Commonly Used Notations

Symbol	Meaning
α	Fine structure constant
a_0	Bohr radius (atomic unit)
a_s	Scattering length
A_e	Electron affinity
$\boldsymbol{B}$	Magnetic field vector
c	Velocity of light
c.c.	Complex conjugate
c.m.	Center of mass
CTMC	Classical trajectory Monte Carlo
Δ	Optical detuning $\omega_0 - \omega$
$\hat{\boldsymbol{\epsilon}}$	Polarization vector
e	Elementary charge
E_h	Hartree energy (atomic unit)
$\boldsymbol{E}$	Electric field vector
${}_2F_1(a, b; c; z)$	Hypergeometric function
h	Planck's constant ($\hbar = h/2\pi$)
H	Hamiltonian
IR	Infrared
I_P	Ionization potential
Im	Imaginary part
$\boldsymbol{k}$	Wave vector (magnitude $k = \lvert\boldsymbol{k}\rvert$)
k_B	Boltzmann constant
k_R	Rutherford coefficient $2\pi\alpha^4 a_0^2 m_e c^2 Z^2$
λ	Wavelength
λ_{dB}	de Broglie wavelength
λ_{mfp}	Mean free path
$\lambdabar$	$\lambda/2\pi$
l, ℓ	Angular momentum quantum number
$\boldsymbol{L}$	Angular momentum vector
$\boldsymbol{\mu}$	Magnetic moment
μ_B	Bohr magneton
m_e	Electron mass
n	Principal quantum number
n_e	Electron density
N	Heavy particle density
ν	Frequency ($\omega = 2\pi\nu$)
ω_p	Plasma frequency
$\wp$	Principal value
$\boldsymbol{p}, \boldsymbol{P}$	Momentum vector
q	Generic charge
$\boldsymbol{r}$	Position vector
r_0	Classical electron radius
R_∞	Rydberg constant (infinite mass)
R_M	Rydberg constant (finite mass)
Re	Real part
T	Temperature
$\mathcal{T}$	Kinetic energy

Commonly Used Notations

Symbol	Meaning
TDHF	Time dependent Hartree–Fock
u	Atomic mass unit
UV	Ultraviolet
v	Velocity
v_B	Bohr velocity (atomic unit)
V	Potential energy
Z	Nuclear charge

Physical Quantities in Atomic Units

Quantity	Unit	Value
Length	a_0	$0.529\,177\,2108\,(18) \times 10^{-10}$ m
Mass	m_e	$0.910\,938\,26\,(16) \times 10^{-30}$ kg
Time	$\hbar/E_h$	$2.418\,884\,326\,505\,(16) \times 10^{-17}$ s
Velocity	$v_B \equiv \alpha c$	$2.187\,691\,2633\,(73) \times 10^{6}$ m s^{-1}
Energy	E_h	$4.359\,744\,17\,(75) \times 10^{-18}$ J
Action	$\hbar$	$1.054\,571\,68\,(18) \times 10^{-34}$ J s
Force	E_h/a_0	$0.823\,872\,25\,(14) \times 10^{-7}$ N
Power	$E_h^2/\hbar$	$0.180\,237\,811\,(31)$ W
Intensity	$\dfrac{E_h^2}{\hbar a_0^2}$	$64.364\,091\,(11) \times 10^{18}$ W m^{-2}
Charge	e	$1.602\,176\,53\,(14) \times 10^{-19}$ C
Electric potential	E_h/e	$27.211\,3845\,(23)$ V
Electric field	$\dfrac{E_h}{ea_0} = \dfrac{\alpha\hbar c}{ea_0^2}$	$0.514\,220\,642\,(44) \times 10^{12}$ V m^{-1}
Magnetic flux density	$\dfrac{E_h}{ea_0\alpha c}$	$2.350\,517\,42\,(20) \times 10^{5}$ T

Atomic Unit Definitions

Length:	$a_0 = \dfrac{4\pi\epsilon_0\hbar^2}{m_e e^2} = \dfrac{\hbar}{\alpha m_e c}$,
Velocity:	$v_B = \dfrac{e^2}{4\pi\epsilon_0\hbar} = \alpha c$,
Time:	$\tau_0 = \dfrac{16\pi^2\epsilon_0^2\hbar^3}{m_e e^4} = \dfrac{\hbar}{\alpha^2 m_e c^2}$,
Energy:	$E_h = \dfrac{e^2}{4\pi\epsilon_0 a_0} = \alpha^2 m_e c^2$,

Energy Conversion Factors

1 eV $= 1.602\,176\,53\,(14) \times 10^{-19}$ J
$= 2.417\,989\,40\,(21) \times 10^{14}$ Hz $\times h$
$= 8065.544\,45\,(69)$ cm$^{-1} \times hc$
$= 3.674\,932\,45\,(31) \times 10^{-2}$ E_h
$= 1.160\,4505\,(20) \times 10^{4}$ K $\times k_B$
$= 96.485\,3383\,(83)$ kJ mol^{-1}

Springer Handbooks
of Atomic, Molecular, and Optical Physics

Springer Handbooks provide a concise compilation of approved key information on methods of research, general principles, and functional relationships in physics and engineering. The world's leading experts in the fields of physics and engineering will be assigned by one or several renowned editors to write the chapters comprising each volume. The content is selected by these experts from Springer sources (books, journals, online content) and other systematic and approved recent publications of physical and technical information.

The volumes will be designed to be useful as readable desk reference books to give a fast and comprehensive overview and easy retrieval of essential reliable key information, including tables, graphs, and bibliographies. References to extensive sources are provided.

Springer Handbook of Atomic, Molecular, and Optical Physics

Gordon W. F. Drake (Ed.)

With CD-ROM, 288 Figures and 111 Tables

Springer

Editor:
Dr. Gordon W. F. Drake
Department of Physics
University of Windsor
Windsor, Ontario N9B 3P4
Canada

Assistant Editor:
Dr. Mark M. Cassar
Department of Physics
University of Windsor
Windsor, Ontario N9B 3P4
Canada

Library of Congress Control Number: 2005931256

ISBN-10: 0-387-20802-X e-ISBN: 0-387-26308-X
ISBN-13: 978-0-387-20802-2 Printed on acid free paper

Printed in Germany.

Production and typesetting: LE-TeX GbR, Leipzig
Handbook coordinator: Dr. W. Skolaut, Heidelberg
Typography, layout and illustrations: schreiberVIS, Seeheim
Cover design: eStudio Calamar Steinen, Barcelona
Cover production: *design&production* GmbH, Heidelberg
Printing and binding: Stürtz GmbH, Würzburg

SPIN 10948934 100/3141/YL 5 4 3 2 1 0

Handbook of Atomic, Molecular, and Optical Physics

Editor

Gordon W. F. Drake
Department of Physics, University of Windsor, Windsor, Ontario, Canada
gdrake@uwindsor.ca

Assistant Editor

Mark M. Cassar
Department of Physics, University of Windsor, Windsor, Ontario, Canada
cassar@uwindsor.ca

Advisory Board

William E. Baylis – Atoms
Department of Physics, University of Windsor, Windsor, Ontario, Canada
baylis@uwindsor.ca

Robert N. Compton – Scattering, Experiment
Oak Ridge National Laboratory, Oak Ridge, Tennessee, USA
ahd@ornl.gov

M. Raymond Flannery – Scattering, Theory
School of Physics, Georgia Institute of Technology, Atlanta, Georgia, USA
flannery@eikonal.physics.gatech.edu

Brian R. Judd – Mathematical Methods
Department of Physics, The Johns Hopkins University, Baltimore, Maryland, USA
juddbr@pha.jhu.edu

Kate P. Kirby – Molecules, Theory
Harvard-Smithsonian Center for Astrophysics, Cambridge, Massachusetts, USA
kirby@cfa.harvard.edu

Pierre Meystre – Optical Physics
Optical Sciences Center, The University of Arizona, Tucson, Arizona, USA
pierre@rhea.opt-sci.arizona.edu

Foreword by Herbert Walther

The Handbook of Atomic, Molecular and Optical (AMO) Physics gives an in-depth survey of the present status of this field of physics. It is an extended version of the first issue to which new and emerging fields have been added. The selection of topics thus traces the recent historic development of AMO physics. The book gives students, scientists, engineers, and other interested people a comprehensive introduction and overview. It combines introductory explanations with descriptions of phenomena, discussions of results achieved, and gives a useful selection of references to allow more detailed studies, making the handbook very suitable as a desktop reference.

AMO physics is an important and basic field of physics. It provided the essential impulse leading to the development of modern physics at the beginning of the last century. We have to remember that at that time not every physicist believed in the existence of atoms and molecules. It was due to Albert Einstein, whose work we commemorate this year with the world year of physics, that this view changed. It was Einstein's microscopic view of molecular motion that led to a way of calculating Avogadro's number and the size of molecules by studying their motion. This work was the basis of his PhD thesis submitted to the University of Zurich in July 1905 and after publication became Einstein's most quoted paper. Furthermore, combining kinetic theory and classical thermodynamics led him to the conclusion that the displacement of a microparticle in Brownian motion varies as the square root of time. The experimental demonstration of this law by Jean Perrin three years later finally afforded striking proof that atoms and molecules are a reality. The energy quantum postulated by Einstein in order to explain the photoelectric effect was the basis for the subsequently initiated development of quantum physics, leading to a revolution in physics and many new applications in science and technology.

The results of AMO physics initiated the development of quantum mechanics and quantum electrodynamics and as a consequence led to a better understanding of the structure of atoms and molecules and their respective interaction with radiation and to the attainment of unprecedented accuracy. AMO physics also influenced the development in other fields of physics, chemistry, astronomy, and biology. It is an astonishing fact that AMO physics constantly went through periods where new phenomena were found, giving rise to an enormous revival of this area. Examples are the maser and laser and their many applications, leading to a better understanding of the basics and the detection of new phenomena, and new possibilities such as laser cooling of atoms, squeezing, and other nonlinear behaviour. Recently, coherent interference effects allowed slow or fast light to be produced. Finally, the achievement of Bose–Einstein condensation in dilute media has opened up a wide range of new phenomena for study. Special quantum phenomena are leading to new applications for transmission of information and for computing. Control of photon emission through specially designed cavities allows controlled and deterministic generation of photons opening the way for a secure information transfer.

Prof. Dr. Herbert Walther

Further new possibilities are emerging, such as the techniques for producing attosecond laser pulses and laser pulses with known and controlled phase relation between the envelope and carrier wave, allowing synthesis of even shorter pulses in a controlled manner. Furthermore, laser pulses may soon be available that are sufficiently intense to allow polarization of the vacuum field. Another interesting development is the generation of artificial atoms, e.g., quantum dots, opening a field where nanotechnology meets atomic physics. It is thus evident that AMO physics is still going strong and will also provide new and interesting opportunities and results in the future.

Preface

The year 2005 has been officially declared by the United Nations to be the International Year of Physics to commemorate the three famous papers of Einstein published in 1905. It is a fitting tribute to the impact of his work that the *Springer Handbook of Atomic, Molecular, and Optical Physics* should be published in coincidence with this event. Virtually all of AMO Physics rests on the foundations established by Einstein in 1905 (including a fourth paper on relativity and his thesis) and his subsequent work. In addition to the theory of relativity, for which he is best known, Einstein ushered in the era of quantum mechanics with his explanation of the photoelectric effect, and he demonstrated the influence of molecular collisions with his explanation of Brownian motion. He also laid the theoretical foundations for all of laser physics with his discovery (in 1917) of the necessity of the process of stimulated emission, and his discussions of the Einstein–Podolsky–Rosen Gedanken experiment (in 1935) led, through Bell's inequalities, to current work on entangled states and quantum information. The past century has been a Golden Age for physics in every sense of the term.

Despite this history of unparalleled progress, the field of AMO Physics continues to advance more rapidly than ever. At the time of publication of an earlier Handbook published by AIP Press in 1996 I wrote "The ever increasing power and versatility of lasers continues to open up new areas for study." Since then, two Nobel Prizes have been awarded for the cooling and trapping of atoms with lasers (Steven Chu, Claude Cohen-Tannoudji, William D. Phillips in 1997), and for the subsequent achievement of Bose–Einstein condensation in a dilute gas of trapped atoms (Eric A. Cornell, Wolfgang Ketterle, Carl E. Wieman in 2001). Although the topic of cooling and trapping was covered in the AIP Handbook, Bose–Einstein condensation was barely mentioned. Since then, the literature has exploded to nearly 2500 papers on Bose–Einstein condensation alone. Similarly, the topics of quantum information and quantum computing barely existed in 1995, and have since become rapidly growing segments of the physics literature. Entirely new topics such as "fast light" and "slow light" have emerged. Techniques for both high precision theory and measurement are opening the possibility to detect a cosmological variation of the fundamental constants with time. All of these topics hold the promise of important engineering and technological applications that come with advances in fundamental science. The more established areas of AMO Physics continue to provide the basic data and broad understanding of a great wealth of underlying processes needed for studies of the environment, and for astrophysics and plasma physics.

Prof. Gordon W. F. Drake

These changes and advances provide more than sufficient justification to prepare a thoroughly revised and updated *Atomic, Molecular and Optical Physics Handbook* for the Springer Handbook Program. The aim is to present the basic ideas, methods, techniques and results of the field at a level that is accessible to graduate students and other researchers new to the field. References are meant to be a guide to the literature, rather than a comprehensive bibliography. Entirely new chapters have been added on Bose–Einstein condensation, quantum information, variations of the fundamental constants, and cavity ring-down spectroscopy. Other chapters have been substantially expanded to include new topics such as fast light and slow light. The intent is to provide a book that will continue to be a valuable resource and source of inspiration for both students and established researchers.

I would like to acknowledge the important role played by the members of the Advisory Board in their continuing support of this project, and I would especially like to acknowledge the talents of Mark Cassar as Assistant Editor. In addition to keeping track of the submissions and corresponding with authors, he read and edited the new material for every chapter to ensure uniformity in style and scientific content, and he composed new material to be added to some of the chapters, as noted in the text.

February 2005 Gordon W. F. Drake

List of Authors

Nigel G. Adams
University of Georgia
Department of Chemistry
Athens, GA 30602-2556, USA
e-mail: *adams@chem.uga.edu*

Miron Ya. Amusia
The Hebrew University
Racah Institute of Physics
Jerusalem, 91904, Israel
e-mail: *amusia@vms.huji.ac.il*

Nils Andersen
University of Copenhagen
Niels Bohr Institute
Universitetsparken 5
Copenhagen, DK-2100, Denmark
e-mail: *noa@fys.ku.dk*

Nigel R. Badnell
University of Strathclyde
Department of Physics
Glasgow, G40NG, United Kingdom
e-mail: *n.r.badnell@strath.ac.uk*

Thomas Bartsch
Georgia Institute of Technology
School of Physics
837 State Street
Atlanta, GA 30332-0430, USA
e-mail: *bartsch@cns.physics.gatech.edu*

Klaus Bartschat
Drake University
Department of Physics and Astronomy
Des Moines, IA 50311, USA
e-mail: *klaus.bartschat@drake.edu*

William E. Baylis
University of Windsor
Department of Physics
Windsor, ON N9B 3P4, Canada
e-mail: *baylis@uwindsor.ca*

Anand K. Bhatia
NASA Goddard Space Flight Center
Laboratory for Astronomy & Solar Physics
Code 681, UV/Optical Astronomy Branch
Greenbelt, MD 20771, USA
e-mail: *anand.k.bhatia@nasa.gov*

Hans Bichsel
University of Washington
Center for Experimental Nuclear Physics and Astrophysics (CENPA)
1211 22nd Avenue East
Seattle, WA 98112-3534, USA
e-mail: *bichsel@npl.washington.edu*

Robert W. Boyd
University of Rochester
Department of Physics and Astronomy
Rochester, NY 14627, USA
e-mail: *boyd@optics.rochester.edu*

John M. Brown
University of Oxford
Physical and Theoretical Chemistry Laboratory
South Parks Road
Oxford, OX1 3QZ, England
e-mail: *john.m.brown@chem.ox.ac.uk*

Henry Buijs
ABB Bomem Inc.
585, Charest Boulevard East
Suite 300
Québec, PQ G1K 9H4, Canada
e-mail: *henry.l.buijs@ca.abb.com*

Philip Burke
The Queen's University of Belfast
Department of Applied Mathematics
and Theoretical Physics
Belfast, Northern Ireland BT7 1NN, UK
e-mail: *p.burke@qub.ac.uk*

Denise Caldwell
National Science Foundation
Physics Division
4201 Wilson Boulevard
Arlington, VA 22230, USA
e-mail: *dcaldwel@nsf.gov*

Mark M. Cassar
University of Windsor
Department of Physics
Windsor, ON N9B 3P4, Canada
e-mail: *cassar@uwindsor.ca*

Kelly Chance
Harvard-Smithsonian Center for Astrophysics
60 Garden Street
Cambridge, MA 02138-1516, USA
e-mail: *kchance@cfa.harvard.edu*

Raymond Y. Chiao
366 Leconte Hall
U.C. Berkeley
Berkeley, CA 94720-7300, USA
e-mail: *chiao@physics.berkeley.edu*

Lew Cocke
Kansas State University
Department of Physics
Manhattan, KS 66506, USA
e-mail: *cocke@phys.ksu.edu*

James S. Cohen
Los Alamos National Laboratory
Atomic and Optical Theory
Los Alamos, NM 87545, USA
e-mail: *cohen@lanl.gov*

Bernd Crasemann
University of Oregon
Department of Physics
Eugene, OR 97403-1274, USA
e-mail: *berndc@uoregon.edu*

David R. Crosley
SRI International
Molecular Physics Laboratory
333 Ravenswood Ave., PS085
Menlo Park, CA 94025-3493, USA
e-mail: *david.crosley@sri.com*

Derrick Crothers
Queen's University Belfast
Department of Applied Mathematics and Theoretical Physics
University Road
Belfast, Northern Ireland BT7 1NN, UK
e-mail: *d.crothers@qub.ac.uk*

Lorenzo J. Curtis
University of Toledo
Department of Physics and Astronomy
2801 West Bancroft Street
Toledo, OH 43606-3390, USA
e-mail: *ljc@physics.utoledo.edu*

Alexander Dalgarno
Harvard-Smithsonian Center for Astrophysics
60 Garden Street
Cambridge, MA 02138, USA
e-mail: *adalgarno@cfa.harvard.edu*

Abigail J. Dobbyn
Max-Planck-Institut für Strömungsforschung
Göttingen, 37073, Germany

Gordon W. F. Drake
University of Windsor
Department of Physics
401 Sunset St.
Windsor, ON N9B 3P4, Canada
e-mail: *gdrake@uwindsor.ca*

Joseph H. Eberly
University of Rochester
Department of Physics and Astronomy and Institute of Optics
Rochester, NY 14627-0171, USA
e-mail: *eberly@pas.rochester.edu*

Guy T. Emery
Bowdoin College
Department of Physics
15 Chestnut Rd.
Brunswick, ME 04011, USA
e-mail: *gemery@bowdoin.edu*

Volker Engel
Universität Würzburg
Institut für Physikalische Chemie
Am Hubland
Würzburg, 97074, Germany
e-mail: *voen@phys-chemie.uni-wuerzburg.de*

Paul Engelking
University of Oregon
Department of Chemistry
and Chemical Physics Institute
Eugene, OR 97403-1253, USA
e-mail: *engelki@uoregon.edu*

Kenneth M. Evenson†

James M. Farrar
University of Rochester
Department of Chemistry
120 Trustee Road
Rochester, NY 14627-0216, USA
e-mail: *farrar@chem.rochester.edu*

Gordon Feldman
The Johns Hopkins University
Department of Physics and Astronomy
Baltimore, MD 21218-2686, USA
e-mail: *gordon.feldman@jhu.edu*

Paul D. Feldman
The Johns Hopkins University
Department of Physics and Astronomy
3400 N. Charles Street
Baltimore, MD 21218-2686, USA
e-mail: *pdf@pha.jhu.edu*

Charlotte F. Fischer
Vanderbilt University
Department of Electrical Engineering
Computer Science
PO BOX 1679, Station B
Nashville, TN 37235, USA
e-mail: *charlotte.f.fischer@vanderbilt.edu*

Victor Flambaum
University of New South Wales
Department of Physics
Sydney, 2052, Australia
e-mail: *v.flambaum@unsw.edu.au*

M. Raymond Flannery
Georgia Institute of Technology
School of Physics
Atlanta, GA 30332-0430, USA
e-mail: *ray.flannery@physics.gatech.edu*

David R. Flower
University of Durham
Department of Physics
South Road
Durham, DH1 3LE, United Kingdom
e-mail: *david.flower@durham.ac.uk*

A. Lewis Ford
Texas A&M University
Department of Physics
College Station, TX 77843-4242, USA
e-mail: *ford@physics.tamu.edu*

Jane L. Fox
Wright State University
Department of Physics
3640 Colonel Glenn Hwy
Dayton, OH 45419, USA
e-mail: *jane.fox@wright.edu*

Matthias Freyberger
Universität Ulm
Abteilung für Quantenphysik
Albert Einstein Allee 11
Ulm, 89069, Germany
e-mail: *matthias.freyberger@uni-ulm.de*

Thomas Fulton
The Johns Hopkins University
The Henry A. Rowland Department
of Physics and Astronomy
Baltimore, MD 21218-2686, USA
e-mail: *Thomas.Fulton@jhu.edu*

Alexander L. Gaeta
Cornell University
Department of Applied and Engineering Physics
Ithaca, NY 14853-3501, USA
e-mail: *a.gaeta@cornell.edu*

Alan Gallagher
JILA, University of Colorado and National Institute of Standards and Technology
Quantum Physics Division
Boulder, CO 80309-0440, USA
e-mail: *alang@jila.colorado.edu*

Thomas F. Gallagher
University of Virginia
Department of Physics
382 McCormick Road
Charlottesville, VA 22904-4714, USA
e-mail: *tfg@virginia.edu*

Muriel Gargaud
Observatoire Aquitain des Sciences de l'Univers
2 Rue de l'Observatoire
33270 Floirac, France
e-mail: *gargaud@obs.u-bordeaux1.fr*

Alan Garscadden
Airforce Research Laboratory
Area B
1950 Fifth Street
Wright Patterson Air Force Base,
OH 45433-7251, USA
e-mail: *alan.garscadden@wpafb.af.mil*

John Glass
British Telecommunications
Solution Design
Riverside Tower (pp RT03-44)
Belfast, Northern Ireland BT1 3BT, UK
e-mail: *john.glass@bt.com*

S. Pedro Goldman
The University of Western Ontario
Department of Physics & Astronomy
London, ON N6A 3K7, Canada
e-mail: *goldman@uwo.ca*

Ian P. Grant
University of Oxford
Mathematical Institute
24/29 St. Giles'
Oxford, OX1 3LB, UK
e-mail: *ipg@maths.ox.ac.uk*

Donald C. Griffin
Rollins College
Department of Physics
1000 Holt Ave.
Winter Park, FL 32789, USA
e-mail: *griffin@vanadium.rollins.edu*

William G. Harter
University of Arkansas
Department of Physics
Fayetteville, AR 72701, USA
e-mail: *wharter@uark.edu*

Carsten Henkel
Universität Potsdam
Institut für Physik
Am Neuen Palais 10
Potsdam, 14469, Germany
e-mail: *carsten.henkel@quantum.physik.uni-potsdam.de*

Eric Herbst
The Ohio State University
Departments of Physics
191 W. Woodruff Ave.
Columbus, OH 43210-1106, USA
e-mail: *herbst@mps.ohio-state.edu*

Robert N. Hill
355 Laurel Avenue
Saint Paul, MN 55102-2107, USA
e-mail: *rnhill@fishnet.com*

David L. Huestis
SRI International
Molecular Physics Laboratory
Menlo Park, CA 94025, USA
e-mail: *david.huestis@sri.com*

Mitio Inokuti
Argonne National Laboratory
Physics Division
9700 South Cass Avenue
Building 203
Argonne, IL 60439, USA
e-mail: *inokuti@anl.gov*

Takeshi Ishihara
University of Tsukuba
Institute of Applied Physics
Ibaraki 305
Tsukuba, 305-8577, Japan

Juha Javanainen
University of Connecticut
Department of Physics
Unit 3046
2152 Hillside Road
Storrs, CT 06269-3046, USA
e-mail: *jj@phys.uconn.edu*

Erik T. Jensen
University of Northern British Columbia
Department of Physics
3333 University Way
Prince George, BC V2N 4Z9, Canada
e-mail: *ejensen@unbc.ca*

Brian R. Judd
The Johns Hopkins University
Department of Physics and Astronomy
3400 North Charles Street
Baltimore, MD 21218, USA
e-mail: *juddbr@pha.jhu.edu*

Alexander A. Kachanov
Research and Development
Picarro, Inc.
480 Oakmead Parkway
Sunnyvale, CA 94085, USA
e-mail: *akachanov@picarro.com*

Isik Kanik
California Institute of Technology
Jet Propulsion Laboratory
Pasadena, CA 91109, USA
e-mail: *isik.kanik@jpl.nasa.gov*

Savely G. Karshenboim
D.I.Mendeleev Institute for Metrology (VNIIM)
Quantum Metrology Department
Moskovsky pr. 19
St. Petersburg, 190005, Russia
e-mail: *sek@mpq.mpg.de*

Kate P. Kirby
Harvard-Smithsonian Center for Astrophysics
60 Garden Street MS-14
Cambridge, MA 02138, USA
e-mail: *kkirby@cfa.havard.edu*

Sir Peter L. Knight
Imperial College London
Department of Physics
Blackett Laboratory
Prince Consort Road
London, SW7 2BW, UK
e-mail: *p.knight@imperial.ac.uk*

Manfred O. Krause
Oak Ridge National Laboratory
125 Baltimore Drive
Oak Ridge, TN 37830, USA
e-mail: *mok@ornl.gov*

Kenneth C. Kulander
Lawrence Livermore National Laboratory
7000 East Ave.
Livermore, CA 94551, USA
e-mail: *kulander@llnl.gov*

Paul G. Kwiat
University of Illinois at Urbana-Champaign
Department of Physics
1110 West Green Street
Urbana, IL 61801-3080, USA
e-mail: *kwiat@uiuc.edu*

Yuan T. Lee
Academia Sinica
Institute of Atomic and Molecular Science
PO BOX 23-166
Taipei, 106, Taiwan

Stephen Lepp
University of Nevada
Department of Physics
4505 Maryland Pkwy
Las Vegas, NV 89154-4002, USA
e-mail: *lepp@unlv.edu*

Maciej Lewenstein
ICFO–Institut de Ciéncies Fotóniques
C. Jordi Ginora 29 Nexus II
Barcelona, 08034, Spain
e-mail: *maciej.lewenstein@icfo.es*

James D. Louck
Los Alamos National Laboratory
Retired Laboratory Fellow
PO BOX 1663
Los Alamos, NM 87545, USA
e-mail: *jimlouck@aol.com*

Joseph H. Macek
University of Tennessee and Oak Ridge National Laboratory
Department of Physics and Astronomy
401 Nielsen Physics Bldg.
Knoxville, TN 37996-1200, USA
e-mail: *jmacek@utk.edu*

Mary L. Mandich
Lucent Technologies Inc.
Bell Laboratories
600 Mountain Avenue
Murray Hill, NJ 07974, USA
e-mail: *mandich@lucent.com*

Edmund J. Mansky
Oak Ridge National Laboratory
Controlled Fusion Atomic Data Center
Oak Ridge, TN 37831, USA
e-mail: *edmundmansky@intdata.com*

Steven T. Manson
Georgia State University
Department of Physics and Astronomy
Atlanta, GA 30303, USA
e-mail: *smanson@gsu.edu*

William C. Martin
National Institute of Standards and Technology
Atomic Physics Division
Gaithersburg, MD 20899-8422, USA
e-mail: *wmartin@nist.gov*

Jim F. McCann
Queen's University Belfast
Dept. of Applied Mathematics
and Theoretical Physics
Belfast, Northern Ireland BT7 1NN, UK
e-mail: *j.f.mccann@qub.ac.uk*

Ronald McCarroll
Université Pierre et Marie Curie
Laboratoire de Chimie Physique
11 rue Pierre et Marie Curie
75231 Paris Cedex 05, France
e-mail: *mccarrol@ccr.jussieu.fr*

Fiona McCausland
Northern Ireland Civil Service
Department of Enterprise Trade and Investment
Massey Avenue
Belfast, Northern Ireland BT4 2JP, UK
e-mail: *fiona.mccausland@detini.gov.uk*

William J. McConkey
University of Windsor
Department of Physics
Windsor, ON N9B 3P4, Canada
e-mail: *mcconk@uwindsor.ca*

Robert P. McEachran
Australian National University
Atomic and Molecular Physics Laboratories
Research School of Physical Sciences
and Engineering
Canberra, ACT 0200, Australia
e-mail: *robert.mceachran@anu.edu.au*

James H. McGuire
Tulane University
Department of Physics
6823 St. Charles Ave.
New Orleans, LA 70118-5698, USA
e-mail: *mcguire@tulane.edu*

Dieter Meschede
Rheinische Friedrich-Wilhelms-Universität Bonn
Institut für Angewandte Physik
Wegelerstraße 8
Bonn, 53115, Germany
e-mail: *meschede@iap.uni-bonn.de*

Pierre Meystre
University of Arizona
Department of Physics
1118 E, 4th Street
Tucson, AZ 85721-0081, USA
e-mail: *meystre@physics.arizona*

Peter W. Milonni
104 Sierra Vista Dr.
Los Alamos, NM 87544, USA
e-mail: *pwm@lanl.gov*

Peter J. Mohr
National Institute of Standards and Technology
Atomic Physics Division
100 Bureau Drive, Stop 8420
Gaithersburg, MD 20899-8420, USA
e-mail: *mohr@nist.gov*

David H. Mordaunt
Max-Planck-Institut für Strömungsforschung
Göttingen, 37073, Germany

John D. Morgan III
University of Delaware
Department of Physics and Astronomy
Newark, DE 19716, USA
e-mail: *jdmorgan@udel.edu*

Michael S. Murillo
Los Alamos National Laboratory
Theoretical Division
PO BOX 1663
Los Alamos, NM 87545, USA
e-mail: *murillo@lanl.gov*

Evgueni E. Nikitin
Technion-Israel Institute of Technology
Department of Chemistry
Haifa, 32000, Israel
e-mail: *nikitin@techunix.technion.ac.il*

Robert F. O'Connell
Louisiana State University
Department of Physics and Astronomy
Baton Rouge, LA 70803-4001, USA
e-mail: *oconnell@phys.lsu.edu*

Francesca O'Rourke
Queen's University Belfast
Department of Applied Mathematics and Theoretical Physics
University Road
Belfast, BT7 1NN, UK
e-mail: *s.orourke@qub.ac.uk*

Ronald E. Olson
University of Missouri-Rolla
Physics Department
Rolla, MO 65409, USA
e-mail: *olson@umr.edu*

Barbara A. Paldus
Skymoon Ventures
3045 Park Boulevard
Palo Alto, CA 94306, USA
e-mail: *bpaldus@skymoonventures.com*

Josef Paldus
University of Waterloo
Department of Applied Mathematics
200 University Avenue West
Waterloo, ON N2L 3G1, Canada
e-mail: *paldus@scienide.uwaterloo.ca*

Gillian Peach
University College London
Department of Physics and Astronomy
London, WC1 E6BT, UK
e-mail: *g.peach@ucl.ac.uk*

Ruth T. Pedlow
Queen's University Belfast
Department of Applied Mathematics and Theoretical Physics
University Road
Belfast, Northern Irland BT7 1NN, UK
e-mail: *r.pedlow@qub.ac.uk*

David J. Pegg
University of Tennessee
Department of Physics
Nielsen Building
Knoxville, TN 37996, USA
e-mail: *djpegg@utk.edu*

Ekkehard Peik
Physikalisch-Technische Bundesanstalt
Bundesallee 100
Braunschweig, 38116, Germany
e-mail: *ekkehard.peik@ptb.de*

Ronald Phaneuf
University of Nevada
Department of Physics
MS-220
Reno, NV 89557-0058, USA
e-mail: *phaneuf@unr.edu*

Michael S. Pindzola
Auburn University
Department of Physics
Auburn, AL 36849, USA
e-mail: *pindzola@physics.auburn.edu*

Eric H. Pinnington
University of Alberta
Department of Physics
Edmonton, AB T6H 0B3, Canada
e-mail: *pinning@phys.ualberta.ca*

Richard C. Powell
University of Arizona
Optical Sciences Center
Tuscon, AZ 85721, USA
e-mail: *rcpowell@email.arizona.edu*

John F. Reading
Texas A&M University
Department of Physics
College Station, TX 77843, USA
e-mail: *reading@physics.tamu.edu*

Jonathan R. Sapirstein
University of Notre Dame
Department of Physics
319 Nieuwland Science
Notre Dame, IN 46556, USA
e-mail: *jsapirst@nd.edu*

Stefan Scheel
Imperial College London
Blackett Laboratory
Prince Consort Road
London, SW7 2BW, UK
e-mail: *s.scheel@imperial.ac.uk*

Axel Schenzle
Ludwig-Maximilians-Universität
Department für Physik
Theresienstraße 37
München, 80333, Germany
e-mail: *axel.schenzle@physik.uni-muenchen.de*

Reinhard Schinke
Max-Planck-Institut für Dynamik &
Selbstorganisation
Bunsenstr. 10
Göttingen, 37073, Germany
e-mail: *rschink@gwdg.de*

Wolfgang P. Schleich
Universität Ulm
Abteilung für Quantenphysik
Albert Einstein Allee 11
Ulm, 89069, Germany
e-mail: *wolfgang.schleich@uni-ulm.de*

David R. Schultz
Oak Ridge National Laboratory
Physics Division
Oak Ridge, TN 37831-6373, USA
e-mail: *schultz@mail.phy.ornl.gov*

Michael Schulz
University of Missouri-Rolla
Physics Department
1870 Miner Circle
Rolla, MO 65409, USA
e-mail: *schulz@umr.edu*

Peter L. Smith
Harvard University
Harvard-Smithsonian Center for Astrophysics
60 Garden Street
Cambridge, MA 02138, USA
e-mail: *plsmith@cfa.havard.edu*

Anthony F. Starace
The University of Nebraska
Department of Physics and Astronomy
116 Brace Laboratory
Lincoln, NE 68588-0111, USA
e-mail: *astarace1@unl.edu*

Glenn Stark
Wellesley College
Department of Physics
106 Central Street
Wellesley, MA 02481, USA
e-mail: *gstark@wellesley.edu*

Allan Stauffer
Department of Physics and Astronomy
York University
4700 Keele Street
Toronto, ON M3J 1P3, Canada
e-mail: *stauffer@yorku.ca*

Aephraim M. Steinberg
University of Toronto
Department of Physics
Toronto, ON M5S 1A7, Canada
e-mail: *steinberg@physics.utoronto.ca*

Stig Stenholm
Royal Institute of Technology
Physics Department
Roslagstullsbacken 21
Stockholm, SE-10691, Sweden
e-mail: *stenholm@atom.kth.se*

Jack C. Straton
Portland State University
University Studies
117P Cramer Hall
Portland, OR 97207, USA

Michael R. Strayer
Oak Ridge National Laboratory
Physics Division
Oak Ridge, TN 37831-6373, USA
e-mail: *strayer@csep2.phy.ornl.gov*

Carlos R. Stroud Jr.
University of Rochester
Institute of Optics
Rochester, NY 14627-0186, USA
e-mail: *stroud@optics.rochester.edu*

Arthur G. Suits
State University of New York
Department of Chemistry
Stony Brook, NY 11794, USA
e-mail: *arthur.suits@sunysb.edu*

Barry N. Taylor
National Institute of Standards and Technology
Atom Physics Division
100 Bureau Drive
Gaithersburg, MD 20899-8401, USA
e-mail: *barry.taylor@nist.gov*

Aaron Temkin
NASA Goddard Space Flight Center
Laboratory for Solar and Space Physics
Solar Physics Branch
Greenbelt, MD 20771, USA
e-mail: *aaron.temkin-1@nasa.gov*

Sandor Trajmar
California Institute of Technology
Jet Propulsion Laboratory
3847 Vineyard Drive
Redwood City, 94063, USA
e-mail: *strajmar@comcast.net*

Elmar Träbert
Ruhr-Universität Bochum
Experimentalphysik III/NB3
Bochum, 44780, Germany
e-mail: *traebert@ep3.rub.de*

Turgay Uzer
Georgia Institute of Technology
School of Physics
837 State Street
Atlanta, GA 30332-0430, USA
e-mail: *turgay.uzer@physics.gatech.edu*

Karl Vogel
Universität Ulm
Abteilung für Quantenphysik
Albert Einstein Allee 11
Ulm, 89069, Germany
e-mail: *karl.vogel@uni-ulm.de*

Jon C. Weisheit
Washington State University
Institute for Shock Physics
PO BOX 64 28 14
Pullman, WA 99164, USA
e-mail: *weisheit@wsu.edu*

Wolfgang L. Wiese
National Institute of Standards and Technology
100 Bureau Drive
Gaithersburg, MD 20899, USA
e-mail: *wiese@nist.gov*

Martin Wilkens
Universität Potsdam
Institut für Physik
Am Neuen Palais 10
Potsdam, 14469, Germany
e-mail: *martin.wilkens@physik.uni-potsdam.de*

David R. Yarkony
The Johns Hopkins University
Department of Chemistry
Baltimore, MD 21218, USA
e-mail: *yarkony@jhu.edu*

Springer Handbook of Atomic, Molecular, and Optical Physics
Organization of the Handbook

Part A gathers together the mathematical methods applicable to a wide class of problems in atomic, molecular, and optical physics. The application of angular momentum theory to quantum mechanics is presented. The basic tenet that isolated physical systems are invariant to rotations of the system is thereby implemented into physical theory. The powerful methods of group theory and second quantization show how simplifications arise if the atomic shell is treated as a basic structural unit. The well established symmetry groups of quantum mechanical Hamiltonians are extended to the larger compact and noncompact dynamical groups. Perturbation theory is introduced as a bridge between an exactly solvable problem and a corresponding real one, allowing approximate solutions of various systems of differential equations. The consistent manner in which the density matrix formalism deals with pure and mixed states is developed, showing how the preparation of an initial state as well as the details regarding the observation of the final state can be treated in a systematic way. The basic computational techniques necessary for accurate and efficient numerical calculations essential to all fields of physics are outlined and a summary of relevant software packages is given. The ever present one-electron solutions of the nonrelativistic Schrödinger equation and the relativistic Dirac equation for the Coulomb potential are then summarized.

Part B presents the main concepts in the theoretical and experimental knowledge of atomic systems, including atomic structure and radiation. Ionization energies for neutral atoms and transition probabilities of selected neutral atoms are tabulated. The computational methods needed for very high precision approximations for helium are summarized. The physical and geometrical significance of simple multipoles is examined. The basic nonrelativistic and relativistic theory of electrons and atoms in external magnetic fields is given. Various properties of Rydberg atoms in external fields and in collisions are investigated. The sources of hyperfine structure in atomic and molecular spectra are outlined, and the resulting energy splittings and isotope shifts given. Precision oscillator strength and lifetime measurements, which provide stringent experimental tests of fundamental atomic structure calculations, are discussed. Ion beam spectroscopy is introduced, and individual applications of ion beam techniques are detailed A basic description of neutral collisional line shapes is given, along with a discussion of radiation transfer in a confined atomic vapor. Many qualitative features of the Thomas–Fermi model are studied and its later outgrowth into general density functional theory delineated. The Hartree–Fock and multiconfiguration Hartree–Fock theories, along with configuration interaction methods, are discussed in detail, and their application to the calculation of various atomic properties presented. Relativistic methods for the calculation of atomic structure for general many-electron atoms are described. A consistent diagrammatic method for calculating the structure of atoms and the characteristics of different atomic processes is given. An outline of the theory of atomic photoionization and the dynamics of the photon–atom collision process is presented. Those kinds of electron correlation that are most important in photoionization are emphasized. The process of autoionization is treated as a quasibound state imbedded in the scattering continuum, and a brief description of the main elements of the theory is given. Green's function techniques are applied to the calculation of higher order corrections to atomic energy levels, and also of transition amplitudes for radiative transitions of atoms. Basic quantum electrodynamic calculations, which are needed to explain small deviations from the solution to the Schrödinger equation in simple systems, are presented. Comparisons of precise measurements and theoretical predictions that provide tests of our knowledge of fundamental physics are made, focussing on several quantitative tests of quantum electrodynamics. Precise measurements of parity nonconserving effects in atoms could lead to possible modifications of the Standard Model, and thus uncover new physics. An approach to this fundamental problem is described. The problem of the possible variation of the fundamental constants with time is discussed in relation to atomic clocks and precision frequency measurements. The most advanced atomic clocks are described, and the current laboratory constraints on these variations are listed.

Part C begins with a discussion of molecular structure from a theoretical/computational perspective using the Born–Oppenheimer approximation as the point of departure. The key role that symmetry considerations play in organizing and simplifying our knowledge of molecular dynamics and spectra is described. The theory of radiative transition probabilities, which determine the intensities of spectral lines, for the rotationally-resolved spectra of certain model molecular systems is summarized. The ways in which molecular photodissociation is studied in the gas phase are outlined. The results presented are particularly relevant to the investigation of combustion and atmospheric reactions. Modern experimental techniques allow the detailed motions of the atomic constituents of a molecule to be resolved as a function of time. A brief description of the basic ideas behind these techniques is given, with an emphasis on gas phase molecules in collision-free conditions. The semiclassical and quantal approaches to nonreactive scattering are outlined. Various quantitative approaches toward a description of the rates of gas phase chemical reactions are presented and then evaluated for their reliability and range of application. Ionic reactions in the gas phase are also considered. Clusters, which are important in many atmospheric and industrial processes, are arranged into six general categories, and then the physics and chemistry common to each category is described. The most important spectroscopic techniques used to study the properties of molecules are presented in detail.

Part D collects together the topics and approaches used in scattering theory. A handy compendium of equations, formulae, and expressions for the classical, quantal, and semiclassical approaches to elastic scattering is given; reactive systems and model potentials are also considered. The dependence of scattering processes on the angular orientation of the reactants and products is discussed through the analysis of scattering experiments which probe atomic collision theories at a fundamental level.

The detailed quantum mechanical techniques available to perform accurate calculations of scattering cross sections from first principles are presented. The theory of elastic, inelastic, and ionizing collisions of electrons with atoms and atomic ions is covered and then extended to include collisions with molecules. The standard scattering theory for electrons is extended to include positron collisions with atomic and molecular systems. Slow collisions of atoms or molecules within the adiabatic approximation are discussed; important deviations from this model are presented in some detail for the low energy case. The main methods in the theoretical treatment of ion-atom and atom–atom collisions are summarized with a focus on intermediate and high collision velocities. The molecular structure and collision dynamics involved in ion–atom charge exchange reactions is studied. Both the perturbative and variational capture theories of the continuum distorted wave model are presented. The Wannier theory for threshold ionization is then developed. Studies of the energy and angular distribution of electrons ejected by the impact of high-velocity atomic or ionic projectiles on atomic targets are overviewed. A useful collection of formulae, expressions, and specific equations that cover the various approaches to electron-ion and ion-ion recombination processes is given. A basic theoretical formulation of dielectronic recombination is described, and its importance in the interpretation of plasma spectral emission is presented. Many of the equations used to study theoretically the collisional properties of both charged and neutral particles with atoms and molecules in Rydberg states are collected together; the primary approximations considered are the impulse approximation, the binary encounter approximation, and the Born approximation. The Thomas mass-transfer process is considered from both a classical and a quantal perspective. Additional features of this process are also discussed. The theoretical background, region of validity, and applications of the classical trajectory Monte Carlo method are then delineated. One-photon processes are discussed and aspects of line broadening directly related to collisions between an emitting, or absorbing, atom and an electron, a neutral atom or an atomic ion are considered.

Part D Scattering Theory

Part E focuses on the experimental aspects of scattering processes. Recent developments in the field of photodetachment are reviewed, with an emphasis on accelerator-based investigations of the photodetachment of atomic negative ions. The theoretical concepts and experimental methods for the scattering of low-energy photons, proceeding primarily through the photoelectric effect, are given. The main photon–atom interaction processes in the intermediate energy range are outlined. The atomic response to inelastic photon scattering is discussed; essential aspects of radiative and radiationless transitions are described in the two-step approximation. Advances such as cold-target recoil-ion momentum spectroscopy are also touched upon. Electron–atom and electron–molecule collision processes, which play a prominent role in a variety of systems, are presented. The discussion is limited to electron collisions with gaseous targets, where single collision conditions prevail, and to low-energy impact processes. The physical principles and experimental methods used to investigate low energy ion–atom collisions are outlined. Inelastic processes which occur in collisions between fast, often highly charged, ions and atoms, are described. A summary of the methods commonly employed in scattering experiments

Part E Scattering Experiment

involving neutral molecules at chemical energies is presented. Applications of single-collision scattering methods to the study of reactive collision dynamics of ionic species with neutral partners are discussed.

Part F presents a coherent collection of the main topics and issues found in quantum optics. Optical physics, which is concerned with the dynamical interactions of atoms and molecules with electromagnetic fields, is first discussed within the context of semiclassical theories, and then extended to a fully quantized version. The theoretical techniques used to describe absorption and emission spectra using density matrix methods are developed. Applications of the dark state in laser physics is briefly mentioned. The basic concepts common to all lasers, such as gain, threshold, and electromagnetic modes of oscillation are described. Recent developments in laser physics, including single-atom lasers, two-photon lasers, and the generation of attosecond pulses are also introduced. The current status of the development of different types of lasers – including nanocavity, quantum-cascade and free-electron lasers – are summarized. The important operational characteristics, such as frequency range and output power, are given for each of the types of lasers described. Nonlinear processes arising from the modifications of the optical properties of a medium due to the passage of intense light beams are discussed. Additional processes that are enabled by the use of ultrashort or ultra-intense laser pulses are presented. The concept of coherent optical transients in atomic and molecular systems reviewed; homogeneous and inhomogeneous relaxation in the theory are properly distinguished. Multiphoton and strong-field processes are given a theoretical description. A discussion of the generation of sub-femtosecond pulses is also included. General and specific theories for the control of atomic motion by light are presented. Various traps used for the cooling and trapping of charged and neutral particles and their applications are discussed. The fundamental physics of dilute quantum degenerate gases is outlined, especially in connection with Bose–Einstein condensation. de Broglie optics, which concerns the propagation of matter waves, is presented with a concentration on the underlying principles and the illustration of these principles. The fundamentals of the quantized electromagnetic field and applications to the broad area of quantum optics are discussed. A detailed description of the changes in the atom–field interaction that take place when the radiation field is modified by the presence of a cavity is given. The basic concepts needed to understand current research, such as the EPR experiment, Bell's inequalities, squeezed states of light, the properties of electromagnetic waves in cavities, and other topics depending on the nonlocality of light are reviewed. Applications to cryptography, tunneling times, and gravity wave detectors are included, along with recent work on "fast light" and "slow light." Correlations and quantum superpositions which can be exploited in quantum information processing and secure communication are delineated. Their link to quantum computing and quantum cryptography is given explicitly.

Part F Quantum Optics

Part G is concerned with the various applications of atomic, molecular, and optical physics. A summary of the processes that take place in photoionized gases, collisionally ionized gases, the diffuse interstellar medium, molecular clouds, circumstellar shells, supernova ejecta, shocked regions, and the early

Part G Applications

Universe are presented. The principal atomic and molecular processes that lead to the observed cometary spectra, as well as the needs for basic atomic and molecular data in the interpretation of these spectra, are focused on. The basic methods used to understand planetary atmospheres are given. The structure of atmospheres and their interaction with solar radiation are detailed, with an emphasis on ionospheres. Atmospheric global change is then studied in terms of the applicable atomic and molecular processes responsible for these changes. A summary of the well-known prescriptions for atomic structure and ionization balance, and a discussion of the modified transition rates for ions in dense plasmas are given. A review of current simulations being used to address a wide array of issues needed to accurately describe atoms in dense plasmas is also presented. The main concepts and processes of the physics and chemistry of the conduction of electricity in ionized gases are described. The physical models and laser diagnostics used to understand combustion systems are presented. Various applications of atomic and molecular physics to phenomena that occur at surfaces are reviewed; particular attention is placed on the application of electron- and photon-atom scattering processes to obtain surface specific structural and spectroscopic information. The effect of finite nuclear size on the electronic energy levels of atoms is also detailed; and conversely, the electronic structure effects in nuclear physics are discussed. A discussion of the concepts needed in the operation of charged particle detectors and in describing radiation effects is introduced. The description is restricted to fast charged particles. The key topics in basic radiation physics are then treated, and illustrative examples are given.

Part G Applications

Contents

Part B Atoms

List of Tables

Table 10.6 Conversion relations between S and A_{ki} for forbidden transitions 192
Table 10.7 Relative strengths for lines of multiplets in LS coupling 193
Table 10.8 Some transitions of the main spectral series of hydrogen 195
Table 10.9 Values of Stark-broadening parameter $\alpha_{1/2}$ of the H_β line of hydrogen (4861 Å) for various temperatures and electron densities 196

11 High Precision Calculations for Helium

Table 11.1 Formulas for the radial integrals $I_0(a,b,c;\alpha,\beta) = \langle r_1^a r_2^b r_{12}^c \, e^{-\alpha r_1 - \beta r_2} \rangle_{rad}$ and $I_0^{log}(a,b,c;\alpha,\beta) = \langle r_1^a r_2^b r_{12}^c \ln r_{12} \, e^{-\alpha r_1 - \beta r_2} \rangle_{rad}$ 203
Table 11.2 Nonrelativistic eigenvalue coefficients ε_0 and ε_1 for helium 205
Table 11.3 Eigenvalue coefficients ε_2 for helium 207
Table 11.4 Values of the reduced electron mass ratio μ/M 207
Table 11.5 Nonrelativistic eigenvalues $E = \varepsilon_0 + (\mu/M)\varepsilon_1 + (\mu/M)^2\varepsilon_2$ for helium-like ions 207
Table 11.6 Expectation values of various operators for He-like ions for the case $M = \infty$ 208
Table 11.7 Total ionization energies for ^{4}He, calculated with $R_M = 3\,289\,391\,006.715$ MHz 210
Table 11.8 QED corrections to the ionization energy included in Table 11.7 for the S- and P-states of helium 211
Table 11.9 Quantum defects for the total energies of helium with the ΔW_n term subtracted (11.54) 212
Table 11.10 Formulas for the hydrogenic expectation value $\langle r^{-j} \rangle \equiv \langle nl | r^{-j} | nl \rangle$ 214
Table 11.11 Oscillator strengths for helium 216
Table 11.12 Singlet–triplet mixing angles for helium 217

13 Atoms in Strong Fields

Table 13.1 Relativistic ground state binding energy $-E_{gs}/Z^2$ and finite nuclear size correction $\delta E_{nuc}/Z^2$ of hydrogenic atoms for various magnetic fields B 230
Table 13.2 Relativistic binding energy $-E_{2S,-1/2}$ for the $2S_{1/2}$ $(m_j = -\frac{1}{2})$ and $-E_{2P,-1/2}$ for the $2P_{1/2}$ $(m_j = -\frac{1}{2})$ excited states of hydrogen in an intense magnetic field B 231
Table 13.3 Relativistic corrections $\delta E = (E - E_{NR})/|E_R|$ to the nonrelativistic energies E_{NR} for the ground state and $n = 2$ excited states of hydrogen in an intense magnetic field B 231
Table 13.4 Relativistic dipole polarizabilities for the ground state of hydrogenic atoms 233

17 Precision Oscillator Strength and Lifetime Measurements

Table 17.1 Measured $np\ ^2P_J$ lifetimes 266

21 Atomic Structure: Multiconfiguration Hartree–Fock Theories

22 Relativistic Atomic Structure

25 Autoionization

27 Quantum Electrodynamics

28 Tests of Fundamental Physics

Part D Scattering Theory

Part E Scattering Experiments

Part F Quantum Optics

92 Radiation Physics

List of Abbreviations

2P/2H	two-particle/two-hole

A

AA	average atom
ACT	activated complex theory
ADDS	angular distribution by Doppler spectroscopy
ADO	average dipole orientation
AES	Auger electron spectroscopy
AI	adiabatic ionization
AL	absorption loss
ALS	advanced light source
AMO	atomic, molecular, and optical
ANDC	arbitrarily normalized decay curve
AO	atomic orbital
AOM	acoustooptic modulator
AS	active space
ASD	atomic spectra database
ASF	atomic state functions
ATI	above threshold ionization
AU	absorbance units

B

BEA	binary encounter approximation
BEC	Bose–Einstein condensate (or condensation)
BF	body-fixed
BI	Bell's inequality
BL	Bethe log
BO	Born–Oppenheimer
BS	Bethe–Salpeter
BW	Brillouin–Wigner

C

CARS	coherent anti-Stokes Raman scattering
CAS	complete active space
CASPT	complete active space perturbation theory
CAUGA	Clifford algebra unitary group approach
CC	coupled cluster
CCA	coupled cluster approximation
CCC	convergent close coupling
CCD	coupled cluster doubles
CCO	coupled-channels optical
CDW	continuum distorted wave
CEAS	cavity enhanced absorption spectroscopy
CES	cavity enhanced spectroscopy
CES	constant energy surface
CETS	cavity enhanced transmission spectroscopy
CFCP	free–free molecular Franck–Condon
CG	Clebsch–Gordan
CH	Clauser–Horne
CI	configuration interaction
CIS	constant ionic state
CL	constant log
CM	center-of-mass
CMA	cylindrical mirror analyzer
COA	classical oscillator approximation
CODATA	Committee on Data for Science and Technology
COIL	chemical-oxygen-iodine
COLTRIMS	cold-target recoil-ion momentum spectroscopy
CP	central potential
CPA	chirped-pulsed-amplification
CQC	classical-quantal coupling
CRDS	cavity ring-down spectroscopy
CSDA	continuous slowing down approximation
CSF	configurational state functions
CTF	common translation factor
CTMC	classical trajectory Monte Carlo
CW	continuous wave
CW-CRDS	continuous-Wave Cavity Ring-Down Spectroscopy
CX	charge exchange
CXO	Chandra X-ray Observatory

D

DB	detailed balance
DCS	differential cross sections
DDCS	doubly differential cross sections
DF	Dirac–Fock
DFB	distributed feedback
DFS	decoherence free subspace
DFT	discrete Fourier transform
DFWM	degenerate four wave mixing
DLR	dielectronic recombination
DODS	different orbitals for different spins
DR	dielectronic recombination
DSPB	distorted wave strong potential Born approximation

E

EA	excitation-autoionization
EBIT	electron beam ion traps
EBS	eikonal Born series

ECP	effective core potential
ECS	exterior complex scaling
EEDF	electron energy distribution functions
EOM	equation of motion
EPR	Einstein–Podolsky–Rosen
ESM	elastic scattering model
ESR	experimental storage ring
EUV	extreme ultraviolet
EW-CRDS	evanescent-wave CRDS
EXAFS	extended X-ray absorption fine structure

F

FBA	first Born approximation
FCPC	full-core plus correlation
FEL	free-electron lasers
FFT	fast Fourier transform
FID	free induction decay
FIR	far-infrared
FM	frequency modulation
FOTOS	first-order theory for oscillator strengths
FS	fine-structure
FT	Fourier transform
FTIR	Fourier transform infrared spectroscopy
FTMS	Fourier transform mass spectrometry
FTS	Fourier transform spectroscopy
FUSE	far ultraviolet spectroscopic explorer
FUV	far ultraviolet
FVP	Fermi virtual photon
FWHM	full width at half maximum

G

GBT	generalized Brillouin's Theorem
GFA	Green's function approach
GGA	generalized gradient approximation
GHZ	Greenberger, Horne, Zeilinger
GI	gauge invariant
GIB	guided ion beam
GOME	global ozone monitoring experiment
GOS	generalized oscillator strength
GPE	Gross–Pitaevskii equation
GRPAE	generalized random phase approximationwith exchange

H

HEDP	high energy-density physics
HF	Hartree–Fock equations
HF	Hellman–Feynman
HG	harmonic generation
HOM	Hong–Ou–Mandel
HREELS	high resolution electron energy loss spectroscopy
HRTOF	H-atom Rydberg time-of-flight
HUM	Hylleraas–Undheim–MacDonald theorem

I

IC	intermediate coupling
ICF	inertial confinement fusion
ICOS	integrated cavity output spectroscopy
ICSLS	international conference on spectral line shapes
IERM	intermediate energy *R*-matrix
IPCC	intergovernmental panel on climate change
IPES	inverse photoemission spectroscopy
IPIR	independent-processes and isolated-resonance
IPM	independent particle model
IPP	impact parameter picture
IR	irreducible representations
IR	infrared
IRI	international reference ionosphere
IRREP	irreducable representation
ISO	infrared space observatory

J

JB	Jeffrey–Born

K

KS	Kohn–Sham
KTA	potassium titanyl arsenate
KTP	potassium titanyl phosphate

L

LA	Lie algebras
LA	linear algebraic
L-CETS	locked cavity enhanced transmission spectroscopy
LDA	local density approximation
LEED	low energy electron diffraction
LER	laser electric resonance
LG	Lie groups
LHC	left-hand circular
LHV	local hidden variable
LIF	laser-induced-fluorescence
LIGO	laser interferometer gravitational-wave observatory
LISA	laser interferometer space antenna
LL	Landau–Lifshitz
LM	Levenberg–Marquardt
LMR	laser magnetic resonance
LPT	laser photodetachment threshold
LRL	Laplace–Runge–Lenz
LTE	local thermodynamic equilibrium
LYP	Lee, Yang, and Parr
LZ	Landau–Zener

M

MBE	molecular beam epitaxy
MCP	microchannel plate
MDAL	minimum detectable absorption loss
MBPT	many-body perturbation theory
MCDHF	multiconfigurational Dirac–Hartree–Fock
MCHF	multiconfiguration Hartree–Fock
MCSCF	multiconfigurational self-consistent field
MEMS	microelectromechanical systems
MFP	mean free path
MIGO	matter–wave interferometric gravitational-wave observatory
MIM	metal-insulator-metal
MKSA	meters, kilograms, seconds, and amperes
MM	Massey–Mohr
MMCDF	multichannel multiconfiguration Dirac–Fock
MO	molecular orbital
MOPA	master oscillator power amplifier
MOT	magneto-optical trap
MOX	molecular orbital X-radiation
MP2	second order Møller–Plesset perturbation theory
MP3	third order Møller–Plesset perturbation theory
MPI	multiphoton ionization
MQDT	multichannel quantum defect theory
MR	multireference
MR-SDCI	multireference singles/doubles configuration interaction
MUV	middle ultraviolet

N

NAR	nonadiabatic region
NEP	noise-equivalent power
NEXAFS	near-edge X-ray absorption fine structure
NDIR	non-dispersive infrared
NFS	nonfine-structure
NICE-OHMS	noise-immune, cavity-enhanced optical heterodyne molecular spectroscopy
NIM	normal incidence monochromator
NIST	National Institute of Standards and Technology
NMR	nuclear magnetic resonance systems
NNS	nearest-neighbor energy level spacings
NR	nonrelativistic
NRQED	NR quantum electrodynamics

O

OAO-2	orbiting astronomical observatory
OB	ordinary Bremsstrahlung
OBE	optical Bloch equations
OBK	Oppenheimer–Brinkman–Kramers
OCP	one-component plasma
OHCE	one-and-a-half centered expansion
OMI	ozone monitoring instrument

P

P-CRDS	pulsed-cavity ringdown spectroscopy
PADDS	angular distribution by Doppler spectroscopy
PAH	polycyclic aromatic hydrocarbon
PBS	polarizing beam splitters
PCDW	projectile continuum distorted wave approximation
PDM	phase diffusion model
PEC	potential energy curves
PES	photoelectron spectroscopy
PES	potential energy surface
PH/HP	particle–hole/hole–particle
PI	photoionization
PID	particle identification
PIMC	path-integral Monte Carlo
PMT	photomultiplier tubes
PNC	parity nonconservation
PPT	positive partial transposes
PR	polarization radiation
PSD	postion senitive detectors
PSS	perturbed stationary state
PT	perturbation theory
PWBA	plane wave Born approximation
PZT	piezo-electric transducer

Q

QCD	quantum chromodynamics
QED	quantum electrodynamics
QIP	quantum information processing
QKD	quantum key distribution
QMC	quantum Monte Carlo
QND	quantum nondemolition
OPO	optical parametric oscillator
QS	quasistatic
QSS	quasi-steady state

R

RATIP	relativistic atomic transition and ionization properties
RDC	ring-down cavity
READI	resonant excitation auto-double ionization
REC	radiative electron capture
REDA	resonant excitation double autoionization
REMPI	resonance-enhanced multiphoton ionization
RES	rotational energy surface
RHC	right-hand circular
RHIC	relativistic heavy ion collider

RIMS	recoil-ion momentum spectroscopy
RMI	relativistic mass increase
RMPS	R-matrix with pseudostates
RNA	Raman–Nath approximation
RPA	random-phase approximation
RPA	retarding potential analyzer
RPAE	random phase approximation with exchange
RR	radiative recombination
RRKM	Rampsberger–Rice–Karplus–Marcus
RSE	radial Schrödinger equation
RSPT	Rayleigh–Schrödinger perturbation theory
RT	Ramsauer–Townsend
RTE	resonant transfer and excitation
RWA	rotating wave approximation

S

SA-MCSCF	state averaged multiconfiguration self-consistent field
SACM	statistical adiabatic channel model
SBS	stimulated Brillouin scattering
SCA	semiclassical approximation
SCF	self-consistent field
SCIAMACHY	scanning imaging absorption spectrometer for atmospheric chartography
SD	spin-dependent
SD	single and double
SDS	singly differential cross section
SDTQ	single, double, triple, quadruple
SE	Schrödinger equation
SEP	stimulated emission pump
SEPE	simultaneous electron photon excitation
SEXAFS	surface extended X-ray absorption fine structure
SF	space-fixed
SI	spin-independent
SIAM	Society for Industrial and Applied Mathematics
SM	submillimeter
SM-FIR	submillimeter far-infrared
SMS	specific mass shift
SOHO	solar and heliospheric observatory
SP	stationary phase
SPA	stationary phase approximations
SQL	standard quantum limit
SQUID	superconducting quantum interference detector
SR	synchrotron radiation
SRS	stimulated Raman scattering
SS	strong-short
STIRAP	stimulated Raman adiabatic passage
STO	Slater type orbital
STP	standard temperature and pressure

T

TCDW	target continuum distorted wave
TDCS	triply differential cross section
TDHF	time-dependent Hartree–Fock
TDS	thermal desorption spectroscopy
TDSE	time dependent Schrödinger equation
TEA	transverse-excitation-atmospheric-pressure
TF	toroidal field
TOF	time-of-flight
TOP	time orbiting potential
TPA	two-photon absorption
TSR	test storage ring
TuFIR	tunable far-infrared

U

UGA	unitary group approach
UHF	unrestricted Hartree–Fock
UPS	ultraviolet photoelectron spectroscopy
UV	ultraviolet
UV-VIS	ultraviolet-visible

V

VASP	Vienna ab-initio simulation package
VCSEL	vertical-cavity surface-emitting laser
VECSEL	vertical external cavity surface-emitting laser
VES	vibrational energy surfaces
VUV	vacuum ultraviolet

W

WCG	Wigner–Clebsch–Gordan
WDM	warm dense matter
WKB	Wentzel, Kramers, Brillouin
WL	weak-long
WMAP	Wilkinson microwave anisotropy probe
WPMD	wavepacket molecular dynamics

X

XPS	X-ray photoelectron spectroscopy

Y

YAG	Yttrium Aluminum Garnet

1. Units and Constants

The currently accepted values for the physical constants are listed in Table 1.1, based on the 2002 CODATA (Committee on Data for Science and Technology) recommendations [1.1]. The quoted values are based on all data available through 31 December 2002, and replace the earlier 1998 CODATA set. Because the uncertainties are correlated, the correlation matrix, given in Table 1.2, must be used in calculating uncertainties for any quantities derived from those tabulated [1.1].

1.1 Electromagnetic Units

The standard electromagnetic units adopted by most scientific journals and elementary texts belong to the *système international* (SI) or rationalized MKSA (meters, kilograms, seconds, and amperes) units. However, many authors working with microscopic phenomena prefer *Gaussian* units, and theoretical physicists often use *Heaviside–Lorentz* (H–L) units. In this Handbook, SI units are used together with *atomic units*. The current section is meant as a reference relating these different systems.

The relations among different sets of units are not simple conversions since the same symbol in different systems can have different physical dimensions. To clarify the meanings of the units, we summarize basic electromagnetic relations for SI, Gaussian, and H–L systems below.

The Coulomb law for the magnitude F of the force acting on each of two static charges q and Q separated by a distance r in a homogeneous medium of permittivity ϵ can be written as

$$F = \frac{1}{4\pi\epsilon}\frac{qQ}{r^2}\,, \tag{1.1}$$

where in a vacuum, ϵ is

$$\epsilon_0 = \begin{cases} \left(\mu_0 c^2\right)^{-1}\,, & \text{SI} \\ (4\pi)^{-1}\,, & \text{Gaussian} \\ 1, & \text{H–L} \end{cases}\,, \tag{1.2}$$

with the closely related permeability of vacuum given by

$$\mu_0 = \begin{cases} 4\pi\times 10^{-7}\,\text{N/A}^2\,, & \text{SI} \\ 4\pi\,, & \text{Gaussian} \\ 1, & \text{H–L} \end{cases}\,. \tag{1.3}$$

(We deviate here from *Jackson* [1.2] who takes $\epsilon_0 = \mu_0 = 1$ in Gaussian units and must introduce additional constants to relate the units. The physically important quantities are the *relative values* $\epsilon_r \equiv \epsilon/\epsilon_0$ and $\mu_r \equiv \mu/\mu_0$, which in traditional Gaussian-unit notation are written without the r subscript.)

Note that ϵ_0 and μ_0 are dimensionless in H–L and Gaussian units, but not in the SI units. Current or electric charge is an independent quantity in the MKSA system but can be expressed in purely mechanical dimensions in the H–L and Gaussian systems. Thus, in Gaussian units, 1 statcoulomb = 1 dyne$^{(1/2)}$ cm, but in SI, even though the ampere is *defined* in terms of the attractive force between thin parallel wires carrying equal currents, there is no mechanical equivalent for the ampere or the coulomb. To establish such an equivalence, one can supplement the SI units by assigning a dimensionless number to ϵ_0 or to μ_0. Gaussian and H–L units arise from two different assignments. The result of assigning a number to ϵ_0 is analogous to the relation $1\,\text{s} = 3\times 10^8$ m established between time and distance units if one sets the speed of light $c = 1$, a convention often used in conjunction with H–L units. (Note that for simplicity, the pure number

Table 1.1 Table of physical constants. Uncertainties are given in parentheses

Quantity	Symbol	Value	Units
Speed of light in vacuum	c	2.997 924 58	10^8 m s^{-1}
Gravitational constant	G	6.6742(10)	10^{-11} m^3 kg^{-1} s^{-2}
Planck constant	h	6.626 0693(11)	10^{-34} J s
	$\hbar = h/2\pi$	1.054 571 68(18)	10^{-34} J s
Elementary charge	e	1.602 176 53(14)	10^{-19} C
		4.803 204 40(42)	10^{-10} esu
Inverse fine structure constant $[4\pi\epsilon_0]\hbar c/e^2$	α^{-1}	137.035 999 11(46)	
Magnetic flux quantum $h/2e$	Φ_0	2.067 833 72(18)	10^{-15} Wb
Atomic mass constant $\frac{1}{12}m\left(^{12}\mathrm{C}\right) = 1$ u	m_u	1.660 538 86(28)	10^{-27} kg
	$m_\mathrm{u}c^2$	931.494 043(80)	MeV
Electron mass	m_e	9.109 3826(16)	10^{-31} kg
		5.485 799 0945(24)	10^{-4} u
Muon mass	m_μ	0.113 428 9264(30)	u
Proton mass	m_p	1.007 276 466 88(13)	u
Neutron mass	m_n	1.008 664 915 60(55)	u
Deuteron mass	m_d	2.013 553 212 70(35)	u
α-particle mass	m_α	4.001 506 179 149(56)	u
Rydberg constant $m_\mathrm{e}c\alpha^2 2h$	R_∞	1.097 373 156 8525(73)	10^7 m^{-1}
	$R_\infty c$	3.289 841 960 360(22)	10^{15} Hz
	$R_\infty hc$	13.605 692 3(1 2)	eV
		2.179 872 09(37)	10^{-18} J
Bohr radius $\alpha/4\pi R_\infty$	a_0	0.529 177 2108(18)	10^{-10} m
Hartree energy $e^2/[4\pi\epsilon_0]a_0 = 2R_\infty hc$	E_h	27.211 3845(23)	eV
	E_h/h	6.579 683 920 721(44)	10^{15} Hz
	E_h/hc	2.194 746 313 705(15)	10^7 m^{-1}
Compton wavelength αa_0	$\bar{\lambda}_\mathrm{C} = \lambda_\mathrm{C}/2\pi$	3.861 592 678(26)	10^{-13} m
Classical electron radius $\alpha^2 a_0$	r_e	2.817 940 325(28)	10^{-15} m
Thomson cross section $8\pi r_\mathrm{e}^2/3$	σ_e	0.665 245 873(13)	10^{-28} m^2
Bohr magneton $[c]e\hbar/2m_\mathrm{e}c$	μ_B	9.274 009 49(80)	10^{-24} J T^{-1}
		5.788 381 804(39)	10^{-5} eV T^{-1}
Electron magnetic moment	$\mu_\mathrm{e}/\mu_\mathrm{B}$	− 1.001 159 652 1859(38)	
Muon magnetic moment	μ_μ/μ_B	− 4.841 970 45(13)	10^{-3}
Proton magnetic moment	$\mu_\mathrm{p}/\mu_\mathrm{B}$	1.521 032 206(15)	10^{-3}
Neutron magnetic moment	$\mu_\mathrm{n}/\mu_\mathrm{B}$	− 1.041 875 63(25)	10^{-3}
Deuteron magnetic moment	$\mu_\mathrm{d}/\mu_\mathrm{B}$	0.466 975 4567(50)	10^{-3}
Electron g factor $-2(1+a_\mathrm{e})$	g_e	− 2.002 319 304 3718(75)	
Muon g factor $-2(1+a_\mu)$	g_μ	− 2.002 331 8396(12)	
Proton gyromagnetic ratio $2\mu_\mathrm{p}/\hbar$	γ_p	2.675 222 05(23)	10^8 s^{-1}T^{-1}
Avogadro constant	N_A	6.022 1415(10)	10^{23} mol^{-1}
Faraday constant $N_\mathrm{A}e$	F	9.648 533 83(83)	10^4 C mol^{-1}
Boltzmann constant R/N_A	k_B	1.380 6505(24)	10^{-23} J K^{-1}
		8.617 343(15)	10^{-5} eV K^{-1}
	$k_\mathrm{B}/E_\mathrm{h}$	3.166 8153(55)	10^{-6} K^{-1}
Molar gas constant	R	8.314 472(15)	J mol^{-1} K^{-1}
Molar volume (ideal gas) RT/P			
$T = 273.15$ K, $P = 101.325$ kPa	V_m	0.022 413 996(39)	m^3 mol^{-1}
$T = 273.15$ K, $P = 100$ kPa	V_m	0.022 710 981(40)	m^3 mol^{-1}

Table 1.1 Table of physical constants. Uncertainties are given in parentheses, cont.

Quantity	Symbol	Value	Units
Stefan–Boltzmann constant $\pi^2 k_B^4/(60\hbar^3 c^2)$	σ	5.670 400(40)	10^{-8} W m^{-2} K^{-4}
First radiation constant $2\pi hc^2$	c_1	3.741 771 38(64)	10^{-16} W m^2
Second radiation constant hc/k_B	c_2	0.014 387 752(25)	m K
Wien displacement law constant $\lambda_{max} T = \frac{c_2}{4.965\,114\,231...}$	b	2.897 7685(51)	10^{-3} m K

Table 1.2 The correlation coefficients of a selected group of constants based on the 2002 CODATA [1.1]

	α	h	e	m_e	N_A	m_e/m_p	F
α	–	–	–	–	–	–	–
h	0.010	–	–	–	–	–	–
e	0.029	1.000	–	–	–	–	–
m_e	−0.029	0.999	0.998	–	–	–	–
N_A	0.029	−0.999	−0.998	−1.000	–	–	–
m_e/m_p	−0.249	−0.002	−0.007	0.007	−0.007	–	–
F	0.087	−0.995	−0.993	−0.998	0.998	−0.022	–

2.997 924 58, equal numerically to the defined speed of light in vacuum in units of 10^8 m/s, is represented by $\dot{3}$.) Thus, although within the Gaussian system, where the assignment $4\pi\epsilon_0 = 1$ is made, it is justified to assert that 1 coulomb *equals* $\dot{3}\times 10^9$ statcoulombs, this is not true in pure SI, where there is no equivalent mechanical unit for charge.

Maxwell's macroscopic equations can be written as

$$\begin{aligned} \lambda\nabla\cdot\boldsymbol{D} &= \rho\,, \\ \lambda c'\nabla\times\boldsymbol{H} - \lambda\frac{\partial\boldsymbol{D}}{\partial t} &= \boldsymbol{j}\,, \\ c'\nabla\times\boldsymbol{E} + \frac{\partial\boldsymbol{B}}{\partial t} &= 0\,, \\ \nabla\cdot\boldsymbol{B} &= 0\,, \end{aligned} \tag{1.4}$$

with the macroscopic field variables related to the polarizations $\boldsymbol{P}$ and $\boldsymbol{M}$ by

$$\begin{aligned} \lambda\boldsymbol{D} &= \epsilon_0\boldsymbol{E} + \boldsymbol{P} = \epsilon\boldsymbol{E} \\ \lambda\boldsymbol{H} &= \mu_0^{-1}\boldsymbol{B} - \boldsymbol{M} = \boldsymbol{B}/\mu \end{aligned} \tag{1.5}$$

(the last equalities for $\boldsymbol{D}$ and $\boldsymbol{H}$ hold only for homogeneous media) and

$$\lambda = \begin{cases} 1, \text{SI} \\ \epsilon_0 = \mu_0^{-1}, \text{Gaussian or H–L}\,, \end{cases} \tag{1.6}$$

$$c' = \begin{cases} 1, \text{SI} \\ c, \text{Gaussian or H–L}\,. \end{cases} \tag{1.7}$$

In Gaussian or H–L units, the fields $\boldsymbol{E}$, $\boldsymbol{B}$, $\boldsymbol{D}$, $\boldsymbol{H}$, and polarizations (dipole moments per unit volume) $\boldsymbol{P}$, $\boldsymbol{M}$ all have the same dimensions, whereas in SI units the microscopic fields $\boldsymbol{E}$ and $\boldsymbol{B}$ have dimensions that are generally distinct from each other as well as from $\boldsymbol{P}$ (or $\boldsymbol{D}$) and $\boldsymbol{M}$ (or $\boldsymbol{H}$), respectively. In all three unit systems, the dimensionless ratio ϵ/ϵ_0 is called the dielectric constant (or relative permittivity) of the medium, and the (dimensionless) fine-structure constant is

$$\alpha = \frac{1}{4\pi\epsilon_0}\frac{e^2}{\hbar c}\,, \tag{1.8}$$

with a numerical value $\alpha^{-1} = 137.035\,999\,11\,(46)$.

In atomic units (Sect. 1.3), the factor $e^2/(4\pi\epsilon_0)$, the electron mass m_e, and $\hbar$, Planck's constant divided by 2π, are all equal to 1. In Gaussian and H–L systems, these conditions determine a numerical value for all electro-mechanical units. Thus in Gaussian units, the electronic charge is $e = 1$, whereas in H–L atomic units $e = \sqrt{4\pi}$. In the SI system, on the other hand, the three conditions $e^2/(4\pi\epsilon_0) = m_e = \hbar = 1$ determine numerical values for mechanical units but not for electromagnetic ones. A complete determination of values requires that ϵ_0 also be assigned a value. The choice most consistent with previous work is to take $e = 1 = 4\pi\epsilon_0$. This choice is made here.

Since a volt is a joule/coulomb and a statvolt is an erg/statcoulomb, 1 volt corresponds to (but is not generally *equal* to, since the physical dimensions may

Table 1.3 Conversion factors for various physical quantities

Quantity	SI units	Gaussian units	Natural H–L units: $\hbar = c = \epsilon_0 = 1$
Length	1 m	$= 10^2$ cm	$= 1$ m
Mass	1 kg	$= 10^3$ g	$\leftrightarrow 2.842\,788\,82(49) \times 10^{42}\ \mathrm{m}^{-1}$
Time	1 s	$= 1$ s	$\leftrightarrow \dot{3} \times 10^8$m
Velocity	$1\ \mathrm{m\,s^{-1}}$	$= 10^2\ \mathrm{cm\,s^{-1}}$	$\leftrightarrow \dot{3}^{-1} \times 10^{-8}$
Energy	$1\ \mathrm{J} = 1\ \mathrm{kg\,m^2\,s^{-2}}$	$= 10^7$ erg	$\leftrightarrow 3.163\,029\,14(54) \times 10^{25}\ \mathrm{m}^{-1}$
Action	1 J s	$= 10^7$ erg s	$\leftrightarrow 0.948\,252\,28(16) \times 10^{34}$
Force	$1\ \mathrm{N} = 1\ \mathrm{J\,m^{-1}}$	$= 10^5$ dyne	$\leftrightarrow 3.163\,029\,14(54) \times 10^{25}\ \mathrm{m}^{-2}$
Power	$1\ \mathrm{W} = 1\ \mathrm{J\,s^{-1}}$	$= 10^7\ \mathrm{erg\,s^{-1}}$	$\leftrightarrow 1.055\,072\,95(18) \times 10^{17}\ \mathrm{m}^{-2}$
Intensity	$1\ \mathrm{W\,m^{-2}}$	$= 10^3\ \mathrm{erg\,cm^{-2}}$	$\leftrightarrow 1.055\,072\,95(18) \times 10^{17}\ \mathrm{m}^{-4}$
Charge	$1\ \mathrm{C} = 1\ \mathrm{A\,s}$	$\leftrightarrow \dot{3} \times 10^9$ statcoul	$\leftrightarrow 1.890\,067\,14(16) \times 10^{18}$
Potential	$1\ \mathrm{V} = 1\ \mathrm{J\,C^{-1}}$	$\leftrightarrow (\dot{3} \times 10^2)^{-1}$ statvolt	$\leftrightarrow 1.673\,500\,94(14) \times 10^{7}\ \mathrm{m}^{-1}$
Electric field	$1\ \mathrm{V\,m^{-1}} = 1\ \mathrm{N\,C^{-1}}$	$\leftrightarrow (\dot{3} \times 10^4)^{-1}$ statvolt cm^{-1}	$\leftrightarrow 1.673\,500\,94(14) \times 10^{7}\ \mathrm{m}^{-2}$
Magnetic field	$1\ \mathrm{T} = 1\ \mathrm{N\,A^{-1}\,m^{-1}}$	$\leftrightarrow 10^4$ gauss	$\leftrightarrow 5.017\,029\,61(43) \times 10^{15}\ \mathrm{m}^{-2}$

differ)

$$\frac{10^7\ \text{erg}}{\dot{3} \times 10^9\ \text{statcoulomb}} = \frac{1}{\dot{3} \times 10^2}\ \text{statvolt}\,. \tag{1.9}$$

In Gaussian units, the unit of magnetic field, namely the Gauss ($\boldsymbol{B}$) or Oersted ($\boldsymbol{H}$) has the same physical size and dimension as the unit of electric field, namely the statvolt/cm, which in turn corresponds to an SI field of $\dot{3} \times 10^4$ V/m. However, the tesla ($1\ \mathrm{T} = 1\ \mathrm{weber/m^2}$), the SI unit of magnetic field $\boldsymbol{B}$ (older texts refer to $\boldsymbol{B}$ as the *magnetic induction*), has the physical dimensions of $\mathrm{V\,s/m^2}$. To find the correspondence to Gaussian units, one must multiply by the speed of light c:

$$1\ \mathrm{T}\,c = \dot{3} \times 10^8\ \mathrm{V/m}\,, \tag{1.10}$$

which corresponds to 10^4 statvolt/cm and hence to 10^4 gauss.

Tables 1.3 and 1.4 related basic mechanical and electromagnetic quantities in the different unit systems. Caution is required both because the same symbol often stands for quantities of different physical dimensions in different systems of units, and because factors of 2π sometimes enter frequencies, depending on whether the units are cycles/s (Hz) or radians/s. The double-headed arrows ($\leftrightarrow$) indicate a correspondence between quantities whose dimensions are not necessarily equal. Thus for example, the force on an electron due to a Gaussian electric field of 1 statvolt/cm is the same as due to an SI electric field of $\dot{3} \times 10^4$ V/m. The correspondences between Gaussian and SI electrostatic quantities become equalities if and only if $4\pi\epsilon_0 = 1$. Thus they are equalities within the Gaussian system but not within the less constrained SI scheme. The SI and Gaussian units of magnetic field have different dimensions unless both ϵ_0 and c are set equal to dimensionless numbers. *Natural H–L units* can be considered SI units supplemented by the conditions $\epsilon_0 = c = \hbar = 1$. They are listed here in units of meters, although eV are also often used: $1\ \mathrm{eV} = 5.067\,731\,04(43) \times 10^6\ \mathrm{m}^{-1} \times \hbar c$. The correspondences may be considered equalities within the natural H–L system but not within SI. Note that the electronic charge in the natural H–L system has the magnitude $e = \sqrt{4\pi\alpha}$. More electromagnetic conversions can be found in *Jackson* [1.2]. The data here are based on the 2002 adjustment by *Mohr* and *Taylor* [1.1].

A few additional energy conversion factors are

$$\begin{aligned}
1\ \mathrm{eV} &= 1.602\,176\,53\,(14) \times 10^{-19}\ \mathrm{J} \\
&= 2.417\,989\,40\,(21) \times 10^{14}\ \mathrm{Hz} \times h \\
&= 8065.544\,45\,(69)\ \mathrm{cm}^{-1} \times hc \\
&= 3.674\,932\,45\,(31) \times 10^{-2} E_\mathrm{h} \\
&= 1.160\,4505\,(20) \times 10^4\ \mathrm{K} \times k_\mathrm{B} \\
&= 96.485\,3383\,(83)\ \mathrm{kJ\,mol^{-1}}
\end{aligned}$$

The basic unit of temperature, the kelvin, is equivalent to about $0.7\ \mathrm{cm}^{-1}$, i. e., the value of the Boltzmann constant k_B expressed in wavenumber units per kelvin is $0.695\,0356(12)\ \mathrm{cm^{-1}K^{-1}}$. Since K is the internationally accepted symbol for the Kelvin [1.3], this suggests that the use of the letter K as a symbol for $1\ \mathrm{cm}^{-1}$ (1 Kayser) should be discontinued.

1.2 Atomic Units

Atomic and molecular calculations based on the Schrödinger equation are most conveniently done in atomic units (a.u.), and then the final result converted to the correct SI units as listed in Table 1.4. In atomic units, $\hbar = m_e = e = 4\pi\epsilon_0 = 1$. The atomic units of length, velocity, time, and energy are then

$$\text{length: } a_0 = \frac{4\pi\epsilon_0\hbar^2}{m_e e^2} = \frac{\hbar}{\alpha m_e c} ,$$

$$\text{velocity: } v_B = \frac{e^2}{4\pi\epsilon_0\hbar} = \alpha c ,$$

$$\text{time: } \tau_0 = \frac{16\pi^2\epsilon_0^2\hbar^3}{m_e e^4} = \frac{\hbar}{\alpha^2 m_e c^2} ,$$

$$\text{energy: } E_h = \frac{e^2}{4\pi\epsilon_0 a_0} = \alpha^2 m_e c^2 ,$$

where, from the definition (1.8), the numerical value of c is $\alpha^{-1} = 137.035\,999\,11(46)$ a.u. For the lowest 1s state of hydrogen (with infinite nuclear mass), a_0 is the Bohr radius, v_B is the Bohr velocity, $2\pi\tau_0$ is the time to complete a Bohr orbit, and E_h (the Hartree energy) is twice the ionization energy. To include the effects of a finite nuclear mass M, one must replace the electron mass m_e by the reduced electron mass $\mu = m_e M/(M + m_e)$.

Atomic energies are often expressed in units of the Rydberg (Ry). The Rydberg for an atom having nuclear mass M is

$$1\ \text{Ry} = R_M = \frac{\mu}{m_e} R_\infty = M(M + m_e)^{-1}\, R_\infty , \tag{1.11}$$

Table 1.4 Physical quantities in atomic units with $\hbar = e = m_e = 4\pi\epsilon_0 = 1$, and $\alpha^{-1} = 137.035\,999\,11(46)$

Quantity	Unit	Value
Length	a_0	$0.529\,177\,2108(18) \times 10^{-10}$ m
Mass	m_e	$0.910\,938\,26(16) \times 10^{-30}$ kg
Time	$\hbar/E_h$	$2.418\,884\,326\,505(16) \times 10^{-17}$ s
Velocity	$v_B \equiv \alpha c$	$2.187\,691\,2633(73) \times 10^{6}$ m s^{-1}
Energy	E_h	$4.359\,744\,17(75) \times 10^{-18}$ J
Action	$\hbar$	$1.054\,571\,68(18) \times 10^{-34}$ J s
Force	E_h/a_0	$0.823\,872\,25(14) \times 10^{-7}$ N
Power	$E_h^2/\hbar$	$0.180\,237\,811(31)$ W
Intensity	$E_h^2/\hbar a_0^2$	$64.364\,091(11) \times 10^{18}$ W m^{-2}
Charge	e	$1.602\,176\,53(14) \times 10^{-19}$ C
Electric potential	E_h/e	$27.211\,3845(23)$ V
Electric field	$E_h/ea_0 = \alpha\hbar c/ea_0^2$	$0.514\,220\,642(44) \times 10^{12}$ V m^{-1}
Magnetic flux density	$E_h/ea_0\alpha c$	$2.350\,517\,42(20) \times 10^{5}$ T

with

$$R_\infty = \frac{m_e c\alpha^2}{2h} = 10\,973\,731.568\,525\,(73)\ \text{m}^{-1} . \tag{1.12}$$

The Rydberg constant R_∞ is thus the limiting value of R_M for infinite nuclear mass, and hcR_∞ is $\frac{1}{2}$ a.u., which is equivalent to 13.605 6923(12) eV.

The energy equivalent of the electron mass, $m_e c^2$, is 0.510 998 918(44) MeV. This energy is a natural unit for relativistic atomic theory. For example, for inner-shell energies in the heaviest elements, the binding energy of the 1s electron in hydrogenic Lr ($Z = 103$) is $0.338\,42\, m_e c^2$.

1.3 Mathematical Constants

A selection of the most important mathematical constants is listed in Table 1.5. More extensive tabulations and formulas can be found in the standard mathematical works [1.4, 5]

1.3.1 Series Summation Formula

The Riemann zeta function defined by $\zeta(n) = \sum_{i=1}^{\infty} i^{-n}$ (Table 1.5) is particularly useful in summing slowly convergent series of the form

$$S = \sum_{i=1}^{\infty} T_i . \tag{1.13}$$

For example, suppose that the series

$$T_i = t_2 i^{-2} + t_3 i^{-3} + \cdots \tag{1.14}$$

for the individual terms in S is rapidly convergent for $i > N$, where N is some suitably large integer.

Then

$$S=\sum_{i=1}^{N} T_i + t_2\zeta^N(2) + t_3\zeta^N(3) + \cdots \,, \tag{1.15}$$

where $\zeta^N(n) = \zeta(n) - \sum_{i=1}^{N} i^{-n}$ is the zeta function with the first N terms subtracted. For N sufficiently large, only the first few t_j coefficients need be known, and they can be adequately estimated by solving the system of equations

$$T_N = t_2 N^{-2} + t_3 N^{-3} + \cdots + t_{k+2} N^{-k-2} \,, \tag{1.16a}$$

$$T_{N-1} = t_2(N-1)^{-2} + t_3(N-1)^{-3} + \cdots + t_{k+2}(N-1)^{-k-2} \,, \tag{1.16b}$$

$$\vdots$$

$$T_{N-k} = t_2(N-k)^{-2} + t_3(N-k)^{-3} + \cdots + t_{k+2}(N-k)^{-k-2} \,, \tag{1.16c}$$

where $k+1 \le N$ is the number of terms retained in (1.14).

Table 1.5 Values of e, π, Euler's constant γ, and the Riemann zeta function $\zeta(n)$

Constant	Value
e	2.718 281 828 459 045 235 360 287 471 352 66
π	3.141 592 653 589 793 238 462 643 383 279 50
$\pi^{1/2}$	1.772 453 850 905 516 027 298 167 483 341 14
γ	0.577 215 664 901 532 860 606 512 090 082 40
$\zeta(2)$	1.644 934 066 848 226 436 472 415 166 646 02
$\zeta(3)$	1.202 056 903 159 594 285 399 738 161 511 45
$\zeta(4)$	1.082 323 233 711 138 191 516 003 696 541 16
$\zeta(5)$	1.036 927 755 143 369 926 331 365 486 457 03
$\zeta(6)$	1.017 343 061 984 449 139 714 517 929 790 92
$\zeta(7)$	1.008 349 277 381 922 826 839 797 549 849 80
$\zeta(8)$	1.004 077 356 197 944 339 378 685 238 508 65
$\zeta(9)$	1.002 008 392 826 082 214 417 852 769 232 41
$\zeta(10)$	1.000 994 575 127 818 085 337 145 958 900 31

References

1.1 P.J. Mohr, B.N. Taylor: Rev. Mod. Phys. **77**, 1 (2005); see also www.physicstoday.org/guide/fundcon.html; all of the values, as well as the correlation coefficients between any two constants, are available online in a searchable database provided by NIST's fundamental constants data center. The internet address is http://physics.nist.gov/constants

1.2 J.D. Jackson: *Classical Electrodynamics*, 3rd edn. (Wiley, New York 1999)

1.3 B.N. Taylor (Ed.): *The International System of Units (SI)*, NIST Spec. Publ. 330 (U.S. Government Printing Office, Washington 2001) p. 7

1.4 M. Abramowitz, I.A. Stegun: *Handbook of Mathematical Functions* (Dover, New York 1965)

1.5 I.S. Gradshteyn, I.M. Ryzhik: *Table of Integrals, Series, and Products* (Academic, New York 1965)

Part A

Mathemat

Part A Mathematical Methods

2 **Angular Momentum Theory**
James D. Louck, Los Alamos, USA

3 **Group Theory for Atomic Shells**
Brian R. Judd, Baltimore, USA

4 **Dynamical Groups**
Josef Paldus, Waterloo, Canada

5 **Perturbation Theory**
Josef Paldus, Waterloo, Canada

6 **Second Quantization**
Brian R. Judd, Baltimore, USA

7 **Density Matrices**
Klaus Bartschat, Des Moines, USA

8 **Computational Techniques**
David R. Schultz, Oak Ridge, USA
Michael R. Strayer, Oak Ridge, USA

9 **Hydrogenic Wave Functions**
Robert N. Hill, Saint Paul, USA

2. Angular Momentum Theory

Angular momentum theory is presented from the viewpoint of the group *SU*(1) of unimodular unitary matrices of order two. This is the basic quantum mechanical rotation group for implementing the consequences of rotational symmetry into isolated complex physical systems, and gives the structure of the angular momentum multiplets of such systems. This entails the study of representation functions of *SU*(2), the Lie algebra of *SU*(2) and copies thereof, and the associated Wigner–Clebsch–Gordan coefficients, Racah coefficients, and 1*n*–*j* coefficients, with an almost boundless set of inter-relations, and presentations of the associated conceptual framework. The relationship to the rotation group in physical 3-space is given in detail. Formulas are often given in a compendium format with brief introductions on their physical and mathematical content. A special effort is made to inter-relate the material to the special functions of mathematics and to the combinatorial foundations of the subject.

Angular momentum theory in its quantum mechanical applications, which is the subject of this section, is the study of the group of 2×2 unitary unimodular matrices and its irreducible representations. It is the mathematics of implementing into physical theory the basic tenet that isolated physical systems are invariant to rotations of the system in physical 3-space, denoted $\boldsymbol{R}^3$, or, equivalently, to the orientation of a Cartesian reference system used to describe the system. That it is the group of 2×2 unimodular matrices that is basic in quantum theory in place of the more obvious group of 3×3 real, orthogonal matrices representing transformations of the coordinates of the constituent particles of the system, or of the reference frame, is a consequence of the Hilbert space structure of the state space of quantum systems and the impossibility of assigning overall phase factors to such states because measurements depend only on the absolute value of transition amplitudes.

The exact relationship between the group $SU(2)$ of 2×2 unimodular unitary matrices and the group $SO(3,\boldsymbol{R})$ of 3×3 real, proper, orthogonal matrices is an important one for keeping the quantum theory of angular momentum, with its numerous conventions and widespread applications across all fields of quantum physics, free of ambiguities. These notations and relations are fixed at the outset.

Presentation of a point in $\boldsymbol{R}^3$:

$$\begin{aligned} \boldsymbol{x} &= \mathrm{col}\,(x_1, x_2, x_3) && \text{column matrix}\,,\\ \boldsymbol{x}^T &= (x_1, x_2, x_3) && \text{row matrix}\,,\\ X &= \begin{pmatrix} x_3 & x_1 - \mathrm{i}x_2 \\ x_1 + \mathrm{i}x_2 & -x_3 \end{pmatrix} \\ &\qquad 2\times2 \text{ traceless Hermitian matrix}\,;\\ &\qquad \text{Cartan's representation}\,. \end{aligned}$$

A one-to-one correspondence between the set $\boldsymbol{R}^3$ of points in 3-space and the set $\boldsymbol{H}^2$ of 2×2 traceless Hermitian matrices is obtained from $x_i = \frac{1}{2}\,\mathrm{Tr}\,(\sigma_i X)$, where the σ_i denote the matrices (Pauli matrices)

$$\sigma_1 = \begin{pmatrix} 0 & 1 \\ 1 & 0 \end{pmatrix}, \quad \sigma_2 = \begin{pmatrix} 0 & -\mathrm{i} \\ \mathrm{i} & 0 \end{pmatrix}, \quad \sigma_3 = \begin{pmatrix} 1 & 0 \\ 0 & -1 \end{pmatrix}. \tag{2.1}$$

Mappings of $\boldsymbol{R}^3$ onto itself:

$$\begin{aligned} \boldsymbol{x} &\to \boldsymbol{x}' = R\boldsymbol{x}\,,\\ X &\to X' = UXU^{\dagger}\,, \end{aligned}$$

where † denotes Hermitian conjugation of a matrix or an operator.

Two-to-one homomorphism of $SU(2)$ onto $SO(3, \boldsymbol{R})$:

$$R_{ij} = R_{ij}(U) = \frac{1}{2}\mathrm{Tr}\left(\sigma_i U \sigma_j U^\dagger\right) , \tag{2.2}$$

$$\begin{pmatrix} \xi \\ \boldsymbol{x}' \end{pmatrix} = \begin{pmatrix} 1 & 0 & 0 & 0 \\ 0 & & & \\ 0 & & R(U) & \\ 0 & & & \end{pmatrix} = \begin{pmatrix} \xi \\ \boldsymbol{x} \end{pmatrix}$$

$$= A^\dagger (U \times U^*) A \begin{pmatrix} \xi \\ \boldsymbol{x} \end{pmatrix} , \tag{2.3}$$

where ξ is an indeterminate, A is the unitary matrix given by

$$A = \frac{1}{\sqrt{2}} \begin{pmatrix} 1 & 0 & 0 & 1 \\ 0 & 1 & -\mathrm{i} & 0 \\ 0 & 1 & \mathrm{i} & 0 \\ 1 & 0 & 0 & -1 \end{pmatrix} ,$$

$U \times U^*$ denotes the matrix direct product, and $*$ denotes complex conjugation. There is a simple unifying theme in almost all the applications. The basic mathematical notions that are implemented over and over again in various contexts are: group action on the underlying coordinates and momenta of the physical system and the corresponding group action in the associated Hilbert space of states; the determination of those subspaces that are mapped irreducibly onto themselves by the group action; the Lie algebra and its actions as derived from the group actions, and conversely; the construction of composite objects from elementary constituents, using the notion of tensor product space and Kronecker products of representations, which are the basic precepts in quantum theory for building complex systems from simpler ones; the reduction of the Kronecker product of irreducible representations into irreducibles with the associated Wigner–Clebsch–Gordan and Racah coefficients determining not only this reduction, but also having a dual role in the construction of the irreducible state spaces themselves; and, finally, the repetition of this process for many-particle systems with the attendant theory of $3n-j$ coefficients. The universality of this methodology may be attributed to being able, in favorable situations, to separate the particular consequences of physical law (e.g., the Coulomb force) from the implications of symmetry imposed on the system by our underlying conceptions of space and time. Empirical models based on symmetry that attempt to identify the more important ingredients underlying observed physical phenomena are also of great importance.

The group actions in complex systems are often modeled after the following examples for the actions of the groups $SO(3, \boldsymbol{R})$ and $SU(2)$ on functions defined over the 2-sphere $S^2 \subset \boldsymbol{R}^3$:

Hilbert space:

$$\boldsymbol{V} = \{f \,|\, f \text{ is a polynomial satisfying } \nabla^2 f(\boldsymbol{x}) = 0\} .$$

Inner or scalar product:

$$(f, f') = \int_{\text{unit sphere}} f^*(\boldsymbol{x}) f'(\boldsymbol{x})\, \mathrm{d}S ,$$

where $f(\boldsymbol{x}) = f(X)$ for $\boldsymbol{x}$ presented in the Cartan matrix form X.

Group actions:

$$(O_R f)(\boldsymbol{x}) = f\left(R^{-1}\boldsymbol{x}\right) , \qquad \text{each } f \in \boldsymbol{V} , \quad \text{each } \boldsymbol{x} \in \boldsymbol{R}^3 ,$$

$$(T_U f)(X) = f\left(U^\dagger X U\right) , \qquad \text{each } f \in \boldsymbol{V} , \quad \text{each } X \in \boldsymbol{H}^2 .$$

Operator properties:

- O_R is a unitary operator on $\boldsymbol{V}$; that is, $(O_R f, O_R f') = (f, f')$.
- T_U is a unitary operator on $\boldsymbol{V}$; that is, $(T_U f, T_U f') = (f, f')$.
- $R \to O_R$ is a unitary representation of $SO(3, \boldsymbol{R})$; that is, $O_{R_1} O_{R_2} = O_{R_1 R_2}$.
- $U \to T_U$ is a unitary representation of $SU(2)$; that is, $T_{U_1} T_{U_2} = T_{U_1 U_2}$.
- $O_{R(U)} = T_U = T_{-U}$ is an operator identity on the space $\boldsymbol{V}$.

One parameter subgroups:

$$U_j(t) = \exp(-\mathrm{i}t\sigma_j/2) , \quad t \in \boldsymbol{R} , \quad j = 1, 2, 3 ;$$

$$R_j(t) = R(U_j(t)) = \exp(-\mathrm{i}tM_j) , \quad t \in \boldsymbol{R} , \quad j = 1, 2, 3 ;$$

where

$$M_1 = \mathrm{i}\begin{pmatrix} 0 & 0 & 0 \\ 0 & 0 & -1 \\ 0 & 1 & 0 \end{pmatrix} , \quad M_2 = \mathrm{i}\begin{pmatrix} 0 & 0 & 1 \\ 0 & 0 & 0 \\ -1 & 0 & 0 \end{pmatrix} ,$$

$$M_3 = \mathrm{i}\begin{pmatrix} 0 & -1 & 0 \\ 1 & 0 & 0 \\ 0 & 0 & 0 \end{pmatrix} . \tag{2.4}$$

Infinitesimal generators:

$$
\begin{aligned}
L_j &= \mathrm{i}(\mathrm{d}O_{R_j(t)}/\mathrm{d}t)_{t=0}\,,\\
L_j &= \mathrm{i}(\mathrm{d}T_{U_j(t)}/\mathrm{d}t)_{t=0}\,,\\
(L_j f)(\boldsymbol{x}) &= -\mathrm{i}\left(x_k\frac{\partial}{\partial x_l} - x_l\frac{\partial}{\partial x_k}\right) f(\boldsymbol{x})\,,\\
&\qquad j, k, l \text{ cyclic in } 1, 2, 3\,.
\end{aligned}
\tag{2.5}
$$

Historically, the algebra of angular momentum came about through the quantum rule of replacing the linear momentum $\boldsymbol{p}$ of a classical point particle, which is located at position $\boldsymbol{r}$, by $\boldsymbol{p} \to -\mathrm{i}\hbar\nabla$, thus replacing the classical angular momentum $\boldsymbol{r}\times\boldsymbol{p}$ about the origin of a chosen Cartesian inertial system by the angular momentum operator:

$$
\boldsymbol{L} = -\mathrm{i}\boldsymbol{r}\times\nabla \quad (\text{in units of } \hbar)\,. \tag{2.6}
$$

The quantal angular momentum properties of this simple one-particle system are then to be inferred from the properties of these operators and their actions in the associated Hilbert space. This remains the method of introducing angular momentum theory in most textbooks because of its simplicity and historical roots. It also leads to focusing the developments of the theory on the algebra of operators in contrast to emphasizing the associated group transformations of the Hilbert space, although the two viewpoints are intimately linked, as illustrated above. Both perspectives will be presented here.

2.1 Orbital Angular Momentum

The model provided by orbital angular momentum operators is the paradigm for standardizing many of the conventions and relations used in more abstract and general treatments. These basic results for the orbital angular momentum operator $\boldsymbol{L} = -\mathrm{i}\boldsymbol{r}\times\nabla$ acting in the vector space $\boldsymbol{V}$ are given in this section both in Cartesian coordinates $\boldsymbol{x} = \mathrm{col}\,(x_1, x_2, x_3)$ and spherical polar coordinates:

$$
\begin{aligned}
&\boldsymbol{x} = (r\sin\theta\cos\phi,\, r\sin\theta\sin\phi,\, r\cos\theta)\,,\\
&0 \le r < \infty\,,\quad 0 \le \phi < 2\pi\,,\quad 0 \le \theta \le \pi\,.
\end{aligned}
$$

2.1.1 Cartesian Representation

Commutation relations:

Cartesian form:

$$
\begin{aligned}
&[L_1, L_2] = \mathrm{i}L_3\,,\quad [L_2, L_3] = \mathrm{i}L_1\,,\\
&[L_3, L_1] = \mathrm{i}L_2\,.
\end{aligned}
$$

Cartan form:

$$
\begin{aligned}
&[L_3, L_+] = L_+\,,\quad [L_3, L_-] = -L_-\,,\\
&[L_+, L_-] = 2L_3\,.
\end{aligned}
$$

Squared orbital angular momentum:

$$
\begin{aligned}
\boldsymbol{L}^2 &= L_1^2 + L_2^2 + L_3^2 = L_-L_+ + L_3(L_3+1)\\
&= L_+L_- + L_3(L_3-1)\\
&= -r^2\nabla^2 + (\boldsymbol{x}\cdot\nabla)^2 + (\boldsymbol{x}\cdot\nabla)\,.
\end{aligned}
$$

$\boldsymbol{L}^2$, L_3 form a complete set of commuting Hermitian operators in $\boldsymbol{V}$ with eigenfunctions

$$
\begin{aligned}
\mathcal{Y}_{lm}(\boldsymbol{x}) = &\left[\frac{2l+1}{4\pi}(l+m)!(l-m)!\right]^{\frac{1}{2}}\\
&\times\sum_k \frac{(-x_1 - \mathrm{i}x_2)^{k+m}(x_1 - \mathrm{i}x_2)^k x_3^{l-m-2k}}{2^{2k+m}(k+m)!k!(l-m-2k)!}\,,
\end{aligned}
$$

where $l = 0, 1, 2, \ldots,;\quad m = l, l-1, \ldots, -l$.

Homogeneous polynomial solutions of Laplace's equation:

$$
\begin{aligned}
&\mathcal{Y}_{lm}(\lambda\boldsymbol{x}) = \lambda^l\mathcal{Y}_{lm}(\boldsymbol{x})\,,\\
&(\boldsymbol{x}\cdot\nabla)\mathcal{Y}_{lm}(\boldsymbol{x}) = l\mathcal{Y}_{lm}(\boldsymbol{x})\,,\\
&\nabla^2\mathcal{Y}_{lm}(\boldsymbol{x}) = 0\,.
\end{aligned}
$$

Complex conjugate:

$$
\mathcal{Y}^*_{lm}(\boldsymbol{x}) = (-1)^m\mathcal{Y}_{l,-m}(\boldsymbol{x})\,.
$$

Action of angular momentum operators:

$$
\begin{aligned}
&L_\pm\mathcal{Y}_{lm}(\boldsymbol{x}) = [(l\mp m)(l\pm m+1)]^{\frac{1}{2}}\mathcal{Y}_{l,m\pm1}(\boldsymbol{x})\,,\\
&L_3\mathcal{Y}_{lm}(\boldsymbol{x}) = m\mathcal{Y}_{lm}(\boldsymbol{x})\,,\\
&\boldsymbol{L}^2\mathcal{Y}_{lm}(\boldsymbol{x}) = l(l+1)\mathcal{Y}_{lm}\,.
\end{aligned}
$$

Highest weight eigenfunction:

$$
\begin{aligned}
&L_+\mathcal{Y}_{ll}(\boldsymbol{x}) = 0\,,\quad L_3\mathcal{Y}_{ll}(\boldsymbol{x}) = l\mathcal{Y}_{ll}(\boldsymbol{x})\,,\\
&\mathcal{Y}_{ll}(\boldsymbol{x}) = \frac{1}{2^l l!}\left(\frac{(2l+1)!}{4\pi}\right)^{\frac{1}{2}}(-x_1 - \mathrm{i}x_2)^l\,.
\end{aligned}
$$

Generation from highest weight:

$$\mathcal{Y}_{lm}(\boldsymbol{x}) = \left(\frac{(l+m)}{(2l)!(l-m)!}\right)^{\frac{1}{2}} L_-^{l-m} Y_{ll}(\boldsymbol{x}) .$$

Relation to Gegenbauer and Jacobi polynomials:

$$\begin{aligned}\mathcal{Y}_{lm}(\boldsymbol{x}) &= r^{l-|m|}\mathcal{Y}_m(x_1,x_2)\\ &\quad\times[(2l+1)(l+m)!(l-m)!/2]^{\frac{1}{2}}\\ &\quad\times H_{l,|m|}(x_3/r) ,\\ H_{l\lambda}(z) &= \frac{(2\lambda)!}{2^\lambda \lambda!}C_{l-\lambda}^{\lambda+\frac{1}{2}}(z) = \frac{(l+\lambda)!}{2^\lambda l!}P_{l-\lambda}^{(\lambda,\lambda)}(z) ,\\ &0 \le \lambda \le l = 0,1,2,\dots ,\end{aligned}$$

where the $\mathcal{Y}_m(x_1,x_2)$ are homogeneous polynomial solutions of degree $|m|$ of Laplace's equation in 2-space, $\boldsymbol{R}^2$:

$$\mathcal{Y}_m(x_1,x_2) = \begin{cases}(-x_1-\mathrm{i}x_2)^m/\sqrt{2\pi} , & m \ge 0 ,\\ (x_1-\mathrm{i}x_2)^{-m}/\sqrt{2\pi} , & m \le 0 .\end{cases}$$

(Section 2.1.2 for the definition of Gegenbauer and Jacobi polynomials.)

Orthogonal group action:

$$(O_R\mathcal{Y}_{lm})(\boldsymbol{x}) = \mathcal{Y}_{lm}(R^{-1}\boldsymbol{x}) = \sum_{m'}\mathcal{D}^l_{m'm}(R)\mathcal{Y}_{lm'}(\boldsymbol{x}) ,$$

where the functions $\mathcal{D}^l_{m'm}(R) = D^l_{m'm}(U(R))$ are defined in Sect. 2.3 for various parametrizations of R.

Unitary group action:

$$\begin{aligned}(T_U\mathcal{Y}_{lm})(X) &= \mathcal{Y}_{lm}(U^\dagger XU)\\ &= \sum_{m'} D^l_{m'm}(U)\mathcal{Y}_{lm'}(X) ,\end{aligned}$$

where the functions $D^l_{m'm}(U)$ are defined in Sects. 2.2 and 2.3.

Orthogonality on the unit sphere:

$$\int_{\text{unit sphere}} \mathcal{Y}^*_{l'm'}(\boldsymbol{x})\mathcal{Y}_{lm}(\boldsymbol{x})\,\mathrm{d}S = \delta_{l'l}\delta_{m'm} .$$

Product of solid harmonics:

$$\begin{aligned}\mathcal{Y}_{k\mu}(\boldsymbol{x})\mathcal{Y}_{lm}(\boldsymbol{x}) &= \sum_{l'}\langle l'||\mathcal{Y}_k||l\rangle C^{lkl'}_{m,\mu,m+\mu}\mathcal{Y}_{l',m+\mu}(\boldsymbol{x})\\ &= \sum_{l'}(l'||\mathcal{Y}_k||l)\begin{pmatrix} l & k & l'\\ m & \mu & -m-\mu\end{pmatrix}\\ &\quad\times(-1)^{l'+m+\mu}\mathcal{Y}_{l',m+\mu}(\boldsymbol{x}) ,\\ \langle l'||\mathcal{Y}_k||l\rangle &= r^{l+k-l'}\left(\frac{(2l+1)(2k+1)}{4\pi(2l'+1)}\right)^{\frac{1}{2}}C^{lkl'}_{000} ,\\ (l'||\mathcal{Y}_k||l) &= r^{l+k-l'}\left(\frac{(2l+1)(2k+1)(2l'+1)}{4\pi}\right)^{\frac{1}{2}}\\ &\quad\times(-1)^{l'}\begin{pmatrix} l & k & l'\\ 0 & 0 & 0\end{pmatrix} ,\end{aligned}$$

where $C^{l_1l_2l}_{m_1m_2m}$ and

$$\begin{pmatrix} l_1 & l_2 & l\\ m_1 & m_2 & -m\end{pmatrix} = \frac{(-1)^{l_1-l_2+m}}{\sqrt{2l+1}}C^{l_1l_2l}_{m_1m_2m}$$

denote Wigner–Clebsch–Gordan coefficients and 3–j coefficients, respectively (Sect. 2.7).

Vector addition theorem for solid harmonics:

$$\begin{aligned}&\mathcal{Y}_{lm}(\boldsymbol{z}+\boldsymbol{z}') =\\ &\quad\sum_{k\mu}\left(\frac{4\pi(2l+1)!}{(2l-2k+1)!(2k+1)!}\right)^{\frac{1}{2}}C^{l-k,k,l}_{m-\mu,\mu,m}\\ &\quad\times\mathcal{Y}_{l-k,m-\mu}(\boldsymbol{z})\mathcal{Y}_{k\mu}(\boldsymbol{z}') , \quad \boldsymbol{z},\boldsymbol{z}' \in \boldsymbol{C}^3 ,\\ &C^{j-k,k,l}_{m-\mu,\mu,m} = \left[\binom{l+m}{k+\mu}\binom{l-m}{k-\mu}\Big/\binom{2l}{2k}\right]^{\frac{1}{2}} .\end{aligned}$$

Rotational invariant in two vectors:

$$\begin{aligned}I_l(\boldsymbol{x},\boldsymbol{y}) &= \frac{1}{2^l}\sum_k(-1)^k\binom{l}{k}\binom{2l-2k}{l}\\ &\quad\times(\boldsymbol{x}\cdot\boldsymbol{y})^{l-2k}(\boldsymbol{x}\cdot\boldsymbol{x})^k(\boldsymbol{y}\cdot\boldsymbol{y})^k\\ &= (\boldsymbol{x}\cdot\boldsymbol{x})^{l/2}(\boldsymbol{y}\cdot\boldsymbol{y})^{l/2}C_l^{(1/2)}(\hat{\boldsymbol{x}}\cdot\hat{\boldsymbol{y}})\\ &= \frac{4\pi}{2l+1}\sum_m(-1)^m\mathcal{Y}_{lm}(\boldsymbol{x})\mathcal{Y}_{l,-m}(\boldsymbol{y}) ,\end{aligned}$$

where $C_l^{(1/2)}(z)$ is a Gegenbauer polynomial (Sect. 2.1.2) and

$$\hat{\boldsymbol{x}} = \boldsymbol{x}/|\boldsymbol{x}| , \quad \hat{\boldsymbol{y}} = \boldsymbol{y}/|\boldsymbol{y}| , \quad \cos\theta = \hat{\boldsymbol{x}}\cdot\hat{\boldsymbol{y}} .$$

Legendre polynomials:

$$P_l(\cos\theta) = \frac{4\pi}{2l+1}\sum_m (-1)^m \mathcal{Y}_{l,-m}(\hat{\boldsymbol{y}})\mathcal{Y}_{lm}(\hat{\boldsymbol{x}}) ,$$

$$\left(\frac{4\pi}{2l+1}\right)^{\frac{1}{2}} \mathcal{Y}_{l0}(\boldsymbol{x})$$
$$= \frac{1}{2^l}\sum_k (-1)^k \binom{l}{k}\binom{2l-2k}{l} x_3^{l-2k}(\boldsymbol{x}\cdot\boldsymbol{x})^k$$
$$= r^l P_l(x_3/r) .$$

Rayleigh plane wave expansion:

$$\mathrm{e}^{\mathrm{i}\boldsymbol{k}\cdot\boldsymbol{x}} = 4\pi \sum_{l=0}^{\infty}\sum_{m=-l}^{l} \mathrm{i}^l j_l(kr)\mathcal{Y}^*_{lm}(\hat{\boldsymbol{k}})\mathcal{Y}_{lm}(\hat{\boldsymbol{x}}) ,$$

$$j_l(kr) = \left(\frac{\pi}{2kr}\right)^{\frac{1}{2}} J_{l+1/2}(kr) .$$

Relations in potential theory:

$$\mathcal{Y}_{lm}(\nabla)\left(\frac{1}{r}\right) = \frac{(-1)^l(2l)!}{2^l l!}\frac{\mathcal{Y}_{lm}(\boldsymbol{x})}{r^{2l+1}} ,$$

$$1/R = \sum_{l=0}^{\infty} I_l(\boldsymbol{x},\boldsymbol{y})/r^{2l+1} ,$$

$$I_l(\boldsymbol{x},\boldsymbol{y})/r^{2l+1} = \frac{(-1)^l}{l!}(\boldsymbol{y}\cdot\nabla)^l\left(\frac{1}{r}\right) .$$

For $\boldsymbol{R}=\boldsymbol{x}-\boldsymbol{y}$, $\quad r=(\boldsymbol{x}\cdot\boldsymbol{x})^{\frac{1}{2}}$, $\quad s=(\boldsymbol{y}\cdot\boldsymbol{y})^{\frac{1}{2}}$,

$$1/R = \sum_l P_l(\cos\theta)\frac{s^l}{r^{l+1}} , \qquad s \le r, \cos\theta = \hat{\boldsymbol{x}}\cdot\hat{\boldsymbol{y}} .$$

Rotational invariants in three vectors:

$$I_{(l_1 l_2 l_3)}\left(\boldsymbol{x}^1,\boldsymbol{x}^2,\boldsymbol{x}^3\right)$$
$$= \frac{(4\pi)^{3/2}}{[(2l_1+1)(2l_2+1)(2l_3+1)]^{\frac{1}{2}}}$$
$$\times \sum_{m_1 m_2 m_3}\begin{pmatrix} l_1 & l_2 & l_3 \\ m_1 & m_2 & m_3\end{pmatrix}$$
$$\times \mathcal{Y}_{l_1 m_1}\left(\boldsymbol{x}^1\right)\mathcal{Y}_{l_2 m_2}\left(\boldsymbol{x}^2\right)\mathcal{Y}_{l_3 m_3}\left(\boldsymbol{x}^3\right) ,$$

where $\begin{pmatrix} l_1 & l_2 & l_3 \\ m_1 & m_2 & m_3\end{pmatrix}$ is a 3–j coefficient (Sect. 2.7).

$$I_{(l_1 l_2 l_3)}\left(\boldsymbol{x}^1,\boldsymbol{x}^2,0\right)$$
$$= \delta_{l_1 l_2}\delta_{l_3 0}(-1)^{l_1} I_{l_1}\left(\boldsymbol{x}^1,\boldsymbol{x}^2\right)/(2l_1+1)^{\frac{1}{2}} .$$

Product law:

$$I_{(l)}(\boldsymbol{x})I_{(k)}(\boldsymbol{x})$$
$$= \sum_{(j)}\left[\prod_{\alpha=1}^{3}(-1)^{j_\alpha}(2j_\alpha+1)\begin{pmatrix} l_\alpha & k_\alpha & j_\alpha \\ 0 & 0 & 0\end{pmatrix}\right.$$
$$\left.\times\left(\boldsymbol{x}^\alpha\cdot\boldsymbol{x}^\alpha\right)^{(l_\alpha+k_\alpha-j_\alpha)/2}\right]\begin{Bmatrix} l_1 & l_2 & l_3 \\ k_1 & k_2 & k_3 \\ j_1 & j_2 & j_3\end{Bmatrix} I_{(j)}(\boldsymbol{x}) ,$$

where $l=(l_1,l_2,l_3)$, etc., $\boldsymbol{x}=\left(\boldsymbol{x}^1,\boldsymbol{x}^2,\boldsymbol{x}^3\right)$.

Coplanar vectors:

$$I_{(l)}\left(\boldsymbol{x}^1,\boldsymbol{x}^2,\alpha\boldsymbol{x}^1+\beta\boldsymbol{x}^2\right)$$
$$= \sum_{kl}\left(\frac{(2l_3+1)!}{(2l_3-2k)!(2k)!}\right)^{\frac{1}{2}}$$
$$\times \alpha^{l_3-k}\beta^k(-1)^{l_1+l_3+k}(2l+1)$$
$$\times \begin{pmatrix} l_3-k & l_1 & l \\ 0 & 0 & 0\end{pmatrix}\begin{pmatrix} k & l_2 & l \\ 0 & 0 & 0\end{pmatrix}\begin{Bmatrix} l_3-k & l_3 & k \\ l_2 & l & l_1\end{Bmatrix}$$
$$\times \left(\boldsymbol{x}^1\cdot\boldsymbol{x}^1\right)^{(l_1+l_3-l-k)/2}\left(\boldsymbol{x}^2\cdot\boldsymbol{x}^2\right)^{(l_2+k-l)/2}$$
$$\times I_l\left(\boldsymbol{x}^1,\boldsymbol{x}^2\right) .$$

The bracket symbols in these relations are 6–j and 9–j coefficients (Sects. 2.9, 2.10).

Cartan's vectors of zero length:

$$\boldsymbol{\alpha} = \left(-z_1^2+z_2^2, -\mathrm{i}\left(z_1^2+z_2^2\right), 2z_1 z_2\right) ,$$
$$\boldsymbol{\alpha}\cdot\boldsymbol{\alpha} = \alpha_1^2+\alpha_2^2+\alpha_3^2 = 0 ,$$
$$\boldsymbol{z} = (z_1, z_2) \in \boldsymbol{C}^2 .$$

Solutions of Laplace's equation using vectors of zero length:

$$\nabla^2(\boldsymbol{\alpha}\cdot\boldsymbol{x})^l = 0 , \quad l = 0, 1, \ldots .$$

Solid harmonics for vectors of zero length:

$$(-1)^m Y_{l,-m}(\boldsymbol{\alpha}) = \frac{(2l)!}{l!}\left(\frac{2l+1}{4\pi}\right)^{\frac{1}{2}} P_{lm}(z_1,z_2) ,$$

$$P_{lm}(z_1,z_2) = \frac{z_1^{l+m} z_2^{l-m}}{\sqrt{(l+m)!(l-m)!}} .$$

Orbital angular momentum operators for vectors of zero length:

$$\boldsymbol{J} = -\mathrm{i}(\boldsymbol{\alpha}\times\nabla_\alpha) ,$$
$$J_+ = z_1\frac{\partial}{\partial z_2} , \qquad J_- = z_2\frac{\partial}{\partial z_1} ,$$
$$J_3 = \frac{1}{2}\left(z_1\frac{\partial}{\partial z_1} - z_2\frac{\partial}{\partial z_2}\right) .$$

Rotational invariant for vectors of zero length:

$$(\boldsymbol{\alpha}\cdot\boldsymbol{x})^l = \left(\frac{4\pi}{2l+1}\right)^{\frac{1}{2}} 2^l l! \sum_m P_{lm}(z)\mathcal{Y}_{lm}(\boldsymbol{x}) .$$

Spinorial invariant under $z^i \to Uz^i (i = 1, 2, 3)$:

$$\sum_{m_1 m_2 m_3} \begin{pmatrix} j_1 & j_2 & j_3 \\ m_1 & m_2 & m_3 \end{pmatrix} P_{j_1 m_1}(z^1) P_{j_2 m_2}(z^2) P_{j_3 m_3}(z^3)$$
$$= [(j_1+j_2+j_3+1)!]^{-1/2}$$
$$\times \frac{(z_{12}^{12})^{j_1+j_2-j_3}(z_{12}^{31})^{j_3+j_1-j_2}(z_{12}^{23})^{j_2+j_3-j_1}}{[(j_1+j_2-j_3)!(j_3+j_1-j_2)!(j_2+j_3-j_1)!]^{\frac{1}{2}}} ,$$
$$z_{12}^{ij} = z_1^i z_2^j - z_1^j z_2^i .$$

This relation is invariant under the transformation

$$z \to Uz = \big((Uz)_1, (Uz)_2\big)$$
$$= (u_{11}z_1 + u_{12}z_2, u_{21}z_1 + u_{22}z_2) ,$$

where $U \in SU(2)$. Transformation properties of vectors of zero length:

$$\boldsymbol{\alpha} \to R\boldsymbol{\alpha}, \quad \boldsymbol{\alpha} = \mathrm{col}(\alpha_1, \alpha_2, \alpha_3) ,$$

where $z \to Uz$ and R is given in terms of U in the beginning of this chapter. Simultaneous eigenvectors of $\boldsymbol{L}^2$ and $\boldsymbol{J}^2$:

$$\boldsymbol{L}^2(\boldsymbol{\alpha}\cdot\boldsymbol{x})^l = l(l+1)(\boldsymbol{\alpha}\cdot\boldsymbol{x})^l , \quad l = 0, 1, \dots ,$$
$$\boldsymbol{J}^2(\boldsymbol{\alpha}\cdot\boldsymbol{x})^l = l(l+1)(\boldsymbol{\alpha}\cdot\boldsymbol{x})^l , \quad l = 0, 1, \dots .$$

2.1.2 Spherical Polar Coordinate Representation

The results given in Sect. 2.1.1 may be presented in any system of coordinates well-defined in terms of Cartesian coordinates. The principal relations for spherical polar coordinates are given in this section, where a vector in $\boldsymbol{R}^3$ is now given in the form

$$\boldsymbol{x} = r\hat{\boldsymbol{x}} = r(\sin\theta\cos\phi, \sin\theta\sin\phi, \cos\theta) ,$$
$$0 \le \theta \le \pi, \quad 0 \le \phi < 2\pi .$$

Orbital angular momentum operators:

$$L_1 = \mathrm{i}\cos\phi\cot\theta\frac{\partial}{\partial\phi} + \mathrm{i}\sin\phi\frac{\partial}{\partial\theta} ,$$
$$L_2 = \mathrm{i}\sin\phi\cot\theta\frac{\partial}{\partial\phi} - \mathrm{i}\cos\phi\frac{\partial}{\partial\theta} ,$$
$$L_3 = -\mathrm{i}\frac{\partial}{\partial\phi} ,$$
$$L_\pm = \mathrm{e}^{\pm\mathrm{i}\phi}\left(\pm\frac{\partial}{\partial\theta} + \mathrm{i}\cot\theta\frac{\partial}{\partial\phi}\right) ,$$
$$\boldsymbol{L}^2 = -\frac{1}{\sin\theta}\frac{\partial}{\partial\theta}\left(\sin\theta\frac{\partial}{\partial\theta}\right) - \frac{1}{\sin^2\theta}\frac{\partial^2}{\partial\phi^2} .$$

Laplacian:

$$\boldsymbol{x}\cdot\nabla = r\frac{\partial}{\partial r} ,$$
$$\nabla^2 = \frac{1}{r^2}\left[\left(r\frac{\partial}{\partial r}\right)^2 + r\frac{\partial}{\partial r} - \boldsymbol{L}^2\right] .$$

Spherical harmonics (solid harmonics on the unit sphere $\boldsymbol{S}^2$):

$$Y_{lm}(\theta,\phi) = (-1)^m\left(\frac{2l+1}{4\pi}(l+m)!(l-m)!\right)^{\frac{1}{2}} \mathrm{e}^{\mathrm{i}m\phi}$$
$$\times \sum_k \frac{(-1)^k(\sin\theta)^{2k+m}(\cos\theta)^{l-2k-m}}{2^{2k+m}(k+m)!k!(l-2k-m)!} .$$

Orthogonality on the unit sphere:

$$\int_0^{2\pi} \mathrm{d}\phi \int_0^{\pi} \mathrm{d}\theta \sin\theta\, Y^*_{l'm'}(\theta,\phi)\, Y_{lm}(\theta,\phi) = \delta_{l'l}\delta_{m'm} .$$

Relation to Legendre, Jacobi, and Gegenbauer polynomials:

$$Y_{lm}(\theta,\phi) = (-1)^m\left(\frac{(2l+1)(l-m)!}{4\pi(l+m)!}\right)^{\frac{1}{2}}$$
$$\times P_l^m(\cos\theta)\,\mathrm{e}^{\mathrm{i}m\phi} ,$$
$$P_l^m(\cos\theta) = \frac{(l+m)!}{l!}\left(\frac{\sin\theta}{2}\right)^m P_{l-m}^{(m,m)}(\cos\theta) .$$

Jacobi polynomials:

$$P_n^{(\alpha,\beta)}(x) = \sum_s \binom{n+\alpha}{s}\binom{n+\beta}{n-s}$$
$$\times \left(\frac{x-1}{2}\right)^{n-s}\left(\frac{x+1}{2}\right)^s ,$$
$$n = 0, 1, \dots ,$$

where α, β are arbitrary parameters and

$$\binom{z}{k} = \begin{cases} z(z-1)\cdots(z-k+1)/k! & \text{for } k = 1, 2, \ldots \\ 1 & \text{for } k = 0 \\ 0 & \text{for } k = -1, -2, \ldots \end{cases}$$

Relations between Jacobi polynomials for $n+\alpha$, $n+\beta$, $n+\alpha+\beta$ nonnegative integers:

$$P_n^{\alpha,\beta}(x) = \frac{(n+\alpha)!(n+\beta)!}{n!(n+\alpha+\beta)!}\left(\frac{x+1}{2}\right)^{-\beta} P_{n+\beta}^{(\alpha,-\beta)}(x)\,,$$

$$P_n^{\alpha,\beta}(x) = \frac{(n+\alpha)!(n+\beta)!}{n!(n+\alpha+\beta)!}\left(\frac{x-1}{2}\right)^{-\alpha} P_{n+\alpha}^{(-\alpha,\beta)}(x)\,,$$

$$P_n^{\alpha,\beta}(x) = \left(\frac{x-1}{2}\right)^{-\alpha}\left(\frac{x+1}{2}\right)^{-\beta} P_{n+\alpha+\beta}^{(-\alpha,-\beta)}(x)\,.$$

Nonstandard form (α arbitrary):

$$P_n^{(\alpha,\alpha)}(x) = \sum_s \frac{(-1)^s(\alpha+s+1)_{n-s}\left(1-x^2\right)^s x^{n-2s}}{2^{2s}s!(n-2s)!}\,,$$

$$(z)_k = z(z+1)\cdots(z+k-1)\,,\quad k = 1, 2, \ldots\,;$$

$$(z)_0 = 1\,.$$

Gegenbauer polynomials ($\alpha > -1/2$):

$$C_n^{(\alpha)}(x) = \frac{(2\alpha)_n}{(\alpha+1/2)_n} P_n^{\left(\alpha-\frac{1}{2},\alpha-\frac{1}{2}\right)}(x) = \sum_s \frac{(-1)^s(\alpha)_{n-s}(2x)^{n-2s}}{s!(n-2s)!}\,.$$

2.2 Abstract Angular Momentum

Abstract angular momentum theory addresses the problem of constructing all finite Hermitian matrices, up to equivalence, that satisfy the same commutation relations

$$[J_1, J_2] = \mathrm{i}J_3\,,\quad [J_2, J_3] = \mathrm{i}J_1\,,\quad [J_3, J_1] = \mathrm{i}J_2 \tag{2.7}$$

as some set of Hermitian operators J_1, J_2, J_3 appropriately defined in some Hilbert space; that is, of constructing all finite Hermitian matrices M_i such that under the correspondence $J_i \to M_i (i = 1, 2, 3)$ the commutation relations are still obeyed. If M_1, M_2, M_3 is such a set of Hermitian matrices, then AM_1A^{-1}, AM_2A^{-1}, AM_3A^{-1}, is another such set, where A is an arbitrary unitary matrix. This defines what is meant by equivalence. The commutation relations (2.7) may also be formulated as:

$$[J_3, J_\pm] = \pm J_\pm\,,\qquad [J_+, J_-] = 2J_3\,,$$
$$J_\pm = J_1 \pm \mathrm{i}J_2\,,\qquad J_+^\dagger = J_-\,. \tag{2.8}$$

The squared angular momentum

$$\boldsymbol{J}^2 = J_1^2 + J_2^2 + J_3^2 = J_-J_+ + J_3(J_3+1) = J_+J_- + J_3(J_3-1) \tag{2.9}$$

commutes with each J_i, and J_3 is, by convention, taken with $\boldsymbol{J}^2$ as a pair of commuting Hermitian operators to be diagonalized.

Examples of matrices satisfying relations (2.7) are provided by $J_i \to \sigma_i/2$ [the 2×2 Hermitian Pauli matrices defined in (2.1)] and $J_i \to M_i$ [the 3×3 matrices defined in (2.4)], these latter matrices being equivalent to those obtained from the matrices of the orbital angular momentum operators for $l = 1$.

One could determine all Hermitian matrices solving (2.7) and (2.8) by using only matrix theory, but it is customary in quantum mechanics to formulate the problem using Hilbert space concepts appropriate to that theory. Thus, one takes the viewpoint that the J_i are linear Hermitian operators with an action defined in a separable Hilbert space $\mathcal{H}$ such that $J_i : \mathcal{H} \to \mathcal{H}$.

One then seeks to decompose the Hilbert space into a direct sum of subspaces that are irreducible with respect to this action; that is, subspaces that cannot be further decomposed as a direct sum of subspaces that all the J_i leave invariant (map vectors in the space into vectors in the space). In this section, the solution of this fundamental problem for angular momentum theory is given. These results set the notation and phase conventions for all of angular momentum theory, in all of its varied realizations, and the relations are therefore sometimes referred to as standard. The method most often used to solve the posed problem is called the method of highest weights.

The solution of this problem is among the most important in quantum theory because of its generality and applicability to a wide range of problems. The space $\mathcal{H}$ can be written as a direct sum

$$\mathcal{H} = \sum_{j=0,\frac{1}{2},1,\ldots} \oplus n_j\mathcal{H}_j\,,\quad \text{each } \mathcal{H}_j \perp \mathcal{H}_{j'}\,,\quad j \neq j'\,, \tag{2.10}$$

in which $\mathcal{H}_j$ denotes a vector space of dimension $2j+1$ that is invariant and irreducible under the action of the set of operators J_i, $i=1,2,3$, and where the direct sum is over all half integers $j=0,\frac{1}{2},1,\ldots$. There may be multiple occurrences, n_j in number, of the same space $\mathcal{H}_j$ for given j, or no such space, $n_j=0$, in the direct sum. Abstractly, in so far as angular momentum properties are concerned, each repeated space $\mathcal{H}_j$ is *identical*. Such spaces may, however, be distinguished by their properties with respect to other physical observables, but not by the angular action of momentum operators themselves. The result, (2.10), applies to any physical system, no matter how complex, in which rotational symmetry, hence $SU(2)$ symmetry, is present, even in situations of higher symmetry where $SU(2)$ is a subgroup. Indeed, the resolution of the terms in (2.10) for various physical systems constitutes "spectroscopy" in the broadest sense.

The characterization of the space $\mathcal{H}_j$ with respect to angular momentum properties is given by the following results, where basis vectors are denoted in the Dirac braket notation.

Orthonormal basis:

$$\{|\,jm\rangle\,|\,m=-j,-j+1,\ldots,j\}\,. \tag{2.11}$$

$$\langle jm'|\,jm\rangle=\delta_{m',m}\,. \tag{2.12}$$

Simultaneous eigenvectors:

$$\boldsymbol{J}^2|\,jm\rangle=j(j+1)|\,jm\rangle\,,\quad J_3|\,jm\rangle=m|\,jm\rangle\,. \tag{2.13}$$

Action of angular momentum operators:

$$\begin{aligned}J_+|\,jm\rangle&=[(j-m)(j+m+1)]^{1/2}|\,jm+1\rangle\,,\\ J_-|\,jm\rangle&=[(j+m)(j-m+1)]^{1/2}|\,jm-1\rangle\,.\end{aligned} \tag{2.14}$$

Defining properties of highest weight vector:

$$J_+|\,jj\rangle=0\,,\quad J_3|\,jj\rangle=j|\,jj\rangle\,.$$

Generation of general vector from highest weight:

$$|\,jm\rangle=\left(\frac{(j+m)!}{(2j)!(j-m)!}\right)^{1/2}J_-^{j-m}|\,jj\rangle\,.$$

Necessary property of lowest weight vector:

$$J_-\,|\,j,-j\rangle=0\,,\quad J_3\,|\,j,-j\rangle=-j\,|\,j,-j\rangle\,.$$

Operator in $\mathcal{H}$ corresponding to a rotation by angle ψ about direction $\hat{\boldsymbol{n}}$ in $\boldsymbol{R}^3$:

$$\begin{aligned}T_{U(\psi,\hat{\boldsymbol{n}})}&=\exp(-\mathrm{i}\psi\hat{\boldsymbol{n}}\cdot\boldsymbol{J})\,,\\ \hat{\boldsymbol{n}}\cdot\hat{\boldsymbol{n}}&=n_1^2+n_2^2+n_3^2=1\,,\\ \hat{\boldsymbol{n}}\cdot\boldsymbol{J}&=n_1J_1+n_2J_2+n_3J_3\,,\\ U(\psi,\hat{\boldsymbol{n}})&=\exp(-\mathrm{i}\psi\hat{\boldsymbol{n}}\cdot\boldsymbol{\sigma}/2)\\ &=\sigma_0\cos\left(\tfrac{1}{2}\psi\right)-\mathrm{i}(\hat{\boldsymbol{n}}\cdot\boldsymbol{\sigma})\sin\left(\tfrac{1}{2}\psi\right)\\ &=\begin{pmatrix}\cos\left(\frac{1}{2}\psi\right)-\mathrm{i}n_3\sin\left(\frac{1}{2}\psi\right) & (-\mathrm{i}n_1-n_2)\sin\left(\frac{1}{2}\psi\right)\\ (-\mathrm{i}n_1+n_2)\sin\left(\frac{1}{2}\psi\right) & \cos\left(\frac{1}{2}\psi\right)+\mathrm{i}n_3\sin\left(\frac{1}{2}\psi\right)\end{pmatrix},\\ &0\le\psi\le 2\pi,\end{aligned} \tag{2.15}$$

where σ_0 denotes the 2 ×2 unit matrix.

Action of $T_{U(\psi,\hat{\boldsymbol{n}})}$ on $\mathcal{H}_j$:

$$T_U|\,jm\rangle=\sum_{m'}D^j_{m'm}(U)|\,jm'\rangle\,, \tag{2.16}$$

in which $U=U(\psi,\hat{\boldsymbol{n}})$ and $D^j_{m'm}(U)$ denotes a homogeneous polynomial of degree $2j$ defined on the elements $u_{ij}=U_{ij}(\psi,\hat{\boldsymbol{n}})$ in row i and column j of the matrix $U(\psi,\hat{\boldsymbol{n}})$ given by (2.15). The explicit form of this polynomial is

$$\begin{aligned}&D^j_{m'm}(U)\\ &=[(j+m)!(j-m)!(j+m')!(j-m')!]^{\frac{1}{2}}\\ &\times\sum_{\boxed{\alpha}}\frac{(u_{11})^{\alpha_{11}}(u_{12})^{\alpha_{12}}(u_{21})^{\alpha_{21}}(u_{22})^{\alpha_{22}}}{\alpha_{11}!\alpha_{12}!\alpha_{21}!\alpha_{22}!}\,.\end{aligned} \tag{2.17}$$

The notation $\boxed{\alpha}$ symbolizes a 2×2 array of nonnegative integers with certain constraints:

$$\begin{array}{|cc|l}\hline \alpha_{11} & \alpha_{12} & j+m'\\ \alpha_{21} & \alpha_{22} & j-m'\,.\\ \hline j+m & j-m & \end{array}$$

In this array the α_{ij} are nonnegative integers subject to the row and column constraints (sums) indicated by the (nonnegative) integers $j\pm m$, $j\pm m'$. Explicitly,

$$\begin{aligned}\alpha_{11}+\alpha_{12}&=j+m'\,, & \alpha_{21}+\alpha_{22}&=j-m'\,,\\ \alpha_{11}+\alpha_{21}&=j+m\,, & \alpha_{12}+\alpha_{22}&=j-m\,.\end{aligned}$$

The summation is over all such arrays. Any one of the α_{ij} may serve as a single summation index if one wishes to

eliminate the redundancy inherent in the square-array notation. The form (2.17) is very useful for obtaining symmetry relations for these polynomials (Sect. 2.3.6).
Unitary property on $\mathcal{H}$:

$$\langle T_U \Psi | T_U \Psi \rangle = \langle \Psi | \Psi \rangle , \quad \text{each } \Psi \in \mathcal{H} .$$

Irreducible unitary matrix representation of $SU(2)$:

$$\begin{aligned}&(D^j(U))_{j-m'+1,j-m+1} = D^j_{m'm}(U) ,\\ &m' = j, j-1, \ldots , -j ; \quad m = j, j-1, \ldots , -j ,\end{aligned} \quad (2.18)$$

denotes the element in row $j-m'+1$ and column $j-m+1$. Then, dimension of $D^j(U) = 2j+1$ and

$$\begin{aligned}&D^j(U)D^j(U') = D^j(UU') ,\\ &U \in SU(2) , \quad U' \in SU(2) ,\\ &(D^j(U))^\dagger = (D^j(U))^{-1} = D^j(U^\dagger) .\end{aligned}$$

Kronecker (direct) product representation:

$$D^{j_1}(U) \times D^{j_2}(U)$$

is a $(2j_1+1)(2j_2+1)$ dimensional reducible representation of $SU(2)$. One can also effect the reduction of this representation into irreducible ones by abstract methods. The results are given in Sect. 2.7.

2.3 Representation Functions

2.3.1 Parametrizations of the Groups $SU(2)$ and $SO(3,R)$

The irreducible representations of the quantal rotation group, $SU(2)$, are among the most important quantities in all of angular momentum theory: These are the unitary matrices of dimension $2j+1$, denoted by $D^j(U)$, where this notation is used to signify that the elements of this matrix, denoted $D^j_{m'm}(U)$, are functions of the elements u_{ij} of the 2×2 unitary unimodular matrix $U \in SU(2)$. It has become standard to enumerate the rows and columns of these matrices in the order $j, j-1, \ldots , -j$ as read from top to bottom down the rows and from left to right across the columns [see also (2.18)]. These matrices may be presented in a variety of parametrizations, all of which are useful. In order to make comparisons between the group $SO(3, \boldsymbol{R})$ and the group $SU(2)$, it is most useful to parametrize these groups so that they are related according to the two-to-one homomorphism given by (2.2).

The general parametrization of the group $SU(2)$ is given in terms of the Euler–Rodrigues parameters corresponding to points belonging to the surface of the unit sphere S^3 in $\boldsymbol{R}^4$,

$$\alpha_0^2+\alpha_1^2+\alpha_2^2+\alpha_3^2 = 1 . \quad (2.19)$$

Each $U \in SU(2)$ can be written in the form:

$$\begin{aligned}U(\alpha_0, \boldsymbol{\alpha}) &= \begin{pmatrix} \alpha_0 - \mathrm{i}\alpha_3 & -\mathrm{i}\alpha_1 - \alpha_2 \\ -\mathrm{i}\alpha_1 + \alpha_2 & \alpha_0 + \mathrm{i}\alpha_3 \end{pmatrix}\\ &= \alpha_0\sigma_0 - \mathrm{i}\boldsymbol{\alpha}\cdot\boldsymbol{\sigma} .\end{aligned} \quad (2.20)$$

The $R \in SO(3, \boldsymbol{R})$ corresponding to this U in the two-to-one homomorphism given by (2.2) is:

$$R(\alpha_0, \boldsymbol{\alpha}) = \begin{pmatrix} \alpha_0^2+\alpha_1^2-\alpha_2^2-\alpha_3^2 & 2\alpha_1\alpha_2-2\alpha_0\alpha_3 & 2\alpha_1\alpha_3+2\alpha_0\alpha_2 \\ 2\alpha_1\alpha_2+2\alpha_0\alpha_3 & \alpha_0^2+\alpha_2^2-\alpha_3^2-\alpha_1^2 & 2\alpha_2\alpha_3-2\alpha_0\alpha_1 \\ 2\alpha_1\alpha_3-2\alpha_0\alpha_2 & 2\alpha_2\alpha_3+2\alpha_0\alpha_1 & \alpha_0^2+\alpha_3^2-\alpha_1^2-\alpha_2^2 \end{pmatrix} . \quad (2.21)$$

The procedure of parametrization is implemented uniformly by first parametrizing the points on the unit sphere S^3 so as to cover the points in S^3 exactly once, thus obtaining a parametrization of each $U \in SU(2)$. Equation (2.21) is then used to obtain the corresponding parametrization of each $R \in SO(3, \boldsymbol{R})$, where one notes that $R(-\alpha_0, -\boldsymbol{\alpha}) = R(\alpha_0, \boldsymbol{\alpha})$. Because of this two-to-one correspondence $\pm U \to R$, the domain of the parameters that cover the unit sphere S^3 exactly once will cover the group $SO(3, \boldsymbol{R})$ exactly twice. This is taken into account uniformly by redefining the domain for $SO(3, \boldsymbol{R})$ so as to cover only the upper hemisphere $(\alpha_0 \geq 0)$ of S^3.

In the active viewpoint (reference frame fixed with points being transformed into new points), an arbitrary vector $\boldsymbol{x} = \mathrm{col}(x_1, x_2, x_3) \in \boldsymbol{R}^3$ is transformed to the new vector $\boldsymbol{x}' = \mathrm{col}(x_1', x_2', x_3')$ by the rule $\boldsymbol{x}' = R\boldsymbol{x}$, or, equivalently, in terms of the Cartan matrix: $X' = UXU^\dagger$. In the passive viewpoint, the basic inertial reference system, which is taken to be a right-handed triad of unit vectors $(\hat{\boldsymbol{e}}_1, \hat{\boldsymbol{e}}_2, \hat{\boldsymbol{e}}_3)$, is transformed by R to a new right-handed triad $(\hat{\boldsymbol{f}}_1, \hat{\boldsymbol{f}}_2, \hat{\boldsymbol{f}}_3)$ by the rule

$$\hat{f}_j = \sum_i R_{ij}\hat{e}_i, \quad i = 1, 2, 3,$$

so that $\hat{e}_i \cdot \hat{f}_j = R_{ij}$. In this viewpoint, the coordinates of one and the same point P undergo a redescription under the change of frame. If the coordinates of P are (x_1, x_2, x_3) relative to the frame $(\hat{e}_1, \hat{e}_2, \hat{e}_3)$ and (x_1', x_2', x_3') relative to the frame $(\hat{f}_1, \hat{f}_2, \hat{f}_3)$, then

$$x_1\hat{e}_1 + x_2\hat{e}_2 + x_3\hat{e}_3 = x_1'\hat{f}_1 + x_2'\hat{f}_2 + x_3'\hat{f}_3,$$

so that $\boldsymbol{x}' = R^T\boldsymbol{x}$.

Rotation about direction $\hat{\boldsymbol{n}} \in S^2$ by positive angle ψ (right-hand rule):

$$(\alpha_0, \boldsymbol{\alpha}) = \left(\cos\frac{1}{2}\psi, \hat{\boldsymbol{n}}\sin\frac{1}{2}\psi\right), \quad 0 \le \psi \le 2\pi,$$

$$U(\psi, \hat{\boldsymbol{n}}) = \exp\left(-\mathrm{i}\frac{1}{2}\psi\hat{\boldsymbol{n}}\cdot\boldsymbol{\sigma}\right) =$$

$$\begin{pmatrix} \cos\frac{1}{2}\psi - \mathrm{i}n_3\sin\frac{1}{2}\psi & (-\mathrm{i}n_1 - n_2)\sin\frac{1}{2}\psi \\ (-\mathrm{i}n_1 + n_2)\sin\frac{1}{2}\psi & \cos\frac{1}{2}\psi + \mathrm{i}n_3\sin\frac{1}{2}\psi \end{pmatrix},$$

$$\begin{aligned} R(\psi, \hat{\boldsymbol{n}}) &= \exp(-\mathrm{i}\psi\hat{\boldsymbol{n}}\cdot\boldsymbol{M}), \quad 0 \le \psi \le \pi \\ &= I_3 - \mathrm{i}\sin\psi(\hat{\boldsymbol{n}}\cdot\boldsymbol{M}) - (\hat{\boldsymbol{n}}\cdot\boldsymbol{M})^2(1-\cos\psi) \\ &= \begin{pmatrix} R_{11} & R_{12} & R_{13} \\ R_{21} & R_{22} & R_{23} \\ R_{31} & R_{32} & R_{33} \end{pmatrix}, \end{aligned}$$

$$\begin{aligned} R_{11} &= n_1^2 + \left(1 - n_1^2\right)\cos\psi, \\ R_{21} &= n_1n_2(1-\cos\psi) + n_3\sin\psi, \\ R_{31} &= n_1n_3(1-\cos\psi) - n_2\sin\psi, \\ R_{12} &= n_1n_2(1-\cos\psi) - n_3\sin\psi, \\ R_{22} &= n_2^2 + \left(1 - n_2^2\right)\cos\psi, \\ R_{32} &= n_2n_3(1-\cos\psi) + n_1\sin\psi, \\ R_{13} &= n_1n_3(1-\cos\psi) + n_2\sin\psi, \\ R_{23} &= n_2n_3(1-\cos\psi) - n_1\sin\psi, \\ R_{33} &= n_3^2 + \left(1 - n_3^2\right)\cos\psi. \end{aligned}$$

The unit vector $\hat{\boldsymbol{n}} \in S^2$ can be further parametrized in terms of the usual spherical polar coordinates:

$$\begin{aligned} &\hat{\boldsymbol{n}} = (\sin\theta\cos\phi, \sin\theta\sin\phi, \cos\theta), \\ &0 \le \theta \le \pi, \quad 0 \le \phi < 2\pi. \end{aligned}$$

Euler angle parametrization:

$$\begin{aligned} U(\alpha\beta\gamma) &= \mathrm{e}^{-\mathrm{i}\alpha\sigma_3/2}\mathrm{e}^{-\mathrm{i}\beta\sigma_2/2}\mathrm{e}^{-\mathrm{i}\gamma\sigma_3/2} \\ &= \begin{pmatrix} \mathrm{e}^{-\mathrm{i}\alpha/2}\cos\left(\frac{1}{2}\beta\right)\mathrm{e}^{-\mathrm{i}\gamma/2} & -\mathrm{e}^{-\mathrm{i}\alpha/2}\sin\left(\frac{1}{2}\beta\right)\mathrm{e}^{\mathrm{i}\gamma/2} \\ \mathrm{e}^{\mathrm{i}\alpha/2}\sin\left(\frac{1}{2}\beta\right)\mathrm{e}^{-\mathrm{i}\gamma/2} & \mathrm{e}^{\mathrm{i}\alpha/2}\cos\left(\frac{1}{2}\beta\right)\mathrm{e}^{\mathrm{i}\gamma/2} \end{pmatrix}, \end{aligned}$$

$$0 \le \alpha < 2\pi, \quad 0 \le \beta \le \pi \quad \text{or } 2\pi \le \beta \le 3\pi,$$
$$0 \le \gamma < 2\pi,$$
$$U(\alpha, \beta + 2\pi, \gamma) = -U(\alpha\beta\gamma);$$

$$\begin{aligned} R(\alpha\beta\gamma) &= \mathrm{e}^{-\mathrm{i}\alpha M_3}\mathrm{e}^{-\mathrm{i}\beta M_2}\mathrm{e}^{-\mathrm{i}\gamma M_3} \\ &= \begin{pmatrix} \cos\alpha & -\sin\alpha & 0 \\ \sin\alpha & \cos\alpha & 0 \\ 0 & 0 & 1 \end{pmatrix}\begin{pmatrix} \cos\beta & 0 & \sin\beta \\ 0 & 1 & 0 \\ -\sin\beta & 0 & \cos\beta \end{pmatrix} \\ &\quad\times\begin{pmatrix} \cos\gamma & -\sin\gamma & 0 \\ \sin\gamma & \cos\gamma & 0 \\ 0 & 0 & 1 \end{pmatrix} \\ &= \begin{pmatrix} \cos\alpha\cos\beta\cos\gamma - \sin\alpha\sin\gamma & -\cos\alpha\cos\beta\sin\gamma - \sin\alpha\cos\gamma & \cos\alpha\sin\beta \\ \sin\alpha\cos\beta\cos\gamma + \cos\alpha\sin\gamma & -\sin\alpha\cos\beta\sin\gamma + \cos\alpha\cos\gamma & \sin\alpha\sin\beta \\ -\sin\beta\cos\gamma & \sin\beta\sin\gamma & \cos\beta \end{pmatrix}, \end{aligned}$$

$$0 \le \alpha < 2\pi, \quad 0 \le \beta \le \pi, \quad 0 \le \gamma < 2\pi.$$

This matrix corresponds to the sequence of frame rotations given by

$$\begin{aligned} &\text{rotate by } \gamma \text{ about } \hat{e}_3 = (0, 0, 1), \\ &\text{rotate by } \beta \text{ about } \hat{e}_2 = (0, 1, 0), \\ &\text{rotate by } \alpha \text{ about } \hat{e}_3 = (0, 0, 1). \end{aligned}$$

Equivalently, it corresponds to the sequence of frame rotations given by

$$\begin{aligned} &\text{rotate by } \alpha \text{ about } \hat{\boldsymbol{n}}_1 = (0, 0, 1), \\ &\text{rotate by } \beta \text{ about } \hat{\boldsymbol{n}}_2 = (-\sin\alpha, \cos\alpha, 0), \\ &\text{rotate by } \gamma \text{ about } \hat{\boldsymbol{n}}_3 = \\ &\qquad (\cos\alpha\sin\beta, \sin\alpha\sin\beta, \cos\beta). \end{aligned}$$

This latter sequence of rotations is depicted in Fig. 2.1 in obtaining the frame $(\hat{f}_1, \hat{f}_2, \hat{f}_3)$ from $(\hat{e}_1, \hat{e}_2, \hat{e}_3)$.

The four complex numbers

$$\begin{aligned} &(a, b, c, d) \\ &= (\alpha_0 + \mathrm{i}\alpha_3, \mathrm{i}\alpha_1 - \alpha_2, \mathrm{i}\alpha_1 + \alpha_2, \alpha_0 - \mathrm{i}\alpha_3) \end{aligned}$$

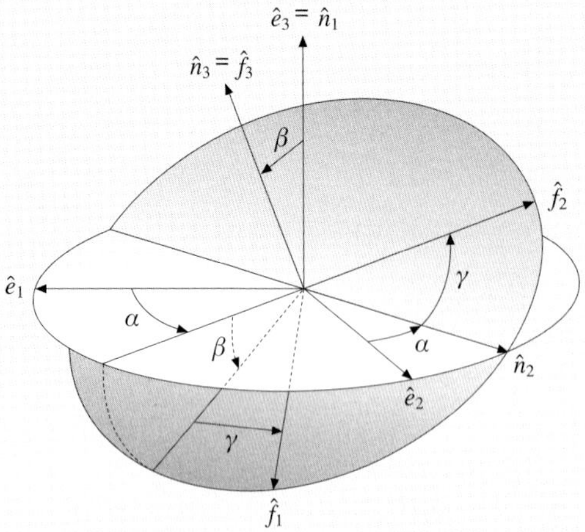

Fig. 2.1 Euler angles. The three Euler angles $(\alpha\beta\gamma)$ are defined by a sequence of three rotations. Reprinted with the permission of Cambridge University Press, after [2.1]

are called the Cayley–Klein parameters, whereas the four real numbers $(\alpha_0, \boldsymbol{\alpha})$ defining a point on the surface of the unit sphere in four-space, $\boldsymbol{S}^3$, are known as the Euler–Rodrigues parameters. The three ratios α_i/α_0 form the homogeneous or symmetric Euler parameters.

2.3.2 Explicit Forms of Representation Functions

The general form of the representation functions is given in its most basic and symmetric form in (2.17). This form applies to every parametrization, it being necessary only to introduce the explicit parametrizations of $U \in SU(2)$ or $R \in SO(3, \boldsymbol{R})$ given in Sect. 2.3.1 to obtain the explicit results given in this section. A choice is also made for the single independent summation parameter in the α-array. The notation for functions is abused by writing

$$
\begin{aligned}
&D^j(\omega) = D^j(U(\omega)) ,\\
&\quad \omega = \text{ set of parameters of } U \in SU(2) .
\end{aligned}
$$

Euler–Rodrigues representation $\left[(\alpha_0, \boldsymbol{\alpha}) \in \boldsymbol{S}^3\right]$:

$$
\begin{aligned}
&D^j_{m'm}(\alpha_0, \boldsymbol{\alpha})\\
&= [(j+m')!(j-m')!(j+m)!(j-m)!]^{\frac{1}{2}}\\
&\quad \times \sum_s \frac{(\alpha_0 - \mathrm{i}\alpha_3)^{j+m-s}(-\mathrm{i}\alpha_1 - \alpha_2)^{m'-m+s}}{(j+m-s)!(m'-m+s)!}\\
&\quad \times \frac{(-\mathrm{i}\alpha_1 + \alpha_2)^s(\alpha_0 + \mathrm{i}\alpha_3)^{j-m'-s}}{s!(j-m'-s)!} . \qquad (2.22)
\end{aligned}
$$

Quaternionic multiplication rule for points on the sphere $\boldsymbol{S}^3$:

$$
\begin{aligned}
(\alpha_0', \boldsymbol{\alpha}')(\alpha_0, \boldsymbol{\alpha}) &= (\alpha_0'', \boldsymbol{\alpha}'') ,\\
\alpha_0'' &= \alpha_0'\alpha_0 - \boldsymbol{\alpha}' \cdot \boldsymbol{\alpha} ,\\
\boldsymbol{\alpha}'' &= \alpha_0'\boldsymbol{\alpha} + \alpha_0\boldsymbol{\alpha}' + \boldsymbol{\alpha}' \times \boldsymbol{\alpha} ;\\
D^j(\alpha_0', \boldsymbol{\alpha}')D^j(\alpha_0, \boldsymbol{\alpha}) &= D^j(\alpha_0'', \boldsymbol{\alpha}'') .
\end{aligned}
$$

The $(\psi, \hat{\boldsymbol{n}})$ parameters:

$$
\alpha_0 = \cos\frac{1}{2}\psi , \quad \boldsymbol{\alpha} = \hat{\boldsymbol{n}}\sin\frac{1}{2}\psi .
$$

Euler angle parametrization:

$$
\begin{aligned}
&D^j_{m'm}(\alpha\beta\gamma) = \mathrm{e}^{-\mathrm{i}m'\alpha} d^j_{m'm}(\beta)\,\mathrm{e}^{-\mathrm{i}m\gamma} ,\\
&d^j_{m'm}(\beta) = \langle jm'|\mathrm{e}^{-\mathrm{i}\beta J_2}|jm\rangle\\
&= [(j+m')!(j-m')!(j+m)!(j-m)!]^{\frac{1}{2}}\\
&\quad \times \sum_s \frac{(-1)^{m'-m+s}\left(\cos\frac{1}{2}\beta\right)^{2j+m-m'-2s}}{(j+m-s)!s!(m'-m+s)!}\\
&\quad \times \frac{\left(\sin\frac{1}{2}\beta\right)^{m'-m+2s}}{(j-m'-s)!} . \qquad (2.23)
\end{aligned}
$$

Explicit matrices:

$$
d^{\frac{1}{2}}(\beta) = \begin{pmatrix} \cos\frac{1}{2}\beta & -\sin\frac{1}{2}\beta \\ \sin\frac{1}{2}\beta & \cos\frac{1}{2}\beta \end{pmatrix} ,
$$

$$
d^1(\beta) = \begin{pmatrix} \frac{1+\cos\beta}{2} & \frac{-\sin\beta}{\sqrt{2}} & \frac{1-\cos\beta}{2} \\ \frac{\sin\beta}{\sqrt{2}} & \cos\beta & \frac{-\sin\beta}{\sqrt{2}} \\ \frac{1-\cos\beta}{2} & \frac{\sin\beta}{\sqrt{2}} & \frac{1+\cos\beta}{2} \end{pmatrix} .
$$

Formal polynomial form (z_{ij} are indeterminates):

$$
\begin{aligned}
D^j_{m'm}(Z) &= [(j+m')!(j-m')!(j+m)!(j-m)!]^{\frac{1}{2}}\\
&\quad \times \sum_{\boxed{\alpha}} \prod_{i,j=1}^{2} (z_{ij})^{\alpha_{ij}}/(\alpha_{ij})! , \qquad (2.24)
\end{aligned}
$$

$$
D^j(Z')D^j(Z) = D^j(Z'Z) .
$$

Boson operator form:

Put $a_i^j = z_{ij}(i, j = 1, 2)$ in (2.24). Let $\bar{a}_i^j$ denote the Hermitian conjugate boson so that

$$
\left[a_l^k, a_i^j\right] = 0 , \quad \left[\bar{a}_l^k, \bar{a}_i^j\right] = 0 , \quad \left[\bar{a}_l^k, a_i^j\right] = \delta^{kj}\delta_{li} .
$$

Then the boson polynomials are orthogonal in the boson inner product:

$$\langle 0 \mid D^{j'}_{\mu'\mu}(\bar{A}) D^{j}_{m'm}(A) \mid 0 \rangle = (2j)!\delta_{j'j}\delta_{\mu'm'}\delta_{\mu m} \,.$$

2.3.3 Relations to Special Functions

Jacobi polynomials (see Sect. 2.1.2):

$$d^{j}_{m'm}(\beta) = \left(\frac{(j+m)!(j-m)!}{(j+m')!(j-m')!}\right)^{\frac{1}{2}} \left(\sin\frac{1}{2}\beta\right)^{m-m'} \times \left(\cos\frac{1}{2}\beta\right)^{m'+m} P^{(m-m',m+m')}_{j-m}(\cos\beta) \,,$$

$$d^{j}_{m'm}(\beta) = (-1)^{m'-m} d^{j}_{-m',-m}(\beta) = (-1)^{m'-m} d^{j}_{mm'}(\beta) = d^{j}_{mm'}(-\beta) \,.$$

Legendre polynomials:

$$D^{l}_{m0}(\beta) = (-1)^m \left(\frac{(l-m)!}{(l+m)!}\right)^{\frac{1}{2}} P^{m}_{l}(\cos\beta) = \left(\frac{(l+m)!}{(l-m)!}\right)^{\frac{1}{2}} P^{-m}_{l}(\cos\beta) \,.$$

Spherical harmonics:

$$Y_{lm}(\beta\alpha) = \left(\frac{2l+1}{4\pi}\right)^{\frac{1}{2}} e^{im\alpha} d^{l}_{m0}(\beta) = \left(\frac{2l+1}{4\pi}\right)^{\frac{1}{2}} D^{l*}_{m0}(\alpha\beta\gamma) \,,$$

$$Y^{*}_{lm}(\beta\alpha) = (-1)^m Y_{l,-m}(\beta\alpha) \,.$$

Gegenbauer polynomials:

$$d^{l}_{m0}(\beta) = (-1)^m [(l+m)!(l-m)!]^{\frac{1}{2}} \times \frac{(2m)!}{m!}\left[\frac{\sin\beta}{2}\right]^m C^{(m+1/2)}_{l-m}(\cos\beta) \,, \quad m \geq 0 \,.$$

Solutions of Laplace's equation in $\boldsymbol{R}^4$ (Sect. 2.5):

$$\nabla^2_4 D^{j}_{m'm}(x_0, \boldsymbol{x}) = 0 \,, \quad (x_0, \boldsymbol{x}) \in \boldsymbol{R}^4 \,,$$

$$\nabla^2_4 = \sum_{\mu=0}^{3} \frac{\partial^2}{\partial x^2_\mu} \,.$$

Replace the Euler–Rodrigues parameters $(\alpha_0, \boldsymbol{\alpha})$ in (2.22) by an arbitrary point $(x_0, \boldsymbol{x}) \in \boldsymbol{R}^4$.

2.3.4 Orthogonality Properties

Inner (scalar) product:

$$(\Psi, \Phi) = \int \mathrm{d}\Omega\, \Psi^*(\boldsymbol{x})\Phi(\boldsymbol{x}) \,,$$

$$\mathrm{d}\Omega = \text{invariant surface measure for } S^3 \,,$$

$$\int_{S^3} \mathrm{d}\Omega = 2\pi^2 \,.$$

Spherical polar coordinate for S^3:

$$(\alpha_0, \boldsymbol{\alpha}) = (\cos\chi, \cos\phi\sin\theta\sin\chi, \sin\phi\sin\theta\sin\chi, \cos\theta\sin\chi) \,,$$

$$0 \leq \theta \leq \pi \,, \quad 0 \leq \phi < 2\pi \,, \quad 0 \leq \chi \leq \pi \,,$$

$$\mathrm{d}\Omega = \mathrm{d}\omega \sin^2\chi \,\mathrm{d}\chi \,,$$

$$\mathrm{d}\omega = \mathrm{d}\phi \sin\theta \,,$$

$$\mathrm{d}\theta = \text{ invariant surface measure for } S^2 \,;$$

$$\int_0^{2\pi} \mathrm{d}\phi \int_0^{\pi} \mathrm{d}\theta \sin\theta \times \int_0^{\pi} \mathrm{d}\chi \sin^2\chi\, D^{j*}_{m'm}(\alpha_0, \boldsymbol{\alpha}) D^{j'}_{\mu'\mu}(\alpha_0, \boldsymbol{\alpha}) = \frac{2\pi^2}{2j+1}\delta_{jj'}\delta_{m'\mu'}\delta_{m\mu} \,.$$

Coordinates $(\psi, \hat{\boldsymbol{n}})$ for S^3:

$$(\alpha_0, \boldsymbol{\alpha}) = \left(\cos\frac{\psi}{2}, \hat{\boldsymbol{n}}\sin\frac{\psi}{2}\right) \,,$$

$$0 \leq \psi \leq 2\pi \,, \quad \hat{\boldsymbol{n}}\cdot\hat{\boldsymbol{n}} = 1 \,,$$

$$\mathrm{d}\Omega = \mathrm{d}S(\hat{\boldsymbol{n}})\sin^2\frac{\psi}{2}\frac{\mathrm{d}\psi}{2} \,,$$

$$\mathrm{d}S(\hat{\boldsymbol{n}}) = \mathrm{d}\omega \quad \text{for } \hat{\boldsymbol{n}} = (\sin\theta\cos\phi, \sin\theta\sin\phi, \cos\theta) \,,$$

$$\int \mathrm{d}S(\hat{\boldsymbol{n}}) \int_0^{2\pi} \frac{\mathrm{d}\psi}{2}\left(\sin\frac{\psi}{2}\right)^2 D^{j*}_{m'm}(\psi, \hat{\boldsymbol{n}}) D^{j'}_{\mu'\mu}(\psi, \hat{\boldsymbol{n}}) = \frac{2\pi^2}{2j+1}\delta_{jj'}\delta_{m'\mu'}\delta_{m\mu} \,,$$

Euler angles for S^3 ($SU(2)$):

$$(\alpha_0, \boldsymbol{\alpha}) = \left(\cos\frac{\beta}{2}\cos\frac{1}{2}(\gamma+\alpha), \sin\frac{\beta}{2}\sin\frac{1}{2}(\gamma-\alpha),\right.$$
$$\left.\sin\frac{\beta}{2}\cos\frac{1}{2}(\gamma-\alpha), \cos\frac{\beta}{2}\sin\frac{1}{2}(\gamma+\alpha)\right),$$
$$\mathrm{d}\Omega = \frac{1}{8}\mathrm{d}\alpha\,\mathrm{d}\gamma\sin\beta\,\mathrm{d}\beta\,, \tag{2.25}$$
$$\frac{1}{8}\int_0^{2\pi}\mathrm{d}\alpha\int_0^{2\pi}\mathrm{d}\gamma\int_0^{\pi}\mathrm{d}\beta\sin\beta D^{j*}_{m'm}(\alpha\beta\gamma)D^{j'}_{\mu'\mu}(\alpha\beta\gamma)$$
$$+\frac{1}{8}\int_0^{2\pi}\mathrm{d}\alpha\int_0^{2\pi}\mathrm{d}\gamma\int_{2\pi}^{3\pi}\mathrm{d}\beta\sin\beta D^{j*}_{m'm}(\alpha\beta\gamma)D^{j'}_{\mu'\mu}(\alpha\beta\gamma)$$
$$=\frac{2\pi^2}{2j+1}\delta_{jj'}\delta_{m'\mu'}\delta_{m\mu}\,. \tag{2.26}$$

Euler angles for hemisphere of S^3 ($SO(3,\boldsymbol{R})$; j' and j both integral):

$$\int_0^{2\pi}\mathrm{d}\alpha\int_0^{2\pi}\mathrm{d}\gamma\int_0^{\pi}\mathrm{d}\beta\sin\beta D^{j*}_{m'm}(\alpha\beta\gamma)D^{j'}_{\mu'\mu}(\alpha\beta\gamma)$$
$$=\frac{8\pi^2}{2j+1}\delta_{jj'}\delta_{m'\mu'}\delta_{m\mu}\,. \tag{2.27}$$

Formal polynomials (2.24):

$$\left(D^j_{m'm}, D^{j'}_{\mu'\mu}\right) = (2j)!\delta_{jj'}\delta_{m'\mu'}\delta_{m\mu}\,,$$

with inner product

$$(P, P') = P^*\left(\frac{\partial}{\partial Z}\right)P'(Z)|_{Z=0}\,,$$

where $P^*\left(\frac{\partial}{\partial Z}\right)$ is the complex conjugate polynomial P^* of P in which each z_{ij} is replaced by $\frac{\partial}{\partial z_{ij}}$.
Boson polynomials:

$$\left\langle D^j_{m'm}\middle| D^{j'}_{\mu'\mu}\right\rangle = (2j)!\delta_{jj'}\delta_{m'\mu'}\delta_{m\mu}\,,$$

with inner product $\langle P|P'\rangle = \langle 0|P^*(\bar{A})P'(A)|0\rangle$.

2.3.5 Recurrence Relations

Many useful relations between the representation functions may be derived as special cases of general relations between these functions and the WCG-coefficients given in Sect. 2.7.1. The simplest of these are obtained from the Kronecker reduction

$$D^j\times D^{\frac{1}{2}} = D^{j+1/2}\oplus D^{j-1/2}\,.$$

Such relations are usually presented in terms of the Euler angle realization of U, leading to the following relations between the functions $d^j_{m',m}(\beta)$:

$$(j-m+1)^{\frac{1}{2}}\cos\left(\frac{1}{2}\beta\right)d^{j+1/2}_{m'-1/2,m-1/2}(\beta)$$
$$+(j+m+1)^{\frac{1}{2}}\sin\left(\frac{1}{2}\beta\right)d^{j+1/2}_{m'-1/2,m+1/2}(\beta)$$
$$=(j-m'+1)^{\frac{1}{2}}d^j_{m'm}(\beta)\,,$$
$$-(j-m+1)^{\frac{1}{2}}\sin\left(\frac{1}{2}\beta\right)d^{j+1/2}_{m'+1/2,m-1/2}(\beta)$$
$$+(j+m+1)^{\frac{1}{2}}\cos\left(\frac{1}{2}\beta\right)d^{j+1/2}_{m'+1/2,m+1/2}(\beta)$$
$$=(j+m'+1)^{\frac{1}{2}}d^j_{m'm}(\beta)\,,$$
$$(j+m)^{\frac{1}{2}}\cos\left(\frac{1}{2}\beta\right)d^{j-1/2}_{m'-1/2,m-1/2}(\beta)$$
$$-(j-m)^{\frac{1}{2}}\sin\left(\frac{1}{2}\beta\right)d^{j-1/2}_{m'-1/2,m+1/2}(\beta)$$
$$=(j+m')^{\frac{1}{2}}d^j_{m'm}(\beta)\,,$$
$$(j+m)^{\frac{1}{2}}\sin\left(\frac{1}{2}\beta\right)d^{j-1/2}_{m'+1/2,m-1/2}(\beta)$$
$$+(j-m)^{\frac{1}{2}}\cos\left(\frac{1}{2}\beta\right)d^{j-1/2}_{m'+1/2,m+1/2}(\beta)$$
$$=(j-m')^{\frac{1}{2}}d^j_{m'm}(\beta)\,.$$

Two useful relations implied by the above are:

$$[(j-m)(j+m+1)]^{\frac{1}{2}}\sin\beta\,d^j_{m',m+1}(\beta)$$
$$+[(j+m)(j-m+1)]^{\frac{1}{2}}\sin\beta\,d^j_{m',m-1}(\beta)$$
$$=2(m\cos\beta-m')d^j_{m'm}(\beta)\,,$$
$$[(j+m)(j-m+1)]^{\frac{1}{2}}d^j_{m',m-1}(\beta)$$
$$+\left[(j+m')(j-m'+1)\right]^{\frac{1}{2}}d^j_{m'-1,m}(\beta)$$
$$=(m-m')\cot\left(\frac{1}{2}\beta\right)d^j_{m'm}(\beta)\,.$$

By considering

$$D^j\times D^1 = D^{j+1}\oplus D^j\oplus D^{j-1}\,,$$

one can also readily derive the matrix elements of the direction cosines specifying the orientation of the body-fixed frame $(\hat{\boldsymbol{f}}_1, \hat{\boldsymbol{f}}_2, \hat{\boldsymbol{f}}_3)$ of a symmetric rotor relative to the inertial frame $(\hat{e}_1, \hat{e}_2, \hat{e}_3)$:

$$\lambda_{\mu,\nu}\Psi^j_{m,m'} = \sum_{j'}\left(\frac{2j+1}{2j'+1}\right)^{\frac{1}{2}} \times C^{j1j'}_{m\mu m+\mu} C^{j1j'}_{m'\nu m'+\nu} \Psi^{j'}_{m+\mu,m'+\nu} ,$$

where the wave functions are those defined for integral j by (2.37), for half-integral j by (2.36), and

$$\begin{aligned}
\lambda_{\mu,\nu} &= \hat{e}_\mu \cdot \hat{f}^*_\nu = \left(D^1_{\mu,\nu}\right)^* , \quad \mu,\nu = -1,0,+1 ;\\
\hat{e}_{+1} &= -(\hat{e}_1 + \mathrm{i}\hat{e}_2)/\sqrt{2} , \qquad \hat{e}_0 = \hat{e}_3 ,\\
\hat{e}_{-1} &= (\hat{e}_1 - \mathrm{i}\hat{e}_2)/\sqrt{2} ,\\
\hat{f}_{+1} &= -(\hat{f}_1 + \mathrm{i}\hat{f}_2)/\sqrt{2} , \qquad \hat{f}_0 = \hat{f}_3 ,\\
\hat{f}_{-1} &= (\hat{f}_1 - \mathrm{i}\hat{f}_2)/\sqrt{2} .
\end{aligned}$$

2.3.6 Symmetry Relations

Symmetry relations for the representation functions $D^j_{m'm}(Z)$ defined by (2.24) are associated with the action of a finite group G on the set $M(2,2)$ of complex 2×2 matrices: $g : M(2,2) \to M(2,2)$, $g \in G$. Equivalently, if $Z \in M(2,2)$ is parametrized by a set Ω of parameters $\omega \in \Omega$ (parameter space), then g may be taken to act directly in the parameter space $g : \Omega \to \Omega$. The action, denoted $\Box$, of a group $G = \{e, g, g', \ldots\}$ (e = identity) on a set $X = \{x, x', \ldots\}$ must satisfy the rules

$$\begin{aligned}
&g : X \to X \quad e\Box x = x, \text{ all } x \in X ,\\
&g'\Box(g\Box x) = (g'g)\Box x, \text{ all } g', g \in G, \text{ all } x \in X .
\end{aligned} \tag{2.28}$$

Using $\cdot$ to denote the action of G on $M(2,2)$ and $\Box$ to denote the action of G on Ω, one has the relation:

$$(g\cdot Z)(\omega) = Z(g^{-1}\Box\omega) .$$

Only finite subgroups G of the unitary group $U(2)$ (group of 2×2 unitary matrices) are considered here: $G \subset U(2)$.

Generally, when G acts on $M(2,2)$, it effects a unitary linear transformation of the set of functions $\{D^j_{m'm}\}$ (j fixed) defined over $Z \in M(2,2)$. For certain groups G, or for some elements of G, a single function $D^j_{\mu'\mu} \in \{D^j_{m'm}\}$ occurs in the transformation, so that

$$\begin{aligned}
&\left(g\cdot D^j_{m'm}\right)(Z) = D^j_{m'm}(g^{-1}\Box Z)\\
&= g_{m'm} D^j_{\mu'\mu}(Z) ,\\
&(\mu'\mu)\\
&\in \{(\lambda' m', \lambda m), (\lambda m, \lambda' m') | \lambda' = \pm 1, \lambda = \pm 1\} ,
\end{aligned} \tag{2.29}$$

where $g_{m'm}$ is a complex number of unit modulus. Relation (2.29) is called a symmetry relation of $D^j_{m'm}$ with respect to g. Usually not all elements in G correspond to symmetry relations. In a symmetry relation, the action of the group is effectively transferred to the discrete quantum labels themselves:

$$\begin{aligned}
g : m' &\to \mu' = m'(g) ,\\
m &\to \mu = m(g) .
\end{aligned} \tag{2.30}$$

In terms of a parametrization Ω of $M(2,2)$, relation (2.29) is written

$$\begin{aligned}
\left(gD^j_{m'm}\right)(\omega) &= D^j_{m'm}(g^{-1}\Box\omega)\\
&= g_{m'm} D^j_{\mu'\mu}(\omega) .
\end{aligned} \tag{2.31}$$

In practice, action symbols such as $\cdot$ and $\Box$ are often dropped in favor of juxtaposition, when the context is clear. Moreover the set of complex matrices $M(2,2)$ may be replaced by $U(2)$ or $SU(2)$ whenever the action conditions (2.28) are satisfied. Relations (2.29–2.31) are illustrated below by examples.

There are several finite subgroups of interest with various group-subgroup relations between them:

1. Pauli group:

   $$\begin{aligned}
   P &= \{\sigma_\mu, -\sigma_\mu, \mathrm{i}\sigma_\mu, -\mathrm{i}\sigma_\mu | \mu = 0,1,2,3\} ,\\
   |P| &= 16 .
   \end{aligned}$$

   Each element of this group is an element of $U(2)$. The action of the group P may therefore be defined on the group $U(2)$ by left and right actions as discussed in Sect. 2.4.1.
2. Symmetric groupS_4:

   $S_4 = \{p | p$ is a permutation of the four Euler–Rodrigues parameters $(\alpha_0, \alpha_1, \alpha_2, \sigma_3)\}$, $|S_4| = 24$. Points in S^3 are mapped to distinct points in S^3; hence, one can take $Z \in SU(2)$, and define the group action directly from $U(\alpha_0, \boldsymbol{\alpha})$ in (2.20). It is simpler, however, to define the action of the group directly on the representation functions (2.22). Not all elements of this group define a symmetry in the sense defined by (2.29) (see below).
3. Abelian group T:

   $$\begin{aligned}
   T &= \{(t_0, t_1, t_2, t_3) | \text{ each } t_\mu = \pm 1\} ,\\
   |T| &= 16 .
   \end{aligned}$$

   Group multiplication is component-wise multiplication and the identity is (1,1,1,1). The action of an element of T is defined directly on the

Euler–Rodrigues parameters by component-wise multiplication, thus mapping points in S^3 to points in S^3; hence, one can take $Z \in SU(2)$. This group is isomorphic to the direct product group $S_2 \times S_2 \times S_2 \times S_2$, S_2 = symmetric group on two distinct objects.

4. Group G:

$$G = \langle \mathcal{R}, \mathcal{C}, \mathcal{T}, \mathcal{K} \rangle , \quad |G| = 32 ,$$

where $\mathcal{R}$, $\mathcal{C}$, $\mathcal{T}$, $\mathcal{K}$ denote the operations of row interchange, column interchange, transposition, and conjugation (see below) of an arbitrary matrix.

$$Z = \begin{pmatrix} a & b \\ c & d \end{pmatrix}$$

The notation $\langle\ \rangle$ designates that the enclosed elements generate G.

It is impossible to give here all the interrelationships among the groups defined in (1)–(4). Instead, some relations are listed as obtained directly from either $D^j_{m'm}(Z)$ defined by (2.24) or $D^j_{m'm}(\alpha_0, \boldsymbol{\alpha})$ defined by (2.22). The actions of the groups T and G defined in (3) and (4) are fully given.

Abelian group T of order 16:

Generators:

$$\begin{aligned}
&T = \langle t_0, t_1, t_2, t_3 \rangle , \quad t_0 = (-1, 1, 1, 1) ,\\
&t_1 = (1, -1, 1, 1) , \quad t_2 = (1, 1, -1, 1) ,\\
&t_3 = (1, 1, 1, -1) .
\end{aligned}$$

Group action:

$$\begin{aligned}
&t \cdot a = (t_0\alpha_0, t_1\alpha_1, t_2\alpha_2, t_3\alpha_3) ,\\
&\text{each } t = (t_0, t_1, t_2, t_3) \in T ,\\
&\text{each } a = (\alpha_0, \alpha_1, \alpha_2, \alpha_3) \in S^3,\\
&(tF)(a) = F(t \cdot a) ,\\
&t_0 D^j_{m'm} = (-1)^{m'-m} D^j_{-m-m'} ,\\
&t_1 D^j_{m'm} = (-1)^{m'-m} D^j_{mm'} ,\\
&t_2 D^j_{m'm} = D^j_{mm'} ,\\
&t_3 D^j_{m'm} = D^j_{-m-m'} .
\end{aligned}$$

Group G of order 32:

Generators:

$$G = \langle \mathcal{R}, \mathcal{C}, \mathcal{T}, \mathcal{K} \rangle ,$$

Generator actions:

$$(\mathcal{R}F)\begin{pmatrix} a & b \\ c & d \end{pmatrix} = F\begin{pmatrix} c & d \\ a & b \end{pmatrix} , \quad \text{row interchange}$$

$$(\mathcal{C}F)\begin{pmatrix} a & b \\ c & d \end{pmatrix} = F\begin{pmatrix} b & a \\ d & c \end{pmatrix} , \quad \text{column interchange}$$

$$(\mathcal{T}F)\begin{pmatrix} a & b \\ c & d \end{pmatrix} = F\begin{pmatrix} a & c \\ b & d \end{pmatrix} , \quad \text{transposition}$$

$$(\mathcal{K}F)\begin{pmatrix} a & b \\ c & d \end{pmatrix} = F\begin{pmatrix} d & -c \\ -b & a \end{pmatrix} , \quad \text{conjugation}$$

Subgroup H:

$$\begin{aligned}
H &= \langle \mathcal{R}, \mathcal{C}, \mathcal{T} \rangle\\
&= \{1, \mathcal{R}, \mathcal{C}, \mathcal{T}, \mathcal{RC} = \mathcal{CR}, \mathcal{TR} = \mathcal{CT}, \mathcal{TC}\\
&= \mathcal{RT}, \mathcal{RCT}\}
\end{aligned}$$

with relations in H given by

$$\begin{aligned}
&\mathcal{R}^2 = \mathcal{C}^2 = \mathcal{T}^2 = 1 ,\\
&\mathcal{TRC} = \mathcal{TCR} = \mathcal{RCT} = \mathcal{CRT} ,\\
&\mathcal{RTC} = \mathcal{CTR} = \mathcal{T} .
\end{aligned}$$

Adjoining the idempotent element $\mathcal{K}$ to H gives the group G of order 32:

$$G = \{H, H\mathcal{K}, H\mathcal{KR}, H\mathcal{KRK}\} .$$

Symmetry relations:

$$\begin{aligned}
&\mathcal{R}D^j_{m'm} = D^j_{-m'm} ,\\
&\mathcal{C}D^j_{m'm} = D^j_{m'-m} ,\\
&\mathcal{T}D^j_{m'm} = D^j_{mm'} ,\\
&\mathcal{K}D^j_{m'm} = (-1)^{m'-m} D^j_{-m'-m} .
\end{aligned} \tag{2.32}$$

These function relations are valid for $D^j_{m'm}$ defined over the arbitrary matrix Z defined by (2.24). They are also true for $Z = U \in SU(2)$, but now the operations $\mathcal{R}$ and $\mathcal{C}$ change the sign of the determinant of the matrix Z so that the transformed matrix no longer belongs to $SU(2)$. It does, however, belong to $U(2)$, the group of all 2×2 unitary matrices. The special irreducible representation functions of $U(2)$ defined by (2.24),

$$D^j_{m'm}(U) , \quad U \in U(2) ,$$

possess each of the 32 symmetries corresponding to the operations in the group G. [There exist other irreducible representations of U(2), involving $\det U$.] The operation $\mathcal{K}$ is closely related to complex conjugation, since for each $U \in U(2)$, $U = (u_{ij})$, one can write

$$U^* = (\det U)^{-1} \begin{pmatrix} u_{22} & -u_{21} \\ -u_{12} & u_{11} \end{pmatrix},$$

$$\begin{aligned}\left(\mathcal{K} D^j_{m'm}\right)(U) &= (\det U)^{2j} D^j_{m'm}(U^*) \\ &= (\det U)^{2j} D^{j*}_{m'm}(U) \\ &= (-1)^{m'-m} D^j_{-m'-m}(U) . \end{aligned} \quad (2.33)$$

Relations (2.32) and (2.33) are valid in an arbitrary parametrization of $U \in U(2)$. In terms of the parametrization

$$U(\chi, \alpha_0, \boldsymbol{\alpha}) = e^{i\chi/2} U(\alpha_0, \boldsymbol{\alpha}) , \quad 0 \le \chi \le 2\pi ,$$

where $U(\alpha_0, \boldsymbol{\alpha}) \in SU(2)$ is the Euler–Rodrigues parametrization, the actions of $\mathcal{R}$, $\mathcal{C}$, $\mathcal{T}$, and $\mathcal{K}$ correspond to the following transformations in parameter space:

$$\begin{aligned}\mathcal{R} :\ & \chi \to \chi + \pi, (\alpha_0, \alpha_1, \alpha_2, \alpha_3) \\ & \to (-\alpha_1, \alpha_0, -\alpha_3, \alpha_2) , \\ \mathcal{C} :\ & \chi \to \chi + \pi, (\alpha_0, \alpha_1, \alpha_2, \alpha_3) \\ & \to (-\alpha_1, \alpha_0, \alpha_3, -\alpha_2) , \\ \mathcal{T} :\ & \chi \to \chi, (\alpha_0, \alpha_1, \alpha_2, \alpha_3) \to (\alpha_0, \alpha_1, -\alpha_2, \alpha_3) , \\ \mathcal{C} :\ & \chi \to \chi, (\alpha_0, \alpha_1, \alpha_2, \alpha_3) \\ & \to (\alpha_0, -\alpha_1, \alpha_2, -\alpha_3) . \end{aligned}$$

The new angle $\chi' = \chi + \pi$ is to be identified with the corresponding point on the unit circle so that these mappings are always in the parameter space, which is the sphere S^3 together with the unit circle for χ. Observe that the following identities hold for functions over $SU(2)$; hence, over $U(2)$:

$$\mathcal{C} = T_{t_1} T_{t_3} , \quad \mathcal{T} = T_{t_2} .$$

Abelian subgroup of S_4:

Generators:

$K = \langle (0,3), (1,2) \rangle$, where $(0,3)$ and $(1,2)$ denote transpositions in S_4, $|K| = 4$.

Group action in parameter space:

$$\begin{aligned}(0,3)(\alpha_0, \alpha_1, \alpha_2, \alpha_3) &= (\alpha_3, \alpha_1, \alpha_2, \alpha_0) , \\ (1,2)(\alpha_0, \alpha_1, \alpha_2, \alpha_3) &= (\alpha_0, \alpha_2, \alpha_1, \alpha_3) .\end{aligned}$$

Symmetry relations:

$$\begin{aligned}(0,3) D^j_{m'm} &= (-i)^{m'+m} D^j_{-m-m'} , \\ (1,2) D^j_{m'm} &= (-i)^{m'-m} D^j_{mm'} .\end{aligned}$$

Diagonal subgroup Σ of the direct product group $P \times P$ (P = Pauli group):

Group elements:

$$\Sigma = \{(\sigma, \sigma) | \sigma \in P\} , \quad |\Sigma| = 16 .$$

Group action:

$$\begin{aligned}&(\sigma, \sigma) : U \to \sigma U \sigma^{\mathrm{T}} , \quad \text{each } \sigma \in P , \\ &[(\sigma, \sigma) F](U) = F(\sigma^{\mathrm{T}} U \sigma) .\end{aligned}$$

Example: $\sigma = i\sigma_2$:

$$\begin{aligned}&(\sigma, \sigma) : (\alpha_0, \alpha_1, \alpha_2, \alpha_3) \to (\alpha_0, -\alpha_1, \alpha_2, -\alpha_3) , \\ &[(\sigma, \sigma) F](\alpha_0, \alpha_1, \alpha_2, \alpha_3) = F(\alpha_0, -\alpha_1, \alpha_2, -\alpha_3) , \\ &(\sigma, \sigma) = t_1 t_2 \text{ on functions over } U(2) .\end{aligned}$$

The relations presented above barely touch on the interrelations among the finite groups introduced in (1)–(4). Symmetry relations (2.32) and (2.33), however, give the symmetries of the $d^j_{m'm}(\beta)$ given in Sect. 2.3.3 in the Euler angle parametrization. In general, it is quite tedious to present the above symmetries in terms of Euler angles, with χ adjoined when necessary, because the Euler angles are not uniquely determined by the points of S^3.

2.4 Group and Lie Algebra Actions

The concept of a group acting on a set is fundamental to applications of group theory to physical problems. Because of the unity that this notion brings to angular momentum theory, it is well worth a brief review in a setting in which a matrix group acts on the set of complex matrices. Thus, let $G \subseteq GL(n, \boldsymbol{C})$ and $H \subseteq GL(m, \boldsymbol{C})$ denote arbitrary subgroups, respectively, of the general linear groups of $n \times n$ and $m \times m$ nonsingular complex matrices, and let $M(n, m)$ denote the set of $n \times m$ complex matrices. A matrix $Z \in M(n, m)$ has row and column entries (z_i^α), $i = 1, 2, \ldots, n$; $\alpha = 1, 2, \ldots, m$.

2.4.1 Matrix Group Actions

Left and right translations of $Z \in M(n, m)$:

$$L_g Z = gZ, \quad \text{each } g \in G, \quad \text{each } Z \in M(n, m),$$
$$R_h Z = Zh^{\mathrm{T}}, \quad \text{each } h \in H, \quad \text{each } Z \in M(n, m).$$

(T denotes matrix transposition.)
Left and right translations commute:

$$Z' = L_g(R_h Z) = R_h(L_g Z), \quad \text{each } g \in G, h \in H,$$
$$Z \in M(n, m).$$

Equivalent form as a transformation on $z \in \boldsymbol{C}^{nm}$:

$$z' = (g \times h)z,$$

where $\times$ denotes the direct product of g and h; the column matrix z (resp., z') is obtained from the columns of Z (resp., Z'), $z^\alpha, \alpha = 1, 2, \ldots, m$, of the $n \times m$ matrix Z as successive entries in a single column vector $z \in \boldsymbol{C}^{nm}$.
Left and right translations in function space:

$$(\mathcal{L}_g f)(Z) = f(g^{\mathrm{T}} Z), \quad \text{each } g \in G,$$
$$(\mathcal{R}_h f)(Z) = f(Zh), \quad \text{each } h \in H,$$

where $f(Z) = f\left(z_i^\alpha\right)$, and the commuting property holds for all well-defined functions f:

$$\mathcal{L}_g(\mathcal{R}_h f) = \mathcal{R}_h(\mathcal{L}_g f).$$

2.4.2 Lie Algebra Actions

Lie algebra of left and right translations:

$$(D_X f)(Z) = \mathrm{i}\frac{\mathrm{d}}{\mathrm{d}t} f\left(\mathrm{e}^{-\mathrm{i}tX^T} Z\right)|_{t=0},$$
$$(D^Y f)(Z) = \mathrm{i}\frac{\mathrm{d}}{\mathrm{d}t} f\left(Z\mathrm{e}^{-\mathrm{i}tY}\right)|_{t=0};$$
$$D_X = \mathrm{Tr}\left(Z^{\mathrm{T}} X \partial/\partial Z\right), \quad \text{each } X \in L(G),$$
$$D^Y = \mathrm{Tr}\left(Y^{\mathrm{T}} Z^{\mathrm{T}} \partial/\partial Z\right), \quad \text{each } Y \in L(H),$$
$$L(G) = \text{Lie algebra of } G,$$
$$L(H) = \text{Lie algebra of } H.$$

Linear derivations:

$$D_{\alpha X + \beta X'} = \alpha D_X + \beta D_{X'}, \quad \alpha, \beta \in \boldsymbol{C},$$
$$[D_X, D_{X'}] = D_{[X, X']},$$

D^Y obeys these same rules.
Commuting property of left and right derivations:

$$\left[D_X, D^Y\right] = 0, \quad X \in L(G), \quad Y \in L(H).$$

Basis set:

$$D_X = \sum_{i,j=1}^{n} x_{ij} D_{ij}, \quad X = (x_{ij}),$$
$$D^Y = \sum_{\alpha,\beta=1}^{m} y^{\alpha\beta} D^{\alpha\beta}, \quad Y = (y^{\alpha\beta}),$$
$$D_{ij} = \sum_{\alpha=1}^{m} z_i^\alpha \frac{\partial}{\partial z_j^\alpha},$$
$$D^{\alpha\beta} = \sum_{i=1}^{n} z_i^\alpha \frac{\partial}{\partial z_i^\beta}.$$

Commutation rules:

$$[D_{ij}, D_{kl}] = \delta_{jk} D_{il} - \delta_{il} D_{kj},$$
$$[D^{\alpha\beta}, D^{\gamma\epsilon}] = \delta^{\beta\gamma} D^{\alpha\epsilon} - \delta^{\alpha\epsilon} D^{\gamma\beta},$$
$$[D_{ij}, D^{\alpha\beta}] = 0,$$

where $i, j, k, l = 1, 2, \ldots, n$ and $\alpha, \beta, \gamma, \epsilon = 1, 2, \ldots, m$. The operator sets $\{D_{ij}\}$ and $\{D^{\alpha\beta}\}$ are realizations of the Weyl generators of $GL(n, \boldsymbol{C})$ and $GL(m, \boldsymbol{C})$, respectively.

2.4.3 Hilbert Spaces

Space of polynomials with inner product:

$$(P, P') = P^*(\partial/\partial Z)P'(Z)|_{Z=0}.$$

Bargmann space of entire functions with inner product:

$$\langle F, F' \rangle = \int F^*(Z)F'(Z)\,\mathrm{d}\mu(Z),$$

$$\mathrm{d}\mu(Z) = \pi^{-nm} \exp\left(-\sum_{i,\alpha} z_i^{\alpha*} z_i^\alpha\right) \prod_{i,\alpha} \mathrm{d}x_i^\alpha\, \mathrm{d}y_i^\alpha,$$

$$z_i^\alpha = x_i^\alpha + \mathrm{i}y_i^\alpha, \; i = 1, 2, \ldots, n; \; \alpha = 1, 2, \ldots, m.$$

Numerical equality of inner products:

$$(P, P') = \langle P, P' \rangle.$$

2.4.4 Relation to Angular Momentum Theory

Spinorial Realization of Sects. 2.4.2 and 2.4.3:

$$G = SU(2), \quad H = (1), \quad Z \in M(2, 1),$$
$$z = \mathrm{col}(z_1, z_2),$$
$X =$ set of 2×2 traceless, Hermitian matrices,
$$(\mathcal{R}_U f)(z) = f(U^{\mathrm{T}} z),$$
$$D_{\sigma_i/2} = (z^{\mathrm{T}} \sigma_i \partial/\partial z)/2,$$

$$J_\pm = D_{\sigma_1/2} \pm \mathrm{i} D_{\sigma_2/2}\,, \quad J_3 = D_{\sigma_3/2}\,,$$
$$J_+ = z_1 \partial/\partial z_2\,, \quad J_- = z_2 \partial/\partial z_1\,,$$
$$J_3 = (1/2)(z_1 \partial/\partial z_1 - z_2 \partial/\partial z_2)\,,$$
$$(P, P') = P^*(\partial/\partial z_1, \partial/\partial z_2) P(z_1, z_2)|_{z_1=z_2=0}\,.$$

Orthonormal basis:

$$P_{jm}(z_1, z_2) = z_1^{j+m} z_2^{j-m} / [(j+m)!(j-m)!]^{\frac{1}{2}}\,,$$
$$j = 0, 1/2, 1, 3/2, \ldots\,; \quad m = j, j-1, \ldots, -j\,.$$

Standard action:

$$\boldsymbol{J}^2 P_{jm}(z) = j(j+1) P_{jm}(z)\,,$$
$$J_3 P_{jm}(z) = m P_{jm}(z)\,,$$
$$J_\pm P_{jm}(z) = [(j \mp m)(j \pm m + 1)]^{\frac{1}{2}} P_{j,m\pm 1}(z)\,.$$

Group transformation:

$$(\mathcal{R}_U P_{jm})(z) = \sum_{m'} D^j_{m'm}(U) P_{jm'}(z)\,,$$

where the representation functions are given by (2.17).

The 2-Spinorial Realization of Sects. 2.4.2 and 2.4.3:

$$G = H = SU(2)\,, \qquad Z \in M(2,2)\,,$$
$$Z = \left[z^1 z^2\right], \qquad z^\alpha = \mathrm{col}\left(z_1^\alpha z_2^\alpha\right),$$

$$X = Y = \text{ set of } 2\times 2 \text{ traceless, Hermitian matrices}\,,$$

$$(\mathcal{R}_U f)(Z) = f(U^{\mathrm{T}} Z)\,, \quad (\mathcal{L}_V f)(Z) = f(ZV)\,,$$
$$U, V \in SU(2)\,,$$
$$D_{\sigma_i/2} = \mathrm{Tr}(Z^{\mathrm{T}} \sigma_i \partial/\partial Z)/2\,,$$
$$D^{\sigma_i/2} = \mathrm{Tr}(\sigma_i Z^{\mathrm{T}} \partial/\partial Z)/2\,.$$

$$M_\pm = D_{\sigma_1/2} \pm \mathrm{i} D_{\sigma_2/2}\,, \qquad M_3 = D_{\sigma_3/2}\,,$$
$$K_\pm = D^{\sigma_1/2} \pm \mathrm{i} D^{\sigma_2/2}\,, \qquad K_3 = D^{\sigma_3/2}\,,$$
$$M_+ = \sum_{\alpha=1}^{2} z_1^\alpha \partial/\partial z_2^\alpha\,, \quad M_- = \sum_{\alpha=1}^{2} z_2^\alpha \partial/\partial z_1^\alpha\,,$$
$$M_3 = \frac{1}{2} \sum_{\alpha=1}^{2} \left(z_1^\alpha \partial/\partial z_1^\alpha - z_2^\alpha \partial/\partial z_2^\alpha\right),$$
$$K_+ = \sum_{i=1}^{2} z_i^1 \partial/\partial z_i^2\,, \quad K_- = \sum_{i=1}^{2} z_i^2 \partial/\partial z_i^1\,,$$
$$K_3 = \frac{1}{2} \sum_{i=1}^{2} \left(z_i^1 \partial/\partial z_i^1 - z_i^2 \partial/\partial z_i^2\right).$$

Mutual commutativity of Lie algebras:

$$[M_i, K_j] = 0\,, \quad i, j = 1, 2, 3\,.$$

Inner product:

$$(P, P') = P^*(Z) P'(\partial/\partial Z)|_{Z=0}\,,$$

Orthogonal basis (2.24):

$$D^j_{mm'}(Z)\,, \quad j = 0, 1/2, 1, 3/2, \ldots\,,$$
$$m = j, j-1, \ldots, -j\,;$$
$$m' = j, j-1, \ldots, -j\,;$$
$$\left(D^j_{mm'}, D^{j'}_{\mu\mu'}\right) = (2j)!\,\delta_{jj'} \delta_{m\mu} \delta_{m'\mu'}\,.$$

Equality of Casimir operators:

$$\boldsymbol{M}^2 = \boldsymbol{K}^2 = M_1^2 + M_2^2 + M_3^2 = K_1^2 + K_2^2 + K_3^2\,.$$

Standard actions:

$$\boldsymbol{M}^2 D^j_{mm'}(Z) = \boldsymbol{K}^2 D^j_{mm'}(Z) = j(j+1) D^j_{mm'}(Z)\,,$$
$$M_3 D^j_{mm'}(Z) = m D^j_{mm'}(Z)\,,$$
$$K_3 D^j_{mm'}(Z) = m' D^j_{mm'}(Z)\,,$$
$$M_\pm D^j_{mm'}(Z) = [(j \mp m)(j \pm m + 1)]^{\frac{1}{2}} D^j_{m\pm 1,m'}(Z)\,,$$
$$K_\pm D^j_{mm'}(Z) = [(j \mp m')(j \pm m' + 1)]^{\frac{1}{2}} \times D^j_{m,m'\pm 1}(Z)\,.$$

Special values:

$$D^j \begin{pmatrix} 1 & 0 \\ 0 & 1 \end{pmatrix} = I_{2j+1} = \text{ unit matrix}\,,$$
$$D^j_{mm'} \begin{pmatrix} z_1 & 0 \\ z_2 & 0 \end{pmatrix} = \delta_{jm'} P_{jm}(z_1, z_2)\,,$$
$$D^j_{mm'} \begin{pmatrix} 0 & z_1 \\ 0 & z_2 \end{pmatrix} = \delta_{jm} P_{jm'}(z_1, z_2)\,,$$
$$D^j_{mm'} \begin{pmatrix} z_1 & 0 \\ 0 & z_2 \end{pmatrix} = \delta_{mm'} z_1^{j+m} z_2^{j-m}\,,$$
$$D^j_{jj}(Z) = \left(z_1^1\right)^{2j}\,.$$

Symmetry relation:

$$\left[D^j(Z)\right]^{\mathrm{T}} = D^j(Z^{\mathrm{T}}) .$$

Generation from highest weight:

$$D^j_{mm'}(Z) = \left(\frac{(j+m)!}{(2j)!(j-m)!} \times \frac{(j+m')!}{(2j)!(j-m')!}\right)^{\frac{1}{2}} \times M_-^{j-m} K_-^{j-m'} D^j_{jj}(Z) .$$

Generating functions:

$$(\boldsymbol{x}^{\mathrm{T}} Z \boldsymbol{y})^{2j}/(2j)! = \sum_{mm'} P_{jm}(\boldsymbol{x}) D^j_{mm'}(Z) P_{jm'}(\boldsymbol{y}) ,$$

$$\exp(t\boldsymbol{x}^{\mathrm{T}} Z \boldsymbol{y}) = \sum_j t^{2j} \sum_{mm'} P_{jm}(\boldsymbol{x}) D^j_{mm'}(Z) P_{jm'}(\boldsymbol{y}) ,$$

$$\boldsymbol{x} = \mathrm{col}(x_1 x_2) , \quad \boldsymbol{y} = \mathrm{col}(y_1 y_2) ,$$

$$Z = (z_i^\alpha) , \quad i, \alpha = 1, 2 ; \quad \text{all indeterminates} .$$

2.5 Differential Operator Realizations of Angular Momentum

Differential operators realizing the standard commutation relations (2.7) and (2.8) can be obtained from the 2-spinorial realizations given in Sect. 2.4.4 by specializing the matrix Z to the appropriate unitary unimodular matrix $U \in SU(2)$ and using the chain rule of elementary calculus. Similarly, one obtains the explicit functions $D^j_{mm'}$ simply by substituting for Z the parametrized U in (2.24). This procedure is used in this section to obtain all the realizations listed. The notations $\boldsymbol{M} = (M_1, M_2, M_3)$ and $\boldsymbol{K} = (K_1, K_2, K_3)$ and the associated $M_\pm$ and $K_\pm$ refer to the differential operators given by the 2-spinorial realization now transformed to the parameters in question.

Euler angles with $Z = U(\alpha\beta\gamma)$ (Sect. 2.3.1):

$$M_3 = \mathrm{i}\partial/\partial\alpha , \quad K_3 = \mathrm{i}\partial/\partial\gamma ,$$

$$\frac{1}{2}\left(\mathrm{e}^{\mathrm{i}\alpha} M_+ - \mathrm{e}^{-\mathrm{i}\alpha} M_-\right) = \frac{1}{2}\left(\mathrm{e}^{-\mathrm{i}\gamma} K_- - \mathrm{e}^{\mathrm{i}\gamma} K_+\right) = \frac{\partial}{\partial\beta} ,$$

$$\frac{1}{2}\left(\mathrm{e}^{\mathrm{i}\alpha} M_+ + \mathrm{e}^{-\mathrm{i}\alpha} M_-\right) = -(\cot\beta)\, M_3 + (\sin\beta)^{-1}\, K_3 ,$$

$$\frac{1}{2}\left(\mathrm{e}^{-\mathrm{i}\gamma} K_- + \mathrm{e}^{\mathrm{i}\gamma} K_+\right) = (\cot\beta)\, K_3 - (\sin\beta)^{-1} M_3 ,$$

$$M_+ = \mathrm{e}^{-\mathrm{i}\alpha}[\partial/\partial\beta - (\cot\beta)\, M_3 + (\sin\beta)^{-1}\, K_3] ,$$

$$M_- = \mathrm{e}^{\mathrm{i}\alpha}[-\partial/\partial\beta - (\cot\beta)\, M_3 + (\sin\beta)^{-1}\, K_3] ,$$

$$K_+ = \mathrm{e}^{-\mathrm{i}\gamma}[-\partial/\partial\beta + (\cot\beta)\, K_3 - (\sin\beta)^{-1}\, M_3] ,$$

$$K_- = \mathrm{e}^{\mathrm{i}\gamma}[\partial/\partial\beta + (\cot\beta)\, K_3 - (\sin\beta)^{-1}\, M_3] .$$

Euler angles with $Z = U^*(\alpha\beta\gamma)$ [replace i by $-$i in the above relations]:

$$M_3 = -\mathrm{i}\partial/\partial\alpha , \quad K_3 = -\mathrm{i}\partial/\partial\gamma , \tag{2.34}$$

$$M_\pm = \mathrm{e}^{\pm\mathrm{i}\alpha}\left[\pm\partial/\partial\beta - (\cot\beta) M_3 + (\sin\beta)^{-1} K_3\right] ,$$

$$K_\pm = \mathrm{e}^{\pm\mathrm{i}\gamma}\left[\mp\partial/\partial\beta + (\cot\beta) K_3 - (\sin\beta)^{-1} M_3\right] .$$

Since $D^j(U^*) = (D^j(U))^*$, which is denoted $D^{j*}(U)$, these operators have the standard action on the complex conjugate functions $D^{j*}_{mm'}(U)$.

Quaternionic variables. $(x_0, \boldsymbol{x}) \in \boldsymbol{R}^4$:

$$(x'_0, \boldsymbol{x}')(x_0, \boldsymbol{x}) = (x'_0 x_0 - \boldsymbol{x}'\cdot\boldsymbol{x},\, x'_0\boldsymbol{x} + x_0\boldsymbol{x}' + \boldsymbol{x}'\times\boldsymbol{x}) ;$$

$$Z = \begin{pmatrix} z_{11} & z_{12} \\ z_{21} & z_{22} \end{pmatrix} = \begin{pmatrix} x_0 - \mathrm{i}x_3 & -\mathrm{i}x_1 - x_2 \\ -\mathrm{i}x_1 + x_2 & x_0 + \mathrm{i}x_3 \end{pmatrix} ;$$

$$\frac{\partial}{\partial Z} = \begin{pmatrix} \partial/\partial z_{11} & \partial/\partial z_{12} \\ \partial/\partial z_{21} & \partial/\partial z_{22} \end{pmatrix} = \frac{1}{2}\begin{pmatrix} \partial/\partial x_0 + \mathrm{i}\partial/\partial x_3 & \mathrm{i}\partial/\partial x_1 - \partial/\partial x_2 \\ \mathrm{i}\partial/\partial x_1 + \partial/\partial x_2 & \partial/\partial x_0 - \mathrm{i}\partial/\partial x_3 \end{pmatrix} ;$$

$$M_i = \frac{1}{2}\mathrm{Tr}\left(Z^{\mathrm{T}}\sigma_i\partial/\partial Z\right) , \quad K_i = \frac{1}{2}\mathrm{Tr}\left(\sigma_i Z^{\mathrm{T}}\partial/\partial Z\right) .$$

(The form of $\partial/\partial Z$ is determined by the requirement $(\partial/\partial z_{ij}) z_{lk} = \delta_{il}\delta_{jk}$; for example, $\frac{1}{2}(\partial/\partial x_0 + \mathrm{i}\partial/\partial x_3)(x_0 - \mathrm{i}x_3) = 1$).

Define the six orbital angular momentum operators in $\boldsymbol{R}^4$ by

$$L_{jk} = -\mathrm{i}(x_j\partial/\partial x_k - x_k\partial/\partial x_j) , \quad j < k = 0, 1, 2, 3 ,$$

which may be written as the orbital angular momentum $\boldsymbol{L}$ in $\boldsymbol{R}^3$ together with the three operators $\boldsymbol{A}$ given by

$$\boldsymbol{L} = -\mathrm{i}\boldsymbol{x}\times\nabla , \quad L_1 = L_{23} ,$$

$$L_2 = L_{31} , \quad L_3 = L_{12} ,$$

$$\boldsymbol{A} = (A_1, A_2, A_3) = (L_{01}, L_{02}, L_{03}) .$$

Then, we have the following relations:

$$K_1 = (L_1 - A_1)/2 , \quad K_2 = (L_2 - A_2)/2 ,$$

$$K_3 = (L_3 - A_3)/2 ;$$

$$M_1 = -(L_1 + A_1)/2 , \quad M_2 = (L_2 + A_2)/2 ,$$

$$M_3 = -(L_3 + A_3)/2 .$$

Commutation rules:

$$\begin{aligned}
[M_j, K_k] &= 0\,, \quad j,k = 1,2,3\,,\\
\boldsymbol{M}\times\boldsymbol{M} &= \mathrm{i}\boldsymbol{M}\,, \quad \boldsymbol{K}\times\boldsymbol{K} = \mathrm{i}\boldsymbol{K}\,,\\
\boldsymbol{L}\times\boldsymbol{L} &= \mathrm{i}\boldsymbol{L}\,, \quad \boldsymbol{A}\times\boldsymbol{A} = \mathrm{i}\boldsymbol{L}\,,\\
[L_j, A_k] &= \mathrm{i}e_{jkl}A_l\,,
\end{aligned}$$

where $e_{jkl} = 1$ for j, k, l an even permutation of $1, 2, 3$; $e_{jkl} = -1$ for an odd permutation of $1, 2, 3$; $e_{jkl} = 0$, otherwise.

The $\boldsymbol{M} = (M_1, M_2, M_3)$ and $\boldsymbol{K} = (K_1, K_2, K_3)$ operators have the standard action given in Sect. 2.2 on the functions $D^j_{mm'}(x_0, \boldsymbol{x})$ defined by (2.22) (Replace α_0 by x_0 and $\boldsymbol{\alpha}$ by $\boldsymbol{x}$). Additional relations:

$$\begin{aligned}
\boldsymbol{K}^2 &= \boldsymbol{M}^2 = -\frac{1}{4}R^2\nabla_4^2 + K_0^2 + K_0\,,\\
R^2 &= x_0^2 + \boldsymbol{x}\cdot\boldsymbol{x}\,,\\
\nabla_4^2 &= \frac{\partial^2}{\partial x_0^2} + \nabla^2 = \sum_\mu \frac{\partial^2}{\partial x_\mu^2}\,,\\
K_0 &= \frac{1}{2}\mathrm{Tr}\left(z^{\mathrm{T}}\frac{\partial}{\partial Z}\right) = \frac{1}{2}\sum_{\mu=0}^{3} x_\mu \frac{\partial}{\partial x_\mu}\,;\\
K_0 D^j_{mm'}(x_0, \boldsymbol{x}) &= jD^j_{mm'}(x_0, \boldsymbol{x})\,,\\
\nabla_4^2 D^j_{mm'}(x_0, \boldsymbol{x}) &= 0\,;
\end{aligned}$$

$$(M_1, -M_2, M_3) = \left(\sum_i R_{i1}K_i, \sum_i R_{i2}K_i, \sum_i R_{i3}K_i\right),$$

$$R_{ij} = \frac{(x_0^2 - \boldsymbol{x}\cdot\boldsymbol{x})\delta_{ij} - 2e_{ijk}x_0x_k + 2x_ix_j}{x_0^2 + \boldsymbol{x}\cdot\boldsymbol{x}}\,,$$

each $(x_0, \boldsymbol{x}) \in \boldsymbol{R}^4$.

The relation $R_{ij} = R_{ij}(x_0, \boldsymbol{x})$ is a mapping of all points of four-space $\boldsymbol{R}^4$ (except the origin) onto the group of proper, orthogonal matrices; for $x_0^2 + \boldsymbol{x}\cdot\boldsymbol{x} = 1$, it is just the Euler–Rodrigues parametrization, (2.21).

The operators $\boldsymbol{R} = (-M_1, M_2, -M_3)$ and $\boldsymbol{K} = (K_1, K_2, K_3)$ have the standard action on $(-1)^{j+m} D^j_{-m,m'}(x_0, \boldsymbol{x})$, so that the orbital angular momentum in $\boldsymbol{R}^3$ is given by the addition $\boldsymbol{L} = \boldsymbol{R} + \boldsymbol{K}$. Thus, one finds:

$$\begin{aligned}
&\sum_{mm'} C^{jjL}_{mm'M}(-1)^{j+m}D^j_{-mm'}(x_0, \boldsymbol{x})\\
&= A_{2j,L}R^{2j-L}\mathcal{Y}_{LM}(\boldsymbol{x})C^{(L+1)}_{2j-L}(x_0/R)\,,\\
&A_{2j,L} = (2\mathrm{i})^L(-1)^{2j}L!\left(\frac{4\pi(2j-L)!}{(2j+L+1)!}\right)^{\frac{1}{2}}.
\end{aligned}$$

2.6 The Symmetric Rotor and Representation Functions

The rigid rotor is an important physical object and its quantum description enters into many physical theories. This description is an application of angular momentum theory with subtleties that need to be made explicit. It is customary to describe the classical rotor in terms of a right-handed triad of unit vectors $(\hat{f}_1, \hat{f}_2, \hat{f}_3)$ fixed in the rotor and constituting a principal axes system located at the center of mass. The instantaneous orientation of this body-fixed frame relative to a right-handed triad of unit vector $(\hat{e}_1, \hat{e}_2, \hat{e}_3)$ specifying an inertial frame, also located at the center of mass, is then given, say, in terms of Euler angles (one could use for this purpose any parametrization of a proper, orthogonal matrix). For Euler angles, the relationship is

$$\hat{f}_j = \sum_i R_{ij}(\alpha\beta\gamma)\hat{e}_i\,. \tag{2.35}$$

The Hamiltonian for the rigid rotor is then of the form

$$H = A\mathcal{P}_1^2 + B\mathcal{P}_2^2 + C\mathcal{P}_3^2\,,$$

where A, B, and C are physical constants related to the reciprocals of the principal moments of inertia, and the angular momenta $\mathcal{P}_j$ $(j = 1, 2, 3)$ are the components of the total angular momentum $\mathcal{J}$ referred to the body-fixed frame:

$$\begin{aligned}
\mathcal{P}_j &= \hat{f}_j\cdot\mathcal{J} = \sum_i R_{ij}(\alpha\beta\gamma)\mathcal{J}_i\,,\\
\mathcal{J} &= \hat{e}_1\mathcal{J}_1 + \hat{e}_2\mathcal{J}_2 + \hat{e}_3\mathcal{J}_3\,.
\end{aligned}$$

For the symmetric rotor (taking $A = B$), the Hamiltonian can be written in the form

$$H = a\boldsymbol{\mathcal{P}}^2 + b\mathcal{P}_3^2\,.$$

It is in the interpretation of this Hamiltonian for quantum mechanics that the subtleties already enter, since the nature of angular momentum components referred to a moving reference system must be treated correctly. Relation (2.35) shows that the body-fixed axes cannot commute with the components of the total angular momentum $\mathcal{J}$ referred to the frame $(\hat{e}_1, \hat{e}_2, \hat{e}_3)$. A position vector $\boldsymbol{x}$ and the orbital angular momentum $\boldsymbol{L}$, with components both referred to an inertial frame, satisfy the commutation relations $[L_j, x_k] = \mathrm{i}e_{jkl}x_l$, and for a rigid body thought of as a collection of point particles rotating together, the same conditions are to be enforced. Relative to the body-fixed frame, the vector $\boldsymbol{x}$ is expressed as

$$\sum_k x_k \hat{\boldsymbol{e}}_k = \sum_h a_h \hat{\boldsymbol{f}}_h , \quad \text{each } a_h = \text{constant} ,$$

$$x_k = \sum_h a_h R_{kh}(\alpha\beta\gamma) .$$

The direction cosines $R_{kh} = R_{kh}(\alpha\beta\gamma) = \hat{\boldsymbol{e}}_k \cdot \hat{\boldsymbol{f}}_h$ and the physical total angular momentum components referred to an inertial frame must satisfy

$$\left[\mathcal{J}_j, R_{kh}\right] = \mathrm{i}e_{jkl}R_{lh} , \quad \text{each } h = 1, 2, 3 ,$$

in complete analogy to $[L_j, x_k] = \mathrm{i}e_{jkl}x_l$.

The description of the angular momentum associated with a symmetric rigid rotor and the angular momentum states is summarized as follows [compare (2.35)]:
Physical total angular momentum $\mathcal{J}$ with components referred to $(\hat{\boldsymbol{e}}_1, \hat{\boldsymbol{e}}_2, \hat{\boldsymbol{e}}_3)$:

$$\mathcal{J}_1 = \mathrm{i}\cos\alpha\cot\beta\frac{\partial}{\partial\alpha} + \mathrm{i}\sin\alpha\frac{\partial}{\partial\beta} - \mathrm{i}\frac{\cos\alpha}{\sin\beta}\frac{\partial}{\partial\gamma} ,$$

$$\mathcal{J}_2 = \mathrm{i}\sin\alpha\cot\beta\frac{\partial}{\partial\alpha} - \mathrm{i}\cos\alpha\frac{\partial}{\partial\beta} - \mathrm{i}\frac{\sin\alpha}{\sin\beta}\frac{\partial}{\partial\gamma} ,$$

$$\mathcal{J}_3 = -\mathrm{i}\frac{\partial}{\partial\alpha} .$$

Physical angular momentum $\mathcal{J}$ with components referred to $\left(\hat{f}_1, \hat{f}_2, \hat{f}_3\right)$:

$$\mathcal{P}_1 = -\mathrm{i}\cos\gamma\cot\beta\frac{\partial}{\partial\gamma} - \mathrm{i}\sin\gamma\frac{\partial}{\partial\beta} + \mathrm{i}\frac{\cos\gamma}{\sin\beta}\frac{\partial}{\partial\alpha} ,$$

$$\mathcal{P}_2 = \mathrm{i}\sin\gamma\cot\beta\frac{\partial}{\partial\gamma} - \mathrm{i}\cos\gamma\frac{\partial}{\partial\beta} - \mathrm{i}\frac{\sin\gamma}{\sin\beta}\frac{\partial}{\partial\alpha} ,$$

$$\mathcal{P}_3 = -\mathrm{i}\frac{\partial}{\partial\gamma} .$$

Standard commutation of the $\mathcal{J}_i$:

$$[\mathcal{J}_i, \mathcal{J}_j] = \mathrm{i}\mathcal{J}_k , \quad i, j, k \text{ cyclic} .$$

$\mathcal{J}_i$ can stand to either side:

$$\mathcal{P}_j = \sum_i R_{ij}(\alpha\beta\gamma)\mathcal{J}_i , \quad \mathcal{J}_i = \sum_i \mathcal{J}_i R_{ij}(\alpha\beta\gamma) .$$

The famous Van Vleck factor of $-\mathrm{i}$ in the commutation of the $\mathcal{P}_i$:

$$[\mathcal{P}_i, \mathcal{P}_j] = -\mathrm{i}\mathcal{P}_k , \quad i, j, k \text{ cyclic} .$$

Mutual commutativity of the $\mathcal{J}_j$ and $\mathcal{P}_i$:

$$[\mathcal{P}_i, \mathcal{J}_j] = 0 , \quad i, j = 1, 2, 3 .$$

Same invariant (squared) total angular momentum:

$$\begin{aligned}\mathcal{P}_1^2 + \mathcal{P}_2^2 + \mathcal{P}_3^2 &= \mathcal{J}_1^2 + \mathcal{J}_2^2 + \mathcal{J}_3^2 = \mathcal{J}^2 \\ &= -\csc^2\beta\left(\frac{\partial^2}{\partial\alpha^2} + \frac{\partial^2}{\partial\gamma^2} - 2\cos\beta\frac{\partial^2}{\partial\alpha\partial\gamma}\right) \\ &\quad - \frac{\partial^2}{\partial\beta^2} - \cot\beta\frac{\partial}{\partial\beta} .\end{aligned}$$

Standard actions:

$$\begin{aligned}\mathcal{J}^2 D^{j*}_{mm'}(\alpha\beta\gamma) &= j(j+1)D^{j*}_{mm'}(\alpha\beta\gamma) , \\ \mathcal{J}_3 D^{j*}_{mm'}(\alpha\beta\gamma) &= m D^{j*}_{mm'}(\alpha\beta\gamma) , \\ \mathcal{P}_3 D^{j*}_{mm'}(\alpha\beta\gamma) &= m' D^{j*}_{mm'}(\alpha\beta\gamma) ; \\ \mathcal{J}_\pm D^{j*}_{mm'}(\alpha\beta\gamma) &= [(j\mp m)(j\pm m+1)]^{\frac{1}{2}} \\ &\quad \times D^{j*}_{m\pm 1,m'}(\alpha\beta\gamma) , \\ (\mathcal{P}_1 - \mathrm{i}\mathcal{P}_2)D^{j*}_{mm'}(\alpha\beta\gamma) &= [(j-m')(j+m'+1)]^{\frac{1}{2}} \\ &\quad \times D^{j*}_{m,m'+1}(\alpha\beta\gamma) , \\ (\mathcal{P}_1 + \mathrm{i}\mathcal{P}_2)D^{j*}_{mm'}(\alpha\beta\gamma) &= [(j+m')(j-m'+1)]^{\frac{1}{2}} \\ &\quad \times D^{j*}_{m,m'-1}(\alpha\beta\gamma) .\end{aligned}$$

Normalized wave functions:

Integral or half-integral j ($SU(2)$ solid body):

$$\left\langle \alpha\beta\gamma \middle| \begin{matrix} j \\ mm' \end{matrix} \right\rangle = \sqrt{\frac{2j+1}{2\pi^2}} D^{j*}_{mm'}(\alpha\beta\gamma) , \tag{2.36}$$

with inner product

$$\langle F | F' \rangle = \int \mathrm{d}\Omega \, \langle \alpha\beta\gamma | F \rangle^* \langle \alpha\beta\gamma | F' \rangle ,$$

where $\mathrm{d}\Omega$ is defined by (2.25) and the integration extends over all α, β, γ given by (2.26).

Integral j (collection of "rigid" point particles):

$$\Psi^j_{mm'}(\alpha\beta\gamma) = \sqrt{\frac{2j+1}{8\pi^2}} D^{j*}_{mm'}(\alpha\beta\gamma) , \tag{2.37}$$

with inner product

$$(\Psi, \Psi') = \int_0^{2\pi} \mathrm{d}\alpha \int_0^{\pi} \mathrm{d}\beta \sin\beta \int_0^{2\pi} \mathrm{d}\gamma \, F^*(\alpha\beta\gamma) F'(\alpha\beta\gamma) .$$

The concept of a solid (impenetrable) body is conceptually distinct from that of a collection of point particles moving collectively together in translation and rotation.

2.7 Wigner–Clebsch–Gordan and 3-j Coefficients

Wigner–Clebsch–Gordan (WCG) coefficients (also called vector coupling coefficients) enter the theory of angular momentum in several ways: (1) as the coefficients in the real, orthogonal matrix that reduces the Kronecker product of two irreducible representations of the quantal rotation group into a direct sum of irreducibles; (2) as the coupling coefficients for constructing basis states of sharp angular momentum in the tensor product space from basis states of sharp angular momentum spanning the two constituent spaces; (3) as purely combinatoric objects in the expansion of a power of a 3×3 determinant; and (4) as coupling coefficients in the algebra of tensor operators. These perspectives are intimately connected, but have a different focus: the first considers the group itself to be primary and views the Lie algebra as the secondary or derived concept; the second considers the Lie algebra and the construction of the vector spaces that carry irreducible representations as primary, and views the representations carried by these spaces as derived quantities; the third is a mathematical construction, at first seeming almost empty of angular momentum concepts, yet the most revealing in showing the symmetry and other properties of the WCG-coefficients; and the fourth is the natural extension of (2) to operators, recognizing that the set of mappings of a vector space into itself is also a vector space. The subject of tensor operator algebra is considered in the next section because of its special importance for physical applications. This section summarizes formulas relating to the first three viewpoints, giving also the explicit mathematical expression of the coefficients in their several forms.

Either viewpoint, (1) or (2), may be taken as an interpretation of the Clebsch–Gordan series, which expresses abstractly the reduction of a Kronecker product of matrices (denoted $\times$) into a direct sum (denoted $\oplus$) of matrices:

$$\begin{aligned}[j_1]\times[j_2] &= \sum_{j=|j_1-j_2|}^{j_1+j_2} \oplus[j] \\ &= [|j_1-j_2|]\oplus[|j_1-j_2|+1]\oplus\cdots\oplus[j_1+j_2]\,.\end{aligned} \tag{2.38}$$

Given two angular momenta $j_1 \in \{0, \frac{1}{2}, 1, \dots\}$ and $j_2 \in \{0, \frac{1}{2}, 1, \dots\}$, the Clebsch–Gordan (CG) series also expresses the rule of addition of two angular momenta:

$$j = j_1+j_2,\ j_1+j_2-1,\ \dots,\ |j_1-j_2|\,.$$

The integers $\epsilon_{j_1 j_2 j}$ defined by

$$\epsilon_{j_1 j_2 j} = \begin{cases} 1\,, & \text{for } j_1, j_2, j \text{ satisfying} \\ & \text{the CG-series rule} \\ 0\,, & \text{otherwise} \end{cases}\,, \tag{2.39}$$

are useful in many relations between angular momentum quantities. The notation $(j_1 j_2 j)$ is used to symbolize the CG-series relation between three angular momentum quantum numbers.

The representation function and Lie algebra interpretations of the CG-series (2.38) are, respectively:

$$C(D^{j_1}\times D^{j_2})C^{\mathrm{T}} = \sum_j \oplus\epsilon_{j_1 j_2 j} D^j\,,$$

$$C\left(J_i^{(j_1)}\times J_i^{(j_2)}\right)C^{\mathrm{T}} = \sum_j \oplus\epsilon_{j_1 j_2 j} J_i^{(j)}\,,$$
$$i = 1, 2, 3\,.$$

The notation $J_i^{(j)}$ denotes the $(2j+1)\times(2j+1)$ matrix with elements

$$J^{(j)}_{m',m} = \langle jm' \mid J_i \mid jm\rangle\,,$$
$$m', m = j, j-1, \dots, -j\,.$$

The elements of the real, orthogonal matrix C of dimension $(2j_1+1)(2j_2+1)$ that effects these reductions are the WCG-coefficients:

$$(C)_{jm;m_1m_2} = C^{j_1 j_2 j}_{m_1 m_2 m}\,.$$

The pairs, (jm) and $(m_1 m_2)$, index rows and columns, respectively, of the matrix C:

$$\begin{aligned}(jm) &: j = j_1+j_2, \dots, |j_1-j_2|\,, \\ &\quad m = j, \dots, -j\,; \\ (m_1m_2) &: m_1 = j_1, \dots, -j_1\,; \\ &\quad m_2 = j_2, \dots, -j_2\,.\end{aligned}$$

Sum rule on projection quantum numbers:

$$C^{j_1 j_2 j}_{m_1 m_2 m} = 0\,, \quad \text{for } m_1+m_2 \neq m\,. \tag{2.40}$$

Clebsch–Gordan series rule on angular momentum quantum numbers:

$$C^{j_1 j_2 j}_{m_1 m_2 m} = 0\,, \text{for } \epsilon_{j_1 j_2 j} = 0\,. \tag{2.41}$$

In presenting formulas that express relations relating to the conceptual framework described above, it is best to use a notation for a WCG-coefficient giving it as an element of an orthogonal matrix. For the expression of symmetry relations, the 3–j coefficient or 3–j symbol notation is most convenient. The following notations are used here:
WCG-coefficient notation:

$$|(j_1 j_2) jm\rangle = \sum_{m_1, m_2} C^{j_1 j_2 j}_{m_1 m_2 m} |j_1 m_1\rangle \otimes |j_2 m_2\rangle \,,$$
$$|j_1 m_1;\, j_2 m_2\rangle = |j_1 m_1\rangle \otimes |j_2 m_2\rangle \,,$$
$$C^{j_1 j_2 j}_{m_1 m_2 m} = \langle j_1 m_1;\, j_2 m_2 | (j_1 j_2) jm\rangle \,,$$
$$\langle (j_1 j_2) j' m' | (j_1 j_2) jm\rangle = \delta_{j' j} \delta_{m' m} \,.$$

The 3–j coefficient notation:

$$\begin{pmatrix} j_1 & j_2 & j \\ m_1 & m_2 & -m \end{pmatrix} = (-1)^{j_1 - j_2 + m} (2j+1)^{-1/2} C^{j_1 j_2 j}_{m_1 m_2 m} \,. \tag{2.42}$$

Orthogonality of WCG-coefficients:
Orthogonality of rows:

$$\sum_{m_1 m_2} C^{j_1 j_2 j}_{m_1 m_2 m} C^{j_1 j_2 j'}_{m_1 m_2 m'} = \delta_{jj'} \delta_{mm'} \,. \tag{2.43}$$

Orthogonality of columns (three forms):

$$\sum_{jm} C^{j_1 j_2 j}_{m_1 m_2 m} C^{j_1 j_2 j}_{m'_1 m'_2 m} = \delta_{m_1 m'_1} \delta_{m_2 m'_2} \,,$$
$$\sum_{j} C^{j_1 j_2 j}_{m_1 m - m_1, m} C^{j_1 j_2 j}_{m'_1, m - m'_1, m} = \delta_{m_1 m'_1} \,,$$
$$\sum_{j} C^{j_1 j_2 j}_{m - m_2, m_2, m} C^{j_1 j_2 j}_{m - m'_2, m'_2, m} = \delta_{m_2 m'_2} \,. \tag{2.44}$$

Orthogonality of 3–j coefficients (symbols):

$$\sum_{m_1 m_2} \begin{pmatrix} j_1 & j_2 & j_3 \\ m_1 & m_2 & m_3 \end{pmatrix} \begin{pmatrix} j_1 & j_2 & j'_3 \\ m_1 & m_2 & m'_3 \end{pmatrix} = \delta_{j_3 j'_3} \delta_{m_3 m'_3} / (2j_3 + 1) \,, \tag{2.45}$$

$$\sum_{j_3 m_3} (2j_3 + 1) \begin{pmatrix} j_1 & j_2 & j_3 \\ m_1 & m_2 & m_3 \end{pmatrix} \begin{pmatrix} j_1 & j_2 & j_3 \\ m'_1 & m'_2 & m_3 \end{pmatrix} = \delta_{m_1 m'_1} \delta_{m_2 m'_2} \,. \tag{2.46}$$

The integers $\epsilon_{j_1 j_2 j}$ ($j_3 = j$) are sometimes included in the orthogonality relations (2.43) and (2.45) to incorporate the extended definition (2.41) of the WCG-coefficients.

2.7.1 Kronecker Product Reduction

Product form:

$$D^{j_1}_{m'_1 m_1}(U) D^{j_2}_{m'_2 m_2}(U) = \sum_j C^{j_1 j_2 j}_{m'_1, m'_2, m'_1 + m'_2} C^{j_1 j_2 j}_{m_1, m_2, m_1 + m_2} \times D^j_{m'_1 + m'_2, m_1 + m_2}(U) \,.$$

Singly coupled form:

$$\sum_{m_1 + m_2 = m} C^{j_1 j_2 j}_{m_1 m_2 m} D^{j_1}_{m'_1 m_1}(U) D^{j_2}_{m'_2 m_2}(U) = C^{j_1 j_2 j}_{m'_1, m'_2, m'_1 + m'_2} D^j_{m'_1 + m'_2, m}(U) \,.$$

Doubly coupled (reduction) form:

$$\sum_{\substack{m'_1 + m'_2 = m' \\ m_1 + m_2 = m}} C^{j_1 j_2 j}_{m'_1 m'_2 m'} C^{j_1 j_2 j}_{m_1 m_2 m} D^{j_1}_{m'_1 m_1}(U) D^{j_2}_{m'_2 m_2}(U) = \delta_{j' j} D^j_{m' m}(U) \,.$$

Integral relation:

$$\int \mathrm{d}\Omega D^{j_1}_{m'_1 m_1}(U) D^{j_2}_{m'_2 m_2}(U) D^{j*}_{m' m_1}(U) = \frac{2\pi^2}{2j+1} C^{j_1 j_2 j}_{m'_1 m'_2 m'} C^{j_1 j_2 j}_{m_1 m_2 m} \,,$$

in any parametrization of $U \in SU(2)$ that covers S^3 exactly once.
Gaunt's integral:

$$\int_0^{2\pi} \mathrm{d}\alpha \int_0^{\pi} \sin\beta \, \mathrm{d}\beta Y^*_{lm}(\beta\alpha) Y_{l_1 m_1}(\beta\alpha) Y_{l_2 m_2}(\beta\alpha) = \left(\frac{(2l_1 + 1)(2l_2 + 1)}{4\pi(2l + 1)} \right)^{1/2} C^{l_1 l_2 l}_{000} C^{l_1 l_2 l}_{m_1 m_2 m'} \,.$$

Integral over three Legendre functions:

$$\int_0^\pi \sin\beta\,\mathrm{d}\beta P_l(\cos\beta)P_{l_1}(\cos\beta)P_{l_2}(\cos\beta) = \frac{2\left(C_{000}^{l_1l_2l}\right)^2}{2l+1}\,.$$

$$C_{000}^{l_1l_2l} = \left(\frac{(2l+1)(l_1+l_2-l)!(l_1-l_2+l)!(-l_1+l_2+l)!}{(l_1+l_2+l+1)!}\right)^{\frac{1}{2}} \times \frac{(-1)^{L-1}L!}{(L-l_1)!(L-l_2)!(L-l)!}\,,$$

$$L=\frac{1}{2}(l_1+l_2+l)\,,$$

for l_1+l_2+l even

$$C_{000}^{l_1l_2l}=0\,,\qquad \text{for } l_1+l_2+l \text{ odd}\,.$$

2.7.2 Tensor Product Space Construction

Orthonormal basis of H_{j_1}:

$$\left\{|j_1m_1\rangle \,\middle|\, m_1=j_1,\,j_1-1,\,\dots,\,-j_1\right\}\,.$$

Orthonormal basis of H_{j_2}:

$$\left\{|j_2m_2\rangle \,\middle|\, m_2=j_2,\,j_2-1,\,\dots,\,-j_2\right\}\,.$$

Uncoupled basis of $H_{j_1}\otimes H_{j_2}$:

$$\left\{|j_1m_1\rangle\otimes|j_2m_2\rangle \,\middle|\, m_1=j_1,\,j_1-1,\,\dots,\,-j_1\,;\; m_2=j_2,\,j_2-1,\,\dots,\,-j_2\right\}.$$

Coupled basis of $H_{j_1}\otimes H_{j_2}$:

$$\left\{|(j_1j_2)jm\rangle \,\middle|\, j=j_1+j_2,\,j_1+j_2-1,\,\dots,\,|j_1-j_2|\,;\; m=j,\,j-1,\,\dots,\,-j\right\}\,,$$

$$|(j_1j_2)jm\rangle=\sum_{m_1m_2}C_{m_1m_2m}^{j_1j_2j}|j_1m_1\rangle\otimes|j_2m_2\rangle\,.$$

Unitary transformations of spaces:

$$T_U|j_1m_1\rangle=\sum_{m_1'}D_{m_1'm_1}^{j_1}(U)\big|j_1m_1'\big\rangle\,,$$

$$m_1=j_1,\,j_1-1,\,\dots,\,-j_1\,,\qquad \text{each } U\in SU(2)\,;$$

$$T_V|j_2m_2\rangle=\sum_{m_2'}D_{m_2'm_2}^{j}(V)\big|j_2m_2'\big\rangle\,,$$

$$m_2=j_2,\,j_2-1,\,\dots,\,-j_2\,,\qquad \text{each } V\in SU(2)\,;$$

$$T_{(U,V)}|j_1m_1\rangle\otimes|j_2m_2\rangle=T_U|j_1m_1\rangle\otimes T_V|j_2m_2\rangle = \sum_{m_1'm_2'}D_{m_1'm_1}^{j_1}(U)D_{m_2'm_2}^{j_2}(V)\big|j_1m_1'\big\rangle\otimes\big|j_2m_2'\big\rangle\,,$$

$$\text{each } U\in SU(2)\,,\quad \text{each } V\in SU(2)\,;$$

$$T_{(U,U)}|(j_1j_2)jm\rangle=\sum_{m'}D_{m'm}^{j}(U)|(j_1j_2)jm'\rangle\,,$$

$$m=j,\,j-1,\,\dots,\,-j;\ \text{each } U\in SU(2)\,.$$

Representation of direct product group $SU(2)\times SU(2)$:

$$T_{(U,V)}T_{(U',V')}=T_{(UU',VV')}\,.$$

Representation of $SU(2)$ as diagonal subgroup of $SU(2)\times SU(2)$:

$$T_{(U,U)}T_{(U',U')}=T_{(UU',UU')}\,,$$

$$T_U=T_{(U,U)}\,.$$

2.7.3 Explicit Forms of WCG-Coefficients

Wigner's form:

$$C_{m_1m_2m}^{j_1j_2j} = \delta(m_1+m_2,m)(2j+1)^{\frac{1}{2}} \times\left(\frac{(j+j_1-j_2)!(j-j_1+j_2)!(j_1+j_2-j)!}{(j+j_1+j_2+1)!}\right)^{\frac{1}{2}} \times\left(\frac{(j+m)!(j-m)!}{(j_1+m_1)!(j_1-m_1)!(j_2+m_2)!(j_2-m_2)!}\right)^{\frac{1}{2}} \times\sum_s\frac{(-1)^{j_2+m_2+s}(j_2+j+m_1-s)!(j_1-m_1+s)!}{s!(j-j_1+j_2-s)!(j+m-s)!(j_1-j_2-m+s)!}\,.$$

Racah's form:

$$C_{m_1m_2m}^{j_1j_2j} = \delta(m_1+m_2,m) \times\left(\frac{(2j+1)(j_1+j_2-j)!}{(j_1+j_2+j+1)!(j+j_1-j_2)!(j+j_2-j_1)!}\right)^{\frac{1}{2}} \times\left(\frac{(j_1-m_1)!(j_2-m_2)!(j-m)!(j+m)!}{(j_1+m_1)!(j_2+m_2)!}\right)^{\frac{1}{2}} \times\sum_t\frac{(-1)^{j_1-m_1+t}(j_1+m_1+t)!(j+j_2-m_1-t)!}{t!(j-m-t)!(j_1-m_1-t)!(j_2-j+m_1+t)!}\,.$$

Van der Waerden's form:

$$C^{j_1 j_2 j}_{m_1 m_2 m} = \delta(m_1+m_2,m) \times \left(\frac{(2j+1)(j_1+j_2-j)!(j+j_1-j_2)!(j+j_2-j_1)!}{(j_1+j_2+j+1)!}\right)^{\frac{1}{2}} \times\left[(j_1+m_1)!(j_1-m_1)!(j_2+m_2)!(j_2-m_2)!\right]^{\frac{1}{2}} \times\left[(j+m)!(j-m)!\right]^{\frac{1}{2}} \times\sum_k(-1)^k\left[k!(j_1+j_2-j-k)!(j_1-m_1-k)! \times(j_2+m_2-k)!(j-j_2+m_1+k)! \times(j-j_1-m_2+k)!\right]^{-1}$$

Regge's formula and its combinatoric structure:

$$(\det A)^k = \sum_{\boxed{\alpha}} A(\alpha) \prod_{i,j=1}^{3} (a_{ij})^{\alpha_{ij}}\,, \quad A=(a_{ij})\,, \tag{2.47}$$

where the summation is over all nonnegative integers α_{ij} that satisfy the row and column sum constraints (2.17) given by

$$\alpha = \begin{array}{|ccc|c} \alpha_{11} & \alpha_{12} & \alpha_{13} & k \\ \alpha_{21} & \alpha_{22} & \alpha_{23} & k \\ \alpha_{31} & \alpha_{32} & \alpha_{33} & k \\ \hline k & k & k & \end{array} \tag{2.48}$$

The coefficients $A(\alpha)$ are constrained sums over multinomial coefficients:

$$A(\alpha) = \sum(-1)^{\phi(K)} \times \begin{pmatrix} k \\ k_{123}, k_{132}, k_{231}, k_{213}, k_{312}, k_{321} \end{pmatrix}\,,$$

where the summation is carried out over all nonnegative integers $k_{i_1 i_2 i_3}$ such that

$$\alpha_{11} = k_{123}+k_{132}\,, \quad \alpha_{12} = k_{231}+k_{213}\,,$$
$$\alpha_{13} = k_{312}+k_{321}\,,$$
$$\alpha_{21} = k_{312}+k_{213}\,, \quad \alpha_{22} = k_{123}+k_{321}\,,$$
$$\alpha_{23} = k_{231}+k_{132}\,,$$
$$\alpha_{31} = k_{231}+k_{321}\,, \quad \alpha_{32} = k_{312}+k_{132}\,,$$
$$\alpha_{33} = k_{123}+k_{213},$$
$$\phi(K) = \sum_{\pi\in A_3} k_\pi = k_{132}+k_{213}+k_{321}\,.$$

The general multinomial coefficient is the integer defined by

$$\begin{pmatrix} k \\ k_1, k_2, \ldots, k_s \end{pmatrix} = k!/k_1!k_2!\cdots k_s!\,, \; k=\sum_i k_i\,.$$

Relation (2.47) generalizes in the obvious way to an $n\times n$ determinant, using the symmetric group S_n and its S_{n-1} subgroups $S^{(j)}_{n-1}$, where j denotes that this is the permutation group on the integers $1, 2, \ldots, n$ with j deleted [2.2].

Regge's formula for the 3–j coefficient is:

$$\begin{pmatrix} j_1 & j_2 & j_3 \\ m_1 & m_2 & m_3 \end{pmatrix} = \delta(m_1+m_2+m_3,0)\left[\prod_{i,j=1}^{3}(\alpha_{ij})!\right]^{\frac{1}{2}} \times\frac{A(\alpha)}{k![(k+1)!]^{\frac{1}{2}}}\,,$$
$$k = j_1+j_2+j_3\,, \tag{2.49}$$
$$\alpha_{11} = j_1+m_1\,, \quad \alpha_{12} = j_2+m_2\,, \quad \alpha_{13} = j_3+m_3\,,$$
$$\alpha_{21} = j_1-m_1\,, \quad \alpha_{22} = j_2-m_2\,, \quad \alpha_{23} = j_3-m_3\,,$$
$$\alpha_{31} = j_2+j_3-j_1\,, \quad \alpha_{32} = j_3+j_1-j_2\,,$$
$$\alpha_{33} = j_1+j_2-j_3\,.$$

Equation (2.49) shows that WCG-coefficients and 3–j coefficients are sums over integers, except for a multiplicative normalization factor.

Schwinger's generating function:

$$\exp(t\det A) = \sum_k \frac{t^k}{k!}\sum_{\boxed{\alpha}} A(\alpha)\prod_{i,j}(a_{ij})^{\alpha_{ij}}\,.$$

The general definition of the ${}_pF_q$ hypergeometric function depending on p numerator parameters, q denominator parameters, and a single variable z is:

$${}_pF_q\left(\begin{matrix} a_1\ldots, a_p \\ b_1\ldots, b_q \end{matrix}; z\right) = \sum_{n=0}^{\infty}\frac{(a_1)_n\cdots(a_p)_n}{(b_1)_n\cdots(b_q)_n}\frac{z^n}{n!}\,,$$
$$(a)_n = a(a+1)\cdots(a+n-1)\,, \quad (a)_0 = 1\,.$$

Such a series is terminating if at least one of the numerator parameters is a negative integer (and all other factors are well-defined). Both WCG-coefficients and Racah 6–j coefficients relate to special series of this type, evaluated at $z=1$. For WCG-coefficients, we have for $\alpha+\beta=\gamma$:

$$C^{abc}_{\alpha\beta\gamma} = [(2c+1)(a+\alpha)!(a-\alpha)!(b+\beta)!(b-\beta)!(c+\gamma)! \times (c-\gamma)!]^{\frac{1}{2}}\,(-1)^{a+b+\gamma+\delta_1}\,\Delta(abc) \times \frac{{}_3F_2\begin{pmatrix}\epsilon_1-\delta_1,\epsilon_2-\delta_1,\epsilon_3-\delta_1\\ \delta_2-\delta_1+1,\delta_3-\delta_1+1\end{pmatrix};1}{(\delta_2-\delta_1)!(\delta_3-\delta_1)!(\delta_1-\epsilon_1)!(\delta_1-\epsilon_2)!(\delta_1-\epsilon_3)!}\,,$$

$\delta_1=\min(a+\alpha+b+\beta, b-\beta+c+\gamma, a+\alpha+c+\gamma)$, $(\delta_1,\delta_2,\delta_3)=$ any permutation of $(a+\alpha+b+\beta, b-\beta+c+\gamma, a+\alpha+c+\gamma)$, after δ_1 is fixed, $(\epsilon_1,\epsilon_2,\epsilon_3)=$ any permutation of $(a+\alpha, b+\alpha+\gamma, c+\gamma)$. A somewhat better form can be found in [2.2].
The quantity

$$\Delta(abc)=\left(\frac{(a+b-c)!(a-b+c)!(-a+b+c)!}{(a+b+c+1)!}\right)^{\frac{1}{2}}.$$

is called a triangle coefficient.

All 72 Regge symmetries are consequences of known properties of the ${}_3F_2$ hypergeometric series.

2.7.4 Symmetries of WCG-Coefficients in 3-*j* Symbol Form

There are 72 known symmetries (up to sign changes) of the 3–j coefficient. There are at least four ways of verifying these symmetries: (1) directly from the van der Waerden form of the coefficients; (2) directly from Regge's generating function; (3) from the known symmetries of the ${}_3F_2$ hypergeometric series; and (4) directly from the symmetries of the representation functions $D^j_{mm'}(U)$. The set of 72 symmetries is succinctly expressed in terms of the coefficient $A(\alpha)$ defined in Sect. 2.8.3 with α_{ij} entries given by (2.48) and (2.49) in which $m_1+m_2+m_3=0$:

$$A\begin{pmatrix} j_1+m_1 & j_2+m_2 & j_3+m_3\\ j_1-m_1 & j_2-m_2 & j_3-m_3\\ j_2+j_3-j_1 & j_3+j_1-j_2 & j_1+j_2-j_3\end{pmatrix}.$$

This coefficient has determinantal symmetry; that is, it is invariant under even permutations of its rows or columns and under transposition, and is multiplied by the factor $(-1)^{j_1+j_2+j_3}$ under odd permutations of its rows or columns. These 72 determinantal operations may be generated from the three operations C_{12}, C_{13}, T consisting of interchange of columns 1 and 2, interchange of columns 1 and 3, and transposition, since the first two operations generate the symmetric group S_3 of permutations of columns, and the symmetric group S_3' of permutations of rows is then given by TS_3T. The transposition T itself generates a group $\{e,T\}$ isomorphic to the symmetric group S_2. Thus, the 72 element determinantal group is the direct product group $S_3\times S_3'\times\{e,T\}$. The three relations between 3–j coefficients corresponding to the generators C_{12}, C_{13}, T are

$$\begin{pmatrix} j_1 & j_2 & j_3\\ m_1 & m_2 & m_3\end{pmatrix} = (-1)^{j_1+j_2+j_3}\begin{pmatrix} j_2 & j_1 & j_3\\ m_2 & m_1 & m_3\end{pmatrix},$$

$$\begin{pmatrix} j_1 & j_2 & j_3\\ m_1 & m_2 & m_3\end{pmatrix} = (-1)^{j_1+j_2+j_3}\begin{pmatrix} j_3 & j_2 & j_1\\ m_3 & m_2 & m_1\end{pmatrix},$$

$$\begin{pmatrix} j_1 & j_2 & j_3\\ m_1 & m_2 & m_3\end{pmatrix} = \begin{pmatrix} \frac{j_1+j_2+m_1+m_2}{2} & \frac{j_1+j_2-m_1-m_2}{2} & j_3\\ \frac{j_1-j_2+m_1-m_2}{2} & \frac{j_1-j_2-m_1+m_2}{2} & j_2-j_1\end{pmatrix}.$$

All 72 relations among 3–j coefficients can be obtained from these 3. The 12 "classical" symmetries of the 3–j symbol

$$\begin{pmatrix} a & b & c\\ \alpha & \beta & \gamma\end{pmatrix}$$

are expressed by:

1. even permutations of the columns leave the coefficient invariant;
2. odd permutations of the columns change the sign by the factor $(-1)^{a+b+c}$;
3. simultaneous sign reversal of the projection quantum numbers changes the sign by $(-1)^{a+b+c}$.

The 72 corresponding symmetries of the WCG-coefficients (up to sign changes and dimensional factors) are best obtained from those of the $3j$-coefficients by using (2.42).

Part A | 2.7

2.7.5 Recurrence Relations

Three-term:

$$[(J+1)(J-2j_1)]^{\frac{1}{2}}\begin{pmatrix} j_1 & j_2 & j_3\\ m_1 & m_2 & m_3\end{pmatrix}$$
$$=[(j_2+m_2)(j_3-m_3)]^{\frac{1}{2}}\begin{pmatrix} j_1 & j_2-\frac{1}{2} & j_3-\frac{1}{2}\\ m_1 & m_2-\frac{1}{2} & m_3+\frac{1}{2}\end{pmatrix}$$
$$-[(j_2-m_2)(j_3+m_3)]^{\frac{1}{2}}\begin{pmatrix} j_1 & j_2-\frac{1}{2} & j_3-\frac{1}{2}\\ m_1 & m_2+\frac{1}{2} & m_3-\frac{1}{2}\end{pmatrix};$$

$$[(J-2j_2)(J+1-2j_3)]^{\frac{1}{2}}\begin{pmatrix} j_1 & j_2 & j_3\\ m_1 & m_2 & m_3\end{pmatrix}$$
$$+[(j_2+m_2+1)(j_3+m_3)]^{\frac{1}{2}}$$
$$\times\begin{pmatrix} j_1 & j_2-\frac{1}{2} & j_3+\frac{1}{2}\\ m_1 & m_2-\frac{1}{2} & m_3+\frac{1}{2}\end{pmatrix}$$
$$+[(j_2-m_2+1)(j_3-m_3)]^{\frac{1}{2}}$$
$$\times\begin{pmatrix} j_1 & j_2-\frac{1}{2} & j_3+\frac{1}{2}\\ m_1 & m_2+\frac{1}{2} & m_3-\frac{1}{2}\end{pmatrix}=0;$$

$$(j_2+m_2)^{\frac{1}{2}}\begin{pmatrix} j_1 & j_2 & j_3\\ m_1 & m & m_3\end{pmatrix}$$
$$=[(j_3-j_1+j_2)(J+1)(j_3-m_3)]^{\frac{1}{2}}$$
$$\times\begin{pmatrix} j_1 & j_2-\frac{1}{2} & j_3-\frac{1}{2}\\ m_1 & m_2-\frac{1}{2} & m_3-\frac{1}{2}\end{pmatrix}$$
$$-[(j_1-j_3+j_2)(J-2j_2+1)(j_3+m_3+1)]^{\frac{1}{2}}$$
$$\times\begin{pmatrix} j_1 & j_2-\frac{1}{2} & j_3+\frac{1}{2}\\ m_1 & m_2-\frac{1}{2} & m_3+\frac{1}{2}\end{pmatrix};$$

$$\begin{pmatrix} j_1 & j_2 & j_3\\ j_2-m_3 & -j_2 & m_3\end{pmatrix}$$
$$=-\left(\frac{2j_2(j_3+m_3)}{(j_3-j_1+j_2)(J+1)}\right)^{\frac{1}{2}}$$
$$\times\begin{pmatrix} j_1 & j_2-\frac{1}{2} & j_3-\frac{1}{2}\\ j_2-m_3 & -j_2+\frac{1}{2} & m_3-\frac{1}{2}\end{pmatrix},$$

$j_3=j_1+j_2,\ j_1+j_2-1,\ldots,\ j_1-j_2+1$

for $j_1\ge j_2$;

$$\begin{pmatrix} j_1 & j_2 & j_3\\ j_2-m_3 & -j_2 & m_3\end{pmatrix}$$
$$=-\left(\frac{2j_2(j_3-m_3+1)}{(j_1-j_3+j_2)(J-2j_2+1)}\right)^{\frac{1}{2}}$$
$$\times\begin{pmatrix} j_1 & j_2-\frac{1}{2} & j_3+\frac{1}{2}\\ j_2-m_3 & -j_2+\frac{1}{2} & m_3-\frac{1}{2}\end{pmatrix},$$

$j_3=j_1+j_2-1,\ j_1+j_2-2,\ldots,\ j_1-j_2$

for $j_1\ge j_2$.

Four-term:

$$[(J+1)(J-2j_1)(J-2j_2)(J-2j_3+1)]^{\frac{1}{2}}$$
$$\times\begin{pmatrix} j_1 & j_2 & j_3\\ m_1 & m_2 & m_3\end{pmatrix}=[(j_2-m_2)(j_2+m_2+1)$$
$$\times(j_3+m_3)(j_3+m_3-1)]^{\frac{1}{2}}$$
$$\times\begin{pmatrix} j_1 & j_2 & j_3-1\\ m_1 & m_2+1 & m_3-1\end{pmatrix}$$
$$-2m_2[(j_3+m_3)(j_3-m_3)]^{\frac{1}{2}}\begin{pmatrix} j_1 & j_2 & j_3-1\\ m_1 & m_2 & m_3\end{pmatrix}$$
$$-[(j_2+m_2)(j_2-m_2+1)(j_3-m_3)(j_3-m_3-1)]^{\frac{1}{2}}$$
$$\times\begin{pmatrix} j_1 & j_2 & j_3-1\\ m_1 & m_2-1 & m_3+1\end{pmatrix}.$$

Five-term:

$$C^{bdf}_{\beta,\delta,\beta+\delta}$$
$$=\left(\frac{(b+d-f)(b+f-d+1)(d-\delta)(f+\beta+\delta+1)}{(2d)(2f+1)(2d)(2f+2)}\right)^{\frac{1}{2}}$$
$$\times C^{bd-1/2f+1/2}_{\beta,\delta+1/2,\beta+\delta+1/2}$$
$$-\left(\frac{(b+d-f)(b+f-d+1)(d+\delta)(f-\beta-\delta+1)}{(2d)(2f+1)(2d)(2f+2)}\right)^{\frac{1}{2}}$$
$$\times C^{bd-1/2f+1/2}_{\beta,\delta-1/2,\beta+\delta-1/2}$$
$$+\left(\frac{(d+f-b)(b+d+f+1)(d+\delta)(f+\beta+\delta)}{(2d)(2f+1)(2d)(2f)}\right)^{\frac{1}{2}}$$
$$\times C^{bd-1/2f-1/2}_{\beta,\delta-1/2,\beta+\delta-1/2}$$
$$+\left(\frac{(d+f-b)(b+d+f+1)(d-\delta)(f-\beta-\delta)}{(2d)(2f+1)(2d)(2f)}\right)^{\frac{1}{2}}$$
$$\times C^{bd-1/2f-1/2}_{\beta,\delta+1/2,\beta+\delta+1/2}.$$

This relation may be used to prove the limit relation (2.50) from the similar recurrence relation (2.84c) for the Racah coefficients.

2.7.6 Limiting Properties and Asymptotic Forms

$$\lim_{a\to\infty} C^{aba+\rho}_{a-\alpha,\beta,a-\alpha+\beta} = \delta_{\rho\beta}\,,$$

$$\lim_{j\to\infty} (-1)^{a+b+2j-\tau}[(2c+1)(2j-2\sigma+1)]^{\frac{1}{2}} \times \begin{Bmatrix} j-\tau & a & j-\sigma \\ b & j & c \end{Bmatrix} = C^{abc}_{\rho\sigma\tau}\,, \tag{2.50}$$

where the brace symbol is a $6\text{–}j$ coefficient (Sect. 2.9).

$$C^{jkj+\Delta}_{m,\mu,m+\mu} \approx (-1)^{\Delta-\mu} D^k_{\mu\Delta}\begin{pmatrix} \cos\frac{1}{2}\beta & \sin\frac{1}{2}\beta \\ -\sin\frac{1}{2}\beta & \cos\frac{1}{2}\beta \end{pmatrix}$$
$$= d^k_{\mu\Delta}(\beta)\,, \quad \text{for large } j\,;$$

$$\cos\frac{1}{2}\beta = \sqrt{\frac{j+m}{2j}}\,, \quad \sin\frac{1}{2}\beta = \sqrt{\frac{j-m}{2j}}\,,$$

$$C^{jkj}_{m0m} \approx P_k(\cos\beta)\,, \quad \text{for large } j\,;$$

$(-1)^k[(2j+1)(2J+1)]^{\frac{1}{2}} W(j, k, J+m, J;\, j, J)$
$\sim P_k(\cos\beta)$,
first for large J, then large j (Sect. 2.9).

2.7.7 WCG-Coefficients as Discretized Representation Functions

$$C^{j_1 j_2 j}_{m_1 m_2 m} = \delta_{m_1+m_2,m}(-1)^{j_1-m_1} \times \left(\frac{(2j+1)(j_1+j_2-j)!}{(j_1+j_2+j+1)!}\right)^{\frac{1}{2}} \times D^j_{m,j_1-j_2}\begin{pmatrix} \sqrt{j_1+m_1} & \sqrt{j_2+m_2} \\ -\sqrt{j_1-m_1} & \sqrt{j_2-m_2}\end{pmatrix}_{\text{symbolic powers}}, \tag{2.51}$$

where in evaluating this result one first substitutes

$$u_{11} = \sqrt{j_1+m_1}\,,$$
$$u_{12} = \sqrt{j_2+m_2}\,,$$
$$u_{21} = -\sqrt{j_1-m_1}\,,$$
$$u_{22} = \sqrt{j_2-m_2}$$

into the form (2.17), followed by the replacement of ordinary powers by generalized powers:

$$(\pm\sqrt{k})^s \to (\pm 1)^s\left(\frac{k!}{(k-s)!}\right)^{\frac{1}{2}}\,.$$

2.8 Tensor Operator Algebra

2.8.1 Conceptual Framework

A tensor operator can be characterized in terms of its algebraic properties with respect to the angular momentum $\boldsymbol{J}$ or in terms of its transformation properties under unitary transformations generated by $\boldsymbol{J}$. Both viewpoints are essential.

A tensor operator $\boldsymbol{T}$ with respect to the group $SU(2)$ is a set of linear operators

$$\boldsymbol{T} = \{T_1, T_2, \ldots, T_n\}\,,$$

where each operator in the set acts in the space $\mathcal{H}$ defined by (2.10) and maps this space into itself $T_i : \mathcal{H} \to \mathcal{H}$, $i = 1, 2, \ldots, n$, and where this set of operators has the following properties with respect to the angular momentum $\boldsymbol{J}$, which acts in the same space $\mathcal{H}$ in the standard way:

1. Commutation relations with respect to the angular momentum $\boldsymbol{J}$:
$$[J_i, T_j] = \sum_{k=1}^{n} t^{(i)}_{kj} T_k\,,$$
where the $t^{(i)}_{kj}$ are scalars (invariants) with respect to $\boldsymbol{J}$.
2. Unitary transformation with respect to $SU(2)$ rotations:
$$\mathrm{e}^{-\mathrm{i}\psi\hat{\boldsymbol{n}}\cdot\boldsymbol{J}} T_i\, \mathrm{e}^{\mathrm{i}\psi\hat{\boldsymbol{n}}\cdot\boldsymbol{J}} = \sum_{j=1}^{n} D_{ji}(U) T_j\,,$$
$$U = U(\psi, \hat{\boldsymbol{n}})\,,$$

where the matrix $D(U)$ is an $n\times n$ unitary matrix representation of $SU(2)$. Reduction of this representation into its irreducible constituents gives the notion of an irreducible tensor operator $\boldsymbol{T}^J$ of rank J. An irreducible tensor operator $\boldsymbol{T}^J$ of rank J is a set of $2J+1$ operators

$$\boldsymbol{T}^J = \left\{T_M^J \mid M = J, J-1, \dots, -J\right\}$$

with the following properties with respect to $SU(2)$:

1. Commutation relations with respect to the angular momentum $\boldsymbol{J}$:
$$\left[J_+, T_M^J\right] = [(J-M)(J+M+1)]^{\frac{1}{2}}\, T_{M+1}^J ,$$
$$\left[J_-, T_M^J\right] = [(J+M)(J-M+1)]^{\frac{1}{2}}\, T_{M-1}^J ,$$
$$\left[J_3, T_M^J\right] = M T_M^J ,$$
$$\sum_i \left[J_i, \left[J_i, T_M^J\right]\right] = J(J+1) T_M^J . \qquad (2.52)$$
2. Generation from highest "weight":
$$T_M^J = \left(\frac{(J+M)!}{(2J)!(J-M)!}\right)^{\frac{1}{2}} \left[J_-, T_J^J\right]_{(J-M)} ,$$
where $[A, B]_{(k)} = [A, [A, B]_{(k-1)}]$, $k = 1, 2, \dots$, with $[A, B]_{(0)} = B$, denotes the k-fold commutator of A with B.
3. Unitary transformation with respect to $SU(2)$ rotations:
$$\mathrm{e}^{-\mathrm{i}\psi\hat{\boldsymbol{n}}\cdot\boldsymbol{J}} T_M^J \mathrm{e}^{\mathrm{i}\psi\hat{\boldsymbol{n}}\cdot\boldsymbol{J}} = \sum_{M'} D_{M'M}^J(U) T_{M'}^J ,\; U = U(\psi, \hat{\boldsymbol{n}}) . \qquad (2.53)$$

Angular momentum operators act in Hilbert spaces by acting linearly on the vectors in such spaces. The concept of a tensor operator generalizes this by replacing the irreducible space H_J by the irreducible tensor $\boldsymbol{T}^J$, and angular momentum operator action on H_J by commutator action on $\boldsymbol{T}^J$, as symbolized, respectively, by

$$\boldsymbol{J} : \{\text{ states }\} \to \{\text{ states }\},$$
$$\{\text{ commutator action of } \boldsymbol{J}\} : \{\text{ tensor operators }\} \to \{\text{tensor operators }\} .$$

Just as exponentiation of the standard generator action (2.13) and (2.14) gives relation (2.16), so does the exponentiation of the commutator action (2.52) give relation (2.53), when one uses the Baker–Campell–Hausdorff identity:

$$\mathrm{e}^{tA} B \mathrm{e}^{-tA} = \sum_k \frac{t^k}{k!} [A, B]_{(k)} .$$

Thus, the linear vector space of states is replaced by the linear vector space of operators. Abstractly, relations (2.13) and (2.52) are identical: only the rule of action and the object of that action has changed.

An example of an irreducible tensor of rank 1 is the angular momentum $\boldsymbol{J}$ itself, which has the special property $\boldsymbol{J} : \mathcal{H}_j \to \mathcal{H}_j$. Thus, relations (2.52) and (2.53) are realized as:

$$T_1^1 = J_{+1} = -(J_1 + \mathrm{i}J_2)/\sqrt{2} ,$$
$$T_0^1 = J_0 = J_3 ,$$
$$T_{-1}^1 = J_{-1} = (J_1 - \mathrm{i}J_2)/\sqrt{2} ;$$
$$\left[J_+, T_\mu^1\right] = [(1-\mu)(2+\mu)]^{\frac{1}{2}}\, T_{\mu+1}^1 ,$$
$$\left[J_-, T_\mu^1\right] = [(1+\mu)(2-\mu)]^{\frac{1}{2}}\, T_{\mu-1}^1 ,$$
$$\left[J_3, T_\mu^1\right] = \mu T_\mu^1 , \quad \mu = 1, 0, -1 ;$$
$$\mathrm{e}^{-\mathrm{i}\psi\hat{\boldsymbol{n}}\cdot\boldsymbol{J}} \boldsymbol{J} \mathrm{e}^{\mathrm{i}\psi\hat{\boldsymbol{n}}\cdot\boldsymbol{J}} = \boldsymbol{J}\cos\psi + \hat{\boldsymbol{n}}(\hat{\boldsymbol{n}}\cdot\boldsymbol{J})(1-\cos\psi) - (\hat{\boldsymbol{n}}\times\boldsymbol{J})\sin\psi ,$$
$$\mathrm{e}^{-\mathrm{i}\psi\hat{\boldsymbol{n}}\cdot\boldsymbol{J}} T_\mu^1 \mathrm{e}^{\mathrm{i}\psi\hat{\boldsymbol{n}}\cdot\boldsymbol{J}} = \sum_\nu D_{\nu\mu}^1(\psi, \hat{\boldsymbol{n}}) T_\nu^1 .$$

2.8.2 Universal Enveloping Algebra of J

The universal enveloping algebra $A(\boldsymbol{J})$ of $\boldsymbol{J}$ is the set of all complex polynomial operators in the components $\boldsymbol{J}_i$ of $\boldsymbol{J}$, or equivalently in (J_+, J_3, J_-). The irreducible tensor operators spanning this algebra are the analogues of the solid harmonics $\mathcal{Y}_{lm}(\boldsymbol{x})$ and are characterized by the following properties: Basis set:

$$\mathcal{T}_k^k = a_k J_+^k , \quad a_k \text{ arbitrary constant} ,$$
$$\mathcal{T}_\mu^k = a_k \left(\frac{(k+\mu)!}{(2k)!(k-\mu)!}\right) \left[J_-, J_+^k\right]_{(k-\mu)} ,$$
$$\mu = k, k-1, \dots, -k ; \quad k = 0, 1, 2, \dots .$$

Standard action with respect to $\boldsymbol{J}$:

$$\left[J_\pm, \mathcal{T}_\mu^k\right] = [(k\mp\mu)(k\pm\mu+1)]^{\frac{1}{2}}\, \mathcal{T}_{\mu\pm1}^k ,$$
$$\left[J_3, \mathcal{T}_\mu^k\right] = \mu \mathcal{T}_\mu^k ,$$
$$\sum_{i=1}^3 \left[J_i, \left[J_i, \mathcal{T}_\mu^k\right]\right] = k(k+1)\mathcal{T}_\mu^k .$$

Unitary transformation:

$$\mathrm{e}^{-\mathrm{i}\psi\hat{\boldsymbol{n}}\cdot\boldsymbol{J}} \mathcal{T}_\mu^k \mathrm{e}^{\mathrm{i}\psi\hat{\boldsymbol{n}}\cdot\boldsymbol{J}} = \sum_\nu D_{\nu\mu}^k(\psi, \hat{\boldsymbol{n}}) \mathcal{T}_\nu^k .$$

2.8.3 Algebra of Irreducible Tensor Operators

Irreducible tensor operators possess, as linear operators acting in the same space, properties 1., 2., and 3. below, and an additional multiplication property 4., which constructs new irreducible tensor operators out of two given ones and is called coupling of irreducible tensor operators. Property 4. extends also to tensor operators acting in the tensor product space associated with kinematically independent systems. It is important that associativity extends to the product (2.54), as well as to the product (2.55). Commutativity in these products is generally invalid. The coupling properties given in 4. and 5. are analogous to the coupling of basis state vectors. The operation of Hermitian conjugation of operators, which is the analogue of complex conjugation of states, is also important, and has the properties presented under 5.

1. Multiplication of an irreducible tensor operator of rank k by a complex number or an invariant with respect to angular momentum $\boldsymbol{J}$ gives an irreducible tensor operator of the same rank.
2. Addition of two irreducible tensor operator of the same rank gives an irreducible tensor of that rank.
3. Ordinary multiplication (juxtaposition) of three irreducible tensor operators is associative, but the multiplication of two is noncommutative, in general.
4. Two irreducible tensor operators $\boldsymbol{S}^{k_1}$ and $\boldsymbol{T}^{k_2}$ of different or the same ranks acting in the same space may be multiplied to obtain new irreducible tensor operators of ranks given by the angular momentum addition rule (Clebsch–Gordan series):

$$\left[\boldsymbol{S}^{k_1} \times \boldsymbol{T}^{k_2}\right]^k_\mu = \sum_{\mu_1,\mu_2} C^{k_1k_2k}_{\mu_1\mu_2\mu} S^{k_1}_{\mu_1} T^{k_2}_{\mu_2}\,, \tag{2.54}$$
$$\mu = k, k-1, \ldots, -k\,;$$
$$\text{rank} = $$
$$k \in \{k_1+k_2, k_1+k_2-1, \ldots, |k_1-k_2|\}\,.$$

The following symbol denotes the irreducible tensor operator with the μ-components (2.54):

$$\left[\boldsymbol{S}^{k_1} \times \boldsymbol{T}^{k_2}\right]^k .$$

5. Two irreducible tensor operators $\boldsymbol{S}^{k_1}$ and $\boldsymbol{T}^{k_2}$ of different or the same ranks acting in different Hilbert spaces, say $\mathcal{H}$ and $\mathcal{K}$, may first be multiplied by the tensor product rule so as to act in the tensor product space $\mathcal{H} \otimes \mathcal{K}$, that is,

$$S^{k_1}_{\mu_1} \otimes T^{k_2}_{\mu_2} : \mathcal{H} \otimes \mathcal{K} \to \mathcal{H} \otimes \mathcal{K}\,,$$

and then coupled to obtain new irreducible tensor operators, acting in the same tensor product space $\mathcal{H} \otimes \mathcal{K}$:

$$\left[\boldsymbol{S}^{k_1} \otimes \boldsymbol{T}^{k_2}\right]^k_\mu = \sum_{\mu,\mu_2} C^{k_1k_2k}_{\mu_1\mu_2\mu} S^{k_1}_{\mu_1} \otimes T^{k_2}_{\mu_2}\,,$$
$$\mu = k, k-1, \ldots, -k\,. \tag{2.55}$$

The following symbol denotes the tensor operator with the μ-components (2.55):

$$\left[\boldsymbol{S}^{k_1} \otimes \boldsymbol{T}^{k_2}\right]^k ,$$
$$k \in \{k_1+k_2, k_1+k_2-1, \ldots, |k_1-k_2|\}\,.$$

6. The conjugate tensor operator to $\boldsymbol{T}^J$, denoted by $\boldsymbol{T}^{J\dagger}$, is the set of operators with components $T^{J\dagger}_M$ defined by

$$\langle j'm'|T^{J\dagger}_M|jm\rangle = \langle jm|T^J_M|j'm'\rangle^* .$$

These components satisfy the following relations:

$$\left[J_\pm, T^{J\dagger}_M\right] = -[(J \pm M)(J \mp M+1)]^{\frac{1}{2}} T^{J\dagger}_{M\mp 1}$$
$$\left[J_3, T^{J\dagger}_M\right] = -MT^{J\dagger}_M\,,$$
$$\sum_i \left[J_i, \left[J_i, T^{J\dagger}_M\right]\right] = J(J+1)T^{J\dagger}_M\,;$$
$$\mathrm{e}^{-\mathrm{i}\psi\hat{\boldsymbol{n}}\cdot\boldsymbol{J}} T^{J\dagger}_M \mathrm{e}^{\mathrm{i}\psi\hat{\boldsymbol{n}}\cdot\boldsymbol{J}} = \sum_{M'} D^{J*}_{M'M}(\psi, \hat{\boldsymbol{n}}) T^{J\dagger}_{M'}\,;$$
$$I^J = \sum_M T^J_M T^{J\dagger}_M = \begin{pmatrix}\text{invariant operator to}\\ SU(2)\text{ rotations}\end{pmatrix},$$
$$\mathrm{e}^{-\mathrm{i}\psi\hat{\boldsymbol{n}}\cdot\boldsymbol{J}} I^J \mathrm{e}^{\mathrm{i}\psi\hat{\boldsymbol{n}}\cdot\boldsymbol{J}} = I^J .$$

An important invariant operator is

$$I^{k_1k_2k} = \sum_{\mu_1\mu_2\mu} C^{k_1k_2k}_{\mu_1\mu_2\mu} T^{k_1}_{\mu_1} T^{k_2}_{\mu_2} T^{k\dagger}_\mu .$$

7. Other definitions of conjugation:

$$T^{\mathrm{T}}_M \to (-1)^{J-M} T^J_{-M}\,, \quad T^J_M \to (-1)^{J+M} T^J_{-M}\,.$$

2.8.4 Wigner–Eckart Theorem

The Wigner–Eckart theorem establishes the form of the matrix elements of an arbitrary irreducible tensor operator:

$$\langle j'm'|T^J_M|jm\rangle = \langle j'\|\boldsymbol{T}^J\|j\rangle C^{jJj'}_{mMm'}$$
$$= \langle j'\|\boldsymbol{T}^J\|j\rangle (-1)^{j+J+m'} \begin{pmatrix} j & J & j' \\ m & M & -m' \end{pmatrix}.$$

Reduced matrix elements with respect to WCG-coefficients:

$$\langle j'\|T^J\|j\rangle = \sum_{\mu M} C^{jJj'}_{\mu M\mu'}\langle j'\mu'|T^J_M|j\mu\rangle ,$$

each $\mu' = j', j'-1, \ldots, -j'$ (the reduced matrix element is independent of μ').

Reduced matrix elements with respect to 3–j coefficients:

$$(j'\|T^J\|j) = (-1)^{2J}\sqrt{2j'+1}\langle j'\|T^J\|j\rangle .$$

Examples of irreducible tensor operators include:

1. The solid harmonics with respect to the orbital angular momentum $\boldsymbol{L}$:

$$\boldsymbol{\mathcal{Y}}^k(\boldsymbol{x}) = \{\mathcal{Y}_{k\mu}(\boldsymbol{x}) : \mu = k, \ldots, -k\} ,$$

$$\mathcal{Y}_{k\mu}|lm\rangle = \sum_{l'} \langle l'\|\boldsymbol{\mathcal{Y}}^k\|l\rangle C^{lkl'}_{m,\mu,m+\mu}|l', m+\mu\rangle ,$$

where

$$\langle \boldsymbol{x} \mid lm\rangle = \mathcal{Y}_{lm}(\boldsymbol{x}) ,$$

$$\langle l'\|\boldsymbol{\mathcal{Y}}^k\|l\rangle = r^{l+k-l'}\left(\frac{(2l+1)(2k+1)}{4\pi(2l'+1)}\right)^{\frac{1}{2}} C^{lkl'}_{000} ,$$

$$\mathcal{Y}_{k\mu}(\boldsymbol{x})\mathcal{Y}_{lm}(\boldsymbol{x}) = \sum_{l'} \langle l'\|\boldsymbol{\mathcal{Y}}^k\|l\rangle C^{lkl'}_{m,\mu,m+\mu}\mathcal{Y}_{l',m+\mu}(\boldsymbol{x}) ,$$

$$\left[\boldsymbol{\mathcal{Y}}^{k_1}(\boldsymbol{x})\otimes\boldsymbol{\mathcal{Y}}^{k_2}(\boldsymbol{x})\right]^k_\mu = \sum_{\mu_1\mu_2} C^{k_1k_2k}_{\mu_1\mu_2\mu}\mathcal{Y}_{k_1\mu_1}(\boldsymbol{x})\mathcal{Y}_{k_2\mu_2}(\boldsymbol{x}) ,$$

$$\left[\boldsymbol{\mathcal{Y}}^{k_1}(\boldsymbol{x})\otimes\boldsymbol{\mathcal{Y}}^{k_2}(\boldsymbol{x})\right]^k_\mu = \langle k\|\boldsymbol{\mathcal{Y}}^{k_1}\|k_2\rangle\mathcal{Y}_{k\mu}(\boldsymbol{x}) .$$

2. The polynomial operator $\mathcal{T}^k$ in the components of $\boldsymbol{J}$ (Sect. 2.8.2):

$$\langle j'm'|\mathcal{T}^k_\mu|jm\rangle = \delta_{j'j}\langle j\|\mathcal{T}^k\|j\rangle C^{jkj}_{m\mu m'} ,$$

$$\langle j\|\mathcal{T}^k\|j\rangle = a_k(-1)^k\left(\frac{(2j+k+1)!k!k!}{(2j+1)(2j-k)!(2k)!}\right)^{\frac{1}{2}} .$$

3. Polynomials in the components of an arbitrary vector operator $\boldsymbol{V}$, which has the defining relations:

$$\left[J_i, V_j\right] = \mathrm{i}e_{ijk}V_k ,$$

$$\left[J_\pm, V_\mu\right] = [(1\mp\mu)(2\pm\mu)]^{\frac{1}{2}}V_{\mu\pm1} ,$$

$$\left[J_3, V_\mu\right] = \mu V_\mu ,$$

$$V_{+1} = -(V_1+\mathrm{i}V_2)/\sqrt{2},\ V_0 = V_3 ,$$

$$V_{-1} = (V_1-\mathrm{i}V_2)/\sqrt{2} .$$

This construction parallels exactly that given in Sect. 2.8.2 upon replacing $\boldsymbol{J}$ by $\boldsymbol{V}$. The explicit form of the resulting polynomials may be quite different since no assumptions are made concerning commutation relations between the components V_i of $\boldsymbol{V}$. The solid harmonics in the gradient operator ∇ constitute an irreducible tensor operator with respect to the orbital angular momentum $\boldsymbol{L}$.

2.8.5 Unit Tensor Operators or Wigner Operators

A unit tensor operator is an irreducible tensor operator $\hat{\boldsymbol{T}}^{J,\Delta}$, indexed not only by the angular momentum quantum number J, but also by an additional label Δ, which specifies that this irreducible tensor operator has reduced matrix elements given by

$$\langle j'\|\hat{\boldsymbol{T}}^{J,\Delta}\|j\rangle = \delta_{j',j+\Delta} .$$

This condition is to be true for all $j = 0, 1/2, 1, \ldots$. There is a unit tensor operator defined for each

$$\Delta = J, J-1, \ldots, -J .$$

The special symbol

$$\left\langle \begin{matrix} & J+\Delta & \\ 2J & & 0 \\ & \bullet & \end{matrix} \right\rangle$$

denotes a unit tensor operator, replacing the boldface symbol $\hat{\boldsymbol{T}}^{J,\Delta}$, while the symbol

$$\left\langle \begin{matrix} & J+\Delta & \\ 2J & & 0 \\ & J+M & \end{matrix} \right\rangle , \quad M = J, J-1, \ldots, -J$$

denotes the components. In the same way that abstract angular momentum $\boldsymbol{J}$ and state vectors $\{|jm\rangle\}$ extract the intrinsic structure of all realizations of angular momentum theory, as given in Sect. 2.2, so does the notion of a unit tensor operator extract the intrinsic structure of the concept of irreducible tensor operator by disregarding the physical content of the theory, which is carried in the structure of the reduced matrix elements. Physical theory is regained from the fact that the unit tensor operators are the basis for arbitrary tensor operators, which is the structural content of the Wigner–Eckart theorem. The concept of a unit tensor operator was introduced by Racah, but it was Biedenharn who recognized the full significance of this concept not only for $SU(2)$, but for all the unitary groups.

All of the content of physical tensor operator theory can be regained from the properties of unit tensor operators or Wigner operators as summarized below:

Notation (double Gel'fand patterns):

$$\left\langle\begin{array}{ccc} & J+\Delta & \\ 2J & & 0 \\ & J+M & \end{array}\right\rangle, \quad \begin{array}{l} M,\,\Delta = J,\, J-1,\,\dots,\,-J \\ 2J = 0,\,1,\,2,\,\dots\,. \end{array}$$

Definition (shift action):

$$\left\langle\begin{array}{ccc} & J+\Delta & \\ 2J & & 0 \\ & J+M & \end{array}\right\rangle |jm\rangle = C^{jJj+\Delta}_{m,M,m+M} \mid j+\Delta, m+M\rangle \tag{2.56}$$

for all $j = 0, \frac{1}{2}, \dots\,;\ m = j, j-1, \dots, -j$.
Conjugation:

$$\begin{aligned} &\left\langle\begin{array}{ccc} & J+\Delta & \\ 2J & & 0 \\ & J+M & \end{array}\right\rangle^{\dagger} |jm\rangle \\ &= C^{j-\Delta Jj}_{m-M,M,m} \mid j-\Delta, m-M\rangle\,. \end{aligned} \tag{2.57}$$

Orthogonality:

$$\sum_M \left\langle\begin{array}{ccc} & J+\Delta' & \\ 2J & & 0 \\ & J+M & \end{array}\right\rangle \left\langle\begin{array}{ccc} & J+\Delta & \\ 2J & & 0 \\ & J+M & \end{array}\right\rangle^{\dagger} = \delta_{\Delta'\Delta} \boldsymbol{I}^J_{\Delta}\,, \tag{2.58}$$

$$\sum_\Delta \left\langle\begin{array}{ccc} & J+\Delta & \\ 2J & & 0 \\ & J+M' & \end{array}\right\rangle^{\dagger} \left\langle\begin{array}{ccc} & J+\Delta & \\ 2J & & 0 \\ & J+M & \end{array}\right\rangle = \delta_{M'M}\,, \tag{2.59}$$

$$\begin{aligned} &\sum_m \langle jm| \left\langle\begin{array}{ccc} & J'+\Delta' & \\ 2J' & & 0 \\ & J'+M' & \end{array}\right\rangle \left\langle\begin{array}{ccc} & J+\Delta & \\ 2J & & 0 \\ & J+M & \end{array}\right\rangle^{\dagger} |jm\rangle \\ &= \frac{2j+1}{2J+1} \delta_{J'J}\delta_{M'M}\delta_{\Delta'\Delta}\,. \end{aligned} \tag{2.60}$$

The invariant operator $\boldsymbol{I}^J_\Delta$ is defined by its action on an arbitrary vector $\psi_j \in \mathcal{H}_j$:

$$\boldsymbol{I}^J_\Delta \psi_j = \epsilon_{j-\Delta,J,j}\psi_j\,.$$

Tensor operator property:

$$\begin{aligned} &\mathrm{e}^{-\mathrm{i}\psi\hat{\boldsymbol{n}}\cdot\boldsymbol{J}} \left\langle\begin{array}{ccc} & J+\Delta & \\ 2J & & 0 \\ & J+M & \end{array}\right\rangle \mathrm{e}^{\mathrm{i}\psi\hat{\boldsymbol{n}}\cdot\boldsymbol{J}} \\ &= \sum_{M'} D^J_{M'M}(\psi,\hat{\boldsymbol{n}}) \left\langle\begin{array}{ccc} & J+\Delta & \\ 2J & & 0 \\ & J+M' & \end{array}\right\rangle\,. \end{aligned} \tag{2.61}$$

Basis property (Wigner–Eckart theorem):

$$\begin{aligned} &T^J_M \mid jm\rangle \\ &= \left(\sum_\Delta \langle j+\Delta \| \boldsymbol{T}^J \| j\rangle \left\langle\begin{array}{ccc} & J+\Delta & \\ 2J & & 0 \\ & J+M & \end{array}\right\rangle\right) \mid jm\rangle\,. \end{aligned} \tag{2.62}$$

Characteristic null space:

The characteristic null space of the Wigner operator defined by (2.56) is the set of irreducible subspaces $\mathcal{H}_j \subset \mathcal{H}$ given be

$$\{\mathcal{H}_j : 2j = 0, 1, \dots, J-\Delta-1\}\,.$$

Coupling law:

$$\begin{aligned} &\sum_{\alpha\beta} C^{abc}_{\alpha\beta\gamma} \left\langle\begin{array}{ccc} & b+\sigma & \\ 2b & & 0 \\ & b+\beta & \end{array}\right\rangle \left\langle\begin{array}{ccc} & a+\rho & \\ 2a & & 0 \\ & a+\alpha & \end{array}\right\rangle \\ &= \boldsymbol{W}^{abc}_{\rho,\sigma,\rho+\sigma} \left\langle\begin{array}{ccc} & c+\rho+\sigma & \\ 2c & & 0 \\ & c+\gamma & \end{array}\right\rangle\,, \end{aligned} \tag{2.63}$$

where $\boldsymbol{W}^{abc}_{\rho\sigma\tau}$ is an invariant operator (commutes with $\boldsymbol{J}$) and is called a Racah invariant. Its relationship to Racah coefficients and 6–j coefficients is given in Sect. 2.9.
Product law:

$$\begin{aligned} &\left\langle\begin{array}{ccc} & b+\sigma & \\ 2b & & 0 \\ & b+\beta & \end{array}\right\rangle \left\langle\begin{array}{ccc} & a+\rho & \\ 2a & & 0 \\ & a+\alpha & \end{array}\right\rangle \\ &= \sum_c \boldsymbol{W}^{abc}_{\rho,\sigma,\rho+\sigma} C^{abc}_{\alpha,\beta,\alpha+\beta} \left\langle\begin{array}{ccc} & c+\rho+\sigma & \\ 2c & & 0 \\ & c+\alpha+\beta & \end{array}\right\rangle\,. \end{aligned} \tag{2.64}$$

Racah invariant:

$$\begin{aligned} \boldsymbol{W}^{abc}_{\rho\sigma\tau} &= \sum_{\alpha\beta\gamma} C^{abc}_{\alpha\beta\gamma} \\ &\times \left\langle\begin{array}{ccc} & b+\sigma & \\ 2b & & 0 \\ & b+\beta & \end{array}\right\rangle \left\langle\begin{array}{ccc} & a+\rho & \\ 2a & & 0 \\ & a+\alpha & \end{array}\right\rangle \left\langle\begin{array}{ccc} & c+\tau & \\ 2c & & 0 \\ & c+\gamma & \end{array}\right\rangle^{\dagger}\,. \end{aligned} \tag{2.65}$$

The notation $\boldsymbol{W}^{abc}_{\rho\sigma\tau}$ for a Racah invariant is designed to "match" that of the WCG-coefficient on the left, the latter being associated with the lower group theoretical labels, for example,

$$\left(\begin{array}{ccc} 2a & & 0 \\ & a+\alpha & \end{array}\right) \to \mid a\alpha\rangle\,,$$

the state vector having a group transformation law under the action of $SU(2)$, and the former with the shift labels of a unit tensor operator,

$$\begin{pmatrix} & \alpha+\rho & \\ 2\alpha & & 0 \end{pmatrix},$$

and having no associated group transformation law. The invariant operator defined by (2.65) has real eigenvalues, hence, is a Hermitian operator,

$$\boldsymbol{W}^{abc\dagger}_{\rho\sigma\tau} = \boldsymbol{W}^{abc}_{\rho\sigma\tau}\,, \tag{2.66}$$

which is diagonal on an arbitrary state vector in $\mathcal{H}_j$ (Sect. 2.9).

The Racah invariant operator does not commute with a unit tensor operator, and it makes a difference whether it is written to the left or to the right of such a unit tensor operator. The convention here writes it to the left.

Relation (2.65) is taken as the definition of $\boldsymbol{W}^{abc}_{\rho\sigma\tau}$ and the following properties all follow from this expression:

Domain of definition:

$$\begin{aligned}
&\boldsymbol{W}^{abc}_{\rho\sigma\tau} : a, b, c \in \{0, 1/2, 1, 3/2, \dots\}\,;\\
&\quad \rho = a, a-1, \dots, -a\\
&\quad \sigma = b, b-1, \dots, -b\\
&\quad \tau = c, c-1, \dots, -c\,;\\
&\boldsymbol{W}^{abc}_{\rho\sigma\tau} = \boldsymbol{0},\ \text{if } \rho+\sigma \neq \tau;\ \text{if } \epsilon_{abc} = 0\,.
\end{aligned}$$

Orthogonality relations:

$$\sum_{\rho\sigma} \boldsymbol{W}^{abc}_{\rho\sigma\tau} \boldsymbol{W}^{abd}_{\rho\sigma\tau'} = \delta_{cd}\delta_{\tau\tau'}\epsilon_{abc}\boldsymbol{I}^{c}_{\tau}\,, \tag{2.67}$$

$$\sum_{c\tau} \boldsymbol{W}^{abc}_{\rho\sigma\tau} \boldsymbol{W}^{abc}_{\rho'\sigma'\tau} = \delta_{\rho\rho'}\delta_{\sigma\sigma'}\boldsymbol{I}^{ab}_{\rho\sigma}\,, \tag{2.68}$$

where the $\boldsymbol{I}$ invariant operators in these expressions have the following eigenvalues on an arbitrary vector $\psi_j \in \mathcal{H}_j$:

$$\begin{aligned}
\boldsymbol{I}^{c}_{\tau}\psi_j &= \epsilon_{j-\tau,c,j}\psi_j\,,\\
\boldsymbol{I}^{ab}_{\rho\sigma}\psi_j &= \epsilon_{j-\sigma-\rho,a,j-\sigma}\epsilon_{j-\sigma,b,j}\psi_j\,.
\end{aligned}$$

The orthogonality relations for Racah invariants parallel exactly those of WCG-coefficients.

Using the orthogonality relations (2.67) for Racah invariants, the following two relations now follow from (2.63) and (2.64), respectively:

WCG and Racah operator coupling:

$$\begin{aligned}
&\sum_{\rho\sigma}\sum_{\alpha\beta} \boldsymbol{W}^{abd}_{\rho,\sigma,\rho+\sigma} C^{abc}_{\alpha,\beta,\alpha+\beta}\\
&\quad\times \left\langle \begin{matrix} & b+\sigma & \\ 2b & & 0 \\ & b+\beta & \end{matrix} \right\rangle \left\langle \begin{matrix} & a+\rho & \\ 2a & & 0 \\ & a+\alpha & \end{matrix} \right\rangle\\
&= \delta_{cd}\epsilon_{abc}\boldsymbol{I}^{d}_{\tau} \left\langle \begin{matrix} & c+\tau & \\ 2c & & 0 \\ & c+\gamma & \end{matrix} \right\rangle\,.
\end{aligned} \tag{2.69}$$

Racah operator coupling of shift patterns:

$$\begin{aligned}
&\sum_{\rho\sigma} \boldsymbol{W}^{abc}_{\rho\sigma\tau} \left\langle \begin{matrix} & b+\sigma & \\ 2b & & 0 \\ & b+\beta & \end{matrix} \right\rangle \left\langle \begin{matrix} & a+\rho & \\ 2a & & 0 \\ & a+\alpha & \end{matrix} \right\rangle\\
&= C^{abc}_{\alpha,\beta,\alpha+\beta} \left\langle \begin{matrix} & c+\tau & \\ 2c & & 0 \\ & c+\alpha+\beta & \end{matrix} \right\rangle\,.
\end{aligned} \tag{2.70}$$

Relations (2.56–2.70) capture the full content of irreducible tensor operator algebra through the concept of unit tensor operators that have only 0 or 1 for their reduced matrix elements. Using the Wigner–Eckart theorem (2.62), the relations between general tensor operators can be reconstructed. Unit tensor operators were invented to exhibit in the most elementary way possible the abstract and intrinsic structure of the irreducible tensor operator algebra, stripping away the details of particular physical applications, thus giving the theory universal application. It accomplishes the same goal for tensor operator theory that the abstract multiplet theory in Sect. 2.2 accomplishes for representation theory.

Physical theory is regained through the concept of reduced matrix element. The coupling rule (2.54) is now transformed to a rule empty of WCG-coefficient content and becomes a rule for coupling of reduced matrix elements using the invariant Racah operators:

$$\begin{aligned}
&\langle (\alpha')\, j' \| [\boldsymbol{S}^{k_1} \times \boldsymbol{T}^{k_2}]^{k} \| (\alpha)\, j \rangle\\
&\quad = (-1)^{k_1+k_2-k} \sum_{(\alpha'')\, j''} W^{k_2 k_1 k}_{j''-j,\, j'-j'',\, j'-j}(j')\\
&\qquad \times \langle (\alpha')\, j' \| \boldsymbol{S}^{k_1} \| (\alpha'')\, j'' \rangle \langle (\alpha'')\, j'' \| \boldsymbol{T}^{k_2} \| (\alpha)\, j \rangle\,.
\end{aligned} \tag{2.71}$$

This coupling rule is invariant to all $SU(2)$ rotations, and reveals the true role of the Racah coefficients and reduced matrix elements in physical theory as invariant objects under $SU(2)$ rotations. It now becomes imperative to understand Racah coefficients as objects free of their original definition in terms of WCG-coefficients.

2.9 Racah Coefficients

Relation (2.65) is taken, initially, as the definition of the Racah coefficient with appropriate adjustments of notations to conform to Racah's W-notation and to Wigner's 6–j notation. Corresponding to each of (2.63–2.65), (2.69, 2.70), there is a corresponding numerical relationship between WCG-coefficients and Racah coefficients. Despite the present day popularity of expressing all such relations in terms of the 3–j and 6–j notation, this temptation is resisted here for this particular set of relations because of their fundamental origins. The relation between the Racah invariant notation and Racah's original W-notation is

$$\begin{aligned}
&\mathbf{W}^{abc}_{\rho\sigma\tau} \mid jm\rangle = \mathbf{W}^{abc}_{\rho\sigma\tau}(j) \mid jm\rangle\,,\\
&\quad \mathbf{W}^{abc}_{\rho\sigma\tau}(j) = 0 \quad \text{if } \tau \neq \rho+\sigma\,, \qquad \text{or } \epsilon_{abc} = 0\,,\\
&\quad \mathbf{W}^{abc}_{\rho\sigma\tau}(j) = [(2c+1)(2j-2\sigma+1)]^{1/2}\\
&\qquad\qquad \times W(j-\tau, a, j, b; j-\sigma, c)\,,\\
&\quad [(2e+1)(2f+1)]^{1/2} W(abcd; ef)\\
&\qquad = W^{bdf}_{e-a,c-e,c-a}(c)\,,
\end{aligned}$$

$W(abcd; ef) = 0$ unless the triples of nonnegative integers and half-integers (abe), (cde), (acf), (bdf) satisfy the triangle conditions.

2.9.1 Basic Relations Between WCG and Racah Coefficients

$$\begin{aligned}
&\sum_{\beta\delta} C^{bdf}_{\beta\delta\gamma} C^{edc}_{\alpha+\beta,\delta,\alpha+\gamma} C^{abe}_{\alpha,\beta,\alpha+\beta}\\
&= [(2e+1)(2f+1)]^{1/2} W(abcd; ef) C^{afc}_{\alpha,\gamma,\alpha+\gamma}\,,\\
&\sum_{f} [(2e+1)(2f+1)]^{1/2} W(abcd; ef)\\
&\qquad \times C^{bdf}_{\beta,\delta,\beta+\delta} C^{afc}_{\alpha,\beta+\delta,\alpha+\beta+\delta}\\
&= C^{edc}_{\alpha+\beta,\delta,\alpha+\beta+\delta} C^{abe}_{\alpha,\beta,\alpha+\beta}\,,\\
&\delta_{cc'} [(2e+1)(2f+1)]^{1/2} W(abcd; ef)\\
&= \sum_{\beta\delta} C^{bdf}_{\beta,\delta,\beta+\delta} C^{edc}_{\gamma-\delta,\delta,\gamma} C^{abe}_{\gamma-\beta-\delta,\beta,\gamma-\delta}\\
&\qquad \times C^{abc'}_{\gamma-\beta-\delta,\beta+\delta,\gamma}\,,\\
&\sum_{\beta\delta e} [(2e+1)(2f+1)]^{1/2} W(abcd; ef)\\
&\qquad \times C^{bdf'}_{\beta\delta\gamma} C^{edc}_{\alpha+\beta,\delta,\alpha+\gamma} C^{abe}_{\alpha,\beta,\alpha+\beta}\\
&= \delta_{ff'} C^{afc}_{\alpha,\gamma,\alpha+\gamma}\,,\\
&\sum_{e} [(2e+1)(2f+1)]^{1/2} W(abcd; ef)\\
&\qquad \times C^{edc}_{\alpha+\beta,\delta,\alpha+\beta+\delta} C^{abe}_{\alpha,\beta,\alpha+\beta}\\
&= C^{bdf}_{\beta,\delta,\beta+\delta} C^{afc}_{\alpha,\beta+\delta,\alpha+\beta+\delta}\,.
\end{aligned}$$

2.9.2 Orthogonality and Explicit Form

Orthogonality relations for Racah coefficients:

$$\begin{aligned}
&\sum_e (2e+1)(2f+1) W(abcd; ef) W(abcd; ef')\\
&\quad = \delta_{ff'} \epsilon_{acf} \epsilon_{bdf}\,,
\end{aligned} \tag{2.72}$$

$$\begin{aligned}
&\sum_f (2e+1)(2f+1) W(abcd; ef) W(abcd; e'f)\\
&\quad = \delta_{ee'} \epsilon_{abe} \epsilon_{cde}\,.
\end{aligned} \tag{2.73}$$

Definition of 6–j coefficients:

$$\begin{Bmatrix} a & b & e \\ d & c & f \end{Bmatrix} = (-1)^{a+b+c+d} W(abcd; ef)\,. \tag{2.74}$$

Orthogonality of 6–j coefficients:

$$\begin{aligned}
&\sum_e (2e+1)(2f+1) \begin{Bmatrix} a & b & e \\ d & c & f \end{Bmatrix} \begin{Bmatrix} a & b & e \\ d & c & f' \end{Bmatrix}\\
&= \delta_{ff'} \epsilon_{acf} \epsilon_{bdf}\,,
\end{aligned} \tag{2.75}$$

$$\begin{aligned}
&\sum_f [(2e+1)(2f+1)] \begin{Bmatrix} a & b & e \\ d & c & f \end{Bmatrix} \begin{Bmatrix} a & b & e' \\ d & c & f \end{Bmatrix}\\
&= \delta_{ee'} \epsilon_{abe} \epsilon_{cde}\,.
\end{aligned} \tag{2.76}$$

Explicit form of Racah coefficients:

$$\begin{aligned}
W(abcd; ef) &= \Delta(abe)\Delta(cde)\Delta(acf)\Delta(bdf)\\
&\times \sum_k \frac{(-1)^{a+b+c+d+k}(k+1)!}{(k-a-b-e)!(k-c-d-e)!}\\
&\times \frac{1}{(k-a-c-f)!(k-b-d-f)!}\\
&\times \frac{1}{(a+b+c+d-k)!}\\
&\times \frac{1}{(a+d+e+f-k)!(b+c+e+f-k)!}\,,
\end{aligned} \tag{2.77}$$

where $\Delta(abc)$ denotes the triangle coefficient, defined for every triple a, b, c of integers and half-odd integers satisfying the triangle conditions by:

$$\begin{aligned}
&\Delta(abc)\\
&= \left(\frac{(a+b-c)!(a-b+c)!(-a+b+c)!}{(a+b+c+1)!}\right)^{\frac{1}{2}}\,.
\end{aligned} \tag{2.78}$$

2.9.3 The Fundamental Identities Between Racah Coefficients

Each of the three relations given in this section is between Racah coefficients alone. Each expresses a fundamental mathematical property. The Biedenharn–Elliott identity is a consequence of the associativity rule for the open product of three irreducible tensor operators; the Racah sum rule is a consequence of the commutativity of a mapping diagram associated with the coupling of three angular momenta; and the triangle coupling rule is a consequence of the associativity of the open product of three symplection polynomials [2.1]. As such, these three relations between Racah coefficients, together with the orthogonality relations, are the building blocks on which is constructed a theory of these coefficients that stands on its own, independent of the WCG-coefficient origins. Indeed, the latter is recovered through the limit relation (2.50).

Biedenharn–Elliott identity:

$$\begin{aligned}&W(a'ab'b;c'e)W(a'ed'd;b'c)\\&=\sum_f(2f+1)W(abcd;ef)W(c'bd'd;b'f)\\&\quad\times W(a'ad'f;c'c)\,,\end{aligned} \tag{2.79a}$$

$$\begin{aligned}&\begin{Bmatrix}a' & a & c'\\ b & b' & e\end{Bmatrix}\begin{Bmatrix}a' & e & b'\\ d & d' & c\end{Bmatrix}\\&=\sum_f(-1)^{\phi}(2f+1)\begin{Bmatrix}a & b & e\\ d & c & f\end{Bmatrix}\begin{Bmatrix}c' & b & b'\\ d & d' & f\end{Bmatrix}\\&\quad\times\begin{Bmatrix}a' & a & c'\\ f & d' & c\end{Bmatrix},\\&\phi=f-e+a'+a+b'+b+c'-c+d'-d\,.\end{aligned} \tag{2.79b}$$

Racah sum rule:

$$\begin{aligned}&\sum_f(-1)^{b+d-f}(2f+1)W(abcd;ef)W(adcb;gf)\\&=(-1)^{e+g-a-c}W(bacd;eg)\,,\end{aligned} \tag{2.80a}$$

$$\begin{aligned}&\sum_f(-1)^{e+g+f}(2f+1)\begin{Bmatrix}a & b & e\\ d & c & f\end{Bmatrix}\begin{Bmatrix}a & d & g\\ b & c & f\end{Bmatrix}\\&=\begin{Bmatrix}b & a & e\\ d & c & g\end{Bmatrix}.\end{aligned} \tag{2.80b}$$

Triangle sum rule:

$$\begin{aligned}&[\Delta(acf)\Delta(bdf)]^{-1}\\&=(2f+1)\sum_e[\Delta(abe)\Delta(cde)]^{-1}W(abcd;ef)\,,\end{aligned} \tag{2.81a}$$

$$\begin{aligned}&(-1)^{a+b+c+d}[\Delta(acf)\Delta(bdf)]^{-1}\\&=(2f+1)\sum_e[\Delta(abe)\Delta(cde)]^{-1}\begin{Bmatrix}abe\\ dcf\end{Bmatrix}.\end{aligned} \tag{2.81b}$$

2.9.4 Schwinger–Bargmann Generating Function and its Combinatorics

Triangles associated with the 6–j symbol $\begin{Bmatrix}j_1 & j_2 & j_3\\ j_4 & j_5 & j_6\end{Bmatrix}$:

$$(j_1j_2j_3)\,,\quad(j_3j_4j_5)\,,\quad(j_1j_5j_6)\,,\quad(j_2j_4j_6)\,.$$

Points in $\boldsymbol{R}^3$ associated with the triangles:

$$\begin{aligned}&(j_1j_2j_3)\to(x_1,x_2,x_3)\,,\ (j_3j_4j_5)\to(y_3,x_4,x_5)\,,\\&(j_1j_5j_6)\to(y_1,y_5,x_6)\,,\ (j_2j_4j_6)\to(y_2,y_4,y_6)\,.\end{aligned}$$

Tetrahedron associated with the points:

The points define the vertices of a general tetrahedron with lines joining each pair of points that share a common subscript, and the lines are labeled by the product of the common coordinates (Fig. 2.2).

Monomial term:

Define the triangle monomial associated with a triangle $(j_aj_bj_c)$ and its associated point (z_a,z_b,z_c) in $\boldsymbol{R}^3$ by

$$(z_a,z_b,z_c)^{(j_aj_bj_c)}=z_a^{j_b+j_c-j_a}z_b^{j_c+j_a-j_b}z_c^{j_a+j_b-j_c}\,. \tag{2.82}$$

Cubic graph (tetrahedral $\boldsymbol{T}_4$) functions:

Interchange the symbols x and y in the coordinates of the vertices of the tetrahedron and define the following polynomials on the vertices and edges of the tetrahedron with this modified labeling.

Vertex function: multiply together the coordinates of each vertex and sum over all such vertices to obtain

$$V_3=y_1y_2y_3+x_3y_4y_5+x_1x_5y_6+x_2x_4x_6\,;$$

Edge function: multiply together the coordinates of a given edge and the opposite edge and sum over all such pairs to obtain

$$E_4=x_1y_1x_4y_4+x_2y_2x_5y_5+x_3y_3x_6y_6\,.$$

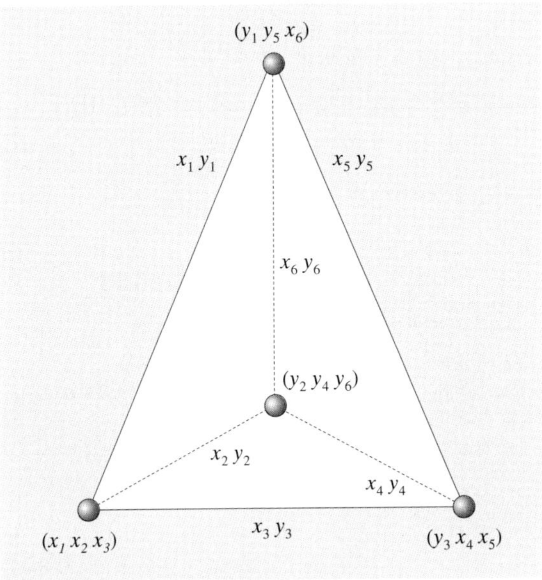

Fig. 2.2 Labeled cubic graph (tetrahedron) associated with $6\text{–}j$ coefficients

Generating function:

$$(1+V_3+E_4)^{-2} = \sum_{\Delta} T(\Delta) Z^{\Delta}\,, \tag{2.83a}$$

$$Z^{\Delta} = (x_1, x_2, x_3)^{(j_1 j_2 j_3)} (y_3, x_4, x_5)^{(j_3 j_4 j_5)} \times (y_1, y_5, x_6)^{(j_1 j_5 j_6)} (y_2, y_4, y_6)^{(j_2 j_4 j_6)}\,, \tag{2.83b}$$

$$\Delta = \begin{bmatrix} (j_1 j_2 j_3) \\ (j_3 j_4 j_5) \\ (j_1 j_5 j_6) \\ (j_2 j_4 j_6) \end{bmatrix};$$

$$T(\Delta) = \sum_k (-1)^k (k+1) \times \binom{k}{k_1, k_2, k_3, k_4, k_5, k_6, k_7}\,, \tag{2.83c}$$

$k_i = k - t_i\,,\quad i = 1, 2, 3, 4\,,$

$k_j = e_{j-4} - k\,,\quad j = 5, 6, 7\,;$

$t_i =$ triangle sum $=$ vertex sum,

$e_j =$ opposite edge sum, in pairs,

$t_1 = j_1 + j_2 + j_3\,,\quad t_2 = j_3 + j_4 + j_5\,,$

$t_3 = j_1 + j_5 + j_6\,,\quad t_4 = j_2 + j_4 + j_6\,,$

$e_1 = (j_2 + j_5) + (j_3 + j_6)\,,$

$e_2 = (j_1 + j_4) + (j_3 + j_6)\,,$

$e_3 = (j_1 + j_4) + (j_2 + j_5)\,.$

The summation in (2.83b) is over the infinite set of all tetrahedra; that is, over the infinite set of arrays Δ having nonnegative integral entries. The $6\text{–}j$ coefficients is then given by

$$\begin{Bmatrix} j_1 & j_2 & j_3 \\ j_4 & j_5 & j_6 \end{Bmatrix} = \frac{T(\Delta)}{\Delta(j_1 j_2 j_3)\Delta(j_1 j_5 j_6)\Delta(j_2 j_4 j_6)\Delta(j_3 j_4 j_5)}\,.$$

Since the factor $T(\Delta)$ is an integer in the expansion (2.83a), this result shows that the $6\text{–}j$ coefficient is an integer, up to the multiplicative triangle coefficient factors.

2.9.5 Symmetries of $6\text{–}j$ Coefficients

There are 144 symmetry relations among the Racah $6\text{–}j$ coefficients. The 24 classical ones, given already by Racah, and corresponding to the tetrahedral point group T_d of rotations-inversions (isomorphic to the symmetric group S_4) mapping the regular tetrahedron onto itself, are realized in the $6\text{–}j$ symbol

$$\begin{Bmatrix} a & b & e \\ d & c & f \end{Bmatrix}$$

as permutations of its columns and the exchange of any pair of letters in the top row with the corresponding pair in the bottom row. Regge discovered the 6-fold increase in symmetry by noting that each term in the summation in (2.77) is invariant not only to the classical 24 symmetries, but also under certain linear transformations of the quantum labels. These symmetries are also implicit in Schwinger's generating function.

The full set, including the original 24 substitutions, of linear transformations of the letters a, b, c, d, e, f thus yields a group of linear transformation isomorphic to $S_4 \times S_3$. The column permutations and row-pair interchanges described above applied to each of the six symbols in the equalities below yield the set of 144 relationships:

$$\begin{Bmatrix} a & b & e \\ d & c & f \end{Bmatrix} = \begin{Bmatrix} a & \frac{b+c+e-f}{2} & \frac{b+e+f-c}{2} \\ d & \frac{b+c+f-e}{2} & \frac{c+e+f-b}{2} \end{Bmatrix}$$

$$= \begin{Bmatrix} \frac{a+d+e-f}{2} & b & \frac{a+e+f-d}{2} \\ \frac{a+d+f-e}{2} & c & \frac{d+e+f-a}{2} \end{Bmatrix}$$

$$= \begin{Bmatrix} \frac{a+b+d-c}{2} & \frac{a+b+c-d}{2} & e \\ \frac{a+c+d-b}{2} & \frac{b+c+d-a}{2} & f \end{Bmatrix}$$

$$= \begin{Bmatrix} \frac{a+b+d-c}{2} & \frac{b+c+e-f}{2} & \frac{a+e+f-d}{2} \\ \frac{a+c+d-b}{2} & \frac{b+c+f-e}{2} & \frac{d+e+f-a}{2} \end{Bmatrix}$$

$$= \begin{Bmatrix} \frac{a+d+e-f}{2} & \frac{a+b+c-d}{2} & \frac{b+e+f-c}{2} \\ \frac{a+d+f-e}{2} & \frac{b+c+d-a}{2} & \frac{c+e+f-b}{2} \end{Bmatrix} .$$

2.9.6 Further Properties

Recurrence relations:
Three-term:

$$\begin{aligned} &[(a+b+e+1)(b+e-a) \\ &\quad \times (c+d+e+1)(d+e-c)]^{1/2} \begin{Bmatrix} a & b & e \\ d & c & f \end{Bmatrix} \\ &= -2e[(b+d+f+1)(b+d-f)]^{1/2} \\ &\quad \times \begin{Bmatrix} a & b-\frac{1}{2} & e-\frac{1}{2} \\ d-\frac{1}{2} & c & f \end{Bmatrix} \\ &\quad + [(a+b-e+1)(a+e-b)(c+d-e+1) \\ &\quad \times (c+e-d)]^{1/2} \begin{Bmatrix} a & b & e-1 \\ d & c & f \end{Bmatrix} , \end{aligned} \tag{2.84a}$$

$$\begin{aligned} &[(a+c+f+1)(c+e-d) \\ &\quad \times (d+e-c+1)(b+d-f+1)]^{1/2} \begin{Bmatrix} a & b & e \\ d & c & f \end{Bmatrix} \\ &= [(a+c-f)(a+e-b) \\ &\quad \times (b+f+d+2)(b+e-a+1)]^{1/2} \\ &\quad \times \begin{Bmatrix} a+\frac{1}{2} & b+\frac{1}{2} & e \\ d+\frac{1}{2} & c-\frac{1}{2} & f \end{Bmatrix} . \\ &\quad + [(c+f-a)(c+e-d)(b-a-c+d+1)]^{1/2} \\ &\quad \times \begin{Bmatrix} a & b & e \\ d+\frac{1}{2} & c-\frac{1}{2} & f-\frac{1}{2} \end{Bmatrix} . \end{aligned} \tag{2.84b}$$

Five-term:

$$\begin{aligned} &(2c+1)(2d)(2f+1) \begin{Bmatrix} a & b & e \\ d & c & f \end{Bmatrix} \\ &= [(b+d-f)(b+f-d+1)(d+e-c) \\ &\quad \times (c+e-d+1)(c+f-a+1)(a+c+f+2)]^{\frac{1}{2}} \\ &\quad \times \begin{Bmatrix} a & b & e \\ d-\frac{1}{2} & c+\frac{1}{2} & f+\frac{1}{2} \end{Bmatrix} + [(b+d-f) \\ &\quad \times (b+f-d+1)(c+d-e)(c+d+e+1) \\ &\quad \times (a+c-f)(a+f-c+1)]^{\frac{1}{2}} \\ &\quad \times \begin{Bmatrix} a & b & e \\ d-\frac{1}{2} & c-\frac{1}{2} & f+\frac{1}{2} \end{Bmatrix} - [(d+f-b) \\ &\quad \times (b+d+f+1)(c+d-e)(c+d+e+1) \\ &\quad \times (c+f-a)(a+c+f+1)]^{\frac{1}{2}} \\ &\quad \times \begin{Bmatrix} a & b & e \\ d-\frac{1}{2} & c-\frac{1}{2} & f-\frac{1}{2} \end{Bmatrix} + [(d+f-b) \\ &\quad \times (b+d+f+1)(d+e-c)(c+e-d+1) \\ &\quad \times (a+f-c)(a+c-f+1)]^{\frac{1}{2}} \\ &\quad \times \begin{Bmatrix} a & b & e \\ d-\frac{1}{2} & c+\frac{1}{2} & f-\frac{1}{2} \end{Bmatrix} . \end{aligned} \tag{2.84c}$$

Relation to hypergeometric series:

$$\begin{aligned} &\begin{Bmatrix} abe \\ dcf \end{Bmatrix} = (-1)^{a+b+c+d} W(abcd; ef) \\ &= \Delta(abe)\Delta(cde)\Delta(acf)\Delta(bdf) \\ &\quad \times \frac{(-1)^{\beta_1}(\beta_1+1)!}{(\beta_2-\beta_1)!(\beta_3-\beta_1)!} \\ &\quad \times \frac{{}_4F_3\begin{pmatrix} \alpha_1-\beta_1, & \alpha_2-\beta_1, & \alpha_3-\beta_1, & \alpha_4-\beta_1 \\ -\beta_1-1, & \beta_2-\beta_1+1, & \beta_3-\beta_1+1, & ;1 \end{pmatrix}}{(\beta_1-\alpha_1)!(\beta_1-\alpha_2)!(\beta_1-\alpha_3)!(\beta_1-\alpha_4)!} , \\ &\beta_1 = \min(a+b+c+d, a+d+e+f, b+c+e+f) , \end{aligned}$$

The parameters β_2 and β_3 are identified in either way with the pair remaining in the 3-tuple

$$(a+b+c+d, a+d+e+f, b+c+e+f)$$

after deleting β_1. The $(\alpha_1, \alpha_2, \alpha_3, \alpha_4)$ may be identified with any permutation of the 4-tuple

$$(a+b+e, c+d+e, a+c+f, b+d+f) .$$

The ${}_4F_3$ series is Saalschützian:

$$\begin{aligned} &1+\sum(\text{numerator parameters}) \\ &= \sum(\text{denominator parameters}) . \end{aligned}$$

2.10 The 9–j Coefficients

2.10.1 Hilbert Space and Tensor Operator Actions

Let $\boldsymbol{T}^a(1)$ and $\boldsymbol{T}^b(2)$ denote irreducible tensor operators of ranks a and b with respect to kinematically independent angular momentum operators $\boldsymbol{J}(1)$ and $\boldsymbol{J}(2)$ that act, respectively, in separable Hilbert spaces $\mathcal{H}(1)$ and $\mathcal{H}(2)$. Let $\mathcal{H}(1)$ and $\mathcal{H}(2)$ be reduced, respectively, into a direct sum of spaces $\mathcal{H}_{j_1}(1)$ and $\mathcal{H}_{j_2}(2)$. The angular momentum $\boldsymbol{J}(1)$ has the standard action on the orthonormal basis $\{|\, j_1 m_1\rangle \mid m_1 = j_1, j_1 - 1, \ldots, -j_1\}$ of $\mathcal{H}_{j_1}(1)$, and $\boldsymbol{J}(2)$ has the standard action on the orthonormal basis $\{|\, j_2 m_2\rangle \mid m_2 = j_2, j_2 - 1, \ldots, -j_2\}$ of $\mathcal{H}_{j_2}(2)$. The irreducible tensor operators $\boldsymbol{T}^a(1)$ and $\boldsymbol{T}^b(2)$ also have the standard actions in their respective Hilbert spaces $\mathcal{H}(1)$ and $\mathcal{H}(2)$, as given by the Wigner–Eckart theorem. The total angular momentum $\boldsymbol{J}$ has the standard action on the coupled orthonormal basis of the tensor product space $\mathcal{H}_{j_1} \otimes \mathcal{H}_{j_2}$:

$$|(j_1 j_2) jm\rangle = \sum_{m_1 m_2} C^{j_1 j_2 j}_{m_1 m_2 m} |\, j_1 m_1\rangle \otimes |\, j_2 m_2\rangle \,. \tag{2.85}$$

The tensor product operator $\boldsymbol{T}^a(1) \otimes \boldsymbol{T}^b(2)$ acts in the tensor product space $\mathcal{H}(1) \otimes \mathcal{H}(2)$ according to the rule:

$$\left(\boldsymbol{T}^a(1) \otimes \boldsymbol{T}^b(2)\right)(|\, j_1 m_1\rangle \otimes |\, j_2 m_2\rangle) = \boldsymbol{T}^a(1)|\, j_1 m_1\rangle \otimes \boldsymbol{T}^b(2)|\, j_2 m_2\rangle \,,$$

so that

$$\left(\boldsymbol{T}^a(1) \otimes \boldsymbol{T}^b(2)\right) |\, (j_1 j_2)\; jm\rangle = \sum_{m_1 m_2} C^{j_1 j_2 j}_{m_1 m_2 m} \boldsymbol{T}^a(1)|\, j_1 m_1\rangle \otimes \boldsymbol{T}^b(2)|\, j_2 m_2\rangle \,. \tag{2.86a}$$

The angular momentum quantities called 9–j coefficients arise when the coupled tensor operators $\boldsymbol{T}^{(ab)c}$ with components γ defined by

$$T^{(ab)c}_{\gamma} = \sum_{\alpha\beta} C^{abc}_{\alpha\beta\gamma} T^a_\alpha(1) \otimes T^b_\beta(2) \,,$$
$$\gamma = c, c-1, \ldots, -c \,, \tag{2.86b}$$

are considered. The quantity $\boldsymbol{T}^{(ab)c}$ is an irreducible tensor operator of rank c with respect to the total angular momentum $\boldsymbol{J}$ for all a, b that yield c under the rule of addition of angular momentum.

2.10.2 9–j Invariant Operators

The entire angular momentum content of relation (2.86b) is captured by taking the irreducible tensor operators $\boldsymbol{T}^a(1)$ and $\boldsymbol{T}^b(2)$ to be unit tensor operators acting in the respective spaces $\mathcal{H}(1)$ and $\mathcal{H}(2)$:

$$\boldsymbol{T}^{(ab)c}_{(\rho\sigma)\gamma} = \sum_{\alpha\beta} C^{abc}_{\alpha\beta\gamma} \left\langle 2a \begin{matrix} \alpha+\rho \\ \\ a+\alpha \end{matrix} 0 \right\rangle_1 \otimes \left\langle 2b \begin{matrix} b+\sigma \\ \\ b+\beta \end{matrix} 0 \right\rangle_2 \,. \tag{2.87}$$

The placement of the unit tensor operators shows in which space they act, so that the additional identification by indices 1 and 2 could be eliminated. For each given $c \in \{0, 1/2, 1, 3/2, 2, \ldots\}$ and all a, b such that the triangle relation (abc) is satisfied, and, for each such pair a, b, all ρ, σ with $\rho \in \{a, a-1, \ldots, -a\}$, $\sigma \in \{b, b-1, \ldots, -b\}$, an irreducible tensor operator of rank c with respect to the total angular momentum $\boldsymbol{J}$ with components γ is defined by (2.87). By the Wigner–Eckart theorem, it must be possible to write

$$\sum_{\alpha\beta} C^{abc}_{\alpha\beta\gamma} \left\langle 2a \begin{matrix} \alpha+\rho \\ \\ a+\alpha \end{matrix} 0 \right\rangle_1 \otimes \left\langle 2b \begin{matrix} b+\sigma \\ \\ b+\beta \end{matrix} 0 \right\rangle_2 = \sum_{\tau} \begin{bmatrix} abc \\ \rho\sigma\tau \end{bmatrix} \left\langle 2c \begin{matrix} c+\tau \\ \\ c+\gamma \end{matrix} 0 \right\rangle \,. \tag{2.88}$$

where: (i) The unit tensor operator on the right-hand side is a irreducible tensor operator with respect to $\boldsymbol{J}$; that is, has the action on the coupled states given by

$$\left\langle 2c \begin{matrix} c+\tau \\ \\ c+\gamma \end{matrix} 0 \right\rangle |(j_1 j_2) jm\rangle = C^{j\,c\;j+\tau}_{m,\gamma,m+\gamma} |(j_1 j_2) j+\tau, m+\gamma\rangle \,; \tag{2.89}$$

and (ii) the symbol $\begin{bmatrix} abc \\ \rho\sigma\tau \end{bmatrix}$ denotes an invariant operator with respect to the total angular momentum $\boldsymbol{J}$. Using the orthogonality of unit tensor operators, we can also write relation (2.88) in the form:

$$\begin{bmatrix} abc \\ \rho\sigma\tau \end{bmatrix} = \sum_{\alpha\beta\gamma} C^{abc}_{\alpha\beta\gamma} \left\langle 2a \begin{matrix} a+\rho \\ \\ a+\alpha \end{matrix} 0 \right\rangle_1$$

$$\otimes \left\langle 2b \begin{matrix} b+\sigma \\ \\ b+\beta \end{matrix} 0 \right\rangle_2 \left\langle 2c \begin{matrix} c+\tau \\ \\ c+\gamma \end{matrix} 0 \right\rangle^\dagger . \tag{2.90}$$

This form is taken as the definition of the 9–j invariant operator. Its eigenvalues in the coupled basis define the 9–j coefficient:

$$\begin{bmatrix} abc \\ \rho\sigma\tau \end{bmatrix} |(j_1 j_2) jm\rangle$$

$$= \langle (j_1+\rho, j_2+\sigma) j+\tau | \left| \begin{bmatrix} abc \\ \rho\sigma\tau \end{bmatrix} \right| |(j_1 j_2) j\rangle$$

$$\times |(j_1 j_2) jm\rangle$$

$$= [(2j+1)(2c+1)(2j_1+2\rho+1)(2j_2+2\sigma+1)]^{\frac{1}{2}}$$

$$\times \begin{Bmatrix} j_1 & j_2 & j \\ a & b & c \\ j_1+\rho & j_2+\sigma & j+\tau \end{Bmatrix} |(j_1 j_2) jm\rangle . \tag{2.91}$$

The 9–j invariant operators play exactly the same role in the tensor product space of two irreducible angular momentum spaces as do the Racah invariants in one such irreducible angular momentum space.

The full content of the coupling law (2.86b) for physical irreducible tensor operators is regained in the coupling law for reduced matrix elements:

$$\langle (\alpha_1'\alpha_2' j_1' j_2') j' \| [\boldsymbol{T}^a(1) \times \boldsymbol{T}^b(2)]^c \| (\alpha_1\alpha_2 j_1 j_2) j\rangle$$

$$= \begin{bmatrix} j_1 & j_2 & j \\ a & b & c \\ j_1' & j_2' & j' \end{bmatrix} \langle (\alpha_1') j_1' \| \boldsymbol{T}^a(1) \| (\alpha_1) j_1\rangle$$

$$\times \langle (\alpha_2') j_2' \| \boldsymbol{T}^b(2) \| (\alpha_2) j_2\rangle ; \tag{2.92a}$$

$$\begin{bmatrix} j_1 & j_2 & j \\ a & b & c \\ j_1' & j_2' & j' \end{bmatrix}$$

$$= \left[(2j_1'+1)(2j_2'+1)(2j+1)(2c+1)\right]^{\frac{1}{2}}$$

$$\times \begin{Bmatrix} j_1 & j_2 & j \\ a & b & c \\ j_1' & j_2' & j' \end{Bmatrix} . \tag{2.92b}$$

2.10.3 Basic Relations Between 9–j Coefficients and 6–j Coefficients

Orthogonality of 9–j coefficients:

$$\sum_{hi} (2c+1)(2f+1)(2h+1)(2i+1)$$

$$\times \begin{Bmatrix} a & b & c \\ d & e & f \\ h & i & j \end{Bmatrix} \begin{Bmatrix} a & b & c' \\ d & e & f' \\ h & i & j \end{Bmatrix} = \delta_{cc'}\delta_{ff'} ,$$

where this relation is to be applied only to triples (abc), (def), (cfj), (abc'), (def'), $(c'f'j)$ for which the triangle conditions hold.

9–j coefficients in terms of 3–j coefficients:

$$\delta_{j_{33}j_{33}'}(2j_{33}+1)^{-1} \begin{Bmatrix} j_{11} & j_{12} & j_{13} \\ j_{21} & j_{22} & j_{23} \\ j_{31} & j_{32} & j_{33} \end{Bmatrix}$$

$$= \sum_{\substack{\text{all } m_{ij} \\ \text{except } m_{33}}} \begin{pmatrix} j_{11} & j_{12} & j_{13} \\ m_{11} & m_{12} & m_{13} \end{pmatrix} \begin{pmatrix} j_{21} & j_{22} & j_{23} \\ m_{21} & m_{22} & m_{23} \end{pmatrix}$$

$$\times \begin{pmatrix} j_{31} & j_{32} & j_{33} \\ m_{31} & m_{32} & m_{33} \end{pmatrix} \begin{pmatrix} j_{11} & j_{21} & j_{31} \\ m_{11} & m_{21} & m_{31} \end{pmatrix}$$

$$\times \begin{pmatrix} j_{12} & j_{22} & j_{32} \\ m_{12} & m_{22} & m_{32} \end{pmatrix} \begin{pmatrix} j_{13} & j_{23} & j_{33}' \\ m_{13} & m_{23} & m_{33} \end{pmatrix} . \tag{2.93}$$

9–j coefficients in terms of 6–j coefficients:

$$\begin{Bmatrix} j_{11} & j_{12} & j_{13} \\ j_{21} & j_{22} & j_{23} \\ j_{31} & j_{32} & j_{33} \end{Bmatrix} = \sum_k (-1)^{2k}(2k+1)$$

$$\times \begin{Bmatrix} j_{11} & j_{21} & j_{31} \\ j_{32} & j_{33} & k \end{Bmatrix} \begin{Bmatrix} j_{12} & j_{22} & j_{32} \\ j_{21} & k & j_{23} \end{Bmatrix}$$

$$\times \begin{Bmatrix} j_{13} & j_{23} & j_{33} \\ k & j_{11} & j_{12} \end{Bmatrix} . \tag{2.94}$$

Basic defining relation for 9–j coefficient from (2.88):

$$(-1)^\phi \begin{Bmatrix} j_{12} & j_{22} & j_{32} \\ j_{11} & j_{21} & j_{31} \\ j_{13} & j_{23} & j_{33} \end{Bmatrix} \begin{pmatrix} j_{31} & j_{32} & j_{33} \\ m_{31} & m_{32} & m_{33} \end{pmatrix}$$

$$= \sum_{\text{all } m_{(1i)} m_{(2i)}} \begin{pmatrix} j_{11} & j_{21} & j_{31} \\ m_{11} & m_{21} & m_{31} \end{pmatrix}$$

$$\times \begin{pmatrix} j_{12} & j_{22} & j_{32} \\ m_{12} & m_{22} & m_{32} \end{pmatrix} \begin{pmatrix} j_{13} & j_{23} & j_{33} \\ m_{13} & m_{23} & m_{33} \end{pmatrix}$$

$$\times \begin{pmatrix} j_{11} & j_{12} & j_{13} \\ m_{11} & m_{12} & m_{13} \end{pmatrix} \begin{pmatrix} j_{21} & j_{22} & j_{23} \\ m_{21} & m_{22} & m_{23} \end{pmatrix},$$

$$\phi = \sum_{kl} j_{kl}\,. \tag{2.95}$$

Additional relations:

$$\sum_{kl} (-1)^{2b+l+h-f}(2k+1)(2l+1) \times \begin{Bmatrix} a & b & c \\ e & d & f \\ k & l & i \end{Bmatrix} \begin{Bmatrix} a & e & k \\ d & b & l \\ g & h & i \end{Bmatrix} = \begin{Bmatrix} a & b & c \\ d & e & f \\ g & h & i \end{Bmatrix},$$

$$\sum_{c} (2c+1) \begin{Bmatrix} a & b & c \\ d & e & f \\ g & h & i \end{Bmatrix} \begin{Bmatrix} a & b & c \\ f & i & j \end{Bmatrix} = (-1)^{2j} \begin{Bmatrix} d & e & f \\ b & j & h \end{Bmatrix} \begin{Bmatrix} g & h & i \\ j & a & d \end{Bmatrix},$$

$$\sum_{klm} (2k+1)(2l+1)(2m+1) \times \begin{Bmatrix} a & b & c \\ d & e & f \\ k & l & m \end{Bmatrix} \begin{Bmatrix} k & l & m \\ a' & b' & c' \\ d' & e' & f' \end{Bmatrix} \times \begin{Bmatrix} a & d & k \\ a' & d' & k' \end{Bmatrix} \begin{Bmatrix} b & e & l \\ b' & e' & l' \end{Bmatrix} \begin{Bmatrix} c & f & m \\ c' & f' & m' \end{Bmatrix} = \begin{Bmatrix} a & b & c \\ d' & e' & f' \\ k' & l' & m' \end{Bmatrix} \begin{Bmatrix} k' & l' & m' \\ a' & b' & c' \\ d & e & f \end{Bmatrix}.$$

2.10.4 Symmetry Relations for 9–j Coefficients and Reduction 6–j Coefficients

The 9–j coefficient

$$\begin{Bmatrix} j_{11} & j_{12} & j_{13} \\ j_{21} & j_{22} & j_{23} \\ j_{31} & j_{32} & j_{33} \end{Bmatrix}$$

is invariant under even permutation of its rows, even permutation of its columns, and under the interchange of rows and columns (matrix transposition). It is multiplied by the factor $(-1)^{\phi}$ (2.95) under odd permutations of its rows or columns. These 72 symmetries are all consequences of the 72 symmetries of the 3–j coefficient in relation (2.93).

Reduction to 6–j coefficients:

$$\begin{Bmatrix} a & b & e \\ c & d & e \\ f & f & 0 \end{Bmatrix} = \begin{Bmatrix} 0 & e & e \\ f & d & b \\ f & c & a \end{Bmatrix} = \begin{Bmatrix} e & 0 & e \\ c & f & a \\ d & f & b \end{Bmatrix}$$

$$= \begin{Bmatrix} f & f & 0 \\ d & c & e \\ b & a & e \end{Bmatrix} = \begin{Bmatrix} f & b & d \\ 0 & e & e \\ f & a & c \end{Bmatrix} = \begin{Bmatrix} a & f & c \\ e & 0 & e \\ b & f & d \end{Bmatrix}$$

$$= \begin{Bmatrix} b & a & e \\ f & f & 0 \\ d & c & e \end{Bmatrix} = \begin{Bmatrix} e & d & c \\ e & b & a \\ 0 & f & f \end{Bmatrix} = \begin{Bmatrix} c & e & d \\ a & e & b \\ f & 0 & f \end{Bmatrix}$$

$$= \frac{(-1)^{b+c+e+f}}{[(2e+1)(2f+1)]^{\frac{1}{2}}} \begin{Bmatrix} abe \\ dcf \end{Bmatrix}.$$

2.10.5 Explicit Algebraic Form of 9–j Coefficients

$$\begin{Bmatrix} a & b & c \\ d & e & f \\ h & i & j \end{Bmatrix} = (1)^{c+f-j} \frac{(dah)(bei)(jhi)}{(def)(bac)(jcf)}$$

$$\times \sum_{xyz} \frac{(-1)^{x+y+z}}{x!y!z!}$$

$$\times \frac{(2f-x)!(2a-z)!}{(2i+1+y)!(a+d+h+1-z)!}$$

$$\times \frac{(d+e-f+x)!(c+j-f+x)!}{(e+f-d-x)!(c+f-j-x)!}$$

$$\times \frac{(e+i-b+y)!(h+i-j+y)!}{(b+e-i-y)!(h+j-i-y)!}$$

$$\times \frac{(b+c-a+z)!}{(a+d-h-z)!(a+c-b-z)!}$$

$$\times \frac{(a+d+j-i-y-z)!}{(d+i-b-f+x+y)!(b+j-a-f+x+z)!},$$

$$(abc) = \left(\frac{(a-b+c)!(a+b-c)!(a+b+c+1)!}{(b+c-a)!}\right)^{\frac{1}{2}}.$$

2.10.6 Racah Operators

A Racah operator is denoted

$$\left\{\begin{matrix} & a+\rho & \\ 2a & & 0 \\ & a+\sigma & \end{matrix}\right\} \qquad \begin{matrix}\rho,\sigma = a, a-1, \dots, -a, \\ 2a = 0, 1, 2, \dots,\end{matrix}$$

and is a special case of the operator defined by (2.87):

$$\left\{\begin{matrix} & a+\rho & \\ 2a & & 0 \\ & a+\sigma & \end{matrix}\right\} |(j_1 j_2) jm\rangle = \left(\frac{(2a+1)(2j_2+1)}{(2j_2+2\sigma+1)}\right)^{\frac{1}{2}} \boldsymbol{T}^{(aa)0}_{(\rho\sigma)0} |(j_1 j_2) jm\rangle \,. \quad (2.96)$$

Thus, a Racah operator is an invariant operator with respect to the total angular momentum $\boldsymbol{J}$. Alternative definitions are:

$$\left\{\begin{matrix} & a+\rho & \\ 2a & & 0 \\ & a+\sigma & \end{matrix}\right\} = (-1)^{a+\sigma} \sum_{\alpha} \left\langle\begin{matrix} & a+\rho & \\ 2a & & 0 \\ & a+\alpha & \end{matrix}\right\rangle \otimes \left\langle\begin{matrix} & a-\sigma & \\ 2a & & 0 \\ & a+\alpha & \end{matrix}\right\rangle^{\dagger} ,$$

$$\begin{aligned}\left\{\begin{matrix} & a+\rho & \\ 2a & & 0 \\ & a+\sigma & \end{matrix}\right\} |(j_1 j_2) jm\rangle &= [(2j_1+2\rho+1)(2j_2+1)]^{\frac{1}{2}} \\ &\quad \times W(j, j_1, j_2+\sigma, a; j_2, j_1+\rho) \\ &\quad \times |(j_1+\rho, j_2+\sigma) jm\rangle\end{aligned}$$

with conjugate

$$\begin{aligned}\left\{\begin{matrix} & a+\rho & \\ 2a & & 0 \\ & a+\sigma & \end{matrix}\right\}^{\dagger} |(j_1 j_2) jm\rangle &= [(2j_1+1)(2j_2-2\sigma+1)]^{\frac{1}{2}} \\ &\quad \times W(j, j_1-\rho, j_2, a; j_2-\sigma, j_1) \\ &\quad \times |(j_1-\rho, j_2-\sigma) jm\rangle \,.\end{aligned}$$

Racah operators satisfy orthogonality relations similar in form to Wigner operators. The open product rule is:

$$\begin{aligned}&\left\{\begin{matrix} & b+\sigma & \\ 2b & & 0 \\ & b+\beta & \end{matrix}\right\} \left\{\begin{matrix} & a+\rho & \\ 2a & & 0 \\ & a+\alpha & \end{matrix}\right\} \\ &= \sum_{c} \overline{\boldsymbol{W}}^{abc}_{\rho,\sigma,\rho+\sigma} \underline{\boldsymbol{W}}^{abc}_{\alpha,\beta,\alpha+\beta} \left\{\begin{matrix} & c+\rho+\sigma & \\ 2c & & 0 \\ & c+\alpha+\beta & \end{matrix}\right\} .\end{aligned} \quad (2.97)$$

In this result $\overline{\boldsymbol{W}}^{abc}_{\rho\sigma\tau}$ and $\underline{\boldsymbol{W}}^{abc}_{\alpha,\beta,\alpha+\beta}$ denote Racah invariants with respect to the angular momenta $\boldsymbol{J}(1)$ and $\boldsymbol{J}(2)$, respectively, so that

$$\begin{aligned}\overline{\boldsymbol{W}}^{abc}_{\rho\sigma\tau} |(j_1 j_2) jm\rangle &= \boldsymbol{W}^{abc}_{\rho\sigma\tau}(j_1) |(j_1 j_2) jm\rangle \,, \\ \underline{\boldsymbol{W}}^{abc}_{\alpha\beta\gamma} |(j_1 j_2) jm\rangle &= \boldsymbol{W}^{abc}_{\alpha\beta\gamma}(j_2) |(j_1 j_2) jm\rangle \,.\end{aligned}$$

The matrix elements of relation (2.97) lead to the Biedenharn–Elliott identity. There are five versions of this relationship in complete analogy to relations (2.63–2.65) and (2.69–2.70) for Wigner operators.

Racah operators are a basis for all invariant operators acting in the tensor product space spanned by the coupled basis vectors (2.85) and are the natural way of formulating interactions in that space. Their algebra is a fascinating study, initiated already in a different guise in the work of *Schwinger* [2.3]. Little use has been made of this concept in physical applications.

Additional relations between Racah coefficients or 6–j coefficients may be derived from the various versions of the rule (2.97) or directly from relation (2.79b) by using the orthogonality relations (2.75). Two of these are:

$$\begin{aligned}&\sum_{e} (-1)^{a+b+e}(2e+1) \\ &\quad \times \begin{Bmatrix} a & b & e \\ d & c & g \end{Bmatrix} \begin{Bmatrix} a' & a & c' \\ b & b' & e \end{Bmatrix} \begin{Bmatrix} a' & e & b' \\ d & d' & c \end{Bmatrix} \\ &\quad = (-1)^{\phi_1} \begin{Bmatrix} c' & b & b' \\ d & d' & g \end{Bmatrix} \begin{Bmatrix} a' & a & c' \\ g & d' & c \end{Bmatrix} , \\ &\phi_1 = g + a' + b' + c' + c + d' + d \,; \\ &\sum_{e,e'} (-1)^{a-c'+e-e'}(2e+1)(2e'+1)(2f+1) \\ &\quad \times \begin{Bmatrix} c' & b & e' \\ d & d' & f \end{Bmatrix} \begin{Bmatrix} a & b & e \\ d & c & g \end{Bmatrix} \begin{Bmatrix} a' & a & c' \\ b & e' & e \end{Bmatrix} \begin{Bmatrix} a' & e & e' \\ d & d' & c \end{Bmatrix} \\ &\quad = \delta_{fg}(-1)^{\phi_2} \begin{Bmatrix} a' & a & c' \\ g & d' & c \end{Bmatrix} , \\ &\phi_2 = g + a' - b + c + d' + d \,.\end{aligned}$$

The W-coefficient form of these relations is obtained by deleting all phase factors and making the substitution (2.74), ignoring the phase factor. There are no phase factors in the corresponding W-coefficient relations.

2.10.7 Schwinger–Wu Generating Function and its Combinatorics

Triangles associated with the 9–j coefficient $\begin{Bmatrix} j_1 j_2 j_3 \\ j_4 j_5 j_6 \\ j_7 j_8 j_9 \end{Bmatrix}$:

$$(j_1 j_2 j_3)\,,\quad (j_4 j_5 j_6)\,,\quad (j_7 j_8 j_9)\,,\quad (j_1 j_4 j_7)\,,$$
$$(j_2 j_5 j_8)\,,\quad (j_3 j_6 j_9)\,.$$

Points in $\boldsymbol{R}^3$ associated with the triangles:

$$(j_1 j_2 j_3) \to (x_1, x_2, x_3),\ (j_4 j_5 j_6) \to (x_4, x_5, x_6)\,,$$
$$(j_7 j_8 j_9) \to (x_7, x_8, x_9),\ (j_1 j_4 j_7) \to (y_1, y_4, y_7)\,,$$
$$(j_2 j_5 j_8) \to (y_2, y_5, y_8),\ (j_3 j_6 j_9) \to (y_3, y_6, y_9)\,.$$

Cubic graph $\boldsymbol{C}_6$ in $\boldsymbol{R}^3$ associated with the points:

The points define the vertices of a cubic graph $\boldsymbol{C}_6$ on six points with lines joining each pair of points that share a common subscript, and the lines are labeled by the products $x_i y_i$, where i is the common subscript (Fig. 2.3).

Cubic graph $\boldsymbol{C}_6$ functions:

Interchange the symbols x and y in the coordinates of the vertices of the cubic graph $\boldsymbol{C}_6$, and define the following polynomials on the vertices and edges of the $\boldsymbol{C}_6$ with this modified labeling:

Vertex function: multiply together the coordinates of each pair of adjacent vertices, divide out the coordinates with a common subscript, and sum over all pairs of vertices to obtain

$$V_4 = y_1 y_2 x_6 x_9 + y_1 y_3 x_5 x_8 + y_2 y_3 x_4 x_7 + y_4 y_5 x_3 x_9 + y_4 y_6 x_2 x_8 + y_5 y_6 x_1 x_7 + y_7 y_8 x_3 x_6 + y_7 y_9 x_2 x_5 + y_8 y_9 x_1 x_4\,.$$

Edge function:

$$E_6 = \det \begin{bmatrix} x_1 y_1 & x_2 y_2 & x_3 y_3 \\ x_4 y_4 & x_5 y_5 & x_6 y_6 \\ x_7 y_7 & x_8 y_8 & x_9 y_9 \end{bmatrix}.$$

Generating function [2.4–6]:

$$(1 - V_4 + E_6)^{-2} = \sum_{\Delta} C(\Delta) Z^{\Delta}\,,$$

$$Z^{\Delta} = \prod_{\text{all vertices}} (z_a, z_b, z_c)^{(j_a j_b j_c)} \quad [\text{see } (2.82,)]\,,$$

$$\Delta = \begin{bmatrix} (j_1 j_2 j_3) \\ (j_4 j_5 j_6) \\ (j_7 j_8 j_9) \\ (j_1 j_4 j_7) \\ (j_2 j_5 j_8) \\ (j_3 j_6 j_9) \end{bmatrix},$$

$$C(\Delta) = \sum_k \sum_{\square} \sum_a (-1)^{k_{10}+k_{11}+k_{12}} (k+1) \times \binom{k}{k_1, \ldots, k_9, k_{10}, \ldots, k_{15}},$$

where summation $\sum_{\square}$ is over all 3×3 square arrays of nonnegative integers $k_j (j = 1, 2, \ldots, 9)$ with fixed row and column sums given by

$$\begin{array}{ccc|c} k_1 & k_2 & k_3 & k - t_1 \\ k_4 & k_5 & k_6 & k - t_2 \\ k_7 & k_8 & k_9 & k - t_3 \\ \hline k - t_4 & k - t_5 & k - t_6 & \end{array}$$

and for each such array the summation $\sum_a$ is over all nonnegative integers a such that the following quantities are nonnegative integers:

$$k_{10} = -a + k_1 - k + j_2 + j_3 + j_4 + j_7\,,$$
$$k_{11} = -a + k_6 - k + j_3 + j_4 + j_5 + j_9\,,$$
$$k_{12} = -a + k_8 - k + j_2 + j_5 + j_7 + j_9\,,$$
$$k_{13} = a + k_5 - k_1 - j_3 + j_6 - j_7 + j_8\,,$$
$$k_{14} = a + k_2 - k_6 + j_1 - j_4 + j_8 - j_9\,,$$
$$k_{15} = a\,.$$

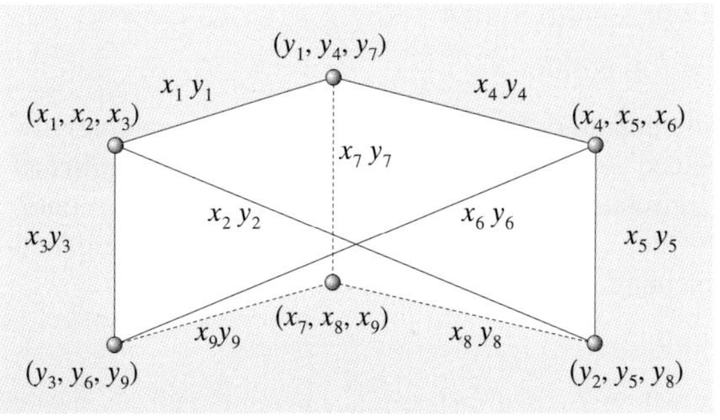

Fig. 2.3 Labeled cubic graph associated with the 9–j coefficient

Note that

$$\sum_{i=10}^{15} k_i = -2k + \sum_{i=1}^{9} j_i \,.$$

The t_i are the following triangle sums:

$$t_1 = j_1 + j_2 + j_3,\ t_2 = j_4 + j_5 + j_6\,,$$
$$t_3 = j_7 + j_8 + j_9\,,$$
$$t_4 = j_1 + j_4 + j_7,\ t_5 = j_2 + j_5 + j_8\,,$$
$$t_6 = j_3 + j_6 + j_9\,.$$

The $9-j$ coefficient is given by

$$\begin{Bmatrix} j_1\, j_2\, j_3 \\ j_4\, j_5\, j_6 \\ j_7\, j_8\, j_9 \end{Bmatrix} = \Delta(j_1 j_2 j_3)\Delta(j_4 j_5 j_6)\Delta(j_7 j_8 j_9) \times \Delta(j_1 j_4 j_7)\Delta(j_2 j_5 j_8)\Delta(j_3 j_6 j_9)C(\Delta)\,.$$

The coefficient $C(\Delta)$ is an integer associated with each cubic graph $\boldsymbol{C}_6$ that counts the number of occurrences of the monomial term Z^{Δ} in the expansion of $(1 - V_4 + E_6)^{-2}$.

2.11 Tensor Spherical Harmonics

Tensor spherical or tensor solid harmonics are special cases of the coupling of two irreducible tensor operators in the tensor product space given in Sect. 2.7.2. They are defined by

$$\mathcal{Y}^{(ls)\,jm} = \sum_{\nu} C^{l\ s\ j}_{m-\nu,\nu,m} \mathcal{Y}_{l,m-\nu} \otimes \xi_\nu$$

and belong to the tensor product space $\mathcal{H}_l \otimes \mathcal{H}'_s$, where the orthonormal bases of the spaces $\mathcal{H}_l$ and $\mathcal{H}'_s$ are:

$$\left\{\mathcal{Y}_{l\mu} : \mu = l, l-1, \ldots, -l\right\}\,,$$
$$\{\xi_\nu : \nu = s, s-1, \ldots, -s\}\,.$$

The orbital angular momentum $\boldsymbol{L}$ has the standard action on the solid harmonics, and a second set of kinematically independent angular momentum operators $\boldsymbol{S}$ has the standard action on the basis set of $\mathcal{H}'_s$.
The total angular momentum is:

$$\boldsymbol{J} = \boldsymbol{L} \otimes \boldsymbol{1}' + \boldsymbol{1} \otimes \boldsymbol{S}\,,$$

The set of vectors

$$\{\ \mathcal{Y}^{(ls)\,jm} : m = j, j-1, \ldots, -j;\ (lsj)$$
$$\text{obey the triangle conditions}\ \}$$

has the following following properties:
Orthogonality:

$$\left\langle \mathcal{Y}^{(l's)\,j'm'}, \mathcal{Y}^{(ls)\,jm}\right\rangle = \sum_{\nu\nu'} C^{l'\ s\ j'}_{m'-\nu',\nu',m'} C^{l\ s\ j}_{m-\nu,\nu,m} (\mathcal{Y}_{l',m'-\nu'}, \mathcal{Y}_{l,m-\nu}) \times (\xi_{\nu'}, \xi_\nu)' = \delta_{j'j}\delta_{l'l}\delta_{m'm}\,,$$

where $\langle\ ,\ \rangle$ denotes the inner product in the space $\mathcal{H}_l \otimes \mathcal{H}'_s$, $(\ ,\ )$ the inner product in $\mathcal{H}_l$, and $(\ ,\ )'$ the inner product in $\mathcal{H}'_s$.
Operator actions:

$$\boldsymbol{J}^2 \mathcal{Y}^{(ls)\,jm} = j(j+1)\mathcal{Y}^{(ls)\,jm}\,,$$
$$J_3 \mathcal{Y}^{(ls)\,jm} = m\mathcal{Y}^{(ls)\,jm}\,,$$
$$(\boldsymbol{L}^2 \otimes \boldsymbol{1}')\mathcal{Y}^{(ls)\,jm} = l(l+1)\mathcal{Y}^{(ls)\,jm}\,,$$
$$(\boldsymbol{1} \otimes \boldsymbol{S}^2)\mathcal{Y}^{(ls)\,jm} = s(s+1)\mathcal{Y}^{(ls)\,jm}\,,$$
$$\boldsymbol{J}^2 = \boldsymbol{L}^2 \otimes \boldsymbol{1}' + \boldsymbol{1} \otimes \boldsymbol{S}^2 + 2\sum_i L_i \otimes S_i\,,$$

$$J_\pm \mathcal{Y}^{(ls)\,jm} = [(j \mp m)(j \pm m + 1)]^{\frac{1}{2}} \mathcal{Y}^{(ls)\,j,m\pm 1}\,.$$

Transformation property under unitary rotations:

$$\exp(-i\psi\hat{\boldsymbol{n}} \cdot \boldsymbol{J})\mathcal{Y}^{(ls)\,jm} = \sum_{m'} D^j_{m'm}(\psi, \hat{\boldsymbol{n}})\mathcal{Y}^{(ls)\,jm'}\,.$$

Special realization:
The eigenvectors ξ_ν are often replaced by column matrices:

$$\xi_\nu = \mathrm{col}(0 \cdots 010 \cdots 0)\,,$$
$$1 \text{ in position } s-\nu+1, \quad \nu = s, s-1, \ldots, -s\,.$$

The operators $\boldsymbol{S} = (S_1, S_2, S_3)$ are correspondingly replaced by their standard $(2s+1) \times (2s+1)$ matrix representations $S_i^{(s)}$. The tensor product of operators becomes a $(2s+1) \times (2s+1)$ matrix containing both operators and numerical matrix elements, e.g.,

$$J_i = L_i I_{2s+1} + S_i^{(s)}\,,$$

in which L_i is a differential operator multiplying the unit matrix, that is, L_i is repeated $2s+1$ times along the diagonal.

2.11.1 Spinor Spherical Harmonics as Matrix Functions

Choose $\xi_{+1/2} = \mathrm{col}(1, 0)$, $\xi_{-1/2} = \mathrm{col}(0, 1)$, and $\boldsymbol{S} = \boldsymbol{\sigma}/2$. The spinor spherical harmonics or Pauli central field spinors are the following, where $j \in \{1/2, 3/2, \dots\}$:

$$\mathcal{Y}^{\left(j-\frac{1}{2},\frac{1}{2}\right)jm} = \begin{pmatrix} \sqrt{\frac{j+m}{2j}}\;\mathcal{Y}_{j-\frac{1}{2},m-\frac{1}{2}} \\ \sqrt{\frac{j-m}{2j}}\;\mathcal{Y}_{j-\frac{1}{2},m+\frac{1}{2}} \end{pmatrix},$$

$$\mathcal{Y}^{\left(j+\frac{1}{2},\frac{1}{2}\right)jm} = \begin{pmatrix} -\sqrt{\frac{j-m+1}{2j+2}}\;\mathcal{Y}_{j+\frac{1}{2},m-\frac{1}{2}} \\ \sqrt{\frac{j+m+1}{2j+2}}\;\mathcal{Y}_{j+\frac{1}{2},m+\frac{1}{2}} \end{pmatrix}.$$

2.11.2 Vector Spherical Harmonics as Matrix Functions

Choose $\xi_{+1} = \mathrm{col}(1, 0, 0)$, $\xi_0 = \mathrm{col}(0, 1, 0)$, $\xi_{-1} = \mathrm{col}(0, 0, 1)$, and $\boldsymbol{S}$ the 3×3 angular momentum matrices given by

$$S_+ = \begin{pmatrix} 0 & \sqrt{2} & 0 \\ 0 & 0 & \sqrt{2} \\ 0 & 0 & 0 \end{pmatrix}, \quad S_- = \begin{pmatrix} 0 & 0 & 0 \\ \sqrt{2} & 0 & 0 \\ 0 & \sqrt{2} & 0 \end{pmatrix},$$

$$S_3 = \begin{pmatrix} 1 & 0 & 0 \\ 0 & 0 & 0 \\ 0 & 0 & -1 \end{pmatrix}.$$

The vector spherical harmonics are the following, where $j \in \{0, 1, 2, \dots\}$:

$$\mathcal{Y}^{(j-1,1)jm} = \begin{pmatrix} \sqrt{\frac{(j+m-1)(j+m)}{2j(2j-1)}}\;\mathcal{Y}_{j-1,m-1} \\ \sqrt{\frac{(j-m)(j+m)}{j(2j-1)}}\;\mathcal{Y}_{j-1,m} \\ \sqrt{\frac{(j-m-1)(j-m)}{2j(2j-1)}}\;\mathcal{Y}_{j-1,m+1} \end{pmatrix},$$

$$\mathcal{Y}^{(j1)jm} = \begin{pmatrix} -\sqrt{\frac{(j+m)(j-m+1)}{2j(j+1)}}\;\mathcal{Y}_{j,m-1} \\ \frac{m}{\sqrt{j(j+1)}}\;\mathcal{Y}_{j,m} \\ \sqrt{\frac{(j-m)(j+m+1)}{2j(j+1)}}\;\mathcal{Y}_{j,m+1} \end{pmatrix},$$

$$\mathcal{Y}^{(j+1,1)jm} = \begin{pmatrix} \sqrt{\frac{(j-m+1)(j-m+2)}{2(j+1)(2j+3)}}\;\mathcal{Y}_{j+1,m-1} \\ -\sqrt{\frac{(j-m+1)(j+m+1)}{(j+1)(2j+3)}}\;\mathcal{Y}_{j+1,m} \\ \sqrt{\frac{(j+m+2)(j+m+1)}{2(j+1)(2j+3)}}\;\mathcal{Y}_{j+1,m+1} \end{pmatrix}.$$

Eigenvalue properties:

$$\begin{aligned} \nabla^2 \mathcal{Y}^{(l1)jm} &= 0\,, \\ \boldsymbol{J}^2 \mathcal{Y}^{(l1)jm} &= j(j+1)\mathcal{Y}^{(l1)jm}\,, \\ J_3 \mathcal{Y}^{(l1)jm} &= m\mathcal{Y}^{(l1)jm}\,, \\ \boldsymbol{L}^2 \mathcal{Y}^{(l1)jm} &= l(l+1)\mathcal{Y}^{(l1)jm}\,, \\ \boldsymbol{S}^2 \mathcal{Y}^{(l1)jm} &= 2\mathcal{Y}^{(l1)jm}\,. \end{aligned}$$

2.11.3 Vector Solid Harmonics as Vector Functions

Vector spherical and solid harmonics can also be defined and their properties presented in terms of the ordinary solid harmonics, using the vectors $\boldsymbol{x}$, ∇, and $\boldsymbol{L}$, and the operations of divergence and curl:

Defining equations:

$$\begin{aligned} \mathcal{Y}^{(l+1,1)lm} &= -[(l+1)(2l+1)]^{-\frac{1}{2}}[(l+1)\boldsymbol{x} \\ &\quad + \mathrm{i}\boldsymbol{x}\times\boldsymbol{L}]\mathcal{Y}_{lm}\,, \\ \mathcal{Y}^{(l1)lm} &= [l(l+1)]^{-1/2}\boldsymbol{L}\mathcal{Y}_{lm}\,, \\ r^2\mathcal{Y}^{(l-1,1)lm} &= -[l(2l+1)]^{-\frac{1}{2}}\times(-l\boldsymbol{x}+\mathrm{i}\boldsymbol{x} \\ &\quad \times\boldsymbol{L})\mathcal{Y}_{lm}\,. \end{aligned}$$

Eigenvalue properties:

$$\begin{aligned} \boldsymbol{J}^2 \mathcal{Y}^{(l1)jm} &= j(j+1)\mathcal{Y}^{(l1)jm}\,, \\ \boldsymbol{L}^2 \mathcal{Y}^{(l1)jm} &= l(l+1)\mathcal{Y}^{(l1)jm}\,, \\ \boldsymbol{S}^2 \mathcal{Y}^{(l1)jm} &= 2\mathcal{Y}^{(l1)jm}\,, \\ J_3 \mathcal{Y}^{(l1)jm} &= m\mathcal{Y}^{(l1)jm}\,, \\ \nabla^2 \mathcal{Y}^{(l1)jm} &= \boldsymbol{0}\,, \\ 2\mathrm{i}\boldsymbol{L}\times\mathcal{Y}^{(l1)jm} &= [j(j+1)-l(l+1)-2]\mathcal{Y}^{(l1)jm}\,. \end{aligned}$$

Orthogonality:

$$\int \mathrm{d}S_{\hat{x}}\,\mathcal{Y}^{(l'1)j'm'*}(\boldsymbol{x})\cdot\mathcal{Y}^{(l1)jm}(\boldsymbol{x}) = \delta_{l'l}\delta_{j'j}\delta_{m'm}r^{l'+l}\,,$$

where the integration is over the unit sphere in $\boldsymbol{R}^3$.

Complex conjugation:

$$\mathcal{Y}^{(l1)jm*} = (-1)^{l+1-j}(-1)^m\mathcal{Y}^{(l1)j,-m}\,.$$

Part A | 2.11

Vector and gradient formulas:

$$x\mathcal{Y}_{lm} = -\left(\frac{l+1}{2l+1}\right)^{\frac{1}{2}} \boldsymbol{\mathcal{Y}}^{(l+1,1)lm} + \left(\frac{l}{2l+1}\right)^{\frac{1}{2}} r^2 \boldsymbol{\mathcal{Y}}^{(l-1,1)lm} ,$$

$$[(l+1)\nabla + \mathrm{i}\nabla\times\boldsymbol{L}](F\mathcal{Y}_{lm}) = -[(l+1)(2l+1)]^{\frac{1}{2}} \left(\frac{1}{r}\frac{\mathrm{d}F}{\mathrm{d}r}\right) \boldsymbol{\mathcal{Y}}^{(l+1,1)lm} ,$$

$$[-l\nabla + \mathrm{i}\nabla\times\boldsymbol{L}](F\mathcal{Y}_{lm}) = -[l(2l+1)]^{\frac{1}{2}} \left[r\frac{\mathrm{d}F}{\mathrm{d}r} + (2l+1)F\right] \boldsymbol{\mathcal{Y}}^{(l-1,1)lm} ,$$

$$\nabla(F\mathcal{Y}_{lm}) = -\left(\frac{l+1}{2l+1}\right)^{\frac{1}{2}} \left(\frac{1}{r}\frac{\mathrm{d}F}{\mathrm{d}r}\right) \boldsymbol{\mathcal{Y}}^{(l+1,1)lm} + \left(\frac{l}{2l+1}\right)^{\frac{1}{2}} \left[r\frac{\mathrm{d}F}{\mathrm{d}r} + (2l+1)F\right] \boldsymbol{\mathcal{Y}}^{(l-1,1)lm} ,$$

$$\mathrm{i}\nabla\times\boldsymbol{L}(F\mathcal{Y}_{lm}) = -l\left(\frac{l+1}{2l+1}\right)^{\frac{1}{2}} \left(\frac{1}{r}\frac{\mathrm{d}F}{\mathrm{d}r}\right) \boldsymbol{\mathcal{Y}}^{(l+1,1)lm} - (l+1)\left(\frac{l}{2l+1}\right)^{\frac{1}{2}} \times \left[r\frac{\mathrm{d}F}{\mathrm{d}r} + (2l+1)F\right] \boldsymbol{\mathcal{Y}}^{(l-1,1)lm} .$$

Curl equations:

$$\mathrm{i}\nabla\times(F\boldsymbol{\mathcal{Y}}^{(l+1,1)lm}) = -\left(\frac{l}{2l+1}\right)^{\frac{1}{2}} \times \left[r\frac{\mathrm{d}F}{\mathrm{d}r} + (2l+3)F\right] \boldsymbol{\mathcal{Y}}^{(l1)lm} ,$$

$$\mathrm{i}\nabla\times(F\boldsymbol{\mathcal{Y}}^{(l1)lm}) = -\left(\frac{l}{2l+1}\right)^{\frac{1}{2}} \times \left(\frac{1}{r}\frac{\mathrm{d}F}{\mathrm{d}r}\right) \boldsymbol{\mathcal{Y}}^{(l+1,1)lm} - \left(\frac{l+1}{2l+1}\right)^{\frac{1}{2}} \times \left[r\frac{\mathrm{d}F}{\mathrm{d}r} + (2l+1)F\right] \boldsymbol{\mathcal{Y}}^{(l-1,1)lm} ,$$

$$\mathrm{i}\nabla\times(F\boldsymbol{\mathcal{Y}}^{(l-1,1)lm}) = -\left(\frac{l+1}{2l+1}\right)^{\frac{1}{2}} \left(\frac{1}{r}\frac{\mathrm{d}F}{\mathrm{d}r}\right) \boldsymbol{\mathcal{Y}}^{(l1)lm} .$$

Divergence equations:

$$\nabla\cdot(F\boldsymbol{\mathcal{Y}}^{(l+1,1)lm}) = -\left(\frac{l+1}{2l+1}\right)^{\frac{1}{2}} \left[r\frac{\mathrm{d}F}{\mathrm{d}r} + (2l+3)F\right] \mathcal{Y}_{lm} ,$$

$$\nabla\cdot(F\boldsymbol{\mathcal{Y}}^{(l1)lm}) = 0 ,$$

$$\nabla\cdot(F\boldsymbol{\mathcal{Y}}^{(l-1,1)lm}) = \left(\frac{l}{2l+1}\right)^{\frac{1}{2}} \left(\frac{1}{r}\frac{\mathrm{d}F}{\mathrm{d}r}\right) \mathcal{Y}_{lm} .$$

Parity property:

$$\boldsymbol{\mathcal{Y}}^{(l+\delta,1)lm}(-\boldsymbol{x}) = (-1)^{l+\delta} \boldsymbol{\mathcal{Y}}^{(l+\delta,1)lm}(\boldsymbol{x}) .$$

Scalar product:

$$\boldsymbol{\mathcal{Y}}^{(l'1)j'm'} \cdot \boldsymbol{\mathcal{Y}}^{(l1)jm} = \sum_{l''} r^{l+l'-l''} \left(\frac{(2j+1)(2j'+1)(2l+1)(2l'+1)}{4\pi(2l''+1)}\right)^{\frac{1}{2}} \times (-1)^{l+j'+l''} C^{ll'l''}_{000} C^{jj'l''}_{m,m',m+m'} \times \begin{Bmatrix} l' & j' & 1 \\ j & l & l'' \end{Bmatrix} \mathcal{Y}_{l'',m+m'} .$$

Cross product:

$$\boldsymbol{\mathcal{Y}}^{(l'1)j'm'} \times \boldsymbol{\mathcal{Y}}^{(l1)jm} = (-\mathrm{i}\sqrt{2}) \sum_{l''j''} r^{l+l'-l''} \times \left(\frac{(2j+1)(2j'+1)(3)(2l+1)(2l'+1)}{4\pi}\right)^{\frac{1}{2}} \times C^{ll'l''}_{000} C^{jj'j''}_{m,m',m+m'} \begin{Bmatrix} l & 1 & j \\ l' & 1 & j' \\ l'' & 1 & j'' \end{Bmatrix} \boldsymbol{\mathcal{Y}}^{(l''1)j'',m+m'} .$$

Conversion to spherical harmonic form:

$$\boldsymbol{\mathcal{Y}}^{(l+\delta,1)lm}(\boldsymbol{x}) = r^{l+\delta} \boldsymbol{\mathcal{Y}}^{(l+\delta,1)lm}(\hat{\boldsymbol{x}}) ,$$

with appropriate modification of F to account for the factor $r^{l+\delta}$.

2.12 Coupling and Recoupling Theory and 3n–j Coefficients

2.12.1 Composite Angular Momentum Systems

An "elementary" angular momentum system is one whose state space can be written as a direct sum of vector spaces $\mathcal{H}_j$ with orthonormal basis

$$\{|\,jm\rangle | m = j, j-1, \ldots, -j\}$$

on which the angular momentum $\boldsymbol{J}$ has the standard action, and which under unitary transformation by

$\exp(-\mathrm{i}\psi\hat{\boldsymbol{n}}\cdot\boldsymbol{J})$ undergoes the standard unitary transformation. A composite angular momentum system is one whose state space is a direct sum of the tensor product spaces $\mathcal{H}_{j_1 j_2\cdots j_n}$ of dimension $\prod_{\alpha=1}^{n}(2j_\alpha+1)$ with orthonormal basis in the tensor product space of the elementary systems given by

$$|\,j_1m_1\rangle\otimes|\,j_2m_2\rangle\otimes\cdots\otimes|\,j_nm_n\rangle\,, \tag{2.98}$$

each $m_\alpha = j_\alpha, j_\alpha-1,\ldots,-j_\alpha$.

The following properties then hold for the composite system:

Independent rotations of the elementary parts:

$$\begin{aligned}&\left\{\exp\left[-\mathrm{i}\psi_1\hat{\boldsymbol{n}}_1\cdot\boldsymbol{J}(1)\right]\otimes\cdots\otimes\exp\left[-\mathrm{i}\psi_n\hat{\boldsymbol{n}}_n\cdot\boldsymbol{J}(n)\right]\right\}\\&\quad\times|\,j_1m_1\rangle\otimes\cdots\otimes j_nm_n\rangle\\&\quad=\exp\left[-\mathrm{i}\psi_1\hat{\boldsymbol{n}}_1\cdot\boldsymbol{J}(1)\right]|\,j_1m_1\rangle\otimes\cdots\\&\qquad\otimes\exp\left[-\mathrm{i}\psi_n\hat{\boldsymbol{n}}_n\cdot\boldsymbol{J}(n)\right]|\,j_nm_n\rangle\\&\quad=\sum_{m_1'\cdots m_n'}\left[D^{j_1}(U_1)\times\cdots\times D^{j_n}(U_n)\right]_{m_1'\cdots m_n';m_1\cdots m_n}\\&\qquad\times\left|j_1m_1'\right\rangle\otimes\cdots\otimes\left|j_nm_n'\right\rangle,\end{aligned} \tag{2.99}$$

$$\begin{aligned}&\left[D^{j_1}(U_1)\times\cdots\times D^{j_n}(U_n)\right]_{m_1'\cdots m_n';m_1\cdots m_n}\\&\quad=D^{j_1}_{m_1'm_1}(U_1)\cdots D^{j_n}_{m_n'm_n}(U_n)\,,\\&U_\alpha=U(\psi_\alpha,\hat{\boldsymbol{n}}_\alpha)\in SU(2)\,,\quad\alpha=1,2,\ldots,n\,.\end{aligned}$$

Multiple Kronecker (direct) product group $SU(2)\times\cdots\times SU(2)$:

Group elements:

$$(U_1,\ldots,U_n)\,,\quad\text{each } U_\alpha\in SU(2)\,.$$

Multiplication rule:

$$(U_1',\ldots,U_n')(U_1,\ldots,U_n)=\left(U_1'U_1,\ldots,U_n'U_n\right).$$

Irreducible representations:

$$D^{j_1}(U_1)\times\cdots\times D^{j_n}(U_n)\,. \tag{2.100}$$

Rotation of the composite system as a unit:

Common rotation:

$$U_1=U_2=\cdots=U_n=U\in SU(2)\,.$$

Diagonal subgroup $SU(2)\subset SU(2)\times\cdots\times SU(2)$:

$$(U,U,\ldots,U),\qquad\text{each } U\in SU(2)\,.$$

Reducible representation of $SU(2)$:

$$D^{j_1}(U)\times\cdots\times D^{j_n}(U)\,. \tag{2.101}$$

Total angular momentum of the composite system:

$$\boldsymbol{J}=\boldsymbol{J}(1)+\boldsymbol{J}(2)+\cdots+\boldsymbol{J}(n)\,,$$

in which the k-th term in the sum is to be interpreted as the tensor product operator:

$$\boldsymbol{I}_1\otimes\cdots\otimes\boldsymbol{J}(k)\otimes\cdots\otimes\boldsymbol{I}_n,\ \boldsymbol{I}_\alpha=\text{ unit operator in }\mathcal{H}_{j_\alpha}\,.$$

The basic problem for composite systems:

The basic problem is to reduce the n-fold direct product representation (2.101) of $SU(2)$ into a direct sum of irreducible representations, or equivalently, to find all subspaces $\mathcal{H}_j\subset\mathcal{H}_{j_1j_2\cdots j_n}$, $j\in\{0,1/2,1,\ldots\}$, with orthonormal bases sets $\{|\,jm\rangle|m=j,j-1,\ldots,-j\}$ on which the total angular momentum $\boldsymbol{J}$ has the standard action.

Form of the solution:

$$\begin{aligned}|\,(j_1j_2\cdots j_n)(k)\,jm\rangle&=\sum_{\substack{\text{all } m_\alpha\\ \sum m_\alpha=m}}C^{j_1j_2\cdots j_nj}_{m_1m_2\cdots m_nm}(k)\\&\quad\times|\,j_1m_1\rangle\otimes|\,j_2m_2\rangle\otimes\cdots\otimes|\,j_nm_n\rangle\,,\end{aligned} \tag{2.102}$$

$m=j,j-1,\ldots,-j$; index set (k) unspecified.

Diagonal operators:

$$\begin{aligned}&\boldsymbol{J}^2(\alpha)=J_1^2(\alpha)+J_2^2(\alpha)+J_3^2(\alpha),\\&\boldsymbol{J}^2(\alpha)\,|\,(j_1j_2\cdots j_n)(k)\,jm\rangle\\&\qquad=j_\alpha(j_\alpha+1)\,|\,(j_1j_2\cdots j_n)(k)\,jm\rangle\,,\\&\qquad\alpha=1,2,\ldots,n\,.\end{aligned} \tag{2.103}$$

Total angular momentum properties imposed:

$$\begin{aligned}&\boldsymbol{J}^2\,|\,(j_1j_2\cdots j_n)(k)\,jm\rangle\\&\quad=j(j+1)\,|\,(j_1j_2\cdots j_n)(k)\,jm\rangle\,,\\&J_3\,|\,(j_1j_2\cdots j_n)(k)\,jm\rangle\\&\quad=m\,|\,(j_1j_2\cdots j_n)(k)\,jm\rangle\,,\\&J_\pm\,|\,(j_1j_2\cdots j_n)(k)\,jm\rangle\\&\quad=[(j\mp m)(j\pm m+1)]^{\frac{1}{2}}\\&\qquad\times|\,(j_1j_2\cdots j_n)(k)\,jm\pm1\rangle\,.\end{aligned} \tag{2.104}$$

Properties of the index set (k):

Reduction of Kronecker product (2.101):

$$\begin{aligned}&D^{j_1}\times D^{j_2}\times\cdots\times D^{j_n}=\sum_j\oplus n_jD^j\,,\\&\prod_\alpha(2j_\alpha+1)=\sum_j n_j(2j+1)\,.\end{aligned} \tag{2.105}$$

For fixed $j_1, j_2, \dots, j_n$, and j, the index set (k) must enumerate exactly n_j perpendicular spaces $\mathcal{H}_j$.
Incompleteness of set of operators:

There are $2n$ commuting Hermitian operators diagonal on the basis (2.98):

$$\left\{ \boldsymbol{J}^2(\alpha), J_3(\alpha) \;\middle|\; \alpha = 1, 2, \dots, n \right\} . \tag{2.106a}$$

There are $n+2$ commuting Hermitian operators diagonal on the basis (2.102):

$$\left\{ \boldsymbol{J}^2, J_3; \boldsymbol{J}^2(\alpha) \;\middle|\; \alpha = 1, 2, \dots, n \right\} . \tag{2.106b}$$

There are $n-2$ additional commuting Hermitian operators, or other rules, required to complete set (2.106b) and determine the indexing set (k).
Basic content of coupling and recoupling theory:

Coupling theory is the study of completing the operator set (2.106b), or the specification of other rules, that uniquely determine the irreducible representation spaces $\mathcal{H}_j$ occurring in the Kronecker product reduction (2.105). Recouping theory is the study of the inter-relations between different methods of effecting this reduction; it is a study of relations between the different ways of spanning the multiplicity space

$$\mathcal{H}_j \oplus \mathcal{H}_j \oplus \dots \oplus \mathcal{H}_j \ (n_j \text{ terms}) .$$

2.12.2 Binary Coupling Theory: Combinatorics

Binary coupling of angular momenta refers to the selecting any pair of angular momentum operators from the set of individual system angular momenta

$$\{\boldsymbol{J}(1), \boldsymbol{J}(2), \dots, \boldsymbol{J}(n)\} ,$$

and carrying out the "addition of angular momenta" for that pair by coupling the corresponding states in the tensor product space by the standard use of $SU(2)$ WCG-coefficients; this is followed by addition of a new pair, which may be a pair distinct from the first pair, or the addition of one new angular momentum to the sum of the first pair, etc. If the order $1, 2, \dots, n$ of the angular momenta is kept fixed in

$$\boldsymbol{J}_1 + \boldsymbol{J}_2 + \dots + \boldsymbol{J}_n , \tag{2.107}$$

one is led to the problem of parentheses. (To avoid misleading parentheses, the notation $\boldsymbol{J}_\alpha = \boldsymbol{J}(\alpha)$ is used in this section.) This is the problem of introducing pairs of parentheses into expression (2.107) that specify the coupling procedure that is to be implemented. The procedure is clear from the following cases for $n = 2, 3$, and 4:

$$\begin{aligned}
n = 2: \ & \boldsymbol{J}_1 + \boldsymbol{J}_2 ; \\
n = 3: \ & (\boldsymbol{J}_1 + \boldsymbol{J}_2) + \boldsymbol{J}_3 , \\
& \boldsymbol{J}_1 + (\boldsymbol{J}_2 + \boldsymbol{J}_3) ; \\
n = 4: \ & (\boldsymbol{J}_1 + \boldsymbol{J}_2) + (\boldsymbol{J}_3 + \boldsymbol{J}_4) , \\
& [(\boldsymbol{J}_1 + \boldsymbol{J}_2) + \boldsymbol{J}_3] + \boldsymbol{J}_4 , \\
& [\boldsymbol{J}_1 + (\boldsymbol{J}_2 + \boldsymbol{J}_3)] + \boldsymbol{J}_4 , \\
& \boldsymbol{J}_1 + [(\boldsymbol{J}_2 + \boldsymbol{J}_3) + \boldsymbol{J}_4] , \\
& \boldsymbol{J}_1 + [\boldsymbol{J}_2 + (\boldsymbol{J}_3 + \boldsymbol{J}_4)] .
\end{aligned}$$

It is customary to use the ordered sequence

$$j_1 j_2 \cdots j_n \tag{2.108}$$

of angular momentum quantum numbers in place of the angular momentum operators in (2.107). Thus, the five placement of parentheses for $n = 4$ becomes:

$$\begin{aligned}
& (j_1 j_2)(j_3 j_4) , \ [(j_1 j_2) j_3] j_4 , \ [j_1 (j_2 j_3)] j_4 , \\
& j_1 [(j_2 j_3) j_4] , \ j_1 [j_2 (j_3 j_4)] .
\end{aligned}$$

(It is also customary to omit the last parentheses pair, which encloses the whole sequence.) A sequence (2.108) into which pairwise insertions of parentheses has been completed is called a binary bracketing of the sequence, and denoted by $(j_1 j_2 \cdots j_n)^B$. This symbol may also be called a coupling symbol. The total number of coupling symbols, that is, the total number of elements a_n in the set

$$\{(j_1 j_2 \cdots j_n)^B | B \text{ is a binary bracketing}\}$$

is given by the Catalan numbers:

$$a_n = \frac{1}{n} \binom{2n-2}{n-1} , \quad n = 2, 3, \dots .$$

Effect of permuting the angular momenta:

Since the position of an individual vector space in the tensor product $\mathcal{H}_{j_1} \otimes \dots \otimes \mathcal{H}_{j_n}$ is kept fixed, the meaning of a permutation of the j_α in the sequence (2.108) corresponding to a given binary bracketing is to permute the positions of the terms in the summation for the total angular momentum, e.g., $(j_1 j_2) j_3 \to (j_3 j_1) j_2$ corresponds to

$$\begin{aligned}
& (\boldsymbol{J}_1 \otimes \boldsymbol{I}_2 \otimes \boldsymbol{I}_3 + \boldsymbol{I}_1 \otimes \boldsymbol{J}_2 \otimes \boldsymbol{I}_3) + \boldsymbol{I}_1 \otimes \boldsymbol{I}_2 \otimes \boldsymbol{J}_3 \\
& = (\boldsymbol{I}_1 \otimes \boldsymbol{I}_2 \otimes \boldsymbol{J}_3 + \boldsymbol{J}_1 \otimes \boldsymbol{I}_2 \otimes \boldsymbol{I}_3) + \boldsymbol{I}_1 \otimes \boldsymbol{J}_2 \otimes \boldsymbol{I}_3 .
\end{aligned}$$

Total number of binary bracketing schemes including permutations:

The number of symbols in the set

$$\left\{ (j_{\alpha_1} j_{\alpha_2} \cdots j_{\alpha_n})^B \;\middle|\; \begin{array}{l} B \text{ is a binary bracketing and} \\ \alpha_1\alpha_2\cdots\alpha_n \text{ is a permutation} \\ \text{of } 1, 2, \ldots, n \end{array} \right\} \tag{2.109}$$

is $c_n = n!a_n = (n)_{n-1} = n(n+1)\cdots(2n-2)$.

Caution: One should not assign numbers to the symbols j_α, since these symbols serve as noncommuting, nonassociative distinct objects in a counting process.

Binary subproducts:

A binary subproduct in the coupling symbol $(j_{\alpha_1} j_{\alpha_2} \cdots j_{\alpha_n})^B$ is the subset of symbols between a given parentheses pair, say, $\{xy\}$. The symbols x and y may themselves contain binary subproducts. Commutation of a binary subproduct is the operation $\{xy\} \to \{yx\}$. For example, the coupling symbol $\{[(j_1 j_2) j_3] j_4\}$ contains three binary subproducts, $\{xy\}$, $[xy]$, (xy).

Equivalence relation:

Two coupling symbols are defined to be equivalent

$$(j_{\alpha_1} j_{\alpha_2} \cdots j_{\alpha_n})^B \sim (j_{\alpha'_1} j_{\alpha'_2} \cdots j_{\alpha'_n})^B$$

if one can be obtained from the other by commutation of the symbols in the binary subproducts. Such commutations change the overall phase of the state vector (2.102) corresponding to a particular coupling symbol, and such states are counted as being the same (equivalent).

Number of inequivalent coupling schemes:

The equivalence relation under commutation of binary subproducts partitions the set (2.109) into equivalence classes, each containing 2^{n-1} elements. There are $d_n = c_n/2^{n-1} = (2n-3)!!$ inequivalent coupling schemes in binary coupling theory. Thus, for $n = 4$, there are $5!! = 5\times3\times1 = 15$ inequivalent binary coupling schemes.

Type of a coupling symbol:

The type of the coupling symbol $(j_{\alpha_1} j_{\alpha_2} \cdots j_{\alpha_n})^B$ is defined to be the symbol obtained by setting all the j_α equal to a common symbol, say, x. Thus, the type of the coupling symbol $\{[(j_1 j_2) j_3] j_4\}$ is $\{[(x^2)x]x\}$.

The Wedderburn–Etherington number b_n gives the number of coupling symbols of distinct types, counting two symbols as equivalent if they are related by commutation of binary subproducts. A closed form of these numbers is not known, although generating functions exist. The first few numbers are:

n	1	2	3	4	5	6	7	8	9	10
b_n	1	1	1	2	3	6	11	23	46	98

There are 15 nontrivial coupling schemes for 4 angular momenta, and they are classified into 2 types, allowing commutation of binary subproducts:

Type $[(x^2)x]x$

$[(j_1 j_2)j_3]j_4$, $[(j_2 j_3)j_1]j_4$, $[(j_3 j_1)j_2]j_4$

$[(j_1 j_2)j_4]j_3$, $[(j_2 j_4)j_1]j_3$, $[(j_4 j_1)j_2]j_3$

$[(j_1 j_3)j_4]j_2$, $[(j_3 j_4)j_1]j_2$, $[(j_4 j_1)j_3]j_2$

$[(j_2 j_3)j_4]j_1$, $[(j_3 j_4)j_2]j_1$, $[(j_4 j_2)j_3]j_1$

Type $(x^2)(x^2)$

$(j_1 j_2)(j_3 j_4)$, $(j_1 j_3)(j_2 j_4)$, $(j_2 j_3)(j_1 j_4)$

2.12.3 Implementation of Binary Couplings

Each binary coupling scheme specifies uniquely a set of intermediate angular momentum operators. For example, the intermediate angular momenta associated with the coupling symbol $[(j_1 j_2) j_3] j_4$ are

$$\boldsymbol{J}(1) + \boldsymbol{J}(2) = \boldsymbol{J}(12)\,, \quad \boldsymbol{J}(12) + \boldsymbol{J}(3) = \boldsymbol{J}(123)\,,$$
$$\boldsymbol{J}(123) + \boldsymbol{J}(4) = \boldsymbol{J}\,,$$

where $\boldsymbol{J}$ is the total angular momentum. Each coupling symbol $(j_{\alpha_1} j_{\alpha_2} \cdots j_{\alpha_n})^B$, defines exactly $n-2$ intermediate angular momentum operators $\boldsymbol{K}(\lambda)$, $\lambda = 1, 2, \ldots, n-2$. The squares of these operators completes the set of operators (2.106b) for each coupling symbol; that is, the states vectors satisfying (2.103–2.104) and the following equations are unique, up to an overall choice of phase factor:

$$\begin{aligned} &\boldsymbol{K}^2(\lambda) \mid (j_{\alpha_1} j_{\alpha_2} \cdots j_{\alpha_n})^B (k_1 k_2 \cdots k_{n-2}) jm\rangle \\ &= k_\lambda(k_\lambda+1)|(j_{\alpha_1} j_{\alpha_2} \cdots j_{\alpha_n})^B (k_1 k_2 \cdots k_{n-2}) jm\rangle\,, \\ &\quad \lambda = 1, 2, \ldots, n-2, \quad n > 2\,. \end{aligned} \tag{2.110}$$

The intermediate angular momentum operators $\boldsymbol{K}^2(\lambda)$ depend, of course, on the choice of binary couplings implicit in the symbol $(j_{\alpha_1} j_{\alpha_2} \cdots j_{\alpha_n})^B$. The vectors have the following properties:

Orthonormal basis of $\mathcal{H}_{j_1}(1) \otimes \cdots \otimes \mathcal{H}_{j_n}(n)$:

$$\langle (j_\alpha)^B (k') jm | (j_\alpha)^B (k) jm\rangle = \prod_\lambda \delta_{k'_\lambda k_\lambda}\,,$$

$$\begin{aligned} (j_\alpha) &= (j_{\alpha_1}, j_{\alpha_2}, \ldots, j_{\alpha_n})\,, \\ (k) &= (k_1, k_2, \ldots, k_{n-2})\,, \\ (k') &= (k'_1, k'_2, \ldots, k'_{n-2})\,. \end{aligned}$$

The range of each k_λ is uniquely determined by the Clebsch–Gordan series and the binary couplings in the

coupling symbol. Together these ranges enumerate exactly the multiplicity n_j of $\mathcal{H}_j$ occurring in the reduction of the multiple Kronecker product.
Uniqueness of state vectors:

$$\begin{aligned}&|\,(j_{\alpha_1} j_{\alpha_2} \cdots j_{\alpha_n})^B (k_1 k_2 \cdots k_{n-2})\, jm\rangle \\ &= \sum_{\sum m_\alpha = m} C \left[\begin{pmatrix} j_{\alpha_1} \cdots j_{\alpha_n} \\ m_{\alpha_1} \cdots m_{\alpha_n} \end{pmatrix}^B \begin{matrix} j \\ m \end{matrix} \right] (k_1, \ldots, k_{n-2}) \\ &\times |\, j_1 m_1\rangle \otimes \cdots \otimes |\, j_n m_n\rangle \,.\end{aligned}$$

In the C-coefficient, the $\begin{pmatrix} j_\alpha \\ m_\alpha \end{pmatrix}$ are paired in the binary bracketing. Each such C-coefficient is a summation over a unique product of $n-1$ $SU(2)$ WCG-coefficients.
Equivalent basis vectors:

$$\begin{aligned}&|\,(j_{\alpha_1} j_{\alpha_2} \cdots j_{\alpha_n})^B (k_1 k_2 \cdots k_{n-2})\, jm\rangle \\ &= \pm\,|\,(j_{\alpha'_1} j_{\alpha'_2} \cdots j_{\alpha'_n})^B (k_1 k_2 \cdots k_{n-2})\, jm\rangle\,,\end{aligned}$$

if and only if $(j_{\alpha_1} j_{\alpha_2} \cdots j_{\alpha_n})^B \sim (j_{\alpha'_1} j_{\alpha'_2} \cdots j_{\alpha'_n})^B$. Inequivalent basis vector are orthonormal in all quantum numbers labeling the state vector.
Recoupling coefficients:

A recoupling coefficient is a transformation coefficient

$$\left\langle (j_\beta)^{B'} \,|\, (l)\, jm \,|\, (j_\alpha)^B (k)\, jm \right\rangle$$

relating any two orthonormal bases of the space $\mathcal{H}_{j_1} \otimes \cdots \otimes \mathcal{H}_{j_n}$, say, the one defined by (2.103, 2.104), and (2.110) for a prescribed coupling scheme corresponding to a bracketing B, and a second one, again defined by these relations but for a different coupling scheme corresponding to a bracketing B'. For example, for $n=3$, there are 3 inequivalent coupling symbols and $\binom{3}{2} = 3$ recoupling coefficients; for $n=4$, there are 15 inequivalent coupling symbols and $\binom{15}{2} = 105$ recoupling coefficients. Each coefficient is, of course, expressible as a sum over products of $2(n-1)$ WCG-coefficients, obtained simply by taking the inner product:

$$\begin{aligned}&\left\langle (j_\beta)^{B'} (l)\, jm \,|\, (j_\alpha)^B (k)\, jm \right\rangle \\ &= \sum_{\sum m_\alpha = m} C \left[\begin{pmatrix} j_{\beta_1} \cdots j_{\beta_n} \\ m_{\beta_1} \cdots m_{\beta_n} \end{pmatrix}^{B'} \begin{matrix} j \\ m \end{matrix} \right] (l) \\ &\times C \left[\begin{pmatrix} j_{\alpha_1} \cdots j_{\alpha_n} \\ m_{\alpha_1} \cdots m_{\alpha_n} \end{pmatrix}^{B} \begin{matrix} j \\ m \end{matrix} \right] (k)\,. \end{aligned} \tag{2.111}$$

The fundamental theorem of binary coupling theory states for inequivalent coupling schemes is:
Each recoupling coefficient is expressible as a sum over products of Racah coefficients, the only other quantities occurring in the summation being phase and dimension factors.
In every instance, the summation over projection quantum numbers in the right-hand side of (2.111) is re-expressible as a sum over Racah coefficients.

2.12.4 Construction of all Transformation Coefficients in Binary Coupling Theory

Augmented notation:

The coupling symbol $(j_{\alpha_1} j_{\alpha_2} \cdots j_{\alpha_n})^B$ contains all information as to how n angular momenta are to be coupled, but is not specific in how the intermediate angular momentum quantum numbers $(k_1 k_2 \cdots k_{n-2})$, are to be matched with the binary couplings implicit in the coupling symbol. For explicit calculations, it is necessary to remedy this deficiency in notation. This may be done by attaching the $n-2$ intermediate angular momentum quantum numbers and the total angular momentum j as subscripts to the $n-1$ parentheses pairs in the coupling symbol. For example, for $(j_1 j_2 j_3 j_4 j_5)^B = \{[(j_1 j_2)(j_3 j_4)] j_5\}$, this results in the replacement

$$\begin{aligned}&\{[(j_1 j_2)(j_3 j_4)] j_5\}(k_1 k_2 k_3) \\ &\quad \to \{[(j_1 j_2)_{k_1} (j_3 j_4)_{k_2}]_{k_3} j_5\}_j\,.\end{aligned}$$

The basic coupling symbol structure is regained simply by ignoring all inferior letters.
Basic rules for commutation and association:

Let x, y, z denote arbitrary disjoint contiguous subcoupling symbols $\{[(x)(y)](z)\}$ contained in the coupling symbol $(j_{\alpha_1} j_{\alpha_2} \cdots j_{\alpha_n})^B$. Let a, b, c denote the intermediate angular momenta associated with addition of the angular momenta represented in x, y, z, respectively, d the angular momentum representing the sum of a and b, and k the sum of d and c. Symbolically, this subcoupling may be presented as

$$\begin{aligned}&\sum \boldsymbol{J}(x) = \boldsymbol{J}(a)\,, \quad \sum \boldsymbol{J}(y) = \boldsymbol{J}(b)\,, \\ &\sum \boldsymbol{J}(z) = \boldsymbol{J}(c)\,, \\ &\quad \boldsymbol{J} = \cdots \{[\boldsymbol{J}(a) + \boldsymbol{J}(b)] + \boldsymbol{J}(c)\} \cdots\,; \\ &\quad \boldsymbol{J}(a) + \boldsymbol{J}(b) = \boldsymbol{J}(d); \quad \boldsymbol{J}(d) + \boldsymbol{J}(c) = \boldsymbol{J}(k)\end{aligned}$$

with augmented coupling symbol

$$(j_{\alpha_1} j_{\alpha_2} \cdots j_{\alpha_n})^B = \cdots \{[(x)_a (y)_b]_d (z)_c\}_k \cdots\,.$$

There are only two basic operations in constructing the recoupling coefficient between any two coupling schemes:

commutation of symbols:

$$(x)_a(y)_b \rightarrow (y)_b(x)_a$$

with the transformation of state vector given by

$$\begin{aligned}&|\cdots[(x)_a(y)_b]_d\cdots\rangle\\&\quad\rightarrow(-1)^{a+b-d}|\cdots[(y)_b(x)_a]_d\cdots\rangle\\&=|\cdots[(x)_a(y)_b]_d\cdots\rangle\,.\end{aligned}$$

Association of symbols:

$$[(x)_a(y)_b](z)_c \rightarrow (x)_a[(y)_b(z)_c]$$

with the transformation of state vector given by

$$\begin{aligned}&|\cdots\{[(x)_a(y)_b]_d(z)_c\}_k\cdots\rangle\\&\quad\rightarrow\sum_e[(2d+1)(2e+1)]^{1/2}W(abkc;\,de)\\&\quad\times|\cdots\{(x)_a[(y)_b(z)_c]_e\}_k\cdots\rangle\\&=|\cdots\{[(x)_a(y)_b]_d(z)_c\}_k\cdots\rangle\,.\end{aligned}$$

The basic result for the calculation of all recoupling coefficients is:
Each pair of coupling schemes for n angular momenta can be brought into coincidence by a series of commutations and associations performed on either of the set of coupling symbols defining the coupling scheme.

In principle, this result gives a method for the construction of all recoupling transformation coefficients and sets the stage for the formulation of still deeper questions arising in recoupling theory, as summarized in Sect. 2.12.5. The following examples illustrate the content of the preceding abstract constructions.
Examples:
WCG-coefficient form:

$$\begin{aligned}&\langle\{[(ab)_e c]_f d\}_g\,|\,[(ac)_h(bd)_k]_g\rangle\\&=\sum_{\alpha+\beta+\gamma+\delta=m}C^{a\,b\,e}_{\alpha,\beta,\alpha+\beta}C^{e\,c\,f}_{\alpha+\beta,\gamma,\alpha+\beta+\gamma}C^{f\,d\,g}_{\alpha+\beta+\gamma,\delta,m}\\&\quad\times C^{a\,c\,h}_{\alpha,\gamma,\alpha+\gamma}C^{b\,d\,k}_{\beta,\delta,\beta+\delta}C^{h\,k\,g}_{\alpha+\gamma,\beta+\delta,m}\,.\end{aligned}$$

6–j coefficient as recoupling coefficient:

$$\begin{aligned}&(ac)(bd)\xrightarrow{\mathrm{R}}[(ac)b]d\xrightarrow{\phi}[b(ac)]d\xrightarrow{\mathrm{R}}[(ba)c]d\\&\xrightarrow{\phi}[(ab)c]d\,,\end{aligned}$$

where ϕ denotes that the communication of symbols effects a phase factor transformation, and R denotes that the associative of symbol effects a Racah coefficient transformation:

$$\begin{aligned}&\langle\{[(ab)_e c]_f d\}_g\,|\,[(ac)_h(bd)_k]_g\rangle\\&=(-1)^{e+h-a-f}[(2f+1)(2k+1)]^{1/2}W(hbgd;\,fk)\\&\quad\times[(2e+1)(2h+1)]^{1/2}W(bafc;\,eh)\,.\end{aligned}$$

9–j coefficient as recoupling coefficient:

$$\begin{aligned}&(ac)(bd)\xrightarrow{\mathrm{R}}[(ac)b]d\xrightarrow{\phi}[b(ac)]d\xrightarrow{\mathrm{R}}[(ba)c]d\\&\xrightarrow{\phi}[(ab)c]d\xrightarrow{\mathrm{R}}(ab)(cd)\,,\\&\langle[(ab)_e(cd)_f]_g\,|\,[(ac)_h(bd)_k]_g\rangle\\&=[(2e+1)(2f+1)(2h+1)(2k+1)]^{\frac{1}{2}}\\&\quad\times\sum_l(-1)^{2l}(2l+1)\begin{Bmatrix}ach\\lbe\end{Bmatrix}\begin{Bmatrix}bdk\\ghl\end{Bmatrix}\begin{Bmatrix}efg\\dlc\end{Bmatrix}\\&=[(2e+1)(2f+1)(2h+1)(2k+1)]^{\frac{1}{2}}\begin{Bmatrix}abe\\cdf\\hkg\end{Bmatrix}\,.\end{aligned}$$

2.12.5 Unsolved Problems in Recoupling Theory

1. Define a route between two coupling symbols for n angular momenta to be any sequence of transpositions and associations that carries one symbol into the other. Each such route then gives rise to a unique expression for the corresponding recoupling coefficient in terms of 6–j coefficients. In general, there are several routes between the same pair of coupling symbols, leading therefore to identities between 6–j coefficients. How many nontrivial routes are there between two given coupling symbols, leading to nontrivial relations between 6–j coefficients (trivial means related by a phase factor)?
2. Only 6–j coefficients arise in all possible couplings of three angular momenta; only 6–j and 9–j coefficients arise in all possible couplings of four angular momenta; in addition to 6–j and 9–j coefficients, two new "classes" of coefficients, called 12–j coefficients of the first and second kind, arise in the coupling of five angular momenta; in addition to 6–j, 9–j, and the two classes of 12–j coefficients, five new classes of 15–j coefficients arise in the coupling of six angular momenta, $\cdots$. What are the classes of 3n—j coefficients? The nonconstructive answer is that a summation over 6–j coefficients arising in the coupling of n angular momenta is of a new class if it cannot be expressed in terms

Part A | 2.12

of previously defined coefficients occurring in the recoupling of $n-1$ or fewer angular momenta.

3. Toward answering the question of classes of $3n$–j coefficients, one is lead into the classification problem of planar cubic graphs. It is known that every $3n$–j coefficient corresponds to a planar cubic graph, but the converse is not true. For small n, the relation between the coupling of n angular momenta, the number of new classes of $3(n-1)$–j coefficients, and the number of nonisomorphic planar cubic graphs on $2(n-1)$ points is:

n	Classes of $3(n-1)-j$ coefficients	Cubic graphs on $2(n-1)$ points
3	1	1
4	1	2
5	2	5
6	5	19
7	18	87
8	84	?
9	576	?

The geometrical object for $n=3$ is a planar graph isomorphic to the tetrahedron in 3-space. The classification of all nonisomorphic cubic graphs on $2(n-1)$ points is an unsolved problem in mathematics, as is the classification of classes for $3(n-1)$–j coefficients.

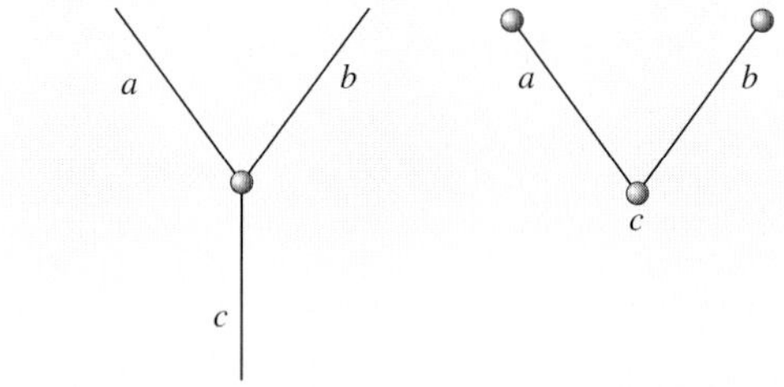

Fig. 2.4 The fundamental triangle $[(ab)c]$ can be realized by lines or points

4. There are (at least) two methods of realizing the basic triangles of angular momentum theory in terms of graphs. The fundamental structural element $[(ab)c]$ is represented either in terms of its points or in terms of its lines (Fig. 2.4):
   The right representation leads to the interpretation of recoupling coefficients as functions defined on pairs of labeled binary trees [2.1]; the left to the diagrams of the Jucys school [2.7, 8]. Either method leads to the relationship of recoupling coefficients to cubic graphs.
5. The approach of classifying $3n-j$ coefficients through the use of unit tensor operator couplings, Racah operators, 9–j invariant operators, and general invariant operators is undeveloped.

2.13 Supplement on Combinatorial Foundations

The quantum theory of angular momentum can be worked out using the abstract postulates of the properties of angular momenta operators and the abstract Hilbert space in which they act. The underlying mathematical apparatus is the Lie algebra of the group $SU(2)$ and multiple copies thereof. An alternative approach is to use special Hilbert spaces that realize all the properties of the abstract postulates and perform calculations within that framework. The framework must be sufficiently rich in structure so as to apply to a manifold of physical situations. This approach has been used often in our treatment; it is an approach that is particularly useful for revealing the combinatorial foundations of quantum angular momentum theory. We illustrate this concretely in this supplementary section. The basic objects are the polynomials defined by (2.24), which we now call $SU(2)$ solid harmonics, where we change the notation slightly by interchanging the role of m and m'.

2.13.1 $SU(2)$ Solid Harmonics

The $SU(2)$ solid harmonics are defined to be the homogeneous polynomials of degree $2j$ in four commuting indeterminates given by

$$D^j_{m\,m'}(Z) = \sqrt{\alpha!\beta!}\sum_{(\alpha:A:\beta)} \frac{Z^A}{A!}\,, \qquad (2.112)$$

in which the indeterminates Z and the nonnegative exponents A are encoded in the matrix arrays

$$Z = \begin{pmatrix} z_{11} & z_{12} \\ z_{21} & z_{22} \end{pmatrix}, \qquad A = \begin{pmatrix} a_{11} & a_{12} \\ a_{21} & a_{22} \end{pmatrix},$$

$$X^A = \prod_{i,j=1}^{2} z_{ij}^{a_{ij}}\,, \qquad A! = \prod_{i,j=1}^{2} (a_{ij})!,$$

$$\alpha! = \alpha_1!\alpha_2!\,, \qquad \beta! = \beta_1!\beta_2!\,.$$

The symbol $(\alpha : A : \beta)$ in (2.112) denotes that the matrix array A of nonnegative integer entries has row and column sums of its a_{ij} entries given in terms of the quantum numbers j, m, m' by

$$a_{11}+a_{12}=\alpha_1=j+m,\ a_{21}+a_{22}=\alpha_2=j-m\,,$$
$$a_{11}+a_{21}=\beta_1=j+m',\ a_{12}+a_{22}=\beta_2=j-m'\,.$$

These $SU(2)$ solid harmonics are among the most important functions in angular momentum theory. Not only do they unify the irreducible representations of $SU(2)$ in any parametrization by the appropriate definition of the indeterminates in terms of generalized coordinates, they also include the popular boson calculus realization of state vectors for quantum mechanical systems, as well as the state vectors for the symmetric rigid rotator.

The realization of the inner product is essential. Physical theory demands an inner product that is given in terms of integrations of wave functions over the variables of the theory, as required by the probabilistic interpretation of wave functions. It is the requirement that realizations of angular momentum operators be Hermitian with respect to the inner product for the spaces being used that assures the orthogonality of functions, so that one is able to take results from one realization of the inner product to another with compatibility of relations. Often, in combinatorial arguments, the inner product plays no direct role.

The nomenclature $SU(2)$ solid harmonics for the polynomials defined by (2.112) is by analogy with the term $SO(3,\boldsymbol{R})$ solid harmonics for the polynomials described in Sect. 2.1.

The polynomials $\mathcal{Y}_{lm}(\boldsymbol{x}),\ \boldsymbol{x}=(x_1,x_2,x_3)\in\boldsymbol{R}^3$ are homogeneous of degree l. The angular momentum operator $\boldsymbol{L}^2$ is given by

$$\boldsymbol{L}^2=-r^2\nabla^2+(\boldsymbol{x}\cdot\nabla)^2+(\boldsymbol{x}\cdot\nabla)\,,$$

which is a sum of two commuting operators $-r^2\nabla^2$ and $(\boldsymbol{x}\cdot\nabla)^2+\boldsymbol{x}\cdot\nabla$, each of which is invariant under orthogonal transformations. The $SO(3,\boldsymbol{R})$ solid harmonics are homogeneous polynomials of degree l that solve $\nabla^2\mathcal{Y}_{lm}(\boldsymbol{x})=0$, so that $\boldsymbol{L}^2\mathcal{Y}_{lm}(\boldsymbol{x})=l(l+1)\mathcal{Y}_{lm}(\boldsymbol{x})$. The component angular momentum operators L_i then have the standard action on these polynomials, and under real, proper, orthogonal transformations give the irreducible representations of the group $SO(3,\boldsymbol{R})$.

The polynomials $D^j_{m\,m'}(Z),\ z=(z_{11},z_{21},z_{12},z_{22})\in\boldsymbol{C}^4$ are homogeneous of degree $2j$. The angular momentum operator $\boldsymbol{J}^2$, with $\boldsymbol{J}=(J_1,J_2,J_3)$, is given by

$$\begin{aligned}\boldsymbol{J}^2&=-(\det Z)\left(\det\frac{\partial}{\partial Z}\right)+J_0(J_0+1)\,,\\ J_0&=\frac{1}{2}z\cdot\boldsymbol{\partial}\,,\\ \boldsymbol{\partial}&=\left(\frac{\partial}{\partial z_{11}},\frac{\partial}{\partial z_{21}},\frac{\partial}{\partial z_{12}},\frac{\partial}{\partial z_{22}}\right)\,,\end{aligned}\tag{2.113}$$

which is a sum of two commuting operators $-(\det Z)\left(\det\frac{\partial}{\partial Z}\right)$ and $J_0(J_0+1)$, each of which is invariant under $SU(2)$ transformations. The $SU(2)$ solid harmonics are homogeneous polynomials of degree $2j$ such that

$$\begin{aligned}\det\frac{\partial}{\partial Z}D^j_{m\,m'}(Z)&=0\,,\\ \boldsymbol{J}^2D^j_{mm'}(Z)&=j(j+1)D^j_{mm'}(Z)\,.\end{aligned}$$

The components of the angular momentum operators $\boldsymbol{J}=(J_1,J_2,J_3)=(M_1,M_2,M_3)$ and $\boldsymbol{K}=(K_1,K_2,K_3)$ then have the standard action on these polynomials as given in Sect. 2.4.4, and under either left or right $SU(2)$ transformations these polynomials give the irreducible representations of the group $SU(2)$.

2.13.2 Combinatorial Definition of Wigner–Clebsch–Gordan Coefficients

The $SU(2)$ solid harmonics have a basic role in the interpretation of WCG-coefficients in combinatorial terms. We recall from Sect. 2.7.2 that the basic abstract Hilbert space coupling rule for compounding two kinematically independent angular momenta with components $\boldsymbol{J}_1=(J_1(1),J_2(1),J_3(1))$ and $\boldsymbol{J}_2=(J_1(2),J_2(2),J_3(2))$ to a total angular momentum $\boldsymbol{J}=(J_1,J_2,J_3)=\boldsymbol{J}_1+\boldsymbol{J}_2$ is

$$\begin{aligned}&|(j_1\,j_2)j\,m\rangle\\ &=\sum_{m_1+m_2=m}C^{j_1\ j_2\ j}_{m_1\ m_2\ m}|j_1\,m_1\rangle\otimes|j_2\,m_2\rangle\,.\end{aligned}\tag{2.114}$$

This relation in abstract Hilbert space is realized explicitly by spinorial polynomials as follows:

$$\begin{aligned}\psi_{(j_1\,j_2)jm}(Z)=&\sum_{m_1+m_2=m}C^{j_1\ j_2\ j}_{m_1\ m_2\ m}\\ &\times\psi_{j_1\,m_1}(z_{11},z_{21})\psi_{j_2\,m_2}(z_{12},z_{22})\,,\end{aligned}\tag{2.115}$$

$$\psi_{(j_1 j_2) jm}(Z) = \sqrt{\frac{2j+1}{(j_1+j_2-j)!(j_1+j_2+j+1)!}} \times (\det Z)^{j_1+j_2-j} D^j_{m, j_1-j_2}(Z)\,, \tag{2.116}$$

$$\psi_{jm}(x, y) = \frac{x^{j+m} y^{j-m}}{\sqrt{(j+m)!(j-m)!}}\,. \tag{2.117}$$

Explicit knowledge of the WCG-coefficients is not needed to prove these relationships. The angular momentum operators

$$\begin{aligned}
J_+(1) &= z_{11}\frac{\partial}{\partial z_{21}}\,, \quad J_-(1) = z_{21}\frac{\partial}{\partial z_{11}}\,,\\
J_3(1) &= \frac{1}{2}\left(z_{11}\frac{\partial}{\partial z_{11}} - z_{21}\frac{\partial}{\partial z_{21}}\right);\\
J_+(2) &= z_{12}\frac{\partial}{\partial z_{22}}\,, \quad J_-(2) = z_{22}\frac{\partial}{\partial z_{12}}\,,\\
J_3(2) &= \frac{1}{2}\left(z_{12}\frac{\partial}{\partial z_{12}} - z_{22}\frac{\partial}{\partial z_{22}}\right)
\end{aligned}$$

are Hermitian in the polynomial inner product defined in Sect. 2.4.3, and have the standard action on the polynomials $\psi_{j_1 m_1}(z_{11}, z_{21})$ and $\psi_{j_2 m_2}(z_{12}, z_{22})$, respectively, which are normalized in the inner product $(\ ,\ )$. The components of total angular momentum operator $\boldsymbol{J} = \boldsymbol{M} = \boldsymbol{J}_1 + \boldsymbol{J}_2$ have the standard action on the polynomials $\psi_{(j_1\, j_2) j\, m}(Z)$, since they have the standard action on the factor $D^j_{m\, j_1 - j_2}(Z)$, as given in Sect. 2.4.4, and $[\boldsymbol{J}, \det X] = \left[\boldsymbol{J}, \det \frac{\partial}{\partial Z}\right] = 0$. Thus, we have

$$\begin{aligned}
\boldsymbol{J}^2 \psi_{(j_1 j_2) jm}(Z) &= j(j+1)\psi_{(j_1 j_2) jm}(Z),\\
J_\pm \psi_{(j_1 j_2) jm}(Z) &= \sqrt{(j \mp m)(j \pm m + 1)}\\
&\quad \times \psi_{(j_1 j_2) jm\pm 1}(Z)\,.
\end{aligned}$$

We also note that the two commuting parts of $\boldsymbol{J}^2$ are diagonal on these functions:

$$\begin{aligned}
&J_0(J_0+1)\psi_{(j_1 j_2) jm}(Z)\\
&\quad = (j_1+j_2)(j_1+j_2+1)\psi_{(j_1 j_2) jm}(Z)\,,\\
&(\det Z)\left(\det \frac{\partial}{\partial X}\right)\psi_{(j_1 j_2) jm}(Z)\\
&\quad = (j_1+j_2-j)(j_1+j_2+j_1+1)\psi_{(j_1 j_2) jm}(Z)\,.
\end{aligned}$$

It is necessary only to verify these properties for the highest weight function $D^j_{j\,j}(Z) = z_{11}^{2j}/\sqrt{(2j)!}$, for which they are seen to hold.

The angular momentum operators $\boldsymbol{K} = (K_1, K_2, K_3)$ defined in Sect. 2.4.4 with components that commute with those of $\boldsymbol{J} = (M_1, M_2, M_3)$ and having $\boldsymbol{K}^2 = \boldsymbol{J}^2$ also have a well-defined action on the functions $\psi_{(j_1\, j_2) j\, m}(Z)$. The action of K_+, K_-, and K_3 on the quantum numbers (j_1, j_2) is to effect the shifts to $\left(j_1 + \frac{1}{2}, j_2 - \frac{1}{2}\right)$, $\left(j_1 - \frac{1}{2}, j_2 + \frac{1}{2}\right)$, and (j_1, j_2), respectively. These actions of Hermitian angular momentum operators satisfying the standard commutation relations $\boldsymbol{K} \times \boldsymbol{K} = \mathrm{i}\boldsymbol{K}$ are quite unusual in that they depend only on the angular momentum quantum numbers j_1, j_2, j themselves, which satisfy the triangle rule, and give further interesting properties of the modified $SU(2)$ solid harmonics $\psi_{(j_1\, j_2) j\, m}(Z)$. We note these properties in full:

$$\begin{aligned}
&\boldsymbol{K}^2 \psi_{(j_1 j_2) jm}(Z) = j(j+1)\psi_{(j_1 j_2) jm}(Z)\,,\\
&K_3 \psi_{(j_1 j_2) jm}(Z) = (j_1 - j_2)\psi_{(j_1 j_2) jm}(Z)\,,\\
&K_+ \psi_{(j_1 j_2) jm}(Z) =\\
&\sqrt{(j - j_1 + j_2)(j + j_1 - j_2 + 1)}\psi_{\left(j_1 + \frac{1}{2} j_2 - \frac{1}{2}\right) jm}(Z)\,,\\
&K_- \psi_{(j_1 j_2) jm}(Z) =\\
&\sqrt{(j + j_1 - j_2)(j - j_1 + j_2 + 1)}\psi_{\left(j_1 - \frac{1}{2} j_2 + \frac{1}{2}\right) jm}(Z)\,.
\end{aligned}$$

These relations play no direct role in our continuing considerations of (2.116) and the determination of the WCG-coefficients, and we do not interpret them further.

The explicit WCG-coefficients are obtained by expanding the 2×2 determinant in (2.116), multiplying this expansion into the D-polynomial, and changing the order of the summation. These operations are most succinctly expressed in terms of the umbral calculus of *Roman* and *Rota* [2.6], using his evaluation operation. The evaluation at y of a divided power $x^k/k!$ of a single indeterminate x to a nonnegative integral power k is defined by

$$\mathrm{eval}_y \frac{x^k}{k!} = \frac{(y)_k}{k!} = \frac{y(y-1)\cdots(y-k+1)}{k!} = \binom{y}{k}\,,$$

where $(y)_k$ is the falling factorial. This definition is extended to products by

$$\mathrm{eval}_{(y_1, y_2, \ldots, y_n)} \prod_{i=1}^{n} \frac{x^{k_i}}{k_i!} = \prod_{i=1}^{n} \mathrm{eval}_{y_i} \frac{x^{k_i}}{k_i!} = \prod_{i=1}^{n} \binom{y_i}{k_i}\,.$$

It is also extended by linearity to sums of such divided powers, multiplied by arbitrary numbers.

The application of these rules to our problem involving four indeterminates gives

$$\frac{(\det X)^n}{n!}\sum_{(\alpha:A:\alpha')}\frac{X^A}{A!}$$
$$=\sum_{(\beta:B:\beta')}\text{eval}_B\left(\frac{(\det X)^n}{n!}\right)\frac{X^B}{B!}, \quad (2.118)$$
$$\beta=(\alpha_1+n,\alpha_2+n),\ \beta'=(\alpha'_1+n,\alpha'_2+n),$$
$$\text{eval}_B\frac{(\det X)^n}{n!}$$
$$=\sum_{k_1+k_2=n}(-1)^{k_2}k_1!k_2!\binom{b_{11}}{k_1}\binom{b_{12}}{k_2}\binom{b_{21}}{k_2}\binom{b_{22}}{k_1}. \quad (2.119)$$

In this result, we do not identify the labels with angular momentum quantum numbers. Relation (2.119) is a purely combinatorial, algebraic identity for arbitrary indeterminates and arbitrary row and column sum constraints on the array A as specified by $\alpha=(\alpha_1,\alpha_2)$ and $\alpha'=(\alpha'_1,\alpha'_2)$. There are no square roots involved.

We now apply relations (2.118–2.119) to the case at hand: $n=j_1+j_2-j$, $\alpha=(j+m,j-m)$, $\alpha'=(j+j_1-j_2,j-j_1+j_2)$, $\beta=(j_1+j_2+m,j_1+j_2-m)$, $\beta'=(2j_1,2j_2)$. This gives the following result for the WCG-coefficients:

$$C^{j_1j_2j}_{m_1m_2m}$$
$$=\sqrt{\frac{(j_1+j_2-j)!(j_1-j_2+j)!(-j_1+j_2+j)!}{(j_1+j_2+j+1)!}}$$
$$\times\sqrt{\frac{(2j+1)(j+m)!(j-m)!}{(j_1+m_1)!(j_1-m_1)!(j_2+m_2)!(j_2-m_2)!}}$$
$$\times\text{eval}_A\frac{(\det X)^{j_1+j_2-j}}{(j_1+j_2-j)!}, \quad (2.120)$$

$$\text{eval}_A\frac{(\det X)^{j_1+j_2-j}}{(j_1+j_2-j)!}$$
$$=\sum_{k_1+k_2=j_1+j_2-j}(-1)^{k_2}k_1!k_2!\binom{j_1+m_1}{k_1}\binom{j_2+m_2}{k_2}$$
$$\times\binom{j_1-m_1}{k_2}\binom{j_2-m_2}{k_1}. \quad (2.121)$$

In summary, we have the following:

Up to multiplicative square-root factors, the WCG-coefficient is the evaluation at the point

$$B=\begin{pmatrix} j_1+m_1 & j_2+m_2\\ j_1-m_1 & j_2-m_2\end{pmatrix}$$

of the divided power

$$\frac{(\det X)^{j_1+j_2-j}}{(j_1+j_2-j)!}$$

of a determinant, which is an integer.

The abstract umbral calculus of Rota thus finds its way, at a basic level, into angular momentum theory. Relation (2.120) is but a rewriting in terms of evaluations of the well-known Van der Waerden form of the WCG-coefficients.

2.13.3 Magic Square Realization of the Addition of Two Angular Momenta

The origin of (2.114), giving the states of total angular momentum by compounding two angular momenta, is usually attributed to properties of the direct sum of two copies of the Lie algebra of the unitary unimodular group $SU(2)$, and to the use of differential operators to realize the Lie algebras and state vectors, as done above. It is an interesting combinatorial result that this structure for adding angular momentum is fully encoded within the properties of magic squares of order 3, and no operators whatsoever are needed, only the condition of being a magic square. We have already noted in Sect. 2.7.4 that Regge observed that the restrictions on the domains of the quantum numbers j_1,m_1,j_2,m_2,j,m are encoded in terms of a magic square A with line-sum $J=j_1+j_2+j$:

$$A=\begin{pmatrix} j_1+m_1 & j_2+m_2 & j-m\\ j_1-m_1 & j_2-m_2 & j+m\\ j_2-j_1+j & j_1-j_2+j & j_1+j_2-j\end{pmatrix}. \quad (2.122)$$

The angular momentum quantum numbers are given in terms of the elements of $A=(a_{ij})_{1\le i,j\le 3}$ by the invertible relations

$$j_1=\frac{1}{2}(a_{11}+a_{21}),\quad j_2=\frac{1}{2}(a_{12}+a_{22}),$$
$$j=\frac{1}{2}(a_{13}+a_{23}),$$
$$m_1=\frac{1}{2}(a_{11}-a_{21}),\quad m_2=\frac{1}{2}(a_{12}-a_{22}),$$
$$m=\frac{1}{2}(a_{23}-a_{13}).$$

It follows from these definitions and the fact that A is a magic square of line-sum J, that the sum rule $m_1+m_2=m$ and the triangle condition are fulfilled.

We use the symbol $\langle j_1, j_2, j\rangle$ to denote any triple j_1, j_2, j of angular momentum quantum numbers that satisfy the triangle conditions, where we note that, if a given triple satisfies the triangle conditions, then all permutations of the triple also satisfy the triangle conditions. The number of magic squares for fixed line-sum J is obtained as follows: Define $\Delta_J = \{\text{all triangles } \langle j_1, j_2, j\rangle \mid j_1 + j_2 + j = J\}$ and $M(j_1, j_2, j) = \{(m_1, m_2)| - j_1 \le m_1 \le j_1;\ -j_2 \le m_2 \le j_2;\ -j \le m_1 + m_2 \le j\}$. Then we have the following identity, which gives the number of angular momentum magic squares with line-sum J:

$$\sum_{\langle j_1, j_2, j\rangle \in \Delta_J} \left|M(j_1, j_2, j)\right| = \binom{J+5}{5} - \binom{J+2}{5} . \tag{2.123}$$

It is nontrivial to effect the summation on the left-hand side of this relation to obtain the right-hand side, but this expression is known from the theory of magic squares *Stanley* [2.9, 10].

Not only can the addition of two angular momenta in quantum theory, with its triangle rule for three angular momentum quantum numbers and its sum rule on the corresponding projection quantum numbers, be codified in magic squares of order 3 and arbitrary line-sum, but also the content of the abstract state vector of (2.114) itself can be so expressed:

$$\left|\frac{1}{2}(a_{11}+a_{21}), \frac{1}{2}(a_{12}+a_{22}); \frac{1}{2}(a_{13}+a_{23}), \frac{1}{2}(a_{23}-a_{13})\right\rangle = \sum_{\begin{pmatrix} a_{11} & a_{12} \\ a_{21} & a_{22}\end{pmatrix}} C(A)\left|\frac{1}{2}(a_{11}+a_{21}), \frac{1}{2}(a_{11}-a_{21})\right\rangle \otimes \left|\frac{1}{2}(a_{12}+a_{22}), \frac{1}{2}(a_{12}-a_{22})\right\rangle ,$$

where the summation is over all subsets

$$\begin{pmatrix} a_{11} & a_{12} \\ a_{21} & a_{22}\end{pmatrix}$$

of the magic square of order 3 such that row 3 and column 3 are held fixed. The coefficients $C(A)$ themselves are the WCG-coefficients, which may be regarded as a function whose domain of definition is the set of all magic squares of order 3. The triangle rule $\langle j_1, j_2, j\rangle$ and the sum rule on (m_1, m_2, m) are implied by the structure of magic squares of order 3. These rich combinatorial footings of angular momentum theory are completed by the observation that the Clebsch–Gordan coefficients themselves are obtained by the Schwinger–Regge generating function given in Sect. 2.7.3 (see [2.2] for the relation to ${}_3F_2$ hypergeometric functions).

2.13.4 MacMahon's and Schwinger's Master Theorems

Generating functions codify the content of many mathematical entities in a unifying, comprehensive way. These functions are very popular in combinatorics, and Schwinger used them extensively in his treatment of angular momentum theory. In this subsection, we present a natural generalization of the $SU(2)$ solid harmonics to a class of polynomials that are homogeneous in n^2 indeterminates. While these polynomials are of interest in their own right, it is their fundamental role in the addition of n kinematically independent angular momenta that motivates their introduction here. They bring an unexpected unity and coherence to angular momentum coupling and recoupling theory [2.11].

We list in compendium format some of the principal results:

Special $U(n)$ solid harmonics:

$$D^k_{\alpha,\beta}(Z) = \sqrt{\alpha!\beta!} \sum_{A \in M(\alpha,\beta)} \frac{Z^A}{A!} , \tag{2.124}$$

$A = (a_{ij})_{1\le i,j\le n}$: matrix of order n in nonnegative integers;

$$A! = \prod_{i,j=1}^{n} a_{ij}! ,\ Z^A = \prod_{i,j=1}^{n} z_{ij}^{a_{ij}} ;$$

where we employ the notations:
$\alpha = (\alpha_1, \alpha_2, \ldots, \alpha_n)$: sequence (composition) of nonnegative integers having the sum k, denoted $\alpha \vdash k$;
$x^\alpha = x_1^{\alpha_1} x_2^{\alpha_2} \cdots x_n^{\alpha_n}$, $\alpha! = \alpha_1!\alpha_2!\cdots\alpha_n!$;
$\beta = (\beta_1, \beta_2, \ldots, \beta_n)$: second composition $\beta \vdash k$;
$M(\alpha, \beta)$, set of all matrices A such that the entries in row i sums to α_i and those in column j to β_j.
The significance of the row-sum vector α is that α_i is the degree of the polynomial in the variables $(z_{i1}, z_{i2}, \ldots, z_{in})$ in row i of Z, with a similar interpretation for β in terms of columns.
Matrix of the $D^k_{\alpha,\beta}(Z)$ polynomials:
The number of compositions of the integer k into n nonnegative parts is given by $\binom{n+k-1}{k}$. The compositions in this set may be linearly ordered by the lexicographical rule $\alpha > \beta$, if the first nonzero part of $\alpha - \beta$ is

positive. The polynomial $D^k_{\alpha,\beta}(Z)$ is then the entry in row α and column β in the matrix $D^k(Z)$ of dimension $\dim D^k(Z) = \binom{n+k-1}{k}$, where, following the convention for $SU(2)$, the rows are labelled from top to bottom by the greatest to the least sequence, and the columns are labelled in the same manner as read from left to right. There is a combinatorial proof by *Chen* and *Louck* [2.12] that these polynomials satisfy the following multiplication rule for arbitrary matrices X and Y:

$$D^k(XY) = D^k(X)D^k(Y) . \tag{2.125}$$

Orthogonality in the inner product (,) defined in Sect. 2.4.3:

$$\left\langle D^k_{\alpha,\beta}, D^k_{\alpha',\beta'} \right\rangle = \delta_{\alpha,\alpha'}\delta_{\beta,\beta'}k! .$$

Value on $Z = \mathrm{diag}(z_1, z_2, \ldots, z_n)$:

$$D^k_{\alpha,\beta}\left[\mathrm{diag}(z_1, z_2, \ldots, z_n)\right] = \delta_{\alpha,\beta}z^\alpha , \tag{2.126}$$
$$D^k(I_n) = I_{\binom{n+k-1}{k}} .$$

Transposition property:

$$D^k(Z^{\mathrm{T}}) = \left[D^k(Z)\right]^{\mathrm{T}} .$$

Special irreducible unitary representations of $U(n)$:

$$D^k(U)D^k(V) = D^k(UV) , \quad \text{all } U, V \in U(n) .$$

Schwinger's Master Theorem: For any two matrices X and Y of order n, the following identities hold:

$$\begin{aligned} &\mathrm{e}^{(\partial_x : X : \partial_y)}\,\mathrm{e}^{(x:Y:y)}\Big|_{x=y=0} \\ &= \sum_{k=0}^{\infty} \sum_{\alpha,\beta \vdash k} D^k_{\alpha,\beta}(X) D^k_{\beta,\alpha}(Y) \\ &= \frac{1}{\det(I - XY)} , \end{aligned} \tag{2.127}$$

$$(x : Z : y) = xZy^{\mathrm{T}} = \sum_{i,j=1}^{n} z_{ij}x_i y_j .$$

MacMahon's Master Theorem: Let X be the diagonal matrix $X = \mathrm{diag}(x_1, x_2, \ldots, x_n)$ and Y a matrix of order n. Then the coefficient of x^α in the expansion of $\frac{1}{\det(I-XY)}$ equals the coefficient of x^α in the product y^α, $y_i = \sum_{j=1}^{n} y_{ij}x_j$, that is,

$$\frac{1}{\det(I - XY)} = \sum_{k=0}^{\infty} \sum_{\alpha \vdash k} D^k_{\alpha,\alpha}(Y)\, x^\alpha . \tag{2.128}$$

Basic Master Theorem: Let Z be a matrix of order n. Then

$$\frac{1}{\det(I - tZ)} = \sum_{k=0}^{\infty} t^k \sum_{\alpha \vdash k} D^k_{\alpha,\alpha}(Z) . \tag{2.129}$$

Schwinger's relation (2.127) follows from the basic relation (2.129) by setting $Z = XY$ and using the multiplication property (2.125); MacMahon's relation then follows from Schwinger's result by setting $X = \mathrm{diag}(x_1, x_2, \ldots, x_n)$ and using property (2.126). Of course, MacMahon's Master Theorem preceded Schwinger's result by many years (see *MacMahon* [2.13]).The unification into the single form by using properties of the $D^k_{\alpha,\beta}(Z)$ polynomials was pointed out in [2.14]. More surprisingly, relation (2.129) was already discovered for the general linear group in 1897 by *Molien* [2.15]; its properties are developed extensively in *Michel* and *Zhilinski* [2.16] in the context of group theory.

For many purposes, it is better in combinatorics to avoid all square roots by using the polynomials

$$L_{\alpha,\beta}(Z) = \sum_{A \in M(\alpha,\beta)} \frac{Z^A}{A!}$$

in place of the $D^k_{\alpha,\beta}(Z)$ defined in (2.124).

2.13.5 The Pfaffian and Double Pfaffian

Schwinger observed that the calculation of $3n-j$ coefficients involves taking the square root $\sqrt{(I - AB)}$, where A and B are skew symmetric (antisymmetric) matrices of order n, but the procedure is rather obscure. The appropriate concepts for taking the square root is that of a Pfaffian and a double Pfaffian, denoted, respectively, by $\mathrm{Pf}(A)$ and $\mathrm{Pf}(A, B)$. The definitions require the concept of a matching of the set of integers $\{1, 2, \ldots, n\}$. A matching of $\{1, 2, \ldots, n\}$ is an unordered set of disjoint subsets $\{i, j\}$ containing two elements. For example, the matchings of $1, 2, 3$ are $\{1, 2\}$, $\{1, 3\}$, and $\{2, 3\}$. We then have the following constructs:
Pfaffian and double Pfaffian of skew symmetric matrices $A = (a_{ij})$ and $B = (b_{ij})$ of order n :

$$\begin{aligned} \mathrm{Pf}(A) = &\sum_{\{i_1,i_2\},\{i_3,i_4\},\ldots,\{i_{n-1},i_n\}} \varepsilon(i_1 i_2 \cdots i_n) \\ &\times a_{i_1,i_2} a_{i_3,i_4} \cdots a_{i_{n-1},i_n} , \end{aligned} \tag{2.130}$$

$$\mathrm{Pf}(A,B) = 1 + \sum_{k\geq 1} \sum_{\substack{\{i_1,i_2\},\{i_3,i_4\},\ldots,\{i_{2k-1},i_{2k}\} \\ \{j_1,j_2\},\{j_3,j_4\},\ldots,\{j_{2k-1},j_{2k}\}}} \varepsilon(i_1 i_2 \cdot i_{2k})\varepsilon(j_1 j_2 \cdots j_{2k}) \times a_{i_1,i_2} a_{i_3,i_4} \cdots a_{i_{2k-1},i_{2k}} \times b_{j_1,j_2} b_{j_3,j_4} \cdots b_{j_{2k-1},j_{2k}} \quad (2.131)$$

where $\{i_1,i_2\},\{i_3,i_4\},\ldots,\{i_{n-1},i_n\}$ is a matching of $\{1,2,\ldots,n\}$,and $\varepsilon(i_1 i_2 \cdots i_n)$ is the sign of the permutation (number of inversions). Similarly, in the double Pfaffian, the 2-subsets are matchings of a subset of $\{1,2,\ldots,n\}$ of even length.

Relations of skew symmetric matrices A, B to Pfaffians:

$$\sqrt{\det A} = \mathrm{Pf}(A)\,; \quad \sqrt{\det(I-AB)} = \mathrm{Pf}(A,B)\,. \quad (2.132)$$

2.13.6 Generating Functions for Coupled Wave Functions and Recoupling Coefficients

This section is a reformulation, nontrivial extension, and interpretation of results found in *Schwinger* [2.3]. We first refine the notation used in Sect. 2.12.3.

Set of triangles in the coupling scheme:

Each coupling scheme, as determined by the bracketing B, has associated with it a unique ordered set of $n-1$ triangles

$$\begin{aligned} \boldsymbol{T}_B(\boldsymbol{j},\boldsymbol{k},j) &= \{\langle a_i,b_i,k_i\rangle | i=1,2,\ldots,n-1\}\,, \\ \boldsymbol{j} &= (j_1,j_2,\ldots,j_n)\,, \\ \boldsymbol{k} &= (k_1,k_2,\ldots,k_{n-2})\,, \\ k_{n-1} &= j\,. \end{aligned}$$

The third part k_i of $\langle a_i,b_i,k_i\rangle$ can always be chosen, without loss of generality, as an intermediate angular momentum $(k_{n-1}=j)$, and the triangles in the set can be ordered by $\langle a_i,b_i,k_i\rangle < \langle a_{i+1},b_{i+1},k_{i+1}\rangle$. The remaining pair of angular momentum labels in the triangle $\langle a_i,b_i,k_i\rangle$ then fall, in general, into four classes: $\langle a_i,b_i,k_i\rangle$ in which a_i can be either a j_r or a k_s, and b_i can be either a $j_{r'}$ or $k_{s'}$. The distribution of the $j's$ and $k's$ among the a_i and b_i is uniquely determined by the bracketing B that defines the coupling scheme.

Clebsch–Gordan coefficients for a given coupling scheme:

$$\begin{pmatrix} \boldsymbol{j} & \boldsymbol{k} & j \\ \boldsymbol{m} & \boldsymbol{q} & m \end{pmatrix}^B = \prod_{\langle a_i,b_i,k_i\rangle \in \boldsymbol{T}_B(\boldsymbol{j},\boldsymbol{k},j)} C^{a_i b_i k_i}_{\alpha_i \beta_i q_i}\,, \quad (2.133)$$

in which the projection quantum numbers α_i and β_i are $m's$ and $q's$ that match the a_i and b_i. In the given coupling scheme determined by the bracketing B, only (j_i,m_i), $i=1,2\ldots,n$; (k_i,q_i), $i=1,2,\ldots,n-2$, and (j,m) appear in the Clebsch–Gordan coefficients. In fact, if one explicitly implements the sum rule on the projection quantum numbers, it is always possible to express the q_i as sums over the m_i and m.

Coupled angular momentum function for n angular momenta:

$$\Psi^B_{(\boldsymbol{j}\,\boldsymbol{k})jm}(\boldsymbol{x},\boldsymbol{y}) = \sum_{\boldsymbol{m}} \begin{pmatrix} \boldsymbol{j} & \boldsymbol{k} & j \\ \boldsymbol{m} & \boldsymbol{q} & m \end{pmatrix}^B \prod_{i=1}^{n} \psi_{j_i m_i}(x_i,y_i)\,. \quad (2.134)$$

$$\begin{pmatrix} \boldsymbol{j} \\ \boldsymbol{m} \end{pmatrix} = \begin{pmatrix} j_1 & j_2 & \cdots & j_n \\ m_1 & m_2 & \cdots & m_n \end{pmatrix}, \quad \begin{pmatrix} \boldsymbol{k} \\ \boldsymbol{q} \end{pmatrix} = \begin{pmatrix} k_1 & k_2 & \cdots & k_{n-2} \\ q_1 & q_2 & \cdots & q_{n-2} \end{pmatrix}. \quad (2.135)$$

$$Z = (z_1 z_2 \ldots z_{n+1}) = \begin{pmatrix} x_1 & x_2 & \ldots & x_{n+1} \\ y_1 & y_2 & \ldots & y_{n+1} \end{pmatrix}. \quad (2.136)$$

Only the first n columns of Z enter into (2.136), but the last column occurs below.

The skew symmetric matrix of a coupling scheme:

The set of triangles $\boldsymbol{T}_B(\boldsymbol{j},\boldsymbol{k},j) = \{\langle a_i,b_i,k_i\rangle | i = 1,2,\ldots,n-1\}$, which is uniquely defined by the bracketing B, can be mapped to a unique skew symmetric matrix of order $n+1$. This mapping is one of the most important results for obtaining generating functions for the coupled wave functions (2.134) and the recoupling coefficients given below. The skew symmetric matrix depends on the bracketing B and the detailed manner in which the $j's$ and $k's$ are distributed among the triangles in $\boldsymbol{T}_B(\boldsymbol{j},\boldsymbol{k},j)$. The rule for constructing the skew symmetric matrix is quite intricate. First, we define a $3\times(n-1)$ matrix T of indeterminates by

$$T = \begin{pmatrix} t_{11} & t_{12} & \cdots & t_{1,n-1} \\ t_{21} & t_{22} & \cdots & t_{2,n-1} \\ t_{31} & t_{32} & \cdots & t_{3,n-1} \end{pmatrix}. \quad (2.137)$$

Second, we associate with each $\langle a_i,b_i,k_i\rangle \in \boldsymbol{T}_B(\boldsymbol{j},\boldsymbol{k},j)$, a triple of indeterminates (u_i,v_i,w_i) as given by

$$\langle a_1, b_1, k_1\rangle \mapsto (u_1, v_1, w_1),\ \text{with } w_1 = t_{21}u_1 + t_{11}v_1,$$
$$\langle a_2, b_2, k_2\rangle \mapsto (u_2, v_2, w_2),\ \text{with } w_2 = t_{22}u_2 + t_{12}v_2,$$
$$\vdots \qquad \vdots$$
$$\langle a_{n-1}, b_{n-1}, k_{n-1}\rangle \mapsto (u_{n-1}, v_{n-1}, w_{n-1}),$$
$$\text{with } w_{n-1} = t_{2,n-1}u_{n-2} + t_{1,n-1}v_{n-1}\,. \qquad (2.138)$$

The indeterminates u_i and v_i are identified as a column $z_i = (x_i, y_i)$ of the $2\times(n+1)$ matrix Z defined by (2.136), or as one of the $w's$ occurring higher in the display (2.138). The distribution rule is in one-to-one correspondence with the distribution of $j's$ and $k's$ in the corresponding triangle. Thus, we have

$$u_i = z_r\,,\quad \text{if } a_i = j_r\,;\quad v_i = z_s\,,\quad \text{if } b_i = j_s\,;$$
$$u_i = z_r\,,\quad \text{if } a_i = j_r\,;\quad v_i = w_s\,,\quad \text{if } b_i = k_s\,;$$
$$u_i = w_r\,,\quad \text{if } a_i = k_r\,;\quad v_i = z_s\,,\quad \text{if } b_i = j_s\,;$$
$$u_i = w_r\,,\quad \text{if } a_i = k_r\,;\quad v_i = w_s\,,\quad \text{if } b_i = k_s\,.$$

The explicit identification of all $j's$ and $k's$ is uniquely determined by the bracketing B. Once this identification has been made, the elements a_{ij}, $i<j$ of the skew symmetric matrix A of order $n+1$ are uniquely determined in terms of the elements of T by equating coefficients of $\det(z_i, z_j) = x_i y_j - x_j y_i$ on the two sides of the form

$$\sum_{1\le i<j\le n+1} a_{ij}\det(z_i, z_j) = \sum_{i=1}^{n-1} t_{3i}\det(u_i, v_i) + \det(w_{n-1}, z_{n+1})\,, \qquad (2.139)$$

where (t_{1i}, t_{2i}, t_{3i}) is the i-th column of the $3\times(n-1)$ matrix T of indeterminates. This relation can be inferred from results given by Schwinger. Since the elements of A are determined as monomials in the elements of T, we sometimes denote A by $A(T)$.

It is useful to illustrate the rule for determining A for $n = 2, 3, 4$:

$n=2$: Triangle: $\langle j_1, j_2, k_1\rangle$:
$w_1 = t_{21}z_1 + t_{11}z_2$

$$\begin{aligned} a_{12}\det(z_1,z_2) + a_{13}\det(z_1,z_3) + a_{23}\det(z_2,z_3) &= t_{31}\det(u_1,v_1) + \det(w_1,z_3)\\ &= t_{31}\det(z_1,z_2) + t_{21}\det(z_1,z_3)\\ &\quad + t_{11}\det(z_2,z_3)\,;\end{aligned}$$

$$a_{12} = t_{31}\,,\quad a_{13} = t_{21}\,,\quad a_{23} = t_{11}\,.$$

$n=3$: Ordered triangles: $\langle j_1, j_2, k_1\rangle$, $\langle k_1, j_3, k_2\rangle$:

$$w_1 = t_{21}u_1 + t_{11}v_1\,,\quad u_1 = z_1,\ v_1 = z_2\,;$$
$$w_2 = t_{22}u_2 + t_{12}v_2\,,\quad u_2 = w_1,\ v_2 = z_3\,.$$
$$\sum_{1\le i<j\le 4} a_{ij}\det(z_i, z_j) = t_{31}\det(u_1,v_1) + t_{32}\det(u_2,v_2) + \det(w_2, z_4)\,;$$

$$a_{12} = t_{31},\ a_{13} = t_{21}t_{32},\ a_{14} = t_{21}t_{22}$$
$$a_{23} = t_{11}t_{32},\ a_{24} = t_{11}t_{22}$$
$$a_{34} = t_{12}$$

$n=4$: Ordered triangles: $\langle j_3, j_1, k_1\rangle$, $\langle j_4, j_2, k_2\rangle$, $\langle k_1, k_2, k_3\rangle$:

$$w_1 = t_{21}u_1 + t_{11}v_1\,,\quad u_1 = z_3\,,\quad v_1 = z_1\,,$$
$$w_2 = t_{22}u_2 + t_{12}v_2\,,\quad u_2 = z_4\,,\quad v_2 = z_2\,,$$
$$w_3 = t_{23}u_3 + t_{13}v_3\,,\quad u_3 = w_1\,,\quad v_3 = w_2\,;$$
$$w_3 = t_{11}t_{23}z_1 + t_{12}t_{13}z_2 + t_{21}t_{23}z_3 + t_{22}t_{13}z_4\,,$$
$$\sum_{1\le i<j\le 5} a_{ij}\det(z_i,z_j) = t_{31}\det(u_1,v_1) + t_{32} \times\det(u_2,v_2) + t_{33}\det(u_3,v_3) + \det(w_3,z_5)\,;$$

$$a_{12} = t_{11}t_{12}t_{33},\ a_{13} = -t_{31},\quad a_{14} = t_{11}t_{22}t_{33},$$
$$a_{23} = -t_{12}t_{21}t_{33},\ a_{24} = -t_{32},$$
$$a_{34} = t_{21}t_{22}t_{33},$$

$$a_{15} = t_{11}t_{23}$$
$$a_{25} = t_{12}t_{13}$$
$$a_{35} = t_{21}t_{23}$$
$$a_{45} = t_{22}t_{13}$$

Triangle monomials:

Let $\langle a, b, c\rangle$ be a triangle of quantum numbers (a, b, c), let (x, y, z) be three indeterminates, and let B denote a binary coupling scheme with the set of triangles $\boldsymbol{T}_B(\boldsymbol{j}, \boldsymbol{k}, j)$:

Elementary triangle monomial:

$$\Phi_{\langle a,b,c\rangle}(x, y, z) = \{abc\}^{-1} x^{b+c-a} y^{a+c-b} z^{a+b-c}\,, \qquad (2.140)$$

$$\{abc\} = \left(\frac{(2c+1)(b+c-a)(a+c-b)!(a+b-c)!}{(a+b+c+1)!}\right)^{\frac{1}{2}}\,.$$

Triangle monomial associated with a given coupling scheme B :

$$\Phi^B_{\boldsymbol{j},\boldsymbol{k},j}(T) = \prod_{\langle a_i,b_i,k_i\rangle\in \boldsymbol{T}_B(\boldsymbol{j},\boldsymbol{k},j)} \Phi_{\langle a_i,b_i,k_i\rangle}(t_{1i}, t_{2i}, t_{3i})\,. \qquad (2.141)$$

Using the definitions introduced above, we can now give the generating functions for the coupled wave functions and the recoupling coefficients for each coupling scheme as determined by the bracketing B.
Generating function for coupled wave functions:

$$\begin{aligned} e^{xA(T)y^T} &= e^{\sum_{1\le i<j\le n+1} a_{i,j}\det(z_i,z_j)} \\ &= \sum_{jk} \Phi^B_{j,k,j}(T) \sum_m (-1)^{j-m} \psi_{j,-m}(x_{n+1},y_{n+1}) \\ &\quad \times \Psi^B_{(jk)jm}(x,y)\,, \end{aligned} \tag{2.142}$$

$$x=(x_1,x_2,\dots,x_{n+1})\,,\qquad y=(y_1,y_2,\dots,y_{n+1})\,.$$

Relation to $U(n+1)$ solid harmonics:

$$e^{xA(T)y^T} = \sum_{k=0}^{\infty} \sum_{\alpha,\beta\vdash k} \frac{x^\alpha}{\sqrt{\alpha!}} D^k_{\alpha,\beta}(A(T)) \frac{y^\beta}{\sqrt{\beta!}}\,. \tag{2.143}$$

Relation of $U(n+1)$ solid harmonics to triangle monomials:

$$D^k_{\alpha,\beta}[A(T)] = (-1)^{j-m} \sum_{k} \begin{pmatrix} j & k & j \\ m & q & m \end{pmatrix}^B \Phi^B_{j,k,j}(T) \tag{2.144}$$

$$\begin{aligned} &\alpha_i = j_i + m_i\,,\quad \beta_i = j_i - m_i \quad i = 1,2,\dots,n\,; \\ &\alpha_{n+1} = j-m\,,\quad \beta_{n+1} = j+m\,; \\ &\sum_i \alpha_i = \sum_i \beta_i = j_1 + j_2 + \dots + j_n + j\,. \end{aligned}$$

The relation between the skew symmetric matrix $A(T)$ of order $n+1$ and the elements of the $3\times(n-1)$ matrix T is that described in relations (2.138).
Generating function for all recoupling coefficients:

$$\begin{aligned} &\frac{1}{[\mathrm{Pf}(A(T),A'(T'))]^2} \\ &= \sum_{j,k,k',j} \Phi^B_{j,k,j}(T)\Phi^{B'}_{j',k',j}(T')\langle j,k,j\|j',k',j\rangle\,, \end{aligned} \tag{2.145}$$

where $\langle j,k,j\,\|\,j',k',j\rangle$ denotes the recoupling coefficient that effects the transformation between the coupling schemes corresponding to the bracketing B and the bracketing B', and where the sequence j' is a permutation of j in accordance with the bracketing B'.
We also note that

$$\begin{aligned} \frac{1}{\mathrm{Pf}(A,A')} &= \frac{1}{\sqrt{\det(I-AA')}} \\ &= \left[1+\sum_{k\ge 1}\sum_{\alpha,\beta\vdash k} D^k_{\alpha,\beta}(A)D^k_{\beta,\alpha}(A')\right]^{\frac{1}{2}}\,, \end{aligned} \tag{2.146}$$

for arbitrary skew symmetric matrices of order n.

Relation (2.145) generates all recoupling coefficients, the trivial ones (those differing by signs) and all the complicated ones, that is, those corresponding to $3n-j$ coefficients. It will also be observed that the expansion of the reciprocal of the double Pfaffian effects an infinite sum in which no radicals occur, which in turn implies that the every recoupling coefficient has the form

$$\begin{aligned} &\langle j,k,j\,\|\,j',k',j\rangle \\ &= \prod_{\langle a_i,b_i,k_i\rangle\in T_B(j,k,j)} \{a_i,b_i,k_i\} \\ &\quad\times \prod_{\langle a'_i,b'_i,k'_i\rangle\in T_{B'}(j',k',j)} \{a'_i,b'_i,k'_i\} \\ &\quad\times I(j,k,j\,\|\,j',k',j)\,, \end{aligned}$$

where $I(j,k,j\,\|\,j',k',j)$ is an integer:
Each recoupling coefficient is an integer multiplied by square-root factors that depend on the triangles associated with the coupling scheme.
Such features can be very useful in the development of algorithms for the calculation of $3n-j$ coefficients, including WCG-coefficients [2.17]. Relation (2.145) should be useful for the classification of $3n-j$ coefficients.

2.14 Tables

Excerpts and Fig. 2.1 are reprinted from *Biedenharn* and *Louck* [2.1] with permission of Cambridge University Press. Tables 2.2–2.4 have been adapted from *Edmonds* [2.18] by permission of Princeton University Press. Thanks are given for this cooperation.

Table 2.1 The solid and spherical harmonics $\mathcal{Y}_{lm}$, and the tensor harmonics $\mathcal{T}^k_\mu$ (labeled by $k=l$ and $\mu=m$) for $l=0$, 1, 2, 3, and 4

l	m	$\sqrt{4\pi}\,\mathcal{Y}_{lm}(\boldsymbol{r})$	$\sqrt{4\pi}\,\mathcal{Y}_{lm}(\theta,\varphi)$	$\mathcal{T}^l_m$
1	±1	$\mp\sqrt{\frac{3}{2}}\,(x\pm\mathrm{i}y)$	$\mp\sqrt{\frac{3}{2}}\,\mathrm{e}^{\pm\mathrm{i}\varphi}\sin\theta$	$\mp\sqrt{2}J_\pm$
	0	$\sqrt{3}\,z$	$\sqrt{3}\cos\theta$	$2J_3$
2	±2	$\frac{1}{2}\sqrt{\frac{15}{2}}\,(x\pm\mathrm{i}y)^2$	$\frac{1}{2}\sqrt{\frac{15}{2}}\,\mathrm{e}^{\pm2\mathrm{i}\varphi}\sin^2\theta$	$\sqrt{6}J_\pm^2$
	±1	$\mp\sqrt{\frac{15}{2}}\,(x\pm\mathrm{i}y)z$	$\mp\sqrt{\frac{15}{2}}\,\mathrm{e}^{\pm\mathrm{i}\varphi}\sin\theta\cos\theta$	$\mp\sqrt{6}J_\pm(2J_3\pm1)$
	0	$\frac{1}{2}\sqrt{5}\,(3z^2-r^2)$	$\frac{1}{2}\sqrt{5}\,(3\cos^2\theta-1)$	$2(3J_3^2-J^2)$
3	±3	$\mp\frac{1}{4}\sqrt{35}\,(x\pm\mathrm{i}y)^3$	$\mp\frac{1}{4}\sqrt{35}\,\mathrm{e}^{\pm3\mathrm{i}\varphi}\sin^3\theta$	$\mp2\sqrt{5}J_\pm^3$
	±2	$\frac{1}{2}\sqrt{\frac{105}{2}}\,(x\pm\mathrm{i}y)^2z$	$\frac{1}{2}\sqrt{\frac{105}{2}}\,\mathrm{e}^{\pm2\mathrm{i}\varphi}\sin^2\theta\cos\theta$	$2\sqrt{30}J_\pm^2(J_3\pm1)$
	±1	$\mp\frac{1}{4}21(x\pm\mathrm{i}y)(5z^2-r^2)$	$\mp\frac{1}{4}\sqrt{21}\,\mathrm{e}^{\pm\mathrm{i}\varphi}\sin\theta(5\cos^2\theta-1)$	$\mp2\sqrt{3}J_\pm(5J_3^2-J^2\pm5J_3+2)$
	0	$\frac{1}{2}\sqrt{7}\,(5z^2-3r^2)z$	$\frac{1}{2}\sqrt{7}\,(5\cos^2\theta-3)\cos\theta$	$4(5J_3^2-3J^2+1)J_3$
4	±4	$\frac{3}{16}\sqrt{70}\,(x\pm\mathrm{i}y)^4$	$\frac{3}{16}\sqrt{70}\,\mathrm{e}^{\pm4\mathrm{i}\varphi}\sin^4\theta$	$\sqrt{70}J_\pm^4$
	±3	$\mp\frac{3}{4}\sqrt{35}\,(x\pm\mathrm{i}y)^3z$	$\mp\frac{3}{4}\sqrt{35}\,\mathrm{e}^{\pm3\mathrm{i}\varphi}\sin^3\theta\cos\theta$	$\mp2\sqrt{35}J_\pm^3(2J_3\pm3)$
	±2	$\frac{3}{8}\sqrt{10}\,(x\pm\mathrm{i}y)^2(7z^2-r^2)$	$\frac{3}{8}\sqrt{10}\,\mathrm{e}^{\pm2\mathrm{i}\varphi}\sin^2\theta(7\cos^2\theta-1)$	$2\sqrt{10}J_\pm^2(7J_3^2-J^2\pm14J_3+9)$
	±1	$\mp\frac{3}{4}\sqrt{5}\,(x\pm\mathrm{i}y)(7z^2-3r^2)z$	$\mp\frac{3}{4}\sqrt{5}\,\mathrm{e}^{\pm\mathrm{i}\varphi}\sin\theta(7\cos^2\theta-3)\cos\theta$	$\mp\sqrt{5}J_\pm(28J_3^3-12J^2J_3\pm42J_3^2$ $-6J^2+38J_3\pm12)$
	0	$\frac{15}{8}\left(7z^4-6z^2r^2+\frac{3}{5}r^4\right)$	$\frac{15}{8}\left(7\cos^4\theta-6\cos^2\theta+\frac{3}{5}\right)$	$70J_3^4-60J^2J_3^2+6(J^2)^2+50J_3^2$ $-12J^2$

Table 2.2 The 3–j coefficients for all M's $=0$, or $J_3=0,\frac{1}{2}$

$$\begin{pmatrix} J_1 & J_2 & J_3 \\ 0 & 0 & 0 \end{pmatrix} = (-1)^{\frac{1}{2}J}\left(\frac{(J_1+J_2-J_3)!(J_1+J_3-J_2)!(J_2+J_3-J_1)!}{(J_1+J_2+J_3+1)!}\right)^{\frac{1}{2}} \times\frac{\left(\frac{1}{2}J\right)!}{\left(\frac{1}{2}J-J_1\right)!\left(\frac{1}{2}J-J_2\right)!\left(\frac{1}{2}J-J_3\right)!}\,,\quad J\ \text{even}$$

$$\begin{pmatrix} J_1 & J_2 & J_3 \\ 0 & 0 & 0 \end{pmatrix} = 0\,,\quad J\ \text{odd, where}\ J=J_1+J_2+J_3$$

$$\begin{pmatrix} J & J & 0 \\ M & -M & 0 \end{pmatrix} = (-1)^{J-M}\frac{1}{(2J+1)^{1/2}}$$

$$\begin{pmatrix} J+\frac{1}{2} & J & \frac{1}{2} \\ M & -M-\frac{1}{2} & \frac{1}{2} \end{pmatrix} = (-1)^{J-M-\frac{1}{2}}\left(\frac{J-M+\frac{1}{2}}{(2J+2)(2J+1)}\right)^{1/2}$$

Table 2.3 The 3–j coefficients for $J_3 = 1, \frac{3}{2}, 2$

$$\begin{pmatrix} J+1 & J & 1 \\ M & -M-1 & 1 \end{pmatrix} = (-1)^{J-M-1}\left(\frac{(J-M)(J-M+1)}{(2J+3)(2J+2)(2J+1)}\right)^{\frac{1}{2}}$$

$$\begin{pmatrix} J+1 & J & 1 \\ M & -M & 0 \end{pmatrix} = (-1)^{J-M-1}\left(\frac{2(J+M+1)(J-M+1)}{(2J+3)(2J+2)(2J+1)}\right)^{\frac{1}{2}}$$

$$\begin{pmatrix} J & J & 1 \\ M & -M-1 & 1 \end{pmatrix} = (-1)^{J-M}\left(\frac{2(J-M)(J+M+1)}{(2J+2)(2J+1)(2J)}\right)^{\frac{1}{2}}$$

$$\begin{pmatrix} J & J & 1 \\ M & -M & 0 \end{pmatrix} = (-1)^{J-M}\frac{M}{[(2J+1)(J+1)J]^{\frac{1}{2}}}$$

$$\begin{pmatrix} J+\frac{3}{2} & J & \frac{3}{2} \\ M & -M-\frac{3}{2} & \frac{3}{2} \end{pmatrix} = (-1)^{J-M+\frac{1}{2}}\left(\frac{(J-M-\frac{1}{2})(J-M+\frac{1}{2})(J-M+\frac{3}{2})}{(2J+4)(2J+3)(2J+2)(2J+1)}\right)^{\frac{1}{2}}$$

$$\begin{pmatrix} J+\frac{3}{2} & J & \frac{3}{2} \\ M & -M-\frac{1}{2} & \frac{1}{2} \end{pmatrix} = (-1)^{J-M+\frac{1}{2}}\left(\frac{3(J-M+\frac{1}{2})(J-M+\frac{3}{2})(J+M+\frac{3}{2})}{(2J+4)(2J+3)(2J+2)(2J+1)}\right)^{\frac{1}{2}}$$

$$\begin{pmatrix} J+\frac{1}{2} & J & \frac{3}{2} \\ M & -M-\frac{3}{2} & \frac{3}{2} \end{pmatrix} = (-1)^{J-M-\frac{1}{2}}\left(\frac{3(J-M-\frac{1}{2})(J-M+\frac{1}{2})(J+M+\frac{3}{2})}{(2J+3)(2J+2)(2J+1)2J}\right)^{\frac{1}{2}}$$

$$\begin{pmatrix} J+\frac{1}{2} & J & \frac{3}{2} \\ M & -M-\frac{1}{2} & \frac{1}{2} \end{pmatrix} = (-1)^{J-M-\frac{1}{2}}\left(\frac{J-M+\frac{1}{2}}{(2J+3)(2J+2)(2J+1)2J}\right)^{\frac{1}{2}}\left(J+3M+\frac{3}{2}\right)$$

$$\begin{pmatrix} J+2 & J & 2 \\ M & -M-2 & 2 \end{pmatrix} = (-1)^{J-M}\left(\frac{(J-M-1)(J-M)(J-M+1)(J-M+2)}{(2J+5)(2J+4)(2J+3)(2J+2)(2J+1)}\right)^{\frac{1}{2}}$$

$$\begin{pmatrix} J+2 & J & 2 \\ M & -M-1 & 1 \end{pmatrix} = (-1)^{J-M}\left(\frac{(J+M+2)(J-M+2)(J-M+1)(J-M)}{(2J+5)(2J+4)(2J+3)(2J+2)(2J+1)}\right)^{\frac{1}{2}}$$

$$\begin{pmatrix} J+2 & J & 2 \\ M & -M & 0 \end{pmatrix} = (-1)^{J-M}\left(\frac{6(J+M+2)(J+M+1)(J-M+2)(J-M+1)}{(2J+5)(2J+4)(2J+3)(2J+2)(2J+1)}\right)^{\frac{1}{2}}$$

$$\begin{pmatrix} J+1 & J & 2 \\ M & -M-2 & 2 \end{pmatrix} = 2(-1)^{J-M+1}\left(\frac{(J-M-1)(J-M)(J-M+1)(J+M+2)}{(2J+4)(2J+3)(2J+2)(2J+1)2J}\right)^{\frac{1}{2}}$$

$$\begin{pmatrix} J+1 & J & 2 \\ M & -M-1 & 1 \end{pmatrix} = (-1)^{J-M+1}2(J+2M+2)\left(\frac{(J-M+1)(J-M)}{(2J+4)(2J+3)(2J+2)(2J+1)2J}\right)^{\frac{1}{2}}$$

$$\begin{pmatrix} J+1 & J & 2 \\ M & -M & 0 \end{pmatrix} = (-1)^{J-M+1}2M\left(\frac{6(J+M+1)(J-M+1)}{(2J+4)(2J+3)(2J+2)(2J+1)2J}\right)^{\frac{1}{2}}$$

$$\begin{pmatrix} J & J & 2 \\ M & -M-2 & 2 \end{pmatrix} = (-1)^{J-M}\left(\frac{6(J-M-1)(J-M)(J+M+1)(J+M+2)}{(2J+3)(2J+2)(2J+1)(2J)(2J-1)}\right)^{\frac{1}{2}}$$

$$\begin{pmatrix} J & J & 2 \\ M & -M-1 & 1 \end{pmatrix} = (-1)^{J-M}(1+2M)\left(\frac{6(J+M+1)(J-M)}{(2J+3)(2J+2)(2J+1)(2J)(2J-1)}\right)^{\frac{1}{2}}$$

$$\begin{pmatrix} J & J & 2 \\ M & -M & 0 \end{pmatrix} = (-1)^{J-M}\frac{2\left[3M^2-J(J+1)\right]}{[(2J+3)(2J+2)(2J+1)(2J)(2J-1)]^{\frac{1}{2}}}$$

Table 2.4 The 6–j coefficients for $d = 0, \frac{1}{2}, 1, \frac{3}{2}, 2$, with $s = a+b+c$

$$\begin{Bmatrix} a & b & c \\ 0 & e & f \end{Bmatrix} = (-1)^s\,[(2b+1)(2c+1)]^{\frac{-1}{2}}\,\delta_{bf}\delta_{ce}$$

$$\begin{Bmatrix} a & b & c \\ \frac{1}{2} & c-\frac{1}{2} & b+\frac{1}{2} \end{Bmatrix} = (-1)^s \left(\frac{(s-2b)(s-2c+1)}{(2b+1)(2b+2)2c(2c+1)} \right)^{\frac{1}{2}}$$

$$\begin{Bmatrix} a & b & c \\ \frac{1}{2} & c-\frac{1}{2} & b-\frac{1}{2} \end{Bmatrix} = (-1)^s \left(\frac{(s+1)(s-2a)}{2b(2b+1)2c(2c+1)} \right)^{\frac{1}{2}}$$

$$\begin{Bmatrix} a & b & c \\ 1 & c-1 & b-1 \end{Bmatrix} = (-1)^s \left(\frac{s(s+1)(s-2a-1)(s-2a)}{(2b-1)2b(2b+1)(2c-1)2c(2c+1)} \right)^{\frac{1}{2}}$$

$$\begin{Bmatrix} a & b & c \\ 1 & c-1 & b \end{Bmatrix} = (-1)^s \left(\frac{2(s+1)(s-2a)(s-2b)(s-2c+1)}{2b(2b+1)(2b+2)(2c-1)2c(2c+1)} \right)^{\frac{1}{2}}$$

$$\begin{Bmatrix} a & b & c \\ 1 & c-1 & b+1 \end{Bmatrix} = (-1)^s \left(\frac{(s-2b-1)(s-2b)(s-2c+1)(s-2c+2)}{(2b+1)(2b+2)(2b+3)(2c-1)2c(2c+1)} \right)^{\frac{1}{2}}$$

$$\begin{Bmatrix} a & b & c \\ 1 & c & b \end{Bmatrix} = (-1)^{s+1} \frac{2\,[b(b+1)+c(c+1)-a(a+1)]}{[2b(2b+1)(2b+2)2c(2c+1)(2c+2)]^{\frac{1}{2}}}\,,$$

$$\begin{Bmatrix} a & b & c \\ \frac{3}{2} & c-\frac{3}{2} & b-\frac{3}{2} \end{Bmatrix} = (-1)^s \left(\frac{(s-1)s(s+1)(s-2a-2)(s-2a-1)(s-2a)}{(2b-2)(2b-1)2b(2b+1)(2c-2)(2c-1)2c(2c+1)} \right)^{\frac{1}{2}}$$

$$\begin{Bmatrix} a & b & c \\ \frac{3}{2} & c-\frac{3}{2} & b-\frac{1}{2} \end{Bmatrix} = (-1)^s \left(\frac{3s(s+1)(s-2a-1)(s-2a)(s-2b)(s-2b+1)}{(2b-1)2b(2b+1)(2b+2)(2c-2)(2c-1)2c(2c+1)} \right)^{\frac{1}{2}}$$

$$\begin{Bmatrix} a & b & c \\ \frac{3}{2} & c-\frac{3}{2} & b+\frac{1}{2} \end{Bmatrix} = (-1)^s \left(\frac{3(s+1)(s-2a)(s-2b-1)(s-2b)(s-2c+1)(s-2c+2)}{2b(2b+1)(2b+2)(2b+3)(2c-2)(2c-1)2c(2c+1)} \right)^{\frac{1}{2}}$$

$$\begin{Bmatrix} a & b & c \\ \frac{3}{2} & c-\frac{3}{2} & b+\frac{3}{2} \end{Bmatrix} = (-1)^s \left(\frac{(s-2b-2)(s-2b-1)(s-2b)(s-2c+1)(s-2c+2)(s-2c+3)}{(2b+1)(2b+2)(2b+3)(2b+4)(2c-2)(2c-1)2c(2c+1)} \right)^{\frac{1}{2}}$$

$$\begin{Bmatrix} a & b & c \\ \frac{3}{2} & c-\frac{1}{2} & b-\frac{1}{2} \end{Bmatrix} = (-1)^s \frac{[2(s-2b)(s-2c)-(s+2)(s-2a-1)]\,[(s+1)(s-2a)]^{\frac{1}{2}}}{[(2b-1)2b(2b+1)(2b+2)(2c-1)2c(2c+1)(2c+2)]^{\frac{1}{2}}}$$

$$\begin{Bmatrix} a & b & c \\ \frac{3}{2} & c-\frac{1}{2} & b+\frac{1}{2} \end{Bmatrix} = (-1)^s \frac{[(s-2b-1)(s-2c)-2(s+2)(s-2a)]\,[(s-2b)(s-2c+1)]^{\frac{1}{2}}}{[2b(2b+1)(2b+2)(2b+3)2c(2c+1)(2c+2)(2c+3)]^{\frac{1}{2}}}\,,$$

$$\begin{Bmatrix} a & b & c \\ 2 & c-2 & b-2 \end{Bmatrix} = (-1)^s \left(\frac{(s-2)(s-1)s(s+1)}{(2b-3)(2b-2)(2b-1)2b(2b+1)} \right)^{\frac{1}{2}} \times \left(\frac{(s-2a-3)(s-2a-2)(s-2a-1)(s-2a)}{(2c-3)(2c-2)(2c-1)2c(2c+1)} \right)^{\frac{1}{2}}$$

$$\begin{Bmatrix} a & b & c \\ 2 & c-2 & b-1 \end{Bmatrix} = (-1)^s\,2 \left(\frac{(s-1)s(s+1)}{(2b-2)(2b-1)2b(2b+1)(2b+2)} \right)^{\frac{1}{2}} \times \left(\frac{(s-2a-2)(s-2a-1)(s-2a)(s-2b)(s-2c+1)}{(2c-3)(2c-2)(2c-1)2c(2c+1)} \right)^{\frac{1}{2}}$$

Table 2.4 The 6–j coefficients for $d = 0, \frac{1}{2}, 1, \frac{3}{2}, 2$, with $s = a+b+c$, cont.

$$\begin{Bmatrix} a & b & c \\ 2 & c-2 & b \end{Bmatrix} = (-1)^s \left(\frac{6s(s+1)(s-2a-1)(s-2a)}{(2b-1)2b(2b+1)(2b+2)(2b+3)}\right)^{\frac{1}{2}} \times \left(\frac{(s-2b)(s-2c+1)(s-2c+2)}{(2c-3)(2c-2)(2c-1)2c(2c+1)}\right)^{\frac{1}{2}}$$

$$\begin{Bmatrix} a & b & c \\ 2 & c-2 & b+1 \end{Bmatrix} = (-1)^s 2 \left(\frac{(s+1)(s-2a)(s-2b-2)(s-2b-1)(s-2b)}{2b(2b+1)(2b+2)(2b+3)(2b+4)}\right)^{\frac{1}{2}} \times \left(\frac{(s-2c+1)(s-2c+2)(s-2c+3)}{(2c-3)(2c-2)(2c-1)2c(2c+1)}\right)^{\frac{1}{2}}$$

$$\begin{Bmatrix} a & b & c \\ 2 & c-2 & b+2 \end{Bmatrix} = (-1)^s \left(\frac{(s-2b-3)(s-2b-2)(s-2b-1)(s-2b)}{(2b+1)(2b+2)(2b+3)(2b+4)(2b+5)}\right)^{1/2} \times \left(\frac{(s-2c+1)(s-2c+2)(s-2c+3)(s-2c+4)}{(2c-3)(2c-2)(2c-1)2c(2c+1)}\right)^{\frac{1}{2}}$$

$$\begin{Bmatrix} a & b & c \\ 2 & c-1 & b-1 \end{Bmatrix} = (-1)^s \frac{4\left[(a+b)(a-b+1)-(c-1)(c-b+1)\right]}{\left[(2b-2)(2b-1)2b(2b+1)(2b+2)\right]^{\frac{1}{2}}} \times \left(\frac{s(s+1)(s-2a-1)(s-2a)}{(2c-2)(2c-1)2c(2c+1)(2c+2)}\right)^{\frac{1}{2}}$$

$$\begin{Bmatrix} a & b & c \\ 2 & c-1 & b \end{Bmatrix} = (-1)^s 2 \frac{\left[(a-b+1)(a-b)-c^2+1\right]}{\left[(2b-1)2b(2b+1)(2b+2)(2b+3)\right]^{\frac{1}{2}}} \times \left(\frac{6(s+1)(s-2a)(s-2b)(s-2c+1)}{(2c-2)(2c-1)2c(2c+1)(2c+2)}\right)^{\frac{1}{2}}$$

$$\begin{Bmatrix} a & b & c \\ 2 & c-1 & b+1 \end{Bmatrix} = (-1)^s \frac{4\left[(a+b+2)(a-b-1)-(c-1)(b+c+2)\right]}{\left[2b(2b+1)(2b+2)(2b+3)(2b+4)\right]^{\frac{1}{2}}} \times \left(\frac{(s-2b-1)(s-2b)(s-2c+1)(s-2c+2)}{(2c-2)(2c-1)2c(2c+1)(2c+2)}\right)^{\frac{1}{2}},$$

$$\begin{Bmatrix} a & b & c \\ 2 & c & b \end{Bmatrix} = (-1)^s \frac{2\left[3X(X-1)-4b(b+1)c(c+1)\right]}{\left[(2b-1)2b(2b+1)(2b+2)(2b+3)\right]^{\frac{1}{2}}} \times \left(\frac{1}{(2c-1)2c(2c+1)(2c+2)(2c+3)}\right)^{\frac{1}{2}},$$

where $X = b(b+1)+c(c+1)-a(a+1)$

References

2.1 L. C. Biedenharn, J. D. Louck: *Encyclopedia of Mathematics and Its Applications*, Vol. 8 & 9, ed. by G.-C. Rota (Addison-Wesley, Reading 1981) presently by (Cambridge Univ. Press, Cambridge)

2.2 J. D. Louck: J. Math. and Math. Sci. **22**, 745 (1999)

2.3 J. Schwinger: On Angular Momentum. U. S. Atomic Energy Commission Report NYO-3071, 1952 (unpublished). In: *Quantum Theory of Angular Momentum*, ed. by L. C. Biedenharn, H. van Dam (Academic, New York 1965) pp. 229–279

2.4 B. R. Judd, G. M. S. Lister: J. Phys. A **20**, 3159 (1987)

2.5 B. R. Judd: *Symmetries in Science*, ed. by B. Gruber, R. S. Millman (Plenum, New York 1980) pp. 151–160

2.6 S. Roman, G.-C. Rota: Adv. in Math. **27**, 95 (1978)

2.7 A. P. Jucys, I. B. Levinson, V. V. Vanagas: *The Theory of Angular Momentum* (Israel Program for Scientific Translation, Jerusalem 1962) (Mathematicheskii apparat teorii momenta kolichestva dvizheniya) Translated from the Russian by A. Sen, A. R. Sen (1962)

2.8 A. P. Jucys, A. A. Bandzaitis: *Angular Momentum Theory in Quantum Physics* (Moksias, Vilnius 1977)
2.9 R. P. Stanley: *Enumerative Combinatorics*, Vol. 1 (Cambridge Univ. Press, Cambridge 1997)
2.10 A. Clebsch: *Theorie der binären algebraischen Formen* (Teubner, Leipzig 1872)
2.11 J. D. Louck, W. Y. C. Chen, H. W. Galbraith: *Symmetry, Structural Properties of Condensed Matter*, ed. by T. Lulek, B. Lulek, A. Wal (World Scientific, Singapore 1999) pp. 112–137
2.12 W. Y. C. Chen, J. D. Louck: Adv. Math. **140**, 207 (1998)
2.13 P. A. MacMahon: *Combinatory Analysis* (Cambridge Univ. Press, Cambridge 1960) (Chelsia Publishing Co., New York, 1960) (Originally published in two volumes by Cambridge Univ. Press, Cambridge, 1915, 1916)
2.14 J. D. Louck: Adv. Appl. Math. **17**, 143 (1996)
2.15 T. Molien: Über die Invarianten der linearen Substitutionsgruppen, Sitzungsber. Konig. Preuss. Akad. Wiss. **52**, 1152 (1897)
2.16 L. Michel, B. I. Zhilinskii: Physics Reports **341**, 11 (2001)
2.17 L. Wei: Comput. Phys. Commun. **120**, 222 (1999)
2.18 A. R. Edmonds: *Angular Momentum in Quantum Mechanics* (Princeton Univ. Press, Princeton 1957)
2.19 E. Cartan: *Thesis* (Paris, Nony 1894) [*Ouevres Complète*, Part 1, pp. 137–287 (Gauthier-Villars, Paris 1952)]
2.20 H. Weyl: *Gruppentheorie und Quantenmechanik* (Hirzel, Leipzig 1928) Translated by H. P. Robertson as *The Theory of Groups and Quantum Mechanics* (Methuen, London 1931)
2.21 P. A. M. Dirac: *The Principles of Quantum Mechanics*, 4th edn. (Oxford Univ. Press, London 1958)
2.22 M. Born, P. Jordan: *Elementare Quantenmechanik* (Springer, Berlin, Heidelberg 1930)
2.23 H. B. G. Casimir: *Thesis, University of Leyden* (Wolters, Groningen 1931) [*Koninkl. Ned. Akad. Wetenschap, Proc.* **34**, 844 (1931)]
2.24 B. L. van der Waerden: *Die gruppentheoretische Methode in der Quantenmechanik* (Springer, Berlin, Heidelberg 1932)
2.25 W. Pauli: *Handbuch der Physik*, Vol. 24, ed. by H. Geiger, K. Scheel (Springer, Berlin, Heidelberg 1933) Chap. 1, pp. 83–272. Later published in *Encyclopedia of Physics*, Vol. 5, Part 1, ed. by S. Flügge (Springer, Berlin, Heidelberg 1958), pp. 45, 46
2.26 H. Weyl: The Structure and Representations of Continuous Groups, Lectures at the Institute for Advanced Study. Princeton, 1934-1935 (unpublished). Notes by R. Brauer
2.27 E. U. Condon, G. H. Shortley: *The Theory of Atomic Spectra* (Cambridge Univ. Press, London 1935)
2.28 H. A. Kramers: *Quantum Mechanics* (North-Holland, Amsterdam 1957) Translation by D. ter Haar of Kramer's monograph published in the *Hand- und Jahrbuch der chemischen Physik* (1937)
2.29 G. Szegö: *Orthogonal Polynomials* (Edwards, Ann Arbor 1948)
2.30 I. M. Gel'fand, Z. Ya. Shapiro: Am. Math. Soc. Transl. **2**, 207 (1956)
2.31 A. Erdelyi, W. Magnus, F. Oberhettinger, G. F. Tricomi: *Higher Transcendental Functions*, Vol. 1 (McGraw-Hill, New York 1953)
2.32 E. P. Wigner: Application of Group Theory to the Special Functions of Mathematical Physics. Lecture notes, 1955 (unpublished)
2.33 H. C. Brinkman: *Applications of Spinor Invariants in Atomic Physics* (North-Holland, Amsterdam 1956)
2.34 M. E. Rose: *Elementary Theory of Angular Momentum* (Wiley, New York 1957)
2.35 U. Fano, G. Racah: *Irreducible Tensorial Sets* (Academic, New York 1959)
2.36 E. P. Wigner: *Group Theory and Its Application to the Quantum Mechanics of Atomic Spectra* (Academic, New York 1959) Translation from the 1931 German edition by J. J. Griffin
2.37 J. C. Slater: *Quantum Theory of Atomic Structure*, Vol. 2 (McGraw-Hill, New York 1960)
2.38 V. Heine: *Group Theory and Quantum Mechanics; An Introduction to Its Present Usage* (Pergamon, New York 1960)
2.39 W. T. Sharp: *Thesis*, Princeton University (1960) (issued as Report AECL-1098, Atomic Energy of Canada, Chalk River, Ontario (1960))
2.40 D. M. Brink, G. R. Satchler: *Angular Momentum* (Oxford Univ. Press, London 1962)
2.41 M. Hamermesh: *Group Theory and Its Applications to Physical Problems* (Addison-Wesley, Reading 1962)
2.42 G. W. Mackey: *The Mathematical Foundations of Quantum Mechanics* (Benjamin, New York 1963)
2.43 A. de-Shalit, I. Talmi: *Nuclear Shell Theory (Pure and Applied Physics Series)*, Vol. 14 (Academic, New York 1963)
2.44 R. P. Feynman: *Feynman Lectures on Physics* (Addison-Wesley, Reading 1963) Chap. 34
2.45 R. Judd: *Operator Techniques in Atomic Spectroscopy* (McGraw-Hill, New York 1963)
2.46 A. S. Davydov: *Quantum Mechanics* (Pergamon, London, Addison-Wesley, Reading 1965) Translation from the Russian of *Kvantovaya Mekhanika* (Moscow, 1963), with revisions and additions by D. ter Haar
2.47 I. M. Gel'fand, R. A. Minlos, Z. Ya. Shapiro: *Representations of the Rotation and Lorentz Groups and Their Applications* (Macmillan, New York 1963) Translated from the Russian by G. Cummins and T. Boddington
2.48 R. Hagedorn: *Selected Topics on Scattering Theory: Part IV, Angular Momentum*, Lectures given at the Max-Planck-Institut für Physik, Munich (1963)
2.49 M. A. Naimark: *Linear Representations of the Lorentz Group* (Pergamon, New York 1964)
2.50 L. C. Biedenharn, H. van Dam: *Quantum Theory of Angular Momentum* (Academic, New York 1965)

2.51 B. L. van der Waerden: *Sources of Quantum Mechanics* (North-Holland, Amsterdam 1967)
2.52 B. R. Judd: *Second Quantization and Atomic Spectroscopy* (Johns Hopkins Press, Baltimore 1967)
2.53 N. Vilenkin: *Special Functions and the Theory of Group Representations*, Vol. 22, Translated from the Russian Am. Math. Soc. Transl. (Amer. Math. Soc., Providence, 1968)
2.54 J. D. Talman: *Special Functions: A Group Theoretic Approach* (Benjamin, New York 1968) Based on E. P. Wigner's lectures (see [2.3])
2.55 B. G. Wybourne: *Symmetry Principles and Atomic Spectroscopy* (Wiley-Interscience, New York 1970)
2.56 E. A. El Baz, B. Castel: *Graphical Methods of Spin Algebras in Atomic, Nuclear, and Particle Physics* (Dekker, New York 1972)
2.57 R. Gilmore: *Lie Groups, Lie Algebras, and Some of Their Applications* (Wiley, New York 1974)
2.58 D. A. Varshalovich, A. N. Moskalev, V. K. Khersonskiĭ: *Quantum Theory of Angular Momentum* (Nauka, Leningrad 1975) (in Russian)
2.59 R. D. Cowan: *The Theory of Atomic Structure and Spectra* (Univ. Calif. Press, Berkeley 1981)
2.60 R. N. Zare: *Angular Momentum* (Wiley-Interscience, New York 1988)
2.61 G. E. Andrews, R. A. Askey, R. Roy: Special Functions. In: *Encyclopidia of Mathematics and Its Applications*, Vol. 71, ed. by G.-C. Rota (Cambridge Univ. Press, Cambridge 1999)
2.62 P. Gordan: *Über das Formensystem binärer Formen* (Teubner, Leipzig 1875)
2.63 W. Heisenberg: Z. Phys. **33**, 879 (1925)
2.64 M. Born, P. Jordan: Z. Phys. **34**, 858 (1925)
2.65 P. A. M. Dirac: Proc. Soc. A **109**, 642 (1925)
2.66 M. Born, W. Heisenberg, P. Jordan: Z. Phys. **35**, 557 (1926)
2.67 W. Pauli: Z. Phys. **36**, 336 (1926)
2.68 E. P. Wigner: Z. Phys. **43**, 624 (1927)
2.69 E. P. Wigner: Z. Phys. **45**, 601 (1927)
2.70 C. Eckart: Rev. Mod. Phys. **2**, 305 (1930)
2.71 E. P. Wigner: Göttinger Nachr., Math.-Phys. **546**, (1932)
2.72 J. H. Van Vleck: Phys. Rev. **47**, 487 (1935)
2.73 E. P. Wigner: On the Matrices which Reduce the Kronecker Products of Representations of S. R. Groups, 1940 (unpublished). In: *Quantum Theory of Angular Momentum*, ed. by L. C. Biedenharn, H. van Dam (Academic, New York 1965) pp. 87–133
2.74 G. Racah: Phys. Rev. **62**, 438 (1942)
2.75 G. Racah: Phys. Rev. **63**, 367 (1943)
2.76 I. M. Gel'fand, M. L. Tseitlin: Dokl. Akad. Nauk SSSR **71**, 825 (1950)
2.77 J. H. Van Vleck: Rev. Mod. Phys. **23**, 213 (1951)
2.78 H. A. Jahn: Proc. R. Soc. A **205**, 192 (1951)
2.79 L. C. Biedenharn, J. M. Blatt, M. E. Rose: Rev. Mod. Phys. **24**, 249 (1952)
2.80 L. C. Biedenharn: *Notes on Multipole Fields*, Lecture notes at Yale University, New Haven 1952 (unpublished)
2.81 J. P. Elliott: Proc. R. Soc. A **218**, 345 (1953)
2.82 L. C. Biedenharn: J. Math. Phys. **31**, 287 (1953)
2.83 H. A. Jahn, J. Hope: Phys. Rev. **93**, 318 (1954)
2.84 T. Regge: Nuovo Cimento **10**, 544 (1958)
2.85 T. Regge: Nuovo Cimento **11**, 116 (1959)
2.86 V. Bargmann: Commun. Pure Appl. Math. **14**, 187 (1961)
2.87 A. Giovannini, D. A. Smith: *Spectroscopic and Group Theoretic Methods in Physics (Racah Memorial Volume)*, ed. by F. Block, S. G. Cohen, A. de-Shalit, S. Sambursky, I. Talmi (Wiley–Interscience, New York 1968) pp. 89–97
2.88 V. Bargmann: Rev. Mod. Phys. **34**, 829 (1962)
2.89 L. Michel: *Lecture Notes in Physics: Group Representations in Mathematics and Physics, Battelle Recontres*, ed. by V. Bargmann (Springer, Berlin, Heidelberg 1970) pp. 36–143
2.90 A. C. T. Wu: J. Math. Phys. **13**, 84 (1972)
2.91 Ya. A. Smorodinskiĭ, L. A. Shelepin: Sov. Phys. Usp. **15**, 1 (1972)
2.92 Ya. A. Smorodinskiĭ, L. A. Shelepin: Usp. Fiz. Nauk **106**, 3 (1972)
2.93 L. A. Shelepin: Invariant algebraic methods and symmetric analysis of cooperative phenomena. In: *Group-Theoretical Methods in Physics*, ed. by D. V. Skobel'tsyn (Fourth Internat. Collog., Nijmegen 1975) pp. 1–109 A special research report translated from the Russian by Consultants Bureau, New York, London.
2.94 Ya. A. Smorodinskiĭ: Sov. Phys. JETP **48**, 403 (1978)
2.95 I. M. Gel'fand, M. I. Graev: Dokl. Math. **33**, 336 (2000) Tranl. from Dokl. Akad. Nauk **372**, 151 (2000)

3. Group Theory for Atomic Shells

The basic elements of the theory of Lie groups and their irreducible representations (IRs) are described. The IRs are used to label the states of an atomic shell and also the components of operators of physical interest. Applications of the generalized Wigner-Eckart theorem lead to relations between matrix elements appearing in different electronic configurations. This is particularly useful in the f shell, where transformations among the seven orbital states of an f electron can be described by the unitary group U(7) and its sequential subgroups SO(7), G_2, and SO(3) with respective IRs [λ], *W*, *U*, and *L*. Extensions to groups that involve electron spin *S* (like Sp(14)) are described, as are groups that do not conserve electron number. The most useful of the latter is the quasispin group whose generators ***Q*** connect states of identical *W*, *U*, *L* and seniority *v* in the f shell. The symmetries of products of objects (states or operators) that themselves possess symmetries are described by the technique of plethysms.

3.1 **Generators** ... 75
- 3.1.1 Group Elements ... 75
- 3.1.2 Conditions on the Structure Constants ... 76
- 3.1.3 Cartan–Weyl Form ... 76
- 3.1.4 Atomic Operators as Generators ... 76

3.2 **Classification of Lie Algebras** ... 76
- 3.2.1 Introduction ... 76
- 3.2.2 The Semisimple Lie Algebras ... 76

3.3 **Irreducible Representations** ... 77
- 3.3.1 Labels ... 77
- 3.3.2 Dimensions ... 77
- 3.3.3 Casimir's Operator ... 77

3.4 **Branching Rules** ... 78
- 3.4.1 Introduction ... 78
- 3.4.2 U(n) ⊃ SU(n) ... 78
- 3.4.3 Canonical Reductions ... 79
- 3.4.4 Other Reductions ... 79

3.5 **Kronecker Products** ... 79
- 3.5.1 Outer Products of Tableaux ... 79
- 3.5.2 Other Outer Products ... 80
- 3.5.3 Plethysms ... 80

3.6 **Atomic States** ... 80
- 3.6.1 Shell Structure ... 80
- 3.6.2 Automorphisms of SO(8) ... 81
- 3.6.3 Hydrogen and Hydrogen-Like Atoms ... 81

3.7 **The Generalized Wigner–Eckart Theorem** ... 82
- 3.7.1 Operators ... 82
- 3.7.2 The Theorem ... 82
- 3.7.3 Calculation of the Isoscalar Factors ... 82
- 3.7.4 Generalizations of Angular Momentum Theory ... 83

3.8 **Checks** ... 83

References ... 84

3.1 Generators

3.1.1 Group Elements

An element S_a of a Lie group $\mathcal{G}$ corresponding to an infinitesimal transformation can be written in the form

$$S_a = 1 + \delta a^\sigma X_\sigma \,, \tag{3.1}$$

where the δa^σ are the infinitesimal parameters and the X_σ are the generators [3.1]. Summation over the repeated Greek index is implied. Transformations corresponding to finite parameters can be found by exponentiation:

$$S_a \to \exp\left(a^1 X_1\right) \exp\left(a^2 X_2\right) \cdots \exp\left(a^r X_r\right) \,. \tag{3.2}$$

The generators necessarily form a Lie algebra, that is, they close under commutation:

$$\left[X_\rho, X_\sigma\right] = c^\tau_{\rho\sigma} X_\tau \,. \tag{3.3}$$

In terms of the structure constants $c^{\tau}_{\rho\sigma}$, the metric tensor is defined as

$$g_{\rho\sigma} = c^{\mu}_{\rho\lambda} c^{\lambda}_{\sigma\mu} \,. \tag{3.4}$$

3.1.2 Conditions on the Structure Constants

For an Abelian group, all the generators commute with one another:

$$c^{\tau}_{\rho\sigma} = 0 \,. \tag{3.5}$$

The operators X_σ, ($\sigma = 1, 2, \ldots, p < r$) form the generators of a subgroup if [3.2]

$$c^{\tau}_{\rho\sigma} = 0 \,, \quad (\rho, \sigma \leq p,\ \tau > p) \,. \tag{3.6}$$

The subgroup is invariant if the stronger condition

$$c^{\tau}_{\rho\sigma} = 0 \,, \quad (\rho \leq p,\ \tau > p) \tag{3.7}$$

is satisfied. A group is *simple* if it contains no invariant subgroup (besides the unit element). A group is *semisimple* if it contains no Abelian invariant subgroup (besides the unit element). A necessary and sufficient condition that a group be semisimple is that

$$\det | g_{\rho\tau} | \neq 0 \,. \tag{3.8}$$

All simple groups are semisimple. For semisimple groups, the inverse tensor $g^{\mu\nu}$ can be formed, thus permitting suffixes to be raised. The quadratic operator

$$C = g^{\rho\sigma} X_\rho X_\sigma \tag{3.9}$$

commutes with all generators of the group and is called Casimir's operator [3.1]. If the generators of a group $\mathcal{G}$ can be broken up into two sets such that each member of one set commutes with all the members of the other, that is, if

$$c^{\tau}_{\rho\sigma} = 0 \,, \quad (\rho \leq p, \sigma > p) \,, \tag{3.10}$$

then the two sets form the generators of two invariant subgroups, $\mathcal{H}$ and $\mathcal{K}$. The group $\mathcal{G}$ is the direct product of $\mathcal{H}$ and $\mathcal{K}$ and is written as $\mathcal{H} \times \mathcal{K}$.

3.1.3 Cartan–Weyl Form

By taking suitable linear combinations H_i and E_α of the generators X_σ, the basic commutation relations (3.3) can be thrown into the so-called Cartan–Weyl form [3.1]

$$[H_i, H_j] = 0 \,, \tag{3.11}$$

$$[H_i, E_\alpha] = \alpha_i E_\alpha \,, \tag{3.12}$$

$$[E_\alpha, E_{-\alpha}] = \alpha^i H_i \,, \tag{3.13}$$

$$[E_\alpha, E_\beta] = N_{\alpha\beta} E_{\alpha+\beta} \,. \tag{3.14}$$

The Roman symbols $i, j, \ldots$ run over an l-dimensional space (the *weight* space of *rank* l) in which the numbers α_i can be visualized as the components of the vectors (called *roots*). The E_α are shift operators, the displacements being specified by the components of $\boldsymbol{\alpha}$. The operator $E_{\alpha+\beta}$ in (3.17) is to be interpreted as 0 if $\boldsymbol{\alpha} + \boldsymbol{\beta}$ is not a root. The coefficient $N_{\alpha\beta}$ depends on the choice of normalization.

3.1.4 Atomic Operators as Generators

The pairs $a^{\dagger}_{\xi} a_\eta$ of creation and annihilation operators for either bosons or fermions, as defined in Sect. 6.1.1 close under commutation and form a Lie algebra. The coupled forms $\boldsymbol{W}^{(\kappa k)}$, defined in Sect. 6.2.2, are often used to play the role of the generators for electrons in an atomic shell.

3.2 Classification of Lie Algebras

3.2.1 Introduction

The semisimple Lie algebras have been classified by *Cartan* [3.3]. They consist of four main classes A_l, B_l, C_l, D_l, and five exceptions G_2, F_4, E_6, E_7, E_8. Each algebra is characterized by an array of roots in the l-dimensional weight space; they are conveniently specified by a set of mutually orthogonal unit vectors $\boldsymbol{e}_i$. The total number of generators (those of type E_α plus the l generators of type H_i) gives the *order* of the algebra.

3.2.2 The Semisimple Lie Algebras

A_l. The roots are conveniently represented by the vectors $\boldsymbol{e}_i - \boldsymbol{e}_j$ ($i, j = 1, 2, \ldots, l+1$). They are all perpendicular to $\Sigma \boldsymbol{e}_k$ and do not extend beyond the l-dimensional weight space. The order of the algebra is $l(l+2)$. The group for which this algebra can serve as a basis is the special unitary group $\mathrm{SU}(l+1)$.

B_l. The roots are $\pm\boldsymbol{e}_i$ and $\pm\boldsymbol{e}_i \pm \boldsymbol{e}_j$ ($i, j = 1, 2, \ldots, l$; $i \neq j$). The order of the algebra is $l(2l+1)$. A cor-

Table 3.1 Generators of the Lie groups for the atomic l shell. The subscripts i and j run over all $4l+2$ states of a single electron

Group	Generators
$SO(8l+5)^a$	$a_i^\dagger a_j^\dagger, a_i^\dagger a_j, a_i a_j, a_i^\dagger, a_j$
$SO(8l+4)^a$	$a_i^\dagger a_j^\dagger, a_i^\dagger a_j, a_i a_j$
$U(4l+2)^b$	$\boldsymbol{W}^{(\kappa k)}$ $(\kappa = 0, 1;\ k = 0, 1, \ldots, 2l)$
$SU(4l+2)^b$	$\boldsymbol{W}^{(\kappa k)}$ (As above, with $\kappa = k = 0$ excluded)
$Sp(4l+2)^c$	$\boldsymbol{W}^{(\kappa k)}$ (As above, with $\kappa + k$ odd)
$U(2l+1)^d$	$\boldsymbol{W}^{(0k)}$ $(k = 0, 1, \ldots, 2l)$
$SU(2l+1)^d$	$\boldsymbol{W}^{(0k)}$ $(k = 1, 2, \ldots, 2l)$
$SO(2l+1)^d$	$\boldsymbol{W}^{(0k)}$ $(k = 1, 3, 5, \ldots, 2l-1)$
G_2^d	$\boldsymbol{W}^{(01)}, \boldsymbol{W}^{(05)}$ (for $l = 3$)
$SO_L(3)^e$	$\boldsymbol{W}^{(01)}$ (or $\boldsymbol{L}$)
$SO_S(3)^e$	$\boldsymbol{W}^{(10)}$ (or $\boldsymbol{S}$)
$U_A(2l+1) \times U_B(2l+1)^f$	$W_{0q}^{(0k)} + W_{0q}^{(1k)}, W_{0q}^{(0k)} - W_{0q}^{(1k)} (k = 0, 1, \ldots, 2l)$
$SO_\lambda(2l+1) \times SO_\mu(2l+1) \times SO_\nu(2l+1) \times SO_\xi(2l+1)^g$	$(\boldsymbol{\theta}^\dagger \boldsymbol{\theta})^{(k)}$ (k odd, $\boldsymbol{\theta} \equiv \boldsymbol{\lambda}, \boldsymbol{\mu}, \boldsymbol{\nu}, \boldsymbol{\xi}$)
$U_\lambda(2^l) \times U_\mu(2^l) \times U_\nu(2^l) \times U_\xi(2^l)^h$	$\boldsymbol{q}_\theta^\dagger \boldsymbol{q}_\theta$ (all components, $\theta \equiv \lambda, \mu, \nu, \xi$)
[a] [3.4,5] [b] [3.1,6] [c] [3.6,7] [d] [3.6] [e] [3.8] [f] [3.9]	[g] [3.10] and (6.69)–(6.72) [h] [3.11]

responding group is the special orthogonal (or rotation) group in $2l+1$ dimensions, $SO(2l+1)$.

C_l. The roots are $\pm 2\boldsymbol{e}_i$ and $\pm \boldsymbol{e}_i \pm \boldsymbol{e}_j$ $(i,\ j = 1, 2, \ldots, l;\ i \neq j)$. The order of the algebra is $l\,(2l+1)$. A corresponding group is the symplectic group in $2l$ dimensions, $Sp(2l)$. A rotation of the roots yields $C_2 = B_2$.

D_l. The roots are $\pm \boldsymbol{e}_i \pm \boldsymbol{e}_j$ $(i, j = 1, 2, \ldots, l\,; i \neq j)$. The order of the algebra is $l(2l-1)$. A corresponding group is the special orthogonal (or rotation) group $SO(2l)$. A rotation of the roots yields $D_3 = A_3$. Also, $D_2 = A_1 \times A_1$.

E_6, E_7, E_8. The roots are given by *Racah* [3.1]. The respective orders are 78, 133, and 248.

F_4. The roots consist of the roots of B_4 together with the 16 vectors $\frac{1}{2}(\pm \boldsymbol{e}_1 \pm \boldsymbol{e}_2 \pm \boldsymbol{e}_3 \pm \boldsymbol{e}_4)$. The order of the algebra is 52.

G_2. The roots consist of the roots of A_2 together with the six vectors $\pm(2\boldsymbol{e}_i - \boldsymbol{e}_j - \boldsymbol{e}_k)$ $(i \neq j \neq k = 1, 2, 3)$. The order of the algebra is 14.

Examples of Lie groups used in atomic shell theory, together with their generators, are given in Table 3.1.

3.3 Irreducible Representations

3.3.1 Labels

If n atomic states of a collection transform among themselves under an arbitrary action of the generators of a group $\mathcal{G}$, then the states form a *representation* of $\mathcal{G}$. The representation is *irreducible* if n' linear combinations of the states cannot be found that also exhibit that property, where $n' < n$. The commuting generators H_i of $\mathcal{G}$ can be simultaneously diagonalized within the n states: their eigenvalues $(m_1, m_2, \ldots, m_l)$ for an eigenstate ψ specify the *weight* of the eigenstate. The weight above is said to be *higher* than $(m_1', m_2', \ldots, m_l')$ if the first non-vanishing term in the sequence $m_1 - m_1',\ m_2 - m_2', \ldots$ is positive. An irreducible representation (IR) of a semisimple group is uniquely specified (to within an equivalence) by its highest weight [3.1], which can therefore be used as a defining label.

3.3.2 Dimensions

The dimensions of the IRs of various groups are expressed in terms of the highest weights and set out in Table 3.2. General algebraic expressions have been given by *Wybourne* [3.12, pp. 137]. Numerical tabulations have been made by Butler in the appendix to another book by *Wybourne* [3.13], and also by *McKay* and *Patera* [3.14]. The latter defines the IRs by specifying the coordinates of the weights with respect to the simple roots of *Dynkin* [3.15].

3.3.3 Casimir's Operator

The eigenvalues of Casimir's operator C, defined in (3.9), can be expressed in terms of the highest weights of an IR [3.1]. A complete algebraic listing for all the semisimple Lie groups has been given by *Wybourne* [3.12, p. 140]. Sometimes Casimir's operator is given in terms of the spherical tensors $\boldsymbol{W}^{(\kappa k)}$, or of their special cases $\boldsymbol{V}^{(k)}\left(=2^{\frac{1}{2}}\boldsymbol{W}^{(0k)}\right)$ for which the single-electron reduced matrix element satisfies

$$(nl\|v^{(k)}\|nl) = (2k+1)^{\frac{1}{2}} . \tag{3.15}$$

The eigenvalues of several operators of that form are given in Table 3.3.

Table 3.2 Dimensions D of the irreducible representations (IR's) of various Lie groups

Group	IR	D
SO(2)	M	1
SO(3)	D_J	$2J+1$
$\mathrm{SO}(4)=\mathrm{SO}_A(3)\times\mathrm{SO}_B(3)$	$D_J\times D_K$	$(2J+1)(2K+1)$
SO(5)	(w_1w_2)	$(w_1+w_2+2)(w_1-w_2+1)(2w_1+3)(2w_2+1)/6$
SO(6)	$(w_1w_2w_3)$	$(w_1-w_2+1)(w_1-w_3+2)(w_2-w_3+1)$ $\times(w_1+w_2+3)(w_1+w_3+2)(w_2+w_3+1)/12$
SO(7)	$(w_1w_2w_3)$	$(w_1+w_2+4)(w_1+w_3+3)(w_2+w_3+2)$ $\times(w_1-w_2+1)(w_1-w_3+2)(w_2-w_3+1)$ $\times(2w_1+5)(2w_2+3)(2w_3+1)/720$
G_2	(u_1u_2)	$(u_1+u_2+3)(u_1+2)(2u_1+u_2+5)(u_1+2u_2+4)$ $\times(u_1-u_2+1)(u_2+1)/120$
SU(3) or U(3)	$[\lambda_1\lambda_2\lambda_3]$	$(\lambda_1-\lambda_2+1)(\lambda_1-\lambda_3+2)(\lambda_2-\lambda_3+1)/2$
SU(4) or U(4)	$[\lambda_1\lambda_2\lambda_3\lambda_4]$	As for $(w_1w_2w_3)$ of SO(6)[a]
Sp(4)	$\langle\sigma_1\sigma_2\rangle$	As for (w_1w_2) of SO(5)[b]
Sp(6)	$\langle\sigma_1\sigma_2\sigma_3\rangle$	$(\sigma_1-\sigma_2+1)(\sigma_1-\sigma_3+2)(\sigma_1+\sigma_2+5)$ $\times(\sigma_1+\sigma_3+4)(\sigma_2+\sigma_3+3)(\sigma_2-\sigma_3+1)$ $\times(\sigma_1+3)(\sigma_2+2)(\sigma_3+1)/720$

[a] Subject to the conditions $w_1=(\lambda_1+\lambda_2-\lambda_3-\lambda_4)/2,\ w_2=(\lambda_1-\lambda_2+\lambda_3-\lambda_4)/2,\ w_3=(\lambda_1-\lambda_2-\lambda_3+\lambda_4)/2$

[b] Subject to the conditions $w_1=(\sigma_1+\sigma_2)/2,\ w_2=(\sigma_1-\sigma_2)/2$

3.4 Branching Rules

3.4.1 Introduction

If a group $\mathcal{H}$ shares some of its generators with a group $\mathcal{G}$, the first can be considered a subgroup of the second. That is, $\mathcal{G}\supset\mathcal{H}$. Many of the groups in Table 3.1 can be put in extended group–subgroup sequences. The IRs of a subgroup that together span an IR of the group constitute a *branching* rule.

3.4.2 U(n) ⊃ SU(n)

The group U(n) differs from SU(n) in that the former contains among its generators a scalar $\left(\text{such as } W^{(00)}\right)$ that, by itself, forms an invariant subgroup. Thus U(n) is not semisimple. The scalar in question commutes with all the generators of the group and so is of type H_i. Its presence enlarges the dimension, l, of the weight space by 1, an extension that can be accommodated by the unit vectors $\boldsymbol{e}_i$ of A_l given in Sect. 3.2.2. The reduction $\mathrm{U}(n)\supset\mathrm{SU}(n)$ leads to the branching rule

$$[\lambda_1\lambda_2\cdots\lambda_n]\to[\lambda_1-a,\lambda_2-a,\cdots,\lambda_n-a] , \tag{3.16}$$

where, in the IR of SU(n) on the right,

$$a=(\lambda_1+\lambda_2+\cdots+\lambda_n)/n . \tag{3.17}$$

Table 3.3 Eigenvalues of Casimir's operator C for groups used in the atomic l shell

Group	IR	Operator	Eigenvalue
$SU(2l+1)$	$[\lambda]^{a}$	$\sum_{k>0}\left(V^{(k)}\right)^2$	$3N+2Nl-\frac{1}{2}N^2-2S(S+1)-N^2/(2l+1)$
$SO(2l+1)$	W^{b}	$\sum_{k_{odd}}\left(V^{(k)}\right)^2$	$\frac{1}{2}\sum_{i=1}^{l} w_i(w_i+1+2l-2i)$
G_2	(u_1u_2)	$\frac{1}{4}\left[\left(V^{(1)}\right)^2+\left(V^{(5)}\right)^2\right]$	$\left(u_1^2+u_2^2+u_1u_2+5u_1+4u_2\right)/12$

[a] Appropriate for terms of l^N with total spin S [3.7], p. 125
[b] Defined by the l weights $(w_1w_2\cdots w_l)$

To avoid fractional weights, the IRs of SU(n) are frequently replaced by those of U(n) for which the λ_i are integers. The weights $\lambda_1, \lambda_2, \ldots$ can be interpreted as the number of cells in successive rows of a *Young Tableau*. When the n states of a single particle are taken as a basis for the IR [10 ... 0] of U(n), thus corresponding to a tableau comprising a single cell, the tableaux comprising N cells can be interpreted in two ways, namely, (1) as an IR of U(n) for a system of N particles, and (2) as an IR of S_N, the finite group of permutations on N objects. A given tableau corresponds to as many permutations as there are ways of entering the numbers 1, 2, ..., N in the cells such that the numbers increase going from left to right along the rows, and from top to bottom down the columns. A tableau possessing cells numbered in this way is called *standard*; it defines a permutation corresponding to a symmetrization with respect to the numbers in the rows, followed by an antisymmetrization with respect to the numbers in the columns [3.16].

3.4.3 Canonical Reductions

A group–subgroup sequence of the type

$$U(n) \supset U(n-1) \supset U(n-2) \supset \cdots \supset U(1) \quad (3.18)$$

is called *canonical* [3.17]. The branching rules for those IRs $\left[\lambda'_1\lambda'_2\cdots\lambda'_{n-1}\right]$ of U($n-1$) contained in $[\lambda_1\lambda_2\cdots\lambda_n]$ of U(n) have been given by *Weyl* [3.18] in terms of the "betweenness" conditions

$$\lambda_1 \geq \lambda'_1 \geq \lambda_2 \geq \lambda'_2 \cdots \geq \lambda'_{n-1} \geq \lambda_n\,. \quad (3.19)$$

The possibility of using the scheme of (3.18) in the theory of complex atomic spectra has been explored by *Harter* and *Patterson* [3.19–21], and by *Drake* and *Schlesinger* [3.22, 23] (see also Sect. 4.3.1).

3.4.4 Other Reductions

The algebraic formulae for $U(n) \supset SO(n)$ and $U(n) \supset Sp(n)$ have been given by *Littlewood* [3.24] and in a rather more accessible form by *Wybourne* [3.13]. Special cases have been tabulated by *Butler* (in Tables C-1 through C-15 in [3.13]). Another set of tables, in which Dynkin's labeling scheme is used, has been given by *McKay* and *Patera* [3.14]. Descriptions of how to apply the mechanics of the mathematics to the Young tableaux that describe the IRs of U(n) can be found in the articles of *Jahn* [3.25] [with particular reference to SO(5)] and *Flowers* [3.26] [for Sp($2j+1$)]. For the atomic l shell, the reductions $SO(2l+1) \supset SO(3)$ and (for f electrons) $SO(7) \supset G_2$ and $G_2 \supset SO(3)$ are important. The sources cited in the previous Section are useful here. It is important to recognize that the embedding of one group in another can often be performed in inequivalent ways, depending on which generators are discarded in the reduction process. Thus the use of $SO(5) \supset SO(3)$ in the atomic d shell involves a different SO(3) group from that derived from the canonical sequence $SO(5) \supset SO(4) \supset SO(3)$.

3.5 Kronecker Products

3.5.1 Outer Products of Tableaux

Consider the tableau $[\lambda_1\lambda_2\cdots\lambda_n]$, where the total number of cells is N. A preliminary definition is required. If among the first r terms of any permutation of the N factors of the product, $x_1{}^{\lambda_1}x_2{}^{\lambda_2}\cdots x_n{}^{\lambda_n}$, the number of times x_1 occurs is $\geq$ the number of times x_2 occurs $\geq$ the number of times x_3 occurs, etc. for all values of r, this permutation is called a *lattice* permutation. The prescription of *Littlewood* [3.24] for finding the tableaux appearing in the Kronecker product of $[\lambda_1\lambda_2\cdots\lambda_n]$ with $[\mu_1\mu_2\cdots\mu_m]$ is as follows. The acceptable tableaux are those that can be built by adding to the tableau $[\lambda_1\lambda_2\cdots\lambda_n]$, μ_1 cells containing the same symbol α, then μ_2 cells containing the same symbol β, etc., subject to two conditions:

1. After the addition of each set of cells labeled by a common symbol we must have a permissible tableau with no two identical symbols in the same column;
2. If the total set of added symbols is read from right to left in the consecutive rows of the final tableau, we obtain a lattice permutation of $\alpha^{\mu_1}\beta^{\mu_2}\gamma^{\mu_3}\cdots$.

Examples of this procedure have been given ([3.24, p. 96], [3.7, p. 136], [3.13, p. 24]). An extensive tabulation involving tableaux with $N < 8$ has been calculated by *Butler* and given by *Wybourne* [3.13, Table B-1].

3.5.2 Other Outer Products

The rules for constructing the Kronecker products for $U(n)$ follow by interpreting the Young tableaux of the previous section as IRs of $U(n)$. The known branching rules for reductions to subgroups enable the Kronecker products for the subgroups to be found. Many examples for $SO(n)$, $Sp(n)$, and G_2 can be found in the book by *Wybourne* ([3.13, Tables D-1 through D-15, and E-4].

3.5.3 Plethysms

Sometimes a particle can be thought of as being composite [as when the six orbital states s + d of a single electron are taken to span the IR [200] of SU(3)]. When the n' component states of a particle form a basis for an IR $[\lambda']$ of $U(n)$ other than [10 ... 0], the process of finding which IRs of $U(n)$ occur for N-particle states whose permutation symmetries are determined by a given Young tableau $[\lambda]$ with N cells is called a *plethysm* [3.24, p. 289] and written as $[\lambda']\otimes[\lambda]$. The special techniques for doing this have been described by *Wybourne* [3.13]. An elementary method, which is often adequate in many cases, runs as follows:

1. Expand $[\lambda']^N$ by repeated use of Table B-1 from [3.13]. The resulting tableaux $[\lambda'']$ are independent of n.
2. Choose a small value of n, and strike out all tableaux from the set $[\lambda'']$ that possess more than n rows [since they are unacceptable as IRs of $U(n)$].
3. Interpret the remaining tableaux $[\lambda'']$ as IRs of $U(n)$ and find their dimensions from Tables A-2 through A-17 of [3.13]. Check that the sum of the dimensions is $(\dim[\lambda'])^N$.
4. Interpret the various tableaux $[\lambda]$ possessing the same number N of cells as IRs of $U(\dim[\lambda'])$, and find their dimensions from [3.13].
5. Match the dimensions of parts (3) and (4), remembering that each tableau $[\lambda]$ occurs as often as the number of its standard forms. This determines the possible ways of assigning the IRs $[\lambda'']$ of $U(n)$ to each $[\lambda]$.
6. Proceed to higher n to remove ambiguities and to include the tableaux struck out in step 2.

This procedure can be extended to calculate the plethysms for other groups. Examples of the type $W\otimes[\lambda]$ and $U\otimes[\lambda]$, where W and U are IRs of SO(7) and G_2, have been given for $[\lambda]\equiv[2]$ and [11] corresponding to the separation of W^2 and U^2 into their symmetric and antisymmetric parts [3.27]. The technique of plethysm is also useful for mixed atomic configurations (Sect. 3.6.1).

3.6 Atomic States

3.6.1 Shell Structure

The 2^{4l+2} states of the l shell span the elementary spinor IR $(\frac{1}{2}\frac{1}{2}\cdots\frac{1}{2})$ of $SO(8l+5)$, which decomposes into the two IRs $(\frac{1}{2}\frac{1}{2}\cdots\pm\frac{1}{2})$ of $SO(8l+4)$, corresponding to an even and an odd number N of electrons [3.4]. The states of l^N span the IR $[1^N0^{4l+2-N}]$ of $U(4l+2)$, corresponding to the antisymmetric Young tableau comprising a single column of N cells. The separation of spin and orbit through the subgroup $U(2)\times U(2l+1)$ yields the tableau products $[\lambda]\times[\tilde{\lambda}]$, where $[\tilde{\lambda}]$ is the tableau obtained by reflecting $[\lambda]$ in a diagonal line [3.1]. The IRs of the subgroup $U(2)\times SO(2l+1)$ are denoted by S and W [3.6]. An alternative way of reaching this subgroup from $U(4l+2)$ involves the intermediary $Sp(4l+2)$, whose IRs $\langle 1^v0^{2l+1-v}\rangle$ possess as a basis the states with seniority v [3.7]. A subgroup of $SO(2l+1)$ is the SO(3) group whose IRs specify L, the total orbital angular momentum.

Alternatives to this classic sequence are provided by the last three groups listed in Table 3.1, together with their respective subgroups. For $\mathrm{U}_A(2l+1)\times\mathrm{U}_B(2l+1)$, the shell is factored by considering spin-up and spin-down electrons as distinct (and statistically independent) particles [3.28]. A further factorization by means of the quasiparticles, θ, leads to four independent spaces. The 2^l states in each space span the elementary spinor $(\frac{1}{2}^l)$ of $\mathrm{SO}_\theta(2l+1)$, which can be regarded as a fictitious particle (or quark), q_θ [3.29].

The standard classification of the states of the d-shell is given in Table 3.4. The component M_Q of the quasispin Q (defined in (6.33–6.35)) is listed, as well as the seniority, $v=2l+1-2Q$, the IRs W of SO(5), and the value of L (as a spectroscopic symbol). Only states in the first half of the shell appear; the classification for the second half is the same as the first except that the signs of M_Q are reversed. A general rule for arbitrary l is exemplified by noting that every W [the IR of SO$(2l+1)$] occurs with two spins (S_1 and S_2) and two quasispins (Q_1 and Q_2) such that $S_1=Q_2$ and $S_2=Q_1$. No duplicated spectroscopic terms appear in Table 3.4. The generators of SO(5) do not commute with the inter-electronic Coulomb interaction; thus the separations effected by SO(5) merely define (to within a phase) a basis. The analog of Table 3.4 has been given

Table 3.4 The states of the d shell

d^N	M_Q	$^{2S+1}[\lambda]$	v	W	L
d^0	$-\frac{5}{2}$	$^1[0]$	0	(00)	S
d^1	-2	$^2[1]$	1	(10)	D
d^2	$-\frac{3}{2}$	$^1[2]$	0	(00)	S
			2	(20)	DG
		$^3[11]$	2	(11)	PF
d^3	-1	$^2[21]$	1	(10)	D
			3	(21)	PDFGH
		$^4[111]$	3	(11)	PF
d^4	$-\frac{1}{2}$	$^1[22]$	0	(00)	S
			2	(20)	DG
			4	(22)	SDFGI
		$^3[211]$	2	(11)	PF
			4	(21)	PDFGH
		$^5[1111]$	4	(10)	D
d^5	0	$^2[221]$	1	(10)	D
			3	(21)	PDFGH
			5	(22)	SDFGI
		$^4[2111]$	3	(11)	PF
			5	(20)	DG
		$^6[11111]$	5	(00)	S

by *Wybourne* [3.30] for the f shell. As *Racah* [3.6] showed, the group G_2 can be used to help distinguish repeated terms, but a few duplications remain. They are distinguished by *Nielson* and *Koster* [3.31] in their tables of spectroscopic coefficients by the letters A and B. The scope for applications of group theory becomes enlarged when the states of a single electron embrace more than one l value. Extensions of the standard model have been made by *Feneuille* [3.32] with particular reference to the configurations $(\mathrm{d}+\mathrm{s})^N$, for which quasiparticles have also been considered [3.33]. The group SU(3) has been used for $(\mathrm{d}+\mathrm{s})^N\mathrm{p}^M$ [3.34]. The mixed configurations $(\mathrm{s}+\mathrm{f})^4$ have found a use in the quark model of the atomic f shell [3.29]. A brief description of this model has been given by *Fano* and *Rao* [3.35].

3.6.2 Automorphisms of SO(8)

The quark structure $s+f$ derives from the SO(3) structure of the elementary spinor $(\frac{1}{2}\frac{1}{2}\frac{1}{2})$ of SO(7). Its eight components span the IR (1000) of SO(8), a group that admits automorphisms [3.36]. This property is exhibited by the existence of the three distinct subgroups SO(7) (Racah's group), SO(7)$'$, and SO(7)$''$, all of which possess the same G_2 and SO(3) as subgoups. A reversal of the relative phase of the s and f quarks takes SO(7) into SO(7)$''$ and vice versa [3.37]. The generators of SO(7)$'$ are the sums of the corresponding generators of SO(7) and SO(7)$''$. The phase reversal between the s and f quarks, when interpreted in terms of electronic states, explains the unexpected simplifications found by *Racah* [3.6] in his equation (87) [3.38], which goes beyond what the Wigner–Eckart theorem for G_2 would predict. Similarly, explanations can be found for some (but not all) proportionalities between blocks of matrix elements of components of the spin–other-orbit interaction for f electrons [3.39]. *Hansen* and *Ven* have given some examples of still unexplained proportionalities [3.40]. The group SO(7)$'$ has proved useful in analyses of the effective three-electron operators used to represent weak configuration interaction in the f shell [3.37].

3.6.3 Hydrogen and Hydrogen-Like Atoms

The nonrelativistic hydrogen atom possesses an SO(4) symmetry associated with the invariance of the Runge–Lenz vector, which indicates the direction of the major axis of the classical elliptic orbit [3.5]. The quantum-mechanical form of this vector can be written in dimensionless units as

$$a = [(l \times p) - (p \times l) + 2Zr/ra_0]/2p_0 , \quad (3.20)$$

where $a_0 = \hbar^2/me^2$ is the Bohr radius, Ze is the nuclear charge, p_0 is related to the principal quantum number n by $p_0 = Z/na_0$, and where the momentum $\boldsymbol{p}$ and angular momentum $\boldsymbol{l}$ of the electron in its orbit are measured in units of $\hbar$. The analysis is best carried out in momentum space [3.41]. The four coordinates to which SO(4) refers can be taken from (9.43–9.46) or directly as kp_x, kp_y, kp_z, and $kp_0(1 - p^2/p_0^2)/2$, where $k = 2p_0/(p^2 + p_0^2)$. The generators of SO(4) are provided by the 6 components of the two mutually commuting vectors $(\boldsymbol{l}+\boldsymbol{a})/2$ and $(\boldsymbol{l}-\boldsymbol{a})/2$, each of which behaves as an angular momentum vector. The equivalence SO(4) = SO(3) × SO(3) corresponds to the isomorphism $D_2 = A_1 \times A_1$ of Sect. 3.2.2.

Hydrogenic eigenfunctions belonging to various energies can be selected to form bases for a number of groups. The inclusion of all the levels up to a given n yields the IR $(n-1, 0)$ of SO(5). Levels of a given l and all n form an infinite basis for an IR of the noncompact group SO(2, 1) [3.42]. All the bound hydrogenic states span an IR of SO(4, 2), as do the states in the continuum [3.43]. Subgroups of SO(4, 2) and their generators have been listed by *Wybourne* [3.12] in his Table 21.2.

To the extent that the central potential of a complex atom resembles the r^{-1} dependence for a bare nucleus, the group SO(4) can be used to label the states [3.44].

3.7 The Generalized Wigner–Eckart Theorem

3.7.1 Operators

All atomic operators involving only the electrons can be built from their creation and annihilation operators. The appropriate group labels for an atomic operator acting on N electrons, each with n relevant component states, reduces to working out the various parts of the Kronecker products $[10\ldots0]^N \times [0\ldots0-1]^N$ of U(n). Subgroups of U(n) can further define these parts, which may be limited by Hermiticity constraints. The group labels for the Coulomb interaction for f electrons were first given by *Racah* [3.6]. Interactions involving electron spin were classified later [3.45–47]. Operators that represent the effects of configuration interaction on the d and f shells have also been studied [3.27, 48–52].

3.7.2 The Theorem

Let the ket, operator T, and bra of a matrix element be labeled by an IR (R_a, R_c, R_b) of a group $\mathcal{G}$, each with a component (i_a, i_c, i_b). Suppose the supplementary labels γ_k are also required to complete the definitions. The generalized Wigner–Eckart theorem is

$$\langle \gamma_a R_a i_a | T(\gamma_c R_c i_c) | \gamma_b R_b i_b \rangle = \sum_\beta A_\beta (\beta R_a i_a | R_b i_b, R_c i_c) , \quad (3.21)$$

where β distinguishes the IRs R_a should they appear more than once in the reduction of the Kronecker product $R_b \times R_c$. The *reduced* matrix element A_β is independent of the i_k [3.6, 8]. The second factor on the right-hand side of (3.21) is a Clebsch–Gordan (CG) coefficient for the group $\mathcal{G}$.

If the specification Ri can be replaced by $R\tau ri$, where r denotes an IR of a subgroup $\mathcal{H}$ of $\mathcal{G}$, and τ is an additional symbol that may be necessary to make the classification unambiguous, the CG coefficient for $\mathcal{G}$ factorizes according to the Racah lemma [3.6]

$$(\beta R_a \tau_a r_a i_a | R_b \tau_b r_b i_b, R_c \tau_c r_c i_c) = \sum_\alpha (\alpha r_a i_a | r_b i_b, r_c i_c)(\beta R_a \tau_a r_a | R_b \tau_b r_b + R_c \tau_c r_c)_\alpha . \quad (3.22)$$

The first factor on the right is a CG coefficient for the group $\mathcal{H}$; the second factor is an *isoscalar* factor [3.53].

3.7.3 Calculation of the Isoscalar Factors

The group $\mathcal{H}$ above is often SO(3), whose Clebsch–Gordan coefficients (and their related $3-j$ symbols) are well-known (Chapt. 2). The principal difficulty in establishing comparable formulae for the isoscalar factors lies in giving algebraic meaning to β. Several methods are available for obtaining numerical results as follows.

Extraction from Tabulated Quantities

If R_b or R_c correspond to the IRs labeling a single electron, the factorization of the known [3.31] coefficients of fractional parentage (cfp) according to formulae of the type [3.6]

$$(\mathrm{d}^N SLv\{|\mathrm{d}^{N-1}S'L'v') = (\mathrm{d}^N Sv\{|\mathrm{d}^{N-1}S'v')(W'L' + (10)\mathrm{d}|WL) \quad (3.23)$$

yields some isoscalar factors. In this example, W and W' are the IRs of SO(5) defined by the triples NSv and $N'S'v'$ (with $N' = N-1$) as in Table 3.4. This approach can be applied to the f shell to give isoscalar factors for SO(7) and G_2. The many-electron cfp of *Donlan* [3.54] and the multielectron cfp of *Velkov* [3.55] further extend the range to IRs R_b and R_c describing many-electron systems. Isoscalar factors found in this way have the advantage that their relative phases as well as the significance of the indices β and τ coincide with current usage.

Evaluation Using Casimir's Operator

Two commuting copies (b and c) are made of the generators of the group $\mathcal{G}$ to form the generators of the direct product $\mathcal{G}_b \times \mathcal{G}_c$ [3.56]. Corresponding generators of $\mathcal{G}_b$ and $\mathcal{G}_c$ are added to give the generators of $\mathcal{G}_a$. Each quadratic operator $(\boldsymbol{T}_a)^2$ appearing in the expression for Casimir's operator C_a for $\mathcal{G}_a$ (as listed in Table 3.3) is written as $(\boldsymbol{T}_b + \boldsymbol{T}_c)^2$. On expanding the expressions of this type, the terms $(\boldsymbol{T}_b)^2$ and $(\boldsymbol{T}_c)^2$ yield Casimir's operators C_b and C_c for $\mathcal{G}_b$ and $\mathcal{G}_c$. Their eigenvalues can be written down in terms of the highest weights of the IRs appearing in the isoscalar factor of (3.22). If the cross products of the type $(\boldsymbol{T}_b \cdot \boldsymbol{T}_c)$ can be evaluated within the uncoupled states $|R_b\tau_b r_b, R_c\tau_c r_c\rangle$, then our knowledge of the eigenvalues of C_a for the coupled states $|\beta R_a \tau_a r_a\rangle$ provides the equations for determining (to within the freedom implied by β) the isoscalar factors relating the uncoupled to the coupled states. The evaluation of the cross products is straightforward when $\mathcal{H} = \mathrm{SO}(3)$, since the relevant 6–$j$ symbols are readily available [3.57]. Examples of this method can be found in the literature [3.48].

3.7.4 Generalizations of Angular Momentum Theory

CG coefficients, n–j symbols, reduced matrix elements, and the entire apparatus of angular momentum theory all have their generalizations to groups other than SO(3). An interchange of two columns of a 3–j symbol has its analog in the interchange of two parts of an isoscalar factor. For IRs W and L of SO($2l+1$) and SO(3), there are two possibilities:

(1) The interchange of the two parts separated by the plus sign, namely,

$$(W_a\tau_a L_a|W_b\tau_b L_b + W_c\tau_c L_c) = (-1)^t (W_a\tau_a L_a|W_c\tau_c L_c + W_b\tau_b L_b)\,, \quad (3.24)$$

where $t = L_a - L_b - L_c + x$, with x dependent on the IRs W only; or,

(2) The *reciprocity* relation of *Racah* [3.6]:

$$(W_a\tau_a L_a|W_b\tau_b L_b + W_c\tau_c L_c) = (-1)^{t'}[(2L_b+1)\dim W_a/(2L_a+1)\dim W_b]^{\frac{1}{2}} \times (W_b\tau_b L_b|W_a\tau_a L_a + W_c\tau_c L_c)\,, \quad (3.25)$$

where $t' = L_a - L_b - L_c + x'$, with x' dependent on the IRs W, but taken to be l by Racah for $W_c = (10\ldots0)$.

Reduced matrix elements in SO(3) can be further reduced by the extraction of isoscalar factors. When W_a occurs once in the decomposition of $W \times W_b$ we have

$$\left(\gamma_a W_a\tau_a L_a\|T^{(WL)}\|\gamma_b W_b\tau_b L_b\right) = [(2L_a+1)/\dim W_a]^{\frac{1}{2}}\left(\gamma_a W_a|||T^{(W)}|||\gamma_b W_b\right) \times (W_b\tau_b L_b + WL|W_a\tau_a L_a)\,. \quad (3.26)$$

Analogs of the n–j symbols are discussed by *Butler* [3.58].

3.8 Checks

The existence of numerical checks is useful when using group theory in atomic physics. The CG coefficients, isoscalar factors, and the various generalizations of the n–j symbols are often calculated in ways that conceal the simplicity and structure of the answer. Practitioners are familiar with several empirical rules:

1. Numbers with different irrationalities, such as $\sqrt{2}$ and $\sqrt{3}$, are never added to one another.
2. The denominators of fractions seldom involve high primes.
3. High primes are uncommon, but when they appear, it is usually in diagonal matrix elements rather than off-diagonal ones.

4. A sum of a number of terms frequently factors in what appears to be an unexpected way, and similar sums often exhibit similar factors.

Guided by these rules, one will find that such errors as do arise occur with phases rather than with magnitudes.

References

3.1 G. Racah: *Group Theory and Spectroscopy*, Springer Tracts in Modern Physics, Vol. 37 (Springer, New York 1965)
3.2 L. P. Eisenhart: *Continuous Groups of Transformations* (Dover, New York 1961)
3.3 E. Cartan: *Sur la Structure des Groupes de Transformations Finis et Continus*, Thesis (Nony, Paris 1894)
3.4 B. R. Judd: *Group Theory and Its Applications*, ed. by E. M. Loebl (Academic, New York 1968)
3.5 E. U. Condon, H. Odabasi: *Atomic Structure* (Cambridge Univ. Press, Cambridge 1980)
3.6 G. Racah: Phys. Rev. **76**, 1352 (1949)
3.7 B. R. Judd: *Operator Techniques in Atomic Spectroscopy* (Princeton Univ. Press, Princeton 1963)
3.8 E. P. Wigner: *Group Theory* (Academic, New York 1959)
3.9 B. R. Judd: Phys. Rev. **162**, 28 (1967)
3.10 L. Armstrong, B. R. Judd: Proc. R. Soc. London Ser. A **315**, 27 and 39 (1970)
3.11 B. R. Judd, G. M. S. Lister: J. Phys. A **25**, 2615 (1992)
3.12 B. G. Wybourne: *Classical Groups for Physicists* (Wiley, New York 1974)
3.13 B. G. Wybourne: *Symmetry Principles and Atomic Spectroscopy* (Wiley, New York 1970)
3.14 W. G. McKay, J. Patera: *Tables of Dimensions, Indices, and Branching Rules for Representations of Simple Lie Algebras* (Dekker, New York 1981)
3.15 E. B. Dynkin: Am. Math. Soc. Transl. Ser. 2 **6**, 245 (1965)
3.16 D. E. Rutherford: *Substitutional Analysis* (Edinburgh Univ. Press, Edinburgh 1948)
3.17 M. Moshinsky: *Group Theory and the Many-Body Problem* (Gordon Breach, New York 1968)
3.18 H. Weyl: *The Theory of Groups and Quantum Mechanics* (Dover, New York undated)
3.19 W. G. Harter: Principles of Symmetry, Dynamics and Spectroscopy **8**, 2819 (1973)
3.20 W. G. Harter: *Principles of Symmetry, Dynamics and Spectroscopy* (Wiley, New York 1993)
3.21 W. G. Harter, C. W. Patterson: *A Unitary Calculus for Electronic Orbitals*, Lect. Notes Phys., Vol. 49 (Springer, Berlin, Heidelberg 1976)
3.22 G. W. F. Drake, M. Schlesinger: Phys. Rev. A **15**, 1990 (1977)
3.23 R. D. Kent, M. Schlesinger: Phys. Rev. A **50**, 186 (1994)
3.24 D. E. Littlewood: *The Theory of Group Characters* (Clarendon, Oxford 1950)
3.25 H. A. Jahn: Proc. R. Soc. London Ser. A **201**, 516 (1950)
3.26 B. H. Flowers: Proc. R. Soc. London Ser. A **212**, 248 (1952)
3.27 B. R. Judd, H. T. Wadzinski: J. Math. Phys. **8**, 2125 (1967)
3.28 C. L. B. Shudeman: J. Franklin Inst. **224**, 501 (1937)
3.29 B. R. Judd, G. M. S. Lister: Phys. Rev. Lett. **67**, 1720 (1991)
3.30 B. G. Wybourne: *Spectroscopic Properties of Rare Earths* (Wiley, New York 1965) p. 15
3.31 C. W. Nielson, G. F. Koster: *Spectroscopic Coefficients for the p^n, d^n, and f^n Configurations* (MIT Press, Cambridge 1963)
3.32 S. Feneuille: J. Phys. (Paris) **28**, 61, 315, 701, and 497 (1967)
3.33 S. Feneuille: J. Phys. (Paris) **30**, 923 (1969)
3.34 S. Feneuille, A. Crubellier, T. Haskell: J. Phys. (Paris) **31**, 25 (1970)
3.35 U. Fano, A. R. P. Rao: *Symmetry Principles in Quantum Physics* (Academic, New York 1996) Sect. 8.3.3
3.36 H. Georgi: *Lie Algebras in Particle Physics* (Benjamin/Cummings, Reading 1982) Chap. XXV
3.37 B. R. Judd: Phys. Rep. **285**, 1 (1997)
3.38 E. Lo, J. E. Hansen, B. R. Judd: J. Phys. B **33**, 819 (2000)
3.39 B. R. Judd, E. Lo: Phys. Rev. Lett. **85**, 948 (2000)
3.40 J. E. Hansen, E. G. Ven: Mol. Phys. **101**, 997 (2003)
3.41 M. J. Englefield: *Group Theory and the Coulomb Problem* (Wiley, New York 1972)
3.42 L. Armstrong: J Phys. (Paris) **31**, 17 (1970)
3.43 C. E. Wulfman: *Group Theory and Its Applications*, Vol. 2, ed. by E.M. Loebl (Academic, New York 1971)
3.44 D. R. Herrick: Adv. Chem. Phys. **52**, 1 (1982)
3.45 A. G. McLellan: Proc. Phys. Soc. London **76**, 419 (1960)
3.46 B. R. Judd: Physica **33**, 174 (1967)
3.47 B. R. Judd, H. M. Crosswhite, H. Crosswhite: Phys. Rev. **169**, 130 (1968)
3.48 B. R. Judd: Phys. Rev. **141**, 4 (1966)
3.49 B. R. Judd, M. A. Suskin: J. Opt. Soc. Am. B **1**, 261 (1984)
3.50 B. R. Judd, R. C. Leavitt: J. Phys. B **19**, 485 (1986)
3.51 R. C. Leavitt: J. Phys. A **20**, 3171 (1987)
3.52 R. C. Leavitt: J. Phys. B **21**, 2363 (1988)
3.53 A. R. Edmonds: Proc. R. Soc. London Ser. A **268**, 567 (1962)
3.54 V. L. Donlan: *Air Force Material Laboratory Report No. AFML-TR-70-249* (Wright-Patterson Air Force Base, Ohio 1970)
3.55 D. D. Velkov: Multi-Electron Coefficients of Fractional Parentage for the p, d, and f Shells. Ph.D. Thesis (The Johns Hopkins University, Baltimore 2000) http://www.pha.jhu.edu/groups/cfp/

3.56 P. Nutter, C. Nielsen: *Fractional parentage coefficients of terms of* f^n, II. Direct Evaluation of Racah's Factored Forms by a Group Theoretical Approach, Technical Memorandum T-133 (Raytheon, Waltham 1963) p. 133

3.57 M. Rotenberg, R. Bivins, N. Metropolis, J. K. Wooten: *The 3-j and 6-j Symbols* (MIT Press, Cambridge 1959)

3.58 P. H. Butler: Phil. Trans. R. Soc. London Ser. A **277**, 545 (1975)

4. Dynamical Groups

The well known *symmetry* (*invariance, degeneracy*) *groups* or *algebras* of quantum mechanical Hamiltonians provide quantum numbers (conservation laws, integrals of motion) for state labeling and the associated selection rules. In addition, it is often advantageous to employ much larger groups, referred to as the *dynamical groups* (*noninvariance groups, dynamical algebras, spectrum generating algebras*), which may or may not be the invariance groups of the studied system [4.1–7]. In all known cases, they are Lie groups (LGs), or rather corresponding Lie algebras (LAs), and one usually requires that all states of interest of a system be contained in a single irreducible representation (irrep). Likewise, one may require that the Hamiltonian be expressible in terms of the Casimir operators of the corresponding universal enveloping algebra [4.8, 9]. In a weaker sense, one regards any group (or corresponding algebra) as a dynamical group if the Hamiltonian can be expressed in terms of its generators [4.10–12]. In nuclear physics, one sometimes distinguishes *exact* (baryon number preserving), *almost exact* (e.g., total isospin), *approximate* (e.g., SU(3) of the "eightfold way") and *model* (e.g., nuclear shell model) *dynamical symmetries* [4.13]. The dynamical groups of interest in atomic and molecular physics can be conveniently classified by their topological characteristic of compactness. Noncompact LGs (LAs) generally arise in simple problems involving an infinite number of bound states, while those involving a finite number of bound states (e.g., molecular vibrations or ab initio models of electronic structure) exploit compact LG's.

We follow the convention of designating Lie groups by capital letters and Lie algebras by lower case letters, e.g., the Lie algebra of the rotation group SO(3) is designated as so(3).

4.1 **Noncompact Dynamical Groups** 87
4.1.1 Realizations of so(2,1) 88
4.1.2 Hydrogenic Realization of so(4,2) 88

4.2 **Hamiltonian Transformation and Simple Applications** 90
4.2.1 *N*-Dimensional Isotropic Harmonic Oscillator 90
4.2.2 *N*-Dimensional Hydrogenic Atom 91
4.2.3 Perturbed Hydrogenic Systems 91

4.3 **Compact Dynamical Groups** 92
4.3.1 Unitary Group and Its Representations 92
4.3.2 Orthogonal Group O(*n*) and Its Representations 93
4.3.3 Clifford Algebras and Spinor Representations 94
4.3.4 Bosonic and Fermionic Realizations of U(*n*) 94
4.3.5 Vibron Model 95
4.3.6 Many-Electron Correlation Problem 96
4.3.7 Clifford Algebra Unitary Group Approach 97
4.3.8 Spin-Dependent Operators 97

References 98

4.1 Noncompact Dynamical Groups

As an illustration we present basic facts concerning LAs that are useful for centrosymmetric Kepler-type problems, their realizations and typical applications. Recall that *a realization* of a LA is a homomorphism associating a concrete set of physically relevant operators with each abstract basis of the given LA. The physical operators we will use are general (intrinsic) position vectors $\boldsymbol{R} = (X_1, X_2, \ldots, X_N)$ in $\mathbb{R}^N$ and their corresponding momenta $\boldsymbol{P} = (P_1, P_2, \ldots, P_N)$, satisfying the basic commutation relations ($\hbar = 1$)

$$\left[X_j, X_k\right] = \left[P_j, P_k\right] = 0, \quad \left[X_j, P_k\right] = \mathrm{i}\delta_{jk} I \,. \tag{4.1}$$

4.1.1 Realizations of so(2,1)

This important LA is a simple noncompact analogue of the well known rotation group LA so(3), (cf. Sects. 2.1 and 3.2). Designating its three generators by T_j ($j = 1, 2, 3$), its structure constants (Sect. 2.1.1 and Sect. 3.1.1) are defined by

$$\begin{aligned} [T_1, T_2] &= \mathrm{i}\gamma T_3 \,, \\ [T_2, T_3] &= \mathrm{i} T_1 \,, \\ [T_3, T_1] &= \mathrm{i} T_2 \,, \end{aligned} \tag{4.2}$$

with $\gamma = -1$, while $\gamma = 1$ gives so(3). Defining the so-called ladder (raising and lowering) operators

$$T_\pm = T_1 \pm \mathrm{i} T_2 \,, \tag{4.3}$$

we also have that

$$\left[T_+, T_-\right] = 2\gamma T_3 \,, \quad \left[T_3, T_\pm\right] = \pm T_3 \,. \tag{4.4}$$

The Casimir operator then has the form

$$T^2 = \gamma\left(T_1^2 + T_2^2\right) + T_3^2 = \gamma T_+ T_- + T_3^2 - T_3 \,. \tag{4.5}$$

With a Hermitian scalar product satisfying $T_j^\dagger = T_j$ ($j = 1, 2, 3$), so that $T_\pm^\dagger = T_\mp$, the unitary irreps (unirreps) carried by the simultaneous eigenstates of T^2 and T_3 have the form (cf. Sects. 2.1.1 and 2.2)

$$\begin{aligned} T^2|kq\rangle &= k(k+1)|kq\rangle \,, \\ T_3|kq\rangle &= q|kq\rangle \,, \\ T_\pm|kq\rangle &= \sqrt{\gamma(k \mp q)(k \pm q + 1)}\,|k, q \pm 1\rangle \,. \end{aligned} \tag{4.6}$$

For so(3), ($\gamma = 1$), only finite dimensional irreps $\mathcal{D}^{(k)}$, $k = 0, 1, 2, \ldots$ with $|q| \le k$ are possible (Sects. 2.2 and 2.3). In contrast, there are no nontrivial finite dimensional unirreps of so(2,1); (for classification, see e.g., [4.2, 14, 15]). The relevant class $\mathcal{D}^+(k)$ of so(2,1) unirreps for bound state problems has a T_3 eigenspectrum bounded from below and is given by $q = -k + \mu$; $\mu = 0, 1, 2, \ldots,$ and $k < 0$ or, equivalently, $\mathcal{D}^+(-k-1)$ with $q = k + 1 + \mu$; $\mu = 0, 1, 2, \ldots$; $k > -1$, since $k_1 = -k - 1$ defines an equivalent unirrep and $k_1(k_1 + 1) = k(k+1)$. There exists a similar class of irreps with the T_3 spectrum bounded from above and two classes (principal and supplementary) with unbounded T_3 spectra (which may be exploited in scattering problems).

For problems involving only central potentials, a useful realization is given in terms of the radial distance $R = |\boldsymbol{R}|$ and the radial momentum

$$\begin{aligned} P_\mathrm{R} &= -\frac{\mathrm{i}}{R}\frac{\partial}{\partial R} R = -\mathrm{i}\left(\frac{\partial}{\partial R} + \frac{1}{R}\right) \\ &= \frac{1}{R}\left(\boldsymbol{R} \cdot \boldsymbol{P} - \mathrm{i} I\right) \,, \end{aligned} \tag{4.7}$$

so that $[R, P_\mathrm{R}] = \mathrm{i}I$. Recall that

$$P^2 = P_\mathrm{R}^2 + \frac{L^2}{R^2} \,, \quad \boldsymbol{L} = \boldsymbol{R} \wedge \boldsymbol{P} \,. \tag{4.8}$$

The general form of the desired so(2,1) realization is [4.2, 14–17]

$$\begin{aligned} \left.\begin{matrix} T_1 \\ T_3 \end{matrix}\right\} &= \frac{1}{2} R^{-\nu}\left(\nu^{-2} R^2 P_\mathrm{R}^2 + \xi \mp R^{2\nu}\right) \,, \\ T_2 &= \frac{1}{2}\left[2\nu^{-1} R P_\mathrm{R} - \mathrm{i}(1 - \nu^{-1}) I\right] \,, \end{aligned} \tag{4.9}$$

where ξ is either a c-number (scalar operator) or an operator which commutes with both R and P_R, and ν is an arbitrary real number.

To interrelate this realization with so(2,1) unirreps $\mathcal{D}^+(k)$ or $\mathcal{D}^+(-k-1)$, we have to establish the connection between the quantum numbers k, q and the parameters ξ and ν. Considering the Casimir operator T^2 in (4.5), we find that in our realization (4.9)

$$T^2 = \xi + \left(1 - \nu^2\right)/4\nu^2 \,, \tag{4.10}$$

so that

$$k = \frac{1}{2}\left(-1 \pm \sqrt{4\xi + \nu^{-2}}\right) \tag{4.11}$$

and

$$q = q_0 + \mu \,, \quad \mu = 0, 1, 2, \ldots \tag{4.12}$$

where

$$q_0 = k + 1 = \frac{1}{2}\left(1 \pm \sqrt{4\xi + \nu^{-2}}\right) \,, \quad k > -1 \,. \tag{4.13}$$

4.1.2 Hydrogenic Realization of so(4,2)

To obtain suitable hydrogenic realizations of so(4,2) it is best to proceed from so(4) (the dynamical symmetry group for the bound states of the nonrelativistic Kepler problem), and merge it with so(2,1) [4.2, 14, 15].

The so(4) LA can be realized either as a direct sum so(4) = so(3) $\oplus$ so(3), or by supplementing so(3) with an appropriately scaled quantum mechanical analogue of the Laplace–Runge–Lenz (LRL) vector (cf. Sect. 3.6.2). In the first case, we use two commuting angular momentum vectors $\boldsymbol{M}$ and $\boldsymbol{N}$ (cf. Sect. 2.5),

$$\begin{aligned} &[M_j, M_k] = \mathrm{i}\varepsilon_{jk\ell} M_\ell\,, \quad [N_j, N_k] = \mathrm{i}\varepsilon_{jk\ell} N_\ell\,, \\ &[M_j, N_k] = 0, \quad (j,k,\ell = 1,2,3) \end{aligned} \tag{4.14}$$

while in the second case we use the components of the total angular momentum vector $\boldsymbol{J}$ and LRL-like vector $\boldsymbol{V}$ with commutation relations

$$\begin{aligned} &[J_j, J_k] = \mathrm{i}\varepsilon_{jk\ell} J_\ell\,, \\ &[V_j, V_k] = \mathrm{i}\sigma\varepsilon_{jk\ell} J_\ell\,, \\ &[J_j, V_k] = \mathrm{i}\varepsilon_{jk\ell} V_\ell\,, \quad (j,k,\ell = 1,2,3)\,, \end{aligned} \tag{4.15}$$

with $\sigma = 1$. For $\sigma = -1$ we obtain so(3,1) (the LA of the homogeneous Lorentz group), which is relevant to the scattering problem of a particle in the Coulomb (or Kepler) potential (see below). For $\sigma = 0$ we get e(3) (the LA of the three-dimensional Euclidean group) [4.18–20]. Note that (4.14) and (4.15) are interrelated by

$$\boldsymbol{M} = \frac{1}{2}(\boldsymbol{J}+\boldsymbol{V})\,, \quad \boldsymbol{N} = \frac{1}{2}(\boldsymbol{J}-\boldsymbol{V})\,, \tag{4.16}$$

so that $\boldsymbol{J} = \boldsymbol{M} \oplus \boldsymbol{N}$ and $\boldsymbol{V} = 2\boldsymbol{M} - \boldsymbol{J}$. The two Casimir operators C_1 and C_2 are

$$\begin{aligned} C_1 &= \sigma J^2 + V^2 \\ &= \sigma J_+ J_- + V_+ V_- + V_3^2 + \sigma J_3(J_3 - 2)\,, \\ C_2 &= (\boldsymbol{V}\cdot\boldsymbol{J}) = (\boldsymbol{J}\cdot\boldsymbol{V}) \\ &= \frac{1}{2}(V_+ J_- + V_- J_+) + V_3 J_3\,, \end{aligned} \tag{4.17}$$

where again

$$X_\pm = X_1 \pm \mathrm{i} X_2\,, \quad X = J \text{ or } V\,. \tag{4.18}$$

For so(3,1) and e(3), only infinite dimensional nontrivial irreps are possible, while for so(4), only finite dimensional ones arise. To get unirreps, we require $\boldsymbol{J}$ and $\boldsymbol{V}$ to be Hermitian. Using $\{J^2, J_3, C_1, C_2\}$ as a complete set of commuting operators for so(4), we label the basis vectors by the four quantum numbers as $|\gamma jm\rangle \equiv |(j_0,\eta)jm\rangle$, so that

$$\begin{aligned} J^2|\gamma jm\rangle &= j(j+1)|\gamma jm\rangle\,, \\ J_3|\gamma jm\rangle &= m|\gamma jm\rangle\,, \\ C_1|\gamma jm\rangle &= \left(j_0^2 - \eta^2 - 1\right)|\gamma jm\rangle\,, \\ C_2|\gamma jm\rangle &= j_0\eta|\gamma jm\rangle\,, \end{aligned} \tag{4.19}$$

with $2|j_0|$ being a nonnegative integer and

$$\begin{aligned} &j = |j_0|,\ |j_0|+1, \ldots, \eta - 1\,; \\ &\eta = |j_0| + k, \quad k = 1,2,\ldots \end{aligned} \tag{4.20}$$

(see, e.g., [4.17] for the action of $J_\pm$, V_3 and $V_\pm$).

To obtain the hydrogenic (or Kepler) realization of so(4), we consider the quantum mechanical analog of the classical LRL vector

$$\begin{aligned} \tilde{\boldsymbol{V}} &= \frac{1}{2}(\boldsymbol{p}\wedge\boldsymbol{L} - \boldsymbol{L}\wedge\boldsymbol{p}) - Zr^{-1}\boldsymbol{r} \\ &= \frac{1}{2}\boldsymbol{r}p^2 - \boldsymbol{p}(\boldsymbol{r}\cdot\boldsymbol{p}) + \boldsymbol{r}H\,, \\ \boldsymbol{L} &= \boldsymbol{r}\wedge\boldsymbol{p}\,, \end{aligned} \tag{4.21}$$

which commutes with the hydrogenic Hamiltonian

$$H = \frac{1}{2}p^2 - \mathcal{Z}r^{-1}\,. \tag{4.22}$$

Note that

$$\begin{aligned} &[\boldsymbol{L}, H] = [\tilde{\boldsymbol{V}}, H] = 0\,, \\ &(\boldsymbol{L}\cdot\tilde{\boldsymbol{V}}) = (\tilde{\boldsymbol{V}}\cdot\boldsymbol{L}) = 0\,, \\ &\quad \tilde{V}^2 = 2H(L^2+1) + \mathcal{Z}^2\,, \end{aligned} \tag{4.23}$$

while the components of $\boldsymbol{L}$ and $\tilde{\boldsymbol{V}}$ satisfy the commutation relations

$$\begin{aligned} &[L_j, L_k] = \mathrm{i}\epsilon_{jk\ell} L_\ell\,, \\ &[L_j, \tilde{V}_k] = \mathrm{i}\epsilon_{jk\ell} \tilde{V}_\ell\,, \\ &[\tilde{V}_j, \tilde{V}_k] = (-2H)\mathrm{i}\epsilon_{jk\ell} L_\ell\,. \end{aligned} \tag{4.24}$$

Thus, restricting ourselves to a specific bound state energy level E_n, we can replace H by E_n and define

$$V_j = (-2E_n)^{-1/2}\tilde{V}_j \quad (j = 1,2,3)\,, \tag{4.25}$$

obtaining the so(4) commutation relations (4.15) (with $\boldsymbol{J}$ replaced by $\boldsymbol{L}$). This is Pauli's hydrogenic realization of so(4) [4.21–23]. [In a similar way we can consider continuum states $E > 0$ and define $\boldsymbol{V} = (2E)^{-1/2}\tilde{\boldsymbol{V}}$, obtaining an so(3,1) realization.] The last identity of (4.23) now becomes

$$V^2 = -(L^2+1) - \mathcal{Z}^2/2E_n\,, \tag{4.26}$$

which immediately implies Bohr's formula, since $V^2 + L^2 = 4M^2 = -1 - \mathcal{Z}^2/2E_n$, so that

$$E_n = -\mathcal{Z}^2/2(2j_1+1)^2 = -\mathcal{Z}^2/2n^2\,, \tag{4.27}$$

where $n = 2j_1 + 1$ and j_1 is the angular momentum quantum number for $\boldsymbol{M}$, (4.16). In terms of the irrep labels (4.20), we have that $j_0 = 0$, $\eta = n$, so that $|\gamma\ell m\rangle = |(0,n)\ell m\rangle \equiv |n\ell m\rangle$, $\ell = 0, 1, \ldots, n-1$.

Using the stepwise merging of so(4) and so(2,1) [adding first T_2 which leads to so(4,1) and subsequently T_1 and T_3], we arrive at the hydrogenic realization of so(4,2) having fifteen generators $\boldsymbol{L}$, $\boldsymbol{A}$, $\boldsymbol{B}$, $\boldsymbol{\Gamma}$, T_1, T_2, T_3, namely (cf. [4.2, 14, 15, 17])

$$\begin{aligned}
\boldsymbol{L} &= \boldsymbol{R} \wedge \boldsymbol{P}\,,\\
\left.\begin{matrix}\boldsymbol{A}\\ \boldsymbol{B}\end{matrix}\right\} &= \frac{1}{2}\boldsymbol{R}P^2 - \boldsymbol{P}(\boldsymbol{R}\cdot\boldsymbol{P}) \mp \boldsymbol{R}\,,\\
\boldsymbol{\Gamma} &= R\boldsymbol{P}\,,\\
\left.\begin{matrix}T_1\\ T_3\end{matrix}\right\} &= \frac{1}{2}\left(RP^2 \mp R\right) = \frac{1}{2}\left(RP_\mathrm{R}^2 + L^2R^{-1} \mp R\right),\\
T_2 &= \boldsymbol{R}\cdot\boldsymbol{P} - \mathrm{i}I = RP_\mathrm{R}\,.
\end{aligned} \tag{4.28}$$

Relabeling these generators by the elements of an antisymmetric 6×6 matrix according to the scheme

$$L_{jk} \leftrightarrow \begin{pmatrix}
0 & L_3 & -L_2 & A_1 & B_1 & \Gamma_1\\
 & 0 & L_1 & A_2 & B_2 & \Gamma_2\\
 & & 0 & A_3 & B_3 & \Gamma_3\\
 & & & 0 & T_2 & T_1\\
 & & & & 0 & T_3\\
 & & & & & 0
\end{pmatrix} \tag{4.29}$$

we can write the commutation relations in the following standard form

$$\begin{aligned}
&\left[L_{jk}, L_{\ell m}\right] =\\
&\quad \mathrm{i}\left(g_{j\ell}L_{km} + g_{km}L_{j\ell} - g_{k\ell}L_{jm} - g_{jm}L_{k\ell}\right),
\end{aligned} \tag{4.30}$$

with the diagonal metric tensor g_{jk} defined by the matrix $G = \mathrm{diag}[1, 1, 1, 1, -1, -1]$. The matrix form (4.29) also implies the subalgebra structure

$$\mathrm{so}(4,2) \supset \mathrm{so}(4,1) \supset \mathrm{so}(4) \supset \mathrm{so}(3)\,, \tag{4.31}$$

with so(4,1) generated by $\boldsymbol{L}$, $\boldsymbol{A}$, $\boldsymbol{B}$ and T_2, and so(4) by $\boldsymbol{L}$, $\boldsymbol{A}$. [$\boldsymbol{L}$, $\boldsymbol{B}$ also generate so(3,1).]

The three independent Casimir operators (quadratic, cubic and quartic) are [4.24], (summation over all indices is implied)

$$\begin{aligned}
Q_2 &= \frac{1}{2}L_{jk}L^{jk}\\
&= L^2 + A^2 - B^2 - \Gamma^2 + T_3^2 - T_1^2 - T_2^2\,,\\
Q_3 &= \frac{1}{48}\varepsilon_{ijk\ell mn}L^{ij}L^{k\ell}L^{mn}\\
&= T_1(\boldsymbol{B}\cdot\boldsymbol{L}) + T_2(\boldsymbol{\Gamma}\cdot\boldsymbol{L}) + T_3(\boldsymbol{A}\cdot\boldsymbol{L})\\
&\quad + \boldsymbol{A}\cdot(\boldsymbol{B}\wedge\boldsymbol{\Gamma})\,,\\
Q_4 &= L_{jk}L^{k\ell}L_{\ell m}L^{mj}\,.
\end{aligned} \tag{4.32}$$

For our hydrogenic realization $Q_2 = -3$, $Q_3 = Q_4 = 0$. Thus, our hydrogenic realization implies a single unirrep of so(4,2) adapted to the chain (4.31).

4.2 Hamiltonian Transformation and Simple Applications

The basic idea is to transform the relevant Schrödinger equation into an eigenvalue problem for one of the operators from the complete set of commuting operators in our realizations, e.g., T_3 for so(2,1). Instead of using a rather involved "tilting" transformation ([4.1, p. 20], and [4.2, 14, 15]), we can rely on a simple scaling transformation [4.16, 25]

$$\boldsymbol{r} = \lambda\boldsymbol{R}\,, \quad \boldsymbol{p} = \lambda^{-1}\boldsymbol{P}\,, \quad r = \lambda R\,, \quad p_\mathrm{r} = \lambda^{-1}P_\mathrm{R}\,, \tag{4.33}$$

where

$$r = \left(\sum_{j=1}^{N} x_j^2\right)^{1/2}, \tag{4.34}$$

and

$$p^2 = p_\mathrm{r}^2 + r^{-2}\left[\frac{1}{4}(N-1)(N-3) + L^2\right], \tag{4.35}$$

with p_r defined analogously to P_R in (4.7); r and p are the physical operators in terms of which the Hamiltonian of the studied system is expressed. Recall that L^2 has eigenvalues [4.26]

$$\ell(\ell + N - 2)\,, \quad \ell = 0, 1, 2, \ldots \quad \text{for} \quad N \geq 2\,, \tag{4.36}$$

and we can set $\ell = 0$ for $N = 1$ (angular momentum term vanishes in one-dimensional case). The units in which $m = e = \hbar = 1$, $c \approx 137$ are used throughout.

4.2.1 N-Dimensional Isotropic Harmonic Oscillator

Considering the Hamiltonian

$$H = \frac{1}{2}p^2 + \frac{1}{2}\omega^2 r^2\,, \tag{4.37}$$

with p^2 in the form (4.35), transforming the corresponding Schrödinger equation using the scaling transformation (4.33) and multiplying by $\frac{1}{4}\lambda^2$, we get for the radial component

$$\frac{1}{2}\left(\frac{1}{4}P_{\mathrm{R}}^2+R^{-2}\xi+\frac{1}{4}\omega^2\lambda^4R^2-\frac{1}{2}\lambda^2E\right)\psi_{\mathrm{R}}(\lambda R)=0\,, \tag{4.38}$$

with

$$\xi=\frac{1}{16}(N-1)(N-3)+\frac{1}{4}\ell(\ell+N-2)\,. \tag{4.39}$$

Choosing λ such that $(\omega/2)^2\lambda^4=1$, we can rewrite (4.38) using the so(2,1) realization (4.9) with $\nu=2$ as

$$\left(T_3-\frac{1}{4}\lambda^2E\right)\psi_{\mathrm{R}}(\lambda R)=0\,. \tag{4.40}$$

Thus, using the second equation of (4.6) we can interrelate $\psi_{\mathrm{R}}(\lambda R)$ with $|kq\rangle$ and set $\frac{1}{4}\lambda^2E=q$, so that

$$E=4q/\lambda^2=2q\omega\,, \tag{4.41}$$

with q given by (4.12) and (4.13), i. e.,

$$q=q_0+\mu\,,\quad \mu=0,1,2,\ldots \tag{4.42}$$

and

$$q_0=k+1=\frac{1}{2}\left[1\pm\left(\ell+\frac{1}{2}N-1\right)\right]\,. \tag{4.43}$$

Now, for $N=1$ we set $\ell=0$ so that $q_0=\frac{1}{4}$ and $q_0=\frac{3}{4}$, yielding for $E=2q\omega$ the values $\left(\frac{1}{2}+2\mu\right)\omega$ and $\left(\frac{1}{2}+2\mu+1\right)\omega$, $\mu=0,1,2,\ldots$. Combining both sets we thus get for $N=1$ the well known result

$$E\equiv E_n=\left(n+\frac{1}{2}\right)\omega\,,\quad n=0,1,2,\ldots\,. \tag{4.44}$$

Similarly, for the general case $N\geq 2$ we choose the upper sign in (4.43) [so that $k>-1$] and get

$$E\equiv E_n=\left(n+\frac{1}{2}N\right)\omega,\quad n=0,1,2,\ldots \tag{4.45}$$

where we identified $(\ell+2\mu)$ with the principal quantum number n.

4.2.2 *N*-Dimensional Hydrogenic Atom

Applying the scaling transformation (4.33) to the hydrogenic Hamiltonian (4.22) in N-dimensions, we get for the radial component (after multiplying from the left by λ^2R)

$$\left[\frac{1}{2}\left(RP_{\mathrm{R}}^2+R^{-1}\xi-2\lambda^2ER\right)-\lambda\mathcal{Z}\right]\psi_{\mathrm{R}}(\lambda R)=0\,, \tag{4.46}$$

where now

$$\xi=\frac{1}{4}(N-1)(N-3)+\ell(\ell+N-2)\,. \tag{4.47}$$

In this case we must set $2\lambda^2E=-1$ and use realization (4.9) with $\nu=1$ to obtain

$$(T_3-\lambda\mathcal{Z})\psi_{\mathrm{R}}(\lambda R)=0\,. \tag{4.48}$$

This immediately implies that

$$\lambda\mathcal{Z}=q \tag{4.49}$$

and

$$q_0=k+1=\frac{1}{2}\left[1\pm(2\ell+N-2)\right]\,. \tag{4.50}$$

Choosing the upper sign [since $\ell\geq 0$ and $k>-1$], so that $q_0=\ell+\frac{1}{2}(N-1)$, and identifying q with the principal quantum number n, we have finally that

$$E\equiv E_n=-\frac{1}{2\lambda^2}=-\frac{\mathcal{Z}^2}{2n^2}\,. \tag{4.51}$$

The N-dimensional relativistic hydrogenic atom can be treated in the same way, using either the Klein–Gordon or Dirac–Coulomb equations [4.2, 14–17].

4.2.3 Perturbed Hydrogenic Systems

The so(4,2) based Lie algebraic formalism can be conveniently exploited to carry out large order perturbation theory (see [4.27–29] and Chapt. 5) for hydrogenic systems described by the Schrödinger equation

$$[H_0+\varepsilon V(\boldsymbol{r})]\,\psi(\boldsymbol{r})=(E_0+\Delta E)\psi(\boldsymbol{r})\,, \tag{4.52}$$

with H_0 given by (4.22) and E_0 by E of (4.51). Applying transformation (4.33), using (4.49), (4.51) and multiplying on the left by λ^2R, we get

$$\left[\frac{1}{2}RP^2-q+\frac{1}{2}R+\varepsilon\lambda^2RV(\lambda\boldsymbol{R})-\lambda^2R\Delta E\right]\Psi(\boldsymbol{R})=0\,, \tag{4.53}$$

where we set $\psi(\lambda\boldsymbol{R})\equiv\Psi(\boldsymbol{R})$. For the important case of a 3-dimensional hydrogenic atom [$N=3$, $\xi=\ell(\ell+1)$, $q\equiv n$] we get using the so(4,2) realization (4.28) [or so(2,1) realization (4.9) with $\nu=1$]

$$(K+\varepsilon W-S\Delta E)\Psi(\boldsymbol{R})=0\,, \tag{4.54}$$

with

$$K = T_3 - n\,,$$
$$W = \lambda^2 R V(\lambda \boldsymbol{R})\,,$$
$$S = \lambda^2 R\,. \quad (4.55)$$

We also have that $\lambda = n/\mathcal{Z}$ and for the ground state case $n = q = 1$. Although (4.54) has the form of a generalized eigenvalue problem requiring perturbation theory formalism with a nonorthogonal basis (where S represents an overlap), T_3 is Hermitian with respect to a $(1/R)$ scalar product, and the required matrix elements can therefore be easily evaluated [4.2, 14, 15, 17, 27–29].

For central field perturbations, $V(\boldsymbol{r}) = V(r)$, the problem reduces to one dimension and since $R = T_3 - T_1$, the so(2,1) hydrogenic realization ($\nu = 1$) can be employed. For problems of a hydrogenic atom in a magnetic field (Zeeman effect) [4.27–30] or a one-electron diatomic ion [4.31], the so(4,2) formalism is required (note, however, that the LoSurdo–Stark effect can also be treated as a one-dimensional problem using parabolic coordinates [4.32]).

The main advantage of the LA approach stems from the fact that the spectrum of T_3 is discrete, so that no integration over continuum states is required. Moreover, the relevant perturbations are closely packed around the diagonal in this representation, so that infinite sums are replaced by small finite sums.

For example, for the LoSurdo–Stark problem when $V(\boldsymbol{r}) = \mathcal{F} z$, where $\mathcal{F}$ designates electric field strength in the z-direction, we get (4.54) with $\varepsilon = \mathcal{F}$ and $W = (n/\mathcal{Z})^3 RZ$, $S = (n/\mathcal{Z})^2 R$. Since both R and Z are easily expressed in terms of so(4,2) generators,

$$Z = B_3 - A_3\,, \quad R = T_3 - T_1\,, \quad (4.56)$$

we can easily compute all the required matrix elements [4.2, 14, 15, 17].

Similarly, considering the Zeeman effect with

$$V(\boldsymbol{r}) = \frac{1}{2}\mathcal{B} L_3 + \frac{1}{8}\mathcal{B}^2\left(r^2 - z^2\right)\,, \quad (4.57)$$

where $\mathcal{B}$ designates magnetic field strength in the z-direction, we have for the ground state when $n = 1$, $\ell = m = 0$ that $\varepsilon = \frac{1}{8}\mathcal{B}^2$, $K = T_3 - 1$, $W = \mathcal{Z}^{-4} R(R^2 - Z^2)$ and $S = \mathcal{Z}^{-2} R$. Again, the matrix elements of W and S are obtained from those of Z and R, (4.56) by matrix multiplication (for tables and programs, see [4.17]).

One can treat one-electron diatomic ions [4.2, 14, 15, 31] and screened Coulomb potentials, including charmonium and harmonium [4.10–12, 17, 33, 34], in a similar way.

Note, finally, that we can also formulate the perturbed problem (4.54) in a standard form not involving the "overlap" by defining the scaling factor as $\lambda = (-2E)^{-1/2}$, where E is now the exact energy $E = E_0 + \Delta E$. Equation (4.54) then becomes

$$(T_3 + \varepsilon W - \lambda \mathcal{Z})\Psi(\boldsymbol{R}) = 0\,, \quad (4.58)$$

with the eigenvalue $\lambda \mathcal{Z}$. In this case any conventional perturbation formalism applies, but the desired energy has to be found from $\lambda \mathcal{Z}$ [4.35].

4.3 Compact Dynamical Groups

Unitary groups U(n) and their LAs often play the role of (compact) dynamical groups since

1. quantum mechanical observables are Hermitian and the LA of U(n) is comprised of Hermitian operators [under the exp(iA) mapping],
2. any compact Lie group is isomorphic to a subgroup of some U(n),
3. "nothing of algebraic import is lost by the unitary restriction" [4.36].

All U(n) irreps have finite dimension and are thus relevant to problems involving a finite number of bound states [4.3–6, 10–12, 36–43].

4.3.1 Unitary Group and Its Representations

The unitary group U(n) has n^2 generators E_{ij} spanning its LA and satisfying the commutation relations

$$\left[E_{ij}, E_{k\ell}\right] = \delta_{jk} E_{i\ell} - \delta_{i\ell} E_{kj} \quad (4.59)$$

and the Hermitian property

$$E_{ij}^{\dagger} = E_{ji}\,. \quad (4.60)$$

They are classified as *raising* ($i < j$), *lowering* ($i > j$) and *weight* ($i = j$) generators according to whether they raise, lower and preserve the weight, respectively. The *weight vector* is a vector of the carrier space of an irrep which is a simultaneous eigenvector of all weight generators E_{ii} of U(n) (comprising its Cartan subalgebra), and the vector $\boldsymbol{m} = (m_1, m_2, \dots, m_n)$ with integer components, consisting of corresponding eigenvalues, is called a *weight*. The *highest weight* $\boldsymbol{m}_n$ (in lexical ordering),

$$\boldsymbol{m}_n = (m_{1n}, m_{2n}, \dots, m_{nn}) , \tag{4.61}$$

with

$$m_{1n} \geq m_{2n} \geq \dots \geq m_{nn} , \tag{4.62}$$

uniquely labels U(n) irreps, $\Gamma(\boldsymbol{m}_n)$, and may be represented by a Young pattern. Subducing $\Gamma(\boldsymbol{m}_r)$ of U(r) to U($r-1$), embedded as U($r-1$) $\oplus$ 1 in U(r), gives [4.41]

$$\Gamma(\boldsymbol{m}_r) \downarrow \mathrm{U}(r-1) = \bigoplus \Gamma(\boldsymbol{m}_{r-1}) , \tag{4.63}$$

where the sum extends over all U($r-1$) weights $\boldsymbol{m}_{r-1} = (m_{1,r-1}, m_{2,r-1}, \dots m_{r-1,r-1})$ satisfying the so-called "betweenness conditions" [4.38]

$$m_{ir} \geq m_{i,r-1} \geq m_{i+1,r} \qquad (i = 1, \dots, r-1) . \tag{4.64}$$

Two irreps $\Gamma(\boldsymbol{m}_n)$ and $\Gamma(\boldsymbol{m}'_n)$ of U(n) yield the same irrep when restricted to SU(n) if $m_i = m'_i + h$, $i = 1, \dots, n$. The SU(n) irreps are thus labeled with highest weights with $m_{nn} = 0$. The dimension of $\Gamma(\boldsymbol{m}_n)$ of U(n) is given by the Weyl dimension formula [4.36]

$$\dim \Gamma(\boldsymbol{m}_n) = \prod_{i<j} (m_{in} - m_{jn} + j - i) \Big/ 1!2! \cdots (n-1)! . \tag{4.65}$$

The U(n) Casimir operators have the form

$$C_k^{\mathrm{U}(n)} = \sum_{i_1, i_2, \dots, i_k = 1}^{n} E_{i_1 i_2} E_{i_2 i_3} \cdots E_{i_{k-1} i_k} E_{i_k i_1} . \tag{4.66}$$

The first order Casimir operator is given by the sum of weight generators and equals the sum of the highest weight components.

Since U(1) is Abelian, the *Gel'fand-Tsetlin* [4.42] canonical chain (Sect. 3.4.3)

$$\mathrm{U}(n) \supset \mathrm{U}(n-1) \supset \dots \supset \mathrm{U}(1) \tag{4.67}$$

can be used to label uniquely the basis vectors of the carrier space of $\Gamma(\boldsymbol{m}_n)$ by triangular Gel'fand tableaux $[m]$ defined by

$$[m] = \begin{bmatrix} \boldsymbol{m}_n \\ \boldsymbol{m}_{n-1} \\ \dots \\ \dots \\ \boldsymbol{m}_2 \\ \boldsymbol{m}_1 \end{bmatrix} = \begin{bmatrix} m_{1n} \quad m_{2n} \quad \cdots\cdots \quad m_{nn} \\ m_{1,n-1} \cdots\cdots m_{n-1,n-1} \\ \dots \\ \dots \\ m_{12}\, m_{22} \\ m_{11} \end{bmatrix} , \tag{4.68}$$

with entries satisfying betweenness conditions (4.64). Matrix representatives of weight generators are diagonal

$$\langle [m'] | E_{ii} | [m] \rangle = \delta_{[m],[m']} \left(\sum_{j=1}^{i} m_{ji} - \sum_{j=1}^{i-1} m_{j,i-1} \right) , \tag{4.69}$$

while those for other generators are rather involved [4.42, 43]. Note that only elementary ($E_{i,i+1}$) raising generators are required since

$$\langle [m'] | E_{ij} | [m] \rangle = \langle [m] | E_{ji} | [m'] \rangle \tag{4.70}$$

and

$$E_{ij} = \left[E_{i,i+1}, E_{i+1,j} \right] . \tag{4.71}$$

In special cases required in applications ([4.39, 40] and Sect. 4.3.4) efficient algorithms exist for the computation of explicit representations.

4.3.2 Orthogonal Group O(n) and Its Representations

Since O(n) is a proper subgroup of U(n), its representation theory has a similar structure. The suitable generators are

$$\begin{aligned} F_{ij} &= E_{ij} - E_{ji} , \quad F_{ji} = -F_{ij} , \\ F_{ii} &= 0 , \qquad\quad F_{ij}^{\dagger} = -F_{ji} \end{aligned} \tag{4.72}$$

and satisfy the commutation relations

$$\left[F_{ij}, F_{k\ell} \right] = \delta_{jk} F_{i\ell} + \delta_{i\ell} F_{jk} - \delta_{ik} F_{j\ell} - \delta_{j\ell} F_{ik} . \tag{4.73}$$

The canonical chain has the form

$$\mathrm{O}(n) \supset \mathrm{O}(n-1) \supset \dots \supset \mathrm{O}(2) . \tag{4.74}$$

The components of the highest weight $\boldsymbol{m}_n$,

$$\boldsymbol{m}_n = (m_{1n}, m_{2n}, \dots, m_{kn}) , \tag{4.75}$$

satisfy the conditions

$$m_{1n} \geq m_{2n} \geq \cdots \geq m_{kn} \geq 0 \quad \text{for} \quad n = 2k+1\,, \tag{4.76}$$

and

$$m_{1n} \geq m_{2n} \geq \cdots \geq |m_{kn}| \quad \text{for} \quad n = 2k\,, \tag{4.77}$$

where m_{in} are simultaneously integers or half-odd integers. The former are referred to as *tensor representations* (since they arise as tensor products of fundamental irreps), while those with half-odd integer components are called *spinor representations*. Note that for $n = 2k$, we have two lowest (mirror-conjugated) spinor representations, namely $\boldsymbol{m}^{(+)} = (\frac{1}{2}, \frac{1}{2}, \ldots, \frac{1}{2})$ and $\boldsymbol{m}^{(-)} = (\frac{1}{2}, \ldots, \frac{1}{2}, -\frac{1}{2})$. Only tensor representations can be labeled by Young tableaux.

Subducing O(n) to O($n-1$), the betweenness conditions (branching rules) have the form

$$m_{in} \geq m_{i,n-1} \geq m_{i+1,n} \quad (i = 1, \ldots, k-1) \tag{4.78}$$

together with

$$m_{k,2k+1} \geq |m_{k,2k}| \tag{4.79}$$

when $n = 2k+1$. The $m_{i,n-1}$ components are integral (half-odd integral) if the m_{in} are integral (half-odd integral). The U(n) $\supset$ O(n) [or SU(n) $\supset$ SO(n)] subduction rules are more involved [4.44].

4.3.3 Clifford Algebras and Spinor Representations

While all reps of U(n) or SL(n) arise as tensor powers of the standard rep, only half of the reps of SO(m) or O(m) arise this way, since SO(m) is not simply connected when $m > 2$. A double covering of SO(m) leads to *spin groups* Spin(m). The best way to proceed is, however, to construct the so-called *Clifford algebras* C_m, whose multiplicative group (consisting of invertible elements) contains a subgroup which provides a double cover of SO(m). The key fact is that C_{2k} is isomorphic with gl(2^k) and C_{2k+1} with gl(2^k) $\oplus$ gl(2^k). The reps of C_m thus provide the required *spinor reps*.

A Clifford algebra C_m is an associative algebra generated by *Clifford numbers* α_i satisfying the anticommutation relations

$$\{\alpha_i, \alpha_j\} = 2\delta_{ij} \quad (i, j = 1, \ldots, m)\,. \tag{4.80}$$

Since $\alpha_i^2 = 1$, $\dim C_m = 2^m$ and a general element of C_m is a product of Clifford numbers $\alpha_1^{\nu_1}\alpha_2^{\nu_2}\cdots\alpha_m^{\nu_m}$ with $\nu_i = 0$ or 1.

To see the relation with so($m+1$), note that

$$F_{0k} = -\frac{1}{2}\mathrm{i}\alpha_k\,, \quad F_{jk} = \frac{1}{4}\left[\alpha_j, \alpha_k\right] = \frac{1}{2}\alpha_j\alpha_k\,, \quad (j \neq k) \tag{4.81}$$

satisfy the commutation relations (4.73).

As an example, C_2 can be realized by Pauli matrices by setting

$$\alpha_1 = \sigma_1 = \begin{pmatrix} 0 & 1 \\ 1 & 0 \end{pmatrix}, \quad \alpha_2 = \sigma_2 = \begin{pmatrix} 0 & \mathrm{i} \\ -\mathrm{i} & 0 \end{pmatrix}. \tag{4.82}$$

Clearly, the four matrices $\mathbf{1}_2, \alpha_1, \alpha_2$ and $\alpha_1\alpha_2$ are linearly independent (note that $\sigma_3 = \mathrm{i}\sigma_1\sigma_2$), so that C_2 is isomorphic to gl(2, $\mathbb{C}$).

Similarly, considering Dirac–Pauli matrices

$$\gamma_0 = \begin{pmatrix} -\mathrm{i}\mathbf{1}_2 & \mathbf{0} \\ \mathbf{0} & -\mathrm{i}\mathbf{1}_2 \end{pmatrix} = \mathrm{i}\gamma_4\,,$$
$$\gamma_k = \begin{pmatrix} \mathbf{0} & \mathrm{i}\sigma_k \\ -\mathrm{i}\sigma_k & \mathbf{0} \end{pmatrix}, \quad (k = 1, 2, 3) \tag{4.83}$$

we have that

$$\{\gamma_i, \gamma_j\} = 2\delta_{ij}, \quad (i, j = 1, \ldots, 4) \tag{4.84}$$

so that γ_i $(i = 1, \ldots, 4)$ or $(i = 0, \ldots, 3)$ represent Clifford numbers for C_4 and $\mathbf{1}_4, \gamma_i, \gamma_i\gamma_j$ $(i < j)$, $\gamma_i\gamma_j\gamma_k$ $(i < j < k)$ and $\gamma_5 \equiv \gamma_1\gamma_2\gamma_3\gamma_4 = i\gamma_0\gamma_1\gamma_2\gamma_3$ form an additive basis for gl(4, $\mathbb{C}$) (the γ_i themselves are said to form a multiplicative basis). For general construction of C_m Clifford numbers in terms of direct products of Pauli matrices see [4.45, 46].

4.3.4 Bosonic and Fermionic Realizations of U(n)

Designating by $b_i^\dagger$ (b_i) the boson creation (annihilation) operators (Sect. 6.1.1) satisfying the commutation relations $\left[b_i, b_j\right] = \left[b_i^\dagger, b_j^\dagger\right] = 0$, $\left[b_i, b_j^\dagger\right] = \delta_{ij}$, we obtain a possible U(n) realization by defining its n^2 generators as follows

$$G_{ij} = b_i^\dagger b_j\,. \tag{4.85}$$

The first order Casimir operator, (4.66) with $k = 1$, then represents the total number operator

$$\hat{N} \equiv C_1^{\mathrm{U}(n)} = \sum_{i=1}^{n} G_{ii} = \sum_{i=1}^{n} b_i^\dagger b_i\,, \tag{4.86}$$

and the physically relevant states, being totally symmetric, carry single row irreps $\Gamma(N\dot{0}) \equiv \Gamma(N0\cdots0)$.

Similarly for fermion creation (annihilation) operators $X_I^\dagger(X_I)$ that are associated with some orthonormal spin orbital set $\{|I\rangle\}$, $I = 1, 2, \ldots, 2n$, and satisfy the anticommutation relations $\{X_I, X_J\} = \{X_I^\dagger, X_J^\dagger\} = 0$, $\{X_I, X_J^\dagger\} = \delta_{IJ}$, the operators

$$e_{IJ} = X_I^\dagger X_J \tag{4.87}$$

again represent the U(2n) generators satisfying (4.59) and (4.60). The first-order Casimir then represents the total number operator $\hat{N} = \sum_I X_I^\dagger X_I$, while the possible physical states are characterized by totally antisymmetric single column irreps $\Gamma(1^N\dot{0}) \equiv \Gamma(11\cdots1\,0\cdots0)$.

4.3.5 Vibron Model

Similar to the unified description of nuclear collective rovibrational states using the interacting boson model [4.47–49], one can build an analogous model for molecular rotation-vibration spectra [4.8]. For diatomics, an appropriate dynamical group is U(4) [4.8, 50–52] and, generally, for rotation-vibration spectra in r-dimensions one requires U($r+1$). For triatomics, the U(4) generating algebra is generalized to U(4) ⊗ U(4), and for the $(k+1)$ atomic molecule to $\mathrm{U}^{(1)}(4) \otimes \cdots \otimes \mathrm{U}^{(k)}(4)$ [4.8,50–52].

For the bosonic realization of U(4), we need four creation ($b_i^\dagger$, $i = 1, \ldots, 4$) and four annihilation (b_i) operators (Sect. 4.3.4). The Hamiltonian may be generally expressed as a multilinear form in terms of boson number preserving products $(b_i^\dagger b_j)$, so that using (4.85) we can write

$$H = h^{(0)} + \sum_{ij} h_{ij}^{(1)} G_{ij} + \frac{1}{2} \sum_{ijk\ell} h_{ijk\ell}^{(2)} G_{ij} G_{k\ell} + \cdots . \tag{4.88}$$

The energy levels (as a function of 0, 1, 2, ... -body matrix elements h^0, $h_{ij}^{(1)}$, $h_{ijk\ell}^{(2)}$, etc.) are then determined by diagonalizing H in an appropriate space, which is conveniently provided by the carrier space of the totally symmetric irrep $\Gamma(N\,0\,0\,0) \equiv \Gamma(N\dot{0})$ of U(4).

The requirement that the resulting states be characterized by angular momentum J and parity P quantum numbers necessitates that the boson operators involved have definite transformation properties under rotations and reflections [4.8]. The boson operators are thus subdivided into the scalar operators $(\sigma^\dagger, \sigma)$, $J = 0$, and vector operators $(\pi_\mu^\dagger, \pi_\mu;\ \mu = 0, \pm1)$, $J = 1$ with parity $P = (-)^J$. All commutators vanish except for

$$[\sigma, \sigma^\dagger] = 1, \quad [\pi_\mu, \pi_{\mu'}^\dagger] = \delta_{\mu\mu'} . \tag{4.89}$$

Since H preserves the total number of vibrons $N = n_\sigma + n_\pi$, the second order Hamiltonian (4.88) within the irrep $\Gamma(N\dot{0})$ can be expressed in terms of four independent parameters (apart from an overall constant) as

$$\begin{aligned} H = {} & e^{(0)} + e^{(1)} \left[\pi^\dagger \times \tilde{\pi}\right]_0^{(0)} \\ & + e_1^{(2)} \left[\left[\pi^\dagger \times \pi^\dagger\right]^{(0)} \times \left[\tilde{\pi} \times \tilde{\pi}\right]^{(0)}\right]_0^{(0)} \\ & + e_2^{(2)} \left[\left[\pi^\dagger \times \pi^\dagger\right]^{(2)} \times \left[\tilde{\pi} \times \tilde{\pi}\right]^{(2)}\right]_0^{(0)} \\ & + e_3^{(2)} \left[\left[\pi^\dagger \times \pi^\dagger\right]^{(0)} \times \left[\tilde{\sigma} \times \tilde{\sigma}\right]^{(0)}\right]^{(0)} \\ & + \left[\left[\sigma^\dagger \times \sigma^\dagger\right]^{(0)} \times \left[\tilde{\pi} \times \tilde{\pi}\right]^{(0)}\right]_0^{(0)} + \cdots , \end{aligned} \tag{4.90}$$

where $\tilde{\sigma} = \sigma$, $\tilde{\pi}_\mu = (-)^{1-\mu}\pi_{-\mu}$ and square brackets indicate the SU(2) couplings.

In special cases the eigenvalue problem for H can be solved analytically, assuming that H can be expressed in terms of Casimir operators of a complete chain of subgroups of U(4) [referred to as dynamical symmetries]. Requiring that the chain contain the physical rotation group O(3), one has two possibilities

$$\begin{aligned} &(I) && \mathrm{U}(4) \supset \mathrm{O}(4) \supset \mathrm{O}(3) \supset \mathrm{O}(2) , \\ &(II) && \mathrm{U}(4) \supset \mathrm{U}(3) \supset \mathrm{O}(3) \supset \mathrm{O}(2) . \end{aligned} \tag{4.91}$$

These imply labels (quantum numbers): N [total vibron number defining a totally symmetric irrep of U(4)], $\omega = N, N-2, N-4, \ldots, 1$ or 0 [defining a totally symmetric irrep of O(4)] and $n_\pi = N, N-1, \ldots, 0$ [defining the U(3) irrep], in addition to the O(3) ⊃ O(2) labels J, M; $|M| \le J$. In terms of these labels one finds for the respective Hamiltonians

$$\begin{aligned} H^{(\mathrm{I})} &= F + A\, C_2^{\mathrm{O}(4)} + B C_2^{\mathrm{O}(3)} , \\ H^{(\mathrm{II})} &= F + \varepsilon C_1^{\mathrm{U}(3)} + \alpha C_2^{\mathrm{U}(3)} + \beta C_2^{\mathrm{O}(3)} , \end{aligned} \tag{4.92}$$

where F, A, B, ε, α, β are free parameters and $C_i^{\mathrm{U}(k)}$, $C_i^{\mathrm{O}(k)}$ are relevant Casimir operators, the following expressions [4.8,50–52] for their eigenvalues

$$\begin{aligned} E^{(\mathrm{I})}(N, \omega, J, M) &= F + A\omega(\omega+2) + BJ(J+1), \\ E^{(\mathrm{II})}(N, n_\pi, J, M) &= F + \epsilon n_\pi + \alpha n_\pi (n_\pi + 3) \\ & \quad + \beta J(J+1) . \end{aligned} \tag{4.93}$$

The limit (I) is appropriate for rigid diatomics and limit (II) for nonrigid ones [4.8, 50–52].

In addition to handling di- and tri-atomic systems, the vibron model was also applied to the overtone spectrum of acetylene [4.53], intramolecular relaxation in benzene and its dimers [4.54, 55], octahedral molecules of the XF_6 type (X = S, W, and U) [4.56], and to linear polyatomics [4.57]. Most recently, the experimental (dispersed fluorescence and stimulated emission pumping) vibrational spectra of H_2O and SO_2 in their ground states, representing typical local-mode and normal-mode molecules, respectively, have been analyzed, including highly excited levels, by relying on the U(2) algebraic effective Hamiltonian approach [4.58–60]. The U(2) algebraic scheme [4.61] also enabled the treatment of Franck–Condon transition intensities [4.62, 63] in rovibronic spectra. The attempts at a similar heuristic phenomenological description of electronic spectra have met so-far with only a limited success [4.64].

4.3.6 Many-Electron Correlation Problem

In atomic and molecular electronic structure calculations one employs a spin-independent model Hamiltonian

$$H = \sum_{i,j} h_{ij} \sum_{\sigma=1}^{2} X_{i\sigma}^{\dagger} X_{j\sigma} + \frac{1}{2} \sum_{i,j,k,\ell} v_{ij,k\ell} \sum_{\sigma,\tau=1}^{2} X_{i\sigma}^{\dagger} X_{j\tau}^{\dagger} X_{\ell\tau} X_{k\sigma} , \quad (4.94)$$

where $X_I^{\dagger} \equiv X_{i\sigma}^{\dagger}$ (X_I) designate the creation (annihilation) operators associated with the orthonormal spin orbitals $|I\rangle \equiv |i\sigma\rangle = |i\rangle \otimes |\sigma\rangle$; $i = 1, \ldots, n$; $\sigma = 1, 2$ $[\sigma = 1, 2$ labeling the spin-up and spin-down eigenstates of $S_z]$, and $h_{ij} = \langle i|\hat{h}|j\rangle$, $v_{ij,k\ell} = \langle i(1)j(2)|\hat{v}|k(1)\ell(2)\rangle$ are the one- and two-electron integrals in the orbital basis $\{|i\rangle\}$. As stated in Sect. 4.3.4, $e_{IJ} \equiv e_{i\sigma,j\tau} = X_{i\sigma}^{\dagger} X_{j\tau}$ may then be regarded as U(2n) generators, and the appropriate U(2n) irrep for N-electron states is $\Gamma\left(1^N \dot{0}\right)$.

Similar to the nuclear many-body problem [4.65], one defines mutually commuting partial traces of spin orbital generators e_{IJ}, (4.87),

$$E_{ij} = \sum_{\sigma=1}^{2} e_{i\sigma,j\sigma} = \sum_{\sigma=1}^{2} X_{i\sigma}^{\dagger} X_{j\sigma} ,$$

$$\mathcal{E}_{\sigma\tau} = \sum_{i=1}^{n} e_{i\sigma,i\tau} = \sum_{i=1}^{n} X_{i\sigma}^{\dagger} X_{i\tau} , \quad (4.95)$$

which again satisfy the unitary group commutation relations (4.59) and property (4.60), and may thus be considered as the generators of the orbital group U(n) and the spin group U(2). The Hamiltonian (4.94) is thus expressible in terms of orbital U(n) generators

$$H = \sum_{i,j} h_{ij} E_{ij} + \frac{1}{2} \sum_{i,j,k,\ell} v_{ij,k\ell} (E_{ik} E_{j\ell} - \delta_{jk} E_{i\ell}) . \quad (4.96)$$

We can thus achieve an automatic spin adaptation by exploiting the chain

$$U(2n) \supset U(n) \otimes U(2) \quad (4.97)$$

and diagonalize H within the carrier space of two-column U(n) irreps $\Gamma(2^a 1^b 0^c) \equiv \Gamma(a, b, c)$ with [4.39, 66]

$$a = \frac{1}{2}N - S , \quad b = 2S ,$$
$$c = n - a - b = n - \frac{1}{2}N - S , \quad (4.98)$$

considering the states of multiplicity $(2S+1)$ involving n orbitals and N electrons. The dimension of each spin-adapted subproblem equals [4.39, 66]

$$\dim \Gamma(2^a 1^b 0^c) = \frac{b+1}{n+1} \binom{n+1}{a} \binom{n+1}{c} , \quad (4.99)$$

where $\binom{m}{n}$ designate binomial coefficients.

Exploiting simplified irrep labeling by triples of integers (a, b, c), (4.98), at each level of the canonical chain (4.67), one achieves more efficient state labeling by replacing Gel'fand tableaux (4.68) by $n \times 3$ *ABC* [4.66] or *Paldus or Gel'fand–Paldus tableaux* [4.40, 67–75]

$$[P] = [a_i b_i c_i] , \quad (4.100)$$

where $a_i + b_i + c_i = i$. Another convenient labeling uses the ternary *step numbers* d_i, $0 \le d_i \le 3$ [4.66–68, 76, 77]

$$d_i = 1 + 2(a_i - a_{i-1}) - (c_i - c_{i-1}) . \quad (4.101)$$

An efficient and transparent representation of this basis can be achieved in terms of *Shavitt graphs* and *distinct row tables* ([4.67, 68], cf. also [4.10–12, 39, 69]). An efficient evaluation of generator matrix representatives, as well as of their products, is formulated in terms of products of *segment values*, whose explicit form has been derived in several different ways [4.10–12, 66–69, 73–75, 77, 78]. Since the dimension (4.99) rapidly increases with n and N, various truncated schemes (limited CI) are often employed. The unitary group formalism that is based either on U(n) or on the universal enveloping algebra of U(n) proved to be of great usefulness in various post-Hartree–Fock approaches to molecular electronic structure [4.79], especially in large-scale CI calculations (in particular in the COLUMBUS Program System [4.80]; see also [24–31] in [4.12]) and in the spin-adapted UGA version of the coupled cluster (CC) method [4.81–83] (cf. Chapter 5; for applications, see [4.84–86]), as well as in various other investigations (e.g. quantum dots [4.87], charge migration in fragmentation of peptide ions [4.88, 89]; see also [4.10–12] for other references).

4.3.7 Clifford Algebra Unitary Group Approach

The Clifford algebra unitary group approach (CAUGA) exploits a realization of the spinor algebra of the rotation group SO($2n+1$) in the covering algebra of U(2^n) to obtain explicit representation matrices for the U(n) [or SO($2n+1$) or SO($2n$)] generators in the basis adapted to the chain [4.90–94]

$$\mathrm{U}(2^n) \supset \mathrm{Spin}(m) \supset \mathrm{SO}(m) \supset \mathrm{U}(n)\,,\quad (m = 2n+1 \text{ or } m = 2n) \tag{4.102}$$

supplemented, if desired, by the canonical chain (4.67) for U(n).

To realize the connection with the fermionic Grassmann algebra generated by the creation $(X_I^\dagger)$ and annihilation (X_I) operators, $I = 1, \dots, 2n$, note that it is isomorphic with the Clifford algebra C_{4n} when we define [4.12, 25]

$$\alpha_I = X_I + X_I^\dagger\,,\quad \alpha_{I+2n} = \mathrm{i}(X_I - X_I^\dagger)\,,\quad (I = 1, \dots, 2n)\,. \tag{4.103}$$

For practical applications, the most important is the final imbedding U(2^n) $\supset$ U(n), (for the role of intermediate groups, see [4.90–92]). All states of an n-orbital model, regardless the electron number N and the total spin S, are contained in a single two-box totally symmetric irrep $\langle 2\dot{0}\rangle$ of U(2^n) [4.93, 94]. To simplify the notation, one employs the one-to-one correspondence between the Clifford algebra monomials, labeled by the occupation numbers $m_i = 0$ or $m_i = 1$ $(i = 1, \dots, n)$, and "multiparticle" single-column U(n) states labeled by

$$p \equiv p\{m_i\} = 2^n - (m_1\, m_2 \cdots m_n)_2\,, \tag{4.104}$$

where the occupation number array $(m_1 \cdots m_n)$ is interpreted as a binary integer, which we then regard as one-box states $|p)$ of U(2^n). The orbital U(n) generators Λ_{ij} may then be expressed as simple linear combinations of U(2^n) generators $E_{pq} = |p)(q|$ with coefficients equal to ±1 [4.93, 94].

Generally, any p-column U(n) irrep is contained at least once in the totally symmetric p-box irrep of U(2^n). For many-electron problems, one thus requires a two-box irrep $\langle 2\dot{0}\rangle$. Any state arising in the U(n) irrep $\Gamma(a, b, c)$ can then be represented as a linear combination of two-box states, labeled by the Weyl tableaux $[i|j] \equiv \boxed{i}\boxed{j}$. In particular, the highest weight state of $\Gamma(a, b, c)$ is represented by $\left[2^c|2^{b+c}\right]$. Once this representation is available, it is straightforward to compute explicit representations of U(n) generators, since E_{pq} act trivially on $[i|j]$ [4.94]. Defining unnormalized states $(i|j)$ as

$$(i|j) = \sqrt{1+\delta_{ij}}\;\;[i|j]\,, \tag{4.105}$$

we have

$$E_{pq}(i|j) = \delta_{qi}(p|j) + \delta_{qj}(i|p)\,. \tag{4.106}$$

The main features of CAUGA may thus be summarized as follows: CAUGA

1. effectively reduces an N-electron problem to a number of two-boson problems;
2. enables an exploitation of an arbitrary coupling scheme (being particularly suited for the valence bond method);
3. can be applied to particle-number nonconserving operators;
4. easily extends to fermions with an arbitrary spin;
5. drastically simplifies evaluation of explicit representations of U(n) generators and of their products;
6. can be exploited in other than shell-model approaches [4.95–101].

4.3.8 Spin-Dependent Operators

The spin-adapted U(n)-based UGA is entirely satisfactory in most investigations of molecular electronic structure. However, when exploring the fine structure in high-resolution spectra, the intersystem crossings, phosphorescent lifetimes, molecular predissociation, spin–orbit interactions in transition metals, and like phenomena, the *explicitly* spin-dependent terms must be included in the Hamiltonian. Since in most cases the total spin S represents a good approximate quantum number, so that the spin-adapted N-electron states render an excellent point of departure, it is necessary to consider the corresponding matrix elements of general spin-orbital U($2n$) generators in terms of which the relevant spin-dependent terms may be expressed. This was first done in the context of the symmetric group and Racah algebra by *Drake* and *Schlesinger* [4.78] and later on in terms of the *Gel'fand–Paldus tableaux* [4.102–106].

In general, the U($2n$) generators $e_{i\sigma,j\tau} \equiv e_{IJ}$ may be resolved into the spin-shift components $e(\pm)_{IJ}$ that increase (+) or decrease (−) the total spin S by one unit and the zero-spin component $e(0)_{IJ}$ that preserves S. The relevant matrix elements can then be expressed in terms of the matrix elements of a single U(n) adjoint tensor operator Δ, which is given by the following second degree polynomial in U(n) generators,

$$\Delta = \boldsymbol{E}(\boldsymbol{E} + N/2 - n - 2), \quad \boldsymbol{E} = \| E_{ij} \| \qquad (4.107)$$

and by the well-known matrix elements of U(2) or SU(2) generators in terms of the pure spin states [4.102, 103] (see also [4.107, 108]). The operator (4.107), referred to as the *Gould–Paldus operator* [4.109], also plays a key role in the determination of reduced density matrices [4.110, 111], and has been recently exploited in the multireference spin-adapted variant of the *density functional theory* [4.109].

References

4.1 A. O. Barut: *Dynamical Groups and Generalized Symmetries in Quantum Theory*, Vols. 1 and 2 (Univ. Canterbury, Christchurch 1971)

4.2 A. Bohm, Y. Ne'eman, A. O. Barut: *Dynamical Groups and Spectrum Generating Algebras* (World Scientific, Singapore 1988)

4.3 R. Gilmore: *Lie Groups, Lie Algebras, and Some of Their Applications* (Wiley, New York 1974)

4.4 B. G. Wybourne: *Classical Groups for Physicists* (Wiley, New York 1974)

4.5 J.-Q. Chen: *Group Representation Theory for Physicists* (World Scientific, Singapore 1989)

4.6 J.-Q. Chen, J. Ping, F. Wang: *Group Representation Theory for Physicists*, 2nd edn. (World Scientific, Singapore 2002)

4.7 O. Castaños, A. Frank, R. Lopez-Peña: J. Phys. A: Math. Gen. **23**, 5141 (1990)

4.8 F. Iachello, R. D. Levine: *Algebraic Theory of Molecules* (Oxford Univ. Press, Oxford 1995)

4.9 A. Frank, P. Van Isacker: *Algebraic Methods in Molecular and Nuclear Structure* (Wiley, New York 1994)

4.10 J. Paldus: *Mathematical Frontiers in Computational Chemical Physics*, ed. by D. G. Truhlar (Springer, Berlin, Heidelberg 1988) pp. 262–299

4.11 I. Shavitt: *Mathematical Frontiers in Computational Chemical Physics*, ed. by D. G. Truhlar (Springer, Berlin, Heidelberg 1988) pp. 300–349

4.12 J. Paldus: *Contemporary Mathematics*, Vol. 160 (AMS, Providence 1994) pp. 209–236

4.13 J. C. Parikh: *Group Symmetries in Nuclear Structure* (Plenum, New York 1978)

4.14 B. G. Adams, J. Čížek, J. Paldus: Int. J. Quantum Chem. **21**, 153 (1982)

4.15 B. G. Adams, J. Čížek, J. Paldus: Adv. Quantum Chem. **19**, 1 (1988)

4.16 J. Čížek, J. Paldus: Int. J. Quantum Chem. **12**, 875 (1977)

4.17 B. G. Adams: *Algebraic Approach to Simple Quantum Systems* (Springer, Berlin, Heidelberg 1994)

4.18 A. Bohm: Nuovo Cimento A **43**, 665 (1966)

4.19 A. Bohm: *Quantum Mechanics* (Springer, New York 1979)

4.20 M. A. Naimark: *Linear Representations of the Lorentz Group* (Pergamon, New York 1964)

4.21 W. Pauli: Z. Phys. **36**, 336 (1926)

4.22 V. Fock: Z. Phys. **98**, 145 (1935)

4.23 A. O. Barut: Phys. Rev. B **135**, 839 (1964)

4.24 A. O. Barut, G. L. Bornzin: J. Math. Phys. **12**, 841 (1971)

4.25 G. A. Zaicev: *Algebraic Problems of Mathematical and Theoretical Physics* (Science Publ. House, Moscow 1974) (in Russian)

4.26 A. Joseph: Rev. Mod. Phys. **39**, 829 (1967)

4.27 M. Bednar: Ann. Phys. (N. Y.) **75**, 305 (1973)

4.28 J. Čížek, E. R. Vrscay: *Group Theoretical Methods in Physics*, ed. by R. Sharp, B. Coleman (Academic, New York 1977) pp. 155–160

4.29 J. Čížek, E. R. Vrscay: Int. J. Quantum Chem. **21**, 27 (1982)

4.30 J. E. Avron, B. G. Adams, J. Čížek, M. Clay, L. Glasser, P. Otto, J. Paldus, E. R. Vrscay: Phys. Rev. Lett. **43**, 691 (1979)

4.31 J. Čížek, M. Clay, J. Paldus: Phys. Rev. A **22**, 793 (1980)

4.32 H. J. Silverstone, B. G. Adams, J. Čížek, P. Otto: Phys. Rev. Lett. **43**, 1498 (1979)

4.33 E. R. Vrscay: Phys. Rev. A **31**, 2054 (1985)

4.34 E. R. Vrscay: Phys. Rev. A **33**, 1433 (1986)

4.35 H. J. Silverstone, R. K. Moats: Phys. Rev. A **23**, 1645 (1981)

4.36 H. Weyl: *Classical Groups* (Princeton Univ. Press, Princeton 1939)

4.37 A. O. Barut, R. Raczka: *Theory of Group Representations and Applications* (Polish Science Publ., Warszawa 1977)

4.38 J. D. Louck: Am. J. Phys. **38**, 3 (1970)

4.39 J. Paldus: *Theoretical Chemistry: Advances and Perspectives*, Vol. 2, ed. by H. Eyring, D. J. Henderson (Academic, New York 1976) pp. 131–290

4.40 F. A. Matsen, R. Pauncz: *The Unitary Group in Quantum Chemistry* (Elsevier, Amsterdam 1986)
4.41 H. Weyl: *The Theory of Groups and Quantum Mechanics* (Dover, New York 1964)
4.42 I. M. Gel'fand, M. L. Tsetlin: Dokl. Akad. Nauk SSSR **71**, 825, 1070 (1950)
4.43 G. E. Baird, L. C. Biedenharn: J. Math. Phys. **4**, 1449 (1963)
4.44 J. Deneen, C. Quesne: J. Phys. A **16**, 2995 (1983)
4.45 H. Boerner: *Representations of Groups*, 2nd edn. (North-Holland, Amsterdam 1970)
4.46 A. Ramakrishnan: *L-Matrix Theory or the Grammar of Dirac Matrices* (Tata McGraw-Hill, India 1972)
4.47 A. Arima, F. Iachello: Phys. Rev. Lett. **35**, 1069 (1975)
4.48 F. Iachello, A. Arima: *The Interacting Boson Model* (Cambridge Univ. Press, Cambridge 1987)
4.49 F. Iachello, P. Van Isacker: *The Interacting Boson–Fermion Model* (Cambridge Univ. Press, Cambridge 2004)
4.50 F. Iachello: Chem. Phys. Lett. **78**, 581 (1981)
4.51 F. Iachello, R. D. Levine: J. Chem. Phys. **77**, 3046 (1982)
4.52 O. S. van Roosmalen, F. Iachello, R. D. Levine, A. E. L. Dieperink: J. Chem. Phys. **79**, 2515 (1983)
4.53 J. Hornos, F. Iachello: J. Chem. Phys. **90**, 5284 (1989)
4.54 F. Iachello, S. Oss: J. Chem. Phys. **99**, 7337 (1993)
4.55 F. Iachello, S. Oss: J. Chem. Phys. **102**, 1141 (1995)
4.56 J.-Q. Chen, F. Iachello, J.-L. Ping: J. Chem. Phys. **104**, 815 (1996)
4.57 T. Sako, D. Aoki, K. Yamanouchi, F. Iachello: J. Chem. Phys. **113**, 6063 (2000)
4.58 T. Sako, K. Yamanouchi, F. Iachello: J. Chem. Phys. **113**, 7292 (2000)
4.59 T. Sako, K. Yamanouchi, F. Iachello: J. Chem. Phys. **114**, 9441 (2001)
4.60 T. Sako, K. Yamanouchi, F. Iachello: J. Chem. Phys. **117**, 1641 (2002)
4.61 F. Iachello, S. Oss: J. Chem. Phys. **104**, 6956 (1996)
4.62 F. Iachello, A. Leviatan, A. Mengoni: J. Chem. Phys. **95**, 1449 (1991)
4.63 T. Müller, P. H. Vaccaro, F. Pèrez-Bernal, F. Iachello: J. Chem. Phys. **111**, 5038 (1999)
4.64 A. Frank, R. Lemus, F. Iachello: J. Chem. Phys. **91**, 29 (1989)
4.65 M. Moshinsky: *Group Theory and the Many-Body Problem* (Gordon Breach, New York 1968)
4.66 J. Paldus: J. Chem. Phys. **61**, 5321 (1974)
4.67 I. Shavitt: Int. J. Quantum Chem. Symp. **11**, 131 (1977)
4.68 I. Shavitt: Int. J. Quantum Chem. Symp. **12**, 5 (1978)
4.69 J. Hinze (Ed.): *The Unitary Group for the Evaluation of Electronic Energy Matrix Elements, Lect. Notes Chem.*, Vol. 22 (Springer, Berlin, Heidelberg 1981)
4.70 R. Pauncz: *Spin Eigenfunctions: Construction and Use* (Plenum, New York 1979) Chap. 9
4.71 S. Wilson: *Electron Correlation in Molecules* (Clarendon, Oxford 1984) Chap. 5
4.72 R. McWeeny: *Methods of Molecular Quantum Mechanics*, 2 edn. (Academic, New York 1989) Chap. 10
4.73 M. D. Gould, G. S. Chandler: Int. J. Quantum Chem. **25**, 553, 603, 1089 (1984)
4.74 M. D. Gould, G. S. Chandler: Int. J. Quantum Chem. **26**, 441 (1984)
4.75 M. D. Gould, G. S. Chandler: Int. J. Quantum Chem. **27**, 878 (1985), Erratum
4.76 J. Paldus: Phys. Rev. A **14**, 1620 (1976)
4.77 J. Paldus, M. J. Boyle: Phys. Scr. **21**, 295 (1980)
4.78 G. W. F. Drake, M. Schlesinger: Phys. Rev. A **15**, 1990 (1977)
4.79 M. A. Robb: Theor. Chem. Acc. **103**, 317 (2000)
4.80 R. Shepard, I. Shavitt, R. M. Pitzer, D. C. Comeau, M. Pepper, H. Lischka, P. G. Szalay, R. Ahlrichs, F. B. Brown, J.-G. Zhao: Int. J. Quantum Chem. Symp. **22**, 149 (1988)
4.81 J. Paldus, B. Jeziorski: Theor. Chim. Acta **86**, 83 (1993)
4.82 X. Li, J. Paldus: J. Chem. Phys. **101**, 8812 (1994)
4.83 B. Jeziorski, J. Paldus, P. Jankowski: Int. J. Quantum Chem. **56**, 129 (1995)
4.84 X. Li, J. Paldus: J. Chem. Phys. **104**, 9555 (1996)
4.85 X. Li, J. Paldus: J. Mol. Struct. (theochem) **527**, 165 (2000)
4.86 P. Jankowski, B. Jeziorski: J. Chem. Phys. **111**, 1857 (1999)
4.87 F. Remacle, R. D. Levin: Chem. Phys. Chem. **2**, 20 (2001)
4.88 F. Remacle, R. D. Levin: J. Chem. Phys. **110**, 5089 (1999)
4.89 F. Remacle, R. D. Levin: J. Phys. Chem. A **104**, 2341 (2000)
4.90 R. S. Nikam, C. R. Sarma: J. Math. Phys. **25**, 1199 (1984)
4.91 C. R. Sarma, J. Paldus: J. Math. Phys. **26**, 1140 (1985)
4.92 M. D. Gould, J. Paldus: J. Math. Phys. **28**, 2304 (1987)
4.93 J. Paldus, C. R. Sarma: J. Chem. Phys. **83**, 5135 (1985)
4.94 J. Paldus, M.-J. Gao, J.-Q. Chen: Phys. Rev. A **35**, 3197 (1987)
4.95 J. Paldus: *Relativistic and Electron Correlation Effects in Molecules and Solids*, NATO ASI Series B, ed. by L. Malli G. (Plenum, New York 1994) pp. 207–282
4.96 J. Paldus, X. Li: *Symmetries in Science VI: From the Rotation Group to Quantum Algebras*, ed. by B. Gruber (Plenum, New York 1993) pp. 573–592
4.97 J. Paldus, J. Planelles: Theor. Chim. Acta **89**, 13 (1994)
4.98 J. Planelles, J. Paldus, X. Li: Theor. Chim. Acta **89**, 33, 59 (1994)
4.99 X. Li, J. Paldus: Int. J. Quantum Chem. **41**, 117 (1992)
4.100 X. Li, J. Paldus: J. Chem. Phys. **102**, 8059 (1995)
4.101 P. Piecuch, X. Li, J. Paldus: Chem. Phys. Lett. **230**, 377 (1994)
4.102 M. D. Gould, J. Paldus: J. Chem. Phys. **92**, 7394 (1990)
4.103 M. D. Gould, J. S. Battle: J. Chem. Phys. **99**, 5961 (1993)
4.104 R. D. Kent, M. Schlesinger: Phys. Rev. A **42**, 1155 (1990)
4.105 R. D. Kent, M. Schlesinger: Phys. Rev. A **50**, 186 (1994)
4.106 R. D. Kent, M. Schlesinger, I. Shavitt: Int. J. Quantum Chem. **41**, 89 (1992)
4.107 M. D. Gould, J. S. Battle: J. Chem. Phys. **98**, 8843 (1993)

4.108 M. D. Gould, J. S. Battle: J. Chem. Phys. **99**, 5983 (1993)
4.109 Y. G. Khait, M. R. Hoffmann: J. Chem. Phys. **120**, 5005 (2004)
4.110 M. D. Gould, J. Paldus, G. S. Chandler: J. Chem. Phys. **93**, 4142 (1990)
4.111 J. Paldus, M. D. Gould: Theor. Chim. Acta **86**, 83 (1993)

5. Perturbation Theory

Perturbation theory (PT) represents one of the bridges that takes us from a simpler, exactly solvable (unperturbed) problem to a corresponding real (perturbed) problem by expressing its solutions as a series expansion in a suitably chosen "small" parameter ε in such a way that the problem reduces to the unperturbed problem when $\varepsilon = 0$. It originated in classical mechanics and eventually developed into an important branch of applied mathematics enabling physicists and engineers to obtain approximate solutions of various systems of differential equations [5.1–4]. For the problems of atomic and molecular structure and dynamics, the perturbed problem is usually given by the time-independent or time-dependent Schrödinger equation [5.5–8].

5.1 **Matrix Perturbation Theory (PT)** 101
5.1.1 Basic Concepts 101
5.1.2 Level-Shift Operators 102
5.1.3 General Formalism 102
5.1.4 Nondegenerate Case 103

5.2 **Time-Independent Perturbation Theory** 103
5.2.1 General Formulation 103
5.2.2 Brillouin–Wigner and Rayleigh–Schrödinger PT (RSPT)... 104
5.2.3 Bracketing Theorem and RSPT..... 104

5.3 **Fermionic Many-Body Perturbation Theory (MBPT)**.................................... 105
5.3.1 Time Independent Wick's Theorem 105
5.3.2 Normal Product Form of PT......... 105
5.3.3 Møller–Plesset and Epstein–Nesbet PT 106
5.3.4 Diagrammatic MBPT................... 107
5.3.5 Vacuum and Wave Function Diagrams 107
5.3.6 Hartree–Fock Diagrams.............. 108
5.3.7 Linked and Connected Cluster Theorems................................ 108
5.3.8 Coupled Cluster Theory.............. 109

5.4 **Time-Dependent Perturbation Theory** ... 111
5.4.1 Evolution Operator PT Expansion 111
5.4.2 Gell–Mann and Low Formula...... 111
5.4.3 Potential Scattering and Quantum Dynamics............ 111
5.4.4 Born Series.............................. 112
5.4.5 Variation of Constants Method.... 112

References 113

5.1 Matrix Perturbation Theory (PT)

A prototype of a time-independent PT considers an eigenvalue problem for the Hamiltonian H of the form

$$H = H_0 + V\,, \quad V = \sum_{i=1}^{\infty} \varepsilon^i V_i\,, \tag{5.1}$$

acting in a (finite-dimensional) Hilbert space $\mathcal{V}_n$, assuming that the spectral resolution of the unperturbed operator H_0 is known; i. e.,

$$H_0 = \sum_i \omega_i P_i\,, \quad P_i P_j = \delta_{ij} P_j\,, \quad \sum_i P_i = I\,, \tag{5.2}$$

where ω_i are distinct eigenvalues of H_0, the P_i form a complete orthonormal set of Hermitian idempotents and I is the identity operator on $\mathcal{V}_n$. The PT problem for H can then be formulated within the Lie algebra $\mathcal{A}$ (see Sect. 3.2) generated by H_0 and V [5.9, 10].

5.1.1 Basic Concepts

Define the diagonal part $\langle X \rangle$ of a general operator $X \in \mathcal{A}$ by

$$\langle X \rangle = \sum_i P_i X P_i\,, \tag{5.3}$$

and recall that the adjoint action of $X \in \mathcal{A}$, $\mathrm{ad}X : \mathcal{A} \to \mathcal{A}$, is defined by

$$\mathrm{ad}\, X(Y) = [X, Y]\,, \quad (\forall\, Y \in \mathcal{A})\,, \tag{5.4}$$

where the square bracket denotes the commutator. The key problem of PT is the 'inversion' of this operation, i. e., the solution of the equation [5.9–11]

$$\mathrm{ad}H_0(X) \equiv [H_0, X] = Y \,. \tag{5.5}$$

Assuming that $\langle Y \rangle = 0$, then

$$X = R(Y) + \langle A \rangle \,, \tag{5.6}$$

where $A \in \mathcal{A}$ is arbitrary and

$$R(Y) = \sum_{i \neq j} \Delta_{ij}^{-1} P_i Y P_j \,, \tag{5.7}$$

with $\Delta_{ij} = \omega_i - \omega_j$, represents the solution of (5.5) with the vanishing diagonal part $\langle R(Y) \rangle = 0$.

5.1.2 Level-Shift Operators

To solve the PT problem for H, (5.1), we search for a unitary level-shift transformation U [5.9, 10], $U^\dagger U = UU^\dagger = I$,

$$UHU^\dagger = U(H_0 + V)U^\dagger = H_0 + E \,, \tag{5.8}$$

where the level-shift operator E satisfies the condition

$$E = \langle E \rangle \,. \tag{5.9}$$

To guarantee the unitarity of U, we express it in the form

$$U = \mathrm{e}^G \,, \quad G^\dagger = -G \,, \quad \langle G \rangle = 0 \,. \tag{5.10}$$

Using the Haussdorff formula

$$\mathrm{e}^A B \mathrm{e}^{-A} = \sum_{k=0}^{\infty} (k!)^{-1} (\,\mathrm{ad}A)^k B \,, \tag{5.11}$$

and defining the operator

$$F = [H_0, G] \,, \tag{5.12}$$

we find from (5.8) that

$$\begin{aligned} E + F &= V + \frac{1}{2}[G, V + E] \\ &\quad + \sum_{k=2}^{\infty} (k!)^{-1} B_k (\,\mathrm{ad}G)^k (V - E) \,, \end{aligned} \tag{5.13}$$

where we used the identity [5.12]

$$\left(\sum_{k=0}^{\infty} \frac{B_k}{k!} X^k \right) \left(\sum_{k=0}^{\infty} \frac{1}{(k+1)!} X^k \right) = I \,, \tag{5.14}$$

and B_k designates the Bernoulli numbers [5.13]

$$\begin{aligned} &B_0 = 1 \,, \quad B_1 = -\frac{1}{2} \,, \quad B_2 = \frac{1}{6} \,, \\ &B_{2k+1} = 0 \quad (k \geq 1) \,, \\ &B_4 = \frac{1}{30} \,, \quad B_6 = \frac{1}{42} \,, \quad \text{etc.} \end{aligned} \tag{5.15}$$

5.1.3 General Formalism

Introducing the PT expansion for relevant operators,

$$X = \sum_{i=1}^{\infty} \varepsilon^i X_i \,, \quad X = E, F, G \,; \quad F_i = [H_0, G_i] \,, \tag{5.16}$$

(5.13) leads to the following system of equations

$$\begin{aligned} &E_1 + F_1 = V_1 \,, \\ &E_2 + F_2 = V_2 + \frac{1}{2}[G_1, V_1 + E_1] \,, \\ &E_3 + F_3 = V_3 + \frac{1}{2}[G_1, V_2 + E_2] \\ &\qquad + \frac{1}{2}[G_2, V_1 + E_1] \\ &\qquad + \frac{1}{12}[G_1, [G_1, V_1 - E_1]] \,, \\ &\text{etc.} \,, \end{aligned} \tag{5.17}$$

which can be solved recursively for E_i and G_i by taking their diagonal part and applying operator R, (5.7), since

$$\begin{aligned} &\langle E_i \rangle = E_i \,, \qquad \langle G_i \rangle = \langle F_i \rangle = 0 \,, \\ &RF_i = G_i \,, \qquad RE_i = 0 \,. \end{aligned} \tag{5.18}$$

We thus get

$$\begin{aligned} &E_1 = \langle V_1 \rangle \,, \\ &E_2 = \langle V_2 \rangle + \frac{1}{2} \langle [RV_1, V_1] \rangle \,, \\ &E_3 = \langle V_3 \rangle + \langle [RV_1, V_2] \rangle \\ &\qquad + \frac{1}{6} \langle [RV_1, [RV_1, 2V_1 + E_1]] \rangle \,, \quad \text{etc.} \,, \end{aligned} \tag{5.19}$$

and

$$\begin{aligned} &G_1 = RV_1 \,, \\ &G_2 = RV_2 + \frac{1}{2} R[RV_1, V_1 + E_1] \,, \quad \text{etc.} \end{aligned} \tag{5.20}$$

Since

$$\begin{aligned} &\langle R(X) \rangle = R\langle X \rangle = 0 \,, \qquad \langle R(X)Y \rangle = -\langle XR(Y) \rangle \,, \\ &R(X\langle Y \rangle) = R(X)\langle Y \rangle, \qquad R(\langle X \rangle Y) = \langle X \rangle R(Y) \,, \end{aligned} \tag{5.21}$$

these relationships can be transformed to a more conventional form

$$\begin{aligned} E_2 &= \langle V_2\rangle - \langle V_1 R V_1\rangle\,,\\ E_3 &= \langle V_3\rangle - \langle V_1 R V_2\rangle - \langle V_2 R V_1\rangle\\ &+ \frac{1}{6}\langle R(V_1)R(V_1)[2V_1+\langle V_1\rangle]\rangle\\ &- \frac{1}{3}\langle R(V_1)[2V_1+\langle V_1\rangle]R(V_1)\rangle\\ &+ \frac{1}{6}\langle[2V_1+\langle V_1\rangle]R(V_1)R(V_1)\rangle\,, \quad \text{etc.} \end{aligned} \tag{5.22}$$

However, in this way certain nonphysical terms arise that exactly cancel when the commutator form is employed (Sect. 5.3.7).

5.1.4 Nondegenerate Case

In the nondegenerate case, when $P_i = |i\rangle\langle i|$, with $|i\rangle$ representing the eigenvector of H_0 associated with the eigenvalue ω_i, the level-shift operator is diagonal and its explicit PT expansion (as well as that for the corresponding eigenvectors) is easily obtained from (5.19) and (5.20). Writing x_{ij} for the matrix element $\langle i|X|j\rangle$, we get

$$\begin{aligned} (e_1)_{ii} &= (v_1)_{ii}\,,\\ (e_2)_{ii} &= (v_2)_{ii} - {\sum_j}' \frac{(v_1)_{ij}(v_1)_{ji}}{\Delta_{ji}}\,,\\ (e_3)_{ii} &= (v_3)_{ii} - {\sum_j}' \frac{(v_1)_{ij}(v_2)_{ji}+(v_2)_{ij}(v_1)_{ji}}{\Delta_{ji}}\\ &+ {\sum_{j,k}}' \frac{(v_1)_{ij}(v_1)_{jk}(v_1)_{ki}}{\Delta_{ji}\Delta_{ki}}\\ &- (v_1)_{ii} {\sum_j}' \frac{(v_1)_{ij}(v_1)_{ji}}{\Delta_{ji}^2}\,,\\ &\text{etc.}, \end{aligned} \tag{5.23}$$

the prime on the summation symbols indicating that the terms with the vanishing denominator are to be deleted.

Note that in contrast to PT expansions which directly expand the level-shift transformation U, $U = 1+\varepsilon U_1 + \varepsilon^2 U_2 + \cdots$, the above Lie algebraic formulation has the advantage that U stays unitary in every order of PT. This is particularly useful in spectroscopic applications, such as line broadening.

5.2 Time-Independent Perturbation Theory

For stationary problems, particularly those arising in atomic and molecular electronic structure studies relying on *ab initio* model Hamiltonians, the PT of Sect. 5.1 can be given a more explicit form which avoids a priori the nonphysical, size inextensive terms [5.6–8, 14, 15].

5.2.1 General Formulation

We wish to find the eigenvalues and eigenvectors of the full (perturbed) problem

$$K|\Psi_i\rangle \equiv (K_0 + W)|\Psi_i\rangle = k_i|\Psi_i\rangle\,, \tag{5.24}$$

assuming we know those of the unperturbed problem

$$K_0|\Phi_i\rangle = \kappa_i|\Phi_i\rangle, \quad \langle\Phi_i|\Phi_j\rangle = \delta_{ij}\,. \tag{5.25}$$

For simplicity, we restrict ourselves to the nondegenerate case ($\kappa_i \neq \kappa_j$ if $i \neq j$) and consider only the first order perturbation [see (5.1), $\varepsilon V_1 \equiv W$, $V_i = 0$ for $i \geq 2$]. Of course, K and K_0 are Hermitian operators acting in a Hilbert space which, in ab initio applications, is finite-dimensional.

Using the intermediate normalization for $|\Psi_i\rangle$,

$$\langle\Psi_i|\Phi_i\rangle = 1\,, \tag{5.26}$$

the asymmetric energy formula gives

$$k_i = \kappa_i + \langle\Phi_i|W|\Psi_i\rangle\,. \tag{5.27}$$

The idempotent Hermitian projectors

$$P_i = |\Phi_i\rangle\langle\Phi_i|,\ Q_i = P_i^{\perp} = 1 - P_i = \sum_{j(\neq i)}|\Phi_j\rangle\langle\Phi_j|\,, \tag{5.28}$$

commute with K_0, so that

$$(\lambda - K_0)Q_i|\Psi_i\rangle = Q_i(\lambda - k_i + W)|\Psi_i\rangle\,, \tag{5.29}$$

λ being an arbitrary scalar (note that we write λI simply as λ). Since the resolvent $(\lambda - K_0)^{-1}$ of K_0 is nonsingular on the orthogonal complement of the ith eigenspace, we get

$$Q_i|\Psi_i\rangle = |\Psi_i\rangle - |\Phi_i\rangle = R_i(\lambda)(\lambda - k_i + W)|\Psi_i\rangle\,, \tag{5.30}$$

where

$$R_i \equiv R_i(\lambda) = (\lambda - K_0)^{-1} Q_i$$
$$= Q_i(\lambda - K_0)^{-1} = \sum_{j(\neq i)} \frac{|\Phi_j\rangle\langle\Phi_j|}{\lambda - \kappa_j}\,, \tag{5.31}$$

assuming ($\lambda \neq \kappa_j$). Iterating this relationship, we get prototypes of the desired PT expansion for $|\Psi_i\rangle$,

$$|\Psi_i\rangle = \sum_{n=0}^{\infty} [R_i(\lambda - k_i + W)]^n |\Phi_i\rangle\,, \tag{5.32}$$

and, from (5.27), for k_i,

$$k_i = \kappa_i + \sum_{n=0}^{\infty} \langle\Phi_i|W[R_i(\lambda - k_i + W)]^n|\Phi_i\rangle\,. \tag{5.33}$$

5.2.2 Brillouin–Wigner and Rayleigh–Schrödinger PT (RSPT)

So far, the parameter λ was arbitrary, as long as $\lambda \neq \kappa_j$ $(j \neq i)$. The following two choices lead to the two basic types of many-body perturbation theory (MBPT):

Brillouin–Wigner (BW) PT
Setting $\lambda = k_i$ gives

$$k_i = \kappa_i + \sum_{n=0}^{\infty} \langle\Phi_i|W\left(R_i^{(\mathrm{BW})} W\right)^n|\Phi_i\rangle\,, \tag{5.34}$$

$$|\Psi_i\rangle = \sum_{n=0}^{\infty} \left(R_i^{(\mathrm{BW})} W\right)^n |\Phi_i\rangle\,, \tag{5.35}$$

where

$$R_i^{(\mathrm{BW})} = \sum_{j(\neq i)} \frac{|\Phi_j\rangle\langle\Phi_j|}{k_i - \kappa_j}\,. \tag{5.36}$$

Rayleigh–Schrödinger (RS) PT
Setting $\lambda = \kappa_i$ gives

$$k_i = \kappa_i + \sum_{n=0}^{\infty} \langle\Phi_i|W\left[R_i^{(\mathrm{RS})}(\kappa_i - k_i + W)\right]^n|\Phi_i\rangle\,, \tag{5.37}$$

$$|\Psi_i\rangle = \sum_{n=0}^{\infty} \left[R_i^{(\mathrm{RS})}(\kappa_i - k_i + W)\right]^n |\Phi_i\rangle\,, \tag{5.38}$$

where now

$$R_i \equiv R_i^{(\mathrm{RS})} = \sum_{j(\neq i)} \frac{|\Phi_j\rangle\langle\Phi_j|}{\kappa_i - \kappa_j}\,. \tag{5.39}$$

The main distinction between these two PTs lies in the fact that the BW form has the exact eigenvalues appearing in the denominators, and thus leads to polynomial expressions for k_i. Although these are not difficult to solve numerically, since the eigenvalues are separated, the resulting energies are never size extensive and thus unusable for extended systems. They are also unsuitable for finite systems when the particle number changes, as in various dissociation processes. From now on, we thus investigate only the RSPT, which yields a fully size-extensive theory.

5.2.3 Bracketing Theorem and RSPT

Expressions (5.37) and (5.38) are not explicit, since they involve the exact eigenvalues k_i on the right-hand side. To achieve an order by order separation, set

$$k_i \equiv k = \sum_{j=0}^{\infty} k^{(j)}\,, \quad |\Psi_i\rangle \equiv |\Psi\rangle = \sum_{j=0}^{\infty} |\Psi^{(j)}\rangle\,, \tag{5.40}$$

where the superscript (j) indicates the jth-order in the perturbation W. We only consider the eigenvalue expressions, since the corresponding eigenvectors are easily recovered from them by removing the bra state and the first interaction W [see (5.37) and (5.38)]. We also simplify the mean value notation writing for a general operator X,

$$\langle X\rangle \equiv \langle\Phi_i|X|\Phi_i\rangle\,. \tag{5.41}$$

Substituting the first expansion (5.40) into (5.37) and collecting the terms of the same order in W, we get

$$\begin{aligned}
k^{(0)} &= \kappa_i\,,\\
k^{(1)} &= \langle W\rangle\,,\\
k^{(2)} &= \langle WRW\rangle\,,\\
k^{(3)} &= \left\langle W(RW)^2\right\rangle - \langle W\rangle\left\langle WR^2W\right\rangle,\\
k^{(4)} &= \left\langle W(RW)^3\right\rangle\\
&\quad - \langle W\rangle\left(\left\langle WR(RW)^2\right\rangle + \left\langle (WR)^2RW\right\rangle\right)\\
&\quad + \langle W\rangle^2\left\langle WR^3W\right\rangle - \langle WRW\rangle\left\langle WR^2W\right\rangle, \quad \text{etc.}
\end{aligned} \tag{5.42}$$

The general expression has the form

$$k^{(n)} = \left\langle W(RW)^{n-1}\right\rangle + \mathcal{R}^{(n)}\,, \tag{5.43}$$

the first term on the right-hand side being referred to as the *principal* nth-order term, while $\mathcal{R}^{(n)}$ designates the so-called *renormalization* terms that are obtained by the *bracketing theorem* [5.14, 16] as follows:

1. Insert the bracketings $\langle\cdots\rangle$ around the W, WRW, ..., $WR\cdots RW$ operator strings of the principal term in all possible ways.
2. Bracketings involving the rightmost and/or the leftmost interaction vanish.
3. The sign of each bracketed term is given by $(-1)^{n_\mathrm{B}}$, where n_B is the number of bracketings.
4. Bracketings within bracketings are allowed, e.g., $\langle WR\langle WR\langle W\rangle RW\rangle RW\rangle = \langle W\rangle\langle WR^2W\rangle^2$.
5. The total number of bracketings (including the principal term) is $(2n-2)!/[n!(n-1)!]$.

5.3 Fermionic Many-Body Perturbation Theory (MBPT)

5.3.1 Time Independent Wick's Theorem

The development of an explicit MBPT formalism is greatly facilitated by the exploitation of the time-independent version of Wick's theorem. This version of the theorem expresses an arbitrary product of creation ($a_\mu^\dagger$) and annihilation (a_μ) operators (see Chapt. 6)as a normal product (relative to $|\Phi_0\rangle$) and as normal products with all possible contractions of these operators [5.14, 15],

$$x_1x_2\cdots x_k = N[x_1x_2\cdots x_k] + \Sigma N[\overbracket{x_1\overbracket{x_2\cdots\cdots}x_k}] ,\quad (x_i = a_{\mu_i}^\dagger \ \text{ or } x_i = a_{\mu_i}) \tag{5.44}$$

where

$$\overbracket{a_\mu^\dagger a_\nu^\dagger} = \overbracket{a_\mu a_\nu} = 0 ,$$
$$\overbracket{a_\mu^\dagger a_\nu} = h(\mu)\delta_{\mu\nu} ,\quad \overbracket{a_\mu a_\nu^\dagger} = p(\mu)\delta_{\mu\nu} , \tag{5.45}$$

and

$$h(\mu) = 1 ,\ p(\mu) = 0 \text{ if } |\mu\rangle \text{ is occupied in } |\Phi_0\rangle \text{ (hole states)},$$
$$h(\mu) = 0 ,\ p(\mu) = 1 \text{ if } |\mu\rangle \text{ is unoccupied in } |\Phi_0\rangle \text{ (particle states)} . \tag{5.46}$$

The N-product with contractions is defined as a product of individual contractions times the N-product of uncontracted operators (defining $N[\emptyset] \equiv 1$ for an empty set) with the sign given by the parity of the permutation reordering the operators into their final order.

Note that the Fermi vacuum mean value of an N-product vanishes unless all operators are contracted. Thus, $\langle x_1x_2\cdots x_k\rangle$ is given by the sum over all possible fully contracted terms (vacuum terms). Similar rules follow for the expressions of the type $(x_1x_2\cdots x_k)|\Phi\rangle$. Moreover, if some operators on the left-hand side of (5.44) are already in the N-product form, all the terms involving contractions between these operators vanish.

5.3.2 Normal Product Form of PT

Consider the eigenvalue problem for a general *ab initio* or semi-empirical electronic Hamiltonian H with one- and two-body components Z and V, namely,

$$H|\Psi_i\rangle = E_i|\Psi_i\rangle ,$$
$$H = Z + V = \sum_i z(i) + \sum_{i<j} v(i, j) , \tag{5.47}$$

and a corresponding unperturbed problem

$$H_0|\Phi_i\rangle = \varepsilon_i|\Phi_i\rangle ,$$
$$H_0 = Z + U ,\quad \langle\Phi_i|\Phi_j\rangle = \delta_{ij} , \tag{5.48}$$

with U representing some approximation to V. In the case that U is also a one-electron operator, $U = \Sigma_i u(i)$, the unperturbed problem (5.48) is separable and reduces to a one-electron problem,

$$(z+u)|\mu\rangle = \omega_\mu|\mu\rangle , \tag{5.49}$$

which is assumed to be solved. Choosing the orthonormal spin orbitals $\{|\mu\rangle\}$ as a basis of the second quantization representation [Chapt. 6, (6.8)], the N-electron solutions of (5.48) can be represented as

$$|\Phi_i\rangle = a_{\mu_1}^\dagger a_{\mu_2}^\dagger \cdots a_{\mu_N}^\dagger|0\rangle , \tag{5.50}$$

$$\varepsilon_i = \sum_{j=1}^{N} \omega_{\mu_j} , \tag{5.51}$$

the state label i representing the occupied spin orbital set $\{\mu_1, \mu_2, \ldots, \mu_N\}$, while the one- and two-body operators take the form

$$X = \sum_{\mu,\nu} \langle\mu|x|\nu\rangle a_\mu^\dagger a_\nu ,\quad X = Z, U;\ x = z, u , \tag{5.52}$$

$$V = \frac{1}{2}\sum_{\mu,\nu,\sigma,\tau} \langle\mu\nu|v|\sigma\tau\rangle a_\mu^\dagger a_\nu^\dagger a_\tau a_\sigma . \tag{5.53}$$

Considering, for simplicity, a nondegenerate ground state $|\Phi\rangle \equiv |\Phi_0\rangle = a_1^\dagger a_2^\dagger \cdots a_N^\dagger |0\rangle$, referred to as a Fermi vacuum, we define the *normal product form* of these operators relative to $|\Phi\rangle$

$$X_N \equiv X - \langle X\rangle = \sum_{\mu,\nu} \langle\mu|x|\nu\rangle N\left[a_\mu^\dagger a_\nu\right], \tag{5.54a}$$

$$(X = Z, U, G; \quad x = z, u, g)$$

$$\begin{aligned} V_N &\equiv V - \langle V\rangle - G_N \\ &= \frac{1}{2} \sum_{\mu,\nu,\sigma,\tau} \langle\mu\nu|v|\sigma\tau\rangle N\left[a_\mu^\dagger a_\nu^\dagger a_\tau a_\sigma\right] \\ &= \frac{1}{4} \sum_{\mu,\nu,\sigma,\tau} \langle\mu\nu|v|\sigma\tau\rangle_A N\left[a_\mu^\dagger a_\nu^\dagger a_\tau a_\sigma\right], \end{aligned} \tag{5.54b}$$

where

$$\langle\mu|g|\nu\rangle = \sum_{\sigma=1}^{N} \langle\mu\sigma|v|\nu\sigma\rangle_A , \tag{5.55}$$

$$\langle\mu\nu|v|\sigma\tau\rangle_A = \langle\mu\nu|v|\sigma\tau\rangle - \langle\mu\nu|v|\tau\sigma\rangle , \tag{5.56}$$

$\langle X\rangle = \langle\Phi|X|\Phi\rangle$, and $N[\cdots]$ designates the normal product relative to $|\Phi\rangle$ [5.14, 15]. (Recall that $N[x_1 x_2 \cdots x_k] = \pm b_{\mu_1}^\dagger \cdots b_{\mu_i}^\dagger b_{\mu_{i+1}} \cdots b_{\mu_k}$, where $x_i = b_{\mu_i}$ or $b_{\mu_i}^\dagger$ are the annihilation and creation operators of the particle-hole formalism relative to $|\Phi\rangle$, i.e., $b_\mu = a_\mu^\dagger$ for $\mu \le N$ and $b_\mu = a_\mu$ for $\mu > N$, the sign being determined by the parity of the permutation $p : j \mapsto \mu_j$.)

Defining

$$K = H - \langle H\rangle, \quad K_0 = H_0 - \langle H_0\rangle = H_0 - \varepsilon_0 , \tag{5.57}$$

we can return to (5.24) and (5.25), where now

$$\begin{aligned} k_i &= E_i - \langle H\rangle, \quad \kappa_i = \varepsilon_i - \varepsilon_0, \\ \varepsilon_0 &= \sum_{\mu=1}^{N} \omega_\mu , \end{aligned} \tag{5.58}$$

and

$$W = K - K_0 = V - U - \langle V - U\rangle . \tag{5.59}$$

With this choice, $\langle W\rangle = 0$, so that for the reference state $|\Phi\rangle$, (5.42) simplify to (we drop the subscript 0 for simplicity)

$$\begin{aligned} k^{(0)} &= 0, \quad k^{(1)} = 0 , \\ k^{(2)} &= \langle WRW\rangle , \\ k^{(3)} &= \langle WRWRW\rangle , \\ k^{(4)} &= \left\langle W(RW)^3\right\rangle - \langle WRW\rangle\left\langle WR^2W\right\rangle, \quad \text{etc.} \end{aligned} \tag{5.60}$$

Note that W is also in the N-product form,

$$W = W_1 + W_2 , \quad W_1 = G_N - U_N , \quad W_2 = V_N . \tag{5.61}$$

5.3.3 Møller–Plesset and Epstein–Nesbet PT

Choosing $U = G$ we have

$$H_0 = Z + G \equiv F , \tag{5.62}$$

so that (5.49) represent Hartree–Fock (HF) equations, and ω_μ and $|\mu\rangle$ the canonical HF orbital energies and spin orbitals, respectively. Since $\langle H\rangle = \sum_{\mu=1}^{N} \left(\langle\mu|z|\mu\rangle + \frac{1}{2}\langle\mu|g|\mu\rangle\right)$ is the HF energy, $k = k_0$ gives directly the ground state correlation energy. (Note, however, that the N-product form of PT eliminates the first-order contribution $k^{(1)} = \langle W\rangle$ in any basis, even when F is not diagonal.) With this choice, $W_1 = 0$, $W = V_N$, and the denominators in (5.39) are given by the differences of HF orbital energies

$$\kappa_0 - \kappa_j = \sum_{i=1}^{\lambda} (\omega_{\mu_i} - \omega_{\nu_i}) \equiv \Delta(\{\mu_i\}; \{\nu_j\}) , \tag{5.63}$$

assuming that $|\Phi_j\rangle$ is a λ-times excited configuration relative to $|\Phi\rangle$ obtained through excitations $\mu_i \to \nu_i$, $i = 1, \ldots, \lambda$. Using the Slater rules (or the second quantization algebra), we can express the second-order contribution in terms of the two-electron integrals and HF orbital energies as

$$k^{(2)} = \frac{1}{2} \sum_{a,b,r,s} \frac{\langle ab|v|rs\rangle\left(\langle rs|v|ab\rangle - \langle rs|v|ba\rangle\right)}{\omega_a + \omega_b - \omega_r - \omega_s} , \tag{5.64}$$

where the summations over a, b (r, s) extend over all occupied (unoccupied) spin orbitals in $|\Phi\rangle$. Obtaining the corresponding higher-order corrections becomes more and more laborious and, beginning with the fourth-order, important cancellations arise between the principal and renormalization terms, even when the N-product form is employed. These will be addressed in Sect. 5.3.7.

The above outlined PT with H_0 given by the HF operator is often referred to as the *Møller–Plesset* PT [5.17] and, when truncated to the n-th order, is designated by the acronym MPn, $n = 2, 3, \ldots$. In this version, the two-electron integrals enter the denominators only through the HF orbital energies. In an alternative, less often employed variant, referred to as the *Epstein–Nesbet* PT [5.18, 19], the whole diagonal part of H is

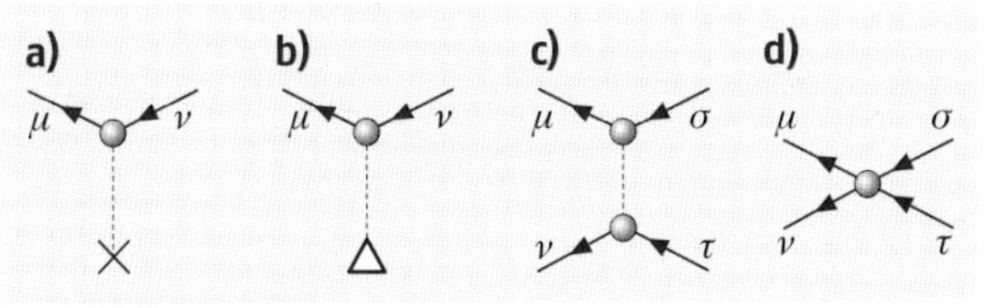

Fig. 5.1a–d Diagrammatic representation of one- and two-electron operators

used as the unperturbed Hamiltonian, i. e.,

$$H_0 = \sum_i \langle \Phi_i | H | \Phi_i \rangle P_i \,. \tag{5.65}$$

With this choice, the denominators are given as differences of the diagonal elements of the configuration interaction matrix.

5.3.4 Diagrammatic MBPT

To facilitate the evaluation of higher order terms, and especially to derive the general properties and characteristics of the MBPT, it is useful to employ a diagrammatic representation [5.6–8, 14, 15]. Representing all the operators in (5.42) and (5.43) or (5.60) in the second quantized form, we have to deal with the reference state (i. e., the Fermi vacuum) mean values of the strings of annihilation and creation operators (or with these strings acting on the reference in the case of a wave function). This is efficiently done using Wick's theorem and its diagrammatic representation via a special form of Feynman diagrams. In this representation we associate with various operators suitable vertices with incident oriented lines representing the creation (outgoing lines) and annihilation (ingoing lines) operators that are involved in their second quantization form. A few typical diagrams representing operators $(-U)$, W_1 and V are shown in Fig. 5.1a, Fig. 5.1b and Fig. 5.1c, Fig. 5.1d, respectively. Using the N-product form of PT with HF orbitals (Sect. 5.3.3), we only need the two-electron operator V or V_N, which can be represented using either non-antisymmetrized vertices (Fig. 5.1c), leading to the *Goldstone diagrams* [5.20], or antisymmetrized vertices (Fig. 5.1d), associated with antisymmetrized two-electron integrals (5.56) and yielding the *Hugenholtz diagrams* [5.21].

5.3.5 Vacuum and Wave Function Diagrams

Applying Wick's theorem to the strings of operators involved, we represent the individual contractions, (5.45), by joining corresponding oriented lines. To obtain a non-

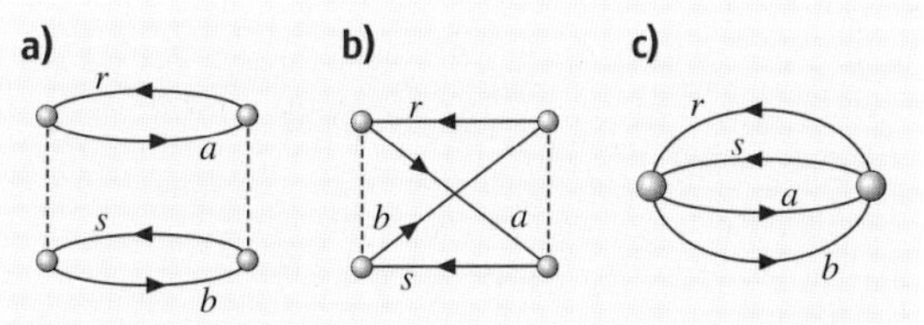

Fig. 5.2a–c The second-order Goldstone (**a**), (**b**) and Hugenholtz (**c**) diagrams

vanishing contribution, only contractions preserving the orientation need be considered [*cf.* (5.45)]. The resulting *internal* lines have either the left–right orientation (*hole* lines) or the right–left one (*particle* lines). Only fully contracted terms, represented by the so-called *vacuum diagrams* (having only internal lines), can contribute to the energy, while those representing wave function contributions have uncontracted or *free* lines extending to the left. When the operators involved are in the N-product form, no contractions of oriented lines issuing from the same vertex are allowed. The projection-like operators R, (5.39), or their powers, lead to the denominators, (5.63), given by the difference of hole and particle orbital energies associated with, respectively, hole and particle lines passing through the interval separating the corresponding two neighboring vertices. Clearly, there must always be at least one pair of such lines lest the denominator vanish. Thus, for example, the second-order contribution $\langle WRW \rangle$ is represented either by the two Goldstone diagrams [5.20] (Fig. 5.2a,b) or by the single Hugenholtz diagram [5.21] (Fig. 5.2c). The rules for the energy (vacuum) diagram evaluation are as follows:

1. Associate appropriate matrix elements with all vertices and form their product. The outgoing (ingoing) lines on each vertex define the bra (ket) states of a given matrix element, and for the Goldstone diagrams, the oriented lines attached to the same node are associated with the same electron number, (e.g., for the leftmost vertex in diagram (a) of Fig. 5.2 we have $\langle ab|\hat{v}|rs\rangle \equiv \langle a(1)b(2)|v|r(1)s(2)\rangle$).
2. Associate a denominator, (5.63), or its appropriate power, with every neighboring pair of vertices (and, for the wave function diagrams, also with the free lines extending to the left of the leftmost vertex; with each pair of such free lines associate also the corresponding pair of particle creation and hole annihilation operators).
3. Sum over all hole and particle labels.
4. Multiply each diagram contribution by the weight factor given by the reciprocal value of the order of

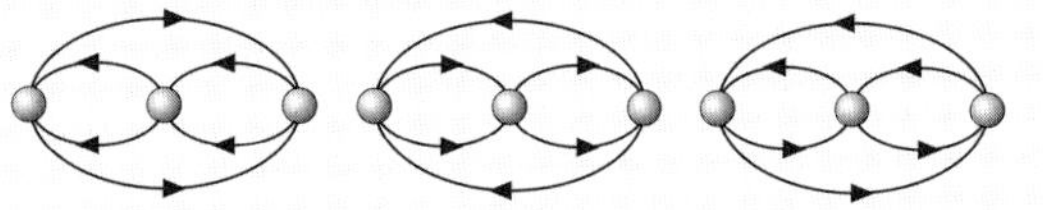

Fig. 5.3 Hugenholtz diagrams for the third-order energy contribution

the group of automorphisms of the diagram (stripped of summation labels) and by the sign $(-1)^{h+\ell}$, where h designates the number of internal hole lines and ℓ gives the number of closed loops of oriented lines (for Hugenholtz diagrams, use any of its Goldstone representatives to determine the correct phase).

Applying these rules to diagrams (a) and (b) of Fig. 5.2 we clearly recover (5.64) or, using the Hugenholtz diagram of Fig. 5.2, the equivalent expression

$$k^{(2)} = \frac{1}{4} \sum_{a,b,r,s} \langle ab|v|rs\rangle_A \langle rs|v|ab\rangle_A \Delta^{-1}(a,b;r,s) . \tag{5.66}$$

The possible third-order Hugenholtz diagrams are shown in Fig. 5.3 with the central vertex involving particle–particle, hole–hole, and particle–hole interaction [5.14, 15].

5.3.6 Hartree–Fock Diagrams

In the general case (non-HF orbitals and/or not normal product form of PT), the one-electron terms, as well as the contractions between operators associated with the same vertex, can occur (the latter are always the hole lines). Representing the W_1 and $(-U)$ operators as shown in Fig. 5.1, the one-body perturbation W_1 represents in fact the three diagrams as shown in Fig. 5.4. The second-order contribution of this type is then represented by the diagrams in Fig. 5.5, which in fact represents nine diagrams which result when each W_1 vertex is replaced by three vertices as shown in Fig. 5.4.

Using HF orbitals, all these terms mutually cancel out as seen above. For this reason, the diagrams involving contractions of lines issuing from the same vertex

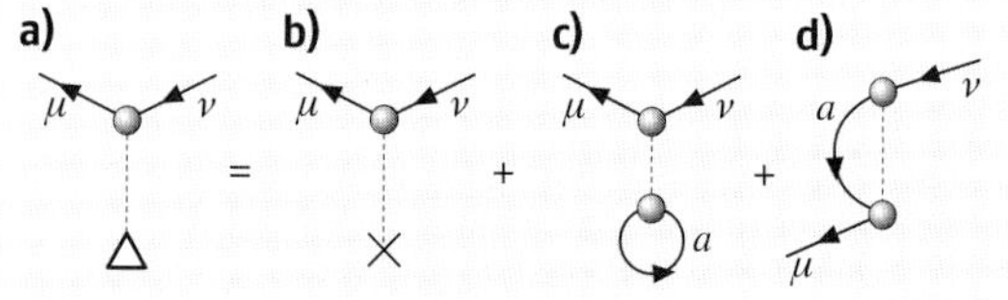

Fig. 5.4a–d Schematic representation of $W_1 = G_N - U_N$

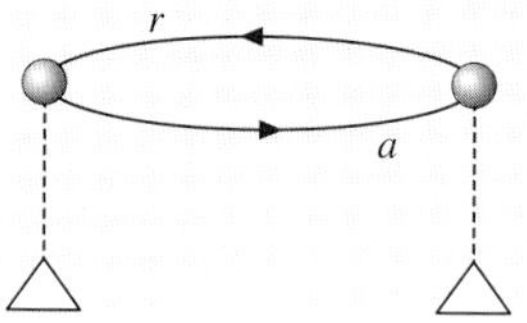

Fig. 5.5 The second-order one-particle contribution

are referred to as *Hartree–Fock diagrams*. Note, however, that even when not employing the canonical HF orbitals, it is convenient to introduce W_1 vertices of the normal product form PT and replace all nine HF-type diagrams by a single diagram of Fig. 5.5 (clearly, this feature provides even greater efficiency in higher orders of PT).

5.3.7 Linked and Connected Cluster Theorems

Using the N-product form of PT, the first nonvanishing renormalization term occurs in the fourth-order [*cf.* (5.60)]. For a system consisting of N noninteracting species, the energy given by this nonphysical term is proportional to N^2, and thus violates the size extensivity of the theory. It was first shown by *Brueckner* [5.22] that in the fourth-order these terms are in fact exactly canceled by the corresponding contributions originating in the principal term. A general proof of this cancellation in an arbitrary order was then given by *Goldstone* [5.20] using the time-dependent PT formalism (Sect. 5.4).

To comprehend this cancellation, consider the fourth-order energy contribution arising from the so-called unlinked diagrams (no such contribution can arise in the second- or the third-order) shown in Fig. 5.6. An *unlinked diagram* is defined as a diagram containing a disconnected vacuum diagram (for the energy diagrams, the terms unlinked and disconnected are synonymous). The numerators associated with both diagrams being identical, we only consider the denominators. Designating the denominator associated with the

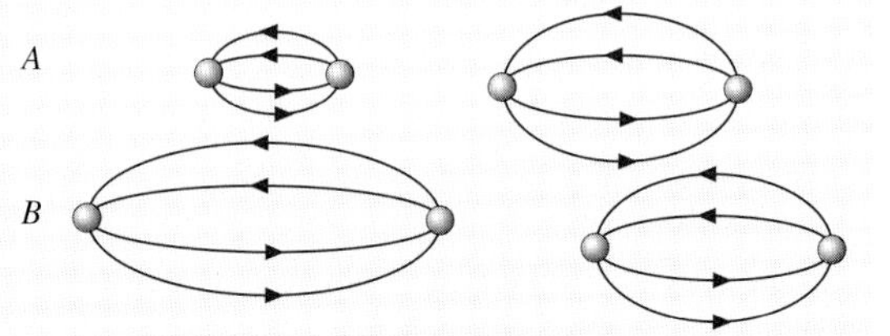

Fig. 5.6 The fourth-order unlinked diagrams

top and the bottom part by A and B, respectively, we find for the overall contribution

$$\frac{1}{B}\cdot\frac{1}{A+B}\cdot\frac{1}{B}+\frac{1}{A}\cdot\frac{1}{A+B}\cdot\frac{1}{B}$$
$$=\left(\frac{1}{B}+\frac{1}{A}\right)\frac{1}{(A+B)B}=\frac{1}{AB^2}\,. \tag{5.67}$$

Thus, the contribution from these terms exactly cancels that from the renormalization term $\langle WRW\rangle\langle WR^2W\rangle$.

Generalizing (5.67), we obtain the *factorization lemma* of *Frantz* and *Mills* [5.23], which implies the cancellation of renormalization terms by the unlinked terms originating from the principal term. This result holds for the energy as well as for the wave function contributions in every order of PT, as ascertained by the *linked cluster theorem*, which states that

$$\Delta E = k = \sum_{n=0}^{\infty}\langle\Phi|W|\Psi^{(n)}\rangle = \sum_{n=0}^{\infty}\langle W(RW)^n\rangle_{\mathrm{L}}\,, \tag{5.68}$$

$$|\Psi\rangle = \sum_{n=0}^{\infty}|\Psi^{(n)}\rangle = \sum_{n=0}^{\infty}\left\{(RW)^n|\Phi\rangle\right\}_{\mathrm{L}}\,, \tag{5.69}$$

where the subscript L indicates that only linked diagrams (or terms) are to be considered. This enables us to obtain general, explicit expressions for the nth-order PT contributions by first constructing all possible linked diagrams involving n vertices and by converting them into the explicit algebraic expressions using the rules of Sect. 5.3.5. Note that linked energy diagrams are always connected, but the linked wave function diagrams are either connected or disconnected, each disconnected component possessing at least one pair of particle–hole free lines extending to the left.

To reveal a deeper structure of the result (5.69), define the *cluster operator* T that generates all connected wave function diagrams,

$$T|\Phi\rangle = \sum_{n=1}^{\infty}\left\{(RW)^n|\Phi\rangle\right\}_{\mathrm{C}}\,, \tag{5.70}$$

the subscript C indicating that only contributions from *connected* diagrams are to be included. Since the general component with r disconnected parts can be shown to be represented by the term $(r!)^{-1}T^r|\Phi\rangle$, the general structure of the exact wave function $|\Psi\rangle$ is given by the *connected cluster theorem*, which states that

$$|\Psi\rangle = \mathrm{e}^T\,|\Phi\rangle\,. \tag{5.71}$$

In other words, the wave operator $\mathcal{W}$ which transforms the unperturbed independent particle model wave function $|\Phi\rangle$ into the exact one according to

$$|\Psi\rangle = \mathcal{W}|\Phi\rangle\,, \tag{5.72}$$

is given by the exponential of the cluster operator T,

$$\mathcal{W} = \mathrm{e}^T\,, \tag{5.73}$$

which in turn is given by the connected wave function diagrams. This is in fact the basis of the coupled cluster methods [5.15, 24–28] (Sect. 5.3.8).

The contributions to T may be further classified by their excitation rank i,

$$T = \sum_{i=1}^{N} T_i\,, \tag{5.74}$$

where T_i designates connected diagrams with i pairs of free particle–hole lines, producing i-times excited components of $|\Psi\rangle$ when acting on $|\Phi\rangle$.

5.3.8 Coupled Cluster Theory

Summing all HF diagrams (Sect. 5.3.6) is equivalent to solving the HF equations. Depending on the average electron density of the system, it may be essential to sum certain types of PT diagrams to infinite order at the post-HF level. A frequently used approach that is capable of recovering a large part of the electronic correlation energy is based on the connected cluster theorem (Sect. 5.3.7), referred to in this context as the exponential cluster Ansatz for the wave operator. Using this Ansatz, one derives a system of energy-independent nonlinear coupled cluster (CC) equations [5.15, 26–28] determining the cluster amplitudes of T. These CC equations can be regarded as recurrence relations generating the MBPT series [5.15], so that by solving these equations one in fact implicitly generates all the MBPT diagrams and sums them to infinite order. Since the solution of the full CC equations is equivalent to the exact solution of the Schrödinger equation, we must – in all practical applications – introduce a suitable truncation scheme, which implies that only diagrams of certain types are summed.

Generally, using the cluster expansion (5.71) in the N-product form of the Schrödinger equation,

$$H_{\mathrm{N}}|\Psi\rangle \equiv (H-\langle H\rangle)|\Psi\rangle = \Delta E|\Psi\rangle\,,$$
$$\Delta E = E - E_0\,, \tag{5.75}$$

premultiplying with the inverse of the wave operator, and using the Hausdorff formula (5.11) yields

$$e^{-T} H_N e^{T} |\Phi\rangle = \sum_{n=0}^{\infty} \frac{[\mathrm{ad}(-T)]^n H_N}{n!} = \Delta E |\Phi\rangle . \tag{5.76}$$

In fact, this expansion terminates, so that using (5.74) and projecting onto $|\Phi\rangle$ we obtain the energy expression

$$\Delta E = \langle H_N T_2 \rangle + \frac{1}{2} \langle H_N T_1^2 \rangle , \tag{5.77}$$

while the projection onto the manifold of excited states $\{|\Phi_i\rangle\}$ relative to $|\Phi\rangle \equiv |\Phi_0\rangle$ gives the system of CC equations

$$\langle \Phi_i | H_N + [H_N, T] + \frac{1}{2}[[H_N, T], T] + \cdots |\Phi\rangle = 0 . \tag{5.78}$$

Approximating, e.g., T by the most important pair cluster component $T \approx T_2$ gives the so-called CCD (coupled clusters with doubles) approximation

$$\langle \Phi_i^{(2)} | H_N + [H_N, T_2] + \frac{1}{2}[[H_N, T_2], T_2] |\Phi\rangle = 0 , \tag{5.79}$$

the superscript (2) indicating pair excitations relative to $|\Phi\rangle$.

Equivalently, (5.77) and (5.78) can be written in the form

$$\Delta E = \langle H_N e^{T} \rangle_C , \tag{5.80}$$

$$\langle \Phi_i | (H_N e^{T})_C | \Phi \rangle = 0 , \tag{5.81}$$

the subscript C again indicating that only connected diagrams are to be considered. The general form of CC equations is

$$a_i + \sum_j b_{ij} t_j + \sum_{j \le k} c_{ijk} t_j t_k + \cdots = 0 , \tag{5.82}$$

where $a_i = \langle \Phi_i | H_N | \Phi_0 \rangle$, $b_{ij} = \langle \Phi_i | H_N | \Phi_j \rangle_C$, $c_{ijk} = \langle \Phi_i | H_N | \Phi_j \otimes \Phi_k \rangle_C$, etc. Writing the diagonal linear term b_{ii} in the form

$$b_{ii} = \Delta_i + b'_{ii} , \tag{5.83}$$

this system can be solved iteratively by rewriting it in the form

$$t_i^{(n+1)} = \Delta_i^{-1} \Big(a_i + b'_{ii} t_i^{(n)} + \sum_j{}' b_{ij} t_j^{(n)} + \sum_{j \le k} c_{ijk} t_j^{(n)} t_k^{(n)} + \cdots \Big) . \tag{5.84}$$

Starting with the zeroth approximation $t_i^{(0)} = 0$, the first iteration is

$$t_i^{(1)} = \Delta_i^{-1} a_i , \tag{5.85}$$

which yields the second-order PT energy when used in (5.77). Clearly, the successive iterations generate higher and higher orders of the PT. At any truncation level, a size extensive result is obtained.

The CC methods belong to the most accurate and often used tools in computations of molecular electronic structure and several general-purpose codes are available for this purpose (for reviews see [5.29–32]). The standard approach truncates the cluster operator (5.74) at the singly (S) and doubly (D) excited level (the CCSD method [5.33]) and is often supplemented by a perturbative account of the triply-excited (T) cluster components [the CCSD(T) method] for greater accuracy [5.34]. To avoid the breakdown of the latter method in quasi-degenerate situations, one can employ one of the renormalized versions of CCSD(T) [5.35]. The CC ansatz (5.71) has also been exploited in the context of the *equation-of-motion* (EOM) and the *linear-response* formalisms, enabling the computation of the excitation energies and of properties other than the energy (dipole and quadrupole moments, polarizabilities, etc., [5.29–32].

At this stage it is important to recall that the above described MBPT and CC approaches pertain to nondegenerate, lowest-lying closed-shell states of a given symmetry species. Although the CC methods are often used even for open-shell states by relying on the unrestricted HF (UHF) reference [of the different-orbitals-for-different-spins (DODS) type], a proper description of such states requires a multireference (MR) generalization based on the *effective Hamiltonian* formalism [5.6, 31, 32, 36–38]. Unfortunately, such a generalization is not unambiguous. The two existing formulations, the so-called *valence universal* [5.6, 37] and *state universal* [5.38] methods, are computationally demanding and often plagued with the intruder state and other problems [5.15, 36]. For these reasons, no general-purpose codes have yet been developed and very few actual applications have been carried out [5.31, 32] (see, however, the recently formulated SU CC approach for general model spaces [5.39, 40]). Nonetheless, the MR CC formalism proved to be very useful in the formulation of the so-called *state selective* or *state specific* approaches (e.g., the reduced MR CCSD method [5.41–46]).

Most recently, the CC approach has been used to handle bosonic-type problems of the vibrational structure in molecular spectra and, generally, multimode dynamics [5.47].

5.4 Time-Dependent Perturbation Theory

5.4.1 Evolution Operator PT Expansion

By introducing the *evolution operator* $U(t, t_0)$

$$|\Psi(t)\rangle = U(t, t_0)|\Psi(t_0)\rangle \,, \tag{5.86}$$

time-dependent Schrödinger equation

$$\mathrm{i}\hbar\frac{\partial}{\partial t}|\Psi(t)\rangle = H|\Psi(t)\rangle \tag{5.87}$$

becomes

$$\mathrm{i}\hbar\frac{\partial}{\partial t}U(t, t_0) = HU(t, t_0) \,. \tag{5.88}$$

Clearly,

$$\begin{aligned} U(t_0, t_0) &= 1 \,, \\ U(t, t_0) &= U(t, t')U(t', t_0) \,, \\ U(t, t_0)^{-1} &= U(t_0, t) = U^{\dagger}(t, t_0) \,. \end{aligned} \tag{5.89}$$

If the Hamiltonian is time independent then

$$U(t, t_0) = \exp\left[-\frac{\mathrm{i}}{\hbar}H(t-t_0)\right] \,. \tag{5.90}$$

In the *interaction picture* (subscript I)

$$|\Psi(t)\rangle_{\mathrm{I}} = \exp\left(\frac{\mathrm{i}}{\hbar}H_0 t\right)|\Psi(t)\rangle \,, \tag{5.91}$$

where now

$$H = H_0 + V \,, \tag{5.92}$$

the Schrödinger equation becomes

$$\mathrm{i}\hbar\frac{\partial}{\partial t}|\Psi(t)\rangle_{\mathrm{I}} = V(t)_{\mathrm{I}}|\Psi(t)\rangle_{\mathrm{I}} \,, \tag{5.93}$$

known as *Tomonaga-Schwinger equation* [5.48]. Analogously, the evolution operator in this picture (we drop the subscript I from now on) satisfies

$$\mathrm{i}\hbar\frac{\partial}{\partial t}U(t, t_0) = V(t)U(t, t_0) \,, \tag{5.94}$$

with the initial condition $U(t_0, t_0) = 1$. This differential equation is equivalent to an integral equation

$$U(t, t_0) = 1 - \frac{\mathrm{i}}{\hbar}\int_{t_0}^{t} V(t_1)U(t_1, t_0)\,\mathrm{d}t_1 \,. \tag{5.95}$$

Iterating we get [5.49, 50]

$$\begin{aligned} U(t, t_0) &= \sum_{n=0}^{\infty}\left(-\frac{\mathrm{i}}{\hbar}\right)^n \\ &\quad \times \int_{t_0}^{t}\mathrm{d}t_1\int_{t_0}^{t_1}\mathrm{d}t_2\cdots\int_{t_0}^{t_{n-1}}\mathrm{d}t_n\,V(t_1)V(t_2)\cdots V(t_n) \\ &= \sum_{n=0}^{\infty}\frac{(-\mathrm{i}/\hbar)^n}{n!} \\ &\quad \times \int_{t_0}^{t}\mathrm{d}t_1\cdots\int_{t_0}^{t}\mathrm{d}t_n\,T[V(t_1)\cdots V(t_n)] \,, \end{aligned} \tag{5.96}$$

where $T[\cdots]$ designates the time-ordering or chronological operator.

5.4.2 Gell-Mann and Low Formula

For a time-independent perturbation, one introduces the so-called *adiabatic switching* by writing

$$H_\alpha(t) = H_0 + \lambda\,\mathrm{e}^{-\alpha|t|}V, \quad \alpha > 0 \tag{5.97}$$

so that $H_\alpha(t \to \pm\infty) = H_0$ and $H_\alpha(t \to 0) = H = H_0 + \lambda V$. Then

$$|\Psi(t)\rangle_{\mathrm{I}} = U_\alpha(t, -\infty|\lambda)|\Phi_0\rangle \,, \tag{5.98}$$

with $U_\alpha(t, -\infty|\lambda)$ obtained with $V_\alpha(t) = \lambda\,\mathrm{e}^{-\alpha|t|}V$ (all in the interaction picture). The desired energy is then given by the Gell–Mann and Low formula [5.51]

$$\Delta E = \lim_{\alpha\to 0+}\mathrm{i}\hbar\alpha\lambda\frac{\partial}{\partial\lambda}\ln\langle\Phi_0|U_\alpha(0, -\infty|\lambda)|\Phi_0\rangle \,, \tag{5.99a}$$

or

$$\Delta E = \frac{1}{2}\lim_{\alpha\to 0+}\mathrm{i}\hbar\alpha\lambda\frac{\partial}{\partial\lambda}\ln\langle\Phi_0|U_\alpha(\infty, -\infty|\lambda)|\Phi_0\rangle \,, \tag{5.99b}$$

which result from the asymmetric energy formula (5.27). One can similarly obtain the perturbation expansion for the one- or two-particle Green functions, e.g.,

$$G_{\mu\nu}(t, t') = \lim_{\alpha\to 0+}\frac{\langle T\{a_\mu(t)a_\nu^{\dagger}(t')U_\alpha(\infty, -\infty|\lambda)\}\rangle}{\langle U_\alpha(\infty, -\infty|\lambda)\rangle} \,, \tag{5.100}$$

with the operators in the interaction representation and the expectation values in the noninteracting ground state $|\Phi_0\rangle$. Analogous expressions result for $G(\boldsymbol{r}t, \boldsymbol{r}'t')$, etc., when the creation and annihilation operators are replaced by the corresponding field operators.

5.4.3 Potential Scattering and Quantum Dynamics

The Schrödinger equation for a free particle of energy $E = \hbar^2 k^2/2m$, moving in the potential $V(\boldsymbol{r})$,

$$\left(\nabla^2 + k^2\right)\psi(k, \boldsymbol{r}) = v(\boldsymbol{r})\psi(k, \boldsymbol{r}) ,$$
$$v(\boldsymbol{r}) = \left(2m/\hbar^2\right)V(\boldsymbol{r}) , \tag{5.101}$$

has the formal solution

$$\psi(k, \boldsymbol{r}) = \Phi(k, \boldsymbol{r}) + \int G_0\left(k, \boldsymbol{r}, \boldsymbol{r}'\right)v\left(\boldsymbol{r}'\right)\psi\left(k, \boldsymbol{r}'\right)\mathrm{d}\boldsymbol{r}' , \tag{5.102}$$

where $\Phi(k, \boldsymbol{r})$ is a solution of the homogeneous equation $[v(\boldsymbol{r}) \equiv 0]$ and $G_0(k, \boldsymbol{r}, \boldsymbol{r}')$ is a classical Green function

$$\left(\nabla^2 + k^2\right)G_0\left(k, \boldsymbol{r}, \boldsymbol{r}'\right) = \delta\left(\boldsymbol{r} - \boldsymbol{r}'\right) . \tag{5.103}$$

For an in-going plane wave $\Phi(k, \boldsymbol{r}) \equiv \Phi_{\boldsymbol{k}_i}(\boldsymbol{r}) = (2\pi)^{3/2}\exp(\mathrm{i}\boldsymbol{k}_i \cdot \boldsymbol{r})$ with the initial wave vector $\boldsymbol{k}_i$ and appropriate asymptotic boundary conditions (outgoing spherical wave with positive phase velocity), when $G_0(k, \boldsymbol{r}, \boldsymbol{r}') \equiv G_0^{(+)}(|\boldsymbol{r} - \boldsymbol{r}'|) = -(4\pi|\boldsymbol{r} - \boldsymbol{r}'|)^{-1}\mathrm{e}^{ik|\boldsymbol{r}-\boldsymbol{r}'|}$, (5.102) is referred to as the *Lippmann–Schwinger equation* [5.52]. It can be equivalently transformed into the integral equation for the Green function

$$G^{(+)}\left(\boldsymbol{r}, \boldsymbol{r}'\right) = G_0^{(+)}\left(\boldsymbol{r}, \boldsymbol{r}'\right) + \int G_0^{(+)}\left(\boldsymbol{r}, \boldsymbol{r}''\right)v\left(\boldsymbol{r}''\right) \times G^{(+)}\left(\boldsymbol{r}'', \boldsymbol{r}'\right)\,\mathrm{d}\boldsymbol{r}'' , \tag{5.104}$$

representing a special case of the Dyson equation.

In the time-dependent case, considering the scattering of a spinless massive particle by a time-dependent potential $V(\boldsymbol{r}, t)$, we get similarly

$$\psi(\boldsymbol{r}, t) = \Phi(\boldsymbol{r}, t) + \int G_0\left(\boldsymbol{r}, \boldsymbol{r}'; t, t'\right)V\left(\boldsymbol{r}', t'\right)\psi\left(\boldsymbol{r}', t'\right)\mathrm{d}\boldsymbol{r}'\,\mathrm{d}t' , \tag{5.105}$$

where the zero-order time-dependent Green function now satisfies the equation

$$\left(\mathrm{i}\hbar\frac{\partial}{\partial t} - H_0\right)G_0\left(\boldsymbol{r}, \boldsymbol{r}'; t, t'\right) = \delta\left(\boldsymbol{r} - \boldsymbol{r}'\right)\delta\left(t - t'\right) . \tag{5.106}$$

Again, for causal propagation one chooses the *time-retarded* or *causal* Green function or propagator $G_0^{(+)}(\boldsymbol{r}, \boldsymbol{r}'; t, t')$.

5.4.4 Born Series

Iteration of (5.105) gives the *Born sequence*

$$\psi_0(\boldsymbol{r}, t) = \Phi(\boldsymbol{r}, t) ,$$

$$\psi_1(\boldsymbol{r}, t) = \Phi(\boldsymbol{r}, t) + \int G_0^{(+)}\left(\boldsymbol{r}, \boldsymbol{r}'; t, t'\right) \times V\left(\boldsymbol{r}', t'\right)\Phi\left(\boldsymbol{r}', t'\right)\mathrm{d}\boldsymbol{r}'\,\mathrm{d}t' , \tag{5.107a}$$

$$\psi_2(\boldsymbol{r}, t) = \Phi(\boldsymbol{r}, t) + \int G_0^{(+)}\left(\boldsymbol{r}, \boldsymbol{r}'; t, t'\right) \times V\left(\boldsymbol{r}', t'\right)\psi_1\left(\boldsymbol{r}', t'\right)\mathrm{d}\boldsymbol{r}'\,\mathrm{d}t' , \tag{5.107b}$$

and, generally

$$\psi_n(\boldsymbol{r}, t) = \Phi(\boldsymbol{r}, t) + \int G_0^{(+)}\left(\boldsymbol{r}, \boldsymbol{r}'; t, t'\right) \times V\left(\boldsymbol{r}', t'\right)\psi_{n-1}\left(\boldsymbol{r}', t'\right)\mathrm{d}\boldsymbol{r}'\,\mathrm{d}t' . \tag{5.108}$$

Summing individual contributions gives the *Born series* for $\psi(\boldsymbol{r}, t) \equiv \psi^{(+)}(\boldsymbol{r}, t)$,

$$\psi(\boldsymbol{r}, t) = \sum_{n=0}^{\infty} \chi_n(\boldsymbol{r}, t) , \tag{5.109}$$

where

$$\chi_0(\boldsymbol{r}, t) = \Phi(\boldsymbol{r}, t) ,$$
$$\chi_n(\boldsymbol{r}, t) = \int \mathcal{G}_n\left(\boldsymbol{r}, \boldsymbol{r}'; t, t'\right)\Phi\left(\boldsymbol{r}', t'\right)\mathrm{d}\boldsymbol{r}'\,\mathrm{d}t' , \tag{5.110}$$

with

$$\mathcal{G}_n\left(\boldsymbol{r}, \boldsymbol{r}'; t, t'\right) = \int \mathcal{G}_1\left(\boldsymbol{r}, \boldsymbol{r}''; t, t''\right) \times \mathcal{G}_{n-1}\left(\boldsymbol{r}'', \boldsymbol{r}'; t'', t'\right)\mathrm{d}\boldsymbol{r}''\,\mathrm{d}t'' , \quad (n > 1)$$
$$\mathcal{G}_1\left(\boldsymbol{r}, \boldsymbol{r}'; t, t'\right) = G_0^{(+)}\left(\boldsymbol{r}, \boldsymbol{r}'; t, t'\right)V\left(\boldsymbol{r}', t'\right) . \tag{5.111}$$

In a similar way we obtain the Born series for the scattering amplitudes or transition matrix elements.

5.4.5 Variation of Constants Method

An alternative way of formulating the time-dependent PT is the method of variation of the constants [5.53, 54]. Start again with the time-dependent Schrödinger equation (5.87) with $H = H_0 + V$, and assume that H_0 is time-independent, while V is a time-dependent perturbation. Designating the eigenvalues and eigenstates of

H_0 by ε_i and $|\Phi_i\rangle$, respectively [*cf.* (5.48)], the general solution of the unperturbed time-dependent Schrödinger equation

$$\mathrm{i}\hbar\frac{\partial}{\partial t}|\Psi_0\rangle = H_0|\Psi_0\rangle \tag{5.112}$$

has the form

$$|\Psi_0\rangle = \sum_j c_j|\Phi_j\rangle \exp\left(-\frac{\mathrm{i}}{\hbar}\varepsilon_j t\right) , \tag{5.113}$$

with c_j representing arbitrary constants, and the sum indicating both the summation over the discrete part and the integration over the continuum part of the spectrum of H_0.

In the spirit of the general variation of constants procedure, write the unknown perturbed wave function $|\Psi(t)\rangle$, (5.87), in the form

$$|\Psi(t)\rangle = \sum_j C_j(t)|\Phi_j\rangle \exp\left(-\frac{\mathrm{i}}{\hbar}\varepsilon_j t\right) , \tag{5.114}$$

where the $C_j(t)$ are now functions of time. Substituting this Ansatz into the time-dependent Schrödinger equation (5.87) gives

$$\dot{C}_j(t) = (\mathrm{i}\hbar)^{-1}\sum_k C_k(t)V_{jk}\exp[(\mathrm{i}/\hbar)\Delta_{jk}t] , \tag{5.115}$$

where

$$\Delta_{jk} = \varepsilon_j - \varepsilon_k , \quad V_{jk} = \langle\Phi_j|V|\Phi_k\rangle . \tag{5.116}$$

Introducing again the 'small' parameter λ by writing the Hamiltonian H in the form

$$H = H_0 + \lambda V(t) , \tag{5.117}$$

and expanding the 'coefficients' $C_j(t)$ in powers of λ,

$$C_j \equiv C_j(t) = \sum_{k=0}^{\infty} C_j^{(k)}(t)\lambda^k , \tag{5.118}$$

gives the system of first order differential equations

$$\dot{C}_j^{(n+1)} = (\mathrm{i}\hbar)^{-1}\sum_k C_k^{(n)}V_{jk}\exp\left[(\mathrm{i}/\hbar)\Delta_{jk}t\right] ,$$
$$n = 0, 1, 2, \ldots , \tag{5.119}$$

with the initial condition $\dot{C}_j^{(0)} = 0$, which implies that $C_j^{(0)}$ are time independent, so that $C_j^{(0)} = c_j$, obtaining (5.113) in the zeroth order. The system (5.119) can be integrated to any prescribed order. For example, if the system is initially in a stationary state $|\Phi_i\rangle$, then set

$$C_j^{(0)} = \begin{cases} \delta_{ji} & \text{for discrete states ,} \\ \delta(j-i) & \text{for continuous states ,} \end{cases} \tag{5.120}$$

so that

$$C_{j(i)}^{(1)}(t) = (\mathrm{i}\hbar)^{-1}\int_{-\infty}^{t} V_{ji}\exp\left[(\mathrm{i}/\hbar)\Delta_{ji}t'\right]\mathrm{d}t' , \tag{5.121}$$

assuming $C_{j(i)}^{(1)}(-\infty) = 0$. Clearly, $|C_{j(i)}^{(1)}(t)|^2$ gives the first order transition probability for the transition from the initial state $|\Phi_i\rangle$ to a particular state $|\Phi_j\rangle$. These in turn will yield the first order differential cross sections [5.55].

References

5.1 T. Kato: *Perturbation Theory for Linear Operators* (Springer, Berlin, Heidelberg 1966)

5.2 H. Baumgärtel: *Analytic Perturbation Theory for Matrices and Operators* (Akademie, Berlin 1984)

5.3 E. J. Hinch: *Perturbation Methods* (Cambridge Univ. Press, Cambridge 1991)

5.4 V. N. Bogaevski, A. Povzner: *Algebraic Methods in Nonlinear Perturbation Theory* (Springer, Berlin, Heidelberg 1991)

5.5 E. M. Corson: *Perturbation Methods in the Quantum Mechanics of n-Electron Systems* (Blackie & Son, London 1951)

5.6 I. Lindgren, J. Morrison: *Atomic Many-Body Theory* (Springer, Berlin, Heidelberg 1982)

5.7 E. K. U. Gross, E. Runge, O. Heinonen: *Many-Particle Theory* (Hilger, New York 1991)

5.8 F. E. Harris, H. J. Monkhorst, D. L. Freeman: *Algebraic and Diagrammatic Methods in Many-Fermion Theory* (Oxford Univ. Press, Oxford 1992)

5.9 H. Primas: Helv. Phys. Acta **34**, 331 (1961)

5.10 H. Primas: Rev. Mod. Phys. **35**, 710 (1963)

5.11 M. Rosenblum: Duke Math. J. **23**, 263 (1956)

5.12 G. Arfken: *Mathematical Methods for Physicists* (Academic, New York 1985) p. 327

5.13 *Encyclopedic Dictionary of Mathematics*, ed. by S. Iyanaga, Y. Kawada (MIT Press, Cambridge 1980) pp. 1494, Appendix B, Table 3

5.14 J. Paldus, J. Čížek: Adv. Quantum Chem. **9**, 105 (1975)

5.15 J. Paldus: *Methods in Computational Molecular Physics*, NATO ASI Series B, Vol. 293, ed. by S. Wilson, G. H. F. Diercksen (Plenum, New York 1992) pp. 99–194

5.16 H. J. Silverstone, T. T. Holloway: J. Chem. Phys. **52**, 1472 (1970)
5.17 C. Møller, M. S. Plesset: Phys. Rev. **46**, 618 (1934)
5.18 P. S. Epstein: Phys. Rev. **28**, 695 (1926)
5.19 R. K. Nesbet: Proc. R. Soc. London A **250**, 312 (1955)
5.20 J. Goldstone: Proc. R. Soc. London A **239**, 267 (1957)
5.21 H. M. Hugenholtz: Physica (Utrecht) **23**, 481 (1957)
5.22 K. A. Brueckner: Phys. Rev. **100**, 36 (1955)
5.23 L. M. Frantz, R. L. Mills: Nucl. Phys. **15**, 16 (1960)
5.24 F. Coester: Nucl. Phys. **7**, 421 (1958)
5.25 F. Coester, H. Kümmel: Nucl. Phys. **17**, 477 (1960)
5.26 J. Čížek: J. Chem. Phys. **45**, 4256 (1966)
5.27 J. Čížek: Adv. Chem. Phys. **14**, 35 (1969)
5.28 J. Paldus, J. Čížek, I. Shavitt: Phys. Rev. A **5**, 50 (1972)
5.29 R. J. Bartlett: *Modern Electronic Structure Theory*, ed. by D. R. Yarkony (World Scientific, Singapore 1995) pp. 47–108, Part I
5.30 R. J. Bartlett (Ed.): *Recent advances in computational chemistry*, Recent Advances in Coupled-Cluster Methods, Vol. 3 (World Scientific, Singapore 1997)
5.31 J. Paldus, X. Li: Adv. Chem. Phys. **110**, 1 (1999)
5.32 J. Paldus: *Handbook of Molecular Physics and Quantum Chemistry*, Vol. 2, Part 3, ed. by S. Wilson (Wiley, Chichester 2003) pp. 272–313
5.33 R. J. Bartlett, G. D. Purvis: Int. J. Quantum Chem. **14**, 561 (1978)
5.34 T. J. Lee, G. E. Scuseria: *Quantum Mechanical Electronic Structure Calculations with Chemical Accuracy*, ed. by S. R. Langhoff (Kluwer, Dordrecht 1995) pp. 47–108
5.35 K. Kowalski, P. Piecuch: J. Chem. Phys. **120**, 1715 (2004)
5.36 J. Paldus: *Relativistic and Electron Correlation Effects in Molecules and Solids*, NATO ASI Series B, Vol. 318, ed. by G. L. Malli (Plenum, New York 1994) pp. 207–282
5.37 I. Lindgren, D. Mukherjee: Phys. Rep. **151**, 93 (1987)
5.38 B. Jeziorski, H. J. Monkhorst: Phys. Rev. A **24**, 1686 (1981)
5.39 X. Li, J. Paldus: J. Chem. Phys. **119**, 5320, 5334, 5343 (2003)
5.40 X. Li, J. Paldus: J. Chem. Phys. **120**, 5890 (2004)
5.41 X. Li, J. Paldus: J. Chem. Phys. **107**, 6257 (1997)
5.42 X. Li, J. Paldus: J. Chem. Phys. **108**, 637 (1998)
5.43 X. Li, J. Paldus: J. Chem. Phys. **110**, 2844 (1999)
5.44 X. Li, J. Paldus: J. Chem. Phys. **113**, 9966 (2000)
5.45 X. Li, J. Paldus: J. Chem. Phys. **118**, 2470 (2003)
5.46 S. Chattopadhyay, D. Pahari, D. Mukherjee, U. S. Mahapatra: J. Chem. Phys. **120**, 5968 (2004)
5.47 O. Christiansen: J. Chem. Phys. **120**, 2149 (2004)
5.48 F. J. Dyson: Phys. Rev. **75**, 486 (1949)
5.49 S. Tomonaga: Prog. Theor. Phys. (Kyoto) **1**, 27 (1946)
5.50 J. Schwinger: Phys. Rev. **74**, 1439 (1948)
5.51 M. Gell-Mann, F. Low: Phys. Rev. **84**, 350 (1951)
5.52 B. A. Lippmann, J. Schwinger: Phys. Rev. **79**, 469 (1950)
5.53 P. A. M. Dirac: Proc. R. Soc. London A **112**, 661 (1926)
5.54 P. A. M. Dirac: Proc. R. Soc. London A **114**, 243 (1926)
5.55 C. J. Joachain: *Quantum Collision Theory* (Elsevier, New York 1975)

6. Second Quantization

In second quantization, the characteristic properties of eigenfunctions are transferred to operators. This approach has the advantage of treating the atomic shell as the basic unit, as opposed to the electron configuration. The creation and annihilation operators allow one to move from configuration to configuration, exposing an intrinsic shell structure. The introduction of coefficients of fractional parentage (cfp) then allows the calculation of the matrix elements of an operator in one configuration to be expressed in terms of those of the same operator in another configuration; hence the matrix elements of an operator in all configurations may be determined from the knowledge of its matrix elements in but one. This can be viewed as an extension of the usual Wigner-Eckart theorem. The basic concepts of quasispin and quasiparticle are also introduced within this context.

6.1 **Basic Properties** 115
6.1.1 Definitions 115
6.1.2 Representation of States 115
6.1.3 Representation of Operators 116

6.2 **Tensors** 116
6.2.1 Construction 116
6.2.2 Coupled Forms 116
6.2.3 Coefficients of Fractional Parentage 117

6.3 **Quasispin** 117
6.3.1 Fermions 117
6.3.2 Bosons 118
6.3.3 Triple Tensors 118
6.3.4 Conjugation 118
6.3.5 Dependence on Electron Number 119
6.3.6 The Half-filled Shell 119

6.4 **Complementarity** 119
6.4.1 Spin–Quasispin Interchange 119
6.4.2 Matrix Elements 119

6.5 **Quasiparticles** 120

References 121

6.1 Basic Properties

6.1.1 Definitions

The creation operator $a_\xi^\dagger$ creates the quantum state ξ. The annihilation (or destruction) operator a_η annihilates the quantum state η. The vacuum (or reference) state $|0\rangle$ satisfies the equation

$$a_\eta|0\rangle = 0\,. \tag{6.1}$$

Bosons satisfy the commutation relations

$$\left[a_\xi^\dagger, a_\eta^\dagger\right] = 0\,, \tag{6.2}$$

$$[a_\xi, a_\eta] = 0\,, \tag{6.3}$$

$$\left[a_\xi, a_\eta^\dagger\right] = \delta(\xi, \eta)\,, \tag{6.4}$$

where $[A, B] \equiv AB - BA$. Fermions satisfy the anticommutation relations

$$\left[a_\xi^\dagger, a_\eta^\dagger\right]_+ = 0\,, \tag{6.5}$$

$$[a_\xi, a_\eta]_+ = 0\,, \tag{6.6}$$

$$\left[a_\xi, a_\eta^\dagger\right]_+ = \delta(\xi, \eta)\,, \tag{6.7}$$

where $[A, B]_+ \equiv AB + BA$.

6.1.2 Representation of States

For an electron in an atom, characterized by the quantum number quartet $(n\;\ell\;m_s\;m_\ell)$, the identification $\xi \equiv (n\;\ell\;m_s\;m_\ell)$ for fermions is made. For normalized Slater determinants $\{\alpha\beta\ldots\nu\}$ characterized by the electron states $\alpha, \beta, \ldots, \nu$, the equivalences

$$a_\alpha^\dagger a_\beta^\dagger \ldots a_\nu^\dagger|0\rangle \equiv \{\alpha\beta\ldots\nu\}\,, \tag{6.8}$$

$$\langle 0|a_\nu \ldots a_\beta a_\alpha \equiv \{\alpha\beta\ldots\nu\}^* \tag{6.9}$$

are valid, where the asterisk denotes the complex conjugate.

Part A | 6

For a normalized boson state $\{\cdots\}$ in which the label ξ appears N_ξ times, the additional factor

$$[N_\alpha!N_\beta!\ldots N_\nu!]^{-\frac{1}{2}} \tag{6.10}$$

must be included on the left-hand sides of the equivalences (6.8) and (6.9).

6.1.3 Representation of Operators

For an operator F, consisting of the sum of operators f_i acting on the single electron i,

$$F \equiv \sum_{\xi,\eta} a_\xi^\dagger \langle\xi|f|\eta\rangle a_\eta \,. \tag{6.11}$$

For an operator G, consisting of the sum of operators g_{ij} acting on the pair of electrons i and j,

$$G \equiv \frac{1}{2} \sum_{\xi,\eta,\zeta,\lambda} a_\xi^\dagger a_\eta^\dagger \langle\xi_1\eta_2|g_{12}|\zeta_1\lambda_2\rangle a_\lambda a_\zeta \,. \tag{6.12}$$

For an N-particle system $|\Psi\rangle$,

$$\sum_\xi a_\xi^\dagger a_\xi |\Psi\rangle = N|\Psi\rangle \,. \tag{6.13}$$

The representations of single-particle and two-particle operators for bosons are identical to those given above for fermions [6.1].

6.2 Tensors

6.2.1 Construction

If the description ξ for a single fermion or boson state includes an angular momentum quantum number t and the corresponding magnetic quantum number m_t, then the $2t+1$ components of a creation operator $a_\sigma^\dagger$, where $\sigma \equiv (t, m_t)$ and $-t \le m_t \le t$, satisfy the commutation relations of *Racah* [6.2] for an irreducible spherical tensor of rank t with respect to the total angular momentum $\boldsymbol{T}$, given by

$$\boldsymbol{T} = \sum a_\xi^\dagger \langle\xi|\boldsymbol{t}|\eta\rangle a_\eta \,. \tag{6.14}$$

That is, with the phase conventions of *Condon* and *Shortley* [6.3],

$$\left[T_z, a_\sigma^\dagger\right] = m_t a_\sigma^\dagger \,, \tag{6.15}$$

$$\left[T_x \pm \mathrm{i}T_y, a_\sigma^\dagger\right] = [t(t+1) - m_t(m_t \pm 1)]^{\frac{1}{2}}\, a_\tau^\dagger \,, \tag{6.16}$$

where $\tau \equiv (t, m_t \pm 1)$.

A spherical tensor $\boldsymbol{a}$ constructed from annihilation operators possesses the components $\tilde{a}_\sigma$, which satisfy

$$\tilde{a}_\sigma = (-1)^p a_\zeta \,, \tag{6.17}$$

with $p = t - m_t$ and $\zeta \equiv (t, -m_t)$.

The $4\ell + 2$ components of the creation operator for an electron in the atomic ℓ shell form a double tensor of rank $\frac{1}{2}$ with respect to the total spin $\boldsymbol{S}$, and rank ℓ with respect to the total angular momentum $\boldsymbol{L}$.

6.2.2 Coupled Forms

Tensors formed from annihilation and creation operators can be coupled by means of the usual rules of angular momentum theory [6.4]. The double tensor defined for electrons in the ℓ shell by

$$\boldsymbol{W}^{(\kappa k)} = -\left(\boldsymbol{a}^\dagger \boldsymbol{a}\right)^{(\kappa k)} \,, \tag{6.18}$$

possesses a rank κ with respect to $\boldsymbol{S}$, and rank k with respect to $\boldsymbol{L}$. Its reduced matrix element, defined here as in (5.4.1) of *Edmonds* [6.4], for a single electron in both the spin and orbital spaces, is given by

$$\left(s\,\ell\,||\,W^{(\kappa k)}\,||\,s\,\ell\right) = [(2\kappa+1)(2k+1)]^{\frac{1}{2}} \,. \tag{6.19}$$

The connections to tensors whose matrix elements have been tabulated [6.5, 6] are

$$\boldsymbol{W}^{(0k)} = [(2k+1)/2]^{\frac{1}{2}} \boldsymbol{U}^{(k)} \,, \tag{6.20}$$

$$\boldsymbol{W}^{(1k)} = [2(2k+1)]^{\frac{1}{2}} \boldsymbol{V}^{(k1)} \,. \tag{6.21}$$

For terms with common spin S, say ψ and ψ',

$$\left(\psi||W^{(0k)}||\psi'\right) = [(2S+1)(2k+1)/2]^{\frac{1}{2}} \left(\psi||U^{(k)}||\psi'\right) . \tag{6.22}$$

This result is obtained because the ranks assigned to the tensors imply that $\boldsymbol{W}^{(0k)}$ is to be reduced with respect to both the spin $\boldsymbol{S}$ and the orbit $\boldsymbol{L}$, while $\boldsymbol{U}^{(k)}$ is to be reduced only with respect to $\boldsymbol{L}$.

The following relations hold for electrons with azimuthal quantum numbers ℓ [6.7]:

$$S = [(2\ell+1)/2]^{\frac{1}{2}}\, W^{(10)}\,, \tag{6.23}$$

$$L = [2\ell(\ell+1)(2\ell+1)/3]^{\frac{1}{2}}\, W^{(01)}\,, \tag{6.24}$$

$$\sum_i \left(s_i C_i^{(2)}\right)^{(1)} = -\left(\frac{\ell(\ell+1)(2\ell+1)}{10(2\ell-1)(2\ell+3)}\right)^{\frac{1}{2}} W^{(12)1}\,, \tag{6.25}$$

$$\sum_i (s_i \cdot \ell_i) = -[\ell(\ell+1)(2\ell+1)/2]\, W^{(11)0}\,, \tag{6.26}$$

where the tensor C^k of *Racah* [6.2] is related to the spherical harmonics by

$$C_q^{(k)} = [4\pi/(2k+1)]^{\frac{1}{2}}\, Y_{kq}\,, \tag{6.27}$$

and where the tensors of the type $W^{(\kappa k)K}$ indicate that the spin and orbital ranks are coupled to a resultant K.

6.2.3 Coefficients of Fractional Parentage

Let ψ and $\bar{\psi}$ denote terms of ℓ^N and ℓ^{N-1} characterized by (S, L) and $(\bar{S}, \bar{L})$. The coefficients of fractional parentage (cfp) $(\psi\{|\bar{\psi})$ of *Racah* [6.8] allow one to calculate an antisymmetrized function ψ by vector-coupling $\bar{\psi}$ to the spin and orbit of the Nth electron:

$$|\psi\rangle = \sum_{\bar{\psi}} |\bar{\psi}, {}^2\ell, SL\rangle \left(\bar{\psi}|\}\psi\right)\,, \tag{6.28}$$

where the sum over $\bar{\psi}$ includes $\bar{S}$, $\bar{L}$, and any other quantum numbers necessary to define the spectroscopic terms of ℓ^{N-1}. The cfp's are given by

$$(\psi||a^\dagger||\bar{\psi}) = (-1)^N [N(2S+1)(2L+1)]^{\frac{1}{2}} \left(\psi\{|\bar{\psi}\right)\,, \tag{6.29}$$

$$(\bar{\psi}||a||\psi) = (-1)^g [N(2S+1)(2L+1)]^{\frac{1}{2}} \left(\bar{\psi}|\}\psi\right)\,, \tag{6.30}$$

where $g = N + \bar{S} + \bar{L} - s - S - \ell - L$. A tabulation for the p, d, and f shells has been given by *Nielson* and *Koster* [6.5].

Two-electron cfp are given by

$$\left(\psi||\left(a^\dagger a^\dagger\right)^{(\kappa k)}||\tilde{\psi}\right) = [N(N-1)(2S+1)(2L+1)]^{\frac{1}{2}} \times \left(\psi\{|\tilde{\psi}, \ell^2(\kappa k)\right)\,, \tag{6.31}$$

where $\tilde{\psi}$ denotes a term of ℓ^{N-2}, and the symbols κ and k stand for the S and the L of a term of ℓ^2. A tabulation for the p, d, and f shells has been given by *Donlan* [6.9]. An extension to all multielectron cfp has been carried out by *Velkov* [6.10].

If, through successive applications of the two-particle operators $(aa)^{(00)}$, a state of ℓ^N can be reduced to ℓ^v, but no further, then v is the seniority number of *Racah* [6.8].

If the ranks s and ℓ of $a^\dagger$ are coupled to $\bar{S}$ and $\bar{L}$ of $\bar{\psi}$, the term

$$\left(a^\dagger|\bar{\psi}\rangle\right)^{(SL)} \tag{6.32}$$

either vanishes, or is a term of ℓ^N characterized by S and L. Such a term is said to possess the *godparent* $\bar{\psi}$. *Redmond* [6.11] has used the notion of godparents to generate an explicit formula for the single particle cfp [6.7].

6.3 Quasispin

6.3.1 Fermions

For electrons, the components $Q_\pm (\equiv Q_x \pm \mathrm{i}Q_y)$ and Q_z of the quasispin Q are defined by [6.7, 12]

$$Q_+ = [(2\ell+1)/2]^{\frac{1}{2}} \left(a^\dagger a^\dagger\right)^{(00)}\,, \tag{6.33}$$

$$Q_- = -[(2\ell+1)/2]^{\frac{1}{2}} (aa)^{(00)}\,, \tag{6.34}$$

$$Q_z = -[(2\ell+1)/8]^{\frac{1}{2}} \left[\left(a^\dagger a\right)^{(00)} + \left(aa^\dagger\right)^{(00)}\right]. \tag{6.35}$$

The term quasispin comes from the fact that the components of Q satisfy the commutation relations of an angular momentum vector. The eigenvalues M_Q of Q_z, for a state of ℓ^N, are given by

$$M_Q = -(2\ell+1-N)/2\,. \tag{6.36}$$

The shift operators Q_+ and Q_- connect states of the ℓ shell possessing the same value of the seniority v of *Racah* [6.8]. A string of such connected states defines the extrema of M_Q, from which it follows that

$$Q = (2\ell+1-v)/2\,. \tag{6.37}$$

Rudzikas has placed special emphasis on quasispin in his reworking of atomic shell theory, and he has also introduced isospin to embrace electrons differing in their principal quantum numbers n [6.13]. Concise tables of one-electron cfp with their quasispin dependence factored out have been given [6.14], as have the algebraic dependences on v and S of two-electron cfp [6.15].

6.3.2 Bosons

For real vibrational modes created by $a_\nu^\dagger$ ($\nu = 1, 2, \ldots, d$), the analogs of (6.33–6.35) are

$$P_+ = -\frac{1}{2}\sum_\nu a_\nu^\dagger a_\nu^\dagger , \tag{6.38}$$

$$P_- = \frac{1}{2}\sum_\nu a_\nu a_\nu , \tag{6.39}$$

$$P_z = \frac{1}{4}\sum_\nu \left(a_\nu^\dagger a_\nu + a_\nu a_\nu^\dagger\right) , \tag{6.40}$$

and $\boldsymbol{P}$ is an angular momentum vector [6.16]. The eigenvalues M_P for an n-boson state are given by

$$M_P = (2n+d)/4 , \tag{6.41}$$

and can therefore be quarter-integral. Successive application of the operator P_+ to a state $|n_0\rangle$, for which $P_-|n_0\rangle = 0$, generates an infinite ladder of states characterized by

$$P = (2n+d-4)/4 . \tag{6.42}$$

6.3.3 Triple Tensors

The creation and annihilation operators $a_\xi^\dagger$ and a_ξ for a given state ξ can be regarded as the two components of a tensor of rank $\frac{1}{2}$ with respect to quasispin (either $\boldsymbol{Q}$ or $\boldsymbol{P}$). For electrons, this leads to triple tensors $\boldsymbol{a}^{(qs\ell)}$ (for which $q = s = \frac{1}{2}$) satisfying

$$a_\lambda^{(qs\ell)} a_\mu^{(qs\ell)} + a_\mu^{(qs\ell)} a_\lambda^{(qs\ell)} = (-1)^{x+1}\delta\left(m_q, -m_q'\right)\delta\left(m_s, -m_s'\right)\delta\left(m_\ell, -m_\ell'\right) , \tag{6.43}$$

where $\lambda \equiv (m_q m_s m_\ell)$, $\mu \equiv (m_q' m_s' m_\ell')$, and $x = q + s + \ell + m_q + m_s + m_\ell$. In terms of the coupled tensor

$$\boldsymbol{X}^{(K\kappa k)} = \left(\boldsymbol{a}^{(qs\ell)}\boldsymbol{a}^{(qs\ell)}\right)^{(K\kappa k)} , \tag{6.44}$$

the angular momenta $\boldsymbol{Q}$, $\boldsymbol{S}$, and $\boldsymbol{L}$ are given by

$$\boldsymbol{Q} = -[(2\ell+1)/4]^{\frac{1}{2}}\boldsymbol{X}^{(100)} , \tag{6.45}$$

$$\boldsymbol{S} = -[(2\ell+1)/4]^{\frac{1}{2}}\boldsymbol{X}^{(010)} , \tag{6.46}$$

$$\boldsymbol{L} = -[\ell(\ell+1)(2\ell+1)/3]^{\frac{1}{2}}\boldsymbol{X}^{(001)} . \tag{6.47}$$

Furthermore, the components of $\boldsymbol{X}^{(K\kappa k)}$ for which $M_K = 0$ are identical to the corresponding components of $2^{\frac{1}{2}}(\boldsymbol{a}^\dagger\boldsymbol{a})^{(\kappa k)}$ when $K+\kappa+k$ is odd; and

$$\boldsymbol{X}^{(K\kappa k)} = -(2\ell+1)^{\frac{1}{2}}\delta(K,0)\delta(\kappa,0)\delta(k,0) \tag{6.48}$$

when $K+\kappa+k$ is even.

6.3.4 Conjugation

Creation and annihilation operators can be interchanged by the operation of the conjugation operator C [6.7, 17]. For electrons in the atomic ℓ shell,

$$C a_\xi^{(qs\ell)} C^{-1} = (-1)^{q-m_q} a_\eta^{(qs\ell)} , \tag{6.49}$$

where $\xi \equiv (m_q m_s m_\ell)$ and $\eta \equiv ((-m_q)m_s m_\ell)$. In terms of the tensors $\boldsymbol{a}^\dagger$ and $\boldsymbol{a}$,

$$C\boldsymbol{a}^\dagger C^{-1} = \boldsymbol{a} , \quad C\boldsymbol{a}C^{-1} = -\boldsymbol{a}^\dagger . \tag{6.50}$$

Furthermore,

$$CX_\lambda^{(K\kappa k)}C^{-1} = (-1)^{K-M_K}X_\mu^{(K\kappa k)} , \tag{6.51}$$

where $\lambda \equiv (M_K M_\kappa M_k)$ and $\mu \equiv [(-M_K)M_\kappa M_k]$, and

$$C\left|QM_Q\right\rangle = (-1)^{Q-M_Q}\left|Q-M_Q\right\rangle . \tag{6.52}$$

Thus, from (6.36), the action of C takes N into $4\ell+2-N$; that is, C interchanges electrons and holes. When the case $\kappa = k = 0$ is excluded, application of (6.51) and (6.52) yields

$$\left(\ell^N\psi\,||\,W^{(\kappa k)}\,||\,\ell^N\psi'\right) = (-1)^y\left(\ell^{4\ell+2-N}\psi||W^{(\kappa k)}||\ell^{4\ell+2-N}\psi'\right) , \tag{6.53}$$

where $y = \kappa + k + \frac{1}{2}(v'-v) + 1$, and where the seniorities v and v' are implied by ψ and ψ'. A similar application to reduced matrix elements of $\boldsymbol{a}^\dagger$ and $\boldsymbol{a}$ gives the following relation between cfp:

$$\left(\ell^{N+1}\psi\{|\ell^N\psi'\right) = (-1)^z\left(\ell^{4\ell+1-N}\psi|\}\ell^{4\ell+2-N}\psi'\right) \times\left(\frac{(4\ell+2-N)(2S'+1)(2L'+1)}{(N+1)(2S+1)(2L+1)}\right)^{\frac{1}{2}} , \tag{6.54}$$

where $z = S + S' - s + L + L' - \ell + \frac{1}{2}(v+v'-1)$. The phases y and z stem from the conventions of angular momentum theory, which enter via quasispin. *Racah* [6.2, 8] did not use this concept, and his phase choices are slightly different from the ones above.

For a Cartesian component Q_u of the quasispin $\boldsymbol{Q}$,

$$CQ_uC^{-1} = -Q_u\,. \tag{6.55}$$

Thus, C is the analog of the time-reversal operator T, for which

$$TL_uT^{-1} = -L_u\,, \tag{6.56}$$

$$TS_uT^{-1} = -S_u\,. \tag{6.57}$$

Both C and T are antiunitary; thus,

$$C\mathrm{i}C^{-1} = -\mathrm{i}\,. \tag{6.58}$$

6.3.5 Dependence on Electron Number

Application of the Wigner–Eckart theorem to matrix elements whose component parts have well-defined quasispin ranks yields the dependence of the matrix elements on the electron number N [6.18, 19]. For $\kappa + k$ even and nonzero, the quasispin rank of $\boldsymbol{W}^{(\kappa k)}$ is 1, and

$$\left(\ell^N\psi||W^{(\kappa k)}||\ell^N\psi'\right) = \tag{6.59}$$
$$\frac{(2\ell+1-N)}{(2\ell+1-v)}\left(\ell^{\,v}\psi\,||\,W^{(\kappa k)}\,||\,\ell^{\,v}\psi'\right).$$

For $\kappa + k$ odd, $\boldsymbol{W}^{(\kappa k)}$ is necessarily a quasispin scalar, and the matrix elements are diagonal with respect to the seniority and independent of N. These properties were first stated in Eqs. (69) and (70) of [6.8].

Application of these ideas to single-electron cfp yields, for states ψ and $\bar{\psi}$ with seniorities v and $v+1$, respectively,

$$\left(\ell^N\psi\{|\ell^{N-1}\bar{\psi}\right) =$$
$$[(N-v)(v+2)/2N]^{\frac{1}{2}}\left(\ell^{\,v+2}\psi|\}\ell^{\,v+1}\bar{\psi}\right). \tag{6.60}$$

6.3.6 The Half-filled Shell

Selection rules for operators of good quasispin rank K, taken between states of the half-filled shell (for which $M_Q = 0$), can be found by inspecting the 3–j symbol

$$\begin{pmatrix} Q & K & Q' \\ 0 & 0 & 0 \end{pmatrix},$$

which appears when the Wigner-Eckart theorem is applied in quasispin space. This 3–j symbol vanishes unless $Q+K+Q'$ is even. An equivalent result can be obtained for $\boldsymbol{W}^{(\kappa k)}$ by referring to (6.53) and insisting that y be even.

6.4 Complementarity

6.4.1 Spin–Quasispin Interchange

The operator R formally interchanges spin and quasispin. The result for the creation and annihilation operators for electrons can be expressed in terms of triple tensors:

$$Ra_\xi^{(qs\ell)}R^{-1} = a_\eta^{(qs\ell)}\,, \tag{6.61}$$

where $\xi \equiv (m_q m_s m_\ell)$ and $\eta \equiv (m_s m_q m_\ell)$. For the tensors $\boldsymbol{X}^{(K\kappa k)}$ defined in (6.44), we get

$$RX_\lambda^{(K\kappa k)}R^{-1} = X_\mu^{(\kappa K k)}\,, \tag{6.62}$$

where $\lambda \equiv (M_K M_\kappa M_k)$ and $\mu \equiv (M_\kappa M_K M_k)$. For states of the ℓ shell,

$$R|\gamma Q M_Q S M_S\rangle = (-1)^t|\gamma S M_S Q M_Q\rangle\,, \tag{6.63}$$

where the quasispin of the ket on the right is S and the spin is Q. The phase factor t depends on S and Q and on phase choices made for the coefficients of fractional parentage. The symbol γ denotes the additional labels necessary to completely define the state in question, including L and M_L.

For every γ, *Racah* [6.20] observed that there are two possible pairs (v_1, S_1) and (v_2, S_2) satisfying

$$v_1 + 2S_2 = v_2 + 2S_1 = 2\ell + 1\,. \tag{6.64}$$

From (6.37) it follows that

$$S_1 = Q_2\,, \qquad S_2 = Q_1\,. \tag{6.65}$$

6.4.2 Matrix Elements

Application of the complementarity operator R to the component parts of a matrix element leads to the equation

$$\langle\gamma Q M_Q S M_S|X_\lambda^{(K\kappa k)}|\gamma' Q' M_Q' S' M_S'\rangle = \tag{6.66}$$
$$(-1)^y\langle\gamma S M_S Q M_Q|X_\mu^{(\kappa K k)}|\gamma' S' M_S' Q' M_Q'\rangle\,,$$

where λ and μ have the same significance as in (6.62), and where y, like t of (6.63), depends on the spins and quasispins but not on the associated magnetic quantum numbers. Equation (6.66) leads to a useful special case when $M_K = M_\kappa = 0$ and the tensors $\boldsymbol{X}$ are converted to those of type $\boldsymbol{W}$, defined in (6.18). The sum $K+\kappa+k$ is taken to be odd, with the scalars $\kappa=k=0$ and $K=k=0$ excluded. Application of the Wigner-Eckart theorem to the spin and orbital spaces yields

$$\frac{\left(\gamma Q M_Q S||W^{(\kappa k)}||\gamma' Q' M_Q' S'\right)}{\left(\gamma S M_S Q||W^{(Kk)}||\gamma' S' M_S' Q'\right)} = (-1)^z \frac{\begin{pmatrix} Q & K & Q' \\ -M_Q & 0 & M_Q \end{pmatrix}}{\begin{pmatrix} S & \kappa & S' \\ -M_S & 0 & M_S \end{pmatrix}}, \tag{6.67}$$

where $z=y+Q-M_Q-S+M_S$. An equivalent form is

$$\frac{\left(\ell^N \gamma v_1 S_1||W^{(\kappa k)}||\ell^N \gamma' v_1' S_1'\right)}{\left(\ell^{N'} \gamma v_2 S_2||W^{(Kk)}||\ell^{N'} \gamma' v_2' S_2'\right)} = (-1)^z \frac{\begin{pmatrix} \frac{1}{2}(2\ell+1-v_1) & K & \frac{1}{2}(2\ell+1-v_1') \\ \frac{1}{2}(2\ell+1-N) & 0 & \frac{1}{2}(N-2\ell-1) \end{pmatrix}}{\begin{pmatrix} \frac{1}{2}(2\ell+1-v_2) & \kappa & \frac{1}{2}(2\ell+1-v_2') \\ \frac{1}{2}(2\ell+1-N') & 0 & \frac{1}{2}(N'-2\ell-1) \end{pmatrix}}, \tag{6.68}$$

where (6.64) is satisfied both for the unprimed and primed quantities.

6.5 Quasiparticles

Sets of linear combinations of the creation and annihilation operators for electrons in the ℓ shell can be constructed such that every member of one set anticommutes with a member of a different set. To preserve the tensorial character of these quasiparticle operators with respect to $\boldsymbol{L}$, it is convenient to define [6.21]

$$\lambda_q^\dagger = 2^{-\frac{1}{2}}\left[a_{\frac{1}{2},q}^\dagger + (-1)^{\ell-q} a_{\frac{1}{2},-q}\right], \tag{6.69}$$

$$\mu_q^\dagger = 2^{-\frac{1}{2}}\left[a_{\frac{1}{2},q}^\dagger - (-1)^{\ell-q} a_{\frac{1}{2},-q}\right], \tag{6.70}$$

$$\nu_q^\dagger = 2^{-\frac{1}{2}}\left[a_{-\frac{1}{2},q}^\dagger + (-1)^{\ell-q} a_{-\frac{1}{2},-q}\right], \tag{6.71}$$

$$\xi_q^\dagger = 2^{-\frac{1}{2}}\left[a_{-\frac{1}{2},q}^\dagger - (-1)^{\ell-q} a_{-\frac{1}{2},-q}\right]. \tag{6.72}$$

The four tensors $\boldsymbol{\theta}^\dagger (\equiv \boldsymbol{\lambda}^\dagger, \boldsymbol{\mu}^\dagger, \boldsymbol{\nu}^\dagger, \text{ or } \boldsymbol{\xi}^\dagger)$ anticommute with each other; the first two act in the spin-up space, the second two in the spin-down space. The tensors $\boldsymbol{\theta}$, whose components $\tilde{\theta}_q$ are defined as in (6.17) with $t=\ell$ and $m_t=q$, are related to their adjoints by the equations

$$\boldsymbol{\lambda}^\dagger = \boldsymbol{\lambda}, \qquad \boldsymbol{\mu}^\dagger = -\boldsymbol{\mu}, \tag{6.73}$$

$$\boldsymbol{\nu}^\dagger = \boldsymbol{\nu}, \qquad \boldsymbol{\xi}^\dagger = -\boldsymbol{\xi}. \tag{6.74}$$

Under the action of the complementarity operator R (see (6.61)) [6.22],

$$R\boldsymbol{\lambda}R^{-1} = \boldsymbol{\lambda}, \qquad R\boldsymbol{\mu}R^{-1} = \boldsymbol{\mu}, \tag{6.75}$$

$$R\boldsymbol{\nu}R^{-1} = \boldsymbol{\nu}, \qquad R\boldsymbol{\xi}R^{-1} = -\boldsymbol{\xi}. \tag{6.76}$$

The tensors $\boldsymbol{\lambda}$, $\boldsymbol{\mu}$, and $\boldsymbol{\nu}$, for a given component q, form a vector with respect to $\boldsymbol{S}+\boldsymbol{Q}$. Every component of $\boldsymbol{\xi}$ is scalar with respect to $\boldsymbol{S}+\boldsymbol{Q}$ [6.23].

The compound quasiparticle operators defined by [6.21]

$$\Theta_q^\dagger = 2^{-\frac{1}{2}}\left[\theta_q^\dagger, \theta_0^\dagger\right], \tag{6.77}$$

where $q>0$ and $\theta \equiv \lambda$, μ, ν, or ξ satisfy the anticommutation relations

$$\left[\Theta_q^\dagger, \Theta_{q'}^\dagger\right]_+ = 0, \tag{6.78}$$

$$\left[\Theta_q, \Theta_{q'}\right]_+ = 0, \tag{6.79}$$

$$\left[\Theta_q^\dagger, \Theta_{q'}\right]_+ = \delta(q,q'), \tag{6.80}$$

for $q, q' > 0$. The $\Theta_q^\dagger$ with $q>0$ can thus be regarded as the creation operators for a fermion quasiparticle with ℓ components.

The connection between the creation and annihilation operators for quasiparticles and for quarks (appearing in the last two rows of Table 3.1) is

$$\boldsymbol{\theta} \to 2^{(\ell-1)/2} \epsilon_\theta \gamma_\theta \left(\boldsymbol{q}_\theta^\dagger \boldsymbol{q}_\theta\right)^{(10\ldots0)}, \tag{6.81}$$

where the γ_θ are Dirac matrices satisfying

$$\gamma_\theta\gamma_\phi + \gamma_\phi\gamma_\theta = 2\delta(\theta,\phi), \tag{6.82}$$

and the ϵ_θ are phases, to some extent dependent on the definitions (6.69–6.72) [6.24]. The superscript $(10\ldots0)$ indicates that $\boldsymbol{q}_\theta^\dagger$ and $\boldsymbol{q}_\theta$ each of which belongs to the elementary spinor $(\frac{1}{2}\frac{1}{2}\ldots\frac{1}{2})$ of $\mathrm{SO}_\theta(2\ell+1)$, are to be coupled to the resultant $(10\ldots0)$, which matches the group label for $\boldsymbol{\theta}$. In the quark model, the $2^{4\ell+2}$ states of the atomic ℓ shell are given by

$$\boldsymbol{q}_\lambda^\dagger \boldsymbol{q}_\mu^\dagger \boldsymbol{q}_\nu^\dagger \boldsymbol{q}_\xi^\dagger |0\rangle_{pp'}, \tag{6.83}$$

where p and p' are parity labels that distinguish the four reference states $|0\rangle$ corresponding to the evenness and oddness of the number of spin-up and spin-down electrons. The scalar nature of ξ (and hence of $\boldsymbol{q}_\xi$) with respect to $\boldsymbol{S}+\boldsymbol{Q}$ can be used to derive relations between spin-orbit matrix elements that go beyond those expected from an application of the Wigner–Eckart theorem [6.25].

References

6.1 E. K. U. Gross, E. Runge, O. Heinonen: *Many-Particle Theory* (Hilger, New York 1991)

6.2 G. Racah: Phys. Rev. **62**, 438 (1942)

6.3 E. U. Condon, G. H. Shortley: *The Theory of Atomic Spectra* (Cambridge Univ. Press, New York 1935)

6.4 A. R. Edmonds: *Angular Momentum in Quantum Mechanics* (Princeton Univ. Press, Princeton 1957)

6.5 C. W. Nielson, G. F. Koster: *Spectroscopic Coefficients for the p^n, d^n, and f^n Configurations* (MIT Press, Cambridge 1963)

6.6 R. Karazija, J. Vizbaraitė, Z. Rudzikas, A. P. Jucys: *Tables for the Calculation of Matrix Elements of Atomic Operators* (Academy Sci. Computing Center, Moscow 1967)

6.7 B. R. Judd: *Second Quantization and Atomic Spectroscopy* (Johns Hopkins, Baltimore 1967)

6.8 G. Racah: Phys. Rev. **63**, 367 (1943)

6.9 V. L. Donlan: *Air Force Materials Laboratory Report No. AFML-TR-70-249* (Wright–Patterson Air Force Base, Ohio 1970)

6.10 D. D. Velkov: Multi-Electron Coefficients of Fractional Parentage for the p, d, and f Shells. Ph.D. Thesis (The Johns Hopkins University, Baltimore 2000) http://www.pha.jhu.edu/groups/cfp/

6.11 P. J. Redmond: Proc. R. Soc. London **A222**, 84 (1954)

6.12 B. H. Flowers, S. Szpikowski: Proc. Phys. Soc. London **84**, 673 (1964)

6.13 Z. Rudzikas: *Theoretical Atomic Spectroscopy* (Cambridge Univ. Press, New York 1997)

6.14 G. Gaigalas, Z. Rudzikas, C. Froese Fischer: At. Data Nucl. Data Tables **70**, 1 (1998)

6.15 B. R. Judd, E. Lo, D. Velkov: Mol. Phys. **98**, 1151 (2000), Table 4

6.16 B. R. Judd: J. Phys. C **14**, 375 (1981)

6.17 J. S. Bell: Nucl. Phys. **12**, 117 (1959)

6.18 H. Watanabe: Prog. Theor. Phys. **32**, 106 (1964)

6.19 R. D. Lawson, M. H. Macfarlane: Nucl. Phys. **66**, 80 (1965)

6.20 G. Racah: Phys. Rev. **76**, 1352 (1949), Table I

6.21 L. Armstrong, B. R. Judd: Proc. R. Soc. London A **315**, 27, 39 (1970)

6.22 B. R. Judd, S. Li: J. Phys. B **22**, 2851 (1989)

6.23 B. R. Judd, G. M. S. Lister, M. A. Suskin: J. Phys. B **19**, 1107 (1986)

6.24 B. R. Judd: Phys. Rep. **285**, 1 (1997)

6.25 B. R. Judd, E. Lo: Phys. Rev. Lett. **85**, 948 (2000)

7. Density Matrices

The density operator was first introduced by J. von *Neumann* [7.1] in 1927 and has since been widely used in quantum statistics. Over the past decades, however, the application of density matrices has spread to many other fields of physics. Density matrices have been used to describe, for example, coherence and correlation phenomena, alignment and orientation and their effect on the polarization of emitted radiation, quantum beat spectroscopy, optical pumping, and scattering processes, particularly when spin-polarized projectiles and/or targets are involved. A thorough introduction to the theory of density matrices and their applications with emphasis on atomic physics can be found in the book by *Blum* [7.2] from which many equations have been extracted for use in this chapter.

7.1 **Basic Formulae** 123
- 7.1.1 Pure States 123
- 7.1.2 Mixed States 124
- 7.1.3 Expectation Values 124
- 7.1.4 The Liouville Equation 124
- 7.1.5 Systems in Thermal Equilibrium .. 125
- 7.1.6 Relaxation Processes 125

7.2 **Spin and Light Polarizations** 125
- 7.2.1 Spin-Polarized Electrons 125
- 7.2.2 Light Polarization 125

7.3 **Atomic Collisions** 126
- 7.3.1 Scattering Amplitudes 126
- 7.3.2 Reduced Density Matrices 126

7.4 **Irreducible Tensor Operators** 127
- 7.4.1 Definition 127
- 7.4.2 Transformation Properties 127
- 7.4.3 Symmetry Properties of State Multipoles 128
- 7.4.4 Orientation and Alignment 128
- 7.4.5 Coupled Systems 129

7.5 **Time Evolution of State Multipoles** 129
- 7.5.1 Perturbation Coefficients 129
- 7.5.2 Quantum Beats 129
- 7.5.3 Time Integration over Quantum Beats 130

7.6 **Examples** 130
- 7.6.1 Generalized *STU*-parameters 130
- 7.6.2 Radiation from Excited States: Stokes Parameters 131

7.7 **Summary** 133

References 133

The main advantage of the density matrix formalism is its ability to deal with pure and mixed states in the same consistent manner. The preparation of the initial state as well as the details regarding the observation of the final state can be treated in a systematic way. In particular, averages over quantum numbers of unpolarized beams in the initial state and incoherent sums over non-observed quantum numbers in the final state can be accounted for via the reduced density matrix. Furthermore, expansion of the density matrix in terms of irreducible tensor operators and the corresponding state multipoles allows for the use of advanced angular momentum techniques, as outlined in Chapts. 2, 3 and 12. More details can be found in two recent textbooks [7.3, 4].

7.1 Basic Formulae

7.1.1 Pure States

Consider a system in a quantum state that is represented by a single wave function $|\Psi\rangle$. The density operator for this situation is defined as

$$\rho = |\Psi\rangle\langle\Psi| \,. \tag{7.1}$$

If $|\Psi\rangle$ is normalized to unity, i.e., if

$$\langle\Psi|\Psi\rangle = 1 \,, \tag{7.2}$$

then

$$\rho^2 = \rho \,. \tag{7.3}$$

Equation (7.3) is the basic equation for identifying *pure* quantum mechanical states represented by a density operator.

Next, consider the expansion of $|\Psi\rangle$ in terms of a complete orthonormal set of basis functions $\{|\Phi_n\rangle\}$, i.e.,

$$|\Psi\rangle = \sum_n c_n |\Phi_n\rangle \,. \quad (7.4)$$

The density operator then becomes

$$\rho = \sum_{n,m} c_n c_m^* |\Phi_n\rangle\langle\Phi_m| = \rho_{nm} |\Phi_n\rangle\langle\Phi_m| \,, \quad (7.5)$$

where the star denotes the complex conjugate quantity. Note that the density matrix elements $\rho_{nm} = \langle\Phi_n|\,\rho\,|\Phi_m\rangle$ depend on the choice of the basis and that the density matrix is Hermitian, i.e.,

$$\rho_{mn}^* = \rho_{nm} \,. \quad (7.6)$$

Finally, if $|\Psi\rangle = |\Phi_i\rangle$ is one of the basis functions, then

$$\rho_{mn} = \delta_{ni}\delta_{mi} \,, \quad (7.7)$$

where δ_{ni} is the Kronecker δ. Hence, the density matrix is diagonal in this representation with only one nonvanishing element.

7.1.2 Mixed States

The above concepts can be extended to treat statistical ensembles of pure quantum states. In the simplest case, such *mixed* states can be represented by a diagonal density matrix of the form

$$\rho = \sum_n w_n |\Psi_n\rangle\langle\Psi_n| \,, \quad (7.8)$$

where the weight w_n is the fraction of systems in the pure quantum state $|\Psi_n\rangle$. The standard normalization for the trace of ρ is

$$\mathrm{Tr}\{\rho\} = \sum_n w_n = 1 \,. \quad (7.9)$$

Since the trace is invariant under unitary transformations of the basis functions, (7.9) also holds if the $|\Psi_n\rangle$ states themselves are expanded in terms of basis functions as in (7.4). For a pure state and the normalization (7.9), one finds in an arbitrary basis

$$\mathrm{Tr}\{\rho\} = \mathrm{Tr}\{\rho^2\} = 1 \,. \quad (7.10)$$

7.1.3 Expectation Values

The density operator contains the maximum available information about a physical system. Consequently, it can be used to calculate expectation values for any operator $\mathcal{A}$ that represents a physical observable. In general,

$$\langle\mathcal{A}\rangle = \mathrm{Tr}\{\mathcal{A}\rho\}/\mathrm{Tr}\{\rho\} \,, \quad (7.11)$$

where $\mathrm{Tr}\{\rho\}$ in the denominator of (7.11) ensures the correct result even for a normalization that is different from (7.9). The invariance of the trace operation ensures the same result – independent of the particular choice of the basis representation.

7.1.4 The Liouville Equation

Suppose (7.8) is valid for a time $t = 0$. If the functions $|\Psi_n(\boldsymbol{r}, t)\rangle$ obey the Schrödinger equation, i.e.

$$\mathrm{i}\frac{\partial}{\partial t}|\Psi_n(\boldsymbol{r}, t)\rangle = H(t)\,|\Psi_n(\boldsymbol{r}, t)\rangle \,, \quad (7.12)$$

the density operator at the time t can be written as

$$\rho(t) = U(t)\,\rho(0)\,U^\dagger(t) \,. \quad (7.13)$$

In (7.13), $U(t)$ is the time evolution operator which relates the wave functions at times $t = 0$ and t according to

$$|\Psi_n(\boldsymbol{r}, t)\rangle = U(t)\,|\Psi_n(\boldsymbol{r}, 0)\rangle \,, \quad (7.14)$$

and $U^\dagger(t)$ denotes its adjoint. Note that

$$U(t) = \mathrm{e}^{-\mathrm{i}Ht} \,, \quad (7.15)$$

if the Hamiltonian H is time-independent.

Differentiation of (7.13) with respect to time and inserting (7.14) into the Schrödinger equation (7.12) yields the equation of motion

$$\mathrm{i}\frac{\partial}{\partial t}\rho(t) = [H(t), \rho(t)] \,, \quad (7.16)$$

where $[\mathcal{A}, \mathcal{B}\,]$ denotes a commutator.

The Liouville equation (7.16) can be used to determine the density matrix and to treat transitions from nonequilibrium to equilibrium states in quantum mechanical systems. Especially for approximate solutions in the presence of small time-dependent perturbation terms in an otherwise time-independent Hamiltonian, i.e., for

$$H(t) = H_0 + V(t) \,, \quad (7.17)$$

the interaction picture is preferably used. The Liouville equation then becomes

$$\mathrm{i}\frac{\partial}{\partial t}\rho_\mathrm{I}(t) = \left[V_\mathrm{I}(t), \rho_\mathrm{I}(t)\right] \,, \quad (7.18)$$

where the subscript I denotes the operator in the interaction picture. In first-order perturbation theory, (7.18) can be integrated to yield

$$\rho_I(t) = \rho_I(0) - i \int_0^t [V_I(\tau), \rho_I(0)] \, d\tau , \tag{7.19}$$

and higher-order terms can be obtained through subsequent iterations.

7.1.5 Systems in Thermal Equilibrium

According to quantum statistics, the density operator for a system which is in thermal equilibrium with a surrounding reservoir $\mathcal{R}$ at a temperature T (canonical ensemble), can be expressed as

$$\rho = \frac{\exp(-\beta H)}{Z} , \tag{7.20}$$

where H is the Hamiltonian, and $\beta = 1/k_B T$ with k_B being the Boltzmann constant. The partition sum

$$Z = \mathrm{Tr}\{\exp(-\beta H)\} , \tag{7.21}$$

ensures the normalization condition (7.9). Expectation values are calculated according to (7.11), and extensions to other types of ensembles are straightforward.

7.1.6 Relaxation Processes

Transitions from nonequilibrium to equilibrium states can also be described within the density matrix formalism. One of the basic problems is to account for irreversibility in the energy (and sometimes particle) exchange between the system of interest, $\mathcal{S}$, and the reservoir, $\mathcal{R}$. This is usually achieved by assuming that the interaction of the system with the reservoir is negligible and, therefore, the density matrix representation for the reservoir at any time t is the same as the representation for $t = 0$.

Another important assumption that is frequently made is the Markov approximation. In this approximation, one assumes that the system "forgets" all knowledge of the past, so that the density matrix elements at the time $t + \Delta t$ depend only on the values of these elements, and their first derivatives, at the time t. When (7.19) is put back into (7.18), the result in the Markov approximation can be rewritten as

$$\frac{\partial}{\partial t} \rho_{\mathcal{S}I}(t) = - i \,\mathrm{Tr}_{\mathcal{R}}\big[V_I(t), \rho_{\mathcal{S}I}(0)\rho_{\mathcal{R}}(0)\big] - \int_0^t d\tau \,\mathrm{Tr}_{\mathcal{R}}\big[V_I(t), [V_I(\tau), \rho_{\mathcal{S}I}(t)\rho_{\mathcal{R}}(0)]\big] , \tag{7.22}$$

where $\mathrm{Tr}_{\mathcal{R}}$ denotes the trace with regard to all variables of the reservoir. Note that the integral over $d\tau$ contains the system density matrix in the interaction picture, $\rho_{\mathcal{S}I}$, at the time t, rather than at all times τ which are integrated over (the Markov approximation), and that the density matrix for the reservoir is taken as $\rho_{\mathcal{R}}(0)$ at all times. For more details, see Chapter 7 of *Blum* [7.2] and references therein.

Equations such as (7.22) are the basis for the master or rate equation approach used, for example, in quantum optics for the theory of lasers and the coupling of atoms to cavity modes. For more details, see Chapts. 68, 69, 70 and 78.

7.2 Spin and Light Polarizations

Density matrices are frequently used to describe the polarization state of spin-polarized particle beams as well as light. The latter can either be emitted from excited atomic or molecular ensembles or can be used, for example, for laser pumping purposes.

7.2.1 Spin-Polarized Electrons

The spin polarization of an electron beam with respect to a given quantization axis $\hat{\boldsymbol{n}}$ is defined as [7.5]

$$P_{\hat{n}} = \frac{N_\uparrow - N_\downarrow}{N_\uparrow + N_\downarrow} , \tag{7.23}$$

where $N_\uparrow (N_\downarrow)$ is the number of electrons with spin up (down) with regard to this axis. An arbitrary polarization state is described by the density matrix

$$\rho = \frac{1}{2} \begin{pmatrix} 1 + P_z & P_x - iP_y \\ P_x + iP_y & 1 - P_z \end{pmatrix} , \tag{7.24}$$

where $P_{x,y,x}$ are the cartesian components of the spin polarization vector. The individual components can be obtained from the density matrix as

$$P_i = \mathrm{Tr}\{\sigma_i \, \rho\} , \tag{7.25}$$

where the σ_i $(i = x, y, z)$ are the standard Pauli spin matrices.

7.2.2 Light Polarization

Another important use of the density matrix formalism is the description of light polarization in terms of the so-called Stokes parameters [7.6]. For a given direction of observation, the general polarization state of light can be fully determined by the measurement of one circular and two independent linear polarizations. Using the notation of Born and Wolf [7.7], the density matrix is given by

$$\rho = \frac{I_{\text{tot}}}{2}\begin{pmatrix} 1-P_3 & P_1 - \mathrm{i}P_2 \\ P_1 + \mathrm{i}P_2 & 1+P_3 \end{pmatrix}, \tag{7.26}$$

where P_1 and P_2 are linear light polarizations while P_3 is the circular polarization (see also Sect. 7.6). In (7.26), the density matrix is normalized in such as way that

$$\mathrm{Tr}\{\rho\} = I_{\text{tot}}, \tag{7.27}$$

where I_{tot} is the total light intensity. Other frequently used names for the various Stokes parameters are

$$P_1 = \quad \eta_3 = M, \tag{7.28}$$

$$P_2 = \quad \eta_1 = C, \tag{7.29}$$

$$P_3 = -\eta_2 = S. \tag{7.30}$$

The Stokes parameters of electric dipole radiation can be related directly to the charge distribution of the emitting atomic ensemble. As discussed in detail in Chapt. 46, one finds, for example,

$$L_\perp = -P_3 \tag{7.31}$$

for the angular momentum transfer perpendicular to the scattering plane in collisional (de-)excitation, and

$$\gamma = \frac{1}{2}\arg\{P_1 + \mathrm{i}P_2\} \tag{7.32}$$

for the alignment angle.

7.3 Atomic Collisions

7.3.1 Scattering Amplitudes

Transitions from an initial state $|J_0 M_0; \boldsymbol{k}_0 m_0\rangle$ to a final state $|J_1 M_1; \boldsymbol{k}_1 m_1\rangle$ are described by scattering amplitudes

$$f(M_1 m_1; M_0 m_0) = \langle J_1 M_1; \boldsymbol{k}_1 m_1 | \mathcal{T} | J_0 M_0; \boldsymbol{k}_0 m_0 \rangle, \tag{7.33}$$

where $\mathcal{T}$ is the transition operator. Furthermore, J_0 (J_1) is the total electronic angular momentum in the initial (final) state of the target and M_0 (M_1) its corresponding z-component, while $\boldsymbol{k}_0$ ($\boldsymbol{k}_1$) is the initial (final) momentum of the projectile and m_0 (m_1) its spin component.

7.3.2 Reduced Density Matrices

While the scattering amplitudes are the central elements in a theoretical description, some restrictions usually need to be taken into account in a practical experiment. The most important ones are: (i) there is no "pure" initial state, and (ii) not all possible quantum numbers are simultaneously determined in the final state. The solution to this problem can be found by using the density matrix formalism. First, the complete density operator after the collision process is given by [7.2]

$$\rho_{\text{out}} = \mathcal{T} \rho_{\text{in}} \mathcal{T}^\dagger, \tag{7.34}$$

where ρ_{in} is the density operator before the collision. The corresponding matrix elements are given by

$$\begin{aligned}(\rho_{\text{out}})^{\boldsymbol{k}_1, M_1' M_1}_{m_1' m_1} = & \sum_{m_0' m_0 M_0' M_0} \rho_{m_0' m_0}\, \rho_{M_0' M_0} \\ & \times f\left(M_1' m_1'; M_0' m_0'\right) \\ & \times f^*\left(M_1 m_1; M_0 m_0\right),\end{aligned} \tag{7.35}$$

where the term $\rho_{m_0' m_0} \rho_{M_0' M_0}$ describes the preparation of the initial state (i). Secondly, "reduced" density matrices account for (ii). For example, if only the scattered projectiles are observed, the corresponding elements of the reduced density matrix are obtained by summing over the atomic quantum numbers as follows:

$$(\rho_{\text{out}})^{\boldsymbol{k}_1}_{m_1' m_1} = \sum_{M_1} (\rho_{\text{out}})^{\boldsymbol{k}_1, M_1 M_1}_{m_1' m_1}. \tag{7.36}$$

The differential cross section for unpolarized projectile and target beams is given by

$$\frac{\mathrm{d}\sigma}{\mathrm{d}\Omega} = C \sum_{m_1} (\rho_{\text{out}})^{\boldsymbol{k}_1}_{m_1 m_1}, \tag{7.37}$$

where C is a constant that depends on the normalization of the continuum waves in a numerical calculation.

On the other hand, if only the atoms are observed (for example, by analyzing the light emitted in optical transitions), the elements

$$(\rho_{\text{out}})_{M_1' M_1} = \int \mathrm{d}^3 \boldsymbol{k}_1 \sum_{m_1} (\rho_{\text{out}})^{\boldsymbol{k}_1, M_1' M_1}_{m_1 m_1} \tag{7.38}$$

determine the integrated Stokes parameters [7.8,9], i.e., the polarization of the emitted light. They contain information about the angular momentum distribution in the excited target ensemble.

Finally, for electron–photon coincidence experiments without spin analysis in the final state, the elements

$$(\rho_{\text{out}})^{k_1}_{M_1' M_1} = \sum_{m_1} (\rho_{\text{out}})^{k_1, M_1' M_1}_{m_1 m_1} \tag{7.39}$$

simultaneously contain information about the projectiles and the target. This information can be extracted by measuring the angle-differential Stokes parameters. In particular, for unpolarized electrons and atoms, the "natural coordinate system", where the quantization axis coincides with the normal to the scattering plane, allows for a simple physical interpretation of the various parameters [7.10] (see Chapt. 46).

The density matrix formalism outlined above is very useful for obtaining a qualitative description of the geometrical and sometimes also of the dynamical symmetries of the collision process [7.11]. Two explicit examples are discussed in Sect. 7.6.

7.4 Irreducible Tensor Operators

The general density matrix theory can be formulated in a very elegant fashion by decomposing the density operator in terms of irreducible components whose matrix elements then become the state multipoles. In such a formulation, full advantage can be taken of the most sophisticated techniques developed in angular momentum algebra (see Chapt. 2). Many explicit examples can be found in [7.3,4].

7.4.1 Definition

The density operator for an ensemble of particles in quantum states labeled as $|JM\rangle$ where J and M are the total angular momentum and its magnetic component, respectively, can be written as

$$\rho = \sum_{J'JM'M} \rho^{J'J}_{M'M} |J'M'\rangle\langle JM| \,, \tag{7.40}$$

where

$$\rho^{J'J}_{M'M} = \langle J'M'|\rho|JM\rangle \tag{7.41}$$

are the matrix elements. (For simplicity, interactions outside the single manifold of momentum states $|JM\rangle$ are neglected). Alternatively, one may write

$$\rho = \sum_{J'JKQ} \left\langle T(J'J)^\dagger_{KQ} \right\rangle T(J'J)_{KQ} \,, \tag{7.42}$$

where the irreducible tensor operators are defined in terms of 3–j symbols as

$$T(J'J)_{KQ} = \sum_{M'M} (-1)^{J'-M'} \sqrt{2K+1} \times \begin{pmatrix} J' & J & K \\ M' & -M & -Q \end{pmatrix} |J'M'\rangle\langle JM| \,, \tag{7.43}$$

and the state multipoles or statistical tensors are given by

$$\left\langle T(J'J)^\dagger_{KQ} \right\rangle = \sum_{M'M} (-1)^{J'-M'} \sqrt{2K+1} \times \begin{pmatrix} J' & J & K \\ M' & -M & -Q \end{pmatrix} \langle J'M'|\rho|JM\rangle \,. \tag{7.44}$$

Hence, the selection rules for the 3–j symbols imply that

$$|J - J'| \le K \le J + J' \,, \tag{7.45}$$

$$M' - M = Q \,. \tag{7.46}$$

Equation (7.44) can be inverted through the orthogonality condition of the 3–j symbols to give

$$\langle J'M'|\rho|JM\rangle = \sum_{KQ} (-1)^{J'-M'} \sqrt{2K+1} \times \begin{pmatrix} J' & J & K \\ M' & -M & -Q \end{pmatrix} \left\langle T(J'J)^\dagger_{KQ} \right\rangle . \tag{7.47}$$

7.4.2 Transformation Properties

Suppose a coordinate system (X_2, Y_2, Z_2) is obtained from another coordinate system (X_1, Y_1, Z_1) through a rotation by a set of three Euler angles (γ, β, α) as defined in *Edmonds* [7.12]. The irreducible tensor operators (7.43) defined in the (X_1, Y_1, Z_1) system are then related to the operators $\langle T(J'J)^\dagger_{KQ}\rangle$ in the (X_2, Y_2, Z_2)

system by

$$T(J'J)_{KQ} = \sum_q T(J'J)_{Kq}\, D(\gamma, \beta, \alpha)^K_{qQ}\,, \quad (7.48)$$

where

$$D(\gamma, \beta, \alpha)^J_{M'M} = e^{iM'\gamma}\, d(\beta)^J_{M'M}\, e^{iM\alpha} \quad (7.49)$$

is a rotation matrix (see Chapt. 2). Note that the rank K of the tensor operator is invariant under such rotations. Similarly,

$$\left\langle T(J'J)^\dagger_{KQ}\right\rangle = \sum_q \left\langle T(J'J)^\dagger_{Kq}\right\rangle D(\gamma, \beta, \alpha)^{K\,*}_{qQ} \quad (7.50)$$

holds for the state multipoles.

The irreducible tensor operators fulfill the orthogonality condition

$$\mathrm{Tr}\left\{T(J'J)_{KQ}\, T(J'J)^\dagger_{K'Q'}\right\} = \delta_{K'K}\,\delta_{Q'Q}\,, \quad (7.51)$$

with

$$T(J'J)_{00} = \frac{1}{\sqrt{2J+1}}\,\delta_{J'J}\,\mathbf{1} \quad (7.52)$$

being proportional to the unit operator $\mathbf{1}$, it follows that all tensor operators have vanishing trace, except for the monopole $T(J'J)_{00}$.

Reduced tensor operators fulfill the Wigner–Eckart theorem (see Sect. 2.8.4)

$$\begin{aligned}&\left\langle J'M'|T(J'J)_{KQ}|JM\right\rangle \\ &= (-1)^{J'-M'}\begin{pmatrix} J' & K & J \\ -M' & Q & M\end{pmatrix} \\ &\quad\times \left\langle J'\|\, T_K \,\|J\right\rangle,\end{aligned} \quad (7.53)$$

where the reduced matrix element is simply given by

$$\left\langle J'\|\, T_K \,\|J\right\rangle = \frac{1}{\sqrt{2K+1}}\,. \quad (7.54)$$

7.4.3 Symmetry Properties of State Multipoles

The Hermiticity condition for the density matrix implies

$$\left\langle T(J'J)^\dagger_{KQ}\right\rangle^* = (-1)^{J'-J+Q}\left\langle T(JJ')^\dagger_{K-Q}\right\rangle, \quad (7.55)$$

which, for sharp angular momentum $J' = J$, yields

$$\left\langle T(J)^\dagger_{KQ}\right\rangle^* = (-1)^Q\left\langle T(J)^\dagger_{K-Q}\right\rangle. \quad (7.56)$$

Hence, the state multipoles $\langle T(J)^\dagger_{K0}\rangle$ are real numbers.

Furthermore, the transformation property (7.50) of the state multipoles imposes restrictions on nonvanishing state multipoles to describe systems with given symmetry properties. In detail, one finds:

1. For spherically symmetric systems,

$$\left\langle T(J'J)^\dagger_{KQ}\right\rangle = \left\langle T(J'J)^\dagger_{KQ}\right\rangle_{\mathrm{rot}} \quad (7.57)$$

for *all* sets of Euler angles. This implies that only the monopole term $\langle T(J)^\dagger_{00}\rangle$ can be different from zero.

2. For axially symmetric systems,

$$\left\langle T(J'J)^\dagger_{KQ}\right\rangle = \left\langle T(J'J)^\dagger_{KQ}\right\rangle_{\mathrm{rot}} \quad (7.58)$$

for *all* Euler angles ϕ that describe a rotation around the z-axis. Since this angle enters via a factor $\exp(-iQ\phi)$ into the general transformation formula (7.50), it follows that only state multipoles with $Q = 0$, i.e., $\langle T(J'J)^\dagger_{K0}\rangle$, can be different from zero in such a situation.

3. For planar symmetric systems with fixed $J' = J$,

$$\left\langle T(J)^\dagger_{KQ}\right\rangle = (-1)^K\left\langle T(J)^\dagger_{KQ}\right\rangle^* \quad (7.59)$$

if the system properties are invariant under reflection in the xz-plane. Hence, state multipoles with even rank K are real numbers, while those with odd rank are purely imaginary in this case.

The above results can be applied immediately to the description of atomic collisions where the incident beam axis is the quantization axis (the so-called "collision system"). For example, impact excitation of unpolarized targets by unpolarized projectiles without observation of the scattered projectiles is symmetric both with regard to rotation around the incident beam axis and with regard to reflection in any plane containing this axis. Consequently, the state multipoles $\langle T(J)^\dagger_{00}\rangle$, $\langle T(J)^\dagger_{20}\rangle$, $\langle T(J)^\dagger_{40}\rangle$, ... fully characterize the atomic ensemble of interest. Using (7.50), similar relationships can be derived for state multipoles defined with regard to other coordinate systems, such as the "natural system" where the quantization axis coincides with the normal vector to the scattering plane (see Chapt. 46).

7.4.4 Orientation and Alignment

From the above discussion, it is apparent that the description of systems that do not exhibit spherical symmetry requires the knowledge of state multipoles with rank $K \neq 0$. Frequently, the multipoles with $K = 1$ and $K = 2$ are determined via the angular correlation and the polarization of radiation emitted from an ensemble of collisionally excited targets. The state multipoles with $K = 1$ are proportional to the spherical components of the angular momentum expectation value and, therefore, give rise to a nonvanishing circular light polarization (see also Sect. 7.6). This corresponds to a sense of rotation or an orientation in the ensemble which is therefore called oriented (see Sect. 46.1).

On the other hand, nonvanishing multipoles with rank $K = 2$ describe the alignment of the system. Some authors, however, use the terms "alignment" or "orientation" synonymously for all nonvanishing state multipoles with ranks $K \neq 0$, thereby describing any system with anisotropic occupation of magnetic sublevels as "aligned" or "oriented". For details on alignment and orientation, see Chapt. 46 and [7.3,4].

7.4.5 Coupled Systems

Tensor operators and state multipoles for coupled systems are constructed as direct products ($\otimes$) of the operators for the individual systems. For example, the density operator for two subsystems in basis states $|L, M_L\rangle$ and $|S, M_S\rangle$ is constructed as [7.2]

$$\rho = \sum_{KQkq} \langle T(L)^\dagger_{KQ} \otimes T(S)^\dagger_{kq}\rangle \left[T(L)_{KQ} \otimes T(S)_{kq}\right] . \tag{7.60}$$

If the two systems are uncorrelated, the state multipoles factor as

$$\langle T(L)^\dagger_{KQ} \otimes T(S)^\dagger_{kq}\rangle = \left\langle T(L)^\dagger_{KQ}\right\rangle \left\langle T(S)^\dagger_{kq}\right\rangle ; \tag{7.61}$$

More generally, irreducible representations of coupled operators can be defined in terms of a 9–j symbol as

$$\begin{aligned} T(J', J)_{K'Q'} &= \sum_{KQkq} \hat{K}\hat{k}\hat{J}\hat{J}' \,(KQ, kq|K'Q') \\ &\quad \times \begin{Bmatrix} K & k & K' \\ L & S & J' \\ L & S & J \end{Bmatrix} T(L)_{KQ} \otimes T(S)_{kq} , \end{aligned} \tag{7.62}$$

where $\hat{x} \equiv \sqrt{2x+1}$, and $(j_1 m_1, j_2 m_2 | j_3 m_3)$ is a standard Clebsch-Gordan coefficient.

7.5 Time Evolution of State Multipoles

7.5.1 Perturbation Coefficients

From the general expansions

$$\rho(t) = \sum_{j'jkq} \left\langle T(j'j;t)^\dagger_{kq}\right\rangle T(j'j)_{kq} \tag{7.63}$$

in terms of irreducible components, together with (7.42) for time $t = 0$ and (7.13) for the time development of the density operator, it follows that

$$\begin{aligned} \left\langle T(j'j;t)^\dagger_{kq}\right\rangle &= \sum_{J'JKQ} \left\langle T(J'J;0)^\dagger_{KQ}\right\rangle \\ &\quad \times G(J'J, j'j;t)^{Qq}_{Kk} , \end{aligned} \tag{7.64}$$

where the perturbation coefficients are defined as

$$\begin{aligned} &G(J'J, j'j;t)^{Qq}_{Kk} \\ &= \mathrm{Tr}\left\{U(t) T(J'J)_{KQ} U(t)^\dagger T(j'j)^\dagger_{kq}\right\} . \end{aligned} \tag{7.65}$$

Hence, these coefficients relate the state multipoles at time t to those at $t = 0$.

7.5.2 Quantum Beats

An important application of the perturbation coefficients is the coherent excitation of several quantum states which subsequently decay by optical transitions. Such an excitation may be performed, for example, in beam-foil experiments or electron–atom collisions where the energy width of the electron beam is too large to resolve the fine structure (or hyperfine structure) of the target states.

Suppose, for instance, that explicitly relativistic effects, such as the spin–orbit interaction between the projectile and the target, can be neglected *during* a collision process between an incident electron and a target atom. In that case, the orbital angular momentum (L) system of the collisionally excited target states may be oriented, depending on the scattering angle of the projectile. On the other hand, the spin (S) system remains unaffected (unpolarized), provided that both the target and the projectile beams are unpolarized. During the lifetime of the excited target states, however, the spin–orbit interaction *within* the target produces an exchange of orientation between the L and the S systems, which results in a net loss of orientation in the L system.

This effect can be observed directly through the intensity and the polarization of the light emitted from the excited target ensemble. The perturbation coefficients for the fine structure interaction are found to be [7.2, 13]

$$G(L;t)_K = \frac{\exp(-\gamma t)}{2S+1} \sum_{J'J} (2J'+1)(2J+1) \times \begin{Bmatrix} L & J' & S \\ J & L & K \end{Bmatrix}^2 \cos\left(\omega_{J'} - \omega_J\right)t\,, \tag{7.66}$$

where $\omega_{J'} - \omega_J$ corresponds to the (angular) frequency difference between the various multiplet states with total electronic angular momenta J' and J, respectively. Also, γ is the natural width of the spectral line; for simplicity, the same lifetime has been assumed in (7.66) for all states of the multiplet.

Note that the perturbation coefficients are independent of the multipole component Q in this case, and that there is no mixing between different multipole ranks K. Similar results can be derived [7.2, 13] for the hyperfine interaction and also to account for the combined effect of fine and hyperfine structure. The cosine terms represent correlation between the signal from different fine structure states, and they lead to oscillations in the intensity as well as the measured Stokes parameters in a time-resolved experiment.

Finally, generalized perturbation coefficients have been derived for the case where both the L and the S systems may be oriented and/or aligned during the collision process [7.14]. This can happen when spin-polarized projectiles and/or target beams are prepared.

7.5.3 Time Integration over Quantum Beats

If the excitation and decay times cannot be resolved in a given experimental setup, the perturbation coefficients need to be integrated over time. As a result, the quantum beats disappear, but a net effect may still be visible through a depolarization of the emitted radiation. For the case of atomic fine structure interaction discussed above, one finds [7.2, 13]

$$\begin{aligned} \bar{G}(L)_K &= \int_0^\infty G(L;t)_K \, dt \\ &= \frac{1}{2S+1} \sum_{J'J} (2J'+1)(2J+1) \\ &\quad \times \begin{Bmatrix} L & J' & S \\ J & L & K \end{Bmatrix}^2 \frac{\gamma}{\gamma^2+\omega_{J'J}^2}\,, \end{aligned} \tag{7.67}$$

where $\omega_{J'J} = \omega_{J'} - \omega_J$. Note that the amount of depolarization depends on the relationship between the fine structure splitting and the natural line width. For $|\omega_{J'J}| \gg \gamma$ (if $J' \neq J$), the terms with $J' = J$ dominate and cause the maximum depolarization; for the opposite case $|\omega_{J'J}| \ll \gamma$, the sum rule for the 6–j symbols can be applied and no depolarization is observed.

Similar depolarizations can be caused through hyperfine structure effects, as well as through external fields. An important example of the latter case is the Hanle effect (see Sect. 17.2.1).

7.6 Examples

In this section, two examples of the reduced density formalism are discussed explicitly. These are: (i) the change of the spin polarization of initially polarized spin-$\frac{1}{2}$ projectiles after scattering from unpolarized targets, and (ii) the Stokes parameters describing the angular distribution and the polarization of light as detected in projectile-photon coincidence experiments after collisional excitation. The recent book by *Andersen* and *Bartschat* [7.4] provides a detailed introduction to these topics, together with a thorough discussion of benchmark studies in the field of electronic and atomic collisions, including extensions to ionization processes, as well as applications in plasma, surface, and nuclear physics. Even more extensive compilations of such studies can be found in a review series dealing with unpolarized electrons colliding with unpolarized targets [7.10], heavy-particle collisions [7.15], and the special role of projectile and target spins in such collisions [7.16].

7.6.1 Generalized *STU*-parameters

For spin-polarized projectile scattering from unpolarized targets, the generalized *STU*-parameters [7.11] contain information about the projectile spin polarization after the collision. These parameters can be expressed in terms of the elements (7.36).

To analyze this problem explicitly, one defines the quantities

$$\langle m_1' m_0';\, m_1 m_0\rangle = \frac{1}{2J_0+1}\sum_{M_1 M_0} f(M_1 m_1';\, M_0 m_0') \times f^*(M_1 m_1;\, M_0 m_0) \qquad (7.68)$$

which contain the *maximum information* that can be obtained from the scattering process, if only the polarization of the projectiles is prepared before the collision and measured thereafter.

Next, the number of independent parameters that can be determined in such an experiment needs to be examined. For spin-$\frac{1}{2}$ particles, there are $2\times2\times2\times2=16$ possible combinations of $\{m_1' m_0';\, m_1 m_0\}$ and, therefore, 16 complex or 32 real parameters (in the most general case of spin-S particles, there would be $(2S+1)^4$ combinations). However, from the definition (7.68) and the Hermiticity of the reduced density matrix contained therein, it follows that

$$\langle m_1' m_0';\, m_1 m_0\rangle = \langle m_1 m_0;\, m_1' m_0'\rangle^* . \qquad (7.69)$$

Furthermore, parity conservation of the interaction or the equivalent reflection invariance with regard to the scattering plane yields the additional relationship [7.11]

$$f(M_1 m_1;\, M_0 m_0) = (-1)^{J_1-M_1+\frac{1}{2}-m_1+J_0-M_0+\frac{1}{2}-m_0} \times \Pi_1 \Pi_0 f(-M_1-m_1;\, -M_0-m_0)\,, \qquad (7.70)$$

where Π_1 and Π_0 are ±1, depending on the parities of the atomic states involved. Hence,

$$\langle m_1' m_0';\, m_1 m_0\rangle = (-1)^{m_1'-m_1+m_0'-m_0} \times \langle -m_1' -m_0';\, -m_1-m_0\rangle . \qquad (7.71)$$

Note that (7.70, 71) hold for the collision frame where the quantization axis ($\hat{z}$) is taken as the incident beam axis and the scattering plane is the xz-plane. Similar formulas can be derived for the natural frame (see Sect. 7.3.2)

Consequently, *eight* independent parameters are sufficient to characterize the reduced spin density matrix of the scattered projectiles. These can be chosen as the *absolute* differential cross section

$$\sigma_u = \frac{1}{2}\sum_{m_1,m_0}\langle m_1 m_0;\, m_1 m_0\rangle \qquad (7.72)$$

for the scattering of unpolarized projectiles from unpolarized targets and the seven *relative* parameters

$$S_A = -\frac{2}{\sigma_u}\,\mathrm{Im}\left\{\left\langle \frac{1}{2}-\frac{1}{2};\, \frac{1}{2}\frac{1}{2}\right\rangle\right\}, \qquad (7.73)$$

$$S_P = -\frac{2}{\sigma_u}\,\mathrm{Im}\left\{\left\langle \frac{1}{2}\frac{1}{2};\, -\frac{1}{2}\frac{1}{2}\right\rangle\right\}, \qquad (7.74)$$

$$T_y = \frac{1}{\sigma_u}\left\{\left\langle -\frac{1}{2}-\frac{1}{2};\, \frac{1}{2}\frac{1}{2}\right\rangle - \left\langle -\frac{1}{2}\frac{1}{2};\, \frac{1}{2}-\frac{1}{2}\right\rangle\right\}, \qquad (7.75)$$

$$T_x = \frac{1}{\sigma_u}\left[\left\langle -\frac{1}{2}-\frac{1}{2};\, \frac{1}{2}\frac{1}{2}\right\rangle + \left\langle -\frac{1}{2}\frac{1}{2};\, \frac{1}{2}-\frac{1}{2}\right\rangle\right], \qquad (7.76)$$

$$T_z = \frac{1}{\sigma_u}\left[\left\langle \frac{1}{2}\frac{1}{2};\, \frac{1}{2}\frac{1}{2}\right\rangle - \left\langle \frac{1}{2}\frac{1}{2};\, -\frac{1}{2}\frac{1}{2}\right\rangle\right], \qquad (7.77)$$

$$U_{xz} = \frac{2}{\sigma_u}\,\mathrm{Re}\left\{\left\langle \frac{1}{2}\frac{1}{2};\, -\frac{1}{2}\frac{1}{2}\right\rangle\right\}, \qquad (7.78)$$

$$U_{zx} = -\frac{2}{\sigma_u}\,\mathrm{Re}\left\{\left\langle \frac{1}{2}-\frac{1}{2};\, \frac{1}{2}\frac{1}{2}\right\rangle\right\}, \qquad (7.79)$$

where $\mathrm{Re}\{x\}$ and $\mathrm{Im}\{x\}$ denote the real and imaginary parts of the complex quantity x, respectively. Note that normalization constants have been omitted in (7.72) to simplify the notation.

Therefore, the most general form for the polarization vector after scattering, $\boldsymbol{P}'$, for an initial polarization vector $\boldsymbol{P}=(P_x, P_y, P_z)$ is given by

$$\frac{(S_P+T_yP_y)\hat{\boldsymbol{y}}+(T_xP_x+U_{xz}P_z)\hat{\boldsymbol{x}}+(T_zP_z-U_{zx}P_x)\hat{\boldsymbol{z}}}{1+S_AP_y}. \qquad (7.80)$$

The physical meaning of the above relation is illustrated in Fig. 7.1.

The following geometries are particularly suitable for the experimental determination of the individual parameters; σ_u and S_P can be measured with unpolarized incident projectiles. A transverse polarization component perpendicular to the scattering plane $(\boldsymbol{P}=P_y\hat{\boldsymbol{y}})$ is needed to obtain S_A and T_y. Finally, the measurement of T_x, U_{zx}, T_z, and U_{xz} requires both transverse $(P_x\hat{\boldsymbol{x}})$ and longitudinal $(P_z\hat{\boldsymbol{z}})$ projectile polarization components in the scattering plane.

7.6.2 Radiation from Excited States: Stokes Parameters

The state multipole description is also widely used for the parametrization of the Stokes parameters that describe the polarization of light emitted in optical decays of excited atomic ensembles. The general case of excitation by spin-polarized projectiles has been treated by

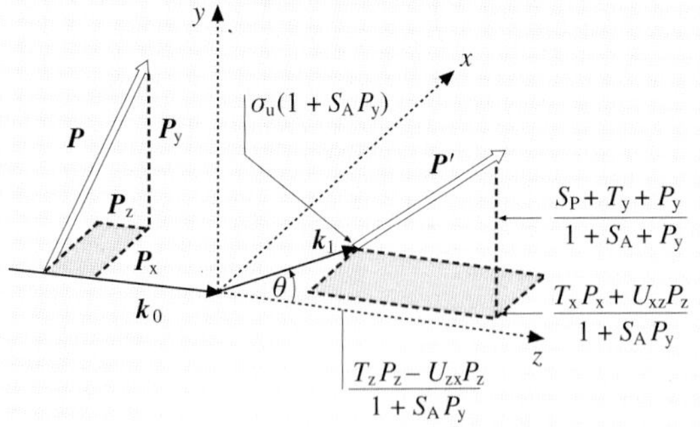

Fig. 7.1 Physical meaning of the generalized *STU*-parameters: the polarization function S_P gives the polarization of an initially unpolarized projectile beam after the collision while the asymmetry function S_A determines a left-right asymmetry in the differential cross section for scattering of a spin-polarized beam. Furthermore, the contraction parameters (T_x, T_y, T_z) describe the change of an initial polarization component along the three cartesian axes while the parameters U_{xz} and U_{zx} determine the rotation of a polarization component in the scattering plane

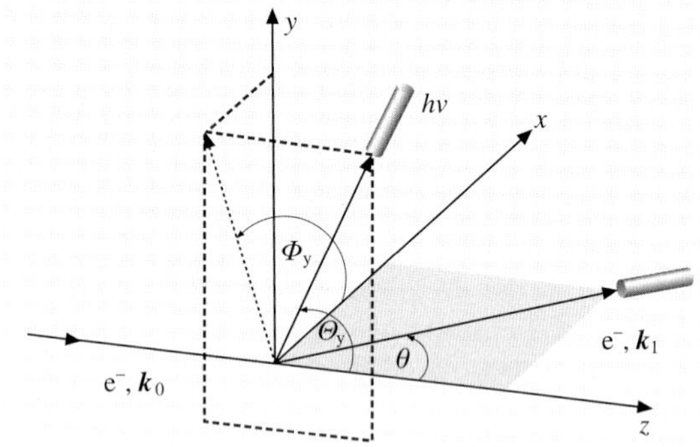

Fig. 7.2 Geometry of electron–photon coincidence experiments

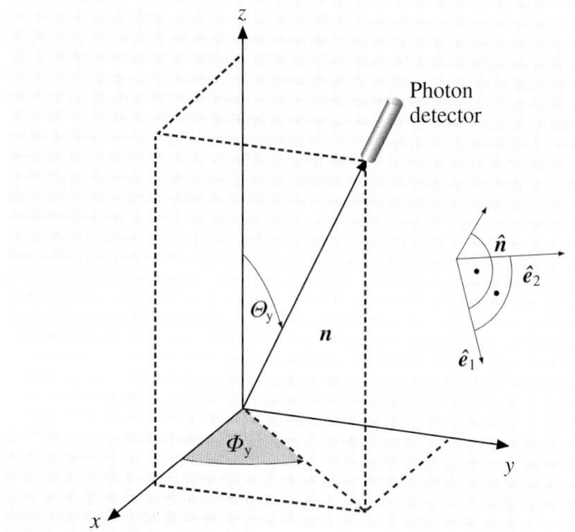

Fig. 7.3 Definition of the Stokes parameters: Photons are observed in a direction $\hat{\boldsymbol{n}}$ with polar angles $(\Theta_\gamma, \Phi_\gamma)$ in the collision system. The three unit vectors $(\hat{\boldsymbol{n}}, \hat{\boldsymbol{e}}_1, \hat{\boldsymbol{e}}_2)$ define the helicity system of the photons, $\hat{\boldsymbol{e}}_1 = (\Theta_\gamma + 90°, \Phi_\gamma)$ lies in the plane spanned by $\hat{\boldsymbol{n}}$ and $\hat{z}$ and is perpendicular to $\hat{\boldsymbol{n}}$ while $\hat{\boldsymbol{e}}_2 = (\Theta_\gamma, \Phi_\gamma + 90°)$ is perpendicular to both $\hat{\boldsymbol{n}}$ and $\hat{\boldsymbol{e}}_1$. In addition to the circular polarization P_3, the linear polarizations P_1 and P_2 are defined with respect to axes in the plane spanned by $\hat{\boldsymbol{e}}_1$ and $\hat{\boldsymbol{e}}_2$. Counting from the direction of $\hat{\boldsymbol{e}}_1$, the axes are located at (0°, 90°) for P_1 and at (45°, 135°) for P_2, respectively

Bartschat and collaborators [7.8]. The basic experimental setup for electron-photon coincidence experiments and the definition of the Stokes parameters are illustrated in Figs. 7.2 and 7.3.

For impact excitation of an atomic state with total electronic angular momentum J and an electric dipole transition to a state with J_f, the photon intensity in a direction $\hat{\boldsymbol{n}} = (\Theta_\gamma, \Phi_\gamma)$ is given by

$$\begin{aligned}
I(\Theta_\gamma, \Phi_\gamma) = C \Bigg[& \frac{2(-1)^{J-J_f}}{3\sqrt{2J+1}} \left\langle T(J)^\dagger_{00} \right\rangle \\
& - \begin{Bmatrix} 1 & 1 & 2 \\ J & J & J_f \end{Bmatrix} \\
& \times \Bigg(\mathrm{Re}\left\{ \left\langle T(J)^\dagger_{22} \right\rangle \right\} \sin^2\Theta_\gamma \cos 2\Phi_\gamma \\
& - \mathrm{Re}\left\{ \left\langle T(J)^\dagger_{21} \right\rangle \right\} \sin 2\Theta_\gamma \cos \Phi_\gamma \\
& + \sqrt{\frac{1}{6}} \left\langle T(J)^\dagger_{20} \right\rangle (3\cos^2\Theta_\gamma - 1) \\
& - \mathrm{Im}\left\{ \left\langle T(J)^\dagger_{22} \right\rangle \right\} \sin^2\Theta_\gamma \sin 2\Phi_\gamma \\
& + \mathrm{Im}\left\{ \left\langle T(J)^\dagger_{21} \right\rangle \right\} \sin 2\Theta_\gamma \sin \Phi_\gamma \Bigg) \Bigg] ,
\end{aligned} \tag{7.81}$$

where

$$C = \frac{e^2\omega^4}{2\pi c^3} \left| \langle J_f \| \boldsymbol{r} \| J \rangle \right|^2 (-1)^{J-J_f} \tag{7.82}$$

is a constant containing the frequency ω of the transition as well as the reduced radial dipole matrix element.

Similarly, the product of the intensity I and the circular light polarization P_3 can be written in terms of state multipoles as

$$
\begin{aligned}
(I\cdot P_3)(\Theta_\gamma,\Phi_\gamma) = &-C\begin{Bmatrix} 1 & 1 & 1 \\ J & J & J_f \end{Bmatrix} \\
&\times\Bigg(\mathrm{Im}\left\{\left\langle T(J)^\dagger_{11}\right\rangle\right\} 2\sin\Theta_\gamma \sin\Phi_\gamma \\
&-\mathrm{Re}\left\{\left\langle T(J)^\dagger_{11}\right\rangle\right\} 2\sin\Theta_\gamma \cos\Phi_\gamma \\
&+\sqrt{2}\left\langle T(J)^\dagger_{10}\right\rangle \cos\Theta_\gamma\Bigg)\,,
\end{aligned}
\tag{7.83}
$$

so that P_3 can be calculated as

$$
P_3(\Theta_\gamma,\Phi_\gamma) = (I\cdot P_3)(\Theta_\gamma,\Phi_\gamma)/I(\Theta_\gamma,\Phi_\gamma)\,.
\tag{7.84}
$$

Note that each state multipole gives rise to a characteristic angular dependence in the formulas for the Stokes parameters, and that perturbation coefficients may need to be applied to deal, for example, with depolarization effects due to internal or external fields. General formulas for $P_1=\eta_3$ and $P_2=\eta_1$ can be found in [7.8] and, for both the natural and the collision systems, in [7.4].

As pointed out before, some of the state multipoles may vanish, depending on the experimental arrangement. A detailed analysis of the information contained in the state multipoles and the generalized Stokes parameters (which are defined for specific values of the projectile spin polarization) has been given by *Andersen* and *Bartschat* [7.4, 17, 18]. They re-analyzed the experiment performed by *Sohn* and *Hanne* [7.19] and showed how the density matrix of the excited atomic ensemble can be determined by a measurement of the generalized Stokes parameters. In some cases, this will allow for the extraction of a complete set of scattering amplitudes for the collision process. Such a "perfect scattering experiment" has been called for by *Bederson* many years ago [7.20] and is now within reach even for fairly complex excitation processes. The most promising cases have been discussed by *Andersen* and *Bartschat* [7.4, 17, 21].

7.7 Summary

The basic formulas dealing with density matrices in quantum mechanics, with particular emphasis on reduced matrix theory and its applications in atomic physics, have been summarized. More details are given in the introductory textbooks by *Blum* [7.2], *Balashov* et al. [7.3], *Andersen* and *Bartschat* [7.4], and the references listed below.

References

7.1 J. von Neumann: Göttinger Nachr. **245** (1927)
7.2 K. Blum: *Density Matrix Theory and Applications* (Plenum, New York 1981)
7.3 V. V. Balashov, A. N. Grum-Grzhimailo, N. M. Kabachnik: *Polarization and Correlation Phenomena in Atomic Collisions. A Practical Theory Course* (Plenum, New York 2000)
7.4 N. Andersen, K. Bartschat: *Polarization, Alignment, and Orientation in Atomic Collisions* (Springer, New York 2001)
7.5 J. Kessler: *Polarized Electrons* (Springer, New York 1985)
7.6 W. E. Baylis, J. Bonenfant, J. Derbyshire, J. Huschilt: Am. J. Phys. **61**, 534 (1993)
7.7 M. Born, E. Wolf: *Principles of Optics* (Pergamon, New York 1970)
7.8 K. Bartschat, K. Blum, G. F. Hanne, J. Kessler: J. Phys. B **14**, 3761 (1981)
7.9 K. Bartschat, K. Blum: Z. Phys. A **304**, 85 (1982)
7.10 N. Andersen, J. W. Gallagher, I. V. Hertel: Phys. Rep. **165**, 1 (1988)
7.11 K. Bartschat: Phys. Rep. **180**, 1 (1989)
7.12 A. R. Edmonds: *Angular Momentum in Quantum Mechanics* (Princeton Univ. Press, Princeton 1957)
7.13 U. Fano, J. H. Macek: Rev. Mod. Phys. **45**, 553 (1973)
7.14 K. Bartschat, H. J. Andrä, K. Blum: Z. Phys. A **314**, 257 (1983)
7.15 N. Andersen, J. T. Broad, E. E. Campbell, J. W. Gallagher, I. V. Hertel: Phys. Rep. **278**, 107 (1997)
7.16 N. Andersen, K. Bartschat, J. T. Broad, I. V. Hertel: Phys. Rep. **279**, 251 (1997)
7.17 N. Andersen, K. Bartschat: Adv. At. Mol. Phys. **36**, 1 (1996)
7.18 N. Andersen, K. Bartschat: J. Phys. B **27**, 3189 (1994); corrigendum: J. Phys. B **29**, 1149 (1996)
7.19 M. Sohn, G. F. Hanne: J. Phys. B **25**, 4627 (1992)
7.20 B. Bederson: Comments At. Mol. Phys. **1**, 41,65 (1969)
7.21 N. Andersen, K. Bartschat: J. Phys. B **30**, 5071 (1997)

8. Computational Techniques

Essential to all fields of physics is the ability to perform numerical computations accurately and efficiently. Whether the specific approach involves perturbation theory, close coupling expansion, solution of classical equations of motion, or fitting and smoothing of data, basic computational techniques such as integration, differentiation, interpolation, matrix and eigenvalue manipulation, Monte Carlo sampling, and solution of differential equations must be among the standard tool kit.

This chapter outlines a portion of this tool kit with the aim of giving guidance and organization to a wide array of computational techniques. Having digested the present overview, the reader is then referred to detailed treatments given in many of the large number of texts existing on numerical analysis and computational techniques [8.1–5], and mathematical physics [8.6–10]. We also summarize, especially in the sections on differential equations and computational linear algebra, the role of software packages readily available to aid in implementing practical solutions.

8.1 **Representation of Functions** 135
8.1.1 Interpolation 135
8.1.2 Fitting 137
8.1.3 Fourier Analysis 139
8.1.4 Approximating Integrals 139
8.1.5 Approximating Derivatives 140

8.2 **Differential and Integral Equations** 141
8.2.1 Ordinary Differential Equations ... 141
8.2.2 Differencing Algorithms for Partial Differential Equations . 143
8.2.3 Variational Methods 144
8.2.4 Finite Elements 144
8.2.5 Integral Equations 146

8.3 **Computational Linear Algebra** 148

8.4 **Monte Carlo Methods** 149
8.4.1 Random Numbers 149
8.4.2 Distributions of Random Numbers 150
8.4.3 Monte Carlo Integration 151

References 151

8.1 Representation of Functions

The ability to represent functions in terms of polynomials or other basic functions is the key to interpolating or fitting data, and to approximating numerically the operations of integration and differentiation. In addition, using methods such as Fourier analysis, knowledge of the properties of functions beyond even their intermediate values, derivatives, and antiderivatives may be determined (e.g., the "spectral" properties).

8.1.1 Interpolation

Given the value of a function $f(x)$ at a set of points $x_1, x_2, \ldots, x_n$, the function is often required at some other values between these abscissae. The process known as *interpolation* seeks to estimate these unknown values by adjusting the parameters of a known function to approximate the local or global behavior of $f(x)$. One of the most useful representations of a function for these purposes utilizes the *algebraic polynomials*, $P_n(x) = a_0 + a_1 x + \cdots + a_n x^n$, where the coefficients are real constants and the exponents are nonnegative integers. The utility stems from the fact that given any continuous function defined on a closed interval, there exists an algebraic polynomial which is as close to that function as desired (Weierstrass Theorem).

One simple application of these polynomials is the power series expansion of the function $f(x)$ about some point, x_0, i. e.,

$$f(x) = \sum_{k=0}^{\infty} a_k (x - x_0)^k . \tag{8.1}$$

A familiar example is the *Taylor expansion* in which the coefficients are given by

$$a_k = \frac{f^{(k)}(x_0)}{k!}\,, \tag{8.2}$$

where $f^{(k)}$ indicates the kth derivative of the function. This form, though quite useful in the derivation of formal techniques, is not very useful for interpolation since it assumes the function and its derivatives are known, and since it is guaranteed to be a good approximation only very near the point x_0 about which the expansion has been made.

Lagrange Interpolation

The polynomial of degree $n-1$ which passes through all n points $[x_1, f(x_1)], [x_2, f(x_2)], \ldots, [x_n, f(x_n)]$ is given by

$$P(x) = \sum_{k=1}^{n} f(x_k) \prod_{i=1, i\neq k}^{n} \frac{x - x_i}{x_k - x_i} \tag{8.3}$$

$$= \sum_{k=1}^{n} f(x_k) L_{nk}(x)\,, \tag{8.4}$$

where $L_{nk}(x)$ are the *Lagrange interpolating polynomials*. Perhaps the most familiar example is that of linear interpolation between the points $[x_1, y_1 \equiv f(x_1)]$ and $[x_2, y_2 \equiv f(x_2)]$, namely,

$$P(x) = \frac{x - x_2}{x_1 - x_2} y_1 + \frac{x - x_1}{x_2 - x_1} y_2\,. \tag{8.5}$$

In practice, it is difficult to estimate the formal error bound for this method, since it depends on knowledge of the $(n+1)$th derivative. Alternatively, one uses *iterated interpolation* in which successively higher order approximations are tried until appropriate agreement is obtained. *Neville's algorithm* defines a recursive procedure to yield an arbitrary order interpolant from polynomials of lower order. This method, and subtle refinements of it, form the basis for most "recommended" polynomial interpolation schemes [8.3].

One important caution to bear in mind is that the more points that are used in constructing the interpolant, and therefore the higher the polynomial order, the greater will be the oscillation in the interpolating function. This highly oscillating polynomial most likely will not correspond more closely to the desired function than polynomials of lower order, and, as a general rule of thumb, fewer than six points should be used.

Cubic Splines

By dividing the interval of interest into a number of subintervals and in each using a polynomial of only modest order, one may avoid the oscillatory nature of high-order (many-point) interpolants. This approach utilizes *piecewise polynomial functions*, the simplest of which is just a linear segment. However, such a straight line approximation has a discontinuous derivative at the data points – a property that one may wish to avoid especially if the derivative of the function is also desired – and which clearly does not provide a smooth interpolant. The solution is therefore to choose the polynomial of lowest order that has enough free parameters (the constants $a_0, a_1, \ldots$) to satisfy the constraints that the function and its derivative are continuous across the subintervals, as well as specifying the derivative at the endpoints x_0 and x_n.

Piecewise cubic polynomials satisfy these constraints, and have a continuous second derivative as well. *Cubic spline interpolation* does not, however, guarantee that the derivatives of the interpolant agree with those of the function at the data points, much less globally. The cubic polynomial in each interval has four undertermined coeffitients,

$$P_i(x) = a_i + b_i(x - x_i) + c_i(x - x_i)^2 + d_i(x - x_i)^3 \tag{8.6}$$

for $i = 0, 1, \ldots, n-1$. Applying the constraints, a system of equations is found which may be solved once the endpoint derivatives are specified. If the second derivatives at the endpoints are set to zero, then the result is termed a *natural spline* and its shape is like that which a long flexible rod would take if forced to pass through all the data points. A *clamped spline* results if the first derivatives are specified at the endpoints, and is usually a better approximation since it incorporates more information about the function (if one has a reasonable way to determine or approximate these first derivatives).

The set of equations in the unknowns, along with the boundary conditions, constitute a *tridiagonal system* or *matrix*, and is therefore amenable to solution by algorithms designed for speed and efficiency for such systems (see Sect. 8.3; [8.1–3]). Other alternatives of potentially significant utility are schemes based on the use of rational functions and orthogonal polynomials.

Rational Function Interpolation

If the function which one seeks to interpolate has one or more poles for real x, then polynomial approximations are not good, and a better method is to use quotients of polynomials, so-called *rational functions*. This occurs since the inverse powers of the dependent variable will fit the region near the pole better if the order is large enough. In fact, if the function is free of poles on the real axis but its analytic continuation in the complex plane has poles, the polynomial approximation may also be poor. It is this property that slows or prevents the convergence of power series. Numerical algorithms very similar to those used to generate iterated polynomial interpolants exist [8.1,3] and can be useful for functions which are not amenable to polynomial interpolation. Rational function interpolation is related to the method of *Padé approximation* used to improve convergence of power series, and which is a rational function analog of Taylor expansion.

Orthogonal Function Interpolation

Interpolation using functions other than the algebraic polynomials can be defined and are often useful. Particularly worthy of mention are schemes based on *orthogonal polynomials* since they play a central role in numerical quadrature. A set of functions $\phi_1(x), \phi_2(x), \ldots, \phi_n(x)$ defined on the interval $[a, b]$ is said to be orthogonal with respect to a weight function $\mathcal{W}(x)$ if the inner product defined by

$$\langle \phi_i | \phi_j \rangle = \int_a^b \phi_i(x)\phi_j(x)\mathcal{W}(x)\,\mathrm{d}x \tag{8.7}$$

is zero for $i \neq j$ and positive for $i = j$. In this case, for any polynomial $P(x)$ of degree at most n, there exists unique constants α_k such that

$$P(x) = \sum_{k=0}^{n} \alpha_k \phi_k(x) . \tag{8.8}$$

Among the more commonly used orthogonal polynomials are *Legendre*, *Laguerre*, and *Chebyshev* polynomials.

Chebyshev Interpolation

The significant advantages of employing a representation of a function in terms of Chebyshev polynomials, $T_k(x)$ [8.4, 6] for tabulations, recurrence formulas, orthogonality properties, etc. of these polynomials), i. e.,

$$f(x) = \sum_{k=0}^{\infty} a_k T_k(x) , \tag{8.9}$$

stems from the fact that (i) the expansion rapidly converges, (ii) the polynomials have a simple form, and (iii) the polynomial approximates very closely the solution of the *minimax* problem. This latter property refers to the requirement that the expansion *mini*mizes the *max*imum magnitude of the error of the approximation. In particular, the Chebyshev series expansion can be truncated so that for a given n it yields the most accurate approximation to the function. Thus, Chebyshev polynomial interpolation is essentially as "good" as one can hope to do. Since these polynomials are defined on the interval $[-1, 1]$, if the endpoints of the interval in question are a and b, the change of variable

$$y = \frac{x - \frac{1}{2}(b+a)}{\frac{1}{2}(b-a)} \tag{8.10}$$

will effect the proper transformation. *Press* et al. [8.3], for example, give convenient and efficient routines for computing the Chebyshev expansion of a function.

8.1.2 Fitting

Fitting of data stands in distinction from interpolation in that the data may have some uncertainty, and therefore, simply determining a polynomial which passes through the points may not yield the best approximation of the underlying function. In fitting, one is concerned with minimizing the deviations of some model function from the data points in an optimal or best fit manner. For example, given a set of data points, even a low-order interpolating polynomial might have significant oscillation, when, in fact, if one accounts for the statistical uncertainties in the data, the best fit may be obtained simply by considering the points to lie on a line.

In addition, most of the traditional methods of assigning this quality of best fit to a particular set of parameters of the model function rely on the assumption that the random deviations are described by a Gaussian (normal) distribution. Results of physical measurements, for example the counting of events, is often closer to a Poisson distribution which tends (not necessarily uniformly) to a Gaussian in the limit of a large number of events, or may even contain "outliers" which lie far outside a Gaussian distribution. In these cases, fitting methods might significantly distort the parameters of the model function in trying to force these different distributions to the Gaussian form. Thus, the *least squares* and *chi-square* fitting procedures discussed below should be used with this caveat in mind. Other techniques, often termed "robust" [8.3, 11], should be used when the distribution is not Gaussian, or replete with outliers.

Least Squares

In this common approach to fitting, we wish to determine the m parameters a_l of some function $f(x; a_1, a_2, \ldots, a_m)$ depending in this example on one variable, x. In particular, we seek to minimize the sum of the squares of the deviations

$$\sum_{k=1}^{n} [y(x_k) - f(x_k; a_1, a_2, \ldots, a_m)]^2 \tag{8.11}$$

by adjusting the parameters, where the $y(x_k)$ are the n data points. In the simplest case, the model function is just a straight line, $f(x; a_1, a_2) = a_1 x + a_2$. Elementary multivariate calculus implies that a minimum occurs if

$$a_1 \sum_{k=1}^{n} x_i^2 + a_2 \sum_{k=1}^{n} x_i = \sum_{k=1}^{n} x_i y_i , \tag{8.12}$$

$$a_1 \sum_{k=1}^{n} x_i + a_2 n = \sum_{k=1}^{n} y_i , \tag{8.13}$$

which are called the *normal equations*. Solution of these equations is straightforward, and an error estimate of the fit can be found [8.3]. In particular, variances may be computed for each parameter, as well as measures of the correlation between uncertainties and an overall estimate of the "goodness of fit" of the data.

Chi-square Fitting

If the data points each have associated with them a different standard deviation, σ_k, the least square principle is modified by minimizing the *chi-square*, defined as

$$\chi^2 \equiv \sum_{k=1}^{n} \left[\frac{y_k - f(x_k; a_1, a_2, \ldots, a_m)}{\sigma_k} \right]^2 . \tag{8.14}$$

Assuming that the uncertainties in the data points are normally distributed, the chi-square value gives a measure of the goodness of fit. If there are n data points and m adjustable parameters, then the probability that χ^2 should exceed a particular value purely by chance is

$$Q = Q\left(\frac{n-m}{2}, \frac{\chi^2}{2} \right) , \tag{8.15}$$

where $Q(a, x) = \Gamma(a, x)/\Gamma(a)$ is the incomplete gamma function. For small values of Q, the deviations of the fit from the data are unlikely to be by chance, and values close to one are indications of better fits. In terms of the chi-square, reasonable fits often have $\chi^2 \approx n - m$.

Other important applications of the chi-square method include simulation and estimating standard deviations. For example, if one has some idea of the actual (i. e., non-Gaussian) distribution of uncertainties of the data points, Monte Carlo simulation can be used to generate a set of test data points subject to this presumed distribution, and the fitting procedure performed on the simulated data set. This allows one to test the accuracy or applicability of the model function chosen. In other situations, if the uncertainties of the data points are unknown, one can assume that they are all equal to some value, say σ, fit using the chi-square procedure, and solve for the value of σ. Thus, some measure of the uncertainty from this statistical point of view can be provided.

General Least Squares

The least squares procedure can be generalized usually by allowing any linear combination of basis functions to determine the model function

$$f(x; a_1, a_2, \ldots, a_m) = \sum_{l=1}^{m} a_l \psi_l(x) . \tag{8.16}$$

The basis functions need not be polynomials. Similarly, the formula for chi-square can be generalized, and normal equations determined through minimization. The equations may be written in compact form by defining a matrix $\boldsymbol{A}$ with elements

$$A_{i,j} = \frac{\psi_j(x_i)}{\sigma_i} , \tag{8.17}$$

and a column vector $\boldsymbol{B}$ with elements

$$B_i = \frac{y_i}{\sigma_i} . \tag{8.18}$$

Then the normal equations are [8.3]

$$\sum_{j=1}^{m} \alpha_{kj} a_j = \beta_k , \tag{8.19}$$

where

$$[\alpha] = \boldsymbol{A}^T \boldsymbol{A} , \quad [\beta] = \boldsymbol{A}^T \boldsymbol{B} , \tag{8.20}$$

and a_j are the adjustable parameters. These equations may be solved using standard methods of computational linear algebra such as Gauss–Jordan elimination. Difficulties involving sensitivity to round-off errors can be avoided by using carefully developed codes to perform this solution [8.3]. We note that elements of the inverse of the matrix $\boldsymbol{\alpha}$ are related to the variances associated with the free parameters and to the covariances relating them.

Statistical Analysis of Data

Data generated by an experiment, or perhaps from a Monte Carlo simulation, have uncertainties due to the statistical, or random, character of the processes by which they are acquired. Therefore, one must be able to describe statistically certain features of the data such as their mean, variance and skewness, and the degree to which correlations exist, either between one portion of the data and another, or between the data and some other standard or model distribution. A very readable introduction to this type analysis has been given by *Young* [8.12], while more comprehensive treatments are also available [8.13].

8.1.3 Fourier Analysis

The *Fourier transform* takes, for example, a function of time, into a function of frequency, or vice versa, namely

$$\tilde{\varphi}(\omega) = \frac{1}{\sqrt{2\pi}} \int_{-\infty}^{\infty} \varphi(t)\,\mathrm{e}^{\mathrm{i}\omega t}\,\mathrm{d}t\,, \tag{8.21}$$

$$\varphi(t) = \frac{1}{\sqrt{2\pi}} \int_{-\infty}^{\infty} \tilde{\varphi}(\omega)\,\mathrm{e}^{-\mathrm{i}\omega t}\,\mathrm{d}\omega\,. \tag{8.22}$$

In this case, the time history of the function $\varphi(t)$ may be termed the "signal" and $\tilde{\varphi}(\omega)$ the "frequency spectrum". Also, if the frequency is related to the energy by $E = \hbar\omega$, one obtains an "energy spectrum" from a signal, and thus the name *spectral methods* for techniques based on the Fourier analysis of signals.

The Fourier transform also defines the relationship between the spatial and momentum representations of wave functions, i. e.,

$$\psi(x) = \frac{1}{\sqrt{2\pi}} \int_{-\infty}^{\infty} \tilde{\psi}(p)\,\mathrm{e}^{\mathrm{i}px}\,\mathrm{d}p\,, \tag{8.23}$$

$$\tilde{\psi}(p) = \frac{1}{\sqrt{2\pi}} \int_{-\infty}^{\infty} \psi(x)\,\mathrm{e}^{-\mathrm{i}px}\,\mathrm{d}x\,. \tag{8.24}$$

Along with the closely related *sine*, *cosine*, and *Laplace transforms*, the Fourier transform is an extraordinarily powerful tool in the representation of functions, spectral analysis, convolution of functions, filtering, and analysis of correlation. Good introductions to these techniques with particular attention to applications in physics can be found in [8.6, 7, 14]. To implement the Fourier transform numerically, the integral tranform pair can be converted to sums

$$\tilde{\varphi}(\omega_j) = \frac{1}{\sqrt{2N2\pi}} \sum_{k=0}^{2N-1} \varphi(t_k)\,\mathrm{e}^{\mathrm{i}\omega_j t_k}\,, \tag{8.25}$$

$$\varphi(t_k) = \frac{1}{\sqrt{2N2\pi}} \sum_{j=0}^{2N-1} \tilde{\varphi}(\omega_j)\,\mathrm{e}^{-\mathrm{i}\omega_j t_k}\,, \tag{8.26}$$

where the functions are "sampled" at $2N$ points. These equations define the *discrete Fourier transform* (DFT). Two cautions in using the DFT are as follows.

First, if a continuous function of time that is sampled at, for simplicity, uniformly spaced intervals, (i. e., $t_{i+1} = t_i + \Delta$), then there is a critical frequency $\omega_\mathrm{c} = \pi/\Delta$, known as the *Nyquist frequency*, which limits the fidelity of the DFT of this function in that it is *aliased*. That is, components outside the frequency range $-\omega_\mathrm{c}$ to ω_c are falsely transformed into this range due to the finite sampling. This effect can be remediated by filtering or windowing techniques. If, however, the function is *bandwidth limited* to frequencies smaller than ω_c, then the DFT does not suffer from this effect, and the signal is completely determined by its samples. Second, implementing the DFT directly from the above equations would require approximately N^2 multiplications to perform the Fourier transform of a function sampled at N points. A variety of *fast Fourier transform* (FFT) algorithms have been developed (e.g., the Danielson–Lanczos and Cooley–Tukey methods) which require only on the order of $(N/2)\log_2 N$ multiplications. Thus, for even moderately large sets of points, the FFT methods are indeed much faster than the direct implementation of the DFT. Issues involved in sampling, aliasing, and selection of algorithms for the FFT are discussed in great detail, for example, in [8.3, 15, 16].

8.1.4 Approximating Integrals

Polynomial Quadrature

Definite integrals may be approximated through a procedure known as numerical quadrature by replacing the integral by an appropriate sum, i. e.,

$$\int_a^b f(x)\,\mathrm{d}x \approx \sum_{k=0}^{n} a_k f(x_k)\,. \tag{8.27}$$

Most formulas for such approximation are based on the interpolating polynomials described in Sect. 8.1.1, especially the Lagrange polynomials, in which case the coefficients a_k are given by

$$a_k = \int_a^b L_{nk}(x_k)\,\mathrm{d}x \,. \tag{8.28}$$

If first or second degree Lagrange polynomials are used with a uniform spacing between the data points, one obtains the *trapezoidal* and *Simpson's rules*, i. e.,

$$\int_a^b f(x)\,\mathrm{d}x \approx \frac{\delta}{2}\left[f(a)+f(b)\right]+\mathcal{O}\left[\delta^3 f^{(2)}(\zeta)\right] , \tag{8.29}$$

$$\int_a^b f(x)\,\mathrm{d}x \approx \frac{\delta}{3}\left[f(a)+4f\left(\frac{\delta}{2}\right)+f(b)\right] + \mathcal{O}\left[\delta^5 f^{(4)}(\zeta)\right] , \tag{8.30}$$

respectively, with $\delta = b-a$, and for some ζ in $[a, b]$.

Other commonly used formulas based on low-order polynomials, and generally referred to as *Newton–Cotes* formulas, are described and discussed in detail in numerical analysis texts [8.1, 2]. Since potentially unwanted rapid oscillations in interpolants may arise, it is generally the case that increasing the order of the quadrature scheme too greatly does not generally improve the accuracy of the approximation. Dividing the interval $[a, b]$ into a number of subintervals and summing the result of application of a low-order formula in each subinterval is usually a much better approach. This procedure, referred to as *composite quadrature*, may be combined with choosing the data points at a nonuniform spacing, decreasing the spacing where the function varies rapidly, and increasing the spacing for economy where the function is smooth to construct an *adaptive quadrature*.

Gaussian Quadrature

If the function whose definite integral is to be approximated can be evaluated explicitly, then the data points (abscissas) can be chosen in a manner in which significantly greater accuracy may be obtained than using Newton–Cotes formulas of equal order. *Gaussian quadrature* is a procedure in which the error in the approximation is minimized owing to this freedom to choose both data points (abscissas) and coefficients. By utilizing orthogonal polynomials and choosing the abscissas at the roots of the polynomials in the interval under consideration, it can be shown that the coefficients may be optimally chosen by solving a simple set of linear equations. Thus, a Gaussian quadrature scheme approximates the definite integral of a function multipled by the weight function appropriate to the orthogonal polynomial being used as

$$\int_a^b \mathcal{W}(x)f(x)\,\mathrm{d}x \approx \sum_{k=1}^{n} a_k f(x_k) , \tag{8.31}$$

where the function is to be evaluated at the abscissas given by the roots of the orthogonal polynomial, x_k. In this case, the coefficients a_k are often referred to as "weights," but should not be confused with the weight function $\mathcal{W}(x)$ (Sect. 8.1.1). Since the Legendre polynomials are orthogonal over the interval $[-1, 1]$ with respect to the weight function $\mathcal{W}(x) \equiv 1$, this equation has a particularly simple form, leading immediately to the *Gauss–Legendre quadrature*. If $f(x)$ contains as a factor the weight function of another of the orthogonal polynomials, the corresponding *Gauss–Laguerre* or *Gauss–Chebyshev* quadrature should be used.

The roots and coefficients have been tabulated [8.4] for many common choices of the orthogonal polynomials (e.g., Legendre, Laguerre, Chebyshev) and for various orders. Simple computer subroutines are also available which conveniently compute them [8.3]. Since the various orthogonal polynomials are defined over different intervals, use of the change of variables such as that given in (8.10) may be required. So, for Gauss–Legendre quadrature we make use of the transformation

$$\int_a^b f(x)\,\mathrm{d}x \approx \frac{(b-a)}{2}\int_{-1}^{1} f\left(\frac{(b-a)y+b+a}{2}\right)\mathrm{d}y \,. \tag{8.32}$$

Other Methods

Especially for multidimensional integrals which can not be reduced analytically to seperable or iterated integrals of lower dimension, *Monte Carlo* integration may provide the only means of finding a good approximation. This method is described in Sect. 8.4.3. Also, a convenient quadrature scheme can easily be devised based on the cubic spline interpolation described in Sect. 8.1.1. since in each subinterval, the definite integral of a cubic polynomial of known coefficients is evident.

8.1.5 Approximating Derivatives

Numerical Differentiation

The calculation of derivatives from a numerical representaion of a function is generally less stable than the calculation of integrals because differentiation tends to enhance fluctuations and worsen the convergence properties of power series. For example, if $f(x)$ is twice continuously differentiable on $[a, b]$, then differentiation of the linear Lagrange interpolation formula (8.5) yields

$$f^{(1)}(x_0) = \frac{f(x_0+\delta) - f(x_0)}{\delta} + \mathcal{O}\left[\delta f^{(2)}(\zeta)\right] \tag{8.33}$$

for some x_0 and ζ in $[a, b]$, where $\delta = b - a$. In the limit $\delta \to 0$, (8.33) coincides with the definition of the derivative. However, in practical calculations with finite precision arithmetic, δ cannot be taken too small because of numerical cancellation in the calculation of $f(a+\delta) - f(a)$.

In practice, increasing the order of the polynomial used decreases the truncation error, but at the expense of increasing round-off error, the upshot being that three- and five-point approximations are usually most useful. Various three- and five-point formulas are given in standard texts [8.2, 4, 17]. Two common five-point formulas (centered and forward/backward) are

$$f^{(1)}(x_0) = \frac{1}{12\delta}\big[f(x_0-2\delta) - 8f(x_0-\delta) + 8f(x_0+\delta) - f(x_0+2\delta)\big] + \mathcal{O}\left[\delta^4 f^{(5)}(\zeta)\right] \tag{8.34}$$

$$f^{(1)}(x_0) = \frac{1}{12\delta}\big[-25f(x_0) + 48f(x_0+\delta) - 36f(x_0+2\delta) + 16f(x_0+3\delta) - 3f(x_0+4\delta)\big] + \mathcal{O}\left[\delta^4 f^{(5)}(\zeta)\right] . \tag{8.35}$$

The second formula is useful for evaluating the derivative at the left or right endpoint of the interval, depending on whether δ is positive or negative, respectively.

Derivatives of Interpolated Functions

An interpolating function can be directly differentiated to obtain the derivative at any desired point. For example, if $f(x) \approx a_0 + a_1 x + a_2 x^2$, then $f^{(1)}(x) = a_1 + 2a_2 x$. However, this approach may fail to give the best approximation to $f^{(1)}(x)$ if the original interpolation was optimized to give the best possible representaion of $f(x)$.

8.2 Differential and Integral Equations

The subject of differential and integral equations is immense in both richness and scope. The discussion here focuses on techniques and algorithms, rather than the formal aspect of the theory. Further information can be found elsewhere under the broad catagories of finite element and finite difference methods. The *Numerov method*, which is particularly useful in integrating the Schrödinger equation, is described in great detail in [8.8].

8.2.1 Ordinary Differential Equations

An *ordinary differential equation* is an equation involving an unknown function and one or more of its derivatives that depends on only one independent variable [8.18]. The *order* of a differential equation is the order of the highest derivative appearing in the equation. A *solution* of a general differential equation of order n,

$$f\left(t, y, \dot{y}, \ldots, y^{(n)}\right) = 0 , \tag{8.36}$$

is a real-valued function $y(t)$ having the following properties: (1) $y(t)$ and its first n derivatives exist, so $y(t)$ and its first $n-1$ derivatives must be continuous, and (2) $y(t)$ satisfies the differential equation for all t. A unique solution requires the specification of n conditions on $y(t)$ and its derivatives. The conditions may be specified as n initial conditions at a single t to give an *initial value problem*, or at the end points of an interval to give a *boundary value problem*.

Consider first solutions to the simple equation

$$\dot{y} = f(t, y) , \qquad y(a) = A . \tag{8.37}$$

The methods discussed below can easily be extended to systems of first-order differential equations and to higher-order differential equations. The methods are referred to as *discrete variable methods* and generate a sequence of approximate values for $y(t)$, $y_1, y_2, y_3, \ldots$ at points $t_1, t_2, t_3, \ldots$. For simplicity, the discussion assumes a constant spacing h between t points. We shall first describe a class of methods known as *one-step* methods [8.19]. They have no memory of the solutions at past times; given y_i, there is a recipe for y_{i+1} that depends only on information at t_i. Errors enter into numerical solutions from two sources. The first is *discretization error* and depends on the method being used. The second is *computational error* which includes such things as round off error.

For a solution on the interval $[a, b]$, let the t points be equally spaced; so for some positive integer n and $h = (b-a)/n$, $t_i = a + ih$, $i = 0, 1, \dots, n$. If $a < b$, h is positive and the integration is forward; if $a > b$, h is negative and the integration is backward. The latter case could occur in solving for the initial point of a solution curve given the terminal point. A general one-step method can then be written in the form

$$y_{i+1} = y_i + h\Delta(t_i, y_i) , \qquad y_0 = y(t_0) , \tag{8.38}$$

where Δ is a function that characterizes the method. Different Δ functions are displayed, giving rise to the Taylor series methods and the Runge–Kutta methods.

Taylor Series Algorithm

To obtain an approximate solution of order p on $[a, b]$, generate the sequence

$$\begin{aligned} y_{i+1} &= y_i + h\left[f(t_i, y_i) + \cdots \right.\\ &\qquad \left. + f^{(p-1)}(t_i, y_i)\frac{h^{p-1}}{p!} \right] ,\\ t_{i+1} &= t_i + h, \quad i = 0, 1, \dots, n-1 \end{aligned} \tag{8.39}$$

where $t_0 = a$ and $y_0 = A$. The Taylor method of order $p = 1$ is known as *Euler's method*:

$$\begin{aligned} y_{i+1} &= y_i + hf(t_i, y_i) ,\\ t_{i+1} &= t_i + h . \end{aligned} \tag{8.40}$$

Taylor series methods can be quite effective if the total derivatives of f are not too difficult to evaluate. Software packages are available that perform exact differentiation, (ADIFOR, MAPLE, MATHEMATICA, etc.) facilitating the use of this approach.

Runge–Kutta Methods

Runge–Kutta methods are designed to approximate Taylor series methods [8.20], but have the advantage of not requiring explicit evaluations of the derivatives of $f(t, y)$. The basic idea is to use a linear combination of values of $f(t, y)$ to approximate $y(t)$. This linear combination is matched up as closely as possible with a Taylor series for $y(t)$ to obtain methods of the highest possible order p. Euler's method is an example using one function evaluation.

To obtain an approximate solution of order $p = 2$, let $h = (b-a)/n$ and generate the sequences

$$\begin{aligned} y_{i+1} &= y_i + h\left[(1-\gamma) f(t_i, y_i) \right.\\ &\qquad \left. + \gamma f\left[t_i + \frac{h}{2\gamma}, y_i + \frac{h}{2\gamma} f(t_i, y_i) \right] \right] ,\\ t_{i+1} &= t_i + h , \qquad i = 0, 1, \dots, n-1 , \end{aligned} \tag{8.41}$$

where $\gamma \neq 0$, $t_0 = a$, $y_0 = A$.

Euler's method is the special case, $\gamma = 0$, and has order 1; the improved Euler method has $\gamma = 1/2$ and the Euler–Cauchy method has $\gamma = 1$.

The Adams–Bashforth and Adams–Moulton Formulas

These formulas furnish important and widely-used examples of multistep methods [8.21]. On reaching a mesh point t_i with approximate solution $y_i \cong y(t_i)$, there are (usually) available approximate solutions $y_{i+1-j} \cong y(t_{i+1-j})$ for $j = 2, 3, \dots, p$. From the differential equation itself, approximations to the derivatives $\dot{y}(t_{i+1-j})$ can be obtained.

An attractive feature of the approach is the form of the underlying polynomial approximation, $P(t)$, to $\dot{y}(t)$ because it can be used to approximate $y(t)$ between mesh points

$$y(t) \cong y_i + \int_{t_i}^{t} P(t)\, dt . \tag{8.42}$$

The lowest-order Adams–Bashforth formula arises from interpolating the single value $f_i = f(t_i, y_i)$ by $P(t)$. The interpolating polynomial is constant so its integration from t_i to t_{i+1} results in $h f(t_i, y_i)$ and the first order Adams–Bashforth formula:

$$y_{i+1} = y_i + hf(t_i, y_i) . \tag{8.43}$$

This is just the forward Euler formula. For constant step size h, the second-order Adams–Bashforth formula is

$$y_{i+1} = y_i + h\left[\left(\frac{3}{2}\right) f(t_i, y_i) - \left(\frac{1}{2}\right) f(t_{i-1}, y_{i-1}) \right] . \tag{8.44}$$

The lowest-order Adams–Moulton formula involves interpolating the single value $f_{i+1} = f(x_{i+1}, y_{i+1})$ and leads to the backward Euler formula

$$y_{i+1} = y_i + hf\,(t_{i+1}, y_{i+1}) , \tag{8.45}$$

which defines y_{i+1} implicitly. From its definition it is clear that it has the same accuracy as the forward Euler method; its advantage is vastly superior stability. The second-order Adams–Moulton method also does not use previously computed solution values; it is called the trapezoidal rule because it generalizes the trapezoidal rule for integrals to differential equations:

$$y_{i+1} = y_i + \frac{h}{2}\left[f(t_{i+1}, y_{i+1}) + f(t_i, y_i) \right] . \tag{8.46}$$

The Adams–Moulton formula of order p is more accurate than the Adams–Bashforth formula of the same order, so that it can use a larger step size; the Adams–Moulton formula is also more stable. A code based on such methods is more complex than a Runge–Kutta code because it must cope with the difficulties of starting the integration and changing the step size. Modern Adams codes attempt to select the most efficient formula at each step, as well as to choose an optimal step size h to achieve a specified accuracy.

8.2.2 Differencing Algorithms for Partial Differential Equations

The modern approach to evolve differencing schemes for most physical problems is based on flux conservation methods [8.22]. One begins by writing the balance equations for a single cell, and subsequently applying quadratures and interpolation formulas. Such approaches have been successful for the full spectrum of hyperbolic, elliptic, and parabolic equations. For simplicity, we begin by discussing systems involving only one space variable.

As a prototype, consider the parabolic equation

$$c\frac{\partial}{\partial t}u(x,t)=\sigma\frac{\partial^2}{\partial x^2}u(x,t)\,, \tag{8.47}$$

where c and σ are constants and $u(x,t)$ is the solution. We begin by establishing a grid of points on the xt-plane with step size h in the x direction and step size k in the t-direction. Let spatial grid points be denoted by $x_n=x_0+nh$ and time grid points by $t_j=t_0+jk$, where n and j are integers and (x_0,t_0) is the origin of the space–time grid. The points ξ_{n-1} and ξ_n are introduced to establish a "control interval". We begin with a conservation statement

$$\int\limits_{\xi_{n-1}}^{\xi_n}\mathrm{d}x\,\left[r(x,t_{j+1})-r(x,t_j)\right]=\int\limits_{t_j}^{t_{j+1}}\mathrm{d}t\,[q(\xi_{n-1},t)-q(\xi_n,t)]\,. \tag{8.48}$$

This equation states that the change in the field density on the interval (ξ_{n-1},ξ_n) from time $t=t_j$ to time $t=t_{j+1}$ is given by the flux into this interval at ξ_{n-1} minus the flux out of the interval at ξ_n from time t_j to time t_{j+1}. This expresses the conservation of material in the case that no sources or sinks are present. We relate the field variable u to the physical variables (the density r and the flux q). We consider the case in which density is assumed to have the form

$$r(x,t)=cu(x,t)+b \tag{8.49}$$

with c and b constants, thus

$$c\int\limits_{\xi_{n-1}}^{\xi_n}\mathrm{d}x\,\left[u(x,t_{j+1})-u(x,t_j)\right]\approx c[u(x_n,t_{j+1})-u(x_n,t_j)]h\,. \tag{8.50}$$

When developing conservation law equations, there are two commonly used strategies for approximating the right-hand-side of (8.48): (i) left end-point quadrature

$$\int\limits_{t_j}^{t_{j+1}}\mathrm{d}t\,\left[q(\xi_{n-1},t)-q(\xi_n,t)\right]\approx\left[q(\xi_{n-1},t_j)-q(\xi_n,t_j)\right]k\,, \tag{8.51}$$

and (ii) right end-point quadrature

$$\int\limits_{t_j}^{t_{j+1}}\mathrm{d}t\,\left[q(\xi_{n-1},t)-q(\xi_n,t)\right]\approx\left[q(\xi_{n-1},t_{j+1})-q(\xi_n,t_{j+1})\right]k\,. \tag{8.52}$$

Combining (8.48) with the respective approximations yields: from (i) an explicit method

$$c\left[u(x_n,t_{j+1})-u(x_n,t_j)\right]h\approx\left[q(\xi_{n-1},t_j)-q(\xi_n,t_j)\right]k\,, \tag{8.53}$$

and from (ii) an implicit method

$$c\left[u(x_n,t_{j+1})-u(x_n,t_j)\right]h\approx\left[q(\xi_{n-1},t_{j+1})-q(\xi_n,t_{j+1})\right]k\,. \tag{8.54}$$

Using centered finite difference formulas to approximate the fluxes at the control points ξ_{n-1} and ξ_n yields

$$q(\xi_{n-1},t_j)=-\sigma\frac{u(x_n,t_j)-u(x_{n-1},t_j)}{h}\,, \tag{8.55}$$

and

$$q(\xi_n,t_j)=-\sigma\frac{u(x_{n+1},t_j)-u(x_n,t_j)}{h} \tag{8.56}$$

where σ is a constant. We also obtain similar formulas for the fluxes at time t_{j+1}.

We have used a lower case u to denote the continuous field variable, $u = u(x, t)$. Note that all of the quadrature and difference formulas involving u are stated as approximate equalities. In each of these approximate equality statements, the amount by which the right side differs from the left side is called the truncation error. If u is a well-behaved function (has enough smooth derivatives), then it can be shown that these truncation errors approach zero as the grid spacings, h and k, approach zero.

If U_n^j denotes the exact solution on the grid, we have from (i) the result

$$c\left(U_n^{j+1} - U_n^j\right) h^2 = \sigma k \left(U_{n-1}^j + U_{n+1}^j - 2U_n^j\right) . \tag{8.57}$$

This is an explicit method since it provides the solution to the difference equation at time t_{j+1}, knowing the values at time t_j.

If we use the numerical approximations (ii) we obtain the result

$$c\left(U_n^{j+1} - U_n^j\right) h^2 = \sigma k \left(U_{n-1}^{j+1} + U_{n+1}^{j+1} - 2U_n^{j+1}\right) . \tag{8.58}$$

Note that this equation defines the solution at time t_{j+1} implicitly, since a system of algebraic equations is required to be satisfied.

8.2.3 Variational Methods

Perhaps the most widely used approximation procedures in AMO physics are the variational methods. We shall outline in detail the Rayleigh–Ritz method [8.23]. This method is limited to boundary value problems which can be formulated in terms of the minimization of a functional $J[u]$. For definiteness we consider the case of a differential operator defined by

$$Lu(\boldsymbol{x}) = f(\boldsymbol{x}) \tag{8.59}$$

with $\boldsymbol{x} = x_i$, $i = 1, 2, 3$ in R, for example, and with $u = 0$ on the boundary of R. The function $f(x)$ is the source. It is assumed that L is always nonsingular and in addition, for the Ritz method L is Hermitian. The real-valued functions u are in the Hilbert space Ω of the operator L. We construct the functional $J[u]$ defined as

$$J[u] = \int_\Omega \mathrm{d}x \, [u(x)Lu(x) - 2u(x)f(x)] . \tag{8.60}$$

The variational ansatz considers a subspace of Ω, Ω_n, spanned by a class of functions $\phi_n(x)$, and we construct the function $u^n \approx u$

$$u^n(x) = \sum_{i=1}^{n} c_i \phi_i(x) . \tag{8.61}$$

We solve for the coefficients c_i by minimizing $J[u^n]$

$$\partial_{c_i} J[u^n] = 0 , \quad i = 1, \ldots , n . \tag{8.62}$$

These equations are simply cast into a set of well-behaved algebraic equations

$$\sum_{j=1}^{n} A_{i,j} c_j = g_i , \quad i = 1, \ldots , n , \tag{8.63}$$

with $A_{i,j} = \int_\Omega \mathrm{d}x \phi_i(x) L \phi_j(x)$, and $g_i = \int_\Omega \mathrm{d}x \phi_i(x) f(x)$. Under very general conditions, the functions u^n converge uniformly to u. The main drawback of the Ritz method is in the assumption of Hermiticity of the operator L. For the *Galerkin Method* we relax this assumption with no other changes. Thus we obtain an identical set of equations, as above with the exception that the function g is no longer symmetric. The convergence of the sequence of solutions u^n to u is no longer guaranteed, unless the operator can be separated into a symmetric part L_0, $L = L_0 + K$ so that $L_0^{-1} K$ is bounded.

8.2.4 Finite Elements

As discussed in Sect. 8.2.2, in the finite difference method for classical partial differential equations, the solution domain is approximated by a grid of uniformly spaced nodes. At each node, the governing differential equation is approximated by an algebraic expression which references adjacent grid points. A system of equations is obtained by evaluating the previous algebraic approximations for each node in the domain. Finally, the system is solved for each value of the dependent variable at each node.

In the finite element method [8.24], the solution domain can be discretized into a number of uniform or nonuniform finite elements that are connected via nodes. The change of the dependent variable with regard to location is approximated within each element by an interpolation function. The interpolation function is defined relative to the values of the variable at the nodes associated with each element. The original boundary value problem is then replaced with an equivalent integral formulation. The interpolation functions are then substituted into the integral equation, integrated, and combined with the results from all other elements in the solution domain.

The results of this procedure can be reformulated into a matrix equation of the form

$$\sum_{j=1}^{n} A_{i,j}c_j = g_i\ , \quad i = 1, \dots, n\ , \tag{8.64}$$

with $A_{i,j} = \int_\Omega \mathrm{d}x\phi_i(x)L\phi_j(x)$, and $g_i = \int_\Omega \mathrm{d}x\phi_i(x) f(x)$ exactly as obtained in Sect. 8.2.3. The only difference arises in the definitions of the support functions $\phi_i(x)$. In general, if these functions are piecewise polynomials on some finite domain, they are called finite elements or splines. Finite elements make it possible to deal in a systematic fashion with regions having curved boundaries of an arbitrary shape. Also, one can systematically estimate the accuracy of the solution in terms of the parameters that label the finite element family, and the solutions are no more difficult to generate than more complex variational methods.

In one space dimension, the simplest finite element family begins with the set of step functions defined by

$$\phi_i(x) = \begin{cases} 1 & x_{i-1} \le x \le x_i \\ 0 & \text{otherwise}\ . \end{cases} \tag{8.65}$$

The use of these simple "hat" functions as a basis provides no advantage over the usual finite difference schemes. However, for certain problems in two or more dimensions, finite element methods have distinct advantages over other methods. Generally, the use of finite elements requires complex, sophisticated computer programs for implementation. The use of higher-order polynomials, commonly called splines, as a basis has been extensively used in atomic and molecular physics. An extensive literature is available [8.25, 26].

We illustrate the use of the finite element method by applying it to the Schrödinger equation. In this case, the linear operator L is $H - E$ where, as usual, E is the energy and the Hamiltonian H is the sum of the kinetic and potential energies, that is, $L = H - E = \mathcal{T} + V - E$ and $Lu(x) = 0$. We define the finite elements through support points, or knots, given by the sequence $\{x_1, x_2, x_3, \dots\}$ which are not necessarily spaced uniformly. Since the "hat" functions have vanishing derivatives, we employ the next more complex basis, i. e., "tent" functions, which are piecewise linear functions given by

$$\phi_i(x) = \begin{cases} \dfrac{x - x_{i-1}}{x_i - x_{i-1}} & x_{i-1} \le x \le x_i \\[2ex] \dfrac{x_{i+1} - x}{x_{i+1} - x_i} & x_i \le x \le x_{i+1} \\[2ex] 0 & \text{otherwise}\ , \end{cases} \tag{8.66}$$

and for which the derivative is given by

$$\frac{\mathrm{d}}{\mathrm{d}x}\phi_i(x) = \begin{cases} \dfrac{1}{x_i - x_{i-1}} & x_{i-1} \le x \le x_i \\[2ex] \dfrac{-1}{x_{i+1} - x_i} & x_i \le x \le x_{i+1} \\[2ex] 0 & \text{otherwise}\ . \end{cases} \tag{8.67}$$

The functions have a maximum value of one at the midpoint of the interval $[x_{i-1}, x_{i+1}]$, with partially overlapping adjacent elements. In fact, the overlaps may be represented by a matrix $\mathbf{O}$ with elements

$$O_{ij} = \int_{-\infty}^{\infty} \mathrm{d}x\ \phi_i(x)\phi_j(x)\ . \tag{8.68}$$

Thus, if $i = j$

$$\begin{aligned} O_{ii} &= \int_{x_{i-1}}^{x_i} \mathrm{d}x\ \frac{(x - x_{i-1})^2}{(x_i - x_{i-1})^2} + \int_{x_i}^{x_{i+1}} \mathrm{d}x\ \frac{(x - x_i)^2}{(x_{i+1} - x_i)^2} \\ &= \frac{1}{3}(x_{i+1} - x_{i-1})\ , \end{aligned} \tag{8.69}$$

if $i = j - 1$

$$\begin{aligned} O_{ij} &= \int_{x_i}^{x_{i+1}} \mathrm{d}x\ \frac{(x - x_i)(x_{i+1} - x)}{(x_{i+1} - x_i)^2} \\ &= \frac{1}{6}(x_{i+1} - x_i)\ , \end{aligned} \tag{8.70}$$

if $i = j + 1$

$$\begin{aligned} O_{ij} &= \int_{x_{i-1}}^{x_i} \mathrm{d}x\ \frac{(x - x_{i-1})(x_i - x)}{(x_i - x_{i-1})^2} \\ &= \frac{1}{6}(x_i - x_{i-1})\ , \end{aligned} \tag{8.71}$$

and $O_{ij} = 0$ otherwise.

The potential energy is represented by the matrix

$$V_{ij} = \int_{-\infty}^{\infty} \mathrm{d}x\ \phi_i(x)V(x)\phi_j(x)\ , \tag{8.72}$$

which may be well approximated by

$$V_{ij} \approx V(x_i) \int_{-\infty}^{\infty} \mathrm{d}x\, \phi_i(x)\phi_j(x) \qquad (8.73)$$
$$= V(x_i) O_{ij}$$

if $x_j - x_i$ is small. The kinetic energy, $\mathcal{T} = -\frac{1}{2}\mathrm{d}^2/\mathrm{d}x^2$, is similarly given by

$$\mathcal{T}_{ij} = -\frac{1}{2}\int_{-\infty}^{\infty} \mathrm{d}x \phi_i(x) \frac{\mathrm{d}^2}{\mathrm{d}x^2}\phi_j(x) , \qquad (8.74)$$

which we compute by integrating by parts since the tent functions have a singular second derivative

$$\mathcal{T}_{ij} = \frac{1}{2}\int_{-\infty}^{\infty} \mathrm{d}x \left(\frac{\mathrm{d}}{\mathrm{d}x}\phi_i(x)\right)\left(\frac{\mathrm{d}}{\mathrm{d}x}\phi_j(x)\right) , \qquad (8.75)$$

which in turn is evalutated to yield

$$\mathcal{T}_{ij} = \begin{cases} \dfrac{x_{i+1} - x_{i-1}}{2(x_i - x_{i-1})(x_{i+1} - x_i)} & i = j \\ \dfrac{1}{2(x_i - x_{i+1})} & i = j-1 \\ \dfrac{1}{2(x_{i-1} - x_i)} & i = j+1 \\ 0 & \text{otherwise} . \end{cases} \qquad (8.76)$$

Finally, since the Hamiltonian matrix is $H_{ij} = \mathcal{T}_{ij} + V_{ij}$, the solution vector $u_i(x)$ may be found by solving the eigenvalue equation

$$[H_{ij} - EO_{ij}]u_i(x) = 0 . \qquad (8.77)$$

8.2.5 Integral Equations

Central to much of practical and formal scattering theory is the integral equation and techniques of its solution. For example, in atomic collision theory, the Schrödinger differential equation

$$[E - H_0(\boldsymbol{r})]\, \psi(\boldsymbol{r}) = V(\boldsymbol{r})\psi(\boldsymbol{r}) \qquad (8.78)$$

where the Hamiltonian $H_0 \equiv -(\hbar^2/2m)\nabla^2 + V_0$ may be solved by exploiting the solution for a delta function source, i. e.,

$$(E - H_0)G(\boldsymbol{r}, \boldsymbol{r}') = \delta(\boldsymbol{r} - \boldsymbol{r}') . \qquad (8.79)$$

In terms of this *Green's function* $G(\boldsymbol{r}, \boldsymbol{r}'$, and any solution $\chi(\boldsymbol{r})$ of the homogeneous equation [i. e. with $V(\boldsymbol{r}) = 0$], the general solution is

$$\psi(\boldsymbol{r}) = \chi(\boldsymbol{r}) + \int \mathrm{d}\boldsymbol{r}'\, G(\boldsymbol{r}, \boldsymbol{r}')V(\boldsymbol{r}')\psi(\boldsymbol{r}') \qquad (8.80)$$

for which, given a choice of the functions $G(\boldsymbol{r}, \boldsymbol{r}')$ and $\chi(\boldsymbol{r})$, particular boundary conditions are determined. This integral equation is the Lippmann–Schwinger equation of potential scattering. Further topics on scattering theory are covered in other chapters (see especially Chapts. 47 to 58) and in standard texts such as those by *Joachain* [8.27], *Rodberg* and *Thaler* [8.28], and *Goldberger* and *Watson* [8.29]. Owing especially to the wide variety of specialized techniques for solving integral equations, we survey briefly only a few of the most widely applied methods.

Integral Transforms

Certain classes of integral equations may be solved using integral transforms such as the Fourier or Laplace transforms. These integral transforms typically have the form

$$f(x) = \int \mathrm{d}x'\, K(x, x')g(x') , \qquad (8.81)$$

where $f(x)$ is the integral transform of $g(x')$ by the kernel $K(x, x')$. Such a pair of functions is the solution of the Schrödinger equation (spatial wave function) and its Fourier transform (momentum representation wave function). *Arfken* [8.6], *Morse* and *Feshbach* [8.9], and *Courant* and *Hilbert* [8.10] give other examples, as well as being excellent references for the application of integral equations and Green's functions in mathematical physics. In their analytic form, these transform methods provide a powerful method of solving integral equations for special cases, and, in addition, they may be implemented by performing the transform numerically.

Power Series Solution

For an equation of the form (in one dimension for simplicity)

$$\psi(r) = \chi(r) + \lambda \int \mathrm{d}r'\, K(r, r')\psi(r') , \qquad (8.82)$$

a solution may be found by iteration. That is, as a first approximation, set $\psi_0(r) = \chi(r)$ so that

$$\psi_1(r) = \chi(r) + \lambda \int \mathrm{d}r'\, K(r, r')\chi(r') . \qquad (8.83)$$

This may be repeated to form a power series solution, i. e.,

$$\psi_n(r) = \sum_{k=0}^{n} \lambda^k I_k(r) , \quad (8.84)$$

where

$$I_0(r) = \chi(r) , \quad (8.85)$$

$$I_1(r) = \int \mathrm{d}r' \, K(r, r')\chi(r') , \quad (8.86)$$

$$I_2(r) = \int \mathrm{d}r'' \int \mathrm{d}r' \, K(r, r')K(r', r'')\chi(r'') , \quad (8.87)$$

$$I_n(r) = \int \mathrm{d}r' \cdots \int \mathrm{d}r^{(n)} \, K(r, r')K(r, r'') \cdots K(r^{(n-1)}, r^{(n)}) . \quad (8.88)$$

If the series converges, then the solution $\psi(r)$ is approached by the expansion. When the Schrödinger equation is cast as an integral equation for scattering in a potential, this iteration scheme leads to the *Born series*, the first term of which is the incident, unperturbed wave, and the second term is usually referred to simply as the *Born approximation*.

Separable Kernels

If the kernel is separable, i. e.,

$$K(r, r') = \sum_{k=1}^{n} f_k(r)g_k(r') , \quad (8.89)$$

where n is finite, then substitution into the prototype integral equation (8.82) yields

$$\psi(r) = \chi(r) + \lambda \sum_{k=1}^{n} f_k(r) \int \mathrm{d}r' \, g(r')\psi(r') . \quad (8.90)$$

Multiplying by $f_k(r)$, integrating over r, and rearranging, yields the set of algebraic equations

$$c_j = b_j + \lambda \sum_{k=1}^{n} a_{jk}c_k , \quad (8.91)$$

where

$$c_k = \int \mathrm{d}r' g_k(r')\psi(r') , \quad (8.92)$$

$$b_k = \int \mathrm{d}r f_k(r)\chi(r) , \quad (8.93)$$

$$a_{jk} = \int \mathrm{d}r g_j(r) f_k(r) , \quad (8.94)$$

or, if $\boldsymbol{c}$ and $\boldsymbol{b}$ denote vectors, and $\boldsymbol{A}$ denotes the matrix of constants a_{jk},

$$\boldsymbol{c} = (\mathbf{1} - \lambda \boldsymbol{A})^{-1}\boldsymbol{b} . \quad (8.95)$$

The eigenvalues are the roots of the determinantal equation. Substituting these into $(\boldsymbol{I} - \lambda \boldsymbol{A})\boldsymbol{c} = 0$ yields the constants c_k which determine the solution of the original equation. This derivation may be found in *Arfken* [8.6], along with an explicit example. Even if the kernel is not exactly separable, if it is approximately so, then this procedure can yield a result which can be substituted into the original equation as a first step in an iterative solution.

Numerical Integration

Perhaps the most straightforward method of solving an integral equation is to apply a numerical integration formula such as Gaussian quadrature. An equation of the form

$$\psi(r) = \int \mathrm{d}r' \, K(r, r')\chi(r') \quad (8.96)$$

can be approximated as

$$\psi(r_j) = \sum_{k=1}^{n} w_k K(r_j, r_k')\chi(r_k) , \quad (8.97)$$

where w_k are quadrature weights, if the kernel is well behaved. However, such an approach is not without pitfalls. In light of the previous subsection, this approach is equivalent to replacing the integral equation by a set of algebraic equations. In this example we have

$$\psi_j = \sum_{k=1}^{n} M_{jk}\chi_k , \quad (8.98)$$

so that the solution of the equation is found by inverting the matrix $\boldsymbol{M}$. Since there is no guarantee that this matrix is not ill-conditioned, the numerical procedure may not produce meaningful results. In particular, only certain classes of integral equations and kernels will lead to stable solutions.

Having only scratched the surface regarding the very rich field of integral equations, the interested reader is encouraged to explore the references given here.

8.3 Computational Linear Algebra

Previous sections of this chapter have dealt with interpolation, differential equations, and related topics. Generally, discretization methodologies lead to classes of algebraic equations. In recent years enormous progress has been made in developing algorithms for solving linear algebraic equations, and many very good books have been written on this topic [8.30]. Furthermore, a large body of numerical software is freely available via an electronic service called *Netlib* (www.netlib.org). In addition to the widely adopted numerical linear algebra packages LAPACK, ScaLAPACK, ARPACK, etc., there are dozens of other libraries, technical reports on various parallel computers and software, test data, facilities to automatically translate Fortran programs to C, bibliographies, names and addresses of scientists and mathematicians, and so on.

Here we discuss methods for solving systems of equations such as

$$\begin{aligned} a_{11}x_1 + a_{12}x_2 + \cdots + a_{1n}x_n &= b_1 \,, \\ a_{21}x_1 + a_{22}x_2 + \cdots + a_{2n}x_n &= b_2 \,, \\ &\vdots \\ a_{m1}x_1 + a_{m2}x_2 + \cdots + a_{mn}x_n &= b_m \,. \end{aligned} \tag{8.99}$$

In these equations a_{ij} and b_i form the set of known quantities, and the x_i must be determined. The solution to these equations can found if they are linearly independent. Numerically, problems can arise due to truncation and roundoff errors that lead to an approximate linear dependence [8.31]. In this case the set of equations are approximately singular and special methods must be invoked. Much of the complexity of modern algorithms comes from minimizing the effects of such errors. For relatively small sets of nonsingular equations, direct methods in which the solution is obtained after a definite number of operations can work well. However, for very large systems iterative techniques are preferable [8.32].

A great many algorithms are available for solving (8.99), depending on the structure of the coefficients. For example, if the matrix of coefficients $\boldsymbol{A}$ is dense, using Gaussian elimination takes $2n^3/3$ operations; if $\boldsymbol{A}$ is also symmetric and positive definite, using the Cholesky algorithm takes a factor of two fewer operations. If $\boldsymbol{A}$ is triangular, i. e., either zero above the diagonal or zero below the diagonal, we can solve the above by simple substitution in only n^2 operations. For example, if $\boldsymbol{A}$ arises from solving certain elliptic partial differential equations, such as Poisson's equation, then $\boldsymbol{Ax} = \boldsymbol{b}$ can be solved using multigrid methods in only n operations.

We shall outline below how to solve (8.99) using elementary Gaussian elimination. More advanced methods, such as *conjugate gradient, generalized minimum residuals*, and the *Lanczos method* are treated elsewhere [8.33].

To solve $\boldsymbol{Ax} = \boldsymbol{b}$, we first use Gaussian elimination to factor the matrix $\boldsymbol{A}$ as $\boldsymbol{PA} = \boldsymbol{LU}$, where $\boldsymbol{L}$ is lower triangular, $\boldsymbol{U}$ is upper triangular, and $\boldsymbol{P}$ is a matrix which permutes the rows of $\boldsymbol{A}$. Then we solve the triangular system $\boldsymbol{Ly} = \boldsymbol{Pb}$ and $\boldsymbol{Ux} = \boldsymbol{y}$. These last two operations are easily performed using standard linear algebra libraries. The factorization $\boldsymbol{PA} = \boldsymbol{LU}$ takes most of the time. Reordering the rows of $\boldsymbol{A}$ with $\boldsymbol{P}$ is called pivoting and is necessary for numerical stability. In the standard partial pivoting scheme, $\boldsymbol{L}$ has ones on its diagonal and other entries bounded in absolute value by one. The simplest version of Gaussian elimination involves adding multiples of one row of $\boldsymbol{A}$ to others to zero out subdiagonal entries, and overwriting $\boldsymbol{A}$ with $\boldsymbol{L}$ and $\boldsymbol{U}$.

We first describe the decomposition of $\boldsymbol{PA}$ into a product of upper and lower triangular matrices,

$$\boldsymbol{A}' = \boldsymbol{LU} \,, \tag{8.100}$$

where the matrix $\boldsymbol{A}'$ is defined by $\boldsymbol{A}' = \boldsymbol{PA}$. A very nice algorithm for pivoting is given in [8.3] and will not be discussed further. Writing out the indices,

$$A'_{ij} = \sum_{k=1}^{\min(i,j)} L_{ik}U_{kj} \,. \tag{8.101}$$

We shall make the choice

$$L_{ii} = 1 \,. \tag{8.102}$$

These equations have the remarkable property that the elements A'_{ij} of each row can be scanned in turn, writing L_{ij} and U_{ij} into the locations A'_{ij} as we go. At each position (i, j), only the current A'_{ij} and already-calculated values of $L_{i'j'}$ and $U_{i'j'}$ are required. To see how this works, consider the first few rows. If $i = 1$,

$$A'_{1j} = U_{1j} \,, \tag{8.103}$$

defining the first row of $\boldsymbol{L}$ and $\boldsymbol{U}$. The U_{1j} are written over the A'_{1j}, which are no longer needed. If $i = 2$,

$$\begin{aligned} A'_{21} &= L_{21}U_{11} \,, && j = 1 \\ A'_{2j} &= L_{21}U_{1j} + U_{2j} \,, && j \geq 2 \,. \end{aligned} \tag{8.104}$$

The first line gives L_{21}, and the second U_{2j}, in terms of existing elements of $\boldsymbol{L}$ and $\boldsymbol{U}$. The U_{2j} and L_{21} are written over the A'_{2j}. (Remember that $L_{ii} = 1$ by definition.) If $i = 3$,

$$\begin{aligned} A'_{31} &= L_{31}U_{11}\,, & j &= 1 \\ A'_{32} &= L_{31}U_{12} + L_{32}U_{22}\,, & j &= 2 \\ A'_{3j} &= L_{31}U_{1j} + L_{32}U_{2j} + U_{3j}\,, & j &\geq 3\,, \end{aligned} \tag{8.105}$$

yielding in turn L_{31},L_{32}, and U_{3j}, which are written over A'_{3j}.

The algorithm should now be clear. At the ith row

$$\begin{aligned} L_{ij} &= U_{jj}^{-1}\left(A'_{ij} - \sum_{k=1}^{j-1} L_{ik}U_{kj}\right), \quad j \leq i-1 \\ U_{ij} &= A'_{ij} - \sum_{k=1}^{i-1} L_{ik}U_{kj}\,, \quad j \geq i\,. \end{aligned} \tag{8.106}$$

We observe from the first line of these equations that the algorithm may run into numerical inaccuracies if any U_{jj} becomes very small. Now $U_{11} = A'_{11}$, while in general $U_{ii} = A'_{ii} - \cdots$. Thus the absolute values of the U_{ii} are maximized if the rows are rearranged so that the absolutely largest elements of $\boldsymbol{A}'$ in each column lie on the diagonal. A little thought shows that the solutions are unchanged by permuting the rows (same equations, different order).

The *LU decomposition* can now be used to solve the system. This relies on the fact that the inversion of a triangular matrix is a simple process of back substitution. We replace ((8.99)) by two systems of equations. Written out in full, the equations for a typical column of $\boldsymbol{y}$ look like

$$\begin{aligned} L_{11}y_1 &= b'_1\,, \\ L_{21}y_1 + L_{22}y_2 &= b'_2\,, \\ L_{31}y_1 + L_{32}y_2 + L_{33}y_3 &= b'_3\,, \\ &\vdots\,, \end{aligned} \tag{8.107}$$

where the vector $\boldsymbol{b}'$ is $\boldsymbol{p}' = \boldsymbol{Pb}$. Thus from successive rows we obtain $y_1, y_2, y_3, \ldots$ in turn

$$\begin{aligned} U_{11}x_1 &= y_1\,, \\ U_{12}x_1 + U_{22}x_2 &= y_2\,, \\ U_{13}x_1 + U_{23}x_2 + U_{33}x_3 &= y_3\,, \\ &\vdots\,, \end{aligned} \tag{8.108}$$

and from successive rows of the latter we obtain $x_1, x_2, x_3, \ldots$ in turn.

Library software also exists for evaluating all the error bounds for dense and band matrices (see discussion of *Netlib* in above). Gaussian elimination with pivoting is almost always numerically stable, so the error bound one expects from solving these equations is of the order of $n\epsilon$, where ϵ is related to the condition number of the matrix $\boldsymbol{A}$. A good discussion of errors and conditioning is given in [8.3].

8.4 Monte Carlo Methods

Owing to the continuing rapid development of computational facilities and the ever-increasing desire to perform *ab initio* calculations, the use of Monte Carlo methods is becoming widespread as a means to evaluate previously intractable multidimensional integrals and to enable complex modeling and simulation. For example, a wide range of applications broadly classified as *Quantum Monte Carlo* have been used to compute, for example, the ground state eigenfunctions of simple molecules. Also, guided random walks have found application in the computation of Green functions, and variables chosen randomly, subject to particular constaints, have been used to mimic the electronic distribution of atoms. The latter application, used in the *classical trajectory Monte Carlo technique* described in Chapt. 58, allows the statistical quasiquantal representation of ion–atom collisions.

Here we summarize the basic tools needed in these methods, and how they may be used to produce specific distributions and make tractable the evaluation of multidimensional integrals with complicated boundaries. Detailed descriptions of these methods can be found in [8.3, 8, 34].

8.4.1 Random Numbers

An essential ingredient of any Monte Carlo procedure is the availability of a computer-generated sequence of random numbers which is not periodic and is free of other significant statistical correlations. Often such numbers are termed *pseudorandom* or *quasirandom*, in distinction to truly random physical processes. While the quality of random number generators supplied with computers has greatly improved over time, it is important to be aware of the potential dangers which can be present. For example, many systems are supplied with a random number generator based on the *linear congruential method.* Typically a sequence of integers $n_1, n_2, n_3, \ldots$ is first produced between 0 and $N-1$ by using the recurrence relation

$$n_{i+1} = (an_i + b) \bmod N\,, \qquad 0 \le i < N-1 \tag{8.109}$$

where a, b, N and the seed value n_0 are positive integers. Real numbers between 0 and (strictly) 1 are then obtained by dividing by N. The period of this sequence is at most N, and depends on the judicious choice of the constants, with N being limited by the wordsize of the computer. A user who is unsure that the character of the random numbers generated on a particular computer platform is proper can perform additional randomizing shuffles or use a portable random number generator, both procedures being described in detail by *Knuth* [8.5] and *Press* et al. [8.3], for example.

8.4.2 Distributions of Random Numbers

Most distributions of random numbers begin with sequences generated uniformly between a lower and an upper limit, and are therefore called *uniform deviates.* However, it is often useful to draw the random numbers from other distributions, such as the Gaussian, Poisson, exponential, gamma, or binomial distributions. These are particularly useful in modeling data or supplying input for an event generator or simulator. In addition, as described below, choosing the random numbers according to some weighting function can signficantly improve the efficiency of integration schemes based on Monte Carlo sampling.

Perhaps the most direct way to produce the required distribution is the *transformation method.* If we have a sequence of uniform deviates x on (0, 1) and wish to find a new sequence y which is distributed with probability given by some function $f(y)$, it can be shown that the required transformation is given by

$$y(x) = \left[\int_0^y f(y)\,\mathrm{d}y\right]^{-1} . \tag{8.110}$$

Evidently, the indefinite integral must be both known and invertible, either analytically or numerically. Since this is seldom the case for distributions of interest, other less direct methods are most often applied. However, even these other methods often rely on the transformation method as one "stage" of the procedure. The transformation method may also be generalized to more than one dimension [8.3].

A more widely applicable approach is the *rejection method,* also known as *von Neumann rejection.* In this case, if one wishes to find a sequence y distributed according to $f(y)$, first choose another function $\tilde{f}(y)$, called the comparison function, which is everywhere greater than $f(y)$ on the desired interval. In addition, a way must exist to generate y according to the comparison function, such as use of the transformation method. Thus, the comparison function must be simpler or better known than the distribution to be found. One simple choice is a constant function which is larger than the maximum value of $f(y)$, but choices which are "closer" to $f(y)$ will be much more efficient.

To proceed, y is generated uniformly according to $\tilde{f}(y)$ and another deviate x is chosen uniformly on (0, 1). One then simply rejects or accepts y depending on whether x is greater than or less than the ratio $f(y)/\tilde{f}(y)$, respectively. The fraction of trial numbers accepted simply depends on the ratio of the area under the desired function to that under the comparison function. Clearly, the efficiency of this scheme depends on how few of the numbers initially generated must be rejected, and therefore on how closely the comparison function approximates the desired distribution. The Lorentzian distribution, for which the inverse definite integral is known (the tangent function), is a good comparison function for a variety of "bell-shaped" distributions such as the Gaussian (normal), Poisson, and gamma distributions.

Especially for distributions which are functions of more than one variable and possess complicated boundaries, the rejection method is impractical and the transformation method simply inapplicable. In the 1950's, a method to generate distributions for such situations was developed and applied in the study of statistical mechanics where multidimensional integrals (e.g., the partition function) must often be solved numerically, and is known as the *Metropolis algorithm.* This procedure, or its variants, has more recently been adopted to aid in the computation of eigenfunctions of complicated Hamiltonians and scattering operators. In essence, the Metropolis method generates a random walk through the space of the dependent variables, and in the limit of a large number of steps in the walk, the points visited approximate the desired distribution.

In its simplest form, the Metropolis method generates this distribution of points by stepping through this space, most frequently taking a step "downhill" but sometimes taking a step "uphill". That is, given a set of coordinates $\boldsymbol{q}$ and a desired distribution function $f(\boldsymbol{q})$, a trial step is taken from the ith configuration $\boldsymbol{q}_i$ to the next, depending on whether the ratio $f(\boldsymbol{q}_i+1)/f(\boldsymbol{q}_i)$ is greater or less than one. If the ratio is greater than one, the step is accepted, but if it is less than one, the step is accepted with a probability given by the ratio.

8.4.3 Monte Carlo Integration

The basic idea of Monte Carlo integration is that if a large number of points is generated uniformly randomly in some n-dimensional space, the number falling inside a given region is proportional to the volume, or definite integral, of the function defining that region. Though this idea is as true in one dimension as it is in n, unless there is a large number ("large" could be as little as three) of dimensions or the boundaries are quite complicated, the numerical quadrature schemes described previously are more accurate and efficient. However, since the Monte Carlo approach is based on just sampling the function at representative points rather than evaluating the function at a large number of finely spaced quadrature points, its advantage for very large problems is apparent.

For simplicity, consider the Monte Carlo method for integrating a function of only one variable; the generalization to n dimensions being straightforward. If we generate N random points uniformly on (a, b), then in the limit of large N the integral is

$$\int_a^b f(x)\,\mathrm{d}x \approx \frac{1}{N}\langle f(x)\rangle \pm \sqrt{\frac{\langle f^2(x)\rangle - \langle f(x)\rangle^2}{N}}\,, \tag{8.111}$$

where

$$\langle f(x)\rangle \equiv \frac{1}{N}\sum_{i=1}^{N} f(x_i) \tag{8.112}$$

is the arithmetic mean. The probable error given is appropriately a statistical one rather than a rigorous error bound and is the one standard error limit. From this one can see that the error decreases as only $N^{1/2}$, more slowly than the rate of decrease for the quadrature schemes based on interpolation. Also, the accuracy is greater for relatively smooth functions, since the Monte Carlo generation of points is unlikely to sample narrowly peaked features of the integrand well. To estimate the integral of a multidimensional function with complicated boundaries, simply find an enclosing volume and generate points uniformly randomly within it. Keeping the enclosing volume as close as possible to the volume of interest miminizes the number of points which fall outside, and therefore increases the efficiency of the procedure.

The Monte Carlo integral is related to techniques for generating random numbers according to prescribed distributions described in Sect. 8.4.2. If we consider a normalized distribution $w(x)$, known as the *weight function*, then with the change of variables defined by

$$y(x) = \int_a^x w(x')\,\mathrm{d}x'\,, \tag{8.113}$$

the Monte Carlo estimate of the integral becomes

$$\int_a^b f(x)\,\mathrm{d}x \approx \frac{1}{N}\left\langle \frac{f[x(y)]}{w[x(y)]}\right\rangle, \tag{8.114}$$

assuming that the transformation is invertible. Choosing $w(x)$ to behave approximately as $f(x)$ allows a more efficient generation of points within the boundaries of the integrand. This occurs since the uniform distribution of points y results in values of x distributed according to w and therefore "close" to f. This procedure, generally termed the *reduction of variance* of the Monte Carlo integration, improves the efficiency of the procedure to the extent that the transformed function f/w can be made smooth, and that the sampled region is as small as possible but still contains the volume to be estimated.

References

8.1 J. Stoer, R. Bulirsch: *Introduction to Numerical Analysis* (Springer, Berlin, Heidelberg 1980)

8.2 R. L. Burden, J. D. Faires, A. C. Reynolds: *Numerical Analysis* (Prindle, Boston 1981)

8.3 W. H. Press, B. P. Flannery, S. A. Teukolsky, W. T. Vetterling: *Numerical Recipes, the Art of Scientific Computing* (Cambridge Univ. Press, Cambridge 1992)

8.4 M. Abramowitz, I. A. Stegun (Eds.): *Handbook of Mathematical Functions*, Applied Mathematics Series, Vol. 55 (National Bureau of Standards, Washington, Dover, New York 1968)

8.5 D. E. Knuth: *The Art of Computer Programming*, Vol. 2 (Addison-Wesley, Reading 1981)
8.6 G. Arfken: *Mathematical Methods for Physicists* (Academic Press, Orlando 1985)
8.7 H. Jeffreys, B. S. Jeffreys: *Methods of Mathematical Physics* (Cambridge Univ. Press, Cambridge 1966)
8.8 S. E. Koonin, D. C. Meredith: *Computational Physics* (Addison-Wesley, Redwood City 1990)
8.9 P. M. Morse, H. Feshbach: *Methods of Theoretical Physics* (McGraw-Hill, New York 1953)
8.10 R. Courant, D. Hilbert: *Methods of Mathematical Physics* (Interscience, New York 1953)
8.11 P. J. Huber: *Robust Statistics* (Wiley, New York 1981)
8.12 H. D. Young: *Statistical Treatment of Experimental Data* (McGraw-Hill, New York 1962)
8.13 P. R. Bevington: *Data Reduction and Error Analysis for the Physical Sciences* (McGraw-Hill, New York 1969)
8.14 D. C. Champeney: *Fourier Transforms and Their Physical Applications* (Academic, New York 1973)
8.15 R. W. Hamming: *Numerical Methods for Scientists and Engineers* (McGraw-Hill, New York 1973)
8.16 D. F. Elliott, K. R. Rao: *Fast Transforms: Algorithms, Analyses, Applications* (Academic, New York 1982)
8.17 D. Zwillinger: *CRC Standard Mathematical Tables and Formulae*, 31st edn. (Chapman & Hall/CRC, New York 2002)
8.18 J. Lambert: *Numerical Methods for Ordinary Differential Equations* (Wiley, New York 1991)
8.19 L. F. Shampine: *Numerical Solution of Ordinary Differential Equations* (Chapman Hall, New York 1994)
8.20 J. Butcher: *The Numerical Analysis of Ordinary Differential Equations: Runge–Kutta and General Linear Methods* (Wiley, New York 1987)
8.21 G. Hall, J. Watt: *Modern Numerical Methods for Ordinary Differential Equations* (Clarendon, Oxford 1976)
8.22 I. Gladwell, R. Wait (Eds.): *A Survey of Numerical Methods for Partial Differential Equations* (Clarendon, Oxford 1979)
8.23 K. Rektorys: *Variational Methods in Mathematics, Science, and Engineering*, 2nd edn. (Reidel, Boston 1980)
8.24 D. Cook, D. S. Malkus, M. E. Plesha: *Concepts and Applications of Finite Element Analysis*, 3rd edn. (Wiley, New York 1989)
8.25 G. Nurnberger: *Approximation by Spline Functions* (Springer, Berlin, Heidelberg 1989)
8.26 C. DeBoor: *Practical Guide to Splines* (Springer, New York 1978)
8.27 C. J. Joachain: *Quantum Collision Theory* (Elsevier, New York 1983)
8.28 L. S. Rodberg, R. M. Thaler: *Introduction to the Quantum Theory of Scattering* (Academic, New York 1967)
8.29 M. L. Goldberger, K. M. Watson: *Collision Theory* (Wiley, New York 1964)
8.30 P. G. Ciarlet: *Introduction to Numerical Linear Algebra and Optimisation* (Cambridge Univ. Press, Cambridge 1989)
8.31 G. Golub, C. Van Loan: *Matrix Computations*, 2nd edn. (Johns Hopkins Univ. Press, Baltimore 1989)
8.32 W. Hackbusch: *Iterative Solution of Large Sparse Systems of Equations* (Springer, New York 1994)
8.33 A. George, J. Liu: *Computer Solution of Large Sparse Positive Definite Systems* (Prentice-Hall, Englewood Cliffs 1981)
8.34 M. H. Kalos, P. A. Whitlock: *The Basics of Monte Carlo Methods* (Wiley, New York 1986)

9. Hydrogenic Wave Functions

This chapter summarizes the solutions of the one-electron nonrelativistic Schrödinger equation, and the one-electron relativistic Dirac equation, for the Coulomb potential. The standard notations and conventions used in the mathematics literature for special functions have been chosen in preference to the notations customarily used in the physics literature whenever there is a conflict. This has been done to facilitate the use of standard reference works such as *Abramowitz* and *Stegun* [9.1], the *Bateman* project [9.2, 3], *Gradshteyn* and *Ryzhik* [9.4], *Jahnke* and *Emde* [9.5], *Luke* [9.6, 7], *Magnus*, *Oberhettinger*, and *Soni* [9.8], *Olver* [9.9], *Szego* [9.10], and the new NIST Digital Library of Mathematical Functions project, which is preparing a hardcover update [9.11] of *Abramowitz* and *Stegun* [9.1] and an online digital library of mathematical functions [9.12]. The section on special functions contains many of the formulas which are needed to check the results quoted in the previous sections, together with a number of other useful formulas. It includes a brief introduction to asymptotic methods.

References to the numerical evaluation of special functions are given.

9.1 **Schrödinger Equation** 153
- 9.1.1 Spherical Coordinates 153
- 9.1.2 Parabolic Coordinates 154
- 9.1.3 Momentum Space 156

9.2 **Dirac Equation** 157

9.3 **The Coulomb Green's Function** 159
- 9.3.1 The Green's Function for the Schrödinger Equation 159
- 9.3.2 The Green's Function for the Dirac Equation 161

9.4 **Special Functions** 162
- 9.4.1 Confluent Hypergeometric Functions 162
- 9.4.2 Laguerre Polynomials 166
- 9.4.3 Gegenbauer Polynomials 169
- 9.4.4 Legendre Functions 169

References 170

9.1 Schrödinger Equation

The nonrelativistic Schrödinger equation for a hydrogenic ion of nuclear charge Z in atomic units is

$$\left(-\frac{1}{2}\nabla^2 - \frac{Z}{r}\right)\psi(\boldsymbol{r}) = E\psi(\boldsymbol{r}) . \qquad (9.1)$$

9.1.1 Spherical Coordinates

The separable solutions of (9.1) in spherical coordinates are

$$\psi(\boldsymbol{r}) = Y_{\ell m}(\theta,\phi) R_\ell(r) , \qquad (9.2)$$

where $Y_{\ell m}(\theta,\phi)$ is a spherical harmonic as defined by *Edmonds* [9.13] and $R_\ell(r)$ is a solution of the radial equation

$$\left[-\frac{1}{2}\left(\frac{\mathrm{d}^2}{\mathrm{d}r^2} + \frac{2}{r}\frac{\mathrm{d}}{\mathrm{d}r} - \frac{\ell(\ell+1)}{r^2}\right) - \frac{Z}{r}\right] R_\ell(r) = ER_\ell(r) . \qquad (9.3)$$

The general solution to (9.3) is

$$R_\ell(r) = r^\ell \exp(\mathrm{i}kr)\left[A\,{}_1F_1(a;c;z) + BU(a,c,z)\right] , \qquad (9.4)$$

where ${}_1F_1$ and U are the regular and irregular solutions of the confluent hypergeometric equation defined in (9.130) and (9.131) below, and

$$k = \sqrt{2E}\,, \tag{9.5}$$
$$a = \ell + 1 - \mathrm{i}k^{-1}Z\,, \tag{9.6}$$
$$c = 2\ell + 2\,, \tag{9.7}$$
$$z = -2\mathrm{i}kr\,. \tag{9.8}$$

A and B are arbitrary constants. The solution given in (9.4) has an $r^{-\ell-1}$ singularity at $r=0$ unless $B=0$ or a is a non-positive integer. The leading term for small r is proportional to r^{ℓ} when $B=0$ and/or a is a non-positive integer. The large r behavior of the solution for (9.4) follows from (9.134), (9.135), and (9.164) below. Bound state solutions, with energy

$$E = -\frac{1}{2}Z^2 n^{-2} \tag{9.9}$$

are obtained when $a = -n + \ell + 1$ where $n > \ell$ is the principal quantum number. The properly normalized bound state solutions are

$$R_{n,\ell}(r) = \frac{2Z}{n^2}\sqrt{\frac{Z(n-\ell-1)!}{(n+\ell)!}}\left(\frac{2Zr}{n}\right)^{\ell} \times \exp(-Zr/n)\, L_{n-\ell-1}^{(2\ell+1)}(2Zr/n)\,, \tag{9.10}$$

where $L_{n-\ell-1}^{(2\ell+1)}$ is the generalized Laguerre polynomial defined in (9.187). The relation in (9.188) shows that ${}_1F_1$ and U are linearly dependent in this case, so that (9.4) is no longer the general solution of (9.3). A linearly independent solution for this case can be obtained by replacing the $L_{n-\ell-1}^{(2\ell+1)}(2Zr/n)$ in (9.10) by the second (irregular) solution $M_{n-\ell-1}^{(2\ell+1)}(2Zr/n)$ of the Laguerre equation [see (9.194), (9.196), and (9.197)]. The first three $R_{n,\ell}$ are

$$R_{1,0}(r) = 2Z^{3/2}\exp(-Zr)\,, \tag{9.11}$$
$$R_{2,0}(r) = \left(\frac{1}{2}Z\right)^{3/2}(2 - Zr)\exp\left(-\frac{1}{2}Zr\right)\,, \tag{9.12}$$
$$R_{2,1}(r) = \left(\frac{1}{2}Z\right)^{3/2}\left(\frac{1}{3}\right)^{1/2} Zr\exp\left(-\frac{1}{2}Zr\right)\,. \tag{9.13}$$

Additional explicit expressions, together with graphs of some of them, can be found in *Pauling* and *Wilson* [9.14].

The $R_{n,\ell}$ can be expanded in powers of $1/n$ [9.15]

$$R_{n,\ell}(r) = -\left(\frac{2Z^2(n+\ell)!}{(n-\ell-1)!\,n^{2\ell+4}}\right)^{1/2} \times r^{-1/2}\sum_{k=0}^{\infty} g_k^{(\ell)}\left[(8Zr)^{1/2}\right] n^{-2k}\,, \tag{9.14}$$

where the functions $g_k^{(\ell)}(z)$ are finite linear combinations of Bessel functions:

$$g_k^{(\ell)}(z) = z^{3k}\sum_{m=0}^{k} a_{k,m}^{(\ell)} J_{2\ell+2m+k+1}(z)\,. \tag{9.15}$$

The coefficients $a_{k,m}^{(\ell)}$ in (9.15) are calculated recursively from

$$a_{k,m}^{(\ell)} = \frac{(2\ell+2m+k+1)}{32(2k+m)(2\ell+m+2k+1)} \times \frac{1}{(2\ell+2m+k-1)} \times \left[(2\ell+2m+k-1)\,a_{k-1,m}^{(\ell)} + 32(k-m+1)(2\ell+m-k)\,a_{k,m-1}^{(\ell)}\right], \tag{9.16}$$

starting with the initial condition

$$a_{0,0}^{(\ell)} = 1\,. \tag{9.17}$$

The expansion (9.14) converges uniformly in r for r in any bounded region of the complex r plane. However, it converges fast enough so that a few terms give a good description of $R_{n,\ell}$ only if r is small. The square root in (9.14) has not been expanded in inverse powers of n because it has a branch point at $1/n = 1/\ell$ which would reduce the radius of convergence of the expansion to $1/\ell$. In some cases, large n expansions of matrix elements can be obtained by inserting (9.14) for $R_{n,\ell}$ and integrating term by term; examples can be found in *Drake* and *Hill* [9.15]. An asymptotic expansion in powers of $1/n$, which is valid from r equal to an arbitrary fixed positive number through the turning point at $r = 2n^2/Z$ out to $r = \infty$, can be assembled from (9.133), (9.166) – (9.181), and (9.188) below.

The $R_{n,\ell}$ are not a complete set because the continuum has been left out. The *Sturmian* functions $\rho_{k,\ell}$, given by

$$\rho_{k,\ell}(\beta; r) = \sqrt{\frac{\beta^3 k!}{\Gamma(k+2\ell+3)}}\,(\beta r)^{\ell}\,\mathrm{e}^{-\beta r/2} \times L_k^{(2\ell+2)}(\beta r)\,, \tag{9.18}$$

do form a complete orthonormal set. The positive constant β, which is independent of k and ℓ, sets the length scale for the basis set (9.18).

9.1.2 Parabolic Coordinates

The Schrödinger equation (9.1) is separable in parabolic coordinates ξ, η, ϕ, which are related to spherical coordinates r, θ, ϕ via

$$\xi = r + z = r\left[1+\cos(\theta)\right] , \tag{9.19}$$

$$\eta = r - z = r\left[1-\cos(\theta)\right] , \tag{9.20}$$

$$\phi = \phi . \tag{9.21}$$

This separability in a second coordinate system is related to the existence of a "hidden" $O(4)$ symmetry, which is also responsible for the degeneracy of the bound states [9.16, 17]. The solutions in parabolic coordinates are particularly convenient for derivations of the Stark effect and the Rutherford scattering cross section. The separable solutions of (9.1) in parabolic coordinates are

$$\psi(\boldsymbol{r}) = \exp(\mathrm{i}m\phi)\, N(\eta)\, \Xi(\xi) , \tag{9.22}$$

where

$$N(\eta) = \eta^{|m|/2}\exp\left(\frac{1}{2}\mathrm{i}k_1\eta\right)\left[A\,{}_1F_1(a_1;c;-\mathrm{i}k_1\eta) + BU(a_1,c,-\mathrm{i}k_1\eta)\right], \tag{9.23}$$

$$\Xi(\xi) = \xi^{|m|/2}\exp\left(\frac{1}{2}\mathrm{i}k_2\xi\right)\left[C\,{}_1F_1(a_2;c;-\mathrm{i}k_2\xi) + DU(a_2,c,-\mathrm{i}k_2\xi)\right], \tag{9.24}$$

with ${}_1F_1$ and U defined in (9.130), (9.131) below, and

$$k_1 = \pm k_2 = \pm\sqrt{2E} , \tag{9.25}$$

$$a_1 = \frac{1}{2}(|m|+1) - \mathrm{i}k_1^{-1}\mu , \tag{9.26}$$

$$a_2 = \frac{1}{2}(|m|+1) - \mathrm{i}k_2^{-1}(Z-\mu) , \tag{9.27}$$

$$c = |m|+1 . \tag{9.28}$$

A, B, C, and D are arbitrary constants; μ is the separation constant. An important special case is the well-known *Coulomb function*

$$\psi_\mathrm{C}(\boldsymbol{r}) = \Gamma\left(1-\mathrm{i}k^{-1}Z\right)\exp\left(\frac{1}{2}\pi k^{-1}Z + \mathrm{i}\boldsymbol{k}\cdot\boldsymbol{r}\right) \times {}_1F_1\left[\mathrm{i}k^{-1}Z; 1; \mathrm{i}(kr-\boldsymbol{k}\cdot\boldsymbol{r})\right] , \tag{9.29}$$

which is obtained by orienting the z-axis in the $\boldsymbol{k}$ direction and taking $m=0$, $-k_1 = k_2 = |\boldsymbol{k}|$, $\mu = Z + \frac{1}{2}\mathrm{i}|\boldsymbol{k}|$. ψ_C is normalized to unit incoming flux [see (9.34) below]. In applications, Z is often replaced by $-Z_1Z_2$, so that the Coulomb potential in (9.1) becomes $+Z_1Z_2/r$. Equation (9.232), the addition theorem for the spherical harmonics ([9.13] p. 63 Eq. 4.6.6), and the $\lambda = c = 1$ special case of (9.163) below can be used to expand ψ_C into an infinite sum of solutions of the form (9.2):

$$\psi_\mathrm{C}(\boldsymbol{r}) = 4\pi\sum_{\ell=0}^{\infty}\sum_{m=-\ell}^{\ell}\frac{\Gamma\left(\ell+1-\mathrm{i}k^{-1}Z\right)}{(2\ell+1)!} \times (-2\mathrm{i}k)^\ell\, \mathrm{e}^{\pi k^{-1}Z/2}\, Y^*_{\ell m}(\theta_k,\phi_k) Y_{\ell m}(\theta,\phi) \times r^\ell \mathrm{e}^{\mathrm{i}kr}\, {}_1F_1\left(\ell+1-\mathrm{i}k^{-1}Z; 2\ell+2; -2\mathrm{i}kr\right) , \tag{9.30}$$

where k, θ_k, and ϕ_k are the spherical coordinates of $\boldsymbol{k}$. ψ_C can be split into an incoming plane wave and an outgoing spherical wave with the aid of (9.134) below:

$$\psi_\mathrm{C}(\boldsymbol{r}) = \psi_\mathrm{in}(\boldsymbol{r}) + \psi_\mathrm{out}(\boldsymbol{r}) , \tag{9.31}$$

where

$$\psi_\mathrm{in}(\boldsymbol{r}) = \exp\left(\mathrm{i}\boldsymbol{k}\cdot\boldsymbol{r} - \frac{1}{2}\pi k^{-1}Z\right) \times U\left[\mathrm{i}k^{-1}Z; 1; \mathrm{i}(kr-\boldsymbol{k}\cdot\boldsymbol{r})\right] , \tag{9.32}$$

$$\psi_\mathrm{out}(\boldsymbol{r}) = -\frac{\Gamma\left(1-\mathrm{i}k^{-1}Z\right)}{\Gamma\left(\mathrm{i}k^{-1}Z\right)}\exp\left(\mathrm{i}kr - \frac{1}{2}\pi k^{-1}Z\right) \times U\left[1-\mathrm{i}k^{-1}Z; 1; -\mathrm{i}(kr-\boldsymbol{k}\cdot\boldsymbol{r})\right] . \tag{9.33}$$

The functions ψ_in and ψ_out can be expanded for $kr - \boldsymbol{k}\cdot\boldsymbol{r}$ large with the aid of (9.164). The result is

$$\psi_\mathrm{in}(\boldsymbol{r}) \sim \exp\left[\mathrm{i}\boldsymbol{k}\cdot\boldsymbol{r} - \mathrm{i}k^{-1}Z\ln(kr-\boldsymbol{k}\cdot\boldsymbol{r})\right] \times \sum_{n=0}^{\infty}\frac{(-\mathrm{i})^n}{n!}\left(\frac{\Gamma\left(\mathrm{i}k^{-1}Z+n\right)}{\Gamma\left(\mathrm{i}k^{-1}Z\right)}\right)^2 \times (kr-\boldsymbol{k}\cdot\boldsymbol{r})^{-n} , \tag{9.34}$$

$$\psi_\mathrm{out}(\boldsymbol{r}) \sim -\frac{\mathrm{i}\Gamma\left(1-\mathrm{i}k^{-1}Z\right)}{\Gamma\left(\mathrm{i}k^{-1}Z\right)(kr-\boldsymbol{k}\cdot\boldsymbol{r})} \times \exp\left[\mathrm{i}kr - \mathrm{i}k^{-1}Z\ln(kr-\boldsymbol{k}\cdot\boldsymbol{r})\right] \times \sum_{n=0}^{\infty}\frac{\mathrm{i}^n}{n!}\left(\frac{\Gamma\left(1-\mathrm{i}k^{-1}Z+n\right)}{\Gamma\left(1-\mathrm{i}k^{-1}Z\right)}\right)^2 \times (kr-\boldsymbol{k}\cdot\boldsymbol{r})^{-n} . \tag{9.35}$$

Because (9.1) is an elliptic partial differential equation, its solutions must be analytic functions of the Cartesian coordinates (except at $r=0$, where the solutions have cusps). The $n=0$ special case of (9.138) shows that ψ_{in} and ψ_{out} are logarithmically singular at $\boldsymbol{k}\cdot\boldsymbol{r}=kr$. Thus ψ_{in} and ψ_{out} are not solutions to (9.1) at $\boldsymbol{k}\cdot\boldsymbol{r}=kr$. The logarithmic singularity cancels when ψ_{in} and ψ_{out} are added to form ψ_{C}, which *is* a solution to (9.1).

Bound state solutions, with energy

$$E=-\frac{1}{2}Z^2\left(n_1+n_2+|m|+1\right)^{-2}, \tag{9.36}$$

are obtained when $a_1=-n_1$ and $a_2=-n_2$ where n_1 and n_2 are non-negative integers. The properly normalized bound state solutions, which can be put into one–one correspondence with the bound state solutions in spherical coordinates, are

$$\begin{aligned}&\psi_{n_1,n_2,m}(\eta,\xi,\phi)\\&=\sqrt{\frac{\beta^{2|m|+4}n_1!n_2!}{2\pi Z\left(n_1+|m|\right)!\left(n_2+|m|\right)!}}\\&\quad\times\exp\left[im\phi-\frac{1}{2}\beta\left(\eta+\xi\right)\right]\\&\quad\times\left(\eta\xi\right)^{|m|/2}L_{n_1}^{(|m|)}\left(\beta\eta\right)L_{n_2}^{(|m|)}\left(\beta\xi\right),\end{aligned} \tag{9.37}$$

where

$$\beta=Z\left(n_1+n_2+|m|+1\right)^{-1}. \tag{9.38}$$

9.1.3 Momentum Space

The nonrelativistic Schrödinger equation (9.1) becomes the integral equation

$$\frac{1}{2}p^2\phi\left(\boldsymbol{p}\right)-\frac{Z}{2\pi^2}\int\frac{\phi\left(\boldsymbol{p}'\right)}{\left(\boldsymbol{p}-\boldsymbol{p}'\right)^2}\mathrm{d}^3\boldsymbol{p}'=E\phi\left(\boldsymbol{p}\right) \tag{9.39}$$

in momentum space. Its solutions are related to the solutions in coordinate space via the Fourier transforms

$$\psi\left(\boldsymbol{r}\right)=\left(2\pi\right)^{-3/2}\int\exp\left(\mathrm{i}\boldsymbol{p}\cdot\boldsymbol{r}\right)\phi\left(\boldsymbol{p}\right)\mathrm{d}^3\boldsymbol{p}, \tag{9.40}$$

$$\phi\left(\boldsymbol{p}\right)=\left(2\pi\right)^{-3/2}\int\exp\left(-\mathrm{i}\boldsymbol{p}\cdot\boldsymbol{r}\right)\psi\left(\boldsymbol{r}\right)\mathrm{d}^3\boldsymbol{r}. \tag{9.41}$$

A trick of *Fock*'s [9.16, 18] can be used to expose the "hidden" $O(4)$ symmetry of hydrogen and construct the bound state solutions to (9.39). Let p, θ_p, ϕ_p and p', θ'_p, ϕ'_p be the spherical coordinates of $\boldsymbol{p}$ and $\boldsymbol{p}'$. Change variables from p, p' to χ, χ' via $p=\sqrt{-2E}\tan\left(\chi/2\right)$ and $p'=\sqrt{-2E}\tan\left(\chi'/2\right)$. This brings (9.39) to the form

$$\begin{aligned}&2\pi^2Z^{-1}\sqrt{-2E}\left[\sec^4(\chi/2)\phi(\boldsymbol{p})\right]\\&=\int\frac{\left[\sec^4(\chi'/2)\phi(\boldsymbol{p}')\right]\sin^2(\chi')\,\mathrm{d}\chi'\sin(\theta_p)\,\mathrm{d}\theta_p\,\mathrm{d}\phi_p}{2-2\left[\cos(\chi)\cos(\chi')+\sin(\chi)\sin(\chi')\cos(\gamma')\right]},\end{aligned} \tag{9.42}$$

where γ' is the angle between $\boldsymbol{p}$ and $\boldsymbol{p}'$. Equation (9.42) is solved by introducing spherical coordinates and spherical harmonics in four dimensions via a natural extension of the procedure used in three dimensions. Going to polar coordinates on x and y yields the cylindrical coordinates r_2, ϕ, z; the further step of going to polar coordinates on r_2 and z yields spherical coordinates r_3, θ, ϕ. If there is a fourth coordinate w, spherical coordinates in four dimensions are obtained via the additional step of going to polar coordinates on r_3 and w. The result is

$$x=r_4\sin\left(\chi\right)\sin\left(\theta\right)\cos\left(\phi\right), \tag{9.43}$$

$$y=r_4\sin\left(\chi\right)\sin\left(\theta\right)\sin\left(\phi\right), \tag{9.44}$$

$$z=r_4\sin\left(\chi\right)\cos\left(\theta\right), \tag{9.45}$$

$$w=r_4\cos\left(\chi\right). \tag{9.46}$$

The volume element, which is easily obtained via the same series of transformations, is

$$\mathrm{d}V=r_4^3\,\mathrm{d}r_4\,\mathrm{d}\Omega_4, \tag{9.47}$$

$$\mathrm{d}\Omega_4=\sin^2\left(\chi\right)\mathrm{d}\chi\sin\left(\theta\right)\mathrm{d}\theta\,\mathrm{d}\phi. \tag{9.48}$$

The four-dimensional spherical harmonics [9.2, Vol. 2, Chap. XI] are

$$\begin{aligned}Y_{n,\ell,m}\left(\chi,\theta,\phi\right)&=2^{\ell+1}\ell!\sqrt{\frac{n\left(n-\ell-1\right)!}{2\pi\left(n+\ell\right)!}}\sin^{\ell}\left(\chi\right)\\&\quad\times C_{n-\ell-1}^{\ell+1}\left[\cos\left(\chi\right)\right]Y_{\ell m}\left(\theta,\phi\right),\end{aligned} \tag{9.49}$$

where $C_{n-\ell-1}^{\ell+1}$ is a Gegenbauer polynomial and $n\geq\ell+1$ is an integer. They have the orthonormality property

$$\begin{aligned}&\int Y_{n,\ell,m}^{*}\left(\chi,\theta,\phi\right)Y_{n',\ell',m'}\left(\chi,\theta,\phi\right)\mathrm{d}\Omega_4\\&=\delta_{n,n'}\delta_{\ell,\ell'}\delta_{m,m'}.\end{aligned} \tag{9.50}$$

Equations (9.229) and (9.230) with $\lambda = 1$, equation (9.231), and the addition theorem for the three dimensional spherical harmonics $Y_{\ell m}$ can be used to show that

$$\begin{aligned}&\Big[1-2\left[\cos(\chi)\cos(\chi')\right.\\&\left.+\sin(\chi)\sin(\chi')\cos(\gamma')\right]t+t^2\Big]^{-1}\\&=\sum_{n=1}^{\infty}\sum_{\ell=0}^{n-1}\sum_{m=-\ell}^{\ell}\frac{2\pi^2}{n}t^{n-1}Y_{n,\ell,m}(\chi,\theta,\phi)\\&\qquad\times Y^*_{n,\ell,m}(\chi',\theta',\phi')\end{aligned}\tag{9.51}$$

holds for $|t| < 1$, where γ' is the angle between $\boldsymbol{p}$ and $\boldsymbol{p}'$. Multiply both sides of (9.51) by $Y_{n,\ell,m}(\chi',\theta',\phi')\,\mathrm{d}\Omega_4'$ (where $\mathrm{d}\Omega_4'$ is $\mathrm{d}\Omega_4$ with χ,θ,ϕ replaced by χ',θ',ϕ') and use the orthogonality relation (9.50). The result can be rearranged to the form

$$\begin{aligned}&2\pi^2 n^{-1}t^{n-1}Y_{n,\ell,m}(\chi,\theta,\phi)=\\&\int\frac{Y_{n,\ell,m}(\chi',\theta',\phi')\sin^2(\chi')\,\mathrm{d}\chi'\sin(\theta)\,\mathrm{d}\theta\,\mathrm{d}\phi}{1-2\left[\cos(\chi)\cos(\chi')+\sin(\chi)\sin(\chi')\cos(\gamma')\right]t+t^2}\,.\end{aligned}\tag{9.52}$$

Analytic continuation can be used to show that (9.52) is valid for all complex t despite the fact that (9.51) is restricted to $|t| < 1$. Comparing the $t = 1$ case of (9.52) with (9.42) shows that $E = -\frac{1}{2}Z^2n^{-2}$ in agreement with (9.9), and that

$$\begin{aligned}\phi(\boldsymbol{p}) = &\left\{\begin{matrix}\text{normalizing}\\ \text{factor}\end{matrix}\right\}\cos^4(\chi/2)\\&\times Y_{n,\ell,m}(\chi,\theta,\phi)\ .\end{aligned}\tag{9.53}$$

Transforming from χ back to p brings these to the form

$$\phi(\boldsymbol{p}) = Y_{\ell m}(\theta_p,\phi_p)\,F_{n,\ell}(p)\ ,\tag{9.54}$$

where the properly normalized radial functions are

$$\begin{aligned}F_{n,\ell}(p) = &\,2^{2\ell+2}n^2\ell!\sqrt{\frac{2(n-\ell-1)!}{\pi Z^3(n+\ell)!}}\left(\frac{np}{Z}\right)^{\ell}\\&\times\frac{Z^{2\ell+4}}{\left(n^2p^2+Z^2\right)^{\ell+2}}\\&\times C_{n-\ell-1}^{\ell+1}\left(\frac{n^2p^2-Z^2}{n^2p^2+Z^2}\right)\,.\end{aligned}\tag{9.55}$$

The first three $F_{n,\ell}$ are

$$F_{1,0}(p) = 4\sqrt{\frac{2}{\pi Z^3}}\frac{Z^4}{\left(p^2+Z^2\right)^2}\ ,\tag{9.56}$$

$$F_{2,0}(p) = \frac{32}{\sqrt{\pi Z^3}}\frac{Z^4\left(4p^2-Z^2\right)}{\left(4p^2+Z^2\right)^3}\ ,\tag{9.57}$$

$$F_{2,1}(p) = \frac{128}{\sqrt{3\pi Z^3}}\frac{Z^5p}{\left(4p^2+Z^2\right)^3}\ .\tag{9.58}$$

The $F_{n,\ell}$ satisfy the integral equation

$$\begin{aligned}&\frac{1}{2}p^2F_{n,\ell}(p)\\&-\frac{Z}{\pi p}\int_0^{\infty}Q_\ell\left(\frac{p^2+p'^2}{2pp'}\right)F_{n,\ell}(p')\,p'\,\mathrm{d}p'\\&=EF_{n,\ell}(p)\ ,\end{aligned}\tag{9.59}$$

which can be obtained by inserting (9.54) in (9.39). Here Q_ℓ is the Legendre function of the second kind, which is defined in (9.233) below.

9.2 Dirac Equation

The relativistic Dirac equation for a hydrogenic ion of nuclear charge Z can be reduced to dimensionless form by using the Compton wavelength $\hbar/(mc)$ for the length scale and the rest mass energy mc^2 for the energy scale. The result is

$$\left(-\mathrm{i}\boldsymbol{\alpha}\cdot\nabla+\beta-\frac{Z\alpha}{r}\right)\psi(\boldsymbol{r}) = E\psi(\boldsymbol{r})\ ,\tag{9.60}$$

where $\alpha = e^2/(\hbar c)$ is the fine structure constant, and $\boldsymbol{\alpha}$, β are the usual Dirac matrices:

$$\boldsymbol{\alpha} = \begin{pmatrix}0 & \boldsymbol{\sigma}\\ \boldsymbol{\sigma} & 0\end{pmatrix},\quad \beta = \begin{pmatrix}I & 0\\ 0 & -I\end{pmatrix}.\tag{9.61}$$

Here $\boldsymbol{\sigma}$ is a vector whose components are the two by two Pauli matrices, and I is the two by two identity matrix given by

$$\sigma_x = \begin{pmatrix} 0 & 1 \\ 1 & 0 \end{pmatrix}, \quad \sigma_y = \begin{pmatrix} 0 & -\mathrm{i} \\ \mathrm{i} & 0 \end{pmatrix},$$
$$\sigma_z = \begin{pmatrix} 1 & 0 \\ 0 & -1 \end{pmatrix}, \quad I = \begin{pmatrix} 1 & 0 \\ 0 & 1 \end{pmatrix}. \tag{9.62}$$

The solutions to (9.60) in spherical coordinates have the form

$$\psi(\boldsymbol{r}) = \begin{pmatrix} G(r)\,\chi_\kappa^m(\theta,\phi) \\ \mathrm{i}F(r)\,\chi_{-\kappa}^m(\theta,\phi) \end{pmatrix}, \tag{9.63}$$

where, for positive energy states, $G(r)$ is the radial part of the large component and $\mathrm{i}F(r)$ is the radial part of the small component. For negative energy states, $G(r)$ is the radial part of the small component and $\mathrm{i}F(r)$ is the radial part of the large component. χ is the two component spinor

$$\chi_\kappa^m = \begin{pmatrix} -\dfrac{\kappa}{|\kappa|}\left(\dfrac{\kappa+\frac{1}{2}-m}{2\kappa+1}\right)^{1/2} Y_{|\kappa+\frac{1}{2}|-\frac{1}{2},m-\frac{1}{2}} \\ \left(\dfrac{\kappa+\frac{1}{2}+m}{2\kappa+1}\right)^{1/2} Y_{|\kappa+\frac{1}{2}|-\frac{1}{2},m+\frac{1}{2}} \end{pmatrix}. \tag{9.64}$$

The relativistic quantum number κ is related to the total angular momentum quantum number j by

$$\kappa = \pm\left(j+\frac{1}{2}\right). \tag{9.65}$$

Because j takes on the values $\frac{1}{2}, \frac{3}{2}, \frac{5}{2}, \ldots$, κ is restricted to the values $\pm1, \pm2, \pm3, \ldots$. The spinor χ_κ^m obeys the useful relations

$$\boldsymbol{\sigma}\cdot\hat{\boldsymbol{r}}\,\chi_\kappa^m = -\,\chi_{-\kappa}^m, \tag{9.66}$$
$$\boldsymbol{\sigma}\cdot\mathbf{L}\,\chi_\kappa^m = -(\kappa+1)\,\chi_\kappa^m, \tag{9.67}$$

where $\hat{\boldsymbol{r}} = \boldsymbol{r}/r$ and $\mathbf{L} = \boldsymbol{r}\times\boldsymbol{p}$ with $\boldsymbol{p} = -\mathrm{i}\nabla$. Equations (9.66), (9.67), and the identity

$$\boldsymbol{\sigma}\cdot\boldsymbol{p} = (\boldsymbol{\sigma}\cdot\hat{\boldsymbol{r}})\left(\hat{\boldsymbol{r}}\cdot\boldsymbol{p} + \frac{\mathrm{i}\boldsymbol{\sigma}\cdot\mathbf{L}}{r}\right) \tag{9.68}$$

can be used to derive the radial equations, which are

$$\left(\frac{\mathrm{d}}{\mathrm{d}r}+\frac{1+\kappa}{r}\right)G(r) - \left(1+E+\frac{Z\alpha}{r}\right)F(r) = 0, \tag{9.69}$$

$$\left(\frac{\mathrm{d}}{\mathrm{d}r}+\frac{1-\kappa}{r}\right)F(r) - \left(1-E-\frac{Z\alpha}{r}\right)G(r) = 0. \tag{9.70}$$

Equations (9.158), (9.159), (9.161), and (9.162) below can be used to show that the general solution to (9.69) and (9.70) is

$$G(r) = r^\gamma \exp(-\lambda r)(1+E)^{1/2}\{A[f_2(r)+f_1(r)] + B[f_4(r)+f_3(r)]\}, \tag{9.71}$$
$$F(r) = r^\gamma \exp(-\lambda r)(1-E)^{1/2}\{A[f_2(r)-f_1(r)] + B[f_4(r)-f_3(r)]\}, \tag{9.72}$$

where

$$f_1(r) = \left(Z\alpha\lambda^{-1}-\kappa\right)\,{}_1F_1(a;c;2\lambda r), \tag{9.73}$$
$$f_2(r) = a\,{}_1F_1(a+1;c;2\lambda r), \tag{9.74}$$
$$f_3(r) = U(a,c,2\lambda r), \tag{9.75}$$
$$f_4(r) = \left(Z\alpha\lambda^{-1}+\kappa\right)U(a+1,c,2\lambda r), \tag{9.76}$$
$$\lambda = (1+E)^{1/2}(1-E)^{1/2}, \tag{9.77}$$
$$\gamma = -1+\left(\kappa^2-Z^2\alpha^2\right)^{1/2}, \tag{9.78}$$
$$a = 1+\gamma-\lambda^{-1}EZ\alpha, \tag{9.79}$$
$$c = 3+2\gamma. \tag{9.80}$$

A and B are arbitrary constants. Because γ is in general not an integer, the solutions have a branch point at $r = 0$, and become infinite at $r = 0$ when $\kappa = \pm1$, which makes γ negative. The solutions for $E < -1$ and $E > +1$ are in the continuum, which implies that one of the factors $(1+E)^{1/2}$, $(1-E)^{1/2}$ is real with the other imaginary. Square integrable solutions, with energy

$$E_{n,\kappa} = \frac{Z}{|Z|}\left(1+\frac{Z^2\alpha^2}{(n+1+\gamma)^2}\right)^{-1/2}, \tag{9.81}$$

are obtained when $a = -n$ where n is a non-negative integer. The properly normalized square integrable solutions are

$$G_{n,\kappa}(r) = C_{n,\kappa}(2\lambda r)^\gamma \exp(-\lambda r)\left(1+E_{n,\kappa}\right)^{1/2} \times\left[g_{n,\kappa}^{(2)}(r)+g_{n,\kappa}^{(1)}(r)\right], \tag{9.82}$$
$$F_{n,\kappa}(r) = C_{n,\kappa}(2\lambda r)^\gamma \exp(-\lambda r)\left(1-E_{n,\kappa}\right)^{1/2} \times\left[g_{n,\kappa}^{(2)}(r)-g_{n,\kappa}^{(1)}(r)\right], \tag{9.83}$$
$$g_{n,\kappa}^{(1)}(r) = \left(Z\alpha\lambda^{-1}-\kappa\right)^{1/2} L_n^{(2+2\gamma)}(2\lambda r), \tag{9.84}$$

$$g^{(2)}_{n,\kappa}(r) = -(n+2+2\gamma)\left(Z\alpha\lambda^{-1}-\kappa\right)^{-1/2} \times L^{(2+2\gamma)}_{n-1}(2\lambda r)\,, \tag{9.85}$$

$$C_{n,\kappa} = \sqrt{\frac{2\lambda^4 n!}{Z\alpha\Gamma(n+3+2\gamma)}}\,. \tag{9.86}$$

When $n=0$, $|Z\alpha\lambda^{-1}| = |\kappa|$, and the value of κ whose sign is the same as the sign of $Z\alpha\lambda^{-1}$ is not permitted. Also, $L^{(2+2\gamma)}_{-1}(2\lambda r)$ is counted as zero, so that $g^{(2)}_{0,\kappa}(r) = 0$.

The eigenvalues and eigenfunctions for the first four states for $Z>0$ will now be written out explicitly in terms of the variable $\rho = Z\alpha r$. For the $1\,S_{1/2}$ ground state, with $n=0$, $j=\frac{1}{2}$, $\kappa=-1$, the formulae are

$$E_{0,-1} = \sqrt{1-Z^2\alpha^2}\,, \tag{9.87}$$

$$G_{0,-1}(r) = \sqrt{\frac{4Z^3\alpha^3\left(1+E_{0,-1}\right)}{\Gamma\left(1+2E_{0,-1}\right)}}\,(2\rho)^{E_{0,-1}-1}\,\mathrm{e}^{-\rho}\,, \tag{9.88}$$

$$F_{0,-1}(r) = -\sqrt{\frac{4Z^3\alpha^3\left(1-E_{0,-1}\right)}{\Gamma\left(1+2E_{0,-1}\right)}} \times (2\rho)^{E_{0,-1}-1}\,\mathrm{e}^{-\rho}\,. \tag{9.89}$$

The formulae for the $2\,S_{1/2}$ excited state, with $n=1$, $j=\frac{1}{2}$, $\kappa=-1$, and for the $2\,P_{1/2}$ excited state, with $n=1$, $j=\frac{1}{2}$, $\kappa=1$, can be written together. They are

$$E_{1,\kappa} = \left(\frac{1}{2}+\frac{1}{2}\sqrt{1-Z^2\alpha^2}\right)^{1/2}\,, \tag{9.90}$$

$$G_{1,\kappa}(r) = \sqrt{\frac{Z^3\alpha^3\left(2E_{1,\kappa}-\kappa\right)\left(1+E_{1,\kappa}\right)}{2E^2_{1,\kappa}\Gamma\left(4E^2_{1,\kappa}+1\right)}} \times \rho_1^{2E^2_{1,\kappa}-2}\,\mathrm{e}^{-\rho_1/2} \times\left[\left(2E_{1,\kappa}-\kappa-1\right)\left(2E_{1,\kappa}+\kappa\right)-\rho_1\right], \tag{9.91}$$

$$F_{1,\kappa}(r) = -\sqrt{\frac{Z^3\alpha^3\left(2E_{1,\kappa}-\kappa\right)\left(1-E_{1,\kappa}\right)}{2E^2_{1,\kappa}\Gamma\left(4E^2_{1,\kappa}+1\right)}} \times \rho_1^{2E^2_{1,\kappa}-2}\,\mathrm{e}^{-\rho_1/2} \times\left[\left(2E_{1,\kappa}-\kappa+1\right)\left(2E_{1,\kappa}+\kappa\right)-\rho_1\right], \tag{9.92}$$

where $\rho_1 = \rho/E_{1,\kappa}$. For the $2\,P_{3/2}$ excited state, with $n=0$, $j=\frac{1}{2}$, $\kappa=-2$, the formulae are

$$E_{0,-2} = \sqrt{1-\frac{1}{4}Z^2\alpha^2}\,, \tag{9.93}$$

$$G_{0,-2}(r) = \sqrt{\frac{Z^3\alpha^3\left(1+E_{0,-2}\right)}{2\Gamma\left(1+4E_{0,-2}\right)}}\,\rho^{2E_{0,-2}-1}\,\mathrm{e}^{-\rho/2}\,, \tag{9.94}$$

$$F_{0,-2}(r) = -\sqrt{\frac{Z^3\alpha^3\left(1-E_{0,-2}\right)}{2\Gamma\left(1+4E_{0,-2}\right)}}\,\rho^{2E_{0,-2}-1}\,\mathrm{e}^{-\rho/2}\,. \tag{9.95}$$

9.3 The Coulomb Green's Function

The abstract Green's operator for a Hamiltonian H is the inverse $G(E) = (H-E)^{-1}$. It is used to write the solution to $(H-E)|\xi\rangle = |\eta\rangle$ in the form $|\xi\rangle = G|\eta\rangle$. It has the spectral representation

$$G(E) = \sum\!\!\!\!\!\!\int_j \frac{1}{E_j-E}|e_j\rangle\langle e_j|\,. \tag{9.96}$$

The sum over j in (9.96) runs over *all* of the spectrum of H, including the continuum. For the bound state part of the spectrum, the numbers E_j and vectors $|e_j\rangle$ are the eigenvalues and eigenvectors of H. For the continuous spectrum, $|e_j\rangle\langle e_j|$ is a projection valued measure [9.19]. The representation (9.96) shows that $G(E)$ has first order poles at the eigenvalues. The reduced Green's operator (also known as the generalized Green's operator), which is the ordinary Green's operator with the singular terms subtracted out, remains finite when E is at an eigenvalue. It can be calulated from

$$G^{(\mathrm{red})}(E_k) = \lim_{E\to E_k}\left\{\frac{\partial}{\partial E}\left[(E-E_k)\,G(E)\right]\right\}\,. \tag{9.97}$$

The coordinate and momentum space representatives of the abstract Green's operator are the Green's functions. The nonrelativistic Coulomb Green's function has been discussed by *Hostler* and *Schwinger* [9.20, 21]. A unified treatment of the Coulomb Green's functions for the Schrödinger and Dirac equations has been given by *Swainson* and *Drake* [9.22]. Reduced Green's functions are discussed in the third of the Swainson–Drake papers, and in the paper of *Hill* and *Huxtable* [9.23].

9.3.1 The Green's Function for the Schrödinger Equation

The Green's function $G^{(S)}$ for the Schrödinger equation (9.98) is a solution of

$$\left(-\frac{1}{2}\nabla^2 - \frac{Z}{r} - E\right) G^{(S)}\left(\boldsymbol{r},\boldsymbol{r}';\,E\right) = \delta\left(\boldsymbol{r}-\boldsymbol{r}'\right) . \tag{9.98}$$

An explicit closed form expression for $G^{(S)}$ is

$$\begin{aligned} G^{(S)}\left(\boldsymbol{r},\boldsymbol{r}';\,E\right) &= \frac{\Gamma(1-\nu)}{2\pi|\boldsymbol{r}-\boldsymbol{r}'|} \\ &\times\left[W_{\nu,\frac{1}{2}}(z_2)\frac{\partial}{\partial z_1}M_{\nu,\frac{1}{2}}(z_1)\right. \\ &\left. - M_{\nu,\frac{1}{2}}(z_1)\frac{\partial}{\partial z_2}W_{\nu,\frac{1}{2}}(z_2)\right] , \end{aligned} \tag{9.99}$$

where $M_{\nu,1/2}$ and $W_{\nu,1/2}$ are the Whittaker functions defined in (9.132) and (9.133) below, and

$$\nu = Z(-2E)^{-1/2} , \tag{9.100}$$

$$z_1 = (-2E)^{1/2}\left(r+r' - |\boldsymbol{r}-\boldsymbol{r}'|\right) , \tag{9.101}$$

$$z_2 = (-2E)^{1/2}\left(r+r' + |\boldsymbol{r}-\boldsymbol{r}'|\right) . \tag{9.102}$$

The branch on which $(-2E)^{1/2}$ is positive should be taken when $E<0$. When $E>0$, the branch which corresponds to incoming (or outgoing) waves at infinity can be selected with the aid of the asymptotic approximation

$$G^{(S)}\left(\boldsymbol{r},\boldsymbol{r}';\,E\right) \approx \frac{\Gamma(1-\nu)}{2\pi|\boldsymbol{r}-\boldsymbol{r}'|} z_2^{\nu}\exp\left(-\tfrac{1}{2}z_2\right) , \tag{9.103}$$

which holds when $z_2 \gg z_1$. This approximation is obtained by using (9.130), (9.132), (9.133), and (9.164) in (9.99). A number of useful expansions for $G^{(S)}$ can be obtained from the integral representation

$$\begin{aligned} &G^{(S)}\left(\boldsymbol{r},\boldsymbol{r}';\,E\right) \\ &= \frac{2Z}{\nu}\int_0^{\infty}\left[\coth\left(\frac{1}{2}\rho\right)\right]^{2\nu}\sinh(\rho) \\ &\times I_0\left\{\nu^{-1}Z\sinh(\rho)\left(2rr'\right)^{1/2}\left[1+\cos(\Theta)\right]\right\} \\ &\times\exp\left[-\nu^{-1}Z\left(r+r'\right)\cosh(\rho)\right]\,\mathrm{d}\rho , \end{aligned} \tag{9.104}$$

where Θ is the angle between $\boldsymbol{r}$ and $\boldsymbol{r}'$. These expansions, and other integral representations, can be found in [9.20–22]. The partial wave expansion of $G^{(S)}$ is

$$\begin{aligned} &G^{(S)}\left(\boldsymbol{r},\boldsymbol{r}';\,E\right) \\ &= \sum_{\ell,m} g_\ell^{(S)}\left(r,r';\,\nu\right) Y_{\ell m}(\theta,\phi)\, Y_{\ell m}^{*}\left(\theta',\phi'\right) . \end{aligned} \tag{9.105}$$

The radial Green's function $g_\ell^{(S)}$ is a solution of the radial equation

$$\begin{aligned} &\left[-\frac{1}{2}\left(\frac{\mathrm{d}^2}{\mathrm{d}r^2}+\frac{2}{r}\frac{\mathrm{d}}{\mathrm{d}r}-\frac{\ell(\ell+1)}{r^2}\right)\right. \\ &\left.-\frac{Z}{r}-E\right] g_\ell^{(S)}\left(r,r';\,\nu\right) = \frac{\delta\left(r-r'\right)}{rr'} . \end{aligned} \tag{9.106}$$

The standard method for calculating the Green's function of a second order ordinary differential equation ([9.24] pp. 354–355) yields

$$\begin{aligned} &g_\ell^{(S)}\left(r,r';\,\nu\right) \\ &= \frac{(2Z)^{2\ell+2}\,\Gamma(\ell+1-\nu)}{(2\ell+1)!\,\nu^{2\ell+1}}\exp\left[-\nu^{-1}Z\left(r+r'\right)\right] \\ &\times\left(rr'\right)^{\ell}\,{}_1F_1\left(\ell+1-\nu;\,2\ell+2;\,2\nu^{-1}Zr_<\right) [3pt] \\ &\times U\left(\ell+1-\nu,\,2\ell+2,\,2\nu^{-1}Zr_>\right) , \end{aligned} \tag{9.107}$$

where $r_<$ is the smaller of the pair r, r' and $r_>$ is the larger of the pair r, r'. Matrix elements of $g_\ell^{(S)}$ can be calculated with the aid of the formula for the double Laplace transform, which is

$$\begin{aligned} &\int_0^{\infty}\mathrm{d}r\int_0^{\infty}\mathrm{d}r'\left(rr'\right)^{\ell+1}\exp\left(-\lambda r-\lambda' r'\right) g_\ell^{(S)}\left(r,r';\,\nu\right) \\ &= \frac{2(2\ell+1)!}{\ell-\nu+1}\left(\frac{\nu}{2Z}\right)^{2\ell+3}\left(\frac{4Z^2}{(\nu\lambda+Z)(\nu\lambda'+Z)}\right)^{2\ell+2} \\ &\times{}_2F_1(2\ell+2,\,\ell-\nu+1;\,\ell-\nu+2;\,1-\zeta) , \end{aligned} \tag{9.108}$$

where

$$\zeta = \frac{2\nu Z\left(\lambda+\lambda'\right)}{(\nu\lambda+Z)(\nu\lambda'+Z)} , \tag{9.109}$$

Matrix elements with respect to Slater orbitals can be calculated from (9.108) by taking derivatives with respect to λ and/or λ' to bring down powers of r and r'. Matrix elements with respect to Laguerre polynomials can be calculated by using (9.108) to evaluate integrals over the generating function (9.199) for the Laguerre polynomial [9.23]. Other methods of calculating matrix elements are discussed in *Swainson* and *Drake* [9.22]. The Green's function $\tilde{G}^{(S)}$ in momentum space is related to the coordinate space Green's function $G^{(S)}$ via the Fourier transforms

$$G^{(S)}(\boldsymbol{r},\boldsymbol{r}';E) = (2\pi)^{-3}\int \exp\left[\mathrm{i}\left(\boldsymbol{p}\cdot\boldsymbol{r} - \boldsymbol{p}'\cdot\boldsymbol{r}'\right)\right] \times \tilde{G}^{(S)}(\boldsymbol{p},\boldsymbol{p}';E)\,\mathrm{d}^3p\,\mathrm{d}^3p' , \tag{9.110}$$

$$\tilde{G}^{(S)}(\boldsymbol{p},\boldsymbol{p}';E) = (2\pi)^{-3}\int \exp\left[-\mathrm{i}\left(\boldsymbol{p}\cdot\boldsymbol{r} - \boldsymbol{p}'\cdot\boldsymbol{r}'\right)\right] \times G^{(S)}(\boldsymbol{r},\boldsymbol{r}';E)\,\mathrm{d}^3r\,\mathrm{d}^3r' . \tag{9.111}$$

The Green's function $\tilde{G}^{(S)}$ is a solution of

$$\left(\frac{1}{2}p^2 - E\right)\tilde{G}^{(S)}(\boldsymbol{p},\boldsymbol{p}';E) - \frac{Z}{2\pi^2}\int \frac{1}{(\boldsymbol{p}-\boldsymbol{p}'')^2} \times \tilde{G}^{(S)}(\boldsymbol{p}'',\boldsymbol{p}';E)\,\mathrm{d}^3p'' = \delta(\boldsymbol{p}-\boldsymbol{p}') . \tag{9.112}$$

An explicit closed form expression for $\tilde{G}^{(S)}$ is

$$\begin{aligned}
&\tilde{G}^{(S)}(\boldsymbol{p},\boldsymbol{p}';E) \\
&= \frac{\delta(\boldsymbol{p}-\boldsymbol{p}')}{\frac{1}{2}p^2 - E} \\
&+ \frac{Z}{2\pi^2|\boldsymbol{p}-\boldsymbol{p}'|^2\left(\frac{1}{2}p^2-E\right)\left[\frac{1}{2}(p')^2-E\right]} \\
&\times\left\{1+\frac{\nu q}{1-\nu}\right. \\
&\times\left[\left(\frac{1-q}{1+q}\right)\,{}_2F_1\left(1,1-\nu;2-\nu;\frac{1-q}{1+q}\right)\right. \\
&\left.\left.-\left(\frac{1+q}{1-q}\right)\,{}_2F_1\left(1,1-\nu;2-\nu;\frac{1+q}{1-q}\right)\right]\right\},
\end{aligned} \tag{9.113}$$

where

$$q = \sqrt{\frac{2E\,|\boldsymbol{p}-\boldsymbol{p}'|^2}{4E^2-4E\,\boldsymbol{p}\cdot\boldsymbol{p}'+(pp')^2}} . \tag{9.114}$$

9.3.2 The Green's Function for the Dirac Equation

The Green's function G_D for the Dirac equation (9.60) is a 4×4 matrix valued solution of

$$\left(-\mathrm{i}\alpha\cdot\nabla+\beta-\frac{Z\alpha}{r}-E\right)G_\mathrm{D}(\boldsymbol{r},\boldsymbol{r}';E) = \delta(\boldsymbol{r}-\boldsymbol{r}')\,I_4 , \tag{9.115}$$

where I_4 is the 4×4 identity matrix. The partial wave expansion of G_D is

$$G_\mathrm{D}(\boldsymbol{r},\boldsymbol{r}';E) = \sum_{\kappa,m}\begin{pmatrix} G_{11}^{\kappa,m} & -\mathrm{i}G_{12}^{\kappa,m} \\ \mathrm{i}G_{21}^{\kappa,m} & G_{22}^{\kappa,m}\end{pmatrix} , \tag{9.116}$$

where

$$G_{11}^{\kappa,m} = \chi_\kappa^m(\theta,\phi)\,\chi_\kappa^{m\dagger}(\theta',\phi')\,g_{11}(r,r';E) , \tag{9.117}$$

$$G_{12}^{\kappa,m} = \chi_\kappa^m(\theta,\phi)\,\chi_{-\kappa}^{m\dagger}(\theta',\phi')\,g_{12}(r,r';E) , \tag{9.118}$$

$$G_{21}^{\kappa,m} = \chi_{-\kappa}^m(\theta,\phi)\,\chi_\kappa^{m\dagger}(\theta',\phi')\,g_{21}(r,r';E) , \tag{9.119}$$

$$G_{22}^{\kappa,m} = \chi_{-\kappa}^m(\theta,\phi)\,\chi_{-\kappa}^{m\dagger}(\theta',\phi')\,g_{22}(r,r';E) . \tag{9.120}$$

The identity

$$\delta(\boldsymbol{r}-\boldsymbol{r}')\,I_4 = \frac{\delta(r-r')}{rr'} \times\sum_{\kappa,m}\begin{pmatrix}\chi_\kappa^m(\theta,\phi)\,\chi_\kappa^{m\dagger}(\theta',\phi') & 0 \\ 0 & \chi_{-\kappa}^m(\theta,\phi)\,\chi_{-\kappa}^{m\dagger}(\theta',\phi')\end{pmatrix} \tag{9.121}$$

can be used to show that the radial functions $g_{jk}(r,r';E)$ satisfy the equation

$$\begin{pmatrix}\left(1-E-\frac{Z\alpha}{r}\right) & -\left(\frac{\mathrm{d}}{\mathrm{d}r}+\frac{1-\kappa}{r}\right) \\ \left(\frac{\mathrm{d}}{\mathrm{d}r}+\frac{1+\kappa}{r}\right) & -\left(1+E+\frac{Z\alpha}{r}\right)\end{pmatrix} \times\begin{pmatrix} g_{11}(r,r';E) & g_{12}(r,r';E) \\ g_{21}(r,r';E) & g_{22}(r,r';E)\end{pmatrix} = \frac{\delta(r-r')}{rr'}\begin{pmatrix}1 & 0\\ 0 & 1\end{pmatrix} . \tag{9.122}$$

The solution to (9.122) is

$$\begin{aligned}
&\begin{pmatrix} g_{11}(r,r';E) & g_{12}(r,r';E) \\ g_{21}(r,r';E) & g_{22}(r,r';E)\end{pmatrix} = \frac{(2\lambda)^{1+2\gamma}\,\Gamma(a)}{\Gamma(3+2\gamma)} \\
&\times\left[\Theta(r'-r)\begin{pmatrix}G_<(r)\\ F_<(r)\end{pmatrix}\begin{pmatrix}G_>(r') & F_>(r')\end{pmatrix}\right. \\
&\left.+\Theta(r-r')\begin{pmatrix}G_>(r)\\ F_>(r)\end{pmatrix}\begin{pmatrix}G_<(r') & F_<(r')\end{pmatrix}\right] ,
\end{aligned} \tag{9.123}$$

where a is defined by (9.79), Θ is the Heaviside unit function, defined by

$$\Theta(x) = \begin{cases} 1, & x > 0\,, \\ \frac{1}{2}, & x = 0\,, \\ 0, & x < 0, \end{cases} \tag{9.124}$$

and the functions $G_<$, $F_<$, $G_>$, and $F_>$ are special cases of the homogeneous solutions (9.71 – 9.80):

$$G_<(r) = r^\gamma \exp(-\lambda r)(1+E)^{1/2}\left[f_2(r) + f_1(r)\right]\,, \tag{9.125}$$

$$F_<(r) = r^\gamma \exp(-\lambda r)(1-E)^{1/2}\left[f_2(r) - f_1(r)\right]\,, \tag{9.126}$$

$$G_>(r) = r^\gamma \exp(-\lambda r)(1+E)^{1/2}\left[f_4(r) + f_3(r)\right]\,, \tag{9.127}$$

$$F_>(r) = r^\gamma \exp(-\lambda r)(1-E)^{1/2}\left[f_4(r) - f_3(r)\right]\,. \tag{9.128}$$

The functions $G_<(r)$ and $F_<(r)$ obey the boundary conditions at $r = 0$. The functions $G_>(r)$ and $F_>(r)$ obey the boundary conditions at $r = \infty$. Integral representations and expansions for the Dirac Green's function can be found in [9.22] and [9.25]. Matrix element evaluation is discussed in [9.22].

9.4 Special Functions

This section contains a brief list of formulae for the special functions which appear in the solutions discussed above. Derivations, and many additional formulae, can be found in the standard reference works listed in the bibliography. For numerically useful approximations and available software packages, see *Olver* et al. [9.12], and *Lozier* and *Olver* [9.26].

9.4.1 Confluent Hypergeometric Functions

The confluent hypergeometric differential equation is

$$\left[z\frac{\mathrm{d}^2}{\mathrm{d}z^2} + (c-z)\frac{\mathrm{d}}{\mathrm{d}z} - a\right]w(z) = 0\,. \tag{9.129}$$

Equation (9.129) has a regular singular point at $r = 0$ with indices 0 and $1-c$ and an irregular singular point at ∞. The regular solution to (9.129) is the confluent hypergeometric function, denoted by ${}_1F_1$ in generalized hypergeometric series notation. It can be defined by the series

$${}_1F_1(a;c;z) = \frac{\Gamma(c)}{\Gamma(a)}\sum_{n=0}^{\infty}\frac{\Gamma(a+n)}{\Gamma(c+n)}\frac{z^n}{n!}\,. \tag{9.130}$$

The series (9.130) for ${}_1F_1$ converges for all finite z if c is not a negative integer or zero. It reduces to a polynomial of degree n in z if $a = -n$ where n is a positive integer and c is not a negative integer or zero. The function ${}_1F_1(a;c;z)$ is denoted by the symbol $M(a,c,z)$ in *Abramowitz* and *Stegun* [9.1], in *Jahnke* and *Emde* [9.5], and in *Olver* [9.9], by ${}_1F_1(a;c;z)$ in both of *Luke's* books [9.6, 7] and in *Magnus* et al. [9.8], and by $\Phi(a,c;z)$ in the *Bateman* project [9.2, 3] and *Gradshteyn* and *Ryzhik* [9.4]. The irregular solution to (9.129) is

$$\begin{aligned} U(a,c,z) &= \frac{\Gamma(1-c)}{\Gamma(1+a-c)}\,{}_1F_1(a;c;z) \\ &\quad + \frac{\Gamma(c-1)}{\Gamma(a)}z^{1-c} \\ &\quad \times {}_1F_1(1+a-c;2-c;z)\,. \end{aligned} \tag{9.131}$$

The function $U(a,c,z)$ is multiple-valued, with principal branch $-\pi < \arg z \leq \pi$. It is denoted by the symbol $U(a,c,z)$ in *Abramowitz* and *Stegun* [9.1], in *Magnus* et al. [9.8], and in *Olver* [9.9], by $\psi(a;c;z)$ in the first of *Luke's* books [9.6], by $U(a;c;z)$ in the second of *Luke's* books [9.7], and by $\Psi(a,c;z)$ in the *Bateman* project [9.2, 3] and *Gradshteyn* and *Ryzhik* [9.4].

The Whittaker functions $M_{\kappa,\mu}$ and $W_{\kappa,\mu}$, which are related to ${}_1F_1$ and U via

$$\begin{aligned} M_{\kappa,\mu}(z) &= \exp\left(-\frac{1}{2}z\right)z^{\mu+\frac{1}{2}} \\ &\quad \times {}_1F_1\left(\mu+\frac{1}{2}-\kappa;2\mu+1;z\right)\,, \end{aligned} \tag{9.132}$$

$$\begin{aligned} W_{\kappa,\mu}(z) &= \exp\left(-\frac{1}{2}z\right)z^{\mu+\frac{1}{2}} \\ &\quad \times U\left(\mu+\frac{1}{2}-\kappa,2\mu+1,z\right)\,, \end{aligned} \tag{9.133}$$

are sometimes used instead of ${}_1F_1$ and U. For numerical evaluation and a program, see [9.27, 28].

The regular solution can be written as a linear combination of irregular solutions via

$$
\begin{aligned}
{}_1F_1\left(a;c;z\right) = {} & \frac{\Gamma\left(c\right)}{\Gamma\left(c-a\right)} \mathrm{e}^{\mathrm{i}\pi\epsilon a} U\left(a,c,z\right) + \frac{\Gamma\left(c\right)}{\Gamma\left(a\right)} \\
& \times \mathrm{e}^{z+\mathrm{i}\pi\epsilon(a-c)} U\left(c-a,c,-z\right) ,
\end{aligned}
\tag{9.134}
$$

$$
\epsilon = \begin{cases} +1\,, & \operatorname{Im} z > 0\,, \\ -1\,, & \operatorname{Im} z < 0\,. \end{cases}
\tag{9.135}
$$

${}_1F_1$ can also be obtained from U as the discontinuity across a branch cut:

$$
\begin{aligned}
& z^{c-1} \exp\left(-z\right) {}_1F_1\left(a;c;z\right) \\
& = \frac{\Gamma\left(1-a\right)\Gamma\left(c\right)}{2\pi\mathrm{i}} \Big[\left(z e^{-\mathrm{i}\pi}\right)^{c-1} U\left(c-a,c,z e^{-\mathrm{i}\pi}\right) \\
& \quad - \left(z e^{\mathrm{i}\pi}\right)^{c-1} U\left(c-a,c,z e^{\mathrm{i}\pi}\right) \Big] .
\end{aligned}
\tag{9.136}
$$

The Wronskian of the two solutions is

$$
\begin{aligned}
& {}_1F_1\left(a;c;z\right) \frac{\mathrm{d}}{\mathrm{d}z} U\left(a,c,z\right) \\
& - U\left(a,c,z\right) \frac{\mathrm{d}}{\mathrm{d}z} {}_1F_1\left(a;c;z\right) \\
& = -\Gamma\left(c\right) z^{-c} \exp\left(z\right) / \Gamma\left(a\right) .
\end{aligned}
\tag{9.137}
$$

A formula for $U\left(a,c,z\right)$ when c is the integer $n+1$ can be obtained by taking the $c \to n+1$ limit of the right-hand side of (9.131) to obtain

$$
\begin{aligned}
U\left(a,n+1,z\right) = {} & \frac{(-1)^{n+1}}{\Gamma\left(a-n\right)} \\
& \times \left[\frac{1}{n!} \ln\left(z\right) {}_1F_1\left(a;n+1;z\right) \right. \\
& \left. + \sum_{k=-n}^{\infty} \frac{\Gamma\left(a+k\right) a_k z^k}{\Gamma\left(a\right)\left(k+n\right)!} \right] ,
\end{aligned}
\tag{9.138}
$$

where

$$
a_k = \begin{cases} (-1)^{k+1}\left(-k-1\right)!\,, & -n \le k \le -1\,, \\ \left[\Psi\left(k+a\right) - \Psi\left(k+1\right) \right. \\ \quad \left. - \Psi\left(k+n+1\right)\right]/k!\,, & k \ge 0\,. \end{cases}
\tag{9.139}
$$

Here Ψ is the logarithmic derivative of the gamma function:

$$
\Psi\left(z\right) = \Gamma'\left(z\right)/\Gamma\left(z\right) .
\tag{9.140}
$$

n is a non-negative integer. When $n=0$, the sum from $k=-n$ to -1 is omitted.

The basic integral representations for ${}_1F_1$ and U are

$$
\begin{aligned}
{}_1F_1\left(a;c;z\right) = {} & \frac{\Gamma\left(c\right)}{\Gamma\left(a\right)\Gamma\left(c-a\right)} \\
& \times \int_0^1 \mathrm{e}^{zt} t^{a-1} \left(1-t\right)^{c-a-1}\,\mathrm{d}t ,
\end{aligned}
\tag{9.141}
$$

$$
U\left(a,c,z\right) = \frac{1}{\Gamma\left(a\right)} \int_0^{\infty} \mathrm{e}^{-zt} t^{a-1} \left(1+t\right)^{c-a-1}\,\mathrm{d}t .
\tag{9.142}
$$

The basic transformation formulae for ${}_1F_1$ and U are

$$
{}_1F_1\left(a;c;z\right) = \mathrm{e}^{z}\, {}_1F_1\left(c-a;c;-z\right) ,
\tag{9.143}
$$

$$
U\left(a,c,z\right) = z^{1-c} U\left(a-c+1,2-c,z\right) .
\tag{9.144}
$$

The recurrence relations among contiguous functions are

$$
\begin{aligned}
& \left(z+2a-c\right) {}_1F_1\left(a;c;z\right) = \left(a-c\right) {}_1F_1\left(a-1;c;z\right) \\
& \quad + a\, {}_1F_1\left(a+1;c;z\right) ,
\end{aligned}
\tag{9.145}
$$

$$
\begin{aligned}
& \left(z+a-1\right) {}_1F_1\left(a;c;z\right) = \left(a-c\right) {}_1F_1\left(a-1;c;z\right) \\
& \quad + \left(c-1\right) {}_1F_1\left(a;c-1;z\right) ,
\end{aligned}
\tag{9.146}
$$

$$
\begin{aligned}
& c\, {}_1F_1\left(a;c;z\right) = c\, {}_1F_1\left(a-1;c;z\right) \\
& \qquad + z\, {}_1F_1\left(a;c+1;z\right) ,
\end{aligned}
\tag{9.147}
$$

$$
\begin{aligned}
& \left(a+1-c\right) {}_1F_1\left(a;c;z\right) = a\, {}_1F_1\left(a+1;c;z\right) \\
& \quad + \left(1-c\right) {}_1F_1\left(a;c-1;z\right) ,
\end{aligned}
\tag{9.148}
$$

$$
\begin{aligned}
& c\left(z+a\right) {}_1F_1\left(a;c;z\right) = a\,c\, {}_1F_1\left(a+1;c;z\right) \\
& \quad + \left(c-a\right) z\, {}_1F_1\left(a;c+1;z\right) ,
\end{aligned}
\tag{9.149}
$$

$$
\begin{aligned}
& c\left(z+c-1\right) {}_1F_1\left(a;c;z\right) = c\left(c-1\right) \\
& \quad \times {}_1F_1\left(a;c-1;z\right) \\
& \quad + \left(c-a\right) {}_1F_1\left(a;c+1;z\right) ,
\end{aligned}
\tag{9.150}
$$

$$
\begin{aligned}
& \left(z+2a-c\right) U\left(a;c;z\right) = U\left(a-1;c;z\right) \\
& \quad + a\left(a-c+1\right) U\left(a+1;c;z\right) ,
\end{aligned}
\tag{9.151}
$$

$$
\begin{aligned}
& \left(z+a-1\right) U\left(a;c;z\right) = U\left(a-1;c;z\right) \\
& \quad + \left(c-a-1\right) U\left(a;c-1;z\right) ,
\end{aligned}
\tag{9.152}
$$

$$
\begin{aligned}
& \left(c-a\right) U\left(a;c;z\right) = -U\left(a-1;c;z\right) \\
& \qquad + zU\left(a;c+1;z\right) ,
\end{aligned}
\tag{9.153}
$$

$$(a+1-c)\,U(a;c;z) = a\,U(a+1;c;z) + U(a;c-1;z)\,, \quad (9.154)$$

$$(z+a)\,U(a;c;z) = a(a-c+1)\,U(a+1;c;z) + zU(a;c+1;z)\,, \quad (9.155)$$

$$(z+c-1)\,U(a;c;z) = (c-a-1)(c-1) \times U(a;c-1;z) + zU(a;c+1;z)\,. \quad (9.156)$$

Useful differentiation formulae include

$$\frac{\mathrm{d}}{\mathrm{d}z}\,{}_1F_1(a;c;z) = a\,c^{-1}\,{}_1F_1(a+1;c+1;z)\,, \quad (9.157)$$

$$\frac{\mathrm{d}}{\mathrm{d}z}\left[z^a\,{}_1F_1(a;c;z)\right] = a\,z^{a-1}\,{}_1F_1(a+1;c;z)\,, \quad (9.158)$$

$$\frac{\mathrm{d}}{\mathrm{d}z}\left[\mathrm{e}^{-z}z^{c-a-1}\,{}_1F_1(a+1;c;z)\right] = (c-a-1)\,\mathrm{e}^{-z}z^{c-a-2}\,{}_1F_1(a;c;z)\,, \quad (9.159)$$

$$\frac{\mathrm{d}}{\mathrm{d}z}U(a,c,z) = -a\,U(a+1,c+1,z)\,, \quad (9.160)$$

$$\frac{\mathrm{d}}{\mathrm{d}z}\left[z^aU(a,c,z)\right] = a(a-c+1)\,z^{a-1}U(a+1,c,z)\,, \quad (9.161)$$

$$\frac{\mathrm{d}}{\mathrm{d}z}\left[\mathrm{e}^{-z}z^{c-a-1}U(a+1,c,z)\right] = -\mathrm{e}^{-z}z^{c-a-2}U(a,c,z)\,. \quad (9.162)$$

An important multiplication theorem is

$${}_1F_1(a;c;z_1z_2) = \sum_{k=0}^{\infty}\frac{\Gamma(a+k)\,\Gamma(\lambda+2k)}{k!\,\Gamma(a)\,\Gamma(\lambda+k)}(-z_1)^k \times {}_2F_1(-k,\lambda+k;c;z_1) \times {}_1F_1(a+k;\lambda+2k+1;z_2)\,. \quad (9.163)$$

The fundamental asymptotic expansion for large z is

$$U(a,c,z) \sim z^{-a}\sum_{n=0}^{\infty}\frac{\Gamma(a+n)\,\Gamma(1+a-c+n)}{n!\,\Gamma(a)\,\Gamma(1+a-c)}(-z)^{-n}\,, \quad -\frac{3}{2}\pi < \arg z < \frac{3}{2}\pi\,. \quad (9.164)$$

The asymptotic expansion of ${}_1F_1$ for large z is obtained by using (9.164) and

$$\exp\left[z+\mathrm{i}\pi\epsilon(a-c)\right]U(c-a,c,-z) \sim \mathrm{e}^z z^{a-c}\sum_{n=0}^{\infty}\frac{\Gamma(c-a+n)\,\Gamma(1-a+n)}{n!\,\Gamma(c-a)\,\Gamma(1-a)}z^{-n}\,, \quad -\frac{5}{2}\pi < \arg z < \frac{5}{2}\pi\,, \quad (9.165)$$

which is a consequence of (9.164), on the right-hand side of (9.134). In the asymptotic expansion (9.165), and in the asymptotic expansion for ${}_1F_1$, the change in the factor $\exp[\mathrm{i}\pi\epsilon(a-c)]$ as $\arg z$ passes through zero is compensated by the phase change which comes from a factor $(-z)^{a-c}$ in the asymptotic expansion of $U(c-a,c,-z)$. The change in the factor $\exp(\mathrm{i}\pi\epsilon a)$ in the first term of (9.134) as $\arg z$ passes through zero is not compensated by any other phase change. However, this discontinuity occurs in a region in which this first term is negligible compared to the second term. This is an example of the Stokes phenomenon [9.29], which occurs because the single-valued function ${}_1F_1$ is being approximated by multiple-valued functions. The large z asymptotic expansion of ${}_1F_1$ is valid for $-\frac{3}{2}\pi < \arg z < \frac{3}{2}\pi$, which is the overlap of the domain of validity of the expansions (9.164) and (9.165).

Uniform asymptotic expansions for the Whittaker functions $M_{\kappa,\mu}$ and $W_{\kappa,\mu}$ introduced in (9.132), (9.133) have been constructed via Olver's method. The following result [9.9], (p. 412, Ex. 7.3), which holds for x positive, κ large and positive, and μ unrestricted, gives the flavor of these approximations:

$$W_{\kappa,\mu}(4\kappa x) = \frac{2^{4/3}\pi^{1/2}\kappa^{\kappa+(1/6)}}{\phi_n(\kappa,\mu)\exp(\kappa)}\left(\frac{x\zeta}{x-1}\right)^{1/4} \times\left\{\mathrm{Ai}\left[(4\kappa)^{2/3}\zeta\right]\sum_{s=0}^{n}\frac{A_s(\zeta)}{(4\kappa)^{2s}} + \frac{\mathrm{Ai}'\left[(4\kappa)^{2/3}\zeta\right]}{(4\kappa)^{2/3}} \times\sum_{s=0}^{n}\frac{B_s(\zeta)}{(4\kappa)^{2s}} + \epsilon_{2n+1,2}(4\kappa,\zeta)\right\}. \quad (9.166)$$

Here Ai is the Airy function, and $\epsilon_{2n+1,2}$ is an error term which tends to zero faster than the last term kept when

$\kappa \to \infty$ with n fixed. ζ is related to x by

$$\frac{4}{3}\zeta^{3/2} = \left(x^2 - x\right)^{1/2} - \ln\left[x^{1/2} + (x-1)^{1/2}\right], \quad x \geq 1\,, \tag{9.167}$$

$$\frac{4}{3}(-\zeta)^{3/2} = \cos^{-1}\left(x^{1/2}\right) - \left(x - x^2\right)^{1/2}, \quad 0 < x \leq 1\,. \tag{9.168}$$

ζ is an analytic function of x in the neighborhood of $x = 1$; conversely, x is an analytic function of ζ in the neighborhood of $\zeta = 0$. The differential form of the relations (9.167), (9.168) is

$$2\zeta^{1/2}\,\mathrm{d}\zeta = x^{-1/2}(x-1)^{1/2}\,\mathrm{d}x \quad x \geq 1\,, \tag{9.169}$$

$$2(-\zeta)^{1/2}\,\mathrm{d}\zeta = x^{-1/2}(1-x)^{1/2}\,\mathrm{d}x \quad 0 < x \leq 1\,. \tag{9.170}$$

A Taylor series expansion of ζ about $x = 1$ is most easily constructed by expanding the $x^{-1/2}$ in (9.169) or (9.170) about $x = 1$ and integrating term by term. The opening terms of such an expansion are

$$\zeta = 2^{-2/3}\left[(x-1) - \frac{1}{5}(x-1)^2 + \frac{17}{175}(x-1)^3\right] + O\left[(x-1)^4\right]. \tag{9.171}$$

The coefficient functions A_s and B_s are calculated recursively from

$$B_s(\zeta) = \frac{1}{2\zeta^{1/2}}\int_0^{\zeta}\left[\psi(\eta)A_s(\eta) - A_s''(\eta)\right]\frac{\mathrm{d}\eta}{\eta^{1/2}}, \quad x \geq 1\,, \tag{9.172}$$

$$B_s(\zeta) = \frac{1}{2(-\zeta)^{1/2}}\int_{\zeta}^{0}\left[\psi(\eta)A_s(\eta) - A_s''(\eta)\right]\frac{\mathrm{d}\eta}{(-\eta)^{1/2}}, \quad 0 < x \leq 1\,, \tag{9.173}$$

$$A_{s+1}(\zeta) = -\frac{1}{2}B_s'(\zeta) + \frac{1}{2}\int \psi(\zeta)B_s(\zeta)\,\mathrm{d}\zeta\,, \tag{9.174}$$

$$\psi(\zeta) = \frac{(4\mu^2-1)\zeta}{x(x-1)} + \frac{(3-8x)\zeta}{4x(x-1)^3} + \frac{5}{16\zeta^2}\,. \tag{9.175}$$

The functions $A_s(\zeta)$, $B_s(\zeta)$, and $\psi(\zeta)$, which appear to be singular at $\zeta = 0$, are actually analytic functions of ζ at $\zeta = 0$. The first few coefficient functions are:

$$A_0(\zeta) = 1\,, \tag{9.176}$$

$$B_0(\zeta) = \frac{1}{4}\zeta^{-1/2}\left(\frac{x}{x-1}\right)^{3/2}\left[\left(8\mu^2 - \frac{1}{2}\right) \times \left(\frac{x}{x-1}\right)^2 - \left(\frac{x}{x-1}\right) + \frac{5}{6}\right] - \frac{5}{48}\zeta^{-2}\,, \tag{9.177}$$

$$A_1(\zeta) = \frac{1}{4}\zeta^{-1}\left[B_0(\zeta) - \psi(\zeta)\right] + \zeta\left[B_0(\zeta)\right]^2\,. \tag{9.178}$$

The function ϕ_n is calculated from

$$\phi_n(\kappa,\mu) = \sum_{s=0}^{n}\frac{A_s(\infty)}{(4\kappa)^{2s}} - \sum_{s=0}^{n-1}\lim_{\zeta\to\infty}\frac{\zeta^{1/2}B_s(\zeta)}{(4\kappa)^{2s+1}}\,. \tag{9.179}$$

The first two ϕ_n are

$$\phi_0(\kappa,\mu) = 1\,, \tag{9.180}$$

$$\phi_1(\kappa,\mu) = 1 - \left(\frac{12\mu^2-1}{24\kappa}\right) + \left(\frac{12\mu^2-1}{24\kappa}\right)^2\,. \tag{9.181}$$

A bound for the error term $\epsilon_{2n+1,2}$ has been given by *Olver* ([9.9] p. 410, Eq. 7.13). The extension to the complex case can be found in *Skovgarrd* [9.30]. Skovgarrd's expansions are in powers of $(4\kappa)^{-1}$ instead of $(4\kappa)^{-2}$, because the factor $1/\phi_n(\kappa,\mu)$ has been expanded out in inverse powers of 4κ in his results; as a consequence, his coefficient functions A_s and B_s differ from Olver's. The corresponding asymptotic expansions for $U\left(\mu + \frac{1}{2} - \kappa, 2\mu+1, 4\kappa z\right)$ and for the Laguerre polynomial $L^{(2\mu)}_{\kappa-\mu-1/2}(4\kappa z)$ can be constructed with the aid of (9.133) and (9.188).

Formulae for matrix element integrals can be obtained by inserting the integral representation (9.141) and interchanging the orders of integration. An example is

$$\int_0^{\infty} z^{\mu}\,{}_1F_1(a_1;c_1;\lambda_1 z)\,{}_1F_1(a_2;c_2;\lambda_2 z)\,\mathrm{d}z = F_2(\mu+1, a_1, a_2, c_1, c_2; -\lambda_1, -\lambda_2)\,. \tag{9.182}$$

Here F_2 is one of the hypergeometric functions of two variables introduced by Appell, which can be defined either by the integral representation ([9.2], Vol. 1, p. 230, Eq. 2)

$$F_2(\alpha,\beta,\beta',\gamma,\gamma';x,y) = \frac{\Gamma(\gamma)\Gamma(\gamma')}{\Gamma(\beta)\Gamma(\beta')\Gamma(\gamma-\beta)\Gamma(\gamma'-\beta')} \times \int_0^1 du \int_0^1 dv\, u^{\beta-1} v^{\beta'-1}(1-u)^{\gamma-\beta-1} \times (1-v)^{\gamma'-\beta'-1}(1-ux-vy)^{-\alpha}, \quad (9.183)$$

or by the series expansion ([9.2], Vol. 1, p. 224, Eq. 7)

$$F_2(\alpha,\beta,\beta',\gamma,\gamma';x,y) = \sum_{m=0}^{\infty}\sum_{n=0}^{\infty} \frac{\Gamma(\alpha+m+n)}{m!n!\Gamma(\alpha)} \times \frac{\Gamma(\beta+m)\Gamma(\beta'+n)\Gamma(\gamma)\Gamma(\gamma')}{\Gamma(\beta)\Gamma(\beta')\Gamma(\gamma+m)\Gamma(\gamma'+n)} x^m y^n, \quad (9.184)$$

which converges for $|x|+|y|<1$. Numerical evaluation of F_2 is particularly convenient in the special cases where it can be expressed in terms of ordinary hypergeometric functions ${}_2F_1$ ([9.2], Vol. 1, Chap. 2), which are easy to calculate, or in terms of elementary functions. The key is the formula ([9.2], Vol. 1, p. 238, Eq. 3)

$$F_2(\alpha,\beta,\beta',\alpha,\alpha;x,y) = (1-x)^{-\beta}(1-y)^{-\beta'} \times {}_2F_1\left(\beta,\beta';\alpha;\frac{xy}{(1-x)(1-y)}\right), \quad (9.185)$$

which shows that it is necessary to get Appell functions F_2 in which the first, fourth, and fifth parameters are equal. This can be achieved by exploiting whatever freedom in the choice of a_1, a_2, c_1, and c_2 is available, and by using identities such as

$$\begin{aligned} &F_2(\alpha+2,\beta,\beta',\alpha+1,\alpha+1;x,y) \\ &= (\alpha+1)^{-1}\beta' y \\ &\quad\times F_2(\alpha+1,\beta,\beta'+1,\alpha+1,\alpha+1;x,y) \\ &\quad+\big[F_2(\alpha+1,\beta,\beta',\alpha+1,\alpha+1;x,y) \\ &\quad+2(\alpha+1)^{-2}\beta\beta' xy \\ &\quad\times F_2(\alpha+2,\beta+1,\beta'+1,\alpha+2,\alpha+2;x,y)\big] \\ &\quad+(\alpha+1)^{-1}\beta x \\ &\quad\times F_2(\alpha+1,\beta+1,\beta',\alpha+1,\alpha+1;x,y), \quad (9.186) \end{aligned}$$

to obtain Appell functions F_2 for which the reduction (9.185) can be used.

9.4.2 Laguerre Polynomials

The Laguerre polynomials $L_n^{(\alpha)}$ are the polynomial solutions of the differential equation

$$\left[z\frac{d^2}{dz^2}+(\alpha+1-z)\frac{d}{dz}+n\right]L_n^{(\alpha)}(z)=0. \quad (9.187)$$

They are a special case of the confluent hypergeometric function:

$$L_n^{(\alpha)}(z) = \frac{\Gamma(n+\alpha+1)}{n!\Gamma(\alpha+1)}\,{}_1F_1(-n;\alpha+1;z) = \frac{(-1)^n}{n!}U(-n,\alpha+1,z), \quad (9.188)$$

and are given explicitly by

$$L_n^{(\alpha)}(z) = \sum_{k=0}^{n}\frac{\Gamma(\alpha+n+1)}{k!(n-k)!\Gamma(\alpha+k+1)}(-z)^k. \quad (9.189)$$

The $L_n^{(\alpha)}$ are sometimes called generalized Laguerre polynomials, because $L_n^{(0)}$, which is often denoted by L_n, is the polynomial introduced by Laguerre. This Laguerre polynomial differs from the "associated Laguerre function" for which the symbol L_q^p (with p and q both integers) is often used in the physics literature. The relation between the two is

$$\left[L_q^p(z)\right]_{\text{physics}} = (-1)^{p+q} q! L_{q-p}^{(p)}(z). \quad (9.190)$$

The first three $L_n^{(\alpha)}$ are

$$L_0^{(\alpha)}(z) = 1, \quad (9.191)$$

$$L_1^{(\alpha)}(z) = \alpha+1-z, \quad (9.192)$$

$$L_2^{(\alpha)}(z) = \frac{1}{2}(\alpha+1)(\alpha+2)-(\alpha+2)z+\frac{1}{2}z^2. \quad (9.193)$$

Equation (9.188) shows that the irregular solution U does not supply a linearly independent second solution. A second solution which remains linearly independent and finite when α is a positive integer is

$$M_n^{(\alpha)}(z) = -\Gamma(\alpha)\big[z^{-\alpha}\,{}_1F_1(-n-\alpha;1-\alpha;z) - \cos(\pi\alpha)\Gamma(1-\alpha)L_n^{(\alpha)}(z)\big]. \quad (9.194)$$

The Wronskian of the two solutions is

$$L_n^{(\alpha)}(z)\frac{d}{dz}M_n^{(\alpha)}(z) - M_n^{(\alpha)}(z)\frac{d}{dz}L_n^{(\alpha)}(z) = \Gamma(n+\alpha+1)z^{-\alpha-1}\exp(z)/n!. \quad (9.195)$$

A formula for $M_n^{(\alpha)}(z)$ when α is a positive integer m can be obtained by taking the $\alpha \to m$ limit of the right-hand side of (9.194) to obtain

$$M_n^{(m)}(z) = \ln(z)\, L_n^{(m)}(z) + \sum_{k=-m}^{\infty} \frac{(n+m)!}{(k+m)!} b_k z^k \,, \tag{9.196}$$

where

$$b_k = \begin{cases} -(-k-1)!/(n-k)!\,, & -m \le k \le -1\,, \\ (-1)^k \left[\Psi(n+1-k) - \Psi(k+1) \right. \\ \quad \left. -\Psi(k+m+1)\right] / \left[k!\,(n-k)!\right]\,, \\ \qquad 0 \le k \le n\,, \\ (-1)^n (k-n-1)!/k!\,, & k \ge n+1\,. \end{cases} \tag{9.197}$$

The Rodrigues formula is

$$L_n^{(\alpha)}(z) = \frac{1}{n!} z^{-\alpha} \exp(z) \frac{\mathrm{d}^n}{\mathrm{d}z^n}\left[z^{n+\alpha}\exp(-z)\right] . \tag{9.198}$$

The generating function is

$$(1-w)^{-\alpha-1} \exp\left(-\frac{wz}{1-w}\right) = \sum_{n=0}^{\infty} L_n^{(\alpha)}(z)\, w^n \,. \tag{9.199}$$

The Christoffel–Darboux formula is

$$\sum_{k=0}^{n} \frac{k!}{\Gamma(k+\alpha+1)} L_k^{(\alpha)}(w)\, L_k^{(\alpha)}(z) = \frac{(n+1)!\left[L_n^{(\alpha)}(w) L_{n+1}^{(\alpha)}(z) - L_{n+1}^{(\alpha)}(w) L_n^{(\alpha)}(z)\right]}{\Gamma(n+\alpha+1)(w-z)} . \tag{9.200}$$

The orthonormality relation is

$$\int_0^\infty x^\alpha \exp(-x)\, L_n^{(\alpha)}(x)\, L_{n'}^{(\alpha)}(x)\, \mathrm{d}x = (n!)^{-1}\Gamma(n+\alpha+1)\,\delta_{n,n'} \,. \tag{9.201}$$

Other useful integration formulae include

$$\int_0^\infty x^{\alpha+1} \exp(-x)\, L_n^{(\alpha)}(x)\, L_{n'}^{(\alpha)}(x)\, \mathrm{d}x = \frac{\Gamma(n+\alpha+1)}{n!}\left[-n\delta_{n,n'+1} + (2n+\alpha+1)\,\delta_{n,n'} - (n+\alpha+1)\,\delta_{n,n'-1}\right] , \tag{9.202}$$

$$\int_0^\infty x^{\alpha-1} \exp(-x)\, L_n^{(\alpha)}(x)\, L_{n'}^{(\alpha)}(x)\, \mathrm{d}x = \frac{\Gamma(n_< + \alpha + 1)}{n_<!\,\alpha} , \quad n_< = \min(n,n') \,, \tag{9.203}$$

$$\begin{aligned} &\int_0^\infty x^{\alpha-2} \exp(-x)\, L_n^{(\alpha)}(x)\, L_{n'}^{(\alpha)}(x)\, \mathrm{d}x \\ &= \frac{\Gamma(n_< + \alpha + 1)}{n_<!\,\alpha(\alpha+2)} \left\{\alpha\left[(\alpha+1)(n+n') + 4\alpha + 5\right] \right. \\ &\quad \left. \times (1-n_<) + 2(n+1)(n'+1) + 2\alpha(n+n')\right\} , \\ &\qquad n_< = \min(n,n') \,. \end{aligned} \tag{9.204}$$

The differentiation and recursion relations are

$$z\frac{\mathrm{d}}{\mathrm{d}z} L_n^{(\alpha)}(z) = n L_n^{(\alpha)}(z) - (n+\alpha)\, L_{n-1}^{(\alpha)}(z) \,, \tag{9.205}$$

$$L_{n+1}^{(\alpha)}(z) = (n+1)^{-1}\left[(2n+\alpha+1-z)\, L_n^{(\alpha)}(z) - (n+\alpha)\, L_{n-1}^{(\alpha)}(z)\right] . \tag{9.206}$$

Additional relations can be obtained as special cases of the relations listed above for the confluent hypergeometric function by using the relation (9.188).

The coefficients c_k in a Laguerre polynomial expansion such as

$$F(x) = \sum_{k=0}^{\infty} \frac{k!}{\Gamma(k+\alpha+1)} c_k \times (\beta x)^{\alpha/2} \exp\left(-\frac{1}{2}\beta x\right) L_k^{(\alpha)}(\beta x) \tag{9.207}$$

are given by the integral

$$c_k = \beta \int_0^\infty F(x)\,(\beta x)^{\alpha/2} \exp\left(-\frac{1}{2}\beta x\right) L_k^{(\alpha)}(\beta x)\, \mathrm{d}x \,. \tag{9.208}$$

The rate of convergence of the expansion (9.207) is determined by the asymptotic behavior of the integral (9.208) for large k. A convenient way of extracting this asymptotic behavior will be described with the special case

$$\alpha = 0\,, \tag{9.209}$$

$$F(x) = (x+c)^{\nu} \exp(-\gamma x)\,, \tag{9.210}$$

as an example. Since the discussion is intended to be illustrative rather than exhaustive, only the case $\gamma < \frac{1}{2}\beta$ will be considered. The method is based on the generating function (9.199). The generating function $g(z)$ for the coefficients c_k is given in general by

$$\begin{aligned} g(z) &= \sum_{k=0}^{\infty} c_k z^n \\ &= \beta^{(2+\alpha)/2} (1-z)^{-\alpha-1}\, G\left[\frac{\beta(1+z)}{2(1-z)}\right], \end{aligned} \tag{9.211}$$

where G is the Laplace transform

$$G(\lambda) = \int_0^{\infty} x^{\alpha/2} F(x) \exp(-\lambda x)\, \mathrm{d}x\,. \tag{9.212}$$

The asymptotic analysis of the expansion coefficient c_k begins with the Cauchy integral for c_k, which is

$$c_k = \frac{1}{2\pi \mathrm{i}} \int_C z^{-k-1} g(z)\, \mathrm{d}z\,, \tag{9.213}$$

where the contour C is a small circle which runs counterclockwise around the origin. The contour C is deformed to give integrals which can be evaluated via standard methods for the asymptotic evaluation of integrals [9.31, 32]. For the example (9.209), (9.210),

$$g(z) = \beta c^{\nu+1} (1-z)^{-1}\, U\left[1, \nu+2, t(z)\right]\,, \tag{9.214}$$

$$t(z) = \frac{(\beta+2\gamma)c + (\beta-2\gamma)cz}{2(1-z)}\,. \tag{9.215}$$

This $g(z)$ has a branch point on the negative real axis at $z = -(\beta+2\gamma)/(\beta-2\gamma)$, and a combination of a branch point and an essential singularity on the positive real axis at $z = 1$. It is convenient to take the associated branch cuts to run from $-\infty$ to $-(\beta+2\gamma)/(\beta-2\gamma)$ on the negative real axis, and from $+1$ to $+\infty$ on the positive real axis. The contour C can be deformed into the sum of two contours which run clockwise around the branch cuts. Then

$$c_k = c_k^{(1)} + c_k^{(2)}\,, \tag{9.216}$$

where $c_k^{(1)}$ is the contribution from a contour which runs clockwise around the branch cut from $-\infty$ to $-(\beta+2\gamma)/(\beta-2\gamma)$ on the negative real axis, and $c_k^{(2)}$ is the contribution from a contour which runs clockwise around the branch cut from $+1$ to $+\infty$ on the positive real axis. The asymptotic behavior of $c_k^{(1)}$ can be extracted in straightforward fashion via the method of *Darboux* ([9.9], pp. 309–315, 321) ([9.32], pp. 116–122). The result is

$$\begin{aligned} c_k^{(1)} &= \left(\frac{2\beta}{\beta+2\gamma}\right)\left(\frac{4\beta}{(\beta+2\gamma)(\beta-2\gamma)}\right)^{\nu} (-1)^k \\ &\quad \times k^{\nu} \left(\frac{\beta-2\gamma}{\beta+2\gamma}\right)^k \left[1 + O\left(k^{-1}\right)\right]\,. \end{aligned} \tag{9.217}$$

The contribution $c_k^{(2)}$ requires a somewhat different strategy. The first step writes it as an integral along the real axis from $+1$ to $+\infty$ of the jump across the branch cut. Evaluating the jump yields

$$\begin{aligned} c_k^{(2)} &= \frac{\beta}{\Gamma(-\nu)} \int_0^{\infty} \mathrm{d}x x^{\nu} \left[\beta + \left(\frac{1}{2}\beta - \gamma\right) x\right]^{-\nu-1} \\ &\quad \times \exp\left[-\left(\frac{1}{2}\beta - \gamma\right) c - (k+1)\ln(1+x) \right. \\ &\quad \left. -\beta c x^{-1}\right]\,. \end{aligned} \tag{9.218}$$

The integrand in (9.218) has a saddle point at $x \approx \left[\beta c/(k+1)\right]^{1/2}$. The asymptotics for $(k+1)c$ large can be extracted via the method of steepest descent (also known as the saddle point method) ([9.9], pp. 136–138), ([9.32], pp. 85–103), [9.31], (see Chapt. 8). However, if $k+1$ is large with $(k+1)c$ small or moderate, the steepest descent approximation breaks down because the saddle point is too close to the branch point of the integrand at $x = 0$. The breakdown can be cured by using the small x approximations $[\beta + (1/2\beta - \gamma)x]^{-\nu-1} \approx \beta^{-\nu-1}$ and $\ln(1+x) \approx x - 1/2(k+1)^{-1}\beta c$ to obtain

$$\begin{aligned} c_k^{(2)} &= \frac{2\beta}{\Gamma(-\nu)} \left(\frac{c}{(k+1)\beta}\right)^{(\nu+1)/2} \exp(\beta c) \\ &\quad \times K_{\nu+1}\left[2\sqrt{(k+1)\beta c}\right]\left[1 + O\left(k^{-1/2}\right)\right]. \end{aligned} \tag{9.219}$$

The function $K_{\nu+1}$ which appears in (9.219) is a modified Bessel function of the third kind in standard notation. The large z approximation $K_{\nu+1}(z) \approx \pi^{1/2} (2z)^{-1/2} \exp(-z)$ can be used to recover the result of a steepest descent approximation to

(9.218), which is

$$c_k^{(2)} = \frac{\pi^{1/2}\beta c^{\nu+1}}{\Gamma(-\nu)}\left[\beta c\left(k+\tfrac{1}{2}\right)\right]^{-(2\nu+3)/4} \times\exp\left\{\gamma c - 2\left[\beta c\left(k+\frac{1}{2}\right)\right]^{1/2}\right\} \times\left[1+O\left(k^{-1}\right)\right] . \quad (9.220)$$

The approximation (9.219) remains valid as $c \to 0$. The approximation (9.220) does not.

The generating function method outlined above has one very nice feature: if the singularity in the complex z-plane which dominates the asymptotics is known, the analysis can be inverted to obtain a "convergence acceleration function" which builds in this singularity, and has no other singularities in the finite complex z plane. The difference between the original $F(x)$ and this convergence acceleration function will have an expansion of the form (9.207) which converges faster than the expansion of $F(x)$. Examples can be found in [9.33–36].

The contributions $c_k^{(1)}$ and $c_k^{(2)}$ exhibit two typical features. The most rapidly varying part, which is

$$[(\beta-2\gamma)/(\beta+2\gamma)]^k \quad (9.221)$$

for $c_k^{(1)}$, and

$$\exp\left\{-2\left[\beta c\left(k+\frac{1}{2}\right)\right]^{1/2}\right\} \quad (9.222)$$

for the steepest descent approximation to $c_k^{(2)}$, is determined by the location of the associated singularity. The next most important part, which is k^ν for $c_k^{(1)}$, and $(k+1/2)^{-(2\nu+3)/4}$ for the steepest descent approximation to $c_k^{(2)}$, is determined by the nature of the singularity (i.e., by the value of ν).

9.4.3 Gegenbauer Polynomials

The Gegenbauer polynomials are a special case of the Jacobi polynomial, and of the hypergeometric function:

$$C_n^\lambda(z) = \frac{\Gamma\left(\lambda+\frac{1}{2}\right)\Gamma(2\lambda+n)}{\Gamma(2\lambda)\,\Gamma\left(\lambda+n+\frac{1}{2}\right)}P_n^{\left(\lambda-\frac{1}{2},\lambda-\frac{1}{2}\right)}(z) = \frac{\Gamma(2\lambda+n)}{n!\,\Gamma(2\lambda)} \times {}_2F_1\left(-n, n+2\lambda; \lambda+\frac{1}{2}; \frac{1}{2}-\frac{1}{2}z\right) , \quad (9.223)$$

and are given explicitly by

$$C_n^\lambda(z) = \sum_{k=0}^{[n/2]}\frac{(-1)^k\,\Gamma(\lambda+n-k)}{k!\,(n-2k)!\,\Gamma(\lambda)}(2z)^{n-2k} . \quad (9.224)$$

The first three C_n^λ are

$$C_0^\lambda(z) = 1 , \quad (9.225)$$

$$C_1^\lambda(z) = 2\lambda z , \quad (9.226)$$

$$C_2^\lambda(z) = 2\lambda(\lambda+1)z^2 - \lambda . \quad (9.227)$$

Additional C_n^λ can be obtained with the aid of the recursion relation

$$C_{n+1}^\lambda(z) = (n+1)^{-1}\left[2(n+\lambda)zC_n^\lambda(z) - (n+2\lambda-1)C_{n-1}^\lambda(z)\right] , \quad (9.228)$$

which is valid for $n \geq 1$. A generating function is

$$\left(1-2zt+t^2\right)^{-\lambda} = \sum_{n=0}^{\infty} C_n^\lambda(z)\,t^n . \quad (9.229)$$

An important addition theorem, which can be used to derive addition theorems for spherical harmonics in any number of dimensions, is

$$C_n^\lambda[\cos(\gamma)] = \frac{\Gamma(2\lambda-1)}{[\Gamma(\lambda)]^2}\sum_{\ell=0}^{n}\frac{2^{2\ell}(2\lambda+2\ell-1)}{\Gamma(n+2\lambda+\ell)} \times\Gamma(n-\ell+1)[\Gamma(\lambda+\ell)]^2 \times\sin^\ell(\chi)\sin^\ell(\chi') \times C_{n-\ell}^{\lambda+\ell}[\cos(\chi)]\,C_{n-\ell}^{\lambda+\ell}[\cos(\chi')] \times C_\ell^{\lambda-\frac{1}{2}}[\cos(\gamma')] , \quad (9.230)$$

$$\cos(\gamma) = \cos(\chi)\cos(\chi') + \sin(\chi)\sin(\chi')\cos(\gamma') . \quad (9.231)$$

9.4.4 Legendre Functions

The Legendre polynomials $P_\ell(z)$ are a special case of the Gegenbauer polynomial, and of the hypergeometric function:

$$P_\ell(z) = C_n^{1/2}(z) = {}_2F_1\left(-n, n+1; 1; \frac{1}{2}-\frac{1}{2}z\right) . \quad (9.232)$$

The second (irregular) solution to the Legendre equation, which appears in (9.43) above, can be defined by the integral representation

$$Q_\ell(z) = 2^{-\ell-1} \int_{-1}^{1} \left(1-t^2\right)^\ell (z-t)^{-\ell-1}\, dt\,. \tag{9.233}$$

The first two Q_ℓ are

$$Q_0(z) = \frac{1}{2}\ln\left(\frac{z+1}{z-1}\right), \tag{9.234}$$

$$Q_1(z) = \frac{1}{2}z\ln\left(\frac{z+1}{z-1}\right) - 1\,, \tag{9.235}$$

Additional Q_ℓ can be obtained with the aid of the recursion relation

$$Q_{\ell+1}(z) = \left[(2\ell+1)\,zQ_\ell(z) - \ell Q_{\ell-1}(z)\right]/(\ell+1)\,, \tag{9.236}$$

which is valid for $\ell \geq 1$.

References

9.1 M. Abramowitz, I. A. Stegun: *Handbook of Mathematical Functions* (Dover, New York 1965)

9.2 A. Erdelyi, W. Magnus, F. Oberhettinger, F. G. Tricomi: *Higher Transcendental Functions*, Vol. 1, 2, 3 (McGraw-Hill, New York 1955) p. 1953

9.3 A. Erdelyi, W. Magnus, F. Oberhettinger, F. G. Tricomi: *Tables of Integral Transforms*, Vol. 1, 2 (McGraw-Hill, New York 1954)

9.4 I. S. Gradshteyn, I. W. Ryzhik: *Tables of Integrals, Series, and Products*, 4th edn. (Academic Press, New York 1965)

9.5 E. Jahnke, F. Emde: *Tables of Functions with Formulae and Curves*, 4th edn. (Dover, New York 1945)

9.6 Y. L. Luke: *The Special Functions and Their Approximations*, Vol. 1, 2 (Academic Press, New York 1969)

9.7 Y. L. Luke: *Mathematical Functions and Their Approximations* (Academic Press, New York 1975)

9.8 W. Magnus, F. Oberhettinger, R. P. Soni: *Formulas and Theorems for the Special Functions of Mathematical Physics*, 3rd edn. (Springer, Berlin, Heidelberg 1966)

9.9 F. W. J. Olver: *Asymptotics and Special Functions* (Academic Press, New York 1974) Reprinted A. K. Peters, Wellesley 1997

9.10 G. Szego: *Orthogonal Polynomials*, Vol. 23, 4th edn. (American Mathematical Society, Providence 1975)

9.11 F. W. J. Olver, D. W. Lozier, R. F. Boisvert, C. W. Clark: *NIST Handbook of Mathematical Functions*, in preparation

9.12 F. W. J. Olver, D. W. Lozier, R. F. Boisvert, C. W. Clark: *NIST Digital Library of Mathematical Functions*, in preparation (see http://dlmf.nist.gov/)

9.13 A. R. Edmonds: *Angular Momentum in Quantum Mechanics* (Princeton Univ. Press, Princeton 1960) Sec. 2.5

9.14 L. Pauling, E. Bright Wilson: *Introduction to Quantum Mechanics With Applications to Chemistry* (McGraw-Hill, New York 1935) Sect. 21

9.15 G. W. F. Drake, R. N. Hill: J. Phys. B **26**, 3159 (1993)

9.16 L. I. Schiff: *Quantum Mechanics*, 3rd edn. (McGraw-Hill, New York 1968) pp. 234–239

9.17 J. D. Talman: *Special Functions, A Group Theoretic Approach* (W. A. Benjamin, New York 1968) pp. 186–188

9.18 V. Fock: Z. Physik **98**, 145 (1935) The relation of the Runge–Lenz vector to the "hidden" *O*(4) symmetry is discussed in Section 3.6.3

9.19 M. Reed, B. Simon: *Methods of Modern Mathematical Physics. I. Functional Analysis* (Academic Press, New York, London 1972) pp. 263–264 Chapters VII The Spectral Theorem and VIII Unbounded Operators. See Theorem VIII.6

9.20 L. Hostler: J. Math. Phys. **5**, 591, 1235 (1964)

9.21 J. Schwinger: J. Math. Phys. **5**, 1606 (1964)

9.22 R. A. Swainson, G. W. F. Drake: J. Phys. A **24**, 79, 95, 1801 (1991)

9.23 R. N. Hill, B. D. Huxtable: J. Math. Phys. **23**, 2365 (1982)

9.24 R. Courant, D. Hilbert: *Methods of Mathematical Physics*, Vol. 1 (Interscience, New York 1953)

9.25 D. J. Hylton: J. Math. Phys. **25**, 1125 (1984)

9.26 D. W. Lozier, F. W. J. Olver: *Mathematics of Computation 1943–1993: A Half-Century of Computational Mathematics*, Proceedings of Symposia in Applied Mathematics Vol. 48, ed. by W. Gautschi (American Mathematical Society, Providence 1994)

9.27 I. J. Thompson, A. R. Barnett: J. Comput. Phys. **64**, 490 (1986)

9.28 I. J. Thompson, A. R. Barnett: Comp. Phys. Commun. **36**, 363 (1985)

9.29 R. E. Meyer: SIAM Rev. **31**, 435 (1989)

9.30 H. Skovgarrd: *Uniform Asymptotic Expansions of Confluent Hypergeometric Functions and Whittaker Functions* (Gjellerups, Copenhagen 1966)

9.31 N. Bleistein, R. A. Handelsman: *Asymptotic Expansions of Integrals* (Holt, Rinehart, & Winston, New York 1975) Reprinted Dover, New York 1986

9.32 R. Wong: *Asymptotic Approximations of Integrals* (Academic Press, San Diego 1989) Reprinted SIAM, Philadelphia 2001
9.33 R. C. Forrey, R. N. Hill: Ann. Phys. **226**, 88–157 (1993)
9.34 R. N. Hill: Phys. Rev. A **51**, 4433 (1995)
9.35 R. C. Forrey, R. N. Hill, R. D. Sharma: Phys. Rev. A **52**, 2948 (1995)
9.36 C. Krauthauser, R. N. Hill: Can. J. Phys. **80**, 181 (2002)

Atoms Part B

Part B Atoms

10 **Atomic Spectroscopy**
William C. Martin, Gaithersburg, USA
Wolfgang L. Wiese, Gaithersburg, USA

11 **High Precision Calculations for Helium**
Gordon W. F. Drake, Windsor, Canada

12 **Atomic Multipoles**
William E. Baylis, Windsor, Canada

13 **Atoms in Strong Fields**
S. Pedro Goldman, London, Canada
Mark M. Cassar, Windsor, Canada

14 **Rydberg Atoms**
Thomas F. Gallagher, Charlottesville, USA

15 **Rydberg Atoms in Strong Static Fields**
Thomas Bartsch, Atlanta, USA
Turgay Uzer, Atlanta, USA

16 **Hyperfine Structure**
Guy T. Emery, Brunswick, USA

17 **Precision Oscillator Strength and Lifetime Measurements**
Lorenzo J. Curtis, Toledo, USA

18 **Spectroscopy of Ions Using Fast Beams and Ion Traps**
Eric H. Pinnington, Edmonton, Canada
Elmar Träbert, Bochum, Germany

19 **Line Shapes and Radiation Transfer**
Alan Gallagher, Boulder, USA

20 **Thomas–Fermi and Other Density-Functional Theories**
John D. Morgan III, Newark, USA

21 **Atomic Structure: Multiconfiguration Hartree–Fock Theories**
Charlotte F. Fischer, Nashville, USA

22 **Relativistic Atomic Structure**
Ian P. Grant, Oxford, UK

23 **Many-Body Theory of Atomic Structure and Processes**
Miron Ya. Amusia, Jerusalem, Israel

24 **Photoionization of Atoms**
Anthony F. Starace, Lincoln, USA

25 **Autoionization**
Aaron Temkin, Greenbelt, USA
Anand K. Bhatia, Greenbelt, USA

26 **Green's Functions of Field Theory**
Gordon Feldman, Baltimore, USA
Thomas Fulton, Baltimore, USA

27 **Quantum Electrodynamics**
Jonathan R. Sapirstein, Notre Dame, USA

28 **Tests of Fundamental Physics**
Peter J. Mohr, Gaithersburg, USA
Barry N. Taylor, Gaithersburg, USA

29 **Parity Nonconserving Effects in Atoms**
Jonathan R. Sapirstein, Notre Dame, USA

30 **Atomic Clocks and Constraints on Variations of Fundamental Constants**
Savely G. Karshenboim, St. Petersburg, Russia
Victor Flambaum, Sydney, Australia
Ekkehard Peik, Braunschweig, Germany

10. Atomic Spectroscopy

This chapter outlines the main concepts of atomic structure, with some emphasis on terminology and notation. Atomic radiation is discussed, in particular the wavelengths, intensities, and shapes of spectral lines, and a few remarks are made regarding continuous spectra. We include updated tabulations of ionization energies for the neutral atoms and transition probabilities for persistent lines of selected neutral atoms. Some sources of additional atomic spectroscopic data are mentioned.

Experimental techniques and the details of atomic theoretical methods are not covered in this chapter; these and a number of other subjects pertinent to atomic spectroscopy are treated in one or more of at least fifteen other chapters in this book.

10.1 **Frequency, Wavenumber, Wavelength** ... 176
10.2 **Atomic States, Shells, and Configurations** 176
10.3 **Hydrogen and Hydrogen-Like Ions** ... 176
10.4 **Alkalis and Alkali-Like Spectra** ... 177
10.5 **Helium and Helium-Like Ions; *LS* Coupling** ... 177
10.6 **Hierarchy of Atomic Structure in *LS* Coupling** ... 177
10.7 **Allowed Terms or Levels for Equivalent Electrons** ... 178
10.7.1 *LS* Coupling ... 178
10.7.2 *jj* Coupling ... 178
10.8 **Notations for Different Coupling Schemes** 179
10.8.1 *LS* Coupling (Russell–Saunders Coupling) ... 179
10.8.2 *jj* Coupling of Equivalent Electrons ... 180
10.8.3 J_1j or J_1J_2 Coupling ... 180
10.8.4 J_1l or J_1L_2 Coupling (J_1K Coupling) ... 180
10.8.5 LS_1 Coupling (*LK* Coupling) ... 181
10.8.6 Coupling Schemes and Term Symbols ... 181
10.9 **Eigenvector Composition of Levels** ... 181
10.10 **Ground Levels and Ionization Energies for the Neutral Atoms** ... 182
10.11 **Zeeman Effect** ... 183
10.12 **Term Series, Quantum Defects, and Spectral-Line Series** ... 184
10.13 **Sequences** ... 185
10.13.1 Isoelectronic Sequence ... 185
10.13.2 Isoionic, Isonuclear, and Homologous Sequences ... 185
10.14 **Spectral Wavelength Ranges, Dispersion of Air** ... 185
10.15 **Wavelength (Frequency) Standards** ... 186
10.16 **Spectral Lines: Selection Rules, Intensities, Transition Probabilities, *f* Values, and Line Strengths** ... 186
10.16.1 Emission Intensities (Transition Probabilities) ... 186
10.16.2 Absorption *f* Values ... 186
10.16.3 Line Strengths ... 186
10.16.4 Relationships Between *A*, *f*, and *S* ... 187
10.16.5 Relationships Between Line and Multiplet Values ... 192
10.16.6 Relative Strengths for Lines of Multiplets in *LS* Coupling ... 193
10.17 **Atomic Lifetimes** ... 194
10.18 **Regularities and Scaling** ... 194
10.18.1 Transitions in Hydrogenic (One-Electron) Species ... 194
10.18.2 Systematic Trends and Regularities in Atoms and Ions with Two or More Electrons ... 194
10.19 **Spectral Line Shapes, Widths, and Shifts** 195
10.19.1 Doppler Broadening ... 195
10.19.2 Pressure Broadening ... 195
10.20 **Spectral Continuum Radiation** ... 196
10.20.1 Hydrogenic Species ... 196
10.20.2 Many-Electron Systems ... 196
10.21 **Sources of Spectroscopic Data** ... 197
References ... 197

10.1 Frequency, Wavenumber, Wavelength

The photon energy due to an electron transition between an upper atomic level k (of energy E_k) and a lower level i is

$$\Delta E = E_k - E_i = h\nu = hc\sigma = hc/\lambda_{\text{vac}} , \qquad (10.1)$$

where ν is the frequency, σ the wavenumber in vacuum, and λ_{vac} the wavelength in vacuum. The most accurate spectroscopic measurements are determinations of transition frequencies, the unit being the Hertz ($1\ \text{Hz} = 1\ \text{s}^{-1}$) or one of its multiples. A measurement of any one of the entities frequency, wavenumber, or wavelength (in vacuum) is an equally accurate determination of the others since the speed of light is exactly defined [10.1]. The most common wavelength units are the nanometer (nm), the Ångström ($1\ \text{Å} = 10^{-1}\ \text{nm}$) and the micrometer (μm). The SI wavenumber unit is the inverse meter, but in practice wavenumbers are usually expressed in inverse centimeters: $1\ \text{cm}^{-1} = 10^2\ \text{m}^{-1}$, equivalent to $2.997\,924\,58 \times 10^4$ MHz. Energy units and conversion factors are further discussed in Chapt. 1.

10.2 Atomic States, Shells, and Configurations

A one-electron atomic *state* is defined by the quantum numbers nlm_lm_s or $nljm_j$, with n and l representing the principal quantum number and the orbital angular momentum quantum number, respectively. The allowed values of n are the positive integers, and $l = 0, 1, \ldots, n-1$. The quantum number j represents the angular momentum obtained by coupling the orbital and spin angular momenta of an electron, i.e., $\boldsymbol{j} = \boldsymbol{l} + \boldsymbol{s}$, so that $j = l \pm 1/2$. The magnetic quantum numbers m_l, m_s, and m_j represent the projections of the corresponding angular momenta along a particular direction; thus, for example, $m_l = -l, -l+1 \cdots l$ and $m_s = \pm 1/2$.

The central field approximation for a many-electron atom leads to wave functions expressed in terms of products of such one-electron states [10.2,3]. Those electrons having the same principal quantum number n belong to the *shell* for that number. Electrons having both the same n value and l value belong to a *subshell*, all electrons in a particular subshell being *equivalent*. The notation for a *configuration* of N equivalent electrons is nl^N, the superscript usually being omitted for $N = 1$. A configuration of several subshells is written as $nl^N n'l'^M \ldots$. The numerical values of l are replaced by letters in writing a configuration, according to the code s, p, d for $l = 0, 1, 2$ and $f, g, h \ldots$ for $l = 3, 4, 5 \ldots$, the letter j being omitted.

The Pauli exclusion principle prohibits atomic states having two electrons with all four quantum numbers the same. Thus the maximum number of equivalent electrons is $2(2l+1)$. A subshell having this number of electrons is *full*, *complete*, or *closed*, and a subshell having a smaller number of electrons is *unfilled*, *incomplete*, or *open*. The $3p^6$ configuration thus represents a full subshell and $3s^2\ 3p^6\ 3d^{10}$ represents a full shell for $n = 3$.

The *parity* of a configuration is *even* or *odd* according to whether $\Sigma_i l_i$ is even or odd, the sum being taken over all electrons (in practice only those in open subshells need be considered).

10.3 Hydrogen and Hydrogen-Like Ions

The quantum numbers n, l, and j are appropriate [10.4]. A particular *level* is denoted either by nl_j or by $nl\ ^2L_J$ with $L = l$ and $J = j$. The latter notation is somewhat redundant for one-electron spectra, but is useful for consistency with more complex structures. The L values are written with the same letter code used for l values, but with capital letters. The *multiplicity* of the L *term* is equal to $2S+1 = 2s+1 = 2$. Written as a superscript, this number expresses the *doublet* character of the structure: each term for $L \geq 1$ has two levels, with $J = L \pm 1/2$, respectively.

The Coulomb interaction between the nucleus and the single electron is dominant, so that the largest energy separations are associated with levels having different n. The hyperfine splitting of the ^{1}H 1s ground level [1420.405 751 766 7(1 0)MHz] results from the interaction of the proton and electron magnetic moments and gives rise to the famous 21 cm line. The separations of the $2n-1$ excited levels having the same n are largely determined by relativistic contributions, including the spin–orbit interaction, with the result that each of the $n-1$ pairs of levels having the same j value is almost degenerate; the separation of the two levels in each pair is mainly due to relatively small Lamb shifts (see Sects. 28.2 and 27.10).

10.4 Alkalis and Alkali-Like Spectra

In the central field approximation there exists no angular-momentum coupling between a closed subshell and an electron outside the subshell, since the net spin and orbital angular momenta of the subshell are both zero. The nlj quantum numbers are, then, again appropriate for a single electron outside closed subshells. However, the electrostatic interactions of this electron with the core electrons and with the nucleus yield a strong l-dependence of the energy levels [10.5]. The differing extent of "core penetration" for ns and np electrons can in some cases, for example, give an energy difference comparable to or exceeding the difference between the np and $(n+1)$p levels. The spin–orbit fine-structure separation between the nl $(l>0)$ levels having $j=l-1/2$ and $l+1/2$, respectively, is relatively small.

10.5 Helium and Helium-Like Ions; *LS* Coupling

The energy structure of the normal $1snl$ configurations is dominated by the electron–nucleus and electron–electron Coulomb contributions [10.4]. In helium and in helium-like ions of the lighter elements, the separations of levels having the same n and having $l=$ s, p, or d are mainly determined by direct and exchange electrostatic interactions between the electrons – the spin–orbit, spin–other orbit, and other relativistic contributions are much smaller. This is the condition for LS coupling, in which:

(a) The orbital angular momenta of the electrons are coupled to give a total orbital angular momentum $\boldsymbol{L}=\Sigma_i \boldsymbol{l}_i$.

(b) The spins of the electrons are coupled to give a total spin $\boldsymbol{S}=\Sigma_i \boldsymbol{s}_i$.

The combination of a particular S value with a particular L value comprises a spectroscopic *term*, the notation for which is ^{2S+1}L. The quantum number $2S+1$ is the *multiplicity* of the term. The $\boldsymbol{S}$ and $\boldsymbol{L}$ vectors are coupled to obtain the total angular momentum, $\boldsymbol{J}=\boldsymbol{S}+\boldsymbol{L}$, for a *level* of the term; the level is denoted as $^{2S+1}L_J$.

The parity is indicated by appended degree symbols on odd parity terms.

For $1snl$ configurations, $\mathrm{L}=l$ and $\mathrm{S}=0$ or 1, i.e., the terms are *singlets* $(\mathrm{S}=0)$ or *triplets* $(\mathrm{S}=1)$. As examples of the He I structure, the ionization energy (energy required to remove one of the 1s electrons in the $1s^2$ ground configuration) is 24.5874 eV, the 1s2s $^3\mathrm{S}-{}^1\mathrm{S}$ separation is 0.7962 eV, the 1s2p $^3\mathrm{P}^\circ-{}^1\mathrm{P}^\circ$ separation is 0.2539 eV, and the 1s2p $^3\mathrm{P}^\circ_2-{}^3\mathrm{P}^\circ_0$ fine-structure spread is only 1.32×10^{-4} eV.

10.6 Hierarchy of Atomic Structure in *LS* Coupling

The centrality of LS coupling in the analysis and theoretical interpretation of atomic spectra has led to the acceptance of notations and nomenclature well adapted to discussions of particular structures and spectra [10.2]. The main elements of the nomenclature are shown in Table 10.1, most of the structural entities having already been defined in the above discussions of simple spectra. The quantum numbers in the table represent a full description for complex configurations, and the accepted names for transitions between the structural elements are also given.

As an example, the Ca I 3d4p $^3\mathrm{D}^\circ_2$ *level* belongs to the $^3\mathrm{D}^\circ$ *term* which, in turn, belongs to the 3d4p $^3(\mathrm{P}^\circ\,\mathrm{D}^\circ\,\mathrm{F}^\circ)$ triplet *triad*. The 3d4p *configuration* also has a $^1(\mathrm{P}^\circ\,\mathrm{D}^\circ\,\mathrm{F}^\circ)$ singlet *triad*. The 3d4s configuration has only *monads*, one ^{1}D and one ^{3}D. The 3d4s $^3\mathrm{D}_2$–3d4p $^3\mathrm{D}^\circ_3$ *line* belongs to the corresponding $^3\mathrm{D}$–$^3\mathrm{D}^\circ$ triplet *multiplet*, and this multiplet belongs to the great Ca I 3d4s ^{3}D–3d4p $^3(\mathrm{P}^\circ\,\mathrm{D}^\circ\,\mathrm{F}^\circ)$ *supermultiplet* of three triplet multiplets discussed by *Russell* and *Saunders* in their classic paper on the alkaline-earth spectra [10.6]. The 3d4s–3d4p *transition*

Table 10.1 Atomic structural hierarchy in LS coupling and names for the groups of all transitions between structural entities

Structural entity	Quantum numbers[a]	Group of all transitions
Configuration	$(n_i l_i)^{N_i}$	Transition array
Polyad	$(n_i l_i)^{N_i}\ \gamma S_1 L_1 nl$ $SL, SL' \cdots$	Supermultiplet
Term	$(n_i l_i)^{N_i}\ \gamma SL$	Multiplet
Level	$(n_i l_i)^{N_i}\ \gamma SLJ$	Line
State	$(n_i l_i)^{N_i}\ \gamma SLJM$	Line component

[a] The configuration may include several open subshells, as indicated by the i subscripts. The letter γ represents any additional quantum numbers, such as ancestral terms, necessary to specify a particular term

array includes both the singlet and triplet supermultiplets, as well as any (LS-forbidden) *intercombination* or *intersystem* lines arising from transitions between levels of the singlet *system* and those of the triplet *system*. The order of the two terms in the transitions as written above, with the lower-energy term on the left, is standard in atomic spectroscopy. Examples of notations for complex configurations are given in Sect. 10.8.

10.7 Allowed Terms or Levels for Equivalent Electrons

10.7.1 *LS* Coupling

The allowed LS terms of a configuration consisting of two nonequivalent groups of electrons are obtained by coupling the $\boldsymbol{S}$ and $\boldsymbol{L}$ vectors of the groups in all possible ways, and the procedure may be extended to any number of such groups. Thus the allowed terms for any configuration can be obtained from a table of the allowed terms for groups of equivalent electrons.

The configuration l^N has more than one allowed term of certain LS types if $l > 1$ and $2 < N < 4l$ (d^3–d^7, f^3–f^{11}, etc.). The recurring terms of a particular LS term type from d^N and f^N configurations are assigned sequential index numbers in the tables of *Nielson* and *Koster* [10.7]; the index numbers stand for additional numbers having group-theoretical significance that serve to differentiate the recurring terms, except for a few terms of f^5 and f^9, f^6 and f^8, and f^7. These remaining terms, which occur only in pairs, are further labeled A or B to indicate Racah's separation of the two terms.

The index numbers of Nielson and Koster are in practice the most frequently used labels for the recurring terms of f^N configurations. Use of their index numbers for the recurring terms of d^N configurations has perhaps the disadvantage of substituting an arbitrary number for a quantum number (the seniority) that itself distinguishes the recurring terms in all cases. The actual value of the seniority number is rarely needed, however, and a consistent notation for the d^N and f^N configurations is desirable. A table of the allowed LS terms of the l^N electrons for $l \leq 3$ is given in [10.8], with all recurring terms having the index numbers of Nielson and Koster as a following on-line integer. The theoretical group labels are also listed. Thus the $d^3\ {}^2D$ term having seniority 3 is designated ${}^2D\,2$, instead of ${}^2_3D$, in this scheme; and the level having $J = 3/2$ is designated ${}^2D_{3/2}2$.

10.7.2 *jj* Coupling

The allowed J values for a group of N equivalent electrons having the same j value, l_j^N, are given in Table 10.2 for $j = 1/2$, $3/2$, $5/2$, and $7/2$ (sufficient

Table 10.2 Allowed J values for l_j^N equivalent electrons (jj) coupling

l_j^N	Allowed J values
$l_{1/2}$	1/2
$l_{1/2}^2$	0
$l_{3/2}$ and $l_{3/2}^3$	3/2
$l_{3/2}^2$	0, 2
$l_{3/2}^4$	0
$l_{5/2}$ and $l_{5/2}^5$	5/2
$l_{5/2}^2$ and $l_{5/2}^4$	0, 2, 4
$l_{5/2}^3$	3/2, 5/2, 9/2
$l_{5/2}^6$	0
$l_{7/2}$ and $l_{7/2}^7$	7/2
$l_{7/2}^2$ and $l_{7/2}^6$	0, 2, 4, 6
$l_{7/2}^3$ and $l_{7/2}^5$	3/2, 5/2, 7/2, 9/2, 11/2, 15/2
$l_{7/2}^4$	0, $_2$2, $_4$2, $_2$4, $_4$4, 5, 6, 8
$l_{7/2}^8$	0

for $l \leq 3$). The $l^4_{7/2}$ group has two allowed levels for each of the J values 2 and 4. The subscripts distinguishing the two levels in each case are the seniority numbers [10.9].

The allowed levels of the configuration nl^N may be obtained by dividing the electrons into sets of two groups $nl^Q_{l+1/2}\, nl^R_{l-1/2}$, $Q+R=N$. The possible sets run from $Q=N-2l$ (or zero if $N<2l$) up to $Q=N$ or $Q=2l+2$, whichever is smaller. The (degenerate) levels for a set with both Q and R nonzero have wave functions defined by the quantum numbers $(\alpha J_1, \beta J_2)J$, with J_1 and J_2 deriving from the Q and R groups, respectively. The symbols α and β represent any additional quantum numbers required to identify levels. The J values of the allowed levels for each $(\alpha J_1, \beta J_2)$ subset are obtained by combining J_1 and J_2 in the usual way.

10.8 Notations for Different Coupling Schemes

In this section we give enough examples to make clear the meaning of the different coupling-scheme notations. Not all the configurations in the examples have been identified experimentally, and some of the examples of a particular coupling scheme given for heuristic purposes may be physically inappropriate. *Cowan* [10.3] describes the physical conditions for the different coupling schemes and gives experimental examples.

10.8.1 *LS* Coupling (Russell–Saunders Coupling)

Some of the examples given below indicate notations bearing on the order of coupling of the electrons

1. $3d^7\ {}^4F_{7/2}$
2. $3d^7\ ({}^4F)4s4p({}^3P^\circ)\ {}^6F^\circ_{9/2}$
3. $4f^7({}^8S^\circ)6s6p^2({}^4P)\ {}^{11}P^\circ_5$
4. $3p^5({}^2P^\circ)3d^2({}^1G)\ {}^2F^\circ_{7/2}$
5. $4f^{10}({}^3K\,2)6s6p({}^1P^\circ)\ {}^3L^\circ_6$
6. $4f^7({}^8S^\circ)5d\ ({}^7D^\circ)6p\ {}^8F_{13/2}$
7. $4f^7({}^8S^\circ)5d\ ({}^9D^\circ)6s\ ({}^8D^\circ)7s\ {}^9D^\circ_5$
8. $4f^7({}^8S^\circ)5d\ ({}^9D^\circ)6s6p({}^3P^\circ)\ {}^{11}F_8$
9. $4f^7({}^8S^\circ)5d^2({}^1G)\ ({}^8G^\circ)6p\ {}^7F_0$
10. $4f\,({}^2F^\circ)\ 5d^2({}^1G)6s\ ({}^2G)\ {}^1P^\circ_1$.

In the second example, the seven 3d electrons couple to give a ^{4}F term, and the 4s and 4p electrons couple to form the ^{3}P° term; the final ^{6}F° term is one of nine possible terms obtained by coupling the ^{4}F grandparent and ^{3}P° parent terms. The next three examples are similar to the second. The meaning of the index number 2 following the ^{3}K symbol in the fifth example is explained in Sect. 10.7.1.

The coupling in example 6 is appropriate if the interaction of the 5d and 4f electrons is sufficiently stronger than the 5d–6p interaction. The ^{7}D° parent term results from coupling the 5d electron to the ^{8}S° grandparent, and the 6p electron is then coupled to the ^{7}D° parent to form the final ^{8}F term. A space is inserted between the 5d electron and the ^{7}D° parent to emphasize that the latter is formed by coupling a term $({}^8S^\circ)$ listed to the left of the space. Example 7 illustrates a similar coupling order carried to a further stage; the ^{8}D° parent term results from the coupling of the 6s electron to the ^{9}D° grandparent.

Example 8 is similar to examples 2–5, but in 8 the first of the two terms that couple to form the final ^{11}F term, i. e., the ^{9}D° term, is itself formed by the coupling of the 5d electron to the ^{8}S° core term. Example 9 shows an ^{8}G° parent term formed by coupling the ^{8}S° and ^{1}G grandparent terms. A space is again used to emphasize that the following $({}^8G^\circ)$ term is formed by the coupling of terms listed before the space.

A different order of coupling is indicated in the final example, the $5d^2\ {}^1G$ term being coupled first to the external 6s electron instead of directly to the 4f core electron. The $4f\,({}^2F^\circ)$ core term is isolated by a space to denote that it is coupled (to the $5d^2({}^1G)6s\ {}^2G$ term) only after the other electrons have been coupled. The notation in this particular case (with a single 4f electron) could be simplified by writing the 4f electron after the ^{2}G term to which it is coupled. It appears more important, however, to retain the convention of giving the core portion of the configuration first.

The notations in examples 1–5 are in the form recommended by *Russell* et al. [10.10], and used in both the Atomic Energy States [10.11] and Atomic Energy Levels [10.8, 12] compilations. The spacings used in the remaining examples allow different orders of coupling of the electrons to be indicated without the use of additional parentheses, brackets, etc.

Some authors assign a short name to each (final) term, so that the configuration can be omitted in tables of classified lines, etc. The most common scheme distinguishes the low terms of a particular SL type by the prefixes $a, b, c, \ldots$, and the high terms by $z, y, x, \ldots$ [10.12].

10.8.2 *jj* Coupling of Equivalent Electrons

This scheme is used, for example, in relativistic calculations. The lower-case j indicates the angular momentum of one electron ($j = l \pm 1/2$) or of each electron in an l_j^N group. Various ways of indicating which of the two possible j values applies to such a group without writing the j-value subscript have been used by different authors; we give the j values explicitly in the examples below. We use the symbols J_i and j to represent total angular momenta.

1. $(6p^2_{1/2})_0$
2. $(6p^2_{1/2}\,6p_{3/2})^\circ_{3/2}$
3. $(6p^2_{1/2}\,6p^2_{3/2})_2$
4. $4d^3_{5/2}\,4d^2_{3/2}\,(9/2,\,2)_{11/2}$

The relatively large spin–orbit interaction of the 6p electrons produces jj-coupling structures for the $6p^2$, $6p^3$, and $6p^4$ ground configurations of neutral Pb, Bi, and Po, respectively; the notations for the ground levels of these atoms are given as the first three examples above. The configuration in the first example shows the notation for equivalent electrons having the same j value l_j^N, in this case two 6p electrons each having $j = 1/2$. A convenient notation for a particular level ($J = 0$) of such a group is also indicated. The second example extends this notation to the case of a $6p^3$ configuration divided into two groups according to the two possible j values. A similar notation is shown for the $6p^4$ level in the third example; this level might also be designated $(6p^{-2}_{3/2})_2$, the negative superscript indicating the two 6p holes. The $(J_1,\,J_2)_J$ term and level notation shown on the right in the fourth example is convenient because each of the two electron groups $4d^3_{5/2}$ and $4d^2_{3/2}$ has more than one allowed total J_i value. The assumed convention is that J_1 applies to the group on the left ($J_1 = 9/2$ for the $4d^3_{5/2}$ group) and J_2 to that on the right.

10.8.3 $J_1 j$ or $J_1 J_2$ Coupling

1. $3d^9(^2D_{5/2})4p_{3/2}\,(5/2,\,3/2)^\circ_3$
2. $4f^{11}(^2H^\circ_{9/2}2)6s6p(^3P^\circ_1)\,(9/2,\,1)_{7/2}$
3. $4f^9(^6H^\circ)5d\,(^7H^\circ_8)6s6p(^3P^\circ_0)\,(8,\,0)_8$
4. $4f^{12}(^3H_6)\,5d(^2D)6s6p(^3P^\circ)\,(^4F^\circ_{3/2})\,(6,\,3/2)^\circ_{13/2}$
5. $5f^4(^5I_4)6d_{3/2}\,(4,\,3/2)_{11/2}7s7p(^1P^\circ_1)\,(11/2,\,1)^\circ_{9/2}$
6. $5f^4_{7/2}5f^5_{5/2}\,(8,\,5/2)^\circ_{21/2}\,7p_{3/2}\,(21/2,\,3/2)_{10}$
7. $5f^3_{7/2}5f^3_{5/2}\,(9/2,\,9/2)_9 7s7p(^3P^\circ_2)\,(9,\,2)^\circ_7$

The first five examples all have core electrons in LS coupling, whereas jj coupling is indicated for the 5f core electrons in the last two examples. Since the J_1 and J_2 values in the final $(J_1,\,J_2)$ term have already been given as subscripts in the configuration, the $(J_1,\,J_2)$ term notations are redundant in all these examples. Unless separation of the configuration and final term designations is desired, as in some data tables, one may obtain a more concise notation by simply enclosing the entire configuration in brackets and adding the final J value as a subscript. Thus, the level in the first example can be designated as $[3d^9(^2D_{5/2})4p_{3/2}]^\circ_3$. If the configuration and coupling order are assumed to be known, still shorter designations may be used; for example, the fourth level above might then be given as $[(^3H_6)\,(^3P^\circ)\,(^4F^\circ_{3/2})]_{13/2}$ or $(^3H_6,\,^3P^\circ,\,^4F^\circ_{3/2})_{13/2}$. Similar economies of notation are of course possible, and often useful, in all coupling schemes.

10.8.4 $J_1 l$ or $J_1 L_2$ Coupling ($J_1 K$ Coupling)

1. $3p^5(^2P^\circ_{1/2})5g\;^2[9/2]^\circ_5$
2. $4f^2(^3H_4)5g\;^2[3]_{5/2}$
3. $4f^{13}(^2F^\circ_{7/2})5d^2(^1D)\;^1[7/2]^\circ_{7/2}$
4. $4f^{13}(^2F^\circ_{5/2})5d6s(^3D)\;^3[9/2]^\circ_{11/2}$

The final terms in the first two examples result from coupling a parent-level $\boldsymbol{J}_1$ to the orbital angular momentum of a 5g electron to obtain a resultant $\boldsymbol{K}$, the K value being enclosed in brackets. The spin of the external electron is then coupled with the $\boldsymbol{K}$ angular momentum to obtain a pair of J values, $J = K \pm 1/2$ (for $K \neq 0$). The multiplicity (2) of such pair terms is usually omitted from the term symbol, but other multiplicities occur in the more general $J_1\,L_2$ coupling (examples 3 and 4). The last two examples are straightforward extensions of $J_1\,l$ coupling, with the $\boldsymbol{L}_2$ and $\boldsymbol{S}_2$ momenta of the "external" term (^{1}D and ^{3}D in examples 3 and 4, respectively) replacing the $\boldsymbol{l}$ and $\boldsymbol{s}$ momenta of a single external electron.

10.8.5 LS_1 Coupling (LK Coupling)

1. $3s^2 3p(^2P°)4f\ G\ ^2[7/2]_3$
2. $3d^7(^4P)4s4p(^3P°)\ D°\ ^3[5/2]^°_{7/2}$

The orbital angular momentum of the core is coupled with the orbital angular momentum of the external electron(s) to give the total orbital angular momentum $\boldsymbol{L}$. The letter symbol for the final L value is listed with the configuration because this angular momentum is then coupled with the spin of the core (S_1) to obtain the resultant $\boldsymbol{K}$ angular momentum of the final term (in brackets). The multiplicity of the $[K]$ term arises from the spin of the external electron(s).

10.8.6 Coupling Schemes and Term Symbols

The coupling schemes outlined above include those now most frequently used in calculations of atomic structure [10.3]. Any term symbol gives the values of two angular momenta that may be coupled to give the total electronic angular momentum of a level (indicated by the J value). For configurations of more than one unfilled subshell, the angular momenta involved in the final coupling derive from two groups of electrons (either group may consist of only one electron). These are often an inner group of coupled electrons and an outer group of coupled electrons, respectively. In any case the quantum numbers for the two groups can be distinguished by subscripts 1 and 2, so that quantum numbers represented by capital letters without subscripts are total quantum numbers for both groups. Thus, the quantum numbers for the two vectors that couple to give the final J are related to the term symbol as follows:

Coupling scheme	Quantum numbers for vectors that couple to give J	Term symbol
LS	$L,\ S$	^{2S+1}L
$J_1 J_2$	$J_1,\ J_2$	$(J_1,\ J_2)$
$J_1 L_2(\rightarrow K)$	$K,\ S_2$	$^{2S_2+1}[K]$
$LS_1(\rightarrow K)$	$K,\ S_2$	$^{2S_2+1}[K]$

10.9 Eigenvector Composition of Levels

The wave functions of levels are often expressed as eigenvectors that are linear combinations of basis states in one of the standard coupling schemes. Thus, the wave function $\Psi(\alpha J)$ for a level labeled αJ might be expressed in terms of LS coupling basis states $\Phi(\gamma SLJ)$:

$$\Psi(\alpha J) = \sum_{\gamma SL} c(\gamma SLJ)\ \Phi(\gamma SLJ)\ . \tag{10.2}$$

The $c(\gamma SLJ)$ are expansion coefficients, and

$$\sum_{\gamma SL} \left| c(\gamma SLJ) \right|^2 = 1\ . \tag{10.3}$$

The squared expansion coefficients for the various γSL terms in the composition of the αJ level are conveniently expressed as percentages, whose sum is 100%. Thus the percentage contributed by the pure Russell–Saunders state γSLJ is equal to $100 \times |c(\gamma SLJ)|^2$. The notation for Russell–Saunders basis states has been used only for concreteness; the eigenvectors may be expressed in any coupling scheme, and the coupling schemes may be different for different configurations included in a single calculation (with configuration interaction). "Intermediate coupling" conditions for a configuration are such that calculations in both LS and jj coupling yield some eigenvectors representing significant mixtures of basis states.

The largest percentage in the composition of a level is called the *purity* of the level in that coupling scheme. The coupling scheme (or combination of coupling schemes if more than one configuration is involved) that results in the largest average purity for all the levels in a calculation is usually best for naming the levels. With regard to any particular calculation, one does well to remember that, as with other calculated quantities, the resulting eigenvectors depend on a specific theoretical model and are subject to the inaccuracies of whatever approximations the model involves.

Theoretical calculations of experimental energy level structures have yielded many eigenvectors having significantly less than 50% purity in any coupling scheme. Since many of the corresponding levels have nevertheless been assigned names by spectroscopists, some caution is advisable in the acceptance of level designations found in the literature.

10.10 Ground Levels and Ionization Energies for the Neutral Atoms

Fortunately, the ground levels of the neutral atoms have reasonably meaningful LS-coupling names, the corresponding eigenvector percentages lying in the range from $\approx 55\%$ to 100%. These names are listed in Table 10.3, except for Pa, U, and Np; the lowest few ground-configuration levels of these atoms comprise better $5f^N(L_1S_1J_1)\,6d_j7s^2\,(J_1j)$ terms than LS-coupling terms. As noted in Sect. 10.8.2, the jj-coupling names given there for the ground levels of Pb, Bi, and Po are more appropriate than the alternative LS-coupling designations in Table 10.3.

The ionization energies in the table are from recent compilations [10.13, 14]. The uncertainties are mainly in the range from less than one to several units in the last decimal place, but a few of the values may be in error by 20 or more units in the final place, i. e., the error could be greater than 0.2 eV. Although no more than four decimal places are given here, values for both the neutral and singly-ionized atoms are given to their full accuracies in [10.14].

Table 10.3 Ground levels and ionization energies for the neutral atoms

Z	Element	Ground configuration [a]	Ground level	Ionization energy (eV)	Z	Element	Ground configuration [a]	Ground level	Ionization energy (eV)
1	H	1s	$^2S_{1/2}$	13.5984	27	Co	[Ar] $3d^7$ $4s^2$	$^4F_{9/2}$	7.8810
2	He	$1s^2$	$^1S_0$	24.5874	28	Ni	[Ar] $3d^8$ $4s^2$	$^3F_4$	7.6398
3	Li	$1s^2$ 2s	$^2S_{1/2}$	5.3917	29	Cu	[Ar] $3d^{10}$ 4s	$^2S_{1/2}$	7.7264
4	Be	$1s^2$ $2s^2$	$^1S_0$	9.3227	30	Zn	[Ar] $3d^{10}$ $4s^2$	$^1S_0$	9.3942
5	B	$1s^2$ $2s^2$ 2p	$^2P^\circ_{1/2}$	8.2980	31	Ga	[Ar] $3d^{10}$ $4s^2$ 4p	$^2P^\circ_{1/2}$	5.9993
6	C	$1s^2$ $2s^2$ $2p^2$	$^3P_0$	11.2603	32	Ge	[Ar] $3d^{10}$ $4s^2$ $4p^2$	$^3P_0$	7.8994
7	N	$1s^2$ $2s^2$ $2p^3$	$^4S^\circ_{3/2}$	14.5341	33	As	[Ar] $3d^{10}$ $4s^2$ $4p^3$	$^4S^\circ_{3/2}$	9.7886
8	O	$1s^2$ $2s^2$ $2p^4$	$^3P_2$	13.6181	34	Se	[Ar] $3d^{10}$ $4s^2$ $4p^4$	$^3P_2$	9.7524
9	F	$1s^2$ $2s^2$ $2p^5$	$^2P^\circ_{3/2}$	17.4228	35	Br	[Ar] $3d^{10}$ $4s^2$ $4p^5$	$^2P^\circ_{3/2}$	11.8138
10	Ne	$1s^2$ $2s^2$ $2p^6$	$^1S_0$	21.5645	36	Kr	[Ar] $3d^{10}$ $4s^2$ $4p^6$	$^1S_0$	13.9996
11	Na	[Ne] 3s	$^2S_{1/2}$	5.1391	37	Rb	[Kr] 5s	$^2S_{1/2}$	4.1771
12	Mg	[Ne] $3s^2$	$^1S_0$	7.6462	38	Sr	[Kr] $5s^2$	$^1S_0$	5.6949
13	Al	[Ne] $3s^2$ 3p	$^2P^\circ_{1/2}$	5.9858	39	Y	[Kr] 4d $5s^2$	$^2D_{3/2}$	6.2173
14	Si	[Ne] $3s^2$ $3p^2$	$^3P_0$	8.1517	40	Zr	[Kr] $4d^2$ $5s^2$	$^3F_2$	6.6339
15	P	[Ne] $3s^2$ $3p^3$	$^4S^\circ_{3/2}$	10.4867	41	Nb	[Kr] $4d^4$ 5s	$^6D_{1/2}$	6.7589
16	S	[Ne] $3s^2$ $3p^4$	$^3P_2$	10.3600	42	Mo	[Kr] $4d^5$ 5s	$^7S_3$	7.0924
17	Cl	[Ne] $3s^2$ $3p^5$	$^2P^\circ_{3/2}$	12.9676	43	Tc	[Kr] $4d^5$ $5s^2$	$^6S_{5/2}$	7.28
18	Ar	[Ne] $3s^2$ $3p^6$	^1S	15.7596	44	Ru	[Kr] $4d^7$ 5s	$^5F_5$	7.3605
19	K	[Ar] 4s	$^2S_{1/2}$	4.3407	45	Rh	[Kr] $4d^8$ 5s	$^4F_{9/2}$	7.4589
20	Ca	[Ar] $4s^2$	$^1S_0$	6.1132	46	Pd	[Kr] $4d^{10}$	$^1S_0$	8.3369
21	Sc	[Ar] 3d $4s^2$	$^2D_{3/2}$	6.5615	47	Ag	[Kr] $4d^{10}$ 5s	$^2S_{1/2}$	7.5762
22	Ti	[Ar] $3d^2$ $4s^2$	$^3F_2$	6.8281	48	Cd	[Kr] $4d^{10}$ $5s^2$	$^1S_0$	8.9938
23	V	[Ar] $3d^3$ $4s^2$	$^4F_{3/2}$	6.7462	49	In	[Kr] $4d^{10}$ $5s^2$ 5p	$^2P^\circ_{1/2}$	5.7864
24	Cr	[Ar] $3d^5$ 4s	$^7S_3$	6.7665	50	Sn	[Kr] $4d^{10}$ $5s^2$ $5p^2$	$^3P_0$	7.3439
25	Mn	[Ar] $3d^5$ $4s^2$	$^6S_{5/2}$	7.4340	51	Sb	[Kr] $4d^{10}$ $5s^2$ $5p^3$	$^4S^\circ_{3/2}$	8.6084
26	Fe	[Ar] $3d^6$ $4s^2$	$^5D_4$	7.9024	52	Te	[Kr] $4d^{10}$ $5s^2$ $5p^4$	$^3P_2$	9.0096

Table 10.3 Ground levels and ionization energies for the neutral atoms, cont.

Z	Element	Ground configuration [a]	Ground level	Ionization energy (eV)	Z	Element	Ground configuration [a]	Ground level	Ionization energy (eV)
53	I	[Kr] $4d^{10}$ $5s^2$ $5p^5$	$^2P^\circ_{3/2}$	10.4513	79	Au	[Xe] $4f^{14}$ $5d^{10}$ 6s	$^2S_{1/2}$	9.2255
54	Xe	[Kr] $4d^{10}$ $5s^2$ $5p^6$	$^1S_0$	12.1298	80	Hg	[Xe] $4f^{14}$ $5d^{10}$ $6s^2$	$^1S_0$	10.4375
55	Cs	[Xe] 6s	$^2S_{1/2}$	3.8939	81	Tl	[Xe] $4f^{14}$ $5d^{10}$ $6s^2$ 6p	$^2P^\circ_{1/2}$	6.1082
56	Ba	[Xe] $6s^2$	$^1S_0$	5.2117	82	Pb	[Xe] $4f^{14}$ $5d^{10}$ $6s^2$ $6p^2$	$^3P_0$	7.4167
57	La	[Xe] 5d $6s^2$	$^2D_{3/2}$	5.5769	83	Bi	[Xe] $4f^{14}$ $5d^{10}$ $6s^2$ $6p^3$	$^4S^\circ_{3/2}$	7.2855
58	Ce	[Xe] 4f 5d $6s^2$	$^1G^\circ_4$	5.5387	84	Po	[Xe] $4f^{14}$ $5d^{10}$ $6s^2$ $6p^4$	$^3P_2$	8.414
59	Pr	[Xe] $4f^3$ $6s^2$	$^4I^\circ_{9/2}$	5.473	85	At	[Xe] $4f^{14}$ $5d^{10}$ $6s^2$ $6p^5$	$^2P^\circ_{3/2}$	
60	Nd	[Xe] $4f^4$ $6s^2$	$^5I_4$	5.5250	86	Rn	[Xe] $4f^{14}$ $5d^{10}$ $6s^2$ $6p^6$	$^1S_0$	10.7485
61	Pm	[Xe] $4f^5$ $6s^2$	$^6H^\circ_{5/2}$	5.582	87	Fr	[Rn] 7s	$^2S_{1/2}$	4.0727
62	Sm	[Xe] $4f^6$ $6s^2$	$^7F_0$	5.6437	88	Ra	[Rn] $7s^2$	$^1S_0$	5.2784
63	Eu	[Xe] $4f^7$ $6s^2$	$^8S^\circ_{7/2}$	5.6704	89	Ac	[Rn] 6d $7s^2$	$^2D_{3/2}$	5.17
64	Gd	[Xe] $4f^7$ 5d $6s^2$	$^9D^\circ_2$	6.1498	90	Th	[Rn] $6d^2$ $7s^2$	$^3F_2$	6.3067
65	Tb	[Xe] $4f^9$ $6s^2$	$^6H^\circ_{15/2}$	5.8638	91	Pa	[Rn] $5f^2$ ($^3H_4$) 6d $7s^2$	$(4, 3/2)_{11/2}$	5.89
66	Dy	[Xe] $4f^{10}$ $6s^2$	$^5I_8$	5.9389	92	U	[Rn] $5f^3$ ($^4I^\circ_{9/2}$) 6d $7s^2$	$(9/2, 3/2)^\circ_6$	6.1941
67	Ho	[Xe] $4f^{11}$ $6s^2$	$^4I^\circ_{15/2}$	6.0215	93	Np	[Rn] $5f^4$ ($^5I_4$) 6d $7s^2$	$(4, 3/2)_{11/2}$	6.2657
68	Er	[Xe] $4f^{12}$ $6s^2$	$^3H_6$	6.1077	94	Pu	[Rn] $5f^6$ $7s^2$	$^7F_0$	6.0260
69	Tm	[Xe] $4f^{13}$ $6s^2$	$^2F^\circ_{7/2}$	6.1843	95	Am	[Rn] $5f^7$ $7s^2$	$^8S^\circ_{7/2}$	5.9738
70	Yb	[Xe] $4f^{14}$ $6s^2$	$^1S_0$	6.2542	96	Cm	[Rn] $5f^7$ 6d $7s^2$	$^9D^\circ_2$	5.9914
71	Lu	[Xe] $4f^{14}$ 5d $6s^2$	$^2D_{3/2}$	5.4259	97	Bk	[Rn] $5f^9$ $7s^2$	$^6H^\circ_{15/2}$	6.1979
72	Hf	[Xe] $4f^{14}$ $5d^2$ $6s^2$	$^3F_2$	6.8251	98	Cf	[Rn] $5f^{10}$ $7s^2$	$^5I_8$	6.2817
73	Ta	[Xe] $4f^{14}$ $5d^3$ $6s^2$	$^4F_{3/2}$	7.5496	99	Es	[Rn] $5f^{11}$ $7s^2$	$^4I^\circ_{15/2}$	6.42
74	W	[Xe] $4f^{14}$ $5d^4$ $6s^2$	$^5D_0$	7.8640	100	Fm	[Rn] $5f^{12}$ $7s^2$	$^3H_6$	6.50
75	Re	[Xe] $4f^{14}$ $5d^5$ $6s^2$	$^6S_{5/2}$	7.8335	101	Md	[Rn] $5f^{13}$ $7s^2$	$^2F^\circ_{7/2}$	6.58
76	Os	[Xe] $4f^{14}$ $5d^6$ $6s^2$	$^5D_4$	8.28	102	No	[Rn] $5f^{14}$ $7s^2$	$^1S_0$	6.65
77	Ir	[Xe] $4f^{14}$ $5d^7$ $6s^2$	$^4F_{9/2}$	9.02	103	Lr	[Rn] $5f^{14}$ $7s^2$ 7p?	$^2P^\circ_{1/2}$?	4.9 ?
78	Pt	[Xe] $4f^{14}$ $5d^9$ 6s	$^3D_3$	8.9588	104	Rf	[Rn] $5f^{14}$ $6d^2$ $7s^2$?	$^3F_2$?	6.0 ?

[a] An element symbol in brackets represents the electrons in the ground configuration of that element

10.11 Zeeman Effect

The Zeeman effect for "weak" magnetic fields (the anomalous Zeeman effect) is of special interest because of the importance of Zeeman data in the analysis and theoretical interpretation of complex spectra. In a weak field, the J value remains a good quantum number although in general a level is split into magnetic sublevels [10.3]. The g value of such a level may be defined by the expression for the energy shift of its magnetic sublevel having magnetic quantum number M, which has one of the $2J+1$ values, $-J, -J+1, \ldots, J$:

$$\Delta E = gM\mu_B B\,. \tag{10.4}$$

The magnetic flux density is B and μ_B is the Bohr magneton ($\mu_B = e\hbar/2m_e$).

The wavenumber shift $\Delta\sigma$ corresponding to this energy shift is

$$\Delta\sigma = gM(0.466\,86B\,\mathrm{cm}^{-1})\,, \tag{10.5}$$

with B representing the numerical value of the magnetic flux density in teslas. The quantity in parentheses, the Lorentz unit, is of the order of 1 or $2\,\mathrm{cm}^{-1}$ for typical flux densities used to obtain Zeeman-effect data with classical spectroscopic methods. Accurate data can be obtained with much smaller fields, of course, by using higher-resolution techniques such as laser spectroscopy. Most of the g values now available for atomic energy levels were derived by application of the above formula (for each of the two combining levels) to measurements of optical Zeeman patterns. A single transverse-Zeeman-effect pattern (two polarizations, resolved components, and sufficiently complete) can yield the J value and the g value for each of the two levels involved.

Neglecting a number of higher-order effects, we can evaluate the g value of a level βJ belonging to a pure LS-coupling term using the formula

$$g_{\beta SLJ} = 1 + \left\{(g_e - 1) \times \frac{J(J+1) - L(L+1) + S(S+1)}{2J(J+1)}\right\}. \tag{10.6}$$

The independence of this expression from any other quantum numbers (represented by β) such as the configuration, etc., is important. The expression is derived from vector coupling formulas by assuming a g value of unity for a pure orbital angular momentum and writing the g value for a pure electron spin as g_e [10.15]. A value of 2 for g_e yields the Landé formula. If the anomalous magnetic moment of the electron is taken into account, the value of g_e is 2.002 3193. "Schwinger" g values obtained with this more accurate value for g_e are given for levels of SL terms in [10.8].

The usefulness of g_{SLJ} values is enhanced by their relation to the g values in intermediate coupling. In the notation used in (10.2) for the wave function of a level βJ in intermediate coupling, the corresponding g value is given by

$$g_{\beta J} = \sum_{\gamma SL} g_{SLJ} |c(\gamma SLJ)|^2\,, \tag{10.7}$$

where the summation is over the same set of quantum numbers as for the wave function. The $g_{\beta J}$ value is thus a weighted average of the Landé g_{SLJ} values, the weighting factors being just the corresponding component percentages.

Formulas for magnetic splitting factors in the J_1J_2 and J_1L_2 coupling schemes are given in [10.8] and [10.15]. Some higher-order effects that must be included in more accurate Zeeman-effect calculations are treated by *Bethe* and *Salpeter* [10.4] and by *Wybourne* [10.15], for example. High precision calculations for helium are given in [10.16]. See also Chapt. 13 and Chapt. 15.

10.12 Term Series, Quantum Defects, and Spectral-Line Series

The Bohr energy levels for hydrogen or for a hydrogenic (one-electron) ion are given by

$$E_n = -\frac{Z^2}{n^2}\,, \tag{10.8}$$

in units of the Rydberg for the appropriate nuclear mass. For a multielectron atom, the deviations of a series of (core)nl levels from hydrogenic E_n values may be due mainly to core penetration by the nl electron (low l-value series), or core polarization by the nl electron (high l-value series), or a combination of the two effects. In either case it can be shown that these deviations can be approximately represented by a constant quantum defect δ_l in the Rydberg formula,

$$E_{nl} = -\frac{Z_c^2}{(n-\delta_l)^2} = -\frac{Z_c^2}{(n^*)^2}\,, \tag{10.9}$$

where Z_c is the charge of the core and $n^* = n - \delta$ is the effective principal quantum number. If the core includes only closed subshells, the E_{nl} values are with respect to a value of zero for the (core)$^1S_0$ level, i. e., the $^1S_0$ level is the limit of the (core)nl series. If the quantities in (10.9) are taken as positive, they represent term values or ionization energies; the term value of the ground level of an atom or ion with respect to the ground level of the next higher ion is thus the principal ionization energy.

If the core has one or more open subshells, the series limit may be the baricenter of the entire core configuration, or any appropriate sub-structure of the core, down to and including a single level. The E_{nl} values refer to the series of corresponding (core)nl structures built on the particular limit structure. The value of the quantum defect depends to some extent on which (core)nl structures are represented by the series formula.

The quantum defect in general also has an energy dependence that must be taken into account if lower members of a series are to be accurately represented by (10.9). For an unperturbed series, this dependence can be expressed by the extended Ritz formula

$$\delta = n - n^* \\ = \delta_0 + \frac{a}{(n-\delta_0)^2} + \frac{b}{(n-\delta_0)^4} + \cdots , \quad (10.10)$$

with $\delta_0, a, b \ldots$ constants for the series (δ_0 being the limit value of the quantum defect for high series members) [10.17]. The value of a is usually positive for core-penetration series and negative for core-polarization series. A discussion of the foundations of the Ritz expansion and application to high precision calculations in helium is given in [10.18].

A spectral-line series results from either emission or absorption transitions involving a common lower level and a series of successive (core)nl upper levels differing only in their n values. The principal series of Na I, $3s\,^2S_{1/2} - np\,^2P^\circ_{1/2,3/2}$ ($n \geq 3$), is an example. The regularity of successive upper term values with increasing n (10.9,10.10) is of course observed in line series; the intervals between successive lines decrease in a regular manner towards higher wavenumbers, and the series of increasing wavenumbers converges towards the term value of the lower level as a limit.

10.13 Sequences

Several types of sequences of elements and/or ionization stages are useful because of regularities in the progressive values of parameters relating to structure and other properties along the sequences. All sequence names may refer either to the atoms and/or ions of the sequence or to their spectra.

10.13.1 Isoelectronic Sequence

A neutral atom and those ions of other elements having the same number of electrons as the atom comprise an isoelectronic sequence. (Note that a negative ion having this number of electrons is a member of the sequence.) An isoelectronic sequence is named according to its neutral member; for example, the Na I isolectronic sequence.

10.13.2 Isoionic, Isonuclear, and Homologous Sequences

An isoionic sequence comprises atoms or ions of different elements having the same charge. Such sequences have probably been most useful along the d- and f-shell rows of the periodic table. Isoionic analyses have also been carried out along p-shell rows, however, and a fine-structure regularity covering spectra of the p-shell atoms throughout the periodic table is known [10.19].

The atom and successive ions of a particular element comprise the isonuclear sequence for that element.

The elements of a particular column and subgroup in the periodic table are homologous. Thus the C, Si, Ge, Sn, and Pb atoms belong to a homologous sequence having np^2 ground configurations (Table 10.3). The singly ionized atoms of these elements comprise another example of a homologous sequence.

10.14 Spectral Wavelength Ranges, Dispersion of Air

The ranges of most interest for optical atomic spectroscopy are:

$\sim$ 2–20 μm	mid-infrared (ir)
700–2000 nm	near ir
400–700 nm	visible
200–400 nm	near ultraviolet (uv)
100–200 nm	vacuum uv or far uv
10–100 nm	extreme uv (euv or xuv)
$<$ 10 nm	soft X-ray, X-ray

The above correspondence of names to ranges should not be taken as exact; the variation as to the extent of some of the named ranges found in the literature is considerable.

Wavelengths in standard air are often tabulated for the region longer than 200 nm. These wavelengths can be related to energy-level differences by conversion to the corresponding (vacuum) wavenumbers or frequencies [10.20, 21].

10.15 Wavelength (Frequency) Standards

In 2001 the Comité International des Poids et Mesures recommended values for optical frequency standards from stabilized lasers using various absorbing atoms, atomic ions, and molecules [10.22]. These frequencies range from 29 054 057 446 579 Hz (10.318 436 884 460 μm; relative standard uncertainty 1.4×10^{-13}) for a transition in OsO_4 to 1 267 402 452 889.92 kHz (236.540 853 549 75 nm; relative standard uncertainty 3.6×10^{-13}) for a transition in the $^{115}In^+$ ion [10.22].

Extensive tables of wavenumbers for molecular transitions in the mid-ir range of 2.3 to 20.5 μm are included in a calibration atlas published in 1991 [10.23]. Wavenumbers of Ar I [10.24] and Ar II [10.25] emission lines having uncertainties as small as $0.0003\,\mathrm{cm}^{-1}$ are included in tables for these spectra covering a broad range from 222 nm to 5.865 μm. Measurements of U and Th lines (575 to 692 nm) suitable for wavenumber calibration at uncertainty levels of $0.0003\,\mathrm{cm}^{-1}$ or $0.0004\,\mathrm{cm}^{-1}$ were reported in [10.26]. Comprehensive tables of lines for U [10.27], Th [10.28], and I_2 [10.29, 30] are useful for calibration at uncertainty levels of 0.002 to $0.003\,\mathrm{cm}^{-1}$, the atlas of the Th spectrum extending down to 278 nm.

A 1974 compilation gives reference wavelengths for some 5400 lines of 38 elements covering the range 1.5 nm to 2.5 μm, with most uncertainties between 10^{-5} and 2×10^{-4} nm [10.31]. The wavelengths for some 1100 Fe lines selected from the Fe/Ne hollow-cathode spectrum have been recommended for reference standards over the range 183 nm to 4.2 μm, with wavenumber uncertainties 0.001 to $0.002\,\mathrm{cm}^{-1}$ [10.32]. Wavelengths for about 3000 vuv and uv lines (110 to 400 nm) from a Pt/Ne hollow-cathode lamp have been determined with uncertainties of 0.0002 nm or less [10.13, 33]. More recent high-accuracy measurements of ultraviolet lines of Fe I, Ge I, Kr II, and Pt I, II include some wavelengths with uncertainties smaller than 10^{-5} nm [10.34]. The wavelengths tabulated for the Kr and Pt lines in [10.34] extend from 171 to 315 nm, and the accuracies of earlier measurements of a number of spectra useful for wavelength calibration are discussed.

10.16 Spectral Lines: Selection Rules, Intensities, Transition Probabilities, *f* Values, and Line Strengths

The selection rules for discrete transitions are given in Table 10.4.

10.16.1 Emission Intensities (Transition Probabilities)

The total power ϵ radiated in a spectral line of frequency ν per unit source volume and per unit solid angle is

$$\epsilon_{\mathrm{line}} = (4\pi)^{-1}\, h\nu\, A_{ki}\, N_k\,, \tag{10.11}$$

where A_{ki} is the atomic transition probability and N_k the number per unit volume (number density) of excited atoms in the upper (initial) level k. For a homogeneous light source of length l and for the optically thin case, where all radiation escapes, the total emitted line intensity (SI quantity: radiance) is

$$I_{\mathrm{line}} = \epsilon_{\mathrm{line}}\, l = \int_0^{+\infty} I(\lambda)\mathrm{d}\lambda = (4\pi)^{-1}\,(hc/\lambda_0)\, A_{ki}\, N_k\, l\,, \tag{10.12}$$

where $I(\lambda)$ is the specific intensity at wavelength λ, and λ_0 the wavelength at line center.

10.16.2 Absorption *f* Values

In absorption, the reduced absorption

$$W(\lambda) = [I(\lambda) - I'(\lambda)]/I(\lambda) \tag{10.13}$$

is used, where $I(\lambda)$ is the incident intensity at wavelength λ, e.g., from a source providing a continuous background, and $I'(\lambda)$ the intensity after passage through the absorbing medium. The reduced line intensity from a homogeneous and optically thin absorbing medium of length l follows as

$$W_{ik} = \int_0^{+\infty} W(\lambda)\,\mathrm{d}\lambda = \frac{e^2}{4\epsilon_0\, m_e\, c^2}\,\lambda_0^2\, N_i\, f_{ik}\, l\,, \tag{10.14}$$

where f_{ik} is the atomic (absorption) oscillator strength (dimensionless).

Table 10.4 Selection rules for discrete transitions

	Electric dipole (E1) ("allowed")	Magnetic dipole (M1) ("forbidden")	Electric quadrupole (E2) ("forbidden")
Rigorous rules	1. $\Delta J = 0, \pm 1$ (except $0 \nleftrightarrow 0$)	$\Delta J = 0, \pm 1$ (except $0 \nleftrightarrow 0$)	$\Delta J = 0, \pm 1, \pm 2$ (except $0 \nleftrightarrow 0$, $1/2 \nleftrightarrow 1/2$, $0 \nleftrightarrow 1$)
	2. $\Delta M = 0, \pm 1$ (except $0 \nleftrightarrow 0$ when $\Delta J = 0$)	$\Delta M = 0, \pm 1$ (except $0 \nleftrightarrow 0$ when $\Delta J = 0$)	$\Delta M = 0, \pm 1, \pm 2$
	3. Parity change	No parity change	No parity change
With negligible configuration interaction	4. One electron jumping, with $\Delta l = \pm 1$, Δn arbitrary	No change in electron configuration; i. e., for all electrons, $\Delta l = 0$, $\Delta n = 0$	No change in electron configuration; *or* one electron jumping with $\Delta l = 0, \pm 2$, Δn arbitrary
For LS coupling only	5. $\Delta S = 0$	$\Delta S = 0$	$\Delta S = 0$
	6. $\Delta L = 0, \pm 1$ (except $0 \nleftrightarrow 0$)	$\Delta L = 0$ $\Delta J = \pm 1$	$\Delta L = 0, \pm 1, \pm 2$ (except $0 \nleftrightarrow 0$, $0 \nleftrightarrow 1$)

10.16.3 Line Strengths

A_{ki} and f_{ik} are the principal atomic quantities related to line intensities. In theoretical work, the *line strength* S is also widely used (see Chapt. 21):

$$S = S(i,k) = S(k,i) = |R_{ik}|^2 , \qquad (10.15)$$

$$R_{ik} = \langle \psi_k | P | \psi_i \rangle , \qquad (10.16)$$

where ψ_i and ψ_k are the initial- and final-state wave functions and R_{ik} is the *transition matrix element* of the appropriate multipole operator P (R_{ik} involves an integration over spatial and spin coordinates of all N electrons of the atom or ion).

10.16.4 Relationships Between A, f, and S

The relationships between A, f, and S for electric dipole (E1, or allowed) transitions in SI units (A in s^{-1}, λ in m, S in $m^2\,C^2$) are

$$A_{ki} = \frac{2\pi e^2}{m_e c \epsilon_0 \lambda^2} \frac{g_i}{g_k} f_{ik} = \frac{16\pi^3}{3h\,\epsilon_0\,\lambda^3\,g_k} S . \qquad (10.17)$$

Numerically, in customary units (A in s^{-1}, λ in Å, S in atomic units),

$$A_{ki} = \frac{6.6702 \times 10^{15}}{\lambda^2} \frac{g_i}{g_k} f_{ik} = \frac{2.0261 \times 10^{18}}{\lambda^3\,g_k} S , \qquad (10.18)$$

and for S and ΔE in atomic units,

$$f_{ik} = \frac{2}{3} (\Delta E / g_i)\, S . \qquad (10.19)$$

g_i and g_k are the statistical weights, which are obtained from the appropriate angular momentum quantum numbers. Thus for the lower (upper) level of a spectral line, $g_{i(k)} = 2J_{i(k)} + 1$ and for the lower (upper) term of a multiplet,

$$\bar{g}_{i(k)} = \sum_{J_{i(k)}} (2J_{i(k)} + 1) = (2L_{i(k)} + 1)\,(2S_{i(k)} + 1) . \qquad (10.20)$$

The A_{ki} values for strong lines of selected elements are given in Table 10.5. For comprehensive numerical

Table 10.5 Wavelengths λ, upper energy levels E_k, statistical weights g_i and g_k of lower and upper levels, and transition probabilities A_{ki} for persistent spectral lines of neutral atoms. Many tabulated lines are resonance lines (marked "g"), where the lower energy level belongs to the ground term

Spectrum	λ[a] (Å)	E_k (cm^{-1})	g_i	g_k	A_{ki} ($10^8\,s^{-1}$)	Accuracy[b]
Ag	3280.7g	30 473	2	4	1.4	B
	3382.9g	29 552	2	2	1.3	B
	5209.1	48 744	2	4	0.75	D
	5465.5	48 764	4	6	0.86	D

Table 10.5 Wavelengths λ, upper energy levels E_k, statistical weights g_i and g_k of lower and upper levels, ... , cont.

Spectrum	λ[a] (Å)	E_k (cm^{-1})	g_i	g_k	A_{ki} ($10^8 s^{-1}$)	Accuracy[b]
Al	3082.2g	32 435	2	4	0.63	C
	3092.7g	32 437	4	6	0.74	C
	3944.0g	25 348	2	2	0.493	C
	3961.5g	25 348	4	2	0.98	C
Ar	1048.2g	95 400	1	3	5.32	B
	4158.6	117 184	5	5	0.0140	B
	3453.5	32 431	10	12	1.1	C+
	4259.4	118 871	3	1	0.0398	B
	7635.1	106 238	5	5	0.245	C
	7948.2	107 132	1	3	0.186	C
	8115.3	105 463	5	7	0.331	C
As	1890.4g	52 898	4	6	2.0	D
	1937.6g	51 610	4	4	2.0	D
	2288.1	54 605	6	4	2.8	D
	2349.8	53 136	4	2	3.1	D
Au	2428.0g	41 174	2	4	1.99	B+
	2676.0g	37 359	2	2	1.64	B+
B	1825.9g	54 767	2	4	1.76	B
	1826.4g	54 767	4	6	2.11	B
	2496.8g	40 040	2	2	0.864	C
	2497.7g	40 040	4	2	1.73	C
Ba	5535.5g	18 060	1	3	1.19	B
	6498.8	24 980	7	7	0.54	D
	7059.9	23 757	7	9	0.50	D
	7280.3	22 947	5	7	0.32	D
Be	2348.6g	42 565	1	3	5.547	AA
	2650.6	59 696	9	9	4.24	AA
Bi	2228.3g	44 865	4	4	0.89	D
	2898.0	45 916	4	2	1.53	C
	2989.0	44 865	4	4	0.55	C
	3067.7g	32 588	4	2	2.07	C
Br	1488.5g	67 184	4	4	1.2	D
	1540.7g	64 907	4	4	1.4	D
	7348.5	78 512	4	6	0.12	D
C	1561.4g	64 087	5	7	1.18	A
	1657.0g	60 393	5	5	2.52	A
	1930.9	61 982	5	3	3.51	B+
	2478.6	61 982	1	3	0.340	B+
Ca	4226.7g	23 652	1	3	2.18	B+
	4302.5	38 552	5	5	1.36	C+
	5588.8	38 259	7	7	0.49	D
	6162.2	31 539	5	3	0.354	C
	6439.1	35 897	7	9	0.53	D
Cd	2288.0g	43 692	1	3	5.3	C
	3466.2	59 498	3	5	1.2	D
	3610.5	59 516	5	7	1.3	D
	5085.8	51 484	5	3	0.56	C

Table 10.5 Wavelengths λ, upper energy levels E_k, statistical weights g_i and g_k of lower and upper levels, cont.

Spectrum	λ [a] (Å)	E_k (cm^{-1})	g_i	g_k	A_{ki} ($10^8 s^{-1}$)	Accuracy [b]
Cl	1347.2g	74 226	4	4	4.19	C
	1351.7g	74 866	2	2	3.23	C
	4526.2	96 313	4	4	0.051	C
	7256.6	85 735	6	4	0.15	C
Co	3405.1	32 842	10	10	1.0	C+
	3453.5	32 431	10	12	1.1	C+
	3502.3	32 028	10	8	0.80	C+
	3569.4	35 451	8	8	1.6	C
Cr	3578.7g	27 935	7	9	1.48	B
	3593.5g	27 820	7	7	1.50	B
	3605.3g	27 729	7	5	1.62	B
	4254.3g	23 499	7	9	0.315	B
	4274.8g	23 386	7	7	0.307	B
	5208.4	26 788	5	7	0.506	B
Cs	3876.1g	25 792	2	4	0.0038	C
	4555.3g	21 946	2	4	0.0188	C
	4593.2g	21 765	2	2	0.0080	C
	8521.1g	11 732	2	4	0.3276	AA
	8943.5g	11 178	2	2	0.287	A
Cu	2178.9g	45 879	2	4	0.913	B
	3247.5g	30 784	2	4	1.39	B
	3274.0g	30 535	2	2	1.37	B
	5218.2	49 942	4	6	0.75	C
F	954.83g	104 731	4	4	5.77	C
	6856.0	116 987	6	8	0.494	C
	7398.7	115 918	6	6	0.285	C+
	7754.7	117 623	4	6	0.382	C+
Fe	3581.2	34 844	11	13	1.02	B+
	3719.9g	26 875	9	11	0.162	B+
	3734.9	33 695	11	11	0.901	B+
	3745.6g	27 395	5	7	0.115	B+
	3859.9g	25 900	9	9	0.0969	B+
	4045.8	36 686	9	9	0.862	B+
Ga	2874.2g	34 782	2	4	1.2	C
	2943.6g	34 788	4	6	1.4	C
	4033.0g	24 789	2	2	0.49	C
	4172.0g	24 789	4	2	0.92	C
Ge	2651.6g	37 702	1	3	0.85	C
	2709.6g	37 452	3	1	2.8	C
	2754.6g	37 702	5	3	1.1	C
	3039.1	40 021	5	3	2.8	C
He	537.03g	186 209	1	3	5.663	AA
	584.33g	171 135	1	3	17.99	AA
	3888.6	185 565	3	9 [c]	0.09475	AA
	4026.2	193 917	9	15 [c]	0.1160	AA
	4471.5	191 445	9	15 [c]	0.2458	AA
	5875.7	186 102	9	15 [c]	0.7070	AA

Table 10.5 Wavelengths λ, upper energy levels E_k, statistical weights g_i and g_k of lower and upper levels, cont.

Spectrum	λ[a] (Å)	E_k (cm^{-1})	g_i	g_k	A_{ki} ($10^8 s^{-1}$)	Accuracy[b]
Hg	2536.5g	39 412	1	3	0.0800	B
	3125.7	71 396	3	5	0.656	B
	4358.3	62 350	3	3	0.557	B
	5460.7	62 350	5	3	0.487	B
I	1782.8g	56 093	4	4	2.71	C
	1830.4g	54 633	4	6	0.16	D
In	3039.4g	32 892	2	6	1.3	D
	3256.1g	32 915	4	4	1.3	D
	4101.8g	24 373	2	2	0.56	C
	4511.3g	24 373	4	2	1.02	C
K	4044.1g	24 720	2	4	0.0124	C
	4047.2g	24 701	2	2	0.0124	C
	7664.9g	13 043	2	4	0.387	B+
	7699.0g	12 985	2	2	0.382	B+
Kr	5570.3	97 919	5	3	0.021	D
	5870.9	97 945	3	5	0.018	D
	7601.5	93 123	5	5	0.31	D
	8112.9	92 294	5	7	0.36	D
Li	3232.7g	30 925	2	6[c]	0.01002	A
	4602.9	36 623	6	10[c]	0.233	B
	6103.6	31 283	6	10[c]	0.6860	AA
	6707.8g	14 904	2	6[c]	0.3691	AA
Mg	2025.8g	49 347	1	3	0.84	D
	2852.1g	35 051	1	3	4.95	B
	4703.0	56 308	3	5	0.255	C
	5183.6	41 197	5	3	0.575	B
Mn	2794.8g	35 770	6	8	3.7	C
	2798.3g	35 726	6	6	3.6	C
	2801.1g	35 690	6	4	3.7	C
	4030.8g	24 820	6	8	0.17	C+
	4033.1g	24 788	6	6	0.165	C+
	4034.4g	24 779	6	4	0.158	C+
N	1199.6g	83 365	4	6	4.01	B+
	1492.6	86 221	6	4	3.13	B+
	4935.1	106 478	4	2	0.0176	B
	7468.3	96 751	6	4	0.193	B+
	8216.3	95 532	6	6	0.223	B+
Na	5890.0g	16 973	2	4	0.611	AA
	5895.9g	16 956	2	2	0.610	AA
	5682.6	34 549	2	4	0.103	C
	8183.3	29 173	2	4	0.453	C
Ne	735.90g	135 889	1	3	6.11	B
	743.72g	134 459	1	3	0.486	B
	5852.5	152 971	3	1	0.682	B
	6402.2	149 657	5	7	0.514	B
	6074.3	150 917	3	1	0.603	B

Table 10.5 Wavelengths λ, upper energy levels E_k, statistical weights g_i and g_k of lower and upper levels, cont.

Spectrum	λ [a] (Å)	E_k (cm^{-1})	g_i	g_k	A_{ki} ($10^8 s^{-1}$)	Accuracy [b]
Ni	3101.6	33 112	5	7	0.63	C+
	3134.1	33 611	3	5	0.73	C+
	3369.6g	29 669	9	7	0.18	C
	3414.8	29 481	7	9	0.55	C
	3524.5	28 569	7	5	1.0	C
	3619.4	31 031	5	7	0.66	C
O	1302.2g	76 795	5	3	3.41	A
	4368.2	99 681	3	9[c]	0.007 58	B
	5436.9	105 019	7	5	0.0180	C+
	7156.7	116 631	5	5	0.505	B
	7771.9	86 631	5	7	0.369	A
P	1775.0g	56 340	4	6	2.17	C
	1782.9g	56 090	4	4	2.14	C
	2136.2	58 174	6	4	2.83	C
	2535.6	58 174	4	4	0.95	C
Pb	2802.0g	46 329	5	7	1.6	D
	2833.1g	35 287	1	3	0.58	D
	3683.5g	34 960	3	1	1.5	D
	4057.8g	35 287	5	3	0.89	D
Rb	4201.8g	23 793	2	4	0.018	C
	4215.5g	23 715	2	2	0.015	C
	7800.3g	12 817	2	4	0.370	B
	7947.6g	12 579	2	2	0.340	B
S	1474.0g	67 843	5	7	1.6	D
	1666.7	69 238	5	5	6.3	C
	1807.3g	55 331	5	3	3.8	C
	4694.1	73 921	5	7	0.0067	D
Sc	3907.5g	25 585	4	6	1.28	C+
	3911.8g	25 725	6	8	1.37	C+
	4020.4g	24 866	4	4	1.65	C+
	4023.7g	25 014	6	6	1.44	C+
Si	2506.9g	39 955	3	5	0.466	C
	2516.1g	39 955	5	5	1.21	C
	2881.6	40 992	5	3	1.89	C
	5006.1	60 962	3	5	0.028	D
	5948.5	57 798	3	5	0.022	D
Sn	2840.0g	38 629	5	5	1.7	D
	3034.1g	34 641	3	1	2.0	D
	3175.1g	34 914	5	3	1.0	D
	3262.3	39 257	5	3	2.7	D
Sr	2428.1g	41 172	1	3	0.17	C
	4607.3g	21 698	1	3	2.01	B
Ti	3642.7g	27 615	7	9	0.774	B
	3653.5g	27 750	9	11	0.754	C+
	3998.6g	25 388	9	9	0.408	B
	4981.7	26 911	11	13	0.660	C+
	5210.4g	19 574	9	9	0.0357	C+

Table 10.5 Wavelengths λ, upper energy levels E_k, statistical weights g_i and g_k of lower and upper levels, cont.

Spectrum	λ [a] (Å)	E_k (cm^{-1})	g_i	g_k	A_{ki} ($10^8\ s^{-1}$)	Accuracy [b]
Tl	2767.9g	36 118	2	4	1.26	C
	3519.2g	36 200	4	6	1.24	C
	3775.7g	26 478	2	2	0.625	B
	5350.5g	26 478	4	2	0.705	B
U	3566.6g	28 650	11	11	0.24	B
	3571.6	38 338	17	15	0.13	C
	3584.9g	27 887	13	15	0.18	B
V	3183.4g	31 541	6	8	2.4	C+
	4111.8	26 738	10	10	1.01	B
	4379.2	25 254	10	12	1.1	C
	4384.7	25 112	8	10	1.1	C
Xe	1192.0g	83 890	1	3	6.2	C
	1295.6g	77 186	1	3	2.5	C
	1469.6g	68 046	1	3	2.8	B
	4671.2	88 470	5	7	0.010	D
	7119.6	92 445	7	9	0.066	D
Zn	2138.6g	46 745	1	3	7.09	B
	3302.6	62 772	3	5	1.2	B
	3345.0	62 777	5	7	1.7	B
	6362.3	62 459	3	5	0.474	C

[a] A "g" following the wavelength indicates that the lower level of the transition belongs to the ground term, i. e., the line is a resonance line. Wavelengths below 2000 Å are in vacuum, and those above 2000 Å are in air

[b] Accuracy estimates pertain to A_{ki} values: AA, uncertainty within 1%; A, within 3%; B, within 10%; C, within 25%; D, within 50%

[c] The A_{ki}, λ, g_i, and g_k are multiplet values; see (10.20) and Sect. 10.16.5

tables of A, f, and S, including forbidden lines, see Sect. 10.21.

Experimental and theoretical methods to determine A, f, or S values as well as atomic lifetimes are discussed in Chapts. 17, 18, and 21.

Conversion relations between S and A_{ki} for the most common forbidden transitions are given in Table 10.6. Oscillator strengths f are not used for forbidden transitions, i. e., magnetic dipole (M1), electric quadrupole (E2), etc.

[Numerical example: For the $1s2p\ ^1P_1^\circ - 1s3d\ ^1D_2$ (allowed) transition in He I at 6678.15 Å: $g_i = 3$; $g_k = 5$; $A_{ki} = 6.38 \times 10^7\ s^{-1}$; $f_{ik} = 0.711$; $S = 46.9\, a_0^2\, e^2$.]

Table 10.6 Conversion relations between S and A_{ki} for forbidden transitions

	SI units [a]	Numerically, in customary units [b]
Electric quadrupole	$A_{ki} = \dfrac{16\pi^5}{15h\,\epsilon_0\lambda^5\,g_k} S$	$A_{ki} = \dfrac{1.1199 \times 10^{18}}{g_k\,\lambda^5} S$
Magnetic dipole	$A_{ki} = \dfrac{16\pi^3\mu_0}{3h\,\lambda^3\,g_k} S$	$A_{ki} = \dfrac{2.697 \times 10^{13}}{g_k\,\lambda^3} S$

[a] A in s^{-1}, λ in m. Electric quadrupole: S in $m^4\,C^2$. Magnetic dipole: S in $J^2\,T^{-2}$

[b] A in s^{-1}, λ in Å. S in atomic units: $a_0^4\,e^2 = 2.013 \times 10^{-79}\ m^4\,C^2$ (electric quadrupole), $e^2h^2/16\pi^2m_e^2 = \mu_B^2 = 8.601 \times 10^{-47}\ J^2\,T^{-2}$ (magnetic dipole). μ_B is the Bohr magneton

10.16.5 Relationships Between Line and Multiplet Values

The relations between the total strength and f value of a multiplet (M) and the corresponding quantities for the lines of the multiplet (allowed transitions) are

$$S_{\mathrm{M}} = \sum S_{\mathrm{line}} , \tag{10.21}$$

$$f_{\mathrm{M}} = \left(\bar{\lambda}\bar{g}_i\right)^{-1} \sum_{J_k, J_i} g_i \lambda(J_i, J_k)\, f(J_i, J_k) . \tag{10.22}$$

$\bar{\lambda}$ is the weighted ("multiplet") wavelength in vacuum:

$$\bar{\lambda} = n\bar{\lambda}_{\mathrm{air}} = hc/\overline{\Delta E} , \tag{10.23}$$

where

$$\overline{\Delta E} = \overline{E_k} - \overline{E_i} = (\bar{g}_k)^{-1} \sum_{J_k} g_k E_k - (\bar{g}_i)^{-1} \sum_{J_i} g_i E_i \tag{10.24}$$

and n is the refractive index of standard air.

10.16.6 Relative Strengths for Lines of Multiplets in *LS* Coupling

Table 10.7 lists relative line strengths for frequently encountered symmetrical (P → P, D → D) and normal (S → P, P → D) multiplets in LS coupling. The strongest, or principal, lines are situated along the main diagonal of the table and are called x_1, x_2, etc. Their

Table 10.7 Relative strengths for lines of multiplets in LS coupling

Normal multiplets S–P, P–D, D–F, etc.

	J_m	J_m-1	J_m-2	J_m-3	J_m-4
J_m-1	x_1	y_1	z_1		
J_m-2		x_2	y_2	z_2	
J_m-3			x_3	y_3	z_3
J_m-4				x_4	y_4
Multiplicity	**1**	**2**	**3**	**4**	**5**
			S–P		
$S_M=$	3	6	9	12	15
x_1	3.00	4.00	5.00	6.00	7.00
y_1		2.00	3.00	4.00	5.00
z_1			1.00	2.00	3.00
			P–P		
$S_M=$	9	18	27	36	45
x_1	9.00	10.00	11.25	12.60	14.00
x_2		4.00	2.25	1.60	1.25
x_3				1.00	2.25
y_1		2.00	3.75	5.40	7.00
y_2			3.00	5.00	6.75
			P–D		
$S_M=$	15	30	45	60	75
x_1	15.00	18.00	21.00	24.00	27.00
x_2		10.00	11.25	12.60	14.00
x_3			5.00	5.00	5.25
y_1		2.00	3.75	5.40	7.00
y_2			3.75	6.40	8.75
y_3				5.60	6.75
z_1			0.25	0.60	1.00
z_2				1.00	2.25
z_3					3.00

Symmetrical multiplets P–P, D–D etc.

	J_m	J_m-1	J_m-2	J_m-3	
J_m	x_1	y_1			
J_m-1	y_1	x_2	y_2		
J_m-2		y_2	x_3	y_3	
J_m-3			y_3	x_4	
Multiplicity	**1**	**2**	**3**	**4**	**5**
			D–D		
$S_M=$	25	50	75	100	125
x_1	25.00	28.00	31.11	34.29	37.50
x_2		18.00	17.36	17.29	17.50
x_3			11.25	8.00	6.25
x_4				5.00	1.25
y_1		2.00	3.89	5.71	7.50
y_2			3.75	7.00	10.00
y_3				5.00	8.75
y_4					5.00
			D–F		
$S_M=$	35	70	105	140	175
x_1	35.00	40.00	45.00	50.00	55.00
x_2		28.00	31.11	34.29	37.50
x_3			21.00	22.40	24.00
x_4				14.00	14.00
x_5					7.00
y_1		2.00	3.89	5.71	7.50
y_2			3.89	7.31	10.50
y_3				5.60	10.00
y_4					7.00
z_1			0.11	0.29	0.50
z_2				0.40	1.00
z_3					1.00

strengths normally diminish along the diagonal. The satellite lines y_n and z_n are usually weaker and deviate more from the LS values than the stronger diagonal lines when departures from LS coupling are encountered. The total multiplet strengths S_M are also listed in Table 10.7. A discussion of their normalization as well as more extensive tables are given in [10.35].

10.17 Atomic Lifetimes

The radiative lifetime τ_k of an atomic level k is related to the sum of transition probabilities to all levels i lower in energy than k:

$$\tau_k = \left(\sum_i A_{ki}\right)^{-1} . \quad (10.25)$$

The *branching* fraction of a particular transition, say to state i', is defined as

$$A_{ki'} / \sum_i A_{ki} = A_{ki'} \tau_k . \quad (10.26)$$

If only one branch (i') exists (or if all other branches may be neglected), one obtains $A_{ki'} \tau_k = 1$, and

$$\tau_k = 1/A_{ki'} . \quad (10.27)$$

Precision lifetime measurement techniques are discussed in Chapts. 17 and 18.

10.18 Regularities and Scaling

10.18.1 Transitions in Hydrogenic (One-Electron) Species

The nonrelativistic *energy* of a hydrogenic transition (10.1, 10) is

$$(\Delta E)_Z = (E_k - E_i)_Z = R_Z hc Z^2 \left(1/n_i^2 - 1/n_k^2\right) . \quad (10.28)$$

Data for some lines of the main spectral series of hydrogen are given in Table 10.8.

Hydrogenic Z Scaling

The spectroscopic quantities for a hydrogenic ion of nuclear charge Z are related to the equivalent quantities in hydrogen ($Z = 1$) as follows (neglecting small differences in the values of R_Z):

$$(\Delta E)_Z = Z^2 (\Delta E)_H , \quad (10.29)$$
$$(\lambda_{vac})_Z = Z^{-2} (\lambda_{vac})_H , \quad (10.30)$$
$$S_Z = Z^{-2} S_H , \quad (10.31)$$
$$f_Z = f_H , \quad (10.32)$$
$$A_Z = Z^4 A_H . \quad (10.33)$$

For large values of Z, roughly $Z > 20$, relativistic corrections become noticeable and must be taken into account.

f-value Trends

f values for high series members (large n' values) of hydrogenic ions decrease according to

$$f(n, l \to n', l \pm 1) \propto (n')^{-3} . \quad (10.34)$$

10.18.2 Systematic Trends and Regularities in Atoms and Ions with Two or More Electrons

Atomic quantities for a given state or transition in an *isoelectronic sequence* may be expressed as power series expansions in Z^{-1}:

$$Z^{-2} E = E_0 + E_1 Z^{-1} + E_2 Z^{-2} + \dots , \quad (10.35)$$
$$Z^2 S = S_0 + S_1 Z^{-1} + S_2 Z^{-2} + \dots , \quad (10.36)$$
$$f = f_0 + f_1 Z^{-1} + f_2 Z^{-2} + \dots , \quad (10.37)$$

where E_0, f_0, and S_0 are hydrogenic quantities. For transitions in which n does not change ($n_i = n_k$), $f_0 = 0$, since states i and k are degenerate.

For equivalent transitions of *homologous atoms*, f values vary gradually. Transitions to be compared in the case of the "alkalis" are [10.36]

$$(nl - n'l')_{Li} \to [(n+1)l - (n'+1)l']_{Na} \to [(n+2)l - (n'+2)l']_{Cu} \to \dots .$$

Complex atomic structures, as well as cases involving strong cancellation in the integrand of the transition integral, generally do not adhere to this regular behavior.

Table 10.8 Some transitions of the main spectral series of hydrogen

Transition	Customary name[a]	λ[b] (Å)	g_i[c]	g_k	A_{ki} ($10^8\,s^{-1}$)	Transition	Customary name[a]	λ[b] (Å)	g_i[c]	g_k	A_{ki} ($10^8\,s^{-1}$)
1–2	(L_α)	1215.67	2	8	4.699	2–6	(H_δ)	4101.73	8	72	9.732 (−3)
1–3	(L_β)	1025.73	2	18	5.575 (−1)[d]	2–7	(H_ϵ)	3970.07	8	98	4.389 (−3)
1–4	(L_γ)	972.537	2	32	1.278 (−1)	3–4	(P_α)	18 751.0	18	32	8.986 (−2)
1–5	(L_δ)	949.743	2	50	4.125 (−2)	3–5	(P_β)	12 818.1	18	50	2.201 (−2)
1–6	(L_ϵ)	937.803	2	72	1.644 (−2)	3–6	(P_γ)	10 938.1	18	72	7.783 (−3)
2–3	(H_α)	6562.80	8	18	4.410 (−1)	3–7	(P_δ)	10 049.4	18	98	3.358 (−3)
2–4	(H_β)	4861.32	8	32	8.419 (−2)	3–8	(P_ϵ)	9545.97	18	128	1.651 (−3)
2–5	(H_γ)	4340.46	8	50	2.530 (−2)						

[a] L_α is often called Lyman α, H_α = Balmer α, P_α = Paschen α, etc.
[b] Wavelengths below 2000 Å are in vacuum; values above 2000 Å are in air
[c] For transitions in hydrogen, $g_{i(k)} = 2(n_{i(k)})^2$, where $n_{i(k)}$, is the principal quantum number of the lower (upper) electron shell
[d] The number in parentheses indicates the power of 10 by which the value has to be multiplied

10.19 Spectral Line Shapes, Widths, and Shifts

Observed spectral lines are always broadened, partly due to the finite resolution of the spectrometer and partly due to intrinsic physical causes. The principal physical causes of spectral line broadening are Doppler and pressure broadening. The theoretical foundations of line broadening are discussed in Chapts. 19 and 59.

10.19.1 Doppler Broadening

Doppler broadening is due to the thermal motion of the emitting atoms or ions. For a Maxwellian velocity distribution, the line shape is *Gaussian*; the full width at half maximum intensity (FWHM) is, in Å,

$$\Delta\lambda^{D}_{1/2} = \left(7.16\times10^{-7}\right)\lambda\,(T/M)^{1/2}\,. \qquad (10.38)$$

T is the temperature of the emitters in K, and M the atomic weight in atomic mass units (amu).

10.19.2 Pressure Broadening

Pressure broadening is due to collisions of the emitters with neighboring particles (see also Chapts. 19 and 59). Shapes are often approximately Lorentzian, i. e., $I(\lambda) \propto \{1+[(\lambda-\lambda_0)/\Delta\lambda_{1/2}]^2\}^{-1}$. In the following formulas, all FWHMs and wavelengths are expressed in Å, particle densities N in cm^{-3}, temperatures T in K, and energies E or I in cm^{-1}.

Resonance broadening (self-broadening) occurs only between identical species and is confined to lines with the upper or lower level having an electric dipole transition (resonance line) to the ground state. The FWHM may be estimated as

$$\Delta\lambda^{R}_{1/2} \simeq 8.6\times10^{-30}(g_i/g_k)^{1/2}\,\lambda^2\lambda_r f_r N_i\,, \qquad (10.39)$$

where λ is the wavelength of the observed line; f_r and λ_r are the oscillator strength and wavelength of the resonance line; g_k and g_i are the statistical weights of its upper and lower levels. N_i is the ground state number density.

For the $1s2p\,^1P^\circ_1 - 1s3d\,^1D_2$ transition in He I [$\lambda = 6678.15$ Å; $\lambda_r\,(1s^2\,^1S_0 - 1s2p\,^1P^\circ_1) = 584.334$ Å; $g_i = 1$; $g_k = 3$; $f_r = 0.2762$] at $N_i = 1\times10^{18}\,cm^{-3}$: $\Delta\lambda^{R}_{1/2} = 0.036$ Å.

Van der Waals broadening arises from the dipole interaction of an excited atom with the induced dipole of a ground state atom. (In the case of foreign gas broadening, both the perturber and the radiator may be in their respective ground states.) An approximate formula for the FWHM, strictly applicable to hydrogen and similar atomic structures only, is

$$\Delta\lambda^{W}_{1/2} \simeq 3.0\times10^{16}\lambda^2\,C_6^{2/5}(T/\mu)^{3/10}N\,, \qquad (10.40)$$

where μ is the atom-perturber reduced mass in units of u, N the perturber density, and C_6 the interaction constant. C_6 may be roughly estimated as follows: $C_6 = C_k - C_i$, with $C_{i(k)} = (9.8\times10^{10})\alpha_d R^2_{i(k)}$ (α_d in cm^3, R^2 in a_0^2). Mean atomic polarizability $\alpha_d \approx (6.7\times10^{-25})(3I_H/4E^*)^2$ cm^3, where I_H is the ionization energy of hydrogen and E^* the energy of the first excited level of the perturber atom. $R^2_{i(k)} \approx 2.5\,[I_H/(I - E_{i(k)})]^2$, where I is the ionization energy of the radiator. Van der Waals broadened lines are red shifted by about one-third the size of the FWHM.

For the $1s2p\,^1P^\circ_1 - 1s3d\,^1D_2$ transition in He I, and with He as perturber: $\lambda = 6678.15$ Å; $I = 198\,311\ cm^{-1}$; $E^* = E_i = 171\,135\ cm^{-1}$; $E_k = 186\,105\ cm^{-1}$; $\mu = 2$. At $T = 15\,000$ K and $N = 1\times10^{18}\ cm^{-3}$: $\Delta\lambda^W_{1/2} = 0.044$ Å.

Stark broadening due to charged perturbers, i. e., ions and electrons, usually dominates resonance and van der Waals broadening in discharges and plasmas. The FWHM for hydrogen lines is

$$\Delta\lambda^{S,H}_{1/2} = \left(2.50\times10^{-9}\right)\alpha_{1/2}\,N_e^{2/3}\,, \qquad (10.41)$$

Table 10.9 Values of Stark-broadening parameter $\alpha_{1/2}$ of the H_β line of hydrogen (4861 Å) for various temperatures and electron densities

T(K)	N_e (cm^{-3})			
	10^{15}	10^{16}	10^{17}	10^{18}
5000	0.0787	0.0808	0.0765	...
10 000	0.0803	0.0840	0.0851	0.0781
20 000	0.0815	0.0860	0.0902	0.0896
30 000	0.0814	0.0860	0.0919	0.0946

where N_e is the electron density. The half-width parameter $\alpha_{1/2}$ for the H_β line at 4861 Å, widely used for plasma diagnostics, is tabulated in Table 10.9 for some typical temperatures and electron densities [10.35]. This reference also contains $\alpha_{1/2}$ parameters for other hydrogen lines, as well as Stark width and shift data for numerous lines of other elements, i. e., neutral atoms and singly charged ions (in the latter, Stark widths and shifts depend linearly on N_e). Other tabulations of complete hydrogen Stark profiles exist.

10.20 Spectral Continuum Radiation

10.20.1 Hydrogenic Species

Precise quantum-mechanical calculations exist only for hydrogenic species. The total power ϵ_{cont} radiated (per unit source volume and per unit solid angle, and expressed in SI units) in the wavelength interval $\Delta\lambda$ is the sum of radiation due to the recombination of a free electron with a bare ion (free–bound transitions) and bremsstrahlung (free–free transitions):

$$\begin{aligned}\epsilon_{cont} = {} & \frac{e^6}{2\pi\epsilon_0^3(6\pi m_e)^{3/2}} N_e N_Z Z^2 \\ & \times \frac{1}{(kT)^{1/2}} \exp\left(-\frac{hc}{\lambda kT}\right)\frac{\Delta\lambda}{\lambda^2} \\ & \times \left\{ \frac{2Z^2 I_H}{kT} \sum_{n\geq(Z^2 I_H \lambda/hc)^{1/2}}^{n'} \frac{\gamma_{fb}}{n^3}\exp\left(\frac{Z^2 I_H}{n^2 kT}\right) \right. \\ & \left. + \bar{\gamma}_{fb}\left[\exp\left(\frac{Z^2 I_H}{(n'+1)^2 kT}\right) - 1\right] + \gamma_{ff} \right\} \end{aligned} \qquad (10.42)$$

where N_e is the electron density, N_Z the number density of hydrogenic (bare) ions of nuclear charge Z, I_H the ionization energy of hydrogen, n' the principal quantum number of the lowest level for which adjacent levels are so close that they approach a continuum and summation over n may be replaced by an integral. (The choice of n' is rather arbitrary; n' as low as 6 is found in the literature.) γ_{fb} and γ_{ff} are the Gaunt factors, which are generally close to unity. (For the higher free-bound continua, starting with $n'+1$, an average Gaunt factor $\bar{\gamma}_{fb}$ is used.) For neutral hydrogen, the recombination continuum forming H^- becomes important, too [10.37].

In the equation above, the value of the constant factor is $6.065\times10^{-55}\ W\,m^4\,J^{1/2}\,sr^{-1}$. [Numerical example: For atomic hydrogen ($Z = 1$), the quantity ϵ_{cont} has the value $2.9\ W\,m^{-3}\,sr^{-1}$ under the following conditions: $\lambda = 3\times10^{-7}$ m; $\Delta\lambda = 1\times10^{-10}$ m; $N_e (= N_{Z=1}) = 1\times10^{21}\ m^{-3}$; $T = 12\,000$ K. The lower limit of the summation index n is 2; the upper limit n' has been taken to be 10. All Gaunt factors γ_{fb}, $\bar{\gamma}_{fb}$, and γ_{ff} have been assumed to be unity.]

10.20.2 Many-Electron Systems

For many-electron systems, only approximate theoretical treatments exist, based on the quantum-defect method (for results of calculations for noble gases, see, e.g., [10.38]). Experimental work is centered on the noble gases [10.39]. Modifications of the continuum by autoionization processes must also be considered.

Near the ionization limit, the f values for bound-bound transitions of a spectral series ($n' \to \infty$) make a smooth connection to the differential oscillator strength distribution $\mathrm{d}f/\mathrm{d}\epsilon$ in the continuum [10.40].

10.21 Sources of Spectroscopic Data

Access to most of the atomic spectroscopic databases currently online is given by links at the Plasma Gate server [10.41]. Extensive data from NIST compilations of atomic wavelengths, energy levels, and transition probabilities are available from the *Atomic Spectra Database* (ASD) at the NIST site [10.13]. Section 10.15 includes additional references for wavelength tables.

References

10.1 B. N. Taylor (Ed.): *The International System of Units (SI)*, NIST Spec. Publ. 330 (U.S. Government Printing Office, Washington 1991) p. 3

10.2 E. U. Condon, G. H. Shortley: *The Theory of Atomic Spectra* (Cambridge Univ. Press, Cambridge 1935)

10.3 R. D. Cowan: *The Theory of Atomic Structure and Spectra* (Univ. of California Press, Berkeley 1981)

10.4 H. A. Bethe, E. E. Salpeter: *Quantum Mechanics of One- and Two-Electron Atoms* (Plenum, New York 1977)

10.5 B. Edlén: *Encyclopedia of Physics*, Vol. 27, ed. by S. Flügge (Springer, Berlin, Heidelberg 1964)

10.6 H. N. Russell, F. A. Saunders: Astrophys. J **61**, 38 (1925)

10.7 C. W. Nielson, G. F. Koster: *Spectroscopic Coefficients for the p^n, d^n, and f^n Configurations* (MIT Press, Cambridge 1963)

10.8 W. C. Martin, R. Zalubas, L. Hagan: *Atomic Energy Levels – The Rare-Earth Elements*, Nat. Stand. Ref. Data Ser., Nat. Bur. Stand. No. 60 (United States Government Printing Office, Washington 1978)

10.9 A. de-Shalit, I. Talmi: *Nuclear Shell Theory* (Academic, New York 1963)

10.10 H. N. Russell, A. G. Shenstone, L. A. Turner: Phys. Rev. **33**, 900 (1929)

10.11 R. F. Bacher, S. Goudsmit: *Atomic Energy States* (McGraw-Hill, New York 1932)

10.12 C. E. Moore: *Atomic Energy Levels*, Nat. Stand. Ref. Data Ser., Nat. Bur. Stand. No. 35 (United States Government Printing Office, Washington 1971)

10.13 W. C. Martin, A. Musgrove, S. Kotochigova, J. E. Sansonetti: *Ground Levels and Ionization Energies for the Neutral Atoms* (version 1.3, 2003) This is one of several online NIST databases referred to in this chapter. The databases are accessible by selecting "Physical Reference Data" at the NIST Physics Laboratory website: http://physics.nist.gov.

10.14 J. E. Sansonetti, W. C. Martin: *Handbook of Basic Atomic Spectroscopic Data*, NIST online database, http://physics.nist.gov/PhysRefData/Handbook. These tables include selected data on wavelengths, energy levels, and transition probabilites for the neutral and singly-ionized atoms of all elements up through einsteinium ($Z = 1-99$)

10.15 B. G. Wybourne: *Spectroscopic Properties of Rare Earths* (Wiley, New York 1965)

10.16 Z. C. Yan, G. W. F. Drake: Phys. Rev. A **50**, R1980 (1980)

10.17 W. C. Martin: J. Opt. Soc. Am. **70**, 784 (1980)

10.18 G. W. F. Drake: Adv. At. Mol. Opt. Phys. **32**, 93 (1994)

10.19 U. Fano, W. C. Martin: *Topics in Modern Physics, A Tribute to E. U. Condon*, ed. by W. E. Brittin, H. Odabasi (Colorado Associated Univ. Press, Colorado 1971) pp. 147–152

10.20 B. Edlén: Metrologia **2**, 71 (1966)

10.21 E. R. Peck, K. Reeder: J. Opt. Soc. Amer. **62**, 958 (1972)

10.22 T. J. Quinn: Metrologia **40**, 103 (2003)

10.23 A. G. Maki, J. S. Wells: *Wavenumber Calibration Tables from Heterodyne Frequency Measurements*, NIST Spec. Publ. 821 (U. S. Government Printing Office, Washington 1991)

10.24 W. Whaling, W. H. C. Anderson, M. T. Carle, J. W. Brault, H. A. Zarem: J. Res. Natl. Inst. Stand. Technol. **107**, 149 (2002)

10.25 W. Whaling, W. H. C. Anderson, M. T. Carle, J. W. Brault, H. A. Zarem: J. Quant. Spectrosc. Radiat. Transfer **53**, 1 (1995)

10.26 C. J. Sansonetti, K.-H. Weber: J. Opt. Soc. Am. B **1**, 361 (1984)

10.27 B. A. Palmer, R. A. Keller, R. Engleman Jr.: Los Alamos National Laboratory Report LA-8251-MS, UC-34a (1980)

10.28 B. A. Palmer, R. Engleman Jr.: Los Alamos National Laboratory Report LA-9615-MS, UC-4 (1983)
10.29 S. Gerstenkorn, P. Luc: *Atlas du Spectre d'Absorption de la Molécule d'Iode entre 14 800–20 000 cm^{-1}* (Editions du CNRS, Paris 1978)
10.30 S. Gerstenkorn, P. Luc: Rev. Phys. Appl. **14**, 791 (1979)
10.31 V. Kaufman, B. Edlén: J. Phys. Chem. Ref. Data **3**, 825 (1974)
10.32 G. Nave, S. Johansson, R. C. M. Learner, A. P. Thorne, J. W. Brault: Astrophys. J. Suppl. Ser. **94**, 221 (1994), and references therein
10.33 J. E. Sansonetti, J. Reader, C. J. Sansonetti, N. Acquista: Atlas of the Spectrum of a Platinum/Neon Hollow-Cathode Lamp in the Region 1130 – 4330 Å, J. Res. Natl. Inst. Stand. Technol. **97**, 1–212 (1992), online database
10.34 G. Nave, C. J. Sansonetti: J. Opt. Soc. Amer. B **21**, 442 (2004)
10.35 A. N. Cox (Ed.): *Allen's Astrophysical Quantities*, 4th edn. (American Inst. Physics Press, Springer, New York 2000)
10.36 A. W. Weiss: J. Quant. Spectrosc. Radiat. Transfer **18**, 481 (1977)
10.37 H. R. Griem: *Spectral Line Broadening by Plasmas* (Academic, New York 1974)
10.38 J. R. Roberts, P. A. Voigt: J. Res. Natl. Bur. Stand. **75**, 291 (1971)
10.39 I. I. Sobelman: *Atomic Spectra and Radiative Transitions*, 2nd edn. (Springer, Berlin, Heidelberg 1992)
10.40 A. T. M. Wilbers, G. M. W. Kroesen, C. J. Timmermans, D. C. Schram: J. Quant. Spectrosc. Radiat. Transfer **45**, 1 (1991)
10.41 Y. Ralchenko: *Databases for Atomic and Plasma Physics*; at site http://plasma-gate.weizmann.ac.il/DBfAPP.html

11. High Precision Calculations for Helium

Exact analytic solutions to the Schrödinger equation are known only for atomic hydrogen, and other equivalent two-body systems (see Chapt. 9). However, very high precision approximations are now available for helium, which are essentially exact for all practical purposes. This chapter summarizes the computational methods and tabulates numerical results for the ground state and several singly excited states. Similar methods can be applied to other three-body problems.

11.1 **The Three-Body Schrödinger Equation** ... 199
11.1.1 Formal Mathematical Properties ... 200
11.2 **Computational Methods** ... 200
11.2.1 Variational Methods ... 200
11.2.2 Construction of Basis Sets ... 201
11.2.3 Calculation of Matrix Elements ... 202
11.2.4 Other Computational Methods ... 205
11.3 **Variational Eigenvalues** ... 205
11.3.1 Expectation Values of Operators and Sum Rules ... 205
11.4 **Total Energies** ... 208
11.4.1 Quantum Defect Extrapolations ... 211
11.4.2 Asymptotic Expansions ... 213
11.5 **Radiative Transitions** ... 215
11.5.1 Basic Formulation ... 215
11.5.2 Oscillator Strength Table ... 216
11.6 **Future Perspectives** ... 218
References ... 218

11.1 The Three-Body Schrödinger Equation

The Schrödinger equation for a three-body system consisting of a nucleus of charge Ze, and mass M, and two electrons of charge $-e$ and mass m_e is

$$\left[\frac{1}{2M}P_{\mathrm{N}}^2+\frac{1}{2m_e}\sum_{i=1}^{2}P_i^2+V(\boldsymbol{R}_{\mathrm{N}},\boldsymbol{R}_i)\right]\Psi=E\Psi\,, \tag{11.1}$$

where $\boldsymbol{P}_i=(\hbar/\mathrm{i})\nabla_i$ and

$$V(\boldsymbol{R}_{\mathrm{N}},\boldsymbol{R}_i)=-\frac{Ze^2}{|\boldsymbol{R}_{\mathrm{N}}-\boldsymbol{R}_1|}-\frac{Ze^2}{|\boldsymbol{R}_{\mathrm{N}}-\boldsymbol{R}_2|}+\frac{e^2}{|\boldsymbol{R}_1-\boldsymbol{R}_2|} \tag{11.2}$$

depends only on the relative particle separations. Since the center of mass (c.m.) is then an ignorable coordinate, it can be eliminated by defining the relative particle coordinates

$$\boldsymbol{r}_i=\boldsymbol{R}_i-\boldsymbol{R}_{\mathrm{N}}$$

to obtain

$$\left[\frac{1}{2\mu}\sum_{i=1}^{2}p_i^2+\frac{1}{M}\boldsymbol{p}_1\cdot\boldsymbol{p}_2+V(\boldsymbol{r}_1,\boldsymbol{r}_2)\right]\Psi=E\Psi\,, \tag{11.3}$$

where $\mu=m_eM/(m_e+M)$ is the electron reduced mass and the term $H_{\mathrm{mp}}=\boldsymbol{p}_1\cdot\boldsymbol{p}_2/M$ is called the mass polarization operator. For computational purposes, it is usual to measure distance in units of $a_\mu=(m_e/\mu)a_0$ and energies in units of $e^2/a_\mu=2(\mu/m_e)R_\infty$ so that (11.3) assumes the dimensionless form

$$\left[-\frac{1}{2}\sum_{i=1}^{2}\nabla_{\rho_i}^2-\frac{\mu}{M}\nabla_{\rho_1}\cdot\nabla_{\rho_2}+V(\boldsymbol{\rho}_1,\boldsymbol{\rho}_2)\right]\Psi=\varepsilon\Psi\,, \tag{11.4}$$

where $\boldsymbol{\rho}_i=\boldsymbol{r}_i/a_\mu$, $\varepsilon=E/(e^2/a_\mu)$, and

$$V(\boldsymbol{\rho}_1,\boldsymbol{\rho}_2)=-\frac{Z}{\rho_1}-\frac{Z}{\rho_2}+\frac{1}{|\boldsymbol{\rho}_1-\boldsymbol{\rho}_2|}\,. \tag{11.5}$$

The limit $\mu/M\to 0$ defines the infinite nuclear mass problem with eigenvalue ε_0 and eigenfunction Ψ_0. If the mass polarization term is treated as a small perturbation, then the total energy assumes the form

$$E=\left[\varepsilon_0+\frac{\mu}{M}\varepsilon_1+\left(\frac{\mu}{M}\right)^2\varepsilon_2+\cdots\right]\frac{\mu}{m_e}\frac{e^2}{a_0}\,, \tag{11.6}$$

where $\varepsilon_1 = -\langle\Psi_0|\nabla_{\rho_1}\cdot\nabla_{\rho_2}|\Psi_0\rangle$ determines the first-order specific mass shift and ε_2 is the second-order coefficient. The common $(\mu/m_e)\varepsilon_0$ mass scaling of all eigenvalues determines the normal mass shift (isotope shift). Since $\mu/m = 1-\mu/M$, the shift is $-(\mu/M)\varepsilon_0$.

11.1.1 Formal Mathematical Properties

Two-Particle Coalescences

The exact nonrelativistic wave function for any many-body system contains discontinuities or cusps in the spherically averaged radial derivative with respect to r_{ij} as $r_{ij}\to 0$, where $r_{ij} = |\boldsymbol{r}_i - \boldsymbol{r}_j|$ is any interparticle coordinate. If the masses and charges are m_i and q_i respectively, then the discontinuities are given by the Kato cusp condition [11.1]

$$\hbar^2\left(\frac{\partial\bar{\Psi}}{\partial r_{ij}}\right)_{r_{ij}=0} = \mu_{ij}q_iq_j\Psi(r_{ij}=0)\,, \tag{11.7}$$

where $\mu_{ij} = m_im_j/(m_i+m_j)$ and $\bar{\Psi}$ denotes the wave function averaged over a sphere centered at $r_{ij}=0$. If Ψ vanishes at $r_{ij}=0$, then its leading dependence on r_{ij} is of the form $r_{ij}^l Y_{lm}(\boldsymbol{r}_{ij})$ for some integer $l>0$ [11.2]. Equation (11.7) applies to any Coulombic system. The electron–nucleus cusp in the wave functions for hydrogen provides a simple example.

Three-Particle Coalescences

Three-particle coalescences are described by the Fock expansion [11.3–6], as recently discussed by *Myers* et al. [11.7]. For the S-states of He-like ions, the expansion has the form

$$\Psi(\boldsymbol{r}_1,\boldsymbol{r}_2) = \sum_{j=0}^{\infty}\sum_{k=0}^{[j/2]}\mathcal{R}^j(\ln\mathcal{R})^k\phi_{j,k}\,, \tag{11.8}$$

where [ ] denotes "greatest integer in", and $\mathcal{R} = (r_1^2+r_2^2)^{1/2}$ is the hyperradius. The leading coefficients are

$$\begin{aligned}\phi_{0,0} &= 1\,,\\ \phi_{1,0} &= -\left(Zr_1+Zr_2-\frac{1}{2}r_{12}\right)\Big/\mathcal{R}\,,\\ \phi_{2,1} &= -2Z\left(\frac{\pi-2}{3\pi}\right)\frac{\boldsymbol{r}_1\cdot\boldsymbol{r}_2}{\mathcal{R}^2}\,.\end{aligned} \tag{11.9}$$

The next term $\phi_{2,0}$ is known in terms of a lengthy expression [11.7–9], but higher terms have not yet been obtained in closed form. The Fock expansion has been proved convergent for all $\mathcal{R}<\frac{1}{2}$ [11.10], and extended to pointwise convergence for all $\mathcal{R}$ [11.11, 12].

Asymptotic Form

The long range behavior of many-electron wave functions has been studied from several points of view [11.13–15]. The basic result of [11.16] is that at large distances, the one-electron density behaves as

$$\rho^{1/2}(r) \approx r^{Z^*/t-1}\mathrm{e}^{-tr}\,, \tag{11.10}$$

where $t=(2I_1)^{1/2}$, I_1 is the first ionization potential (in a.u.), and $Z^* = Z-N+1$ is the screened nuclear charge seen by the outer most electron. For hydrogenic systems with principal quantum number n, $I_1 = (Z^*)^2/2n^2$.

11.2 Computational Methods

11.2.1 Variational Methods

Most high precision calculations for the bound states of three-body systems such as helium are based on the Rayleigh–Ritz variational principle. For any normalizable trial function Ψ_{tr}, the quantity

$$E_{\text{tr}} = \frac{\langle\Psi_{\text{tr}}|H|\Psi_{\text{tr}}\rangle}{\langle\Psi_{\text{tr}}|\Psi_{\text{tr}}\rangle} \tag{11.11}$$

satisfies the inequality $E_{\text{tr}}\ge E_1$, where E_1 is the true ground state energy. Thus E_{tr} is an upper bound to E_1. The inequality is easily proved by expanding Ψ_{tr} in the complete basis set of eigenfunctions Ψ_1, Ψ_2, Ψ_3, $\cdots$ of H with eigenvalues $E_1<E_2<E_3<\cdots$, so that

$$\Psi_{\text{tr}} = \sum_{i=1}^{\infty}c_i\Psi_i\,, \tag{11.12}$$

where the c_i are expansion coefficients. This can always be done in principle, even though the exact Ψ_i are not actually known. If Ψ_{tr} is normalized so that $\langle\Psi_{\text{tr}}|\Psi_{\text{tr}}\rangle = 1$, then $\sum_{i=1}^{\infty}|c_i|^2 = 1$ and

$$\begin{aligned}E_{\text{tr}} &= |c_1|^2E_1+|c_2|^2E_2+|c_3|^2E_3+\cdots\\ &= E_1+|c_2|^2(E_2-E_1)+|c_3|^2(E_3-E_1)+\cdots\\ &\ge E_1\,,\end{aligned} \tag{11.13}$$

which proves the theorem.

The basic idea of variational calculations then is to write Ψ_{tr} in some arbitrarily chosen mathematical form with variational parameters (subject to normalizability and boundary conditions at the origin and infinity), and then adjust the parameters to obtain the minimum value of E_{tr}.

The minimization problem for the case of *linear* variational coefficients can be solved algebraically. For example, let

$$\chi_p(\alpha,\beta) = r_1^i r_2^j r_{12}^k \, e^{-\alpha r_1 - \beta r_2} \tag{11.14}$$

denote the members of a basis set, where p is an index labeling distinct triplets of nonnegative integer values for the powers $\{i, j, k\}$, and α, β are (for the moment) fixed constants determining the distance scale.

If Ψ_{tr} is expanded in the form

$$\Psi_{tr} = \sum_{p=1}^{N} c_p \chi_p(\alpha,\beta)\,, \tag{11.15}$$

then the solution to the system of equations $\partial E_{tr}/\partial c_p = 0$, $p = 1, \ldots, N$, is exactly equivalent to solving the N-dimensional generalized eigenvalue problem

$$\mathbf{H}\boldsymbol{c} = \lambda \mathbf{O}\boldsymbol{c}\,, \tag{11.16}$$

where $\boldsymbol{c}$ is a column vector of coefficients c_p; and $\mathbf{H}$ and $\mathbf{O}$ have matrix elements $H_{pq} = \langle \chi_p | H | \chi_q \rangle$ and $O_{pq} = \langle \chi_p | \chi_q \rangle$. There are N eigenvalues $\lambda_1, \lambda_2, \ldots \lambda_N$, of which the lowest is an upper bound to E_1.

Extension to Excited States

By the Hylleraas–Undheim–MacDonald (HUM) theorem [11.17, 18], the remaining eigenvalues $\lambda_2, \lambda_3, \ldots$ are also upper bounds to the exact energies $E_2, E_3, \ldots$, provided that the spectrum is bounded from below. The HUM theorem is a consequence of the matrix eigenvalue interleaving theorem, which states that as the dimensions of $\mathbf{H}$ and $\mathbf{O}$ are progressively increased by adding an extra row and column, the N old eigenvalues λ_p fall between the $N+1$ new ones. Consequently, as illustrated in Fig. 11.1, all eigenvalues numbered from the bottom up must move inexorably downward as N is increased. Since the exact spectrum of bound states is obtained in the limit $N \to \infty$, no λ_p can cross the corresponding exact E_p on its way down. Thus $\lambda_p \geq E_p$ for every finite N.

11.2.2 Construction of Basis Sets

Since the Schrödinger equation (11.4) is not separable in the electron coordinates, basis sets which incorporate

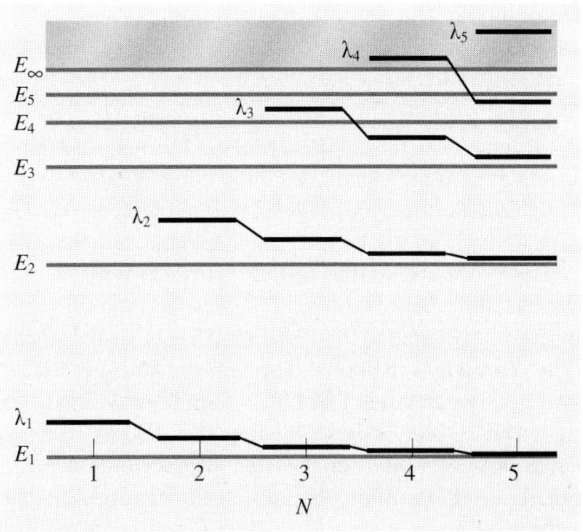

Fig. 11.1 Diagram illustrating the Hylleraas–Undheim–MacDonald Theorem. The λ_p, $p = 1, \ldots, N$ are the variational eigenvalues for an N-dimensional basis set, and the E_i are the exact eigenvalues of H. The highest λ_p lie in the continuous spectrum of H

the $r_{12} = |\boldsymbol{r}_1 - \boldsymbol{r}_2|$ interelectron coordinate are most efficient. The necessity for r_{12} terms also follows from the Fock expansion (11.8). A basis set constructed from terms of the form (11.14) is called a Hylleraas basis set [11.19, 20]. (The basis set is often expressed in terms of the equivalent variables $s = r_1 + r_2$, $t = r_1 - r_2$, $u = r_{12}$.)

With $\chi_p(\alpha,\beta)$ defined as in (11.14), the general form for a state of total angular momentum L is

$$\Psi_{tr} = \sum_{l_1=0}^{[L/2]} \sum_{p} C_{p,l_1} \chi_p(\alpha,\beta) r_1^{l_1} r_2^{l_2} \mathcal{Y}^M_{l_1 L-l_1 L}(\hat{\boldsymbol{r}}_1, \hat{\boldsymbol{r}}_2) \pm \text{exchange}\,, \tag{11.17}$$

where

$$\mathcal{Y}^M_{l_1 l_2 L}(\hat{\boldsymbol{r}}_1, \hat{\boldsymbol{r}}_2) = \sum_{m_1,m_2} Y_{l_1 m_1}(\hat{\boldsymbol{r}}_1) Y_{l_2 m_2}(\hat{\boldsymbol{r}}_2) \times \langle l_1 l_2 m_1 m_2 | LM \rangle \tag{11.18}$$

is the vector coupled product of angular momenta l_1, l_2 for the two electrons. The sum over p in (11.17) typically includes all terms in (11.14) with $i + j + k \leq \Omega$, where Ω is an integer determining a so-called *Pekeris shell* of terms, and the exchange term denotes the interchange of r_1 and r_2 with (+) for singlet states and (−) for triplet states. Convergence is studied by progressively increasing Ω. The number of terms is

$$N = \frac{1}{6}(\Omega+1)(\Omega+2)(\Omega+3) .$$

Basis sets of this type were used by many authors, culminating in the extensive high precision calculations of *Pekeris* and coworkers [11.21] for low-lying states, using as many as 1078 terms. Their accuracy is not easily surpassed because of the rapid growth of N with Ω, and because of numerical linear dependence in the basis set for large Ω. Recently, their accuracy has been surpassed by two principal methods. The first explicitly includes powers of logarithmic and half-integral terms in χ_p, as suggested by the Fock expansion [11.22–25]. This is particularly effective for S-states. The second focuses directly on the multiple distance scales required for an accurate representation of the wave function by writing the trial function in terms of the double basis set [11.26]

$$\Psi_{\text{tr}} = \sum_{l_1=0}^{[L/2]} \sum_{p} \left[C^{(1)}_{p,l_1} \chi_p(\alpha_1, \beta_1) + C^{(2)}_{p,l_1} \chi_p(\alpha_2, \beta_2) \right] \times r_1^{l_1} r_2^{l_2} \mathcal{Y}_{l_1 l_2 L}(\boldsymbol{r}_1, \boldsymbol{r}_2) \pm \text{exchange} , \tag{11.19}$$

where each $\chi_p(\alpha, \beta)$ is of the form (11.14), but with different values for the distance scales α_1, β_1 and α_2, β_2 in the two sets of terms. They are determined by a complete minimization of E_{tr} with respect to all four parameters, producing a natural division of the basis set into an asymptotic sector and a close-range correlation sector. The method produces a dramatic improvement in accuracy for higher-lying Rydberg states (where variational methods typically deteriorate rapidly in accuracy) and is also effective for low-lying S-states [11.27–29]. Nonrelativistic energies accurate to 1 part in 10^{16} are obtainable with modest computing resources.

Another version of the variational method is the quasi-random (or stochastic) method in which nonlinear exponential parameters for all three of r_1, r_2, and r_{12} are chosen at random from certain specified intervals [11.30, 31]. The method is remarkably accurate and efficient for low-lying states, but subject to severe roundoff error.

11.2.3 Calculation of Matrix Elements

The three-body problem has the unique advantage that the full six-dimensional volume element (in the c.m. frame) can be transformed to the product of a three-dimensional angular integral (ang) and a three-dimensional radial integral (rad) over r_1, r_2, and r_{12}. The transformation is

$$\iint \mathrm{d}\boldsymbol{r}_1\, \mathrm{d}\boldsymbol{r}_2 = \int_0^{2\pi} \mathrm{d}\phi \int_0^{2\pi} \mathrm{d}\varphi_1 \int_0^{\pi} \sin\theta_1\, \mathrm{d}\theta_1 \times \int_0^{\infty} r_1\, \mathrm{d}r_1 \int_0^{\infty} r_2\, \mathrm{d}r_2 \int_{|r_1-r_2|}^{r_1+r_2} r_{12}\, \mathrm{d}r_{12} , \tag{11.20}$$

where θ_1, φ_1 are the polar angles of $\boldsymbol{r}_1$ and ϕ is the angle of rotation of the triangle formed by $\boldsymbol{r}_1$, $\boldsymbol{r}_2$, and $\boldsymbol{r}_{12}$ about the $\boldsymbol{r}_1$ direction. The polar angles θ_2, φ_2 are then dependent variables. The basic angular integral is

$$\langle Y^*_{l_1 m_1}(\theta_1, \varphi_1) Y_{l_2 m_2}(\theta_2, \varphi_2) \rangle_{\text{ang}} = 2\pi \delta_{l_1 l_2} \delta_{m_1 m_2} P_{l_1}(\cos\theta) , \tag{11.21}$$

where $\cos\theta \equiv \hat{\boldsymbol{r}}_1 \cdot \hat{\boldsymbol{r}}_2$ denotes the radial function

$$\cos\theta = \frac{r_1^2 + r_2^2 - r_{12}^2}{2 r_1 r_2} , \tag{11.22}$$

and $P_l(\cos\theta)$ is a Legendre polynomial. The angular integral over vector-coupled spherical harmonics is [11.32]

$$\left\langle \mathcal{Y}^{M'*}_{l_1' l_2' L'}(\hat{\boldsymbol{r}}_1, \hat{\boldsymbol{r}}_2)\, \mathcal{Y}^{M}_{l_1 l_2 L}(\hat{\boldsymbol{r}}_1, \hat{\boldsymbol{r}}_2) \right\rangle_{\text{ang}} = \delta_{L,L'} \delta_{M,M'} \sum_{\Lambda} C_\Lambda P_\Lambda(\cos\theta) , \tag{11.23}$$

where

$$C_\Lambda = \frac{1}{2}[(2l_1+1)(2l_1'+1)(2l_2+1)(2l_2'+1)]^{1/2} \times (-1)^{L+\Lambda}(2\Lambda+1) \times \begin{pmatrix} l_1' & l_1 & \Lambda \\ 0 & 0 & 0 \end{pmatrix} \begin{pmatrix} l_2' & l_2 & \Lambda \\ 0 & 0 & 0 \end{pmatrix} \begin{Bmatrix} L & l_1 & l_2 \\ \Lambda & l_2' & l_1' \end{Bmatrix} , \tag{11.24}$$

and the sum over Λ includes all nonvanishing terms. This can be extended to general matrix elements of tensor operators by further vector coupling [11.32].

Radial Integrals

Table 11.1 lists formulas for the radial integrals arising from matrix elements of H, as well as those from the Breit interaction (see Sect. 21.1). Although they can all be written in closed form, some have been expressed as infinite series in order to achieve good numerical stability. The exceptions are formulas 5 and 10 in the Table, which became unstable as $\alpha \to \beta$. More elaborate

Table 11.1 Formulas for the radial integrals $I_0(a,b,c;\alpha,\beta) = \langle r_1^a r_2^b r_{12}^c\, \mathrm{e}^{-\alpha r_1-\beta r_2}\rangle_{\mathrm{rad}}$ and $I_0^{\log}(a,b,c;\alpha,\beta) = \langle r_1^a r_2^b r_{12}^c \ln r_{12}\, \mathrm{e}^{-\alpha r_1-\beta r_2}\rangle_{\mathrm{rad}}$; $\psi(n) = -\gamma + \sum_{k=1}^{n-1} k^{-1}$ is the digamma function, ${}_2F_1(a,b;c;z)$ is the hypergeometric function, and $s = a+b+c+5$. Except as noted, the formulas apply for $a \ge -1$, $b \ge -1$, $c \ge -1$

1.	$I_0(-2,-2,-1;\alpha,\beta)$	$= \frac{2}{\alpha}\ln\left(\frac{\alpha+\beta}{\beta}\right)+\frac{2}{\beta}\ln\left(\frac{\alpha+\beta}{\alpha}\right)$
2.	$I_0(a,b,c;\alpha,\beta)$	$= \frac{2}{c+2}\sum_{i=0}^{[(c+1)/2]}\binom{c+2}{2i+1}\left[F_{a+2i+2,\,b+c-2i+2}(\alpha,\beta)+F_{b+2i+2,\,a+c-2i+2}(\beta,\alpha)\right]$ $(c\ge -1,\ s\ge 0)$ where $F_{p,q}(\alpha,\beta)=\begin{cases} \frac{q!}{(\alpha+\beta)^{p+1}\beta^{q+1}}\sum_{j=0}^{q}\frac{(p+j)!}{j!}\left(\frac{\beta}{\alpha+\beta}\right)^j & q\ge 0,\ p\ge 0\\ \frac{p!}{\alpha^{p+q+2}}\sum_{j=p+q+1}^{\infty}\frac{j!}{(j-q)!}\left(\frac{\alpha}{\alpha+\beta}\right)^{j+1} & q<0,\ p\ge 0\\ 0\ \text{[a]} & p<0\end{cases}$
3.	$I_0(a,b,c;\alpha,\alpha)$	$= \frac{2^{c+3}s!}{(c+2)(2\alpha)^{s+1}}\left[\sum_{j=0}^{a+1}\binom{a+1}{j}(b+1)!\left(\frac{j!}{(j+b+2)!}-\frac{(j+c+2)!}{(j+b+c+4)!}\right)+(a\leftrightarrow b)\right]$
5.	$I_0(a,b,-2;\alpha,\beta)$	$= \frac{(a+1)!}{\alpha^{a+2}}\sum_{j=0}^{a+1}\frac{(b+1+j)!}{j!}\left(\frac{\alpha^j}{(\alpha+\beta)^{b+2+j}}-\frac{(-\alpha)^j}{(\beta-\alpha)^{b+2+j}}\right) \times\left[\ln\left(\frac{2\alpha}{\alpha+\beta}\right)-\psi(a+2-j)+\psi(1)\right]+\binom{a\leftrightarrow b}{\alpha\leftrightarrow\beta}$
6.	$I_0(a,b,-2;\alpha,\alpha)$	$= \frac{2s!(a+1)!(b+1)!}{(2\alpha)^{s+1}}\left[\sum_{j=0}^{a+1}\frac{\psi(s+1-j)-\psi(a+2-j)}{j!(s-j)!}+(a\leftrightarrow b)\right]$
7.	$I_0(-1,-1,-3;\alpha,\beta)$	$= \frac{2(\beta\ln\beta-\alpha\ln\alpha)}{\alpha^2-\beta^2}+\frac{2[\psi(2)-\ln\epsilon]}{\alpha+\beta}$ [b]
8.	$I_0(a,b,-3;\alpha,\beta)$	$= \left[\frac{(a+1)!}{\alpha^{a+1}(\alpha+\beta)^{b+2}}\sum_{j=0}^{a}\frac{(b+1+j)!}{j!(a+1-j)}\left(\frac{\alpha}{\alpha+\beta}\right)^j+\binom{a\leftrightarrow b}{\alpha\leftrightarrow\beta}\right]-\frac{s!\left[\ln(\alpha\beta\epsilon^2)-2\psi(2)\right]}{(\alpha+\beta)^{s+1}} -\frac{(a+1)!(b+1)!}{(s+1)\alpha^{a+2}\beta^{b+1}}\,{}_2F_1\left(a+2,1;s+2;\frac{\alpha-\beta}{\alpha}\right),\quad a\ge -1,\ b\ge -1$
9.	$I_0^{\log}(-1,-1,c;\alpha,\beta)$	$= \frac{2(c+1)!}{\alpha^2-\beta^2}\left(\frac{\ln\alpha-\psi(c+2)}{\alpha^{c+2}}-\frac{\ln\beta-\psi(c+2)}{\beta^{c+2}}\right)$
10.	$I_0^{\log}(a,b,c;\alpha,\beta)$	$= \frac{(a+1)!}{(c+2)\alpha^{a+c+4}}\sum_{j=0}^{a+1}\frac{(b+1+j)!(a+c+3-j)!}{j!(a+1-j)!}\left(\frac{\alpha^j}{(\alpha+\beta)^{b+2+j}}-\frac{(-\alpha)^j}{(\beta-\alpha)^{b+2+j}}\right) \times\left[-\psi(a+c+4-j)+\frac{1}{c+2}+\ln\alpha\right]+\binom{a\leftrightarrow b}{\alpha\leftrightarrow\beta}$
11.	$I_0^{\log}(a,b,c;\alpha,\alpha)$	$= \frac{2^{c+3}s!(b+1)!}{(c+2)(2\alpha)^{s+1}}\sum_{j=0}^{a+1}\binom{a+1}{j}\left\{\left(\frac{j!}{(j+b+2)!}-\frac{(j+c+2)!}{(j+b+c+4)!}\right)\left[\psi(s+1)-\ln\alpha-\frac{1}{c+2}\right] +\frac{(a+c+3-j)!}{(s+1-j)!}\left[\psi(s+1-j)-\psi(a+c+4-j)\right]\right\}+(a\leftrightarrow b)$

[a] Terms with $p<0$ represent divergent parts which cancel from convergent differences between integrals with the same α and β

[b] ϵ is the radius of an infinitesimal sphere about $r_{12}=0$ which is omitted from the range of integration

techniques for these are discussed in [11.33]. Other cases can be derived by use of the formula

$$\left\langle r_1^{-1} r_2^{-1} f(r_{12})\, \mathrm{e}^{-\alpha r_1-\beta r_2} \right\rangle_{\text{rad}} = \frac{2}{\alpha^2-\beta^2} \int_0^\infty \left(\mathrm{e}^{-\beta r} - \mathrm{e}^{-\alpha r}\right) r f(r)\, \mathrm{d}r\,, \tag{11.25}$$

and then differentiating or integrating with respect to α or β to raise or lower the powers of r_1 and r_2.

Total Integral

The angular integral (11.23) combined with the radial integrals from Table 11.1 yields the total integral

$$\left\langle \mathcal{Y}_{l_1 l_2 L}^{M*}\, \mathcal{Y}_{l_1' l_2' L}^{M}\, f(a,b,c;\alpha,\beta) \right\rangle = \sum_\Lambda C_\Lambda I_\Lambda(a,b,c;\alpha,\beta)\,, \tag{11.26}$$

where

$$I_\Lambda(a,b,c;\alpha,\beta) = \langle f(a,b,c;\alpha,\beta) P_\Lambda(\cos\theta)\rangle_{\text{rad}}\,,$$
$$f(a,b,c;\alpha,\beta) = r_1^a\, r_2^b r_{12}^c\, \mathrm{e}^{-\alpha r_1-\beta r_2}\,.$$

Starting from I_0 and I_1, the general I_Λ can be efficiently calculated from the recursion relations [11.32]

$$\begin{aligned} I_{\Lambda+1}&(a,b,c;\alpha,\beta) \\ &= \frac{2\Lambda+1}{c+2} I_\Lambda(a-1,b-1,c+2;\alpha,\beta) \\ &\quad + I_{\Lambda-1}(a,b,c;\alpha,\beta), \quad c\neq -2 \end{aligned} \tag{11.27}$$

$$\begin{aligned} I_{\Lambda+1}&(a,b,-2;\alpha,\beta) \\ &= (2\Lambda+1) I_\Lambda^{\log}(a-1,b-1,0;\alpha,\beta) \\ &\quad + I_{\Lambda-1}(a,b,-2;\alpha,\beta), \quad c=-2 \end{aligned} \tag{11.28}$$

where

$$\begin{aligned} I_\Lambda^{\log}&(a,b,c;\alpha,\beta) \\ &= \langle f(a,b,c,\alpha,\beta) \ln r_{12} P_\Lambda(\cos\theta)\rangle_{\text{rad}}\,. \end{aligned}$$

The $I_\Lambda^{\log}$ integrals follow the recursion relation

$$\begin{aligned} I_{\Lambda+1}^{\log}&(a,b,c;\alpha,\beta) \\ &= \frac{(2\Lambda+1)}{c+2}\left[I_\Lambda^{\log}(a-1,b-1,c+2;\alpha,\beta) \right. \\ &\quad \left. - \frac{1}{c+2} I_\Lambda(a-1,b-1,c+2;\alpha,\beta) \right] \\ &\quad + I_{\Lambda-1}^{\log}(a,b,c;\alpha,\beta)\,. \end{aligned} \tag{11.29}$$

Hamiltonian Matrix Elements

The general form of the Laplacian operator in terms of r_1, r_2, r_{12} variables is

$$\begin{aligned} \nabla_1^2 &= \frac{1}{r_1^2}\frac{\partial}{\partial r_1}\left(r_1^2 \frac{\partial}{\partial r_1}\right) + \frac{1}{r^2}\frac{\partial}{\partial r}\left(r^2\frac{\partial}{\partial r}\right) - \frac{l_1^2}{r_1^2} \\ &\quad + \frac{2(r_1 - r_2\cos\theta)}{r}\frac{\partial^2}{\partial r_1 \partial r} - 2\left(\nabla_1^Y \cdot \boldsymbol{r}_2\right)\frac{1}{r}\frac{\partial}{\partial r}\,, \end{aligned} \tag{11.30}$$

and similarly for ∇_2^2 with subscripts 1 and 2 interchanged. The term ∇_1^Y is understood to act only on the $\mathcal{Y}_{l_1 l_2 L}^M(\hat{\boldsymbol{r}}_1, \hat{\boldsymbol{r}}_2)$ part of the wave function. This term in ∇_1^2 can be easily evaluated by means of the effective operator replacement

$$\begin{aligned} &\langle \mathcal{Y}_{l_1' l_2' L'}^{M*}\, \nabla_1^Y \cdot \boldsymbol{r}_2\, \mathcal{Y}_{l_1 l_2 L}^M \rangle_{\text{ang}} \frac{1}{r_{12}}\frac{\partial g(r_{12})}{\partial r_{12}} \\ &\quad\to \frac{g(r_{12})}{2 r_1 r_2} \sum_\Lambda \tilde{C}_\Lambda P_\Lambda(\cos\theta) \end{aligned} \tag{11.31}$$

for the angular part of the total integral, where

$$\tilde{C}_\Lambda = [l_1'(l_1'+1) - l_1(l_1+1) - \Lambda(\Lambda+1)] C_\Lambda\,. \tag{11.32}$$

The replacement (11.31) becomes an equality after radial integration with any function $g(r_{12})$ in the integrand. The matrix elements of H between arbitrary basis functions defined by

$$\chi = r_1^a\, r_2^b r_{12}^c\, \mathrm{e}^{-\alpha r_1 - \beta r_2} \mathcal{Y}_{l_1 l_2 L}^M(\hat{\boldsymbol{r}}_1, \hat{\boldsymbol{r}}_2)\,,$$
$$\chi' = r_1^{a'} r_2^{b'} r_{12}^{c'}\, \mathrm{e}^{-\alpha' r_1 - \beta' r_2} \mathcal{Y}_{l_1' l_2' L}^M(\hat{\boldsymbol{r}}_1, \hat{\boldsymbol{r}}_2)\,,$$

can then be written in the explicitly Hermitian form (for infinite nuclear mass)

$$\begin{aligned} \langle \chi' | H | \chi \rangle &= \frac{1}{8}\sum_\Lambda C_\Lambda \sum_{i=0}^{2} \left[A_i^{(1)} \right. \\ &\quad \times I_\Lambda(a_+ - i, b_+, c_+; \alpha_+, \beta_+) \\ &\quad + A_i^{(2)} I_\Lambda(a_+, b_+ - i, c_+; \alpha_+, \beta_+) \\ &\quad \left. + A_i^{(3)} I_\Lambda(a_+, b_+, c_+ - i; \alpha_+, \beta_+) \right] \end{aligned} \tag{11.33}$$

where $a_\pm = a' \pm a$, $\alpha_\pm = \alpha' \pm \alpha$ etc., and

$$A_0^{(1)} = -\alpha_+^2 - \alpha_-^2 + 2\alpha_-\alpha_+(c_-/c_+)\,,$$

$$
\begin{aligned}
A_1^{(1)} &= 2\{\alpha_+(a_+ + 2) + \alpha_- a_- \\
&\quad - [\alpha_+ a_- + \alpha_-(a_+ + 2)](c_-/c_+)\} - 8Z\,, \\
A_2^{(1)} &= -a_+^2 - a_-^2 - 2a_+ + 2a_-(a_+ + 1)(c_-/c_+) \\
&\quad + 2l_1(l_1+1)(1 - c_-/c_+) \\
&\quad + 2l_1'(l_1'+1)(1 + c_-/c_+)\,, \\
A_0^{(3)} &= 0\,, \qquad A_1^{(3)} = 8\,, \\
A_2^{(3)} &= 2(c_+^2 - c_-^2)\,,
\end{aligned}
$$

with $(c_-/c_+) = 0$ for $c_+ = 0$. The $A_i^{(2)}$ are defined similarly to $A_i^{(1)}$ with the replacements $a \to b$, $\alpha \to \beta$, $l_1 \to l_2$. The overlap integral is

$$
\langle \chi' | \chi \rangle = \sum_\Lambda C_\Lambda I_\Lambda(a_+, b_+, c_+; \alpha_+, \beta_+)\,. \tag{11.34}
$$

11.2.4 Other Computational Methods

Although not yet at the same level of accuracy as variational methods, certain nonvariational methods, such as finite element methods [11.34], solutions to the Faddeev equations [11.35], and the correlated-function hyperspherical-harmonic method [11.36], have their own advantages of flexibility and/or generality. A characteristic feature of these methods is that they provide direct numerical solutions to the three-body problem which in principle converge pointwise to the exact solution, rather than depending upon a globally optimized solution. Other methods particularly suited to doubly-excited states are discussed in Chapt. 25

11.3 Variational Eigenvalues

High precision variational eigenvalues are available for all states of helium up to $n = 10$ and $L = 7$ [11.27–29]. The nonrelativistic values of ε_0, ε_1 and ε_2 [see (11.6)] are listed in Table 11.2 and Table 11.3. The ε_0 are the eigenvalues for infinite nuclear mass, and ε_1 and ε_2, together with (11.6) give the finite mass corrections for the isotopes ^{3}He and ^{4}He. The values of μ/M can be calculated from

$$
\frac{\mu}{M} = \left(\frac{M_\mathrm{A}}{5.485\,799\,110\,(12) \times 10^{-4}} - N + 1 \right)^{-1} \tag{11.35}
$$

where M_A is the atomic mass (in amu, see [11.37] for a tabulation. For high precision work, the helium electronic binding energy of 8.48×10^{-8}amu should be added to M_A.) and N is the number of electrons. For ^{4}He, one can use directly the accurately known value of m_e/m_α to calculate $\mu/M = 1/(m_\alpha/m_\mathrm{e} + 1)$. Values of μ/M for the first several isotopes are listed in Table 11.4, and the corresponding energy coefficients for the $1s^2\,{}^1S$ ground state are given in Table 11.5.

11.3.1 Expectation Values of Operators and Sum Rules

Expectation values for various powers of the radial coordinates, together with operators appearing in the Breit interaction, are listed in Table 11.6 for the ground state of helium and He-like ions. Included are all terms required to calculate $\langle V^2 \rangle$, and the oscillator strength sum rules [11.38]

$$
S(-1) = \frac{2}{3}\langle (\boldsymbol{r}_1 + \boldsymbol{r}_2)^2 \rangle\,, \tag{11.36a}
$$

$$
S(0) = 2\,, \tag{11.36b}
$$

$$
S(1) = -\frac{4}{3}(\varepsilon_0 - \varepsilon_1)\,, \tag{11.36c}
$$

$$
S(2) = \frac{2\pi Z}{3}\langle \delta(\boldsymbol{r}_1) + \delta(\boldsymbol{r}_2) \rangle\,, \tag{11.36d}
$$

where $S(k) = \sum_n [\varepsilon_0(n^1\mathrm{P}) - \varepsilon_0(1^1\mathrm{S})]^k f_{0n}$, with energies in a.u., and f_{0n} is the $1^1\mathrm{S} - n^1\mathrm{P}$ oscillator strength (see Sect. 11.5.1).

Table 11.2 Nonrelativistic eigenvalue coefficients ε_0 and ε_1 for helium

State	$\varepsilon_0(n\,^1L)$	$\varepsilon_1(n\,^1L)$	$\varepsilon_0(n\,^3L)$	$\varepsilon_1(n\,^3L)$
1S	−2.903 724 377 034 1195	0.159 069 475 085 84	–	–
2S	−2.145 974 046 054 419(6)	0.009 503 864 419 28	−2.175 229 378 236 791 30	0.007 442 130 706 04
2P	−2.123 843 086 498 093(2)	0.046 044 524 937(1)	−2.133 164 190 779 273(5)	−0.064 572 425 024(4)
3S	−2.061 271 989 740 911(5)	0.002 630 567 0977(1)	−2.068 689 067 472 457 19	0.001 896 211 617 81
3P	−2.055 146 362 091 94(3)	0.014 548 047 097(1)	−2.058 081 084 274 28(4)	−0.018 369 001 636(2)
3D	−2.055 620 732 852 246(6)	−0.000 249 399 9921(1)	−2.055 636 309 453 261(4)	0.000 025 322 839(1)

Table 11.2 Nonrelativistic eigenvalue coefficients ε_0 and ε_1 for helium, cont.

State	$\varepsilon_0(n\,^1L)$	$\varepsilon_1(n\,^1L)$	$\varepsilon_0(n\,^3L)$	$\varepsilon_1(n\,^3L)$
4S	−2.033 586 717 030 72(1)	0.001 073 641 2266(1)	−2.036 512 083 098 236 30(2)	0.000 742 661 516 18
4P	−2.031 069 650 450 24(3)	0.006 254 923 5543(1)	−2.032 324 354 296 62(2)	−0.007 555 178 98(1)
4D	−2.031 279 846 178 687(7)	−0.000 129 175 1887(8)	−2.031 288 847 501 795(3)	0.000 029 442 651(2)
4F	−2.031 255 144 381 749(1)	−0.000 010 024 2694(2)	−2.031 255 168 403 2456(6)	−0.000 009 669 6396
5S	−2.021 176 851 574 363(5)	0.000 538 860 3605(1)	−2.022 618 872 302 312 27(1)	0.000 363 697 136 49
5P	−2.019 905 989 900 83(2)	0.003 230 021 84(2)	−2.020 551 187 256 25(1)	−0.003 810 911 035(1)
5D	−2.020 015 836 159 984(4)	−0.000 071 883 131(6)	−2.020 021 027 446 911(5)	0.000 019 568 85(1)
5F	−2.020 002 937 158 7427(5)	−0.000 005 704 2946(4)	−2.020 002 957 377 3694(4)	−0.000 005 406 4900(5)
5G	−2.020 000 710 898 584 71(1)	−0.000 001 404 4136	−2.020 000 710 925 343 92(1)	−0.000 001 404 0013
6S	−2.014 563 098 446 60(1)	0.000 307 704 277(1)	−2.015 377 452 992 862 19(3)	0.000 204 329 479 10
6P	−2.013 833 979 671 73(2)	0.001 878 058 536(1)	−2.014 207 958 773 74(1)	−0.002 184 346 463(1)
6D	−2.013 898 227 424 286(5)	−0.000 043 412 2689(9)	−2.013 901 415 453 792(7)	0.000 012 742 22(3)
6F	−2.013 890 683 815 5497(3)	−0.000 003 482 257(7)	−2.013 890 698 348 5320(2)	−0.000 003 268 4586(8)
6G	−2.013 889 345 387 313 22(3)	−0.000 000 898 5799(7)	−2.013 889 345 416 952 96(3)	−0.000 000 898 1237(7)
6H	−2.013 889 034 754 279 72	−0.000 000 290 3471	−2.013 889 034 754 301 55	−0.000 000 290 3467
7S	−2.010 625 776 210 87(2)	0.000 191 925 025(1)	−2.011 129 919 527 626 21(4)	0.000 125 981 736 89
7P	−2.010 169 314 529 35(2)	0.001 186 152 30(1)	−2.010 404 960 007 94(2)	−0.001 366 5008(3)
7D	−2.010 210 028 457 98(1)	−0.000 028 027 840(2)	−2.010 212 105 955 595(2)	0.000 008 563 121(3)
7F	−2.010 205 248 074 013(1)	−0.000 002 262 00(4)	−2.010 205 258 374 865(1)	−0.000 002 110 58(3)
7G	−2.010 204 386 224 772 55(7)	−0.000 000 598 3963(3)	−2.010 204 386 250 217 93(6)	−0.000 000 598 005(1)
7H	−2.010 204 182 806 482 04(2)	−0.000 000 201 0978	−2.010 204 182 806 512 04(1)	−0.000 000 201 0973
7I	−2.010 204 120 606 191 32	−0.000 000 077 7755	−2.010 204 120 606 191 340	−0.000 000 077 7755
8S	−2.008 093 622 105 61(4)	0.000 127 650 436(1)	−2.008 427 122 064 721 42(6)	0.000 083 070 552 34
8P	−2.007 789 127 133 22(2)	0.000 796 195 83(5)	−2.007 947 013 771 12(1)	−0.000 911 0535(3)
8D	−2.007 816 512 563 811(7)	−0.000 019 076 181(1)	−2.007 817 934 711 706(3)	0.000 005 971 1234(3)
8F	−2.007 813 297 115 0141(6)	−0.000 001 545 48(1)	−2.007 813 304 535 0908(5)	−0.000 001 436 452(2)
8G	−2.007 812 711 494 0241(1)	−0.000 000 415 0040(1)	−2.007 812 711 514 424 82(9)	−0.000 000 414 6904
8H	−2.007 812 571 828 655 81(1)	−0.000 000 142 6492(3)	−2.007 812 571 828 685 73(1)	−0.000 000 142 6487(2)
8I	−2.007 812 528 549 584 59	−0.000 000 056 9359	−2.007 812 528 549 584 61	−0.000 000 056 9359
8K	−2.007 812 512 570 229 31	−0.000 000 025 1113	−2.007 812 512 570 229 306	−0.000 000 025 1113
9S	−2.006 369 553 107 85(3)	0.000 089 149 6387(7)	−2.006 601 516 715 010 67(3)	0.000 057 628 311 52
9P	−2.006 156 384 652 86(5)	0.000 559 978 028(2)	−2.006 267 267 366 41(4)	−0.000 637 531 359(6)
9D	−2.006 175 671 437 641(6)	−0.000 013 542 185(3)	−2.006 176 684 884 697(2)	0.000 004 306 538(6)
9F	−2.006 173 406 897 3246(8)	−0.000 001 099 9671(3)	−2.006 173 412 365 0430(7)	−0.000 001 019 651(2)
9G	−2.006 172 991 627 5863(2)	−0.000 000 298 2672(1)	−2.006 172 991 643 6650(3)	−0.000 000 298 0198(1)
9H	−2.006 172 891 903 619 14(2)	−0.000 000 104 0022	−2.006 172 891 903 645 88(2)	−0.000 000 104 0019
9I	−2.006 172 860 732 382 57	−0.000 000 042 3136	−2.006 172 860 732 382 60	−0.000 000 042 3136(1)
9K	−2.006 172 849 096 329 78	−0.000 000 019 1516	−2.006 172 849 096 329 780	−0.000 000 019 1516
10S	−2.005 142 991 748 00(8)	0.000 064 697 214(3)	−2.005 310 794 915 6113(2)	0.000 041 598 811 52
10P	−2.004 987 983 802 22(4)	0.000 408 649 4263	−2.005 068 805 4978(1)	−0.000 463 433 718(8)
10D	−2.005 002 071 654 250(6)	−0.000 009 947 5060(6)	−2.005 002 818 080 232(8)	0.000 003 198 298(8)
10F	−2.005 000 417 564 6682(9)	−0.000 000 809 442(9)	−2.005 000 421 686 6036(7)	−0.000 000 748 9264(2)
10G	−2.005 000 112 764 3180(3)	−0.000 000 220 982(2)	−2.005 000 112 777 0031(4)	−0.000 000 220 785(3)
10H	−2.005 000 039 214 394 52(2)	−0.000 000 077 8067	−2.005 000 039 214 417 41(2)	−0.000 000 077 8062
10I	−2.005 000 016 086 516 19	−0.000 000 032 0590(1)	−2.005 000 016 086 516 22	−0.000 000 032 0589(2)
10K	−2.005 000 007 388 375 88	−0.000 000 014 7514	−2.005 000 007 388 375 88	−0.000 000 014 7514

Table 11.3 Eigenvalue coefficients ε_2 for helium

State	$\varepsilon_2(n\,^1L)$	$\varepsilon_2(n\,^3L)$
1S	−0.470 391 870(1)	−
2S	−0.135 276 864(1)	−0.057 495 8479(2)
2P	−0.168 271 22(7)	−0.204 959 88(1)
3S	−0.058 599 3124(4)	−0.040 455 8505(5)
3P	−0.066 047 859(3)	−0.070 292 710(2)
3D	−0.057 201 299(9)	−0.054 737 73(1)
4S	−0.032 522 293(2)	−0.025 628 6338(1)
4P	−0.035 159 71(6)	−0.036 129 973(2)
4D	−0.032 150 91(2)	−0.030 747 891(7)
4F	−0.031 274 336(4)	−0.031 277 9921(3)
5S	−0.020 647 26(9)	−0.017 322 734 96
5P	−0.021 8476(3)	−0.022 166 61(9)
5D	−0.020 5101(2)	−0.019 7062(2)
5F	−0.020 013 498(6)	−0.020 016 561(4)
5G	−0.020 003 5608	−0.020 003 5646
6S	−0.014 261 796(4)	−0.012 411 3991(3)
6P	−0.014 902 86(9)	−0.015 033 58(5)
6D	−0.014 1994(2)	−0.013 707 27(1)
6F	−0.013 896 984(2)	−0.013 899 22(3)
6G	−0.013 891 179(6)	−0.013 891 184(8)
6H	−0.013 889 6191	−0.013 889 6190
7S	−0.010 4382(2)	−0.009 304 4433(3)
7P	−0.010 8186(2)	−0.010 879(2)
7D	−0.010 405 09(3)	−0.010 085 212(1)
7F	−0.010 2092(3)	−0.010 2107(3)
7G	−0.010 205 61(5)	−0.010 205 61(5)
7H	−0.010 204 590(2)	−0.010 204 587(2)
7I	−0.010 204 2767	−0.010 204 2768
8S	−0.007 968 944(3)	−0.007 224 7705(3)
8P	−0.008 2117(5)	−0.008 2487(6)
8D	−0.007 9507(4)	−0.007 731 59(2)

Table 11.3 Eigenvalue coefficients ε_2 for helium, cont.

State	$\varepsilon_2(n\,^1L)$	$\varepsilon_2(n\,^3L)$
8F	−0.007 8159(3)	−0.007 8170(2)
8G	−0.007 813 563(1)	−0.007 813 568(3)
8H	−0.007 812 855(4)	−0.007 812 859(5)
8I	−0.007 812 6429	−0.007 812 6429
8K	−0.007 812 5630	−0.007 812 5630
9S	−0.006 282 5136(1)	−0.005 768 0285(1)
9P	−0.006 4457(2)	−0.006 464 9369(1)
9D	−0.006 270 99(7)	−0.006 1152(1)
9F	−0.006 175 20(1)	−0.006 176 0254(7)
9G	−0.006 173 5796(1)	−0.006 173 592(4)
9H	−0.006 173 104(2)	−0.006 173 101(2)
9I	−0.006 172 9459(1)	−0.006 172 9460(2)
9K	−0.006 172 8876	−0.006 172 8876
10S	−0.005 079 8362(8)	−0.004 709 4530(1)
10P	−0.005 197(1)	−0.005 2067(1)
10D	−0.005 0724(4)	−0.004 9580(8)
10F	−0.005 001 76(2)	−0.005 002 386(2)
10G	−0.005 000 55(2)	−0.005 000 55(2)
10H	−0.005 000 1935(2)	−0.005 000 1935(1)
10I	−0.005 000 0803(4)	−0.005 000 081(1)
10K	−0.005 000 0369	−0.005 000 0368

Table 11.4 Values of the reduced electron mass ratio μ/M

Isotope	$\mu/M \times 10^4$
^{1}H	5.443 205 771(12)
^{2}D	2.723 695 064(6)
^{3}He	1.819 212 075(4)
^{4}He	1.370 745 641(3)
^{6}Li	0.912 167 61(8)
^{7}Li	0.782 020 21(6)
^{9}Be	0.608 820 45(3)

Table 11.5 Nonrelativistic eigenvalues $E = \varepsilon_0 + (\mu/M)\varepsilon_1 + (\mu/M)^2\varepsilon_2$ for helium-like ions (in units of e^2/a_μ)

Atom	$\varepsilon_0(1\,^1S)$	$\varepsilon_1(1\,^1S)$	$\varepsilon_2(1\,^1S)$
H$^-$	−0.527 751 016 544 377	0.032 879 781 852 30	−0.059 779 492 64(1)
He	−2.903 724 377 034 1195	0.159 069 475 085 84	−0.470 391 870(1)
Li$^+$	−7.279 913 412 669 3059	0.288 975 786 393 99	−1.277 369 3776(2)
Be^{++}	−13.655 566 238 423 5867	0.420 520 303 439 44	−2.491 572 8581(1)

Table 11.6 Expectation values of various operators for He-like ions for the case $M = \infty$ (in a.u.)

Quantity	H^-	He	Li^+	Be^{++}
$\langle r_1^2 \rangle$	11.913 699 678 05(6)	1.193 482 995 019	0.446 279 011 201	0.232 067 315 531
$\langle r_{12}^2 \rangle$	25.202 025 2912(1)	2.516 439 312 833	0.927 064 803 063	0.477 946 525 143
$\langle \boldsymbol{r}_1 \cdot \boldsymbol{r}_2 \rangle$	−0.687 312 967 569	−0.064 736 661 398	−0.017 253 390 330	−0.006 905 947 040
$\langle r_1 \rangle$	2.710 178 278 444(1)	0.929 472 294 874	0.572 774 149 971	0.414 283 328 006
$\langle r_{12} \rangle$	4.412 694 497 992(2)	1.422 070 255 566	0.862 315 375 456	0.618 756 314 066
$\langle 1/r_1 \rangle$	0.683 261 767 652	1.688 316 800 717	2.687 924 397 413	3.687 750 406 344
$\langle 1/r_{12} \rangle$	0.311 021 502 214	0.945 818 448 800	1.567 719 559 137	2.190 870 773 906
$\langle 1/r_1^2 \rangle$	1.116 662 8246(1)	6.017 408 8670(1)	14.927 623 7214(2)	27.840 105 671 33(2)
$\langle 1/r_{12}^2 \rangle$	0.155 104 152 58(3)	1.464 770 923 350(1)	4.082 232 787 55(2)	8.028 801 781 824(1)
$\langle 1/r_1 r_2 \rangle$	0.382 627 890 340	2.708 655 474 480	7.011 874 111 824(1)	13.313 954 940 144(1)
$\langle 1/r_1 r_{12} \rangle$	0.253 077 567 065	1.920 943 921 900	5.069 790 932 379	9.717 071 116 528
$\langle \delta(\boldsymbol{r}_1) \rangle$	0.164 552 872 86(3)	1.810 429 318 49(3)	6.852 009 4344(1)	17.198 172 544 74(3)
$\langle \delta(\boldsymbol{r}_{12}) \rangle$	0.002 737 9923(3)	0.106 345 3712(2)	0.533 722 5371(9)	1.522 895 3541(2)
$\langle p^4 \rangle$	2.462 558 614(3)	54.088 067 230(2)	310.547 150 179(6)	1047.278 491 476(2)
$\langle H_{oo} \rangle / \alpha^2$	−0.008 875 022 10(1)	−0.139 094 690 556(1)	−0.427 991 611 178(9)	−0.878 768 694 709(1)

11.4 Total Energies

As discussed in Chapts. 21 and 27, relativistic and QED corrections must be added to the nonrelativistic eigenvalues of Sect. 11.3 before a meaningful comparison with measured transition frequencies can be made. The corrections are discussed in detail in [11.27, 28, 39].

The terms in order of decreasing size are:

1. *Relativistic corrections of $O(\alpha^2)$*

$$\begin{aligned} H_{\text{rel}} &= H_{\text{NFS}} + H_{\text{FS}} \,, \\ H_{\text{NFS}} &= H_{\text{mass}} + H_{\text{D}} + H_{\text{ssc}} + H_{\text{oo}} \,, \\ H_{\text{FS}} &= H_{\text{so}} + H_{\text{soo}} + H_{\text{ss}} \,. \end{aligned}$$

The various nonfine-structure (NFS) and fine-structure (FS) terms are defined in Sect. 21.1. The off-diagonal matrix elements of H_{FS} mix states of different spin and cause a break-down of LS-coupling.

2. *Anomalous magnetic moment corrections of $O(\alpha^3)$*
The general FS matrix elements between states with spins S and S' due to the anomalous magnetic moment a_e are (see Sect. 27.4)

$$\begin{aligned} \langle \gamma S | H_{\text{FS}}^{\text{anom}} | \gamma' S' \rangle &= 2a_\text{e} \Big\langle \gamma S | H_{\text{so}} + \frac{2}{3} \delta_{S,S'} H_{\text{soo}} \\ &\quad + \left(1 + \frac{1}{2} a_\text{e}\right) H_{\text{ss}} | \gamma' S' \Big\rangle \,, \end{aligned} \tag{11.37}$$

where $a_\text{e} = (g_\text{e} - 2)/2 = \alpha/(2\pi) - 0.328\,479\alpha^2 + \cdots$.

3. *QED corrections of $O(\alpha^3)$*
The lowest order QED corrections (including NFS anomalous magnetic moment terms) can be written in the form $\Delta_{L,1} + \Delta_{L,2}$, where

$$\begin{aligned} \Delta E_{L,1} &= \frac{4}{3} Z\alpha^3 \left(\ln(Z\alpha)^{-2} + \frac{19}{30} - \ln k_0 \right) \\ &\quad \times \langle \delta(\boldsymbol{r}_1) + \delta(\boldsymbol{r}_2) \rangle \end{aligned} \tag{11.38}$$

$$\begin{aligned} \Delta E_{L,2} &= \alpha^3 \left(\frac{89}{15} + \frac{14}{3} \ln\alpha - \frac{20}{3} \boldsymbol{s}_1 \cdot \boldsymbol{s}_2 \right) \\ &\quad \times \langle \delta(\boldsymbol{r}_{12}) \rangle - \frac{14}{3} \alpha^3 Q \,, \end{aligned} \tag{11.39}$$

$\ln k_0$ is the two-electron Bethe logarithm defined by (27.86) and Q is the matrix element defined by (27.83). For a highly excited $1snl$ state, $\Delta E_{L,2} \to 0$, $\langle \delta(\boldsymbol{r}_1) + \delta(\boldsymbol{r}_2) \rangle \to Z^3/\pi$, $\ln k_0 \to \ln k_0(1s) = 2.984\,128\,555$, and $\Delta E_{L,1}$ reduces to the Lamb shift of the 1s core state (see Sects. 28.3.4 and 28.3.5). Thus $\Delta E_{L,1}$ represents the electron–nucleus part of the QED shift with the factor of Z^3/π replaced by the correct electron density at the nucleus. Accurate values of $\ln k_0$ for two-electron atoms and ions are tabulated in [11.40].

For the low-lying S-states and P-states of helium [11.31],

$$\ln k_0(1\,^1\mathrm{S}) = 2.983\,865\,861\,, \quad (11.40a)$$

$$\ln k_0(2\,^1\mathrm{S}) = 2.980\,118\,365\,, \quad (11.40b)$$

$$\ln k_0(2\,^3\mathrm{S}) = 2.977\,742\,459\,, \quad (11.40c)$$

$$\ln k_0(2\,^1\mathrm{P}) = 2.983\,803\,377\,, \quad (11.40d)$$

$$\ln k_0(2\,^3\mathrm{P}) = 2.983\,690\,995\,. \quad (11.40e)$$

For a $1snl$ state with large l, the asymptotic expansion [11.41,42]

$$\ln k_0(1snl) \approx \ln k_0(1s) + \frac{1}{n^3}\left(\frac{Z-1}{Z}\right)^4 \ln k_0(nl) + 0.316\,205(6)Z^{-6}\langle r^{-4}\rangle_{nl} + \Delta\beta(1snl) \quad (11.41)$$

becomes essentially exact. Here $\ln k_0(nl)$ is the one-electron Bethe logarithm [11.43] and

$$\langle r^{-4}\rangle_{nl} = \frac{16(Z-1)^4[3n^2 - l(l+1)]}{(2l-1)2l(2l+1)(2l+2)(2l+3)}\,. \quad (11.42)$$

The correction $\Delta\beta(1snl)$ for higher order terms is

$$\Delta\beta(1snl\,^1\mathrm{L}) = 95.8(8)\langle r^{-6}\rangle - 845(19)\langle r^{-7}\rangle + 1406(50)\langle r^{-8}\rangle \quad (11.43)$$

$$\Delta\beta(1snl\,^3\mathrm{L}) = 95.1(9)\langle r^{-6}\rangle - 841(23)\langle r^{-7}\rangle + 1584(60)\langle r^{-8}\rangle\,. \quad (11.44)$$

For example, for the 1s4f ^{1}F state, $\beta(4\,^1\mathrm{F}) = 2.984\,127\,1493(3)$.
For higher Z, $1/Z$ expansions of [11.40] should be used.

4. *Relativistic finite mass corrections of $O(\alpha^2\mu/M)$*
Relativistic finite mass corrections come from two sources. First, a transformation to relative coordinates as in (11.3) is applied to the pairwise Breit interactions among the three particles, generating the new terms [11.44,45]

$$\Delta = \Delta_{\mathrm{oo}} + \Delta_{\mathrm{so}} + 2\frac{m_\mathrm{e}}{M}H_{\mathrm{so}}$$

where

$$\Delta_{\mathrm{oo}} = \frac{-Z\alpha^2 m_\mathrm{e}}{2M}\sum_{i,j}\frac{1}{r_i}\left[\boldsymbol{p}_j\cdot\boldsymbol{p}_i + \hat{\boldsymbol{r}}_i\cdot(\hat{\boldsymbol{r}}_i\cdot\boldsymbol{p}_j)\boldsymbol{p}_i\right]\,, \quad (11.45)$$

$$\Delta_{\mathrm{so}} = \frac{Z\alpha^2 m_\mathrm{e}}{M}\sum_{i\neq j}\frac{1}{r_i^3}\boldsymbol{r}_i\times\boldsymbol{p}_j\cdot\boldsymbol{s}_i\,. \quad (11.46)$$

Second, the mass polarization term H_{mp} in (11.3) generates second-order cross-terms between H_{mp} and H_{rel}. If the wave functions are calculated by solving (11.4) in scaled atomic units, the H_{mp} correction is then automatically included to all orders and the mass-corrected relativistic energy shift is (in units of e^2/a_0)

$$\Delta E_{\mathrm{rel}} = \left(\frac{\mu}{m_\mathrm{e}}\right)^3 \left\langle \left(\frac{\mu}{m_\mathrm{e}}\right) H_{\mathrm{mass}} + H_\mathrm{D} + H_{\mathrm{ssc}} + H_{\mathrm{oo}} + \Delta_{\mathrm{oo}} + \left(1 + \frac{2m_\mathrm{e}}{M}\right) H_{\mathrm{so}} + H_{\mathrm{soo}} + H_{\mathrm{ss}} + \Delta_{\mathrm{so}} \right\rangle \quad (11.47)$$

with $\mu/m_\mathrm{e} = 1 - \mu/M$. The difference $\Delta E_{\mathrm{rel}} - \langle H_{\mathrm{rel}}\rangle_\infty$ calculated for infinite nuclear mass is the relativistic finite mass correction.

5. *Higher-order corrections*
Spin-dependent terms of $O(\alpha^4)$ are known in their entirety, and have recently been calculated to high precision [11.46]. Nonrelativistic operators for the spin-independent part have recently been derived and calculated for the 1s2s $^3\mathrm{S}_1$ state [11.47] and 1s^2 $^1\mathrm{S}_0$ state [11.48]. The dominant electron–nucleus part is known from the one-electron Lamb shift to be

$$\Delta E'_{L,1} = Z\alpha^3\left[\pi Z\alpha\left(\frac{427}{96} - 2\ln 2\right) + 0.538\,931\frac{\alpha}{\pi}\right] \times \langle\delta(\boldsymbol{r}_1) + \delta(\boldsymbol{r}_2)\rangle + O(\alpha^5) \quad (11.48)$$

and the electron–electron logarithmic part is [11.49]

$$\Delta E'_{L,2} = \pi\alpha^4 \ln\alpha^{-1}\langle\delta(\boldsymbol{r}_{12})\rangle\,. \quad (11.49)$$

As an example, $\Delta E'_{L,1}$ contributes -50.336 MHz and -88.267 MHz to the $2\,^1\mathrm{P}$–$2\,^1\mathrm{S}$ and $2\,^3\mathrm{P}_\mathrm{J}$–$2\,^3\mathrm{S}_1$ transition frequencies respectively, while the differences between theory and experiment are ≈ 1 MHz and ≈ -7 MHz for the two cases (see [11.39]). Thus, two-electron corrections (for example, relativistic corrections of relative order $Z\alpha$ to $\Delta E_{L,2}$) are evidently small.
Table 11.7 lists the calculated ionization energies for all states of helium up to $n = 10$ and $L = 7$. For the D-states and beyond, the uncertainties are sufficiently small that these states can be taken as known points of reference in the interpretation of experimental transition frequencies. However,

long-range Casimir–Polder corrections [11.50–52] are not included since they still lack experimental confirmation [11.53]. The QED shifts are the largest for the S- and P-states. The contributions from $\Delta E_{L,1} + \Delta E'_{L,1}$ and $\Delta E_{L,2} + \Delta E'_{L,2}$ for these states are listed separately in Table 11.8. Applications to isotope shifts and measurements of the nuclear radius are discussed in Sects. 16.2 and 90.1

Table 11.7 Total ionization energies for ^{4}He, calculated with $R_M = 3\,289\,391\,006.715$ MHz

State	$E(n\,^1L_L)$	$E(n\,^3L_{L-1})$	$E(n\,^3L_L)$	$E(n\,^3L_{L+1})$
1S	5 945 204 223.(91)			
2S	960 332 041.(25)			1 152 842 741.2(6)
2P	814 709 150.(9)	876 078 642.(17)	876 108 265.(17)	876 110 558.(17)
3S	403 096 132.(8)			451 903 472.(8)
3P	362 787 968.(3)	382 109 902.(6)	382 118 017.(6)	382 118 676.(6)
3D	365 917 749.018(5)	366 018 892.97(1)	366 020 218.086(8)	366 020 293.415(2)
4S	220 960 311.(3)			240 210 377.(3)
4P	204 397 211.(1)	212 658 040.(3)	212 661 348.(2)	212 661 617.7(4)
4D	205 783 935.816(3)	205 842 547.918(6)	205 843 103.149(4)	205 843 139.171(1)
4F	205 620 797.145	205 621 029.602(1)	205 621 502.019(1)	205 621 287.974
5S	139 318 258.(2)			148 807 312.(2)
5P	130 955 541.8(7)	135 203 443.(2)	135 205 105.(1)	135 205 240.8(2)
5D	131 680 211.938(2)	131 714 043.938(3)	131 714 327.498(2)	131 714 346.719(1)
5F	131 595 041.501	131 595 195.235(1)	131 595 419.741(1)	131 595 327.454
5G	131 580 320.1329(1)	131 580 370.9465(2)	131 580 529.5188(2)	131 580 446.4606(1)
6S	95 807 682.0(9)			101 166 442.3(9)
6P	91 009 810.5(4)	93 472 041.5(9)	93 472 992.5(7)	93 473 070.0(1)
6D	91 433 655.841(1)	91 454 440.605(2)	91 454 604.486(1)	91 454 615.8316(5)
6F	91 383 852.0310(2)	91 383 954.3008(5)	91 384 078.8996(5)	91 384 030.7936(2)
6G	91 374 997.961 01(7)	91 375 027.4177(2)	91 375 119.1361(2)	91 375 071.113 69(7)
6H	91 372 940.612 32(3)	91 372 961.813 55(7)	91 373 021.530 29(7)	91 372 990.226 74(3)
7S	69 904 819.7(6)			73 222 269.3(6)
7P	66 901 127.5(2)	68 452 586.6(6)	68 453 180.9(4)	68 453 229.30(8)
7D	67 169 717.1562(6)	67 183 264.590(1)	67 183 367.7091(9)	67 183 374.9339(3)
7F	67 138 158.5571(1)	67 138 228.5582(3)	67 138 305.0654(3)	67 138 276.7195(1)
7G	67 132 455.947 62(6)	67 132 474.5216(1)	67 132 532.2572(1)	67 132 502.036 82(5)
7H	67 131 109.015 31(2)	67 131 122.366 81(5)	67 131 159.972 34(5)	67 131 140.259 24(2)
7I	67 130 692.480 04(1)	67 130 702.489 54(2)	67 130 728.915 04(2)	67 130 715.088 42(1)
8S	53 246 283.1(4)			55 440 834.1(4)
8P	51 242 587.4(2)	52 282 092.0(4)	52 282 488.0(3)	52 282 520.19(5)
8D	51 423 248.1412(4)	51 432 523.2471(8)	51 432 592.2921(6)	51 432 597.1660(2)
8F	51 402 021.6289(1)	51 402 071.1099(2)	51 402 121.5341(2)	51 402 103.3700(1)
8G	51 398 146.238 04(6)	51 398 158.6930(1)	51 398 197.3600(1)	51 398 177.125 22(7)
8H	51 397 221.579 33(2)	51 397 230.523 93(4)	51 397 255.716 51(4)	51 397 242.510 26(2)
8I	51 396 931.943 12(1)	51 396 938.648 74(2)	51 396 956.351 73(2)	51 396 947.088 95(1)
8K	51 396 822.734 66	51 396 827.929 60(1)	51 396 841.052 96(1)	51 396 834.203 39
9S	41 903 979.2(3)			43 430 382.9(3)
9P	40 501 246.4(1)	41 231 283.2(3)	41 231 560.2(2)	41 231 582.71(4)

Table 11.7 Total ionization energies for ^{4}He, calculated with $R_M = 3\,289\,391\,006.715$ MHz, cont.

State	$E(n\,^1L_L)$	$E(n\,^3L_{L-1})$	$E(n\,^3L_L)$	$E(n\,^3L_{L+1})$
9D	40 628 480.2670(3)	40 635 090.4472(6)	40 635 138.9215(4)	40 635 142.3606(2)
9F	40 613 531.5089(1)	40 613 567.5551(2)	40 613 602.5777(2)	40 613 590.2103(1)
9G	40 610 783.198 70(5)	40 610 791.952 31(8)	40 610 819.103 53(8)	40 610 804.897 45(5)
9H	40 610 123.035 88(2)	40 610 129.318 02(3)	40 610 147.011 47(3)	40 610 137.736 30(2)
9I	40 609 914.516 79(1)	40 609 919.226 38(2)	40 609 931.659 73(2)	40 609 925.154 18(1)
9K	40 609 835.097 59	40 609 838.746 17(1)	40 609 847.963 12(1)	40 609 843.152 44
10S	33 834 679.6(2)			34 938 883.9(2)
10P	32 814 665.30(8)	33 346 784.3(2)	33 346 985.6(1)	33 347 001.97(3)
10D	32 907 601.9150(2)	32 912 470.7559(4)	32 912 506.0839(3)	32 912 508.5992(1)
10F	32 896 683.0965(1)	32 896 710.0670(1)	32 896 735.3970(1)	32 896 726.5815(1)
10G	32 894 665.770 94(3)	32 894 672.155 70(5)	32 894 691.945 68(5)	32 894 681.592 48(3)
10H	32 894 178.909 63(1)	32 894 183.489 34(2)	32 894 196.387 81(2)	32 894 189.626 22(1)
10I	32 894 024.241 08(1)	32 894 027.674 38(1)	32 894 036.738 28(1)	32 894 031.995 73(1)
10K	32 893 964.927 04	32 893 967.586 86(1)	32 893 974.306 01(1)	32 893 970.799 02

Table 11.8 QED corrections to the ionization energy included in Table 11.7 for the S- and P-states of helium (in MHz)

	$\Delta E_{L,1} + \Delta E'_{L,1}$		$\Delta E_{L,2} + \Delta E'_{L,2}$	
State	Singlet	Triplet	Singlet	Triplet
1S	−45 409.		4173.[a]	
2S	−3134.4	−4098.7	327.865	39.883[a]
3S	−858.34	−1030.29	91.258	8.468
4S	−349.09	−402.29	37.303	3.203
5S	−174.93	−196.80	18.735	1.544
6S	−99.807	−110.505	10.702	0.861
7S	−62.221	−68.113	6.677	0.528
8S	−41.369	−44.904	4.441	0.347
9S	−28.885	−31.147	3.102	0.240
10S	−20.959	−22.482	2.251	0.173
2P	−103.6	1208.7	62.608	45.502
3P	−35.13	344.96	19.559	12.376
4P	−15.15	142.33	8.413	5.035
5P	−7.816	71.911	4.348	2.529
6P	−4.540	41.256	2.529	1.446
7P	−2.866	25.824	1.598	0.904
8P	−1.923	17.223	1.073	0.602
9P	−1.352	12.055	0.755	0.421
10P	−0.987	8.764	0.551	0.306

[a] Includes additional contributions of −4 MHz for the $1\,^1$S state [11.48] and 3.00(1) MHz for the $2\,^3$S state [11.47] due to electron–electron terms of $O(\alpha^4)R_\infty$

11.4.1 Quantum Defect Extrapolations

As discussed in Sect. 14.1, the ionization energies of an isolated Rydberg series of states can be expressed in the form

$$W_n = R_M(Z-1)^2/n^{*2}\,, \tag{11.50}$$

where $Z-1$ is the screened nuclear charge and n^* is the effective principal quantum number defined by an iterative solution to the equation

$$n^* = n - \delta(n^*)\,, \tag{11.51}$$

where $\delta(n^*)$ is the *quantum defect* defined by the Ritz expansion

$$\delta(n^*) = \delta_0 + \frac{\delta_2}{(n-\delta)^2} + \frac{\delta_4}{(n-\delta)^4} + \cdots \tag{11.52}$$

with constant coefficients δ_i. The absence of *odd* terms in this series is a special property of the eigenvalues of Hamiltonians of the form $H_C + V$, where H_C is a pure one-electron Coulomb Hamiltonian, and V is a local, short-range, spherically symmetric potential of arbitrary strength (see [11.54] for further discussion). For the Rydberg states of helium, odd terms must be included in the Ritz expansion (11.52) due to relativistic and mass polarization corrections, but they can be removed again by first adjusting the energies according to

$$W'_n = W_n - \Delta W_n\,, \tag{11.53}$$

where, to sufficient accuracy [11.54] [see discussion following (11.66)]

$$\Delta W_n = R_M \left\{ \frac{-3\alpha^2(Z-1)^4}{4n^4} + \left(\frac{\mu}{M}\right)^2 \frac{(Z-1)^2}{n^2} \times \left[1 + \frac{5}{6}(\alpha Z)^2\right] \right\} , \qquad (11.54)$$

with $Z = 2$ for helium. The quantum defect parameters listed in Table 11.9 provide accurate extrapolations to higher-lying Rydberg states, with

$$W_n = R_M / n^{*2} + \Delta W_n \,. \qquad (11.55)$$

Table 11.9 Quantum defects for the total energies of helium with the ΔW_n term subtracted (11.54)

δ_i	Value			
	$^1S_0$			$^3S_1$
δ_0	0.139 718 064 86(21)			0.296 656 487 71(75)
δ_2	0.027 835 737(18)			0.038 296 666(59)
δ_4	0.016 792 29(41)			0.007 5131(12)
δ_6	−0.001 4590(31)			−0.004 5476(79)
δ_8	0.002 9227(65)			0.002 180(14)
	$^1P_1$	$^3P_0$	$^3P_1$	$^3P_2$
δ_0	−0.012 141 803 603(64)	0.068 328 002 51(27)	0.068 357 857 65(27)	0.068 360 283 79(23)
δ_2	0.007 519 0804(59)	−0.018 641 975(24)	−0.018 630 462(24)	−0.018 629 228(21)
δ_4	0.013 977 80(15)	−0.012 331 65(57)	−0.012 330 40(57)	−0.012 332 75(51)
δ_6	0.004 8373(12)	−0.007 9515(45)	−0.007 9512(45)	−0.007 9527(41)
δ_8	0.001 2283(29)	−0.005 448(10)	−0.005 450(10)	−0.005 451(9)
	$^1D_2$	$^3D_1$	$^3D_2$	$^3D_3$
δ_0	0.002 113 378 464(49)	0.002 885 580 281(22)	0.002 890 941 493(25)	0.002 891 328 825(26)
δ_2	−0.003 090 0510(58)	−0.006 357 6012(27)	−0.006 357 1836(30)	−0.006 357 7040(33)
δ_4	0.000 008 27(22)	0.000 336 67(11)	0.000 337 77(11)	0.000 336 70(13)
δ_6	−0.000 3094(31)	0.000 8394(16)	0.000 8392(16)	0.000 8395(18)
δ_8	−0.000 401(14)	0.000 3798(72)	0.000 4323(75)	0.000 3811(83)
	$^1F_3$	$^3F_2$	$^3F_3$	$^3F_4$
δ_0	0.000 440 294 26(62)	0.000 444 869 89(22)	0.000 448 594 83(28)	0.000 447 379 27(21)
δ_2	−0.001 689 446(65)	−0.001 739 275(24)	−0.001 727 232(30)	−0.001 739 217(23)
δ_4	−0.000 1183(20)	0.000 104 76(76)	0.000 1524(9)	0.000 104 78(71)
δ_6	0.000 326(18)	0.000 0337(69)	−0.000 2486(83)	0.000 0331(64)
	$^1G_4$	$^3G_3$	$^3G_4$	$^3G_5$
δ_0	0.000 124 734 490(79)	0.000 125 707 43(12)	0.000 128 713 16(10)	0.000 127 141 67(11)
δ_2	−0.000 796 230(12)	−0.000 796 498(19)	−0.000 796 246(15)	−0.000 796 484(17)
δ_4	−0.000 012 05(53)	−0.000 009 80(81)	−0.000 011 89(66)	−0.000 009 85(75)
δ_6	−0.000 0136(69)	−0.000 019(11)	−0.000 0141(85)	−0.000 019(10)
	$^1H_5$	$^3H_4$	$^3H_5$	$^3H_6$
δ_0	0.000 047 100 899(61)	0.000 047 797 067(43)	0.000 049 757 614(51)	0.000 048 729 846(45)
δ_2	−0.000 433 2277(84)	−0.000 433 2322(55)	−0.000 433 2274(65)	−0.000 433 2281(57)
δ_4	−0.000 008 14(26)	−0.000 008 07(16)	−0.000 008 13(19)	−0.000 008 10(16)
	$^1I_6$	$^3I_5$	$^3I_6$	$^3I_7$
δ_0	0.000 021 868 881(17)	0.000 022 390 759(20)	0.000 023 768 483(14)	0.000 023 047 609(26)
δ_2	−0.000 261 0673(22)	−0.000 261 0680(28)	−0.000 261 0662(18)	−0.000 261 0672(35)
δ_4	−0.000 004 048(67)	−0.000 004 042(87)	−0.000 004 076(58)	−0.000 004 04(11)

11.4.2 Asymptotic Expansions

The asymptotic expansion method [11.39, 55] rapidly increases in accuracy with increasing angular momentum L of the Rydberg electron, and can be used to high precision for $L \geq 7$. The method is based on a model in which:

1. the Rydberg electron, treated as a distinguishable particle, moves in the field of the core consisting of the He nucleus and a tightly bound 1s electron.
2. the core, as characterized by its various multipole moments, is perturbed by the electric field of the Rydberg electron.

A systematic perturbation expansion yields an asymptotic series of the form

$$\varepsilon_0^{nL} = -2 - \frac{(Z-1)^2}{2n^2} + a_0 \sum_{j=4}^{N} A_j \langle r^{-j} \rangle_{nL} \frac{e^2}{a_0}, \quad (11.56)$$

where the expectation value $\langle r^{-j} \rangle_{nL}$ is calculated with respect to the hydrogenic nL-electron wave function [11.56] and the series is truncated at the upper limit N where the series begins diverging. The leading coefficients A_j are

$$A_4 = -\frac{1}{2}\alpha_1, \quad A_5 = 0, \quad A_6 = -\frac{1}{2}(\alpha_2 - 6\beta_1),$$

where α_k is the 2^k-pole polarizability of the hydrogenic core and β_k is a nonadiabatic correction. The exact hydrogenic values are

$$\alpha_1 = \frac{9a_0^3}{2Z^4}, \quad \alpha_2 = \frac{15a_0^5}{Z^6}, \quad \beta_1 = \frac{43a_0^5}{8Z^6}.$$

All terms are known up to A_{10} (see [11.39, 55] for detailed results). The expansions for the terms ε_0, ε_1, and ε_2 in (11.6) for helium are

$$\begin{aligned}\varepsilon_0^{nL} = & -2 - \frac{1}{2n^2} - \frac{9}{64}\langle r^{-4}\rangle \\ & + \frac{69}{512}\langle r^{-6}\rangle + \frac{3833}{15\,360}\langle r^{-7}\rangle \\ & - \left(\frac{55\,923}{65\,536} + \frac{957L(L+1)}{10\,240}\right)\langle r^{-8}\rangle \\ & - \frac{908\,185}{688\,128}\langle r^{-9}\rangle \\ & + \left(\frac{3\,824\,925}{1\,048\,576} + \frac{33\,275L(L+1)}{28\,672}\right)\langle r^{-10}\rangle \\ & + \mathrm{e}^{(1,1)} - \frac{23}{20}\mathrm{e}^{(1,2)}, \end{aligned} \quad (11.57)$$

$$\begin{aligned}\varepsilon_1^{nL} = & -\frac{9}{32}\langle r^{-4}\rangle + \frac{249}{256}\langle r^{-6}\rangle + \frac{319}{3840}\langle r^{-7}\rangle \\ & - \left(\frac{34\,659}{16\,384} + \frac{957L(L+1)}{5120}\right)\langle r^{-8}\rangle \\ & - \frac{14\,419}{3072}\langle r^{-9}\rangle \\ & + \left(\frac{6\,413\,781}{262\,144} + \frac{24\,155L(L+1)}{8192}\right)\langle r^{-10}\rangle \\ & + 4\mathrm{e}^{(1,1)} - \frac{53}{5}\mathrm{e}^{(1,2)}, \end{aligned} \quad (11.58)$$

$$\begin{aligned}\varepsilon_2^{nL} = & -\frac{1}{2n^2} - \frac{45}{64}\langle r^{-4}\rangle + \frac{165}{512}\langle r^{-6}\rangle + \frac{2555}{3072}\langle r^{-7}\rangle \\ & - \left(\frac{268\,485}{32\,768} + \frac{957L(L+1)}{2048}\right)\langle r^{-8}\rangle \\ & + \frac{598\,909}{172\,032}\langle r^{-9}\rangle \\ & + \left(\frac{3\,907\,923}{524\,288} + \frac{629\,515L(L+1)}{114\,688}\right)\langle r^{-10}\rangle \\ & + 14\mathrm{e}^{(1,1)} - \frac{251}{10}\mathrm{e}^{(1,2)}. \end{aligned} \quad (11.59)$$

The terms $\mathrm{e}^{(1,1)}$ and $\mathrm{e}^{(1,2)}$ are second-order dipole–dipole and dipole–quadrupole perturbation corrections. Defining $f_p^L = (L+p)!/(L-p)!$, they are given by

$$\begin{aligned}\mathrm{e}^{(i,j)} = & -\frac{(2-\delta_{j,k})2^{2i+2j+1}(2L-2i)!(2L-2j)!}{n^3(2L+2i+1)!(2L+2j+1)!} \\ & \times \left(\frac{2^{2i+2j}(2L-2i-2j)!A^{(i,j)}}{n^{2i+2j+2}(2L+2i+2j+1)!} \right. \\ & \left. + \frac{B^{(i,j)}}{n^{2i+2j+1}}\right) \end{aligned} \quad (11.60)$$

with

$$\begin{aligned} A^{(1,1)} = & \, 3n^2(3n^2 - 2f_1)(f_1 - 2)(45 + 623f_1^L \\ & + 3640f_2^L + 560f_3^L), \\ B^{(1,1)} = & \, (9n^2 - 7f_1^L)(3n^2 - f_1^L), \\ A^{(1,2)} = & -21n^6(94\,500 + 122\,850f_1^L \\ & - 1\,126\,125f_2^L - 18\,931\,770f_3^L \\ & - 11\,171\,160f_4^L - 1\,029\,600f_5^L \\ & - 18\,304f_6^L) \\ & - 15n^4(94\,500 - 444\,150f_1^L \\ & + 7\,747\,425f_2^L + 337\,931\,880f_3^L \\ & + 375\,290\,190f_4^L + 66\,518\,760f_5^L \end{aligned}$$

$$
\begin{aligned}
&+2\,880\,416 f_6^L + 29\,568 f_7^L)\\
&+ 9n^2 f_1^L (90\,300 - 177\,450 f_1^L\\
&+1\,738\,450 f_2^L + 133\,125\,575 f_3^L\\
&+160\,040\,870 f_4^L + 29\,322\,216 f_5^L\\
&+1\,293\,600 f_6^L + 13\,440 f_7^L)\\
&+ 2 f_1^L f_2^L f_3^L (45 + 252 f_1^L - 1680 f_2^L\\
&-2240 f_3^L)\,,\\
B^{(1,2)} &= 315n^6 + 125n^4(3 - 5 f_1^L)\\
&- 7n^2 f_1^L (43 - 39 f_1^L) - 27 f_1^L f_2^L\,.
\end{aligned}
$$

The accuracy of the expansion for the ε_0, ε_1, and ε_2 can be reliably estimated to be one-half of the last $\langle r^{-j}\rangle$ term included in the sum. Formulas for the $\langle r^{-j}\rangle$ are given in Table 11.10.

The asymptotic formulas for the NFS relativistic corrections are [11.39, 57]

$$
\begin{aligned}
\langle H_{\mathrm{mass}} + H_{\mathrm{D}}\rangle \to &-\frac{\alpha^2 Z^4}{8} + h_1(nL) + \chi_1(nL)\\
&+ \frac{(Z\alpha)^2}{2}\left(\frac{14}{3Z^4}\langle r^{-4}\rangle\right.\\
&\left.-\frac{5041}{240Z^6}\langle r^{-6}\rangle\right)
\end{aligned}
\tag{11.61}
$$

$$
\begin{aligned}
\langle H_{\mathrm{oo}}\rangle \to \frac{\alpha^2}{Z^2}&\left(\langle r^{-4}\rangle + \frac{3(Z-1)}{Z^2}\langle r^{-5}\rangle\right.\\
&\left.-\frac{3(f_1^L+8)}{4Z^2}\langle r^{-6}\rangle\right)\,,
\end{aligned}
\tag{11.62}
$$

where

$$
h_1(nL) = \frac{\alpha^2(Z-1)^4}{2n^3}\left(\frac{3}{4n} - \frac{1}{L+\frac{1}{2}}\right)
\tag{11.63}
$$

is the leading one-electron Dirac energy and

$$
\begin{aligned}
\chi_1(nL) = \frac{\alpha^2\alpha_1}{2}&\left\{3\left(\frac{Z-1}{n}\right)^2\langle r^{-4}\rangle\right.\\
&-(Z-1)\langle r^{-5}\rangle - \frac{4(2L-2)!}{(2L+3)!}\\
&\times\left[4\left(\frac{Z-1}{n}\right)^6\left(n + \frac{9n^2 - 5 f_1^L}{2L+1}\right)\right.\\
&\left.\left.+(Z-1)^2\left(\frac{40 f_2^L + 70 f_1^L - 3}{2L+1}\right)\langle r^{-4}\rangle\right]\right\}
\end{aligned}
\tag{11.64}
$$

is the correction due to the dipole perturbation of the Rydberg electron. The relativistic recoil terms due to mass polarization are

$$
\begin{aligned}
&\langle H_{\mathrm{mass}} + H_{\mathrm{D}}\rangle_{\mathrm{RR}}\\
&\to \frac{\mu}{M}\left[\frac{22(Z\alpha)^2(Z-1)}{9Z^4}\langle r^{-4}\rangle + 2(Z-1)\chi_1(nL)\right]\\
&\quad+\left(\frac{\mu}{M}\right)^2\left[-\frac{5}{12}\left(\frac{\alpha Z(Z-1)}{n}\right)^2 + 4h_1(nL)\right]\,,
\end{aligned}
\tag{11.65}
$$

$$
\begin{aligned}
&\langle H_{\mathrm{oo}}\rangle_{\mathrm{RR}} + \langle \Delta_{\mathrm{oo}}\rangle\\
&\to -\frac{\alpha^2\mu}{M}\left[Z^4 + \frac{(Z-1)^4}{n^3}\left(\frac{1}{n} - \frac{3}{2L+1}\right)\right.\\
&\qquad\left.-\frac{25[1+13f(Z)]}{16Z^2}\langle r^{-4}\rangle\right]\,,
\end{aligned}
\tag{11.66}
$$

with $f(Z) \simeq 1 + (Z-2)/6$. The $-(5/12)[\alpha Z(Z-1)/n]^2$ term in (11.65) is the dominant contribution in helium for $L \geq 4$. It is included in (11.54) for ΔW_n, along with the leading $1/n^2$ term from (11.59), and the $1/n^4$ term from (11.63). The complete relativistic finite

Table 11.10 Formulas for the hydrogenic expectation value $\langle r^{-j}\rangle \equiv \langle nl|r^{-j}|nl\rangle$ in terms of

$$
G_p^{nl} = \frac{2^p Z^p (2l-p+2)!}{n^{p+1}(2l+p-1)!}\,, \quad f_p^l = \frac{(l+p)!}{(l-p)!}\,.
$$

j	$\langle r^{-j}\rangle$ (a_0)
2	$\frac{1}{2}G_2^{nl}$
3	nG_3^{nl}
4	$G_4^{nl}(3n^2 - f_1^l)$
5	$2G_5^{nl}\left[5n^3 - n(3f_1^l - 1)\right]$
6	$G_6^{nl}\left[35n^4 - 5n^2(6f_1^l - 5) + 3f_2^l\right]$
7	$2G_7^{nl}\left[63n^5 - 35n^3(2f_1^l - 3) + n(15f_2^l - 20f_1^l + 12)\right]$
8	$G_8^{nl}\left[462n^6 - 210n^4(3f_1^l - 7) + 42n^2(5f_2^l - 15f_1^l + 14) - 10f_3^l\right]$
9	$2G_9^{nl}\left[858n^7 - 462n^5(3f_1^l - 10) + 42n^3(15f_2^l - 75f_1^l + 101) - 2n(35f_3^l - 105f_2^l + 252f_1^l - 180)\right]$
10	$G_{10}^{nl}\left[6435n^8 - 6006n^6(2f_1^l - 9) + 1155n^4(6f_2^l - 44f_1^l + 81) - 6n^2(210f_3^l - 1365f_2^l + 4648f_1^l - 4566) + 35f_4^l\right]$

mass correction includes also the mass-scaling terms $-(\mu/M)\langle 4H_{\text{mass}}+3H_{\text{D}}+3H_{\text{oo}}\rangle$ obtained by expanding μ/m_{e} in (11.47). The $\langle\delta(\boldsymbol{r}_1)\rangle$ term is [11.58]

$$\pi\langle\delta(\boldsymbol{r}_1)\rangle \to \frac{Z^3}{2}-\frac{31}{4Z^3}\langle r^{-4}\rangle+\frac{1447}{32Z^5}\langle r^{-7}\rangle - \frac{-31(Z-1)}{2Z^3}\left(\frac{\mu}{M}\right)\langle r^{-4}\rangle+\cdots . \tag{11.67}$$

$\langle\delta(\boldsymbol{r}_{12})\rangle$ vanishes exponentially as $1/n^{2L+4}$ with increasing L. The complete asymptotic expressions for the FS matrix elements are summarized by the formulas

$$\begin{aligned}&\langle nL\,^3L_J|H_{\text{FS}}|nL\,^3L_J\rangle\\&\to T_{nL}(J)\{Z-3+2S_L(J)+2a_{\text{e}}[Z-2\\&\quad+(2+a_{\text{e}})S_L(J)]+(\mu/M)[2-4S_L()]\}\,,\end{aligned} \tag{11.68}$$

$$\begin{aligned}&\langle nL\,^3L_J|H_{\text{FS}}|nL\,^1L_J\rangle\\&\to T_{nL}(L)(Z+1+2a_{\text{e}}Z-2\mu/M)\sqrt{L(L+1)}\,,\end{aligned} \tag{11.69}$$

where

$$T_{nL}(J)=\begin{cases}-\alpha^2(L+1)\langle r^{-3}\rangle/4 & J=L-1\,,\\ -\alpha^2\langle r^{-3}\rangle/4 & J=L\,,\\ \alpha^2L\langle r^{-3}\rangle/4 & J=L+1\,,\end{cases}$$

$$S_L(J)=\begin{cases}1 & J=L\,,\\ \pm1/(2J+1) & J=L\pm1\,.\end{cases}$$

The asymptotic form for the QED term $\Delta E_{L,1}$ follows from (11.38) with the use of (11.67) for $\langle\delta(\boldsymbol{r}_1)\rangle$ and (11.41) for $\ln k_0$. The electron–electron part is

$$\Delta E_{L,2}\to-\frac{7\alpha^3}{6\pi}\left(\langle r^{-3}\rangle+\frac{3}{Z^2}\langle r^{-5}\rangle\right). \tag{11.70}$$

With the use of the formulas in this section, the variationally calculated ionization energies for the K-states ($L=7$) in Table 11.7 can be reproduced to within ±20 Hz. For $L>7$, the uncertainty becomes less than 1 Hz, up to the Casimir-Polder retardation effects which have not been included.

11.5 Radiative Transitions

11.5.1 Basic Formulation

In a semiclassical picture, the interaction Hamiltonian with the radiation field is obtained by making the minimal coupling replacements

$$\begin{aligned}\boldsymbol{P}_{\text{N}}&\to\boldsymbol{P}_{\text{N}}-\frac{Ze}{c}\boldsymbol{A}(\boldsymbol{R}_{\text{N}})\\ \boldsymbol{P}_i&\to\boldsymbol{P}_i+\frac{e}{c}\boldsymbol{A}_i(\boldsymbol{R}_i)\end{aligned} \tag{11.71}$$

in (11.1), where

$$\boldsymbol{A}(\boldsymbol{R})=c\left(\frac{2\pi\hbar}{\omega\mathcal{V}}\right)^{1/2}\hat{\boldsymbol{\epsilon}}\,\mathrm{e}^{\mathrm{i}\boldsymbol{k}\cdot\boldsymbol{R}} \tag{11.72}$$

is the time-independent part of the vector potential $\boldsymbol{A}(\boldsymbol{r},t)=\boldsymbol{A}(\boldsymbol{r})\mathrm{e}^{-\mathrm{i}\omega t}+\text{c.c}$ for a photon of frequency ω, wave vector $\boldsymbol{k}$, and polarization $\hat{\boldsymbol{\epsilon}}\perp\boldsymbol{k}$ normalized to unit photon energy $\hbar\omega$ in volume $\mathcal{V}$. The linear coupling terms then yield

$$H_{\text{int}}=-\frac{Ze}{Mc}\boldsymbol{P}_{\text{N}}\cdot\boldsymbol{A}(\boldsymbol{R}_{\text{N}})+\frac{e}{m_{\text{e}}c}\sum_{i=1}^{2}\boldsymbol{P}_i\cdot\boldsymbol{A}(\boldsymbol{R}_i)\,, \tag{11.73}$$

and from Fermi's Golden Rule, the decay rate for spontaneous emission from state γ to γ' is

$$w_{\gamma,\gamma'}\,\mathrm{d}\Omega=\frac{2\pi}{\hbar}|\langle\gamma|H_{\text{int}}|\gamma'\rangle|^2\rho_f\,, \tag{11.74}$$

where $\rho_f=\mathcal{V}\omega^2\,\mathrm{d}\Omega/(2\pi c)^3\hbar$ is the number of photon states with polarization $\hat{\boldsymbol{\epsilon}}$ per unit energy and solid angle in the normalization volume $\mathcal{V}$. In the long wavelength and electric dipole approximations, the factor $\mathrm{e}^{\mathrm{i}\boldsymbol{k}\cdot\boldsymbol{R}}$ in (11.72) is replaced by unity. After integrating over angles $\mathrm{d}\Omega$ and summing over polarizations $\hat{\boldsymbol{\epsilon}}$, the decay rate reduces to

$$w_{\gamma,\gamma'}=\frac{4}{3}\alpha\omega_{\gamma,\gamma'}|\langle\gamma|\boldsymbol{Q}_p|\gamma'\rangle|^2\,, \tag{11.75}$$

where $\omega_{\gamma',\gamma}$ is the transition frequency and $\boldsymbol{Q}_p$ is the velocity form of the transition operator

$$\boldsymbol{Q}_p=-\frac{Z}{Mc}\boldsymbol{P}_{\text{N}}+\frac{1}{m_{\text{e}}c}\sum_{i=1}^{N}\boldsymbol{P}_i \tag{11.76}$$

for the general case of N electrons. From the commutator $[H_0,\boldsymbol{Q}_r/\hbar\omega_{\gamma,\gamma'}]=\boldsymbol{Q}_p$, where H_0 is the field-free Hamiltonian in (11.1), the equivalent length form is

$$\boldsymbol{Q}_r=-\frac{i}{c}\omega_{\gamma,\gamma'}\left(Z\boldsymbol{R}_{\text{N}}-\sum_{i=1}^{N}\boldsymbol{R}_i\right). \tag{11.77}$$

After transforming to c.m. plus relative coordinates in parallel with (11.3), the dipole transition operators become

$$\boldsymbol{Q}_p = \frac{Z_p}{m_e c}\sum_{i=1}^{N} \boldsymbol{p}_i\,, \quad \boldsymbol{Q}_r = \frac{i\omega_{\gamma,\gamma'}}{c} Z_r \sum_{i=1}^{N} \boldsymbol{r}_i\,, \tag{11.78}$$

with

$$Z_p = \frac{Zm_e + M}{M}\,, \quad Z_r = \frac{Zm_e + M}{Nm_e + M}\,,$$

and H_0 now contains the H_{mp} term. If (11.3) is solved exactly for the states $|\gamma\rangle$ and $|\gamma'\rangle$, then the identity

$$\langle\gamma|\boldsymbol{Q}_p|\gamma'\rangle = \langle\gamma|\boldsymbol{Q}_r|\gamma'\rangle \tag{11.79}$$

is satisfied to all orders in m_e/M. For a neutral atom, $N = Z$ and $Z_r = 1$. If the oscillator strength is defined by

$$\begin{aligned} f_{\gamma',\gamma} &= \frac{2m_e\omega_{\gamma',\gamma}}{3\hbar}\left(\frac{Z_r}{Z_p}\right)\left|\left\langle\gamma'\left|\sum_{i=1}^{N}\boldsymbol{r}_i\right|\gamma\right\rangle\right|^2 \\ &= \frac{2}{3m_e\hbar\omega_{\gamma',\gamma}}\left(\frac{Z_p}{Z_r}\right)\left|\left\langle\gamma'\left|\sum_{i=1}^{N}\boldsymbol{p}_i\right|\gamma\right\rangle\right|^2 \end{aligned} \tag{11.80}$$

then the sum rule $\sum_{\gamma'} f_{\gamma',\gamma} = N$ remains valid, independent of m_e/M. The decay rate, summed over final states and averaged over initial states, is

$$\bar{w}_{\gamma,\gamma'} = -\frac{2\alpha\hbar\omega^2_{\gamma,\gamma'}}{m_e c^2} Z_p Z_r \bar{f}_{\gamma,\gamma'}\,, \tag{11.81}$$

where $\bar{f}_{\gamma,\gamma'} = -(g'_\gamma/g_\gamma)\bar{f}_{\gamma',\gamma}$ is the (negative) oscillator strength for photon emission, and g'_γ, g_γ are the statistical weights of the states.

11.5.2 Oscillator Strength Table

Table 11.11 provides arrays of nonrelativistic oscillator strengths among various states of helium, including the effects of finite nuclear mass as a separate factor. In the absence of mass polarization, the correction factor would be $(1+\mu/M)^{-1} \simeq 1-\mu/M$. Mass polarization effects are particularly strong for P-states, and for transitions with $\Delta n = 0$.

The largest relativistic correction comes from singlet–triplet mixing between states with the same n, L, and J (e.g. $3\,^1D_2$ and $3\,^3D_2$) due to $\boldsymbol{H}_{FS}$. The wave functions obtained by diagonalizing the 2×2 matrices

Table 11.11 Oscillator strengths for helium. The factor in brackets gives the finite mass correction, with $y = \mu/M$

	$1\,^1S$	$2\,^1S$	$3\,^1S$	$4\,^1S$
$2\,^1P$	$0.276\,1647(1-2.282y)$	$0.376\,4403(1+1.255y)$	$-0.145\,4703(1+1.351y)$	$-0.025\,8703(1+0.885y)$
$3\,^1P$	$0.073\,4349(1-1.789y)$	$0.151\,3417(1-3.971y)$	$0.626\,1931(1+1.234y)$	$-0.307\,5074(1+1.097y)$
$4\,^1P$	$0.029\,8629(1-1.583y)$	$0.049\,1549(1-3.235y)$	$0.143\,8889(1-4.650y)$	$0.858\,0214(1+1.205y)$
$5\,^1P$	$0.015\,0393(1-1.474y)$	$0.022\,3377(1-2.967y)$	$0.050\,4714(1-3.764y)$	$0.146\,2869(1-5.080y)$
$6\,^1P$	$0.008\,6277(1-1.407y)$	$0.012\,1340(1-2.829y)$	$0.024\,1835(1-3.444y)$	$0.052\,7562(1-4.105y)$
$7\,^1P$	$0.005\,4054(1-1.362y)$	$0.007\,3596(1-2.75y)$	$0.013\,6794(1-3.279y)$	$0.025\,8918(1-3.75y)$
	$2\,^3S$	$3\,^3S$	$4\,^3S$	$5\,^3S$
$2\,^3P$	$0.539\,0861(1-3.185y)$	$-0.208\,5359(1-3.773y)$	$-0.031\,7208(1-2.819y)$	$-0.011\,3409(1-2.609y)$
$3\,^3P$	$0.064\,4612(1+5.552y)$	$0.890\,8513(1-2.967y)$	$-0.435\,6711(1-3.362y)$	$-0.067\,6073(1-2.359y)$
$4\,^3P$	$0.025\,7689(1+3.886y)$	$0.050\,0833(1+7.505y)$	$1.215\,2630(1-2.878y)$	$-0.668\,3003(1-3.185y)$
$5\,^3P$	$0.012\,4906(1+3.332y)$	$0.022\,9141(1+5.209y)$	$0.044\,2305(1+9.009y)$	$1.530\,6287(1-2.827y)$
$6\,^3P$	$0.006\,9822(1+3.063y)$	$0.011\,9933(1+4.460y)$	$0.021\,6301(1+6.198y)$	$0.041\,5177(1+10.215y)$
$7\,^3P$	$0.004\,2990(1+2.908y)$	$0.007\,0772(1+4.092y)$	$0.011\,7754(1+5.292y)$	$0.021\,1003(1+6.981y)$
	$2\,^1P$	$3\,^1P$	$4\,^1P$	$5\,^1P$
$3\,^1D$	$0.710\,1641(1-0.281y)$	$-0.021\,1401(1+29.947y)$	$-0.015\,3034(1-6.680y)$	$-0.003\,1128(1-6.27y)$
$4\,^1D$	$0.120\,2704(1-1.307y)$	$0.648\,1049(1+0.435y)$	$-0.040\,0610(1+29.183y)$	$-0.039\,2932(1-6.163y)$
$5\,^1D$	$0.043\,2576(1-1.681y)$	$0.141\,3027(1-0.566y)$	$0.647\,6679(1+0.817y)$	$-0.057\,3258(1+28.903y)$
$6\,^1D$	$0.020\,9485(1-1.866y)$	$0.056\,2766(1-0.936y)$	$0.152\,8104(1-0.170y)$	$0.669\,8361(1+1.056y)$
$7\,^1D$	$0.011\,8970(1-1.975y)$	$0.028\,8961(1-1.127y)$	$0.063\,5953(1-0.538y)$	$0.163\,0272(1+0.082y)$
$8\,^1D$	$0.007\,4645(1-2.046y)$	$0.017\,0777(1-1.241y)$	$0.033\,6403(1-0.731y)$	$0.069\,3063(1-0.26y)$

Table 11.11 Oscillator strengths for helium. The factor in brackets gives the finite mass correction, with $y = \mu/M$, cont.

	$2\,^3$P	$3\,^3$P	$4\,^3$P	$5\,^3$P
$3\,^3$D	0.610 2252(1 − 2.029y)	0.112 1004(1 + 6.653y)	−0.036 9592(1 + 3.292y)	−0.006 9009(1 + 2.678y)
$4\,^3$D	0.122 8469(1 − 1.001y)	0.477 5938(1 − 3.059y)	0.200 9498(1 + 6.368y)	−0.088 3017(1 + 2.939y)
$5\,^3$D	0.047 0071(1 − 0.631y)	0.124 5532(1 − 2.019y)	0.438 3888(1 − 3.607y)	0.280 0558(1 + 6.225y)
$6\,^3$D	0.023 4692(1 − 0.449y)	0.053 0093(1 − 1.631y)	0.123 9414(1 − 2.555y)	0.429 4411(1 − 3.961y)
$7\,^3$D	0.013 5638(1 − 0.346y)	0.028 1587(1 − 1.432y)	0.055 2332(1 − 2.153y)	0.125 2389(1 − 2.904y)
$8\,^3$D	0.008 6047(1 − 0.280y)	0.016 9809(1 − 1.315y)	0.030 2853(1 − 1.94y)	0.057 0589(1 − 2.498y)
	$3\,^1$D	$4\,^1$D	$5\,^1$D	$6\,^1$D
$4\,^1$F	1.015 0829(1 − 1.010y)	0.002 4920(1 + 3.833y)	−0.012 6968(1 − 0.888y)	−0.002 2631(1 − 0.890y)
$5\,^1$F	0.156 8808(1 − 0.993y)	0.886 1343(1 − 1.023y)	0.004 6467(1 + 4.139y)	−0.033 2539(1 − 0.893y)
$6\,^1$F	0.054 0508(1 − 0.984y)	0.186 0576(1 − 1.001y)	0.839 1374(1 − 1.031y)	0.006 6028(1 + 4.302y)
$7\,^1$F	0.025 6799(1 − 0.978y)	0.072 3229(1 − 0.994y)	0.196 3692(1 − 1.014y)	0.826 9464(1 − 1.039y)
$8\,^1$F	0.014 4782(1 − 0.978y)	0.036 6627(1 − 0.987y)	0.080 7847(1 − 1.003y)	0.203 1182(1 − 1.019y)
$9\,^1$F	0.009 0730(1 − 0.977y)	0.021 5401(1 − 0.975y)	0.042 4256(1 − 1.000y)	0.086 0955(1 − 1.01y)
	$3\,^3$D	$4\,^3$D	$5\,^3$D	$6\,^3$D
$4\,^3$F	1.014 3389(1 − 0.997y)	0.003 3992(1 − 2.166y)	−0.012 8084(1 − 1.042y)	−0.002 2830(1 − 1.044y)
$5\,^3$F	0.156 9831(1 − 1.004y)	0.884 5767(1 − 0.991y)	0.006 5121(1 − 2.387y)	−0.033 5369(1 − 1.043y)
$6\,^3$F	0.054 1179(1 − 1.006y)	0.186 0264(1 − 1.003y)	0.837 0221(1 − 0.988y)	0.009 3836(1 − 2.499y)
$7\,^3$F	0.025 7201(1 − 1.008y)	0.072 3579(1 − 1.003y)	0.196 2031(1 − 0.996y)	0.824 4031(1 − 0.984y)
$8\,^3$F	0.014 5037(1 − 1.009y)	0.036 6936(1 − 1.004y)	0.080 7712(1 − 1.00y)	0.202 8407(1 − 0.993y)
$9\,^3$F	0.009 0903(1 − 1.008y)	0.021 5632(1 − 1.011y)	0.042 4344(1 − 0.99y)	0.086 0373(1 − 0.99y)

$\boldsymbol{H}_0 + \boldsymbol{H}_{\rm NFS} + \boldsymbol{H}_{\rm FS}$ are then

$$\Psi(n\,^1L_L) = \Psi_0(n\,^1L_L)\cos\theta + \Psi_0(n\,^3L_L)\sin\theta$$
$$\Psi(n\,^3L_L) = -\Psi_0(n\,^1L_L)\sin\theta + \Psi_0(n\,^3L_L)\cos\theta\,.$$

Values of $\sin\theta$ are listed in Table 11.12. The corrected oscillator strengths $\tilde{f}_{\gamma,\gamma'}$ for the singlet (s) and triplet (t) components of a $\gamma \to \gamma'$ transition can then be calculated from the values in Table 11.11 according to

$$\tilde{f}^{\rm ss}_{\gamma,\gamma'} = \omega^{\rm ss}_{\gamma,\gamma'}\Big(X^{\rm ss}_{\gamma,\gamma'}\cos\theta_\gamma\cos\theta_{\gamma'} + X^{\rm tt}_{\gamma,\gamma'}\sin\theta_\gamma\sin\theta_{\gamma'}\Big)^2\,,$$

$$\tilde{f}^{\rm tt}_{\gamma,\gamma'} = \omega^{\rm tt}_{\gamma,\gamma'}\Big(X^{\rm ss}_{\gamma,\gamma'}\sin\theta_\gamma\sin\theta_{\gamma'} + X^{\rm tt}_{\gamma,\gamma'}\cos\theta_\gamma\cos\theta_{\gamma'}\Big)^2\,,$$

$$\tilde{f}^{\rm st}_{\gamma,\gamma'} = \omega^{\rm st}_{\gamma,\gamma'}\Big(X^{\rm ss}_{\gamma,\gamma'}\cos\theta_\gamma\sin\theta_{\gamma'} - X^{\rm tt}_{\gamma,\gamma'}\sin\theta_\gamma\cos\theta_{\gamma'}\Big)^2\,,$$

$$\tilde{f}^{\rm ts}_{\gamma,\gamma'} = \omega^{\rm ts}_{\gamma,\gamma'}\Big(X^{\rm ss}_{\gamma,\gamma'}\sin\theta_\gamma\cos\theta_{\gamma'} - X^{\rm tt}_{\gamma,\gamma'}\cos\theta_\gamma\sin\theta_{\gamma'}\Big)^2\,,$$

Table 11.12 Singlet–triplet mixing angles for helium

State	$\sin\theta$	State	$\sin\theta$	State	$\sin\theta$
2P	0.000 2783				
3P	0.000 2558	3D	0.015 6095		
4P	0.000 2498	4D	0.011 3960	4F	0.604 1024
5P	0.000 2473	5D	0.010 1143	5F	0.549 9291
6P	0.000 2460	6D	0.009 5289	6F	0.518 0737
7P	0.000 2452	7D	0.009 2067	7F	0.498 4184
8P	0.000 2447	8D	0.009 0087	8F	0.485 5768
9P	0.000 2444	9D	0.008 8777	9F	0.476 7620
10P	0.000 2442	10D	0.008 7862	10F	0.470 4595
5G	0.693 4752				
6G	0.693 1996	6H	0.696 2385		
7G	0.692 9889	7H	0.696 2377	7I	0.697 9315
8G	0.692 8356	8H	0.696 2372	8I	0.697 9315
9G	0.692 7195	9H	0.696 2374	9I	0.697 9316
10G	0.692 6329	10H	0.696 2353	10I	0.697 9316
8K	0.699 1671				
9K	0.699 1671	9L	0.700 1089		
10K	0.699 1671	10L	0.700 1089	10M	0.700 8507

where $X^{ss}_{\gamma,\gamma'} = (f^{ss}_{\gamma,\gamma'}/\omega^{ss}_{\gamma,\gamma'})^{1/2}$, and similarly for $X^{tt}_{\gamma,\gamma'}$. From (11.80), $X_{\gamma,\gamma'}$ is proportional to the dipole *length* form of the transition operator, for which there are no spin-dependent relativistic corrections [11.59, 60]. The mixing corrections are particularly significant for D–F and F–G transitions, where intermidiate coupling prevails. The two-state approximation becomes increasingly accurate with increasing L, but for P-states, where $\sin\theta$ is small, states with $n' \neq n$ must also be included [11.61].

11.6 Future Perspectives

The variational calculations, together with quantum defect extrapolations for high n and asymptotic expansions for high L, provide essentially exact results for the entire singly-excited spectrum of helium. In this sense, helium joins hydrogen as a fundamental atomic system. The dominant uncertainties arise from two-electron QED effects beyond the current realm of standard atomic physics. Transition frequencies from the $1s2s\ {}^1S_0$ state are now known to better than ± 0.5 MHz (± 1.8 parts in 10^9) [11.62, 63], and the fine structure intervals in the $1s2p\ {}^3P$ state have been measured to an accuracy exceeding 1 kHz [11.64]. Comparisons with theory [11.65, 66] hold the promise of determining an "atomic physics" value for the fine structure constant. Transition frequencies among the $n = 10$ states are known even more accurately [11.53]. Recent progress in the use of isotope the shift to deduce the size of the nucleus from the nuclear volume effect (see Sect. 90.1) has attracted a great deal of attention, especially for neutron-rich "halo" nuclei such as ^{6}He and ^{11}Li. [11.67, 68]

Part B | 11

References

11.1 T. Kato: Commun. Pure Appl. Math. **10**, 151 (1957)
11.2 M. Hoffman-Ostenhoff, T. Hoffmann-Ostenhoff, H. Stremnitzer: Phys. Rev. Lett. **68**, 3857 (1992)
11.3 V. A. Fock: Izv. Akad. Nauk SSSR, Ser. Fiz. **18**, 161 (1954)
11.4 V. A. Fock: D. Kngl. Norske Videnskab. Selsk. Forh. **31**, 138 (1958)
11.5 V. A. Fock: D. Kngl. Norske Videnskab. Selsk. Forh. **31**, 145 (1958)
11.6 G. B. Sochilin: Int. J. Quantum Chem. **3**, 297 (1969)
11.7 C. R. Myers, C. J. Umriger, J. P. Sethna, J. D. Morgan III: Phys. Rev. A **44**, 5537 (1991)
11.8 J. E. Gottschalk, E. N. Maslen: J. Phys. A **20**, 2781 (1987)
11.9 K. McIsaac, E. N. Maslen: Int. J. Quantum Chem. **31**, 361 (1987)
11.10 J. H. Macek: Phys. Rev. **160**, 170 (1967)
11.11 J. Leray: *Trends and Applications of Pure Mathematics to Mechanics*, Lect. Notes Phys., Vol. 195, ed. by P. G. Ciarlet, M. Roseau (Springer, Berlin, Heidelberg 1984) pp. 235–247
11.12 J. D. Morgan III: Theoret. Chem. Acta **69**, 181 (1986)
11.13 P. Deift, W. Hunziker, B. Simon, E. Vock: Commun. Math. Phys. **64**, 1 (1978)
11.14 J. D. Morgan III: J. Phys. A **10**, L91 (1977)
11.15 M. Hoffman-Ostenhoff, T. Hoffmann-Ostenhoff: Phys. Rev. A **16**, 1782 (1977)
11.16 R. Ahlrichs, T. Hoffman-Ostenhoff, M. Hoffmann-Ostenhoff, J. D. Morgan III: Phys. Rev. A **23**, 2107 (1981)
11.17 E. A. Hylleraas, B. Undheim: Z. Phys. **65**, 759 (1930)
11.18 J. K. L. MacDonald: Phys. Rev. **43**, 830 (1933)
11.19 E. A. Hylleraas: Z. Phys. **48**, 469 (1928)
11.20 E. A. Hylleraas: Z. Phys. **54**, 347 (1929)
11.21 Y. Accad, C. L. Pekeris, B. Schiff: Phys. Rev. A **4**, 516 (1971)
11.22 K. Frankowski, C. L. Pekeris: Phys. Rev. **146**, 46 (1966)
11.23 K. Frankowski, C. L. Pekeris: Phys. Rev. **150**, 366(E) (1966)
11.24 D. E. Freund, B. D. Huxtable, J. D. Morgan III: Phys. Rev. A **29**, 980 (1984)
11.25 A. J. Thakkar, T. Koga: Phys. Rev. A **50**, 854 (1994)
11.26 G. W. F. Drake: Nucl. Instrum. Methods Phys. Res. Sect B **31**, 7 (1988)
11.27 G. W. F. Drake, Z.-C. Yan: Phys. Rev. A **46**, 2378 (1992)
11.28 G. W. F. Drake: *Long Range Casimir Forces: Theory and Recent Experiments in Atomic Systems*, ed. by F. S. Levin, D. A. Micha (Plenum Press, New York 1993) p. 107
11.29 G. W. F. Drake, Z.-C. Yan: Chem. Phys. Lett. **229**, 486 (1994)
11.30 A. M. Frolov, V. H. Smith: J. Phys. B **37**, 2917 (2004)
11.31 V. Korobov: Phys. Rev. A **69**, 0545012 (2004) The Bethe logarithms calculated by Korobov include an additional $\ln Z^2$ in their definition
11.32 G. W. F. Drake: Phys. Rev. A **18**, 820 (1978)
11.33 Z.-C. Yan, G. W. F. Drake: Can. J. Phys. **72**, 822 (1994)

11.34 J. Ackermann: Phys. Rev. A **52**, 1968 (1995) and earlier references therein
11.35 C.-Y. Hu, A. A. Kvitsinsky, J. S. Cohen: J. Phys. B **28**, 3629 (1995) and earlier references therein
11.36 R. Krivec, V. B. Mandelzweig, K. Varga: Phys. Rev. **61**, 062503 (2000) and earlier references therein
11.37 A. H. Wapstra, G. Audi: Nucl. Phys. A **432**, 1 (1985)
11.38 A. Dalgarno, N. Lynn: Proc. Phys. Soc. (London) A **70**, 802 (1957)
11.39 G. W. F. Drake: Adv. At. Mol. Opt. Phys. **31**, 1 (1993)
11.40 G. W. F. Drake, S. P. Goldman: Can. J. Phys. **77**, 835 (1999)
11.41 S. P. Goldman, G. W. F. Drake: Phys. Rev. Lett. **68**, 1683 (1992)
11.42 G. W. F. Drake: Phys. Scr. **T95**, 22 (2001)
11.43 G. W. F. Drake, R. A. Swainson: Phys. Rev. A **41**, 1243 (1990)
11.44 A. P. Stone: Proc. Phys. Soc. (London); **77**, 786 (1961)
11.45 A. P. Stone: Proc. Phys. Soc. (London); **81**, 868 (1963)
11.46 G. W. F. Drake: Phys. Rev. Lett. **74**, 4791 (1995)
11.47 K. Pachucki: Phys. Rev. Lett. **84**, 4561 (2000)
11.48 V. Korobov, A. Yelkhovsky: Phys. Rev. Lett. **87**, 193003 (2001)
11.49 G.W.F. Drake, I.B. Khriplovich, A.I. Milstein, A.S. Yelkhovsky: Phys. Rev. A **48**, R15 (1993)
11.50 J. F. Babb, L. Spruch: Phys. Rev. A **38**, 13 (1988)
11.51 C.-K. Au: Phys. Rev. A **39**, 2789 (1989)
11.52 C.-K. Au, M. A. Mesa: Phys. Rev. A **41**, 2848 (1990)
11.53 C.H. Storry, N.E. Rothery, E.A. Hessels: Phys. Rev. Lett. **75**, 3249 (1995) and earlier references therein
11.54 G. W. F. Drake: Adv. At. Mol. Opt. Phys. **32**, 93 (1994)
11.55 R. J. Drachman: Phys. Rev. A **47**, 694 (1993)
11.56 G. W. F. Drake, R. A. Swainson: Phys. Rev. A **42**, 1123 (1990)
11.57 E. A. Hessels: Phys. Rev. A **46**, 5389 (1992)
11.58 G. W. F. Drake: Phys. Rev. A **45**, 70 (1992)
11.59 G. W. F. Drake: Phys. Rev. A; J. Phys. B **5**, 1979 (1972)
11.60 G. W. F. Drake: J. Phys. B **9**, L169 (1976)
11.61 G. W. F. Drake: Phys. Rev. A **19**, 1387 (1979)
11.62 C. J. Sansonetti, J. D. Gillaspy: Phys. Rev. A **45**, R1 (1992)
11.63 W. Lichten, D. Shiner, Z.-X. Zhou: Phys. Rev. A **43**, 1663 (1991)
11.64 M. C. George, L. D. Lombardi, E. A. Hessels: Phys. Rev. Lett. **87**, 173002 (2001)
11.65 G. W. F. Drake: Can. J. Phys. **80**, 1195 (2002)
11.66 K. Panchucki, J. Sapirstein: J. Phys. B **36**, 803 (2003) and earlier references therein
11.67 L.-B. Wang, P. Müller, K. Bailey, G.W.F. Drake, J. P. Greene, D. Henderson, R. J. Holt, R. V. F. Janssens, c. L. Jiang, Z.-T. Lu, T. P. O'Connor, R. C. Pardo, M. Paul, K. E. Rehm, J. P. Schiffer, X. D. Tang: Phys. Rev. Lett. **93**, 142501 (2004)
11.68 G. Ewald, W. Nörtershäuser, A. Dax, S. Göte, R. Kirchner, H.J. Kluge, Th. Kühl, R. Sanchez, A. Wojtaszek, B.A. Bushaw, G.W.F. Drake, Z.-C. Yan, C. Zimmermann: Phys. Rev. Lett. **93**, 113002 (2004)

12. Atomic Multipoles

Often symmetries in the experiment limit the number of nonvanishing multipoles, and frequently only populations proportional to diagonal elements of the density matrix are significant. This chapter studies the physical and geometrical significance of simple multipoles and examines whether symmetries allow a complete characterization of the ensemble with state populations. More thorough treatments can be found elsewhere [12.1].

12.1 **Polarization and Multipoles** 222

12.2 **The Density Matrix in Liouville Space** 222

12.3 **Diagonal Representation: State Populations** 224

12.4 **Interaction with Light** 224

12.5 **Extensions** 225

References 226

A typical atomic experiment involves the preparation of an atomic or molecular ensemble, its perturbation by a combination of collisions and external fields, and the characterization of the perturbed system through detection of emitted or scattered particles or quanta. The ensemble can be described by its density matrix, whose full specification generally requires elements between every pair of states in a complete basis set, and whose elements generally depend on spatial position, velocity, and time. The time development of such an ensemble may depend on interactions among the atoms of the ensemble with each other, with external fields, and with external perturbers. Most problems burdened by this much detail are intractable, but fortunately, significant simplifications can usually be made.

In this chapter, only ensembles of homogeneously excited, independent atoms are considered. The spatial and velocity distribution of the atoms is thus assumed to be independent of the quantum-state distribution, and the total density matrix can be factored into the product of a phase-space distribution with a state density matrix that is independent of position and velocity. The assumption is severe and prevents us from treating cases of radiation transfer (see Chapt. 19) where the degree of excitation varies as a function of position or problems of velocity-selective laser excitation. Nevertheless, many experiments in atomic physics do use ensembles that are well-described by our assumption, and even in more complex cases, our model can often serve as a starting point for analysis. Typical of the experiments for which our approach is well suited are collisionally induced polarization relaxation experiments in a gas cell, either of an excited state or of an optically pumped ground state.

In addition, we shall for the most part focus on a few isolated manifolds, usually a ground-state manifold and one or two excited-state manifolds. Each manifold, we assume, can be described by a given eigenvalue J of the total angular momentum. The qualifier "isolated" means that coherences between manifolds oscillate rapidly in time compared to other processes and can be ignored. These assumptions permit a great simplification and frequently allow a decoupling of the equations of motion, as shown below. The density matrix within each state manifold can be described by multipole moments, which are coefficients of the expansion of the density matrix in irreducible tensor operators, as described in Chapt. 7. Each state multipole is associated with a physical electric or magnetic moment in the atom.

The amount of information which can be imparted to, or obtained from, an atomic ensemble is limited by the finite number of accessible multipole moments.The nature of that information may depend on the time evolution of the multipoles in external fields and their relaxation through collisions. The ideal experiment is a *complete* measurement that determines all the multipoles. Often *over-complete* experiments can be designed, but there is little reason for determining redundant information unless consistency checks are important.

12.1 Polarization and Multipoles

An ensemble is said to be *polarized* if there is a nonstatistical distribution of atoms in magnetic sublevels. The type of polarization can be described by the *multipoles* that exist. In many experiments, symmetries in the preparation or detection of the ensemble limit the detectable polarizations to ones caused by different sublevel populations. One can then employ the relatively simple *population* (or *diagonal*) representation. The more general case is discussed in the next section.

As an introduction to the concepts, consider an atomic manifold of total angular momentum 3/2, for example an excited alkali $P_{3/2}$ state. There are four magnetic sublevels $|m\rangle$ with $m = \pm 3/2, \pm 1/2$ so that state populations are given by a four-dimensional population–density vector

$$N = \begin{pmatrix} N_{3/2} \\ N_{1/2} \\ N_{-1/2} \\ N_{-3/2} \end{pmatrix} , \tag{12.1}$$

where N_m is the density of atoms in state $|m\rangle$. If there are no external fields, the time evolution of $\boldsymbol{N}$ is given by

$$\dot{N} = \mathcal{S} - \Gamma N - \boldsymbol{\gamma} \cdot \boldsymbol{N} , \tag{12.2}$$

where $\mathcal{S}$ is the source vector and gives the excitation rate to the various sublevels, Γ is the radiative decay rate, assumed the same for all levels, and $\boldsymbol{\gamma}$ is the collisional matrix, whose elements $-\gamma_{mn}$ give the collisional transition rate from n to m.

The equation of motion is most easily solved if the basis used for the state space exploits the symmetry. Define the "spherical" orthonormal basis of vectors

$$\hat{T}_0 = \frac{1}{2}\begin{pmatrix} 1 \\ 1 \\ 1 \\ 1 \end{pmatrix} , \qquad \hat{T}_1 = \frac{1}{\sqrt{20}}\begin{pmatrix} 3 \\ 1 \\ -1 \\ -3 \end{pmatrix} , \tag{12.3}$$

$$\hat{T}_2 = \frac{1}{2}\begin{pmatrix} 1 \\ -1 \\ -1 \\ 1 \end{pmatrix} , \qquad \hat{T}_3 = \frac{1}{\sqrt{20}}\begin{pmatrix} 1 \\ -3 \\ 3 \\ -1 \end{pmatrix} . \tag{12.4}$$

(These relations follow from the $\boldsymbol{T}_{LM}$ with $M = 0$. See Sects. 12.2 and 12.3). The population vector $\boldsymbol{N}$ can be expanded

$$N = \sum_{L=0}^{4} n_L \hat{T}_L , \tag{12.5}$$

where the coefficients $n_L = \boldsymbol{N} \cdot \hat{\boldsymbol{T}}_L$ contain all the knowable information about the population distribution. The coefficient n_L is called the $2L$ multipole moment of the ensemble. In particular, $n_0 = \frac{1}{2}\sum_m N_m$ is twice the population of the entire ensemble; $n_1 = 5^{-1/2}\sum_m m N_m = 2n_0 \langle J_z \rangle / \sqrt{5}$ is the "orientation" (associated with a physical magnetic-dipole moment) of the system, $n_2 = \frac{1}{3} n_0 \langle 3J_z^2 - J^2 \rangle$ is the "alignment" (electric quadrupole moment), and $n_3 = (2n_0/\sqrt{45})\langle J_z(5J_z + 1 - 3\boldsymbol{J}^2)\rangle$ is the octupole moment of $\boldsymbol{N}$. The source vector $\mathcal{S}$ is readily expanded in the same basis.

If the collisions are isotropic, as is usually approximately the case in gas-cell experiments, then the collision matrix $\boldsymbol{\gamma}$ is invariant under rotations and

$$\boldsymbol{\gamma} \cdot \hat{\boldsymbol{T}}_L = \gamma_L \hat{\boldsymbol{T}}_L . \tag{12.6}$$

The equation of motion (12.2) is separated into four uncoupled scalar equations:

$$\dot{n}_L = \mathcal{S}_L - \Gamma n_L - \gamma_L n_L . \tag{12.7}$$

If the source term $\mathcal{S}_L = \mathcal{S} \cdot \hat{\boldsymbol{T}}_L$ is constant, (12.7) have the simple solutions

$$n_L(t) = n_L(\infty) + [n_L(0) - n_L(\infty)] \exp[-(\Gamma + \gamma_L)t] , \tag{12.8}$$

with the steady-state value

$$n_L(\infty) = \mathcal{S}_L / (\Gamma + \gamma_L) . \tag{12.9}$$

The detection of such multipoles is discussed at the end of Sect. 12.4.

12.2 The Density Matrix in Liouville Space

If in the example of the last section the state basis is changed, say by rotating the axis of quantization, then populations are generally no longer sufficient to characterize the ensemble. In the new basis, the ensemble will have been prepared in a coherent superposition of states. An adequate description for the quantum state of an ensemble in a manifold of n quantum states requires an $n \times n$ density matrix $\boldsymbol{\rho}(t)$ as discussed in Chapt. 7.

The density matrix may be considered a vector in a space, called *Liouville space*, of n^2 dimensions. A convenient set of complex orthonormal basis vectors in this space is given by the irreducible tensor operators $\boldsymbol{T}_{LM}$:

$$\boldsymbol{T}^*_{LM} \cdot \boldsymbol{T}_{L'M'} := \mathrm{Tr}\left\{\boldsymbol{T}^\dagger_{LM}\boldsymbol{T}_{L'M'}\right\} = \delta_{LL'}\delta_{MM'} , \tag{12.10}$$

where the trace (Tr) expression is appropriate for the usual $n \times n$ matrix representations of the $\boldsymbol{T}_{LM}$. Every quantum operator on the manifold can be expanded in the basis $\{\boldsymbol{T}_{LM}\}$, and in particular the density matrix has an expansion

$$\boldsymbol{\rho}(t) = \sum_{LM} \rho_{LM}(t)\,\boldsymbol{T}_{LM} , \tag{12.11}$$

with time-dependent coefficients

$$\begin{aligned}\rho_{LM}(t) &= \boldsymbol{T}^*_{LM} \cdot \boldsymbol{\rho}(t) := Tr\left[\boldsymbol{T}^\dagger_{LM}\boldsymbol{\rho}(t)\right]\\ &= N\left\langle \boldsymbol{T}^\dagger_{LM}\right\rangle \end{aligned} \tag{12.12}$$

that are identified with the multipole moments of the ensemble. Here, $\langle\cdots\rangle$ indicates the average value over the ensemble and $N = \mathrm{Tr}\{\boldsymbol{\rho}\}$ is the normalization of the density matrix $\boldsymbol{\rho}$, which is often set equal to unity but is sometimes more conveniently set equal to the total density of atoms in the manifold.

The $\boldsymbol{T}_{LM}$ are defined to transform under rotations as spherical harmonics Y_{LM} [12.2, 3]. The expansion (12.11) splits the density matrix into basis vectors $\boldsymbol{T}_{LM}$ which contain the geometric information and scalar coefficients (the multipole moments) $\rho_{LM}(t)$ which contain the physical information about state distributions.

The time dependence of $\boldsymbol{\rho}(t)$ is given by the Schrödinger equation and in the interaction picture takes the form

$$\mathrm{i}\hbar\dot{\boldsymbol{\rho}} = [\boldsymbol{V}, \boldsymbol{\rho}] , \tag{12.13}$$

where $\boldsymbol{V}$ is the interaction due to external fields (including radiation) and collisions. When $\boldsymbol{V}$ is expanded and second-order perturbation theory is applied, the time evolution (12.13) can be expressed in a form analogous to (12.2) but with an extra term [12.1]:

$$\dot{\boldsymbol{\rho}} = \mathcal{S} - \boldsymbol{\Gamma}\cdot\boldsymbol{\rho} - \boldsymbol{\gamma}\cdot\boldsymbol{\rho} - \mathrm{i}\boldsymbol{L}_0\cdot\boldsymbol{\rho} , \tag{12.14}$$

where $\boldsymbol{L}_0$ is the *Liouville operator* arising from the interaction $\boldsymbol{V}_0$ with slowly varying external fields. Its matrix elements with $\boldsymbol{\rho}$ are

$$(\boldsymbol{L}_0\cdot\boldsymbol{\rho})_{mn} = \hbar^{-1}[\boldsymbol{V}_0, \boldsymbol{\rho}]_{mn} = (\omega_m - \omega_n)\,\rho_{mn} , \tag{12.15}$$

where $\omega_m - \omega_n$ are the field-induced frequency splittings between the states.

For example, in a weak magnetic field $B_0\hat{z}$ oriented along the quantization axis $\hat{z}$, $\boldsymbol{V}_0 = \omega_0 J_z$ where $\omega_0 = g_J\mu_B B_0/\hbar$. Since $[J_z, \boldsymbol{T}_{LM}] = M\hbar\boldsymbol{T}_{LM}$ (see Chapt. 7), then assuming that the collisions are isotropic and that the radiative decay rate is the same for all states of the manifold, the expansion of $\boldsymbol{\rho}$ in $\boldsymbol{T}_{LM}$ decouples the equation of motion (12.14) into multipoles:

$$\dot{\rho}_{LM} = \mathcal{S}_{LM} - \Gamma\rho_{LM} - \gamma_L\rho_{LM} - iM\omega_0\rho_{LM} , \tag{12.16}$$

with the solution

$$\begin{aligned}\rho_{LM}(t) &= \rho_{LM}(\infty) + [\rho_{LM}(0) - \rho_{LM}(\infty)]\\ &\quad\times\exp[-(\Gamma + \gamma_L + iM\omega_0)\,t]\end{aligned} \tag{12.17}$$

comprising a transient part that decays as it precesses at the angular rate ω_0 plus a steady-state part that predicts the collisionally broadened Hanle effect (see Chapt. 17):

$$\rho_{LM}(\infty) = \frac{\mathcal{S}_{LM}}{\Gamma + \gamma_L + \mathrm{i}M\omega_0} . \tag{12.18}$$

If $M = 0$, the solutions reduce to the field-independent ones given in (12.9). On the other hand, any stray field that is not aligned with an induced multipole moment can rotate the moment and possibly cause systematic measurement errors if not taken into account. Measurements of γ_L in a gas cell at temperature T are frequently used to determine thermally averaged multipole-relaxation cross sections Q_L, defined by

$$\gamma_L = N\bar{v}Q_L , \tag{12.19}$$

where N is the density of perturbers, and the mean relative velocity is $\bar{v} = 2\pi^{-1/2}(2kT/\mu)^{1/2}$, with μ the reduced mass of the polarized-atom/perturber system. Many semiclassical and full quantum calculations of such cross sections have been made, both for atoms [12.1, 4] and for molecules [12.5].

12.3 Diagonal Representation: State Populations

An ensemble is said to be in a *coherent state* when one or more of the off-diagonal elements of the density matrix do not vanish. In a manifold of angular momentum substates of a single total angular momentum j, the ensemble has coherence if and only if it is not axially symmetric, that is, if and only if $[\boldsymbol{\rho}, J_z] \neq 0$. In an ensemble lacking coherence, or one in which any coherence that is present does not affect the observation, only the $N = 2j+1$ diagonal elements of the $N \times N$ density matrix, those that represent state populations, need be considered. In terms of multipoles, only the N elements ρ_{L0} then play a role; the ρ_{LM} elements with $M \neq 0$ can be safely ignored. Such a *diagonal representation* is valid whenever (1) the time evolution of the system is axially symmetric and (2) either the preparation of the system or its detection is axially symmetric.

12.4 Interaction with Light

A polarized system can be prepared in a variety of ways, including beam splitting in external fields, collisional excitation (including beam-foil excitation of fast beams), or radiative excitation with directed and/or polarized light. Several options for the detection of multipoles also exist, including the measurement of the anisotropy and/or polarization of scattered particles or photons. In this section, the interaction of the atomic ensemble with dipole radiation is considered.

It is convenient to define a *detection operator* for electric dipole radiation of polarization $\hat{\boldsymbol{\epsilon}}$ by

$$\boldsymbol{D}_\lambda\left(\hat{\boldsymbol{\epsilon}}\right) = \sum_\mu \hat{\boldsymbol{\epsilon}} \cdot \boldsymbol{d} \left|\lambda\mu\right\rangle \left\langle\lambda\mu\right| \boldsymbol{d} \cdot \hat{\boldsymbol{\epsilon}}^* \tag{12.20}$$

for decay to the state λ. Here, $\boldsymbol{d} = \sum_k e\boldsymbol{r}_k$ is the dipole-moment operator of the atom and μ is a magnetic sublevel of λ. The detection operator is a vector in Liouville space and can be expanded in irreducible tensor operators according to

$$\boldsymbol{D}_\lambda\left(\hat{\boldsymbol{\epsilon}}\right) = \sum_{LM} D_{\lambda LM}\left(\hat{\boldsymbol{\epsilon}}\right) \boldsymbol{T}_{LM} , \tag{12.21}$$

where

$$\begin{aligned} D_{\lambda LM}\left(\hat{\boldsymbol{\epsilon}}\right) &= \boldsymbol{T}^*_{LM} \cdot \boldsymbol{D}_\lambda\left(\hat{\boldsymbol{\epsilon}}\right) \\ &= B_L\left(\lambda\right) \Phi_{LM}\left(\hat{\boldsymbol{\epsilon}}\right) , \end{aligned} \tag{12.22}$$

with the dynamics contained in

$$B_L\left(\lambda\right) = \left(-1\right)^{\lambda+L+j+1} \left|d_{j\lambda}\right|^2 \begin{Bmatrix} 1 & L & 1 \\ j & \lambda & j \end{Bmatrix} , \tag{12.23}$$

and the angular dependence in

$$\begin{aligned} \Phi_{LM}\left(\hat{\boldsymbol{\epsilon}}\right) = {}& \left(2L+1\right)^{1/2} \sum_{rs} \hat{\boldsymbol{\epsilon}} \cdot \hat{\boldsymbol{r}} \left(\hat{\boldsymbol{\epsilon}} \cdot \hat{\boldsymbol{s}}\right)^* \\ & \times \left(-1\right)^{1-s} \begin{pmatrix} 1 & L & 1 \\ r & \lambda & -s \end{pmatrix} . \end{aligned} \tag{12.24}$$

Here, $\hat{\boldsymbol{r}}$ and $\hat{\boldsymbol{s}}$ range over the unit spherical tensors of rank one, namely $\pm\hat{\boldsymbol{1}} = \mp\left(\hat{\boldsymbol{x}} \pm \mathrm{i}\hat{\boldsymbol{y}}\right)/\sqrt{2}$ and $\hat{\boldsymbol{0}} = \hat{\boldsymbol{z}}$ (see also *Omont* [12.6], who gives tables of Φ_{LM}), and $d_{j\lambda}$ is a reduced matrix element of $\boldsymbol{d}$.

The intensity of radiation of polarization $\hat{\boldsymbol{\epsilon}}'$ emitted by an excited ensemble of atoms with state density matrix $\boldsymbol{\rho}$ in its radiative decay to level λ is

$$\begin{aligned} \boldsymbol{\rho}^* \cdot \boldsymbol{D}_\lambda\left(\hat{\boldsymbol{\epsilon}}'\right) &= \sum_{m,n,\mu} \left\langle\lambda\mu\right| \boldsymbol{d} \cdot \hat{\boldsymbol{\epsilon}}'^* \left|m\right\rangle \rho_{mn} \left\langle n\right| \boldsymbol{d} \cdot \hat{\boldsymbol{\epsilon}}' \left|\lambda\mu\right\rangle \\ &= \sum_{LM} \rho_{LM} D_{\lambda LM} , \end{aligned} \tag{12.25}$$

and by selection of polarization or spatial distribution, individual multipole components $D_{\lambda LM}$ and hence ρ_{LM} can be determined. The source terms $\mathcal{S}_{LM}$ excited by electric-dipole radiation from an isotropic ground state are given by a linear combination of $D_{\lambda LM}$:

$$\mathcal{S}_{LM} = \left(2\pi\right)^2 \sum_{\lambda,\hat{\epsilon}} u_{\hat{\epsilon}}\left(\lambda\right) D_{\lambda LM}\left(\hat{\boldsymbol{\epsilon}}\right) , \tag{12.26}$$

where $u_{\hat{\epsilon}}(\lambda)$ is the energy density of exciting radiation with polarization vector $\hat{\boldsymbol{\epsilon}}$ per unit energy. From the $3-j$ symbol in (12.24), only the $L = 0, 1$, and 2 components of $\boldsymbol{\rho}$ are observable unless the splitting of the Zeeman sublevels of λ are spectroscopically resolved. Similarly, if the ground state is unpolarized and its Zeeman sublevels unresolved, then only $L = 0, 1$, and 2 components of $\boldsymbol{\rho}$ can be excited. To excite higher-order multipoles, the ground state can be polarized by optical pumping [12.7].

As an example, consider the axially symmetric system discussed in Sect. 12.1 with $j = 3/2$ in the excited state and $\lambda = 1/2$ in the ground state. Only the four diagonal elements of $\boldsymbol{D}(\hat{\boldsymbol{\epsilon}}')$,

$$D_m\left(\hat{\boldsymbol{\epsilon}}'\right) = \sum_\mu \left|\left\langle jm\left|\hat{\boldsymbol{\epsilon}}'\cdot\boldsymbol{d}\right|\lambda\mu\right\rangle\right|^2$$
$$= \left|d_{j\lambda}\right|^2 \sum_{\mu r} \left|\hat{\boldsymbol{\epsilon}}'\cdot\hat{\boldsymbol{r}}\right|^2 \begin{pmatrix} \lambda & 1 & j \\ -\mu & r & m \end{pmatrix}^2 ,$$

are needed for the $2j+1$ values of $m=-j,\ldots,j$, where μ ranges over the $2\lambda+1$ values from $-\lambda$ to λ and $r=\pm1,0$. One finds directly

$$\boldsymbol{D}\left(\hat{\boldsymbol{\epsilon}}'\right) = \frac{\left|d_{j\lambda}\right|^2}{12}\left[\left|\hat{\boldsymbol{\epsilon}}'\cdot-\hat{\boldsymbol{1}}\right|^2\begin{pmatrix}3\\1\\0\\0\end{pmatrix} + \left|\hat{\boldsymbol{\epsilon}}'\cdot\hat{\boldsymbol{0}}\right|^2\begin{pmatrix}0\\2\\2\\0\end{pmatrix} + \left|\hat{\boldsymbol{\epsilon}}'\cdot\hat{\boldsymbol{1}}\right|^2\begin{pmatrix}0\\0\\1\\3\end{pmatrix}\right] . \tag{12.27}$$

The detected signal of polarization $\hat{\boldsymbol{\epsilon}}'$ is proportional to $\boldsymbol{N}\cdot\boldsymbol{D}(\hat{\boldsymbol{\epsilon}}')$, where $\boldsymbol{N}$ is the population–density vector (12.1). Some of the state polarization can be monitored by looking at polarized dipole radiation. If the emitted light is observed as it propagates perpendicular to the polarization vector, the intensities are given to within an overall constant of proportionality by

$$I_{\sigma\pm} = \boldsymbol{D}\left(\pm\hat{\boldsymbol{1}}\right)\cdot\boldsymbol{N}, \quad I_\pi = \boldsymbol{D}\left(\hat{\boldsymbol{0}}\right)\cdot\boldsymbol{N} , \tag{12.28}$$

$$I_\sigma = \boldsymbol{D}\left(\hat{\boldsymbol{\xi}}\right)\cdot\boldsymbol{N} = \frac{1}{2}\left(I_{\sigma+}+I_{\sigma-}\right) , \tag{12.29}$$

where $\hat{\xi}$ is any real unit vector (linear polarization vector) in the xy-plane. The intensities of polarized radiation may also be given in terms of Stokes parameters [12.8] (see also Chapt. 7). By expanding $\boldsymbol{N}$ in the spherical basis vectors $\boldsymbol{T}_L$ as in (12.5), one obtains a detected signal

$$\boldsymbol{N}\cdot\boldsymbol{D}\left(\hat{\boldsymbol{\epsilon}}'\right) = \sum_{L=0}^{2j+1} n_L D_{\lambda L0} \tag{12.30}$$

$$= \frac{\left|d_{j\lambda}\right|^2}{12}\left[2n_0 + \sqrt{5}\left(\left|\hat{\boldsymbol{\epsilon}}'\cdot\hat{\boldsymbol{1}}\right|^2 - \left|\hat{\boldsymbol{\epsilon}}'\cdot-\hat{\boldsymbol{1}}\right|^2\right)n_1 + \left(3\left|\hat{\boldsymbol{\epsilon}}'\cdot\hat{\boldsymbol{0}}\right|^2-1\right)n_2\right] . \tag{12.31}$$

Thus there is no way to monitor the octupole polarization n_3 from the dipole radiation since $D_{\lambda LM}=0$ for $L>2$. To monitor the orientation n_1, the circular polarization must be measured

$$\frac{\sqrt{5}}{3}\frac{n_1}{n_0} = \frac{I_{\sigma+}-I_{\sigma-}}{2I_\sigma+I_\pi} , \tag{12.32}$$

since the coefficient $\boldsymbol{D}_{\lambda L0}$ vanishes for any linear polarization. Conversely, from

$$\frac{n_2}{2n_0} = \frac{I_\sigma - I_\pi}{2I_\sigma+I_\pi} , \tag{12.33}$$

alignment is given by linear polarization. However, the denominator in both cases is different from that common in polarization measurements, and indeed, the ratio $(I_{\sigma+}-I_{\sigma-})/(I_{\sigma+}+I_{\sigma-})$ is *not* proportional to the orientation but contains a contribution from the alignment, and $(I_\sigma-I_\pi)/(I_\sigma+I_\pi)$ is not linear in the alignment. The expressions for other total angular momenta j are the same except for the numerical coefficients on the LHS of (12.32) and (12.33). Finally, if the desire is to measure the total excited-state population n_0 without any polarization component (which might rotate in stray external fields), one can choose linear polarization at an angle to make $D_{\lambda20}$ disappear, namely at the "magic angle" $\theta=\arccos 3^{-1/2}=54.74$ degrees. Of course, the same angle may be chosen to excite an unpolarized population.

12.5 Extensions

Although the discussion here has focused on cell experiments in which collisional perturbations are isotropic and state manifolds with a given total angular momentum j are well isolated, the concept of state multipoles is also useful in many applications where the collision symmetry is lower and where states with different j values interact, possibly due to the presence of fine and hyperfine structure [12.1]. Applications have been made in electron–atom collisions (Chapt. 7), atom–atom collisions [12.9], and atom–molecule collisions [12.5].

References

12.1 W. E. Baylis: Part. B, Chap. 28. In: *Progress in Atomic Physics*, ed. by W. Hanle, H. Kleinpoppen (Plenum, New York 1977)

12.2 U. Fano: Rev. Mod. Phys. **29**, 74 (1957)

12.3 U. Fano, G. Racah: *Irreducible Tensorial Sets* (Academic, New York 1967)

12.4 E. L. Lewis: Phys. Rep. **58**, 1 (1980)

12.5 W. E. Baylis, J. Pascale, F. Rossi: Phys. Rev. A **36**, 4212 (1987)

12.6 A. Omont: Prog. Quantum Electron. **5**, 69 (1977)

12.7 W. Happer: Rev. Mod. Phys. **44**, 169 (1972)

12.8 W. E. Baylis, J. Bonenfant, J. Derbyshire, J. Huschilt: Am. J. Phys. **61**, 534 (1993)

12.9 E. E. Nikitin, S. Ya. Umanski: *Theory of Slow Atomic Collisions* (Springer, Berlin, Heidelberg 1984)

13. Atoms in Strong Fields

Interest in the effect that electric and magnetic fields have on the internal structure of atoms is as old as quantum mechanics itself. In practical terms, an atom's spectrum acts as its signature, and so it is important to understand how electric and magnetic fields alter this characteristic. In this chapter, a summary of the basic nonrelativistic and relativistic theory of electrons and atoms in external magnetic fields is given. Extensions to the case of very strong fields are then introduced for both types of fields.

13.1 **Electron in a Uniform Magnetic Field** 227
  13.1.1 Nonrelativistic Theory 227
  13.1.2 Relativistic Theory 228

13.2 **Atoms in Uniform Magnetic Fields** 228
  13.2.1 Anomalous Zeeman Effect 228
  13.2.2 Normal Zeeman Effect 229
  13.2.3 Paschen–Back Effect 229

13.3 **Atoms in Very Strong Magnetic Fields** 230

13.4 **Atoms in Electric Fields** 231
  13.4.1 Stark Ionization 231
  13.4.2 Linear Stark Effect 231
  13.4.3 Quadratic Stark Effect 232
  13.4.4 Other Stark Corrections 232

13.5 **Recent Developments** 233

References 234

13.1 Electron in a Uniform Magnetic Field

13.1.1 Nonrelativistic Theory

The nonrelativistic Hamiltonian (in Gaussian units) for an electron in an external field $\boldsymbol{A}$ is [13.1]

$$H=\frac{1}{2m}\left(\boldsymbol{p}-\frac{e}{c}\boldsymbol{A}\right)^2-\boldsymbol{\mu}\cdot\boldsymbol{B}+eV\,, \tag{13.1}$$

where $\boldsymbol{A}$ is the vector potential and V is the scalar potential. The second term in (13.1) must be included to account for the interaction of the electron magnetic moment with an external magnetic field. The potentials $\boldsymbol{A}$ and V are only defined to within a gauge transformation [13.2]:

$$\boldsymbol{A}'=\boldsymbol{A}+\nabla f\,,\quad V'=V-\frac{1}{c}\frac{\partial f}{\partial t}\,. \tag{13.2}$$

We choose the gauge $\nabla\cdot\boldsymbol{A}=0$ in which the momentum operator $\boldsymbol{p}=\mathrm{i}\hbar\nabla$ and vector potential $\boldsymbol{A}$ commute. $\boldsymbol{B}$ is a constant uniform magnetic field with vector potential

$$\boldsymbol{A}=\frac{1}{2}\boldsymbol{B}\times\boldsymbol{r}\,. \tag{13.3}$$

$\boldsymbol{\mu}$ is the magnetic moment of the electron:

$$\boldsymbol{\mu}=\frac{1}{2}g_\mathrm{e}\mu_\mathrm{B}\boldsymbol{\sigma}\,, \tag{13.4}$$

where the σ_i are the Pauli spin matrices, $\mu_\mathrm{B}=e\hbar/2m_\mathrm{e}c$ is the Bohr magneton and g_e is the electron g-factor which accounts for the anomalous magnetic moment of the electron (see Sect. 27.4).

Consider now the case of a free electron in a constant uniform field in the z-direction with no scalar potential, i. e., $V=0$ in (13.1). This case also describes an atom in the limit of strong magnetic fields such that the Coulomb interactions are negligible. In this case, which is different from (13.3), we have

$$\boldsymbol{B}=B\hat{z}\text{ with }\begin{cases}A_x=-By\\ A_y=A_x=0\end{cases}. \tag{13.5}$$

For this field, the operators μ_z, p_x and p_z commute with the Hamiltonian and are therefore conserved. Calling their respective eigenvalues μ_z, p_x and p_z, with $-\infty\leq p_x,\,p_y\leq\infty$, the eigenstates are written as

$$\psi=\mathrm{e}^{\mathrm{i}\left(p_xx+p_zz\right)/\hbar}\varphi(y)\,. \tag{13.6}$$

Calling $y_0=-cp_x/eB$, φ satisfies

$$-\frac{\hbar^2}{2m}\frac{\mathrm{d}^2}{\mathrm{d}y^2}\varphi+\frac{1}{2}m\omega_\mathrm{B}^2\left(y-y_0\right)^2\varphi=\left(E+\mu_zB-\frac{p_z^2}{2m}\right)\varphi\,, \tag{13.7}$$

which is the Schrödinger equation for a one-dimensional harmonic oscillator with angular frequency $\omega_B = |e|B/mc$, the cyclotron frequency of the electron. The solutions to (13.7) give the eigenstates for an electron in an external homogeneous magnetic field. They are called *Landau levels* [13.1] with energies given by

$$E_n = \left(n + \frac{1}{2} + m_s\right)\hbar\omega_B + \frac{p_z^2}{2m} , \tag{13.8}$$

where $\hbar m_s = \pm\hbar/2$ is the eigenvalue of the z-component of the spin operator s, and eigenfunctions given by

$$\varphi_n(y) = \frac{1}{\sqrt{\pi^{1/2} a_B 2^n n!}} \times \exp\left(-\frac{(y-y_0)^2}{2a_B^2}\right) H_n\left(\frac{y-y_0}{a_B}\right) , \tag{13.9}$$

where $a_B = \sqrt{\hbar/m\omega_B}$ and the H_n are Hermite polynomials [13.3].

13.1.2 Relativistic Theory

The relativistic analog of (13.1) is given by the Dirac Hamiltonian

$$H_D = c\boldsymbol{\alpha} \cdot \left(\boldsymbol{p} - \frac{e}{c}\boldsymbol{A}\right) + \beta mc^2 + eV , \tag{13.10}$$

where α and β are the 4×4 matrices defined by (9.61). The eigenfunctions of the Dirac Hamiltonian (13.10) are written as four-dimensional spinors:

$$\psi_D = \begin{pmatrix} \psi^{(1)} \\ \psi^{(2)} \\ \psi^{(3)} \\ \psi^{(4)} \end{pmatrix} \equiv \begin{pmatrix} \phi_1 \\ \phi_{-1} \\ \chi_1 \\ \chi_{-1} \end{pmatrix} \equiv \begin{pmatrix} \phi \\ \chi \end{pmatrix} , \tag{13.11}$$

where ϕ and χ are two-dimensional (Pauli) spinors and ϕ_i and χ_i are functions of the coordinates of the electron.

In the case $V = 0$, the lower component χ can be eliminated in the eigenvalue equation $H_D\psi = E\psi$ to obtain a Hamiltonian equation for ϕ only, similar to (13.1) [13.4]. For a field defined by (13.5) one writes ϕ in the form (as in the nonrelativistic case)

$$\phi = e^{i(p_x x + p_z z)/\hbar} f(y) . \tag{13.12}$$

Here, $f(y)$ satisfies the equation

$$-\frac{\hbar^2}{2m}\frac{d^2}{dy^2}\varphi + \frac{1}{2}m\omega_B^2(y-y_0)^2\varphi = \left(\frac{E^2 - m^2c^4}{2mc^2} + \mu_z B - \frac{p_z^2}{2m}\right)\varphi , \tag{13.13}$$

which reduces to the nonrelativistic form (13.7) in the limit $E \to mc^2$. In (13.13), the term involving μ is not included arbitrarily as in (13.1) but is a consequence of $H_D\phi$, which predicts the value $g = 2$. The value g_e used in (13.4) includes radiative corrections to the Dirac value.

From (13.13) we obtain

$$E_n^2 = 2mc^2\left[\frac{mc^2}{2} + \left(n + \frac{1}{2} + m_s\right)\hbar\omega_B + \frac{p_z^2}{2m}\right] , \tag{13.14}$$

and the eigenfunctions $\psi_{m_s}^D$ are

$$\psi_{1/2}^D = N \begin{pmatrix} \frac{E+mc^2}{c}\psi_n \\ 0 \\ p_z\psi_n \\ -\frac{i\hbar}{a_B}\sqrt{2n}\,\psi_{n-1} \end{pmatrix} ,$$

$$\psi_{-1/2}^D = N \begin{pmatrix} 0 \\ \frac{E+mc^2}{c}\psi_n \\ \frac{i\hbar}{a_B}\sqrt{2(n+1)}\,\psi_{n+1} \\ -p_z\psi_n \end{pmatrix} , \tag{13.15}$$

where ψ_n is defined as in (13.6) and (13.9), and N is a normalization constant.

13.2 Atoms in Uniform Magnetic Fields

13.2.1 Anomalous Zeeman Effect

Consider now the nonrelativistic Hamiltonian for a one-electron atom in the presence of an external magnetic field $\boldsymbol{B}$. The one-electron Hamiltonian can be written as [13.5]

$$H = \frac{1}{2m}\boldsymbol{p}^2 - \frac{1}{4\pi\epsilon_0}\frac{Ze^2}{r} + \xi(r)\boldsymbol{L}\cdot\boldsymbol{S} + \frac{\mu_B}{\hbar}(\boldsymbol{L} + g_e\boldsymbol{S})\cdot\boldsymbol{B} + \frac{e^2}{8mc^2}(\boldsymbol{B}\times\boldsymbol{r})^2 , \tag{13.16}$$

where

$$\xi(r) = \frac{1}{2m^2c^2}\frac{Ze^2}{4\pi\epsilon_0}\frac{1}{r^3} . \quad (13.17)$$

The anomalous Zeeman effect corresponds to the case of *weak* magnetic fields such that the magnetic interaction is small compared with the $\boldsymbol{L}\cdot\boldsymbol{S}$ spin-orbit term. The energy shifts are obtained from a perturbation of (13.16) with $\boldsymbol{B}=0$. The unperturbed wave functions are eigenfunctions of $\boldsymbol{L}^2$, $\boldsymbol{S}^2$, $\boldsymbol{J}^2$ and J_z, with $\boldsymbol{J}=\boldsymbol{L}+\boldsymbol{S}$, but are not eigenfunctions of L_z or S_z. The energy levels with given values of l, s and j split in the presence of a field defined by (13.5) according to

$$\Delta E_{m_j} = g\mu_B B m_j , \quad (13.18)$$

where g is the *Landé splitting factor* given by

$$g = g_l \frac{j(j+1)-s(s+1)+l(l+1)}{2j(j+1)} + g_e \frac{j(j+1)+s(s+1)-l(l+1)}{2j(j+1)} , \quad (13.19)$$

where g_e is defined by (13.4), and $g_l = 1 - m_e/M$ to lowest order in m_e/M, where m_e is the electron mass and M is the nuclear mass. To a first approximation, it is often sufficient to take $g_e = 2$ (the Dirac value) and $g_l = 1$. In the case of a many-electron atom, j, l and s are replaced by the *total* angular momenta J, L and S. In the one-electron case (neglecting corrections to g_l or g_e) the g-value is simply

$$g \approx \frac{j+\frac{1}{2}}{l+\frac{1}{2}} , \quad (13.20)$$

which shows that the splitting of the $j=l+\frac{1}{2}$ levels is larger than that of the $j=l-\frac{1}{2}$ levels. The selection rules for the splitting of spectral lines are $\delta m_j = 0$ for components polarized parallel to the field (π components), and $\delta m_j = \pm 1$ for those perpendicular to the field (σ components).

13.2.2 Normal Zeeman Effect

For moderately strong fields up to $B \sim 10^4$ T, the quadratic $(\boldsymbol{B}\times\boldsymbol{r})^2$ term in (13.16) can be neglected. If the spin-orbit interaction term is also neglected, then the energies relative to the field-free eigenvalues E_n are given by

$$E_n(B) = E_n + \mu_B B(m_l + 2m_s) . \quad (13.21)$$

The selection rules for transitions are $\Delta m_s = 0$ and $\Delta m_l = 0, \pm 1$. The transition energy of a spectral line is split into three components, the *Lorentz triplet*

$$\Delta E_n = \Delta E_{n,0} + (\Delta m_l)\hbar\omega_L , \quad (13.22)$$

where $\Delta E_{n,0}$ is the transition energy in the absence of a field, and $\omega_L = |e|B/2m$ is the Larmor frequency. Transitions with $\Delta m_l = 0$ produce the π line at the unshifted transition energy; transitions with $\Delta m_l = \pm 1$ produce the shifted σ lines. Lorentz triplets can be observed in many-electron atoms in which the total spin is zero.

13.2.3 Paschen–Back Effect

We add now to the results of the last section the first-order perturbation caused by the spin-orbit term. The calculation can be performed in closed form for hydrogenic atoms, for which the contribution to the energy of the level n is [13.6]

$$\begin{aligned} \Delta E_n &= 0 , \quad \text{for } l = 0 \\ \Delta E_n &= \frac{\alpha^4 Z^4 \mu c^2}{2n^3} m_l m_s \\ &\times \left[l\left(l+\frac{1}{2}\right)(l+1) \right] , \quad \text{for } l \neq 0 \end{aligned} \quad (13.23)$$

where μ is the reduced mass, $\mu = mM/(m+M)$. A general expression of the relativistic solution for the Paschen–Back effect can be written for the one-electron case in the Pauli approximation in which the eigenstates are given in terms of ϕ in (13.11). Call $\phi^0_\pm$ the eigenfunctions with no external field corresponding to $j = l \pm \frac{1}{2}$ with unperturbed energies $E^0_\pm$ respectively. In terms of the dimensionless variables

$$\begin{aligned} \eta &= \frac{\mu_B B}{E_+ - E_-} , \quad \epsilon = \sqrt{1+\eta\frac{4m_j}{2l+1}+\eta^2} , \\ \gamma &= \frac{1}{\epsilon}\left(1+\eta\frac{4m_j}{2l+1}\right) , \end{aligned} \quad (13.24)$$

the energies $E_\pm$ and wave functions $\phi_\pm$ of the states, in the presence of the field B, are

$$E_\pm = \frac{E^0_+}{2}(1 \pm \epsilon + 2m_j\eta) + \frac{E^0_-}{2}(1 \mp \epsilon - 2m_j\eta) \quad (13.25)$$

and

$$\phi_\pm = \pm\sqrt{\frac{1\pm\gamma}{2}}\,\phi^0_+ + \sqrt{\frac{1\mp\gamma}{2}}\,\phi^0_- . \quad (13.26)$$

13.3 Atoms in Very Strong Magnetic Fields

The case of *very* strong magnetic fields (i. e., $B > 10^4$ T), such as those encountered at the surface of neutron stars, is also called the *quadratic Zeeman effect*, as the last term in (13.16) is dominant. In this range, perturbation calculations fail to yield good results as the field is too large, and even at fields of the order $B \sim 10^7$ T the Landau high B approximation of (13.8) and (13.9) is not adequate.

Very accurate calculations have been performed using variational finite basis set techniques for both the relativistic Dirac and nonrelativistic Schrödinger Hamiltonians. The calculations use the following relativistic basis set [13.7] that includes nuclear size effects (R is the nuclear size) and contains both asymptotic limits, the Coulomb limit for $B = 0$ and the Landau limit for $B \to \infty$:

$$\psi_{nl}^{(k,\nu)} = \begin{cases} r^{q_k - 1 + 2n} \mathrm{e}^{-a_{n\nu}^{(k)} r^2 - \beta\rho^2} \Omega_k & r \le R \\ b_{n\nu}^{(k)} r^{\gamma_\nu - 1 + n} \mathrm{e}^{-\lambda r - \beta\rho^2} \Omega_k & r > R \end{cases} \tag{13.27}$$

with

$$\Omega_k = (\cos\theta)^{l - |m_k|} (\sin\theta)^{|m_k|} \mathrm{e}^{\mathrm{i} m_k \phi} \omega_k , \tag{13.28}$$

where

$$\begin{aligned} &n = 0, 1, \ldots, N_r, \quad k = 1, 2, 3, 4, \quad \nu = 1, 2, 3, 4, \\ &q_1 = q_2 = k_0, \quad q_3 = q_4 = k_0' , \\ &m_k = \mu - \sigma_k/2, \quad \sigma_1 = \sigma_3 = 1, \quad \sigma_2 = \sigma_4 = -1 . \end{aligned}$$

Here, k refers to the component $\psi^{(k)}$ in (13.11), and λ and β are variational parameters. The power of r at the origin is given by

$$k_0 = \begin{cases} |\kappa| & \text{if } \kappa < 0 \\ |\kappa| + 1 & \text{if } \kappa > 0 , \end{cases} \tag{13.29}$$

$$k_0' = \begin{cases} |\kappa| + 1 & \text{if } \kappa < 0 \\ |\kappa| & \text{if } \kappa > 0 , \end{cases} \tag{13.30}$$

The index ν refers to the two regular and two irregular solutions for $r > R$ that match the corresponding powers at the origin k_0 and k_0'.

$$\gamma_1 = \gamma_0, \ \gamma_2 = \gamma_0 + 1, \ \gamma_3 = -\gamma_0, \ \gamma_4 = -\gamma_0 + 1 , \tag{13.31}$$

$$\gamma_0 = \sqrt{\kappa^2 - (\alpha Z)^2} , \quad \kappa = \mp\left(\nu \pm \frac{1}{2}\right) + \frac{1}{2} , \tag{13.32}$$

$$\omega_1 = \begin{pmatrix} \vartheta_1 \\ 0 \end{pmatrix}, \ \omega_2 = \begin{pmatrix} \vartheta_{-1} \\ 0 \end{pmatrix}, \quad \omega_3 = \begin{pmatrix} 0 \\ i\vartheta_{-1} \end{pmatrix}, \ \omega_4 = \begin{pmatrix} 0 \\ i\vartheta_{-1} \end{pmatrix}, \tag{13.33}$$

where ϑ_k is a two-component Pauli spinor: $\boldsymbol{\sigma}_z \vartheta_k = k\,\vartheta_i$. For even (odd) parity states, the value of l for the large components ($k = 1, 2$) is an even (odd) number greater than or equal to $|m_k|$ up to $2N_\theta$ (for even parity) or $2N_\theta + 1$ (for odd parity), while for the small components ($k = 3, 4$) it is an odd (even) number greater than or equal to $|m_k|$ up to $2N_\theta + 1$ (for even parity) or $2N_\theta$ (for odd parity), since the small component has a different nonrelativistic parity than the large component. The coefficients $a_{n\nu}^{(k)}$ and $b_{n\nu}^{(k)}$ are determined by the continuity condition of the basis functions and their first derivatives at R. For a point nucleus, the section $r \le R$ is omitted; for a nonrelativistic calculation, take $\alpha = 0$ in the basis set.

Table 13.1 presents relativistic (Dirac) energies for the ground state of one-electron atoms. Values for a point nucleus and finite nuclear size corrections are given. Table 13.2 presents the relativistic energies for $n = 2$

Table 13.1 Relativistic ground state binding energy $-E_{gs}/Z^2$ and finite nuclear size correction $\delta E_{nuc}/Z^2$ (in a.u.) of hydrogenic atoms for various magnetic fields B (in units of 2.35×10^5 T). δE_{nuc} should be added to E_{gs}

Z	B	$-E_{gs}/Z^2$	$\delta E_{nuc}/Z^2$
1	0	0.500 006 656 597 483 75	$1.557\,86 \times 10^{-10}$
1	10^{-5}	0.500 011 656 4837	1.5579×10^{-10}
1	10^{-2}	0.504 981 572 360	1.5580×10^{-10}
1	10^{-1}	0.547 532 408 3429	1.5718×10^{-10}
1	2	1.022 218 0290	3.23×10^{-10}
1	10	1.747 800 68	1.182×10^{-9}
1	20	2.215 400 91	2.360×10^{-9}
1	200	4.727 1233	3.032×10^{-8}
1	500	6.257 0326	8.778×10^{-8}
20	0	0.502 691 308 407 5098	1.3372×10^{-6}
20	1	0.503 930 867 05	1.34×10^{-6}
20	10	0.514 950 248	1.3×10^{-6}
20	100	0.612 377 94	1.4×10^{-6}
40	0	0.511 129 686 143	1.1878×10^{-5}
92	0	0.574 338 140 7377	8.4155×10^{-4}
92	1	0.574 386 987	8.4155×10^{-4}

Table 13.2 Relativistic binding energy $-E_{2S,-1/2}$ for the $2S_{1/2}$ $(m_j = -\frac{1}{2})$ and $-E_{2P,-1/2}$ for the $2P_{1/2}$ $(m_j = -\frac{1}{2})$ excited states of hydrogen (in a.u.) in an intense magnetic field B (in units of 2.35×10^5 T)

B	$-E_{2S,-1/2}$	$-E_{2P,-1/2}$
10^{-6}	0.125 002 580 164	0.125 002 283 074
10^{-5}		0.125 006 104 950
10^{-4}	0.125 052 044 95	0.125 050 967 92
10^{-3}		0.125 499 4694
10^{-2}	0.129 653 6428	0.129 851 3642
0.05	0.142 018 956	
0.1	0.148 091 7386	0.162 411 0524
0.2	0.148 989 58	
0.5	0.150 810 15	
1	0.160 471 07	0.260 009 34
10	0.208 955 91	0.382 663 18
100	0.256 191	0.463 6641

excited states of hydrogen with the (negligible) finite nuclear size correction included.

Table 13.3 Relativistic corrections $\delta E = (E - E_{NR})/|E_R|$ to the nonrelativistic energies E_{NR} for the ground state and $n = 2$ excited states of hydrogen in an intense magnetic field B (in units of 2.35×10^5 T). The numbers in brackets denote powers of 10

B	δE_{gs}	B	$\delta E_{2S,-1/2}$	$\delta E_{2P,-1/2}$
0.1	−1.08[−5]	1[−6]	−1.66[−5]	−1.43[−5]
1	−5.21[−6]	1[−4]	−1.66[−5]	−7.86[−6]
2	−4.03[−6]	1[−3]		−7.72[−6]
3	−3.48[−6]	1[−2]	−1.60[−5]	−7.30[−6]
20	−1.09[−6]	0.05	−1.57[−5]	
200	4.61[−6]	0.1	−1.74[−6]	−6.00[−6]
500	8.81[−6]	1	−1.3[−5]	−1.05[−5]
2000	1.85[−5]	10	−2.0[−5]	−3.48[−5]
5000	2.78[−5]	100	−3.9[−5]	−1.0[−4]

Table 13.3, which displays the relativistic corrections of the energies of the previous two tables, presents one of the most interesting relativistic results: the change in sign of the relativistic correction of the energy of the ground state at $B \sim 10^7$ T.

13.4 Atoms in Electric Fields

13.4.1 Stark Ionization

An external electric field $\boldsymbol{F}$ introduces the perturbing potential

$$V = -\boldsymbol{d} \cdot \boldsymbol{F} , \tag{13.34}$$

where

$$\boldsymbol{d} = \sum_i q_i \boldsymbol{r}_i \tag{13.35}$$

is the dipole moment of the atom, and i runs over all electrons in the atom. In the case of strong external electric fields, bound states do not exist because the atom ionizes. Consider a hydrogenic atom in a static electric field

$$\boldsymbol{F} = F\hat{z} . \tag{13.36}$$

The total potential acting on the electron is then

$$V_{tot}(\boldsymbol{r}) = -\frac{e^2 Z}{4\pi\epsilon_0} \frac{1}{r} + eFz . \tag{13.37}$$

Consider the z-dependence of this potential. Call $\rho = \sqrt{x^2 + y^2}$ and $v(z, \rho) = V(x, y, z)$. Unlike the Coulomb case in which $v_{Coul}(\pm\infty, \rho) = 0$ resulting in an infinite number of bound states, now $v(\pm\infty, \rho) = \pm\infty$ and v has a local maximum. On the z axis, this maximum occurs at $z_{max} = -\sqrt{Z|e|/(4\pi\epsilon_0 F)}$ for which $v(z_{max}, 0) = 0$. There is then a potential barrier through which the electron can tunnel, i. e., there are no bound states any longer but *resonances*. The potential barrier is shallower the stronger the field; the well can contain a smaller number of bound states and ionization occurs.

13.4.2 Linear Stark Effect

The electric field (13.36) produces a dipole potential

$$V_F = eFz = eF\,r\sqrt{\frac{4\pi}{3}}\,Y_{10}(\hat{\boldsymbol{r}}) , \tag{13.38}$$

which does not preserve parity. A first-order perturbation calculation for the energy

$$E_n^{(1)} = \langle n\,|V_F|\,n\rangle \tag{13.39}$$

yields null results unless the unperturbed states are degenerate with states of opposite parity.

In the remainder of this chapter, atomic units will be used. Final results for energies can be multiplied by $2R_\infty hc$ to translate to SI or other units. The calculation can be carried out in detail for the case of hydrogenic atoms [13.8]. In this case it is convenient to work in parabolic coordinates: φ denotes the usual angle in the xy-plane, and

$$\begin{aligned}\xi &= r + z\,,\\ \eta &= r - z\,.\end{aligned} \tag{13.40}$$

The Hamiltonian for a hydrogenic atom with a field $V_F = \frac{1}{2} F(\xi - \eta)$ from (13.38) is

$$(\xi + \eta)H = \xi h_+(\xi) + \eta h_-(\eta)\,. \tag{13.41}$$

The wave function is written in the form

$$\Psi(\xi, \eta, \varphi) = \frac{1}{\sqrt{2\pi Z}} \psi_+(\xi)\psi_-(\eta)\mathrm{e}^{\mathrm{i}m_l\varphi}\,, \tag{13.42}$$

with the $\psi_\pm$ satisfying

$$h_\pm(x)\psi_\pm(x) = E\psi_\pm(x)\,, \tag{13.43}$$

where $x = \xi$ for ψ_+ and $x = \eta$ for ψ_-, and

$$h_\pm(x) = -\frac{2}{x}\frac{\mathrm{d}}{\mathrm{d}x}\left(x\frac{\mathrm{d}}{\mathrm{d}x}\right) - \frac{2Z_\pm}{x} + \frac{m_l^2}{2x^2} \mp \frac{1}{2}Fx\,, \tag{13.44}$$

with

$$Z = Z_+ + Z_-\,. \tag{13.45}$$

Using the notation

$$\begin{aligned}\epsilon &= \sqrt{-2E}\,,\\ n_\pm &= Z_\pm/\epsilon - \frac{1}{2}(|m_l| + 1)\,,\\ n &= n_+ + n_- + |m_l| + 1\,,\\ n_+, N_- &= 0, 1, \ldots, n - |m_l| + 1\,,\\ |m_l| &= 0, 1, 2, \ldots, n-1\,,\\ \delta n &= n_+ - n_-\,,\end{aligned} \tag{13.46}$$

where n is the principal quantum number, the unperturbed eigenfunctions are

$$\psi_\pm(x) = \frac{\epsilon\, n_\pm!^{\frac{1}{2}}}{(n_\pm + |m_l|)!^{\frac{1}{2}}}\,\mathrm{e}^{-\frac{1}{2}\epsilon x}(\epsilon x)^{\frac{1}{2}|m_l|} L_{n_\pm}^{(|m_l|)}(\epsilon x)\,, \tag{13.47}$$

where the $L_b^{(a)}$ are generalized Laguerre polynomials (Sect. 9.4.2). The zero-order eigenvalues are

$$Z_\pm^{(0)} = \left(n_\pm + \frac{m_l + 1}{2}\right)\epsilon\,. \tag{13.48}$$

The first-order perturbation yields

$$\begin{aligned}Z_\pm^{(1)} = \pm\frac{1}{4}\frac{F}{\epsilon^2}\big[&6n\pm(n_\pm\\ &+ m_l + 1) + m_l(m_l + 3) + 2\big]\,.\end{aligned} \tag{13.49}$$

From these

$$\epsilon = \frac{Z}{n} - \frac{3}{2F\left(\frac{n}{Z}\right)^2\delta_n}\,, \tag{13.50}$$

and to first order in F,

$$\begin{aligned}E &= -\frac{1}{2}\epsilon^2 \approx E^{(0)} + E^{(1)}\,,\\ E^{(0)} &= -\frac{1}{2}\frac{Z^2}{n^2}\,,\\ E^{(1)} &= \frac{3}{2}\frac{F}{Z}n\,\delta_n\,.\end{aligned} \tag{13.51}$$

13.4.3 Quadratic Stark Effect

A perturbation linear in the field F yields no contribution to nondegenerate states (e.g., the ground state $n_+ = n_- = m = 0$; $n = 1$). In this case, the lowest order contribution comes from the *quadratic Stark effect*, the contribution of order F^2. The quadratic perturbation to a level $E_n^{(0)}$ caused by a general electric field $\boldsymbol{F}$ can be written in terms of the symmetric tensor α_{ij}^n as

$$E_n^{(2)} = -\frac{1}{2}\alpha_{ij}^n F_i F_j\,, \tag{13.52}$$

with

$$\alpha_{ij}^n = -2\sum_{\substack{m\\ m\neq n}} \frac{\langle n\,|d_i|\,m\rangle\langle m\,|d_j|\,n\rangle}{E_n - E_m}\,, \tag{13.53}$$

where d_i is defined in (13.35).

For a field (13.36),

$$\Delta E_n = -\frac{1}{2}\alpha^n F^2\,, \tag{13.54}$$

where

$$\alpha^n \equiv \alpha_{zz}^n = -2e^2\sum_{\substack{m\\ m\neq n}} \frac{|\langle n\,|z|\,m\rangle|^2}{E_n - E_m}\,. \tag{13.55}$$

In terms of (13.46), a general nonrelativistic expression for the *dipole polarizability* of hydrogenic ions is [13.9]

$$\alpha^n = \frac{a_0^3 n^4}{8Z^4}\left(17n^2 - 3\delta_n^2 - 9m_l^2 + 19\right)\,. \tag{13.56}$$

For the ground state of hydrogenic atoms,

$$\alpha^{n=1} = \frac{9a_0^3}{2Z^4}\,. \tag{13.57}$$

Table 13.4 lists the *relativistic* values for the ground state polarizability $\alpha_{\mathrm{rel}}^{n=1}$, obtained by calculating (13.55) using relativistic variational basis sets [13.10]. The values are interpolated by

$$\alpha^{n=1} = \frac{a_0^3}{Z^4}\left[\frac{9}{2} - \frac{14}{3}(\alpha Z)^2 + 0.53983(\alpha Z)^4\right] . \tag{13.58}$$

13.4.4 Other Stark Corrections

Third Order Corrections

For the energy correction cubic in the external field (13.36), one obtains [13.9]

$$E^{(3)} = \frac{3}{32} F^3 \left(\frac{n}{Z}\right)^7 \times \delta_n \left(23n^2 - \delta_n^2 + 11m_l^2 + 39\right) . \tag{13.59}$$

Relativistic Linear Stark-Shift of the Fine Structure of Hydrogen

For a Stark effect small relative to the fine structure, the degenerate levels corresponding to the same value of j split according to

$$\delta_m E_{nj} = \frac{3}{4}\sqrt{n^2 - \left(j + \frac{1}{2}\right)^2} \frac{nm}{j(j+1)} F . \tag{13.60}$$

Other Stark Corrections in Hydrogen

The expectation value of the delta function, is, in a.u. [13.11],

$$2\pi \langle 1s|\delta(\boldsymbol{r})|1s\rangle = 2 - 31F^2 . \tag{13.61}$$

Table 13.4 Relativistic dipole polarizabilities for the ground state of hydrogenic atoms

Z	$\alpha_{\text{rel}}^{n=1} Z^4 / (a_0^3)$
1	4.499 7515
5	4.493 7883
10	4.475 1644
20	4.400 8376
30	4.277 5621
40	4.106 2474
50	3.888 1792
60	3.625 0295
70	3.318 8659
80	2.972 1524
90	2.587 7205
100	2.168 6483

For the Bethe logarithm β defined by

$$\beta_{1s} = \frac{\sum_n |\langle 1s\,|\boldsymbol{p}|\,n\rangle|^2 (E_n - E_{1s}) \ln|E_n - E_{1s}|}{\sum_n |\langle 1s\,|\boldsymbol{p}|\,n\rangle|^2 (E_n - E_{1s})} , \tag{13.62}$$

the result is [13.12]

$$\beta_{1s} = 2.290\,981\,375\,205\,552\,301 + 0.316\,205(6)\,F^2 . \tag{13.63}$$

These results are useful in calculating an asymptotic expansion for the two-electron Bethe logarithm [13.13].

13.5 Recent Developments

The drastic change of an atom's internal structure in the presence of external electric and magnetic fields is shown most clearly through the changes induced in its spectral features. Of these features, avoided crossings are a distinctive example. Recent work in this area by *Férez* and *Dehesa* [13.14] has suggested the use of Shannon's information entropy [13.15], defined by

$$S = -\int \rho(\boldsymbol{r}) \ln \rho(\boldsymbol{r}) \mathrm{d}\boldsymbol{r} , \tag{13.64}$$

where $\rho(\boldsymbol{r}) = |\psi(\boldsymbol{r})|^2$, as an indicator or predictor of such irregular features of atomic spectra. By studying some excited states of hydrogen in parallel fields it was shown that, for the states involved, a marked confinement of the electron cloud and an information-theoretic exchange occurs when the magnetic field strength is adjusted adiabatically through the region of an avoided crossing. The field strengths studied are characteristic of compact astronomical objects, such as white dwarfs and neutron stars.

Although the effects of strong magnetic fields on the structure and dynamics of hydrogen have been known for some time, knowledge of the helium atom in such fields has only recently become sufficient for comparison with astrophysical observations [13.16–18]. As one example of their importance, such studies have proven critical in showing the presence of helium in the atmospheres of certain magnetic white dwarfs [13.19].

In recent years, the increased sophistication and resolution of observation techniques has not only increased the number of known astronomical objects, but also motivated the study of the effects of strong fields on heavier atoms [13.20].

Another interesting area of current research concerns the relationship between quantum mechanics and classically chaotic systems. For these studies, Rubidium Rydberg atoms are an ideal system since laboratory fields can easily push the atom to the strong-field limit [13.21–23].

For a very useful review of various topics up to 1998 see [13.24]; a more concise review, concerning the electronic structure of atoms, molecules, and bulk matter, including some properties of dense plasma, in strong fields, is given in [13.25].

References

13.1 L. D. Landau, E. M. Lifshitz: *Quantum Mechanics (Course of Theoretical Physics)*, Vol. 3 (Pergamon, Oxford 1977) p. 456
13.2 L. D. Landau, E. M. Lifshitz: *The Classical Theory of Fields (Course of Theoretical Physics)*, Vol. 2 (Pergamon, Oxford 1975) p. 49
13.3 A. Messiah: *Quantum Mechanics* (Wiley, New York 1999) p. 491
13.4 C. Itzykson, J.-B. Zuber: *Quantum Field Theory* (McGraw-Hill, New York 1980) p. 67
13.5 H. A. Bethe, E. Salpeter: *Quantum Mechanics of One- and Two-electron Atoms* (Plenum, New York 1977) p. 208
13.6 H. A. Bethe, E. Salpeter: *Quantum Mechanics of One- and Two-electron Atoms* (Plenum, New York 1977) p. 211
13.7 Z. Chen, S. P. Goldman: Phys. Rev. A **48**, 1107 (1993)
13.8 H. A. Bethe, E. Salpeter: *Quantum Mechanics of One- and Two-electron Atoms* (Plenum, New York 1977) p. 229
13.9 H. A. Bethe, E. Salpeter: *Quantum Mechanics of One- and Two-electron Atoms* (Plenum, New York 1977) p. 233
13.10 G. W. F. Drake, S. P. Goldman: Phys. Rev. A **23**, 2093 (1981)
13.11 G. W. F. Drake: Phys. Rev. A **45**, 70 (1992)
13.12 S. P. Goldman: Phys. Rev. A **50**, 3039 (1994)
13.13 S. P. Goldman, G. W. F. Drake: Phys. Rev. Lett. **68**, 1683 (1992)
13.14 R. González-Férez, J. S. Dehesa: Phys. Rev. Lett. **91**, 113001 (2003)
13.15 C. E. Shannon: Bell Syst. Tech. J. **27**, 623 (1948)
13.16 W. Becken, P. Schmelcher, F. K. Diakonos: J. Phys. B. **32**, 1557 (1999)
13.17 W. Becken, P. Schmelcher: Phys. Rev. A **63**, 053412 (2001)
13.18 W. Becken, P. Schmelcher: Phys. Rev. A **65**, 033416 (2002)
13.19 S. Jordan, P. Schmelcher, W. Becken, W. Schweizer: Astron. Astrophys. **336**, 33 (1998)
13.20 P. Schmelcher: private communication
13.21 J. von Milczewski, T. Uzer: *Atoms and Molecules in Strong External Fields, edited by P. Schmelcher and W. Schweizer* (Springer, Berlin, Heidelberg 1998) p. 199
13.22 J. Main, G. Wunner: *Atoms and Molecules in Strong External Fields, edited by P. Schmelcher and W. Schweizer* (Springer, Berlin, Heidelberg 1998) p. 223
13.23 J. R. Guest, G. Raithel: Phys. Rev. A **68**, 052502 (2003)
13.24 P. Schmelcher, W. Schweizer (Eds.): *Atoms and Molecules in Strong External Fields* (Springer, Berlin 1998)
13.25 D. Lai: Ref. Mod. Phys **73**, 629 (2001)

14. Rydberg Atoms

Rydberg atoms are those in which the valence electron is in a state of high principal quantum number n. They are of historical interest since the observation of Rydberg series helped in the initial unraveling of atomic spectroscopy [14.1]. Since the 1970s, these atoms have been studied mostly for two reasons. First, Rydberg states are at the border between bound states and the continuum, and any process which can result in either excited bound states or ions and free electrons usually leads to the production of Rydberg states. Second, the exaggerated properties of Rydberg atoms allow experiments to be done which would be difficult or impossible with normal atoms.

14.1 **Wave Functions and Quantum Defect Theory** 235

14.2 **Optical Excitation and Radiative Lifetimes** 237

14.3 **Electric Fields** 238

14.4 **Magnetic Fields** 241

14.5 **Microwave Fields** 242

14.6 **Collisions** 243

14.7 **Autoionizing Rydberg States** 244

References 245

14.1 Wave Functions and Quantum Defect Theory

Many of the properties of Rydberg atoms can be calculated accurately using quantum defect theory, which is easily understood by starting with the H atom [14.2]. We shall use atomic units, as discussed in Sect. 1.2. The Schrödinger equation for the motion of the electron in a H atom in spherical co-ordinates is

$$\left(-\frac{1}{2}\nabla^2-\frac{1}{r}\right)\Psi(r,\theta,\phi)=E\Psi(r,\theta,\phi)\,, \tag{14.1}$$

where E is the energy, r is the distance between the electron and the proton, and θ and ϕ are the polar and azimuthal angles of the electron's position. Equation (14.1) can be separated, and its solution expressed as the product

$$\Psi(r,\theta,\phi)=R(r)Y_{\ell m}(\theta,\phi)\,, \tag{14.2}$$

where ℓ and m are the orbital and azimuthal-orbital angular momentum (i. e., magnetic) quantum numbers and $Y_{\ell m}(\theta,\phi)$ is a normalized spherical harmonic. $R(r)$ satisfies the radial equation

$$\frac{\mathrm{d}^2R(r)}{\mathrm{d}r^2}+\frac{2\,\mathrm{d}R(r)}{r\,\mathrm{d}r}+2ER(r)+\frac{2R(r)}{r}=\frac{\ell(\ell+1)R}{r^2}\,, \tag{14.3}$$

which has the two physically interesting solutions

$$R(r)=\frac{f(\ell,E,r)}{r}\,, \tag{14.4}$$

$$R(r)=\frac{g(\ell,E,r)}{r}\,. \tag{14.5}$$

The f and g functions are the regular and irregular Coulomb functions which are the solutions to a variant of (14.3). As $r\to 0$ they have the forms [14.3]

$$f(\ell,E,r)\propto r^{\ell+1}\,, \tag{14.6}$$

$$g(\ell,E,r)\propto r^{-\ell}\,, \tag{14.7}$$

irrespective of whether E is positive or negative. As $r\to\infty$, for $E>0$ the f and g functions are sine and cosine waves, i. e., there is a phase shift of $\pi/2$ between them. For $E<0$ it is useful to introduce ν, defined by $E=-1/2\nu^2$, and for $E<0$ as $r\to\infty$

$$f=u(\ell,\nu,r)\sin\pi\nu-v(\ell,\nu,r)\mathrm{e}^{\mathrm{i}\pi\nu}\,, \tag{14.8}$$

$$g=-u(\ell,\nu,r)\cos\pi\nu+v(\ell,\nu,r)\mathrm{e}^{\mathrm{i}\pi(\nu+1/2)}\,, \tag{14.9}$$

where u and v are exponentially increasing and decreasing functions of r. As $r\to\infty$, $u\to\infty$ and $v\to 0$.

Part B | 14

Requiring that the wave function be square integrable means that as $r \to 0$ only the f function is allowed. Equation (14.8) shows that the $r \to \infty$ boundary condition requires that $\sin \pi \nu$ be zero or ν an integer n, leading to the hydrogenic Bohr formula for the energies:

$$E = -\frac{1}{2n^2} \,. \tag{14.10}$$

The classical turning point of an s wave occurs at $r = 2n^2$, and the expectation values of positive powers of r reflect the location of the outer turning point, i.e.,

$$\left\langle r^k \right\rangle \approx n^{2k} \,. \tag{14.11}$$

The expectation values of negative powers of r are determined by the properties of the wave function at small r. The normalization constant of the radial wave function scales as $n^{-3/2}$, so that $R(r) \propto n^{-3/2} r^{\ell+1}$ for small r. Accordingly, the expectation values of negative powers of r, except r^{-1}, and any properties which depend on the small r part of the wave function, scale as n^{-3}. Using the properties of the wave function and the energies, the n-scaling of the properties of Rydberg atoms can be determined.

The primary reason for introducing the Coulomb waves instead of the more common Hermite polynominal solution for the radial function is to set the stage for single channel quantum defect theory, which enables us to calculate the wave functions and properties of one valence electron atoms such as Na. The simplest picture of an Na Rydberg atom is an electron orbiting a positively charged Na^+ core consisting of 10 electrons and a nucleus of charge +11. The ten electrons are assumed to be frozen in place with spherical symmetry about the nucleus, so their charge cloud is not polarized by the outer valence electron, although the valence electron can penetrate the ten-electron cloud. When the electron penetrates the charge cloud of the core electrons, it sees a potential well deeper than $-1/r$ due to the decreased shielding of the +11 nuclear charge. For Na and other alkali atoms, we assume that there is a radius r_c such that for $r < r_c$ the potential is deeper than $-1/r$, and for $r > r_c$ it is equal to $-1/r$. As a result of the deeper potential at $r < r_c$, the radial wave function is pulled into the core in Na, relative to H, as shown in Fig. 14.1. For $r \geq r_c$, the potential is a Coulomb $-1/r$ potential, and $R(r)$ is a solution of (14.3) which can be expressed

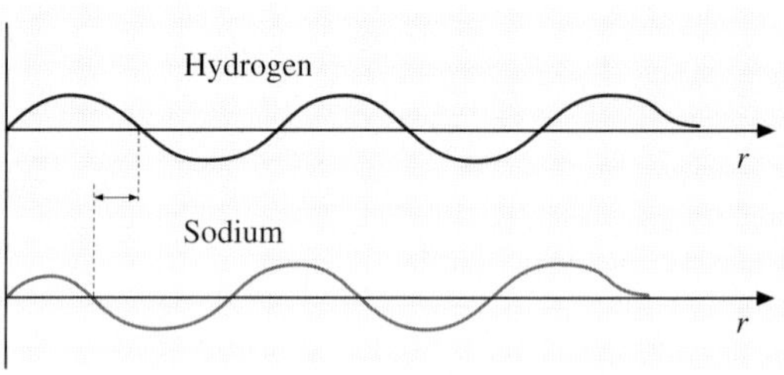

Fig. 14.1 Radial wave functions for H and Na showing that the Na wave function is pulled in toward the ionic core

as

$$R(r) = \left[f(\ell, \nu, r) \cos \tau_\ell - g(\ell, \nu, r) \sin \tau_\ell \right] / r \,, \tag{14.12}$$

where τ_ℓ is the radial phase shift.

Near the ionization limit, $E \sim 0$, and as a result, the kinetic energy of the Rydberg electron is greater than $1/r_c$ (~ 10 eV) when $r < r_c$. As a result, changes in E of 0.10 eV, the $n = 10$ binding energy, do not appreciably alter the phase shift τ_ℓ, and we can assume τ_ℓ to be independent of E. The ℓ dependence of τ_ℓ arises because the centrifugal $\ell(\ell+1)/r$ term in (14.3) excludes the Rydberg electron from the region of the core in states of high ℓ. Applying the $r \to \infty$ boundary condition to the wave function of (14.12) leads to the requirement that the coefficient of u vanish, i.e.,

$$\cos \tau_\ell \sin(\pi \nu) + \sin \tau_\ell \cos(\pi \nu) = 0 \,, \tag{14.13}$$

which implies that $\sin(\pi\nu + \tau_\ell) = 0$ or $\nu = n - \tau_\ell/\pi$. Usually τ_ℓ/π is written as δ_ℓ and termed the quantum defect, and the energies of members of the $n\ell$ series are written as

$$E = -\frac{1}{2(n-\delta_\ell)^2} = -\frac{1}{2n^{*2}} \,, \tag{14.14}$$

where $n^* = n - \delta_\ell$ is often termed the effective quantum number (see also Sect. 11.4.1).

Knowledge of the quantum defect δ_ℓ of a series of ℓ states determines their energies, and it is a straightforward matter to calculate the Coulomb wave function specified in (14.12) using a Numerov algorithm [14.4, 5]. This procedure gives wave functions valid for $r \geq r_c$, which can be used to calculate many of the properties of Rydberg atoms with great accuracy. The effect of core penetration on the energies is easily seen in the energy level diagram of Fig. 14.2. The Na $\ell \geq 2$ states have the same energies as hydro-

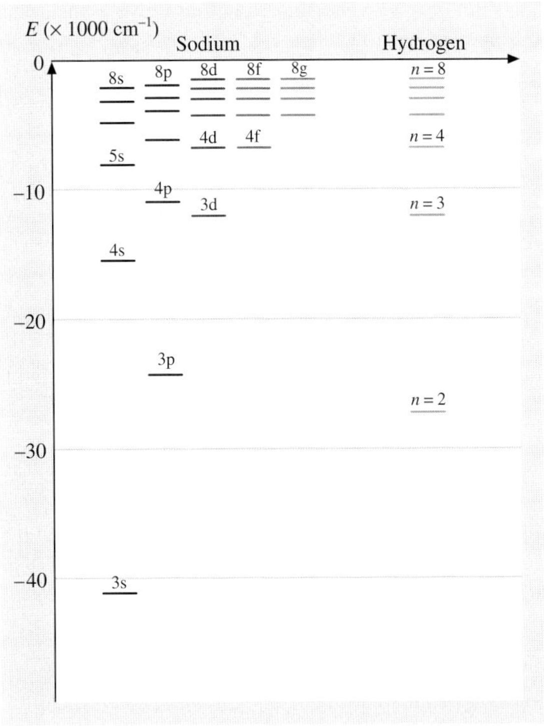

Fig. 14.2 Energy levels of Na and H

gen, while the s and p states, with quantum defects of 1.35 and 0.85 respectively, lie far below the hydrogenic energies.

Although it is impossible to discern in Fig. 14.2, the Na $\ell \geq 2$ states also lie below the hydrogenic energies. For these states it is not core penetration, but core polarization which is responsible for the shift to lower energy. Contrary to our earlier assumption that the outer electron does not affect the inner electrons if $r > r_c$, the outer electron polarizes the inner electron cloud even when $r > r_c$, and the energies of even the high ℓ states fall below the hydrogenic energies. The leading term in the polarization energy is due to the dipole polarizability of the core, α_d. For high ℓ states it gives a quantum defect of [14.6]

$$\delta_\ell = \frac{3\alpha_d}{4\ell^5} . \tag{14.15}$$

Quantum defects due primarily to core polarization rarely exceed 10^{-2}, while those due to core penetration are often greater than one.

14.2 Optical Excitation and Radiative Lifetimes

Optical excitation of the Rydberg states from the ground state or any other low lying state is the continuation of the photoionization cross section σ_{PI} below the ionization limit. The photoionization cross section, discussed more extensively in Chapt. 24, is approximately constant at the limit. Above and below the limit the average photoabsorption cross section is the same, as evidenced by the fact that a discontinuity is not evident in an absorption spectrum, i. e., it is not possible to see where the unresolved Rydberg states end and the continuum begins. Nonetheless, below the limit the cross section is structured by the Δn spacing of $1/n^3$ between adjacent members of the Rydberg series. In any experiment, there is a finite resolution $\Delta\omega$ with which the Rydberg states can be excited. It can arise, for example, from the Doppler width or a laser linewidth. This resolution determines the cross section σ_n for exciting the Rydberg state of principal quantum number n. Explicitly, σ_n is given by

$$\sigma_n = \frac{\sigma_{PI}}{n^3 \Delta\omega} . \tag{14.16}$$

A typical value for σ_{PI} is $10^{-18}\,\text{cm}^2$. For a resolution $\Delta\omega = 1\,\text{cm}^{-1}$ $(6 \times 10^{-6}\,\text{a.u.})$ the cross section for exciting an $n = 20$ atom is $3 \times 10^{-17}\,\text{cm}^2$.

From the wave functions of the Rydberg states, we can also derive the n^{-3} dependence of the photoexcitation cross section. The dipole matrix element from the ground state to a Rydberg state only involves the part of the Rydberg state wave function near the core. At small r, the Rydberg wave function only depends on n through the n^{-3} normalization factor, and as a result, the squared dipole matrix element between the ground state and the Rydberg state and the cross section both have an n^{-3} dependence.

Radiative decay, which is covered in Chapt. 17, is, to some extent, the reverse of optical excitation. The general expression for the spontaneous transition rate from the $n\ell$ state to the $n'\ell'$ state is the Einstein A coefficient, given by [14.2]

$$A_{n\ell,n'\ell'} = \frac{4}{3}\mu^2_{n\ell,n'\ell'}\omega^3_{n\ell,n'\ell'}\frac{\alpha^3 g_>}{2g_n + 1} , \tag{14.17}$$

where $\mu_{n\ell,n'\ell'}$ and $\omega_{n\ell,n',\ell'}$ are the electric dipole matrix elements and frequencies of the $n\ell \to n'\ell'$ transitions, g_n and $g_{n'}$ are the degeneracies of the $n\ell$ and $n'\ell'$ states, and $g_>$ is the greater of g_n and $g_{n'}$. The lifetime $\tau_{n\ell}$ of the $n\ell$ state is obtained by summing the decay rates to all possible lower energy states. Explicitly,

$$\frac{1}{\tau_{n\ell}} = \sum_{n'\ell'} A_{n\ell,n'\ell'} \,. \tag{14.18}$$

Due to the ω^3 factor in (14.17), the highest frequency transition usually contributes most heavily to the total radiative decay rate, and the dominant decay is likely to be the lowest lying state possible. For low-ℓ Rydberg states, the lowest lying ℓ' states are bound by orders of magnitude more than the Rydberg states, and the frequency of the decay is nearly independent of n. Only the squared dipole moment depends on n, as n^{-3}, because of the normalization of the Rydberg wave function at the core. Consequently, for low-ℓ states,

$$\tau_{n\ell} \propto n^3 \,. \tag{14.19}$$

As a typical example, the 10f state in H has a lifetime of 1.08 μs [14.7].

The highest ℓ states, with $\ell = n-1$, have radiative lifetimes with a completely different n dependence. The only possible transitions are $n \to n-1$, with frequency $1/n^3$. In this case the dipole moments reflect the large size of both the n and $n-1$ states and have the n^2 scaling of the orbital radius. Using (14.17) for $\ell = n-1$ leads to

$$\tau_{n(n-1)} \propto n^5 \,. \tag{14.20}$$

Another useful lifetime, τ_n, is that corresponding to the average decay rate of all ℓ, m states of the same n. It scales as $n^{4.5}$ [14.2, 8].

Equation (14.17) describes spontaneous decay to lower lying states driven by the vacuum. At room temperature, 300 K, there are many thermal photons at the frequencies of the $n \to n \pm 1$ transitions of Rydberg states for $n \geq 10$, and these photons drive transitions to higher and lower states [14.9]. A convenient way of describing blackbody radiation is in terms of the photon occupation number $\bar{n}$, given by

$$\bar{n} = \frac{1}{e^{\omega/kT} - 1} \,. \tag{14.21}$$

The stimulated emission or absorption rate $K_{n\ell,n'\ell'}$ from state $n\ell$ to state $n'\ell'$ is given by

$$K_{n\ell,n'\ell'} = \frac{4}{3}\mu_{n\ell,n'\ell'}\omega^3_{n\ell,n'\ell'}\frac{\alpha^3 \bar{n} g_>}{2g_n + 1} \,. \tag{14.22}$$

Summing these rates over n' and ℓ' gives the total blackbody decay rate $1/\tau_n^{\mathrm{bb}}$. Explicitly,

$$\frac{1}{\tau_{n\ell}^{\mathrm{bb}}} = \sum_{n'\ell'} K_{n\ell,n'\ell'} \,. \tag{14.23}$$

The resulting lifetime $\tau_{n\ell}^{\mathrm{T}}$ at any given temperature is given by

$$\frac{1}{\tau_{n\ell}^{\mathrm{T}}} = 1/\tau_{n\ell} + 1/\tau_{n\ell}^{\mathrm{bb}} \,. \tag{14.24}$$

For low-ℓ states with $10 < n < 20$, blackbody radiation produces a 10% decrease in the lifetimes, but for high-ℓ states of the same n, it reduces the lifetimes by a factor of ten. Since $1/\tau_{n\ell}^{\mathrm{bb}} \propto n^{-2}$, this term must dominate normal spontaneous emission at high n.

The above discussion of spontaneous and stimulated transitions is based on the implicit assumption that the atoms are in free space. If the atoms are in a cavity, which introduces structure into the blackbody and vacuum fields, the transition rates are significantly altered [14.10]. These alterations are described in Chapt. 79. If the cavity is tuned to a resonance, it increases the transition rate by the finesse of the cavity (approximately the Q for low-order modes). On the other hand, if the cavity is tuned between resonances, the transition rate is suppressed by a similar factor.

14.3 Electric Fields

As a starting point, consider the H atom in a static electric field $\mathcal{E}$ in the z-direction, and focus on the states of principal quantum number n. The field couples ℓ and $\ell \pm 1$ states of the same m by the electric dipole matrix elements. Since the states all have a common zero field energy of $-1/2n^2$, and the off-diagonal Hamiltonian matrix elements are all proportional to $\mathcal{E}$, the eigenstates are field-independent linear combinations of the zero field ℓ states of the same m, and the energy shifts from $-1/2n^2$ are linear in $\mathcal{E}$. In this first-order approximation, the energies are given by [14.2]

$$E = -\frac{1}{2n^2} + \frac{3}{2}(n_1 - n_2)n\mathcal{E} \,, \tag{14.25}$$

where n_1 and n_2 are parabolic quantum numbers (see Sect. 9.1.2) which satisfy

$$n_1 + n_2 + |m| + 1 = n\,. \tag{14.26}$$

Consider the $m = 0$ states as an example. The $n_1 - n_2 = n - 1$ state is shifted up in energy by $\frac{3}{2}n(n-1)\mathcal{E}$ and is called the extreme blue Stark state, and the $n_2 - n_1 = n - 1$ state is shifted down in energy by $\frac{3}{2}n(n-1)\mathcal{E}$ and is called the extreme red Stark state. These two states have large permanent dipole moments, and in the red (blue) state the electron spends most of its time on the downfield (upfield) side of the proton as shown in Fig. 14.3, a plot of the potential along the z-axis. We have here ignored the electric dipole couplings to other n states, which introduce small second order Stark shifts to lower energy. As implied by (14.26), states of higher m have smaller shifts. In particular, the circular $m = \ell = n - 1$ state has no first order shift since there are no degenerate states to which it is coupled by the field.

The Stark effect in other atoms is similar, but not identical to that observed in H. This point is shown by Fig. 14.4, a plot of the energies of the Na $m = 0$ levels near $n = 20$. The energy levels are similar to those of H in that most of the levels exhibit apparently linear Stark shifts from the zero field energy of the high-ℓ states. The differences, however, are twofold. First, the levels from s and p states with nonzero quantum defects join the manifold of Stark states at some nonzero field, given

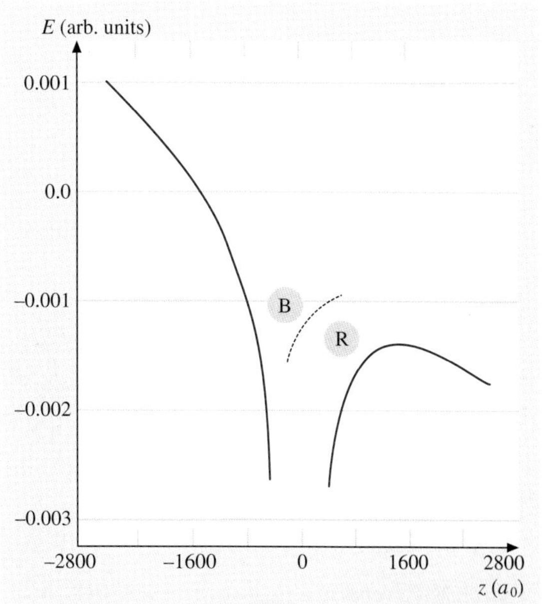

Fig. 14.3 Combined Coulomb–Stark potential along the z-axis when a field of 5×10^{-7} a.u. (2700 V/cm) is applied in the z-direction (*solid*). The extreme red state (R) is near the saddlepoint, and the extreme blue (B) state is held on the upfield side of the atom by an effective potential (*dashed*) roughly analogous to a centrifugal potential

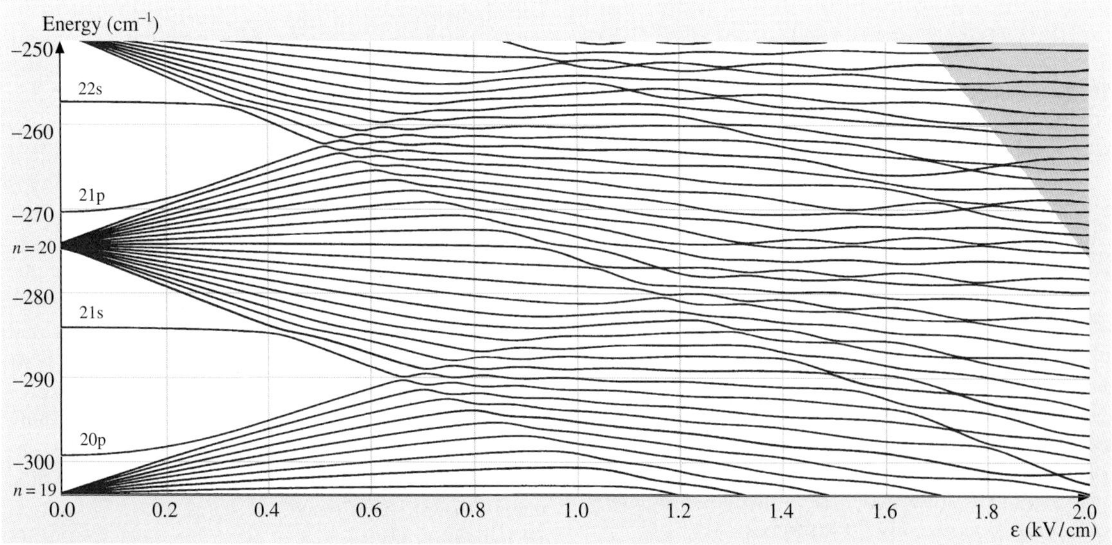

Fig. 14.4 Energies of Na $m = 0$ levels of $n \approx 20$ as a function of electric field. The *shaded region* is above the classical ionization limit

approximately by [14.4]

$$\mathcal{E} = \frac{2\delta'_\ell}{3n^5}\,, \tag{14.27}$$

where δ'_ℓ is the magnitude of the difference between δ_ℓ and the nearest integer. Second, there are avoided crossings between the blue $n = 20$ and red $n = 21$ Stark states. In H these states would cross, but in Na they do not because of the finite sized Na^+ core, which also leads to the nonzero quantum defects of the ns and np states. This point, and other related points, are described in Chapt. 15.

Field ionization is both intrinsically interesting and of great practical importance for the detection of Rydberg atoms [14.11]. The simplest picture of field ionization can be understood with the help of Fig. 14.4. The potential along the z-axis of an atom in a field $\mathcal{E}$ in the z-direction is given by

$$V = -\frac{1}{r} - \mathcal{E}z\,. \tag{14.28}$$

If an atom has an energy E relative to the zero field limit, it can ionize classically if the energy E lies above the saddle point in the potential. The required field is given by

$$\mathcal{E} = \frac{E^2}{4}\,. \tag{14.29}$$

Ignoring the Stark shifts and using $E = -1/2n^2$ yields the expression

$$\mathcal{E} = \frac{1}{16n^4}\,. \tag{14.30}$$

The H atom ionizes classically as described above, or by quantum mechanical tunneling which occurs at slightly lower fields. Since the tunneling rates increase exponentially with field strength, typically an order of magnitude for a 3% change in the field, specifying the classical ionization field is a good approximation to the field which gives an ionization rate of practical interest. The red and blue states of H ionize at very different fields, as shown by Fig. 14.5, a plot of the $m = 0$ Stark states out to the fields at which the ionization rates are $10^6\,s^{-1}$ [14.12]. First, note the crossing of the levels of different n mentioned earlier. Second, note that the red states ionize at lower fields than do the blue states, in spite of the fact that they are lower in energy. In the red states, the electron is close to the saddle point of the potential of Fig. 14.3, and it ionizes according to (14.29). If the Stark shift of the extreme red state to lower energy is taken into account, (14.30) becomes

$$\mathcal{E} = \frac{1}{9n^4}\,. \tag{14.31}$$

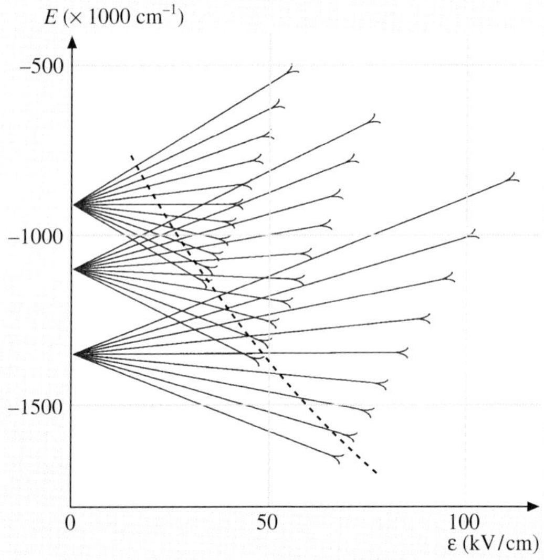

Fig. 14.5 Energies of H $m = 0$ levels of $n = 9$, 10, and 11 as functions of electric field. The widths of the levels due to ionization broaden exponentially with fields, and the onset of the broadening indicated is at an ionization rate of $10^6\,s^{-1}$. The *broken line* indicating the classical ionization limit, $\mathcal{E} = E^2/4$ passes near the points at which the extreme red states ionize

In the blue state the electron is held on the upfield side of the atom by an effective potential roughly analogous to a centrifugal potential, as shown by Fig. 14.3. At the same field the blue state's energy is lower relative to the saddle point of its potential, shown by the broken line of Fig. 14.3, than is the energy of the red state relative to the saddle point of its potential, given by (14.28) and shown by the solid line of Fig. 14.3. As shown by the broken line of Fig. 14.5, the classical ionization limit of (14.29) is simply a line connecting the ionization fields of the extreme red Stark states. All other states are stable above the classical ionization limit. In the Na atom, ionization of $m = 0$ states occurs in a qualitatively different fashion [14.12]. Due to the finite size of the Na^+ core, there are avoided crossings between the blue and red Stark states of different n, as is shown by Fig. 14.4. In the region above the classical ionization limit, shown by the shaded region of Fig. 14.4, the same coupling between hydrogenically stable blue states and the degenerate red continua leads to autoionization of the blue states [14.13]. As a result, all states above the classical ionization limit ionize at experimentally significant rates. In higher m states, the core coupling is smaller, and the behavior is more similar to H.

Field ionization is commonly used to detect Rydberg atoms in a state selective manner. Experiments are most often conducted at or near zero field, and afterwards the field is increased in order to ionize the atoms. Exactly how the atoms pass from the low field to the high ionizing field is quite important. The passage can be adiabatic, diabatic or anything in between. The selectivity is best if the passage is purely adiabatic or purely diabatic, for in these two cases unique paths are followed.

In zero field, optical excitation from a ground s state leads only to final np states. In the presence of an electric field, all the Stark states are optically accessible, because they all have some p character. The fact that all the Stark states are optically accessible from the ground state allows the population of arbitrary ℓ states of nonhydrogenic atoms by a technique called Stark switching [14.6, 14]. In any atom other than H, the ℓ states are nondegenerate in zero field, and each of them is adiabatically connected to one, and only one, high field Stark state, as shown by Fig. 14.4. If one of the Stark states is excited with a laser and the field reduced to zero adiabatically, the atoms are left in a single zero field ℓ state.

In zero field, the photoionization cross section is structureless. However, in an electric field, it exhibits obvious structure, sometimes termed strong field mixing resonances. Specifically, when ground state s atoms are exposed to light polarized parallel to the static field, an oscillatory structure is observed in the cross section, even above the zero field ionization limit [14.15]. The origin of the structure can be understood with the aid of a simple classical picture [14.16, 17]. The electrons ejected in the downfield direction can simply leave the atom, while the electrons ejected in the upfield direction are reflected back across the ionic core and also leave the atom in the downfield direction. The wave packets corresponding to these two classical trajectories are added, and they can interfere constructively or destructively at the ionic core depending on the phase accumulation of the reflected wave packet. Since the phase depends on the energy, there is an oscillation in the photoexcitation spectrum. This model suggests that no oscillations should be observed for light polarized perpendicular to the static field, and none are. The oscillations can also be thought of as arising from the remnants of quasistable extreme blue Stark states which have been shifted above the ionization limit, and, using this approach or a WKB approach, one can show that the spacing between the oscillations at the zero field limit is $\Delta E = \mathcal{E}^{3/4}$ [14.18, 19].

The initial photoexcitation experiments were done using narrow bandwidth lasers, so that the time dependence of the classical pictures was not explicitly observed. Using mode locked lasers it has been possible to create a variety of Rydberg wave packets [14.20, 21] and observe, in effect, the classical motion of an electron in an atom. Of particular interest, it has been possible to directly observe the time delay of the ejection of electrons subsequent to excitation in an electric field [14.22].

14.4 Magnetic Fields

To first order, the energy shift of a Rydberg atom due to a magnetic field B (the Zeeman effect) is proportional to the angular momentum of the atom. Since the states optically accessible from the ground state have low angular momenta, the energy shifts are the same as those of low-lying atomic states. In contrast, the second order diamagnetic energy shifts are proportional to the area of the Rydberg electron's orbit and scale as B^2n^4 [14.23]. The diamagnetic interaction mixes the ℓ states, allowing all to be excited from the ground state, and produces large shifts to higher energies. The energy levels as a function of magnetic field are reminiscent of the Stark energy levels shown in Fig. 14.5, differing in that the energy shifts are quadratic in the magnetic field.

One of the most striking phenomena in magnetic fields is the existence of quasi-Landau resonances, spaced by $\Delta E = 3\hbar B/2$, in the photoionization cross section above the ionization limit [14.24]. The origin of this structure is similar to the origin of the strong field mixing resonances observed in electric fields. An electron ejected in the plane perpendicular to the B fields is launched into a circular orbit and returns to the ionic core. The returning wave packet can be in or out of phase with the one leaving the ionic core, and thus, can interfere constructively or destructively with it. While the electron motion in the xy-plane is bound, motion in the z-direction is unaffected by the magnetic field and is unbounded above the ionization limit, leading to resonances of substantial width. The Coulomb potential does provide some binding in the z-direction and allows the existence of quasistable three-dimensional orbits [14.25].

14.5 Microwave Fields

Strong microwave fields have been used to drive multiphoton transitions between Rydberg states and to ionize them. Here we restrict our attention to ionization. Ionization by both linearly and circularly polarized fields has been explored with both H and other atoms.

Hydrogen atoms have been studied with linearly polarized fields of frequencies up to 36 GHz [14.26]. When the microwave frequency $\omega \ll 1/n^3$, ionization of $m = 0$ states occurs at a field of $\mathcal{E} = 1/9n^4(E^2/4)$, which is the field at which the extreme red Stark state is ionized by a static field. Due to the second-order Stark effect, the blue and red shifted states are not quite mirror images of each other, and when the microwave field reverses, transitions between Stark states occur. There is a rapid mixing of the Stark states of the same n and m by a microwave field, and all of them are ionized at the same microwave field amplitude, $\mathcal{E} = 1/9n^4$. Important points are that no change in n occurs and the ionization field is the same as the static field required for ionization of the extreme red Stark state. As ω approaches $1/n^3$, the field falls below $1/9n^4$ due to Δn transitions to higher lying states, allowing ionization at lower fields. This form of ionization can be well described as the transition to the classically chaotic regime [14.27]. For $\omega > 1/n^3$ the ionization field is more or less constant, and for $\omega > 1/2n^2$ the process becomes photoionization.

The ionization of nonhydrogenic atoms by linearly polarized fields has also been investigated at frequencies of up to 30 GHz, but the result is very different from the hydrogenic result. For $\omega \ll 1/n^3$ and low m, ionization occurs at a field of $\mathcal{E} \approx 1/3n^5$ [14.28]. This is the field at which the $m = 0$ extreme blue and red Stark states of principal quantum number n and $n+1$ have their avoided crossing. For $n = 20$ this field is $\approx 500\,\mathrm{V/cm}$, as shown by Fig. 14.4. How ionization occurs can be understood with a simple model based on a time-varying electric field. As the microwave field oscillates in time, atoms follow the Stark states of Na shown in Fig. 14.4. Even with very small field amplitudes, transitions between the Stark states of the same n are quite rapid because of the zero field avoided crossings. If the field reaches $1/3n^5$, the avoided crossing between the extreme red n and blue $n+1$ state is reached, and an atom in the blue n Stark state can make a Landau–Zener transition to the red $n+1$ Stark state. Since the analogous red–blue avoided crossings between higher lying states occur at lower fields, once an atom has made the $n \to n+1$ transition it rapidly makes a succession of transitions through higher n states to a state which is itself ionized by the field.

The Landau–Zener description given above is somewhat oversimplified in that we have ignored the coherence between field cycles. When it is included, we see that the transitions between levels are resonant multiphoton transitions. While the resonant character is obscured by the presence of many overlapping resonances, the coherence substantially increases the $n \to n+1$ transition probability even when $\mathcal{E} < 1/3n^5$. The fields required for ionization calculated using this model are lower than $1/3n^5$, in agreement with the experimental observations. Nonhydrogenic Na states of high m behave like H, because no states with significant quantum defects are included, and the $n \to n+1$ avoided crossings are vanishingly small.

Experiments on ionization of alkali atoms by circularly polarized fields of frequency ω show that for $\omega \ll 1/n^3$, a field amplitude of $\mathcal{E} = 1/16n^4$ is required for ionization [14.29]. This field is the same as the static field required. In a frame rotating with frequency ω, the circularly polarized field is stationary and cannot induce transitions, so this result is not surprising. On the other hand, when the problem is transformed to the rotating frame, the potential of (14.28) is replaced by

$$V = -\frac{1}{r} - \mathcal{E}x - \frac{\omega^2\rho^2}{2}\,, \tag{14.32}$$

where $\rho^2 = x^2 + y^2$, and we have assumed the field to be in the x-direction in the rotating frame. This potential has a saddle point below $\mathcal{E} = 1/16n^4$ [14.30]. As n or ω is raised so that $\omega \to 1/n^3$, the experimentally observed field falls below $1/16n^4$, but not so fast as implied by (14.32). Equation (14.32) is based solely on energy considerations, and ionization at the threshold field implied by (14.32) requires that the electron escape over the saddle point in the rotating frame at nearly zero velocity. For this to happen, when ω approaches $1/n^3$, more than n units of angular momentum must be transferred to the electron, which is unlikely. Models based on a restriction of the angular momentum transferred from the field to the Rydberg electron are in better agreement with the experimental results. Small deviations of a few percent from circular polarization allow ionization at fields as low as $\mathcal{E} = 1/3n^5$. This sensitivity can be understood as follows. In the rotating frame, a field with slightly elliptical polarization appears to be a large static field with a superimposed oscillating field at frequency 2ω. The oscillating field drives transitions to states of higher energy, allowing ionization at fields less than $\mathcal{E} = 1/16n^4$.

In the regime in which $\omega > 1/n^3$, microwave ionization of nonhydrogenic atoms is essentially the same as it is in H [14.31]. In this regime, the microwave field couples states differing in n by more than one, and the pressure or absence of quantum defects is not so important. Consequently, only for $\omega > 1/n^3$ is the microwave ionization of H and other atoms different.

14.6 Collisions

Since Rydberg atoms are large, with geometric cross sections proportional to n^4, one might expect the cross sections for collisions to be correspondingly large. In fact, such is often not the case. A useful way of understanding collisions of neutral atoms and molecules with Rydberg atoms is to imagine an atom or molecule M passing through the electron cloud of an Na Rydberg atom. There are three interactions

$$e^- - Na^+ , \quad e^- - M , \quad M - Na^+ . \tag{14.33}$$

The long range e^-–Na^+ interaction determines the energy levels of the Na atom. The short range of the e^-–M and M–Na^+ interactions makes it likely that only one will be important at any given time. This approximation, termed the binary encounter approximation, is described in Chapt. 56. The M–Na^+ interaction can only lead to cross sections of $\approx 10-100\,\text{Å}^2$. On the other hand, since the electron can be anywhere in the cloud, the cross sections due to the e^-–M interaction can be as large as the geometric cross section of the Rydberg atom. Accordingly, we focus on the e^-–M interaction.

Consider a thermal collision between M and an Na Rydberg atom. Typically, M passes through the electron cloud slowly compared with the velocity of the Rydberg electron, and it is the e^-–M scattering which determines what happens in the M–Na collision, as first pointed out by *Fermi* [14.32]. First consider the case where M is an atom. There are no energetically accessible states of atom M which can be excited by the low energy electron, so the scattering must be elastic. The electron can transfer very little kinetic energy to M, but the direction of the electron's motion can change. With this thought in mind, we can see that only the collisional mixing of nearly degenerate ℓ states of the same n has very large cross sections. The ℓ-mixing cross sections are approximately geometric at low n [14.33]. If the M atom comes anywhere into the Rydberg orbit, scattering into a different ℓ state occurs. At high n, the cross section decreases, because the probability distribution of the Rydberg electron becomes too dilute, and it becomes increasingly likely that the M atom will pass through the Rydberg electron's orbit without encountering the electron. The n at which the peak ℓ-mixing cross section occurs increases with the electron scattering length of the atom. While ℓ-mixing cross sections are large, n changing cross sections are small $(\approx 100\,\text{Å}^2)$ since they cannot occur when the Rydberg electron is anywhere close to the outer turning point of its orbit [14.34].

If M is a molecule, there are likely to be energetically accessible vibrational and rotational transitions which can provide energy to or accept energy from the Rydberg electron, and this possibility increases the likelihood of n changing collisions with Rydberg atoms [14.11]. Electronic energy from the Rydberg atom must be resonantly transferred to rotation or vibration in the molecule. In heavy or complex molecules, the presence of many rotational-vibrational states tends to obscure the resonant character of the transfer, but in several light systems the collisional resonances have been observed clearly [14.11].

Using the large Stark shifts of Rydberg atoms it is possible to tune the levels so that resonant energy transfer between two colliding atoms can occur [14.35] by the resonant dipole–dipole coupling,

$$Vd = \frac{\mu_1 \mu_2}{R^3} . \tag{14.34}$$

Here μ_1 and μ_2 are the dipole matrix elements of the upward and downward transitions in the two atoms, and R is their separation. At room temperature, this process leads to enormous cross sections, substantially in excess of the geometric cross sections. At the low temperatures (300 μK) attainable using cold atoms, the atoms do not move, and therefore cannot collide. However, resonant dipole–dipole energy transfer is still observed due to the static dipole–dipole interactions of not two, but many atoms [14.36, 37].

Since Rydberg atoms are easily perturbed by electric fields, it is hardly a surprise that collisions of charged particles with Rydberg atoms have large cross sections. In cold Rydberg atom samples, these large cross sections can lead to the spontaneous evolution to a plasma, since the macroscopic positive charge of the cold ions can trap any liberated electrons, leading to impact ionization for a large part of the Rydberg atom sample [14.38, 39].

14.7 Autoionizing Rydberg States

The bound Rydberg atoms considered thus far are formed by adding the Rydberg electron to the ground state of the ionic core. It could equally well be added to an excited state of the core [14.40]. Figure 14.6 shows the energy levels of the ground 5s state of Sr^+ and the excited 5p state. Adding an $n\ell$ electron to the 5s state yields the bound Sr $5sn\ell$ state, and adding it to the excited 5p state gives the doubly excited $5pn\ell$ state, which is coupled by the Coulomb interaction to the degenerate $5s\epsilon\ell'$ continuum. The $5pn\ell$ state autoionizes at the rate $\Gamma_{n\ell}$ given by [14.41]

$$\Gamma_{n\ell} = 2\pi |\langle 5pn\ell | V | 5s\epsilon\ell' \rangle|^2 , \tag{14.35}$$

where V denotes the Coulomb coupling between the nominally bound $5pn\ell$ state and the $5s\epsilon\ell'$ continuum. A more general description of autoionization can be found in Chapt. 25.

A simple picture, based on superelastic electron scattering from the Sr^+ 5p state, gives the scaling of the autoionization rates of (14.35) with n and ℓ. The $n\ell$ Rydberg electron is in an elliptical orbit, and each time it comes near the core it has an n-independent probability γ_ℓ of scattering superelastically from the Sr^+ 5p ion, leaving the core in the 5s state and gaining enough energy to escape from the Coulomb potential of the Sr^+ core. The autoionization rate of the $5pn\ell$ state is obtained by multiplying γ_ℓ by the orbital frequency of the $n\ell$ state, $1/n^3$ to obtain

$$\Gamma_{n\ell} = \frac{\gamma_\ell}{n^3} . \tag{14.36}$$

Equation (14.36) displays the n dependence of the autoionization rate explicitly and the ℓ dependence through γ_ℓ. As ℓ increases, the closest approach of the Rydberg electron to the Sr^+ is at a larger orbital radius, so that superelastic scattering becomes progressively less probable, and γ_ℓ decreases rapidly with increasing ℓ. The simple picture of autoionization given above implies a finite probability of autoionization each time the $n\ell$ electron passes the ionic core, so the probability of an atom's remaining in the autoionizing state should resemble stair steps [14.42], which can be directly observed using mode locked laser excitation and detection [14.43].

To a first approximation, the Sr $5pn\ell$ states can be described by the independent electron picture used above, but in states converging to higher lying states of Sr^+, the independent electron picture fails. Consider the Sr^+ $\ell \geq 4$ states of $n > 5$. They are essentially degenerate, and the field due to an outer Rydberg electron converts the zero field ℓ states to superpositions much like Stark states. The outer electron polarizes the Sr^+ core, so that the outer electron is in a potential due to a charge and a dipole, and the resulting dipole states of the outer electron display a qualitatively different excitation spectrum than do states such as the $5pn\ell$ states, which are well described by an independent particle picture [14.44]. When both electrons are excited to very high-lying states, with the outer electron in a state of relatively low ℓ, the classical orbits of the two electrons cross. Time domain measurements, made using wave packets, show that in this case autoionization is likely to occur in the first orbit of the outer electron [14.45].

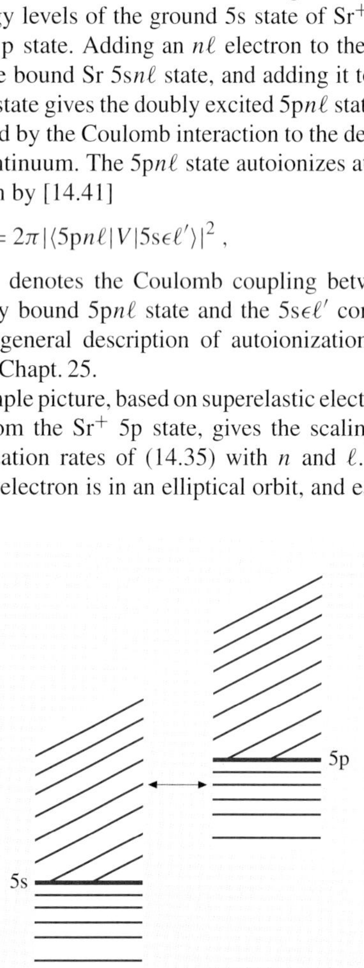

Fig. 14.6 Sr^+ 5s and 5p states (—), the Rydberg states of Sr converging to these two ionic states are shown by (—), the continuum above the two ionic levels (///). The $5pn\ell$ states are coupled to the $5s\epsilon\ell'$ continua and autoionize

References

14.1 H. E. White: *Introduction to Atomic Spectra* (McGraw-Hill, New York 1934)
14.2 H. A. Bethe, E. A. Salpeter: *Quantum Mechanics of One and Two Electron Atoms* (Academic, New York 1975)
14.3 U. Fano: Phys. Rev. A **2**, 353 (1970)
14.4 M. L. Zimmerman, M. G. Littman, M. M. Kash, D. Kleppner: Phys. Rev. A **20**, 2251 (1979)
14.5 S. A. Bhatti, C. L. Cromer, W. E. Cooke: Phys. Rev. A **24**, 161 (1981)
14.6 R. R. Freeman, D. Kleppner: Phys. Rev. A **14**, 1614 (1976)
14.7 A. Lindgard, S. E. Nielsen: At. Data Nucl. Data Tables **19**, 534 (1977)
14.8 E. S. Chang: Phys. Rev. A **31**, 495 (1985)
14.9 W. E. Cooke, T. F. Gallagher: Phys. Rev. A **21**, 588 (1980)
14.10 S. Haroche, J. M. Raimond: Radiative properties of Rydberg states in resonant cavities. In: *Advances in Atomic and Molecular Physics*, Vol. 20, ed. by D. Bates, B. Bederson (Academic, New York 1985)
14.11 F. B. Dunning, R. F. Stebbings: Experimental studies of thermal-energy collisions of Rydberg atoms with molecules. In: *Rydberg States of Atoms with Molecules*, ed. by R. F. Stebbings, F. B. Dunning (Cambridge Univ. Press, Cambridge 1983)
14.12 D. S. Bailey, J. R. Hiskes, A. C. Riviere: Nucl. Fusion **5**, 41 (1965)
14.13 M. G. Littman, M. M. Kash, D. Kleppner: Phys. Rev. Lett. **41**, 103 (1978)
14.14 R. R. Jones, T. F. Gallagher: Phys. Rev. A **38**, 2946 (1988)
14.15 R. R. Freeman, N. P. Economou, G. C. Bjorklund, K. T. Lu: Phys. Rev. Lett. **41**, 1463 (1978)
14.16 W. P. Reinhardt: J. Phys. B **16**, 635 (1983)
14.17 J. Gao, J. B. Delos, M. C. Baruch: Phys. Rev. A **46**, 1449 (1992)
14.18 T. F. Gallagher: *Rydberg Atoms* (Cambridge Univ. Press, Cambridge 1994)
14.19 A. R. P. Rau: J. Phys. B **12**, L193 (1979)
14.20 R. R. Jones, L. D. Noordam: Adv. At. Mol. Opt. Phys. **38**, 1 (1997)
14.21 G. Alber, P. Zoller: Phys. Rep. **199**, 231 (1991)
14.22 J. B. M. Warntjes, C. Wesdorp, F. Robicheaux, L. D. Noordam: Phys. Rev. Lett. **83**, 512 (1999)
14.23 D. Kleppner, M. G. Littman, M. L. Zimmerman: Rydberg atoms in strong fields. In: *Rydberg States of Atoms and Molecules*, ed. by R. F. Stebbings, F. B. Dunning (Cambridge Univ. Press, Cambridge 1983)
14.24 W. R. S. Garton, F. S. Tomkins: Astrophys. J. **158**, 839 (1969)
14.25 A. Holle, J. Main, G. Wiebusch, H. Rottke, K. H. Welge: Phys. Rev. Lett. **61**, 161 (1988)
14.26 B. E. Sauer, M. R. W. Bellerman, P. M. Koch: Phys. Rev. Lett. **68**, 1633 (1992)
14.27 R. V. Jensen, S. M. Susskind, M. M. Sanders: Phys. Rept. **201**, 1 (1991)
14.28 P. Pillet, H. B. van Linden van der Heuvell, W. W. Smith, R. Kachru, N. H. Tran, T. F. Gallagher: Phys. Rev. A **30**, 280 (1984)
14.29 P. Fu, T. J. Scholz, J. M. Hettema, T. F. Gallagher: Phys. Rev. Lett. **64**, 511 (1990)
14.30 M. Nauenberg: Phys. Rev. Lett. **64**, 2731 (1990)
14.31 A. Krug, A. Buchleitner: Phys. Rev. A **66**, 053416 (2002)
14.32 E. Fermi: Nuovo Cimento **11**, 157 (1934)
14.33 T. F. Gallagher, S. A. Edelstein, R. M. Hill: Phys. Rev. A **15**, 1945 (1977)
14.34 F. Gounand, J. Berlande: Experimental studies of the interaction of Rydberg atoms with atomic species at thermal energies. In: *Rydberg States of Atoms and Molecules*, ed. by R. F. Stebbings, F. B. Dunning (Cambridge Univ. Press, Cambridge 1983)
14.35 T. F. Gallagher: Phys. Rept. **210**, 319 (1992)
14.36 I. Mourachko, D. Comparat, F. de Tomasi, A. Fioretti, P. Nosbaum, V. M. Akulin, P. Pillet: Phys. Rev. Lett. **80**, 253 (1998)
14.37 W. R. Anderson, J. R. Veale, T. F. Gallagher: Phys. Rev. Lett. **80**, 249 (1998)
14.38 M. P. Robinson, B. Laburthe-Tolra, M. W. Noel, T. F. Gallagher, P. Pillet: Phys. Rev. Lett. **85**, 4466 (2000)
14.39 S. K. Dutta, J. R. Guest, D. Feldbaum, A. Walz-Flannigan, G. Raithel: Phys. Rev. Lett. **86**, 3993 (2001)
14.40 T. F. Gallagher: J. Opt. Soc. Am. B **4**, 794 (1987)
14.41 U. Fano: Phys. Rev. **124**, 1866 (1961)
14.42 X. Wang, W. E. Cooke: Phys. Rev. Lett. **67**, 696 (1991)
14.43 S. N. Pisharody, R. R. Jones: Phys. Rev. A **65**, 033418 (2002)
14.44 U. Eichmann, V. Lange, W. Sandner: Phys. Rev. Lett. **68**, 21 (1992)
14.45 S. N. Pisharody, R. R. Jones: Science **303**, 813 (2004)

15. Rydberg Atoms in Strong Static Fields

Confronting classical and quantum mechanics in systems whose classical motion is chaotic is one of the fundamental problems of physics, as evidenced by the enormous outpouring of research during the last three decades [15.1, 2]. Highly excited Rydberg atoms in external fields [15.3] play a prominent role in this quest because they are the best known examples of quantum systems whose classical counterpart is chaotic. For a wide variety of field configurations and field strengths, their spectra can be measured to high precision. At the same time, since their Hamiltonians are known analytically, they are equally amenable to accurate theoretical investigations using either classical or quantum mechanics.

This chapter is restricted to a description of Rydberg atoms in strong static fields. Related information on atoms in strong fields can be found in Chapt. 13 of this Handbook, on Rydberg atoms in Chapt. 14, and on the interaction of atoms with strong laser fields in Chapt. 74.

15.1 **Scaled-Energy Spectroscopy** 248
15.2 **Closed-Orbit Theory** 248
15.3 **Classical and Quantum Chaos** 249
15.3.1 Magnetic Field 249
15.3.2 Parallel Electric and Magnetic Fields 250
15.3.3 Crossed Electric and Magnetic Fields 250
15.4 **Nuclear-Mass Effects** 251
References 251

Different configurations of external fields have been studied: (i) an electric field, which in hydrogen leads to integrable classical dynamics [15.4, 5] (ii) a magnetic field, which produces a transition from regular to chaotic classical dynamics and which sparked the interest in Rydberg atoms as prototype examples for the study of the quantum-classical correspondence [15.6–15] and references therein (iii) parallel electric and magnetic fields [15.16, 17] (iv) crossed electric and magnetic fields which break all continuous symmetries of the unperturbed atom and thus allow one to study the transition from regularity to chaos in three coupled degrees of freedom [15.18–21] and references therein.

The hydrogen atom is the prototype example for states with a single highly excited electron under the influence of strong external fields. For an electron in the hydrogen ground state, the influence of external electric or magnetic fields becomes comparable to that of the nuclear Coulomb field when the field strengths are in the order of the atomic units of electric field strength, $F_0 = e/(4\pi\varepsilon_0 a_0^2) = 5.142\,206\,42\,(44) \times 10^{11}$ V/m, or magnetic field strength, $B_0 = \hbar/\left(ea_0^2\right) = 2.350\,517\,42\,(20) \times 10^5$ T, which is far beyond experimental reach. However, the relative importance of the external fields scales with the principal quantum number n as n^4F and n^3B, so that for highly excited atoms, laboratory fields can easily be "strong". Atomic units will be used throughout this chapter.

In a non-hydrogenic atom, the influence of the inner-shell electrons can be summarized by means of a short-range effective core potential or a set of quantum defects [15.22]. For laboratory field strengths, the core is too small to be appreciably influenced by the external fields. For this reason, the field-free quantum defects can be used to model core effects even in the presence of external fields [15.23].

15.1 Scaled-Energy Spectroscopy

The Hamiltonian for a hydrogen atom in a $\hat{z}$-directed magnetic field and an electric field of arbitrary orientation is

$$H = \frac{\boldsymbol{p}^2}{2} - \frac{1}{r} + \frac{1}{2} BL_z + \frac{1}{8} B^2\rho^2 + \boldsymbol{r}\cdot\boldsymbol{F} = E\,, \tag{15.1}$$

where $\rho^2 = x^2 + y^2$ and L_z is the angular momentum component along the magnetic field axis. The dynamics depends on three parameters: the field strengths F and B and the energy E. We can reduce the number of independent parameters to two if we exploit a scaling property of the Hamiltonian: In terms of the scaled quantities

$$\tilde{\boldsymbol{r}} = w^{-2}\boldsymbol{r}\,, \qquad \tilde{\boldsymbol{p}} = w\boldsymbol{p}$$
$$\tilde{E} = w^2 E\,, \qquad \tilde{F} = w^4 F \tag{15.2}$$

with the scaling parameter

$$w = B^{-1/3}\,, \tag{15.3}$$

the scaled Hamiltonian reads

$$\tilde{H} = \frac{\tilde{\boldsymbol{p}}^2}{2} - \frac{1}{\tilde{r}} + \frac{1}{2}\tilde{L}_z + \frac{1}{8}\tilde{\rho}^2 + \tilde{\boldsymbol{r}}\cdot\tilde{\boldsymbol{F}} = \tilde{E}\,. \tag{15.4}$$

The scaled dynamics thus depends only on two parameters, the scaled energy $\tilde{E}$ and the scaled electric field strength $\tilde{F}$. Instead of the above scaling with the magnetic field strength, which is the most common one, equivalent scaling prescriptions with the electric field strength or the energy can be used [15.24].

The way of recording an atomic spectrum that is most suitable for the investigation of quantum-classical correspondence is scaled-energy spectroscopy. A scaled spectrum consists of a list of eigenvalues w_n of the scaling parameter (15.3) characterizing the quantum states for a given scaled energy $\tilde{E}$ and scaled electric field strength $\tilde{F}$. It offers the advantage that the underlying classical dynamics does not change across the spectrum, which makes the spectrum more easily accessible to a semiclassical interpretation (Sect. 15.2). For this reason, scaled-energy spectroscopy has been adopted in numerous experimental [15.4, 5, 8, 18] and theoretical investigations.

To obtain a theoretical description of a scaled spectrum, the Schrödinger equation must be rewritten in terms of the scaling parameter w. In the case of a single external field, either electric or magnetic, this procedure leads to a generalized eigenvalue equation for the scaling parameter w [15.25]. In the presence of both electric and magnetic fields, the scaled spectrum is described by a quadratic eigenvalue equation that has become tractable only recently [15.26].

In a non-hydrogenic atom, the extent of the core imposes an absolute length scale and thus breaks the scaling symmetry. However, if the extent of the Rydberg electron's orbital is large, the size of the core can be neglected and the scaling behavior is restored. This renders scaled-energy spectroscopy a useful concept also for non-hydrogenic atoms [15.4, 5, 18].

15.2 Closed-Orbit Theory

Among the most remarkable effects strong external fields produce in Rydberg atoms are the Quasi-Landau oscillations: Close to the ionization limit, the photoabsorption spectrum of Ba I in a magnetic field shows regular oscillations [15.6]. This phenomenon was given a convincing interpretation by *Starace* [15.27] and embedded by *Du* and *Delos* [15.28, 29] and *Bogomolny* [15.30] into the general framework of closed-orbit theory, which has since become the central interpretative tool for a description of Rydberg spectra in external fields [15.4, 5, 8, 18]). Recently, it has also been used for the computation of Rydberg spectra [15.31, 32].

Closed-orbit theory represents an atomic photoabsorption spectrum as a superposition of regular oscillations, each of which is related to a closed orbit of the underlying classical dynamics, i. e., to an orbit that starts and ends at the position of the nucleus. The period of an oscillation is given by the return time of the associated closed orbit (divided by $\hbar$), the amplitude is determined on the one hand by the initial state and the polarization of the exciting photon, and on the other hand by the stability of the closed orbit. Once the initial state is specified, both can therefore be calculated within classical mechanics.

A spectrum that contains contributions from many closed orbits can look enormously complicated. Its Fourier transform, on the other hand, consists of a series of isolated peaks that can be identified with the contributions of individual closed orbits. The Fourier transform thus provides the means to identify the crucial dynamics underlying a complicated spectrum. If a spectrum is recorded at constant external field strength, however, this analysis is inhibited by the fact that the oscillations are not strictly harmonic because both the return time of a closed orbit and the recurrence amplitude associated with it vary across the spectrum. This is the principal reason why scaled-energy spectroscopy (see Sect. 15.1) is customarily used. In a scaled spectrum, the period of an oscillation is given by the scaled action of the corresponding orbit and is fixed across the spectrum.

Although initially devised for atoms in magnetic fields [15.28–30], closed-orbit theory is equally applicable to atoms in electric [15.33] as well as parallel [15.34] or crossed [15.34, 35] electric and magnetic fields. In the case of non-hydrogenic atoms, the influence of the ionic core can be modelled either by means of an effective classical potential [15.35, 36] or in terms of quantum defects [15.37]. Recently, closed-orbit theory has even been applied to the spectra of simple molecules in external fields [15.38].

Since a non-hydrogenic core is much smaller than the extent of a closed orbit (which is comparable to the size of the atomic Rydberg state), it does not appreciably modify the shape of the orbit. The peaks observed in a hydrogen spectrum are therefore also observed in the corresponding spectrum of a non-hydrogenic atom, although their strengths may be altered considerably (core shadowing) [15.37]. The principal effect of a core is to scatter the electron returning along one closed orbit into the initial direction of another, so that concatenations of hydrogenic closed orbits appear in the spectrum [15.37]. For this reason, the closed orbits of the hydrogen atom in external fields are the crucial ingredient for the interpretation of any Rydberg spectrum. They have been systematically studied for hydrogen in magnetic [15.39, 40] as well as electric [15.41] and crossed [15.42, 43] fields.

15.3 Classical and Quantum Chaos

In the absence of external electric and magnetic fields, the classical dynamics described by the atomic Hamiltonian (15.1) is integrable and completely degenerate [15.44]. When external fields are present, a transition to classical chaos can be observed whose details depend on the precise field configuration (see below). It is characterized by the break-up of invariant tori and the appearance of irregular regions in the classical phase space.

Chaos, as understood in classical mechanics, does not exist in closed quantum systems [15.45]. Nevertheless, in the dynamics of a quantum system clear indications of regularity or chaos in the underlying classical system can be found [15.2]. Most prominent among them is the statistical distribution of nearest-neighbor energy level spacings (NNS). In a classically chaotic system, energy levels show avoided crossings. Level repulsion is statistically reflected by NNS following a Wigner distribution

$$P(S) = \frac{\pi}{2} S \mathrm{e}^{-\pi S/4} \tag{15.5}$$

that restricts small spacings. On the other hand, integrable systems possess a complete set of quantum numbers, so that levels are allowed to cross. This gives rise to a Poissonian NNS distribution

$$P(S) = \mathrm{e}^{-S} \tag{15.6}$$

that favors small spacings. For mixed regular-chaotic systems, a transition from a Poisson to a Wigner NNS distribution is found [15.14].

15.3.1 Magnetic Field

An atom exposed to a magnetic field possesses rotational symmetry around the magnetic field axis, which leads in classical mechanics to the conservation of the angular-momentum component L_z along the field axis and in quantum mechanics to a good magnetic quantum number $m = L_z$. The dynamics of the rotation coordinate can thus be separated, leaving two coupled degrees of freedom.

The quantum numbers l and n that characterize the pure hydrogen states both break down in a magnetic field. However, the field affects them differently: Whereas l breaks down extensively even for small fields (l-mixing region), the breakdown of n is only achieved through considerably stronger fields or higher energies (n-mixing region). Since chaotic dynamics can only exist in at least two degrees of freedom, a single n-manifold of electronic energy levels does not have enough degrees of freedom to support chaos. Chaos can develop only when different n-manifolds mix (intermanifold chaos). The regular dynamics that prevails as long as n is approximately conserved is reflected in the existence of a second adiabatic constant of motion (apart from the energy), which is given by [15.46, 47]

$$\Lambda = 4A^2 - 5A_z^2 , \quad (15.7)$$

in terms of the Runge–Lenz vector

$$A = \frac{1}{\sqrt{-2E}}\left[\frac{1}{2}(p\times L - L\times p) - \frac{r}{r}\right], \quad (15.8)$$

and is conserved to second order in the magnetic field strength.

As the magnetic field strength increases, corresponding to an increase in the scaled energy from $-\infty$ toward zero, the classical dynamics changes from regular to almost entirely chaotic [15.13, 14]. For positive scaled energies above $\tilde{E}_c = 0.328\,782\ldots$, completely hyperbolic dynamics is reached [15.48]. In step with the onset of classical chaos, the quantum NNS distribution changes from a Poisson to a Wigner distribution [15.13, 14].

15.3.2 Parallel Electric and Magnetic Fields

In parallel fields, as in a pure magnetic field, an atom retains rotational symmetry around the field axis. It therefore shows a similar transition from regular dynamics to intermanifold chaos at scaled energies $\tilde{E} \approx 0$. At small field strengths, a second-order adiabatic invariant akin to (15.7) is given by [15.49, 50]

$$\Lambda_\beta = 4A^2 - 5(A_z - \beta)^2 + 5\beta^2 , \quad (15.9)$$

where the parameter

$$\beta = \frac{12}{5}\frac{F}{n^2 B^2} \quad (15.10)$$

measures the relative strengths of the electric and magnetic fields.

15.3.3 Crossed Electric and Magnetic Fields

In nonaligned electric and magnetic fields, the rotational symmetry of the field-free atom is broken completely, so that all three degrees of freedom are coupled. The angular momentum quantum numbers l and m break down extensively even at small field strengths; the principal quantum number n follows only gradually. Even when n is approximately conserved, however, in the crossed-fields atom two coupled degrees of freedom remain. They allow the occurrence of chaotic dynamics within a single n-manifold (intramanifold chaos) [15.51, 52].

The intramanifold dynamics can conveniently be described in terms of the vectors [15.53]

$$I_1 = \frac{1}{2}(L + A) ,$$
$$I_2 = \frac{1}{2}(L - A) , \quad (15.11)$$

that obey independent angular momentum Poisson bracket (or, in quantum mechanics, commutator) relations. For fixed n, I_1 and I_2 are restricted to the spheres

$$I_1^2 = I_2^2 = \frac{n^2}{4} . \quad (15.12)$$

They span a four-dimensional space that is a convenient representation of the intramanifold phase space.

Within a given n manifold, the position vector r can be replaced with $-\frac{3}{2}nA$ [15.54]. Using this replacement, we can rewrite the contributions to the Hamiltonian (15.1) that are linear in the field strengths as

$$H_{\text{lin}} = \omega_1 \cdot I_1 + \omega_2 \cdot I_2 \quad (15.13)$$

with the constant vectors

$$\omega_1 = \frac{1}{2}(B - 3nF) ,$$
$$\omega_2 = \frac{1}{2}(B + 3nF) . \quad (15.14)$$

The first-order Hamiltonian H_{lin} describes a precession of the vectors I_1 and I_2 around ω_1 and ω_2, respectively, and preserves the integrability of the dynamics. Intramanifold chaos arises only if the quadratic contribution to the Hamiltonian (15.1) is taken into account. It can be detected either in classical mechanics [15.51, 52] or in quantum mechanics via its imprint on the intramanifold NNS distribution [15.52].

The properties of the crossed-fields hydrogen atom above the ionization threshold provide an example of a chaotic scattering system. Classically, chaotic ionization manifests itself in a fractal dependence of the electron escape time on the initial conditions [15.55]. Experimentally, a distinction has been made between "prompt" electrons that ionize fast, and "delayed" electrons that ionize only after more than 100 ns [15.20]. The latter can be interpreted as electrons that undergo chaotic scattering and circle the nucleus many times before they escape. A detailed classical model of chaotic scattering was presented in [15.56]. In quantum mechanics, chaotic scattering can be identified through the occurrence of Ericson fluctuations in the above-threshold spectrum [15.55].

15.4 Nuclear-Mass Effects

So far, only the relative motion of the electron with respect to the ionic core has been described. This is appropriate if the nucleus can be assumed to be infinitely heavy and thus not to take part in the motion. To include the effects of a finite nuclear mass, the description must start from the coupled two-body Hamiltonian and then work toward a separation of the internal dynamics from the center-of-mass (CM) motion.

It turns out that in the presence of a magnetic field, unlike the field-free two-body problem, a complete separation of the relative and CM motions is impossible. Instead, only a pseudo-separation can be achieved, where the relative and CM motions remain coupled through a new constant of motion called the pseudomomentum $\boldsymbol{K}$ [15.57]. This coupling introduces a number of novel effects into the dynamics (see [15.58] for a detailed discussion).

The influence of the CM motion on the internal dynamics is twofold: on the one hand, the motion of the atom in the magnetic field causes an induced electric field (motional Stark effect). On the other hand, the kinetic energy of the CM motion gives rise to an additional confining potential for the internal motion that could, in principle, locate the electron at a large distance from the nucleus, and produce atomic states with a huge dipole moment.

Conversely, the motion of the CM is driven by the internal motion, most strongly so in the case of vanishing pseudomomentum. It thus reflects the transition from regular to chaotic internal dynamics: A regular internal motion leads to a regular CM motion, whereas chaotic internal dynamics, for $\boldsymbol{K} = 0$, give rise to a classical diffusion of the CM.

References

15.1 M. C. Gutzwiller: *Chaos in Classical and Quantum Mechanics* (Springer, Berlin, Heidelberg 1990)

15.2 F. Haake: *Quantum Signatures of Chaos*, 2nd edn. (Springer, Berlin, Heidelberg 2000)

15.3 T. F. Gallagher: *Rydberg Atoms* (Cambridge Univ. Press, Cambridge 1994)

15.4 A. Kips, W. Vassen, W. Hogervorst: Phys. Rev. A **59**, 2948 (1999)

15.5 R. V. Jensen, H. Flores-Rueda, J. D. Wright, M. L. Keeler, T. J. Morgan: Phys. Rev. A **62**, 053410 (2000)

15.6 W. R. S. Garton, F. S. Tomkins: Astroph. J. **185**, 839 (1969)

15.7 K. T. Lu, F. S. Tomkins, W. R. S. Garton: Proc. Roy. Soc. London Ser. A **362**, 421 (1978)

15.8 J. Main, G. Wiebusch, K. Welge, J. Shaw, J. B. Delos: Phys. Rev. A **49**, 847 (1994)

15.9 R. J. Elliott, G. Droungas, J.-P. Connerade: J. Phys. B **28**, L537 (1995)

15.10 R. J. Elliott, G. Droungas, J.-P. Connerade, X.-H. He, K. T. Taylor: J. Phys. B **29**, 3341 (1996)

15.11 J. C. Gay: The structure of Rydberg atoms in strong static fields. In: *NATO Advanced Study Institute Series B: Physics*, Vol. 143, ed. by J. P. Briand (Plenum Press, New York 1986) pp. 107–152

15.12 J. C. Gay: Hydrogenic systems in electric and magnetic fields. In: *The spectrum of atomic hydrogen: Advances*, ed. by G. W. Series (World Scientific, Singapore 1988) pp. 367–446

15.13 H. Friedrich, D. Wintgen: Phys. Rep. **183**, 37 (1989)

15.14 H. Hasegawa, M. Robnik, G. Wunner: Prog. Theor. Phys. Suppl. **98**, 198 (1989)

15.15 D. Delande: Chaos in atomic and molecular physics. In: *Chaos and Quantum Physics, Session LII of Les Houches*, ed. by M. J. Giannoni, A. Voros, J. Zinn-Justin (North-Holland, Amsterdam 1991) pp. 665–726

15.16 M. Courtney, H. Jiao, N. Spellmeyer, D. Kleppner, D. Gao, J. B. Delos: Phys. Rev. Lett. **74**, 1538 (1995)

15.17 I. Seipp, K. T. Taylor, W. Schweizer: J. Phys. B **29**, 1 (1996)

15.18 G. Raithel, M. Fauth, H. Walther: Phys. Rev. A **44**, 1898 (1991)

15.19 J.-P. Connerade, M.-S. Zhan, J. Rao, K. T. Taylor: J. Phys. B **32**, 2351 (1999)

15.20 S. Freund, R. Ubert, E. Flöthmann, K. Welge, D. M Wang, J. B Delos: Phys. Rev. A **65**, 053408 (2002)

15.21 T. Uzer: Phys. Scr. **90**, 176 (2001)

15.22 H. Friedrich: *Theoretical Atomic Physics* (Springer, Berlin, Heidelberg 1998)

15.23 J. Rao, K. T. Taylor: J. Phys. B **30**, 3627 (1997)

15.24 H. Friedrich: Scaling properties for atoms in external fields. In: *Atoms and Molecules in Strong External Fields*, ed. by P. Schmelcher, W. Schweizer (Plenum Press, New York 1998) pp. 153–167

15.25 T. S. Monteiro, G. Wunner: Phys. Rev. Lett. **65**, 1100 (1990)

15.26 J. Rao, K. T. Taylor: J. Phys. B **35**, 2627 (2002)

15.27 A. F. Starace: J. Phys. B **6**, 585 (1973)

15.28 M. L. Du, J. B. Delos: Phys. Rev. A **38**, 1896 (1988)

15.29 M. L. Du, J. B. Delos: Phys. Rev. A **38**, 1913 (1988)
15.30 E. B. Bogomolny: Sov. Phys. JETP **69**, 275 (1989)
15.31 J. Main, G. Wunner: Phys. Rev. A **59**, R2548 (1999)
15.32 T. Bartsch, J. Main, G. Wunner: J. Phys. B **36**, 1231 (2003)
15.33 J. Gao, J. B. Delos, M. Baruch: Phys. Rev. A **46**, 1449 (1992)
15.34 J.-M. Mao, K. A. Rapelje, S. J. Blodgett-Ford, J. B. Delos: Phys. Rev. A **48**, 2117 (1996)
15.35 K. Weibert, J. Main, G. Wunner: Ann. Phys. (NY) **268**, 172 (1998)
15.36 B. Hüpper, J. Main, G. Wunner: Phys. Rev. A **53**, 744 (1996)
15.37 P. A. Dando, T. S. Monteiro, D. Delande, K. T. Taylor: Phys. Rev. A **54**, 127 (1996)
15.38 A. Matzkin, T. S. Monteiro: Phys. Rev. Lett. **87**, 143002 (2001)
15.39 M. A. Al-Laithy, P. F. O'Mahony, K. T. Taylor: J. Phys. B **19**, L773 (1986)
15.40 J. Main, A. Holle, G. Wiebusch, K. H. Welge: Z. Phys. D **6**, 295 (1987)
15.41 J. Gao, J. B. Delos: Phys. Rev. A **49**, 869 (1994)
15.42 D. M. Wang, J. B. Delos: Phys. Rev. A **63**, 043409 (2001)
15.43 T. Bartsch, J. Main, G. Wunner: Phys. Rev. A **67**, 063410 (2003)
15.44 H. Goldstein: *Classical Mechanics* (Addison-Wesley, Reading 1965)
15.45 M. V. Berry: Phys. Scr. **40**, 335 (1989)
15.46 E. A. Solov'ev: Sov. Phys. JETP **55**, 1017 (1982)
15.47 D. R. Herrick: Phys. Rev. A **26**, 232 (1982)
15.48 K. T. Hansen: Phys. Rev. E **51**, 1838 (1995)
15.49 P. Cacciani, E. Luc-Koenig, J. Pinard, C. Thomas, S. Liberman: J. Phys. B **21**, 3499 (1988) and references therein
15.50 P. A. Braun: Rev. Mod. Phys. **65**, 115 (1993) and references therein
15.51 J. von Milczewski, T. Uzer: Phys. Rev. E **55**, 6540 (1997)
15.52 J. Main, M. Schwacke, G. Wunner: Phys. Rev. A **57**, 1149 (1998)
15.53 M. J. Englefield: *Group Theory and the Coulomb Problem* (Wiley-Interscience, New York 1972)
15.54 W. Pauli: Z. Phys. **36**, 339 (1926)
15.55 J. Main, G. Wunner: J. Phys. B **27**, 2835 (1994)
15.56 C. Jaffé, D. Farrelly, T. Uzer: Phys. Rev. Lett. **84**, 610 (2000)
15.57 J. E. Avron, I. W. Herbst, B. Simon: Ann. Phys. (NY) **114**, 431 (1978)
15.58 P. Schmelcher, L. S. Cederbaum: Two interacting charged particles in strong static fields. A variety of two-body phenomenon. In: *Structure and Bonding*, Vol. 86, ed. by L. S. Cederbaum, K. C. Kulander, N. H. March (Springer, Berlin 1997) pp. 27–62

16. Hyperfine Structure

Hyperfine structure in atomic and molecular spectra is a result of the interaction between electronic degrees of freedom and nuclear properties other than the dominant one, the nuclear Coulomb field. It includes splittings of energy levels (and thus of spectral lines) from magnetic dipole and electric quadrupole interactions (and higher multipoles, on occasion). Isotope shifts are experimentally entangled with hyperfine structure, and the so-called field effect in the isotope shift can be naturally included as part of hyperfine structure. Studies of hyperfine structure can be used to probe nuclear properties, but they are an equally important probe of the structure of atomic systems, providing especially good tests of atomic wave functions near the nucleus. There are also isotope shifts owing to the mass differences between different nuclear species, and the study of these shifts provides useful atomic information, especially about correlations between electrons.

Hyperfine effects are usually small and often, but not always, it is sufficient to consider only diagonal matrix elements for the atomic or molecular system and for the nuclear system. In some cases, however, matrix elements off-diagonal in the atomic space, even though small, can be of importance; one possible result is to cause admixtures sufficient to make normally *forbidden* transitions possible.

16.1 **Splittings and Intensities** 254
16.1.1 Angular Momentum Coupling 254
16.1.2 Energy Splittings 254
16.1.3 Intensities 255

16.2 **Isotope Shifts** 256
16.2.1 Normal Mass Shift 256
16.2.2 Specific Mass Shift 256
16.2.3 Field Shift 256
16.2.4 Separation of Mass Shift and Field Shift 257

16.3 **Hyperfine Structure** 258
16.3.1 Electric Multipoles 258
16.3.2 Magnetic Multipoles 258
16.3.3 Hyperfine Anomalies 259

References 259

In the diagonal case, one can picture each electron undergoing elastic scattering from the nucleus and returning to its initial bound state. As pointed out by *Casimir* [16.1, 2], however, the *internal conversion* of nuclear gamma-ray transitions involves the inelastic down-scattering from an excited nuclear state to a lower one as an electron goes from an initial bound state to the continuum. By further conversion of bound to continuum states, one sees the connection with electron scattering from the nucleus – elastic, inelastic, and break-up. Hyperfine structure of outer-shell electronic states is at the low momentum-transfer end of this chain of related processes.

Some of the standard textbooks which discuss hyperfine structure are [16.3–10] and a few newer texts [16.11–14]. Especially relevant are [16.15–19] and the conference proceedings [16.20–22].

The study of hyperfine structure in free atoms, ions, and molecules is part of the more extensive research area of hyperfine interactions, which includes the study of atoms and molecules in matter, both at rest, for example as part of the structure of a solid, and moving, such as ions moving through condensed or gaseous matter. This more general subject also includes the ways in which atomic electrons shield the nucleus, or antishield it, from external or collective fields. Thus nuclear magnetic resonance, nuclear quadrupole resonance, electron-nuclear double resonance, recoilless nuclear absorption and emission, nuclear orientation, production of polarized beams, and many other widely used techniques, are intimately connected with hyperfine effects.

Though hyperfine effects are ordinarily small in electronic systems, they can become much larger in "exotic" atoms: those with a heavier lepton or hadron as the "light" particle. Hyperfine effects are typically related to light particle density at the nucleus, or to expectation values of r^{-3}, and thus scale as the cube of the light particle mass. The study of muonic atoms has contributed importantly to knowledge of the nuclear charge distribution [16.23–26]. There has been considerable interest in pionic atoms, where the strong interaction also contributes to hyperfine structure (e.g., [16.27, 28]), and also in kaonic, antiprotonic, and other hadronic "atoms" [16.29, 30]. See especially [16.31] for recent work on antiprotonic helium.

Some other examples of interaction between atomic and nuclear degrees of freedom are discussed in Chapt. 90.

16.1 Splittings and Intensities

16.1.1 Angular Momentum Coupling

When the nuclear system is in an isotropic environment, each nuclear state β has a definite value of nuclear angular momentum $I_\beta \hbar$, where the possible values of I_β are related to the number of nucleons (protons plus neutrons) in the same way as those for J_α are related to the number of electrons in electronic state $|\alpha\rangle$. The nuclear operators, eigenstates, and eigenvalues are related to each other in the same way as for atomic angular momentum by

$$I^2|\beta\rangle = I_\beta(I_\beta+1)|\beta\rangle\,,$$
$$I_z|\beta\rangle = M_\beta|\beta\rangle\,, \tag{16.1}$$

in units with $\hbar = 1$. Shift operators move the system from one M-value to another, as for the atomic system (see Chapt. 2), and the operator $\boldsymbol{I}$ is the generator of rotations. When the combined atomic-nuclear system is considered, in an isotropic environment, it is the total angular momentum of the combined system defined by

$$\boldsymbol{F} = \boldsymbol{J} + \boldsymbol{I}\,, \tag{16.2}$$

that has definite values. The state of the combined system can be labeled by γ, so that

$$F^2|\gamma\rangle = F_\gamma(F_\gamma+1)|\gamma\rangle\,,$$
$$F_z|\gamma\rangle = M_\gamma|\gamma\rangle\,. \tag{16.3}$$

The shift operators are defined as before, and it is now $\boldsymbol{F}$ that is the generator of rotations of the (combined) system, or of the coordinate frame to which the system is referred.

By the rules of combining angular momenta, the possible values of the quantum number F are separated by integer steps and run from an upper limit of $J_\alpha + I_\beta$ to a lower limit of $|J_\alpha - I_\beta|$. The number of possible eigenvalues F is the smaller of $2J_\alpha + 1$ and $2I_\beta + 1$. Experimental values of the nuclear quantum number I may be found in a number of compilations [16.32–34].

16.1.2 Energy Splittings

Electromagnetic interactions between atomic electrons and the nucleus can be expanded in a multipole series

$$H_{\mathrm{eN}} = \sum_k \boldsymbol{T}^k(\mathrm{N}) \cdot \boldsymbol{T}^k(\mathrm{e})\,,$$
$$\equiv \sum_{k,q} (-1)^i T^k_q(\mathrm{N}) T^k_{-q}(\mathrm{e}) \tag{16.4}$$

where $\boldsymbol{T}^k(\mathrm{N})$ is an irreducible tensor operator of rank k operating in the nuclear space, and similarly $\boldsymbol{T}^k(\mathrm{e})$ operates in the space of the electrons. Since one is taking diagonal matrix elements (in the nuclear space, at least) in states that are to a very good approximation eigenstates of the parity operator, only even electric multipoles (E0, E2, etc.) and odd magnetic multipoles (M1, M3, etc.) contribute to the series. The effects of the parity nonconserving weak interaction are considered in Chapt. 29.

The term with $k = 0$ contributes directly to the structure (and fine structure) of atomic systems, and its dominant contributions come from the external r^{-1} electrostatic field of the nucleus. The *hyperfine* Hamiltonian is defined by subtracting that term to obtain

$$H_{\mathrm{hfs}} = \sum_i \left[\boldsymbol{T}^0(\mathrm{N})\boldsymbol{T}^0(i) - \left(-Ze^2/r_i\right)\right]$$
$$+ \sum_{k=1} \boldsymbol{T}^k(\mathrm{N}) \cdot \boldsymbol{T}^k(\mathrm{e})\,, \tag{16.5}$$

where Z is the nuclear charge number. The difference between the Ze^2/r term(s) and the full monopole term is called the *field effect* or *finite nuclear size effect* in the isotope shift, and the remaining terms contribute dipole ($k = 1$), quadrupole ($k = 2$), and higher multipoles in hyperfine structure.

Since the hyperfine Hamiltonian can be expressed as a multipole expansion, its contributions to the pattern of energy levels for the various F values in a given

J_α, I_β multiplet in first-order perturbation theory can be described relatively simply in terms of 3–j and 6–j symbols. The contribution of the term which is the scalar product of electron and nuclear operators of multipole k is

$$\begin{aligned}&\Delta E_k(JIF, JIF)\\&= (-)^{J+I+F}\begin{Bmatrix} J & J & k \\ I & I & F \end{Bmatrix}\\&\times\left[\begin{pmatrix} J & k & J \\ J & 0 & -J \end{pmatrix}\begin{pmatrix} I & k & I \\ I & 0 & -I \end{pmatrix}\right]^{-1} A_k\,, \quad (16.6)\end{aligned}$$

where for $k \geq 1$,

$$A_k = \langle JJ|\boldsymbol{T}^k(\mathrm{e})|JJ\rangle \cdot \langle II|\boldsymbol{T}^k(\mathrm{N})|II\rangle\,. \quad (16.7)$$

The commonly used hfs coefficients A, B, etc., are related to the A_k by

$$A = A_1/IJ, \quad B = 4A_2, \quad C = A_3, \quad D = A_4\,. \quad (16.8)$$

The isotope shift A_0 is the matrix element of the reduced monopole operator.

The pattern of the splitting depends on the total angular momentum F wholly through the 6–j symbol. Since for $k = 0$ the value of the 6–j symbol is independent of F, the monopole term shifts all levels of the hyperfine multiplet equally, independent of the value of F.

The F-dependence of the dipole contribution can be found from the fact that the same 6–j symbol would appear for any scalar product of $k = 1$ operators, for example $\boldsymbol{J} \cdot \boldsymbol{I}$. But in this product space, with J, I, and F all good quantum numbers, the diagonal matrix elements of $\boldsymbol{J} \cdot \boldsymbol{I}$ are just

$$\langle \boldsymbol{J} \cdot \boldsymbol{I} \rangle = \frac{1}{2}[F(F+1) - J(J+1) - I(I+1)]\,, \quad (16.9)$$

so that

$$\begin{aligned}&\Delta E_1(JIF, JIF)\\&= \frac{1}{2}A[F(F+1) - J(J+1) - I(I+1)]\,, \quad (16.10)\end{aligned}$$

where, in terms of reduced matrix elements according to the convention of *Brink* and *Satchler* ([16.35, p. 152]), (the first version given in Sect. 2.8.4)

$$\begin{aligned}A = {}&[J(J+1)]^{-1/2}\langle J\|\boldsymbol{T}^1(\mathrm{e})\|J\rangle[I(I+1)]^{-1/2}\\&\times\langle I\|\boldsymbol{T}^1(\mathrm{N})\|I\rangle\,. \quad (16.11)\end{aligned}$$

A is the magnetic dipole hyperfine structure constant for the atomic level J and nuclear state I. M1 hfs shows the same pattern of splittings as spin-orbit fine structure, described sometimes as the *Landé interval rule*.

Electric quadrupole hfs is described by the quadrupole hyperfine structure constant B. If we define the quantity $K = [F(F+1) - J(J+1) - I(I+1)]$, then

$$\begin{aligned}&\Delta E_2(JIF, JIF)\\&= \frac{\left(\frac{1}{4}B\right)[3K(K+1)/2 - 2J(J+1)I(I+1)]}{J(2J-1)I(2I-1)}\,. \quad (16.12)\end{aligned}$$

The constant B is related to the tensor operators by

$$\begin{aligned}\frac{1}{4}B = {}&[J(2J-1)/(J+1)(2J+3)]^{-1/2}\\&\times\langle J\|\boldsymbol{T}^2(\mathrm{e})\|J\rangle\\&\times[I(2I-1)/(I+1)(2I+3)]^{-1/2}\\&\times\langle I\|\boldsymbol{T}^2(\mathrm{N})\|I\rangle\,, \quad (16.13)\end{aligned}$$

For higher multipoles, see [16.36].

The multipole expansion is important because it is valid for relativistic as well as nonrelativistic situations, and for nuclear penetration effects (hyperfine anomalies discussed in Sect. 16.3.3) as well as for normal hyperfine structure. Its limitation comes from its nature as a first-order diagonal perturbation. Off-diagonal contributions, even when small, can perturb the pattern, but, more importantly, can lead to misleading values for the A_k coefficients, including the isotope shift.

16.1.3 Intensities

When hyperfine structure is observed as a splitting in an optical transition between different atomic levels, there are relations between the intensities of the components. The general rule for reduced matrix elements of a tensor operator operating in the first part of a coupled space is ([16.35, p. 152])

$$\begin{aligned}&\langle JIF\|Q^\lambda(\mathrm{e})\|J'IF'\rangle\\&= (-1)^{\lambda+I+F'+J}(2F'+1)^{1/2}\begin{Bmatrix} F & F' & \lambda \\ J' & J & I \end{Bmatrix}\\&\times(2J+1)^{1/2}\langle J\|Q^\lambda(\mathrm{e})\|J'\rangle\,. \quad (16.14)\end{aligned}$$

For a dipole transition ($\lambda = 1$) connecting atomic states J and J', with fixed nuclear spin I, the line strength $S_{FF'}$ of the hyperfine component connecting F and F' is related to the line strength $S_{JJ'}$ by

$$S_{FF'} = (2F+1)(2F'+1)\begin{Bmatrix} F & F' & 1 \\ J' & J & I \end{Bmatrix}^2 S_{JJ'}\,. \quad (16.15)$$

16.2 Isotope Shifts

Two distinct mechanisms contribute to isotope shifts in atomic energy levels and transition energies. First, there are shifts due to the different mass values of the isotopes; these *mass shifts* can again be separated into two kinds, the normal mass shift and the specific mass shift. Second, there are shifts due to different nuclear charge distributions in different isotopes. Shifts of this sort are called *field shifts*, and can be considered to be the monopole part of the hyperfine interaction.

The usual convention is to describe an isotope shift in a transition as positive when the line frequency is greater for the heavier isotope.

16.2.1 Normal Mass Shift

The *normal mass shift* occurs already for one-electron atoms, where the energy scale in the c.m. frame for the electron-nucleus system is directly proportional to the reduced mass of the system, $\mu = mM/(M+m)$, where m is the mass of the electron and M that of the nucleus. The two natural limits are those of an infinitely heavy nucleus, where $\mu = m$, and positronium, where $\mu = \frac{1}{2}m$. The normal mass shift applies to all levels of all atomic systems. The fractional shift in frequency between isotopes of mass M_H and M_L is given by

$$(\nu_H - \nu_L)/\nu_H = m(M_H - M_L)/M_H(M_L + m)\,, \tag{16.16}$$

or

$$(\nu_H - \nu_L)/\nu_L = m(M_H - M_L)/(M_H + m)M_L\,. \tag{16.17}$$

The normal mass shift between deuterium and protonic hydrogen is $(\nu_{\mathrm{D}} - \nu_{\mathrm{H}})/\nu_{\mathrm{H}} = 2.721\times10^{-4}$, amounting to $4.15\,\mathrm{cm}^{-1}$ (0.179 nm) for Balmer α. It decreases rapidly for heavy elements, with $\Delta\nu/\nu \approx (A_H - A_L)/(1823A_HA_L)$, where the A-values are the atomic mass numbers; for the pair ^{208}Pb–^{206}Pb, $\Delta\nu/\nu$ is then 2.56×10^{-8}, corresponding to a wavelength shift of 0.000 014 nm for a line at 550 nm.

16.2.2 Specific Mass Shift

When the system has more than two particles the situation is more complicated; in particular, the center of mass of any particular electron and the nucleus is no longer the center of mass of the whole system. When c.m. motion is removed from the Hamiltonian, there remains a set of *mass polarization* terms

$$H_{\mathrm{mp}} = \frac{1}{M}\sum_{j<k} \boldsymbol{p}_j \cdot \boldsymbol{p}_k\,, \tag{16.18}$$

where M is the nuclear mass and $\boldsymbol{p}_j$ is the momentum of the jth electron. The matrix elements of these terms can be strongly state-dependent, and the difference in their contributions for isotopes of different mass is called the *specific mass shift* (SMS). For a transition $a \to b$, the lowest-order SMS between isotopes A and A' is

$$\Delta\nu(a, b; A, A') = K^{a,b}(M_A - M_{A'})/M_AM_{A'}\,, \tag{16.19}$$

where

$$K^{a,b} = \langle a|\textstyle\sum_{j<k}\boldsymbol{p}_j\cdot\boldsymbol{p}_k|a\rangle - \langle b|\textstyle\sum_{j<k}\boldsymbol{p}_j\cdot\boldsymbol{p}_k|b\rangle\,. \tag{16.20}$$

It was earlier thought that the SMS, like the normal mass shift, was always very small for heavy atoms, but that has turned out not to be the case.

Since the operator $\boldsymbol{p}_1\cdot\boldsymbol{p}_2$ resembles the product of two dipole transition operators, the matrix elements of H_{mp} vanish between simple product-type wave functions unless allowed by dipole selection rules. For example, in the $1sn\ell\ ^1L$ and ^3L states of helium, the only nonvanishing diagonal terms are the $\ell = 1$ exchange terms $\pm\langle 1s(1)np(2)|\boldsymbol{p}_1\cdot\boldsymbol{p}_2|np(2)1s(1)\rangle$, with $(-)$ for singlet states and $(+)$ for triplet states. The resulting Hughes–Eckart level shift [16.37] is positive for singlets and negative for triplets. For other states, $\langle H_{\mathrm{mp}}\rangle$ acquires a nonzero value due to electron correlation (configuration mixing) effects, but the resulting level shifts are correspondingly smaller. Detailed tabulations for many states of helium, including second-order corrections, are given by *Drake* and *Yan* [16.38].

For other atoms, the diagonal matrix elements of H_{mp} within a single electron configuration are weighted by the same coefficients as those that weight the Slater exchange integral G^1. Relativistic corrections have been given by *Stone* [16.39, 40]. *Bauche* and *Champeau* [16.11] provide a useful discussion of the SMS, as does *King* [16.16].

Among recent results, we mention high-precision experimental work on the difference in the splittings $2^3\mathrm{S}_1$–$2^3\mathrm{P}_J$ between ^{4}He and ^{3}He [16.41–43], which can be compared with theoretical results [16.43, 44], and an extensive multi-isotope study in Sm II, in which isotope-shift results are used to deduce structure information for a number of levels [16.45].

16.2.3 Field Shift

The field shift is due to different electric monopole interactions within the nucleus for different nuclei. Since the field shift is state-dependent, it contributes to the isotope shift of electronic transition frequencies. When field shifts occur between different isomeric levels of the same nuclear species, they are called *isomer shifts*.

Following *Seltzer* [16.46], (see also the summary in *Fricke* et al.[16.47]), a level shift can be written as

$$\Delta E = -e^2 \int [\rho_N(A) - \rho_N(A')] \times \left[r_N^{-1} \int \rho_e \, d\tau_e - \int r_e^{-1} \rho_e \, d\tau_e \right] d\tau_N . \tag{16.21}$$

The electronic factor inside the second bracket can be fitted as an even power series in r_N, starting with r_N^{2j+1}. The result is that the field shift for a transition a is

$$\Delta\nu(a, AA') = F^a \lambda^{AA'} , \tag{16.22}$$

with

$$\lambda^{AA'} = \delta\langle r^2\rangle_{AA'} + (C_2/C_1)\delta\langle r^4\rangle_{AA'} + (C_3/C_1)\delta\langle r^6\rangle_{AA'} + \cdots , \tag{16.23}$$

where $\delta\langle r^k\rangle_{AA'} = \langle r^k\rangle_A - \langle r^k\rangle_{A'}$. The term F^a is an electronic factor proportional to the change in electron density at the nucleus between the initial and final electronic states; in the simplest perturbation approach F^a is equal to $-(4\pi/6)Ze^2\Delta\rho_e(0)$, where $\Delta\rho_e(0)$ is the electron density at the nucleus in the lower atomic state minus that in the upper (see Sect. 90.1). The ratios C_2/C_1 etc., which weight the higher even moments, depend only on Z and not on the particular transition; values are tabulated in [16.46, 48, 49], as are F-factors for K X-ray transitions. This approach has been generalized and reformulated by *Blundell* et al. [16.49].

Measurements of isotope shifts of optical lines have long played an important role in the determination of nuclear rms radii. A more comprehensive picture of nuclear charge distributions can be found by combining optical isotope shift results with those for X-ray lines and for transitions in muonic atoms, together with results of elastic electron scattering from nuclei; a compilation of the results of such an analysis is to be found in *Fricke* et al. [16.47].

16.2.4 Separation of Mass Shift and Field Shift

From the measured isotope shift for a spectral line a between a pair of isotopes A and A', one can define a *residual isotope shift* by subtracting the normal mass shift. The residual shift is the sum of two terms, each with an electronic factor and a nuclear factor

$$\Delta\nu_{a,AA'} = K^a(M_A - M_{A'})/M_A M_{A'} + F^a\lambda^{AA'} . \tag{16.24}$$

If the field shift is assumed to be negligible, the electronic factor K^a can be extracted. Similarly, if the SMS is assumed to be negligible, from the ratios $\Delta\nu(a, A_1A_2)/\Delta\nu(a, A_3A_4)$, ratios of $\lambda(A_1A_2)/\lambda(A_3A_4)$ can be obtained, and the ratios $\Delta\nu(a, A_1A_2)/\Delta\nu(b, A_1A_2)$ give ratios of the electronic factors F^a/F^b.

If both contributions have to be considered, one can gain information from a *King* plot [16.16, 50] in which a *modified residual shift* $\Delta\nu'(a, AA')$ for line a is defined as

$$\Delta\nu'(a, AA') = \Delta\nu(a, AA')M_A M_{A'}/(M_A - M_{A'}) . \tag{16.25}$$

(Some authors modify the whole shift, without subtracting the normal mass shift; since the isotopic mass dependence is the same for the two mass-shift contributions, the separation proceeds as before, but with an altered definition of the constant K.) For each isotope pair AA', this defines a point in the King plot whose ordinate is, e.g., $\Delta\nu'(a, AA')$ and whose abscissa is $\Delta\nu'(b, AA')$. If the assumed additivity is valid, one finds a linear relationship between the modified shifts for line a and those for another line b according to

$$\Delta\nu'(a, AA') = \left(F^a/F^b\right)\Delta\nu'(b, AA') + K^a - K^b\left(F^a/F^b\right) . \tag{16.26}$$

From the slope of the line one has the ratio of F-factors for the two lines, and from the intercept a relation between the K-factors. A sometimes useful variant of the King plot involves choosing a pair of isotopes, say B and B' as a reference pair [16.16, 47]. A *reduced shift* $\Delta\nu''(a, AA')$ is then defined as

$$\Delta\nu''(a, AA') = \frac{\Delta\nu(a, AA')M_A M_{A'}(M_B - M_{B'})}{(M_A - M_{A'})M_B M_{B'}} , \tag{16.27}$$

and $\Delta\nu''(a)$ is plotted vs. $\Delta\nu''(b)$.

It is not in general possible to extract F- and K-values for individual transitions from optical isotope shifts alone, unless one has at least one calculated or otherwise reliably known value. Information about nuclear charge distributions, however, and isotopic changes in them, is available from other sources, including X-ray isotope shifts, muonic X-rays, and elastic electron scattering from nuclei. Calculated electronic factors for K X-ray transitions can be expected to be more reliable than those for optical transitions. The effects of finite nuclear size are much bigger in muonic than electronic atoms, and screening effects much smaller. Combined analyses of all four types of data lead to the best current knowledge of changes in nuclear charge parameters, and thus contribute to knowledge of the electronic factors F^a and K^a for optical transitions [16.47].

16.3 Hyperfine Structure

We discuss first *normal* hyperfine structure, in which terms coming from penetration of electrons into the nuclear charge and current distributions can be neglected; such contributions, the *anomalous* hyperfine structure, are briefly discussed in Sect. 16.3.3. See also Sect. 21.7.2.

16.3.1 Electric Multipoles

The electric quadrupole tensor operators, when penetration is neglected, are

$$T^2(N) = \rho_N(\boldsymbol{r}_N) r_N^2 \boldsymbol{C}^{(2)} , \quad (16.28)$$

where ρ_N is the nuclear charge density, and

$$T^2(e_i) = -e r_i^{-3} \boldsymbol{C}^{(2)}(i) . \quad (16.29)$$

The matrix element of $T^2(\mathrm{N})$ that occurs in the expansion coefficient A_2 is just $e/2$ times the nuclear quadrupole moment Q:

$$eQ = 2\langle II | T^2(\mathrm{N}) | II \rangle . \quad (16.30)$$

Experimental values of Q are tabulated by *Raghavan* [16.32] and more recently discussed by *Pyykkö* [16.51]. The matrix element of $T^2(\mathrm{e})$, summed over all electrons, also has a simple semiclassical interpretation [16.7];

$$\begin{aligned}\langle JJ | T^2(\mathrm{e}) | JJ \rangle &= (e/2) q_J \\ &= (1/2) \langle \partial^2 V / \partial z^2 \rangle_{JJ} ,\end{aligned} \quad (16.31)$$

where V is the electrostatic potential at the nucleus due to the electrons. The quadrupole splitting constant can thus be expressed as

$$B = 4A_2 = e^2 Q q_J = eQ \langle \partial^2 V / \partial z^2 \rangle_{JJ} . \quad (16.32)$$

Because the matrix elements of spherical harmonics in lsj-coupled states are independent of l (aside from the parity requirement that $l + l' + k$ be even), the one-electron matrix elements of $T^2(\mathrm{e})$ are [16.36]

$$\begin{aligned}&\langle lsj \| T^2(\mathrm{e}) \| l'sj' \rangle \\ &= -e \langle lsj \| \boldsymbol{C}^{(2)} \| l'sj' \rangle \int r^{-3} (gg' + ff') \, \mathrm{d}r ,\end{aligned} \quad (16.33)$$

where g and f are the large and small r-multiplied Dirac radial functions, and the right-hand reduced matrix element is equal to $(-1)^{k-1} C^{jkj'}_{\frac{1}{2}0\frac{1}{2}}$ with $k = 2$ ([16.35, p. 153]).

Electric hexadecapole hyperfine structure follows similar rules [16.36]. While there is considerable indirect evidence that many nuclear states have nonzero E4 moments [16.52], their effect on hyperfine structure has been identified only occasionally [16.53, 54].

16.3.2 Magnetic Multipoles

For the magnetic multipole modes, when penetration effects are neglected, and when the the nuclear current is taken to be a sum of the orbital-current and spin-current contributions of individual nucleons, the nuclear tensor operator can be written as the sum

$$\begin{aligned}T^k(\mathrm{N}) = &\ (e\hbar/2m_{\mathrm{p}}c) \sum_n [k(2k-1)]^{1/2} r_n^{k-1} \\ &\times \Big[2g_{ln}(k+1)^{-1} \left(\boldsymbol{C}^{(k-1)} \boldsymbol{L}_n \right)^{(k)} \\ &+ g_{sn} \left(\boldsymbol{C}^{(k-1)} \boldsymbol{S}_n \right)^{(k)} \Big] ,\end{aligned} \quad (16.34)$$

where $e\hbar/2m_{\mathrm{p}}c$ is the nuclear magneton, and in the extreme single-nucleon model the orbital g-factors g_{ln} are 1 for protons and 0 for neutrons, while the spin g-factors g_{sn} are 5.585 694 701(56) for protons and $-$3.826 085 46(90) for neutrons. Similarly, for electron i, the atomic Mk operator can be written

$$T^k(\mathrm{e}_i) = ie[(k+1)/k]^{1/2} r_i^{-k-1} \left(\boldsymbol{\alpha}_i \boldsymbol{C}_i^{(k)} \right)^{(k)} . \quad (16.35)$$

For the magnetic dipole case, the nuclear operator of (16.29) becomes

$$T^1(\mathrm{N}) = (e\hbar/2m_\mathrm{p}c)\sum_n (g_{ln}\boldsymbol{L}_n + g_{sn}\boldsymbol{S}_n)\,. \quad (16.36)$$

The nuclear currents, however, are more complicated in the strongly interacting and relatively dense environment of real nuclear systems; renormalized g-factors can be used to take some such effects into account. For the quenching of spin matrix elements, see *Castel* and *Towner* [16.55]; for the M1 mode, the effective g-factor is typically in the neighborhood of 0.5 or 0.6 of the free-space value. It is also sometimes useful to consider effective orbital g-factors, but the effective value is smaller, with $g_\mathrm{eff}/g_\mathrm{free} = 1.05-1.1$ [16.56]. The matrix element of $T^1(\mathrm{N})$ is the nuclear magnetic dipole moment

$$\mu = \langle II|T^1(\mathrm{N})|II\rangle\,. \quad (16.37)$$

Experimental values of μ are tabulated by *Raghavan* [16.32].

In the semiclassical picture where $\Delta E = -\boldsymbol{\mu}\cdot\boldsymbol{B}(0)$, the contributions to $\boldsymbol{B}(0)$ from the orbital and spin magnetism of the ith electron are

$$\boldsymbol{B}_l(0) = -2\mu_0 r_i^{-3}\boldsymbol{l}_i\,, \quad (16.38)$$

$$\boldsymbol{B}_s(0) = 2\mu_0 r_i^{-3}(10)^{1/2}\left[s_i\boldsymbol{C}_i^{(2)}\right]^{(1)}\,, \quad (16.39)$$

while the "contact" term for s-electrons is

$$\boldsymbol{B}_\mathrm{c}(0) = -(16\pi/3)\mu_0\rho_i(0)\boldsymbol{s}\,. \quad (16.40)$$

The one-electron matrix elements of $T^1(\mathrm{e})$ are [16.36]

$$\langle lsj\|T^1(\mathrm{e})\|l's j'\rangle = -e(\kappa+\kappa')\langle lsj\|\boldsymbol{C}^{(1)}\|l'_x s j'\rangle \times \int r^{-2}\left(fg'+gf'\right)\mathrm{d}r\,, \quad (16.41)$$

where for a given combination lsj, $\kappa = (l-j)(2j+1)$, and l_x is the opposite-parity orbital quantum number of the small Dirac component, $l_x = l+1$ for $j = l+1/2$, $l_x = l-1$ for $j = l-1/2$. Recent high precision calculations for the hyperfine structure of helium and lithium, including relativistic corrections and second-order effects, have recently been done by *Pachucki* [16.57, 58].

Magnetic octupole hyperfine structure [16.59, 60] follows similar rules [16.36, 61]. Systems studied in recent years include Eu I excited states [16.62].

16.3.3 Hyperfine Anomalies

When the electron density is nonzero inside the nucleus, the interaction with different kinds of nuclear current density is in general different; there is, for example, a sensitivity to differing radial distributions of spin and orbital currents. The result is that the ratio of A-values for two isotopes is not necessarily the same as the ratio of nuclear g-factors [16.63]. The anomaly for two nuclear species a and b is characterized by

$${}^a\Delta^b = (A_a/A_b)(g_a/g_b) - 1\,. \quad (16.42)$$

Δ is seldom larger than a few per cent. The theory for the M1 mode was worked out by *Bohr* and *Weisskopf* [16.64]; there are relatively recent reviews of theory and experiment [16.65, 66], and an especially clear exposition of the semiclassical limit [16.67]. Recent experimental results include transitions in isotopes of La II [16.68], and the ground state of hydrogenlike thallium [16.69]. Anomalies are expected to be considerably smaller for the electric mode.

References

16.1 H. B. G. Casimir: *On the Interaction Between Atomic Nuclei and Electrons* (Teyler's Tweede Genootschap, Haarlem 1963)

16.2 H. B. G. Casimir: *On the Interaction Between Atomic Nuclei and Electrons* (Freeman, San Francisco 1936)

16.3 L. Pauling, S. Goudsmit: *The Structure of Line Spectra* (McGraw Hill, New York 1930)

16.4 G. Herzberg: *Atomic Spectra and Atomic Structure*, 1st and 2nd edn. (Prentice-Hall and Dover, New York 1937 and 1944)

16.5 G. Herzberg: *Molecular Spectra and Molecular Structure, I. Spectra of Diatomic Molecules*, 2nd edn. (Van Nostrand, Princeton 1950)

16.6 H. Kopfermann: *Nuclear Moments* (Academic, New York 1958)

16.7 N. F. Ramsey: *Molecular Beams* (Clarendon, Oxford 1956)

16.8 J. C. Slater: *Quantum Theory of Atomic Structure*, Vol. II (McGraw-Hill, New York 1960)

16.9 I. I. Sobelman: *Introduction to the Theory of Atomic Spectra* (Pergamon, London 1972)

16.10 H. G. Kuhn: *Atomic Spectra*, 1st and 2nd edn. (Academic, New York 1962 and 1969)

16.11 A. Corney: *Atomic and Laser Spectroscopy* (Clarendon, Oxford 1977)
16.12 M. Weissbluth: *Atoms and Molecules* (Academic, New York 1978)
16.13 I. I. Sobelman: *Atomic Spectra and Radiative Transitions* (Springer, Berlin, Heidelberg 1979)
16.14 H. Haken, H. C. Wolf: *Atomic and Quantum Physics* (Springer, Berlin, Heidelberg 1984)
16.15 L. Armstrong, Jr.: *Theory of the Hyperfine Structure of Free Atoms* (Wiley, New York 1971)
16.16 W. H. King: *Isotope Shifts in Atomic Spectra* (Plenum, New York 1984)
16.17 I. Lindgren, J. Morrison: *Atomic Many-Body Theory* (Springer, Berlin, Heidelberg 1982)
16.18 J. Bauche, R.-J. Champeau: Adv. At. Mol. Opt. Phys. **12**, 39 (1976)
16.19 K. Heilig, A. Steudel: New developments of classical optical spectroscopy. In: *Progress in Atomic Spectroscopy, Part A*, ed. by W. Hanle, H. Kleinpoppen (Plenum, New York 1978) pp. 263–328
16.20 V. W. Hughes, B. Bederson, V. W. Cohen, F. M. J. Pichanik (Eds.): *Atomic Physics* (Plenum, New York 1969)articles by H. M. Foley, pp. 509–522, and H. H. Stroke, pp. 523–550
16.21 J. C. Zorn, R. R. Lewis, M. K. Weiss (Eds.): *Atomic Physics*, AIP Conf. Proc. 233 (American Institute of Physics, New York 1991)
16.22 H. Walther, T. W. Hänsch, B. Neizert: *Atomic Physics*, AIP Conf. Proc. 275 (American Institute of Physics, New York 1993)
16.23 J. Hüfner, F. Scheck, C. S. Wu: Muon Phys. **1**, 201 (1977)
16.24 R. C. Barrett: Muon Phys. **1**, 309–322 (1977)
16.25 R. C. Barrett, D. Jackson: *Nuclear Sizes and Structure* (Oxford Univ. Press, Oxford 1977)
16.26 K. Pachucki: Phys. Rev. A **63**, 032508 (2001)
16.27 G. Backenstoss: Ann. Rev. Nucl. Sci. **20**, 467 (1970)
16.28 J. Konijn: An improved parametrization of the optical potential for pionic atoms. In: *Pions in Nuclei*, ed. by E. Oset, M. J. Vicente Vacas, C. Garcia Recio (World Scientific, Singapore 1992) p. 303
16.29 C. J. Batty: Nucl. Phys. A **508**, 89C (1990)
16.30 G. Backenstoss: Comtemp. Phys. **30**, 433 (1989)
16.31 V. I. Korobov, D. Bakalov: J. Phys. B **34**, L519 (2001)
16.32 P. Raghavan: At. Data Nucl. Data Tables **42**, 189 (1989)
16.33 M. J. Martin, J. K. Tuli (Eds.): *Nuclear Data Sheets* (Academic, New York 1995) Nos. 74–76; each issue contains a guide to the most recent compilation for each atomic mass number
16.34 C. M. Lederer, V. S. Shirley: *Table of Isotopes*, 7th edn. (Wiley, New York 1978)
16.35 D. M. Brink, G. R. Satchler: *Angular Momentum*, 2nd edn. (Clarendon Press, Oxford 1968)
16.36 C. Schwartz: Phys. Rev. **97**, 380 (1955)
16.37 D. S. Hughes, C. Eckart: Phys. Rev. **36**, 694 (1930)
16.38 G. W. F. Drake, Z.-C. Yan: Phys. Rev. A **46**, 2378 (1992)
16.39 A. P. Stone: Proc. Phys. Soc. Lond. **77**, 786 (1961)
16.40 A. P. Stone: Proc. Phys. Soc. Lond. **81**, 868 (1963)
16.41 J. R. Lawall. Ping Zhao, F. M. Pipkin: Phys. Rev. Lett. **66**, 592 (1991)
16.42 D. Shiner, R. Dixson, V. Vedantham: Phys. Rev. Lett. **74**, 3553 (1995)
16.43 F. Marin, F. Minardi, F. S. Pavone, G. W. F. Drake: Z. Phys. D **32**, 285 (1995)
16.44 G. W. F. Drake: High-precision calculations for the Rydberg states of helium. In: *Long Range Casimir Forces: Theory and Recent Experiments on Atomic Systems*, ed. by F. S. Levin, D. A. Micha (Plenum, New York 1993) pp. 196–199 For isotope shifts in Li^+, see also [16.70]
16.45 P. Villemoes et al.: Phys. Rev. A **51**, 2838 (1995)
16.46 E. C. Seltzer: Phys. Rev. **188**, 1916 (1969)
16.47 G. Fricke et al.: At. Data Nucl. Data Tables **60**, 177 (1995)
16.48 F. Boehm, P. L. Lee: At. Data Nucl. Data Tables **37**, 455 (1987)
16.49 S. A. Blundell et al.: J. Phys. B **20**, 3663 (1987)
16.50 W. H. King: J. Opt. Soc. Am. **53**, 638 (1963)
16.51 P. Pyykkö: Z. Naturforsch. **47a**, 189 (1992)
16.52 R. F. Casten: *Nuclear Structure from a Simple Perspective* (Oxford Univ. Press, Oxford 1990) pp. 296–300
16.53 W. Dankwort, J. Ferch, H. Gebauer: Z. Phys. **267**, 229 (1974)
16.54 O. Becker et al.: Phys. Rev. A **48**, 3546 (1993)
16.55 B. Castel, I. S. Towner: *Modern Theories of Nuclear Moments* (Clarendon Press, Oxford 1990)
16.56 T. Yamazaki et al.: Phys. Rev. Lett. **25**, 547 (1970)
16.57 K. Pachucki: J. Phys. B **34**, 3357 (2001)
16.58 K. Pachucki: Phys. Rev. A **66**, 062501 (2002)
16.59 V. Jaccarino, J. G. King, R. A. Satten, H. H. Stroke: Phys. Rev. **94**, 1798 (1954)
16.60 P. Kusch, T. G. Eck: Phys. Rev. **94**, 1799 (1954)
16.61 M. Mizushima: *Quantum Mechanics of Atomic Spectra and Atomic Structure* (Benjamin, New York 1970) Sect. 9–10
16.62 W. J. Childs: Phys. Rev. A **44**, 1523 (1991)
16.63 F. Bitter: Phys. Rev. **76**, 150 (1949)
16.64 A. Bohr, V. F. Weisskopf: Phys. Rev. **77**, 94 (1950)
16.65 S. Büttgenbach: Hyperfine Interactions **20**, 1 (1984)
16.66 G. Savard, G. Werth: Ann. Rev. Nucl. Part. Sci. **50**, 119 (2000)
16.67 R. A. Sorensen: Am. J. Phys. **35**, 1078 (1967)
16.68 H. Iimura et al.: Phys. Rev. C **68**, 054328 (2003)
16.69 P. Beiersdorfer et al.: Nucl. Instrum. Methods Physics Research B **205**, 62 (2003)
16.70 E. Riis, A. G. Sinclair, O. Paulsen, G. W. F. Drake, W. R. C. Rowley, A. P. Levick: Phys. Rev. A **49**, 207 (1994)

17. Precision Oscillator Strength and Lifetime Measurements

The accuracy of oscillator strength and lifetime measurements has improved greatly in the past twenty years. Nevertheless, these high accuracies have been achieved for only a restricted number of lines belonging to a few elements and ionization stages [17.1]. Large numbers of precision measurements must still be made as improved experimental oscillator strengths are needed, both as tests of theoretical concepts, and for diagnostics and engineering applications.

17.1 **Oscillator Strengths** 262
17.1.1 Absorption and Dispersion Measurements 262
17.1.2 Emission Measurements 263
17.1.3 Combined Absorption, Emission and Lifetime Measurements 263
17.1.4 Branching Ratios in Highly Ionized Atoms 264

17.2 **Lifetimes** 264
17.2.1 The Hanle Effect 265
17.2.2 Time-Resolved Decay Measurements 265
17.2.3 Other Methods 267
17.2.4 Multiplexed Detection 267

References 268

A spectral line arising from a radiative transition between atomic states i and k is characterized by its wavelength λ_{ik}, its intensity and its shape. In the limit of free atoms, the intensity per atom is determined by the emission transition rate A_{ik} or absorption oscillator strength f_{ki}, and the shape by the natural width $\Gamma_i = \hbar/\tau_i$, where τ_i is the lifetime of the excited state. While classical spectroscopic methods provide precise wavelength measurements (1 part in 10^8 or better; Chapt. 10), it has only recently been possible to measure oscillator strengths and lifetimes to better than a few percent, as discussed in this chapter. Examples of applications which require an accurate knowledge of these quantities are the interpretation of astrophysical data (Chapt. 82), atmospheric physics (Chapt. 84), combustion (Chapt. 88), the modeling and diagnosis of thermonuclear plasmas (Chapt. 86), nonlinear optics (Chapt. 72), isotope separation (Chapt. 16), and the development of new types of lasers (Chapt. 71).

Precision measurements of oscillator strengths and lifetimes also provide stringent tests of atomic structure calculations. These quantities are very sensitive to the wave functions and the approximations used, particularly in cases where electron correlations (Sect. 23.2.1) and relativistic effects (Sect. 22.1) are significant. They provide experimental tests of fundamental theory; for example, of quantum electrodynamic corrections (Sect. 27.2), and of the nonconservation of parity predicted by the unified electro-weak theory (Sect. 29.1).

In applications where only modest precision is required, semi-empirical parameterizations involving quantum defects, charge screening and polarization allow a few precise measurements to be extrapolated along isoelectronic, homologous, isoionic and Rydberg sequences. Similar methods have been applied to the atomic energy levels themselves (Sect. 10.13). They allow one to produce a very large data base of moderate precision [17.2].

The spontaneous transition rate A_{ik} (Sect. 10.16.2) is the probability per unit time (s^{-1}) for an atom in any one of the g_i states of the energy level i to make a transition to any of the g_k states of the level k. The lifetime τ_i is then given by $1/\tau_i = \sum_k A_{ik}$. The branching fraction for the kth channel of the decay of level i is defined as $F_B = \tau_i A_{ik}$, and the branching ratio (Sect. 10.17) between two decay channels is $R_B = A_{ik}/A_{ij}$. Emission and absorption rate constants differ by a factor of λ_{ik}^2, and the absorption oscillator strength f_{ki} (Sect. 10.17) is defined by

$$g_k f_{ki} = C\lambda_{ik}^2 g_i A_{ik} , \tag{17.1}$$

where $C = (32\pi^3 \alpha a_0^2 \text{Ry})^{-1} = 1.499\,19 \times 10^{-14}\,\text{nm}^{-2}\,\text{s}$. Because of this relationship, the words transition rate and oscillator strength will be used almost interchangeably.

17.1 Oscillator Strengths

Oscillator strengths can be determined directly through absolute emission, absorption, or dispersion measurements, or through the combined measurement of branching ratios and lifetimes. Direct measurements compare different transitions at the same time, and require sample equilibrium, a knowledge of the absolute number density, and an absolute intensity measurement. These are in contrast to time-resolved lifetime measurements, which compare relative intensities from the same transition at different times, and require no absolute measurements. However, lifetime measurements yield oscillator strengths directly only in cases where a single decay transition channel exists, such as the lowest excited level in an atom or ion. Thus, combined measurements of lifetimes and branching ratios are often used where high precision values for oscillator strengths are required. Both absorption and dispersion measurements involve the number density of the lower level of the transition, whereas emission measurements involve that of the upper level.

17.1.1 Absorption and Dispersion Measurements

Absorption measurements involve placing a sample of atoms (for example, in a gas cell, an atomic beam, an arc, a shock tube, or within the vapor column in a furnace) between a continuous light source and a spectrometer. For an isolated spectral line, the absorption cross section for a beam of photons of frequency ν passing through the sample is

$$\sigma_{ik}(\nu) = \pi\alpha(\hbar/m)g(\nu)f_{ik} , \tag{17.2}$$

where $g(\nu)$ is the spectral distribution function per unit frequency normalized so that $\int_0^\infty g(\nu)\,\mathrm{d}\nu = 1$. If there are N atoms per unit volume, the absorption coefficient is $k_\nu = N\sigma(\nu)$. The integrated intensity lost after passing through a distance L of the sample is

$$\begin{aligned}\int_0^\infty \Delta I(\nu)\,\mathrm{d}\nu &= I_0 \int_0^\infty \left(1-\exp^{-\sigma_{ik}(\nu)NL}\right)\mathrm{d}\nu \\ &\simeq I_0\pi\alpha(\hbar/m)NLf_{ik} ,\end{aligned} \tag{17.3}$$

where the second line applies if the sample is optically thin. Otherwise, the integral can be calculated directly, if $g(\nu)$ is known, to determine f_{ik} by the curve-of-growth method.

The Furnace Method

High precision absorption measurements have been achieved by *Blackwell* and co-workers [17.3], who have used the furnace method to study the astrophysically important neutral iron spectrum. These measurements have quoted accuracies of 0.5% on a relative scale, and 2.5% on an absolute scale. This accuracy was obtained through the use of a stable and isothermal furnace, low-noise spectral intensity recording techniques, and two identical high resolution spectrometers for the simultaneous recording of pairs of absorption lines. By selecting successive line pairs of a suitable oscillator strength ratio and adjusting the temperature and vapor pressure in the furnace, a large dynamic range of oscillator strengths could be covered. Recently, corroborative studies of the uncertainties quoted in these measurements have been undertaken, including tests that are coupled to other methods that use combinations of lifetime and branching ratio measurements [17.3].

The Hook Method

The absorption measurements described above determine the oscillator strength from the line intensity. An alternative absorptive approach is the anomalous dispersion or "hook" method, which determines the oscillator strength from the index of refraction at wavelengths near the edge of an absorption line [17.1]. The advantages of this method are its large dynamic range, its insensitivity to the line shape, and the fact that it does not saturate. The absorbing gas is placed in one arm of an interferometer and a compensator is placed in the other arm. This leads to the formation of oblique interference fringes with two characteristic hooks symmetric about the center of an absorption line. The oscillator strength is determined by the wavelength separation W between the hooks. Defining $K = \lambda_{ik}N_\mathrm{f}$, where N_f is the number of fringes per unit wavelength, then

$$f_{ik} = \frac{\pi K W^2}{\lambda_{ik}^3\alpha^2 a_0 NL} \tag{17.4}$$

where a_0 is the Bohr radius.

Synchrotron Radiation

Storage ring synchrotron radiation facilities now provide a source of continuum radiation that can extend the wavelength range of absorption measurements. A technique [17.4] has been developed and applied that utilizes a hollow cathode discharge as an absorbing sample, synchrotron radiation as a continuum source, and a CCD array for vacuum ultraviolet (VUV) detection. Here relative oscillator strengths were obtained using two separate detection systems. However, VUV calibration standards are presently lacking (Sect. 17.1.4), and will be required to obtain independent tests (Sect. 17.1.3) of sets of branching fractions.

17.1.2 Emission Measurements

Emission methods use, for example, a hollow cathode, wall-stabilized arc, or shock tube to excite the source. Recent developments in the methods of Fourier transform spectrometry (FTS) [17.5, 6] offer several advantages over grating spectroscopy, as discussed next.

Fourier Transform Spectrometry

As opposed to the dispersive nature of a grating spectrograph, FTS uses interference effects from a Michelson interferometer. All radiation admitted to the spectrometer is thus incident on the detectors at all times, but different wavelengths are distinguished by their spatial modulation frequencies. The interferogram from all sinusoidal signals is sampled at a prescribed interval of path length, and compared with a laser of known frequency following the same optical path. The spectrum is recovered from the interferogram by means of a fast Fourier transform (Sect. 8.1.3). Thus, for an FTS instrument, the spectral range is determined by the sampling step size, whereas the resolution depends on the maximum path difference (this is in contrast to a grating spectrometer where the sampling step determines the resolution, and the scan length determines the spectral range).

The FTS method provides a number of attractive features in precision oscillator strength measurements. The axial symmetry and the replacement of the slit by a larger aperture can provide a throughput that is two orders of magnitude greater than that of a grating instrument of the same resolution. The precision and reproducibility are determined by the laser standard and the linearity of the wavenumber scale. The resolution can be increased as necessary to resolve a specific source line. The superior resolution allows blending and self-absorption to be more readily detected. The spectral range is limited only by detectors and filters. Recently, uv FTS instruments have been constructed that operate from the visible down to 175 nm. Since all wavelengths are observed at all times, errors from drifts in source conditions during scanning are reduced.

Hollow Cathode Lamps

Hollow cathode discharge lamps are used extensively for emission branching fraction measurements [17.5]. These lamps can generate an emission spectrum of essentially any element, and the relatively low collision rates result in line profiles that are narrow and primarily Doppler broadened. The narrow line width is a major advantage when studying line-rich spectra. The low collision rates also imply that the discharges are far from local thermodynamic equilibrium (LTE). However, this does not affect the determination of branching fractions, where only the relative strengths of lines from a common upper level are measured. These measurements can then be put on an absolute basis if lifetime measurements are available for all of the upper levels.

17.1.3 Combined Absorption, Emission and Lifetime Measurements

By combining measurements obtained in emission with those obtained in absorption (or dispersion) to obtain branching ratios, and then incorporating lifetime measurements, it is possible to use a scheme that requires no knowledge of level populations [17.1]. The scheme was originally proposed by Ladenburg in 1933, and its various modern implementations are known as "leap-frogging," "linkage" and "bow ties".

The principle of "leap-frogging" or "linkage" is illustrated in Fig. 17.1a. The decay of level 1 is unbranched, so the $1 \to 2$ transition rate is assumed known (kn) from a lifetime measurement. This is used to specify the $2 \to 3$ oscillator strength using relative absorption (ab) measurements for the $2 \to 1 : 2 \to 3$ oscillator strength ratio (and appropriate factors of the wavelengths and degeneracies). In a similar manner, this is subsequently combined with relative emission (em) measurements of the $3 \to 2 : 3 \to 4$ branching ratio to determine the $3 \to 4$ transition rate.

The principle of "bow ties" is illustrated in Fig. 17.1b. The two branching ratios for $1 \to 2 : 1 \to 4$ and for $3 \to 2 : 3 \to 4$ are measured in emission (em). The two oscillator strength ratios for $2 \to 1 : 2 \to 3$ and for $4 \to 1 : 4 \to 3$ are measured in absorption (ab). After correction for wavelength and degeneracy factors between f and A values, these relationships are combined into a quantity known as the "bow tie ratio," which would be unity for ideally accurate measurements. A significant deviation from unity can be used to trace the observations that are in error. Figure 17.1b is a "simple bow tie" connecting two upper and two lower levels with four transitions. Higher order sets of measurements can

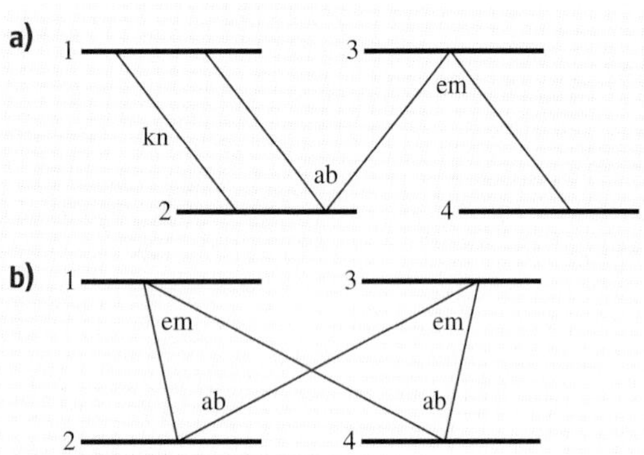

Fig. 17.1a,b Illustration of methods for the determination of ratios of oscillator strengths (ab = absorption, em = emission, kn = known)

be similarly coupled; for example, a set of transitions between three lower levels and three upper levels can be coupled by nine simple bow ties.

In cases where both lifetime and complete branching ratio measurements exist for the same upper level, then the branching ratios can be normalized to branching fractions divided by the lifetime to obtain absolute transition probabilities.

17.1.4 Branching Ratios in Highly Ionized Atoms

In highly ionized atoms, many measurements of lifetimes in the 1–5% accuracy range have been made by ANDC (Sect. 17.2.2) analysis of beam–foil measurements. However, little work exists for the precision measurement of branching ratios in highly ionized atoms, where an intensity calibration of the detection equipment is particularly difficult. In beam-foil excitation, for example, Doppler broadening and Doppler shifts, polarization due to anisotropic excitation, and the short wavelength (≤ 200 nm) nature of the radiation are not well-suited for use with standard techniques for calibrating grating spectrometer and detection systems by the use of calibrated lamps. Calibrations have been carried out using synchrotron radiation, or through the use of previously known branching ratios.

One way in which lifetime measurements are used to determine branching ratios involves precision studies of the lifetimes of the individual fine structure components of a multiplet decay. If one fine structure level has a decay channel that is not available to the other levels (for example, a spin-changing transition or an autoionization mode made possible by a J-dependent intermediate coupling) then the transition rate of the extra channel can be determined by differential lifetime measurements.

The use of Si(Li) detectors in the measurement of very short wavelength radiation in highly ionized atoms also offers possibilities for branching ratio measurements. Since these devices can specify the photon energy from pulse height information without the need for a spectrometer, they can be calibrated for detection efficiency as a function of energy. This type of detection was recently used to determine the branching ratio of the magnetic dipole channel to the two-photon decay channel in the 2s $2S_{\frac{1}{2}}$ state in one-electron krypton [17.7].

Although radiometric calibration standards are available for $\lambda > 280$ nm, technical challenges exist for their extension to shorter wavelengths. The urgent need for these standards and their desired characteristics have been discussed by *Lawler* et al. [17.4], and an operational prescription for combining sources of uncertainty in their specification has been presented by *Sikström* et al. [17.8]. A semi-empirical method for obtaining line intensity standards in the VUV has also been proposed [17.9, 10] that uses intermediate coupling (IC) amplitudes deduced from measured energy level data to obtain intensity ratios. This utilizes the observation that the ns^2np^2 and $ns^2np(n+1)$s configurations in the Si, Ge and Sn isoelectronic sequences exhibit negligible configuration interaction, hence the intensity ratios within their transition arrays are accurately prescribed by the IC amplitudes. For the neutral atoms the transitions occur in the visible region, and measurements confirm the validity of the empirical values. Thus, isoelectronic extensions can yield VUV standards.

17.2 Lifetimes

The total transition rate summed over all decay channels can be measured either through frequency-resolved studies of the level width, or through time-resolved studies of the level lifetime. In order to determine the natural linewidth in a field-free spectroscopic measurement, either the lifetime must be very short, or the Doppler, pressure, and instrumental broadenings must be made very small. Line widths have been determined using

Fabry–Perot spectrometry at very low temperatures and pressures, and in beam–foil studies of radiative transitions in which the lower level decays very rapidly via autoionization. The linewidth can also be determined through the use of the phase shift method. Here modulated excitation is applied to the source, thus producing similarly modulated emitted radiation, and the width can be specified from the phase shift between the two signals. Other methods involve resonance fluorescence techniques, where sub-Doppler widths are obtained because the width of the exciting radiation selects a subset of particle motions within the sample. Resonance fluorescence techniques that can be used to determine level widths include zero-field level crossing (the Hanle effect), high-field level crossing, and double optical resonance methods, but the Hanle effect is the most common.

17.2.1 The Hanle Effect

In its most commonly used form, the Hanle effect makes use of polarized resonance radiation to excite atoms in the presence of a known variable magnetic field. The magnetic substates of the sample are anisotropically excited, and the subsequent radiation possesses a preferred angular distribution. By applying the magnetic field in a direction perpendicular to the anisotropy, the angular distribution is made to precess, producing oscillations in the radiation observed at a fixed angle. At infinite precessional frequency the intensity would be proportional to the instantaneous average angular intensity, but at finite precessional frequency it depends upon the decay that has occurred during each quarter rotation. Measured as a function of magnetic field, the emitted intensity has a Lorentzian shape centered about zero field with a width that depends on the lifetime and g-factor of the level [17.2].

17.2.2 Time-Resolved Decay Measurements

The most direct method for the experimental determination of level lifetimes is through the time-resolved measurement of the free decay of the fluorescence radiation following a cutoff of the source of excitation. An important factor limiting the accuracy is the repopulation of the level of interest by cascade transitions from higher-lying levels. For this reason, decay curve measurements fall into two classes: those that involve selective excitation of the level of interest, thus eliminating cascading altogether; and those that use correlations between cascade connected decays to account for the effects of cascades.

Selective Excitation

Lifetime measurements accurate to within a few parts in 10^3 have been obtained through selective excitation produced when appropriately tuned laser light is incident on a gas cell or a thermal beam, or on a fast ion beam (Sect. 18.1). With a gas cell or thermal beam, the timing is obtained by a pulsed laser and delayed coincidence detection. With a fast ion beam, time-of-flight methods are used. In either case, after removal of the background, the decay curve of intensity vs. time is a single exponential, and the lifetime is obtained from its semilogarithmic slope. Laser-excited time-of-flight studies were first carried out by observing the optical decay of the ion in flight following excitation using a laser beam which crossed the ion beam. In these studies, the laser light was tuned to the frequency of the desired absorption transition either through the use of a tunable dye laser, or by varying the angle of intersection to exploit the Doppler effect. Recent measurements have utilized diode laser excitation in this geometry [17.11]. A number of adaptations of this technique have been developed in which the laser and ion beams are made to be collinear, and are switched into and out of resonance within a segment of the beam by use of the Doppler effect [17.12]. The collinear geometry can provide a longer excitation region and less scattering of laser light into the detector than occurs in the crossed beam geometry. In one adaptation [17.13], excitation occurs within an electrostatic velocity switch, and the time resolution is obtained by physically moving the velocity switch. In another adaptation [17.12], the ion beam is accelerated with a spatially varying voltage ramp. The resonance region is moved relative to a fixed detector by time-sweeping either the laser tuning or the ramp voltage.

While these selective excitation methods totally eliminate the effects of cascade repopulation, they are generally limited to levels in neutral and singly ionized atoms that can be accessed from the ground state by strongly absorptive E1 transitions, and the selectivity itself is a limitation. Many very precise measurements have been made by these techniques, but they have primarily involved $\Delta n = 0$ resonance transitions in neutral alkali atoms and singly ionized alkali-like ions. A summary of these measurements is given in Table 17.1, and a comparison of these values with theoretical calculations is given in [17.14].

Table 17.1 Measured $np\ ^2P_J$ lifetimes

Atom	n	J	τ (ns)
Li	2	1/2	27.20(20)[a], 27.29(4)[b]
Na	3	1/2	16.38(8)[c], 16.40(3)[b], 16.30(2)[d]
	3	3/2	16.36(2)[c], 16.25(2)[d]
Mg^+	3	1/2	3.854(30)[d]
	3	3/2	3.810(40)[e]
Ca^+	4	1/2	7.07(7)[f], 7.098(20)[g]
	4	3/2	6.87(6)[f], 6.924(19)[g]
Cu	4	1/2	7.27(6)[h]
	4	3/2	7.17(6)[h]
Sr^+	5	1/2	7.47(7)[i]
	5	3/2	6.69(7)[i]
Ag	5	1/2	7.408(32)[j]
	5	3/2	6.791(19)[j]
Cd^+	5	1/2	3.11(3)[k]
	5	3/2	2.77(6)[k]
Cs	6	3/2	30.55(27)[l]
Ba^+	6	1/2	7.92(8)[i]
	6	3/2	6.312(16)[m]

[a][17.17] [b][17.18] [c][17.19] [d][17.20] [e][17.21] [f][17.22] [g][17.23] [h][17.24] [i][17.25] [j][17.26] [k][17.27] [l][17.11] [m][17.28]

Advances have recently been made in the measurement of transition probabilities of long-lived metastable levels. Through a laser probing technique [17.15] using an ion storage ring, an extremely long lifetime (28 s) has been measured [17.16]. The metastable level is populated in the ion source and preserved in the storage ring until it is destructively probed at variable times after excitation. The laser light is collinearly merged with the beam and tuned to a transition that promotes the population of the metastable level to a higher unstable level. The fluorescence subsequently emitted by this unstable level gives a relative measure of the population of the metastable level. Thus, by varying the delay time between the excitation and the laser probing, a decay curve of the population remaining is obtained.

Nonselective Excitation

Much more general access can be obtained by nonselective excitation methods, such as pulsed electron beam bombardment of a gas cell or gas jet, or in-flight excitation of a fast ion beam by a thin foil. Pulsed electron beam excitation can be achieved either through use of a suppressor grid, or by repetitive high frequency deflection of the beam across a slit so as to chop the beam. Particularly in the case of weak lines, the high frequency deflection technique offers the advantages of high current and sharp cut-off times. The high currents yield high light levels, so that high resolution spectroscopic methods can be used to eliminate the effects of line blending. Pulsed electron excitation methods are well suited to measurements in neutral and near neutral ions (although for very long lifetimes in ionized species, the decay curves can be distorted if particles escape from the viewing volume through the Coulomb explosion effect [17.29]). However, for highly ionized atoms, the only generally applicable method is thin foil excitation of a fast ion beam. Nonselective excitation techniques can also be applied to measurements such as the phase shift method [17.30] and the Hanle effect [17.31], in which case cascade repopulation also can become a serious problem. However, most of the attempts to eliminate or account for cascade effects have occurred in decay curve studies.

In beam–foil studies, the excitation is created in the dense solid environment of the foil, after which the ions emerge into a field free, collision free, high vacuum region downstream from the foil. A time-resolved decay curve is obtained by translating the foil upstream or downstream relative to the detection apparatus. The beam is a very tenuous plasma, which has both advantages and disadvantages. The low density avoids the effects of collisional de-excitation and radiation trapping, but also produces relatively low light levels. This requires fast optical systems with a corresponding reduction in wavelength dispersion, and care must be taken to avoid blending of these Doppler broadened lines. Methods have been developed by which grating monochromators can be refocussed to a moving light source, thus utilizing the angular dependence of the Doppler effect to narrow and enhance the lines.

In these nonselective excitation methods, the decay curve involves a sum of many exponentials, one corresponding to the primary level, and one to each level that cascades (either directly or indirectly) into it. Decay exponentials do not comprise an orthogonal set of functions, and the representation of an infinite sum by a finite sum through curve fitting methods (Sect. 8.1.2) can lead to large errors. Fortunately, alternative methods to exponential curve fitting exist, which permit the accurate extraction of lifetimes to be made from correlated sets of nonselectively populated decay curves.

ANDC Method

Precision lifetime values have been extracted from cascade-affected decay curves by a technique known as the arbitrarily normalized decay curve (ANDC) method (Sect. 18.1.1, [17.32]), which exploits dynamical correlations among the cascade-related decay curves. These correlations arise from the rate equation that connects the population of a given level to those of the levels that cascade directly into it. The instantaneous population of each level n is, to within constant factors ξ_n involving the transition probabilities and detection efficiencies, proportional to the intensity of radiation $I_n(t)$ emitted in any convenient decay branch. In terms of these intensities, the population equation for the level n can be written in terms of its direct cascades from levels i as

$$\frac{\mathrm{d}I_n}{\mathrm{d}t}(t_\mathrm{p}) = \sum_i \xi_i I_i(t_\mathrm{p}) - I_n(t_\mathrm{p})/\tau_n \,. \tag{17.5}$$

Thus, if all decay curves are measured at the same discrete intervals of time t_p, the population equation provides a separate independent linear relationship among these measured decay curves for each value of t_p, with common constant coefficients given by the lifetime τ_n and the normalization parameters ξ_i. Although the sum over cascades is formally unbounded, the dominant effects of cascading from highly excited states are often accounted for by indirect cascading through the lower states, in which case the sum can be truncated after only a few terms. ANDC analysis consists of using this equation to relate the measured $I_k(t_\mathrm{p})$ (using numerical differentiation or integration) to determine τ_k and the ξ_i through a linear regression. If all significant direct cascades have been included, the goodness-of-fit will be uniform for all time subregions, indicating reliability. If important cascades have been omitted or blends are present, the fit will vary over time subregions, indicating a failure of the analysis. Very rugged algorithms have been developed [17.33] that permit accurate lifetimes to be extracted even in cases where statistical fluctuations are substantial, and studies of the propagation and correlation of errors have been made. Clearly the ANDC method is most easily applied to systems for which repopulation effects are dominated by a small number of cascade channels. Further applications are discussed in Sect. 18.1.1.

17.2.3 Other Methods

Coincidence measurements provide another method for elimination of cascade effects. While the low count rates and correspondingly high accidental rates make the application of these methods difficult for optical spectra, the use of Si(Li) detectors for the very short wavelength emission in very highly ionized systems offers new possibilities [17.34] for these measurements.

Another method of accounting for cascading involves the combined use of a laser and beam–foil excitation [17.35, 36]. The beam–foil excitation provides a source of ions in excited states, and a chopped laser is used to stimulate transitions between two excited states. By subtracting the decay curves obtained with laser on and with laser off, the cascade-free, laser-produced portion of the decay curve is obtained.

17.2.4 Multiplexed Detection

Recently, the detection efficiency and reliability of beam–foil measurements have been improved through the use of position sensitive detectors (PSD), which permits measurement of decay curves as a function both of wavelength and of time since excitation. The PSD is mounted at the exit focus of the analyzing monochromator, where it records all lines within a given wavelength interval simultaneously, including reference lines with the same Doppler shifts. Decay curves can be constructed by integrating over a line profile, and that profile can be examined for exponential content to eliminate blending. The time dependent backgrounds underlying the decay curves are directly available from the neighboring channels.

The use of multiplexed detection greatly enhances the data collection efficiency, and causes many possible systematic errors to cancel in differential measurements. Effects such as fluctuations in the beam current, degradation of the foil, divergence of the beam, etc., affect all decay curves in the same way.

The most accurate measurement [17.37] made by beam–foil excitation ($\pm 0.26\%$) used a type of multiplexed detection in which two spectrometers simultaneously viewed the decays of the $^1\mathrm{P}_1$ and $^3\mathrm{P}_1$ levels of the 1s3p configuration in neutral helium. The $^1\mathrm{P}_1$ emission exhibited the desired multi-exponential decay curve (with only weak cascading), whereas the $^3\mathrm{P}_1$ emission exhibited a zero field quantum beat pattern superimposed on its decay because of the anisotropic excitation of that level. The quantum beats provided an in-beam time base calibration which permitted this high precision.

References

17.1 M. C. E. Huber, R. J. Sandeman: Rep. Prog. Phys. **49**, 397 (1986)

17.2 L. J. Curtis: *Atomic Structure and Lifetimes: A Conceptual Approach* (Cambridge Univ. Press, Cambridge 2003)

17.3 D. E. Blackwell: *Atomic Spectra and Oscillator Strengths for Astrophysics and Fusion Research*, ed. by J. E. Hansen (North-Holland, Amsterdam 1990) p. 160

17.4 J. E. Lawler, S. D. Bergeson, J. A. Feldchak, K. L. Mullman: Phys. Scr. **T83**, 11 (1999)

17.5 J. E. Lawler, S. D. Bergeson, R. C. Wamsley: Phys. Scr. **T47**, 29 (1993)

17.6 A. Thorne, U. Litzén, S. Johansson: *Spectrophysics: Principles and Applications* (Springer, Berlin, Heidelberg 1999)

17.7 S. Cheng, H. G. Berry, R. W. Dunford, D. S. Gemmell, E. P. Kanter, B. J. Zabransky, A. E. Livingston, L. J. Curtis, J. Bailey, J. A. Nolen Jr.: Phys. Rev. A **47**, 903 (1993)

17.8 C. M. Sikström, H. Nilsson, U. Litzén, H. Lundberg: J. Quant. Spectrosc. Radiat. Transfer **74**, 355 (2002)

17.9 L. J. Curtis: J. Phys. B **31**, L769 (1998)

17.10 L. J. Curtis: J. Phys. B **33**, L259 (2000)

17.11 C. E. Tanner, A. E. Livingston, R. J. Rafac, F. G. Serpa, K. W. Kukla, H. G. Berry, L. Young, C. A. Kurtz: Phys. Rev. Lett. **69**, 2765 (1992)

17.12 J. Jin, D. A. Church: Phys. Rev. **47**, 132 (1993)

17.13 H. Winter, M. L. Gaillard: Z. Phys. **A281**, 311 (1977)

17.14 L. J. Curtis: Phys. Scr. **48**, 599 (1992)

17.15 J. Lidberg, A. Al-Kahlili, L. O. Norlin, P. Royen, X. Tordoir, S. Mannervik: Nucl. Instrum. Meth. Phys. Res. B **152**, 157 (1999)

17.16 H. Hartman, D. Rostohar, A. Derkatch, P. Lundin, P. Schef, S. Johansson, H. Lundberg, S. Mannervik, L.-O. Norlin, P. Royan: J. Phys. B **36**, L197 (2003)

17.17 J. Carlsson, L. Sturesson: Z. Phys. **D14**, 281 (1989)

17.18 A. Gaupp, P. Kuske, H. J. Andrä: Phys. Rev. A **26**, 3351 (1982)

17.19 J. Carlsson: Z. Phys. **D9**, 147 (1988)

17.20 U. Volz, M. Majerus, H. Liebel, A. Schmitt, H. Schmoranzer: Phys. Rev. Lett. **76**, 2862 (1996)

17.21 W. Ansbacher, Y. Li, E. H. Pinnington: Phys. Lett. A **139**, 165 (1989)

17.22 R. N. Gosselin, E. H. Pinnington, W. Ansbacher: Nucl. Instrum. Meth. Phys. Res. **B31**, 305 (1988)

17.23 J. Jin, D. A. Church: Phys. Rev. Lett. **70**, 3213 (1993)

17.24 J. Carlsson, L. Sturesson, S. Svanberg: Z. Phys. **D11**, 287 (1989)

17.25 P. Kuske, N. Kirchner, W. Wittmann, H. J. Andrä, D. Kaiser: Phys. Lett. A **64**, 377 (1978)

17.26 J. Carlsson, P. Jönsson, L. Sturesson: Z. Phys. **D16**, 87 (1990)

17.27 E. H. Pinnington, J. J. van Hunen, R. N. Gosselin, B. Guo, R. W. Berends: Phys. Scr. **49**, 331 (1994)

17.28 H. J. Andrä: *Beam–Foil Spectroscopy*, Vol. 2, ed. by I. A. Sellin, D. J. Pegg (Plenum, New York 1976) p. 835

17.29 L. J. Curtis, P. Erman: J. Opt. Soc. Am. **67**, 1218 (1977)

17.30 L. J. Curtis, W. H. Smith: Phys. Rev. A **9**, 1537 (1974)

17.31 M. Dufay: Nucl. Instrum. Meth. **110**, 79 (1973)

17.32 L. J. Curtis: *Beam–Foil Spectroscopy*, ed. by S. Bashkin (Springer, Berlin, Heidelberg 1976) Chap. 3, pp. 63–109

17.33 L. Engström: Nucl. Instrum. Meth. **202**, 369 (1982)

17.34 R. W. Dunford, H. G. Berry, S. Cheng, E. P. Kanter, C. Kurtz, B. J. Zabransky, A. E. Livingston, L. J. Curtis: Phys. Rev. A **48**, 1929 (1993)

17.35 H. Harde, G. Guthöhrlein: Phys. Rev. A **10**, 1488 (1974)

17.36 Y. Baudinet-Robinet, H.-P. Garnir, P.-D. Dumont, J. Résimont: Phys. Rev. A **42**, 1080 (1990)

17.37 G. Astner, L. J. Curtis, L. Liljeby, S. Mannervik, I. Martinson: Z. Phys. **279**, 1 (1976)

18. Spectroscopy of Ions Using Fast Beams and Ion Traps

A knowledge of the spectra of ionized atoms is of importance in many fields. A wide variety of light sources are available for the study of such spectra. In recent years, techniques coming under the broad headings of fast beams and ion traps have been used extensively for such studies. This chapter will consider the advantages each technique has for particular applications.

18.1 **Spectroscopy Using Fast Ion Beams** 269
18.1.1 Beam–Foil Spectroscopy............ 269
18.1.2 Beam–Gas Spectroscopy............ 270
18.1.3 Beam–Laser Spectroscopy.......... 271
18.1.4 Other Techniques of Ion-Beam Spectroscopy.......... 272

18.2 **Spectroscopy Using Ion Traps** 272
18.2.1 Electron Beam Ion Traps............ 273
18.2.2 Heavy-Ion Storage Rings........... 275

References 277

Part B | 18

18.1 Spectroscopy Using Fast Ion Beams

A beam of ionized atoms has several advantages as a spectroscopic source. Unlike arcs, sparks, and high temperature plasmas, the ions can be studied in an environment that is free of electric and magnetic fields and relatively free of interparticle collisions [18.1]. Standard accelerator techniques can be used to produce a well-collimated, mass-analyzed beam of ions having a low velocity spread. In principle, virtually any charge state of any element, isotopically pure if required, can be obtained. Finally, the well-defined velocity of the ions permits the study of processes evolving in time in terms of their spatial evolution along the beam. This is particularly important in the case of lifetime measurements.

18.1.1 Beam–Foil Spectroscopy

A beam of ions passing through a thin (50–200 nm) foil emerges in a range of ionization states, with the mean charge state increasing with the incident energy [18.2]. Thin foils made from a light element, usually carbon, are used to minimize particle scattering and energy straggling. Thus, a F^+ beam of 0.5 MeV that enters a carbon foil emerges with a mean charge of about $+2e$, while a Xe ion beam at 180 MeV emerges with a mean charge of about $+29e$. The outer electrons are distributed over many different states, most of which then decay to lower states by photon emission as the ions move away from the foil. The beam-foil interaction is a highly nonselective excitation process, which is an advantage for spectroscopic studies but causes a problem for lifetime measurements. (Methods for tackling this problem are discussed in Sect. 17.2.2.) A major disadvantage of the beam-foil light source is its low intensity; consequently, scanning spectrometers have usually been equipped with photon-counting detectors. In recent years, however, position sensitive detectors have become available, which permit the simultaneous recording of information over a wide spectral range, resulting in a greatly improved detection efficiency (Sect. 44.4).

The intrinsic properties of the beam-foil interaction are important in understanding its usefulness for spectroscopic studies. The electrons of the moving ion are shielded from the travelling ion core as the ion passes through the foil and are then recaptured into a statistical distribution of outer states. The probability that more than one electron in a given ion will be captured in an excited state is high relative to other light sources, and hence the technique has been used extensively to study doubly- and multiply-excited states [18.3] (Sect. 64.1). The interaction also favours the production of high-L Rydberg states [18.4] (Chapt. 14). At low incident ion energies, electron capture can give a downstream beam containing neutral and even negative ions. The first observation of photon emission between bound states in a negative ion was achieved using beam-foil excitation [18.5]. A more recent example of multiple excitation in a negative ion is the identification in lithium-like He^- of the $2p^3\ ^4S^o$ state in which all three electrons are excited [18.6,7]. The observation of the same triply excited level in neutral Li has resulted in one of the most precise (40 ppm) wavelength measurements in beam-foil spectroscopy [18.8]. The time-resolution inherent in the beam-foil source can also be used to aid in the identification of transitions from long-lived states, such as intercombination transitions [18.9] (Sect. 10.16). Here the beam foil spectrum is first recorded close to the foil and then far downstream (a few mm to a few cm); "far" in this context means that the short-lived states have had time to decay. The intercombination (or other long-lived) transitions are then easily identified by their relatively strong intensity in the (overall much weaker) downstream spectrum. This has been a decisive step in identifying laboratory lines with lines in the solar corona [18.10, 11]. A recent example of beam-foil spectroscopic analysis is given in [18.12].

One problem with using fast ions as a light source for precision wavelength measurements is the inevitable Doppler broadening of the spectral lines. However, this can be removed by appropriately refocusing the spectrometer [18.13, 14]. Furthermore, since the Doppler width varies with the wavelength, and the line width of a given grating spectrometer used on a fast ion beam usually is dominated by spectrometer geometry, not by diffraction, Doppler broadening often becomes less important in measurements at shorter wavelengths, as in the UV or EUV. This is the primary range of emission from the more highly-ionized atoms, and that is where beam-foil spectroscopy really comes into its own. For example, the leading terms omitted in calculations of the wavelength of the $1s2s\,^3S - 1s2p\,^3P^o$ transition in He-like ions [18.15] scale as Z^4, with the result that measurements made with higher-Z ions need not have as high a precision for a meaningful test of the calculation as would be required for a low-Z ion. The beam-foil measurement for the leading ($J = 1$ to $J = 2$) component in Ni^{26+} [18.16] has an uncertainty of 0.02%, which is about equal to that in the calculation. The two values agree well within this limit. (Naturally, measurements made using the beam-laser techniques discussed in Sect. 18.1.3 yield results with a much higher precision, but such measurements are restricted to low-Z ions because of the excitation energy steps and transition wavelengths involved; the theoretical uncertainties in this case are also much smaller.)

Turning now to beam-foil lifetime measurements, here the intensity of a given transition is studied as a function of the time that has elapsed since excitation, usually by stepping the foil upstream. Because of the nonselective nature of the excitation, the possibility exists that the state being studied is itself being repopulated by higher-lying states, resulting in a decay curve that consists of the sum of exponential terms, one term corresponding to the primary level being studied and the other terms to higher-lying levels involved in repopulating that level. The analysis of such decay curves can be problematic. Several computer routines have been developed to tackle this problem, such as DISCRETE [18.17] and HOMER [18.18]. One useful trick here is to record the decay curve for each of the major transitions repopulating a given primary level and then include the lifetimes obtained for those transitions as fixed parameters in fitting the decay curves for that primary level. A more rigorous method to include the decay data from the repopulating transitions in the analysis of the primary lifetime is the ANDC technique described in detail in Sect. 17.2.2. An additional problem may arise in the measurement of very short lifetimes, which tend to be associated with short-wavelength transitions for which no lenses are available to focus the beam at the spectrometer entrance slit. The observation region is then defined by the spectrometer aperture and extends for a finite length along the beam that can be comparable to, or even greater than, the decay length for that transition. Here it is necessary to fit the decay curve including the vignetted region around the foil [18.19, 20]. Lifetimes in the picosecond regime have been successfully measured in this way.

18.1.2 Beam-Gas Spectroscopy

While the use of a gas target in place of a foil has the obvious advantage that it cannot break, the loss of a tightly-localized excitation region means that the fine time-resolution of the beam-foil light source is largely lost. The beam-gas source, however, has two main advantages. First, the passage of a beam of fast, highly-stripped ions through a neutral gas results in the production of recoil ions (Sect. 65.2) that are moving very slowly relative to the beam ions, thus reducing the Doppler broadening problem mentioned earlier. Secondly, it is possible to study details of the interaction between the gas and beam ions, such as charge-exchange reactions, as they occur, rather than merely observing their consequences. Such experiments have experienced a resurgence with the advent of the ECR ion source [18.21, 22]. A recent example of such work is given in [18.23].

When a fast beam of highly-charged ions passes through a neutral gas, it leaves a trail of ionized gas atoms in its wake. These ions recoil from the interaction region with relatively low velocities ($v/c = 10^{-4}$–10^{-5}).

Furthermore, the recoil velocities tend to be restricted to a narrow range of angles approximately perpendicular to the beam direction. Hence, observation of the radiation emitted by the recoil ions transverse to their motion, i. e., along a direction parallel to the beam, gives a very low Doppler width, $\Delta\lambda_D/\lambda$ being typically 10^{-6}, the limit imposed by the thermal motion of the target gas atoms. Recoil ion spectroscopy is therefore a useful procedure for precision wavelength measurements for highly-stripped ions, such as He-like Ar^{16+} recoil ions produced by a beam of 2 GeV U^{70+} ions [18.24]. The energies and charge states of the recoil ions may be determined using standard time-of-flight techniques [18.25], and detecting the recoil ions in coincidence with their progenitor ions yields information on the dependence of the recoil energy on the details of the ionizing collision [18.26]. In later developments, the differentially pumped gas target has first been replaced by a supersonic jet target which reduces the thermal motion of the target particles, and then by a cold atom sample in an atom trap; replacing optical detection by position-sensitive fast-timing detectors for all collision products, the technique of COLTRIMS (cold target recoil ion momentum spectroscopy) can now be employed to study the momentum distribution of the collision partners (developed largely by the groups of H. Schmidt-Böcking (Frankfurt) and C. L. Cocke (Manhattan, Kansas)).

Measurements made on the projectile ions themselves also yield useful information about electron-capture processes. The strength of such a process is described in terms of its cross section (Sect. 63.1.2). Recent work has shifted from measurements of the total cross section for electron capture to more detailed studies of the individual capture channels [18.27]. Such studies provide much more stringent tests of theoretical models of ion–atom charge transfer processes. They often involve such techniques as spectroscopy of the optical radiation or of the Auger electrons (Sect. 25.1.1) emitted by the ions after electron exchange. This topic is covered in Chaps. 51, 64, and 65. A further example is the study by *Prior* et al. [18.28] of the angular distribution of Auger electrons emitted by doubly-excited states formed in hydrogen-like projectile ions with an energy of 40 keV, following double-electron capture from target helium atoms. They found that significant alignment of the magnetic substates of the projectile ions can result from electron capture. Such anisotropies in the Auger electron emission demonstrate the danger of using single-angle measurements to determine cross sections.

18.1.3 Beam-Laser Spectroscopy

As for the beam-foil source, excitation of a beam of fast ions by a transverse tuneable laser produces the localized excitation required for high temporal resolution. Now, however, the excitation is highly selective, permitting the population of just a single level. The laser-induced-fluorescence (LIF) signal as a function of the distance along the beam from the excitation region is therefore described by a single exponential decay, for which an exact analysis with rigorous error bounds is possible. The restriction to levels that can be accessed by electric dipole (E1) transitions from the ground and metastable levels may be overcome by combining laser excitation with a nonselective mode of excitation, as in beam-gas-laser [18.29,30] or beam-foil-laser [18.31] measurements. A discussion of precision lifetime measurements using laser excitation of a fast beam is given in Sect. 17.2.2. Here the discussion will be limited to precision optical spectroscopy. Examples of precision beam-laser wavelength measurements may be found in [18.32, 33]. A further example is the measurement of the spin-forbidden 1s2s $^1S_0$–1s2p $^3P_1$ interval in N^{5+} [18.34], where the experimental value, 986.321 (7) cm^{-1} is in agreement with the calculated value, 986.58 (30) cm^{-1} [18.34]. An example of a similar measurement in a molecular ion is given in [18.35], while a recent study of the hyperfine structure in a rare-earth ion is given in [18.36].

A major aim in beam-laser spectroscopy is to minimize the width of the LIF signal. The instrumental linewidth of the laser itself can be reduced to below 1 kHz, so that the width of the LIF signal is usually dominated by the velocity spread of the ions and by the divergences of the ion and laser beams. The effects of beam divergence can be minimized by using a collinear geometry, in which the ion and laser beams are parallel. If the angle between the ion and laser beams is θ, the Doppler-shifted laser frequency, as measured in the ion's rest frame, is $f_L(1-\beta\cos\theta)$, where f_L is the laser frequency and $\beta = v/c$ and is much less than unity. Hence the range in frequency resulting from a beam divergence $\Delta\theta$ is given by $f_L\beta\sin\theta\Delta\theta$, which tends to zero as θ tends to zero. One disadvantage of the collinear geometry is that, if the laser is brought into resonance with an atomic transition by adjusting the ion velocity and/or the laser frequency, the LIF signal is produced over the entire overlap region between the ion and laser beams. The resonance can be restricted to a desired region by setting the ion velocity to be slightly off resonance. The velocity can then be adjusted locally for resonance with the laser by passing the ion beam through a Faraday cage electrode to which an adjustable voltage is applied [18.37, 38]. The width of the resonance signal here is usually dominated by the spread in the ion velocity, $c\,\Delta\beta$, usually arising in the ion source being used. The width resulting from a given $\Delta\beta$ can therefore be reduced by using a higher ion energy. This is known as kinematic compression. In terms of the ion energy E, the range in the ion energy ΔE and the ion mass M, the Doppler width of the LIF signal is given by $f_L \Delta E/(2Mc^2 E)^{1/2}$, and thus decreases as E is increased.

A more significant improvement in frequency resolution is made possible by including rf resonance in a laser double-resonance experiment. Here the ions are brought into resonance with an off resonance laser using two separate Faraday cage electrodes. The first resonance depletes the population in the ion state from which excitation occurs, thus weakening the second resonance signal. An rf field is then applied to the ions between the two electrodes. Tuning the frequency of this field over the region that corresponds to fine- or hyperfine-structure intervals in the ion can then repopulate the state from which laser excitation occurs, thus re-establishing the second laser LIF resonance signal. The width of the resonance signal is now determined by the *transit-time broadening* that results from the finite time spent by the ions in the rf field. For example, in the experiments by *Sen* et al. [18.38] with a beam of $^{131}Eu^+$ ions at 1.35 keV, the width of the laser-rf double-resonance signal was 59 kHz, compared with a width of 45 MHz obtained using a single LIF resonance.

18.1.4 Other Techniques of Ion-Beam Spectroscopy

Ion beams find uses in many other applications, three of the main areas involving storage rings (discussed in Sect. 18.2.2), merged beams, and studies of the ion-surface interaction at grazing incidence. Merged beam experiments usually study recombination processes involving electrons and atomic or molecular ions (Sect. 54.1 regarding recombination processes). The advantage of using merged beams is that the time development of the processes may be studied spatially, while maintaining a low relative velocity between the ions and the electrons. This permits measurements at the low energies of importance in studies of Rydberg state formation and in some astrophysical applications. A very different type of experiment studies the ion-surface interaction using a well-collimated ion beam at grazing incidence on a clean, flat surface. Such experiments have revealed very large atomic orientations [18.39]. This orientation can be passed on to the nuclei of the atoms via the hyperfine interaction, thus providing a source of oriented nuclei.

18.2 Spectroscopy Using Ion Traps

A basic purpose of ion traps is to confine ions to the field of view of detectors for time intervals that are longer than the radiative lifetimes of long-lived atomic levels of possible interest. At thermal energies, the ion velocities are large enough to leave a typical detection zone within microseconds. Electrostatic (Kingdon), magnetic (Penning), and radiofrequency (Paul) traps have served for this task for decades (Chapt. 75), with recent additions to the armory by electrostatic mirrors of various shapes [18.40,41]. Two trap varieties of particular interest for spectroscopy, the electron beam ion trap and the heavy-ion storage ring, will be treated in Sects. 18.2.1 and 18.2.2, respectively.

Collisions with the neutral atoms and molecules of the residual gas cause charge exchange, and thus loss of the ion species. Therefore, an ultrahigh vacuum is of primary importance. Over the last three decades the figure of merit has moved from pressures of about 10^{-8} mbar to about 10^{-11} mbar. Further improvements can be expected from working with traps at liquid helium temperature; in fact, even ion traps as large as an ion storage ring at Aarhus (Sect. 18.2.1) have been cooled considerably to vary both the vacuum and the amount of blackbody radiation experienced by ions in weakly bound states.

The energy steps in multiply charged ions are regularly larger than what is available from lasers, at least for excitation from the ground state. Hence, single-ion trapping and laser spectroscopic investigation are rarely an option for these ions; many ions are needed to provide a sufficiently strong emission signal. The production of quantities of multiply charged ions used to be achieved by electron bombardment of a dilute gas inside the trap volume, or by ablation from a surface. Evidently, this is detrimental to any subsequent measurements, since the residual gas is still present. Precision work like mass spectrometry that exploits the ion cyclotron motion of stored ions, or detailed studies of the radiative processes (including the effects of the interrogating laser field) in ions nowadays employ a sequence of ion traps. In a first trap, the ions of interest are produced and possibly cooled by laser light or other mechanisms, and then, by applying electric fields, the ions of interest are moved to a second trap that works under better vacuum conditions or that can be more finely tuned. In the same sense, a heavy-ion storage ring is being fed by an isotopically pure, charge-state selected ion beam. Any loss of ions, measured by whatever means, is thus proportional to the loss of the ion species of interest.

18.2.1 Electron Beam Ion Traps

Electron beam ion traps (EBIT) make use of the attractive potential of a high-density electron beam, as well as of the space charge compensation that is provided by the electron beam to any ion cloud already trapped. Most electron beam ion traps generate the high-current density electron beam by feeding the beam from an electron gun into a magnetic field that then compresses and guides the beam. In most cases, superconducting magnets with fields of 3 to 8 T are being used, and current densities of the order of 10^4 A/cm^2 are reached. This high current density corresponds to an electron density of the order of 10^{11} /cm^3. This low-density environment, roughly comparable to tokamak discharges, is one of the factors that renders the electron beam ion trap a very interesting device for laboratory astrophysics. The magnetic field helps to confine any ion cloud that is produced from the residual gas (or gas bled in) or from injected low charge ions. However, the ions could move away along the field lines, if they were not stopped by potential barriers provided by electrically charged drift tubes. Obviously, the basic design is the same as that of a Penning trap, with the permanent electron beam added. In fact, EBIT with the electron beam on has been said to operate in electronic trapping mode [18.42]; while the same device with the electron beam off ("magnetic trapping mode") still works as a Penning trap. This option of producing an ion cloud with intense electron bombardment and then studying the ions without the electron beam present is the basis for a variety of experiments on charge exchange (CX) reactions and long-lived excited levels [18.43] (see below).

The first working electron beam ion trap, EBIT-I, has been set up at Livermore [18.44,45]. The successful operation instigated an upgrade to SuperEBIT, the first such machine that was able to completely ionize all naturally occurring elements [18.46]. Based mostly on the Livermore design, some 8 to 10 EBITs are now either running or under construction around the world. The EBIT operating principles have, for example, been described by *Currell* [18.47].

Ionization of ions trapped in the combination of electrical fields proceeds as long as the electron beam energy is high enough to overcome the ionization potential. Thus, the highest charge state can be pre-selected by the appropriate choice of the electron beam energy. The technical effort required to reach, for example, bare uranium in SuperEBIT is much smaller than in an ion accelerator. In both cases the ionization is achieved by frequent energetic collisions of ions with electrons. In SuperEBIT, the ions are (practically) stationary, and the design energy of SuperEBIT, 250 keV, is enough to remove even the last electron of uranium. At a heavy-ion accelerator, the electrons are stationary (in a foil target), and the ions are fast. Consequently an ion energy per nucleon that is higher by the proton/electron mass ratio is required – some 500 MeV/amu. Such energetic ion beams are only available in a few large accelerator laboratories, whereas an electron beam ion trap with its auxiliary equipment fits into an office-sized laboratory space. Of course, there are experiments that need specific properties of either fast ion beams or stationary ions, so both types of devices have their specific merit.

The ions in an EBIT are not only stationary in the sense that they are localized in a cloud, and moving either way along the magnetic field with the same probability, but their energy (temperature) can also be controlled by the height of the potential barriers. The voltages on the confining drift tubes are usually chosen to be few hundred volts. This makes for barrier potentials $+qeU$ (charge state q, elementary charge e, voltage U) that are higher for highly charged ions than for low-charge state ions. This not only benefits the confinement of highly charged ions directly; light ions (residual gas or purposely bled in gases) may become fully ionized by the collisions with the electron beam and by charge exchange, but they still have a larger chance to evaporate from the trap and thus they cool the remaining ion cloud. Under typical conditions, the ion cloud may have a temperature of a few keV. This can be lowered by introducing a cooling gas and by lowering the potential barriers. With Cs^{45+} in the trap, this has been demonstrated by reducing the (thermal) Doppler spread of X-ray emission lines until it was smaller than the natural line width of the emitter, thus yielding a measurement of femtosecond level lifetimes from highly resolved X-ray spectra [18.48].

The other good level lifetime range of an EBIT reaches from a few microseconds (limited by practical switching issues) to many milliseconds. Under direct optical (X-ray, EUV, visible) observation of a spectral feature, the electron beam is used to produce an ion cloud with ions of a desired charge state. When the electron beam is stopped, all direct excitation and prompt emission ceases. Any later photon signal relates to delayed emission from long-lived levels, or from excitation by charge transfer collisions (highly charged ions capturing electrons from the residual gas atoms). CX is an important loss mechanism, and the CX signal also serves as a monitor of the number of ions remaining in the trap. Owing to the excellent vacuum in cryogenic EBITs, trapping times of many seconds, if not minutes have been observed [18.43]. The ion loss rate is the major correction to the apparent decay time of the delayed photon signal. For atomic level lifetimes of a few milliseconds and less, this correction is small (a few percent or less). EBIT lifetime measurements that take this correction into account yield results that agree with those from heavy-ion storage rings [18.49] and that are, at uncertainties of 0.5% and less, remarkably consistent with theory in a case for which the theory can do very well (see Fig. 18.1). This observation can be turned around and interpreted as a demonstration of the reliability of the experimental techniques that then can be applied to more complex cases in which theory evidently has problems (for examples, see [18.50, 51]).

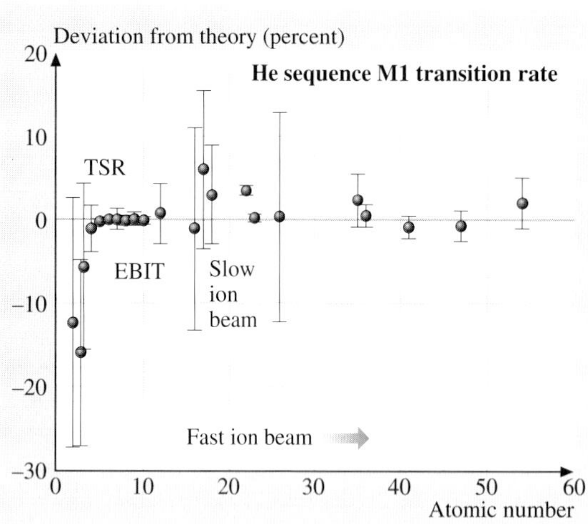

Fig. 18.1 The lifetime of the 1s2s $^3S_1$ level in the He isoelectronic sequence varies by 15 orders of magnitude from He (about 6000 s) to Xe^{52+} (a few picoseconds). The figure shows the deviation of selected experimental results from calculations, and indicates the dominant experimental techniques for the various lifetime ranges. The consistency of experimental results from a heavy-ion storage ring (TSR Heidelberg) and an electron beam ion trap (Livermore EBIT-II) with theory in the low-Z range is impressive. (After [18.50]).

An EBIT is an excellent light source for precision spectroscopy of highly charged ions because it not only gives access to all charge states of all elements, but it does so at low particle densities. Of particular interest to precision spectroscopy have been the ions with a single valence electron (Li, Na, Cu isoelectronic sequences) that are rather amenable to calculation. Such spectra of ions up to $Z = 60$ or 70 have been measured at low-density plasmas like the tokamak, and they have been found to agree well with theory. Higher charge states were then reached in laser-produced plasmas (Sect. 44.1.2), but the wavelength results seemed to deviate from the theoretical trend that had supported the tokamak results. There also were (very few) high-nuclear charge Z data from fast ion beams. Electron beam ion trap data have now confirmed that the trend of the tokamak data was correct and that theory [including second-order quantum electrodynamics (QED) contributions] provides a good description of the $n = 3$ (Na sequence) [18.52] and $n = 4$ levels (Cu sequence) [18.53] up to $Z = 92$, within the 40 ppm error margin of the EBIT results for the heaviest Cu-like ions. The agreement with theory for the Zn-like ions of the same elements is much poorer. As the QED contributions are rather similar, this is a problem of theory with the computational treatment of two (and more) valence electrons in the same shell.

The atomic systems purportedly under best theoretical control are H-like ions, with only one electron in total. In high-Z H-like ions, the lines of highest interest are in the hard X-ray range ($n = 1{-}2$), or are severely lifetime-broadened (2s–2p) and in the EUV. One line is in the visible (for a number of ions), where it is accessible to high-resolution spectroscopy, and not even notably broadened, because its upper level has a millisecond-range lifetime: the M1 transition between the hyperfine levels of the ground state. For two isotopes (of Bi and Pb), this transition has been induced by laser radiation in a heavy-ion storage ring ([18.54], see below), and for 5 isotopes (of Re, Ho, and Tl) emission spectroscopy at the SuperEBIT electron beam ion trap [18.55] was successful. Using the ion cloud (with its cross section largely determined by the electron beam diameter of about

70 μm) as a light source without a further entrance slit, a high-efficiency transmission grating spectrometer, and a position-sensitive detector, lines from two Tl isotopes were recorded simultaneously [18.55]. The differential effects of the two nuclei yielded additional information on the charge distribution in the nucleus, on top of the magnetic moment distribution that makes up the dominant effect. These results are required for a better interpretation of parity nonconservation (PNC; Chapt. 29) measurements of neutral Tl atoms, in which the nuclear magnetic moments so far have been treated only as a point dipole. Corresponding measurements are also being done on Li-like ions, and again the most precise data so far (on Bi) have been obtained at SuperEBIT.

One of the pertinent problems in astrophysics is the cataloguing of spectral lines in all spectral ranges, both for identifying spectral features and thus learning about the composition of a light source, and for modeling of the light source in order to understand its "operating conditions". Even data bases that claim "practical completeness", however, are found to be grossly incomplete in the EUV and soft-X-ray ranges that have been opened to high-resolution observations by the grating spectrometers on board the *Chandra* and *XMM-Newton* spacecrafts. Here observations at electron beam ion traps fill in much of the needed data, at comparable quality. Moreover, an EBIT as a kind of analog computer does more than provide correct line positions: the spectra show line ratios from a light source with known electron energy (and the option of known temperature by simulating a Maxwellian energy distribution [18.56]) and particle density. This serves both as a check on collisional-radiative models and as an immediate data resource for astrophysics.

Most X-ray data from EBITs have been collected using solid state detectors [Si(Li), Ge] that offer high detection efficiency (large solid angle) and signal timing on the microsecond scale, but feature poor spectral resolution. High-resolution instruments like crystal spectrometers are necessary to analyze spectra in any detail; equipped with position-sensitive detectors, such instruments do much better than scanning spectrometers in terms of data collection rate and calibration. However, they suffer from the low diffraction efficiency of the crystals. Recently a new device, the microcalorimeter, has started to show its interesting properties. In these devices, small absorbers at mK-temperatures show a measurable temperature increase when absorbing an X-ray photon, and the signal is proportional to the deposited photon energy [18.57]. The linewidth of the best devices is below 10 eV (not as good as a crystal spectrometer, but much better than a traditional solid state diode), the sensor pixels can be grouped in arrays to make for a larger area and for cross references among pixels (which helps with calibration and with the rejection of cosmic ray events), and the signal processing is fast enough to permit time resolution on the millisecond range. A spaceflight engineering spare has been used at the Livermore EBITs to study, for example, soft-X-rays of light elements as seen from CX near comets [18.58], or measure the (10 ms) time constant of the M3 decay in a Ni-like ion (Xe^{26+}) [18.59].

Last, but not least, the well defined and adjustable electron beam energy permits detailed studies of the interaction of fast electrons with highly charged ions. This includes, for example, the spectroscopy of dielectronic recombination (DR) resonances, or the exploration of radiative recombination (RR) (Chapt. 55). These processes can be investigated up to the highest ion charges, where relativistic and QED contributions (like the Generalized Breit interaction) matter.

18.2.2 Heavy-Ion Storage Rings

With foil-excited ion beams and long level-lifetimes, decay curves spread out along the beam, and the signal from a given width of the field of view may drop to the detector background level. The decay lengths of microsecond lifetime levels are on the order of 10 m, and those of millisecond levels are tens of kilometers. In such cases it is advantageous to curve the beam line around on itself, forming a storage ring, in which the ions pass in front of the detector over and over again. Heavy-ion storage rings need excellent vacuum conditions (10^{-11} mbar and better) to reach storage times of seconds, minutes, or hours, depending on the electronic structure of the stored ions and on the ion beam energy. The dominant loss processes are electron capture and loss, small angle scattering, and large angle (inelastic) scattering as with any fast ion beams [18.60], but these are aggravated here by the much longer path lengths.

Storage rings (for example, TSR Heidelberg, ASTRID Aarhus, CRYRING Stockholm, ESR Darmstadt), with magnetic dipoles and quadrupoles for beam transport and focusing, sort the stored particles by their momentum. The typical ion beam energies range from a few dozen keV total to hundreds of MeV per nucleon (ESR). Electrostatic storage rings (ELISA Aarhus and more under construction) have only electrical fields and select by particle energy, usually below about 100 keV (for a review, see [18.61]). They are more suitable for low-charge heavy ions and ion molecules, including biomolecules, than the magnetic rings that can handle very fast particles. Magnetic storage rings usually have electron cooler sections in which a "cool" electron beam (with a low longitudinal velocity spread from kinematic compression) of almost the same velocity as the circulating ion beam is merged with the latter for a path of a few meters (and is then deflected out again). By scattering among electrons and ions, the momentum spread of the electrons (small) and ions (larger) equilibrates, leaving the ion beam with a narrower momentum distribution and thus cooled. Cooling, which typically takes a few seconds, improves the storage behaviour of the ion beam and the energy resolution of, for example, dielectronic recombination (DR) studies (Chapt. 55). For these, the same electron cooler is now tuned to provide electrons at a well defined but different velocity. Thus the electron cooler can serve as an electron target, without the complications of a foil target in beam-foil spectroscopy. The difference velocity can be chosen from a wide range, including zero. Extremely low energy collisions are being investigated for the study of DR and for the recombination of molecular radicals. When a beam of molecular ions is injected, storage is long enough to let some of the internal degrees of freedom relax, and then a beam of molecular ions can be extracted that are closer to their ground state.

A cooled ion beam, with its narrow velocity distribution, is also of interest for laser-ion interaction studies, offering higher resolution and better signal. Laser-assisted electron capture in the electron cooler, as well as laser spectroscopy on high-lying levels populated by DR, are possible. One of the problems with precision wavelength measurements involving fast ions is, as always the accurate determination of the velocity, as a step towards determining Doppler corrections. At TSR Heidelberg, a beam of Li^+ ions was subjected to a laser beam from ahead, tuned to one of the 2s–2p transitions. A second laser beam from behind probed the position of the Lamb dip in the velocity distribution and thus assured that it would meet the same velocity group of the multi-MeV stored ions. Accurate off-line calibrations of the laser frequencies then permitted a test of the Doppler formula to a relative precision of 2.2×10^{-7} [18.62].

The Doppler shift determination in any observation of fast ions requires accurate angle measurements. These are nontrivial, because the detection efficiency of any finite size spectrometer or extended detector may be non-uniform as a function of position or angle. One technique calls for segmented ("granular") X-ray detectors, the strips of which are calibrated individually [18.63]. Relativity changes the emission pattern seen in the laboratory rest frame to one that favours forward emission. This is beneficial for zero-degree spectroscopy, that is, an observation along the ion beam path. At ESR Darmstadt, bare ions captured an electron in the electron cooler section, and the resulting X-rays were detected from straight ahead (behind the next dipole magnet that deflected the ion beam). This geometry maximizes the Doppler shift, but minimizes the uncertainty relating to geometry. Also, after electron capture, the ion in the bending magnet section follows a trajectory that differs from that of the unperturbed ions. The ion can be detected and, in coincidence with the X-ray detector, make for a very clean and charge-specific spectrum. Similar coincidence measurements make it possible to use a low-pressure gas jet target in a high-energy ion storage ring, evaluating only coincidences of X-ray photons and charge-changing events [18.63]. Fast ions (energetic enough to achieve the desired charge state) can also be decelerated in a storage ring, which helps to do systematic checks of the Doppler effect and to work at lower Doppler shift [18.63].

As mentioned in Sect. 18.2.1, laser-resonance techniques have been used to find the ground state hyperfine transition in two H-like heavy isotopes. In one of them, the lifetime was also measured, by recording the fluorescence decay from the ion beam after switching off the laser [18.54]. For lower charge states, one can exploit the excitation that ions carry into the ring from their production in the ion source, or from stripping processes in the injector accelerator [18.64]. Lifetimes from half a millisecond to several seconds have been measured this way by passive observation [18.50], with an accuracy of better than 0.2% in favourable cases. Other techniques use excitation by DR in the ring [18.65], or laser probing of the remaining metastable level population so that fluorescence is emitted near a photomultiplier detector [18.66]. With a stored beam of negative ions, even blackbody radiation may be sufficient to photodetach the weakly bound last electron; the ensuing neutral atom is not deflected at the next bending section and leaves the ring to be detected. All in all, lifetime measurements at storage rings reach from 10 μs to about 1 min.

References

18.1 I. Martinson: Rep. Prog. Phys. **52**, 157 (1989)
18.2 K. Shima, T. Mikumo, H. Tawara: At. Data Nucl. Data Tables **34**, 357 (1986)
18.3 T. Andersen, S. Mannervik: Comments At. Mol. Phys. **16**, 185 (1985)
18.4 F. G. Serpa, A. E. Livingston: Phys. Rev. A **43**, 6447 (1991)
18.5 C. F. Bunge: Phys. Rev. Lett. **44**, 1450 (1980)
18.6 E. J. Knystautas: Phys. Rev. Lett. **69**, 2635 (1992)
18.7 E. Träbert, P. H. Heckmann, J. Doerfert, J. Granzow: J. Phys. B: At. Mol. Opt. Phys. **25**, L353 (1992)
18.8 S. Mannervik, R. T. Short, D. Sonnek, E. Träbert, G. Möller, V. Lodwig, P. H. Heckmann, J. H. Blanke, K. Brand: Phys. Rev. A **39**, 3964 (1989)
18.9 E. Träbert: Physica Scripta **48**, 699 (1993)
18.10 E. Träbert, R. Hutton, I. Martinson: Mon. Not. R. Astron. Soc. **227**, 27 (1987)
18.11 E. Träbert: Mon. Not. R. Astron. Soc. **297**, 399 (1998)
18.12 C. Jupén, P. Bengtsson, L. Engström, A. E. Livingston: Physica Scripta **64**, 329 (2001)
18.13 J. O. Stoner, J. A. Leavitt: Opt. Acta **20**, 435 (1973)
18.14 K.-E. Bergkvist: A high-intensity method for beam-foil spectroscopy, with retained spatial resolution along the beam. In: *Beam Foil Spectroscopy*, ed. by I. A. Sellin, D. J. Pegg (Plenum, New York 1976) p. 719
18.15 G. W. F. Drake: Can. J. Phys. **66**, 586 (1988)
18.16 A. S. Zacharias, A. E. Livingston, Y. N. Lu, R. F. Ward, H. G. Berry, R. W. Dunford: Nucl. Instrum. Methods Phys. Res. B **31**, 41 (1988)
18.17 S. W. Provencher: J. Chem. Phys. **64**, 2772 (1976)
18.18 D. J. G. Irwin, A. E. Livingston: Comput. Phys. Commun. **7**, 95 (1974)
18.19 P. H. Heckmann, E. Träbert, H. Winter, F. Hannebauer, H. H. Bukow, H. von Buttlar: Phys. Lett. A **57**, 126 (1976)
18.20 E. H. Pinnington, W. Ansbacher, J. A. Kernahan, Z.-Q. Ge, A. S. Inamdar: Nucl. Instrum. Methods Phys. Res. B **31**, 206 (1988)
18.21 R. Geller: App. Phys. Lett. **16**, 40 (1970)
18.22 C. M. Lyneis, T. A. Antaya: Rev. Sci. Instrum. **61**, 221 (1990)
18.23 S. J. Smith, J. A. Lorenzo, S. S. Tayal, A. Chutjian: Phys. Rev. A **68**, 062708 (2003)
18.24 J. A. Laming, J. D. Silver: Phys. Lett. A **123**, 395 (1987)
18.25 G. P. Grandin, D. Hennecart, X. Hussin, D. Lecler, I. Lesteven-Vaisse, D. Lisfi: Europhys. Lett. **6**, 683 (1988)
18.26 J. C. Levin, R. T. Short, C.-S. O, H. Cederquist, S. B. Elston, J. P. Gibbons, I. A. Sellin, H. Schmidt-Böcking: Phys. Rev. A **36**, 1649 (1987)
18.27 M. Barat: Nucl. Instrum. Methods Phys. Res. B **9**, 364 (1985)
18.28 M. H. Prior, R. A. Holt, D. Schneider, K. L. Randall, R. Hutton: Phys. Rev. A **48**, 1964 (1993)
18.29 D. Schulze-Hagenest, H. Harde, W. Brand, W. Demtröder: Z. Phys. A **282**, 149 (1977)
18.30 H. Schmoranzer, U. Volz: Physica Scripta **T47**, 42 (1993)
18.31 Y. Baudinet-Robinet, P.-D. Dumont, H.-P. Garnir, A. El. Himdy: Phys. Rev. A **40**, 6321 (1989)
18.32 T. J. Scholl, R. Cameron, S. D. Rosner, L. Zhang, R. A. Holt, C. J. Sansonetti, J. D. Gillaspy: Phys. Rev. Lett. **71**, 2188 (1993)
18.33 T. P. Dinneen, N. Berrah-Mansour, H. G. Berry, L. Young, R. C. Pardo: Phys. Rev. Lett. **66**, 2859 (1991)
18.34 E. G. Myers, J. K. Thompson, E. P. Gavathas, N. R. Clausen, J. D. Silver, D. J. H. Howie: Phys. Rev. Lett. **75**, 3637 (1995)
18.35 T. J. Scholl, S. D. Rosner, R. A. Holt: Can. J. Phys. **76**, 39 (1998)
18.36 R. C. Rivest, M. R. Izawa, S. D. Rosner, T. J. Scholl, G. Wu, R. A. Holt: Can. J. Phys. **80**, 557 (2002)
18.37 M. L. Gaillard, D. J. Pegg, C. R. Bingham, H. K. Carter, R. L. Mlekodaj, J. D. Cole: Phys. Rev. A **26**, 1975 (1982)
18.38 A. Sen, W. J. Childs, L. S. Goodman: Nucl. Instrum. Methods Phys. Res. B **31**, 324 (1988)
18.39 H. J. Andrä, R. Fröhling, H. J. Plöhn, J. D. Silver: Phys. Rev. Lett. **37**, 1212 (1976)
18.40 D. Zajfman, O. Heber, L. Vejby-Christensen, I. Ben-Itzhak, M. Rappaport, R. Fishman, M. Dahan: Phys. Rev. A **55**, 1577 (1997)
18.41 H. T. Schmidt, H. Cederquist, J. Jensen, A. Fardi: Nucl. Instrum. Methods Phys. Res. B **173**, 523 (2001)
18.42 P. Beiersdorfer, L. Schweikhard, J. Crespo López-Urrutia, K. Widmann: Rev. Sci. Instrum. **67**, 3818 (1996)
18.43 L. Schweikhard, P. Beiersdorfer, E. Träbert: Non-neutral plasma physics IV: Proc. 2001 Int. Workshop on Non-neutral Plasmas, San Diego (CA, USA), edited by F. Anderegg, C. F. Driscoll, L. Schweikhard, Am. Inst. Phys. Conf. Proc. **606**, 174 (2002)
18.44 M. A. Levine, R. E. Marrs, J. N. Bardsley, P. Beiersdorfer, C. L. Bennett, M. H. Chen, T. Cowan, D. Dietrich, J. R. Henderson, D. A. Knapp, A. Osterheld, B. M. Penetrante, M. B. Schneider, J. H. Scofield: Nucl. Instrum. Methods B **43**, 431 (1989)
18.45 M. A. Levine, R. E. Marrs, J. R. Henderson, D. A. Knapp, M. B. Schneider: Physica Scripta T **22**, 157 (1988)
18.46 R. E. Marrs, S. R. Elliott, D. A. Knapp: Phys. Rev. Lett. **72**, 4082 (1994)
18.47 F. J. Currell: The Physics of Electron Beam Ion Traps. In: *Trapping highly charged ions: Fundamentals and applications*, ed. by J. Gillaspy (Nova Science, Commack, N.Y. 2001) p. 3

18.48 P. Beiersdorfer, A. L. Osterheld, V. Decaux, K. Widmann: Phys. Rev. Lett. **71**, 2196 (1993)
18.49 E. Träbert, P. Beiersdorfer, G. Gwinner, E. H. Pinnington, A. Wolf: Phys. Rev. A **66**, 052507 (2002)
18.50 E. Träbert: Can. J. Phys. **80**, 1481 (2002)
18.51 P. Beiersdorfer, E. Träbert, E. H. Pinnington: Astrophys. J. **587**, 836 (2003)
18.52 P. Beiersdorfer, E. Träbert, H. Chen, M.-H. Chen, M. J. May, A. L. Osterheld: Phys. Rev. A **67**, 052103 (2003)
18.53 E. Träbert, P. Beiersdorfer, H. Chen: Phys. Rev. A **70**, 032506 (2004)
18.54 I. Klaft, S. Borneis, T. Engel, B. Fricke, R. Grieser, G. Huber, T. Kühl, D. Marx, R. Neumann, S. Schröder, P. Seelig, L. Völker: Phys. Rev. Lett. **73**, 2425 (1994)
18.55 P. Beiersdorfer, S. B. Utter, K. L. Wong, J. R. Crespo López-Urrutia, J. A. Britten, H. Chen, C. L. Harris, R. S. Thoe, D. B. Thorn, E. Träbert, M. G. H. Gustavsson, C. Forssén: Phys. Rev. A **64**, 032506 (2001)
18.56 D. W. Savin, B. Beck, P. Beiersdorfer, S. M. Kahn, G. V. Brown, M. F. Gu, D. A. Liedahl, J. H. Scofield: Phys. Scr. T **80**, 312 (1999)
18.57 F. S. Porter, K. Gendreau, K. Boyce, A. Szymkowiak, R. Kelley, S. Stahle, J. Gygax, R. Brekosky, P. Beiersdorfer, G. V. Brown, S. Kahn: http://phonon.gsfc.nasa.gov/pubs/spie00-ppt.pdf
18.58 P. Beiersdorfer, K. R. Boyce, G. V. Brown, H. Chen, S. M. Kahn, R. L. Kelley, M. May, R. L. Olson, F. S. Porter, C. K. Stahle, W. A. Tillotson: Science **300**, 1558 (2003)
18.59 E. Träbert, P. Beiersdorfer et al.: (work in progress)
18.60 I. S. Dmitriev, V. S. Nikolaev, Ya. A. Teplova: Phys. Lett. **26A**, 122 (1968)
18.61 L. H. Andersen, O. Heber, D. Zajfman: J. Phys. B **37**, R57 (2004)
18.62 G. Saathoff, S. Karpuk, U. Eisenbarth, G. Huber, S. Krohn, S. Reinhardt, D. Schwalm, A. Wolf, G. Gwinner: Phys. Rev. Lett. **91**, 190403 (2003)
18.63 Th. Stöhlker, P. H. Mokler, F. Bosch, R. W. Dunford, F. Franzke, O. Klepper, C. Kozhuharov, T. Ludziejewski, F. Nolden, H. Reich, P. Rymuza, Z. Stachura, M. Steck, P. Swiat, A. Warczak: Phys. Rev. Lett. **85**, 3109 (2000)
18.64 J. Doerfert, E. Träbert, A. Wolf, D. Schwalm, O. Uwira: Phys. Rev. Lett. **78**, 4355 (1997)
18.65 H. T. Schmidt, P. Forck, M. Grieser, D. Habs, J. Kenntner, G. Miersch, R. Repnow, U. Schramm, T. Schüssler, D. Schwalm, A. Wolf: Phys. Rev. Lett. **72**, 1616 (1994)
18.66 S. Mannervik: Physica Scripta T **105**, 67 (2003)

19. Line Shapes and Radiation Transfer

The shapes of collisionally broadened atomic lines is a topic almost as old as Fraunhofer's discovery of the existence of discrete lines. *Lorentz* provided the first quantitative theory in 1906 [19.1], and *Weisskopf* advanced this to the *impact theory* by 1933 [19.2]. *Holtsmark* [19.3], *Kuhn* [19.4] and *Margenau* [19.5] meanwhile developed the *quasistatic* or *statistical* theory which describes the line wing, and *Jablonski* put this on a quantum mechanical footing in the context of free–free molecular radiation [19.6,7]. By the 1940s, satellite bands in the line wings, and a variety of high and low pressure line shapes and broadening rates had been measured. Initial confusion regarding the validity of the contrasting *impact* versus *static* approaches was largely resolved by unified treatments of the Fourier integral theory [19.8–13]. *Baranger* then provided a quantum basis for the impact theory, including level degeneracies [19.14]. Descriptions can be found in a variety of reviews and references therein, including [19.2,5,15–20]. The broadening of molecular lines involves the additional complication of rotationally nonadiabatic collisions; this was initially addressed by *Anderson* [19.12,13] and later with great thoroughness by *van Kranendonk* [19.21]. This chapter and most of the above theories are concerned with neutral atomic gases, which is sometimes called *pressure broadening*. In plasmas, electron, ion, and neutral collisions all contribute to the line shapes and strengths; thus the emitted lines provide a powerful diagnostic of plasma conditions. Neither molecular nor plasma broadening will be covered here; the latter is reviewed in [19.17–20, 22], and in Chapts. 59 and 48.

19.1 **Collisional Line Shapes** 279
- 19.1.1 Voigt Line Shape 279
- 19.1.2 Interaction Potentials 280
- 19.1.3 Classical Oscillator Approximation 280
- 19.1.4 Impact Approximation 281
- 19.1.5 Examples: Line Core 282
- 19.1.6 Δ and γ_c Characteristics 284
- 19.1.7 Quasistatic Approximation 284
- 19.1.8 Satellites 285
- 19.1.9 Bound States and Other Quantum Effects 286
- 19.1.10 Einstein *A* and *B* Coefficients 286

19.2 **Radiation Trapping** 287
- 19.2.1 Holstein–Biberman Theory 287
- 19.2.2 Additional Factors 289
- 19.2.3 Measurements 290

References 292

19.1 Collisional Line Shapes

The neutral-gas theories described above generally used phenomenological or long range forms of the atomic and molecular interactions, and most measurements were not sufficiently detailed to test the validity of these parameterizations or the theoretical approximations. A great deal of the work since mid century has been directed towards obtaining more realistic and accurate descriptions of these interactions and of the full line shapes. In addition, many new types of observations have stimulated variations on the basic theories and descriptions. This includes topics such as collision-induced forbidden transitions, satellite shapes, spectral and polarization redistribution, orientation and alignment effects, Doppler-free spectroscopy and very low temperature collisions. Most of these topics are beyond the scope of this brief and basic description of neutral collisional line shapes.

19.1.1 Voigt Line Shape

An atomic (or molecular or ionic) line has an intrinsic Lorentzian shape that reflects the Fourier transform of the exponentially decaying spontaneous emission. For a spontaneous decay rate Γ, the fullwidth at half maximum (FWHM) is $\Delta\omega = \Gamma$. Due to the Maxwellian distribution of atomic velocities v in a thermal vapor, and the Doppler shift $\Delta\omega_D = v/\lambda\!\!\!^-$, the resonance frequencies of atoms in the laboratory frame have a Gaussian distribution. The Doppler width of this distribution is approximately $\omega_0 u/c$, where ω_0 is the resonant frequency and u is the mean thermal velocity. The full line shape is thus a convolution of the natural Lorentzian with the thermal Gaussian; a Voigt profile. In the presence of collisions, the line from each atom broadens, shifts and becomes asymmetric, and this is normally convoluted with the Gaussian velocity distribution to obtain the complete line shape. Collisions may also cause weak or dipole-forbidden transitions to become stronger as well as broader.

19.1.2 Interaction Potentials

Theories of collisional line shapes consider an ensemble average of collisional interactions. For atomic gases, the description of each individual collision or interaction generally starts with a molecular model for a pair of interacting atoms, since the Born–Oppenheimer approximation is appropriate for thermal atomic collisions (electron velocities $\gg$ nuclear velocities) [19.23]. The radiative transition then occurs between adiabatic

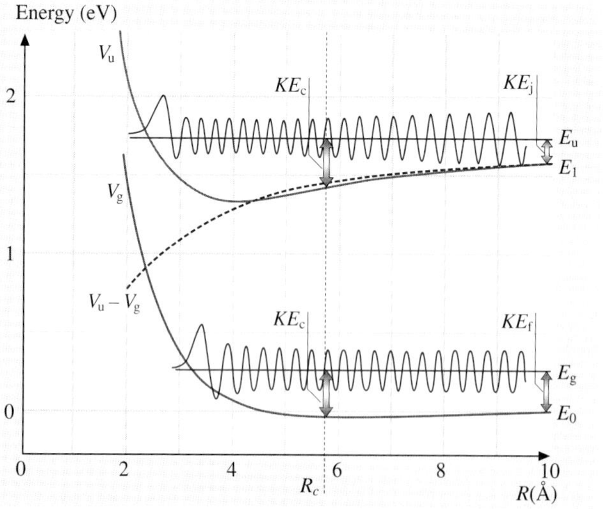

Fig. 19.1 Diagrammatic representation of a free–free molecular radiative transition and the classical Franck–Condon principle. Adiabatic potentials (V_u and V_g), the difference potential equal to the classical transition frequency (*dashed*), nuclear kinetic energies and wave functions $\Phi_u(R)$ and $\Phi_g(R)$ are indicated. The position R_C is the classical radiation position for the initial (E_u) and final (E_g) energies shown

electronic molecular states that separate to the atomic states of the transition under consideration. The notation $V_u(R)$ denotes the electronic energy of the upper state, and $V_g(R)$ the lower state, where R is the internuclear separation, and the total atomic energies are E_1 and E_0 with $E_1 - E_0 = \hbar\omega_0$ (Fig. 19.1). The next simplification is to assume that the atom-pair statistically branch into adiabatic motion along each of the molecular states associated with the initial atomic state, and radiation to each of the final states is summed independently, assuming they are also completely adiabatic. This ignores an inevitable nonadiabatic coupling between the states as their energy separation decreases to zero at large R, but as discussed below this has been shown to have a very minor effect on line broadening. The single-collision problem then reduces to calculating the spectrum of a molecular transition between each upper and lower pair of adiabatic states, for each initial state of internuclear motion and all possible final motions, and summing these weighted by the rate of initial collisional motions. However, such calculations are only necessary to elucidate particular quantum features, because the *classical oscillator approximation, impact approximation, quasistatic approximation*, and *classical Franck–Condon principle* provide major conceptual and calculational simplifications. These are presented in the context of free–free transitions in the following sections.

19.1.3 Classical Oscillator Approximation

Consider a free–free molecular radiative transition of energy $\hbar\omega = E_u - E_g$ between upper (u) and lower (g) states of total energy E_u and E_g, as shown in Fig. 19.1. Each elastic scattering state is the product of an electronic adiabatic state $\phi(r, R)$ and a nuclear motion state $\Phi(R)$ of the molecule. The electronic states u and g have effective potentials

$$V_q^e = V_q + l(l+1)/R^2 , \tag{19.1}$$

where $q = $ u or g. Examples of such $V(R)$ and $\Phi(R)$ for the case of $l = 0$ are shown in Fig. 19.1. The electric dipole radiation operator that couples these states is normally a weak perturbation that does not alter these potentials or wave functions. For definiteness assume the atom is initially in the upper state, and that an atom pair approach with kinetic energy $\mathcal{T}_i$ and separate with $\mathcal{T}_f$, as indicated. The intensity I, or transition probability, is proportional to the squared matrix element of the dipole operator $e\boldsymbol{r} = e\Sigma_i \boldsymbol{r}_i$ between initial and final states:

$$I \propto \left| \int \mathrm{d}^3 R \int \mathrm{d}^3 \boldsymbol{r} \Phi_\mathrm{u}(R)^* \phi_\mathrm{u}(\boldsymbol{r}, R)^* e \boldsymbol{r} \phi_\mathrm{g}(\boldsymbol{r}, R) \Phi_\mathrm{g}(R) \right|^2$$
$$= \mu^2 \left| \int \mathrm{d}^3 R \Phi_\mathrm{u}(R)^* \Phi_\mathrm{g}(R) \right|^2 , \tag{19.2}$$

where the last factor is the Franck–Condon factor, μ is the dipole moment, here assumed to be the atomic value independent of R, and $\mathrm{d}^3 R = 4\pi R^2 \mathrm{d}R$ for diatomics. The $\Phi(R)$ can be represented, in the WKB approximation, as

$$R\Phi_q = k_q^{-1/2} \exp\left[-\mathrm{i} \int_0^\infty \mathrm{d}R k_q(R) \right] , \tag{19.3}$$

where

$$k_q = \left\{ 2M\left[E_q - V_q^\mathrm{e}(R) \right] \right\}^{1/2} / \hbar \tag{19.4}$$

and M is the reduced mass of the atom pair ($\hbar^2 k^2 / 2M$ is the kinetic energy in Fig. 19.1).

The integrand in the second line of (19.2) then contains $\exp[\mathrm{i} \int \mathrm{d}^3 R (k_\mathrm{u} - k_\mathrm{g})]$. If one multiplies $k_\mathrm{u} - k_\mathrm{g}$ by $(k_\mathrm{u} + k_\mathrm{g})/(k_\mathrm{u} + k_\mathrm{g})$, the exponent becomes

$$\frac{2M(\omega - \omega_\mathrm{c})}{\hbar(k_\mathrm{u} + k_\mathrm{g})} ,$$

and if one further defines a variable $t(R)$ by $\mathrm{d}t = \mathrm{d}R / v(R)$ with $v(R) = \hbar(k_\mathrm{u} + k_\mathrm{g})/2M$, the exponent becomes $\mathrm{i} \int \omega_\mathrm{c}(t')\,\mathrm{d}t' - \mathrm{i}\omega t$, where

$$\omega_\mathrm{c}(R) = [V_\mathrm{u}(R) - V_\mathrm{g}(R)]/\hbar \tag{19.5}$$

and $R(t)$ is the classical trajectory in the average potential. Thus, the Franck–Condon factor reduces to the squared Fourier transform of $\exp[\mathrm{i} \int^t \omega_\mathrm{c}(t')\,\mathrm{d}t']$, where $\hbar\omega_\mathrm{c}(R) = V_\mathrm{u}(R) - V_\mathrm{g}(R)$ (Fig. 19.1). This is the *classical oscillator approximation* (COA), in which the atoms are considered to be a classical oscillator at $\omega_\mathrm{c}(R)$ during the collisional interaction, and $R(t)$ is the classical orbit.

The $\sqrt{k}$ in the denominator of the WKB wave function leads to $1/v(R)$ in the nuclear density at R, again corresponding to classical motion. This COA is used as the starting point in many line shape calculations. It is apparent that the COA breaks down when the initial- and final-state orbits are significantly different, or when the Born–Oppenheimer or WKB approximation is not valid. However, the COA is normally a very good approximation for thermal atomic vapors when $h(\omega - \omega_0) \ll k_\mathrm{B} T$, since most orbits are then straight lines within the R region that yields the observed radiation.

One further conceptual and mathematical simplification is to subtract ω_0 from $\omega_\mathrm{c}(t')$ and from ω by defining $\Delta\omega_\mathrm{c} = \omega_\mathrm{c} - \omega_0$ and $\Delta\omega = \omega - \omega_0$. The Franck–Condon factor in (19.2) then becomes

$$I(\omega) = \left(\frac{4\mu^2 \omega^4}{3c^3} \right) \times \left| \int \mathrm{d}t \exp\left[\mathrm{i} \int^t \Delta\omega_\mathrm{c}(t')\,\mathrm{d}t' - \mathrm{i}\Delta\omega t \right] \right|^2_\mathrm{av} , \tag{19.6}$$

where the prefactors that yield the correct intensity are explained in [19.12, 13] (Sect. 10.6). The average is, in general, over a sequence of collisions leading to an integral over collision velocity and impact parameter, weighted by the rate of occurrence.

In the next section, the character of the broadened line core is obtained by the traditional method of evaluating (19.6) using the impact approximation.

19.1.4 Impact Approximation

Atomic collisions in thermal vapors typically occur in a time interval $\tau_\mathrm{c} < 1$ ps, whereas the time between significant collisions is much longer at vapor pressures below a few atmospheres. Thus, one can consider the atom as radiating its unperturbed frequency ω_0 most of the time, but occasionally suffering a rapid, strong perturbation. If the duration of a collision is τ_c, then $\phi(t) = \int^t \Delta\omega_\mathrm{c}(t')\,\mathrm{d}t'$ in the exponent of (19.6) undergoes a net phase shift $\Delta\theta$ in a time τ_c, and is constant between collisions. If $\Delta\omega \ll 1/\tau_\mathrm{c}$, then one can approximate this phase shift as instantaneous; this is the essential assumption of the *impact approximation*. When the factor $\exp[\mathrm{i}\Delta\theta(b, v)]$, which then occurs in (19.6) is averaged over collision rates with velocity v and impact parameter b this leads to $(\gamma_\mathrm{c} - \mathrm{i}\delta)t$ in place of $\phi(t)$, where γ_c and δ are given by [19.15]

$$\gamma_\mathrm{c} = \int_0^\infty n(v) v\,\mathrm{d}v \int_0^\infty 2\pi b [1 - \cos\Delta\theta(b, v)]\,\mathrm{d}b ,$$
$$\delta = \int_0^\infty n(v) v\,\mathrm{d}v \int_0^\infty 2\pi b \sin\Delta\theta(b, v)\,\mathrm{d}b .$$

Including the spontaneous emission dipole decay rate $\exp(-\Gamma t/2)$, the Fourier integral produces the normalized Lorentzian line shape

$$I(\omega) = \frac{(\Gamma/2+\gamma_c)}{\pi\left[(\Gamma/2+\gamma_c)^2+(\Delta\omega-\delta)^2\right]} \,. \tag{19.7}$$

The full line shape is then a Voigt profile with a Lorentzian-component half width $\Gamma + 2\gamma_c$ and a shift of δ from ω_0. The shift can be understood as the perturbed oscillator frequency advancing relative to ω_0 by the average value of $\sin \Delta\theta$, or alternately as indicating the direction of the frequency shift of $V_u - V_g$. The Lorentz approximation corresponds to taking $1 - \cos \Delta\theta = 1$ or 0, with the former representing a collision. This gives $\gamma = 1/\Delta t_c$, where Δt_c is the average time between strong collisions.

The b for which $\Delta\theta(b, \langle v\rangle) = 1$ is called the Weisskopf radius (R_W). The collisional line width $2\gamma_c$ can be thought of as a collisional rate $n\langle v\rangle Q$, where $Q = \pi R_W^2$. The importance of R_W is that γ_c and δ result primarily from atomic interactions in the region $R \approx R_W$. Since the size of an electronic wave function increases with increasing excitation energy, $|\Delta V(R)| = |V(R) - V(\infty)|$ is normally larger for the upper state of a transition. (High Rydberg states can be an exception.) Since $\Delta\theta = 1$ for $b = R_W$ and $\Delta\theta \approx (\Delta V/\hbar)\tau_c$, then $\Delta V(R_W) \approx \hbar/\tau_c$. For a typical $\tau_c = 1$ ps this implies $\Delta V(R_W)/hc \approx 5\,\mathrm{cm}^{-1}$, which is a relatively weak, long range interaction. Thus, reasonable approximations to the necessary $V(R)$ can often be obtained from atomic perturbation theory. Simple expressions for γ_c and δ are obtained for $V_i - V_f = A/R^n$ with $n > 2$ [19.24]. The case of $n = 3$ corresponds to the resonant interaction between identical atoms, while $n = 6$ corresponds to the van der Waals interaction, which holds at long range for foreign gas interactions. This result for one pair of adiabatic states must be averaged over all of the pairs that separate to each atomic state.

Higher order approximations obtain an asymmetric line, rather than a pure Lorentzian, due to the asymmetry of $\Delta V(R)$ [19.24]. In the line core this asymmetry appears as a multiplicative correction $1 + \Delta\omega/D$, where $|D| \approx 1/\tau_c$ [19.25], and with increasing detuning it increases until at $|\Delta\omega| > 1/\tau_c$ the static wing approaches the quasistatic limit described in Sect. 19.1.7. If $\Delta V(R)$ changes monotonically with decreasing R, only one side of the line has a static contribution and the other, antistatic side falls off exponentially at $|\Delta\omega| > \tau_c$ [19.11]. However, this situation is seldom observed, as more than one difference potential generally contributes and there is usually a static contribution on both sides of the line. Another factor that produces a small divergence from the Voigt profile is the velocity dependence of the shift and width. When combined with the higher velocities of atoms emitting or absorbing in the Doppler wing, this produces an asymmetry in the Doppler wings [19.26, 27].

19.1.5 Examples: Line Core

It is possible to deconvolve a Voigt line shape to separate the Doppler and Lorentzian components, and thereby deduce broadenings of considerably less than the Doppler width ([19.28] and references therein). However, the broadening is most easily observed at perturber densities where the collisional broadening exceeds the Doppler broadening. Such a pressure-dependent line shape is shown in Fig. 19.2, for a range of perturber density n_P such that the broadening exceeds the Doppler width and hyperfine structure, yet the line core is described by the impact theory [19.29]. In Fig. 19.2 the normalized line intensity has been divided by n_P; as the line wings are proportional to n_P they are constant in such a plot, while the line center broadens and shifts with increasing n_P. For this case of fairly heavy atoms, $(2\pi c\tau_c)^{-1} = \Delta k_c$ corresponds to $\approx 0.5\,\mathrm{cm}^{-1}$, and the line becomes asymmetric and non-Lorentzian beyond $\approx 1\,\mathrm{cm}^{-1}$ (Fig. 19.3a); the red wing intensity falls more slowly and the blue wing more rapidly than $(\Delta\omega)^{-2}$. This behavior is typical for most heavy perturbers, and is

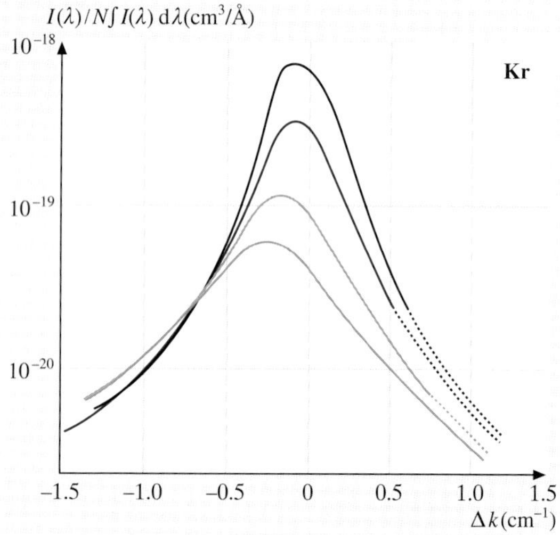

Fig. 19.2 Normalized line shape of the Rb $5P_{3/2} - 5S_{1/2}$ transition broadened by Kr, for Kr densities of 4.5, 9, 18, and $27 \times 10^{18}\,\mathrm{cm}^{-3}$ (*top* to *bottom*). Hyperfine structure and instrumental resolution cause $\approx 0.3\,\mathrm{cm}^{-1}$ of the broadening shown

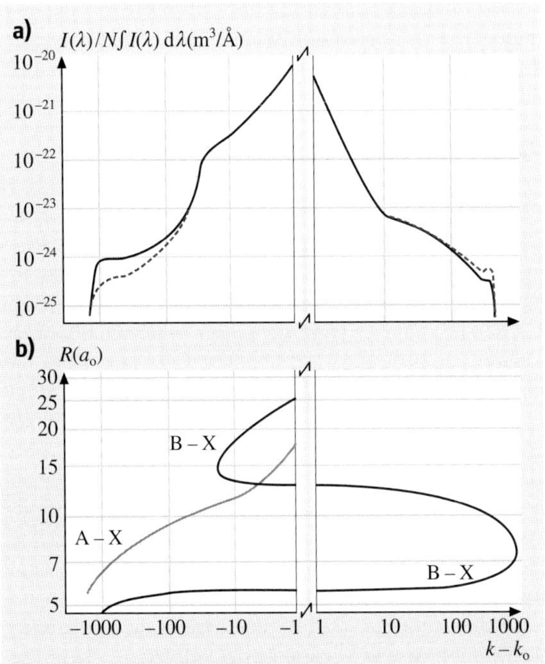

Fig. 19.3a,b Normalized intensity in the wings of the Rb $5P_{3/2}-5S_{1/2}$ transition broadened by Kr, in frequency units of $k=1/\lambda$. The measured spectrum in **(a)** is from [19.29–31]. The *solid line* is at 310 K and the *dashed line* at 540 K. The difference potentials corresponding to the A, B and X states of Rb – Kr, taken from [19.32], are shown in **(b)**

attributed to a long range attractive V_u which dominates $V_u - V_g$.

For the lowest n_P shown in Fig. 19.2, a convolution with the Doppler, hyperfine and instrumental broadenings showed that the line is essentially a symmetric Lorentzian for $|\Delta k| < \Delta k_c$ [19.29]. However, at the highest density, the half-height point is beginning to fall outside of Δk_c; the impact approximation is marginally valid for describing the half width of the line at this density. Most early experiments were done at more than 10 times this density [19.15]; most of the line-core was at $|\Delta k| > \Delta k_c$ and describable by the static theory (Sect. 19.1.7) rather than the impact approximation. The impact approximation was also not valid under these conditions because collisions overlap in time. These very broad lines are well represented by the multiple-perturber, static theories that assume scalarly additive perturber interactions [19.5, 8]. This transition between an impact and quasistatic line core, and to multiple perturber interactions, occurs at lower pressures for transitions to higher states, as the interactions have a longer range. In addition, nearby intensity peaks or satellites often occur, and strongly affect the line as pressure increases. An example calculation, based on an interpretation of measured spectra [19.16], is shown in Fig. 19.4. This shows how a line with a satellite feature progressively broadens and finally blends with the satellite as n_P increases.

With the advent of saturated-absorption (Doppler free) spectroscopy, collisional line broadening can be measured at much lower densities, where $2\gamma_c \ll \Delta\omega_D$. In principle, this can allow measurement of line broadenings and shifts, although with a complication that affects the line shape; the same collisions that produce optical phase shifts also change the atomic velocity. These velocity changes have a minor effect outside the Doppler envelope where high pressure measurements are normally made, but they are quite important in saturated absorption line shapes. This affects primarily the low intensity wings of the line, so it does not prevent measuring the broadening and shift of the nearly Lorentzian core.

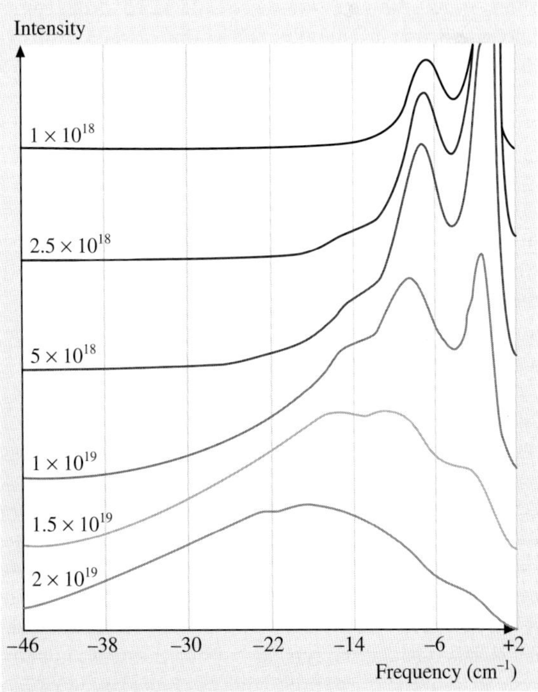

Fig. 19.4 Calculated line shapes of the $Cs(9P_{1/2}-6S_{1/2})$ line broadened by Xe at the densities indicated (from [19.16]). The assumed interaction is based on measured line shapes, but data corresponding to the calculated conditions are not available

Details can be found in [19.33] and references therein. Two-photon absorption yields Doppler free lines that are not affected by velocity changing collisions. This provides the most exacting test of line shapes. These narrow lines are precisely Lorentzian, with a broadening that reflects the upper state interaction since this is usually much stronger than that of the ground state. The technique has been used to measure the broadening of two photon transitions to many excited states [19.34,35].

19.1.6 Δ and γ_c Characteristics

Since 1970, neutral broadening has generally been measured in the $|\Delta\omega| < 1/\tau_c$ region where the impact approximation and (19.7) is valid. Measurements through 1982 are tabulated in [19.16], and through 1992 in [19.36]. More recent measurements are tabulated in the NIST Reference Data bibliography, which is accessible (free) at the web site http://physics.nist.gov/PhysRefData. These involve primarily metal vapor resonance lines broadened by noble gases. For collisions with the heavier (more polarizable) gases, the sizes of these measured broadening rate coefficients generally fall within a factor of 10 range, and approximately fit the prediction of (19.7) with $\Delta V_u(R) - \Delta V_g(R) = C_6 R^{-6}$ [19.37] and C_6 given by a simple effective quantum number formula. This occurs because the potentials are fairly close to van der Waals for $R \geq R_W$ and the broadening is insensitive to details of the potentials at $R < R_W$ since $\cos\Delta\theta$ in (19.7) averages to ≈ 0 for the closer (strong) collisions. It also occurs because the full quantum solution for broadening by a van der Waals interaction, with Zeeman degeneracies, yields nearly the same result as the above single-level theory with an average C_6 [19.38].

For the heavy, more polarizable perturbers, the excited state interactions are attractive and red shifts occur, but the measured shifts have a very poor correlation with the van der Waals prediction. As b decreases and $\Delta\theta(b, v)$ increases, $\sin\Delta\theta$ oscillates and major cancellations occur in the average of $\sin\Delta\theta$ in (19.7). The shift is therefore only a fraction of the broadening and is very sensitive to the interaction throughout the region $R \approx R_W$. This often differs considerably from the van der Waals form, at the typical $\approx 5\,\mathrm{cm}^{-1}$ interaction energy at R_W. The shape of the red wing just beyond $\Delta\omega_c$ also frequently fails to fit that expected for a van der Waals interaction [19.29]. Thus, the often good agreement of γ_c with van der Waals numbers is not a reliable indicator of the actual $V(R)$ in the relevant R region, even for heavy noble gas perturbers.

For He and sometimes Ne perturbers, a repulsive interaction due to charge overlap normally dominates at $R \approx R_W$, causing a blue shift as well as a larger broadening than the van der Waals prediction.

19.1.7 Quasistatic Approximation

The impact approximation is valid for $|\Delta\omega| \ll 1/\tau_c$, where the $1/\tau_c$ is typically $1-10\,\mathrm{cm}^{-1}$. For larger $|\Delta\omega|$ the line shape becomes asymmetric, with higher intensity on the wing corresponding to the long-range $V_u(R) - V_g(R)$. At large detunings where $\Delta\omega \gg 1/\tau_c$ a major simplification occurs. The COA describes the interacting atom pair as an oscillator of frequency $\omega_c(R) = [V_u(R) - V_g(R)]/\hbar$ when at separation R. Since R is time dependent during the classical orbit, ω_c is as well and the Fourier spectrum is broadened relative to the simple distribution of $\omega_c(t)$ that occurs during the orbit. But if the motion is sufficiently slow, the intensity at ω reduces to the probability of finding the atom pair at the appropriate $R(\omega = \omega_c)$. The spectrum then reduces, at low pressure, to the probability distribution of pair separations R, subject to (19.5) between R and ω. This is the binary quasistatic, static, or statistical spectrum, which accurately describes most line wings for $|\Delta\omega| > 1/\tau_c$. When the pressure is large enough to yield a significant probability of one perturber at $R \leq R_c$, multiple-perturber interactions must also be considered as in [19.5].

This intuitive deduction of the statistical spectrum from the COA [19.4] can also be obtained more formally from (19.6) by expanding the exponent about the time during a collision when $\omega_c(t) = \omega$. Alternatively, it follows directly from (19.2) using WKB wave functions to evaluate free–free molecular Franck–Condon factors [19.6, 7]. This result is identical to the classical Franck–Condon principle (CFCP), originally established in the context of bound–bound molecular radiation.

The CFCP yields important insights for all molecular radiation. Again consider (19.2) with the substitution of the WKB wave functions ϕ_q, given below it. Examples of ϕ_u and ϕ_g are given in Fig. 19.1. For large detunings $\omega - \omega_0$, as shown in Fig. 19.1, the integrand on the right side oscillates rapidly everywhere except at the stationary phase point R_c, where $k_u = k_g$. As a consequence, the dominant contribution to the integral occurs at R_c and one can consider the transition to be localized at R_c. Since $k_u(R_c) = k_g(R_c)$, $\mathcal{T}_u(R_c) = \mathcal{T}_g(R_c) = \mathcal{T}_c$ also holds, and as can be seen in Fig. 19.1 it then follows that $\hbar\omega = V_u(R_c) - V_g(R_c)$. Thus, radiation at frequency ω "occurs" when the atoms are near R_c, where the electronic state energies differ by $\hbar\omega$. Note that this holds for all initial kinetic energies and angular momenta, as long as the conditions for validity of the Born–Oppenheimer and WKB approximations hold for the initial and final nuclear motions. This is the CFCP, which is equivalent to the classical-oscillator model for radiation at large detunings from the atomic transition.

Another insight evident from Fig. 19.1 is that the photon energy associated with the frequency difference $\omega - \omega_c$ is supplied by nuclear kinetic energy $\hbar(\omega - \omega_0) = \mathcal{T}_i - \mathcal{T}_f$. This transformation of nuclear into electronic energy takes place as the nuclei move from large R to R_c on one $V(R)$ and back to large R on the other.

If an absorbing or emitting atom interacts as $V_i(R)$ with a density n_P of perturbers in a vapor of temperature T, the probability of a perturber at separation $R \to R + \mathrm{d}R$ is $n_P 4\pi R^2 \exp[-V_i(R)/k_B T]\,\mathrm{d}R$ if the interatomic motion is in equilibrium. Inverting (19.5) for $R(\omega)$ yields $\mathrm{d}R = \mathrm{d}\omega/(\mathrm{d}\omega/\mathrm{d}R)$, and this pair of relations yields the (single perturber) quasistatic (QS) spectrum

$$I(\omega) = N n_P \Gamma 4\pi R(\omega)^2 \times \exp[-V_i(\omega)/k_B T]/[\mathrm{d}\omega(R)/\mathrm{d}R]\,, \quad (19.8)$$

where N is the radiator density and $I(\omega)$ the radiation per unit volume and frequency interval.

Figure 19.3a gives an example of far wing emission line shapes versus photon energy in units of cm^{-1}, for the Kr broadened Rb D2 line for which $\Delta k_c \approx 0.7\,\mathrm{cm}^{-1}$. These data are normalized by dividing by perturber density, so they are independent of perturber density for the density region of the experiment. The excited state produces two $V_u(R)$, called the A and B states, while the ground state produces one $V_g(R)$, called the X state. Each of these potentials has a single minimum at long range and is strongly repulsive at close range, but the well depths and positions are very different [19.32]. This causes the complex forms of $V_u(R) - V_g(R)$ that are shown in Fig. 19.3b. There I have plotted $\ln(R)$ vertically and $\ln(\Delta V(R)/hc)$ horizontally, where ΔV refers to $V_A - V_X$ and $V_B - V_X$. The right side of (19.8) can also be written as the exponential and constant factors times $\mathrm{d}R^3(\omega)/\mathrm{d}\omega$. Since $\ln[R^3(\omega)] \propto \ln[R(\omega)]$, the static spectrum at $\Delta k = \Delta\omega/2\pi c$ is proportional to the slopes of the curves in Fig. 19.3b, divided by $|\Delta k|$ due to the $\ln(\Delta k)$ horizontal axis. One can see qualitatively that the overall spectrum follows such a relation to the lines in Fig. 19.3b; in fact in most spectral regions this relation is quantitatively accurate.

The temperature dependence in Fig. 19.3a corresponds to the exponential factors in (19.8) [19.30]. At large R, both $\Delta V(R)$ are attractive, and this causes a large intensity on the negative Δk (red) wing. However, once $|\Delta k|$ exceeds $\approx 20\,\mathrm{cm}^{-1}$, where $V_B - V_X$ reverses direction, the red wing intensity drops rapidly. This extremum in $\Delta V(R)$ causes a satellite at $\approx -20\,\mathrm{cm}^{-1}$, although it is spread out by the finite collision speed and does not cause a distinct peak in the spectrum. Satellite features are discussed in more detail in the next paragraph. The antistatic blue wing drops rapidly for several decades, then suddenly flattens beyond $\approx 10\,\mathrm{cm}^{-1}$ due to the positive portion of $V_B - V_X$ at small R. The remaining blue wing is the B–X band, and has a satellite at $\approx 350\,\mathrm{cm}^{-1}$ as $V_B - V_X$ passes through another extremum. The theory predicts this at $800\,\mathrm{cm}^{-1}$, but clearly represents all the basic aspects correctly. This satellite is also spread out by finite collision speed, but a definite intensity peak remains. The red wing beyond $\approx 50\,\mathrm{cm}^{-1}$ is the A–X band. The feature near $-1000\,\mathrm{cm}^{-1}$ is due to the exponential factor in (19.8), not an extremum in $\Delta V(R)$; the feature essentially disappears if the normalized intensity is extrapolated to infinite temperature.

19.1.8 Satellites

In regions of the wing where the intensity falls slowly with increasing frequency, motional broadening of the static spectrum is not noticeable and the static spectrum is a good approximation. However, if $\Delta V(R)$, or equivalently $\omega_c(R)$, has an extremum at some R_S, the denominator of (19.8) is zero at $\omega(R_S) = \omega_S$. This produces a local maximum, or satellite, in the far wing intensity, as seen in Fig. 19.3a at $350\,\mathrm{cm}^{-1}$. If one expands $\omega(R)$ in a Taylor series about $R = R_S$ this produces in (19.8) a square root divergence of finite area, with no intensity beyond ω_S. The area under this feature is meaningful, but not its shape; the quasistatic assumption is clearly not valid for such sharp features. The more accurate satellite shape is obtained by returning to (19.2) and expanding $V_u(R)$, $V_g(R)$ and the WKB wave functions about R_S, or using (19.6) with $\omega_c(t)$ expanded about $t(R_S)$. *Sando* and *Wormhoudt* used the former method to obtain a universal satellite shape [19.39]. *Szudy* and *Baylis* improved the expansion to yield a smooth transition to the quasistatic spectrum at smaller detunings [19.37]. This result is nearly the same as Sando et al. in the spectral region of the satellite, but it more accurately connects to the adjacent static line wing. Intensity undulations between the satellite and the line occur in this calculation; these arise from alternating constructive and destructive interference between two contributions to the same frequency from $R > R_S$ and $R < R_S$. This can not be seen in the low resolution of Fig. 19.3a, but such undulations are seen near the $350\,cm^{-1}$ satellite [19.31]. At antistatic detunings beyond ω_S, which are not quasistatically allowed, the calculated intensities decay exponentially. This is also observed experimentally [19.31] and is the same behavior predicted for the antistatic wing of a line [19.11].

At higher perturber densities and closer to the line core, corresponding to larger R interactions, the multiple perturber probability distribution must be included. If the interactions are additive, this leads to a secondary satellite at twice the detuning of the low pressure satellite, as seen in Fig. 19.4.

The wings of a collisionally broadened atomic line are molecular radiation. In the context of molecular bound state spectroscopy, a satellite is a "head of band heads," corresponding to a frequency region where bound–bound band heads congregate. This occurs, of course, at the classical satellite frequency and when $V_u(R) - V_g(R)$ has an extremum [19.40].

An extremum in $\Delta V(R)$ is the most common cause of satellites, but similar looking features can occur for other reasons. Forbidden bands often appear in the wings of forbidden atomic transitions, due to an increase in the transition dipole moment $\mu(R)$ resulting from the collisional interaction. These are described, in the QS approximation, by (19.8) with Γ replaced by $\Gamma(R)$. If $\Gamma(R)$ increases rapidly with decreasing R, the intensity increases as ω moves into the far wing until the dR^3 and exponential factors cause a net decrease at small R. This leads to forbidden bands far from the atomic frequency, such as those in [19.41]. In some cases, a collision-induced feature also appears at the frequency of the forbidden transition. The shapes of such features, which also include radiative collisions, in which both atoms change state, are calculated and reviewed in [19.42].

A variety of related line shape phenomena has been investigated, including the relation between absorbed and emitted wavelengths (spectral redistribution), the dependence of fluorescence polarization on absorbed wavelength (polarization redistribution), and high power effects. Some references regarding these phenomena are [19.43–47].

19.1.9 Bound States and Other Quantum Effects

The validity of the QS spectrum requires the validity of the WKB approximation in the initial and final state, but it is not restricted to free–free molecular transitions. In fact, the equilibrium probability distribution in (19.8) must include bound states in an attractive $V_i(R)$. The QS spectrum describes the *average* behavior of bound–free and bound–bound molecular bands, as well as the free–free radiation implied by the above method of derivation. The quantum character is expressed in the discrete bound–bound lines that make up this average, and in *Condon oscillations*, where the intensity oscillates about the average $I_{QS}(\omega)$. The latter occur as oscillations in Franck–Condon factors in the bound–bound case, and as smooth oscillations in bound–free spectra and low resolution bound–bound spectra. An additional quantum feature occurs in regions of the spectrum dominated by classical turning points, usually at the far edge of a line wing where the intensity is dropping rapidly. There, quantum tunneling past the edge of the classically allowed region spreads the spectrum. Yet another is the energy $h\omega_0/2$ of the ground vibrational state, which effectively adds to $k_B T$ in (19.8) for attractive V_i. All of these quantum features become more pronounced as the reduced mass decreases; examples and details can be found in [19.40, 48–50].

19.1.10 Einstein *A* and *B* Coefficients

The relationship between absorption coefficient $B_{12}(\omega)$, stimulated emission coefficient $B_{21}(\omega)$ and spontaneous emission coefficient $A_{21}(\omega)$ are given by the Einstein relations; $A_{21}/B_{21} = 8\pi h \lambda^{-3}$ and $B_{21}/B_{12} = g_1/g_2$. These relations are most familiar for atomic lines, but if they are referred to the density of absorbers $dN_g/d\omega$ and emitters $dN_u/d\omega$ that emit or absorb at ω, then they also apply to the wings of lines, i. e.,

$$k(\omega) = B_{12}(\omega)\left(\frac{2\pi h}{\lambda}\right)\left(\frac{g_u}{g_g}\right)\frac{dN_g}{d\omega} = \frac{1}{4}\lambda^2 A_{21}(\omega)\left(\frac{g_u}{g_g}\right)\frac{dN_g}{d\omega}\,, \tag{19.9}$$

$$g(\omega) = \frac{1}{4}\lambda^2 A_{21}(\omega)\frac{dN_u}{d\omega}\,, \tag{19.10}$$

$$I(\omega) = \hbar\omega A_{21}(\omega)\frac{dN_u}{d\omega}\,. \tag{19.11}$$

Here $k(\omega)$ is the absorption coefficient due to lower state atoms, $g(\omega)$ is the stimulated emission coefficient and $I(\omega)$ the spontaneous emission due to excited state population, and g_u and g_g are the statistical weights of the atomic states. For absorbing atoms of density N_g and perturber density n_P, the QS approximation with equilibrated internuclear motion sets

$$\frac{dN_g}{d\omega} = N_g n_P 4\pi R^2 \frac{dR}{d\omega} \exp\left(-\frac{V_g}{k_B T}\right) , \quad (19.12)$$

and equivalently for a radiating atom density N_u with perturber interaction V_u. Normally most of the radiation, and $dN_g/d\omega$, is concentrated at the atomic line, so integrating over $d\omega$ near the line leads to the relations

$$\int_0^\infty k(\omega)\,d\omega = \frac{1}{2}\lambda^2 A_{21} N_g \left(\frac{g_u}{g_g}\right) , \quad (19.13)$$

$$\int_0^\infty I(\omega)\,d\omega = A_{21} N_u \left(\frac{h\omega}{2\pi}\right) , \text{ etc.} \quad (19.14)$$

Note that

$$\frac{g(\omega)/N_u}{k(\omega)/N_g} \propto \exp\left(-\frac{\hbar\omega}{k_B T}\right) ; \quad (19.15)$$

if N_u/N_g is also in equilibrium at T, this yields the correct equilibrium relation between $k(\omega)$, $g(\omega)$, $I(\omega)$, and a black body spectrum. While these relations are much more general than the QS theory, the latter provides a helpful conceptual basis. The above expressions in terms of spontaneous emission thus cover all cases.

19.2 Radiation Trapping

Atoms and ions efficiently absorb their own resonance radiation, and their emission can be reabsorbed before escaping a vapor. Molecules are less efficient absorbers, since each electronic transition branches into multiple-line bands, but interesting effects result if such reabsorption occurs. This emission and reabsorption process is fundamental to the formation of stellar lines, where it is called *radiation transfer*, and to confined vapors and plasmas where it is also called *radiation diffusion* or *trapping*. Fraunhofer's observation of dark lines in the stellar spectrum result from this radiation transfer process. Highly sophisticated treatments of line formation in inhomogeneous and nonequilibrium plasmas containing many species [19.19, 20] also apply to laboratory plasmas, but the simplifications inherent in a one- or two-element, confined plasma with cylindrical or planar symmetry leads to easier treatments. This sections discusses only a uniform density and temperature, confined atomic vapor.

The flourescent lamp in which 254 nm mercury radiation diffuses to the walls and excites a phosphor, provides a prime example of radiation trapping. Its improvement motivated the seminal *Biberman* [19.51, 52] and *Holstein* [19.53, 54] theories, continuing through modern theory and experiment that is particularly relevant to electrodeless and compact lamps. Dense clouds of cold, trapped atoms are also influenced by radiation trapping. Reference [19.55] provides and excellent overview of this topic, which we will not discuss here. The effect of radiation trapping on the *polarization* of flourescent radiation played a major role in developing a correct understanding of the coherent response of atoms to radiation. This is reviewed in [19.44], and will not be covered here. *Molisch* and *Oehry* [19.56] have provided a detailed discussion of research on radiation transport up to 1998.

19.2.1 Holstein–Biberman Theory

An atom in a dense vapor may be excited by externally applied radiation plus the fluorescence from other excited atoms within the vapor, and it will decay by spontaneous emission (neglecting stimulated emission). This is expressed by the Holstein–Biberman equation

$$dn(\boldsymbol{r},t)/dt = S(\boldsymbol{r},t) + \gamma \int_{\text{vol}} K(\boldsymbol{r}-\boldsymbol{r}')n(\boldsymbol{r}',t)\,d^3\boldsymbol{r}' - \gamma n(\boldsymbol{r},t) , \quad (19.16)$$

where $n(\boldsymbol{r},t)$ is the excited state density at position $\boldsymbol{r}$, $S(\boldsymbol{r},t)$ is the excitation rate due to externally applied radiation, γ is the spontaneous emission rate, the kernel $K(\boldsymbol{r}-\boldsymbol{r}')$ is the probability of a reabsorption at $\boldsymbol{r}$ due to fluorescence by an atom at $\boldsymbol{r}'$, and the integral is over the vapor filled volume [19.51–54]. Since $K(\boldsymbol{r},\boldsymbol{r}')$ is assumed the same for all excited atoms, this contains an implicit assumption that all atoms emit the same fully redistributed spectrum. The solution of this linear integral equation, subject to boundary values at the vapor boundary, can be expressed as a sum over an orthogonal set of solutions $n(\boldsymbol{r},t)_i = n(\boldsymbol{r})_i \exp(-g_i\gamma t)$ of the homogeneous equation

$$n(\boldsymbol{r},t) = \sum_{i=1}^{\infty} a(t)_i n(\boldsymbol{r})_i , \quad (19.17)$$

where, if $S(\boldsymbol{r},t) = S(\boldsymbol{r})f(t)$, then $a(t)_i = \bar{a}_i \int_{-\infty}^{t} f(t') \times \exp[-g_i\gamma(t-t')]\,\mathrm{d}t'$ and $\bar{a}_i = \int S(\boldsymbol{r})n(\boldsymbol{r})_i\,\mathrm{d}^3\boldsymbol{r}$. Here $n(\boldsymbol{r},t)_i$ is the ith normal mode and $g_i\gamma$ is the decay rate of this mode, as it would decay without change in its shape $n(\boldsymbol{r})_i$ from a pulse of excitation.

Two shapes of vapor regions have been studied in detail: an infinitely long cylinder of radius R and the region between two infinite parallel plates with separation L. The first three symmetric modes of the latter slab geometry are shown with unit height in Fig. 19.5. A spatial integration over the normalized $i = 1$ or fundamental mode yields 1 and all others integrate to zero, so $a(t)_1$ equals the total excited state population. g_1 is the escape probability; i. e., the probability of photon escape averaged over the fundamental mode distribution of emitters $n(\boldsymbol{r})_1$. Since $n(\boldsymbol{r},t)$ must be everywhere positive, the negative contributions of the higher order modes only reduce the density in some regions. The g_i can vary from 0 to 1 and increase with increasing i, so that higher order modes die out faster after pulsed excitation. The ratios of decay rates is $g_i : g_3 : g_5 = 1 : 3.7 : 6.4$ for the symmetric slab modes shown in Fig. 19.5. For steady state excitation, (19.17) yields $a(t)_i = \bar{a}_i/g_i\gamma$, so the lower order modes are more heavily weighted because they decay more slowly. The fundamental mode decay rate $g_1\gamma$ is of primary interest in most situations, and we will now discuss its properties.

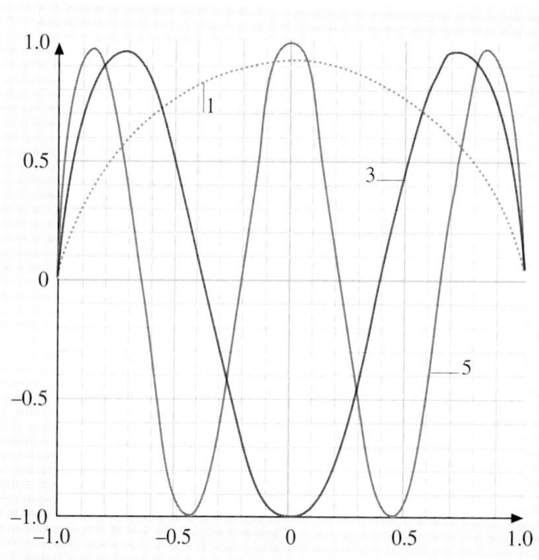

Fig. 19.5 The first three symmetric eigenfunctions ($j = 0.2, 4$) of radiation trapping between slab windows, for a Doppler line profile, from [19.57–59]. The windows are at ± 1

The kernel $K(x)$ is the probability of fluorescence transport over a distance x followed by reabsorption, averaged over the emitted frequency distribution. It is conceptually useful to express it in terms of the spectrally averaged transmission $T(x)$

$$K(x) = \frac{1}{4\pi x^2}\frac{\mathrm{d}T(x)}{\mathrm{d}x}\,, \tag{19.18}$$

$$T(x) = \int_0^\infty \mathcal{L}(\omega)\exp[-k(\omega)x]\,\mathrm{d}\omega\,,$$

where $\mathcal{L}(\omega)$ is the emission line shape normalized to unit area, and $x = |\boldsymbol{r} - \boldsymbol{r}'|$. If one assumes that the fluorescence frequency of an atom does not depend on the frequency it absorbed (i. e., complete spectral redistribution), this leads to $k(\omega) = \kappa\mathcal{L}(\omega)$, where $\kappa = (\lambda^2/8\pi)(g_u/g_g)n\Gamma$ and g_u and g_g are statistical weights. This simplification applies under most conditions and will be used here; its range of validity and more accurate treatments are discussed below.

The transmission factor $\mathcal{L}(\omega)$ and the integrand of (19.18) are shown in Fig. 19.6, for a Gaussian line shape and several values of k_0x, where k_0 is the line center absorption coefficient. At small k_0x, the transmitted spectrum is similar to $\mathcal{L}(\omega)$; for these conditions $T(x) \simeq \exp(-k_{av}x)$, where $k_{av} \simeq 0.7k_0$ is the average attenuation. For $k_0x > 5$, the transmission is small except at the edges of the line. The transmitted radiation is then predominately in the ω region near ω_1, defined by $k_0x\mathcal{L}(\omega_1) = 1$. Since the integrand is sharply peaked near ω_1, this leads to simple analytic forms for $T(x)$. In

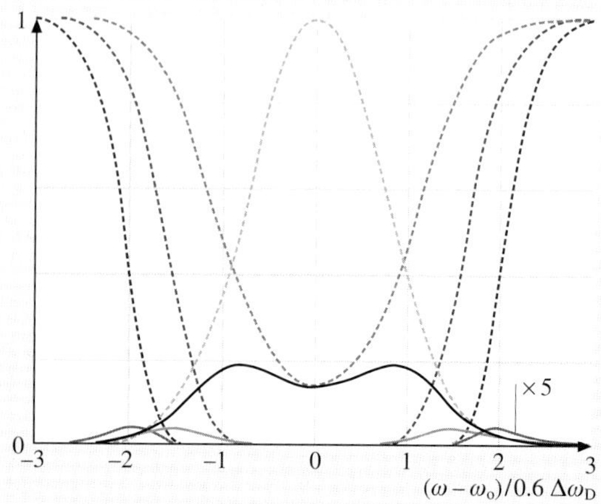

Fig. 19.6 Gaussian emission spectrum $\mathcal{L}(\omega)$ (*short-dash line*), transmissions $T(\omega)$ (*long-dash lines*), and transmitted intensities (*solid lines*) for $k_0x = 2$, 10, and 50

the large $k_0 x$ limit, $T(x) \simeq [k_0 x(\pi \ln k_0 x)^{1/2}]^{-1}$ in the Gaussian case, and $T(x) \simeq (\pi k_0 x)^{-1/2}$ for a Lorentzian line shape. These asymptotic forms of $T(x)$ are compared with the exact $T(x)$ in Fig. 19.7; $T(x)$ follows the asymptotic formulas for $k_0 x > 5$ and 10, respectively. $T(x)$ for several Voigt line shapes is also shown in Fig. 19.7; these follow the Gaussian $T(x)$ at smaller $k_0 x$, then rise above as ω_1 moves into the Lorentzian wing.

For radiative escape from a cell, transmission over distances near the cell dimension (R or L) is most important, since transport over this distance often escapes the vapor and transport over much smaller distances does not have much effect. The escape probability g_1, averaged over the fundamental mode distribution, is close to $T(R)$ or $T(L/2)$, while the higher order modes are related to the same asymptotic forms of $T(x)$ at smaller distances. Thus, in the large $k_0 L$ slab case,

$$g_i = \begin{cases} \dfrac{G_i}{k_0 L(\frac{1}{2}\ln k_0 L)^{1/2}} & \text{Gaussian line} \quad (19.19\text{a}) \\ \dfrac{G_1'}{(k_0 L)^{1/2}} & \text{Lorentzian line,} \quad (19.19\text{b}) \end{cases}$$

with $G_1 = 1.03$ and $G_1' \simeq 0.65$. For an infinite cylinder, the same equations hold with $L/2 \to R$ and slightly larger G_i values. Exact G_i and G_1' values can be found in [19.57–59].

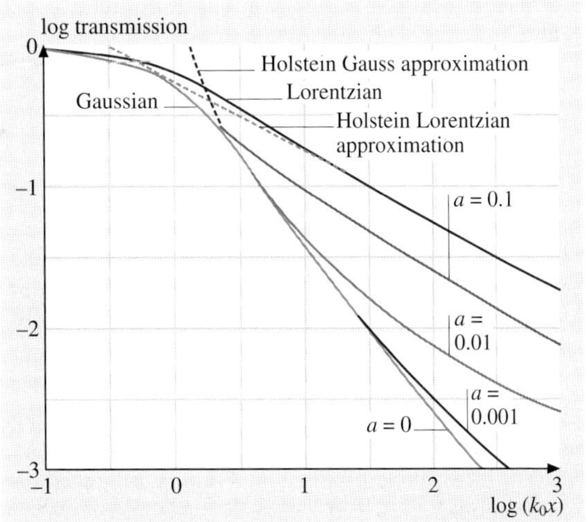

Fig. 19.7 Transmission $T(x)$ versus distance in units of $k_0 x$, for Voigt line shapes with the a parameters indicated, where $a = (\ln 2)^{1/2}\Delta\omega_{\text{Lor}}/\Delta\omega_{\text{Gauss}}$. The Gaussian limit corresponds to $a = 0$ and the Lorentzian limit to $a = 1$. The Holstein, large $k_0 x$, approximations are also indicated

19.2.2 Additional Factors

As noted above, the line shape of a two-level atom in a thermal vapor is a Voigt shape; a convolution of a Lorentzian of width $\Gamma + 2\gamma_c$ with a Gaussian of width $\Delta\omega_D$. In most cases, $\Delta\omega_D \gg \Gamma$, so in the absence of a buffer gas the line shape is nearly Gaussian at low density (n). As a result, ω_1 is in the Gaussian region of the line at low density and g_1 behaves similarly to the Gaussian transmission in Fig. 19.7 with x replaced by the confinement dimension. k_0 is proportional to n, so from (19.19a) g_1 is approximately inversely proportional to n for $k_0 L > 5$. As n increases, ω_1 moves further into the wing of the line, and when ω_1 reaches the Lorentzian tail of the Voigt line profile a transition to (19.19b) occurs, where k_0 corresponds to a purely Lorentzian line. (That the core of the line does not have a Lorentzian shape does not matter, since the fraction of emission well into the Lorentzian wing is nearly the same as that of a pure Lorentzian line.)

In the absence of a collision, a two level atom reradiates in its rest frame the same frequency it absorbed. Thermal motion redistributes this coherent scattering frequency within the Doppler envelope when the emission and absorption are in different directions, but it does not transfer it into the natural Lorentzian wing outside the Doppler envelope. This leads to the property that an atomic vapor will scatter frequencies in the natural wing, but will not emit in this wing unless it absorbed there or is excited by or during a collision. With line broadening collisions, a fraction $\Gamma/(\Gamma + 2\gamma_c)$ of optical attenuation is coherently scattered and a fraction $2\gamma_c/(\Gamma + 2\gamma_c)$ is redistributed into "incoherent" emission with a Lorentzian spectrum of width $\Gamma + 2\gamma_c$ centered at $\omega_0 + \delta$ in the reference frame of the moving atom. This redistributed emission can escape in the Lorentzian wing of the Voigt line. In this radiation transport problem, the consequence is that (19.19b) with $k_0 = n(\lambda^2/2\pi)(\Gamma/\gamma_c)(g_u/g_g)$, corresponding to a Lorentzian with $\Gamma_{\text{Total}} = 2\gamma_c$ not $\Gamma + 2\gamma_c$, provides the best approximation to g_1 in the density region where ω_1 is in the Lorentzian wing of the line. Since $k_0 \propto n/\gamma_c$ and in the absence of a buffer gas $\gamma_c = k_c n$, where k_c is the rate coefficient for self broadening collisions, g_1 becomes independent of n. In fact, $k_c \propto \Gamma$ as well, so g_1 is also independent of Γ. For the case of a $J = 0$ ground state and a $J = 1$ excited state, $g_1 = 0.21(\lambda/L)^{1/2}$; the broadening coefficient for other cases can be found in [19.60]. If the broadening is due to a buffer gas, $\gamma_c = k_c n_B$ in (19.19b) yields

$$g_1 \propto \left(\frac{n_B}{n}\right)^{1/2} ; \qquad (19.20)$$

this has been studied in [19.64].

Post et al. have numerically evaluated g_1 for all values of $k_0 L$ for slab and cylinder geometries, by integrating the radiative escape probability $g(z)$ over the fundamental mode distribution $N(z)$, where z is the position between the windows [19.65]. To obtain $g(z)$ they integrate over the angular distribution of the emission, using $T(x)$ from the exact line shape. Thus all features of the calculation correspond to the Holstein–Biberman theory for an isolated line without approximation. As will now be discussed real atomic vapors are generally not that simple.

Many atomic "lines" have multiple components due to hyperfine structure and isotope shifts; some components are isolated while others are separated by less than a Doppler line width and overlap. The absorption line shape then becomes a weighted sum over components, each with an equivalent Voigt shape. In a high density vapor or a plasma, collisions will usually distribute the excited state population between the isotopes and hyperfine states in proportion to their isotopic fraction and statistical weight. The emission line shape $\mathcal{L}(\omega)$ is then a similarly weighted distribution over components. Since radiation only escapes in the wings of a line component at high $k_0 L$, overlapping components act almost as a single component. If the line has M isolated components, the right-hand side of (19.19a) and (19.19b) become sums over the fraction f_j of the intensity in the j component times the escape probability for that component. The latter is obtained, for large $k_0 L$, by replacing k_0 with $k_0 f_j$ in (19.19a) and (19.19b). The net result, after summing over components, is an increase in g_i by a factor of $\approx M$ in the Gaussian case and $\approx M^{1/2}$ in the Lorentzian. This approximation was obtained by Holstein in the context of the Hg 254 nm radiation under conditions appropriate to the fluorescent lamp [19.66]. Walsh made a more detailed study of these overlapping components [19.67], and the dependence of g_1 on the ratio of line separation to Doppler width is also given in [19.63].

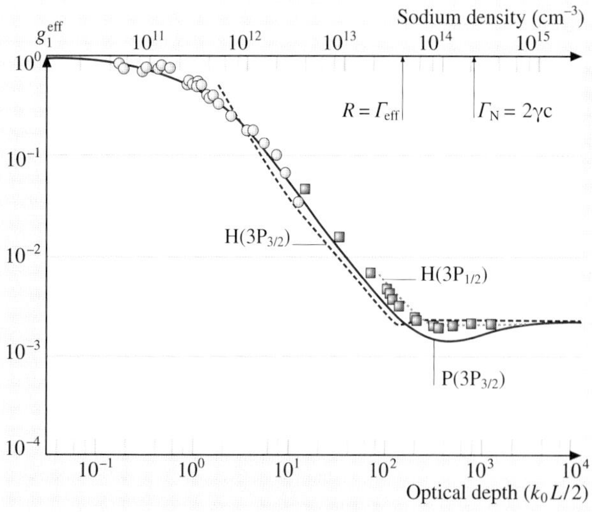

Fig. 19.8 Radiative escape probability g_1 for Na vapor excited to the $3P_{3/2}$ state, for a slab geometry. The Holstein approximation for the $3P_{3/2} - 3S_{1/2}$ (D2) line and the $3P_{1/2} - 3S_{1/2}$ (D1) line are indicated as *dashed lines*. The Post-type calculation of [19.61] for the D2 line is indicated as a *solid line*. *Solid squares* are data from [19.62], and *open circles* are data from [19.63]. The effective escape probability corresponds to the D2 line rate at low densities but a combination of D1 and D2 at high densities

19.2.3 Measurements

The overall behavior of g_1 versus n is shown in Fig. 19.8 for the $Na(3P_{3/2})$ or D2 resonance line in pure Na vapor [19.62, 63]. In this type of experiment the fundamental mode decay rate is established by a combination of optimally exciting that spatial mode and of waiting until the fluorescence decay is exponential in time after termination of the excitation. A transition to approximately $1/n$ behavior, corresponding to (19.19a), is seen to occur at $k_0 L/2 \approx 5$. At $k_0 L/2 \approx 100$ the transition to n^0 behavior, corresponding to a self-broadened Lorentzian line in (19.19b), can be seen. The behavior at $k_0 L < 5$ fits the Milne diffusion theory [19.68] as well as the Post et al. theory shown as a solid line; this is also similar to $T(L/2)$, as seen in Fig. 19.7. For $5 < k_0 L/2 < 100$, the behavior is similar to (19.19a) (dashed line), but the Post et al. theory (solid line) is $\approx 20\%$ higher due to the inclusion of the Na hyperfine structure (hfs splitting $\simeq$ Doppler width). For $k_0 L/2 > 1000$, the Post theory converges to the Holstein–Lorentzian-line result with $\Gamma_{Total} = 2\gamma_c$.

The experiment is complicated in the $50 < k_0 L/2 < 500$ region by fine structure mixing [19.62]. The $3P_{3/2}$ state was excited, but at high densities, collisions populate the $3P_{1/2}$ state, which has a smaller g_1 than the $3P_{3/2}$ state (Fig. 19.8). At low densities, $g_1^{eff} = g_1(3P_{3/2})$, and at high densities these states are statistically populated and $g_1^{\mathrm{eff}} = \frac{1}{3} g_1(3\mathrm{P}_{1/2}) + \frac{2}{3} g_1(3\mathrm{P}_{3/2})$. The transition density where the fine structure mixing rate R equals Γ_{eff} is indicated in Fig. 19.8. The theory is also complicated in this intermediate $k_0 L$ region by the necessity of including incomplete frequency redistribution [19.65]; this leads to the dip in g_1 near $k_0 L \approx 500$. While the overall behavior of the data in Fig. 19.8 is consistent with the Post et al. theory, there is $\approx 30\%$ systematic discrepancy at $k_0 L/2 = 10-100$ and the dip near 500 is not seen. Part of this difference probably results from the experimental geometry, which was between a slab and a cylinder of radius $R = L/2$; g_1 for the cylinder is 17% larger than the slab value used in Fig. 19.8.

The fundamental mode decay rate has also been measured for the Hg 254 nm [19.69] and 185 nm [19.65] lines, for the Ne resonance line [19.70] and for the Ar resonance line [19.71]. The Hg measurements are complicated by multiple isotopes and hyperfine structure, producing a mixture of partially overlapping and isolated lines combined with density-dependent uncertainties in excited state populations of the various isotopes. Serious efforts to model and measure these effects have been made [19.65, 67, 69, 72]. The Ne and Ar measurements have similar complications, as will now be discussed.

In essence, g_1 behaves like the Gaussian $T(x = L/2)$ in Fig. 19.7 until n is large enough for ω_1 to approach the collision induced Lorentzian wing of the Voigt line. g_1 then decreases more slowly since the line wing does not fall off as rapidly as a Gaussian. With continued increase in n, ω_1 moves further into the Lorentzian wing, a broader spectral region escapes and g_1 reaches a minimum. Finally, when the entire escaping spectral region is Lorentzian, g_1 reaches the constant value described above. Independent and detailed treatments of this density region, including incomplete frequency redistribution, predict a dip in g_1 as seen in Fig. 19.8 [19.65, 71, 73, 74]. However, this has not been clearly confirmed experimentally. In Fig. 19.8 this dip occurs where fine structure mixing also occurs, and in addition the data are higher than the calculations throughout this n region. *Post* et al. [19.65] did observe such a dip for the Hg (149 nm) resonance line, but the data do not fit the calculation at other densities; hyperfine and isotopic structure within the line cause major complications. This long-standing issue has finally been clarified by *Menningen* and *Lawler* [19.75], who measured the decay of the Hg (185 nm) resonance line following laser excitation. They observed a clear dip in g_1 due to incomplete redistribution. They also carried out sophisticated Monte Carlo simulations, obtaining g_1 values that compared favorably with the measurements. By extending the simulations over a large range of a parameter space, they constructed an analytic formula for g_1 of a single-component line in cylindrical geometry [19.76]. This formula includes effects of incomplete frequency redistribution and varying ratios of Doppler broadening, radiative broadening and collisional broadening, so that it can be applied to any resonance line. *Payne* et al. [19.71] did not observe the predicted dip for the Ar resonance line; again a minor isotope with an isolated line occurs and could be very important at these high optical depths. *Phelps* [19.70] reported such a dip for the Ne 74.3 nm resonance line, but with rather large uncertainties due to the necessity of correcting for other collisional effects. Again there are isotopes with isolated lines that may have effected the data. Thus, experiments have verified the essential aspects of the above theories, but quantitative agreement in all aspects has not yet been achieved.

The fact that the escaping radiation is concentrated in the wings of the line, near the unity optical depth point ω_1, is reflected in the emitted spectrum. Calculated examples are shown in [19.74]; the Gaussian case looks somewhat like the transmitted spectra in Fig. 19.6 for $x \approx L/4$. These spectra, and all results described so far, are calculated assuming no motion of the atoms. This is appropriate in the central region of the vapor, because the distance moved in an excited state lifetime ($L_v = v/\Gamma$) is much smaller than L. In fact, resonant collisions between excited and ground state atoms further limits the distance an excited atom moves in one direction before transfer of excitation. However, near the window or wall of the container, atomic motion will cause wall collisions of excited atoms and loss of radiation. This loss will be primarily within the Doppler core of the line, since these frequencies can only escape if emitted near the vapor edge. This loss depends on the excited state density in the neighborhood of the wall, and can be significant if $L_v > 1/k_0$. The excited atom density near the wall must be self consistent with the radiation transport and wall quenching. This situation has been modeled and studied experimentally ([19.77] and references therein).

Additional aspects of radiation trapping, such as higher-order spatial modes and non-uniform absorber distributions, can be significant in lighting plasmas (and trapped atom clouds). Propagator function techniques have been developed for modeling radiation transport when the excitation has unusual temporal or spatial character [19.78, 79]. Non-uniform absorber spatial distributions can be particularly important at high power densities, and have been considered in [19.80].

References

19.1 H. A. Lorentz: Proc. Akad. Wet. (Amsterdam) **8**, 591 (1906)
19.2 V. Weisskopf: Phys. Z. **34**, 1 (1933)
19.3 J. Holtsmark: Ann. Phys. (Leipzig) **58**, 577 (1919)
19.4 H. G. Kuhn: Philos. Mag. **18**, 987 (1934)
19.5 H. Margenau, W. W. Watson: Rev. Mod. Phys. **8**, 22 (1936)
19.6 A. Jablonski: Z. Physik **70**, 723 (1931)
19.7 A. Jablonski: Acta Phys. Polon. **27**, 49 (1965)
19.8 E. Lindholm: Ark. Fys. A **32**, 1 (1945)
19.9 H. M. Foley: Phys. Rev. **69**, 616 (1946)
19.10 H. M. Foley: Phys. Rev. **73**, 259 (1948)
19.11 T. Holstein: Phys. Rev. **79**, 744 (1950)
19.12 P. W. Anderson: Phys. Rev. **76**, 647 (1949)
19.13 P. W. Anderson: Phys. Rev. **86**, 809 (1952)
19.14 M. Baranger: Phys. Rev. **112**, 855 (1958)
19.15 S-Y. Chen, M. Takeo: Rev. Mod. Phys. **29**, 20 (1957)
19.16 N. Allard, J. Kielkopf: Rev. Mod. Phys. **54**, 1103 (1982)
19.17 H. Griem: *Plasma Spectroscopy* (McGraw Hill, New York 1964)
19.18 H. Griem: *Spectral Line Broadening by Plasmas* (Academic, New York 1974)
19.19 J. Jeffries: *Spectral Line Formation* (Blaisdell, Waltham 1968)
19.20 D. Mihalas: *Stellar Atmospheres* (Freeman, San Francisco 1970)
19.21 J. van Kranendonk: Cnd. J. Phys. **46**, 1173 (1968)
19.22 J. Cooper: Rep. Prog. Phys. **29**, 35 (1966)
19.23 J. Simons: *Energetic Principles of Chemical Reactions* (Jones Bartlett, Boston 1983)
19.24 J. Szudy, W. E. Baylis: J. Quant. Spectrosc. Radiat. Trans. **17**, 681 (1977)
19.25 R. E. Walkup, A. Spielfiedel, D. E. Pritchard: Phys. Rev. Lett. **45**, 986 (1980)
19.26 J. Ward, J. Cooper, E. W. Smith: J. Quant. Spectrosc. Radiat. Trans. **14**, 555 (1974)
19.27 D. G. McCarten, N. Lwin: J. Phys. B **10**, 17 (1977)
19.28 D. N. Stacey, R. C. Thompson: Acta Phys. Polon. A **54**, 833 (1978)
19.29 Ch. Ottinger, R. Scheps, G. W. York, A. Gallagher: Phys. Rev. A **11**, 1815 (1975)
19.30 D. Drummond, A. Gallagher: J. Chem. Phys. **60**, 3426 (1974)
19.31 C. G. Carrington, A. Gallagher: Phys. Rev. A **10**, 1464 (1974)
19.32 J. Pascale, J. Vandeplanque: J. Chem. Phys. **60**, 2278 (1974)
19.33 M. J. O'Callaghan, J. Cooper: Phys. Rev. A **39**, 6206 (1989)
19.34 B. P. Stoicheff, E. Weinberger: Phys. Rev. Lett. **44**, 733 (1980)
19.35 K. H. Weber, K. Niemax: Z. Phys. A **307**, 13 (1982)
19.36 J. R. Fuhr, A. Lesage: *Bibliography of Atomic Line Shapes and Shifts*, NIST Special Publication 366 (U. S. Gov't. Printing Office, Washington 1993), Suppl. 4
19.37 J. Szudy, W. E. Baylis: J. Quant. Spectrosc. Radiat. Trans. **15**, 641 (1975)
19.38 D. N. Stacey, J. Cooper: J. Quant. Spectrosc. Radiat. Trans. **11**, 1271 (1971)
19.39 K. M. Sando, J. Wormhoudt: Phys. Rev. A **7**, 1889 (1973)
19.40 L. K. Lam, M. M. Hessel, A. Gallagher: J. Chem. Phys. **66**, 3550 (1977)
19.41 A. Tam, G. Moe, W. Park, W. Happer: Phys. Rev. Lett. **35**, 85 (1975)
19.42 T. Holstein, A. Gallagher: Phys. Rev. A **16**, 2413 (1977)
19.43 J. L. Carlston, A. Szoke: J. Phys. B **9**, L231 (1976)
19.44 A. Omont: Prog. Quant. Electr. **5**, 69 (1977)
19.45 J. Light, A. Szoke: Phys. Rev. A **15**, 1029 (1977)
19.46 K. Burnett, J. Cooper, R. J. Ballagh, E. W. Smith: Phys. Rev. A **22**, 2005 (1980)
19.47 A. Streater, J. Cooper, W. J. Sandle: J. Quant. Spectrosc. Radiat. Trans. **37**, 151 (1987)
19.48 J. Tellinghausen: J. Mol. Spectrosc. **103**, 455 (1984)
19.49 F. H. Mies: J. Chem. Phys. **48**, 482 (1968)
19.50 C. G. Carrington, D. Drummond, A. V. Phelps, A. Gallagher: Chem. Phys. Lett. **22**, 511 (1973)
19.51 L. M. Biberman: J. Exp. Theor. Phys. U. S. S. R. **17**, 416 (1947)
19.52 L. M. Biberman: J. Exp. Theor. Phys. U. S. S. R. **59**, 659 (1948)
19.53 T. Holstein: Phys. Rev. **72**, 1212 (1947)
19.54 T. Holstein: Phys. Rev. **83**, 1159 (1951)
19.55 A. Fioretti, A. F. Molisch, J. H. Müller, P. Verkerk, M. Allegrini: Opt. Commun. **149**, 415 (1998)
19.56 A. F. Molisch, B. P. Oehry: *Radiation Trapping in Atomic Vapours* (Clarendon, Oxford 1998)
19.57 C. van Trigt: Phys. Rev. **181**, 97 (1969)
19.58 C. van Trigt: Phys. Rev. A **4**, 1303 (1971)
19.59 C. van Trigt: Phys. Rev. **13**, 726 (1976)
19.60 C. G. Carrington, D. N. Stacey, J. Cooper: J. Phys. B **6**, 417 (1973)
19.61 J. Huennekins, T. Colbert: J. Quant. Spectrosc. Radiat. Trans. **41**, 439 (1989)
19.62 J. Huennekins, A. Gallagher: Phys. Rev. A **28**, 238 (1983)
19.63 T. Colbert, J. Huennekens: Phys. Rev. **41**, 6145 (1990)
19.64 J. Huennekins, H. J. Park, T. Colbert, S. C. McClain: Phys. Rev. **35**, 2892 (1987)
19.65 H. A. Post, P. van de Weijer, R. M. M. Cremers: Phys. Rev. A **33**, 2003 (1986)
19.66 T. Holstein: Phys. Rev. **83**, 1159 (1951)
19.67 P. J. Walsh: Phys. Rev. **116**, 511 (1959)
19.68 E. Milne: J. Math. Soc. (London) **1**, 40 (1926)
19.69 T. Holstein, D. Alpert, A. O. McCoubrey: Phys. Rev. **85**, 985 (1952)
19.70 A. V. Phelps: Phys. Rev. **114**, 1011 (1959)

19.71 M. G. Payne, J. E. Talmage, G. S. Hurst, E. B. Wagner: Phys. Rev. A **9**, 1050 (1974)
19.72 J. B. Anderson, J. Maya, M. W. Grossman, R. Lagushenko, J. F. Weymouth: Phys. Rev. A **3**, 2986 (1985)
19.73 G. J. Parker, W. N. G. Hitchon, J. E. Lawler: J. Phys. B **26**, 4643 (1993)
19.74 C. van Trigt: Phys. Rev. A **1**, 1298 (1970)
19.75 K. L. Menningen, J. E. Lawler: J. Appl. Phys. **88**, 3190 (2000)
19.76 J. E. Lawler, J. J. Curry, G. G. Lister: J. Phys. D **33**, 252 (2000)
19.77 A. Zajonc, A. V. Phelps: Phys. Rev. A **23**, 2479 (1981)
19.78 J. E. Lawler, G. J. Parker, W. N. G. Hitchon: J. Quant. Spectrosc. Radiat. Trans. **49**, 627 (1993)
19.79 G. J. Parker, W. N. G. Hitchon, J. E. Lawler: J. Phys. B **26**, 4643 (1993)
19.80 J. J. Curry, J. E. Lawler, G. G. Lister: J. Appl. Phys. **86**, 731 (1999)

20. Thomas–Fermi and Other Density-Functional Theories

The key idea in Thomas–Fermi theory and its generalizations is the replacement of complicated terms in the kinetic energy and electron–electron repulsion energy contributions to the total energy by relatively simple functionals of the electron density ρ. This chapter first describes Thomas–Fermi theory, and then its various generalizations which attempt to correct, with varying success, some of its deficiencies. It concludes with an overview of the Hohenberg-Kohn and Kohn-Sham density functional theories.

20.1 **Thomas–Fermi Theory and Its Extensions** 296
20.1.1 Thomas–Fermi Theory 296
20.1.2 Thomas–Fermi–von Weizsäcker Theory 298
20.1.3 Thomas–Fermi–Dirac Theory 299
20.1.4 Thomas–Fermi–von Weizsäcker-Dirac Theory 299
20.1.5 Thomas–Fermi Theory with Different Spin Densities 300
20.2 **Nonrelativistic Energies of Heavy Atoms** 300
20.3 **General Density Functional Theory** 301
20.3.1 The Hohenberg–Kohn Theorem for the One-Electron Density 301
20.3.2 The Kohn–Sham Method for Including Exchange and Correlation Corrections 302
20.3.3 Density Functional Theory for Excited States 303
20.3.4 Relativistic and Quantum Field Theoretic Density Functional Theory 303
20.4 **Recent Developments** 303
References 304

In the early years of quantum physics, *Thomas* [20.1] and *Fermi* [20.2–5] independently invented a simplified theory, subsequently known as *Thomas–Fermi theory*, to describe nonrelativistically an atom or atomic ion with a large nuclear charge Z and a large number of electrons N. Many qualitative features of this model can be studied analytically, and the precise solution can be found by solving numerically a nonlinear ordinary differential equation. *Lenz* [20.6] demonstrated that this equation for the electrostatic potential could be derived from a variational expression for the energy as a functional of the electron density. Refinements to Thomas–Fermi theory include a term in the energy functional to account for electron exchange effects introduced by *Dirac* [20.7], and nonlocal gradient corrections to the kinetic energy introduced by *von Weizsäcker* [20.8].

Although the Hartree–Fock method or other more elaborate techniques for calculating electronic structure now provide much more accurate results (Chapts. 21, 22, and 23), Thomas–Fermi theory provides quick estimates and global insight into the total energy and other properties of a heavy atom or ion. A rigorous analysis of Thomas–Fermi theory by *Lieb* and *Simon* [20.9, 10] showed that it is asymptotically exact in that it yields the correct leading asymptotic behavior, for both the total nonrelativistic energy and the electronic density, in the limit as both Z and N tend to infinity, with the ratio Z/N fixed. (In a real atom, of course, relativistic and other effects become increasingly important as Z increases.) However, Thomas–Fermi theory has the property that molecules do not bind, as first noted by *Sheldon* [20.11] and proved by *Teller* [20.12]. That the interatomic potential energy curve for a homonuclear diatomic molecule is purely repulsive was demonstrated by *Balàzs* [20.13]. This 'no binding' property of clusters of atoms was used by *Lieb* and *Thirring* [20.14] to prove the *stability of matter*, in the sense that as the number of particles increases, the total nonrelativistic energy decreases only linearly rather than as a higher power of the number of particles, as it would if electrons were bosons rather than fermions. Lieb went on to explore the mathematical structure of the modifications of the Thomas–Fermi model when gradient terms (von Weizsäcker) and/or exchange (Dirac) terms are included [20.15, 16]. A review article by *Spruch* [20.17] explicates the linkage between long-developed physical intuition and the mathematically rigorous results obtained in the 1970's and 1980's. The older literature was reviewed by *Gombás* [20.18, 19] and by *March* [20.20].

An outgrowth of Thomas–Fermi theory is the general density functional theory initiated by *Hohenberg* and *Kohn* [20.21] and by *Kohn* and *Sham* [20.22], as discussed in Sect. 20.3 of this chapter.

20.1 Thomas–Fermi Theory

20.1.1 Thomas–Fermi Theory

In a D-dimensional Euclidean space, the expectation value of the electronic kinetic energy operator in a quantum state ψ can be approximated by

$$\frac{\hbar^2}{m_e}2\pi^2\frac{D}{D+2}\left(\frac{D}{2\Omega_D}\right)^{2/D}\int\rho^{(D+2)/D}(\boldsymbol{r})\,\mathrm{d}^D r\,, \quad (20.1)$$

where

$$\Omega_D = D\pi^{D/2}/\Gamma(1+D/2) \quad (20.2)$$

is the surface area of a unit hypersphere in D dimensions [20.17, p. 176]. These expressions can easily be derived by considering the energy levels of a system of a large number of noninteracting fermions confined to a D-dimensional box. Specialization to the physically interesting case of $D = 3$ yields the well-known expression

$$\frac{\hbar^2}{m_e}2\pi^2\frac{3}{5}\left(\frac{3}{2\Omega_3}\right)^{2/3}\int\rho^{5/3}(\boldsymbol{r})\,\mathrm{d}\boldsymbol{r}\,, \quad (20.3)$$

where

$$\Omega_3 = 3\pi^{3/2}/\Gamma(1+3/2) = 4\pi\,. \quad (20.4)$$

The electron–nucleus attraction energy in a three-dimensional space is given exactly by

$$\int\rho(\boldsymbol{r})V(\boldsymbol{r})\,\mathrm{d}\boldsymbol{r}\,, \quad (20.5)$$

where $V(\boldsymbol{r})$ is the Coulomb potential due to a single nucleus ($V(\boldsymbol{r}) = -Z/r$) or to several nuclei [$V(\boldsymbol{r}) = -\sum_i Z_i/|\boldsymbol{r}-\boldsymbol{R}_i|$]. The electron–electron Coulomb repulsion energy in a three-dimensional space is approximated by

$$\frac{1}{2}\int\frac{\rho(\boldsymbol{r})\rho(\boldsymbol{r}')}{|\boldsymbol{r}-\boldsymbol{r}'|}\,\mathrm{d}\boldsymbol{r}\,\mathrm{d}\boldsymbol{r}'\,, \quad (20.6)$$

which tends to overestimate the actual repulsion energy because it includes the self-energy of the densities of individual electrons. This is, however, a higher-order effect for a system with a large number of electrons concentrated in a small region of space. As was suggested by *Fermi* and *Amaldi* [20.23], this overestimation can be approximately corrected for an atom with N electrons by multiplying this term by the ratio of the number of ordered pairs of different electrons to the total number of ordered pairs

$$\frac{N(N-1)}{N^2} = 1 - \frac{1}{N}\,. \quad (20.7)$$

This is approximately correct for an atom, with many electrons concentrated close together, but it would still be an overestimate for a diffuse system, such as one composed of N electrons and N protons separated by large distances of $O(R)$, for which the ground-state electron–electron repulsion term should be proportional to $\frac{1}{2}N(N-1)/R$ rather than to N times a constant of $O(1)$. For this reason the Fermi–Amaldi correction, which complicates the mathematical analysis without eliminating the unphysical overestimation of the electron–electron repulsion term, is not usually included. It is evident that the treatment of both the electronic kinetic energy term and the electron–electron repulsion energy term depends on the assumption that the number N of electrons (actually, the number of electrons *per atom*) is large. Hence the Thomas–Fermi model is sometimes called the *statistical model* of an atom.

The three contributions to the total energy are now added together and one seeks to minimize their sum, the Lenz functional [20.6]

$$E[\rho] = \frac{\hbar^2}{m_e}2\pi^2\frac{3}{5}\left(\frac{3}{2\Omega_3}\right)^{2/3}\int\rho^{5/3}(\boldsymbol{r})\,\mathrm{d}\boldsymbol{r} + \int\rho(\boldsymbol{r})V(\boldsymbol{r})\,\mathrm{d}\boldsymbol{r} + \frac{1}{2}\int\frac{\rho(\boldsymbol{r})\rho(\boldsymbol{r}')}{|\boldsymbol{r}-\boldsymbol{r}'|}\,\mathrm{d}\boldsymbol{r}\,\mathrm{d}\boldsymbol{r}'\,, \quad (20.8)$$

over all admissible densities ρ. The mathematical question now arises: what is an admissible density? The answer was provided by *Lieb* and *Simon* [20.9, 10]: a density for which both

$$\int\rho(\boldsymbol{r})\,\mathrm{d}\boldsymbol{r}\,, \quad (20.9)$$

the total number of electrons, and

$$\int \rho^{5/3}(\boldsymbol{r})\,\mathrm{d}\boldsymbol{r}\,, \tag{20.10}$$

which is proportional to the estimate of their kinetic energy, are finite, automatically yields finite values of the other terms in the expression for the energy. As Lieb and Simon proved, the minimization of this functional over all such densities yields a well-determined result.

Carrying out the variation of $E[\rho]$ with respect to ρ yields the Euler–Lagrange equation

$$\begin{aligned}0 = &\frac{\hbar^2}{m_\mathrm{e}} 2\pi^2 \left(\frac{3}{2\Omega_3}\right)^{2/3} \rho^{2/3}(\boldsymbol{r})\\ &+ V(\boldsymbol{r}) + \int \frac{\rho(\boldsymbol{r}')}{|\boldsymbol{r}-\boldsymbol{r}'|}\,\mathrm{d}\boldsymbol{r}'\,.\end{aligned} \tag{20.11}$$

The sum of the last two terms is of course the negative of the total electrostatic potential $\phi(\boldsymbol{r})$, so one sees that in Thomas–Fermi theory the density is proportional to the 3/2-power of the potential. To simplify subsequent manipulations, let

$$\frac{\hbar^2}{m_\mathrm{e}} 2\pi^2 \left(\frac{3}{2\Omega_3}\right)^{2/3} = \frac{\hbar^2}{2m_\mathrm{e}}\left(3\pi^2\right)^{2/3} = \gamma_\mathrm{p}\,, \tag{20.12}$$

so that

$$\gamma_\mathrm{p}\rho^{2/3}(\boldsymbol{r}) = \phi(\boldsymbol{r})\,. \tag{20.13}$$

By Poisson's theorem,

$$-\nabla^2\phi = 4\pi\left[\sum_i Z_i\delta^{(3)}(\boldsymbol{r}-\boldsymbol{R}_i) - \rho(\boldsymbol{r})\right], \tag{20.14}$$

and from (20.13) one has $\rho = \gamma_\mathrm{p}^{-3/2}\phi^{3/2}$, so from the integral equation for the electronic density ρ one obtains the differential equation

$$-\nabla^2\phi = 4\pi\left[\sum_i Z_i\delta^{(3)}(\boldsymbol{r}-\boldsymbol{R}_i) - \gamma_\mathrm{p}^{-3/2}\phi^{3/2}\right] \tag{20.15}$$

for the potential ϕ. In the case of a single nucleus, the usual separation of variables in spherical polar coordinates yields for ϕ the ordinary differential equation

$$\frac{1}{r}\frac{\mathrm{d}^2}{\mathrm{d}r^2}(r\phi) = 4\pi\,\gamma_\mathrm{p}^{-3/2}\,\phi^{3/2}\,, \tag{20.16}$$

whose similarity to Emden's equation, which Eddington had used to study the internal constitution of stars, was recognized by *Milne* [20.24]. The numerical solution of this equation with the appropriate boundary conditions at $r=0$ and $r=\infty$ was extensively discussed by *Baker* [20.25], and accurate solutions tabulated by *Tal* and *Levy* [20.26]. The numerical solution determines that the total energy of a neutral atom is

$$\begin{aligned}E &= -3.678\,745\,21\ldots\ \gamma_\mathrm{p}^{-1}\,Z^{7/3}\\ &= -1.537\,490\,24\ldots\ Z^{7/3}\,\mathrm{Ry}\,.\end{aligned} \tag{20.17}$$

Another possibility is to do the constrained minimization over all densities which obey

$$\int \rho(\boldsymbol{r})\,\mathrm{d}\boldsymbol{r} = \lambda\,, \tag{20.18}$$

where λ is the number of electrons, which for purposes of mathematical analysis is allowed to be nonintegral. One then introduces a Lagrange multiplier $-\mu$, the *chemical potential*, to correspond with this constraint, and thereby obtains the Euler–Lagrange equation

$$\begin{aligned}0 = &\frac{\hbar^2}{m_\mathrm{e}} 2\pi^2 \left(\frac{3}{2\Omega_3}\right)^{2/3} \rho^{2/3}(\boldsymbol{r})\\ &+ V(\boldsymbol{r}) + \int \frac{\rho(\boldsymbol{r}')}{|\boldsymbol{r}-\boldsymbol{r}'|}\,\mathrm{d}\boldsymbol{r}' + \mu\,,\end{aligned} \tag{20.19}$$

which holds wherever ρ is positive. As was shown by *Lieb* and *Simon* [20.9, 10], this procedure too is well-defined. The analogue of (20.13), the relationship between the density and the electrostatic potential for the neutral atom, is now

$$\gamma_\mathrm{p}\rho^{2/3}(\boldsymbol{r}) = [\phi(\boldsymbol{r}) - \mu]_+\,, \tag{20.20}$$

where $[f]_+ = \max(f, 0)$. The corresponding differential equation for the potential ϕ is

$$\begin{aligned}-\nabla^2\phi = 4\pi\Bigg[&\sum_i Z_i\delta^{(3)}(\boldsymbol{r}-\boldsymbol{R}_i)\\ &-\gamma_\mathrm{p}^{-3/2}[\phi(\boldsymbol{r})-\mu]_+^{3/2}\Bigg]\,.\end{aligned} \tag{20.21}$$

Lieb and Simon rigorously proved a large number of results concerning the solution of the Thomas–Fermi model. When $V(\boldsymbol{r})$ is a sum of Coulomb potentials arising from a set of nuclei of positive charges Z_i, with $\sum_i Z_i = Z$, then the energy $E(\lambda)$ is a continuous, monotonically decreasing function of λ for $0 \le \lambda \le Z$, and its derivative $\mathrm{d}E/\mathrm{d}\lambda$ is the chemical potential $-\mu(\lambda)$, which vanishes at $\lambda = Z$. For λ in this range, there is a unique minimizing density ρ, whereas for $\lambda > Z$ there is no unique minimizing ρ, since one can place arbitrarily large clumps of charge with arbitrarily low energy arbitrarily far away from the nuclei. In the atomic case, with a single nucleus, $\rho(\boldsymbol{r})$ is a spherically symmetric monotonically decreasing function of r.

Moreover, for an atom or atomic ion, the Thomas–Fermi density obeys the *virial theorem*

$$2\langle \mathcal{T} \rangle = -\langle V \rangle = -2E \,, \tag{20.22}$$

and for a neutral atom the electronic kinetic energy, electron–nucleus attraction energy, and electron–electron repulsion energy terms in the expression for the total energy satisfy the ratios $3 : -7 : 1$.

It is straightforward to examine the behavior of the Thomas–Fermi density ρ in the limit as either $r \to 0$ or $r \to \infty$. For large r, the electron density vanishes identically outside a sphere of finite radius for a positive ion. For a neutral atom, the ordinary differential equation (20.16) for the potential ϕ can be analyzed to show that

$$\phi(r) \simeq \gamma_\mathrm{p}\,(3\gamma_\mathrm{p}/\pi)^2\, r^{-4} \,, \tag{20.23}$$

from which it follows that

$$\rho(r) \simeq (3\gamma_\mathrm{p}/\pi)^3\, r^{-6} \,, \tag{20.24}$$

independent of Z. This implies that as $Z \to \infty$, a neutral atom described by the Thomas–Fermi model has a finite size defined in terms of a radius within which all but a fixed amount of electronic probability density is located. For example, if one defines the size of an atom as that value of r_a for which

$$\int\limits_{|\boldsymbol{r}|\geq r_\mathrm{a}} \rho(\boldsymbol{r})\, \mathrm{d}\boldsymbol{r} = \frac{1}{2} \,, \tag{20.25}$$

one finds that in the large-Z limit

$$r_\mathrm{a} = \left(\frac{8\pi}{3}\right)^{1/3} \frac{3\gamma_\mathrm{p}}{\pi} \,. \tag{20.26}$$

In atomic units, $\gamma_\mathrm{p} = \frac{1}{2}(3\pi^2)^{2/3}$, and

$$r_\mathrm{a} = (9\pi)^{2/3}\, a_0 \simeq 9.3\, a_0 \,, \tag{20.27}$$

which is about what one would expect for a 'real' nonrelativistic atom with a large nuclear charge Z. On the other hand, the characteristic distance scale in Thomas–Fermi theory, defined as the 'average' value of r, or in terms of a radius within which a fixed *fraction* of electronic probability density is located, is proportional to $Z^{-1/3}$, which shrinks to 0 as $Z \to \infty$. The resolution of this paradox is that outside the typical 'core' scale of distance set by $Z^{-1/3}$, within which most of the electron density is located, there resides in the 'mantle' region a fraction of electrons proportional to $Z^{2/3}/Z = Z^{-1/3}$, and almost all of these are concentrated within a sphere of radius of about $10\, a_0$.

Moving deeper into the core and approaching the nucleus, the $-Z/r$ singularity in the electron–nucleus Coulomb potential dominates the smeared-out electron–electron potential, so one readily finds that

$$\rho(r) \simeq \left(\frac{Z}{\gamma_\mathrm{p}\, r}\right)^{3/2} \,. \tag{20.28}$$

This singularity is integrable but unphysical, since it arises from the approximation of the local kinetic energy by $\rho^{5/3}$, which breaks down where ρ is rapidly varying on a length scale proportional to $1/Z$. In a 'real' nonrelativistic heavy atom governed by the Schrödinger equation, the actual electron density at the nucleus is finite, being proportional to Z^3. This unphysical singularity in the electron density in Thomas–Fermi theory can be eliminated by adding a gradient correction to the Thomas–Fermi kinetic energy term.

20.1.2 Thomas–Fermi–von Weizsäcker Theory

The semiclassical approximation (20.3) for the quantum kinetic energy in terms of a power of the density is capable of improvement, particularly in regions of space where the density is rapidly varying. The incorporation of such corrections leads to a gradient expansion for the kinetic energy [20.27]. The leading correction is of the form

$$\frac{\hbar^2}{2m} \int \left|\left(\nabla \rho^{1/2}\right)(\boldsymbol{r})\right|^2 \mathrm{d}\boldsymbol{r} \,. \tag{20.29}$$

Addition of such a term to the Thomas–Fermi expression for the kinetic energy yields a theory which avoids many of the unphysical features of ordinary Thomas–Fermi theory at very short and moderately large distances. The more important points, as rigorously proved in Lieb's review article [20.15, 16], are as follows. The leading features of the energy are unchanged; for large Z the energy $E(Z)$ of a neutral atom or atomic ion is still proportional to $Z^{7/3}$, but now there enter higher-order corrections arising from the gradient terms of order $Z^{7/3}\, Z^{-1/3} = Z^2$ and higher powers of $Z^{-1/3}$. The maximum number of electrons which can be bound by an atom of nuclear charge Z is no longer exactly Z, but a slightly larger number; thus Thomas–Fermi–von Weizsäcker theory allows for the formation of negatively charged atomic ions. It was further proven by

Benguria and *Lieb* [20.28] that in the Thomas–Fermi–von Weizsäcker model a neutral atom can bind at most one extra electron, and that a neutral molecule can bind at most as many extra electrons as it has nuclei.

The effect on the electronic density ρ is more profound. While the general shape and properties of ρ in the 'core' and 'mantle' regions is unchanged, the fact that $\nabla\rho^{1/2}$ need not, and in general does not, vanish when ρ vanishes on some surface implies that for a positive ion ρ no longer vanishes outside of a sphere, as it does in the case of Thomas–Fermi theory, but instead extends over all space. For positive ions, neutral atoms, and negative ions alike, 'differential inequality' techniques [20.29] can be used to show that $\rho(r)$ decays exponentially for large r, with the constant in the exponential proportional to $\mu^{1/2}(\lambda)$. For small r, the gradient terms dominate the energy expression, so one finds that the electronic density no longer diverges as $r \to 0$, but instead tends to a finite limit, with a first derivative which obeys a relation analogous to the Kato cusp condition [20.30] (see Sect. 11.1.1).

The study of molecules within the Thomas–Fermi–von Weizsäcker model involves several subtleties and pitfalls, which can lead to physical absurdities. Since two neutral atoms with different nuclear charges will in general have different chemical potentials, a pair of such atoms placed a long distance apart will *spontaneously ionize*, with a small amount of electric charge being transferred from one to the other until the chemical potentials of the positively charged ion and the negatively charged ion become equal. The result is a long-range Coulomb attraction between them [20.31]. This phenomenon does *not* occur in the real world, since the amount of electric charge which can be transferred is quantized in units of $-e$, and it is empirically true that the smallest atomic ionization potential exceeds the largest atomic electron affinity. For two neutral atoms with the same nuclear charges, the situation is more subtle. Nonetheless, a careful analysis shows that in this case too, though no spontaneous ionization occurs, there is a long-range attractive interaction between them arising from the overlap of the exponentially small tails of the electron clouds. Since electron correlation is not included in this model, it could not be expected to describe attractive van der Waals forces.

In summary, the Thomas–Fermi–von Weizsäcker model yields a more realistic picture of a single atom than does the Thomas–Fermi model. However, it does not provide a useful picture for understanding the interaction between atoms at large distances. These kinds of unphysical features provide a glimpse into the complicated nature of the universal density functional, which must include terms which rigorously suppress an unphysical feature like spontaneous ionization of a distant pair of heteronuclear atoms [20.32, 33]. It is evident from the mathematical properties of Thomas–Fermi–von Weizsäcker theory and related models that a density functional which 'fixes up' the Thomas–Fermi expression simply by adding a few gradient terms and/or simple exchange terms and the like must still differ in important ways from the universal density functional, particularly for properties of extended systems.

20.1.3 Thomas–Fermi–Dirac Theory

The effect of the exchange of electrons can be approximated, following *Dirac* [20.7], by including in the Thomas–Fermi energy functional an expression of the form

$$-\frac{1}{4\pi^3}\left(3\pi^2\right)^{4/3}\int \rho^{4/3}(\boldsymbol{r})\,\mathrm{d}\boldsymbol{r}\,. \tag{20.30}$$

Minimization of the resulting Thomas–Fermi–Dirac energy functional over all admissible densities ρ whose integral is λ yields a well-defined $E(\lambda)$, which has the correct behavior for $\lambda \le Z$, and it has been shown that for an atom the exchange correction to the energy is of order $Z^{5/3}$. However, this model exhibits unphysical behavior for $\lambda > Z$, because one can obtain a completely artificial lowering of the energy by placing many small clumps of electrons a large distance from the nucleus, for which the negative $\int \rho^{4/3}\,\mathrm{d}\boldsymbol{r}$ term dominates the energy expression [20.15, 16]. At the conclusion of his original article, Dirac clearly stated that the correction he had derived, although giving a better approximation in the interior of an atom, gives "a meaningless result for the outside of the atom" [20.7]. It is therefore clear that any physically reasonable theory must somehow profoundly modify this correction in the region where the electronic density is very small.

20.1.4 Thomas–Fermi–von Weizsäcker–Dirac Theory

One can also include the Dirac exchange correction in the Thomas–Fermi–von Weizsäcker energy functional. In this case, however, the mathematical foundations of the theory are still incomplete ([20.15, 16, pp. 638–9]). Nonetheless, it is clear that this theory too suffers from the unphysical lowering of the energy by small clumps of electrons at large distances from the nucleus.

In summary, one can say that the inclusion of Dirac's exchange correction in its most straightforward form leads to an improvement of energies for positive ions or neutral atoms, but to unphysical behavior for systems where the charge of the electrons exceeds the nuclear charge, in line with Dirac's own observations on the limitations of his correction [20.7]. We see here again a manifestation of how complicated must be the behavior of the true universal density functional.

20.1.5 Thomas–Fermi Theory with Different Spin Densities

As was remarked by *Lieb* and *Simon* [20.10], it is possible to consider a variant of Thomas–Fermi theory with a pair of spin densities ρ_α and ρ_β for the spin-up and spin-down electrons, with the two adding together to produce the total electronic density ρ. This theory has been rigorously formulated and analyzed by *Goldstein* and *Rieder* [20.34]. Because the problem is nonlinear, the mathematical complications are substantial, and the theory is not a trivial extension of ordinary Thomas–Fermi theory. *Goldstein* and *Rieder* first considered the case where the total number of electrons of each type of spin is specified in advance [20.35]. There is no mathematical obstacle to constructing such a spin-polarized Thomas–Fermi theory, but it does not yield the kind of spontaneous spin-polarization that one observes in the ground states of many real quantum mechanical atoms and molecules, which is not surprising in view of the fact that such spin-polarization in accord with Hund's first rule arises from exchange and correlation effects not included in this simple functional. However, in the case where the electronic spins (but *not* their currents) are coupled to an external magnetic field, the ground state is naturally spin-polarized [20.34].

20.2 Nonrelativistic Energies of Heavy Atoms

Thomas–Fermi theory suggests that (20.17) provides the leading term in a power series expansion for the nonrelativistic energy of a neutral atom of the form

$$E(Z) = -\left(c_7 Z^{7/3} + c_6 Z^{6/3} + c_5 Z^{5/3} + \cdots\right) \tag{20.31}$$

with $c_7 = 1.537\,490\,24\ldots$ Ry, $c_6 = -1$ Ry, and $c_5 \simeq 0.5398$ Ry. The c_6 term was first calculated by *Scott* [20.36] from the observation that it arises from the energy of the innermost electrons for which the electron–electron interaction can be neglected. The difference between the exact and Thomas–Fermi energies for this case of noninteracting electrons yields the correct c_6 [20.17, 37]. A mathematically nonrigorous but physically insightful justification of the Scott correction was provided in 1980 by *Schwinger* [20.38]. This result has now been rigorously proved, with upper and lower bounds coinciding [20.39–43].

The c_5 term is much more subtle, since it arises from a combination of effects from the exchange interaction and from the bulk motion of electrons in the Thomas–Fermi potential. A general analytic procedure devised by *Schwinger* [20.44] yields the above value, in good agreement with a much earlier estimate by *March* and *Plaskett* [20.45].

A numerical check of these results, based on a fit to Hartree–Fock calculations for Z up to 290 with correlation corrections, yielded the values [20.46] $c_5 = 0.55 \pm 0.02$ Ry and $c_4 \simeq 0$. It seems likely that, because of shell structure, the terms c_4 and beyond have an oscillatory dependence on Z [20.47]. The oscillatory structure and other refinements of Thomas–Fermi theory are considered in a series of papers by *Englert* and *Schwinger* [20.48–50].

Iantchenko, *Lieb*, and *Siedentop* [20.51] have proven Lieb's 'strong Scott conjecture' that for small r, the rescaled density for the exact quantum system converges to the sum of the densities of the bound noninteracting hydrogenic orbitals; the properties of this function were explored by *Heilman* and *Lieb* [20.52]. *Fefferman* and *Seco* [20.53] have rigorously proved the correctness of Schwinger's procedure for calculating not just the $O(Z^{6/3})$ Scott correction but also the $O(Z^{5/3})$ exchange term. Their full proof includes a demonstration that the Hartree–Fock energy agrees with the exact quantum energy through $O(Z^{5/3})$, with an error of smaller order [20.54]. Numerous auxiliary theorems and lemmas are published in [20.55–58]. Progress toward obtaining higher-order oscillatory terms is described in [20.59–63]. The analytical evaluation of accurate approximations to the energy of a heavy atom, or at least of the contributions to that energy of all but the few outermost electronic orbitals, would be of particularly great value if it led to the construction of more accurate and better justified pseudopotentials [20.64–67] for describing the valence orbitals.

20.3 General Density Functional Theory

The literature on general density functional theory and its applications is enormous, so any bibliography must be selective. The reader interested in learning more could begin by consulting a number of review articles [20.68–73], collections of articles [20.74–76], and conference proceedings [20.77–87], and the recent textbooks by *Parr* and *Yang* [20.88] and by *Dreizler* and *Gross* [20.89].

20.3.1 The Hohenberg–Kohn Theorem for the One-Electron Density

In 1964 *Hohenberg* and *Kohn* [20.21] argued that there exists a universal density functional $F[\rho]$, independent of the external potential $V(\boldsymbol{r})$, such that minimization of the sum

$$F[\rho] + \int \rho(\boldsymbol{r}) V(\boldsymbol{r}) \, \mathrm{d}\boldsymbol{r} \,, \tag{20.32}$$

subject to the constraint

$$\int \rho(\boldsymbol{r}) \, \mathrm{d}\boldsymbol{r} = N \text{ (a positive integer)} \,, \tag{20.33}$$

yields the ground state energy of a quantum-mechanical N-electron system moving in this external potential. However, Hohenberg and Kohn's 'theorem' is like a mathematical 'existence theorem'; no procedure exists to calculate explicitly this unknown universal functional, which surely is extremely complicated if it can be written down at all in closed form. (E. Bright Wilson, however, defined it, implicitly and whimsically, as follows: "Take the ground-state density and integrate it to find the total number of electrons. Find the cusps in the density to locate all nuclei, and then use the cusp condition – that the radial derivative of the density at the cusp is minus twice the nuclear charge density at each cusp – to determine the charges on each nucleus. Finally solve Schrödinger's equation for the ground-state density or any other property that is desired" (paraphrased by B. I. Dunlap, in [20.83, p. 3], from J. W. D. Connolly).)

Moreover, Hohenberg and Kohn glossed over two problems: it is not clear a priori that every well-behaved ρ is derivable from a well-behaved properly antisymmetric many-electron wave function (the so-called n-representability problem, since n was used by Hohenberg and Kohn to represent the density of electrons), and it is also not clear a priori that every well-behaved density ρ can be derived from a quantum-mechanical many-electron wave function ψ which is the properly antisymmetric ground-state wave function for a system of electrons moving in some external potential $V(\boldsymbol{r})$ (the so-called v-representability problem, since Hohenberg and Kohn used v in place of V).

The n-representability problem was solved by *Gilbert* [20.90] and by *Harriman* [20.91], who gave a prescription for starting from an arbitrary well-behaved ρ and from it constructing a many-electron wave function ψ which generated that ρ [20.33,92]. The v-representability problem is much more formidable, as demonstrated by the discovery that there are well-behaved densities ρ which are *not* the ground-state densities for any fermionic system in an external potential V [20.92, 93]. Following Percus' definition of a universal kinetic energy functional for independent fermion systems [20.94], *Levy* [20.95] proposed to circumvent this v-representability problem by modifying the definition of $F[\rho]$ so that instead of being defined in terms of densities which might not be v-representable, it is defined as

$$F[\rho] = \min \, [\psi, (T+V)\psi] \,, \tag{20.34}$$

with the minimum being taken over all properly antisymmetric normalized ψ's which yield that ρ.

A great deal of effort has been devoted to trying to find approximate representations of the universal functional $F[\rho]$. One route is mathematical, and features a careful exploration of the abstract properties which $F[\rho]$ must have. Another route is numerical, and can be characterized as involving the guessing of some *ansatz* with a general resemblance to Thomas–Fermi–von Weizsäcker–Dirac theory, with some flexible parameters which are determined by least-squares fitting of the energies resulting from insertion of Hartree–Fock densities into the trial functional to theoretical Hartree–Fock energies, or the like. If, however, the basic *ansatz* exhibits unphysical features in the case of negatively charged ions or heteronuclear molecules, it is not likely that the optimization of parameters in that *ansatz* will get one closer to the true universal density functional. In the opinion of this writer, significant progress in density functional theory based solely upon the one-electron density is likely to require a major revolution in our mathematical understanding of this field, with a useful procedure made explicit for constructing progressively better approximations to the universal density functional, which, like π or other transcendental numbers, probably will never be written down exactly in closed form. Moreover, the numerical solution of the highly nonlinear Euler–Lagrange equations for a very complicated density functional is likely to require large amounts of computer time, as well as problems with landing in local minima of the energy.

20.3.2 The Kohn–Sham Method for Including Exchange and Correlation Corrections

Density functional theory posed solely in terms of the one-electron density and based upon the Hohenberg–Kohn variational principle provides no general procedure for accurately calculating relatively small energy differences such as excitation energies, ionization potentials, electron affinities, or the binding energies of molecules. There is, however, a powerful method inspired by the Hohenberg–Kohn variational principle, which has been used with great success in the calculation of such quantities. This is the Kohn–Sham variational method [20.22].

The key idea in the Kohn–Sham variational method is to replace the nonlocal exchange term in the Hartree–Fock equations with an exchange-correlation potential, which at least in principle can be used to determine energies exactly. The oldest, simplest, and most common *ansatz* used for the exchange-correlation potential involves the local density approximation (LDA), in which one assumes that the exchange-correlation potential for the actual system under study has the same functional form as does the exchange-correlation potential for a uniform interacting gas of electrons. If the density is not too small or not too rapidly varying, the exchange part of this potential can be approximated by $\rho^{1/3}$, which appears in Dirac's first-order approximation for exchange energies, with a systematic procedure for deriving higher-order corrections in a gradient expansion. The correlation part of this potential is accurately known from *Ceperley* and *Alder*'s quantum Monte Carlo calculation of the properties of the uniform electron gas [20.96]. One therefore retains the important features of the quantum theory based on wave functions, with a determinantal approximation to ψ, while approximately including exchange and correlation effects through a simply computable effective potential. Higher corrections, which are important for quantitative accuracy, can be incorporated by taking account of the variation of ρ by means of a gradient expansion [20.27] involving $\nabla\rho$ and higher derivatives [20.89, Chapt. 7], thus yielding a generalized gradient approximation (GGA) for the exchange-correlation potential.

The Kohn–Sham procedure has become the backbone for the vast majority of accurate calculations of the electronic structure of solids [20.72, 86]. In the 1990's, motivated by *Becke*'s work on constructing simple gradient-corrected exchange potentials [20.97–103], and incorporating the Lee, Yang, and Parr (LYP) expression for the correlation potential [20.104] derived from *Colle* and *Salvetti*'s correlation-energy formula [20.105–107], the Kohn–Sham method is finding increasing application in efficiently estimating relatively small energy differences of relevance to chemistry [20.108–110] (However, Becke's gradient-corrected exchange potential does *not* have the correct $1/r$ behavior at large r, as was observed by several authors [20.111–113]). For definitive results, however, one must still resort to an ab initio theory which at least in principle converges toward the correct result.

The generation of improved generalized gradient approximations has recently become a growth industry, with increasingly many proposals of increasingly greater complexity [20.97, 104, 113–125]. Inevitably, some expressions work better for some properties than for others. It is found that usually most of the errors in the long-range tails of the exchange and correlation potentials tend to cancel each other, thus leading to better overall energies than one could reasonably expect [20.126]. Under these circumstances, it is important to have benchmarks for testing the accuracy of the various approximations. Such comparisons have been carried out for two important sets of two-electron systems [20.127–129]:

1. a pair of electrons moving in harmonic potential wells and coupled by the Coulomb repulsion, which yields an exactly solvable system;
2. helium-like ions of variable nuclear charge Z, for which extremely accurate energies and wave functions are available which take account of the behavior of the exact but unknown wave function in the vicinity of all two-particle coalescences and the three-particle coalescence.

The results indicate that the approximate exchange-correlation potentials differ quite considerably from the true exchange-correlation potentials, thus indicating the need for further analytical work in understanding how to design accurate exchange-correlation potentials, and for devising tests of exchange-correlation potentials for larger atoms and for molecules.

Another important way of testing the validity of various approximate exchange and correlation potentials is checking whether they obey inequalities imposed by such general properties as scaling and the Hellmann–Feynman theorem. Such general tests have been devised by *Levy* and his co-workers [20.130–139], who have found that many of the commonly used approximate potentials violate general inequalities which must be obeyed by the exact potential. These abstract results are helpful in designing potentials which should be better approximations to the true potential.

20.3.3 Density Functional Theory for Excited States

The Hohenberg–Kohn theorem and the Kohn–Sham method were originally formulated in terms of the ground electronic state. These techniques can be extended to calculate the ground state of a given symmetry [20.140], but that leaves unresolved the issue of using density functional theory to calculate the energies of excited states for a given symmetry. Using the Rayleigh–Ritz principle for ensembles, general abstract procedures for generalizing density functional theory to excited state calculations have been formulated by *Theophilou* [20.141] and by several other workers [20.142–151]. Unfortunately, the errors typically seem to be much larger than for ground-state density functional theory.

20.3.4 Relativistic and Quantum Field Theoretic Density Functional Theory

At a formal level, one can discuss the development of density functional theory for a relativistic system of electrons. For an overview of this challenging subject, see the discussions by *Dreizler* and *Gross* [20.89, Chapt. 8] and by *Dreizler* [20.152]. Much of the formalism carries over, but no good way has yet been found of incorporating vacuum polarization corrections.

20.4 Recent Developments

During the last eight years there has continued to be exponentially growing interest in applications of density functional theory of the Kohn-Sham variety to atoms and molecules, especially those of chemical relevance which are too large for accurate ab initio electronic structure calculations. The awarding of the 1998 Nobel Prize in Chemistry to Walter Kohn and John A. Pople recognised their individual contributions to this increasingly important field. Their Nobel lectures were published the following year in the *Reviews of Modern Physics* [20.153, 154].

Since a comprehensive summary of the wide-ranging developments in density functional theory during the past decade is not feasible within the limited space available for this supplementary section, I will briefly cite some of the most extensive surveys of various aspects of this field that have appeared since 1995.

Many aspects of density functional theory were reviewed in four consecutive volumes of *Topics in Current Chemistry* published in 1996 [20.155], and in 1999 an entire volume of *Advances in Quantum Chemistry* was devoted to density functional theory [20.156]. This has also been the subject of several conference proceedings [20.157–159] and introductory textbooks [20.160, 161]. Developments in time-dependent density functional theory for chemical systems have been surveyed in two very recent review articles [20.162, 163].

Although the locality of DFT was proved for a large class of functionals [20.164–166], this issue has come under recent dispute. The question that has been raised is whether there exists an exact Thomas-Fermi model for non-interacting electrons. If such an exact model does not exist, as it is a direct consequence of the Hohenberg-Kohn theorem, then DFT would be incomplete.

Nesbet [20.167–171] has argued that such a theory would be inconsistent with the Pauli exclusion principle for atoms of more than two electrons (or for a two electron atom where both electrons are in the same spin state). The contention is that if only the total electron density were normalized (which corresponds to only one Lagrange multiplier), as in the TF model, then no shell structure can exist; hence such a system would violate the exclusion principle.

A counter-example has recently been constructed by *Lindgren* and *Salomonson* [20.172] showing that shell structure can indeed be generated through a single Lagrange multiplier. In addition, they have verified numerically that a local Kohn-Sham potential can reproduce to high accuracy the many-body electron density and the 2s eigenvalue for the 1s2s ^{3}S state of neutral helium.

References

20.1 L. H. Thomas: Proc. Camb. Philos. Soc. **23**, 542 (1927)
20.2 E. Fermi: Rend. Accad. Naz. Lincei **6**, 602 (1927)
20.3 E. Fermi: Rend. Accad. Naz. Lincei **7**, 342 (1928)
20.4 E. Fermi: Z. Phys. **48**, 73 (1928)
20.5 E. Fermi: Z. Phys. **49**, 550 (1928)
20.6 W. Lenz: Z. Phys. **77**, 713 (1932)
20.7 P. A. M. Dirac: Proc. Camb. Philos. Soc. **26**, 376 (1930)
20.8 C. F. von Weizsäcker: Z. Phys. **96**, 431 (1935)
20.9 E. H. Lieb, B. Simon: Phys. Rev. Lett. **31**, 681 (1973)
20.10 E. H. Lieb, B. Simon: Adv. Math. **23**, 22 (1977)
20.11 J. W. Sheldon: Phys. Rev. **99**, 1291 (1955)
20.12 E. Teller: Rev. Mod. Phys. **34**, 627 (1962)
20.13 N. Balàzs: Phys. Rev. **156**, 42 (1967)
20.14 E. H. Lieb, W. Thirring: Phys. Rev. Lett. **35**, 687 (1975)
20.15 E. H. Lieb: Rev. Mod. Phys. **53**, 603 (1981)
20.16 E. H. Lieb: Rev. Mod. Phys., **54**, 311(E) (1982)
20.17 L. Spruch: Rev. Mod. Phys. **63**, 151 (1991)
20.18 P. Gombás: *Die statistische Theorie des Atoms und ihre Anwendungen* (Springer, Berlin, Heidelberg 1949)
20.19 P. Gombás: Statistische Behandlung des Atoms. In: *Atome II*, Handbuch der Physik, Vol. 36, ed. by S. Flügge (Springer, Berlin, Heidelberg 1956) pp. 109–231
20.20 N. H. March: The Thomas–Fermi approximation in quantum mechanics, Adv. Phys. **6**, 1–98 (1957)
20.21 P. Hohenberg, W. Kohn: Phys. Rev. B **136**, 864 (1964)
20.22 W. Kohn, L. J. Sham: Phys. Rev. A **140**, 1133 (1965)
20.23 E. Fermi, E. Amaldi: Mem. Accad. d'Italia **6**, 119 (1934)
20.24 E. A. Milne: Proc. Camb. Philos. Soc. **23**, 794 (1927)
20.25 E. B. Baker: Phys. Rev. **36**, 630 (1930)
20.26 Y. Tal, M. Levy: Phys. Rev. A **23**, 408 (1981)
20.27 C. H. Hodges: Can. J. Phys. **51**, 1428 (1973)
20.28 R. Benguria, E. H. Lieb: J. Phys. B **18**, 1045 (1985)
20.29 R. Ahlrichs, T. Hoffmann-Ostenhof, M. Hoffmann-Ostenhof, J. D. Morgan III: Phys. Rev. A **23**, 2107 (1981) and references therein
20.30 T. Kato: Commun. Pure Appl. Math. **10**, 151 (1957)
20.31 J. P. Perdew, R. G. Parr, M. Levy, J. L. Balduz Jr.: Phys. Rev. Lett. **49**, 1691 (1982)
20.32 M. Levy, J. P. Perdew: In: *Density Functional Methods in Physics*, NATO ASI Series B, Vol. 123, ed. by R. M. Dreizler, J. da Providência (Plenum, New York 1985) pp. 11–30
20.33 E. H. Lieb: In: *Density Functional Methods in Physics*, NATO ASI Series B, Vol. 123, ed. by R. M. Dreizler, J. da Providência (Plenum, New York 1985) pp. 31–80
20.34 J. A. Goldstein, G. R. Rieder: J. Math. Phys. **32**, 2907 (1991)
20.35 J. A. Goldstein, G. R. Rieder: J. Math. Phys. **29**, 709 (1988)
20.36 J. M. C. Scott: Philos. Mag. **43**, 859 (1952)
20.37 E. H. Lieb: Rev. Mod. Phys. **48**, 553 (1976)
20.38 J. Schwinger: Phys. Rev. A **22**, 1827 (1980)
20.39 H. Siedentop, R. Weikard: Commun. Math. Phys. **112**, 471 (1987)
20.40 H. Siedentop, R. Weikard: Invent. Math. **97**, 159 (1989)
20.41 W. Hughes: Adv. Math. **79**, 213 (1990)
20.42 V. Bach: Rep. Math. Phys. **28**, 213 (1989)
20.43 V. J. Ivrii, I. M. Sigal: Annals of Math. **138**, 243 (1993)
20.44 J. Schwinger: Phys. Rev. A **24**, 2353 (1981)
20.45 N. H. March, J. S. Plaskett: Proc. Roy. Soc. Lond. A **235**, 419 (1956)
20.46 R. Shakeshaft, L. Spruch, J. Mann: J. Phys. B **14**, L121 (1981)
20.47 R. Shakeshaft, L. Spruch: Phys. Rev. A **23**, 2118 (1981)
20.48 B.-G. Englert, J. Schwinger: Phys. Rev. A **26**, 2322 (1982)
20.49 B.-G. Englert, J. Schwinger: Phys. Rev. A **29**, 2331, 2339, 2353 (1984)
20.50 B.-G. Englert, J. Schwinger: Phys. Rev. A **32**, 28, 36, 47 (1985)
20.51 A. Iantchenko, E. H. Lieb, H. Siedentop: J. Reine Angew. Math. **472**, 177 (1996)
20.52 O. J. Heilman, E. H. Lieb: Phys. Rev. A **52**, 3628 (1995)
20.53 C. L. Fefferman, L. A. Seco: Bull. Amer. Math. Soc. **23**, 525 (1990)
20.54 C. Fefferman, L. A. Seco: Adv. Math. **107**, 1 (1994)
20.55 C. Fefferman, L. Seco: Adv. Math. **95**, 145 (1992)
20.56 C. Fefferman, L. A. Seco: Revista Matemática Iberoamericana **9**, 409 (1993)
20.57 C. Fefferman, L. Seco: Adv. Math. **107**, 187 (1994)
20.58 C. Fefferman, L. Seco: Adv. Math. **108**, 263 (1994)
20.59 A. Cordoba, C. Fefferman, L. Seco: Proc. Natl. Acad. Sci. **91**, 5776 (1994)
20.60 A. Cordoba, C. Fefferman, L. A. Seco: Revista Matématica Iberoamericana **11**, 165 (1995)
20.61 C. Fefferman, L. Seco: Adv. Math. **111**, 88 (1995)
20.62 C. Fefferman, L. Seco: Adv. Math. **119**, 26 (1996)
20.63 C. Fefferman: Adv. Math. **124**, 100 (1996)
20.64 V. Heine: Solid State Phys. **24**, 1 (1970)
20.65 M. L. Cohen, V. Heine: Solid State Phys. **24**, 37 (1970)
20.66 V. Heine, D. Weaire: Solid State Phys. **24**, 249 (1970)
20.67 M. C. Payne, M. P. Teter, D. C. Allan, T. A. Arias, J. D. Joannopoulos: Rev. Mod. Phys. **64**, 1045 (1992)
20.68 A. K. Rajagopal: Adv. Chem. Phys. **41**, 59 (1980)
20.69 A. S. Bamzai, B. M. Deb: Rev. Mod. Phys. **53**, 95, 593(E) (1981)
20.70 R. G. Parr: Ann. Rev. Phys. Chem. **34**, 631 (1983)
20.71 J. Callaway, N. H. March: Solid State Phys. **38**, 135 (1984)
20.72 R. O. Jones, O. Gunnarsson: Rev. Mod. Phys. **61**, 689 (1989)
20.73 S. B. Trickey: In: *Conceptual trends in quantum chemistry*, Vol. 1, ed. by E. S. Kryachko, J. L. Calais (Kluwer, Amsterdam 1994) pp. 87–100

20.74 S. Lundqvist, N. H. March (Eds.): *Theory of the Inhomogeneous Electron Gas* (Plenum, New York 1983)
20.75 N. H. March, B. M. Deb (Eds.): *The Single-Particle Density in Physics and Chemistry* (Academic, New York 1987)
20.76 S. B. Trickey (Ed.): *Density Functional Theory of Many-Fermion Systems*, Advances in Quantum Chemistry, Vol. 21 (Academic, New York 1990)
20.77 J. Keller, J. L. Gázquez: In: *Density Functional Theory*, Lecture Notes in Physics, Vol. 187 (Springer, Berlin, Heidelberg 1983)
20.78 D. Langreth, H. Suhl (Eds.): *Many-Body Phenomena at Surfaces* (Academic, New York 1984)
20.79 J. P. Dahl, J. Avery (Eds.): *Local Density Approximations in Quantum Chemistry and Solid State Physics* (Plenum, New York 1984)
20.80 P. Phariseau, W. M. Temmermann (Eds.): *The Electronic Structure of Complex Systems*, NATO ASI Series B, Vol. 113 (Plenum, New York 1984)
20.81 R. M. Dreizler, J. da Providência (Eds.): *Density Functional Methods in Physics*, NATO ASI Series B, Vol. 123 (Plenum, New York 1985)
20.82 R. Erdahl, V. H. Smith Jr. (Eds.): *Density Matrices and Density Functionals* (Reidel, Dordrecht 1987)
20.83 J. K. Labanowski, J. W. Andzelm (Eds.): *Density Functional Methods in Chemistry* (Springer, New York 1991)
20.84 P. J. Grout, A. B. Lidiard (Eds.): Density Functional Theory and its Applications. To celebrate the 65th Birthday of Norman H. March, Philos. Mag. B **69**, 725–1074 (1994)
20.85 E. S. Kryachko, E. V. Ludeña: *Energy Density Functional Theory of Many-Electron Systems* (Kluwer, Dordrecht 1990)
20.86 E. K. U. Gross, R. M. Dreizler (Eds.): *Density Functional Theory*, NATO ASI Series B, Vol. 337 (Plenum, New York 1995)
20.87 A. Goursot, C. Mijoule, N. Russo (Eds.): CECAM Workshop, Theoret. Chim. Acta **91**, 111–266 (1995)
20.88 R. G. Parr, W. Yang: *Density-Functional Theory of Atoms and Molecules* (Clarendon, Oxford 1989)
20.89 R. M. Dreizler, E. K. U. Gross: *Density Functional Theory: An Approach to the Quantum Many-Body Problem* (Springer, Berlin, Heidelberg 1990)
20.90 T. L. Gilbert: Phys. Rev. B **12**, 2111 (1975)
20.91 J. E. Harriman: Phys. Rev. A **24**, 680 (1981)
20.92 E. H. Lieb: In: *Physics as Natural Philosophy: Essays in Honor of Laszlo Tisza on his 75th Birthday*, ed. by H. Feshbach, A. Shimony (MIT, Cambridge 1982) pp. 111–149; reprinted in [20.173]
20.93 M. Levy: Phys. Rev. A **26**, 1200 (1982)
20.94 J. K. Percus: Intern. J. Quantum Chem. **13**, 89 (1978)
20.95 M. Levy: Proc. Natl. Acad. Sci. USA **76**, 6062 (1979)
20.96 D. M. Ceperley, B. J. Alder: Phys. Rev. Lett. **45**, 566 (1980)
20.97 A. D. Becke: Phys. Rev. A **38**, 3098 (1988)
20.98 A. D. Becke: Intern. J. Quantum Chem. Symp. **23**, 599 (1989)
20.99 A. D. Becke: J. Chem. Phys. **96**, 2155 (1992)
20.100 A. D. Becke: J. Chem. Phys. **97**, 9173 (1992)
20.101 A. D. Becke: J. Chem. Phys. **98**, 5648 (1993)
20.102 A. D. Becke: J. Chem. Phys. **98**, 1372 (1993)
20.103 R. M. Dickson, A. D. Becke: J. Chem. Phys. **99**, 3898 (1993)
20.104 C. Lee, W. Yang, R. G. Parr: Phys. Rev. B **37**, 785 (1988)
20.105 R. Colle, O. Salvetti: Theoret. Chim. Acta **37**, 329 (1975)
20.106 R. Colle, O. Salvetti: Theoret. Chim. Acta **53**, 55 (1979)
20.107 R. Colle, O. Salvetti: J. Chem. Phys. **79**, 1404 (1983)
20.108 B. G. Johnson, P. M. W. Gill, J. A. Pople: J. Chem. Phys. **98**, 5612 (1993)
20.109 C. W. Murray, N. C. Handy, R. D. Amos: J. Chem. Phys. **98**, 7145 (1993)
20.110 G. J. Laming, V. Termath, N. C. Handy: J. Chem. Phys. **99**, 8765 (1993)
20.111 G. Ortiz, P. Ballone: Phys. Rev. B **43**, 6376 (1991)
20.112 C. Lee, Z. Zhou: Phys. Rev. A **44**, 1536 (1991)
20.113 E. Engel, J. A. Chevary, L. D. Macdonald, S. H. Vosko: Z. Phys. D **23**, 7 (1992)
20.114 S. H. Vosko, L. Wilk, M. Nusair: Can. J. Phys. 1200 (1980)
20.115 D. C. Langreth, M. J. Mehl: Phys. Rev. Lett. **47**, 446 (1981)
20.116 D. C. Langreth, M. J. Mehl: Phys. Rev. B **28**, 1809 (1983)
20.117 D. C. Langreth, M. J. Mehl: Phys. Rev. B **29**, 2310(E) (1984)
20.118 J. P. Perdew, Y. Wang: Phys. Rev. B **33**, 8800 (1986)
20.119 J. P. Perdew: Phys. Rev. B **33**, 8822 (1986)
20.120 J. P. Perdew: Phys. Rev. B **34**, 7406(E) (1986)
20.121 A. E. DePristo, J. D. Kress: J. Chem. Phys. **86**, 1425 (1987)
20.122 L. C. Wilson, M. Levy: Phys. Rev. B **41**, 12930 (1990)
20.123 J. P. Perdew: Physica B **172**, 1 (1991)
20.124 J. P. Perdew, Y. Wang: Phys. Rev. B **45**, 13244 (1992)
20.125 D. J. Lacks, R. G. Gordon: Phys. Rev. A **47**, 4681 (1993)
20.126 J. P. Perdew: Int. J. Quantum Chem. Symp. **27**, 93 (1993)
20.127 C. Filippi, C. J. Umrigar, M. Taut: J. Chem. Phys. **100**, 1290 (1994)
20.128 C. J. Umrigar, X. Gonze: Phys. Rev. A **50**, 3827 (1994)
20.129 C. J. Umrigar: High Performance Computing and its Application to the Physical Sciences. In: *Proc. of the Mardi Gras '93 Conference*, ed. by D. A. Browne, J. Callaway, J. P. Drayer, R. W. Haymaker, R. K. Kalia, J. E. Tohline, P. Vashishta (World Scientific, Singapore 1993) pp. 43–59
20.130 M. Levy, J. P. Perdew: Phys. Rev. A **32**, 2010 (1985)
20.131 M. Levy, H. Ou-Yang: Phys. Rev. A **38**, 625 (1988)
20.132 M. Levy: In: *Density Functional Theory of Many-Fermion Systems*, Advances in Quantum Chemistry,

Vol. 21, ed. by S. B. Trickey (Academic, New York 1990) pp. 69–95
20.133 M. Levy: In: *Density Functional Methods in Chemistry*, ed. by J. K. Labanowski, J. W. Andzelm (Springer, New York 1991) pp. 175–193
20.134 A. Görling, M. Levy: Phys. Rev. A **45**, 1509 (1992)
20.135 Q. Zhao, M. Levy, R. G. Parr: Phys. Rev. A **47**, 918 (1993)
20.136 A. Görling, M. Levy: Phys. Rev. B **47**, 13105 (1993)
20.137 M. Levy, J. P. Perdew: Phys. Rev. B **48**, 11638 (1993)
20.138 A. Görling, M. Levy: Phys. Rev. A **50**, 196 (1994)
20.139 M. Levy, A. Görling: Philos. Mag. B **69**, 763 (1994)
20.140 O. Gunnarson, B. J. Lundqvist: Phys. Rev. B **13**, 4274 (1976)
20.141 A. K. Theophilou: J. Phys. C **12**, 5419 (1979)
20.142 N. Hadjisavvas, A. Theophilou: Phys. Rev. A **30**, 2183 (1984)
20.143 N. Hadjisavvas, A. K. Theophilou: Philos. Mag. B **69**, 771 (1994)
20.144 J. Katriel: J. Phys. C **13**, L375 (1980)
20.145 A. Theophilou: Phys. Rev. A **32**, 720 (1985)
20.146 W. Kohn: Phys. Rev. A **34**, 737 (1986)
20.147 E. K. U. Gross, L. N. Oliveira, W. Kohn: Phys. Rev. A **37**, 2805, 2809 (1988)
20.148 L. N. Oliveira, E. K. U. Gross, W. Kohn: Phys. Rev. A **37**, 2821 (1988) and earlier references therein
20.149 H. Englisch, H. Fieseler, A. Haufe: Phys. Rev. A **37**, 4570 (1988)
20.150 L. N. Oliveira: In: *Density Functional Theory of Many-Fermion Systems*, Advances in Quantum Chemistry, Vol. 21, ed. by S. B. Trickey (Academic, New York 1990) pp. 135–154
20.151 L. N. Oliveira, E. K. U. Gross, W. Kohn: Int. J. Quantum Chem. Symp. **24**, 707 (1990)
20.152 R. M. Dreizler: Phys. Scr. **T46**, 167 (1993)
20.153 W. Kohn: Rev. Mod. Phys. **71**, 1253 (1999)
20.154 J. A. Pople: Rev. Mod. Phys. **71**, 1267 (1999)
20.155 R. F. Nalewajski (Ed.): *Functionals and Effective Potentials* I; *Relativistic and Time Dependent Extensions* II; *Interpretation, Atoms, Molecules and Clusters* III; *Theory of Chemical Reactivity* IV, Topics in Current Chemistry, Vol. 180–183 (Springer, New York 1996)
20.156 J. M. Seminario (Ed.): *Density Functional Theory*, Advances in Quantum Chemistry, Vol. 33 (Academic, New York 1999)
20.157 B. B. Laird, R. B. Ross, T. Ziegler: *Chemical Applications of Density-Functional Theory*, ACS Symposium Series, Vol. 629 (American Chemical Society, Washington 1996)
20.158 J. F. Dobson, G. Vignale, M. P. Das (Eds.): *Electronic Density Functional Theory: Recent Progress and New Directions* (Plenum, New York 1998)
20.159 D. Joubert (Ed.): *Density Functionals: Theory and Applications* (Springer, New York 1998)
20.160 W. Koch, M. C. Holthausen: *A Chemist's Guide to Density Functional Theory* (Wiley, New York 2001)
20.161 C. Fiolhais, F. Nogueira, M. Marques: *A Primer in Density Functional Theory*, Lecture Notes in Physics, Vol. 620 (Springer, New York 2003)
20.162 M. A. L. Marques, E. K. U. Gross: Annu. Rev. Phys. Chem. **55**, 427 (2004)
20.163 F. Furche, K. Burke: Ann. Rep. Comput. Chem. **1**, 19 (2004)
20.164 H. Englisch, R. Englisch: Phys. Status Solidi B **123**, 711 (1984)
20.165 H. Englisch, R. Englisch: Phys. Status Solidi B **124**, 373 (1984)
20.166 R. van Leeuwen: Adv. Quantum Chem. **43**, 25 (2003)
20.167 R. K. Nesbet: Phys. Rev. A **58**, R12 (1998)
20.168 R. K. Nesbet: Phys. Rev. A **65**, 010502(R) (2001)
20.169 R. K. Nesbet: Adv. Quantum Chem. **43**, 1 (2003)
20.170 R. K. Nesbet: e-print physics/0309120.
20.171 R. K. Nesbet: e-print physics/0309121.
20.172 I. Lindgren, S. Salomonson: Phys. Rev. A **70**, 032509 (2004)
20.173 E. H. Lieb: Intern. J. Quantum Chem. **24**, 243 (1983)

21. Atomic Structure: Multiconfiguration Hartree–Fock Theories

This chapter outlines variational methods for the determination of wave functions either in nonrelativistic LS or relativistic LSJ theory. The emphasis is on Hartree–Fock and multiconfiguration Hartree–Fock theory though configuration interaction methods are also mentioned. Some results from the application of these methods to a number of atomic properties are presented.

21.1 Hamiltonians: Schrödinger and Breit–Pauli 307
21.2 Wave Functions: LS and LSJ Coupling 308
21.3 Variational Principle 309
21.4 Hartree–Fock Theory 309
21.4.1 Diagonal Energy Parameters and Koopmans' Theorem 311
21.4.2 The Fixed-Core Hartree–Fock Approximation 311
21.4.3 Brillouin's Theorem 311
21.4.4 Properties of Hartree–Fock Functions 312
21.5 Multiconfiguration Hartree–Fock Theory 313
21.5.1 Z-Dependent Theory 313
21.5.2 The MCHF Approximation 314
21.5.3 Systematic Methods 315
21.5.4 Excited States 316
21.5.5 Autoionizing States 316
21.6 Configuration Interaction Methods 316
21.7 Atomic Properties 318
21.7.1 Isotope Effects 318
21.7.2 Hyperfine Effects 319
21.7.3 Metastable States and Lifetimes 320
21.7.4 Transition Probabilities 321
21.7.5 Electron Affinities 321
21.8 Summary 322
References 322

21.1 Hamiltonians: Schrödinger and Breit–Pauli

The state of a many-electron system is described by a wave function Ψ that is the solution of a partial differential equation (called the wave equation),

$$(H-E)\Psi=0\,, \tag{21.1}$$

where H is the Hamiltonian operator for the system and E the total energy. The operator H depends on the system (atomic, molecular, solid-state, etc.) as well as the underlying quantum mechanical formalism (nonrelativistic, Breit–Pauli, Dirac–Coulomb, or Dirac–Breit, etc.). In atomic systems, the Hamiltonian of the nonrelativistic Schrödinger equation is (in atomic units)

$$H_{\text{nr}}=-\frac{1}{2}\sum_{i=1}^{N}\left(\nabla_i^2+\frac{2Z}{r_i}\right)+\sum_{i<j}\frac{1}{r_{ij}}\,. \tag{21.2}$$

Here Z is the nuclear charge of the atom with N electrons, r_i is the distance of electron i from the nucleus, and r_{ij} is the distance between electron i and electron j. This equation was derived under the assumption of a point-nucleus of infinite mass. The term $2Z/r$ represents the nuclear attraction and $1/r_{ij}$ the inter-electron repulsion. The operator H_{nr} has both a discrete and continuous spectrum: for the former, $\Psi(\boldsymbol{r}_1,\boldsymbol{r}_2,\ldots,\boldsymbol{r}_N)$ has a probability interpretation and consequently must be square integrable. In the Breit–Pauli approximation, the Hamiltonian is extended to include relativistic corrections up to relative order $(\alpha Z)^2$. It is convenient to write the Breit–Pauli Hamiltonian as the sum [21.1]

$$H_{\text{BP}}=H_{\text{nr}}+H_{\text{rel}}\,, \tag{21.3}$$

where H_{rel} represents the relativistic contributions. The latter may again be subdivided into nonfine-structure (NFS) and fine structure (FS) contributions:

$$H_{\text{rel}}=H_{\text{NFS}}+H_{\text{FS}}\,. \tag{21.4}$$

The NFS contributions

$$H_{\text{NFS}} = H_{\text{mass}} + H_{\text{D}} + H_{\text{ssc}} + H_{\text{oo}} \quad (21.5)$$

shift nonrelativistic energy levels without splitting the levels. The mass-velocity term

$$H_{\text{mass}} = -\frac{\alpha^2}{8} \sum_i \nabla_i^4 \quad (21.6)$$

corrects for the variation of mass with velocity; the one- and two-body Darwin terms

$$H_{\text{D}} = -\frac{\alpha^2 Z}{8} \sum_i \nabla_i^2 r_i^{-1} + \frac{\alpha^2}{4} \sum_{i<j} \nabla_i^2 r_{ij}^{-1} \quad (21.7)$$

are the corrections of the one-electron Dirac equation due to the retardation of the electromagnetic field produced by an electron; the spin–spin contact term

$$H_{\text{ssc}} = -\frac{8\pi\alpha^2}{3} \sum_{i<j} (\boldsymbol{s}_i \cdot \boldsymbol{s}_j) \delta(\boldsymbol{r}_{ij}) \quad (21.8)$$

accounts for the interaction of the spin magnetic moments of two electrons occupying the same space; the orbit–orbit interaction

$$H_{\text{oo}} = -\frac{\alpha^2}{2} \sum_{i<j} \left(\frac{\boldsymbol{p}_i \cdot \boldsymbol{p}_j}{r_{ij}} + \frac{\boldsymbol{r}_{ij} (\boldsymbol{r}_{ij} \cdot \boldsymbol{p}_i) \cdot \boldsymbol{p}_j}{r_{ij}^3} \right) \quad (21.9)$$

accounts for the interaction of two orbital moments.

The FS contributions

$$H_{\text{FS}} = H_{\text{so}} + H_{\text{soo}} + H_{\text{ss}} \,, \quad (21.10)$$

split the nonrelativistic energy levels into a series of closely-spaced fine structure levels. The nuclear spin–orbit interaction

$$H_{\text{so}} = \frac{\alpha^2 Z}{2} \sum_i \frac{1}{r_i^3} (\boldsymbol{l}_i \cdot \boldsymbol{s}_i) \,, \quad (21.11)$$

represents the interaction of the spin and angular magnetic moments of an electron in the field of the nucleus. The spin–other-orbit term

$$H_{\text{soo}} = -\frac{\alpha^2}{2} \sum_{i \neq j} \left(\frac{\boldsymbol{r}_{ij}}{r_{ij}^3} \times \boldsymbol{p}_i \right) \cdot (\boldsymbol{s}_i + 2\boldsymbol{s}_j) \,, \quad (21.12)$$

and the spin–spin term

$$H_{\text{ss}} = \alpha^2 \sum_{i<j} \frac{1}{r_{ij}^3} \left[\boldsymbol{s}_i \cdot \boldsymbol{s}_j - \frac{3}{r_{ij}^2} (\boldsymbol{s}_i \cdot \boldsymbol{r}_{ij})(\boldsymbol{s}_j \cdot \boldsymbol{r}_{ij}) \right] , \quad (21.13)$$

arise from spin-dependent interactions with the other electrons in the system.

21.2 Wave Functions: LS and LSJ Coupling

In the configuration interaction model, the approximate wave function Ψ for a many-electron system is expanded in terms of configuration state functions (CSF).

The assignment of nl quantum numbers to electrons specifies a configuration, often written as $(n_1 l_1)^{q_1} (n_2 l_2)^{q_2} \cdots (n_m l_m)^{q_m}$, where q_i is the occupation of subshell $(n_i l_i)$. Associated with each subshell are one-electron spin-orbitals

$$\phi(r, \theta, \varphi, \sigma) = (1/r) P_{nl}(r) Y_{lm_l}(\theta, \varphi) \chi_{m_s}(\sigma) \,,$$

where $P_{nl}(r)$ is the radial function, $Y_{lm_l}(\theta, \varphi)$ a spherical harmonic, and $\chi_{m_s}(\sigma)$ a spinor. Each CSF is a linear combination of products of one-electron spin-orbitals, one for each electron in the system, such that the sum is an eigenfunction of the total angular momenta operators L^2, L_z and the total spin operators S^2, S_z. It can be considered to be a product of radial factors, one for each electron, an angular and a spin factor obtained by vector coupling methods. It also is required to be antisymmetric with respect to the interchange of any pair of electron co-ordinates. Often, the specification of the configuration and the final LS quantum numbers is sufficient to define the configuration state, but this is not always the case. Additional information about the order of coupling or the seniority of a subshell of equivalent electrons may be needed. Let γ specify the configuration information and any additional information about coupling to uniquely specify the configuration state function denoted by $\Phi(\gamma LS)$.

The wave function for a many-electron system is usually labeled in the same manner as a CSF and generally designates the largest component. Thus, in the LS approximation,

$$\Psi(\gamma LS) = \sum_{\alpha=1}^{M} c_\alpha \Phi(\gamma_\alpha LS) \,.$$

However, cases are known where the configuration states are so highly mixed that no dominant component can be found. Then the assignment is made using other criteria. Clearly no two states should have the same label.

In the LSJ scheme, the angular and spin momenta are coupled to form an eigenstate of the total momenta J^2, J_z. The label often still includes an LS designation, as in $2p^{2\,3}P_2$, but only the subscript J is a good quantum number. Thus,

$$\Psi(\gamma LSJ) = \sum_{\alpha=1}^{M} c_\alpha \Phi(\gamma_\alpha L_\alpha S_\alpha J) . \tag{21.14}$$

21.3 Variational Principle

Variational theory shows the equivalence between solutions of the wave equation, $(H-E)\psi = 0$, and stationary solutions of a functional. For bound states where approximate solutions Ψ are restricted to a square integrable subspace, say $\tilde{\mathcal{H}}$, the best solutions are those for which the energy functional

$$\mathcal{E}(\Psi) = \langle \Psi | H | \Psi \rangle / \langle \Psi | \Psi \rangle \tag{21.15}$$

is stationary. The condition $\delta\mathcal{E}(\Psi) = 0$ leads to

$$\langle \delta\Psi | H - E | \Psi \rangle = 0, \quad \forall\, \delta\Psi \in \tilde{\mathcal{H}}, \quad E = \mathcal{E}(\Psi) . \tag{21.16}$$

Several results readily follow. The eigenvalues of H are bounded from below. Let $E_0 \le E_1 \le \cdots$. Then

$$E_0 \le \mathcal{E}(\Psi), \quad \forall\, \Psi \in \tilde{\mathcal{H}} . \tag{21.17}$$

Consequently, for any approximate wave function, the computed energy is an upper bound to the exact lowest eigenvalue. By the Hylleraas–Undheim–MacDonald theorem (see Sect. 11.3.1) the computed excited states are also upper bounds to the exact eigenvalues, provided that the correct number of states lies below.

21.4 Hartree–Fock Theory

In the Hartree–Fock (HF) approximation, the approximate wave function consists of only one configuration state function. The radial function of each spin–orbital is assumed to depend only on the nl quantum numbers. These are determined using the variational principle and the nonrelativistic Schrödinger Hamiltonian.

The energy functional can be written as an energy expression for the matrix element $\langle \Phi(\gamma LS) | H | \Phi(\gamma LS) \rangle$. Racah algebra may be used to evaluate the spin–angular contributions, resulting in two types of radial integrals:

One-Body

Let $\mathcal{L}$ be the differential operator

$$\mathcal{L} = \frac{d^2}{dr^2} + \frac{2Z}{r} - \frac{\ell(\ell+1)}{r^2} . \tag{21.18}$$

Then,

$$I(nl, n'l') = -\frac{1}{2} \int_0^\infty P(nl; r) \times \mathcal{L} P(n'l'; r)\, dr . \tag{21.19}$$

Two-Body

The other integrals arise from the multipole expansion of the two-electron part

$$\frac{1}{r_{12}} = \sum_k \frac{r_<^k}{r_>^{k+1}} P^k(\cos\theta) , \tag{21.20}$$

where θ is the angle between the vectors $\boldsymbol{r}_1$ and $\boldsymbol{r}_2$, and $r_<, r_>$ are the lesser and greater of r_1, r_2, respectively. In general, let a, b, c, d be four nl quantum numbers, two from the left (bra) and two from the right (ket) CSF. Then

$$R^k(ab, cd) = \int_0^\infty \int_0^\infty P(a; r_1) P(b; r_2) \times \frac{r_<^k}{r_>^{k+1}} P(c; r_1) P(d; r_2)\, dr_1\, dr_2 , \tag{21.21}$$

which is called a Slater integral. It has the symmetries

$$R^k(ad, cb) \equiv R^k(cb, ad) \equiv R^k(cd, ab) \equiv R^k(ab, cd) .$$

In the Hartree–Fock approximation, the Slater integrals that occur depend on only two sets of quantum numbers. These special cases are denoted separately as

$$F^k(a,b) \equiv R^k(ab,ab) \text{ and}$$
$$G^k(a,b) \equiv R^k(ab,ba)\,. \tag{21.22}$$

The former is the direct interaction between a pair of orbitals whereas the latter arises from the exchange operator.

The energy expression may be written as

$$\begin{aligned}\mathcal{E}(\gamma LS) = \sum_i^m q_i \Bigg[& I(n_il_i, n_il_i) \\ & + \frac{q_i-1}{2}\sum_{k=0}^{2l_i} f_k(i,i) F^k(n_il_i, n_il_i)\Bigg] \\ & + \sum_{j<i} q_iq_j \Bigg[\sum_{k=0}^{2\min(l_i,l_j)} f_k(i,i) F^k(n_il_i, n_jl_j) \\ & + \sum_{k=|l_i-l_j|}^{l_i+l_j} g_k(i,j) G^k(n_il_i, n_jl_j)\Bigg]\,.\end{aligned} \tag{21.23}$$

In general, the coefficients $f_k(i,j)$ and $g_k(i,j)$ depend not only on the configuration, but also on the coupling. An extremely useful concept, introduced by Slater, is the "average energy of a configuration", $\mathcal{E}(\text{av})$. This is a weighted average of all possible LS terms, where the weighting factor is $(2L+1)(2S+1)$. In this case the coefficients have simple formulas that depend on the configuration:

$$\begin{aligned} f_k(i,i) &= 1,\, k=0\,, \\ &= -\left(\frac{2l_i+1}{4l_i+1}\right)\begin{pmatrix} l_i & k & l_i \\ 0 & 0 & 0\end{pmatrix}^2,\, k>0 \\ f_k(i,j) &= 1,\, i\neq j \text{ and } k=0\,, \\ &= 0,\, i\neq j \text{ and } k>0\,, \\ g_k(i,j) &= -\left(\frac{2l_i+1}{4l_i+1}\right)\begin{pmatrix} l_i & k & l_j \\ 0 & 0 & 0\end{pmatrix}^2\,. \end{aligned} \tag{21.24}$$

Let $\Im(P_a)$ be an integral that depends on P_a. Then $\delta\Im$ is defined as the first-order term of $\Im(P_a+\delta P_a)-\Im(P_a)$. To derive the first-order variation of the F^k and G^k integrals (and R^k in general) it is convenient to replace the variables (r_1, r_2) by (r, s) and introduce the function

$$\begin{aligned} Y^k(ab;r) &= r\int_0^\infty \frac{r_<^k}{r_>^{k+1}} P(a;s)P(b;s)\,\mathrm{d}s \\ &= \int_0^r \left(\frac{s}{r}\right)^k P(a;s)P(b;s)\,\mathrm{d}s \\ &\quad + \int_r^\infty \left(\frac{r}{s}\right)^{k+1} P(a;s)P(b;s)\,\mathrm{d}s \end{aligned} \tag{21.25}$$

so that

$$F^k(a,b) = \int_0^\infty P^2(a;r)\left(\frac{1}{r}\right) Y^k(bb;r)\,\mathrm{d}r \tag{21.26}$$

and

$$G^k(a,b) = \int_0^\infty P(a;r)P(b;r)\left(\frac{1}{r}\right) Y^k(ab;r)\,\mathrm{d}r\,. \tag{21.27}$$

Then the variations are

$$\begin{aligned} \delta I(a,b) &= -\frac{1}{2}(1+\delta_{a,b})\int_0^\infty \delta P(b;r)\mathcal{L}P(b;r)\,\mathrm{d}r\,, \\ \delta F^k(a,b) &= 2(1+\delta_{a,b})\int_0^\infty \delta P(a;r)P(a;r) \\ &\quad \times \left(\frac{1}{r}\right) Y^k(bb;r)\,\mathrm{d}r\,, \\ \delta G^k(a,b) &= 2\int_0^\infty \delta P(a;r)P(b;r)\left(\frac{1}{r}\right) Y^k(ab;r)\,\mathrm{d}r\,. \end{aligned} \tag{21.28}$$

The part of the expression that depends on $P(n_il_i;r)$, for example, is the negative of the removal energy of the entire $(n_il_i)^{q_i}$ subshell, say $-\bar{\mathcal{E}}[(n_il_i)^{q_i}]$. The stationary condition for a Hartree–Fock solution applies to this expression, but since the variations must be constrained in order to satisfy orthonormality assumptions, Lagrange multipliers λ_{ij} need to be introduced. The stationary condition applies to the functional

$$\begin{aligned}\mathcal{F}[P(n_il_i)] = & -\bar{\mathcal{E}}\left[(n_il_i)^{q_i}\right] \\ & + \sum_j \delta_{l_i,l_j}\lambda_{ij}\langle P(n_il_i)|P(n_jl_j)\rangle\,. \end{aligned} \tag{21.29}$$

Applying the variational conditions to each of the integrals, and dividing by $-q_i$, we get the equation

$$\left(\frac{d^2}{dr^2}+\frac{2}{r}[Z-Y(n_il_i;r)]-\frac{l(l+1)}{r^2}-\varepsilon_{ii}\right)\times P(n_il_i;r)=\frac{2}{r}X(n_il_i;r)+\sum_{j\neq i}\delta_{l_i,l_j}\varepsilon_{ij}P(n_jl_j;r)\,, \quad (21.30)$$

where

$$Y(n_il_i;r)=(q_i-1)\sum_k f_k(i,i)Y^k(n_il_in_il_i;r)+\sum_{j\neq i}q_jY^k(n_jl_jn_jl_j;r)\,,$$
$$X(n_il_i;r)=\sum_{j\neq i}q_j\sum_k g_k(i,j)Y^k(n_il_in_jl_j;r)\times P(n_jl_j;r)\,. \quad (21.31)$$

21.4.1 Diagonal Energy Parameters and Koopmans' Theorem

The diagonal (ε_{ii}) and off-diagonal (ε_{ij}) energy parameters are related to the Lagrange multipliers by $\varepsilon_{ii}=2\lambda_{ii}/q_i$ and $\varepsilon_{ij}=\lambda_{ij}/q_i$. In fact,

$$\varepsilon_{ii}=\frac{2}{q_i}\bar{E}\big[(n_il_i)^{q_i}\big]-(q_i-1)\sum_k f_k(i,i)F^k(n_il_i,n_il_i) \quad (21.32)$$

where $\bar{E}[(n_il_i)^{q_i}]$ is the Hartree–Fock value of the removal energy functional $\bar{\mathcal{E}}[(n_il_i)^{q_i}]$. In the special case where $q_i=1$, ε_{ii} is twice the removal energy, or ionization energy. This is often referred to as Koopmans' Theorem; but, as discussed in Sect. 21.4.3, if a rotation of the radial basis leaves the wave function unchanged while transforming the matrix of energy parameters (ε_{ij}), the removal energies are extreme values obtained by setting the off-diagonal energy parameters to zero. For multiply occupied shells, $\varepsilon_{ii}/2$ can be interpreted as an average removal energy, with a correction arising from the self-interaction.

21.4.2 The Fixed-Core Hartree–Fock Approximation

The above derivation has assumed that the solution is stationary with respect to *all* allowed variations. In practice, it may be convenient to assume that certain radial functions are "fixed" or "frozen". In other words, these radial functions are assumed to be given. Such approximations are often made for core orbitals and so, this is called a fixed-core HF approximation.

21.4.3 Brillouin's Theorem

The Hartree–Fock approximation has some special properties not possessed by other single configuration approximations. One such property is referred to as satisfying Brillouin's theorem, though, in complex systems with multiple open shells of the same symmetry, Brillouin's theorem is not always obeyed.

Let $\Phi^{HF}(\gamma LS)$ be a Hartree–Fock configuration state, where γ denotes the configuration and coupling scheme. With $\Phi^{HF}(\gamma LS)$ are associated the m Hartree–Fock radial functions $P^{HF}(n_1l_1;r)$, $P^{HF}(n_2l_2;r)$, $\ldots$, $P^{HF}(n_ml_m;r)$. These radial functions define the occupied orbitals. To this set may be added virtual orbitals that maintain the necessary orthonormality conditions. Let one of the radial functions (nl) be replaced by another $(n'l)$, either occupied or virtual, without any change in the coupling of the spin–angular factor. Let the resulting function be denoted by $F(nl\to n'l)$.

The perturbation of the Hartree–Fock radial function, $P(nl;r)\to P^{HF}(nl;r)+\epsilon P(n'l;r)$ induces a perturbation $\Phi^{HF}(\gamma LS)\to\Phi^{HF}(\gamma LS)+\epsilon F(nl\to n'l)$. But the Hartree–Fock energy is stationary with respect to such variations and so,

$$\langle\Phi^{HF}(\gamma LS)\mid H\mid F(nl\to n'l)\rangle=0\,. \quad (21.33)$$

If the function $F(nl\to n'l)$ is a CSF for a configuration γ^*, or proportional to one, then Brillouin's theorem is said to hold between the two configuration states. When $n'l$ is a virtual orbital, it may happen that $F(nl\to n'l)$ is a linear combination of CSFs, as in the $2p\to3p$ replacement from $2p^3\,{}^2P$, yielding a linear combination of $\{2p^2(^1S)3p,\ 2p^2(^3P)3p,\ 2p(^1D)3p\}$, the linear combination being determined by coefficients of fractional parentage. Thus, Brillouin's theorem will not hold for any of the three individual configuration states in the above equation, only for the linear combination.

When perturbations are constrained by orthogonality conditions between occupied orbitals, the perturbation is of the form of a rotation, where both are perturbed simultaneously,

$$P(nl;r)\to P(nl;r)+\epsilon P(n'l;r)\,,$$
$$P(n'l;r)\to P(n'l;r)-\epsilon P(nl;r)\,. \quad (21.34)$$

Then the perturbation has the form $F(nl\to n'l,n'l\to -nl)$. For $1s^22s\,{}^2S$, the simultaneous perturbations,

Part B | 21.4

$F(1s \to 2s, 2s \to -1s)$, lead to a linear combination of $\{1s2s^{2\,2}S,\ 1s^{3\,2}S\}$. The CSF for the $1s^{3\,2}S$ is identically zero by antisymmetry, and so Brillouin's theorem holds for the lithium-like ground state. In the $1s2s\,^3S$ state, neither the $1s \to 2s$ nor the $2s \to 1s$ substitutions are allowed; in fact, it can be shown that for these states, Brillouin's theorem holds for all mono-excited configurations. The same is not true for $1s2s\,^1S$ where the simultaneous perturbations lead to the condition

$$\left\langle \Phi^{\mathrm{HF}}(1s2s\,^1S)|H|\left[\Phi\left(1s^2\right)-\Phi\left(2s^2\right)\right]\right\rangle = 0\,. \tag{21.35}$$

Thus, Brillouin's theorem is not obeyed for either the $\Phi\left(1s^2\right)$ or $\Phi\left(2s^2\right)$ CSF in an HF calculation for 1s2s.

The importance of Brillouin's theorem lies in the fact that certain interactions have already been included to first order. This has the consequence that certain classes of diagrams can be omitted in many-body perturbation theory [21.2].

21.4.4 Properties of Hartree–Fock Functions

Term Dependence

The radial distribution for a given nl orbital may depend significantly on the LS term. A well known example is the $1s^22s2p$ configuration in Be which may couple to form either a ^3P or ^1P term. The energy expression differs only in the exchange interaction, $\pm(1/3)G^1(2s, 2p)$, where the $+$ refers to ^1P and the $-$ to ^3P. Clearly, the energies of these two terms differ. What is not quite as obvious is the extent to which the $P(2p)$ radial functions differ for the two states. The most affected orbital is the one that is least tightly bound, which in this case is the 2p orbital. Figure 21.1 shows the two radial functions. The ^1P orbital is far more diffuse (not as localized) as the one for ^3P. Such a change in an orbital is called LS term dependence.

Orbital Collapse

Another phenomenon, called *orbital collapse*, occurs when an orbital rapidly contracts as a function of the energy. This could be an LS dependent effect, but it can also occur along an isoelectronic sequence. This effect is most noticeable in the high-l orbitals. In hydrogen, the mean radius of an orbital is $\langle r\rangle = (1/2)[3n^2 - l(l+1)]\,a_0$. Thus, the higher-$l$ orbitals are more contracted; but in neutral systems, the high-l orbitals have a higher energy and are more diffuse. This is due, in part, to the $l(l+1)/r^2$ angular momentum barrier that appears in the definition of the $\mathcal{L}$ operator.

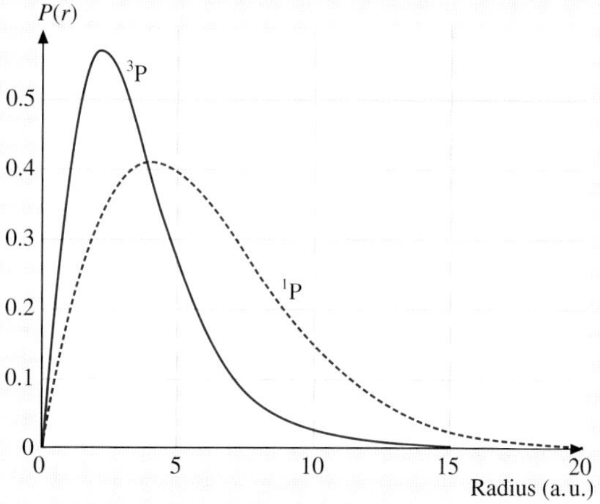

Fig. 21.1 A comparison of the 2p Hartree–Fock radial functions for the $1s2p\ ^{1,3}P$ states of Be

In the Hartree model, it is possible for $V(r) + l(l+1)/r^2$ to have two wells: an inner well and an outer shallow well. As the lowest eigenfunction changes rapidly from the outer well to the inner well, as Z changes, orbital collapse is said to occur.

Quantum Defects and Rydberg Series

Spectra of atoms often exhibit phenomena associated with a Rydberg series of states where one of the electrons is in an nl orbital, with n assuming a sequence of values. An example is the $1s^22snd\,^3D$ series in Be, $n = 3, 4, 5, \ldots$. For such a series, an useful concept is that of a quantum defect parameter δ. In hydrogen, the ionization energy (I_P) in atomic units is $1/(2n^2)$. In complex neutral systems, the effective charge would be the same at large r. As n increases, the mean radius becomes larger and the probability of the electron being in the hydrogen-like potential increases. Thus, one could define an effective quantum number, $n^* = n - \delta$, such that

$$I_P(nl) = (1/2)/(n-\delta)^2\,. \tag{21.36}$$

Table 21.1 The effective quantum number and quantum defect parameters of the $2snd$ Rydberg series in Be

	^3D		^1D	
n	n^*	$\delta(nl)$	n^*	$\delta(nl)$
3	2.968	0.032	3.014	−0.014
4	3.960	0.040	4.012	−0.012
5	4.957	0.043	5.013	−0.013
6	5.955	0.045	6.013	−0.013

For ionized systems, the equation must be modified to $I_P(nl) = (1/2)[(Z - N + 1)/(n - \delta)]^2$.

Often, this parameter is defined with respect to observed data, but it can also be used to evaluate Hartree–Fock energies, where $I_P = \varepsilon_{nl,nl}/2$, so that $\varepsilon_{nl,nl} = [(Z - N + 1)/(n - \delta)]^2$. Table 21.1 shows the effective quantum number and quantum defect for the Hartree–Fock $2snd$ ^{3}D and ^{1}D orbitals in Be as a function of n. For the triplet part the quantum defect is positive whereas for the singlet it is negative. This is the effect of exchange. Note that as n increases the quantum defect becomes constant. This observation is often the basis for determining ionization potentials from observed data.

21.5 Multiconfiguration Hartree–Fock Theory

The Hartree–Fock method predicts many atomic properties remarkably well; but when analyzed carefully, systematic discrepancies can be observed. Consider the ionization potentials tabulated in Table 21.2 compared with the observed values. In these calculations, the energy of the ion was computed using the same radial functions as for the atom. Thus, no "relaxation" effects were included.

The observed data include other effects as well, such as relativistic effects, finite mass and volume of the nucleus, but these are small for light atoms. For these systems, the largest source of discrepancy arises from the fact that the Hartree–Fock solution is an independent particle approximation to the exact solution of Schrödinger's equation. Neglected entirely is the notion of "correlation in the motion of the electrons"; each electron is assumed to move independently in a field determined by the other electrons. For this reason, the error in the energy was defined by Löwdin [21.3], to be the correlation energy, that is,

$$E^{\text{corr}} = E^{\text{exact}} - E^{\text{HF}} . \tag{21.37}$$

In this definition, E^{exact} is not the observed energy – it is the exact solution of Schrödinger's equation which itself is based on a number of assumptions.

21.5.1 Z-Dependent Theory

An indication of the important correlation corrections can be obtained from a perturbation theory study of the exact wave function. In the following section, we follow closely the approach taken by *Layzer* et al. [21.4] in the study of the Z-dependent structure of the total energy.

Let us introduce a new scaled length, $\rho = Zr$. Then the Hamiltonian becomes

$$H = Z^2\left(H_0 + Z^{-1}V\right) , \tag{21.38}$$

where

$$H_0 = -\frac{1}{2}\sum_i \left(\nabla_i^2 + \frac{2}{\rho}\right) , \tag{21.39}$$

$$V = \sum_{i>j} \frac{1}{\rho_{ij}} , \tag{21.40}$$

and Schrödinger's equation becomes

$$\left(H_0 + Z^{-1}V\right)\psi = \left(Z^{-2}E\right)\psi . \tag{21.41}$$

With $Z^{-1}V$ regarded as a perturbation, the expansions of ψ and E in the powers of Z^{-1} are

$$\psi = \psi_0 + Z^{-1}\psi_1 + Z^{-2}\psi_2 + \cdots , \tag{21.42}$$

in the ρ unit of length, and

$$E = Z^2\left(E_0 + Z^{-1}E_1 + Z^{-2}E_2 + Z^{-3}E_3 + \cdots\right) . \tag{21.43}$$

The zero-order equation is

$$H_0\psi_0 = E_0\psi_0 . \tag{21.44}$$

The solutions of this equation are products of hydrogenic orbitals.

Table 21.2 Observed and Hartree–Fock ionization potentials for the ground states of neutral atoms, in eV. (See also Table 10.3.)

Atom	Obs.	HF	Diff.
Li	5.39	5.34	0.05
Be	9.32	8.42	0.90
B	8.30	8.43	−0.13
C	11.26	11.79	−0.53
N	14.53	15.44	−0.91
O	13.62	14.45	−0.85
F	17.42	18.62	−1.20
Ne	21.56	23.14	−1.58

Let $|(nl)\nu LS\rangle$ be a configuration state function constructed by vector coupling methods from products of hydrogenic orbitals. Here (nl) represents a set of N quantum numbers $(n_1l_1, n_2l_2, \ldots, n_Nl_N)$ and ν any additional quantum numbers such as the coupling scheme or seniority needed to distinguish the different configuration states. Then,

$$H_0|(nl)\nu LS\rangle = E_0|(nl)\nu LS\rangle\ , \tag{21.45}$$

$$E_0 = -\frac{1}{2}\sum_i \frac{1}{n_i^2}\ . \tag{21.46}$$

Since E_0 is independent of the l_i, different configurations may have the same E_0; that is, E_0 is degenerate. According to first-order perturbation theory for degenerate states, ψ_0 then is a linear combination of the degenerate configuration state functions $|(nl')\nu' LS\rangle$ with the same set of principal quantum numbers n_i and parity π. The coefficients are components of an eigenvector of the interaction matrix $\langle(nl')\nu' LS|V|(nl)\nu LS\rangle$, and E_1 is the corresponding eigenvalue. This is the set of configurations referred to as the *complex* by *Layzer* [21.5] and denoted by the quantum numbers $(n)\pi LS$.

The zero-order wave function ψ_0 describes the many-electron system in a general way. It can be shown that the square of the expansion coefficients of ψ_0 over the degenerate set of configuration states can be interpreted as a probability that the many-electron system is in that configuration state, that ψ_1 is then a weighted linear combination of first-order corrections to each such configuration state. Let us now assume the nondegenerate case where $\psi_0 = \Phi(\gamma LS)$. The configurations interacting with γLS are of two types: those that differ by a single electron (single substitution S) and those that differ by two electrons (double substitution D). The former can be further subdivided into three categories:
(i) Those that differ from γLS by one principal quantum number but retain the same spin–angular coupling. These configuration states are part of *radial correlation*.
(ii) Those that differ by one principal quantum number but differ in their coupling. If the only change is the coupling of the spins, the configuration states are part of *spin-polarization*.
(iii) Those that differ in the angular momentum of one electron and are accompanied by a change in angular coupling of the configuration state and possibly also the spin coupling. The latter represent *orbital-polarization*.

The sums over intermediate states involve infinite sums. In practice, the set of orbitals is finite. In the nondegenerate case, these orbitals can be divided into occupied orbitals and unoccupied, or virtual, orbitals depending on whether or not they occur in the *reference configuration* that defines γLS. Single and double (SD) replacements of occupied orbitals by other occupied or virtual orbitals generate the set of configurations that interact with ψ_0. Consider the $1s^22s$ ground state of Li and the {1s, 2s, 3s, 4s, 2p, 3p, 4p, 3d, 4d, 4f} set of orbitals. The 1s and 2s orbitals are occupied and all the other orbitals are virtual orbitals, vl. The set of replacements can then be classified as follows:

Replacement	Configuration	Type of correlation
$1s \to 2s$	$1s2s^2$	Radial and spin-polarization
$2s \to vs$	$1s^2vs$	Radial
$1s \to vs$	$1svs(^1S)2s$	Radial
	$1svs(^3S)2s$	Spin-polarization
$1s2s \to vlv'l$	$1svlv'l$	Core-polarization
$1s^2 \to vlv'l$	$2svlv'l$	Core

The above discussion has considered only the Z-dependence of the wave function, but the notion can readily be extended to other properties. For example, in transition studies, the dipole transition matrix element decreases as $1/Z$, whereas the transition energy increases linearly with Z for $\Delta n = 0$ transitions, and quadratically as Z^2 otherwise. A first-order theory for oscillator strengths (FOTOS) [21.6] is based on similar concepts.

21.5.2 The MCHF Approximation

In the multiconfiguration Hartree–Fock (MCHF) method, the wave function is approximated by a linear combination of orthogonal configuration states so that

$$\Psi(\gamma LS) = \sum_i^m c_i \Phi(\gamma_i LS)\ , \tag{21.47}$$

where

$$\sum_i^m c_i^2 = 1\ .$$

Then the energy expression becomes

$$\mathcal{E}\left[\Psi(\gamma LS)\right] = \sum_i^m \sum_j^m c_i c_j H_{ij} \tag{21.48}$$

where

$$H_{ij} = \left\langle \Phi(\gamma_i LS)\middle|H\middle|\Phi(\gamma_j LS)\right\rangle\ . \tag{21.49}$$

Because $H_{ij} = H_{ji}$, the sum over i, j may be limited to the diagonals and the lower part of the matrix $\boldsymbol{H} = (H_{ij})$, called the *interaction* matrix. Let $\boldsymbol{c} = (c_i)$ be a column vector of the expansion coefficients, also called *mixing coefficients*. Then the energy of the system is

$$E = \boldsymbol{c}^{\mathrm{T}} \boldsymbol{H} \boldsymbol{c} . \tag{21.50}$$

Let $\boldsymbol{P}$ be the column vector of radial functions, $(P_a, P_b, \dots)^{\mathrm{T}}$. Since the interaction matrix elements depend on the radial functions, it is clear that the energy functional depends on both $\boldsymbol{P}$ and $\boldsymbol{c}$.

In deriving the MCHF equations, the energy needs to be expressed in terms of the radial functions and $\boldsymbol{c}$. From the theory of angular momenta, it follows that

$$H_{ij} = \sum_{ab} q_{ab}^{ij} I(a,b) + \sum_{abcd;k} v_{abcd;k}^{ij} R^k(ab,cd) , \tag{21.51}$$

where the sum over ab or $abcd$ covers the occupied orbitals of configuration states i and j.

Substituting into the energy expression (21.48), and interchanging the order of summation, we get

$$\mathcal{E}(\Psi) = \sum_{ab} q_{ab} I(a,b) + \sum_{abcd;k} v_{abcd;k} R^k(ab,cd) , \tag{21.52}$$

where

$$q_{ab} = \sum_i \sum_j c_i c_j q_{ab}^{ij} \quad \text{and}$$

$$v_{abcd;k} = \sum_i \sum_j c_i c_j v_{abcd;k}^{ij} .$$

In this form, the energy is expressed as a list of integrals and their contribution to the energy – a form suitable for the derivation of the MCHF radial equations.

As in the derivation of the Hartree–Fock equations, the stationary principle must be applied to a functional that includes Lagrange multipliers for all the constraints. Thus,

$$\mathcal{F}(\boldsymbol{P},\boldsymbol{c}) = \mathcal{E}(\Psi) + \sum_{a<b} \delta_{\ell_a,\ell_b} \lambda_{ab} \langle a|b\rangle - E \sum_i c_i^2 . \tag{21.53}$$

In deriving the stationary conditions with respect to variations in c_i, the most convenient form for $\mathcal{E}(\Psi)$ is (21.48), which leads to the secular equation

$$\boldsymbol{H}\boldsymbol{c} = E\boldsymbol{c} . \tag{21.54}$$

Thus, the Lagrange multiplier E is the total energy of the system.

The requirement of a stationary condition with respect to variations in the radial functions leads to a system of equations with exactly the same form as the Hartree–Fock equations except: (i) the occupation numbers q_{aa} are not integers but rather *expected* occupation numbers, and (ii) the function $X(r)$ arises not only from the exchange of electrons within a configuration state, but also from interactions between configuration states.

A solution of the MCHF problem requires the simultaneous solution of the secular equation and the variational radial equations. When the latter are assumed to be given, then only the secular problem needs to be solved and the problem is called a *configuration interaction* (CI) problem. If any radial function is optimized, the calculation is called a *multi-configuration Hartree–Fock* (MCHF) calculation. The iterative procedure for its solution is the MCHF-SCF method, details of which can be found in [21.7].

21.5.3 Systematic Methods

For MCHF calculations with only a few configuration states, the latter must be carefully chosen. According to first-order perturbation theory, the expansion coefficient is $c_i = \langle \Phi_i|H|\Phi_0\rangle/(E_0 - E_i)$. Clearly, configurations near in energy to the state under investigation are candidates for a strong interaction, but it is also important to remember the numerator in this expression. Generally, the latter is large if the electrons in the two configurations occupy the same region of space. Thus in the ground state of beryllium, the strongest mixing is with $2p^{2\,1}S$, even though the energy of the latter is far removed from the energy of $2s^{2\,1}S$. In highly ionized systems, the complex identifies the configurations that might interact strongly, but for neutral systems, there are many exceptions, particularly in atoms such as the transition metals where the 4s and 4p subshells may be filled before the 3d subshell.

For large scale computation, it is desirable to define the configuration states systematically in terms of an active set (AS) of orbitals. Several notations are commonly used for this set. A very simple designation is "the $n = 4$ active set" in that all orbitals up to and including those with principal quantum number $n = 4$ are included. Another notation common in quantum chemistry is one where the number of orbitals of each symmetry is specified, as in 4s3p2d1f.

Closely associated with this set is the set of configuration states that can be generated. The latter is referred to as the complete active space (CAS). Often the CAS is defined relative to a set of closed shells (or subshells) common to all configurations. For example, the complex for the $3s^23p^{23}P$ ground state of Si, is a neon-like core, $1s^22s^22p^6$, coupled to the $\{N = 4, \pi =$ even, $^3P\}$ CAS of the active set {3s, 3p, 3d}. If the AS is extended to {3s, 3p, 3d, 4s, 4p, 4d, 4f}, then all single, double, triple, quadruple (SDTQ) replacements from the outer four electrons are generated. Of course, the number of CSF increase rapidly, both with the size of the active set and with the number of electrons N defining the CAS. For this reason, other models may be used, such as multi-reference singles and doubles (MR–SD) modeled on the results of Z-dependent perturbation theory. These multi-reference functions may be extended to include all important contributors to a wave function. Calculations in which correlation is restricted to a few outer electrons is called *valence correlation*.

A consequence of a CAS expansion is that the MCHF problem is over determined: a rotation of radial functions of the same symmetry merely transforms the CSFs and the expansion coefficients. The wave function and energy do not change. When similar situations occur in Hartree–Fock theory, Koopmans' theorem requires that the lagrange multiplier associated with the rotation be set to zero. In MCHF calculations, the degree of freedom is usually used to eliminate a CSF or, more precisely, to determine that solution for which a specific CSF has a zero expansion coefficient. A generalized Brillouin's Theorem (GBT) then holds. In the case of the helium ground state (or any ns^2 pair function), applying GBT leads to the *natural orbital expansion* of the form $\{1s^2, 2s^2, 3s^2, \ldots, 2p^2, 3p^2, \ldots, 3d^2, \ldots\}$ in which all CSF differ by two electrons. For other symmetries, such as $1s2p\,^3P$, a *reduced form* [21.8] can also be defined in which all CSF differ by two electrons, but now involves different sets of orthonormal radial functions for the different partial waves as in $\{1s2p_1, 2s3p_1, 3s4p_1, \ldots, 2p_23d_1, 3p_24d_1, \ldots\}$. Such expansions yield the fastest rate of convergence, but are difficult to apply in complex systems. For a history of Brillouin's theorem and its use in solving multiconfiguration self-consistent field problems see [21.9].

21.5.4 Excited States

For ground states or atomic states lowest in their symmetry, the variational procedure is a minimization procedure, and consequently any approximate energy is an upper bound. For all others the energies are stationary. Such calculations may be difficult. One of the most difficult has been the HF calculation for $1s2s\,^1S$ and, once obtained is disappointing since the energy is too low. Excited state calculations become minimization problems through the use of the Hylleraas–Undheim–MacDonald theorem (Sect. 11.3.1). Consider a CAS calculation over the active set {1s, 2s} for which the CSFs are $\{1s^2, 1s2s, 2s^2\}$. In determining orbitals, we have a degree of freedom so one CSF may be removed. Selecting $2s^2$ has the consequence that the eigenvalue for the 1s2s state to be determined is now the second eigenvalue, and hence an upper bound to the exact.

21.5.5 Autoionizing States

The MCHF variational method may be applied to core excited states imbedded in the continuum, provided that certain CSFs with filled shells are omitted. An example is $1s2s2p^2P$ of Lithium. An MCHF calculation omitting $1s^2np$ configuration states represents the localized charge distribution of a state in the continuum. However, the resulting energy is not guaranteed to be an upper bound to the exact energy.

The saddle-point variational method, as described in Chapt. 25 is a minimax method for such states.

21.6 Configuration Interaction Methods

Configuration interaction (CI) methods differ from variational methods like MCHF in that the radial functions are assumed to be fixed, and hence known in advance. There are several situations where these methods can be used effectively:

MCHF with Breit–Pauli

Since the Breit–Pauli operators are valid only as first order perturbations, variational calculations for the Breit–Pauli Hamiltonian are not justified. Instead, MCHF calculations are performed to provide a basis for LSJ wave functions of (21.14). The expansion coefficients are obtained as a CI calculation. Usually such expansions are a concatenation of the expansions of all the LS terms of a configuration and possibly some close lying configurations. Most Breit–Pauli codes require a single orthonormal basis. In order to have an orbital basis that simultaneously describes the correlation of all these LS terms from a systematic procedure, the MCHF derivation described earlier has been extended to derive systems of coupled equations for linear combinations of energy functionals, one for each LS term and eigenstate. With this extension, it has been possible to perform "spectrum" calculations in which all LSJ levels up to a certain point in the spectrum are included. This requires a balanced correlation approach so that the energy differences with respect to the ground state is in good agreement with the observed excitation energy. Once wave functions have been obtained for each level, transition probabilities can be computed. With all E1 transitions between these levels and some E2/M1 transitions, the lifetimes of levels can be obtained [21.10].

Full-Core Methods

The full-core plus correlation method (FCPC) [21.16] is a configuration interaction method. It is mostly used for an atomic system in which a well defined "core" is present. For example, in a system such as the three-elec tron $1s^2nl$ or the four-electron $1s^2nln'l'$, the system has a 1s1s core. In this case, the wave function for an N-electron system is written as

$$\begin{aligned}\Psi(1,2,3,\ldots,N) &= \\ &\mathcal{A}\phi_{1s1s}(1,2)\sum_i C_i\Phi_i(3,4,\ldots,N) \\ &+\mathcal{A}\sum_j D_j\psi_j(1,2,3,\ldots,N)\,,\end{aligned}$$

where $\phi_{1s1s}(1,2)$, the wave function of the core, is used as a single term, and $\mathcal{A}$ denotes antisymmetrization. The relaxation of the core and other correlation effects are accounted for by the last term in this equation. The correlation effect of the 1s1s core is, in general, very strong and it is difficult to fully account for this effect in a conventional N-electron wave function. If inaccurate results are obtained from such a wave function, it may be difficult to distinguish between errors coming from the core part or from the other parts. In the FCPC, the correlation effect in $\phi_{1s1s}(1,2)$ can be precalculated to a desired accuracy such that ψ_j no longer contains the contribution from the unperturbed core. In physical processes such as ionization, optical transitions and others, the 1s1s core wave function in the final state does not change much from that of the initial state. Using the same $\phi_{1s1s}(1,2)$ for the core may minimize possible errors due to the inaccuracy of the core wave function. The application of this method is not limited to systems with 1s1s cores. For example, for the lithium-like $1s2snp\ ^4P^o$, for $n \geq 4$, the $1s2s\ ^3S$ can also be considered as an appropriate "core".

From a computational view point, the FCPC wave function has the advantage that it reduces the matrix size of the secular equation substantially. This drastically reduces the memory and CPU time needed on a computer. This method has been quite successful in getting accurate results for four-electron systems as shown in Table 21.3. See [21.13] for more details.

Table 21.3 Comparison of theoretical and experimental energies for Be $1s^22s^2\ ^1S$ in hartrees. All theoretical values include some form of extrapolation

Ref.	Method	$-E_{NR}$	$-E_{rel}$
[21.11]	FEM MCHF	14.667 37	14.669 67
[21.12]	MCHF	14.667 315	
[21.13]	Full-core CI	14.667 3492	14.669 6774
[21.14]	Semi-empirical	14.667 353	
[21.15]	Experiment		14.669 6759

Slater Type Orbitals

The essential characteristic of a radial function can be well represented by the expansion

$$P_{nl}(r) = \sum_j c_{jnl}\phi_{jnl}(r)\,,$$

where

$$\phi_{jnl}(r) = \left(\frac{(2\zeta_{jnl})^{2I_{jnl}+1}}{(2I_{jnl})}\right)^{1/2} r^{I_{jnl}}\exp(-\zeta_{jnl}r)$$

is a *Slater type orbital* (STO). Optimized sets of parameters have been tabulated for many Hartree–Fock wave functions and these may be used to represent the core [21.17]. Others may be added and selected orbitals *exponent optimized* (only the exponent is varied) so as to augment the basis. This method has been used effectively by A. Hibbert as implemented in the *Configuration Interaction Version 3* program [21.18].

Spline Basis

The analytic basis methods described in the previous section have some similarities with MCHF in that linear combinations of STOs first represent orbitals which then define the CSFs. These bases result in extensive cancellation as shown by *Hansen* et al. [21.19]. An expansion in B-spline basis functions, $B_{i,k}(r)$, which are a basis for a piecewise polynomial subspace (see Chapt. 8), provide a more flexible basis with very little cancellation in this representation.

By solving a radial equation with a well chosen potential using the spline Galerkin method that leads to a matrix eigenvalue problem, a complete set of orbitals for a piecewise polynomial space can be obtained. The resulting orthogonal orbital basis may then be used in

CI calculations. Such methods have been reviewed by *Hansen* et al. [21.19]. Methods that deal directly with the primitive B-spline basis have been applied to the study of Rydberg series [21.20], though orthogonality to target orbitals was still required. More recently, by using non-orthogonal theory, an Rmatrix-CI method has been developed that leads to a generalized eigenvalue problem. As in the close-coupling approximation, the wave function is expressed in terms of one or more "targets," each coupled to a Rydberg orbital, and an arbitrary number of pseudo-states. Each Rydberg orbital is expressed as a linear combination of B-splines but there is no requirement of orthogonality of the Rydberg orbital to the orbitals defining the target function [21.21]

21.7 Atomic Properties

The discussion so far has concentrated entirely on determining accurate wave functions based on expressions for the energy. Energy levels are a by-product of such calculations, but once a wave function is known a number of atomic properties can be evaluated.

21.7.1 Isotope Effects

The isotope shift observed in atomic transitions can be separated into a mass shift and a field shift. The mass shift is due to differences in the nuclear mass of the isotopes, and is the dominant effect for light atoms. The volume shift arises from the finite volume of the nuclear charge distribution, and is important for heavy atoms. From a physical point of view the field shift is the more interesting, since it yields information about differences in the nuclear charge distribution between the isotopes.

The isotope shift in a transition is given as the difference between the shift for the upper and lower level. The individual shifts are often large, but cancel, and therefore it is necessary to calculate them very accurately in order to get a reliable value for the difference. The energy shifts are evaluated in first-order perturbation theory with wave functions obtained from the zero-order Schrödinger Hamiltonian, H_0.

Mass Shift

Up to this point, the nuclear mass has been taken to be infinite. If instead it has a finite value M, an N-electron atom turns into an $(N+1)$-particle dynamical system. A transformation to center of mass $\boldsymbol{R}$ plus relative coordinates $\boldsymbol{r}'_i = \boldsymbol{r}_i - \boldsymbol{r}_{\text{nuc}}$ yields the transformed Hamiltonian [21.22]

$$H = \frac{P^2}{2M_{\text{tot}}} + \sum_{i=1}^{N} \frac{p_i'^2}{2\mu} + \sum_{i<j}^{N} \frac{\boldsymbol{p}'_i \cdot \boldsymbol{p}'_j}{M} + V(\boldsymbol{r}'_i) , \tag{21.55}$$

where $M_{\text{tot}} = M + Nm$ is the total mass and $\mu = Mm/(M+m)$ is the reduced mass. The first term is the kinetic energy of the center of mass, which can be neglected if $\boldsymbol{R}$ is an ignorable coordinate. For the remaining terms, introduce the scaled distances $\boldsymbol{\rho}_i = \boldsymbol{r}'_i/a_\mu$, where $a_\mu = (m/\mu)a_0$ is the scaled Bohr radius. If the potential is entirely Coulombic, then H assumes the form

$$H = -\frac{1}{2}\sum_{i=1}^{N} \nabla^2_{\rho_i} - \frac{\mu}{M}\sum_{i<j}^{N} \nabla_{\rho_i} \cdot \nabla_{\rho_j} + V(\rho_i) \tag{21.56}$$

in units of e^2/a_μ. This can be written in the form $H = H_0 + H_{\text{MP}}$, where H_0 includes the first and last terms of (21.56), and

$$H_{\text{MP}} = -\frac{\mu}{M}\sum_{i<j}^{N} \nabla_{\rho_i} \cdot \nabla_{\rho_j} \quad (e^2/a_\mu) \tag{21.57}$$

is the additional *mass polarization* term.

Equation (21.56) gives rise to two kinds of mass shifts. First, since H is identical to the infinite mass Hamiltonian, all its eigenvalues E_0 are multiplied by $a_0/a_\mu = \mu/m$, resulting in the *normal mass shift* (nms)

$$\Delta E_{\text{nms}} = -(\mu/M)E_0 . \tag{21.58}$$

Second, if H_{MP} is treated as a first order perturbation, then the resulting *specific mass shift* (sms) is

$$\Delta E_{\text{sms}} = \langle \Psi_0 \mid H_{\text{MP}} \mid \Psi_0 \rangle \quad (e^2/a_\mu) \tag{21.59}$$

where Ψ_0 is an eigenvector of H_0. The specific mass shift parameter S is defined by

$$S = \left\langle \Psi_0 \mid -\sum_{i<j}^{N} \nabla_{\rho_i} \cdot \nabla_{\rho_j} \mid \Psi_0 \right\rangle . \tag{21.60}$$

Field Shift

The field shift of an atomic energy level is due to the extended nuclear charge distribution. The field inside the nucleus deviates from the Coulomb field of a point charge, and this is reflected in the calculated levels. For light atoms, the resulting energy correction to the level E_0 is expressed in terms of the nonrelativistic electron probability at the origin, $|\Psi(0)|^2$:

$$E_{\text{fs}} = \frac{2\pi}{3} Z\langle r_{\text{N}}^2\rangle |\Psi(0)|^2 \,, \tag{21.61}$$

where $\langle r_{\text{N}}^2\rangle$ is the mean square radius of the nucleus. For heavier atoms ($Z > 10$) it becomes necessary to include a relativistic correction factor. Numerical values of this correction factor can be found in [21.23].

Table 21.4 shows the convergence of an MCHF calculation for the specific mass shift parameter and the electron density at the nucleus of the ground state of Boron II as the active set increases. The calculated $^{11}\text{B} - {}^{10}\text{B}$ isotope shift is 13.3 mÅ with an estimated uncertainty of 1%. The size of the isotope shift is similar to the limit of resolution of the Goddard High Resolution Spectrograph aboard the Hubble Space Telescope [21.24].

21.7.2 Hyperfine Effects

The magnetic hyperfine structure (hfs) is due to an interaction between the magnetic field generated by the electrons and the nuclear magnetic dipole moment. The interaction couples the nuclear and electronic angular momenta to a total momentum $\boldsymbol{F} = \boldsymbol{I} + \boldsymbol{J}$, and the interaction energy can be written as the expectation value of a scalar product between an electronic and a nuclear tensor operator [21.25] (see Chapt. 16)

$$W_{M1}(J) = \langle \gamma_I \gamma_J IJFM_F \mid \boldsymbol{T}^{(1)} \cdot \boldsymbol{M}^{(1)} \mid \gamma_I \gamma_J IJFM_F\rangle \,. \tag{21.62}$$

The nuclear operator $\boldsymbol{M}^{(1)}$ is related to the scalar magnetic dipole moment, μ_I, according to

$$\langle \gamma_I II \mid M_0^{(1)} \mid \gamma_I II\rangle = \mu_I \,. \tag{21.63}$$

Table 21.4 Specific mass shift parameter and electron density at the nucleus as a function of the active set

Active set	S (a.u.)	$\lvert\psi(0)\rvert^2$
HF	0.000 00	72.629
2s 1p	−0.020 17	72.452
3s 2p 1d	0.625 18	72.490
4s 3p 2d 1f	0.624 81	72.497
5s 4p 3d 2f 1g	0.601 69	72.501
6s 5p 4d 3f 2g 1h	0.598 03	72.503
7s 6p 5d 4f 3g 2h 1i	0.597 09	72.504

The magnetic dipole moments are known quantities, obtained with high accuracy from experiments. For a recent tabulation see [21.26].

The electronic operator is

$$\boldsymbol{T}^{(1)} = \frac{\alpha^2}{2}\sum_{i=1}^{N}\Big\{ 2\boldsymbol{l}^{(1)}(i) r_i^{-3} - g_s\sqrt{10}\big[\boldsymbol{C}^{(2)}(i)\times \boldsymbol{s}^{(1)}(i)\big]^{(1)} r_i^{-3} + g_s\frac{8}{3}\pi\delta(\boldsymbol{r}_i)\boldsymbol{s}^{(1)}(i)\Big\} \,, \tag{21.64}$$

where $g_s = 2.002\,3193$ is the electron spin g-factor, $\delta(\boldsymbol{r})$ the three-dimensional delta function and $\boldsymbol{C}_q^{(k)} = \sqrt{4\pi/(2k+1)}Y_{kq}$, with Y_{kq} being a normalized spherical harmonic. The first term of the electronic operator represents the magnetic field generated by the orbiting electric charges and is called the orbital term. The second term represents the field generated by the orbiting magnetic dipole moments, which are coupled to the spin of the electrons. This is the spin-dipole term. The last term represents the contact interaction between the nuclear magnetic dipole moment and the electron magnetic moment. It is called the Fermi contact term and contributes only for s-electrons.

By recoupling I and J, the interaction energy can be rewritten as

$$W_{M1}(J) = (-1)^{I+J-F}W(IJIJ;\,F1) \times \langle \gamma_J J\|\boldsymbol{T}^{(1)}\|\gamma_J J\rangle\langle \gamma_I I\|\boldsymbol{M}^{(1)}\|\gamma_I I\rangle \,, \tag{21.65}$$

where $W(IJIJ;\,F1)$ is a W coefficient of Racah. When the magnetic dipole interaction constant

$$A_J = \frac{\mu_I}{I}\frac{1}{[J(J+1)(2J+1)]^{\frac{1}{2}}}\langle \gamma_J J\|\boldsymbol{T}^{(1)}\|\gamma_J J\rangle \tag{21.66}$$

is introduced, the energy is given by

$$W_{M1}(J) = \frac{1}{2}A_J C \,, \tag{21.67}$$

where $C = F(F+1) - J(J+1) - I(I+1)$. In theoretical studies, the A-factor is often given as a linear combination of the hyperfine parameters

$$a_l = \langle \gamma LSM_L M_S \mid \sum_i l_0^{(1)}(i) r_i^{-3} \mid \gamma LSM_L M_S\rangle \,, \tag{21.68}$$

$$a_d = \langle \gamma LSM_L M_S | \sum_i 2C_0^{(2)}(i)s_0^{(1)}(i)r_i^{-3} | \gamma LSM_L M_S\rangle , \quad (21.69)$$

$$a_c = \langle \gamma LSM_L M_S | \sum_i 2s_0^{(1)}(i)r_i^{-2}\delta(r_i) | \gamma LSM_L M_S\rangle , \quad (21.70)$$

where $M_L = L$ and $M_S = S$. These parameters correspond to the orbital, spin-dipole and Fermi contact term of the electronic operator.

The electric hyperfine structure is due to the interaction between the electric field gradient produced by the electrons and the nonspherical charge distribution of the nucleus. The interaction energy is

$$W_{E2}(J) = \langle \gamma_I \gamma_J IJFM_F | \boldsymbol{T}^{(2)} \cdot \boldsymbol{M}^{(2)} | \gamma_I \gamma_J IJFM_F\rangle , \quad (21.71)$$

where the nuclear operator $\boldsymbol{M}^{(2)}$ is related to the scalar electric quadrupole moment, Q, according to

$$\langle \gamma_I II | M_0^{(2)} | \gamma_I II\rangle = \frac{Q}{2} . \quad (21.72)$$

The electronic operator is

$$\boldsymbol{T}^{(2)} = -\sum_{i=1}^{N} \boldsymbol{C}^{(2)}(i)r_i^{-3} , \quad (21.73)$$

and represents the electric field gradient. By introducing the electric quadrupole interaction constant B,

$$B_J = 2Q\left(\frac{J(2J-1)}{(J+1)(2J+1)(2J+3)}\right)^{\frac{1}{2}} \times \langle \gamma_J J \| \boldsymbol{T}^{(2)} \| \gamma_J J\rangle , \quad (21.74)$$

the interaction energy can be written as

$$W_{E2}(J) = B_J \frac{\frac{3}{4}C(C+1) - I(I+1)J(J+1)}{2I(2I-1)J(2J-1)} . \quad (21.75)$$

In many cases the electric hyperfine interaction is weaker than the magnetic and manifests itself as a small deviation from the Landé interval rule for the magnetic hfs. If the electronic part of the interaction can be calculated accurately, a value of the electric quadrupole moment Q, which is a difficult quantity to measure with direct nuclear techniques, can be deduced from the measured B-factor. A recent tabulation of nuclear quadrupole moments is given in [21.27].

Table 21.5 shows the convergence of an MCHF active space calculation for two different isotopes for the $1s^2 2s2p\,^1P$ state of B II [21.24]. Some oscillations are observed since each new "layer" of orbitals may localize in different regions of space.

Table 21.5 MCHF Hyperfine constants (in MHz) for the $1s^2 2s2p\ ^1P$ state of B II

	^{10}B		^{11}B	
Active set	A_1	B_1	A_1	B_1
HF	60.06	8.338	179.36	4.001
2s 1p	60.22	8.360	179.83	4.011
3s 2p 1d	60.98	8.193	182.11	3.932
4s 3p 2d 1f	60.05	7.677	179.34	3.684
5s 4p 3d 2f 1g	60.48	7.764	180.62	3.725
6s 5p 4d 3f 2g 1h	60.85	8.052	181.71	3.864
7s 6p 5d 4f 3g 2h 1i	60.81	8.002	181.60	3.840

21.7.3 Metastable States and Lifetimes

States above an ionization threshold may decay via a radiationless transition to a continuum. When the interaction with the continuum is spin-forbidden the state is metastable. The $nsnp^{2\,4}P$ of negative ions, for example, decay through Breit–Pauli interactions with $nsnp^2$ doublets. The latter, in turn interact with continuum states, thus opening a decay channel. Such metastable states may be treated as bound states.

The foundation for the theory of autoionization was laid down by Fano [21.28], where he developed a configuration interaction (CI) theory for autoionization. Let $\Psi_b(N;\gamma LS)$ be a normalized, multiconfiguration component of a discrete perturbor for an N-electron system, in which all orbitals are bound orbitals, decaying exponentially for large r. Let Ψ_k be an asymptotically normalized continuum component of the wave function, also for an N-electron system, at energy E, of the form,

$$\Psi_k(N;E\gamma' LS) = |\Psi_b(N-1;\beta\tilde{L}\tilde{S}) \cdot \phi(kl)LS\rangle , \quad (21.76)$$

where $\Psi_b(N-1;\beta\tilde{L}\tilde{S})$ is a bound MCHF wave function for the $(N-1)$-electron target system. Then, in the LS-coupling scheme, the width of the autoionizing state is given by the "Golden Rule"

$$\Gamma = 2\pi V_{E_0}^2 , \quad (21.77)$$

where

$$V_{E_0} = \langle \Psi_b(N;\gamma LS)|H - E_0|\Psi_k(N;E_0\gamma' LS)\rangle . \quad (21.78)$$

A similar formula can be derived for the LSJ-scheme and the Breit–Pauli Hamiltonian. In the above equation,

$E_0 = \langle \Psi_b(N;\gamma LS)|H|\Psi_b(N;\gamma LS)\rangle$. The energy of the core, or target, is $E_{\text{target}} = \langle \Psi_b(N-1;\beta\tilde{L}\tilde{S})|H|\Psi_b(N-1;\beta\tilde{L}\tilde{S})\rangle$, where H in the latter equation is the Hamiltonian for an $(N-1)$-electron system.

The wave function for the continuum component is assumed to have only one unknown, namely $\phi(kl) = P_{kl}(r)|ls\rangle$, the one-particle continuum function, where $|ls\rangle$ is the known spin–angular part. The radial equation for $P_{kl}(r)$ has exactly the same form as (21.26), except that $\varepsilon_{nl,nl} = -k^2$, where $E_0 = E_{\text{target}} + k^2/2$. The radial function can be obtained iteratively using an SCF procedure from outward integration.

One of the more accurate calculations of a lifetime is that for He^- $1s2s2p\,^4P_{5/2}$ by *Miecznik* et al. [21.29]. In this case, a lifetime of $(345\pm10)\,\mu s$ was found and compared with a recent experimental value of $(350\pm15)\,\mu s$ [21.30]. In this LSJ state, the ^4P interacts with the $1s^2kf\,^2F$. It was found that correlation in the target of the continuum orbital modified the lifetime. In calculations like these, it is always a question whether orthogonality conditions should be applied [as in projection operator formalism (see Chapt. 25)]. Some theorems relating to this question have been published by *Brage* et al. [21.31].

The position and widths of autoionizing resonances can also be determined from the study of photoionization or photodetachment using a spline Galerkin method together with inverse iteration. No boundary condition need be applied nor is there an inner and outer region. Resonance properties are obtained from a fitting of the cross-section [21.32]. Non-orthogonal, spline-based R-matrix methods with an inner and outer region have also been developed where there is no need of the "Buttle correction" and, at the same time, the non-orthogonality eliminates the need of certain pseudo-states [21.33]. For an extensive review of the application of splines in atomic and molecular physics, see [21.34].

21.7.4 Transition Probabilities

The most fundamental quantity for the probability of a transition from an initial state i to a final state f is the reduced matrix element related to the line strength by

$$S^{1/2} = \langle \Psi_i||O||\Psi_f\rangle\,, \tag{21.79}$$

where O is the transition operator. In the case of the electric dipole transition, there are two frequently used forms: the length form $O = \sum_j \boldsymbol{r}_j$, and the velocity form, $O = \sum_j \nabla_j/E_{if}$, where $E_{if} = E_f - E_i$. For exact *non-relativistic* wave functions, the two forms are equivalent, but for approximate wave functions, the matrix elements in general differ. Thus, the computation of the line strength and the oscillator strength, or f-value, where

$$f = (2/3)E_{if}S/[(2S_i+1)(2L_i+1)]\,,$$

forms a critical test of the wave function in nonrelativistic theory and also the model describing a many-electron system. The same operators are often also used in Breit–Pauli calculations. In this case, the velocity form of the operator has neglected some terms of order $(\alpha Z)^2$ and the length value is preferred.

Some well-known discrepancies between theory and experiment existed for more than a decade for the resonance transitions of Li and Na. For the nonrelativistic 2s–2p transition in Li, a full-core CI [21.35] calculation produced f-values of (0.747 04, 0.747 04, 0.753 78) for the length-, velocity-, and acceleration form, respectively. When relativistic corrections were included, the value changed to 0.747 15, in agreement with a number of theories, tabulated to only four decimal places. A fast beam-laser experiment by *Gaupp* et al. [21.36], yielded a value of 0.7416 ± 0.0012 but this value was revised in 1996 by a beam-gas-laser experiment in perfect agreement with theory [21.37]. In the case of the resonance transition in Na, when theory included correlation in the core as well as core-polarization and some relativistic effects, results were in agreement with an almost simultaneous cascade of new experimental values [21.38].

For MCHF calculations of transition data, an important consideration is that the matrix element is between two different states. For independently optimized wave functions, the orbitals of the initial and final states are not orthonormal, as assumed when Racah algebra techniques are used to evaluate the transition matrix element. Through the use of biorthogonal transformations, the orbitals and the coefficients of expansion of the wave functions of the initial and final state can be transformed efficiently so that standard Racah algebra techniques may be applied [21.39]. Table 21.6 shows the convergence of an MCHF calculation for the ground state of Boron from independently optimized wave functions.

21.7.5 Electron Affinities

By definition, the electron affinity is $A_e = E(A^-) - E(A)$. Thus it is the energy difference between Hamiltonians differing by one electron. Correlation plays a very important role in the binding of the extra electron in the negative ion. It has been known for a long time that the alkali metals have a positive electron affinity, but only recently has it been found, theoretically [21.40] and experimentally [21.41], that some of the alkaline earths may also be able to bind an extra electron. The d-electrons need to play a strong role, so Be and Mg, do not have a positive A_e, but according to the most recent experimental measurement, the electron affinity for Ca is 18 meV [21.42]. A calculation based on the spline methods and using a model potential with adjustable parameters to describe the core, obtained a value of 17.7 meV [21.43] in close agreement with experiment.

Atomic systems such as Ca are often thought of as two-electron systems and indeed, for qualitative descriptions, many observations can be explained. A number of physical effects need to be considered when predicting such electron affinities:
(i) Valence correlation is crucial for obtaining binding.
(ii) Intershell correlation between the valence electrons and the core (core polarization) is also important. The first electron may polarize the core considerably, but this is reduced by the second electron since the two avoid each other dynamically and prefer to be on opposite sides of the core.
(iii) Core rearrangement, which occurs because of the presence of one or more outer electrons, and is particularly large if any of these penetrate the core. In the case of Ca^+, the fixed-core Hartree–Fock energy of 3d ^2D state is 300 meV higher if computed in the fixed potential the Ca^{+2} ion compared with a fully variational calculation!
(iv) Intracore exclusion effects due to the presence of an extra valence electron which reduces the correlation of the core.
(v) Relativistic shift effects, which are present in observed levels and are particularly important for s-electrons.

Table 21.6 Convergence of transition data for the $1s^22s^22p\,^2P^o \rightarrow 1s^22s2p^2\,^2D$ transition in Boron with increasing active set

N	gf_l	gf_v	S_l	ΔE (cm^{-1})
3	0.6876	0.8156	2.5534	53 197
4	0.2456	0.2696	0.9959	48 720
5	0.2625	0.2695	1.0705	48 440
6	0.2891	0.2866	1.1868	48 125
7	0.2928	0.2900	1.2036	48 051
Expt. [a]	0.28(02)			47 857

[a] [21.44]

Model potential methods attempt to capture all but valence correlation in a potential so that calculations for Calcium, for example, can proceed as though for a two-electron system. A review of various theoretical approaches, many of which include different effects, may be found in [21.45].

For small systems such as Li, the electron affinity has been computed [21.46] to experimental accuracy [21.47] of (0.6176 ± 0.0002) eV. In neutral oxygen, it has been found that there is an isotope effect on the electron affinity [21.48].

21.8 Summary

More comprehensive treatments of atomic structure may be found [21.49, 50]. An atomic structure package is available for many of the calculations described here [21.51]. A review of the application of systematic procedures to the prediction of atomic properties has been published [21.52, 53].

References

21.1 R. Glass, A. Hibbert: Comput. Phys. Commun. **16**, 19 (1978)
21.2 I. Lindgren, J. Morrison: *Atomic Many-Body Theory* (Springer, Berlin, Heidelberg, New York 1982)
21.3 P.-O. Löwdin: Phys. Rev. **97**, 1509 (1955)
21.4 D. Layzer, Z. Horák, M. N. Lewis, D. P. Thompson: Ann. Phys. (N. Y.) **29**, 101 (1964)
21.5 D. Layzer: Ann. Phys. (N. Y.) **8**, 271 (1959)
21.6 C. A. Nicolaides, D. R. Beck: Chem. Phys. Lett. **35**, 79 (1975)
21.7 C. Froese Fischer: Comput. Phys. Rep. **3**, 273 (1986)
21.8 C. Froese Fischer: J. Comput. Phys. **13**, 502 (1973)
21.9 M. R. Godefroid, J. Lievin, J.-Y. Metz: Int. J. Quantum Chem. **XL**, 243 (1991)
21.10 G. Tachiev, C. Froese Fischer: J. Phys. B **32**, 5805 (1999)
21.11 S. A. Alexander, J. Olsen, P. Öster, H. M. Quiney, S. Salomonson, D. Sundholm: Phys. Rev. A **43**, 3355 (1991)
21.12 C. Froese Fischer: J. Phys. B **26**, 855 (1993)
21.13 K. T. Chung, X.-W. Zhu, Z.-W. Wang: Phys. Rev. A **47**, 1740 (1993)

21.14 E. Lindroth, H. Persson, S. Salomonson, A.-M. Mårtensson-Pendrill: Phys. Rev. A **45**, 1493 (1992)
21.15 R. L. Kelly: J. Phys. Chem. Ref. Data **16**, 1371–1678 (1987), Suppl. 1
21.16 K. T. Chung: *Many-Body Theory of Atomic Structure and Photoionization*, ed. by T. N. Chang (World Scientific, Singapore 1993) p. 83
21.17 E. Clementi, C. Roetti: At. Data Nucl. Data Tables **14**, 177 (1974)
21.18 A. Hibbert: Comput. Phys. Commun. **9**, 141 (1975)
21.19 M. Bentley, H.W. van der Hart, M. Landtman, G. M. S. Lister, Y.-T. Shen, N. Vaeck: Phys. Scr. **T47**, 7 (1993)
21.20 T. Brage, C. Froese Fischer: Phys. Scr. **49**, 651 (1994)
21.21 O. Zatsarinny, C. Froese Fischer: J. Phys. B **35**, 4669 (2002)
21.22 D. S. Hughes, C. Eckart: Phys. Rev. **36**, 694 (1930)
21.23 W. H. King: *Isotope Shifts in Atomic Spectra* (Plenum, New York 1984)
21.24 P. Jönsson, S. G. Johansson, C. Froese Fischer: Astrophys. J. **429**, L45 (1994)
21.25 I. Lindgren, A. Rosén: Case Stud. At. Phys. 1772 **4**, 93 (1974)
21.26 P. Raghavan: At. Data Nucl. Data Tables **42**, 189 (1989)
21.27 P. Pyykkö: Z. Naturforsch. **47a**, 189 (1992)
21.28 U. Fano: Phys. Rev. **129**, 1866 (1961)
21.29 G. Miecznik, T. Brage, C. Froese Fischer: Phys. Rev. A **47**, 3718 (1993)
21.30 T. Andersen, L. H. Andersen, P. Balling, H. K. Haugen, P. Hvelplund, W. W. Smith, K. Taulbjerg: Phys. Rev. A **47**, 890 (1993)
21.31 T. Brage, C. Froese Fischer, N. Vaeck: J. Phys. B **26**, 621 (1993)
21.32 J. Xi, C. Froese Fischer: Phys. Rev. A **53**, 3169 (1996)
21.33 O. Zatsarinny, C. Froese Fischer: J. Phys. B **35**, 4161 (2002)
21.34 H. Bachau, E. Cormier, J. E. Hansen, F. Martin: Rep. Prog. Phys. **64**, 1815 (2001)
21.35 K. T. Chung: Proceedings of the international conference on highly charged ions, Manhattan Kansas, 1992. In: *AIP Conference Proceedings #274* (AIP, New York 1993)
21.36 A. Gaupp, P. Kuske, H. J. Andrä: Phys. Rev. A **26**, 3351 (1982)
21.37 U. Volz, H. Schmoranzer: Phys. Scripta **T65**, 48 (1996)
21.38 P. Jönsson, A. Ynnerman, C. Froese Fischer, M. R. Godefroid, J. Olsesn: Phys. Rev. A **53**, 4021 (1996)
21.39 J. Olsen, M. R. Godefroid, P. Jönsson, P. Å. Malmqvist, C. Froese Fischer: Phys. Rev. E **52**, 449 (1995)
21.40 C. Froese Fischer, J. B. Lagowski, S. H. Vosko: Phys. Rev. Lett. **59**, 2263 (1987)
21.41 D. J. Pegg, J. S. Thompson, R. N. Compton, G. D. Alton: Phys. Rev. Lett. **59**, 2267 (1989)
21.42 K. W. McLaughlin, D. W. Duquette: Phys. Rev. Lett. **72**, 1176 (1994)
21.43 H. W. van der Hart, C. Laughlin, J. E. Hansen: Phys. Rev. Lett. **71**, 1506 (1993)
21.44 T. R. O'Brian, J. E. Lawler: Astron. Astrophys. **255**, 420 (1992)
21.45 C. Froese Fischer, T. Brage: Can. J. Phys. **70**, 1283 (1992)
21.46 C. Froese Fischer: J. Phys. B **26**, 855 (1993)
21.47 J. Dellwo, Y. Liu, D. J. Pegg, G. D. Alton: Phys. Rev. A **45**, 1544 (1992)
21.48 M. R. Godefroid, C. Froese Fischer: Phys. Rev. A **60**, R2637 (1999)
21.49 I. I. Sobel'man: *Introduction to the Theory of Atomic Spectra* (Pergamon, Oxford 1972)
21.50 R. D. Cowan: *The Theory of Atomic Structure and Spectra* (Univ. of California Press, Berkeley 1981)
21.51 C. Froese Fischer: Computer Phys. Commun. **64**, 369 (1991)
21.52 C. Froese Fischer, P. Jönsson: Comput. Phys. Commun. **84**, 37 (1994)
21.53 C. Froese Fischer, T. Brage, P. Jönsson: *Computational Atomic Structure: An MCHF approach* (IOP, Bristol 1997)

22. Relativistic Atomic Structure

Relativistic quantum mechanics is required for the description of atoms and molecules whenever their orbital electrons probe regions of space with high potential energy near the atomic nuclei. Primary effects of a relativistic description include changes to spatial and momentum distributions; spin–orbit interactions; quantum electrodynamic corrections such as the Lamb shift; and vacuum polarization. Secondary effects in many-electron systems arise from shielding of the outer electrons by the distributions of electrons in penetrating orbitals; they change orbital binding energies and dimensions and so modify the order in which atomic shells are filled in the lower rows of the Periodic Table.

Relativistic atomic and molecular structure theory can be regarded as a simplification of the fundamental description provided by quantum electrodynamics (QED). This treats the atom or molecule as an assembly of electrons and atomic nuclei interacting primarily by exchanging photons. This model is far too difficult and general for most purposes, and simplifications are required. The most important of these is the representation of the nuclei as classical charge distributions, or even as point particles. Since the motion of the nuclei is usually slow relative to the electrons, it is often adequate to treat the nuclear motion nonrelativistically, or even to start from nuclei in fixed positions, correcting subsequently for nuclear motion.

The emphasis in this chapter is on relativistic methods for the calculation of atomic structure for general many-electron atoms based on an effective Hamiltonian derived from QED in the manner sketched in Sect. 22.2 below. An understanding of the Dirac equation, its solutions and their numerical approximation, is essential material for studying many-electron systems, just as the corresponding properties of the Schrödinger equation underpin Chapt. 21. We shall use atomic units throughout. Where it aids interpretation we shall, however, insert factors of c, m_e and $\hbar$. In these units, the velocity of light, c, has the numerical value $\alpha^{-1} = 137.035\,999\,11(46)$, where α is the fine structure constant.

22.1 **Mathematical Preliminaries** ... 326
- 22.1.1 Relativistic Notation: Minkowski Space-Time ... 326
- 22.1.2 Lorentz Transformations ... 326
- 22.1.3 Classification of Lorentz Transformations ... 326
- 22.1.4 Contravariant and Covariant Vectors ... 327
- 22.1.5 Poincaré Transformations ... 327

22.2 **Dirac's Equation** ... 328
- 22.2.1 Characterization of Dirac States ... 328
- 22.2.2 The Charge-Current 4-Vector ... 328

22.3 **QED: Relativistic Atomic and Molecular Structure** ... 329
- 22.3.1 The QED Equations of Motion ... 329
- 22.3.2 The Quantized Electron–Positron Field ... 329
- 22.3.3 Quantized Electromagnetic Field ... 330
- 22.3.4 QED Perturbation Theory ... 331
- 22.3.5 Propagators ... 333
- 22.3.6 Effective Interaction of Electrons ... 333

22.4 **Many-Body Theory For Atoms** ... 334
- 22.4.1 Effective Hamiltonians ... 335
- 22.4.2 Nonrelativistic Limit: Breit–Pauli Hamiltonian ... 335
- 22.4.3 Perturbation Theory: Nondegenerate Case ... 335
- 22.4.4 Perturbation Theory: Open-Shell Case ... 336
- 22.4.5 Perturbation Theory: Algorithms ... 337

22.5 **Spherical Symmetry** ... 337
- 22.5.1 Eigenstates of Angular Momentum ... 337
- 22.5.2 Eigenstates of Dirac Hamiltonian in Spherical Coordinates ... 338
- 22.5.3 Radial Amplitudes ... 340
- 22.5.4 Square Integrable Solutions ... 341
- 22.5.5 Hydrogenic Solutions ... 342
- 22.5.6 The Free Electron Problem in Spherical Coordinates ... 343

22.6 **Numerical Approximation of Central Field Dirac Equations** 344
22.6.1 Finite Differences 344
22.6.2 Expansion Methods 345
22.6.3 Catalogue of Basis Sets for Atomic Calculations 347
22.7 **Many-Body Calculations** 350
22.7.1 Atomic States 350
22.7.2 Slater Determinants 350
22.7.3 Configurational States 350
22.7.4 CSF Expansion 350
22.7.5 Matrix Element Construction 350
22.7.6 Dirac–Hartree–Fock and Other Theories 351
22.7.7 Radiative Corrections 353
22.7.8 Radiative Processes 353
22.8 **Recent Developments** 354
22.8.1 Technical Advances 354
22.8.2 Software for Relativistic Atomic Structure and Properties 354
References 355

22.1 Mathematical Preliminaries

22.1.1 Relativistic Notation: Minkowski Space-Time

An *event* in Minkowski space-time is defined by a 4-vector $x = \{x^\mu\}$ $(\mu = 0, 1, 2, 3)$ where $x^0 = ct$ is the time coordinate and x^1, x^2, x^3 are Cartesian coordinates in 3-space. The bilinear form (The Einstein suffix convention, in which repeated pairs of Greek subscripts are assumed to be summed over all values 0, 1, 2, 3, will be used where necessary in this chapter.)

$$(x, y) = x^\mu g_{\mu\nu} y^\nu \,, \tag{22.1}$$

in which

$$g = (g_{\mu\nu}) = (g^{\mu\nu}) = \begin{pmatrix} 1 & 0 & 0 & 0 \\ 0 & -1 & 0 & 0 \\ 0 & 0 & -1 & 0 \\ 0 & 0 & 0 & -1 \end{pmatrix} \tag{22.2}$$

are called *metric coefficients*, defines the metric of Minkowski space.

22.1.2 Lorentz Transformations

Lorentz transformations are defined as linear mappings Λ such that

$$(\Lambda x, \Lambda y) = (x, y) \tag{22.3}$$

so that

$$g_{\mu\nu} = \Lambda^\rho{}_\mu \, g_{\rho\sigma} \, \Lambda^\sigma{}_\nu \,. \tag{22.4}$$

This furnishes 10 equations connecting the 16 components of Λ; at most 6 components can be regarded as independent parameters. The (infinite) set of Λ matrices forms a regular matrix group (with respect to matrix multiplication) called the Lorentz group, $\mathcal{L}$, designated O(3,1) [22.1, 2].

22.1.3 Classification of Lorentz Transformations

Rotations
Lorentz transformations with matrices of the form

$$\Lambda = \begin{pmatrix} 1 & \mathbf{0}^\top \\ \mathbf{0} & \boldsymbol{R} \end{pmatrix} , \tag{22.5}$$

where $\boldsymbol{R} \in \mathrm{SO}(3)$ is an orthogonal 3×3 matrix with determinant $+1$, and $\mathbf{0}$ is a null three dimensional column vector, correspond to three-dimensional space rotations. They form a group isomorphic to SO(3).

Boosts
Lorentz transformations with matrices of the form

$$\Lambda = \begin{pmatrix} \gamma(v) & \gamma(v)\boldsymbol{v}^\top \\ \gamma(v)\boldsymbol{v} & \boldsymbol{I}_3 + (\gamma(v) - 1)\boldsymbol{n}\boldsymbol{n}^\top \end{pmatrix} , \tag{22.6}$$

with $\boldsymbol{v} = v\boldsymbol{n}$ a three dimensional column vector, $|\boldsymbol{n}| = 1$, $v = |\boldsymbol{v}|$ and $\gamma(v) = (1 - v^2/c^2)^{-1/2}$, are called *boosts*. The matrix Λ describes an 'active' transformation from an inertial frame in which a free classical particle is at rest to another inertial frame in which its velocity is $\boldsymbol{v}$.

Boosts form a submanifold of $\mathcal{L}$ though they do not in general form a subgroup. However, the set of boosts in a fixed direction $\boldsymbol{n}$ form a one-parameter subgroup.

Discrete Transformations
The matrices

$$P = \begin{pmatrix} 1 & \mathbf{0}^\top \\ \mathbf{0} & -\boldsymbol{I}_3 \end{pmatrix} , \quad T = \begin{pmatrix} -1 & \mathbf{0}^\top \\ \mathbf{0} & \boldsymbol{I}_3 \end{pmatrix} \quad \text{with} \quad PT = -\boldsymbol{I}_4 \tag{22.7}$$

are called discrete Lorentz transformations and form a subgroup of the Lorentz group along with the identity I_4. The matrix P performs *space* or *parity inversion*; the matrix T performs *time reversal*.

Infinitesimal Lorentz Transformations

The proper Lorentz transformations close to the identity are of particular importance: they have the form

$$\begin{aligned} \Lambda^{\mu}_{\nu} &= \delta^{\mu}_{\nu} + \epsilon\omega^{\mu}_{\nu} + \cdots , \\ (\Lambda^{-1})^{\mu}_{\nu} &= \delta^{\mu}_{\nu} - \epsilon\omega^{\mu}_{\nu} + \cdots , \end{aligned} \tag{22.8}$$

where

$$\omega_{\mu\nu} = -\omega_{\nu\mu}$$

and ϵ is infinitesimal. The *infinitesimal generators*, components $\omega_{\mu\nu}$, can be treated as quantum mechanical observables: see Sect. 22.2.1.

The Lorentz Group

The Lorentz group $\mathcal{L}$ is a Lie group with a six-dimensional group manifold which has four connected components, namely

$$\mathcal{L}^{\uparrow}_{+} \equiv \left\{ \Lambda \in \mathcal{L} \,|\, \Lambda^{0}{}_{0} \geq 1, \det\Lambda = +1 \right\} , \tag{22.9}$$

$$\mathcal{L}^{\uparrow}_{-} \equiv \left\{ \Lambda \in \mathcal{L} \,|\, \Lambda^{0}{}_{0} \geq 1, \det\Lambda = -1 \right\} = P\mathcal{L}^{\uparrow}_{+} , \tag{22.10}$$

$$\mathcal{L}^{\downarrow}_{+} \equiv \left\{ \Lambda \in \mathcal{L} \,|\, \Lambda^{0}{}_{0} \leq 1, \det\Lambda = -1 \right\} = T\mathcal{L}^{\uparrow}_{+} , \tag{22.11}$$

$$\mathcal{L}^{\downarrow}_{+} \equiv \left\{ \Lambda \in \mathcal{L} \,|\, \Lambda^{0}{}_{0} \leq 1, \det\Lambda = +1 \right\} = PT\mathcal{L}^{\uparrow}_{+} . \tag{22.12}$$

The connected component $\mathcal{L}^{\uparrow}_{+}$ containing the identity is a Lie subgroup of $\mathcal{L}$ called the *proper Lorentz group*. All its group elements can be obtained from boosts and rotations. It is not simply connected because the subgroup of rotations is not simply connected. The group is also noncompact as the subset of boosts is homeomorphic to $\mathcal{R}^3$.

These topological properties of $\mathcal{L}^{\uparrow}_{+}$ are essential for understanding the properties of relativistic wave equations. In particular the multiple connectedness forces the introduction of spinor representations, and to the appearance of half-integer angular momenta or spin.

22.1.4 Contravariant and Covariant Vectors

Contravariant 4-vectors (such as events x) transform according to the rule

$$a^{\mu} \mapsto a^{\mu\prime} = \Lambda^{\mu}{}_{\nu} a^{\nu} . \tag{22.13}$$

Covariant 4-vectors can be formed by writing

$$a_{\mu} = g_{\mu\nu} a^{\nu} , \tag{22.14}$$

so that

$$a^{\mu} a_{\mu} = a^{\mu} g_{\mu\nu} a^{\nu} = (a, a) \tag{22.15}$$

is invariant with respect to Lorentz transformations. Similarly, we can construct a contravariant 4-vector from a covariant one by writing

$$a^{\mu} = g^{\mu\nu} a_{\nu} . \tag{22.16}$$

The transformation law for covariant vectors is therefore

$$a_{\mu} \mapsto a'_{\mu} = [\Lambda^{-1}]^{\nu}{}_{\mu} a_{\nu} . \tag{22.17}$$

The most important example of a covariant vector is the 4-momentum operator

$$p_{\mu} = \mathrm{i}\frac{\partial}{\partial x^{\nu}} \quad \mu = 0, 1, 2, 3 . \tag{22.18}$$

From this we derive the *contravariant* 4-momentum operator with components p^{μ} by writing

$$\begin{aligned} p^{\mu} &= g^{\mu\nu} p_{\nu} \\ &= \left(\mathrm{i}\frac{\partial}{\partial x^0}, -\mathrm{i}\frac{\partial}{\partial x^1}, -\mathrm{i}\frac{\partial}{\partial x^2}, -\mathrm{i}\frac{\partial}{\partial x^3} \right) , \end{aligned} \tag{22.19}$$

in agreement with nonrelativistic expressions.

22.1.5 Poincaré Transformations

More generally, a Poincaré transformation is obtained by combining Lorentz transformations and space-time translations:

$$\Pi(x) = \Lambda x + a . \tag{22.20}$$

The set of all Poincaré transformations, $\Pi = (a, \Lambda)$, with the composition law

$$(a_1, \Lambda_1)(a_2, \Lambda_2) = (a_1 + \Lambda_1 a_2, \Lambda_1 \Lambda_2) , \tag{22.21}$$

also forms a group, $\mathcal{P}$.

Properties of the Lorentz and Poincaré groups will be introduced as needed. For a concise account of their properties see [22.3]. For more detail on relativistic quantum mechanics in general see textbooks such as [22.3, 4].

22.2 Dirac's Equation

We present Dirac's equation for an electron in a classical electromagnetic field defined by the 4-potential $A_\mu(x)$:

Covariant Form

$$\left\{\gamma^\mu\left[p_\mu - eA_\mu(x)\right] - m_{\rm e}c\right\}\psi(x) = 0\,. \qquad (22.22)$$

where

$\gamma^\mu(\mu = 0, 1, 2, 3)$, are 4×4 matrices.

$\psi(x)$ is a 4-component spinor wave function.

Here, and elsewhere in this chapter, identity matrices are omitted when it is safe to do so.

Dirac Gamma Matrices

- Anticommutation relations:

$$\gamma^\mu\gamma^\nu + \gamma^\nu\gamma^\mu = 2g^{\mu\nu}\,.$$

- Standard representation:

$$\gamma^0 = \begin{pmatrix}1 & 0\\ 0 & -1\end{pmatrix}$$

$$\gamma^i = \begin{pmatrix}0 & \sigma^i\\ -\sigma^i & 0\end{pmatrix} \quad \mathrm{i} = 1, 2, 3\,,$$

where $\sigma_i(i = 1, 2, 3)$ are Pauli matrices [22.1–4].

Noncovariant Form

In the majority of atomic structure calculations, a frame of reference is chosen in which the nuclear center is taken to be fixed at the origin. In this case it is convenient to write Dirac's equation in noncovariant form. Then functions of

$$x = (x^0, \boldsymbol{x})\,,$$

where $x^0 = ct$, can be regarded as functions of the time t and the position 3-vector $\boldsymbol{x}$, so that (22.22) is replaced by

$$\mathrm{i}\frac{\partial}{\partial t}\psi(\boldsymbol{x}, t) = \hat{\boldsymbol{h}}_{\rm D}\psi(\boldsymbol{x}, t) \qquad (22.23)$$

where the scalar and 3-vector potentials are defined by

$$\phi(\boldsymbol{x}, t) = cA^0(x)\,,$$
$$\boldsymbol{A}(\boldsymbol{x}, t) = \left[A^1(x), A^2(x), A^3(x)\right]\,, \qquad (22.24)$$

and

$$\hat{\boldsymbol{h}}_{\rm D} = \left\{c\boldsymbol{\alpha}\cdot\left[\boldsymbol{p} - e\boldsymbol{A}(\boldsymbol{x}, t)\right] + e\phi(\boldsymbol{x}, t) + \beta m_{\rm e}c^2\right\} \qquad (22.25)$$

defines the Dirac Hamiltonian. The matrices $\boldsymbol{\alpha}$, with Cartesian components $(\alpha^1, \alpha^2, \alpha^3)$, and β, have the standard representation

$$\beta = \gamma^0 = \begin{pmatrix}1 & 0\\ 0 & -1\end{pmatrix} \qquad (22.26)$$

$$\alpha^i = \gamma^0\gamma^i = \begin{pmatrix}0 & \sigma^i\\ \sigma^i & 0\end{pmatrix} \quad i = 1, 2, 3\,. \qquad (22.27)$$

22.2.1 Characterization of Dirac States

The solutions of Dirac's equation span representations of the Lorentz and Poincaré groups, whose infinitesimal generators can be identified with physical observables. The Lorentz group algebra has 10 independent self-adjoint infinitesimal generators: these can be taken to be the components p^μ of the four-momentum (which generate displacements in each of the four coordinate directions); the three generators, J_i, of rotations about each coordinate axis in space; and the pseudovector w_μ. The irreducible representations can be characterized by invariants

$$(p, p) = m_e^2c^2\,, \qquad (22.28)$$

$$(w, w) = -m_e^2c^2\boldsymbol{s}^2 = -\frac{3}{4}m_e^2c^2\,, \qquad (22.29)$$

where p is the momentum four-vector and $\boldsymbol{s}$ is a 3-vector defined in terms of Pauli matrices by

$$s_i = \frac{1}{2}\sigma_i\,, \quad i = 1, 2, 3\,.$$

which can be interpreted as the electronic angular momentum (intrinsic spin) in its rest frame. For more detail see [22.3] and the original papers [22.5, 6].

22.2.2 The Charge-Current 4-Vector

Dirac's equation (22.22) is covariant with respect to Lorentz (22.3) and Poincaré (22.20) transformations, provided that there exists a nonsingular 4×4 matrix $S(\Lambda)$

with the property

$$\psi'(x) = S(\Lambda)\psi\left[\Lambda^{-1}(x-a)\right] . \quad (22.30)$$

The matrices $S(\Lambda)$ are characterized by the equation

$$S^{-1}(\Lambda)\gamma^{\lambda}S(\Lambda) = \Lambda^{\lambda}{}_{\mu}\gamma^{\mu} . \quad (22.31)$$

The most important observable expression required in this chapter is the *charge–current four-vector*

$$j^{\mu} = ec\overline{\psi}(x)\gamma^{\mu}\psi(x) , \quad (22.32)$$

where the Dirac adjoint is defined by

$$\overline{\psi}(x) = \psi^{\dagger}(x)\gamma^{0} , \quad (22.33)$$

and the dagger denotes spinor conjugation and transposition. Since

$$\begin{aligned}\overline{\psi}'(x) &= \overline{\psi}[\Lambda^{-1}(x-a)]\gamma^{0}S(\Lambda)^{\dagger}\gamma^{0}\\ &= \overline{\psi}[\Lambda^{-1}(x-a)]S^{-1}(\Lambda) ,\end{aligned}$$

$j^{\mu}(x)$ transforms as a 4-vector

$$j^{\mu\prime}(x) = \Lambda^{\mu}{}_{\nu}j^{\nu}(x)$$

by virtue of (22.31). The component $j^{0}(x)$ can be interpreted as a multiple of the *charge density* $\rho(x)$,

$$j^{0}(x) = ec\rho(x) = ec\overline{\psi}(x)\gamma^{0}\psi(x) = ec\psi^{\dagger}(x)\psi(x) \quad (22.34)$$

and the space-like components as the *current density*

$$j^{i}(x) = ec\overline{\psi}(x)\gamma^{i}\psi(x) = ec\psi^{\dagger}(x)\alpha^{i}\psi(x) . \quad (22.35)$$

The charge–current density satisfies a continuity equation, which in noncovariant form reads

$$\frac{\partial\rho(x)}{\partial t} + \sum_{i=1}^{3}\frac{\partial j^{i}(x)}{\partial x^{i}} = 0 ,$$

or, in covariant notation,

$$\partial_{\mu}j^{\mu} = 0 . \quad (22.36)$$

This is readily proved by using the Dirac equation (22.22) and its Dirac adjoint. Equation (22.36) is clearly invariant under Poincaré transformations, and this yields the important property that *electric charge* is conserved in Dirac theory.

22.3 QED: Relativistic Atomic and Molecular Structure

22.3.1 The QED Equations of Motion

The conventional starting point [22.7–10] for deriving equations of motion in quantum electrodynamics (QED) is a Lagrangian density of the form

$$\mathcal{L}(x) = \mathcal{L}_{\mathrm{em}}(x) + \mathcal{L}_{\mathrm{e}}(x) + \mathcal{L}_{\mathrm{int}}(x) . \quad (22.37)$$

The first term is the Lagrangian density for the free electromagnetic field, $F^{\mu\nu}(x)$,

$$\mathcal{L}_{\mathrm{em}}(x) = -\frac{1}{4}F^{\mu\nu}F_{\mu\nu} , \quad (22.38)$$

the second term is the Lagrangian density for the electron–positron field in the presence of the external potential $A^{\mu}_{\mathrm{ext}}(x)$,

$$\mathcal{L}_{\mathrm{e}}(x) = \bar{\psi}(x)\left\{\gamma_{\mu}\left[p^{\mu} - eA^{\mu}_{\mathrm{ext}}(x)\right] - m_{\mathrm{e}}c\right\}\psi(x) . \quad (22.39)$$

We assume that the electromagnetic fields are expressible in terms of the four-potentials by

$$F^{\mu\nu} = \partial^{\mu}A^{\nu}_{\mathrm{tot}} - \partial^{\nu}A^{\mu}_{\mathrm{tot}} ,$$

where

$$A^{\mu}_{\mathrm{tot}}(x) = A^{\mu}_{\mathrm{ext}}(x) + A^{\mu}(x)$$

is the sum of a four-potential $A^{\mu}_{\mathrm{ext}}(x)$ describing the fields generated by classical external charge–current distributions, and a quantized field $A^{\mu}(x)$ which through

$$\mathcal{L}_{\mathrm{int}}(x) = -j_{\mu}(x)A^{\mu}(x) , \quad (22.40)$$

accounts for the interaction between the uncoupled electrons and the radiation field. The field equations deduced from (22.37) are

$$\begin{aligned}\left\{\gamma_{\mu}\left[p^{\mu} - A^{\mu}_{\mathrm{ext}}(x)\right] - m_{\mathrm{e}}c\right\}\psi(x) &= \gamma_{\mu}(x)\psi(x)A^{\mu}(x)\\ \partial_{\mu}F^{\mu\nu}(x) &= j^{\nu}(x) , \quad (22.41)\end{aligned}$$

and clearly exhibit the coupling between the fields.

Quantum electrodynamics requires the solution of the system (22.41) when $A^{\mu}(x)$, $\psi(x)$ and its adjoint $\bar{\psi}(x)$ are quantized fields. This formulation is purely formal: it ignores all questions of zero-point energies, normal ordering of operators, choice of gauge associated with the quantized photon field, or the need to include (infinite) counterterms to render the theory finite.

22.3.2 The Quantized Electron–Positron Field

Furry's bound interaction picture of QED [22.7, 11] exploits the fact that a one-electron model is often a good starting point for a more accurate calculation of atomic or molecular properties. The electrons are described by a field operator

$$\psi(x) = \sum_{E_m > E_F} a_m \psi_m(x) + \sum_{E_n < E_F} b_n^\dagger \psi_n(x) , \tag{22.42}$$

where $E_F \geq -mc^2$ is a "Fermi level" separating the states describing electrons (bound and continuum) from the positron states (lower continuum) in the chosen time-independent model potential $V(\boldsymbol{r})$. Equation (22.42) is written as if the spectrum were entirely discrete, as in finite matrix models; more generally, this must be replaced by integrals over the continuum states together with a sum over the bound states. We assume that the amplitudes $\psi_m(x)$ are orthonormalized (which can be achieved, for example, by enclosing the system in a finite box). The operators a_m and $a_m^\dagger$ respectively annihilate and create electrons, and b_n and $b_n^\dagger$ perform the same role for vacancies in the "negative energy" states, which we interpret as antiparticles (positrons). These operators satisfy the *anticommutation* rules (see Sect. 6.1.1)

$$\begin{aligned} \left[a_m, a_{m'}^\dagger\right] &= \delta_{m,m'} , \\ \left[b_n, b_{n'}^\dagger\right] &= \delta_{n,n'} , \end{aligned} \tag{22.43}$$

where $[a, b] = ab + ba$. All other anticommutators vanish. The operator representing the number of electrons in state m is then

$$N_m = a_m^\dagger a_m , \tag{22.44}$$

having the eigenvalues 0 or 1; the states of a system of noninteracting electrons and positrons can therefore be labeled by listing the occupation numbers, 0 or 1 of the one-electron states participating.

We define the *vacuum state* as the (reference) state $|0\rangle$ in which $N_m = N_n = 0$ for all m, n, so that

$$a_m|0\rangle = b_n|0\rangle = 0 . \tag{22.45}$$

The operator representing the total number of particles is given by

$$N = \int \psi^\dagger(x)\psi(x)\,\mathrm{d}\boldsymbol{x} = \sum_{E_m > E_F} N_m + \sum_{E_n < E_F} (1 - N_n) .$$

This is not quite satisfactory: $N = \sum_{E_n < E_F} 1$ is infinite for the vacuum state, as are the total charge and energy of the vacuum.

These infinite "zero-point" values can be eliminated by introducing *normal ordered operators*. A product of annihilation and creation operators is in normal order if it is rearranged so that all annihilation operators are to the right of all creation operators. Such a product has a null value in the vacuum state. In performing the rearrangement, each anticommutator is treated as if it were zero. We denote normal ordering by placing the operators between colons. Thus $: a_m^\dagger a_m := a_m^\dagger a_m$ whilst $: b_n b_n^\dagger : = -b_n^\dagger b_n$. This means that if we *redefine* N by

$$N = \int : \psi^\dagger(x)\psi(x) : \mathrm{d}\boldsymbol{x} = \sum_{E_m > E_F} N_m - \sum_{E_n < E_F} N_n , \tag{22.46}$$

then $\langle 0|N|0\rangle = 0$. The same trick eliminates the infinity from the total energy of the vacuum;

$$\begin{aligned} H_0 &= \int : \psi^\dagger(x)\hat{\boldsymbol{h}}_D\psi(x) : \mathrm{d}\boldsymbol{x} \\ &= \sum_{E_m > E_F} N_m E_m - \sum_{E_n < E_F} N_n E_n , \end{aligned} \tag{22.47}$$

so that $\langle 0|H_0|0\rangle = 0$.

The current density operator is given by the commutator of two field variables

$$j^\mu(x) = -\frac{1}{2} ec \left[\bar{\psi}(x)\gamma^\mu, \psi(x)\right] , \tag{22.48}$$

where the Dirac adjoint, $\bar{\psi}(x)$ is defined by (22.33). The definition (22.48) differs from (22.32) by expressing the total current as the difference between the electron (negatively charged) and positron (positively charged) currents. We can write

$$j^\mu(x) =: j^\mu(x) : + ec \operatorname{Tr}\left[\gamma^\mu S_F(x, x)\right] , \tag{22.49}$$

where $S_F(x, y)$ is the Feynman causal propagator, defined below. Since $\langle 0| : j^\mu(x) : |0\rangle = 0$, the last term in (22.49) is the *vacuum polarization current* due to the asymmetry between positive and negative energy states induced by the external field. From this, the net charge of the system is

$$\begin{aligned} Q &= \frac{1}{c} \int j^0(x)\,\mathrm{d}^3x \\ &= -e\left(\sum_{E_m > E_F} N_m - \sum_{E_n < E_F} N_n\right) + Q_{\text{vac}} . \end{aligned} \tag{22.50}$$

Q_{vac} is the total charge of the vacuum, which vanishes for free electrons, but is finite in the presence of an external field (the phenomenon of *vacuum polarization*). Note that whilst Q is conserved for all processes, the total number of particles need not be; it is always possible to add virtual states incorporating electron–positron pairs without changing Q.

22.3.3 Quantized Electromagnetic Field

The four-potential of the quantized electromagnetic field can be expressed in terms of a spectral expansion over the field modes in, for example, plane waves

$$A_\mu(x) = \int \frac{d^3k}{2(2\pi)^3 k_0} \sum_{\lambda=0}^{3} \Big[q^{(\lambda)}(k)\epsilon_\mu^{(\lambda)}(k)\,e^{-ik\cdot x} + q^{(\lambda)\dagger}(k)\epsilon_\mu^{(\lambda)*}(k)\,e^{ik\cdot x} \Big] . \tag{22.51}$$

The vectors $\epsilon_\mu^{(\lambda)}(k)$ describe the polarization modes; there are four linearly independent vectors, which may be assumed real, for each k on the positive light cone. Two of these ($\lambda = 1, 2$) can be chosen perpendicular to the photon momentum $\boldsymbol{k}$ (*transverse polarization*); one component ($\lambda = 3$) along $\boldsymbol{k}$ (*longitudinal polarization*); and the final component ($\lambda = 0$) is time-like (*scalar polarization*). The operators $q^{(\lambda)}(k)$ and $q^{(\lambda)\dagger}(k)$ describe respectively photon absorption and emission. They satisfy commutation relations

$$\begin{aligned} \Big[q^{(\lambda)}(k), q^{(\lambda')\dagger}(k')\Big] &= \delta_{\lambda,\lambda'}\delta_{k,k'},\ \lambda, \lambda' = 1, 2, 3 \\ \Big[q^{(0)}(k), q^{(0)\dagger}(k')\Big] &= -\delta_{k,k'}\ ; \end{aligned} \tag{22.52}$$

all other commutators vanish. The photon vacuum state, $|0\rangle_\gamma$, is such that

$$q^{(\lambda)}(k)|0\rangle_\gamma = 0 . \tag{22.53}$$

Further details may be found in the texts [22.7–10].

22.3.4 QED Perturbation Theory

The textbook perturbation theory of QED, see for example [22.7,8,10,12] and other works, has been adapted for applications to relativistic atomic and molecular structure and is also the source of methods of nonrelativistic many-body perturbation theory (MBPT). We offer a brief sketch emphasizing details not found in the standard texts.

The Perturbation Expansion

The Lagrangian approach leads to an interaction Hamiltonian

$$H_I = -\int j^\mu(x)A_\mu(x)\,d\boldsymbol{x} . \tag{22.54}$$

In the interaction representation, this gives an equation of motion

$$i\hbar\partial_t|\Phi(t)\rangle = H_I(t)|\Phi(t)\rangle , \tag{22.55}$$

where $|\Phi(t)\rangle$ is the QED state vector, and

$$H_I(t) = \exp(iH_0t/\hbar)H_I\exp(-iH_0t/\hbar) ,$$

where $H_0 = H_{em} + H_e$ is the Hamiltonian for the uncoupled photon and electron–positron fields. If $S(t,t')$ is the *time development* operator such that

$$|\Phi(t)\rangle = S(t,t')|\Phi(t')\rangle ,$$

then

$$i\hbar\partial_t S(t,t') = H_I(t)S(t,t') .$$

The equivalent integral equation, incorporating the inital condition $S(t,t) = 1$,

$$S(t,t_0) = 1 - \frac{i}{\hbar}\int_{t_0}^{t} H_I(t_1)S(t_1,t_0)\,d_1t , \tag{22.56}$$

can be solved iteratively, giving

$$S(t,t_0) = 1 + \sum_{n=1}^{\infty} S^{(n)}(t,t_0) , \tag{22.57}$$

where

$$S^{(n)}(t,t_0) = (-i/\hbar)^n \int_{t_0}^{t} dt_1 \int_{t_0}^{t_1} dt_2 \cdots \int_{t_0}^{t_{n-1}} dt_n \times H_I(t_1)H_I(t_2)\cdots H_I(t_n) .$$

This can be put into a more symmetric form by using *time-ordered* operators. Define the T-product of two operators by

$$T\Big[A(t_1)B(t_2)\Big] = \begin{cases} A(t_1)B(t_2) , & t_1 > t_2 \\ \pm B(t_2)A(t_1) , & t_2 > t_1 \end{cases} \tag{22.58}$$

where the *positive* sign refers to the product of photon operators and the *negative* sign to electrons. Then

$$S^{(n)}(t,t_0) = \frac{(-i/\hbar)^n}{n!} \int_{t_0}^{t} dt_1 \int_{t_0}^{t} dt_2 \cdots \int_{t_0}^{t} dt_n \times T\Big[H_I(t_1)H_I(t_2)\cdots H_I(t_n)\Big] . \tag{22.59}$$

The operator $S(t,t')$ relates the state vector at time t to the state vector at some earlier time $t' < t$. Its matrix elements therefore give the transition amplitudes for different processes, for example the emission or absorption of radiation by a system, or the outcome of scattering of a projectile from a target. The techniques for extracting cross-sections and other observable quantities from the S-operator are described at length in the texts [22.7, 8, 10, 12].

Although the use of normal ordering means that the charge and mass of the reference state, the vacuum, is zero, it fails to remove other infinities due to the occurrence of divergent integrals. The method of extracting finite quantities from this theory involves *renormalization* of the charge and mass of the electron. We shall refer especially to [22.10, Chapt. 8] for a detailed discussion. The most difficult technical problems are posed by *mass renormalization.* Formally, we modify the interaction Hamiltonian to read

$$j^{\mu}(x)A_{\mu}(x) - \delta M(x)\,,$$

where $\delta M(x)$ is the *mass renormalization operator*

$$\delta M(x) = \frac{1}{2}\delta m\left[\bar{\psi}(x), \psi(x)\right]$$

where δm is infinite.

A further problem is that electrons in a many-electron atom or molecule move in a potential which is quite unlike that of the bare nucleus. It is therefore useful to introduce a local *mean field* potential, say $U(x)$, representing some sort of average interaction with the rest of the electron charge distribution, so that the zero-order orbitals satisfy

$$\left[c\boldsymbol{\alpha}\cdot\boldsymbol{p} + \beta c^2 + V_{\text{nuc}}(\boldsymbol{x}) + U(\boldsymbol{x}) - E_m\right]\psi_m(\boldsymbol{x}) = 0\,. \tag{22.60}$$

With this starting point, the interaction Hamiltonian becomes

$$H_{\text{I}}(x) = H_{\text{I}}^{(1)}(x) + H_{\text{I}}^{(2)}(x)\,, \tag{22.61}$$

where

$$H_{\text{I}}^{(1)}(x) = -U(x),\; H_{\text{I}}^{(2)}(x) = j^{\mu}(x)A_{\mu}(x) - \delta M(x)\,,$$

and the electron current is defined in terms of the mean field orbitals of (22.60). The expression $H_{\text{I}}^{(2)}(x)$ is sometimes referred to as a *fluctuation potential.* The term $j^{\mu}(x)A_{\mu}(x)$ is proportional to the electron charge, e, which serves as an ordering parameter for perturbation expansions.

Effective Interactions

Although the S-matrix formalism provides in principle a complete computational scheme for many-electron systems, it is generally too cumbersome for practical use, and approximations are necessary. Usually, this is a matter of selecting a subset of dominant contributions to the perturbation series depending on the application. We are faced with the evaluation of T-products of the form

$$T\left[\phi(t_1)\phi(t_2)\cdots\phi(t_n)\right]$$

which is done using Wick's Theorem [22.10, p. 25].

In the simplest case,

$$T\left[\phi(t_1)\phi(t_2)\right] = \;:\phi(t_1)\phi(t_2):\; + \langle 0 \mid T\left[\phi(t_1)\phi(t_2)\right] \mid 0\rangle\,. \tag{22.62}$$

The vacuum expectation value is called a *contraction.* More generally, we have

$$\begin{aligned} &T\left[\phi(t_1)\phi(t_2)\cdots\phi(t_n)\right] \\ &= \;:\phi(t_1)\phi(t_2)\cdots\phi(t_n): \\ &+\left\{\langle 0|T[\phi(t_1)\phi(t_2)]|0\rangle :\phi(t_3)\cdots\phi(t_n): \right. \\ &\left. + \text{permutations}\right\} \\ &+\left\{\langle 0|T\left[\phi(t_1)\phi(t_2)\right]|0\rangle\langle 0|T[\phi(t_3)\phi(t_4)]|0\rangle \right. \\ &\left. \times :\phi(t_5)\cdots\phi(t_n): + \text{permutations}\right\}\cdots\,. \end{aligned}$$

This result has the effect that a T-product with an odd number of factors vanishes. A rigorous statement can be found in all standard texts; each term in the expansion gives rise to a Feynman diagram which can be interpreted as the amplitude of a physical process. As an example, consider the simple but important case

$$S^{(2)} = \frac{(\mathrm{i}e)^2}{2!}T\left[j^{\mu}(x)A_{\mu}(x).j^{\nu}(y)A_{\nu}(y)\right]. \tag{22.63}$$

One of the terms (there are others) found by using Wick's Theorem is

$$j^{\mu}(x)j^{\nu}(y)\langle 0|T\left[A_{\mu}(x)A_{\nu}(y)\right]|0\rangle\,.$$

We see that this involves the contraction of two photon amplitudes

$$-\frac{1}{2}D_{F\mu\nu}(x-y) = \langle 0 \mid T\left[A_{\mu}(x)A_{\nu}(y)\right] \mid 0\rangle\,,$$

which plays the role of a *propagator* (22.69): it relates the photon amplitudes at two space-time points x, y.

With the introduction of a spectral expansion for the electron current (22.48), the contribution to the energy of the system becomes

$$\frac{1}{2}\sum_{pqrs} : a_p^\dagger a_q^\dagger a_s a_r : \langle pq|V|rs\rangle \,, \tag{22.64}$$

which can be interpreted, in the familiar language of ordinary quantum mechanics, as the energy of two electrons due to the electron–electron interaction V which is directly related to the photon propagator.

22.3.5 Propagators

Propagators relate field variables at different space-time points. Here we briefly define those most often needed in atomic and molecular physics.

Electrons

Define Feynman's causal propagator for the electron–positron field by the contraction

$$S_F(x_2, x_1) = \langle 0|T\left[\psi(x_2)\bar{\psi}(x_1)\right]|0\rangle \,. \tag{22.65}$$

This has a spectral decomposition of the form

$$S_F(x_2, x_1) = \begin{cases} \sum_{E_m > E_F} \psi_m(x_2)\bar{\psi}_m(x_1) & t_1 > t_2 \,, \\ -\sum_{E_n < E_F} \psi_n(x_2)\bar{\psi}_n(x_1) & t_1 < t_2 \,, \end{cases} \tag{22.66}$$

which ensures that positive energy solutions are propagated forwards in time, and negative energy solutions backwards in time in accordance with the antiparticle interpretation of the negative energy states. By noting that the stationary state solutions $\psi_m(x)$ have time dependence $\exp(-\mathrm{i}E_m t)$, we can write (22.66) in the form

$$\begin{aligned} S_F(x_2, x_1) &= \frac{1}{2\pi \mathrm{i}} \int_{-\infty}^{\infty} \sum_n \frac{\psi_n(\boldsymbol{x}_2)\bar{\psi}_n(\boldsymbol{x}_1)}{E_n - z(1+\mathrm{i}\delta)} \mathrm{e}^{-\mathrm{i}z(t_2-t_1)} \mathrm{d}z \\ &= \frac{1}{2\pi \mathrm{i}} \int_{-\infty}^{\infty} G(\boldsymbol{x}_2, \boldsymbol{x}_1, z)\gamma^0 \mathrm{e}^{-\mathrm{i}z(t_2-t_1)} \,\mathrm{d}z \,, \end{aligned} \tag{22.67}$$

where δ is a small positive number, the sum over n includes *the whole spectrum*, and where the Green's function $G(\boldsymbol{x}_2, \boldsymbol{x}_1, z)$, in the specific case in which the potential of the external field $a^\mu(x)$ has only a scalar time-independent part, $V_{\mathrm{nuc}}(\boldsymbol{x})$, satisfies

$$\begin{aligned} &\left[c\boldsymbol{\alpha}\cdot\boldsymbol{p} + \beta c^2 + (V_{\mathrm{nuc}}\boldsymbol{x}) - z\right] G(\boldsymbol{x}, \boldsymbol{y}, z) \\ &= \delta^{(3)}(\boldsymbol{x}_2 - \boldsymbol{x}_1) \,. \end{aligned} \tag{22.68}$$

$G(\boldsymbol{x}_2, \boldsymbol{x}_1, z)$ is a meromorphic function of the complex variable z with branch points at $z = \pm c^2$, and cuts along the real axis (c^2, ∞) and $(-\infty, -c^2)$. The poles lie on the segment $(-c^2, c^2)$ at the bound eigenvalues of the Dirac Hamiltonian for this potential.

Photons

The photon propagator $D_{F\mu\nu}(x_2 - x_1)$ is constructed in a similar manner:

$$-\frac{1}{2} D_{F\mu\nu}(x_2 - x_1) = \langle 0\left|T\left[A_\mu(x_2)A_\nu(x_1)\right]\right|0\rangle \,, \tag{22.69}$$

where μ_0 is the permeability of the vacuum. This has the integral representation

$$\begin{aligned} D_{F\mu\nu}(x_2 - x_1) &= g_{\mu\nu} D_F(x_2 - x_1) \\ &= -g_{\mu\nu}\frac{\mathrm{i}}{(2\pi)^4} \int \mathrm{d}^4 q \, D(q^2) \,, \end{aligned} \tag{22.70}$$

where

$$D(q^2) = \frac{1}{q^2 + \mathrm{i}\delta} \,,$$

and δ is a small positive number. This is not unique, as the four-potentials depend on the choice of gauge; for details see [22.8, Sect. 77]. The various forms for the electron–electron interaction given below express such gauge choices.

22.3.6 Effective Interaction of Electrons

The expression (22.63) can be viewed in several ways: it is the interaction of the current density $j^\mu(x)$ at the space-time point x with the four-potential due to the current $j^\mu(y)$; the interaction of the current density $j^\mu(y)$ with the four-potential due to the current $j^\mu(x)$; or, as is commonly assumed in nonrelativistic atomic theory, the effective interaction between two charge density distributions, as represented by (22.64). In terms of the corresponding Feynman diagram, it can be thought of as the energy due to the exchange of a virtual photon.

The form of V depends on the choice of gauge potential, as follows.

Feynman Gauge

$$\langle pq|V|rs\rangle = \int\int \psi_p^\dagger(\boldsymbol{x})\psi_r(\boldsymbol{x})v_{sq}^{\mathrm{F}}(\boldsymbol{x},\boldsymbol{y})\psi_q^\dagger(\boldsymbol{y})\psi_s(\boldsymbol{y})\,\mathrm{d}^3x\,\mathrm{d}^3y \tag{22.71}$$

where

$$v_{sq}^{\mathrm{F}}(\boldsymbol{x},\boldsymbol{y}) = \frac{\mathrm{e}^{\mathrm{i}\omega_{sq}R}}{R}(1-\boldsymbol{\alpha}_x\cdot\boldsymbol{\alpha}_y)\,, \tag{22.72}$$

with

$$\boldsymbol{R}=\boldsymbol{x}-\boldsymbol{y},\quad R=|\boldsymbol{R}|,\quad \omega_{sq}=\frac{E_s-E_q}{c}\,.$$

This interaction gives both a real and an imaginary contribution to the energy; only the former is usually taken into account in structure calculations. Since the orbital indices are dummy variables, it is usual to symmetrize the interaction kernel by writing

$$\bar{v}^{\mathrm{F}}(\boldsymbol{x},\boldsymbol{y}) = \frac{1}{2}\left[v_{sq}^{\mathrm{F}}(\boldsymbol{x},\boldsymbol{y})+v_{rp}^{\mathrm{F}}(\boldsymbol{x},\boldsymbol{y})\right],$$

which places the orbitals on an equal footing.

Coulomb Gauge

Here the Feynman propagator is replaced by that for the Coulomb gauge, giving

$$v_{sq}^{T}(\boldsymbol{x},\boldsymbol{y}) = \frac{\mathrm{e}^{\mathrm{i}\omega_{sq}R}}{R} - \left[\boldsymbol{\alpha}_x\cdot\boldsymbol{\alpha}_y\frac{\mathrm{e}^{\mathrm{i}\omega_{sq}R}}{R} + (\boldsymbol{\alpha}_x\cdot\nabla)(\boldsymbol{\alpha}_y\cdot\nabla)\frac{\mathrm{e}^{\mathrm{i}\omega_{rp}R}-1}{\omega_{rp}^2R}\right] \tag{22.73}$$

in which the operator ∇ involves differentiation with respect to $\boldsymbol{R}$.

Symmetrization is also used with this interaction.

Breit Operator

The *low frequency limit*, $\omega_{rp}\to 0$, $\omega_{sq}\to 0$, is known as the Breit interaction:

$$\lim_{\omega_{rp},\omega_{sq}\to 0} v_{sq}^{T}(\boldsymbol{x},\boldsymbol{y}) = \frac{1}{R}+v^{\mathrm{B}}(\boldsymbol{R})\,, \tag{22.74}$$

where

$$v^{\mathrm{B}}(\boldsymbol{R}) = -\frac{1}{2R}\left(\boldsymbol{\alpha}_x\cdot\boldsymbol{\alpha}_y+\frac{\boldsymbol{\alpha}_x\cdot\boldsymbol{R}\,\boldsymbol{\alpha}_y\cdot\boldsymbol{R}}{R^2}\right).$$

Gaunt Operator

This is a further approximation in which $v^{\mathrm{B}}(\boldsymbol{R})$ is replaced by

$$v^{\mathrm{G}}(\boldsymbol{R}) = -\frac{\boldsymbol{\alpha}_x\cdot\boldsymbol{\alpha}_y}{R}\,, \tag{22.75}$$

the residual part of the Breit interaction being neglected.

Comments

The choice of gauge should not influence the predictions of QED for atomic and molecular structure when the perturbation series is summed to convergence, so that it should not matter if the unapproximated effective operators are taken in Feynman or Coulomb gauge. However, this need not be true at each order of perturbation. It has been shown that the results are equivalent, order by order, if the orbitals have been defined in a *local* potential, but not otherwise. There have also been suggestions that the Feynman operator introduces spurious terms in lower orders of perturbation that are canceled in higher orders [22.13]. For this reason, most structure calculations have used Coulomb gauge.

It is often argued, following *Bethe* and *Salpeter* [22.14, Sect. 38], that the Breit interaction should only be used in first order perturbation theory. The reason is the approximation $\omega\to 0$; however, this approximation is quite adequate for many applications in which the dominant interactions involve only small energy differences.

22.4 Many-Body Theory For Atoms

The relativistic theory of atomic structure can be viewed as a simplification of the QED approach using an effective Hamiltonian operator in which the Dirac electrons interact through the effective electron–electron interaction of Sect. 22.3.6. This approach retains the dominant terms from the perturbation solution; those that are omitted are small and can, with sufficient trouble, be taken into account perturbatively [22.10, 15]. In particular, radiative correction terms requiring renormalization are explicitly omitted, and their effects incorporated at a later stage. Once a model has been chosen, the techniques and methods used for practical calculations acquire a close resemblance to those of the nonrelativistic theory described, for example in Chapt. 21.

22.4.1 Effective Hamiltonians

The models which are closest to QED are those in which the full electron–electron interaction is included, usually in Coulomb gauge. We define a Fock space Hamiltonian

$$H_{\rm DCB} = H_0 + H_1 + H_2 \qquad (22.76)$$

where, as in (22.47),

$$H_0 = \sum_{E_m > E_{\rm F}} N_m E_m - \sum_{E_n < E_{\rm F}} N_n E_n \,, \qquad (22.77)$$

in which the states are those determined with respect to a mean-field central potential $U(\boldsymbol{x})$ as in (22.60)

$$\left[c\boldsymbol{\alpha}\cdot\boldsymbol{p} + \beta c^2 + V_{\rm nuc}(\boldsymbol{x}) + U(\boldsymbol{x}) - E_m\right]\psi_m(\boldsymbol{x}) = 0 \,,$$

and

$$H_1 = -\sum_{pq} : a_p^\dagger a_q : \langle p|U(\boldsymbol{x})|q\rangle$$

$$H_2 = \frac{1}{2}\sum_{pqrs} : a_p^\dagger a_q^\dagger a_r a_s : \langle pq|V|sr\rangle \,.$$

Here the sums run only over states p with $E_p > E_{\rm F}$; this means that states with $E_p < E_{\rm F}$ are treated as inert.

The models are named according to the choice of V from Sect. 22.5.3.

Dirac–Coulomb–Breit Models

These incorporate the *full* Coulomb gauge operator (22.73) or the less accurate Breit operator (22.74). The fully retarded operator is usually taken in the symmetrized form. The Gaunt operator (22.75) is sometimes considered as an approximation to the Breit operator.

Dirac–Coulomb Models

The electron–electron interaction is simply taken to be the static $1/R$ potential. Note that although the equations are relativistic, the choices of electron–nucleus interaction all implicitly restrict these models to a frame in which the nuclei are fixed in space. The full electron–electron interaction is gauge invariant; however, it is common to start from the Dirac–Coulomb operator, in which case the gauge invariance is lost. Since radiative transition rates are sensitive to loss of gauge invariance [22.16] the choice of potential in (22.76) can make a big difference. Such choices may also affect the rate of convergence in correlation calculations in which the relativistic parts of the electron–electron interaction are treated as a second, independent, perturbation.

22.4.2 Nonrelativistic Limit: Breit–Pauli Hamiltonian

The nonrelativistic limit of the Dirac–Coulomb–Breit Hamiltonian is described in Chapt. 21. The derivation is given in many texts, for example [22.8, 10, 14], and in principle involves the following steps:

1. Express the relativistic 4-spinor in terms of nonrelativistic Pauli 2-spinors of the form (see Sect. 21.2)

   $$\phi_{nlm_l,m_s}(\boldsymbol{x}) = \text{const.}\frac{P_{nl}(r)}{r} Y_{lm_l}(\theta,\phi)\chi_{m_s}(\sigma) \,,$$

   where χ_{m_s} is a 2-component eigenvector of the spin operator $\boldsymbol{s}$ to lowest order in $1/c$.
2. Extract effective operators to order $1/c^2$.

Thus the Breit–Pauli Hamiltonian is written as the sum of terms of Sect. 21.2 which can be correlated with specific parts of the parent relativistic operator:

1. One-body terms originate from the Dirac Hamiltonian: they are $H_{\rm mass}$ (21.5), the one-body part of $H_{\rm Darwin}$ (21.7) and the spin–orbit couplings $H_{\rm so}$ (21.11) and $H_{\rm soo}$ (21.12). The forms given in these equations assume that the electron interacts with a point-charge nucleus and only require the Coulomb part of the electron–electron interaction.
2. Two-body terms, including the two-body parts of $H_{\rm Darwin}$ (21.7), the spin–spin contact term $H_{\rm ssc}$ (21.8), the orbit–orbit term $H_{\rm oo}$ (21.9) and the spin–spin term $H_{\rm ss}$ (21.13) originate from the Breit interaction.

22.4.3 Perturbation Theory: Nondegenerate Case

We give a brief resumé of the Rayleigh–Schrödinger perturbation theory following *Lindgren* [22.17]. The material presented here supplements the general discussion of perturbation theory in Chapt. 5. First consider the simplest case with a nondegenerate reference state Φ belonging to the Hilbert space $\mathcal{H}$ satisfying

$$H_0|\Phi\rangle = E_0|\Phi\rangle \,, \qquad (22.78)$$

which is a first approximation to the solution of the full problem

$$H|\Psi\rangle = E|\Psi\rangle, \quad H = H_0 + V \,. \qquad (22.79)$$

Next, introduce a projection operator P such that

$$P = |\Phi\rangle\langle\Phi|, \quad P|\Phi\rangle = |\Phi\rangle \,,$$

and its complement $Q = 1 - P$, projecting onto the complementary subspace $\mathcal{H} \setminus \{\Phi\}$. With the *intermediate normalization*

$$\langle\Phi|\Psi\rangle = \langle\Phi|\Phi\rangle = 1 ,$$

it follows that

$$P|\Psi\rangle = |\Phi\rangle\langle\Phi|\Psi\rangle = |\Phi\rangle ,$$

so that the perturbed wave function can be decomposed into two parts:

$$|\Psi\rangle = (P + Q)|\Psi\rangle = |\Phi\rangle + Q|\Psi\rangle .$$

Thus, with intermediate normalization,

$$E = \langle\Phi|H|\Psi\rangle = E_0 + \langle\Phi|V|\Psi\rangle .$$

We now use this decomposition to write (22.79) in the form

$$(E_0 - H_0)|\Psi\rangle = (V - \Delta E)|\Psi\rangle , \tag{22.80}$$

where $\Delta E = \langle\Phi|V|\Psi\rangle$. Thus

$$(E_0 - H_0)Q|\Psi\rangle = Q(V - \Delta E)|\Psi\rangle . \tag{22.81}$$

Introduce the *resolvent operator*

$$R = \frac{Q}{E_0 - H_0} , \tag{22.82}$$

which is well-defined except on $\{\Phi\}$. Then the perturbation contribution to the wave function is

$$Q|\Psi\rangle = R(V - \Delta E)|\Psi\rangle = R(V|\Psi\rangle - |\Psi\rangle\langle\Phi|V|\Psi\rangle) .$$

The *Rayleigh–Schrödinger* perturbation expansion can now be written

$$|\Psi\rangle = |\Phi\rangle + \left|\Psi^{(1)}\right\rangle + \left|\Psi^{(2)}\right\rangle + \cdots$$
$$E = E_0 + E^{(1)} + E^{(2)} + \cdots$$

The contributions are ordered by the number of occurrences of V, the leading terms being

$$\left|\Psi^{(1)}\right\rangle = RV|\Phi\rangle ,$$
$$\left|\Psi^{(2)}\right\rangle = \left(RVRV - R^2VPV\right)|\Phi\rangle ,$$

and so on. The corresponding contribution to the energy can then be found from

$$E^{(n)} = \langle\Phi|V\left|\Psi^{(n-1)}\right\rangle .$$

22.4.4 Perturbation Theory: Open-Shell Case

Consider now the case in which there are several unperturbed states, $\left|\Phi^{(a)}\right\rangle$, $a = 1, 2, \ldots, d$, having the same energy E_0, which span a d-dimensional linear subspace (the *model space*) $\mathcal{M} \subset \mathcal{H}$, so that

$$H_0\left|\Phi^{(a)}\right\rangle = E_0\left|\Phi^{(a)}\right\rangle, \quad a = 1, 2, \ldots, d .$$

Let P be the projector onto $\mathcal{M}$, and Q onto the orthogonal subspace $\mathcal{M}^{\perp}$.

The perturbed states $\left|\Psi^{(a)}\right\rangle$, $a = 1, 2, \ldots, d$ are related to the unperturbed states by the *wave operator* Ω,

$$\left|\Psi^{(a)}\right\rangle = \Omega\left|\Phi^{(a)}\right\rangle, \quad a = 1, 2, \ldots, d .$$

The *effective Hamiltonian*, H_{eff}, is defined so that

$$H_{\text{eff}}\left|\Phi^{(a)}\right\rangle = E^{(a)}\left|\Phi^{(a)}\right\rangle ,$$

and thus

$$\Omega H_{\text{eff}}\left|\Phi^{(a)}\right\rangle = E^{(a)}\left|\Psi^{(a)}\right\rangle = H\Omega\left|\Phi^{(a)}\right\rangle .$$

Thus on the domain $\mathcal{M}$ we can write an operator equation

$$\Omega H_{\text{eff}} P = H\Omega P , \tag{22.83}$$

known as the Bloch equation. We now partition H_{eff} so that

$$H_{\text{eff}} P = (H_0 + V_{\text{eff}})P ,$$

enabling a reformulation of (22.83) as the commutator equation

$$[\Omega, H_0]P = (V\Omega - \Omega V_{\text{eff}})P . \tag{22.84}$$

With the intermediate normalization convention of Sect. 22.4.3, this becomes

$$\left|\Psi^{(a)}\right\rangle = P\left|\Phi^{(a)}\right\rangle$$

so that $P\Omega P = P$ and

$$H_{\text{eff}} P = PH\Omega P , \quad V_{\text{eff}} P = PV\Omega P .$$

Then (22.84) can be put in the final form

$$[\Omega, H_0]P = (V\Omega - \Omega PV\Omega)P . \tag{22.85}$$

The general Rayleigh–Schrödinger perturbation expansion can now be generated by expanding the wave operator order by order

$$\Omega = 1 + \Omega^{(1)} + \Omega^{(2)} + \cdots ,$$

and inserting into (22.85), resulting in a hierarchy of equations

$$\left[\Omega^{(1)}, H_0\right]P = (V - PV)P = QVP\,,$$
$$\left[\Omega^{(2)}, H_0\right]P = \left(QV\Omega^{(1)} - \Omega^{(1)}PV\right)P\,,$$

and so on, with $H^{(n)}_{\text{eff}} = PV\Omega^{(n-1)}$.

22.4.5 Perturbation Theory: Algorithms

The techniques of QED perturbation theory of Sect. 22.3.4 can be utilized to give computable expressions for perturbation calculations order by order. They exploit the second quantized representation of operators of Sect. 22.4.1 along with the use of diagrams to express the contributions to the wave operator and the energy as sums over virtual states. The use of Wick's theorem to reduce products of normally-ordered operators, and the *linked-diagram* or *linked-cluster* theorem are explained in *Lindgren*'s article [22.17] and Chapt. 5. Further references and discussion of features which can exploit vector-processing and parallel-processing computer architectures may be found in [22.18].

The theory can also be recast so as to sum certain classes of terms to completion. This depends on the possibility of expressing the wave operator as a normally ordered exponential operator

$$\Omega = \{\exp S\} = 1 + \{S\} + \frac{1}{2!}\{S^2\} + \cdots\,,$$

where the normally ordered operator S is known as the *cluster* operator. Expanding S order by order leads to the *coupled cluster expansion* (see also Chapts. 5 and 27).

22.5 Spherical Symmetry

A popular starting point for most calculations in atomic and molecular structure is the independent particle central field approximation. This assumes that the electrons move independently in a potential field of the form

$$A_0(x) = \frac{1}{c}\phi(r)\,, \quad r = |\boldsymbol{x}|\,;$$
$$A_i(x) = 0\,, \quad i = 1, 2, 3\,. \tag{22.86}$$

Clearly $\phi(r)$ is left unchanged by any rotation about the origin, $r = 0$, but transforms as the component $A_0(x)$ of a 4-vector under other types of Lorentz and Poincaré transformation such as boosts or translations. However, solutions in central potentials of this form have a simple form which is convenient for further calculation.

With this restriction on the 4-potential, Dirac's Hamiltonian becomes

$$\hat{\boldsymbol{h}}_{\text{D}} = \left\{c\boldsymbol{\alpha}\cdot\boldsymbol{p} + e\phi(r) + \beta m_e c^2\right\}. \tag{22.87}$$

Consider stationary solutions with energy E satisfying

$$\hat{\boldsymbol{h}}_{\text{D}}\psi_E(x) = E\psi_E(x)\,.$$

Since $\hat{\boldsymbol{h}}_{\text{D}}$ is invariant with respect to rotation about $r = 0$, it commutes with the generators J_1, J_2, J_3 mentioned in Sect. 22.1.1, corresponding to components of the total angular momentum $\boldsymbol{j}$ of the electron, usually decomposed into an orbital part $\boldsymbol{l}$ and a spin part $\boldsymbol{s}$,

$$\boldsymbol{j} = \boldsymbol{l} + \boldsymbol{s} \tag{22.88}$$

where

$$l_j = \mathrm{i}\epsilon_{jkl}x_k\partial_l\,, \quad j = 1, 2, 3$$
$$s_j = \frac{1}{2}\epsilon_{jkl}\sigma_{kl}\,, \quad j = 1, 2, 3\,.$$

22.5.1 Eigenstates of Angular Momentum

We can construct simultaneous eigenstates of the operators $\boldsymbol{j}^2$ and j_3 by using the product representation $\mathcal{D}^{(l)} \times \mathcal{D}^{(1/2)}$ of the rotation group SO(3), which is reducible to the Clebsch–Gordan sum of two irreps

$$\mathcal{D}^{(l+1/2)} \oplus \mathcal{D}^{(l-1/2)}\,. \tag{22.89}$$

We construct a basis for each irrep from products of basis vectors for $\mathcal{D}^{(1/2)}$ and $\mathcal{D}^{(l)}$ respectively. $\mathcal{D}^{(1/2)}$ is a 2-dimensional representation spanned by the simultaneous eigenstates ϕ_σ of $\boldsymbol{s}^2$ and s_3

$$\boldsymbol{s}^2\,\phi_\sigma = \frac{3}{4}\phi_\sigma\,, \quad s_3\,\phi_\sigma = \sigma\phi_\sigma\,, \quad \sigma = \pm\frac{1}{2}\,,$$

for which we can use 2-rowed vectors

$$\phi_{1/2} = \begin{pmatrix}1\\0\end{pmatrix}, \quad \phi_{-1/2} = \begin{pmatrix}0\\1\end{pmatrix}.$$

The representation $\mathcal{D}^{(l)}$ is $(2l+1)$-dimensional; its basis vectors can be taken to be the spherical harmonics

$$\left\{Y_l^m(\theta, \varphi) \mid m = -l, -l+1, \ldots, l\right\},$$

so that

$$l^2\, Y_l^m(\theta,\varphi) = l(l+1)\hbar^2 Y_l^m(\theta,\varphi)\,,$$
$$l_3\, Y_l^m(\theta,\varphi) = m\hbar\, Y_l^m(\theta,\varphi)\,.$$

We shall assume that spherical harmonics satisfy the standard relations

$$l_\pm Y_l^m(\theta,\varphi) = [l(l+1)-m(m\pm 1)]^{1/2}\hbar Y_l^{m\pm 1}(\theta,\varphi)\,,$$

where $l_\pm = l_1 \pm l_2$, so that

$$\begin{aligned}
Y_l^m(\theta,\varphi) &= \left(\frac{2l+1}{4\pi}\right)^{1/2} C_l^m(\theta,\varphi)\,,\\
C_l^m(\theta,\varphi) &= (-1)^m \left(\frac{(l-m)!}{(l+m)!}\right)^{1/2} P_l^m(\theta)\,\mathrm{e}^{im\varphi}\,,\\
&\quad \text{if } m \geq 0\,,\\
C_l^{-m}(\theta,\varphi) &= (-1)^m C_l^m(\theta,\varphi)^*\,.
\end{aligned} \tag{22.90}$$

Basis functions for the representations $\mathcal{D}^j$ with $j = l \pm \frac{1}{2}$ have the form (The order of coupling is significant, and great confusion results from a mixing of conventions. Here we couple in the order l, s, j. The same spin-angle functions are obtained if we use the order s, l, j but there is a phase difference $(-1)^{l-j+1/2} = (-1)^{(1-a)/2}$. You have been warned!)

$$\begin{aligned}
&\chi_{j,m,a}(\theta,\varphi)\\
&= \sum_\sigma \left\langle l, m-\sigma, \frac{1}{2}, \sigma \,\middle|\, l, \frac{1}{2}, j, m\right\rangle Y_l^{m-\sigma}(\theta,\varphi)\phi_\sigma
\end{aligned} \tag{22.91}$$

where $\langle l, m-\sigma, \frac{1}{2}, \sigma \,|\, l, \frac{1}{2}, j, m\rangle$ is a Clebsch–Gordan coefficient with

$$l = j - \frac{1}{2}a, \quad a = \pm 1,$$
$$m = -j, -j+1, \ldots, j-1, j\,.$$

Inserting explicit expressions for the Clebsch–Gordan coefficients gives

$$\chi_{j,m,-1}(\theta,\varphi) = \begin{pmatrix} -\left(\frac{j+1-m}{2j+2}\right)^{1/2} Y_{j+1/2}^{m-1/2}(\theta,\varphi) \\ \left(\frac{j+1+m}{2j+2}\right)^{1/2} Y_{j+1/2}^{m+1/2}(\theta,\varphi) \end{pmatrix},$$
$$\chi_{j,m,1}(\theta,\varphi) = \begin{pmatrix} \left(\frac{j+m}{2j}\right)^{1/2} Y_{j-1/2}^{m-1/2}(\theta,\varphi) \\ \left(\frac{j-m}{2j}\right)^{1/2} Y_{j-1/2}^{m+1/2}(\theta,\varphi) \end{pmatrix}. \tag{22.92}$$

The vectors (22.92) satisfy

$$\begin{aligned}
&\boldsymbol{j}^2\chi_{j,m,a} = j(j+1)\chi_{j,m,a}, \quad \boldsymbol{s}^2\chi_{j,m,a} = \frac{3}{4}\chi_{j,m,a}\,,\\
&\boldsymbol{l}^2\chi_{j,m,a} = l(l+1)\chi_{j,m,a}, \quad l = j - \frac{1}{2}a, \quad a = \pm 1\,.
\end{aligned} \tag{22.93}$$

The parity of the angular part is given by $(-1)^l$, with the two possibilities distinguished by means of the operator

$$K' = -(\boldsymbol{j}^2 - \boldsymbol{l}^2 - \boldsymbol{s}^2 + 1) = -(2\boldsymbol{s}\cdot\boldsymbol{l} + 1) \tag{22.94}$$

so that

$$K'\chi_{j,m,a} = k'\chi_{j,m,a}, \quad k' = -\left(j + \frac{1}{2}\right)a, \quad a = \pm 1\,.$$

The basis vectors are orthonormal on the unit sphere with respect to the inner product

$$\begin{aligned}
&(\chi_{j',m',a'} \mid \chi_{jma})\\
&= \int\!\!\int \chi^\dagger_{j',m',a'}(\theta,\varphi)\chi_{j,m,a}(\theta,\varphi)\sin\theta\,\mathrm{d}\theta\,\mathrm{d}\varphi\\
&= \delta_{j',j}\delta_{m',m}\delta_{a',a}\,.
\end{aligned} \tag{22.95}$$

22.5.2 Eigenstates of Dirac Hamiltonian in Spherical Coordinates

Eigenstates of Dirac's Hamiltonian (22.87) in spherical coordinates with a spherically symmetric potential $V(r) = e\phi(r)$,

$$\hat{\boldsymbol{h}}_\mathrm{D}\psi_E(\boldsymbol{r}) = E\psi_E(\boldsymbol{r})\,, \tag{22.96}$$

are also simultaneous eigenstates of $\boldsymbol{j}^2$, of j_3 and of the operator

$$K = \begin{pmatrix} K' & 0 \\ 0 & -K' \end{pmatrix}, \tag{22.97}$$

where K' is defined in (22.94) above. Denote the corresponding eigenvalues by j, m and κ, where

$$\kappa = \pm\left(j + \frac{1}{2}\right). \tag{22.98}$$

Then the simultaneous eigenstates take the form

$$\psi_{E\kappa m}(\boldsymbol{r}) = \frac{1}{r}\begin{pmatrix} P_{E\kappa}(r)\chi_{\kappa,m}(\theta,\varphi) \\ \mathrm{i}Q_{E\kappa}(r)\chi_{-\kappa,m}(\theta,\varphi) \end{pmatrix}, \tag{22.99}$$

where $\kappa = -(j+1/2)a$ is the eigenvalue of K', and the notation $\chi_{\kappa,m}$ replaces the notation $\chi_{j,m,a}$ used previously in (22.91). The factor i in the lower component ensures that, at least for bound states, the radial amplitudes $P_{E\kappa}(r)$, $Q_{E\kappa}(r)$ can be chosen to be real. This decomposition into radial and angular factors exploits the identity

$$\sigma \cdot p\left[\frac{F(r)}{r}\chi_{\kappa,m}(\theta,\varphi)\right] = \mathrm{i}\hbar\frac{1}{r}\left(\frac{\mathrm{d}F}{\mathrm{d}r}+\frac{\kappa F}{r}\right)\chi_{-\kappa,m}(\theta,\varphi) \quad (22.100)$$

and gives a reduced eigenvalue equation

$$\begin{pmatrix} mc^2-E+V & -c\left(\frac{\mathrm{d}}{\mathrm{d}r}-\frac{\kappa}{r}\right) \\ c\left(\frac{\mathrm{d}}{\mathrm{d}r}+\frac{\kappa}{r}\right) & -mc^2-E+V \end{pmatrix}\begin{pmatrix} P_{E\kappa}(r) \\ Q_{E\kappa}(r)\end{pmatrix} = 0\,. \quad (22.101)$$

Angular Density Distributions

It is a remarkable fact that the angular density distribution

$$A_{\kappa,m}(\theta,\varphi) = \chi_{\kappa,m}(\theta,\varphi)^{\dagger}\chi_{\kappa,m}(\theta,\varphi)\,, \quad (22.102)$$

where $m=-j,-j+1,\ldots,j-1,j$, is *independent of the sign of* κ; the equivalence of

$$A_{j+1/2,m}(\theta,\varphi) = \frac{1}{4\pi}\frac{(j-m)!}{(j+m)!} \times\left[(j-m+1)^2\left|P^{m-1/2}_{j+1/2}(\mu)\right|^2+\left|P^{m+1/2}_{j+1/2}(\mu)\right|^2\right],$$

and

$$A_{-(j+1/2),m}(\theta,\varphi) = \frac{1}{4\pi}\frac{(j-m)!}{(j+m)!} \times\left[(j+m)^2\left|P^{m-1/2}_{j-1/2}(\mu)\right|^2+\left|P^{m+1/2}_{j-1/2}(\mu)\right|^2\right],$$

where $\mu=\cos\theta$, was demonstrated by *Hartree* [22.19].

Angular densities for the lowest $|\kappa|$ values are given in Table 22.1. The corresponding nonrelativistic angular densities

$$A_{l,m}(\theta,\varphi)_{nr} = \left|Y^m_l(\theta,\varphi)\right|^2 = \frac{2l+1}{4\pi}\frac{(l-m)!}{(l+m)!}\left|P^{|m|}_l(\mu)\right|^2;$$

are listed in Table 22.2.

Radial Density Distributions

The probability density distribution $\rho_{E\kappa m}(\boldsymbol{r})$ associated with the stationary state (22.99) is given by

$$\rho_{E,\kappa,m}(\boldsymbol{r}) = \frac{1}{r^2}\Big[|P_{E,\kappa}(r)|^2A_{\kappa,m}(\theta,\varphi) + |Q_{E,\kappa}(r)|^2A_{-\kappa,m}(\theta,\varphi)\Big]\,. \quad (22.103)$$

Table 22.1 Relativistic angular density functions

$\lvert\kappa\rvert$	$\lvert m\rvert$	$4\pi\, A_{\lvert\kappa\rvert,m}(\theta,\varphi)$
1	$\frac{1}{2}$	1
2	$\frac{3}{2}$	$\frac{3}{2}\sin^2\theta$
	$\frac{1}{2}$	$\frac{1}{2}(1+3\cos^2\theta)$
3	$\frac{5}{2}$	$\frac{15}{8}\sin^4\theta$
	$\frac{3}{2}$	$\frac{3}{8}\sin^2\theta(1+15\cos^2\theta)$
	$\frac{1}{2}$	$\frac{3}{4}(3\cos^2\theta-1)^2+3\sin^2\theta\cos^2\theta$

Table 22.2 Nonrelativistic angular density functions

l	$\lvert m\rvert$	$4\pi\, A_{l,m}(\theta,\varphi)_{nr}$
0	0	1
1	1	$\frac{3}{2}\sin^2\theta$
	0	$3\cos^2\theta$
2	2	$\frac{15}{8}\sin^4\theta$
	1	$\frac{15}{2}\sin^2\theta\cos^2\theta$
	0	$\frac{5}{4}(3\cos^2\theta-1)^2$

Since $A_{\kappa,m}$ does not depend on the sign of κ, the angular part can be factored so that

$$\rho_{E,\kappa,m}(\boldsymbol{r}) = \frac{D_{E,\kappa}(r)}{r^2}A_{|\kappa|,m}(\theta,\varphi)\,,$$

where

$$D_{E,\kappa}(r) = \left[|P_{E,\kappa}(r)|^2+|Q_{E,\kappa}(r)|^2\right] \quad (22.104)$$

defines the *radial density* distribution.

Subshells in j–j Coupling

The notion of a *subshell* depends on the observation that the set $\{\psi_{E,\kappa,m}, m=-j,\ldots,j\}$ have a common radial density distribution. The simplest atomic model is one in which the electrons move independently in a mean field central potential. Since

$$\sum_{m=-j}^{j}\rho_{E,\kappa,m}(\boldsymbol{r}) = \frac{2j+1}{4\pi}\frac{D_{E,\kappa}(r)}{r^2}\,, \quad (22.105)$$

a state of $2j+1$ independent electrons, with one in each member of the set $\{\psi_{E,\kappa,m}, m=-j,\ldots,j\}$, has a spherically symmetric probability density. If E belongs to the point spectrum of the Hamiltonian, then (22.105) gives a distribution localized in r, and we refer to the states $\{E,\kappa,m\}$, $m=-j,\ldots,j$ as belonging to the subshell $\{E,\kappa\}$.

The notations in use for Dirac central field states are set out in Table 22.3. Here l is associated with the orbital angular quantum number of the upper pair of

Table 22.3 Spectroscopic labels and angular quantum numbers

Label:	s	$\bar{p}$	p	$\bar{d}$	d	$\bar{f}$	f
$\kappa = -(j+\frac{1}{2})a$	-1	$+1$	-2	$+2$	-3	$+3$	-4
$j = l + \frac{1}{2}a$	$\frac{1}{2}$	$\frac{1}{2}$	$\frac{3}{2}$	$\frac{3}{2}$	$\frac{5}{2}$	$\frac{5}{2}$	$\frac{7}{2}$
a	1	-1	1	-1	1	-1	1
$l = j - \frac{1}{2}a$	0	1	1	2	2	3	3
$\bar{l} = j + \frac{1}{2}a$	1	0	2	1	3	2	4

components and $\bar{l}$ with the lower pair. Note the useful equivalence

$$\kappa(\kappa+1) = l(l+1)\,.$$

Defining $\bar{\kappa} := -\kappa$ we have also $\bar{\kappa}(\bar{\kappa}+1) = \bar{l}(\bar{l}+1)$.

22.5.3 Radial Amplitudes

Textbooks on quantum electrodynamics usually contain extensive discussions of the formalism associated with the Dirac equation but rarely go beyond the treatment of the hydrogen atom Chapt. 10. *Greiner*'s textbook [22.4] is an honorable exception, with many worked examples. A more exhaustive list of problems in which exact solutions are known is contained in [22.20]; it is particularly rich in detail about equations of motion and Green's functions in external electromagnetic fields of various configurations; coherent states of relativistic particles; charged particles in quantized plane wave fields. It also incorporates discussion of extensions of the Dirac equations due to Pauli which include explicit interaction terms arising from anomalous magnetic or electric moments.

Atoms and molecules with more than one electron are not soluble analytically so that numerical models are needed to make predictions. The solutions are sensitive to boundary conditions on which we focus in this section. For large r, solutions of (22.101) can be found proportional to $\exp(\pm\lambda r)$, where

$$\lambda = +\sqrt{c^2 - E^2/c^2}\,. \tag{22.106}$$

Thus λ is *real* when $-c^2 \le E \le c^2$, and *pure imaginary* otherwise.

Singular Point at $r = 0$

Singularities of the nuclear potential near $r = 0$ have a major influence on the nature of solutions of the Dirac equation. Suppose that the potential has the form

$$V(r) = -\frac{Z(r)}{r}\,, \tag{22.107}$$

so that $Z(r)$ is the effective charge seen by an electron at radius r from the nuclear center. The dependence of $Z(r)$ on r may reflect the finite size of the nuclear charge distribution, so far treated as a point, or the screening due to the environment. Assume that $Z(r)$ can be expanded in a power series of the form

$$Z(r) = Z_0 + Z_1 r + Z_2 r^2 + \cdots \tag{22.108}$$

in a neighborhood of $r = 0$. This property characterizes a number of well-used models

1. *Point nucleus*: $Z_0 \neq 0$; $Z_n = 0$, $\quad n > 0$.
2. *Uniform nuclear charge distribution*:

$$V(r) = \begin{cases} -\dfrac{3Z}{2a}\left(1 - \dfrac{r^2}{3a^2}\right), & 0 \le r \le a\,, \\ -\dfrac{Z}{r}\,, & r > a\,. \end{cases} \tag{22.109}$$

   This gives the expansion $Z_0 = -3Z/2a$, $Z_1 = 0$, $Z_2 = +Z/2a^3$, $Z_n = 0$ for $n > 2$ when $r \le a$.
3. *Fermi distribution*: The nuclear charge density has the form

$$\rho_{\text{nuc}}(r) = \frac{\rho_0}{1 + \exp[(r-a)/d]}\,,$$

   where ρ_0 is chosen so that the total charge on the nucleus is Z.

Other nuclear models, reflecting the density distributions deduced from nuclear scattering experiments, can be found in the literature.

Series Solutions Near $r = 0$

Any solution for the radial amplitudes of Dirac's equation in a central potential

$$u(r) = \begin{pmatrix} P(r) \\ Q(r) \end{pmatrix}\,, \tag{22.110}$$

with radial density

$$D(r) = P^2(r) + Q^2(r)\,,$$

can be expanded in a power series near the singular point at $r = 0$ in the form

$$u(r) = r^\gamma\left(u_0 + u_1 r + u_2 r^2 + \cdots\right)\,, \tag{22.111}$$

where

$$u_k = \begin{pmatrix} p_k \\ q_k \end{pmatrix}\,, \quad k = 1, 2, \ldots$$

and γ, p_k, q_k are constants which depend on the nuclear potential model.

Point Nuclear Models

For a Coulomb singularity, $Z_0 \neq 0$, the leading coefficients satisfy

$$-Z_0 p_0 + c(\kappa - \gamma) q_0 = 0 \,, \quad c(\kappa + \gamma) p_0 - Z_0 q_0 = 0 \,, \tag{22.112}$$

so that

$$\gamma = \pm\sqrt{\kappa^2 - \frac{Z_0^2}{c^2}} \,, \quad \frac{q_0}{p_0} = \frac{Z_0}{c(\kappa - \gamma)} = \frac{c(\kappa + \gamma)}{Z_0} \,. \tag{22.113}$$

Finite Nuclear Models

Finite nuclear models, for which $Z_0 = 0$, have no singularity in the potential at $r = 0$. The indicial equation (22.113) reduces to $\gamma = \pm|\kappa|$, so that for $\kappa < 0$,

$$P(r) = p_0 r^{l+1} + O\left(r^{l+3}\right) \,, \tag{22.114}$$

$$Q(r) = q_1 r^{l+2} + O\left(r^{l+4}\right) \,, \tag{22.115}$$

with

$$q_1/p_0 = \left(E - mc^2 + Z_1\right)/[c(2l+3)] \,, \quad q_0 = p_1 = 0 \,,$$

and for $\kappa \geq 1$,

$$P(r) = p_1 r^{l+1} + O\left(r^{l+3}\right) \,, \tag{22.116}$$

$$Q(r) = q_0 r^{l} + O\left(r^{l+2}\right) \,, \tag{22.117}$$

with

$$p_1/q_0 = -\left(E - mc^2 + Z_1\right)/[c(2l+1)] \,, \quad p_0 = q_1 = 0 \,.$$

In both cases the solutions consist of either purely *even* powers or purely *odd* powers of r, contrasting strongly with the point nucleus case, where both even and odd powers are present in the series expansion.

The Nonrelativistic Limit

For a solution linked to a nonrelativistic state with orbital angular momentum l, one expects the nonrelativistic limit

$$P(r) = O\left(r^{l+1}\right), \quad c \to \infty \,.$$

The limiting behavior reveals some significant features.

Finite nuclear models.
The behavior is entirely regular:

$$P(r) = O\left(r^{l+1}\right), \quad Q(r) = O\left(c^{-1}\right) \to 0 \,.$$

Point nuclear models.
Since

$$\gamma = |\kappa| - \frac{Z^2}{2c^2|\kappa|} + \cdots \,,$$

(22.113) shows that the leading coefficient p_0 *vanishes* in the limit so that,

$$P(r) \approx p_1 r^{l+1}\left[1 + O\left(r^2\right)\right], \quad \text{when } \kappa \geq 1,\ l = \kappa \,. \tag{22.118}$$

All higher powers of *odd* relative order vanish in the limit for both components. The behavior in the case $\kappa < 0$ is entirely regular.

22.5.4 Square Integrable Solutions

Square integrable solutions require $\int D_{E,\kappa}(r)\,\mathrm{d}r$ to be finite; since the solutions are smooth, except possibly near the singular endpoints $r \to 0$ and $r \to \infty$, we focus on the behavior at the endpoints:

$r \to \infty$
For *real* values of λ the condition

$$\int_R^\infty D_{E,\kappa}(r)\,\mathrm{d}r < \infty \,, \quad 0 < R < \infty \,,$$

requires that $P_{E\kappa}(r)$, $Q_{E\kappa}(r)$ are proportional to $\exp(-\lambda r)$ with $\lambda > 0$.

This means that bound states can only exist when E lies in the interval $-c^2 \leq E \leq c^2$. Outside this interval solutions are necessarily of scattering type and so

$$\int_R^\infty D_{E,\kappa}(r)\,\mathrm{d}r$$

diverges when $|E| > c^2$.

$r \to 0$
This limit requires

$$\int_0^{R'} D_{E,\kappa}(r)\,\mathrm{d}r < \infty, \quad R' > 0 \,.$$

Since $D_{E,\kappa}(r) \sim r^{\pm 2\gamma}$ as $r \to 0$, this condition holds when $\pm\gamma > -\frac{1}{2}$. Only the solution with $\gamma > 0$ satisfies the condition when $|\gamma| > \frac{1}{2}$, or $Z < \alpha^{-1}\sqrt{\kappa^2 - 1/4}$, and the solution with $\gamma < 0$ must be disregarded. This corresponds to the *limit point case* of a second-order differential operator [22.21]. In the special case $|\kappa| = 1$ or $j = \frac{1}{2}$ this limits Z to be smaller than $c\sqrt{3}/2 \approx 118.6$. For $Z > c\sqrt{3}/2$, both solutions are square integrable near the origin (the *limit circle case*) and the differential operator is no longer essentially self-adjoint.

The Coulomb potential must have a finite expectation for any physically acceptable solution, so that we also require

$$\int_0^{R'} D_{E,\kappa}(r)\frac{\mathrm{d}r}{r} < \infty, \qquad R' > 0\,.$$

This is always satisfied by the solution with $\gamma > 0$ for all $|Z| < \alpha^{-1}|\kappa|$, but not by the solution with $\gamma < 0$. Imposing this condition restores essential self-adjointness (on a restricted domain) for $118 < Z \le 137$.

22.5.5 Hydrogenic Solutions

The wave functions for hydrogenic solutions of Dirac's equation are presented in Sect. 22.8.2. Here we note some properties of hydrogenic solutions that reveal dynamical effects of relativity in the absence of screening by orbital electrons. In this case $Z_0 = Z$, $Z_n = 0$, $n > 0$. When $-c^2 < E < c^2$ we have bound states. The parameter λ, (22.106), can conveniently be written

$$\lambda = Z/N\,, \tag{22.119}$$

so that rearranging (22.106) gives

$$E = +c^2\sqrt{1 - \frac{Z^2}{N^2c^2}}\,, \tag{22.120}$$

essentially equivalent to Sommerfeld's fine structure formula. In the formal nonrelativistic limit, $c \to \infty$, we have

$$E = c^2 - \frac{Z^2}{2N^2} + O(1/c^2)$$

so that N is closely related to the principal quantum number, n, appearing in the Rydberg formula. As in Sect. 22.8.2, we write $\rho = 2\lambda r$.

Define the *inner quantum number*

$$n_\mathrm{r} = -a = -\gamma + \frac{NE}{c^2}, \qquad n_\mathrm{r} = 0, 1, 2, \ldots\,.$$

Substitute for E from (22.120) to get

$$\begin{aligned} N &= \left[(n_\mathrm{r}+\gamma)^2 + \alpha^2 Z^2\right]^{1/2} \\ &= \left[n^2 - 2n_\mathrm{r}(|\kappa| - \gamma)\right]^{1/2}\,, \end{aligned} \tag{22.121}$$

where $n = n_\mathrm{r} + |\kappa|$ is the *principal quantum number*, the exact equivalent of the principal quantum number of the nonrelativistic state to which the Dirac solution reduces in the limit $c \to \infty$. With this notation, the radial amplitudes for bound hydrogenic states are

$$\begin{aligned} &P_{E\kappa}(r) \\ &= \mathcal{N}_{E\kappa}(c+E/c)^{1/2}\rho^\gamma \mathrm{e}^{-\rho/2}\big[-n_\mathrm{r}M(-n_\mathrm{r}+1, \\ &\quad 2\gamma+1;\rho) + (N-\kappa)M(-n_\mathrm{r}, 2\gamma+1;\rho)\big]\,, \end{aligned} \tag{22.122}$$

$$\begin{aligned} &Q_{E\kappa}(r) \\ &= \mathcal{N}_{E\kappa}(c-E/c)^{1/2}\rho^\gamma \mathrm{e}^{-\rho/2}\big[-n_\mathrm{r}M(-n_\mathrm{r}+1, \\ &\quad 2\gamma+1;\rho) - (N-\kappa)M(-n_\mathrm{r}, 2\gamma+1;\rho)\big]\,, \end{aligned} \tag{22.123}$$

where

$$\mathcal{N}_{E\kappa} = \left(\frac{\alpha Z}{2N^2(N-\kappa)} \cdot \frac{\Gamma(2\gamma+n_\mathrm{r}+1)}{n_\mathrm{r}!\,[\Gamma(2\gamma+1)]^2}\right)^{1/2}$$

is the normalization constant. For definitions of the confluent hypergeometric functions $M(a,b;c;z)$ see [22.22, Sect. 13.1].

Table 22.4 lists expectation values of simple powers of the radial variable $\rho = 2Zr/N$ from [22.23]

Table 22.4 Radial moments $\langle\rho^s\rangle$

s	Nonrelativistic	Relativistic[a]
2	$2n^2[5n^2+1 - 3l(l+1)]$	$2\left[N^2(5N^2-2\kappa^2)R^2(N) + N^2(1-\gamma^2) - 3\kappa N^2 R(N)\right]$
1	$3n^2 - l(l+1)$	$-\kappa + (3N^2-\kappa^2)R(N)$
0	1	1
−1	$\dfrac{1}{2n^2}$	$\dfrac{n\gamma + (\lvert\kappa\rvert-\gamma)\lvert\kappa\rvert}{2\gamma N^3}$
−2	$\dfrac{1}{2n^3(2l+1)}$	$\dfrac{\kappa^2 R(N)}{2\gamma^2 N^3(2\gamma - \mathrm{sgn}\kappa)}$
−3	$\dfrac{1}{4n^3 l(l+1)(2l+1)}$	$\dfrac{N^2 + 2\gamma^2\kappa^2 - 3N^2\kappa R(N)}{4N^5\gamma(\gamma^2-1)(4\gamma^2-1)}$

[a] $R(N) = \sqrt{1 - Z^2/N^2c^2}$

and [22.24]. Simple algebra, using the inequalities $\gamma < |\kappa|$ and $N < n$, yields the inequality

$$\langle \rho^s \rangle_{n\kappa} < \langle \rho^s \rangle_{nl}, \quad s > 0 ;$$

the inequality is reversed for $s < 0$. In the same way, it is easy to deduce that relativistic hydrogenic eigenvalues lie below the nonrelativistic eigenvalues

$$\epsilon_{n\kappa} < \epsilon_{nl} .$$

Thus, in the absence of screening, Dirac orbitals both *contract* and are *stabilized* with respect to their nonrelativistic counterparts. The relativistic and nonrelativistic expectation values approach each other as the *relativistic coupling constant*, $Z/c = \alpha Z \to 0$. This formal *nonrelativistic limit* is approached as $\alpha \to 0$ or $c \to \infty$, in which the speed of light is regarded as infinite.

22.5.6 The Free Electron Problem in Spherical Coordinates

The radial equation (22.101) for the free electron ($V(r) = 0$) gives a pair of first order ordinary differential equations

$$(mc^2 - E) P_{E\kappa}(r) = c\left(\frac{\mathrm{d}}{\mathrm{d}r} - \frac{\kappa}{r}\right) Q_{E\kappa}(r) ,$$
$$c\left(\frac{\mathrm{d}}{\mathrm{d}r} + \frac{\kappa}{r}\right) P_{E\kappa}(r) = \left(mc^2 + E\right) Q_{E\kappa}(r) , \quad (22.124)$$

from which we deduce that

$$\frac{\mathrm{d}^2 P_{E\kappa}(r)}{\mathrm{d}r^2} + \left(p^2 - \frac{\kappa(\kappa+1)}{r^2}\right) P_{E\kappa}(r) = 0 ,$$
$$\frac{\mathrm{d}^2 Q_{E\kappa}(r)}{\mathrm{d}r^2} + \left(p^2 - \frac{\bar{\kappa}(\bar{\kappa}+1)}{r^2}\right) Q_{E\kappa}(r) = 0 , \quad (22.125)$$

where $p^2 = m^2c^2 - E^2/c^2 = \boldsymbol{p}.\,\boldsymbol{p}$ and the angular quantum numbers κ and $\bar{\kappa}$ are associated respectively with the upper and lower components. These are defining equations of Riccati–Bessel functions [22.22, Sect. 10.1.1] of orders l and $\bar{l}$ respectively, where

$$\kappa(\kappa+1) = l(l+1), \quad \bar{\kappa}(\bar{\kappa}+1) = \bar{l}(\bar{l}+1) .$$

Thus the solutions of (22.125) are functions of the variable $x = pr$ of the form

$$P_{E\kappa}(r) = Ax f_l(x), \quad Q_{E\kappa}(r) = Bx f_{\bar{l}}(x) ,$$

where the ratio of A and B is determined by (22.124) and where $f_l(x)$ is a spherical Bessel function of the first, second or third kind [22.22, Sect. 10.1.1]. Thus

$$P_{E\kappa}(r) = \mathcal{N}\left(\frac{E + mc^2}{\pi E}\right)^{1/2} x f_l(x) ,$$
$$Q_{E\kappa}(r) = \mathcal{N}\,\mathrm{sgn}(\kappa)\left(\frac{E - mc^2}{\pi E}\right)^{1/2} x f_{\bar{l}}(x) . \quad (22.126)$$

Equations (22.124) require that Riccati–Bessel solutions of the same type be chosen for both components. The possibilities are:

Standing Waves

The two solutions of the same type are $f_l(x) = j_l(x)$, $f_l(x) = y_l(x)$. The $j_l(x)$ are bounded everywhere, including the singular points $x = 0$, $x \to \infty$ and have zeros of order l at $x = 0$. The $y_l(x)$ are bounded at infinity but have poles of order $l+1$ at $x = 0$.

Progressive Waves

The spherical Hankel functions (functions of the third kind) are linear combinations

$$h_l^{(1)}(x) = j_l(x) + \mathrm{i}y_l(x), \quad h_l^{(2)}(x) = j_l(x) - \mathrm{i}y_l(x) .$$

Recalling that p is *real* if and only if $|E| > mc^2$, we see that $h_l^{(1)}(x)$, $h_l^{(2)}(x)$ are bounded as $x \to \infty$ and have poles of order $l+1$ at $x = 0$. Notice that when $|E| < mc^2$, which does not occur for a free particle, p becomes pure imaginary and no solution exists which is finite at both singular points.

The normalization constant $\mathcal{N}$ can be determined by using the well-known result

$$\int_0^\infty j_l(pr) j_l(p'r) r^2 \,\mathrm{d}r = \frac{\pi}{2p^2} \delta(p - p') .$$

The choice $\mathcal{N} = 1$ ensures that

$$\int_0^\infty \left[P^\dagger_{E\kappa}(r) P_{E'\kappa}(r) + Q^\dagger_{E\kappa}(r) Q_{E'\kappa}(r)\right] \mathrm{d}r = \delta(p - p') .$$

Noting that

$$\delta(E - E') = \left|\frac{\mathrm{d}p}{\mathrm{d}E}\right| \delta(p - p') ,$$

and $\mathrm{d}p/\mathrm{d}E = c^2 p/E$ gives

$$\int_0^\infty \left[P^\dagger_{E\kappa}(r) P_{E'\kappa}(r) + Q^\dagger_{E\kappa}(r) Q_{E'\kappa}(r)\right] \mathrm{d}r = \delta(E - E') .$$

when $\mathcal{N} = \left(|E|/c^2 p\right)^{1/2}$.

22.6 Numerical Approximation of Central Field Dirac Equations

The main drive for understanding methods of numerical approximation of solutions of Dirac's equation comes from their application to many-electron systems. Approximate wave functions for atomic or molecular states are usually constructed from products of one-electron orbitals, and their determination exploits knowledge gained from the treatment of one-electron problems. Whilst the numerical methods described here are strictly one-electron in character, extension to many-electron problems is relatively straightforward.

22.6.1 Finite Differences

The numerical approximation of eigensolutions of the first order system of differential equations (22.101)

$$E\begin{pmatrix} P_{E\kappa}(r) \\ Q_{E\kappa}(r) \end{pmatrix} = \begin{pmatrix} mc^2+V(r) & -c\left(\frac{\mathrm{d}}{\mathrm{d}r}-\frac{\kappa}{r}\right) \\ c\left(\frac{\mathrm{d}}{\mathrm{d}r}+\frac{\kappa}{r}\right) & -mc^2+V(r) \end{pmatrix}\begin{pmatrix} P_{E\kappa}(r) \\ Q_{E\kappa}(r) \end{pmatrix} \tag{22.127}$$

can be achieved by more or less standard finite difference methods given in texts such as [22.25]. For states in either continuum, $E > mc^2$ or $E < -mc^2$, the calculation is completely specified as an *initial value problem* for a prescribed value of E starting from power series solutions in the neighborhood of $r = 0$. Solutions of this sort exist for all values of (complex) E except at the bound eigensolutions in the gap $-mc^2 < E < mc^2$. For bound states, the calculation becomes that of a *two-point boundary value problem* in which the eigenvalue E has to be determined iteratively along with the numerical solution. We concentrate on the latter, which is more involved.

It is convenient to write

$$\epsilon_{n\kappa} = E_{n\kappa} - mc^2 , \tag{22.128}$$

so that ϵ approaches the nonrelativistic eigenvalue in the limit $c \to \infty$. For the one-electron problem, (22.101) can be written in the general form

$$J\frac{\mathrm{d}u}{\mathrm{d}s} + \frac{1}{c}\frac{\mathrm{d}r}{\mathrm{d}s}\left[r\epsilon + W(s)\right]u(s) = \chi(s)\frac{\mathrm{d}r}{\mathrm{d}s} , \tag{22.129}$$

where $u(s)$ and $\chi(s)$ are two-component vectors, such that

$$u(s) = \begin{pmatrix} P(s) \\ Q(s) \end{pmatrix} , \quad J = \begin{pmatrix} 0 & 1 \\ -1 & 0 \end{pmatrix} ,$$

$$W(s) = \begin{pmatrix} -rV(r) & -c\kappa \\ -c\kappa & 2rc^2 - rV(r) \end{pmatrix} ,$$

and $r(s)$ is a smooth differentiable function of a new independent variable s. This facilitates the use of a uniform grid for s mapping onto a suitable nonuniform grid for r. Common choices are

$$r_n = r_0\,\mathrm{e}^{s_n} , \quad s_n = nh , \quad n = 0, 1, 2, \ldots , N ,$$

for suitable values of the parameters r_0 and h, and

$$A\,r_n + \log\left(1 + \frac{r_n}{r_0}\right) = s_n , \quad n = 0, 1, 2, \ldots , N ,$$

where A is a constant, chosen so that the spacing in r_n increases exponentially for small values of n and approaches a constant for large values of n. The exponentially increasing spacing is appropriate for tightly bound solutions, but a nearly linear spacing is advisable to ensure numerical stability in the tails of extended and continuum solutions.

The most convenient numerical algorithm involves *double shooting* from $s_0 = 0$ and $s_N = Nh$ towards an intermediate *join* point $s = Jh$, adjusting ϵ until the trial solutions have the right number of nodes and have left- and right-limits at $s = Jh$ which agree to a pre-set tolerance (commonly about 1 part in 10^8).

The *deferred correction* method [22.26, 27] allows the precision of the numerical approximation to be improved as the iteration converges. Consider the simplest implicit linear difference scheme for the first order system

$$\frac{\mathrm{d}y}{\mathrm{d}s} = F\left[y(s), s\right] ,$$

based on the trapezoidal rule of quadrature, is

$$z_{j+1} - z_j = \frac{1}{2}h(F_{j+1} + F_j) , \tag{22.130}$$

which has a local truncation error $O(h^2)$. The precision can be improved, at the expense of increasing the computational cost per iterative cycle, by adding higher order difference terms to the right-hand side in (22.130). Use of the trial solution from the previous cycle leaves the stability properties of (22.130) are unaltered, but the converged solution has much higher accuracy.

To apply this to the Dirac system, write $f(s) = \mathrm{d}r/\mathrm{d}s$ and

$$A_j^{\pm} = J \pm \frac{h}{2c} f(s_j) W(s_j) .$$

Also consider a slightly generalized problem in which $V(r)$ is replaced by a discretized potential $U_j^{(\nu)}$ that may change from one iteration to the next as in a self-consistent field calculation. The first iteration is

$$\begin{aligned}&\boldsymbol{A}_{j+1}^{+(0)}\,U_{j+1}^{(1)}-\boldsymbol{A}_{j}^{-(0)}\,U_{j}^{(1)}\\&\quad+\epsilon^{(0)}\frac{h}{2c}\left[r_{j+1}f(s_{j+1})U_{j+1}^{(1)}+r_j f(s_j)U_j^{(1)}\right]\\&=\frac{1}{2}h\left[f(s_{j+1})\chi(s_{j+1})^{(0)}+f(s_j)\chi(s_j)^{(0)}\right],\end{aligned} \tag{22.131}$$

where superscript 0 refers to initial estimates and superscript 1 to the result of the first iteration. On the $(\nu+1)$-th iteration, we solve

$$\begin{aligned}&\boldsymbol{A}_{j+1}^{+(\nu)}\,U_{j+1}^{(\nu+1)}-\boldsymbol{A}_{j}^{-(\nu)}\,U_{j}^{(\nu+1)}\\&\quad+\epsilon^{(\nu)}\frac{h}{2c}\left[r_{j+1}f(s_{j+1})U_{j+1}^{(\nu+1)}+r_j f(s_j)U_j^{(\nu+1)}\right]\\&=\frac{1}{2}h\left[f(s_{j+1})\chi(s_{j+1})^{(\nu)}+f(s_j)\chi(s_j)^{(\nu)}\right]\\&\quad+\frac{1}{12}\delta^3U_{j+1/2}^{(\nu)}+\cdots,\end{aligned} \tag{22.132}$$

where $\delta^3\,U_{j+1/2}^{(\nu)}$ is the central-difference correction of order 3 [22.22, Sect. 25.1.2]. Higher order difference corrections (at least to order 5) are included in modern codes to improve the accuracy and numerical stability of weakly bound solutions. This deferred correction algorithm can be shown to converge asymptotically to the required solution of the differential system with a local truncation error of order $O(h^{2p+2})$ when difference corrections of order $2p+1$ are employed [22.28].

22.6.2 Expansion Methods

Methods of solving the Dirac equation which represent the one-electron wave function as a linear combination of sets of square integrable functions (basis sets) have become popular in the last 10 years. Simple and rigorous criteria for choosing effective basis sets for this purpose are now available, and classes of functions that satisfy these criteria are known. Consequently, cheap and accurate calculations of the electronic structure of atoms and molecules are now a practical possibility.

Finite difference algorithms generate eigensolutions one at a time. Basis set methods replace the differential operator $\hat{\boldsymbol{h}}_\mathrm{D}$ of (22.87) with a finite symmetric (in some cases complex Hermitian) matrix of dimension $2N$. The spectrum of this operator, which is of course a pure point spectrum, consists of three pieces: N eigensolutions with $E<-mc^2$ $(\epsilon<-2mc^2)$ representing the eigenstates of the lower continuum; $N_\mathrm{b}<N$ eigensolutions in the gap $-mc^2<E<mc^2$ $(-2mc^2<\epsilon<0)$ corresponding to bound states; and $N-N_\mathrm{b}$ eigensolutions with $E>mc^2$ $(\epsilon>0)$ representing the eigenstates of the upper continuum. For properly chosen basis sets, the approximation properties of bound state eigensolutions are similar to those of the equivalent nonrelativistic eigensolutions. Solutions at continuum energies have the correct behavior near $r=0$, but their amplitudes decrease exponentially like bound state solutions at large values of r. The criteria on which this description rests are as follows:

A. The eigenstates of $\hat{\boldsymbol{h}}_\mathrm{D}$ are 4-component central field spinors whose components are coupled. The basis functions should therefore also consist of 4-component spinors of the form

$$\Phi_{\kappa m}(\boldsymbol{r})=\frac{1}{r}\begin{bmatrix}f_\kappa^L(r)\chi_{\kappa,m}(\theta,\varphi)\\ i f_\kappa^S(r)\chi_{-\kappa,m}(\theta,\varphi)\end{bmatrix}. \tag{22.133}$$

B. The spinor basis functions should, as far as practicable, satisfy the boundary conditions near $r=0$ of Sect. 22.5.3. They should also be square integrable at infinity.
C. Acceptable spinor basis functions should satisfy the relation

$$\mathrm{i}\hbar\frac{f_\kappa^S(r)}{r}\chi_{-\kappa,m}(\theta,\varphi)\to\boldsymbol{\sigma}\cdot\boldsymbol{p}\,\frac{f_\kappa^L(r)}{r}\chi_{\kappa,m}(\theta,\varphi) \tag{22.134}$$

in the *nonrelativistic limit*, $c\to\infty$.
D. Acceptable spinor basis functions must have *finite expectation values* of component operators of $\hat{\boldsymbol{h}}_\mathrm{D}$, namely $\boldsymbol{\alpha}\cdot\boldsymbol{p}$, β and $V(r)$.

Finite Basis Set Formalism

Assume that each solution of the target problem is approximated as a linear combination

$$\psi_{\kappa m}(\boldsymbol{r})=\frac{1}{r}\begin{pmatrix}\sum_j c_{\kappa j}^L f_{\kappa j}^L(r)\chi_{\kappa,m}(\theta,\varphi)\\ \mathrm{i}\sum_j c_{\kappa j}^S f_{\kappa j}^S(r)\chi_{-\kappa,m}(\theta,\varphi)\end{pmatrix}, \tag{22.135}$$

where $c_{\kappa j}^L, c_{\kappa j}^S$ $j=1\cdots N$, are arbitrary constants, so that each j-term on the right-hand side has the form (22.133). This enables us to construct a *Rayleigh quotient*

$$W[\psi]=\frac{\langle\psi|\hat{\boldsymbol{h}}_\mathrm{D}|\psi\rangle}{\langle\psi|\psi\rangle}, \tag{22.136}$$

where both $\langle\psi|\hat{\boldsymbol{h}}_{\mathrm{D}}|\psi\rangle$ and $\langle\psi|\psi\rangle$ are quadratic expressions in the expansion coefficients c_j^L, c_j^S. By requiring that $W[\psi]$ shall be stationary with respect to arbitrary variations in the expansion coefficients, we arrive at the matrix eigenvalue equation

$$f_\kappa \begin{pmatrix} \boldsymbol{c}_\kappa^L \\ \boldsymbol{c}_\kappa^S \end{pmatrix} = \epsilon \begin{pmatrix} \boldsymbol{S}_\kappa^{LL} & 0 \\ 0 & \boldsymbol{S}_\kappa^{SS} \end{pmatrix} \begin{pmatrix} \boldsymbol{c}_\kappa^L \\ \boldsymbol{c}_\kappa^S \end{pmatrix}, \tag{22.137}$$

where the matrix Hamiltonian is denoted by

$$f_\kappa = \begin{pmatrix} \boldsymbol{V}_\kappa^{LL} & c\boldsymbol{\Pi}_\kappa^{LS} \\ c\boldsymbol{\Pi}_\kappa^{SL} & \boldsymbol{V}_\kappa^{SS} - 2mc^2\boldsymbol{S}_\kappa^{SS} \end{pmatrix},$$

$\boldsymbol{c}_\kappa^L, \boldsymbol{c}_\kappa^S$ are N-vectors, and $\boldsymbol{V}_\kappa^{LL}$, $\boldsymbol{V}_\kappa^{SS}$, $\boldsymbol{S}_\kappa^{LL}$, $\boldsymbol{S}_\kappa^{SS}$, $\boldsymbol{\Pi}_\kappa^{LS}$ and $\boldsymbol{\Pi}_\kappa^{SL}$ are all $N \times N$ matrices. Using superscripts T to denote either of the letters L, S, the elements of the matrices are defined by

$$S_{\kappa ij}^{TT} = \int_0^\infty f_{i\kappa}^{T*}(r) f_{j\kappa}^T(r)\,\mathrm{d}r\,, \tag{22.138}$$

$$V_{\kappa ij}^{TT} = \int_0^\infty f_{i\kappa}^{T*}(r) V(r) f_{j\kappa}^T(r)\,\mathrm{d}r\,, \tag{22.139}$$

and

$$\Pi_{\kappa ij}^{LS} = \int_0^\infty f_{i\kappa}^{L*}(r)\left(-\frac{\mathrm{d}}{\mathrm{d}r} + \frac{\kappa}{r}\right) f_{j\kappa}^S(r)\,\mathrm{d}r\,, \tag{22.140}$$

$$\Pi_{\kappa ij}^{SL} = \int_0^\infty f_{i\kappa}^{S*}(r)\left(\frac{\mathrm{d}}{\mathrm{d}r} + \frac{\kappa}{r}\right) f_{j\kappa}^L(r)\,\mathrm{d}r\,. \tag{22.141}$$

If $f_{i\kappa}^L(r)$ and $f_{i\kappa}^S(r)$ vanish at both $r = 0$ and $r \to \infty$, then a simple integration by parts shows that $\boldsymbol{\Pi}_\kappa^{LS}$ and $\boldsymbol{\Pi}_\kappa^{SL}$ are Hermitian conjugate matrices.

Physically Acceptable Basis Sets

The four criteria described above are exploited in the following way:

A. The structure of (22.133) ensures (i) that the upper and lower components have properly matched angular behavior. It also emphasizes that the radial parts are part of a spinor structure which should be kept intact when making approximations.
B. The nuclear singularity drives the dynamics of the electronic motion. It is therefore important that approximate trial solutions should have the correct analytic character as defined in Sects. 22.5.3 and 22.5.4. An expansion of $f_{i\kappa}^L(r)$ and $f_{i\kappa}^S(r)$ at $r = 0$ must reproduce this analytic behavior *exactly* if the approximation is to be physically reliable. The boundary conditions are part of the definition of a quantum mechanical operator; changing them gives a different operator with a different eigenvalue spectrum, so that trial functions which do not satisfy the boundary conditions of the physical problem cannot reproduce the physical solution. The behavior as $r \to \infty$ is less crucial. Provided a bound wavefunction is well approximated over the region containing most of the electron density, the results are insensitive to many choices.
C. The correct reduction of the Dirac equation to Schrödinger's equation in the nonrelativistic limit (for example see [22.4, p. 97]) depends upon the operator identity

$$\boldsymbol{p}^2 = (\boldsymbol{\sigma}\cdot\boldsymbol{p})(\boldsymbol{\sigma}\cdot\boldsymbol{p})\,.$$

In the basis set formalism, the matrix equivalent of this equation is

$$T_{lij} = \frac{1}{2}\sum_{k=1}^{N} \Pi_{\kappa ik}^{LS} \Pi_{\kappa kj}^{SL}\,, \tag{22.142}$$

where

$$T_{lij} = \int_0^\infty f_{i\kappa}^{L*}(r)\frac{1}{2}\left(-\frac{\mathrm{d}^2}{\mathrm{d}r^2} + \frac{l(l+1)}{r^2}\right) f_{j\kappa}^L(r)\,\mathrm{d}r$$

is the ij-element of the nonrelativistic radial kinetic energy matrix. *This is not true in general unless criterion C holds* [22.29, 30]. The criterion can only be satisfied by matched pairs of functions $f_{i\kappa}^L(r)$, $f_{i\kappa}^S(r)$, ruling out all choices of basis set in which large and small components are not matched in pairs. Another way of viewing this result is to observe that for a general basis set, the sum over intermediate states in (22.142) is necessarily incomplete. The Hermitian conjugacy property, $\Pi_{\kappa ij}^{LS} = \Pi_{\kappa ji}^{SL}$ ensures that the omitted terms give real and non-negative contributions. Thus all other choices of basis set cause (22.142) to *underestimate* the nonrelativistic kinetic energy [22.29] and to give spuriously large relativistic energy corrections.

We emphasize that (22.134) need only be true in the limit $c \to \infty$; however, basis sets used for finite values of c should be smooth functions of c^{-1} as $c \to \infty$

so that the equality

$$i\hbar \frac{f_\kappa^S(r)}{r} \chi_{-\kappa,m}(\theta,\varphi) = \boldsymbol{\sigma}\cdot\boldsymbol{p}\, \frac{f_\kappa^L(r)}{r} \chi_{\kappa,m}(\theta,\varphi) \tag{22.143}$$

holds *in the limit*.

D. This ensures that the basis functions are in the domain of the Dirac operator; the meaning of this statement can be made precise in a functional analytic discussion such as in [22.3]. Some implications for the finite basis set approach are given in the author's paper [22.15, pp. 235–253], which discusses the convergence of expectation values of operators for approximate Dirac wavefunctions obtained by this method. Here the main importance is that a (possibly singular) multiplicative operator $V(r)$ (say, $-Z/r$) has $N\times N$ matrices $\boldsymbol{V}_\kappa^{LL}$, $\boldsymbol{V}_\kappa^{SS}$ with *finite* elements. This must be true both for exact solutions and for approximations if the wave functions are to represent physical states. In particular, both matrices must have a *lowest eigenvalue* $V_{\min}^{(N)}$ say. Consider now the quantity $(\psi|\hat{\boldsymbol{h}}_{\mathrm{D}}(\lambda)|\psi)$, where

$$\hat{\boldsymbol{h}}_{\mathrm{D}}(\lambda) = \left(c\boldsymbol{\alpha}\cdot\boldsymbol{p} + \beta m_{\mathrm{e}}c^2\right) + \lambda V(r)\,.$$

With $\lambda = 0$ we have a free Dirac particle with a two-branched continuous spectrum $E > mc^2$ and $E < -mc^2$. A negative definite $V(r)$ has always $(\psi|V(r)|\psi) > V_{\min}$; clearly,

$$V_{\min}^{(N)} \geq V_{\min} > -2mc^2\,, \tag{22.144}$$

for all values of N. So if we increase λ from 0 to 1, the eigenvalues of trial solutions corresponding to eigenvalues in the upper continuum at $\lambda = 0$ will be smoothly decreasing functions of λ bounded below by $V_{\min}$ for all values of N. It follows that *the upper set of eigenvalues has a fixed lower bound* in the gap $\left(-mc^2, mc^2\right)$ for each finite matrix approximation. If the basis set satisfies suitable completeness criteria in an appropriate Hilbert space as $N\to\infty$ (see [22.15, pp. 235–253], [22.31] for more details) we see that, if (22.144) holds for all values of N, the infinite sequence $\{E_{N+i}^{(N)},\ N = N_0, N_0+1, \ldots\}$ of eigenvalues approximating the ith bound state has a finite lower bound, and therefore, by the completeness of the real numbers, it must have a limit point E_i in the bound state gap $\left(-mc^2, mc^2\right)$. Thus Rayleigh–Ritz approximations for Dirac's Hamiltonian converge in the same fashion as the corresponding nonrelativistic Rayleigh–Ritz approximations [22.30, 31].

22.6.3 Catalogue of Basis Sets for Atomic Calculations

A. L-Spinors:

L-spinors [22.31] are related to Dirac hydrogenic functions in much the same way as Sturmian functions [22.32, 33] are related to Schrödinger hydrogenic functions (Sect. 22.3). They are solutions of the differential equation system

$$\begin{pmatrix} \frac{1}{2} - \frac{\alpha_{n_r\kappa} Z\mu^2}{cx} & -\frac{\mathrm{d}}{\mathrm{d}x} + \frac{\kappa}{x} \\ \frac{\mathrm{d}}{\mathrm{d}x} + \frac{\kappa}{x} & -\frac{1}{2} - \frac{Z}{\alpha_{n_r\kappa}\mu^2 cx} \end{pmatrix} \begin{pmatrix} \frac{f_{n_r\kappa}^L(x)}{\mu} \\ \mu f_{n_r\kappa}^S(x) \end{pmatrix} = 0\,, \tag{22.145}$$

where $x = 2\lambda r$ is a scaled radial coordinate, with fixed λ which can be related to an energy parameter $E_0^R = c^2\sqrt{1-\lambda^2/c^2}$, and μ^2, a root of the equation $\mu^4 - 2c\mu^2/\lambda + 1 = 0$, is given by

$$\mu^2 = \frac{c}{\lambda}\left(1 + E_0^R/c^2\right)\,. \tag{22.146}$$

This choice ensures that $f_{n_r\kappa}^L(x)$ tends smoothly to the corresponding Coulomb Sturmian in the nonrelativistic limit $c\to\infty$ [22.31]. L^2 boundary conditions are satisfied if $\alpha_{n_r\kappa} = N_{n_r\kappa}\lambda/Z$; when $\alpha_{n_r\kappa} = 1$, then $f_{n_r\kappa}^L(x)$, $f_{n_r\kappa}^S(x)$ respectively coincide with the Dirac–Coulomb eigenfunctions $P_{n\kappa}(r)$ and $Q_{n\kappa}(r)$ having principal quantum number $n = n_r + |\kappa|$. The explicit form for L-spinors, in terms of Laguerre polynomials (see Sect. 9.3.2), $L_{n_r}^{(2\gamma)}(x)$, is

$$f_{\kappa,n_r}^{(L)}(x) = \mathcal{N}_{n_r,\kappa}\, x^\gamma \mathrm{e}^{-x/2}\left[-(1-\delta_{n_r,0})L_{n_r-1}^{(2\gamma)}(x) + \frac{(N_{n_r,\kappa}-\kappa)}{(n_r+2\gamma)} L_{n_r}^{(2\gamma)}(x)\right], \tag{22.147}$$

$$f_{\kappa,n_r}^{(S)}(x) = \mathcal{N}_{n_r,\kappa}\, x^\gamma \mathrm{e}^{-x/2}\left[-(1-\delta_{n_r,0})L_{n_r-1}^{(2\gamma)}(x) - \frac{(N_{n_r,\kappa}-\kappa)}{(n_r+2\gamma)} L_{n_r}^{(2\gamma)}(x)\right], \tag{22.148}$$

where

$$\mathcal{N}_{n_r\kappa} = \left(\frac{n_r!\,(2\gamma+n_r)}{2N_{n_r\kappa}(N_{n_r\kappa}-\kappa)\,\Gamma(2\gamma+n_r)}\right)^{1/2} \tag{22.149}$$

is chosen so that the diagonal elements g_{n_r,n_r}^κ of the Gram (or overlap) matrix are unity for both large and small components. Both Gram matrices are tri-diagonal with non-zero off-diagonal elements

$$g^{(\kappa)}_{n_r,(n_r+1)} = g^{(\kappa)}_{(n_r+1),n_r} = \frac{\eta^T}{2}\left(\frac{(n_r+1)(2\gamma+n_r+1)(N_{n_r\kappa}-\kappa)}{N_{n_r\kappa}N_{(n_r+1),\kappa}(N_{(n_r+1),\kappa}-\kappa)}\right)^{1/2}, \tag{22.150}$$

where $T = L, S$, $\eta^L = -1$ and $\eta^S = +1$. This convention facilitates the construction of the blocks of the matrix Hamiltonian (22.137), which are banded when the operators are the powers r^n, $n > -1$. The properties of Laguerre polynomials ensure that the matrix of the Coulomb potential is diagonal. For a full discussion of L-spinors, their orthogonality and completeness properties, and applications to hydrogenic atoms see [22.31].

L-spinors are most useful for hydrogenic problems, either for isolated atoms or for atoms in strong electromagnetic fields (see Chapt. 13). The equivalent nonrelativistic Coulomb Sturmians have for a long time been used to study the Zeeman effect on high Rydberg levels, especially in the region where chaotic behavior is expected [22.34] (see Chapt. 15).

B. S-Spinors:
S-spinors have the functional form of *the most nearly nodeless* L-spinors characterized by the minimal value of n_r, and can be viewed as the relativistic analogues of Slater functions (STOs). When κ is negative, take $n_r = 0$, so that

$$f^{(L)}_{\kappa,0}(x) = -f^{(S)}_{\kappa,0}(x) = \mathcal{N}_{\kappa,0}\, x^\gamma \exp(-x/2)\frac{N_{0,\kappa}-\kappa}{2\gamma}L^{(2\gamma)}_0(x)\,.$$

When κ is positive, we must take $n_r = 1$, and then

$$f^{(L)}_{\kappa,1}(x) = \mathcal{N}_{1,\kappa}\, x^\gamma\, e^{-x/2} \times\left[-L^{(2\gamma)}_0(x) + \frac{N_{1,\kappa}-\kappa}{1+2\gamma}L^{(2\gamma)}_1(x)\right],$$

$$f^{(S)}_{\kappa,1}(x) = \mathcal{N}_{1,\kappa}\, x^\gamma\, e^{-x/2} \times\left[-L^{(2\gamma)}_0(x) - \frac{N_{1,\kappa}-\kappa}{1+2\gamma}L^{(2\gamma)}_1(x)\right].$$

These can be simplified by inserting the explicit expressions $L^{(2\gamma)}_0(x) = 1$, $L^{(2\gamma)}_1(x) = 2\gamma+1-x$. We define a *set of S-spinors* with exponents $\{\lambda_m, m = 1, 2, \dots, N\}$ by rewriting the above in the form

$$f^{(T)}_m(r) = A^T g_m(\gamma, r) + B^T g_m(\gamma+1, r)\,, \tag{22.151}$$

where $T = L, S$, $g_m(\theta, r) = r^\theta\, e^{-\lambda_m r}$,

$$\begin{aligned} &A^L = A^S = 1,\ B^L = B^S = 0 \quad \text{for } \kappa < 0\,,\\ &\left.\begin{aligned} A^L &= \frac{(\kappa+1-N_{1,\kappa})(2\gamma+1)}{2(N_{1,\kappa}-\kappa)}\\ A^S &= \frac{(\kappa-1-N_{1,\kappa})(2\gamma+1)}{2(N_{1,\kappa}-\kappa)}\\ B^L &= B^S = 1 \end{aligned}\right\} \quad \text{for } \kappa > 0\,, \end{aligned} \tag{22.152}$$

and

$$\gamma = \sqrt{\kappa^2 - Z^2/c^2}, \quad N_{1,\kappa} = \sqrt{\kappa^2+2\gamma+1}\,.$$

The choice of the set of positive real exponents $\{\lambda_m, m = 1, 2, \dots, N\}$, must be such as to assure Rayleigh–Ritz convergence [22.15, pp. 235–253)] and to maximize the rate at which it is attained. In particular, if *one* particular exponent is chosen to have the value $\lambda_m = Z/N_{n_r,\kappa}$, then the corresponding S-spinor is a true hydrogenic solution. In this case the trial solution is exact. Clearly, S-spinors inherit desirable properties of L-spinors and, in particular, satisfy criteria A–D.

All elements of the matrix Hamiltonian of the Dirac hydrogenic problem can be expressed in terms of Euler's integral for the gamma function [22.22, Sect. 6.1.1]:

$$\Gamma(z) = k^z \int_0^\infty t^{z-1} e^{-kt}\, dt\,, \quad (\Re z > 0\,,\ \Re k > 0)$$

and are therefore readily written down and evaluated. The effectiveness of this method depends upon the choice of exponent set: see D below. We refer to calculations using this scheme for many-electron systems in Sect. 22.7.

C. G-Spinors:
The G-spinors are the relativistic analogues of nonrelativistic spherical Gaussians (SGTO), popular in quantum chemistry for studying the electronic structure of atoms and molecules. They satisfy the relativistic boundary conditions for a *finite size* nuclear charge density distribution at $r = 0$, and are therefore the most convenient for relativistic molecular electronic structure calculations. They are defined so that (22.143) holds for finite c as well as in the nonrelativistic limit, which is equivalent to

$$f^{(S)}_m(r) = \text{const.}\left(\frac{d}{dr} + \frac{\kappa}{r}\right) f^{(L)}_m(r)\,. \tag{22.153}$$

Thus, if

$$f_m^{(L)}(r) = \mathcal{N}_{l,m}^{(L)}\, r^{l+1} e^{-\lambda_m r^2}\,, \qquad (22.154)$$

$$f_m^{(S)}(r) = \mathcal{N}_{l,m}^{(S)}\left[(\kappa+l+1)r^l - 2\lambda_m r^{l+2}\right] e^{-\lambda_m r^2}\,. \qquad (22.155)$$

Note that the leading term in (22.155) vanishes when $\kappa < 0$, so that the radial amplitude $r^{-1} f_m^{(S)}(r)$ is never singular, even in the s-state case when $\kappa = -1$, $l = 0$.

D. Exponent Sets for S- and G-Spinors:

Quantum chemists are familiar with the use of nonrelativistic STO and GTO basis sets, and there are extensive collections of optimized exponents which permit economical calculations for atomic and molecular calculations [22.35–37]. These sets are a good starting point for relativistic calculations also. By and large, the compilations ignore mathematical completeness, which although desirable is unattainable in practice. However, basis sets can almost always be constructed to give adequate numerical precision for most purposes.

An effective alternative to optimization, especially for atoms, is to use geometrical sequences $\{\lambda_m\}$ of the form

$$\lambda_m = \alpha_N \beta_N^{m-1}\,, \qquad m = 1, 2, \ldots, N\,, \qquad (22.156)$$

which depend upon just two parameters α_N, β_N. A convenient way to do this is to find a pair α_{N_0}, β_{N_0} for small N_0, say $N_0 = 9$, in a cheap and simple nonrelativistic calculation and then to increase N systematically using relations such as

$$\frac{\alpha_N}{\alpha_{N_0}} = \left(\frac{\beta_N - 1}{\beta_{N_0} - 1}\right)^a\,, \quad \text{or} \quad \frac{\ln \beta_N}{\ln \beta_{N_0}} = \left(\frac{N_0}{N}\right)^b\,,$$

where a, b are positive constants. Experience shows that no linear dependence problems (caused by ill-conditioning of the $\boldsymbol{S}^T$ matrices) are encountered when $\beta_N > 1.2$ for S-spinors, with N up to about 30, or $\beta_N > 1.5$ for G-spinors with N up to about 50.

E. Other Types of Analytic Basis Sets; Variational Collapse:

The earliest work with atoms [22.38,39] used STO functions of the form $\{r^\gamma \exp(-\lambda_m r),\ m = 1, \ldots N\}$ for both large and small components, whilst *Kagawa* [22.40,41] used integer powers instead of the noninteger γ. *Drake* and *Goldman* [22.42] used functions of the form $\{r^{\gamma+i} \exp(-\lambda r),\ i = 0, \ldots N-1\}$. For hydrogenic problems, these worked well for negative κ states, but gave a single spurious eigenvalue for positive κ, which could be simply deleted from the basis set. Various test calculations are included in the review article [22.43, Sect. IV]. Other attempts to use GTOs in the early 1980's led to problems interpreted as a failure of the Rayleigh–Ritz method because of the presence of "negative energy states" with a spectrum unbounded below: so-called "variational collapse". It is clear that all these approaches fail to observe three, and sometimes all, of the four criteria for acceptable basis sets. They are incapable of satisfying the physical boundary conditions, and it is therefore hardly surprising that they give unphysical spectra.

Several procedures have been advocated to overcome the problem, of which the two most popular are "kinetic balance" and projection operators. Kinetic balance, suggested by *Lee* and *McLean* [22.44], advocates augmenting a GTO basis, common to both large and small components, with additional functions to "balance the set kinetically". This appears to "fix up" the problem for the upper spectrum, but introduces spurious states, mainly in the lower part of the spectrum, as well as increasing the size of the small component basis set. There is no rigorous nonrelativistic limit, and no mathematical proof of convergence such as that guaranteed by criteria A–D. A model with spurious negative energy states cannot furnish a consistent physical interpretation of negative energy solutions as positron states, expected of a proper relativistic theory.

If "variational collapse" is attributed to the absence of a lower bound to the Dirac spectrum as a whole, the idea of introducing a projection operator to eliminate collapse seems attractive. This is easy to do for free electrons, where the operators

$$\Lambda^{\pm} = \frac{\hat{h}_D \pm E}{2|E|}\,,$$

select positive/negative energy solutions. Unfortunately, this cannot be done in the presence of a potential except by an approximation which complicates calculations and reduces the efficiency of algorithms. The "negative energy sea" also depends upon the choice of potential; perturbing the potential (as long as it does not change the domain of the Hamiltonian) induces a unitary transformation taking one set of eigenstates into another which inevitably mixes the old positive and negative energy states. For example a relativistic calculation on a hydrogenic atom in which the nuclear charge is perturbed gives incorrect answers if the negative energy contribution to the perturbation series is omitted [22.45].

In any event, the finite matrix eigensolutions include both positive energy and negative energy states. It is therefore a simple matter to exclude the negative energy states if their contribution is expected to be negligible; this is the *no virtual-pair approximation*. The negative energy solutions are inert spectators for most atomic processes, just as are those positive energy solutions which lie deep in the atomic core. It is easy to go beyond the no virtual-pair approximation if the physical problem demands it.

Finite Element Methods:
Johnson et al. [22.46, 47], following earlier work on relativistic ion–ion collisions by *Bottcher* and *Strayer* [22.48], popularized the use of a basis of B-splines in relativistic atomic calculations. See Sect. 8.1.1. The method has mainly been of use in relativistic many-body calculations on the spectra of heavy ions. See Sect. 21.6 for spline-Galerkin representations in nonrelativistic atomic structure, such as [22.49]. *Parpia* and *Fischer* explored the spline-Galerkin approach for the Dirac equation [22.50], but this method has not been extended so far to relativistic many-electron atoms.

22.7 Many-Body Calculations

22.7.1 Atomic States

The construction of atomic many-electron wavefunctions from products of central field Dirac orbitals is employed to simplify the algorithms for calculating electronic structures and properties. This can be either in the context of expansions in Slater determinants of the traditional type, or by use of Racah algebra. A complete description of the methods of the latter sort used in popular computer codes is found in [22.27, Sect. 2].

22.7.2 Slater Determinants

An antisymmetric state of N independent electrons in configuration space can be constructed in the form

$$\begin{aligned}&\{\alpha_1, \alpha_2, \ldots, \alpha_N\} \\ &= \langle \boldsymbol{x}_1, \boldsymbol{x}_2, \ldots, \boldsymbol{x}_n | a^\dagger_{\alpha_1} a^\dagger_{\alpha_2} \cdots a^\dagger_{\alpha_N} |0\rangle \\ &= \frac{1}{N!}\begin{vmatrix} \psi_{\alpha_1}(\boldsymbol{x}_1) & \psi_{\alpha_2}(\boldsymbol{x}_1) & \cdots & \psi_{\alpha_N}(\boldsymbol{x}_1) \\ \psi_{\alpha_1}(\boldsymbol{x}_2) & \psi_{\alpha_2}(\boldsymbol{x}_2) & \cdots & \psi_{\alpha_N}(\boldsymbol{x}_2) \\ \cdots & \cdots & \cdots & \cdots \\ \psi_{\alpha_1}(\boldsymbol{x}_N) & \psi_{\alpha_2}(\boldsymbol{x}_N) & \cdots & \psi_{\alpha_N}(\boldsymbol{x}_N) \end{vmatrix}\end{aligned} \tag{22.157}$$

This *Slater determinant* is an antisymmetric eigenfunction of H_0 corresponding to the energy $\sum E_{\alpha_n}$ and of the angular momentum projection $J_3 = \sum j_{3,\alpha_n}$ corresponding to the eigenvalue $M_3 = \sum m_{3,\alpha_n}$. Defining the parity of a Dirac electron orbital as that of its upper component, $(-1)^{l_{\alpha_n}}$, we see that this has parity $\Pi(-1)^{l_{\alpha_n}}$.

22.7.3 Configurational States

Configurational state functions (CSF) having specified total angular momentum J and parity Π can be constructed by vector addition of the individual angular momenta: $\boldsymbol{J} = \sum \boldsymbol{j}_{\alpha_n}$. We write such states as

$$\phi(\gamma JM) = \sum_{\{m_{\alpha_n}\}} \langle \gamma JM | m_{\alpha_1}, m_{\alpha_2} \ldots, m_{\alpha_N} \rangle \times \{\alpha_1, \alpha_2, \ldots, \alpha_N\}\,, \tag{22.158}$$

where $\langle \gamma JM | m_{\alpha_1}, m_{\alpha_2} \ldots, m_{\alpha_N} \rangle$ is a generalized Clebsch–Gordon coefficient, and γ defines the angular momentum coupling scheme.

A list of orbital quantum numbers, $\{\alpha_1, \alpha_2, \ldots, \alpha_N\}$ defines an electron *configuration*. If the configuration belongs to a single subshell, then the states share a common set of labels $\{n, \kappa\}$ where n is the principal quantum number. In $j-j$ coupling, the *α-subshell states* of N_α equivalent electrons can therefore be identified (we can suppress the projection M_α and the parity Π_α) by the labeling α^{N_α}, γ_α, J_α, where γ_α distinguishes degenerate states of the same J_α. For $j-j$ coupling, such labels are needed only for $j \geq \frac{5}{2}$; the *seniority scheme*, [22.27, Sects. 2.3, 2.4)], provides a complete classification for $j < \frac{9}{2}$. A list of states of configurations j^N, classified in terms of the *seniority number* v and of *total angular momentum* J, appears in Table 22.5.

22.7.4 CSF Expansion

Atomic state functions (ASF) are linear superpositions of CSF's, of the form

$$\Psi(\gamma\Pi J) = \sum_{\alpha=1}^{N} c_\alpha \phi(\gamma_\alpha J)\,, \tag{22.159}$$

where c_α are a set of (normally) real coefficients. These coefficients are usually chosen so that $\Psi(\gamma\Pi J)$ is an

Table 22.5 j^N configurational states in the seniority scheme. The multiplicity of each unresolved degenerate state is indicated by a superscript

j	N	v	J
$\frac{1}{2}$	0, 2	0	0
	1	0	$\frac{1}{2}$
$\frac{3}{2}$	0, 4	0	0
	1, 3	1	$\frac{3}{2}$
	2	0	0
	2	2	2
$\frac{5}{2}$	0, 6	0	0
	1, 5	1	$\frac{5}{2}$
	2, 4	0	0
		2	2, 4
	3	1	$\frac{5}{2}$
		3	$\frac{3}{2}, \frac{9}{2}$
$\frac{7}{2}$	0, 8	0	0
	1, 7	1	
	2, 6	0	0
		2	2, 4, 6
	3, 5	1	$\frac{7}{2}$
		3	$\frac{3}{2}, \frac{5}{2}, \frac{9}{2}, \frac{11}{2}, \frac{15}{2}$
	4	0	0
		2	2, 4, 6
		4	2, 4, 5, 8
$\frac{9}{2}$	0, 10	0	0
	1, 9	1	$\frac{9}{2}$
	2, 8	0	0
		2	2, 4, 6, 8
	3, 7	1	$\frac{9}{2}$
		3	$\frac{3}{2}, \frac{5}{2}, \frac{7}{2}, \frac{9}{2}, \frac{11}{2}, \frac{13}{2}, \frac{15}{2}, \frac{17}{2}, \frac{21}{2}$
	4, 6	0	0
		2	2, 4, 6, 8
		4	$0, 2, 3, 4^2, 5, 6^2, 7, 8, 9, 10, 12$
	5	1	$\frac{9}{2}$
		3	$\frac{3}{2}, \frac{5}{2}, \frac{7}{2}, \frac{9}{2}, \frac{11}{2}, \frac{13}{2}, \frac{15}{2}, \frac{17}{2}, \frac{21}{2}$
	$\frac{7}{2}$	5	$\frac{3}{2}, \frac{5}{2}, \frac{7}{2}, \frac{9}{2}, \frac{11}{2}, \frac{13}{2}, \frac{15}{2}, \frac{17}{2}, \frac{19}{2}, \frac{25}{2}$

eigenstate of the many-electron Hamiltonian matrix in a finite subspace of CSF's.

22.7.5 Matrix Element Construction

A full presentation of the reduction of matrix elements between CSF's to computable form is beyond the scope of this chapter. There are two approaches: one is based on expanding all CSF's and ASF's in Slater determinants, whilst the other exploits the properties of central field orbital spinors. The principles underlying the first are straightforward and may be found in atomic physics texts and review articles such as [22.26, 27].

The use of second quantization and diagrammatic methods of the quantum theory of angular momentum provides a powerful means of reducing matrix elements between atomic CSF's to a linear combination of radial integrals in a systematic way. The method, which is fully explained in [22.27], leads to a complete classification of matrix element expressions for all the one- and two-electron operators treated in this chapter. A full implementation within the $j-j$ coupling seniority scheme is available in various versions of the GRASP code [22.51–53].

22.7.6 Dirac–Hartree–Fock and Other Theories

The notation above echoes that of the nonrelativistic theory of Chapt. 21, and it is possible to proceed along similar lines.

Dirac–Hartree–Fock Theory

Dirac–Hartree–Fock theory works exactly as described in Sect. 21.4; relativistic counterparts of Koopmans' theorem, fixed-core approximations, Brillouin's theorem are easy to obtain. The properties of Dirac–Hartree–Fock functions closely resemble those of Hartree–Fock functions, though allowance must be made for the fact that, for example, np orbitals (with $\kappa = -2$, $j = \frac{3}{2}$) and $n\bar{p}$ orbitals (with $\kappa = +1$, $j = \frac{1}{2}$) have different spatial distributions as a consequence of the dynamical and indirect effects of relativity. For further insight see [22.23, 26].

Most such calculations are currently made with updated versions of the codes of *Desclaux* [22.54] or *Grant* [22.51–53] which rely on finite difference methods resting on the techniques of Sect. 22.6.1. Further details may be found in the code descriptions.

Finite Matrix Methods for Atoms and Molecules

In view of the rapid pace of development of finite matrix methods, especially for the treatment of relativistic molecular electronic structure in the Born–Oppenheimer (fixed nucleus) approximation, it seems appropriate to give a brief outline of the extension of the one-electron equations of Sect. 22.4.2 to the many-electron case.

The method of approximation generalizes the one-body approximation scheme of Sect. 22.6.2 to the many-body problem based on the effective Hamiltonian of

Sect. 22.4.1. This leads to an energy functional of the form

$$E = E_0 + E_1 \tag{22.160}$$

where E_0 is the expected value of H_0 (22.76) and E_1 the expected value of H_1 for the finite basis many-body trial function. This leads to matrix Dirac–Fock equations of the form

$$\mathbf{FX} = \mathbf{ESX}. \tag{22.161}$$

In general, the Fock matrix $\mathbf{F}$ is a sum of several matrices

$$\mathbf{F} = \mathbf{f} + \mathbf{g} + \mathbf{b}, \tag{22.162}$$

where, for each symmetry κ and nuclear center, A, of the molecule, $\mathbf{f}$ can be partitioned into blocks

$$f_{A\kappa} = \begin{pmatrix} V_{A\kappa}^{LL} & c\Pi_{A\kappa}^{LS} \\ c\Pi_{A\kappa}^{SL} & V_{A\kappa}^{SS} - 2mc^2 S_{A\kappa}^{SS} \end{pmatrix}. \tag{22.163}$$

The matrix

$$\mathbf{g} = \begin{pmatrix} \mathbf{J}^{LL} - \mathbf{K}^{LL} & -\mathbf{K}^{LS} \\ -\mathbf{K}^{SL} & \mathbf{J}^{SS} - \mathbf{K}^{SS} \end{pmatrix} \tag{22.164}$$

is the matrix of the Coulomb repulsion part of the electron–electron interaction and

$$\mathbf{b} = \begin{pmatrix} \mathbf{B}^{LL} & \mathbf{B}^{LS} \\ \mathbf{B}^{SL} & \mathbf{B}^{SS} \end{pmatrix}. \tag{22.165}$$

is the matrix of the Breit interaction.

In the atomic (one nuclear center) case, following [22.55], these matrices can also be blocked by symmetry κ. Using superscripts T to label the L or S components, and the notation $\bar{T}$ to denote the complementary label: $\bar{T} = S$ when $T = L$ or $\bar{T} = L$ when $T = S$, then the direct Coulomb part $\mathbf{J}_\kappa^{TT}$ has matrix elements

$$J_{\kappa pq}^{TT} = \sum_{\kappa' rs}(2j'+1)\left(D_{\kappa' rs}^{TT} J_{\kappa pq,\kappa' rs}^{0,TTTT} + D_{\kappa' rs}^{\bar{T}\bar{T}} J_{\kappa pq,\kappa' rs}^{0,TT\bar{T}\bar{T}}\right), \tag{22.166}$$

whilst the exchange part $\mathbf{K}_\kappa^{TT'}$ has the form

$$K_{\kappa pq}^{TT'} = \sum_{\kappa' rs}\sum_{\nu}(2j'+1)b_\nu(jj')D_{\kappa' rs}^{TT'} K_{\kappa pq,\kappa' rs}^{\nu,TT'TT'}, \tag{22.167}$$

where TT' denotes any combination of component labels. Here $\mathbf{D}_\kappa^{TT'}$ is a density matrix with elements

$$D_{\kappa pq}^{TT'} = c_{\kappa p}^{T*} c_{\kappa q}^{T'}, \tag{22.168}$$

where $c_{\kappa p}^T$ are the expansion coefficients. The Breit interaction matrices have the similar form

$$B_{\kappa pq}^{TT} = \sum_{\kappa' rs}\sum_{\nu}(2j'+1)e_\nu(jj')D_{\kappa' rs}^{\bar{T}\bar{T}} K_{\kappa pq,\kappa' rs}^{\nu,TT\bar{T}\bar{T}}, \tag{22.169}$$

and

$$B_{\kappa pq}^{T\bar{T}} = \sum_{\kappa' rs}\sum_{\nu}(2j'+1)D_{\kappa' rs}^{\bar{T}T}$$
$$\times \left[d_\nu(\kappa\kappa') K_{\kappa pq,\kappa' rs}^{\nu,T\bar{T}\bar{T}T} + g_\nu(\kappa\kappa') M_{\kappa pq,\kappa' rs}^{\nu,T\bar{T}\bar{T}T}\right]. \tag{22.170}$$

The matrix elements are constructed from standard radial integrals

$$J_{\kappa pq,\kappa' rs}^{\nu,TTT'T'} = \int_0^\infty \int_0^\infty f_{\kappa p}^T(r_1) f_{\kappa q}^T(r_1) U_\nu(r_1, r_2)$$
$$\times f_{\kappa' r}^{T'}(r_2) f_{\kappa' s}^{T'}(r_2)\, dr_2\, dr_1 \tag{22.171}$$

where

$$U_\nu(r_1, r_2) = \begin{cases} r_1^\nu / r_2^{\nu+1} & \text{for } r_1 < r_2, \\ r_2^\nu / r_1^{\nu+1} & \text{for } r_1 > r_2. \end{cases}$$

Similarly

$$K_{\kappa pq,\kappa' rs}^{\nu,TT'TT'} = J_{\kappa p,\kappa' r,\kappa q,\kappa' s}^{\nu,TTT'T'} \tag{22.172}$$

and

$$M_{\kappa pq,\kappa' rs}^{\nu,T\bar{T}\bar{T}T} = \int_0^\infty \int_{r_1}^\infty f_{\kappa p}^T(r_1) f_{\kappa' r}^{\bar{T}}(r_1) U_\nu(r_1, r_2)$$
$$\times f_{\kappa q}^{\bar{T}}(r_2) f_{\kappa' s}^T(r_2)\, dr_2\, dr_1. \tag{22.173}$$

Further details about the coefficients $b_\nu(jj')$, $e_\nu(jj')$, $d_\nu(\kappa\kappa')$ and $g_\nu(\kappa\kappa')$ may be found in [22.55].

This formalism has been implemented for closed shell atoms with both S-spinors and G-spinors [22.55]. Computational aspects of calculating the radial integrals using S-spinors are discussed in [22.56, 57], and can be adapted with relatively small modifications to G-spinor basis sets. As yet, there have been relatively few applications by comparison with codes based on finite difference methods, but the potential can be gauged from papers such as [22.55, 58–61], which deal with Dirac–Fock and Dirac–Fock–Breit calculations, many-body perturbation theory and coupled-cluster schemes.

G-spinor basis sets provide the most promising technique for application to the electronic structure of molecules; computer codes are under active development.

Electron Correlation in Atomic Calculations

Here we use the term *correlation* to denote methods which go beyond the single determinant approximation of Dirac–Hartree–Fock theory. These include *configuration interaction* schemes, in which each ASF is represented as a linear combination of CSF's built from previously determined orbital spinors and *multiconfiguration Dirac–Fock* calculations in which the orbitals are optimized simultaneously. Calculations representative of state of the art techniques will be found in [22.62, 63].

Many-body perturbation theory calculations and coupled-cluster calculations are not well suited to calculations with finite difference codes, because of the expense of calculating more than a limited orbital basis and all the matrix elements required. Calculations based on finite matrix methods enable this sort of calculation to be done more economically. Some justification for the use of finite matrix methods in relativistic many-body theory is given in [22.15, pp. 235–253].

The relativistic version of quantum defect theory [22.64, 65] also gives insight into the competing roles of relativistic dynamics and screening in atoms. Compared with nonrelativistic quantum defect theory, it has been under-used.

22.7.7 Radiative Corrections

The term "radiative corrections" is usually interpreted to mean QED contributions to energies, expectation values or rates of atomic or molecular processes that arise from interaction between the electron–positron and photon fields, apart from those directly attributable to the nonrelativistic Coulomb interaction. This includes the relativistic and retardation effects embodied in the effective interaction between electrons as well as contributions from processes that are not so included. We consider two such processes, the electron self-energy and the vacuum polarization, which involve interactions of the same formal order as those giving rise to the covariant electron–electron interaction discussed above, but which are formally infinite. These are the lowest order processes requiring renormalization. See [22.7, 8, 10, 15] for more details.

Electron Self-Energy

For a one-electron system, the renormalized expression for the self-energy of an electron in the state a in Feynman gauge is

$$\begin{aligned}\Delta E_a = \lim_{\Lambda\to\infty} \mathcal{R}\Big[& -\mathrm{i}\alpha\pi . mc^2 \int \bar{\psi}_a(x_2)\gamma^\mu S_{\mathrm{F}}(x_2,x_1) \\ & \times \gamma^\nu \psi_a(x_1) g_{\mu\nu} D_{\mathrm{F}}^{\Lambda}(x_2-x_1)\,\mathrm{d}^3x_2\,\mathrm{d}^3x_1 \\ & \times \mathrm{d}(t_2-t_1) - \delta m(\Lambda)\langle\psi_a|\beta|\psi_a\rangle\Big]\,,\end{aligned} \tag{22.174}$$

where

$$\delta m(\Lambda) = \frac{\alpha}{\pi} mc^2 \left[\frac{3}{4}\ln(\Lambda^2) + \frac{3}{8}\right].$$

This represents the contribution from virtual processes involving the exchange of a single photon. The photon propagator has been modified to give the photon and effective mass Λ, so that the denominator of $D(q^2)$ (22.69) becomes $q^2 - \Lambda^2 + \mathrm{i}\delta$. The two parts of this formula diverge as $\Lambda \to \infty$, though the limit of their difference is finite. This makes calculation difficult and expensive. There are several approaches:

1. For atomic number $Z \lesssim 20$, an expansion in powers of the electron–nucleus coupling parameter $\alpha Z = Z/c$ is satisfactory.
2. At larger atomic numbers an expansion in αZ evidently fails to converge, and nonperturbative methods must be sought. This too is computationally difficult and expensive. The results for hydrogenic ions have been tabulated [22.66] for atomic numbers in the range $1 \leq Z \leq 100$. (See [22.10, Chapt. 2] for an up-to-date summary biased towards applications to the spectroscopy of highly-ionized atoms.)
3. Processes involving more than one virtual photon are hard to calculate, and have mostly been ignored. See [22.10] for references.

Vacuum Polarization

The contribution of vacuum polarization is next in order of importance in the list of radiative corrections in atoms. As shown by (22.49), the nuclear potential generates a current in the vacuum that is responsible for a short-range screening of the nuclear charge. This can be represented as a local perturbing potential which is easy to take into account [22.67–69].

22.7.8 Radiative Processes

The operator $j^\mu(x)A_\mu(x)$ which occurs in the interaction Hamiltonian (22.61) describes processes in which the number of photons present can increase or decrease by one. The Fock space operator may be written

$$H_{\text{int}} = \sum_{a,b} \sum_{\rho} \left[a_a^\dagger a_b q_\rho^\dagger M_{ab}^{(\rho)\dagger}(t) + a_a^\dagger a_b q_\rho M_{ab}^{(\rho)}(t) \right] , \tag{22.175}$$

where the first set of terms in the sum represents emission of a photon in the mode labeled ρ and the second to absorption of a photon by the same initial state. The operators a_a and $a_a^\dagger$ are anticommuting annihilation and creation operators of electrons, whilst q_ρ and $q_\rho^\dagger$ are *commuting* annihilation and creation operators of photons. If ω denotes photon frequency, then

$$M_{ab}^{(\rho)\dagger}(t) = M_{ab}^{(\rho)} \mathrm{e}^{\mathrm{i}(E_a - E_b + \omega)t} ,$$

$$M_{ab}^{(\rho)}(t) = M_{ab}^{(\rho)} \mathrm{e}^{\mathrm{i}(E_a - E_b - \omega)t} ,$$

where

$$M_{ab}^{(\rho)} = \left(\frac{\omega}{\pi c}\right)^{1/2} \int \psi_a^\dagger(\boldsymbol{x}) \left[\Phi^{(\rho)}(\boldsymbol{x}) + c\boldsymbol{\alpha} \cdot \boldsymbol{A}^{(\rho)}(\boldsymbol{x}) \right] \times \psi_b(\boldsymbol{x})\, \mathrm{d}^3 x$$

is the transition amplitude. For a discussion of this expression including the effect of gauge transformations on the computed amplitudes, the elimination of angular coordinates for atomic central field orbitals and connection with the nonrelativistic limit, see [22.10, 16, 27].

22.8 Recent Developments

22.8.1 Technical Advances

Relativistic atomic structure continues to develop to meet modern demands for high quality calculations on many-electron atoms. The computing power now available makes it possible to carry out multi-configurational Dirac–Hartree–Fock (MCDHF) or configuration interaction (CI) calculations on a scale unimaginable when this chapter was first drafted. Some of the software currently available is surveyed below.

On the theoretical side, there have been new technical applications of tensor operator theory. Whilst the approach initiated by *Fano* [22.27, 70] continues to be the basis on which many relativistic and nonrelativistic calculations are based, recent work aims to simplify the calculation, not only by exploiting second quantization techniques and the coupling of tensor operators, but by better utilization of quasispin methods [22.71–74]. A new jj-coupling package along these lines [22.75] has been constructed for evaluation of fractional parentage coefficients, reduced fractional parentage coefficients (in which the dependence on particle number is extracted as a quasispin $3j$-symbol), matrix elements of unit tensors T^k and double tensor operators $W^{k_q k_j}$, from which to construct many-particle matrix elements of physical operators. *Fritzsche* et al. [22.76–80] have recently published utilities which exploit the capabilities of the Maple computer algebra system to evaluate Racah algebra expressions.

22.8.2 Software for Relativistic Atomic Structure and Properties

Many software packages for relativistic atomic physics calculations can now be downloaded from the internet. The earliest codes, which generate many-electron wavefunctions and bound energy levels, taking account of the full relativistic electron–electron interaction and QED corrections, of *Desclaux* [22.54] and *Grant* et al. [22.51], though now much modified, are still in use, as is the code of *Chernysheva* and *Yakhontov* [22.81]. These codes can use various (MC)DHF and CI procedures, albeit with not more than a few hundred CSF. A more recent version of Grant et al.'s code appeared in 1989 [22.52] and GRASP92 embodied major changes to the user interface and to file-handling to permit calculations with very large CSF sets [22.53]. Most earlier calculations were of the AL or EAL type, in which a large number of states are treated together using a common orbital set. These are cheap and work well for highly ionized, few-electron systems but the results only have modest accuracy. More accurate treatment of electron correlation requires MCDHF calculations on single levels (OL calculations) or small groups of fine structure levels (EOL calculations). The CSF sets are chosen through some active space (AS) procedure as in nonrelativistic MCHF [22.82]; complete active spaces (CAS) are often too large for practical use, so that the AS must be restricted in some way, for example by using only SD (single and double) replacements from the reference CSF set. With such large CSF basis sets it is not practical or desirable to diagonalize the complete Hamiltonian matrix, and *Davidson*'s version [22.83, 84] of the Lanczos algorithm, as implemented by *Stathopoulos* and *Fischer* [22.85], is used in GRASP92 to construct the small number of eigenvalues and eigenvectors of physical importance.

This approach generally gives highly accurate wavefunctions and energy levels for a small number of atomic states. Each state is determined in a separate SCF calculation, and therefore has *its own set of orbitals*. The GRASP software for calculating radiative transition probabilities was based on the assumption that initial and final states of a transition are described by *the same orbital set*. Most if this machinery can still be used by way of a procedure to express sets of non-orthogonal orbitals as a biorthonormal system [22.86]. An adaptation for GRASP92 was used, for example, to calculate radiative transition probabilities for lines of the C III spectrum [22.87] and the oscillator strengths of the $n\mathrm{d}\,^2\mathrm{D}_{3/2} - (n+1)\mathrm{p}\,^2\mathrm{P}^0_{1/2,3/2}$ lines in Lu ($n = 5$) and Lw ($n = 6$) which are very sensitive to correlation effects [22.88]. These two calculations involved CSF sets of order 300,000. Desclaux's code, which uses an expansion of the many-electron wavefunction in determinantal wavefunctions rather than the Fano approach using jj-coupled CSFs, has similarly been modernized [22.89]; its capabilities are rather similar to those of GRASP. There is no published description.

GRASP92 has been enhanced recently with new utilities to calculate hyperfine interactions [22.90–92] and isotope shifts [22.93]. *Fritzsche* et al. have developed a new suite of programs, RATIP (an acronym for Relativistic Atomic Transition and Ionization Properties), which uses MCDHF wavefunctions from GRASP92 to study a range of atomic properties [22.94, 95]. Like Desclaux's package, this expresses jj-coupled symmetry functions in terms of Slater determinants [22.96] and also provides the relevant utilities for coefficients of fractional parentage and the calculation of angular coefficients. The package supports CI calculations of ASF and energy levels taking account of the Breit interaction and QED estimates. A new utility [22.97] permits calculation of relaxed orbital radiative transition probabilities and lifetimes within the RATIP framework. The code generates continuum orbitals, which enable calculation of Auger energies, relative intensities and angular distributions, and should also enable calculation of photoionization cross-sections and angular distributions. The papers cited contain information on how to obtain the programs, many of which are also obtainable from the Computer Physics Communications International Program Library [22.98].

References

22.1 B. G. Wybourne: *Classical Groups for Physicists* (Wiley, New York 1974)

22.2 J. P. Elliott, P. G. Dawber: *Symmetry in Physics* (Macmillan, Basingstoke, London 1979)

22.3 B. Thaller: *The Dirac Equation* (Springer, Berlin, Heidelberg 1992)

22.4 W. Greiner: *Relativistic Quantum Mechanics* (Springer, Berlin, Heidelberg 1990)

22.5 L. L. Foldy: Phys. Rev. **102**, 568 (1956)

22.6 Yu. M. Shirokov: Sov. Phys. JETP **6**, 568,919,929 (1958)

22.7 S. S. Schweber: *Introduction to Relativistic Quantum Field Theory* (Harper Row, New York 1964)

22.8 V. B. Berestetskii, E. M. Lifshitz, L. P. Pitaevskii: *Relativistic Quantum Theory* (Pergamon, Oxford 1971)

22.9 C. Itzykson, J.-B. Zuber: *Quantum Field Theory* (McGraw-Hill, New York 1980)

22.10 L. N. Labzowsky, G. L. Klimchitskaya, Yu. Yu. Dmitriev: *Relativistic Effects in the Spectra of Atomic Systems* (Institute of Physics Publishing, Bristol 1993)

22.11 W. H. Furry: Phys. Rev. **81**, 115 (1951)

22.12 I. P. Grant, H. M. Quiney: Adv. At. Mol. Phys. **23**, 37 (1988)

22.13 I. Lindgren: J. Phys. B **23**, 1085 (1990)

22.14 H. A. Bethe, E. E. Salpeter: *Quantum Mechanics of One-, and Two-Electron Systems* (Springer, Berlin, Heidelberg 1957)

22.15 W. R. Johnson, P. J. Mohr, J. Sucher (Eds.): *Relativistic, Quantum Electrodynamic, and Weak Interaction Effects in Atoms* (American Insitute of Physics, New York 1989)

22.16 I. P. Grant: J. Phys. B **7**, 1458 (1974)

22.17 I. Lindgren in 22.15, pp. 3–27

22.18 D. J. Baker, D. Moncrieff, S. Wilson: Vector processing and parallel processing in many-body perturbation theory calculations for electron correlation effects in atoms and molecules. In: *Supercomputational Science*, ed. by R. G. Evans, S. Wilson (Plenum Press, New York 1990) pp. 201–209

22.19 D. R. Hartree: Proc. Camb. Phil. Soc. **25**, 225 (1929)

22.20 V. G. Bagrov, D. M. Gitman: *Exact Solutions of Relativistic Wave Equations* (Kluwer Academic, Dordrecht 1990)
22.21 E. A. Coddington, N. Levinson: *Theory of Ordinary Differential Equations* (McGraw-Hill, New York 1955)
22.22 M. Abramowitz, I. A. Stegun: *Handbook of Mathematical Functions* (Dover, New York 1965)
22.23 V. M. Burke, I. P. Grant: Proc. Phys. Soc. **90**, 297 (1967)
22.24 J. Kobus, J. Karwowski, W. Jaskolski: Phys. Rev. A **20**, 3347 (1987)
22.25 G. Hall, J. M. Watt: *Modern Numerical Methods for Ordinary Differential Equations* (Clarendon Press, Oxford 1976)
22.26 I. P. Grant: Adv. Phys. **19**, 747 (1970)
22.27 I. P. Grant: *Methods in Computational Chemistry*, Vol. 2 (Clarendon Press, Oxford 1976)
22.28 I. P. Grant: Phys. Scr. **21**, 443 (1980)
22.29 K. G. Dyall, I. P. Grant, S. Wilson: J. Phys. B **17**, 493 (1984)
22.30 I. P. Grant: J. Phys. B **19**, 3187 (1986)
22.31 I. P. Grant, H. M. Quiney: Phys. Rev. A **62**, 022508 (2000)
22.32 M. Rotenberg: Ann. Phys. **19**, 262 (1962)
22.33 M. Rotenberg: Adv. At. Mol. Phys. **6**, 233 (1970)
22.34 C. W. Clark, K. T. Taylor: J. Phys. B **15**, 1175 (1982)
22.35 P. Čarsky, M. Urban: *Ab Initio Calculations. Methods, and Applications in Chemistry* (Springer, Berlin, Heidelberg 1980)
22.36 S. Huzinaga, J. Andzelm, M. Klobukowski, E. Radzio-Andselm, Y. Sakai, H. Tatewaki (Eds.): *Gaussian Basis Sets for Molecular Calculations* (Elsevier, Amsterdam 1984)
22.37 R. Poirier, R. Kari, I. G. Csizmadia: *Handbook of Gaussian Basis Sets* (Elsevier, Amsterdam 1985)
22.38 Y.-K. Kim: Phys. Rev. **154**, 17 (1967)
22.39 Y.-K. Kim: Phys. Rev. A **159**, 190 (1967)
22.40 T. Kagawa: Phys. Rev. A **12**, 2245 (1975)
22.41 T. Kagawa: Phys. Rev. **22**, 2340 (1980)
22.42 G. W. F. Drake, S. P. Goldman: Phys. Rev. A **23**, 2093 (1981)
22.43 G. W. F. Drake, S. P. Goldman: Adv. At. Mol. Phys. **25**, 393 (1988)
22.44 Y. S. Lee, A. D. McLean: J. Chem. Phys. **76**, 735 (1982)
22.45 H. M. Quiney, I. P. Grant, S. Wilson: J. Phys. B **18**, 2805 (1985)
22.46 W. R. Johnson, S. A. Blundell, J. Sapirstein: Phys. Rev. A **37**, 307 (1988)
22.47 W. R. Johnson, S. A. Blundell, J. Sapirstein: Phys. Rev. A **40**, 2233 (1988)
22.48 C. Bottcher, M. R. Strayer: Pair production at GeV/u energies. In: *Atomic Theory Workshop on Relativistic and QED Effects in Heavy Atoms*, AIP Conference Proceedings No 136 (AIP, New York 1985) pp. 268–298
22.49 T. Brage, C. F. Fischer: Phys. Scr. **49**, 651 (1994)
22.50 C. F. Fischer, F. A. Parpia: Phys. Lett. A **179**, 198 (1993)
22.51 I. P. Grant, B. J. McKenzie, P. H. Norrington, D. F. Mayers, N. C. Pyper: Comput. Phys. Commun. **21**, 207 (1980)
22.52 K. G. Dyall, I. P. Grant, C. T. Johnson, F. A. Parpia, E. P. Plummer: Comput. Phys. Commun. **55**, 425 (1989)
22.53 F. A. Parpia, C. F. Fischer, I. P. Grant: Comput. Phys. Commun. **94**, 249 (1996)
22.54 J. P. Desclaux: Comput. Phys. Commun. **9**, 31 (1975) see also **13**, 71 (1977)
22.55 H. M. Quiney, I. P. Grant, S. Wilson: J. Phys. B **20**, 1413 (1987)
22.56 H. M. Quiney: Relativistic atomic structure calculations I: Basic theory and the finite basis set approximation. In: *Supercomputational Science*, ed. by R. G. Evans, S. Wilson (Plenum Press, New York 1990) pp. 159–184
22.57 H. M. Quiney: Relativistic atomic structure calculations II: Computational aspects of the finite basis set method. In: *Supercomputational Science*, ed. by R. G. Evans, S. Wilson (Plenum Press, New York 1990) pp. 185–200
22.58 S. Wilson: Relativistic molecular structure calculations. In: *Methods in Computational Chemistry*, Vol. 2, ed. by S. Wilson (Plenum Press, New York 1988) p. 73
22.59 H. M. Quiney: Relativistic many-body perturbation theory. In: *Methods in Computational Chemistry*, Vol. 2, ed. by S. Wilson (Plenum Press, New York 1988) p. 227
22.60 H. M. Quiney, I. P. Grant, S. Wilson: On the relativistic many-body perturbation theory of atomic and molecular electronic structure. In: *Many-Body Methods in Quantum Chemistry*, Lecture Notes in Chemistry, Vol. 52, ed. by U. Kaldor (Springer, Berlin, Heidelberg 1989) pp. 307–344
22.61 H. M. Quiney, I. P. Grant, S. Wilson: J. Phys. B **23**, L271 (1990)
22.62 S. Fritzsche, I. P. Grant: Phys. Scr. **50**, 473 (1994)
22.63 A. Ynnerman, C. F. Fischer: Phys. Rev. A **51**, 2020 (1995)
22.64 W. R. Johnson, K. T. Cheng: J. Phys. B **12**, 863 (1979)
22.65 C. M. Lee, W. R. Johnson: Phys. Rev. A **22**, 979 (1980)
22.66 W. R. Johnson, G. Soff: At. Data Nucl. Data Tables **33**, 405 (1985)
22.67 E. A. Uehling: Phys. Rev. **48**, 55 (1935)
22.68 E. H. Wichmann, N. M. Kroll: Phys. Rev. **101**, 843 (1956)
22.69 L. W. Fullerton, G. A. Rinker: Phys. Rev. A **13**, 1283 (1976)
22.70 U. Fano: Phys. Rev. A **140**, A67 (1965)
22.71 Z. B. Rudzikas: *Theoretical Atomic Spectroscopy* (Cambridge Univ. Press, Cambridge 1997)
22.72 G. Gaigalas, Z. B. Rudzikas, C. Froese Fischer: J. Phys. B **30**, 3747 (1997)
22.73 G. Gaigalas: Lithuanian J. Phys. **39**, 80 (1999)

22.74 J. Kaniauskas, Z. B. Rudzikas: J. Phys. B **13**, 3521 (1980)
22.75 G. Gaigalas, S. Fritzsche: Comput. Phys. Commun. **134**, 86 (2001)
22.76 S. Fritzsche: Comput. Phys. Commun. **103**, 51 (1997)
22.77 S. Fritzsche, S. Varga, D. Geschke, B. Fricke: Comput. Phys. Commun. **111**, 167 (1998)
22.78 G. Gaigalas, S. Fritzsche, B. Fricke: Comput. Phys. Commun. **135**, 219 (2001)
22.79 T. Inghoff, S. Fritzsche, B. Fricke: Comput. Phys. Commun. **139**, 297 (2001)
22.80 S. Fritzsche, T. Inghoff, T. Bastug, M. Tomaselli: Comput. Phys. Commun. **139**, 314 (2001)
22.81 L. V. Chernysheva, V. L. Yakhontov: Comput. Phys. Commun. **119**, 232 (1999)
22.82 C. F. Fischer, T. Brage, P. Jönsson: *Computational Atomic Structure. An MCHF Approach* (Institute of Physics Publishing, Bristol, Philadelphia 1997)
22.83 E. R. Davidson: J. Comput. Phys. **17**, 87 (1975)
22.84 E. R. Davidson: Comput. Phys. Commun. **53**, 49 (1989)
22.85 A. Stathopoulos, C. F. Fischer: Comput. Phys. Commun. **79**, 268 (1994)
22.86 J. Olsen, M. R. Godefroid, P. Jönsson, P. Å. Malmqvist, C. F. Fischer: Phys. Rev. E **52**, 4499 (1995)
22.87 P. Jönsson, C. F. Fischer: Phys. Rev. A **57**, 4967 (1998)
22.88 Y Zou, C. F. Fischer: Phys. Rev. Lett. **88**, 183001 (2002)
22.89 J. P. Desclaux, private communication, December 2003
22.90 J. Bieroń, P. Jönsson, C. F. Fischer: Phys. Rev. A **53**, 1 (1995)
22.91 J. Bieroń, I. P. Grant, C. F. Fischer: Phys. Rev. A **58**, 4401 (1998)
22.92 J. Bieroń, C. F. Fischer, I. P. Grant: Phys. Rev. A **59**, 4295 (1999)
22.93 P. Jönsson, C. F. Fischer: Comput. Phys. Commun. **100**, 81 (1997)
22.94 S. Fritzsche: J. Elec. Spec. Rel. Phenom. **114–116**, 1155 (2001)
22.95 S. Fritzsche: Phys. Scr. T **100**, 37 (2002)
22.96 S. Fritzsche, J. Anton: Comput. Phys. Commun. **124**, 353 (2000)
22.97 S. Fritzsche, C. F. Fischer, C. Z. Dong: Comput. Phys. Commun. **124**, 340 (2000)
22.98 The CPC Program Library home page is at http://www.cpc.cs.qub.ac.uk/cpc/

23. Many-Body Theory of Atomic Structure and Processes

All atoms except hydrogen are many-body systems, in which the interelectron interaction plays an important or even decisive role. The aim of this chapter is to describe a consistent method for calculating the structure of atoms and the characteristics of different atomic processes, by applying perturbation theory to take into account the interelectron interaction. This method involves drawing a characteristic diagram based on the structure or process. This is then used to create an analytical expression to the lowest order in the interelectron interaction. Higher-order corrections are subsequently generated.

This technique was invented about half a century ago in quantum electrodynamics by *Feynman* [23.1], then modified and adjusted for multiparticle systems by a number of authors. Its application to atomic structure and atomic processes required further modifications, which were initiated at the end of the fifties (see, e.g., [23.2]) and later. The corresponding technique was successfully applied to the calculation of a wide variety of characteristics and processes in many papers and several review articles [23.3, 4]. The increasing amount of experimental data available has led to improved accuracy for this technique, so that it can be applied to current problems considering not only atoms and ions, both positive and negative [23.5–8], but also molecules [23.9], clusters [23.10] and fullerenes.

23.1 **Diagrammatic Technique** 360
23.1.1 Basic Elements 360
23.1.2 Construction Principles for Diagrams 360
23.1.3 Correspondence Rules 362
23.1.4 Higher-Order Corrections and Summation of Sequences 363

23.2 **Calculation of Atomic Properties** 365
23.2.1 Electron Correlations in Ground State Properties 365
23.2.2 Characteristics of One-Particle States 366
23.2.3 Electron Scattering 367
23.2.4 Two-Electron and Two-Vacancy States 369
23.2.5 Electron–Vacancy States 370
23.2.6 Photoionization in RPAE and Beyond 371
23.2.7 Photon Emission and Bremsstrahlung 374

23.3 **Concluding Remarks** 375

References 376

The elements of the diagrammatic technique, which form a convenient and simple "language", are given together with the rules for creating "sentences" using basic "words". A kind of "dictionary" helps to translate diagrammatic "sentences" into analytical expressions suitable for calculations.

An essential part of the program is to learn how the simplest approximation can be improved, and what are the mechanisms and processes connected with, and responsible for, higher-order corrections.

When the diagrammatic technique of many-body theory is used, it is unnecessary to be restricted to a finite number of lowest-order terms in the interelectron interaction. On the contrary, some infinite sequences may be taken into account. The sum of all many-body diagrams is completely equivalent to the many-particle Schrödinger equation. Therefore, taking all of them into account is just as complicated as solving the corresponding equation. Compared with other approaches, the diagrammatic technique can easily uncover hidden approximations and transparently demonstrate possible sources of corrections.

23.1 Diagrammatic Technique

23.1.1 Basic Elements

Each physical atomic process (or a process with participation of a molecule, cluster or fullerenes) involves an electronic interaction with a projectile or external field, in general time-dependent, or a mutual interelectronic interaction. By convention, the ground state of the atom (if it is not degenerate) is regarded as the vacuum state. Then the simplest process in this target is excitation of an electron to an unoccupied level, leaving behind a vacancy. The basic elements of a diagram are

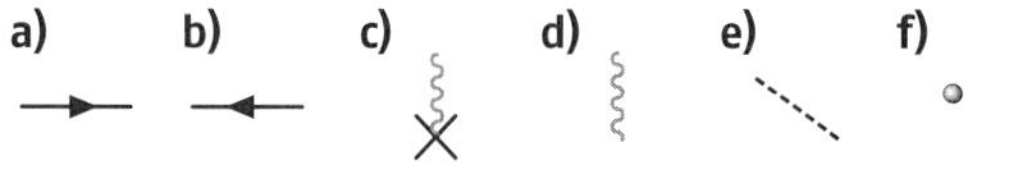

(23.1)

where (a) with an arrow directed to the right represents an electron excited to a vacant level; (b) with an arrow directed to the left represents a vacancy; (c) with a cross represents the static Coulomb interaction; (d) represents the interelectron Coulomb interaction; (e) represents interaction with a time-dependent external field, usually electromagnetic; and (f) represents the very act of interaction.

The elements (23.1a–f) in combination can describe the following real or virtual basic processes

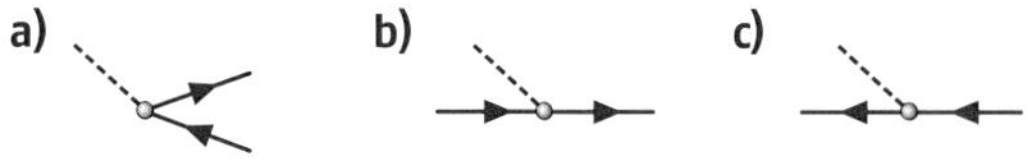

(23.2)

which represent (a) photon absorption by the vacuum with electron–vacancy pair creation; (b) electron excitation; and (c) vacancy excitation. Diagrams (23.2) depict processes as developing in time, shown increasing from left to right. A vacancy can be thought of as an antiparticle to the electron, moving backward in time. The time-reverse of processes (23.2) represent processes of photon emission due to annihilation of an electron–vacancy pair, vacancy transition, and electron inelastic scattering, respectively.

A static [for example, Coulomb (23.1c)] field can virtually create an electron–vacancy pair, or affect the moving electron or vacancy, as shown in the following diagrams:

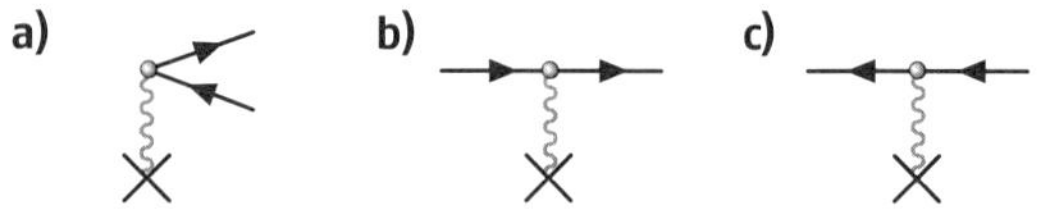

(23.3)

Just as for (23.2), diagrams (23.3) have their time-reversed counterparts.

Inclusion of interelectron interaction leads to a number of processes of which some examples are

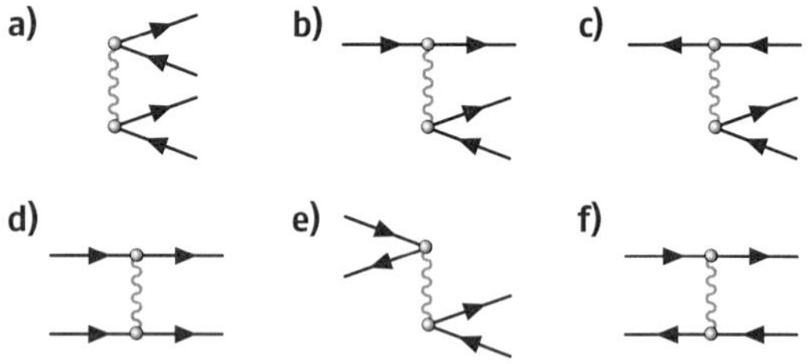

(23.4)

Here, (23.4a) describes creation of two electron–vacancy pairs, (23.4b) represents the simplest picture of electron inelastic scattering, (23.4c) depicts vacancy decay with electron–vacancy pair creation, (23.4d) stands for electron–electron scattering, (23.4e) represents a process which can be called electron–vacancy annihilation and creation, while (23.4f) shows electron–vacancy scattering.

23.1.2 Construction Principles for Diagrams

The foundations of diagrammatic techniques are discussed in a number of books such as [23.11]. This chapter presents recipes for the construction and evaluation of diagrams corresponding to various atomic processes [23.12].

The basic procedure is to connect the initial and final states of the atom, drawn at the left and right sides of the diagram, using any of the elements in (23.1). In doing so, the following rules apply:

1. At each dot (23.1f), only three lines can meet: wavy (or dashed) and electron–vacancy.
2. A vacancy cannot be transformed into an electron or vice versa.

3. Electrons and vacancies can be created only pairwise from the vacuum.
4. Only linked diagrams are allowed; i. e. only those having no parts entirely disconnected from one another.

The simplest or initial approximation to a process is represented by a diagram which includes the lowest possible number of elements (23.1–23.4). Higher-order corrections can be derived by including additional elements of interaction with the static field of the nucleus (23.1c) and between electrons and/or vacancies (23.1d).

As an illustration of the method, consider the following three processes:

1. one-electron photoionization – the initial state is a photon while the simplest final state is an electron–vacancy pair. They can be combined together giving the basic diagram (23.2a).
2. elastic electron scattering – the initial and final states are single electrons. To describe the simplest scattering process, the interaction with the Coulomb field must be taken into account, leading to (23.3b). To account for interelectron interaction, the simplest element

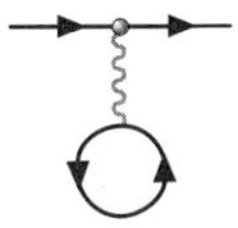

(23.5)

must be introduced. It is a modification of (23.4b) accounting for the interaction of an incoming electron with all target electrons individually, not altering their states. This is emphasized by the loop in which the same vacancy leaving the lower dot reenters it. Indistinguishability of all electrons as fermions is taken into account by permutation of the electron (vacancy) line ends, as illustrated in the following diagrams:

(23.6)

Diagram (23.6a) is obtained from (23.5) by permutation of the electron lines on one side of the interelectron interaction. Diagram (23.6b) is equivalent to (23.6a), but is simpler to draw.

3. inelastic electron scattering – the initial state is a single electron. For the final state we choose one with two electrons and therefore a single vacancy. The simplest diagram in this case is given by (23.4b).

To illustrate the description of the ground state characteristics, consider the contributions to the ground state energy of an atom. If this state is not degenerate, its potential energy is given by vacuum diagrams which have no free lines in the initial or final states. The simplest vacuum diagrams are

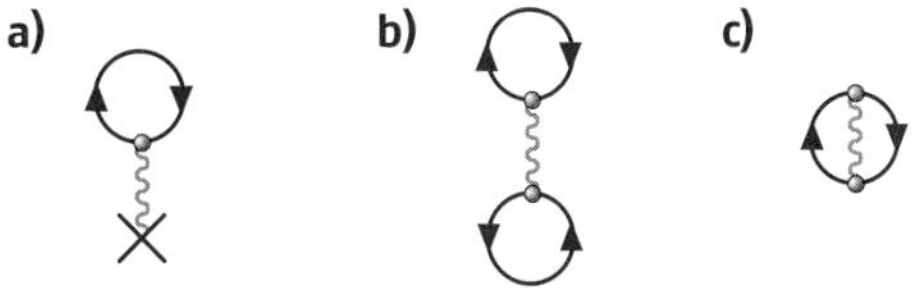

(23.7)

Higher-order corrections to all these diagrams can be obtained by adding elements such as a static external field (23.3) or interelectron interaction (23.4) without changing the initial and final states of the processes. The lowest-order processes are represented by (23.2a), (23.5), (23.6), (23.4b), and (23.7). There are many corrections even in the next order of interaction, either with an external field or with electrons or vacancies. To illustrate, only one correction to each process will be presented:

1. Simple photoionization (23.2a) may be combined with (23.4e) to obtain

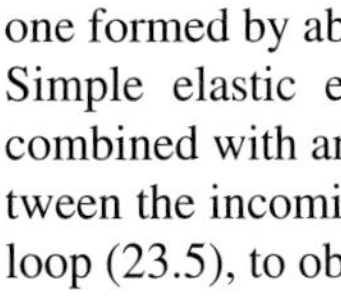

(23.8)

This describes the effect of the creation of another electron–vacancy pair, after annihilation of the first one formed by absorption of the initial photon.
2. Simple elastic electron scattering (23.5) can be combined with an extra interaction term (23.4f) between the incoming electron and the vacancy of the loop (23.5), to obtain

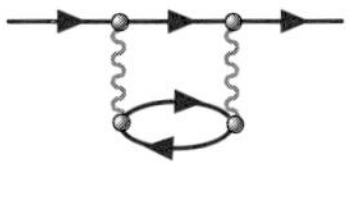

(23.9)

3. Simple inelastic electron scattering (23.4b) can also be combined with (23.4f), accounting for the interaction of an electron and vacancy created in the lowest-order process, to obtain

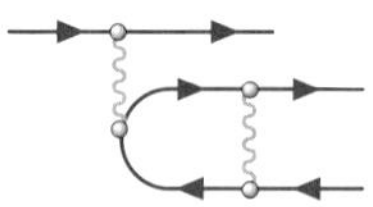

(23.10)

4. The ground state energy term (23.7b) can be combined with (23.7b) and the element (23.3c) of the interaction between the vacancy in the (23.7b) loops and the static field to obtain

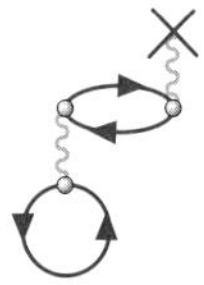

(23.11)

Higher-order corrections can be constructed step by step by introducing further elements of interaction. In some cases, classes of diagrams may be taken into account up to infinite order by solving closed systems of integral or differential equations.

23.1.3 Correspondence Rules

These rules describe how to obtain an analytical expression corresponding to a given diagram. One starts by choosing a zero-order approximation which can be that of independent electrons moving in the Coulomb field of an atomic nucleus. Atoms with completely occupied shells, or subshells having a non-degenerate ground state, can be chosen as the vacuum. Electron (vacancy) states are characterized in this case by the quantum numbers n, ℓ, m_ℓ, and $\sigma = \pm 1/2$.

The first correspondence rule is to substitute a matrix element for each interaction:

$$\begin{aligned} &\text{Diagram (23.2)} \rightarrow \langle p|W|q\rangle\,, \\ &\text{Diagram (23.3)} \rightarrow \langle p|U|q\rangle\,, \\ &\text{Diagram (23.4)} \rightarrow \langle pt|V|qs\rangle\,, \end{aligned} \quad (23.12)$$

where W is the interaction potential of an electron with the external time-dependent field, U is the interaction potential of an electron (vacancy) with an external static field, for example that of the nucleus, and V is the Coulomb interelectron interaction. Each of the letters p, q, t, s represents a full set of n, ℓ, m_ℓ, σ quantum numbers. Vacancy states are below (and include) the highest occupied energy level, called the Fermi level, so that $p \leq F$. Electron states are above the Fermi level so that $q > F$. Thus diagram (23.2a) is represented by $\langle p|W|q\rangle$ with $p \leq F$ and $q > F$.

Apart from initial and final states, each diagram can have sections, i. e., intervals between successive interactions. For instance (23.9) and (23.10) each have one section. Each section is represented by an inverse energy denominator $\varepsilon_{\mathrm{d}}^{-1}$. It includes the sum over all vacancy energies $\sum_{\mathrm{vac}} \varepsilon_i$ minus the sum of the electron energies $\sum_{\mathrm{el}} \varepsilon_n$ to which the entrance energy E of the diagram (e.g. $\hbar\omega$ for a time-dependent field) must be added:

$$\varepsilon_{\mathrm{d}}^{-1} = \left(\sum_{\mathrm{vac}} \varepsilon_i - \sum_{\mathrm{el}} \varepsilon_n + E\right)^{-1}\,. \quad (23.13)$$

The second correspondence rule is to identify sections and write down their energy denominators. After attributing to each electron (vacancy) line a letter, denoting its state, the analytical expression for a diagram is given by

$$\begin{aligned} &\text{Analytical Expression} \\ &= \text{(the product of all interaction matrix elements)} \\ &\quad \times \text{(all energy denominators)}^{-1} \\ &\quad \times (-1)^L \text{ summed over all intermediate} \\ &\qquad \text{electron and vacancy states,} \end{aligned} \quad (23.14)$$

where L is equal to the sum of the total number of vacancy lines and closed vacancy or electron–vacancy loops.

Although electrons are fermions, the summation in (23.14) has no additional restrictions caused by the Pauli principle. It runs over all electron ($> F$) and vacancy ($\leq F$) states, including those where two or more electrons (or vacancies) are in the same state. The correspondence rules (23.12), (23.13), and (23.14) can be illustrated by giving as examples the analytical expressions of two diagrams (23.8) and (23.9).

Attributing letters denoting electron and vacancy states, diagram (23.8) becomes

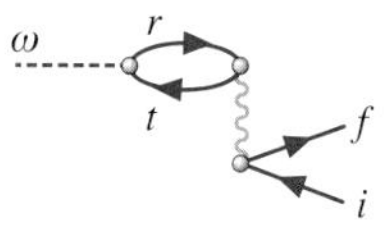

(23.15)

According to (23.12–23.14), the analytical formula

$$A_{if}(\omega) = \sum_{r>F, t\le F}\!\!\!\!\!\!\!\!\int \frac{\langle t|W|r\rangle\langle ri|V|tf\rangle}{\varepsilon_t - \varepsilon_r + \omega}(-1)^{2+1} \tag{23.16}$$

is obtained. The symbol $\sum\!\!\!\!\!\!\int$ includes summation over discrete levels and integration. In (23.16), the intermediate state is $r > F$, $t \le F$ and the diagram has two vacancies (t and i) and one loop rt. Integration must be performed over those states r which belong to the continuum.

Assigning letters denoting states, (23.9) appears as

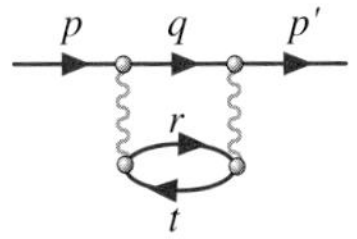

(23.17)

where $p, q, p', r > F$, while $t \le F$. According to (23.14),

$$\Delta E = \sum_{r,q>F; t\le F}\!\!\!\!\!\!\!\!\int \frac{\langle pt|V|qr\rangle\langle qr|V|p't\rangle}{\varepsilon_t - \varepsilon_q - \varepsilon_r + \varepsilon_p}(-1)^{1+1}\,, \tag{23.18}$$

where the intermediate states are $q, r > F$ and $t \le F$. It has one vacancy and one electron–vacancy loop rt.

An intermediate state in a diagram can be real or virtual. It is real if the energy conservation law can be fulfilled, i. e. if for some values of the section energy the following relation holds:

$$E = \sum_{\text{el}} \varepsilon_n - \sum_{\text{vac}} \varepsilon_i\,. \tag{23.19}$$

If (23.19) can be fulfilled, a prescription for avoiding the singularity in (23.13) is to substitute the expression $\varepsilon_{\text{d}}^{-1}Q$, where

$$\begin{aligned} Q &= \lim_{\eta\to 0}\left(E - \sum_{\text{el}} \varepsilon_n + \sum_{\text{vac}} \varepsilon_i + \text{i}\eta\right)^{-1} \\ &= P\left(E - \sum_{\text{el}} \varepsilon_n + \sum_{\text{vac}} \varepsilon_{\text{i}}\right)^{-1} \\ &\quad - \text{i}\pi\delta\left(E - \sum_{\text{el}} \varepsilon_n + \sum_{\text{vac}} \varepsilon_{\text{i}}\right), \end{aligned} \tag{23.20}$$

for $\varepsilon_{\text{d}}^{-1}$. Here P denotes that the principal value is to be taken on integration over intermediate state energies. The result of (23.20) can thus be complex.

An intermediate state is virtual if the energy conservation law (23.19) is violated for all values of the section energy. In general, the bigger the virtuality, i. e. the difference $E - \sum_{\text{el}} \varepsilon_n + \sum_{\text{vac}} \varepsilon_i$, the smaller the contribution to the amplitude of the process.

23.1.4 Higher-Order Corrections and Summation of Sequences

An important feature of the diagrammatic technique is the convenience in constructing higher-order corrections and in the summation of infinite sequences of diagrams. According to (23.13), each new interaction line leads to an additional interaction matrix element, extra energy denominator and summation over new intermediate states.

An important example of infinite summation is that of determining the one-electron states. The interaction with the nucleus (23.3b) and with atomic electrons (23.5) and (23.6) is not small and must be taken into account non-perturbative; i. e., these elements must be iterated infinitely. To simplify the drawing, only the element (23.5) is repeated, leading to the diagrammatic equation:

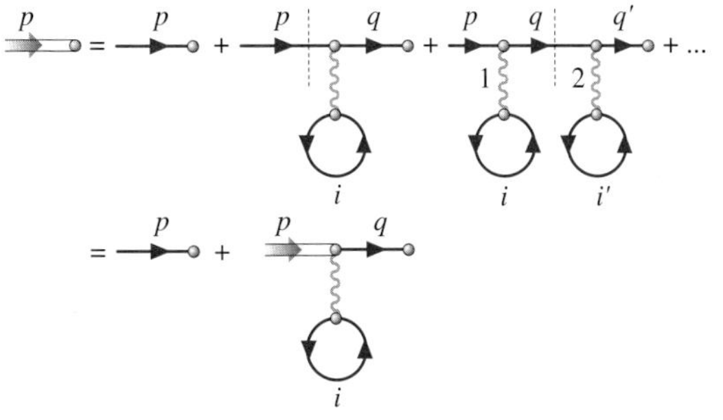

(23.21)

Indeed, everything in the infinite sum which is in front of the dashed line repeats the infinite sum itself, thus leading to a closed equation of the form

$$\langle\tilde{p}| = \langle p| + \sum_{q>F}\!\!\!\!\!\!\!\int \langle\tilde{p}i|V|qi\rangle \frac{1}{-\varepsilon_q + \varepsilon_p}\langle q|\,. \tag{23.22}$$

The two interactions leading to (23.21) can be permuted, so that the interaction 1 can be after 2. This leads to extension of the sum to include states with $q \leq F$. As a result, the summation in (23.22) must be performed over all states q.

Interaction with the nucleus and the other electrons affects also the occupied (or vacancy) states i in (23.21) and therefore the latter must be modified by inserting the elements (23.3b), (23.5), and (23.6) into them. Here again, the vacancy line in (23.5) and (23.6) must be modified by including the corrections (23.3b), (23.5), (23.6) and so on. Finally, the diagrammatic equation

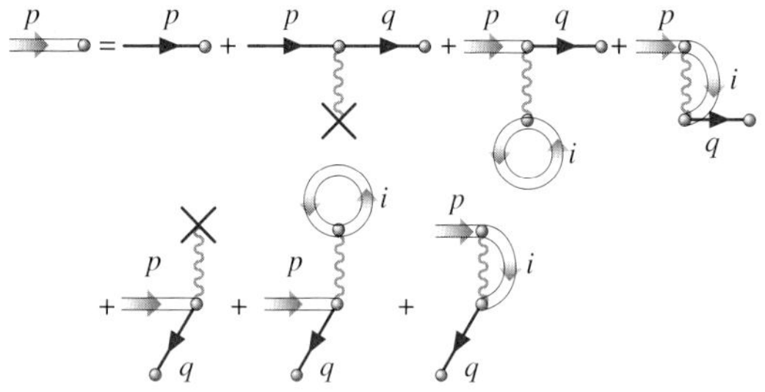

(23.23)

is obtained. The doubled line for i emphasizes that the vacancy wave function is determined by an equation similar to (23.21). The corresponding analytical equation looks like (23.22), but includes also the Coulomb interaction with the nucleus and the exchange interaction with other atomic electrons. The summation over q in this equation is extended over all q, not only $q > F$. Multiplying the corresponding equation by $(\hat{H}_0 - \varepsilon_p)$ from the right (atomic units are used in this chapter: $e = m_e = \hbar = 1$), where $\hat{H}_0 = -\nabla^2/2$, and using the completeness of the functions $\sum_q |q\rangle\langle q| = \delta(\boldsymbol{r} - \boldsymbol{r}')$, results in the equation

$$\left[-\frac{\nabla^2}{2} - \frac{Z}{r} + \sum_{i \leq F} \int \frac{d\boldsymbol{r}'}{|\boldsymbol{r}' - \boldsymbol{r}|} |\phi_i(\boldsymbol{r}')|^2 - \varepsilon_p\right] \phi_p(\boldsymbol{r}) = \sum_{i \leq F} \int \frac{d\boldsymbol{r}'}{|\boldsymbol{r}' - \boldsymbol{r}|} \phi_i^*(\boldsymbol{r}')\phi_p(\boldsymbol{r}')\phi_i(\boldsymbol{r}) \quad (23.24)$$

for the electron wave function $\phi_p(\boldsymbol{r})$. Here $\phi_i(\boldsymbol{r})$ are wave functions determined by equations similar to (23.24). These are the Hartree–Fock (HF) equations.

HF includes a part for interelectron interaction matrix elements, namely that given by (23.5) and (23.6). The rest is called the residual interaction, and its inclusion leads beyond the HF frame, accounting for correlations.

When a perturbative approach is used, it is essential to define the zero-order approximation. In this chapter, and very often in the literature, the Hartree–Fock approximation is used in this role. To simplify the drawing of diagrams, from now on single (rather than double) lines will represent electrons (vacancies), whose wave functions are determined in the HF approximation by (23.24). Obviously, in this case elements (23.3a), (23.5) and (23.6) should not be added to any other diagrams.

The procedure used in deriving (23.21) and (23.23) is in fact more general. Let us separate all diagrams describing elastic scattering which do not include a single one-electron or one-vacancy state as intermediate. Depicting their total contribution by a square, the precise one-particle state is determined by an infinite sequence of iterative diagrams which can be summed, similarly to (23.21), by

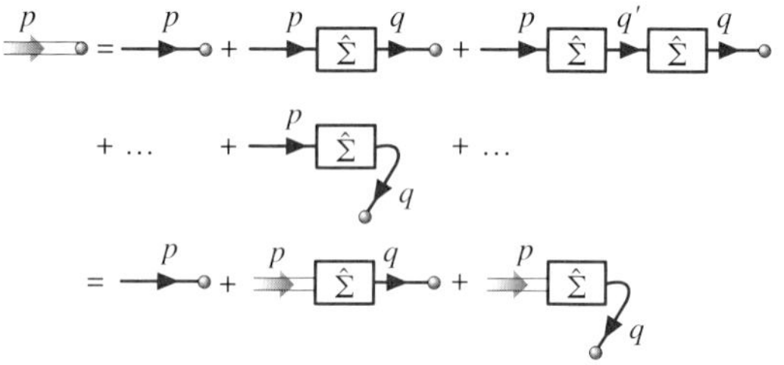

(23.25)

Here the single line stands for an HF state. Using the correspondence rule (23.14), an analytical equation similar to the Schrödinger equation can be derived with the operator $\hat{\boldsymbol{\Sigma}}$ playing the role of an external potential. The essential difference is, however, that this "potential" depends in principle upon the energy and state of the particle. The same kind of iterative procedure leading to (23.21) or (23.23) will be used several times in this chapter.

Other zero-order approximations can be chosen. But then diagrams with corrections of the type (23.3a) must be included, with the external static field potential equal to the difference between the HF and the chosen one.

To calculate the numerical value of a given diagram or a sequence of diagrams one needs to know, according to the description given above, the matrix elements of external fields and interelectron interactions obtained with the help of one-electron HF wave functions. The required calculational procedures are described in [23.13].

23.2 Calculation of Atomic Properties

23.2.1 Electron Correlations in Ground State Properties

A major advantage of the diagrammatic technique in many-body theory is that it is usually unnecessary to know the total wave function of the atom. On the contrary, only actively participating electrons or vacancies appear in a diagram. The HF zero-order approximation for one-electron and one-vacancy wave functions is used in what follows. All atomic characteristics and cross sections for atomic processes calculated with HF form the one-electron approximation. Everything beyond the HF frame, i. e., caused by residual interaction, are called correlation corrections or correlations. They can be calculated using the many-body perturbation theory (MBPT) [23.3], random phase approximation (RPA) [23.14] and random phase approximation with exchange (RPAE) or its generalized version GRPAE [23.12, 13].

The simplest diagrammatic expression for the correlation energy is given by the two diagrams

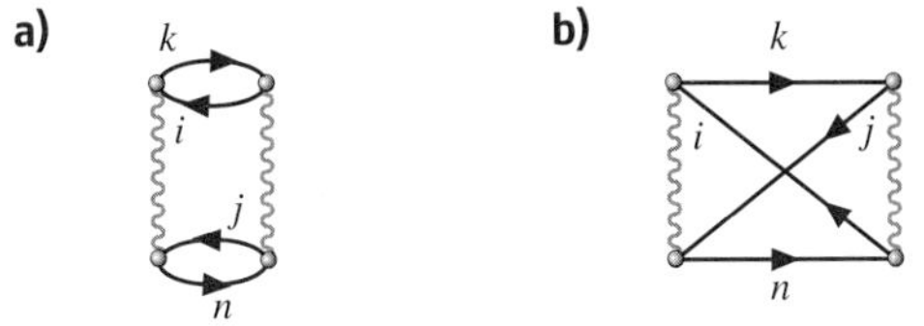

(23.26)

The analytical expression $\Delta E^{(2)}_{\text{corr}}$ for (23.26a) is

$$\Delta E^{(2)}_{\text{corr}} = \sum_{k,n>F;i,j\leq F}\!\!\!\!\!\!\!\!\int \frac{\langle ij|V|kn\rangle\langle kn|V|ij\rangle}{\varepsilon_i+\varepsilon_j-\varepsilon_k-\varepsilon_n} . \qquad (23.27)$$

The analytical expression for (23.26b) differs from (23.27) by the sign and an exchange matrix element $\langle kn|V|ji\rangle$ instead of a direct $\langle kn|V|ij\rangle$ one. The contribution (23.26) overestimates the correlation energy by $\approx 10\%$.

Diagrams (23.26) can also be used to describe the interaction potential of two atoms, designated A and B. Let the ki states belong to A and nj to B. At large distances R between the atoms, the contribution of (23.26b) is exponentially small. Because the vacancies i and j are located inside atoms A and B respectively, the interelectron potential $V = |\boldsymbol{r}_{\text{A}} - \boldsymbol{r}_{\text{B}} + \boldsymbol{R}|^{-1}$ at large distances $R \gg R_{\text{A,B}}$, ($R_{\text{A,B}}$ are atomic radii), can be expanded as a series in powers of R^{-1}. The first term giving a non-vanishing contribution to (23.27) is $V \simeq R^{-3}[(\boldsymbol{r}_{\text{A}}\cdot\boldsymbol{r}_{\text{B}}) - 3(\boldsymbol{r}_{\text{A}}\cdot\boldsymbol{n})(\boldsymbol{r}_{\text{B}}\cdot\boldsymbol{n})]$, $\boldsymbol{n}$ being the unit vector in the direction of $\boldsymbol{R}$. Substituted into (23.26), this potential leads to the expression

$$U(R) = -\frac{C_6}{R^6} \qquad (23.28)$$

for the interatomic potential [23.15], where

$$C_6 \approx \sum_{k,n>F;i,j\leq F}\!\!\!\!\!\!\!\!\int \frac{|\langle i|\boldsymbol{r}|k\rangle|^2|\langle j|\boldsymbol{r}|n\rangle|^2}{(\varepsilon_i+\varepsilon_j-\varepsilon_k-\varepsilon_n)} . \qquad (23.29)$$

Calculations [23.16] show that the inclusion of higher-order corrections is important for obtaining accurate values for ΔE_{corr} and C_6. However, to improve accuracy by taking into account the corrections to diagrams (23.26) requires considerable effort. Indeed, there are several types of corrections to (23.26) such as (i) screening of the Coulomb interelectron interaction by the electron–vacancy excitations; (ii) interaction between vacancies ij; (iii) interaction between electrons and vacancies $ki(nj)$ $(kj(ni))$; and (iv) interaction between electrons kn. Corrections to the HF field itself which acts upon electrons k, n and vacancies i, j are discussed in Sect. 23.2.3.

Screening of the Coulomb interelectron interaction is very important, and in many cases must be taken into account non-perturbative. The simplest way to do this is to use RPA, which defines the effective interelectron interaction $\tilde{\Gamma}$ as a solution of an integral equation, shown diagrammatically by

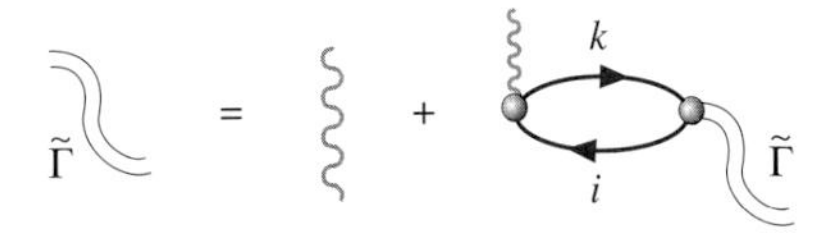

(23.30)

If V in (23.27) is replaced by $\tilde{\Gamma}$, an expression for ΔE_{corr} in RPA can be derived.

Exchange is very important in atoms and molecules, so diagram (23.30) can be modified to include this effect, thus leading to the effective interaction Γ in RPAE [23.12, 13, 16]:

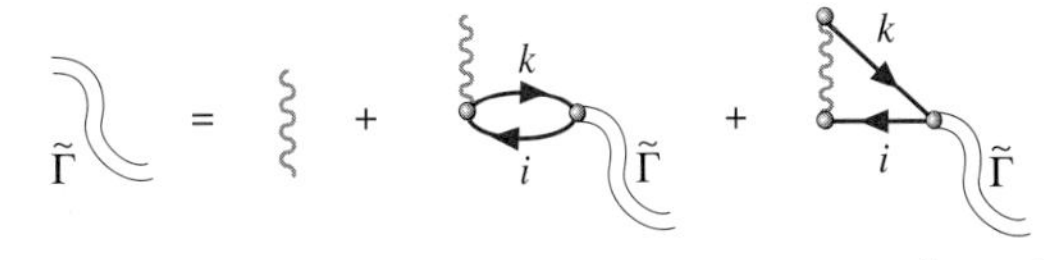

(23.31)

Replacing V in (23.27) by Γ gives a rather accurate expression for ΔE_{corr} in RPAE. Taking into account screening also affects the long-range interatomic interaction considerably by altering the constant C_6 in (23.28).

The ground state energy of an atom or molecule is modified by an external field. For a not too intense electromagnetic field, the simplest correction to the ground state energy is given by the diagrams

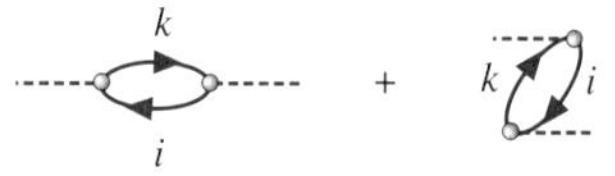

(23.32)

Considering a dipole external field, its interaction with the atomic electrons is given by $W = \sum_{i \le F} \boldsymbol{E} \cdot \boldsymbol{r}_i$, $\boldsymbol{E}$ being the strength of the field. The ground state energy shift is given by $\Delta E = -\alpha(\omega) E^2/2$, where $\alpha(\omega)$ is the dynamical dipole polarizability and ω is the frequency of the field. According to (23.32), $\alpha(\omega)$ is determined by

$$\alpha(\omega) = \sum_{k>F;\, i \le F} \frac{2|\langle i|z|k\rangle|^2 (\varepsilon_k - \varepsilon_i)}{(\varepsilon_k - \varepsilon_i)^2 - \omega^2}\,, \tag{23.33}$$

where z is a component of the vector $\boldsymbol{r}$.

RPAE corrections to $\alpha(\omega)$ are discussed in Sect. 23.2.5 in connection with the photoionization process.

Non-dipole polarizabilities of other multipolarities can be obtained in the lowest order of interelectron interaction using (23.32) with a properly chosen interaction operator between the electromagnetic field and an electron, instead of $W = \sum_{i \le F} \boldsymbol{E} \cdot \boldsymbol{r}_i$.

23.2.2 Characteristics of One-Particle States

A single vacancy or electron can propagate from one instant of interaction to another, as described to zero order by elements (23.1b) [or (23.1a)] with dots (23.1f) at the ends. This line represents an HF one-particle state with a given angular momentum, spin, and total momentum. Accounting for virtual or real atomic excitations leads, for a vacancy, to a diagram similar to (23.25) but with oppositely directed arrows. Because the interaction with these excitations is usually much smaller than the energy distance between shells, in the sum over q only the term $q = i$, i being the considered vacancy state, need be taken into account. Interaction with the vacuum leaves the angular momentum, spin,and total momentum unaltered. It can however change the energy, and lead to a finite lifetime for a vacancy state.

Analytically, the vacancy propagation in the HF approximation is described by the one-particle HF Green's function G^{HF}:

$$G_i^{\mathrm{HF}}(E) = 1/(\varepsilon_i - E)\,. \tag{23.34}$$

Solving (23.25) for a vacancy i with only Σ_{ii} terms included gives

$$G_i(E) = 1/[\varepsilon_i + \Sigma_{ii}(E) - E]\,. \tag{23.35}$$

The pole in $G(E)$ which determines the vacancy energy is shifted from $E = \varepsilon_i$ to $E_i = \varepsilon_i + \Sigma_{ii}(E_i)$. The quantity $\Sigma_{ii}(E)$ is called the self-energy, and is in general a complex function of energy, its imaginary part determining the lifetime of the vacancy i. Near E_i, (23.35) can be written in the form

$$G_i(E) \approx F_i/\left[\varepsilon_i + \Sigma_{ii}(E_i) - E\right] = F_i/(E_i - E)\,, \tag{23.36}$$

where

$$F_i = \left(1 - \left.\frac{\partial \Sigma_{ii}(E)}{\partial E}\right|_{E=E_i}\right)^{-1} \tag{23.37}$$

is called the *spectroscopic factor*. It characterizes the probability for more complicated configurations to be admixed into a single vacancy state i [23.12].

An important problem is to calculate the self-energy part $\Sigma(E)$. The first nonzero contributions are

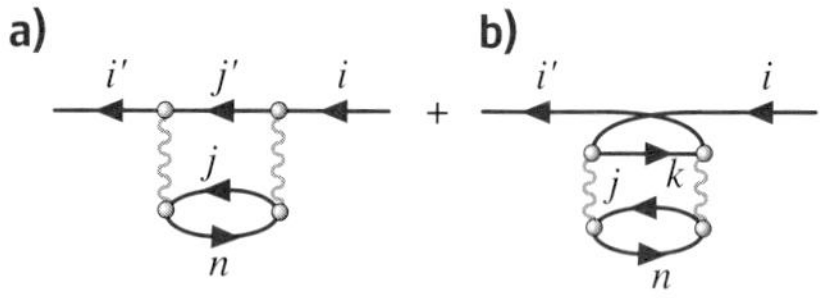

+ exchange terms.

(23.38)

Specific calculations [23.16] demonstrate that if the intermediate electron states n [in (23.38a)] and kn [in (23.38b)] are found in the field of vacancies jj' and $ii'j$, the diagrams (23.38) are able to reproduce the values of the correlation energy shift with about 5% accuracy. For outer subshell vacancies, the contributions (23.38a) and (23.38b) are almost equally important, to a large extent cancelling each other. For example, (23.38a) shifts the outer 3p vacancy in Ar to lower binding energies by 0.1 Ry, while the contribution of (23.38b) is −0.074 Ry. The total value 0.026 Ry is small and close to the experimental one, which is 0.01 Ry. For inner vacancies, (23.38a) is dominant because the intermediate states in (23.38b) have large virtualities and are therefore small. The main contribution to the sum over j' comes from the term $j' = i' = i$, which gives for the energy shift of level i

$$\Delta\varepsilon_i = \Sigma_{ii}^{(2)}(\varepsilon_i) = \sum\!\!\!\!\!\!\!\int_{n>F;j\leq F} \frac{|\langle n|r^{-1}|j\rangle|^2}{\varepsilon_n - \varepsilon_j} . \quad (23.39)$$

The value (23.39) is positive. Most important higher-order corrections will be included if V in (23.38a) is replaced by Γ from (23.31).

The physical meaning of diagram (23.38a) is transparent: it accounts for configuration mixing of one vacancy i and "two vacancies jj' – one electron n" states in the lowest order in the interelectron interaction. Diagram (23.38b) is not as transparent, and for $i = i'$ its intermediate state appears to violate the Pauli principle. However, as noted in connection with (23.14), the Pauli principle should not be considered as a restriction in constructing intermediate states.

Diagram (23.38a) and its exchange can have an imaginary part, which gives the probability of Auger decay $\gamma_i^{(\mathrm{A})}$, calculated to the lowest order in the interelectron interaction. For (23.38a) one has

$$\begin{aligned}\gamma_i^{(\mathrm{A})} &= \mathrm{Im}\,[\Sigma_{ii}(E_i)] \\ &= 2\pi \sum\!\!\!\!\!\!\!\int_{j,j'\leq F;n>F} |\langle in|V|jj'\rangle|^2 \delta(\varepsilon_j + \varepsilon_{j'} - \varepsilon_n - \varepsilon_i) .\end{aligned} \quad (23.40)$$

The width $\gamma_i^{(\mathrm{A})}$ is usually much smaller than $\mathrm{Re}[\Sigma_{ii}(E_i)]$, but there are several exceptional cases with abnormally large Auger widths, among which the most impressive is the 4p-vacancy in Xe with its $\gamma_{4\mathrm{p}} \approx 10\,\mathrm{eV}$.

Higher-order corrections include those which are taken into account when V in (23.40) is replaced by Γ from (23.31). The others include jj' vacancy–vacancy interaction, the interaction between vacancies jj' and the electron n and so on. As noted above, all of these can be obtained step by step by inserting the elements (23.4) into (23.38). To select the most important corrections, a physical idea and/or experience are necessary. For instance, if the energy transferred in the decay process $\Delta\varepsilon = \varepsilon_{j'} - \varepsilon_j$ is close to some threshold energies of atomic intermediate or outer shells, corrections which include virtual excitation of this shell must be taken into account.

The contribution to the spectroscopic factor from (23.38a) is given according to (23.37) by

$$F_i^{(2)} \equiv \left(1 + \sum\!\!\!\!\!\!\!\int_{j,j'\leq F;n>F} \frac{|\langle in|V|j'j\rangle|^2}{(\varepsilon_j + \varepsilon_{j'} - \varepsilon_n - \varepsilon_i)^2}\right)^{-1} . \quad (23.41)$$

Generally, for any Fermi particle, $F_i \leq 1$ [23.17] because there cannot be more than one particle in a given state. Note that the integrand in (23.41) is the lowest-order admixture of the $jj'n$ state to a pure one-vacancy state i. A small F value means strong mixing. For atoms, F_i is usually close to 1, but there are exceptions where F is small. For example, $F_{5\mathrm{s}}$ in Xe is about 0.33 [23.12].

The operator $\Sigma_{ij(n)}(\varepsilon)$ has non-diagonal matrix elements, which leads to admixture of other one-vacancy j or one-electron n states to the vacancy i. A measure of this admixture is given by the ratio $\Sigma_{ij(n)}(\varepsilon_i)/(\varepsilon_{j(n)} - \varepsilon_i)$.

In higher orders, decay processes more complex than those described by the imaginary part of (23.38a) become possible. For example, this could be a two-electron Auger decay in which the transition energy is distributed between two outgoing electrons. An example of the lowest-order diagram for this process is

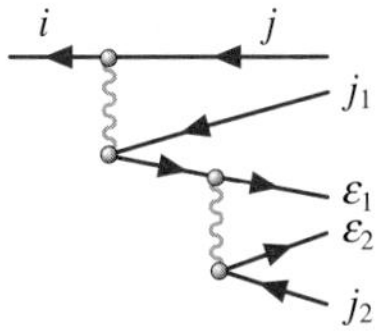

(23.42)

This is one of those diagrams which describe the mixing of a pure one-vacancy state with a quite complex configuration $jj_1j_2\varepsilon_1\varepsilon_2$.

23.2.3 Electron Scattering

Propagation of an electron in a discrete level or in a scattering state can be described in the same way as for a vacancy. The electron wave function is determined by (23.25). Using the correspondence rule (23.14), (23.25) can be expressed analytically in the form

$$\begin{aligned}&\left[-\frac{\nabla^2}{2} - \frac{Z}{r} + \sum_{i\leq F}\int \frac{\mathrm{d}\boldsymbol{r}'}{|\boldsymbol{r}'-\boldsymbol{r}|}|\phi_i(\boldsymbol{r}')|^2 - \varepsilon\right]\psi_\varepsilon(\boldsymbol{r}) \\ &= \sum_{i\leq F}\int \frac{\mathrm{d}\boldsymbol{r}'}{|\boldsymbol{r}'-\boldsymbol{r}|}\phi_i^*(\boldsymbol{r}')\psi_\varepsilon(\boldsymbol{r}')\phi_i(\boldsymbol{r}) \\ &\quad + \int \hat{\Sigma}(\boldsymbol{r},\boldsymbol{r}',\varepsilon)\psi_\varepsilon(\boldsymbol{r}')\mathrm{d}\boldsymbol{r}' .\end{aligned} \quad (23.43)$$

The terms with the Coulomb interelectron interaction $|\boldsymbol{r}' - \boldsymbol{r}|^{-1}$ determine the Hartree–Fock self-consistent potential. The last term in (23.43) represents the nonlocal energy dependent polarization interaction of the continuous spectrum electron with the target atom. Although

(23.43) resembles the ordinary Schrödinger equation, because of the energy dependence of the self-energy part $\hat{\Sigma}(\boldsymbol{r}, \boldsymbol{r}', \varepsilon)$, it is not the same. Consequently, the wave function $\psi_\varepsilon(\boldsymbol{r})$, often called a Dyson orbital, must be normalized according to the condition

$$(\psi_{\varepsilon'}|\psi_\varepsilon) = F_\varepsilon \delta(\varepsilon' - \varepsilon) , \qquad (23.44)$$

where

$$F_\varepsilon = \left[1 - \left(\varepsilon \left| \frac{\partial \hat{\Sigma}(\boldsymbol{r}, \boldsymbol{r}', E)}{\partial E} \right| \varepsilon \right) \bigg|_{E=\varepsilon} \right]^{-1} . \qquad (23.45)$$

This is different from that for ordinary wave functions by the factor $F_\varepsilon < 1$. In F_ε, the matrix element of the operator $\partial \hat{\Sigma}(\boldsymbol{r}, \boldsymbol{r}', E)/\partial E$ is calculated between states $\psi_\varepsilon(\boldsymbol{r})$. It is seen that F_ε is the same spectroscopic factor as determined by (23.37), but for a continuous spectrum or an excited electron state.

The object described by the wave function $\psi_\varepsilon(\boldsymbol{r})$ differs from an individual electron because it can be unstable, and its state is mixed with those of more complicated configurations, such as "two electrons" $k'k''$ – one vacancy j''. This object is called a *quasi-electron*.

Equation (23.43) also determines the energies of electrons in discrete levels which are shifted from their HF values. Contrary to the case of deep vacancies, it is impossible to predict the sign of the energy shift without detailed calculations.

For incoming electron energies ε higher than the target ionization threshold, the operator $\hat{\Sigma}(\boldsymbol{r}, \boldsymbol{r}', \varepsilon)$ acquires an imaginary part, thus becoming an optical potential for the projectile. The additional elastic scattering phase shifts from their HF values can be expressed via matrix elements of $\hat{\Sigma}(r, r', \varepsilon)$ between the wave functions $\phi^*_\varepsilon(r)$ and $\psi_\varepsilon(r)$; but to find numbers for these phase shifts, the self-energy part or polarization interaction $\hat{\Sigma}(\boldsymbol{r}, \boldsymbol{r}', \varepsilon)$ must be calculated. It appears that the second-order projectile–target interactions

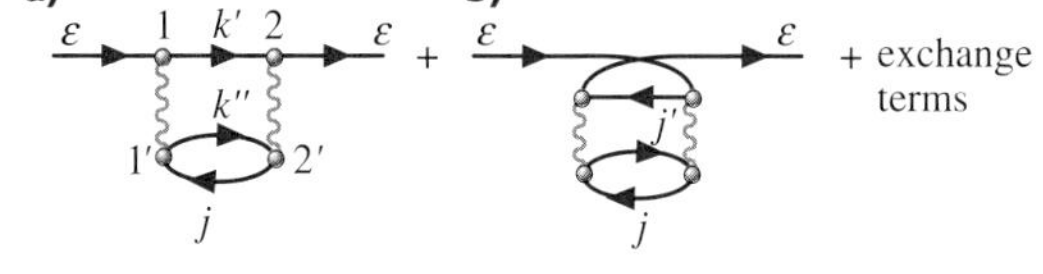

(23.46)

provide a reasonably good approximation [23.10, 12].

The expression for $\hat{\Sigma}(\boldsymbol{r}, \boldsymbol{r}', \varepsilon)$ simplifies at distances far from the target. Only (23.46a) contributes in this region, while other terms are exponentially small. Expanding the Coulomb interelectron interactions in (23.46a) in powers of $r_{1'}/r_1 \ll 1$, $r_{2'}/r_2 \ll 1$ gives

$$\Sigma(\boldsymbol{r}, \boldsymbol{r}', \varepsilon) = -\delta(\boldsymbol{r} - \boldsymbol{r}') \frac{\alpha^{\mathrm{HF}}(0)}{2r^4} , \qquad r, r' \to \infty \qquad (23.47)$$

where $\alpha^{\mathrm{HF}}(0)$ is the static ($\omega = 0$) dipole polarizability determined by (23.33).

Accounting for each additional interaction between the projectile and target increases the power of r in the denominator of (23.47). Most important is the interaction between target electrons. By including this, the asymptotic expression (23.47) is also modified, where instead of $\alpha^{\mathrm{HF}}(0)$, the RPAE polarizability $\alpha^{\mathrm{RPAE}}(0)$ appears. If the target–projectile interelectron interaction is taken into account to second order as in (23.46) while all the rest is included exactly, the expression for $\hat{\Sigma}(\boldsymbol{r}, \boldsymbol{r}', \varepsilon)$ for $r, r' \to \infty$ is still given by (23.46), but with the exact static polarizability.

Many experimental results for low energy electron scattering by noble gases, alkalis, and alkaline earths agree well with calculations of elastic scattering phase shifts obtained by solving equation (23.42) in which $\hat{\Sigma}(\boldsymbol{r}, \boldsymbol{r}', \varepsilon)$ is given by (23.46) with RPAE corrections taken into account. Total cross sections are reproduced with an accuracy as high as several percent, including the Ramsauer minimum region. This is illustrated in Fig. 23.1 for the $e^- + Xe$ case [23.18]. This approximation is also reasonably good in describing the angular distributions. As the projectile energy ε increases, the contribution of (23.46) decreases rapidly.

The approach presented here applies to other incoming particles, such as for instance positrons e^+.

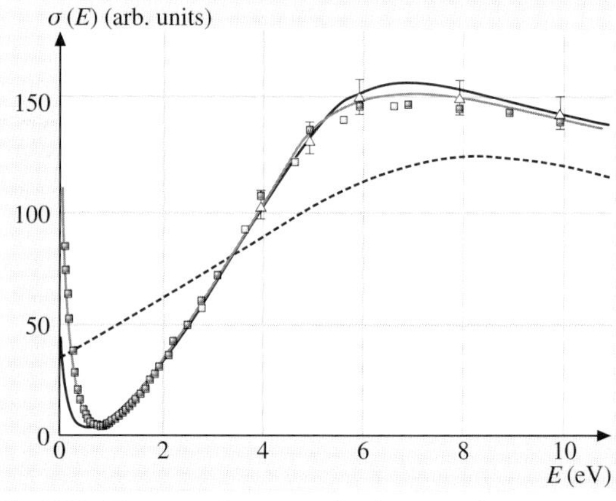

Fig. 23.1 Electron–Xe atom elastic scattering cross section [23.12]. *Solid line*: including polarization interaction; *dashed line*: HF; *dash-dotted line*: experiment

The exchange between e^+ and target electrons must be omitted by discarding (23.6) and all but (23.46a) terms in the polarization interaction $\hat{\Sigma}(\boldsymbol{r}, \boldsymbol{r}', \varepsilon)$. The incoming positron in its intermediate state k' [see (23.46a)] interacts strongly with the virtually excited electron k'', forming a positronium-like object. Corresponding diagrams are obtained by inserting elements (23.4d) for the e^+–e^- interaction into (23.46a). Summation of the infinite sequence of such diagrams corresponds to substituting the product $\phi_{k'}(\boldsymbol{r}_1)\phi_{k''}(\boldsymbol{r}_2)$ in the intermediate state of (23.46a) by the exact e^+e^- wave function in the field of a target with a vacancy j. Such a program is very complicated, and a simplification has proved to be satisfactory. Only the positronium (Ps) binding is taken into account by subtracting its binding energy in the denominator of (23.13). This is equivalent to adding the Ps ionization potential I_{Ps} to ε in (23.46) [23.19]. It enhances the polarization interaction and leads to an interesting qualitative feature: the possibility of alteration of the sign of the interaction (23.46). Indeed, instead of $\alpha(0)$, the corresponding expression for e^+-atom collision includes $\alpha(I_{\mathrm{Ps}})$. For alkalis, the binding energy is less than I_{Ps}, and for the energy region $\hbar\omega > I_{\mathrm{Ps}}$ the polarizability (23.33) is negative, while $\alpha(0) > 0$. This leads to a repulsive polarization interaction, rather than the usual attractive one. This difference affects the cross section qualitatively [23.20].

Another, more complicated, approximation substitutes $\phi_{k'}(\boldsymbol{r}_1)\phi_{k''}(\boldsymbol{r}_2)$ by the product of the precise Ps wave function $\psi_{\mathrm{Ps}}(|\boldsymbol{r}_1 - \boldsymbol{r}_2|)$ and the wave function of the free motion of the Ps center of gravity [23.21]

According to the diagrams (23.23), (23.43) describes the target with an additional electron. If the target is a neutral atom, solution of (23.43) with discrete energy values describes negative ion states; both ground and excited.

Again, as in Sect. 23.2.2, diagrams (23.46) with RPAE corrections form a reasonably good starting point for calculating the negative ion binding energies, even in cases when this binding is comparatively small, as in alkaline earth negative ions [23.22]. The inclusion of only the outer shell polarizability (j is a vacancy in the outer shell) leads to overbinding of the additional electron forming the negative ion. Only the inclusion of screening due to inner shell excitations yields good agreement with experiment. For instance, recent measurements of Ca^- affinity [23.23] give about 20 meV, while the calculations without inner shell excitations give about 50 meV [23.22]. Their inclusion must considerably reduce the theoretical value.

23.2.4 Two-Electron and Two-Vacancy States

One can construct a diagrammatic equation for the wave function of a two-electron or a two-vacancy state by separating all diagrams describing two-electron (two-vacancy) scattering which do not include these states as intermediate, and denoting their total contribution by a circle. Then the exact two-electron (vacancy) state is determined by the infinite sequence of diagrams

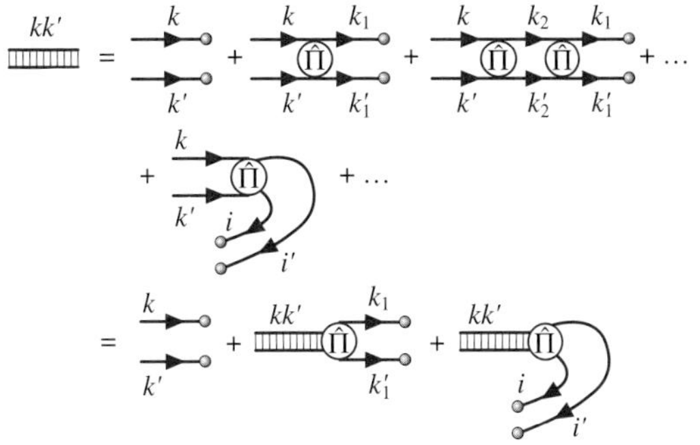

(23.48)

The analytic equation for two electrons in an atom interacting with each other can be written in the form

$$\left(\hat{H}_1^{\mathrm{HF}} + \hat{\Sigma}_1 + \hat{H}_2^{\mathrm{HF}} + \hat{\Sigma}_2 + \hat{Q}\hat{\Pi}_{12} - \varepsilon\right) \times \psi_{12}(\boldsymbol{r}_1 \cdot \boldsymbol{r}_2) = 0\,. \tag{23.49}$$

Here $\hat{H}_{1(2)}^{\mathrm{HF}}$ is the HF part of the one-particle Hamiltonian in (23.24), $\hat{\Pi}_{12}$ is the effective interelectron interaction and $\hat{Q}$ is the projection operator

$$\hat{Q} = 1 - n_1 - n_2\,, \tag{23.50}$$

with $n_{1(2)}$ being the Fermi step function, $n_{1(2)} = 1$ for $1(2) \le F$ and $n_{1(2)} = 0$ for $1(2) > F$. The function $n_{1(2)}$ thus eliminates contributions of vacant states. The operator $\hat{Q}$ takes into account the fact that propagation of two electrons (or, more precisely, quasi-electrons due to the presence of $\hat{\Sigma}$) takes place in a system of other particles which occupy all levels up to the Fermi level. The presence of $\hat{Q}$ makes (23.49) essentially different from a simple two-electron Schrödinger equation: $\hat{Q}$ requires that after each act of interaction described by $\hat{\Pi}_{12}$, both participants either remain electrons or become vacancies.

Diagrams (23.48) describe the states of two electrons outside the closed shell core, or electron scattering by an atom with one electron outside the closed shells. In general, $\hat{\Pi}_{12}$ is nonlocal, dependent upon ε, and can have an imaginary part. The lowest-order approximation to $\hat{\Pi}_{12}$ is $V = |\boldsymbol{r}_1 - \boldsymbol{r}_2|^{-1}$. Equation (23.49) must be solved in order to obtain, for example, the excitation spectrum of two electrons in atoms with two electrons outside closed subshells, such as Ca. Instead of V, the interaction Γ from (23.31) can be used, which would account for screening due to virtual excitation of inner shell electrons. For low level two-electron excitations, the energy dependence of $\hat{\Pi}_{12}$ can be neglected.

The same type of equation can be obtained for two-vacancy states, describing their energies, decay widths, and structure due to configuration mixing with more complex states. For inner and intermediate shell vacancies, however, the corresponding corrections can be taken into account perturbatively, and the screening by outer shells is not essential. The admixture of outer shell excitation with inner vacancies is important, leading to satellites of the main spectral lines.

The interaction between vacancies leads to correlation two-vacancy decay processes in which the energy is carried away by a single electron or photon. Some diagrams exemplifying these processes in the lowest possible order of interaction are

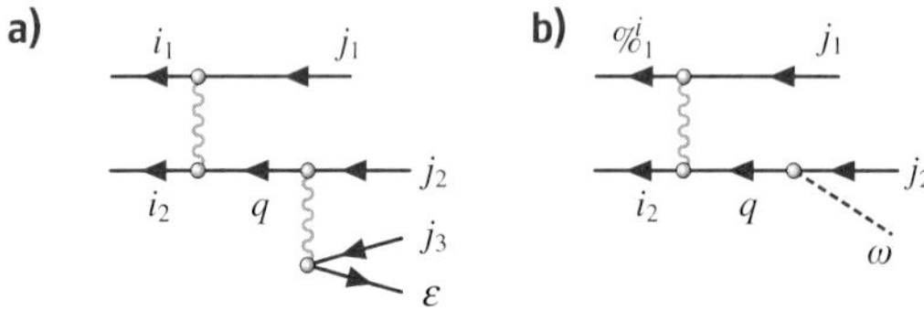

(23.51)

Even in this order, there are several diagrams giving together the amplitude of the correlation Auger (23.51a) and radiative (23.51b) decay. For inner shells, a specific feature of such processes is that the released energy is about twice that for a single vacancy decay. Of course, in these cases the decay probability is relatively small. It is not, however, necessarily much smaller than that of individual vacancies if the energies of the intermediate and initial states are close to each other.

The presence of residual two-body forces leads to effective multiparticle interactions. The simplest diagram presenting a three-electron interaction is given by

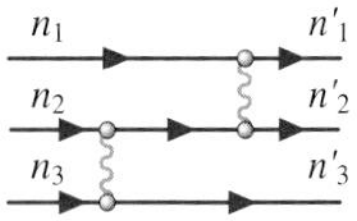

(23.52)

The role of multi-particle interactions in atomic structure is far from being clear. Diagrams similar to (23.52) are important if it is of interest to calculate the energy levels of atoms with three or more electrons (or vacancies) outside of closed shells. This is a very complicated calculation because even two electrons in the Coulomb field of the nucleus is a difficult three-body problem.

23.2.5 Electron–Vacancy States

The one-electron–one-vacancy state is the simplest excitation of a closed shell system under the action of an external time-dependent field, represented by (23.2a). Beginning with electrons and vacancies described in the HF approximation, the result of residual interactions leads to excitations of more complex states, including those with two or more electron–vacancy pairs. The interaction can also lead to a single electron–vacancy pair. Let us concentrate on the latter case and separate all diagrams describing electron–vacancy interaction which do not include these states as intermediate, and denote them by a circle. Then the exact electron–vacancy state is determined by the infinite sequence of diagrams

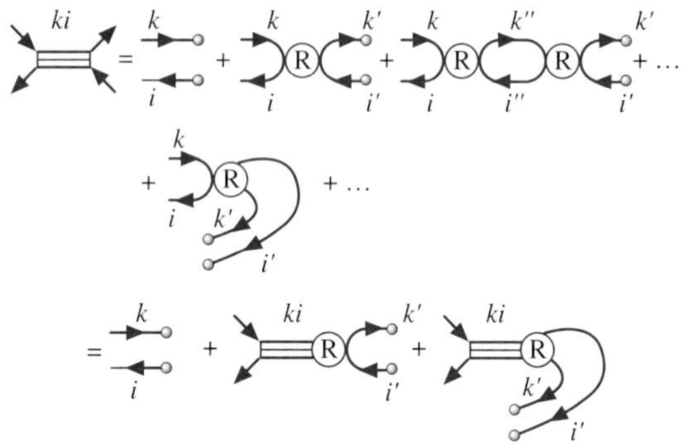

(23.53)

Contrary to the electron–electron case in Sect. 23.2.4, the analytical expression corresponding to (23.53) cannot be represented as a Schrödinger-type equation. Indeed, being symmetric under time reversal, (23.53) leads to an equation depending, unlike (23.49), upon the second power of the electron–vacancy energy ω. Note that R in (23.53) and $\hat{\Pi}$ in (23.46) are different.

It is necessary to solve (23.53) when calculating the photoabsorption amplitude, which can then be represented as

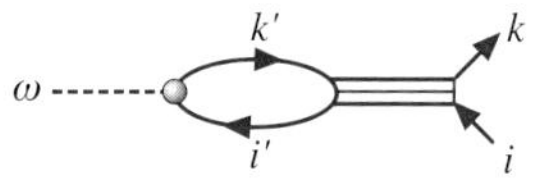

(23.54)

The amplitude for elastic photon scattering is also expressed via the exact electron–vacancy state, determined

by (23.53) to be

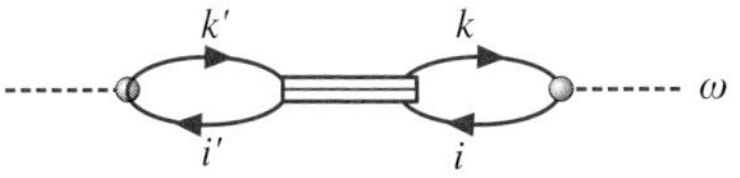

(23.55)

The other case where it is necessary to solve (23.53) is the scattering of electrons, both elastic and inelastic, by atoms with a vacancy in their outer shell, such as the halogens.

The simplest approximation to R is given by the Coulomb interaction to lowest order (23.4e) and (23.4f). With such R, the sequence of diagrams (23.53) is the same as (23.31) and thus forms the RPAE, which is often used to describe photoionization and other atomic processes. Both terms (23.4e) and (23.4f) contribute to the electron–vacancy in (23.53) only if the external field is spin independent, such as an ordinary photon. For a magnetic interaction, which is proportional to spin, the term (23.4e) does not contribute.

23.2.6 Photoionization in RPAE and Beyond

In RPAE, the photoionization amplitude $\langle k|D(\omega)|i\rangle$ is determined by solving an integral equation obtained from (23.53) and (23.54) using the correspondence rule (23.14) [23.12]:

$$\begin{aligned}&\langle k|D(\omega)|i\rangle\\&=\langle k|d|i\rangle\\&+\sum_{k'>F;i'\leq F}\!\!\!\!\!\!\!\!\!\int\;\left(\frac{\langle k'|D(\omega)|i'\rangle\langle ki'|V|ik'-k'i\rangle}{\omega-\varepsilon_{k'}+\varepsilon_{i'}+i\eta}\right.\\&\left.-\frac{\langle i'|D(\omega)|k'\rangle\langle kk'|V|ii'-i'i\rangle}{\omega-\varepsilon_{k'}-\varepsilon_{i'}-i\eta}\right),\end{aligned} \quad (23.56)$$

where d is the dipole operator describing the photon–electron interaction and $V=|\boldsymbol{r}_1-\boldsymbol{r}_2|^{-1}$. To obtain the RPAE photoionization cross section, the usual expression (24.19) must be multiplied by the square modulus of the ratio $\langle k|D(\omega)|i\rangle/\langle k|d|i\rangle$, with $\omega=\varepsilon_k+I_i$. The RPAE corrections described by the term in square brackets in (23.56) are very large for outer and intermediate electron shells. Through the sum over i', if only terms with the same energies $\varepsilon_{i'}=\varepsilon_i$ are included, (23.56) accounts for intra-shell correlations. By adding terms with $\varepsilon_{i'}\neq\varepsilon_i$, the effect of inter-shell correlations is taken into account.

For atoms, (23.56) has to be solved numerically, but can be presented in a symbolical operator form that creates the possibility of qualitative analyzes of its solutions:

$$D(\omega)=d+D(\omega)\chi(\omega)U\,, \quad (23.57)$$

where U is a combination of the direct V_d and exchange V_e Coulomb interelectron potentials, $U=V_\mathrm{d}-V_\mathrm{e}$, $\chi(\omega)=\chi_1(\omega)+\chi_2(\omega)$, $\chi_1(\omega)=1/(\omega-\omega'+i\eta)$ and $\chi_2(\omega)=1/(\omega+\omega')$, with ω' being the excitation energy of the virtual electron–vacancy state. Using Γ from (23.31), one can present $D(\omega)$ as

$$D(\omega)=d+d\chi(\omega)\Gamma(\omega)\,. \quad (23.58)$$

Equation (23.57) allows a rather simple, also symbolic, solution

$$D(\omega)=d/[1-\chi(\omega)U]\,. \quad (23.59)$$

If the denominator in (23.59) has a solution Ω determined by the equation

$$1-\chi(\Omega)U=0\,, \quad (23.60)$$

at $\Omega>I$, where I is the atomic ionization potential, then the cross section has a powerful maximum called a giant resonance with energy Ω.

A giant resonance is of a collective nature, in the sense that it appears to be due to coherent virtual excitation of all electrons of at least one considered multi-electron subshell.

These intra-shell correlations are most important for multielectron shells with large photoionization cross sections. Their inclusion leads to a quantitative description of the above mentioned giant resonances – huge maxima in the photoionization cross sections. An example is the $4d^{10}$ photoionization cross section of Xe shown in Fig. 23.2 [23.12], where satisfactory agreement with experiment is demonstrated.

It appears that all RPAE intra-shell time-forward diagrams, such as that on the first line in (23.53), and the first term in brackets in (23.56) with $\varepsilon_{i'}=\varepsilon_i$, may be taken into account by the matrix element $\langle\tilde{k}|d|i\rangle$. This is the one-electron approximation, but with the function $\tilde{\phi}_k(\boldsymbol{r})$ calculated in the term-dependent HF approximation [23.12]. Term-dependency means that only the total angular momentum and spin and their projections for the electron–vacancy pair are conserved, being equal to that of the incoming photon. The individual values for the electron and vacancy angular momentum and spin are not considered to be good quantum numbers. Thus the term-dependent HF includes a large fraction of RPAE correlations.

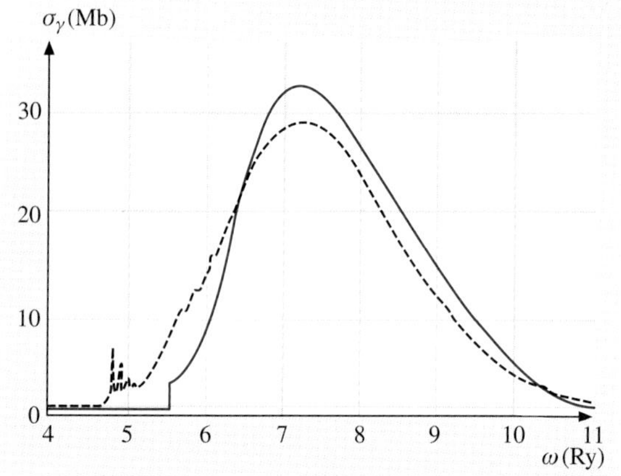

Fig. 23.2 Photoabsorption in the vicinity of the $4d^{10}$ subshell threshold in Xe [23.7]. *Solid line*: RPAE; *dashed line*: experiment

RPAE permitted to the prediction of interference resonances. To describe them, let us consider a situation in which the direct HF amplitude d_s is small, while there are other electrons with large photoionization amplitude D_b, $D_b(\omega) \gg d_s$. Then, from (23.57) one has

$$D_s(\omega) \approx d_s + D_b(\omega)\chi(\omega)$$
$$U_{bs} \approx D_b(\omega)\chi(\omega)U_{bs} \gg d_s \tag{23.61}$$

if the inter-transition interaction U_{bs} is not too small. The enhancement of the photoionization amplitude described by (23.61) manifests itself as a resonance in the partial cross section of s electrons photoionization. Very often the term $D_b(\omega)\chi(\omega)U_{bs}$ and d_s are of opposite sign, so that the total amplitude acquires two minima, along with an extra maximum, thus forming a rather complicated structure in the partial cross section that was named *interference* or *correlation* resonance.

Usually, these resonances are manifestations of inter-shell correlations. These are taken into account if the sum over i' in (23.56) includes terms with $\varepsilon_{i'} \neq \varepsilon_i$. An example is the $5s^2$ subshell in Xe, which is strongly affected by the outer $5p^6$ and inner $4d^{10}$ neighboring electrons. Due to this interaction, the $5s^2$ cross section is completely altered, as illustrated in Fig. 23.3 [23.12]. The RPAE results predict a qualitative feature of the experimental data, namely the formation of a maximum and two minima in the cross section. The second minimum is not seen in Fig. 23.3, since it lies at considerably higher ω.

RPAE is able to describe a number of other effects, such as *giant autoionizational* resonance (decay of a powerful discrete excitation into a continuum, with which the excitation interacts strongly), *continuous spectrum autoionization* (modification of a broad continuous spectrum excitation due to its strong interaction with a narrow continuum that happens in negative ions [23.22]) and *quadrupole giant resonances* [23.24].

Above we concentrated on dipole Giant resonances. Quadrupole amplitudes in RPAE are determined by an equation similar to (23.57):

$$Q(\omega) = q + Q(\omega)\chi(\omega)U \,, \tag{23.62}$$

where q is the quadrupole amplitude in HF approximation.

Giant quadrupole resonance was found in excitations of $4d^6$ electrons in Xe [23.25]. Its direct observation in photoabsorption is almost impossible, since the corresponding cross section is very small due to the inclusion of the extra factor $\alpha^2 = 1/c^2 \approx 10^{-4}$ as compared to the dipole cross section. However, the quadrupole amplitude leads to noticeable corrections to the angular distributions of photoelectrons where their relative contribution is considerably bigger.

Note that the amplitude of electron elastic scattering on an atom with a vacancy is expressed in RPAE via Γ given by (23.31) [23.26].

For the inner or deep intermediate shells, RPAE proves to be insufficient. First, screening of the Coulomb interaction between the outgoing or virtually excited electron and the vacancy [see (23.4f)] must be taken into account. This can be done by replacing V by Γ

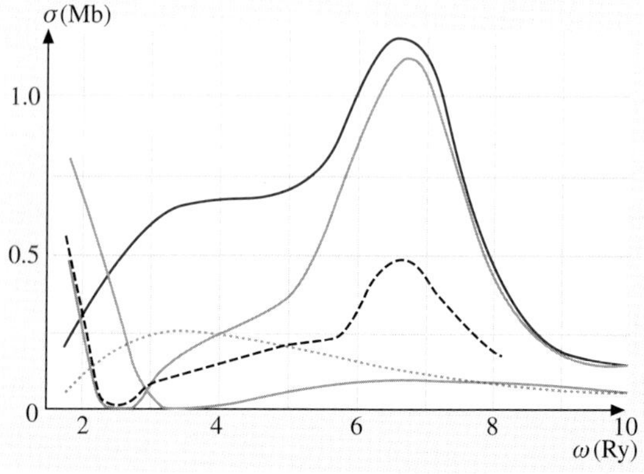

Fig. 23.3 Photoionization of $5s^2$ electrons in Xe [23.7]. *Solid line*: RPAE with effects of $5p^6$ and $4d^{10}$ included; *dashed line*: $5s^2$ electrons only; *dash-dotted line*: with effect of $5p^6$ electrons; *dash-double-dotted line*: with effect of $4d^{10}$ electrons; *dotted line*: experiment

from (23.31). The ionization potential (or the energy of the vacancy i) must also be corrected, which requires inclusion of at least the contribution from the first term of (23.38). It has been demonstrated [23.27] that the screening of the electron–vacancy interaction can be taken into account by calculating the wave function of the virtually excited or outgoing electron in the self-consistent HF field of an ion instead of that of a neutral atom. A method which uses only these one-particle wave functions in (23.54) (23.56) is called the generalized RPAE or GRPAE. The use of this approximation considerably improves the agreement with experiment near the intermediate shell thresholds, decreasing there the cross section value and shifting its maximum to higher energies. GRPAE permitted to disclose *intra-doublet resonances* that results from interaction of electrons belonging to two components of the spin-orbit doublet, e.g., $3d_{3/2}$ and $3d_{5/2}$ in Xe, Cs and Ba atoms [23.24, 28].

RPAE and GRPAE corrections affect not only the cross sections but also characteristics of the photoelectron angular distribution, i. e., dipole and non-dipole angular anisotropy parameters [23.12, 16, 24]. As an example, Fig. 23.4a presents the partial cross sections [23.28] while Fig. 23.4b depicts the dipole anisotropy parameter β [23.24] for $3d_{5/2}$ and $3d_{3/2}$ electrons in Cs. The effect of intra-doublet resonance – an additional maximum in the $3d_{5/2}$ cross section under the action of $3d_{3/2}$ electrons – is clearly seen.

Figure 23.5 depicts the non-dipole angular anisotropy parameter γ_{5s} for 5s electrons in Xe [23.29]. The parameter γ_{ns} (in Fig. 23.5, $n = 5$) is given by the simple formula [23.16]

$$\gamma_{ns}(\omega) = 6[|Q_{ns}(\omega)|/|D_{ns}(\omega)|]\cos(\Delta_q - \Delta_d)\,, \tag{23.63}$$

where $Q_{ns}(D_{ns})$ are the RPAE (GRPAE) quadrupole (dipole) photoionization amplitudes and $\Delta_q(\Delta_d)$ are their phases. Thus, $\gamma_{ns}(\omega)$ is sensitive to the presence of interference, dipole, and quadrupole resonances. The latter is presented by a small but noticeable maximum on the high energy slope of the huge maximum, caused by the presence of the giant dipole resonance. The variation of γ_{5s} near the 5s threshold is determined by the resonant behavior of $\cos(\Delta_q - \Delta_d)$, called *phase resonance* [23.24].

Close to inner shell thresholds, the Auger decay of a deep vacancy must be taken into account. Due to decay, the photoelectron instantly finds itself in the field of at least two vacancies instead of one, leading

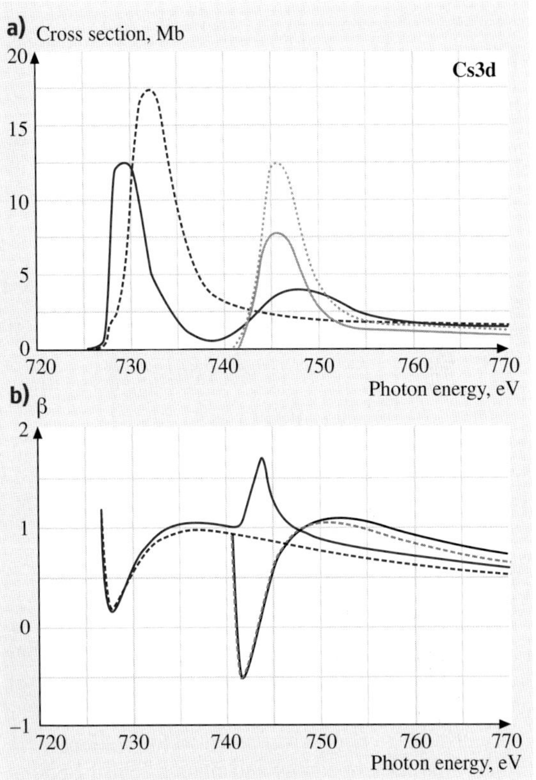

Fig. 23.4a,b Intra-doublet resonance in $3d^{10}$ Cs. (**a**) Partial photoionization cross sections $\sigma_{5/2}$ and $\sigma_{3/2}$ [23.28], (**b**) Dipole angular anisotropy parameters $\beta_{5/2}$ and $\beta_{3/2}$. *Solid line*: data for 5/2 with account of 3/2; *Dash-dotted line*: data for 5/2 without account of 3/2; *Dotted line*: data for 3/2 with account of 5/2; *Dashed line*: data for 3/2 without account of 5/2

to considerable growth of the threshold cross section. Diagrammatically, the effect of decay may be described by

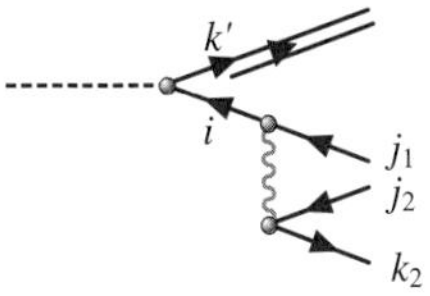

(23.64)

Here, the double line emphasizes that starting from the instant of decay, the photoelectron moves in the field $j_1 j_2$ of double instead of a single i vacancy. For inner vacancies, this is a strong effect which can lead even to recapture of the photoelectron into some of the discrete levels in the field of the double vacancy $j_1 j_2$.

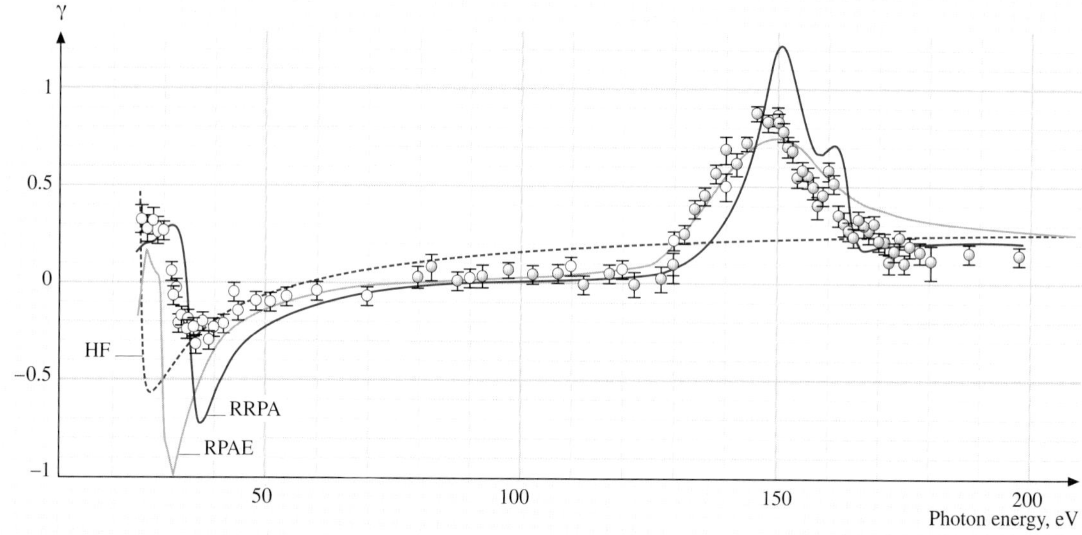

Fig. 23.5 Nondipole anisotropy parameter $\gamma_{5s}(\omega)$ of $5s^2$ electrons in Xe [23.29]

So-called Post-collision interactions in photoionization can be taken into account by this diagram when the Auger electron is much faster than the photoelectron [23.30]. If their speed is of the same order, their mutual Coulomb repulsion must be accounted for, leading to additional alteration of energy and angular redistributions.

A photoelectron can excite or knock out another atomic electron. To lowest order in the residual interelectron interaction, this process can be represented by

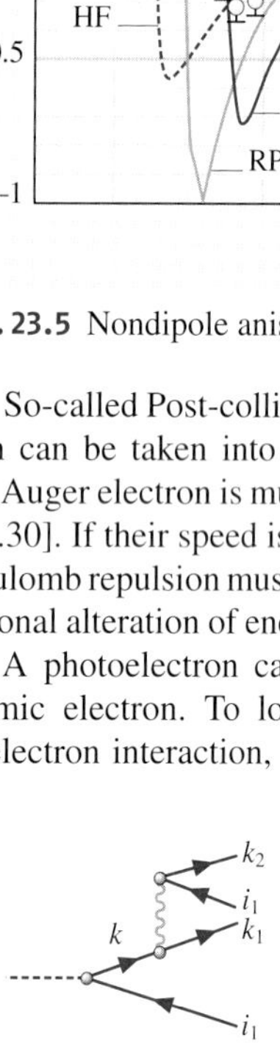

(23.65)

While the formation of an initial electron–vacancy ki_1 pair requires RPAE or GRPAE for its description, the second step of (23.65) can be reasonably well reproduced by the lowest-order term in V. It appears that process (23.65) has a high probability for not too fast photoelectrons, changing considerably the cross section [23.31] for ionization by creating a vacancy i_1.

A comparatively simple diagram

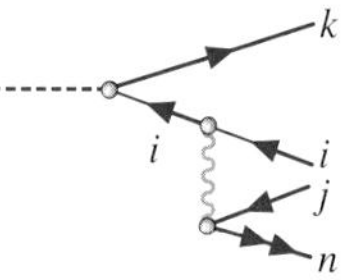

(23.66)

describes the photoionization process in which a more complicated state is created in the ion than a single vacancy, e.g. a state with two vacancies and one electron. Here the doubled arrow indicates that the electron is in a discrete level n.

Diagrams (23.64–23.66) present some corrections which mix electron–vacancy and two-electron–two-vacancy configurations. Each additional interaction line increases the number of possible physical processes considerably. With growth of the number of particles actively participating in a process, the calculational difficulties increase enormously. However, this is not a shortcoming of the diagrammatic approach, but a specific feature of more and more complex physical processes.

23.2.7 Photon Emission and Bremsstrahlung

The amplitude of photon emission in lowest order is given by the time-reverse of (23.2c):

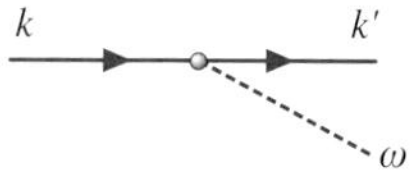

(23.67)

This diagram represents ordinary Bremsstrahlung; i.e. a process of projectile deceleration in the field of the target. If the target has internal structure as atoms do, it can be really or virtually excited during the collision process. The simplest excitation means creation of an electron–vacancy pair. The annihilation of this pair results via the time-reverse of (23.2a) in photon emission. The process thus looks like

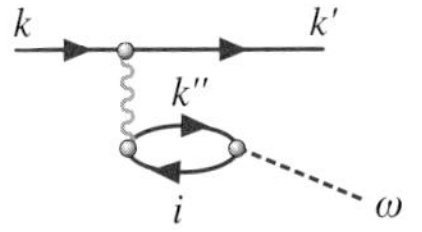

(23.68)

To obtain the total Bremsstrahlung amplitude, the terms (23.67) and (23.68) must be summed. The polarization radiation (PR) created by the mechanism (23.68) has a number of features which are different from the ordinary Bremsstrahlung (OB) represented by (23.67). The intensity is proportional to $1/M_p^2$ for OB, where M_p is the projectile mass, and the spectrum, at least for high ε_k, is proportional to $1/\omega$. On the other hand, the PR intensity is almost completely independent of M_p and its frequency dependence is quite complex, being determined by the target polarizability $\alpha(\omega)$ [23.32]. PR is most important for frequencies ω of the order of and higher than the target's ionization potential. At sufficiently large distances and for neutral targets, PR starts to predominate over OB. Close to discrete excitations of the k' electron, the contribution (23.68) becomes resonantly enhanced.

Higher-order corrections are important in the PR amplitude. First, the Coulomb interaction V in (23.68) must be replaced by Γ from (23.31).

The analytical expression for the total Bremsstrahlung amplitude, including RPAE corrections to (23.68), is given by the expression

$$\langle k|A(\omega)|k'\rangle = \langle k|d|k'\rangle + \sum\!\!\!\!\!\!\int_{k''>F;i\leq F} \langle ki|V|k'k''\rangle \times \lim_{\eta\to+0} \frac{2(\varepsilon_{k''}-\varepsilon_i)}{\omega^2-(\varepsilon_{k''}-\varepsilon_i)^2+\mathrm{i}\omega\eta} \langle i|D(\omega)|k''\rangle\,. \tag{23.69}$$

To derive the Bremsstrahlung spectrum, the usual general expression must be multiplied by the square modulus of the ratio $\langle k|A(\omega)|k'\rangle / \langle k|d|k'\rangle$. If the incoming electron is slow, corrections (23.46) also become important. The intermediate state in (23.68) includes two electrons k' and k, and a vacancy i. The extent of interaction between them could be considerable.

An important feature of PR is that it is nonzero even if the projectile is neutral, but is able to polarize the target. For example, it leads to emission of continuous spectrum radiation in atom–atom collisions, whose intensity for frequencies of the order of the ionization potentials is close to that in electron–atom collisions.

The second and higher orders in the residual interaction involve processes more complicated than (23.68), for instance those which include simultaneous photon emission and target excitation (ionization) [23.33].

23.3 Concluding Remarks

It is most convenient to apply diagrammatic techniques to closed shell atoms whose ground state is nondegenerate. Degeneracy means that some of the energy denominators (23.13) become zero with nonzero statistical weight. All such contributions must be summed to eliminate this degeneracy. This leads to strong mixing of some states. For example, the energy required for electron–vacancy transitions jn [see (23.38)] within an open shell is zero, thus leading to strong mixing of i and ijn states. If a pair with zero excitation energy has nonzero angular momentum, taking into account the mixing within such a pair destroys angular momentum as a characteristic of a one-vacancy state. This makes all calculations much more complicated, reflecting a specific feature of the degenerate physical system.

In using the diagrams and formulas presented above, it is essential that the interelectron and electron–nucleus interactions be purely potential. Inclusion of retardation and spin-dependence in the interparticle interaction makes the calculations much more complicated. These parts of the interaction appear as relativistic corrections. They are comparatively small in all but the heaviest atoms, and can be taken into account perturbatively. Beyond lowest order, these additional interactions are strongly altered when virtual excitations of electron–vacancy pairs and the Coulomb interaction between them is taken into account. An example is given by the sequence of diagrams

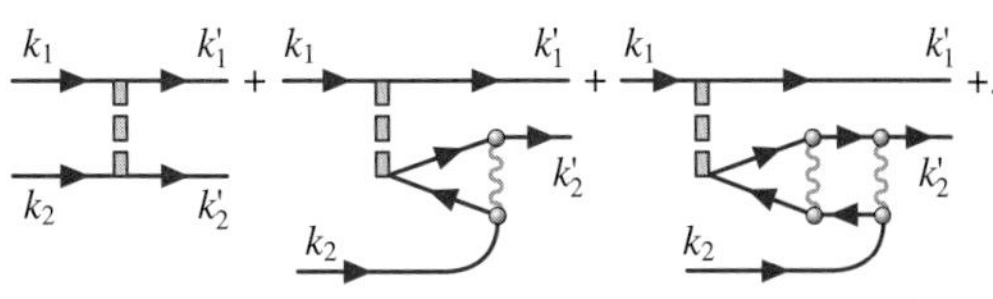

(23.70)

where the heavy dashed line stands for the spin-dependent interelectron interaction. Note that here there are no electron–vacancy loops as in (23.28) because the Coulomb interaction is unable to affect the electron spin and thus to transfer spin excitations.

The same kind of diagram describes the one-particle field acting upon an electron or vacancy due to the presence of spin-orbit interaction or weak interaction between electrons and the nucleus. For instance, the effective weak potential includes contributions from the sequence

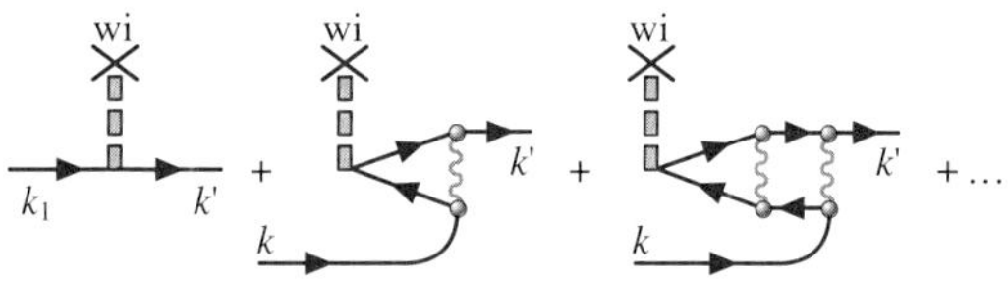

(23.71)

This is another example demonstrating how flexible and convenient the many-body approach is for considering different processes and interactions.

References

23.1 R. P. Feynman: *Quantum Electrodynamics* (Benjamin, New York 1961)
23.2 H. A. Bethe, J. Goldstone: Proc. R. Soc. London A **238**, 551 (1957)
23.3 H. P. Kelly: *Advances in Theoretical Physics*, Vol. 2, ed. by K. A. Brueckner (Academic Press, New York 1968) pp. 75–169
23.4 M. Ya. Amusia: *Many-body Effects in Electron Atomic Shells* (A. F. Ioffe Physical-Technical Institute Publications, Leningrad 1968) pp. 1–144 in Russian
23.5 M. Ya. Amusia: *X-Ray and Inner-Shell Processes*, AIP Conf. Proc. 389, ed. by R. L. Johnson, H. Schmidt-Böking, B. F. Sonntag (AIP Press, Woodbury, New York 1997) pp. 415–430
23.6 M. Ya. Amusia, J.-P. Connerade: Rep. Prog. Phys. **63**, 41 (2000)
23.7 M. Ya. Amusia: Phys. Essays **13**, 444 (2000)
23.8 M. Ya. Amusia: *The Physics of Ionized Gases*, ed. by N. Konjevich, Z. L. Petrovich, G. Malovich (Institute of Physics, Belgrade, Yugoslavia 2001) pp. 19–40
23.9 N. A. Cherepkov, S. K. Semenov, Y. Hikosaka, K. Ito, S. Motoki, A. Yagishita: Phys. Rev. Lett. **84**, 250 (2000)
23.10 A. N. Ipatov, V. K. Ivanov, B. D. Agap'ev, W. Ekardt: W. J. Phys. B: At. Mol. Opt. Phys. **31**, 925 (1998)
23.11 N. H. March, W. H. Young, S. Sampanthar: *The Many-Body Problem in Quantum Mechanics* (Cambridge Univ. Press, Cambridge 1967)
23.12 M. Ya. Amusia: *Atomic Photoeffect* (Plenum Press, New York 1990)
23.13 M. Ya. Amusia, L. V. Chernysheva: *Computation of Atomic Processes* (Institute of Physics Publishing, Bristol-Philadelphia 1997)
23.14 D. Pines: *The Many-Body Problem* (Benjamin, New York 1961)
23.15 L. D. Landau, E. M. Lifshits: *Quantum Mechanics* (Pergamon Press, Oxford 1965) pp. 319–322
23.16 M. Ya. Amusia, N. A. Cherepkov: Case Studies At. Phys. **5**, 47 (1975)
23.17 A. B. Migdal: *Theory of Finite Fermi-Systems and Applications to Atomic Nuclei* (Interscience, New York 1967)
23.18 W. R. Johnson, C. Guet: Phys. Rev. A **49**, 1041 (1994)
23.19 M. Ya. Amusia, N. A. Cherepkov, L. V. Chernysheva: JETP **124**, 1 (2003)
23.20 S. Zhou, S. P. Parikh, W. E. Kauppila, C. K. Kwan, D. Lin, A. Surdutovich, T. S. Stein: Phys. Rev. Lett. **73**, 236 (1994)
23.21 G. F. Gribakin, W. A. King: Can. J. Phys. **74**, 449 (1996)
23.22 V. K. Ivanov: J. Phys. B: At. Mol. Opt. Phys. **32**, R67 (1999)
23.23 C. W. Walter, J. R. Peterson: Phys. Rev. Lett. **68**, 2281 (1992)
23.24 M. Ya. Amusia: Radiation Physics and Chemistry **70**, 237 (2004)
23.25 W. Johnson, K. Cheng: Phys. Rev. A **63**, 022504 (2001)
23.26 M. Ya. Amusia, V. A. Sosnivker, N. A. Cherepkov, L. V. Chernysheva, S. I. Sheftel: J. Tech. Phys. (USSR Acad. Sci.) **60**, 1 (1990) in Russian
23.27 M. Ya. Amusia: *Photoionization in VUV and Soft X-Ray Frequency Regions*, ed. by U. Becker, D. Shirley (Plenum Press, New York 1996) pp. 1–46
23.28 M. Ya. Amusia, L. V. Chernysheva, S. T. Manson, A. Z. Msezane, V. Radoevich: Phys. Rev. Lett. **88**, 093002 (2002)
23.29 O. Hemmers, R. Guillemin, E. P. Kanter, B. Krassig, D. W. Lindle, S. H. Southworth , R. Wehlitz, J. Baker, A. Hudson, M. Lotrakul, D. Rolles, W. C. Stolte, I. C. Tran, A. Wolska, S. W. Yu, M. Y. Amusia, K. T. Cheng, L. V. Chernysheva, W. R. Johnson, S. T. Manson: Phys. Rev. Lett. **91**, 053002 (2003)
23.30 M. Ya. Amusia, M. Yu. Kuchiev, S. A. Sheinerman: J. Exp. Theor. Phys. **76**(2), 470 (1979)

23.31 M. Ya. Amusia, G. F. Gribakin, K. L. Tsemekhaman, V. L. Tsemekhaman: J. Phys. B **23**, 393 (1990)
23.32 M. Ya. Amusia: Phys. Rep. **162**, 249 (1988)
23.33 V. N. Tsitovich, I. M. Oiringel (Eds.): *Polarizational Radiation of Particles and Atoms* (Plenum Press, New York 1992)

24. Photoionization of Atoms

This chapter outlines the theory of atomic photoionization, and the dynamics of the photon–atom collision process. Those kinds of electron correlation that are most important in photoionization are emphasized, although many qualitative features can be understood within a central field model. The particle–hole type of electron correlations are discussed, as they are by far the most important for describing the single photoionization of atoms near ionization thresholds. Detailed reviews of atomic photoionization are presented in [24.1] and [24.2]. Current activities and interests are well-described in two recent books [24.3, 4]. Other related topics covered in this volume are experimental studies of photon interactions at both low and high energies in Chapts. 61 and 62, photodetachment in Chapt. 60, theoretical descriptions of electron correlations in Chapt. 23, autoionization in Chapt. 25, and multiphoton processes in Chapt. 74.

24.1 **General Considerations** 379
24.1.1 The Interaction Hamiltonian 379
24.1.2 Alternative Forms for the Transition Matrix Element 380
24.1.3 Selection Rules for Electric Dipole Transitions 381
24.1.4 Boundary Conditions on the Final State Wave Function 381
24.1.5 Photoionization Cross Sections 382

24.2 **An Independent Electron Model** 382
24.2.1 Central Potential Model 382
24.2.2 High Energy Behavior 383
24.2.3 Near Threshold Behavior 383

24.3 **Particle–Hole Interaction Effects** 384
24.3.1 Intrachannel Interactions 384
24.3.2 Virtual Double Excitations 384
24.3.3 Interchannel Interactions 385
24.3.4 Photoionization of Ar 385

24.4 **Theoretical Methods for Photoionization** 386
24.4.1 Calculational Methods 386
24.4.2 Other Interaction Effects 387

24.5 **Recent Developments** 387

24.6 **Future Directions** 388

References 388

24.1 General Considerations

24.1.1 The Interaction Hamiltonian

Consider an N-electron atom with nuclear charge Z. In the nonrelativistic approximation, it is described by the Hamiltonian

$$H=\sum_{i=1}^{N}\left(\frac{{p_i}^2}{2m}-\frac{Ze^2}{r_i}\right)+\sum_{i>j=1}^{N}\frac{e^2}{|\boldsymbol{r}_i-\boldsymbol{r}_j|}\,. \tag{24.1}$$

The one-electron terms in brackets describe the kinetic and potential energy of each electron in the Coulomb field of the nucleus; the second set of terms describe the repulsive electrostatic potential energy between electron pairs. The interaction of this atom with external electromagnetic radiation is described by the additional terms obtained upon replacing $\boldsymbol{p}_i$ by $\boldsymbol{p}_i+(|e|/c)\boldsymbol{A}(\boldsymbol{r}_i,t)$, where $\boldsymbol{A}(\boldsymbol{r}_i,t)$ is the vector potential for the radiation. The interaction Hamiltonian is thus

$$H_{\text{int}}=\sum_{i=1}^{N}\left\{\frac{+|e|}{2mc}[\boldsymbol{p}_i\cdot\boldsymbol{A}(\boldsymbol{r}_i,t)+\boldsymbol{A}(\boldsymbol{r}_i,t)\cdot\boldsymbol{p}_i]\right.$$
$$\left.+\frac{e^2}{2mc^2}|\boldsymbol{A}(\boldsymbol{r}_i,t)|^2\right\}\,. \tag{24.2}$$

Under the most common circumstance of single-photon ionization of an outer-subshell electron, the interaction Hamiltonian in (24.2) may be simplified considerably. First, the third term in (24.2) may be dropped, as it introduces two-photon processes (since it is of second order in $\boldsymbol{A}$). In any case, it is small compared with single photon processes since it is of second order in the coupling constant $|e|/c$. Second, we choose the Coulomb gauge for $\boldsymbol{A}$, which fixes the divergence of $\boldsymbol{A}$ as $\nabla\cdot\boldsymbol{A}=0$. $\boldsymbol{A}$ thus describes a transverse radiation field. Furthermore $\boldsymbol{p}$ and $\boldsymbol{A}$ now commute and hence the first and second terms in (24.2) may be combined. Third, we introduce the following form for $\boldsymbol{A}$:

$$\boldsymbol{A}(\boldsymbol{r}_i, t) = \left(\frac{2\pi c^2\hbar}{\omega V}\right)^{\frac{1}{2}} \hat{\boldsymbol{\epsilon}} \mathrm{e}^{\mathrm{i}(\boldsymbol{k}\cdot\boldsymbol{r}_i-\omega t)} . \tag{24.3}$$

This classical expression for $\boldsymbol{A}$ may be shown [24.5] to give photoabsorption transition rates that are in agreement with those obtained using the quantum theory of radiation. Here $\boldsymbol{k}$ and ω are the wave vector and angular frequency of the incident radiation, $\hat{\boldsymbol{\epsilon}}$ is its polarization unit vector, and V is the spatial volume. Fourth, the *electric dipole* (E1) approximation, in which $\exp[\mathrm{i}(\boldsymbol{k}\cdot\boldsymbol{r}_i)]$ is replaced by unity, is usually appropriate. The radii r_i of the atomic electrons are usually of order 1 Å. Thus for $\lambda \gg 100$ Å, $|\boldsymbol{k}\cdot\boldsymbol{r}_i| \ll 1$. Now $\lambda \gg 100$ Å corresponds to photon energies $\hbar\omega \ll 124$ eV. For outer atomic subshells, most of the photoabsorption occurs for much smaller photon energies, thus validating the use of the E1 approximation. (This approximation cannot be used uncritically, however. For example, photoionization of excited atoms (which have large radii), photoionization of inner subshells (which requires the use of short wavelength radiation), and calculation of differential cross sections or other measurable quantities that are sensitive to the overlap of electric dipole and higher multipole amplitudes all require that the validity of the electric dipole approximation be checked.) Use of all of the above conventions and approximations allows the reduction of H_{int} in (24.2) to the simplified form

$$H_{\text{int}} = \frac{+|e|}{mc}\left(\frac{2\pi c^2\hbar}{\omega V}\right)^{\frac{1}{2}} \sum_{i=1}^{N} \hat{\boldsymbol{\epsilon}}\cdot\boldsymbol{p}_i \exp(-\mathrm{i}\omega t) . \tag{24.4}$$

H_{int} thus has the form of a harmonically time-dependent perturbation. According to time-dependent perturbation theory, the photoionization cross section is proportional to the absolute square of the matrix element of (24.4) between the initial and final electronic states described by the atomic Hamiltonian in (24.1). Atomic units, in which $|e| = m = \hbar = 1$, are used in what follows.

24.1.2 Alternative Forms for the Transition Matrix Element

The matrix element of (24.4) is proportional to the matrix element of the momentum operator $\sum_i \boldsymbol{p}_i$. Alternative expressions for this matrix element may be obtained from the following operator equations involving commutators of the exact atomic Hamiltonian in (24.1):

$$\sum_{i=1}^{N} \boldsymbol{p}_i = -\mathrm{i}\left[\sum_{i=1}^{N} \boldsymbol{r}_i, H\right] , \tag{24.5}$$

$$\left[\sum_{i=1}^{N} \boldsymbol{p}_i, H\right] = -\mathrm{i}\sum_{i=1}^{N} \frac{Z\boldsymbol{r}_i}{r_i^3} . \tag{24.6}$$

Matrix elements of (24.5) and (24.6) between eigenstates $\langle\psi_0|$ and $|\psi_f\rangle$ of H having energies E_0 and E_f respectively give

$$\langle\psi_0|\sum_{i=1}^{N} \boldsymbol{p}_i|\psi_f\rangle = -\mathrm{i}\omega\langle\psi_0|\sum_{i=1}^{N} \boldsymbol{r}_i|\psi_f\rangle , \tag{24.7}$$

$$\langle\psi_0|\sum_{i=1}^{N} \boldsymbol{p}_i|\psi_f\rangle = \frac{-\mathrm{i}}{\omega}\langle\psi_0|\sum_{i=1}^{N} \frac{Z\boldsymbol{r}_i}{r_i^3}|\psi_f\rangle , \tag{24.8}$$

where $\omega = E_f - E_0$. Matrix elements of $\sum_{i=1}^{N} \boldsymbol{p}_i$, $\sum_{i=1}^{N} \boldsymbol{r}_i$, and $\sum_{i=1}^{N} Z\boldsymbol{r}_i/r_i^3$ are known as the "velocity," "length," and "acceleration" forms of the E1 matrix element.

Equality of the matrix elements in (24.7) and (24.8) does not hold when approximate eigenstates of H are used [24.6]. In such a case, qualitative considerations may help to determine which form is most reliable. For example, the length form tends to emphasize the large r part of the approximate wave functions, the acceleration form tends to emphasize the small r part of the wave functions, and the velocity form tends to emphasize intermediate values of r.

If instead of employing approximate eigenstates of the exact H, one employs exact eigenstates of an approximate N-electron Hamiltonian, then inequality of the matrix elements in (24.7) and (24.8) is a measure of the nonlocality of the potential in the approximate Hamiltonian [24.7, 8]. The exchange part of the Hartree–Fock potential is an example of such a nonlocal potential. Nonlocal potentials are also implicitly introduced in configuration interaction calculations employing a finite number of configurations [24.7, 8]. One may eliminate the ambiguity of which form of the E1 transition operator to use by requiring that the Schrödinger equation be gauge invariant. Only the length form is consistent with such gauge invariance [24.7, 8].

However, equality of the alternative forms of the transition operator does not necessarily imply high accuracy. For example, they are exactly equal when one uses an approximate local potential to describe the N-electron atom, as in a central potential model, even though the accuracy is often poor. The length and velocity forms are also exactly equal in the random phase approximation [24.9], which does generally give accurate cross sections for single photoionization of closed shell atoms. No general prescription exists, however, for ensuring that the length and velocity matrix elements are equal at each level of approximation to the N-electron Hamiltonian.

24.1.3 Selection Rules for Electric Dipole Transitions

If one ignores relativistic interactions, then a general atomic photoionization process may be described in LS-coupling as follows:

$$\begin{aligned}&\mathcal{A}(L, S, M_L, M_S, \pi_{\mathcal{A}}) + \gamma(\pi_\gamma, \ell_\gamma, m_\gamma)\\&\longrightarrow \mathcal{A}^+(\bar{L}\bar{S}\pi_{\mathcal{A}^+})\varepsilon\ell(L', S', M_{L'}, M_{S'})\,. \end{aligned} \quad (24.9)$$

Here the atom $\mathcal{A}$ is ionized by the photon γ to produce a photoelectron with kinetic energy ε and orbital angular momentum ℓ. The photoelectron is coupled to the ion $\mathcal{A}^+$ with total orbital and spin angular momenta L' and S'. In the electric dipole approximation, the photon may be regarded as having odd parity, i. e., $\pi_\gamma = -1$, and unit angular momentum, i. e., $\ell_\gamma = 1$. This is obvious from (24.7) and (24.8), where the E1 operator is seen to be a vector operator. The component m_γ of the photon in the E1 approximation is ± 1 for right or left circularly polarized light and 0 for linearly polarized light. (The z axis is taken as $\hat{\boldsymbol{k}}$ in the case of circularly polarized light and as $\hat{\boldsymbol{\epsilon}}$ in the case of linearly polarized light, where $\boldsymbol{k}$ and $\hat{\boldsymbol{\epsilon}}$ are defined in (24.3).) Angular momentum and parity selection rules for the E1 transition in (24.9) imply the following relations between the initial and final state quantum numbers:

$$L' = L \oplus 1 = \bar{L} \oplus \ell\,, \quad (24.10)$$

$$M_{L'} = M_L + m_\gamma = M_{\bar{L}} + m_\ell\,, \quad (24.11)$$

$$S' = S = \bar{S} \oplus \frac{1}{2}\,, \quad (24.12)$$

$$M_{S'} = M_S = M_{\bar{S}} + m_s\,, \quad (24.13)$$

$$\pi_{\mathcal{A}}\pi_{\mathcal{A}^+} = (-1)^{\ell+1}\,. \quad (24.14)$$

Equation (24.14) follows from the parity $(-1)^\ell$ of the photoelectron. The direct sum symbol $\oplus$ denotes the vector addition of $\boldsymbol{A}$ and $\boldsymbol{B}$ i.e, $\boldsymbol{A} \oplus \boldsymbol{B} = A+B, A+B-1, \dots, |A-B|$.

In (24.9), the quantum numbers $\alpha \equiv \bar{L}, \bar{S}, \pi_{\mathcal{A}^+}, \ell, L', S', M_{L'}, M_{S'}$ (plus any other quantum numbers needed to specify uniquely the state of the ion $\mathcal{A}^+$) define a final state *channel*. All final states that differ only in the photoelectron energy ε belong to the same channel. The quantum numbers L', S', $M_{L'}$, $M_{S'}$, and $\pi_{\text{tot}} = (-1)^\ell \pi_{\mathcal{A}^+}$ are the only good quantum numbers for the final states. Thus the Hamiltonian (24.1) mixes final state channels having the same angular momentum and parity quantum numbers but differing quantum numbers for the ion and the photoelectron; i. e., differing $\bar{L}$, $\bar{S}$, π_{A^+}, and ℓ but the same L', S', $M_{L'}$, $M_{S'}$ and $(-1)^\ell \pi_{\mathcal{A}^+}$.

24.1.4 Boundary Conditions on the Final State Wave Function

Photoionization calculations obtain final state wave functions satisfying the asymptotic boundary condition that the photoelectron is ionized in channel α. This boundary condition is expressed as

$$\begin{aligned}&\psi^-_{\alpha E}(\boldsymbol{r}_1 s_1, \dots, \boldsymbol{r}_N s_N)\\&\underset{r_N\to\infty}{\longrightarrow} \theta_\alpha(\boldsymbol{r}_1 s_1, \dots, \hat{\boldsymbol{r}}_N s_N)\frac{1}{\mathrm{i}(2\pi k_\alpha)^{\frac{1}{2}}}\frac{1}{r_N}\mathrm{e}^{\mathrm{i}\Delta_\alpha}\\&-\sum_{\alpha'}\theta_{\alpha'}(\boldsymbol{r}_1 s_1, \dots, \hat{\boldsymbol{r}}_N s_N)\frac{1}{\mathrm{i}(2\pi k_{\alpha'})^{\frac{1}{2}}}\frac{1}{r_N}\mathrm{e}^{-\mathrm{i}\Delta_{\alpha'}}S^\dagger_{\alpha'\alpha}\,,\end{aligned} \quad (24.15)$$

where the phase appropriate for a Coulomb field is

$$\Delta_\alpha \equiv k_\alpha r_N - \frac{1}{2}\pi\ell_\alpha + \frac{1}{k_\alpha}\log 2k_\alpha r_N + \sigma_{\ell_\alpha}\,. \quad (24.16)$$

The minus superscript on the wave function in (24.15) indicates an "incoming wave" normalization: i. e., asymptotically $\psi^-_{\alpha E}$ has outgoing spherical Coulomb waves only in channel α, while there are incoming spherical Coulomb waves in all channels. $S^\dagger_{\alpha'\alpha}$ is the Hermitian conjugate of the S-matrix of scattering theory, θ_α indicates the coupled wave function of the ion and the angular and spin parts of the photoelectron wave function, k_α is the photoelectron momentum in channel α and ℓ_α is its orbital angular momentum, and σ_{ℓ_α} in (24.16) is the Coulomb phase shift.

While one calculates channel functions $\psi^-_{\alpha E}$, experimentally one measures photoelectrons which asymptotically have well-defined linear momenta $\boldsymbol{k}_\alpha$ and well-defined spin states $m_{\frac{1}{2}}$, and ions in well-defined states $\bar{\alpha} \equiv \bar{L}\bar{S}M_{\bar{L}}M_{\bar{S}}$. The wave function appropriate for this experimental situation is related to the channel functions by uncoupling the ionic and electronic orbital and spin angular momenta and projecting the photoelectron angular momentum states ℓ_α, m_α onto the direction $\hat{\boldsymbol{k}}_\alpha$ by means of the spherical harmonic $Y^\star_{\ell_\alpha m_\alpha}(k_\alpha)$. This relation is [24.1]:

$$\begin{aligned}
&\psi^-_{\bar{\alpha}\boldsymbol{k}_\alpha}(\boldsymbol{r}_1 s_1, \ldots, \boldsymbol{r}_N s_N) \\
&= \sum_{\ell_\alpha m_\alpha} \frac{\mathrm{i}^{\ell_\alpha} \exp(-\mathrm{i}\sigma_{\ell_\alpha})}{k_\alpha^{\frac{1}{2}}} Y^\star_{\ell_\alpha m_\alpha}(\hat{\boldsymbol{k}}_\alpha) \\
&\times \sum_{LM_L SM_S} \langle \bar{L} M_{\bar{L}} \ell_\alpha m_\alpha | L M_L \rangle \\
&\times \langle \bar{S} M_{\bar{S}} \tfrac{1}{2} m_{\frac{1}{2}} | S M_S \rangle \psi^-_{\alpha E}(\boldsymbol{r}_1 s_1, \ldots, \boldsymbol{r}_N s_N) \,, \quad (24.17)
\end{aligned}$$

where the coefficients in brackets are Clebsch–Gordon coefficients. This wave function is normalized to a delta function in momentum space, i. e.,

$$\int \left(\psi^-_{\bar{\alpha}\boldsymbol{k}_\alpha}\right)^\dagger \psi^-_{\bar{\alpha}'\boldsymbol{k}_{\alpha'}} \mathrm{d}^3\boldsymbol{r} = \delta_{\bar{\alpha}\bar{\alpha}'} \delta(\boldsymbol{k}_\alpha - \boldsymbol{k}_{\alpha'}) \,. \quad (24.18)$$

The factors $\mathrm{i}^{\ell_\alpha} \exp(-\mathrm{i}\sigma_{\ell_\alpha}) k_\alpha^{-\frac{1}{2}}$ ensure that for large r_N (24.17) represents a Coulomb wave (with momentum $\boldsymbol{k}_\alpha$) times the ionic wave function for the state $\bar{\alpha}$ plus a sum of terms representing incoming spherical waves. Thus only the ionic term $\bar{\alpha}$ has an outgoing wave. One uses the wave function in (24.17) to calculate the angular distribution of photoelectrons.

24.1.5 Photoionization Cross Sections

If one writes H_{int} in (24.4) as $H_{\text{int}}(t) = H_{\text{int}}(0)\,\mathrm{e}^{-\mathrm{i}\omega t}$, then from first order time-dependent perturbation theory, the transition rate for transition from an initial state with energy E_0 and wave function ψ_0 to a final state with total energy E_f and wave function $\psi^-_{\bar{\alpha}\boldsymbol{k}_\alpha}$ is

$$\begin{aligned}
\mathrm{d}W_{\boldsymbol{k}_\alpha} &= 2\pi |\langle \psi_0 | H_{\text{int}}(0) | \psi^-_{\bar{\alpha}\boldsymbol{k}_\alpha} \rangle|^2 \\
&\times \delta(E_f - E_0 - \omega) k_\alpha^2 \, \mathrm{d}k_\alpha \, \mathrm{d}\Omega(\hat{\boldsymbol{k}}_\alpha) \,. \quad (24.19)
\end{aligned}$$

The delta function expresses energy conservation and the last factors on the right are the phase space factors for the photoelectron. Dividing the transition rate by the incident photon current density c/V, integrating over $\mathrm{d}k_\alpha$, and inserting $H_{\text{int}}(0)$, the differential photoionization cross section is

$$\frac{\mathrm{d}\sigma_{\bar{\alpha}}}{\mathrm{d}\Omega} = \frac{4\pi^2}{c} \frac{k_\alpha}{\omega} \left| \hat{\boldsymbol{\epsilon}} \cdot \langle \psi_0 | \sum_{i=1}^{N} \boldsymbol{p}_i | \psi^-_{\bar{\alpha}\boldsymbol{k}_\alpha} \rangle \right|^2 . \quad (24.20)$$

Implicit in (24.19) and (24.20) is an average over initial magnetic quantum numbers $M_{L_0} M_{S_0}$ and a sum over final magnetic quantum numbers $M_{\bar{L}} M_{\bar{S}} m_{\frac{1}{2}}$. The length form of (24.20) is obtained by replacing each $\boldsymbol{p}_i$ by $\omega \boldsymbol{r}_i$ (24.7).

Substitution of the final state wave function (24.17) in (24.20) permits one to carry out the numerous summations over magnetic quantum numbers and obtain the form

$$\frac{\mathrm{d}\sigma_{\bar{\alpha}}}{\mathrm{d}\Omega} = \frac{\sigma_{\bar{\alpha}}}{4\pi} [1 + \beta P_2(\cos\theta)] \quad (24.21)$$

for the differential cross section [24.10]. Here $\sigma_{\bar{\alpha}}$ is the partial cross section for leaving the ion in the state $\bar{\alpha}$, β is the asymmetry parameter [24.11], $P_2(\cos\theta) = \frac{3}{2}\cos^2\theta - \frac{1}{2}$, and θ indicates the direction of the outgoing photoelectron with respect to the polarization vector $\hat{\boldsymbol{\epsilon}}$ of the incident light. The form of (24.21) follows in the electric dipole approximation from general symmetry principles, provided that the target atom is unpolarized [24.12]. The partial cross section is given in terms of reduced E1 matrix elements involving the channel functions in (24.15) by

$$\sigma_{\bar{\alpha}} = \frac{4\pi^2}{3c} \omega [L]^{-1} \sum_{\ell_\alpha L'} \left| \langle \psi_0 \| \sum_{i=1}^{N} r_i^{[1]} \| \psi^-_{\alpha E} \rangle \right|^2 . \quad (24.22)$$

The β parameter has a much more complicated expression involving interference between different reduced dipole amplitudes [24.1]. Thus measurement of β provides information on the relative phases of the alternative final state channel wave functions, whereas the partial cross-section in (24.22) does not. From the requirement that the differential cross section in (24.21) be positive, one sees that $-1 \leq \beta \leq +2$.

24.2 An Independent Electron Model

The many-body wave functions ψ_0 and $\psi^-_{\alpha E}$ are usually expressed in terms of a basis of independent electron wave functions. Key qualitative features of photoionization cross sections can often be interpreted in terms of the overlaps of initial and final state one electron radial wave functions [24.1, 13]. The simplest independent electron representation of the atom, the central potential model, proves useful for this purpose.

24.2.1 Central Potential Model

In the central potential (CP) model the exact H in (24.1) is approximated by a sum of single-particle terms describing the independent motion of each electron in a central potential $V(r)$:

$$H_{\mathrm{CP}} = \sum_{i=1}^{N} \left[\frac{p_i{}^2}{2m} + V(r_i) \right] . \tag{24.23}$$

The potential $V(r)$ must describe the nuclear attraction and the electron–electron repulsion as well as possible and must satisfy the boundary conditions

$$V(r) \underset{r\to 0}{\longrightarrow} -Z/r \quad \text{and} \quad V(r) \underset{r\to\infty}{\longrightarrow} -1/r \tag{24.24}$$

in the case of a neutral atom. H_{CP} is separable in spherical coordinates and its eigenstates can be written as Slater determinants of one-electron orbitals of the form $r^{-1} P_{n\ell} Y_{\ell m}(\Omega)$ for bound orbitals and of the form $r^{-1} P_{\varepsilon\ell}(r) Y_{\ell m}(\Omega)$ for continuum orbitals. The one-electron radial wave functions satisfy

$$\frac{\mathrm{d}^2 P_{\varepsilon\ell}(r)}{\mathrm{d}r^2} + 2\left[\varepsilon - V(r) - \frac{\ell(\ell+1)}{2r^2} \right] P_{\varepsilon\ell}(r) = 0 , \tag{24.25}$$

subject to the boundary condition $P_{\varepsilon\ell}(0) = 0$, and similarly for the discrete orbitals $P_{n\ell}(r)$. *Hermann* and *Skillman* [24.14] have tabulated a widely used central potential for each element in the periodic table as well as radial wave functions for each occupied orbital in the ground state of each element.

24.2.2 High Energy Behavior

The hydrogen atom cross section, which is nonzero at threshold and decreases monotonically with increasing photon energy, serves as a model for inner-shell photoionization cross sections in the X-ray photon energy range. A sharp onset at threshold followed by a monotonic decrease above threshold is precisely the behavior seen in X-ray photoabsorption measurements. A simple hydrogenic approximation at high energies may be justified theoretically as follows: (1) Since a free electron cannot absorb a photon (because of kinematical considerations), at high photon energies one expects the more strongly bound inner electrons to be preferentially ionized as compared with the outer electrons. (2) Since the $P_{n\ell}(r)$ for an inner electron is concentrated in a very small range of r, one expects the integrand of the radial dipole matrix element to be negligible except for those values of r where $P_{n\ell}(r)$ is greatest. (3) Thus it is only necessary to approximate the atomic potential locally, e.g., by means of a screened Coulomb potential

$$V_{n\ell}(r) = -\left(\frac{Z - s_{n\ell}}{r} \right) + V_{n\ell}{}^{\mathrm{o}} \tag{24.26}$$

appropriate for the $n\ell$ orbital. Here $s_{n\ell}$ is the "inner-screening" parameter, which accounts for the screening of the nuclear charge by the other atomic electrons, and $V^{\mathrm{o}}_{n\ell}$ is the "outer-screening" parameter, which accounts for the lowering of the $n\ell$ electrons' binding energy due to repulsion between the outer electrons and the photoelectron as the latter leaves the atom. The potential in (24.26) predicts hydrogen-like photoionization cross sections for inner-shell electrons with onsets determined by the outer-screening parameters $V^{\mathrm{o}}_{n\ell}$.

Use of more accurate atomic central potentials in place of the screened hydrogenic potential in (24.26) generally enables one to obtain photoionization cross sections below the keV photon energy region to within 10% of the experimental results [24.15]. For $\ell > 0$ subshells and photon energies in the keV region and above, the independent particle model becomes increasingly inadequate owing to coupling with nearby ns-subshells, which generally have larger partial cross sections at high photon energies [24.16]. For high, but still nonrelativistic photon energies, i. e., $\omega \ll mc^2$, the energy dependence of the cross section for the $n\ell$ subshell within the independent particle model is [24.17]

$$\sigma_{n\ell} \sim \omega^{-\ell-\frac{7}{2}} . \tag{24.27}$$

However, when interchannel interactions are taken into account, the asymptotic energy dependence for subshells having $\ell > 0$ becomes independent of ℓ [24.18]:

$$\sigma_{n\ell} \sim \omega^{-\frac{9}{2}} (\ell > 0) . \tag{24.28}$$

This result stems from coupling of the $\ell > 0$ photoionization channels with nearby s-subshell channels.

24.2.3 Near Threshold Behavior

For photons in the vuv energy region, i. e., near the outer-subshell ionization thresholds, the photoionization cross sections for subshells with $\ell \geq 1$ frequently have distinctly nonhydrogenic behavior. The cross section, instead of decreasing monotonically as for hydrogen, rises above threshold to a maximum (the so called *delayed maximum* above threshold). Then it decreases to a minimum (the *Cooper minimum* [24.19, 20]) and rises to a second maximum. Finally the cross section decreases monotonically at high energies in accordance with hydrogenic behavior. Such nonhydrogenic behavior may be interpreted as due either to an effective potential barrier or to a zero in the radial dipole matrix element. We examine each of these effects in turn.

The *delayed maximum* above outer subshell ionization thresholds of heavy atoms (i. e., $Z \gtrsim 18$) is due to an effective potential barrier seen by $\ell = 2$ and $\ell = 3$ photoelectrons in the region of the outer edge of the atom (24.25). This effective potential lowers the probability of photoelectron escape until the photoelectrons have enough excess energy to surmount the barrier. Such behavior is nonhydrogenic. Furthermore, in cases where an inner subshell with $\ell = 2$ or 3 is being filled as Z increases (as in the transition metals, the lanthanides and the actinides) there is a double well potential. This double well has profound effects on the 3p-subshell spectra of the transition metals, the 4d-subshell spectra of the lanthanides, and the 5d-subshell spectra of the actinides, as well as on atoms with Z just below those of these series of elements [24.1, 21, 22].

Cross section minima arise due to a change in sign of the radial dipole transition matrix element in a particular channel [24.23, 24]. Rules for predicting their occurrence were developed by *Cooper* [24.19,20]. Studies of their occurrence in photoionization from excited states [24.25], in high Z atoms [24.26], and in relativistic approximation [24.27] have been carried out. Only recently has a proof been given [24.28] that such minima do not occur in atomic hydrogen spectra. For other elements, there are further rules on when and how many minima may occur [24.29–31].

Often within such minima, one can observe effects of weak interactions that are otherwise obscured. Relativistic and weak correlation effects on the asymmetry parameter β for s-subshells is a notable example [24.32]. *Wang* et al. [24.33] have also emphasized that near such minima in the E1 amplitudes, one cannot ignore the effects of quadrupole and higher corrections to the differential cross section. Central potential model calculations [24.33] show that quadrupole corrections can be as large as 10% of the E1 cross section at such cross section minima, even for low photon energies.

24.3 Particle–Hole Interaction Effects

The experimental photoionization cross sections for the outer subshells of the noble gases (The noble gases have played a prominent role in the development of the theory of photoionization for two reasons. These were among the first elements studied by experimentalists with synchrotron radiation beginning in the 1960's. Also, their closed-shell, spherically symmetric ground states simplified the theoretical analysis of their cross sections.) near the ionization thresholds can be understood in terms of interactions between the photoelectron, the residual ion, and the photon field which are called, in many-body theory language, "particle–hole" interactions (see Chapt. 47). These may be described as interactions in which two electrons either excite or deexcite each other out of or into their initial subshell locations in the unexcited atom. To analyze the effects of these interactions on the cross section, it is convenient to classify them into three categories: intrachannel, virtual double excitation, and interchannel. These alternative kinds of particle–hole interactions are illustrated in Fig. 24.1 using both many-body perturbation theory (MBPT) diagrams and more "physical" scattering pictures. We discuss each of these types of interaction in turn.

24.3.1 Intrachannel Interactions

The MBPT diagram for this interaction is shown on the left in Fig. 24.1a; on the right a slightly more pictorial description of this interaction is shown. The wiggly line indicates a photon, which is absorbed by the atom in such a way that an electron is excited out of the $n\ell$ subshell. During the escape of this excited electron, it collides or interacts with another electron from the same subshell in such a way that the second electron absorbs all the energy imparted to the atom by the photon; the first electron is de-excited back to its original location in the $n\ell$ subshell. For closed-shell atoms, the photoionization process leads to a $^1P_1$ final state in which the intrachannel interaction is strongly repulsive. This interaction tends to broaden cross section maxima and push them to higher photon energies as compared with the results of central potential model calculations.

Intrachannel interaction effects are taken into account automatically when the correct Hartree-Fock (HF) basis set is employed in which the photoelectron sees a net Coulomb field due to the residual ion and is coupled to the ion to form the appropriate total orbital L and

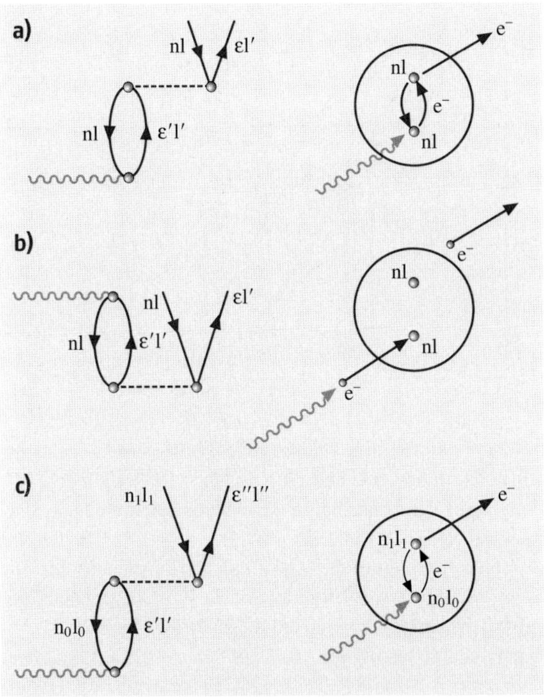

Fig. 24.1a–c MBPT diagrams (*left*) and scattering pictures (*right*) for three kinds of particle–hole interaction: (**a**) intrachannel scattering following photoabsorption; (**b**) photoabsorption by a virtual doubly-excited state; (**c**) interchannel scattering following photoabsorption

spin S angular momenta. Any other basis set requires explicit treatment of intrachannel interactions.

24.3.2 Virtual Double Excitations

The MBPT diagram for this type of interaction is shown on the left in Fig. 24.1b. Topologically, this diagram is the same as that on the left in Fig. 24.1a. In fact, the radial parts of the two matrix elements are identical; only the angular factors differ. A more pictorial description of this interaction is shown on the right of Fig. 24.1b. The ground state of the atom before photoabsorption is shown to have two electrons virtually excited out of the $n\ell$ subshell. In absorbing the photon, one of these electrons is de-excited to its original location in the $n\ell$ subshell, while the other electron in ionized. These virtual double excitations imply a more diffuse atom than in central-potential or HF models, with the effect that the overly repulsive intrachannel interactions are weakened, leading to cross sections for noble gas atoms that are in very good agreement with experiment with the exception that resonance features are not predicted.

24.3.3 Interchannel Interactions

The interchannel interaction shown in Fig. 24.1c is important, particularly for s subshells. This interaction has the same form as the intrachannel interaction shown in Fig. 24.1a, except now when an electron is photoexcited out of the $n_0\ell_0$ subshell, it collides or interacts with an electron in a different subshell – the $n_1\ell_1$ subshell. This interaction causes the second electron to be ionized, and the first electron to fall back into its original location in the $n_0\ell_0$ subshell.

Interchannel interaction effects are usually very conspicuous features of photoionization cross sections. When the interacting channels have partial photoionization cross sections which differ greatly in magnitude, one finds that the calculated cross section for the weaker channel is completely dominated by its interaction with the stronger channel. At the same time, it is often a safe approximation to ignore the effect of weak channels on stronger channels. In addition, when the interacting channels have differing binding energies, their interchannel interactions lead to resonance structure in the channel with lower binding energy (arising from its coupling to the Rydberg series in the channel with higher binding energy).

At high photon energies, s-subshell partial cross sections dominate over $\ell > 0$ subshell partial cross sections [(24.27), (24.28)]. Hence interchannel interactions of $\ell > 0$ subshells with nearby s-subshells change independent particle model predictions significantly. In particular, as noted in Sect. 24.3.2, such interactions can drastically change the magnitudes of the $\ell > 0$ partial cross sections [24.16] as well as their asymptotic energy behavior [24.18].

24.3.4 Photoionization of Ar

An example of both the qualitative features exhibited by photoionization cross sections in the vuv energy region and of the ability of theory to calculate photoionization cross reactions is provided by photoionization of the $n = 3$ subshell of argon, i. e.,

$$\begin{aligned} \mathrm{Ar}3s^2 3p^6 + \gamma &\to \mathrm{Ar}^+ 3s^2 3p^5 + e^- \\ &\to \mathrm{Ar}^+ 3s3p^6 + e^- . \end{aligned} \qquad (24.29)$$

Figure 24.2 shows the MBPT calculation of *Kelly* and *Simons* [24.34], which includes both intrachannel and interchannel interactions as well as the effect of virtual double excitations. The cross section is in excellent agreement with experiment [24.35, 36], even to the extent of describing the resonance behavior due to discrete members of the 3s → εp channel.

Figure 24.2 illustrates most of the features of photoionization cross sections described so far. First, the cross section rises to a delayed maximum just above the threshold because of the potential barrier seen by photoelectrons from the 3p subshell having $\ell = 2$. For photon energies in the range of 45–50 eV, the calculated cross section goes through a minimum because of a change in sign of the 3p → εd radial dipole amplitude. The HFL and HFV calculations include the strongly repulsive intrachannel interactions in the ^{1}P final-state channels and calculate the transition amplitude using the length (L) and velocity (V) form respectively for the electric dipole transition operator (24.7). With respect to the results of central potential model calculations, the HFL and HFV results have lower and broader maxima at higher energies. They also disagree with each other by a factor of two! Inclusion of virtual double excitations results in length and velocity results that agree to within 10% with each other and with experiment, except that the resonance structures are not reproduced. Finally, taking into account the interchannel interactions, one obtains the length and velocity form results shown in Fig. 24.2 by dash-dot and dashed curves respectively. Agreement with experiment is excellent and the observed resonances are well-reproduced.

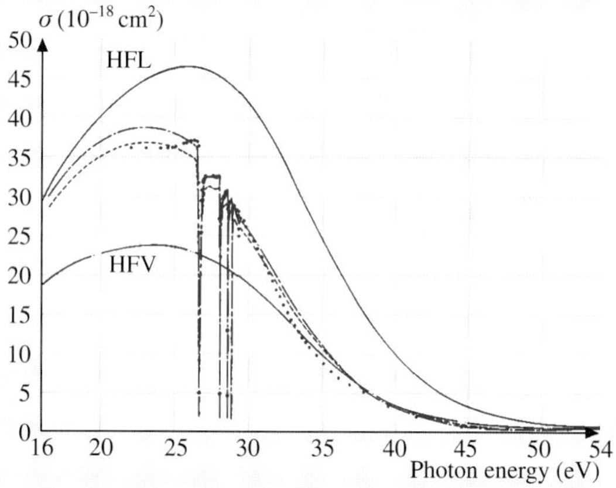

Fig. 24.2 Photoionization cross section for the 3p and 3s subshells of Ar. HFL and HFV indicate the length and velocity results obtained using HF orbitals calculated in a $^1P_1$ potential. *Dot-dash* and *dashed lines* represent the length and velocity results of the MBPT calculation of *Kelly* and *Simons* [24.34]. Only the four lowest 3s → np resonances are shown; the series converges to the 3s threshold at 29.24 eV. Experimental results are those of *Samson* [24.35] above 37 eV and of *Madden* et al. [24.36] below 37 eV (After [24.34])

24.4 Theoretical Methods for Photoionization

24.4.1 Calculational Methods

Most of the ab initio methods for the calculation of photoionization cross sections (e.g., the MBPT method [24.37], the close-coupling (CC) method [24.38], the R-matrix method [24.39, 40], the random phase approximation (RPA) method [24.9], the relativistic RPA method [24.41], the transition matrix method [24.42,43], the multiconfiguration Hartree-Fock (MCHF) method [24.44–46], etc.) have successfully calculated outer p-subshell photoionization cross sections of the noble gases by treating in their alternative ways the key interactions described above, i. e., the particle–hole interactions. In general, these methods all treat both intrachannel and interchannel interactions to infinite order and differ only in their treatment of ground state correlations. (The exception is MBPT, which often treats interchannel interactions between weak and strong channels only to first or second order.) These methods therefore stand in contrast to central potential model calculations, which do not treat any of the particle–hole interactions, and single-channel term-dependent HF calculations, which treat only the intrachannel interactions. The key point is that selection of the interactions that are included in a particular calculation is more important than the method by which such interactions are handled.

Treatment of photoionization of atoms other than the noble gases presents additional challenges for theory. For example, elements such as the alkaline earths, which have s^2 outer subshells, require careful treatment of electron pair excitations in both initial and final states. Open shell atoms have many more ionization thresholds than do the noble gases. Treatment of the resultant rich resonance structures typically relies heavily on quantum defect theory [24.46] (see Chapt. 32). All the methods listed above can be used to treat elements other than the noble gases, but a method which has come to prominence because of the excellent results it obtains for both alkaline earth and open-shell atoms is the eigenchannel R-matrix method [24.47].

24.4.2 Other Interaction Effects

A number of interactions, not of the particle–hole type, lead to conspicuous effects in localized energy regions. When treating photoionization in such energy regions, one must be careful to choose a theoretical method which is appropriate. Among the interactions which may be important are the following:

Relativistic and Spin-Dependent Interactions

The fact that $j = \ell - \frac{1}{2}$ electrons are contracted more than $j = \ell + \frac{1}{2}$ electrons at small distances has an enormous effect on the location of cross section minima in heavy elements [24.15, 48]. It may explain the large observed differences in the profiles of a resonance decaying to final states that differ only in their fine structure quantum numbers [24.49].

Inner-Shell Vacancy Rearrangement

Inner-shell vacancies often result in significant production of satellite structures in photoelectron spectra. Calculations for inner subshell partial photoionization cross sections are often substantially larger than results of photoelectron measurements [24.50–52]. This difference is attributed to such satellite production, which is often not treated in theoretical calculations.

Polarization and Relaxation Effects

Negative ion photodetachment cross sections often exhibit strong effects of core polarization near threshold. These effects can be treated semi-empirically, resulting in excellent agreement between theory and experiment [24.53]. Even for inner shell photoionization cross sections of heavy elements, ab initio theories do not reproduce measurements near threshold without the inclusion of polarization and relaxation effects [24.54, 55].

An Example

The calculation of the energy dependence of the asymmetry parameter β for the 5s subshell of xenon requires the theoretical treatment of all of the above effects. In the absence of relativistic interactions, β for Xe 5s would have the energy-independent value of two. Deviations of β from two are therefore an indication of the presence of these relativistic interactions. The greatest deviation of β from two occurs in the localized energy region where the partial photoionization cross section for the 5s subshell has a minimum. In this region, however, relativistic calculations show larger deviations from two than are observed experimentally. Inner shell rearrangement and relaxation effects play an important role [24.56, 57] and must be included to achieve good agreement with experiment.

24.5 Recent Developments

One of the most intensively studied areas in atomic photoionization in recent years has been the double photoionization of the helium atom. Extensive sets of experimental measurements for the two electron angular distributions (i. e., the triply differential cross sections) have provided stringent tests for various theoretical models and their treatments of electron correlations. A number of excellent reviews of this field have been published recently [24.58–60].

Another intensively studied area has been the analysis and measurement of non-dipole effects in photoionization, which were first observed in the X-ray region [24.61] but have been found to be significant even in the vuv photon energy region [24.62, 63]. In general, these effects stem from interference between electric quadrupole and the (usual) electric dipole transition amplitudes in differential cross sections (for a recent review, see [24.64]). Besides asymmetries in the photoelectron angular distributions, non-dipole effects lead also to new features for spin-resolved measurements [24.65, 66], and for the case of polarized atoms [24.67]. Recently, non-dipole effects have been predicted to be significant also in double photoionization of helium at relatively low photon energies [24.68].

Finally, both experimental and theoretical studies of ionic species have flourished over the past decade. In particular, photodetachment of negative ions near excited atomic thresholds provides an opportunity to study correlated, three-body Coulomb states unencumbered by Rydberg series. Only with the advent of powerful computer workstations have theorists been able to carry out numerical calculations for such high, doubly excited states with spectroscopic accuracy. Following experiments for photodetachment of H^- with excitations of atomic levels in H($n > 2$), theorists developed propensity rules for identifying and characterizing the dominant photodetachment channels [24.69–71]. More recently, experimental and theoretical interest shifted to the negative alkali ions (e.g., Li^- and Na^-), which for low photon energies have outer electron detachment spectra grossly resembling that of H^-. However, the negative alkali spectra contain clear signatures of propensity-rule-forbidden states that become increasingly prominent as the atomic number increases (owing to the nonhydrogenic inner electron cores). A brief review of low energy negative alkali photodetachment is given in [24.72]. Among the more general features brought to light by these studies is the mirroring of resonance profiles in alternative partial cross sections, which appears to be a very general phenomenon common to photodetachment and photoionization processes involving highly excited residual atoms or ions [24.73].

Recently, high energy (K-shell) photodetachment of the negative ions Li^- and He^- (resulting in two electron ionization) has been studied both experimentally and theoretically [24.74, 75]. These studies represent the first results for inner shell photodetachment. There is general agreement between theory and experiment well above the K edge, but the theoretical cross sections at the K edge are significantly higher than the experimental measurements. The latter discrepancy is now understood as arising from recapture of the low-energy detached electron following Auger decay of the inner-shell vacancy, which when taken into account theoretically has been shown to provide results that agree with experiment [24.76, 77]. Also, the first experimental data together with theoretical analyses were recently presented for photoionization of ground and metastable positive ions (O^+ and Sc^{++}) [24.78, 79]. With the advent of data for photoionization of positive ions it now becomes possible (using the principle of detailed balance) to make connections to data for electron–ion photo-recombination cross sections [24.79].

24.6 Future Directions

The construction of high brightness synchrotron light sources and the increasing use of lasers are providing the means to study atomic photoionization processes at an unparalleled level of detail. The synchrotrons generally produce photons in the soft X-ray and X-ray regions. Thus, inner shell vacancy production and decay, satellite production, and multiple ionization phenomena are all being increasingly studied. Laser sources are allowing production of atoms in tailored initial states. Studies of ions, both negativeand positive, in well-specified states are also increasingly being carried out. Thus, photoionization of excited atoms and ions and, in particular, complete measurements of particular photoionization processes, are now possible. Recent collections of short review papers provide references to these topics [24.3, 4]. In addition, two recent reviews of experimental results for noble gas atom photoionization [24.80] and for metal atom photoionization [24.81] also provide valuable information on the current state of the correspondingtheoretical results.

References

24.1 A. F. Starace: *Handbuch der Physik*, Vol. XXXI, ed. by W. Mehlhorn (Springer, Berlin, Heidelberg 1982) pp. 1–121
24.2 M. Ya. Amusia: *Atomic Photoeffect* (Plenum, New York 1990)
24.3 T. N. Chang (Ed.): *Many-Body Theory of Atomic Structure and Photoionization* (World Scientific, Singapore 1993)
24.4 U. Becker, D. A. Shirley: *VUV and Soft X-Ray Photoionization Studies* (Plenum, New York 1994)
24.5 J. J. Sakurai: *Advanced Quantum Mechanics* (Addison-Wesley, Reading 1967) p. 39
24.6 S. Chandrasekhar: Astrophys. J. **102**, 223 (1945)
24.7 A. F. Starace: Phys. Rev. A **3**, 1242 (1971)
24.8 A. F. Starace: Phys. Rev. A **8**, 1141 (1973)
24.9 M. Ya. Amusia, N. A. Cherepkov: Case Studies in At. Phys. **5**, 47 (1975)
24.10 J. M. Blatt, L. C. Biedenharn: Rev. Mod. Phys. **24**, 258 (1952)
24.11 See Sect. 7 of [24.1]
24.12 C. N. Yang: Phys. Rev. **74**, 764 (1948)
24.13 See, e.g., (9.6)–(9.15) of for expressions for the reduced matrix elements in (24.22) in which all angular integrations have been carried out and the results are expressed in terms of one-electron radial dipole matrix elements

24.14 F. Hermann, S. Skillman: *Atomic Structure Calculations* (Prentice-Hall, Englewood Cliffs 1963)
24.15 R. H. Pratt, A. Ron, H. K. Tseng: Rev. Mod. Phys. **45**, 273 (1973)
24.16 E. W. B. Dias, H. S. Chakraborty, P. C. Deshmukh, S. T. Manson, O. Hemmers, P. Glans, D. L. Hansen, H. Wang, S. B. Whitfield, D. W. Lindle, R. Wehlitz, J. C. Levin, I. A. Sellin, R. C. C. Perera: Phys. Rev. Lett. **78**, 4553 (1997)
24.17 H. A. Bethe, E. E. Salpeter: *Quantum Mechanics of One- and Two-Electron Atoms* (Springer, Berlin, Heidelberg 1957), Sects. 69–71
24.18 M. Ya. Amusia, N. B. Avdonina, E. G. Drukarev, S. T. Manson, R. H. Pratt: Phys. Rev. Lett. **85**, 4703 (2000)
24.19 J. W. Cooper: Phys. Rev. **128**, 681 (1962)
24.20 U. Fano, J. W. Cooper: Rev. Mod. Phys. **40**, 441 (1968)
24.21 , pp. 50–55, and references therein
24.22 J. P. Connerade, J. M. Estiva, R. C. Karnatak (Eds.): *Giant Resonances in Atoms, Molecules, and Solids* (Plenum, New York 1987) and references therein
24.23 D. R. Bates: Mon. Not. R. Astron. Soc. **106**, 432 (1946)
24.24 M. J. Seaton: Proc. Roy. Soc. A **208**, 418 (1951)
24.25 A. Z. Msezane, S. T. Manson: Phys. Rev. Lett. **48**, 473 (1982)
24.26 Y. S. Kim, R. H. Pratt, A. Ron: Phys. Rev. A **24**, 1626 (1981)
24.27 Y. S. Kim, A. Ron, R. H. Pratt, B. R. Tambe, S. T. Manson: Phys. Rev. Lett. **46**, 1326 (1981)
24.28 S. D. Oh, R. H. Pratt: Phys. Rev. A **34**, 2486 (1986)
24.29 R. H. Pratt, R. Y. Yin, X. Liang: Phys. Rev. A **35**, 1450 (1987)
24.30 R. Y. Yin, R. H. Pratt: **35**, 1149 (1987)
24.31 R. Y. Yin, R. H. Pratt: **35**, 1154 (1987)
24.32 S. T. Manson, A. F. Starace: Rev. Mod. Phys. **54**, 389 (1982)
24.33 M. S. Wang, Y. S. Kim, R. H. Pratt, A. Ron: Phys. Rev. A **25**, 857 (1982)
24.34 H. P. Kelly, R. L. Simons: Phys. Rev. Lett. **30**, 529 (1973)
24.35 J. A. R. Samson: Adv. At. Mol. Phys. **2**, 177 (1966)
24.36 R. P. Madden, D. L. Ederer, K. Codling: Phys. Rev. **177**, 136 (1969)
24.37 H. P. Kelly: *Photoionization and Other Probes of Many-Electron Interactions*, ed. by F. J. Wuilleumier (Plenum, New York 1976) pp. 83–109
24.38 P. G. Burke, M. J. Seaton: Methods Comput. Phys. **10**, 1 (1971)
24.39 P. G. Burke, W. D. Robb: Adv. At. Mol. Phys. **11**, 143 (1975)
24.40 P. G. Burke, W. D. Robb: *Electronic and Atomic Collisions: Invited Papers and Progress Reports*, ed. by G. Watel (North Holland, Amsterdam 1978) pp. 201–280
24.41 W. R. Johnson, C. D. Lin, K. T. Cheng, C. M. Lee: Phys. Scr. **21**, 409 (1980)
24.42 T. N. Chang, U. Fano: Phys. Rev. A **13**, 263 (1976)
24.43 T. N. Chang, U. Fano: Phys. Rev. A **13**, 282 (1976)
24.44 J. R. Swanson, L. Armstrong, Jr.: Phys. Rev. A **15**, 661 (1977)
24.45 J. R. Swanson, L. Armstrong, Jr.: Phys. Rev. A **16**, 1117 (1977)
24.46 M. J. Seaton: Rep. Prog. Phys. **46**, 167 (1983)
24.47 C. H. Greene: *Fundamental Processes of Atomic Dynamics*, ed. by J. S. Briggs, H. Kleinpoppen, H. O. Lutz (Plenum, New York 1988) pp. 105–127 and references therein
24.48 D. W. Lindle, T. A. Ferrett, P. A. Heiman, D. A. Shirley: Phys. Rev. A **37**, 3808 (1988)
24.49 M. Krause, F. Cerrina, A. Fahlman, T. A. Carlson: Phys. Rev. Lett. **51**, 2093 (1983)
24.50 J. B. West, P. R. Woodruff, K. Codling, R. G. Houlgate: J. Phys. B **9**, 407 (1976)
24.51 M. Y. Adam, F. Wuilleumier, N. Sandner, V. Schmidt, G. Wendin: J. Phys. (Paris) **39**, 129 (1978)
24.52 U. Becker, T. Prescher, E. Schmidt, B. Sonntag, H.-E. Wetzel: Phys. Rev. A **33**, 3891 (1986)
24.53 K. T. Taylor, D. W. Norcross: Phys. Rev. A **34**, 3878 (1986)
24.54 M. Ya. Amusia: *Atomic Physics 5*, ed. by R. Marrus, M. Prior, H. Shugart (Plenum, New York 1977) pp. 537–565
24.55 W. Jitschin, U. Werner, G. Materlik, G. D. Doolen: Phys. Rev. A **35**, 5038 (1987)
24.56 G. Wendin, A. F. Starace: Phys. Rev. A **28**, 3143 (1983)
24.57 J. Tulkki: Phys. Rev. Lett. **62**, 2817 (1989)
24.58 J. S. Briggs, V. Schmidt: J. Phys. B **33**, R1 (2000)
24.59 G. C. King, L. Avaldi: J. Phys. B **33**, R215 (2000)
24.60 J. Berakdar, H. Klar: Phys. Rep. **340**, 474 (2001)
24.61 B. Krässig, M. Jung, M. S. Gemmel, E. P. Kanter, T. LeBrun, S. H. Southworth, L. Young: Phys. Rev. Lett. **75**, 4736 (1995)
24.62 O. Hemmers, R. Guillemin, E. P. Kanter, B. Krässig, D. W. Lindle, S. H. Southworth, R. Wehlitz, J. Baker, A. Hudson, M. Lotrakul, D. Rolles, W. C. Stolte, I. C. Tran, A. Wolska, S. W. Yu, M. Ya. Amusia, K. T. Cheng, L. V. Chernysheva, W. R. Johnson, S. T. Manson: Phys. Rev. Lett. **91**, 053002 (2003)
24.63 E. P. Kanter, B. Krässig, S. H. Southworth, R. Guillemin, O. Hemmers, D. W. Lindle, R. Wehlitz, M. Ya. Amusia, L. V. Chernysheva, N. L. S. Martin: Phys. Rev. A **68**, 012714 (2003)
24.64 D. W. Lindle, O. Hemmers: J. Elec. Spectros. Rel. Phen. **100**, 297 (1999)(Special Issue, October)
24.65 N. A. Cherepkov, S. K. Semenov: J. Phys. B **34**, L211 (2001)
24.66 T. Khalil, B. Schmidtke, M. Drescher, N. Müller, U. Heinzmann: Phys. Rev. Lett. **89**, 053001 (2002)
24.67 A. N. Grum-Grzhimailo: J. Phys. B **34**, L359 (2001)
24.68 A. Y. Istomin, N. L. Manakov, A. V. Meremianin, A. F. Starace: Phys. Rev. Lett. **92**, 062002 (2004)
24.69 H. R. Sadeghpour, C. H. Greene: Phys. Rev. Lett. **65**, 313 (1990)
24.70 J. M. Rost, J. S. Briggs, J. M. Feagin: Phys. Rev. Lett. **66**, 1642 (1991)

24.71 H. R. Sadeghpour, C. H. Greene, M. Cavagnero: Phys. Rev. A **45**, 1587 (1992)

24.72 A. F. Starace: *Novel Doubly Excited States Produced in Negative Ion Photodetachment*, ed. by F. Aumayr, H. Winter (World Scientific, Singapore 1998) pp. 107–116

24.73 C. N. Liu, A. F. Starace: Phys. Rev. A **59**, R1731 (1999)

24.74 N. Berrah, J. D. Bozek, A. A. Wills, G. Turri, L. L. Zhou, S. T. Manson, G. Akerman, B. Rude, N. D. Gibson, C. W. Walter, L. VoKy, A. Hibbert, S. M. Ferguson: Phys. Rev. Lett. **87**, 253002 (2001)

24.75 N. Berrah, J. D. Bozek, G. Turri, G. Akerman, B. Rude, H. L. Zhou, S. T. Manson: Phys. Rev. Lett. **88**, 093001 (2002)

24.76 J. L. Sanz-Vicario, E. Lindroth, N. Brandefelt: Phys. Rev. A **66**, 052713 (2002)

24.77 T. W. Gorczyca, O. Zatsarinny, H. L. Zhou, S. T. Manson, Z. Felfli, A. Z. Msezane: Phys. Rev. A **68**, 050703 (R) (2003)

24.78 A. M. Covington, A. Aguilar, I. R. Covington, M. Gharaibeh, C. A. Shirley, R. A. Phaneuf, I. Alvarez, C. Cisneros, G. Hinojosa, J. D. Bozek, I. Dominguez, M. M. Sant'Anna, A. S. Schlachter, N. Berrah, S. N. Nahar, B. M. McLaughlin: Phys. Rev. Lett. **87**, 243002 (2001)

24.79 S. Schippers, A. Müller, S. Ricz, M. E. Bannister, G. H. Dunn, J. Bozek, A. S. Schlacter, G. Hinojosa, C. Cisneros, A. Aguilar, A. M. Covington, M. F. Gharaibeh, R. A. Phaneuf: Phys. Rev. Lett. **89**, 193002 (2002)

24.80 V. Schmidt: Rep. Prog. Phys. **55**, 1483 (1992)

24.81 B. Sonntag, P. Zimmerman: Rep. Prog. Phys. **55**, 911 (1992)

25. Autoionization

The phenomenon of autoionization, or more particularly the autoionization state itself, is treated for the most part in this chapter as a bound state. The process is rigorously a part of the scattering continuum (Chapt. 47), but, due mostly to the work of *Feshbach* [25.1], a rigorous formulation can be established whereby the main element of the theory can be made into a bound state problem with the scattering elements built around it. The major constituent of both these features is accomplished with projection operators. A brief description of the above elements of the theory, centered around projection operators, is the aim of this chapter [25.2], although some additional methods are discussed in Sect. 25.5. Rydberg units are used unless otherwise noted.

25.1 **Introduction** 391
25.1.1 Auger Effect 391
25.1.2 Autoionization, Autodetachment, and Radiative Decay 391
25.1.3 Formation, Scattering, and Resonances 391

25.2 **The Projection Operator Formalism** 392
25.2.1 The Optical Potential 392
25.2.2 Expansion of V_{op}: The QHQ Problem 392

25.3 **Forms of P and Q** 393
25.3.1 The Feshbach Form 393
25.3.2 Reduction for the $N=1$ Target 394
25.3.3 Alternative Projection and Projection-Like Operators 394

25.4 **Width, Shift, and Shape Parameter** 394
25.4.1 Width and Shift 394
25.4.2 Shape Parameter 395
25.4.3 Relation to Breit–Wigner Parameters 396

25.5 **Other Calculational Methods** 396
25.5.1 Complex Rotation Method 396
25.5.2 Pseudopotential Method 397

25.6 **Related Topics** 398

References 399

25.1 Introduction

25.1.1 Auger Effect

Autoionization falls within the general class of phenomena known as the Auger effect. In the Auger effect an atomic system "seemingly" (Sect. 25.1.3) spontaneously decays into a partition of its constituent parts.

25.1.2 Autoionization, Autodetachment, and Radiative Decay

If the initial, composite system is neutral, or positively charged, and its constituent decay particles are an electron and the residual ion, then the process is called *autoionization*. If the original system is a negative ion, so that the residual heavy particle system is neutral, then the process is technically called *autodetachment*; for the most part, the physics and the mathematical treatment are the same.

It is also possible, before electron emission takes place, that the system will alternatively decay radiatively to an autoionization state of lower energy, or a true bound state of the composite system. The latter process is called *radiative stabilization* (which is a basic part of dielectronic recombination (Chapt. 55)).

25.1.3 Formation, Scattering, and Resonances

Autoionizing states are formed by scattering processes and photoabsorption. These are the inverses of the autoionization and photon emission processes by which they can decay. In the scattering process, formation of the autoionization state corresponds to a resonance in the scattering cross section (Chapt. 47). Autoionization is the process which corresponds to the decay of the resonance. The decay of the resonance (autoionization) is then seen to be the last half of the resonant scattering process. Strictly speaking therefore, the resonant or autoionization state, although it may be long-lived, is not completely stationary, and that is the reason that the word "seemingly" was used to describe the Auger process. The compound system can also be formed by absorption of photons impinging on a bound state (usually the ground state) of the compound system, in which case the autionization state shows up as a "line" in the absorption spectrum.

25.2 The Projection Operator Formalism

In the energy domain where the Schrödinger equation (SE)

$$H\Psi = E\Psi \tag{25.1}$$

describes scattering, the wave function does not vanish at infinity, i. e.,

$$\lim_{r_i \to \infty} \Psi \neq 0 , \tag{25.2}$$

where r_i is the radial coordinate of ith electron.

The basic idea of the projection operator formalism [25.1] is to define projection operators P and Q which separate Ψ into scattering-like ($P\Psi$) and quadratically integrable ($Q\Psi$) parts:

$$\Psi = P\Psi + Q\Psi . \tag{25.3}$$

Implicit in (25.3) are

$$\begin{aligned} &\text{completeness: } P + Q = 1 , \\ &\text{idempotency: } P^2 = P , \quad Q^2 = Q , \\ &\text{orthogonality: } PQ = 0 , \end{aligned} \tag{25.4}$$

and the asymptotic properties

$$\lim_{r_i \to \infty} \begin{cases} P\Psi = \lim\limits_{r_i \to \infty} \Psi \\ Q\Psi = 0 \end{cases} . \tag{25.5}$$

25.2.1 The Optical Potential

Straightforward manipulation of (25.1) and (25.3) leads to an important relation between $Q\Psi$ and $P\Psi$:

$$Q\Psi = (E - QHQ)^{-1} QHP\Psi . \tag{25.6}$$

From (25.6) a basic equation for $P\Psi$ (which, by virtue of (25.5), contains all the scattering information) emerges:

$$(PHP + V_{\text{op}} - E)P\Psi = 0 . \tag{25.7}$$

The most significant part of (25.7) is the optical potential V_{op} given by

$$V_{\text{op}} = PHQ(E - QHQ)^{-1} QHP . \tag{25.8}$$

V_{op} is a nonlocal potential; the most incisive way to give it meaning is to the define the QHQ problem.

25.2.2 Expansion of V_{op}: The QHQ Problem

The following eigenvalue problem constitutes the heart of the projection operator formalism:

$$QHQ\Phi_n = \mathcal{E}_n \Phi_n . \tag{25.9}$$

For calculational purposes, it is best to recast (25.9) in the variational form

$$\delta\left(\frac{\langle \Phi QHQ\Phi \rangle}{\langle \Phi Q\Phi \rangle}\right) = 0 . \tag{25.10}$$

This equation may yield a discrete plus a continuous spectrum in an energy domain where the SE has only a continuous spectrum. Moreover, if the Q operator is appropriately chosen, then the discrete eigenvalues $\mathcal{E}_n$ are close to the desired class of many-body resonances, called variously Feshbach resonances, core-excited [25.3], or doubly-excited states. In terms of these solutions, V_{op} has the expansion

$$V_{\text{op}} = \sum\!\!\!\!\!\!\int PHQ|\Phi_n\rangle(E - \mathcal{E}_n)^{-1}\langle\Phi_n|QHP . \tag{25.11}$$

25.3 Forms of P and Q

25.3.1 The Feshbach Form

Projection operators are not unique. *Feshbach* [25.1] has sketched a derivation of a "robust" projection operator for the general N-electron target system. Robust means that $Q\Phi$ is devoid of open channels. The complete expression for P, including inelastic channels has been derived in [25.5]

$$P = \sum_{i=1}^{N+1} \left[\boldsymbol{\psi}(r^{(i)})\rangle \cdot \langle \boldsymbol{\psi}(r^{(i)}) + \sum_{\lambda_\alpha}{}' \frac{\boldsymbol{v}_\alpha(r_i) \cdot \boldsymbol{\psi}(r^{(i)})\rangle\langle \boldsymbol{v}_\alpha(r_i) \cdot \boldsymbol{\psi}(r^{(i)})}{\lambda_\alpha - 1} \right], \tag{25.12}$$

where the prime on the second summation means that terms with $\lambda_\alpha = 1$ are to be omitted, r_i denotes the radial coordinate of electron i, and $r^{(i)}$ stands for the angular and spin coordinates of electron i plus all coordinates of the remaining N electrons. Thus $r^{(i)}$ indicates the totality of all coordinates of the $(N+1)$ electrons *except* the radial coordinate of the ith electron r_i. $\boldsymbol{\psi}(r^{(i)})$ is the vector channel wave function in which the angular momentum and spin of the electron i are coupled to the target in state ν ($\nu = 0, \ldots, \nu_{\max}$). A component of a vector labels the inelastic channel, and dot products represent sums over channels. The Q operator is then made explicit by completeness: $Q = 1 - P$, (25.4).

The α-indexed quantities in (25.12) arise from exchange; they are the eigensolutions of the integral eigenvalue problem

$$\boldsymbol{v}_\alpha(r_i) = \lambda_\alpha \langle \mathbf{K}(r_i|r_j) \cdot \boldsymbol{v}_\alpha(r_j)\rangle_{r_j} . \tag{25.13}$$

Here, $\mathbf{K}(r_i|r_j)$ is a matrix with components

$$K_{\mu\nu}(r_i|r_j) \propto N\big\langle \psi_\mu(r^{(i)})\psi_\nu(r^{(j)})\big\rangle_{r^{(ij)}} , \tag{25.14}$$

and $r^{(ij)}$ indicates that all variables, except r_i and r_j are integrated over. In the inelastic regime, therefore, the $[\boldsymbol{v}_\alpha, \lambda_\alpha]$ are not associated with specific channels, but rather with the totality of open channels. This means that every component of $\boldsymbol{v}_\alpha$ is associated with all inelastic channels [25.5]. The $\boldsymbol{v}_\alpha$ are orthogonal, and can be normalized so that $\langle \boldsymbol{v}_\alpha \cdot \boldsymbol{v}_\beta\rangle = \delta_{\alpha\beta}$. The λ_α obey several sum rules [25.4], of which the most useful is

$$\sum_{\alpha=1}^{n_\lambda} (\lambda_\alpha)^{-1} = \sum_{\nu=0}^{\nu_{\max}} \langle K_{\nu\nu}(r|r)\rangle_r , \tag{25.15}$$

where n_λ is the number of eigenvalues of (25.13). The v_α can be accurately calculated by use of a variational principle [25.4]. A test of (25.15) for the lowest $He^-(^2S)$ resonance, using Hylleraas functions to construct Q in the evaluation of QHQ, is shown in Table 25.1 [25.4], and results for the resonance position are compared in Table 25.2.

Table 25.1 Test of sum rule (25.15) for the lowest He^- $(1s2s^2\,{}^2S)$ autodetachment state [25.4]. Projection operators are based on a 4 term Hylleraas ϕ_0 and the variational form of v_α given below. Values of other constants are given in [25.4]

$$\phi_0 \propto (1 + C_1 r_1 + C_2 r_2 + C_{12} r_{12}) \exp(-\gamma_1 r_1 - \gamma_2 r_2) + (r_1 \leftrightarrow r_2) ,$$

$$v_\alpha \propto \left(c_{11}^{(\alpha)} + c_{12}^{(\alpha)} r\right) \exp(-\gamma_1 r) + \left(c_{21}^{(\alpha)} + c_{22}^{(\alpha)} r\right) \exp(-\gamma_2 r) ,$$

and the variational eigenvalues are [25.4]

$$\lambda_1 = 1.009\,453 \quad \lambda_2 = 232.8540$$
$$\lambda_3 = 80\,101.08 \quad \lambda_4 = 4\,817\,341$$

Summation	Value (Ry^{-1})
$\sum_{\alpha=1}^{4} (\lambda_\alpha)^{-1}$	0.994 9425
$\langle K(r\|r)\rangle_r$	0.994 9514

Table 25.2 Comparison of methods for calculating the energy of the lowest He^- $(1s2s^2\,{}^2S)$ autodetachment state. The QHQ results are denoted $(\hat{\mathcal{E}} - E_0)_{\text{Quasi}}$ for the quasi-projection method and $(\mathcal{E} - E_0)_{\text{Complete}}$ for the complete projection method [25.4]. The entries labeled "Other results" give the full resonant energy. Units are eV

Target	$(\hat{\mathcal{E}} - E_0)_{\text{Quasi}}$	$(\mathcal{E} - E_0)^{a}_{\text{Complete}}$
Closed shell	19.366	19.593
Open shell	19.385	19.666
$\|1s1s'\| + 2p^2$	19.388	19.615
4 term Hylleraas	19.381	19.496
10 term Hylleraas	19.379	19.504
Other results 19.402[b], 19.376[c], 19.367[d], 19.367 ± 0.007[e]		

[a] Values obtained using $R_\infty = 13.605\,698$ eV and $E_0 = -79.0151$ eV [25.6]
[b] Complex rotation method; *Junker* and *Huang* [25.7]
[c] Hole-projection complex-rotation; *Davis* and *Chung* [25.8]
[d] Hermitian-representation complex-rotation; [25.9]
[e] Experiment; *Brunt* et al. [25.10]

25.3.2 Reduction for the $N = 1$ Target

Explicit rigorous P and Q operators of the above type are only possible for $N = 1$ (i.e. hydrogenic) targets. In that case, spin can be easily eliminated by using spatially symmetric or antisymmetric wave functions. In the elastic region, P and Q reduce to [25.11, 12]

$$P = P_1 + P_2 - P_1 P_2 ; \quad Q = 1 - P_1 - P_2 + P_1 P_2 . \tag{25.16}$$

Here the $P_i = \phi(r_i)\rangle\langle\phi(r_i)$ are purely spatial projectors. Forms for the inelastic continuum are easily generalized [25.13].

There have been many calculations of QHQ for two-electron systems ($N = 1$), starting with fundamental work of *O'Malley* and *Geltman* [25.13]. A small sample is given in Table 25.3. They are given for their historical importance, demonstrating for the first time the convergence of eigenvalues to well defined values in the continuum. All eigenvalues (below the $n = 2$ threshold, in this case) correspond to resonances.

Table 25.3 Energies $\mathcal{E}_s$ of the He(2s2p ^{1}P^0) autoionization states below He$^+$ ($n = 2$) threshold from the variational calculations of *O'Malley* and *Geltman* [25.13]. Units are Ry. N is the number of terms in the trial function

N	$s = 1$	$s = 2$	$s = 3$
9	−1.377 08	−1.178 92	−1.097 16
15	−1.380 44	−1.183 12	−1.104 32
20	−1.382 16	−1.183 48	−1.108 28
25	−1.383 16	−1.190 00	−1.111 88

25.3.3 Alternative Projection and Projection-Like Operators

Two alternative methods based on the idea of projection are available: quasi-projectors and hole projection operators.

Quasi-projectors [25.14] relax the condition of idempotency, but still maintain a discrete spectrum, which is in a one-to-one correspondence with resonances, with a predeterminable number of exceptions [25.2].

Hole projection operators have proven to be a more practical and effective approach [25.15]. The method uses one-particle (say, hydrogenic) orbitals, $\phi_n(q;\boldsymbol{r})$, to build holes via projectors, $[1 - \sum_n \phi_n\rangle\langle\phi_n]$, operating on the $(N+1)$-particle wave function. The Rayleigh–Ritz functional is minimized with respect to the parameters in Φ, but it is maximized with respect to q, the nonlinear parameter in all the ϕ_n. In a model case, this *minimax* procedure has been shown [25.15] to optimize the eigenvalues to describe resonance energies; many calculations since then [25.16] have verified the minimax criteria in many-electron systems. More recently, the technique has been combined with complex rotation, so as to enable calculation of other resonant quantities [25.17]. Remarkably accurate results have been obtained.

Finally, hole projectors are ideally suited for inner-shell vacancy states of many-electron systems if high accuracy is required [25.2]. Effectively, this amounts to a reliable method for optimizing parameters of a hole orbital to be used in an Auger transition integral for filling such a vacancy, although that method has apparently never been used (Chapt. 62).

25.4 Width, Shift, and Shape Parameter

25.4.1 Width and Shift

Here one requires $P\Psi$ as well as $Q\Psi$. The former is obtained from a "nonresonant continuum," defined as the scattering solution of

$$\left[PHP + V_s^{(\mathrm{nr})} - E\right] P\Psi_s^{(\mathrm{nr})} = 0 , \tag{25.17}$$

where the nonresonant potential,

$$V_s^{(\mathrm{nr})} = V_{\mathrm{op}} - PHQ\Phi_s\rangle(E - \mathcal{E}_s)^{-1}\langle\Phi_s QHP , \tag{25.18}$$

excludes the resonant state s from the optical potential. In terms of $P\Psi_s^{(\mathrm{nr})}$, whose phase shift, η_0, is smooth in the vicinity of $E \approx \mathcal{E}_s$, a solution of the complete problem, (25.7), can be constructed [25.13] with a phase shift $\eta_0 + \eta_{\mathrm{r}}$, where the additional phase shift

$$\eta_{\mathrm{r}} = \arctan\left(\frac{\Gamma/2}{(\mathcal{E}_s + \Delta_s) - E}\right) \tag{25.19}$$

exhibits typical resonant behavior ($0 < \eta_{\mathrm{r}} < \pi$). From (25.19) it is clear that the "true" position of the resonance is

$$E_s(E) = \mathcal{E}_s + \Delta_s(E) . \tag{25.20}$$

The width Γ_s and shift Δ_s are given by [25.13]:

$$\Gamma_s(E) = 2k|\langle\Psi_s^{(\mathrm{nr})}(E)\, PHQ\, \Phi_s\rangle|^2 , \tag{25.21}$$

$$\Delta_s(E) = \langle\Phi_s\, QHP\, G_P(E)\, PHQ\Phi_s\rangle , \tag{25.22}$$

where k is the scattering momentum (i. e. $k^2 = E - E_0$), and G_P is the Green's function associated with (25.17); G_P can be simplified from the form given in [25.13] [(2.28) of first article of [25.2]].

Equation (25.20) is an implicit equation for the energy at which the resonance occurs. It can be solved graphically [25.18], and that energy defines the Feshbach resonant energy E_F, which differs (very slightly) from the Breit–Wigner energy (Sect. 25.4.3 and Fig. 25.1).

25.4.2 Shape Parameter

The shape of an isolated radiative transition between an autoionization state and some other state can be described by Γ, E, and an additional parameter q_s, often called the shape parameter [25.19]. The ratio of transition probabality (in, say, absorption from an initial state labeled $i\rangle$) through the resonant state to its nonresonant value, parametrized in its Fano form [25.19] on the left-hand side of (25.23), can be equated to its meaning in Feshbach terms on the right-hand side of (25.23):

$$\frac{(e_s+q_s)^2}{1+e_s^2} = \frac{|\langle P\Psi + Q\Psi|T|i\rangle|^2}{|\langle P\Psi_s^{(\mathrm{nr})} + Q\Psi_s^{(\mathrm{nr})}|T|i\rangle|^2}\,, \tag{25.23}$$

where e_s is the scaled energy

$$e_s = (E - E_s)/(\Gamma_s/2) \tag{25.24}$$

and T is a radiative transition operator (25.25).

To analyze (25.23) in the Feshbach formalism, the key point [25.13] is to recognize that the bras on the right-hand side must include P as well as Q parts of the respective wave functions, as is indicated in (25.23). That is because the T operator is a perturbation and not part of the dynamical problem. With T in length form

$$T \propto \sum_{j=1}^{N+1} z_j\,, \tag{25.25}$$

the rhs of (25.23) can be calculated by noting that $P\Psi_s$ and $P\Psi_s^{(\mathrm{nr})}$ can be, in principle, determined from (25.7) and (25.17). $Q\Psi_s$ is then derived from $P\Psi_s$ using (25.6). But $Q\Psi_s^{(\mathrm{nr})}$ excludes the sth term from the right-hand side of (25.6):

$$Q\Psi_s^{(\mathrm{nr})} = \sum_{n\neq s} \frac{\Phi_n\rangle\langle\Phi_n QHP\Psi^{(\mathrm{nr})}\rangle}{E - \mathcal{E}_n}\,. \tag{25.26}$$

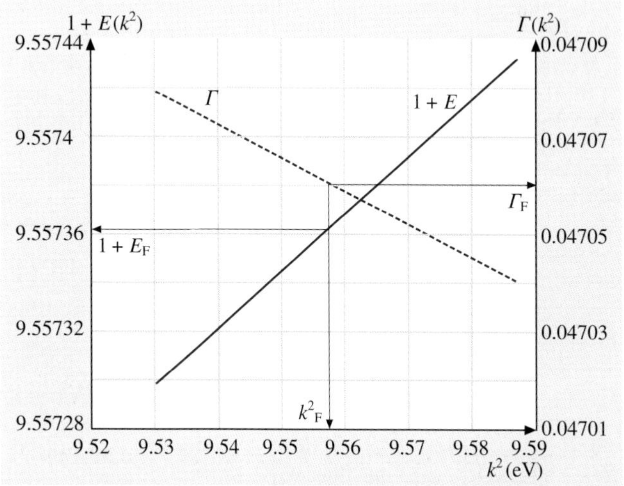

Fig. 25.1 Precision calculation of $H^-(^1S)$ resonance. *Solid curve*: $E+1$ vs. k^2, where $E = \mathcal{E} + \Delta$, (25.20). k_F^2 is that value of k^2 at which $E+1 = k^2$. *Dashed curve* is Γ vs. k^2 from (25.21), and $\Gamma_F = \Gamma\left(k_F^2\right)$. *Curves* are from calculations of [25.18]; results are $E_F = 9.557\,35$ eV and $\Gamma_F = 0.047\,0605$ eV. Applying corrections, (25.31) gives finally $E_{BW} = E_F + O(10^{-6})$, $\Gamma_{BW} = 0.047\,17$ eV

With use of these relations in the rhs of (25.23) an explicit formula for q_s was derived in [25.2]; it is given by

$$q_s = \frac{\langle\Phi_s\tilde{Q}|T|i\rangle}{k\langle i|T|\Psi_s^{(\mathrm{nr})}\rangle\langle\Psi_s^{(\mathrm{nr})}|PHQ\Phi_s\rangle}\,, \tag{25.27}$$

where

$$\tilde{Q} = Q + QHPG_s + \sum_{n\neq s} \frac{QHPG_sHQ|\Phi_n\rangle\langle\Phi_n}{E - \mathcal{E}_n}\,. \tag{25.28}$$

The Green's function G_s in (25.28) is the one associated with (25.17). It can be expanded in terms of the eigensolutions of (25.17), but its spectrum may have a discrete as well as continuous part, in which case,

$$G_s = \sum_\nu{}' \frac{P\Psi_\nu^{(\mathrm{nr})}\rangle\langle P\Psi_\nu^{(\mathrm{nr})}}{E_s - E_\nu} + \frac{\wp}{\pi}\int \frac{P\Psi_s^{(\mathrm{nr})}(E')\rangle\langle P\Psi_s^{(\mathrm{nr})}(E')\sqrt{E'}\,\mathrm{d}E'}{E_s - E'}\,. \tag{25.29}$$

The sum over ν refers to the discrete part of the spectrum of (25.17) (if there is one), and $\wp$ denotes a principal value integral over the continuum solutions. The latter are always assumed to be normalized as

Table 25.4 Comparison of high precision calculations with experiment for the resonance parameters of the He(^{1}P^0) resonances below the $n = 2$ threshold. For photoabsorption, the appropriate Rydberg constant is $R_M = R_\infty/(1 + m/M)$ [25.20]. The value used here is $R_M = 13.603\,833$ eV, and $E_0 = -79.0151$ eV [25.6]

Quantity	Units	Calculation [a] $s = 1$	$s = 2$	Experiment $s = 1$	$s = 2$
$\mathcal{E}_s$	Ry	$-1.385\,7895$	$-1.194\,182$		
Δ	eV	$-0.007\,13$	$0.000\,6202$		
$E_s - E_0$	eV	60.1444	62.7587	60.133 ± 0.015 [b] 60.151 ± 0.0103 [c]	62.756 ± 0.01 [b]
Γ	eV	0.0369 [d]	$0.000\,1165$	0.038 ± 0.004 [b] 0.038 ± 0.002 [c]	
q_s		-2.849 [e]	-4.606 [e]	-2.80 ± 0.25 [b] -2.55 ± 0.16 [c]	

[a] *Bhatia* and *Temkin* [25.20], except as noted
[b] *Madden* and *Codling* [25.21]
[c] *Morgan* and *Ederer* [25.22]
[d] *Bhatia*, *Burke*, and *Temkin* [25.23]
[e] *Bhatia* and *Temkin* [25.24]

plane waves (not energy normalized) throughout this chapter.

Equation (25.27) is a nontrivial example of what can be done with the projection operator technique. Not only does it allow very accurate calculations ([25.17, 20] and footnote e of Table 25.4), but it provides a theoretical incisiveness which far exceeds all previous resonance formalisms.

A formula for the resonant scattering cross section can be derived which is of the same form as the left-hand side of (25.23); however, in that case, the parameter corresponding to q_s is related to the nonresonant phase shift [25.25], and has no quantitative relationship to the above shape parameter (q_s).

25.4.3 Relation to Breit–Wigner Parameters

Inferring resonance parameters from experimental data is generally done by fitting to resonance formulae in which the resonance parameters are assumed to be energy independent. A phase shift for example would be inferred by assuming

$$\eta(E) = \delta_0(E) + \arctan\left(\frac{\Gamma_{\mathrm{BW}}/2}{E_{\mathrm{BW}} - E}\right) . \tag{25.30}$$

The relation between the above Breit–Wigner parameters and those of the Feshbach theory has been derived in lowest order by *Drachman* [25.26]:

$$\begin{aligned} E_{\mathrm{BW}} &= E_{\mathrm{F}} - (1/4)\Gamma_{\mathrm{F}}(\mathrm{d}\Gamma_{\mathrm{F}}/\mathrm{d}E)_{E=E_{\mathrm{F}}} , \\ \Gamma_{\mathrm{BW}} &= \Gamma_{\mathrm{F}}(1 + \mathrm{d}\Delta_{\mathrm{F}}/\mathrm{d}E)_{E=E_{\mathrm{F}}} , \\ \delta_0(E) &= \eta_0(E) - (1/2)\mathrm{d}\Gamma_{\mathrm{F}}/\mathrm{d}E . \end{aligned} \tag{25.31}$$

In only one precision calculation (for the lowest ^{1}S resonance in electron–hydrogen scattering) have these differences, thus far, been evaluated [25.18]. A précis is given in Fig. 25.1.

25.5 Other Calculational Methods

We now briefly review two calculational methods used for basic applications in autoionization of few body systems: (a) complex rotation and (b) a pseudopotential method.

25.5.1 Complex Rotation Method

Complex rotation, which is based on a theorem of *Balslev* and *Combes* [25.27], has been extensively applied with great accuracy. Two additional basic systems to be mentioned here are H^- and Ps^- (Ps = positronium). In complex rotation the particle distances are multiplied by a common phase factor

$$r_i \to r_i\,e^{i\theta}\,. \tag{25.32}$$

Under this replacement the Hamiltonian undergoes the transformation

$$H \to T\,e^{-2i\theta} + V\,e^{-i\theta} \tag{25.33}$$

(only Coulomb interactions are assumed). A real variational wave function Φ is used (for the applications here, they are of Hylleraas form, multiplied by rotational harmonics of symmetric Euler angles of the desired angular momentum, parity, and spin [25.31]). The functional

$$[E] = \frac{\langle\Phi|H|\Phi\rangle}{\langle\Phi|\Phi\rangle} = \frac{\langle\Phi|T|\Phi\rangle\,e^{-2i\theta} + \langle\Phi|V|\Phi\rangle\,e^{-i\theta}}{\langle\Phi|\Phi\rangle} \tag{25.34}$$

is then evaluated. Minimization of (25.34) with respect to the linear parameters, for a given value of θ, is carried out in the usual way, but by virtue of the complex dependence on rotation angle the matrix elements H_{ij} in the matrix eigenvalue equation

$$\det|H_{ij}(\theta) - E\Delta_{ij}| = 0 \tag{25.35}$$

are complex. Thus the solution of (25.35) gives rise to complex eigenvalues $E_\lambda = E_\lambda(\theta)$. For a given λ, the optimum θ is the one for which $E_\lambda(\theta)$ is effectively stationary as a function of θ [25.29]. Note that no projection operators are used: the real part of E_λ corresponds to the Breit–Wigner (i. e., experimental) position of the resonance, and $\mathrm{Im}(E_\lambda) = \Gamma_{BW}/2$, where Γ_{BW} corresponds to the Breit–Wigner width of the resonance. These parameters thus include the full Feshbach values plus corrections (25.31).

Using Hylleraas wave functions with up to 1230 terms in the complex rotation method, resonance parameters have been obtained for resonance states of H^- below the $n = 2$ and 3 thresholds of H which compare very well with those obtained using the projection-operator, R-matrix and close-coupling methods. Results for the $^1D^e$ states of H^- are given in the Table 25.5. Similar calculations for Ps^- have been carried out [25.32]. The complex rotation method has been applied to the autoionization states of many different systems including muonic systems [25.33], as well as to study the combined effect of electric field and spin-orbit interaction on resonance parameters [25.34].

25.5.2 Pseudopotential Method

The second method that is included in this section is done so for the reason that it represents a rather different idea for the calculation of autoionization rather than being a more elaborate application of methodologies that are already known, with results too numerous to be referenced here. The method, described as a pseudopotential approach, was introduced by *Martin* et al. [25.35]. An effective Hamiltonian H_{eff} is defined as

$$H_{eff} = H + MP\,, \tag{25.36}$$

where M is a scalar parameter (i. e., a number), which will be taken to be very large, multiplying the P operator, (25.16). [Applications have thus far been restricted to one-electron targets and resonances below $n = 3$ excited state.] In practice, one minimizes the expectation value of H_{eff}, i. e.,

$$\delta\left(\frac{\langle\Psi_v|H_{eff}|\Psi_v\rangle}{\langle\Psi_v\Psi_v\rangle}\right) = 0\,, \tag{25.37}$$

using an arbitrary, quadratically integrable, variational function Ψ_v.

In order to understand the nature of the spectrum that arises from this variation, we imagine Ψ_v divided into its P and Q space components:

$$\Psi_v = Q\Psi_v + P\Psi_v = \Psi_v^Q + \Psi_v^P\,. \tag{25.38}$$

Table 25.5 Comparison of resonance parameters (in eV) obtained from different methods for calculating $^1D^e$ states in H^-

Threshold n	Complex-cordinate rotation [25.29]		*R*-matrix [25.28]		Feshbach projection [25.20] (25.31)	
	E	Γ	E	Γ	E_F	Γ_F
2	10.124 36	0.008 62	10.1252	0.008 81	10.1244	0.010
3	11.811 02	0.045 12	11.810 97 [a]	0.044 49 [a]		

[a] Close coupling (18-state), [25.30]

The expectation value $\langle\Psi_\mathrm{v}|H_\mathrm{eff}|\Psi_\mathrm{v}\rangle$ is written in matrix form as

$$\langle\Psi_\mathrm{v}|H_\mathrm{eff}|\Psi_\mathrm{v}\rangle = \left\langle \left(\Psi_\mathrm{v}^Q\ \Psi_\mathrm{v}^P\right) \begin{pmatrix} H_{QQ} & H_{QP} \\ H_{PQ} & H_{PP}+M \end{pmatrix} \begin{pmatrix} \Psi_\mathrm{v}^Q \\ \Psi_\mathrm{v}^P \end{pmatrix} \right\rangle . \tag{25.39}$$

The eigenvalue problem resulting from (25.37) reduces to finding the eigenvalues of the determinant

$$\det \begin{pmatrix} \langle H_{QQ}\rangle - \lambda & \langle H_{QP}\rangle \\ \langle H_{PQ}\rangle & \langle H_{PP}\rangle + M - \lambda \end{pmatrix} = 0 . \tag{25.40}$$

Note that only the bottom right component contains the term M. As a result, in the limit of large M, the eigenvalues, which can readily be solved for from (25.39), are

$$\lim_{M\to\mathrm{large}} \lambda = \begin{cases} M + \langle H_{PP}\rangle \\ \langle H_{QQ}\rangle \end{cases} . \tag{25.41}$$

The lower eigenvalue is the desired Feshbach resonant energy $\mathcal{E}_\mathrm{F} = \langle H_{QQ}\rangle$. The width is calculated from [with our normalization, (25.21)]

$$\Gamma = 2k|\langle\Psi_\lambda|H_\mathrm{eff}|\chi_\mathrm{E}\rangle|^2 \tag{25.42}$$

where χ_E is the solution of the exchange approximation

$$(H_{PP} - E)\chi_\mathrm{E} = 0 \tag{25.43}$$

It is emphasized that this method only calculates the Feshbach energy; thus the shifts are not included. On the other hand the method uses no projection operators in calculating the matrix elements of H, and only the matrix elements of P by itself occur. This is much easier than a standard QHQ calculation (Sect. 25.2.2).

In practice, the matrix in (25.39) will expand to an $N\times N$ matrix, where N is the number of linear parameters in Ψ_v, and (if one uses a Hylleraas form of Ψ_v, for example) the matrix in (25.39) will not overtly divide itself into the simple form of this heuristic exposition pictured in (25.39) or (25.40). Nevertheless, the conclusion holds; in detail, the eigenvalue spectrum will span a range of values with those below the threshold, appropriate to the P operator being used, corresponding to real resonances, and the largest eigenvalue will approach the value of M used in the specific calculation.

A sample of results for the He(^{1}P) resonances below the $n=2$ threshold of He^+ taken from [25.35], with some comparisons, is given in Table 25.6. Note that the value of $\mathcal{E}_\mathrm{F}$ of the second resonance in the *Martin* et al. [25.35] calculation is lower than the rigorous QHQ calculation [25.20]. It is believed that this may be due to the residual M dependence of H_eff.

Table 25.6 Resonance energies $\mathcal{E}_\mathrm{F}$ (Ry) and widths (eV) for ^{1}P states of He below $n=2$ threshold (-1 Ry) of He^+

State	*Martin* et al. [25.35]		*Lipsky* and *Conneely* [25.36]		*Bhatia* and *Temkin* [25.20, 24]	
	Position	Width	Position	Width	Position	Width
1	−1.384 00	0.0382	−1.376 72	0.0341	−1.385 79	0.0363
2	−1.194 60	0.000 146	−1.193 12	0.000 131	−1.194 18	0.000 106
3	−1.127 52	0.000 860	−1.125 84	0.007 27	−1.127 72	0.0090

25.6 Related Topics

This chapter is necessarily of limited scope. Within the projection operator formalism, overlapping resonance theory [25.37] is discussed in Sect. 47.1.3. Recent calculations [25.38] have shown that such effects, when present, can induce significant alteration of isolated resonance results. Other prominent items not included are stabilization methods [25.39] and hyperspherical coordinate methods [25.40, 41]. The latter methods have the appealing property of presenting energies as a function of the hyperradius, $R = \left(\sum_i r_i^2\right)^{1/2}$, which look like potential energy curves of diatomic molecules as a function of the intermolecular separation, which is also usually denoted by R. The molecular structure analogy has also been used to uncover additional (approximate) symmetries with corresponding quantum labels [25.40, 41]. They thus have a global character not present in the foregoing methods. On a purely quantitative level, however, they are not generally as accurate as methods based on the projection operator or complex rotation formalism.

There are many other areas in which autoionization can play an important role, such as satellite line formation [25.42], inner-shell ionization [25.43], to mention only a few (Chapt. 62 and [25.44]). In addition, significant application of the phenomena associated with autoionization to diagnostics of astrophysical [25.45] and fusion [25.46] plasmas, for example, shows that autoionization has considerable applied utility.

References

25.1 H. Feshbach: Ann. Phys. (N. Y.) **19**, 287 (1962)

25.2 For a more complete review of much of this material: A. Temkin, A. K. Bhatia: *Autoionization, Recent Developments and Applications*, ed. by A. Temkin (Plenum, New York, 1985), pp. 1 ff, 35 ff

25.3 H. S. Taylor: Adv. Chem. Phys. **18**, 91 (1970)

25.4 A. Berk, A. Temkin: Phys. Rev. A **32**, 3196 (1985)

25.5 A. Temkin, A. K. Bhatia: Phys. Rev. A **31**, 1259 (1985)

25.6 C. L. Pekeris: Phys. Rev. **112**, 1649 (1958)

25.7 B. R. Junker, C. L. Huang: Phys. Rev. A **18**, 313 (1978)

25.8 B. F. Davis, K. Chung: Phys. Rev. A **29**, 313, 2437 (1978)

25.9 M. Bylicki: J. Phys. B **24**, 413 (1991)

25.10 J. Brunt, G. King, F. Read: J. Phys. B **10**, 433 (1977)

25.11 Y. Hahn, T. F. O'Malley, L. Spruch: Phys. Rev. **128**, 932 (1962)

25.12 Y. Hahn: Ann. Phys. (N. Y.) **58**, 137 (1960)

25.13 T. F. O'Malley, S. Geltman: Phys. Rev. A **137**, 1344 (1965)

25.14 A. Temkin, A. K. Bhatia, J. N. Bardsley: Phys. Rev. A **7**, 1633 (1972)

25.15 K. T. Chung: Phys. Rev. A **20**, 1743 (1979)

25.16 K. T. Chung, B. F. Davis: Hole-projection method for calculating Feshbach resonances and inner shell vacancies. In: *Autoionization, Recent Developments and Applications*, ed. by A. Temkin (Plenum, New York 1985) p. 73

25.17 K. T. Chung, B. F. Davis: Phys. Rev. A **26**, 3278 (1982)

25.18 Y.-K. Ho, A. K. Bhatia, A. Temkin: Phys. Rev. A **15**, 1432 (1977)

25.19 U. Fano: Phys. Rev. **124**, 1866 (1960)

25.20 A. K. Bhatia, A. Temkin: Phys. Rev. A **11**, 2018 (1975)

25.21 R. P. Madden, K. Codling: Astrophs. J. **141**, 364 (1965)

25.22 H. Morgan, D. Ederer: Phys. Rev. A **29**, 1901 (1984)

25.23 A. K. Bhatia, P. G. Burke, A. Temkin: Phys. Rev. A **8**, 21 (1973)

25.24 A. K. Bhatia, A. Temkin: Phys. Rev. A **29**, 1895 (1984)

25.25 K. Smith: *The Calculation of Atomic Collision Processes* (Wiley-Interscience, New York 1971) p. 47

25.26 R. J. Drachman: Phys. Rev. A **15**, 1432 (1977)

25.27 E. B. Balslev, J. W. Combes: Comm. Math. Phys. **22**, 280 (1971)

25.28 T. Scholz, P. Scott, P. G. Burke: J. Phys. B **21**, L139 (1988)

25.29 A. K. Bhatia, Y. K. Ho: Phys. Rev. A **41**, 504 (1990)

25.30 J. Callaway: Phys. Rev. A **26**, 199 (1982)

25.31 A. K. Bhatia, A. Temkin: Rev. Mod. Phys. **36**, 1050 (1964)

25.32 A. K. Bhatia, Y. K. Ho: Phys. Rev. A **42**, 1119 (1990)

25.33 C.-Y. Hu, A. K. Bhatia: Muon Catalyzed Fusion **5/6**, 439 (1990/1991)

25.34 I. A. Ivanov, Y. K. Ho: Phys. Rev. A **68**, 033410 (2003)

25.35 F. Martin, O. Mo, A. Riera, M. Yañez: Europhys. Lett. **4**, 799 (1987)

25.36 L. Lipsky, M. J. Conneely: Phys. Rev. A **14**, 2193 (1976)

25.37 H. Feshbach: Ann. Phys. (N. Y.) **43**, 410 (1967)

25.38 D. C. Griffin, M. S. Pindzola, F. Robicheaux, T. W. Gorczyca, N. R. Badnell: Phys. Rev. Lett. **72**, 3491 (1994)

25.39 V. A. Mandelshtam, T. R. Ravuri, H. S. Taylor: J. Chem. Phys. **101**, 8792 (1994)

25.40 C.-D. Lin (Ed.): Classifications and properties of doubly excited states of atoms. In: *Review of Fundamental Processes and Applications of Atoms and Ions* (World Scientific, Singapore 1993) p. 24

25.41 D. R. Herrick, O. Sinanoglu: Phys. A **11**, 97 (1975)

25.42 F. Bely-Debau, A. H. Gabriel, S. Valonte: Mon. Not. Roy. Astron. Soc. **186**, 305 (1979)

25.43 C. J. Powell: Inner shell ionization cross sections. In: *Electron Impact Ionization*, ed. by T. D. Mark, G. H. Dunn (Springer, Berlin, Heidelberg 1985) Chap. 6

25.44 R. D. Cowan: *The Theory of Atomic Structure and Spectra* (Univ. of California Press, Berkeley 1981)

25.45 G. Doschek: Diagnostics of solar and astrophysical plasmas dependent on autoionization. In: *Autoionization, Recent Developments and Applications*, ed. by A. Temkin, A. K. Bhatia (Plenum, New York 1985) p. 171, op. cit. in [25.2]

25.46 M. Finkenthal: Atomic processes responsible for XUV emisssion. In: *AIP Conference Proceedings # 206; Atomic Processes in Plasmas, 1990*, ed. by Y.-K. Kim, R. E. Elton (American Institute of Physics, New York 1990) p. 95

26. Green's Functions of Field Theory

26.1 **The Two-Point Green's Function** 402

26.2 **The Four-Point Green's Function** 405

26.3 **Radiative Transitions** 406

26.4 **Radiative Corrections** 408

References 411

The discussion in this chapter is restricted to Green's function techniques as applied to problems in atomic physics, specifically to the calculation of higher order (correlation, Breit, as well as radiative) corrections to energy levels, and also of transition amplitudes for radiative transitions of atoms which are gauge invariant (GI) at every level of approximation.

Green's function techniques were first applied to many-electron atoms in 1971 as specific instances of the use of field theory techniques in many-particle problems [26.1, 2]. They initially provided alternative derivations of known approximations such as the Hartree–Fock (HF) approximation and random phase approximation (RPA). The starting point for the derivations was a nonrelativistic (NR) field-theoretical effective Hamiltonian for the system, which involved the nucleus–electron potential and only Coulomb interactions between electrons.

It subsequently became apparent that it was possible to formulate the problem more generally, with the full quantum electrodynamic (QED) Hamiltonian as a starting point [26.3, 4]. Thus, the theory contains both relativity, and virtual and real transverse photons, as well as the Coulomb interaction between the electrons. The relativistic (R) approximate equations, such as the Dirac–Fock (DF) and relativistic random phase approximation (RRPA) are the natural outcomes of the formalism, and the NR results can be viewed as approximations to these R cases. Moreover, the Green's function approach (GFA) now provides a means of carrying out programs involving systematic approximations of successively higher and higher accuracy. The GFA provides a framework which allows one to make corrections to results obtained in the DF approximation or in the coupled cluster approximation (CCA) and many-body perturbation theory (MBPT), including magnetic (Breit type) interactions [26.5, 6]. It ensures that there is neither double counting nor omission of contributions. For radiative transitions, the formalism allows for a systematic treatment of such effects which is gauge independent that at any given level of approximation [26.7–10]. (It should be noted that there is a subtle difference between gauge invariance and gauge independence. The first refers to the transition amplitude and the second to quantities which are directly observable experimentally.) Finally, the GFA is numerically implementable. Renormalization can be carried out for radiative corrections, resulting in finite and calculable expressions [26.11–13]. The integro-differential equations (i. e., the Dyson equations) needed to calculate energies or transition amplitudes in nontrivial approximations are also soluble [26.8, 9, 14].

In quantum field theory, Green's functions are defined in terms of vacuum expectation values of products of field operators. While this restriction can be relaxed, expectation values must still be taken for a nondegenerate state. As a practical matter, this requirement ultimately restricts one to consider atoms with electron numbers associated only with closed shells or subshells, and those with closed shells or subshells plus or minus one or two electrons. The corresponding Green's functions considered here are the two- and four-point functions for energy levels, and the three- and five-point functions for transition amplitudes (leading to oscillator strengths). The restriction in electron numbers is clearly not a severe one. It allows one to cover many atomic species.

Starting from relativistic QED, the electron field operator $\psi(\boldsymbol{r}, t)$ written in the Heisenberg picture, satisfies equal-time anticommutation relations

$$\left[\psi(\boldsymbol{r}, t), \psi(\boldsymbol{r}', t)\right] = \left[\psi^\dagger(\boldsymbol{r}, t), \psi^\dagger(\boldsymbol{r}', t)\right] = 0\,, \quad (26.1)$$

$$\left[\psi(\boldsymbol{r}, t), \psi^\dagger(\boldsymbol{r}', t)\right] = \delta^3(\boldsymbol{r} - \boldsymbol{r}')\,. \quad (26.2)$$

(Spinor labels are suppressed.)

The time-translation operator, acting on any Heisenberg operator $\mathcal{O}(\boldsymbol{r}, t) = \{\psi(\boldsymbol{r}, t), \psi^\dagger(\boldsymbol{r}, t), j_\mu(\boldsymbol{r}, t), \ldots\}$, is

$$\mathcal{O}(\boldsymbol{r}, t) = \mathrm{e}^{\mathrm{i}\mathcal{H}t}\mathcal{O}(\boldsymbol{r}, 0)\,\mathrm{e}^{-\mathrm{i}\mathcal{H}t}, \tag{26.3}$$

where $\mathcal{H}$ is the full QED Hamiltonian.

Notations and Definitions

For brevity, plainface numbers are used to denote both a coordinate and time, while boldface numbers denote a coordinate vector alone. For example,

$$\begin{aligned} 1 &\equiv (\mathbf{1}, t_1) \equiv (\boldsymbol{r}_1, t_1)\,, \\ \mathrm{d}^3\mathbf{1} &\equiv \mathrm{d}^3\boldsymbol{r}_1\,, \\ \mathrm{d}^4 1 &\equiv \mathrm{d}^3\boldsymbol{r}_1\,\mathrm{d}t_1\,. \end{aligned} \tag{26.4}$$

For radiative transitions, $\mathcal{M}_{fi}^{N'}(k_0)$ denotes the transition amplitude for the emission of a single photon of energy k_0 for an N'-electron atom from an initial energy $E_i^{N'}$ to a final energy $E_f^{N'}$, where

$$\mathcal{M}_{fi}^{N'}(k_0) \equiv (2\pi)^4\delta\big(E_f^{N'} - E_i^{N'} - k_0\big)M_{fi}^{N'}(k_0)\,. \tag{26.5}$$

The notation on the right-hand side of (26.5) is somewhat redundant, since k_0 is taken to be $k_0 = E_f^{N'} - E_i^{N'}$.

The current density operators at $t = 0$ are given respectively by

$$j_\mu(\boldsymbol{r}, 0) = \begin{cases} \dfrac{1}{2}\displaystyle\int \mathrm{d}^3\mathbf{1}\,\mathrm{d}^3\mathbf{2}\left[\psi^\dagger(\mathbf{1}, 0), u_\mu^{\mathrm{R}}(\mathbf{12}; \boldsymbol{r})\psi(\mathbf{2}, 0)\right], & \text{for R} \\ \displaystyle\int \mathrm{d}^3\mathbf{1}\,\mathrm{d}^3\mathbf{2}\,\psi^\dagger(\mathbf{1}, 0)u_\mu^{\mathrm{NR}}(\mathbf{12}; \boldsymbol{r})\psi(\mathbf{2}, 0)\,, & \text{for NR} \end{cases} \tag{26.6}$$

with

$$u_\mu(\mathbf{12}; \boldsymbol{r}) = \begin{cases} e\alpha_\mu\delta^3(\mathbf{1} - \boldsymbol{r})\delta^3(\mathbf{2} - \boldsymbol{r})\,, & \text{for R} \\ e\left[1, \left(\dfrac{1}{2\mathrm{i}m}\right)(\nabla_2 - \nabla_1)\delta^3(\mathbf{1} - \boldsymbol{r})\delta^3(\mathbf{2} - \boldsymbol{r})\right], & \text{for NR}\,, \end{cases} \tag{26.7}$$

where $\alpha_\mu = (\boldsymbol{\alpha}, 1)$, and the components of $\boldsymbol{\alpha}$ are the usual Dirac matrices.

The corresponding charge operators are defined as

$$\mathcal{Q} = \begin{cases} \dfrac{1}{2}e\displaystyle\int \mathrm{d}^3\mathbf{1}\left[\psi^\dagger(\mathbf{1}, 0), \psi(\mathbf{1}, 0)\right], & \text{for R} \\ e\displaystyle\int \mathrm{d}^3\mathbf{1}\,\psi^\dagger(\mathbf{1}, 0)\psi(\mathbf{1}, 0)\,, & \text{for NR}\,. \end{cases} \tag{26.8}$$

The transition amplitude for a photon of polarization four-vector $\epsilon^\mu(k)$, momentum $\boldsymbol{k}$, energy $k_0 (k_0 = |\boldsymbol{k}|)$, and photon attachment point $\boldsymbol{r}$ is

$$M_{fi}^{N'}(k_0) = \int \mathrm{d}^3\boldsymbol{r}\,\frac{\mathrm{e}^{\mathrm{i}\boldsymbol{k}\cdot\boldsymbol{r}}}{\sqrt{2k_0}}M_{fi}^{N'}(\boldsymbol{r}; k_0)\,. \tag{26.9}$$

In terms of the current density operator, $M_{fi}^{N'}(\boldsymbol{r}; k_0)$ can be written as

$$\begin{aligned} M_{fi}^{N'}(\boldsymbol{r}; k_0) &= \left\langle {}^{N'}_{f}\right| \epsilon^\mu(k) j_\mu(\boldsymbol{r}, 0) \left| {}^{N'}_{i}\right\rangle \\ &\equiv \left\langle {}^{N'}_{f}\right| j^k(\boldsymbol{r}, 0) \left| {}^{N'}_{i}\right\rangle, \end{aligned} \tag{26.10}$$

where $\left| {}^{N'}_{n}\right\rangle$ is a state of leptonic charge number N', with N' corresponding to an atom of N' electrons, with total energy $E_n^{N'}$. The term *lepton charge* is used to refer to the charge of electrons and positrons.

In the dipole approximation, $\mathrm{e}^{\mathrm{i}\boldsymbol{k}\cdot\boldsymbol{r}} \approx 1$, $\epsilon^\mu j_\mu$ contains the quantity

$$\epsilon^\mu(k)u_\mu(\mathbf{12}; \boldsymbol{r}) \equiv \lambda_k(\boldsymbol{r})\delta(\mathbf{1} - \boldsymbol{r})\delta(\mathbf{2} - \boldsymbol{r})\,, \tag{26.11}$$

where, in the radiation gauge,

$$\lambda_k(\boldsymbol{r}) = \begin{cases} e\boldsymbol{\epsilon}(k)\cdot\boldsymbol{\alpha}\,, & \text{velocity form} \\ \mathrm{i}ek_0\boldsymbol{\epsilon}(k)\cdot\boldsymbol{r}\,, & \text{length form}\,. \end{cases} \tag{26.12}$$

26.1 The Two-Point Green's Function

The two-point Green's function, or one-body propagator [26.3, 4], for a system of lepton charge Ne is defined as the expectation value of a time-ordered product

$$G^N(1, 1') \equiv -\mathrm{i}\left\langle {}^{N}_{0}\right| T\left[\psi(1)\psi^\dagger(1')\right]\left| {}^{N}_{0}\right\rangle, \tag{26.13}$$

where $\left| {}^{N}_{0}\right\rangle$ is the ground state of leptonic charge number N, with N corresponding to an atom with electron number N in a filled shell or subshell ($N = 2, 4, 8, \ldots$).

The relative time $(t_1 - t_1')$ Fourier transform of (26.13) yields the spectral representation of

$G^N(\mathbf{1},\mathbf{1}')$:

$$G^N_\omega = \sum\!\!\!\!\!\!\int_j \frac{|u^j\rangle\langle u^j|}{\omega-\varepsilon^j+\mathrm{i}\eta} + \sum\!\!\!\!\!\!\int_\zeta \frac{|v^\zeta\rangle\langle v^\zeta|}{\omega-\varepsilon^\zeta+\mathrm{i}\eta}\,,\quad \eta=0^+\,, \tag{26.14}$$

where

$$G^N_\omega(\mathbf{1},\mathbf{1}') \equiv \langle \mathbf{1}|G^N_\omega|\mathbf{1}'\rangle,\; u^j(\boldsymbol{r}) \equiv \left\langle \boldsymbol{r}\,\middle|\,u^j\right\rangle\,,\ \text{etc.}\,, \tag{26.15}$$

and where the symbol $\sum\!\!\!\!\!\!\int$ denotes a summation over discrete and integration over continuous states. The two terms in (26.14) are obtained by counting each time-ordering separately in (26.13), introducing a complete set of intermediate states and using the time translation operator (26.3). The functions $u^j(\boldsymbol{r})$ and $v^\zeta(\boldsymbol{r})$ are defined by

$$\begin{aligned} u^j(\boldsymbol{r}) &\equiv \left\langle {}^N_0 \,\middle|\, \psi(\boldsymbol{r},0) \,\middle|\, {}^{N+1}_{\;\;j} \right\rangle\,,\\ \left\langle {}^N_0 \,\middle|\, \psi(\boldsymbol{r},t) \,\middle|\, {}^{N+1}_{\;\;j} \right\rangle &= \mathrm{e}^{\mathrm{i}(E^N_0-E^{N+1}_j)t} u^j(\boldsymbol{r})\,, \end{aligned} \tag{26.16}$$

and

$$\begin{aligned} v^\zeta(\boldsymbol{r}) &\equiv \left\langle {}^{N-1}_{\;\;\zeta} \,\middle|\, \psi(\boldsymbol{r},0) \,\middle|\, {}^N_0 \right\rangle\,,\\ \left\langle {}^{N-1}_{\;\;\zeta} \,\middle|\, \psi(\boldsymbol{r},t) \,\middle|\, {}^N_0 \right\rangle &= \mathrm{e}^{\mathrm{i}(E^{N-1}_\zeta-E^N_0)t} v^\zeta(\boldsymbol{r})\,. \end{aligned} \tag{26.17}$$

Here, $\left|{}^N_0\right\rangle$ is the ground state of an atom of lepton charge number N (corresponding to a nondegenerate state – a closed shell or subshell) of energy E^N_0, $\left|{}^{N+1}_{\;\;j}\right\rangle$ an atomic state of total energy E^{N+1}_j and leptonic charge number $N+1$, and $\left|{}^{N-1}_{\;\;\zeta}\right\rangle$ is the state of energy E^{N-1}_ζ and leptonic charge number $N-1$. These several states satisfy the eigenvalue equations

$$\begin{aligned} \mathcal{H}\left|{}^N_0\right\rangle &= E^N_0\left|{}^N_0\right\rangle\,,\\ \mathcal{H}\left|{}^{N+1}_{\;\;j}\right\rangle &= E^{N+1}_j\left|{}^{N+1}_{\;\;j}\right\rangle\,,\\ \mathcal{H}\left|{}^{N-1}_{\;\;\zeta}\right\rangle &= E^{N-1}_\zeta\left|{}^{N-1}_{\;\;\zeta}\right\rangle\,. \end{aligned} \tag{26.18}$$

The energy parameters ε^j and ε^ζ are defined by

$$\varepsilon^j \equiv E^{N+1}_j - E^N_0\,, \tag{26.19}$$

$$\varepsilon^\zeta \equiv E^N_0 - E^{N-1}_\zeta\,. \tag{26.20}$$

Equations (26.19) and (26.20) are generalizations of Koopmans's theorem [26.15] (see Sect. 21.4.1). In the DF approximation, the state $\left|{}^{N+1}_{\;\;j}\right\rangle$ can be thought of as N effective particles (electrons) making up the core, plus one valence electron, with energy label j. The atom can also be an isoelectronic ion with nuclear charge number Z. In this (DF) approximation, the state $\left|{}^{N-1}_{\;\;\zeta}\right\rangle$ is one of two possible types. It can have N independent electrons making up the core, with one of the core electrons missing, or equivalently the N electron core with one electron–hole, with energy label ζ. There is a finite number N of such hole states, which shall be labeled a. There are no other states in the NR (HF) case. In the R (DF) case, the second group of states $\left|{}^{N-1}_{\;\;\zeta}\right\rangle$ can also describe an atom with a core of N electrons, plus one positron. Its continuum energy label, will be taken as $\zeta=\bar{\ell}$. This energy will appear with a negative sign in (26.14).

The second step in leading to an explicit $G^N(1,1')$ is the generation of a Dyson equation which it satisfies. In both the DF and HF approximations, the Dyson equation can be obtained through a successive series of steps [26.11]. Working in the Coulomb gauge (see Sect. 27.2), begin with the Coulomb interaction between electrons and neglect the exchange of transverse virtual photons. The Dyson equation for the resulting two-point function is then expanded as a power series in the electron–electron (ee) interaction

$$V_{\mathrm{ee}} \equiv V = \frac{\alpha}{|\boldsymbol{x}-\boldsymbol{y}|}\,, \tag{26.21}$$

resulting in an infinite set of Feynman diagrams ($\alpha=e^2/4\pi$ in rationalized mks units). As a next approximation, consider only diagrams involving single Coulomb exchanges and their iterates ("ladders"). That is, set aside for later consideration nonladder Feynman diagrams of two or more Coulomb photons and their iterates (e.g., two crossed Coulomb photon lines). A summation of the infinite set of these remaining terms generates an equation for a propagator labeled $G^N_{\omega,\mathrm{Coul}}$, which contains Coulomb radiative corrections in its kernel. Finally, modify the kernel by isolating these radiative corrections (self energy and vacuum polarization) by constructing a spectral representation of $G^{\text{“0”}}_\omega$ which mimics that for the usual QED propagator, which is a vacuum (rather than an N-lepton ground state) expectation value. This requires shifting the poles corresponding to core energies of the atom from the upper to the lower complex ω plane. The shifting of poles is accomplished by means of the equation

$$\frac{1}{x-\mathrm{i}\eta} - \frac{1}{x+\mathrm{i}\eta} = 2\pi\mathrm{i}\delta(x)\,, \tag{26.22}$$

which, when used to shift the poles of core electrons in the ω plane, gives

$$G^N_{\omega,\mathrm{Coul}} = 2\pi \mathrm{i} \sum_{a=1}^{N} \left|v^a\right\rangle \left\langle v^a\right| \delta(\omega - \varepsilon^a) + G^{\text{“}0\text{”}}_{\omega} . \tag{26.23}$$

It is the first term on the right-hand side of this equation which occurs in the kernel of (26.24) below and generates the DF approximation.

Some of the terms set aside in the course of this sequence of approximations are reconsidered in Sect. 26.4 to obtain more accurate energy results. What remains at the end of this sequence is an approximation to G^N_ω of (26.14), designated as $G^N_{\omega,\mathrm{DF}}$, and which satisfies the self-consistent Dyson equation

$$G^N_{\omega,\mathrm{DF}} = g_\omega + g_\omega \Sigma^N_{\mathrm{DF}} G^N_{\omega,\mathrm{DF}} , \tag{26.24}$$

where $\sum^N_{\mathrm{DF}}$ is the kernel defined in (26.27). For the R (DF) case,

$$g_\omega^{-1} = \omega - h^{\mathrm{R}}(\boldsymbol{r}),\ h^{\mathrm{R}}(\boldsymbol{r}) = \boldsymbol{\alpha} \cdot \boldsymbol{p} + m\beta - Z\alpha/r , \tag{26.25}$$

and the corresponding h^{NR}, used in the HF case, is

$$h^{\mathrm{NR}}(\boldsymbol{r}) = \frac{\boldsymbol{p}^2}{2m} - \frac{Z\alpha}{r} . \tag{26.26}$$

The function g_ω in (26.25) is the R or NR Coulomb Green's function. It is the solution of the corresponding inhomogeneous c-number Schrödinger or Dirac equation. It is a single-particle equation which has had a long history of specific treatments (see Chapt. 9 and [26.16–19]). It is the semiclassical limit of the two-point propagator we consider here when only the c-number nuclear Coulomb potential is kept and all other (q-number) interactions are turned off.

In the DF and HF approximations in (26.14), we replace ε^j by $e_j > 0$, $\left|u^j\right\rangle$ by $|j\rangle$ (the valence energies and states of an atom with $N+1$ electrons and a frozen relaxed core of N electrons), ε^ζ by $e_a > 0$, $|u^\zeta\rangle$ by $|a\rangle$, for the N discrete electron core states, and ε^ζ by $e_{\bar{\ell}} < 0$, $|u^\zeta\rangle$ by $|\bar{\ell}\rangle$, for the continuum of negative energy states (which do not appear in the HF approximation). The kernel Σ^N_{DF} contains *only* core states $|a\rangle$, in *both* the DF and HF approximations, and is given by the sum over states

$$\begin{aligned}\left\langle m \middle| \Sigma^N_{\mathrm{DF}} \middle| n\right\rangle &= \alpha \sum_{a=1}^{N} \int \mathrm{d}^3\boldsymbol{x}\,\mathrm{d}^3\boldsymbol{y} \frac{1}{|\boldsymbol{x}-\boldsymbol{y}|} \langle m|\boldsymbol{x}\rangle \langle a|\boldsymbol{y}\rangle \\ &\quad \times (\langle \boldsymbol{x}|n\rangle \langle \boldsymbol{y}|a\rangle - \langle \boldsymbol{x}|a\rangle \langle \boldsymbol{y}|n\rangle) \\ &\equiv \sum_{a=1}^{N} \left[\left(\begin{smallmatrix} m \\ a \end{smallmatrix} |V| \begin{smallmatrix} n \\ a \end{smallmatrix}\right) - \left(\begin{smallmatrix} m \\ a \end{smallmatrix} |V| \begin{smallmatrix} a \\ n \end{smallmatrix}\right) \right] \\ &\equiv \sum_{a=1}^{N} \left[\begin{smallmatrix} m \\ a \end{smallmatrix} |V| \begin{smallmatrix} n \\ a \end{smallmatrix}\right]\end{aligned} \tag{26.27}$$

for arbitrary states m and n. The first term in brackets is the Hartree term and the second is the electron exchange term in the DF (HF) approximation. The DF (HF) equation is the homogeneous equation corresponding to the inhomogeneous (26.24). Thus,

$$\left(e_n - h - \Sigma^N_{\mathrm{DF}}\right)|n\rangle = 0, \qquad \left\langle n|n'\right\rangle = \delta_{nn'} . \tag{26.28}$$

The states $|n\rangle$ (valence, core, and negative energy states) are orthonormal and complete. In the coordinate basis, (26.28) takes the more familiar form

$$[e_n - h(\boldsymbol{x})] \langle \boldsymbol{x}|n\rangle - \left\langle \boldsymbol{x} \middle| \Sigma^N_{\mathrm{DF}} \middle| n\right\rangle = 0 , \tag{26.29}$$

where

$$\begin{aligned}\left\langle \boldsymbol{x} \middle| \Sigma^N_{\mathrm{DF}} \middle| n\right\rangle &= \sum\!\!\!\!\!\!\int_m \sum_{a=1}^{N} \langle \boldsymbol{x}|m\rangle \left[\begin{smallmatrix} m \\ a \end{smallmatrix} |V| \begin{smallmatrix} n \\ a \end{smallmatrix}\right] \\ &\equiv \sum_{a=1}^{N} \left[\begin{smallmatrix} x \\ a \end{smallmatrix} |V| \begin{smallmatrix} n \\ a \end{smallmatrix}\right] \\ &\equiv \int V_{\mathrm{DF}}(\boldsymbol{x},\boldsymbol{y})\,\mathrm{d}^3\boldsymbol{y} \langle \boldsymbol{y}|n\rangle .\end{aligned} \tag{26.30}$$

One can also generate equations corresponding to higher approximations than DF using the same approach. For example, one can obtain a Brueckner equation [26.20],

$$\left[e_n - h - \Sigma^N_{\mathrm{DF}} - \Sigma^N_{\mathrm{B}}(e_n)\right] |n\rangle = 0 , \tag{26.31}$$

and the states now satisfy the orthonormality condition

$$\begin{aligned}&\lim_{\omega \to e_n} \int \mathrm{d}^3\boldsymbol{x}\,\mathrm{d}^3\boldsymbol{y}\, \langle n|\boldsymbol{x}\rangle \left\langle \boldsymbol{y}|n'\right\rangle \\ &\quad \times \left(\frac{\left\{(\omega - e_n)\delta^3(\boldsymbol{x}-\boldsymbol{y}) - \left\langle \boldsymbol{x} \middle| \left\{\Sigma^N_{\mathrm{B}}(\omega) - \Sigma^N_{\mathrm{B}}(e_n)\right\} \middle| \boldsymbol{y}\right\rangle\right\}}{\omega - e_n} \right) \\ &= \delta_{nn'} ,\end{aligned} \tag{26.32}$$

where the energy-dependent kernel $\Sigma_{\mathrm{B}}(e_n)$ arises from irreducible Feynman diagrams involving two Coulomb photons. The kernel is given by

$$\begin{aligned}&\left\langle m \middle| \Sigma^N_{\mathrm{B}}(e_n) \middle| n\right\rangle \\ &= \sum\!\!\!\!\!\!\int_{a,i,j} \frac{1/2}{e_n + e_a - e_i - e_j} \left[\begin{smallmatrix} m \\ a \end{smallmatrix} \middle| V \middle| \begin{smallmatrix} i \\ j \end{smallmatrix}\right] \left[\begin{smallmatrix} i \\ j \end{smallmatrix} \middle| V \middle| \begin{smallmatrix} n \\ a \end{smallmatrix}\right]\end{aligned}$$

$$+\sum_{a,b,i}\!\!\!\!\!\!\!\!\int \frac{1/2}{e_n+e_i-e_a-e_b}\left[\begin{smallmatrix}m\\i\end{smallmatrix}\middle|V\middle|\begin{smallmatrix}a\\b\end{smallmatrix}\right]\left[\begin{smallmatrix}a\\b\end{smallmatrix}\middle|V\middle|\begin{smallmatrix}n\\i\end{smallmatrix}\right]. \tag{26.33}$$

$\sum\!\!\!\!\!\int$ involves summations over core states (a, b), and summation and integration over valence states (i, j). In perturbation theory, the lowest order contribution of $\Sigma_B(e_n)$ is an ee correlation term.

26.2 The Four-Point Green's Function

The four-point Green's function [26.4, 8, 11], or two-body propagator, is defined as

$$G^N(12, 1'2') \equiv -\left\langle \begin{smallmatrix}N\\0\end{smallmatrix}\middle| T\left[\psi(1)\psi^\dagger(1')\psi(2)\psi^\dagger(2')\right]\middle|\begin{smallmatrix}N\\0\end{smallmatrix}\right\rangle . \tag{26.34}$$

In order to avoid unnecessary complication, only simple ladders of $V_{ee} \equiv V$ are considered. There are 4! possible time orderings. Of these, there are four groups of four with $t_1, t_{1'} > t_2, t_{2'}$, $t_1, t_{1'} < t_2, t_{2'}$, $t_1, t_2 > t_{1'}, t_{2'}$, and $t_1, t_2 < t_{1'}, t_{2'}$, corresponding to the particle–hole (PH/HP) and the two-particle/two-hole (2P/2H) cases for the first eight and last eight time orderings, respectively. For each of these cases, introduce a total time and relative time variable defined by

$$T = \frac{1}{2}(t_1+t_{1'}),\ T' = \frac{1}{2}(t_2+t_{2'}),$$
$$t = t_1 - t_{1'},\ t' = t_2 - t_{2'}, \tag{26.35}$$

for the first eight cases (PH/HP) and

$$T = \frac{1}{2}(t_1+t_2),\ T' = \frac{1}{2}(t_{1'}+t_{2'}),$$
$$t = t_1 - t_2,\ t' = t_{1'} - t_{2'}, \tag{26.36}$$

for the last eight cases (2P/2H). For a particular set of eight time orderings, a time translation with respect to the relevant c.m. time t or t', followed by a separate Fourier transformation for each case with respect to $T - T'$ (with integration variable $d\Omega$), yields contributions with poles in the separately defined Ω-planes at

$$\omega_{(i\alpha)} = E^N_{(i\alpha)} - E^N_0,\ \text{for PH} \tag{26.37a}$$
$$-\omega_{(i\alpha)} = E^N_0 - E^N_{(i\alpha)},\ \text{for HP} \tag{26.37b}$$

$$\omega_{(ij)} = E^{N+2}_{(ij)} - E^N_0,\ \text{for 2P} \tag{26.38a}$$
$$\omega_{(ab)} = E^N_0 - E^{N-2}_{(ab)},\ \text{for 2H}. \tag{26.38b}$$

Equations (26.37a,b) parallel (26.19) and (26.20) as generalizations of Koopmans' theorem. The spectral representation is of a form similar to (26.14), with wave functions corresponding to (26.16) and (26.17) given by

$$\chi_{(ia)}(11') \equiv \left\langle \mathbf{1}\middle|\chi_{(ia)}(t)\middle|\mathbf{1}'\right\rangle e^{-i\omega_{(ia)}T}$$
$$= \left\langle \begin{smallmatrix}N\\0\end{smallmatrix}\middle| T\left[\psi(1)\psi^\dagger(1')\right]\middle|\begin{smallmatrix}N\\(ia)\end{smallmatrix}\right\rangle \tag{26.39}$$

for the PH/HP case and

$$\varphi_{(ij)}(12) \equiv \langle \mathbf{12}|\,\varphi_{(ij)}(t)\rangle e^{-i\omega_{(ij)}T}$$
$$= \left\langle \begin{smallmatrix}N\\0\end{smallmatrix}\middle| T\left[\psi(1)\psi(2)\right]\middle|\begin{smallmatrix}N+2\\(ij)\end{smallmatrix}\right\rangle , \tag{26.40}$$

$$\gamma_{(ab)}(1'2') \equiv \langle \mathbf{1'2'}|\,\gamma_{(ab)}(t')\rangle e^{-i\omega_{(ab)}T'}$$
$$= \left\langle \begin{smallmatrix}N-2\\(ab)\end{smallmatrix}\middle| T\left[\psi(1')\psi(2')\right]\middle|\begin{smallmatrix}N\\0\end{smallmatrix}\right\rangle , \tag{26.41}$$

for the 2P and 2H cases, respectively. The antisymmetry under exchange follows from the definitions (26.40) and (26.41):

$$\zeta_{(\tau)}(12) = -\zeta_{(\tau)}(21),\quad \zeta_{(\tau)} = \varphi_{(ij)}, \gamma_{(ab)}, \tag{26.42}$$

where it is understood that the suppressed spinor indices are interchanged as well as the coordinate and time variables.

The three amplitudes defined above satisfy Bethe–Salpeter (BS) equations. The PH/HP case is the analog of the positronium atom and the 2P case is analogous to He. For the case of Coulomb ladder exchanges, to which we have restricted ourselves, these BS equations can be reduced to simpler ones, with the relative time set equal to zero (the Salpeter equation [26.21]). The corresponding BS wave functions in the DF (HF) basis (rather than the coordinate basis), are for PH/HP,

$$\left\langle m\middle|\bar{\chi}_{(ia)}\middle|n\right\rangle \equiv \int d^3\mathbf{1}\,d^3\mathbf{1}'\langle m|\mathbf{1}\rangle\langle \mathbf{1}|\chi_{(ia)}(0)|\mathbf{1}'\rangle\langle \mathbf{1}'|n\rangle , \tag{26.43}$$

with $m = k$ and $n = c$, or $m = c$ and $n = k$, and for 2P/2H,

$$\left\langle mn\middle|\bar{\zeta}_{(\tau)}\right\rangle \equiv \int d^3\mathbf{1}\,d^3\mathbf{2}\langle m|\mathbf{1}\rangle\langle n|\mathbf{2}\rangle\left\langle \mathbf{12}\middle|\zeta_{(\tau)}(0)\right\rangle , \tag{26.44}$$

where $\zeta_{(\tau)} = \varphi_{(ij)}, \gamma_{(ab)}$. The states $|i\rangle$, $|j\rangle$, $|a\rangle$, $|b\rangle$, $|m\rangle$, and $|n\rangle$ label one-particle DF (HF) eigenkets and

e_i, e_j, e_a, e_b, e_m, and e_n, the corresponding eigenvalues.

In the PH/HP case, the BS equation is

$$-(\omega_{(ia)} - e_k + e_c)\langle i\,|\bar{\chi}_{(ia)}|\,a\rangle = \sum\!\!\!\!\!\!\int \left\{ \left[\begin{smallmatrix}k\\b\end{smallmatrix}\middle|V\middle|\begin{smallmatrix}j\\c\end{smallmatrix}\right]\langle j\,|\bar{\chi}_{(ia)}|\,b\rangle + \left[\begin{smallmatrix}k\\j\end{smallmatrix}\middle|V\middle|\begin{smallmatrix}b\\c\end{smallmatrix}\right]\langle b\,|\bar{\chi}_{(ia)}|\,j\rangle \right\},$$
$$(\omega_{(ia)} - e_c + e_k)\langle a\,|\bar{\chi}_{(ia)}|\,i\rangle = \sum\!\!\!\!\!\!\int_{j,b} \left\{ \left[\begin{smallmatrix}c\\b\end{smallmatrix}\middle|V\middle|\begin{smallmatrix}j\\k\end{smallmatrix}\right]\langle j\,|\bar{\chi}_{(ia)}|\,b\rangle + \left[\begin{smallmatrix}c\\j\end{smallmatrix}\middle|V\middle|\begin{smallmatrix}b\\k\end{smallmatrix}\right]\langle b\,|\bar{\chi}_{(ia)}|\,j\rangle \right\}. \quad (26.45)$$

In the 2P/2H case, the coupled pairs of BS equations are

$$-(\omega_{(\tau)} - e_c - e_d)\left\langle cd|\bar{\zeta}_{(\tau)}\right\rangle = \sum\!\!\!\!\!\!\int_{m>n} \left[\begin{smallmatrix}c\\d\end{smallmatrix}\middle|V\middle|\begin{smallmatrix}m\\n\end{smallmatrix}\right]\left\langle mn\,|\bar{\zeta}_{(\tau)}\right\rangle, \quad (26.46)$$
$$(\omega_{(\tau)} - e_k - e_\ell)\left\langle k\ell|\bar{\zeta}_{(\tau)}\right\rangle = \sum\!\!\!\!\!\!\int_{m>n} \left[\begin{smallmatrix}k\\\ell\end{smallmatrix}\middle|V\middle|\begin{smallmatrix}m\\n\end{smallmatrix}\right]\left\langle mn\,|\bar{\zeta}_{(\tau)}\right\rangle,$$

with m and n in these equations labeling either both core, or both valence states.

The wave functions also satisfy the additional conditions:

$$\langle c\,|\bar{\chi}_{(ia)}|\,d\rangle = \langle k\,|\bar{\chi}_{(ia)}|\,\ell\rangle = 0, \quad \text{PH/HP (no 2P/2H terms)} \quad (26.47)$$
$$\langle ck\,|\bar{\varphi}_{(ij)}\rangle = \langle ck\,|\bar{\gamma}_{(ab)}\rangle = 0, \quad \text{2P/2H (no PH/HP terms)} \quad (26.48)$$

The single indices a, b, c, d refer to core and i, j, k, ℓ to valence DF (HF) states. The BS wave functions satisfy the orthonormality conditions

$$\sum_{k,c}\left(\langle k\,|\bar{\chi}_{(ia)}|\,c\rangle\langle c\,|\bar{\chi}_{(ia)'}|\,k\rangle + \langle c\,|\bar{\chi}_{(ia)}|\,k\rangle\langle k\,|\bar{\chi}_{(ia)'}|\,c\rangle\right) = \delta_{(ia)(ia)'}, \quad (26.49)$$

for the PH/HP case, and

$$\pm\left(\sum_{c>d}\left\langle\bar{\zeta}_{(\tau)}\middle|\,cd\right\rangle\left\langle cd\,\middle|\bar{\zeta}_{(\tau')}\right\rangle - \sum_{k>\ell}\left\langle\bar{\zeta}_{(\tau)}\middle|\,k\ell\right\rangle\left\langle k\ell\,\middle|\bar{\zeta}_{(\tau')}\right\rangle\right) = \delta_{(\tau)(\tau')}, \quad (26.50)$$

where $+(-)$ corresponds to the 2H(2P) case and $(\tau) = (ab)$ or (ij).

The PH/HP case in the GFA, involving Coulomb ladders for the ee interaction, is just the R and NR RPA [26.2]. The labels c, d should also refer to antiparticles, but the contributions of the integrals from these terms to (26.45) and (26.46) are negligible.

26.3 Radiative Transitions

For the majority of applications, one begins with the function $\Gamma^N(12;3)$, the reducible three-point vertex:

$$\Gamma^N(12;3) = -\left\langle\begin{smallmatrix}N\\0\end{smallmatrix}\middle|T\left[\psi(1)j^k(3)\psi^\dagger(2)\right]\middle|\begin{smallmatrix}N\\0\end{smallmatrix}\right\rangle. \quad (26.51)$$

The usual strategy is followed. The spectral representation serves to identify the functions of ultimate interest. (Energies are not relevant in this case.) One then generates Dyson equations in the chosen approximation by summing an infinite series of perturbation terms.

There are 3! time orderings in (26.51). As with the four-point function, not all of them are subsequently useful. The two useful cases are $t_1 > t_3 > t_2$, and $t_2 > t_3 > t_1$. To obtain a spectral representation for these two cases, one first carries out a time translation of t_3, using the operator $\exp(\mathrm{i}\mathcal{H}t_3)$ of (26.3), so that $t_1 \to \tau_1 = t_1 - t_3$, $t_2 \to \tau_2 = t_2 - t_3$. One then introduces complete sets of intermediate states. The functions defined in connection with the two-point function in (26.16) and (26.17) will now appear, as well as the radiative transition amplitude, defined in (26.10). If one next carries out a separate time translation of τ_1 and τ_2 and Fourier-transforms the resulting expressions, one obtains (with **3** replaced by $\boldsymbol{r}$)

$$\Gamma^N(\boldsymbol{r};\omega_1\omega_2) = \sum\!\!\!\!\!\!\int_{j\ell} \frac{|u^j\rangle M^{N+1}_{j\ell}(\boldsymbol{r})\langle u^\ell|}{(\omega_1 - \varepsilon_j + \mathrm{i}\eta)(\omega_2 - \varepsilon_\ell + \mathrm{i}\eta)} + \sum\!\!\!\!\!\!\int_{\zeta\chi} \frac{|v^\zeta\rangle M^{N-1}_{\zeta\chi}(\boldsymbol{r})\langle v^\chi|}{(\omega_1 - \varepsilon_\zeta - \mathrm{i}\eta)(\omega_2 - \varepsilon_\chi - \mathrm{i}\eta)} + \cdots. \quad (26.52)$$

We next define the three-point irreducible electron vertex Λ^N from the reducible vertex Γ^N (the electron "legs" in Γ^N are missing in Λ^N) as

$$\left\langle \mathbf{1} \left| \Gamma^N(\mathbf{r};\omega_1\omega_2) \right| \mathbf{2} \right\rangle \equiv \int\int d^3\mathbf{1}' d^3\mathbf{2}' \left\langle \mathbf{1} \left| G^N_{\omega_1} \right| \mathbf{1}' \right\rangle \left\langle \mathbf{1}' \left| \Lambda^N(\mathbf{r};\omega_1\omega_2) \right| \mathbf{2}' \right\rangle \times \left\langle \mathbf{2}' \left| G^N_{\omega_2} \right| \mathbf{2} \right\rangle , \quad (26.53)$$

and pick out the residues of the ω_1 and ω_2 poles at specific energies ε_m and ε_n. [$\Lambda^N(\mathbf{r};\omega_1\omega_2)$ has no such poles.] With the equivalent of the DF(HF) approximation to the Dyson equations (26.57) below, the corresponding kets form an orthonormal set, and scalar products are calculated with respect to $\langle m|$ and $|n\rangle$. The second term on the right-hand side of (26.52) refers to hole states and is of less interest than the first term describing 1P–1P transitions. From the first term, the transition matrix element in terms of Λ^N is

$$M^{N+1}_{fi}(\mathbf{r};k_0) = \left\langle f \left| \Lambda^N(\mathbf{r};e_f e_i) \right| i \right\rangle . \quad (26.54)$$

Generation of a Dyson Equation

The approximation of only Coulomb "ladder" ee interactions in the two-point Green's function produces an infinite set of Feynman diagrams to which one end of a single transverse photon line is attached in all possible ways. A resummation of these diagrams generates the $G^N_{\omega,\text{Coul}}$ functions, which are approximated by the DF (HF) propagators written in their spectral form. The scalar products leading to (26.54) can then be taken.

A further simplification results because of the Coulomb ladder approximation, which is the same as in the four-point Green's function case: the relative time can be set equal to zero, which corresponds to integrating over the relative frequency, $\omega = \omega_1 - \omega_2$. With the total frequency defined as $\Omega = \frac{1}{2}(\omega_1 + \omega_2)$, and the definitions

$$\bar{\Lambda}^N(\mathbf{r},\Omega) = \frac{1}{2\pi}\int d\omega \Lambda^N\left(\mathbf{r};\omega + \frac{1}{2}\Omega, \omega - \frac{1}{2}\Omega\right),$$

$$\left\langle m \left| \bar{\Lambda}^N(\Omega) \right| n \right\rangle = \int d^3\mathbf{r} \frac{e^{i\mathbf{k}\cdot\mathbf{r}}}{\sqrt{2k_0}} \left\langle m \left| \bar{\Lambda}^N(\mathbf{r},\Omega) \right| n \right\rangle ,$$

$$\Omega_{mn} = e_n - e_m = k_0 , \quad (26.55)$$

with a relation similar to (26.55) for the quantity $\lambda_k(r)$ defined in (26.11), i. e.,

$$\langle m|\lambda(k_0)|n\rangle = \int d^3\mathbf{r} \frac{e^{i\mathbf{k}\cdot\mathbf{r}}}{\sqrt{2k_0}} \langle m|\lambda_k(r)|n\rangle , \quad (26.56)$$

the matrix elements in the DF (HF) basis are [26.14, 20]

$$\left\langle m \left| \bar{\Lambda}^N(k_0) \right| n \right\rangle = \langle m|\lambda(k_0)|n\rangle + \sum\!\!\!\!\!\!\int_{aj} \left\{ \frac{1}{e_a - e_j + k_0} \left[{m \atop j} \middle| V \middle| {n \atop a} \right] \left\langle a \left| \bar{\Lambda}^N(k_0) \right| j \right\rangle + \frac{1}{e_a - e_j - k_0} \left[{m \atop a} \middle| V \middle| {n \atop j} \right] \left\langle j \left| \bar{\Lambda}^N(k_0) \right| a \right\rangle \right\} . \quad (26.57)$$

As discussed in Sect. 26.1, the label a should include not just hole states but also negative energy states, which have been neglected in (26.57). Note also that only PH or HP matrix elements appear on the right-hand side of (26.57). Therefore, a closed set of inhomogeneous linear algebraic equations for $\langle b|\bar{\Lambda}^N(k_0)|\ell\rangle$ and $\langle \ell|\bar{\Lambda}^N(k_0)|b\rangle$ results from setting $m = b, n = \ell$ or $m = \ell$, $n = b$, respectively, in (26.57). These equations can be solved numerically, and the resulting $\langle a|\bar{\Lambda}^N(k_0)|j\rangle$ and $\langle j|\bar{\Lambda}^N(k_0)|a\rangle$ substituted in (26.57) to obtain the final result

$$M^{N+1}_{fi}(k_0) \equiv \left\langle f \left| \bar{\Lambda}^N(k_0) \right| i \right\rangle . \quad (26.58)$$

An integro-differential equation form, which provides the option of choosing alternative numerical techniques [26.14], is obtained from (26.57), after some rearrangement and passage to a coordinate basis. Defining $m_{fi}(k_0)$ from $\lambda(k_0)$ in analogy with the definition of $M^{N+1}_{fi}(k_0)$ from $\bar{\Lambda}^N(k_0)$ in (26.58), one has

$$M^{N+1}_{fi}(k_0) = m_{fi}(k_0) + \sum_a (\langle a|\lambda(k_0)|A^-\rangle + \langle A^+|\lambda(k_0)|a\rangle) , \quad (26.59)$$

where

$$[h(\mathbf{r}) \mp k_0 - e_a]\langle \mathbf{r}|A^\pm\rangle + \sum_a \left[{\mathbf{r} \atop a} \middle| V \middle| {A\pm \atop a} \right] = -\sum\!\!\!\!\!\!\int_j \langle \mathbf{r}|j\rangle\langle j| \left[v_\pm + \mathcal{V}_\pm \right] |a\rangle , \quad (26.60)$$

$$\langle j|v_+|a\rangle = \left[{j \atop i} \middle| V \middle| {a \atop f} \right], \quad \langle j|v_-|a\rangle = \left[{j \atop f} \middle| V \middle| {a \atop i} \right] , \quad (26.61)$$

$$\langle j|\mathcal{V}_\pm|a\rangle = \sum_b \left\{ \left[{j \atop b} \middle| V \middle| {a \atop B_\pm} \right] + \left[{j \atop B_\mp} \middle| V \middle| {a \atop b} \right] \right\} . \quad (26.62)$$

The three-point Green's function, as can be seen from this summary, describes transition amplitudes for radiative transitions between two valence states of atoms with closed shells (subshells) plus one electron (the 1P case). In the Coulomb ladder approximation, closely related to the DF equation, the photon vertex $\langle \mathbf{1} | \Lambda^N(\boldsymbol{r}; \omega_1\omega_2) | \mathbf{2} \rangle$ of (26.53), is nonlocal in space, rather than the local vertex $\langle \mathbf{1} | \lambda_k(\boldsymbol{r}) | \mathbf{2} \rangle$ of (26.11) [as follows from the factor of $\delta^3(\mathbf{1}-\boldsymbol{r})\delta^3(\mathbf{2}-\boldsymbol{r})$]. The presence of these additional nonlocal contributions to $M_{fi}^{N+1}(k_0)$ is made apparent in (26.59). The $m_{fi}(k_0)$ term is the contribution of the local vertex. Of course, if one knew the *exact* $N+1$ electron wave functions and energies, only the local vertex would enter and a GI result would be obtained. However, the length and velocity versions of (26.12) are equal and GI even in the approximation just discussed [26.3, 7]. Gauge invariance is an *essential* constraint on radiative transition amplitudes. It has been proven for general gauges not only in the present approximation, but also in the ones discussed below [26.8, 10], all of them arising in the GFA. Since the effective potential in the DF (or HF) approximation is nonlocal, the effective current must also be nonlocal in order to maintain the GI.

The somewhat more complicated Dyson equation satisfied by the nonlocal vertex corresponding to the Brueckner approximation has also been generated [26.22], and put in a numerically implementable form.

An alternative approximation [26.3, 7] for transition matrix elements, proposed earlier [26.2] than the one just discussed, is based on the RPA [our PH/HP case, with wave function $\bar{\chi}_{(fa)}$ the solution of (26.45)]. Reference to (26.6), (26.7), (26.10) and (26.11) gives immediately

$$M_{fa}^N(\boldsymbol{r}; \omega_{(fa)}) = -\lambda_k(\boldsymbol{r}) \langle \boldsymbol{r} | \bar{\chi}_{(fa)} | \boldsymbol{r} \rangle \,, \qquad (26.63)$$

where $k_0 = k = \omega_{(fa)}$ and N corresponds to atoms with a closed shell or subshell of electrons.

Aside from being a different approximation to transition amplitudes than (26.54), (26.63) covers a different set of cases than does (26.54), since the corresponding initial states i in (26.54) are restricted to core states a in (26.63). Thus, for example, the case $N=2$ in (26.54) can describe transitions between any two valence states of Li. The corresponding case for (26.63) is $N=4$, but only transitions to a higher level, originating in the 1s or 2s level of Li, can be described by the formalism.

Finally, we shall discuss radiative transitions [26.8] for 2P atoms (closed shell/subshell plus two valence electrons). The general case involves a five-point nonlocal vertex. However, in the ladder approximation, this reduces to transition matrix elements which contain combinations of DF, 2P BS wave functions, and three-point functions. The final expressions are

$$\begin{aligned} &M_{fi}^{N+2}(\boldsymbol{r}; k_0) \\ &= \left\langle \bar{\zeta}_{(fp)} \left| \left(\frac{\bar{\Lambda}^N(\boldsymbol{r}, k_0)}{k_0 - \Delta H} V - V \frac{\bar{\Lambda}^N(\boldsymbol{r}, k_0)}{k_0 - \Delta H} \right) \right| \bar{\zeta}_{(iq)} \right\rangle . \end{aligned} \qquad (26.64)$$

The expressions appearing in this equation are, in more detail, in the DF basis:

$$\left\langle \begin{matrix} m \\ n \end{matrix} \left| \frac{\bar{\Lambda}^N(\boldsymbol{r}, k_0)}{k_0 - \Delta H} \right| \begin{matrix} q \\ s \end{matrix} \right\rangle \equiv \frac{\langle m | \bar{\Lambda}^N(\boldsymbol{r}, k_0) | q \rangle \delta_{ns}}{k_0 + e_q - e_m} \,, \qquad (26.65)$$

where m, n, q, s label DF states (P or H), $\left|\begin{smallmatrix} q \\ s \end{smallmatrix}\right\rangle \equiv |qs\rangle \equiv |q\rangle|s\rangle$; (26.65) serves to define ΔH, as a difference of two DF Hamiltonians, the argument $\boldsymbol{r}$ of $\bar{\Lambda}$ refers to the electron which emits the photon, and

$$\begin{aligned} \left\langle \begin{smallmatrix} m \\ n \end{smallmatrix} \middle| V \middle| \bar{\zeta}_{(\tau)} \right\rangle &= \sum\!\!\!\!\!\!\!\int_{cd} \left\langle \begin{smallmatrix} m \\ n \end{smallmatrix} \middle| V \middle| \begin{smallmatrix} c \\ d \end{smallmatrix} \right\rangle \left\langle cd \middle| \bar{\zeta}_{(\tau)} \right\rangle \\ &+ \sum\!\!\!\!\!\!\!\int_{k\ell} \left\langle \begin{smallmatrix} m \\ n \end{smallmatrix} \middle| V \middle| \begin{smallmatrix} k \\ \ell \end{smallmatrix} \right\rangle \left\langle k\ell \middle| \bar{\zeta}_{(\tau)} \right\rangle . \end{aligned} \qquad (26.66)$$

In (26.66) we used the fact that, in the DF basis, $|\bar{\zeta}_{(\tau)}\rangle$ only has 2P or 2H components (26.46, 26.48).

26.4 Radiative Corrections

This section summarizes radiative corrections for 1P/1H atoms, starting from the two-point Green's function. The Dyson equations are generated and solved perturbatively for the energy to order $\alpha^5 m$ (α^3 a.u.). The perturbation theory starts from the DF solution as the zero-order one.

As done for the three-point function in Sect. 26.3, the Dyson equation for radiative corrections involving a single transverse photon is generated by expanding the two-point propagator to all orders in the Coulomb ladder approximation, inserting a transverse virtual photon in all possible ways, and then resumming. The resulting integral equation [26.23] is quite complicated, containing even the three-point vertex in its inhomogeneous term, as well as a mass counter term to eliminate divergences.

It is sufficient for the present purposes to expand these vertices through first order matrix elements of $V_{ee} \equiv V$, but to exclude matrix elements of type $\left[{}^{m}_{a}\,|V|{}^{n}_{a}\right]$ (where a is the label for core states), since these are already included in the DF approximation [see (26.27)].

Other radiative corrections are also generated, which include Coulomb photons. Two of these corrections, involving only such photons, have already been referred to in the text between (26.21) and (26.23). Finally, to order $\alpha^5 m$, corrections involving two transverse photons must be included.

Systematic application of the pole-shifting process in (26.23) to the one- and two-transverse-photon and the Coulomb-photon expressions yields two types of terms: first, photon exchange (of one transverse photon or of two-photons of either kind, Coulomb or transverse) between core and valence states; and second, self-energy and vacuum polarization contributions. Among the photon exchange terms, one can identify those contributions included in other approaches (at least for a few special cases, such as those arising from a single transverse photon interaction, and from two Coulomb interactions and Coulomb–Breit interactions with positive energy intermediate electron states). These are electron correlation terms which are included as part of a MBPT calculation [26.5], or one which involves consideration of an infinite subset of MBPT terms [26.6]. They need not be re-evaluated. The terms not covered by calculations of the type in [26.5, 6], which are of $O(m\alpha^5 c^2)$, involve retardation in Coulomb-transverse photon exchange, negative energy intermediate electron states for two Coulomb and Coulomb-transverse photon exchanges, two transverse photon exchange, self-energy and vacuum polarization terms, as well as anomalous magnetic moment corrections.

After lengthy calculation, one obtains final results in finite analytic form, which are numerically implementable. The results given below are obtained after further approximations. First, the remnants of the original integral equation are solved iteratively. Second, the self-energy terms are calculated in a joint expansion [26.24–26] in α and αZ, and are thus only valid for low Z in isoelectronic sequences. Finally, there are characteristic logarithmic terms generated by low virtual photon momenta in the self-energy contributions. A standard approach is to scale these with a factor of $(\alpha Z)^2$ to obtain Bethe log (BL) terms as constants independent of α, together with constant log (CL) terms of the form $\ln(\alpha Z)^2$. The BL terms are independent of Z for hydrogenic ions, and remain nearly so for other atoms. BL and CL terms are associated with both nuclear and ee Coulomb potentials. The ee CL terms arising from core and valence self-energy terms are canceled exactly by those coming from Coulomb and transverse photon exchange. Thus, the numerous and rather complicated corresponding ee BL terms, which also are *individually* small compared to their associated CL terms, should additionally almost cancel, and are therefore neglected.

One finally obtains (with all state labels denoting principal and orbital quantum numbers as well as spin indices: $n \equiv (n, \ell; m_s)$)

$$\Delta E(n, \ell) = Z|\langle \mathbf{0}|n\rangle|^2 F(n, \ell)\,, \tag{26.67}$$

in a.u., where $\langle \mathbf{0}|n\rangle \equiv \langle \mathbf{0}|n, 0\rangle$ is the s-state HF wave function at the origin of coordinates, and serves to define an effective (shielded) nuclear charge $Z_{n,\text{eff}}$:

$$\langle \boldsymbol{r}|n, \ell\rangle \equiv \Psi_{n,\ell}(\boldsymbol{r}),\; |\langle \mathbf{0}|n\rangle|^2 \equiv \frac{1}{\pi n^3}(Z_{n,\text{eff}}\alpha)^3\,. \tag{26.68}$$

(Just as in hydrogen, for which the final expressions for the radiative corrections require NR and not R wave functions and energies, so in this case we use HF (NR) and not DF (R) quantities.) $F(n, \ell)$ consists of a valence or hole contribution $F_{\text{v,h}}$, a core term $F_{\text{core}}(n, \ell)$, and one due to photon exchange between electrons $F_{\text{ee}}(n, \ell)$:

$$F(n, \ell) = F_{\text{v,h}}(n, \ell) + F_{\text{core}}(n, \ell) + F_{\text{ee}}(n, \ell)\,, \tag{26.69}$$

$$F_{\text{v,h}}(n, \ell) = \frac{4}{3}\left[N(Z)\delta_{\ell 0} + L(n\ell) + U^{\text{so}}_{\text{v,h}}(n\ell)\right]\,, \tag{26.70}$$

$$F_{\text{core}}(n, \ell) = \frac{4}{3Z}\big[N(Z)\rho(n\ell) + L_{\text{core}}(n\ell) + U(n\ell) + U^{\text{so}}_{\text{core}}(n\ell)\big]\,, \tag{26.71}$$

$$F_{\text{ee}}(n, \ell) = \frac{4}{3Z}\left[E(Z)K^C(n\ell) + K^L(n\ell)\right]\,, \tag{26.72}$$

with $U^{\text{so}}_{\text{v,h}}(n\ell)$ and $U^{\text{so}}_{\text{core}}(n\ell)$ being the spin-orbit terms,

$$N(Z) = \ln\frac{1}{(Z\alpha)^2} + \frac{19}{30} + Z\alpha C_5\,, \tag{26.73}$$

$$C_5 = 3\pi\left(1 + \frac{11}{128} - \frac{1}{2}\ln 2\right)\,, \tag{26.74}$$

$$E(Z) = -\frac{7}{2}\ln\frac{1}{Z\alpha} + \frac{59}{20} - \frac{9}{8}\pi\,, \tag{26.75}$$

$$L(n\ell) = \frac{\mathrm{i}/4\pi}{|\langle \mathbf{0}|n\rangle|^2}\left\langle n, \ell\left\|\left[\boldsymbol{p}\cdot\mathcal{L}_{\text{B}}(n\ell)\frac{\boldsymbol{r}}{r^3} - \frac{\boldsymbol{r}}{r^3}\mathcal{L}_{\text{B}}(n\ell)\cdot\boldsymbol{p}\right]\right\| n, \ell\right\rangle\,, \tag{26.76}$$

$$\mathcal{L}_{\text{B}}(n\ell) \equiv \ln\frac{Z^2}{2|e_n - H|}\,, \tag{26.77}$$

Part B | 26.4

H the HF Hamiltonian, $e_n \equiv e(n\ell)$ the HF energy,

$$U_{\mathrm{v,h}}^{\mathrm{so}}(n\ell) = \frac{3/16\pi}{|\langle \mathbf{0}|n\rangle|^2}\left\langle n\left|\frac{1}{r^3}\right|n\right\rangle C_\ell\,, \qquad (26.78)$$

where

$$C_\ell = \begin{cases} \ell & j=\ell+\frac{1}{2} \\ -(\ell+1) & j=\ell-\frac{1}{2} \end{cases}\,, \qquad (26.79)$$

$$\rho(n\ell) = \frac{\mathrm{i}/4\pi}{|\langle \mathbf{0}|n\rangle|^2}\sum_a \sum\!\!\!\!\!\!\int_{jk}\left\{\frac{(e_k-e_j)(e_a-2e_k+e_j)}{e_a-e_j}\left[{n \atop a}\middle|V\middle|{n \atop j}\right]\langle j|\boldsymbol{r}|k\rangle\cdot\langle k|\boldsymbol{p}|a\rangle+\text{c.c.}\right\}\,, \qquad (26.80)$$

$$L_{\mathrm{core}}(n\ell) = \frac{\mathrm{i}/4\pi}{|\langle \mathbf{0}|n\rangle|^2}\sum_a \sum\!\!\!\!\!\!\int_{jk}\left\{\frac{(e_k-e_j)(e_a-2e_k+e_j)}{e_a-e_j}\left[{n \atop a}\middle|V\middle|{n \atop j}\right]\ln\frac{Z^2}{2|e_a-e_k|}\langle j|\boldsymbol{r}|k\rangle\cdot\langle k|\boldsymbol{p}|a\rangle+\text{c.c.}\right\}\,, \qquad (26.81)$$

$$U_{(1)}(n\ell) = \frac{1/2\pi}{|\langle \mathbf{0}|n\rangle|^2}\sum_a\left\{\left\langle a\left|\boldsymbol{p}^2\right|a\right\rangle\left[{n \atop a}\middle|V\middle|{n \atop a}\right]-\left[{n \atop (a)'}\middle|V\middle|{n \atop (a)'}\right]-\left(\left[{(b)' \atop a}\middle|V\middle|{a \atop (b)'}\right]+\text{c.c.}\right)\delta_{nb}\right\}\,, \qquad (26.82)$$

$$U_{(2)}(n\ell) = \frac{\mathrm{i}/2\pi}{|\langle \mathbf{0}|n\rangle|^2}\sum_a \sum\!\!\!\!\!\!\int_{k<q}\left\{\frac{(2e_a-e_k-e_q)}{e_q-e_k}\left[{n \atop q}\middle|V\middle|{n \atop k}\right]\left(\ln\left|\frac{e_a-e_q}{e_a-e_k}\right|+2\right)\langle a|\boldsymbol{p}|q\rangle\cdot\langle k|\boldsymbol{p}|a\rangle+\text{c.c.}\right\}\,, \qquad (26.83)$$

$$U_{\mathrm{core}}^{\mathrm{so}}(n\ell) = \frac{3/16\pi}{|\langle \mathbf{0}|n\rangle|^2}\sum\!\!\!\!\!\!\int_{aj}\left\{\frac{\delta_{\ell_a\ell_j}}{e_a-e_j}\left[{n \atop a}\middle|V\middle|{n \atop j}\right]\left\langle j\left|\frac{1}{r^3}\right|a\right\rangle C_\ell+\text{c.c.}\right\}\,, \qquad (26.84)$$

$$K^C(n\ell) = \frac{1}{|\langle \mathbf{0}|n\rangle|^2}\sum_a\left[{n \atop a}\middle|\delta^3(\boldsymbol{x}-\boldsymbol{y})\middle|{n \atop a}\right]\,, \qquad (26.85)$$

$$K^L(n\ell) = \frac{7/2}{|\langle \mathbf{0}|n\rangle|^2|}\sum_a\left[{n \atop a}\middle|\frac{1}{4\pi}\nabla^2\left(\frac{\ln(Z|\boldsymbol{x}-\boldsymbol{y}|)+\gamma}{|\boldsymbol{x}-\boldsymbol{y}|}\right)\middle|{n \atop a}\right]\,, \qquad (26.86)$$

where

$$U(n\ell) \equiv U_{(1)}(n\ell)+U_{(2)}(n\ell)\,, \qquad (26.87)$$

$$\langle \boldsymbol{r}|(a)'\rangle \equiv \frac{\mathrm{d}}{\mathrm{d}r}\langle \boldsymbol{r}|a\rangle\,, \qquad (26.88)$$

and γ is the Euler–Mascheroni constant. As in (26.27), x refers to the top row and y to the bottom row in the two-row expression in (26.86).

$L(n\ell)$, which is associated with the valence electron and the nuclear Coulomb potential, has the form of a hydrogenic BL, except that it is calculated with HF wave functions and energies. $L_{\mathrm{core}}(n\ell)$ comes from BL terms associated with core electrons.

The expressions appearing in (26.80), (26.81), and (26.83) originate from the double commutator $[\boldsymbol{p}\cdot[\boldsymbol{p},H]]$, which is then approximated by using the commutator $[\boldsymbol{r},H]\approx \mathrm{i}\boldsymbol{p}$. This is not an exact result because the ee exchange term in the HF potential is neglected. One thus obtains [26.13]

$$\langle m\,|[\boldsymbol{p}\cdot[\boldsymbol{p},H]]|\,n\rangle \approx \mathrm{i}(e_n-2e_k+e_m)(e_k-e_n)\,\langle m\,|\boldsymbol{p}|\,k\rangle\cdot\langle k\,|\boldsymbol{r}|\,n\rangle\,. \qquad (26.89)$$

A less accurate approximation is [26.12]

$$4\pi Z\delta^3(\boldsymbol{r}) \approx [\boldsymbol{p}\cdot[\boldsymbol{p},H]]\,. \qquad (26.90)$$

Using the left-hand side of this equation instead of the approximation (26.89) to the right-hand side would lead to only s-state a, j and p-state k contributions in (26.80) and alternative forms of (26.81) and (26.83).

The major part of the contribution to the energy due to radiative corrections comes from $F_{\mathrm{v,h}}(n,0)$ (s-states), and numerical tests indicate that the principal effect in this term comes from the renormalization of the electron density at the coordinate origin due to electron shielding.

This reduction occurs because the shielded wave function is more spread out than the unshielded one. Thus, one can extend the above results semi-empirically to high Z (for which $F_{\text{core}}(n,\ell)$ and $F_{\text{ee}}(n,\ell)$ play a much smaller role because of their $1/Z$ dependence) by using the hydrogenic results [26.27] for high Z, i. e., replacing Z^4 for hydrogen by $Z(Z_{n,\text{eff}})^3$ for 1P/1H atoms (where $Z_{n,\text{eff}}$ is the shielded nuclear charge number, defined in (26.68), and Z is the unshielded nuclear charge number). The results obtained in this way are competitive with other evaluations of ΔE_n [26.28].

Finally, there is a correction to $F(n,\ell)$ for smaller Z when the integral equations for the self-energy and vacuum polarization contributions are solved more accurately than by iteration. These corrections play little role in s-state energies, but are somewhat more important in p-states for the hard core case (closed shells, $N=2$, 10, etc.). They are expected to provide a much more significant contribution in the soft core case (closed *sub*shells, $N=4$, 12, etc.).

The correction is based on the simpler approximation given in (26.90) of the more accurate (26.89). It is given by

$$\delta F(n,\ell)=\frac{1}{Z\langle\mathbf{0}|n\rangle^2}\sum_a \sum\!\!\!\!\!\!\int_j \left\{\frac{\left[\begin{smallmatrix}n\\a\end{smallmatrix}\middle|V\middle|\begin{smallmatrix}n\\j\end{smallmatrix}\right]}{e_a-e_j}\left[\langle j|\mathcal{E}|a\rangle - Z\mathcal{F}^{(0)}_{\text{v,h}}\langle j|\mathbf{0}\rangle\langle\mathbf{0}|a\rangle\right]+\text{c.c.}\right\}, \tag{26.91}$$

where

$$\langle j|\mathcal{E}|a\rangle = Z\mathcal{F}^{(0)}_{\text{v,h}}\langle j|\mathbf{0}\rangle\langle\mathbf{0}|a\rangle + \sum_b \sum\!\!\!\!\!\!\int_k \left(\frac{\left[\begin{smallmatrix}j\\b\end{smallmatrix}\middle|V\middle|\begin{smallmatrix}a\\k\end{smallmatrix}\right]}{e_b-e_k}\left[\langle k|\mathcal{E}|b\rangle+\text{c.c.}\right]\right), \tag{26.92}$$

$$\mathcal{F}^{(o)}_{\text{v,h}}=\frac{4}{3}\{N(Z)+L\}\,. \tag{26.93}$$

The inhomogeneous term in (26.92) appears only for states $|j\rangle$ and $|a\rangle$ which are s-states. L in (26.93) is taken to be a constant, an approximation sufficient for the desired accuracy, and reflects the fact that $L(n,0)$ for s-states is essentially constant as a function of radial quantum number, and is approximately the same for hydrogen and the HF approximation.

In order to obtain the correction $\delta F(n,\ell)$ of (26.91), it is necessary first to solve the coupled inhomogeneous linear equations of (26.92). While the sum $\sum_a$ over core states is always over a finite number of discrete states, the symbol $\sum\!\!\!\!\!\!\int$ denotes an infinite sum over discrete bound valence states and an integral over the continuum of such states. Indeed, expressions of this type occur throughout the GFA. They can be dealt with by the use of finite basis techniques, for example the B-spline approach [26.29, 30].

References

26.1 A. L. Fetter, J. D. Walecka: *Quantum Theory of Many Particle Systems* (McGraw-Hill, New York 1971)

26.2 Gy. 1Csanak, H. S. Taylor, R. Yaris: Adv. Atom. Mol. Phys. **7**, 287 (1971)

26.3 G. Feldman, T. Fulton: Ann. Phys. (N. Y.) **172**, 40 (1986)

26.4 G. Feldman, T. Fulton: Ann. Phys. (N. Y.) **179**, 20 (1987)

26.5 W. R. Johnson, S. A. Blundell, J. Sapirstein: Phys. Rev. A **37**, 2764 (1988)

26.6 W. R. Johnson, S. A. Blundell, Z. W. Liu, J. Sapirstein: Phys. Rev. A **40**, 2233 (1989)

26.7 G. Feldman, T. Fulton: Ann. Phys. (N. Y.) **152**, 376 (1984)

26.8 S.-S. Liaw, G. Feldman, T. Fulton: Phys. Rev. A **38**, 5985 (1988)

26.9 S.-S. Liaw: Phys. Rev. A **47**, 1726 (1993)

26.10 S.-S. Liaw, F.-Y. Chiou: Phys. Rev. A **49**, 2435 (1994)

26.11 G. Feldman, T. Fulton: Ann. Phys. (N. Y.) **201**, 193 (1990)

26.12 G. Feldman, T. Fulton, J. Ingham: Ann. Phys. (N. Y.) **219**, 1 (1992)

26.13 A. Devoto, G. Feldman, T. Fulton: Ann. Phys. (N. Y.) **232**, 88 (1994)

26.14 T. Fulton, W. R. Johnson: Phys. Rev. **A34**, 1686 (1986)

26.15 T. H. Koopmans: Physica **1**, 104 (1933)

26.16 L. Hostler: J. Math. Phys. **5**, 1234 (1964)

26.17 R. A. Swainson, G. W. F. Drake: J. Phys. A: Math. Gen. **24**, 1801 (1991)

26.18 J. Schwinger: J. Math. Phys. **5**, 1606 (1964)

26.19 R. A. Swainson, G. W. F. Drake: J. Phys. A: Math. Gen. **24**, 95 (1991)

26.20 S.-S. Liaw: Phys. Rev. A **48**, 3555 (1993)

26.21 E. E. Salpeter: Phys. Rev. **87**, 328 (1952)

26.22 S.-S. Liaw, F.-Y. Chiou: Phys. Rev. A **49**, 2435 (1994), Eq. (15)

26.23 G. Feldman, T. Fulton: Ann. Phys. (N. Y.) **201**, 193 (1990), Fig. 6

26.24 J. B. French, V. F. Weisskopf: Phys. Rev. **75**, 1240 (1949)

26.25 G.W. Erickson, D.R. Yennie: Ann. Phys. (N. Y.) **35**, 271 (1965)

26.26 G.W. Erickson, D.R. Yennie: Ann. Phys. (N. Y.) **35**, 447 (1965)

26.27 W.R. Johnson, G. Soff: At. Data Nucl. Data Tables **33**, 405 (1985)

26.28 K.T. Cheng, W.R. Johnson, J. Sapirstein: Phys. Rev. Lett. **66**, 2960 (1991)

26.29 C. de Boor: *A Practical Guide to Splines* (Springer, Berlin, Heidelberg 1987)

26.30 W.R. Johnson, J. Sapirstein: Phys. Rev. Lett. **57**, 1126 (1986)

27. Quantum Electrodynamics

Quantum Electrodynamics (QED) is the underlying theory of atomic and molecular physics. Despite this generality, it is not necessary to use the full theory in most atomic physics problems. This is because in the nonrelativistic limit QED reduces to the Schrödinger equation, and the extra physics in QED is in general quite small, being suppressed by powers of the fine structure constant α. Given the difficulty of solving the Schrödinger equation with high accuracy in most atomic physics situations, these small corrections can usually be neglected. The theory is however needed to explain small deviations from the solution to the Schrödinger equation in simple systems, in particular a single electron in a constant magnetic field and few-electron atoms. Larger deviations occur for highly charged ions, and also for high-energy scattering of electrons and photons. We note that a rather extensive review of QED is available [27.1], and refer the reader interested in more details to that work. In addition, comparison with experiment is made by *Mohr* in Chapt. 28, and thus is done here only in selected cases.

27.1 **Covariant Perturbation Theory** 413
27.2 **Renormalization Theory and Gauge Choices** 414
27.3 **Tests of QED in Lepton Scattering** 416
27.4 **Electron and Muon *g* Factors** 416
27.5 **Recoil Corrections** 418
27.6 **Fine Structure** 420
27.7 **Hyperfine Structure** 421
27.7.1 Muonium Hyperfine Splitting 421
27.7.2 Hydrogen Hyperfine Splitting 422
27.8 **Orthopositronium Decay Rate** 422
27.9 **Precision Tests of QED in Neutral Helium** 423
27.10 **QED in Highly Charged One-Electron Ions** 424
27.11 **QED in Highly Charged Many-Electron Ions** 425
References 427

27.1 Covariant Perturbation Theory

QED combines relativity, electromagnetism, and quantum mechanics. As the first two theories were well understood when quantum mechanics was formulated, the development of the the fundamental equations of QED (after Dirac's introduction of his relativistic equation for the electron) took place rapidly, being in place in 1928 [27.2, 3]. However, it was recognized almost immediately that when higher order perturbation theory was considered, infinities associated with short wavelengths, known as ultraviolet divergences, were present, and that this apparently predicted infinite shifts in spectral lines. These difficulties were not overcome for two decades, but at that time improvements in calculational technology coupled with an understanding that the infinities could be grouped into renormalizations of the electron's mass, charge, and wave function and the photon's wave function, led to the modern form of QED.

A central object in this form of the theory is the S-matrix. To introduce it, we start with the Schrödinger equation,

$$\mathrm{i}\hbar\frac{\partial}{\partial t}\Psi(t) = (H_0 + H_\mathrm{I})\Psi(t) \tag{27.1}$$

where H_0 is the Hamiltonian of free electrons and photons (although this can be easily generalized to include external potentials such as a nuclear Coulomb field), and H_I the electromagnetic interaction between them. These Hamiltonians follow from the Lagrangian density $\mathcal{L} = \mathcal{L}_0 + \mathcal{L}_1$, where

$$\begin{aligned}\mathcal{L}_0 =& \bar{\psi}_0(\boldsymbol{x},t)\left(\gamma_\mu p^\mu - m_0\right)\psi_0(\boldsymbol{x},t)\\ &-\frac{1}{4}F_{0\mu\nu}(\boldsymbol{x},t)F_0^{\mu\nu}(\boldsymbol{x},t)\end{aligned} \tag{27.2}$$

and

$$\mathcal{L}_\mathrm{I} = -e_0\bar{\psi}_0(\boldsymbol{x},t)\gamma_\mu\psi_0(\boldsymbol{x},t)A_0^\mu(\boldsymbol{x},t)\,. \tag{27.3}$$

The 0 subscripts in the above emphasize that the fields and couplings are unrenormalized: renormalization will be discussed in the next section. In addition $\bar{\psi} p^\mu \psi$ is understood to represent $\frac{1}{2}\mathrm{i}\bar{\psi}(\partial^\mu \psi) - \frac{1}{2}\mathrm{i}(\partial^\mu \bar{\psi})\psi$, and gauge fixing terms have been suppressed. By making the unitary transformation

$$\Psi(t) \equiv \mathrm{e}^{-\mathrm{i}H_0 t/\hbar}\Phi(t) , \tag{27.4}$$

which transforms from the Schrödinger to the interaction representation, and further defining the U matrix through

$$\Phi(t) = U(t, -\infty)\Phi(-\infty) , \tag{27.5}$$

one finds an equation for this matrix

$$\mathrm{i}\hbar\frac{\partial}{\partial t}U(t, -\infty) = \hat{H}_{\mathrm{I}}(t)U(t, -\infty) \tag{27.6}$$

where

$$\hat{H}_{\mathrm{I}}(t) = \mathrm{e}^{\mathrm{i}H_0 t/\hbar} H_{\mathrm{I}}\, \mathrm{e}^{-\mathrm{i}H_0 t/\hbar} . \tag{27.7}$$

Solving this equation iteratively then gives for the S-matrix, defined as $S = U(\infty, -\infty)$,

$$S = \sum_{n=0}^{\infty}(-\mathrm{i})^n \frac{1}{n!}\int_{-\infty}^{\infty} \mathrm{d}t_1 \cdots \int_{-\infty}^{\infty} \mathrm{d}t_n \times T\left[\hat{H}_{\mathrm{I}}(t_1), \cdots \hat{H}_{\mathrm{I}}(t_n)\right] , \tag{27.8}$$

where T is the time ordering operator. An initial state consisting of free electrons, positrons, and photons will then have an amplitude to scatter into a final state with different momenta and perhaps different numbers of particles given by the S-matrix. This amplitude can be calculated using Wick's theorem, and the result conveniently represented by Feynman diagrams. Lowest order results of this procedure (tree approximation) describe processes such as electron scattering, electron–positron annihilation, etc. to fairly high precision. However, as mentioned above, when higher terms in the perturbation expansion are considered, diagrams containing closed loops are encountered that are formally infinite, and a renormalization program must be introduced.

27.2 Renormalization Theory and Gauge Choices

Before we discuss renormalization theory, we mention that QED has the same freedom to choose gauge as classical electromagnetism. We will discuss four gauges that have been used in QED calculations, though there is of course an arbitrary number. All of these gauges can be defined through the photon propagator in momentum space. If this is defined by

$$\int \mathrm{d}^4 x\, \mathrm{e}^{-\mathrm{i}k\cdot x}\langle 0|T[A_\mu(x)A_\nu(0)]|0\rangle \equiv -\mathrm{i}\frac{G_{\mu\nu}(k)}{k^2} \tag{27.9}$$

the Coulomb gauge is defined by

$$\begin{aligned} G_{00} &= -\frac{k^2}{\boldsymbol{k}^2} , \\ G_{ij} &= -\left(\delta_{ij} - \frac{k_i k_j}{\boldsymbol{k}^2}\right) , \\ G_{i0} &= G_{0i} = 0 . \end{aligned} \tag{27.10}$$

While this gauge is particularly physical, with the G_{00} part directly corresponding to the instantaneous Coulomb interaction and the G_{ij} to magnetic interactions, it is relatively difficult to work with. For ease of calculation, the covariant gauges, defined by

$$G_{\mu\nu} = g_{\mu\nu} + \beta\frac{k_\mu k_\nu}{k^2} \tag{27.11}$$

are useful, particularly the case $\beta = 0$, the Feynman gauge. Other values of β are $\beta = 2$, the Yennie gauge [27.4], and $\beta = -1$, the Landau gauge. The former has the advantage of controlling infrared divergences, and the latter of controlling ultraviolet divergences.

Two of these infinities are first encountered when the self-energy of a free electron is calculated in second order perturbation theory. In order to deal with finite quantities, we first must introduce a device to regulate the high-frequency range of the integrations. This can be done in a number of ways, among them Pauli–Villars regularization [27.5]. In this method one modifies the photon propagator to

$$1/q^2 \to 1/\left(q^2 - \lambda^2\right) - 1/\left(q^2 - \Lambda^2\right) , \tag{27.12}$$

where λ is a photon mass that regulates infrared divergences and Λ an ultraviolet cutoff mass. In this case the self-energy operator is represented by the Feynman diagram of Fig. 27.1a, and is, using Feynman gauge,

$$\Sigma(p) = -\mathrm{i}e^2 \int \frac{\mathrm{d}^4 k}{(2\pi)^4}\gamma_\mu \frac{1}{\not{p} - \not{k} - m_0}\gamma^\mu \times \left(\frac{1}{k^2 - \lambda^2} - \frac{1}{k^2 - \Lambda^2}\right) . \tag{27.13}$$

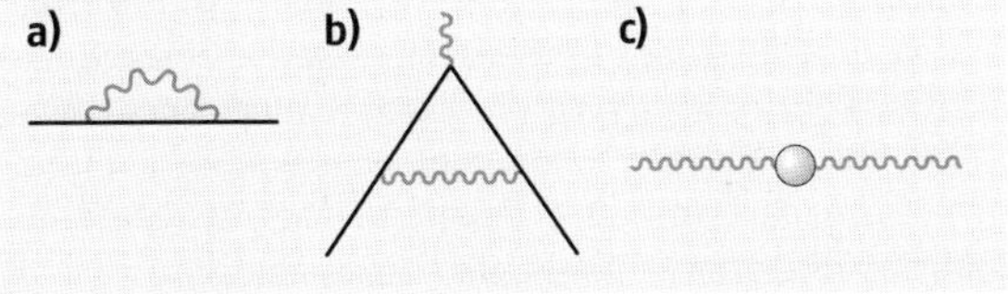

Fig. 27.1a–c Ultraviolet divergent one-loop Feynman diagrams

Combining the two denominators together with a Feynman parameter and carrying out the integration over k then leads to

$$\Sigma(p) = \delta m^{(2)} + B^{(2)}(\not p - m_0) + \Sigma_F(p) , \quad (27.14)$$

where

$$\delta m^{(2)} = \frac{3\alpha m_0}{2\pi}[\ln(\Lambda/m_0) + 1/4] \quad (27.15)$$

and

$$B^{(2)} = -\frac{\alpha}{2\pi}[\ln(\Lambda/m_0) + 2\ln(\lambda/m_0) + 9/4] . \quad (27.16)$$

Σ_F will not be given here, but the important point is that it is ultraviolet finite. Thus the ultraviolet infinities of QED are isolated in the first two terms, which are of a very simple structure.

The next infinity is connected with the vertex function of Fig. 27.1b. This is defined by the equation

$$\Gamma_\mu = -\mathrm{i}e_0^3 \int \frac{\mathrm{d}^4k}{(2\pi)^4} \gamma_\rho \frac{1}{\not p - \not k - m} \gamma_\mu \times \frac{1}{\not p' - \not k - m} \gamma^\rho \left(\frac{1}{k^2 - \lambda^2} - \frac{1}{k^2 - \Lambda^2} \right) . \quad (27.17)$$

While Γ_μ is a fairly complicated function, its ultraviolet divergent part is simply a multiple of γ_μ. When the electron momenta p and p' are equal and on shell, this integral can be evaluated to be

$$\Gamma_\mu = e_0 \gamma_\mu L^{(2)} \quad (27.18)$$

where

$$L^{(2)} = \frac{\alpha}{2\pi}[\ln(\Lambda/m) + 2\ln(\lambda/m) + 9/4] . \quad (27.19)$$

The fact that $L^{(2)} = -B^{(2)}$ is a consequence of the Ward identity [27.6].

The final infinity of second-order QED arises from the vacuum polarization diagram of Fig. 27.1c. The associated integral is

$$\Pi_{\mu\nu}(p) = -\mathrm{i}e^2 \int \frac{\mathrm{d}^4k}{(2\pi)^4} \mathrm{Tr} \times \left(\gamma_\mu \frac{1}{\not p + \not k - m_0} \gamma_\nu \frac{1}{\not k - m_0} \right) . \quad (27.20)$$

In this case the integral cannot be regulated as before, since there is no photon propagator to modify. Instead one can subtract a similar integral with the electron mass replaced by M. Vacuum polarization is particularly sensitive to ultraviolet divergences, since the nominal order of the divergence is quadratic. However, gauge invariance requires $\Pi_{\mu\nu}(p)$ to have the structure $(p^2 g_{\mu\nu} - p_\mu p_\nu)\Pi(p^2)$, and if one considers only the $p_\mu p_\nu$ part of the vacuum polarization integral, the ultraviolet divergence is only logarithmic. This divergence is independent of the photon momentum p, and one can write

$$\Pi(p^2) = C^{(2)} + \Pi_{\text{finite}}(p^2) \quad (27.21)$$

where

$$C^{(2)} = \frac{\alpha}{3\pi} \ln\left(M^2/m_0^2\right) \quad (27.22)$$

and

$$\Pi_{\text{finite}}(p^2) = -\frac{2\alpha}{\pi} \int_0^1 \mathrm{d}x\, x(1-x) \times \ln\left[1 - x(1-x)p^2/m_0^2\right] . \quad (27.23)$$

The four infinite quantities encountered in second order perturbation theory are modified by higher order corrections, but no new divergent structures arise. The basic idea of renormalization is to note that these structures are already present in the lowest order Lagrangian. We now make the following definitions:

$$m = m_0 + \delta m , \quad (27.24)$$

$$\psi(x) = Z_2^{-1/2} \psi_0(x) , \quad (27.25)$$

$$A^\mu(x) = Z_3^{-1/2} A_0^\mu(x) , \quad (27.26)$$

and

$$e = e_0 Z_1^{-1} Z_2 Z_3^{1/2} . \quad (27.27)$$

These correspond to an additive renormalization of the electron mass and multiplicative renormalizations of the electron and photon wave functions and the electron charge. Rewriting the original bare Lagrangian in terms of these renormalized quantities then gives that Lagrangian without the 0 subscripts, plus the following counterterms:

$$\mathcal{L}_{CT1} = Z_2 \delta m \bar\psi(\boldsymbol{x}, t)\psi(\boldsymbol{x}, t) , \quad (27.28)$$

$$\mathcal{L}_{CT2} = -e(Z_1 - 1)\bar\psi(\boldsymbol{x}, t)\gamma_\mu \psi(\boldsymbol{x}, t) A^\mu(\boldsymbol{x}, t) , \quad (27.29)$$

$$\mathcal{L}_{CT3} = -\frac{1}{4}(Z_3 - 1) F_{\mu\nu}(\boldsymbol{x}, t) F^{\mu\nu}(\boldsymbol{x}, t) , \quad (27.30)$$

and

$$\mathcal{L}_{CT4} = (Z_2 - 1)\bar{\psi}(\boldsymbol{x}, t)\left(\gamma_\mu p^\mu - m\right)\psi(\boldsymbol{x}, t) \,. \tag{27.31}$$

By choosing $Z_2 = 1 + B^{(2)}$, $Z_1 = 1 - L^{(2)}$, $\delta m = \delta m^{(2)}$, and $Z_3 = 1 - C^{(2)}$, these counterterms will precisely cancel the previously encountered divergences in second order. At this point we identify m and e as the experimentally determined mass and charge of the electron: as long as these are used, the radiative corrections discussed above have no effect for free electrons. However, when an electron undergoes scattering or is in the presence of an external magnetic or nuclear Coulomb field, the finite terms no longer vanish, and give rise to small corrections. We now turn to a discussion of these corrections.

27.3 Tests of QED in Lepton Scattering

The highest energy tests of QED come from scattering experiments at accelerators. While the dominant part of QED corrections for all the other tests discussed in this chapter involves electron and photon propagators close to the mass shell, scattering experiments involve propagators very far off the mass shell, which allows tests of the theory at very small distances. It is standard to parameterize possible deviations from the predictions of QED at these small distances by the introduction of form factors of the form

$$F(q^2) = 1 - \frac{q^2}{q^2 - \Lambda^2} \tag{27.32}$$

where q is photon momentum at an electron–photon vertex. In QED Λ is infinite and this form factor is unity even at very high q^2, but this can be tested in various scattering experiments. For example, Bhabha scattering, $e^+ e^- \to e^+ e^-$, has been accurately measured at high center of mass energy, $\sqrt{s} = 34.8\,\text{GeV}$, at TASSO [27.1, 7]. To compare with QED, very sizable radiative corrections must be carefully calculated, and at these energies electroweak effects involving the Z boson, while small, must also be considered. Although the accuracy of the experiments is not high compared with atomic physics measurements, being at the percent level, the good agreement with QED that is found allows lower limits on the cutoff $\Lambda > 500\,\text{GeV}$ to be placed. This corresponds to distances of under 10^{-16} cm. It is of interest to compare this sensitivity with that available from atomic physics tests. The change in the photon propagator given above corresponds to a potential $e^{-\Lambda r}/r$. This would lead to an energy shift of a 2s electron in hydrogen of $46/\Lambda^2$ kHz with Λ in units of GeV. Thus even 1 kHz accuracy in the Lamb shift would only restrict $\Lambda > 7\,\text{GeV}$.

27.4 Electron and Muon *g* Factors

One of the successes of the Dirac equation is the prediction $g = 2$ for the electron. The leading correction to this result coming from QED is the Schwinger correction [27.8],

$$g = 2\left(1 + \frac{\alpha}{2\pi}\right) . \tag{27.33}$$

While in principle this is an external field problem, because of the weakness of laboratory magnetic fields, the correction can be related to Feynman diagrams with free propagators. To see the weakness we note that $eB/m_e^2 = 2.3 \times 10^{-14}\,\text{B}$ (Gauss). In extremely intense magnetic fields such as can be encountered in astrophysical situations, a bound state approach [27.9] should be used both for calculating energy shifts and the imaginary part of these shifts, which describe synchrotron radiation. An example of the more precise approach is Demeur's formula for the (real) energy shift of an electron in the lowest energy level

$$\begin{aligned}\Delta E = \frac{m\alpha}{2\pi}\Bigg[& -\frac{eB}{2m^2} + \left(\frac{eB}{m^2}\right)^2 \left(\frac{4}{3}\ln\frac{m^2}{2eB} - \frac{13}{18}\right) \\ & + \left(\frac{eB}{m^2}\right)^3 \left(\frac{14}{3}\ln\frac{m^2}{2eB} - \frac{32}{5}\ln 2 + \frac{83}{90}\right) \\ & + \cdots\Bigg] .\end{aligned} \tag{27.34}$$

The second term could actually be seen at the present level of precision, but is spin-independent, and the third term is negligible. An interesting feature of the experiment is the effect of the conducting cavity, which must be understood to extract the correct value of $g - 2$ [27.1, 10].

After the initial verification of the Schwinger correction, experiments of increasing precision culminating in the Penning trap measurements in Washington [27.11] have stimulated advances in theoretical calculations. These involve the evaluation of constants C_i defined by

$$a_e = C_1 \frac{\alpha}{\pi} + C_2 \left(\frac{\alpha}{\pi}\right)^2 + C_3 \left(\frac{\alpha}{\pi}\right)^3 + C_4 \left(\frac{\alpha}{\pi}\right)^4 + \cdots \quad (27.35)$$

where $a_e = (g-2)/2$ is the anomalous magnetic moment of the electron. The computational effort involved in computing the coefficients C_i increases very rapidly with i, and the four loop calculation is the largest QED calculation ever carried out. The situation with regard to these calculations is as follows. After the calculation of the Schwinger correction, the next step was the evaluation of the seven Feynman diagrams of Fig. 27.2. A feature of the one-loop calculation, that it is ultraviolet finite, is no longer present at this level, and renormalizations of the self-mass, vertex, and wave function must be performed, although the latter two cancel by Ward's identity. When the calculation is carried out in Feynman gauge, each graph has an infrared divergence that must be regulated in some fashion, for example by giving the photon a small mass λ. This calculation was first correctly carried out by *Sommerfield* [27.12] and *Petermann* [27.13]. The result is

$$C_2 = \frac{197}{144} + \frac{\pi^2}{6}(0.5 - 3\ln 2) + 0.75\zeta(3) \quad (27.36)$$

where $\zeta(3) = 1.202\,05\ldots$ is the Riemann zeta function of argument 3. While each individual diagram is infrared divergent in Feynman gauge, *Adkins* [27.14] has used Yennie gauge [27.4] to recalculate the effect, free of infrared divergent terms. His results are given in Table 27.1.

The vacuum polarization graph of Fig. 27.2e plays an interesting role. While the result for C_2 given above includes only the case where the intermediate particles in the loop are electrons, that loop can also involve any charged particle, such as the muon, pion, or tau. However, because all these particles are much heavier than the electron, which sets the energy scale of the Feynman integral, their effect is suppressed by the square of the mass ratio of the electron to their mass, and are thus quite small. Specifically, in units of 10^{-12}, the muon loop contributes 2.80, the tau loop 0.01, and hadrons 1.6(2). These act to increase C_2 by 0.000 000 82. We note in passing that the effect of the weak interactions enters in one loop, and contributes 0.05×10^{-12} to the electron $g-2$ value and $1\,95(10) \times 10^{-11}$ for the muon.

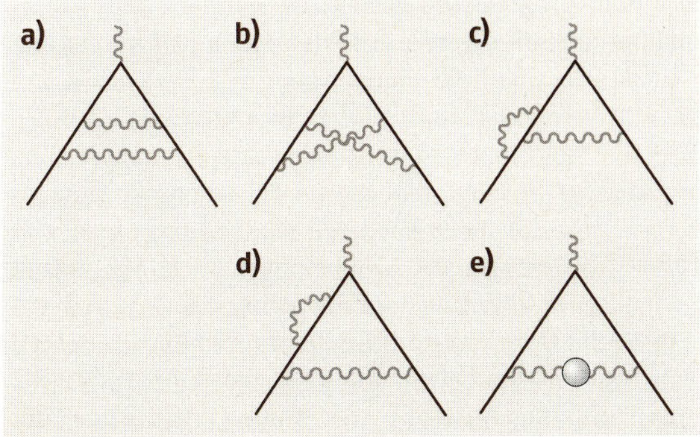

Fig. 27.2a–e Two-loop Feynman diagrams contributing to $g-2$

Table 27.1 Contributions to of C_2 in Yennie gauge

Graph	Value
a	$-\frac{1}{2}B^{(2)} - \frac{3}{16}$
b	$\frac{5}{4}\zeta(3) - \frac{5}{6}\pi^2 \ln 2 + \frac{5}{12}\pi^2 + \frac{7}{12}$
c	$-B^{(2)} - \frac{1}{2}\zeta(3) + \frac{1}{3}\pi^2 \ln 2 - \frac{29}{24}$
d	$B^{(2)} - \frac{9}{8}$
e	$-\frac{1}{3}\pi^2 + \frac{119}{36}$
Counterterm	$\frac{1}{2}B^{(2)}$

When the anomalous magnetic moment of the muon is considered the situation changes significantly. Firstly, the contribution of electrons in the loop is enhanced, since they are now relatively light particles, and the vacuum polarization loop behaves as a logarithm of the mass ratio. Specifically, the difference between the muon and electron $g-2$ factors behaves as [27.15]

$$a_\mu - a_e = 1.094\left(\frac{\alpha}{\pi}\right)^2 + 22.9\left(\frac{\alpha}{\pi}\right)^3 + 132.7\left(\frac{\alpha}{\pi}\right)^4 + \cdots \quad (27.37)$$

These large coefficients arise primarily because the vacuum polarization loops involving electrons change the effective value of α to $\alpha/(1 - 2\alpha/3\pi \ln m_\mu/m_e)$. Note however that such logarithmic terms also arise from other sources, most notably the light-by-light scattering graphs that enter first in third order.

The second important change in the muon case is the significant role of strongly interacting particles in the loop. Fortunately, while our present inability to carry out high accuracy calculations of the strong interactions could in principle interfere with the interpretation of the muon $g-2$ as a QED test, the bulk of this contribution can be related to the experimentally available cross-section for $e^+ e^-$ annihilation into hadrons [27.16]. It is also possible to use τ decay to determine the contribution [27.17]. At present the two methods are not in agreement, and this situation will have to be resolved before a possible discrepancy between theory and the most recent experiment [27.18] can be interpreted as indicating new physics. Specifically, if the τ data is used a 1.4 standard deviation difference exists, but if the $e^+ e^-$ data is used the discrepancy increases to 2.7 standard deviations.

The calculation of C_3 involves 72 Feynman graphs, although they can be grouped together into a smaller number of gauge invariant sets. As discussed in more detail in Chapt. 28, C_3 is now known analytically, removing an important source of numerical uncertainty. The evaluation of such high-order graphs requires an intricate set of subtractions to lead to finite answers, and provides a practical demonstration of the renormalizability of QED. The result for C_3 is

$$C_3 = 1.181\,241\,456\ldots \tag{27.38}$$

precision measurements of recoil in cesium [27.20, 21], which has achieved 7 ppb and has the potential of reach-

Finally, the very large scale calculation of C_4, which is almost completely numerical, has been carried out by *Kinoshita* and *Lindquist* [27.19]. When their result,

$$C_4 = -1.509\,8(384)\,, \tag{27.39}$$

is compared with experiment, agreement is found, but the largest source of error is the uncertainty in the fine structure constant as determined from solid state physics. However, if one instead assumes the validity of QED, the situation can be turned around to determine a QED value of the fine structure constant. This is done by combining the experimental result [27.11]

$$\begin{aligned} a_{e^-} &= 1\,159\,652\,188.4(4.3)\times 10^{-12} \\ a_{e^+} &= 1\,159\,652\,187.9(4.3)\times 10^{-12} \end{aligned} \tag{27.40}$$

with the previous formulas for the C coefficients (including the small vacuum polarization corrections for C_2 discussed above). The result is

$$\alpha^{-1}_{\mathrm{QED}} = 137.035\,992\,22(51)(48) \tag{27.41}$$

where the errors come from experiment and C_4 respectively. It would be of great interest to have another QED determination of the fine structure constant of comparable accuracy: a very promising approach involves ing 1 ppb. Another way to determine α involves the fine structure of helium, as will be discussed below.

27.5 Recoil Corrections

Because of the smallness of the ratio of the electron mass to most nuclear masses, a reasonable approximation to atoms is to treat the nucleus as a source of a classical Coulomb and magnetic dipole field, which corresponds to the use of *Furry* representation [27.22]. In addition, the leading correction to this approximation can be accounted for by using the reduced mass in place of the electron mass. However, to calculate higher order terms consistently, one must treat a one-electron atom as a two-body system and an N-electron atom as a $N+1$-body system. Shortly after the development of modern QED, a number of workers [27.23–25] developed two-body equations by considering the Green's function for electron–nucleus scattering, with the original equation known as the Bethe–Salpeter equation. Considered as a function of the total energy in the c.m. frame, this Green's function has poles at bound state energies. In practice an approximate Green's function is considered that has poles at either the Schrödinger or the Dirac energies, with reduced mass built in to some degree. Then a perturbation theory is set up that allows corrections of higher order in α and m_e/m_N to be calculated in a systematic way.

While the treatment of recoil using various versions of the Bethe–Salpeter equation gives correct answers, its implementation is quite complicated. In recent years enormous progress has been made using the very different approach of effective field theory. One of the earliest uses of effective field theory was in QED [27.26], and we will refer to it as NRQED (nonrelativistic QED). Effective field theories have been used in many different areas of physics, and are useful whenever physics at one scale can be treated separately from physics at a widely different scale. Atomic physics is an ideal place for the use of effective field theory, as, for example, the scale of the Bohr radius, the basic atomic physics length scale, is separated from the electron Compton wavelength, which is characteristic of most QED effects, by two orders of magnitude. NRQED applies to both recoil and non-recoil QED corrections, and always has as its starting point the well-understood Schrödinger equation. Relativistic effects, such as relativistic mass increase, magnetic interactions, or the Darwin term, are then included perturbatively. This requires cutoff methods to deal with higher orders of perturbation theory, where those operators lead to ultraviolet singular results. In addition, the short distance physics of QED is accounted for by adding in perturbations that involve delta functions or derivatives of delta functions, with coefficients determined by what is called a matching procedure, in which scattering calculations in full QED and in the effective field theory are forced to agree. This approach, which unfortunately has not yet been treated at textbook level, has had a very great impact on higher order QED calculations, with a number of higher order effects calculated in recent years using the technique.

To illustrate the method, we follow a treatment of *Pachucki* [27.27] in which the Dirac energies of hydrogen in the non-recoil limit are calculated to order $m\alpha^6$. While he treated the general case, for simplicity we consider here only the ground state energy, which has the Taylor expansion

$$E = mc^2\left[1-\frac{1}{2}(Z\alpha)^2-\frac{1}{8}(Z\alpha)^4-\frac{1}{16}(Z\alpha)^6\right], \tag{27.42}$$

where we follow the convention of allowing for a general nuclear charge Z. The fine structure, as is well known, can be derived from perturbations associated with the relativistic mass increase and the Darwin term, with the spin–orbit interaction not contributing for s-states. While the contribution of these perturbations is finite in lowest order perturbation theory, they give rise to singularities when treated in second order. To see this, we give the momentum space version of second order perturbation theory,

$$E^{(2)} = \int \frac{\mathrm{d}^3p\,\mathrm{d}^3k\,\mathrm{d}^3l\,\mathrm{d}^3q}{(2\pi)^9}\phi_0(\boldsymbol{p})V(\boldsymbol{p},\boldsymbol{k}) \times G_{\mathrm{R}}(\boldsymbol{k},\boldsymbol{l})V(\boldsymbol{l},\boldsymbol{q})\phi_0(\boldsymbol{q}), \tag{27.43}$$

where G_{R} is the reduced Coulomb Green's function. It can be expanded in terms of a free term, a one-potential term, and a many-potential term. The strongest singularities are associated with the free term,

$$G^0_{\mathrm{R}}(\boldsymbol{k},\boldsymbol{l}) = -\frac{(2\pi)^3 2m\delta^3(\boldsymbol{k}-\boldsymbol{l})}{k^2+\gamma^2}, \tag{27.44}$$

where $\gamma = mZ\alpha$. The Darwin term in momentum space is simply $V_{\mathrm{D}} = \pi Z\alpha/2m^2$, and it is simple to see that when both V's in the expression for $E^{(2)}$ are Darwin terms and the free part of the Green's function is used a linearly divergent integral results. This can be regulated by imposing a cutoff Λ on the magnitude of all momenta in the integral, in which case a simple calculation gives

$$E^{(2)}(\mathrm{DD0}) = -\frac{\Lambda(Z\alpha)^5}{4\pi}+\frac{m(Z\alpha)^6}{8}. \tag{27.45}$$

Linear divergences also exist when two relativistic mass increase (RMI) terms or Darwin-RMI cross terms are considered, but these terms happen to cancel.

To get a finite answer, the contribution of operators of intrinsic order $m(Z\alpha)^6$ must be considered. These operators can be obtained from consideration of the Bethe–Salpeter equation, but a great simplification of NRQED is the fact that they can also be obtained by considering free-particle scattering, where the complications of the bound state problem are not present. We illustrate this by considering electron scattering in a Coulomb potential created by a stationary charge $Z|e|$. In the Dirac theory this is given by

$$\begin{aligned}&\frac{-4\pi Z\alpha}{|\boldsymbol{p}_2-\boldsymbol{p}_1|^2}\bar{\psi}(\boldsymbol{p}_2)\gamma_0\psi(\boldsymbol{p}_1)\\ &= \frac{-4\pi Z\alpha}{|\boldsymbol{p}_2-\boldsymbol{p}_1|^2}\left(1-\frac{|\boldsymbol{p}_2-\boldsymbol{p}_1|^2}{8m^2}\right.\\ &\quad\left.+\frac{6|\boldsymbol{p}_2-\boldsymbol{p}_1|^2\left(p_1^2+p_2^2\right)+5\left(p_2^2-p_1^2\right)^2}{128m^4}\right).\end{aligned} \tag{27.46}$$

In the above we have carried out a Taylor expansion to fourth order in the electron momenta and dropped spin–orbit terms. While this coincides with Schrödinger theory in lowest order, a set of extra terms exists in the Dirac case. To account for these, we modify the nonrelativistic theory by adding extra operators that make the theories agree. If the Taylor expansion is stopped in order p^2, we recognize the Darwin term already treated above, but to go to order $m(Z\alpha)^6$ the last terms must be treated. They are again linearly divergent, as are two other terms – one associated with the next term in RMI,

$$V_{\mathrm{RMI}} = -(2\pi)^3\delta(\boldsymbol{k}-\boldsymbol{l})\left(\frac{|\boldsymbol{k}|^4}{8m^3}-\frac{|\boldsymbol{k}|^6}{16m^5}\right) \tag{27.47}$$

and the other a term that can either be derived with a Foldy–Wouthuysen transformation or by comparing two-Coulomb photon scattering in the Dirac and Schrödinger theories. When combined, first order perturbation theory for these operators cancels out the linear divergence found above arising from second order perturbation theory, along with a logarithmic singularity, leaving a finite answer in agreement with the Dirac theory.

27.6 Fine Structure

The fine structure of hydrogenic atoms in the non-recoil limit is correctly described by the Dirac equation, which gives for a state of principal quantum number n and spin j $E_{nj} = m_e f(n, j)$, with

$$f(n, j) = \left(1 + \frac{(Z\alpha)^2}{(n-\beta)^2}\right)^{-1/2} . \quad (27.48)$$

Here

$$\beta \equiv j + \frac{1}{2} - \sqrt{\left(j+\frac{1}{2}\right)^2 - (Z\alpha)^2} , \quad (27.49)$$

which gives the expansion

$$E_{nj} = m\left\{1 - \frac{(Z\alpha)^2}{2n^2} - \frac{(Z\alpha)^4}{2n^3}\left(\frac{1}{j+\frac{1}{2}} - \frac{3}{4n}\right) + \dots\right\} \quad (27.50)$$

With the use of the Bethe–Salpeter equation or NRQED one can include recoil corrections exactly to order $m(Z\alpha)^4$, and to this order

$$E = M + m_r[f(n, j) - 1] - \frac{m_r^2}{2M}[f(n, j) - 1]^2 + \frac{(Z\alpha)^4 m_r^3}{2n^3 m_N^2}\left(\frac{1}{j+\frac{1}{2}} - \frac{1}{l+\frac{1}{2}}\right)(1-\delta_{l0}) \quad (27.51)$$

The last term in this expression leads to a slight breaking of the degeneracy of the $2s_{1/2}$ and $2p_{1/2}$ states: this splitting is referred to as the Lamb shift. A larger breaking arises from the finite size of the nucleus, which in the nonrelativistic limit shifts s-states by

$$\Delta E_n(\text{finite size}) = \frac{2}{3n^3}(Z\alpha)^4 m_r^3 \langle r^2 \rangle . \quad (27.52)$$

A major issue for the Lamb shift in hydrogen is the disagreement between measurements at Stanford [27.28] and Mainz [27.29] of the charge radius of the proton. As discussed in Mohr's chapter, the 18 kHz difference creates difficulties in interpreting the experimental status of the Lamb shift. While assuming the validity of QED and completing all calculations that enter at the few kHz level may allow one to determine the proton charge radius independently, it is clearly highly desirable that a definitive electron-scattering experiment be carried out. A promising alternative resolution to the problem may come from measurements on muonic hydrogen undergoing at PSI. Because the scale of such atoms is two orders of magnitude smaller than hydrogen, the effect of proton size is greatly enhanced, and a measurement of the Lamb shift would then allow a very precise measurement of the proton charge radius.

The largest correction to the Lamb shift comes from the self-energy and vacuum polarization graphs previously introduced in Sect. 27.2, with the understanding that the free electron propagators are replaced by Dirac–Coulomb propagators. The self-energy diagram in Feynman gauge gives an energy shift

$$\Delta E_n(\text{SE}) = -\mathrm{i}e^2 \int \mathrm{d}^3 r\, \mathrm{d}^3 r' \times \int \frac{\mathrm{d}^4 k}{(2\pi)^4} \frac{\mathrm{e}^{\mathrm{i}\boldsymbol{k}\cdot(\boldsymbol{r}-\boldsymbol{r}')}}{k^2 + i\epsilon} \bar{\psi}_n(\boldsymbol{r}) \times \gamma_\mu S_F(\boldsymbol{r}, \boldsymbol{r}'; E_n - k_0)\gamma^\mu \psi_n(\boldsymbol{r}') \quad (27.53)$$

and the vacuum polarization a shift

$$\Delta E_n(\text{VP}) = -\mathrm{i}e^2 \int \mathrm{d}^3 r \int \mathrm{d}^3 r' \times \int \frac{\mathrm{d}^3 k}{(2\pi)^3} \int \frac{\mathrm{d}E}{2\pi} \bar{\psi}_n(\boldsymbol{r}) \times \gamma_0 \psi_n(\boldsymbol{r}) \frac{\mathrm{e}^{\mathrm{i}\boldsymbol{k}\cdot(\boldsymbol{r}-\boldsymbol{r}')}}{\boldsymbol{k}^2} \mathrm{Tr}\left[\gamma^0 S_F(\boldsymbol{r}', \boldsymbol{r}'; E)\right] . \quad (27.54)$$

In both cases the counterterms discussed above are understood to be added. It is conventional to pull out the overall behavior in Z, α, and the principal quantum number n as follows:

$$\Delta E_n(\text{SE}) + \Delta E_n(\text{VP}) \equiv m\alpha \frac{(Z\alpha)^4}{\pi n^3} F_n(Z\alpha) . \quad (27.55)$$

The evaluation of the function $F_n(Z\alpha)$ is quite difficult. At low Z, a perturbative expansion to order $(Z\alpha)^2$ is available, and is expressed as

$$F_n(Z\alpha) = A_{40} + A_{41}\ln(Z\alpha)^{-2} + (Z\alpha)A_{50} + (Z\alpha)^2\left[A_{62}\ln^2(Z\alpha)^{-2} + A_{61}\ln(Z\alpha)^{-2} + A_{60}\right] + \cdots \quad (27.56)$$

The constants A can be found in *Mohr*'s Chapt. 28. A_{40} contains a constant that must be obtained numerically, known as the Bethe logarithm. It can be defined from a nonrelativistic limit of the Coulomb gauge version of (27.53), in which one carries out the integral over k_0 with Cauchy's theorem, keeping only the photon pole, makes the dipole approximation, and replaces α with $\boldsymbol{p}/m$. In that limit one has

$$\Delta E_n^{\mathrm{NR}} = \frac{2e^2}{3m^2} \int_0^{\Lambda} \frac{\mathrm{d}^3k}{2\omega(2\pi)^3} \sum_m \frac{|\boldsymbol{p}_{nm}|^2}{E_n - \omega - E_m}, \tag{27.57}$$

where $\boldsymbol{p}_{nm} \equiv \langle n|\boldsymbol{p}|m\rangle$. Because negative energy states are not present, this expression diverges linearly with Λ. The Bethe logarithm is defined by subtracting $-1/\omega$ from the above denominator and carrying out the d^3k integration, giving

$$\Delta E_n^{\mathrm{NR}} = \frac{e^2}{6m^2\pi^2} \sum_m |\boldsymbol{p}_{nm}|^2 (E_m - E_n) \ln \frac{\Lambda}{|E_m - E_n|} \equiv \frac{4m\alpha(Z\alpha)^4}{3\pi n^3} \ln\left(\frac{\Lambda}{k_0(n)\mathrm{Ry}}\right). \tag{27.58}$$

Elegant high accuracy determinations of these quantities using Schwinger's Coulomb Green's function have been made [27.30, 31]. Reference [27.32] contains an extensive tabulation.

We next discuss recoil corrections to the Lamb shift. One recoil effect is the modification of the A constants by a factor $(m_\mathrm{r}/m_\mathrm{e})^3$ (although part of A_{40} is modified instead by $(m_\mathrm{r}/m_\mathrm{e})^2$). Another correction, similar to the Lamb shift but involving a photon exchanged between the electron and nucleus was first derived by *Salpeter* [27.25], and is given by

$$\Delta E_n = \frac{m_\mathrm{r}^3}{m_\mathrm{e} m_N} \frac{(Z\alpha)^5}{\pi n^3} \left[\frac{2}{3}\delta_{l0} \ln\left(\frac{1}{Z\alpha}\right) - \frac{8}{3}\ln k_0(n) - \frac{7}{3}a_n - \frac{1}{9}\delta_{l0} - \frac{2}{m_N^2 - m_\mathrm{e}^2}\delta_{l0} \times \left(m_N^2 \ln\frac{m_\mathrm{e}}{m_\mathrm{r}} - m_\mathrm{e}^2 \ln\frac{m_N}{m_\mathrm{r}}\right)\right], \tag{27.59}$$

where $a_{1\mathrm{s}} = -3 - \ln 4$, $a_{2\mathrm{s}} = -9/2$, and $a_{2\mathrm{p}} = -13/3$. Present research is concerned with corrections of order $\alpha^6 m_\mathrm{e}^2/m_N$.

Finally we discuss the two-loop Lamb shift. The leading contribution is [27.33–35]

$$\Delta E_n(\text{two-loop}) = \frac{m\alpha^2(Z\alpha)^4}{\pi^2 n^3} H_n(Z\alpha) \tag{27.60}$$

where for s-states

$$H_n = -\frac{4358}{1296} - \frac{10}{27}\pi^2 + \frac{3}{2}\pi^2 \ln 2 - \frac{9}{4}\zeta(3) \tag{27.61}$$

and for non-s states

$$H_n = \left[\frac{197}{72} + \frac{\pi^2}{6} - \pi^2 \ln 2 + \frac{3}{2}\zeta(3)\right] \frac{C_{lj}}{2(2l+1)} \tag{27.62}$$

where $C_{lj} = 2(j-l)/(j+1/2)$. The $Z\alpha$ corrections to this quantity have been carried out [27.36, 37], and considerable progress has been made in calculating terms of order $(Z\alpha)^2$ [27.38], though a complete calculation of the constant in that order has not yet been carried out.

27.7 Hyperfine Structure

27.7.1 Muonium Hyperfine Splitting

After the electron $g-2$, the best test of QED is afforded by ground state muonium hyperfine splitting. This is dominated by the Fermi splitting, which is given by

$$E_\mathrm{F} = \frac{16}{3}\alpha^2 \frac{m_\mathrm{r}^3}{m_\mathrm{e}^2 m_\mu} c R_\infty \tag{27.63}$$

where m_r is the reduced mass. The leading correction to the Fermi splitting involves the electron and muon anomalous magnetic moment, which act to increase the splitting by a factor $(1+a_\mathrm{e})(1+a_\mu)$. It is convenient to split the remaining corrections into non-recoil and recoil parts. The non-recoil terms are given by

$$\Delta\nu_{\text{non-recoil}} = (1+a_\mu)\left\{1 + a_\mathrm{e} + \frac{3}{2}(Z\alpha)^2 + \alpha(Z\alpha)\left(\ln 2 - \frac{5}{2}\right) - \frac{8\alpha(Z\alpha)^2}{3\pi}\ln(Z\alpha) \times \left[\ln(Z\alpha) - \ln 4 + \frac{281}{480}\right] + \frac{\alpha(Z\alpha)^2}{\pi}[14.88\,(0.29)] + \frac{\alpha^2(Z\alpha)}{\pi} D_1\right\} E_\mathrm{F}. \tag{27.64}$$

A recent development has been the complete evaluation of D_1, which is a binding correction to the two-loop $g-2$ contribution, by *Kinoshita* and *Nio* [27.39], who found $D_1 = 0.82(4)$. Note also the constant 14.88 has changed from the previous value of 15.38 because of a new calculation of the vacuum polarization component, as discussed further in Mohr's chapter. Subtracting $\Delta\nu_{\text{non-recoil}}$ from experiment leaves 795 kHz to be accounted for by recoil. The present state of recoil corrections is given by

$$\begin{aligned}\Delta\nu_{\text{recoil}} = \Bigg\{ & -\frac{3Z\alpha}{\pi}\frac{m_e m_\mu}{m_\mu^2 - m_e^2}\ln\frac{m_\mu}{m_e} + \frac{(Z\alpha)^2 m_r^2}{m_e m_\mu} \\ & \times\left[-2\ln(Z\alpha) - 8\ln 2 + \frac{65}{18}\right] \\ & + \frac{\alpha(Z\alpha)}{\pi^2}\frac{m_e}{m_\mu}\Bigg[-2\ln^2\frac{m_\mu}{m_e} + \frac{13}{12}\ln\frac{m_\mu}{m_e} \\ & + \frac{21}{2}\zeta(3) + \frac{\pi^2}{6} + \frac{35}{9} + 2.15(14) \\ & + \frac{\alpha}{\pi}\left(-\frac{4}{3}\ln^3\frac{m_\mu}{m_e} + \frac{4}{3}\ln^2\frac{m_\mu}{m_e}\right)\Bigg]\Bigg\} E_F\,. \end{aligned} \tag{27.65}$$

The full power of modern forms of the Bethe–Salpeter equation was needed for the evaluation of the second term [27.40]. While these terms account for just the needed 795 kHz mentioned above, the relatively large uncertainty in the muon mass leads to an uncertainty in the Fermi splitting of 1.3 kHz, and further progress in muonium hyperfine splitting will need this uncertainty to be reduced.

27.7.2 Hydrogen Hyperfine Splitting

The situation in hydrogen hyperfine splitting is quite different from the muonium case because of the structure of the proton. The recoil corrections in (27.65) that involve $\ln m_\mu/m_e$ involve the offshell muon propagator. Because the muon is a pointlike particle, there is no uncertainty in the calculation. However, when it is replaced by a proton, two strong-interaction problems arise that limit the theoretical accuracy that can be attained. The first arises from the fact that the charge distribution of the proton modifies the hyperfine splitting. The fractional correction can be expressed in terms of the electric and magnetic form factors of the proton as follows:

$$\begin{aligned}\delta_p(\text{Zemach}) = & \frac{2\alpha m_e}{\pi^2}\int\frac{d^3p}{p^4} \\ & \times\left(\frac{G_E(-p^2)G_M(-p^2)}{1+\kappa} - 1\right),\end{aligned} \tag{27.66}$$

where κ is the anomalous magnetic moment of the proton. It contributes -38.72 ppm, with an error of 0.5 ppm coming from the uncertainty in the form factors. The second correction, δ_{recoil}, replaces the first term in $\Delta\nu_{\text{recoil}}$ discussed in muonium hfs. The logarithm of the ratio of the muon and electron masses in that expression arises from high internal momenta. In the case of hydrogen hfs, this is sensitive to details of the proton structure. The most recent evaluation of this quantity [27.41] gives a 5.68 ppm effect, which is smaller than the Zemach correction only because of cancellation of individual terms of comparable size. When these corrections are added to the the other QED corrections, theory agrees with experiment at under the 1 ppm level, with hadronic uncertainties of about 0.5 ppm. This limits the size of hadronic effects arising from the polarizability of the proton, which is known to be bounded in magnitude by 4 ppm from inelastic electron scattering data [27.42]. As with the case of the Lamb shift in hydrogen, further progress will require better understanding from either an experimental or theoretical approach of properties of the proton.

27.8 Orthopositronium Decay Rate

The decay rate of orthopositronium has been measured with increasing accuracy over the years, and has reached the level where corrections of order α^2 to the lowest order decay rate need to be considered. The most recent experiment [27.43] has determined

$$\Gamma_{\text{experiment}} = 7.0404(10)(8)\,\mu\text{s}^{-1}\,. \tag{27.67}$$

The lowest order rate is

$$\Gamma_0 = 2\frac{\pi^2-9}{9\pi}m\alpha^6 = 7.211\,1670(1)\,\mu\text{s}^{-1}\,, \tag{27.68}$$

so that radiative corrections must be quite large, about 2.4%, to account for the difference. These radiative corrections are conventionally written as

$$\Gamma_{\text{theory}} = \left[1 + A\frac{\alpha}{\pi} + \frac{\alpha^2}{3}\ln\alpha + B\left(\frac{\alpha}{\pi}\right)^2 - \frac{3\alpha^3}{2\pi}\ln^2\alpha + C\frac{\alpha^3}{\pi}\ln\alpha + D\left(\frac{\alpha}{\pi}\right)^3 + \ldots\right]\Gamma_0 . \quad (27.69)$$

The constant A was first correctly calculated in [27.44]. Analytic and numeric improvements to this work [27.45] have led to the present value of

$$A = -10.286\,606 \pm 0.000\,010 , \quad (27.70)$$

which accounts for all but 0.1% of the difference between theory and experiment. The logarithmic term of order $\alpha^2\Gamma_0$ was determined by *Caswell* and *Lepage* [27.46] and the leading logarithmic term of order α^3 by *Karshenboim* [27.47, 48]. The coefficient of the nonleading logarithmic term in order α^3, C, has recently been determined, but is numerically insignificant. For quite some time a discrepancy appeared to be present between theory and experiment, because a set of experiments in both powders and vacuum appeared to require a very large coefficient B. However, when this constant was finally calculated [27.49], it was found to be

$$B = 45.06(26) , \quad (27.71)$$

which is too small to explain the discrepancy. However, another experiment [27.50] was consistent with theory, and further work at Michigan led to the above result, which is in complete agreement with theory: there is at present no discrepancy in the decay rate of orthopositronium. We note also progress in the theory of the decay rate of parapositronium [27.51, 52], which is also consistent with experiment.

27.9 Precision Tests of QED in Neutral Helium

The Bethe–Salpeter equation mentioned above was developed in terms of an expansion of free electron and nuclear propagators. However, if these propagators are replaced with two electron propagators in an external Coulomb field, the formalism, while now more complicated, allows a rigorous treatment of helium in the non-recoil limit. This was carried out by *Araki* [27.53] and *Sucher* [27.54], and allowed theoretical predictions up to order $m\alpha^5$. However, the actual calculations are far more difficult because of the more complicated propagators. Firstly, the Schrödinger equation cannot be solved analytically, but rather numerically. However, the use of sophisticated Hylleras basis sets [27.55, 56], or the more recently introduced method of random exponents [27.57], leads to accuracies far beyond experimental precision. Secondly, the fine structure cannot be calculated analytically, but instead involves expectation values of operators that are somewhat singular: again the high-quality basis sets just mentioned allow the precision evaluation of this quantity [27.58, 59]. A third difficulty is the evaluation of the Lamb shift. In helium the analog of the Bethe logarithm discussed in connection with hydrogen is much more difficult to evaluate. However, significant progress has been made in recent years, with one approach using *Schwartz*'s [27.60] idea of using an integral representation of the numerator of the Bethe logarithm [27.61, 62]. An even more accurate and computationally simpler approach has been developed by *Drake* and *Goldman* [27.63].

Most of the leading QED corrections in helium can be expressed in terms of expectation values of various operators $\langle O\rangle$, where

$$\langle O\rangle \equiv \int \mathrm{d}^3r_1\,\mathrm{d}^3r_2\,\phi_0^*(\boldsymbol{r}_1,\boldsymbol{r}_2)O(\boldsymbol{r}_1,\boldsymbol{r}_2)\phi_0(\boldsymbol{r}_1,\boldsymbol{r}_2) . \quad (27.72)$$

Explicitly, one has

$$\begin{aligned} H^{(4)} &= \frac{\alpha}{\pi}\langle H_{\text{so}}\rangle + \frac{\alpha}{2\pi}\langle H_{\text{soo}}\rangle + \frac{\alpha}{\pi}\langle H_{\text{ss}}\rangle \\ &+ \left(\frac{89}{15} + \frac{14}{3}\ln\alpha - \frac{20}{3}\boldsymbol{s}_1\cdot\boldsymbol{s}_2\right)\frac{\alpha^2}{m^2}\langle\delta(\boldsymbol{r}_{12})\rangle \\ &+ \left(\frac{76}{45} - \frac{8}{3}\ln 2\alpha^2\right)\frac{\alpha^2}{m^2} \\ &\times[\langle Z\delta(\boldsymbol{r}_1) + Z\delta(\boldsymbol{r}_2)\rangle] \\ &- \frac{14}{3}\frac{\alpha^2}{m^2}Q - \frac{2\alpha}{3\pi m^2}M . \end{aligned} \quad (27.73)$$

Here

$$H_{\text{so}} = \frac{\alpha}{4m^2}\left\{\frac{Z}{r_1^3}\boldsymbol{\sigma}_1\cdot\boldsymbol{L}_1 + \frac{Z}{r_2^3}\boldsymbol{\sigma}_2\cdot\boldsymbol{L}_2 - \frac{1}{r_{12}^3}[\boldsymbol{\sigma}_1\cdot(\boldsymbol{r}_{12}\times\boldsymbol{p}_1)\boldsymbol{\sigma}_2\cdot(\boldsymbol{r}_{12}\times\boldsymbol{p}_2)]\right\} , \quad (27.74)$$

$$H_{\mathrm{soo}} = \frac{\alpha}{2m^2}\left[\boldsymbol{\sigma}_1 \cdot \left(\frac{\boldsymbol{r}_{12}}{r_{12}^3} \times \boldsymbol{p}_2\right) + \boldsymbol{\sigma}_2 \cdot \left(\frac{\boldsymbol{r}_{12}}{r_{12}^3} \times \boldsymbol{p}_1\right)\right] \tag{27.75}$$

and

$$H_{\mathrm{ss}} = \frac{\alpha}{4m^2}\left\{\left[\frac{\boldsymbol{\sigma}_1 \cdot \boldsymbol{\sigma}_2 - 3\boldsymbol{\sigma}_1 \cdot \hat{\boldsymbol{r}}_{12}\boldsymbol{\sigma}_2 \cdot \hat{\boldsymbol{r}}_{12}}{r_{12}^3}\right] - \frac{8\pi}{3}\boldsymbol{\sigma}_1 \cdot \boldsymbol{\sigma}_2 \delta^3(\boldsymbol{r}_{12})\right\} . \tag{27.76}$$

In the above, units in which $\hbar = c = 1$ are used. Then p and $1/r$ are of order $m\alpha$, and $H^{(4)}$ is of order $m\alpha^5$. The quantity Q is defined by

$$Q = \frac{1}{4\pi}\lim_{a\to 0}\left\langle \frac{1}{r_{12}^3(a)} + 4\pi\delta(\boldsymbol{r}_{12})[\gamma + \ln(a)]\right\rangle , \tag{27.77}$$

where $r_{12}(a)$ vanishes when $r_{12} < a$ and M is given by

$$M = \sum_n \left(E_n^0 - E^0\right)|\boldsymbol{P}_n|^2 \ln \frac{\left(E_n^0 - E^0\right)}{m\alpha^2} \tag{27.78}$$

where

$$\boldsymbol{P}_n = \int \mathrm{d}^3 r_1\, \mathrm{d}^3 r_2 \phi_n^*(\boldsymbol{r}_1, \boldsymbol{r}_2)(\boldsymbol{p}_1 + \boldsymbol{p}_2)\phi_0(\boldsymbol{r}_1, \boldsymbol{r}_2) . \tag{27.79}$$

M can be written in terms of an expectation value if we define the Bethe logarithm $\ln k_0$ through

$$M \equiv \sum_n \left(E_n^0 - E^0\right)|\boldsymbol{P}_n|^2 \ln \frac{k_0}{m\alpha^2} . \tag{27.80}$$

Calculations of the next order, that is, those of order $m\alpha^6$, were carried out for triplet P states by *Douglas* and *Kroll* [27.64] some time ago, but extension of the method to S states has only recently become possible through the use of NRQED techniques: we note in particular a calculation of the $2^3\mathrm{S}_1$ state by *Pachucki* [27.65] and of the ground state by *Korobov* and *Yelkhovsky* [27.66]. An outstanding problem in QED is the completion of the extension of the fine structure calculation to order $m\alpha^7$. This latter calculation should allow an extraction of the fine structure constant from helium competitive to that obtained from electron $g-2$ [27.67]. Work on this problem using both a Bethe–Salpeter formalism [27.68] and NRQED [27.69] is ongoing.

27.10 QED in Highly Charged One-Electron Ions

Because the Lamb shift scales as Z^4 and energy levels as Z^2, the relative importance of this effect increases as one goes out along an isoelectronic sequence. In addition, the approach used in the previous section, which relies on an expansion in powers of $Z\alpha$, becomes inappropriate, and requires methods that do not use such an expansion. Such methods were introduced by *Wichmann* and *Kroll* [27.70] for vacuum polarization. They were extended to the more difficult self-energy calculation by *Brown* and others [27.71], and the first correct calculations using the method were carried out by *Desiderio* and *Johnson* [27.72]. The basic idea is first to carry out the d^3k integration in (27.53) analytically, leaving an integration over d^3r, d^3r', and k_0 of a product of the electron and photon propagators in coordinate space. If one then makes a partial wave expansion of these propagators, the angle integrations can be carried out, and the self-energy becomes a sum over partial waves of an integral over two radial coordinates and the photon energy, which integral is carried out numerically. While simple in principle, this method is numerically awkward because the parts of the self-energy that are ultraviolet divergent are not well described in coordinate space. To solve this problem, the parts of the self-energy that have this sensitivity, which are associated with the part of the electron propagator in which the electron propagates freely, or with a single interaction, can be separated out. These subtractions allowed the first high-accuracy calculation of the self-energy in Coulomb potentials by *Mohr* [27.73–75]. Later *Blundell* and *Snyderman* [27.76] used such subtractions in a purely numeric approach that allowed the treatment of non-Coulomb potentials, which must be used for the many-electron ions discussed in the next section. While the most accurate calculations have been carried out for the case of the Coulomb potential, the finite size of the nucleus cannot be neglected, particularly at high Z. This issue has recently been studied by several groups [27.77, 78], and is now well understood. The experimental status of QED in one-electron ions is discussed by Mohr in this volume. An important recent development has been the calculation of the two-loop Lamb shift using exact electron propagators. While so far carried out only for the ground state in the ranges $Z > 40$ [27.79] it should be straightforward, though computationally intensive, to extend the calculations both to excited states and to lower values of Z.

27.11 QED in Highly Charged Many-Electron Ions

The previous applications of QED have been concentrated on simple systems, the most complex being helium. However, slight modifications of the S-matrix approach to QED described in Sect. 27.1 allow the treatment of ions with any number of electrons. In practice, this does not necessarily allow progress, since diagrams of arbitrarily high order contribute at the level of 1 a.u. in the many-body problem. However, when the nuclear Coulomb field dominates the Coulomb fields of the electrons, a small parameter is available, the quantity $1/Z$. Physically this simply reflects the fact the the energy of repulsion of any pair of electrons is of the order of Z a.u., while the energy of attraction to the nucleus scales as Z^2 a.u.. This $1/Z$ expansion has been studied empirically in the context of many-body perturbation theory, and it can be seen that while diagrams involving two-photon interactions are important, those involving three photons are highly suppressed. Thus the possibility exists of putting the QED of highly charged ions on the same footing as, for example, the electron $g-2$, in the sense that evaluation of a limited number of Feynman diagrams will allow precision tests of QED. However, while in principle one can start with the original Furry representation, in practice, particularly when larger numbers of electrons are present, it is better to start with non-Coulomb potentials that build in a major part of the electron screening present in the ion. For example, if one considers sodiumlike platinum ($Z=78$), the valence electron should see a nuclear charge of around 68 because of the ten 'core' electrons. One potential that builds in this property is the so-called 'core-Hartree' potential [27.80], which is defined so that

$$V_{\rm CH}(r) = V_{\rm nuc}(r) + \sum_a (2j_a+1) v_0(a,a;r)\,, \tag{27.81}$$

where

$$v_0(a,a;r) \equiv \int_0^\infty \mathrm{d}r' \frac{1}{r_>}\left[g_a^2(r') + f_a^2(r')\right] \tag{27.82}$$

with g_a and f_a the upper and lower radial components of the Dirac wavefunction for core state a. For sodium-like systems, a ranges over the 1s, 2s, $2\mathrm{p}_{1/2}$, and $2\mathrm{p}_{3/2}$ states. These states are solved for self consistently. The resulting potential gives results close to a full Dirac–Fock potential, but has the advantage of being local, which makes the connection with QED transparent. Because as $r\to\infty$, $v_0(a,a;r)\to 1/r$, the long range behavior of this potential has the physically expected limit. It is very simple to modify the Furry representation to include this potential: one simply writes

$$H_0 = \int \mathrm{d}^3 r \psi^\dagger(\boldsymbol{r})[\alpha\cdot\boldsymbol{p} + \beta m + V_{\rm CH}(r)]\psi(\boldsymbol{r}) \tag{27.83}$$

and

$$\begin{aligned} H_{\rm I} = &\int \mathrm{d}^3 r \psi^\dagger(\boldsymbol{r})[V_{\rm nuc}(r) - V_{\rm CH}(r)]\psi(\boldsymbol{r}) \\ &- \mathrm{e}\int \mathrm{d}^3 r \psi^\dagger(\boldsymbol{r})\boldsymbol{\alpha}\cdot\boldsymbol{A}(\boldsymbol{r})\psi(\boldsymbol{r}) \\ &+ \frac{\alpha}{2}\int \frac{\mathrm{d}^3 r\,\mathrm{d}^3 r'}{|\boldsymbol{r}-\boldsymbol{r}'|}\psi^\dagger(\boldsymbol{r})\psi(\boldsymbol{r})\psi^\dagger(\boldsymbol{r}')\psi(\boldsymbol{r}')\,. \end{aligned} \tag{27.84}$$

The only difference with the QED discussed in the previous section is the first term of $H_{\rm I}$, which acts in a manner similar to the self-mass counterterm, though it is of course finite. However, when more than one electron is present, new types of Feynman diagram are encountered: a representative set of them is given in Fig. 27.3. There is an interesting connection between these diagrams and the many-body perturbation theory (MBPT) method of solving the many-electron Schrödinger equation. In this latter method the N-electron Hamiltonian is written $H = H_0 + V_C$, where

$$H_0 = \sum_{i=1}^{i=N}\left[\boldsymbol{\alpha}_i\cdot\boldsymbol{p}_i + \beta_i m - \frac{Z\alpha}{r_i} + U(r_i)\right] \tag{27.85}$$

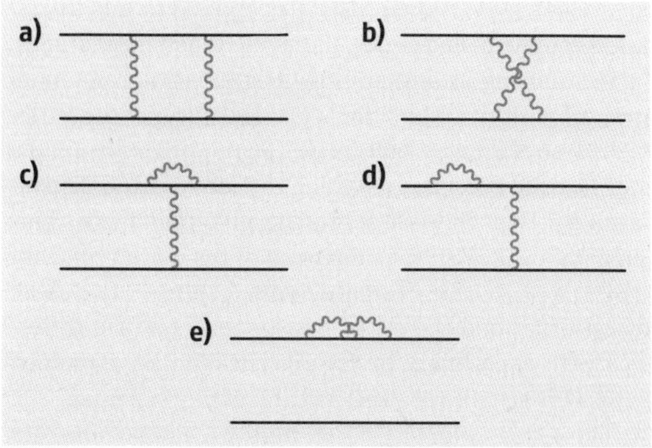

Fig. 27.3a–e Two-photon diagrams contributing to energy levels of highly charged ions

and

$$V_C = \frac{\alpha}{2} \sum_{ij} \frac{1}{|\boldsymbol{r}_i - \boldsymbol{r}_j|} - \sum_{i=1}^{i=N} U(r_i) \,. \tag{27.86}$$

Care must be taken when working with the relativistic version of the N-electron Schrödinger equation because of negative energy states, a point that has been emphasized by *Sucher* [27.81]. This problem shows up in MBPT in second order, where the associated energy shift for the ground state of an alkali atom with valence electron v in the case when U is the Hartree–Fock potential is

$$E^{(2)} = -\sum_{amn} \frac{g_{vamn}(g_{mnva} - g_{mnav})}{\epsilon_m + \epsilon_n - \epsilon_a - \epsilon_v} + \sum_{abm} \frac{g_{abmv}(g_{mvab} - g_{mvba})}{\epsilon_m + \epsilon_v - \epsilon_a - \epsilon_b} \,. \tag{27.87}$$

Here a and b range over the occupied core states, and m and n over excited states including in principle negative energy states. However, if they are included, the energy denominator in the first term of the above expression can vanish when m is a positive energy state and n negative energy, and vice versa. This problem can be avoided by restricting sums over excited states to positive energy states, but this procedure must be justified in the framework of QED. This can indeed be done [27.82], with the result that the use of MBPT with the positive energy restriction is justified, but negative energy state effects enter in a well-defined way, giving effects on the order of the Lamb shift.

To illustrate the present status of QED in highly charged many-electron ions, we consider the spectrum of sodium-like platinum. The transition energy between $3p_{3/2}$ and $3s_{1/2}$ has been measured [27.83] to be 653.44(7) eV. When MBPT is applied to this transition through second order, the answer depends slightly on the starting potential. The Hartree–Fock potential gives 659.56 eV, while the core-Hartree potential described above gives 659.59 eV. A modification of the core-Hartree potential in which the factor $2j_a + 1$ is reduced by 1 for the last core state give 659.57 eV. This spread of 0.03 eV gives a measure of the convergence of MBPT: it should be compared with a spread of 12 eV in lowest order and 0.4 eV in first order. Thus the difference with experiment of 6.1 eV that is to be explained by QED is reliable at the 0.1 eV level.

The QED calculations in highly charged ions are not as advanced as in the QED tests in simple neutral atoms. As discussed above, only in the last decade have calculations of one-loop radiative corrections in non-Coulomb potentials begun to be carried out. In the case of the core-Hartree potential they give exactly the 6.1 eV required to explain the difference between MBPT and experiment. This is actually somewhat unfortunate, since the set of two-photon Feynman diagrams comprising the two-loop Lamb shift mentioned above, some of which are shown in Fig. 27.3, should be detectable. Specifically, scaling arguments indicate that they should contribute at the few tenths of an eV level, and it is likely the good agreement with experiment found above involves fortuitous calculations. The graph of Fig. 27.3a plays a particularly interesting role, since it can be shown to 'contain' $E^{(2)}$. Specifically, when both photons are Coulomb photons and the electron propagators are written in terms of a spectral representation, the fourth component of the internal momentum integration can be carried out with contour methods. When both propagators involve positive energy states, $E^{(2)}$ from MBPT is precisely reproduced, including of course the previously introduced ad hoc rule of having only positive energy states in intermediate sums. When one is positive and the other negative the undefined term discussed above could arise, but since in this case both poles are on the same side of the axis, the contribution vanishes. However, when both are negative, a contribution outside of MBPT, part of the QED effect, arises. It is of order $Z^3\alpha^3$ a.u., in other words $1/Z$ of the leading Lamb shift. For sodium-like platinum, this should enter at the 0.1 eV level. For this reason, to be sensitive to corrections beyond the one-loop Lamb shift, ever more precise experiments along with more sophisticated MBPT calculations are required.

In recent years considerable progress has been made in the evaluation of 'two-photon' physics in lithium-like ions. This physics involves not only the two-loop Lamb shift mentioned in the previous section, but also vertex corrections to one-photon exchange and a full QED treatment of the correlation effects discussed in the previous paragraph. These effects have been included in recent calculations [27.80, 84], and the extension to sodiumlike and other alkalilike ions should be straightforward. This latter development would be desirable, as new data on sodiumlike and copperlike uranium of extremely high accuracy has become available.

As this progress is made, the interesting prospect arises of extending the high-accuracy QED tests in simple neutral systems discussed in this review to more complicated many-electron ions, and eventually to all the neutral atoms in the periodic table.

References

27.1 T. Kinoshita (Ed.): *Quantum Electrodynamics* (World Scientific, Singapore 1990)
27.2 J. Schwinger: *Selected Papers on Quantum Electrodynamics* (Dover, New York 1958)
27.3 S.S. Schweber: *QED and the Men Who Made It* (Princeton Univ. Press, Princeton, New Jersey 1994)
27.4 H. Fried, D.R. Yennie: Phys. Rev. **112**, 1391 (1958)
27.5 W. Pauli, F. Villars: Rev. Mod. Phys. **21**, 434 (1949)
27.6 J.C. Ward: Phys. Rev. **73**, 416 (1948)
27.7 This experiment and others are discussed by H.-U. Martyn in [27.1]
27.8 J. Schwinger: Phys. Rev. **73**, 416 (1948)
27.9 W.-Y. Tsai, A. Yildiz: Phys. Rev. D **8**, 3446 (1973)
27.10 G. Gabrielse, J. Tan, and L. S. Brown: in [27.1]
27.11 R.S. Van Dyck Jr., P.B. Schwinberg, H.G. Dehmelt: Phys. Rev. Lett. **59**, 26 (1987)
27.12 C.M. Sommerfield: Ann. Phys. (N.Y.) **5**, 26 (1958)
27.13 A. Petermann: Helv. Phys. Acta **30**, 407 (1957)
27.14 G.S. Adkins: Phys. Rev. D **39**, 3798 (1989)
27.15 T. Kinoshita, M. Nio: arXiv.hep-ph/0402206 (2004)
27.16 T. Kinoshita, B. Nizic, Y. Okamoto: Phys. Rev. D **31**, 2108 (1985)
27.17 M. Davier, S. Eidelman, A. Höcker, Z. Zhang: Eur. Phys. J. **C 31**, 503 (2003)
27.18 G.W. Bennett et al.: Phys. Rev. Lett. **92**, 161802 (2004)
27.19 T. Kinoshita, W.B. Lindquist: Phys. Rev. D **42**, 636 (1990)
27.20 A. Wicht, J.M. Hensley, E. Sarajlic, S. Chu: Phys. Scr. **T102**, 82 (2002)
27.21 W. Furry: Phys. Rev. **107**, 1448 (1957)
27.22 W.H. Furry: Phys. Rev. **81**, 115 (1951)
27.23 J. Schwinger: Proc. Nat. Acad. Sci. USA **37**, 452, 455 (1951)
27.24 E.E. Salpeter, H.A. Bethe: Phys. Rev. **84**, 1232 (1951)
27.25 E.E. Salpeter: Phys. Rev. **87**, 328 (1952)
27.26 W.E. Caswell, G.P. Lepage: Phys. Lett. B **167**, 437 (1986)
27.27 K. Pachucki: Phys. Rev. A **56**, 297 (1997)
27.28 L.N. Hand, D.J. Miller, R. Wilson: Rev. Mod. Phys. **35**, 335 (1963)
27.29 G.G. Simon, Ch. Schmidt, F. Borkowski, V.H. Walther: Nucl. Phys. A **333**, 381 (1980)
27.30 M. Lieber: Phys. Rev. **174**, 2037 (1968)
27.31 R.W. Huff: Phys. Rev. **186**, 1367 (1969)
27.32 G.W.F. Drake: Phys. Rev. A **41**, 1243 (1990)
27.33 T.W. Appelquist, S.J. Brodsky: Phys. Rev. A **2**, 2293 (1970)
27.34 B.E. Lautrup, A. Peterman, E. deRafael: Phys. Lett. B **31**, 577 (1970)
27.35 R. Barbieri, J.A. Mignaco, E. Remiddi: Nuovo Cimento Lett. **3**, 588 (1970)
27.36 K. Pachucki: Phys. Rev. Lett. **72**, 3154 (1994)
27.37 M.I. Eides, V.A. Shelyuto: JETP Letters **61**, 478 (1995)
27.38 U.D. Jentschura, K. Pachucki: J. Phys. A **35**, 1927 (2002)
27.39 T. Kinoshita, M. Nio: Phys. Rev. Lett. **72**, 3803 (1994)
27.40 G.T. Bodwin, D.R. Yennie, M.A. Gregorio: Rev. Mod. Phys. **57**, 723 (1985)
27.41 G.T. Bodwin, D.R. Yennie: Phys. Rev. D **37**, 498 (1988)
27.42 V.W. Hughes, J. Kuti: Ann. Rev. Nucl. Part. Sci. **33**, 611 (1983)
27.43 R.S. Vallery, P.W. Zitzewitz, D. Gidley: Phys. Rev. Lett. **90**, 203402 (2003)
27.44 W.E. Caswell, G.P. Lepage, J. Sapirstein: Phys. Rev. Lett. **38**, 488 (1977)
27.45 G.S. Adkins: Phys. Rev. Lett **76**, 4903 (1996)
27.46 W.E. Caswell, G.P. Lepage: Phys. Rev. A **20**, 36 (1979)
27.47 S.G. Karshenboim: Zh. Eksp. Teor. Fiz. **103**, 1105 (1993)
27.48 S.G. Karshenboim: JETP **76**, 541 (1993)
27.49 G.S. Adkins, RN. Fell, J. Sapirstein: Annals of Physics **295**, 136 (2002)
27.50 S. Asai, S. Orito, N. Shinohara: Phys. Lett. B **357**, 475 (1995)
27.51 A. Czarnecki, K. Melnikov, A. Yelkhovsky: Phys. Rev. A **61**, 052502 (2000)
27.52 G.S. Adkins, NM. McGovern, J. Sapirstein: Phys. Rev. A **68**, 032512 (2003)
27.53 H. Araki: Prog. Theor. Phys. **17**, 619 (1957)
27.54 J. Sucher: Phys. Rev. **109**, 1010 (1958)
27.55 See the contribution by Drake in this book
27.56 J.D. Morgan III: High precision calculation of helium atom energy levels. In: *AIP Conference Proceedings #189, Relativistic, Quantum Electrodynamic, and Weak Interaction Effects in Atoms, 1988*, ed. by W. Johnson, P. Mohr, J. Sucher (American Institute of Physics, New York 1988) p.123
27.57 V.I. Korobov: Phys. Rev. A **66**, 024501 (2002)
27.58 G.W.F. Drake, Z.-C. Yan: Phys. Rev. A **46**, 2378 (1992)
27.59 F. Drake G. W. Adv. Mol. Opt. Phys. **31**, 1 (1993)
27.60 C. Schwartz: Phys. Rev. **123**, 1700 (1961)
27.61 V.I. Korobov, S.V. Korobov: Phys. Rev. A **59**, 3394 (1999)
27.62 J.D. Baker, R.C. Forrey, M. Jesiorska, J.D. Morgan III: aiXiv:physics/0002005(2000)
27.63 G.W.F. Drake, S.P. Goldman: Can. J. Phys. **77**, 835 (1999)
27.64 M. Douglas, N. Kroll: Ann. Phys. (N.Y.) **82**, 89 (1974)
27.65 K. Pachucki: Phys. Rev. Lett. **84**, 4561 (2000)
27.66 V.I. Korobov, A. Yelkhovsky: Phys. Rev. Lett. **87**, 123003 (2001)
27.67 G.W.F. Drake: Can. J. Phys. **80**, 1195 (2002)

27.68 T. Zhang: Phys. Rev. A **53**, 3896 (1996)
27.69 K. Pachucki, J. Sapirstein: J. Phys. B **35**, 1783 (2002)
27.70 E. H. Wichmann, N. M. Kroll: Phys. Rev. **101**, 843 (1956)
27.71 G. E. Brown, J. S. Kanger, G. W. Schaefer: Proc. R. Soc. London Ser. A **251**, 92 (1959)
27.72 A. M. Desiderio, W. R. Johnson: Phys. Rev. A **3**, 1267 (1971)
27.73 J. Mohr P. Ann. Phys. (N.Y.) **88**, 26 (1974)
27.74 P. J. Mohr: Ann. Phys. (N.Y.) **88**, 52 (1974) higher accuracy results are given for $n=1$ and $n=2$ states see in [27.85]
27.75 P. J. Mohr, Y. K. Kim: Phys. Rev. A **45**, 2727 (1992)
27.76 S. A. Blundell, N. Snyderman: Phys. Rev. A **44**, R1427 (1991)
27.77 P. J. Mohr, G. Soff: Phys. Rev. Lett. **70**, 158 (1993)
27.78 K. T. Cheng, W. R. Johnson, J. Sapirstein: Phys. Rev. A **47**, 1817 (1993)
27.79 V. A. Yerokhin, P. Indelicato, V. M. Shabaev: Phys. Rev. Lett. **91**, 073001 (2003)
27.80 J. Sapirstein, K. T. Cheng: Phys. Rev. A **64**, 022502 (2001)
27.81 J. Sucher: Phys. Rev. A **22**, 348 (1980)
27.82 J. Sapirstein: Phys. Scr. T **46**, 52 (1993)
27.83 T. E. Cowan, C. L. Bennett, D. D. Dietrich, J. Bixler, C. J. Hailey, J. R. Henderson, D. A. Knapp, M. A. Levine, R. E. Marrs, M. B. Schneider: Phys. Rev. Lett. **66**, 1150 (1991)
27.84 V. A. Yerokhin, A. N. Artemyev, V. M. Shabaev, M. M. Sysak, O. M. Zherebtsov, G. Soff: Phys. Rev. A **60**, 3522 (1999)
27.85 P. J. Mohr: Phys. Rev. A **46**, 4421 (1992)

28. Tests of Fundamental Physics

This chapter describes comparisons of precise measurements and theoretical predictions that provide tests of our knowledge of fundamental physics. The focus is on several quantitative tests of quantum electrodynamics (QED).

The basic formulation of the theory of QED and calculational methods are discussed in Chapt. 27. Here, only the end results of calculations are collected, numerically evaluated, and compared with the corresponding experiments. (All quoted uncertainties are meant to be approximately at the one standard deviation level.)

It should be remarked that QED theory and the fundamental constants that are employed in evaluating the theoretical expressions are intimately linked. Fundamental constants are discussed in Chapt. 1. Values of the constants needed for comparison of theory and experiment are generally determined by other comparisons of theory and experiment, so that only the consistency of a set of tests is checked. However, the fact that this overall consistency is maintained at a high level of precision and over a broad range of phenomena provides confidence that QED is sound despite mathematical shortcomings in its formulation.

28.1 **Electron g-Factor Anomaly** 429

28.2 **Electron g-Factor in $^{12}C^{5+}$ and $^{16}O^{7+}$** 432

28.3 **Hydrogen and Deuterium Atoms** 437
- 28.3.1 Dirac Eigenvalue 437
- 28.3.2 Relativistic Recoil 438
- 28.3.3 Nuclear Polarization 439
- 28.3.4 Self Energy 439
- 28.3.5 Vacuum Polarization.................. 440
- 28.3.6 Two-Photon Corrections............. 441
- 28.3.7 Three-Photon Corrections........... 442
- 28.3.8 Finite Nuclear Size...................... 443
- 28.3.9 Nuclear-Size Correction to Self Energy and Vacuum Polarization . 443
- 28.3.10 Radiative-Recoil Corrections....... 444
- 28.3.11 Nucleus Self Energy 444
- 28.3.12 Total Energy and Uncertainty...... 444
- 28.3.13 Transition Frequencies Between Levels with $n=2$ 445

References 445

Recent reviews that cover topics in this chapter are given in [28.1, 2]. Much of the content in this chapter is reprinted from [28.3]. Values of the fundamental constants used in calculations in this chapter are the 2002 CODATA recommended values [28.3].

28.1 Electron g-Factor Anomaly

The magnetic moment of any of the three charged leptons (e, μ, τ) is written as

$$\boldsymbol{\mu} = g\frac{e}{2m}\boldsymbol{s}\,, \tag{28.1}$$

where g is the g-factor of the particle, m is its mass, and $\boldsymbol{s}$ is its spin. In (28.1), e is the elementary charge and is positive. For the negatively charged leptons (e^-, μ^-, and τ^-) g is negative, and for the corresponding antiparticles (e^+, μ^+, and τ^+) g is positive. CPT invariance implies that the masses and absolute values of the g-factors are the same for each particle–antiparticle pair.

These leptons have eigenvalues of spin projection $s_z = \pm\hbar/2$, and in the case of the electron and positron it is conventional to write, based on (28.1),

$$\mu_e = \frac{g_e}{2}\mu_B\,, \tag{28.2}$$

where $\mu_B = e\hbar/(2m_e)$ is the Bohr magneton.

For nucleons or nuclei with spin $\boldsymbol{I}$, the magnetic moment can be written as

$$\boldsymbol{\mu} = g\frac{e}{2m_p}\boldsymbol{I}\,, \tag{28.3}$$

or

$$\mu = g\mu_N i \,. \tag{28.4}$$

In (28.4), $\mu_N = e\hbar/(2m_p)$ is the nuclear magneton, defined in analogy with the Bohr magneton, and i is the spin quantum number of the nucleus defined by $\boldsymbol{I}^2 = i(i+1)\hbar^2$ and $I_z = -i\hbar, \ldots, (i-1)\hbar, i\hbar$, where I_z is the spin projection. However, in some publications moments of nucleons are expressed in terms of the Bohr magneton with a corresponding change in the definition of the g-factor.

One of the most precise tests of QED is the comparison of theory and experiment for the electron magnetic moment anomaly; the current status of this comparison is given in this section.

The electron magnetic moment is proportional to

$$g_e = -2(1+a_e)\,, \tag{28.5}$$

where the anomaly a_e characterizes the deviation of the g-factor from the Dirac value of $g_e(\text{Dirac}) = -2$. Measurement of the anomaly by the University of Washington group has yielded values for the electron and positron given respectively by [28.4]

$$a_{e^-}(\text{exp}) = 1\,159\,652\,188.4\,(4.3)\times 10^{-12}\,, \tag{28.6}$$

$$a_{e^+}(\text{exp}) = 1\,159\,652\,187.9\,(4.3)\times 10^{-12}\,, \tag{28.7}$$

which yield the mean value

$$a_e(\text{exp}) = 1\,159\,652\,188.3\,(4.2)\times 10^{-12}\,, \tag{28.8}$$

based on the analysis described in [28.1]. This analysis assumes that CPT invariance holds for the electron–positron system.

The same experiment has provided a precision comparison of the electron and positron g-factors

$$\left|\frac{g_{e^-}}{g_{e^+}}\right| = 1 + (0.5 \pm 2.1)\times 10^{-12}\,, \tag{28.9}$$

which provides a test of CPT invariance of the electron–positron system.

The theoretical expression for a_e may be written as

$$a_e(\text{th}) = a_e(\text{QED}) + a_e(\text{weak}) + a_e(\text{had})\,, \tag{28.10}$$

where the terms denoted by QED, weak, and had account for the purely quantum electrodynamic, predominantly electroweak, and predominantly hadronic (that is, strong interaction) contributions to a_e, respectively. The QED contribution may be written as [28.5]

$$a_e(\text{QED}) = A_1 + A_2(m_e/m_\mu) + A_2(m_e/m_\tau) + A_3(m_e/m_\mu, m_e/m_\tau)\,. \tag{28.11}$$

The term A_1 is mass independent and the other terms are functions of the indicated mass ratios. For these terms the lepton in the numerator of the mass ratio is the particle under consideration, while the lepton in the denominator of the ratio is the virtual particle that is the source of the vacuum polarization that gives rise to the term.

Each of the four terms on the right-hand side of (28.11) is expressed as a power series in the fine-structure constant α:

$$A_i = A_i^{(2)}\left(\frac{\alpha}{\pi}\right) + A_i^{(4)}\left(\frac{\alpha}{\pi}\right)^2 + A_i^{(6)}\left(\frac{\alpha}{\pi}\right)^3 + A_i^{(8)}\left(\frac{\alpha}{\pi}\right)^4 + \cdots\,. \tag{28.12}$$

The fine-structure constant α is proportional to the square of the elementary charge e, and the order of a term containing $(\alpha/\pi)^n$ is $2n$ and its coefficient is called the $2n$th-order coefficient.

The second-order coefficient is known exactly, and the fourth- and sixth-order coefficients are known analytically in terms of readily evaluated functions:

$$A_1^{(2)} = \frac{1}{2} \tag{28.13}$$

$$A_1^{(4)} = -0.328\,478\,965\,579\ldots \tag{28.14}$$

$$A_1^{(6)} = 1.181\,241\,456\ldots\,. \tag{28.15}$$

A total of 891 Feynman diagrams give rise to the eighth-order coefficient $A_1^{(8)}$, and only a few of these are known analytically. However, in an effort begun in the 1970s, *Kinoshita* and collaborators have calculated $A_1^{(8)}$ numerically (for a summary of some of this work see [28.6, 7]). The value of $A_1^{(8)}$ used in the 1998 CODATA adjustment of the fundamental constants was $-1.5098\,(384)$ [28.1]. Recently an error in the program employed in the evaluation of a gauge-invariant 18 diagram subset of the 891 diagrams was discovered in the course of carrying out an independent calculation to check this value [28.8]. The corrected program together with improved precision in the numerical integration for all diagrams leads to the tentative value $A_1^{(8)} = -1.7366\,(60)$ [28.9], where the shift from the earlier value is predominately due to the correction of the error. As a result of this recent work, *Kinoshita* and *Nio* [28.8] report that the integrals from all 891 Feynman diagrams have now been verified by independent calculation and/or checked by analytic comparison with lower-order integrals. Nevertheless, because the precision of the numerical evaluation of some integrals is still being improved and a closer examination is being made of the uncertainty of the numerical evaluation of other integrals, we retain the uncertainty estimate of the earlier reported value of $A_1^{(8)}$. This gives

$$A_1^{(8)} = -1.7366\,(384)\,. \tag{28.16}$$

The 0.0384 standard uncertainty of $A_1^{(8)}$ contributes a standard uncertainty to a_e(th) of $0.96\times10^{-9}a_e$, which may be compared to the $3.7\times10^{-9}a_e$ uncertainty of the experimental value (28.8). We also note that work is in progress on analytic calculations of eighth-order integrals. See, for example, *Laporta* [28.10] and *Mastrolia* and *Remiddi* [28.11].

Little is known about the tenth-order coefficient $A_1^{(10)}$ and higher-order coefficients. To evaluate the contribution to the uncertainty of a_e(th) due to lack of knowledge of $A_1^{(10)}$, we follow [28.1] to obtain $A_1^{(10)} = 0.0(3.8)$. Because the 3.8 standard uncertainty of $A_1^{(10)}$ contributes a standard uncertainty component to a_e(th) of only $0.22\times10^{-9}a_e$, the uncertainty contributions to a_e(th) from all other higher-order coefficients are assumed to be negligible.

The mass-dependent coefficients of possible interest and corresponding contributions to a_e(th), based on the 2002 CODATA recommended values of the mass ratios [28.3], are

$$A_2^{(4)}(m_e/m_\mu) = 5.197\,386\,70\,(27)\times10^{-7} \rightarrow 2.418\times10^{-9}a_e\,, \tag{28.17}$$

$$A_2^{(4)}(m_e/m_\tau) = 1.837\,63\,(60)\times10^{-9} \rightarrow 0.009\times10^{-9}a_e\,, \tag{28.18}$$

$$A_2^{(6)}(m_e/m_\mu) = -7.373\,941\,58\,(28)\times10^{-6} \rightarrow -0.080\times10^{-9}a_e\,, \tag{28.19}$$

$$A_2^{(6)}(m_e/m_\tau) = -6.5819\,(19)\times10^{-8} \rightarrow -0.001\times10^{-9}a_e\,, \tag{28.20}$$

where the standard uncertainties of the coefficients are due to the uncertainties of the mass ratios. However, the contributions are so small that the uncertainties of the mass ratios are negligible. It may also be noted that the contributions from $A_3^{(6)}(m_e/m_\mu, m_e/m_\tau)$ and all higher-order mass-dependent terms are negligible as well.

For the electroweak contribution we have

$$a_e(\text{weak}) = 0.0297\,(5)\times10^{-12} = 0.0256\,(5)\times10^{-9}a_e\,, \tag{28.21}$$

as in [28.1].

The hadronic contribution is

$$a_e(\text{had}) = 1.671\,(19)\times10^{-12} = 1.441\,(17)\times10^{-9}a_e\,, \tag{28.22}$$

and is the sum of the following three contributions: $a_e^{(4)}(\text{had}) = 1.875\,(18)\times10^{-12}$ obtained by *Davier* and *Höcker* [28.12]; $a_e^{(6a)}(\text{had}) = -0.225\,(5)\times10^{-12}$ given by *Krause* [28.13]; and $a_e^{(\gamma\gamma)}(\text{had}) = 0.0210\,(36)\times10^{-12}$ obtained by multiplying the corresponding result for the muon given in [28.3] by the factor $(m_e/m_\mu)^2$, since $a_e^{(\gamma\gamma)}(\text{had})$ is assumed to vary approximately as m_μ^2. The contribution a_e(had), although larger than a_e(weak), is not yet of major significance.

Since the dependence on α of any contribution other than a_e(QED) is negligible, we obtain a convenient form for the function by combining terms in a_e(QED) that have like powers of α/π. This leads to the following summary of the above results:

$$a_e(\text{th}) = a_e(\text{QED}) + a_e(\text{weak}) + a_e(\text{had})\,, \tag{28.23}$$

where

$$a_e(\text{QED}) = C_e^{(2)}\left(\frac{\alpha}{\pi}\right) + C_e^{(4)}\left(\frac{\alpha}{\pi}\right)^2 + C_e^{(6)}\left(\frac{\alpha}{\pi}\right)^3 + C_e^{(8)}\left(\frac{\alpha}{\pi}\right)^4 + C_e^{(10)}\left(\frac{\alpha}{\pi}\right)^5 + \cdots\,, \tag{28.24}$$

with

$$\begin{aligned} C_e^{(2)} &= 0.5\,,\\ C_e^{(4)} &= -0.328\,478\,444\,00\,,\\ C_e^{(6)} &= 1.181\,234\,017\,,\\ C_e^{(8)} &= -1.7366\,(384)\,,\\ C_e^{(10)} &= 0.0\,(3.8)\,, \end{aligned} \tag{28.25}$$

and where

$$a_e(\text{weak}) = 0.030\,(1)\times10^{-12} \tag{28.26}$$

and

$$a_e(\text{had}) = 1.671\,(19)\times10^{-12}\,. \tag{28.27}$$

The standard uncertainty of a_e(th) from the uncertainties of the terms listed above, other than that due to α, is

$$u[a_e(\text{th})] = 1.15\times10^{-12} = 0.99\times10^{-9}a_e\,, \tag{28.28}$$

and is dominated by the uncertainty of the coefficient $C_e^{(8)}$.

We define an additive correction δ_e to a_e(th) to account for the lack of exact knowledge of a_e(th), and hence the complete theoretical expression for the electron anomaly is

$$a_e(\alpha, \delta_e) = a_e(\text{th}) + \delta_e \,, \tag{28.29}$$

where all the uncertainty is associated with δ_e. The theoretical estimate of δ_e is zero and its standard uncertainty is $u[a_e(\text{th})]$:

$$\delta_e = 0.0\,(1.1) \times 10^{-12} \,. \tag{28.30}$$

Equating the theoretical expression with $a_e(\text{exp})$ given in (28.8) yields

$$\alpha^{-1}(a_e) = 137.035\,998\,80\,(52) \,. \tag{28.31}$$

The uncertainty of $a_e(\text{th})$ is significantly smaller than the uncertainty of $a_e(\text{exp})$, and thus the uncertainty of this inferred value of α is determined mainly by the uncertainty of $a_e(\text{exp})$. This result has the smallest uncertainty of any value of alpha currently available.

This result compares favorably with the value

$$\alpha^{-1}(\text{recoil}) = 137.036\,0001\,(11) \tag{28.32}$$

derived from the atomic recoil frequency shift of photons absorbed and emitted by cesium, as reviewed in [28.3].

28.2 Electron g-Factor in $^{12}C^{5+}$ and $^{16}O^{7+}$

For a ground-state hydrogenic ion $^{A}X^{(Z-1)+}$ with mass number A, atomic number (proton number) Z, nuclear spin quantum number $i = 0$, and g-factor $g_{e^-}(^{A}X^{(Z-1)+})$ in an applied magnetic flux density B, the ratio of the electron's spin-flip (often called precession) frequency $f_s = |g_{e^-}(^{A}X^{(Z-1)+})|(e\hbar/2m_e)B/h$ to the cyclotron frequency of the ion $f_c = (Z-1)\,eB/2\pi m(^{A}X^{(Z-1)+})$ in the same magnetic flux density is

$$\frac{f_s(^{A}X^{(Z-1)+})}{f_c(^{A}X^{(Z-1)+})} = -\frac{g_{e^-}(^{A}X^{(Z-1)+})}{2(Z-1)} \frac{A_r(^{A}X^{(Z-1)+})}{A_r(e)} \,, \tag{28.33}$$

where $A_r(X)$ is the relative atomic mass of particle X. If the frequency ratio f_s/f_c is determined experimentally with high accuracy, and $A_r(^{A}X^{(Z-1)+})$ of the ion is also accurately known, then this expression can be used to determine an accurate value of $A_r(e)$, assuming the bound-state electron g-factor can be calculated from QED theory with sufficient accuracy; or the g-factor can be determined if $A_r(e)$ is accurately known from another experiment. In fact, a broad program involving workers from a number of European laboratories has been underway since about the mid-1990s to measure the frequency ratio and calculate the g-factor for different ions, most notably (to date) $^{12}C^{5+}$ and $^{16}O^{7+}$. The measurements themselves are being performed at the GSI (Gesellschaft für Schwerionenforschung, Darmstadt, Germany) by GSI and University of Mainz researchers. Values reported are [28.14–16]

$$\frac{f_s(^{12}C^{5+})}{f_c(^{12}C^{5+})} = 4376.210\,4989\,(23) \tag{28.34}$$

and [28.16–18]

$$\frac{f_s(^{16}O^{7+})}{f_c(^{16}O^{7+})} = 4164.376\,1836\,(31) \,. \tag{28.35}$$

It should be noted that these two frequency ratios are correlated. Based on the detailed uncertainty budget of the two results [28.16], we find

$$r\left(\frac{f_s(^{12}C^{5+})}{f_c(^{12}C^{5+})}, \frac{f_s(^{16}O^{7+})}{f_c(^{16}O^{7+})}\right) = 0.035 \tag{28.36}$$

for the correlation coefficient.

We next consider the g-factor in (28.33) for an electron in the 1S state of hydrogen-like carbon 12 (atomic number $Z = 6$, nuclear spin quantum number $i = 0$) or in the 1S state of hydrogen-like oxygen 16 (atomic number $Z = 8$, nuclear spin quantum number $i = 0$) within the framework of relativistic bound-state theory. The measured quantity is the transition frequency between the two Zeeman levels of the atom in an externally applied magnetic field.

The energy of a free electron with spin projection s_z in a magnetic flux density B in the z direction is

$$E_{s_z} = -g_{e^-}\frac{e}{2m_e} s_z B \,, \tag{28.37}$$

and hence the spin-flip energy difference is

$$\Delta E = -g_{e^-}\mu_B B \,. \tag{28.38}$$

(In keeping with the definition of the g-factor in Sect. 28.1 the quantity g_{e^-} is negative.) The analogous expressions for the ions considered here are

$$\Delta E_b(X) = -g_{e^-}(X)\mu_B B \,, \tag{28.39}$$

which defines the bound-state electron g-factor in the case where there is no nuclear spin, and where X is either $^{12}C^{5+}$ or $^{16}O^{7+}$.

The main theoretical contributions to $g_{e^-}(X)$ can be categorized as follows:

- Dirac (relativistic) value g_D;
- radiative corrections Δg_{rad};
- recoil corrections Δg_{rec};
- nuclear size corrections Δg_{ns}.

Thus we write

$$g_{e^-}(X) = g_D + \Delta g_{rad} + \Delta g_{rec} + \Delta g_{ns} + \cdots \,, \tag{28.40}$$

where terms accounting for other effects are assumed to be negligible at the current level of uncertainty of the relevant experiments (relative standard uncertainty $u_r \approx 6\times10^{-10}$). These theoretical contributions are discussed in the following paragraphs; numerical results are summarized in Tables 28.1, 28.2.

Breit [28.19] obtained the exact value

$$\begin{aligned} g_D &= -\frac{2}{3}\left[1+2\sqrt{1-(Z\alpha)^2}\right] \\ &= -2\left[1-\frac{1}{3}(Z\alpha)^2-\frac{1}{12}(Z\alpha)^4 \right. \\ &\quad \left. -\frac{1}{24}(Z\alpha)^6+\cdots\right] \end{aligned} \tag{28.41}$$

from the Dirac equation for an electron in the field of a fixed point charge of magnitude Ze, where the only uncertainty is that due to the uncertainty in α.

The radiative corrections may be written as

$$\begin{aligned} \Delta g_{rad} = -2\Big[& C_e^{(2)}(Z\alpha)\left(\frac{\alpha}{\pi}\right) \\ & + C_e^{(4)}(Z\alpha)\left(\frac{\alpha}{\pi}\right)^2+\cdots\Big] \,, \end{aligned} \tag{28.42}$$

where the coefficients $C_e^{(2n)}(Z\alpha)$, corresponding to n virtual photons, are slowly varying functions of $Z\alpha$. These coefficients are defined in direct analogy with the corresponding coefficients for the free electron $C_e^{(2n)}$ given in Sect. 28.1 so that

$$\lim_{Z\alpha\to 0} C_e^{(2n)}(Z\alpha) = C_e^{(2n)} \,. \tag{28.43}$$

The coefficient $C_e^{(2)}(Z\alpha)$ has been calculated to second order in $Z\alpha$ by *Grotch* [28.20] who finds

$$\begin{aligned} C_e^{(2)}(Z\alpha) &= C_e^{(2)} + \frac{1}{12}(Z\alpha)^2+\cdots \\ &= \frac{1}{2}+\frac{1}{12}(Z\alpha)^2+\cdots \,. \end{aligned} \tag{28.44}$$

This result has been confirmed by *Faustov* and *Close* [28.21] and *Osborn* [28.22], as well as by others.

The terms listed in (28.44) do not provide a value of $C_e^{(2)}(Z\alpha)$ which is sufficiently accurate at the level of uncertainty of the current experimental results. However, *Yerokhin* [28.23, 24] have recently calculated numerically the self-energy contribution $C_{e,SE}^{(2)}(Z\alpha)$ to the coefficient to all orders in $Z\alpha$ over a wide range of Z. These results are in general agreement with, but are more accurate than, the earlier results of *Beier* et al. [28.25] and *Beier* [28.26]. Other calculations of the self energy have been carried out by *Persson* [28.27]; *Blundell* et al. [28.28]; and *Goidenko* [28.29]. For $Z=6$ and $Z=8$ the calculation of *Yerokhin* et al. [28.23] gives

$$\begin{aligned} C_{e,SE}^{(2)}(6\alpha) &= 0.500\,183\,609\,(19) \\ C_{e,SE}^{(2)}(8\alpha) &= 0.500\,349\,291\,(19) \,, \end{aligned} \tag{28.45}$$

where we have converted their quoted result to conform with our notation convention, taking into account the value of α employed in their calculation.

Table 28.1 Theoretical contributions and total for the g-factor of the electron in hydrogenic carbon 12 based on the 2002 recommended values of the constants

Contribution	Value	Source (Eq.)
Dirac g_D	−1.998 721 354 39 (1)	(28.41)
$\Delta g_{SE}^{(2)}$	−0.002 323 672 45 (9)	(28.45)
$\Delta g_{VP}^{(2)}$	0.000 000 008 51	(28.49)
$\Delta g^{(4)}$	0.000 003 545 74 (16)	(28.53)
$\Delta g^{(6)}$	−0.000 000 029 62	(28.54)
$\Delta g^{(8)}$	0.000 000 000 10	(28.55)
Δg_{rec}	−0.000 000 087 64 (1)	(28.56, 28.58)
Δg_{ns}	−0.000 000 000 41	(28.60)
$g_{e^-}(^{12}C^{5+})$	−2.001 041 590 16 (18)	(28.61)

Table 28.2 Theoretical contributions and total for the g-factor of the electron in hydrogenic oxygen 16 based on the 2002 recommended values of the constants

Contribution	Value	Source (Eq.)
Dirac g_D	−1.997 726 003 06 (2)	(28.41)
$\Delta g_{SE}^{(2)}$	−0.002 324 442 15 (9)	(28.45)
$\Delta g_{VP}^{(2)}$	0.000 000 026 38	(28.49)
$\Delta g^{(4)}$	0.000 003 546 62 (42)	(28.53)
$\Delta g^{(6)}$	−0.000 000 029 62	(28.54)
$\Delta g^{(8)}$	0.000 000 000 10	(28.55)
Δg_{rec}	−0.000 000 117 02 (1)	(28.56, 28.58)
Δg_{ns}	−0.000 000 001 56 (1)	(28.60)
$g_{e^-}(^{16}O^{7+})$	−2.000 047 020 31 (43)	(28.61)

The lowest-order vacuum-polarization correction is conveniently considered as consisting of two parts. In one the vacuum polarization loop modifies the interaction between the bound electron and the Coulomb field of the nucleus, and in the other the loop modifies the interaction between the bound electron and the external magnetic field. The first part, sometimes called the "wave function" correction, has been calculated numerically by *Beier* et al. [28.25], with the result (in our notation)

$$C^{(2)}_{\mathrm{e,VPwf}}(6\alpha) = -0.000\,001\,840\,3431\,(43)\,,$$
$$C^{(2)}_{\mathrm{e,VPwf}}(8\alpha) = -0.000\,005\,712\,028\,(26)\,. \quad (28.46)$$

Each of these values is the sum of the Uehling potential contribution and the higher-order Wichmann–Kroll contribution, which were calculated separately.

The values in (28.46) are consistent with the result of an evaluation of the correction in powers of $Z\alpha$. Terms to order $(\alpha/\pi)(Z\alpha)^7$ have been calculated for the Uehling potential contribution [28.30–32]; and an estimate of the leading order $(\alpha/\pi)(Z\alpha)^6$ term of the Wichmann–Kroll contribution has been given by *Karshenboim* et al. [28.32] based on a prescription of *Karshenboim* [28.30]. To the level of uncertainty of interest here, the values from the power series are the same as the numerical values in (28.46). (Note that for the Wichmann–Kroll term, the agreement between the power-series results and the numerical results is improved by an order of magnitude if an additional term in the power series for the energy level [28.33] used in Karshenboim's prescription is included.)

For the second part of the lowest-order vacuum polarization correction, sometimes called the "potential" correction, *Beier* et al. [28.25] found that the Uehling potential contribution is zero. They also calculated the Wichmann–Kroll contribution numerically over a wide range of Z. Their value at low Z is very small and only an uncertainty estimate of 3×10^{-10} in g is given because of poor convergence of the partial wave expansion. The reduction in uncertainty (by a factor of 30 for carbon) employed by *Beier* et al. [28.14] for this term, based on the assumption that it is of the order of $(\alpha/\pi)(Z\alpha)^7$, is not considered here, because the reference quoted for this estimate [28.32] does not explicitly discuss this term. *Yerokhin* et al. [28.23] obtained numerical values for this contribution for carbon and oxygen by a least-squares fit to the values of *Beier* et al. [28.25] at higher Z.

Subsequently, *Karshenboim* and *Milstein* [28.34] analytically calculated the Wichmann–Kroll contribution to the potential correction to lowest order in $Z\alpha$. Their result in our notation is

$$C^{(2)}_{\mathrm{e,VPp}}(Z\alpha) = \frac{7\pi}{432}(Z\alpha)^5 + \cdots\,. \quad (28.47)$$

This result, together with the numerical values from *Beier* [28.26], yields

$$C^{(2)}_{\mathrm{e,VPp}}(6\alpha) = 0.000\,000\,007\,9595\,(69)\,,$$
$$C^{(2)}_{\mathrm{e,VPp}}(8\alpha) = 0.000\,000\,033\,235\,(29)\,, \quad (28.48)$$

which are used in the present analysis. We obtained these results by fitting a function of the form $\left[a+bZ\alpha+c(Z\alpha)^2\right](Z\alpha)^5$ to the point in (28.47) and two values of the complete function calculated by [28.26] (separated by about 10 calculated values) and evaluating the fitted function at $Z=6$ or 8. This was done for a range of pairs of points from [28.26], and the results in (28.48) are the apparent limit of the values as the lower Z member of the pair used in the fit approaches either 6 or 8 as appropriate. (This general approach is described in more detail in [28.35].)

The total one-photon vacuum polarization coefficients are given by the sum of (28.46) and (28.48):

$$\begin{aligned} C^{(2)}_{\mathrm{e,VP}}(6\alpha) &= C^{(2)}_{\mathrm{e,VPwf}}(6\alpha) + C^{(2)}_{\mathrm{e,VPp}}(6\alpha) \\ &= -0.000\,001\,832\,384\,(11)\,; \\ C^{(2)}_{\mathrm{e,VP}}(8\alpha) &= C^{(2)}_{\mathrm{e,VPwf}}(8\alpha) + C^{(2)}_{\mathrm{e,VPp}}(8\alpha) \\ &= -0.000\,005\,678\,793\,(55)\,. \end{aligned} \quad (28.49)$$

The total for the one-photon coefficient $C^{(2)}_{\mathrm{e}}(Z\alpha)$, given by the sum of (28.45) and (28.49), is

$$\begin{aligned} C^{(2)}_{\mathrm{e}}(6\alpha) &= C^{(2)}_{\mathrm{e,SE}}(6\alpha) + C^{(2)}_{\mathrm{e,VP}}(6\alpha) \\ &= 0.500\,181\,777\,(19)\,, \\ C^{(2)}_{\mathrm{e}}(8\alpha) &= C^{(2)}_{\mathrm{e,SE}}(8\alpha) + C^{(2)}_{\mathrm{e,VP}}(8\alpha) \\ &= 0.500\,343\,613\,(19)\,, \end{aligned} \quad (28.50)$$

where in this case, following *Beier* et al. [28.25], the uncertainty is simply the sum of the individual uncertainties in (28.45) and (28.49). The total one-photon contribution $\Delta g^{(2)}$ to the g-factor is thus

$$\begin{aligned} \Delta g^{(2)} &= -2C^{(2)}_{\mathrm{e}}(Z\alpha)\left(\frac{\alpha}{\pi}\right) \\ &= -0.002\,323\,663\,93\,(9) \quad \text{for } Z=6 \\ &= -0.002\,324\,415\,77\,(9) \quad \text{for } Z=8\,. \end{aligned} \quad (28.51)$$

The separate one-photon self energy and vacuum polarization contributions to the g-factor are given in Tables 28.1, 28.2.

Evaluations by *Eides* and *Grotch* [28.36] using the Bargmann–Michel–Telegdi equation and by *Czarnecki* et al. [28.37] using an effective potential approach yield

$$C_e^{(2n)}(Z\alpha) = C_e^{(2n)}\left(1 + \frac{(Z\alpha)^2}{6} + \cdots\right) \quad (28.52)$$

as the leading binding correction to the free electron coefficients $C_e^{(2n)}$ for any n. For $n = 1$, this result was already known, as is evident from (28.44). We include this correction for the two-photon term, that is, for $n = 2$, which gives

$$\begin{aligned} C_e^{(4)}(Z\alpha) &= C_e^{(4)}\left(1 + \frac{(Z\alpha)^2}{6} + \cdots\right) \\ &= -0.328\,583\,(14) \quad \text{for } Z = 6 \\ &= -0.328\,665\,(39) \quad \text{for } Z = 8\,, \end{aligned} \quad (28.53)$$

here $C_e^{(4)} = -0.328\,478\,444\ldots$. The uncertainty is due to uncalculated terms and is obtained by assuming that the unknown higher-order terms for $n = 2$, represented by the dots in (28.52), are the same as the higher-order terms for $n = 1$ as can be deduced by comparing the numerical results given in (28.50) to (28.44). This is the same general approach as that employed by *Beier* et al. [28.14].

The three-photon term is calculated in a similar way but the uncertainty due to uncalculated higher-order terms is negligible:

$$\begin{aligned} C_e^{(6)}(Z\alpha) &= C_e^{(6)}\left(1 + \frac{(Z\alpha)^2}{6} + \cdots\right) \\ &= 1.1816\ldots \quad \text{for } Z = 6 \\ &= 1.1819\ldots \quad \text{for } Z = 8\,, \end{aligned} \quad (28.54)$$

where $C_e^{(6)} = 1.181\,234\ldots$. For the four-photon correction, at the level of uncertainty of current interest, only the free-electron coefficient is necessary:

$$C_e^{(8)}(Z\alpha) \approx C_e^{(8)} = -1.7366\,(384)\,. \quad (28.55)$$

The preceding corrections Δg_D and Δg_{rad} are based on the approximation that the nucleus of the hydrogenic atom has an infinite mass. The recoil correction to the bound-state g-factor associated with the finite mass of the nucleus is denoted by Δg_{rec}, which we write here as the sum $\Delta g_{rec}^{(0)} + \Delta g_{rec}^{(2)}$ corresponding to terms that are zero- and first-order in α/π, respectively. For $\Delta g_{rec}^{(0)}$, we have

$$\begin{aligned} \Delta g_{rec}^{(0)} &= \left[-(Z\alpha)^2 + \frac{(Z\alpha)^4}{3\left[1+\sqrt{1-(Z\alpha)^2}\right]^2} - (Z\alpha)^5 P(Z\alpha) \right] \frac{m_e}{m_N} + \mathcal{O}\left(\frac{m_e}{m_N}\right)^2 \\ &= -0.000\,000\,087\,71\,(1)\ldots \text{ for } Z = 6 \\ &= -0.000\,000\,117\,11\,(1)\ldots \text{ for } Z = 8\,, \end{aligned} \quad (28.56)$$

where m_N is the mass of the nucleus. The mass ratios, obtained from the 2002 CODATA adjustment, are $m_e/m(^{12}C^{6+}) = 0.000\,045\,7275\ldots$ and $m_e/m(^{16}O^{8+}) = 0.000\,034\,3065\ldots$. In (28.56), the first term in the brackets was calculated by *Grotch* [28.38]. Shortly thereafter, this term and higher-order terms were obtained by *Grotch* [28.39], *Hegstrom* [28.40], *Faustov* [28.21], *Close* and *Osborn* [28.22], and *Grotch* and *Hegstrom* [28.41] (see also *Hegstrom* [28.42] and *Grotch* [28.20]). The second and third terms in the brackets were calculated by *Shabaev* and *Yerokhin* [28.43] based on the formulation of *Shabaev* [28.44] (see also *Yelkhovski* [28.45]). Shabaev and Yerokhin have numerically evaluated the function $P(Z\alpha)$ over a wide range of Z, with the result $P(6\alpha) = 10.493\,95\,(1)$ for hydrogenic carbon and $P(8\alpha) = 9.300\,18\,(1)$ for hydrogenic oxygen.

An additional term of the order of the mass ratio squared has been considered by various authors. Earlier calculations of this term for atoms with a spin one-half nucleus, such as muonium, have been done by *Close* and *Osborn* [28.22] and *Grotch* and *Hegstrom* [28.41] (see also *Eides* and *Grotch* [28.36]). Their result for this term is

$$(1+Z)(Z\alpha)^2\left(\frac{m_e}{m_N}\right)^2 . \quad (28.57)$$

Eides and *Grotch* [28.36], *Eides* [28.46] find that this correction to the g-factor is independent of the spin of the nucleus, so (28.57) gives the correction for carbon and oxygen, as well as atoms with a spin one-half nucleus. On the other hand, *Martynenko* and *Faustov* [28.47, 48] find that the correction of this order depends on the spin of the nucleus and give a result with the factor $1 + Z$ replaced by $Z/3$ for a spin zero nucleus. In view of this discrepancy, we include a contribution to $\Delta g_{\text{rec}}^{(0)}$ in (28.56) that is the average of the two quoted results with an uncertainty of half of the difference between them.

For $\Delta g_{\text{rec}}^{(2)}$, we have

$$\begin{aligned}\Delta g_{\text{rec}}^{(2)} &= \frac{\alpha}{\pi}\frac{(Z\alpha)^2}{3}\frac{m_e}{m_N}+\cdots \\ &= 0.000\,000\,000\,06\ldots \quad \text{for } Z=6 \\ &= 0.000\,000\,000\,09\ldots \quad \text{for } Z=8\,, \qquad (28.58)\end{aligned}$$

There is a small correction to the bound-state g-factor due to the finite size of the nucleus:

$$\Delta g_{\text{ns}} = \frac{8}{3}(Z\alpha)^4\left(\frac{R_N}{\lambda\!\!\!^{-}_C}\right)^2+\cdots\,, \qquad (28.59)$$

where R_N is the bound-state nuclear rms charge radius and $\lambda\!\!\!^{-}_C$ is the Compton wavelength of the electron divided by 2π. In (28.59), the term shown is the nonrelativistic approximation given by *Karshenboim* [28.30]. This term and the dominant relativistic correction have been calculated by *Glazov* and *Shabaev* [28.49]. We take $R_N = 2.4705\,(23)$ fm and $R_N = 2.6995\,(68)$ from the compilation of *Angeli* [28.50] for the values of the ^{12}C and ^{16}O nuclear radii, respectively, which, based on *Glazov* and *Shabaev* [28.49], yields

$$\begin{aligned}\Delta g_{\text{ns}} &= -0.000\,000\,000\,41 \quad \text{for } {}^{12}\text{C}\,, \\ \Delta g_{\text{ns}} &= -0.000\,000\,001\,56\,(1) \quad \text{for } {}^{16}\text{O}\,. \qquad (28.60)\end{aligned}$$

The theoretical value for the g-factor of the electron in hydrogenic carbon 12 or oxygen 16 is the sum of the individual contributions discussed above and summarized in Tables 28.1 and 28.2:

$$\begin{aligned}g_{e^-}\left({}^{12}\text{C}^{5+}\right) &= -2.001\,041\,590\,16\,(18) \\ g_{e^-}\left({}^{16}\text{O}^{7+}\right) &= -2.000\,047\,020\,31\,(43)\,. \qquad (28.61)\end{aligned}$$

We define $g_C(\text{th})$ to be the sum of g_D as given in (28.41), the term $-2(\alpha/\pi)C_e^{(2)}$, and the numerical values of the remaining terms in (28.40) as given in Table 28.1 without the uncertainties. The standard uncertainty of $g_C(\text{th})$ from the uncertainties of these latter terms is

$$u[g_C(\text{th})] = 1.8\times10^{-10} = 9.0\times10^{-11}|g_C(\text{th})|\,. \qquad (28.62)$$

The uncertainty in $g_C(\text{th})$ due to the uncertainty in α enters primarily through the functional dependence of g_D and the term $-2(\alpha/\pi)C_e^{(2)}$ on α. Therefore this particular component of uncertainty is not explicitly included in $u[g_C(\text{th})]$, but it is included in Tables 28.1 and 28.2. To take the uncertainty $u[g_C(\text{th})]$ into account we employ as the theoretical expression for the g-factor

$$g_C(\alpha,\delta_C) = g_C(\text{th})+\delta_C\,, \qquad (28.63)$$

where the input value of the additive correction δ_C is taken to be zero and its standard uncertainty is $u[g_C(\text{th})]$:

$$\delta_C = 0.0\,(1.8)\times10^{-10}\,. \qquad (28.64)$$

Analogous considerations apply for the g-factor in oxygen:

$$u[g_O(\text{th})] = 4.3\times10^{-10} = 2.2\times10^{-10}|g_O(\text{th})| \qquad (28.65)$$

$$g_O(\alpha,\delta_O) = g_O(\text{th})+\delta_O \qquad (28.66)$$

$$\delta_O = 0.0\,(4.3)\times10^{-10}\,. \qquad (28.67)$$

Since the uncertainties of the theoretical values of the carbon and oxygen g-factors arise primarily from the same sources, the quantities δ_C and δ_O are highly correlated. Their covariance is

$$u(\delta_C,\delta_O) = 741\times10^{-22}\,, \qquad (28.68)$$

which corresponds to a correlation coefficient of $r(\delta_C,\delta_O) = 0.95$.

The theoretical value of the ratio of the two g-factors, which is relevant to the following discussion, is

$$\frac{g_{e^-}\left({}^{12}\text{C}^{5+}\right)}{g_{e^-}\left({}^{16}\text{O}^{7+}\right)} = 1.000\,497\,273\,23\,(13)\,, \qquad (28.69)$$

where the covariance is taken into account in calculating the uncertainty, and for this purpose includes the contribution due to the uncertainty in α.

Finally, we consider evaluation of the mass ratio in (28.33) by applying the relation for the relative atomic mass $A_r\left({}^AX\right)$ of a neutral atom ${}^AX$ in terms of the relative atomic mass of an ion of the atom formed by the removal of n electrons, which is given by

$$\begin{aligned}A_r\left({}^AX\right) &= A_r\left({}^AX^{n+}\right)+nA_r(\text{e}) \\ &\quad - \frac{E_b\left({}^AX\right)-E_b\left({}^AX^{n+}\right)}{m_uc^2}\,. \qquad (28.70)\end{aligned}$$

Here A is the mass number, Z is the atomic number (proton number), $E_b\left({}^AX\right)/m_uc^2$ is the relative-atomic-mass equivalent of the total binding energy of the Z electrons of the atom, $E_b\left({}^AX^{n+}\right)/m_uc^2$ is the relative-atomic-mass-equivalent of the binding energy of the ${}^AX^{n+}$ ion, and m_u is the atomic mass constant.

From (28.33) and (28.70), we have

$$\frac{f_s\left({}^{12}\mathrm{C}^{5+}\right)}{f_c\left({}^{12}\mathrm{C}^{5+}\right)} = -\frac{g_{e^-}\left({}^{12}\mathrm{C}^{5+}\right)}{10A_r(e)} \times \left[12 - 5A_r(e) + \frac{E_b\left({}^{12}\mathrm{C}\right) - E_b\left({}^{12}\mathrm{C}^{5+}\right)}{m_u c^2}\right], \quad (28.71)$$

which is the equation for the ${}^{12}\mathrm{C}^{5+}$ frequency-ratio. Evaluation of this expression using the result for $f_s({}^{12}\mathrm{C}^{5+})/f_c({}^{12}\mathrm{C}^{5+})$ in (28.34), the theoretical result for $g_{e^-}({}^{12}\mathrm{C}^{5+})$ in Table 28.1, and the relevant binding energies from [28.3], yields

$$A_r(e) = 0.000\,548\,579\,909\,31\,(29)\,, \quad (28.72)$$

a result that is consistent with the University of Washington result [28.51], but has about a factor of four smaller uncertainty.

Similarly, we have

$$\frac{f_s\left({}^{16}\mathrm{O}^{7+}\right)}{f_c\left({}^{16}\mathrm{O}^{7+}\right)} = -\frac{g_{e^-}\left({}^{16}\mathrm{O}^{7+}\right)}{14A_r(e)} A_r\left({}^{16}\mathrm{O}^{7+}\right) \quad (28.73)$$

with

$$A_r\left({}^{16}\mathrm{O}\right) = A_r\left({}^{16}\mathrm{O}^{7+}\right) + 7A_r(e) - \frac{E_b\left({}^{16}\mathrm{O}\right) - E_b\left({}^{16}\mathrm{O}^{7+}\right)}{m_u c^2}, \quad (28.74)$$

which are the equations for the oxygen frequency ratio and $A_r({}^{16}\mathrm{O})$, respectively. The first expression, evaluated using the result for $f_s({}^{16}\mathrm{O}^{7+})/f_c({}^{16}\mathrm{O}^{7+})$ in (28.35) and the theoretical result for $g_{e^-}({}^{16}\mathrm{O}^{7+})$ in Table 28.2, in combination with the second expression, evaluated using the value of $A_r({}^{16}\mathrm{O})$ from the *University of Washington group* [28.52, 53] and the relevant binding energies from [28.3], yields

$$A_r(e) = 0.000\,548\,579\,909\,57\,(43)\,, \quad (28.75)$$

a value that is consistent with both the *University of Washington* value [28.51] and the value in (28.72) obtained from $f_s({}^{12}\mathrm{C}^{5+})/f_c({}^{12}\mathrm{C}^{5+})$.

As a consistency test, it is of interest to compare the experimental and theoretical values of the ratio of $g_{e^-}({}^{12}\mathrm{C}^{5+})$ to $g_{e^-}({}^{16}\mathrm{O}^{7+})$ [28.54]. The main reason is that the experimental value of the ratio is only weakly dependent on the value of $A_r(e)$. The theoretical value of the ratio is given in (28.69) and takes into account the covariance of the two theoretical values. The experimental value of the ratio can be obtained by combining (28.34) to (28.36), (28.71), (28.73), and (28.74), and using the 2002 CODATA recommended value for $A_r(e)$. (Because of the weak dependence of the experimental ratio on $A_r(e)$, the value used is not at all critical.) The result is

$$\frac{g_{e^-}\left({}^{12}\mathrm{C}^{5+}\right)}{g_{e^-}\left({}^{16}\mathrm{O}^{7+}\right)} = 1.000\,497\,273\,70\,(90)\,, \quad (28.76)$$

in agreement with the theoretical value in (28.69).

28.3 Hydrogen and Deuterium Atoms

This section gives a brief survey of the theory of the energy levels of hydrogen and deuterium relevant to measurements of transition frequencies. Although information to completely determine the theoretical values for the energy levels is provided, results that are included in [28.1, 2] are given with minimal discussion, and the emphasis is on recent results. For brevity, reference to most historical works is not included.

The theoretical data provided here are confined to that needed to evaluate the theoretical values of the precisely measured transition frequencies in hydrogen and deuterium summarized in [28.3].

It should be noted that the theoretical values of the energy levels of different states of hydrogen and deuterium are highly correlated. For example, for S states, the uncalculated terms are primarily of the form of an unknown common constant divided by n^3. This fact is taken into account by calculating covariances between energy levels in addition to the uncertainties of the individual levels as discussed in detail in Sect. 28.3.12. To provide the information needed to calculate the covariances, where necessary we distinguish between components of uncertainty that are proportional to $1/n^3$, denoted by u_0, and components of uncertainty that are essentially random functions of n, denoted by u_n.

Theoretical values of the energy levels of hydrogen and deuterium atoms are determined mainly by the Dirac eigenvalue, QED effects such as self energy and vacuum polarization, and nuclear size and motion effects. We consider each of these contributions in turn.

28.3.1 Dirac Eigenvalue

The binding energy of an electron in a static Coulomb field (the external electric field of a point nucleus of charge Ze with infinite mass) is determined predominantly by the Dirac eigenvalue

$$E_D = \left(1+\frac{(Z\alpha)^2}{(n-\delta)^2}\right)^{-1/2} m_e c^2 , \tag{28.77}$$

where n is the principal quantum number,

$$\delta = |\kappa| - \left[\kappa^2 - (Z\alpha)^2\right]^{1/2} , \tag{28.78}$$

and κ is the angular momentum-parity quantum number ($\kappa = -1, 1, -2, 2, -3$ for $S_{1/2}$, $P_{1/2}$, $P_{3/2}$, $D_{3/2}$, and $D_{5/2}$ states, respectively). States with the same principal quantum number n and angular momentum quantum number $j = |\kappa| - \frac{1}{2}$ have degenerate eigenvalues. The nonrelativistic orbital angular momentum is given by $l = \left|\kappa + \frac{1}{2}\right| - \frac{1}{2}$. (Although we are interested only in the case where the nuclear charge is e, we retain the atomic number Z in order to indicate the nature of various terms.)

Corrections to the Dirac eigenvalue that approximately take into account the finite mass of the nucleus m_N are included in the more general expression for atomic energy levels, which replaces (28.77) [28.55,56]:

$$\begin{aligned} E_M &= Mc^2 + [f(n,j)-1]m_r c^2 \\ &\quad - [f(n,j)-1]^2 \frac{m_r^2 c^2}{2M} \\ &\quad + \frac{1-\delta_{l0}}{\kappa(2l+1)} \frac{(Z\alpha)^4 m_r^3 c^2}{2n^3 m_N^2} + \cdots , \end{aligned} \tag{28.79}$$

where

$$f(n,j) = \left[1+\frac{(Z\alpha)^2}{(n-\delta)^2}\right]^{-1/2} , \tag{28.80}$$

$M = m_e + m_N$, and $m_r = m_e m_N/(m_e + m_N)$ is the reduced mass.

28.3.2 Relativistic Recoil

Relativistic corrections to (28.79) associated with motion of the nucleus are considered relativistic-recoil corrections. The leading term, to lowest order in $Z\alpha$ and all orders in m_e/m_N, is [28.56, 57]

$$\begin{aligned} E_S &= \frac{m_r^3}{m_e^2 m_N} \frac{(Z\alpha)^5}{\pi n^3} m_e c^2 \\ &\quad \times \left\{ \frac{1}{3}\delta_{l0} \ln(Z\alpha)^{-2} - \frac{8}{3}\ln k_0(n,l) - \frac{1}{9}\delta_{l0} - \frac{7}{3}a_n \right. \\ &\quad \left. - \frac{2}{m_N^2 - m_e^2}\delta_{l0}\left[m_N^2 \ln\left(\frac{m_e}{m_r}\right) - m_e^2 \ln\left(\frac{m_N}{m_r}\right)\right]\right\} , \end{aligned} \tag{28.81}$$

where

$$\begin{aligned} a_n &= -2\left[\ln\left(\frac{2}{n}\right) + \sum_{i=1}^{n}\frac{1}{i} + 1 - \frac{1}{2n}\right]\delta_{l0} \\ &\quad + \frac{1-\delta_{l0}}{l(l+1)(2l+1)} . \end{aligned} \tag{28.82}$$

To lowest order in the mass ratio, higher-order corrections in $Z\alpha$ have been extensively investigated; the contribution of the next two orders in $Z\alpha$ can be written as

$$\begin{aligned} E_R &= \frac{m_e}{m_N}\frac{(Z\alpha)^6}{n^3} m_e c^2 \\ &\quad \times \left[D_{60} + D_{72} Z\alpha \ln^2 (Z\alpha)^{-2} + \cdots\right] , \end{aligned} \tag{28.83}$$

where for $nS_{1/2}$ states [28.58, 59]

$$D_{60} = 4\ln 2 - \frac{7}{2} \tag{28.84}$$

and for states with $l \geq 1$ [28.60–62]

$$D_{60} = \left(3 - \frac{l(l+1)}{n^2}\right)\frac{2}{(4l^2-1)(2l+3)} . \tag{28.85}$$

(As usual, the first subscript on the coefficient refers to the power of $Z\alpha$ and the second subscript to the power of $\ln(Z\alpha)^{-2}$.) The next coefficient in (28.83) has been calculated recently with the result [28.63, 64]

$$D_{72} = -\frac{11}{60\pi}\delta_{l0} . \tag{28.86}$$

The relativistic recoil correction used here is based on (28.81) to (28.86). Numerical values for the complete contribution of (28.83) to all orders in $Z\alpha$ have been obtained by *Shabaev* et al. [28.65]. While these results are in general agreement with the values given by the power series expressions, the difference between them for S states is about three times larger than expected (based on the uncertainty quoted by *Shabaev* et al. [28.65] and the estimated uncertainty of the truncated power series which is taken to be one-half the contribution of the term proportional to D_{72}, as suggested by *Eides* et al. [28.2]). This difference is not critical, and we allow for the ambiguity by assigning an uncertainty for S states of 10% of the contribution given by (28.83). This is sufficiently large that the power series value is consistent with the numerical all-order calculated value. For the states with $l \geq 1$, we assign an uncertainty of 1% of the contribution in (28.83). The covariances of the theoretical values are calculated by assuming that the uncertainties are predominately due to uncalculated terms proportional to $(m_e/m_N)/n^3$.

28.3.3 Nuclear Polarization

Another effect involving specific properties of the nucleus, in addition to relativistic recoil, is nuclear polarization. It arises from interactions between the electron and nucleus in which the nucleus is excited from the ground state to virtual higher states.

For hydrogen, the result that we use for the nuclear polarization is [28.66]

$$E_P(H) = -0.070\,(13)h\frac{\delta_{l0}}{n^3}\text{kHz}\,. \tag{28.87}$$

Larger values for this correction have been reported by *Roedenfelder* [28.67], *Martynenko* and *Faustov* [28.68], but apparently they are based on an incorrect formulation of the dispersion relations [28.2, 66].

For deuterium, to a good approximation, the polarizability of the nucleus is the sum of the proton polarizability, the neutron polarizability [28.69], and the dominant nuclear structure polarizability [28.70], with the total given by

$$E_P(D) = -21.37\,(8)h\frac{\delta_{l0}}{n^3}\text{kHz}\,. \tag{28.88}$$

We assume that this effect is negligible in states of higher l.

28.3.4 Self Energy

The second order (in e, first order in α) level shift due to the one-photon electron self energy, the lowest-order radiative correction, is given by

$$E_{SE}^{(2)} = \frac{\alpha}{\pi}\frac{(Z\alpha)^4}{n^3}F(Z\alpha)m_ec^2\,, \tag{28.89}$$

where

$$\begin{aligned} F(Z\alpha) = {} & A_{41}\ln(Z\alpha)^{-2} + A_{40} + A_{50}(Z\alpha) \\ & + A_{62}(Z\alpha)^2\ln^2(Z\alpha)^{-2} \\ & + A_{61}(Z\alpha)^2\ln(Z\alpha)^{-2} \\ & + G_{SE}(Z\alpha)(Z\alpha)^2\,, \end{aligned} \tag{28.90}$$

with [28.71]

$$\begin{aligned} A_{41} &= \frac{4}{3}\delta_{l0} \\ A_{40} &= -\frac{4}{3}\ln k_0(n,l) + \frac{10}{9}\delta_{l0} \\ &\quad - \frac{1}{2\kappa(2l+1)}(1-\delta_{l0}) \\ A_{50} &= \left(\frac{139}{32} - 2\ln 2\right)\pi\delta_{l0} \\ A_{62} &= -\delta_{l0} \\ A_{61} &= \left[4\left(1+\frac{1}{2}+\cdots+\frac{1}{n}\right) + \frac{28}{3}\ln 2 - 4\ln n \right. \\ &\quad \left. - \frac{601}{180} - \frac{77}{45n^2}\right]\delta_{l0} \\ &\quad + \left(1-\frac{1}{n^2}\right)\left(\frac{2}{15}+\frac{1}{3}\delta_{j\frac{1}{2}}\right)\delta_{l1} \\ &\quad + \frac{96n^2 - 32l(l+1)}{3n^2(2l-1)(2l)(2l+1)(2l+2)(2l+3)} \\ &\quad \times (1-\delta_{l0})\,. \end{aligned} \tag{28.91}$$

Selected Bethe logarithms $\ln k_0(n,l)$ that appear in (28.91) are given in Table 28.3 [28.72].

The function $G_{SE}(Z\alpha)$ in (28.90) gives the higher-order contribution (in $Z\alpha$) to the self energy, and values for $G_{SE}(\alpha)$ are listed in Table 28.4. For the states with $n = 1$ and $n = 2$, the values in the table are based on direct numerical evaluations by *Jentschura* et al. [28.73, 74], and the values for the 3S and 4S states are from *Jentschura* and *Mohr* [28.75]. The remaining values of $G_{SE}(\alpha)$ are based on the

Table 28.3 Relevant Bethe logarithms $\ln k_0(n,l)$

n	S	P	D
1	2.984 128 556		
2	2.811 769 893	−0.030 016 709	
3	2.767 663 612		
4	2.749 811 840	−0.041 954 895	−0.006 740 939
6	2.735 664 207		−0.008 147 204
8	2.730 267 261		−0.008 785 043
12			−0.009 342 954

Table 28.4 Values of the function $G_{\mathrm{SE}}(\alpha)$

n	$\mathrm{S}_{1/2}$	$\mathrm{P}_{1/2}$	$\mathrm{P}_{3/2}$	$\mathrm{D}_{3/2}$	$\mathrm{D}_{5/2}$
1	− 30.290 24 (2)				
2	− 31.185 15 (9)	− 0.9735 (2)	− 0.4865 (2)		
3	− 31.0477 (9)				
4	− 30.912 (4)	− 1.165 (2)	− 0.611 (2)		0.031 (1)
6	− 30.82 (8)				0.034 (2)
8	− 30.80 (9)			0.008 (5)	0.034 (2)
12				0.009 (5)	0.035 (2)

low-Z limit of this function, $G_{\mathrm{SE}}(0) = A_{60}$, in the cases where it is known, together with extrapolations of the results of complete numerical calculations of $F(Z\alpha)$ (28.90) at higher Z [28.76, 77]. There is a long history of calculations of A_{60} [28.2], leading up to the accurate values of A_{60} for the 1S and 2S states obtained by *Pachucki* [28.73, 78–80]. Values for P and D states have been reported subsequently by *Jentschura* and *Pachucki* [28.62], *Jentschura* et al. [28.81, 82]. Extensive numerical evaluations of $F(Z\alpha)$ at higher Z, which in turn yield values for $G_{\mathrm{SE}}(Z\alpha)$, have been done by *Mohr* [28.83], *Mohr* and *Kim* [28.84], *Indelicato* and *Mohr* [28.85], *Le Bigot* [28.86].

The dominant effect of the finite mass of the nucleus on the self energy correction is taken into account by multiplying each term of $F(Z\alpha)$ by the reduced-mass factor $(m_{\mathrm{r}}/m_{\mathrm{e}})^3$, except that the magnetic moment term $-1/[2\kappa(2l+1)]$ in A_{40} is instead multiplied by the factor $(m_{\mathrm{r}}/m_{\mathrm{e}})^2$. In addition, the argument $(Z\alpha)^{-2}$ of the logarithms is replaced by $(m_{\mathrm{e}}/m_{\mathrm{r}})(Z\alpha)^{-2}$ [28.56].

The uncertainty of the self energy contribution to a given level arises entirely from the uncertainty of $G_{\mathrm{SE}}(\alpha)$ listed in Table 28.4 and is taken to be entirely of type u_n.

28.3.5 Vacuum Polarization

The second-order vacuum-polarization level shift, due to the creation of a virtual electron–positron pair in the exchange of photons between the electron and the nucleus, is

$$E^{(2)}_{\mathrm{VP}} = \frac{\alpha}{\pi}\frac{(Z\alpha)^4}{n^3} H(Z\alpha) m_{\mathrm{e}} c^2 , \tag{28.92}$$

where the function $H(Z\alpha)$ is divided into the part corresponding to the Uehling potential, denoted here by $H^{(1)}(Z\alpha)$, and the higher-order remainder $H^{(\mathrm{R})}(Z\alpha) = H^{(3)}(Z\alpha) + H^{(5)}(Z\alpha) + \cdots$, where the superscript denotes the order in powers of the external field. The individual terms are expanded in a power series in $Z\alpha$ as

$$\begin{aligned} H^{(1)}(Z\alpha) &= V_{40} + V_{50}(Z\alpha) + V_{61}(Z\alpha)^2 \ln(Z\alpha)^{-2} \\ &\quad + G^{(1)}_{\mathrm{VP}}(Z\alpha)(Z\alpha)^2 , \end{aligned} \tag{28.93}$$

$$H^{(\mathrm{R})}(Z\alpha) = G^{(\mathrm{R})}_{\mathrm{VP}}(Z\alpha)(Z\alpha)^2 , \tag{28.94}$$

with

$$\begin{aligned} V_{40} &= -\frac{4}{15}\delta_{l0} , \\ V_{50} &= \frac{5}{48}\pi\delta_{l0} , \\ V_{61} &= -\frac{2}{15}\delta_{l0} . \end{aligned} \tag{28.95}$$

The part $G^{(1)}_{\mathrm{VP}}(Z\alpha)$ arises from the Uehling potential, and is readily calculated numerically [28.76, 87]; values are given in Table 28.5. The higher-order remainder $G^{(\mathrm{R})}_{\mathrm{VP}}(Z\alpha)$ has been considered by *Wichmann* and *Kroll*, and the leading terms in powers of $Z\alpha$ are [28.33,88,89]

$$\begin{aligned} G^{(\mathrm{R})}_{\mathrm{VP}}(Z\alpha) &= \left(\frac{19}{45} - \frac{\pi^2}{27}\right)\delta_{l0} \\ &\quad + \left(\frac{1}{16} - \frac{31\pi^2}{2880}\right)\pi(Z\alpha)\delta_{l0} + \cdots . \end{aligned} \tag{28.96}$$

Higher-order terms omitted from (28.96) are negligible.

In a manner similar to that for the self energy, the leading effect of the finite mass of the nucleus is taken into account by multiplying (28.92) by the factor $(m_{\mathrm{r}}/m_{\mathrm{e}})^3$ and including a multiplicative factor of $(m_{\mathrm{e}}/m_{\mathrm{r}})$ in the argument of the logarithm in (28.93).

There is also a second-order vacuum polarization level shift due to the creation of virtual particle pairs other than the $\mathrm{e}^-\mathrm{e}^+$ pair. The predominant contribution for nS states arises from $\mu^+\mu^-$, with the leading term being [28.90, 91]

$$E^{(2)}_{\mu\mathrm{VP}} = \frac{\alpha}{\pi}\frac{(Z\alpha)^4}{n^3}\left(-\frac{4}{15}\right)\left(\frac{m_{\mathrm{e}}}{m_{\mu}}\right)^2\left(\frac{m_{\mathrm{r}}}{m_{\mathrm{e}}}\right)^3 m_{\mathrm{e}} c^2 . \tag{28.97}$$

Table 28.5 Values of the function $G_{\text{VP}}^{(1)}(\alpha)$

n	$S_{1/2}$	$P_{1/2}$	$P_{3/2}$	$D_{3/2}$	$D_{5/2}$
1	−0.618 724				
2	−0.808 872	−0.064 006	−0.014 132		
3	−0.814 530				
4	−0.806 579	−0.080 007	−0.017 666		−0.000 000
6	−0.791 450				−0.000 000
8	−0.781 197			−0.000 000	−0.000 000
12				−0.000 000	−0.000 000

The next order term in the contribution of muon vacuum polarization to nS states is of relative order $Z\alpha m_e/m_\mu$ and is therefore negligible. The analogous contribution $E_{\tau\text{VP}}^{(2)}$ from $\tau^+\tau^-$ (−18 Hz for the 1S state) is also negligible at the level of uncertainty of current interest.

For the hadronic vacuum polarization contribution, we take the result given by *Friar* et al. [28.92] that utilizes all available e^+e^- scattering data:

$$E_{\text{had VP}}^{(2)} = 0.671\,(15) E_{\mu\text{VP}}^{(2)} , \tag{28.98}$$

where the uncertainty is of type u_0.

The muonic and hadronic vacuum polarization contributions are negligible for P and D states.

28.3.6 Two-Photon Corrections

Corrections from two virtual photons, of order α^2, have been calculated as a power series in $Z\alpha$:

$$E^{(4)} = \left(\frac{\alpha}{\pi}\right)^2 \frac{(Z\alpha)^4}{n^3} m_e c^2 F^{(4)}(Z\alpha) , \tag{28.99}$$

where

$$\begin{aligned} F^{(4)}(Z\alpha) &= B_{40} + B_{50}(Z\alpha) + B_{63}(Z\alpha)^2 \ln^3(Z\alpha)^{-2} \\ &\quad + B_{62}(Z\alpha)^2 \ln^2(Z\alpha)^{-2} \\ &\quad + B_{61}(Z\alpha)^2 \ln(Z\alpha)^{-2} + B_{60}(Z\alpha)^2 \\ &\quad + \cdots . \end{aligned} \tag{28.100}$$

The leading term B_{40} is well known:

$$\begin{aligned} B_{40} &= \left[\frac{3\pi^2}{2}\ln 2 - \frac{10\pi^2}{27} - \frac{2179}{648} - \frac{9}{4}\zeta(3)\right]\delta_{l0} \\ &\quad + \left[\frac{\pi^2 \ln 2}{2} - \frac{\pi^2}{12} - \frac{197}{144} - \frac{3\zeta(3)}{4}\right]\frac{1-\delta_{l0}}{\kappa(2l+1)} . \end{aligned} \tag{28.101}$$

The second term has been calculated by *Eides* and *Shelyuto* [28.93], *Pachucki* [28.94], *Eides* et al. [28.90], *Pachucki* [28.95] with the result

$$B_{50} = -21.5561\,(31)\delta_{l0} . \tag{28.102}$$

The next coefficient, as obtained by *Karshenboim* [28.96], *Yerokhin* [28.97], *Manohar* and *Stewart* [28.98], *Pachucki* [28.99], is

$$B_{63} = -\frac{8}{27}\delta_{l0} . \tag{28.103}$$

For S states the coefficient B_{62} has been found to be

$$\begin{aligned} B_{62} = \frac{16}{9}\bigg[&\frac{71}{60} - \ln 2 + \gamma + \psi(n) \\ &- \ln n - \frac{1}{n} + \frac{1}{4n^2}\bigg] , \end{aligned} \tag{28.104}$$

where $\gamma = 0.577\ldots$ is Euler's constant and ψ is the psi function [28.100]. The difference $B_{62}(1) - B_{62}(n)$ was calculated by *Karshenboim* [28.101] and confirmed by *Pachucki* [28.99] who also calculated the n-independent additive constant. For P states the calculated value is [28.101]

$$B_{62} = \frac{4}{27}\frac{n^2-1}{n^2} . \tag{28.105}$$

This result has been confirmed by *Jentschura* and *Nándori* [28.102] who also show that for D and higher angular momentum states $B_{62} = 0$.

The single-logarithm coefficient B_{61} for S states has been given as [28.99]

$$\begin{aligned} B_{61} = &\frac{39751}{10800} + \frac{4N(n)}{3} + \frac{55\pi^2}{27} - \frac{616\ln 2}{135} \\ &+ \frac{3\pi^2 \ln 2}{4} + \frac{40\ln^2 2}{9} - \frac{9\zeta(3)}{8} \\ &+ \left(\frac{304}{135} - \frac{32\ln 2}{9}\right) \\ &\times \left[\frac{3}{4} + \gamma + \psi(n) - \ln n - \frac{1}{n} + \frac{1}{4n^2}\right] , \end{aligned} \tag{28.106}$$

where $N(n)$ is a term that was numerically evaluated for the 1S state by *Pachucki* [28.99]. *Jentschura* [28.103] has evaluated $N(n)$ for excited S states with $n=2$ to $n=8$, has made an improved evaluation for $n=1$, and has given an approximate fit to the calculated results in order to extend them to higher n. Values of the function $N(n)$ for the states of interest here are given in Table 28.6. There are no results yet for P or D states for B_{61}. Based on the relative magnitude of A_{61} for the S, P, and D states, we take as uncertainties $u_n(B_{61}) = 5.0$ for P states and $u_n(B_{61}) = 0.5$ for D states.

Recent work indicates that there may be an additional contribution to B_{61} and/or B_{60} [28.104, 105]. The effect of such a contribution would be to change the S-state energy levels by an amount that is likely to be less than half the uncertainty of the nuclear size correction due to uncertainty in the rms radius of the nucleus.

The two-loop Bethe logarithm b_L, which is expected to be the dominant part of the no-log term B_{60}, has been calculated for the 1S and 2S states by *Pachucki* and *Jentschura* [28.106] who obtained

$$b_L = -81.4\,(3) \quad \text{1S state} \tag{28.107a}$$

$$b_L = -66.6\,(3) \quad \text{2S state}\,. \tag{28.107b}$$

An additional contribution for S states,

$$b_M = \frac{10}{9}N\,, \tag{28.108}$$

was derived by *Pachucki* [28.99], where N is given in Table 28.7 as a function of the state n. These contributions can be combined to obtain an estimate for the coefficient B_{60} for S states:

$$B_{60} = b_L + \frac{10}{9}N + \cdots\,, \tag{28.109}$$

where the dots represent uncalculated contributions to B_{60} which are at the relative level of 15% [28.106]. In order to obtain an approximate value for B_{60} for S states with $n \geq 3$, we employ a simple extrapolation formula,

$$b_L = a + \frac{b}{n}\,, \tag{28.110}$$

with a and b fitted to the 1S and 2S values of b_L, and we include a component of uncertainty $u_0(b_L) = 5.0$. The results for b_L, along with the total estimated values of B_{60} for S states, are given in Table 28.7. For P states, there is a calculation of fine-structure differences [28.107], but because of the uncertainty in B_{61} for P states, we do not include this result. We assume that for both the P and D states, the uncertainty attributed to B_{61} is sufficiently large to account for the uncertainty in B_{60} and higher-order terms as well.

As in the case of the order α self-energy and vacuum-polarization contributions, the dominant effect of the finite mass of the nucleus is taken into account by multiplying each term of the two-photon contribution by the reduced-mass factor $(m_r/m_e)^3$, except that the magnetic moment term, the second line of (28.101), is instead multiplied by the factor $(m_r/m_e)^2$. In addition, the argument $(Z\alpha)^{-2}$ of the logarithms is replaced by $(m_e/m_r)(Z\alpha)^{-2}$.

28.3.7 Three-Photon Corrections

The leading contribution from three virtual photons is assumed to have the form

$$E^{(6)} = \left(\frac{\alpha}{\pi}\right)^3 \frac{(Z\alpha)^4}{n^3} m_e c^2 \left[C_{40} + C_{50}(Z\alpha) + \cdots\right]\,, \tag{28.111}$$

in analogy with (28.99) for two photons. The level shifts of order $(\alpha/\pi)^3 (Z\alpha)^4 m_e c^2$ that contribute to C_{40} can be characterized as the sum of a self-energy correction, a magnetic-moment correction, and a vacuum polarization correction. The self-energy correction arises from the slope of the Dirac form factor, and it has recently been calculated by *Melnikov* and *Ritbergen* [28.108]

Table 28.6 Values of N

n	N
1	17.855 672 (1)
2	12.032 209 (1)
3	10.449 810 (1)
4	9.722 413 (1)
6	9.031 832 (1)
8	8.697 639 (1)

Table 28.7 Values of b_L and B_{60}

n	b_L	B_{60}
1	−81.4 (3)	−61.6 (9.2)
2	−66.6 (3)	−53.2 (8.0)
3	−61.7 (5.0)	−50.1 (9.0)
4	−59.2 (5.0)	−48.4 (8.8)
6	−56.7 (5.0)	−46.7 (8.6)
8	−55.5 (5.0)	−45.8 (8.5)

who obtained

$$E_{\mathrm{SE}}^{(6)} = \left(\frac{\alpha}{\pi}\right)^3 \frac{(Z\alpha)^4}{n^3} m_e c^2 \left(-\frac{868a_4}{9} + \frac{25\zeta(5)}{2} - \frac{17\pi^2\zeta(3)}{6} - \frac{2929\zeta(3)}{72} - \frac{217\ln^4 2}{54} - \frac{103\pi^2\ln^2 2}{270} + \frac{41671\pi^2\ln 2}{540} + \frac{3899\pi^4}{6480} - \frac{454979\pi^2}{9720} - \frac{77513}{46656}\right)\delta_{l0}\,, \qquad (28.112)$$

where ζ is the Riemann zeta function and $a_4 = \sum_{n=1}^{\infty} 1/(2^n n^4) = 0.517\,479\,061\ldots$. The magnetic-moment correction comes from the known three-loop electron anomalous magnetic moment [28.109], and is given by

$$E_{\mathrm{MM}}^{(6)} = \left(\frac{\alpha}{\pi}\right)^3 \frac{(Z\alpha)^4}{n^3} m_e c^2 \left(-\frac{100a_4}{3} + \frac{215\zeta(5)}{24} - \frac{83\pi^2\zeta(3)}{72} - \frac{139\zeta(3)}{18} - \frac{25\ln^4 2}{18} + \frac{25\pi^2\ln^2 2}{18} + \frac{298\pi^2\ln 2}{9} + \frac{239\pi^4}{2160} - \frac{17101\pi^2}{810} - \frac{28259}{5184}\right)\frac{1}{\kappa(2l+1)}\,, \qquad (28.113)$$

and the vacuum-polarization correction is [28.110, 111]

$$E_{\mathrm{VP}}^{(6)} = \left(\frac{\alpha}{\pi}\right)^2 \frac{(Z\alpha)^4}{n^3} m_e c^2 \left(-\frac{8135\zeta(3)}{2304} + \frac{4\pi^2\ln 2}{15} - \frac{23\pi^2}{90} + \frac{325805}{93312}\right)\delta_{l0}\,. \qquad (28.114)$$

The total for C_{40} is

$$C_{40} = \left(-\frac{568a_4}{9} + \frac{85\zeta(5)}{24} - \frac{121\pi^2\zeta(3)}{72} - \frac{84071\zeta(3)}{2304} - \frac{71\ln^4 2}{27} - \frac{239\pi^2\ln^2 2}{135} + \frac{4787\pi^2\ln 2}{108} + \frac{1591\pi^4}{3240} - \frac{252251\pi^2}{9720} + \frac{679441}{93312}\right)\delta_{l0} + \left(-\frac{100a_4}{3} + \frac{215\zeta(5)}{24} - \frac{83\pi^2\zeta(3)}{72} - \frac{139\zeta(3)}{18} - \frac{25\ln^4 2}{18} + \frac{25\pi^2\ln^2 2}{18} + \frac{298\pi^2\ln 2}{9} + \frac{239\pi^4}{2160} - \frac{17101\pi^2}{810} - \frac{28259}{5184}\right)\frac{1-\delta_{l0}}{\kappa(2l+1)}\,. \qquad (28.115)$$

An uncertainty in the three-photon correction is assigned by taking $u_0(C_{50}) = 30\delta_{l0}$ and $u_n(C_{63}) = 1$, where C_{63} is defined by the usual convention.

The dominant effect of the finite mass of the nucleus is taken into account by multiplying C_{40} in (28.115) by the reduced-mass factor $(m_r/m_e)^3$ for $l = 0$ or by the factor $(m_r/m_e)^2$ for $l \neq 0$.

The contribution from four photons is expected to be of order

$$\left(\frac{\alpha}{\pi}\right)^4 \frac{(Z\alpha)^4}{n^3} m_e c^2\,, \qquad (28.116)$$

which is about 10 Hz for the 1S state and is negligible at the level of uncertainty of current interest.

28.3.8 Finite Nuclear Size

At low Z, the leading contribution due to the finite size of the nucleus is

$$E_{\mathrm{NS}}^{(0)} = \mathcal{E}_{\mathrm{NS}}\delta_{l0}\,, \qquad (28.117)$$

with

$$\mathcal{E}_{\mathrm{NS}} = \frac{2}{3}\left(\frac{m_r}{m_e}\right)^3 \frac{(Z\alpha)^2}{n^3} m_e c^2 \left(\frac{Z\alpha R_N}{\lambda\!\!\!^{-}_C}\right)^2\,, \qquad (28.118)$$

where R_N is the bound-state root-mean-square (rms) charge radius of the nucleus and $\lambda\!\!\!^{-}_C$ is the Compton wavelength of the electron divided by 2π. The leading higher-order contributions have been examined by *Friar* [28.112], *Friar* and *Payne* [28.113], *Karshenboim* [28.114] (see also *Mohr* [28.115], *Borisoglebski* and *Trofimenko* [28.89]). The expressions that we employ to evaluate the nuclear size correction are the same as those discussed in more detail in [28.1].

For S states the leading and next-order corrections are given by

$$E_{\mathrm{NS}} = \mathcal{E}_{\mathrm{NS}}\left\{1 - C_\eta \frac{m_r}{m_e}\frac{R_N}{\lambda\!\!\!^{-}_C} Z\alpha - \left[\ln\left(\frac{m_r}{m_e}\frac{R_N}{\lambda\!\!\!^{-}_C}\frac{Z\alpha}{n}\right) + \psi(n) + \gamma - \frac{(5n+9)(n-1)}{4n^2} - C_\theta\right](Z\alpha)^2\right\}\,, \qquad (28.119)$$

where C_η and C_θ are constants that depend on the details of the assumed charge distribution in the nucleus. The values used here are $C_\eta = 1.7\,(1)$ and $C_\theta = 0.47\,(4)$ for hydrogen or $C_\eta = 2.0\,(1)$ and $C_\theta = 0.38\,(4)$ for deuterium.

For the $P_{1/2}$ states in hydrogen the leading term is

$$E_{\mathrm{NS}} = \mathcal{E}_{\mathrm{NS}}\frac{(Z\alpha)^2(n^2-1)}{4n^2}\,. \qquad (28.120)$$

For $P_{3/2}$ states and D states the nuclear-size contribution is negligible.

28.3.9 Nuclear-Size Correction to Self Energy and Vacuum Polarization

In addition to the direct effect of finite nuclear size on energy levels, its effect on the self energy and vacuum polarization contributions must also be considered. This same correction is sometimes called the radiative correction to the nuclear-size effect.

For the self energy, the additional contribution due to the finite size of the nucleus is [28.116–119]

$$E_{\mathrm{NSE}} = \left(4\ln 2 - \frac{23}{4}\right)\alpha(Z\alpha)\mathcal{E}_{\mathrm{NS}}\delta_{l0}\,, \qquad (28.121)$$

and for the vacuum polarization it is [28.117, 120, 121]

$$E_{\mathrm{NVP}} = \frac{3}{4}\alpha(Z\alpha)\mathcal{E}_{\mathrm{NS}}\delta_{l0}\,. \qquad (28.122)$$

For the self-energy term, higher-order size corrections for S states [28.118] and size corrections for P states have been calculated [28.103, 122], but these corrections are negligible for the current work, and are not included. The D-state corrections are assumed to be negligible.

28.3.10 Radiative-Recoil Corrections

The dominant effect of nuclear motion on the self energy and vacuum polarization has been taken into account by including appropriate reduced-mass factors. The additional contributions beyond this prescription are termed radiative-recoil effects with leading terms given by

$$\begin{aligned} E_{\mathrm{RR}} = {} & \frac{m_{\mathrm{r}}^3}{m_{\mathrm{e}}^2 m_{\mathrm{N}}} \frac{\alpha(Z\alpha)^5}{\pi^2 n^3} m_{\mathrm{e}}c^2\delta_{l0} \\ & \times \left[6\zeta(3) - 2\pi^2\ln 2 + \frac{35\pi^2}{36} - \frac{448}{27}\right. \\ & \left. + \frac{2}{3}\pi(Z\alpha)\ln^2(Z\alpha)^{-2} + \cdots\right]. \end{aligned} \qquad (28.123)$$

The leading constant term in (28.123) is the sum of the analytic result for the electron-line contribution [28.123, 124] and the vacuum-polarization contribution [28.125, 126]. The log-squared term has been calculated by *Pachucki* and *Karshenboim* [28.63] and *Melnikov* and *Yelkhovski* [28.64].

For the uncertainty, we take a term of order $(Z\alpha)\ln(Z\alpha)^{-2}$ relative to the square brackets in (28.123) with numerical coefficients 10 for u_0 and 1 for u_n. These coefficients are roughly what one would expect for the higher-order uncalculated terms.

28.3.11 Nucleus Self Energy

An additional contribution due to the self energy of the nucleus has been given by *Pachucki* [28.126]:

$$\begin{aligned} E_{\mathrm{SEN}} = {} & \frac{4Z^2\alpha(Z\alpha)^4}{3\pi n^3} \frac{m_{\mathrm{r}}^3}{m_{\mathrm{N}}^2} c^2 \\ & \times \left[\ln\left(\frac{m_{\mathrm{N}}}{m_{\mathrm{r}}(Z\alpha)^2}\right)\delta_{l0} - \ln k_0(n,l)\right]. \end{aligned} \qquad (28.124)$$

This correction has also been examined by *Eides* et al. [28.2], who consider how it is modified by the effect of structure of the proton. The structure effect leads to an additional model-dependent constant in the square brackets in (28.124).

To evaluate the nucleus self-energy correction, we use (28.124) and assign an uncertainty u_0 that corresponds to an additive constant of 0.5 in the square brackets for S states. For P and D states, the correction is small and its uncertainty, compared to other uncertainties, is negligible.

28.3.12 Total Energy and Uncertainty

The total energy E_{nLj}^X of a particular level (where $L = \mathrm{S}, \mathrm{P}, \ldots$ and $X = \mathrm{H}, \mathrm{D}, \ldots$) is the sum of the various contributions listed above plus an additive correction δ_{nLj}^X that accounts for the uncertainty in the theoretical expression for E_{nLj}^X. Our theoretical estimate of the value of δ_{nLj}^X for a particular level is zero with a standard uncertainty of $u(\delta_{nLj}^X)$ equal to the square root of the sum of the squares (rss) of the individual uncertainties of the contributions, since, as they are defined above, the contributions to the energy of a given level are independent. (Components of uncertainty associated with the fundamental constants are not explicitly shown here.) Thus we have for the square of the uncertainty, or variance, of a particular level

$$u^2(\delta_{nLj}^X) = \sum_i \frac{u_{0i}^2(XLj) + u_{ni}^2(XLj)}{n^6}\,, \qquad (28.125)$$

where the individual values $u_{0i}(XLj)/n^3$ and $u_{ni}(XLj)/n^3$ are the components of uncertainty from each of the contributions, labeled by i, discussed above. [The factors of $1/n^3$ are isolated so that $u_{0i}(XLj)$ is explicitly independent of n.]

Table 28.8 Measured transition frequencies ν in hydrogen

Authors	Laboratory	Frequency interval	Reported value ν/kHz
Hagley and *Pipkin* 28.127	Harvard Univ.	$\nu_H(2S_{1/2}-2P_{3/2})$	9 911 200 (12)
Lundeen and *Pipkin* 28.128	Harvard Univ.	$\nu_H(2P_{1/2}-2S_{1/2})$	1 057 845.0 (9.0)
Newton et al. 28.129	Univ. Sussex	$\nu_H(2P_{1/2}-2S_{1/2})$	1 057 862 (20)

The covariance of any two δs follows from (F7) of Appendix F of [28.1]. For a given isotope X, we have

$$u\left(\delta^X_{n_1Lj}, \delta^X_{n_2Lj}\right) = \sum_i \frac{u^2_{0i}(XLj)}{(n_1n_2)^3}\,, \qquad (28.126)$$

which follows from the fact that $u(u_{0i}, u_{ni}) = 0$ and $u\left(u_{n_1i}, u_{n_2i}\right) = 0$ for $n_1 \neq n_2$. We also set

$$u\left(\delta^X_{n_1L_1j_1}, \delta^X_{n_2L_2j_2}\right) = 0\,, \qquad (28.127)$$

if $L_1 \neq L_2$ or $j_1 \neq j_2$. For covariances between δs for hydrogen and deuterium, we have for states of the same n

$$u\left(\delta^H_{nLj}, \delta^D_{nLj}\right) = \sum_{i=i_c} \frac{u_{0i}(\mathrm{H}Lj)u_{0i}(\mathrm{D}Lj) + u_{ni}(\mathrm{H}Lj)u_{ni}(\mathrm{D}Lj)}{n^6}\,, \qquad (28.128)$$

and for $n_1 \neq n_2$

$$u\left(\delta^H_{n_1Lj}, \delta^D_{n_2Lj}\right) = \sum_{i=i_c} \frac{u_{0i}(\mathrm{H}Lj)u_{0i}(\mathrm{D}Lj)}{(n_1n_2)^3}\,, \qquad (28.129)$$

where the summation is over the uncertainties common to hydrogen and deuterium. In most cases, the uncertainties can in fact be viewed as common except for a known multiplicative factor that contains all of the mass dependence. We assume

$$u\left(\delta^H_{n_1L_1j_1}, \delta^D_{n_2L_2j_2}\right) = 0\,, \qquad (28.130)$$

if $L_1 \neq L_2$ or $j_1 \neq j_2$. These covariances correspond to correlation coefficients as large as 0.991.

Since the transitions between levels are measured in frequency units (Hz), in order to apply the above equations for the energy level contributions we divide the theoretical expression for the energy difference ΔE of the transition by the Planck constant h to convert it to a frequency. Further, we replace the group of constants $\alpha^2 m_e c^2/2h$ in $\Delta E/h$ by cR_∞ to calculate the uncertainties.

28.3.13 Transition Frequencies Between Levels with $n = 2$

As an indication of the consistency of the theory summarized above and the experimental data, we list values of the transition frequencies between levels with $n = 2$ in hydrogen. These results are based on a variation of the 2002 least-squares adjustment in which the measurements of these particular transitions are not included [28.1]. The calculated values are

$$\begin{aligned} \nu_H(2P_{1/2} - 2S_{1/2}) &= 1\,057\,844.5\,(2.6)\ \mathrm{kHz}\,,\\ \nu_H(2S_{1/2} - 2P_{3/2}) &= 9\,911\,197.1\,(2.6)\ \mathrm{kHz}\,,\\ \nu_H(2P_{1/2} - 2P_{3/2}) &= 10\,969\,041.57\,(89)\ \mathrm{kHz}\,. \end{aligned} \qquad (28.131)$$

These results compare favorably with the most recent experimental values given in Table 28.8.

In addition, in He^+, a recent experimental value of the Lamb shift is $\mathcal{S} = 14\,041.13\,(17)$ MHz [28.130], and the current theoretical value is $\mathcal{S} = 14\,041.474\,(42)$ MHz [28.131].

Part B | 28

References

28.1 P. J. Mohr, B. N. Taylor: Rev. Mod. Phys. **72**, 351 (2000)

28.2 M. I. Eides, H. Grotch, V. A. Shelyuto: Phys. Rep. **342**, 63 (2001)

28.3 P. J. Mohr, B. N. Taylor: Rev. Mod. Phys. **77**, 1 (2005)

28.4 R. S. Van Dyck Jr., P. B. Schwinberg, H. G. Dehmelt: Phys. Rev. Lett. **59**, 26 (1987)

28.5 T. Kinoshita, B. Nižić, Y. Okamoto: Phys. Rev. D **41**, 593 (1990)

28.6 T. Kinoshita: Present status of $g-2$ of electron and muon. In: *The Hydrogen Atom: Precision Physics of Simple Atomic Systems*, ed. by S. G. Karshenboim, F. S. Pavone, G. F. Bassani, M. Inguscio, T. W. Hänsch (Springer, Berlin 2001)

28.7 T. Kinoshita: IEEE Trans. Instrum. Meas. **50**, 568 (2001)

28.8 T. Kinoshita, M. Nio: Phys. Rev. Lett. **90**, 021803 (2003)

28.9 T. Kinoshita: Anomalous magnetic moment of the electron and the fine structure constant. In: *An Isolated Atomic Particle at Rest in Free Space*, ed. by E. N. Fortson, E. M. Henley (Al-

pha Science International, Harrow, UK 2004) pp. 77–90
28.10 S. Laporta: Phys. Lett. B **523**, 95 (2001)
28.11 P. Mastrolia, E. Remiddi: Precise evaluation of the electron $g-2$ at 4 loops: The algebraic way. In: *The Hydrogen Atom: Precision Physics of Simple Atomic Systems*, ed. by S. G. Karshenboim, F. S. Pavone, G. F. Bassani, M. Inguscio, T. W. Hänsch (Springer, Berlin 2001) pp. 776–783
28.12 M. Davier, A. Höcker: Phys. Lett. B **435**, 427 (1998)
28.13 B. Krause: Phys. Lett. B **390**, 392 (1997)
28.14 T. Beier, H. Häffner, N. Hermanspahn, S. G. Karshenboim, H.-J. Kluge, W. Quint, S. Stahl, J. Verdú, G. Werth: Phys. Rev. Lett. **88**, 011603 (2002)
28.15 H. Häffner, T. Beier, S. Djekić, N. Hermanspahn, H.-J. Kluge, W. Quint, S. Stahl, J. Verdú, T. Valenzuela, G. Werth: Eur. Phys. J. D **22**, 163 (2003)
28.16 G. Werth: private communication (2003)
28.17 J. L. Verdú, T. Beier, S. Djekic, H. Häffner, H.-J. Kluge, W. Quint, T. Valenzuela, G. Werth: Can. J. Phys. **80**, 1233 (2002)
28.18 J. Verdú, T. Beier, S. Djekić, H. Häffner, H.-J. Kluge, W. Quint, T. Valenzuela, M. Vogel, G. Werth: J. Phys. B **36**, 655 (2003)
28.19 G. Breit: Nature (London) **122**, 649 (1928)
28.20 H. Grotch: Phys. Rev. Lett. **24**, 39 (1970)
28.21 R. Faustov: Phys. Lett. B **33**, 422 (1970)
28.22 F. E. Close, H. Osborn: Phys. Lett. B **34**, 400 (1971)
28.23 V. A. Yerokhin, P. Indelicato, V. M. Shabaev: Phys. Rev. Lett. **89**, 143001 (2002)
28.24 V. A. Yerokhin, P. Indelicato, V. M. Shabaev: Can. J. Phys. **80**, 1249 (2002)
28.25 T. Beier, I. Lindgren, H. Persson, S. Salomonson, P. Sunnergren, H. Häffner, N. Hermanspahn: Phys. Rev. A **62**, 032510 (2000)
28.26 T. Beier: Phys. Rep. **339**, 79 (2000)
28.27 H. Persson, S. Salomonson, P. Sunnergren, I. Lindgren: Phys. Rev. A **56**, R2499 (1997)
28.28 S. A. Blundell, K. T. Cheng, J. Sapirstein: Phys. Rev. A **55**, 1857 (1997)
28.29 I. Goidenko, L. Labzowsky, G. Plunien, G. Soff: Phys. Rev. A **66**, 032115 (2002)
28.30 S. G. Karshenboim: Phys. Lett. A **266**, 380 (2000)
28.31 S. G. Karshenboim, V. G. Ivanov, V. M. Shabaev: Can. J. Phys. **79**, 81 (2001)
28.32 S. G. Karshenboim, V. G. Ivanov, V. M. Shabaev: Zh. Eksp. Teor. Fiz. **120**, 546 (2001) [JETP **93**, 477 (2001)]
28.33 P. J. Mohr: Lamb shift in hydrogen-like ions. In: *Beam-Foil Spectroscopy*, Vol. 1, ed. by I. A. Sellin, D. J. Pegg (Plenum Press, New York 1975) pp. 89–96
28.34 S. G. Karshenboim, A. I. Milstein: Phys. Lett. B **549**, 321 (2002)
28.35 E.-O. Le Bigot, U. D. Jentschura, P. J. Mohr, P. Indelicato, G. Soff: Phys. Rev. A **68**, 042101 (2003)
28.36 M. I. Eides, H. Grotch: Ann. Phys. (N.Y.) **260**, 191 (1997)
28.37 A. Czarnecki, K. Melnikov, A. Yelkhovsky: Phys. Rev. A **63**, 012509 (2001)
28.38 H. Grotch: Phys. Rev. A **2**, 1605 (1970)
28.39 H. Grotch: Electromagnetic interactions of hydrogenic atoms: Correction to g factors. In: *Precision Measurement and Fundamental Constants*, ed. by D. N. Langenberg, B. N. Taylor (NBS Spec. Pub. 343, US Government Printing Office, Washington, DC 1971) pp. 421–425
28.40 R. A. Hegstrom: $1S_{1/2}$ bound state corrections to the electron and proton g factors for atomic hydrogen. In: *Precision Measurement and Fundamental Constants*, ed. by D. N. Langenberg, B. N. Taylor (NBS Spec. Pub. 343, US Government Printing Office, Washington, DC 1971) pp. 417–420
28.41 H. Grotch, R. A. Hegstrom: Phys. Rev. A **4**, 59 (1971)
28.42 R. A. Hegstrom: Phys. Rev. **184**, 17 (1969)
28.43 V. M. Shabaev, V. A. Yerokhin: Phys. Rev. Lett. **88**, 091801 (2002)
28.44 V. M. Shabaev: Phys. Rev. A **64**, 052104 (2001)
28.45 A. Yelkhovsky: hep-ph/0108091 (2001)
28.46 M. Eides: private communication (2002)
28.47 A. P. Martynenko, R. N. Faustov: Yad. Fiz. **65**, 297 (2002) [Phys. At. Nucl. **65**, 271 (2002)]
28.48 A. P. Martynenko, R. N. Faustov: Zh. Eksp. Teor. Fiz. **120**, 539 (2001) [JETP **93**, 471 (2001)]
28.49 D. A. Glazov, V. M. Shabaev: Phys. Lett. A **297**, 408 (2002)
28.50 I. Angeli: Heavy Ion Phys. **8**, 23 (1998)
28.51 D. L. Farnham, R. S. Van Dyck, Jr., P. B. Schwinberg: Phys. Rev. Lett. **75**, 3598 (1995)
28.52 R. S. Van Dyck, Jr., S. L. Zafonte, P. B. Schwinberg: Hyp. Int. **132**, 163 (2001)
28.53 R. S. Van Dyck, Jr.: private communication (2003)
28.54 S. G. Karshenboim, V. G. Ivanov: Can. J. Phys. **80**, 1305 (2002)
28.55 W. A. Barker, F. N. Glover: Phys. Rev. **99**, 317 (1955)
28.56 J. R. Sapirstein, D. R. Yennie: In: *Quantum Electrodynamics*, ed. by T. Kinoshita (World Scientific, Singapore 1990) Chap. 12, pp. 560–672
28.57 G. W. Erickson: J. Phys. Chem. Ref. Data **6**, 831 (1977)
28.58 K. Pachucki, H. Grotch: Phys. Rev. A **51**, 1854 (1995)
28.59 M. I. Eides, H. Grotch: Phys. Rev. A **55**, 3351 (1997)
28.60 E. A. Golosov, A. S. Elkhovskiĭ, A. I. Mil'shteĭn, I. B. Khriplovich: Zh. Eksp. Teor. Fiz. **107**, 393 (1995) [JETP **80**, 208 (1995)]
28.61 A. S. Elkhovskiĭ: Zh. Eksp. Teor. Fiz. **110**, 431 (1996) [JETP **83**, 230 (1996)]
28.62 U. Jentschura, K. Pachucki: Phys. Rev. A **54**, 1853 (1996)
28.63 K. Pachucki, S. G. Karshenboim: Phys. Rev. A **60**, 2792 (1999)
28.64 K. Melnikov, A. Yelkhovsky: Phys. Lett. B **458**, 143 (1999)
28.65 V. M. Shabaev, A. N. Artemyev, T. Beier, G. Soff: J. Phys. B **31**, L337 (1998)

28.66 I. B. Khriplovich, R. A. Sen'kov: Phys. Lett. B **481**, 447 (2000)
28.67 R. Rosenfelder: Phys. Lett. B **463**, 317 (1999)
28.68 A. P. Martynenko, R. N. Faustov: Yad. Fiz. **63**, 915 (2000) [Phys. At. Nucl. **63**, 845 (2000)]
28.69 I. B. Khriplovich, R. A. Sen'kov: Phys. Lett. A **249**, 474 (1998)
28.70 J. L. Friar, G. L. Payne: Phys. Rev. C **56**, 619 (1997)
28.71 G.W. Erickson, D. R. Yennie: Ann. Phys. (N.Y.) **35**, 271 (1965)
28.72 G. W. F. Drake, R. A. Swainson: Phys. Rev. A **41**, 1243 (1990)
28.73 U. D. Jentschura, P. J. Mohr, G. Soff: Phys. Rev. Lett. **82**, 53 (1999)
28.74 U. D. Jentschura, P. J. Mohr, G. Soff: Phys. Rev. A **63**, 042512 (2001)
28.75 U. D. Jentschura, P. J. Mohr: Phys. Rev. A **69**, 064103 (2004)
28.76 S. Kotochigova, P. J. Mohr, B. N. Taylor: Can. J. Phys. **80**, 1373 (2002)
28.77 E.-O. Le Bigot, U. D. Jentschura, P. J. Mohr, P. Indelicato: private communication (2003)
28.78 K. Pachucki: Phys. Rev. A **46**, 648 (1992)
28.79 K. Pachucki: quoted in [28.73], 1999
28.80 K. Pachucki: Ann. Phys. (N.Y.) **226**, 1 (1993)
28.81 U. D. Jentschura, G. Soff, P. J. Mohr: Phys. Rev. A **56**, 1739 (1997)
28.82 U. D. Jentschura, E.-O. Le Bigot, P. J. Mohr, P. Indelicato, G. Soff: Phys. Rev. Lett. **90**, 163001 (2003)
28.83 P. J. Mohr: Phys. Rev. A **46**, 4421 (1992)
28.84 P. J. Mohr, Y.-K. Kim: Phys. Rev. A **45**, 2727 (1992)
28.85 P. Indelicato, P. J. Mohr: Phys. Rev. A **58**, 165 (1998)
28.86 E.-O. Le Bigot: QED dans les ions á un et deux électrons: états trés excités ou quasi-dégénérés. Ph.D. Thesis (University of Paris VI, Paris 2001)
28.87 P. J. Mohr: Phys. Rev. A **26**, 2338 (1982)
28.88 E. H. Wichmann, N. M. Kroll: Phys. Rev. **101**, 843 (1956)
28.89 P. J. Mohr: At. Data. Nucl. Data Tables **29**, 453 (1983)
28.90 M. I. Eides, V. A. Shelyuto: Phys. Rev. A **52**, 954 (1995)
28.91 S. G. Karshenboim: J. Phys. B **28**, L77 (1995)
28.92 J. L. Friar, J. Martorell, D. W. L. Sprung: Phys. Rev. A **59**, 4061 (1999)
28.93 K. Pachucki: Phys. Rev. A **48**, 2609 (1993)
28.94 M. I. Eides, H. Grotch, V. A. Shelyuto: Phys. Rev. A **55**, 2447 (1997)
28.95 K. Pachucki: Phys. Rev. Lett. **72**, 3154 (1994)
28.96 S. G. Karshenboim: Zh. Eksp. Teor. Fiz. **103**, 1105 (1993) [JETP **76**, 541 (1993)]
28.97 V. A. Yerokhin: Phys. Rev. A **62**, 012508 (2000)
28.98 A. V. Manohar, I. W. Stewart: Phys. Rev. Lett. **85**, 2248 (2000)
28.99 K. Pachucki: Phys. Rev. A **63**, 042503 (2001)
28.100 M. Abramowitz, I. A. Stegun: *Handbook of Mathematical Functions* (Dover Publications, Inc., New York, NY 1965)
28.101 S. G. Karshenboim: J. Phys. B **29**, L29 (1996)
28.102 U. D. Jentschura, I. Nándori: Phys. Rev. A **66**, 022114 (2002)
28.103 U. D. Jentschura: J. Phys. A **36**, L229 (2003)
28.104 K. Pachucki: private communication (2004)
28.105 V. A. Yerokhin, P. Indelicato, and V. M. Shabaev: hep-ph/0409048, 2004
28.106 K. Pachucki, U. D. Jentschura: Phys. Rev. Lett. **91**, 113005 (2003)
28.107 U. D. Jentschura, K. Pachucki: J. Phys. A **35**, 1927 (2002)
28.108 K. Melnikov, T. V. Ritbergen: Phys. Rev. Lett. **84**, 1673 (2000)
28.109 S. Laporta, E. Remiddi: Phys. Lett. B **379**, 283 (1996)
28.110 P. A. Baikov, D. J. Broadhurst: Three-loop QED vacuum polarization and the four-loop muon anomalous magnetic moment. In: *New Computing Techniques in Physics Research IV. International Workshop on Software Engineering and Artificial Intelligence for High Energy and Nuclear Physics*, ed. by B. Denby, D. Perret-Gallix (World Scientific, Singapore 1995) pp. 167–172
28.111 M. I. Eides, H. Grotch: Phys. Rev. A **52**, 3360 (1995)
28.112 J. L. Friar: Ann. Phys. (N.Y.) **122**, 151 (1979)
28.113 J. L. Friar, G. L. Payne: Phys. Rev. A **56**, 5173 (1997)
28.114 S. G. Karshenboim: Z. Phys. D **39**, 109 (1997)
28.115 L. A. Borisoglebsky, E. E. Trofimenko: Phys. Lett. B **81**, 175 (1979)
28.116 K. Pachucki: Phys. Rev. A **48**, 120 (1993)
28.117 M. I. Eides, H. Grotch: Phys. Rev. A **56**, R2507 (1997)
28.118 A. I. Milstein, O. P. Sushkov, I. S. Terekhov: Phys. Rev. Lett. **89**, 283003 (2002)
28.119 A. I. Milstein, O. P. Sushkov, I. S. Terekhov: Phys. Rev. A **67**, 062103 (2003)
28.120 J. L. Friar: Z. Phys. A **292**, 1 (1979)
28.121 D. J. Hylton: Phys. Rev. A **32**, 1303 (1985)
28.122 A. I. Milstein, O. P. Sushkov, I. S. Terekhov: Phys. Rev. A **67**, 062111 (2003)
28.123 M. I. Eides, H. Grotch, V. A. Shelyuto: Phys. Rev. A **63**, 052509 (2001)
28.124 A. Czarnecki, K. Melnikov: Phys. Rev. Lett. **87**, 013001 (2001)
28.125 M. I. Eides, H. Grotch: Phys. Rev. A **52**, 1757 (1995)
28.126 K. Pachucki: Phys. Rev. A **52**, 1079 (1995)
28.127 E. W. Hagley, F. M. Pipkin: Phys. Rev. Lett. **72**, 1172 (1994)
28.128 S. R. Lundeen, F. M. Pipkin: Metrologia **22**, 9(1986)
28.129 G. Newton, D. A. Andrews, P. J. Unsworth: Philos. Trans. R. Soc. London, Ser. A **290**, 373 (1979)
28.130 A. van Wijngaarden, F. Holuj, G. W. F. Drake: Phys. Rev. A **63**, 012505 (2001)
28.131 U. D. Jentschura, G. W. F. Drake: Can. J. Phys. **82**, 103 (2004)

29. Parity Nonconserving Effects in Atoms

Until 1957 the invariance of the laws of physics under the process of parity inversion was assumed to hold. The concept of this invariance has its origin in atomic physics, where Laporte introduced it to explain certain aspects of the iron spectrum. The underlying theory of atomic structure, quantum electrodynamics (QED), is an example of a theory that has this invariance. However, it was shown in 1957 that weak interaction processes are not invariant under parity inversion. At that time the only known weak interactions were charge changing and thus did not affect the spectrum of a stable atom. However, if charge conserving, or neutral weak interactions exist, these can lead to parity nonconserving (PNC) processes in atoms. These include electric dipole transitions between states of the same parity and optical rotation.

29.1 **The Standard Model** ... 450
29.2 **PNC in Cesium** ... 451
29.3 **Many-Body Perturbation Theory** ... 451
29.4 **PNC Calculations** ... 452
29.5 **Recent Developments** ... 453
29.6 **Comparison with Experiment** ... 453
References ... 454

It is now well established that such interactions do exist, and are mediated by the Z boson. Unfortunately, because of the extremely small ratio of the energy scale of atoms to the Z mass, PNC processes in atoms must be correspondingly small. In 1974, however, the *Bouchiats* [29.1], in a paper that laid the foundation for the field, showed that parity nonconserving (PNC) transitions in heavy atoms with atomic number Z were enhanced by a factor of Z^3. While still very small, this effect has been observed in a variety of heavy atoms, specifically cesium ($Z = 55$) [29.2–4], thallium ($Z = 81$) [29.5,6], lead ($Z = 82$) [29.7], and bismuth ($Z = 83$) [29.8]. The accurate calculation of the electronic structure of such atoms is of course a very challenging atomic physics problem, and such calculations must be carried out before the experiments can be interpreted in terms of particle physics. This problem is not present for hydrogen, but experimental problems have impeded progress in this direction.

At the time of the Bouchiats's paper, the question of whether or not neutral currents existed was an important problem in weak interaction physics, and even a qualitative observation of atomic PNC would have been of great interest. However, now that neutral currents are well established, what now has become the most important aspect of atomic PNC is the precise measurement of the effect. This is because of the present intense interest in possible modifications of the standard model. While this model, described in more detail in the next section, is extremely successful, it is incomplete in some ways, in particular, in the way particle masses are treated. In the standard model, for example, the mystery of the fermion mass spectrum is simply transferred to another mystery of the sizes of Yukawa coupling constants that couple Higgs fields to fermions. It is hoped that some deviation from the standard model will be uncovered, generally referred to as 'new physics', that will provide a hint as to a more satisfactory theory. At the present time, since all standard model predictions are in agreement with experiment at the few percent level, the hoped for deviation will be small; and for this reason high precision experiments combined with calculations of similar accuracy are needed. For this reason atomic PNC, while apparently a somewhat esoteric branch of atomic physics, actually is involved with one of the most central problems of the field, the accurate solution of the Schrödinger equation for many-electron atoms. We will describe an approach to this fundamental problem below.

Another weak interaction effect important in atomic physics has to do with CP noninvariance. While the weak interactions are not invariant under parity inversion (P), until 1964 it appeared that they were invariant under the simultaneous operations of P and charge conjugation (C), or CP. In 1964, however, measurements of the neutral K meson system showed that CP invariance is slightly broken. While this does not directly affect atomic physics, it has been suggested that this still incompletely explained breaking in the quark sector may have a counterpart in the lepton sector, which could have, as a consequence, a nonvanishing electron electric dipole moment (edm). This opens the possibility of using precise atomic experiments to search for such an effect. In this case, as with 'ordinary' PNC, a calculation of atomic structure is necessary. The requirements of precision are not as important in this case, as the simple detection of a non-zero atomic edm would be a discovery of importance comparable to the detection of CP noninvariance in the kaon system. Because of limitations of space, we do not discuss this interesting field further, referring the interested reader to the review of the subject given in [29.9].

29.1 The Standard Model

The Standard Model is the present theory of the strong, weak, and electromagnetic interactions. We are concerned with the part of the standard model that unifies the weak and electromagnetic interactions, which is referred to as the electroweak theory. This theory involves the interactions of leptons and quarks with four vector bosons, the photon (A), a heavy neutral partner of the photon (Z), and two charged bosons (W^+, W^-). Interactions involving the photon conserve parity, while interactions between the other bosons do not. The neutral currents affecting atomic PNC are described by the following Lagrangian:

$$L = -e\sum_i q_i\bar{\psi}_i\gamma_\mu\psi_i A^\mu - \frac{e}{\sin 2\theta_W} \times\sum_i \bar{\psi}_i\gamma_\mu(V_i - A_i\gamma_5)\psi_i Z^\mu , \tag{29.1}$$

where i ranges over three types of fermions, the electron $\left(q_e = -1,\ V_e = -\frac{1}{2} + 2\sin^2\theta_W,\ A_e = -\frac{1}{2}\right)$, the up quark $\left(q_u = \frac{2}{3},\ V_u = \frac{1}{2} - \frac{4}{3}\sin^2\theta_W,\ A_u = \frac{1}{2}\right)$, and the down quark $\left(q_d = -\frac{1}{3},\ V_d = -\frac{1}{2} + \frac{2}{3}\sin^2\theta_W,\ A_d = -\frac{1}{2}\right)$. The weak angle θ_W is a fundamental parameter of the standard model and is discussed further below.

The physics in the above Lagrangian that leads to atomic PNC is the exchange of a virtual Z, either between a quark in the nucleus and an electron, or between two electrons, though calculations of the latter effect show it to be negligible [29.10]. PNC arises when the Z matrix element is a vector on the nucleus and an axial vector on the electron ($V_N A_e$), or vice versa ($A_N V_e$). The dominant PNC contribution comes from the former case, because all the quarks contribute coherently. The latter case, however, cannot be neglected and is discussed further below. Because the nuclear current is of vector nature, one can introduce a conserved charge, the *weak charge* Q_W,

$$Q_W = 2Z(2V_u + V_d) + 2N(V_u + 2V_d) . \tag{29.2}$$

Here Z is the number of protons and N the number of neutrons. Putting in the above values of V_u and V_d then gives

$$Q_W = Z\left(1 - 4\sin^2\theta_W\right) - N . \tag{29.3}$$

It would appear that the dependence of Q_W on the weak angle would allow its determination from atomic PNC. While this can be done, it is important to note that the above discussion has been at the tree level. One of the most important features of the electroweak theory is that radiative corrections can be calculated. These corrections enter at the percent level, and must be included for an accurate interpretation of experiments sensitive to electroweak effects. Until recently, of the parameters affecting the electroweak theory only the fine structure constant and the muon decay rate were known with high precision, leaving the weak angle a free parameter. However, now that the Z mass has been measured with very high precision, it is now possible to use that measurement to determine θ_W, and to then predict a large set of radiatively corrected electroweak processes, including atomic PNC [29.11]. While there is still some sensitivity to the top quark and Higgs masses, as discussed further below, using the Z mass allows the prediction for cesium of, assuming a Higgs mass of 100 GeV,

$$Q_W = -73.20(0.13) . \tag{29.4}$$

Therefore, the importance of atomic PNC for fundamental physics is not so much its ability to determine the weak angle, but rather the fact that the standard model makes a definite prediction for the size of the effect, and any disagreement indicates new physics; and conversely, agreement to within a given precision puts limits on new physics.

29.2 PNC in Cesium

We concentrate on PNC in cesium because it is the simplest atom in which the effect has been measured. This is because it consists of a single electron outside a closed xenonlike core which is relatively unpolarizable. This should be contrasted with, for example, thallium. While thallium nominally also consists of one $6p_{1/2}$ electron outside a closed core, part of that core is a filled $6s^2_{1/2}$ subshell. It is quite easy to polarize the outer subshell, so that one really has three electrons outside a closed core. This leads to distinctly poorer convergence properties of many-body perturbation theory, the theoretical method used for these calculations, and consequently less accurate atomic theory predictions. Similar considerations apply to lead and bismuth. The PNC transition in cesium that has been studied is $6s_{1/2} \rightarrow 7s_{1/2}$, and accurate measurements have been made in *Paris* [29.2] and *Boulder* [29.3, 4]. Cesium is a 55 electron atom with a nucleus consisting of 78 neutrons and 55 protons with nuclear spin $I = 7/2$. The total angular momentum of atomic s-states is then $F = 3$ or $F = 4$. Both of the transitions, $6s_{1/2}(F = 4) \rightarrow 7s_{1/2}(F = 3)$ and $6s_{1/2}(F = 3) \rightarrow 7s_{1/2}(F = 4)$, have been measured, allowing the isolation of PNC effects that depend on the spin of the nucleus. The bulk of atomic PNC comes from the timelike contribution of the $(V_N A_e)$ exchange discussed in the previous section, and can be described by the effective Hamiltonian

$$H_W = \frac{G_F}{\sqrt{8}} Q_W \rho_{nuc}(\boldsymbol{r}) \gamma_5 . \tag{29.5}$$

Here $\rho_{nuc}(\boldsymbol{r})$ is a weighted average of the neutron and proton distributions in the nucleus, which leads to nuclear structure uncertainties that will be discussed below. Using this Hamiltonian, a large scale calculation [29.10] leads to the prediction for the nuclear-spin-independent part of the PNC transition

$$E_{PNC} = -0.905(9) \times 10^{-11} \mathrm{i}|e|a_0(-Q_W/N) . \tag{29.6}$$

Here the unknown Q_W has been factored out and divided by its approximate value $-N$. This result is in good agreement with independent MBPT calculations of the Novosibirsk [29.12] and Göteborg [29.13] groups. When this is compared with the experimental measurement [29.3, 4]

$$E^{exp}_{PNC} = -0.8374(67) \times 10^{-11} \mathrm{i}|e|a_0 , \tag{29.7}$$

there results a prediction for Q_W of

$$Q_W = -72.17(0.58)[0.72] , \tag{29.8}$$

where the first error is experimental and the second theoretical. The spacelike part of $(V_N A_e)$ exchange and the timelike part of $(A_N V_e)$ exchange are negligible, but the spacelike part of the latter gives a nuclear spin-dependent effect that will be discussed below. Also discussed below is an interesting nuclear physics source of PNC known as the *anapole* moment [29.14], which arises from photon exchange with weak radiative corrections on the nuclear vertex. This effect enters at the several percent level, but in a way that can be subtracted out as will be described below.

29.3 Many-Body Perturbation Theory

While there are a variety of ways of solving the many-electron Schrödinger equation, the most accurate treatments of cesium PNC are based on many-body perturbation theory (MBPT). In MBPT, the Hamiltonian is broken up into $H = H_0 + V_C$, where

$$H_0 = \sum_{i=1}^{N} \left[\boldsymbol{\alpha}_i \cdot \boldsymbol{p}_i + \beta_i m + U(r_i) \right] \tag{29.9}$$

and

$$V_C = \frac{1}{2} \sum_{ij} \frac{\alpha}{|\boldsymbol{r}_i - \boldsymbol{r}_j|} - \sum_{i=1}^{N} \left[\frac{Z\alpha}{r_i} + U(r_i) \right] . \tag{29.10}$$

The starting potential $U(r)$ is generally chosen to be a frozen core Hartree–Fock (HF) potential. This model of the atom is rather inaccurate: valence removal energies disagree with experiment by about 10%, and matrix elements of the hyperfine operator by about 40%. Thus it is essential for accurate calculations to include the effects of V_C as fully as possible. MBPT proceeds by expanding the many-body wave function and the energy in powers of V_C. While going to second order in V_C improves agreement with experiment to the percent level for energies and the few percent level for matrix elements [29.15], more powerful methods that sum infinite classes of MBPT contributions are needed for higher accuracy, and we turn to a description of these *all order* methods.

If the frozen-core HF wave function is denoted by Ψ_0, all-order *singles–doubles* methods add to it a correction $\Delta\Psi$ with either one or two orbitals in the HF wave function excited:

$$\delta\Psi = \Bigg(\sum_{am} \rho_{ma} a_m^\dagger a_a + \sum_{abmn} \rho_{mnab} a_m^\dagger a_n^\dagger a_a a_b + \sum_m \rho_{mv} a_m^\dagger a_v + \sum_{amn} \rho_{mnav} a_m^\dagger a_n^\dagger a_a a_v \Bigg) \Psi_0 . \tag{29.11}$$

The terms on the first line of (29.11) describe single and double excitations of the closed core, while those on the second line describe single and double excitations of the atom where the valence orbital is also excited. Substituting (29.11) into the Schrödinger equation, one obtains a set of coupled equations for the expansion coefficients that can be found in [29.16]. The first and second iterations of the equations for the expansion coefficients leads to results that are identical to first- and second-order perturbation theory. In third-order perturbation theory, terms associated with triple excitations contribute to the energy. These terms have no counterpart in the iterative solution to the equations under consideration.

To account for such terms, it is necessary to add to $\Delta\Psi$ a triple-excitation correction of the specific form:

$$\Delta\Psi = \Bigg(\sum_{abcmnr} \rho_{mnrabc} a_n^\dagger a_m^\dagger a_r^\dagger a_a a_b a_c + \sum_{abmnr} \rho_{mnrabv} a_n^\dagger a_m^\dagger a_r^\dagger a_a a_b a_v \Bigg) \Psi_0 . \tag{29.12}$$

Such a term enters in two ways. First, there are a set of equations giving the triple-excitation coefficients in terms of the single-, double-, and triple-excitation coefficients. Second, the triple-excitation coefficients enter on the right-hand side of the equations for the single- and double-excitation coefficients. Solving for the triple-excitation coefficients in terms of the singles and doubles (ignoring the triples on the right-hand sides of these equations), one can use them on the right-hand sides of the equations for the singles. This procedure leads to equations which, when iterated to third order, include all of the terms from MBPT. However, the triples also modify the the doubles equation, and this more computationally demanding step has not yet been implemented. As the effects of this modification enter first in fourth-order MBPT, the calculation is complete through third order, but still misses some fourth-order contributions. Nevertheless, the method is accurate enough to predict PNC to 1%. Greater accuracy is expected when the doubles equation is modified, which should lead to a calculation complete through fourth order, and work on this problem is in progress.

29.4 PNC Calculations

The above discussion of MBPT allows the calculation on cesium to be accurate to a few tenths of percent for energies and under one percent for 'ordinary' matrix elements [29.16], and thus should allow a calculation of similar accuracy for PNC. To calculate such transitions, one modifies H_0 by adding the weak-interaction h_W to the HF potential. This approach leads to a generalization of the single-particle states in which each state acquires an opposite-parity admixture, that is,

$$\phi_k \to \phi_k + \tilde{\phi}_k . \tag{29.13}$$

Thus, for example, each $s_{1/2}$ orbital will pick up a small $p_{1/2}$ state admixture. One can then calculate a PNC transition along the lines of a parity allowed transition calculation, simply replacing each orbital occuring in the MBPT formulas in turn with its opposite parity admixture. Details of this approach can be found in [29.10].

There are a number of smaller effects that must be considered when the 1% level of accuracy is reached. The first stems from the fact that the Z couples to the density of up and down quarks, which in turn is related to the density of protons and neutrons in the cesium nucleus. While experimental data is available on the charge distribution [29.17], the neutron distribution is available only theoretically. While this uncertainty can be shown to be unimportant for Cs^{133} [29.10], there has been interest in looking at PNC in several different isotopes with the aim of taking ratios and greatly reducing atomic physics uncertainties. However, the neutron distribution in different nuclei is more uncertain, and taking the ratio enhances the nuclear physics uncertainty. This issue has been addressed recently by *Chen* and *Vogel* [29.18], who estimate the nuclear physics uncertainties to be smaller than the anticipated experimental error.

The next small effect is the interaction between the nuclear axial-vector current and the electron vector current from Z exchange. In the limit of nonrelativistic nucleon motion, this interaction is given by the spin-dependent Hamiltonian

$$h_{\mathrm{W}}^{(2)} = -\frac{G}{\sqrt{2}} K_2 \frac{\kappa - 1/2}{I(I+1)} \boldsymbol{\alpha} \cdot \boldsymbol{I} \rho(r) \,. \tag{29.14}$$

Here, $\kappa = 4$, $I = 7/2$ and $K_2 \simeq -0.05$ for the valence proton of ^{133}Cs. Additionally, parity violation in the nucleus leads to a parity-violating nuclear moment, the anapole moment mentioned above, that couples electromagnetically to the atomic electrons. The anapole–electron interaction is described by a Hamiltonian similar to (29.14), with K_2 replaced by $-\kappa/(\kappa - 1/2)K_a$, where $K_a = 0.24 - 0.33$ is determined from nuclear model calculations [29.19]. Linear combinations of amplitudes for different $F \to F'$ transitions can be used to isolate either the spin-dependent or spin-independent parts of the interaction. The final smaller effects are Z exchange between the electrons, which as mentioned above turns out to be negligible, and the inclusion of the Breit interaction. While the original estimate of the latter was a small 0.2% effect, included in (29.6), this turned out to be an underestimate, one of a number of unexpectedly large contributions found in recent years, as discussed in the next section.

29.5 Recent Developments

The Breit interaction was reexamined by *Derevianko* [29.20] and found to be significantly larger than the 0.2% mentioned above. The 0.6% includes the effects of negative energy states, which are enhanced in the context of PNC. This was followed by the discovery of a number of effects that contributed at the several tenths of a percent level, with theory and experiment going into and out of agreement with one another as each effect was found. At present the situation has stabilized, and, as will be discussed in the next section, no discrepancy with the standard model is present. In the past eight years, the principal developments have been the discovery of a surprisingly large two-loop radiative correction associated with vacuum polarziation [29.21,22], recalculation of the atomic physics part of the calculation [29.23], which, however, only reduced the theoretical error bar, an independent measurement of the polarizability of cesium [29.24], and the discovery that the self-energy radiative correction, which in lowest order is $-\alpha/2\pi$, and is included in the radiative corrections to Q_{W} presented in [29.11], is strongly enhanced by binding corrections starting at order $Z\alpha$ [29.25–27]. When taken together, these different effects tend to cancel, so the qualitative picture of agreement with the standard model remains correct.

29.6 Comparison with Experiment

We can now make use of the above analysis to extract the value of the weak charge Q_{W} from experiment. The PNC amplitudes measured by *Wieman* et al. [29.3, 4] are

$$\frac{\Im(E_{\mathrm{PNC}})}{\beta} = \begin{cases} -1.6349(80) & F = 4 \to 3 \\ -1.5576(77) & F = 3 \to 4 \end{cases} \tag{29.15}$$

in units of mV/cm. The quantity β is the vector part of the Stark induced polarizability for the 6s → 7s transition in cesium. This quantity has also been calculated with an accuracy of better than 1% using the all-order techniques outlined above, giving $\beta = 27.00(20)a_0^3$. We note that this quantity can also be obtained from experiment, but at present two recent experiments [29.3, 4, 24] are in disagreement, so we use the number obtained from theory. Eliminating the spin-dependence from (29.15), by taking an appropriate linear combination and using the theoretical value for β, one finds

$$\Im\left(E_{\mathrm{PNC}}^{\mathrm{exp}}\right) = -0.8252(184)[61]10^{-11}|e|a_0 \,, \tag{29.16}$$

where the the first error is from experiment and the second from theory. Combining this result with our calculation of the spin-independent amplitude given in (29.6), we obtain

$$Q_{\mathrm{W}} = -72.17(0.58)[0.72] \,. \tag{29.17}$$

Alternatively, taking the opposite linear combination to eliminate the spin-independent terms in (29.15), we obtain the value

$$K_a - (\kappa - 1/2)/\kappa K_2 = 0.72 \pm 0.10 \qquad (29.18)$$

for the constant governing the spin-dependent interaction.

Radiative corrections to the weak charge Q_W incorporating a parameterization of new physics beyond the standard model have been worked out by *Marciano* and *Rosner* [29.11], who find

$$Q_W(^{133}Cs_{55}) = -73.20 - 0.8S - 0.005T \pm 0.13 , \qquad (29.19)$$

assuming the values $m_t = 140$ GeV for the top quark mass and $m_H = 100$ GeV for the Higgs particle mass. The parameters S and T in (29.19) are associated partly with deviations of the top quark and Higgs masses from their assumed values and partly with new physics beyond the standard model. In the absence of new physics, the small factor multiplying T makes this prediction very insensitive to the top quark mass. Unfortunately, both the experimental and theoretical errors are presently too large to make atomic PNC in cesium a precision test of the standard model. However, there are two features of cesium PNC that even at the present accuracy lead to particle physics implications. The first is the fact that large positive values of S, such as can arise in technicolor theories [29.28], will lead to disagreement of theory and experiment in cesium PNC. The second is the effect of extra Z bosons, which is not accounted for in (29.19). Exchange of new Z's can be shown to be strongly constrained by atomic PNC [29.29]. Perhaps more interesting is the possibility of having entirely new physics that has not been thought of. Since new physics affects different weak interaction tests differently, it is important to have as many such tests as possible. The value of atomic PNC tests will increase when the next stage of accuracy is reached, at which time atomic physics will have a significant role in precision tests of the standard model.

References

29.1 M. A. Bouchiat, C. C. Bouchiat: J. Phys. (Paris) **35**, 899 (1974)
29.2 M. A. Bouchiat et al.: J. Phys. (Paris) **47**, 1709 (1986)
29.3 S. C. Bennett, C. E. Wieman: Phys. Rev. Lett. **82**, 2484 (1999)
29.4 C. S. Wood et al.: Science **275**, 1759 (1997)
29.5 P. S. Drell, E. D. Commins: Phys. Rev. Lett. **53**, 968 (1984)
29.6 T. M. Wolfenden, P. E. G. Baird, P. G. H. Sandars: Europhys. Lett. **15**, 731 (1991)
29.7 T. P. Emmons, J. M. Reeves, E. N. Fortson: Phys. Rev. Lett. **51**, 2089 (1983)
29.8 M. J. D. Macpherson, K. P. Zetie, R. B. Warrington, D. N. Stacey, J. P. Hoare: Phys. Rev. Lett. **67**, 2784 (1991)
29.9 W. Bernreuther, M. Suzuki: Rev. Mod. Phys. **63**, 313 (1991)
29.10 S. A. Blundell, J. Sapirstein, W. R. Johnson: Phys. Rev. D **45**, 1602 (1992)
29.11 W. Marciano, J. Rosner: Phys. Rev. Lett. **65**, 2963 (1990)
29.12 V. A. Dzuba, V. V. Flambaum, O. P. Sushkov: Phys. Lett. A **141**, 147 (1989)
29.13 A. C. Hartley, E. Lindroth, A.-M. Mårtensson-Pendrill: J. Phys. B **23**, 3417 (1990)
29.14 Ya. Zel'dovich: Zh. Eksp. Teor. Fiz. **33**, 1531 (1957) [Sov. Phys. JETP **6**, 1184 (1958)]
29.15 W. R. Johnson, M. Idrees, J. Sapirstein: Phys. Rev. A **35**, 3218 (1987)
29.16 S. A. Blundell, W. R. Johnson, J. Sapirstein: Phys. Rev. A **43**, 3407 (1991)
29.17 R. Engfer et al.: At. Data Nucl. Data Tables **14**, 479 (1974)
29.18 B. Q. Chen, P. Vogel: Phys. Rev. C **48**, 1392 (1993)
29.19 V. V. Flambaum, I. B. Khriplovich, O. P. Sushkov: Phys. Lett. **146B**, 367 (1984)
29.20 A. Derevianko: Phys. Rev. Lett. **85**, 1618 (2000)
29.21 W. R. Johnson, I. Bednyakov, G. Soff: Phys. Rev. Lett. **87**, 233001 (2001)
29.22 W. R. Johnson, I. Bednyakov, G. Soff: Phys. Rev. Lett. **88**, 079903 (2002)
29.23 V. A. Dzuba, V. V. Flambaum, J. S. M. Ginges: Phys. Rev. **66**, 076013 (2002)
29.24 A. A. Vasilyev, I. M. Savukov, M. S. Safronova, H. G. Berry: Phys. Rev. A **66**, 020101 (2002)
29.25 M. Yu. Kuchiev: J. Phys. B **35**, L503 (2002)
29.26 A. I. Milstein, O. P. Sushkov, I. S. Terekhov: Phys. Rev. Lett. **89**, 283003 (2002)
29.27 J. Sapirstein, K. Pachucki, A. Veitia, K. T. Cheng: Phys. Rev. A **67**, 052110 (2003)
29.28 M. E. Peskin, T. Takeuchi: Phys. Rev. Lett. **65**, 964 (1990)
29.29 P. Langacker, M. Luo: Phys. Rev. D **45**, 278 (1992)

30. Atomic Clocks and Constraints on Variations of Fundamental Constants

Fundamental constants play an important role in modern physics, being landmarks that designate different areas. We call them constants, however, as long as we only consider minor variations with the cosmological time/space scale, their constancy is an experimental fact rather than a basic theoretical principle. Modern theories unifying gravity with electromagnetic, weak, and strong interactions, or even the developing quantum gravity itself often suggest such variations.

Many parameters that we call *fundamental constants*, such as the electron charge and mass [30.1, 2], are actually not truly fundamental constants but effective parameters which are affected by renormalization or the presence of matter [30.3]. Living in a changing universe we cannot expect that matter will affect these parameters the same way during any given cosmological epoch. An example is the inflationary model of the universe which states that in a very early epoch the universe experienced a phase transition which, in particular, changed a vacuum average of the so-called Higgs field which determines the electron mass. The latter was zero before this transition and reached a value close or equal to the present value after the transition.

The problem of variations of constants has many facets and here we discuss aspects related to atomic clocks and precision frequency measurements. Other related topics may be found in [30.4].

30.1 **Atomic Clocks and Frequency Standards.** 456
30.1.1 Caesium Atomic Fountain 456
30.1.2 Single-Ion Trap 457
30.1.3 Laser-Cooled Neutral Atoms 457
30.1.4 Two-Photon Transitions and Doppler-Free Spectroscopy... 458
30.1.5 Optical Frequency Measurements 458
30.1.6 Limitations on Frequency Variations 458

30.2 **Atomic Spectra and their Dependence on the Fundamental Constants** 459
30.2.1 The Spectrum of Hydrogen and Nonrelativistic Atoms 459
30.2.2 Hyperfine Structure and the Schmidt Model.............. 459
30.2.3 Atomic Spectra: Relativistic Corrections 460

30.3 **Laboratory Constraints on Time the Variations of the Fundamental Constants** 460
30.3.1 Constraints from Absolute Optical Measurements 460
30.3.2 Constraints from Microwave Clocks 461
30.3.3 Model-Dependent Constraints 461

30.4 **Summary** 462

References 462

Laboratory searches for a possible time variation of fundamental physical constants currently consist of two important parts: (i) one has to measure a certain physical quantity at two different moments of time that are separated by at least a few years; (ii) one has to be able to interpret the result in terms of fundamental constants. The latter is a strong requirement for a cross comparison of different results.

The measurements which may be performed most accurately are frequency measurements; and thus, frequency standards or atomic clocks will be involved in most of the laboratory searches. Frequency metrology has shown great progress in the last decade and will continue to do so for some time. The constraints on the variations of the fundamental constants obtained in this manner are, so far, somewhat weaker than those from other methods (astrophysics, geochemistry), but still competitive with them. In contrast to other methods, however, frequency measurements allow a very clear interpretation of the final results and a transparent evaluation procedure, making them less vulnerable to systematic errors. While there is still potential for improvement, the basic details of the method have been recently fixed.

The most advanced atomic clocks are discussed in Sect. 30.1. They are realized with many-electron atoms and their frequency cannot be interpreted in terms of fundamental constants. However, a much simpler problem needs to be solved: to interpret their variation in terms of fundamental constants. This idea is discussed in Sect. 30.2. The current laboratory constraints on the variations of the fundamental constants are summarized in Sect. 30.3.

30.1 Atomic Clocks and Frequency Standards

Frequency standards are important tools for precision measurements and serve various purposes which, in turn, have different requirements that must be satisfied. In particular, it is not necessary for a frequency standard to reproduce a frequency which is related to a certain atomic transition although it may be expressed in its terms. A well known example is the hydrogen maser, where the frequency is affected by the wall shift which may vary with time [30.5]. For the study of time variations of fundamental constants it is necessary to use standards similar to a primary caesium clock. In this case, any deviation of its frequency from the unperturbed atomic transition frequency should be known (within a known uncertainty) because this is a necessary requirement for being a 'primary' standard.

From the point of view of fundamental physics, the hydrogen maser is an artefact quite similar, in a sense, to the prototype of the kilogram held at the Bureau International des Poids et Mesures (BIPM) in Paris. Both artefacts are somehow related to fundamental constants (e.g., the mass of the prototype can be expressed in terms of the nucleon masses and their number) but they also have a kind of residual classical-physics flexibility which allows their properties to change. In contrast, standards similar to the caesium clock have a frequency (or other property) that is determined by a certain natural constant which is not flexible, being of pure quantum origin. It may change only if the fundamental constants are changing.

In Sect. 30.3, results obtained with caesium and rubidium fountains, a hydrogen beam, ultracold calcium clouds, and trapped ions of ytterbium and mercury are discussed. While caesium and rubidium clocks operate in the radio frequency domain, most of the other standards listed above rely on optical transitions.

30.1.1 Caesium Atomic Fountain

Caesium clocks are the most accurate primary standards for time and frequency [30.6]. The hyperfine splitting frequency between the $F = 3$ and $F = 4$ levels of the $^2S_{1/2}$ ground state of the ^{133}Cs atom at 9.192 GHz has been used for the definition of the SI second since 1967. In a so-called caesium fountain (Fig. 30.1), a dilute cloud of laser cooled caesium atoms at a temperature of about 1 μK is launched upwards to initiate a free parabolic flight with an apogee at about 1 m above the cooling zone. A microwave cavity is mounted near the lower endpoints of the parabola and is traversed by the atoms twice – once during ascent, once during descent – so that Ramsey's method of interrogation with separated oscillatory fields [30.5] can be realized. The total interrogation time being on the order of 0.5 s, a resonance linewidth of 1 Hz is achieved, about a factor of 100 narrower than in traditional devices using a thermal atomic beam from an oven. Selection and detection

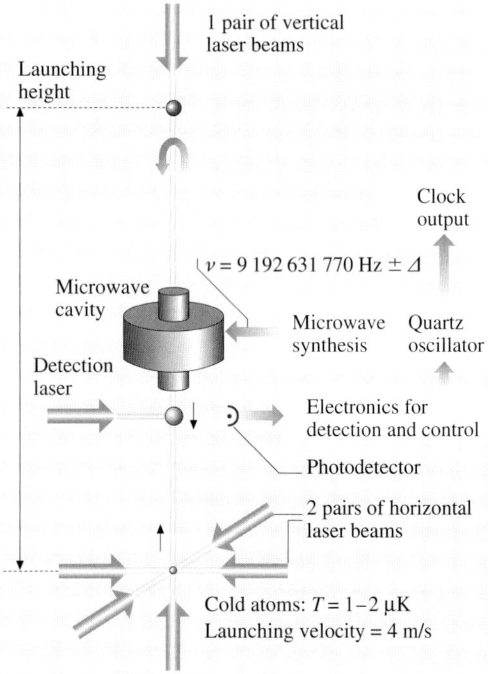

Fig. 30.1 Schematic of an atomic fountain clock

of the hyperfine state is performed via optical pumping and laser induced resonance fluorescence. In a carefully controlled setup, a relative uncertainty slightly below 1×10^{-15} can be reached in the realization of the resonance frequency of the unperturbed Cs atom. The averaging time that is required to reach this level of uncertainty is on the order of 10^4 s. One limiting effect that contributes significantly to the systematic uncertainty of the caesium fountain is the frequency shift due to cold collisions between the atoms. In this respect, a fountain frequency standard based on the ground state hyperfine frequency of the ^{87}Rb atom at about 6.835 GHz is more favorable, since its collisional shift is lower by more than a factor of 50 for the same atomic density. With the caesium frequency being fixed by definition in the SI system, the ^{87}Rb frequency is therefore presently the most precisely measured atomic transition frequency [30.7].

30.1.2 Single-Ion Trap

An alternative to interrogating atoms in free flight, and a possibility to obtain practically unlimited interaction time, is to store them in a trap. Ions are well suited because they carry electric charge and can be trapped in radio frequency ion traps (Paul traps [30.8]) that provide confinement around a field-free saddle point of an electric quadrupole potential. This ensures that the internal level structure is only minimally perturbed by the trap. Combined with laser cooling it is possible to reach the so-called Lamb–Dicke regime where the linear Doppler shift is eliminated. A single ion, trapped in an ultrahigh vacuum is conceptually a very simple system that allows good control of systematic frequency shifts [30.9]. The use of the much higher, optical reference frequency allows one to obtain a stability that is superior to microwave frequency standards, although only a single ion is used to obtain a correction signal for the reference oscillator.

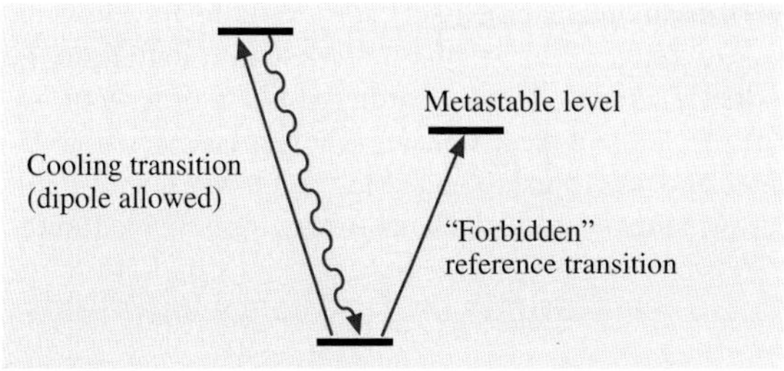

Fig. 30.2 Double resonance scheme applied in single-ion-trap frequency standards

A number of possible reference optical transitions with a natural linewidth of the order of 1 Hz and below are available in different ions, such as Yb^+ [30.10] and Hg^+ [30.11, 12]. These ions possess a useful level system, where both a dipole-allowed transition and a forbidden reference transition of the optical clock can be driven with two different lasers from the ground state (Fig. 30.2). The dipole transition is used for laser cooling and for the optical detection of the ion via its resonance fluorescence. If a second laser excites the ion to the metastable upper level of the reference transition, the fluorescence disappears and every single excitation can thus be detected with practically hundred percent efficiency as a dark period in the fluorescence signal.

Using these techniques and a femtosecond laser frequency comb generator (see Sect. 30.1.5) for the link to primary caesium clocks, the absolute frequencies of the transitions $^2S_{1/2} \to {}^2D_{5/2}$ in $^{199}Hg^+$ at 1065 THz and $^2S_{1/2} \to {}^2D_{3/2}$ in $^{171}Yb^+$ at 688 THz have been measured with relative uncertainties of only 9×10^{-15}. It is believed that single-ion optical frequency standards offer the potential to ultimately reach the 10^{-18} level of relative accuracy.

A similar double resonance technique can be employed if the reference transition is in the microwave domain and a number of accurate measurements of hyperfine structure intervals in trapped ions has been performed. In particular, the HFS interval in $^{171}Yb^+$ has been measured several times [30.13–15] and can be used to obtain constraints on temporal variations.

30.1.3 Laser-Cooled Neutral Atoms

Optical frequency standards have been developed with free laser-cooled neutral atoms, most notably of the alkaline-earth elements that possess narrow intercombination transitions. The atoms are collected in a magneto-optical trap, are then released and interogated by a sequence of laser pulses to realize a frequency-sensitive Ramsey–Bordé atom interferometer [30.16]. Of these systems, the one based on the $^1S_0 \to {}^3P_1$ intercombination line of ^{40}Ca at 657 nm has reached the lowest relative uncertainty so far (about 2×10^{-14}) [30.11, 12, 17, 18]. Limiting factors in the uncertainty of these standards are the residual linear Doppler effect and phase front curvature of the laser beams that excite the ballistically expanding atom cloud. It has therefore been proposed to confine the atoms in an optical lattice, i. e., in the array of interference maxima produced by several intersecting, red-detuned laser beams [30.19]. The detuning of the trapping laser could be chosen such that the light shift it produces in the ground and excited state of the reference transition are equal, and therefore it would produce no shift of the reference frequency. This approach is presently being investigated and may be applied to the very narrow (mHz natural linewidth) $^1S_0 \rightarrow {}^3P_0$ transitions in neutral strontium, ytterbium, or mercury.

30.1.4 Two-Photon Transitions and Doppler-Free Spectroscopy

The linear Doppler shift of an absorption resonance can also be avoided if a two-photon excitation is induced by two counterpropagating laser beams. A prominent example that has been studied with high precision is the two-photon excitation of the $^1S \rightarrow {}^2S$ transition in atomic hydrogen. The precise measurement of this frequency is of importance for the determination of the Rydberg constant and as a test of quantum electrodynamics (QED). Hydrogen atoms are cooled by collisions in a cryogenic nozzle and interact with a standing laser-wave of 243 nm wavelength inside a resonator. Since the atoms are not as cold as in laser cooled samples, a correction for the second order Doppler effect is performed. The laser excitation is interrupted periodically and the excited atoms are detected in a time resolved manner so that their velocity can be examined. An accuracy of about 2×10^{-14} has been obtained in absolute frequency measurements with a transportable caesium fountain [30.20, 21].

30.1.5 Optical Frequency Measurements

In recent years, the progress in stability and accuracy of optical frequency standards has been impressive; and there is belief that in the future an optical clock may supersede the microwave clocks because the optical oscillators offer a much higher number of periods in a given time. In addition, some systematic effects, such as the Zeeman effect, have an absolute order of magnitude that does not scale with the transition frequency, and consequently is relatively less important at higher transition frequencies. A long-standing problem, however, was the precise conversion of an optical frequency to the microwave domain, where frequencies can be counted electronically in order to establish a time scale or can easily be compared in a phase coherent way.

This problem has recently been solved by the so-called femtosecond laser frequency comb generator [30.22]. Briefly, a mode-locked femtosecond laser

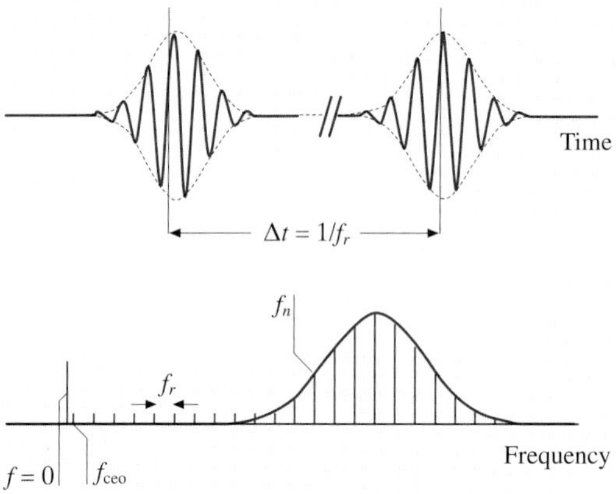

Fig. 30.3 Frequency comb generated from femtosecond laser pulses

produces, in the frequency domain, a comb of equally spaced optical frequencies f_n (Fig. 30.3) that can be written as $f_n = n f_r + f_{ceo}$ (with $f_{ceo} < f_r$), where f_r is the pulse repetition rate of the laser, the mode number n is a large integer (of order 10^5), and f_{ceo} (carrier-envelope-offset) is a shift of the whole comb that is produced by group velocity dispersion in the laser. The repetition rate f_r can easily be measured with a fast photodiode. In order to determine f_{ceo}, the comb is broadened in a nonlinear medium so that it covers at least one octave. Now the second harmonic of mode n from the "red" wing of the spectrum, at frequency $2(n f_r + f_{ceo})$, can be mixed with mode $2n$ from the "blue" wing, at frequency $2n f_r + f_{ceo}$, and f_{ceo} is obtained as a difference frequency. In this way, the precise relation between the two microwave frequencies f_r and f_{ceo} and the numerous optical frequencies f_n is known. The setup can now be used for an absolute optical frequency measurement by referencing f_r and f_{ceo} to a microwave standard and recording the beat note between the optical frequency f_o to be measured and the closest comb frequency f_n. Vice versa, the setup may work as an optical clockwork, for example, by adjusting f_{ceo} to zero and by stabilizing one comb line f_n to f_o so that f_r is now an exact subharmonic to order n of f_o. The precision of these transfer schemes has been investigated and was found to be so high that it will not limit the performance of optical clocks for the foreseeable future.

30.1.6 Limitations on Frequency Variations

The frequency standards described above have been successfully developed and their accuracy has been improved

Table 30.1 Limits on possible time variation of frequencies of different transitions in SI units. Here $\delta f/f$ is the fractional uncertainty of the most accurate measurement of the frequency f

Atom, transition	f GHz	$\delta f/f$ (10^{-15})	$\Delta f/\Delta t$ (Hz/yr)	Ref.
H, Opt	2 466 061	14	-8 ± 16	[30.20, 21]
Ca, Opt	455 986	13	-4 ± 5	[30.17, 18]
Rb, HFS	6.835	1	$(0 \pm 5) \times 10^{-6}$	[30.7]
Yb^+, Opt	688 359	9	-1 ± 3	[30.23]
Yb^+, HFS	12.642	73	$(4 \pm 4) \times 10^{-4}$	[30.13–15]
Hg^+, Opt	1 064 721	9	0 ± 7	[30.11, 12]

in the last decade. This progress, as a consequence, has led to certain constraints on the possible variations of the fundamental constants. Considering frequency variations, one has to have in mind that not only the numerical value but also the units may vary. For this reason, one needs to deal with dimensionless quantities which are unit-independent. During the last decade, a number of transition frequencies were measured in the corresponding SI unit, the hertz (see Table 30.1). These dimensional results are actually related to dimensionless quantities since a frequency measurement in SI is a measurement with respect to the caesium hyperfine interval

$$\{f\} = 9\,192\,631\,770 \cdot \frac{f}{f_{\mathrm{HFS}}(\mathrm{Cs})}\,, \tag{30.1}$$

where $\{f\}$ stands for the numerical value of the frequency f. (Most absolute frequency measurements have been realized as a direct comparison with a primary caesium standard.) In Sect. 30.3, in order to simplify notation, this symbol for the numerical value is dropped.

30.2 Atomic Spectra and their Dependence on the Fundamental Constants

30.2.1 The Spectrum of Hydrogen and Nonrelativistic Atoms

The hydrogen atom is the simplest atom and one can easily calculate the leading contribution to different kinds of transitions in its spectrum [30.24, 25], such as the gross, fine, and hyperfine structure. The scaling behavior of these contributions with the values of the Rydberg constant R_∞, the fine structure constant α, and the magnetic moments of proton and Bohr magneton is clear. The results for some typical hydrogenic transitions are

$$f(2\mathrm{p} \rightarrow 1\mathrm{s}) \simeq \frac{3}{4}\, cR_\infty\,,$$

$$f\left(2\mathrm{p}_{3/2} - 2\mathrm{p}_{1/2}\right) \simeq \frac{1}{16}\alpha^2\, cR_\infty\,,$$

$$f_{\mathrm{HFS}}(1\mathrm{s}) \simeq \frac{4}{3}\,\alpha^2\, \frac{\mu_{\mathrm{p}}}{\mu_{\mathrm{B}}}\, cR_\infty\,. \tag{30.2}$$

In the nonrelativistic approximation, the basic frequencies and the fine and hyperfine structure intervals of all atomic spectra have a similar dependence on the fundamental constants. The presence of a few electrons and a nuclear charge of $Z \neq 1$ makes theory more complicated and introduces certain multiplicative numbers but involves no new parameters. The importance of this scaling for a search for the variations was first pointed out in [30.26] and was applied to astrophysical data. Similar results may be presented for molecular transitions (electronic, vibrational, rotational and hyperfine) [30.27], however, up to now no measurement with molecules has been performed at a level of accuracy that is competitive with atomic transitions. They have been used only in a search for variations of constants in astrophysical observations [30.28].

30.2.2 Hyperfine Structure and the Schmidt Model

The atomic hyperfine structure

$$f_{\mathrm{NR}}(\mathrm{HFS}) = \mathrm{const}\,\alpha^2\, \frac{\mu}{\mu_{\mathrm{B}}}\, cR_\infty \tag{30.3}$$

involves nuclear magnetic moments μ which are different for different nuclei; thus, a comparison of the constraints on the variations of nuclear magnetic moments has a reduced value. To compare them, one may apply the Schmidt model [30.3, 29], which predicts all the magnetic moments of nuclei with an odd number of nucleons (odd value of atomic number A) in terms of the proton and neutron g-factors, g_{p} and g_{n}, respectively, and the nuclear magneton only. Unfortunately, the uncertainty of the calculation within the Schmidt model is quite high (usually from 10% to 50%). The Schmidt model, being a kind of ab initio model, only allows for improvements which, unfortunately, involve some effective phenomenological parameters. This would not really improve the situation, but return us to the case where there are too many possibly varying independent parameters. A comparison of the Schmidt values to the actual data is presented for caesium, rubidium, and ytterbium in Table 30.2.

30.2.3 Atomic Spectra: Relativistic Corrections

A theory based on the leading nonrelativistic approximation may not be accurate enough. Any atomic frequency can be presented as

$$f = f_{\mathrm{NR}}\, F_{\mathrm{rel}}(\alpha)\,, \qquad (30.4)$$

where the first (nonrelativistic) factor is determined by a scaling similar to the hydrogenic transitions (30.2). The second factor stands for relativistic corrections which vanish at $\alpha = 0$; and thus, $F_{\mathrm{rel}}(0) = 1$.

The importance of relativistic corrections for the hyperfine structure was first emphasized in [30.30]. Relativistic many-body calculations for various transitions have been performed in [30.31–36]. A typical accuracy is about 10%. Some results are summarized in Tables 30.2 and 30.3, where we list the relative sensitivity of the relativistic factors F_{rel} to changes in α,

$$\kappa = \frac{\partial \ln F_{\mathrm{rel}}}{\partial \ln \alpha}\,. \qquad (30.5)$$

Note that the relativistic corrections in heavy atoms are proportional to $(Z\alpha)^2$ because of the singularity of relativistic operators. Due to this, the corrections rapidly increase with the nuclear charge Z.

The signs and magnitudes of κ are explained by a simple estimate of the relativistic correction. For example, an approximate expression for the relativistic correction factor for the hyperfine structure of an s-wave electron in an alkali-like atom is [30.30]

$$F_{\mathrm{rel}}(\alpha) = \frac{1}{\sqrt{1-(Z\alpha)^2}} \cdot \frac{1}{1-(4/3)(Z\alpha)^2} \simeq 1 + \frac{11}{6}(Z\alpha)^2\,.$$

Table 30.2 Magnetic moments and relativistic corrections for atoms involved in microwave standards. The relativistic sensitivity κ is defined in Sect. 30.2.3. Here μ is an actual value of the nuclear magnetic moment, μ_{N} is the nuclear magneton, and μ_{S} stands for the Schmidt value of the nuclear magnetic moment; the nucleon g factors are $g_{\mathrm{p}}/2 \simeq 2.79$ and $g_{\mathrm{n}}/2 \simeq -1.91$

Z	Atom	μ/μ_{N}	$\mu_{\mathrm{S}}/\mu_{\mathrm{N}}$	μ/μ_{S}	κ
37	^{87}Rb	2.75	$g_{\mathrm{p}}/2+1$	0.74	0.34
55	^{133}Cs	2.58	$7/18\cdot(10-g_{\mathrm{p}})$	1.50	0.83
70	^{171}Yb$^+$	0.49	$-g_{\mathrm{n}}/6$	0.77	1.5

Table 30.3 Limits on possible time variation of the frequencies of different transitions and their sensitivity to variations in α due to relativistic corrections

Atom, transition	$\partial f/f\partial t$	κ
H, 1s – 2s	$-3.2(63)\times10^{-15}\ \mathrm{yr}^{-1}$	0.00
^{40}Ca, $^1\mathrm{S}_0-{}^3\mathrm{P}_1$	$-8(11)\times10^{-15}\ \mathrm{yr}^{-1}$	0.03
^{171}Yb$^+$, $^2\mathrm{S}_{1/2}-{}^2\mathrm{D}_{3/2}$	$-1.2(44)\times10^{-15}\ \mathrm{yr}^{-1}$	0.9
^{199}Hg$^+$, $^2\mathrm{S}_{1/2}-{}^2\mathrm{D}_{5/2}$	$-0.2(70)\times10^{-15}\ \mathrm{yr}^{-1}$	−3.2

A similar rough estimation for the energy levels may be performed for the gross structure:

$$E = -\frac{Z_a^2 mc^2\alpha^2}{2n_*^2}\left(1+\frac{(Z\alpha)^2}{n_*}\frac{1}{j+1/2}\right)\,. \qquad (30.6)$$

Here j is the electron angular momentum, n_* is the effective value of the principle quantum number (which determines the nonrelativistic energy of the electron), and Z_a is the charge "seen" by the valence electron – it is 1 for neutral atoms, 2 for singly charged ions, etc. This equation tells us that κ, for the excitation of the electron from the orbital j to the orbital j', has a different sign for $j > j'$ and $j < j'$. The difference of sign between the sensitivities of the ytterbium and mercury transitions in Table 30.3 reflects the fact that in Yb$^+$ a 6s-electron is excited to the empty 5d-shell, while in Hg$^+$ a hole is created in the filled 5d-shell if the electron is excited to the 6s-shell.

30.3 Laboratory Constraints on Time the Variations of the Fundamental Constants

Logarithmic derivatives (30.5) appear since we are looking for a variation of the constants in relative units. In other words, we are interested in a determination of, e.g., $\Delta\alpha/\alpha\Delta t$ while the input data of interest are related to $\Delta f/f\Delta t$. Their relation takes the form

$$\frac{\partial \ln f}{\partial t} = \frac{\partial \ln f_{\mathrm{NR}}}{\partial t} + \kappa\frac{\partial \ln \alpha}{\partial t}\,. \qquad (30.7)$$

If one compares transitions of the same type – gross structure, fine structure – the first term cancels.

30.3.1 Constraints from Absolute Optical Measurements

Absolute frequency measurements offer the possibility to compare a number of optical transitions with frequencies f_{NR}, which scale as cR_∞, with the caesium hyperfine structure. One can rewrite (30.7) as

$$\frac{\partial \ln f_{opt}}{\partial t} = \frac{\partial \ln cR_\infty}{\partial t} + \kappa \frac{\partial \ln \alpha}{\partial t}, \quad (30.8)$$

where dimensional quantities, such as frequency and the Rydberg constant, are stated in SI units (30.1). This equation may be used in different ways. For example, in Fig. 30.4 we plot experimental data for $\partial \ln f_{opt}/\partial t$ as a function of the sensitivity κ and derive a model-independent constraint on the variation of the fine structure constant

$$\frac{\partial \ln \alpha}{\partial t} = \left(-0.3 \pm 2.0 \times 10^{-15}\right) \text{y}^{-1} \quad (30.9)$$

and the numerical value of the Rydberg frequency cR_∞ (Table 30.4) in the SI unit of Hertz. The latter is of great metrological importance, being related to a common drift of optical clocks with respect to a caesium clock, i. e., to the definition of the SI second. The SI definition of the metre is unpractical and so, in practice, the optical wavelengths of reference lines calibrated against the caesium standard are used to determine the SI metre [30.37].

The constraints on the variations of α and cR_∞ are correlated and the standard uncertainty ellipse, defined as

$$\sum_i \frac{1}{u_i^2} \left(\frac{\partial \ln f_i}{\partial t} - \frac{\partial ln \text{Ry}}{\partial t} - \kappa_i \frac{\partial \ln \alpha}{\partial t} \right)^2 = 1 + \chi^2_{min},$$

is presented in Fig. 30.5. Here we sum over all available data: $\partial (\ln f_i) / \partial t$ is the central value of the observed drift rate, u_i its 1σ uncertainty, and χ^2_{min} the minimized χ^2 of the fit.

Table 30.4 Model-independent laboratory constraints on the possible time variations of natural constants

X	$\partial \ln X / \partial t$
α	$(-0.3 \pm 2.0) \times 10^{-15}\,\text{yr}^{-1}$
$\{cR_\infty\}$	$(-2.1 \pm 3.1) \times 10^{-15}\,\text{yr}^{-1}$
μ_{Cs}/μ_B	$(3.0 \pm 6.8) \times 10^{-15}\,\text{yr}^{-1}$
μ_{Rb}/μ_{Cs}	$(-0.2 \pm 1.2) \times 10^{-15}\,\text{yr}^{-1}$
μ_{Yb}/μ_{Cs}	$(3 \pm 3) \times 10^{-14}\,\text{yr}^{-1}$

The numerical value of the Rydberg constant, from the point of view of fundamental physics, can be expressed in terms of the caesium hyperfine interval in atomic units and its variation may be expressed in terms of the variations of α and μ_{Cs}/μ_B. A constraint for the latter is presented in Table 30.4.

30.3.2 Constraints from Microwave Clocks

A model-independent comparison of different HFS transitions is not simple because their nonrelativistic contributions f_{NR} are not the same, but involve different magnetic moments. Applying (30.9) to experimental data, one can obtain constraints on the relative variations of the magnetic moments of Rb, Cs, and Yb (Table 30.4).

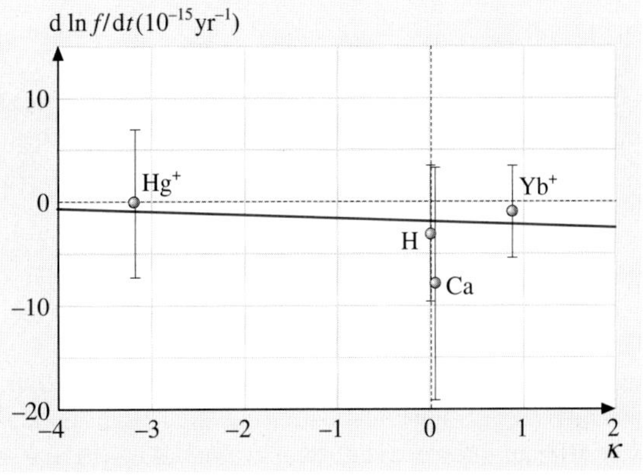

Fig. 30.4 Frequency variations versus their sensitivity κ.

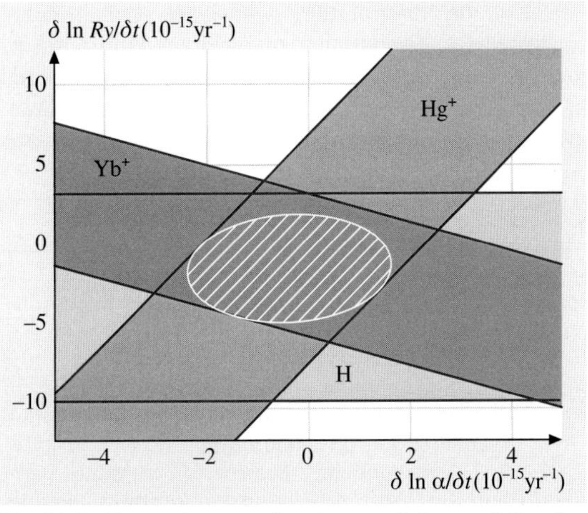

Fig. 30.5 Constraints on the time variations of the fine structure constant α and the numerical value of the Rydberg constant. The preliminary data on Ca are not included

30.3.3 Model-Dependent Constraints

In order to gain information on constants more fundamental than the nuclear magnetic moments, any further evaluation of the experimental data should involve the Schmidt model, which is far from perfect. Model-dependent constraints are summarized in Table 30.5.

The nucleon g factors, in their turn, depend on a dimensionless fundamental constant m_q/Λ_{QCD}, where m_q is the quark mass and Λ_{QCD} is the quantum chromodynamic (QCD) scale. A study of this dependence may supply us with deep insight into the possible variations of the more fundamental properties of Nature (see [30.31,32] for details). This approach is promising, but its accuracy needs to be better understood.

Table 30.5 Model-dependent laboratory constraints on possible time variations of fundamental constants. The uncertainties here do not include uncertainties from the application of the Schmidt model

X	$\partial \ln X/\partial t$
m_e/m_p	$(2.9 \pm 6.2) \times 10^{-15}\,\mathrm{yr}^{-1}$
μ_p/μ_e	$(2.9 \pm 5.8) \times 10^{-15}\,\mathrm{yr}^{-1}$
g_p	$(-0.1 \pm 0.5) \times 10^{-15}\,\mathrm{yr}^{-1}$
g_n	$(3 \pm 3) \times 10^{-14}\,\mathrm{yr}^{-1}$

30.4 Summary

The results collected in Tables 30.4 and 30.5 are competitive with data from other searches and have a more reliable interpretation. The results from astrophysical searches and the study of the samarium resonance from Oklo data claim higher sensitivity (see, e.g., [30.4]), however, they are more difficult to interpret. We have, for example, not assumed any hierarchy in variation rates or that some constants stay fixed while others vary, as it is done in the study of the position of the Oklo resonance. The evaluation presented here is transparent, and any particular calculation or measurement can be checked. In contrast, the astrophysical data show significant results only after an intensive statistical evaluation.

The laboratory searches involving atomic clocks have definitely shown progress and in a few years we expect an increase in the accuracy of these clocks, an increase in the number of different kinds of frequency standards (e.g., optical Sr, Sr^+, In^+ standards and a microwave Hg^+ standard are being tried now), and indeed an increase in the time separation between accurate experiments, since it is now typically only 2–3 years. An optical clock based on a nuclear transition in Th-229 is also under consideration [30.38]. Such a clock would offer different sensitivity to systematic effects, as well as to variations of different fundamental constants.

Laboratory searches are not necessarily limited to experiments with metrological accuracy. An example of a high-sensitivity search with a relatively low accuracy is the study of the dysprosium atom for a determination of the splitting between the $4f^{10}5d\,6s$ and $4f^{9}5d^{2}6s$ states, which offers a great sensitivity value of $\kappa \simeq 5.7 \times 10^8$ [30.36].

Variations of constants on the cosmological time scale can be expected but the magnitude, as well as other details, is unclear. Because of a broad range of options there is a need for the development of as many different searches as possible, and the laboratory search for variations is an attractive opportunity to open up a way that could lead to new physics.

References

30.1 W. E. Baylis, G. W. F. Drake: Chapt. 1 in this Handbook.

30.2 P. J. Mohr, B. N. Taylor: Rev. Mod. Phys. **77**, 1 (2005)

30.3 S. G. Karshenboim: Time and space variation of Fundamental constants: Motivation and laboratory search. Eprints physics/0306180;
S. G. Karshenboim: Search for possible variation of the fine structure constant. Eprints physics/0311080. To be published.

30.4 S. G. Karshenboim, E. Peik (Eds.): *Astrophysics, Clocks and Fundamental Constants*, Lect. Notes in Phys., Vol. 648 (Springer, Berlin, Heidelberg 2004)

30.5 N. F. Ramsey: Rev. Mod. Phys. **62**, 541 (1990)

30.6 A. Bauch, H. R. Telle: Rep. Prog. Phys. **65**, 789 (2002)

30.7 H. Marion, F. Pereira Dos Santos, M. Abgrall, S. Zhang, Y. Sortais, S. Bize, I. Maksimovic, D. Calonico, J. Gruenert, C. Mandache, P. Lemonde, G. Santarelli, Ph. Laurent, A. Clairon, C. Salomon: Phys. Rev. Lett. **90**, 150801 (2003)

30.8 W. Paul: Rev. Mod. Phys. **62**, 531 (1990) See also: J. Javanainen, Chapt. 75 in this Handbook.

30.9 H. Dehmelt: IEEE Trans. Instrum. Meas. **31**, 83 (1982)

30.10 J. Stenger, C. Tamm, N. Haverkamp, S. Weyers, H. R. Telle: Opt. Lett. **26**, 1589 (2001)

30.11 T. Udem, S. A. Diddams, K. R. Vogel, C. W. Oates, E. A. Curtis, W. D. Lee, W. M. Itano, R. E. Drullinger, J. C. Bergquist, L. Hollberg: Phys. Rev. Lett. **86**, 4996 (2001)

30.12 S. Bize, S. A. Diddams, U. Tanaka, C. E. Tanner, W. H. Oskay, R. E. Drullinger, T. E. Parker, T. P. Heavner, S. R. Jefferts, L. Hollberg, W. M. Itano, D. J. Wineland, J. C. Bergquist: Phys. Rev. Lett. **90**, 150802 (2003)

30.13 P. T. H. Fisk, M. J. Sellars, M. A. Lawn, C. Coles: IEEE Trans. UFFC **44**, 344 (1997)

30.14 P. T. Fisk: Rep. Prog. Phys. **60**, 761 (1997)

30.15 R. B. Warrington, P. T. H. Fisk, M. J. Wouters, M. A. Lawn: *Proceedings of the 6th Symposium Frequency Standards and Metrology* (World Scientific, Singapore 2002) p. 297

30.16 C. J. Bordé: Phys. Lett. A **140**, 10 (1989)

30.17 G. Wilpers, T. Binnewies, C. Degenhardt, U. Sterr, J. Helmcke, F. Riehle: Phys. Rev. Lett. **89**, 230801 (2002)

30.18 F. Riehle, C. Degenhardt, Ch. Lisdat, G. Wilpers, H. Schnatz, T. Binnewies, H. Stoehr, U. Sterr: An optical frequency standard with cold and ultra-cold calcium atoms. In: *Astrophysics, Clocks and Fundamental Constants*, Lect. Notes in Phys., Vol. 648, ed. by S. G. Karshenboim, E. Peik (Springer, Berlin, Heidelberg 2004) p. 229

30.19 H. Katori, M. Takamoto, V. G. Pal'chikov, V. D. Ovsiannikov: Phys. Rev. Lett. **91**, 173005 (2003)

30.20 M. Niering, R. Holzwarth, J. Reichert, P. Pokasov, T. Udem, M. Weitz, T. W. Hänsch, P. Lemonde, G. Santarelli, M. Abgrall, P. Laurent, C. Salomon, A. Clairon: Phys. Rev. Lett. **84**, 5496 (2000)

30.21 M. Fischer, N. Kolachevsky, M. Zimmermann, R. Holzwarth, Th. Udem, T. W. Hänsch, M. Abgrall, J. Grünert, I. Maksimovic, S. Bize, H. Marion, F. Pereira Dos Santos, P. Lemonde, G. Santarelli, P. Laurent, A. Clairon, C. Salomon, M. Haas, U. D. Jentschura, C. H. Keitel: Phys. Rev. Lett. **92**, 230802 (2004)

30.22 T. Udem, J. Reichert, R. Holzwarth, S. Diddams, D. Jones, J. Ye, S. Cundiff, T. W. Hänsch, J. Hall: A new type of frequency chain and its application to fundamental frequency metrology. In: *The Hydrogen Atom: Precision Physics of Simple Atomic Systems*, Lect. Notes in Phys., Vol. 570, ed. by S. G. Karshenboim, F. S. Pavone, F. Bassani, M. Inguscio, T. W. Hänsch (Springer, Berlin, Heidelberg 2001) p. 229

30.23 E. Peik, B. Lipphardt, H. Schnatz, T. Schneider, Chr. Tamm, S. G. Karshenboim: Phys. Rev. Lett. **93**, 170801 (2004)

30.24 J. Sapirstein, Chapt. 27 in this Handbook

30.25 P. Mohr, Chapt. 28 in this Handbook

30.26 M. P. Savedoff: Nature **178**, 688 (1956)

30.27 R. I. Thompson: Astrophys. Lett. **16**, 3 (1975)

30.28 D. A. Varshalovich, A. V. Ivanchik, A. V. Orlov, A. Y. Potekhin, P. Petitjean: Current status of the problem of cosmological variability of fundamental physical constants. In: *Precision Physics of Simple Atomic Systems*, Lect. Notes in Phys., Vol. 627, ed. by S. G. Karshenboim, V. B. Smirnov (Springer, Berlin, Heidelberg 2003) p. 199

30.29 S. G. Karshenboim: Can. J. Phys. **78**, 639 (2000)

30.30 J. D. Prestage, R. L. Tjoelker, L. Maleki: Phys. Rev. Lett. **74**, 3511 (1995)

30.31 V. V. Flambaum: Limits on temporal variation of fine structure constant, quark masses and strong interaction from atomic clock experiments. In: *Laser Spectroscopy*, ed. by P. Hannaford, A. Sidorov, H. Bachor, K. Baldwin (World Scientific, New Jersey, London, Singapore, Shanghai, Taipei, Chennai 2004) p. 49

30.32 V. V. Flambaum, L. B. Leinweber, A. W. Thomas, R. D. Young: Phys. Rev. **69**, 115006 (2004)

30.33 V. A. Dzuba, V. V. Flambaum, J. K. Webb: Phys. Rev. Lett. **82**, 888 (1999)

30.34 V. A. Dzuba, V. V. Flambaum, J. K. Webb: Phys. Rev. A **59**, 230 (1999)

30.35 V. A. Dzuba, V. V. Flambaum: Phys. Rev. A **61**, 034502 (2001)

30.36 V. A. Dzuba, V. V. Flambaum, M. V. Marchenko: Phys. Rev. A **68**, 022506 (2003)

30.37 T. J. Quinn: Metrologia **40**, 103 (2003)

30.38 E. Peik, Chr. Tamm: Europhys. Lett. **61**, 181 (2003)

Part C Molecules

Part C Molecules

31 **Molecular Structure**
David R. Yarkony, Baltimore, USA

32 **Molecular Symmetry and Dynamics**
William G. Harter, Fayetteville, USA

33 **Radiative Transition Probabilities**
David L. Huestis, Menlo Park, USA

34 **Molecular Photodissociation**
Abigail J. Dobbyn, Göttingen, Germany
David H. Mordaunt, Göttingen, Germany
Reinhard Schinke, Göttingen, Germany

35 **Time-Resolved Molecular Dynamics**
Volker Engel, Würzburg, Germany

36 **Nonreactive Scattering**
David R. Flower, Durham, United Kingdom

37 **Gas Phase Reactions**
Eric Herbst, Columbus, USA

38 **Gas Phase Ionic Reactions**
Nigel G. Adams, Athens, USA

39 **Clusters**
Mary L. Mandich, Murray Hill, USA

40 **Infrared Spectroscopy**
Henry Buijs, Québec, Canada

41 **Laser Spectroscopy in the Submillimeter and Far-Infrared Regions**
Kenneth M. Evenson†
John M. Brown, Oxford, England

42 **Spectroscopic Techniques: Lasers**
Paul Engelking, Eugene, USA

43 **Spectroscopic Techniques: Cavity-Enhanced Methods**
Barbara A. Paldus, Palo Alto, USA
Alexander A. Kachanov, Sunnyvale, USA

44 **Spectroscopic Techniques: Ultraviolet**
Glenn Stark, Wellesley, USA
Peter L. Smith, Cambridge, USA

31. Molecular Structure

Molecular structure is a reflection of the Born–Oppenheimer separation of electronic and nuclear motion, which is in turn a consequence of the large difference between the electron and nuclear masses. One consequence of this separation is the concept of a potential energy surface for nuclear motion created by the faster moving electrons. Corollaries include equilibrium structures, transition states, and reaction paths which are the foundation of the description of molecular structure and reactivity. However the Born–Oppenheimer approximation is not uniformly applicable and its breakdown results in perturbations in molecular spectra, radiationless decay, and nonadiabatic chemical reactions.

There are many issues that can be addressed in a discussion of molecular structure, including the structure and bonding of individual classes of molecules, computational and/or experimental techniques used to determine or infer molecular structure, the accuracy of those methods, etc. In an effort to provide a broad view of the essential aspects of molecular structure, this Chapter considers issues in molecular structure from a theoretical/computational perspective using the Born–Oppenheimer approximation as the point of origin. Rather than providing a compendium of results, this chapter will explain how issues in molecular structure are investigated and how the questions that can be addressed reflect the available methodology. Even with these restrictions the scope of this topic remains enormous and precludes a detailed presentation of any one issue. Thus the abbreviated discussions in this work are supplemented by ample references to the literature.

Several aspects of potential energy surfaces and their relation to molecular structure will be considered: (i) the electronic structure techniques used to determine a single point on a potential energy surface, and the interactions that couple the electronic states in question, (ii) the local properties of potential energy surfaces, in particular equilibrium structures and rovibrational levels that provide the link to experimental inferences concerning molecular structure, (iii) global chemistry deduced from potential energy surfaces including reaction mechanisms and reaction paths and (iv) phenomena resulting from the nonadiabatic interactions that couple potential energy surfaces.

31.1 **Concepts** 468
31.1.1 Nonadiabatic Ansatz: Born–Oppenheimer Approximation 468
31.1.2 Born–Oppenheimer Potential Energy Surfaces and Their Topology 469
31.1.3 Classification of Interstate Couplings: Adiabatic and Diabatic Bases 469
31.1.4 Surfaces of Intersection of Potential Energy Surfaces 470

31.2 **Characterization of Potential Energy Surfaces** 470
31.2.1 The Self-Consistent Field (SCF) Method 471
31.2.2 Electron Correlation: Wave Function Based Methods 472
31.2.3 Electron Correlation: Density Functional Theory 475
31.2.4 Weakly Interacting Systems 476

31.3 **Intersurface Interactions: Perturbations** 476
31.3.1 Derivative Couplings 476
31.3.2 Breit–Pauli Interactions 477
31.3.3 Surfaces of Intersection 479

31.4 **Nuclear Motion** 480
31.4.1 General Considerations 480
31.4.2 Rotational-Vibrational Structure . 481
31.4.3 Coupling of Electronic and Rotational Angular Momentum in Weakly Interacting 482
31.4.4 Reaction Path 483

31.5 **Reaction Mechanisms: A Spin-Forbidden Chemical Reaction** 484

31.6 **Recent Developments** 486

References 486

31.1 Concepts

31.1.1 Nonadiabatic Ansatz: Born–Oppenheimer Approximation

Basic Quantities

The total Hamiltonian for electronic and nuclear motion in the space fixed coordinate frame, in atomic units, is

$$\begin{aligned} H^{\mathrm{eN}}(\boldsymbol{r},\boldsymbol{R}) &= \sum_{\alpha=1}^{N} \frac{-1}{2M_\alpha}\nabla_\alpha^2 + H^{\mathrm{e}}(\boldsymbol{r};\boldsymbol{R}) \\ &\equiv T^{\mathrm{nuc}} + H^{\mathrm{e}}(\boldsymbol{r};\boldsymbol{R})\,, \end{aligned} \tag{31.1}$$

where $\boldsymbol{R}$ denotes the $3N$ nuclear coordinates, with $\boldsymbol{R} = (\boldsymbol{R}_1, \boldsymbol{R}_2, \ldots, \boldsymbol{R}_N)$, $\boldsymbol{R}_i = (X_i, Y_i, Z_i)$, $\boldsymbol{r}$ denotes the $3M$ electronic coordinates 2using similar conventions, T^{nuc} is the nuclear kinetic energy operator and $H^{\mathrm{e}}(\boldsymbol{r},\boldsymbol{R})$ is the total electronic Hamiltonian taken as

$$H^{\mathrm{e}}(\boldsymbol{r};\boldsymbol{R}) = H^0(\boldsymbol{r};\boldsymbol{R}) + H^{\mathrm{rel}}(\boldsymbol{r};\boldsymbol{R})\,. \tag{31.2}$$

Here $H^0(\boldsymbol{r};\boldsymbol{R})$ is the nonrelativistic Born–Oppenheimer Hamiltonian

$$\begin{aligned} H^0(\boldsymbol{r};\boldsymbol{R}) &= -\frac{1}{2}\sum_{i=1}^{M}\nabla_i^2 - \sum_{K,i}\frac{Z_K}{|\boldsymbol{R}_k - \boldsymbol{r}_i|} \\ &+ \frac{1}{2}\sum_{i\neq j}^{M}\frac{1}{|\boldsymbol{r}_i - \boldsymbol{r}_j|} + \frac{1}{2}\sum_{K\neq L}^{N}\frac{Z_K Z_L}{|\boldsymbol{R}_K - \boldsymbol{R}_L|}\,, \end{aligned} \tag{31.3}$$

and $H^{\mathrm{rel}}(\boldsymbol{r};\boldsymbol{R})$ is the relativistic contribution to the electronic Hamiltonian, for light atoms conventionally treated within the Breit–Pauli approximation [31.1] and discussed further in Sect. 31.3.2. Note that in (31.3) the nuclear kinetic energy is absent, so that only the $\boldsymbol{r}_i$ are dynamical variables in (31.2) and (31.3). Using the Born–Huang approach [31.2] to the Born–Oppenheimer approximation, the total wave function $\Psi_L^{\mathrm{eN}}(\boldsymbol{r},\boldsymbol{R})$ is expanded in a basis of electronic states, $\Psi_I^{\mathrm{e}}(\boldsymbol{r};\boldsymbol{R})$. The total wave function for the system thus has the form

$$\Psi_L^{\mathrm{eN}}(\boldsymbol{r},\boldsymbol{R}) = \sum_I \Psi_I^{\mathrm{e}}(\boldsymbol{r};\boldsymbol{R})\beta_I^L(\boldsymbol{R})\,. \tag{31.4}$$

Equation (31.4) is valid for any complete set of electronic states depending parametrically on nuclear coordinates. The parametrical dependence of $\Psi_I^{\mathrm{e}}(\boldsymbol{r};\boldsymbol{R})$ on $\boldsymbol{R}$, denoted by the semicolon, reflects the fact that the $\boldsymbol{R}$ are not dynamical variables in (31.2) and (31.3). As a practical matter it is necessary to make a particular choice of $\Psi_I^{\mathrm{e}}(\boldsymbol{r};\boldsymbol{R})$ in order to limit the size of the expansion in electronic states. An adiabatic electronic state is an eigenfunction of $H^{\mathrm{e}}(\boldsymbol{r};\boldsymbol{R})$ for fixed $\boldsymbol{R}$. A particularly useful choice of adiabatic state, denoted by $\Psi_I^0(\boldsymbol{r};\boldsymbol{R})$, employs $H^0(\boldsymbol{r};\boldsymbol{R})$ rather than the full $H^{\mathrm{e}}(\boldsymbol{r},\boldsymbol{R})$. These electronic wave functions satisfy

$$H^0(\boldsymbol{r};\boldsymbol{R})\Psi_I^0(\boldsymbol{r};\boldsymbol{R}) = E_I^0(\boldsymbol{R})\Psi_I^0(\boldsymbol{r};\boldsymbol{R})\,. \tag{31.5}$$

The $\Psi_I^0(\boldsymbol{r};\boldsymbol{R})$ are determined up to a geometry dependent phase. This phase is usually chosen such that the $\Psi_I^0(\boldsymbol{r};\boldsymbol{R})$ are real. This assumption is acceptable except in the situation where there is a conical intersection on the potential energy surface in question. In that case, the real-valued $\Psi_I^0(\boldsymbol{r};\boldsymbol{R})$ changes sign when a closed loop surrounding the conical intersection point is traversed, that is, $\Psi_I^0(\boldsymbol{r};\boldsymbol{R})$ is not single-valued [31.3]. This geometric or *Berry* [31.4] phase condition has consequences in such phenomena as the dynamic Jahn–Teller effect [31.5, 6] but will not be addressed in detail in this review.

As a consequence of the parametric dependence of $\Psi_I^0(\boldsymbol{r};\boldsymbol{R})$ on $\boldsymbol{R}$, $E_I^0(\boldsymbol{R})$ becomes a function of $\boldsymbol{R}$ and is referred to as the nonrelativistic Born–Oppenheimer potential energy surface for reasons discussed below. Approaches based on (31.5) are most appropriate for molecular systems with only light atoms. However with the use of pseudopotential techniques [31.7], formally equivalent approaches can be developed for heavier systems (Sect. 31.3.2).

Inserting (31.4) into the time independent Schrödinger equation $\left[H^{\mathrm{eN}}(\boldsymbol{r};\boldsymbol{R}) - E\right]\Psi^{\mathrm{eN}}(\boldsymbol{r};\boldsymbol{R}) = 0$ and taking the inner product with the electronic basis state $\Psi_I^0(\boldsymbol{r};\boldsymbol{R})$ gives the following system of coupled equations for the rovibronic functions $\beta_I^L(\boldsymbol{R})$ [31.8, 9]:

$$\begin{aligned} &\left[T^{\mathrm{nuc}} + E_I^0(\boldsymbol{R}) - E\right]\beta_I(\boldsymbol{R}) \\ &= -\sum_J \Bigg(\tilde{K}^{IJ}(\boldsymbol{R}) + H_{IJ}^{\mathrm{BP}}(\boldsymbol{R}) \\ &\quad - \sum_{\alpha=1}^{N}\left\{\frac{1}{2M_\alpha}\left[\boldsymbol{f}_\alpha^{IJ}(\boldsymbol{R})\cdot\nabla_\alpha + \nabla_\alpha\cdot\boldsymbol{f}_\alpha^{IJ}(\boldsymbol{R})\right]\right\}\Bigg) \\ &\quad \times \beta_J(\boldsymbol{R})\,, \end{aligned} \tag{31.6}$$

where the state label L on β_I^L has been suppressed, and

$$\boldsymbol{f}_\alpha^{IJ}(\boldsymbol{R}) \equiv \left[f_{X_\alpha}^{IJ}(\boldsymbol{R}), f_{Y_\alpha}^{IJ}(\boldsymbol{R}), f_{Z_\alpha}^{IJ}(\boldsymbol{R})\right]\,, \tag{31.7}$$

$$\tilde{K}^{JI}(\boldsymbol{R}) = \sum_{W,\alpha}\frac{1}{2M_\alpha}\tilde{k}_{W_\alpha W_\alpha}^{JI}(\boldsymbol{R})\,, \tag{31.8}$$

$$f_{W_\alpha}^{JI}(\boldsymbol{R}) = \left\langle \Psi_J^0(\boldsymbol{r};\boldsymbol{R}) \middle| \frac{\partial}{\partial W_\alpha}\Psi_I^0(\boldsymbol{r};\boldsymbol{R})\right\rangle_{\boldsymbol{r}}\,, \tag{31.9}$$

$$\tilde{k}^{JI}_{W_\alpha W'_\beta} = \left\langle \frac{\partial}{\partial W_\alpha} \Psi^0_J(\boldsymbol{r};\boldsymbol{R}) \middle| \frac{\partial}{\partial W'_\beta} \Psi^0_I(\boldsymbol{r};\boldsymbol{R}) \right\rangle_{\boldsymbol{r}} , \tag{31.10}$$

with $W = X, Y, Z, \alpha = 1 \cdots N$. The subscript $\boldsymbol{r}$ on the matrix elements in (31.9–31.10) denotes integration over all electronic coordinates and the final ∇_α in (31.6) acts on both $f^{IJ}_\alpha(\boldsymbol{R})$ and $\beta_J(\boldsymbol{R})$. While this representation may not be optimal for a treatment of the nuclear dynamics, since it employs a space fixed frame representation, it is adequate for the present development as it clearly displays the origins of the Born–Oppenheimer approximation and its breakdown.

Modifications to the Nonrelativistic Born–Oppenheimer Potential Energy Surface

When the interstate couplings on the right-hand side of (31.6) can be neglected, the nuclear motion is governed by the effective potential given by $V^I(\boldsymbol{R}) \equiv E^{\mathrm{e}}_I(\boldsymbol{R}) + \tilde{K}^{II}(\boldsymbol{R}) \equiv E^0_I(\boldsymbol{R}) + H^{\mathrm{BP}}_{II}(\boldsymbol{R}) + \tilde{K}^{II}(\boldsymbol{R})$ and the Born–Oppenheimer approximation is valid. Here it is assumed that $f^{II}_\alpha(\boldsymbol{R}) = 0$, a sufficient condition for which is that the $\Psi^0_I(\boldsymbol{r};\boldsymbol{R})$ are chosen real-valued [see discussion following (31.5)]. $E^0_I(\boldsymbol{R})$ is generally the principal $\boldsymbol{R}$-dependent contribution to $V^I(\boldsymbol{R})$. $H^{\mathrm{BP}}_{II}(\boldsymbol{R})$ is referred to as the relativistic contribution to the nonrelativistic Born–Oppenheimer potential energy surface and is discussed further in Sect. 31.3.2. The contribution $\tilde{K}^{II}(\boldsymbol{R})$ is referred to as the adiabatic correction. Unlike $E^0_I(\boldsymbol{R})$ and $H^{\mathrm{BP}}_{IJ}(\boldsymbol{R})$, which are independent of mass, $\tilde{K}^{II}(\boldsymbol{R})$ is mass dependent (31.8 and 31.10). It has been computed from first principles [31.10–12] and inferred from experiments [31.13, 14] principally for diatomic systems.

31.1.2 Born–Oppenheimer Potential Energy Surfaces and Their Topology

$E^0_I(\boldsymbol{R})$ is the main focus of this chapter. There have been several recent reviews of ground state ($I = 1$) potential energy surfaces [31.15, 16].

Equilibrium structures (stable or metastable species) represent local minima on $E^0_I(\boldsymbol{R})$, that is $g^I_{W_\alpha}(\boldsymbol{R}) \equiv \partial E^0_I(\boldsymbol{R})/\partial W_\alpha = 0$ for all W_α at $\boldsymbol{R} = \boldsymbol{R}^{\mathrm{e}}$. There may be several equilibrium structures for a given set of nuclei; for example the atoms H, C and N form stable molecules HCN and HNC. Saddle points or transition states, extrema on $E^0_I(\boldsymbol{R})$ with one negative eigenvalue of the *Hessian matrix* $\boldsymbol{F}^I(\boldsymbol{R})$, whose elements are defined by $F^I_{W_\alpha W_\beta}(\boldsymbol{R}) \equiv \partial^2 E^0_I(\boldsymbol{R})/\partial W_\alpha \partial W_\beta$, represent mountain passes separating the various equilibrium structures and the asymptotes – values of $\boldsymbol{R}$ corresponding to isolated molecular fragments. This situation is illustrated in Fig. 31.1.

Reaction paths [31.17] connect the asymptotes, minima, and saddle points. They can best be defined as the steepest descent paths from a transition state structure down to local minima [31.17], although methods for walking uphill (shallowest ascent path) [31.18] also exist. Unlike equilibrium structures and saddle points which are independent of the choice of coordinate system used to represent $\boldsymbol{R}$, the reaction path is coordinate system dependent. The intrinsic reaction coordinate [31.17], the reaction coordinate in mass weighted cartesian coordinates $\boldsymbol{R}_i \to \boldsymbol{R}_i M_i$, is most frequently used (Sect. 31.4.4)

These features of the potential energy surfaces and their determination will be discussed further below. Their determination consists of two parts: (i) the level of treatment, i. e., the electronic structure technique used to determine $E^0_I(\boldsymbol{R})$ and (ii) the characterization of the features of $E^0_I(\boldsymbol{R})$, i. e., minima, saddle points, and reaction paths [31.19,20] at the level of treatment chosen. In this chapter, only fully ab initio levels of treatment are considered [31.21,22].

31.1.3 Classification of Interstate Couplings: Adiabatic and Diabatic Bases

Intersurface couplings, shown on the right-hand side of (31.6), result in the breakdown of the single potential energy surface Born–Oppenheimer approximation. Within the Born–Huang ansatz such a nonadiabatic process is interpreted in terms of motion on more than one Born–Oppenheimer potential energy surface [31.8]. From (31.6), it is seen that there are three types of matrix elements coupling the electronic states, $\tilde{K}^{IJ}(\boldsymbol{R})$, $H^{\mathrm{BP}_{IJ}}(\boldsymbol{R})$ and $f^{IJ}(\boldsymbol{R})$. Two of these couplings, $\tilde{K}^{IJ}(\boldsymbol{R})$

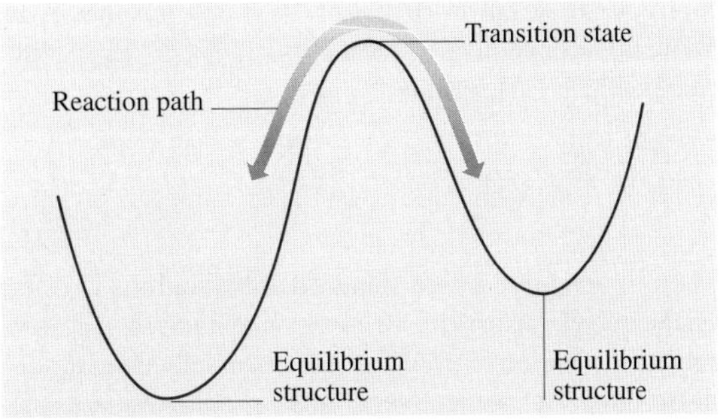

Fig. 31.1 Schematic representation of a transition state and reaction path connecting two minima

and $f^{IJ}(\boldsymbol{R})$ arise from the nuclear kinetic energy operator T^{nuc}, while the third coupling arises from the Breit–Pauli interaction H^{BP}. This classification of the intersurface interactions is a consequence of the choice of the adiabatic electronic states through (31.5). Other choices of $\Psi_I^0(\boldsymbol{r};\boldsymbol{R})$, most notably the diabatic electronic states [31.23–27], are possible.

In the adiabatic state basis, $H^0(\boldsymbol{r};\boldsymbol{R})$ is diagonal and in the absence of relativistic effects, intersurface couplings originate exclusively from the derivative coupling terms in (31.9) and (31.10). The diabatic basis seeks to 'transfer' the intersurface coupling from these derivative operators to a potential term analogous in form to $H_{IJ}^{\text{BP}}(\boldsymbol{R})$. The diabatic basis, $\Psi_I^{0-d}(\boldsymbol{r};\boldsymbol{R})$, is a unitary transformation of the adiabatic electronic basis, defined such that [31.24]

$$f_{R_\alpha}^{JI,d} \equiv \left\langle \Psi_J^{0-d}(\boldsymbol{r};\boldsymbol{R}) \middle| \frac{\partial}{\partial R_\alpha} \Psi_I^{0-d}(\boldsymbol{r};\boldsymbol{R}) \right\rangle_r = 0 \tag{31.11}$$

where R_α is an internal coordinate. For polyatomic systems ($N > 2$), rigorous diabatic bases do not exist [31.27] and approximate diabatic bases are sought [31.27–30]. A discussion of this issue can be found in [31.31].

31.1.4 Surfaces of Intersection of Potential Energy Surfaces

The intersurface couplings are most effective in promoting a nonadiabatic process when $|V^I(\boldsymbol{R}) - V^J(\boldsymbol{R})|$ is small for some J on the right-hand side of (31.6), so that

$$\begin{aligned} &E_I^e(\boldsymbol{R}) + \tilde{K}^{II}(\boldsymbol{R}) - \left[E_J^e(\boldsymbol{R}) + \tilde{K}^{JJ}(\boldsymbol{R})\right] \\ &\equiv \Delta E_{IJ}^0(\boldsymbol{R}) + \Delta E_{IJ}^{\text{BP}}(\boldsymbol{R}) + \Delta K_{IJ}(\boldsymbol{R}) \end{aligned} \tag{31.12}$$

is small. Here

$$\Delta E_{IJ}^0(\boldsymbol{R}) \equiv E_I^0(\boldsymbol{R}) - E_J^0(\boldsymbol{R});\,, \tag{31.13}$$

$$\Delta K_{IJ}(\boldsymbol{R}) \equiv \tilde{K}^{II}(\boldsymbol{R}) - \tilde{K}^{JJ}(\boldsymbol{R})\,, \tag{31.14}$$

$$\Delta E_{IJ}^{\text{BP}}(\boldsymbol{R}) \equiv H_{II}^{\text{BP}}(\boldsymbol{R}) - H_{JJ}^{\text{BP}}(\boldsymbol{R})\,. \tag{31.15}$$

Since in general $\Delta K_{IJ}(\boldsymbol{R})$ is quite small, and for the low atomic number systems considered here H_{II}^{BP} is itself small, we are led to consider regions of nuclear coordinate space for which the magnitude of $\Delta E_{IJ}^0(\boldsymbol{R})$ is small, referred to as avoided intersections when $\Delta E_{IJ}^0(\boldsymbol{R}) \neq 0$ and as conical and Renner-type [31.32] intersections when $\Delta E_{IJ}^0(\boldsymbol{R}) = 0$. The set of $\boldsymbol{R}$ for which $\Delta E_{IJ}^0(\boldsymbol{R}) = 0$, the noncrossing rule, was first discussed in 1929 by *von Neumann* and *Wigner* [31.33]. The rule states that for diatomic systems, crossings between two potential energy curves of the same symmetry are not possible (actually, are extremely rare), whereas for polyatomic systems, potential energy surface intersections are allowed. Mathematically the noncrossing rule is expressed in terms of the dimensions of a surface, the surface of intersection, on which a set of conditions can be satisfied [31.3, 34]. For electronic states of the same symmetry (neglecting electron spin degeneracy), the dimension of the surface of conical intersection is $K-2$, where K is the number of internal nuclear degrees of freedom. For states of different spin symmetry, the dimension of the surface of intersection is $K-1$.

31.2 Characterization of Potential Energy Surfaces

This section is concerned with the determination of the approximate solution of (31.5). For this purpose it is frequently convenient to re-express $H^0(\boldsymbol{r};\boldsymbol{R})$ in second quantized [31.35] form as

$$H^0(\boldsymbol{r};\boldsymbol{R}) = \sum_{i,j} h_{ij} a_i^\dagger a_j + \frac{1}{2} \sum_{i,j,k,l} (il|jk) a_i^\dagger a_j^\dagger a_k a_l\,, \tag{31.16}$$

where h_{ij} and $(il|jk)$ are standard abbreviations [31.36] for the one electron (kinetic energy and nuclear–electron attraction) and two electron (electron–electron repulsion) integrals respectively, and $a_i^\dagger$ and a_i are the fermion creation and destruction operators. The integrals and the creation and destruction operators are defined in terms of a basis of one electron functions, usually molecular spin-orbitals $\phi_i\gamma_i$, $\gamma_i = \alpha$ or β. Thus

$$\begin{aligned} h_{ij}(\boldsymbol{R}) = \Big\langle \phi_i(\boldsymbol{r}_k)\gamma_i \Big| &-\frac{1}{2}\nabla_k^2 \\ &- \sum_K \frac{Z_K}{|\boldsymbol{R}_K - \boldsymbol{r}_k|} \Big| \phi_j(\boldsymbol{r}_k)\gamma_j \Big\rangle_{\boldsymbol{r}_k} \end{aligned} \tag{31.17}$$

$$\begin{aligned} (ij|kl) = \Big\langle \phi_i(\boldsymbol{r}_m)\gamma_i \phi_j(\boldsymbol{r}_m)\gamma_j \Big| &\frac{1}{|\boldsymbol{r}_m - \boldsymbol{r}_n|} \\ &\times \Big| \phi_k(\boldsymbol{r}_n)\gamma_k \phi_l(\boldsymbol{r}_n)\gamma_l \Big\rangle_{\boldsymbol{r}_m,\boldsymbol{r}_n}\,. \end{aligned} \tag{31.18}$$

Equation (31.18) is referred to as the Mulliken notation [31.37] for a two-electron integral. It differs from the frequently used [31.36] bra-ket representation $\langle ij|kl\rangle = (ik|jl)$. The molecular orbitals $\phi_i(\boldsymbol{r}_j;\boldsymbol{R})$ are in turn expanded in terms of a basis set of atom-centered functions [31.38], $\chi(\boldsymbol{r}_j;\boldsymbol{R})$, with $\phi_i(\boldsymbol{r}_j;\boldsymbol{R}) = \sum_{P=1}^{L} \chi_P(\boldsymbol{r}_j;\boldsymbol{R}) T_{P,i}(\boldsymbol{R})$. Note that here we have distinguished between a spin-orbital $\phi_i(\boldsymbol{r}_j;\boldsymbol{R})\gamma_i$ and a molecular orbital $\phi_i(\boldsymbol{r}_j;\boldsymbol{R})$, but we will follow the usual convention of allowing the functions in (31.17) and (31.18) to be molecular orbitals or spin-orbitals with the use being clear from the context.

Since $H^0(\boldsymbol{r};\boldsymbol{R})$ is independent of electron spin, it can also be written in terms of the $\tilde{E}_{pq}$, spin-averaged excitation operators [31.39], where $\tilde{E}_{pq} = \sum_\gamma a^\dagger_{p\gamma} a_{q\gamma}$, p, q label orbitals and $\gamma = \alpha, \beta$ so that

$$H^0(\boldsymbol{r};\boldsymbol{R}) = \sum_{i,j} h_{ij}\tilde{E}_{ij} + \frac{1}{2}\sum_{i,j,k,l}(ij|kl)\tilde{E}_{ij}\tilde{E}_{kl} - \delta_{jk}\tilde{E}_{il}\,. \tag{31.19}$$

The $\tilde{E}_{pq}$ satisfy $\tilde{E}_{ij}, \tilde{E}_{kl}] = \tilde{E}_{il}\delta_{jk} - \tilde{E}_{kj}\delta_{il}$ as a consequence of the commutation relations for the $a_i^\dagger$ and a_i and are generators of the unitary group [31.40], (Chapt. 4). This observation allows the powerful machinery of the unitary group to be applied to the evaluation of $E_I^0(\boldsymbol{R})$ [31.41] (Sect. 31.1.2).

The determination of $E_I^0(\boldsymbol{R})$ involves two steps; (i) the determination of the molecular orbitals $\phi_i(\boldsymbol{r}_j;\boldsymbol{R})$, and (ii) sometimes simultaneously, the determination of the approximate wave function $\Psi_I^0(\boldsymbol{r};\boldsymbol{R})$.

Wave functions discussed in this section are constructed from functions that are antisymmetrized products of spin-orbitals $\phi_i(\boldsymbol{r}_j;\boldsymbol{R})\gamma_i$

$$\Psi_I^0(\boldsymbol{r};\boldsymbol{R}) = \mathcal{A}\prod_{i=1}^{M}\phi_i(\boldsymbol{r}_j;\boldsymbol{R})\gamma_i\,, \tag{31.20}$$

where $\gamma_i = \alpha$ or β. The antisymmetrizer $\mathcal{A}$ is required to take into account the Fermi statistics of the electrons. The wave function in (31.20) is referred to as a Slater determinant and is an eigenfunction of M_s, the z-component of total electron spin. A linear combination of Slater determinants that is also an eigenfunction of S^2 is referred to as a configuration state function (CSF) [31.42]. Equation (31.20) corresponds to a particular distribution of electrons in the orbitals $\phi_i(\boldsymbol{r}_j;\boldsymbol{R})$. This distribution is referred to as an electron configuration.

In principle, it is possible to learn everything about the topology of a potential energy surface from the pointwise evaluation of $E_I^0(\boldsymbol{R})$. In practice however, the determination of topological features, including location of equilibrium structures, saddle points and reaction paths, characterization of quadratic and anharmonic force fields [31.43], and even the evaluation of derivative couplings [31.44], has benefitted immensely from techniques in which the energy gradient $\boldsymbol{g}^I(\boldsymbol{R})$ or higher derivatives are determined directly from knowledge of the wave function at the $\boldsymbol{R}$ in question. These analytic derivative techniques, i. e., techniques that do not use divided difference differentiation, have been actively developed since their first introduction [31.45].

The techniques most commonly used to calculate $E_I^0(\boldsymbol{R})$ and its derivatives are described below.

31.2.1 The Self-Consistent Field (SCF) Method

The SCF Energy

In this most basic treatment, the electronic wave function $\Psi_I^{0-\mathrm{SCF}}$ is taken as a single Slater determinant or a single CSF. The spin-orbitals are determined to give an extremum of the energy functional $\langle\Psi_I^{0-\mathrm{SCF}}(\boldsymbol{r};\boldsymbol{R})|H^0(\boldsymbol{r};\boldsymbol{R})|\Psi_I^{0-\mathrm{SCF}}(\boldsymbol{r};\boldsymbol{R})\rangle_r$. This gives rise to the self-consistent field conditions. For restricted closed shell wave functions [31.46], defined by the conditions

$$\Psi_I^{0-\mathrm{RSCF}}(\boldsymbol{r};\boldsymbol{R}) = \mathcal{A}\prod_{i=1}^{L}\phi_{2i}(\boldsymbol{r}_{2i};\boldsymbol{R})\alpha\phi_{2i+1}(\boldsymbol{r}_{2i+1};\boldsymbol{R})\beta \tag{31.21}$$

with $\phi_{2i}(\boldsymbol{r}_j;\boldsymbol{R}) = \phi_{2i-1}(\boldsymbol{r}_j;\boldsymbol{R})$, $i = 1, \ldots, L$ and $2L = M$, the SCF equations are

$$\varepsilon_{ri} = h_{ri} + \sum_{j\in\{j_\mathrm{occ}\}} 2(ir|jj) - (ij|rj) = 0 \tag{31.22}$$

for ϕ_i, ϕ_r occupied and virtual orbitals, respectively. Here, occupied orbitals j_occ are those occurring in the orbital product in (31.21), and the remainder of the orbitals are referred to as virtual orbitals. The solution to (31.22) is referred to as the canonical SCF solution, provided the orbitals satisfy the SCF equations

$$F\phi_i = \lambda_i\phi_i\,, \tag{31.23}$$

where the Fock operator F is defined by $F_{ri} = \varepsilon_{ri}$ and $\lambda_i = \varepsilon_{ii}$. The corresponding electronic energy is $E_I^0(\boldsymbol{R}) \equiv E_I^{0-\mathrm{SCF}}(\boldsymbol{R}) = \sum_{j \in \{j_{\mathrm{occ}}\}} (\varepsilon_{jj} + h_{jj})$.

SCF Energy Derivatives

The derivative of $E_i^{0-\mathrm{SCF}}(\boldsymbol{R})$ can be expanded through third order using analytic gradient techniques [31.47, 48], and through fourth order using divided differences and more recently using analytic derivative techniques [31.43]. The ability to evaluate the energy derivatives or force field through fourth order is important in the determination of vibrational properties of molecules, as discussed in Sect. 31.4.2.

Direct SCF

The range of molecular systems accessible to treatment at the SCF level has been expanded considerably by the introduction of the direct SCF methods [31.49]. In this method, none of the two-electron integrals in the χ basis $(pq|rs)$ are stored during the iterative solution of (31.22) and (31.23). Since the number of such integrals grows as $L^4/8$, direct SCF procedures avoid the 'L^4-storage bottleneck', enabling SCF wave functions to be determined for systems with more than 100 atoms and basis sets with more than 1000 functions. (If all the $L^4/8$ integrals were stored on disk this basis would require ≈ 1000 gigabytes).

31.2.2 Electron Correlation: Wave Function Based Methods

Wave functions more accurate than SCF wave functions are obtained by including the effects of electron correlation. The correlation energy is defined as $E_I^{\mathrm{corr}}(\boldsymbol{R}) \equiv E_I^0(\boldsymbol{R}) - E_I^{0-\mathrm{SCF}}(\boldsymbol{R})$. Methods for the determination of $E_I^{\mathrm{corr}}(\boldsymbol{R})$ are commonly classified as single reference or mulitreference methods. In single reference methods, an SCF wave function $\Psi_I^{0-\mathrm{SCF}}(\boldsymbol{r}; \boldsymbol{R})$ given by (31.20) is improved. In multireference methods, the starting point is the space spanned by a set of terms like (31.20). This space is referred to as the reference space and the functions in the space as the reference configurations. Sometimes the molecular orbitals $\boldsymbol{\phi}(\boldsymbol{r}_i; \boldsymbol{R})$ for the reference space are the SCF orbitals of a single reference configuration calculation, but more often the $\boldsymbol{\phi}(\boldsymbol{r}_i; \boldsymbol{R})$ are chosen to satisfy multiconfigurational self-consistent field (MCSCF) equations [31.50–53].

Single Reference Methods

Second Order Møller–Plesset Perturbation Theory (MP2). In this approach [31.54], the solutions of the SCF equations are used to determine $E_I^0(\boldsymbol{R})$ to second order as

$$E_I^{0-\mathrm{MP2}}(\boldsymbol{R}) = E_I^{0-\mathrm{SCF}}(\boldsymbol{R}) - \frac{1}{4} \sum_{a,b,i,j} \frac{(ab||ij)^2}{\lambda_a + \lambda_b - \lambda_i - \lambda_j} , \tag{31.24}$$

where i, j denote occupied orbitals, a, b denote virtual orbitals, and

$$(ab||ij) = (ai|bj) - (aj|bi) = \langle ab|ij\rangle - \langle ab|ji\rangle .$$

Both gradients of $E_I^{0-\mathrm{MP2}}(\boldsymbol{R})$ and direct implementations of MP2 [31.37, 55] are currently available. Higher order perturbation theories, also in common use, are discussed in the context of coupled cluster methods which are considered next.

Coupled Cluster Method. Perhaps the most reliable method currently available for characterizing near equilibrium properties of moderately sized molecules is the coupled cluster approach [31.56–60] (Chapt. 5). In this approach, the exact wave function is written as a unitary transformation of a reference wave function $\Phi_0(\boldsymbol{r}; \boldsymbol{R})$ [usually, but not necessarily, the SCF wave function $\Phi_0(\boldsymbol{r}; \boldsymbol{R}) = \Psi_I^{0-\mathrm{SCF}}(\boldsymbol{r}; \boldsymbol{R})$],

$$\Psi_I^{0-\mathrm{CC}} = \exp(T)\Phi_0 . \tag{31.25}$$

where T is an excitation operator defined as

$$T = \sum_p T_p ; \tag{31.26}$$

with

$$T_p = \frac{1}{p!^2} \sum_{i,j,k,\ldots,a,b,c} t_{ijk\cdots}^{abc\cdots} a_a^\dagger a_b^\dagger a_c^\dagger a_i a_j a_k , \tag{31.27}$$

where i, j, k denote occupied orbitals in Φ_0 and a, b, c denote virtual orbitals. Thus

$$\Psi_I^{0-\mathrm{CC}} = \Phi_0 + \sum_{i,a} t_i^a \Phi_i^a + \frac{1}{4} \sum_{i,j,a,b} t_{ij}^{ab} \Phi_{ij}^{ab} + \cdots , \tag{31.28}$$

where Φ_i^a, Φ_{ij}^{ab} are single and double excitations from Φ_0, and it is the amplitudes t that must be determined. Wave functions of the form in (31.25–31.28) have the important property referred to as size consistency [31.60] or size extensivity [31.59]. Consider N noninteracting helium atoms. The electronic wave function for the ith helium atom in the natural orbital basis [31.61] can be written as $\Psi_i = [1 + T_2(i)]\phi_0(i)$, so that the N atom electronic wave function becomes

$$\Psi = \prod_i \Psi_i = \prod_i [1 + T_2(i)]\Phi_0$$
$$= \left[\exp \sum_i T_2(i)\right] \Phi_0 = \exp(T_2)\Phi_0 \qquad (31.29)$$

Thus this exponential type of solution scales properly with the number of particles.

To obtain the amplitudes $\boldsymbol{t}$, define $H_N = H^0 - \langle\Phi_0|H^0|\Phi_0\rangle$, $E_I^{0-\text{CC}} = \Delta E_I^{0-\text{CC}} + \langle\Phi_0|H^0|\Phi_0\rangle \equiv \Delta E_I^{0-\text{CC}} + E_I^{0-0}$, $P = |\Phi_0\rangle\langle\Phi_0|$ and $Q = 1 - P$. Then, inserting (31.25–31.28) into (31.5) gives the energy (31.30) and amplitude (31.31) equations, which define the coupled cluster approach:

$$\Delta E_I^{0-\text{CC}} = \langle\Phi_0|\tilde{H}_N|\Phi_0\rangle \,, \qquad (31.30)$$
$$\langle\Phi_{ijk\cdots}^{abc\cdots}|\tilde{H}_N|\Phi_0\rangle = 0 \text{ for all } \Phi_{ijk\cdots}^{abc\cdots} \,, \qquad (31.31)$$

where

$$\tilde{H}_N = \exp(-T)H_N\exp(T)$$
$$= -E_I^{0-0} + \exp(-T)H^0\exp(T) \,. \qquad (31.32)$$

To appreciate the nature of these equations, consider the approximation $T = T_2$ constructed from SCF orbitals [31.36, 59], referred to as the coupled cluster doubles (CCD) level. At this level the energy equation becomes

$$E_I^{0-\text{CCD}} = \langle\Phi_0|\exp(-T_2)H^0\exp(T_2)|\Phi_0\rangle$$
$$= \langle\Phi_0|H^0T_2|\Phi_0\rangle + E_I^{0-0}$$
$$= E_I^{0-0} + \frac{1}{4}\sum_{i,j,r,s} \Phi_0 H^0 a_r^\dagger a_s^\dagger a_i a_j \Phi_0 t_{ij}^{rs}$$
$$= E_I^{0-0} + \frac{1}{4}\sum_{i,j,r,s} (rs||ij)t_{ij}^{rs} \qquad (31.33)$$

for i, j occupied and r, s virtual orbitals. The amplitude equations become

$$\langle\Phi_{ij}^{rs}|\exp(-T_2)H^0\exp(T_2)|\Phi_0\rangle$$
$$= \langle\Phi_{ij}^{rs}|\left(1 - T_2 + \frac{1}{2}T_2^2\right)H^0$$
$$\times\left(1 + T_2 + \frac{1}{2}T_2^2\right)|\Phi_0\rangle$$
$$= 0 \qquad (31.34)$$

which reduces (after considerable commutator algebra) [31.36] to

$$\begin{aligned}
&(\lambda_r + \lambda_s - \lambda_i - \lambda_j)t_{ij}^{rs} = \\
&\quad -(rs||ij) - \sum_{p>q}(rs||pq)t_{ij}^{pq} - \sum_{k>l}(kl||ij)t_{kl}^{rs} \\
&\quad + \sum_{k,p}\left[(ks||jp)t_{ik}^{rp} - (kr||jp)t_{ik}^{sp}\right. \\
&\quad \left. - (ks||ip)t_{jk}^{rp} + (kr||ip)t_{jk}^{sp}\right] \\
&\quad - \sum_{k>l;p>q}(kl||pq)\left[t_{ij}^{pq}t_{kl}^{rs} - 2\left(t_{ij}^{rp}t_{kl}^{sq} + t_{ij}^{sq}t_{kl}^{rp}\right)\right. \\
&\quad \left. - 2\left(t_{ik}^{rs}t_{jl}^{pq} + t_{ik}^{pq}t_{jl}^{rs}\right) + 4\left(t_{ik}^{rp}t_{jl}^{sq} + t_{ik}^{sq}t_{jl}^{rp}\right)\right] \,.
\end{aligned} \qquad (31.35)$$

The solution to (31.35) is obtained iteratively. The first iteration gives $t_{ij}^{rs} = -(rs||ij)/(\lambda_r + \lambda_s - \lambda_i - \lambda_j)$, which when inserted into (31.33) gives (31.24), i. e., the MP2 result. If the quadratic terms in (31.35) are neglected, then the second iteration gives the third order Møller–Plesset (MP3) result [31.36]. The result of iterating (31.35) to convergence gives the CCD result.

Of the levels of coupled cluster treatments in current use, those including T_1, T_2, T_3, and T_4 in (31.25) provide the most reliable results [31.62–66].

Multireference Methods

Single reference methods provide probably the most powerful tools for treating near-equilibrium properties of ground electronic state systems. In other instances, such as the study of electronically excited states, determination of global potential energy surfaces and for systems with multiple open shells such as diradicals, multireference techniques are found to be extremely useful. In the multireference techniques discussed below, the wave function is written as

$$\Psi_I^{0-\text{MRF}}(\boldsymbol{r}; \boldsymbol{R}) = \sum_\alpha c_\alpha^{\text{I}}\psi_\alpha(\boldsymbol{r}; \boldsymbol{R}) \qquad (31.36)$$

where $\psi_\alpha(\boldsymbol{r}; \boldsymbol{R})$ is a CSF. This expansion is usually referred to as a configuration interaction (CI) expansion, and the coefficients $\boldsymbol{c}^{\text{I}}(\boldsymbol{R})$ are referred to as CI coefficients [31.42]. Since (31.36) does not involve the exponential ansatz, it is not automatically size extensive. In the description of these wave functions, it is useful to generalize the notion of occupied and virtual orbitals to inactive, active, and virtual orbitals, where, referring to the CSF's defining the reference space, inactive orbitals are fully occupied in all CSFs, virtual orbitals are not occupied in any CSF and the active orbitals are (partially) occupied in at least one CSF.

Most multireference techniques begin with the determination of a multiconfigurational self-consistent field (MCSCF) wave function or state averaged MCSCF (SA-MCSCF) wave function [31.50–53, 67]. This approach is capable of describing the internal or static correlation energy, the part of the correlation energy that leads to sizeable separation of two electrons in a pair, and near-degeneracy effects. A particularly robust type of MCSCF wave function, the complete active space (CAS) [31.68–70] wave function, includes in (31.36) all CSFs arising from the distribution of the available electrons among the active orbitals. Note that a CAS wave function is size consistent.

The remaining part of the correlation energy, the dynamic correlation, describes the two-electron cusp, i. e., the regions of space for which two electrons experience the singularity in the Coulomb potential. Empirically, it has been shown [31.71] that wave functions capable of providing chemically accurate descriptions of the dynamic correlation can be obtained by augmenting the MCSCF or reference wave function with all CSFs that differ by at most two molecular orbitals, double excitations (31.28) from those of the reference space. Such wave functions are generally referred to as multireference single and double excitation configuration interaction (MR-SDCI) wave functions. First (second) order wave functions [31.72] include all single (single and double) excitations relative to a CAS reference space.

Multiconfigurational Self-Consistent Field (MCSCF) Theory. In the MCSCF approximation, a wave function $\Psi_I^{0-\mathrm{MCSCF}}(\boldsymbol{r};\boldsymbol{R}) = \sum_\alpha c_\alpha^I \psi_\alpha(\boldsymbol{r};\boldsymbol{R})$ of the form in (31.36) is to be determined. In this procedure, both the molecular orbitals $\phi(\boldsymbol{r}_j;\boldsymbol{R})$ and $\boldsymbol{c}^I(\boldsymbol{R})$ are determined from the requirement that

$$E_I^{0-\mathrm{MCSCF}} = \left\langle \Psi_I^{0-\mathrm{MCSCF}}(\boldsymbol{r};\boldsymbol{R}) \middle| H^0 \middle| \Psi_I^{0-\mathrm{MCSCF}}(\boldsymbol{r};\boldsymbol{R}) \right\rangle_r , \tag{31.37}$$

be a minimum. In a popular variant, the state averaged MCSCF (SA-MCSCF) [31.50–53] procedure, the average energy functional

$$E^{0-\mathrm{SA-MCSCF}} = \sum_{I=2}^{K} w_I \left\langle \Psi_I^{0-\mathrm{MCSCF}}(\boldsymbol{r};\boldsymbol{R}) \middle| H^0 \middle| \times \Psi_I^{0-\mathrm{MCSCF}}(\boldsymbol{r};\boldsymbol{R}) \right\rangle_r , \tag{31.38}$$

where the weight vector $\boldsymbol{w} = (w_1, w_2, \ldots, w_K)$ has only positive elements, is minimized. This procedure should be compared with the multireference CI approach described below in which a predetermined set of molecular orbitals is used.

The optimum molecular orbitals and CI coeffients can be written as unitary transformations of an initial set of such quantities, i. e.,

$$\Psi_I^{0-\mathrm{MCSCF}} = \exp(\gamma)\exp(\Delta)\Phi_I^0 , \tag{31.39}$$

where

$$\boldsymbol{\gamma} = \sum_{i,s} \gamma_{s,i} a_s^\dagger a_i , \quad \Delta = \sum_{n,l} \Delta_{l,n} \left|\Phi_l^0\right\rangle\!\left\langle\Phi_n^0\right| , \tag{31.40}$$

and $\boldsymbol{\gamma}$, Δ are general anti-Hermitian matrices. Since $m_{i,j} = -m_{j,i} \equiv m_{ij}$ for $\boldsymbol{m} = \boldsymbol{\gamma}$, Δ, the upper triangle of $\boldsymbol{m}$ forms a vector $\vec{\boldsymbol{m}}$ that enumerates the independent parameters of $\boldsymbol{m}$. The MCSCF or SA-MCSCF equations can be succinctly formulated by inserting (31.39) into (31.37) or (31.38), expanding the commutators to second order and requiring $\partial E/\partial m_{ij} = 0$ [31.36]. The result provides a system of Newton-Raphson equations that can be solved iteratively for the $\boldsymbol{\gamma}$, $\boldsymbol{\Delta}$ [31.52, 53, 67, 73].

Multireference Configuration Interaction Theory: The MR-SDCI Method. In multireference configuration interaction theory, the wave function is again of the form $\Psi_I^{0-\mathrm{MRCI}}(\boldsymbol{r};\boldsymbol{R}) = \sum_\alpha c_\alpha^I(\boldsymbol{R})\psi_\alpha(\boldsymbol{r};\boldsymbol{R})$, but now CSFs involving the large space of virtual orbitals are included. In this approach the CI coefficients are found for a predetermined set of molecular orbitals. The CSF expansions at the MR-SDCI level become quite large (1–10 million CSFs is routine), and even larger expansions are tractable using specialized methods. The $\boldsymbol{c}^I$ satisfy the usual matrix equation

$$\left(\boldsymbol{H}^0 - E_I^0\right)\boldsymbol{c}^I = 0 , \tag{31.41}$$

where

$$H_{\alpha\beta}^0 = \langle \psi_\alpha | H^0 | \psi_\beta \rangle = \sum_{i,j} A_{ij}^{\alpha\beta} h_{ij} + \sum_{i,j,k,l} A_{ijkl}^{\alpha\beta} (ij|kl) . \tag{31.42}$$

It is the computationally elegant solution of (31.41) for large expansions that is the essence of modern MR-SDCI methods.

Because of the large dimension of the CSF space, it is not possible (or even desirable) to find all the solutions of (31.41). The few lowest eigenstates and eigenvalues can be found [31.74] using an iterative direct CI procedure [31.75] in which a subspace is generated sequentially from the residual defined at the kth iteration by

$$\sigma_\mu^{(k)} = \sum_\nu \left(H_{\mu\nu}^0 - E^{(k-1)}\delta_{\mu\nu} \right) c_\nu^{(k-1)} , \qquad (31.43)$$

where for simplicity of notation, the state index I is suppressed. The computationally demanding step in this procedure is the efficient evaluation of the $A_{ijkl}^{\alpha\beta} = \langle \psi_\alpha | \tilde{E}_{ij} \tilde{E}_{kl} - \delta_{jk} \tilde{E}_{il} | \psi_\beta \rangle$. Key to the efficiency of this evaluation is the factorization formally achieved by

$$\langle \psi_\alpha | \tilde{E}_{ij} \tilde{E}_{kl} | \psi_\beta \rangle = \sum_m \langle \psi_\alpha | \tilde{E}_{ij} | \psi_m \rangle \langle \psi_m | \tilde{E}_{kl} | \psi_\beta \rangle . \qquad (31.44)$$

Using unitary group techniques, this apparently intractable summation can be used to express the $A_{ijkl}^{\mu\nu}$ as a simple finite product [31.76].

Contracted CI and Complete Active Space Perturbation Theory (CASPT2). The direct approach outlined above makes treatment of large MR-SDCI expansions possible. However, as the size of the reference space grows, the CSF space in the MR-SDCI expansion may become intractably large, particularly if the full second-order wave function is used. To avoid this bottleneck, the reference CSFs may be selected from the active space, and perturbation theory may be used to select CSFs involving orbitals in the virtual space [31.42]. The use of selection procedures complicates the implementation of 'direct' techniques although recently progress in selected direct CI procedures has been reported [31.77–79]. Alternatively, new techniques have been developed that avoid this selection procedure. In these approaches, the MC-SCF wave function itself is used as the reference wave function for CI or pertubation theory techniques. The use of a reference wave function rather than a reference space considerably reduces the size of the CSF space to be handled. In this approach, one of the principal computational complications is that the excited functions are not necessarily mutually orthogonal. Two computational procedures currently in wide use, known as contracted CI [31.80–83] and CASPT2 [31.84, 85], are based on this approach.

The CASPT2 method is a computationally efficient variant of second order perturbation theory in which the reference wave function is a CAS-MCSCF wave function and thus may itself contain tens of thousands of CSFs. In this case the full multireference CI problem would be intractable owing to the large space of double excitations. A similar approach is adopted in the contracted CI method, in that the excitations are defined in terms of a general MCSCF reference wave function $\Psi_I^{0-\mathrm{MRF}}$ rather than the reference space as in the MR-SDCI methods described above.

31.2.3 Electron Correlation: Density Functional Theory

The approaches in Sect. 31.1.1 and 31.1.2 can be referred to as wave function based approaches in the sense that determination of $E_I^0(\boldsymbol{R})$ is accompanied by the determination of the corresponding electronic wave function $\Psi_I^0(\boldsymbol{r};\boldsymbol{R})$. An alternative approach is known as density functional theory (DFT) [31.86]. The ultimate goal of DFT is the determination of total densities and energies without the determination of wave functions, as in the Thomas–Fermi approximation. DFT is based on the Hohenberg–Kohn Theorem [31.87], which states that the total electronic density can be considered to be the independent variable in a multi-electron theory (see also Chapt. 20). Computationally viable approaches exploit the Kohn and Sham formulation [31.88], which introduces molecular orbitals as an intermediate device.

The essential features of the Kohn–Sham (KS) theory [31.86] are as follows [31.89]. Assume that the real N-electron system for a particular arrangement of the nuclei $\boldsymbol{R}$ has a total electron density $\rho(\boldsymbol{r};\boldsymbol{R})$. Consider a system of N independent noninteracting electrons subject to a one-body potential V_0 with total density $\rho^0(\boldsymbol{r};\boldsymbol{R})$ such that $\rho(\boldsymbol{r};\boldsymbol{R}) = \rho^0(\boldsymbol{r};\boldsymbol{R})$. The corresponding independent particle orbitals, the Kohn–Sham orbitals $\phi_i^{\mathrm{KS}}(\boldsymbol{r}_j;\boldsymbol{R})$, satisfy a Hartree–Fock-like equation

$$\left(-\frac{1}{2}\nabla^2 + V_0 - \varepsilon_i \right) \phi_i^{\mathrm{KS}} = 0 \qquad (31.45)$$

with

$$\rho^0(\boldsymbol{r};\boldsymbol{R}) = \sum_i |\phi_i^{\mathrm{KS}}|^2 . \qquad (31.46)$$

The relation between the energy of the ideal system and that of the true system $E^{0-\mathrm{DFT}}(\boldsymbol{R})$ is obtained from the adiabatic connection formula [31.89].

In order to determine $E^{0-\mathrm{DFT}}(\boldsymbol{R})$, functions ϕ_i^{KS} are required, which in turn means that V_0, the Kohn–Sham noninteracting one-body potential, must be determined. V_0 is written as (31.3)

$$V_0 = V_{\mathrm{e-N}} + V_{\mathrm{Coul}} + V_{\mathrm{xc}} \qquad (31.47)$$

where V_{Coul} is the Coulomb interaction corresponding to the electron density, $V_{\mathrm{e-N}}$ is the electron-nuclear attraction interaction and V_{xc} is the exchange-correlation density. The effect of (31.47) is to isolate from V_0 the straightforward contributions V_{Coul} and V_{e-N}, and transfers our ignorance to the remaining portion V_{xc}. Although, by the Hohenberg–Kohn theorem V_{xc} must exist, its determination remains the challenge of modern density functional theory, which currently uses approximate functional forms. These approximate treatments of V_{xc} are quite useful in practice, and for large systems, DFT offers a promising alternative to wave function based methods.

Through (31.45), the KS approach is formally similar to SCF theory, although the Kohn–Sham theory is in principle exact. This formal similarity is exploited in the evaluation of the derivative $E^{0-\mathrm{DFT}}(\boldsymbol{R})$ [31.90].

31.2.4 Weakly Interacting Systems

When attempting to describe weak chemical interactions such as van der Waals or dispersion forces, the techniques outlined in Sect. 31.2.2 must be modified somewhat. These modifications arise from the finiteness, and hence incompleteness, of the basis (the set χ) used to describe the molecular orbitals. Assume that the interaction of two molecules A and B is to be determined. Consider the description of molecule A as the distance between A and B decreases from infinity. The atom centered basis functions on molecule B augment those on molecule A, lowering its energy, independent of any physical interaction. This computational artifact serves to overestimate the interaction energy, and is known as the basis set superposition error [31.91,92]. In chemically bonded systems where interaction energies are large, it is of negligible importance. However in weakly bonded systems for which the interaction energies may be on the order of 10 to 100 cm^{-1}, the basis set superpostion error can be significant.

The basis set superposition error can be reduced by the counterpoise correction [31.93,94]. In this approach, the interaction energy is evaluated directly as

$$E_I^{0-\mathrm{int}}(\boldsymbol{R}) = E_I^0(\boldsymbol{R}) - \left[E_I^{0-A}(\boldsymbol{R}) + E_I^{0-B}(\boldsymbol{R})\right] , \tag{31.48}$$

where $E_I^{0-A}(\boldsymbol{R})$ and $E_I^{0-B}(\boldsymbol{R})$, the energies of A and B respectively, are evaluated in the full basis.

31.3 Intersurface Interactions: Perturbations

The existence of interstate interactions can lead to 'unexpected' shifts in spectral lines as well as predissociation of the states themselves [31.95]. These situations are illustrated in Fig. 31.2a,b which present the 1, 2, 3 $^3\Pi_g$ potential energy curves for Al_2 and the corresponding derivative couplings $f^{IJ}(\boldsymbol{R})$ respectively. The derivative couplings were evaluated using the method described in Sect. 31.3.1. In this molecule, derivative couplings between the 2, 3 $^3\Pi_g$ states are responsible for the perturbations in the vibrational levels of the bound 2, 3 $^3\Pi_g$ states. Derivative couplings of these states with the 1 $^3\Pi_g$ state causes predissociation of all levels in the 2, 3 $^3\Pi_g$ manifold [31.96].

Section 31.1 describes two classes of interstate matrix elements that can lead to these nonadiabatic phenomena, the derivative coupling matrix elements in (31.7–31.10) and H_{IJ}^{rel}, which is usually treated for low Z systems within the Breit–Pauli approximation. An illustration of a nonadiabatic process induced by H^{rel} is provided in Sect. 31.5. A key issue in the treatment of the electronic structure aspects of these phenomena is the reliable evaluation of the interstate matrix elements. The interstate interactions are usually of most interest in regions of nuclear coordinate space far removed from the equlibrium nuclear configuration. Thus it is desirable to evaluate these interactions using multireference CI wave functions which (Sect. 31.1) are well suited for use in these regions of coordinate space. The evaluation of nonadiabatic interactions, based on SA-MCSCF/CI wave functions, is discussed below.

31.3.1 Derivative Couplings

From (31.9–31.11), two classes of matrix elements are required, $\tilde{k}_{W_\alpha W_\beta}^{JI}(\boldsymbol{R})$ and $f_{W_\alpha}^{JI}(\boldsymbol{R})$. In fact, techniques to evaluate both of these exist [31.44]. However, it is common to approximate $\tilde{k}_{W_\alpha W_\beta}^{JI}(\boldsymbol{R})$ by

$$\begin{aligned}\tilde{k}_{W_\alpha W'_\beta}^{JI}(\boldsymbol{R}) &= \sum_M \left\langle \frac{\partial}{\partial W_\alpha} \Psi_J^0(\boldsymbol{r};\boldsymbol{R}) \middle| \Psi_M^0(\boldsymbol{r};\boldsymbol{R}) \right\rangle_r \\ &\quad \times \left\langle \Psi_M^0(\boldsymbol{r};\boldsymbol{R}) \middle| \frac{\partial}{\partial W'_\beta} \Psi_I^0(\boldsymbol{r};\boldsymbol{R}) \right\rangle_r \\ &= \sum_M f_{W_\alpha}^{MJ} f_{W'_\beta}^{MI} .\end{aligned} \tag{31.49}$$

with the (in principle infinite) summation over states truncated to reflect only the states explicitly treated in the

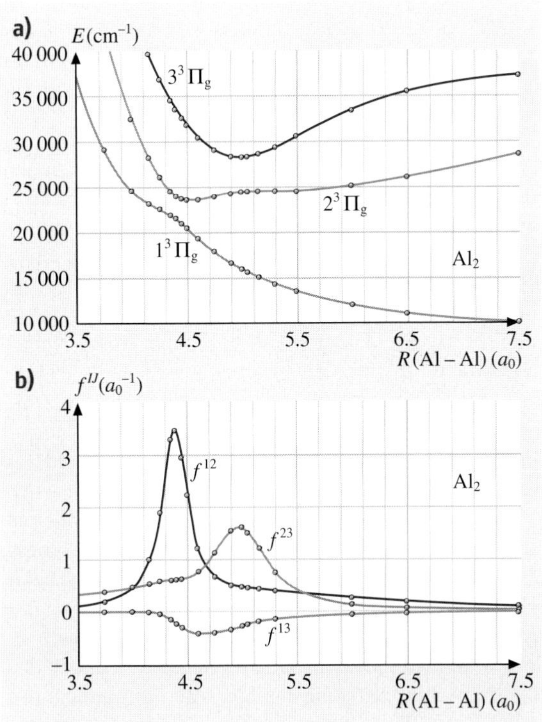

Fig. 31.2 **(a)** Adiabatic potential energy curves for the 1, 2, 3 $^3\Pi_g$ states of Al_2 from [31.96]. **(b)** Derivative couplings $F^{IJ}(\boldsymbol{R})$ for $(I, J) = 1, 3\,^3\Pi_g$, $1, 2\,^3\Pi_g$ and $2, 3\,^3\Pi_g$ from [31.96]

nonadiabatic dynamics. Thus it is sufficient to discuss the determination of $f^{JI}_{W_\alpha}(\boldsymbol{R})$.

The use of analytic derivative theory [31.97, 98] greatly improves the computational efficiency of evaluating $f^{JI}_{W_\alpha}(\boldsymbol{R})$ relative to the earlier divided difference techniques [31.99]. The key ideas are given below. In this presentation, the standard, real-valued normalization is used. Additional contributions owing to the geometric phase, if required, must be evaluated separately as they do not follow from the electronic Schrödinger equation at a single point [31.100].

Differentiation of the $\Psi^0_I(\boldsymbol{r};\boldsymbol{R})$ defined in (31.36) gives:

$$\frac{\partial}{\partial W_\alpha}\Psi^0_I(\boldsymbol{r};\boldsymbol{R}) = \sum_\lambda \left\{ \left[\frac{\partial}{\partial W_\alpha} c^I_\lambda(\boldsymbol{R})\right]\psi_\lambda(\boldsymbol{r};\boldsymbol{R}) + c^I_\lambda(\boldsymbol{R})\left[\frac{\partial}{\partial W_\alpha}\psi_\lambda(\boldsymbol{r};\boldsymbol{R})\right]\right\} . \tag{31.50}$$

Thus $f^{JI}_{W_\alpha}(\boldsymbol{R})$, consists of two terms

$$f^{JI}_{W_\alpha}(\boldsymbol{R}) = {}^{\mathrm{CI}}f^{JI}_{W_\alpha}(\boldsymbol{R}) + {}^{\mathrm{CSF}}f^{JI}_{W_\alpha}(\boldsymbol{R}) , \tag{31.51}$$

where the CI contribution is given by

$${}^{\mathrm{CI}}f^{JI}_{W_\alpha}(\boldsymbol{R}) = \sum_\lambda c^J_\lambda(\boldsymbol{R})\left[\frac{\partial}{\partial W_\alpha} c^I_\lambda(\boldsymbol{R})\right] , \tag{31.52}$$

and the CSF contribution has the form

$${}^{\mathrm{CSF}}f^{JI}_{W_\alpha}(\boldsymbol{R}) = \sum_{\lambda,\mu} c^J_\lambda(\boldsymbol{R})\left\langle \psi_\lambda(\boldsymbol{r};\boldsymbol{R}) \middle| \frac{\partial}{\partial W_\alpha}\psi^I_\mu(\boldsymbol{r};\boldsymbol{R})\right\rangle_{\boldsymbol{r}} c^I_\mu(\boldsymbol{R}) . \tag{31.53}$$

Evaluation of ${}^{\mathrm{CSF}}f^{JI}_{W_\alpha}(\boldsymbol{R})$ is straightforward using analytical derivative techniques [31.101], so that the remainder of the discussion focusses on ${}^{\mathrm{CI}}f^{JI}_{W_\alpha}(\boldsymbol{R})$. From (31.52), it would appear that the derivative of the CI coefficients $\partial/\partial W_\alpha c^I_\lambda(\boldsymbol{R}) \equiv V^I_{W_\alpha,\lambda}(\boldsymbol{R})$ would be required to evaluate ${}^{\mathrm{CI}}f^{JI}_{W_\alpha}(\boldsymbol{R})$. This is quite costly and is in fact not necessary, since only the projection onto the state $\Psi^0_J(\boldsymbol{r};\boldsymbol{R})$ is required. Equation (31.52) for ${}^{\mathrm{CI}}f^{JI}_{W_\alpha}(\boldsymbol{R})$ can be recast in a form similar to that of $\boldsymbol{g}^I(\boldsymbol{R})$. This transformation of (31.52) enables the explicit determination of $\boldsymbol{V}^I_{W_\alpha}(\boldsymbol{R})$ to be avoided, and is the key to the efficient use of analytic gradient techniques in the evaluation of $\boldsymbol{f}^{JI}(\boldsymbol{R})$.

Differentiating (31.41) with respect to W_α gives

$$\left[\boldsymbol{H}^0 - E^0_I(\boldsymbol{R})\right]\boldsymbol{V}^I_{W_\alpha}(\boldsymbol{R}) = -\left\{\frac{\partial}{\partial W_\alpha}\left[\boldsymbol{H}^0 - E^0_I(\boldsymbol{R})\right]\right\}\boldsymbol{c}^I(\boldsymbol{R}) . \tag{31.54}$$

Taking the inner product of (31.54) with $\boldsymbol{c}^J(\boldsymbol{R})$ gives

$${}^{\mathrm{CI}}f^{JI}_{W_\alpha}(\boldsymbol{R}) \equiv \boldsymbol{c}^J(\boldsymbol{R})^\dagger \boldsymbol{V}^I_{W_\alpha}(\boldsymbol{R}) \tag{31.55}$$

$$= \Delta E^0_{IJ}(\boldsymbol{R})^{-1}\boldsymbol{c}^J(\boldsymbol{R})^\dagger \frac{\partial \boldsymbol{H}^0(\boldsymbol{R})}{\partial W_\alpha}\boldsymbol{c}^I(\boldsymbol{R}) \tag{31.56}$$

$$\equiv \Delta E^0_{IJ}(\boldsymbol{R})^{-1} h^{JI}_{W_\alpha}(\boldsymbol{R}) . \tag{31.57}$$

Observe that (31.56) and (31.57) are not the Hellmann–Feynman theorem [31.101, 102] (Chapt. 51), to which they bear a formal resemblance, since it is not the Hamiltonian operator $H^0(\boldsymbol{r};\boldsymbol{R})$ but rather the Hamiltonian matrix $\boldsymbol{H}^0(\boldsymbol{R})$ that is being differentiated. Since the energy gradient has the form [31.44]

$$\frac{\partial}{\partial W_\alpha}E^0_I(\boldsymbol{R}) = \boldsymbol{c}^I(\boldsymbol{R})^\dagger \frac{\partial \boldsymbol{H}^0(\boldsymbol{R})}{\partial W_\alpha}\boldsymbol{c}^I(\boldsymbol{R}) , \tag{31.58}$$

its relation to ${}^{\text{CI}}f^{JI}_{W_\alpha}(\boldsymbol{R})$ is clear. This identification is the key step in the evaluation of ${}^{\text{CI}}f^{JI}_{W_\alpha}(\boldsymbol{R})$ using analytic gradient techniques [31.97, 98].

31.3.2 Breit–Pauli Interactions

For light systems, it is possible to introduce relativistic effects using the Breit–Pauli approximation [31.1], in which the four component Dirac description of a single electron is replaced by a two component (α, β) description. The $H^{\text{e}}(\boldsymbol{r}; \boldsymbol{R})$ becomes [31.1, 103]

$$H^{\text{e}}(\boldsymbol{r}; \boldsymbol{R}) = H^0(\boldsymbol{r}; \boldsymbol{R}) + H^{\text{rel}}(\boldsymbol{r}; \boldsymbol{R}) , \tag{31.59}$$

where, in parallel with (31.4–31.13) for the atomic case, the relativistic correction $H^{\text{rel}} = \sum_{k=1}^{3} H^k(\boldsymbol{r}; \boldsymbol{R})$ can be divided into spin-dependent (SD) and spin-independent (SI) parts, plus an external field interaction term H^{ext}. These are given by

$$H^1 = H^{\text{SD}} \equiv H^{\text{so}} + H^{\text{soo}} + H^{\text{ss}} , \tag{31.60}$$

$$H^2 = H^{\text{SI}} \equiv H^{\text{mass}} + H^{\text{D}} + H^{\text{ssc}} + H^{\text{oo}} , \tag{31.61}$$

$$H^3 = H^{\text{ext}} , \tag{31.62}$$

where, in atomic units,

$$H^{\text{so}} = \frac{\alpha^2}{2} \sum_{K,i} \frac{Z_K(\boldsymbol{r}_i - \boldsymbol{R}_K) \times \boldsymbol{p}_i \cdot \boldsymbol{s}_i}{|\boldsymbol{r}_i - \boldsymbol{R}_K|^3} , \tag{31.63}$$

$$H^{\text{soo}} = -\frac{\alpha^2}{2} \sum_{i \neq j} \frac{\boldsymbol{r}_{ij} \times \boldsymbol{p}_i \cdot (\boldsymbol{s}_i + 2\boldsymbol{s}_j)}{|\boldsymbol{r}_{ij}|^3} , \tag{31.64}$$

$$H^{\text{ss}} = \alpha^2 \sum_{i<j} \left[\frac{\boldsymbol{s}_i \cdot \boldsymbol{s}_j}{|\boldsymbol{r}_{ij}|^3} - \frac{3(\boldsymbol{r}_{ij} \cdot \boldsymbol{s}_i)(\boldsymbol{r}_{ij} \cdot \boldsymbol{s}_j)}{|\boldsymbol{r}_{ij}|^5} \right] , \tag{31.65}$$

$$H^{\text{mass}} = -\frac{\alpha^2}{8} \sum_i p_i^4 , \tag{31.66}$$

$$H^{\text{D}} = -\frac{\alpha^2}{8} \left[\sum_{K,i} Z_K \nabla_i^2 |\boldsymbol{r}_i - \boldsymbol{R}_K|^{-1} - \sum_{i \neq j} \nabla_i^2 |\boldsymbol{r}_{ij}|^{-1} \right] , \tag{31.67}$$

$$H^{\text{ssc}} = -\frac{8\pi\alpha^2}{3} \sum_{i<j} (\boldsymbol{s}_i \cdot \boldsymbol{s}_j) \delta(\boldsymbol{r}_{ij}) , \tag{31.68}$$

$$H^{\text{oo}} = -\frac{\alpha^2}{4} \left(\sum_{j \neq i} \frac{\boldsymbol{p}_i \cdot \boldsymbol{p}_j}{|\boldsymbol{r}_{ij}|} - \frac{\boldsymbol{r}_{ij} \cdot [\boldsymbol{r}_{ij} \cdot \boldsymbol{p}_j] \boldsymbol{p}_i}{|\boldsymbol{r}_{ij}|^3} \right) , \tag{31.69}$$

$$H^{\text{ext}} = \frac{\alpha^2}{2} \sum_i \boldsymbol{E}(\boldsymbol{r}_i) \times \boldsymbol{p}_i \cdot \boldsymbol{s}_i + \frac{i\alpha^2}{4} \sum_i \boldsymbol{E}(\boldsymbol{r}_i) \cdot \boldsymbol{p}_i + 2\mu \sum_i \mathcal{H}(\boldsymbol{r}_i) \cdot \boldsymbol{s}_i , \tag{31.70}$$

$\boldsymbol{E}(\boldsymbol{r})$ and $\mathcal{H}(r)$ are electric and magnetic fields, and $\mu = e\hbar/(2m_{\text{e}})$ is the Bohr magneton. The physical significance of these terms is discussed is Sect. 21.1. One of the most important consequences of these relativistic effects is that total electron spin is no longer a good quantum number as it is in $H^0(\boldsymbol{r}; \boldsymbol{R})$. The term H^{SD} couple states corresponding to distinct eigenvalues of S^2 and lead to the nonadiabatic effects that are the subject of this section.

The Breit–Pauli approximation is most useful for light atoms, but the approximation breaks down when Z becomes large [31.104–106]. One of the principal effects omitted in a treatment which includes only H^{SD} is the relativistic contraction of the molecular orbitals [31.107] due to the mass-velocity operator (H^{mass}), an effect whose importance increases with Z. Several approaches exist which attempt to correct this situation while retaining the spirit and simplifications of the Breit–Pauli approximation. The first of these is the relativistic effective core potential (ECP) approximation [31.108–111]. In this approach, the results of an atomic Dirac–Fock calculation [31.112] are used to replace innermost or core electrons of a given atom with (i) an effective one electron potential that modifies the electron-nuclear attraction term in H^0 and (ii) an effective one electron spin-orbit operator, so that

$$H^{\text{e}}(\boldsymbol{r}; \boldsymbol{R}) \to H^{\text{e-ECP}}(\boldsymbol{r}; \boldsymbol{R}) = H^{0\text{-ECP}}(\boldsymbol{r}; \boldsymbol{R}) + H^{\text{so-ECP}}(\boldsymbol{r}; \boldsymbol{R}) . \tag{31.71}$$

The formal similarity between $H^{\text{e-ECP}}$ and $H^{\text{e}}(\boldsymbol{r}; \boldsymbol{R})$ results in a similar phenomenological interpretation of relativistically induced nonadiabatic processes. Applications of this approach have been reviewed [31.107, 113].

A second approach includes all electrons explicitly, and uses H^{SD} as defined above. The relativistic contraction of the core electrons is included by using a variational one-component spin-free approximation [31.114, 115] to the no-pair Hamiltonian [31.116] at the orbital optimization stage. The variational nature of the approximation provides advantages over the use of the H^{mass} term. Applications of this approach to the spectra of CuH and NiH have been reported [31.117, 118].

In order to evaluate $H^{SD}_{IJ}(\boldsymbol{R})$, it is necessary to specify the molecular orbitals to be used to construct the $\Psi^0_I(\boldsymbol{r};\boldsymbol{R})$. The choice of molecular orbitals is dictated by the following considerations. Matrix elements of H^{SD} between different states are required. The molecular orbitals appropriate for one state may not be appropriate for the description of the second state. Two approaches are available to handle this situation. In one approach, distinct sets of (mutually nonorthogonal) molecular orbitals are used to describe each state [31.119]. This permits a more compact description of the spaces in question. However in this case one is required to evaluate the matrix element of a two electron operator H^{soo} and/or H^{ss} in a nonorthogonal molecular orbital basis, an imposing computational task. This significantly limits the size of the CSF space which is tractable. The alternative approach is to use a common orthonormal basis balanced between the two spaces in question, and to use larger CSF spaces [31.120–122]. The use of a common orthonormal basis decreases significantly the computational effort required to evaluate the matrix elements. A symbolic matrix element method [31.123] has been applied to H^{SD}, as described in the review [31.124].

31.3.3 Surfaces of Intersection

The preceding subsections have considered what must be calculated in order to characterize an electronically nonadiabatic process. As noted in Sect. 31.1, it is also necessary to consider where in nuclear coordinate space electronic nonadiabaticity is important. Nonadiabatic processes are important in regions of close approach of the potential energy surfaces with regions of surface intersections being of preeminent interest. Until recently, these surfaces of intersection were determined by indirect methods, i.e., the potential energy surfaces were characterized, and then the surface of intersection was determined. This made the determination of these surfaces of intersection a computationally daunting task. However, computational advances have made it possible to determine these surfaces of intersection directly, i.e., without prior determination of the individual potential energy surfaces. This point is discussed next.

A point on the surface of conical intersection of two states of the same symmetry, subject to a set of geometric equality constraints of the form $C_i(\boldsymbol{R})=0,\ i=1,\ldots,m$, is determined from the Newton–Raphson equations [31.125]

$$-\begin{pmatrix} \boldsymbol{Q}^{IJ}(\boldsymbol{R},\boldsymbol{\xi},\boldsymbol{\lambda}) & \boldsymbol{g}^{IJ}(\boldsymbol{R}) & \boldsymbol{h}^{IJ}(\boldsymbol{R}) & \boldsymbol{k}(\boldsymbol{R}) \\ \boldsymbol{g}^{IJ}(\boldsymbol{R})^\dagger & 0 & 0 & \boldsymbol{0} \\ \boldsymbol{h}^{IJ}(\boldsymbol{R})^\dagger & 0 & 0 & \boldsymbol{0} \\ \boldsymbol{k}(\boldsymbol{R})^\dagger & \boldsymbol{0}^\dagger & \boldsymbol{0}^\dagger & \boldsymbol{0} \end{pmatrix}\begin{pmatrix}\delta\boldsymbol{R}\\ \delta\xi_1\\ \delta\xi_2\\ \delta\boldsymbol{\lambda}\end{pmatrix} = \begin{pmatrix} \boldsymbol{g}^I(\boldsymbol{R})+\xi_1\boldsymbol{g}^{IJ}(\boldsymbol{R})+\xi_2\boldsymbol{h}^{IJ}(\boldsymbol{R})+\sum_{i=1}^m\lambda_i\boldsymbol{k}^i(\boldsymbol{R}) \\ \Delta E_{IJ}(\boldsymbol{R}) \\ 0 \\ \boldsymbol{C}(\boldsymbol{R}) \end{pmatrix} \tag{31.72}$$

where $\delta\boldsymbol{R}=\boldsymbol{R}'-\boldsymbol{R}$, $\delta\boldsymbol{\lambda}=\boldsymbol{\lambda}'-\boldsymbol{\lambda}$, $\delta\boldsymbol{\xi}=\boldsymbol{\xi}'-\boldsymbol{\xi}$, $g^{IJ}_\alpha(\boldsymbol{R})\equiv\partial\Delta E_{IJ}(\boldsymbol{R})/\partial R_\alpha$, $k^i_\alpha(\boldsymbol{R})\equiv\partial C_i(\boldsymbol{R})/\partial R_\alpha$, $h^{IJ}_\alpha(\boldsymbol{R})\equiv\boldsymbol{c}^{I\dagger}(\boldsymbol{R})\boldsymbol{H}^0(\boldsymbol{R})/\partial R_\alpha\boldsymbol{c}^J(\boldsymbol{R})$,$\boldsymbol{\xi}$ and $\boldsymbol{\lambda}$ are Lagrange multipliers, and $\boldsymbol{Q}^{IJ}(\boldsymbol{R},\boldsymbol{\xi},\boldsymbol{\lambda})$ is a matrix of second derivatives [31.125]. For two states of different symmetry, the analogue of (31.72) is used with the terms related to ξ_2 omitted [31.126]. The excellent performance of this algorithm has been documented [31.125, 127].

Equations (31.72) can be motivated as follows. $\boldsymbol{R}^c$ is sought so that $E^0_I(\boldsymbol{R})$ is minimized subject to the constraints $E^0_I(\boldsymbol{R})=E^0_J(\boldsymbol{R})$ and$\boldsymbol{C}(\boldsymbol{R})=0$. The key is to impose the first of these constraints, noting that at each step in the Newton–Raphson procedure,$\boldsymbol{c}^I(\boldsymbol{R})$ and$\boldsymbol{c}^J(\boldsymbol{R})$ are eigenvectors. Equation (31.72) are the Newton-Raphson equations corresponding to the Lagrangian function [31.128]

$$\begin{aligned}L_{IJ}(\boldsymbol{R},\boldsymbol{\xi},\boldsymbol{\lambda}) &= E_I(\boldsymbol{R})+\xi_1\Delta E_{IJ}(\boldsymbol{R})\\ &\quad+\xi_2H_{IJ}(\boldsymbol{R})+\sum_{k=1}^M\lambda_kC_k(\boldsymbol{R})\,,\end{aligned} \tag{31.73}$$

provided that the gradient of $H^0_{IJ}(\boldsymbol{R})$ is interpreted as a change in $\boldsymbol{H}^0(\boldsymbol{R})$ within the subspace spanned by $\boldsymbol{c}^I(\boldsymbol{R})$ and $\boldsymbol{c}^J(\boldsymbol{R})$. This can be derived from quasidegenerate perturbation theory and understood as follows. Assume for convenience that $m=0$, i.e., there are no geometrical constraints.

Consider an $\boldsymbol{R}$ for which (31.72) is not satisfied. The 2×2 matrix $\boldsymbol{H}(\boldsymbol{R})$ with matrix elements $\boldsymbol{H}_{KL}(\boldsymbol{R})$, $K,L\in\{I,J\}$ becomes at $\boldsymbol{R}+\delta\boldsymbol{R}$

$$\begin{aligned}&H_{IJ}(\boldsymbol{R}+\delta\boldsymbol{R})\\ &=\boldsymbol{c}^{I\dagger}(\boldsymbol{R})\left(\boldsymbol{H}(\boldsymbol{R})+\sum_\alpha\frac{\partial\boldsymbol{H}(\boldsymbol{R})}{\partial R_\alpha}\delta R_\alpha\right)\boldsymbol{c}^J(\boldsymbol{R})\,,\end{aligned} \tag{31.74}$$

$\boldsymbol{H}(\boldsymbol{R})$ is diagonal but nondegenerate at $\boldsymbol{R}$, i. e.,

$$\boldsymbol{H}(\boldsymbol{R}) = \begin{pmatrix} E_I^0(\boldsymbol{R}) - E_J^0(\boldsymbol{R}) & 0 \\ 0 & 0 \end{pmatrix} . \qquad (31.75)$$

From (31.74) at $\boldsymbol{R}^c = \boldsymbol{R} + \delta\boldsymbol{R}$, $\boldsymbol{H}(\boldsymbol{R}^c)$ becomes, to first order, ignoring an irrelevant uniform shift of the diagonal elements,

$$\boldsymbol{H}(\boldsymbol{R}^c) = \begin{pmatrix} \Delta E_{IJ}(\boldsymbol{R}) + \boldsymbol{g}^{IJ\dagger} \cdot \delta\boldsymbol{R} & \boldsymbol{h}^{IJ\dagger} \cdot \delta\boldsymbol{R} \\ \boldsymbol{h}^{IJ\dagger} \cdot \delta\boldsymbol{R} & 0 \end{pmatrix} . \qquad (31.76)$$

Thus (31.72) is seen to be the requirement that $\boldsymbol{H}(\boldsymbol{R}^c)$ has degenerate eigenvalues, with eigenvectors $\boldsymbol{c}^I(\boldsymbol{R})$ and $\boldsymbol{c}^J(\boldsymbol{R})$. When (31.72) have been solved, $\boldsymbol{H}_{IJ}(\boldsymbol{R}^c)$ is diagonal and degenerate, $E_I(\boldsymbol{R}^c)$ has been minimized, and $\boldsymbol{g}^I(\boldsymbol{R}^c) = 0$, except along directions contained in the two dimensional subspace spanned by $\boldsymbol{g}^{IJ}(\boldsymbol{R}^c)$ and $\boldsymbol{h}^{IJ}(\boldsymbol{R}^c)$. From (31.76), it is these two directions which lift the degeneracy of $\boldsymbol{H}(\boldsymbol{R}^c)$. Equation (31.72) is also relevant to the reaction path in a nonadiabatic process as discussed in Sect. 31.4. A discussion of these points from an alternative perspective has been presented by *Radazos* et al. [31.129].

A solution to (31.72) is referred to as a conical intersection, although rigorously degenerate states cannot be obtained from a numerical procedure. If required, the existence of a conical intersection can be rigorously established by showing that the adiabatic wave functions undergo a change of sign when transported around a closed loop containing $\boldsymbol{R}^c$ in the plane defined by $\boldsymbol{g}^{IJ}(\boldsymbol{R}^c)$ and $\boldsymbol{h}^{IJ}(\boldsymbol{R}^c)$. This is the geometric or Berry phase criterion [31.130, 131]. It has been explicitly demonstrated for a conical intersection in O_3 by *Ruedenberg* and coworkers [31.132].

31.4 Nuclear Motion

31.4.1 General Considerations

The determination of the rovibrational spectrum of polyatomic systems from first principles is a problem of primary importance since it allows the determination of molecular forces and structure from spectral data. In the adiabiatic case, the solution of (31.6)

$$\left[T^{\text{nuc}}(\boldsymbol{R}) + E_I^0(\boldsymbol{Q}) - E_K\right]\beta_K(\boldsymbol{R}) = 0 \qquad (31.77)$$

is sought, where $\boldsymbol{Q}$ denotes a set of internal nuclear coordinates. The reliability of the solution of (31.77) reflects the accuracy of the Born–Oppenheimer potential energy surface $E_I^0(\boldsymbol{Q})$ appearing in that equation. The methods for determining $E_I^0(\boldsymbol{Q})$ were discussed in Sect. 31.2. Conversely, the determination of the E_K from spectroscopic measurements can be used to infer information concerning $E_I^0(\boldsymbol{Q})$. In either case, the accurate solution of (31.77) is requisite and this section is concerned with its solution.

The operator in (31.77) has a continuous spectrum since T^{nuc} includes translations of the nuclear center of mass (cm). An operator with a discrete spectrum is obtained by replacing the Hamiltonian in (31.77) with one in which the translation of the nuclear c.m. has been eliminated by transforming to the c.m. frame. In the center of mass frame, (31.77) has the general form [31.133]

$$\left[T^{\text{vr}}(\boldsymbol{\Omega}, \boldsymbol{Q}) + T^{\text{vib}}(\boldsymbol{Q}) + E_I^0(\boldsymbol{Q}) - E_K\right]\beta(\boldsymbol{\Omega}, \mathbf{Q}) = 0 \qquad (31.78)$$

where $\boldsymbol{\Omega}$ are three rotational coordinates and T^{vr} and T^{vib} are the rotational and vibrational kinetic energy operators. In $T^{\text{vr}}(\boldsymbol{\Omega}, \mathbf{Q})$ all the complexity associated with the coupling of nuclear and electronic angular momentum is buried. The determination of the appropriate form for (31.78) is by no means straightforward [31.134, 135], and treatment of the effects of multiple angular momentum is a complex problem in angular momentum algebra.

In diatomic systems, the vibrational problem involves only one internal coordinate and is straightforward. Angular momentum coupling is usually treated using Hund's case (a), (b), etc., or an intermediate case approach [31.95] with Van Vleck's reversed angular momentum commutation relations being helpful in analyzing the coupled angular momentum problem [31.134].

Polyatomic systems introduce new complications, since in addition to the increased dimensionality of the vibrational problem, internal and rotational coordinates interconvert for colinear arrangements of the nuclei; this is particularly relevant in triatomic systems. In addition, since the $C_{\infty v}$ point group has doubly degenerate representations at collinear geometries, electronic state degeneracies may arise (the Renner–Teller effect) [31.32,136,137], further complicating the analysis. Techniques associated with the description of triatomic systems involving coupled angular momentum and electronic degeneracy are illustrated in Sect. 31.4.3.

More generally, (31.78) can be rewritten as

$$\left[T^{\mathrm{vr}}(\boldsymbol{\Omega}, \boldsymbol{Q}^{\mathrm{e}}) + T^{\mathrm{vib}}(\boldsymbol{Q}) + E_I^{\mathrm{sep}}(\boldsymbol{Q}) - E_K\right]\beta(\boldsymbol{R}) = -\Delta T^{\mathrm{vr}}(\boldsymbol{\Omega}, \boldsymbol{Q})\beta(\boldsymbol{R}) - \Delta E_I^0(\boldsymbol{Q}) , \quad (31.79)$$

where

$$\Delta T^{\mathrm{vr}}(\boldsymbol{\Omega}, \boldsymbol{Q}) \equiv T^{\mathrm{vr}}(\boldsymbol{\Omega}, \boldsymbol{Q}) - T^{\mathrm{vr}}(\boldsymbol{\Omega}, \boldsymbol{Q}^{e}) , \quad (31.80)$$

$$\Delta E_I^0(\boldsymbol{Q}) \equiv E_I^0(\boldsymbol{Q}) - E_I^{\mathrm{sep}}(\boldsymbol{Q}) , \quad (31.81)$$

$\boldsymbol{Q}^{\mathrm{e}}$ is an equilibrium structure, and $E_I^{\mathrm{sep}}(\boldsymbol{Q})$ is a separable function of the normal coordinates. In this case, the solution to the left-hand side of (31.79) can be factorized into a rotational part and a vibrational part [31.138] according to

$$\psi_I^J = \sum_K k_K^J D_{K0}^J(\theta, \chi) \prod_i \alpha_{l_i}(Q_i) \quad (31.82)$$

where D_{K0}^J is a symmetric top wave function [31.139], and $\alpha_j(Q_i)$ is a pure vibrational wave function, e.g., harmonic oscillator function, for the ith internal normal coordinate. It is in this approximation that the notion of a molecule rotating and vibrating about a fixed molecular structure is achieved. In this case, E_K is just the sum of the pure rotational and vibrational energies. These rigid rotator-vibrator solutions then form the basis for the inclusion of the effects of the right-hand side using, for example, perturbation theory [31.140]. They can also be used as a basis for a nonperturbative treatment (Sect. 31.4.2).

The classic treatments of (31.77) [31.141–144] have been periodically revisited [31.134, 145]. In these investigations, the contributions of electronic angular momentum to this Hamiltonian were frequently suppressed [31.144] so that these treatments are appropriate to totally symmetric electronic states. *Van Vleck* [31.134] has shown how the effects of electronic angular momentum can be incorporated into these treatments.

For nonlinear polyatomic systems, the Hamiltonian in (31.78) can be transformed to the Eckart or body fixed frame in which the reference axis is a body fixed axis oriented along the principal moments of inertia [31.146], although other choices of the body fixed axes are possible [31.147]. The principal moments of inertia represent the eigenvectors of the inertial tensor matrix I_{ij}, $i, j = x, y, z$, where [31.144]

$$I_{xx} = \sum_i M_i(\boldsymbol{R}_i^{\mathrm{e}} \cdot \boldsymbol{R}_i^{\mathrm{e}} - X_i^{\mathrm{e}} X_i^{\mathrm{e}}) , \quad (31.83)$$

$$I_{xy} = -\sum_i M_i(X_i^{\mathrm{e}} Y_i^{\mathrm{e}}) , \quad (31.84)$$

and cyclic permutations, and $\boldsymbol{R}^{\mathrm{e}}$ denotes an equilibrium structure. The Hamiltonian determined by this procedure is referred to as the Watson Hamiltonian [31.145], and is widely used in discussing the rovibrational spectrum of nonlinear polyatomic molecules in singlet electronic states. Treatments of the Watson Hamiltonian are mentioned in Sect. 31.4.2. For linear molecules, an alternative treatment is required since there is one fewer overall rotational coordinate and one more internal coordinate [31.138].

31.4.2 Rotational-Vibrational Structure

Various approaches to the solution of (31.79) exist. The principal issues, which are interrelated, are (i) the range of nuclear configurations over which $E_I^0(\boldsymbol{Q})$ is known, (ii) the coordinate system used to express the internal coordinates, and (iii) particularly in larger systems, the number of modes or internal coordinates retained in the calculations (Sect. 31.4.3). With regard to point (i), two approaches are currently in use. The force field method uses a power series expansion of $E_I^0(\boldsymbol{Q})$ about $E_I^0(\boldsymbol{Q}^{\mathrm{e}})$, that is (using the Einstein summation convention)

$$E_I^0(\boldsymbol{Q}) = E_I^0(\boldsymbol{Q}^{\mathrm{e}}) + \frac{1}{2}\frac{\partial^2 E_I^0(\boldsymbol{Q}^{\mathrm{e}})}{\partial Q_i \partial Q_j}\Delta Q_i \Delta Q_j + \frac{1}{6}\frac{\partial^3 E_I^0(\boldsymbol{Q}^{\mathrm{e}})}{\partial Q_i \partial Q_j \partial Q_k}\Delta Q_i \Delta Q_j \Delta Q_k \quad (31.85)$$

together with perturbation theory to determine spectroscopic constants. In this approach, the force fields [the partial derivatives in (31.85) are usually evaluated directly with the aid of analytic gradient techniques (Sect. 31.2). Since the expansion of $E_I^0(\boldsymbol{Q})$ is truncated, the results are not independent of the coordinate system used. For example, significant differences in the description of Fermi-resonance parameters [31.148, 149] in rectilinear and curvilinear coordinates [31.150] have been reported.

Alternatively, $E_I^0(\boldsymbol{Q})$ can be represented by a grid of points around $\boldsymbol{Q}^{\mathrm{e}}$. In the most reliable calculations reported to date, $E_I^0(\boldsymbol{Q})$ is determined using the coupled cluster techniques discussed in Sect. 31.2. Then (31.79) can be solved in a basis analogous to

ψ_l^J in (31.82). This approach is frequently referred to as the vibrational CI problem [31.151]. The reliability of the results depends to a considerable extent on the basis functions and coordinate system used to describe the problem. Considerable success has been reported for a technique in which the Watson Hamiltonian and harmonic oscillator functions are used to solve (31.79) [31.151, 152].

31.4.3 Coupling of Electronic and Rotational Angular Momentum in Weakly Interacting

An understanding of the molecular structure of the weakly bound compounds of noble gas atoms and diatomic molecules provides important insights into the nature of chemical bonding. The inference of structural data from spectroscopic observations is an important aspect of this problem. In this subsection, a theoretical framework for understanding the spectroscopy of these systems is outlined as an illustration of the methods used to treat coupling of electronic and nuclear angular momentum in weakly interacting triatomic molecular systems. Detailed discussion of this class of problems can be found in [31.153–155].

As an example, consider the rovibronic structure of a noble gas-diatom complex Rg-AB, in either its $^1\Sigma^+$ or $^1\Pi$ states (which may be closely spaced). Note that degeneracy of the $^1\Pi$ state will only persist for collinear geometries of the triatom system. The rovibrational wave functions can be expanded in a product basis of functions describing (i) the rovibronic structure of AB and (ii) the relative motion (vibrational and end-over-end rotation) of Rg and AB. Since total angular momentum J and its space fixed projections M are good quantum numbers, the rovibronic wave function can be expanded as

$$\psi^{JM} = \frac{1}{R} \sum_{vj\Omega l\varepsilon} C^{JM}_{j\Omega l\varepsilon v}(R) \times |\psi_{\mathrm{Rg}}\rangle |v_{Ij}\rangle |Ij\Omega\varepsilon, l;\, JM\rangle \,, \quad (31.86)$$

where ψ_{Rg} is the ground state wave function for the noble gas atom, l denotes the angular momentum associated with the end-over-end motion, and v, j, ε and I denote respectively the vibrational, angular momentum (electronic + rotational), e/f symmetry index and state label of the electronic state of the AB molecule in Hund's case (a) basis. The angular momentum coupling algebra noted above is reflected in the definition of $|Ij\Omega\varepsilon, l;\, JM\rangle$, which is

$$|Ij\Omega\varepsilon, l;\, JM\rangle = \sum_{m_j m_l} \langle jm_j lm_l | JM\rangle Y_{lm_l}(\theta,\phi)\psi_{Ijm_j\Omega\varepsilon}(\beta,\alpha)\,, \quad (31.87)$$

where θ, ϕ and β, α are the polar and azimuthal angles for the line connecting the noble gas atom to the center of mass of the diatom and the diatom axis, respectively, and $\langle\cdots|\cdots\rangle$ is a Clebsch–Gordan coefficient [31.156] (Chapt. 2). The $\psi_{Ijm_j\Omega\varepsilon}$ are defined in turn by [31.95]

$$\psi_{Ijm_j\Omega\varepsilon} = \left(\frac{2J+1}{8\pi}\right)^{1/2} \times \left[D^j_{m_j,\Omega}(\alpha,\beta,0)^*|I^{2S+1}\Lambda^+(\Sigma)\rangle + \varepsilon D^j_{m_j,-\Omega}(\alpha,\beta,0)^*|I^{2S+1}\Lambda^-(-\Sigma)\rangle\right] \quad (31.88)$$

or

$$\psi_{Ijm_j\Omega\varepsilon} = \left(\frac{2J+1}{4\pi}\right)^{1/2} \times D^j_{m_j,0}(\alpha,\beta,0)^*|I^{2S+1}\Lambda(\Sigma)\rangle \quad (31.89)$$

when $\Lambda = \Sigma = 0$, where the electronic state has term symbol $I^{2S+1}\Lambda$ with $\Omega = \Lambda + \Sigma$.

The $\boldsymbol{C}^{JM}(\boldsymbol{R})$ satisfy the usual close coupled equations [31.157, 158] Sect. 47.1.1

$$\left[\left(\frac{-1}{2\mu}\frac{d^2}{dR^2} + \frac{l(l+1)}{2\mu R^2}\right)\boldsymbol{I} - \frac{\boldsymbol{k}^2}{2\mu} + \boldsymbol{V}(r)\right]\boldsymbol{C}^{JM}(r) = 0\,, \quad (31.90)$$

where $l(l+1)$ and $\boldsymbol{k}^2/2\mu$ designate the diagonal matrices of orbital angular momentum and asymptotic net scattering energy of individual channels. $\boldsymbol{V}(R)$ represents the matrix elements of $H^{\mathrm{e}}(\boldsymbol{r};\boldsymbol{R})$ in the vibronic basis defined in (31.86–31.89). It is built from terms of the form

$$\left\langle Y_{lm} D^{j\,*}_{m_j\Omega} v_{Ij}\right|\left\langle \psi_{\mathrm{Rg}} I^{2S+1}\Lambda\right| H^{\mathrm{e}}(\boldsymbol{r};\,\boldsymbol{R}) \times \left|\psi_{\mathrm{Rg}} I'^{2S'+1}\Lambda'\right\rangle_{\boldsymbol{r}} \left|v_{I'j'} Y_{l'm'} D^{j'\,*}_{m'_j\Omega'}\right\rangle \equiv \left\langle Y_{lm} D^{j\,*}_{m_j\Omega} v_{Ij}\right| H^{\mathrm{e}}_{II'}(R,\tilde{\beta},r)\left| Y_{l'm'} D^{j'\,*}_{m'_j\Omega'} v_{I'j'}\right\rangle . \quad (31.91)$$

The angle $\tilde{\beta}$ is the polar angle of the diatom with respect to the atom-diatom axis. The angular integrations on the right-hand side of (31.91) are accomplished by expanding the angular dependence of $H^{e}_{II'}$ in d-matrices and using angular momentum coupling algebra [31.158]. $H^{e}_{II'}$ gives the interaction potential of the ground state noble gas atom and the $I^{2S+1}\Lambda$ state of AB. It represents the potential coupling in the diabatic basis $\psi_{\mathrm{Rg}}|I^{2S+1}\Lambda\rangle$. These matrix elements are derived from the corresponding adiabatic potential energy surfaces, determined using the techniques of Sect. 31.2. Since the van der Waals interaction is weak, the counterpoised method discussed in Sect. 31.2.4 should be used. When only one electronic state of a given symmetry is involved (point group C_s in the case of a triatom), the adiabatic and diabatic potentials are taken as equal. This is the case if, for example, only the $^1\Pi$ state of AB is considered. When more than one electronic state of a given sysmmetry is involved, for example, when both $^1\Pi$ and $^1\Sigma^+$ electronic states of AB are included, the adiabatic potentials are used to determine the diabatic state potentials using an approximate adiabatic $\rightarrow$ diabatic state transformation [31.159, 160].

The solution to (31.90) can be obtained by either variational methods or by direct integration [31.153]. Results for Ar-OH [31.161], Ar-BH [31.155], and Ar-CH [31.162], based on potential energy surfaces determined from multireference contracted CI wave functions, are encouraging. In these studies, the r-dependence of the potential energy surfaces was neglected.

31.4.4 Reaction Path

The rovibrational spectrum reflects the molecular structure in the vicinity of an equilibrium structure of the molecule. However a chemical reaction samples a much broader range of molecular structures. The evolution of molecular structure can be characterized by the reaction path, which represents a minimum energy path along the potential energy surface connecting the reactants and products. The reaction path Hamiltonian [31.163, 164] enables dynamical aspects of a chemical reaction to be inferred from a characterization of the reaction path.

The reaction path $\boldsymbol{R}(s)$ on potential energy surface $E^0_I(\boldsymbol{R})$ is the curve in mass weighted Cartesian coordinates given parametrically in terms of the arc length s and is defined as that solution to the differential equation

$$\frac{\mathrm{d}\boldsymbol{R}}{ds} = \boldsymbol{g}^I \Big/ \sqrt{\boldsymbol{g}^{I\dagger}\boldsymbol{g}^I} \equiv \boldsymbol{R}'(s) \qquad (31.92)$$

which approaches the saddle point from below [31.17, 165], where the arc length is given by

$$\mathrm{d}s^2 = \sum_{i=1}^{N} \mathrm{d}R_i^2 \, . \qquad (31.93)$$

At each point s, $\boldsymbol{R}(s)$ represents the molecular structure in terms of $3N$ mass weighted Cartesian coordinates. (In this subsection, the notation suppresses the difference between Cartesian and mass weighted Cartesian coordinates.)

The reaction path can be obtained by integrating (31.92). Integration begins at the saddle point along the direction of the single negative eigenvalue of $\boldsymbol{F}^I$, which is related to the reaction path curvature $\mathrm{d}\boldsymbol{R}^2/\mathrm{d}s^2$, by

$$\frac{\mathrm{d}\boldsymbol{R}^2}{\mathrm{d}s^2} \equiv \frac{\mathrm{d}\boldsymbol{R}'}{\mathrm{d}s} = \left[\boldsymbol{F}^I\boldsymbol{R}' - \left(\boldsymbol{R}'^{\dagger}\boldsymbol{F}^I\boldsymbol{R}'\right)\boldsymbol{R}'\right] \Big/ \sqrt{\boldsymbol{g}^{I\dagger}\boldsymbol{g}^I} \, . \qquad (31.94)$$

Thus integration of (31.92) gives the steepest descent path from the saddle point. This integration requires, in principle, only knowledge of the energy gradient, $\boldsymbol{g}^I(\boldsymbol{R})$, which is readily available using analytic gradient techniques (see Sect. 31.2). This approach works quite well when the reaction path is used only to investigate the topology of a potential energy surface. However if quantities such as the curvature are desired, procedures that follow the reaction path more closely are required. Such procedures require higher derivatives of the energy, as fully discussed in [31.165].

In a nonadiabatic process, the reaction path necessarily involves more than one potential energy surface. In the case of nonadiabatic reactions involving avoided intersections, the reaction path is difficult to define since the propensity for an intersurface transition spreads out over a range of nuclear coordinates. However, for nonadiabatic reactions that proceed though a surface of intersection of two potential energy surfaces, the reaction path passes through the minimum energy point on the surface of intersection of the two potential energy surfaces [31.166]. An example of such a process is provided in Sect. 31.5. The minimum energy point on the surface of intersection satisfies (31.72) with no geometric constraints. For a spin-forbidden reaction, the solution of (31.72) yields $\boldsymbol{g}^I\boldsymbol{R} + \xi_1\boldsymbol{g}^{IJ}(\boldsymbol{R}) = 0$, so that $\boldsymbol{g}^I(\boldsymbol{R}) = \xi_1/(1+\xi_1)\boldsymbol{g}^J(\boldsymbol{R})$, i. e., the gradients on the two surfaces are either parallel or anti-parallel depending on the value of ξ_1. The relationship between the gradients in the vicinity of an intersection of two states of the same symmetry is not as simple. The wave functions are not differentiable at a point of intersection. In the vicinity of the point of intersection, the condition $\boldsymbol{g}^I(\boldsymbol{R}) + \xi_1 \boldsymbol{g}^{IJ}(\boldsymbol{R}) + \xi_2 \boldsymbol{h}^{IJ}(\boldsymbol{R}) \approx 0$ holds, so that $\boldsymbol{g}^I(\boldsymbol{R})$ and $\boldsymbol{g}^J(\boldsymbol{R})$ will not be simply related. Finally, note that the adiabatic correction, defined in Sect. 31.1, is large (and positive) in the vicinity of the conical intersection, so that the nuclear wave function exhibits a node at this point.

31.5 Reaction Mechanisms: A Spin-Forbidden Chemical Reaction

This section presents an example to illustrate how the ideas outlined in the preceding sections can be used to obtain a mechanistic description of a chemical reaction which could not be deduced on intuitive grounds. Considered here is the ground state reaction

$$\begin{aligned} &\mathrm{CH}(\mathrm{X}\,^2\Pi) + \mathrm{N_2}(\mathrm{X}\,^1\Sigma_g^+) \to \mathrm{HCN_2}(1\,^2\mathrm{A}'') \\ &\to \mathrm{HCN_2}(1\,4\mathrm{A}'') \to \mathrm{HCN}(\mathrm{X}\,^1\Sigma^+) + \mathrm{N}(^4\mathrm{S})\,, \end{aligned} \tag{31.95}$$

which is of considerable importance in the chemistry of planetary atmospheres [31.168, 169] and hydrocarbon flames, having been suggested as the initial step in the production of prompt or Fenimore NO [31.169–172]. However, because the reaction is spin-forbidden, its importance has been questioned.

The reaction will be interpreted in terms of the intermediate complex model [31.173, 174]. In this model, the reaction can occur despite a small probability for intersystem crossing by repeatedly traversing the 2A″–4A″ surface of intersection. The repeated traversals result from a local minimum on the 2A″ potential energy surface. This mechanism is operative for the spin-forbidden oxygen quenching reactions [31.173, 174]

$$\begin{aligned} &\mathrm{O}(^1\mathrm{D}) + \mathrm{N_2}(\mathrm{X}\,^1\Sigma_g^+) \to \mathrm{N_2O}^* \\ &\qquad \to \mathrm{O}(^3\mathrm{P}) + \mathrm{N_2}(\mathrm{X}^1\Sigma_g^+)\,, \end{aligned} \tag{31.96}$$

$$\begin{aligned} &\mathrm{O}(^1\mathrm{D}) + \mathrm{CO}(\mathrm{X}\,^1\Sigma^+) \to \mathrm{CO}_2^* \\ &\qquad \to \mathrm{O}(^3\mathrm{P}) + \mathrm{CO}(\mathrm{X}\,^1\Sigma^+)\,, \end{aligned} \tag{31.97}$$

which are important in the chemistry of the atmosphere.

To demonstrate the feasibility of reaction (31.95) as an intermediate complex assisted chemical reaction it is necessary to (i) determine the energetically relevant portion of the doublet-quartet surface of intersection, (ii) determine the spin-orbit interaction coupling the doublet and quartet surfaces, (iii) characterize a local minimum on the doublet surface from which the surface of intersection is accessible, (iv) characterize a path from the reactant channel to this region, and (v) determine the exit channel path on the quartet surface. The results of electronic structure calculations at the MR-CI level, addressing these points [31.167, 175–178] are summarized below.

Local Extrema on the 2A″–4A″ Surface of Intersection. Points in the 2A″–4A″ surface of intersection were determined from the solution to (31.72) [31.126, 175]. The minimum energy point on the surface of intersection, labelled $\mathcal{C}_{\mathrm{mex}}^{C_{2v}}$, has approximate C_{2v} symmetry, and is shown in Fig. 31.3. It is the region of this crossing point that must be accessed from the doublet surface.

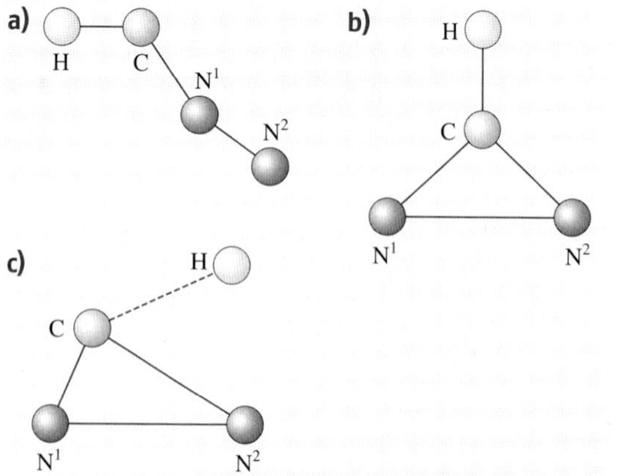

Fig. 31.3a–c Key bond distances in Å for **(a)** $R_{\mathrm{min}}^{\mathrm{dative}}(\mathrm{CH}) = 1.082$, $R_{\mathrm{min}}^{\mathrm{dative}}(\mathrm{CN^1}) = 1.340$, $R_{\mathrm{min}}^{\mathrm{dative}}(\mathrm{N^1N^2}) = 1.143$; **(b)** $R_{\mathrm{min}}^{C_{2v}}(\mathrm{CH}) = 1.071$, $R_{\mathrm{min}}^{C_{2v}}(\mathrm{CN^1}) = 1.311$; $R_{\mathrm{min}}^{C_{2v}}(\mathrm{N^1N^2}) = 1.735$; $R_{\mathrm{mex}}^{C_{2v}}(\mathrm{CH}) = 1.076$, $R_{\mathrm{mex}}^{C_{2v}}(\mathrm{CN^1}) = 1.3$, $R_{\mathrm{mex}}^{C_{2v}}(\mathrm{N^1N^2}) = 2.221$; **(c)** $R_{\mathrm{TS}}^{c}(\mathrm{CH}) = 1.093$, $R_{\mathrm{TS}}^{c}(\mathrm{CN^1}) = 1.515$, $R_{\mathrm{TS}}^{c}(\mathrm{N^1N^2}) = 1.214$. The notation is as defined in Sect. 31.5. The *dashed line* in structure **(c)** indicates that H is out of the plane of the paper. **(a)**, **(b)** after [31.124], **(c)** after [31.167]

The Spin-Orbit Interaction. In the double group corresponding to C_s symmetry, the $4A''$ and $2A''$ wave functions each carry degenerate irreducible representations, Kramer's doublets [31.179]. The following pairs of nonrelativistic zeroth-order wave functions can be used to span these degenerate representations (i) $\Psi^0[2A''(\frac{1}{2})]$, $\Psi^0[2A''(-\frac{1}{2})]$, (ii) $i\Psi^0[4A''(\frac{1}{2})]$, $i\Psi^0[4A''(-\frac{1}{2})]$, and (iii) $i\Psi^0[4A''(\frac{3}{2})]$, $i\Psi^0[4A''(-\frac{3}{2})]$, where the M_s value has been given parenthetically. In this case, all nonvanishing matrix elements connecting the components of the $4A''$ and $2A''$ states can be expressed in terms of the single real-valued matrix element [31.139]

$$
\begin{aligned}
&H^{\text{so}}(4A'', 2A'') \\
&\equiv \left\langle i\Psi^0\left[4A''\left(\frac{3}{2}\right)\right]\middle|H^{\text{so}}\middle|\Psi^0\left[2A''\left(\frac{1}{2}\right)\right]\right\rangle , \qquad (31.98)
\end{aligned}
$$

which was found to be $\approx 12.5\,\text{cm}^{-1}$ [31.126, 175] in the vicinity of the minimum energy crossing point determined above.

Local Extrema on the $2A''$ Potential Energy Surface. Two local minima on the $2A''$ potential energy surface have been found [31.175–177]. The local minimum configuration labeled $\mathfrak{C}_{\text{min}}^{\text{dative}}$ pictured in Fig. 31.3 represents a datively bonded structure in which N_2 donates a pair of electrons to the empty $CH(1\pi)$ orbital. This can be thought of as a reactant channel structure, since the N–N and C–H bond lengths are similar to those in the isolated molecules. This point on the $2A''$ potential energy surface is stable by 20.2 kcal/mol relative to the doublet asymptote in the reactant channel. The second minimum on the $2A''$ potential energy surface is also pictured in Fig. 31.3 and is seen to have C_{2v} symmetry. This configuration point, denoted $\mathfrak{C}_{\text{min}}^{C_{2v}}$, is stable by 22.3 kcal/mol relative to the doublet asymptote in the reactant channel. Note that $\mathfrak{C}_{\text{min}}^{C_{2v}}$ and $\mathfrak{C}_{\text{mex}}^{C_{2v}}$ differ only in the length of the N–N bond.

Reaction Path. From the above data it might appear that both the $\mathfrak{C}_{\text{min}}^{C_{2v}}$ and $\mathfrak{C}_{\text{min}}^{\text{dative}}$ structures are involved in the mechanism of reaction (31.95). However, the energy of the transition state connecting these two minima is quite large [31.177], so that the reaction path avoids $\mathfrak{C}_{\text{min}}^{\text{dative}}$ [31.167]. The true reaction path involves $\mathfrak{C}_{\text{min}}^{C_{2v}}$, $\mathfrak{C}_{\text{mex}}^{C_{2v}}$, and a reactant channel transition state $\mathfrak{C}_{\text{TS}}^{c}$ that connects $\mathfrak{C}_{\text{min}}^{C_{2v}}$ and the reactant channel species $CH+N_2$. $\mathfrak{C}_{\text{TS}}^{c}$ is also pictured

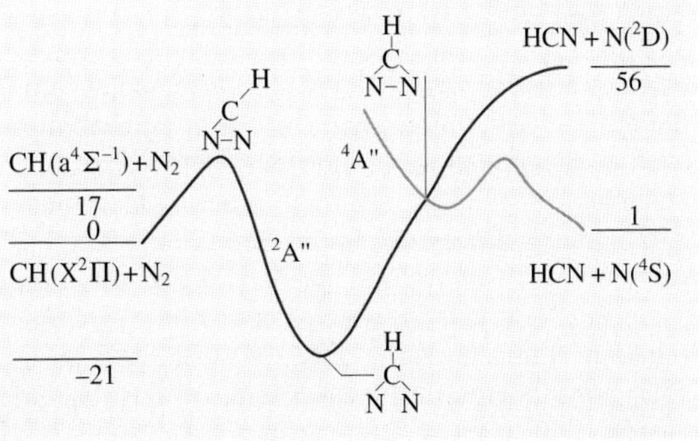

Fig. 31.4 Reaction path for $CH(X^2\Pi)+N_2(X^1\Sigma_g^+) \rightarrow HCN(X^1\Sigma^+)+N(^4S)$ after [31.167]. Energy scale, which is approximate, is in kcal/mol

in Fig. 31.3. It involves the nonplanar cis- approach of HC to N_2 [31.167]. The total reaction path is presented in Fig. 31.4 [31.178]. Note that following intersystem crossing, the system encounters a final transition state before proceeding to the spin-forbidden products.

From this qualitative analysis, a bimodal picture of the reaction of ground state CH with N_2 emerges. $CH(^2\Pi)$ can be removed by either (i) forming a complex $\mathfrak{C}_{\text{min}}^{\text{dative}}$ which is stabilized by a third body collision or (ii) forming the spin-forbidden products with the reaction mechanism indicated in Fig. 31.4. This mechanism has been deduced entirely from the computational results cited above. The key computational findings responsible for this mechanism are (i) the existence and geometrical relationship between $\mathfrak{C}_{\text{min}}^{C_{2v}}$ and $\mathfrak{C}_{\text{mex}}^{C_{2v}}$, (ii) the large barrier separating $\mathfrak{C}_{\text{min}}^{\text{dative}}$ and $\mathfrak{C}_{\text{min}}^{C_{2v}}$, and (iii) the entrance channel path across $\mathfrak{C}_{\text{TS}}^{c}$ to $\mathfrak{C}_{\text{min}}^{C_{2v}}$.

Since there is apparently no or little barrier to forming $\mathfrak{C}_{\text{min}}^{\text{dative}}$ [31.177], at low temperatures CH will be removed exclusively by this process at a rate with a small or negative temperature dependence, indicating a barrierless reaction. At higher temperatures, the rate constant is expected to exhibit a more usual Arrhenius type behavior as the spin-forbidden channel can be accessed [31.180]. These expectations are consistent with previous low temperature measurements [31.169, 181] and high temperature shock tube results [31.172, 182, 183]. The above reaction path for the spin-forbidden reaction has been used in a two internal coordinate model for reaction (31.95) [31.180] that also supports the proposed description of this reaction.

31.6 Recent Developments

Added by Mark M. Cassar. If the motion of nuclei within a molecule causes more than one Born-Oppenheimer potential energy surface to intersect, then nonadiabatic transitions become an essential process in molecular dynamics. These transitions are at the heart of many biological processes, such as light harvesting by plants [31.184, 185], which depends on electronic excitation transfer, and a host of processes in the upper atmosphere. In recent years, it has become apparent that conical intersections (see Sect. 31.3.3 for a definition) of two states of the same symmetry are a fundamental element of electronically nonadiabatic phenomena [31.186]. As an example, conical intersections can facilitate transitions from upper to lower electronic states. Transitions of this nature provide efficient pathways to lower electronic states, and are thus amenable to experimental study (e.g., through dissociation products of photoexcitation [31.187]).

Although points of conical intersection, which are not individual but form seams, can be located in a straightfoward manner, finding points of minimum energy has proved more difficult. This difficulty has been attributed to the erratic behavior of the parameters used in the search algorithm. This seemingly intrinsic problem which would preclude extrapolation – a consequence of the singular nature of the intersection itself – has been avoided through the use of extrapolatable functions [31.188]. This recent work introduces functions that vary smoothly (i. e., are well-behaved) as one moves along the intersection seam, and hence allow the use of extrapolation techniques.

The study of conical intersections has also been extended to include three-state intersections and first-order relativistic effects. Accidental conical intersections (i. e., those not required by symmetry arguments) of three states of the same symmetry, which may provide an efficient mechanism for radiationless decay, have been shown to exist [31.189]. Spin-orbit effects are essential for the proper description of the nuclear dynamics in molecules containing an odd number of electrons [31.190]. The topography of conical intersections changes in this case, requiring the development of new algorithms [31.191].

References

31.1 H. A. Bethe, E. E. Salpeter: *Quantum Mechanics of One and Two Electron Atoms* (Plenum/Rosetta, New York 1977)

31.2 M. Born, K. Huang: *Dynamical Theory of Crystal Lattices* (Oxford Univ. Press, Oxford 1954)

31.3 G. Herzberg, H. C. Longuet-Higgins: Disc. Faraday Soc. **35**, 77 (1963)

31.4 M. V. Berry: Proc. R. Soc. London **A392**, 45 (1984)

31.5 H. C. Longuet-Higgins: Adv. Spectrosc. **2**, 429 (1961)

31.6 W. H. Gerber, E. Schumacher: J. Chem. Phys. **69**, 1692 (1978)

31.7 L. R. Kahn, P. Baybutt, D. G. Truhlar: J. Chem. Phys. **65**, 3826 (1976)

31.8 J. C. Tully: In: *Modern Theoretical Chemistry*, Vol. 2, ed. by W. H. Miller (Plenum, New York 1976) p. 217

31.9 B. C. Garrett, D. G. Truhlar: *Theoretical Chemistry Advances and Perspectives* (Academic, New York 1981)

31.10 D. M. Bishop, L. M. Cheung: J. Chem. Phys. **78**, 1396 (1983)

31.11 D. M. Bishop, L. M. Cheung: J. Chem. Phys. **80**, 4341 (1984)

31.12 J. O. Jensen, D. R. Yarkony: J. Chem. Phys. **89**, 3853 (1988)

31.13 C. R. Vidal, W. C. Stwalley: J. Chem. Phys. **77**, 883 (1982)

31.14 Y. C. Chen, D. R. Harding, W. C. Stwalley, C. R. Vidal: J. Chem. Phys. **85**, 2436 (1986)

31.15 J. N. Murrell, S. Carter, S. C. Farantos, P. Huxley, A. J. C. Varandas: *Molecular Potential Energy Surfaces* (Wiley, New York 1984)

31.16 D. G. Truhlar, R. Steckler, M. S. Gordon: Chem. Rev. **87**, 217 (1987)

31.17 K. Fukui: Acc. Chem. Res. **14**, 363 (1981)

31.18 C. J. Cerjan, W. H. Miller: J. Chem. Phys. **75**, 2800 (1981)

31.19 H. B. Schlegel: Adv. Chem. Phys. **67**, 249 (1987)

31.20 H. B. Schlegel: In: *Modern Electronic Structure Theory, Part 1*, ed. by D. R. Yarkony (World Scientific, Singapore 1995)

31.21 H. F. Schaefer (Ed.): *Modern Theoretical Chemistry* (Plenum, New York 1976)

31.22 D. R. Yarkony (Ed.): *Modern Electronic Structure Theory* (World Scientific, Singapore 1995)

31.23 A. D. McLachlan: Mol. Phys. **4**, 417 (1961)

31.24 F. T. Smith: Phys. Rev. **179**, 111 (1969)

31.25 W. Lichten: Phys. Rev. **131**, 339 (1963)

31.26 W. Lichten: Phys. Rev. **164**, 131 (1967)

31.27 C. A. Mead, D. G. Truhlar: J. Chem. Phys. **77**, 6090 (1982)

31.28 M. Baer: Chem. Phys. Lett. **35**, 112 (1975)

31.29 M. Baer: Chem. Phys. **15**, 49 (1976)

31.30 T. Pacher, C. A. Mead, L. S. Cederbaum, H. Koppel: J. Chem.Phys. **91**, 7057 (1989)
31.31 V. Sidis: In: *State-Selected and State-to-State Ion-Molecule Reaction Dynamics Part 2, Theory*, Vol. 82, ed. by M. Baer, C. Y. Ng (Wiley, New York 1992) p. 73
31.32 R. Renner: Z. Phys **92**, 172 (1934)
31.33 J. von Neumann, E. Wigner: Z.. Phys **30**, 467 (1929)
31.34 C. A. Mead: J. Chem. Phys. **70**, 2276 (1979)
31.35 A. Szabo, N. S. Ostlund: *Modern Quantum Chemistry: Introduction to Advanced Electronic Structure Theory* (McMillan, New York 1982)
31.36 P. Jørgensen, J. Simons: *Second Quantization Based Methods in Quantum Chemistry* (Academic, New York 1981)
31.37 J. Almlöf: In: *Modern Electronic Structure Theory, Part 1*, ed. by D. R. Yarkony (World Scientific, Singapore 1995) p. 1
31.38 T. H. Dunning, P. J. Hay: In: *Modern Theoretical Chemistry*, Vol. 3, ed. by H. F. Schaefer (Plenum, New York 1976)
31.39 B. O. Roos, P. Linse, P. E. M. Siegbahn, M. R. A. Blomberg: Chem. Phys **66**, 197 (1982)
31.40 M. Moshinsky: *Group Theory and the Many Body Problem* (Gordon and Breach, New York 1968)
31.41 J. Hinze: *The Unitary Group*, Lecture Notes in Chem. (Springer, Berlin, Heidelberg 1979) p. 22
31.42 I. Shavitt: In: *Modern Theoretical Chemistry*, Vol. 3, ed. by H./,F. Schaefer (Plenum, New York 1976) p. 189
31.43 Y. Yamaguchi, Y. Osamura, J. D. Goddard, H. F. Schaefer: *A New Dimension to Quantum Chemistry: Analytic Derivative Methods in ab initio Molecular Electronic Structure Theory* (Oxford Univ. Press, Oxford 1994)
31.44 B. H. Lengsfield, D. R. Yarkony: In: *State-Selected and State to State Ion-Molecule Reaction Dynamics: Part 2, Theory*, Vol. 82, ed. by M. Baer, C. Ng (Wiley, New York 1992)
31.45 P. Pulay: Mol. Phys. **17**, 197 (1969)
31.46 C. C. J. Roothaan: Rev. Mod. Phys. **23**, 69 (1951)
31.47 J. F. Gaw, Y. Yamaguchi, H. F. Schaefer: J. Chem. Phys. **81**, 6395 (1984)
31.48 J. F. Gaw, Y. Yamaguchi, H. F. Schaefer, N. C. Handy: J. Chem.Phys. **85**, 5132 (1986)
31.49 J. Almlöf: J. Comput. Chem. **3**, 385 (1982)
31.50 K. Docken, J. Hinze: J. Chem. Phys. **57**, 4928 (1972)
31.51 J. Hinze: J. Chem. Phys. **59**, 6424 (1973)
31.52 H. Werner, W. Meyer: J. Chem. Phys. **74**, 5794 (1981)
31.53 B. H. Lengsfield: J. Chem. Phys. **77**, 4073 (1982)
31.54 C. Møller, M. S. Plesset: Phys. Rev. **46**, 618 (1934)
31.55 J. Almlöf: Chem. Phys. Lett. **181**, 319 (1991)
31.56 J. Cizek: J. Chem. Phys. **45**, 4256 (1966)
31.57 J. Cizek: Adv. Chem. Phys. **14**, 35 (1969)
31.58 J. Paldus, J. Cizek: In: *Energy Structure and Reactivity*, ed. by D. Smith, W. B. McRae (Wiley, New York 1973) p. 35
31.59 R. J. Bartlett, G. D. Purvis III: Int. J. Quantum Chem. Symp. **14**, 561 (1978)
31.60 J. A. Pople, R. Krishnan, H. B. Schlegel, J. S. Binkley: Int. J. Quantum Chem. **14**, 545 (1978)
31.61 P. O. Löwdin: Phys. Rev. **97**, 1474 (1955)
31.62 R. J. Bartlett, J. D. Watts, S. A. Kucharski, J. Noga: Chem. Phys. Lett. **165**, 513 (1990)
31.63 S. A. Kucharski, R. J. Bartlett: Chem. Phys. Lett. **158**, 550 (1989)
31.64 K. Raghavachari, G. W. Trucks, J. A. Pople, M. Head-Gordon: Chem. Phys. Lett. **157**, 479 (1989)
31.65 J. D. Watts, J. Gauss, R. J. Bartlett: J. Chem. Phys. **98**, 8718 (1993)
31.66 R. J. Bartlett: In: *Modern Electronic Structure Theory*, ed. by D. R. Yarkony (World Scientific, Singapore 1995) p. 2
31.67 R. N. Diffenderfer, D. R. Yarkony: J. Chem. Phys. **86**, 5098 (1982)
31.68 B. O. Roos, P. R. Taylor, P. E. M. Siegbahn: Chem. Phys. **48**, 157 (1980)
31.69 B. O. Roos: Int. J. Quantum Chem. Symp. **14**, 175 (1980)
31.70 P. Siegbahn, A. Heiberg, B. Roos, B. Levy: Phys. Scr. **21**, 323 (1980)
31.71 C. W. Bauschlicher, S. R. Langhoff, P. R. Taylor: Adv. Chem. Phys. **77**, 103 (1990)
31.72 H. J. Silverstone, O. Sinanoglu: J. Chem. Phys. **44**, 1899 (1966)
31.73 E. Dalgaard, P. Jørgensen: J. Chem. Phys. **69**, 3833 (1978)
31.74 E. R. Davidson: J. Comput. Phys. **17**, 87 (1975)
31.75 B. O. Roos: Chem. Phys. Lett. **15**, 153 (1972)
31.76 P. E. M. Siegbahn: J. Chem. Phys. **72**, 1647 (1980)
31.77 R. J. Harrison: J. Chem. Phys. **94**, 5021 (1991)
31.78 R. Caballol, J. P. Malrieu: Chem. Phys. Lett. **188**, 543 (1992)
31.79 A. Povill, J. Rubio, F. Illas: Theo. Chim. Acta **82**, 229 (1992)
31.80 P. J. Knowles, H. Werner: Chem. Phys. Lett. **115**, 259 (1985)
31.81 P. J. Knowles, H. Werner: Chem. Phys. Lett. **145**, 514 (1988)
31.82 H. Werner, P. J. Knowles: J. Chem. Phys. **82**, 5053 (1985)
31.83 H. Werner, P. J. Knowles: J. Chem. Phys. **89**, 5803 (1988)
31.84 K. Andersson, P. Å. Malmqvist, B. O. Roos, A. J. Sadlej, K. Wolinski: J. Phys. Chem. **94**, 5483 (1990)
31.85 K. Andersson, P. Å. Malmqvist, B. O. Roos: J. Chem. Phys. **96**, 1218 (1992)
31.86 R. G. Parr, W. Yang: *Density-Functional Theory of Atoms and Molecules* (Oxford, New York 1989)
31.87 P. Hohenberg, W. Kohn: Phys. Rev. B **136**, 864 (1964)
31.88 W. Kohn, L. J. Sham: Phys. Rev. A **140**, 1133 (1965)
31.89 A. D. Becke: In: *Modern Electronic Structure Theory, Part 2*, ed. by D. R. Yarkony (World Scientific, Singapore 1995)

31.90 P. Pula: In: *Modern Electronic Structure Theory, Part 2*, ed. by D. R. Yarkony (World Scientific, Singapore 1995)
31.91 B. Liu, A. D. McLean: J. Chem. Phys. **59**, 4557 (1973)
31.92 B. Liu, A. D. McLean: J. Chem. Phys. **91**, 2348 (1989)
31.93 S. F. Boys, F. Bernardi: Mol. Phys. **19**, 553 (1970)
31.94 M. Gutowski, J. van Lenthe, J. C. G. M. van Duijneveldt, F. B. van Duijneveldt: J. Chem. Phys. **98**, 4728 (1993)
31.95 H. Lefebvre-Brion, R. W. Field: *Perturbations in the Spectra of Diatomic Molecules* (Academic, New York 1986)
31.96 S. Han, H. Hettema, D. R. Yarkony: J. Chem. Phys. **102**, 1955 (1995)
31.97 B. H. Lengsfield, P. Saxe, D. R. Yarkony: J. Chem. Phys. **81**, 4549 (1984)
31.98 P. Saxe, B. H. Lengsfield, D. R. Yarkony: Chem. Phys. Lett. **113**, 159 (1985)
31.99 R. J. Buenker, G. Hirsch, S. D. Peyerimhoff, P. J. Bruna, J. Romelt, M. Bettendorff, C. Petrongolo: In: *Current Aspects of Quantum Chemistry 1981*, Vol. 21, ed. by R. Carbo (Elsevier, New York 1982) p. 81
31.100 C. A. Mead: J. Chem. Phys. **78**, 807 (1993)
31.101 J. Hellmann: *Einführung in die Quantenchemie* (Deuticke, Leipzig 1937)
31.102 R. P. Feynman: Phys. Rev. **56**, 340 (1939)
31.103 S. R. Langhoff, C. W. Kern: In: *Modern Theoretical Chemistry*, Vol. 4, ed. by H. F. Schaefer (Plenum, New York 1977) p. 381
31.104 M. Blume, R. E. Watson: Proc. R. Soc. London A **270**, 127 (1962)
31.105 M. Blume, R. E. Watson: Proc. R. Soc. London A **271**, 565 (1963)
31.106 S. Fraga, K. M. S. Saxena, B. W. N. Lo: At. Data **3**, 323 (1971)
31.107 W. C. Ermler, R. B. Ross, P. A. Christiansen: *Advances in Quantum Chemistry*, Vol. 19 (Academic, New York 1988) p. 139
31.108 W. C. Ermler, Y. S. Lee, P. A. Christiansen, K. S. Pitzer: Chem. Phys. Lett. **81**, 70 (1981)
31.109 Y. S. Lee, W. C. Erlmer, K. S. Pitzer: J. Chem. Phys. **67**, 5861 (1977)
31.110 R. M. Pitzer, N. W. Winter: J. Chem. Phys. **92**, 3061 (1988)
31.111 W. J. Stevens, M. Krauss: J. Chem. Phys. **76**, 3834 (1982)
31.112 J. P. Desclaux: Comp. Phys. Comm. **9**, 31 (1975)
31.113 K. Balasubramanian: J. Phys. Chem. **93**, 6585 (1989)
31.114 B. A. Hess: Phys. Rev. A **33**, 3742 (1986)
31.115 B. A. Hess, P. Chandra: Phys. Scr. **36**, 412 (1987)
31.116 J. Sucher: Phys. Rev. A **22**, 348 (1980)
31.117 C. M. Marian: J. Chem. Phys. **93**, 1176 (1990)
31.118 C. M. Marian: J. Chem. Phys. **94**, 5574 (1991)
31.119 T. R. Furlani, H. F. King: J. Chem. Phys. **82**, 5577 (1985)
31.120 B. A. Hess, R. J. Buenker, C. M. Marian, S. D. Peyerimhoff: Chem. Phys. **71**, 79 (1982)
31.121 D. R. Yarkony: J. Chem. Phys. **84**, 2075 (1986)
31.122 J. O. Jensen, D. R. Yarkony: Chem. Phys. Lett. **141**, 391 (1987)
31.123 B. Liu, M. Yoshimine: J. Chem. Phys. **74**, 612 (1981)
31.124 D. R. Yarkony: Int. Rev. Phys. Chem. **11**, 195 (1992)
31.125 M. R. Manaa, D. R. Yarkony: J. Chem. Phys. **99**, 5251 (1993)
31.126 D. R. Yarkony: J. Chem. Phys. **97**, 4407 (1993)
31.127 D. R. Yarkony: J. Chem. Phys. **100**, 3639 (1994)
31.128 R. Fletcher: *Practical Methods of Optimization*, Vol. 2 (Wiley, New York Vol 1981)
31.129 I. N. Radazos, M. A. Robb, M. A. Bernardi, M. Olivucci: Chem. Phys. Lett. **197**, 217 (1992)
31.130 G. Herzberg, H. C. Longuet-Higgins: Disc. Faraday Soc. **35**, 77 (1963)
31.131 M. V. Berry: Proc. R. Soc. London A **392**, 45 (1984)
31.132 S. Xantheas, S. T. Elbert, K. Ruedenberg: J. Chem. Phys. **93**, 7519 (1990)
31.133 J. Tennyson, S. Miller, B. T. Sutcliffe: J. Chem. Soc. Faraday Trans. **84**, 1295 (1988)
31.134 J. H. Van Vleck: Rev. Mod. Phys. **23**, 213 (1951)
31.135 M. Mizushima: *The Theory of Rotating Diatomic Molecules* (Wiley, New York 1975)
31.136 T. J. Lee, D. J. Fox, H. F. Schaefer, R. M. Pitzer: J. Chem.Phys. **81**, 356 (1984)
31.137 R. N. Dixon: Mol. Phys. **54**, 333 (1985)
31.138 R. J. Whitehead, N. C. Handy: J. Mol. Spec. **55**, 356 (1975)
31.139 M. E. Rose: *Elementary Theory of Angular Momentum* (Wiley, New York 1957)
31.140 I. M. Mills: In: *Molecular Spectroscopy: Modern Research*, ed. by K. N. Rao, C. W. Mathews (Academic, New York 1970) p. 115
31.141 C. Eckart: Phys. Rev. **47**, 552 (1935)
31.142 E. B. Wilson, Jr., J. B. Howard: J. Chem. Phys. **4**, 260 (1936)
31.143 B. T. Darling, D. M. Dennison: Phys. Rev. **57**, 128 (1940)
31.144 E. B. Wilson, J. C. Decius, P. C. Cross: *Molecular Vibrations* (McGraw-Hill, New York 1955)
31.145 J. K. G. Watson: Mol. Phys. **15**, 479 (1968)
31.146 R. N. Zare: *Angular Momentum* (Wiley, New York 1988)
31.147 B. T. Sutcliffe, J. Tennyson: Mol. Phys. **58**, 1053 (1986)
31.148 J. Segall, R. N. Zare, H. R. Dubal, M. Lewerenz, M. Quack: J. Chem. Phys. **86**, 634 (1987)
31.149 W. H. Green, W. D. Lawrence, C. B. Moore: J. Chem. Phys. **86**, 6000 (1987)
31.150 R. D. Amos, J. G. Gaw, N. C. Handy, S. Carter: J. Chem. Soc. Faraday Trans. **84**, 1247 (1988)
31.151 P. Botschwina: Chem. Phys. **68**, 41 (1982)
31.152 P. Botschwina: J. Chem. Soc. Faraday Trans. **84**, 1263 (1988)

31.153 J. M. Hutson: In: *Advances in Molecular Vibrations and Collision Dynamics*, Vol. 1A, ed. by J. M. Bowman, M. A. Ratner (JAI Press, Greenwich 1991)
31.154 M. L. Dubernet, D. Flower, J. M. Hutson: J. Chem. Phys. **94**, 7602 (1991)
31.155 M. H. Alexander, S. Gregurick, P. J. Dagdigian: J. Chem. Phys. **101**, 2887 (1994)
31.156 D. M. Brink, G. R. Satchler: *Angular Momentum* (Clarendon, Oxford 1968)
31.157 A. Arthurs, A. Dalgarno: Proc. Roy. Soc. London A **256**, 540 (1960)
31.158 M. Alexander: Chem. Phys. **92**, 337 (1985)
31.159 H. Werner, B. Follmeg, M. H. Alexander: J. Chem. Phys. **89**, 3139 (1988)
31.160 H. Werner, B. Follmeg, M. H. Alexander, D. Lemoine: J. Chem. Phys. **91**, 5425 (1989)
31.161 C. Chakravarty, D. Clary: J. Chem. Phys. **94**, 4149 (1991)
31.162 M. H. Alexander, S. Gregurick, P. J. Dagdigian, G. W. Lemire, M. J. McQuaid, R. C. Sausa: J. Chem. Phys. **101**, 4547 (1994)
31.163 W. H. Miller: In: *The Theory of Chemical Reaction Dynamics*, ed. by D. C. Clary (Reidel, Dordrecht 1986)
31.164 W. H. Miller, N. Handy, J. Adams: J. Chem. Phys. **72**, 99 (1980)
31.165 M. Page, J. W. McIver: J. Chem. Phys. **88**, 922 (1988)
31.166 S. Kato, R. L. Jaffe, A. Komornicki, K. Morokuma: J. Chem. Phys. **78**, 4567 (1983)
31.167 S. P. Walch: Chem. Phys. Lett. **208**, 214 (1993)
31.168 D. F. Strobel: Planet. Space Sci. **30**, 839 (1982)
31.169 M. R. Berman, M. C. Lin: J. Phys. Chem **87**, 3933 (1983)
31.170 C. P. Fenimore: *13th International Symposium on Combustion* (The Combustion Institute, Pittsburgh 1971) p. 373
31.171 J. Blauwens, B. Smets, J. Peeters: *16th International Symposium on Combustion* (The Combustion Institute, Pittsburgh 1977) p. 1055
31.172 A. J. Dean, R. K. Hanson, C. T. Bowman: *23rd International Symposium on Combustion* (The Combustion Institute, Pittsburgh 1990) p. 259
31.173 J. C. Tully: J. Chem. Phys. **61**, 61 (1974)
31.174 G. E. Zahr, R. K. Preston, W. H. Miller: J. Chem. Phys **62**, 1127 (1975)
31.175 M. R. Manaa, D. R. Yarkony: J. Chem. Phys. **95**, 1808 (1991)
31.176 M. R. Manaa, D. R. Yarkony: Chem. Phys. Lett. **188**, 352 (1992)
31.177 J. M. L. Martin, P. R. Taylor: Chem. Phys. Lett. **209**, 143 (1993)
31.178 T. Seideman, S. P. Walch: J. Chem. Phys. **101**, 3656 (1994)
31.179 M. Tinkham: *Group Theory and Quantum Mechanics* (McGraw-Hill, New York 1964)
31.180 T. Seideman: J. Chem. Phys. **101**, 3662 (1994)
31.181 K. H. Becker, B. Engelhardt, H. Geiger, R. Kurtenbach, G. Schrey, P. Wissen: Chem. Phys. Lett. **154**, 342 (1992)
31.182 D. Lindackers, M. Burmeister, P. Roth: *23rd International Symposium on Combustion* (Combustion Institute, Pittsburgh 1990) p. 251
31.183 D. Lindackers, M. Burmeister, P. Roth: Combustion and Flame **81**, 251 (1990)
31.184 X. Hu, K. Schulten: Phys. Tod. **50**, 28 (1997)
31.185 W. Weber, V. Helms, A. McMCammon, P. W. Langhoff: Proc. Natl. Acad. Sci. U.S.A. **96**, 6177 (1999)
31.186 D. R. Yarkony: J. Phys. Chem. A **105**, 6277 (2001)
31.187 D. R. Yarkony: J. Chem. Phys. **121**, 628 (2004)
31.188 D. R. Yarkony: J. Phys. Chem. A **108**, 3200 (2004)
31.189 S. Matsika, D. R. Yarkony: J. Chem. Phys. **117**, 6907 (2002)
31.190 S. Matsika, D. R. Yarkony: J. Phys. Chem. B **106**, 8108 (2002)
31.191 S. Matsika, D. R. Yarkony: J. Chem. Phys. **115**, 5066 (2001)

32. Molecular Symmetry and Dynamics

Molecules are aggregates of two or more nuclei bound by at least one electron. The nuclei of most stable molecules can be imagined to be points in a more or less rigid body whose relative positions are constrained by an electronic bonding potential. This potential depends strongly upon the electronic state as described in Chapt. 31. Most of this discussion is about stable molecules in their electronic ground state. In Sect. 32.6 some comments are made about molecules with excited, or "loose", parts.

32.1 **Dynamics and Spectra of Molecular Rotors** 491
32.1.1 Rigid Rotors 492
32.1.2 Molecular States Inside and Out .. 492
32.1.3 Rigid Asymmetric Rotor Eigensolutions and Dynamics 493

32.2 **Rotational Energy Surfaces and Semiclassical Rotational Dynamics** .. 494

32.3 **Symmetry of Molecular Rotors** 498
32.3.1 Asymmetric Rotor Symmetry Analysis 498

32.4 **Tetrahedral-Octahedral Rotational Dynamics and Spectra** 499
32.4.1 Semirigid Octahedral Rotors and Centrifugal Tensor Hamiltonians 499
32.4.2 Octahedral and Tetrahedral Rotational Energy Surfaces 500
32.4.3 Octahedral and Tetrahedral Rotational Fine Structure 500
32.4.4 Octahedral Superfine Structure ... 502

32.5 **High Resolution Rovibrational Structure** 503
32.5.1 Tetrahedral Nuclear Hyperfine Structure 505
32.5.2 Superhyperfine Structure and Spontaneous Symmetry Breaking 505
32.5.3 Extreme Molecular Symmetry Effects 506

32.6 **Composite Rotors and Multiple RES** 507
32.6.1 3D-Rotor and 2D-Oscillator Analogy 509
32.6.2 Gyro-Rotors and 2D-Local Mode Analogy 510
32.6.3 Multiple Gyro-Rotor RES and Eigensurfaces 511

References 512

32.1 Dynamics and Spectra of Molecular Rotors

Motions that stretch or compress the bonds are called vibrational motions, and give rise to spectral resonances in the infrared region of the spectrum. Typical fundamental vibrational quanta (ν_0) lie between $80\,\mathrm{cm}^{-1}$ (the lowest $GeBr_4$ mode) and $3020\,\mathrm{cm}^{-1}$ (the highest CH_4 mode). (A $1000\,\mathrm{cm}^{-1}$ wave has a wavelength of $10\,\mu\mathrm{m}$ and a frequency defined by the speed of light: 29.979 2458 THz.) Vibrational amplitudes are usually tiny since zero-point motions or vibrations involving one or two quanta ($\nu = 0, 1, 2, \dots$) are constrained by the steep bonding potential to less than a few percent of the bond lengths, but high overtones may lead to dissociation, i. e., molecular breakup.

Overall rotation of molecules in free space is unconstrained, and gives rise to far-infrared or microwave pure rotational transitions or sidebands on top of vibrational spectra. Typical rotational quanta ($2B$) lie between $0.18\,\mathrm{cm}^{-1}$ (5.4 GHz) for SF_6 and $10.6\,\mathrm{cm}^{-1}$ for CH_4. Individual molecules are free to rotate or translate as a whole while undergoing tiny but usually rapid vibrations. Vibrating molecules may be thought of as tumbling collections of masses held together by 'springs' (the electronic vibrational potential or force field), and are called semirigid rotors. The coupling of rotational and vibrational motion is called rovibrational coupling and includes centrifugal and Coriolis coupling, which will be introduced in Sect. 32.6.

This discussion of molecular dynamics and spectra mainly involves molecular rotation and properties of rotationally excited molecules, particularly those with high rotational quantum number $J = 10$–200. However, the discussion also applies to molecules in excited vibrational states, and even certain cases of molecules in excited electronic states. The analysis of vibronic (vibrational-electronic), rovibrational (rotation–vibration), or rovibronic (all three) types of excitation can be very complicated [32.1–5] and is beyond the scope of this article, but these problems can all benefit from the elementary considerations described here.

32.1.1 Rigid Rotors

As a first approximation, and for the purposes of discussing basic molecular dynamics and spectra, one may ignore vibrations and model stable molecules as 'stick-and-ball' structures or rigid rotors. Then the Hamiltonian has just three terms:

$$H = AJ_x^2 + BJ_y^2 + CJ_z^2 \,. \tag{32.1}$$

Here $\{J_x, J_y, J_z\}$ are rotational angular momentum operators, and the rotational constants are half the inverses of the principal moments of inertia I_α of the body:

$$A = \frac{1}{2I_x}\,, \quad B = \frac{1}{2I_y}\,, \quad C = \frac{1}{2I_z}\,. \tag{32.2}$$

This implies that the J-coordinate system being used is a special one fixed to the rotor's body and aligned to its principal axes, an elementary body, or Eckart, frame.

Many molecules, particularly all diatomic molecules, have two of these rotational constants equal, say, $A = B$. Such rotors are called symmetric tops, and their Hamiltonian can be written in terms of the square of the total angular momentum $\boldsymbol{J}\cdot\boldsymbol{J}$ and one other body component J_z as

$$\begin{aligned} H &= BJ_x^2 + BJ_y^2 + CJ_z^2 \\ &= BJ_x^2 + BJ_y^2 + BJ_z^2 + (C-B)J_z^2 \\ &= B\boldsymbol{J}\cdot\boldsymbol{J} + (C-B)J_z^2 \end{aligned} \tag{32.3}$$

This gives a simple formula for the symmetric top rotational energy levels in terms of the quantum numbers J for the total angular momentum and K for the body z-component:

$$E(J,K) = BJ(J+1) + (C-B)K^2 \tag{32.4}$$

However, this eigenvalue formula may be a little too simple, since it hides the structure of the eigenstates or eigenfunctions. Indeed, the full Schrödinger angular differential equation based upon the Hamiltonian (32.1) is more lengthy. One should remember that H is written in a rotating body coordinate frame that must be connected to a star-fixed, or laboratory, frame in order to get the full theory.

32.1.2 Molecular States Inside and Out

Rotor angular momentum eigenfunctions can be expressed as continuous linear combinations of rotor angular position states $|\alpha\beta\gamma\rangle$ defined by Euler angles of the lab azimuth α, the polar angle β of body z-axis, and the body azimuth, or 'gauge twist', γ. The eigenfunctions are,

$$\left|\begin{matrix} J \\ MK \end{matrix}\right\rangle = \frac{\sqrt{2J+1}}{8\pi^2} \int_0^{2\pi} \mathrm{d}\alpha \int_0^{\pi} \sin\beta\, \mathrm{d}\beta \times \int_0^{2\pi} \mathrm{d}\gamma\, D^{J}_{MK}{}^{*}(\alpha\beta\gamma)|\alpha\beta\gamma\rangle \,, \tag{32.5}$$

where the rotor wave functions $D^{J}_{MK}{}^{*}$ are just the conjugates of the Wigner rotation matrices described in Sect. 32.3.1, and row and column indices M and K, respectively, are the lab and body components of the angular momentum [32.5–7].

An important feature of polyatomic molecules is that their angular momentum states have two kinds of azimuthal quantum numbers. In addition to the usual lab component of momentum M associated with the lab coordinate α (α and β are usually labeled φ and ϑ), there is a body component K associated with the Euler coordinate γ, the body azimuthal angle of the laboratory Z-axis relative to the body z-axis.

The physics of atomic or diatomic angular momentum states has no internal or "body" structure, so the quantum number K is always zero. Unless one sets $K = 0$, the energy formula (32.4) blows up for a point particle because z-inertia for a point is zero and C is infinite. Also, the dimension of the angular momentum state multiplet of a given J is larger than the usual $(2J+1)$ found in atomic or diatomic molecular physics. In polyatomic rotors, the number of states for each J is $(2J+1)^2$, since both quantum numbers M and K range between -J and $+J$.

A further important feature is that the molecular rotor wave functions contain, as a special ($K = 0$) case, all the usual atomic spherical harmonics Y_m^ℓ complete with correct normalization and phase, since

$$\sqrt{4\pi}Y_m^\ell(\varphi\vartheta) = D_{m0}^\ell(\varphi\vartheta\cdot)^*\sqrt{2\ell+1}\,. \qquad (32.6)$$

This is part of a powerful symmetry principle: group representations are quantum wave functions, and symmetry analysis is an extension of Fourier analysis – not just for translations as in Fourier's original work, but for any group of symmetry operations. The usual Fourier coefficients e^{ikx} are replaced by the D functions in the rotational Fourier transform embodied by (32.5).

Molecular rotational analysis displays another important but little known aspect of symmetry analysis in general. For every group of symmetry operations, such as the external lab-based rotations familiar to atomic physics, there is an independent dual group of internal or body-based operations. The external symmetry of the environment or laboratory is independent of the internal symmetry of the molecular body, and all the operations of one commute with all those of the other. The molecular rotation group is thus written as an outer product $R(3)_{\mathrm{LAB}}\otimes R(3)_{\mathrm{BODY}}$ of the external and internal parts, and the degeneracy associated with this group's representations for a single J is $(2J+1)^2$ as mentioned above. It is a special $\otimes$-product, however, since the J-number is shared.

The inversion or parity operator $\boldsymbol{I}(\boldsymbol{r}\to-\boldsymbol{r})$ can be defined to be the same for both lab and body frames. Including $\boldsymbol{I}$ with the rotational group $R(3)$ gives the orthogonal group $O(3) = R(3)\otimes\{\boldsymbol{1},\boldsymbol{I}\}$. If parity is conserved (e.g., no weak neutral currents), the fundamental molecular orthogonal group is $O(3)_{\mathrm{LAB}}\otimes O(3)_{\mathrm{BODY}}$.

How this symmetry breaks down and which levels split depends upon both the perturbative laboratory environment and the internal molecular structure. A spherical top Hamiltonian is (32.1) with $A = B = C$. This has a full $O(3)_{\mathrm{BODY}}$ (spherical) symmetry since it is just $B\boldsymbol{J}\cdot\boldsymbol{J}$. Given that the rotor is in an $O(3)_{\mathrm{LAB}}$ laboratory (empty space), the original symmetry $O(3)_{\mathrm{LAB}}\otimes O(3)_{\mathrm{BODY}}$ remains intact and the $(2J+1)^2$ degeneracy is to be expected. However, a symmetric rotor in a lab vacuum has its internal symmetry broken down to $O(2)_{\mathrm{BODY}}$ if $A = B \neq C$, and the energies given by (32.4) consist of internal quantum singlets for $K = 0$ and $\pm K$ doublets for $K \neq 0$. But each of these levels still has a lab degeneracy of $(2J+1)$ if $O(3)_{\mathrm{LAB}}$ is still in effect. So the $(2J+1)^2$ level degeneracies are each split into multiplets of degeneracy $(2J+1)$ and $2(2J+1)$ for $K = 0$ and $K \neq 0$, respectively. The resulting levels are often labeled Σ, Π, Δ, Φ, Γ, ... in a somewhat inappropriate analogy with the atomic s, p, d, f, g, ... labels of Bohr model electronic orbitals.

Only by perturbing the lab environment can one reduce the $O(3)_{\mathrm{LAB}}$ symmetry and split the M degeneracies. For example, a uniform electric field would reduce the $O(3)_{\mathrm{LAB}}$ to an $O(2)_{\mathrm{LAB}}$, giving Stark splittings which consist of external quantum singlets for $M = 0$ and $\pm M$ doublets for $M \neq 0$. A uniform magnetic field would reduce the $O(3)_{\mathrm{LAB}}$ to an $R(2)_{\mathrm{LAB}}$, giving Zeeman splittings into external quantum singlets for each M. The analogy between atomic external field splitting and internal molecular rotational structure splitting is sometimes a useful one and will be used later.

32.1.3 Rigid Asymmetric Rotor Eigensolutions and Dynamics

The general case for the rigid rotor Hamiltonian (32.1) has three unequal principal moments of inertia ($A \neq B \neq C$). This is called the rigid asymmetric top Hamiltonian, and provides a first approximation for modeling rotation of low symmetry molecules, such as H_20. Also, a number of properties of its eigensolutions are shared by more complicated systems. The dynamics of an asymmetric top is quite remarkable, as demonstrated by tossing a tennis racquet in the air, flat side up. The corresponding quantum behavior of such a molecule is also nontrivial.

Given the total angular momentum J, one may construct a $(2J+1)$-dimensional matrix representation of H using standard matrix elements of the angular momentum operators J_x, J_y, and J_z, as given in Chapt. 2. The H matrix connects states with $(2J+1)$-different body quantum numbers $K(-J \leq K \leq J)$, but the matrix is independent of the lab quantum numbers M, so there are $(2J+1)$ identical H matrices; one for each value of the lab quantum number $M(-J \leq K \leq J)$.

A plot of the 21 eigenvalues of (32.1) for $J = 10$ is shown in Fig. 32.1. Here, the constants are set to $A = 0.2\,\mathrm{cm}^{-1}$ and $C = 0.6\,\mathrm{cm}^{-1}$, while B is varied between $B = A$, which corresponds to a prolate symmetric top (an elongated cylindrical object) and $B = C$, which corresponds to an oblate symmetric top (a flattened cylindrical object or discus). For all B values between those of A and C, the object is asymmetric.

The left hand end $\left(A = B = 0.2\,\mathrm{cm}^{-1},\ C = 0.6\,\mathrm{cm}^{-1}\right)$ of the plot in Fig. 32.1 corresponds to a prolate symmetric top. The symmetric top level spectrum is given by (32.4). It consists of a lowest singlet state corresponding to $K = 0$ and an ascending quadratic ladder of doublets corresponding to $K = \pm1, \pm2, \dots, \pm J$.

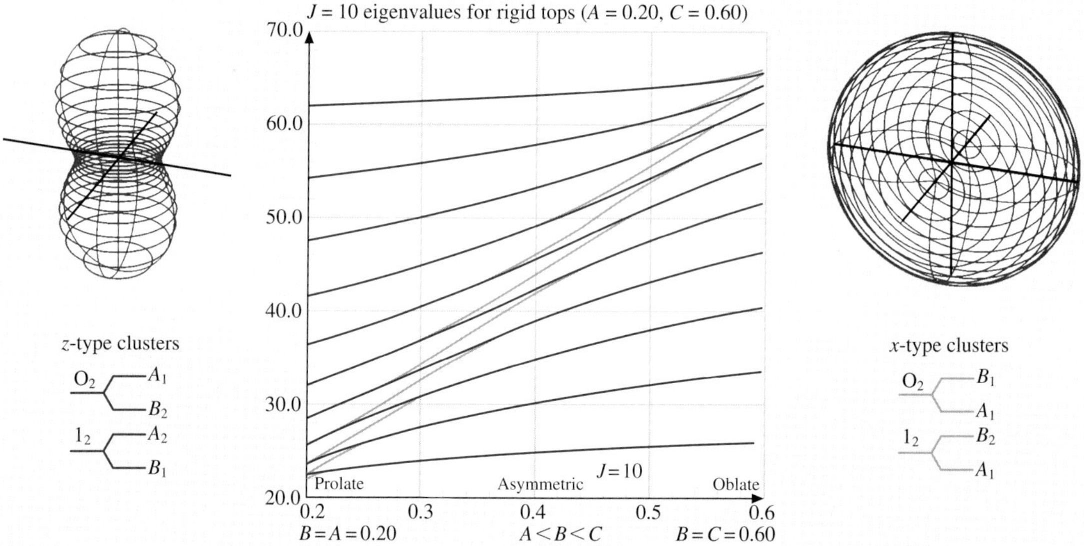

Fig. 32.1 $J = 10$ eigenvalue plot for symmetric rigid rotors. ($A = 0.2$, $C = 0.6\,\mathrm{cm}^{-1}$ $A < B < C$). Prolate and oblate RE surfaces are shown

The right hand end $\left(A = 0.2\,\mathrm{cm}^{-1}, B = C = 0.6\,\mathrm{cm}^{-1}\right)$ of the plot corresponds to an oblate symmetric top with a descending quadratic ladder of levels, the $K = 0$ level being highest. Also, the internal K-axis of quantization switches from the body z-axis for $\left(A = B = 0.2\,\mathrm{cm}^{-1}, C = 0.6\,\mathrm{cm}^{-1}\right)$ to the body x-axis for $\left(A = 0.2\,\mathrm{cm}^{-1}, B = C = 0.6\,\mathrm{cm}^{-1}\right)$. Note that the lab M-degeneracy is invisible here, but exists nevertheless.

For intermediate values of B, one has an asymmetric top level structure and, strictly speaking, no single axis of quantization. As a result, the eigenlevel spectrum is quite different. A detailed display of asymmetric top levels for the case $\left(A = 0.2\,\mathrm{cm}^{-1}, B = 0.4\,\mathrm{cm}^{-1}, C = 0.6\,\mathrm{cm}^{-1}\right)$ is given at the bottom of Fig. 32.2. They are shown to correspond to semiclassical orbits discussed in Sect. 32.2. This example is the most asymmetric top, since parameter B has a value midway between the symmetric top limits of $B = A$ and $B = C$.

The twenty-one $J = 10$ asymmetric top levels are arranged into roughly ten asymmetry doublets and one singlet. This resembles the symmetric top levels except that doublets are split by varying amounts, and the singlet is isolated from the other levels in the middle of the band instead of being crowded at the top or bottom. The doublet splittings are magnified in circles drawn next to the levels, and these indicate that the splitting decreases quasi-exponentially with each doublet's separation from the central singlet.

An asymmetric doublet splitting is also called superfine structure and can be viewed as the result of a dynamic tunneling process in a semiclassical model of rotation [32.8–10]. Such a model clarifies the classical-quantum correspondence for polyatomic rovibrational dynamics in general. It can also help to derive simple approximations for eigenvalues and eigenvectors.

32.2 Rotational Energy Surfaces and Semiclassical Rotational Dynamics

A semiclassical model of molecular rotation can be based upon what is called a rotational energy surface (RES) [32.7–14]. Examples of RES for an asymmetric top are shown in Fig. 32.2, and for prolate and oblate symmetric tops in Fig. 32.1. Each surface is a radial plot of the classical energy derived from the Hamiltonian (32.1) as a function of the polar direction of the classical angular momentum $\boldsymbol{J}$-vector in the body frame. The magnitude of $\boldsymbol{J}$ is fixed for each surface. Note that the $\boldsymbol{J}$-vector in the lab frame is a classical constant of the motion if there are no external perturbations. However, $\boldsymbol{J}$ may gyrate considerably in the moving body frame,

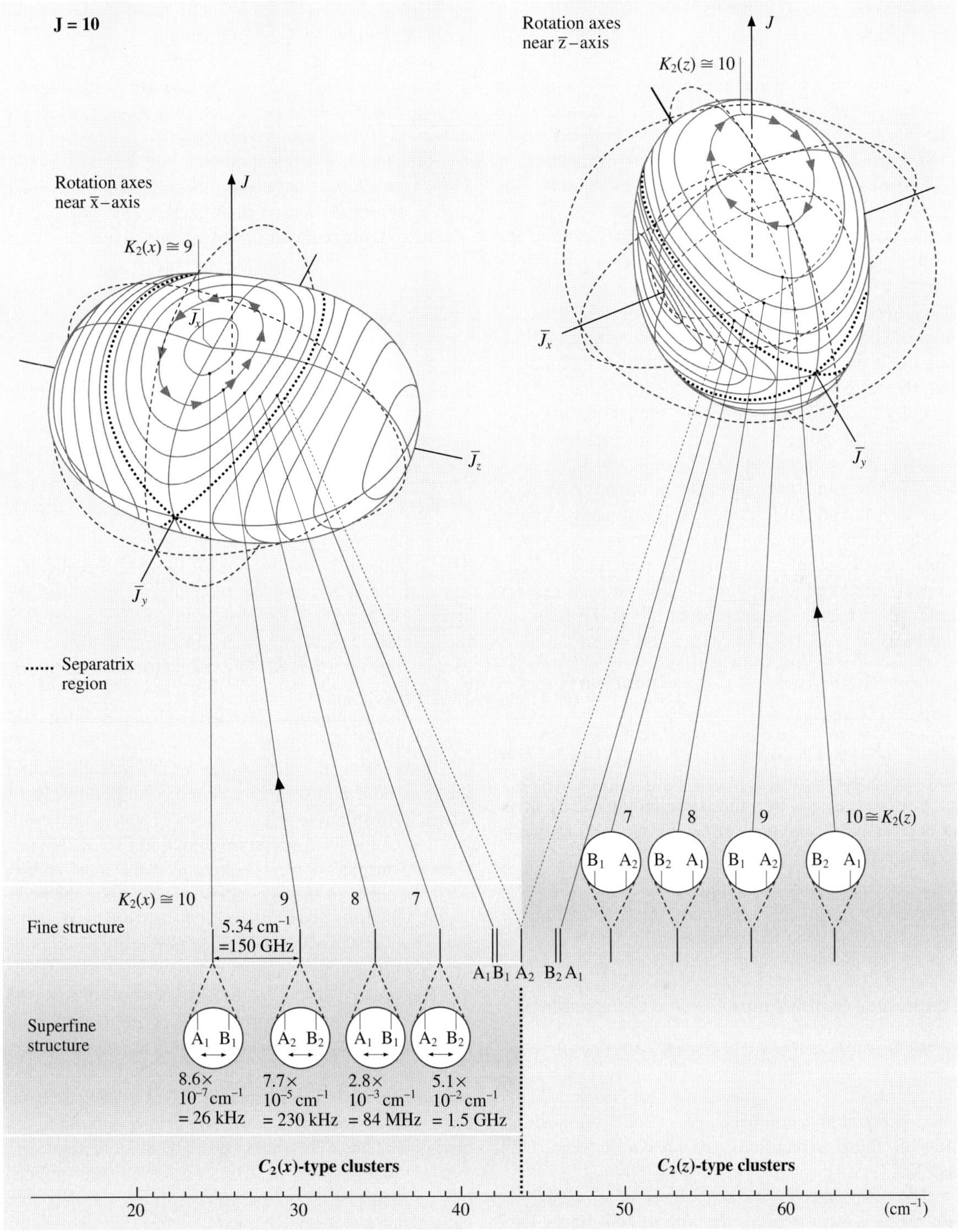

Fig. 32.2 $J = 10$ rotational energy surface and related level spectrum for an asymmetric rigid rotator $\left(A = 0.2,\ B = 0.4,\ C = 0.6\,\mathrm{cm}^{-1}\right)$

but its magnitude $|\boldsymbol{J}|$ stays the same in all frames for free rotation.

An RES differs from what is called a constant energy surface (CES), which is obtained by simply plotting $E = H = \text{const.}$ in $\boldsymbol{J}$-space using (32.1). A rigid rotor CES is an ellipsoid covering a range of $|\boldsymbol{J}|$ values at a single energy. An RE surface, on the other hand, is a spherical harmonic plot at a single $|\boldsymbol{J}|$ value for a range of energies. The latter is more appropriate for spectroscopic studies of fine structure, since one value of the rotational quantum number J corresponds to a multiplet of energy levels or transitions. An RES also shows loci of high and low energy rotations. Also, it has roughly the same shape as the body it represents, i. e., an RES is long in the direction that the corresponding molecule is long (but vice-versa for CES).

For a freely rotating molecule, the laboratory components of the classical total angular momentum $\boldsymbol{J}$ are constant. If one chooses to let $\boldsymbol{J}$ define the lab Z-axis, then the direction of the $\boldsymbol{J}$-vector in the body frame is given by polar and body azimuthal coordinates β and γ, which are the second and third Euler angles, respectively. (It is conventional to use the negatives $-\beta$ and $-\gamma$ as polar coordinates, but this will not be necessary here.) Then the body components of the $\boldsymbol{J}$-vector are written as

$$\left(J_x = |J| \sin\beta \cos\gamma\,, \quad J_y = |J| \sin\beta \sin\gamma\,, \quad J_z = |J| \cos\beta\right), \tag{32.7}$$

where the magnitude of the quantum value $|J| = \sqrt{J(J+1)} \cong J + \frac{1}{2}$.

Substituting this into the Hamiltonian (32.1) gives an expression for the general rigid rotor RES radius in polar coordinates:

$$E(\beta,\gamma) = \langle H \rangle = J(J+1) \times \left[\sin^2\beta \left(A\cos^2\gamma + B\sin^2\gamma\right) + C\cos^2\beta\right]. \tag{32.8}$$

The prolate symmetric top ($A = B < C$) expression

$$E(\beta) = \langle H \rangle = J(J+1)\left[B + (C - B)\cos^2\beta\right] \tag{32.9}$$

is independent of azimuthal angle γ. The 3-dimensional plots of these expressions are shown in Figs. 32.1 and 32.2.

The RES have topography lines of constant energy ($E = \text{const.}$) that are the intersection of an RES (constant $|\boldsymbol{J}|$) with spheres of constant energy. The topography lines are allowed classical paths of the angular momentum $\boldsymbol{J}$-vector in the body frame, since these paths conserve both energy and momentum.

The trajectories in these figures are special ones. They are the quantizing trajectories for total angular momentum $J = 10$. For the prolate symmetric top, the quantizing trajectories have integral values for the body z-component K of angular momentum. According to the Dirac vector model, angular momentum vectors trace out a cone of altitude K and slant height $|J| = \sqrt{J(J+1)}$. The quantizing polar angles Θ^J_k are given by

$$\Theta^J_K = \cos^{-1}\frac{K}{\sqrt{J(J+1)}} \quad (K = J, J-1, \cdots, -J)\,. \tag{32.10}$$

These are the latitude angles of the paths on the RES in Fig. 32.1 for $K = 10, 9, 8, \ldots, -10$. (For the oblate RES, the angles are relative to the x-axis.) If $\beta = \left(\Theta^J_K\right)$ is substituted into the symmetric top RES (32.9), the result is

$$E\left(\Theta^J_K\right) = J(J+1)B + (C - B)\,K^2\,, \tag{32.11}$$

which is precisely the symmetric top eigenvalue (32.4). The quantizing paths are circles lying at the intersections of the Dirac angular momentum cones and the RES. The angle $\left(\Theta^J_K\right)$ is a measure of the angular momentum uncertainty ΔJ_x or ΔJ_y transverse to the z-axis of quantization. Clearly, $K = \pm J$ states have minimum uncertainty.

For the asymmetric top, the classical paths that conserve both $|J|$ and E are one of two types. First, there are those pairs of equal-energy orbits that go around the hills on the plus or minus end of the body z-axis, which correspond to the $\pm K$ pairs of levels in the upper half of the level spectrum drawn in Fig. 32.2. Then there are the pairs of levels belonging to the equal-energy orbits in either of the two valleys surrounding the body x-axis, which are associated with the pairs of levels in the lower half of the level spectrum. Different eigensolutions occupy different geography.

The upper pairs of paths are seen to be distorted versions of the prolate top orbits seen on the left-hand side of Fig. 32.1, while the lower pairs are distorted versions of the oblate top orbits seen on the right-hand side of Fig. 32.1. The distortion makes J_z deviate from a constant K-value and corresponds to K-mixing in the quantum states. This also shows that more than one axis of quantization must be considered; the prolate-like paths are based on the z-axis, while the oblate-like paths belong to the body x-axis.

The two types of orbits, x and y, are separated by what is called a separatrix curve, which crosses the saddle points on either side of the body y-axis. In the example shown in Fig. 32.2, the separatrix is associated with a single level which separates the upper and lower energy doublets. The doublets that are closer to the separatrix level are split more than those which are farther away. Apart from the splitting, the energy levels can be obtained by generalized Bohr quantization of the classical paths on the RES. The quantization condition is,

$$\int J_z \, d\gamma = K , \tag{32.12a}$$

where

$$J_z = \sqrt{\frac{J(J+1)\left(C\cos^2\gamma + B\sin^2\gamma\right) - E}{\left(C\cos^2\gamma + B\sin^2\gamma\right) - A}} \tag{32.12b}$$

follows from (32.7) and (32.8). The resulting E_K-values are obtained by iteration.

The doublet, or superfine, splitting is a quantum effect which may be associated with tunneling between orbits that would have had equal energies E_K in the purely classical or semiclassical model. Approximate tunneling rates are obtained from integrals over the saddle point between each pair of equal-energy quantizing paths. The K-th rate, or amplitude, is,

$$S_K = \nu_K \, e^{-P_K} , \tag{32.13}$$

where

$$P_K = \tag{32.14}$$
$$i \int_{\gamma-}^{\gamma+} d\gamma \sqrt{\frac{J(J+1)\left(C\cos^2\gamma + B\sin^2\gamma\right) - E_K}{\left(C\cos^2\gamma + B\sin^2\gamma\right) - A}}$$

is the saddle path integral between the points of closest approach, $\gamma+$ and $\gamma-$, and ν_K is the classical precession frequency or quantum level spacing around energy level E_K. Since there are two tunneling paths, the amplitude S_K is doubled in a tunneling Hamiltoni an matrix for the K-th semiclassical doublet of z and $-z = \bar{z}$ paths:

$$\langle H\rangle_K = \begin{pmatrix} E_K & 2S_K \\ 2S_K & E_K \end{pmatrix} \begin{matrix} |z\rangle \\ |\bar{z}\rangle \end{matrix} . \tag{32.15}$$

The resulting tunneling energy eigensolutions are given in Table 32.1.

A- or B-states correspond to symmetric and antisymmetric combinations of waves localized on the two semiclassical paths. Rotational symmetry is considered in Sect. 32.3.

The total doublet splitting is $4S_K$, and decreases exponentially with the saddle path integral (32.14). The superfine A–B splittings in Fig. 32.2 are seen to range from several GHz near the separatrix down to only 26 kHz for the highest-K doublets at the band edges.

Meanwhile, the typical interdoublet level spacing or classical precessional frequency is about 150 GHz for the $J = 10$ levels shown in Fig. 32.2. This K-level spacing is called rotational fine structure splitting, and is also present in the symmetric top case. (The superfine splitting of the symmetric top doublets is exactly zero, since they have $O(2)_{BODY}$ symmetry if $A = B$ or $B = C$. In this case, all tunneling amplitudes cancel.)

The classical precession of $\boldsymbol{J}$ in the body frame follows a "left-hand rule" similar to what meteorologists use to determine Northern Hemisphere cyclonic rotation. A left "thumbs-down" or "low" has counterclockwise precession as does an oblate rotor valley, but a prolate RES "high" supports clockwise motion just like a weather "high".

Finally, consider the spacing between adjacent J-levels, which is called rotational structure, in a spectrum. This spacing is

$$E(J, K) - E(J-1, K) = 2BJ , \tag{32.16}$$

according to the symmetric top energy formulas (32.4). For the example just treated, $2BJ$ is about $10\,\text{cm}^{-1}$ or 300 GHz. This corresponds to the actual rotation frequency of the body. It is the only kind of rotational dynamics or spectrum that is possible for a simple diatomic rotor. A diatomic molecule, however, can have internal electronic or nuclear spin rotation, which gives an additional fine structure as discussed later [32.1,6,15].

To summarize, polyatomic molecules can be expected to exhibit all three types of rotational motion and spectra (from faster to slower): rotational, precessional, and precessional tunneling. These are related to three kinds of spectral structure (from coarser to finer spectra): rotational structure, fine structure, and superfine structure, respectively. Again, this neglects internal rotational and spin effects, which can have abnormally strong rotational resonance coupling due to the superfine structure [32.9,16]. Examples of this are discussed at the end of this chapter.

Table 32.1 Tunneling energy eigensolutions

Eigenvectors	$\lvert z\rangle$	$\lvert\bar{z}\rangle$	Eigenvalues
$\lvert A\rangle$	1	1	$E^A(K) = E_K + 2S_K$
$\lvert B\rangle$	1	-1	$E^B(K) = E_K - 2S_K$

32.3 Symmetry of Molecular Rotors

Molecular rotational symmetry is most easily introduced using examples of rigid rotors. Molecular rotor structures may have more or less internal molecular symmetry, depending on how their nuclei are positioned relative to one another in the body frame. A molecule's rotational symmetry is described by one of the elementary rotational point symmetry groups. These are the n-fold axial cyclic groups C_n and polygonal dihedral groups $D_n (n = 1, 2, \ldots)$, the tetrahedral group T, the cubic-octahedral group O, or the icosahedral group Y. All other point groups, such as C_{nv}, T_d, and O_h, are a combination of an elementary point group with the inversion operation $\boldsymbol{I}(\boldsymbol{r} \to -\boldsymbol{r})$. Each of these groups consist of operations which leave at least one point (origin) of a structure fixed while mapping identical atoms or nuclei into each other in such a way that the appearance of the structure is unchanged. The point groups are subgroups of the nuclear permutation groups [32.17].

In other words, molecular symmetry is based upon one of the most fundamental properties of atomic physics: the absolute identity of all atoms or, more precisely, nuclei of a given atomic number Z and mass number A. It is the identity of the so-called 'elementary' electronic and nucleonic constituent particles that underlies the symmetry.

The Pauli principle states that all half-integer spin particles are antisymmetrized with every other one of their kind in the universe. The Pauli–Fermi antisymmetrization principle and the related Bose–Einstein symmetrization principle determine much of molecular symmetry and dynamics, just as the Pauli exclusion principle is fundamental to electronic structure.

32.3.1 Asymmetric Rotor Symmetry Analysis

For an asymmetric rigid rotor, any rotation which interchanges $x-$, $y-$, or z-axes of the body cannot possibly be a symmetry, since all three axes are assumed to have different inertial constants. This restricts one to consider only 180° rotations about the body axes, and these are the elements of the rotor groups C_2 and D_2.

The two symmetry types for C_2 are even (denoted A or 0_2) and odd (denoted B or 1_2) with respect to a 180° rotation. For D_2, which is just $C_2 \otimes C_2$, the four symmetry types are even-even (denoted A_1), even-odd (denoted A_2), odd-even (denoted B_1), and odd-odd (denoted B_2) with respect to 180° rotations about the y- and x-axes, respectively. (The z-symmetry is determined by a product of the other two since $\boldsymbol{R}_z = \boldsymbol{R}_x\ \boldsymbol{R}_y$.) This is summarized in the character Tables 32.2 and 32.3.

The RES for the rigid rotor shown in Fig. 32.2 is invariant under 180° rotations about each of the three body axes. Therefore, its Hamiltonian symmetry is D_2 and its quantum eigenlevels must correspond to one of the four types listed under D_2 in Table 32.3. The D_2 symmetry labels are called rotational (or in general rovibronic) species of the molecular state. The species label the symmetry of a quantum wave function associated with a pair of C_2 symmetric semiclassical paths.

The classical $\boldsymbol{J}$-paths come in D_2 symmetric pairs, but each individual classical $\boldsymbol{J}$-path on the rigid rotor RES has a C_2 symmetry which is a subgroup of D_2. Each path in the valley around the x-axis is invariant under just the 180° rotation around the x-axis. This is $C_2(x)$ symmetry. The other member of its pair that goes around the negative x-axis also has this local $C_2(x)$ symmetry. The combined pair of paths has the full D_2 symmetry but classical mechanics does not permit occupation of two separate paths. Multiple path occupation is a completely quantum effect.

Similarly, each individual $\boldsymbol{J}$-path on the hill around the z-axis is invariant under just the 180° rotation around the z-axis, so it has $C_2(z)$ symmetry as does the equivalent path around the negative z-axis. Only the separatrix has the full D_2 symmetry, since its pairs are linked up on the y-axis to form the boundary between the x and z paths. No $\boldsymbol{J}$-paths encircle the unstable y-axis since it is a saddle point.

Each classical $\boldsymbol{J}$-path near the x- or z-axis belongs to a particular K-value through the semiclassical quantization conditions (32.12). Depending upon whether the K-value is even (0_2) or odd (1_2), the corresponding K-doublet is correlated with a pair of D_2 species as shown in the columns of the correlation tables in Fig. 32.3. These three correlation tables give the axial

C_2	$\boldsymbol{1}$	$\boldsymbol{R}$
A	1	1
B	1	−1

Table 32.2 Character table for symmetry group C_2

Table 32.3 Character table for symmetry group D_2

D_2	$\boldsymbol{1}$	$\boldsymbol{R}_x$	$\boldsymbol{R}_y$	$\boldsymbol{R}_z$
A_1	1	1	1	1
A_2	1	−1	1	−1
B_1	1	1	−1	−1
B_2	1	−1	−1	1

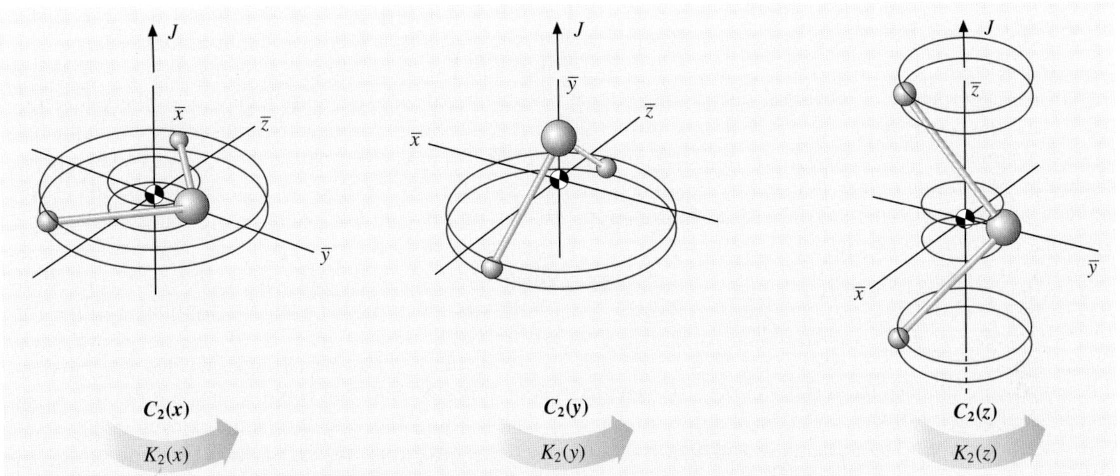

$C_2(x)$

	0_2	I_2
A_1	|	•
A_2	•	|
B_1	|	•
B_2	•	|

$C_2(y)$

	0_2	I_2
A_1	|	•
A_2	|	•
B_1	•	|
B_2	•	|

$C_2(z)$

	0_2	I_2
A_1	|	•
A_2	•	|
B_1	•	|
B_2	|	•

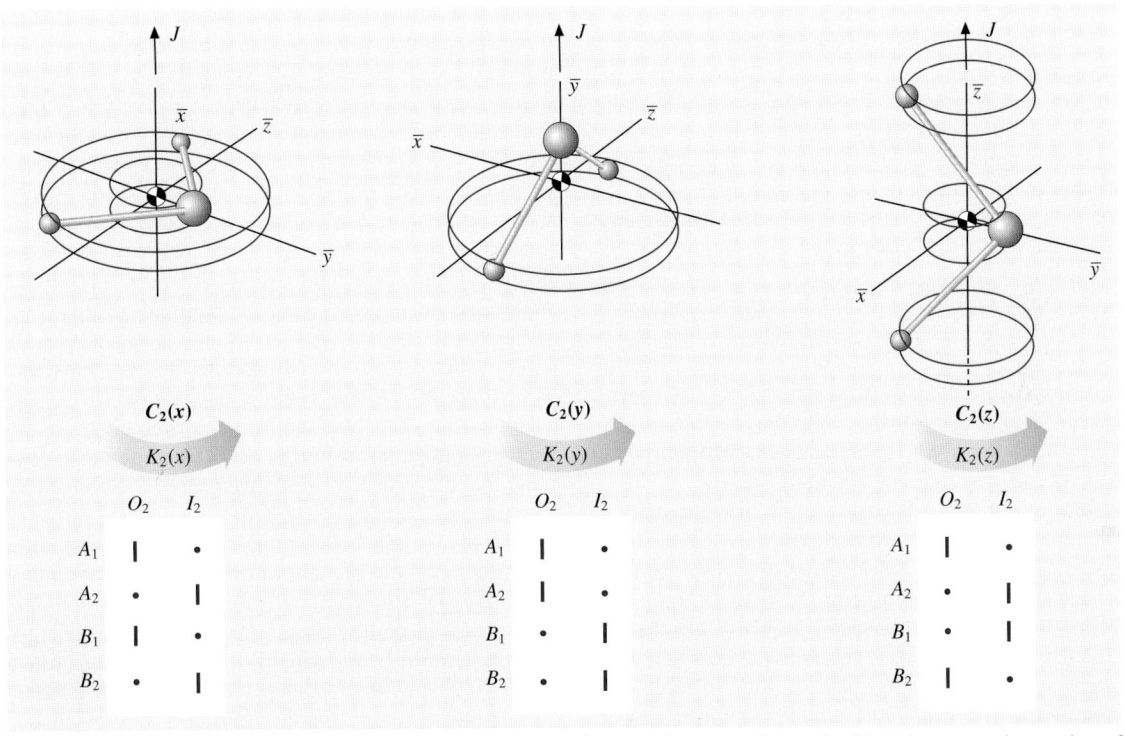

Fig. 32.3 Tables of correlations between D_2 symmetry species and the even (0_2) and odd (I_2) symmetric species of subgroups $C_2(x)$, $C_2(y)$, and $C_2(z)$

180° rotational symmetry of each D_2 species for rotation near each body axis x, y and z, respectively, but only the stable rotation axes x and z support stable path doublets for this Hamiltonian (32.1).

For example, consider the $K = 10$ paths which lie lowest in the x-axis valleys. Since $K = 10$ is even (0_2), it is correlated with an A_1 and B_1 superfine doublet [see the 0_2 column of the $C_2(x)$ table]. On the high end near the z-axis hilltop, $K = 10$ gives rise to an A_1 and B_2 doublet [see the 0_2 column of the $C_2(z)$ table]. All the doublets in Fig. 32.2 may be assigned in this way.

32.4 Tetrahedral–Octahedral Rotational Dynamics and Spectra

The highest symmetry rigid rotor is the *spherical top* for which the three inertial constants are equal ($A = B = C$). The spherical top Hamiltonian

$$H = B\boldsymbol{J}\cdot\boldsymbol{J}$$

has the full $R(3)_{\text{LAB}} \otimes R(3)_{\text{BODY}}$ symmetry. With inversion parity, the symmetry is $O(3)_{\text{LAB}} \otimes O(3)_{\text{BODY}}$. In any case, the J-levels are $(2J+1)^2$-fold degenerate. The resulting $BJ(J+1)$ energy expression is the first approximation for molecules which have regular polyhedral symmetry of, for example, a tetrahedron (CF_4), cube (C_6H_6), octahedron (SF_6), dodecahedron or icosahedron ($C_{20}H_{20}$, $B_{12}H_{12}$, or C_{60}). Rigid regular polyhedra have isotropic or equal inertial constants and rotate just like they were perfectly spherical distributions of mass.

However, no molecule can really have spherical $O(3)_{\text{BODY}}$ symmetry; even molecules of the highest symmetry contain a finite number of nuclear mass points, and therefore have a finite internal point symmetry. Evidence of octahedral or tetrahedral symmetry shows up in fine structure splittings analogous to those for asymmetric tops. However, spherical top fine structure is due to symmetry breaking caused by anisotropic or tensor rotational distortion. To discuss this, one needs to consider what are called semirigid rotors.

32.4.1 Semirigid Octahedral Rotors and Centrifugal Tensor Hamiltonians

The lowest order tensor centrifugal distortion perturbation has the same form for both tetrahedral and octahedral molecules. It is simply a sum of fourth powers of angular momentum operators given in the third term below. The first two terms are the scalar rotor energy and scalar centrifugal energy.

$$H = B\,|J|^2 + D\,|J|^4 + 10t_{044} \times \left[J_x^4 + J_y^4 + J_z^4 - (3/5)J^4 \right] . \tag{32.17}$$

The tensor term includes the scalar $(3/5)J^4$ to preserve the center of gravity of the tensor level splitting. This type of semirigid rotor Hamiltonian was first used in the study of methane (CH_4) spectra [32.18].

The scalar terms do not reduce the symmetry or split the levels. The tensor term (t_{044}) breaks the molecular symmetry from $O(3)_{\mathrm{LAB}} \otimes\ O(3)_{\mathrm{BODY}}$ to the lower symmetry subgroup $O(3)_{\mathrm{LAB}} \otimes T_{d\mathrm{BODY}}$, or $O(3)_{\mathrm{LAB}} \otimes O_{h\mathrm{BODY}}$, and splits the $(2J+1)^2$-fold degeneracy into intricate fine structure patterns which are analogous to cubic crystal field splitting of atomic orbitals. The first calculations of the tensor spectrum were done by direct numerical diagonalization [32.18–21]. As a result, many of the subtle symmetry properties were missed. The semiclassical analysis [32.22] described in the following Sections exposes these properties.

32.4.2 Octahedral and Tetrahedral Rotational Energy Surfaces

By substituting in (32.7) and plotting the energy as a function of body polar angles β and γ, an RES is obtained, two views of which are shown in Fig. 32.4. Here the tensor term is exaggerated in order to exhibit the topography clearly. (In ($n = 0$) SF_6, the t_{044} coefficient is only about 5.44 Hz, while the rotational constant is $B = 0.09\,\mathrm{cm}^{-1}$. The t_{244} coefficient of ($n = 1$) SF_6 is much greater.)

A positive tensor coefficient ($t_{a44} > 0$) gives an octahedral shaped RES, as shown in Fig. 32.4. This is appropriate for octahedral molecules since they are least susceptible to distortion by rotations around the x-, y-, and z-axes containing the strong radial bonds. Thus the rotational energy is highest for a $\boldsymbol{J}$-vector near one of six body axes $(\pm 1, 0, 0)$, $(0, \pm 1, 0)$, or $(0, 0, \pm 1)$, i. e., one of the six RES hills in Fig. 32.4.

However, if the $\boldsymbol{J}$-vector is set in any of the eight interaxial directions $(\pm 1, \pm 1, \pm 1)$, the centrifugal force will more easily bend the weaker angular bonds, raise the molecular inertia, and lower the rotational energy. This accounts for the eight valleys on the RES in Fig. 32.4.

A negative tensor coefficient ($t_{a44} < 0$) gives a cubic shaped RES. This is usually appropriate for cubic and tetrahedral molecules, since they are most susceptible to distortion by rotations around the x-, y-, and z-axes which lie between the strong radial bonds on the cubic diagonals. Instead of six hills and eight valleys, one finds six valleys and eight hills on the cubic RES. Both freon CF_4 and cubane C_8H_8 are examples of this type of topology.

Note that a semirigid tetrahedral rotor may have the same form of rotational Hamiltonian and RE surface as a cubic rotor. The four tetrahedral atomic sites are in the same directions as four of the eight cubic sites. The other four cubic sites form an inverted tetrahedron of the same shape.

If only tetrahedral symmetry were required, the Hamiltonian could contain a third order tensor of the form $J_x J_y J_z$. However, pure rotational Hamiltonians must also satisfy time-reversal symmetry: the energy for each $\boldsymbol{J}$ must be the same as for $-\boldsymbol{J}$, and thus rotational sense should not matter. This symmetry excludes all odd powers of $\boldsymbol{J}$. Simple rotor RES have inversion symmetry even if their molecules do not. Compound rotors containing spins or other rotors may have "lopsided" pairs of RES as shown in Sect. 32.6.

32.4.3 Octahedral and Tetrahedral Rotational Fine Structure

An example of rotational fine structure for angular momentum quantum number $J = 30$ is shown in Fig. 32.4. The levels consist mainly of clusters of levels belonging to the octahedral symmetry species A_1, A_2, E, T_1, or T_2. The characters of these species are given in Table 32.4. (The tetrahedral T_d group has a similar table where T_1 and T_2 are often labeled F_1 and F_2).

The first column gives the dimension or degeneracy of each species; A_1, A_2, are singlets, E is a doublet,

Table 32.4 Character table for symmetry group O

O	0°	120°	180°	90°	180°
A_1	1	O 1	1	1	1
A_2	1	1	1	−1	−1
E	2	−1	2	0	0
T_1	3	0	−1	1	−1
T_2	3	0	−1	−1	1

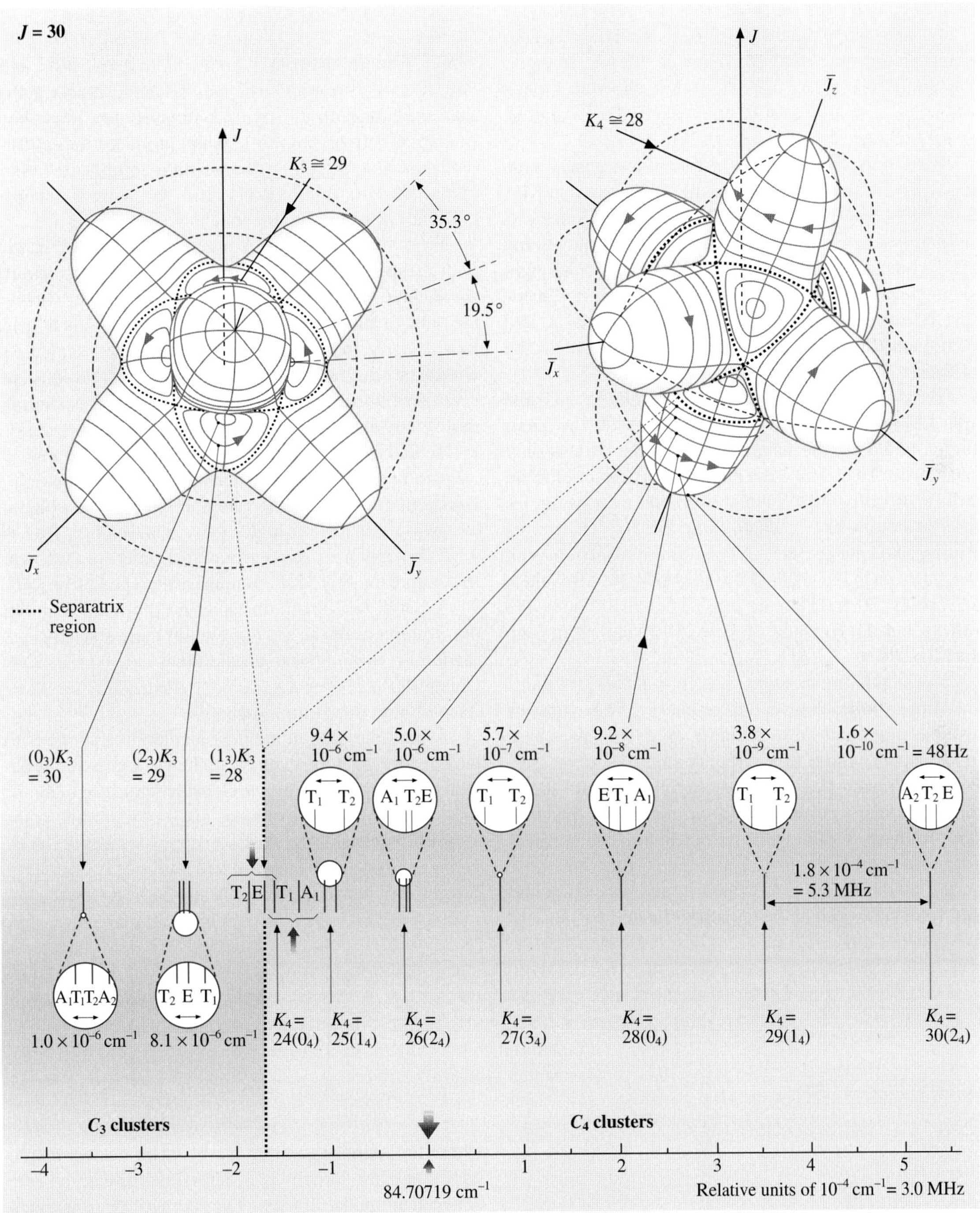

Fig. 32.4 $J = 10$ rotational energy surface related level spectrum for a semirigid octahedral or thetrahedral rotor

while T_1 and T_2 are triplets. These species form two clusters (A_1, T_1, T_2, A_2) and (T_2, E, T_1) on the low end of the spectrum and six clusters (T_1, T_2), (A_2, T_2, E), (T_1, T_2), (E, T_1, A_1), (T_1, T_2), and (A_2, T_2, E) on the upper part of the spectrum. (See the right-hand side of Fig. 32.4). Note that the total dimension or (near) degeneracy for each of the two lower clusters is eight: $(1+3+3+1)$ and $(3+2+3)$, while the upper clusters each have a six-fold (near) degeneracy: $(3+3)$, $(1+3+2)$, etc.

Each of the two lower eight-fold clusters can be associated with semiclassical quantizing paths in an RES valley as shown in Fig. 32.4. The eight-fold dimension or (near) degeneracy occurs because each quantizing path is repeated eight times – once in each of the eight identical valleys. Similarly, the six-fold cluster dimension occurs because there are six identical hills, and each quantizing path is repeated six times around the surface.

The majority of the paths lie on the hills because the hills are bigger than the valleys. The hills subtend a half angle of 35.3° to the separatrix, while the valleys only have 19.5°. To estimate the number of paths or clusters in hills or valleys, the angular momentum cone angles for $J = 30$ may be calculated using (32.10). The results are displayed in Fig. 32.5. The results are consistent with the spectrum in Fig. 32.4. Only the two highest K-values of $K = 29, 30$ have cones small enough to fit in the valleys, but the six states of $K = 25–30$ can all fit onto the hills.

The angular momentum cone formula also provides an estimate for each level cluster energy. The estimates become more and more accurate as K increases (approaching J), while the uncertainty angle Θ_K^J decreases. Paths for higher K are more nearly circular and therefore more nearly correspond to symmetric top quantum states of pure K. The paths on octahedral RE surfaces are more nearly circular for a given K than are those on the asymmetric top RE surface, and so the octahedral rotor states can be better approximated by those of a symmetric top.

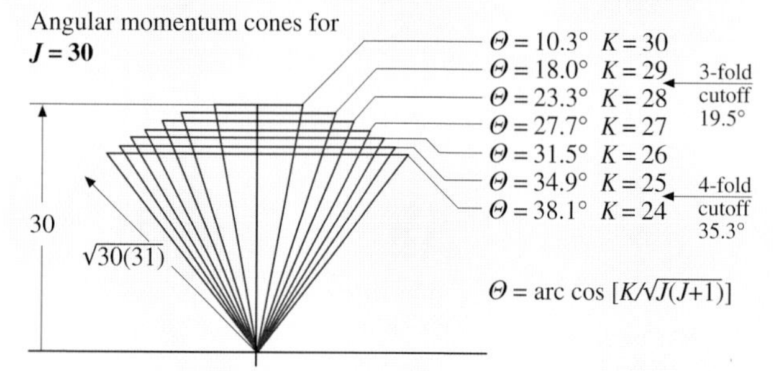

Fig. 32.5 $J = 30$ angular momentum cone half angles and octahedral cutoffs

32.4.4 Octahedral Superfine Structure

The octahedral RES has many more local hills and valleys and corresponding types of semiclassical paths than are found on the rigid asymmetric top RES. The tunneling between multiple paths produces an octahedral superfine structure that is more complicated than the asymmetric top doublets. Still, the same symmetry correlations and tunneling mechanics may be used.

First, the octahedral symmetry must be correlated with the local symmetry of the paths on the hills and in the valleys. The hill paths have a C_4 symmetry while the valley paths have a local C_3 symmetry. This is seen most clearly for the low-K paths near the separatrix which are less circular. The C_3 and C_4 correlations are given in Fig. 32.6 with a sketch of the corresponding molecular rotation for each type of path.

To find the octahedral species associated with a $K_3 = 30$ path in a C_3 valley one notes that 30 is 0 modulo 3. Hence the desired species are found in the 0_3 column of the C_3 correlation table: (A_1, A_2, T_1, T_2). This is what appears (not necessarily in that order) in the lower left corner of Fig. 32.4. Similarly, the species (A_2, E, T_2) for a $K_4 = 30$ path on top of a C_4 hill are found in the 2_4 column of the C_4 correlation table since 30 is 2 modulo 4; these appear on the other side of Fig. 32.4. Clusters (T_1, T_2) for $K_4 = 29$ and (A_1, E, T_1) for $K_4 = 28$ are found in a similar manner.

A multiple path tunneling calculation analogous to the one for rigid rotors can be applied to approximate octahedral superfine splittings. Consider the cluster (A_1, E, T_1) for $K_4 = 28$, for example. Six C_4-symmetric paths located on octahedral vertices on opposite sides of the x-, y-, and z-axes may be labeled $\{|x\rangle, |\bar{x}\rangle, |y\rangle, |\bar{y}\rangle, |z\rangle, |\bar{z}\rangle\}$. A tunneling matrix between the six paths follows:

$$\langle H\rangle_{K_4=28} = \begin{pmatrix} H & 0 & S & S & S & S \\ 0 & H & S & S & S & S \\ S & S & H & 0 & S & S \\ S & S & 0 & H & S & S \\ S & S & S & S & H & 0 \\ S & S & S & S & 0 & H \end{pmatrix} \begin{matrix} |x\rangle \\ |\bar{x}\rangle \\ |y\rangle \\ |\bar{y}\rangle \\ |z\rangle \\ |\bar{z}\rangle \end{matrix} , \tag{32.18}$$

where the tunneling amplitude between nearest neighbor octahedral vertices is S, but is assumed to be zero between antipodal vertices. The eigenvectors and eigenvalues for this matrix are given in the Table 32.5.

This predicts that the triplet (T_1) level should fall between the singlet (A_1) and the doublet (E) levels and the singlet-triplet spacing $(4S)$ should be twice the split-

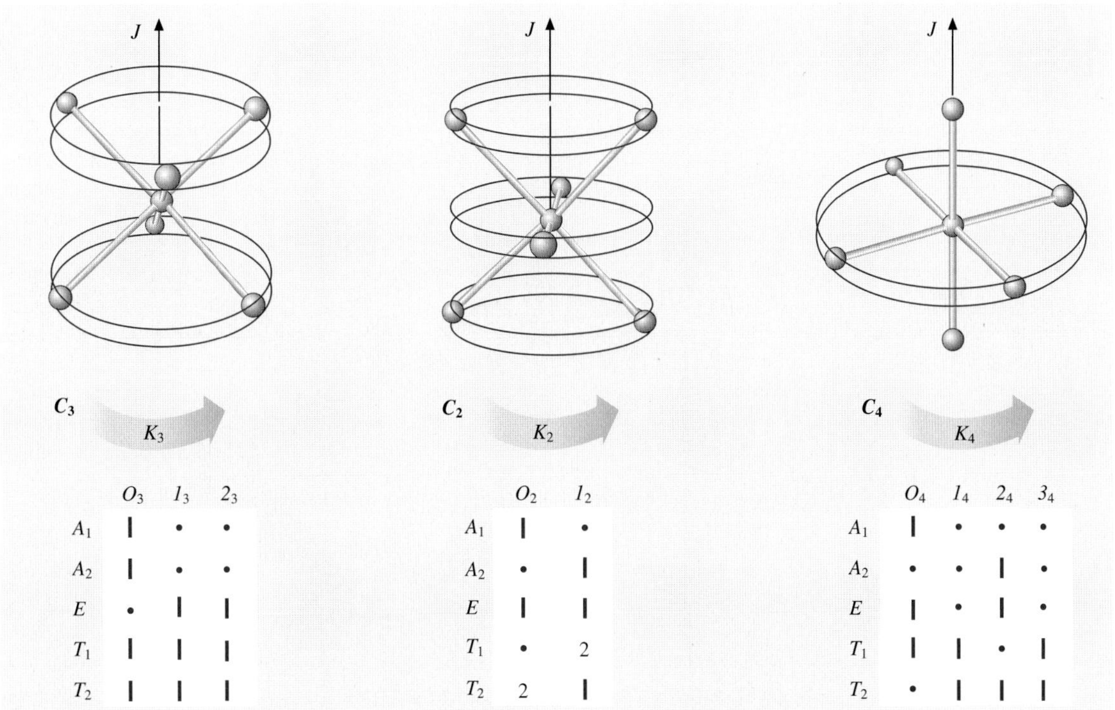

Fig. 32.6 Tables of correlations between 0 symmetry species and the cyclic axial symmetry species (K_p means K mod p) of subgroups C_3, C_2 and C_4

ting (-2S) between the triplet and doublet. This 2 : 1 ratio is observed in the (E, T_1, A_1) and (A_2, T_2, E) clusters which can be resolved and also in numerical calculation [32.18–21].

The tunneling amplitudes can be calculated by a separatrix path integral analogous to the asymmetric top formula (32.13) [32.10, 11]. As shown in Fig. 32.4, the tunneling rates or superfine splittings near the separatrix are ~ 1 MHz, which is only slightly slower than the classical precessional frequency. But as K approaches J on the hilltops, the tunneling rate slows down to a few Hz.

Table 32.5 Eigenvectors and eigenvalues of the tunneling matrix for the (A_1, E, T_1) cluster with $K = 28$

Eigenvector	$\|x\rangle$	$\|\bar{x}\rangle$	$\|y\rangle$	$\|\bar{y}\rangle$	$\|z\rangle$	$\|\bar{z}\rangle$	Eigenvalue
$\sqrt{6}\,\|A_1\rangle =$	1	1	1	1	1	1	$E^{A_1} = H+4S$
$\sqrt{12}\,\|E,1\rangle =$	2	2	-1	-1	-1	-1	$E^{E} = H-2S$
$2\,\|E,2\rangle =$	0	0	1	1	-1	-1	
$\sqrt{2}\,\|T_1,1\rangle =$	1	-1	0	0	0	0	$E^{T_1} = H$
$\sqrt{2}\,\|T_1,2\rangle =$	0	0	1	-1	0	0	
$\sqrt{2}\,\|T_1,3\rangle =$	0	0	0	0	1	-1	

32.5 High Resolution Rovibrational Structure

A display of spectral hierarchy for higher and higher resolution is shown in Fig. 32.7 for the 630 cm^{-1} or 16 μm bands of CF_4. This will serve to summarize the possible rovibrational spectral structures and place them in a larger context. The ν_4 resonance in part (a) corresponds to a dipole active $n_4 = 0 \rightarrow 1$ vibrational

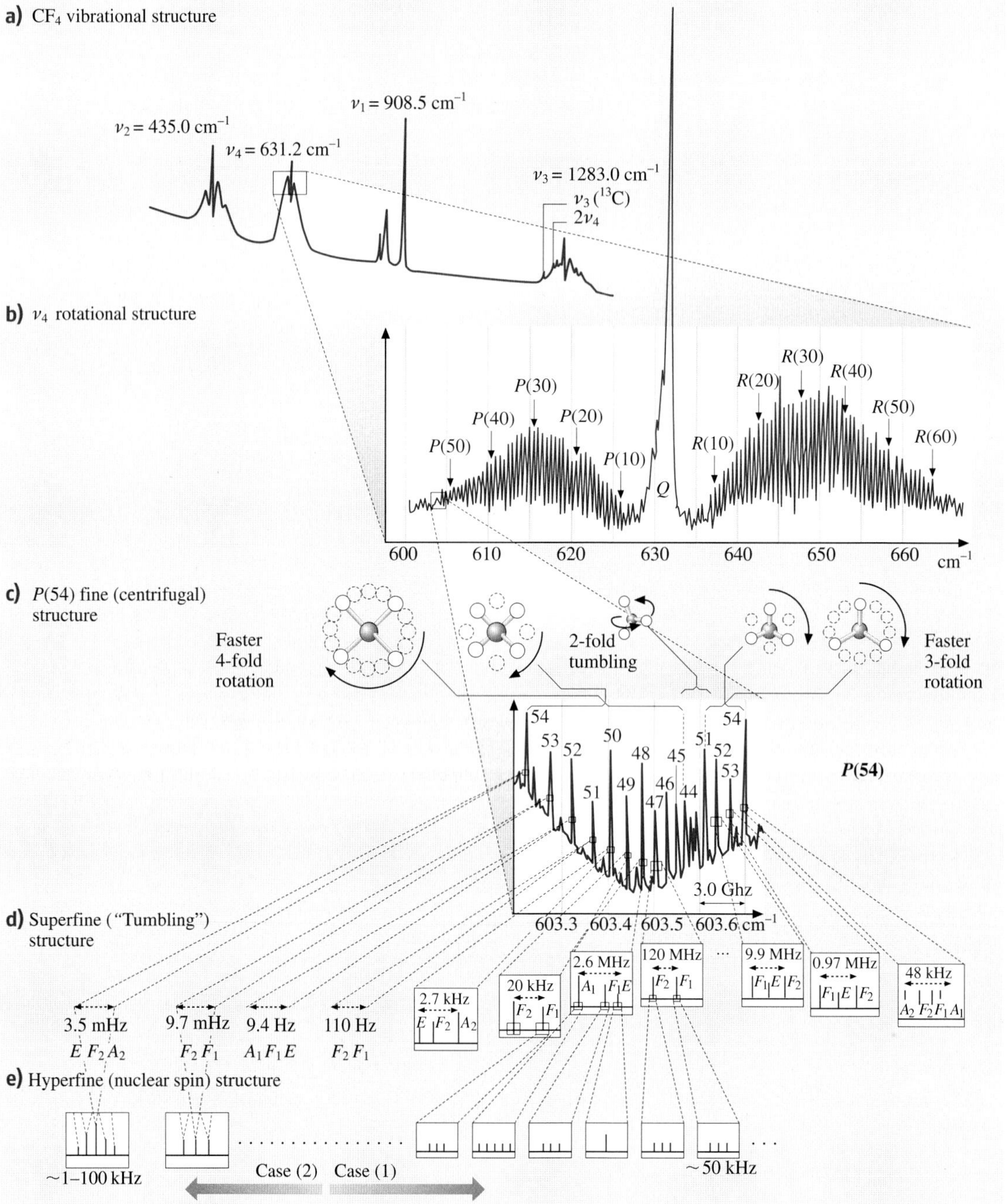

Fig. 32.7a–e Rovibrational structure in the 630 cm^{-1} or 16 μm bands of CF_4 [32.16]. **(a)** Vibrational resonances and band profiles. (Raman spectra from [32.23]). **(b)** Rotational P, Q, and R band structur corresponding to $J \to J-1$, $J \to J+1$ transitions. (FTIR spectra from [32.24]). **(c)** P(54) rotational fine structure due to rotation–vibration coupling and angular momentum precessional motion. (Laser diode spectra from [32.25]). **(d)** Superfine structure due to precessional tunneling [32.26]. **(e)** Hyperfine structure due to nuclear spin precession [32.26]

transition, and is just one of many vibrational structures to study. The $P(54)$ sideband resonance in part (b) corresponds to a $(J = 54) \to (J-1)$ rotational transition, and is just one of hundreds of rotational structures to study within the ν_4 bands.

Each band is something like a Russian doll; it contains structure within structure within structure down to the resolution of few tens of Hz. Examples of rotational fine and superfine structures described in Sect. 32.4 are shown in Fig. 32.7c, d, but even more resolution is needed to see the hyperfine structure in Fig. 32.7e. Such extremely high resolution has been reached with a CO_2 saturation absorption spectrometer [32.27, 28]. The 10 μm bands of SF_6 and SiF_4 have been studied in this manner, the latter being similar to CF_4 [32.26].

32.5.1 Tetrahedral Nuclear Hyperfine Structure

High resolution spectral studies of SiF_4 showed unanticipated effects involving the four fluorine nuclear spin and magnetic moments and their associated hyperfine states. First, the Pauli principle restricts the nuclear spin multiplicity associated with each of the rotational symmetry species in much the same way that atomic $L-S$ coupled states ${}^{2S+1}L$ have certain spin multiplicities $(2S+1)$ allowed for a given orbital L species involving two or more equivalent electrons. Second, since superfine splittings can easily be tiny, different spin species can end up close enough that hyperfine interactions, however small, can cause strongly resonant mixing of the normally inviolate species. Finally, a pure and simple form of spontaneous symmetry breaking is observed in which otherwise equivalent nuclei fall into different subsets due to quantum rotor dynamics.

Connecting nuclear spin to rotational species is done by correlating the full permutation symmetry (S_n for XY_n molecules) with the full molecular rotation and parity symmetry $[O(3)_{\mathrm{LAB}} \otimes T_{d\mathrm{BODY}}$ for CF_4 molecules or $O(3)_{\mathrm{LAB}} \otimes O_{h\mathrm{BODY}}$ and for $SF_6]$. For four spin-1/2 nuclei, the Pauli principle allows a spin of $I = 2$ and a spin multiplicity of five $(2I+1=5)$ for (J^+, A_2) or (J^-, A_1) species, but excludes (J^-, A_2) or (J^+, A_1) species altogether. The Pauli allowed spin for (J^+, T_1) or (J^-, T_2) species is $I = 1$ with a multiplicity of three, but there are no allowed (J^+, T_2) or (J^-, T_1) species. Finally, both (J^+, E) and (J^-, E) belong to singlet spin $I = 0$ and are singlet partners to an inversion doublet. (None of the other species can have both + and − parity.)

The E inversion doublet is analogous to the doublet in NH_3 which is responsible for the ammonia maser. However, NH_3-type inversion is not feasible in CF_4 or SiF_4, and so the splitting of the E doublet in these molecules is due to hyperfine resonance [32.9, 16, 23].

The Pauli analysis gives the number of hyperfine lines that each species would exhibit if it were isolated and resolved, as shown in the center of Fig. 32.7e. The rotational singlets A_1 and A_2 have five lines each, the rotational triplets T_1 and T_2 are spin triplets, and the rotational doublet E is a spin singlet but an inversion doublet. If the hyperfine structure of a given species A_1, A_2, T_1, T_2, or E is not resolved, then their line heights are proportional to their total spin weights of 5, 5, 3, 3, and 2, respectively.

If the unresolved species are clustered, then the total spin weights of each add to give a characteristic cluster line height. The line heights of the C_4 clusters (T_1, T_2), (A_2, T_2, E), (T_1, T_2), (E, T_1, A_1) are 6, 10, 6, 10, respectively. The line heights of the C_3 clusters (A_1, T_1, T_2, A_2), (T_1, E, T_2), (T_1, E, T_2) are 16, 8, 8, respectively. This is roughly what is seen in the $P(54)$ spectrum in Fig. 32.7c.

32.5.2 Superhyperfine Structure and Spontaneous Symmetry Breaking

The superfine cluster splittings ($2S$, $4S$, etc.) are proportional to the $\boldsymbol{J}$-precessional tunneling or 'tumbling' rates between equivalent C_3 or C_4 symmetry axes, and they decrease with increasing K_3 or K_4. At some point, the superfine splittings decrease to less than the hyperfine splittings which are actually increasing with K. The resulting collision of superfine and hyperfine structure has been called superhyperfine structure or Case 2 clusters. The following is a rough sketch of the phenomenology of this very complex effect, using the results of *Pfister* [32.26].

As long as the tunneling rates are > 1 MHz, the nuclear spins will tend to average over spherical top motion. The spins couple into states of good total nuclear spin I, which in turn couple weakly with the overall angular momentum and with well defined rovibrational species A_1, A_2, T_1, T_2, or E as described above. The resulting coupling is called Case 1, and is analogous to LS coupling in atoms.

Stick figures for two examples of spectra observed by *Pfister* [32.26] are shown in Fig. 32.8a and b. The first Case 1 cluster, shown in (a), is a C_4 type (0_4) cluster (A_1, T_1, E), which was solved in Table 32.6. The other Case 1 cluster, shown in (b), is a C_3 type $(\pm 1_3)$ cluster (T_1, E, T_2) (recall the C_3 correlations in Fig. 32.3). They are

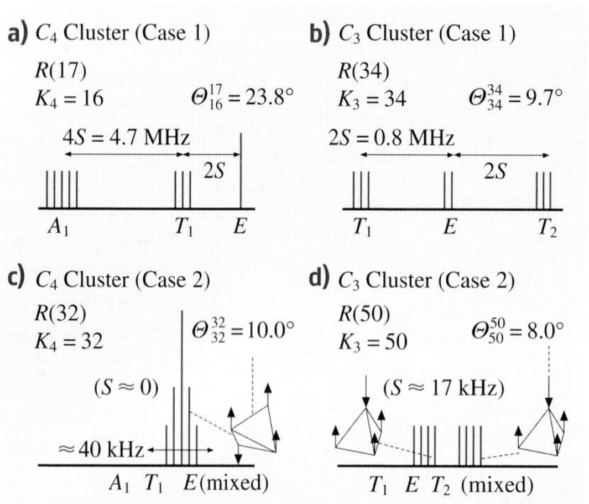

Fig. 32.8a–d Stick sketches for example of superfine and hyperfine spectral structure found by *Pfister* [32.26]; **(a)**,**(b)** Case 1 clusters (high tunneling amplitude S); **(c)**,**(d)** Case 2 clusters (low tunneling amplitude S)

similar to the corresponding sketches in Fig. 32.7e. One notable difference is that the inversion doublet shows little or no splitting in the (A_1, T_1, E) cluster, but does split in the (T_1, E, T_2) cluster.

When the tunneling rates fall below 10 or 20 kHz, the angular momentum can remain near a particular C_3 or C_4 symmetry axis for a time longer than the nuclear spin precession rates. Spin precession rates and the corresponding hyperfine splittings are ≈ 50 kHz, and increase with K. Hence, there is plenty of time for each of the nuclear spins to align or anti-align with the C_3 or C_4 symmetry axes of rotation. This is called Case 2 coupling, and the resulting spectrum resembles that of an NMR scan of the nuclei, but here the magnetic field is provided by the molecule's own body frame rotation.

If SiF_4 rotates uniformly about one C_4 symmetry axis, then all four F nuclei occupy equivalent positions at the same average distance from the rotation axis and experience the same local magnetic fields. The molecule can be thought of as a paired diatomic F_2–F_2 rotor with each one symmetrized or antisymmetrized so as to make the whole state symmetric. Table 32.6 shows the spin-$^1/_2$ base states arranged horizontally according to the total projection I_z of nuclear spins on the C_4 axis. Horizontal arrays ($\uparrow\downarrow$) of spins denote symmetric states, while vertical arrays ($\updownarrow$) denote antisymmetric spin states.

The hyperfine energy is approximately proportional to the projection I_z. The resulting spectrum

Table 32.6 Spin $-\frac{1}{2}$ basis states for SiF_4 rotating about a C_4 symmetry axis

$I_z = 2$	$I_z = 1$	$I_z = 0$	$I_z = -1$	$I_z = -2$
		$\left\lvert\begin{smallmatrix}\uparrow & \uparrow\\ \downarrow & \downarrow\end{smallmatrix}\right\rangle$		
		$\lvert\uparrow\downarrow\ \uparrow\downarrow\rangle$		
	$\lvert\uparrow\downarrow\ \uparrow\uparrow\rangle$	$\lvert\downarrow\downarrow\ \uparrow\uparrow\rangle$	$\lvert\downarrow\downarrow\ \uparrow\downarrow\rangle$	
$\lvert\uparrow\uparrow\ \uparrow\uparrow\rangle$	$\lvert\uparrow\uparrow\ \uparrow\downarrow\rangle$	$\lvert\uparrow\uparrow\ \downarrow\downarrow\rangle$	$\lvert\uparrow\downarrow\ \downarrow\downarrow\rangle$	$\lvert\downarrow\downarrow\ \downarrow\downarrow\rangle$

is (1, 2, 4, 2, 1)-degenerate pyramid of equally spaced lines as shown in Fig. 32.8c. Four spin-1/2 states without symmetry restrictions would give the standard binomial (1, 4, 6, 4, 1)-degeneracy seen in NMR spectra.

If the molecule settles upon C_3 symmetry axes of rotation, the situation is markedly different. The four nuclei no longer occupy equivalent positions. One nucleus sits on the rotation axis, while the other three nuclei occupy equivalent off-axis positions. The off-axis nuclei experience a different local magnetic field than the single on-axis nucleus (Fig. 32.8d). From the spectrum, it appears that the spin-up to spin-down energy difference is much greater for the lone on-axis nucleus than for the three equatorial nuclei, whose spin states form the energy quartet $\{\lvert\uparrow\uparrow\uparrow\rangle, \lvert\uparrow\uparrow\downarrow\rangle, \lvert\uparrow\downarrow\downarrow\rangle, \lvert\downarrow\downarrow\downarrow\rangle\}$. The on-axis nucleus has an energy doublet with a large splitting, so that the four nuclei together give a doublet of quartets as shown in the figure.

If the off-axis nuclei had experienced the greatest splitting, then the spectrum would have been a quartet of doublets instead of a doublet of quartets. Something like this does occur in the SF_6 superhyperfine structure, which shows a quintet of triplets for a Case-2 C_4-symmetry cluster. For either one of these molecules, it is remarkable how different the rovibrational 'chemical shifts' can become for equivalent symmetry sites. The result is a microscopic example of spontaneous symmetry breaking.

32.5.3 Extreme Molecular Symmetry Effects

The most common high symmetry molecules belong to either the tetrahedral T_d or cubic-octahedral O groups. Until the recent discovery of fullerenes and the structure of virus coats, the occurrence of molecular point groups of icosahedral symmetry was thought to be rare or nonexistent in nature [32.24, 25].

For an extreme example of symmetry breaking effects, consider the Buckminsterfullerene or Buckyball molecule C_{60} which has the highest possible molecular point symmetry Y_h. A semiclassical approach to rotational symmetry and dynamics is useful here since the rotational quantum constant is so small for the fullerenes (for C_{60} $2B = 0.0056\,\mathrm{cm}^{-1}$ or 168 MHz) [32.29–31].

Since there are two isotopes ^{12}C (nuclear spin 0) and ^{13}C (nuclear spin 1/2) it is possible to have a Bose-symmetric molecule ($^{12}C_{60}$), or Fermi-symmetric molecule ($^{13}C_{60}$), or many broken-symmetry combinations ($^{12}C_x{}^{13}C_{60-x}$). The most likely combination is $^{12}C_{59}{}^{13}C$, which has no rotational symmetry at all, only one reflection plane. This may be the most extreme example of molecular isotopic symmetry breaking; it goes from the highest possible symmetry Y_h to one of the lowest, C_h.

The Fermi-symmetric molecule $^{13}C_{60}$ has ten times as many rotating spin-1/2 nuclei as SF_6, and 2^{10} times as many hyperfine states, or about 1.15×10^{18} spin states distributed among 10 symmetry species [32.32]. In contrast,the Bose-symmetric molecule $^{12}C_{60}$ has only one spin symmetry species allowed by the Bose exclusion principle: A_{1g}. It provides an even more extreme example of Bose exclusion than the $Os^{16}O_4$ molecule. In all, 119 of the 120 Y_h rovibrational symmetry states are Bose-excluded, giving $^{12}C_{60}$ an extraordinarily sparse rotational structure. However, it only takes the addition of a single neutron to make $^{12}C_{59}{}^{13}C$. Then all the excluded rovibrational states must return!

32.6 Composite Rotors and Multiple RES

So far, the discussion has focused on Hamiltonians and RES involving functions of even mulipolarity, i. e., constant ($k=0$), quadrupole ($k=2$), hexadecapole ($k=4$), while ignoring odd functions, i. e., dipole ($k=1$), octupole ($k=3$), for reasons of time-reversal symmetry. However, for composite "rotor-rotors" any mulitpolarity is possible, and the dipole is of primary utility.

A composite rotor is one composed of two or more objects with more or less independent angular momenta. This could be a molecule with attached methyl (CH_3) "gyro" or "pinwheel" sub-rotors, a system of considerable biological interest. It could be a molecule with a vibration or "phonon" excitation that couples strongly to rotation. Also, any nuclear or electronic spin with significant coupling may be regarded as an elementary sub-rotor. The classical analogy is a spacecraft with gyros on board.

A rotor–rotor Hamiltonian has the general interaction form

$$H_{\text{rotor }R+S} = H_{\text{rotor}R} + H_{\text{rotor}S} + V_{RS}\,. \tag{32.19}$$

A useful approximation assumes that rotor S, the "gyro", is fastened to the frame of rotor R, so that the interaction V_{RS} becomes a constraint, does no work, and is thus assumed to be zero. An asymmetric top with body-fixed spin has the Hamiltonian

$$\begin{aligned} H_{R+S(\text{Body-fixed})} &= A\boldsymbol{R}_x^2 + B\boldsymbol{R}_y^2 + C\boldsymbol{R}_z^2 + H_{\text{rotor}S} \\ &\quad + (\sim 0)\,, \end{aligned} \tag{32.20a}$$

which is a modified version of (32.1). The total angular momentum of the system is a conserved vector$\boldsymbol{J} = \boldsymbol{R}+\boldsymbol{S}$ in the lab-frame and a conserved magnitude $|\boldsymbol{J}|$ in the rotor-R body frame. So we use $\boldsymbol{R} = \boldsymbol{J} - \boldsymbol{S}$ in place of $\boldsymbol{R}$:

$$\begin{aligned} H_{R,S(\text{fixed})} &= A\,(\boldsymbol{J}_x - \boldsymbol{S}_x)^2 + B\left(\boldsymbol{J}_y - \boldsymbol{S}_y\right)^2 \\ &\quad + C\,(\boldsymbol{J}_z - \boldsymbol{S}_z)^2 + H_{\text{rotor}S} \\ &= A\boldsymbol{J}_x^2 + B\boldsymbol{J}_y^2 + C\boldsymbol{J}_z^2 - 2A\boldsymbol{J}_x\boldsymbol{S}_x \\ &\quad - 2B\boldsymbol{J}_y\boldsymbol{S}_y - 2C\boldsymbol{J}_z\boldsymbol{S}_z + H'_{\text{rotor}S}\,. \end{aligned} \tag{32.20b}$$

The gyro spin components $\boldsymbol{S}_a$ are first treated as constant classical parameters S_a:

$$\begin{aligned} H_{R,S(\text{fixed})} &= \text{const.}\ 1 - 2AS_x\boldsymbol{J}_x - 2BS_y\boldsymbol{J}_y - 2CS_z\boldsymbol{J}_z \\ &\quad + A\boldsymbol{J}_x^2 + B\boldsymbol{J}_y^2 + C\boldsymbol{J}_z^2 \\ &= M_0\boldsymbol{T}_0^0 + \sum\nolimits_d D_d\boldsymbol{T}_d^1 + \sum\nolimits_q Q_q\boldsymbol{T}_q^2\,. \end{aligned} \tag{32.20c}$$

This is a simple Hamiltonian multipole tensor operator expansion having here just a monopole $\boldsymbol{T}_0^0$ term, three dipole $\boldsymbol{T}_a^1$ terms, and two quadrupole $\boldsymbol{T}_q^2$ terms. Figure 32.9 shows these three tensor terms, where each graph is a radial plot of a spherical harmonic function $Y_q^k(\phi, \Phi)$ representing a tensor operator $\boldsymbol{T}_q^k$. The tensor components are

$$\boldsymbol{T}_0^0 = \frac{\boldsymbol{J}_x^2 + \boldsymbol{J}_y^2 + \boldsymbol{J}_z^2}{3} \tag{32.21a}$$

$$\begin{aligned} \boldsymbol{T}_x^1 &= \boldsymbol{J}_x = \frac{\boldsymbol{T}_{+1}^1 + \boldsymbol{T}_{-1}^1}{\sqrt{2}} \\ \boldsymbol{T}_y^1 &= \boldsymbol{J}_y = \frac{\boldsymbol{T}_{+1}^1 - \boldsymbol{T}_{-1}^1}{\mathrm{i}\sqrt{2}} \\ \boldsymbol{T}_z^1 &= \boldsymbol{J}_z = \boldsymbol{T}_0^1 \end{aligned} \tag{32.21b}$$

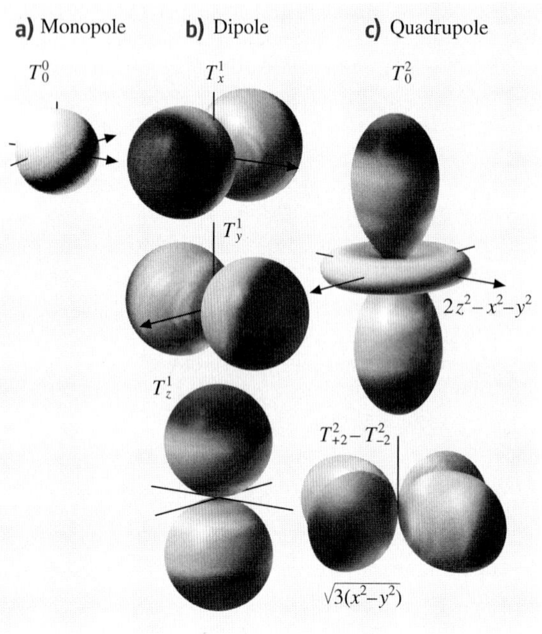

Fig. 32.9a–c The six lowest order RES components needed to describe rigid gyro-motors

$$\begin{aligned} \boldsymbol{T}_{zz}^2 &= \frac{2\boldsymbol{J}_z^2 - \boldsymbol{J}_x^2 - \boldsymbol{J}_y^2}{2} = \boldsymbol{T}_0^2 \\ \boldsymbol{T}_{x^2-y^2}^2 &= \boldsymbol{J}_x^2 - \boldsymbol{J}_y^2 = \frac{2\left(\boldsymbol{T}_2^2 - \boldsymbol{T}_{-2}^2\right)}{\sqrt{6}} \end{aligned} \tag{32.21c}$$

The constant coefficients or moments indicate the strength of each multipole symmetry:

$$M_0 = A + B + C + 3H'_{\text{rotor}_S} \tag{32.22a}$$

$$\begin{aligned} D_x &= -2AS_x , \\ D_y &= -2BS_y , \\ D_z &= -2CS_z \end{aligned} \tag{32.22b}$$

$$\begin{aligned} Q_{zz} &= (2C - A - B)/6 \\ Q_{x^2-y^2} &= (A - B)/2 \end{aligned} \tag{32.22c}$$

The scalar monopole RES (a) is a sphere, the vector dipole RES (b) are bi-spheres pointing along Cartesian axes, and the RES (c) resemble quadrupole antenna patterns. Also, Fig. 32.9a–c plot the six s, p, and d Bohr–Schrödinger orbitals that are analogs for the six octahedral J-tunneling states listed in Table 32.5.

The asymmetric and symmetric rotor Hamiltonians (32.1) and (32.1) are combinations of a monopole (32.21a), which by itself makes a spherical rotor, and varying amounts of the two quadrupole terms (32.21c) to give the rigid rotor RES pictured in Figs. 32.1 and Fig. 32.2. The Q coefficients in (32.22c) are both zero for a spherical top ($A = B = C$), but only one is zero for a symmetric top ($A = B$).

Combining the monopole (32.21a) with the dipole terms (32.21b) gives the gyro-rotor Hamiltonian (32.20b) for a spherical rotor ($A = B = C$):

$$H = \text{const} + BJ^2 - g\mu \boldsymbol{S} \cdot \boldsymbol{J} , \tag{32.23}$$

where $-g\mu = 2A = 2B = 2C$. This Hamiltonian resembles a dipole potential $-\boldsymbol{m} \cdot \boldsymbol{B}$ for a magnetic moment $\boldsymbol{m} = g\boldsymbol{J}$ that precesses clockwise around a lab-fixed magnetic field $\boldsymbol{B} = \mu \boldsymbol{S}$. (The PE is least for $\boldsymbol{J}$ along $\boldsymbol{S}$.)

The Hamiltonian (32.23) is a simple example of Coriolis rotational energy. It is least for $\boldsymbol{J}$ along $\boldsymbol{S}$, where $|\boldsymbol{R}| = |\boldsymbol{J} - \boldsymbol{S}|$ and the rotor kinetic energy BR^2 are least. (Magnitudes $|J|$ and $|S|$ are constant here.) The spherical rotor-gyro RES in Fig. 32.10 has a minimum along the body-axis $+\boldsymbol{S}$ and a maximum along $-\boldsymbol{S}$, where BR^2 is greatest.

As is the case for the rigid solid rotors in Figs. 32.1 and Fig. 32.2, the RES topography lines determine the precession $\boldsymbol{J}$-paths in the body frame, wherein gyro-$\boldsymbol{S}$ is fixed, as shown in Fig. 32.10. The left-hand rule gives the sense of the $\boldsymbol{J}$-precession in the body $\boldsymbol{S}$-frame, i. e., all $\boldsymbol{J}$ precess counterclockwise relative to the "low" on the $+\boldsymbol{S}$-axis, or clockwise relative to the "high" on the $-\boldsymbol{S}$-axis. In the lab, $\boldsymbol{S}$ precesses in a clockwise manner around a fixed $\boldsymbol{J}$.

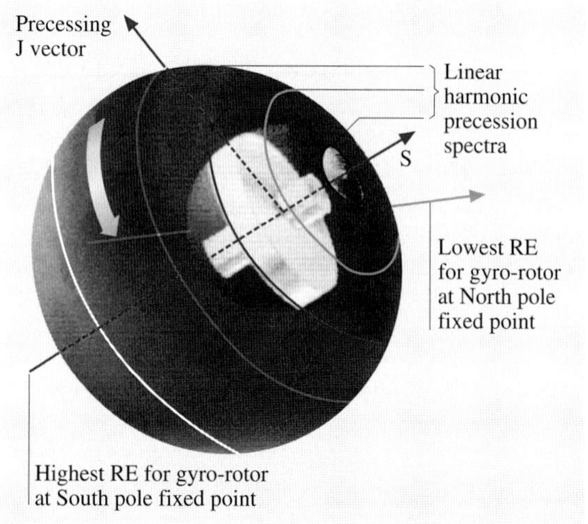

Fig. 32.10 The spherical gyro-rotor RES is a cardioid of revolution around gyro spin $\boldsymbol{S}$

Gyro-RES differ from solid rotor RES, which have two opposite "highs" and/or two opposite "lows" separated by saddle fixed points where the precessional flow direction reverses, as seen in Fig. 32.2. The gyro-RES in Fig. 32.10 has no saddle fixed points, and thus has only one "high" and one direction of flow with the same harmonic precession frequency for all $\boldsymbol{J}$-vectors between the high $+\boldsymbol{S}$ and low $-\boldsymbol{S}$-axes. This is because the spectrum of the gyro-rotor Hamiltonian (32.23) is harmonic, or linear, in the K:

$$\left\langle {}^{J}_{K} \middle| H \middle| {}^{J}_{K} \right\rangle = \text{const.} + BJ(J+1) - 2BK . \quad (32.24)$$

In contrast, even the symmetric rigid rotor spectrum (32.4) is quadratic in K. Other rotors shown in Figs. 32.2 and Fig. 32.4 have levels that have an even more nonlinear spacing.

32.6.1 3D-Rotor and 2D-Oscillator Analogy

Linear levels are usually associated with harmonic oscillators not rotors, but the gyro-rotor's linear spectrum highlights a 160-year-old analogy between the motions of 3D rotors and 2D vibrations [32.33–45]. Stokes [32.35] first described 2D electric vibration or optical polarization, by a 3D vector that became known as the Stokes vector $\boldsymbol{S}$, and later as the "spin" $\boldsymbol{S}$. The Stokes spin was based on Hamilton's quaternions q_μ [32.33,34]. The Pauli spinors $\sigma_\mu = \mathrm{i}q_\mu$ [32.36] were defined, 83 years later, as components of a general 2D Hermitian matrix H. Spinors square to the unit matrix $\left(\sigma_\mu^2 = \boldsymbol{1} = \sigma_0\right)$, while quaternions square to $-\boldsymbol{1}$. The 3D Hamiltonian is

$$H = \begin{pmatrix} A & B - \mathrm{i}C \\ B + \mathrm{i}C & D \end{pmatrix} = \frac{A+D}{2}\sigma_0 + \frac{A-D}{2}\sigma_\mathrm{A} + B\sigma_\mathrm{B} + C\sigma_\mathrm{C} , \quad (32.25)$$

where

$$\sigma_0 = \begin{pmatrix} 1 & 0 \\ 0 & 1 \end{pmatrix}, \quad \sigma_\mathrm{A} = \begin{pmatrix} 1 & 0 \\ 0 & -1 \end{pmatrix}, \quad \sigma_\mathrm{B} = \begin{pmatrix} 0 & 1 \\ 1 & 0 \end{pmatrix}, \quad \sigma_\mathrm{C} = \begin{pmatrix} 0 & -\mathrm{i} \\ i & 0 \end{pmatrix}.$$

The 3D-component labels $\frac{A-D}{2}$ (Asymmetric-diagonal), B (Bilateral-balanced), and C (Circular-Coriolis) are ABC mnemonics for Pauli's z, x, and y, respectively. The 2D operator H has a $\boldsymbol{1} + \boldsymbol{S} \cdot \boldsymbol{J}$ form of the Coriolis coupling Hamiltonian (32.23):

$$\begin{aligned} H &= S_0\boldsymbol{1} + S_\mathrm{A}J_\mathrm{A} + S_\mathrm{B}J_\mathrm{B} + S_\mathrm{C}J_\mathrm{C} \\ &= S_0 J_0 + \boldsymbol{S} \cdot \boldsymbol{J} , \end{aligned} \quad (32.26)$$

where

$$J_0 = \boldsymbol{1}, \quad J_A = \frac{\sigma_\mathrm{A}}{2}, \quad J_\mathrm{B} = \frac{\sigma_\mathrm{B}}{2}, \quad J_\mathrm{C} = \frac{\sigma_\mathrm{C}}{2},$$

and

$$S_0 = (A+\mathrm{D})/2 , \quad S_A = (A - \mathrm{D}) , \quad S_\mathrm{B} = 2\mathrm{B} , \quad S_\mathrm{C} = 2\mathrm{C} .$$

The elementary 2D-oscillator ladder operators $a^\dagger$, and a make the 2D-3D theory more powerful. This is known as the Jordan–Schwinger map [32.37–39] between 2D oscillation and 3D rotation. In terms of the ladder operators

$$\begin{aligned} &\boldsymbol{J}_0 = \boldsymbol{N} = a_1^\dagger a_1 + a_2^\dagger a_2 , \ \boldsymbol{J}_\mathrm{A} = \frac{1}{2}\left(a_1^\dagger a_1 - a_2^\dagger a_2\right) , \\ &\boldsymbol{J}_\mathrm{B} = \frac{1}{2}\left(a_1^\dagger a_2 + a_2^\dagger a_1\right) , \ \boldsymbol{J}_\mathrm{C} = \frac{-\mathrm{i}}{2}\left(a_1^\dagger a_2 - a_2^\dagger a_1\right) . \end{aligned} \quad (32.27)$$

where

$$a_1^\dagger a_1 = \begin{pmatrix} 1 & 0 \\ 0 & 0 \end{pmatrix}, \quad a_1^\dagger a_2 = \begin{pmatrix} 0 & 1 \\ 0 & 0 \end{pmatrix}, \quad a_2^\dagger a_1 = \begin{pmatrix} 0 & 0 \\ 1 & 0 \end{pmatrix}, \quad a_2^\dagger a_2 = \begin{pmatrix} 0 & 0 \\ 0 & 1 \end{pmatrix}.$$

The $a^\dagger a$-algebra gives Schwinger's 3D angular momentum raising and lowering operators $\boldsymbol{J}_+ = \boldsymbol{J}_\mathrm{B} + \mathrm{i}\boldsymbol{J}_\mathrm{C} = a_1^\dagger a_2$ and $\boldsymbol{J}_- = \boldsymbol{J}_\mathrm{B} - \mathrm{i}\boldsymbol{J}_\mathrm{C} = a_2^\dagger a_1$, where in two dimensions 1 and 2 are spin-up $(+\hbar/2)$ and spin-down $(-\hbar/2)$, instead of the x-and y-polarized states envisioned by Stokes.

The angular 3D ladder operation is replaced by a simpler 2D oscillator operation:

$$\begin{aligned} &\boldsymbol{J}_+ |n_1 n_2\rangle = a_1{}^\dagger a_2 \, |n_1 n_2\rangle = \\ &\quad \sqrt{n_1+1}\sqrt{n_2}\, |n_1+1, n_2-1\rangle , \\ &\boldsymbol{J}_- |n_1 n_2\rangle = a_2^\dagger a_1 \, |n_1 n_2\rangle = \\ &\quad \sqrt{n_1}\sqrt{n_2+1}\, |n_1-1, n_2+1\rangle . \end{aligned} \quad (32.28)$$

The 2D oscillator states are labeled by the total number $N = (n_1 + n_2)$ of quanta and the net quantum population $\Delta N = (n_1 - n_2)$. The 3D angular momentum states $\left| {}^{J}_{K} \right\rangle$ are labeled by the total momentum $J = N/2 = (n_1+n_2)/2$ and the z-component

$K = \Delta N/2 = (n_1 - -n_2)/2$, just half (or $\eta/2$) of N and ΔN.

$$|n_1, n_2\rangle = \frac{\left(a_1^\dagger\right)^{n_1}\left(a_2^\dagger\right)^{n_2}}{\sqrt{n_1! n_2!}}|0,0\rangle =$$

$$\left|{J \atop K}\right\rangle = \frac{\left(a_1^\dagger\right)^{J+K}\left(a_2^\dagger\right)^{J-K}}{\sqrt{(J+K)!\,(J-K)!}}|0,0\rangle\ , \tag{32.29}$$

where

$$n_1 = J + K\ , \quad n_2 = J - K\ .$$

From this *Schwinger* [32.38] rederived the Wigner matrices $D^J_{MK}(\alpha\beta\gamma)$, which appear in (32.5) and (32.6), and the Wigner–Eckart or Clebsch–Gordan matrix values. This helps clarify the approximation of these values by (J, K)−cone levels around RES hills or valleys [recall (32.10) and (32.11)], since

$$\left\langle {J' \atop K} \middle| T^k_0 \middle| {J \atop K} \right\rangle = C^{kJJ}_{0KK}\langle J\,\|k\|\,J\rangle \sim D^J_{JK}\left(\Theta^J_K\right)\ .$$

32.6.2 Gyro-Rotors and 2D-Local Mode Analogy

The 2D–3D analogy provides insight into spin [32.40–42] and rovibrational dynamics [32.40–45], as well as having computational value. Consider extending a single 2D-oscillator-rotor analogy in the Stokes model to a model of two 1D oscillators with coordinates $x_1 = x$ and $x_2 = y$.

Identical side-by-side oscillators have bilateral B-symmetry. The Hamiltonian H_{B} commutes with the matrices σ_{B} (+45° mirror reflection of axes $\pm x \leftrightarrows \pm y$) and $-\sigma_{\text{B}}$ (−45° mirror reflection of axes $\mp x \leftrightarrows \pm y$), both of which switch oscillators. A first-order bilateral Hamiltonian is $H_{\text{B}} = 2B\sigma_{\text{B}}$. This is analogous to a gyro rotor $\boldsymbol{T}^1_x$ with $\boldsymbol{S}$ along the B-axis, as shown in Fig. 32.11a. (The added unit operator $\boldsymbol{T}^0_0$ shifts levels, but does not affect eigenstates.)

The eigenstates of H_{B} are the symmetric and antisymmetric normal modes that belong to the fixed points on the $\boldsymbol{S}$-vector or $\pm B$-axes of the 3D Stokes space. If instead, the $\boldsymbol{S}$-vector lies on the A-axis, the Hamiltonian is an asymmetric diagonal $H_{\text{A}} = 2A\sigma_{\text{A}}$ matrix. From (32.25) we see that the operator σ_{A} reflects y into $-y$ but leaves x alone, so that the eigenvectors of H_{A} are localized on the x-oscillator or the y-oscillator, but not on both. Such motions are local modes, but they are not modes of H_{B} since it does not commute with H_{A}.

If the vector $\boldsymbol{J}$ is on the $+A$-axis (local x-mode), the Hamiltonian H_B rotates $\boldsymbol{J}$ to the $-C$-axis, then to the $-A$-axis (local y-mode), then to the $+C$-axis, and then back to the $+A$-axis. This $\boldsymbol{J}$-path is the equator of Fig. 32.11a. The $\pm C$-axes label circular polarization with right and left chirality, respectively. Twice during a B-beat, $\boldsymbol{J}$ passes the $\pm C$-axes, where one vibrator's phase is 90° ahead and resonantly pumping

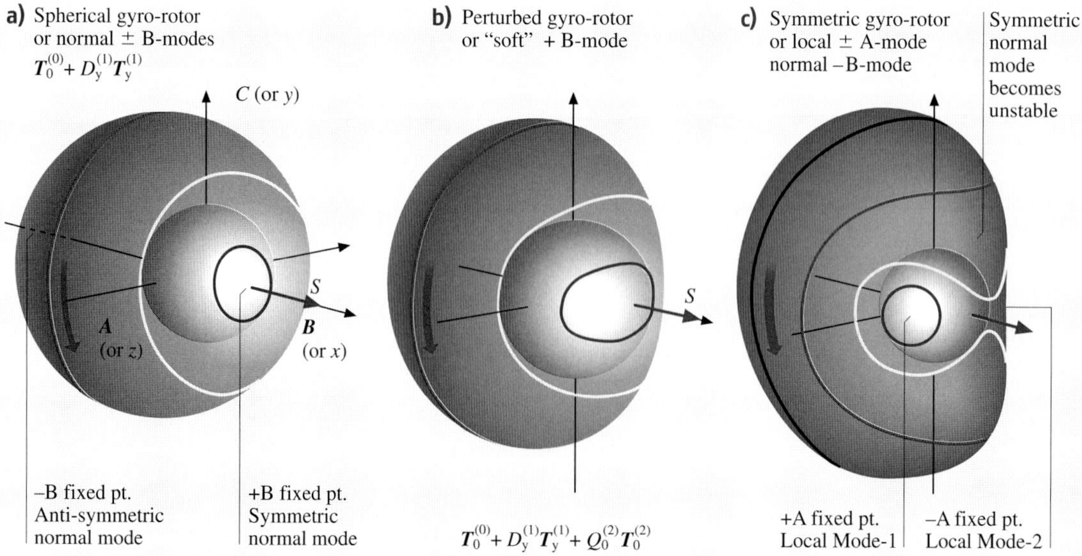

Fig. 32.11a–c A spherical gyro-rotor becomes a symmetric gyro-rotor by adding T_0^2

the other. Such bilateral beat and resonant transfer is disrupted by adding anharmonic T_0^2 or $T_{\pm 2}^2$ terms to the B-symmetry terms T_x^1 and T_0^0. Adding T_0^2 causes B-circles in Fig. 32.11(a) to distort near the B-axis, as shown in Fig. 32.11b–c.

In molecular rotation theory, the T_0^2 and T_0^0 terms comprise the initial unperturbed Hamiltonian (32.3) of a symmetric top, while the gyro terms T_q^1 are viewed as perturbations in (32.20), due to an "on-board" gyro rotor. For vibration theory, the T_q^1 terms make up a normal-mode Hamiltonian, and the T_0^2 term is viewed as an anharmonic perturbation.

The effect of T_0^2, seen in Fig. 32.11c, is to replace the stable fixed point $+B$ (representing the (+)-normal mode) by a saddle point as B bifurcates (splits) into a pair of fixed points that head toward the $\pm A$-axes. So one normal mode dies and begets two stable local modes, wherein one mass may keep its energy, and not lose it to the other through the usual B-beating process. (The A-modes become anharmonically detuned.)

Pairs of classical modes, each localized on different sides of the RES in Fig. 32.11, are analogous to the asymmetric top $\pm K$-precession pairs in Fig. 32.2 with degenerate energy in a classical RES picture. The quantum-tunneling Hamiltonian (32.15) splits each trajectory pair into a superfine doublet with (±)-eigenstates sharing both RES paths, as seen in Table 32.1. The quantum gyro-spin doublets also share $\pm J$ components both up and down the A-axis, as seen in Fig. 32.11c.

32.6.3 Multiple Gyro-Rotor RES and Eigensurfaces

While simple quantum rotors delocalize J to multiple RES paths, a gyro-rotor J may delocalize to multiple paths and surfaces. Gyro-rotor RES vary with S, and if S is a quantum spin, the possibility arises for a distribution over multiple RES [32.46, 47]. A simple quantum theory of S allows both $+S$ and $-S$ at once. The RES for each is plotted one on top of the other, as in Fig. 32.12a, while component RES are shown in Fig. 32.12b for $+S$ and in Fig. 32.12c for $-S$. An energy sphere is shown intersecting an RES pair for an asymmetric gyro-rotor. If the spin S is set to zero, the pair of RES collapses into a rigid asymmetric top RES, shown in Fig. 32.2, having angular inversion (time-reversal $J \to -J$) and D_{2h} reflection symmetry. The composite RES in Fig. 32.12a has inversion symmetry, but lacks reflection symmetry. Its parts in Fig. 32.12b and c have neither inversion nor reflection symmetry if gyro-spins $\pm S$ are off-axis.

The gyro-rotor Hamiltonian (32.20) allows tunneling or mixing of multiple RES. A two-state spin$-1/2$ gyro-spin model has a $2 \otimes 2$ Hamiltonian matrix and two base-RES:

$$H_{\text{gyro}} = M_0 J \cdot J + D_x S_x J_x + D_y S_y J_y + D_z S_z J_z + Q_{xx} J_x^2 + Q_{yy} J_y^2 + Q_{zz} J_z^2 \quad (32.30)$$

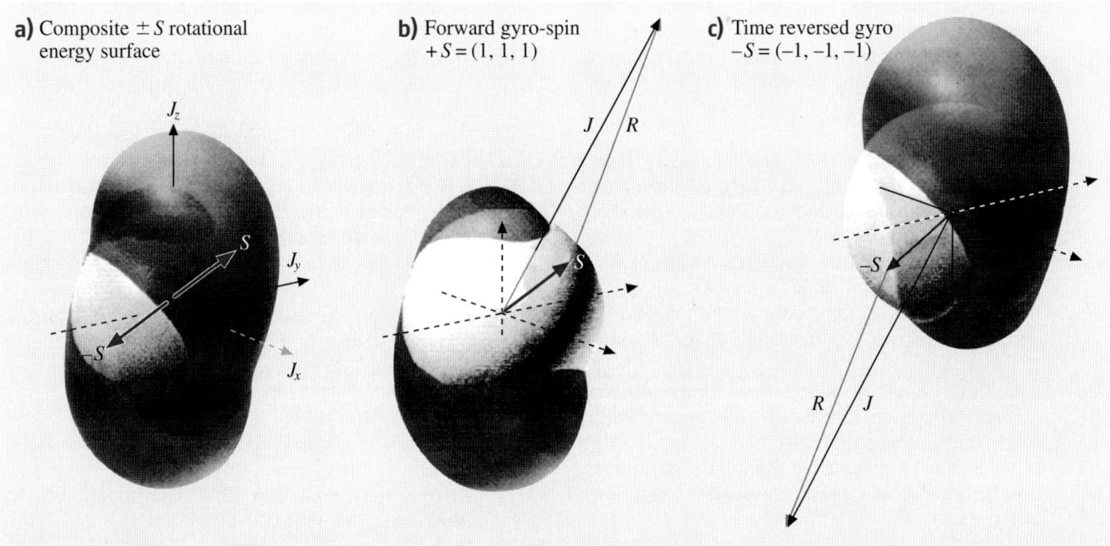

Fig. 32.12a–c Asymmetric gyro-rotor RES (classical body-fixed-spin case); **(a)** Composite $\pm S$; **(b)** Forward spin $\pm S$; **(c)** Reversed spin $-S$

As in (32.7), $\boldsymbol{J}$ is approximated by classical vector components in the body frame:

$$\left(J_x = |J| \sin\beta \cos\gamma\ , \quad J_y = |J| \sin\beta \sin\gamma\ , \right.$$
$$\left. J_z = |J| \cos\beta \right)\ . \tag{32.31a}$$

But the gyro-spin S uses its quantum representation $\boldsymbol{S} = |S|\boldsymbol{\sigma}/2 = \sqrt{3}\boldsymbol{\sigma}/2$ from (32.25):

$$\begin{aligned}\langle H_{\text{gyro}}\rangle &= M_0 J^2 + Q_{xx} J_x^2 + Q_{yy} J_y^2 + Q_{zz} J_z^2 \\ &\quad + D_x |S| \sigma_x J_x + D_y |S| \sigma_y J_y + D_z |S| \sigma_z J_z \\ &= \begin{pmatrix} h(J) + D_z |S| J_z & |S| \left(D_x J_x - \mathrm{i} D_y J_y\right) \\ |S| \left(D_x J_x + \mathrm{i} D_y J_y\right) & h(J) - D_z |S| J_z \end{pmatrix} \\ &= \begin{pmatrix} h(J) + d_z \cos\beta & \left(d_x \cos\gamma - \mathrm{i} d_y \sin\gamma\right) \\ & \times \sin\beta \\ \left(d_x \cos\gamma + \mathrm{i} d_y \sin\gamma\right) & h(J) - d_z \cos\beta \\ \times \sin\beta & \end{pmatrix} ,\end{aligned} \tag{32.31b}$$

where

$$h(J) = M_0 J^2 + Q_{xx} J_x^2 + Q_{yy} J_y^2 + Q_{zz} J_z^2$$

and

$$d_\mu = D_\mu |S| |J|\ . \tag{32.31c}$$

The dynamics generated by Hamiltonian approximations such as (32.31b) are analogous to other semiclassical approximations, such as the Maxwell–Bloch model of an atom in a cavity. Their solutions are very complicated and often chaotic. The classical variable ($\boldsymbol{J}$ in this case) follows phase contours on a changing RES that depends on the instantaneous expectation values of the quantum variables ($\boldsymbol{S}$ in this case), which in turn vary according to the instantaneous classical variables.

In spite of this complexity, semiclassical spectra may be approximated using RES pairs obtained from eigenvalues of a $2 \otimes 2$ matrix such as (32.31b) for each classical angular orientation ($\beta\gamma$) of the J-vector in the body frame [32.46, 47]. The results are pairs of surfaces roughly like those in Fig. 32.12a, but without the intersection lines. The Wigner non-crossing effect prevents degeneracy, except at isolated points.

Near-crossing RES are the rotational equivalent of near-crossing vibrational-potential energy surfaces (VES) described in treatments of *Jahn–Teller* effects [32.48, 49]. The classical, semiclassical, and quantum theory for such loosely-bound or fluxional systems is still inits infancy, but is potentially a very rich source of new effects.

References

32.1 G. Herzberg: *Molecular Spectra and Structure: Vol. I, Spectra of Diatomic Molecules* (Van-Norstrand-Reinhold, New York 1950)
32.2 G. Herzberg: *Molecular Spectra and Structure: Vol. II, Infrared and Raman Spectra of Polyatomic Molecules* (Van-Norstrand-Reinhold, New York 1945)
32.3 G. Herzberg: *Molecular Spectra and Structure: Vol. III, Electronic Structure of Polyatomic Molecules* (Van-Norstrand-Reinhold, New York 1966)
32.4 F. B. Wilson, V. C. Decius, P. C. Cross: *Molecular Vibrations* (McGraw Hill, New York 1955)
32.5 D. Papousek, M. R. Aliev: *Molecular Vibrational-Rotational Spectra, Studies Phys. Theor. Chem. 17* (Elsevier, Amsterdam 1982)
32.6 R. N. Zare: *Angular Momentum: Understanding Spatial Aspects in Chemistry and Physics* (Wiley Interscience, New York 1988)
32.7 W. G. Harter: *Principles of Symmetry, Dynamics, and Spectroscopy* (Wiley Interscience, New York 1993)
32.8 W. G. Harter, C. W. Patterson: J. Math. Phys. **20**, 1453 (1979)
32.9 W. G. Harter: Phys. Rev. A **24**, 192 (1981)
32.10 W. G. Harter, C. W. Patterson: J. Chem. Phys. **80**, 4241 (1984)
32.11 W. G. Harter: Comp. Phys. Rep. **8**, 319 (1988)
32.12 D. A. Sadovskii, B. I. Zhilinskii: Mol. Phys **65**, 109 (1988)
32.13 D. A. Sadovskii, B. I. Zhilinskii: Phys. Rev. A **47**, 2653 (1993)
32.14 I. M. Pavlichenkov: Phys. Rep. **226**, 173 (1993)
32.15 W. G. Harter, C. W. Patterson, F. J. daPaixao: Rev. Mod. Phys. **50**, 37 (1978)
32.16 W. G. Harter, C. W. Patterson: Phys. Rev. A **19**, 2277 (1979)
32.17 P. R. Bunker: *Molecular Symmetry and Spectroscopy* (Academic, New York 1979)
32.18 K. T. Hecht: J. Mol. Spectrosc. **5**, 355 (1960)
32.19 K. R. Lea, M. J. M. Leask, W. P. Wolf: J. Phys. Chem. Solids **23**, 1381 (1962)
32.20 A. J. Dorney, J. K. G. Watson: J. Mol. Spectrosc. **42**, 1 (1972)
32.21 K. Fox, H. W. Galbraith, B. J. Krohn, J. D. Louck: Phys. Rev. A **15**, 1363 (1977)
32.22 W. G. Harter, C. W. Patterson: J. Chem. Phys. **66**, 4872 (1977)

32.23 R. J. Butcher, Ch. Chardonnet, Ch. Bordé: Phys. Rev. Lett., **70**, 2698 (1993)
32.24 H.W. Kroto, J. R. Heath, S. C. O'Brian, R. F. Curl, R. E. Smalley: Nature **318**, 162 (1985)
32.25 W. Kratschmer, W. D. Lamb, K. Fostiropolous, D. R. Huffman: Nature **347**, 354 (1990)
32.26 O. Pfister: *Etude esperimentale et theorique des interactions hyperfines dans la bande de vibration* ν_3 *de la molecule* $^{28}SiF_4$ (Dissertation, Univ., Paris-Nord 1993)
32.27 J. Bordé, Ch. J. Bordé: Chem. Phys. **71**, 417 (1982)
32.28 Ch. Bordé, J. Bordé, Ch. Breant, Ch. Chardonnet, A. Vanlerberghe, Ch. Salomon: *Laser Spectroscopy VII* (Springer, Berlin, Heidelberg 1985) p. 95
32.29 W. G. Harter, D. E. Weeks: Chem. Phys. Lett. **132**, 187 (1986)
32.30 D. E. Weeks, W. G. Harter: Chem. Phys. Lett. **144**, 366 (1988)
32.31 D. E. Weeks, W. G. Harter: Chem. Phys. Lett. **176**, 209 (1991)
32.32 W. G. Harter, T. C. Reimer: Chem. Phys. Lett. **194**, 230 (1992)
32.33 W. R. Hamilton: Proc. R. Irish Acad. **II**, 424 (1844)
32.34 W. R. Hamilton: Phi. Mag. **25**, 489 (1844)
32.35 G. Stokes: Proc. R. Soc. London **11**, 547 (1862)
32.36 W. Pauli: Z. Phys. **37**, 601 (1927)
32.37 P. Jordan: Z. Phys. **94**, 531 (1935)
32.38 J. Schwinger: *Quantum Theory of Angular Momentum*, ed. by L. C. Biedenharn, H. van Dam (Academic, New York 1965) p. 229
32.39 L. C. Biedenharn, J. D. Louck: *Angular Momentum in Quantum Physics*, Encyclopedia of Mathematics, Vol 8, ed. by G. C. Rota (Addison Wesley, Reading, Massachusetts 1981) p. 212
32.40 I. I. Rabi, N. F. Ramsey, J. Schwinger: Rev. Mod. Phys. **26**, 167 (1954)
32.41 R. P. Feynman, F. I. Vernon, R. W. Helwarth: J. Appl. Phys. **28**, 49 (1957)
32.42 W. G. Harter, N. dos Santos: Am. J. Phys. **46**, 251 (1978)
32.43 K. K. Lehmann: J. Chem. Phys. **79**, 1098 (1983)
32.44 W. G. Harter: J. Chem. Phys. **85**, 5560 (1986)
32.45 Z. Li, L. Xiao, M. E. Kellman: J. Chem. Phys. **92**, 2251 (1990)
32.46 W. G. Harter: Comp. Phys. Rep. **8**, 319 (1988), see pp. 378–85
32.47 J. Ortigoso, I. Kleiner, J. T. Hougen: J. Chem. Phys. **110**, 11688 (1999)
32.48 H. A. Jahn, E. Teller: Proc. R. Soc. London **A161**, 220 (1937)
32.49 H. A. Jahn, E. Teller: Proc. R. Soc. London **A164**, 117 (1938)

33. Radiative Transition Probabilities

This chapter summarizes the theory of radiative transition probabilities or intensities for rotationally-resolved (high-resolution) molecular spectra. A combined treatment of diatomic, linear, symmetric-top, and asymmetric-top molecules is based on angular momentum relations. Generality and symmetry relations are emphasized. The *energy-intensity* model is founded in a rotating-frame basis-set expansion of the wave functions, Hamiltonians, and transition operators. The intensities of the various rotational branches are calculated from a small number of transition-moment matrix elements, whose relative values can be assumed from the supposed nature of the transition, or inferred by fitting experimental intensities.

33.1 **Overview** ... 515
33.1.1 Intensity versus Line-Position Spectroscopy ... 515

33.2 **Molecular Wave Functions in the Rotating Frame** ... 516
33.2.1 Symmetries of the Exact Wave Function ... 516
33.2.2 Rotation Matrices ... 517
33.2.3 Transformation of Ordinary Objects into the Rotating Frame ... 517

33.3 **The Energy-Intensity Model** ... 518
33.3.1 States, Levels, and Components ... 518
33.3.2 The Basis Set and Matrix Hamiltonian ... 518
33.3.3 Fitting Experimental Energies ... 520
33.3.4 The Transition Moment Matrix ... 520
33.3.5 Fitting Experimental Intensities ... 520

33.4 **Selection Rules** ... 521
33.4.1 Symmetry Types ... 521
33.4.2 Rotational Branches and Parity ... 521
33.4.3 Nuclear Spin, Spatial Symmetry, and Statistics ... 522
33.4.4 Electron Orbital and Spin Angular Momenta ... 523

33.5 **Absorption Cross Sections and Radiative Lifetimes** ... 524
33.5.1 Radiation Relations ... 524
33.5.2 Transition Moments ... 524

33.6 **Vibrational Band Strengths** ... 525
33.6.1 Franck–Condon Factors ... 525
33.6.2 Vibrational Transitions ... 526

33.7 **Rotational Branch Strengths** ... 526
33.7.1 Branch Structure and Transition Type ... 526
33.7.2 Hönl–London Factors ... 527
33.7.3 Sum Rules ... 528
33.7.4 Hund's Cases ... 528
33.7.5 Symmetric Tops ... 530
33.7.6 Asymmetric Tops ... 530

33.8 **Forbidden Transitions** ... 530
33.8.1 Spin-Changing Transitions ... 530
33.8.2 Orbitally-Forbidden Transitions ... 531

33.9 **Recent Developments** ... 531

References ... 532

33.1 Overview

33.1.1 Intensity versus Line-Position Spectroscopy

The fact that atoms and molecules absorb and emit radiation with propensities that vary with wavelength is the origin of the field called spectroscopy. The relatively sharp intensity maxima are interpreted as corresponding to transitions between discrete states or energy levels. The frequencies or energies of these transitions are used as the primary source of information about the internal structure of the atom or molecule. Line positions can be measured with very high precision (1 ppm or better). Excellent calibration standards have been developed. The quality of these experimental data has attracted extensive analytical and theoretical effort. Sophisticated parametrized models have been developed in which the smallest shifts from the expected line positions can be used to identify perturbations or other subtle effects.

For molecules, knowledge of the strengths of these transitions is far less well developed. One reason is that quantitative experimental data on rotationally-resolved absorption cross sections and emission intensities are much rarer and the experiments themselves are much more difficult to calibrate. Few measurements claim a precision better than 1% and agreement within 10% of measurements in different laboratories is typically viewed as good. This situation is undesirable because most applications of molecular spectroscopy are in fact measurements of intensity. In many cases the strengths of absorptions or emissions are used to infer gas composition, temperature, time evolution, or other environmental conditions. In other examples the actual absorption and emission is the primary interest. Among the most important of these are atmospheric absorption of solar radiation and the greenhouse effect.

33.2 Molecular Wave Functions in the Rotating Frame

33.2.1 Symmetries of the Exact Wave Function

The exact total wave function for any isolated molecule with well-defined energy and total angular momentum can be expressed in a basis-set expansion over configurations with well-defined internal quantum numbers,

$$\Phi_{\text{exact}} = \Phi_{\text{trans}} \sum_{\alpha\beta\gamma\delta\epsilon} C_{\alpha\beta\gamma\delta\epsilon}\, \Phi^{\alpha}_{\text{rot}}\, \Phi^{\beta}_{\text{vib}}\, \Phi^{\gamma}_{\text{elec}}\, \Phi^{\delta}_{\text{espin}}\, \Phi^{\epsilon}_{\text{nspin}}\,. \tag{33.1}$$

In principle, the coefficients $C_{\alpha\beta\gamma\delta\epsilon}$ can be found only by diagonalizing the exact Hamiltonian. In practice one attempts to find a sufficiently good approximation, containing only a few terms, with coefficients chosen by diagonalizing an approximate or model Hamiltonian. This is the basis of the energy–intensity model developed in Sect. 33.3. As discussed by *Longuet-Higgins* [33.1] and *Bunker* [33.2], there are only six true symmetries of the exact Hamiltonian of an isolated molecule:

1. translation of the center of mass;
2. permutation of electrons;
3. permutation of identical nuclei;
4. time reversal or momentum reversal;
5. inversion of all particles through the center of mass;
6. rotation about space-fixed axes.

Of these, only the symmetries numbered 5 and 6 give quantum numbers (parity and the total angular momentum F) that are both rigorous and useful spectroscopic labels of the states of the molecule. The other symmetries are convenient for simplifying the description of the molecular wave function, for the evaluation of relations between matrix elements, and for classification of molecular states according to approximate symmetries.

The first symmetry, translation of the center of mass, allows the choice of a coordinate system referenced to the center of mass, and suppression of the portion of the wave function describing motion through space (as long as the molecule does not dissociate).

Symmetry number 2, exchange of electrons, does not directly provide any labels or quantum numbers, since the Fermi–Dirac statistics of electrons require that all wave functions must be antisymmetric. However, it provides considerable information about the probable electronic states since it controls whether molecular orbitals can be doubly or only singly occupied. For most (low-Z) molecules, each state will have a nearly well-defined value of electron spin: singlet or triplet for example. Admixture of other spin values usually can be treated as a perturbation. These points will be elaborated in Sects. 33.4.4 and 33.7.4.

Permutation of identical nuclei, symmetry number 3, also gives an identical quantum number to all the states of the molecule (± 1 depending on the character of the permutation and on whether nuclei with integral or half-integral spin are being permuted). It supplies little direct information about the energy separations between the states of the molecule. On the other hand, many molecules have identical nuclei in geometrically or dynamically equivalent positions. The existence of spatial symmetry, for nonplanar molecules, is really the same thing as permutational symmetry. Consequently, nuclear permutation, combined with inversion (symmetry number 5), is the basis for naming the states according to the approximate spatial symmetry group of the molecular frame and vibrational motion. These concepts will be explored in Sect. 33.4.3.

Symmetry number 4, time reversal, is both subtle and simple. In the absence of external magnetic fields the Hamiltonian for a molecule will contain only even combinations of angular momentum operators, e.g., $F_\alpha F_\beta$, $F_\alpha L_\beta$, or $F_\alpha S_\beta$. Thus changing the signs of all the angular momenta should result in an equivalent wave function. This will require that matrix elements retain the same absolute value when the angular quantum numbers are reversed, leading in general to complex conjugation [33.3].

Spatial inversion, symmetry number 5, is always an allowed operation for any molecule, even if it appears to lack internal inversion symmetry. This operation can be considered as a symmetry of the spherically-symmetric laboratory in which the molecule resides. If the molecule is linear, triatomic, rigid with a plane of symmetry, or is nonrigid with accessible vibrational or tunneling modes that correspond to plane reflections, inversion symmetry divides the states of the molecule into two classes, called parities. Perturbations can occur only between states of the same parity. For optical transitions, the change in parity of the states must match the parity of the operator. Otherwise, reflection of the molecule in a plane will interchange inconvertible optical isomers. Such optical isomers are energetically degenerate, so in all cases, inversion through the center of mass remains a valid symmetry of the rotating molecule. However, the separation of the states into two kinds does not provide any selection rules. The two parity classes are perfectly degenerate, thus there is always an allowed level with the correct parity either for perturbations or for optical transitions.

33.2.2 Rotation Matrices

The final symmetry, rotation about the center of mass, restricts the discussion to states with well-defined laboratory angular momentum, and to re-expression of the exact wave function by changing variables from *laboratory* coordinates to *body-fixed* or internal coordinates, and introducing the Euler angles relating these two coordinate systems,

$$\begin{aligned}\Phi_{\text{exact}}^{F,M_F}(\text{lab}) &= \sum_{K_F} \Phi_{\text{rot}}^{F,M_F K_F}(\text{Euler angles}) \\ &\quad \times \Phi_{\text{vesn}}^{(F,K_F)}(\text{internal, spins}) . \end{aligned} \tag{33.2}$$

Here F is the total angular momentum of the molecule, including vibrational, mechanical-rotation, electron-orbital, electron-spin, and nuclear-spin contributions. M_F and K_F are the projections of F in the laboratory and body-fixed frames, respectively. In the majority of cases, the magnitude of nuclear hyperfine interactions is sufficiently small that its influence can be ignored when analyzing wave functions and computing energies. Thus the quantum numbers J, M_J, and K_J, or just J, M, and K can be used.

Explanation is postponed of how the body-fixed frame is to be selected, but for any choice, the wave function for rotation of the entire molecule can be expressed using a rotation matrix [33.4, 5]

$$\begin{aligned}&\Phi_{\text{rot}}^{F,M_F K_F}(\text{Euler angles}) \\ &= \left(\frac{2F+1}{8\pi^2}\right)^{1/2} D_{M_F K_F}^{*F}(\phi,\theta,\chi) . \end{aligned} \tag{33.3}$$

For diatomics, *Zare* [33.5] suggests multiplying by $(2\pi)^{1/2}$ and setting $\chi = 0$. The internal wave function for the vibrational, electronic, electron-spin, and nuclear-spin degrees of freedom [thus the label (vesn)] can be thought of as the partial summation

$$\begin{aligned}&\Phi_{\text{vesn}}^{(F,K_F)}(\text{internal, spins}) \\ &= \sum_{\beta\gamma\delta\epsilon} C_{(FK_F)\beta\gamma\delta\epsilon} \Phi_{\text{vib}}^{\beta} \Phi_{\text{elec}}^{\gamma} \Phi_{\text{espin}}^{\delta} \Phi_{\text{nspin}}^{\epsilon} , \end{aligned} \tag{33.4}$$

expressed in the internal or rotated coordinate system. Note that the FK_F designation is only a parametric label. The rotational wave function has been absorbed into the rotation matrix.

33.2.3 Transformation of Ordinary Objects into the Rotating Frame

The assumption of rotational symmetry allows re-expression of matrix elements between total wave functions as a sum of matrix elements between internal wave functions. For example, the tensor operator $T^{(L)}$ belonging to the L representation of the rotation group, can be written in the rotating frame as [33.5–8]

$$T_p^{(L)}(\text{lab}) = \sum_q D_{pq}^{*L}(\phi\theta\chi) T_q^{(L)}(\text{body}) , \tag{33.5}$$

and can be used to evaluate matrix elements that might represent radiative transitions:

$$\begin{aligned}&\left\langle \psi^{F',M_F''+p} \middle| T_p^{(L)}(\text{lab}) \middle| \Phi^{F'',M_F''} \right\rangle \\ &= \left(\frac{(2F''+1)}{(2F'+1)}\right)^{1/2} \left\langle F''M_F'', Lp \middle| F'M_F''+p \right\rangle \\ &\quad \times \sum_{qK_F''} \left\langle F''K_F'', Lq \middle| F'K_F''+q \right\rangle \\ &\quad \times \left\langle \psi_{\text{vesn}}^{(F',K_F''+q)} \middle| T_q^{(L)}(\text{body}) \middle| \Phi_{\text{vesn}}^{(F'',K_F'')} \right\rangle , \end{aligned} \tag{33.6}$$

where $\langle F''M_F'', Lp|F'M_F''+p\rangle$ and $\langle F''K_F'', Lq|F'K_F''+q\rangle$ are Clebsch–Gordan coefficients that vanish if $|F'-F''| > L$, $|M_F''+p| > F'$, or $|K_F''+q| > F'$.

This transformation forms the basis for the derivation of rotational branch strengths (Sect. 33.7.2) and for the description of electron motions that are weakly coupled to the molecular frame (Sect. 33.7.4).

33.3 The Energy–Intensity Model

33.3.1 States, Levels, and Components

The previous section introduced the concept of representing the wave function of a molecule as a product of five simpler wave functions:

$$\psi \approx \psi_{\text{elec}}\psi_{\text{vib}}\psi_{\text{rot}}\psi_{\text{espin}}\psi_{\text{nspin}} . \quad (33.7)$$

This construction yields a similar separation of the Hamiltonian,

$$H \approx H_{\text{elec}} + H_{\text{vib}} + H_{\text{rot–fs}} + H_{\text{hf}} , \quad (33.8)$$

and representations of the energies as sums of contributions,

$$E \approx T_{\text{e}} + G_{\text{v}} + F_{\text{c}}(J) , \quad (33.9)$$

and absorption or emission transition strengths as products,

$$I \approx I_{\text{elec}} I_{\text{vib}} I_{\text{rot–fs}} I_{\text{hf}} . \quad (33.10)$$

Whatever theoretical arguments might favor such a separation, the real impetus is the empirical observation that most molecular absorption and emission spectra exhibit recognizable patterns arising from the dissimilar magnitudes of the energies associated with these five degrees of freedom. Separation of the wave function and the Hamiltonian into these four or five contributions facilitates the assignment of molecular spectra, in addition to suggesting models with parameters that can be adjusted to quantitatively represent the observed spectra.

Most states of molecules are dominated by a single set of electronic and vibrational quantum numbers. Electronic states are often well separated. With each electronic state is associated a potential energy surface, the energy at the minimum being labeled T_{e}. Motion of nuclei within this potential generates various levels corresponding to different vibrational quantum numbers, following regular patterns or progressions in energy, summarized by a small number of parameters called vibrational frequencies. The quantity G_{v} represents the energy of the the vibrational level above the potential minimum. For each vibrational level, a progression of rotational levels is expected. For linear molecules in electronic states without electronic angular momentum (i.e., $^1\Sigma$ states) the rotational energies are also reproduced by a few rotational constants.

For more complicated molecules and electronic states, i.e., most cases, there are multiple energetically distinct levels with the same value of J (in addition to the $2J+1$ orientational degeneracy of each level). These multiple levels all share the same nominal quantum numbers (additional analysis may subdivide them into parity or permutational symmetry types). These sublevels are called "components" with energies expressed by the notation $F_{\text{c}}(J)$. The quantity N_{c}, the number of components expected, reflects the assignment of the nature of the vibronic state. For linear molecules there is a limited number of components corresponding to the various orientations of electron spin and orbital angular momentum. For example, a $^2\Pi$ electronic state will have four components for each value of J (except for $J = 1/2$, where there are only two components). For nonlinear molecules the number of components increases with J, proportional to $2J+1$, corresponding to various possible projections of the total angular momentum onto the tumbling molecular frame.

The conclusion of this analysis is that a basis set be chosen, over which a model rotational and fine-structure Hamiltonian can be expressed. The wave functions then become vectors of numbers. A priori, only the form of the matrix elements and their dependence on J and body–frame projection (K or Ω) are known. Little is known in advance about how strong the interactions are in any given molecule. Thus one tends to write the Hamiltonian with parameters that are to be determined by fitting the observed energy levels.

Similarly, the choice of the basis sets for the upper and lower states specifies the overall form of the matrix of transition moments between the basis functions. The transition can be chosen to be of a simple standard form, for example, parallel or perpendicular, with only one unknown parameter representing the overall strength of the transition. Alternatively, the transition matrix elements can be considered to be independently adjustable, within the symmetry restrictions that are required (time reversal) or assumed (spatial symmetry).

33.3.2 The Basis Set and Matrix Hamiltonian

For linear molecules it is convenient to choose a basis set labeled by the projections of orbital and spin angular momenta in the body–frame coordinate system, represented symbolically by

$$|\Lambda\Sigma;\,JM\Omega\rangle=\left(\frac{(2J+1)}{8\pi^2}\right)^{1/2}D^{*J}_{M\Omega}|\Lambda\Sigma\rangle\,, \tag{33.11}$$

where $\Omega=\Lambda+\Sigma$. This is called the Hund's case (a) basis set, which is an accurate representation in a single term if the body–frame angular momenta are nearly conserved. This is true if the spin-orbit interaction is larger than the separation between rotational levels. Under all circumstances, this basis set facilitates construction of the matrix Hamiltonian and representation of sources of transition probability [33.9].

One parametrization for the spin–rotation Hamiltonian is provided by *Brown* et al. [33.10–12]:

$$\begin{aligned}&\boldsymbol{H}_{\text{spin-rot}}\\&=T_e+G_v+B_vN^2-D_vN^4\\&\quad+\frac{1}{2}\left[A_v+A_{D_v}N^2,\,L_zS_z\right]_+\\&\quad+(\gamma_v+\gamma_{D_v}N^2)\boldsymbol{N}\cdot\boldsymbol{S}\\&\quad+\frac{1}{3}\left[\lambda_v+\lambda_{D_v}N^2,\,3S_z^2-\boldsymbol{S}^2\right]_+\\&\quad+\eta_vL_zS_z\left[S_z^2-\frac{1}{5}(3S^2-1)\right]\\&\quad-\frac{1}{4}\left[o_v+o_{D_v}N^2,\,\Lambda_+^2S_-^2+\Lambda_-^2S_+^2\right]_+\\&\quad+\frac{1}{4}\left[p_v+p_{D_v}N^2,\,\Lambda_+^2S_-N_-+\Lambda_-^2S_+N_+\right]_+\\&\quad+\frac{1}{4}\left[q_v+q_{D_v}N^2,\,\Lambda_+^2N_-^2+\Lambda_-^2N_+^2\right]_+\,,\end{aligned} \tag{33.12}$$

where $[x,y]_+$ is the anticommutator $(xy+yx)$, and $\boldsymbol{N}=\boldsymbol{J}-\boldsymbol{S}$. *Zare* et al. [33.13] provide an alternative parametrization, with different interpretations of the *spectroscopic constants* (B, D, A, γ, λ, etc.) because they multiply different symbolic operators. One significant difference is that *Zare* et al. use the "mechanical angular momentum" $\boldsymbol{R}=\boldsymbol{J}-\boldsymbol{L}-\boldsymbol{S}$ as the expansion operator, rather than $\boldsymbol{N}$, which might be called the "spinless angular momentum." These differences mean that care must be taken in attempting to construct simulated spectra from published constants. In spite of much discussion in the literature, there is little theoretical foundation for preferring one parametrization over another, as long as the observed levels are accurately fit. In a number of cases naive assumptions about the origin of certain types of interactions have been overturned. For example, the spin–spin interaction, represented by the constant λ, is often dominated by level shifts due to off-diagonal spin-orbit perturbations [33.14].

For polyatomic molecules, a suitable basis set for expansion can be chosen to have a similar form [33.8, 15, 16]

$$|l\Lambda\Sigma;\,JMK\rangle=\left(\frac{(2J+1)}{8\pi^2}\right)^{1/2}D^{*J}_{MK}|l\Lambda\Sigma\rangle\,, \tag{33.13}$$

where Λ and Σ represent the projections of the electron- orbital (L) and spin (S) angular momenta, and l represents the projection of the vibrational angular momentum (p for degenerate vibrational modes). This is the symmetric top basis set. Generalizing the work of *Watson* [33.17, 18], the parametrized Hamiltonian might be written in a form such as

$$\begin{aligned}\boldsymbol{H}_{\text{rot}}=\sum h^{\alpha\beta\gamma\delta\epsilon}_{\zeta\eta\theta l}\Big\{&(\boldsymbol{J}^2)^{\alpha}(\boldsymbol{J}_z)^{2\beta}(\boldsymbol{J}_+^{2\gamma}+\boldsymbol{J}_-^{2\gamma})\\&\times(\boldsymbol{J}\cdot\boldsymbol{p})^{\delta}(\boldsymbol{J}\cdot\boldsymbol{L})^{\epsilon}(\boldsymbol{J}\cdot\boldsymbol{S})^{\zeta}\\&\times(\boldsymbol{p}\cdot\boldsymbol{L})^{\eta}(\boldsymbol{p}\cdot\boldsymbol{S})^{\theta}(\boldsymbol{L}\cdot\boldsymbol{S})^{l}\Big\}\,,\end{aligned} \tag{33.14}$$

where the {} indicates that an appropriately symmetric combination be constructed with anticommutators.

For both linear and nonlinear molecules, it is convenient to use the Wang transformation [33.19] to combine basis functions with opposite sense of rotation: for diatomics

$$\frac{1}{\sqrt{2}}\left[||\Lambda|,\,\Sigma;J,\,M,\,\Omega\rangle\pm|-|\Lambda|,-\Sigma;\,J,\,M,-\Omega\rangle\right]\,; \tag{33.15}$$

and for polyatomics

$$\begin{aligned}\frac{1}{\sqrt{2}}&[|l,\,\Lambda,\,\Sigma;\,J,\,M,\,K\rangle\\&\pm|-l,-\Lambda,-\Sigma;\,J,\,M,-K\rangle]\,.\end{aligned} \tag{33.16}$$

For diatomic molecules, these combinations can be assigned the parity $\pm(-1)^{J+S+s}$, where $s=1$ for Σ^- states, and 0 otherwise [33.13, 20]. For symmetric top molecules, each term is to be accompanied by the appropriate hidden nuclear-spin basis function [33.8, 21].

For asymmetric top molecules, the Wang transformation divides the basis functions into four symmetry classes $E^{\pm}$ and $O^{\pm}$ according to the combining sign and whether K is even or odd. The eigenstates are often labeled by two projection quantum numbers called K_{-1} and K_1. Assuming that $A \geq B \geq C$, the asymmetry parameter

$$\kappa = \frac{(2B - A - C)}{(A - C)} \tag{33.17}$$

ranges from -1 for a prolate symmetric top ($B = C$) to 1 for an oblate symmetric top ($B = A$). A, B, and C are the rotational constants, or reciprocals of the moments of inertia, about the three principal top axes. Each asymmetric top level can be correlated with specific symmetric top levels (i. e., K-values) in the two limits. The prolate limiting K-value is called K_{-1} and the oblate limit is called K_1 (i. e., $\kappa = \pm 1$). Note that the symmetric-top principal axes rotate by 90° during this correlation. The eigenstates are given additional symmetry names (ee,eo,oe,oo) according to whether K_{-1} and K_1 are even or odd. *Papousek* and *Aliev* [33.18] discuss the relations between the ($E^{\pm}$, $O^{\pm}$) and (ee,eo,oe,oo) labeling schemes.

33.3.3 Fitting Experimental Energies

Having chosen a basis set and model Hamiltonian for both the upper and lower levels, the observed transition energies can be used to infer the numerical values of the constants that best fit the spectrum. The following quotation provides a good description of the process:

> *The calculational procedure logically divides into three steps: (1) The matrix elements of the upper and lower state Hamiltonians are calculated for each J value using initial values of the adjustable molecular constants; (2) both Hamiltonians are numerically diagonalized and the resulting sets of eigenvalues are used to construct a set of calculated line positions; and (3) from a least-squares fit of the calculated to the observed line positions, an improved set of molecular constants is generated. This nonlinear least-squares procedure is repeated until a satisfactory set of molecular constants is obtained.*

This quotation is taken from the article by *Zare* et al. [33.13] in which they describe the basis for the LINFIT computer program, one of the first to accomplish direct extraction of constants from diatomic spectral line positions based on numerically diagonalized Hamiltonians.

33.3.4 The Transition Moment Matrix

Diagonalization of the model Hamiltonians for the upper and lower states yields vector wave functions that can be used for calculating matrix elements, especially those needed to evaluate radiative transition probabilities. The wave functions for diatomic molecules have the form

$$\begin{aligned} \psi'_{J'M'c'} &= \sum_{\Lambda'\Sigma'} b^{J'c'}_{\Lambda'\Sigma'} |\Lambda'\Sigma';\, J'M'\Omega'\rangle' \,, \\ \Phi''_{J''M''c''} &= \sum_{\Lambda''\Sigma''} a^{J''c''}_{\Lambda''\Sigma''} |\Lambda''\Sigma'';\, J''M''\Omega''\rangle'' \,. \end{aligned} \tag{33.18}$$

Section 33.2.3 expresses matrix elements of laboratory-frame operators in terms of matrix elements in the rotating body-fixed frame. Terms of the form

$$\mu_{K'K''} = \langle \psi^{(J'K')} | T^{(L)}_q(\text{body}) | \Phi^{(J''K'')} \rangle \theta(-q) \tag{33.19}$$

need to be evaluated. These terms are multiplied by zero if $K' \neq K'' + q$. In the diatomic basis set these become

$$\mu_{\Lambda'\Sigma'\Lambda''\Sigma''} = {}'\langle \Lambda'\Sigma' | T^{(L)}_q(\text{body}) | \Lambda''\Sigma'' \rangle'' \theta(-q) \,, \tag{33.20}$$

where $\theta(-q)$ is a phase factor described in Sect. 33.7.2. Only a few of these matrix elements are independent and nonzero. For electric dipole transitions ($L = 1$), time-reversal and inversion-symmetry can be used to establish the relation

$$\mu_{-\Omega'-\Omega''} = \eta(-1)^{\Omega'-\Omega''} \mu_{\Omega'\Omega''} \,. \tag{33.21}$$

The sign of $\eta = \pm 1$ is determined by the overall character of the electronic transition, and is related to the classification of levels into e- and f-parity types and to the determination of which components are involved in the rotational branches (P, Q, and R). These concepts are elaborated in Sect. 33.4.

33.3.5 Fitting Experimental Intensities

For allowed transitions in linear molecules and symmetric tops, only one independent parameter is normally expected in the transition moment matrix. Thus no additional information is available from fitting the experimental rotational branch strengths (assuming the energy–intensity model is adequate). In diatomic molecules, the intensities of different vibrational bands can be used to infer the internuclear-distance dependence of the electronic transition moment (for example, see *Luque* and *Crosley* [33.22]).

For forbidden transitions and allowed transitions in asymmetric tops, more than one independent parameter is expected. The intensity of a single given rotational line can be expressed in the form

$$I^{\text{line}} = \left| \sum_{K'K''} \mu_{K'K''} Z_{K'K''}(\text{line}) \right|^2 , \quad (33.22)$$

where $Z_{K'K''}(\text{line})$ can be calculated in advance from the energies (wave functions) and quantum numbers alone, using the formulas in Sect. 33.7.2. Nonlinear least-squares fitting can be used to derive the best intensity parameters [33.23–26], analysis of which can help characterize the nature of the transition, and identify the sources of transition probability.

33.4 Selection Rules

33.4.1 Symmetry Types

Selection rules are guidelines for identifying which transitions are expected to be strong and which are expected to be weak. These rules are based on classifying rovibronic levels into labeled symmetry types. Some symmetry distinctions are effectively exact: such as total angular momentum F, or laboratory-inversion parity. Others are approximate, derived from estimates that certain matrix elements are expected to be much larger than others. The most important of these are based on electron spin (for light molecules) and geometrical point-group symmetry (for relatively rigid polyatomics). In actual fact, no transition is completely forbidden. The multipole nature of electromagnetic radiation (electric-dipole, magnetic-dipole, electric-quadrupole, etc.) implies that any change in angular momentum or parity is possible in principle. Practical interest emphasizes identification of the origin of the strongest source of transition probability, and estimation of the strengths of the weak transitions relative to the stronger ones. The result is a collection of *propensity rules* using *selection rules* as tools of estimation.

Basis functions for expansion of the wave functions for the upper and lower states were chosen in Sect. 33.3.2. The first step in the symmetry classification of rovibronic levels consists of identifying various linear combinations of basis functions that block-diagonalize the exact or approximate Hamiltonians. Symmetry-type names are then assigned to these linear combinations based on the value of F or J and knowledge of the symmetry properties of the underlying vibrational and electronic states. Thus each eigenfunction or rovibronic level consists of an expansion over only one of the kinds of linear combination, and the level can be assigned a specific symmetry type.

Similarly, the basis-set expansion leads to a matrix representation of the possible transitions. Spin- and spatial-symmetry arguments establish relationships between these transition matrix elements, and provide estimates of which are much smaller than the others. Each combination of upper- and lower-state symmetry types results in a specific pattern of rotational branches. The most important patterns are ΔJ even (Q-branches) or odd (P- and R-branches), and intensity alternation for consecutive values of J (nuclear spin statistics).

33.4.2 Rotational Branches and Parity

The symmetry of time or momentum reversal implies that changing the signs of all the angular momenta should result in an equivalent wave function. For example, the phase convention

$$\begin{aligned} &\Phi_{\text{vesn}}^{(F,-K_F)}(\text{internal, spins}) \\ &= (-1)^{-F+K_F} \Phi_{\text{vesn}}^{*(F,K_F)}(\text{internal}, -\text{spins}) \quad (33.23) \end{aligned}$$

can be chosen to establish that the relative phases of matrix elements of the Hamiltonian can be taken as

$$\begin{aligned} &\left\langle D^{*F}_{M_F-K'_F} \Phi_{\text{vesn}}^{(F,-K'_F)} \middle| H \middle| D^{*F}_{M_F-K''_F} \Phi_{\text{vesn}}^{(F,-K''_F)} \right\rangle \\ &= \left\langle D^{*F}_{M_F K'_F} \Phi_{\text{vesn}}^{(F,K'_F)} \middle| H \middle| D^{*F}_{M_F K''_F} \Phi_{\text{vesn}}^{(F,K''_F)} \right\rangle^* . \quad (33.24) \end{aligned}$$

The formula for matrix elements of optical transition operators can also be reanalyzed,

$$\begin{aligned} &\left\langle \psi^{F',M''_F+p} \middle| T_p^{(L)}(\text{lab}) \middle| \Phi^{F'',M''_F} \right\rangle \quad (33.25) \\ &= \left[\frac{(2F''+1)}{(2F'+1)} \right]^{1/2} \left\langle F''M''_F, Lp \middle| F'M''_F+p \right\rangle \\ &\times \frac{1}{2} \sum_{qK''} \left\langle F''K''_F, Lq \middle| F'K''_F+q \right\rangle \\ &\times \left\{ \left\langle \psi_{\text{vesn}}^{(F',K''_F+q)} \middle| T_q^{(L)}(\text{body}) \middle| \Phi_{\text{vesn}}^{(F'',K''_F)} \right\rangle \right. \\ &+ (-1)^{F'+L-F''} \\ &\times \left. \left\langle \psi_{\text{vesn}}^{(F',-K''_F-q)} \middle| T_{-q}^{(L)}(\text{body}) \middle| \Phi_{\text{vesn}}^{(F'',-K''_F)} \right\rangle \right\} , \end{aligned}$$

to establish that all contributions are purely real or purely imaginary [33.27]. Since such transition matrix elements will be used as absolute squares, they can be treated as if they were purely real.

Parity classification of molecular states according to inversion through the center of mass is important for establishing which transitions are electric-dipole allowed and which states can perturb each other. As discussed by *Larsson* [33.4], such classification is not without subtlety and opportunity for confusion. Inversion of the laboratory spatial coordinates (i_{sp}, also called E^* [33.2]) is equivalent to a reflection σ of the molecule-fixed electronic and nuclear coordinates in an arbitrary plane followed by rotation of the molecular frame by 180° about an axis through the origin and perpendicular to the reflection plane (if F is half-integral special care must be taken about the sense of rotation). It follows that

$$\begin{aligned} & i_{sp}\Phi^{F,M_F}_{exact}(\text{lab}) \\ & = E^*\Phi^{F,M_F}_{exact}(\text{lab}) \\ & = \eta(-1)^{F-\gamma}\Phi^{F,M_F}_{exact}(\text{lab}) \\ & = \left[\frac{(2F+1)}{8\pi^2}\right]^{1/2} \\ & \quad \times \sum_{K_F}(-1)^{F-K_F} D^{*F}_{M_F-K_F}\sigma_{xz}\Phi^{(F,K_F)}_{vesn} \end{aligned} \tag{33.26}$$

and

$$\sigma_{xz}\Phi^{(F,K_F)}_{vesn} = \eta(-1)^{K_F-\gamma}\Phi^{(F,-K_F)}_{vesn}\,, \tag{33.27}$$

where $\gamma = 0$ or $1/2$ for integral or half-integral F, respectively, and η is the parity label for the state, having values of ± 1. In linear molecules, levels with $\eta = +1$ are called e-levels while those with $\eta = -1$ are called f-levels [33.28, 29].

Inversion symmetry can be combined with time reversal to establish that all matrix elements of the Hamiltonian can be taken to be real [33.27]. The wave function can also be expressed in the form of the Wang transformation [33.19], uniting the $\pm K_F$ components,

$$\begin{aligned} & \Phi^{F,M_F}_{exact}(\text{lab}) \\ & = \left(\frac{(2F+1)}{8\pi^2}\right)^{1/2} \sum_{K_F\geq 0}\left[D^{*F}_{M_F K_F}\Phi^{(F,K_F)}_{vesn}\right. \\ & \quad \left. + \eta(-1)^{-K_F+\gamma} D^{*F}_{M_F-K_F}\sigma_{xz}\Phi^{(F,K_F)}_{vesn}\right] \\ & \quad \times\left[2(1+\delta_{K_F 0})\right]^{-1/2}\,. \end{aligned} \tag{33.28}$$

If the molecule is rigid and has a plane of symmetry, or is nonrigid with accessible vibrational or tunneling modes that correspond to plane reflections, inversion symmetry divides the states of the molecule into two classes, according to the sign of η. Perturbations can occur only for $\Delta F = 0$ and $\eta'\eta'' = +1$. For optical transitions, the change in parity of the states must match the parity of the operator. Odd operators (e.g., electric-dipole) require $\eta'\eta''(-1)^{\Delta F} = -1$. Even operators (e.g., magnetic-dipole and electric-quadrupole) require $\eta'\eta''(-1)^{\Delta F} = +1$.

33.4.3 Nuclear Spin, Spatial Symmetry, and Statistics

For most molecules, the coupling of nuclear spin with the electron-spin, electron-orbital, and frame-rotational angular momenta is sufficiently weak that treatment of the energetics of hyperfine interactions can be postponed. The first-order effect of nuclear spin is that rovibronic wave functions for molecules containing identical nuclei must be combined with appropriate nuclear spin wave functions in order to obtain the necessary Fermi–Dirac or Bose–Einstein nuclear permutation symmetry. For many molecules, there exist combinations of nuclear permutations that correspond to combinations of frame rotations, laboratory inversions, and feasible vibrational motions (the rotational wave function makes a contribution because renumbering the nuclei requires a reanalysis of the Euler angles). For rigid molecules, these permutations (possibly including inversion) can be used to generate the point symmetry group of the molecule. For fluxional molecules, with multiple energetically equivalent nuclear configurations, a rather large "molecular symmetry group" can result, one that may not correspond to any ordinary point group [33.1, 2].

In the discussion immediately following, consider the case of N occurrences of one kind of nucleus, the others being unique (e.g., PD_3). The treatment can easily be extended to the case of multiple kinds of identical nuclei (e.g., C_2H_6). The exact wave function can be rearranged into a sum over products of the form

$$\Phi_{exact} = \sum_{a,b}\Phi^{(a)}_{rves}\Phi^{(b)}_{nspin}\,, \tag{33.29}$$

where $\Phi^{(a)}_{rves}$ is a rovibronic wave function belonging to the $\Gamma(a)$ representation of the symmetric group S_N of permutations over N objects, and $\Phi^{(b)}_{nspin}$ is a nuclear spin wave function, belonging to the $\Gamma(b)$ representation of S_N. In order to obtain the correct permutation symmetries for the overall wave function, the only terms that can appear in this sum are those for which the direct product $\Gamma(a)\otimes\Gamma(b)$ contains the symmetric or antisymmetric representation, for bosons or fermions, respectively.

Assumption of negligible hyperfine interactions allows evaluation of matrix elements of the form

$$\left\langle \Phi^{(c)}_{\text{rves}} \Phi^{(d)}_{\text{nspin}} \middle| H \middle| \Phi^{(a)}_{\text{rves}} \Phi^{(b)}_{\text{nspin}} \right\rangle = \left\langle \Phi^{(c)}_{\text{rves}} \middle| H \middle| \Phi^{(a)}_{\text{rves}} \right\rangle \left\langle \Phi^{(d)}_{\text{nspin}} \middle| \Phi^{(b)}_{\text{nspin}} \right\rangle , \tag{33.30}$$

which vanishes unless $\Gamma(a) = \Gamma(c)$ and $\Gamma(b) = \Gamma(d)$. Thus the nearly-exact wave function can be written as a sum over products where the rovibronic and nuclear-spin factors correspond to basis functions from single known representations:

$$\Phi_{\text{exact}} \approx \sum_{a,b} \Phi^{(a)}_{\text{rves}} \Phi^{(b)}_{\text{nspin}} , \tag{33.31}$$

with $\Gamma(a) = \Gamma_{\text{rves}}$ and $\Gamma(b) = \Gamma_{\text{nspin}}$. This divides the states of the molecule into a number of noninteracting symmetry classes, labeled by the representations of the symmetric group. In the absence of hyperfine interactions, optical transitions are possible only within a certain symmetry class.

Thus the existence of spatial or dynamical symmetry implies that each rovibronic wave function transforms according to a particular representation of a subgroup of the permutation-inversion group (called CNPI by *Bunker* [33.2]). Each representation includes only specific values of nuclear spin, corresponding to the permutational properties of the nuclear spin wave functions. The most important effect of this analysis is to assign statistical weights or relative intensities to the different symmetry types. For example, the symmetry group for NH_3 is $D_{3\text{h}}$ (including umbrella inversion), with representations A_1', A_2', A_1'', A_2'', E', and E''. The A_1' and A_1'' representations must be combined with the (nonexistent) antisymmetric spin function, yielding a statistical weight of 0. Similarly, A_2' and A_2'' combine with the symmetric $I = 3/2$ spin function, with a statistical weight of 4 (i. e., $2I + 1$). E' and E'' combine with the nonsymmetric $I = 1/2$ spin functions, with a statistical weight of 2. This material is discussed from various viewpoints in numerous articles and text books, of which only a few can be cited here [33.1, 2, 8, 18, 21, 30–36]. See Chapt. 32 for additional details and examples.

Although this analysis appears rather complicated, the selection rules that result are actually the same, at least in simple cases, as the ones that are derivable from simpler ideas. For example, for a ${}^1\Sigma_g^+$ lower state, even J levels are permutation symmetric and have parity $+1$, while odd J levels are permutation antisymmetric and have parity -1. For a ${}^1\Sigma_u^+$ upper state, even J levels are permutation antisymmetric and have parity $+1$, while odd J levels are permutation symmetric and have parity -1. Both the parity selection rule and permutation-symmetry selection rule independently require that $\Delta J = \pm 1$ for electric-dipole transitions. Similarly, that the permanent dipole moment of a symmetric-top molecule must lie along the body-fixed axis replicates the $\Delta K = 0$ selection rule for pure-rotation transitions provided by permutational symmetry arguments. This means that when simulating absorption and emission spectra, the nuclear-spin wave function can usually be ignored. The intensity alternation imposed by spin-statistics can be represented by multiplying each wave function by the appropriate factor, for example, 0 or $[(2I+1)/(2i+1)^N]^{1/2}$, where I is the total nuclear spin, and i is the spin of one of the N equivalent nuclei.

Group theory remains vital for understanding the relative strengths of vibrational transitions in polyatomics (see *Cotton* [33.37], for example, and Sect. 33.6.2) and becomes very interesting as interaction between vibration and rotation increases. For the purposes of this discussion, the most important issue is identification of which transition-moment matrix elements $\mu_{K'K''}$ vanish and which are related by symmetry.

33.4.4 Electron Orbital and Spin Angular Momenta

For all molecules, the strongest transitions tend to be those that conserve electron spin. The zero-order transition-moment matrix is diagonal both in total spin and in spin-projection onto the body–frame axis. In the $|\Lambda\Sigma\rangle$ basis set for linear molecules this is expressed by

$$\mu_{\Lambda'\Sigma'\Lambda''\Sigma''} = \mu_{\Lambda'\Lambda''} \delta_{\Sigma'\Sigma''} . \tag{33.32}$$

The transition-moment tensor operator $T_q^{(L)}$(body) can connect basis functions that differ in Λ by at most L. Allowed electric-dipole transitions thus satisfy

$$\mu_{\Lambda'\Lambda''} = 0, \quad \text{for } |\Lambda' - \Lambda''| > 1 . \tag{33.33}$$

In the usual case that the upper and lower states each consist of only a single value of $|\Lambda|$, there is only one independent, nonzero, matrix element $\mu_{|\Lambda'|,|\Lambda''|}$ with

$$\mu_{-|\Lambda'|,-|\Lambda''|} = \eta(-1)^{|\Lambda'|-|\Lambda''|} \mu_{|\Lambda'|,|\Lambda''|} . \tag{33.34}$$

See Sect. 33.8 for a discussion of spin-forbidden and orbitally-forbidden transitions. Similar arguments and phase relationships can be developed for polyatomic molecules with nonzero electron spin or degenerate vibrational or electronic states.

33.5 Absorption Cross Sections and Radiative Lifetimes

33.5.1 Radiation Relations

Among the most important radiation relations is the connection between the absorption cross section and the rate of spontaneous emission. *Einstein* [33.38] introduced his A and B coefficients to describe the rates of absorption and emission of radiation of a collection of two-level atoms or molecules in equilibrium with a radiation field at the same temperature. The discussion here follows that of *Condon* and *Shortley* [33.39], *Penner* [33.40], *Thorne* [33.41], and *Steinfeld* [33.42] (with corrections). Also see Chapts. 10, 17, and 68 of this Handbook. The number of absorption events per unit volume per unit time is written as

$$N_l B_{lu} \rho(\nu) . \tag{33.35}$$

While the rate of emission is

$$N_u B_{ul} \rho(\nu) + N_u A_{ul} . \tag{33.36}$$

In thermal equilibrium, the radiative energy density is given by the Planck blackbody law

$$\rho(\nu) = \left(\frac{8\pi h \nu^3}{c^3}\right)\left(e^{h\nu/kT} - 1\right)^{-1} , \tag{33.37}$$

and the ground and excited state densities satisfy a Boltzmann relationship,

$$\frac{N_u}{N_l} = \left(\frac{g_u}{g_l}\right) e^{-h\nu/kT} , \tag{33.38}$$

where g_u and g_l are the degeneracies of the upper and lower states. The requirement that the rates of absorption and emission must be equal leads to the relations

$$A_{ul} = \left(\frac{8\pi h \nu^3}{c^3}\right) B_{ul} = \left(\frac{8\pi h \nu^3}{c^3}\right)\left(\frac{g_l}{g_u}\right) B_{lu} . \tag{33.39}$$

Numerical values of the B coefficients can be derived from the optical absorption cross section, and thus

$$\begin{aligned} A_{ul} &= \left(\frac{8\pi \nu^2}{c^2}\right)\left(\frac{g_l}{g_u}\right)\int \sigma_{\text{abs}}(\nu)\,d\nu \\ &= \left(\frac{8\pi \nu^2}{c^2}\right)\int \sigma_{\text{se}}(\nu)\,d\nu . \end{aligned} \tag{33.40}$$

Finally, the expression for the absorption oscillator strength is

$$\begin{aligned} f_{\text{abs}} &= (4\pi\epsilon_0)\left(\frac{mc^3}{8\pi^2\nu^2 e^2}\right)\left(\frac{g_u}{g_l}\right) A_{ul} \\ &= (4\pi\epsilon_0)\left(\frac{mc}{\pi e^2}\right)\int \sigma_{\text{abs}}(\nu)\,d\nu . \end{aligned} \tag{33.41}$$

The emission oscillator strength is simply related to that for absorption: $f_{\text{em}} = -(g_l/g_u) f_{\text{abs}}$. The oscillator strength offers considerable advantages as a means of reporting and comparing the strengths of radiative transitions. It is dimensionless, obeys the simple sum rule (for electric-dipole transitions)

$$\sum_u f_{ul} = \text{number of electrons} , \tag{33.42}$$

and is directly derivable from an experimental absorption cross section even before the assignment of the upper level has been determined (i. e., before its degeneracy is known).

33.5.2 Transition Moments

In many cases, the intention is to construct model quantum mechanical wave functions for the two states involved in the transition under study. In addition, ab initio electronic wave functions and matrix elements may be available (see Chapt. 31). Quantum mechanics suggests the following expression for the Einstein A coefficient (see Sect. 11.5.1):

$$\begin{aligned} A_{ul} = &\left(\frac{64\pi^4\nu^3}{3hc^3}\right)\left(\frac{1}{4\pi\epsilon_0}\right)\left(\frac{1}{g_u}\right) \\ &\times \sum_{u',l'',p} \left|\langle \psi'_{u'} | e r_p | \psi''_{l''} \rangle\right|^2 . \end{aligned} \tag{33.43}$$

The summation is over all three optical polarization directions p (i. e., r_p runs over x, y, and z in the lab frame), all degenerate components l'' of the lower state (i. e., g_l of them), and all degenerate components u' of the upper state (i. e., g_u of them). This triple sum is also called the *line strength* S_{ul}. Division by the upper-level degeneracy corrects for the fact that the transitions should be averaged rather than summed over the initial levels.

In practice, choosing the appropriate degeneracy to divide by is a question of some ambiguity. For atoms, it is sufficient to understand how the individual matrix elements and the line strength were calculated. For example, *Bethe* and *Salpeter* [33.43] use a degeneracy of $(2L+1)$ for Schrödinger wave functions for the hydrogen atom, and $(2J+1)$ for Dirac wave functions.

For molecules with internal angular momentum, i. e., everything other than $^1\Sigma$ states of linear molecules, the situation is much more complicated. For electric-dipole allowed transitions in light molecules, ab initio transition moments are calculated in a body-fixed coordinate system, ignoring spin, and not summed over anything.

For diatomic molecules, following the work of *Whiting* et al. [33.44, 45], the transition probability from a single upper-state component $(J'c')$ to a single lower-state component $(J''c'')$ is written as

$$A_{J'c'J''c''} = \left(\frac{64\pi^4\nu^3}{3hc^3}\right)\left(\frac{1}{4\pi\epsilon_0}\right) \times \left(\frac{1}{2J'+1}\right) q_{v'v''}|R_e|^2 S^{c'c''}_{J'J''} . \quad (33.44)$$

In this formula $q_{v'v''}|R_e|^2$ represents the rotationless contribution to the transition moment, symbolically represented as a product of a vibrational overlap ($q_{v'v''}$, i. e., a Franck–Condon factor) and an electronic-only component $|R_e|^2$ (Sect. 33.6.1). All of the rotational complexity is absorbed into the rotational-branch strength factor $S^{c'c''}_{J'J''}$ (Sects. 33.7.2 and 33.7.3). The issue to be addressed here is how to divide numerical factors between $|R_e|^2$ and $S^{c'c''}_{J'J''}$. One approach is to construct an estimate for the rotationless transition probability

$$A_{v'v''} = \left(\frac{1}{N'_c}\right) \sum_{c'J''c''} A_{J'c'J''c''} , \quad (33.45)$$

where N'_c is the number of internal spin-orbit components of the upper state. *Whiting* et al. suggest that $S^{c'c''}_{J'J''}$ be normalized such that for spin-allowed transitions

$$\sum_{c'J''c''} S^{c'c''}_{J'J''} = (2-\delta_{0,\Lambda'}\delta_{0,\Lambda''})(2S+1)(2J'+1) . \quad (33.46)$$

The first factor is 1 for Σ–Σ transitions, and 2 for all others. The final factor is replaced by $(2J''+1)$ if the sum is over J' instead of J''. For spin-forbidden transitions the following is a plausible extension of this sum rule,

$$\sum_{c'J''c''} S^{c'c''}_{J'J''} = \max(N'_c, N''_c)(2J'+1) . \quad (33.47)$$

Section 33.7.3 provides a corresponding sum rule for polyatomic molecules. This normalization yields

$$A_{v'v''} = \left(\frac{64\pi^4\nu^3}{3hc^3}\right)\left(\frac{1}{4\pi\epsilon_0}\right) \times \left(\frac{\max(N'_c, N''_c)}{N'_c}\right) q_{v'v''}|R_e|^2 \quad (33.48)$$

and for spin-allowed transitions, the simple spin-free expressions for the electronic transition moments:

$$|R_e|^2 = |\langle\Lambda|ez|\Lambda\rangle|^2 \quad (33.49)$$

for parallel transitions and

$$|R_e|^2 = \left|\langle\Lambda+1|e\frac{1}{\sqrt{2}}(x+iy)|\Lambda\rangle\right|^2 \quad (33.50)$$

for perpendicular transitions.

33.6 Vibrational Band Strengths

33.6.1 Franck–Condon Factors

The Born–Oppenheimer separation of electron and nuclear motion suggests that during an optical transition between different electronic states the nuclei should change neither their position nor momentum. This concept was developed from semiclassical arguments by *Franck* [33.46] and justified quantum mechanically by *Condon* [33.47]. Following *Herzberg* [33.48] and *Steinfeld* [33.42] the vibronic transition moment can be written as

$$\begin{aligned}\mu_{v'v''} &= \langle\psi'_e\psi'_{v'}|\mu|\psi''_e\psi''_{v''}\rangle ,\\ &= \int dR\, \psi^{*\prime}_{v'}(R)\psi''_{v''}(R)\\ &\quad\times \int dr\, \psi^{*\prime}_e(r,R)\psi''_e(r,R)\mu(r,R) ,\\ &= \int dR\, \psi^{*\prime}_{v'}(R)\psi''_{v''}(R)\mu(R) .\end{aligned} \quad (33.51)$$

If the R-dependence of $\mu(R)$ is sufficiently weak, it can be factored out to obtain

$$\mu_{v'v''} = R_e \int dR\, \psi^{*\prime}_{v'}(R)\psi''_{v''}(R) , \quad (33.52)$$

where R_e is called the electronic transition moment. The transition probability is proportional to the square of the above, which is usually written as

$$I \approx q_{v'v''} R_e^2 , \quad (33.53)$$

where

$$q_{v'v''} = \left|\int dR\, \psi^{*\prime}_{v'}(R)\psi''_{v''}(R)\right|^2 , \quad (33.54)$$

the square of the overlap between initial and final vibrational wave functions, is called the Franck–Condon factor.

The Franck–Condon factors satisfy the sum rule

$$\sum_{v'} q_{v'v''} = \sum_{v''} q_{v'v''} = 1 \quad (33.55)$$

provided the summations include the continuum vibrational wave functions above the dissociation limits. The Franck–Condon approach also can be used to calculate intensities and cross sections for bound–free [33.49] and free–free [33.50] emission and absorption.

In some cases the variation of $\mu(R)$ is significant. Calculation of the effect on the intensities can usually be handled by the *r-centroid* method in which the expression

$$\bar{r}_{v'v''} = \frac{\langle \psi'_{v'} | R | \psi''_{v''} \rangle}{\langle \psi'_{v'} | \psi''_{v''} \rangle} \tag{33.56}$$

is used to calculate an effective internuclear distance for the transition. The transition strength is then proportional to $q_{v'v'}|\mu(\bar{r}_{v'v''})|^2$. An advantage of this formulation is that the vibrational overlaps can be calculated from energy information only, before the transition moment function is known and before the transition strengths are investigated experimentally. The quantitative accuracy of the r-centroid method for transition moments that are not linear in the internuclear distance, has been addressed by a considerable literature, which has been summarized by *McCallum* [33.51].

A second complication arises from the fact that the vibrational wave functions themselves depend parametrically on the rotational angular momentum. Calculation of rotationally-dependent Franck–Condon factors is described by *Dwivedi* et al. [33.52] who also discuss the r-centroid method.

33.6.2 Vibrational Transitions

Vibrational transitions derive their strength from the variation of the "permanent" dipole moment of the molecule as a function of geometry or internuclear coordinates. As described by several authors [33.31, 32, 34, 53] one can expand the dipole moment as a power series in the internal-Cartesian or normal-mode coordinates

$$M_{xyz}(Q) = M^0_{xyz} + \sum_i \left(\frac{\partial M_{xyz}}{\partial Q_i} \right) Q_i + \cdots \tag{33.57}$$

and calculate intensities from a formula like

$$I \approx \left| \langle \psi_{v'_1} \psi_{v'_2} \cdots | M(\boldsymbol{Q}) | \psi_{v''_1} \psi_{v''_2} \cdots \rangle \right|^2 . \tag{33.58}$$

For homonuclear diatomic molecules, the dipole moment vanishes identically, so there is no rovibrational spectrum. The dipole moment for heteronuclear diatomics is often close to linear in the internuclear distance. The harmonic oscillator model suggests that transitions with $\Delta v = \pm 1$ are the strongest, with intensities approximated by

$$I_{v+1,v} \approx \left| \frac{\mathrm{d}M}{\mathrm{d}R} \right|^2 (v+1) . \tag{33.59}$$

Overtone bands, i. e., with $|\Delta v| > 1$, are observed, as dramatically illustrated by the $\Delta v = 4, 5$ emissions from the OH radical observed from the Earth's night sky [33.54].

For polyatomic molecules, overtone and combination bands are often quite strong. The presence or absence of which is used to establish the symmetries of the vibrational modes. In general, it is difficult to construct quantitative vibrational intensity formulas with only a few parameters that can be inferred experimentally.

33.7 Rotational Branch Strengths

33.7.1 Branch Structure and Transition Type

The overall rotational structure of a molecular transition is determined by the relative values and phases of the body–frame transition-moment matrix elements, the relative values and phases of coefficients in the expansion of the upper-state and lower-state component wave functions over the angular-momentum-projection basis functions, the energy separations between the components, and the relative values and phases of the vector-coupling coefficients. In simple cases, each lower component might be connected to only a single upper component [Hund's case (b) or symmetric tops], or $\Delta J = \pm 1$ (P- and R-branches) may dominate over $\Delta J = 0$ (Q-branches).

For diatomic molecules, symmetry arguments are used to divide the components into the two parity classes e and f. For electric dipole transitions, the selection rules from Sect. 33.4.2 imply that

$$\begin{aligned} &\left(N'_e N''_e + N'_f N''_f\right) && P\text{- and } R\text{-branches} \\ &\left(N'_e N''_f + N'_f N''_e\right) && Q\text{-branches} \end{aligned} \tag{33.60}$$

are expected, where N_e and N_f indicate the number of components of each parity class (N_e and N_f differ by no more than one). Rotational branches are labeled with the notation ${}^{\Delta R}\Delta J_{c'c''}$, using symbols P for -1, Q for 0, and R for $+1$. ΔR indicates the "apparent" change in mechanical rotational angular momentum (i. e., energy) [see Hund's case (b) in Sect. 33.7.4] and ΔJ indicates the "actual" change (i. e., quantum mechanical). The labels $c'c''$ indicate the components involved. Thus a ${}^{P}Q_{21}$ branch is expected to be "red shaded" ($\Delta R = -1$) with $\Delta J = 0$ involving the second component of the upper state and the first (lowest) component of the lower state. The notation ${}^{P}Q_{21}(J)$ [or sometimes ${}^{P}Q_{21}(R = N)$ for Σ lower states] identifies an individual rotational line and specifies the rotational quantum number of the lower state involved in the transition. If the upper- and lower-state component numbers are the same, one of them may be dropped. Thus P_{22} is sometimes written as P_2.

For symmetric top molecules, the rotational branches are labeled ΔJ_K (e.g., P_1). For asymmetric tops, the branches are labeled by $\Delta J_{\Delta K_{-1},\Delta K_1}$, where K_{-1} and K_1 are the prolate- and oblate-limit angular momenta projections (described in Sect. 33.3.2).

The transition dipole moment commonly lies parallel or perpendicular to the body–frame axis. In the former case, $\mu_{K'K''}$ vanishes for $K' \neq K''$, and in the latter for $K' = K''$. Thus parallel bands correspond to $\Delta K = 0$ transitions, while perpendicular bands have $\Delta K = \pm 1$. As enforced by the vector-coupling coefficients or Hönl–London factors described below, for low values of K (e.g., diatomics), $\Delta K = 0$ implies strong $\Delta J = \pm 1$ (P and R) branches and weak $\Delta J = 0$ (Q) branches. On the other hand, $\Delta K = \pm 1$ leads to Q-branches that are approximately twice as strong as either the P- or R-branches.

33.7.2 Hönl–London Factors

The matrix model Hamiltonians for the upper and lower states have been diagonalized, yielding the wave functions

$$\psi_{c'}^{J'M'} = \sum_{K'} b_{J'c'}^{K'} \left(\frac{(2J'+1)}{8\pi^2} \right)^{1/2} \times D_{M'K'}^{*J'}(\text{lab})\, \chi_{K'}(\text{body}) \qquad (33.61)$$

and

$$\Phi_{c''}^{J''M''} = \sum_{K''} a_{J''c''}^{K''} \left(\frac{(2J''+1)}{8\pi^2} \right)^{1/2} \times D_{M''K''}^{*J''}(\text{lab})\, \xi_{K''}(\text{body}) . \qquad (33.62)$$

In these expressions, the designations K' and K'' are slightly symbolic. They represent the body–frame projection of the total angular momentum and also a running index over basis functions. For complicated cases, more than one basis function can have a given value of K.

Following Sect. 33.2.3, the rotational branch strength is then written as

$$\begin{aligned} S_{J'J''}^{c'c''} &= \sum_{pM'M''} \left|\langle \psi_{c'}^{J'M'} | T_p^{(L)}(\text{lab}) | \Phi_{c''}^{J''M''} \rangle\right|^2 \\ &= (2J''+1) \left| \sum_{q'K'K''} b_{J'c'}^{*K'} a_{J''c''}^{K''} \langle \chi_{K'} | T_q^{(L)}(\text{body}) | \xi_{K''} \rangle \right. \\ &\quad \left. \times \langle J''K'', Lq | J'K' \rangle \right|^2 \end{aligned} \qquad (33.63)$$

or

$$S_{J'J''}^{c'c''} = \left| \sum_{K'K''} b_{J'c'}^{*K'} a_{J''c''}^{K''} \mu_{K'K''} \zeta(J', K', J'', K'') \right|^2 , \qquad (33.64)$$

where

$$\mu_{K'K''} = \langle \chi_{K'} | T_q^{(L)}(\text{body}) | \xi_{K''} \rangle \theta(K'' - K') \qquad (33.65)$$

is the body–frame transition-moment matrix introduced in Sect. 33.3.4, with relative values that are hypothesized based on interpretation of the nature of the transition, calculated from ab initio wave functions, or inferred by fitting the observed rotational branch strengths. The Clebsch–Gordan expression

$$\begin{aligned} &\zeta(J', K', J'', K'') \\ &= (2J''+1)^{1/2} \langle J''K'', LK' - K'' | J'K' \rangle \\ &\quad \times \theta(K'' - K')\theta(J' - J'') \end{aligned} \qquad (33.66)$$

represents the transformation of the radiation field from the laboratory-frame to the body–frame, also related to the "direction cosines" used by many authors. The additional phase factors

$$\theta(k) = \text{sgn}(k) = \begin{cases} +1 & k \geq 0 \\ -1 & k < 0 \end{cases} \qquad (33.67)$$

have been included here to make the signs and symmetry relations of $\mu_{K'K''}$ and $\zeta(J', K', J'', K'')$ agree with those already in use [33.9, 44, 55, 56]. They are related to the choice of the leading signs when T_+ and T_- are expressed as $\pm(T_x + \mathrm{i}T_y)$ and $\pm(T_x - \mathrm{i}T_y)$. Their inclusion has no effect for spin-allowed transitions with only one source of transition probability, e.g., purely parallel or perpendicular.

In the usual case of electric-dipole (or magnetic-dipole) radiation (i. e., $L=1$), ζ^2 is the well-known Hönl–London factor [33.57]. $\zeta(J', K', J'', K'')$ is a real, signed quantity: negative for $\Delta J \Delta K > 0$; or $\Delta J = \Delta K = 0$ and $K < 0$; otherwise positive [33.9].

Setting $L=2$ provides intensity formulas for electric-quadrupole [33.58], *Raman* [33.59], and two-photon [33.60, 61] transitions. Additional Rayleigh-like terms can appear for $K' = K'' \neq 0$. *Halpern* et al. [33.61] also give formulas for three-photon transitions in diatomics, expressed in terms of Clebsch–Gordan coefficients with $L=3$ and $L=1$ (for $|\Delta\Omega| \leq 1$).

33.7.3 Sum Rules

The orthonormality relations for component eigenvectors

$$\sum_{c'} b^{*K'}_{J'c'} b^{K}_{J'c'} = \delta_{K',K} \,, \tag{33.68}$$

$$\sum_{c''} a^{*K''}_{J''c''} a^{K}_{J''c''} = \delta_{K'',K} \tag{33.69}$$

can be used to construct the sum rule

$$\sum_{c'c''} S^{c'c''}_{J'J''} = \sum_{K'K''} \left|\mu_{K'K''}\zeta(J', K', J'', K'')\right|^2 . \tag{33.70}$$

Finally, the orthonormality relations of the Clebsch–Gordan coefficients result in

$$\sum_{J'c'c''} S^{c'c''}_{J'J''} = (2J''+1) \sum_{K'K''} \left|\mu_{K'K''}\right|^2 , \tag{33.71}$$

$$\sum_{J''c'c''} S^{c'c''}_{J'J''} = (2J'+1) \sum_{K'K''} \left|\mu_{K'K''}\right|^2 . \tag{33.72}$$

As discussed in Sect. 33.5.2, it is convenient to have the $\mu_{K'K''}$ matrix elements consist of numbers that represent the nature of the transition but not its strength, the latter being expressed by the "vibrational" ($q_{v'v''}$) and "electronic" (R_e) contributions. Following Sect. 33.5.2, for diatomic molecules, the "orientational" part $\mu_{K'K''}$ is taken to have a fixed sum rule

$$\sum_{K'K''} |\mu_{K'K''}|^2 = \max(N'_c, N''_c) \,, \tag{33.73}$$

where N'_c is the number of components (K' values, or basis functions) for the upper state, and N''_c is the number of components in the lower state. For polyatomic molecules, the sum rule can be written as

$$(2J+1)^2 (\mu_a^2 + \mu_b^2 + \mu_c^2) \max(N'_c, N''_c) \,, \tag{33.74}$$

where

$$\mu_a^2 + \mu_b^2 + \mu_c^2 = |\mu_+|^2 + |\mu_-|^2 + |\mu_0|^2 = 1 \tag{33.75}$$

and N'_c and N''_c are the numbers of spin-electronic-vibrational components in the upper and lower states, respectively. Also see *Whiting* et al. [33.44, 45] and *Brown* et al. [33.7].

33.7.4 Hund's Cases

In diatomic molecules, several limiting cases are useful as short-hand or first-approximation concepts for classification of energy levels and rotational branch strengths. These are called the Hund's cases [33.62–64]. They are distinguished by the extent to which the electron orbital and spin angular momenta are rigidly attached to the tumbling molecular frame, i. e., whether Λ, Σ, and S are good quantum numbers. Hund's cases are discussed in many journal articles and in every textbook dealing with the rotational structure of diatomic spectra. An appealing recent description is provided by *Nikitin* and *Zare* [33.65].

In most works, the emphasis has been on finding a favorable zero-order approximation for perturbation expansion of energy levels. The advance of precision measurement of transition energies and the availability of sophisticated parametrical matrix models and fast computers on which to realize them, has reduced the importance of Hund's cases for actual computations. In particular, the need to derive and implement numerous explicit energy and intensity formulas leads to unfortunate transcription errors. Nevertheless, they remain of value for qualitative and pedagogical understanding, especially for estimates of the relative intensities of rotational branches.

Hund's case (a) describes the situation in which Λ and Σ are separately well-defined. This is a common case in which the separation between electronic states, i. e., different values of $|\Lambda|$, is larger than the spin-orbit interaction, which in turn, is larger than the separation between rotational levels. At low J, there are $(2S+1)$ pairs of nearly-degenerate energy levels separated from each other by the spin orbit constant: $E \approx A\Lambda\Sigma +$

$BJ(J+1)$. The wave functions are of the form

$$\psi^{JM}_{\Omega\pm} = \frac{1}{\sqrt{2}}\left(\frac{(2J+1)}{8\pi^2}\right)^{1/2} \times\left[D^{*J}_{M\Lambda+\Sigma}|\Lambda,\,\Sigma\rangle \pm D^{*J}_{M-\Lambda-\Sigma}|-\Lambda,\,-\Sigma\rangle\right] = \frac{1}{\sqrt{2}}[|\Lambda,\,\Sigma;\,J,\,M,\,\Omega\rangle \pm|-\Lambda,\,-\Sigma;\,J,\,M,\,-\Omega\rangle]\,, \quad (33.76)$$

with only two nonzero expansion coefficients a^K_{Jc}. If both the upper and lower states are well described by Hund's case (a), then each lower component is optically connected only to the upper components with the same value of $|\Sigma|$. Then

$$S^{\Omega'\Omega''}_{J'J''} = (2J''+1)\left|\langle J''\Omega'',\,1\Omega'-\Omega|J'\Omega'\rangle\right|^2\,, \quad (33.77)$$

with $\Omega'' = \Lambda''+\Sigma$ and $\Omega' = \Lambda'+\Sigma$.

Hund's case (b) indicates that Λ is well-defined, but spin-orbit coupling is weak. The components correspond to well-defined values of $\boldsymbol{N} = \boldsymbol{J}-\boldsymbol{S}$, ranging from $|J-S|$ to $(J+S)$, with energies approximated by $E \approx BN(N+1)$, and wave functions of the form

$$\psi^{JM}_{N\pm} = \sum_{\Sigma}(-1)^{S+\Sigma}\langle J-\Omega,\,S\Sigma|N-\Lambda\rangle \times\frac{1}{\sqrt{2}}[|\Lambda,\,\Sigma;\,J,\,M,\,\Omega\rangle \pm|-\Lambda,\,-\Sigma;\,J,\,M,\,-\Omega\rangle]\,. \quad (33.78)$$

This equation is derivable from Mizushima's equation (2-3-26) [33.14] and Zare's equations (2.8), (2.26), and (3.105) [33.5], using the lab-to-body transformation

$$|SM_S\rangle(\text{lab}) = \sum_{\Sigma} D^{*S}_{M_S\Sigma}(\phi\theta\chi)|S\Sigma\rangle(\text{body})\,. \quad (33.79)$$

It disagrees with Judd's problem 9.1 [33.66] by a phase factor $(-1)^{J+2S+\Sigma-N}$ but agrees with Mizushima's expansion of a $^3\Pi$ state [33.14, p. 287] if the Clebsch–Gordan coefficients are taken from *Condon* and *Shortley* [33.39, p. 76].

If both the upper and lower states are well described by Hund's case (b), these wave functions can be substituted into the general rotational-branch strength equations above. Following *Edmonds* [33.67, (6.2.8) and (6.2.13), and Table 5] yields the square of a product of Clebsch–Gordan and Racah coefficients

$$S^{N'N''}_{J'J''} = (2J''+1)(2J'+1)(2N''+1) \times\left|\langle N''\Lambda'',\,1\Lambda'-\Lambda''|N'\Lambda'\rangle \times W(N',\,J'',\,N'',\,J';\,S,\,1)\right|^2\,. \quad (33.80)$$

The Clebsch–Gordan coefficient enforces the case (b) selection rule $\Delta N = 0, \pm1$, while the Racah coefficient provides the $\Delta J = \Delta N$ propensity rule, which becomes more precise as N increases. A similar propensity rule, $\Delta F = \Delta J$, is common for transitions between hyperfine components (see also *Femenias* [33.68]).

Hund's case (c) corresponds to the situation in which spin-orbit coupling is so strong that each level described by the projection Ω actually consists of multiple values of $|\Lambda|$ (e.g., mixing of Σ and Π states) or multiple values of S (e.g., mixing of singlet and triplet spins). This limiting case is formally similar to Hund's case (a), but no assumptions can be made about the relative magnitudes of transition-moment matrix elements $\mu_{\Omega'\Omega''}$. Any of which can be nonzero for $|\Delta\Omega| \le 1$, for example

$$S^{\Omega'_+\Omega''_\pm}_{J'J''} = \frac{1}{4}\big|\mu_{\Omega'\Omega''}\zeta(J',\,\Omega',\,J'',\,\Omega'') \pm\mu_{\Omega'-\Omega''}\zeta(J',\,\Omega',\,J'',\,-\Omega'') +\mu_{-\Omega'\Omega''}\zeta(J',\,-\Omega',\,J'',\,\Omega'') \pm\mu_{-\Omega'-\Omega''}\zeta(J',\,-\Omega',\,J'',\,-\Omega'')\big|^2\,. \quad (33.81)$$

The symmetry (sign) relations between $\mu_{\Omega'\Omega''}\zeta(J',\,\Omega',\,J'',\,\Omega'')$ and $\mu_{-\Omega'-\Omega''}\zeta(J',\,-\Omega',\,J'',\,-\Omega'')$ determine whether this transition occurs only for $\Delta J = \pm1$ (P- and R-branches) or only for $\Delta J = 0$ (Q-branches).

The interest and complexity of Hund's case (c) were exemplified by a seminal work by *Kopp* and *Hougen* [33.69], who considered $\Omega' = 1/2$, $\Omega'' = 1/2$ transitions, under the assumption that both states could consist of arbitrary mixtures of $^2\Sigma_{1/2}$ and $^2\Pi_{1/2}$ character. Each of the six rotational branches shows constructive or destructive interference of parallel ($\Delta\Omega = 0$) and perpendicular ($\Delta\Omega = \pm1$) contributions. Hund's case (c) also describes spin-orbit mixing collisions [33.70] or dissociation to specific spin-orbit limits [33.71–73].

Hund's case (d) arises in the investigation of Rydberg series [33.74], in which the separation between Σ and Π from the same orbital configuration approach each other as the principal quantum number (n) increases. Spin-orbit coupling between these projections also diminishes. The eigenfunction components correspond to well-defined values of $\boldsymbol{R} = \boldsymbol{J}-\boldsymbol{L}$, ranging from $|J-L|$ to $(J+L)$, with energies approximated by $E \approx BR(R+1)$, and wave functions of the form

$$\psi_{RL}^{JM} = \sum_{\Lambda} (-1)^{L+\Lambda} \langle J-\Lambda, L\Lambda | R0 \rangle | L\Lambda; JM\Lambda \rangle \,. \tag{33.82}$$

Carroll [33.74] used intensity formulas provided by *Kovacs* [33.58] to analyze, by spectral simulation, the 4p–15p $\left({}^1\Sigma_u^+ \text{ and } {}^1\Pi_u\right)$ Rydberg states of N_2 excited from the ground $X\,{}^1\Sigma_g^+$. Three strong Q-form branches survive, correspondingto $R' = J''$, two arising from $\Pi \leftarrow \Sigma$ and one from $\Sigma \leftarrow \Sigma$ case (a) branches. The remaining O-form ($R' = J'' - 2$) and S-form ($R' = J'' + 2$) branches fade rapidly as n increases. With the phase conventions used here, this situation corresponds to body–frame transition matrix elements satisfying

$$\mu_{00} = \mu_{-10} = -\mu_{10} \,. \tag{33.83}$$

In the opposite case, corresponding to a parity change of the parent-ion core [33.73], two of the Q-form branches are extinguished, while one Q-form, one O-form, and one S-form branch remain. The transition matrix elements would satisfy

$$\mu_{00} = 0 \,, \quad \mu_{-10} = \mu_{10} \,. \tag{33.84}$$

Hund's case (d) polyatomics are also known [33.8].

Hund's case (e) would correspond to a situation in which $\boldsymbol{L}$ and $\boldsymbol{S}$ are strongly coupled to each other, but neither is strongly coupled to the internuclear axis. No examples are known for bound states of molecules.

33.7.5 Symmetric Tops

For transitions between nondegenerate vibronic states, the transition moment must lie along the principal top axis, leading to the selection rule $\Delta K = 0$. Otherwise, Hougen's convenient quantum number [33.33] $G = \Lambda + l - K$, provides the selection rule $\Delta G = 0, \pm n$ (for an n-fold major symmetry axis) (Sects. Section 33.3.2, Section 33.4.3, and Chapt. 32). Transitions with $\Delta G = \pm n$ are much weaker than those with $\Delta G = 0$ and are not calculable from a simple formula. Branch intensities can be calculated with the Hönl–London formulas of Sect. 33.7.2.

33.7.6 Asymmetric Tops

In general, no assumptions can be made about the orientation of the transition moment. The vector representations (μ_x, μ_y, μ_z), (μ_a, μ_b, μ_c), and $(\mu_0, \mu_{+1}, \mu_{-1})$ can have any combination of independent nonzero values. It is common that one of the (μ_a, μ_b, μ_c) values is significantly larger than the others, especially for planar molecules with a two-fold symmetry axis. In this case one obtains a type A, B, or C band, if μ_a, μ_b, μ_c dominates, respectively [33.31, 75]. The tradition of analytic calculation of line strengths from explicit representations of wave functions and transition moments leads to formulas of considerable complexity, with somewhat restrictive assumptions [33.35, 36, 76]. In the more general notation of Sect. 33.7.2, the rotational line strength can be written as

$$S_{J'J''}^{\tau'\tau''} = \Bigg| \sum_{K'K''} b_{J'\tau'}^{*K'} a_{J''\tau''}^{K''} \left(\mu_0 \delta_{K'K''} + \mu_+ \delta_{K'K''+1} \right. \\ \left. + \mu_- \delta_{K'K''-1}\right) \zeta\left(J', K', J'', K''\right) \Bigg|^2 \,, \tag{33.85}$$

where

$$\mu_0 = \mu_c \,, \quad |\mu_+| = |\mu_-| \,, \\ |\mu_+|^2 + |\mu_-|^2 = \mu_a^2 + \mu_b^2 \,. \tag{33.86}$$

Papousek and *Aliev* [33.18] and *Zare* [33.5] follow the present formulation, but with somewhat less generality with respect to wave function expansion coefficients or transition moment components.

33.8 Forbidden Transitions

33.8.1 Spin-Changing Transitions

The formalism presented above permits simulation of any allowed transition or forbidden transitions mediated by spin-orbit or spin-spin perturbations, or any perturbation that is diagonal in Ω. For spin-allowed transitions, the transition moment matrix is taken to be diagonal in and independent of the spin projection, so that

$$\mu_{\Lambda'\Sigma'\Lambda''\Sigma''} = \mu_{\Lambda'\Lambda''} \delta_{\Sigma'\Sigma''} \,. \tag{33.87}$$

For forbidden transitions, or complicated Hund's case (c) mixings, the transition moment matrix elements can be considered as independent variable parameters, limited only by the symmetry constraint

$$\mu_{-\Omega'-\Omega''} = \eta(-1)^{\Omega'-\Omega''}\mu_{\Omega'\Omega''}\,, \tag{33.88}$$

and the fact that terms with $|\Delta\Omega| > 1$ will be multiplied by zero. Alternatively, a specific set of candidate perturbers can be selected, and the Λ- and Σ-dependence of their contributions to the transition-moment matrix evaluated explicitly. For example, first-order spin-orbit mixing would lead to terms of the form

$$\begin{aligned}&\mu^{S'S''}_{\Lambda'\Sigma'\Lambda''\Sigma''}\\&=\sum_{\Lambda}\Bigg[\mu^{S'S'}_{\Lambda'\Sigma'\Lambda\Sigma'}\frac{\langle S'\Lambda\Sigma'|H_{SO}|S''\Lambda''\Sigma''\rangle}{\Delta E''}\\&\quad+\left(\frac{\langle S'\Lambda'\Sigma'|H_{SO}|S''\Lambda\Sigma''\rangle}{\Delta E'}\right)\mu^{S''S''}_{\Lambda\Sigma''\Lambda''\Sigma''}\Bigg]\,.\end{aligned} \tag{33.89}$$

However, care must be taken in reducing these matrix elements using the Wigner–Eckart theorem, for example, following *Lefebvre-Brion* and *Field* [33.55], in order to satisfy the $\Delta\Omega = 0$ requirement for matrix elements of the rotationless Hamiltonian.

33.8.2 Orbitally-Forbidden Transitions

Even if the upper and lower states share the same value of electron spin, the transition may still be forbidden. The change in orbital angular momentum may be too large, $|\Delta\Lambda| > 1$, or a change in reflection parity, $\Sigma^- \to \Sigma^+$, may cause the zero-order transition matrix elements to vanish. Spin-orbit mixing with other $^{2S+1}\Lambda$ states, as described above, is usually the largest source of transition probability. In addition, terms in the Hamiltonian of the form $\boldsymbol{J}\cdot\boldsymbol{L}$ lead to contributions to the transition strength that increase with J, and that may mix-in higher values of Ω than were present in the zero-order $\Lambda\Sigma$ basis set for the upper and lower states. This situation can be represented by generalizing the formula from Sect. 33.7.2, following *Huestis* et al. [33.23],

$$\begin{aligned}S^{c'c''}_{J'J''} &= \Bigg|\sum_{K'K''} b^{*K'}_{J'c'}a^{K''}_{J''c''}\\&\quad\times\sum_{i=-1}^{1}\mu^{(i)}_{K'K''}\zeta^{(i)}(J',K',J'',K'')\Bigg|^2\,,\end{aligned} \tag{33.90}$$

where $\mu^{(0)}_{K'K''}$ is the rotationless contribution ($\mu_{K'K''}$ from Sect. 33.7.2) and $\mu^{(\pm1)}_{K'K''}$ are the new rotation-assisted terms. The new reflection-symmetry rule is

$$\mu^{(i)}_{-K'-K''} = \eta(-1)^{K'-K''+i}\mu^{(-i)}_{K'K''}\,. \tag{33.91}$$

The revised square-root Hönl–London factors are

$$\zeta^{(0)}(J',K',J'',K'') = \zeta(J',K',J'',K'') \tag{33.92}$$

(from Sect. 33.7.2) and

$$\begin{aligned}&\zeta^{(\pm1)}(J',K',J'',K'')\\&=\frac{1}{2}\{[J'(J'+1)-K'(K'\mp1)]^{1/2}\\&\quad\times\zeta(J',K'\mp1,J'',K'')\\&\quad+[J''(J''+1)-K''(K''\pm1)]^{1/2}\\&\quad\times\zeta(J',K',J'',K''\pm1)\}\,.\end{aligned} \tag{33.93}$$

As in Sect. 33.7.2 the symbols K' and K'' represent $\Lambda'\Sigma'$ and $\Lambda''\Sigma''$ when used as labels, and Ω' and Ω'' when used as numbers (a distinction that is relevant only when $S \geq |\Lambda|$ and $\Lambda \neq 0$). This formulation is more symmetric than that proposed by *Huestis* et al. [33.23], in that it explicitly allows for either the upper or lower state to be mixed by rotation (of significance only for low J and $\Delta\Lambda > 1$).

33.9 Recent Developments

Added by Mark M. Cassar. Astronomical sky spectra are important for an understanding of processes both in Earth's and other terrestrial environments. These spectra are the background spectra obtained through the slit of a spectrometer while excluding the object of primary interest to the astronomer – the star, galaxy, etc. The sky spectrum is subsequently subtracted from the object spectrum so that the final product contains no emissions from extraneous sources – nightglow, zodiacal light, and the light of other stellar objects. This operation then leaves the astronomer with purer astronomical spectra, which can then be compared to theoretical transition probability calculations to identify emission sources. This procedure has recently been used to identify the atomic oxygen green line in the Venus night airglow [33.77], which relied on an understanding of molecular oxygen emissions. In addition, interpretation of the intensities of molecular oxygen emissions also furthers the understanding of the elementary processes occurring in the Earth's atmosphere [33.78].

Two recent studies have focused on the radiative properties of the CaN and $^{39}K^{85}Rb$ molecules. In the former study, the radiative transition probabilities and lifetimes for the $A^4\Pi - X^4\Sigma^-$ and $B^4\Sigma^- - X^4\Sigma^-$ band systems were calculated [33.79]. These results will in turn facilitate future spectroscopic studies of CaN showing the essential interplay between theory and experiment, which is required for a deeper understanding of these processes. (Radiative properties are sensitive to electronic coupling schemes and to configuration interaction, and thus present an important testing ground for theoretical models [33.80, 81].) The second study provides quantitative estimates for the radiative cooling of heteronuclear translationally ultracold molecules [33.82, 83]. By calculating the radiative transition probabilities for $^{39}K^{85}Rb$, which lead to the radiative lifetime through the total Einstein A coefficient, it has been shown that under appropriate laboratory conditions such a cooling process is not relevant [33.84].

References

33.1 H. C. Longuet-Higgins: Mol. Phys. **6**, 445 (1963)
33.2 P. R. Bunker: *Molecular Symmetry and Spectroscopy* (Academic, New York 1979)
33.3 E. P. Wigner: *Group Theory* (Academic, New York 1959)
33.4 M. Larsson: Phys. Scr. **23**, 835 (1981)
33.5 R. N. Zare: *Angular Momentum* (Wiley, New York 1988)
33.6 E. S. Chang, U. Fano: Phys. Rev. A **6**, 173 (1972)
33.7 J. M. Brown, B. J. Howard, C. M. L. Kerr: J. Mol. Spectrosc. **60**, 433 (1976)
33.8 H. Helm, L. J. Lembo, P. C. Cosby, D. L. Huestis: Photoionization and dissociation of the triatomic hydrogen molecule. In: *Fundamentals of Laser Interactions II*, ed. by F. Ehlotzky (Springer, Berlin, Heidelberg 1989)
33.9 J. T. Hougen: *The Calculation of Rotational Energy Levels and Rotational Line Intensities in Diatomic Molecules*, NBS Monograph 115 (U.S. Government Printing Office, Washington, DC 1970)
33.10 J. M. Brown, E. A. Colbourn, J. K. G. Watson, F. D. Wayne: J. Mol. Spectrosc. **74**, 294 (1979)
33.11 J. M. Brown, A. J. Merer: J. Mol. Spectrosc. **74**, 488 (1979)
33.12 J. M. Brown, D. J. Milton, J. K. G. Watson, R. N. Zare, D. L. Albritton, M. Horani, J. Rostas: J. Mol. Spectrosc. **90**, 139 (1981)
33.13 R. N. Zare, A. L. Schmeltekopf, W. J. Harrop, D. L. Albritton: J. Mol. Spectrosc. **46**, 37 (1973)
33.14 M. Mizushima: *The Theory of Rotating Diatomic Molecules* (Wiley, New York 1975)
33.15 G. Herzberg: *Molecular Spectra and Molecular Structure III. Electronic Spectra and Electronic Structure of Polyatomic Molecules* (Van Nostrand, New York 1966)
33.16 L. J. Lembo, H. Helm, D. L. Huestis: J. Chem. Phys. **90**, 5299 (1989)
33.17 J. K. G. Watson: Aspects of quartic and sextic centrifugal effects on rotational energy levels. In: *Vibrational Spectra and Structure*, Vol. 6, ed. by J. Durig (Elsevier, Amsterdam 1977)
33.18 D. Papousek, M. R. Aliev: *Molecular Vibrational-Rotational Spectra* (Elsevier, Amsterdam 1982)
33.19 S. C. Wang: Phys. Rev. **34**, 243 (1929)
33.20 H. Helm, P. C. Cosby, R. P. Saxon, D. L. Huestis: J. Chem. Phys. **76**, 2516 (1982)
33.21 C. H. Townes, A. L. Schawlow: *Microwave Spectroscopy* (McGraw Hill, New York 1955) p. 1975 (reprinted, Dover, New York)
33.22 J. Luque, D. R. Crosley: J. Quant. Spectrosc. Radiat. Transfer **53**, 189 (1995)
33.23 D. L. Huestis, R. A. Copeland, K. Knutsen, T. G. Slanger, R. T. Jongma, M. G. H. Boogaarts, G. Meijer: Can. J. Phys. **72**, 1109 (1994)
33.24 T. G. Slanger, D. L. Huestis: J. Chem. Phys. **78**, 2274 (1983)
33.25 C. M. L. Kerr, J. K. G. Watson: Can. J. Phys. **64**, 36 (1986)
33.26 M. J. Dyer, G. W. Faris, P. C. Cosby, D. L. Huestis, T. G. Slanger: Chem. Phys. **171**, 237 (1993)
33.27 C. Di Lauro, F. Lattanzi, G. Graner: Molec. Phys. **71**, 1285 (1990)
33.28 R. L. Kronig: *Band Spectra and Molecular Structure* (Cambridge Univ. Press, London 1930)
33.29 J. M. Brown, J. T. Hougen, K.-P. Huber, J. W. C. Johns, I. Kopp, H. Lefebvre-Brion, A. J. Merer, D. A. Ramsey, J. Rostas, R. N. Zare: J. Mol. Spectrosc. **55**, 500 (1975)
33.30 E. B. Wilson Jr.: J. Chem. Phys. **3**, 276 (1935)
33.31 G. Herzberg: *Molecular Spectra and Molecular Structure II. Infrared and Raman Spectra of Polyatomic Molecules* (Van Nostrand, New York 1945)
33.32 E. B. Wilson Jr., J. C. Decius, P. C. Cross: *Molecular Vibrations* (McGraw Hill, New York 1955)
33.33 J. T. Hougen: J. Chem. Phys. **37**, 1433 (1962)
33.34 H. C. Allen, P. C. Cross: *Molecular Vib-Rotors* (Wiley, New York 1963)

33.35 J. E. Wollrab: *Rotational Spectra and Molecular Structure* (Academic, New York 1967)
33.36 H. W. Kroto: *Molecular Rotation Spectra* (Wiley, London 1975) (reprinted, Dover, New York 1992)
33.37 F. A. Cotton: *Chemical Applications of Group Theory*, 2nd edn. (Wiley, New York 1971)
33.38 A. Einstein: Physik. Z. **18**, 121 (1917)
33.39 E. U. Condon, G. H. Shortley: *The Theory of Atomic Spectra* (Cambridge Univ. Press, London 1935) (reprinted 1967)
33.40 S. S. Penner: *Quantitative Molecular Spectroscopy and Gas Emissivities* (Addison-Wesley, Reading 1995)
33.41 A. P. Thorne: *Spectrophysics* (Chapman Hall, London 1974)
33.42 J. I. Steinfeld: *Molecules and Radiation* (MIT, Cambridge 1978)
33.43 H. A. Bethe, E. E. Salpeter: *Quantum Mechanics of One- and Two-Electron Atoms* (Springer, Berlin, Heidelberg 1957)
33.44 E. E. Whiting, R. W. Nicholls: Astrophys. J. Suppl. Series **235**, 27 (1974)
33.45 E. E. Whiting, A. Schadee, J. B. Tatum, J. T. Hougen, R. W. Nicholls: J. Mol. Spectrosc. **80**, 249 (1980)
33.46 J. Franck: Trans. Faraday Soc. **21**, 536 (1925)
33.47 E. U. Condon: Phys. Rev. **32**, 858 (1928)
33.48 G. Herzberg: *Molecular Spectra and Molecular Structure I. Spectra of Diatomic Molecules* (Van Nostrand, New York 1950)
33.49 J. G. Winans, E. C. G. Stueckelberg: Proc. Nat. Acad. Sci. **14**, 867 (1928)
33.50 R. E. M. Hedges, D. L. Drummond, A. Gallagher: Phys. Rev. A **6**, 1519 (1972)
33.51 J. C. McCallum: J. Quant. Spectrosc. Radiat. Transfer **21**, 563 (1979)
33.52 P. H. Dwivedi, D. Branch, J. H. Huffaker, R. A. Bell: Astrophys. J. Suppl. **36**, 573 (1978)
33.53 G. W. King: *Spectroscopy and Molecular Structure* (Holt, Rinehart and Winston, New York 1964)
33.54 A. B. Meinel: Astrophys. J. **111**, 555 (1950)
33.55 H. Lefebvre-Brion, R. W. Field: *Perturbations in the Spectra of Diatomic Molecules* (Academic, New York 1986)
33.56 D. L. Huestis: *DIATOM Spectral Simulation Computer Program, Version 7.0* (SRI International, Menlo Park, CA 1994)
33.57 H. Hönl, F. London: Z. Phys. **33**, 803 (1925)
33.58 I. Kovacs: *Rotational Structure in the Spectra of Diatomic Molecules* (Elsevier, New York 1969)
33.59 A. Weber: High resolution Raman studies of gases. In: *The Raman Effect*, Vol. 2, ed. by A. Anderson (Dekker, New York 1973)
33.60 K. Chen, E. S. Yeung: J. Chem. Phys. **69**, 43 (1978)
33.61 J. B. Halpern, H. Zacharias, R. Wallenstein: J. Mol. Spectrosc. **79**, 1 (1980)
33.62 F. Hund: Z. Phys. **36**, 657 (1926)
33.63 F. Hund: Z. Phys. **40**, 742 (1927)
33.64 F. Hund: Z. Phys. **42**, 93 (1927)
33.65 E. E. Nikitin, R. N. Zare: Mol. Phys. **82**, 85 (1994)
33.66 B. R. Judd: *Angular Momentum Theory for Diatomic Molecules* (Academic, New York 1975)
33.67 A. R. Edmonds: *Angular Momentum in Quantum Mechanics*, ed. by 2nd (Princeton Univ. Press, Princeton 1960) (reprinted with corrections 1974)
33.68 J. L. Femenias: Phys. Rev. A **15**, 1625 (1977)
33.69 I. Kopp, J. T. Hougen: Can. J. Phys. **45**, 2581 (1967)
33.70 A. P. Hickman, D. L. Huestis, R. P. Saxon: J. Chem. Phys. **96**, 2099 (1992)
33.71 H. Helm, P. C. Cosby, D. L. Huestis: J. Chem. Phys. **73**, 2629 (1980)
33.72 H. Helm, P. C. Cosby, D. L. Huestis: Phys. Rev. A **30**, 851 (1984)
33.73 P. C. Cosby, D. L. Huestis: J. Chem. Phys. **97**, 6108 (1992)
33.74 P. K. Carroll: J. Chem. Phys. **58**, 3597 (1973)
33.75 H. H. Nielsen: Phys. Rev. **38**, 1432 (1931)
33.76 P. C. Cross, R. M. Hainer, G. W. King: J. Chem. Phys. **12**, 210 (1944)
33.77 T. G. Slanger, P. C. Crosby, D. L. Huestis, T. A. Bida: Science **291**, 463 (2001)
33.78 T. G. Slanger, R. A. Copeland: Chem. Rev. **103**, 4731 (2003)
33.79 M. Peligrini, O. Roberto-Neto, F. B. C. Machado: Chem. Phys. Lett. **375**, 9 (2003)
33.80 E. Biémont, H. P. Garnir, P. Palmeri, P. Quinet, Z. S. Li, Z. O. Zhang, S. Svanberg: Phys. Rev. A **64**, 022503 (2001)
33.81 H. L. Xu, A. Persson, S. Svanberg, K. Blagoev, G. Malcheva, V. Pentchev, E. Biémont, J. Campos, M. Ortiz, R. Mayo: Phys. Rev. A **70**, 042508 (1970)
33.82 W. C. Stwalley, H. Wang: J. Mol. Spectrosc. **195**, 194 (1999)
33.83 J. T. Bahns, P. L. Gould, W. C. Stwalley: Adv. At. Mol. Opt. Phys. **42**, 171 (2000)
33.84 W. T. Zemke, W. C. Stwalley: J. Chem. Phys. **120**, 88 (2004)

34. Molecular Photodissociation

Molecular photodissociation is the photoinitiated fragmentation of a bound molecule [34.1]. The purpose of this chapter is to outline the ways in which molecular photodissociation is studied in the gas phase [34.2]. The results are particularly relevant to the investigation of the species involved in combustion and atmospheric reactions [34.3].

34.1 **Observables** 537
34.1.1 Scalar Properties 537
34.1.2 Vector Correlations 537

34.2 **Experimental Techniques** 539

34.3 **Theoretical Techniques** 540

34.4 **Concepts in Dissociation** 541
34.4.1 Direct Dissociation 541
34.4.2 Vibrational Predissociation 542
34.4.3 Electronic Predissociation 542

34.5 **Recent Developments** 543

34.6 **Summary** 544

References 545

Conceptually the photodissociation process can be divided into three stages. During the first stage the molecule absorbs a photon and is promoted to an excited state. This is generally an excited electronic state, but can be a highly excited vibrational state in the ground electronic state. In the second stage, the transient complex evolves through a series of *transition states*, until finally, in the third stage, the molecule enters the *exit channel* and dissociates into the products. Schematically, this might be represented, for a triatomic molecule ABC (see Fig. 34.1), as

$$\mathrm{ABC} + \hbar\omega \to (\mathrm{ABC})^{\ddagger} \to \mathrm{AB}(v, j) + \mathrm{C}\,. \qquad (34.1)$$

In the case of the triatomic molecule represented here, the dissociation involves the transformation of one of the vibrational modes to a translational, or dissociative, mode, another vibrational mode (the bending) to rotational motion of the products (j), whilst the third vibrational mode is preserved (v).

When the molecule is promoted to an electronic state which has a purely repulsive potential energy surface (PES), it undergoes very rapid dissociation, often in less than one vibrational period. This is called *direct dissociation*. However, the dissociation of the transient complex can be delayed, taking place over many vibrational periods. This is called *indirect dissociation*, or predissociation, and has been divided into three different categories [34.4], though as with the division between direct and indirect, this is sometimes somewhat arbitrary.

Vibrational Predissociation (Herzberg Type II)

In this case, the transient complex is on a vibrationally adiabatic potential energy surface (this is an effective potential for the molecule when it is in a particular vibrational state v) which is not dissociative, or which has a barrier to dissociation. Therefore, to dissociate it must either tunnel through the barrier, which is the only possibility for $v = 0$, or undergo a nonadiabatic transition to a lower vibrational state, thereby transferring energy from a vibrational degree of freedom to the dissociative mode. This process of energy exchange is commonly called the intramolecular redistribution of vibrational energy (IVR).

Rotational Predissociation (Herzberg Type III)

In this case, the transient complex is on a nondissociative rotationally adiabatic potential energy surface. Therefore, in a similar manner to vibrational predissociation, if it is to dissociate it must undergo a nonadiabatic transition to a lower rotational state, thereby transferring energy from rotation to the dissociative mode.

Electronic Predissociation (Herzberg Type I)

In this case, the PES of the electronic state of the transient complex is not dissociative at the given energy, and in order to dissociate the molecule must undergo a nonadiabatic transition to a second dissociative electronic state. This involves the coupling of nuclear and electronic motion and therefore leads to

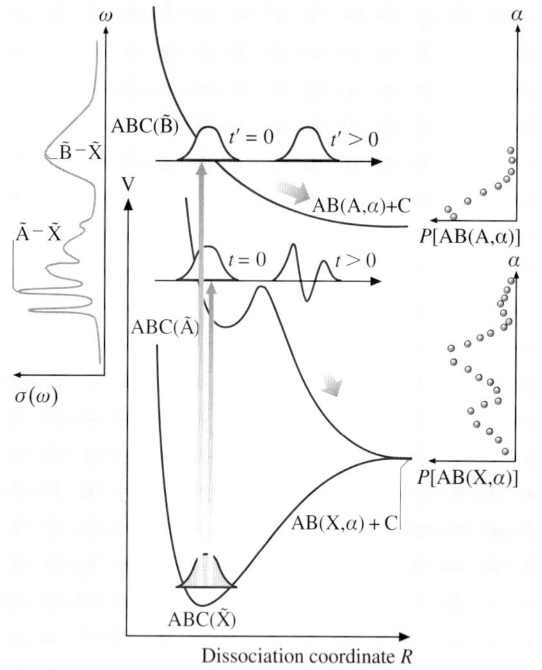

Fig. 34.1 Schematic representation of the uv photodissociation of a triatomic molecule ABC into products AB(α) and C, illustrating the total absorption cross section $\sigma(\omega)$, evolution of the molecular wavepacket, and asymptotic product state distributions $P(\alpha)$, for direct and indirect dissociations on $\tilde{B}$ and $\tilde{A}$ state PESs respectively

a break down of the Born–Oppenheimer (BO) approximation. There are two main types of electronic predissociation. In the first case, there is only a very small coupling, and no actual crossing, between two different electronic states, and the transition between the two is driven by the very high density of vibrational states on the second electronic state. This is called internal conversion for spin-allowed processes, and intersystem crossing for spin-forbidden processes. In the second case, the transition between the electronic states is driven by strong coupling. This coupling can be vibronic (vibrational-electronic) in nature, e.g., for the Renner–Teller and Jahn–Teller effects, or purely electronic, as in the case of a *conical intersection*.

Selection Rules

Two sets of selection rules apply to photodissociation. The first set governs the allowed states to which the molecule can be promoted by the photon. These selection rules are simply those for bound-state spectroscopy (Sect. 33.4). Note in particular the selection rule $\Delta J = 0, \pm 1$. This has important practical implications since it means that a molecule which is initially rotationally cold remains so after absorption of a photon. Thus, those observables which are averaged over J will have a clear structure experimentally, and will be easier to calculate theoretically. This is in contrast to scattering experiments which in general involve a summation over many J states (Chapt. 36).

The second set of selection rules governs the dissociation process. The transient complex, or prepared (p) state, will undergo transitions to a final (f) vibrational, rotational or electronic state in order to dissociate; these transitions have their own set of selection rules. As for all selection rules, these are determined on the basis of symmetry. For a total wave function Ψ and a perturbation function W, which consists of the coupling terms or neglected terms in the Hamiltonian, $\int \Psi_p^* W \Psi_f \, d\tau$ must be nonzero for a transition to take place. As W forms a part of the Hamiltonian, it is totally symmetric, and therefore the integral is nonzero only if the prepared and final state irreducible representations are equal, $\boldsymbol{\Gamma}_p = \boldsymbol{\Gamma}_f$. If there is a transition to an excited electronic state, the point groups of the initial and final states are often not the same, in which case the point group formed by the joint elements of symmetry is used, or, in the case where there is no stable geometry for one of the states, the symmetry of the potential is used. For a diatomic molecule these selection rules are given in Table 49.2.

If the motion can be separated into vibrational, rotational, and electronic parts, so that $\Psi = \Psi^{\mathrm{v}} \Psi^{\mathrm{r}} \Psi^{\mathrm{e}}$ and $W = W^{\mathrm{v}} + W^{\mathrm{r}} + W^{\mathrm{e}}$, it is then possible to derive three separate selection rules: $\boldsymbol{\Gamma}_p^{\mathrm{r}} = \boldsymbol{\Gamma}_f^{\mathrm{r}}$ for the rotational motion, i.e. conservation of internal angular momentum; $\boldsymbol{\Gamma}_p^{\mathrm{e}} = \boldsymbol{\Gamma}_f^{\mathrm{e}}$ for the electronic motion; and $\boldsymbol{\Gamma}_p^{\mathrm{v}} = \boldsymbol{\Gamma}_f^{\mathrm{v}}$ for the vibrational motion. Since the final vibrational state is in the continuum, in practice all vibrational species ($\boldsymbol{\Gamma}_f^{\mathrm{v}}$) are available at a given energy, so that the vibrational selection rule is not significantly restrictive. This separation is not possible in the case of electronic predissociation occurring through the Renner–Teller or Jahn–Teller effect, where it is necessary to consider the vibronic species of the initial and final states.

34.1 Observables

Fundamental to any study of photodissociation is the measurement or calculation of the characteristic properties, or observables, of the reaction, from which the underlying dynamics of the fragmentation process can be inferred.

34.1.1 Scalar Properties

The absorption cross section $\sigma(\omega)$ is a measure of the capacity of the molecule to absorb photons with frequency ω. It is analogous to the line intensity in bound-state spectroscopy. Assuming that the light-matter interaction is weak (Chapt. 68), and that the light pulse is on for a long time, the absorption cross section is given by

$$\sigma(\omega) \propto \omega_{fi} \left| \langle \Psi_f | \boldsymbol{E} \cdot \hat{\boldsymbol{\mu}} | \Psi_i \rangle \right|^2 , \tag{34.2}$$

where Ψ_i and Ψ_f represent the initial and final states, whose energies differ by $\hbar\omega_{fi}$; $\boldsymbol{E}$ is a unit vector in the direction of the polarization of the electric field, and $\hat{\boldsymbol{\mu}}$ is the electric dipole operator of the molecule. Assuming the Born–Oppenheimer separation of electronic and nuclear motion, (34.2) can be rewritten as

$$\sigma(\omega) \propto \omega_{fi} \left| \langle \Psi_f^{\mathrm{rv}} | \mu_{fi} | \Psi_i^{\mathrm{rv}} \rangle \right|^2 , \tag{34.3}$$

where the electronic transition dipole moment μ_{fi} equals $\langle \Psi_f^{\mathrm{e}} | \boldsymbol{E} \cdot \hat{\boldsymbol{\mu}} | \Psi_i^{\mathrm{e}} \rangle$, and is in general dependent on the internal coordinates of the molecule. The superscripts (r, v) will be dropped from now on, and Ψ will refer to the wave function for the internal coordinates of the molecule.

The absorption cross section reflects not only the nature of the transient complex but also its evolution through the transition states. For direct dissociation the absorption cross section is usually very broad and structureless. In contrast, the absorption cross section for predissociation is structured, containing lines which are normally Lorentzian in shape, and whose widths Γ are related to the lifetime of the transient complex at that energy by $\tau = \hbar/\Gamma$.

The partial photodissociation cross sections $\sigma(\omega, \alpha)$ are a measure of the capacity of the molecule to absorb photons with frequency ω *and* to yield products in quantum state α. They are defined by

$$\sigma(\omega, \alpha) \propto \omega_{fi} \left| \langle \Psi_f^{\alpha} | \mu_{fi} | \Psi_i \rangle \right|^2 , \tag{34.4}$$

where Ψ_f^{α} is the final wave function for the products in the quantum state α. The partial cross sections for direct dissociation are broad and featureless. For predissociation, similar structures are seen in the partial cross sections as in the absorption cross section. The absorption or total cross section is given by the sum of the partial cross sections over all final product states:

$$\sigma(\omega) = \sum_{\alpha} \sigma(\omega, \alpha) . \tag{34.5}$$

The rotational and vibrational product distributions $P(\omega, \alpha)$ provide information about the amount of product formed by a photon with frequency ω in a particular rotational or vibrational state α. These are related to the partial cross sections by $P(\omega, \alpha) = \sigma(\omega, \alpha)/\sigma(\omega)$. The rotational and vibrational product distributions reflect the nature of the transient complex as it enters the exit channel, as well as the dynamics in the exit channel.

The branching ratios for different chemical species produced in photodissociation are defined as the fraction of the total number of parent molecules that produce the particular species of interest. In (34.1), the molecule ABC dissociated into AB + C. It might equally well have dissociated to A + BC, or indeed A + B + C. It is clear then that there may be several different reaction schemes, or channels, for the photodissociation of one particular molecule. Thus, the branching ratio for forming AB is the yield of this first channel divided by the total dissociation yield into all possible channels. Further, it would sometimes be possible to produce AB, or any of the other chemical species, in various electronic, vibrational or/and rotational quantum states. In this case, the branching ratio for forming AB(α) is the yield of AB in the specific quantum state α divided by the total yield of AB; however, this describes the branching into only this particular reaction channel.

For the reaction scheme represented in (34.1), the quantum yields of the products AB and C are the same. However, for example, in the reaction

$$\mathrm{ABC_2} + \hbar\omega \rightarrow (\mathrm{ABC_2})^{\ddagger} \rightarrow \mathrm{AB}(v, j) + 2\mathrm{C} , \tag{34.6}$$

the quantum yield of C is twice that for AB. In general, the quantum yield of a particular product fragment for one reaction channel is the ratio of the number of fragments formed to the number of photons absorbed. However, it is again possible for a molecule to dissociate into various different reaction channels. In such a case, to obtain the overall quantum yield for a particular product, the quantum yield for each reaction channel must be summed over all the available reaction channels, taking into account the branching ratios for the channels.

34.1.2 Vector Correlations

Photodissociation is by its very nature an anisotropic process, as can be seen from (34.2). The operator $\hat{\mu}$ defines a specific axis in the molecular body-fixed frame of reference. At the instant of photoexcitation, $\hat{\mu}$ is preferentially aligned parallel to the polarization of the electric field $\boldsymbol{E}$ in the *external laboratory space-fixed* frame of reference. Hence, $\boldsymbol{E}$ defines a specific axis, and thus cylindrical symmetry in the body-fixed frame. If fragmentation occurs on a time scale which is short compared with overall rotation of the excited complex, this correlation persists between the body-fixed frame and the space-fixed frame, and a wealth of information can be obtained. However, rotation of the transient complex prior to fragmentation serves to degrade this symmetry in the external body-fixed frame.

Three vectors fully describe the photodissociation process for both the parent molecule and the products: (i) $\hat{\mu}$ in the body-fixed frame (and hence $\boldsymbol{E}$, in the space-fixed frame, at the instant of photoexcitation); (ii) $\boldsymbol{v}$, the recoil velocity of the products; and (iii) $\boldsymbol{j}$, the rotational angular momenta of the fragments. Vector correlations can exist between all of these vectors [34.5].

The most commonly observed is the angular distribution of the photofragments $I(\theta, \alpha)$, i. e. the relation

$$I(\theta, \alpha) \propto \frac{1}{4\pi}\left[1 + \beta(\alpha) P_2(\cos\theta)\right] \qquad (34.7)$$

between $\boldsymbol{v}$ and $\boldsymbol{E}$. $P_2(x)$ is the second-order Legendre polynomial and θ is the angle between $\boldsymbol{v}$ and $\boldsymbol{E}$. The *anisotropy parameter* $\beta(\alpha)$ ranges between -1 for a perpendicular transition and $+2$ for a parallel transition. Thus, measuring the angular distribution of the fragments provides information about the type of electronic transition and hence the electronic symmetry of the excited state [34.6]. If the alignment between the body-fixed and space-fixed frames is destroyed, the angular distribution becomes *isotropic* and $\beta(\alpha) = 0$. The anisotropy parameter depends on the product channel α.

A second vector correlation concerns the direction of $\boldsymbol{j}$ with respect to $\boldsymbol{E}$. Fragmentation generates rotational motion in the nuclear plane: for a perpendicular transition this is perpendicular to the plane containing the atoms, leading to the projection of $\boldsymbol{j}$ being preferentially aligned parallel to $\hat{\mu}$, and thus $\boldsymbol{E}$ in the space-fixed frame. For a parallel transition, the opposite would be true, i. e. $\boldsymbol{j}$ would be aligned

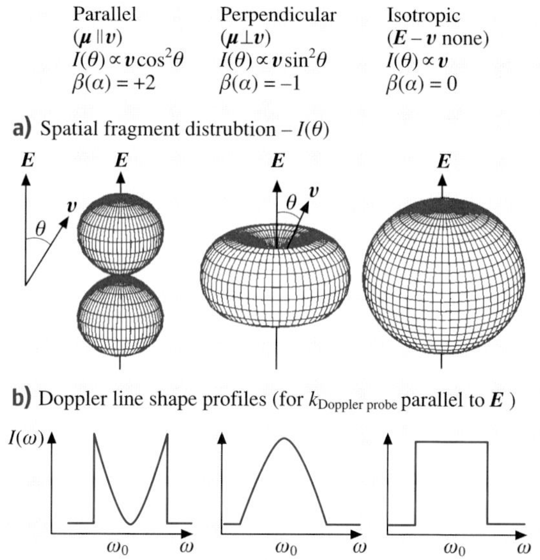

Fig. 34.2 Spatial recoil anisotropies and Doppler line shape profiles, for parallel and perpendicular transitions compared with an isotropic distribution

in the plane perpendicular to $\hat{\mu}$. The alignment of $\boldsymbol{j}$ leads to polarized emission/absorption depending on whether molecules are created in an electronically excited/ground state. Therefore the orientation of the product polarization with respect to the original photolysis polarization $\boldsymbol{E}$ also yields information about the symmetry of the electronic states involved in dissociation.

The final association in this series is independent of the space-fixed frame, since $\boldsymbol{v}$ and $\boldsymbol{j}$ are both defined in the body-fixed frame. Unlike the two previous correlations, a long lifetime does not destroy the alignment, as it is not established until the bond breaks and the two fragments recoil. For a tetratomic (or larger) molecule there are, in principle, two possible sources of product rotational excitation: bending motion in a plane of the molecule producing fragments with $\boldsymbol{v}$ perpendicular to $\boldsymbol{j}$; or torsional motion leading to fragment rotation out of the plane. A prime example of this is the distinction between frisbee and propeller type motion of the two OH fragments in the dissociation of hydrogen peroxide [34.7, 8]. Measurement of only the scalar properties *cannot* discriminate between these two possibilities, highlighting the additional information that can be gained about the bond rupture and the exit channel dynamics by the study of vector correlations (Fig. 34.2).

34.2 Experimental Techniques

Early photochemical experiments used broad white-light continuum sources and large diffractometers [34.4]. However, it has been the development of lasers in combination with molecular-beam techniques that has dramatically increased the understanding of photodissociation processes. The ever increasing spectral and time resolutions of lasers, in addition to the power and range of wavelengths available, have made it possible to excite molecules selectively and with high efficiency. This has enabled state-specific preparation of the parent molecule, study of time evolution, as well as the measurement of the scalar and vector properties of the asymptotic products [34.9].

Specification of the Initial State

A room temperature sample of a gas will have a Boltzmann distribution over rotational states. Molecular beam techniques provide an improved specification of the initial angular momenta in the parent molecule by isentropically cooling its internal rotational energy [34.10]. Full quantum state specification can be achieved by various two-photon excitation schemes, e.g., stimulated emission pump (SEP) spectroscopy and vibrationally mediated dissociation. SEP is commonly used to study dissociation on the ground state PES; the molecules are excited to a stable upper electronic state, stimulated emission back to the ground state prepares a *single* quasibound state. Vibrationally mediated dissociation provides information about both ground and upper electronic states; the molecules are excited to a stable intermediate vibrational level on the ground state and further excitation promotes this fully defined wave function to an upper dissociative electronic state [34.11].

Detailed Measurement of the Absorption Cross Section

UV and VUV electronic spectroscopy has proven a very powerful tool for examining the interaction of a photon with a parent molecule (Chapt. 69). Absorption cross sections are typically measured by scanning the frequency domain and monitoring either the intensity of radiation absorbed or the flux of product molecules produced. State specific detection of the product flux yields the partial cross section $\sigma(\omega, \alpha)$. Explicit measurement of $\sigma(\omega)$ is a direct application of the Beer–Lambert law (Sect. 69.1) and thus depends on the length of the optical cavity: cavity ring-down spectroscopy with multiple passes through a cell provides an effective cell length of several tens of kilometers.

Evolution of the Transient Complex

The evolution of the molecular wavepacket can be probed by time-resolved spectroscopy, as discussed in Chapt. 35 and [34.12]. Real-time analysis of the molecular wavepacket provides a direct insight into the forces acting during molecular photodissociation. This type of time-resolved spectroscopy and the energy-resolved spectroscopy described above are mutually exclusive due to the time-energy uncertainty principle (Sect. 80.3.1).

Asymptotic Properties

The vast majority of photodissociation studies determine asymptotic properties of the dissociation process, measuring either internal energy, recoil velocity, or angular distributions of the dissociation products. The product state distributions are usually explicitly probed by laser-induced fluorescence (LIF), resonance-enhanced multiphoton ionization (REMPI) spectroscopy, or coherent Raman scattering with the relative populations of the products obtained via line intensities.

The distribution and anisotropy of the recoil velocities are measured using Doppler spectroscopy or time-of-flight (TOF) techniques [34.13]. Doppler spectroscopy uses the Doppler-broadening of lines in the LIF or REMPI excitation spectra (Sect. 69.5); the profile of the line shape also depends on the recoil anisotropy of the probed species (Fig. 34.2). However, many important molecular fragments are not amenable to spectroscopic detection. Thus, though lacking the ultimate state-specificity of spectroscopy, TOF techniques by virtue of their general applicability provide an appealing alternative route to determine the product state distributions. In TOF techniques, the time is recorded for photofragments to recoil a known distance from the interaction region to a detector. Due to total energy and momentum conservation, the translational energy distribution of a fragment state specifically detected directly implies the internal energy distribution of the other partner product. In less favorable cases where this is not possible, *coincidence* detection schemes are employed to define the partition between translational and internal energies. Rotation of the detection axis with respect to the polarization of the photolysis laser yields the recoil anisotropy.

Advances in spatially resolved detection schemes are now providing an improved measure of vector correlations.

Mass spectrometry can be used to measure branching ratios and quantum yields, which can also be obtained from the techniques described above.

34.3 Theoretical Techniques

The calculation of the observables of photodissociation can be carried out using quantum mechanics either in the time-independent or the time-dependent frame, as well as using classical mechanics [34.1]. Theoretical studies have contributed greatly to the understanding of photodissociation processes, as they provide the ability not only to calculate the observables, but also, through the knowledge of the wave function, to view directly the dissociation dynamics. This has enabled the inference of the underlying dynamics from the observables of the reaction to be more precisely established.

Due to computer limitations, the majority of quantum mechanical studies currently treat fully only three degrees of freedom, and thus have mainly concentrated on triatomic molecules. Jacobi coordinates are usually used, with the appropriate set for the dissociation of ABC into AB and C as follows: R, denoting the distance between the atom C and the centre of mass of the AB fragment; r, denoting the internal vibration coordinate of AB; and γ, denoting the bending angle between R and r.

The initial state Ψ_i is generally taken to be a single bound state. It is obtained either by the solution of the Schrödinger equation at a particular energy (Sect. 31.4), or simply by taking a product of three Gaussians in the three coordinates, with the parameters of the Gaussians being determined from spectroscopic information on the ground state.

Further, to calculate the observables it is also necessary to have information about μ_{fi}. However, this is often assumed to be a constant, i. e. independent of the internal coordinates of the molecule. The *Franck–Condon principle* assumes that the nuclear geometry changes after the electronic transition, and not during it. Therefore a molecule, with a particular geometry, will, when promoted by the photon to the excited electronic state, be centred around the same geometry, which is thus referred to as the Franck–Condon region, or point.

To carry out any dynamical calculations it is necessary to have PESs for the electronic states involved. These are usually obtained from *ab initio* calculations which are described in Chapt. 31. The accuracy of the PES surface largely determines the accuracy of the results obtained, as the PES essentially determines the dynamics of the fragmentation process.

In the *time-independent* approach, a solution of the time-independent Schrödinger equation

$$(\hat{\boldsymbol{H}} - E)\Psi^{\alpha} = 0\,, \qquad (34.8)$$

is sought for a specific total energy $\boldsymbol{E}$ subject to appropriate boundary conditions at infinite product separation. There are many different approaches to solving this problem, but they can be broadly separated into two groups: scattering methods and $\mathcal{L}^2$ methods. The scattering methods involve the solution of the coupled channel equations described in Chapt. 36. These can be solved directly to yield the wave functions, which can then be used to calculate the observables, or can be solved indirectly to provide similar information. The use of $\mathcal{L}^2$ methods, which attempt to expand Ψ in a finite basis set, is not directly applicable since the wave functions are in the continuum and spread out to infinite distances in the R coordinate. Thus various modifications have been introduced in order to take this into account. The most important of these use variational principles [34.14], such as that due to *Kohn* [34.15]. In the Kohn variational principle the wave function in the inner or interaction region is expanded in a finite $\mathcal{L}^2$ basis. However, in the outer region, the wave function is expanded in an energy dependent basis of outgoing and incoming waves, which are approximate solutions of the coupled channel equations. Other methods which can sometimes be used to indirectly extract information about the observables are stabilization [34.16] and complex scaling [34.17].

In the *time-dependent* approach [34.18], one solves the time-dependent Schrödinger equation

$$\mathrm{i}\hbar\frac{\partial}{\partial t}\Phi(t) = \hat{\boldsymbol{H}}\Phi(t) \qquad (34.9)$$

for the wavepacket $\Phi(t)$ with initial condition $\Phi(0) = \Psi_i$, i. e. it is assumed that the molecule is *vertically* promoted by an infinitely short pulse to the electronic state under consideration. The wavepacket is a coherent superposition of stationary wave functions Ψ^{α} (Chapt. 35), and since it comprises many of the stationary states, it contains all the information necessary to characterize the dissociation (see Fig. 34.3).

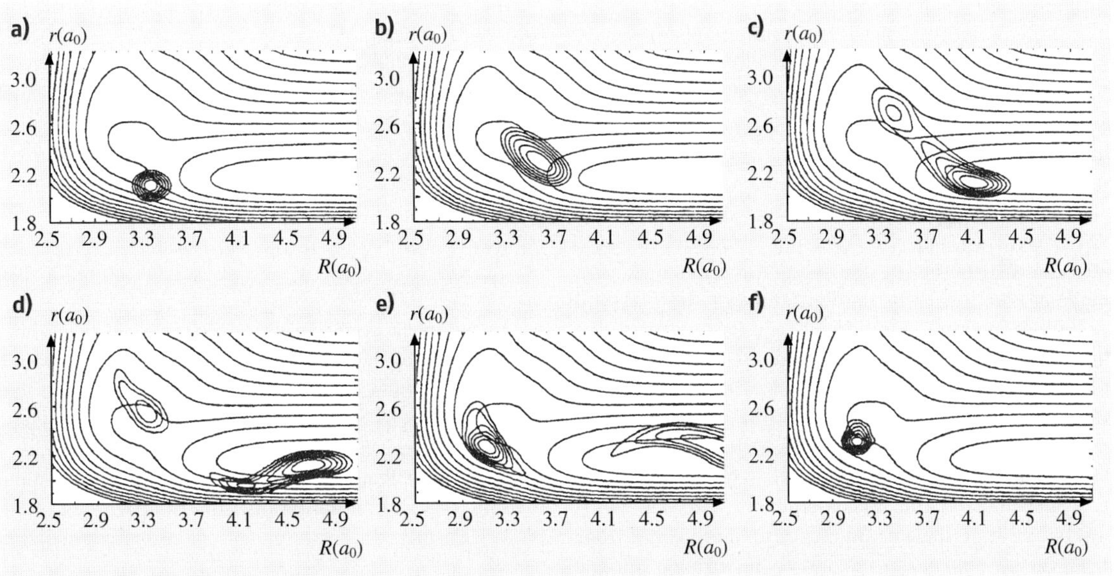

Fig. 34.3 Time evolution of a wavepacket in the dissociation of FNO in the S_1 state

The total absorption cross section is given by

$$\sigma(\omega) \propto \int_{-\infty}^{+\infty} \mathrm{d}t\, S(t)\, \mathrm{e}^{-\mathrm{i}\omega t}\,, \tag{34.10}$$

where the autocorrelation function $S(t)$ is defined as

$$S(t) = \langle \Phi(0) \mid \Phi(t) \rangle\,. \tag{34.11}$$

The autocorrelation function reflects the motion of the wavepacket, and therefore is a convenient means for visualizing the molecular dynamics. The individual partial cross sections can be obtained in the limit $t \to \infty$ by projection of the wavepacket onto the stationary wave functions of the products, i. e. plane waves in the dissociation coordinate and vibrational-rotational wave functions for the free products.

34.4 Concepts in Dissociation

There has been substantial experimental and theoretical work to elucidate the processes involved in photodissociation from knowledge of the observables, and much progress has been made. In this section, an attempt is made to present some of the simpler ideas which have emerged [34.1].

34.4.1 Direct Dissociation

Direct dissociation is the very fast rupture of a bond after a molecule has been promoted to an electronic state which has a purely repulsive PES. A very clear picture of this process can be obtained from wavepacket calculations: the wavepacket which is placed on the repulsive surface moves directly down the PES and into the exit channel. The autocorrelation function decays from one to zero in a short time and does not show any recurrences, i. e. oscillations in the autocorrelation function. The absorption cross section $\sigma(\omega)$, which is the Fourier transform of the autocorrelation function (34.10), is therefore a very broad Gaussian with no structure. The breadth of $\sigma(\omega)$ is inversely proportional to the width of the autocorrelation function and can, using simple classical pictures, be taken to be approximately proportional to the steepness of the potential at the Franck–Condon point. The partial cross sections have a similar structure to the total cross section, though they have differing intensities and are shifted relative to each other on the energy scale.

The product distributions can be predicted using simple classical pictures. These methods can be divided into two groups, depending on the extent of the excitation/de-excitation, or coupling, in the exit channel. If there is very little excitation/de-excitation in the exit channel, the rotational and vibrational product distributions are best described using Franck–Condon mapping. Another model which gives good results for the rotational distributions is the impact parameter, or impulsive, model. If excitation/de-excitation in the exit channel is not negligible, the product distributions are best described using the *reflection principle*. This relates the distribution of the initial wavepacket in γ to the final rotational distributions through the classical excitation function. Similarly, the distribution of the initial wavepacket in space is related to the final vibrational distributions through another classical excitation function.

Even for a purely repulsive PES, the potential may be very flat in the Franck–Condon region so that the molecule may be able to undergo one internal vibration before it dissociates. In this case, there is a diffuse structure in $\sigma(\omega)$, associated with recurrences in the autocorrelation function. The spacing of the structures in $\sigma(\omega)$ are related to the period of the internal vibrations by $\Delta E = 2\pi\hbar/T$. Diffuse structures in $\sigma(\omega)$ have also been linked to unstable periodic orbits.

34.4.2 Vibrational Predissociation

Vibrational predissociation is dissociation delayed due to the trapping of the energy of the molecule in modes orthogonal to the dissociation coordinate. It can be explained very clearly in the time-independent picture as resonances (sometimes known as Feshbach resonances), which are simply extensions of the bound states into the continuum. $\sigma(\omega)$ in this case consists of a series of Lorentzian lines, whose width is inversely proportional to the lifetime of the resonance. In the case where the internal modes of the molecule are not strongly coupled to each other or to the dissociation mode, these resonance states can often be assigned, with the number of quanta in each mode being specified. In this case, the lines in $\sigma(\omega)$ form a series of progressions. The widths of these lines, and thus the lifetimes of the resonance states, often show trends relating the lifetimes to the assignment. This is called *mode-specificity*. In the case that the system is strongly mixed, it is not possible to make an assignment of the resonances, and the lifetimes show strong fluctuations. This is called *statistical state-specificity* [34.19].

The resonances can also be seen in the time-dependent picture, where the autocorrelation function shows many recurrences with periods T, depending on the fundamental frequencies of the internal modes $\omega = 2\pi/T$ (Chapt. 35).

The partial cross sections for vibrational predissociation also consist of Lorentzian lines, with positions and widths exactly as for the total cross sections, but with differing intensities. The partial widths, which describe the rate of dissociation into each product channel, are given by

$$\Gamma_\alpha = \Gamma \frac{\sigma(\omega, \alpha)}{\sigma_{\text{tot}}(\omega)} . \tag{34.12}$$

In the weak coupling case, simple pictures can be used to describe the product distributions. The rotational product distributions can be explained using again the reflection principle; but in this case, instead of considering the distribution of the initial wave function in γ, the distribution of the wave function at the transition state is used. The vibrational product distributions can often be well described by examining vibrationally adiabatic curves.

In the case that the modes are strongly coupled, the simple models break down. It is then sometimes possible to use statistical models to describe both the rates and the product distributions. One example of these unimolecular-statistical theories is the Ramsperger–Rice–Karplus–Marcus (RRKM) theory, which is widely used for the description of unimolecular dissociations [34.20]. Another example is phase space theory (PST) [34.21, 22] which is often used to calculate the product distributions for reactions which have no barrier. The quantum mechanical results fluctuate about these average values. These fluctuations, which can be considered as being independent of the system, can be described well by the predications of random matrix theory [34.23].

34.4.3 Electronic Predissociation

Nonadiabatic transitions between two or more electronic states are a common phenomenon in photodissociation [34.24] as well as in other chemical reactions (Chapt. 49). Such transitions can result in the production of both electronically excited and ground state fragments.

Adiabatic molecular PESs can vary in complex fashions. Many of these contortions arise from avoided and real crossings of the surfaces, and in all such cases, the physical and chemical understanding is greatly facilitated by expressing the adiabats in terms of the diabatic states (Chapt. 31). The electronic diabatic states are chosen to simplify the structure of the electronic wave functions by incorporating the off-diagonal or coupling elements as a pure potential energy term, rather than as a kinetic term, or a mixture of both.

Under these conditions, the BO approximation is inadequate since there is coupling between the different adiabatic states, and the electronic and nuclear motion cannot be separated. Therefore the solution of the time-dependent Schrödinger equation (Sect. 34.3) requires the set of coupled equations

$$i\hbar\frac{\partial}{\partial t}\begin{pmatrix}\Psi_1\\ \Psi_2\end{pmatrix}=\begin{pmatrix}V_{11}+T_{11} & V_{12}\\ V_{12} & V_{22}+T_{22}\end{pmatrix}\begin{pmatrix}\Psi_1\\ \Psi_2\end{pmatrix}, \tag{34.13}$$

to be propagated, where the coupling between the diabatic surfaces is in the potential (V) and not the kinetic (T) terms. The wavepacket evolves on both diabatic (or adiabatic) surfaces, and shows a complicated motion in moving between the two surfaces. The coupling between the nuclear and electronic motion can be thought of as resulting in the nuclear motion forcing the transfer of a valence electron to another molecular orbital. Since the efficiency of this transfer is greatest when the orbitals are degenerate, the crossings of the wavepacket between the PESs are generally localized around their degeneracies. Finally, the wavepacket moves out on the adiabatic surfaces towards the products, with which they are correlated.

34.5 Recent Developments

In recent years, the field of photodissociation has seen a number of intriguing applications and comparisons between detailed experimental data and high-quality *ab initio* calculations. These applications have become feasible mainly because of the possibility to construct accurate potential energy surfaces from first principle electronic structure calculations. Cases in which the fragmentation proceeds via two or several electronic states have been especially concentrated on [34.26]. In these cases the Born–Oppenheimer approximation is not valid and the coupling between electronic and nuclear degrees of freedom is essential (Sect. 34.4.3). A nice example is the photodissociation of water in the second absorption band. Since water has only 10 electrons, highly accurate potential energy surfaces have been calculated theoretically, and these have been used in extensive dynamics calculations – including motion on three potential energy surfaces [34.27]. The agreement between the calculated and the measured absorption cross section at room temperature is outstanding [34.28]. From the elaborate analysis of product state distributions (rotational, vibrational, and electronic), many details about the coupled motion on several potential energy surfaces have been learned [34.29, 30]. The electronic density of water is small, and therefore the photodissociation can be treated on a nearly exact level. For other triatomic molecules, with more electrons and a higher density of electronic states, this is generally not feasible. An important example is ozone, which plays a vital role in the atmosphere. The electronic structure of O_3 is illustrated in Fig. 34.4, where many spin-allowed as well as spin-forbidden fragmentation pathways are

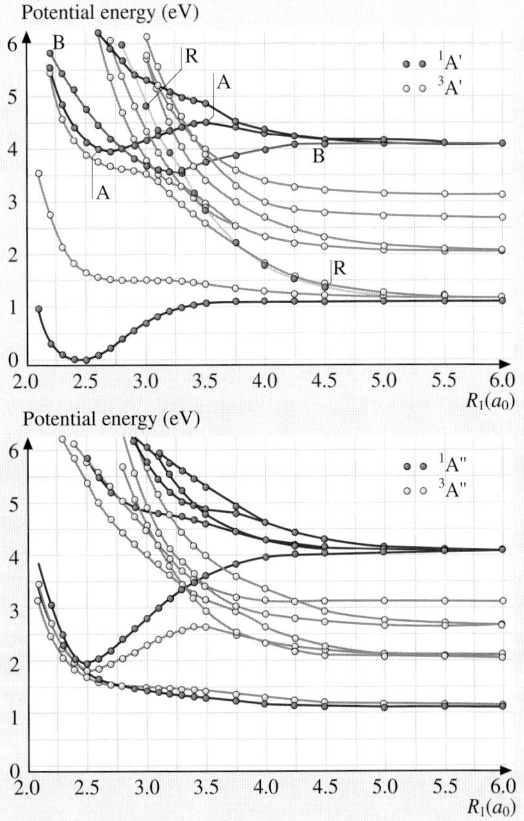

Fig. 34.4 Electronic structure of ozone. Shown are cuts through the potential energy surfaces for the singlet and triplet states. After [34.25]

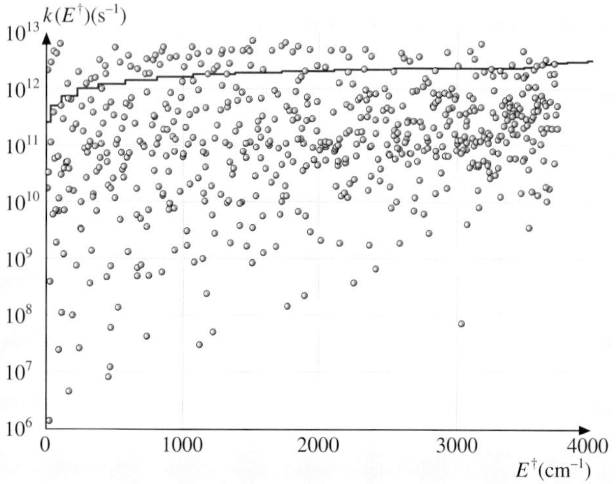

Fig. 34.5 Overview of the calculated dissociation rates of HOCl as a function of the excess energy. The *solid line* is the prediction of a statistical model. After [34.32]

seen [34.25]. The photodissociation of ozone in the uv range has been the target of many experimental studies [34.31]. The interpretation of the many experimental results on the basis of realistic potential energy surfaces is a great challenge for theoretical chemistry.

Photodissociation studies are particularly rewarding if the lifetime in the excited electronic state is long because then the absorption spectrum shows well resolved lines (resonances), the widths of which are inversely related to the state-specific lifetimes [34.33]. A typical situation is the excitation of a particular vibrational-rotational state in a bound electronic state which can decay only via coupling to a dissociative electronic state. The lifetime then reflects the coupling of this state to the continuum of the dissociative state (predissociation). An example is the photodissociation of HCO [34.34, 35]; in this case, the upper and the lower state are coupled by Renner–Teller coupling. A similar example is the photodissociation of HNO. For this molecule, the lower state has a deep potential well which supports long-lived states in its own continuum. The mixing between the quasi-bound states of the upper state with the resonance states of the lower state leads to interesting behavior in the lifetime as a function of the rotational quantum number [34.36] (resonance between resonances).

Resonances are also prominent features of ground state potential energy surfaces; they are the continuation of the true bound states into the continuum (Sect. 34.4.2). Since resonances determine the kinetics of chemical reactions, they are usually studied in the framework of *unimolecular dissociations* or *unimolecular reactions* [34.37]. On the other hand, these resonance states can be excited by photons, and therefore it is meaningful to discuss them also in the context of photodissociation. In the last few years, numerical methods have been developed to efficiently calculate the resonance parameters [34.38–40]; see [34.37] for a comprehensive overview. Several triatomic molecules with dramatically different intramolecular dynamics have been investigated. The main observation is a strong fluctuation in the resonance lifetimes over several orders of magnitude, even for molecules whose classical dynamics is chaotic, such as NO_2 [34.41]. Figure 34.5 shows the results for HOCl $\rightarrow$ HO + Cl [34.32, 42]. The large fluctuations of the lifetimes (or dissociation rates) are believed to affect the fall-off behavior of recombination rate coefficients [34.43].

The concept of first calculating a potential energy surface as function of all coordinates and then performing dynamics calculations is suitable only for triatomics. For molecules with more than four atoms it is not applicable, simply because of the rapidly increasing number of degrees of freedom. For larger molecules, *direct dynamics* simulations, in which the methodology of classical trajectory simulations is coupled directly to electronic structure calculations, are the method of choice [34.44, 45]. In these simulations, the derivatives of the potential, which are required for the numerical integration of the equations of motion, are obtained directly from electronic structure theory without the need for an analytic potential energy surface. An important application of direct dynamics is the study of post-transition state intramolecular and unimolecular dynamics. When the dissociation proceeds through a transition state, it may be sufficient to start trajectories at the transition state and to follow them into the product channels [34.46, 47]

34.6 Summary

Photodissociation of polyatomic molecules is an ideal field for studying the details of molecular dynamics. The primary goal of the experimental and theoretical approaches (Sects. 34.2, 34.3) is to understand the connection between the observables (Sect. 34.1) and the underlying chemical dynamics (Sect. 34.4).

Once this connection has been established, it is possible to have a detailed understanding of the dissociation dynamics, transition state geometries, and the PESs which ultimately govern the molecular chemical reactivity. The interplay between powerful experimental and theoretical techniques has enabled this goal to be realized for many photodissociation reactions.

References

34.1 R. Schinke: *Photodissociation Dynamics* (Cambridge Univ. Press, Cambridge 1993)
34.2 H. Okabe: *Photochemistry of Small Molecules* (Wiley, New York 1978)
34.3 R. P. Wayne: *Chemistry of Atmospheres*, 2nd edn. (Oxford Univ. Press, New York 1991)
34.4 G. Herzberg: *Molecular Spectra and Molecular Structure III, Electronic Spectra and Electronic Structure of Polyatomic Molecules* (Van Nostrand, New York 1967)
34.5 G. E. Hall, P. L. Houston: Ann. Rev. Phys. Chem. **40**, 375 (1989)
34.6 R. N. Zare: *Angular Momentum* (Wiley, New York 1988)
34.7 A. U. Grunewald, K.-H. Gericke, F. J. Comes: J. Chem. Phys. **87**, 5709 (1987)
34.8 A. U. Grunewald, K.-H. Gericke, F. J. Comes: J. Chem. Phys. **89**, 345 (1988)
34.9 M. N. R. Ashfold, J. E. Baggott (Eds.): *Molecular Photodissociation Dynamics* (Royal Society of Chemistry, London 1987)
34.10 G. Scoles (Ed.): *Atomic and Molecular Beams Methods* (Oxford Univ. Press, New York 1988)
34.11 F. F. Crim: Ann. Rev. Phys. Chem. **44**, 397 (1993)
34.12 J. Manz, L. Wöste (Eds.): *Femtosecond Chemistry* (VCH, Weinheim 1995)
34.13 M. N. R. Ashfold, I. R. Lambert, D. H. Mordaunt, G. P. Morley, C. M. Western: J. Phys. Chem. **96**, 2938 (1992)
34.14 R. K. Nesbet: *Variational Methods in Electron–Atom Scattering Theory* (Plenum, New York 1980)
34.15 J. Z. H. Zhang, W. H. Miller: J. Phys. Chem. **92**, 1811 (1990)
34.16 R. Lefebvre: J. Phys. Chem. **89**, 4201 (1985)
34.17 W. P. Reinhardt: Ann. Rev. Phys. Chem. **33**, 223 (1982)
34.18 R. Kosloff: J. Phys. Chem. **92**, 2087 (1988)
34.19 W. L. Hase, S. Cho, D. Lu, K. N. Swamy: Chem. Phys. **139**, 1 (1989)
34.20 P. J. Robinson, K. A. Holbrook: *Unimolecular Reactions* (Wiley, London 1972)
34.21 P. Pechukas, J. C. Light: J. Chem. Phys. **42**, 3285 (1965)
34.22 P. Pechukas, J. C. Light, C. Rankin: J. Chem. Phys. **44**, 794 (1966)
34.23 T. A. Brody, J. Flores, J. B. French, P. A. Mello, A. Pandey, S. S. M. Wong: Rev. Mod. Phys. **53**, 385 (1981)
34.24 R. N. Dixon: Chem. Soc. Rev. **23**, 375 (1994)
34.25 H. Zhu, Z.-W. Qu, M. Tashiro, R. Schinke: Chem. Phys. Lett. **384**, 45 (2004)
34.26 R. Schinke: Quantum mechanical studies of photodissociation dynamics using accurate global potential energy surfaces. In: *Conical Intersections*, ed. by W. Domcke, D. R. Yarkony, H. Köppel (World Scientific, Singapore 2004)
34.27 R. van Harrevelt, M. C. van Hemert: J. Chem. Phys. **112**, 5777 (2000)
34.28 B.-M. Cheng, C.-Y. Chung, M. Bahou, Y.-P. Lee, L. C. Lee, R. van Harrevelt, M. C. van Hemert: J. Chem. Phys. **120**, 224 (2004)
34.29 J. H. Fillion, R. van Harrevelt, J. Ruiz, M. Castillejo, A. H. Zanganeh, J. L. Lemaire, M. C. van Hemert, F. Rostas: J. Phys. Chem. A **105**, 11414 (2001)
34.30 S. A. Harich, X. F. Yang, X. Yang, R. van Harrevelt, M. C. van Hemert: Phys. Rev. Lett. **87**, 263001 (2001)
34.31 Y. Matsumi, M. Kawasaki: Chem. Rev. **103**, 4767 (2003)
34.32 J. Hauschildt, J. Weiß, C. Beck, S. Yu. Grebenshchikov, R. Düren, R. Schinke, J. Koput: Chem. Phys. Lett. **300**, 569 (1999)
34.33 R. Schinke, H.-M. Keller, M. Stumpf, A. J. Dobbyn: J. Phys. B **28**, 3081 (1995)
34.34 D. W. Neyer, P. L. Houston: The HCO potential energy surface; probes using molecular scattering and photodissociation. In: *The Chemical Dynamics and Kinetics of Small Radicals*, ed. by K. Liu, A. Wagner (World Scientific, Singapore 1994)
34.35 J. Weiß, R. Schinke, V. A. Mandelshtam: J. Chem. Phys. **113**, 4588 (2000)
34.36 J. Weiß, R. Schinke: J. Chem. Phys. **115**, 3173 (2001)
34.37 S. Yu. Grebenshchikov, R. Schinke, W. L. Hase: State-specific dynamics of unimolecular dissociation. In: *Unimolecular Kinetics*, ed. by N. Green (Elsevier, Amsterdam 2003)
34.38 N. Moiseyev: Phys. Rep. **302**, 211 (1998)
34.39 V. A. Mandelshtam, H. S. Taylor: J. Chem. Phys. **102**, 7390 (1995)
34.40 V. A. Mandelshtam, H. S. Taylor: J. Chem. Phys. **106**, 5085 (1997)

34.41 B. Kirmse, B. Abel, D. Schwarzer, S.Yu. Grebenshchikov, R. Schinke: J. Phys. Chem. A **104**, 10374 (2000)

34.42 S. Skokov, J. M. Bowman: J. Chem. Phys. **110**, 9789 (1999)

34.43 H. Hippler, N. Krasteva, F. Striebel: Phys. Chem. Chem. Phys. **6**, 3383 (2004)

34.44 L. Sun, W. L. Hase: Born–Oppenheimer direct dynamics classical trajectory simulations. In: *Review in Computational Chemistry*, Vol. 19, ed. by K. B. Lipkowitz, R. Larter, T. R. Cundari (Wiley-VCH, Hoboken, NJ 2003)

34.45 W. L. Hase, K. Song, M. S. Gordon: Comp. Sci. Engineer. **5**, 36 (2003)

34.46 K. Bolton, H. B. Schlegel, W. L. Hase, K. Song: Phys. Chem. Chem. Phys. **1**, 999 (1999)

34.47 W. Chen, W. L. Hase, H. B. Schlegel: Chem. Phys. Lett. **228**, 436 (1994)

35. Time-Resolved Molecular Dynamics

Time-resolved experiments have been performed on a diversity of molecular systems. Applications of spectroscopic techniques which work in the time-domain range from the detection of simple vibrational motion of a diatomic molecule to the direct determination of relaxation times in polyatomic molecules in a liquid environment, or the recording of isomerization processes in biomolecules. The underlying principles of these experiments are more or less the same. In this chapter, a brief description of the basic ideas of transient spectroscopy is given with the emphasis on gas-phase molecules under collision-free conditions, as are usually provided in a molecular beam.

For the development of ultrashort pulse techniques and their application to areas as different as optical engineering, solid state physics or biology see the series of conference proceedings in [35.1–14]. For special applications to molecular physics and chemistry consult [35.15–22] and the review in [35.23].

Besides an overall rotation and the translational motion of the molecular center of mass, nuclei within a molecule possess vibrational degrees of freedom. The real-time detection of internal vibrational dynamics is discussed in what follows, but the considerations apply equally well to the case of a fragmentation process where the dissociation dynamics is resolved.

35.1 **Pump–Probe Experiments** 548

35.2 **Theoretical Description** 548

35.3 **Applications** 550
- 35.3.1 Internal Vibrational Dynamics of Diatomic Molecules in the Gas Phase 550
- 35.3.2 Elementary Gas-Phase Chemical Reactions 550
- 35.3.3 Molecular Dynamics in Liquid and Solid Surroundings 551

35.4 **Recent Developments** 551
- 35.4.1 Faster Dynamics 551
- 35.4.2 X-Ray Pulses 551
- 35.4.3 Time-Resolved Diffraction 551
- 35.4.4 Dynamics and Control 552

References 552

Classically, a molecule can be imagined to consist of atoms connected by springs, each spring representing a chemical bond. The atoms vibrate, performing a periodic oscillation around an equilibrium position. Quantum mechanically, the molecule has discrete vibrational states φ_n, where n represents a set of vibrational quantum numbers and ϵ_n are the corresponding eigenenergies. A vibrational period T_{vib} is calculated from the energy spacing as:

$$T_{\text{vib}} = \frac{2\pi\hbar}{\epsilon_{n+1} - \epsilon_n} . \tag{35.1}$$

Typical vibrational periods for smaller molecules are of the order of several hundred femtoseconds.

Traditional high resolution spectroscopy uses laser pulses with a temporal width T_P much larger than any of the time-scales on which the internal molecular motion takes place:

$$T_\text{P} \gg T_{\text{vib}} . \tag{35.2}$$

In the energy domain, this means that the spectral width of the laser pulse is small enough to excite a single eigenstate φ_n. The time-evolution of this state is that of a stationary state

$$\psi(t) = a_n \, e^{-i\epsilon_n t/\hbar} \varphi_n , \tag{35.3}$$

where a_n is a complex number determined by the particular preparation process and the molecular properties.

To detect molecular motion within a time-resolved measurement, pulses with a width T_P smaller than the vibrational period have to be used:

$$T_\text{P} < T_{\text{vib}} . \tag{35.4}$$

Thus, to resolve vibrational dynamics of smaller molecules, one needs ultrashort pulses in the femtosecond regime. Because a short pulse has a broad spectral distribution it is possible to excite several eigenstates simultaneously: a coherent superposition of the functions φ_n is prepared. The resulting wave function is a vibrational wave packet of the form

$$\psi(t) = \sum_n a_n \, \mathrm{e}^{-\mathrm{i}\epsilon_n t/\hbar} \varphi_n \,. \tag{35.5}$$

Due to the time-dependent phase factors, the average position of the wave packet changes as a function of time. The principle of time-resolved molecular dynamics is to create such wave packets within a molecule and follow their motion in time.

Pulses in the femtosecond regime are now commercially available. Schemes have been developed to use these pulses to detect wave-packet dynamics. This is, of course, not limited to bound state motion but applies as well to fragmentation processes. In the latter case, the sum in (35.5) has to be replaced by an integral over continuum functions. The condition (35.4) then reads $T_\mathrm{P} < T_\mathrm{F}$, where T_F is the time during which the atoms separate and do not interact with each other afterwards. It is then possible to monitor the breaking of a chemical bond.

Typical experimental schemes are described in Sect. 35.1, and a theoretical treatment which closely follows the experimental procedure is outlined in Sect. 35.2. Applications are listed in Sect. 35.3 and a brief overview highlighting recent developments is presented in Sect. 35.4.

35.1 Pump–Probe Experiments

The experimental setup for a time-resolved measurement commonly uses two ultra-short laser pulses which are delayed with respect to each other. Therefore, a single pulse is produced and split. The resulting two pulses are delayed by sending them along different paths until they reach the molecular sample. The temporal difference is adjusted by variation of the path length. The first pulse (pump) excites the molecule and prepares a wave packet, i. e., a coherent superposition of molecular eigenstates. The second pulse (probe) interacts with the molecular sample after a defined delay-time and prepares the system in yet another state. A signal is then measured as a function of the delay-time between the pulses. The idea is that, because the wave packet is located in different spatial regions at different times, the signal, in general, depends on the delay-time.

There are various detection schemes. If the probe-pulse prepares the molecule in an energetically higher electronic state, fluorescence can be measured. The pump–probe signal then consists of the total fluorescence yield, recorded as a function of the delay-time between the pulses. Because this signal is proportional to the population created in the state from which the radiative decay takes place, the signal reflects how effective the probe-pulse excitation occurs which, in turn, depends on the wave packet dynamics in the intermediate state accessed by the pump-pulse excitation. It is also possible to detect the transient absorption spectrum of the system which is subject to the probe-pulse excitation. If the parameters of the probe-pulse are chosen such that the molecule is ionized, a total ion signal or a photoelectron spectrum can be detected as a function of the pulse delay. Another technique involves time-resolved four-wave mixing schemes, where coherently emitted radiation is employed to track the system dynamics.

35.2 Theoretical Description

There are several theoretical approaches to describe the experiments which have been outlined in Sect. 35.1. They are based on classical, quantum mechanical or semi-classical descriptions [35.24, 25]. Under the conditions specified at the beginning of this chapter, the most straightforward approach is to solve the time-dependent Schrödinger equation for the field coupled nuclear motion in different electronic states. Here, we outline this approach when only three electronic states $|i\rangle (i = 1, 2, 3)$ participate in the excitation process and only the states $(|i\rangle,\ |i \pm 1\rangle)$ are coupled. Then, the nuclear wave function $\psi(r)$, where r denotes the nuclear co-ordinates, consists of three components ψ_i. The equation of motion for the nuclei reads

$$\begin{pmatrix} H_1 & W_{12} & 0 \\ W_{12} & H_2 & W_{23} \\ 0 & W_{23} & H_3 \end{pmatrix} \psi(r,t) = \mathrm{i}\hbar \frac{\partial}{\partial t} \psi(r,t) \,, \qquad (35.6)$$

with the field-matter interaction

$$W_{ji}(r,t) = -\mu_{ji}(r)\{f(t-T_1)\cos[\omega_1(t-T_1)] + f(t-T_2)\cos[\omega_2(t-T_2)]\} \,. \qquad (35.7)$$

Here, H_i denotes the nuclear Hamiltonian within the electronic state $|i\rangle$. The components $\psi_i(r,t)$ of the wave function are coupled by the dipole interaction terms $W_{ji}(r,t)$ containing the projection of the (classical) electric field vector on the transition dipole moment $\mu_{ji}(r)$ connecting states $|i\rangle$ and $|j\rangle$. The field envelope function $f(t)$ is assumed to be the same for pump- and probe-pulses having frequencies ω_1, ω_2, respectively. The pulses interact with the molecule around times T_1, T_2 so that the time-delay is $T = T_2 - T_1$.

Treating a pump-probe fluorescence set-up as an example, we may assume that the population in the state prepared by the probe-pulse is proportional to the total fluorescence signal. Then it is sufficient to determine the wave function $\psi_3(r,t)$ for a fixed pump-probe delay and obtain the signal from its norm. To do so it is, in principle, possible to integrate equation (35.6) numerically with a given initial condition. This indeed is necessary if the fields are of high intensity. In most cases, however, when the interest is in the molecular dynamics rather than in non-linear effects induced by high power laser pulses, a perturbative approach is of conceptual and technical advantage.

Except in the case when the two pulses have temporal overlap, the pump-process and probe-process may be separated. In first-order perturbation theory, the wave function created by the pump-pulse in state $|2\rangle$ is

$$\psi_2(r,0) = \frac{1}{\mathrm{i}\hbar} \int_{-\infty}^{+\infty} \mathrm{d}t U_2(-t) W_{21}(r,t) U_1(t) \varphi_{1,n}(r) \,, \qquad (35.8)$$

where time $t = 0$ refers to the end of the pump-pulse interaction. The initial (stationary) vibrational state with vibrational quantum numbers n and eigenenergy ϵ_n is denoted as $\varphi_{1,n}$ and U_i is the propagator in electronic state $|i\rangle$.

Expanding ψ_2 in the set of vibrational eigenfunctions $\{\varphi_{2,m}\}$ with eigenenergies E_m in electronic state $|2\rangle$ yields:

$$\psi_2(r,0) = \sum_m \varphi_{2,m}(r)\, c_{mn}\, I_{mn}(\omega_1) \qquad (35.9)$$

with the overlap integrals

$$c_{mn} = \int \mathrm{d}r \varphi_{2,m}(r) \mu_{21}(r) \varphi_{1,n}(r) \qquad (35.10)$$

and the time integrals

$$I_{mn}(\omega_1) = \frac{1}{2} \int_{-\infty}^{\infty} \mathrm{d}t f(t)\, \mathrm{e}^{\mathrm{i}(E_m - \epsilon_n - \hbar\omega_1)t/\hbar} \,, \qquad (35.11)$$

where we have set $T_1 = 0$ and replaced (within the 'rotating wave approximation') the cosine-term in (35.7) by $\frac{1}{2}\exp(-\mathrm{i}\omega_1 t)$.

From (35.9–35.11) it is clear that ψ_2 is a vibrational wave packet. The weights of the states which build the packet are determined by products of the overlap integrals and the Fourier-transform of the pulse envelope $f(t)$ taken with respect to the energy $E_m - \epsilon_n - \hbar\omega_1$. In the limit of an infinite long pulse, I_{mn} becomes proportional to a δ-function and, for resonance excitation, only a single vibrational state is excited.

Once the pump-pulse no longer interacts with the system, the packet propagates unperturbed. This motion is detected in a time-resolved experiment by exposing the molecule to a probe-pulse at time T, thereby inducing a transition to state $|3\rangle$. As above, the wave function in state $|3\rangle$ can be written as

$$\psi_3(r,T) = \frac{1}{\mathrm{i}\hbar} \int_{-\infty}^{\infty} \mathrm{d}t U_3(-t) W_{32}(r,t) U_2(t'-T) \psi_2(r,T) \,. \qquad (35.12)$$

There is an essential difference between the wave functions ψ_2 and ψ_3. The wave packet created by the pump-pulse results from a stationary initial wave function, whereas ψ_3, prepared by the probe-pulse, contains the wave packet ψ_2 as initial state. The latter packet changes its position and is centered around different distances r for different delay times T.

Because, during the short time the probe-pulse interacts with the molecule, the heavy nuclei do not move essentially, one may, to a good approximation, neglect the kinetic energy part of the propagators U_i in (35.12). This approximation yields

$$\psi_3(r,T) \approx \mu_{32}(r)\psi_2(r,T) \times \int_{-\infty}^{\infty} \mathrm{d}t\, f(t)\, \mathrm{e}^{\mathrm{i}(V_3(r)-V_2(r)-\hbar\omega_2)t/\hbar}, \quad (35.13)$$

where V_i is the potential energy in electronic state $|i\rangle$. The time integral has its largest modulus at the point $r = r_M$, where the equation

$$V_3(r) - V_2(r) = \hbar\omega_2 \quad (35.14)$$

holds. Assuming that the wave packet is localized, this implies that the norm of ψ_3, and thus the population in state $|3\rangle$, is maximal at those times T when the center of the moving wave packet ψ_2 is located around r_M, where the difference potential $V_3(r) - V_2(r)$ equals the photon energy $\hbar\omega_2$. This consideration shows the central idea of a pump–probe scheme: for a fixed frequency ω_2 it is possible to detect the wave packet each time it reaches the region around the position r_M because then the pump–probe signal exhibits a maximum. If ω_2 is changed, the packet is detected when it visits a different position r_M.

35.3 Applications

It is not possible to give a complete account of the numerous studies of molecular dynamics in the time-domain here. Only selected examples are listed in what follows, so that the cited work is to be taken as typical, being far from a complete compilation.

35.3.1 Internal Vibrational Dynamics of Diatomic Molecules in the Gas Phase

Prototype experiments investigated the simplest case of molecular motion: the vibration of a diatomic molecule. The change of the bond-length within an electronically excited state could be detected in real-time for molecules such as I_2 [35.26] or Na_2 [35.27]. It was shown that not only the vibrational periods but also the Born–Oppenheimer bound-state potentials [35.27, 28] and coordinate dependent transition dipole moments [35.29] can be constructed from the data. In this connection, the phenomena of wave packet dispersion and revival, i. e., the spreading and re-focusing of an initially localized wave packet, was documented [35.26, 30]. The interesting phenomenon of fractional revival, where a wave packet splits into two or several parts was also verified experimentally [35.31].

By using time-resolved CARS (coherent-antistokes-raman-scattering) spectroscopy, the particular temporal arrangement of three time-delayed pulses allows one to monitor electronic ground- and excited-state rotational/vibrational dynamics within a single experiment [35.32].

Employing time-resolved photoelectron spectroscopy [35.33, 34], the time-evolution of probability densities could be directly mapped into the kinetic energy distributions of the photoelectrons, thus obtaining a one-to-one picture of quantum mechanical wave packet dynamics [35.35, 36].

35.3.2 Elementary Gas-Phase Chemical Reactions

The study of gas phase chemical reactions in real-time has been pioneered by *Zewail* and coworkers [35.23]. It is of primary interest to chemistry to monitor how chemical bonds are formed and broken.

In the case of a direct bond rupture, the time-scale to be resolved is much shorter than in cases where long-lived resonances exist. In a first femtosecond experiment, the ICN molecule was prepared in an electronically excited state with a repulsive potential energy surface. and free CN was detected in a pump–probe arrangement [35.37]. By changing the probe-laser frequency, the atom-molecule separation could be recorded in time and for different inter-nuclear distances r_M.

The decay of a quasi-bound complex via electronic predissociation was monitored. From the time-signal the nonadiabatic coupling between a bound and a dissociative electronic state could be extracted for the NaI molecule [35.38]. Another type of indirect decay was studied in the OClO molecule which fragments via different decay channels with characteristic time scales [35.39]. The latter are determined by the coupling of internal nuclear degrees of freedom on one hand and the coupling between electronic and nuclear motion on the other.

As an example of a reaction triggered by multi-photon absorption, we mention the multiple fragmentation of $Fe(CO)_5$. This prototype organometallic compound was studied, and how the various CO-ligands emerge from the complex as a function of time [35.40] was determined.

35.3.3 Molecular Dynamics in Liquid and Solid Surroundings

One main application of time-resolved measurements is the study of molecules embedded in a surrounding. For example, transient spectroscopy has been applied to study diatomic molecules in matrices [35.41, 42], zeolites [35.43], and also nanotubes [35.44], thereby revealing information about coherent dynamics, nonadiabatic transitions and energy transfer mechanisms in a solid environment.

Characteristics of relaxation processes were determined for, e.g., I_2 in rare-gas environments which allowed one to monitor the gas-phase to liquid state transition dynamics [35.45–47]. Another important phenomenon is that of the caging of a molecule in a liquid. This effect describes the fragmentation and recombination of a molecule, the latter being induced by the cage consisting of the surrounding molecules. This effect has been detected within a time-resolved measurement [35.48].

As other prominent examples we mention proton-transfer processes in liquids [35.49], the dynamics of hydrogen bonds [35.50], and also time-dependent processes in carotenoids which have been reviewed recently [35.51].

35.4 Recent Developments

The rapid advance in laser technology and the development of sophisticated experiments have led to a revolution in the spectroscopy with ultrashort pulses. Experimental techniques used for time-resolved measurements are evolving with incredible speed. In what follows we list some recent developments which will, along with other efforts, be the basis of future applications.

35.4.1 Faster Dynamics

According to (35.1) and (35.4), the possibility of resolving molecular dynamics is determined by the relation between the average energy spacing in a system and the laser-pulse duration. Thus, if the aim is to observe the dynamics of electrons taking place within a few femtosecond, shorter pulses are necessary. The production of attosecond pulses has been reported [35.52]. Although such pulses are far from being used in an experiment needing a reasonable repetition rate, the first steps towards a sub-femtosecond time-resolution have been taken.

35.4.2 X-Ray Pulses

Much effort is spent in the generation of ultrashort pulses having wavelengths reaching into the X-ray regime [35.53, 54].

Concerning molecular motion, several prospects are emerging. Firstly, using a high photon-energy pump-pulse, it is possible to prepare core-excited states of a molecule such that the following dynamics could be detected with a time-delayed optical probe pulsc. As a second scenario, a uv-pump pulse initiating a chemical reaction, could be followed by a time-delayed X-ray pulse which then is able to selectively excite core levels of the atoms involved in the process. Thus, a signal would track the transient chemical shift of an atom undergoing a molecular re-arrangement process. Here, the first experimental results of this kind have been reported. For example, employing soft X-ray laser pulses, it was possible to monitor the photo-dissociation of the Br_2 molecule [35.55]. Also, it became possible to synchronize an optical pulse from a laser source with an X-ray pulse from a storage ring [35.56] to monitor the temporal changes of an oxidation state of Ru bound in a transition metal complex [35.57].

35.4.3 Time-Resolved Diffraction

Diffraction experiments are of tremendous importance for structural analysis. Geometrical changes of molecules or molecular decomposition could, in principle, be followed using ultrashort light or particle pulses. Concerning time-resolved diffraction using electromagnetic waves, much success has been reported in recent years [35.58]. The applications, to date, are in material science. For example, it is now possible to follow phase transitions in time [35.59]. Concerning the application of time-resolved X-ray diffraction to molecules, the experimental difficulties are numerous. However, first experiments on a chemical reaction in solutions have been reported [35.60].

A promising technique employs not electromagnetic light but pulses of electrons (UED = ultrafast electron diffraction) [35.61]. Such pulses have been used in monitoring, e.g., the structural dynamics of pyridine molecules [35.62].

35.4.4 Dynamics and Control

As has been discussed throughout this chapter, molecular motion can be traced with the help of time-resolved spectroscopy. It is now possible to go one step further and use laser pulses to control molecular motion. Employing pulse shapers [35.63] which allow for the amplitude- and phase-modulation of a given input field, it was shown that, e.g., the branching ratio of photo-products in a photochemical reaction can be influenced by modulating the excitation pulses using feedback algorithms [35.64]. The field of laser control has been recently reviewed extensively [35.65–67].

References

35.1 C. V. Shank, E. P. Ippen, S. L. Shapiro (Eds.): *Picosecond Phenomena, Ultrafast Phenomena*, Springer Ser. Chem. Phys. I-X, Vol. 4 (Springer, Berlin, Heidelberg 1978)

35.2 R. M. Hochstrasser, W. Kaiser, C. V. Shank (Eds.): *Picosecond Phenomena, Ultrafast Phenomena*, Springer Ser. Chem. Phys. I-X, Vol. 14 (Springer, Berlin, Heidelberg 1980)

35.3 K. B. Eisenthal, R. M. Hochstrasser, W. Kaiser, A. Laubereau (Eds.): *Picosecond Phenomena, Ultrafast Phenomena*, Springer Ser. Chem. Phys. I-X, Vol. 23 (Springer, Berlin, Heidelberg 1982)

35.4 D. H. Auston, K. B. Eisenthal (Eds.): *Picosecond Phenomena, Ultrafast Phenomena*, Springer Ser. Chem. Phys. I-X, Vol. 38 (Springer, Berlin, Heidelberg 1984)

35.5 G. R. Fleming, A. E. Siegman (Eds.): *Picosecond Phenomena, Ultrafast Phenomena*, Springer Ser. Chem. Phys. I-X, Vol. 46 (Springer, Berlin, Heidelberg 1986)

35.6 T. Yajima, K. Yoshihara, C. B. Harris, S. Shionoya (Eds.): *Picosecond Phenomena, Ultrafast Phenomena*, Springer Ser. Chem. Phys. I-X, Vol. 48 (Springer, Berlin, Heidelberg 1988)

35.7 C. B. Harris, E. P. Ippen, G. A. Mourou, A. H. Zewail (Eds.): *Picosecond Phenomena, Ultrafast Phenomena*, Springer Ser. Chem. Phys. I-X, Vol. 53 (Springer, Berlin, Heidelberg 1990)

35.8 J.-L. Martin, A. Migus, G. A. Mourou, A. H. Zewail (Eds.): *Picosecond Phenomena, Ultrafast Phenomena*, Springer Ser. Chem. Phys. I-X, Vol. 55 (Springer, Berlin, Heidelberg 1993)

35.9 P. F. Barbara, W. H. Knox, G. A. Mourou, A. H. Zewail (Eds.): *Picosecond Phenomena, Ultrafast Phenomena*, Springer Ser. Chem. Phys. I-X, Vol. 60 (Springer, Berlin, Heidelberg 1994)

35.10 P. F. Barbara, J. G. Fujimoto, W. H. Knox, W. Zinth (Eds.): *Picosecond Phenomena, Ultrafast Phenomena*, Springer Ser. Chem. Phys. I-X, Vol. 62 (Springer, Berlin, Heidelberg 1996)

35.11 T. Elsaesser, J. G. Fujimoto, D. A. Wiersma, W. Zinth (Eds.): *Picosecond Phenomena, Ultrafast Phenomena*, Springer Ser. Chem. Phys. I-X, Vol. 63 (Springer, Berlin, Heidelberg 1998)

35.12 T. Elsaesser, S. Mukamel, M. M. Murnane, N. F. Scherer (Eds.): *Picosecond Phenomena, Ultrafast Phenomena*, Springer Ser. Chem. Phys. I-X, Vol. 66 (Springer, Berlin, Heidelberg 2001)

35.13 R. D. Miller, M. M. Murnane, N. F. Scherer, A. M. Weiner (Eds.): *Picosecond Phenomena, Ultrafast Phenomena*, Springer Ser. Chem. Phys. I-X, Vol. 71 (Springer, Berlin, Heidelberg 2003)

35.14 T. Kobayashi, T. Okada, K. A. Nelson, S. DeSilvestri (Eds.): *Picosecond Phenomena, Ultrafast Phenomena*, Springer Ser. Chem. Phys. I-X, Vol. 79 (Springer, Berlin, Heidelberg 2005)

35.15 W. Kaiser (Ed.): *Ultrashort Laser Pulses* (Springer, Berlin, Heidelberg 1993)

35.16 A. H. Zewail: *Femtochemistry*, Vol. I, II (World Scientific, Singapore 1994)

35.17 L. Wöste, J. Manz (Eds.): *Femtosecond Chemistry*, Vol. I, II (VCH, Weinheim 1995)

35.18 M. Chergui (Ed.): *Femtochemistry - Ultrafast Chemical and Physical Processes in Molecular Systems* (World Scientific, Singapore 1996)

35.19 V. Sundström (Ed.): *Femtochemistry and Femtobiology: Ultrafast Reaction Dynamics at Atomic-Scale Resolution* (Imperial College Press, London 1996)

35.20 F. D. Schryver, S. DeFeyter, G. Schweitzer (Eds.): *Femtochemistry* (Wiley-VCH, Weinheim 2001)

35.21 A. Douhal, J. Santamaria (Eds.): *Femtochemistry and Femtobiology* (World Scientific, Singapore 2002)

35.22 M. Martin, J. T. Hynes (Eds.): *Femtochemistry and Femtobiology: Ultrafast Events in Molecular Science* (Elsevier, Oxford 2004)

35.23 A. H. Zewail: Angew. Chem., Int. Ed. Engl. **39**, 2586 (2000)

35.24 S. Mukamel: *Principles of Nonlinear Optical Spectroscopy* (Oxford Univ. Press, New York 1995)

35.25 G. Stock, W. Domcke: Adv. Chem. Phys. **100**, 1 (1997)

35.26 R. M. Bowman, M. Dantus, A. H. Zewail: Chem. Phys. Lett. **161**, 297 (1989)

35.27 T. Baumert, G. Gerber: Adv. At. Molec. Opt. Phys. **35**, 163 (1995)

35.28 M. Gruebele, A. H. Zewail: J. Chem. Phys. **98**, 883 (1993)

35.29 M. Wollenhaupt, A. Assion, O. Bazhan, Ch. Horn, D. Liese, C. Sarpe-Tudoran, M. Winter, T. Baumert: Chem. Phys. Lett. **3,76**, 457 (2003)
35.30 T. Baumert, V. Engel, C. Röttermann, W.T. Strunz, G. Gerber: Chem. Phys. Lett. **191**, 639 (1992)
35.31 M.J.J. Vrakking, D.M. Villeneuve, A. Stolow: Phys. Rev. A **54**, 1 (1996)
35.32 A. Materny, T. Chen, M. Schmitt, T. Siebert, A. Vierheilig, V. Engel, W. Kiefer: Appl. Phys. B **71**, 299 (2000)
35.33 D.M. Neumark: Ann. Rev. Phys. Chem. **52**, 255 (2001)
35.34 A. Stolow: Ann. Rev. Phys. Chem. **54**, 89 (2003)
35.35 A. Assion, M. Geisler, J. Helbing, V. Seyfried, T. Baumert: Phys. Rev. A **54**, R4605 (1996)
35.36 T. Frohnmeyer, M. Hofmann, M. Strehle, T. Baumert: Chem. Phys. Lett. **312**, 447 (1999)
35.37 M.J. Rosker, M.D.A.H. Zewail: J. Chem. Phys. **89**, 6113 (1988)
35.38 T.S. Rose, M.J. Rosker, A.H. Zewail: J. Chem. Phys. **91**, 7415 (1989)
35.39 T. Baumert, J.L. Herek, A.H. Zewail: J. Chem. Phys. **99**, 4430 (1993)
35.40 L. Bañares, T. Baumert, M. Bergt, B. Kiefer, G. Gerber: J. Chem. Phys. **108**, 5799 (1997)
35.41 M. Bargheer, N. Schwentner: Phys. Rev. Lett. **91**, 085504 (2003)
35.42 M. Karavitis, V.A. Apkarian: J. Chem. Phys. **120**, 292 (2004)
35.43 V.A. Ermoshin, G. Flachenecker, A. Materny, V. Engel: J. Chem. Phys. **114**, 8132 (2001)
35.44 A. Douhal: Chem. Rev. **104**, 1955 (2004)
35.45 C. Lienau, A.H. Zewail: J. Phys. Chem. **100**, 18629 (1996)
35.46 Q. Liu, C. Wan, A.H. Zewail: J. Phys. Chem. **100**, 18666 (1996)
35.47 A. Materny, C. Lienau, A.H. Zewail: J. Phys. Chem. **100**, 18650 (1996)
35.48 C. Wan, M. Gupta, J.S. Baskin, Z.H. Kim, A.H. Zewail: J. Chem. Phys. **106**, 4353 (1997)
35.49 S. Lochbrunner, A.J. Wurzer, E. Riedle: J. Phys. Chem. A **107**, 10580 (2003)
35.50 E.T.J. Nibbering, T. Elsaesser: Chem. Rev. **104**, 1887 (2004)
35.51 T. Polivka, V. Sundström: Chem. Rev. **104**, 2021 (2004)
35.52 A. Baltuška, T. Udem, M. Uiberacker, M. Hentschel, E. Goulielmakis, C. Gohle, R. Holzwarth, V.S. Yakovlev, A. Scrinzi, T.W. Hänsch, F. Krausz: Nature **421**, 611 (2003)
35.53 T. Brabec, F. Krausz: Rev. Mod. Phys. **72**, 545 (2000)
35.54 A. Rousse, C. Rischel, J.-C. Gauthier: Rev. Mod. Phys. **73**, 17 (2001)
35.55 L. Nugent-Glandorf, M. Scheer, D.A. Samuels, A.M. Mulhiesen, E.R. Grant, X. Yang, V.M. Bierbaum, S.R. Leone: Phys. Rev. Lett. **87**, 193002 (2001)
35.56 C. Bressler, M. Chergui: Chem. Rev. **104**, 1781 (2004)
35.57 M. Saes, C. Bressler, R. Abela, D. Grolimund, S.L. Johnson, P.A. Heimann, M. Chergui: Phys. Rev. Lett. **90**, 047403 (2003)
35.58 J.R. Helliwell, P.M. Rentzepis (Eds.): *Time-resolved Diffraction* (Clarendon Press, Oxford 1997)
35.59 D. von der Linde, K. Sokolowski-Tinten: J. Mod. Opt. **50**, 683 (2003)
35.60 R. Neutze, R. Wouts, S. Techert, J. Davidsson, M. Kocsis, A. Kirrander, F. Schotte, M. Wulff: Phys. Rev. Lett. **87**, 195508 (2001)
35.61 R. Srinivasan, V. Lobastov, C.-Y. Ruan, A. Zewail: Helvetica Chimica Acta **86**, 1763 (2003)
35.62 V.A. Lobastov, R. Srinivasan, B.M. Goodson, C.-Y. Ruan, J.S. Feenstra, A.H. Zewail: J. Phys. Chem. A **105**, 11159 (2001)
35.63 A.M. Weiner: Rev. Sci. Instrum. **71**, 1929 (2000)
35.64 A. Assion, T. Baumert, M. Bergt, T. Brixner, V. Seyfried, M. Strehle, G. Gerber: Science **282**, 919 (1998)
35.65 T. Brixner, N.H. Damrauer, G. Gerber: Adv. At. Molec. Opt. Phys. **46**, 1 (2001)
35.66 T. Brixner, G. Gerber: Chem. Phys. Chem. **4**, 418 (2003)
35.67 M. Dantus, V.V. Lozovoy: Chem. Rev. **104**, 1813 (2004)

36. Nonreactive Scattering

The basic formulations of nonreactive scattering are presented in the sections to follow. The semiclassical and quantal approaches to this problem are outlined. Specific symmetries, and their closely related conservation laws, which reduce the complexity of computation are discussed, along with the usual coordinate systems used to express the necessary scattering equations. The chapter ends with prescriptions for determining the various matrix elements needed for a given calculation.

36.1 Definitions 555
36.2 Semiclassical Method 556
36.3 Quantal Method 556
36.4 Symmetries and Conservation Laws 557
36.5 Coordinate Systems 557
36.6 Scattering Equations 558
36.7 Matrix Elements 558
36.7.1 Centrifugal Potential 558
36.7.2 Interaction Potential 559
References 560

36.1 Definitions

The cross section σ for a transition from state j' to j is defined classically as

$$\sigma(j \leftarrow j') = 2\pi \int_0^\infty P_b(j \leftarrow j') b \, \mathrm{d}b \tag{36.1}$$

where P_b is the transition probability for impact parameter b (Fig. 36.1). The impact parameter is related to the relative angular momentum quantum number, ℓ, by

$$\begin{aligned} 2\mu E b^2 &= \ell(\ell+1) \\ &= k^2 b^2 \end{aligned} \tag{36.2}$$

where k is the wave number at relative collision energy E and μ is the reduced mass; atomic units ($e = m_e = \hbar = 1$) are used throughout. Differentiating (36.2) and setting $\mathrm{d}\ell = 1$ in the quantal limit,

$$b \, \mathrm{d}b = \frac{2\ell+1}{2k^2},$$

whence the quantum mechanical equivalent of (36.1) may be obtained

$$\sigma(j \leftarrow j') = \frac{\pi}{k_{j'}^2} \sum_\ell (2\ell+1) P_\ell(j \leftarrow j'), \tag{36.3}$$

$k_{j'}$ being the wave number in the initial channel. If the initial state is degenerate and the $\omega_{j'}$ degenerate substates are labelled by Ω', then

$$P_\ell(j \leftarrow j') = \frac{1}{\omega_{j'}} \sum_{\Omega',\Omega} |T_\ell(j\Omega, j'\Omega')|^2, \tag{36.4}$$

where T_ℓ is an element of the transmission matrix $\boldsymbol{T}$, which contains all the information on the scattering event. The scattering matrix $\boldsymbol{S}$ is related to the $\boldsymbol{T}$ matrix by

$$\boldsymbol{S} = \boldsymbol{1} - \boldsymbol{T} \tag{36.5}$$

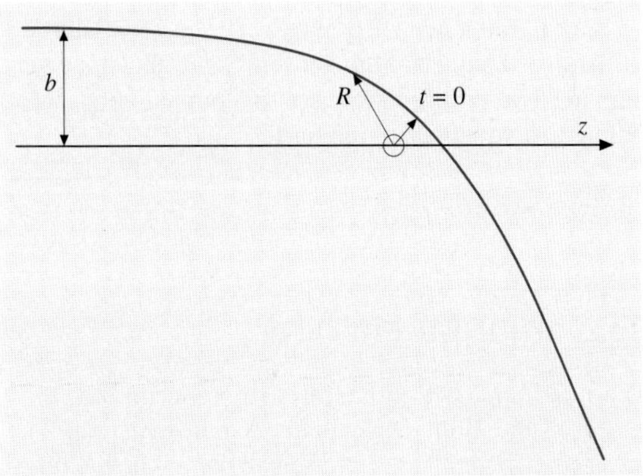

Fig. 36.1 Classical scattering by a fixed scattering center. The trajectory is symmetric about the point of closest approach, which is the time origin

and thence to the reactance matrix $\boldsymbol{K}$ through

$$S = \frac{1 + \mathrm{i}\boldsymbol{K}}{1 - \mathrm{i}\boldsymbol{K}} . \tag{36.6}$$

The elements of the $\boldsymbol{K}$ matrix are real. Conservation of the incident flux of particles requires that

$$\sum_{j\Omega} |S(j\Omega, j'\Omega')|^2 = 1 . \tag{36.7}$$

Micro-reversibility (time-reversal symmetry) implies that the $\boldsymbol{S}$ matrix is symmetric,

$$S(j\Omega, j'\Omega') = S(j'\Omega', j\Omega) ,$$

and hence

$$\sigma(j \leftarrow j')k_{j'}^2\omega_{j'} = \sigma(j' \leftarrow j)k_j^2\omega_j . \tag{36.8}$$

The thermally averaged (Maxwellian) rate coefficient is

$$\langle\sigma v\rangle_{j\leftarrow j'} = \left(\frac{8k_\mathrm{B}T}{\pi\mu}\right)^{\frac{1}{2}} \int_0^\infty x_{j'}\sigma(j \leftarrow j')\,\mathrm{e}^{-x_{j'}}\,\mathrm{d}x_{j'} , \tag{36.9}$$

where $x_{j'} = \mu v_{j'}^2/2k_\mathrm{B}T$, $v_{j'}$ being the relative collision velocity; k_B is Boltzmann's constant, and $(8k_\mathrm{B}T/\pi\mu)^{\frac{1}{2}}$ may be identified with the mean thermal velocity at temperature T. From (36.8) and (36.9),

$$\begin{aligned}&\langle\sigma v\rangle_{j\leftarrow j'}\omega_{j'}\exp(-\epsilon_{j'}/k_\mathrm{B}T)\\ &= \langle\sigma v\rangle_{j'\leftarrow j}\omega_j\exp(-\epsilon_j/k_\mathrm{B}T) ,\end{aligned} \tag{36.10}$$

where $\epsilon_{j'}, \epsilon_j$ denote the energies of the states j', j with respect to the reference level. Equation (36.10) relates the rate coefficients of inverse transitions to their relative degeneracies and excitation energies.

36.2 Semiclassical Method

If the collision dynamics are formulated using elements of both classical and quantum mechanics, then the method is called "semiclassical". Thus, the relative motion might be described by a classical trajectory, whilst internal degrees of freedom (rotation, vibration, ...) are quantized. The de Broglie wavelength associated with the linear motion of a proton attains the Bohr radius at a collision energy of 0.29 eV, and a classical trajectory is a good approximation at much higher energies.

The introduction of a classical trajectory leads to a time-dependent Hamiltonian, and Schrödinger's equation may be reduced to a set of coupled, first-order differential equations of the form

$$\mathrm{i}\dot{\boldsymbol{a}} = \boldsymbol{V}\boldsymbol{a} , \tag{36.11}$$

where $\boldsymbol{a}$ is a column vector of transition amplitudes and $\dot{\boldsymbol{a}}$ is the vector of their time derivatives. The square matrix $\boldsymbol{V}$ is the interaction matrix, whose elements incorporate oscillatory factors of the type $\exp[(\epsilon_{j'} - \epsilon_j)t]$. The trajectory is taken to be symmetric about the point of closest approach, $t = 0$, and $\boldsymbol{a}$ is initialized through

$$a_j \to \delta_{j'j} \quad \text{as } t \to -\infty.$$

The differential equations (36.11) may be integrated numerically, and the transition probabilities derived from

$$P_b(j \leftarrow j') \to |a_j(b)|^2 \quad \text{as } t \to \infty.$$

The cross sections are given by (36.1).

36.3 Quantal Method

The usual approach to nonreactive scattering [36.1] is based on the Born–Oppenheimer approximation. The interaction potential V is taken to be a known function of the *nuclear* coordinates, the electronic energy being minimized for each geometry and separation of target and projectile. The total Hamiltonian may then be written

$$\begin{aligned}H = &-\frac{1}{2\mu}\nabla_R^2 + h_1(\boldsymbol{x}_1) + h_2(\boldsymbol{x}_2)\\ &+ V(\boldsymbol{R}, \boldsymbol{x}_1, \boldsymbol{x}_2) ,\end{aligned} \tag{36.12}$$

where the first term represents the relative kinetic energy of the target and projectile, and h_1, h_2 are functions of the intramolecular nuclear coordinates $\boldsymbol{x}_1, \boldsymbol{x}_2$. The eigenfunctions ψ_1, ψ_2 of h_1, h_2 describe the vibrational and rotational motions of the isolated molecules and form a basis for expanding the total wave function Ψ of the system:

$$\Psi(\boldsymbol{R}, \boldsymbol{x}_1, \boldsymbol{x}_2) = \sum_{\alpha_1\alpha_2\ell m} \frac{F(\alpha_1\alpha_2\ell m|R)}{R} \times Y_{\ell m}(\Theta, \Phi)\psi_1(\alpha_1|\boldsymbol{x}_1)\psi_2(\alpha_2|\boldsymbol{x}_2) . \tag{36.13}$$

In (36.13), α_1, α_2 denote the sets of quantum numbers required to specify the states of the isolated molecules, and $Y_{\ell m}(\Theta, \Phi)$ is a spherical harmonic function of the angular coordinates of the intermolecular vector $\boldsymbol{R}$. $F(\alpha_1\alpha_2\ell m|R)$ are R-dependent expansion coefficients which are solutions of the Schrödinger equation

$$(H - E)\Psi = 0 . \tag{36.14}$$

These solutions may be arranged as the columns of a square matrix, $\boldsymbol{F}(R)$, in which each column is labelled by a different initial scattering state. The radial functions must satisfy the physical boundary conditions,

$$\boldsymbol{F}(R) \to \boldsymbol{0} \quad \text{as} \quad R \to 0$$
$$\boldsymbol{F}(R) \to \boldsymbol{J}(R)\boldsymbol{A} - \boldsymbol{N}(R)\boldsymbol{B} \quad \text{as} \quad R \to \infty ,$$

where $\boldsymbol{J}$ and $\boldsymbol{N}$ are diagonal matrices whose non-vanishing elements are given by

$$J_{ii} = k_i^{\frac{1}{2}} R j_\ell(k_i R) \tag{36.15}$$

$$N_{ii} = k_i^{\frac{1}{2}} R n_\ell(k_i R) , \tag{36.16}$$

and j_ℓ, n_ℓ are spherical Bessel functions of the first, second kinds; k_i is the wave number in channel i (a given set of values of the quantum numbers $\alpha_1\alpha_2\ell m$). The reactance matrix,

$$\boldsymbol{K} = \boldsymbol{B}\boldsymbol{A}^{-1} , \tag{36.17}$$

yields the state-to-state cross sections.

36.4 Symmetries and Conservation Laws

The expansion of the total wave function, (36.13), does not explicitly incorporate the invariance of the Hamiltonian under an arbitrary rotation in space or reflection in the coordinate origin. The Hamiltonian commutes with $\boldsymbol{J}^2$, where $\boldsymbol{J}$ is the total angular momentum, and any component of $\boldsymbol{J}$, J_x, J_y, or J_z, although these components do not commute amongst themselves. Eigenfunctions of H may, therefore, be chosen to be simultaneous eigenfunctions of $\boldsymbol{J}^2$ and (conventionally) J_z, with eigenvalues $J(J+1)$ and M. If $\boldsymbol{j}_1$ and $\boldsymbol{j}_2$ are the angular momenta of the isolated molecules, then

$$\boldsymbol{j}_{12} = \boldsymbol{j}_1 + \boldsymbol{j}_2$$

is their resultant and

$$\boldsymbol{J} = \boldsymbol{j}_{12} + \boldsymbol{\ell} ,$$

where $\boldsymbol{\ell}$ denotes the relative angular momentum of the two molecules. The coupling of the angular momenta to the multipolar expansion of the electromagnetic field gives rise to collisional selection rules.

The parity operator, P, reflects the coordinates in the origin. Because two successive operations with P restore the original values of the coordinates, the corresponding eigenvalue p satisfies the equation $p^2 = 1$, or $p = \pm 1$. For electromagnetic interactions, the commutation of P and H implies conservation of the parity of the system.

If one takes advantage of the conservation laws associated with these symmetries of the system, substantial savings in computing time can be made: only one value of the total angular momentum and of the parity need to be considered simultaneously.

36.5 Coordinate Systems

The natural choice of coordinate system in which to express the interaction potential, $V(\boldsymbol{R}, \boldsymbol{x}_1, \boldsymbol{x}_2)$, is a body-fixed (BF) system in which the z-axis coincides with the direction of the intermolecular vector $\boldsymbol{R} = (R, \Theta, \Phi)$. A rotation of the space-fixed (SF) coordinate system through the Euler angles $(\Phi, \Theta, 0)$ generates such a BF frame. The intramolecular coordinates $\boldsymbol{x}_1, \boldsymbol{x}_2$ must then be expressed relative to the BF frame, as must the Laplacian operator ∇_R^2 which appears in the expression for the total Hamiltonian, (36.12). The latter may be written as

$$\nabla_R^2 = \frac{1}{R}\frac{\partial^2}{\partial R^2}R - \frac{\ell^2}{R^2}$$
$$= \frac{1}{R}\frac{\partial^2}{\partial R^2}R - \frac{(\boldsymbol{J}-\boldsymbol{j}_{12})^2}{R^2}\,, \qquad (36.18)$$

which are the forms suitable for calculations in SF and BF coordinates, respectively.

A unitary transformation relates the normalized eigenfunctions of a given parity in SF and BF coordinates. In Dirac notation,

$$|j_{12}\ell JM\rangle_{\rm SF}$$
$$= \sum_{\bar\Omega} |j_{12}\bar\Omega\epsilon JM\rangle_{\rm BF}\langle j_{12}\bar\Omega\epsilon JM|j_{12}\ell JM\rangle\,, \qquad (36.19)$$

where $\bar\Omega = |\Omega|$ and Ω is the projection of $\boldsymbol{J}$ on the BF z-axis. As the projection of $\boldsymbol{\ell}$ on the intermolecular axis is zero, Ω is also the projection of $\boldsymbol{j}_{12}$ on the BF z-axis. The absolute value of Ω appears in the transformation because $|j_{12} \pm \Omega JM\rangle$ are *not* eigenfunctions of the parity operator P, whereas the linear combinations

$$|j_{12}\bar\Omega\epsilon JM\rangle = \frac{|j_{12}\Omega JM\rangle + \epsilon|j_{12} - \Omega JM\rangle}{\left[2\left(1+\delta_{\bar\Omega 0}\right)\right]^{\frac12}}$$

($\epsilon = \pm 1$) are eigenfunctions of P. The factor $[2(1+\delta_{\bar\Omega 0})]^{\frac12}$ ensures the correct normalization of these functions. The elements of the matrix which performs the unitary transformation (36.19) are

$$\langle j_{12}\bar\Omega\epsilon JM|j_{12}\ell JM\rangle$$
$$= \left(\frac{2(2\ell+1)}{(1+\delta_{\bar\Omega 0})(2J+1)}\right)^{\frac12} C^{j_{12}\ell J}_{\bar\Omega 0\bar\Omega}\,, \qquad (36.20)$$

where $C^{j_{12}\ell J}_{\bar\Omega 0\bar\Omega}$ is a Clebsch–Gordan coefficient.

36.6 Scattering Equations

Schrödinger's equation for the scattering system may be reduced to a set of coupled, ordinary, second-order differential equations for the radial functions $F(R)$. Expressed in matrix form, they become

$$\left[\mathbf{1}\frac{\mathrm{d}^2}{\mathrm{d}R^2} + \boldsymbol{W}(R)\right]\boldsymbol{F}(R) = 0\,. \qquad (36.21)$$

There exists a set of equations (36.21) for each value of the total angular momentum J and parity p (Sect. 36.4). The matrix $\boldsymbol{W}$ may be written

$$\boldsymbol{W}(R) = \boldsymbol{k}^2 - 2\mu \boldsymbol{V}_{\rm eff}\,, \qquad (36.22)$$

where $\boldsymbol{k}^2$ is a diagonal matrix whose non-vanishing elements are

$$k^2_{\alpha_1\alpha_2} = 2\mu\left(E - \epsilon_{\alpha_1} - \epsilon_{\alpha_2}\right)\,. \qquad (36.23)$$

$k_{\alpha_1\alpha_2}$ is the wave number at infinite separation ($R \to \infty$) when the molecules are in eigenstates α_1, α_2 with eigenenergies $\epsilon_{\alpha_1}, \epsilon_{\alpha_2}$. $\boldsymbol{V}_{\rm eff}$ is the matrix of the effective potential,

$$V_{\rm eff} = V(\boldsymbol{R}, \boldsymbol{x}_1, \boldsymbol{x}_2) + \frac{\boldsymbol{\ell}^2}{2\mu R^2}\,, \qquad (36.24)$$

in which V is the interaction potential and $\boldsymbol{\ell}^2/(2\mu R^2)$ may be identified with the centrifugal potential. There exist standard computer codes for solving equations of the form (36.21) [36.2, 3].

36.7 Matrix Elements

36.7.1 Centrifugal Potential

When evaluated in the SF frame, the matrix of the centrifugal potential is diagonal, with non-vanishing elements $\ell(\ell+1)/(2\mu R^2)$. In the BF frame, the diagonal elements are [see (36.18)]

$$\langle j_{12}\bar\Omega\epsilon JM|(\boldsymbol{J}-\boldsymbol{j}_{12})^2/2\mu R^2|j_{12}\bar\Omega\epsilon JM\rangle$$
$$= \left[J(J+1) + j_{12}(j_{12}+1) - 2\bar\Omega^2\right]/(2\mu R^2)\,, \qquad (36.25)$$

and, in addition, there are off-diagonal elements

$$\langle j_{12}\bar\Omega\epsilon JM|(\boldsymbol{J}-\boldsymbol{j}_{12})^2/2\mu R^2|j_{12}\bar\Omega\pm 1\epsilon JM\rangle$$
$$= -\left\{\left(1+\delta_{\bar\Omega 0}\right)\left(1+\delta_{\bar\Omega\pm 1,0}\right)\right.$$
$$\times[J(J+1) - \bar\Omega(\bar\Omega\pm 1)]$$
$$\left.\times[j_{12}(j_{12}+1) - \bar\Omega(\bar\Omega\pm 1)]\right\}^{\frac12}/(2\mu R^2)\,. \qquad (36.26)$$

The matrix elements (36.26), which are off-diagonal in $\bar\Omega$, are associated with the rotation in space of the BF coordinate system (Coriolis coupling). In the "coupled states" approximation [36.4], the off-diagonal elements (36.26) are neglected, and the diagonal elements (36.25) are often replaced by their SF equivalent form $\ell(\ell+1)/(2\mu R^2)$. The matrix of the interaction potential, on the other hand, continues to be evaluated in BF coordinates. The net effect of these approximations is to ignore the rotation of the BF frame in the course of the collision, and the associated dynamical terms.

36.7.2 Interaction Potential

The interaction potential is usually expressed and computed in BF coordinates. For the purposes of the analysis, it is convenient to derive a multipolar expansion of the potential from a least-squares fit to the original data points. If the collision calculations are being done in the BF frame, the potential matrix elements may be evaluated directly. However, if the SF frame is to be used, the potential expansion must first be transformed into SF form.

Consider the interaction between a symmetric top, such as ammonia (NH_3), and a rigid rotor, such as H_2 [36.5, 6]. This complicated example serves as a paradigm, and simpler cases may be derived by reduction. In the BF frame, the potential can be expanded as

$$V(R,\hat{\omega}'_1,\hat{\boldsymbol{r}}'_2)=\sum_{\substack{\lambda_1\lambda_2\\ \mu\nu}} v_{\lambda_1\lambda_2\mu\nu}(R) \times \mathcal{D}^{\lambda_1}_{\mu\nu}(\hat{\omega}'_1)Y_{\lambda_2-\nu}(\hat{\boldsymbol{r}}'_2)\,, \tag{36.27}$$

where the Euler angles $\hat{\omega}'_1=(\phi'_1,\theta'_1,\psi'_1)$ determine the orientation of the principal axes of the top and $\hat{\boldsymbol{r}}'_2=(\theta'_2,\phi'_2)$ the orientation of the axis of the rotor (Fig. 36.2). Expanding the rotation matrix element $\mathcal{D}^{\lambda_1}_{\mu\nu}$ and the spherical harmonic $Y_{\lambda_2-\nu}$, (36.27) becomes

$$V(R,\hat{\omega}'_1,\hat{\boldsymbol{r}}'_2)=\sum_{\substack{\lambda_1\lambda_2\\ \mu\nu}} v_{\lambda_1\lambda_2\mu\nu}(R)\left(\frac{2\lambda_2+1}{4\pi}\right)^{\frac{1}{2}} \times \mathrm{e}^{\mathrm{i}\mu\psi'_1}d^{\lambda_1}_{\mu\nu}(\theta'_1)\,\mathrm{e}^{\mathrm{i}\nu\phi'_1}d^{\lambda_2}_{0-\nu}(\theta'_2)\,\mathrm{e}^{-\mathrm{i}\nu\phi'_2} \tag{36.28}$$

where

$$d^j_{m'm}(\beta)=\langle jm'|\exp(\mathrm{i}\beta J_y)|jm\rangle \tag{36.29}$$

according to the definition in [36.7]. We note that the $d^j_{m'm}$ are real. The combination of positive and negative values of the index ν in (36.28) reflects the invariance of the potential under rotations about the intermolecular axis (alternatively, its dependence on the *difference* between ϕ'_1 and ϕ'_2). In SF coordinates, the potential becomes

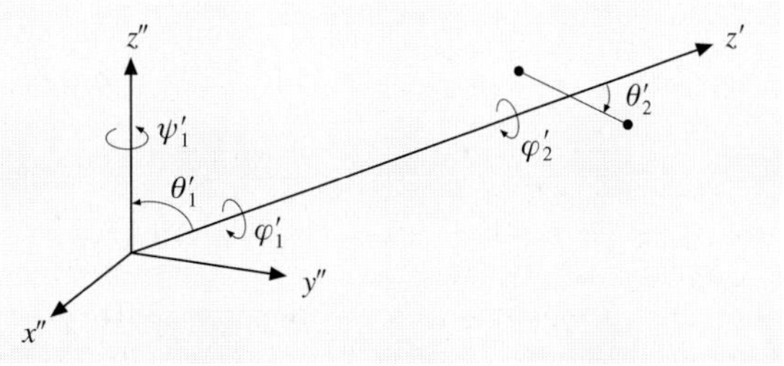

Fig. 36.2 The body fixed coordinate system: scattering of a rigid rotor and a symmetric top

$$V(\boldsymbol{R},\hat{\omega}_1,\hat{\boldsymbol{r}}_2) =\sum_{\substack{\lambda_1\lambda_2\\ \lambda\mu}} v_{\lambda_1\lambda_2\lambda\mu}(R) \times \sum_{m_1m_2} C^{\lambda_1\lambda_2\lambda}_{m_1m_2m}\mathcal{D}^{\lambda_1}_{\mu m_1}(\hat{\omega}_1)Y_{\lambda_2m_2}(\hat{\boldsymbol{r}}_2)Y^*_{\lambda m}(\hat{\boldsymbol{R}})\,. \tag{36.30}$$

The expansion coefficients in the SF and BF frames are related by

$$v_{\lambda_1\lambda_2\lambda\mu}(R) =\left(\frac{4\pi}{2\lambda+1}\right)^{\frac{1}{2}}\sum_{\nu\geq 0} C^{\lambda_1\lambda_2\lambda}_{\nu-\nu 0}(1+\delta_{\nu 0})^{-1} \times\left[v_{\lambda_1\lambda_2\mu\nu}(R)+(-1)^{\lambda_1+\lambda_2+\lambda}v_{\lambda_1\lambda_2\mu-\nu}(R)\right]. \tag{36.31}$$

The matrix elements of the potential given by (36.30) are

$$\begin{aligned}&\langle j_1kj_2j_{12}\ell JM|V(\boldsymbol{R},\hat{\omega}_1,\hat{\boldsymbol{r}}_2)|j'_1k'j'_2j'_{12}\ell' JM\rangle\\ &=\sum_{\substack{\lambda_1\lambda_2\\ \lambda\mu}} v_{\lambda_1\lambda_2\lambda\mu}(R)(-1)^{j'_1+j'_2+j_{12}+k'-J}\left(\frac{2\lambda+1}{4\pi}\right)\\ &\times\left[(2j_1+1)(2j_2+1)(2j_{12}+1)(2\ell+1)(2\lambda_2+1)\right.\\ &\left.\times(2\ell'+1)(2j'_{12}+1)(2j'_2+1)(2j'_1+1)\right]^{\frac{1}{2}}\\ &\times\begin{pmatrix}\lambda&\ell&\ell'\\0&0&0\end{pmatrix}\begin{pmatrix}\lambda_2&j_2&j'_2\\0&0&0\end{pmatrix}\begin{pmatrix}\lambda_1&j_1&j'_1\\\mu&k&-k'\end{pmatrix}\\ &\times\begin{Bmatrix}\ell'&\ell&\lambda\\j_{12}&j'_{12}&J\end{Bmatrix}\begin{Bmatrix}j_{12}&j_2&j_1\\j'_{12}&j'_2&j'_1\\\lambda&\lambda_2&\lambda_1\end{Bmatrix}\end{aligned} \tag{36.32}$$

where k, k' denote the projection of j_1, j_1' on the symmetry axis of the symmetric top; $\begin{pmatrix} \cdots \\ \cdots \end{pmatrix}$ is a Wigner 3j-, $\begin{Bmatrix} \cdots \\ \cdots \end{Bmatrix}$ a 6j-, and $\begin{Bmatrix} \cdots \\ \cdots \\ \cdots \end{Bmatrix}$ a 9j-coefficient. The potential matrix elements for other important cases may be derived from (36.32). Rigid rotor-rigid rotor [36.8] obtains when $k = \mu = 0$, and atom-rigid rotor when, additionally, $j_1 = \lambda_1 = 0$. References to numerical results for specific systems have been compiled in Appendix 2 of [36.9].

In order to exploit conservation of the parity of the total system, it is necessary to use symmetry adapted rotational eigenfunctions

$$\left|j_1\bar{k}m\eta\right\rangle = \frac{|j_1 km\rangle + \eta|j_1 -km\rangle}{\left[2(1+\delta_{\bar{k}0})\right]^{\frac{1}{2}}}\,, \tag{36.33}$$

where $\bar{k}$ is the absolute magnitude of the projection of $\boldsymbol{j}_1$ on the symmetry axis of the top. The parity of the total wave function $\left|j_1\bar{k}\eta j_2 j_{12}\ell JM\right\rangle$ is then $p = \eta(-1)^{j_1+j_2+\ell+\bar{k}}$ and is conserved during the collision. This fact enables the coupled equations to be separated into two non-interacting parity blocks. Because the orientation of the SF z-axis is arbitrary, the matrix elements (36.32) must be independent of M.

The rotational eigenfunctions of an asymmetric top, such as water (H_2O) or formaldehyde (H_2CO), may be written as linear combinations of the symmetric top functions of (36.33):

$$|j_1\tau m\rangle = \sum_{\bar{k}} \left|j_1\bar{k}m\eta\right\rangle\!\left\langle j_1\bar{k}m\eta\middle| j_1\tau m\right\rangle , \tag{36.34}$$

where the expansion coefficients $\left\langle j_1\bar{k}m\eta\middle| j_1\tau m\right\rangle$ are labelled by $-j_1 \le \tau \le j_1$. The parity (inversion symmetry) of these functions is $\eta(-1)^{j_1+\bar{k}}$. Because protons are fermions, the total (rotational and spin) nuclear wave function must be antisymmetric under exchange of the two protons in H_2O or in H_2CO. Proton exchange is equivalent to a rotation through π about the symmetry axis of the molecule. The rotational functions (36.34) are eigenfunctions of this operator with eigenvalues $(-1)^{\bar{k}}$. As the ortho nuclear spin function is symmetric under proton exchange, and the para function is antisymmetric, it follows that rotational states with $\bar{k}$ odd are ortho states, whereas those with $\bar{k}$ even are para states. Conservation of the parity $\eta(-1)^{j_1+\bar{k}}$ then implies that η is *either* 1 *or* -1 in the summation on the right hand side of (36.34). Thus, the index τ implies $\bar{k} =$ even or $\bar{k} =$ odd and $\eta = 1$ or $\eta = -1$. The molecular internal Hamiltonian matrix has four non-interacting diagonal blocks, corresponding to the four possible combinations of $\bar{k}$ and η [36.10, 11].

Recent studies of the rotational excitation of H_2O by H_2 [36.12, 13] show that, at low energies, the cross sections can be much larger for collisions with ortho-H_2 (j_2 odd) than with para-H_2 (j_2 even). The long range interaction between the (large) dipole moment of H_2O and the quadrupole moment of H_2 – absent for para-H_2 in its ground state, $j_2 = 0$ – is responsible for this difference in behavior.

References

36.1 R. B. Bernstein: *Atom–Molecule Collision Theory* (Plenum, New York 1979)

36.2 J. M. Hutson, S. Green: *MOLSCAT Computer Code Version 12, distributed by CCP6* (UK Science, Engineering Research Council, Daresbury Laboratory 1993)

36.3 D. R. Flower, G. Bourhis, J.-M. Launay: Comput. Phys. Comm. **131**, 187 (2000)

36.4 A. S. Dickinson: Comput. Phys. Commun. **17**, 51 (1979)

36.5 A. Offer, D. R. Flower: J. Chem. Soc. Faraday Trans. **86**, 1659 (1990)

36.6 C. Rist, M. H. Alexander, P. Valiron: J. Chem. Phys. **98**, 4662 (1993)

36.7 A. R. Edmonds: *Angular Momentum in Quantum Mechanics* (Princeton Univ. Press, Princeton 1974)

36.8 J. Schaefer, W. Meyer: J. Chem. Phys. **70**, 344 (1979)

36.9 D. R. Flower: *Molecular Collisions in the Interstellar Medium* (Cambridge Univ. Press, Cambridge 1990)

36.10 B. J. Garrison, W. A. Lester, W. H. Miller: J. Chem. Phys. **65**, 2193 (1976)

36.11 T. R. Phillips, S. Maluendes, S. Green: J. Chem. Phys. **102**, 6024 (1995)

36.12 M.-L. Dubernet, A. Grosjean: Astron. Astrophys. **390**, 793 (2002)

36.13 A. Grosjean, M.-L. Dubernet, C. Ceccarelli: Astron. Astrophys. **408**, 1197 (2003)

37. Gas Phase Reactions

The rates of gas phase chemical processes can generally be described by rate laws in which the rate of formation of products or disappearance of reactants is related to the product of the concentrations of reactants raised to various powers [37.1]. Rate laws are deterministic expressions that are usually accurate even though they are used to represent a stochastic reality. Rate equations may fail in the limit of small numbers of reacting particles, where both fluctuations and discrete aspects are important. Exceptions to the reliability of rate equations have been found recently in a variety of fields including surface chemistry on small particles [37.2]. Moreover, both Monte Carlo and master equation methods can be used in their place, given sufficient computing power [37.3–6].

37.1 **Normal Bimolecular Reactions** 563
37.1.1 Capture Theories 563
37.1.2 Phase Space Theories 565
37.1.3 Short-Range Barriers 566
37.1.4 Complexes Followed by Barriers 568
37.1.5 The Role of Tunneling 569

37.2 **Association Reactions** 570
37.2.1 Radiative Stabilization 570
37.2.2 Complex Formation and Dissociation 571
37.2.3 Competition with Exoergic Channels 572

37.3 **Concluding Remarks** 572

References 573

As an example of a rate law, consider the rate of disappearance of reactant A in a gas mixture containing species A, B, and C. The rate law for this disappearance can be expressed by the equation

$$\mathrm{d}[A]/\mathrm{d}t = -k[A]^a[B]^b[C]^c\,, \qquad (37.1)$$

where the symbols [ ] refer to concentration, and the rate coefficient k is dependent on temperature, and possibly other parameters such as the total gas density. The above law is not the most general that can be envisaged. For example, if species A reacts via more than one set of processes, more than one negative term on the right-hand side of the equation will be needed. In addition, if the reverse reaction to form A from products is appreciable, a positive term must also be included. At equilibrium, the rate of change of reactant must be zero.

The relation above does not necessarily refer to one chemical reaction, but to a succession of elementary reactions known collectively as a mechanism. The most elementary reaction is a simple bimolecular process with a second order rate law of the type

$$\mathrm{d}[C]/\mathrm{d}t = -\mathrm{d}[A]/\mathrm{d}t = k[A][B]\,, \qquad (37.2)$$

where two species A and B collide to form products, one of which can be labeled C. In this law, the rate coefficient k is related simply to the reaction cross section σ via the equation

$$k = \sigma v\,, \qquad (37.3)$$

where v is the relative velocity of reactants. The rate coefficient in a bimolecular process has units of volume per time, typically $\mathrm{cm}^3\,\mathrm{s}^{-1}$. In a thermal system, the rate coefficient $k(T)$ is averaged over all degrees of freedom of the reactants, both internal and translational. The most specific rate coefficient is termed a state-to-state coefficient and refers to a process in which reactant A in quantum state a reacts with reactant B in quantum state b at a specific translational energy $\mathcal{T}$ to form products in specific states separating with a specific translational energy [37.7].

Although normal bimolecular processes produce more than one product, it is also possible for the two reactants A and B to stick together if sufficient energy is released in the form of a photon:

$$A + B \longrightarrow AB + h\nu\,. \qquad (37.4)$$

Such a process is called radiative association [37.8], and has mainly been studied for ion-molecule systems, i.e., reactions in which one of the two reactants is an ion. The process of radiative association is normally thought to occur in two steps. The first step produces a collision complex AB^*, which is a molecule existing in a transitory fashion above its dissociation limit:

$$A + B \longrightarrow AB^* . \tag{37.5}$$

Once formed, the complex, which is often thought of as an ergodic entity retaining memory only of its total energy and angular momentum, can either redissociate into reactants, or emit a photon of sufficient energy to stabilize itself:

$$AB^* \longrightarrow AB + h\nu . \tag{37.6}$$

Since redissociation of the complex is generally more rapid than radiative emission, radiative association rate coefficients are normally small. Emission of one infrared photon is sufficient to achieve stabilization, but stabilization of a complex via electronic emission may be a more rapid mechanism if suitable electronic states exist [37.8,9]. The lifetime of the complex against redissociation into reactants is a strong direct function of the binding energy of the species and the number of atoms it possesses.

In addition to bimolecular processes, two other types of reactions often referred to as elementary are unimolecular and termolecular reactions. Although complex reaction mechanisms are sometimes divided into unimolecular and termolecular steps, these are not strictly elementary because they can be subdivided into a series of bimolecular steps. In a unimolecular reaction [37.10,11], a molecule A is destroyed at sufficiently high gas density by a process that has the seemingly simple first order rate law

$$\mathrm{d}[A]/\mathrm{d}t = -k[A] . \tag{37.7}$$

At low pressures, on the other hand, the rate law is second order. A simplified series of events, called the Lindemann mechanism, explains these limiting cases by invoking the activation of species A by strong inelastic collisions with bath gas M to form an activated complex A^* which can either be deactivated by inelastic collisions or spontaneously decompose, since it possesses sufficient energy to do so. The steps are written as

$$A + M \rightleftharpoons A^* + M , \tag{37.8}$$

$$A^* \longrightarrow B + C , \tag{37.9}$$

where the spontaneous destruction of A^* can be thought of as a truly elementary unimolecular reaction, akin to spontaneous emission of radiation. Studies of spontaneous dissociation occupy an important place in gas phase reaction theory, and are discussed in Sects. 37.1.2 and 37.1.4 in the context of dissociation of intermediate complexes. If the rate coefficients for the forward and reverse processes in (37.8) are labeled k_1 and k_{-1} respectively, and the rate coefficient for spontaneous dissociation of A^* into products is labeled $k_2(\mathrm{s}^{-1})$, application of the steady-state principle to the concentration of the activated complex, namely,

$$\mathrm{d}[A^*]/\mathrm{d}t = k_1[A][M] - k_{-1}[A^*][M] - k_2[A^*] \approx 0 \tag{37.10}$$

leads to the rate law

$$-\mathrm{d}[A]/\mathrm{d}t = k'[A][M] , \tag{37.11}$$

where

$$k' = k_1k_2/\left(k_{-1}[M] + k_2\right) . \tag{37.12}$$

At low pressures, $k_{-1}[M] \ll k_2$ and $k' \approx k_1$ so that a second order rate law prevails. At high pressures, $k_{-1}[M] \gg k_2$ and $k' \approx k_1k_2/k_{-1}[M]$ so that the rate law becomes first order:

$$-\mathrm{d}[A]/\mathrm{d}t \approx (k_1/k_{-1})\, k_2[A] . \tag{37.13}$$

Since both the activation and deactivation of complexes occur stepwise, rather than in single strong collisions, reality is far more complex than the Lindemann mechanism [37.12], and, at the highest degree of detail, master equation treatments are needed, especially for intermediate pressures [37.13].

In a termolecular reaction, which is actually the inverse of a unimolecular process [37.13], two species B and C collide to form a collision complex A^*, which can be regarded as the activated complex of stable species A. The collision complex can be stabilized by subsequent strong collisions with other species:

$$A^* + M \longrightarrow A + M , \tag{37.14}$$

or can redissociate into reactants. If the complex formation step occurs with a bimolecular rate coefficient k_{-2}, and k_{-1} and k_2 refer to complex stabilization and dissociation as in the unimolecular case, the rate law for the termolecular process can be written

$$\mathrm{d}[A]/\mathrm{d}t = k'[B][C][M] , \tag{37.15}$$

where

$$k' = k_{-2}k_{-1}/\left(k_{-1}[M] + k_2\right) . \tag{37.16}$$

At low pressures, $k' \approx k_{-2}k_{-1}/k_2$ and the rate law is third order. At high pressures, when every activated complex is deactivated, the rate law becomes second order since $k' \approx k_{-2}/[M]$. Actually, at very low pressures radiative stabilization of the complex dominates and the rate law once again becomes second order. As in the unimolecular case, the complex does not really undergo strong inelastic collisions, so that reality is once again more complex than pictured here [37.13]. For detailed theories of thermal reactions, there is the additional problem in (37.12) and (37.16) of deciding when thermalization of the partial rate coefficients should be undertaken. In reality, the partial rate coefficients refer to reactions with specific amounts of energy and angular momentum, and should not be thermally averaged before incorporation into the equations for k'.

37.1 Normal Bimolecular Reactions

The rates of elementary (bimolecular) chemical reactions are governed by Born–Oppenheimer potential surfaces, which contain electronic energies and nuclear-nuclear repulsions. Reactions can be classified as exoergic or endoergic depending upon whether the 0 K energies of the products lie below or above those of the reactants, respectively. In the simplest types of exoergic chemical reactions, the potential energy flows downhill from reactants to products, or flows downhill from reactants to a global minimum (the reaction complex) after which it flows uphill to products. More commonly, the morphology of the potential surface is such that after some long-range attraction, the potential rises as old chemical bonds are broken before falling as new bonds are formed. Generally, there is a minimum energy pathway through the region of large potential referred to as the reaction coordinate. The system is said to traverse a transition state barrier, which refers to the configuration of atoms at which a potential saddle point occurs. The height of this transition state barrier is related to the activation energy barrier E_a in the classical Arrhenius rate law

$$k(T) = A(T)\exp\left[-E_a/(k_B T)\right], \tag{37.17}$$

for rate coefficients of bimolecular reactions which contain short-range barriers [37.1]. In the Arrhenius expression, k_B is the Boltzmann constant, and the pre-exponential factor $A(T)$ can be related to the form of the long-range potential, or to an equilibrium coefficient between the transition state and reactants (see the discussion of activated complex theory in Sect. 37.1.3). Although fits of experimental data over short temperature ranges often assume the pre-exponential factor to be totally independent of temperature, theories show that this is not strictly true in most instances. A more serious problem with the expression undoubtedly occurs at low temperatures since tunneling will clearly lead to deviations.

Although, in principle, it is possible to calculate reaction cross sections and rate coefficients via the quantum theory of scattering, in practice few systems have been studied by this technique given the immense computational effort required [37.14, 15]. Another set of approaches, which has been used to study a large variety of systems, is known as classical molecular dynamics (see Chapt. 58). In these approaches, the atoms move classically on the quantum mechanically generated Born–Oppenheimer potential surfaces. For many reactions, however, neither technique is applicable, and a variety of simpler approaches has been developed, using capture and statistical approximations; these approaches will be emphasized here. Indeed, the use of an ergodic complex in our preliminary discussion of association and unimolecular reactions above presages the use of statistical approximations. An excellent high-level review article on many of the topics covered here has been written by *Troe* [37.16].

37.1.1 Capture Theories

For reactions that do not possess a potential energy barrier at short-range, it is tempting to apply long-range capture theories between structureless particles to calculate the reaction rate coefficient. Such theories assume that (a) all hard collisions lead to reaction, and (b) hard collisions occur for all partial waves up to a maximum impact parameter b_{max} or relative angular momentum quantum number L_{max}. This is normally defined so that the reactant translational energy $\mathcal{T}_{AB}$ is just sufficient to overcome a centrifugal barrier. The centrifugal barrier produces a long-range maximum in the effective potential energy function V_{eff} given by the relation

$$V_{eff}(r,b) = V(r) + \mathcal{T}_{AB}b^2/r^2, \tag{37.18}$$

where r is the separation between reactants and $\mathcal{T}_{AB}b^2/r^2$ is the angular kinetic energy. If the reactants overcome the centrifugal barrier, they will spiral in towards each other in the absence of short-range repulsive forces. The approach has had its most notable success for exoergic reactions between ions and nonpolar neutral molecules [37.17]. The long-range potential in this situation is simply (in cgs-esu units)

$$V(r) = -e^2\alpha_d/2r^4\,, \quad (37.19)$$

where α_d is the polarizability of the neutral reactant. This potential leads to the Langevin rate coefficient

$$k = v\pi b_{max}^2 = 2\pi e(\alpha_d/\mu)^{1/2}\,, \quad (37.20)$$

where μ is the reduced mass of the reactants. The theory leads to a temperature-independent rate coefficient with magnitude $\approx 10^{-9}\,cm^3\,s^{-1}$. Numerous experiments from above room temperature to below 30 K confirm its validity for the majority of ion–molecule reactions, which appear rarely to possess potential energy surfaces with short-range barriers [37.18, 19].

An analogous central force potential for structureless neutral–neutral reactants is the van der Waals or Lennard–Jones attraction

$$V(r) = -C_6/r^6\,, \quad (37.21)$$

where C_6 can be defined simply in terms of the ionization potentials and polarizabilities of the reactants [37.20,21]. The rate coefficient obtained using a capture theory with this long-range potential is given, after translational thermal averaging, by the equation [37.21]

$$k(T) = 8.56\,C_6^{1/3}\mu^{-1/2}(k_B T)^{1/6}\,, \quad (37.22)$$

where all quantities are in cgs units. Equation (37.22) leads to estimates for the rate coefficient at room temperature in the range $\approx 10^{-10}$–$10^{-9}\,cm^3\,s^{-1}$. Unlike the situation for ion–molecule reactions, this estimate has not received much attention mainly because most neutral–neutral reactions involve activation energy. Even for those systems without activation energy, the approximation appears to lead to rate coefficients that are too large by at least a factor of a few [37.21]. In place of the result of (37.22), kineticists often use the simple hard-sphere model with atomic dimensions for the reaction cross section. The hard-sphere model leads to a temperature dependence of $T^{1/2}$, which is in disagreement with a whole series of new experiments by *Smith*, *Rowe*, and co-workers on fast neutral–neutral reactions down to temperatures near 10 K [37.22].

Both long-range potentials considered above are isotropic in nature. A variety of capture theories [37.23, 24] have been developed which take angular degrees of freedom into account. For ion–molecule systems in which the neutral species has a permanent dipole moment μ_d, the long-range potential becomes

$$V(r,\theta) = -\frac{e^2\alpha_d}{2r^4} - \frac{e\mu_d\cos\theta}{r^2}\,, \quad (37.23)$$

where θ is the angle between the radius vector from the charged species to the center-of-mass of the dipolar species and the dipole vector. Adiabatic effective (centrifugal) potential energy curves can be defined at any given fixed intermolecular separation r by diagonalizing angular kinetic energy and potential energy matrices using a suitable basis set. If an atomic ion is reacting with a linear neutral, a suitable basis set would consist of spherical harmonics for the rotation of the linear molecule (angular momentum j) as well as rotation matrices for the relative motion (angular momentum L) between the species. The eigenvalues of a matrix with fixed total angular momentum J then correspond to the effective radial potential energy functions V_{eff} at each r. The L, j labeling of the potential curves can be ascertained by starting the calculation at sufficiently large r so that L and j are reasonably good quantum numbers. Additional approximations, such as the centrifugal sudden approximation, can be made to facilitate the calculation [37.25]. Making the adiabatic assumption that transitions between these potentials are not allowed at long range, one obtains an adiabatic capture rate coefficient for each initial value of j and translational (radial kinetic) energy $\mathcal{T}_{AB}$ from the criterion

$$V_{eff}(j, L_{max}, R) = \mathcal{T}_{AB}\,, \quad (37.24)$$

where R is the separation corresponding to the maximum effective potential. Unlike the Langevin approach, the rate coefficients thermalized over translation for each j show an inverse dependence on temperature, with the $j = 0$ state showing the most severe dependence since the dipole is essentially "locked" for this state. Thermal averaging over j typically leads to rate coefficients with an inverse dependence on temperature between $T^{-1/2}$ and T^{-1} [37.26]. Rate coefficients as large as $10^{-7}\,cm^3\,s^{-1}$ can be obtained as the temperature approaches 10 K. Although most ion–dipole reactions seem to obey capture theory models at low temperature, there are exceptions.

An alternative approach to ion–dipole reactions using the classical concept of adiabatic invariants has been developed [37.27]. In addition, variants to the capture theory discussed here have been formulated. A statistical adiabatic approach has been developed [37.23], in which adiabatic effective potential maxima are not used to define capture cross sections, but to define rate coefficients via the activated complex (transition state) theory discussed in Sect. 37.1.3, which posits thermal equilibrium between reactants and the species existing at the potential maxima. One advantage of this simplification is that it permits the adiabatic treatment to be more easily extended to complex geometries, especially if perturbation approximations are utilized. Other approaches involve the variational principle; an upper bound to the capture rate coefficient can be determined within the transition state, or bottleneck approach, which defines the transition state through minimization of the number of available vibrational/rotational states at the bottleneck [37.28] (see Sect. 37.1.3).

Nonspherical capture theories have also been used [37.29, 30] to study rapid neutral–neutral reactions. The role of atomic fine structure at low temperatures is an especially interesting application; reactions involving atomic C and O with a variety of reactants have been studied. Use of an electrostatic potential shows that the reactivity of C or O atoms in their $^3P_0$ states with dipolar species is minimal. This is particularly important for atomic carbon since at low temperatures it lies primarily in its ground $^3P_0$ state. The choice of an electrostatic potential has been disputed [37.21] because experimental results for C–hydrocarbon reactions at room temperature are best understood if the long-range potential is dispersive (Lennard–Jones) in character rather than electrostatic.

Reactions between radicals in $^2\Pi$ states (e.g. OH) and $^2\Sigma$ states (e.g. CN) and stable molecules have also been considered, especially at low temperature, with long range potentials that contain both electrostatic and dispersion terms. The results can be compared with new low temperature experimental results on the $CN-O_2$ reaction, but the temperature dependence is not matched by theory if the dispersion term is included [37.30, 31]. In general, even the most recent capture theories are not as reliable as those for ion–molecule systems [37.32]. Rapid neutral–neutral reactions can also be treated by transition state theories (see below) [37.33] or, rarely, by detailed quantum mechanical means [37.34].

The last five years have witnessed a burgeoning interest in rapid neutral–neutral reactions at low temperature studied with the so-called CRESU technique (an acronym for Cinétique de Réaction en Ecoulement Supersonique Uniforme) [37.22, 35–37]. The new data should provoke new attempts at theoretical understanding.

37.1.2 Phase Space Theories

Capture theories tell us neither the products of reaction, if several sets of exoergic products are available, nor the distributions of quantum states of the products. The simplest approach to these questions for reactions with barrierless potentials is to make a statistical approximation – all detailed outcomes being equally probable as long as energy and angular momentum are conserved. Such a result requires strong coupling at short range. The most prominent treatment along these lines is referred to as phase space theory [37.38]. In this theory, the cross section σ for a reaction between two species A and B with angular momentum quantum numbers J_A and J_B and in specific vibrational-electronic states colliding with asymptotic translational kinetic energy $\mathcal{T}_{AB}$ to form products C and D in specific vibrational-electronic states with angular momentum quantum numbers J_C and J_D is

$$\sigma(J_A, J_B \to J_C, J_D) = \frac{\pi\hbar^2}{2\mu\mathcal{T}_{AB}} \sum_{L_i, J}(2L_i+1) \times P(J_A, J_B, L_i \to J) P'(J \to J_C, J_D)\,, \quad (37.25)$$

where J is the total angular momentum of the combined system, L_i is the initial relative angular momentum of reactants, P is the probability that the angular momenta of the reactants add vectorially to form J, P' is the probability that the combined system with angular momentum J dissociates into the particular final state of C and D, μ is the reduced mass of reactants, and the summation is over the allowable ranges of initial relative angular momentum and total angular momentum quantum numbers. P' is equal to the sum over the final relative angular momentum L_f, of angular momentum allowed $(J \to L_f, J_C, J_D)$ combinations leading to the specific product state divided by the sum of like combinations for all energetically accessible product and reactant states. The ranges of initial and final relative angular momenta are given by appropriate capture models (e.g. Langevin, ion–dipole, Lennard–Jones) as well as angular momentum triangular rules. This procedure involves the implicit assumption that strong coupling does not occur at long range; rather, adiabatic effective potentials can be assumed for initial and final states. The state-to-state rate coefficient is simply the cross section multiplied by the relative velocity of the two reactants. Summation over all product states, as well as thermal averaging over the reactant state distributions and the translational energy distribution, can all be undertaken. Strategies for summing/integrating over rotational/vibrational degrees of freedom have been given [37.39]. As can be seen by writing out the expression for P', the phase space formula for the state-to-state cross section in (37.25) obeys microscopic reversibility:

$$\begin{aligned} &2\mu\mathcal{T}_{AB}\sigma(J_A, J_B \to J_C, J_D) \\ &= 2\mu'\mathcal{T}'_{AB}\sigma(J_A, J_B \leftarrow J_C, J_D) \end{aligned} \quad (37.26)$$

where the reduced mass and asymptotic translational energy of the products are denoted by primes. Thus, thermalization of the forward and backward rates leads to detailed balancing.

As a detailed theory for prediction of reaction products and their state distributions, phase space theory and its variants have had mixed success. It is true that the theory correctly predicts that exoergic reactions occurring on barrierless potential surfaces proceed on every strong collision. It is also true, however, that the theory is generally not useful in predicting the branching ratios among several sets of exoergic products because potential surfaces do not often show barrierless pathways for more than one set of products. When applied to product state distributions, the theory yields statistical results, so that population inversions in nondegenerate degrees of freedom (such as the vibrations of a diatomic molecule) are not replicated. With respect to more thermalized entities, such as total cross sections vs. collision energy for endoergic reactions [37.40], the theory can be quite successful, especially when the potential surface involves a deep minimum or intermediate complex. In this instance, the strong coupling hypothesis comes closest to actuality. A deep well is also associated with a high density of vibrational states in the quasicontinuum above the dissociation limit of the molecular structure defined by the potential minimum [37.1,41,42]. Such a high density renders direct (specific) dynamical processess less likely.

A useful variant of phase space theory if complex lifetimes are needed [37.43] (see Sects.37.1.4 and 37.2) is based on a unimolecular decay theory of *Klots* [37.44]. The reaction is considered to proceed via a capture cross section to form the intermediate complex, which then can dissociate into all available reactant and product states consistent with conservation of energy and angular momentum. The complex dissociation rate k_{uni} into a specific state can be obtained via the principle of microscopic reversibility in terms of the capture cross section from that state, obtained with $P' = 1$ in (37.25), to form the complex. In particular, if a complex with total angular momentum J can dissociate into one state of reactants A and B separating with translational energy $\mathcal{T}_{AB}$, then

$$k_{\text{uni}} = \rho_{\text{vib}}^{-1} g(J)^{-1} g(J_A) g(J_B) 2\mu\mathcal{T}_{AB}\sigma(J_A, J_B \to J)\,, \quad (37.27)$$

where ρ_{vib} is the density of complex vibrational states obtained via a variety of prescriptions [37.1,41,42], and g is the rotational degeneracy. A cross section analogous to that in (37.25) can be formulated in terms of capture to form the complex multiplied by the complex dissociation rate into a particular state divided by the total (summed) dissociation rate. Note that the dissociation rate of the complex is proportional to ρ_{vib}^{-1}. Since ρ_{vib} is a strong function of the well depth, long-lived complexes are associated with large well depths (>1 eV). The Klots form for k_{uni} is especially useful for ion–molecule systems, where the cross section for complex formation can be assumed to be Langevin or ion–dipole. The concept of a long-range potential is less useful for most neutral–neutral systems, so that unimolecular rate coefficients for unstable entities are normally obtained quite differently in terms of ρ_{vib} at the transition state (Sect. 37.1.4).

For smaller reaction systems, especially those involving ions, it is quite common for the electronic states of reactants to correlate with more than one potential surface of the combined system, although typically only one surface leads to energetically accessible products. In such instances, it is normal to assume statistical partitioning, although fine structure effects can complicate this picture. A generalization of phase space theory to account for multiple potential surfaces has been proposed for the $C^+ + H_2$ reaction [37.45].

Phase space theory has been used to predict product branching ratios for dissociative recombination reactions between polyatomic positive ions and electrons [37.46]; its success here has been limited at best.

37.1.3 Short-Range Barriers

Most reactions involving neutral molecules, as well as a minority of ion–molecule reactions, possess potential surfaces with short-range barriers. The experimental rate coefficients of these systems over selected temperature ranges are normally fit to the Arrhenius expression (37.17).

The simplest theoretical method of taking short-range potential barriers into account is the line-of-centers approach, which resembles the capture theories previously discussed [37.7]. In this crude approximation, it is assumed that structureless reactants colliding with impact parameter b along a repulsive potential must reach some minimum distance, d, for reaction to occur. If the potential energy in the absence of angular momentum at d is E_0, this condition implies that the asymptotic translational energy of reactants $\mathcal{T}_{AB}$ must exceed the sum of E_0 and the centrifugal energy $\mathcal{T}_{AB}b^2/d^2$, which in turn yields a maximum impact parameter b_{max}. Thermal averaging of the rate coefficient $k = v\pi b_{\text{max}}^2$ over a Maxwell–Boltzmann distribution then yields

$$k(T) = \pi d^2 (8k_BT/\pi\mu)^{1/2} \exp(-E_0/k_BT) , \tag{37.28}$$

which bears some resemblance to the Arrhenius expression if one equates the parameter E_0 with the activation energy. Assuming $T = 300\,\text{K}$, $\mu = 10\,\text{u}$, and $d = 1\,\text{Å}$, the pre-exponential factor is $2.5 \times 10^{-11}\,\text{cm}^3\,\text{s}^{-1}$, which lies in a typical range.

The simple line-of-centers approach reduces the problem to that of structureless particles. The standard method of including all degrees of freedom is to use canonical ensemble statistical mechanics and to imagine that the transition state is in equilibrium with reactants. In the transition state, one of the vibrational degrees of freedom of a normal polyatomic molecule is replaced by a degree of freedom (the reaction coordinate) along which the potential is a maximum (with a corresponding imaginary frequency of vibration). The reaction coordinate is treated as a separable translation, so that the reaction rate coefficient can be envisaged as the equilibrium coefficient between transition state (minus one coordinate) and reactants multiplied by the (average) speed of the transition state structure over the saddle point in the potential energy surface. The canonical result for $k(T)$ is given by [37.1, 7]

$$k(T) = \frac{k_BT}{h} \frac{q_{AB^\dagger}}{q_A q_B} \exp(-E_0/k_BT) \tag{37.29}$$

for reactants A and B, where E_0 is the energy difference between the transition state and the reactants referred to zero-point levels, the $\dagger$ refers to the transition state, and the q are partition functions per unit volume. The partition functions can be factored into products representing electronic, vibrational, rotational, and translational degrees of freedom [37.47]. This formulation for $k(T)$ is known as the activated complex theory (ACT); a more appropriate name would be transition state theory since the term activated complex is also used to refer to an unstable state of a molecule in a deep potential well. The rate coefficient can also be written in terms of the thermodynamic parameters $\Delta H^\dagger$ and $\Delta S^\dagger$ [37.1].

Both the size and temperature dependence of the pre-exponential factor depend critically on the characteristics of transition state and reactants. Originally, ACT theory was used mainly to fit transition state characteristics to rate data. Increasingly accurate ab initio calculations of potential surfaces now allow purely theoretical determinations of $k(T)$ [37.48].

In addition to the assumption of a separable reaction coordinate that can be treated as a translation, several other assumptions have gone into the derivation of the ACT rate coefficient. First, molecules at the saddle point configuration (the transition state) have been arbitrarily chosen to be in equilibrium with reactants. More recently, the transition state assumption has been generalized to refer to that portion of the potential surface in which the reaction flux or number of states is a minimum [37.49], whether or not this occurs at a well-defined saddle point. Loose transition states can thus be defined even if there is no barrier along the reaction coordinate [37.50]. The procedure can be undertaken for transition states as a function of temperature, energy, or, in its most detailed version, energy and angular momentum. The variational theorem shows rate coefficients determined in this way to be upper bounds to the true rate coefficients, the more detailed procedures leading to the better bounds [37.23, 49, 50]. Secondly, the implicit assumption is made that translation along the reaction coordinate at the transition state structure invariably leads to products. Sometimes a transmission coefficient κ is used as an unknown factor in the expression for $k(T)$ to account for the possibility that translation leads back to reactants instead. Thirdly, the assumption is made that two-body collisions can lead to canonical statistical equilibrium. This assumption becomes worse as temperature declines, because at low temperatures the long-range centrifugal constraints on angular momentum become more significant [37.8]. The influence of the long-range potential on ACT is contained in the statistical adiabatic theory discussed earlier [37.23]. Here loose transition states are defined as the maxima of effective adiabatic potential curves, and (37.29) must be modified since, as in detailed variational transition state theory, the transition states are themselves dependent on energy and angular momentum. A related problem is how to treat angular momentum constraints at low temperatures for potential surfaces which also contain a large short-range barrier or tight transition state. One approach is discussed in Sect. 37.1.4.

Rate coefficients derived from ACT do not indicate what the reaction products or sets of products might be. This information can be obtained from ab initio studies of the potential surface if different transition states lead to different products.

37.1.4 Complexes Followed by Barriers

Since long-range forces are always attractive, it makes sense to consider theories in which attraction and repulsion occur sequentially. For ion–molecule systems, this is especially important because many potential surfaces are monotonically attractive from long range to formation of a deep minimum at short range, but possess transition state barriers in their exit channels which are not large enough to prevent reaction but which affect the reaction dynamics. In addition, there are ion–molecule systems with potential surfaces closer to the norm for neutral–neutral species; in these systems there is only a weak long-range minimum followed by a short-range transition state barrier with energy above that of reactants. For the former type and, more arguably, for the latter type of potential surface, one can assume that the reaction proceeds through initial formation of an ergodic complex, followed by dissociation of the complex back into reactants or over the transition state barrier. If the reactants are labeled A and B, the complex AB^*, and the transition state $AB^\dagger$, the mechanism is

$$A + B \rightleftharpoons AB^* \longrightarrow AB^\dagger \longrightarrow \text{Product}\,, \tag{37.30}$$

which leads to the steady-state rate law

$$\mathrm{d}[A]/\mathrm{d}t = -k[A][B]\,, \tag{37.31}$$

$$k = \frac{k_{\mathrm{cf}}}{k_{\mathrm{cd}} + k_{\mathrm{cd'}}} k_{\mathrm{cd'}}\,, \tag{37.32}$$

where k_{cf}, k_{cd}, and $k_{\mathrm{cd'}}$ are the rate coefficients for complex formation, redissociation into reactants, and dissociation into products over the transition state, respectively. One statistical approach to such systems is to use a capture theory for complex formation, to use the Klots formulation of complex redissociation into reactants, and to use a different theory for the unimolecular dissociation of the complex into products. Transition state theories of unimolecular complex decay have been studied for many years [37.1]. The current version is called RRKM theory, after the four authors *Rice*, *Ramsperger*, *Kassel*, and *Marcus* [37.10–13]. This theory builds upon the earlier RRK approach, in which random intramolecular vibrational energy transfer leads to large amounts of energy in the bond to be broken. An alternative hypothesis, in which amplitudes of well-defined normal modes add up to extend the bond to be broken past a certain amount, is now discredited [37.51]. In the RRKM approach, which is perfectly analogous to ACT theory, an equilibrium is envisaged between the activated complex and transition state species. The main use of the theory is in thermal unimolecular decay where it explains complex activation, deactivation, and random unimolecular dissociation as a function of gas density. At high density in a thermal environment, an equilibrium among complex, activated complex, and transition state leads to a thermal unimolecular decay rate (37.13)

$$k_{\mathrm{RRKM}}(T) = \frac{k_{\mathrm{B}}T}{h}\frac{q_{AB^\dagger}}{q_{AB}} \exp(-E_0/k_{\mathrm{B}}T)\,, \tag{37.33}$$

which is perfectly analogous to the ACT result. At lower pressures, the activated complex is not in thermal equilibrium with the stable complex and the details of complex activation and deactivation are important [37.12, 13].

In the microcanonical formulation of RRKM theory, the dissociation rate coefficient k_{RRKM} as a function of (activated) complex total energy E and angular momentum J is given by

$$k_{\mathrm{RRKM}}(E, J) = \frac{N^\dagger[E - E_0 - E_{\mathrm{rot}}(J)]}{h\rho^*[E - E_{\mathrm{rot}}(J)]}\,, \tag{37.34}$$

where $N^\dagger$ refers to the total number of vibrational states of the transition state from its minimum allowable saddle point energy E_0 through E, and ρ^* refers to the density of vibrational states of the complex at energy E. For both the transition state and the complex, the available vibrational energy is the total available energy minus the rotational energy $E_{\mathrm{rot}}(J)$, which is a function of the angular momentum. The energy not used for vibration and rotation in the transition state is considered to belong to the separable reaction coordinate. The most common expressions for the number and density of vibrational states are the semiclassical empirical values [37.1, 41, 42]; direct counting schemes also exist [37.52]. Both empirical and direct counting refer to states representing a bath of harmonic oscillators; anharmonic effects are rarely treated.

A theory of reaction rates for the mechanism in (37.30) incorporating a capture theory, and the Klots and RRKM unimolecular decays has been applied successfully to a variety of ion–molecule reactions in competition with association channels for which the reaction proceeds via a deep well and an exit channel barrier (see Sect. 37.2.3) [37.53–55]. Some authors prefer to use RRKM theory to describe redissociation of the complex into reactants [37.13, 56] by means of a loose transition state [37.57], which can be defined via a variety of variational and other methods [37.50]. Some recent work on unimolecular decay of polyatomic ions and neutrals following excitation indicates the superiority of the variational RRKM approach to unimolecular decay [37.58], especially if there is a significant amount of energy. Both the Klots and RRKM approaches to complex redissociation have problems associated with them: the Klots expression obeys microscopic reversibility but requires both a capture cross section and the assumption that the phase space formulation of the probability of individual quantum states is accurate; the RRKM expression does not obviously obey microscopic reversibility and, unless a variational calculation is performed, can create a somewhat artificial transition state.

Reaction mechanisms with more than one potential barrier can also be treated via a combination of ACT, capture, and unimolecular approaches [37.59, 60].

37.1.5 The Role of Tunneling

The role of tunneling in bimolecular and unimolecular reactions grows in importance as the temperature is lowered and hopping over potential barriers becomes more difficult. A simple one-dimensional correction for tunneling in the ACT expression for bimolecular rate coefficients, obtained by *Wigner* [37.1, 61], appears to be reasonable if the tunneling mechanism is not dominant. This correction $\Gamma > 1$ is simply the quantum correction to the partition function for the reaction coordinate:

$$\Gamma \approx 1 - \frac{(h\nu_i)^2}{24(k_B T)^2} + \cdots , \quad (37.35)$$

where ν_i is the (imaginary) harmonic frequency at the saddle point. Improved corrections have also been developed [37.62, 63] and, for the $H + H_2$ reaction, tested versus accurate quantum calculations [37.64]. At very low temperatures, such as those found in the interstellar medium, tunneling corrections are likely to be very large and to require proper multidimensional treatments.

A one-dimensional tunneling correction to the RRKM expression for the microcanonical unimolecular decay rate coefficient also exists [37.65]. The effective potential representing the reaction coordinate at the saddle point is assumed to be an Eckart barrier. The probability of tunneling for each vibrational state of the transition state is computed, and this probability takes the place of simply counting the state in the standard formula for k_{RRKM}. In particular, $N^\dagger(E - E_0 - E_{\text{rot}})$ in the equation for $k_{\text{RRKM}}(E, J)$ is replaced by

$$N^\dagger_{\text{QM}}(E - E_0 - E_{\text{rot}}) = \sum_n P(E - E_0 - E_{\text{rot}} - \epsilon_n) , \quad (37.36)$$

where ϵ_n is the vibrational energy of state n of the transition state with respect to its zero point energy, and the sum is over all states n for which the energy in the reaction coordinate (the energy in parentheses on the right-hand-side) is negative, but not so negative that the classical energy in the reaction coordinate lies below the minimum of the complex potential well. In general, only a few vibrational states of the transition state need be considered for the tunneling correction.

This tunneling correction has been incorporated into statistical theories for the rate of reactions that proceed via complexes and transition state barriers [37.60]. It has been used recently in a calculation of the slow ion–molecule reaction

$$NH_3^+ + H_2 \longrightarrow NH_4^+ + H , \quad (37.37)$$

which successfully reproduces the observation that, despite initially decreasing as the temperature is reduced below 300 K, the reaction rate coefficient begins to increase as the temperature is decreased below 100 K [37.66]. The parameters for the calculation were obtained from an ab initio calculation of the potential surface, which shows the system to proceed through a weakly-bound long-range complex before encountering a transition state with a rather small barrier of ≈ 0.25 eV. Although the dynamical theory is not quantitative (presumably because of the one-dimensional tunneling approximation), it does reproduce the isotope effects seen when the reactants $NH_3^+ + D_2$ and $ND_3^+ + H_2$ are used, definitively showing that tunneling is the cause. The actual increase in rate at very low temperature comes from the fact that the tunneling rate is less dependent on temperature than the dissociation rate of the complex into reactants. Similar calculations have been performed for the analogous ion–molecule reaction

$$C_2H_2^+ + H_2 \longrightarrow C_2H_3^+ + H \quad (37.38)$$

to explain a similar observation, although in this latter instance there is still a controversy concerning whether or not the reaction is truly exoergic [37.67]. It is interesting to speculate on whether similar effects can be detected for analogous neutral radical–H_2 reactants with moderate activation energy barriers, such as

$$CCH + H_2 \longrightarrow C_2H_2 + H . \quad (37.39)$$

This system possesses a transition state of rather moderate energy [37.48], but the only long-range complex is presumably the very weakly-bound van der Waals structure. Unlike the ion–molecule case, there are very few experimental studies of low temperature neutral–neutral reactions. Studies of pressure broadening at very low temperatures do reveal, however, the strong influence of the van der Waals bond [37.68].

37.2 Association Reactions

Association reactions have been studied for both ion–molecule and neutral–neutral systems. For the ion–molecule case, association processes can be investigated in great detail at low densities near or in the radiative association regime because ions can be stored in low pressure traps. Review articles [37.69,70] describe such experiments for small and large ions, respectively. The results of many higher pressure studies of termolecular ion–molecule association, especially those undertaken in the SIFT apparatus, are also available [37.71]. In general, below the high pressure limit, the rate of association reactions without activation energy is known to increase with (a) an increase in the size of the reactants, (b) an increase in the bond energy of the product species, and (c) a decrease in temperature. These trends are all reproduced by statistical theories in which association proceeds by the formation of a complex followed by radiative and/or collisional stabilization [37.8]. More limited data for termolecular neutral–neutral reactions, with and without short-range potential barriers, have also been reviewed [37.72]. Although radiative stabilization can proceed via a single photon, collisional stabilization proceeds most probably in a stepwise fashion rather than in a single strong collision. The most detailed statistical theories incorporating multistep collisional stabilization use master equation techniques; in general it is preferable to solve the inverse problem of unimolecular dissociation via detailed RRKM theory and then invoke detailed balance [37.13,56]. Such theories are more successful than strong collision approaches, especially in considering the dependence of ternary association rates on pressure. This dependence can be especially difficult to treat when association competes with an exoergic channel which does not necessarily dominate because of a barrier in the exit channel.

37.2.1 Radiative Stabilization

The problem of radiative association in the absence of competitive exoergic channels is in principle much more simple. In addition to the rates of complex formation and dissociation, one needs only the rate of stabilization via spontaneous emission. If the complex abundance is at steady state, the rate law for radiative association of species A and B is simply

$$\mathrm{d}[A]/\mathrm{d}t = -k_{\mathrm{ra}}[A][B]\,, \tag{37.40}$$

$$k_{\mathrm{ra}} = \frac{k_{\mathrm{cf}}}{k_{\mathrm{cd}}+k_{\mathrm{rad}}}k_{\mathrm{rad}}\,, \tag{37.41}$$

where k_{ra} represents the rate coefficient for radiative association and k_{rad} refers to the rate coefficient for radiative stabilization. If ternary association is considered in addition, the rate coefficient for radiative association is the low pressure limit of a more complex rate expression.

Radiative stabilization, which is normally considered to proceed via emission of a single infrared photon, has been studied in some detail [37.73–75]. If the unstable complex is imagined to be a collection of harmonic oscillators, each vibrational state i, which is best regarded as a Feshbach resonance, can be expressed as a set of occupation numbers $n_1^i, n_2^i, \ldots, \left[n^i\right]$ for the assorted modes. If it is further assumed that upon formation, the complex has a probability P_i of being formed in state i, the rate of single-photon vibrational emission $k_{\mathrm{rad}}\left(\mathrm{s}^{-1}\right)$ is then given by

$$k_{\mathrm{rad}} = \sum_{i,j} P_i A_{i\to j} = \sum_{[n^i],[n^j]} P_{[n^i]} A_{[n^i]\to[n^j]}\,, \tag{37.42}$$

where the sum is over initial states i and final states j, and A is the Einstein coefficient for spontaneous emission. With some algebraic manipulation and the assumption that dipole selection rules apply, the result simplifies to

$$k_{\mathrm{rad}} = \sum_{k=1}^{s}\sum_{n_k} P_{n_k}^k n_k A_{1\to 0}^k\,, \tag{37.43}$$

where the index k refers to each of the s normal modes, the index n_k to the occupation number of mode k, and the Einstein A coefficient, which here refers to the fundamental transition of mode k, can be expressed in terms of the absorption intensity (see Chapt. 10). The probability $P^k_{n_k}$ that mode k is excited to state n_k is given by the statistical formula

$$P^k_{n_k} = \rho'_{\text{vib}}(E_{\text{vib}} - n_k h\nu_k)/\rho_{\text{vib}}(E_{\text{vib}}) , \quad (37.44)$$

where ρ_{vib} is the vibrational density of states and ρ'_{vib} is the reduced density of states when n_k quanta are assigned to mode k. The role of overtone and combination contributions to the radiative emission rate has also been investigated [37.74], as have a canonical approximation to the microcanonical formulation discussed here [37.75] and the small additional effect of stabilization by sequential emission [37.73]. With the use of the classical approximation to the standard vibrational density of state functions [37.41,42], (37.43) for k_{rad} reduces to a linear function of the vibrational energy:

$$k_{\text{rad}} = \frac{E_{\text{vib}}}{s} \sum_{k=1}^{s} \frac{A^k_{1\to 0}}{h\nu_k} . \quad (37.45)$$

The constant of proportionality between k_{rad} and E_{vib} depends strongly on the fundamental intensities; many of these can be obtained from infrared absorption spectra, although there is normally insufficient information for all of the active modes in polyatomic molecules, especially ions. Reasonable estimates, as well as limited experimental data, show that k_{rad} varies between 10 and $1000\,\text{s}^{-1}$ for vibrational energies up to a few eV [37.69, 70, 73–75]. The radiative stabilization rate typically decreases slightly with increasing molecular size due to the inclusion of infrared inactive modes of vibration. There also appears to be a distinction between ions and neutral species, with rates for ions apparently faster.

Since spontaneous emission rates between electronic states are often far more rapid than infrared vibrational rates, the possibility that radiative stabilization can proceed via electronic emission has been examined [37.9]. The well-studied association between C^+ and H_2 appears to proceed via electronic relaxation, with a radiative stabilization rate one to two orders of magnitude larger than would be the case for vibrational emission. Whether or not the association between CH_3^+ and H_2 proceeds via an electronically excited complex has been debated for some time; the answer is apparently negative [37.69].

37.2.2 Complex Formation and Dissociation

The formation and dissociation of the complex can be studied with a variety of statistical approximations other than what is discussed in this section. In particular, complex dissociation has been treated by a thermal RRKM approach [37.70, 75], an energy but not angular momentum specific RRKM approach [37.73], variational transition state theory, statistical adiabatic theory, and flexible transition state theory [37.72]. We first consider a simple thermal approximation for systems without activation energy incorporating microscopic reversibility [37.8]. In the limit that the complex redissociation into reactants is much more rapid than radiative stabilization, the rate coefficient k_{ra} for radiative association in the thermal model is

$$k_{\text{ra}} = K(T) k_{\text{rad}} , \quad (37.46)$$

where $K(T)$ is the ratio between k_{cf} and k_{cd} in a canonical ensemble, and the radiative stabilization rate has been assumed to be independent of temperature. The equilibrium coefficient $K(T)$ between complex and reactants can be rewritten in terms of partition functions in the standard way [37.47]. However, the partition function of the complex is best calculated via the equation

$$q_{AB^*} = \int_0^\infty \rho_{\text{vr}}(E + D_0) \exp(-E/k_B T)\, \mathrm{d}E ,$$
$$\approx \rho_{\text{vr}}(D_0) k_B T , \quad D_0 \gg k_B T , \quad (37.47)$$

where ρ_{vr} is the density of vibrational-rotational states [37.41, 42], E is the energy of reactants, and D_0 is the bond energy of the complex. One immediate prediction of this approach (deriving mainly from the rotational partition functions of the reactants) is that k_{ra} possesses a strong inverse dependence on temperature since

$$K(T) \propto T^{-(r_A + r_B + 1)/2} , \quad (37.48)$$

where r refers to the number of rotational degrees of freedom (two for a linear molecule and three for a nonlinear one). It also predicts a strong dependence on well depth and size of the complex since the density of complex states is a strong function of both these parameters. Although the thermal model agrees with the strong inverse temperature dependence of the coefficients for both radiative and, more commonly, ternary ion–molecule association [37.8, 71] (in which case, the thermal model with strong collisions replaces k_{rad} with a collisional rate coefficient), the absolute rate coefficients calculated tend to range up to an order of magnitude too high. Given the large range of values exhibited by radiative association rate coefficients $(10^{-20}–10^{-9}\,\text{cm}^3\,\text{s}^{-1})$, this might not seem too large a problem. It has been shown, however, that the thermal model is deficient, especially at low temperatures because thermal equilibrium cannot be achieved by two-body collisions [37.8]. A revised approach, called the modified thermal theory, replaces the rovibrational density of complex states in the thermal model with a vibrational density of states coupled with sums over the allowable ranges of complex angular momentum achievable from the specific capture model assumed [37.8]. The result is a somewhat lessened inverse temperature dependence and somewhat smaller rate coefficients, especially at low temperatures. Both modifications result in better agreement with experiment [37.8].

The modified thermal approach considers structureless reactants; full consideration of the internal states of the reactants is achieved via the Klots version of phase space theory, in which k_{cf} and k_{cd} in (37.41) are the capture and unimolecular rates discussed in Sect. 37.1.2. The phase space treatment reduces to the modified thermal treatment for small reactant angular momentum if the possibility of saturation $[k_{cd}(E, J) < k_{rad}]$ may be ignored. Both conditions are normally met. The phase space approach has also been used for ion–molecule ternary association reactions, but here is coupled with the strong collision hypothesis and must be regarded as inferior to the more detailed RRKM calculations with master equation treatments for collisional stabilization [37.13, 56].

37.2.3 Competition with Exoergic Channels

There have been numerous reactions reported (especially ion–molecule reactions) in which exoergic reaction channels compete with association channels, both radiative (at low pressure) and ternary (at higher pressures). One view of such competition is that it occurs via a sequential mechanism in which the reactants form a long-lived collision complex which either redissociates, dissociates into exoergic products, or is stabilized [37.53, 54]. The rate coefficient for competitive radiative association is then given by the steady-state expression

$$k_{ra} = \frac{k_{cf}}{k_{cd} + k_{cd'} + k_{rad}} k_{rad} , \tag{37.49}$$

where $k_{cd'}$ refers to complex dissociation into products. The additional and normally large $k_{cd'}$ term in the denominator means that the thermal and modified thermal theories cannot, in general, be used. The phase space treatment for this mechanism [37.53, 54, 76, 77] shows that association cannot compete with exoergic channels unless there is a significant barrier in the exit channel which considerably slows the dissociation rate of the complex into products, especially for partial waves of high angular momentum. Such barriers tend to be large enough to slow dissociation down, but not large enough to require tunneling. Although the results of phase space calculations are in good agreement with experiment for a variety of competitive systems [37.69, 76], they are once again inferior to RRKM calculations with master equation collisional deactivation when the competition involves collisional stabilization of the complex [37.13, 53, 54, 56].

Another mechanism exists for competition between association and normal exoergic channels [37.69]. Ab initio studies show that there is a parallel type of competition in which the product and association channels occur on different portions of the potential surface, with a branching at long range in the entrance channel. For example, the competing reactions [37.69, 78]

$$C_2H_2^+ + H_2 \longrightarrow C_2H_3^+ + H , \tag{37.50}$$

$$C_2H_2^+ + H_2 \longrightarrow C_2H_4^+ , \tag{37.51}$$

occur via distinct pathways. The former reaction is a possibly endoergic direct process in which the molecular hydrogen attacks perpendicularly, leading to the cyclic form of the $C_2H_3^+$ ion, whereas the association reaction occurs via the deep well of the ethylene ion. The competition in the analogous $C_3H^+ + H_2$ system is not currently well understood, with both parallel and series mechanisms suggested [37.69, 79].

37.3 Concluding Remarks

While the study of chemical reaction dynamics and kinetics is a relatively mature area of investigation, many challenges remain. A central one is the development of an efficient, fully quantum mechanical method for evaluating $k(T)$ that can be applied to a range of systems [37.80]. A second set of challenges comes in the intrinsically non-adiabatic nature of chemical reactions. As a system moves from reactants to products, there is a dramatic change in the electronic wave function. In fact the very existence of a transition state reflects the avoided crossing of two potential energy surfaces. There is experimental and theoretical evidence that the fact that the dynamics takes place on multiple potential surfaces has a large effect on the gas-phase state-to-state dynamics, and eventually the rate coefficients. Finally, as the evaluation of the electronic structure of atoms and molecules becomes more easily accomplished through the use of sophisticated computer packages, researchers will be able to ask more detailed questions about the relationships between the topology of potential surfaces and the corresponding rate coefficients than was possible as recently as ten years ago [37.81].

References

37.1 H. S. Johnston: *Gas Phase Reaction Rate Theory* (Ronald, New York 1966)

37.2 T. Stantcheva, V. I. Shematovich, E. Herbst: Astron. Astrophys. **391**, 1069 (2002)

37.3 D. T. Gillespie: J. Comp. Phys. **22**, 403 (1976)

37.4 S. B. Charnley: Astrophys. J. **509**, L121 (1998)

37.5 O. Biham, I. Furman, V. Pirronello, G. Vidali: Astrophys. J. **553**, 595 (2001)

37.6 N. J. B. Green, T. Toniazzo, M. J. Pilling, D. P. Ruffle, N. Bell, T. W. Hartquist: Astron. Astrophys. **375**, 1111 (2001)

37.7 R. D. Levine, R. B. Bernstein: *Molecular Reaction Dynamics* (Oxford Univ. Press, New York 1974)

37.8 D. R. Bates, E. Herbst: *Rate Coefficients in Astrochemistry*, ed. by T. J. Millar, D. A. Williams (Kluwer, Dordrecht 1988) p. 17

37.9 E. Herbst, D. R. Bates: Astrophys. J. **329**, 410 (1988)

37.10 P. J. Robinson, K. A. Holbrook: *Unimolecular Reactions* (Wiley-Interscience, New York 1972)

37.11 W. Forst: *Theory of Unimolecular Reactions* (Academic, New York 1973)

37.12 J. Troe: J. Phys. Chem. **83**, 114 (1979)

37.13 R. G. Gilbert, S. C. Smith: *Theory of Unimolecular and Recombination Reactions* (Blackwell, Oxford 1990)

37.14 W. Jakubetz, D. Sokolovski, J. N. L. Connor, G. C. Schatz: J. Chem. Phys. **97**, 6451 (1992)

37.15 S. M. Auerbach, W. H. Miller: J. Chem. Phys. **98**, 6917 (1993)

37.16 J. Troe: Adv. Chem. Phys. **LXXXII**, Part 2, 485 (1992)

37.17 T. Su, M. T. Bowers: *Gas Phase Ion Chemistry*, Vol. 1, ed. by M. T. Bowers (Academic, New York 1979) p. 83

37.18 B. R. Rowe: *Rate Coefficients in Astrochemistry*, ed. by T. J. Millar, D. A. Williams (Kluwer, Dordrecht 1988) p. 135

37.19 V. G. Anicich, W. T. Huntress: Astrophys. J. Suppl. **62**, 553 (1986)

37.20 J. O. Hirschfelder, C. F. Curtiss, R. B. Bird: *Molecular Theory of Gases and Liquids* (Wiley, New York 1954)

37.21 D. C. Clary, N. Halder, D. Husain, M. Kabir: Astrophys. J. **422**, 416 (1994)

37.22 I. W. M. Smith, E. Herbst, Q. Chang: Mon. Not. R. Astron. Soc. **350**, 323 (2004)

37.23 S. C. Smith, J. Troe: J. Chem. Phys. **97**, 8820 (1992)

37.24 D. C. Clary: *Rate Coefficients in Astrochemistry*, ed. by T. J. Millar, D. A. Williams (Wiley, New York 1954) p. 1

37.25 D. C. Clary: J. Chem. Phys. **91**, 1718 (1987)

37.26 N. G. Adams, D. Smith, D. C. Clary: Astrophys. J. **196**, L31 (1985)

37.27 D. R. Bates, W. L. Morgan: J. Chem. Phys. **87**, 2611 (1987)

37.28 S. C. Smith, M. J. McEwan, R. G. Gilbert: J. Phys. Chem. **93**, 8142 (1989)

37.29 M. M. Graff: Astrophys. J. **339**, 239 (1989)

37.30 D. C. Clary, T. S. Stoecklin, A. G. Wickham: J. Chem. Soc. Faraday Trans. **89**, 2185 (1993)

37.31 B. R. Rowe, A. Canosa, I. R. Sims: J. Chem. Soc. Faraday Trans. **89**, 2193 (1993)

37.32 E. I. Dashevskaya, A. I. Maergoiz, J. Troe, I. Litvin, E. E. Nikitin: J. Chem. Phys. **118**, 7313 (2003)

37.33 Y. Georgievskii, S. J. Klippenstein: J. Chem. Phys. **118**, 5442 (2003)

37.34 D. C. Clary, E. Buonomo I. R. Sims, I. W. M. Smith, W. D. Geppert, C. Naulin, M. Costes, L. Cartechini, P. Casavecchia: J. Phys. Chem. **A106**, 5541 (2002)

37.35 D. Chastaing, P. L. James, I. R. Sims, I. W. M. Smith: J. Chem. Soc. Faraday Discussions **109**, 165 (1998)

37.36 B. R. Rowe, C. Rebrion-Rowe, A. Canosa: *Astrochemistry: from Molecular Clouds to Planetary Systems*, ed. by Y. C. Minh, E. F. van Dishoeck (Sheridan Books, Chelsea, Michigan 2000) p. 237

37.37 D. Chastaing, S. D. Le Picard, I. R. Sims, I. W. M. Smith: Astron. Astrophys. **365**, 241 (2001)

37.38 J. C. Light: Disc. Faraday Soc. **44**, 14 (1967)

37.39 W. J. Chesnavich, M. T. Bowers: *Gas Phase Ion Chemistry*, ed. by M. T. Bowers (Academic, New York 1979) p. 119

37.40 D. Gerlich: J. Chem. Phys. **90**, 3574 (1989)

37.41 G. Z. Whitten, B. S. Rabinovitch: J. Chem. Phys. **38**, 2466 (1963)

37.42 G. Z. Whitten, B. S. Rabinovitch: J. Chem. Phys. **41**, 1883 (1964)

37.43 L. M. Bass, W. J. Chesnavich, M. T. Bowers: J. Am. Chem. Soc. **101**, 5493 (1979)

37.44 C. E. Klots: J. Phys. Chem. **75**, 1526 (1971)

37.45 E. Herbst, S. K. Knudson: Chem. Phys. **55**, 293 (1981)

37.46 E. T. Galloway, E. Herbst: Astrophys. J. **376**, 531 (1991)

37.47 T. L. Hill: *An Introduction to Statistical Thermodynamics* (Addison-Wesley, Reading 1960)

37.48 L. B. Harding, G. C. Schatz, R. A. Chiles: J. Chem. Phys. **76**, 5172 (1982)

37.49 D. G. Truhlar, B. C. Garrett: Acc. Chem. Res. **13**, 440 (1980)
37.50 W. L. Hase, D. M. Wardlaw: *Bimolecular Collisions*, ed. by M. N. R. Ashfold, J. E. Baggott (Royal Society of Chemistry, London 1989) p. 171
37.51 N. B. Slater: *Theory of Unimolecular Reactions* (Cornell Univ. Press, Ithaca 1959)
37.52 S. E. Stein, B. S. Rabinovitch: J. Chem. Phys. **58**, 2438 (1973)
37.53 L. M. Bass, R. D. Cates, M. F. Jarrold, N. J. Kirchner, M. T. Bowers: J. Am. Chem. Soc. **105**, 7024 (1983)
37.54 E. Herbst: J. Chem. Phys. **82**, 4017 (1985)
37.55 S. Wlodek, D. K. Bohme, E. Herbst: Mon. Not. R. Astron. Soc. **242**, 674 (1990)
37.56 S. C. Smith, P. F. Wilson, P. Sudkeaw, R. G. A. R. MacLagan, M. J. McEwan, V. G. Anicich, W. T. Huntress: J. Chem. Phys. **98**, 1944 (1993)
37.57 S. C. Smith: J. Chem. Phys. **97**, 2406 (1993)
37.58 S. J. Klippenstein, J. D. Faulk, R. C. Dunbar: J. Chem. Phys. **98**, 243 (1993)
37.59 Y. Chen, A. Rauk, E. Tschuikow-Roux: J. Phys. Chem. **95**, 9900 (1991)
37.60 M. J. Frost, P. Sharkey, I. W. M. Smith: J. Phys. Chem. **97**, 12254 (1993)
37.61 E. P. Wigner: Z. Phys. Chem. **B19**, 203 (1932)
37.62 T. N. Truong, D. G. Truhlar: J. Chem. Phys. **93**, 1761 (1990)
37.63 T. N. Truong, D. G. Truhlar: J. Chem. Phys. **97**, 8820 (1993)
37.64 T. Takayanagi, N. Masaki, K. Nakamura, M. Okamota, S. Sato, G. C. Schatz: J. Chem. Phys. **86**, 6133 (1987)
37.65 W. H. Miller: J. Am. Chem. Soc. **101**, 6810 (1979)
37.66 E. Herbst, D. J. DeFrees, D. Talbi, F. Pauzat, W. Koch, A. D. McLean: J. Chem. Phys. **94**, 7842 (1991)
37.67 D. Smith, J. Glosik, V. Skalsky, P. Spanel, W. Lindinger: Int. J. Mass Spectrom. Ion Proc. **129**, 145 (1993)
37.68 D. C. Flatin, J. J. Holton, M. M. Beaky, T. M. Goyette, F. C. DeLucia: J. Mol. Spectrosc. **164**, 425 (1994)
37.69 D. Gerlich, S. Horning: Chem. Rev. **92**, 1509 (1992)
37.70 R. C. Dunbar: *Unimolecular and Bimolecular Ion-Molecule Reaction Dynamics*, ed. by C. Y. Ng, T. Baer, I. Powis (Wiley, New York 1994) p. 270
37.71 N. G. Adams, D. Smith: *Reactions of Small Transient Species: Kinetics and Energetics*, ed. by A. Fonijn, M. A. A. Clyne (Academic, New York 1983) p. 311
37.72 J. W. Davies, M. J. Pilling: *Bimolecular Collisions*, ed. by M. N. R. Ashfold, J. E. Baggott (Royal Society of Chemistry, London 1989) p. 105
37.73 J. R. Barker: J. Phys. Chem. **96**, 7361 (1992)
37.74 E. Herbst: Chem. Phys. **65**, 185 (1982)
37.75 R. C. Dunbar: Int. J. Mass Spectrom. Ion Proc. **100**, 423 (1990)
37.76 E. Herbst, M. J. McEwan: Astron. Astrophys. **229**, 201 (1990)
37.77 E. Herbst, R. C. Dunbar: Mon. Not. R. Astron. Soc. **253**, 341 (1991)
37.78 S. A. Maluendes, A. D. McLean, E. Herbst: Chem. Phys. Lett. **217**, 571 (1994)
37.79 S. A. Maluendes, A. D. McLean, K. Yamashita, E. Herbst: J. Chem.Phys. **99**, 2812 (1993)
37.80 W. H. Miller: Acc. Chem. Res. **26**, 174 (1993)
37.81 S. K. Klippenstein, L. B. Harding: J. Phys. Chem. A **103**, 9388 (1999)

38. Gas Phase Ionic Reactions

Ionic reactions in the gas phase is a broad field encompassing a multitude of interactions between ions (both positively and negatively charged), electrons and neutrals. These can be as simple as the transfer of an electron between molecules, or can be complex with considerable bond breaking, reforming and rearrangement. Reactions of the general type

$$A^{+,-} + B^{-,n} \rightarrow \text{products}\,, \quad (38.1)$$

illustrating this diversity, are given in Table 38.1. The reaction process can be considered as consisting of three parts: (a) the initial interaction, in which the colliding particles, $A^{+,-}$ and $B^{-,n}$ are drawn together by an attractive potential, (b) the reaction intermediate and transition state, in which reactants are transformed into products and (c) the weakening interaction as the product particles separate. Interactions can also occur where only elastic scattering is involved and these are considered in Chapt. 67.

38.1 **Overview** 575
38.2 **Reaction Energetics** 576
38.3 **Chemical Kinetics** 578
38.4 **Reaction Processes** 578
38.4.1 Binary Ion–Neutral Reactions 579
38.4.2 Ternary Ion–Molecule Reactions.. 581
38.5 **Electron Attachment** 582
38.6 **Recombination** 583
38.6.1 Electron–Ion Recombination 583
38.6.2 Ion–Ion Recombination (Mutual Neutralization)............. 584
References 585

38.1 Overview

Part (a) of the interaction is readily studied theoretically by classical mechanics in terms of reactant particle motions controlled by the interaction potential $u(r)$ between the particles, where r is the interparticle separation. For ion–neutral interactions, this can take the form (i) of an attractive ion-polarization or induced dipole potential [38.1] $u(r) = -\alpha_d q^2/2r^4$, where α_d is the polarizability of the neutral and q is the charge on the ion; anisotropy in the polarizability can also be taken into account [38.2], (ii) of an ion-permanent dipole potential [38.3] with $u(r) = -q\mu_D \cos\theta/r^2$, where μ_D is the permanent dipole of the neutral and θ is the angle that the dipole makes with r, and (iii) of an ion-quadrupole potential [38.4] with $u(r) = -Qq(3\cos^2\theta - 1)/2r^3$, where Q is the quadrupole moment and θ is the angle the quadrupole axis makes with r. Such capture theories are considered in Chapt. 37. Other interaction potentials can be considered, but these are of lesser significance. Coulombic interaction potentials $u(r) = -q_1 q_2/r$, where q_1 and q_2 are the charges on the interacting particles, are appropriate for positive ion–negative ion recombination (more correctly termed mutual neutralization) and electron–ion recombination. For processes involving electrons, i. e., electron attachment and electron–ion recombination, the wave nature of the electron also has to be considered [38.5, 6].

In addition to providing a means of bringing the reactant particles together, the attractive interaction potential has another function. In part (b) of the reaction mechanism, where considerable rearrangement of the atoms in the colliding species may occur (i. e., in the intermediate complex and transition state), there may be barriers to reaction. Here, the kinetic energy gained from the interaction potential is available in the intermediate complex for overcoming such energy barriers (very much less interaction energy is available in neutral-neutral reactions and the effects of energy barriers are very much more evident [38.7]). Of course, this energy has to be reconverted to potential energy as the product particles separate, and is therefore not available to drive the overall reaction. However, the amount of energy returned may differ from that initially converted to kinetic energy since

Table 38.1 Examples illustrating the range of ionic reactions that can occur in the gas phase

Reaction Process	Reaction Type
$Ar^+ + O_2 \rightarrow O_2^+ + Ar$	Nondissociative charge transfer/charge exchange
$O_2^- + NO_2 \rightarrow NO_2^- + O_2$	
$He^+ + O_2 \rightarrow O^+ + O + He$	Dissociative charge transfer
C_{60}^{3+} + Corannulene → Corannulene^{2+} + C_{60}^+	Multiple charge transfer
$O^+ + N_2 \rightarrow NO^+ + N$	Ion/atom interchange or atom abstraction
$O^- + CH_4 \rightarrow OH^- + CH_3$	Atom abstraction
$H_3O^+ + HCN \rightarrow H_2CN^+ + H_2O$	Proton transfer
$OH^- + HCN \rightarrow CN^- + H_2O$	
$OH^- + H \rightarrow H_2O + e^-$	Associative detachment
$C^+ + C_2H_2 \rightarrow C_3H^+ + H$	Atom insertion
$CH_3^+ + H_2 + M \rightarrow CH_5^+ + M$	Ternary collisional association
$Cl^- + BCl_3 + M \rightarrow BCl_4^- + M$	
$He_2^+ + e^- \rightarrow He_2(2He) + h\nu$	Radiative electron–ion recombination
$HCO^+ + e^- \rightarrow H + CO$	Dissociative electron–ion recombination
$NO^+ + NO_2^- \rightarrow NO + NO_2$	Ion–ion recombination (mutual neutralization)
$e^- + CCl_4 \rightarrow Cl^- + CCl_3$	Dissociative electron attachment
$e^- + C_7F_{14} \rightarrow C_7F_{14}^-$	Nondissociative electron attachment
$e^- + O_2 + M \rightarrow O_2^- + M$	Ternary collisionally stabilized attachment
$^{13}C^+ + {}^{12}CO \rightleftharpoons {}^{12}C^+ + {}^{13}CO$	Isotope exchange
$OD^- + NH_3 \rightleftharpoons OH^- + NH_2D$	

the polarizability/dipole moment of the product neutral(s) may differ from those of the reactant neutrals. Generally, the final part (c) of the interaction has little influence on the magnitude of the reaction rate coefficient or on the product distribution once the products have significantly separated (i. e., beyond the range at which processes such as long range electron transfer can occur). Thus, to a major degree, part (b) of the mechanism and the reaction energetics determine the products of the reaction. Theories which address this part of the interaction are also discussed in Chapt. 37 and these have met with mixed success. Thus, at present, experimental measurements are providing a more definitive understanding of reactions and their mechanisms.

Also, the situation is not in general clear as to the form of the thermodynamic energy that governs whether a reaction will proceed spontaneously, i. e., whether it is controlled by the enthalpy change ΔH in the reaction or by the Gibbs' free energy change $\Delta G = \Delta H - T\Delta S$, where ΔS is the entropy change in the reaction [38.8]. This is discussed in more detail in Sect. 38.2.

Rate coefficients and product distributions for these reaction processes are important in all ionized media where molecular species exist, such as interstellar gas clouds [38.9], planetary atmospheres [38.10] (including that of the Earth [38.11, 12]), comets (Chapt. 83), the space shuttle environment [38.13], laser plasmas [38.14], plasmas used to etch semiconductors [38.15], hydrocarbon flames [38.16], etc.

38.2 Reaction Energetics

The availability of sufficient energy is a primary consideration for determining whether a reaction can proceed spontaneously. Criteria for determining whether energy is given out or absorbed in a reaction are (i) exo- or endoergicity, ΔE representing the internal energy change involved in a single interaction, (ii) exo- or endothermicity for an ensemble of particles in thermal equilibrium as defined by the enthalpy change per mole, ΔH, in the reaction and (iii) exergonic or endergonic as defined by the Gibbs' free energy change, ΔG [38.17]. ΔE is related to ΔH at temperature T by

$$\Delta H_T \approx -N\Delta E_T \,, \qquad (38.2)$$

where N is Avogadro's number. ΔE is usually deduced from bond energies [38.18–20], ionization potentials [38.21], electron affinities [38.22, 23], proton affinities [38.22, 24, 25], gas phase basicities [38.22, 24–26], etc. ΔH_T^0 is determined from the heats of formation $H_{\mathrm{f},T}^0$ by [38.27]

$$\Delta H_T^0 = \sum_{\text{products}} H_{\mathrm{f},T}^0 - \sum_{\text{reactants}} H_{\mathrm{f},T}^0 , \tag{38.3}$$

where the superscript 0 refers to the standard state [38.17] of the reactants and products [for example, see (38.1)]. ΔG_T^0 is determined from ΔH_T^0 and the entropy change ΔS_T^0 by

$$\Delta G_T^0 = \Delta H_T^0 - T \Delta S_T^0 , \tag{38.4}$$

where

$$\Delta S_T^0 = \sum_{\text{products}} S_T^0 - \sum_{\text{reactants}} S_T^0 , \tag{38.5}$$

and the S_T^0 are the standard entropies [38.17]. In cases where all of the $H_{\mathrm{f},T}^0$ and S_T^0 are not available, they can often be deduced by constructing other reactions involving the species of interest and other species for which the required thermodynamic information is known ([38.17] discusses the details of ways in which this can be achieved). Alternatively, the magnitude of a thermodynamic parameter can be calculated using equilibrium statistical thermodynamics, if the energies of all of the occupied molecular energy levels are known [38.28]. For studies where all reacting particles of a given type have the same energy, such as in beam/beam interactions where cross sections $\sigma(E)$ are measured, then ΔE is most appropriate. Alternatively, for reactions involving an ensemble of particles in thermal equilibrium at a temperature T, such as those studied in high pressure mass spectrometer ion sources or afterglows where rate coefficients $k(T)$ are measured, ΔH and ΔG are more appropriate. (The terms "rate coefficient" and "rate constant" are used interchangeably in the literature.) $\sigma(E)$ and $k(T)$ are directly related via

$$k(T) = \int_0^\infty v(E) f(E) \sigma(E)\, \mathrm{d}E , \tag{38.6}$$

where $v(E)$ is the relative speed of the reactants and $f(E)$ is the Maxwell–Boltzmann energy distribution. Whether ΔH or ΔG is more important for determining reaction spontaneity depends on the degree of interaction of the reacting systems with the surroundings during the course of the whole reaction process. For example, if the reactions occur at low pressure such that the reaction time is much less than the collision time with the background gas, there will be no interactions with the surroundings, and the only energy available in the reactions will be ΔH. Here, ΔS can only determine the probability that the intermediate complex dissociates forward to products or back to reactants. At the other extreme, if the reactions are conducted at high pressure such that the reaction species are always in thermal equilibrium with the surroundings (the limiting case of this is reaction in solution), then the additional energy $T\Delta S$ is available, and ΔG determines whether the reactions are spontaneous. Obviously, for intermediate pressures, there is a varying degree of contact with the surroundings, and which of ΔH and ΔG is most applicable is more obscure. The definition of surrounding is also somewhat loose, since in the present context it represents anything that can provide a source of energy during the reaction (for example, vibrational modes of the reacting species that are not involved in the interaction and which could be cooled as the reaction proceeds). Usually, the $T\Delta S$ term is not sufficiently large that ΔH and ΔG have different signs so that confusion about the spontaneity of reactions often does not arise. Recently, however, reactions have been discovered which appear to proceed spontaneously even though ΔH is positive, and there has been considerable discussion of this [38.8].

Eventually, all reaction processes will reach equilibrium as defined by

$$\Delta G = -RT \ln K , \tag{38.7}$$

where

$$K = \left(p_C^c\, p_D^d\right) / \left(p_A^a\, p_B^b\right) \tag{38.8}$$

is the equilibrium constant for the reaction

$$aA + bB \rightleftharpoons cC + dD . \tag{38.9}$$

The subscripts to the pressures p denote the components A to D, and the superscripts represent the stoichiometries [38.17,28]. The equilibrium constant K also equals $k_\mathrm{f}/k_\mathrm{r}$, the ratio of the forward to reverse rate coefficients, obtained for thermalized particles at temperature T. Thus, from measurements of k or the equilibrium constant at a series of temperatures, ΔH and ΔS of the reaction can be separately obtained.

38.3 Chemical Kinetics

The rate coefficient k for the reaction process

$$A+B \rightarrow C+D \tag{38.10}$$

is defined by the rate equation [38.29]

$$\frac{d[A]}{dt} = \frac{d[B]}{dt} = -k[A][B] , \tag{38.11}$$

where the square parentheses represent the concentrations of the enclosed species, and the units of k are typically $cm^3\,molec^{-1}\,s^{-1}$ (often $molec^{-1}$ is not written explicitly in the units). These k are deduced from the time variation of specific concentrations of thermalized reactants under a variety of conditions. For ion–neutral reactions, with A being the positive or negative ion species, the situation can usually be achieved in the laboratory where $[B] \gg [A]$, i. e., a pseudo first-order reaction. Simple integration yields

$$[A]_t = [A]_0 \exp(-k\,[B]_0\, t) , \tag{38.12}$$

where the subscript 0 indicates that this concentration is time invariant. Similar circumstances apply to binary electron attachment reactions, where $[A]=[e]$, the electron number density (the symbol β is often used here in place of k). If the reaction proceeds by association that is stabilized by collision with a third body M then

$$A+B+M \rightarrow AB+M , \tag{38.13}$$

and the solution is

$$[A]_t = [A]_0 \exp(-k\,[B]_0[M]_0\, t) \tag{38.14}$$

with k now having units of $cm^6\,molec^{-2}\,s^{-1}$. Note that situations exist where binary and ternary reactions can occur simultaneously.

If $[B] \not\gg [A]$, the solution of (38.11) is different. For the special case $[B]=[A]$, integration yields

$$1/[A]_t - 1/[A]_0 = kt , \tag{38.15}$$

and similarly for $[B]$. This situation is usually achieved in the laboratory for the determination of k for electron–ion and ion–ion recombination (the symbols α_e and α_i respectively are often used to replace k here). In many applications where several processes contribute to changes in $[A]$ and $[B]$, and where the simple limits are not applicable, the situation has to be analyzed numerically.

38.4 Reaction Processes

Recently, a nine volume series, concerned with all aspects of mass spectrometry, is being published of which Volume 1 deals with theoretical and experimental aspects of ionic reactions in great detail [38.30]. The reader is recommended to first consult the present chapter to get an overview of the reaction processes, and to go to this other text if more detail is required.

The wide variety of possible reaction processes is listed in Table 38.1 by specific examples. However, the reaction mechanisms illustrated are completely general. Magnitudes of the rate coefficients for binary interactions between charged and neutral particles, vary from $10^{-7}\,cm^3\,s^{-1}$ for electron attachment [38.31,32] to 10^{-8} or $10^{-9}\,cm^3\,s^{-1}$ for unit efficiency ion–molecule reactions [38.33–36]. The efficiency is defined as the ratio of the measured k to the theoretical collisional value, i. e., that determined using the appropriate interaction potential. Ion–molecule reactions involving molecules with large permanent dipoles (for example, HCN at 2.98 D and HCl at 1.08 D [38.37]), can have $k > 10^{-7}\,cm^3\,s^{-1}$ at low T [38.37–39] due to locking of the dipole along the line joining the reactant species, thereby maximizing the strength of the interaction. Rate coefficients can be much smaller (by orders of magnitude) than these upper limits if the efficiency of part (b) of the mechanism is small. The part of the mechanism after collision occurs is treated by transition state theory in terms of the partition function of the transition state; where the transition state is, and its partition function, relates to the number of ways that the available energy can be distributed in the transition state [38.40]. More details of the theories are given in Chapt. 37 and [38.41].

For dissociative electron–ion recombination, the upper limits on k are generally much larger as a consequence of the long range Coulombic interaction potential, varying from $\approx 10^{-7}\,cm^3\,s^{-1}$ for diatomic ions to $> 10^{-6}\,cm^3\,s^{-1}$ for more polyatomic species [38.42–45]. For ion–ion recombination, the rate coefficients are about an order of magnitude smaller than these values because of the larger mass of the negative ion relative to the electron, and thus the smaller interaction velocity (the σ of the two processes are, in fact, similar) [38.46]. Electron–ion recombinations that are radiatively stabilized generally have a small k, as small as $\approx 10^{-12}\,cm^3\,s^{-1}$ [38.47], because of the large magnitude of the radiative lifetime of the reaction intermediate compared with the time for autoionization. For most reaction types, experimental rate coefficients are available at room temperature, and for some, the temperature and/or energy dependencies have also been determined [38.31, 32, 43, 44, 46, 48]. Generally, only the ion products are identified using mass spectrometry, and the product neutrals are not determined (unless the energetics allow only one neutral product), and the states of excitation of the products are not identified.

For ternary processes involving association, k varies from totally saturated at the binary collision limit to very small ($\approx 10^{-32}\,\mathrm{cm^6\,s^{-1}}$) for ion–neutral reactions [38.33, 34] and similarly for electron attachment [38.49]. Little is known about collision stabilized recombination [38.50]. Each type of reaction process has its own characteristic behavior and dependence on temperature. Rate coefficients, temperature dependencies, and product distributions (where available) have been tabulated for ion–neutral reactions [38.33, 34, 51] and electron attachment [38.31, 32]. Less data are available for electron–ion and ion–ion recombination, so no attempts have been made to compile these. Data are available in [38.6, 42–44, 46].

38.4.1 Binary Ion–Neutral Reactions

Over the years, these reaction processes have been reviewed several times, but from different perspectives. Most recently, studies of positive ion–molecules in flow tubes [38.52] and negative ion–molecule reactions from an organic mechanistic viewpoint [38.53] have been discussed.

Charge Transfer and Charge Exchange

These processes involve the exchange of an electron which can occur at relatively large interparticle separations (i. e., up to $\approx 6\,\text{Å}$ [38.54]). Thus, in principle, k values can be larger than the collision limiting value. Such reactions are, in general, relatively fast, although there are some notable exceptions (for example, $He^+ + H_2, k = 1\times10^{-13}\,\mathrm{cm^3\,s^{-1}}$; $Ne^+ + H_2, k < 2\times10^{-14}$ and $Ne^+ + N_2, k = 1.1\times10^{-13}$ [38.33]). Attempts have been made to relate the efficiency of charge transfer to the Franck–Condon overlap between the neutral reactant and the product ion, with mixed success [38.55]. More energy is generally available in the positive ion reactions than the negative ion reactions since ionization potentials are much larger than electron affinities, and thus more dissociative products would be expected, as is usually observed. The reaction

$$He^+ + N_2 \to \begin{cases} N^+ + N + He + 0.28\,\mathrm{eV} \\ N_2^+ + He + 9.00\,\mathrm{eV}\,, \end{cases}$$

is particularly interesting and has been studied in considerable detail. The product distribution slightly favors N^+ (60%) rather than N_2^+ [38.56], with the N^+ channel becoming even more important with increasing vibrational temperature of the N_2 [38.57]. Spectroscopic emission studies [38.58] have shown that a significant fraction ($\geq 5\%$) of the reactions proceed by charge transfer into the $N_2^+\ (C\,^2\Sigma_g^+)$ state followed by the radiative decay

$$N_2^+(C\,^2\Sigma_u^+) \to N_2^+(X\,^2\Sigma_g^+) + h\nu\,. \qquad (38.16)$$

This channel competes with the predissociation

$$N_2^+(C\,^2\Sigma_u^+) \to N^+ + N\,. \qquad (38.17)$$

A new mechanism has recently been observed in which charge transfer occurs in parallel with chemi-ionization (or, equivalently, electron detachment):

$$He^+ + C_{60} \to C_{60}^{2+} + e^- + He\,. \qquad (38.18)$$

Such a process is, of course, only energetically possible for high recombination energy ions like He^+ and Ne^+ [38.59]. Also, multiply charged C_{60} (C_{60}^{3+}) has been seen to undergo two electron transfer with Corannulene and some Polycyclic Aromatic Hydrocarbons (PAH's), generally in parallel with a whole series of other reaction channels [38.60].

Proton Transfer

Where proton transfer is significantly exoergic, as determined by the difference in the proton affinities of the reactant and product neutrals, reaction usually proceeds at the collisional rate [38.61]. If the reaction is close to thermoneutral, then the amount of phase space available when the intermediate complex dissociates to products is similar to that available when it dissociates back to the reactants, and the k approximates to one-half the collisional value. For proton transfer reactions which are not highly exo- or endoergic, k_f and k_r can be measured and thus ΔG determined [using (38.7)]. If ΔS can be determined in some way, or if k_f/k_r can be determined as a function of T, then ΔH can be deduced and used to obtain the proton affinity difference (ΔS can also be determined in the latter case). This has been used to construct proton affinity scales [38.24, 25]. Care is required in such studies to ensure that no vibrational excitation remains in the reactant ion due to its formation, and that the identities (i. e., isomeric forms) of the reactant and product ions are known (e.g. whether the ion is HCO^+ or the higher energy form COH^+ [38.62]).

Often the individual reaction processes do not occur in isolation. For example, in the reactions of CH_4^+ with COS, H_2S, NH_3, H_2CO and CH_3OH, both charge transfer and proton transfer are energetically possible, and both channels are observed. Here, the least exoergic channel is favored in all cases [38.63]. Proton transfer also occurs in negative ion reactions, for example those of OH^- and NH_2^- giving H_2O and NH_3 respectively (e.g. see Table 38.1), and these types of reactions are production sources for many negative ions [38.53, 64].

Ion–Atom Interchange and Atom Abstraction

In many simple cases, these two processes are the same, for example

$$Ar^+ + H_2 \rightarrow ArH^+ + H\,, \tag{38.19}$$

but not so for the more complicated molecules (e.g., O^- + CH_4; see Table 38.1). Such reactions, when exoergic, usually occur close to the collisional rate although somewhat slower. Series of such reactions with H_2 occur in the interstellar medium and are responsible for producing the hydrogenation in many of the species observed there, for example, in CH_3^+ production from CH^+, NH_4^+ from NH^+ and H_3O^+ from OH^+ (see Chapt. 82). The reaction,

$$NH_3^+ + H_2 \rightarrow NH_4^+ + H\,, \tag{38.20}$$

is particularly interesting. At temperatures greater than 300 K, the reaction shows an activation energy barrier of 2 kcal mol^{-1}. The rate coefficient k decreases with decreasing T, reaching a minimum of $2\times10^{-13}\,\mathrm{cm^3\,s^{-1}}$ at 80 K and then increases at lower T due to tunneling through the barrier [38.48]. This behavior occurs because, at the higher temperatures, the lifetime of the intermediate complex is not sufficient for significant tunneling to occur, but there is sufficient energy in the reacting species to overcome the barrier. At lower temperatures, there is not sufficient energy available to overcome the barrier, but the lifetime of the intermediate complex becomes long enough for significant tunneling to occur [38.65]. This explanation has been substantiated by the isotopic studies of the reactions NH_3^+ + D_2, and ND_3^+ + H_2 and D_2 [38.66].

For some reactions, isotopic labeling has been used to identify the reaction mechanism. For example, the reaction

$$O^+ + O_2 \rightarrow O_2^+ + O \tag{38.21}$$

has been shown to be predominantly charge transfer, rather than ion–atom interchange, by labeling the ion as $^{18}O^+$ [38.67, 68].

Associative Detachment

For negative ions, an additional process is possible for which there is no equivalent positive ion analog (although there is the related process of chemi-ionization, i. e., $AB + C \rightarrow ABC^+ + e^-$). In this process, the negative ion and the neutral associate, and the association is stabilized by the ejection of the electron (see Table 38.1). Such reactions can only occur when the electron detachment energy is less than the energy of the bond that is produced. Therefore, these reactions usually involve radical species which produce stable molecules. k values are usually an appreciable fraction of the collisional values [38.69, 70]. Infrared emissions have been detected from a series of these reactions [38.71], for example O^- with CO and F^-, Cl^- and CN^- with H, and show that the reactions populate the highest vibrational levels that are energetically accessible.

Other Binary Ion–Molecule Reaction Channels

As reactant species become more polyatomic, there is a greater variety of reaction processes that can occur, and these often occur in parallel. The processes that occur are too numerous to list so a few examples will have to suffice. Reactions that occur in isolation are insertion reactions of the type [38.72]

$$C^+ + C_nH_m \rightarrow C_{n+1}H_{m-x}^+ + xH \tag{38.22}$$

and

$$S^+ + C_nH_m \rightarrow H_{m-1}C_nS^+ + H\,. \tag{38.23}$$

Multiple channels are very evident in reactions of ions produced from species with large ionization potentials and small proton affinities, for example, NH^+ [38.73],

$$NH^+ + CH_3NH_2 \rightarrow \begin{cases} H_4CN^+ + NH_2 \\ CH_3NH_2^+ + NH\ \text{(charge tsfr.)} \\ CH_3NH_3^+ + N\ \text{(proton tsfr.)} \\ H_2CN^+ + (NH_3 + H) \\ H_3CN^+ + NH_3\,. \end{cases}$$

In addition to charge and proton transfer, other channels requiring more rearrangement are evident.

Isotopically Labeled Reactants

A great deal can be learned about reaction mechanisms in significantly exoergic ion–neutral reactions by isotopic labeling. Some such reactions have been mentioned above, but a particularly graphic example is the reaction

$$CH_4^+ + CH_4 \rightarrow CH_5^+ + CH_3 , \qquad (38.24)$$

which was assumed to be either proton transfer and/or H-atom abstraction, depending on the interaction energy. By studying the reaction with both the ion and the neutral reactants separately deuterated, the reaction was shown to be much more complex, with the product ions being CH_4D^+ (10%), $CH_3D_2^+$ (22%), $CH_2D_3^+$ (43%) and CHD_4^+ (25%) for the reaction of CH_4^+ with CD_4 [38.74]. The reaction clearly proceeds via a long-lived intermediate in which there is a large degree of isotope mixing before unimolecular decomposition to products.

A further class of isotopic reactions are those for which the exoergicity is provided only by the different zero point energies of the reactants and products. Many such reactions have been studied. Examples of the various types are: (i) The symmetrical charge transfer

$$^{15}N_2^+ + {}^{14}N_2 \rightarrow {}^{14}N_2^+ + {}^{15}N_2 , \qquad (38.25)$$

which proceeds in both directions at more than half the collisional rate. This implies that the reaction proceeds via long-range charge transfer, a conclusion substantiated by the fact that no mixed product $^{14}N\,^{15}N^+$ is produced [38.67]. (ii) Proton transfer reactions exemplified by $^{14}N_2H^+ + {}^{15}N_2$ and $H\,^{12}CO^+ + {}^{13}CO$, which are both exoergic by about 1 meV [38.48]. Some isotopic scrambling has been observed in the latter reaction by using the double isotopic substitution [38.75],

$$H^{12}C^{18}O^+ + {}^{13}C^{16}O \rightarrow \begin{cases} H^{13}C^{16}O^+ + {}^{12}C^{18}O \\ (\geq 90\%) \\ H^{13}C^{18}O^+ + {}^{12}C^{16}O \\ (\leq 10\%) . \end{cases}$$

Other examples where considerable isotope scrambling occurs are H/D exchange in the reaction systems $H_3^+ + H_2$, $CH_3^+ + H_2$, $H_3O^+ + H_2O$, $CH_5^+ + CH_4$, etc. [38.48]. Such scrambling reactions are believed to proceed via proton bound dimer intermediates (e.g., $H_2O \cdots H^+ \cdots D_2O$). (iii) Isotope exchange reactions such as $^{13}C^+ + {}^{12}CO \rightarrow {}^{12}C^+ + {}^{13}CO$ have also been studied. For this reaction, both k_f and k_r increase with decreasing T due to an increase in the intermediate complex lifetime; k_r decreases at lower T due to the endoergicity in this reaction direction [38.74].

Isotope exchange is also observed to occur with negative ions, viz.

$$DO^- + MH \rightarrow HO^- + MD , \qquad (38.26)$$

and is a common process. Particularly significant is the comparison where MH is H_2 and NH_3. These species have very similar gas phase acidities, and thus the exoergicities of the reactions are similar. However, NH_3 has a larger polarizability than H_2 and, in addition, has a permanent dipole moment. The observation that the k for the NH_3 reaction is a factor of ten larger than that for the H_2 reaction is explained as being due to the stronger interaction potential which makes more energy available in the intermediate complex and facilitates H/D exchange [38.64].

Temperature Dependencies of Binary Reactions

When the k values at room temperature are close to the collisional value, appreciably exoergic reactions generally exhibit little temperature dependence. For slower reactions, a significant inverse temperature dependence is observed, being $\approx T^{-1/2}$ for several reactions. This behavior is rationalized as a decreasing lifetime of the intermediate complex with increasing temperature, thus allowing less time for reaction [38.48].

38.4.2 Ternary Ion–Molecule Reactions

In cases where binary channels are not energetically possible, association reactions can still occur, both for positive and negative ions. The reactions can be as simple as

$$He^+ + He + He \rightarrow He_2^+ + He , \qquad (38.27)$$

to associations more complex than those listed in Table 38.1. Rate coefficients k_3 at room temperature vary from $1 \times 10^{-31}\,cm^6\,s^{-1}$ for reactions such as (38.27) to in excess of $1 \times 10^{-25}\,cm^6\,s^{-1}$, this upper limit being due to experimental constraints rather than the reaction itself. The mechanism usually postulated is

$$A^\pm + B \rightleftharpoons \left(AB^\pm\right)^* , \qquad (38.28)$$

$$\left(AB^\pm\right)^* + M \rightarrow AB^\pm + M^* \qquad (38.29)$$

where M stabilizes the excited intermediate, $\left(AB^\pm\right)^*$ by removing energy in the collision. For this type of reaction, there is a considerable dependence of k on T.

Statistical theory predicts (see Chapt. 37),

$$k_3 \propto T^{-(\ell/2+\delta)} , \quad (38.30)$$

where ℓ is the number of rotational degrees of freedom in the separated reactants, and δ is a parameter attributed to the temperature variation of the collision efficiency of the stabilizing third body M. Experimentally, many association reactions exhibit such power law dependencies of the k_3 [38.48, 51] with δ being small (≈ 0.2 or 0.3) for a helium third body. Also, some evidence exists for a contribution of vibrational degrees of freedom to the temperature coefficient in cases where the vibrational levels in the reactant neutral are closely spaced [38.76].

For complex reactant species, lifetimes of intermediate complexes become very long so that, in the higher pressure experiments (see below), all intermediate complexes are stabilized, and thus the reaction is independent of the pressure of M. This form of "saturation" can be eliminated in many cases by using low pressure experiments, so that the time between collisions is long, and normal ternary kinetics (38.14) are restored. Under these conditions, sometimes there is still a pressure independent component to the k which is postulated as being due to radiative stabilization of the intermediate, viz.

$$\left(AB^{\pm}\right)^{*} \rightarrow AB^{\pm} + h\nu . \quad (38.31)$$

An example of this is the reaction [38.77]

$$CH_3^+ + CH_3CN \rightarrow CH_3^+ \cdot CH_3CN + h\nu . \quad (38.32)$$

The stabilizing photon is often considered to be in the infrared due to a vibrational transition within the ground electronic state (with a radiative lifetime of $\approx 10^{-2}$ to 10^{-3} s). Note that only a photon with $h\nu > 3/2 k_B T$ is required for the complex (although still vibrationally excited) to be stable against unimolecular dissociation. However, there are cases where it is believed that an electronically excited state is accessed [38.78, 79]. Examples are

$$C^+ + H_2 \rightarrow CH_2^+ + h\nu \quad (38.33)$$

and

$$Cl^- + BCl_3 \rightarrow BCl_4^- + h\nu . \quad (38.34)$$

As yet, radiative association has not been observed directly (i. e., by the detection of the emitted photon), however, the kinetic evidence is strong for the existence of this process.

For the more rapid collisional associations, competition with binary channels is possible, for example

$$CH_3^+ + NH_3 \rightarrow \begin{cases} H_4CN^+ + H_2 & (70\%) \\ NH_4^+ + CH_2 & (10\%) \\ \xrightarrow{He} CH_3^+ \cdot NH_3 + He & (20\%) \end{cases}$$

where the percentages refer to product abundances at a He pressure of 0.2 Torr [38.80].

38.5 Electron Attachment

These reactions have generally been studied by different techniques from those used to examine ion–neutral reactions and also by different workers [38.31, 32]. Analogous to the ion–neutral situation, there are dissociative processes,

$$e^- + CCl_4 \rightarrow Cl^- + CCl_3 , \quad (38.35)$$

and nondissociative processes,

$$e^- + C_6F_6 \rightarrow C_6F_6^- , \quad (38.36)$$

the former being stabilized by dissociation, and the latter occurring because of the long lifetime of the intermediate against autodetachment with eventual collisional stabilization (cf. saturation in ion–neutral association reactions; Sect. 38.4.2). Measured rate coefficients vary in the range 1×10^{-7} to the smallest measurable value of $\approx 1\times10^{-12}\,cm^3\,s^{-1}$ [38.31, 32]. The upper limit on k for this process is determined, from a consideration of the electron de Broglie wavelength $\lambdabar = \lambda/2\pi$, to be $\approx 5\times10^{-7}(300/T)^{1/2} cm^3\,s^{-1}$ [38.81], or, more rigorously,

$$k(E) = \sum_i \left(\sigma/\pi\lambdabar^2\right)_i / [h\rho(E)] , \quad (38.37)$$

where $\rho(E)$ is the density of states and

$$\sigma/\pi\lambdabar^2 \approx 1 - \exp\left[-\left(4\gamma^2\mathcal{T}\right)^{1/2}\right] , \quad (38.38)$$

with $\gamma^2 R_\infty = (2\mu/m)^2 \alpha_d / a_0^3$ [38.5]; σ is the collision cross section, $\mathcal{T}$ the kinetic energy, α_d the polarizability, a_0 the Bohr radius, μ the reduced mass, and i represents the available product channels. Where k is less than the upper limit value, this is usually due to activation energy barriers, and k shows Arrhenius behavior

$$k(T) = k_0 \exp\left(-\Delta E / k_B T\right) , \quad (38.39)$$

where k_B is the Boltzmann constant, with the barrier height ΔE being in the range 0 to ≈ 300 meV [38.32]. If the attachment is nondissociative, and the lifetime of the intermediate complex against autoionization is small compared with the collision time (as occurs for less complex species), the reaction exhibits normal ternary behavior, for example,

$$e^- + O_2 + M \to O_2^- + M . \quad (38.40)$$

Some interesting mechanisms have been identified for attachment. Different product ions have been observed in experiments conducted at very different pressures, for example, in the reaction

$$C_6F_5I + e^- \to C_6F_5^- + I \quad (\geq 95\%) \quad (38.41)$$

$$\to C_6F_5I^- \quad (\leq 5\%) . \quad (38.42)$$

at high pressures (≈ 1 Torr), the products, as indicated, were observed with an association channel [see (38.42)] being detected [38.82]. This association channel was not seen in low pressure experiments [38.83]. Such behavior is explained in terms of the relative magnitudes of the autodetachment lifetime and the time between collisions. When the former time is larger, the association product will dominate, whereas if the converse is true, dissociation products dominate. Thus, the different product distributions, rather than indicating discrepancies, yield information about the autodetachment lifetime.

In attachments to the Br-containing compounds, CF_2Br_2, $CFBr_3$, and CF_2BrCF_2Br and CH_2BrCH_2Br, in which the Br atoms are on different carbon atoms, up to a 20% Br_2^- product is observed [38.84]. This shows that atoms in product molecules can come from spatially separated parts of the reactant molecule. Thus, considerable distortion and rearrangement must occur in the intermediate complex.

38.6 Recombination

This process has been much less studied than ion–neutral reactions with only ≈ 100 reactions being studied [38.44] compared to about 10 000 in the former case [38.33]. Also, relatively little information is available concerning the products since these are neutral and very much more difficult to detect than ions.

38.6.1 Electron–Ion Recombination

Electron–ion recombination can proceed by series of mechanisms, which have been discussed in detail recently [38.85], and these are included in Table 38.1. In brief, these are radiative recombination, where the neutralized ion (excited into the continuum) is stabilized by radiation emission, dielectronic recombination, which is similar to radiative recombination, except that there is a double electron excitation into the continuum, collisionally stabilized recombination, where the intermediate is stabilized by collisions with electrons or a heavy third body (combined collisional radiative recombination is also possible), and dissociative recombination. Dissociative recombination can obviously only occur when the recombining ion is molecular, however, when it occurs it is usually several orders of magnitude faster than the other recombination processes. Thus, only dissociative recombination will be discussed here.

Rate coefficients for dissociative electron–ion recombination at room temperature vary from $> 1\times10^{-6}$ for polyatomic ions to $\approx 1\times10^{-7}$ cm^3 s^{-1} for diatomic ions [38.42, 43]. For large k, there is little temperature dependence, temperature dependencies being more marked for the slower reactions. Two mechanisms were initially proposed: (i) the direct mechanism [38.86] where the neutralized ion undergoes a radiationless transition to a repulsive potential curve on which dissociation to products occurs, and (ii) the indirect mechanism [38.87] where the neutralized ion initially transfers to a Rydberg state and then undergoes a radiationless transition to the repulsive curve. For the former process, the theoretical temperature dependence is $T_e^{-0.5}$, while it is $T_v^{-1}\,T_T^{-1.5}$ for the latter, where the subscripts T and e refer to the thermal ion vibrational and electron temperatures respectively [38.6]. There is no reason why these two processes cannot occur in parallel, although it is not straightfoward to determine the relative contributions [38.87]. Fortunately, both processes are automatically included in multichannel quantum defect theory (MQDT) [38.88]. Experimental T-dependencies for reactions which are not close to the collisional limit have power law dependencies in the range ≈ 0.7 to 1.5 [38.43, 89, 90], i. e., between the theoretical predictions. In cases where there is detailed temperature data, e.g., for some hydrocarbon ions $(CH_5^+, C_2H_3^+, C_2H_5^+, C_3H_7^+, C_6H_7^+)$ [38.89, 90], the dependence changes from the lower dependence at low temperature to the higher dependence at higher temperature.

Less is known about the products. Detailed theory carried out for the diatomic ions O_2^+, NO^+ and N_2^+, and to a large degree HCO^+, are in general agreement with experiment [38.91, 92]. For the more polyatomic ions, the theory is not yet sufficiently quantitative [38.93, 94] and reliance is placed on experiment. H-atom contributions to the product distributions for ten ions (N_2H^+, HCO^+, HCO_2^+, N_2OH^+, $OCSH^+$, H_2CN^+, H_3O^+, H_3S^+, NH_4^+ and CH_5^+) vary from ≈ 20% for $OCSH^+$ to 120% for CH_5^+ [38.95]. OH is a substantial product (30 to 65%) in the above reactions where it is energetically possible [38.96] (i. e., excluding HCO^+), except in the case of $OCSH^+$, perhaps indicating that the proton is exclusively on the S-atom rather than the O-atom (i. e., in the lowest energy form [38.97, 98]). More complete product distributions are now being obtained, initially using flowing afterglows (O_2H^+, HCO_2^+, H_2O^+, H_3O^+) [38.99], but more recently using storage rings (including H_3^+, CH_2^+, H_2O^+, NH_2^+, H_3O^+, CH_3^+, NH_4^+, CH_5^+) [38.100]. These studies are showing that fragmentation to give three products is a common, indeed dominant, mechanism [38.101]. Dissociative recombination has recently benn reviewed by [38.102].

Electron–ion recombination is an energetic process and electronically excited states can also be populated. Vibrational population distributions for these states can readily be determined if Einstein *A* coefficients are known for the observed transitions ($I \propto A[*]$, where I is the photon intensity, and $[*]$ is the number density of the excited state). The N_2 ($B\,^3\Pi_g$) state and the CO ($a\,^3\Pi_r$) state vibrational population produced in the recombinations of N_2H^+, HCO^+, HCO_2^+ and CO_2^+ have been determined [38.103, 104]. Possible mechanisms [38.105] have been suggested for this vibrational excitation: (i) the impulsive force on the molecular fragments as the neutralized ion rapidly dissociates, and (ii) the Franck–Condon overlap between the wave functions of the molecular products (in various vibrational states) and the wave function of this particular fragment in the neutralized ion before it dissociates. Theory underestimates the populations of the higher levels, but does predict the small observed oscillation in the occupancy of the various vibrational levels in the CO ($a\,^3\Pi_r$) state generated in the recombination of HCO^+ [38.106].

The theories of recombination discussed above assume favorable potential curve crossings, however, it has been shown both experimentally [38.107, 108] and theoretically [38.109] that such are not necessary when quantum tunneling can occur. Experimental evidence has been obtained in the recombinations of N_2H^+ and N_2D^+ where the populations of the $\nu' = 6$ vibrational level of the N_2 electronically excited ($B\,^3\Pi_g$) state is greatly enhanced (≈ 6 at 100 K) for N_2H^+ over N_2D^+. This level is resonant with the $\nu = 0$ vibrational level of the recombining ion, making tunneling more facile; and H atom tunneling is further enhanced because of the smaller mass [38.108].

38.6.2 Ion–Ion Recombination (Mutual Neutralization)

Less information is available on this process than on electron–ion recombination; a detailed review has recently been published [38.110]. k values vary from about 4×10^{-8} to $1\times10^{-7}\,cm^3\,s^{-1}$ at room temperature [38.44, 46] with power law temperature dependencies of ≈ 0.4 for the only two systems that have been studied as a function of T ($NO^+ + NO_2^-$ and $NH_4^+ + Cl^-$) [38.46]. This is consistent with theory [38.111] in which the Landau–Zener approximation is used to determine the probability of a crossing between the Coulombic ion–ion attractive potential and the potentials of the neutral products. This transition occurs by an electron transfer (the optimum distance for such a transfer when a favorable crossing exists is about 10 Å). Theory gives that

$$k \approx 2\left(v^2 Q\right)\left(\mu / 2\pi k_B T\right)^{1/2} , \tag{38.43}$$

where Q is the cross section

$$Q = \pi R_c^2\left[1 + (R_c E)^{-1}\right] . \tag{38.44}$$

R_c is the crossing distance, E is the interaction energy and v is the relative velocity at the crossing point [38.111]. Once neutralized by electron transfer, the products continue undeflected with the velocity gained from the Coulombic field. Photon emission has been detected from a series of excited state products. Following the early detection [38.112] of NO ($A\,^2\Sigma^+$) emissions from $NO^+ + NO_2^-$, a variety of NO emissions have been detected from NO^+ recombinations with molecular (SF_6^-, $C_6F_6^-$, and $C_6F_5CH_3^-$) and atomic (Cl^- and I^-) negative ions [38.113, 114] and He_2 emissions from He_2^+ recombinations with $C_6F_6^-$ and $C_6F_5Cl^-$ [38.113]. These emissions were interpreted in terms of long-range electron transfer [38.112], however, in the recombination of Kr^+ and Xe^+ with SF_6^-, KrF, and XeF, excimer emissions have been seen [38.113], suggesting an intimate encounter in these cases. Also, few data are available on the effects of pressure. For the reactions $SF_3^+ + SF_5^-$ and $NO^+ + NO_2^-$, k increases by about a factor of 4 between 1 and 8 Torr [38.46, 50].

References

38.1 G. Gioumousis, D. P. Stevenson: J. Chem. Phys. **29**, 294 (1958)

38.2 L. Bass, T. Su, M. T. Bowers: Int. J. Mass Spectrom. Ion. Phys. **28**, 389 (1978)

38.3 T. Su, M. T. Bowers: Classical ion–molecule collision theory. In: *Gas Phase Ion Chemistry*, Vol. 1, ed. by M. T. Bowers (Academic, New York 1979) p. 83

38.4 T. Su, M. T. Bowers: Int. J. Mass Spectrom. Ion Phys. **25**, 1 (1975)

38.5 C. E. Klots: Chem. Phys. Lett. **38**, 61 (1976)

38.6 J. N. Bardsley, M. A. Biondi: Adv. At. Mol. Phys. **6**, 1 (1970)

38.7 I. W. M. Smith: Experimental measurements of the rate constants for neutral–neutral reactions. In: *Rate Coefficients in Astrochemistry*, ed. by T. J. Millar, D. A. Williams (Kluwer, Dordrecht 1988) p. 103

38.8 M. J. Henchman: Entropy driven reactions: Summary of the panel discussion. In: *Structure/Reactivity and Thermochemistry*, ed. by P. Ausloos, S. G. Lias (Reidel, Dordrecht 1987) p. 381

38.9 A. Dalgarno: J. Chem. Soc. Farad. Trans. **89**, 2111 (1993)

38.10 S. K. Atreya: *Atmospheres and Ionospheres of the Outer Planets and their Satellites* (Springer, Berlin, Heidelberg 1986)

38.11 E. E. Ferguson, F. C. Fehsenfeld, D. L. Albritton: Ion chemistry of the Earth's atmosphere. In: *Gas Phase Ion Chemistry*, Vol. 1, ed. by M. T. Bowers (Academic, New York 1979) p. 45

38.12 D. Smith, N. G. Adams: Topics Curr. Chem. **89**, 1 (1980)

38.13 E. Murad, S. Lai: J. Geophys. Res. **91**, 13745 (1986)

38.14 E. W. McDaniel, W. L. Nighan: *Applied Atomic Collision Physics* (Academic, New York 1982)

38.15 G. Turban: Pure Appl. Chem. **56**, 215 (1984)

38.16 S. Williams et al.: J. Phys. Chem. A **104**, 10336 (2000)

38.17 I. M. Klotz, R. M. Rosenberg: *Chemical Thermodynamics* (Krieger, Malabar 1991)

38.18 R. C. West (Ed.): *CRC Handbook of Chemistry and Physics*, 69th edn. (CRC, Boca Raton 1988)

38.19 K. P. Huber, G. Herzberg: *Molecular Spectra and Molecular Structure, IV Constants of Diatomic Molecules* (Van Nostrand, New York 1979)

38.20 G. Herzberg: *Molecular Spectra and Molecular Structure, III Electronic Spectra and Electronic Structure of Polyatomic Molecules* (Van Nostrand, New Jersey 1967)

38.21 H. M. Rosenstock, K. Draxl, B. W. Steiner, J. T. Herron: J. Phys. Chem. Ref. Data **6**, I1 (1977)

38.22 NIST Positive and Negative Ion Energetics Database, Standard Reference Data Program (Gaithersburg, Maryland, 1990)

38.23 B. K. Janousek, J. I. Brauman: Electron affinities. In: *Gas Phase Ion Chemistry*, Vol. 2, ed. by M. T. Bowers (Academic, New York 1979) p. 53

38.24 S. G. Lias, J. F. Liebman, R. D. Levin: J. Phys. Chem. Ref. Data **13**, 695 (1984)

38.25 S. G. Lias, J. E. Bartmess, J. F. Liebman, J. L. Holmes, R. D. Levin, W. G. Mallard: J. Phys. Chem. Ref. Data **17**, Suppl. 1 (1988)

38.26 J. E. Bartmess, R. T. McIver: The gas phase acidity scale. In: *Gas Phase Ion Chemistry*, Vol. 2, ed. by M. T. Bowers (Academic, New York 1979) p. 87

38.27 M. W. Chase: NIST-JANAF Thermochemical Tables, 4th Edn., J. Phys. Chem. Ref. Data **9** (1998)

38.28 D. A. McQuarrie: *Statistical Mechanics* (Harper-Collins, New York 1976)

38.29 J. I. Steinfeld, J. S. Francisco, W. L. Hase: *Chemical Kinetics and Dynamics* (Prentice-Hall, Englewood 1989) p. 1

38.30 P. B. Armentrout: *The Encyclopedia of Mass Spectrometry*, Vol. 1 (Elsevier, Amsterdam 2003)

38.31 L. G. Christophorou, D. L. McCorkle, A. A. Christodoulides: Electron attachment processes. In: *Electron-Molecule Interactions and their Applications*, Vol. 1, ed. by L. G. Christophorou (Academic, Orlando 1984) p. 477

38.32 D. Smith, N. G. Adams: Studies of plasma reaction processes using a flowing afterglow/langmuir probe apparatus. In: *Swarms of Ions and Electrons in Gases*, ed. by W. Lindinger, T. D. Mark, F. Howorka (Springer, Vienna 1984) p. 284

38.33 Y. Ikezoe, S. Matsuoka, M. Takebe, A. A. Viggiano: *Gas Phase Ion-Molecule Reaction Rate Constants through 1986* (Maruzen, Tokyo 1987)

38.34 V. G. Anicich: J. Phys. Chem. Ref. Data **22**, 1469 (1993)

38.35 V. Anicich: Astrophys. J. **84**, 215 (1993)

38.36 V. Anicich: *An Index of the Literature for Bimolecular Gas Phase Cation-Molecule Reaction Kinetics*, JPL Publication 03-19 (JPL, Pasadena, California November, 2003)

38.37 D. C. Clary, D. Smith, N. G. Adams: Chem. Phys. Lett. **119**, 320 (1985)

38.38 D. C. Clary: Mol. Phys. **54**, 605 (1985)

38.39 N. G. Adams, D. Smith, D. C. Clary: Ap. J. **296**, L31 (1985)

38.40 D. A. McQuarrie: *Statistical Mechanics* (Harper & Row, New York 1976)

38.41 D. R. Ridge: Ion–molecule collision theory. In: *The Encyclopedia of Mass Spectrometry*, Vol. 1, ed. by P. B. Armentrout (Elsevier, Amsterdam 2003) p. 1

38.42 R. Johnsen: Int. J. Mass Specrom. Ion Proc. **81**, 67 (1987)

38.43 N. G. Adams, D. Smith: Laboratory studies of dissociative recombination and mutual neutralization and their relevance to interstellar chemistry. In: *Rate Coefficients in Astrochemistry*, ed. by T. J. Millar, D. A. Williams (Kluwer, Dordrecht 1988) p. 173

38.44 J. B. A. Mitchell: Phys. Rep. **186**, 216 (1990)

38.45 J. B. A. Mitchell, C. Rebrion-Rowe: Int. Rev. Phys. Chem. **16**, 201 (1997)

38.46 D. Smith, N. G. Adams: Studies of ion–ion recombination using flowing afterglow plasmas. In: *Physics of Ion–Ion and Electron–Ion Collisions*, ed. by F. Brouillard, J. W. McGowan (Plenum, New York 1983) p. 501

38.47 D. R. Bates, A. Dalgarno: Electronic recombination. In: *Atomic and Molecular Processes*, ed. by D. R. Bates (Academic, London 1962) p. 245

38.48 N. G. Adams, D. Smith: Ion–molecule reactions at low temperatures. In: *Reactions of Small Transient Species*, ed. by A. Fontijn (Academic Press, London 1983) p. 311

38.49 H. W. S. Massey: *Negative Ions* (Cambridge Univ. Press, Cambridge 1976)

38.50 D. Smith, N. G. Adams: Geophys. Res. Lett. **9**, 1085 (1982)

38.51 M. Meot-Ner: Temperature and pressure effects in the kinetics ion–molecule reactions. In: *Gas Phase Ion Chemistry*, Vol. 1, ed. by M. T. Bowers (Academic, New York 1979) p. 198

38.52 D. K. Bohme: Int. J. Mass Spectrom. **200**, 97 (2000)

38.53 C. H. DePuy: Int. J. Mass Spectrom. **200**, 79 (2000)

38.54 D. Smith, N. G. Adams, E. Alge, H. Villinger, W. Lindinger: J. Phys. B **13**, 2787 (1980)

38.55 J. B. Lauderslager, W. T. Huntress, M. T. Bowers: J. Chem. Phys. **61**, 4600 (1974)

38.56 N. G. Adams, D. Smith: J. Phys. B **9**, 1439 (1976)

38.57 A. L. Schmeltekopf, E. E. Ferguson, F. C. Fehsenfeld: J. Chem. Phys. **48**, 2966 (1968)

38.58 E. C. Inn: Planet. Space Sci. **15**, 19 (1967)

38.59 G. Javahery, S. Petrie, J. Wang, D. K. Bohme: Chem. Phys. Lett. **195**, 7 (1992)

38.60 G. Javahery et al.: Org. Mass Spectrom. **20**, 1005 (1993)

38.61 D. K. Bohme: The kinetics and energetics of proton transfer. In: *Interactions between Ions and Molecules*, ed. by P. Ausloos (Plenum, New York 1975) p. 489

38.62 M. J. McEwan: Flow tube studies of small isomeric ions. In: *Advances in Gas Phase Ion Chemistry*, Vol. 1, ed. by N. G. Adams, L. M. Babcock (JAI Press, Greenwich 1992) p. 1

38.63 N. G. Adams, D. Smith: Chem. Phys. Lett. **54**, 530 (1978)

38.64 C. H. DePuy: SIFT-drift studies of anions. In: *Ionic Processes in the Gas Phase*, ed. by M. A. Almoster Ferreira (Reidel, Dortrecht 1984) p. 227

38.65 S. E. Barlow, G. H. Dunn: Int. J. Mass Spectrom. Ion Proc. **80**, 227 (1987)

38.66 N. G. Adams, D. Smith: Int. J. Mass Spectrom. Ion Proc. **61**, 133 (1984)

38.67 F. C. Fehsenfeld, D. L. Albritton, D. L. Bush, P. G. Fornier, T. R. Govers, J. Fourier: J. Chem. Phys. **61**, 2150 (1974)

38.68 I. Dotan: Chem. Phys. Lett. **75**, 509 (1980)

38.69 F. C. Fehsenfeld: Associative detachment. In: *Interactions between Ions and Molecules*, ed. by P. Ausloos (Plenum, New York 1975) p. 387

38.70 A. A. Viggiano, J. F. Paulson: Reactions of negative ions. In: *Swarms of Ions and Electrons in Gases*, ed. by W. Lindinger, T. D. Mark, F. Howorka (Springer, Vienna 1984) p. 218

38.71 V. M. Bierbaum, G. B. Ellison, S. R. Leone: Flowing afterglow studies of ion reaction dynamics using infrared chemiluminescence and laser-induced fluorescence. In: *Gas Phase Ion Chemistry*, Vol. 3, ed. by M. T. Bowers (Academic, New York 1984) p. 1

38.72 D. Smith, N. G. Adams, E. E. Ferguson: Interstellar ion chemistry: Laboratory studies. In: *Molecular Astrophysics*, ed. by T. W. Hartquist (Cambridge Univ. Press, Cambridge 1990) p. 181

38.73 N. G. Adams, D. Smith, J. F. Paulson: J. Chem. Phys. **72**, 288 (1980)

38.74 D. Smith, N. G. Adams: Isotope exchange in ion–molecule reactions. In: *Ionic Processes in the Gas Phase*, ed. by M. A. Almoster Ferreira (Reidel, Dordrecht 1984) p. 41

38.75 D. Smith, N. G. Adams: Ap. J. **242**, 424 (1980)

38.76 A. A. Viggiano, J. F. Paulson: J. Phys. Chem. **95**, 10719 (1991)

38.77 E. Herbst, M. J. McEwan: Astron. Astrophys. **229**, 201 (1990)

38.78 E. Herbst, J. G. Schubert, P. R. Certain: Ap. J. **213**, 696 (1977)

38.79 C. R. Herd, L. M. Babcock: J. Phys. Chem. **93**, 245 (1989)

38.80 D. Smith, N. G. Adams: Chem. Phys. Lett. **54**, 535 (1978)

38.81 J. M. Warman, M. C. Sauer: Int. J. Radiat. Chem. **3**, 273 (1971)

38.82 C. R. Herd, N. G. Adams, D. Smith: Int. J. Mass Spectrom. Ion Proc. **87**, 331 (1989)

38.83 W. T. Naff, R. N. Compton, C. D. Cooper: J. Chem. Phys. **54**, 212 (1971)

38.84 D. Smith, C. R. Herd, N. G. Adams, J. F. Paulson: Int. J. Mass Spectrom. Ion Proc. **96**, 341 (1990)

38.85 N. G. Adams, L. M. Babcock, J. L. McLain: Electron–ion recombination. In: *The Encyclopedia of Mass Spectrometry*, Vol. 1, ed. by P. B. Armentrout (Elsevier, Amsterdam 2003) p. 542

38.86 D. R. Bates: Phys. Rev. **78**, 492 (1950)

38.87 J. N. Bardsley: J. Phys. B **1**, 349 (1968)

38.88 A. Giusti: J. Phys. B **13**, 3867 (1980)

38.89 J. L. McLain, V. Poterya, C. D. Molek, N. G. Adams, L. M. Babcock: J. Phys. Chem. A **108**, 6704 (2004)

38.90 A. Ehlerding et al.: J. Phys. Chem. A **107**, 2179 (2003)

38.91 S. L. Guberman: Electron–ion continuum–continuum mixing in dissociative recombination. In: *Dissociative Recombination: Theory, Experiment and Applications*, ed. by B. R. Rowe, J. B. A. Mitchell, A. Canosa (Plenum, New York 1993) p. 47

38.92 D. Talbi, Y. Ellinger: A theoretical study of the HCO^+ and HCS^+ electronic dissociative recombi-

nations. In: *Dissociative Recombination: Theory, Experiment and Applications*, ed. by B. R. Rowe, J. B. A. Mitchell, A. Canosa (Plenum, New York 1993) p. 59
38.93 D. R. Bates: Ap. J. **344**, 531 (1989)
38.94 E. T. Galloway, E. Herbst: Ap. J. **376**, 531 (1991)
38.95 N. G. Adams, C. R. Herd, M. Geoghegan, D. Smith, A. Canosa, B. R. Rowe, J. L. Queffelec, M. Morlais: J. Chem. Phys. **94**, 4852 (1991)
38.96 C. R. Herd, N. G. Adams, D. Smith: Ap. J. **349**, 388 (1990)
38.97 P. G. Jasien, W. J. Stevens: J. Chem. Phys. **83**, 2984 (1985)
38.98 M. Scarlett, P. R. Taylor: Chem. Phys. **101**, 17 (1986)
38.99 N. G. Adams: Adv. Gas Phase Ion Chem. **1**, 272 (1992)
38.100 M. Larsson: Dissociative electron–ion recombination studies using ion synchrotrons. In: *Photoionization and Photodetachment*, Vol. II, ed. by C.-Y. Ng (World Scientific, Singapore 2000) p. 693
38.101 S. Datz: J. Phys. Chem. A **105**, 2369 (2001)
38.102 N. G. Adams, V. Poterya, L. M. Babcock: Mass Spectrom. Revs. (2005) (in press)
38.103 N. G. Adams, L. M. Babcock: J. Phys. Chem. **98**, 4564 (1994)
38.104 M. Skrzypkowski, T. Gougousi, M. F. Golde: Spectroscopic emissions from the recombination of N_2O^+, N_2OH^+/HN_2O^+, CO_2^+, CO_2H^+, HCO^+/COH^+, H_2O^+, NO_2^+, HNO^+ and LIF measurements of the H atom yield from H_3^+. In: *Dissociative Recombination: Theory, Experiment and Applications*, Vol. IV, ed. by M. Larsson, J. B. A. Mitchell, I. F. Schneider (World Scientific, Singapore 2000) p. 200
38.105 D. R. Bates: Mon. Not. R. Astron. Soc. **263**, 369 (1993)
38.106 N. G. Adams, L. M. Babcock: Ap. J. **434**, 184 (1994)
38.107 J. M. Butler, L. M. Babcock, N. G. Adams: Mol. Phys. **91**, 81 (1997)
38.108 V. Poterya, J. L. McLain, L. M. Babcock, N. G. Adams: J. Phys. Chem. A (2005) (in press)
38.109 D. R. Bates: Adv. Atom. Mol. Opt. Phys. **34**, 427 (1994)
38.110 N. G. Adams, L. M. Babcock, C. D. Molek: Ion–ion recombination. In: *The Encyclopedia of Mass Spectrometry*, Vol. 1, ed. by P. B. Armentrout (Elsevier, Amsterdam 2003) p. 555
38.111 R. E. Olsen: J. Chem. Phys. **56**, 2979 (1972)
38.112 D. Smith, N. G. Adams, M. J. Church: J. Phys. B **11**, 4041 (1978)
38.113 M. Tsuji: Adv. Gas Phase Ion Chem. **4**, 137 (2001)
38.114 P. Spanel, D. Smith: Chem. Phys. Lett. **258**, 477 (1996)

39. Clusters

Clusters are small aggregates of atoms or molecules which are transitional forms of matter between atoms or molecules and their corresponding bulk forms. Just as this definition spans an incredibly broad range of clusters from, say, He_2 to Na_{10000}, so do the properties of these clusters span a broad range. This chapter attempts to bring order to this diverse cluster kingdom by first sorting them into six general categories. Within each category, the physics and chemistry of the more or less similar cluster species are described. Particular emphasis is placed on the unique properties of clusters owing to their finite size and finite lattice.

This chapter summarizes one of the youngest topics in this volume. Much of what is known is highly qualitative and has not yet been assembled into overarching tables or equations. Thus, this review is best regarded as a progress report on the current knowledge in this rapidly advancing field. Many of the concepts and the language used to discuss clusters are derived from condensed matter physics. The nature of these clusters impels such descriptions.

39.1 **Metal Clusters** ... 590
39.1.1 Geometric Structures ... 590
39.1.2 Electronic and Magnetic Properties ... 590
39.1.3 Chemical Properties ... 592
39.1.4 Stable Metal Cluster Molecules and Metallocarbohedrenes ... 593

39.2 **Carbon Clusters** ... 593
39.2.1 Small Carbon Clusters ... 594
39.2.2 Fullerenes ... 594
39.2.3 Giant Carbon Clusters: Tubes, Capsules, Onions, Russian Dolls, Papier Mâché ... 595

39.3 **Ionic Clusters** ... 596
39.3.1 Geometric Structures ... 596
39.3.2 Electronic and Chemical Properties ... 596

39.4 **Semiconductor Clusters** ... 597
39.4.1 Silicon and Germanium Clusters ... 597
39.4.2 Group III–V and Group II–VI Semiconductor Clusters ... 598

39.5 **Noble Gas Clusters** ... 599
39.5.1 Geometric Structures ... 599
39.5.2 Electronic Properties ... 600
39.5.3 Doped Noble Gas Clusters ... 600
39.5.4 Helium Clusters ... 601

39.6 **Molecular Clusters** ... 602
39.6.1 Geometric Structures and Phase Dynamics ... 602
39.6.2 Electronic Properties: Charge Solvation ... 602

39.7 **Recent Developments** ... 603

References ... 604

Clusters discussed in this chapter are isolated species composed primarily of a single type of atom or molecule. Most of these clusters are highly reactive and can only be made and studied under rarified conditions such as in a molecular beam. In keeping with the definition of clusters given above, this chapter will not attempt to cover the truly vast literature on atomic and molecular dimers and trimers. A great many of these species have been thoroughly characterized, however they are better described as molecules rather than "tiny clusters". Stabilized clusters in the form of Zintl ions, colloids, and nanoparticles have also been made. While it is impossible to resist mentioning these clusters throughout this chapter, they are more appropriate to condensed matter physics. Finally, experimental and theoretical techniques for forming and studying clusters will not be covered; excellent reviews of these methods are available elsewhere [39.1–6].

39.1 Metal Clusters

Clusters of a wide variety of refractory and nonrefractory metals have been made and studied. These include clusters of alkali metals, transition metals, coinage metals, main group metals, and lanthanides. Clusters of alkali metals, particularly sodium, are the most well studied and understood. The emerging picture of alkali clusters is that they behave much like quasifree electron metal spheres. Thus, approximations such as the jellium model used for describing bulk metals explain a large number of alkali metal properties quite well [39.2]. Other metallic clusters, such as some main group metal and noble metal clusters, can also be understood within the jellium model. Nonetheless, there are examples of metallic clusters which deviate significantly from these simple models, particularly among the transition metal clusters.

39.1.1 Geometric Structures

The ground state geometry is known accurately for only a handful of small metal clusters. Examples include lithium and sodium clusters containing up to nine atoms, where electronic spectra are compared with accurate ab initio quantum chemical calculations to deduce their structures [39.5–7]. Such an approach is not a general one, either experimentally or theoretically. As the size and/or atomic valency of the metal cluster increases, the number of possible ground state structures grows enormously. Interpretation of experimental spectra requires theoretical guidance, which means that each one of these structures must be calculated using an accurate electronic structure calculational method. Thorough theoretical investigation is only practical with current computational tools for clusters containing up to about thirteen atoms. Methods such as molecular dynamics combined with density functional calculations have been used to speed up the process of finding and comparing various isomers [39.4, 8]. Such approaches are still restricted by available computational power to clusters with relatively few atoms and valence electrons per atom. Metal clusters of atoms with higher valency, e.g., mid-row transition metals, have proved to be quite difficult to treat accurately [39.9]. The large electron correlation problems inherent in these clusters must be treated semi-empirically, leading to large uncertainties in the relative energies between isomers of different spin and geometry.

Despite these difficulties, approximate geometries are known for many metal clusters. For cluster sizes greater than, roughly, several tens of atoms, there is strong experimental evidence that spherical close packed geometries, particularly the Mackay filled icosahedra, predominate the geometric structures [39.7, 10, 11] (The Mackay filled icosahedra contain concentric closed icosahedral shells of atoms plus one central atom. These structures are pentagonally symmetric and occur every n atoms by $n = 13, 55, \cdots 1/3(2n+1)(5n^2+5n+3)$ [39.4]). These geometries have been deduced by a variety of means. For example, abundance mass spectra of metal clusters up to sizes containing thousands of atoms show intensity enhancements, termed *magic numbers*, at each cluster size which corresponds to a complete Mackay icosahedron. Saturation coverages of transition metal clusters with various reagents can be explained in terms of covering the icosahedral faces of these clusters. The persistence of icosahedral structures to large cluster sizes raises the question about where the crossover to the metallic packing occurs. For example, a spherical piece of an fcc metal would have cuboctohedral symmetry rather than noncrystalline icosahedral symmetry. Ultrafine metal particles containing 10^4 to 10^5 atoms typically have bulk crystalline geometries [39.12]. Little theoretical or experimental data currently exist to resolve the question of when the bulk crystalline structure emerges [39.10]. The point of crossover must involve kinetic as well as energetic factors. Although icosahedral packing is seen in beam experiments on metal clusters containing hundreds to thousands of atoms, high resolution electron microscopy experiments on supported metal particles in this size range show fluctuating structures that can rapidly evolve between icosahedral, cuboctohedral, and other crystalline arrangements, depending on cluster temperature [39.13].

39.1.2 Electronic and Magnetic Properties

A number of electronic properties have been measured for metallic clusters, particularly alkali metal and noble metal clusters. These include ionization potentials, electronic affinities, polarizabilities, and photoabsorption cross sections.

The spherical jellium approximation is generally a good model for these properties in clusters of alkali and noble metals [39.2, 7]. This model treats the valence electrons of the cluster as a delocalized sea of electrons smeared over a uniform spherical background of ionic cores. The energy levels of such a jellium sphere can be calculated by confining the jellium electrons to a three dimensional potential. A suitable form of this potential yields level spacings that are given by principal and angular momentum quantum numbers (n) and (l). There are no restrictions on l for a given n, and the degeneracy of a given l level is $2(2l+1)$. The levels order as 1s, 1p, 1d, 2s, 1f, 2p, 1g, 2d, 3s, $\cdots$. The electron configuration of a cluster is given by filling successive energy levels (termed shells) with the available valence electrons. Special stability occurs for those clusters with a closed shell configuration. This stabilization can be seen in the experimental measurements of the dependences of total binding energies, ionization potentials, electron affinities, and polarizabilities on cluster size [39.2, 7, 10, 11, 13, 14]. The potential used to calculate the energy levels within the jellium model has been formulated to include elliptical distortions of open shell clusters [39.7]. This refinement has been successful in describing the fine structures of these trends. Despite the usefulness of the jellium model, it is not applicable for very small or very large clusters. In many smaller clusters containing up to a few tens of atoms, electron-ionic lattice interactions cannot be neglected, and the spherical approximation is poor. For these clusters, accurate quantum chemical calculations must be used to determine the electronic structure. Alternatively, for clusters larger than roughly 500 to 3000 atoms (depending on which clusters and the cluster temperature), the level spacings become increasingly continuous, blurring the shell structure. When this occurs, other effects are seen to dominate the trends in cluster stability, particularly the stability arising from completing a geometric shell of atoms on the cluster [39.7, 10, 11, 15, 16].

The convergence of cluster electronic structure to bulk metal electronic structure has been seen in the evolution of metal cluster ionization potentials (I_P) and electron affinities (A_e) as a function of cluster size. A good approximation for the overall trend in the I_P as a function of cluster size (N is the number of atoms) is given by the electrostatic model for the work function of a classical conducting sphere [39.2]

$$I_{\mathrm{P},N} = W_\mathrm{B} + A\frac{e^2}{R_N}\,, \qquad (39.1)$$

where W_B is the work function of the bulk metal, e is the electron charge, and R is the cluster radius, which is often set proportional to $N^{1/3}$. A is a constant which is found experimentally to have values of about 0.3 to 0.5; the theoretically derived value for A depends on the model used [39.2, 11]. A similar form of (39.1), where A is replaced by $(A-1)$, describes the A_e of a negatively charged cluster. Experimentally, the I_P and A_e of alkali metal, noble metal, and some main group metal clusters behave as described by (39.1) with shell structure superimposed on the overall trend [39.2, 6, 14]. Equation (39.1) predicts a smooth convergence of the work function to the bulk value with increasing cluster size. This has been seen experimentally, such as in copper cluster valence and inner shell A_e which extrapolate smoothly to the corresponding bulk values [39.17]. In mercury clusters, the transition to bulk metallic behavior occurs more abruptly and clusters with less than ≈ 17 atoms appear to behave as nonmetals [39.18]. Transition metal clusters often deviate strongly from the trend in (39.1) [39.10].

Electronic spectra are available for a number of metal clusters including alkali, noble metal, transition metal, and aluminum clusters [39.7, 10, 11]. Small alkali clusters exhibit rich spectra in the visible. Many of these spectra have been assigned using accurate electronic structure calculations as described above. Visible spectra of larger alkali metal clusters and other metal clusters are typified by giant resonances with cross sections reaching values as large as 2000 Å^2 [39.6, 7]. In the absence of detailed electronic structure information, these spectra have been assigned using comparisons with bulk metal spectra. In particular, the giant resonances are assigned to collective excitations of the cluster valence electrons, in an analogy to bulk metal plasmon resonances. Theoretical treatments of these giant resonances for clusters have been derived from classical treatments of conducting spheres driven by an external electromagnetic field [39.7, 10]. While the blue shift of the resonance frequency with increasing cluster size is well predicted by the classical models, other details are less well described, such as the magnitude of the shift and the width of the resonance. Clusters which are known to be nonspherical from other measurements exhibit multiple resonance peaks; these have yet to be quantitatively described by theory. Transition metal clusters also show evidence of collective excitations; however, the magnitudes of these absorptions are 2 to 5 times larger than predicted by classical models [39.19]. An intriguing *non sequitur* occurs when the classical models are extended down to describe the resonances of small alkali clusters. This is illustrated by the example of Na_8, which exhibits a large single resonance in the photoabsorption spectrum. At first glance, the classical jellium model does fine: it predicts a spherical closed shell cluster which should exhibit a single resonance. Yet, a more thorough examination of the time dependence of this feature has revealed that it consists of four overlapping absorptions [39.10, 20]. These multiple absorptions clearly must arise from one-electron excitations, not a collective all-electron excitation [39.5]. Further work is needed to weave together the two disparate pictures of collective versus one electron excitations in metal clusters.

Inner shell electrons of some metal clusters have been probed spectroscopically. Whereas excitations of the delocalized valence electrons primarily reflect the entire cluster environment, excitations of the localized inner shell electrons reflect the atomic environment. In mercury clusters, excitation of the 5d core electrons reveals a transition from insulating clusters at small sizes to more metal-like clusters with increasing s-p hybridization typical of bulk mercury [39.21]. Inner core electron spectra of copper and antimony clusters also reveal details regarding the evolution of the cluster lattice structure as a function of size [39.11, 17]. Such valuable information can be obtained from inner shell electron spectra of metal clusters that more experiments are warranted.

Magnetic moments have been measured and described theoretically for a range of transition metal, rare earth, and Group III metal clusters [39.5, 7, 10, 22–24]. Examples of clusters which exhibit a positive magnetic moment include cobalt, iron, nickel, gadolinium, and terbium clusters. Clusters of metals such as vanadium, palladium, chromium, and aluminum are observed to be diamagnetic. The effective moment per atom, μ, in magnetic clusters is greater than in the bulk because of the lower average coordination number in clusters [39.24]. As the cluster size increases, the surface to volume atom ratio decreases and μ converges to the bulk value. The size of these magnetic clusters is smaller than the critical domain size for bulk ferromagnetism, thus they are best described as paramagnetic with a single moment given by $N\mu$, where N is the number of atoms in the cluster. Stern–Gerlach deflection data have been obtained to measure μ. A critical parameter in interpreting these data is the so-called blocking temperature T_B, the temperature at which the cluster moment unlocks from the cluster axes and orients thermally in an external field. For clusters with temperatures $T > T_B$, the observed effective moment μ_{eff} for paramagnetic clusters is given by [39.24]

$$\mu_{eff} = \mu \left[\coth\left(\frac{N\mu H}{k_B T} \right) - \frac{k_B T}{N\mu H} \right] , \tag{39.2}$$

where k_B is the Boltzmann constant, and H is the magnetic field strength. Medium size iron and cobalt clusters behave well according to this classically derived expression. Deviations from this model are observed in internally cold clusters and rare earth clusters which can be explained by partial and complete locking of the magnetic moment to the cluster lattice. Also, in very large clusters, $N\mu$ becomes sufficiently large that alignment overwhelms thermal statistical behavior.

39.1.3 Chemical Properties

The chemistry of both charged and neutral metal clusters has been studied, particularly for clusters of transition metals and Group III metals [39.6, 7, 25, 26]. Both chemisorption and physisorption are seen, depending on the type of metal cluster and reagent. In many cases, the observed chemistry is quite similar to that observed for the corresponding bulk metal. For example, platinum clusters dehydrogenate hydrocarbons, hydrogen chemisorbs on transition metal clusters with the exception of coinage metal clusters, and oxygen reacts readily with aluminum and iron clusters.

Chemisorption reactions of metal clusters have been seen with a wide variety of reagents. Products, reaction rates, and activation barriers have been measured as a function of cluster size. With a few notable exceptions, such as the reactions of hydrogen with nickel and aluminum clusters, none of these reactions are understood at the microscopic level [39.10, 11]. Macroscopically, some correlations have been seen between reaction rates and other measured cluster properties. For example, some cluster reactions show evidence that the shell structure affects reaction rates, with open shell clusters having much higher reaction rate and lower activation barriers than clusters with closed shells. In other cases, the pattern of cluster reaction rates with cluster size correlates with the cluster I_P's. These correlations have been used to infer the essential cluster-reagent interaction which governs the reactivity for a given set of clusters with a given reagent. For example, open shell clusters and clusters with low I_P's favor reactions which involve electron donation from the cluster to the reagent at a critical point along the reaction coordinate. Such generalities must be made with caution, however, since they do not hold up well over a broad range of cluster sizes, compositions, and reagents. Even a single cluster size can exhibit complex reactivity: isomers with different reactivity have been observed for niobium clusters and, in some cases, these isomers have been interconverted by annealing. Of course, complexity is to be expected, given the richness and diversity of chemistry that is known to occur for different metal systems in various electronic and geometric environments.

Geometric structures of transition metal clusters have been studied using physisorption reactions. Weakly bound adsorbates such as hydrogen, water, and ammonia show strong saturation behavior in their uptake by metal clusters [39.10,11,27,28]. Trends in the saturation coverage with cluster size yield information about the type of binding site and total number of binding sites for a given cluster size. Additional information regarding the nature of the adsorbate site is often available from studies of the corresponding physisorption on bulk metal surfaces. This knowledge is used to sort through possible cluster structures and deduce which geometries exhibit the correct number and type of adsorbate binding sites. For example, saturation coverages of iron, cobalt, and nickel clusters correlate with icosahedrally packed geometries over a wide range of sizes. Adsorbate binding energies have also been determined in some cases [39.27, 28].

39.1.4 Stable Metal Cluster Molecules and Metallocarbohedrenes

Thus far, this section has focussed entirely on the properties of metal clusters isolated in the gas phase. This discussion would not be complete, however, without mentioning that a number of metal cluster molecules have been made which are sufficiently stable that they can be made in quantity and bottled [39.7, 10, 11, 29, 30]! Also, recently a new class of metal-carbon clusters has been discovered, termed metallocarbohedrenes, that are believed to be sufficiently stable and abundant that they can be made in bulk. The brief overview of these isolatable clusters given below is not meant to be complete but is intended to introduce the large and impressive body of work in this field. Interested readers are encouraged to consult the reviews cited and references therein.

Metal cluster molecules consist of a metal cluster core surrounded by a stabilizing ligand shell. A large number of such metal cluster molecules has been made, including some with metal cores as large as 300 platinum or 561 palladium atoms. In many cases, crystals of single size clusters have been made with an exactly known number of metal atoms in the core. The availability of macroscopic samples of these clusters has made it possible to measure a number of their properties. Exact structures are known for many metal cluster molecules from X-ray crystallography. Electronic and magnetic properties have been determined which reveal the development of metallic behavior within the metal core. The ligand shell is found to interact strongly with the metal atoms on the surface of the metal core. This outer shell of metal atoms does not behave as a surface of metal atoms with free valence electrons such as is found on the surface of a bulk metal. Undoubtedly, such strong interactions between surface metal atoms and coordinating ligands are necessary to make a cluster which is sufficiently stable to be isolated and crystallized without coalescing. The core of atoms inside the outer metal shell of atoms does not appear to be greatly perturbed by the coordinating ligands. Studies of the electronic and magnetic properties of the core show the onset of metallic properties as a function of particle size and atomic packing.

A distinct class of transition metal-carbon clusters has been recently been found which have been termed metallocarbohedrenes [39.31]. Within this class, the clusters M_8C_{12} ($M = $ Ti, V, and Zr) are particularly abundant and stable. Extensive electronic structure calculations show that the structure of M_8C_{12} is metallic and should be viewed as a distorted M_8 cube where each face is decorated with a C_2 dimer [39.32]. This structure differs greatly from the corresponding bulk metal carbides, which have cubic rock salt crystalline forms. Furthermore, the metal carbide cluster formation conditions can be adjusted to yield cubic fragments of the bulk. The related metal nitride clusters are only observed to form cubic structures. Theoretical calculations on the metallocarbohedrene and cubic forms of these metal carbides and nitrides show that they are comparable in energy for the carbides, but that the cubic structures are much more stable in the metal nitride clusters [39.33]. Despite their apparent stability, none of the metallocarbohedrenes has yet been isolated and purified.

39.2 Carbon Clusters

The discovery of the especially stable, spherical cluster of sixty carbon atoms, named buckminsterfullerene, has ignited an intense research effort in carbon clusters [39.7, 8, 10, 11, 34–39]. The family of pure carbon clusters that have been made extends from the dimer all the way up to tubular shaped clusters containing thousands of atoms. Several of these clusters, most notably C_{60} and C_{70}, have been isolated as a single size in macroscopic quantities [39.40]. The fascinating properties of bulk materials made of pure and doped C_{60} and C_{70} are outside the scope of this review [39.8, 10, 11, 37, 39, 41].

Carbon clusters can be roughly grouped into three distinct classes. The first consists primarily of small carbon clusters which have linear or ring geometries.

Hollow spheres (termed fullerenes) appear at about 28 carbon atoms and persist up to at least several hundred atoms. Very large carbon clusters containing hundreds to thousands of atoms assume various forms, including onion-like structures of concentric spheres and hollow tubes.

39.2.1 Small Carbon Clusters

Theoretical and experimental studies have found that rings and linear chains are the most stable configurations for small carbon clusters. Below about ten atoms, linear cumulenes are the most stable structures for odd numbered carbon clusters and ionic carbon clusters [39.13, 26, 42, 43]. For the even numbered neutral carbon clusters, C_4, C_6, and C_8, however, cyclic and linear geometries are nearly isoenergetic [39.7, 11, 13, 42]. Various theoretical and experimental studies have yielded conflicting results as to which geometry is more stable for each of these three clusters [39.7,42,44]. Much of the controversy over the experimental evidence probably arises because the two structures are so close in energy that either isomer or both may be present, depending on the preparation conditions. By way of proof, direct experimental data have been found for coexisting linear and cyclic isomers of C^+_{7-9} and C_{11} [39.26, 44]. Also, some experimental probes are not equally sensitive to the signature from a linear versus a cyclic geometry.

At ten carbon atoms, a distinct transition occurs from linear to cyclic structures [39.26, 43, 44]. Starting at this size, the additional bonding stabilization accrued by joining the two ends of the linear chain overcomes the strain energy resulting from ring closure. Stable monocyclic ring structures are observed for carbon clusters over a surprisingly wide range of sizes, even persisting into the size range where three dimensional fullerenes appear. Bicyclic rings, higher order polycyclic rings, and graphitic fragments also occur in this size range but many of these configurations appear to be metastable [39.43,45].

A_e's and I_P's have been measured for these small carbon clusters. Dramatic effects appear in the size dependence of the A_e's as a result of the changeover from linear to ring structures [39.44]. A_e's of the chain structures are noticeably higher than for ring structures for carbon clusters containing similar numbers of atoms. Distinct odd-even alternations in the A_e's are also seen, plus there is evidence for aromatic stabilization according to the $4n+2$ rule. No break in the I_P's is seen across the structural transition, but the trend in the I_P's also follows the $4n+2$ rule expected for aromaticity.

Chemical reactions have been observed for both neutral and charged small carbon clusters, particularly for the cations [39.7, 26, 43]. In general, the linear cationic clusters, containing fewer than ten atoms, react readily with a variety of reagents and exhibit reactivity typical of carbenes. The ring shaped cationic clusters with ten or more atoms are much less reactive, and often show no detectable reaction with reagents that react efficiently with the smaller carbon clusters. It is this differential reactivity which has revealed the presence of coexisting linear and cyclic isomers. Evidence for polyacetylene versus cumulene structure in the small linear chains can also be seen in the cluster reaction patterns.

39.2.2 Fullerenes

Fullerenes make up a class of carbon clusters with closed hollow carbon atom cage morphologies. The term "*fullerene*" was inspired by the geodesic domes of architect R. Buckminster Fuller, and has come into widespread usage despite its nonstandard nomenclature. The fullerene cage network is composed of interlocking rings of sp^2 hybridized carbon atoms, where every carbon atom is bonded to three other carbon atoms. Five and six member rings predominate in the cages. Highly strained four and smaller membered rings are unfavorable. Seven and higher membered rings usually can facilely rearrange within the network to form five and six membered rings. Delocalization of the π electrons over the cage contributes significantly to the stabilization of the fullerenes.

Assuming at least five-membered rings, the smallest possible fullerene is C_{20}, consisting of twelve pentagons. However, the observed fullerenes all contain about 30 atoms or more. Larger fullerenes are formed by joining together pentagons, hexagons, and heptagons, with the pentagons providing the curvature necessary to close the cage. In fullerene structures composed entirely of pentagons, hexagons, and heptagons, it can be rigorously shown from Euler's theorem that, $N_5 = N_7 + 12$, where N_5 is the number of pentagons and N_7 the number of heptagons. By far the most abundant fullerene that is seen is C_{60}. There are 1812 possible isomers for C_{60} but only one forms in great abundance [39.37], which has each of its 12 pentagons isolated and surrounded by its 20 hexagons. The resulting molecule, resembling a soccer ball, belongs to the highest point symmetry group I_h, where all the carbon atoms are equivalent [39.34–37]. The *isolated pentagon rule* arises out of minimizing strain and maximizing resonance energy and is a powerful tool for predicting the most stable fullerene isomers [39.37]. Besides C_{60}, a number of other fullerenes have been seen, most notably C_{70} which is the next larger fullerene that can be constructed using the isolated pentagon rule. The largest proven fullerene structure contains 84 carbon atoms. It is possible to construct increasingly larger fullerene cages containing hundreds of carbon atoms. Such giant fullerenes have yet to be conclusively verified in experiments, however evidence for their existence has been seen in bulk samples containing fullerene mixtures and in gas phase abundances of carbon clusters [39.36, 37, 39].

A number of the fullerenes has been made and purified as a single size in macroscopic quantities [39.37, 39, 40]. This has enabled their geometries to be determined quite accurately using a wide variety of spectroscopic and theoretical techniques. Even structures of fullerenes such as C_{76}, C_{78}, and C_{84} which occur as isomeric mixtures have been elucidated. The electronic structures of many of these fullerenes have been calculated and compared with experiment [39.37, 38].

Yet to be resolved is the issue of how these low entropy fullerene structures can form so efficiently in carbon vapor. One proposal is that cup-like prefullerenes form first which add new carbon moieties to close the cage. Other suggestions include coalescence of small carbon rings such as C_6 or C_{10}, or stacking of intermediate size rings on a small "seed" ring such as C_{10}, or folding up of large defective graphitic fragments to form closed hollow spheroids [39.8, 10, 36, 37]. Some intriguing insights into this open question are provided by experiments on nonfullerene metastable forms of carbon clusters in the fullerene size regime [39.43]. These species can be made in the gas phase and are found to have bicyclic, graphitic, and other polycyclic ring shapes which are not in the form of partial fullerene cages. Upon annealing, some of these ring forms are seen to convert into the fullerenes, which suggests that they are important intermediates for forming the fullerene cage. Similar polycyclic ring shapes are observed during the initial "melting" of fullerenes in molecular dynamics simulations [39.46].

Since the fullerenes are hollow, much attention has focussed on putting something inside. A wide variety of noble gas atoms and metal atoms has been successfully loaded into fullerenes, forming a family of so-called "endohedral" complexes [39.36,37,39,47–51]. Noble gas atoms encapsulated in fullerenes appear to assume central positions within the cage and do not perturb the overall electronic structure of the fullerene [39.11, 47, 52]. Endohedral complexes where the encapsulated species is a metal atom, called *metallofullerenes*, encompass a wide range of metals (M) and fullerene sizes from $U@C_{28}$ to $Sc_3@C_{82}$ (@ denotes that the metal atom is inside the fullerene). Separation and purification of a macroscopic amount has been achieved for several of these [39.48, 49, 51]. Theoretical and experimental data indicate that the endohedral metallofullerenes have considerable charge transfer from metal atom(s) to the fullerene cage [39.37,47,48,51,52]. This interaction affects the oxidation states of the metal atom(s) and the cage and, in some cases, causes the metal atom(s) to locate off-center. It is yet to be generally understood why, within the range of known fullerene sizes, only some form endohedral complexes, particularly C_{28}, C_{60}, C_{70}, C_{74}, C_{80}, C_{82}, and C_{84}. The $M@C_{28}$ endohedral complex is especially intriguing because it appears that C_{28} itself does not form as an empty fullerene.

The unsaturated surfaces of fullerenes undergo a broad range of chemical reactions. The availability of macroscopic samples of fullerenes, especially C_{60} and C_{70}, has enabled the preparation of numerous fullerene derivatives. With their high electron affinities, fullerenes readily form a wide variety of charge transfer compounds. These include the so-called *exohedral* complexes, where metal atoms are attached to the outside of the cage. Particularly stable exohedral metal-fullerene clusters have been observed which have one alkaline earth atom decorating each ring of the cage [39.53]. The unsaturated double bonds of the fullerenes can be functionalized with reagents such as halogens, aromatics, and alcohols [39.37, 39]. Some of the bulk forms of these complexes and derivatives exhibit amazing properties such as superconductivity at temperatures as high as 33 K seen in C_{60} films doped with alkali or alkaline earth metals [39.51]. References [39.37, 39, 51] contain recent reviews.

39.2.3 Giant Carbon Clusters: Tubes, Capsules, Onions, Russian Dolls, Papier Mâché ...

Giant clusters of pure carbon have been found which have tubular, capsular, and spherical shapes [39.8, 10, 37, 51, 54]. These clusters occur both as single entities and with multiple concentric layers. The basic structure consists of a spiralled or rolled graphitic sheet made up of hexagonal rings of sp^2 hybridized carbon atoms. Just as in the fullerenes, pentagonal rings provide the curvature required to form a ball or to cap the ends of tube shapes. Negative curvature has also been seen which is believed to result from heptagonal rings. Carbon tubes and capsules are formed prolifically in the same carbon arcs which produce fullerenes. Typical diameters range from about 5 to 100 nm, depending on the inner diameter and number of layers, and lengths of up to several microns are seen. For concentric or spiralled structures, the average interlayer spacing is 3.4 to 3.5 Å which is slightly larger than in crystalline graphite. Although tubes and capsules are the most commonly seen morphologies, these are observed to convert to layered spherical "onion" structures under intense irradiation [39.8]. This suggests that collapsed onion structures are more stable. Currently, there exists some debate over whether the multilayer tubes or capsules consist of layers of complete shells within shells, such as in a "Russian doll", or whether these layers are so highly defective that the overall structures are best described as "papier-mâché" consisting of numerous overlying graphitic fragments [39.54].

Little experimental detail is available on the electronic or mechanical properties of these tubes. Theoretical calculations have been performed on the electronic properties of perfect tubes constructed in various ways [39.37, 51]. Numerous configurations of tubes are possible, depending on the tube diameter and the overall screw axis formed by the rows of carbon hexagons wound around the tube waist. The geometrical arrangement strongly affects the electronic properties of a tube. Appropriate choices of diameter and tilt angle of the screw axis yield tubes which are metallic or semiconducting. Tubes and capsules with perfect lattice arrangements are calculated to be extremely stiff, forming the strongest carbon fibers known. Almost no data exist on the chemical properties of carbon tubes, capsules, and onions. While they are stable enough to isolate in air, reaction occurs with O_2 and CO_2 at high temperatures which destroy the tubes [39.51]. Finally, as for the fullerenes, tubes, capsules, and onions have a hollow cavity which can be filled. These "nanocapsules" have been successfully loaded with lead and gold atoms, as well as crystalline metal carbide particles [39.51].

39.3 Ionic Clusters

A growing body of experimental and theoretical evidence on alkali halide and alkaline earth oxide clusters shows that these tiny clusters of ionic materials are ionically bound. Most of these clusters have highly ordered crystal structures even at very small sizes. Thus, these clusters offer excellent systems for studying electron localization on finite-sized crystalline lattices.

39.3.1 Geometric Structures

Most ionic clusters of alkali halides and alkaline earth oxides assume the cubic rock salt lattice typical of bulk sodium chloride [39.6, 7, 10, 13, 55]. This lattice is favored not only for clusters of rock salt cubic solids such as NaCl and NaF, but also for other clusters such as Cs_xI_y where the bulk form has the cesium chloride cubic crystal structure. Exceptions occur when the cluster size is smaller than, roughly, a unit cube, or when the cluster anions and cations differ greatly in ionic radius, such as in lithium bromide clusters (even though solid LiBr has the NaCl crystal structure) [39.10]. The bulk rock salt lattice is retained even for clusters where the number of anions and cations are unequal, as long as the deviation from stoichiometry is not too great. In clusters with a large excess of alkali atoms, the extra metal atoms segregate to a face of the cluster and form a metallic overlayer [39.55].

Maximization of ionic interactions and minimization of surface energy leads to cuboid crystal morphologies for these clusters which have as many (1 0 0) faces as possible. Particularly stable clusters occur for sizes where the cuboid lattice is completely filled and has nearly equal numbers of ions on all sides [39.55]. Excess electrons or holes also play a crucial role in the stability of ionic clusters. For example, while the $[Na_{14}Cl_{13}]^+$ cluster forms a particularly stable cube (with 9 atoms on each square face and one atom in the center), its counterpart $[Na_{13}Cl_{14}]^+$ does not because it has fewer holes than the available electrons; however, the anion $[Na_{13}Cl_{14}]^-$ does form a particularly stable cube [39.11, 55].

39.3.2 Electronic and Chemical Properties

The I_P's, A_e's, photoabsorption spectra and photoabsorption cross sections have all been measured for ionic clusters, particularly alkali halide $[M_jX_k]$ clusters. Understanding the role of the coulombic interactions, particularly excess electrons, in these clusters is key to understanding the trends in their electronic properties with cluster size, composition, and overall charge. Excess electrons are known to localize in at least four distinct ways: at anion vacancies, in weakly bound surface states, on specific alkali metal ions, or in cation-anion pair dipole fields [39.11, 55, 56].

First, there is a large class of ionic clusters which contain equal numbers of electrons and holes, such as clusters in the series $\left[M_{j+1}X_j\right]^+$. I_P and A_e of clusters within this class reflect the overall stability associated with forming the *perfect* filled cuboid lattices described in Sect. 39.3.1. Such perfect clusters do not chemisorb polar molecules, whereas the *imperfect* clusters readily do [39.55]. The lowest energy absorptions for clusters in this class result from charge transfer excitations, just as in perfect ionic solids. Spectra obtained for $\left[Cs_{j+1}I_j\right]^+$ clusters show large cross sections for absorption, consistent with charge transfer, and features which converge towards the bulk for the perfectly cubic cluster, $[Cs_{14}I_{13}]^+$ [39.57].

Other ionic clusters do not contain equal numbers of electrons and holes; clusters with excess electrons have been the most studied. One such class consists of clusters with at least one anionic vacancy and one or more excess electrons. In this class, e.g., $[Na_{14}Cl_{12}]^+$, the excess electron localizes at the lattice site of the missing anion. This is conceptually similar to an F-center in an ionic bulk crystal. Enhanced stabilization is observed for those size clusters within this class where the excess electron sitting at an anionic site yields a filled cuboid lattice. Nonetheless, because the localized electron has a large zero-point energy, its binding energy is much less than an anion at the same site. This is reflected in both the first and second I_P's, as well as the A_e's of this class of clusters [39.55]. Just as for F-centers in ionic crystals, clusters with an excess electron localized on an anionic vacancy have strong optical absorptions at energies well below the charge transfer bands [39.55].

The next class of ionic clusters with excess electrons consists of perfect cuboid ionic clusters which contain one or more excess electrons, but do not have a defect binding site such as an anionic vacancy or an excess metal atom. In these clusters, e.g., $Na_{14}Cl_{13}$, the electron is quite weakly bound in a surface state which primarily involves the surface metal cations. These clusters have particularly low electron binding energies; for example, the I_P of ≈ 1.9 eV determined for $Na_{14}F_{13}$ is the lowest I_P measured for any compound [39.55].

In the other two known classes of ionic clusters with excess electrons, the extra electrons have been calculated to localize on a metal cation or in a dipole potential well [39.58]. Some evidence for these forms of localization has been seen in photoabsorption, I_P's and A_e's [39.10, 11, 55]. For example, the photoelectron spectra of some $\left[Na_{j+1}Cl_j\right]^-$ clusters show that the two excess electrons are singlet-coupled, and localized either at an anion vacancy, or at a single Na site so that they behave as a Na^- anion loosely bound to a neutral $\left[Na_jCl_j\right]$ cluster. The spectral behavior of other clusters within this same series, however, suggests they have the two excess electrons in a triplet coupled state, forming the analog of a bipolaron in a solid [39.10].

39.4 Semiconductor Clusters

Semiconductor clusters make up a class of clusters where, by analogy with bulk semiconductors, covalent forces are expected to dominate electronic and geometric structure. Silicon clusters are by far the most studied of the semiconductor clusters. Some information is available on germanium clusters and compound clusters made of Groups III and V atoms or Groups II and IV atoms. Also, there are the well known stable molecules of bulk semiconductors such as the P_4 tetrahedron and various sulfur rings; these have been reviewed in detail elsewhere [39.3]. Clusters of other possible semiconductors have been made, but little data beyond their nascent distributions are available [39.3]. There is also a growing body of data on silicon, III–V, and, especially II–VI semiconductor clusters in the nanometer size regime where bulk samples of stabilized forms of these clusters have been made and isolated. The crystalline and electronic structure properties, particularly quantum confinement effects, are reviewed in [39.10, 11, 59–61].

39.4.1 Silicon and Germanium Clusters

The geometries of small silicon clusters depart radically from microcrystalline fragments of the bulk silicon diamond lattice. The structures of silicon clusters containing up to about 13 atoms have been studied extensively both experimentally and theoretically [39.62]. These structures are more compact and, starting at Si_7 which has a pentagonal bipyramidal structure, have higher coordination than silicon in the bulk lattice. Multiple isomers of similar energy also appear starting at about Si_{10}. For example, the tetracapped trigonal prism and the symmetric tetracapped octahedron structures of

Si_{10} are nearly isoenergetic; microcrystalline fragments such as the adamantane form of Si_{10} are much higher in energy. The geometric structures of larger silicon clusters are less well known. The gross shapes of clusters containing up to about 60 silicon atoms have been found experimentally to undergo a transition between Si_{20} and Si_{30} from increasingly elongated structures to more spherical structures [39.26]. In the transition region, both prolate and oblate isomers are observed for a single cluster size. Multiple isomers over a wide range of silicon cluster sizes have been repeatedly observed in various experiments; it appears that somewhat different sets of isomers can be produced depending on the cluster formation conditions [39.26]. Elucidating the ground state structures of these larger clusters theoretically is, in general, an intractable problem and requires simplifying approaches. Use of semiempirical quantum mechanical techniques or silicon interaction potentials derived from bulk silicon, however, has led to unsatisfactory results, which suggests that the silicon atoms in silicon clusters are strongly reconstructed away from the usual sp^3 silicon atom environment [39.63, 64]. Some consensus has recently emerged that larger silicon clusters consist of internal silicon atoms strongly interconnected to a surrounding cage that has been described as a buckled fullerene [39.63, 65, 66]. Such a silicon cage must not be construed as a true fullerene, however, since silicon does not form the strong double bonds which stabilize the interconnected carbon rings of the fullerenes. Within this scenario, the shape change is believed to occur at the point where the cage begins to contain one or more internal atoms which provide the additional bonding needed to stabilize a spherical geometry.

I_P's and A_e's, cohesive energies, and photoabsorption spectra have been measured for silicon clusters containing up to several hundred atoms. Silicon cluster I_P's start near the Si atom I_P, fall abruptly between 20 and 22 silicon atoms where the shape change has been observed, and then slowly converge towards the bulk work function. This convergence is apparently quite gradual since little change is seen between Si_{100} and Si_{200} [39.67]. A_e's are only known for silicon clusters up to $\approx$15 atoms and reflect the large structural changes which occur in this small size regime [39.3]. Indication of a structural shape change is not observed in either the trends in the cohesive energies or the electronic spectra for silicon clusters. The cohesive energies increase smoothly with increasing size and exceed the cohesive energy of bulk silicon by 10–20% [39.68]. Silicon clusters exhibit strong sharp absorption spectra in the near UV. A common set of absorption features appears at $\approx Si_{15}$ and persists up to at least Si_{70} [39.69]. This suggests that the strong absorptions arise from localized Si–Si bond excitations, and that these obscure the delocalized excitations which are more sensitive to silicon cluster structure. The most unusual aspect of these spectra is that the common signature for these minute clusters which are strongly reconstructed from the bulk is, nonetheless, strikingly similar to the spectrum of bulk diamond-lattice crystalline silicon.

Silicon cluster ions are observed to react with a variety of reagents [39.3, 26, 68]. For example, silicon clusters chemisorb small organic molecules such as ethylene, and inorganic molecules such as O_2, NH_3, and XeF_2. Some of these reactions are sufficiently exothermic that they cause cluster fragmentation, or loss of small neutral fragments. The chemistry of silicon clusters often bears a close relationship to that known for bulk silicon surfaces; however, the clusters are often much less reactive by as much as several orders of magnitude. Silicon cluster reactions reveal the presence of numerous isomers which differ in their reaction rate for a given reagent. Isomerization has been induced by adding sufficient thermal energy to some clusters; the resulting more stable cluster form may or may not be more reactive than the higher energy isomer. This is particularly well illustrated in reactions of oblate versus prolate isomers where the more reactive isomer varies depending on cluster size [39.26]. In general, differential chemistry is not a useful predictor of cluster geometry *per se*, and appears to correlate more readily with other factors such as the number and type of dangling bonds available for reaction.

Much less is known about germanium clusters. The most stable structures that have been calculated for germanium clusters are quite similar to those for silicon clusters, although the overall binding energies are lower for germanium clusters [39.3, 7]. Available data such as A_e's and photoelectron spectra for germanium clusters also underscore their similarities to silicon clusters in general [39.3]. The chemistry of germanium clusters has not been reported.

39.4.2 Group III–V and Group II–VI Semiconductor Clusters

Geometric structures have been calculated for some of the smaller III–V and II–VI clusters, particularly aluminum phosphide, gallium arsenide, and magnesium sulfide clusters [39.3, 11, 70–72]. As for silicon clusters, these clusters differ significantly from microcrystalline fragments snipped from their corresponding bulk crystalline forms. Electronegativity differences between the two constituent atoms play a major role in determining and stabilizing these structures. This is manifested in several ways. The bonding arrangements in the most stable structures have, in general, alternating electropositive and electronegative atoms. In those clusters such as III–V clusters where electronegativity differences are smaller, covalent interactions predominate ionic interactions and the energetics of the various geometric structures are similar to those of the covalent silicon clusters. In II–VI clusters, electronegativity differences are much larger, and these clusters have structures more comparable to those seen in ionic clusters where ionic interactions are maximized.

Some data are available on the electronic structures of III–V clusters. A strong even-odd alternation occurs in both the I_P's and A_e's of gallium arsenide clusters where those clusters having a total even number of atoms have higher I_P's and lower A_e's than neighboring odd-numbered clusters [39.3,73]. This has been explained by electronic structure calculations which find that, in general, the odd clusters are triplets while the even clusters of gallium arsenide are singlets [39.70–72]. Electronic absorption spectra have also been recorded for indium phosphide clusters which exhibit strong differences with cluster size and stoichiometry [39.74]. These spectra are also consistent with odd-numbered In_xP_y clusters having open-shell configurations, and even-numbered In_xP_y clusters having closed-shell singlet ground states, even for clusters which are considerably off-stoichiometry. A strikingly similar strong continuum-like absorption appears in the blue end of spectra for the even-numbered In_xP_y clusters. The onset of these bsorptions lies close to the bulk indium phosphide band gap. The overall spectral behavior, however, is quantitatively similar to the absorptions seen in semiconductor glasses.

Little is known about the chemical properties of these clusters, with gallium arsenide clusters being the only ones studied. Hydrogen chloride is observed ubiquitously to etch $[Ga_xAs_y]^-$; however, multiple isomers are seen with differing degrees of reactivity [39.3]. Chemisorption of ammonia on $[Ga_xAs_y]^+$ has also been seen with the highest rates of reactivity occurring for the stoichiometric ($x = y$) clusters [39.75].

39.5 Noble Gas Clusters

Clusters have been made of all of the stable noble gases including helium. As a general class, these clusters are the most weakly bound of all clusters, and are held together only by van der Waals forces. These interactions are well understood theoretically, and thus noble gas clusters are excellent model systems for studying a variety of structural and electronic effects at finite sizes. Helium clusters behave much as quantum liquids, and are treated separately in the discussion below.

39.5.1 Geometric Structures

The overall evolution of noble gas cluster structures as a function of size is well established [39.4,6,11,76,77]. Icosahedral packing dominates at small cluster sizes. This packing involves polyicosahedral structures for the small cluster sizes, and different noble gas clusters exhibit somewhat different preferred arrangements. Starting at 100 to 250 atoms (depending on which noble gas), formation of closed-shell Mackay icosahedra dominates the structures [39.78]. Finally, at somewhere in the range 600 to 6000 atoms, the structures cross over to the close-packed fcc arrangement of the noble gas solids.

The increased coordination characteristic of the Mackay icosahedra explains why this morphology is adopted at small cluster sizes. With the addition of each successive icosahedral shell however, this five-fold symmetric packing becomes increasingly strained as a result of both atom-atom radial compression and tangential dilation. At some size, icosahedral structures are no longer more stable than fcc structures, and the noble gas clusters undergo a phase change. Currently, the critical size required for fcc packing is a matter of some dispute. Experimental evidence for this transformation at ≈ 600 atoms comes from electron diffraction data [39.76]. The spectral signatures of molecules doped in noble gas clusters suggest however, that this transition occurs at ≈ 2000 atoms [39.77]. Theoretical studies find a range of critical sizes, depending on the treatment used, but typically favor an even larger size regime [39.4,11,13,77].

Knowledge of accurate pair-wise interaction potentials for noble gases has enabled extensive simulations of the physical properties of noble gas clusters [39.4,6,7,79–81]. Phenomena such as specific heat, bond length and coordination number fluctuations, phase equilibria, dynamical freezing and melting, isomerization, and solvation have been explored using various molecular dynamics simulations. Numerous effects peculiar to the finite sizes and proliferation of isomers in small noble gas clusters are observed. For example, the melting temperature decreases significantly with an overall decrease in cluster size. Internal diffusion rates depend strongly on cluster size and the stability of certain favored structures such as the complete Mackay icosahedra. A well-studied finite size effect is the dependence of the mean energy per particle in a cluster versus the particle temperature (the so-called *caloric curve*) in the region where the particle is observed to "melt." Unlike in the bulk, calculations find hysteresis in the melting transition of small noble gas clusters where solid-like and liquid-like forms are observed to coexist. Such coexistence is quite size-dependent. It becomes more pronounced when the clusters are rapidly heated or cooled, and appears to be a finite time averaging effect [39.4, 6, 7].

39.5.2 Electronic Properties

The fragility of noble gas clusters has hampered measurements of their I_P's and A_e's. Extensive fragmentation is known to accompany ionization, making it difficult to establish the size of the ionized parent cluster. Nonetheless, I_P's have been measured for cluster sizes containing up to several tens of atoms [39.82]. Measured I_P's agree with theoretical predictions which are based on the assumption that the ionizing state within the cluster is a dimer, trimer, or higher n-mer cationic core [39.6, 82]. Absorption profiles of noble gas cations provide supporting evidence for the delocalization of the positive charge over a n-mer unit within the cluster, although consensus has yet to be reached on the size of this cationic cluster within the cluster [39.6, 11].

Several distinct types of electronic excitations are observed in absorption spectra of neutral noble gas clusters [39.6, 7, 11, 83]. Clusters containing less than ≈ 30 atoms exhibit broad absorptions near the atomic resonance lines. These absorptions are molecular-like, but have not been described in detail above the dimer. In addition, broad continuum absorptions assigned to Rydberg excitations are observed at these small sizes. Bulk-like excitations corresponding to surface and bulk excitons in solid noble gases emerge for clusters larger than ≈ 50 atoms. The profiles of these excitonic absorptions indicate that they arise from delocalized excitations analogous to Wannier excitons in bulk solids. These cluster exciton absorptions are blue-shifted relative to the bulk as a result of quantum confinement of the exciton within the small cluster. Another type of excitonic excitation has also been reported which appears to be a localized excitonic state whose character is sensitive to the cluster structure. Relaxation of cluster excited states is accompanied by extensive fragmentation [39.7]. Details of the relaxation processes differ according to noble gas, cluster size and cluster structure. Again, these differences can be traced to finite size effects in these noble gas clusters [39.4].

39.5.3 Doped Noble Gas Clusters

Various chromophores have been added to noble gas clusters as a microscopic variation of matrix isolation [39.3, 4, 7, 10, 11, 77, 84]. A wide range of chromophores have been studied including other noble gas atoms, metal atoms, small polyatomic molecules such as SF_6, CH_3F, and HCl, and a variety of organic molecules such as benzene, carbazole, and naphthalene. Comparisons of the electronic and vibrational spectra of these guest molecules with their spectra in noble gas liquids or solid matrices have revealed information such as noble gas cluster structure, solvation effects, and solute diffusion.

A crucial issue in understanding the spectral signatures of doped noble gas clusters is the location of the solute in or on the noble gas cluster. Chromophores well-embedded in a noble gas cluster behave as molecules surrounded by a dielectric medium, which is most correctly viewed as both imperfect and finite sized [39.4, 7, 84, 85]. The evolution of the spectral changes with cluster size reflect the interplay of repulsive forces and collective dielectric effects as the cluster builds the solvent shell around the solute core. Both theoretical and experimental data illustrate this evolution and show how, once the first few solvent shells are established, the spectrum converges asymptotically to the matrix isolation value. For smaller clusters, the effects of too few or incomplete solvation shells appear as shifts and/or broadening of the spectral lines relative to the bulk [39.3, 4]. Differential solvation of the ground and excited states causes spectral shifts which are typically to lower energy since the excited state is usually more stabilized than is the ground state. Spectral broadening results from multiple isomers. Broadened electronic spectra have also been explained as a signature of cluster melting, although this interpretation has been questioned [39.3, 4, 84, 85]. Some infrared spectra of solute molecules have also yielded information on the nature of the solute binding site, revealing that the solvating noble gas cluster undergoes an icosahedral to fcc phase transition in a critical size regime [39.77].

Not all dopant species are found well-solvated, or *wetted* in the cluster interior. Both electronic and infrared dopant spectra indicate the existence of chromophores bound to the cluster surface. In some cases, the chromophore is ubiquitously *nonwetting*, such as for SF_6–Xe_n and SiF_4–Ar_n, and resides exclusively on the cluster surface [39.77]. In other systems, e.g., carbazole–Ar_n and CF_3Cl–Ar_n, both wetting and nonwetting are observed [39.7,77,84]. Systems such as these have enabled measurements of solute diffusion rates into or out of the cluster. Finally, wetting–nonwetting transitions have been observed which depend on cluster size and/or cluster temperature [39.7, 84]. Understanding this wide variety of behavior requires modeling cluster solute-solvent structures (in both ground and excited states in order to interpret electronic spectra) as a function of size and temperature. To date, most of the theoretical simulations have focussed on smaller clusters at a given temperature [39.4, 7, 77, 84, 85]. Nonetheless, theoretical models have provided qualitative and quantitative explanations for some of the observed spectral shifts, and have shown that the propensity for wetting/nonwetting can be related to the degree to which maximum possible coordination within the solvent is attainable.

39.5.4 Helium Clusters

Quantum effects play a dominant role in helium clusters since they consist of very weakly interacting small mass particles. Statistical effects are also expected since ^3He is a fermion and ^4He is a boson. Overall, helium clusters are believed to be in fluid-like or superfluid-like states with highly fluctional structures [39.86].

Neutral helium clusters have been made experimentally in sizes ranging from the dimer up to 10^6 atoms [39.4, 11, 77]. The diatomic He_2, has just been detected recently and is found to have a binding energy of only $\approx 10^{-7}$ eV with an average internuclear distance of ≈ 55 Å [39.87,88]. Evidence for magic numbers in the helium cluster abundances is seen; however, the sizes of these especially stable clusters differ substantially from what is seen in the other noble gas clusters [39.4]. Helium cluster structures have been theoretically investigated using fully quantum mechanical treatments [39.4, 86]. These studies find that ^4He_n clusters should be bound at all sizes, but that a minimum number of atoms, ≈ 30, are required for ^3He_n clusters to be stable. The latter prediction has not yet been verified experimentally, owing to the expense of ^3He, and the difficulties of ascertaining the true size of helium clusters since they easily boil off atoms upon ionization. Calculations reveal that the packing of helium clusters is highly delocalized with no evidence of icosahedral morphology. The structures derived have been found to be extremely sensitive to temperature and total angular momentum. Cluster binding energies and densities increase smoothly and monotonically with increasing size, approaching bulk behavior at about 300 atoms. Thus, helium clusters do not exhibit any especially stable sizes (magic numbers) such as the heavier noble gases do. In short, helium clusters behave as liquid-like quantum fluids, and may even be superfluid-like at the temperatures required to stabilize them in beam experiments. Based on these calculations, it appears that the magic numbers observed experimentally in abundance spectra pertain to the more strongly bound ionized helium clusters, rather than the neutral clusters which are the subject of this discussion.

Electronic absorption spectra have recently been recorded for clusters containing ≈ 50–10^6 helium atoms [39.89]. Broad strong absorption bands are observed which do not behave like the Wannier exciton bands seen for the heavier noble gas clusters, nor are they well-described by the Frenkel excitonic model. Note that although the Wannier and Frenkel exciton models were originally developed for solids with translational symmetry, they are also good descriptions for excitations in liquid noble gases [39.4]. At this writing, the helium cluster electronic spectra cannot be compared with absorption spectra of liquid or solid helium because these latter spectra have not yet been measured! With this limitation and a current lack of sufficient theoretical guidance, the helium cluster spectra have not yet been thoroughly interpreted. Theoretical calculations are available which describe the collective excitations in helium clusters on the electronic ground state surface [39.86]. Spectra in this energy regime have not yet been recorded.

Experimentally, it has been quite easy to dope helium clusters with various atomic and molecular species such as other noble gas atoms, oxygen, and SF_6 [39.4, 11, 77]. Electrons, however, are not well-solvated and negatively charged helium clusters only appear for sizes containing more than 10^5–10^6 atoms [39.11]. Impurity species provide a spectroscopic probe for studying the properties of the solvating helium cluster. Infrared spectra of helium clusters doped with SF_6 show that the impurity molecule resides on the cluster surface, in contradiction of theoretical calculations which predict that it should be found inside a helium cluster. Understanding this discrepancy has stimulated further theoretical investigations that reveal the dramatic structural effects which can occur when angular momentum is added to these very fragile clusters during dopant pick-up [39.90].

39.6 Molecular Clusters

Molecular clusters are in many ways similar to noble gas clusters. Both are weakly bound, and the interactions between the particles which make up the cluster can usually be described to a good approximation as the sum of pairwise interactions. The molecule constituency adds complexity, however, which is reflected in the diversity seen amongst these clusters. Molecular clusters provide model systems for studying solvation, including electron solvation, nucleation, and phase transitions at finite sizes. Furthermore, properties which are often difficult to study in the bulk, such as phase transition dynamics, are much more amenable to study in molecular clusters.

Numerous molecular clusters have been made and studied. By far the most work has focussed on the smaller molecular clusters such as molecular dimers, trimers and tetramers. A number of outstanding reviews are available on this subset of molecular clusters [39.3,4,6,76,91,92].

39.6.1 Geometric Structures and Phase Dynamics

In contrast to noble gas clusters, molecular clusters typically assume bulk phase structures at relatively small sizes. Thus, the range of condensed phases of molecular solids is reflected in the variety of packing structures found in molecular clusters. At one extreme are clusters of small nearly-spherical molecules such CH_4 and N_2 which form clusters closely connected to noble gas clusters [39.76]. Small clusters of these molecules are packed in polyicosahedral arrangements at small sizes, and cross over to fcc structures at sizes containing several thousand atoms. At another extreme are clusters such as water clusters where stronger hydrogen bonding forces lead to the formation of crystalline networks such as in a diamond cubic phase [39.76]. Some molecular clusters appear to be liquid-like with thermal motion wiping out any persistent periodic order. The liquid-like structure observed in clusters of molecules such as benzene and various hydrocarbons may freeze into a crystalline or amorphous state at sufficiently low temperatures.

Some molecular clusters have been found to exhibit a panoply of phases, depending on size and temperature. For example, not only do clusters of TeF_6 assemble into the bulk body-centered cubic and orthorhombic phases, they can also be made in other phases including two trigonal forms, a rhombohedral phase, and two monoclinic phases [39.93, 94]. Similarly complex phase formation has been found in other clusters of other hexafluoride molecules such as SF_6, SeF_6, MoF_6, and WF_6.

Phase transitions in molecular clusters have been examined both experimentally and theoretically using molecular dynamics. Transition temperatures are found to depend on cluster size, and span a much broader range than in the corresponding bulk phase transitions. Large rates for nucleation and phase changes are seen in molecular clusters. Once the critical nucleus, which initiates the phase change, forms in the interior of the cluster, the remainder of the cluster transforms to the new phase extremely rapidly. Critical nucleus sizes depend on the cluster size and are typically quite small, e.g., for TeF_6 clusters, they contain only a few dozen molecules for clusters consisting of a few hundred molecules [39.10, 93, 94]. A key finding is that phase transitions in molecular clusters appear to violate *Ostwald's step rule*, which states that equilibrium systems must pass through all intermediate-energy stable phases during a transition from a higher energy to a lower energy phase [39.10]. This noncompliance apparently occurs because of the speed at which molecular clusters undergo phase transitions: intermediate phases simply do not have time to form.

39.6.2 Electronic Properties: Charge Solvation

I_P's, A_e's, and some spectral data are available for larger molecular clusters. Nearly all of the spectra measured pertain to clusters containing less than ten molecules [39.3, 4, 6, 76, 91, 92]. Precise molecular orientations within many of these clusters have been derived from detailed analyses of these spectra. Such information has not yet emerged from the few spectra that have been measured for larger molecular clusters [39.3, 6]. Alternatively, measurements of electronic properties of larger clusters have generally targeted a different issue: solvation of excess positive or negative charge in a restricted bulk-like system.

Charged molecular clusters provide model systems for studying the mechanism of charge stabilization within the confines of a finite system. Excess positive charge appears to be highly localized in molecular clusters, residing on a small unit containing at most a few monomers. This positive core is surrounded and stabilized by overlying shells of molecules. For example, $(CO_2)_n^+$ clusters behave as the dimer cation $(CO_2)_2^+$, surrounded by the remaining CO_2 molecules [39.4]. Positively charged solutes such as alkali cations have also been introduced to study solvation of positive charge within molecular clusters [39.10]. For those dopants which have much lower ionization potentials than the solvating molecules, the positive charge remains strongly localized on the impurity. The cluster molecules are observed to build up solvation shells around this central impurity. The first few solvation shells are strongly affected by the positive charge, which has a decreasing influence with each successive layer. The structure of such a cluster reflects the accommodation between the geometry of the molecules influenced by the central positive charge, and those in the outer solvation layers where intermolecular forces predominate [39.10, 95]. Evidence for intracluster reactions in positively charged clusters has also been seen, such as proton transfer in cationic water or ammonia clusters [39.3, 11].

Whereas virtually every molecular cluster containing as little as two molecules exists stably as a positive ion, the same cannot be said for the negative cluster ions. Measurements of the minimum number of molecules required within the cluster to support an additional electron show that, while the dimeric $(HCl)_2^-$, $(SO_2)_2^-$, and $(H_2O)_2^-$ clusters are stable, ≈ 35 or ≈ 41 molecule-clusters of ND_3 or NH_3, respectively, are required to stabilize an electron [39.4]. Spectroscopic studies show evidence for internal and external solvation of the excess electron in these anionic clusters.

Charge solvation in water clusters has been examined in some detail. Of particular interest is the $(H_2O)_{20}$ cluster which, from experimental and theoretical data, appears to form an especially stable, well-defined clathrate cage. The interior of this cage is large enough that it can and does hold various cations such as NH_4^+, H_3O^+, and alkali ions [39.10]. This cage is not observed to surround negative ions or electrons, which instead, reside on the surface [39.10]. This is a general result for excess negative charge in water clusters in this size regime where excess charge can be external to the cluster. Excess electrons in water clusters are found on the cluster surface for small cluster sizes up to 60 to 70 water molecules. Above this size, the electron resides in the cluster interior, and behaves analogously to hydrated electrons in bulk water [39.4, 7].

39.7 Recent Developments

Added by Mark M. Cassar. Experimental and theoretical work on clusters has continued to be an active and rapidly growing area of research over the past decade. This Section provides a non-exhaustive snapshot of some recent work in the vast field of cluster science.

The original focus on the scalable properties of clusters (concerning a smooth transition from small particles to bulk matter) has now extended to include important non-scalable properties. These properties, particularly at the nanoscale level, have enormous potential for technological application [39.96–98]. Studies aimed at understanding the underlying atomic structure of noble-metal clusters and nanoparticles, which is the first step toward their controlled use in future nanotechnologies, e.g., catalysis, labeling, or photonics, have been carried out [39.99]. Ab initio all-electron molecular-orbital calculations for small ($n = 7-11$) and medium ($n = 12-20$) silicon clusters Si_n have been performed in order to study their structure and relative stability [39.100, 101]; such calculations are important in determining the scalability of present day semiconductor technology.

Other interesting work has been done on the electron transfer properties of metal clusters that could act as conducting bridges between molecular wires [39.102]; and in the role that tetramanganese clusters play in one of the active photosynthesis sites (photosystem II) in green plants, and in certain bacteria and algae [39.103].

The reader is referred to various reviews that can be found in the literature: for time-resolved photoelectron spectroscopy (TRPES) of clusters, which allows the dynamics along the entire reaction coordinate to be followed, see [39.104]; for ultrafast dynamics in cluster systems and atomic clusters, see [39.105, 106]; for small carbon clusters, important in the chemistry of carbon stars, comets, interstellar molecular clouds, and hydrocarbon flames, see [39.107]; for the relation between electronic structure, atomic structure and magnetism of clusters of transition elements, see [39.97].

References

39.1 Special issue on gas phase clusters, Chem. Rev. **86**, 375 (1986)

39.2 W. A. deHeer, W. D. Knight, M. Y. Chou, M. L. Cohen: Electronic shell structure and metal clusters. In: *Solid State Physics*, Vol. 40, ed. by H. Ehrenreich, D. Turnbull (Academic, New York 1987)

39.3 E. R. Bernstein (Ed.): *Atomic and Molecular Clusters* (Elsevier, New York 1990)

39.4 G. Scoles (Ed.): *The Chemical Physics of Atomic and Molecular Clusters* (North-Holland, New York 1990)

39.5 V. Bonacic-Koutecky, P. Fantucci, J. Koutecky: Special issue on gas phase clusters, Chem. Rev. **91**, 1035 (1991)

39.6 H. Haberland (Ed.): *Clusters of Atoms and Molecules* (Springer, Berlin 1994)

39.7 Proceedings of the 5th international meeting on small particles and inorganic clusters, Z. Phys. D. **19–20** (1991)

39.8 V. Kumar, T. P. Martin, E. Tosatti (Eds.): *Clusters and Fullerenes* (World Scientific, River Edge 1993)

39.9 M. D. Morse: Chem. Rev. **86**, 1049 (1986)

39.10 Proceedings of the 6th international meeting on small particles and inorganic clusters, Z. Phys. D. **26–26S** (1993)

39.11 P. Jena, S. N. Khanna, B. K. Rao (Eds.): *Physics and Chemistry of Finite Systems: From Clusters to Crystals*, Vol. 1 & 2 (Kluwer, Netherlands 1992)

39.12 C. G. Granqvist, R. A. Buhrman: J. Appl. Phys. **47**, 2200 (1976)

39.13 Proceedings of the 4th international meeting on small particles and inorganic clusters, Z. Phys. D. **12** (1989)

39.14 M. A. Duncan (Ed.): *Advances in Metal and Semiconductor Clusters*, Vol. 2 (JAI Press, Greenwith 1994)

39.15 S. Bjornholm, J. Borggreen, O. Echt, K. Hansen, J. Pedersen, H. D. Rasmussen: Phys. Rev. Lett. **65**, 1627 (1990)

39.16 T.P. Martin, T. Bergmann, H. Gohlich, T. Lange: Chem. Phys. Lett. **172**, 209 (1990)

39.17 O. Cheshnovsky, K.J. Taylor, J. Conceicao, R.E. Smalley: Phys. Rev. Lett. **64**, 1785 (1990)

39.18 K. Rademann: Ber. Bunsenges. Phys. Chem. **93**, 653 (1989)

39.19 M. B. Knickelbein, W. J. C. Menezes: Phys. Rev. Lett. **69**, 1046 (1992)

39.20 T. Baumert, C. Rottgermann, C. Rothenfusser, R. Thalweiser, V. Weiss, G. Gerber: Phys. Rev. Lett. **69**, 1512 (1992)

39.21 C. Bréchignac, M. Broyer, Ph. Cahuzac, G. Delacretaz, P. Labastie, J. P. Wolf, L. Wöste: Phys. Rev. Lett. **60**, 275 (1988)

39.22 J. P. Bucher, D. C. Douglass, L. A. Bloomfield: Phys. Rev. Lett. **66**, 3052 (1991)

39.23 I. M. L. Billas, J. A. Becker, A. Chatelain, W. A. deHeer: Phys. Rev. Lett. **71**, 4067 (1993)

39.24 S. N. Khanna, S. Linderoth: Phys. Rev. Lett. **67**, 742 (1991)

39.25 D. M. Cox, K. C. Reichmann, D. J. Trevor, A. Kaldor: J. Chem. Phys. **88**, 111 (1988) and references therein

39.26 D. C. Parent, S. L. Anderson: Chem. Rev. **92**, 1541 (1992)

39.27 E. K. Parks, T. D. Klots, B. J. Winter, S. J. Riley: J. Chem. Phys. **99**, 5831 (1993)

39.28 E. K. Parks, S. J. Riley: J. Chem. Phys. **99**, 5898 (1993)

39.29 T. Schmid: Chem. Rev. **922**, 1709 (1992)

39.30 L. J. de Jongh, H. B. Brom, J. M. van Ruitenbeek, R. C. Thiel, G. Schmid, G. Longoni, A. Ceriotti, R. E. Benfield, R. Zanoni: Physical and chemical properties of high nuclearity metal-cluster compounds: Model systems for small metal particles. In: *Cluster Models for Surface and Bulk Phenomena*, ed. by G. Pacchioni, P. S. Bragus, F. Parmigiani (Plenum, New York 1992)

39.31 B. C. Buo, K. P. Kerns, A. W. Castleman: Science **255**, 1411 (1992)

39.32 H. Chen, M. Feyereisen, X.P. Long, G. Fitzgerald: Phys. Rev. Lett. **71**, 1732 (1993)

39.33 B. V. Reddy, S. N. Khanna: Chem. Phys. Lett. **209**, 104 (1993)

39.34 H. W. Kroto, J. R. Heath, S. C. O'Brien, R. F. Curl, R. E. Smalley: Nature **318**, 162 (1985)

39.35 W. R. Creasy: Fullerene Sci. Tech. **1**, 23 (1993)

39.36 H. W. Kroto, A. W. Allaf, S. P. Balm: Chem. Rev. **91**, 1213 (1991)

39.37 W.E. Billups, M. A. Ciufolini (Eds.): *Buckminsterfullerenes* (VCH, New York 1993)

39.38 J. Cioslowski: *Electronic Structure Calculations on Fullerenes and Their Derivatives (Topics in Physical Chemistry)* (Oxford University Press, Oxford 1995)

39.39 Special issue on Buckminsterfullerenes, Acc. Chem. Res. **25** (1992)

39.40 W. Kratschmer, L. D. Lamb, K. Fostiropoulos, D. R. Huffman: Nature **347**, 354 (1990)

39.41 A.F. Hebard, M. J. Rosseinsky, R. C. Haddon, D. W. Murphy, S. H. Glarum, T. T. M. Palstra, A. P. Ramirez, A. R. Kortan: Nature **350**, 600 (1991)

39.42 J. R. Heath, R. J. Saykally: In: *On Clusters and Clustering*, ed. by P. J. Reynolds (North-Holland, Amsterdam 1993) p. 7

39.43 G. von Helden, M.-T. Hsu, N. Gotts, M. T. Bowers: J. Phys. Chem. **97**, 8182 (1993)

39.44 W. Weltner, R. J. Van Zee: Chem. Rev. **89**, 1713 (1989)
39.45 Special issue on fullerenes, carbon, and metal-carbon clusters, Int. J. Mass Spectrom. Ion Processes **138** (1994)
39.46 S. G. Kim, D. Tomanek: Phys. Rev. Lett. **72**, 2418 (1994)
39.47 D. S. Bethune, R. D. Johnson, R. J. Salem, M. S. de Vries, C. S. Yannoni: Nature **366**, 123 (1993)
39.48 X.-D. Wang, T. Hazhizume, Q. Xue, H. Shinohara, Y. Saito, Y. Nishina, T. Sakurai: Japan. J. Appl. Phys. **32**, L147 (1993)
39.49 S. Hino, H. Takahashi, K. Iwasaki, K. Matsumoto, T. Miyazaki, S.Hasegawa, K. Kikuchi, Y. Achiba: Phys. Rev. Lett. **71**, 4261 (1993)
39.50 M. Saunders, H. A. Jimenez-Vazquez, R.J. Cross, S. Mroczkowski, M. L. Gross, D. E. Giblin, R. J. Poreda: J. Am. Chem. Soc. **116**, 2193 (1994)
39.51 H. Ehrenreich, F. Spaepen (Eds.): *Fullerenes in Solid State Physics*, Vol. 48 (Academic, New York 1994)
39.52 J. Cioslowski: Ab initio electronic structure calculations on endohedral complexes of the C_{60} cluster. In: *Spectroscopic and Computational Studies of Supramolecular Systems*, Vol. 40, ed. by J. E. D. Davies (Kluwer, Netherlands 1992)
39.53 U. Zimmermann, N. Malinowski, U. Naher, S. Frank, T. P. Martin: Phys. Rev. Lett. **72**, 3542 (1994)
39.54 O. Zhou, R. M. Fleming, D. W. Murphy, C. H. Chen, R. C. Haddon, A. P.Ramirez, S. H. Glarum: Science **263**, 1744 (1994)
39.55 R. L. Whetten: Acc. Chem. Res. **26**, 49 (1993)
39.56 R. N. Barnett, U. Landman, D. Scharf, J. Jortner: Acc. Chem. Res. **22**, 350 (1989)
39.57 X. Li, R. L. Whetten: J. Chem. Phys. **98**, 6170 (1993)
39.58 G. Rajagopal, R. N. Barnett, U. Landman: Phys. Rev. Lett. **67**, 727 (1991)
39.59 M. L. Steigerwald, L. E. Brus: Acc. Chem. Res. **23**, 183 (1990)
39.60 Y. Wang: Acc. Chem. Res. **24**, 133 (1991)
39.61 L. Brus: Adv. Mater. **5**, 286 (1993)
39.62 K. Raghavachari, L. A. Curtiss: Accurate theoretical Studies of small elemental clusters. In: *Quantum Mechanical Electronic Structure Calculations With Chemical Accuracy*, ed. by R. S. Langhoff (Kluwer, Netherlands 1994)
39.63 K. Raghavachari: Phase Transitions **24–26**, 61 (1990)
39.64 N. Bingelli, J. L. Martins, J. R. Chelikowsky: Phys. Rev. Lett. **68**, 2956 (1992)
39.65 E. Kaxiras, K. Jackson: Phys. Rev. Lett. **71**, 727 (1993)
39.66 U. Röthlisberger, W. Andreoni, M. Parrinello: Phys. Rev. Lett. **72**, 665 (1994)
39.67 K. Fuke, K. Tsukamoto, F. Misaizu, M. Sanekata: J. Chem. Phys. **99**, 7807 (1993)
39.68 M. F. Jarrold: Science **252**, 1085 (1991)
39.69 K.-D. Rinnen, M.L. Mandich: Phys. Rev. Lett. **69**, 1823 (1992)
39.70 M. A. Al-Laham, K. Raghavachari: J. Chem. Phys. **98**, 8770 (1993)
39.71 L. Lou, P. Nordlander, R. E. Smalley: J. Chem. Phys. **97**, 1858 (1992)
39.72 R. M. Graves, G. E. Scuseria: J. Chem. Phys. **95**, 6602 (1991)
39.73 C. Jin, K. J. Taylor, J. Conceicao, R. E. Smalley: Chem. Phys. Lett. **175**, 17 (1990)
39.74 K.-D. Rinnen, K.D. Kolenbrander, A. M. DeSantolo, M. L. Mandich: J. Chem. Phys. **96**, 4088 (1992)
39.75 L. Wang, L. P. F. Chibante, F. K. Tittel, R. F. Curl, R. E. Smalley: Chem. Phys. Lett. **172**, 335 (1990)
39.76 Special issue on gas phase clusters, Chem. Rev. **86** (1986)
39.77 S. Goyal, D. L. Schutt, G. Scoles: Acc. Chem. Res. **26**, 123 (1993)
39.78 W. Miehle, O. Kandler, T. Leisner: J. Chem. Phys. **91**, 5940 (1989)
39.79 D. J. Wales: Mol. Phys. **78**, 151 (1993)
39.80 R.S. Berry: Chem. Rev. **93**, 2379 (1993)
39.81 H. Matsuoka, T. Hirokawa, M. Matsui, M. Doyama: Phys. Rev. Lett. **69**, 297 (1992)
39.82 W. Kamke, J. de Vries, J. Krauss, E. Kaiser, B. Kamke, I. V. Hertel: Z. Phys. D **14**, 339 (1989)
39.83 J. Wormer, M. Joppien, G. Zimmerer, T. Moller: Phys. Rev. Lett. **67**, 2053 (1991)
39.84 S. Leutwyler, J. Bosiger: Chem. Rev. **90**, 189 (1990)
39.85 J. E. Adams, R. M. Stratt: J. Chem. Phys. **99**, 789 (1993)
39.86 K.B. Whaley: Int. Rev. Phys. Chem. **13**, 41 (1994)
39.87 F. Luo, G. C. McBane, G. Kim, C. F. Giese, W. R. Gentry: J. Chem. Phys. **98**, 3564 (1993)
39.88 J.A. Anderson, C.A. Traynor, B. M. Boghosian: J. Chem. Phys. **99**, 345 (1993)
39.89 M. Joppien, R. Karbach, T. Moller: Phys. Rev. Lett. **71**, 2654 (1993)
39.90 M. A. McMahon, R. N. Barnett, K. B. Whaley: J. Chem. Phys. **99**, 8816 (1993)
39.91 E. J. Bieske, J. P. Maier: Chem. Rev. **93**, 2603 (1993)
39.92 A. McIlroy, D. J. Nesbitt: Adv. Mol. Vib. Collision Dyn. A **1**, 109 (1991)
39.93 S. Xu, L. S. Bartell: J. Phys. Chem. **97**, 13544 (1993)
39.94 S. Xu, L. S. Bartell: J. Phys. Chem. **97**, 13550 (1993)
39.95 T. J. Selegue, N. Moe, J. A. Draves, J. M. Lisy: J. Chem. Phys. **96**, 7268 (1992)
39.96 J. Jortner: Faraday Discuss. **108**, 1 (1997)
39.97 J. A. Alonso: Chem. Rev. **100**, 637 (2000)
39.98 G. Meloni, M. J. Ferguson, S. Sheehan, D. M. Neumark: Chem. Phys. Lett. **399**, 389 (2004)
39.99 H. Häkkinen, M. Moseler, O. Kostkos, N. Morgner, M. Hoffmann, B. v. Issendorff: Phys. Rev. Lett. **93**, 093401 (2004)

39.100 X. L. Zhu, X. C. Zeng: J. Chem. Phys. **118**, 3558 (2003)
39.101 X. L. Zhu, X. C. Zeng, Y. A. Lei, B. Pan: J. Chem. Phys. **120**, 8985 (2004)
39.102 J. H. K. Yip, J. Wu, K.-Y. Wong, K. P. Ho, C. S. Pun, J. J. Vittal: J. Chin. Chem. Soc. **51**, 1245 (2004)
39.103 S. Mukhopadhyay, S. K. Mandal, S. Bhaduri, W. H. Armstrong: Chem. Rev. **104**, 2981 (2004)
39.104 D. M. Neumark: Annu. Rev. Phys. Chem. **52**, 255 (2001)
39.105 T. E. Dermota, Q. Zhong, A. W. Castleman, Jr.: Chem. Rev. **104**, 1862 (2004)
39.106 V. Bonačić-Koutecký, R. Mitrić: Chem. Rev. **104**, 1861 (2004)
39.107 A. Van Orden, R. J. Saykelly: Chem. Rev. **98**, 2313 (1998)

40. Infrared Spectroscopy

Infrared spectroscopy consists of the measurement of interactions of waves of the infrared (IR) part of the electromagnetic spectrum with matter. The IR spectrum starts just beyond the red part of the visible spectrum at a wavelength $\lambda = 700$ nm and extends to the microwave region at $\lambda = 0.1$ cm. Electromagnetic waves are generally described in terms of their frequency ν in Hz. In IR spectroscopy it is common practice however to use the spatial frequency $\sigma = \nu/c$. These are called wavenumbers and have units of cm^{-1}. In this way the near, mid and far IR spectrum spans the frequencies from 14 300 cm^{-1} to 10 cm^{-1}. The interactions observed in the IR spectrum involve principally the energies associated with molecular structure change. Infrared spectroscopy is therefore useful for molecular structure elucidation and the identification and quantification of different molecular species in a sample [40.1].

The most common IR analysis of a sample is by IR *absorption* spectroscopy. This involves transmitting a beam of intense IR radiation through the sample and observing the distribution of wavenumbers absorbed by the molecules. Molecules in a sample may also be studied by IR *emission* spectroscopy simply by observing specific wavenumbers being emitted by virtue of the nonzero absolute temperature of the sample. Finally, radiation *reflected* from a smooth surface of a solid sample also provides information about the molecular structure of the material by virtue of the anomalous dispersion associated with absorption bands.

40.1 **Intensities of Infrared Radiation** 607

40.2 **Sources for IR Absorption Spectroscopy** .. 608

40.3 **Source, Spectrometer, Sample and Detector Relationship** 608

40.4 **Simplified Principle of FTIR Spectroscopy** 608
- 40.4.1 Interferogram Generation: The Michelson Interferometer 609
- 40.4.2 Description of Wavefront Interference with Time Delay 609
- 40.4.3 The Operation of Spectrum Determination 610

40.5 **Optical Aspects of FTIR Technology** 611

40.6 **The Scanning Michelson Interferometer** 612

40.7 **Recent Developments** 613

40.8 **Conclusion** 613

References 613

40.1 Intensities of Infrared Radiation

For strong interactions of electromagnetic waves with matter, the emitted and absorbed intensities are governed by Planck's radiation law in addition to the emissivity and absorptivity of the material. Planck's radiation law for thermal radiation from an ideal black body is

$$P_{\mathrm{bb}}(\sigma)\,\mathrm{d}\sigma = \frac{c_1\sigma^3\,\mathrm{d}\sigma}{\exp(h\sigma/k_\mathrm{B}T)+1}\,, \qquad (40.1)$$

where c_1 is a proportionality constant. Depending on the definition of c_1, $P_{\mathrm{bb}}(\sigma)$ may represent a radiation density per unit spectral interval (cm^{-1}) in a cavity at temperature T in $\mathrm{ergs/cm^3}$, or an energy flux emitted from a surface in $\mathrm{W/cm^2}$ steradians. At frequencies low compared with $h/k_\mathrm{B}T$, the energy distribution increases with σ^2 and is approximately proportional to T at a given σ. At high frequency, the energy distribution falls off exponentially. In the near IR, a high temperature is required to emit radiation. Room temperature objects emit strongest in the 300 to 600 cm^{-1} region, and emit negligible energy above 3000 cm^{-1}. Materials cooled to liquid nitrogen temperature (77 K) only emit below $100\,\text{cm}^{-1}$, while materials cooled to liquid helium temperature (4.2 K) only emit below $20\,\text{cm}^{-1}$. In contrast to visible spectroscopy, IR absorption spectroscopyis complicated by emission of IR radiation from the sample and the surrounding environment.

40.2 Sources for IR Absorption Spectroscopy

A silicon carbide element electrically heated to 1400 K provides a strong continuum of IR radiation over a major part of the IR spectrum. It is commonly used as a source of radiation for IR absorption spectroscopy. For near IR spectroscopy, a tungsten filament lamp operated at 2800 K provides a strong continuum all the way up to the visible part of the spectrum; it is not useful below $3000\,\text{cm}^{-1}$ because of absorption by the glass or quartz envelope. Various electrically heated ceramic elements, such as the Nernst glower and high temperature carbon rods, have been devised to serve as IR sources.

40.3 Source, Spectrometer, Sample and Detector Relationship

Since a sample at room temperature emits IR radiation in the mid IR, it is important to distinguish between transmitted radiation used in the determination of its absorption spectrum and its emission spectrum. By employing an intense IR beam, the effect of emission is minimized. Further distinction is achieved by encoding the IR beam before it impinges on the sample.

With classical grating or prism spectrometers, the source radiation is chopped by means of a mechanical chopper before it passes through the sample. The IR detector is provided with a means of synchronously decoding the chopped signal, thereby eliminating the emitted spectrum. Often the chopper is arranged such that it alternately switches between an empty reference beam and the sample. The logarithm of the ratio of the demodulated sample and reference spectra provides the absorption spectrum directly.

In *Fourier Transform infrared (FTIR) Spectroscopy*, a scanning Michelson interferometer provides the encoding function directly and no chopper is required. The interferometer is commonly placed before the sample so that it does not encode the thermally emitted radiation of the sample. Ratio recording of a sample against an empty beam is not common in FTIR. Instead, the absorption spectrum is obtained by sequentially recording the spectra of the sample and of the empty beam (sample removed), and computing the logarithm of the ratio numerically.

If it is not convenient to place the sample after the scanning Michelson interferometer, the absorption spectrum of a sample placed in front of the interferometer can be deduced by subtracting the separately recorded emission spectrum from the combined transmission plus emission spectrum.

Infrared emission and reflection spectroscopy form the basis for remote sensing. Solid and gaseous (cloud) objects may be identified and quantified by direct observation of their IR spectra at a distance. Gaseous clouds reflect poorly, providing only transmitted or emitted IR radiation. Their emission spectrum is contrasted directly with the spectrum of the scene or object beyond the cloud. With a background at lower temperature than the gas cloud, the gas spectrum appears in emission, while with a warmer background the gas spectrum appears in absorption.

Only a few solid materials transmit IR radiation over a substantial thickness. Remote sensing of solid objects relates therefore to surface emission and (diffuse) reflection of IR radiation from the surrounding environment.

40.4 Simplified Principle of FTIR Spectroscopy

In FTIR spectroscopy, the spectrum of a beam of incident IR radiation is obtained by first generating and recording an interferogram with a scanning Michelson interferometer. Subsequently the interferogram is inverted by means of a cosine Fourier transform into the spectrum.

40.4.1 Interferogram Generation: The Michelson Interferometer

The scanning Michelson interferometer shown in Fig. 40.1 consists of a beam splitter, which is a substrate with a dielectric coating such that 50% of an incident beam is reflected and the remaining 50% is transmitted, and two plane mirrors (M_1 and M_2), one or both of which are translated along the direction of the beam. After splitting, the two equal amplitude wave fronts are propagated along different optical paths. The mirrors at the end of each path return the wavefronts to the beamsplitter, which then acts as a wavefront combiner. Because of their common coherent origin, the wavefronts interfere with one another when they combine. The state of interference is varied by scanning one or both of the mirrors such that there is a variable time delay between the two separated beams. The resulting intensity variation of the combined output beam as a function of relative time delay is the interferogram.

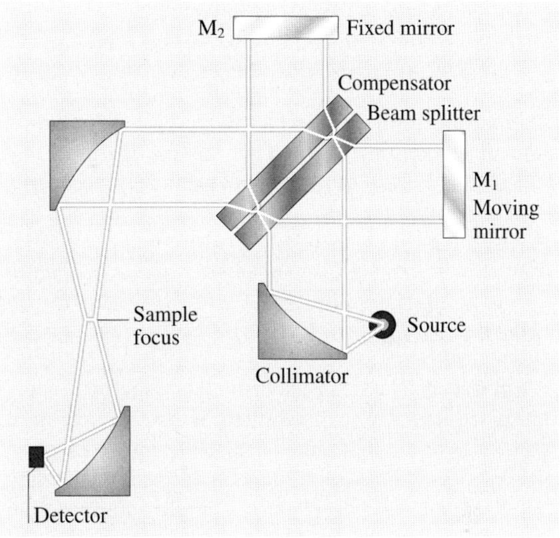

Fig. 40.1 The Michelson interferometer

40.4.2 Description of Wavefront Interference with Time Delay

The intensity $I_0(\nu)$ at frequency ν of a plane wave in space is given by the expectation value of its electric field vector $\boldsymbol{E}(\nu, t) = \mathcal{E}(\nu)\exp(\mathrm{i}2\pi\nu t)$ according to

$$I_0(\nu) = \langle \boldsymbol{E}(\nu,t)|\boldsymbol{E}(\nu,t)\rangle = \mathcal{E}(\nu)^2 . \tag{40.2}$$

The intensity at the output of the interferometer due to an incident intensity $I_0(\nu)$ is given by $I(\nu, \delta)$, where δ is the time delay between the two wavefronts which have been propagated along two different paths, and

$$\begin{aligned} I(\nu,\delta) &= \langle \boldsymbol{E}(\nu,t){+}\boldsymbol{E}(\nu,t{+}\delta)|\boldsymbol{E}(\nu,t){+}\boldsymbol{E}(\nu,t+\delta)\rangle \\ &= \frac{1}{4}\mathcal{E}(\nu)^2\left(2+\mathrm{e}^{-\mathrm{i}2\pi\nu\delta}+\mathrm{e}^{\mathrm{i}2\pi\nu\delta}\right) \\ &= \frac{1}{2}I_0(\nu)[1+\cos(2\pi\nu\delta)] . \end{aligned} \tag{40.3}$$

As can be seen, the output intensity of a single frequency source at the output of an ideal scanning Michelson interferometer fluctuates sinusoidally between zero and the input intensity $I_0(\nu)$ as the time delay δ between the separated wavefronts is varied by means of scanning one of the mirrors.

The quantity δ is related to mirror displacement x with respect to equal distance of the mirrors from the beamsplitter by

$$\delta = 2x\cos\theta/c , \tag{40.4}$$

where θ is the angle between the wavefront and the optical axis of the spectrometer, and the optical axis is the normal to each plane mirror M_1 and M_2. From this, (40.3) becomes

$$I_0(\nu, x) = \frac{1}{2}I_0(\nu)\{1+\cos[2\pi\nu(2x\cos\theta/c)]\} , \tag{40.5}$$

or, using $\sigma = \nu/c$,

$$I_0(\sigma, x) = \frac{1}{2}I_0(\sigma)\{1+\cos[2\pi\sigma(2x\cos\theta)]\} . \tag{40.6}$$

Thus the output intensity fluctuates at frequency $\sigma' = 2\sigma\cos\theta$ as a function of the mirror displacement x.

The incident intensity generally consists of a distribution of intensities over many frequencies $S(\nu)\,\mathrm{d}\nu$ with integrated intensity

$$I_0 = \int S(\nu)\,\mathrm{d}\nu . \tag{40.7}$$

For this case, the output intensity of an ideal scanning Michelson interferometer is given by

$$I_0(\delta) = \frac{1}{2}\int S(\nu)[1+\cos(2\pi\nu\delta)]\,\mathrm{d}\nu . \tag{40.8}$$

The second term on the right side of (40.8) has the form of the cosine Fourier transform of the spectrum.

By rearrangement of (40.8), it is given by

$$\int S(\nu)\cos(2\pi\nu\delta)\,\mathrm{d}\nu = 2I_0(\delta) - I_0\,. \tag{40.9}$$

The constant term I_0 provides no useful information about the spectrum. The inverse cosine Fourier transform of $2I_0(\delta)$ results in the spectrum $S(\nu)$ according to

$$S(\nu) = \int 2I_0(\delta)\cos(2\pi\nu\delta)\,\mathrm{d}\delta\,, \tag{40.10}$$

$$S(\sigma) = \int 2I_0(x)\cos\left(2\pi\sigma' x\right)\,\mathrm{d}x\,. \tag{40.11}$$

Contrary to classical spectrometers, where the spectrum is sequentially scanned, there is no segregation of frequencies of the input intensity. All frequencies in the source are modulated simultaneously by the scanning Michelson interferometer into a single interferogram signal. This multiplex mode of spectrum determination was first exploited by *Felgett* [40.2]. It contributes to a large advantage in sensitivity compared with other spectrometers, and is referred to as the Felgett or multiplex advantage.

40.4.3 The Operation of Spectrum Determination

The output interferogram is detected by an IR detector which converts the intensity variations $I_0(x)$ as a function of different mirror positions x into an electrical signal. Continuous determination of the inverse cosine Fourier transform of the evolving interferogram requires continuous multiplication of the signal by cosine functions with all the different frequencies of the spectrum and integrating these products. Continuous Fourier analysis with a bank of narrow band filters has been implemented both in analog and digital form in early versions of FTIRs [40.3].

It is, however, far more practical to capture the interferogram signal in numeric form, using an analog to digital converter, store it in computer memory, and compute the Fourier transform numerically after the mirror displacement range has been covered.

The numerical representation of the interferogram is determined at known intervals of mirror displacement Δx. The computed spectrum is then determined at regular intervals of spatial frequency $\Delta\sigma$ by the discrete cosine Fourier transform

$$S(j\Delta\sigma) = \sum_n 2I_0(n\Delta x)\cos(2\pi jn\Delta\sigma\Delta x)\,. \tag{40.12}$$

The sampling interval Δx of optical path difference determines the extent of the numerically computed spectrum. A higher density of sampling permits a wider spectral range to be determined, up to

$$\sigma_{\max} = \frac{1}{2\Delta x}\,. \tag{40.13}$$

Beyond $\sigma_{\max}$, the spectrum repeats in reverse order, and beyond $2\sigma_{\max}$, the spectrum repeats as is. This is called spectral aliasing, and results from the incomplete knowledge of the full interferogram function between the discrete numeric representation. In order to insure that the numeric representation of the interferogram describes the continuous function uniquely, it is important to band limit the interferogram information to the range 0 to $\sigma_{\max}$ by means of optical and electrical filtering.

Conversely, the higher the density of sampling in the spectral domain, the longer the interferogram needs to be. For a wavenumber range $\Delta\sigma$, the length is

$$x_{\max} = \frac{1}{2\Delta\sigma}\,. \tag{40.14}$$

From the orthogonality property of the discrete cosine Fourier transform, unique linearly independent information can occur only at spectral intervals equal to or greater than the sampling interval. Hence the sampling interval in the spectrum is related to the achievable resolution. The full width at half maximum of the representation of a single frequency in the spectrum is $1.2\Delta\sigma$. This factor applies for the case where the single cosine wave interferogram has been abruptly truncated at the end of the mirror scan. The lineshape function for this case is quite oscillatory due to the abrupt termination of the interferogram signal at the end of the scan. This is not always satisfactory, and frequently the interferogram is modified by a windowing or apodization function to make the lineshape more localized and monotonic. Apodization always results in an increase in the full width at half maximum.

A particularly convenient and accurate way to establish the sampling intervals of the interferogram is the use of a single frequency laser directed coaxially with the source radiation through the scanning Michelson interferometer. The intensity of the laser at the output of the interferometer is a highly consistent cosine wave with one cycle per change in mirror movement of one half wavelength of the laser light.

40.5 Optical Aspects of FTIR Technology

The description of wavefront interference developed in Sect. 40.4.2 applies to interference of plane wave fronts only. A plane wave of IR radiation is obtained at the output of a collimator optics having an IR point source at its focus. In practice a point source has insignificant intensity. A finite size source, which may be represented by a distribution of point sources in the focal plane of the collimator, provides a distribution of plane wavefronts with different angles of propagation through the scanning Michelson interferometer.

As shown in (40.6), this distribution of angles results in a distribution of modulation frequencies as a function of mirror displacement x of the output intensity for a given IR wavenumber. This is illustrated in Fig. 40.2 which shows, for a single frequency IR source, the distribution of output intensity modulation frequencies for an ideal point source on the optical axis, a circularly symmetric distribution of uniform intensity about the optical axis, and finally a circularly symmetric distribution positioned slightly off the axis of the collimator.

As a result of illuminating the scanning Michelson interferometer with a finite size source, a single IR frequency source is observed having a distribution of modulation frequencies in its interferogram signal. This distribution limits the ability to resolve two closely spaced IR frequencies and determines the optical resolution limit of the FTIR: The larger the extent of the source, the more restricted the resolution becomes. The ratio $\sigma/\Delta\sigma$ is the resolving power R of the scanning Michelson interferometer and, based on (40.6), is given by

$$R = 1/(1 - \cos\theta_m) \,, \qquad (40.15)$$

where θ_m is the maximum off-axis angle of illumination. For small θ_m, $\cos\theta_m \approx 1 - \frac{1}{2}\theta_m^2$, so that

$$R \sim 2/\theta_m^2 \,. \qquad (40.16)$$

From (40.14), a resolution limit is also imposed by the maximum length of the interferogram recorded. The interferogram length dependent resolution is constant for all spectral regions, while the optical resolution is proportional to the spectral frequency. Both resolution limits combine to give the overall resolution of an FTIR. At low resolution, the available throughput is so high that the optical resolution is often negligible compared with the length resolution. At high resolution, throughput is at a premium and the optical resolution is often closely matched to the length resolution at the frequency of interest.

To insure a symmetrical frequency distribution, the area integrated illumination must increase as $\sin\theta$, which is approximately linear for small angles, up to its maximum θ_m, as shown with the centered circular illumination in Fig. 40.2. Any deviation from this, such as off-axis positioning of the circle, noncircular shapes, or a poorly focused centered symmetrical circle, will result in a gradual roll-off of the distribution on the low frequency side only [40.4]. This results in an asymmetric spectral line shape.

For a collimator of given focal length and for a given resolving power, it is easily shown that the area of the source, or the stop that delineates it, is much larger than the slit area for classical grating spectrometers. This is particularly the case at high resolving powers.

This throughput advantage was first pointed out by *Jacquinot* [40.5], and plays an important role in the large sensitivity advantage of FTIR. The stop that delineates the source extent for a scanning Michelson interferometer is often referred to as the Jacquinot stop or the field of view stop.

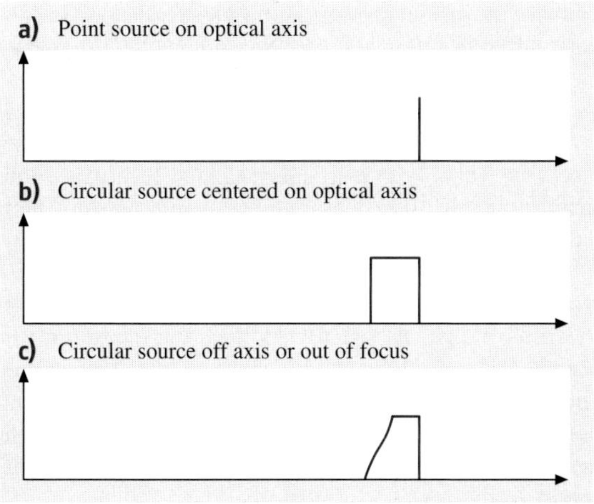

Fig. 40.2 Distribution of interferogram modulation frequencies for a single optical frequency source

40.6 The Scanning Michelson Interferometer

Optical throughput is not only determined by the area of the field of view stop, but also the solid angle subtended by the rays traversing this area. The solid angle Ω of rays traversing a Jacquinot stop positioned in the focal plane of an input collimator is given by the ratio of the interferometer beam area divided by the square of the focal length of the collimator.

For a given collimator focal length, the area of the Jacquinot stop is inversely proportional to the resolving power. To maintain equal throughput, the area of the interferometer optics should be increased as resolving power is increased in order to offset the decrease in the Jacquinot stop area.

It is common to construct interferometers with 2.5 cm diameter optics for resolving powers up to 5000, 5.0 cm diameter optics for resolving powers up to 40 000 and 7.5 cm diameter optics for resolving powers up to 1 000 000.

In order to obtain a uniform state of interference across the entire beam of the interferometer, the beamsplitter substrate and the two mirrors must be flat to within a small fraction of the wavelength used. Also, these elements must be oriented correctly so that the optical path difference error across the beam is less than a small fraction of the wavelength.

Figure 40.1 shows two substrates at the beamsplitter position. One of the substrates supports the beamsplitting coating, while the companion substrate of precisely the same thickness acts as a compensating element to insure identical optical paths through the two arms of the interferometer. To avoid secondary interference effects, both beamsplitter and compensator substrates are normally wedged. The direction of the wedges of the two substrates must be aligned again to insure symmetry in both arms of the interferometer.

The maintenance of a very close orientational alignment tolerance of the two mirrors with respect to the beamsplitter in a stable manner over time and while scanning one of the mirrors is the greatest challenge of interferometer design and is also the greatest weakness of FTIR.

In early models of FTIRs, alignment was maintained by means of a stable mechanical structure and a highly precise linear air bearing for the scanning mirror. Satisfactory operation required a stable environment and frequent alignment tuning and could be achieved for mirror displacements of only several centimeters, thus limiting the maximum resolution.

Different techniques have been developed to overcome this weakness in FTIR. The two most prominent are (1) Dynamic alignment of the interferometer, where optical alignment is servo-controlled using the reference laser not only for mirror displacement control but also for mirror orientation control, and (2) the use of cube corner mirrors in place of the flat mirrors in the interferometer. Dynamic alignment has the advantage of retaining a high degree of simplicity in the optical design of the interferometer. On the other hand it is more complex electronically. Cube corners have the property of always reflecting light 180° to incident light independent of orientation. Cube corners always insure wavefront parallelism at the point of recombination of the two beams in the interferometer. Cube corners lack a defined optical axis. In a cube corner interferometer, the optical axis is defined as the direction in which the wavefronts undergo zero shear.

The scanning Michelson Interferometer is normally provided with a drive mechanism to displace one of the mirrors precisely parallel to its initial position and at uniform velocity. The uniform velocity translates the mirror displacement dependent intensities into time dependent intensities. This facilitates signal processing electronics. In some measurement scenarios however, where the sample spectrum may vary with time, it is undesirable to deal with the multifrequency time-varying intensities of the interferogram signal. In this case it is preferable to scan the moving mirror in a stepwise mode, where the mirror is momentarily stationary at the time of signal measurement and then advanced rapidly to the next position. The mirror scan velocity v can be varied so that electrical signals can have different frequencies for the same optical frequencies:

$$f = \sigma/2v \,. \tag{40.17}$$

The scan velocity is normally selected to provide the most favorable frequency regime for the detector and electronics used and for the mechanical capabilities of the mirror drive. Typically the velocity range is from 0.05 cm/s to 4 cm/s, putting the frequencies in the audio range. At these velocities, the measurement scan may be completed in a shorter time than is desired for signal averaging purposes. In this case it is common to repeat the scan a number of times and add the results together: this is called co-adding of scans.

40.7 Recent Developments

Added by Mark M. Cassar. Infrared spectromicroscopy, which combines the well-established technique of Fourier transform infrared (FTIR) spectroscopy with a microscope, has been one of the main developments in this area in the past decade. Recently, array detectors have made infrared imaging practical and quick. The brightness attainable in an IR spectromicroscope has also been enhanced through the use of a synchrotron radiation (SR) source, allowing the source beam to be focused to a spot with a diameter $\leq 10\,\mu m$ [40.6]. This also allows, because of the high signal-to-noise ratio, the measurement of dilute sample concentrations. One important application of SR-FTIR microscopy is to study the effects of various stimuli on biomolecules in order to understand how diseases start and spread [40.7, 8].

A testament to the robustness and versatility of FTIR is given by its potential inclusion in future Mars expeditions [40.9].

40.8 Conclusion

The modern technique of Fourier transform IR spectroscopy has evolved rapidly from its beginnings in the early 1950s to being the dominant technique of IR spectroscopy in many diverse disciplines. The Aspen Conference on Fourier Spectroscopy in 1970 was the watershed for FTIR, where many fundamental issues of the technique were treated [40.10]. A practical book by *Griffith* and *de Haseth* gives many examples of applications of FTIR [40.11].

FTIR is a powerful technique for infrared spectroscopy. It has a large sensitivity advantage over conventional dispersive spectrometers because of the efficient multiplexing of all the spectral elements and the greater throughput of the Jacquinot stop. It can be used efficiently for resolving powers from less than 1000 up to 1 000 000. FTIR combines techniques of optical interferometry, laser metrology and digital signal processing.

The computation of the cosine Fourier transform was initially a daunting task. Today, with highly efficient factoring algorithms brought to FTIR by *Forman* [40.12], and the widespread availability of inexpensive high performance personal computers, the computation time for a Fourier transform is only a fraction of a second per 1000 data points in the interferogram.

A traditional weakness of FTIR is the severe demand of alignment stability and accuracy of the scanning Michelson interferometer. Newer optical designs and control procedures have largely overcome this weakness. Now FTIR can be justified not only for its high sensitivity but also for its high degree of reproducibility and stability, permitting demanding applications in quantitative IR analysis.

References

40.1 N. B. Colthup, L. H. Daly, S. E. Wiberley: *Introduction to Infrared, Raman Spectroscopy* (Academic, New York 1990)

40.2 P. B. Fellgett: J. Phys. Radium **19**, 187, 237 (1958)

40.3 J. E. Hoffman, G. A. Vanasse: J. Phys. (Paris) **28**, C2:79 (1967)

40.4 P. Saarinen, J. Kauppinnen: Appl. Opt. **31**, 2353 (1992)

40.5 P. Jacquinot: J. Opt. Soc. Am. **44**, 761 (1954)

40.6 M. C. Martin: Synchrotron Radiation News **15**, 10 (2002)

40.7 H.-Y. N. Holman, M. C. Martin, W. McKinney: Spectroscopy – An International Journal **17**, 139 (2003)

40.8 H.-Y. N. Holman, K. A. Bjornsted, M. P. McNamara, M. C. Martin, W. R. McKinney, EA. Blakely: J. Biomed. Opt. **7**, 417 (2002)

40.9 MS. Anderson, J. M. Andringa, R. W. Carlson, P. Conrad, W. Hartford, M. Shafer, A. Soto, A. I. Tsapin, J. P. Dybwad, W. Wadsworth, K. Hand: Rev. Sci. Instr. **76**, 034101 (2005)

40.10 G. A. Vanasse, A. T. Stair, D. J. Baker (Eds.): Aspen International Conference on Fourier Spectroscopy, 1970, AFCRL-71-0019

40.11 P. R. Griffith, J. A. de Haseth: *Fourier Transform Infrared Spectrometry*, J. Biomed. Opt., Vol. 83 (Wiley Interscience, New York 1986)

40.12 M. L. Forman: J. Opt. Soc. Am. **56**, 978 (1966)

41. Laser Spectroscopy in the Submillimeter and Far-Infrared Regions

Recent technical developments in sub-millimeter and far-infrared laser spectroscopy are described. This includes new laser sources, both side-band and difference-frequency generation. An experiment which uses fixed-frequency far-infrared lasers to study open-shell molecules (free radicals) is described; the technique is known as laser magnetic resonance (LMR). Sub-millimeter and far-infrared laser spectroscopies are finding expensive use in the detection and monitoring of molecules in astrophysical sources and in the earth's atmosphere.

41.1 **Experimental Techniques using Coherent SM-FIR Radiation** 616
41.1.1 Tunable FIR Spectroscopy with CO_2 Laser Difference Generation in a MIM Diode 617
41.1.2 Laser Magnetic Resonance 618
41.1.3 TuFIR and LMR Detectors 619

41.2 **Submillimeter and FIR Astronomy** 620

41.3 **Upper Atmospheric Studies** 620

References 621

Research in the submillimeter and far-infrared (SM-FIR) regions of the electromagnetic spectrum (1000 to 150 μm, 0.3 to 2.0 THz; and 150 to 20 μm, 2.0 to 15 THz, respectively) had been relatively inactive until about 30 years ago. Three events were responsible for enhanced activity in this part of the electromagnetic spectrum: the discovery of far-infrared (FIR) lasers [41.1], the development of background-limited detectors [41.2], and the invention of the FIR Fourier transform (FT) spectrometer [41.3]. Following these developments major discoveries have taken place in laboratory spectroscopic studies [41.4, 5], in astronomical observations [41.6], and in spectroscopic studies of our upper atmosphere [41.7].

Rotational transition frequencies of light molecules (such as hydrides) lie in this region, and the associated electric dipole transitions are especially strong at these frequencies. In fact they are 10 000 times stronger than at microwave frequencies because they are 100 times typical microwave frequencies and their peak absorptivities depend approximately on the square of the frequency. Fine structure transitions of atoms and molecules also lie in this region; however, they are much weaker, magnetic dipole transitions. The observation of fine structure spectra is very important in determining atomic concentrations in astronomical and atmospheric sources and for determining the local physical conditions. Bending frequencies of larger molecules also lie in this region, but their transitions are not as strong as rotational transitions (typically a factor of 10^3 weaker).

Spectral accuracy has been increased by several orders of magnitude with the extension of direct frequency measurement metrology into the SM-FIR region [41.8]. Transitions whose frequencies have been measured (including absorptions and laser emissions) are useful wavelength calibration sources ($\lambda_{vac} = c/\nu$) for FT spectrometers. FIR spectra of a series of rotational transitions have been measured in CO [41.9], HCl [41.10], HF [41.11], and CH_3OH [41.12] to be used for FT calibration standards. These lines are ten to a hundred times more accurately measured than can be realized in present state-of-the-art Fourier transform spectrometers; thus, they are excellent calibration standards. High-accuracy and high-resolution spectroscopy has permitted the spectroscopic assignment of the SM-FIR lasing transitions themselves [41.13] resulting in a much better understanding of the lasing process.

Astronomical spectroscopy in this region [41.6] may be in emission or absorption and is performed using either interferometric [41.14] (wavelength-based), or heterodyne (frequency-based) [41.15] techniques to resolve the individual spectral features.

Most high resolution spectra of our upper atmosphere have been taken with FT spectrometers flown above the heavily absorbing water vapor region in the lower atmosphere. Emission lines are generally observed in these spectrometers [41.7].

41.1 Experimental Techniques using Coherent SM-FIR Radiation

The earliest sources of coherent SM radiation came from harmonics of klystron radiation generated in point-contact semiconductor diodes [41.16]. Spectroscopically useful powers up to about one terahertz are produced. This technique is being replaced by electronic oscillators which oscillate to over one terahertz [41.17]. The group at Cologne University, Germany [41.18] has been particularly successful with this approach. Spectroscopy above this frequency generally is performed with either lasers or FT spectrometers (see Chapt. 40).

Laser techniques use either tunable radiation synthesized from the radiation of other lasers, or fixed-frequency SM-FIR laser radiation and tuning of the transition frequency of the species by an electric or magnetic field. Spectroscopy with tunable far-infrared radiation is called TuFIR spectroscopy. Spectroscopy with fixed frequency lasers is called either laser electric resonance (LER) or laser magnetic resonance (LMR). LMR is applicable only to paramagnetic species and is noteworthy for its extreme sensitivity. LMR spectroscopy has been more widely applied than LER, and is discussed in this chapter.

Tunable SM-FIR radiation has been generated either by adding microwave sidebands to radiation from a SM-FIR laser [41.19] or by using a pair of higher frequency lasers and generating the frequency difference [41.20]. The sideband technique was first reported by *Dymanus* [41.19] and uses a Schottky diode as the mixing element. It has been used up to 4.25 THz and produces a few microwatts of radiation [41.21]. Groups at Berkley, California [41.22] and Cologne, Germany [41.23] have developed this technique to good effect. The CO_2 laser

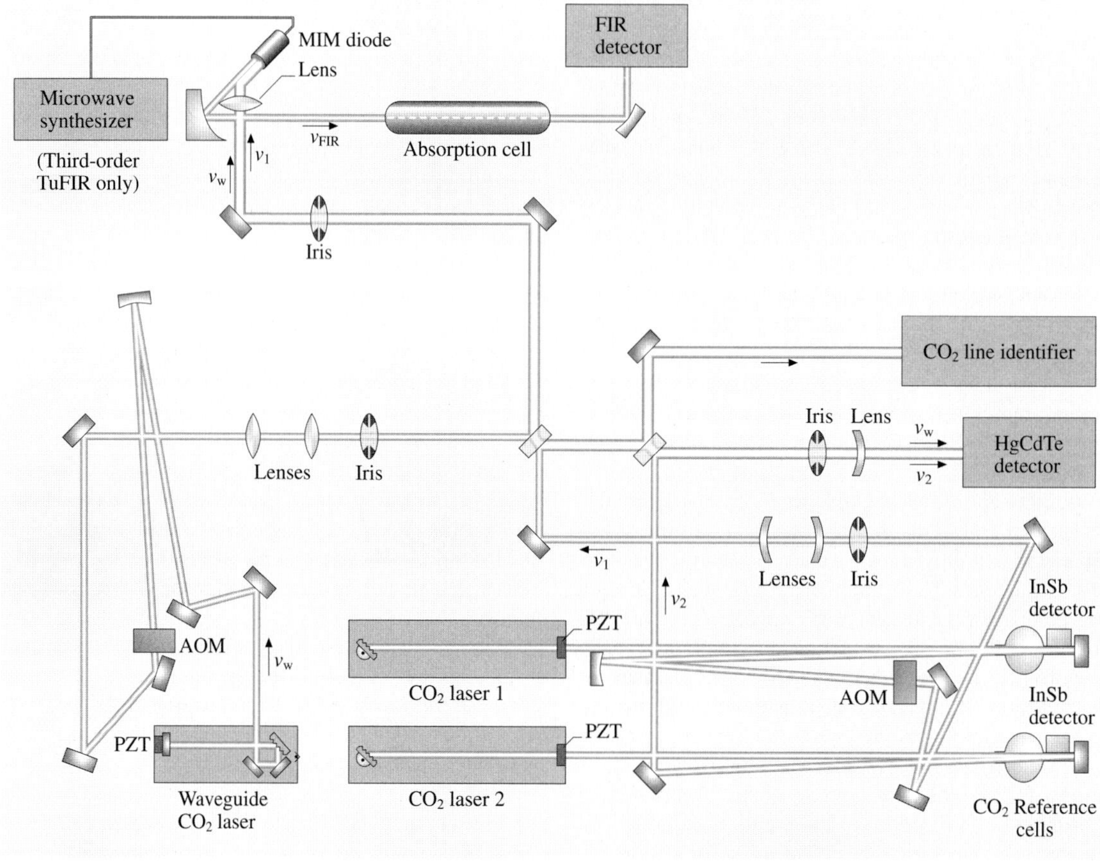

Fig. 41.1 Tunable far-infrared (TuFIR) spectrometer for second- or third-order operation using CO_2 laser difference-frequency generation in the MIM diode

frequency-difference technique with difference generation in the metal-insulator-metal (MIM) diode covers the FIR region out to over 6 THz and produces about 0.1 μW. This technique uses fluorescence-stabilized CO_2 lasers whose frequencies have been directly measured, and it is about two orders of magnitude more accurate than the sideband technique. However, it is somewhat less sensitive because of the decreased power available. There are several review articles on the laser sideband technique [41.21, 24], and only the laser difference technique is described here.

41.1.1 Tunable FIR Spectroscopy with CO_2 Laser Difference Generation in a MIM Diode

There are two different ways of generating FIR radiation using a pair of CO_2 lasers and the MIM diode. One is by second-order generation, in which tunability is achieved by using a tunable waveguide CO_2 laser as one of the CO_2 lasers; it is operated at about 8 kPa (60 Torr) and is tunable by about ±120 MHz. The second technique uses third-order generation, in which tunable microwave sidebands are added to the difference frequency of the two CO_2 lasers. The complete spectrometer which can be operated in either second or third order is shown in Fig. 41.1.

The FIR frequencies generated are:

$$\text{second-order: } \nu_{\text{fir}} = \left| \nu_{1,CO_2} - \nu_{W,CO_2} \right| \tag{41.1a}$$

$$\text{third-order: } \nu_{\text{fir}} = \left| \nu_{1,CO_2} - \nu_{W,CO_2} \right| \pm \nu_{\text{mw}} \tag{41.1b}$$

where ν_{fir} is the tunable FIR radiation, ν_{1,CO_2} and ν_{W,CO_2} are the CO_2 laser frequencies, and ν_{mw} is the microwave frequency.

Different MIM diodes are used for the two different orders: in second-order, a tungsten whisker contacts a nickel base and the normal oxide layer on nickel serves as the insulating barrier; in third-order, a cobalt base with its natural cobalt oxide layer is substituted for the nickel base. The third-order cobalt diodes produce about one third as much FIR radiation in each sideband as there is in the second-order difference; hence third-order generation is not quite as sensitive as second-order. The much larger tunability, however, makes it considerably easier to use.

The common isotope of CO_2 is used in both the waveguide laser and in laser 2; in laser 1, one of four isotopic species is used. Ninety percent of all frequencies from 0.3 to 4.5 THz can be synthesized; the coverage decreases between 4.5 and 6.3 THz. Ninety megahertz acoustooptic modulators (AOMs) are used in the output beams of the two CO_2 lasers which irradiate the MIM diode; they increase the frequency coverage by an additional 180 MHz and isolate the CO_2 lasers from the MIM diode. This isolation decreases amplitude noise in the generated FIR radiation, caused by the feedback to the CO_2 laser from the MIM diode, by an order of magnitude; hence, the spectrometer sensitivity increases by an order of magnitude.

The radiations from laser 1 and the waveguide laser are focused on the MIM diode. Laser 2 serves as a frequency reference for the waveguide laser; the two lasers beat with each other in the HgCdTe detector and a servosystem offset-locks the waveguide laser to laser 2. Lasers 1 and 2 are frequency modulated using piezoelectric drivers on the end mirrors, and they are servoed to the line center of 4.3 μm saturated fluorescence signals obtained from the external low-pressure CO_2 reference cells. In both second and third order, the CO_2 reference lasers are stabilized to line center with an uncertainty of 10 kHz. The overall uncertainty in the FIR frequency due to two lasers is thus $\sqrt{2} \times 10$ or 14 kHz. This number was determined experimentally in a measurement of the rotational frequencies of CO out to $J = 38$ (4.3 THz), with the analysis of the data set determining the molecular constants [41.9].

The CO_2 radiation is focused by a 25 mm focal length lens onto the conically sharpened tip of the 25 μm diameter tungsten whisker. The FIR radiation is emitted from the 0.1 to 3 mm long whisker in a conical long-wire antenna pattern. Then it is collimated to a polarized beam by a corner reflector [41.25] and a 30 mm focal length off-axis segment of a parabolic mirror.

The largest uncertainty in the measurement of a transition frequency comes from finding the centers of the Doppler broadened lines. This is about 0.05 of the line width for lines observed with a signal-to-noise ratio of 50 or better and corresponds to about 0.05 to 0.5 MHz. A linefitting program [41.26] improves the line center determinations by nearly an order of magnitude. The FIR radiation is frequency modulated due to a 0.5 to 6 MHz frequency modulation of CO_2 laser 1; this modulation is at a rate of 1 kHz. The FIR detector and lock-in amplifier detect at this modulation rate; hence the derivatives of the absorptions are recorded.

Absorption cells from 0.5 to 3.5 m in length with diameters ranging from 19 to 30 mm have been used in the spectrometer. The cells have either glass, copper, or Teflon walls and have polyethylene or polypropylene windows at each end. These long absorption cells lend themselves naturally to electrical discharges for the study of molecular ions. A measurement of the HCO^+ line at 1 THz exhibited a signal-to-noise ratio of 100 : 1 using a 1 s time constant; this is the same signal-to-noise ratio that was obtained using the laser sideband technique.

The instrumental resolution of the spectrometer is limited by the combined frequency fluctuation from each CO_2 laser (about 15 kHz). This is less than any Doppler-limited line width and, therefore, does not limit the resolution. The entire data system is computerized to facilitate the data recording and optimize the data handling.

Improvements in this TuFIR technique may come from either better diodes or detection schemes. The nonlinearity of the MIM diode is extremely small, and conversion efficiencies could be much larger if a more efficient diode is discovered. Differential detection (which requires more sensitive detectors) could also significantly improve the sensitivity and permit the detection of weaker lines.

41.1.2 Laser Magnetic Resonance

Laser magnetic resonance (LMR) is performed by magnetically tuning the transition frequencies of paramagnetic species into coincidence with the fixed frequency radiation from a laser. LMR is a type of Zeeman spectroscopy [41.27], and its chief forte is its extreme sensitivity. Approximately one hundred species have been observed in the SM-FIR region. These include atoms, diatomics (especially hydrides), polyatomics, ions, metastables, metastable ions [41.28], and many "first observations" of free radicals. These observations are summarized in several review articles on LMR [41.29–31].

The FIR LMR spectrometer at NIST, Boulder is shown in Fig. 41.2. This spectrometer has an intracavity paramagnetic sample in a variable magnetic field which is labeled "sample volume" in the figure. A regular 38 cm EPR magnet with a 7.5 cm air gap is used. The laser cavity is divided into two parts by a Brewster angle polypropylene beam splitter about 12 μm thick. A CO_2 laser optically pumps the FIR laser gas. This pump beam makes many nearly perpendicular passes reflected by the walls of the gold-lined Pyrex tube. A 45° coupler serves to couple out the FIR radiation from the cavity to the helium-cooled detector. Mirrors each with an 89 cm radius of curvature are used in the nearly confocal geometry of the 94 cm cavity. One of the gold-coated end mirrors is moved with a micrometer to tune the cavity

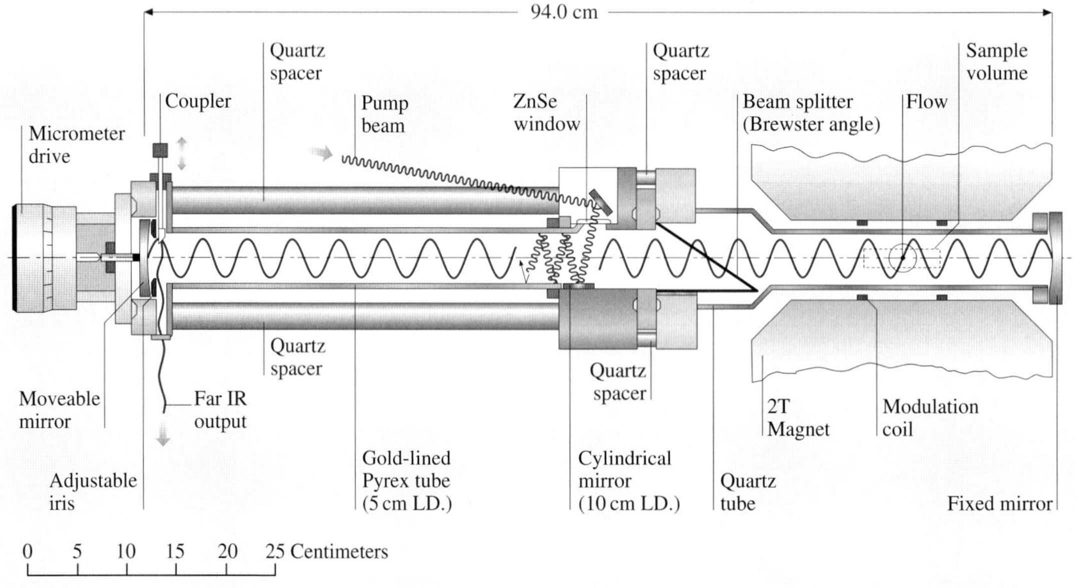

Fig. 41.2 40 μm to 1000 μm Far-infrared laser magnetic resonance (LMR) spectrometer using an optically pumped FIR laser

to resonance and also to determine the oscillating wavelength by moving the end mirror several half-wavelengths. The beam splitter is rotatable so that the polarization of the laser can be varied with respect to the magnetic field. Quartz spacers are used to minimize the thermal expansion of the cavity. This LMR spectrometer oscillates at wavelengths between 40 and 1000 μm. The technology of FIR lasers has been reviewed by *Douglas* [41.32].

The LMR technique requires a close coincidence (typically within 20 GHz) between the absorption line and the frequency of the laser. Zero field frequencies which are 100 times more accurate than those obtained from optical spectra are obtained from the analysis of the Zeeman spectra observed (i. e. using the laser frequency and the magnetic field values). The accuracy is within 1 or 2 MHz and has permitted the far infrared astronomical observation of many of these species.

Atomic FIR LMR spectra are due to fine structure transitions and are magnetic dipole transitions; hence, they are several orders of magnitude weaker than electric dipole rotational spectra of molecules. The production of a sufficient atomic number density can be difficult; however, "atomic flames" have been very effective sources, and the high sensitivity of LMR has been the most successful spectroscopic technique for measuring these transitions. The atoms O, C, metastable Mg, S, Si, Fe, Al, N^+, C^+, P^+, Fe^+, and F^+ have been measured by FIR LMR. Atomic Zeeman spectra are relatively simple to analyze, and fine structure frequencies accurate to within 1 MHz can be determined.

In a FIR LMR spectrometer the sample is inside the laser cavity; hence, sub-Doppler line widths can be observed by operating the sample at low pressure and observing saturation dips in the signals. This has permitted the resolution of proton hyperfine structure in a number of hydrides. The observation of the resolved hyperfine structure is useful in the identification of the species involved and also yields accurate values of the hyperfine splittings.

Rotational spectra of most hydrides lie in this spectral region and many of the hydrides, such as OH, CH, SH, and NH have been observed. Others, such as CaH, MnH, TiH, and ZnH are excellent candidates for LMR studies. Ions are much more difficult because of their low concentrations; however, the use of a special microwave discharge operating in the magnetic field has proved to be an extremely productive source of ions for LMR studies [41.28].

The spectra of a number of polyatomic species has been observed by LMR: for example, NO_2, HO_2, HCO, PH_2, CH_2, NH_2, AsH_2, HO_2, HS_2, CH_3O, and CCH. The spectra of these polyatomic species are more difficult to analyze, resulting in a somewhat less accurate prediction of the zero-field frequencies. Spectra of the extremely elusive CH_2 and CD_2 radicals have finally yielded to analysis with the simultaneous observation of FIR LMR spectra [41.33] and IR LMR spectra [41.34]. The data yield rotational constants which predict the ground state rotational transition frequencies and permit the determination of the singlet–triplet splitting in that molecule [41.35].

In the last ten years, there has been an experimental push to higher frequencies in FIR spectroscopy. In the LMR spectrometer (shown in Fig. 41.2), this has been achieved principally by reducing the internal diameter of the gold-lined tube in the pumping region to 19 mm so as to increase the overlap between the laser gain medium and the FIR radiation field in the laser cavity. As a result, many new laser lines have been discovered; the system operates down to below 40 μm. This has enabled molecules with larger fine-structure intervals to be studied (e.g., FO and HF^+ [41.36, 37]) and has also led to the first detection of vibration-rotation transitions in molecules with low-frequency bending vibrations (CCN, HCCN, FeH_2) [41.38–40].

The sensitivity of TuFIR is only about 1% of that of laser magnetic resonance, but it is difficult to compare the two because the sample can be very large in the TuFIR spectrometer and is limited to about 2 cm^3 in the LMR spectrometer.

41.1.3 TuFIR and LMR Detectors

Four different detectors have been used in the TuFIR and LMR spectrometers: (1) an InSb 4 K, liquid ^{4}He-cooled, hot-electron bolometer, operating from 0.3 to 0.6 THz with a noise equivalent power (NEP) of about $10^{-13}\,W/\sqrt{Hz}$; (2) a gallium doped germanium bolometer, cooled to the lambda point of liquid-helium, operating from 0.6 to 8.5 THz with an NEP of about $10^{-13}\,W/\sqrt{Hz}$; (3) a similar, but liquid ^{3}He-cooled bolometer, with a NEP two orders of magnitude smaller; and (4) an unstressed Ge:Ga photoconductor, cooled to 4 K, with an NEP of $10^{-14}\,W/\sqrt{Hz}$, operating from 2.5 to 8.5 THz. The optimization of detector systems for this new technique has been responsible for a significant improvement in sensitivity of these spectrometers.

41.2 Submillimeter and FIR Astronomy

Submillimeter astronomy is reviewed in [41.6]. Many atomic species (both neutrals and ions) and more than 100 molecules have been detected in interstellar space; many were first observed in the submillimeter and FIR region. In the microwave and submillimeter region, radio (heterodyne) techniques are employed in the receivers. At higher frequencies, interferometric techniques are used [41.14]; however, heterodyne techniques recently have been employed at these higher frequencies and have resulted in the detection of CO [41.41] and OH [41.42] at 2.5 THz and a search for the IR band of methylene [41.43] at 30 THz. Submillimeter radio astronomy observatories (at high altitudes above much of the water vapor absorptions in our atmosphere) use laboratory determined frequencies. For example, the Kuiper Airborne Observatory and the submillimeter and FIR telescopes on Mauna Kea, Hawaii require frequencies accurate to about 1 MHz in their heterodyne receivers. For heterodyne detection, fixed frequency FIR gas lasers generally serve as local oscillators, and the frequencies of these lasers must be known.

The discovery and direct frequency measurement of FIR laser lines has continued apace since the last major review of this topic was published in 1986 [41.44]. Almost 1000 lines were listed in that publication; the present number is at least twice this. Two main factors have fuelled this progress. First, the design of CO_2 lasers has continued to develop so that now a single laser can produce 275 individual transitions from the regular, hot [41.45], and sequence [41.46] bands of CO_2 with power levels from 2 to 30 W. Secondly, many new, short-wavelength laser lines have been discovered in a cavity specially designed to promote them. As a result, many high frequency lines have now been characterized, several in the range of 26 to 45 μm. The main lasing molecules used in this work are CH_3OH [41.47–50] and hydrazine, N_2H_4 [41.51].

Recent observations of FIR transitions in molecules in the interstellar medium have been made from satellite-based platforms such as the Infrared Space Observatory (ISO) [41.52] using Fourier Transform methods. Attention has been paid to the measurement of less abundant isotopic forms, such as ^{18}OH and ^{17}OH, because of the information that they provide about star formation processes.

There are four main goals of laboratory SM-FIR spectroscopy which serve the needs of the SM-FIR radio astronomy field:

1. to provide accurate frequencies of SM-FIR species for their detection,
2. to find new far infrared active species,
3. to measure the frequencies of FIR species which can be used to calibrate Fourier transform spectrometers, and
4. to measure frequencies of far-infrared lasers for use as local oscillators in radio astronomy receivers (and to be used in the analysis of laboratory LMR data).

41.3 Upper Atmospheric Studies

A very impressive set of SM-FIR spectra of our upper atmosphere has been observed using balloon-borne FT spectrometers flown at altitudes where the lines are narrow and the spectrometer is above the "black", heavily absorbing water vapor transitions [41.7]. A number of very important species with strong SM-FIR spectra have been observed. They include: O_2, H_2O, NO, ClO, OH, HO_2, O_3, and O. Numerous lines have not yet been identified. It is difficult to calibrate these instruments absolutely because the spectra come from the species emitting at temperatures of about 200 K, and from an indeterminate path length. The high sensitivity of LMR might provide an alternate way of measuring the paramagnetic species in our upper atmosphere by flying an LMR spectrometer to high altitudes. A light-weight solenoid magnet could be used to increase the path length. Absolute concentrations of the paramagnetic species such as OH, HO_2, NO, and O could be obtained.

References

41.1 A.Crocker, H. A. Gebbie, M. F. Kimmitt, L. E. S. Mathias: Nature **201**, 250 (1964)

41.2 E. E. Haller: Infrared Phys. Technol. **35**, 127–146 (1994)

41.3 J. Strong, G. Vanasse: J. Opt. Soc. Amer. **49**, 884 (1959)

41.4 K. M. Evenson, H. P. Broida, J. S. Wells, R. J. Mahler: Phys. Rev. Lett. **21**, 1038 (1968)

41.5 B. F. J. Zuidberg, A. Dymanus: Appl. Phys. Lett. **32**, 367 (1978)

41.6 T. G. Phillips: Techniques of submillimeter Astronomy. In: *Millimetre and Submillimetre Astronomy*, ed. by W. B. Burton (Kluwer, Doordrecht 1988) p.1

41.7 W. A. Traub, K. V. Chance, D. G. Johnson, K. W. Jucks: Proc. Soc. Photo-Opt. Instrum. Engrs. **1491**, 298 (1991)

41.8 D. A. Jennings, K. M. Evenson, D. J. E. Knight: Proc. IEEE **74**, 168 (1986)

41.9 T. D. Varberg, K. M. Evenson: IEEE Trans. Instrum. Meas. **42**, 412 (1993)

41.10 I. G. Nolt, J. V. Radostitz, G. DiLonardo, K. M. Evenson, D. A. Jennings, K. R. Leopold, M. D. Vanek, L. R. Zink, A. Hinz, K. V. Chance: J. Mol. Spectrosc. **125**, 274–287 (1987)

41.11 D. A. Jennings, K. M. Evenson, L. R. Zink, C. Demuynck, J. L. Destombes, B. Lemoine: J. Mol. Spectrosc. **122**, 477 (1987)

41.12 F. Matsushima, K. M. Evenson, L. R. Zink: J. Mol. Spectrosc. **164**, 517 (1994)

41.13 L.-H. Xu, R. M. Lees, K. M. Evenson, C.-C. Chou, J.-T. Shy, E. C. C. Vasconcellos: Can. J. Phys. **72**, 1155 (1994)

41.14 J. W. V. Storey, D. M. Watson, C. H. Townes: Int. J. Infrared Millimeter Waves **1**, 15 (1980)

41.15 H.-P. Röser: Infrared Phys. **32**, 385 (1991)

41.16 P. Helminger, J. K. Messer, F. C. DeLucia: Appl. Phys. Lett. **42**, 309 (1983)

41.17 S. P. Belov, L. I. Gershstein, A. F. Krupnov, A. V. Maslovsky, V. Spirko, D. Papousek: J. Mol. Spectrosc. **84**, 288 (1980)

41.18 G. Winnewisser: Vib. Spectrosc. **8**, 241 (1995)

41.19 B. F. J. Zuidberg, A. Dymanus: Appl. Phys. Lett. **32**, 367 (1978)

41.20 K. M. Evenson, D. A. Jennings, F. R. Petersen: Appl. Phys. Lett. **44**, 576 (1984)

41.21 P. Verhoeve, E. Zwart, M. Versluis, M. Drabbels, J. J. ter Meulen, W. Leo Meerts, A. Dymanus: Rev. Sci. Instrum. **61**, 1612 (1990)

41.22 R. C. Cohen, K. L. Busarow, K. B. Laughlin, G. A. Blake, M. Havenith, Y. T. Lee, R. J. Saykally: J. Chem. Phys. **89**, 4494 (1988)

41.23 F. Lewen, E. Michael, R. Gendriesch, J. Stutzki, G. Winnewisser: J. Molec. Spectrosc. **183**, 207 (1997)

41.24 G. A. Blake, K. B. Laughlin, R. C. Cohen, K. L. Busarow, D.-H. Gwo, C. A. Schmuttenmaer, D. W. Steyert, R. J. Saykally: Rev. Sci. Instrum. **62**, 1693, 1701 (1991)

41.25 E. N. Grossman: Infrared Phys. **29**, 875 (1989)

41.26 K. V. Chance, D. A. Jennings, K. M. Evenson, M. D. Vanek, I. G. Nolt, J. V. Radostitz, K. Park: J. Mol. Spectrosc. **146**, 375 (1991)

41.27 J. S. Wells, K. M. Evenson: Rev. Sci. Instrum. **41**, 226 (1970)

41.28 T. D. Varberg, K. M. Evenson, J. M. Brown: J. Chem. Phys. **100**, 2487 (1994)

41.29 K. M. Evenson: Faraday Discussions of The Royal Society of Chemistry **71**, 7 (1981)

41.30 D. K. Russell: Laser magnetic resonance spectroscopy. In: *Electron Spin Resonance*, Specialist Periodical Reports, Vol. 16, ed. by M. Symons (The Royal Sociey of Chemistry, London 1990) p. 64

41.31 A. I. Chichinin: Magnetic Resonance: Laser Magnetic Resonance. In: *Encyclopedia of Spectroscopy and Spectrometry*, ed. by J. Linton, G. Tranter, J. Holmes (Academic, London 2000) pp. 1133–1140

41.32 N. G. Douglas: *Millimetre and Submillimetre Wavelength Lasers* (Springer, Berlin, Heidelberg 1990)

41.33 T. J. Sears, P. R. Bunker, A. R. W. McKellar, K. M. Evenson, D. A. Jennings, J. M. Brown: J. Chem. Phys. **77**, 5348 (1982)

41.34 T. J. Sears, P. R. Bunker, A. R. W. McKellar: J. Chem. Phys. **75**, 4731 (1981)

41.35 A. R. W. McKellar, P. R. Bunker, T. J. Sears, K. M. Evenson, R. J. Saykally, S. R. Langhoff: J. Chem. Phys. **79**, 5251 (1983)

41.36 F. Tamassia, J. M. Brown, K. M. Evenson: J. Chem. Phys. **110**, 7273 (1999)

41.37 M. D. Allen, K. M. Evenson, J. M. Brown: J. Mol. Spectrosc. **227**, 13 (2004)

41.38 M. D. Allen, K. M. Evenson, D. A. Gillett, J. M. Brown: J. Mol. Spectrosc. **201**, 18 (2000)

41.39 M. D. Allen, K. M. Evenson, J. M. Brown: J. Mol. Spectrosc. **209**, 143 (2001)

41.40 H. Körsgen, K. M. Evenson, J. M. Brown: J. Chem. Phys. **107**, 1025 (1997)

41.41 R. T. Boreiko, A. L. Betz: Astrophys. J. **346**, L97 (1989)

41.42 A. L. Betz, R. T. Boreiko: Astrophys. J. **346**, L101 (1989)

41.43 D. M. Goldhaber, A. L. Betz, J. J. Ottusch: Astrophys. J. **314**, 356 (1987)

41.44 M. Inguscio, G. Moruzzi, K. M. Evenson, D. A. Jennings: J. Appl. Phys. **60**, R161 (1986), Table 5

41.45 A. G. Maki, C.-C. Chou, K. M. Evenson, L. R. Zink, J.-T. Shy: J. Mol. Spectrosc. **167**, 211 (1994)

41.46 C.-C. Chou, A. G. Maki, S. J. Tochitsky, J.-T. Shy, K. M. Evenson, L. R. Zink: J. Mol. Spectrosc. **172**, 233 (1995)
41.47 S. C. Zerbetto, E. C. C. Vasconcellos: Int. J. IR & mmwaves **15**, 889 (1994)
41.48 D. Pereira, E. M. Telles, F. Strumia: Int. J. IR & mmwaves **15**, 1 (1994)
41.49 S. C. Zerbetto, L. R. Zink, K. M. Evenson, E. C. C. Vasconcellos: Int. J. IR & mmwaves **17**, 1049 (1996)
41.50 E. M. Telles, H. Odashima, L. R. Zink, K. M. Evenson: J. Mol. Spectrosc. **195**, 360 (1999)
41.51 E. C. C. Vascocellos, S. C. Zerbetto, L. R. Zink, K. M. Evenson: J. Opt. Soc. Am. B **15**, 1839 (1998)
41.52 C. Ceccarelli, J.-P. Baluteau, M. Walmsley, B. M. Swinyard, E. Caux, S. D. Sidher, P. Cox, C. Gry, M. Kessler, T. Prusti: Astron. Astrophys. **383**, 603 (2002)

42. Spectroscopic Techniques: Lasers

As a primary research tool, the laser plays a fundamental role in the spectroscopic study of atomic and molecular systems. This Chapter describes the basic operating principles, configurations, and characteristic parameters of lasers. Laser designs are discussed and then the details of the interaction of the laser light with matter delineated. The reader is also referred to Chapts. 70 (Laser Priciples) and 71 (Types of Lasers) for further information.

42.1 **Laser Basics** ... 623
42.1.1 Stimulated Emission ... 623
42.1.2 Laser Configurations ... 623
42.1.3 Gain ... 623
42.1.4 Laser Light ... 624

42.2 **Laser Designs** ... 625
42.2.1 Cavities ... 625
42.2.2 Pumping ... 626

42.3 **Interaction of Laser Light with Matter** ... 628
42.3.1 Linear Absorption ... 628
42.3.2 Multiphoton Absorption ... 628
42.3.3 Level Shifts ... 629
42.3.4 Hole Burning ... 629
42.3.5 Nonlinear Optics ... 629
42.3.6 Raman Scattering ... 630

42.4 **Recent Developments** ... 630

References ... 631

42.1 Laser Basics

42.1.1 Stimulated Emission

A cross section σ_{21} for absorption of radiation by a lower state 1 engenders a balancing cross section σ_{12} for emission stimulated by radiation interacting with an upper state 2. Detailed balance relates these two cross sections according to

$$g_2\sigma_{12} = g_1\sigma_{21} , \quad (42.1)$$

where g_1 and g_2 are the statistical degeneracies of their respective states [42.1].

For a collection of emitting and absorbing states with densities n_2 and n_1, amplification may occur when $n_2\sigma_{12} > n_1\sigma_{21}$, which leads to a requirement for an *inversion* of the state populations:

$$n_2/n_1 > g_2/g_1 . \quad (42.2)$$

The rate of spontaneous emission at frequency ν can be modeled itself by stimulated emission induced by a noise source of the magnitude of the density of states $\rho(\nu)$

$$\gamma_{12}(\nu) = \sigma_{12}(\nu)\rho(\nu)/c = \sigma_{12}(\nu)8\pi\lambda^{-2} . \quad (42.3)$$

42.1.2 Laser Configurations

A practical laser combines a population inversion with a means for controlling the radiation.

The basic laser source is the laser oscillator, an amplifier possessing positive feedback. The usual form is simply a piece of active gain medium placed inside a resonant optical cavity (Fig. 42.1). Tunability is produced if the resonant cavity is frequency selective and adjustable (Fig. 42.2). Many laser sources use an amplifier after the oscillator.

42.1.3 Gain

The fundamental gain per pass is given by

$$G = J/J_0 = \exp[(\kappa - \mu)L] , \quad (42.4)$$

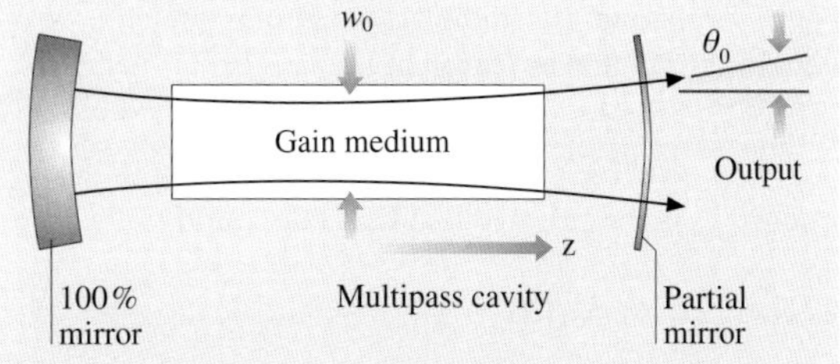

Fig. 42.1 Simple laser oscillator and beam parameters. Distances z are generally measured from the minimum beam waist w_0. The beam appears with a far-field divergence angle θ_0

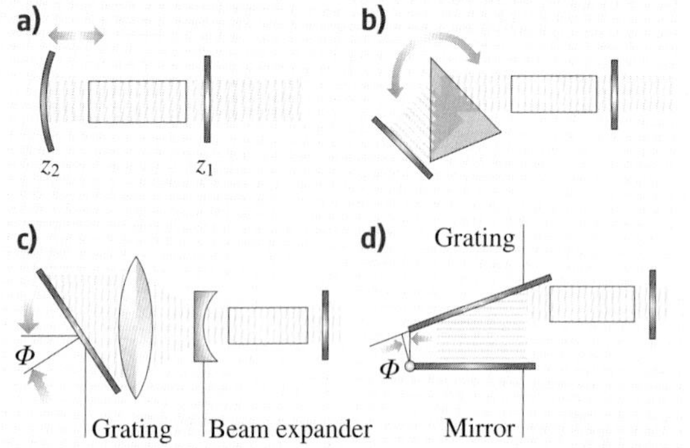

Fig. 42.2a–d Tunable laser oscillator geometries. **(a)** Fabry–Perot: tuning is usually done by changing the cavity length, although changing the index of refraction by changing the temperature or current is common with solid state laser diodes. **(b)** Littrow prism line selector: typical of atomic ion lasers capable of multiple line output. **(c)** Littrow grating tuning: common in pulsed dye lasers with high gain (> 10) per pass. Telescope increases resolution by filling, and reducing angular divergence at the grating. **(d)** Grazing incidence, mirror tuned, grating mount

where J/J_0 is the ratio of light output to input, κ is the gain coefficient, μ is the nonradiative loss rate, and L is the path length. The gain coefficient

$$\kappa = n^* \sigma_{12} \quad (42.5)$$

depends upon the net inversion n^*,

$$n^* = n_2 - (g_1/g_2) n_1 \,. \quad (42.6)$$

If λ is the wavelength of radiation, F_{12} is the emission line shape function normalized over frequency ν, τ_2 is the lifetime of the transition, and f_{12} is the branching ratio for the upper state to undergo this transition, then the stimulated emission cross section is

$$\sigma_{12} = \frac{\lambda^2 f_{12} F_{12}(\nu)}{8\pi\tau_2} \,. \quad (42.7)$$

For a Lorentzian lifetime-broadened line, the cross section for stimulated emission at the line center becomes

$$\sigma_{12} = \frac{\lambda^2 f_{12}}{4\pi^2 \Gamma_{12} \tau_2} \,, \quad (42.8)$$

where Γ_{12} is the full width at half maximum of the line.

42.1.4 Laser Light

Lasers are inherently bright sources of radiation: the radiation field within a practical laser must be high enough for stimulated emission to compete with spontaneous emission. The effective source of spontaneous fluctuations approximates that of the density of states. In terms of the beam parameters photon flux J per solid angle Ω, and frequency ν, this is

$$\frac{\mathrm{d}^2 J}{\mathrm{d}\Omega\,\mathrm{d}\nu} = \frac{2\nu^2}{\epsilon_{\mathrm{r}} c^2} \,. \quad (42.9)$$

Beam *quality* is given by the product of the angular divergence times the beam width. Highest beam quality is associated with diffraction limited light emitted from a Gaussian spot. For circular laser beams traveling in the z-direction, this corresponds to a solution of the electromagnetic wave equation

$$u(r, \phi, z) = \psi(r, z) \exp(-\mathrm{i}kz) \,, \quad (42.10)$$

where u is a polarization component of the field. For high values of $k = 2\pi/\lambda$, corresponding to short wavelength, the adiabatic radial solution is also Gaussian:

$$\psi(r, z) = \exp\left\{-\mathrm{i}\left[P + kr^2/(2q)\right]\right\} \,, \quad (42.11)$$

where the complex phase shift P, beam parameter q, and beam radius w are functions of z:

$$\begin{aligned} P(z) &= -\mathrm{i}\ln\left[1 - \mathrm{i}\lambda z/\left(\pi w_0^2\right)\right] \\ &= -\mathrm{i}\ln\sqrt{1 + \left[\lambda z/\left(\pi w_0^2\right)\right]^2} \\ &\quad - \tan^{-1}\left[\lambda z/\left(\pi w_0^2\right)\right] \,, \end{aligned} \quad (42.12)$$

$$q(z) = \mathrm{i}\pi w_0^2/\lambda + z \,, \quad (42.13)$$

$$w^2(z) = w_0^2 \left[1 + \left(\frac{\lambda z}{\pi w_0^2}\right)^2\right] \,. \quad (42.14)$$

Here, w_0 is the beam waist parameter, the minimum width of the Gaussian beam at a focused spot. For Gaussian beams, the product of the minimum beam waist and beam divergence angle θ_0 is given by

$$\theta_0 w_0 = \lambda/\pi \,. \quad (42.15)$$

The beam waist and divergence follow optical imaging according to paraxial ray theory.

Higher order circular modes with p radial nodes and l angular node planes are specified by multiplying (42.11) and (42.12) by angular and radial factors to obtain

$$\psi_{pl}(r, \phi, z) = \left(\sqrt{2}r/w\right)^l L_p^l\left(2r^2/w^2\right) \mathrm{e}^{\mathrm{i}l\phi} \psi(r, z) \,, \quad (42.16)$$

$$P_{pl}(z) = (2p + l + 1) P(z) \,. \quad (42.17)$$

Here, the functions $L_p^l(x)$ are the Laguerre polynomials as defined in Sect. 9.4.2. The radial phase shifts produce a wave front curvature of effective radius

$$R = z + \left[(2p+l+1)/2 + w_0^2\pi/\lambda^2\right] w_0^2\pi/z . \quad (42.18)$$

Modes with the same values of $2p+l$ have identical axial and radial phase shifts. The two polarization components of the electromagnetic field double the degeneracies of all modes considered here. Often these degeneracies are split in practice by optical inhomogeneities of the medium through which they pass. More details can be found in the summary of *Kogelnick* and *Li* [42.2], or in the texts by *Verdeyen* [42.3] or *Svelto* [42.4].

Some applications require knowledge of the electric field in additionto the flux density J. For purely sinusoidal single mode beams, the rms field is

$$\langle E\rangle = \left(\frac{h\nu J}{c\epsilon_0}\right)^{1/2} . \quad (42.19)$$

Nonlinear effects are often expressed in terms of powers of the field by

$$\langle E^n\rangle = 2^{(n-1)/2}\langle E\rangle^n \quad (42.20)$$

for single mode and multimode radiation of random frequency spacings. For m equally spaced modes, this is increased by $m!/(m-n)!$.

42.2 Laser Designs

42.2.1 Cavities

The simple Fabry–Perot cavity consists of two spherical mirrors facing one another. The surfaces are chosen to be constant phase surfaces for the desired modes (42.18). Stability criteria are shown in Fig. 42.3. A cavity is stable when initial angles θ and displacements r of paraxial rays transform during a round trip into θ' and r' satisfying

$$-2 < \frac{\partial\theta'}{\partial\theta} + \frac{\partial r'}{\partial r} < 2 . \quad (42.21)$$

At frequencies for which the round-trip phase change per passage

$$\delta(\nu) = 2\pi(z_2 - z_1)\nu/c + 2\left[P_{pl}(z_2) - P_{pl}(z_1)\right] \approx 2\pi(z_2 - z_1)/\lambda + \pi(2p+l+1) \quad (42.22)$$

is an integer multiple of 2π, the phases from different passages interfere constructively, giving longitudinal modes. (Here $P_{pl}(z_i)$ gives the additional phase shift for higher transverse modes at mirrors $i = 1, 2$.)

For a particular radial mode structure in an empty Fabry–Perot cavity, the ratio of the maximum cavity decay time for these standing waves to the minimum cavity decay time for frequencies between longitudinal modes is

$$(1+r)^2/(1-r)^2 . \quad (42.23)$$

Here r is the reflectivity of the end mirrors; for cavities with mirrors having different reflectivities, one may use the square root of the product of their reflectivities. A simple Fabry–Perot cavity may be tuned by changing

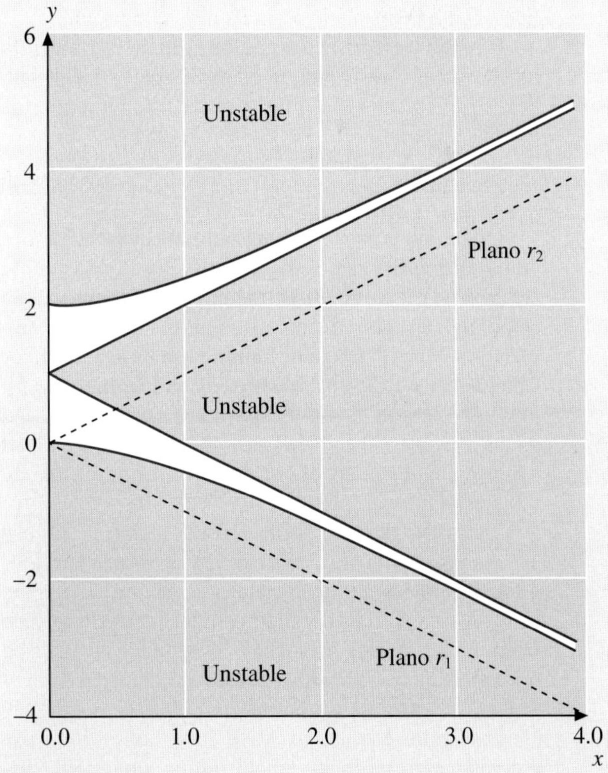

Fig. 42.3 Stability parameters for simple two-mirror laser cavities of length L and mirror radii of curvature R_1 and R_2. Here, $x = \frac{1}{2}(L/R_1 - L/R_2)$ is the mean curvature difference of the two mirrors; $y = \frac{1}{2}(L/R_1 + L/R_2)$ is the mean curvature of the two mirrors. Cavities with parameters in the unshaded region are stable

the cavity length or by changing the index of refraction of the cavity material. Since both of these may be properties of temperature, temperature tuning may be possible. The index of refraction of a material may also be sensitive to the intensity of excitation. Diode lasers consisting of a semiconductor die with polished, reflecting faces are often tuned by changing temperature and pumping current.

For lasers with a dispersive optical element within the cavity, highest selectivity is obtained when the light has low angular divergence and impinges upon the dispersive element as nearly plane waves. A beam expander may reduce the angular spread while simultaneously increasing the beam width. For a grating used as a mirror in a Littrow mount, the dispersion equation is

$$\Delta\lambda = (d/n)\cos\phi\,\Delta\phi\,, \tag{42.24}$$

where n is the diffraction order, d is the distance between lines, and ϕ is the angle of incidence off normal.

Cavities with prisms or gratings can be conveniently tuned by rotating the angle of the dispersive element. For a grating used as a mirror in a Littrow mount, the grating equation is

$$\lambda = (2d/n)\sin\phi\,. \tag{42.25}$$

Often, more than one longitudinal cavity mode operates within the selected frequency band, and tuning consists of "hopping" from mode to mode, rather than smoothly sweeping a single line across a band of frequencies. Smooth frequency tuning can be achieved, for example, in a design by *Wallenstein* and *Hänsch* [42.5], in which the grating and Fabry–Perot cavity are placed together inside a chamber. The whole laser is then tuned by changing the index of refraction of the gas inside by varying its pressure.

The current trend with pulsed lasers is to use a very lossy, short oscillator cavity in which the longitudinal modes are nearly absent, making up for the cavity losses with a very high gain lasing medium. The front, output mirror of such a cavity may actually be *anti*reflection coated, with a reflectivity of only a percent or less.

Low gain, continuous wave (cw) lasers often use a combination of cavity length tuning along with dispersive element rotation, such as a prism or Lyot filter. Often, lasing on a traveling wave in a ring configuration is used to avoid longitudinal modes, as illustrated in Fig. 42.4. Some commonly used gain media are listed in Tables 42.1 and 42.2.

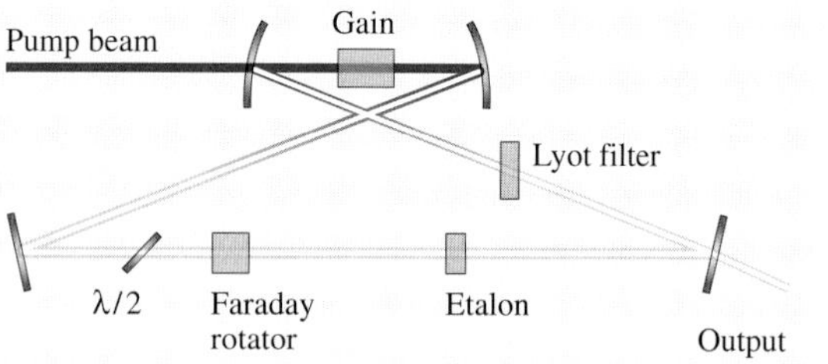

Fig. 42.4 Ring laser. The Faraday rotator and half-wave plate permit circulation of the cavity fields in only one direction

42.2.2 Pumping

Many methods, including electrical discharges and flashlamps, have been used to pump the gain media of lasers. Generally, the best pump is another laser.

Two notable pump lasers have dominated the field of tunable, visible lasers: pulsed neodymium YAG, frequency doubled to ≈ 503 nm, and cw Ar II ion at 514.5 nm. Both are extremely effective at exciting the highly efficient rhodamine class dyes in the red–yellow portion of the visible spectrum. The typical pump beam of a Nd/YAG laser enters the amplifying dye cell transversely along one on the faces of the cell. The typical cw pump beam enters the dye almost collinearly with the laser axis.

By temporarily "spoiling" the Q by making the laser cavity lossy, lasing can be held off until the gain medium stores a greater energy density than the minimum required for lasing. Rapidly switching off the loss mechanism releases this energy in one giant pulse. An electronic optical shutter, such as a Pockels cell, or a saturable dye inside the laser cavity, designed to photobleach from the spontaneous emission just before the laser reaches threshold, are commonly used.

Periodically spoiling the laser gain or the cavity Q at the period of a round trip produces intense, short pulses. Viewed from the frequency domain, the phases of individual longitudinal modes are "locked" together to produce a light packet circulating at the frequency of the reciprocal of the mode spacing. Extremely short pulses (< 1 ps) can be produced by mode locking. Practically, mode locking can be achieved by using a thin, saturable absorber near one of the cavity mirrors [42.6], or by acoustically modulating an optical element in the cavity – even an end mirror itself. One of the best ways of mode locking is to pump a short lifetime gain medium, such as a dye, with mode locked laser light, such as from a mode locked argon ion laser [42.7].

Table 42.1 Fixed frequency lasers

Laser	Wavelength (nm)	Excitation method
ArCl	≈170 (band)	Pulsed, gas discharge
ArF	≈193	Pulsed, gas discharge
KrCl	≈222	Pulsed, gas discharge
KrF	≈248	Pulsed, gas discharge
XeBr	≈282	Pulsed, gas discharge
XeCl	≈308	Pulsed, gas discharge
XeF	≈351, 353	Pulsed, gas discharge
N_2	≈337.1 (other bands)	Pulsed (1-10 ns), gasdischarge
Ar^+	488.0, 514.5 (454.4, 457.9, 465.8, 472.7, 476.5, 501.7, 528.7)	cw, gas discharge
Ar^{+2}	351.1, 363.8	cw, gas discharge, high magnetic field
Kr^+	568.2, 647.1, (476.2, 520.8, 530.9)	cw, gas discharge
Kr^{+2}	350.7, 356.4, 406.7	cw, gas discharge, high magnetic field
Ne/He	632.8, 1152.3, 3390 (others)	cw, gas discharge
Cr^{+3}-Ruby	694.3	Pulsed, flashlamp
Nd^{+3}-YAG	≈1065, ≈1300	Pulsed: flashlamp; cw: lamp, LED or laser diode
Nd^{+3}-glass(various)	≈920, ≈1060, ≈1370	Pulsed, flashlamp
Yb^{+3}-glass	≈1060	Pulsed, flashlamp
Er^{+3}-glass(various)	≈1540, ≈1536, ≈1543, ≈1550	Pulsed: flashlamp; cw: lamp, LED or laser diode
Xe	3507	cw gas discharge
CO_2	≈10 600 (lines)	cw: gas discharge or gas dynamic
Molecular vibration and rotation	1×10^5–3×10^6 (numerous lines)	IR laser (CO_2, CO, ⋯)

Table 42.2 Approximate tuning ranges for tunable lasers

Laser	Wavelengths (nm)	Notes
Dye solution	<330 – >1200 [≈10 each dye]	Pulsed: laser, flashlamp; cw: laser
Alexandrite	700–820	Pulsed: lamp, laser; cw: laser
Ti^{+3}/Sapphire	680–1100	Pulsed: laser, lamp; cw: laser
GaAs	840–900	pn junction
InGaAlP/GaAs	630–700 [1–10]	pn junction
GaAsP	550–880	pn junction
AlGaAs, AlGaAs/GaAs	≈820 [1–20], 720–880	pn junction, temp. & current tuning
InP	≈900	pn junction, temp. & current tuning
GaInAs	906–3100	pn junction, temp. & current tuning
InGaAlAs/GaAs	800–1100	pn junction, temp. & current tuning
InPAs	900–4000 [1–50]	pn junction, temp. & current tuning
InGaAsP/InP	1200–1650	pn junction, temp. & current tuning
InAs	≈3100	pn junction, temp. & current tuning
InSb	≈5200	pn junction, temp. & current tuning
PbS	≈4300	pn junction, temp. & current tuning
PbTe	≈6500	pn junction, temp. & current tuning
PbSe	≈8500	pn junction, temp. & current tuning
PbSnSeTe "Lead Salt Diode"	4500–15 000 [1–50]	pn junction, temp. & current tuning

Table 42.2 Approximate tuning ranges for tunable lasers, cont.

Laser	Wavelengths (nm)	Notes
Color Centers: (F2+)/LiF	800–1040	cw: laser, Ar, Kr
/NaF	900–1050	cw: laser, Ar, Kr
/KCl:Tl	1400–1700	cw: laser, Ar, Kr, Nd/YAG
/KF	1300–1400	cw: laser, Ar, Kr, Nd/YAG
/NaCl	1400–1600	cw: laser, Ar, Kr, Nd/YAG
/KCl	1600–1700	cw: laser, Ar, Kr, Nd/YAG
/KBr	1700–1900	cw: laser, Ar, Kr, Nd/YAG
/KCl:Li	2500–2900	cw: laser, Ar, Kr, Nd/YAG
/KCl:Na	1600–1950	cw: laser, Ar, Kr, Nd/YAG

42.3 Interaction of Laser Light with Matter

42.3.1 Linear Absorption

The absorption cross section σ_{fi} for transition to the final state $|f\rangle$, when integrated over frequency ν, is given theoretically by the leading first-order perturbation of the initial state $|i\rangle$ in the electric dipole approximation

$$\int \sigma_{fi}(\nu)\,\mathrm{d}\nu = 4\pi^2\alpha\bar{\nu}\left|\langle f|\sum_e \boldsymbol{r}_e \cdot \hat{\boldsymbol{\epsilon}}|i\rangle\right|^2 , \qquad (42.26)$$

where the sum is over all charges e at distances r_e, $\hat{\epsilon}$ is a unit polarization vector, and $\bar{\nu}$ is the averaged transition energy. Averaged over orientations and summed over possible final states, each electron contributes to the total integrated absorption cross section one electron oscillator, $\pi r_0 c \approx 0.03\,\mathrm{cm}^2\,\mathrm{s}^{-1}$; here, r_0 is the classical electron radius. Electronic absorption bands typically contain an oscillator strength $f = 0.01$–0.5 of an "electron oscillator", while weaker vibrational transitions have $f = 10^{-6}$–10^{-4} in each band.

For narrow lines with radiation of broader band width $\Delta\nu$ at flux density J, the linear absorption rate constant can be usefully estimated as $(\pi r_0 c) fJ/\Delta\nu$.

42.3.2 Multiphoton Absorption

Second-order perturbation theory gives the theoretical two-photon contribution to the absorption. An absorption cross section $\sigma^{(2)} J_1$ for a photon of frequency ν_2 is induced by an off-resonance monochromatic field of frequency ν_1 and photon flux density J_1. When integrated over the frequency of the second photon, the second-order cross section can be related to the dipole matrix elements

$$\int \sigma_{fi}^{(2)}(\nu_1, \nu_2) J_1\,\mathrm{d}\nu_2 = \frac{4\pi^2\alpha^2}{\nu_1\bar{\nu}_2} J_1 \times \left|\sum_m \frac{\nu_{fm}\nu_{mi}\langle r_{fm}\rangle\langle r_{mi}\rangle}{\nu_{mi} - \nu_1}\right|^2 \qquad (42.27)$$

where

$$\langle r_{fm}\rangle = \langle f|\sum_e \boldsymbol{r}_e \cdot \hat{\boldsymbol{\epsilon}}_2|m\rangle ,$$
$$\langle r_{mi}\rangle = \langle m|\sum_e \boldsymbol{r}_e \cdot \hat{\boldsymbol{\epsilon}}_1|i\rangle , \qquad (42.28)$$

and the sum is taken over intermediate states $|m\rangle$ having frequencies ν_{mi} and ν_{fm} for transitions to the initial and final states. The energy for the overall transition comes from two photons; hence the two-photon resonance condition $\nu_{fi} = \nu_1 + \nu_2$. A special case often occurs when only one radiation frequency is used: then $\nu_1 = \nu_2$, and two-photon resonance is achieved when the energy of the transition corresponds to twice the frequency of the radiation field.

This integrated, induced cross section is roughly of the order $\alpha(\pi c r_0)^2 J_1/(\nu_1\bar{\nu}_2)$. Two-photon absorption becomes comparable to the one-photon absorption with off-resonance fields on the order of

$$J_1 \approx \frac{\nu_1\nu_2}{\alpha\pi c r_0} , \qquad (42.29)$$

or about 10^{21} photons $\mathrm{s}^{-1}\,\mathrm{cm}^{-2}$ ($10^{17}\,\mathrm{W\,m}^{-2}\,\mathrm{s}^{-1}$ for typical green light).

Typical dipole-allowed molecular multiphoton absorption cross sections are $\approx 10^{-58}\,\mathrm{m^4\,s}$ for two photons and $10^{-94}\,\mathrm{m^6\,s^2}$ for three photons.

Multiphoton absorption was one of the first effects explored with lasers. Two-photon absorption was first reported for inorganic crystals containing europium ions [42.8]. Blue and ultraviolet fluorescence appeared in the interaction of red ruby laser light with organic compounds [42.9–12]. Others observed two-photon absorption directly [42.13, 14]. Selection rules for multiphoton absorption are summarized by *McClain* [42.15]; recent work is reviewed by *Ashfold* and *Howe* [42.16].

Highly excited states may subsequently ionize in the intense fields in a multiphoton ionization (MPI) process. The ionization signal is often detected in a proportional ionization cell. A low pressure cell containing the vapor of a transition metal organometallic, such as iron carbonyl or ferrocene which photodissociates to give the metal atom, may be used for wavelength calibration by MPI.

42.3.3 Level Shifts

High radiation power causes resonant frequencies to shift, responding to the ac Stark effect [42.17, 18]. Even moderate fields, tuned near resonance, interact strongly with an atom or small molecule, which undergoes rapid excitation and stimulated emission at the characteristic power dependent Rabi frequency,

$$\nu_{\mathrm{Rabi}} = \frac{2\pi g_1}{g_1 + g_2}\sigma_{21} J\,. \qquad (42.30)$$

Fluorescence from such an interacting system has a characteristic "head and shoulders" spectrum best understood as radiation at the fundamental frequency amplitude modulated at the Rabi frequency [42.19].

42.3.4 Hole Burning

Radiation at a particular frequency generally moves population out of states that absorb that radiation. Molecules that interact resonantly with the radiation may also spontaneously emit at different frequencies, thereby ending in nonabsorbing states. This optical pumping effect can make resonance features in an absorption spectrum disappear at high power levels [42.20]. Depletion may appear as a "hole" in the absorption spectrum [42.21]. Recently, interest has shifted to permanent "hole burning" as a method of information storage in materials.

Hole burning is the basis for Doppler-free Lamb-dip spectroscopy, in which only absorbing atoms or molecules having little or no velocity component along the axis of two counterpropagating beams are temporarily depleted [42.22]. This technique is commonly used for laser frequency stabilization, such as with the iodine-locked He–Ne laser.

42.3.5 Nonlinear Optics

Multiplying

Nonlinear susceptibility of an optical medium can generate radiation at frequencies which are multiples of the frequency of laser radiation passing through. Phenomenologically, the second order polarization

$$\boldsymbol{P}_{2\nu} = \epsilon_0 \chi^{[2]} \boldsymbol{E}_\nu^2 \qquad (42.31)$$

is given in terms of the second order nonlinear susceptibility $\chi^{[2]}$, a third rank tensor, and the electric field at the fundamental frequency (see Chapt. 72). This nonlinear susceptibility can range from 0.5–5 pm/V for typical materials used for frequency doubling. Typical materials and their use are reviewed by *Bordui* and *Fejer* [42.23].

For a nonlinear process occurring over a length l in a cylindrical region with Gaussian waist w_0 with polarization P_0 on axis, the far field flux is given by

$$J(R,\theta) = \frac{\pi^4 \nu^3 n^3 w_0^4 P_0^2}{4hc^3 \epsilon_0 R^2}\left(\frac{\sin^2[(k_0 - k\cos\theta)l/2]}{(k_0 - k\cos\theta)^2}\right) \times \exp\left[-k^2 w_0^2 \sin^2(\theta)/2\right]. \qquad (42.32)$$

Here, n is the index of refraction and k the propagation constant for the induced radiation, while k_0 is that for the induced polarization, the vector sum of those of the original radiation. When phase matched, $k_0 - k\cos\theta = 0$, and the term in the brackets maximizes to $(l/2)^2$.

The greatest difficulty is in selecting materials which can be phase matched such that the relative phases of the fundamental and overtone radiation propagate together through the material; otherwise radiation at the higher frequency generated at different places inside the material destructively interfere. Phase matching is usually achieved by either angle tuning of a birefringent crystal, or by temperature tuning.

Only materials without a center of inversion in their crystal structures have a second order nonlinear susceptibility. All materials, including gases, will have a third order nonlinear susceptibility $\chi^{[3]}$. This can be used to generate third harmonics, especially in the vacuum ultraviolet (VUV), where doubling materials are not available [42.24, 25].

Mixing

The same materials that permit frequency doubling and tripling also allow 3-wave and 4-wave frequency mixing. The frequency matching conditions are, respectively,

$$\nu_1 \pm \nu_2 = \nu_3 ,$$
$$\nu_1 \pm \nu_2 \pm \nu_3 = \nu_4 . \quad (42.33)$$

Tunable UV radiation may be generated by adding the frequencies of a fixed and tunable visible outputs. Tunable IR has been obtained by differencing fixed and tunable visible lasers. Mixing of radiation from an Ar ion laser with that from a tuned R6G dye laser in lithium iodate to produce tunable 2200–4600 nm radiation is noteworthy.

Optical Parametric Oscillator

It is possible to reverse 3-wave mixing, generating two frequencies whose sum is that of the input radiation. In parametric generation, the output frequencies are given by the phase match conditions. Both the desired frequency and the secondary "idler" frequency must be allowed to build up in the nonlinear medium. The idler radiation is not present initially, but results from the frequency mixing process itself. The process has many of the characteristics of a laser oscillator, including that of a gain threshold. This makes tuning of an optical parametric oscillator similar to that of a laser, but with more degrees of freedom: now oscillation at two frequencies must be attained simultaneously, along with the correct *phase matching* of the nonlinear material [42.26].

42.3.6 Raman Scattering

It is possible to have one or more of the fields in a mixing process belong to just polarization, rather than radiation. The frequency additive case of multiphoton absorption has already been consisdered; the frequency subtractive case is Raman scattering.

Incoherent Raman Scattering

Radiation at a higher frequency can excite a lower frequency vibration or rotation within a material, with appearance of radiation at the frequency of the incident radiation minus that of the absorption. The integrated cross section for this effect is given by (42.27). However, in this case the cross section is for *emission* of the second photon. The rate of spontaneous Raman emission is obtained by multiplying (42.27) by the spectral flux density of the zero-point field, $8\pi\nu^2/\epsilon_r c$. Typical vibrational Raman cross sections for transparent molecules are about $10^{-34}\,\mathrm{m^2\,sr^{-1}}$ in the blue–green (488 nm) [42.27].

Alternatively, energy can be extracted from an excited state, with the inelastically scattered photon departing with the sum of the incident frequency and that of the deexcitation. The term "anti-Stokes" distinguishes it from the more usual "Stokes" type of Raman scattering.

Coherent Raman Emission

The integrated cross section for emission stimulated at the Raman frequency is given directly by (42.27).

A fourth-order mixing process that is less susceptible to saturation involves coherent anti-Stokes Raman scattering (CARS). Two beams excite the material: the difference of their frequencies corresponds to an excitation of the material. Stimulated Raman scattering excites the material, which then deexcites through an anti-Stokes process, giving rise to a third, higher frequency radiation field. The phase is determinate, and the radiation leaves the region of scattering as a beam [42.28].

If the incident radiation induces a Raman process over a sufficiently long path, the stimulated Raman process can be used for gain at both the Stokes and anti-Stokes frequencies. Since spontaneous Raman processes are proportional to the integrated cross section, while gain in the stimulated Raman process is proportional to the peak cross section, simple materials with sharp, simple line spectra are most suitable for Raman gain media. While Raman lasers have been produced using vibrational excitation of organic liquids, currently the most important technical application is for Raman shifting the output of lasers, tunable or otherwise, to frequencies which otherwise would be inaccessible. The high pressure H_2 Raman shifter produces, at low powers, beams consisting almost entirely of well separated, sharp lines shifted by $4160\,\mathrm{cm^{-1}}$ from the pump beam; at high powers, a series of Stokes and anti-Stokes bands appear, each separated by $4160\,\mathrm{cm^{-1}}$ from each other.

42.4 Recent Developments

One of the most exciting advancements in the past decade in laser physics has been the generation of optical frequency combs; and, more specifically, their applicability in the domain of high-resolution laser spectroscopy. Basically, through a superposition process of many continuous wave modes, a short train of frequency spikes may be produced from a mode-locked laser [42.29] (see also Sect. 30.1.5). These spikes are equally spaced and are referred to as a frequency comb. The frequency ω_n of the nth cavity mode may be expressed as

$$\omega_n = n\omega_r + \omega' , \tag{42.34}$$

where ω_r is characteristic of the laser and ω' is a frequency offset due to the difference between the phase and group velocity of the superposed waves.

The microwave frequencies ω_r and ω' are determined through the use of nonlinear optics. Once these two parameters are determined, any unknown optical frequency ω_o may be measured by recording the beat frequency between it and the closest comb frequency. This technique gives experimenters a high-precision method for the spectroscopic determination of such fundamental quantities as the fine structure constant, the Rydberg constant, and the Lamb shift [42.30, 31].

Text and references updated by Mark M. Cassar

References

42.1 A. Einstein: Phys. Z. **18**, 121 (1917)
42.2 H. Kogelnik, T. Li: Proc. IEEE **54**, 1312 (1966)
42.3 J. T. Verdeyen: *Laser Electronics*, 3rd edn. (Prentice Hall, Englewood Cliffs 1994)
42.4 O. Svelto: *Principles of Lasers* (Plenum, New York 1976)
42.5 R. Wallenstein, T. W. Hänsch: Opt. Commun. **14**, 353 (1975)
42.6 E. P. Ippen, C. V. Shank, A. Dienes: Appl. Phys. Lett. **21**, 348 (1972)
42.7 R. K. Jain, C. P. Ausschnitt: Opt. Lett. **2**, 117 (1978)
42.8 W. Kaiser, C. G. B. Garrett: Phys. Rev. Lett. **7**, 229 (1961)
42.9 W. L. Peticolas, J. P. Goldsborough, K. E. Rieckhoff: Phys. Rev. Lett. **10**, 4345 (1963)
42.10 W. L. Peticolas, K. E. Reickhoff: J. Chem. Phys. **39**, 1347 (1963)
42.11 S. Singh, B. P. Stoicheff: J. Chem. Phys. **38**, 2032 (1963)
42.12 S. Z. Weisz, A. B. Zahlen, J. Gilreath, R. C. Jarnagin, M. Silver: J. Chem. Phys. **41**, 3491 (1964)
42.13 J. J. Hopfield, J. M. Warlock, K. Park: Phys. Rev. Lett. **11**, 414 (1963)
42.14 B. Staginnus, D. Frölich, T. Caps: Rev. Sci. Instrum. **39**, 1129 (1968)
42.15 W. M. McClain: Acc. Chem. Res. **7**, 129 (1974)
42.16 M. N. Ashfold, J. D. Howe: Ann. Rev. Phys. Chem. **45**, 57 (1994)
42.17 H. J. Carmichael, D. F. Walls: J. Phys. B **9**, 1199 (1976)
42.18 L.-P. Li, B.-X. Yang, P. M. Johnson: J. Opt. Soc. Am. **2**, 748 (1985)
42.19 B. R. Mallow: Phys. Rev. **188**, 1969 (1969)
42.20 L. J. Rothberg, D. P. Gerrity, V. Vaida: J. Chem. Phys. **75**, 4403 (1981)
42.21 S. Völker: Ann. Rev. Phys. Chem. **40**, 499 (1989)
42.22 W. R. Bennett Jr.: Phys. Rev. **126**, 580 (1962)
42.23 P. F. Bordui, M. F. Martin: Ann. Rev. Mater. Sci. **23**, 321 (1993)
42.24 A. Kung: Opt. Lett. **8**, 24 (1983)
42.25 J. Bokor, P. Bucksbaum, R. Freeman: Opt. Lett. **8**, 217 (1983)
42.26 J. Falk, J. M. Yarbourough, E. O. Ammann: IEEE J. Quantum Elec. **7**, 359 (1971)
42.27 C. M. Penney, R. L. St. Peters, M. Lapp: J. Opt. Soc. Am. **64**, 712 (1974)
42.28 P. R. Regnier, J. P. Taran: Appl. Phys. Lett. **23**, 240 (1973)
42.29 T. Udem, R. Holzwarth, M. Zimmerman, C. Goble, T. Hänsch: Topics Appl. Phys. **95**, 295 (2004)
42.30 J. M. Hensley: A precision measurement of the fine structure constant. Ph.D. Thesis (Stanford University, Stanford 2001)
42.31 B. de Beauvoir, C. Schwob, O. Acef, L. Jozefowski, L. Hilico, F. Nez, L. Julien, A. Clairon: Eur. Phys. J. D **12**, 61 (2000)

43. Spectroscopic Techniques: Cavity-Enhanced Methods

Cavity enhanced spectroscopy (CES) methodology provides a much higher degree of sensitivity than that available from conventional absorption spectrometers. The aim of this chapter is to present the fundamentals of the method, and the various modifications and extensions that have been developed. In order to set the stage, the limitations of traditional absorption spectrometers are first discussed, followed by a description of cavity ring-down spectroscopy (CRDS), the most popular CES embodiment. A few other well-known CES approaches are also described in detail. The chapter concludes with a discussion of recent work on extending CRDS to the study of liquids and solids.

43.1 **Limitations of Traditional Absorption Spectrometers** 633
43.2 **Cavity Ring-Down Spectroscopy** 634
 43.2.1 Pulsed Cavity Ring-Down Spectroscopy 634
 43.2.2 Continuous-Wave Cavity Ring-Down Spectroscopy (CW-CRDS) 635
43.3 **Cavity Enhanced Spectroscopy**............... 636
 43.3.1 Cavity Enhanced Transmission Spectroscopy (CETS)..................... 637
 43.3.2 Locked Cavity Enhanced Transmission Spectroscopy (L-CETS).................................. 638
43.4 **Extensions to Solids and Liquids** 639
References 640

43.1 Limitations of Traditional Absorption Spectrometers

An absorption spectrometer measures the difference in intensity between the incident light intensity I_0 and the transmitted light intensity $I(x, \lambda)$. Beer's law relates the absorbed light to the sample absorption $\alpha(\lambda)$

$$I(x, \lambda) = I_0\,\mathrm{e}^{-\alpha(\lambda)x}\,, \qquad (43.1)$$

where λ is wavelength, and x is path length. Absorption is related to concentration C through the extinction coefficient $\varepsilon(\lambda)$ namely $\alpha(\lambda) = C\varepsilon(\lambda)$. Typically, a spectral feature, called an absorption peak, of the target species is measured in order to obtain its concentration. The performance of a spectrometer has two figures of merit: sensitivity and selectivity.

Sensitivity is the smallest detectable change in one centimeter of path length that a spectrometer can measure during one second. If many measurements can be made within a second, averaging may be used to further improve (by a factor of the square root of the number of measurements or the square root of the data acquisition rate) the achievable sensitivity. Sensitivity has units of $\mathrm{cm^{-1}\,Hz^{-1/2}}$. Sensitivity can also be quantified using the minimum detectable absorption loss (MDAL), i. e., the normalized standard deviation of the smallest detectable change in absorption (units of $\mathrm{cm^{-1}}$). Equation (43.1) shows that the sensitivity of a spectrometer depends not only on the light path length through the sample, but also on the intensity noise of the light source.

Selectivity is the ability of a spectrometer to distinguish between two different species absorbing at similar wavelengths. The instrument must be able to resolve the different spectral lines. Thus, selectivity depends on spectral resolution, which has units of frequency (MHz), wavelength (nm), or wavenumbers ($\mathrm{cm^{-1}}$).

Traditional spectrometers, such as non-dispersive infrared (NDIR), and Fourier Transform infrared (FTIR), use incoherent thermal light sources. For both techniques, the physical length of their sample chamber limits their sensitivity. Some devices try to fold the light path several times through the sample chamber in order to improve sensitivity, but this approach encounters physical size and mechanical stability limitations. Typical MDAL are in the $10^{-5}\,\mathrm{cm^{-1}}$ range. These instruments therefore rely on measuring the strongest absorption transitions available, which are found in the mid-infrared range (3 to 12 μm). Often, however, the strongest transitions overlap with features of other species found in the sample mixture. The instrument performance becomes a sensitivity-selectivity tradeoff.

Laser-based optical detection methods, called tunable diode laser absorption spectroscopy (TDLAS), circumvent some of these problems by exploiting the coherent nature of laser light. A tunable continuous-wave laser source brings two benefits:

- A narrow linewidth, which allows high spectral resolution scans to be performed, and
- Low spatial beam divergence, which allows it to be folded hundreds, if not thousands of times.

By transmitting laser light having a small beam size over long distances, multi-pass cells can be designed to achieve up to a kilometer of path length enhancement. Multi-pass cell laser spectroscopy systems have demonstrated MDAL down to $10^{-9}\,\mathrm{cm}^{-1}$. However, such instruments still remain limited by laser intensity fluctuations and interference fringes. Moreover, standard multi-pass techniques do not provide an absolute optical loss measurement.

43.2 Cavity Ring-Down Spectroscopy

Cavity ring-down spectroscopy (CRDS) is a more recently developed TDLAS approach that replaces a multi-pass cell with a stable optical resonator, called the ring-down cavity (RDC). CRDS is based on the principle of measuring the rate of decay of light intensity inside the RDC. The transmitted wave decays exponentially in time. The decay rate is proportional to the total optical losses inside the RDC.

In a typical CRDS setup, light from a laser is first injected into the RDC, and is then interrupted. The circulating light inside the RDC is both scattered and transmitted by the mirrors on every round-trip, and can be monitored using a photodetector placed behind one of the cavity mirrors. The decay constant, also called the ring-down time τ is then measured as a function of laser wavelength to obtain a spectrum of the cavity optical losses. Detailed mathematical treatments of CRDS can be found in [43.1]. A simple derivation is presented here.

For a given wavelength λ the transmitted light $I(t, \lambda)$ from the RDC is given by

$$I(t,\lambda) = I_0\,\mathrm{e}^{-\frac{t}{\tau(\lambda)}}\,, \tag{43.2}$$

where I_0 is the transmitted light at the time the light source is shut off, and $\tau(\lambda)$ is the ring-down time constant. The total optical loss inside the cavity is $L(\lambda) = [c\tau(\lambda)]^{-1}\, l_{\mathrm{rt}}$ where c is the speed of light. The total optical loss comprises the empty cavity optical loss and the sample optical loss. CRDS provides an absolute measurement of these optical losses. The empty cavity (round-trip) optical loss $L_{\mathrm{empty}}(\lambda)$ comprises the scattering and transmission losses of the mirrors. In general, better mirrors provide lower empty cavity losses and higher sensitivity. The sample (round-trip) optical loss is $A(\lambda) = \alpha(\lambda) l_{\mathrm{rt}}$ where l_{rt} is the cavity round-trip length, and is simply the difference between total cavity losses and empty cavity losses, namely $A(\lambda) = L(\lambda) - L_{\mathrm{empty}}(\lambda)$. Once the absorption spectrum $\alpha(\lambda)$ of the sample has been measured, then the sample concentration can be readily computed using the absorption cross section and lineshape parameters.

The MDAL for a CRDS system is defined by

$$\alpha_{\min} = \frac{1}{l_{\mathrm{eff}}}\left(\frac{\Delta\tau}{\tau}\right)\,, \tag{43.3}$$

where $(\Delta\tau/\tau)$ is called the shot-to-shot noise of the system. The effective path length of a CRDS measurement is $l_{\mathrm{eff}} = c\tau$. For typical RDC mirrors having a reflectivity of 99.995%, and scattering losses of less than 0.0005%, the path length enhancement can exceed 20 000. For a 20 cm long sample cell, the effective path length is 8 km, which exceeds the best performance of multi-pass spectroscopy by a factor of three, based on effective path length alone. A good CRDS system can achieve a shot to shot variation of 0.03%, leading to a MDAL of $4\times10^{-10}\,\mathrm{cm}^{-1}$. Note also that the CRDS measurement is not dependent on either the initial intensity of the light inside the cavity, provided that the signal has a sufficient signal to noise ratio at the detector, or on the physical sample path length like traditional absorption spectroscopy. Moreover, CRDS can use laser sources having narrow linewidths and achieving high spectral resolution.

CRDS can resolve all three limitations of absorption spectroscopy, namely sensitivity, selectivity, and dependence on intensity noise of the light source [43.1]. CRDS has been implemented using many different approaches. This chapter will discuss several commonly used variations on the CRDS technique.

43.2.1 Pulsed Cavity Ring-Down Spectroscopy

Early implementations of CRDS used pulsed lasers sources (P-CRDS) [43.2]. A typical P-CRDS setup

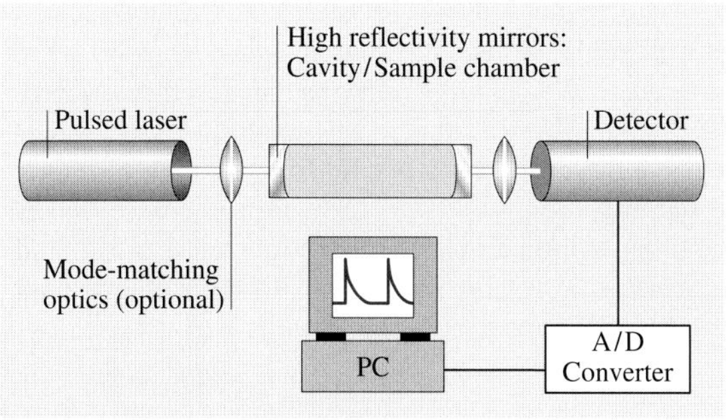

Fig. 43.1 Typical P-CRDS setup

is shown in Fig. 43.1. Today, CRDS has been implemented in the broadest possible range of wavelengths, from the UV (216 nm) to the mid-infrared (10 μm). Because of its experimental simplicity, P-CRDS has become a widespread tool for chemists and spectroscopists, finding applications in the measurement of predissociation dynamics, photolysis products, radiative lifetimes, aerosols or soot detection, temporal imaging, overtone vibrational spectroscopy, and kinetic studies. Typical P-CRDS sensitivity is $1\times10^{-9}\,\text{cm}^{-1}\text{Hz}^{-1/2}$.

P-CRDS methods have also been combined with other detection methodologies. Variations on P-CRDS include:

- Fourier-transform (FT) P-CRDS [43.3], where an RDC is excited with a broadband source and time-resolved FT scans of the RDC output waveform are taken. Inversion of the interferogram then produces time-dependent ring-down waveforms at all resolved frequencies within the source wavelength range.
- Polarization P-CRDS [43.4], where the spectral splitting induced by the magnetic field is observed from the difference of the ring-down spectra of the two orthogonal polarizations.
- Pulse-stacked P-CRDS [43.5], where the length of the RDC is set so that pulses from a very high repetition rate pulsed source coherently add together, which increases the effective cavity light throughput to yield improved detection of the ring-down waveform.

Spectral resolution and sensitivity of P-CRDS are, however, limited by the use of short-pulse lasers. The effects of pulsed laser bandwidth on spectral resolution of P-CRDS have been studied extensively and have led to system designs where only a single longitudinal and transverse mode of the RDC is excited [43.6]. However, single-mode P-CRDS is difficult to implement experimentally, and still has limited sensitivity because the laser pulse is substantially attenuated by the RDC mirrors at the cavity output. The requirement for improving CRDS sensitivity by reducing the variability in the measurement of the decay constant from shot to shot with single-mode excitation, and improving the light transmission through the cavity to increase the signal to noise ratio of the decay waveform on the detector, provided the catalyst to implement continuous wave (CW) lasers in CRDS.

43.2.2 Continuous-Wave Cavity Ring-Down Spectroscopy (CW-CRDS)

CW lasers have narrow linewidths (< 50 MHz) and can be tuned in small spectral increments (< 50 MHz) to achieve high spectral resolution with excellent wavelength reproducibility. Moreover, owing to their narrow linewidths, they are better suited for efficiently coupling into a single mode of a high finesse RDC, thereby reducing shot-to-shot variations in measured ring-down decay constant. Furthermore, they can be directly modulated, thereby allowing higher data repetition rates, leading in turn to improved sensitivity.

The first efforts in CW-CRDS involved optically locking a laser diode to a high-finesse cavity, but the performance was limited because the laser diode would drift and lock to different RDC modes. The use of sufficient optical isolation and an external optical switch resolved this issue. The most common approach used today is to sweep a RDC mode through the emission profile of a diode laser, and shut the laser off with an acousto-optic modulator (AOM) (Fig. 43.2) when sufficient light is injected into the cavity [43.7]. Numerous variations on this approach exist. For example, CW-CRDS can be implemented by rapidly sweeping the cavity mode into and

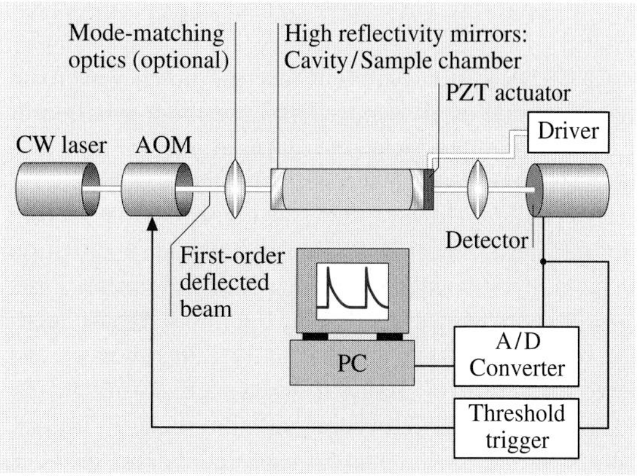

Fig. 43.2 Typical CW-CRDS setup

Part C | 43.2

out of resonance with the laser. The simplest approach is to sweep the laser wavelength into and out of resonance with a cavity mode and directly modulate the laser source current [43.8]. The most popular CW lasers used today are distributed feedback (DFB) diode lasers.

In most CW-CRDS embodiments, the ring-down cavities are linear, i. e., consist of two mirrors. Ring resonators (e.g., triangular or bow-tie cavities) can also be used. Ring cavities provide the benefits of minimizing optical feedback and eliminating the need for extensive isolation or frequency shifting of the laser source. Moreover, ring resonators break the frequency degeneracy between cavity modes having orthogonal polarizations, effectively creating two coupled resonators having high and low finesse, respectively. The low finesse cavity can be used to lock the laser to a single cavity mode, while the high finesse cavity can be used for CRDS. The AOM acts as both a frequency-shifter and an on-off switch. This method was used to demonstrate the highest CRDS sensitivity achieved to date, namely $1.0 \times 10^{-12}\,\mathrm{cm}^{-1}\mathrm{Hz}^{-1/2}$ [43.9].

Prism-based, rather than mirror-based cavities have also been used [43.10]. Such cavities comprise two prisms whose intra-cavity facing angles are at Brewster's angle, and whose extra-cavity facing angles are such that total internal reflection occurs, and results in unit reflectivity. One of the prisms has a curved facet to create a stable optical resonator. The purpose of this design is to extend operation over a much broader range of wavelengths than can be achieved using dielectric coatings. Currently, high reflectivity mirrors are limited in bandwidth to about $\pm 15\%$ of their center wavelength.

CW-CRDS has been applied over a wide range of wavelengths. It has been used for medical breath analysis, trace gas detection in environmental and process control applications, and isotopic analysis. In the near-infrared, CW-CRDS systems achieve sensitivities of $10^{-11}\,\mathrm{cm}^{-1}\mathrm{Hz}^{-1/2}$, and a concentration measurement repeatability of 1 part in 5000. Similar performance in the mid-infrared ($3\,\mu\mathrm{m}$) resulted in the detection of parts-per-trillion ethylene concentrations [43.11].

Extensions of the basic CW-CRDS technique have also been developed:

- Phase-shifted CW-CRDS [43.12]: the phase shift accrued by a sinusoidally modulated CW laser is measured for both an empty RDC and one having a sample. The concentration is deduced from these two measurements.
- Heterodyne CW-CRDS [43.13]: enhances the power in the ring-down decay waveform by mixing with a local oscillator. For example, the local oscillator can be the orthogonal polarization used to lock the laser to the RDC, or can be the reflected signal from the RDC when laser is frequency-shifted (by the local oscillator frequency) off resonance from the RDC mode. Heterodyne CRDS can approach the shot-noise limit.
- V-cavity CW-CRDS [43.14]: a three-mirror V-shaped RDC is exploited to achieve consistent repetitive optical locking of a DFB diode laser to a single cavity mode, thereby significantly enhancing light throughput.

CW-CRDS is now maturing to a level of robustness and reliability that it can be commercially deployed in industrial applications, where the sensitivity requirements can no longer be met by FTIR, NDIR, or gas chromatography.

43.3 Cavity Enhanced Spectroscopy

Most CRDS systems capture the ring-down waveform using a digital oscilloscope. The Levenberg–Marquardt (LM) method produces the optimal fit, so that LM methods have become the de facto "gold standard" in CRDS [43.15]. However, the LM algorithm can require multiple iterations, thereby limiting the data acquisition times to several hundred Hz. Fast-fitting algorithms that closely approximate the LM fit, but allow data acquisition rates up to 10 kHz have recently gained widespread deployment [43.16]. Today, the data acquisition rates are no longer limited by the back-end numerical fit. Rather, the speed of CRDS systems is limited by the speed of laser modulation itself.

Cavity enhanced spectroscopy (CES) was developed in an effort to simplify CRDS and eliminate the requirement for digitization of a time-domain signal and laser modulation. CES has many different implementations. All CES methods are based on the principle that the build up of intracavity power, and hence cavity throughput, depends on intracavity losses, which include absorption by a sample. CES involves measuring the steady-state transmission through a cavity as a function of wavelength in order to determine changes in integrated transmitted intensity caused by the absorbing species.

For a cavity having two mirrors of intensity reflectivity, R, and length, L, the effective steady-state path length is $L_{eff} = L/(1-R)$. If an absorbing gas is present, the reflectivity will be "reduced" by the Beer–Lambert factor $e^{-\alpha L}$ namely $R' = R e^{-\alpha L}$ so that one can effectively relate the ratio R'/R to the Beer–Lambert ratio I/I_0 for a single pass. The steady state transmitted laser intensity for such a cavity is given by $I = I_L C_p T [2(1-R)]^{-1}$ where T is the intensity transmissivity I_L is the laser intensity, and C_p is the coupling efficiency in the cavity mode. Thus, for an absorbing gas, the change in transmitted intensity at a given wavelength is:

$$\Delta I/I = GA\,(1+GA)^{-1}\;, \qquad (43.4)$$

where $A = 1 - e^{-\alpha L}$ and $G = R/(1-R)$ so that for $\alpha L \ll 1$, $(\Delta I/I) \approx GA \approx G\alpha L$. This latter relation has been interpreted as a linear response in absorption loss, multiplied by an effective cavity "gain" of $R/(1-R)$, i. e., the absorbance is measured over the effective length of the cavity which corresponds to the number of cavity passes occurring within the cavity ring-down time constant. Note that the transmitted power level will be attenuated by the mirror transmissivity, T, so that these methods are limited by laser power and detector sensitivity.

Because CES techniques measure transmitted light intensity, they are no longer immune to laser intensity noise. Furthermore, when the absorbance becomes comparable to the cavity loss, the sensitivity improvement of CES saturates, as the sample absorption begins to dominate the effective number of cavity passes. Note that for this case, the effective path length becomes a function of the sample concentration, which underlines another limitation of CES: this technique is not independent of the cavity length and hence depends on cavity alignment. Moreover, CES systems are not self-calibrated to the species extinction coefficient, and therefore require calibration against a known sample concentration, or against the absolute cavity loss, often measured using CRDS. Finally, it should be noted that the rate of data collection in CES is limited by the RDC time constant, because the cavity acts like a single-pole, low pass filter having a 3 dB frequency of $(2\pi\tau)^{-1}$ which can range from 5 to 50 kHz. Unlike CRDS, CES does not require fast digitization of the decay waveform followed by a non-linear fit, so that the data acquisition hardware can be a much slower, less expensive A/D converter and spectral data can be acquired almost instantaneously.

Five distinguishable variants of CES have been developed and are discussed. The first three methods, called cavity enhanced transmission spectroscopy (CETS), find their origins in CRDS. These methods are cavity enhanced absorption spectroscopy (CEAS) [43.17], integrated cavity output spectroscopy (ICOS) [43.18], and off-axis ICOS [43.19]. For these three approaches, the laser intensity is no longer interrupted to observe a "ring-down event", although the path length enhancing properties of the RDC remain. More sensitive CES methods involve locking the laser to the cavity mode resonance. These will be referred to as locked cavity enhanced transmission spectroscopy and have two variations: locked cavity enhanced transmission spectroscopy (L-CETS) [43.20] and noise-immune, cavity-enhanced optical heterodyne molecular spectroscopy (NICE-OHMS) [43.21].

43.3.1 Cavity Enhanced Transmission Spectroscopy (CETS)

CETS has been implemented using several variations, all of which are based on measuring the time-integrated transmission through a high finesse RDC as function of wavelength. As stated earlier, the transmitted light provides an effectively enhanced path length to any sample absorption inside the cavity. The transmitted light intensity in all CETS approaches is a small fraction [about $(1-R)$] of the incident intensity, which reduces the signal to noise on detection, so that averaging is required. All CETS approaches are also dependent on laser intensity noise and sample path length.

The trade-off in using a high finesse cavity is that in steady state, its transmission is a non-uniform function of wavelength, and consists of a series of sharp cavity mode peaks, namely the transverse and longitudinal modes. This transmission pattern repeats itself periodically every free spectral range (FSR). The density of the mode spacing is a function of the cavity design: round-trip length and mirror radius of curvature. The quality of mode matching between the laser and the RDC determines the number of modes into which light can couple efficiently.

CEAS is the simplest CETS approach: the laser, coupled through a RDC, is tuned in wavelength over the absorption feature of interest, and the integrated cavity transmission is measured as a function of wavelength [43.17]. The cavity length is free-running (neither modulated nor locked to the laser). In order to minimize the non-uniformity of cavity transmission, CEAS exploits cavity geometries such that the inherent mode structure is as dense as possible. No mode-matching is employed, so that laser light is coupled into as many modes as possible. The laser is scanned multiple times over the cavity modes in order to time average over the unstabilized cavity length. Extensive averaging can be required to achieve reasonable performance. The residual mode structure of the cavity can be significant and produces intensity modulation in the spectrum. Typically, the laser linewidth is larger than the individual cavity mode resonances, so that the output can be very noisy. CEAS does appear to have mechanical stability advantages, in that a cavity length change or deformation is in fact desirable to randomize the excited modes over wavelength. Typical sensitivities of CEAS range from are $5\times10^{-7}\,\mathrm{cm^{-1}Hz^{-1/2}}$.

ICOS tries to achieve uniform transmission through the RDC by systematically disrupting the cavity mode resonances, in order to recover the frequency-averaged response of the cavity as a function of wavelength [43.18]. RDC length modulation was implemented by either moving one of the cavity mirrors using a piezo-electric transducer (PZT), or by slightly modulating the angle of injection in the cavity using a PZT-driven mirror mount. When the laser is scanned over the desired wavelength range with only the RDC modulation sweeping the modes (5 to 10 free spectral ranges), the resulting absorption spectra show an intensity modulation of about 10%. This intensity modulation results from a periodic non-uniformity in the mirror movement at the turning points of the PZT modulation, where the increased overlap time between the cavity mode and laser produces a higher transmitted light intensity. In order to eliminate this intensity modulation, the laser wavelength is frequency modulated simultaneously. The typical sensitivity of ICOS approach is about $2\times10^{-7}\,\mathrm{cm^{-1}Hz^{-1/2}}$.

A third approach, called off-axis ICOS, resolves the cavity transmission uniformity problem of CEAS, and eliminates the need for modulation of ICOS. In off-axis ICOS, shown in Fig. 43.3, the RDC is aligned so as to generate a set of spatially separated reflections within the cavity that eventually satisfy the re-entrant condition [43.19]. Furthermore, the mirrors are slightly astigmatized, which results in Lissajous spot patterns on the mirrors. The combination of this alignment and slight astigmatism significantly reduces the RDC mode degeneracy. As a result, the cavity mode spacing can become smaller than the cavity mode width, the cavity transmission appears to become largely frequency independent, or "white". Thus, the RDC provides path length enhancement, without introducing intensity noise in the transmitted spectrum. However, increasing the number of spots on the mirrors usually requires larger mirrors (e.g., 2 inches for off-axis ICOS versus 0.5 inches for CRDS), which in turn results in a larger cavity (sample) volume. In cases of low sample flow, sample chamber filling times can limit the measurement time. Moreover, the system is now dependent on the mechanical stability of the cavity and the laser to cavity alignment. The best off-axis ICOS sensitivity achieved to date is $3\times10^{-11}\,\mathrm{cm^{-1}Hz^{-1/2}}$ and is comparable to CW-CRDS.

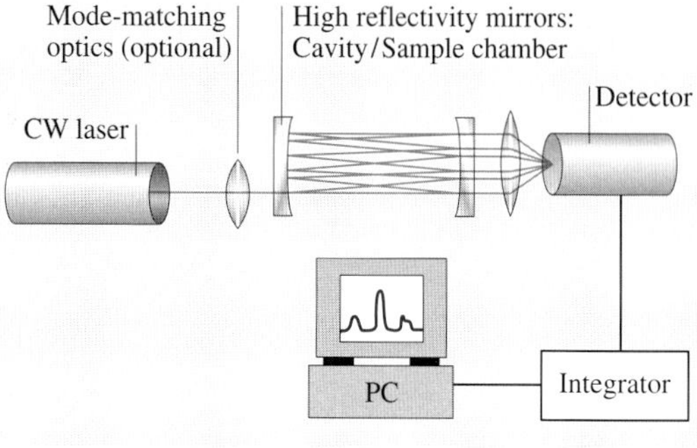

Fig. 43.3 Typical off-axis ICOS setup

43.3.2 Locked Cavity Enhanced Transmission Spectroscopy (L-CETS)

An alternative approach (L-CETS) achieves higher sensitivity by locking the laser frequency to a single cavity resonance and then scanning the cavity over the absorption feature of interest [43.20]. In this case, the cavity throughput becomes uniform across the entire spectral scan, and the cavity transmission increases to the maximum theoretical value, leading to better signal to noise at the detector than CETS approaches. Laser wavelength jitter at high frequencies, which cannot be compensated by the locking control loops, will lead to increased noise in the spectral scans. In order to lock the laser and cavity together robustly, the laser linewidth typically cannot exceed the cavity mode resonance linewidth, so that for a high finesse sample cavity, the laser choice can be limited. Sensitivities of $10^{-11}\,\mathrm{cm^{-1}Hz^{-1/2}}$ have been demonstrated using L-CETS.

The NICE-OHMS technique was developed to overcome the dependence of L-CETS on laser wavelength jitter [43.21]. NICE-OHMS combines the benefits of frequency modulation (FM) spectroscopy with the path length enhancements of a high finesse optical resonator. The technique is called "noise immune", because it does not depend on the quality of the laser and cavity lock. Effectively, it is immune to laser frequency noise although residual amplitude modulation can reduce the performance. NICE-OHMS sensitivity depends on the transmitted laser power, the efficiency and bandwidth of the photodetector, and the FM modulation index. NICE-OHMS has exploited cavities having a finesse of 100 000 to achieve sensitivities of $1 \times 10^{-14}\,\mathrm{cm}^{-1}\mathrm{Hz}^{-1/2}$. NICE-OHMS holds the world record in detection sensitivity of all cavity enhanced techniques.

In a NICE-OHMS experiment, phase modulation of the laser produces side bands that are set to equal the free spectral range of the high finesse resonator. Because the sidebands are transmitted by the cavity in the same manner as the carrier, any small wavelength fluctuations in the laser or small optical phase shifts of the transmitted carrier that contribute to noise in the transmitted intensity will appear identically in the sidebands. After demodulation, this noise will cancel out. Thus, the transmitted carrier and sidebands are an accurate representation of the carrier and sidebands impinging on the input mirror of the cavity. The carrier laser frequency is locked to the peak of the optical cavity mode and tracks this mode if the cavity length is changed in order to produce a spectral scan. The sidebands are detected and demodulated as in conventional FM spectroscopy. The key to NICE-OHMS is that the noise level can approach the intrinsic shot noise of the laser at the FSR frequency (namely hundreds of MHz).

If there is no sample in the cavity, the transmitted sidebands will cancel after demodulation because they have opposite phases. No signal will be observed. If there is a sample in the cavity, the sidebands will experience different transmission amplitudes, so that demodulation will produce a signal proportional to the difference in absorption between the sideband frequencies. In addition, the absorption feature produces a phase shift that pulls the carrier cavity mode frequency, so that the sidebands become detuned from the mode peak and acquire a phase shift on transmission. This sideband phase shift contributes to the demodulated signal.

43.4 Extensions to Solids and Liquids

Thus far, the discussion of cavity-enhanced techniques has addressed traditional optical resonators formed using high reflectivity dielectric mirrors that encompass gas samples. Extensions of cavity enhanced methods to liquid and solid media has required additional innovation, specifically in the optical cavities used.

Evanescent-wave CRDS (EW-CRDS) exploits the fact that total internal reflection allows probing the surface layer of a sample in contact with a prism. The simplest configuration places a Brewster prism inside a linear ring-down cavity. The prism folds the cavity beam path by 90 degrees, thereby producing one point in the prism having total internal reflection [43.22]. The evanescent wave produced by this internal reflection can be used to probe liquid or solid samples. Another possible embodiment is a ring cavity having its optical path inside a multi-faceted polygon, where at each facet, total internal reflection occurs [43.23]. Light is coupled into and out from the polygon by photon tunneling – which effectively controls the overall finesse of the cavity. The absorbing sample material can then be placed on any one or more of the polygon facets, and its detection is done using the polygon's evanescent waves.

Fiber cavities are attractive because they can extend the use of CRDS into harsh environments, can probe liquids as well as gases, and can be used to measure pressure [43.24]. The high reflectivity mirrors on each end of the fiber can be either dielectrically coated, as in a traditional ring-down cavity design, or can consist of Fiber Bragg gratings. Fiber-based CRDS has also been applied to detection in liquid media via a fiber loop cavity wherein a liquid sample replaced the index matching fluid in the gap between fibers at the connector splice [43.25]. This approach produced a 100-fold enhancement over linear detection.

More direct approaches to measuring liquid samples involved confining the liquid samples within a more traditional RDC. The most direct method involves placing a liquid directly into the cavity, but this approach results in very limited sensitivity. By placing a Brewster cell filled with liquid sample in a RDC, the sensitivity of the P-CRDS can be slightly improved. By matching the Brewster angles to the refractive indices of the adjacent media (the outside angle matches the air-filled RDC, while the inner angle matches the index of the liquid), the sensitivity can be dramatically improved [43.26]. The peak-to-peak baseline noise level of such a P-CRDS system was 1.0×10^{-5} absorbance units (AU), rivaling the best available commercial ultraviolet-visible (UV-VIS) direct absorption detectors. The performance remained limited by the excitation of multiple cavity modes A CW-CRDS system using the same angle-matched Brewster cell [43.27] improved the peak-to-peak baseline noise by a factor of 50 to 2×10^{-7} AU. This CRDS detector outperformed the best commercially available UV-VIS detector by a factor of 30, again illustrating the potential for CRDS to replace standard absorption spectroscopy techniques.

Solid samples have also been placed inside linear CRDS cavities for characterization. Mostly, P-CRDS are used. Examples of applications include characterization of C-60 films, thin-film coatings, and silicon wafers.

It is anticipated that CRDS will eventually reach sensitivities for liquid and thin film samples that are comparable to those achieved in gases. CRDS is also expected to find commercial applications in high performance liquid chromatography, thin film characterization, and biological detection. Combination techniques of CRDS and fluorescence, or CRDS and Raman spectroscopy, may also be on the not-so-distant horizon, where CRDS provides the quantification, while the complementary technique provides identification [43.28].

References

43.1 K. W. Busch (Ed.): *Cavity Ringdown Spectroscopy: An Ultratrace Absorption Measurement Technique*, ACS Symp. Ser., Vol. 720 (Oxford Univ. Press, Washington 1997)

43.2 A. O'Keefe, D. A. G. Deacon: Rev. Sci. Instrum. **59**, 2544 (1988)

43.3 R. Engeln, G. Meijer: Rev. Sci. Instrum. **67**, 2708 (1996)

43.4 R. Engeln, G. Bierden, E. van den Berg, G. Meijer: J. Chem. Phys. **107**, 4458 (1997)

43.5 E. R. Crosson, P. Haar, G. A. Marcus, H. A. Schwettman, B. A. Paldus, T. G. Spence, R. N. Zare: Rev. Sci. Instrum. **70**, 4 (1999)

43.6 R. D. van Zee, J. T. Hodges, J. P. Looney: Appl. Opt. **38**, 3951 (1999)

43.7 D. Romanini, A. A. Kachanov, N. Sadeghi, F. Stoeckel: Chem. Phys. Lett. **270**, 538 (1997)

43.8 B. A. Paldus, R. N. Zare: U.S. Patent #6 466 322

43.9 T. G. Spence, C. C. Harb, B. A. Paldus, R. N. Zare, B. Willke, R. L. Byer: Rev. Sci. Instrum. **71**, 347 (2000)

43.10 K. K. Lehmann, P. Rabinowitz: U.S. Patent #5 973 864

43.11 F. Kuhnemann, F. Muller, B. von Basun, D. Halmer, A. Popp, S. Schiller, P. Hering, M. Murtz: SPIE Proc **5337**, paper 18 (SPIE, Bellingham 2004)

43.12 R. Engeln, G. Von Helden, G. Berden, G. Meijer: Chem. Phys. Lett. **262**, 105 (1996)

43.13 M. D. Levenson, B. A. Paldus, T. G. Spence, C. C. Harb, J. S. Harris, R. N. Zare: Chem. Phys. Lett. **290**, 335 (1998)

43.14 D. Romanini, A. A. Kachanov, J. Morville, M. Chenevier: EOS/SPIE International Symp. on Ind. Lasers Inspection Envirosense, **3821**, 94 (SPIE, Bellingham 1999)

43.15 H. Naus, I. H. M. van Stokkum, W. Hogervorst, W. Ubachs: Appl. Opt. **40**, 4416 (2001)

43.16 D. Halmer, G. von Basum, P. Hering, M. Murtz: Rev. Sci. Instrum. **75**, 2187 (2004)

43.17 R. Engeln, G. Berden, E. vandenBerg, G. Meijer: J. Chem. Phys. **107**, 4458 (1997)

43.18 A. O'Keefe, J. J. Scherer, J. B. Paul: Chem. Phys. Lett. **307**, 343 (1999)

43.19 J. B. Paul, L. Lapson, J. G. Anderson: Appl. Opt. **40**, 4904 (2001)

43.20 L. Gianfrani, R. W. Fox, L. Hollberg: J. Opt. Soc. Am. B **16**, 2247 (1999)

43.21 J. Ye, L. S. Ma, J. L. Hall: J. Opt. Soc. Am. B **15**, 6 (1998)

43.22 A. C. R. Pipino, J. W. Hudgens, R. E. Huie: Rev. Sci. Instrum. **68**, 2978 (1997)

43.23 A. C. R. Pipino: Phys. Rev. Lett. **83**, 3093 (1999)

43.24 T. von Lerber, M. W. Sigrist: Appl. Opt. **41**, 3567 (2002) and EP00121314.9

43.25 R. S. Brown, I. Kozin, Z. Tong, R. D. Oleschuk, H. P. Loock: J. Chem. Phys. **117**, 10444 (2002)

43.26 K. Snyder, R. N. Zare: Anal. Chem. **75**, 3086 (2003)

43.27 K. Snyder, R. N. Zare, A. A. Kachanov, B. A. Paldus: Anal. Chem. **77**, 1177 (2005)

43.28 B. A. Richmann, A. A. Kachanov, B. A. Paldus, A. W. Strawa: Opt. Express **13**, 3376 (2005)

44. Spectroscopic Techniques: Ultraviolet

The design of UV spectroscopic experiments and apparatus must conform to the constraints associated with producing, dispersing, and detecting UV photons. In this chapter, we review the instrumentation available for UV spectroscopy, concentrating on the VUV, where special instrumentation is necessary. Recent advances are stressed, particularly in the areas of synchrotron radiation and the production of VUV laser light.

The most inclusive, up-to-date review of vacuum ultraviolet techniques is Vacuum Ultraviolet Spectroscopy I & II (1998), edited by *Samson* and *Ederer* [44.1]. This text significantly expands on Samson's classic, authoritative review of VUV instrumentation, first published in 1967 [44.2]. There are many comprehensive reviews of VUV light sources, spectrometers, and detectors, which are referenced in the appropriate sections of this chapter.

44.1 **Light Sources** ... 642
44.1.1 Synchrotron Radiation ... 642
44.1.2 Laser-Produced Plasmas ... 643
44.1.3 Arcs, Sparks, and Discharges ... 644
44.1.4 Supercontinuum Radiation ... 644

44.2 **VUV Lasers** ... 645

44.3 **Spectrometers** ... 647
44.3.1 Grating Spectrometers ... 647
44.3.2 Fourier Transform Spectrometers ... 648

44.4 **Detectors** ... 648

44.5 **Optical Materials** ... 651

References ... 652

The ultraviolet (UV) spectral region extends from the short wavelength side of the visible, about 400 nm, to the long wavelength side of the X-ray region at approximately 10 nm. Ultraviolet photon energies range from about 3 to 120 eV. Because laboratory air at STP does not transmit below about 200 nm, the UV region is conventionally subdivided into the near ultraviolet, wavelengths $\lambda > 200$ nm, and the vacuum ultraviolet (VUV), $\lambda < 200$ nm. Further subdivisions are widespread, but not uniformly adopted. The term middle ultraviolet (MUV) is sometimes used to designate the 200 to 300 nm wavelength region, with the term near UV used only for 300 to 400 nm. VUV wavelengths are often divided into the far ultraviolet (FUV), 100 to 200 nm, and the extreme ultraviolet (EUV), 10 to 100 nm. In this review, we will simply refer to the UV (10 to 400 nm), the near UV (200 to 400 nm), and the VUV (10 to 200 nm).

The absorption of UV radiation by atoms and molecules involves transitions to highly excited discrete and continuum levels. The relevant atomic and molecular physics of highly excited states is discussed in Chapts. 14, 24, 25, and 61. Ultraviolet absorption initiates many important photochemical reactions; applications in astrophysics, aeronomy, and combustion are reviewed in Chapts. 82, 84, and 88. The experimental techniques used at ultraviolet wavelengths include high-resolution spectroscopic measurements of wavelengths and line shapes (Chapt. 10) , absolute photoabsorption oscillator strength and radiative lifetime measurements (Chapt. 17), fluorescence spectroscopy, and energy, angle, and spin-resolved photoelectron spectroscopy (Chapt. 61).

44.1 Light Sources

44.1.1 Synchrotron Radiation

Synchrotron radiation (SR) facilities provide intense continuum radiation in the visible, UV, and soft X-ray spectral regions. The radiation is emitted by relativistic electrons (or positrons) accelerated by the Lorentz force of a magnetic field. In addition to its continuum nature, synchrotron radiation is characterized by high degrees of collimation and polarization. Advances in specialized magnet designs, e.g., wigglers and undulators, that dramatically increase spectral brightness at selectable wavelengths, have stimulated continuing growth in synchrotron-radiation-based spectroscopic research. References [44.3–13] give general reviews on the nature of synchrotron radiation and on specific SR-based experimental techniques.

SR facilities use either a linear accelerator, a microtron, or a synchrotron to accelerate electrons or positrons to relativistic energies (0.1 to 10 GeV). The particles are then injected into a storage ring, where a series of bending magnets steers them in a closed orbit. The particles in the storage ring radiate energy, which is replaced by radio frequency accelerating cavities that are part of the ring. These accelerating systems produce well-defined, regularly spaced bunches of stored particles, so that radiation from SR facilities is characterized by sub-nanosecond pulses at 1 to 10 MHz repetition rates.

Typical storage ring currents are on the order of 100 to 1 A. The current decreases continuously because of collisions with residual gas molecules and electron–electron (or positron–positron) scattering. Beam lifetimes in most facilities range from a few to about 20 hours. The longest lifetimes are achieved by positron beams, which repel any ions produced in residual gas collisions. Storage ring pressures must be in the 10^{-10} Torr range in order to reduce collisions to acceptable levels; such pressure requirements can restrict the types of measurements performed at SR facilities.

The spectral properties of synchrotron radiation can be derived from Larmor's formula for an accelerating relativistic charged particle [44.4, 14–18]. The total power P radiated by a beam of current I and energy E in a bending magnet field B is

$$P = 0.0265 E^3 I B \text{ kW} , \qquad (44.1)$$

where, throughout this chapter, I is in mA, E is in GeV, and B is in Tesla. For example, a 300 mA beam of 1 GeV electrons radiates about 8 kW when traversing the field of a typical 1.0 T bending magnet. One-half of the total power is radiated above the critical wavelength, λ_c, where

$$\lambda_c = 1.9/\left(BE^2\right) \text{ nm} . \qquad (44.2)$$

For the example just considered, $\lambda_c = 1.9$ nm.

The power radiated by, and the flux from, synchrotron radiation sources is often expressed per fractional (or percent) bandwidth. For example the flux Φ, in units of photons $\left(\text{cm}^2\text{s}\right)^{-1}$, at 10 nm (124 eV) per 1% bandwidth is the flux in a 0.1 nm band at 10 nm, which is the same as that in a 1.24 eV energy band. The radiated power per fractional bandwidth peaks at $\lambda \approx 0.75\,\lambda_c$.

The flux per fractional bandwidth reaches a broad maximum at $\lambda \approx 3.3\,\lambda_c$ and decreases slowly at longer wavelengths, with a limiting behavior that is proportional to $\lambda^{-1/3}$ for $\lambda \gg \lambda_c$. Thus, although the photon flux typically peaks in the soft X-ray region, a high flux is also present throughout the VUV wavelength range. Below λ_c, the flux drops rapidly, with little usable radiation at wavelengths below about $0.1\,\lambda_c$.

The angular distribution of the radiated light is sharply peaked in the instantaneous direction of the beam of radiating particles. Practically all the radiation is emitted into an opening half-angle θ that at $\lambda = \lambda_c$ is equal to $1/\gamma$, where $\gamma = E/mc^2$; i. e.,

$$\theta = 1/\gamma = 5.11 \times 10^{-4}/E \text{ rad} . \qquad (44.3)$$

Radiation from beams with energies of a few GeV has a divergence somewhat less than 1 mrad at λ_c. The divergence increases at longer wavelengths, with

$$\theta \approx (\gamma)^{-1}(\lambda/\lambda_c)^{1/3} \text{ rad} . \qquad (44.4)$$

Radiation from bending magnets maintains this narrow divergence only in the vertical plane; in the orbital plane, the radiation is emitted in an opening angle equal to the bending angle of the magnet. Typical beam cross sections are 0.01 to 1 mm^2. Thus, the extremely high spectral brightness, i. e., photon flux per unit solid angle and % bandwidth, of synchrotron sources reflects both the very low divergence of the radiation and the small source size. A typical brightness associated with bending magnet radiation is on the order of 10^{12} photons $(\text{s mm}^2\text{ mrad}^2\ 0.1\%\text{ bandwidth})^{-1}$.

Synchrotron radiation has well-characterized polarization properties. Within the orbital plane the radiation is completely linearly polarized with the electric field vector in the orbital plane; above and below the orbital plane the radiation is elliptically polarized, with the degree of polarization dependent on both wavelength and viewing angle. Undulator designs incorporating either crossed planar magnets or a helical array of magnets can produce circularly polarized light (e.g., [44.19]), allowing for the development of circular dichroism spectroscopies in the vacuum ultraviolet [44.20].

The pulsed nature of the radiation at synchrotron facilities is ideal for sub-nanosecond time-resolved spectroscopies [44.21–25]. Electron bunch lengths l_b are typically a few centimeters; radiation is observed from each bunch for a time interval approximately equal to $l_b/c \approx 100$ ps. The time for one full orbit is determined by the size of the storage ring and varies from about 20 ns at the smallest rings to 500 to 1000 ns at the larger rings. The number of bunches is typically on the order of a few hundred, resulting in pulse repetition rates from a few MHz to about 500 MHz.

Many synchrotron facilities incorporate straight sections between bending magnets to accommodate insertion devices such as wigglers and undulators. These are linear arrays of magnets with alternating polarities that cause the beam trajectory to oscillate, although no net displacement of the beam occurs. The oscillations produce synchrotron radiation that is characterized by extremely high brightness and that can be spectrally tailored to individual experiments [44.4, 13, 17, 26, 27].

Wigglers and undulators are distinguished by the value of their deflection parameter K, which is a measure of the maximum bending angle of the stored particle beam in units of θ, the angular divergence of the emitted radiation. The deflection parameter K is determined by the peak strength of the alternating magnetic field B_0 and its spatial period x_B by

$$K = 0.934 B_0 x_B \,, \qquad (44.5)$$

with x_B in cm.

In the wiggler regime, $K \gg 1$; i. e., the angle at which the magnets deflect the electron beam is large compared with θ. The resulting spectrum resembles that from a bending magnet, but is brighter by a factor of N_m, the number of magnet periods (usually 10 to 100). Wigglers are generally used as wavelength shifters; magnetic fields larger than those available in conventional bending magnets decrease the critical wavelength of the spectral distribution and shift the overall spectrum to shorter wavelengths.

In the undulator regime, $K \ll 1$; i. e., the angle at which the electrons are deflected is close to θ and radiation from individual oscillations adds coherently at certain resonant wavelengths, producing a gain of N_m^2. Undulator radiation is characterized by sharp peaks at a fundamental wavelength λ_1 and its odd harmonics. λ_1 is determined by the beam energy and x_B:

$$\lambda_1 \approx 0.1\, x_B / \left(2\gamma^2\right) = 1.3\, x_B E^{-2} \text{ nm} \,, \qquad (44.6)$$

where x_B is in cm. For example, the fundamental wavelength is 6.5 nm for a 1 GeV beam traversing an undulator with a spatial period of 5 cm. The fractional bandwidth of the peaks in an undulator spectrum is $\approx 1/N_m$. Unlike the radiation from a bending magnet, the angular divergence of undulator radiation is sharply peaked both vertically and in the orbital plane; the brightness of undulator radiation is $\approx 10^{18}$ to 10^{19} photons/(s mm^2 mrad^2 0.1% bandwidth), about six orders of magnitude greater than bending magnet radiation. The wavelength of the fundamental is tunable; the strength of the peak magnetic field is altered via changes in the size of the gaps between the poles of the magnets.

44.1.2 Laser-Produced Plasmas

When the output of a high-power pulsed laser is focused onto a solid target, a short-lived, high-temperature ($T \approx 50$ to 100 eV), high-density $\left(n_e \approx 10^{21}\,\text{cm}^{-3}\right)$ plasma is created. The radiation from certain target materials, particularly the rare earths ($57 \le Z \le 71$) and neighboring metals on the periodic table, produces a strong VUV quasicontinuum that is essentially free of discrete lines [44.28, 29]. The continua are most intense in the 4 to 30 nm region but often extend to about 180 nm. A review of laser-produced plasmas and their applications in the VUV can be found in [44.30].

The primary mechanisms responsible for the continua are recombination radiation (ionization stages up to $\approx$ 16 are attained) and bremsstrahlung. The absence of discrete features in rare earth plasmas is thought to be caused by the extreme complexity of rare earth atomic energy level structures; individual emission lines are blended into an apparent continuum [44.31]. The necessary peak laser powers, which are in the range of 10^{10} to 10^{11} W cm^{-2} [44.32, 33], are easily achieved with most commercially available Q-switched lasers. The continuum pulse duration is comparable to the duration of the laser pulse. Peak output intensities are significantly greater than those from other pulsed table-top continuum sources such as spark discharges; synchrotron sources produce much higher average intensities [44.33]. The target material is usually a cylindrical rod that is rotated to present a fresh surface for each laser shot. References [44.32, 34–38] present and discuss the spectral characteristics of several laser-produced plasma sources. Atomic photoabsorption techniques based on laser-produced plasma sources are described in [44.33, 39–41].

44.1.3 Arcs, Sparks, and Discharges

Several laboratory sources of VUV line and continuum radiation are based on gas discharges, high-pressure arcs, and low-pressure and vacuum sparks. While these well-established radiation sources are not as intense as synchrotron radiation and laser-produced plasmas, the traditional sources have the advantages of being portable and inexpensive. Reviews can be found in [44.42, 43]

H_2 and D_2 Discharges

Direct current discharges (approximately 100 to 500 mA) through 1 to 2 Torr of H_2 or D_2 generate continuum radiation from about 165 to 350 nm [44.2, 44]. The continuum is produced by transitions from the bound $1s\sigma\, 2s\sigma\, a^3\Sigma_g^+$ state to the repulsive $1s\sigma\, 2p\sigma\, b^3\Sigma_u^+$ state [44.45]; the very steep potential curve of the lower state results in the extended nature of the continuum. The photon flux per unit wavelength from such lamps peaks at about 185 nm; below 165 nm the many-line spectra of H_2 or D_2 begin to dominate.

Ar Mini-Arc

A high-current (20 to 50 A) wall-stabilized arc discharge through 1 to 2 atm of Ar, known as a mini-arc, produces a continuum associated with recombination radiation. The continuum extends from the near UV down to approximately 110 nm [44.44, 46]. The stability and reproducibility of the output of the mini-arc has led to its use as a secondary radiometric standard in the VUV [44.46, 47].

Noble Gas Discharges

High voltage (10 kV), mildly condensed, repetitive (5 kHz) discharges through the noble gases produce continua associated with molecular transitions between bound excited states and the very weakly bound ground states of the noble gas dimers [44.2, 48, 49]. These continua begin at the first resonance line of the atomic species and extend a few tens of nanometers to longer wavelengths. For example, the onset of the He continuum is approximately 59 nm, while the useful continuum extends from 65 to 95 nm with a peak intensity at about 80 nm. Discharges through He, Ar, Kr, and Xe cover the 65 to 180 nm region. Optimum discharge pressures vary from about 40 Torr for He up to a few hundred Torr for Kr and Xe. Differential pumping is required when using these discharge sources at wavelengths shortward of 105 nm, the short wavelength transmission limit of window materials (see Sect. 44.5).

Flash Discharges and Vacuum Sparks

Flash discharges of approximately 1 μs duration through low-pressure (approximately 0.02 Torr) gases in ceramic or glass capillaries yield useful continua down to about 30 nm [44.2, 50]. The continua are produced by passage of the discharge through sputtered wall materials and are independent of the carrier gas. The BRV source [44.51], a pulsed vacuum spark with extremely low inductance, utilizes a high-Z anode (e.g., W or U) to generate a smooth continuum down to about 10 nm. The short pulse duration (approximately 50 ns) [44.52] and well-defined triggering make this source useful for transient absorption studies at short wavelengths.

Line Radiation Sources

Hollow cathode discharges [44.53, 54], which are easy and inexpensive to construct and operate, are the most common line-emission sources used in UV spectroscopy. Low power (approximately 5 W) lamps designed to produce the spectra of about 70 elements are commercially available for spectrochemical and atomic absorption spectroscopy. Line widths are narrow, so blends are avoided. Most manufacturers can provide lamps with special windows for use at VUV wavelengths (Sect. 44.5).

High-current, differentially-pumped hollow cathode sources, which have been developed for use as radiometric standards in the 40 to 125 nm wavelength range [44.55], can also be used as line sources. At shorter wavelengths, e.g., 5 to 40 nm, the Penning discharge has proved to be a useful source of line radiation [44.43, 56, 57]. Electron-beam excitation sources [44.43] produce stable and reproducible line emission spectra and find use as VUV calibration standards [44.54, 58]. Electron-beam ion trap (EBIT) sources [44.59, 60] generate VUV and X-ray spectra of highly ionized atomic species.

44.1.4 Supercontinuum Radiation

Continua extending throughout the near UV wavelength range to below 200 nm can be produced by focusing femtosecond laser pulses in both liquids and noble gases. This supercontinuum generation [44.61] allows for time-resolved, sub-picosecond spectroscopy of photochemical reactions [44.62, 63]. The supercontinuum arises from a number of competing nonlinear processes [44.64–66]. Access to the UV has been demonstrated both by focusing 308 nm laser pulses in water [44.67] and by focusing 248 nm laser pulses in high pressure (10 to 40 atm) cells of Ne, Ar, and Kr [44.68]. The continuum produced using H_2O is particularly broad, extending from 200 to 600 nm. The neon continuum extends into the VUV to at least 187 nm.

44.2 VUV Lasers

Advances in laser technologies have led to an impressive increase in spectroscopically useful VUV laser sources. Recent reviews can be found in [44.69, 70]. The following summary of VUV lasers does not cover the rapidly developing fields of X-ray and soft X-ray lasers [44.71, 72] or the development of free electron lasers operating in the VUV ([44.73, 74] and references therein).

Primary Lasers

Pulsed lasing at VUV wavelengths has been achieved in a number of media, including the molecular gases H_2, CO, and F_2, excimer systems (Ar_2, Kr_2, Xe_2, ArF), and Auger-pumped noble gases (Xe, Kr) [44.75]. The fundamental barrier to producing stimulated emission in the VUV is the ν^3 dependence of the spontaneous emission rate, which necessitates very high pump powers to create and maintain population inversions. Existing VUV primary lasers have generally not been widely used in spectroscopic applications because of either a lack of tunability or unacceptably broad line widths.

Pulsed lasing in CO, H_2, and F_2 occurs between rovibronic levels of a high-lying electronic state and excited vibrational levels of the ground electronic state. The lasing output consists of discrete lines and is not tunable. The CO and H_2 outputs [44.76, 77] span relatively broad regions (181 to 197 nm and 110 to 164 nm, respectively), while the F_2 laser output consists of two or three closely-spaced lines at 157 nm [44.78]. Pulse energies are in the 1 to 100 μJ range for CO and H_2 lasers. Commercially available F_2 lasers offer pulse energies of about 50 mJ at 50 Hz repetition rates.

Noble-gas and noble-gas-halide excimer lasers operate via bound-upper-level to continuum-lower-level transitions. (Retrospective reviews of excimer laser technologies can be found in [44.79, 80].) The resulting gain profiles are broad and the outputs are typically tunable over about 2 nm. ArF excimer systems (193 nm) are pumped by high pressure gas discharges; lasers with pulse energies of $\approx$ 20 to 40 mJ are commercially available. Xe_2 (173 nm), Kr_2 (146 nm), and Ar_2 (126 nm) lasers [44.81–83] produce comparable pulse energies but require electron beam excitation. Auger-pumped lasers achieve a population inversion through the ejection of an inner-shell electron from a neutral atom to produce a highly excited singly ionized species, followed by Auger decay to excited states of the doubly ionized species. The soft X-ray (approximately 100 eV) pump photons often originate from a laser-produced plasma. The lasing output is nontunable. Lasing in Xe (108.9 nm) and Kr (90.7 nm) [44.84, 85] has been achieved with output pulses in the range of a few μJ, and Auger-pumped lasing in Zn (at approximately 130 nm) has also been reported [44.86].

Nonlinear Techniques

In the past thirty years, developments in the techniques of nonlinear optics have extended the range of tunable, pulsed coherent radiation to VUV wavelengths. Nonlinear frequency conversion techniques, such as stimulated anti-Stokes Raman scattering, harmonic generation, and sum- and difference-frequency mixing (see Chapt. 72) , can produce narrow-bandwidth radiation spanning the entire VUV. The generated VUV radiation has the spectral and spatial characteristics of the input laser radiation. Pulses of 10^{10} to 10^{12} photons are commonly generated with bandwidths on the order of $0.1\,\mathrm{cm}^{-1}$ (2×10^{-4} nm). Many reviews are available on this subject [44.75, 87–95]. The standard nonlinear techniques for producing tunable coherent radiation with $\lambda < 200$ nm are summarized below.

Second harmonic generation in nonlinear optical crystals is a well-established method for generating tunable laser light in the near UV (Chapt. 42). At the shortest wavelengths, second harmonic generation in β-barium borate (BBO) produces usable outputs down to approximately 205 nm, below which the phase matching requirement between the fundamental and the second harmonic cannot be met. Sum-frequency mixing ($\omega = \omega_1 + \omega_2$) in BBO [44.96, 97], where ω_1 is typically the Nd:YAG fundamental (1064 nm) and ω_2 is tunable UV light, extends the useful range of BBO to approximately 190 nm, where the crystal begins to exhibit strong absorption.

Stimulated anti-Stokes Raman scattering of UV laser light in molecular gases (e.g., H_2, N_2, CH_4) is used to generate coherent radiation down to about 120 nm, although low conversion efficiencies often restrict the usable output to $\lambda > 160$ nm. The scattering is a four-wave mixing process resulting in a series of output frequencies shifted from the pump laser frequency by multiples of the vibrational splitting of the ground electronic state of the gas [44.98, 99]. Because high anti-Stokes orders are not efficiently generated, H_2, with its large ground state vibrational splitting of 4155 cm^{-1}, is often the gas of choice. The experimental requirements are relatively minimal; a tunable visible or UV source (e.g., a frequency-doubled dye laser) and a high pressure gas cell (about 2 to 10 atm of H_2 in a 1 m cell). Conversion efficiencies of a few percent are reported for the first few anti-Stokes orders in H_2 with pump laser energies of about 10 to 50 mJ [44.100, 101]; efficiencies then drop continuously, reaching about 10^{-5} to 10^{-6} for the highest orders ($n = 9$ to 13) reported. Stokes seeding techniques [44.102] have been shown to increase efficiencies for the highest anti-Stokes orders by as much as a factor of 100.

Third harmonic generation ($\omega = 3\omega_1$), sum-frequency mixing ($\omega = 2\omega_1 + \omega_2$), and difference-frequency mixing ($\omega = 2\omega_1 - \omega_2$) in appropriate noble gases and metal vapors are the most broadly applicable methods for producing tunable VUV light. These processes result from the presence of the third-order nonlinear term in the expansion of the induced macroscopic polarization of the gas as a power series in the electric field (Chapt. 72). (The second-order term, which is responsible for frequency doubling in crystals, is zero for isotropic media such as gases.) The power generated in third-order effects is proportional to

$$N^2 \left|\chi^{(3)}\right|^2 P_1^2 P_2 F \,, \tag{44.7}$$

where N is the number density of the gas, $\chi^{(3)}$ is the third-order nonlinear susceptibility, P_1 and P_2 are the input laser powers at ω_1 and ω_2, and the factor F describes the phase matching between the generated VUV light and the induced polarization [44.88, 89]. For third harmonic generation, $P_1^2 P_2$ is replaced by P_1^3.

The most critical constraint in third-order frequency conversion is the phase matching requirement; a comprehensive treatment is presented in [44.103]. The factor F is a function of the product $b\Delta k$, where b is the confocal beam parameter of the focused input radiation and Δk is the phase mismatch between the generated VUV light and the input radiation:

$$\Delta k = k - (2k_1 + k_2) \,. \tag{44.8}$$

where k_i is the wave vector of the radiation with frequency ω_i. In the standard case of tight focusing ($b \ll L$, where L is the linear dimension of the gas cell) for sum-frequency mixing and third harmonic generation processes, F is nonzero only for $\Delta k < 0$. Therefore, these techniques are applicable only in spectral regions where the gas exhibits negative dispersion. In contrast, for difference-frequency mixing, the factor F is nonzero in regions of both positive and negative dispersion, and wider tuning ranges are generally possible.

Conversion efficiencies for sum- and difference-frequency mixing schemes are usually in the range of 10^{-7} to 10^{-4} with input peak laser powers of 1 to 10 MW. Resonant enhancement of $\chi^{(3)}$, achieved by tuning the input radiation to transition frequencies in the gas, dramatically increases conversion efficiencies (by factors of about 10^4 and allows for much lower (≈ 1 kW) input laser powers. Most commonly, one of the incident frequencies is tuned to an allowed two-photon transition, with the second input frequency providing subsequent tunability in the VUV; resonant methods therefore require two tunable inputs.

Metal vapors (e.g., Mg, Zn, Hg) are negatively dispersive over fairly broad regions of the VUV between 85 and 200 nm and are consequently used for sum-frequency mixing and third harmonic generation. Conversion efficiencies are further enhanced in these vapors by three photon resonances ($2\omega_1 + \omega_2$) with the ionization continuum or broad autoionizing features [44.91]. One experimental drawback is the complexity associated with generating the metal vapors in ovens or heat pipes.

Noble gases are generally less suited for sum-frequency mixing because they exhibit negative dispersion over fairly limited spectral ranges in the VUV. However, they do provide an experimentally simple medium for difference-frequency mixing schemes. In particular, Xe, Kr, and Ar (as well as H_2) have been used in resonant and nonresonant difference-frequency mixing schemes to produce tunable radiation over the 100 to 200 nm region [44.104–110]. The noble gases are also used for sum-frequency mixing and third harmonic generation. Kr, Ar, and Ne are used for the generation of tunable radiation down to 65 nm [44.104, 111–114]. Below the LiF transmission cutoff (105 nm) the gases are introduced as pulsed jets. *Hollenstein* et al. [44.115] describe an ultra-narrow bandwidth $\left(\approx 0.008\ \mathrm{cm}^{-1}\right)$ system, utilizing resonance-enhanced sum-frequency mixing in rare gases, capable of producing coherent radiation between 73 and 124 nm. A representative summary of third-order frequency conversion schemes is presented in Table 44.1

Higher-order frequency conversion techniques have been used to generate both fixed frequency and tunable coherent radiation shortward of 70 nm. For example, radiation at 53.2 nm and 38.0 nm has been produced via fifth and seventh harmonic generation in He, using the 266.1 nm Nd:YAG fourth harmonic [44.116]. Tunable radiation at 58 nm has been produced through fifth harmonic generation in C_2H_2 using the frequency-doubled output of a dye laser [44.117, 118]. A system with continuous tunability over the 40 to 100 nm region with sub-cm^{-1} resolution has been developed using high-order harmonic generation in Ar and Kr [44.119].

44.3 Spectrometers

44.3.1 Grating Spectrometers

The design and characteristics of VUV grating spectrometers and monochromators are reviewed by many authors [44.2, 50, 129–134]. Two basic types of spectrometer are used in the VUV; normal incidence instruments for 200 nm > λ > 30 nm and grazing incidence instruments for 50 nm > λ > 2 nm. The particulars of these two types are largely dictated by the low reflectivities of both metal and dielectric surfaces in the VUV.

Concave gratings are used almost exclusively in VUV spectrometers. Such gratings provide both dispersion and focusing, thus eliminating the need for additional mirrors and their associated reflection losses. Most VUV spectrometers make use of the focusing properties of the Rowland circle, which is tangent to the grating at its center, lies in a plane perpendicular to the grating grooves, and has a diameter equal to the radius of curvature of the grating [44.2, 135]. A source on a horizontal Rowland circle is focused horozontally by the grating to a location also on the circle. The dispersion introduced by the grating results in a focused, diffracted spectrum lying on the Rowland circle. Almost all normal incidence and grazing incidence VUV instruments are designed with the entrance and exit slits (or photographic plate) lying on, or nearly on, the Rowland circle.

The image formed by a concave grating is not stigmatic, i. e., the vertical focus does not coincide with the horizontal focus. Hence, a point source is imaged into a vertical line on the Rowland circle [44.2, 131, 136]. Astigmatism is particularly severe at grazing incidence angles, resulting in both loss of signal (the image of the entrance slit being larger in extent than the exit slit) and loss of resolution (the image of the entrance slit is curved in the dispersion direction). Aspherical concave gratings (e.g., toroidal gratings) reduce the astigmatism associated with conventional spherical gratings ([44.136] and references in [44.137]). The most important recent advances in concave grating production are the use of interference techniques to produce holographic gratings [44.137–139] and the development of variable line spacing gratings [44.140, 141]. Interference techniques eliminate the periodic irregularities in conventionally ruled gratings that lead to spectroscopic "ghosts", reduce the level of scattered light by significant amounts, allow for very high groove densities (e.g., 4800 mm^{-1}), and can be relatively easily applied to aspheric surfaces. Mechanically-ruled variable line spacing gratings correct for spherical aberrations and allow for relatively simple focusing and scanning designs in EUV spectrometers and monochromators [44.141].

Table 44.1 Representative third-order frequency conversion schemes for generation of tunable coherent VUV light

Medium	λ (nm)	Process	Ref.
Sr	165–200	res. diff. mixing	[44.120]
	178–196	res. sum mixing	[44.121]
Mg	121–174	res. sum mixing	[44.91]
Zn	106–140	res. sum mixing	[44.122]
Hg	142–182	nonres. tripling	[44.88]
	117–122	res. sum mixing	[44.123]
	104–108	res. sum mixing	[44.124]
	85–125	res. sum mixing	[44.125]
Xe	160–206	nonres. diff. mixing	[44.105]
	140–147	nonres. tripling	[44.126]
	113–119	nonres. sum mixing	[44.105]
Kr	117–150	res. diff. mixing	[44.127]
	110–116	nonres. sum mixing	[44.105]
	120–124	nonres. tripling	[44.128]
	127–180	res. diff. mixing	[44.104]
	72–83	res. sum mixing	[44.104]
Ar	102–124	res. diff. mixing	[44.108]
	97–105	nonres. tripling	[44.111]
Ne	72–74	nonres. tripling	[44.112]

Normal incidence grating designs are appropriate for wavelengths greater than about 30 to 40 nm. The most common types include the Eagle, Wadsworth, and Seya–Namioka designs [44.129, 131, 132]. Eagle mounts, with entrance and exit slits on the Rowland circle and approximately equal angles of incidence and reflection ($< 10°$), can be either in-plane or off-plane. Photographic resolving powers $\lambda/\delta\lambda$ of more than 2.5×10^5 and photoelectric resolving powers of ≈ 1 to 2×10^5 have been achieved in the 100 nm region with 6.65 m Eagle mount instruments [44.142–145]. In the Wadsworth mount, the light source is at a large distance from the grating and no entrance slit is required. This design is appropriate for collimated light sources such as synchrotron radiation. It has the advantages of high throughput and minimal astigmatism [44.2, 130]. The Seya–Namioka mount is a Rowland circle instrument in which the angle subtended by the fixed entrance and exit slits relative to the center of the grating is approximately 70°. The spectrometer remains in good focus for small grating rotations, resulting in a simple scanning mechanism and thus an inexpensive design [44.2]. Seya–Namioka instruments provide high throughput at moderate spectral resolutions (typically 0.02 to 0.05 nm) [44.130].

The normal incidence reflectance of all standard metal coatings is no greater than a few percent for $\lambda < 30$ nm [44.131]; in this wavelength region grazing incidence instruments take advantage of the total reflection of photons at extreme grazing angles. There is a sharp reflectance cutoff at photon energies above a characteristic energy that typically limits the use of grazing incidence optics to $\lambda \gtrsim 1$ nm. Grazing incidence designs, tailored to the constraints of synchrotron radiation facilities, are described in [44.4, 130, 133]. Astigmatism becomes severe at grazing angles; interferometrically produced aspherical gratings and mechanically ruled variable line spacing gratings are used to reduce this aberration [44.139, 141]. The resolving power of a grazing incidence instrument is generally lower than that of a comparably sized normal incidence instrument, in large part because of the decreased effective width of the grating at grazing incidence [44.2]. The largest grazing incidence instruments achieve resolving powers of about 10^5 in the 5 to 40 nm region [44.146, 147].

A number of nonstandard instrument designs are reported in the literature, often for use in VUV astronomy and aeronomy applications. These novel designs take advantage of the properties of conical diffraction [44.148–150] and dual-grating crossed dispersion or echelle mounts [44.151, 152] to reduce the effects of aberrations. Reference [44.129] reviews many specialized VUV spectrometer designs.

44.3.2 Fourier Transform Spectrometers

Fourier transform spectroscopy (FTS) is a well-established technique for high-resolution emission and absorption spectroscopy at infrared and visible wavelengths. The technique has been extended into the near UV and VUV regions, where FTS is characterized by the large optical throughput, high spectral resolution, and accurate linear wavenumber scale available at longer wavelengths [44.153]. However, the multiplex advantage of FTS is not realized at UV wavelengths because the signal-to-noise ratio is photon-noise limited rather than detector limited. A review of interferometric techniques in the VUV can be found in [44.154].

A scanning Michelson interferometer with a fused silica beamsplitter has achieved a resolving power of 1.8×10^6 at wavelengths down to 178 nm [44.153]. The same interferometer design, with a MgF_2 beamsplitter, has achieved a resolving power of 8.5×10^5 at wavelengths shorter than 140 nm [44.155]. This spectral resolution is significantly better than that realized by the best VUV grating instruments.

At shorter wavelengths, diffraction gratings can be used as division-of-amplitude beamsplitters to create all-reflecting interference spectrometers for VUV wavelengths. This technique has been used to create spatial heterodyned spectrometers for astronomical applications [44.156–158]. The interferometer mirrors are not moved in such instruments; the interferogram is viewed by an array detector. The design of a wave-front-division interferometer for wavelengths down to 60 nm is described in [44.159].

44.4 Detectors

There is a wide variety of photon detectors with useful response and sensitivity at VUV wavelengths. With the exception of those involving the photo-ionization of gases, VUV detectors are based on the same underlying principles as their counterparts in the visible and infrared regions – the common detection schemes are initiated by surface photoemission, electron–hole pair creation in semiconducting materials, or chemical changes in photographic emulsions. The details of the design of VUV detectors, and the constraints on their use, are most often determined by the low levels of transmission of VUV light through suitable window and semiconducting materials. Nevertheless, VUV detectors with single photon counting sensitivities and/or imaging capabilities are widely available. Reviews of VUV detectors can be found in [44.1, 2, 50, 160–163]. References [44.164–166] review the operating characteristics and calibration of VUV detector standards. A more general summary of photon detector technology is presented in [44.167, 168]. Below we summarize and compare the most common VUV detection schemes, including the use of photographic plates, photomultiplier tubes and vacuum photodiodes, multichannel plate detectors, silicon photodiodes, charge-coupled devices, and ionization chambers.

Photographic Plates

Photographic plates have long been used at VUV wavelengths for spectroscopic measurements, and they are still sometimes the detector of choice for both spectroscopic surveys and high resolution wavelength measurements. They have the advantages of being imaging detectors with very high spatial resolution; spectral line positions can be determined with an uncertainty of about 1 μm. Plates also have multiplexing and very flexible integrating capabilities, being insensitive to fluctuations in signal intensities during the period of exposure. However, when compared with photoelectric detectors, photographic plates have the disadvantages of limited dynamic range and a nonlinear response, which makes radiometric measurements very problematic when plates are used. The gelatin base used in standard photographic emulsions strongly absorbs VUV radiation, so special plates with no gelatin base are required for VUV work [44.2].

Photomultiplier Tubes

Photomultiplier tubes (PMTs) [44.169] are used through the VUV for single-photon-counting, nonimaging, applications. Dark count rates are low $\left(\text{about } 1\,\text{s}^{-1}\right)$ for solar-blind tubes with $1\,\text{cm}^2$ photocathodes; pulse rise times are about 1 to 10 ns ([44.170] for a general review of amplifying detectors in the VUV). In the 105 to 200 nm region, useful PMT window materials are fused silica, MgF_2, and LiF; sapphire is also used in environments where ionizing radiation, which causes sapphire to fluoresce, is not present. In the windowless region of the VUV wavelength range ($2\,\text{nm} \leq \lambda \leq 105\,\text{nm}$), two options are available: the PMT may be operated bare, i. e., without a window, or a fluorescent coating can be deposited on the window to down-convert the VUV light to longer wavelengths.

Peak quantum efficiencies for VUV photomultiplier tubes are about 15 to 20%. Some of the most useful coatings for photocathodes in the VUV are CsI, KBr, and other alkali halides. These materials combine high VUV quantum efficiencies with a solar blind response – their relatively high photoelectric work functions result in long wavelength cutoffs between 150 and 300 nm. Solar blind PMTs have low dark count rates and have minimal response to stray ambient light.

Bare PMTs, although used in some applications, are constrained by the degradation of many photocathode and dynode surfaces upon exposure to air and humidity. Metal surfaces such as tungsten and aluminum/Al_2O_3 [44.162] must be used rather than the higher efficiency alkali halide coated surfaces. Sodium salicylate is the most commonly used VUV-visible conversion phosphor; others include liumogen, terphenyl, and coronene [44.2, 171, 172]. The conversion efficiency of sodium salicylate (peak fluorescence about 430 nm) is relatively constant for VUV light from 30 to 200 nm. It displays some aging effects, which may be associated with contaminants such as oil vapor in the associated vacuum system [44.2]. The fluorescent decay time of sodium salicylate is ≈ 10 ns [44.2].

Vacuum Photodiodes

Vacuum photodiodes consist of a photocathode and an anode, with no amplifying dynode chain. The current leaving the photocathode is measured. (A review of vacuum photodiode designs and performance can be found in [44.173].) Because of their stability, spatial uniformity, and portability, vacuum photodiodes are used as radiometric transfer-standard detectors throughout much of the VUV [44.162, 174, 175]. There is no gain in vacuum photodiodes, so minimum signal levels that can be accurately measured are on the order of 10^6 photons/s. For fast timing applications, vacuum photodiodes have been designed with risetimes as short as about 60 ps [44.160].

Microchannel Plates

The microchannel plate (MCP) detector is a photoemissive array detector that combines the single-photon counting sensitivity of a PMT with high resolution imaging capability [44.161–163, 170, 176, 177]. An MCP consists of an array of semiconducting glass channels with diameters of about 10 to 25 μm and length-to-diameter ratios of about 50 : 1. Electrons ejected from the front surface of an MCP via the photoelectric effect are accelerated through the channels; repeated collisions with the channel walls result in an amplification of the charge by about 10^6. The exiting charge clouds are detected by a variety of position-sensing anode structures.

Bare MCPs have quantum efficiencies of about 10% for $\lambda \leq 100$ nm and a long wavelength cutoff about 120 nm [44.178]. Alkali halide coatings increase the quantum efficiency to about 20% and extend MCP sensitivities toward longer wavelengths [44.178–180]. Feedback instabilities produced by positive ions created in the channels during the electron cloud amplification are minimized through two common channel geometries; the chevron configuration, where two or more straight-channel MCPs oriented at different angles are cascaded [44.176], or a configuration with one set of curved channels [44.181, 182].

Readout schemes for determining the position of individual detected photons rely on either direct detection of the resulting electron cloud, or conversion of the electron cloud into visible photons via a phosphor to produce an optical image [44.170, 183]. Direct detection schemes include centroid-detecting anodes such as the wedge and strip design [44.178, 184], cross strip anodes [44.185]. Other designs are described in [44.186, 187]. Discrete anode arrays that digitally locate event positions include the MAMA (multianode microchannel array) [44.188, 189] and the CODACON systems [44.190]. Spatial resolutions of 15 μm are possible.

Silicon Photodiodes

Broad use of silicon photodiodes at VUV wavelengths has traditionally been limited by the strong absorption of VUV photons in the outer SiO_2 passivation layer that covers the p–n junction of these devices. Standard silicon detectors are sensitive throughout the infrared and visible regions and also in the soft X-ray ($\lambda < 2$ nm) and X-ray regions. Significant improvements in silicon photodiode sensitivities in the VUV are realized by thinning the SiO_2 passivation region to thicknesses of about 5 to 10 nm. References [44.173, 191] review VUV semiconductor photodiode designs; references [44.192, 193] describe the development of non-silicon-based photodiodes, e.g., wide bandgap materials such as GaN, for this spectral range. References [44.194–197] report devices with a quantum efficiency of 120% (electron–hole pairs per incident photon) at 100 nm. The development of such photodiodes with appreciable VUV sensitivity and good temporal stability has led to their use as radiometric transfer standards [44.198, 199]. Because silicon photodiodes respond strongly to radiation throughout the visible, ir, and X-ray regions, a method for rejection of stray light is essential for their effective use in the VUV. Reference [44.200] reports on the use of thin-film filters, deposited on the photodiode surface, to restrict the bandpass of the radiation impinging on the diode to selected VUV wavelengths. Wide bandgap semiconductors reject visible light with their natural solar-blind response [44.199].

Charge-Coupled Devices

CCDs are widely used for low light level imaging applications throughout the visible and near-ir regions. In the VUV, charge-coupled devices suffer from strong absorption both by surface gate structures and by the inactive passivation layer [44.170]. Two approaches are employed to overcome these limitations: the CCD front surface is overcoated with a photon down-converting phosphor [44.201, 202], or a thinned and surface-treated CCD is back-illuminated with VUV light [44.203, 204]. Such techniques have resulted in CCD VUV quantum efficiencies that rival those of photoemissive devices such as microchannel plates. A comparison of CCD and MCP performance in the VUV is given by [44.178]; they conclude that MCPs are most appropriate for low light-level imaging where photon counting is applicable, while the larger dynamic range and readout capabilities of CCDs make them suitable for higher light level applications.

Ionization Chambers

In gas ionization chambers, the impact of VUV photons with gas atoms or molecules produces electron–ion pairs which are then collected by the application of an appropriate voltage [44.205]. The photoionization efficiency and the detection efficiency approach 100% for photon energies above the ionization threshold of the gas [44.2, 162]. With appropriate gain, ionization chambers can detect individual photons (e.g., the Geiger counter); these chambers require relatively high gas pressures (100 to 2000 Torr) and therefore cannot be used in the windowless region of the VUV. The double ionization chamber described by [44.206] serves as a primary standard detector from 5 to 100 nm [44.162].

44.5 Optical Materials

The design of VUV instrumentation is dictated in large part by the constraints of VUV optical materials. Transmission through bulk materials is limited to $\lambda \gtrsim 105$ nm, the short wavelength transmission limit of LiF. Normal incidence reflectance from metal surfaces and coatings decreases dramatically at short wavelengths; polarizers and narrow-band interference filters are relatively difficult to produce because of the lack of materials with suitable optical constants. Below we briefly summarize the properties and uses of VUV optical materials. Comprehensive discussions can be found in [44.1,2,50].

Windows

Few bulk materials transmit light below 200 nm; none transmit light from about 2 nm to 105 nm. In the 105 to 200 nm wavelength region, the most common window materials are synthetic fused quartz or suprasil (with a short λ transmission cutoff about 160 nm), sapphire (145 nm), CaF_2 (125 nm) MgF_2 (112 nm), and LiF (105 nm). Compilations of the transmission characteristics of these and other materials are presented in [44.2, 50, 131, 207, 208]. Single crystal windows with dimensions up to 10 cm can be obtained for most of these materials. Sapphire fluoresces when struck by ionizing radiation and is therefore not suitable for certain environments.

Capillaries

In the windowless region, 2 to 105 nm, it is standard practice to use differentially pumped slits or circular apertures to isolate different pressure regions, e.g., to isolate an absorption cell from a vacuum spectrometer. This technique has the drawback of requiring large pumping capacities and limiting optical throughputs. Thin films (see below) can be used below 105 nm as windows, but they provide limited transmission in narrow bandpasses and cannot support large pressure differentials. Reference [44.209] describes a differentially pumped MCP capillary array as an alternative window in the VUV. The capillaries are typically ≈ 3 mm in length and have diameters of 10 to 100 μm. Measured optical transmissions are in the 20 to 50% range throughout the VUV; tradeoffs must be considered between optical throughput and gas conductance. The main limitation to an MCP window is its small angular aperture.

Thin Films

Below 105 nm, thin (about 10 to 200 nm) metallic films are used as transmission filters with bandpasses of approximately 10 to 50 nm. Details of filter properties are given in [44.208, 210]. A review of thin film filter preparation techniques is presented in [44.211]. Reference [44.212] describes composite metallic filters (e.g., Al/Ti/C, Ti/Sb/Al), developed for the Extreme Ultraviolet Explorer satellite, with transmission bandpasses of about 10 to 20 nm and high rejection at wavelengths of strong VUV geocoronal lines.

Coatings

Above 120 nm, the principal broadband reflector for VUV wavelengths is Al with a thin protective overcoat of MgF_2, having a normal incidence reflectance of ≈ 80 to 85% [44.2, 213]. The normal incidence reflectivities of all materials drop dramatically below about 100 nm. A compilation of coating reflectivities at normal and grazing incidence is presented in [44.214]. Materials with the highest reflectivities include Os, Pt, Au, and Ir, with reflectivities of 15 to 30% from 30 to 110 nm [44.2, 215]. [44.216] reviews the preparation of VUV reflectance coatings for diffraction gratings. Reference [44.217] reports grazing incidence reflection coefficients for Rh, Os, Pt, and Au from 5 to 30 nm; Rh has the highest reflectivity in this region.

Interference Filters and Multilayer Coatings

The development of interference filters for VUV wavelengths has been limited by the availability of coating materials with both high transmission and appreciable range of refractive differentials [44.218]. Multilayer dielectric reflectors thus require many layers to compensate for the small reflectivities at the material interfaces. The theory of multilayer reflecting optics and optic designs are reviewed in [44.219]. Reference [44.215] evaluates the commercially available VUV reflectance filters, antireflection coatings, and neutral density filters as of 1983. Developments as of the earlys 1990s are described in [44.220–223]. More recent advances are presented in [44.224–227].

Normal incidence optics with multilayer reflection coatings can be used in the nominally grazing incidence spectral region below 30 nm [44.228–231]. Reference [44.232] summarizes synthesis procedures and models of multilayer structures. References [44.233–238] describe normal incidence gratings coated with Mo/Si multilayers. The performances of other multilayer coatings are described in [44.239–242].

Polarizers

The production and detection of linearly polarized light in the VUV is discussed in [44.2, 131, 243–247]. Above about 112 nm, transmission polarizers, based on the birefringence of MgF_2, are employed [44.248, 249]. Reflection polarizers must be used below 112 nm. An analysis of double-reflection circular polarizers is given in [44.250].

References

44.1 J. A. R. Samson, D. L. Ederer (Eds.): *Vacuum Ultraviolet Spectroscopy I & II* (Academic, San Diego 1998).
44.2 J. A. R. Samson: *Techniques of Vacuum Ultraviolet Spectroscopy* (Wiley, New York 1967)
44.3 G. S. Brown, D. E. Moncton (Eds.): *Handbook on Synchrotron Radiation*, Vol. 3 (North Holland, Amsterdam 1991)
44.4 G. Margaritondo: *Introduction to Synchrotron Radiation* (Oxford Univ. Press, New York 1988)
44.5 G. V. Marr (Ed.): *Handbook on Synchrotron Radiation*, Vol. 2 (North Holland, Amsterdam 1987)
44.6 V. Schmidt: Rep. Prog. Phys. **55**, 1483 (1992)
44.7 H. Winick: Synchrotron radiation. In: *Physics of Particle Accelerators* (American Institute of Physics, New York 1989) p. 2138
44.8 H. Winick: *Synchrotron Radiation Sources: A Primer* (World Scientific, Singapore 1994)
44.9 S. L. Hulbert, G. P. Williams: Synchrotron radiation sources. In: *Vacuum Ultraviolet Spectroscopy I*, ed. by J. A. R. Samson, D. L. Ederer (Academic, San Diego 1998) p. 1
44.10 H. Winick: J. Synch. Rad. **5**, 168 (1998)
44.11 H. Wiedemann: *Synchrotron Radiation* (Springer, Berlin, Heidelberg 2003)
44.12 V. A. Bordovitsyn (Ed.): *Synchrotron Radiation Theory and its Development* (World Scientific, Singapore 1999)
44.13 F. Ciocci, G. Dattoli, A. Torre, A. Renieri: *Insertion Devices for Synchrotron Radiation and Free Electron Lasers* (World Scientific, Singapore 2000)
44.14 E.-E. Koch, D. E. Eastman, Y. Farge: Synchrotron radiation – a powerful tool in science. In: *Handbook on Synchrotron Radiation*, Vol. 1, ed. by E.-E. Koch (North Holland, Amsterdam 1983) p. 1
44.15 S. Krinsky, M. L. Perlman, R. E. Watson: Characteristics of synchrotron radiation and of its sources. In: *Handbook on Synchrotron Radiation*, Vol. 1, ed. by E.-E. Koch (North Holland, Amsterdam 1983) p. 65
44.16 I. H. Munro, G. V. Marr: Synchrotron radiation sources. In: *Handbook on Synchrotron Radiation*, Vol. 2, ed. by G. V. Marr (North Holland, Amsterdam 1987) p. 1
44.17 K. Wille: Rep. Prog. Phys. **54**, 1005 (1991)
44.18 H. Winick: Properties of synchrotron radiation. In: *Synchrotron Radiation Research*, ed. by H. Winick, S. Doniach (Plenum, New York 1980) p. 11
44.19 L. Nahon, F. Polack, B. Lagarde, R. Thissen, C. Alcaraz, O. Dutuit, K. Ito: Nucl. Instr. Meth. **467**, 453 (2001)
44.20 B. A. Wallace: J. Synchr. Rad. **7**, 289 (2000)
44.21 T. Möller, G. Zimmerer: Phys. Scr. T **17**, 177 (1987)
44.22 I. H. Munro, N. Schwentner: Nucl. Instrum. Methods **208**, 819 (1983)
44.23 V. Rehn: Nucl. Instrum. Methods **177**, 193 (1980)
44.24 B. Craseman: Can. J. Phys. **76**, 251 (1998)
44.25 D. W. Lindle, O. A. Hemmers: J. Alloys & Compounds **328**, 27 (2001)
44.26 G. S. Brown, W. Lavender: Synchrotron-radiation spectra. In: *Handbook on Synchrotron Radiation*, Vol. 3, ed. by G. S. Brown, D. E. Moncton (North Holland, Amsterdam 1991) p. 37
44.27 P. Elleaume: Rev. Sci. Instrum. **63**, 321 (1992)
44.28 P. K. Carroll, E. T. Kennedy, G. O'Sullivan: Opt. Lett. **2**, 72 (1978)
44.29 P. K. Carroll, E. T. Kennedy, G. O'Sullivan: Appl. Opt. **19**, 1454 (1980)
44.30 M. Richardson: Laser-produced plasmas. In: *Vacuum Ultraviolet Spectroscopy I*, ed. by J. A. Samson, D. L. Ederer (Academic, San Diego 1998) p. 83
44.31 G. O'Sullivan: J. Phys. B **16**, 3291 (1983)
44.32 J. M. Bridges, C. L. Cromer, T. J. McIlrath: Appl. Opt. **25**, 2208 (1986)
44.33 J. T. Costello, J. P. Mosnier, E. T. Kennedy, P. K. Carroll, G. O'Sullivan: Phys. Scr. T **34**, 77 (1991)
44.34 P. Gohil, V. Kaufman, T. J. McIlrath: Appl. Opt. **25**, 2039 (1986)
44.35 F. B. Orth, K. Ueda, T. J. McIlrath, M. L. Ginter: Appl. Opt. **25**, 2215 (1986)
44.36 K. Eidmann, T. Kishimoto: Appl. Phys. Lett. **49**, 377 (1986)
44.37 O. Meighan, A. Gray, J. P. Mosnier, W. Whitty, J. T. Costello, C. L. S. Lewis, A. MacPhee, R. Allott, I. C. E. Turcu, A. Lamb: Appl. Phys. Lett. **70**, 1497 (1997)
44.38 D. Giulietti, L. A. Gizza: Rivista del Nuovo Cim. **21**, 1 (1998)
44.39 E. T. Kennedy, J. T. Costello, J. P. Mosnier, A. A. Cafolla, M. Collins, L. Kiernan, U. Koble, M. H. Sayyard, M. Shaw, B. F. Sonntag, R. Barchewitz: Opt. Eng. **33**, 3984 (1994)
44.40 E. T. Kennedy, J. T. Costello, A. Gray, C. McGuinness, J. P. Mossnier, P. van Kampen: J. Electr. Spectrosc. Rel. Phen. **103**, 161 (1999)

44.41 O. Meighan, C. Danson, L. Dardis, C. L. S. Lewis, A. MacPhee, C. McGuinness, R. O'Rourke, W. Shaikh, I. C. E. Turcu, J. T. Costello: J. Phys. B **33**, 1159 (2000)
44.42 J. R. Roberts: Glow discharges and wall-stabilized arcs. In: *Vacuum Ultraviolet Spectroscopy I*, ed. by J. A. Samson, D. L. Ederer (Academic, San Diego 1998) p. 37
44.43 M. Kühne: Hollow cathodes and Penning discharges. In: *Vacuum Ultraviolet Spectroscopy I*, ed. by J. A. Samson, D. L. Ederer (Academic, San Diego 1998) p. 65
44.44 A. P. Thorne, U. Litzen, S. Johansson: *Spectrophysics: Principles and Applications* (Springer, Berlin 1999)
44.45 K. P. Huber, G. Herzberg: *Constants of Diatomic Molecules* (Van Nostrand, New York 1979)
44.46 J. M. Bridges, W. R.. Ott: Appl. Opt. **16**, 367 (1977)
44.47 J. Z. Klose, J. M. Bridges, W. R. Ott: J. Res. Nat. Bur. Stand. **93**, 21 (1988)
44.48 Y. Tanaka: J. Opt. Soc. Am. **45**, 710 (1955)
44.49 Y. Tanaka, A. S. Jursa, F. J. LeBlanc: J. Opt. Soc. Am. **48**, 304 (1958)
44.50 A. N. Zaidel, E. Ya. Schreider: *Vacuum Ultraviolet Spectroscopy* (Humphrey, Ann Arbor 1970)
44.51 G. Balloffet, J. Romand, B. Vodar: C. R. Acad. Sci. **252**, 4139 (1961)
44.52 T. B. Lucatorto, T. J. McIlrath, G. Mehlman: Appl. Opt. **18**, 2916 (1979)
44.53 M. E. Pillow: Spectrochem. Acta. B **36**, 821 (1981)
44.54 M. Kühne: Radiometric characterization of VUV sources. In: *Vacuum Ultraviolet Spectroscopy I*, ed. by J. A. Samson, D. L. Ederer (Academic, San Diego 1998) p. 119
44.55 J. Hollandt, M. Kühne, B. Wende: Appl. Opt. **33**, 68 (1994)
44.56 D. S. Finley, S. Bowyer, F. Paresce, R. F. Malina: Appl. Opt. **18**, 649 (1979)
44.57 C. Heise, J. Hollandt, R. Kling, M. Kock, M. Kühne: Appl. Opt. **33**, 5111 (1994)
44.58 J. S. Risley, W. B. Westerveld: Appl. Opt. **28**, 389 (1989)
44.59 R. E. Marrs, M. A. Levine, D. A. Knapp, J. R. Henderson: Phys. Rev. Lett. **60**, 1715 (1988)
44.60 P. Beiersdorfer: Ann. Rev. Astr. Astrophys. **41**, 343 (2003)
44.61 R. R. Alfano, S. L. Shapiro: Phys. Rev. Lett. **24**, 592 (1970)
44.62 J. H. Glownia, J. A. Misewich, P. P. Sorokin: In: *The Supercontinuum Laser Source*, ed. by R. R. Alfano (Springer, New York 1989) p. 337
44.63 J. H. Glownia, J. A. Misewich, P. P. Sorokin: J. Chem. Phys. **92**, 3335 (1990)
44.64 R. R. Alfano: *The Supercontinuum Laser Source* (Springer, New York 1989)
44.65 P. B. Corkum, C. Rolland: In: *The Supercontinuum Laser Source*, ed. by R. R. Alfano (Springer, New York 1989) p. 318
44.66 Y. R. Shen, G. Yang: In: *The Supercontinuum Laser Source*, ed. by R. R. Alfano (Springer, New York 1989) p. 1
44.67 G. Rodriguez, J. P. Roberts, A. J. Taylor: Opt. Lett. **19**, 1146 (1994)
44.68 T. R. Gosnell, A. J. Taylor, D. P. Greene: Opt. Lett. **15**, 130 (1990)
44.69 P. Jaeglé: Vacuum ultraviolet lasers. In: *Vacuum Ultraviolet Spectroscopy I*, ed. by J. A. Samson, D. L. Ederer (Academic, San Diego 1998) p. 101
44.70 P. Misra, M. A. Dubinskii: *Ultraviolet Spectroscopy and UV Lasers* (Marcel Dekker, New York 2002)
44.71 H. Daido: Rep. Prog. Phys. **65**, 1513 (2002)
44.72 J. J. Rocca: Rev. Sci. Instrum. **70**, 3799 (1999)
44.73 V. Ayvazyan et al.: Phys. Rev. Lett **88**, 104802 (2002)
44.74 G. R. Neil, L. Merminga: Rev. Mod. Phys. **74**, 685 (2002)
44.75 S. M. Hooker, C. E. Webb: Prog. Quant. Electron. **18**, 227 (1994)
44.76 R. W. Dreyfus, R. T. Hodgson: Phys. Rev. A **9**, 2635 (1974)
44.77 R. T. Hodgson: J. Chem. Phys. **55**, 5378 (1971)
44.78 C. J. Sansonetti, J. Reader, K. Vogler: Appl. Opt. **40**, 1974 (2001)
44.79 J. G. Eden: IEEE J. Sel. Top. Quant. Electron. **6**, 1051 (2000)
44.80 J. J. Ewing: IEEE J. Sel. Top. Quant. Electron. **6**, 1061 (2000)
44.81 P. W. Hoff, J. C. Swingle, C. K. Rhodes: Appl. Phys. Lett. **23**, 245 (1973)
44.82 W. M. Hughes, J. Shannon, R. Hunter: Appl. Phys. Lett. **24**, 488 (1974)
44.83 M. H. R. Hutchinson: Appl. Opt. **19**, 3883 (1980)
44.84 H. C. Kapteyn, R. W. Lee, R. W. Falcone: Phys. Rev. Lett. **57**, 2939 (1986)
44.85 H. C. Kapteyn, R. W. Falcone: Phys. Rev. A **37**, 2033 (1988)
44.86 D. J. Walker, C. P. J. Barty, G. Y. Yin, J. F. Young, S. E. Harris: Opt. Lett. **12**, 894 (1987)
44.87 M. N. R. Ashfold, J. D. Prince: Contemp. Phys. **29**, 125 (1988)
44.88 R. Hilbig, G. Hilber, A. Lago, B. Wolff, R. Wallenstein: Comments At. Mol. Phys. **18**, 157 (1986)
44.89 W. Jamroz, B. P. Stoicheff: Prog. Opt. **XX**, 327 (1983)
44.90 B. P. Stoicheff: *Frontiers of Laser Spectroscopy of Gases*, ed. by A. C. Alves, J. M. Brown, J. M. Hollas (Kluwer, Dordrecht 1988) p. 63
44.91 B. P. Stoicheff: *Frontiers in Laser Spectroscopy*, ed. by T. W. Hänsch, M. Inguscio (North Holland, Amsterdam 1994) p. 105
44.92 C. R. Vidal: *Tunable Lasers*, ed. by L. F. Mollenauer, J. C. White (Springer, Heidelberg 1987) p. 57
44.93 S. C. Wallace, R. H. Huebner: *Photophysics and Photochemistry in the Vacuum Ultraviolet*, ed. by S. P. McGlynn, G. L. Findley (Reidel, Dordrecht 1985) p. 185
44.94 H. F. Döbele: Plasma Sources Sci. Tech. **4**, 224 (1995)

44.95 R.H. Lipson, S.S. Dimov, P. Wang, Y.J. Shi, D.M. Mao, X.K. Hu, J. Vanstone: Instrum. Sci. Tech. **28**, 85 (2000)
44.96 W. Mückenheim, P. Lokai, B. Burghardt, D. Basting: Appl. Phys. B **45**, 259 (1988)
44.97 G.C. Bhar, U. Chatterjee, A.M. Rudra, P. Kumbhakar: Quant. Electron. **29**, 800 (1999)
44.98 A. Penzkofer, A. Laubereau, W. Kaiser: Prog. Quant. Electron. **6**, 55 (1979)
44.99 Y.R. Shen: *The Principles of Nonlinear Optics* (Wiley-Interscience, New York 1984)
44.100 Y. Huo, K. Shimizu, T. Yagi: J. Appl. Phys. **72**, 3258 (1992)
44.101 H. Schomburg, H.F. Döbele, B. Rückle: Appl. Phys. B **30**, 131 (1983)
44.102 A. Goehlich, U. Czarnetzki, H.F. Döbele: Appl. Opt. **37**, 8453 (1998)
44.103 G.J. Bjorklund: IEEE J. Quantum Electron. **11**, 289 (1975)
44.104 G. Hilber, A. Lago, R. Wallenstein: J. Opt. Soc. Am. B **4**, 1753 (1987)
44.105 R. Hilbig, R. Wallenstein: Appl. Opt. **21**, 913 (1982)
44.106 B. Wellegehausen, H. Welling, C. Momma, M. Feuerhake, K. Massavi, H. Eichmann: Opt. Quant. Electron. **28**, 267 (1996)
44.107 N. Melikechi, S. Gangopadhyay, E.E. Eyler: Appl. Opt. **36**, 7776 (1997)
44.108 A. Nazarkin, G. Korn, O. Kittlemann, J. Ringling, I.V. Hertel: Phys. Rev. A **56**, 671 (1997)
44.109 G.W. Faris, S.A. Meyer, M.J. Dyer, M.J. Banks: J. Opt. Soc. Am. B **17**, 1856 (2000)
44.110 M. Wittmann, M.T. Wick, O. Steinkellner, V. Stert. Pl Farmanara, W. Radloff, G. Korn, I.V. Hertel: Opt. Comm. **173**, 323 (2000)
44.111 R. Hilbig, R. Wallenstein: Opt. Comm. **44**, 283 (1983)
44.112 R. Hilbig, A. Lago, R. Wallenstein: Opt. Comm. **49**, 297 (1984)
44.113 E. Cromwell, T. Trickl, Y.T. Lee, A.H. Kung: Rev. Sci. Instrum. **60**, 2888 (1989)
44.114 H. Palm, F. Merkt: Appl. Phys. Lett. **73**, 157 (1998)
44.115 U. Hollenstein, H. Palm, F. Merkt: Rev. Sci. Instrum. **71**, 4023 (2000)
44.116 J. Reintjes, C.Y. She, R.C. Eckardt: IEEE J. Quantum Electron. **14**, 581 (1978)
44.117 K.S.E. Eikema, W. Ubachs, W. Vassen, W. Hogervorst: Phys. Rev. Lett. **71**, 1690 (1993)
44.118 K.S.E. Eikema, W. Ubachs, W. Vassen, W. Hogervorst: Phys. Rev. A **55**, 1866 (1997)
44.119 F. Brandi, D. Neshev, W. Ubachs: Phys. Rev. Lett. **91**, 163901 (2003)
44.120 K. Yamanouchi, S. Tsuchiya: J. Phys. B **28**, 133 (1995)
44.121 R.T. Hodgson, P.P. Sorokin, J.J. Wynne: Phys. Rev. Lett. **32**, 343 (1974)
44.122 W. Jamroz, P.E. LaRocque, B.P. Stoicheff: Opt. Lett. **7**, 148 (1982)
44.123 R. Mahon, F.S. Tomkins: IEEE J. Quantum Electron. **18**, 913 (1982)
44.124 C.H. Kwon, H.L. Kim, M.S. Kim: Rev. Sci. Instrum. **74**, 2939 (2003)
44.125 P.R. Herman, B.P. Stoicheff: Opt. Lett. **10**, 502 (1985)
44.126 R. Hilbig, R. Wallenstein: IEEE J. Quantum Electron. **17**, 1566 (1981)
44.127 C.E.M. Strauss, D.J. Funk: Opt. Lett. **16**, 1192 (1991)
44.128 D. Cotter: Opt. Comm. **31**, 397 (1979)
44.129 W.R. Hunter: Diffraction gratings and mountings for the vacuum ultraviolet spectral region. In: *Spectrometric Techniques*, Vol. IV, ed. by G. Vanasse (Academic, New York 1985) p. 63
44.130 R.L. Johnson: Grating monochromators and optics for the VUV and soft X-ray region. In: *Handbook of Synchrotron Radiation*, Vol. 1, ed. by E.E. Koch (North-Holland, Amsterdam 1983) p. 173
44.131 J.A.R. Samson: Far ultraviolet region. In: *Methods of Experimental Physics*, Vol. 13, ed. by D. Williams (Academic, New York 1976) p. 204
44.132 M. Koike: Normal incidence monochromators. In: *Vacuum Ultraviolet Spectroscopy II*, ed. by J.A. Samson, D.L. Ederer (Academic, San Diego 1998) p. 1
44.133 H.A. Padmore, M.R. Howells, W.R. McKinney: Grazing incidence monochromators. In: *Vacuum Ultraviolet Spectroscopy II*, ed. by J.A. Samson, D.L. Ederer (Academic, San Diego 1998) p. 21
44.134 S. Singh: Opt. Laser Tech. **31**, 195 (1999)
44.135 E.W. Palmer, M.C. Hutley, A. Franks, J.F. Verrill, B. Gale: Rep. Prog. Phys. **38**, 975 (1975)
44.136 J.H. Underwood: Imaging properties and aberrations of spherical optics and non-spherical optics. In: *Vacuum Ultraviolet Spectroscopy I*, ed. by J.A. Samson, D.L. Ederer (Academic, San Diego 1998) p. 145
44.137 M.C. Hutley: *Diffraction Gratings* (Academic, New York 1982)
44.138 G. Schmahl, D. Rudolf: Prog. Opt. **XIV**, 197 (1976)
44.139 T. Namioka: Diffraction gratings. In: *Vacuum Ultraviolet Spectroscopy I*, ed. by J.A. Samson, D.L. Ederer (Academic, San Diego 1998) p. 347
44.140 M.C. Hettrick: Appl. Opt. **24**, 1251 (1985)
44.141 J.H. Underwood: Spectrograph and monochromators using varied line spacing gratings. In: *Vacuum Ultraviolet Spectroscopy II*, ed. by J.A. Samson, D.L. Ederer (Academic, San Diego 1998) p. 55
44.142 K. Ito, T. Namioka, Y. Morioka, T. Sasaki, H. Noda, K. Goto, T. Katayama, M. Koike: Appl. Opt. **25**, 837 (1986)
44.143 K. Ito, T. Namioka: Rev. Sci. Instr. **60**, 1573 (1989)
44.144 K. Yoshino, D.E. Freeman, W.H. Parkinson: Appl. Opt. **19**, 66 (1980)
44.145 L. Nahon, C. Alcaraz, J.L. Marlats, B. Lagarde, F. Polack, R. Thissen, D. Lepere, K. Ito: Rev. Sci. Instrum. **72**, 1320 (2001)
44.146 J. Nordgren, L. Pettersson, L. Selander, S. Griep, C. Nordling, K. Siegbahn: Phys. Scr. **20**, 623 (1979)

44.147 J. Reader, N. Acquista: J. Opt. Soc. Am. **69**, 1285 (1979)
44.148 W. C. Cash: Appl. Opt. **21**, 710 (1982)
44.149 M. Neviere, D. Maystre, W. R. Hunter: J. Opt. Soc. Am. **68**, 1106 (1978)
44.150 P. Lemaire: Appl. Opt. **30**, 1294 (1991)
44.151 W. C. Cash: Appl. Opt. **22**, 3971 (1983)
44.152 M. C. Hettrick, S. Bowyer: Appl. Opt. **22**, 3921 (1983)
44.153 A. P. Thorne, C. J. Harris, I. Wynne-Jones, R. C. M. Learner, G. Cox: J. Phys. E **20**, 54 (1987)
44.154 A. P. Thorne, M. R. Howells: Interferometric spectrometers. In: *Vacuum Ultraviolet Spectroscopy II*, ed. by J. A. Samson, D. L. Ederer (Academic, San Diego 1998) p. 73
44.155 A. P. Thorne, G. Cox, P. L. Smith, W. H. Parkinson: Proc. SPIE **2282**, 58 (1994)
44.156 S. Chakrabarti, D. M. Cotton, J. S. Vickers, B. C. Bush: Appl. Opt. **33**, 2596 (1994)
44.157 J. Harlander, R. J. Reynolds, F. L. Roesler: Astrophys. J. **396**, 730 (1992)
44.158 R. A. Kruger, L. W. Anderson, F. L. Roesler: J. Opt. Soc. Am. **62**, 938 (1972)
44.159 N. De Oliveira, D. Joyeux, D. Phalippou: Surf. Rev. Lett. **9**, 655 (2002)
44.160 C. B. Johnson: Proc. SPIE **1243**, 2 (1990)
44.161 O. H. W. Siegmund, M. A. Gummin, J. Stock, D. Marsh: *Photon Detectors for Space Instrumentation*, ed. by T. D. Guyenne, J. Hunt (European Space Agency, Paris 1992) p. 356
44.162 J. G. Timothy, R. P. Madden: Photon detectors for the ultraviolet and X-ray region. In: *Handbook on Synchrotron Radiation*, Vol. 1, ed. by E.-E. Koch (North-Holland, Amsterdam 1983) p. 315
44.163 J. G. Timothy: *Photoelectronic Imaging Devices*, ed. by B. L. Morgan (I. O. P., Bristol 1991) p. 85
44.164 M. L. Furst, R. M. Graves, L. R. Canfield, R. E. Vest: Rev. Sci. Instrum. **66**, 2257 (1995)
44.165 P. S. Shaw, T. C. Larason, R. Gupta, K. R. Lykke: Rev. Sci. Instrum. **73**, 1625 (2002)
44.166 G. Ulm: Metrologia **40**, 101 (2003)
44.167 J. H. Moore, C. C. Davis, M. A. Coplan: *Building Scientific Apparatus: A Practical Guide to Design and Construction*, 2nd edn. (Addison-Wesley, Reading 1989)
44.168 G. Rieke: *Detection of Light* (Univ. Cambridge, Cambridge 2003)
44.169 R. W. Engstrom: *Photomultiplier Handbook* (RCA Electro Optics and Devices, Lancaster 1980)
44.170 O. H. W. Siegmund: Photodiode detectors. In: *Vacuum Ultraviolet Spectroscopy II*, ed. by J. A. Samson, D. L. Ederer (Academic, San Diego 1998) p. 139
44.171 J. B. Birks: In: *The Physics and Chemistry of the Organic Solid State*, ed. by D. Fox, M. M. Labes, A. Weissberger (Interscience, New York 1965) pp. 434–509
44.172 J. A. R. Samson, G. N. Haddad: J. Opt. Soc. Am. **64**, 1346 (1974)
44.173 L. R. Canfield: Gas detectors. In: *Vacuum Ultraviolet Spectroscopy II*, ed. by J. A. Samson, D. L. Ederer (Academic, San Diego 1998) p. 117
44.174 E. B. Saloman: Nucl. Instrum. Meth. **172**, 79 (1980)
44.175 L. R. Canfield, N. Swanson: J. Res. Nat. Bur. Stand. **92**, 97 (1987)
44.176 J. L. Wiza: Nucl. Instrum. Meth. **162**, 587 (1979)
44.177 O. H. W. Siegmund, A. S. Tremsin, J. V. Vallerga: Nucl. Instrum. Meth. A **510**, 185 (2003)
44.178 J. V. Vallerga, M. Lampton: Proc. SPIE **868**, 25 (1987)
44.179 O. H. W. Siegmund, E. Everman, J. V. Vallerga, J. Sokolowski, M. Lampton: Appl. Opt. **26**, 3607 (1987)
44.180 O. H. W. Siegmund, E. Everman, J. V. Vallerga, S. Labov, J. Bixler, M. Lampton: Proc. SPIE **687**, 117 (1987)
44.181 J. G. Timothy, R. L. Bybee: Rev. Sci. Instrum. **48**, 292 (1977)
44.182 J. G. Timothy: Rev. Sci. Instrum. **52**, 1131 (1981)
44.183 C. L. Cromer, J. M. Bridges, J. R. Roberts, T. B. Lucatorto: Appl. Opt. **24**, 2996 (1985)
44.184 C. Martin, P. Jelinsky, M. Lampton, R. F. Malina: Rev. Sci. Instrum. **52**, 1067 (1981)
44.185 O. H. W. Siegmund, A. S. Tremsin, J. V. Vallerga, J. Hull: IEEE Trans. Nucl. Sci. **48**, 430 (2001)
44.186 J. S. Vickers, S. Chakrabarti: Rev. Sci. Instrum. **70**, 2912 (1999)
44.187 J. S. Lapington, K. Rees: Nucl. Instrum. Meth. A **477**, 273 (2002)
44.188 J. G. Timothy, G. H. Mount, R. L. Bybee: SPIE Space Opt. **183**, 169 (1979)
44.189 J. G. Timothy, R. L. Bybee: Proc. SPIE **687**, 109 (1986)
44.190 W. E. McClintock, C. A. Barth, R. E. Steele, G. M. Lawrence, J. G. Timothy: Appl. Opt. **21**, 3071 (1982)
44.191 M. Razeghi, A. Rogalski: J. Appl. Phys. **79**, 7433 (1996)
44.192 E. Monroy, F. Omnes, F. Calle: Semiconductor Sci. Tech. **18**, R33 (2003)
44.193 A. Motogaito, H. Watanabe, K. Hiramatsu, K. Fukui, Y. Hamamura, K. Tadatomo: Physica Status Solidi A **200**, 147 (2003)
44.194 L. R. Canfield, J. Kerner, R. Korde: Appl. Opt. **28**, 3940 (1989)
44.195 L. R. Canfield, J. Kerner, R. Korde: Proc. SPIE **1344**, 372 (1990)
44.196 R. Korde, L. R. Canfield, B. Wallis: Proc. SPIE **932**, 153 (1988)
44.197 R. Korde, J. S. Cable, L. R. Canfield: IEEE Trans. Nucl. Sci. **40**, 1655 (1993)
44.198 E. M. Gullikson, R. Korde, L. R. Canfield, R. E. Vest: J. Electron Spectrosc. Rel. Phenom. **80**, 313 (1996)
44.199 R. E. Vest, B. Hertog, P. Chow: Metrologia **40**, S141 (2003)
44.200 L. R. Canfield, R. Vest, T. N. Woods, R. Korde: Proc. SPIE **2282**, 31 (1994)
44.201 P. F. Morrisey, S. R. McCandliss, P. D. Feldman, S. D. Friedman: Appl. Opt. **33**, 2534 (1994)

44.202 J. Janesick, T. Elliot, R. Winzenread, J. Pinter, R. Dyck: SPIE **2415**, 2 (1995)
44.203 J. Janesick, D. Campbell, T. Elliot, T. Daud: Opt. Eng. **26**, 853 (1987)
44.204 R. A. Stern, R. C. Catura, R. Kimble, A. F. Davidsen, M. Winzenread, M. M. Blouke, R. Hayes, D. M. Walton, J. L. Culhane: Opt. Eng. **26**, 875 (1987)
44.205 J. B. West: *Vacuum Ultraviolet Spectroscopy II*, ed. by J. A. Samson, D. L. Ederer (Academic, San Diego 1998) p. 107
44.206 J. A. R. Samson: J. Opt. Soc. Am. **54**, 6 (1964)
44.207 E. D. Palik: *Handbook of Optical Constants of Solids* (Academic, New York 1985)
44.208 W. R. Hunterin: *Vacuum Ultraviolet Spectroscopy I*, ed. by J. A. Samson, D. L. Ederer (Academic, San Diego 1998) p. 305
44.209 T. B. Lucatorto, T. J. McIlrath, J. R. Roberts: Appl. Opt. **18**, 2505 (1979)
44.210 F. R. Powell, P. W. Vedder, J. F. Lindblom, S. F. Powell: Opt. Eng. **26**, 614 (1990)
44.211 W. R. Hunter: In: *Physics of Thin Films*, Vol. 7, ed. by G. Hass, M. H. Francombe, R. W. Hoffman (Academic, New York 1973) pp. 43–114
44.212 P. W. Vedder, J. V. Vallerga, O. H. W. Siegmund, J. Gibson, J. Hull: SPIE **1159**, 392 (1989)
44.213 J. I. Larruquert, R. A. M. Keski-Kuha: Opt. Comm. **215**, 93 (2003)
44.214 W. R. Hunterin: *Vacuum Ultraviolet Spectroscopy I*, ed. by J. A. Samson, D. L. Ederer (Academic, San Diego 1998) p. 205
44.215 B. K. Flint: Adv. Space. Res. **2**, 135 (1983)
44.216 W. R. Hunter: *Physics of Thin Films*, Vol. 11, ed. by G. Hass, M. H. Francombe (Academic, New York 1980) pp. 1–34
44.217 M. C. Hettrick, S. A. Flint, J. Edelstein: Appl. Opt. **24**, 3682 (1985)
44.218 M. Zukic, D. G. Torr, J. F. Spann, M. R. Torr: Appl. Opt. **29**, 4284 (1990)
44.219 E. Spillerin: *Vacuum Ultraviolet Spectroscopy I*, ed. by J. A. Samson, D. L. Ederer (Academic, San Diego 1998) p. 271
44.220 M. Zukic, D. G. Torr, J. F. Spann, M. R. Torr: Appl. Opt. **29**, 4293 (1990)
44.221 M. Zukic, D. G. Torr: VUV thin films. In: *Thin Films*, ed. by K. H. Guenther (Springer, Berlin 1991) Chap. 7
44.222 M. Zukic, D. G. Torr: Appl. Opt. **31**, 1588 (1992)
44.223 M. Zukic, D. G. Torr, J. Kim, J. F. Spann, M. R. Torr: Opt. Eng. **32**, 3069 (1993)
44.224 Y. A. Uspenskii, V. E. Levashov, A. V. Vinogradov, A. I. Fedorenko, V. V. Kondratenko, Y. P. Pershin, E. N. Zubarev, V. Y. Fedotov: Opt. Lett. **23**, 771 (1998)
44.225 J. I. Larruquert, R. A. M. Kedki-Kuha: Appl. Opt. **38**, 1231 (1999)
44.226 J. I. Larruquert, R. A. M. Kedki-Kuha: Appl. Opt. **40**, 1126 (2001)
44.227 J. I. Larruquert, R. A. M. Kedki-Kuha: Appl. Opt. **41**, 5398 (2002)
44.228 T. W. Barbee, S. Mrowka, M. C. Hettrick: Appl. Opt. **24**, 883 (1985)
44.229 J. V. Bixler, T. W. Barbee, D. D. Dietrich: Proc. Soc. Photo-Opt. Instrum. Eng. **1160**, 648 (1989)
44.230 E. Spiller: Appl. Opt. **15**, 2333 (1976)
44.231 P. P. Naulleau, J. A. Liddle, E. H. Anderson, E. M. Gullikson, P. Mirkarimi, F. Salmassi, E. Spiller: Opt. Commun. **229**, 109 (2004)
44.232 T. W. Barbee: Phys. Scr. T **31**, 147 (1990)
44.233 J. F. Seely, C. M. Brown: Appl. Opt. **32**, 6288 (1993)
44.234 J. M. Slaughter, D. W. Schulze, C. R. Hills, A. Mirone, R. Stalia, R. N. Watts, C. Tarrio, T. B. Lucatorto, M. Krumrey, T. B. Mueller, C. M. Falco: J. Appl. Phys. **76**, 2144 (1994)
44.235 D. G. Stearns, R. S. Rosen, S. P. Vernon: Appl. Opt. **32**, 6952 (1993)
44.236 M. Toyoda, N. Miyata, M. Yanagihara, M. Watanabe: Jap. J. Appl. Phys. **37**, 2066 (1998)
44.237 C. Montcalm, R. F. Grabner, R. M. Hudyma, M. A. Schmidt, E. Spiller, C. C. Walton, M. Wedowski, J. A. Folta: Appl. Opt. **41**, 3262 (2002)
44.238 E. Spiller, S. L. Baker, P. B. Mirkarimi, V. Sperry, E. M. Gullikson, D. G. Stearns: Appl. Opt. **42**, 4049 (2003)
44.239 R. J. Thomas, R. A. M. Keski-Kuha, W. M. Neupert, C. E. Condor, J. S. Gunn: Appl. Opt. **30**, 2245 (1991)
44.240 M. Grigonis, E. J. Knystautas: Appl. Opt. **36**, 2839 (1997)
44.241 Y. Hotta, M. Furudate, M. Yamamoto, M. Watanabe: Surf. Rev. Lett. **9**, 571 (2002)
44.242 J. F. Seely, Y. A. Uspenskii, Y. P. Pershin, V. V. Kondratenko, A. V. Vinogradov: Appl. Opt. **41**, 1846 (2002)
44.243 W. R. Hunter: Polarization. In: *Vacuum Ultraciolet Spectroscopy I*, ed. by J. A. Samson, D. L. Ederer (Academic, San Diego 1998) p. 227
44.244 S. Chwirot, J. Slevin: Meas. Sci. Tech. **4**, 1305 (1993)
44.245 P. Zetner, A. Pradhan, W. B. Westerveld, J. W. McConkey: Appl. Opt. **22**, 2210 (1983)
44.246 P. Zetner, K. Becker, W. B. Westerveld, J. W. McConkey: Appl. Opt. **23**, 3184 (1984)
44.247 L. Museur, C. Olivero, D. Reidel, M. C. Castex: Appl. Phys. B **70**, 499 (2000)
44.248 R. Hippler, M. Faust, R. Wolf, H. Kleinpoppen, H. O. Lutz: Phys. Rev. A **31**, 1399 (1985)
44.249 H. Winter, H. W. Ortjohann: Rev. Sci. Instrum. **58**, 359 (1987)
44.250 W. B. Westerveld, K. Becker, P. W. Zetner, J. J. Orr, J. W. McConkey: Appl. Opt. **24**, 2256 (1985)

Part D Scattering

Part D Scattering Theory

45 Elastic Scattering: Classical, Quantal, and Semiclassical
M. Raymond Flannery, Atlanta, USA

46 Orientation and Alignment in Atomic and Molecular Collisions
Nils Andersen, Copenhagen, Denmark

47 Electron–Atom, Electron–Ion, and Electron–Molecule Collisions
Philip Burke, Belfast, UK

48 Positron Collisions
Robert P. McEachran, Canberra, Australia
Allan Stauffer, Toronto, Canada

49 Adiabatic and Diabatic Collision Processes at Low Energies
Evgueni E. Nikitin, Haifa, Israel

50 Ion–Atom and Atom–Atom Collisions
A. Lewis Ford, College Station, USA
John F. Reading, College Station, USA

51 Ion–Atom Charge Transfer Reactions at Low Energies
Muriel Gargaud, Floirac, France
Ronald McCarroll, Paris Cedex 05, France

52 Continuum Distorted Wave and Wannier Methods
Derrick Crothers, Belfast, UK
Fiona McCausland, Belfast, UK
John Glass, Belfast, UK
Jim F. McCann, Belfast, UK
Francesca O'Rourke, Belfast, UK
Ruth T. Pedlow, Belfast, UK

53 Ionization in High Energy Ion–Atom Collisions
Joseph H. Macek, Knoxville, USA
Steven T. Manson, Atlanta, USA

54 Electron–Ion and Ion–Ion Recombination
M. Raymond Flannery, Atlanta, USA

55 Dielectronic Recombination
Michael S. Pindzola, Auburn, USA
Donald C. Griffin, Winter Park, USA
Nigel R. Badnell, Glasgow, United Kingdom

56 Rydberg Collisions: Binary Encounter, Born and Impulse Approximations
Edmund J. Mansky, Oak Ridge, USA

57 Mass Transfer at High Energies: Thomas Peak
James H. McGuire, New Orleans, USA
Jack C. Straton, Portland, USA
Takeshi Ishihara, Tsukuba, Japan

58 Classical Trajectory and Monte Carlo Techniques
Ronald E. Olson, Rolla, USA

59 Collisional Broadening of Spectral Lines
Gillian Peach, London, UK

45. Elastic Scattering: Classical, Quantal, and Semiclassical

Scattering cross sections determine the rates at which gas phase processes and chemical reactions happen, whether in the atmosphere, or in an industrial reactor. This chapter provides a handy compendium of equations, formulae, and expressions for the classical, quantal, and semiclassical approaches to elastic scattering. Reactive systems and model potentials are also considered.

45.1 **Classical Scattering Formulae** ... 659
45.1.1 Deflection Functions ... 660
45.1.2 Elastic Scattering Cross Section ... 661
45.1.3 Center-of-Mass to Laboratory Coordinate Conversion ... 662
45.1.4 Glory and Rainbow Scattering ... 662
45.1.5 Orbiting and Spiraling Collisions ... 662
45.1.6 Quantities Derived from Classical Scattering ... 663
45.1.7 Collision Action ... 663

45.2 **Quantal Scattering Formulae** ... 664
45.2.1 Basic Formulae ... 664
45.2.2 Identical Particles: Symmetry Oscillations ... 666
45.2.3 Partial Wave Expansion ... 667
45.2.4 Scattering Length and Effective Range ... 668
45.2.5 Logarithmic Derivatives ... 670
45.2.6 Coulomb Scattering ... 671
45.2.7 Resonance Scattering ... 671
45.2.8 Integral Equation for Phase Shift. 673
45.2.9 Variable Phase Method ... 673
45.2.10 General Amplitudes ... 674

45.3 **Semiclassical Scattering Formulae** ... 675
45.3.1 Scattering Amplitude: Exact Poisson Sum Representation ... 675
45.3.2 Semiclassical Procedure ... 675
45.3.3 Semiclassical Amplitudes: Integral Representation ... 676
45.3.4 Semiclassical Amplitudes and Cross Sections ... 677
45.3.5 Diffraction and Glory Amplitudes 679
45.3.6 Small-Angle (Diffraction) Scattering ... 680
45.3.7 Small-Angle (Glory) Scattering ... 681
45.3.8 Oscillations in Elastic Scattering ... 683

45.4 **Elastic Scattering in Reactive Systems** ... 683
45.4.1 Quantal Elastic, Absorption and Total Cross Sections ... 683

45.5 **Results for Model Potentials** ... 684
45.5.1 Born Amplitudes and Cross Sections for Model Potentials ... 689

References ... 689

45.1 Classical Scattering Formulae

Central Field. The total energy $E > 0$ and orbital angular momentum L of relative planar motion are conserved. For a particle of mass M with coordinates (R, ψ), a symmetric potential $V(R)$, and asymptotic speed v,

$$E = \frac{p^2(R)}{2M} + V(R) + \frac{L^2}{2MR^2} = \frac{1}{2}Mv^2 = \text{constant}\,, \tag{45.1}$$

$$\begin{aligned} L^2 &= (2ME)b^2 = M^2v^2b^2 \\ &= \left[MR^2(t)\dot{\psi}(t)\right]^2 = \text{constant}\,. \end{aligned} \tag{45.2}$$

Equation (45.2) implies constant areal velocity.

Radial momentum $p(R)$:

$$p(R;\, E, b) = (2ME)^{1/2}\left(1 - \frac{V(R)}{E} - \frac{b^2}{R^2}\right)^{1/2}\,. \tag{45.3}$$

Effective potential:

$$V_{\text{eff}}(R) = V(R) + \frac{L^2}{2MR^2} = V(R) + \frac{b^2}{R^2}E\,. \tag{45.4}$$

The turning points $R_i(E, b)$ are the roots of $E = V_{\text{eff}}(R)$. The smallest R_i is the distance of closest approach $R_c(E, b)$ at the pericenter. The maximum impact parameter b_X and angular momentum L_X for approach to within a distance R_X are

$$b_X^2 = R_X^2 \left[1 - V(R_X)/E\right] , \quad (45.5a)$$

$$L_X^2 = 2MR_X^2 \left[E - V(R_X)\right] . \quad (45.5b)$$

The trajectory is defined by $R = R(t; E, b)$, $\psi = \psi(t; E, \mathrm{b})$, where R is the distance from the scattering center O and ψ is measured with respect to the apse line OA joining O to the pericenter R_c. Taking $t(R_c) = 0$, (45.1) implies

$$t(R) = \left(\frac{M}{2E}\right)^{1/2} \int_{R_c}^{R} \left(1 - \frac{V(R)}{E} - \frac{b^2}{R^2}\right)^{1/2} \mathrm{d}R , \quad (45.6)$$

which implicitly provides $R = R(t; E, b)$.

$$\psi(t; E, b) = \frac{L}{M} \int_0^t R^{-2}(t)\, \mathrm{d}t$$

$$= \frac{\left(2MEb^2\right)^{1/2}}{M} \int_0^t R^{-2}(t)\, \mathrm{d}t. \quad (45.7)$$

Orbit integral ($0 \le \psi \le \pi$): ψ is symmetrical about and measured from the apse line joining O and R_c.

$$\psi(R; E, b) = b \int_{R_c}^{R} \frac{\mathrm{d}R/R^2}{\left[1 - V(R)/E - b^2/R^2\right]^{1/2}} \quad (45.8)$$

$$= -\frac{\partial}{\partial b} \int_{R_c}^{R} \left(1 - \frac{V(R)}{E} - \frac{b^2}{R^2}\right)^{1/2} \mathrm{d}R . \quad (45.9)$$

For large b and/or small $V(R)/E \ll 1$, (45.9) reduces to

$$\psi(R; E, b) = \frac{\pi}{2} - \sin^{-1}\frac{b}{R} + \frac{1}{2E}\frac{\partial}{\partial b} \times \int_b^R \frac{V(R)\, \mathrm{d}R}{\left[1 - b^2/R^2\right]^{1/2}} , \quad (45.10)$$

$$\psi(R \to \infty; E, b) = \frac{\pi}{2} + \frac{b}{2E} \times \int_b^\infty \left(\frac{\mathrm{d}V}{\mathrm{d}R}\right) \frac{\mathrm{d}R}{\left(R^2 - b^2\right)^{1/2}} . \quad (45.11a)$$

For a straight-line path, $R^2 = b^2 + Z^2$, where Z is the distance along the scattering axis, and

$$\psi(R \to \infty; E, b) = \frac{\pi}{2} + \frac{1}{4E}\frac{\partial}{\partial b} \int_{-\infty}^{\infty} V(b, Z)\, \mathrm{d}Z . \quad (45.11b)$$

45.1.1 Deflection Functions

The deflection function $\chi(E, b)$, $(-\infty \le \chi < \pi)$, is defined to be $\chi(E, b) = \pi - 2\psi(R \to \infty; E, b)$. Then

$$\chi(E, b) = \pi - 2b \int_{R_c}^{\infty} \frac{\mathrm{d}R/R^2}{\left[1 - V(R)/E - b^2/R^2\right]^{1/2}} , \quad (45.12)$$

$$= \pi - 2 \times \int_0^1 \left(\left\{\frac{1 - V(R_c/x)/E}{1 - V(R_c)/E}\right\} - x^2\right)^{-1/2} \mathrm{d}x . \quad (45.13)$$

An expression which avoids spurious divergences is

$$\chi(E, b) = \pi + 2\frac{\partial}{\partial b} \times \int_{R_c}^{\infty} \left[1 - V(R)/E - b^2/R^2\right]^{1/2} \mathrm{d}R . \quad (45.14)$$

Small-angle scattering, $V(R_c)/E \ll 1, b \sim R_c$:

$$\chi(E, b) = \left(\frac{R_c}{E}\right) \int_{R_c}^{\infty} \frac{\left[V(R_c) - V(R)\right] R\, \mathrm{d}R}{\left(R^2 - R_c^2\right)^{3/2}} \quad (45.15a)$$

$$= \frac{1}{E} \int_0^1 \frac{\left[V(R_c) - V(R_c/x)\right] \mathrm{d}x}{\left(1 - x^2\right)^{3/2}} , \quad (45.15b)$$

where $x = R_c/R$. From (45.10)–(45.11b), other forms are

$$\chi(E, b) = -\frac{1}{E}\frac{\partial}{\partial b} \int_b^\infty \frac{V(R)\, \mathrm{d}R}{\left(1 - b^2/R^2\right)^{1/2}} \quad (45.16a)$$

$$= -\frac{b}{E} \int_b^\infty \left(\frac{\mathrm{d}V}{\mathrm{d}R}\right) \frac{\mathrm{d}R}{\left(R^2 - b^2\right)^{1/2}} \quad (45.16b)$$

$$= -\frac{1}{2E}\frac{\partial}{\partial b} \int_{-\infty}^{\infty} V(b, Z)\, \mathrm{d}Z . \quad (45.16c)$$

For straight-line paths $R^2 = b^2 + v^2 t^2$, (45.12) yields the impulse-momentum result

$$\chi(E, b) = (Mv)^{-1} \int_{-\infty}^{\infty} F_\perp(t)\, dt = \Delta p_\perp / p_\infty , \tag{45.17}$$

where $\Delta p_\perp$ is the momentum transferred perpendicular to the incident direction and $F_\perp = -(\partial V/\partial R)(b/R)$ is the impulsive force causing scattering. Special cases are

Head-on collisions ($b = 0$):	$\chi(E, 0) = \pi$.
Overall repulsion:	$0 < \chi \le \pi$.
Overall attraction:	$-\infty \le \chi \le 0$.
Forward glory:	$\chi = -2n\pi$.
Backward glory:	$\chi = -(2n-1)\pi$, $n = 0, 1, 2, \ldots$.
Rainbow scattering:	$(d\chi/db) = 0$ at $\chi_r < 0$.
Deflection range:	$\chi_r \le \chi \le \pi$.
Orbiting collisions:	cf. (45.34).
Diffraction scattering:	$\chi \to 0$ as $b \to \infty$.

The scattered particle may wind or spiral many times around ($\chi \to -\infty$) the scattering center. The experimentally observed quantity is the scattering angle θ ($0 \le \theta \le \pi$) which is associated with various deflections

$$\chi_i = +\theta, -\theta, -2\pi \pm \theta, -4\pi \pm \theta, \ldots \quad (i = 1, 2, \ldots n)$$

resulting from n different impact parameters b_i.

Gauss–Mehler Quadrature Evaluation of the Deflection Function.

$$\chi(E, b) = \pi \left[1 - \left(\frac{b}{R_c} \right) \frac{1}{n} \sum_{j=1}^{n} a_k g(a_j) \right] , \tag{45.18}$$

where

$$a_k = \cos\left(\frac{2j-1}{4n} \pi \right) , \quad k = n + 1 - j , \text{ and}$$

$$g(x) = \left[1 - V(R_c/x)/E - b^2 x^2 / R_c^2 \right]^{-1/2} , \quad 0 \le x \le 1 .$$

45.1.2 Elastic Scattering Cross Section

Differential cross section:

$$\frac{d\sigma(\theta, E)}{d\Omega} \equiv I(\theta; E) \equiv \sigma(\theta; E) ,$$

$$\sigma(\theta, E) = \sum_{i=1}^{n} \left| \frac{b_i\, db_i}{d(\cos \chi_i)} \right| = \sum_{i=1}^{n} I_i(\theta) . \tag{45.19}$$

Integral cross section for scattering by angles $\theta \ge \theta_0$:

$$\sigma_0(E) = 2\pi \int_{\theta_0}^{\pi} I(\theta; E)\, d(\cos\theta) = 2\pi \int_0^{b_0(\theta_0)} b\, db , \tag{45.20}$$

where θ_0 results from one $b_0 = b(\theta_0)$. When θ_0 results from three impact parameters b_1, b_2, b_3, for example, then

$$\sigma_0(E) = 2\pi \int_0^{b_1} b\, db + 2\pi \int_{b_2}^{b_3} b\, db = \pi \left(b_1^2 + b_3^2 - b_2^2 \right) . \tag{45.21}$$

Diffusion (momentum-transfer) cross section:

$$\sigma_d(E) = 2\pi \int_0^{\pi} [1 - \cos\theta(E, b)]\, I(\theta)\, d(\cos\theta) , \tag{45.22a}$$

$$= 2\pi \int_0^{\infty} [1 - \cos\theta(E, b)]\, b\, db , \tag{45.22b}$$

$$= 4\pi \int_0^{\infty} \sin^2 \left[\frac{1}{2} \theta(E, b) \right] b\, db . \tag{45.22c}$$

Viscosity cross section:

$$\sigma_v(E) = 2\pi \int_0^{\pi} \left[1 - \cos^2\theta(E, b) \right] I(\theta)\, d(\cos\theta) , \tag{45.23a}$$

$$= 2\pi \int_0^{\infty} \left[1 - \cos^2\theta(E, b) \right] b\, db , \tag{45.23b}$$

$$= 2\pi \int_0^{\infty} \left[\sin^2\theta(E, b) \right] b\, db . \tag{45.23c}$$

Small-Angle Diffraction Scattering. The small-angle scattering regime is defined by the conditions $V(R_c)/E \ll 1$, $b \geq R_c$, where $\theta = |\chi|$. The main contribution to $d\sigma/d\Omega$ for small-angle scattering arises from the asymptotic branch of the deflection function χ at large impact parameters b, and is primarily determined by the long-range (attractive) part of the potential $V(R)$ (Sect. 45.3.6).

Large-Angle Scattering. The main contribution to $d\sigma/d\Omega$ for large-angle scattering arises from the positive branch of χ at small b and is mainly determined by the repulsive part of the potential.

45.1.3 Center-of-Mass to Laboratory Coordinate Conversion

Let ψ_1, ψ_2 be the angles for scattering and recoil, respectively, of the projectile by a target initially at rest in the lab frame. Then

$$\sigma_1(\psi_1)\,d\Omega_1 = \sigma_2(\psi_2)\,d\Omega_2 = \sigma_{cm}^{(1,2)}(\theta)\,d\Omega_{cm}\,, \quad 0 \leq \theta \leq \pi\,; \tag{45.24}$$

$$\sigma_{cm}^{(2)}(\theta,\phi) = \sigma_{cm}^{(1)}(\pi-\theta,\phi+\pi)\,. \tag{45.25}$$

(A) Two-body elastic scattering process without conversion of translational kinetic energy into internal energy: $(1)+(2) \to (1)+(2)$.

$$\sigma_1(\psi_1) = \sigma_{cm}(\theta)\frac{\left(1+2x\cos\theta+x^2\right)^{3/2}}{|1+x\cos\theta|}\,; \tag{45.26}$$

$$\sigma_2(\psi_2) = \sigma_{cm}(\theta)\left|4\sin\frac{1}{2}\theta\right|\,;$$

$$\psi_2 = \frac{1}{2}(\pi-\theta)\,, \quad 0 \leq \psi_2 \leq \frac{1}{2}\pi\,; \tag{45.27}$$

$$\tan\psi_1 = \frac{\sin\theta}{(x+\cos\theta)}\,, \quad x = M_1/M_2\,. \tag{45.28}$$

$M_1 > M_2$: As $0 \leq \theta \leq \theta_c = \cos^{-1}(-M_2/M_1)$,
$0 \leq \psi_1 \to \psi_1^{max} = \sin^{-1}(M_2/M_1) < \frac{1}{2}\pi$.
As $\theta_c \leq \theta \to \pi$, $\psi_1^{max} \leq \psi_1 \to 0$.
θ is a double-valued function of ψ;

$M_1 = M_2$: $\sigma_1(\psi_1) = (4\cos\psi_1)\sigma_{cm}(\theta = 2\psi_1)$,
$0 \leq \psi_1 \leq \frac{1}{2}\pi$, $\psi_1+\psi_2 = \frac{1}{2}\pi$;
no backscattering.

$M_1 \ll M_2$: $\sigma_1(\psi_1) = \sigma_{cm}(\theta = \psi_1)$;
lab and cm frames identical.

(B) Two-body elastic scattering process with conversion of translational kinetic energy into internal energy: $(1)+(2) \to (3)+(4)$. For conversion of internal energy ε_i so that kinetic energy of relative motion (in the cm frame) increases from E_i to $E_f = E_i + \varepsilon_i$. For $j = 3, 4$,

$$\sigma_j(\psi_j) = \sigma_{cm}(\theta)\frac{\left[1+2x_j\cos\theta+x_j^2\right]^{3/2}}{\left|1+x_j\cos\theta\right|}\,, \tag{45.29}$$

$$x_3 = \left(\frac{M_1M_3E_i}{M_2M_4E_f}\right)^{1/2}\,,$$

$$x_4 = -\left(\frac{M_1M_4E_i}{M_2M_3E_f}\right)^{1/2}\,,$$

$$\tan\psi_3 = \frac{\sin\theta}{(x_3+\cos\theta)}\,, \quad \tan\psi_4 = \frac{\sin\theta}{(|x_4|-\cos\theta)}\,.$$

45.1.4 Glory and Rainbow Scattering

Glory. The deflection function χ passes through $-2n\pi$ (forward glory) or $-(2n+1)\pi$ (backward glory) at finite impact parameters b_g. Then $\sin\theta \to 0$ as $\theta \to \theta_g$ so that classical cross section diverges as

$$\sigma(E,\theta) = \left(\frac{2b_g}{\sin\theta}\right)\left|\frac{db}{d\chi}\right|_g \quad \text{as } \theta \to \theta_g\,. \tag{45.30}$$

Rainbow. The deflection function χ passes through a negative minimum at $b = b_r$; $(d\chi/db)_r \to 0$ so that

$$\chi(b) = \chi(b_r) + \omega_r(b-b_r)^2\,, \tag{45.31}$$

$$\omega_r = \frac{1}{2}\left(\frac{d^2\chi}{db^2}\right)_{b_r} > 0\,. \tag{45.32}$$

The classical cross section diverges as

$$\sigma(E,\theta) = \frac{b_r}{2\sin\theta}\left[\omega_r(\theta_r-\theta)\right]^{-1/2}\,, \quad \theta < \theta_r\,, \tag{45.33}$$

and is augmented by the contribution from the positive branch of $\chi(b)$.

45.1.5 Orbiting and Spiraling Collisions

Attractive interactions $V(R) = -C/R^n$ $(n \geq 2)$ can support quasibound states with positive energy within the angular momentum barrier. Particles with $b < b_0$ spiral towards the scattering center. Those with $b = b_0$ are in unstable circular orbits of radius R_0. The radius R_0 is determined from the two conditions

$$\left(\frac{\mathrm{d}V_{\mathrm{eff}}}{\mathrm{d}R}\right)_{R_0}=0\,,\quad E=V_{\mathrm{eff}}(R_0)\,,\tag{45.34}$$

which, when combined, yields

$$E=V_{\mathrm{eff}}(R_0)=V(R_0)+\frac{1}{2}R_0\left(\frac{\mathrm{d}V}{\mathrm{d}R}\right)_{R_0}\,.\tag{45.35}$$

The angular momentum L_0 of the circular orbit is

$$L_0^2=(2ME)b_0^2=2MR_0^2\left[E-V(R_0)\right]\,.\tag{45.36}$$

Thus $b_0^2=R_0^2F$, where

$$F=1-\frac{V(R_0)}{E}=\frac{1}{2}\left(\frac{R_0}{E}\right)\left(\frac{\mathrm{d}V}{\mathrm{d}R}\right)_{R_0}\tag{45.37}$$

is the *focusing factor*. The orbiting and spiraling cross section is then

$$\sigma_{\mathrm{orb}}(E)=\pi b_0^2=\pi R_0^2F\,.\tag{45.38}$$

45.1.6 Quantities Derived from Classical Scattering

The semiclassical phase $\eta(E,b)\equiv\eta(E,\lambda)$, with $\lambda=\left(\ell+\frac{1}{2}\right)=kb$ is a function of b or λ. The quantities $p(R)\equiv p(R;\,E,L)$ and $p_0(R)\equiv p_0(R;\,E,L)$ are radial momenta in the presence and absence of the potential $V(R)$, respectively.

$$\eta^{\mathrm{SC}}(E,b)=\frac{1}{\hbar}\left[\int_{R_c}^{\infty}p(R)\,\mathrm{d}R-\int_{b}^{\infty}p_0(R)\,\mathrm{d}R\right]\tag{45.39}$$

$$=k\int_{R_c}^{\infty}\left[1-V/E-b^2/R^2\right]^{1/2}\mathrm{d}R$$

$$-k\int_{b}^{\infty}\left[1-b^2/R^2\right]^{1/2}\mathrm{d}R\,.\tag{45.40}$$

Asymptotic speed v: $E=\frac{1}{2}Mv^2=\hbar^2k^2/2M$, $k/2E=1/\hbar v$.

Jeffrey–Born phase function:
For small V/E and $b\sim R_c$,

$$\eta_{\mathrm{JB}}(E,b)=-\frac{k}{2E}\int_{b}^{\infty}\frac{V(R)\,\mathrm{d}R}{\left(1-b^2/R^2\right)^{1/2}}\,.\tag{45.41}$$

Eikonal phase function:
For small V/E and a linear trajectory $R^2=b^2+Z^2$,

$$\eta_{\mathrm{E}}(E,\boldsymbol{b})=-\frac{k}{4E}\int_{-\infty}^{\infty}V(\boldsymbol{b},Z)\,\mathrm{d}Z\,.\tag{45.42}$$

Semiclassical cross sections:

$$\sigma(E)=8\pi\int_0^{\infty}\left[\sin^2\eta(E,b)\right]b\,\mathrm{d}b\,,\tag{45.43}$$

$$=\left(8\pi/k^2\right)\int_0^{\infty}\sin^2\eta(E,\lambda)\lambda\,\mathrm{d}\lambda\,.\tag{45.44}$$

Landau–Lifshitz cross section:

$$\sigma_{\mathrm{LL}}(E)=8\pi\int_0^{\infty}\left[\sin^2\eta_{\mathrm{JB}}(E,b)\right]b\,\mathrm{d}b\,.\tag{45.45}$$

Massey–Mohr cross section:

$$\eta(E,b_0)=\frac{1}{2}\,,\quad\left\langle\sin^2\eta(E,b<b_0)\right\rangle=\frac{1}{2}\,,\tag{45.46}$$

$$\sigma_{\mathrm{MM}}(E)=2\pi b_0^2+8\pi\int_{b_0}^{\infty}\eta_{\mathrm{JB}}^2(E,b)b\,\mathrm{d}b\,.\tag{45.47}$$

Schiff cross section:

$$\sigma_{c(E)}=4\int_{-\infty}^{\infty}\mathrm{d}X\int_{-\infty}^{\infty}\mathrm{d}Y\left[\sin^2\eta_{\mathrm{E}}(X,Y)\right]\,.\tag{45.48}$$

This reduces to (45.45) for spherical $V(\boldsymbol{R})$; $b\equiv(x,y)$.

Random-phase approximation (RPA):
For angle α,

$$4\pi\int_0^{\infty}P(b)\sin^2\alpha(b)b\,\mathrm{d}b=2\pi\int_0^{b_c}P(b)b\,\mathrm{d}b\,,\tag{45.49}$$

where $\alpha(b_c)=1/\pi$.

Collision delay time function:

$$\tau(E,\lambda)=2\hbar\frac{\partial\eta(E,\lambda)}{\partial E}=\frac{2}{v}\frac{\partial\eta_\ell(k)}{\partial k}\,.\tag{45.50}$$

Deflection-angle phase function relation:

$$\chi(E,\lambda)=\frac{2}{k}\frac{\partial\eta(E,b)}{\partial b}=2\frac{\partial\eta(E,\lambda)}{\partial\lambda}\,.\tag{45.51}$$

45.1.7 Collision Action

The classical collision action along a classical path with deflection $\chi = \chi(E, L)$, measured relative to the action along the path of the undeflected particle, is

$$S^{C}(E, L; \chi) = S_R(E, L) - L\chi(E, L) \tag{45.52a}$$
$$= 2\eta^{SC}(E, L)\hbar - L\chi . \tag{45.52b}$$

Radial component of collision action S^C:

$$S_R(E, L) = 2\int_{R_c(E,L)}^{\infty} p(R)\,dR - 2\int_{b(E,L)}^{\infty} p_0(R)\,dR \tag{45.53a}$$
$$= 2\eta^{SC}(E, L)\hbar \tag{45.53b}$$

Collision delay time function:

$$\tau(E, L) = 2M\left(\int_{R_c}^{\infty}\frac{dR}{p(R)} - \int_{b}^{\infty}\frac{dR}{p_0(R)}\right) \tag{45.54a}$$
$$= \left(\frac{\partial S_R}{\partial E}\right)_L . \tag{45.54b}$$

Deflection angle function:

$$\chi(E, L) = \pi - 2L\int_{R_c}^{\infty}\frac{dR/R^2}{p(R)} \tag{45.55a}$$
$$= \left(\frac{\partial S_R}{\partial L}\right)_E . \tag{45.55b}$$

Radial collision action change:

$$dS_R = \tau(E, L)\,dE + \chi(E, L)\,dL = 2\hbar\,d\eta^{SC} \tag{45.56}$$

45.2 Quantal Scattering Formulae

The basic quantity in *quantal* elastic scattering is the complex scattering amplitude $f(E, \theta)$, expressed in terms of the phase shifts $\eta_\ell(E)$ associated with scattering of the ℓ th partial wave. Derived quantities are the diagonal elements of the scattering matrix S, transition matrix T and reactance matrix K.

Reduced Energy: $k^2 = \left(2M/\hbar^2\right) E$.
Reduced Potential: $U(R) = \left(2M/\hbar^2\right) V(R)$.

45.2.1 Basic Formulae

Wave function: As $R \to \infty$,

$$\Psi(\boldsymbol{R}) \to \exp(ikZ) + \frac{1}{R} f(\theta)\exp(ikR) \tag{45.57}$$

for symmetric interactions $V = V(R)$.
Elastic scattering DCS:

$$\frac{d\sigma}{d\Omega} = I(\theta) = |f(\theta)|^2 . \tag{45.58}$$

Scattering, transition and reactance matrix elements in terms of η_ℓ:

$$S_\ell(k) = \exp(2i\eta_\ell) , \tag{45.59a}$$
$$T_\ell(k) = \sin\eta_\ell \exp(i\eta_\ell) , \tag{45.59b}$$
$$K_\ell(k) = \tan\eta_\ell . \tag{45.59c}$$

Scattering amplitudes $f(\theta)$:

$$f(\theta) = \frac{1}{2ik}\sum_{\ell=0}^{\infty}(2\ell+1)\left[\exp(2i\eta_\ell) - 1\right]P_\ell(\cos\theta)$$
$$= \sum_{\ell=0}^{\infty} f_\ell(\theta) , \tag{45.60a}$$

$$f(\theta) = \frac{1}{2ik}\sum_{\ell=0}^{\infty}(2\ell+1)\left[S_\ell(k) - 1\right]P_\ell(\cos\theta) , \tag{45.60b}$$

$$f(\theta) = \frac{1}{k}\sum_{\ell=0}^{\infty}(2\ell+1)T_\ell(k)P_\ell(\cos\theta) , \tag{45.60c}$$

$$f(\theta) = \frac{1}{2ik}\sum_{\ell=0}^{\infty} S_\ell(k)P_\ell(\cos\theta) , \quad \theta \neq 0 , \tag{45.60d}$$
$$= \frac{1}{k}\sum_{\ell=0}^{\infty}(2\ell+1)T_\ell(k), \quad \theta = 0 . \tag{45.60e}$$

Integral cross sections $\sigma(E)$:

$$\sigma(E) = 2\pi\int_0^{\pi} I(\theta)\,d(\cos\theta) , \tag{45.61a}$$
$$= \frac{4\pi}{k^2}\sum_{\ell=0}^{\infty}(2\ell+1)\sin^2\eta_\ell . \tag{45.61b}$$

Optical theorem:

$$\sigma(E) = (4\pi/k)\,\mathrm{Im}[f(0)]\,. \tag{45.62}$$

Partial cross sections $\sigma_\ell(E)$:

$$\sigma(E) = \sum_{\ell=0}^{\infty} \sigma_\ell(E) \tag{45.63}$$

$$\sigma_\ell(E) = \frac{4\pi}{k^2}(2\ell+1)\sin^2\eta_\ell \tag{45.64a}$$

$$= \frac{4\pi}{k^2}(2\ell+1)|T_\ell|^2$$

$$= \frac{2\pi}{k}(2\ell+1)\,[1-\mathrm{Re}(S_\ell)]\,. \tag{45.64b}$$

Upper limit:

$$\sigma_\ell(E) \le \left(4\pi/k^2\right)(2\ell+1)\,. \tag{45.65}$$

Unitarity, flux conservation, η_ℓ is real:

$$|S_\ell|^2 = 1\,,\quad |T_\ell|^2 = \mathrm{Im}(T_\ell)\,. \tag{45.66}$$

Differential cross sections (DCS):

$$\frac{\mathrm{d}\sigma(E,\theta)}{\mathrm{d}\Omega} = |f(\theta)|^2 = I(\theta) = A(\theta)^2 + B(\theta)^2\,, \tag{45.67}$$

$$A(\theta) = \mathrm{Re}[f(\theta)]$$

$$= \frac{1}{2k}\sum_{\ell=0}^{\infty}(2\ell+1)\sin 2\eta_\ell\, P_\ell(\cos\theta)\,,$$

$$B(\theta) = \mathrm{Im}[f(\theta)]$$

$$= \frac{1}{2k}\sum_{\ell=0}^{\infty}(2\ell+1)\,[1-\cos 2\eta_\ell]\, P_\ell(\cos\theta)\,.$$

$$\int A(\theta)^2\,\mathrm{d}\Omega = \frac{4\pi}{k^2}\sum_{\ell=0}^{\infty}(2\ell+1)\sin^2\eta_\ell\cos^2\eta_\ell\,, \tag{45.68a}$$

$$\int B(\theta)^2\,\mathrm{d}\Omega = \frac{4\pi}{k^2}\sum_{\ell=0}^{\infty}(2\ell+1)\sin^4\eta_\ell\,, \tag{45.68b}$$

$$\frac{\mathrm{d}\sigma(E,\theta)}{\mathrm{d}\Omega} = \frac{1}{k^2}\sum_{L=0}^{\infty} a_L(E) P_L(\cos\theta)\,, \tag{45.69a}$$

$$a_L = \sum_{\ell=0}^{\infty}\sum_{\ell'=|\ell-L|}^{\ell+L}(2\ell+1)(2\ell'+1)$$

$$(\ell\ell'00\mid\ell\ell'L0)^2$$

$$\times\sin\eta_\ell\sin\eta_{\ell'}\cos(\eta_\ell-\eta_{\ell'})\,, \tag{45.69b}$$

where $(\ell\ell'mm'\mid\ell\ell'LM)$ are the Clebsch–Gordan Coefficients.

Three-term expansion:

$$\frac{\mathrm{d}\sigma(E,\theta)}{\mathrm{d}\Omega} = \frac{1}{k^2}\Bigg[\left(a_0-\frac{1}{2}a_2\right)+a_1\cos\theta + \frac{3}{2}a_2\cos^2\theta\Bigg] \tag{45.70}$$

$$a_0(E) = \sum_{\ell=0}^{\infty}(2\ell+1)\sin^2\eta_\ell\,, \tag{45.71a}$$

$$a_1(E) = 6\sum_{\ell=0}^{\infty}(\ell+1)\sin\eta_\ell\sin\eta_{\ell+1} \times\cos(\eta_{\ell+1}-\eta_\ell)\,, \tag{45.71b}$$

$$a_2(E) = 5\sum_{\ell=0}^{\infty}\Big[b_\ell\sin^2\eta_\ell + c_\ell\sin\eta_\ell\sin\eta_{\ell+2} \times\cos(\eta_{\ell+2}-\eta_\ell)\Big]\,. \tag{45.71c}$$

with coefficients

$$b_\ell = \frac{\ell(\ell+1)(2\ell+1)}{(2\ell-1)(2\ell+3)}\,, \tag{45.72a}$$

$$c_\ell = \frac{3(\ell+1)(\ell+2)}{(2\ell+3)}\,. \tag{45.72b}$$

S, P wave ($\ell = 0, 1$) net contribution:

$$\frac{\mathrm{d}\sigma}{\mathrm{d}\Omega} = \frac{1}{k^2}\Big\{\sin^2\eta_0 + [6\sin\eta_0\sin\eta_1\cos(\eta_1-\eta_0)] \times\cos\theta + 9\sin^2\eta_1\cos^2\theta\Big\}\,, \tag{45.73}$$

$$\sigma(E) = \frac{4\pi}{k^2}\left(\sin^2\eta_0 + 3\sin^2\eta_1\right)\,. \tag{45.74}$$

For pure S-wave scattering, the DCS is isotropic. For pure P-wave scattering, the DCS is symmetric about $\theta = \pi/2$, where it vanishes; the DCS rises to equal maxima at $\theta = 0, \pi$. For combined S- and P-wave scattering, the DCS is asymmetric with forward-backward asymmetry.

Transport cross sections ($n \ge 1$):

$$\sigma^{(n)}(E) = 2\pi\left(1-\frac{1+(-1)^n}{2(n+1)}\right)^{-1} \times\int_0^\pi\left(1-\cos^n\theta\right) I(\theta)\,\mathrm{d}(\cos\theta)\,. \tag{45.75}$$

The diffusion and viscosity cross sections (45.22a) and (45.23a) are given by the transport cross sections $\sigma^{(1)}$ and $\frac{2}{3}\sigma^{(2)}$, respectively.

$$\sigma^{(1)}(E) = \frac{4\pi}{k^2}\sum_{\ell=0}^{\infty}(\ell+1)\sin^2(\eta_\ell - \eta_{\ell+1}) , \tag{45.76a}$$

$$\sigma^{(2)}(E) = \frac{4\pi}{k^2}\left(\frac{3}{2}\right)\sum_{\ell=0}^{\infty}\frac{(\ell+1)(\ell+2)}{(2\ell+3)} \times \sin^2(\eta_\ell - \eta_{\ell+2}) , \tag{45.76b}$$

$$\sigma^{(3)}(E) = \frac{4\pi}{k^2}\sum_{\ell=0}^{\infty}\frac{(\ell+1)}{(2\ell+5)} \times\left[\frac{(\ell+2)(\ell+3)}{(2\ell+3)}\sin^2(\eta_\ell - \eta_{\ell+3}) + \frac{3\left(\ell^2+2\ell-1\right)}{(2\ell-1)}\sin^2(\eta_\ell - \eta_{\ell+1})\right] , \tag{45.76c}$$

$$\sigma^{(4)}(E) = \frac{4\pi}{k^2}\left(\frac{5}{4}\right)\sum_{\ell=0}^{\infty}\frac{(\ell+1)(\ell+2)}{(2\ell+3)(2\ell+7)} \times\left[\frac{(\ell+3)(\ell+4)}{(2\ell+5)}\sin^2(\eta_\ell - \eta_{\ell+4}) + \frac{2\left(2\ell^2+6\ell-3\right)}{(2\ell-1)}\sin^2(\eta_\ell - \eta_{\ell+2})\right] . \tag{45.76d}$$

Collision integrals: Averages of $\sigma^{(n)}(E)$ over a Maxwellian distribution at temperature T are

$$\Omega^{(n,s)}(T) = \left[(s+1)!(kT)^{s+2}\right]^{-1}\int_0^{\infty}\sigma^{(n)}(E) \times \exp(-E/kT)E^{s+1}\,\mathrm{d}E . \tag{45.77}$$

Normalization factors are chosen so that the above expressions for $\sigma^{(n)}$ and $\Omega^{(n,s)}$ reduce to πd^2 for classical rigid spheres of diameter d.

Mobility: The mobility K of ions of charge e in a gas of density N is given by the Chapman–Enskog formula

$$K = \frac{3e}{8N}\left(\frac{\pi}{2MkT}\right)^{1/2}\left[\Omega^{(1,1)}(T)\right]^{-1} . \tag{45.78}$$

Phase shifts η_ℓ can be determined from the numerical solution of the radial Schrödinger equation (45.93), from an integral equation (45.179b), from solving a nonlinear first-order differential equation (45.179a), from Logarithmic Derviatives (Sect. 45.2.5) or from variational techniques (Sect. 45.2.4).

45.2.2 Identical Particles: Symmetry Oscillations

Colliding Particles, each with spin **s**, *in a Total Spin* $\mathbf{S_t}$ *Resolved State in the Range* $(0 \to 2\mathbf{s})$. Particle interchange: $\Psi(\boldsymbol{R}) = (-1)^{S_t}\Psi(-\boldsymbol{R})$

$$I_{A,S}(\theta) = \frac{1}{2}|f(\theta) \mp f(\pi-\theta)|^2 , \tag{45.79}$$

$$\Psi_{A,S}(\boldsymbol{R}) \to \left[\exp(ikZ) \mp \exp(-ikZ)\right] + \frac{1}{R}\left[f(\theta) \mp f(\pi-\theta)\right]\exp(ikR) , \tag{45.80}$$

$$I_{A,S}(\theta) = \frac{1}{4k^2}\left|\sum_{\ell=0}^{\infty}\omega_\ell(2\ell+1)\left[\exp 2i\eta_\ell - 1\right] \times P_\ell(\cos\theta)\right|^2 , \tag{45.81}$$

$$\sigma_{A,S}(E) = \frac{4\pi}{k^2}\sum_{\ell=0}^{\infty}\omega_\ell(2\ell+1)\sin^2\eta_\ell , \tag{45.82}$$

where A and S denote antisymmetric and symmetric wave functions (with respect to particle interchange) for collisions of identical particles with odd and even total spin S_t

A: S_t odd $\omega_\ell = 0$ (ℓ even); $\omega_\ell = 2$ (ℓ odd);
S: S_t even$\omega_\ell = 2$ (ℓ even); $\omega_\ell = 0$ (ℓ odd).

Spin-States $\mathbf{S_t}$ *Unresolved*. S/A combination:

$$I(\theta) = g_A I_A(\theta) + g_S I_S(\theta) , \tag{45.83}$$

$$\sigma(E) = g_A\sigma_A(E) + g_S\sigma_S(E) , \tag{45.84}$$

where g_A and g_S are the fractions of states with odd and even total spins $S_t = 0, 1, 2, \ldots , 2s$. For Fermions (F) with half integer spin s, and Bosons (B) with integral spin s

F: $g_A = (s+1)/(2s+1)$,$g_S = s/(2s+1)$,
B: $g_A = s/(2s+1)$, $g_S = (s+1)/(2s+1)$,

so that (45.83) and (45.84) have the alternative forms

$$I(\mathrm{F}) = |f(\theta)|^2 + |f(\pi-\theta)|^2 - \mathcal{I} , \tag{45.85a}$$

$$\sigma(\mathrm{F}) = \frac{1}{2}\left[\sigma_S + \sigma_A\right] - \frac{1}{2}\left[\sigma_S - \sigma_A\right]/(2s+1) , \tag{45.85b}$$

$$I(\mathrm{B}) = |f(\theta)|^2 + |f(\pi-\theta)|^2 + \mathcal{I} , \tag{45.85c}$$

$$\sigma(\mathrm{B}) = \frac{1}{2}\left[\sigma_S + \sigma_A\right] + \frac{1}{2}\left[\sigma_S - \sigma_A\right]/(2s+1) , \tag{45.85d}$$

where the interference term is

$$\mathcal{I} = \left(\frac{2}{2s+1}\right) \operatorname{Re}\left[f(\theta) f^*(\pi-\theta)\right] . \quad (45.86)$$

Example: For fermions with spin 1/2,

$$\sigma(E) = \frac{2\pi}{k^2}\left[\sum_{\ell=\text{even}}^{\infty} (2\ell+1)\sin^2\eta_\ell + 3\sum_{\ell=\text{odd}}^{\infty} (2\ell+1)\sin^2\eta_\ell\right] . \quad (45.87)$$

Symmetry oscillations originate from the interference between unscattered incident particles in the forward ($\theta=0$) direction and backward scattered particles ($\theta=\pi$, $\ell=0$). Symmetry oscillations are sensitive to the repulsive wall of the interaction.

Resonant Charge Transfer and Transport Cross Sections for $\mathbf{A}^+$*–*$\mathbf{A}$ *Collisions.* The phase shifts for elastic scattering by the gerade (g) and ungerade (u) potentials of A_2^+ are, respectively, η_ℓ^g and η_ℓ^u. The charge transfer (X) and transport cross sections are

$$\sigma_X(E) = \frac{\pi}{k^2}\sum_{\ell=0}^{\infty}(2\ell+1)\sin^2\left(\eta_\ell^g-\eta_\ell^u\right) , \quad (45.88a)$$

$$\sigma_{A,S}^{(1)}(E) = \frac{4\pi}{k^2}\sum_{\ell=0}^{\infty}(\ell+1)\sin^2(\beta_\ell-\beta_{\ell+1}) , \quad (45.88b)$$

$$\sigma_{A,S}^{(2)}(E) = \frac{4\pi}{k^2}\left(\frac{3}{2}\right)\sum_{\ell=0}^{\infty}\frac{(\ell+1)(\ell+2)}{(2\ell+3)} \times \sin^2(\beta_\ell-\beta_{\ell+2}) ; \quad (45.88c)$$

A: $\beta_\ell = \eta_\ell^g$ (ℓ even), or η_ℓ^u (ℓ odd) ,
S: $\beta_\ell = \eta_\ell^u$ (ℓ even), or η_ℓ^g (ℓ odd) .

$\sigma_{A,S}^{(1)}$ contains (g/u) interference; $\sigma_{A,S}^{(2)}$ does not. When nuclear spin degeneracy is acknowledged, the cross sections $\sigma_{A,S}$ are summed according to (45.85b) or (45.85d).

Since there is no coupling between molecular states of different electronic angular momentum, the scattering by the ${}^2\Sigma_{g,u}$ pair and the ${}^2\Pi_{g,u}$ pair of Ne_2^+ potentials (for example) is independent and

$$\sigma_X(E) = \frac{1}{3}\sigma_\Sigma(E) + \frac{2}{3}\sigma_\Pi(E) . \quad (45.89)$$

Singlet–Triplet Spin Flip Cross Section.

$$\sigma_{ST}(E) = \frac{\pi}{k^2}\sum_{\ell=0}^{\infty}(2\ell+1)\sin^2\left(\eta_\ell^s-\eta_\ell^t\right) , \quad (45.90)$$

where $\eta_\ell^{S,T}$ are the phase shifts for individual scattering by the singlet and triplet potentials, respectively.

45.2.3 Partial Wave Expansion

$$\Psi(\mathbf{R}) = \frac{1}{kR}\sum_{\ell=0}^{\infty} A_\ell v_\ell(kR) P_\ell(\cos\theta) , \quad (45.91)$$

$$A_\ell = \mathrm{i}^\ell(2\ell+1)\exp(\mathrm{i}\eta_\ell) . \quad (45.92)$$

Radial Schrödinger equation (RSE):

$$\frac{\mathrm{d}^2 v_\ell}{\mathrm{d}R^2} + \left[k^2 - U(R) - \frac{\ell(\ell+1)}{R^2}\right] v_\ell(R) = 0 \quad (45.93)$$

where v_ℓ is normalized so that

$$v_\ell(R) \overset{R>R_0}{=} \cos\eta_\ell F_\ell(kR) + \sin\eta_\ell G_\ell(kR) \quad (45.94)$$
$$\to \sin\left(kR - \frac{1}{2}\ell\pi + \eta_\ell\right) \text{ as } R\to\infty .$$

The regular (nonsingular) solution (zero at $R=0$) of the field-free RSE (45.93) with $U(R)=0$ is

$$F_\ell(kR) = (kR) j_\ell(kR) = \left(\frac{1}{2}\pi kR\right)^{1/2} J_{\ell+1/2}(kR) \quad (45.95)$$

$$\to \begin{cases} (kR)^{\ell+1}/(2\ell+1)!!, & R\to 0 \\ \sin\left(kR - \frac{1}{2}\ell\pi\right), & R\to\infty, \end{cases} \quad (45.96)$$

where j_ℓ is the spherical Bessel function. Equation (45.91) with $v_\ell = F_\ell$ is the partial-wave expansion for the incident plane-wave $\exp(\mathrm{i}kZ)$.

The irregular solution (divergent at $R=0$) of the field-free RSE is

$$G_\ell(kR) = -(kR) n_\ell(kR) = \left(\frac{1}{2}\pi kR\right)^{1/2} J_{-(\ell+1/2)}(kR) \quad (45.97)$$

$$\to \begin{cases} (2\ell-1)!!/(kR)^\ell, & R\to 0 \\ \cos\left(kR - \frac{1}{2}\ell\pi\right), & R\to\infty, \end{cases} \quad (45.98)$$

where n_ℓ is the spherical Neumann function.

The full asymptotic scattering solution is the combination (45.94) of the regular and irregular solutions. The mixture depends upon:

Forms of Normalization for v_ℓ. In (45.91), possible choices of normalization are:

(a) $$A_\ell = \mathrm{i}^\ell (2\ell+1) \exp \mathrm{i}\eta_\ell \,, \tag{45.99a}$$

$$v_\ell(R) \sim \sin\left(kR - \frac{1}{2}\ell\pi + \eta_\ell\right) ; \tag{45.99b}$$

(b) $$A_\ell = \mathrm{i}^\ell (2\ell+1) \cos \eta_\ell \,, \tag{45.100a}$$

$$v_\ell(R) \sim \sin\left(kR - \frac{1}{2}\ell\pi\right) + K_\ell \cos\left(kR - \frac{1}{2}\ell\pi\right) ; \tag{45.100b}$$

(c) $$A_\ell = \mathrm{i}^\ell (2\ell+1) \,, \tag{45.101a}$$

$$v_\ell(R) \sim \sin\left(kR - \frac{1}{2}\ell\pi\right) + T_\ell e^{\mathrm{i}(kR-\ell\pi/2)} ; \tag{45.101b}$$

(d) $$A_\ell = \frac{1}{2}\mathrm{i}^{\ell+1}(2\ell+1) \,, \tag{45.102a}$$

$$v_\ell(R) \sim \mathrm{e}^{-\mathrm{i}(kR-\ell\pi/2)} - S_\ell \mathrm{e}^{\mathrm{i}(kR-\ell\pi/2)} ; \tag{45.102b}$$

$$S_\ell = 1 + 2\mathrm{i}T_\ell ; \qquad K_\ell = T_\ell/(1+\mathrm{i}T_\ell) ; \tag{45.103}$$

$$T_\ell = \sin\eta_\ell \,\mathrm{e}^{\mathrm{i}\eta_\ell} ; \quad 1+\mathrm{i}T_\ell = \cos\eta_\ell \,\mathrm{e}^{\mathrm{i}\eta_\ell} \,. \tag{45.104}$$

Significance of η_ℓ, K_ℓ, T_ℓ, and S_ℓ. The effect of scattering is therefore to: (1) introduce a phase shift η_ℓ in (45.99b) to the regular standing wave, (2) leave the regular standing wave alone and introduce either an irregular standing wave of real amplitude K_ℓ in (45.100b) or, a spherical outgoing wave of amplitude T_ℓ in (45.101b), and (3) to convert in (45.102b) an incoming spherical wave of unit amplitude to an outgoing spherical wave of amplitude S_ℓ.

Levinson's Theorem. A local potential $U(R)$ can support n_ℓ bound states of angular momentum ℓ and energy E_n such that

$$\lim_{k\to 0} \eta_0(k) = \begin{cases} n_0\pi \,, & E_n < 0 \\ \left(n_0 + \frac{1}{2}\right)\pi \,, & E_{n+1} = 0 \,, \end{cases} \tag{45.105}$$

$$\lim_{k\to 0} \eta_\ell(k) = n_\ell \pi \,, \quad \ell > 0 \,. \tag{45.106}$$

45.2.4 Scattering Length and Effective Range

Blatt–Jackson Effective Range Formula. For short-range potentials,

$$k \cot \eta_0 = -\frac{1}{a_\mathrm{s}} + \frac{1}{2} R_\mathrm{e} k^2 + \mathcal{O}\left(k^4\right) . \tag{45.107}$$

Effective range:

$$R_\mathrm{e} = 2 \int_0^\infty \left[u_0^2(R) - v_0^2(R)\right] \mathrm{d}R \,, \tag{45.108}$$

where $u_0 = \sin(kR+\eta_0)/\sin\eta_0$ is the $k=0$ limit of the potential-free $\ell = 0$ radial wave function and normalized so that $u_0(R)$ goes to unity as $k \to 0$. The potential distorted $\ell = 0$ radial function v_0 is normalized at large R to $u_0(R)$. The effective range is a measure of the distance over which v_0 differs from u_0. The outside factor of 2 in (45.108) is chosen such that $R_\mathrm{e} = a$ for a square well of range a.

Scattering length:

$$a_\mathrm{s} = -\lim_{k\to 0} f(\theta) \,. \tag{45.109}$$

Relation with $\mathbf{k} \to 0$ Elastic Cross Section.

$$\sigma(k\to 0) = \frac{4\pi}{k^2}\sin^2\eta_0 = 4\pi a_\mathrm{s}^2 \left[\left(1 - \frac{1}{2}k^2 a_\mathrm{s} R_\mathrm{e}\right)^2 + k^2 a_\mathrm{s}^2\right]^{-1} \tag{45.110}$$

$$\sim 4\pi a_\mathrm{s}^2 \left[1 + a_\mathrm{s} k^2 (R_\mathrm{e} - a_\mathrm{s})\right] \tag{45.111}$$

Relation with Bound Levels. If a $\ell = 0$ bound level of energy $E_n = -\hbar^2 k_n^2/2M$ lies sufficiently near the dissociation limit the effective range and scattering lengths, R_e and a_s, respectively, are related by,

$$-\frac{1}{a_\mathrm{s}} = -k_n + \frac{1}{2} R_\mathrm{e} k_n^2 + \cdots \tag{45.112}$$

Wigner Causality Condition. If η_ℓ provides the dominant contribution to $f(\theta)$ then

$$\frac{\partial \eta_\ell(k)}{\partial k} \geq -a_\mathrm{s} \tag{45.113}$$

where a_s is the scattering length ($\ell = 0$) and is a measure of the range of the interaction.

Effective Range Formulae. The Blatt–Jackson formula must be modified [45.1–5] for long-range interactions as follows.

(1) Modified Coulomb potential: $V(R) \sim Z_1 Z_2 e^2/R$

$$2(K/a_0) = -(1/a_s) + \frac{1}{2} R_e k^2 \tag{45.114}$$

$$K = \frac{\pi \cot \eta_0}{e^{2\pi\alpha} - 1} - \ln \beta - 0.5772 + \beta^2 \sum_{n=1}^{\infty} \left[n \left(n^2 + \beta^2 \right) \right]^{-1} \tag{45.115}$$

where $\beta = Z_1 Z_2 e^2/\hbar v = Z_1 Z_2/(k a_0)$.

(2) Polarization potential: $V(R) = -\alpha_d e^2/2R^4$,

$$\tan \eta_0 = -a_s k - \frac{\pi}{3} C_4 k^2 - \frac{4}{3} C_4 a_s k^3 \ln(k a_0) + D k^3 + F k^4 , \tag{45.116}$$

$$\tan \eta_1 = \frac{\pi}{15} C_4 k^2 - a_s^{(1)} k^3 , \tag{45.117}$$

$$\tan \eta_\ell = \frac{\pi C_4 k^2}{(2\ell+3)(2\ell+1)(2\ell-1)} + \mathcal{O}\left(k^{2\ell+1}\right) , \tag{45.118}$$

for $\ell > 1$, where

$$C_4 = \frac{2M}{\hbar^2} \left(\frac{\alpha_d e^2}{2} \right) = \left(\frac{\alpha_d}{a_0} \right) \left(\frac{M}{m_e} \right) . \tag{45.119}$$

Example: e^-–Ar low energy collisions: The values

$a_s = -1.459 a_0$; $D = 68.93 a_0^3$
$a_s^{(1)} = 8.69 a_0^3$; $F = -97 a_0^4$

provide an accurate fit to recent measurements [45.6] of (45.76a) for the diffusion cross section σ_d.

(3) Van der Waals potential: $V(R) = -C/R^6$

$$k \cot \eta_0 = -\frac{1}{a_s} + \frac{1}{2} R_e k^2 - \frac{\pi}{15 a_s^2} \left(\frac{2MC}{\hbar^2} \right) k^3 - \frac{4}{15 a_s} \left(\frac{2MC}{\hbar^2} \right) k^4 \ln(k a_0) + \mathcal{O}\left(k^4\right) . \tag{45.120}$$

e–Atom Collisions with Polarization Attraction. As $k \to 0$, the differential cross section is

$$\frac{d\sigma}{d\Omega} = a_s^2 \left[1 + \frac{C_4}{a_s} k \sin \frac{\theta}{2} + \frac{8}{3} C_4 k^2 \ln(k a_0) + \cdots \right] \tag{45.121}$$

and the elastic and diffusion cross sections are

$$\sigma(k \to 0) = 4\pi a_s^2 \left[1 + \frac{2\pi C_4 k}{3 a_s} + \frac{8}{3} C_4 k^2 \ln(k a_0) + \cdots \right] \tag{45.122}$$

$$\sigma_d(k \to 0) = 4\pi a_s^2 \left[1 + \frac{4\pi C_4 k}{5 a_s} + \frac{8}{3} C_4 k^2 \ln(k a_0) + \cdots \right] \tag{45.123}$$

For e^-–noble gas collisions, the scattering lengths are

	He	Ne	Ar	Kr	Xe
a_s (a_0)	1.19	0.24	-1.459	-3.7	-6.5

For atoms with $a_s < 0$, a Ramsauer–Townsend minimum appears in both σ and σ_d at low energies, provided that scattering from higher partial waves can be neglected, because from (45.116), $\eta_0 \simeq 0$ at $k = -3a_s/\pi C_4$.

Semiclassical Scattering Lengths. For heavy particle collisions, $\eta^{SC}(E \to 0, b)$ tends to

$$\eta_0^{SC} = \left(\frac{2M}{\hbar^2} \right)^{1/2} \int_{R_0}^{\infty} |V(R)|^{1/2} \, dR . \tag{45.124}$$

(a) Hard-core + well:

$$V(R) = \begin{cases} \infty, & R < R_0 \\ -V_0, & R_0 \le R < R_1 \\ 0, & R_1 < R , \end{cases}$$

$$a_s = \left[1 - \tan \eta_0^{SC}/(k R_1) \right] R_1 , \tag{45.125}$$

$$\eta_0^{SC} = k(R_1 + R_0) , \quad k^2 = 2MV_0/\hbar^2 . \tag{45.126}$$

The phase-averaged scattering length is $\langle a_s \rangle = R_1$.

(b) Hard-core + power-law ($n > 2$):

$$V(R) = \begin{cases} \infty , & R < R_0 \\ \pm C/R^n , & R > R_0 \end{cases} . \tag{45.127}$$

Repulsion (+): with $\gamma^2 = 2MC/\hbar^2$,

$$a_s^{(+)} = \left(\frac{\gamma}{n-2} \right)^{2/(n-2)} \Gamma\left(\frac{n-3}{n-2} \right) / \Gamma\left(\frac{n-1}{n-2} \right) . \tag{45.128}$$

Attraction (−): with $\theta_n = \pi/(n-2)$,

$$a_s^{(-)} = a_s^{(+)}\left[1 - \tan\theta_n \tan\left(\eta_0^{SC} - \frac{1}{2}\theta_n\right)\right]\cos\theta_n\,, \tag{45.129}$$

$$\eta_0^{SC} = \gamma \int_{R_0}^{\infty} R^{-n/2}\,\mathrm{d}R = \frac{2\gamma R_0^{1-n/2}}{n-2}\,, \tag{45.130}$$

$$\langle a_s^{(-)}\rangle = a_s^{(+)}\cos\theta_n\,. \tag{45.131}$$

Number of bound states:

$$N_b = \mathrm{int}\left\{\frac{1}{\pi}\left[\eta_0^{SC} - \frac{1}{2}(n-1)\theta_n\right]\right\} + 1\,, \tag{45.132}$$

where $\mathrm{int}(x)$ denotes the largest integer of the real argument x. For integer x, $a_s^{(-)}$ is infinite and a new bound state appears at zero energy.

45.2.5 Logarithmic Derivatives

Phase shift calculations can be based on the logarithmic derivative at $R=a$ separating internal and external regions. Two equivalent forms using sets (R_ℓ, j_ℓ, n_ℓ) or (v_ℓ, F_ℓ, G_ℓ) when $R_\ell = v_\ell/kR$ are

$$K_\ell(k) = \frac{k j_\ell'(ka) - \gamma_\ell(k) j_\ell(ka)}{k n_\ell'(ka) - \gamma_\ell(k) n_\ell(ka)} = \tan\eta_\ell\,, \tag{45.133a}$$

$$= -\frac{k F_\ell'(ka) - L_\ell(k) F_\ell(ka)}{k G_\ell'(ka) - L_\ell(k) G_\ell(ka)}\,, \tag{45.133b}$$

where the logarithmic derivative of the internal solution at $R=a$ appropriate to either set, is

$$\gamma_\ell = \left(R_\ell^{-1}\,\mathrm{d}R_\ell/\mathrm{d}R\right)_{R=a}\,, \tag{45.134}$$

or alternatively, $L_\ell = [v_\ell^{-1} dv_\ell/\mathrm{d}R]_{R=a}$. The primes denote differentiation with respect to the argument, i. e.

$$B_\ell'(ka) = \left(\frac{\mathrm{d}B_\ell(x)}{\mathrm{d}x}\right)_{x=ka} = \frac{1}{k}\left(\frac{\mathrm{d}B_\ell(kR)}{\mathrm{d}R}\right)_{R=a}\,, \tag{45.135}$$

where B_ℓ denotes the functions F_ℓ, G_ℓ, j_ℓ, and n_ℓ.

Decomposition of the S-Matrix Element.

$$S_\ell(k) = \mathrm{e}^{2\mathrm{i}\eta_\ell} = \left(\frac{\gamma_\ell - (r_\ell - \mathrm{i}s_\ell)}{\gamma_\ell - (r_\ell + \mathrm{i}s_\ell)}\right)\mathrm{e}^{2\mathrm{i}\eta_\ell^H}\,, \tag{45.136}$$

where

$$\eta_\ell^H(k) = -\frac{j_\ell(ka) - \mathrm{i}n_\ell(ka)}{j_\ell(ka) + \mathrm{i}n_\ell(ka)}\,, \tag{45.137}$$

$$r_\ell + \mathrm{i}s_\ell = k\left(\frac{j_\ell'(ka) + \mathrm{i}n_\ell'(ka)}{j_\ell(ka) + \mathrm{i}n_\ell(ka)}\right)\,. \tag{45.138}$$

Decomposition of the Phase Shift.

$$\eta_\ell = \eta_\ell^H + \delta_\ell\,, \tag{45.139}$$

where η_ℓ^H is determined by (45.137), and where δ_ℓ is determined by

$$\tan\delta_\ell = \frac{s_\ell}{\gamma_\ell - r_\ell}\,, \tag{45.140}$$

which depends on the shape of U via the logarithmic derivative γ_ℓ of (45.134), and can vary rapidly with k, thereby giving rise to resonances.

Examples. (1) Hard sphere: if $V(R) = \infty$ for $R < a$, and $V(R) = 0$ for $R > a$, then $\gamma_\ell = \infty$, and

$$K_\ell^{(HS)} = \tan\eta_\ell^{(HS)}(k) = \frac{j_\ell(ka)}{n_\ell(ka)} \tag{45.141}$$

$$\to \begin{cases} -(ka)^{2\ell+1}/\left[(2\ell+1)!!(2\ell-1)!!\right], & ka \ll 1 \\ -\tan\left(ka - \frac{1}{2}\ell\pi\right), & ka \gg 1 \end{cases} \tag{45.142}$$

$$S_\ell^{(HS)} = \exp\left(2\mathrm{i}\eta_\ell^{(HS)}\right) = -\frac{j_\ell(ka) - \mathrm{i}n_\ell(ka)}{j_\ell(ka) + \mathrm{i}n_\ell(ka)} = 1 + 2\mathrm{i}T_\ell^{(HS)}\,. \tag{45.143}$$

The phase shift η_ℓ^H in the decomposition (45.139) is therefore identified as $\eta_\ell^{(HS)}$ for hard sphere scattering.

$$\sigma(E \to 0) = 4\pi a^2\,. \tag{45.144}$$

Diffraction pattern: As $E \to \infty$,

$$\frac{\mathrm{d}\sigma}{\mathrm{d}\Omega} \to \frac{1}{4}a^2\left[1 + \cot^2\left(\frac{1}{2}\theta\right) J_1^2(ka\sin\theta)\right]\,, \tag{45.145}$$

$$\sigma(E) \to 2\pi a^2\,. \tag{45.146}$$

Classical hard sphere scattering and diffraction about the sharp edge each contribute πa^2 to σ.

(2) Spherical Well: if $U(R) = -U_0$ for $R \le a$, and $U(R) = 0$ for $R > a$, then

$$\gamma_\ell(k) = \kappa\frac{j_\ell'(\kappa a)}{j_\ell(\kappa a)}; \qquad \kappa^2 = U_0 + k^2 \equiv k_0^2 + k^2\,. \tag{45.147}$$

S-wave ($\ell = 0$) properties:

$$\eta_0(k) = -ka + \tan^{-1}[(k/\kappa)\tan\kappa a]\ . \qquad (45.148)$$

As $k \to 0$, $\sigma_0(E) \to 4\pi a_s^2$, where the scattering length is

$$a_s = [1 - \tan(k_0 a)/(k_0 a)]\,a\ . \qquad (45.149)$$

For a shallow well $k_0 a \ll 1$: $\sigma_0(E) = (4\pi/9)U_0^2 a^6$, which agrees with the Born result (45.171) as $k \to 0$.

The condition for $\ell = 0$ bound state with energy $E_n = -(\hbar^2 k_n^2/2M)$ is

$$k_n \tan\kappa' a = -\kappa', \quad \kappa'^2 = k_0^2 - k_n^2\ . \qquad (45.150)$$

As the well is further deepened, $\sigma_0(E)$ oscillates between zero, where $\tan k_0 a = k_0 a$, and ∞, where $k_0 a = n\pi/2$, the condition both for appearance of a new level n at energy E and for $a_0 \to \infty$. In the neighborhood of these infinite resonances,

$$\sigma_0(E) = \frac{4\pi}{k^2 + \kappa'^2}\ , \qquad (45.151)$$

where $\kappa' = \kappa/\tan\kappa a$.

45.2.6 Coulomb Scattering

Direct solution of RSE (45.93) yields

$$v_\ell \sim \sin\left(kR - \frac{1}{2}\ell\pi + \eta_\ell^{(C)} - \beta\ln 2kR\right)\ , \qquad (45.152)$$

$$\beta = Z_1 Z_2 e^2/\hbar v = Z_1 Z_2/(k a_0)\ . \qquad (45.153)$$

Coulomb phase shift:

$$\eta_\ell^{(C)} = \arg\Gamma(\ell+1+\mathrm{i}\beta) = \mathrm{Im}\,[\ln\Gamma(\ell+1+\mathrm{i}\beta)]\ . \qquad (45.154)$$

Coulomb S-matrix element:

$$S_\ell^{(C)} = \exp\left(2\mathrm{i}\eta_\ell^{(C)}\right) = \frac{\Gamma(\ell+1+\mathrm{i}\beta)}{\Gamma(\ell+1-\mathrm{i}\beta)}\ . \qquad (45.155)$$

Coulomb scattering amplitude:

$$f_C(\theta) = -\frac{\beta\exp\left[2\mathrm{i}\eta_\ell^{(C)} - \mathrm{i}\beta\ln\left(\sin^2\frac{1}{2}\theta\right)\right]}{2k\sin^2\frac{1}{2}\theta}\ . \qquad (45.156)$$

Coulomb differential cross section:

$$\frac{d\sigma}{d\Omega} = \frac{\beta^2}{4k^2\sin^4\frac{1}{2}\theta} = \frac{Z_1^2 Z_2^2 e^4}{16E^2}\csc^4\frac{1}{2}\theta\ , \qquad (45.157)$$

which is the Rutherford scattering cross section.

Mott Formula. For the Coulomb scattering of two identical particles: From (45.85a) and (45.85c)

(a) spin-zero bosons (e.g. ^{4}He–^{4}He):

$$\frac{d\sigma}{d\Omega} = \frac{\beta^2}{4k^2}\left(\csc^4\frac{1}{2}\theta + \sec^4\frac{1}{2}\theta + 2\csc^2\frac{1}{2}\theta\sec^2\frac{1}{2}\theta\cos\Gamma\right)\ , \qquad (45.158)$$

(b) spin-$\frac{1}{2}$ fermions (e.g. H^+–H^+, $e^\pm$–$e^\pm$)

$$\frac{d\sigma}{d\Omega} = \frac{\beta^2}{4k^2}\left(\csc^4\frac{1}{2}\theta + \sec^4\frac{1}{2}\theta - \csc^2\frac{1}{2}\theta\sec^2\frac{1}{2}\theta\cos\Gamma\right)\ , \qquad (45.159)$$

where $\Gamma = 2\beta\ln\left(\tan\frac{1}{2}\theta\right)$.

45.2.7 Resonance Scattering

Zero-Energy Broad Resonances. The spherical well example (45.147) serves to illustrate broad resonances. When the well depth U_0 is strong enough to accomodate the (n_0+1) th energy level at zero energy, the bound state condition (45.150) implies that $\eta_0(k \to 0) = (n_0+1)\pi$, illustrating Levinson's theorem (45.105). As k increases, η_0 generally decreases through either $(2n-1)\pi/2$, or $(n-1)\pi$, where σ_0 has, respectively, a maximum value $4\pi/k^2$ and a minimum value of zero. If the phase shifts η_ℓ for $\ell > 0$ are small, then a nonzero minimum value in $\sigma(E)$ is evident. This is the Ramsauer–Townsend minimum manifest when the potential is just strong enough to introduce one or more wavelengths into the well with no observable scattering. Since the rate of decrease of η_0 cannot be arbitrarily rapid, (45.105), broad resonances will be exhibited in contrast to narrow (Breit–Wigner) resonances when η_ℓ increases rapidly through $(2n-1)\pi/2$ over a small energy range ΔE.

Narrow Resonances. The general decomposition (45.139) can be used to analyze narrow resonances. When γ_ℓ varies rapidly within an energy width Γ about a resonance energy E_r then δ_ℓ increases through odd multiples of $\pi/2$ and

$$\delta_\ell = \delta_\ell^r = \tan^{-1}\frac{\Gamma}{2(E_r - E)}\ , \qquad (45.160)$$

so that (45.60a) with (45.59a) and (45.59b) is

$$f_\ell = \frac{(2\ell+1)}{k}\left(T_\ell^{(HS)} + S_\ell^{(HS)}\frac{\Gamma/2}{E_r - E - \frac{1}{2}\Gamma}\right) \times P_\ell(\cos\theta)\ . \qquad (45.161)$$

Breit–Wigner Formula. For a pure resonance with no background phase shift, $S_\ell^{(\mathrm{HS})} = 1$ and the cross section has the Lorentz shape

$$\sigma(E) = \frac{4\pi(2\ell+1)}{k^2} \times \left(\frac{\Gamma^2/4}{(E-E_\mathrm{r})^2 + \Gamma^2/4} \right) . \tag{45.162}$$

Shape Resonances. (Also called quasibound state or tunneling resonances). At very low impact energies near or below the energy threshold for orbiting, sharp spikes superimposed on the glory oscillations may be evident in the E-dependance of $\sigma(E)$. These are due to quasibound states with positive energy $E_{n\ell}$ supported by the effective potential $V(r) + L^2/\left(2Mr^2\right)$. In heavy particle collisions, quasibound levels are the continuation of the bound rotational levels to positive energies $E_{n\ell} > 0$.

Systems in quasibound states $(n\ell)$, with nonresonant eigenenergy $E_{n\ell}$ and phase shift $\eta_\ell^{(0)}$, have

$$E = E_{n\ell} - \frac{\mathrm{i}}{2}\Gamma_{n\ell} , \tag{45.163a}$$

$$\eta_\ell = \eta_\ell^{(0)} + \eta_\ell^{(\mathrm{r})} , \tag{45.163b}$$

$$\eta_\ell = \eta_\ell^{(0)} + \tan^{-1} \frac{\Gamma_{n\ell}}{2(E_{n\ell} - E)} ; \tag{45.163c}$$

$$S_\ell(E) = \mathrm{e}^{2\mathrm{i}\eta_\ell} = \left(\frac{E - E_{n\ell} - \frac{\mathrm{i}}{2}\Gamma_{n\ell}}{E - E_{n\ell} + \frac{\mathrm{i}}{2}\Gamma_{n\ell}} \right) \mathrm{e}^{2\mathrm{i}\eta_\ell^{(0)}} . \tag{45.164}$$

The phase shift $\eta_\ell^{(\mathrm{r})}$ increases by π as E increases through $E_{n\ell}$ at a rate determined by $\Gamma_{n\ell}$. The dominant amplitude shifts from the external to the quasibound internal regions at $E = E_{n\ell}$.

Partial-Wave Scattering Amplitude.

$$f_\ell(\theta) = \frac{1}{2\mathrm{i}k}(2\ell+1)\exp\left(2\mathrm{i}\eta_\ell^0\right) P_\ell(\cos\theta) \times \left(1 - \frac{\mathrm{i}\Gamma_{n\ell}}{(E - E_{n\ell} + \frac{\mathrm{i}}{2}\Gamma_{n\ell})} \right) ; \tag{45.165}$$

$=$ background potential scattering amplitude
$+$ resonance scattering amplitude.

The partial-wave cross section is composed of the following: potential resonance and interference contributions to $\sigma_\ell = |f_\ell(\theta)|^2$:

$$\sigma_\ell = \frac{4\pi}{k^2}(2\ell+1)\left(\sin^2\eta_\ell^{(0)} + \frac{\Gamma_{n\ell}^2 \cos 2\eta_\ell^{(0)} + 2\Gamma_{n\ell}(E_{n\ell} - E)\sin 2\eta_\ell^{(0)}}{4(E - E_{n\ell})^2 + \Gamma_{n\ell}^2} \right) \tag{45.166a}$$

$$= \frac{4\pi}{k^2}(2\ell+1)\Bigg[\sin^2\eta_\ell^{(0)} \times \left(\frac{(E-E_{n\ell})^2}{(E-E_{n\ell})^2 + (\Gamma_{n\ell}/2)^2} \right) + \cos^2\eta_\ell^{(0)} \left(\frac{(\Gamma_{n\ell}/2)^2}{(E-E_{n\ell})^2 + (\Gamma_{n\ell}/2)^2} \right) + \sin 2\eta_\ell^{(0)} \left(\frac{(\Gamma_{n\ell}/2)(E_{n\ell} - E)}{(E-E_{n\ell})^2 + (\Gamma_{n\ell}/2)^2} \right) \Bigg] , \tag{45.166b}$$

$$\sigma(E) = \sum_{\ell=0}^{\infty} \sigma_\ell = \sigma_0(E) + \sigma_{\mathrm{res}}(E) . \tag{45.166c}$$

Resonance Shapes. Resonance shapes depend on the value of the background phase shift $\eta_\ell^{(0)}$. The case $\eta_\ell^{(0)} = 0$ gives a Lorentz line shape through the Breit–Wigner formula

$$\sigma_\ell = \frac{4\pi(2\ell+1)}{k^2} \frac{\Gamma_{n\ell}^2/4}{(E-E_{n\ell})^2 + \Gamma_{n\ell}^2/4} . \tag{45.167}$$

The other cases from (45.166b) are

$$\begin{aligned}
&\eta_\ell^{(0)} = n\pi && \text{large positive spike;}\\
&\eta_\ell^{(0)} = \left(n + \tfrac{1}{2}\right)\pi && \text{large negative spike;}\\
&n\pi < \eta_\ell^{(0)} < \left(n + \tfrac{1}{2}\right)\pi && \text{positive then negative;}\\
&\left(n + \tfrac{1}{2}\right)\pi < \eta_\ell^{(0)} < (n+1)\pi && \text{negative then positive.}
\end{aligned} \tag{45.168}$$

Time Delay.

$$\tau = 2\hbar \left(\frac{\partial \eta_\ell}{\partial E} \right)_\ell = 2\hbar \left(\frac{\partial \eta_\ell^{(0)}}{\partial E} \right) + \frac{4\hbar}{\Gamma_{n\ell}} . \tag{45.169}$$

The time for capture into quasibound levels is $\tau_\mathrm{c} = 4\hbar/\Gamma_{n\ell}$, and the capture frequency is $\nu_\mathrm{c} = \Gamma_{n\ell}/4\hbar$.

45.2.8 Integral Equation for Phase Shift

$$\sin\eta_\ell = -\frac{1}{k}\int_0^\infty F_\ell(kR)U(R)v_\ell^{(A)}(kR)\,\mathrm{d}R\,, \tag{45.170a}$$

$$K_\ell = \tan\eta_\ell = -\frac{1}{k}\int_0^\infty F_\ell(kR)U(R)v_\ell^{(B)}(kR)\,\mathrm{d}R\,, \tag{45.170b}$$

$$T_\ell = \mathrm{e}^{\mathrm{i}\eta_\ell}\sin\eta_\ell = -\frac{1}{k}\int_0^\infty F_\ell(kR)U(R)v_\ell^{(C)}(kR)\,\mathrm{d}R\,, \tag{45.170c}$$

$$S_\ell - 1 = \exp(2\mathrm{i}\eta_\ell) - 1 \tag{45.170d}$$

$$= -\frac{1}{k}\int_0^\infty F_\ell(kR)U(R)v_\ell^{(D)}(kR)\,\mathrm{d}R\,, \tag{45.170e}$$

where $v_\ell^{(A)}$, $v_\ell^{(B)}$, $v_\ell^{(C)}$, $v_\ell^{(D)}$ are so chosen to have asymptotes prescribed by (45.99b)–(45.102b), respectively.

Born Approximation for Phase Shifts. Set $v_\ell = F_\ell$ in (45.170a)–(45.170d) to obtain

$$\tan\eta_\ell^B(k) = -k\int_0^\infty U(R)\,[j_\ell(kR)]^2\,R^2\,\mathrm{d}R\,. \tag{45.171}$$

For $\lambda = (\ell+1/2) \gg ka$, substitute

$$\left\langle k^2R^2\,[j_\ell(kR)]^2\right\rangle = \frac{1}{2}\left(1-\lambda^2/k^2R^2\right)^{-1/2}\,. \tag{45.172}$$

For the Jeffrey–Born (JB) phase shift function, $\ell \gg ka$,

$$\tan\eta_{\mathrm{JB}}(\lambda) = -\frac{k}{2E}\int_{\lambda/k}^\infty \frac{V(R)\,\mathrm{d}R}{\left[1-\lambda^2/(kR)^2\right]^{1/2}}\,, \tag{45.173}$$

which agrees with (45.41) since $bk = \ell + \frac{1}{2} = \lambda$. For linear trajectories $R^2 = b^2 + Z^2$, the eikonal phase (45.42) is recovered.

Born S-Wave Phase Shift.

$$\tan\eta_0^{\mathrm{B}}(k) = -\frac{1}{k}\int_0^\infty U(R)\sin^2(kR)\,\mathrm{d}R \tag{45.174}$$

Examples: (i) $U = U_0\dfrac{\mathrm{e}^{-\alpha R}}{R}$, (ii) $U = \dfrac{U_0}{\left(R^2+R_0^2\right)^2}$;

$$\text{(i)}\tan\eta_0^{\mathrm{B}} = -\frac{U_0}{4k}\ln\left(1+4k^2/\alpha^2\right)\,, \tag{45.175}$$

$$\text{(ii)}\tan\eta_0^{\mathrm{B}} = -\frac{\pi U_0}{4kR_0^3}\left[1-(1+2kR_0)\,\mathrm{e}^{-2kR_0}\right]. \tag{45.176}$$

Born Phase Shifts (Large ℓ). For $\ell \gg ka$,

$$\tan\eta_\ell^{\mathrm{B}} = -\frac{k^{2\ell+1}}{[(2\ell+1)!!]^2}\int_0^\infty U(R)R^{2\ell+2}\,\mathrm{d}R\,, \tag{45.177}$$

valid only for finite range interactions $U(R>a) = 0$. Example: $U = -U_0$, $R \le a$ and $U = 0$, $R > a$.

$$\tan\eta_\ell^{\mathrm{B}}(\ell \gg ka) = U_0a^2\frac{(ka)^{2\ell+1}}{[(2\ell+1)!!]^2\,(2\ell+3)}\,. \tag{45.178}$$

For $\ell \gg ka$, $\eta_{\ell+1}/\eta_\ell \sim (ka/2\ell)^2$.

45.2.9 Variable Phase Method

The phase function $\eta_\ell(R)$ is defined to be the scattering phase shift produced by the part of the potential $V(R)$ contained within a sphere of radius R. It satisfies the nonlinear differential equation for $\eta_\ell(R)$

$$\frac{\mathrm{d}\eta_\ell}{\mathrm{d}R} = -kR^2U(R)\,[\cos\eta_\ell(R)j_\ell(kR) - \sin\eta_\ell(R)n_\ell(kR)]^2\,. \tag{45.179a}$$

The corresponding integral equation for $\eta_\ell(R)$ is

$$\eta_\ell(R) = -k\int_0^R \left[\cos\eta_\ell(R)j_\ell(kR) - \sin\eta_\ell(R)n_\ell(kR)\right]^2 U(R)R^2\,\mathrm{d}R\,. \tag{45.179b}$$

The Born approximation (45.171) is recovered by substituting $\eta_\ell = 0$ on the RHS of (45.179b) as $R \to \infty$.

45.2.10 General Amplitudes

For a general potential $V(\boldsymbol{R})$, define the reduced potential $U(\boldsymbol{R}) = \left(2M/\hbar^2\right) V(\boldsymbol{R})$. The plane wave scattering states are

$$\phi_{\boldsymbol{k}}(\boldsymbol{R}) = \exp(\mathrm{i}\boldsymbol{k} \cdot \boldsymbol{R}) = \phi\_k^*(\boldsymbol{R}) , \qquad (45.180)$$

and the full scattering solutions have the form

$$\Psi_{\boldsymbol{k}}^{(\pm)}(\boldsymbol{R}) \sim \phi_{\boldsymbol{k}}(\boldsymbol{R}) + \frac{f(\boldsymbol{k}, \boldsymbol{k}')}{R} \exp(\pm \mathrm{i}kR) , \qquad (45.181)$$

where the scattering amplitude is

$$f(\boldsymbol{k}, \boldsymbol{k}') = -\frac{1}{4\pi} \left\langle \phi_{\boldsymbol{k}'}(\boldsymbol{R}) \mid U(\boldsymbol{R}) \mid \Psi_{\boldsymbol{k}}^{(+)}(\boldsymbol{R}) \right\rangle \qquad (45.182a)$$

$$= -\frac{1}{4\pi} \left\langle \Psi_{\boldsymbol{k}'}^{(-)}(\boldsymbol{R}) \mid U(\boldsymbol{R}) \mid \phi_{\boldsymbol{k}}(\boldsymbol{R}) \right\rangle \qquad (45.182b)$$

$$\equiv -\frac{1}{4\pi} \left\langle \phi_{\boldsymbol{k}'}(\boldsymbol{R}) \,|T|\, \phi_{\boldsymbol{k}}(\boldsymbol{R}) \right\rangle . \qquad (45.182c)$$

The last equation defines the T-matrix element.

First Born Approximation.
Set $\Psi_{\boldsymbol{k}}^{+} = \phi_{\boldsymbol{k}}$ in (45.182a). Then

$$f_{\mathrm{B}}(K) = -\frac{1}{4\pi} \int U(\boldsymbol{R}) \exp(\mathrm{i}\boldsymbol{K} \cdot \boldsymbol{R})\, \mathrm{d}\boldsymbol{R} , \qquad (45.183)$$

where the momentum change is $\boldsymbol{K} = \boldsymbol{k} - \boldsymbol{k}'$, and $K = 2k \sin \frac{1}{2}\theta$. For a symmetric potential,

$$f_{\mathrm{B}}(K) = -\int \frac{\sin KR}{KR} U(R) R^2 \,\mathrm{d}R . \qquad (45.184)$$

Connection with partial wave analysis:

$$\frac{\sin KR}{KR} = \sum_{\ell=0}^{\infty} (2\ell+1) \left[j_\ell(kR)\right]^2 P_\ell(\cos\theta) . \qquad (45.185)$$

which is consistent with (45.171).

The static e$^-$–atom scattering potential and Born scattering amplitude are

$$V(R) = -\frac{Ze^2}{R} + e^2 \int \frac{|\psi(\boldsymbol{r})|^2 \,\mathrm{d}\boldsymbol{r}}{|\boldsymbol{R} - \boldsymbol{r}|} , \qquad (45.186)$$

$$f_{\mathrm{B}}(K) = \frac{2Me^2}{\hbar^2} \frac{[Z - F(K)]}{K^2} . \qquad (45.187)$$

where the *elastic form factor* is

$$F(K) = \int |\psi(\boldsymbol{r})|^2 \exp(\mathrm{i}\boldsymbol{K} \cdot \boldsymbol{r})\, \mathrm{d}\boldsymbol{r} .$$

For a pure Coulomb field, $F(K) = 0$ and $\sigma_{\mathrm{B}}(\theta, E) = |f_{\mathrm{B}}(K)|^2$ reduces to (45.157).

Two Potential Formulae. For scattering from the combined potential $U(\boldsymbol{R}) = U_0(\boldsymbol{R}) + U_1(\boldsymbol{R})$,

$$f(\boldsymbol{k}, \boldsymbol{k}') = -\frac{1}{4\pi} \Big[\left\langle \phi_{\boldsymbol{k}'}(\boldsymbol{R}) \mid U_0(\boldsymbol{R}) \mid \chi_{\boldsymbol{k}}^{(+)}(\boldsymbol{R}) \right\rangle + \left\langle \chi_{\boldsymbol{k}'}^{(-)}(\boldsymbol{R}) \mid U_1(\boldsymbol{R}) \mid \Psi_{\boldsymbol{k}}^{(+)} \right\rangle \Big] , \qquad (45.188a)$$

where $\chi_{\boldsymbol{k}}^{(\pm)}(\boldsymbol{R})$ and $\Psi_{\boldsymbol{k}}^{(\pm)}(\boldsymbol{R})$ are full solutions for scattering by V_0 and $V_0 + V_1$, respectively. For symmetric interactions,

$$f(\theta) = \frac{1}{k} \sum_{\ell=0}^{\infty} (2\ell+1) \left(T_\ell^{(0)} + T_\ell^{(1)} \right) P_\ell(\cos\theta) , \qquad (45.189)$$

$$T_\ell^{(0)} = \exp\left(\mathrm{i}\eta_\ell^{(0)}\right) \sin \eta_\ell^{(0)} \qquad (45.190a)$$

$$= -\frac{1}{k} \int_0^\infty \mathrm{d}R \left[F_\ell(R) U_0(R) u_\ell(R)\right] , \qquad (45.190b)$$

$$T_\ell^{(1)} = \exp\left(2\mathrm{i}\eta_\ell^{(0)}\right) \exp\left(\mathrm{i}\eta_\ell^{(1)}\right) \sin \eta_\ell^{(1)} , \qquad (45.191a)$$

$$T_\ell^{(1)} = -\frac{1}{k} \int_0^\infty \mathrm{d}R \left[u_\ell(R) U_1(R) v_\ell(R)\right] , \qquad (45.191b)$$

where u_ℓ and v_ℓ are the radial wave functions in (45.93), with phase-shifts $\eta_\ell^{(0)}$ and $\eta_\ell = \eta_\ell^{(0)} + \eta_\ell^{(1)}$, for scattering by V_0 and $V_0 + V_1$, respectively, normalized according to (45.101b).

Distorted-Wave Approximation.

$$\Psi_{\boldsymbol{k}}^{(+)}(\boldsymbol{R}) \sim \chi_{\boldsymbol{k}}^{(+)}(\boldsymbol{R})$$

$$f(\boldsymbol{k}, \boldsymbol{k}') = -\frac{1}{4\pi} \Big[\left\langle \phi_{\boldsymbol{k}'}(\boldsymbol{R}) \mid U_0(\boldsymbol{R}) \mid \chi_{\boldsymbol{k}}^{(+)}(\boldsymbol{R}) \right\rangle + \left\langle \chi_{\boldsymbol{k}'}^{(-)}(\boldsymbol{R}) \mid U_1(\boldsymbol{R}) \mid \chi_{\boldsymbol{k}}^{(+)}(\boldsymbol{R}) \right\rangle \Big] . \qquad (45.192)$$

45.3 Semiclassical Scattering Formulae

45.3.1 Scattering Amplitude: Exact Poisson Sum Representation

Poisson Sum Formula. With $\lambda = \ell + \frac{1}{2}$,

$$\sum_{\ell=0}^{\infty} F\left(\ell + \frac{1}{2}\right) = \sum_{m=-\infty}^{\infty} (-1)^m \int_0^{\infty} F(\lambda) \mathrm{e}^{\mathrm{i}2m\pi\lambda} \, \mathrm{d}\lambda \,. \tag{45.193}$$

When applied to (45.60b),

$$f(\theta) = (\mathrm{i}k)^{-1} \sum_{m=-\infty}^{\infty} (-1)^m \int_0^{\infty} \lambda \left(\mathrm{e}^{2\mathrm{i}\eta(\lambda)} - 1\right) \times P_{\lambda-\frac{1}{2}}(\cos\theta) \mathrm{e}^{\mathrm{i}2m\pi\lambda} \, \mathrm{d}\lambda \,, \tag{45.194}$$

where $\eta(\lambda)$ and $P_{\lambda-1/2} \equiv P(\lambda, \theta)$ are now phase functions and Legendre functions of the continuous variable λ, being interpolated from discrete to continuous ℓ. This infinite-sum-of-integrals representation for $f(\theta)$ is in principle exact. It is the appropriate technique for conversion from a sum over (quantal) discrete values of a variable to a continuous integration over that variable which classically can assume any value. The particular merit here is that the index m labels the classical paths that have encircled the (attractive) scattering center m times, and that the terms with $m < 0$ have no regions of stationary phase (SP). For deflections χ in the range $-\pi < \chi < \pi$, the only SP contribution is the $m = 0$ term.

45.3.2 Semiclassical Procedure

Semiclassical analysis [45.7–9] involves reducing (45.194) by the three approximations represented by cases A to C below. Since the integrands can oscillate very rapidly over large regions of λ, the main contributions to the integrals arise from points λ_i of stationary phase of each integrand. The amplitude can then be evaluated by the method of stationary phase, the basis of semiclassical analysis.

A. Legendre Function Asymptotic Expansions

Main range: $\sin\theta > \lambda^{-1}$, θ not within λ^{-1} of zero or π.

$$P_\ell(\cos\theta) = (2/(\pi\lambda\sin\theta))^{1/2} \cos(\lambda\theta - \pi/4) \,. \tag{45.195}$$

Forward formula: θ within λ^{-1} of zero.

$$P_\ell(\cos\theta) = [\theta/\sin\theta]^{1/2} J_0(\lambda\theta) \,, \tag{45.196a}$$

$$J_0(\lambda\theta) \frac{1}{\pi} \int_0^{\pi} \mathrm{e}^{-\mathrm{i}\lambda\theta\cos\phi} \, \mathrm{d}\phi \,. \tag{45.196b}$$

Backward formula: θ within λ^{-1} of π.

$$P_\ell(\cos\theta) = \left(\frac{\pi - \theta}{\sin\theta}\right)^{1/2} \times J_0[\lambda(\pi - \theta)] \, \mathrm{e}^{-\mathrm{i}\pi(\lambda - 1/2)} \,. \tag{45.197}$$

Equations (45.195–45.197) are useful for analysis of caustics (rainbows), diffraction and forward and backward glories, respectively. Also, a useful identity is

$$\sum_{\ell=0}^{\infty} (2\ell + 1) P_\ell(\cos\theta) = \begin{cases} 4\delta(1 - \cos\theta), & \theta > 0 \\ 0 & \theta = 0 \end{cases}$$

where $\delta(x)$ is the Dirac delta function.

B. *JWKB* Phase Shift Functions

$$\eta(\lambda) = \int_{R_c}^{\infty} k_\lambda(R) \, \mathrm{d}R - \int_{\lambda/k}^{\infty} \left(k^2 - \frac{\lambda^2}{R^2}\right)^{1/2} \mathrm{d}R \tag{45.198a}$$

$$= \lim_{R \to \infty} \left[\int_{R_c}^{\infty} k_\ell(R') \, \mathrm{d}R' - kR\right] + \frac{1}{2}\lambda\pi \,. \tag{45.198b}$$

Local wave number:

$$k_\lambda^2(R) = k^2 - U(R) - \lambda^2/R^2 \,. \tag{45.199}$$

The *Langer modification*:

$$b = \frac{\sqrt{\ell(\ell+1)}}{k} = \frac{\ell + 1/2}{k} = \frac{\lambda}{k} \,. \tag{45.200}$$

Useful identity: As $R \to \infty$,

$$\sin\left[\int_{\lambda/k}^{R} \left(k^2 - \lambda^2/R^2\right)^{1/2} \mathrm{d}R + \frac{\pi}{4}\right] \to \sin\left(kR - \frac{1}{2}\ell\pi\right) \tag{45.201}$$

JWKB phase functions are valid when variation of the potential over the local wavenumber $k_\lambda^{-1}(R)$ is a small fraction of the available kinetic energy $E - V(R)$. Many wavelengths can then be accomodated within a range ΔR for a characteristic potential change ΔV. The classical method is valid when $(1/k)(\mathrm{d}V/\mathrm{d}R) \ll (E - V)$.

Phase–Deflection Function Relation.

$$\chi(\lambda) = 2\frac{\partial\eta(\lambda)}{\partial\lambda} . \tag{45.202}$$

C. Stationary Phase Approximations (SPA) to Generic Integrals

$$A^{\pm}(\theta) = \int_{-\infty}^{\infty} g(\theta;\lambda)\exp\left[\pm\mathrm{i}\gamma(\theta;\lambda)\right]\mathrm{d}\lambda \tag{45.203}$$

for parametric θ. In cases where the phase function γ has two stationary points, a phase minimum γ_1 at λ_1 and a phase maximum γ_2 at λ_2, then $\gamma_i' = 0$, $\gamma_1'' > 0$, $\gamma_2'' < 0$ where $\gamma_i' = (\mathrm{d}\gamma/\mathrm{d}\lambda)$ at λ_i and $\gamma_i'' = (\mathrm{d}^2\gamma/\mathrm{d}\lambda^2)$ for $i = 1, 2$. Since g is real, $A^- = (A^+)^*$, $g_i(\theta) = g(\theta, \lambda_i)$.

Uniform Airy result.

$$\begin{aligned} A^{+}(\theta) &= a_1(\theta)\mathrm{e}^{\mathrm{i}(\gamma_1+\pi/4)}F^*(\gamma_{21}) \\ &\quad + a_2(\theta)\mathrm{e}^{\mathrm{i}(\gamma_2-\pi/4)}F(\gamma_{21}) , \end{aligned} \tag{45.204}$$

$$a_i(\theta) = \left[2\pi/|\gamma_i''|\right]^{1/2} g_i(\theta) , \quad i = 1, 2 , \tag{45.205}$$

$$\begin{aligned} \gamma_{21}(\theta) &= \gamma_2 - \gamma_1 \\ &\equiv \frac{4}{3}|z(\theta)|^{3/2} > 0 , \end{aligned} \tag{45.206}$$

$$\begin{aligned} F[\gamma_{21}(\theta)] &= \left[z^{1/4}\mathrm{Ai}(-z) + \mathrm{i}z^{-1/4}\mathrm{Ai}'(-z)\right]\sqrt{\pi} \\ &\quad \times \mathrm{e}^{-\mathrm{i}(\gamma_{21}/2-\pi/4)} , \end{aligned} \tag{45.207}$$

where Ai and Ai′ are the Airy function and its derivative.

This result holds for all separations $(\lambda_2 - \lambda_1)$ in location of stationary phases including a caustic (or rainbow), which is a point of inflection in γ, i. e. $\gamma_1 = \gamma_2$, $\gamma_i' = 0 = \gamma_i''$. An equivalent expression is [45.9].

$$\begin{aligned} A^{+}(\theta) &= \Big[(a_1 + a_2)z^{1/4}\mathrm{Ai}(-z) \\ &\quad - \mathrm{i}(a_1 - a_2)z^{-1/4}\mathrm{Ai}'(-z)\Big]\sqrt{\pi}\,\mathrm{e}^{\mathrm{i}\bar{\gamma}} , \end{aligned} \tag{45.208}$$

where the mean phase is $\bar{\gamma} = \frac{1}{2}(\gamma_1 + \gamma_2)$. The first form (45.204) is useful for analysis of widely separated regions of stationary phase when $\gamma_{21} \gg 0$ and $F \to 1$. The equivalent second form (45.208) is valuable in the neighborhood of caustics or rainbows when the stationary phase regions coalesce as $a_1 \to a_2$.

Primitive result. For widely separated regions λ_1 and λ_2, $F \to 1$ and

$$A^{\pm}(\epsilon) = \left[a_1(\epsilon) \mp \mathrm{i}a_2(\epsilon)\mathrm{e}^{\pm\mathrm{i}\gamma_{21}}\right]\mathrm{e}^{\pm\mathrm{i}(\gamma_1+\pi/4)} , \tag{45.209a}$$

$$\begin{aligned} A^{\pm}(\epsilon) &= a_1(\epsilon)\exp\left[\pm\mathrm{i}\left(\gamma_1 + \frac{\pi}{4}\right)\right] \\ &\quad + a_2(\epsilon)\exp\left[\pm\mathrm{i}\left(\gamma_2 - \frac{\pi}{4}\right)\right] . \end{aligned} \tag{45.209b}$$

Note that the minimum phase γ_1 is increased by $\pi/4$ and the maximum phase γ_2 is reduced by $\pi/4$.

Transitional Airy Result. In the neighborhood of a caustic or rainbow where $\gamma'' = 0$, at the inflection point $\lambda_1 = \lambda_2 = \lambda_r$, then

$$A^{\pm}(\theta) = 2\pi\left|\frac{2}{\gamma'''(\lambda_r)}\right|^{1/3} g(\theta;\lambda_r)\mathrm{Ai}(-z)\,\mathrm{e}^{\pm\mathrm{i}\gamma(\theta;\lambda_r)} \tag{45.210}$$

$$z = \left|\frac{2}{\gamma'''(\lambda_r)}\right|^{1/3}\gamma'(\theta;\lambda_r) . \tag{45.211}$$

Only over a very small angular range does this result agree in practice with the uniform result (45.204), which uniformly connects (45.209a) and (45.210). These stationary-phase formulae are not only applicable to integrals involving (λ, θ) but also to (t, E) and (R, p) combinations which occur in the Method of Variation of Constants and in Franck–Condon overlaps of vibrational wave functions, respectively.

45.3.3 Semiclassical Amplitudes: Integral Representation

A. Off-Axis Scattering: $\sin\theta > \lambda^{-1}$

Except in the forward and backward directions, (45.194) with (45.195) reduces to

$$\begin{aligned} f(\theta) &= -\frac{1}{k(2\pi\sin\theta)^{1/2}}\sum_{m=-\infty}^{\infty}(-1)^m \\ &\quad \times \int_0^{\infty}\mathrm{d}\lambda\,\lambda^{1/2} \\ &\quad \times \left(\mathrm{e}^{\mathrm{i}\Delta^{+}(\lambda,m)} - \mathrm{e}^{\mathrm{i}\Delta^{-}(\lambda;m)}\right) , \end{aligned} \tag{45.212}$$

$$\begin{aligned} \Delta^{\pm}(\lambda;m) &= 2\eta(\lambda) + 2m\pi\lambda \pm \lambda\theta \pm \pi/4 \quad &(45.213) \\ &\equiv S^{\mathrm{C}} \pm \pi/4 , &(45.214) \end{aligned}$$

where S^{C} is the classical action (45.52b) divided by $\hbar$.

The stationary phase condition $\mathrm{d}\Delta^{\pm}/\mathrm{d}\lambda = 0$ yields the deflection function χ to scattering angle θ relation

$$\chi(\lambda_i) = \mp\theta - 2m\pi\,, \tag{45.215}$$

where λ_i are points of stationary phase (SP). Since $\pi \geq \chi \geq -\infty$, integrals with $m < 0$ have no SP's and vanish under the SPA. For cases involving no orbiting (where $\chi \to -\infty$) and when $\pi > \chi > -\pi$, then integrals with $m > 0$ also vanish under SPA so that the only remaining contribution from $m = 0$ to (45.214) is

$$f(\theta) = -\frac{1}{k(2\pi\sin\theta)^{1/2}} \times \int_0^\infty \lambda^{1/2}\left[\mathrm{e}^{\mathrm{i}\Delta^+(\lambda)} - \mathrm{e}^{\mathrm{i}\Delta^-(\lambda)}\right]\mathrm{d}\lambda\,, \tag{45.216}$$

$$\Delta^{\pm}(\lambda) = 2\eta(\lambda) \pm \lambda\theta \pm \pi/4\,. \tag{45.217}$$

The attractive branch Δ^+ contributes only negative deflections and the repulsive branch Δ^- contributes only positive deflections and has one SP point at λ_1 where Δ^- is maximum.

Rainbow angle θ_r: $(\Delta^+)''_{\lambda_r} = 0$, so that $\chi'(\lambda_r) = 0$ where $\chi(\lambda_r) < 0$ has reached its most negative value.

$\theta < \theta_r$: Δ^+ has two SP points $\lambda_{2,3}$;
a maximum at λ_2 and
a minimum at λ_3.

$\theta = \theta_r$: $\lambda_2 = \lambda_3$: SP's coalesce.

$\theta > \theta_r$: no classical attractive scattering.
Δ^+ has no SP points.

B. Forward Amplitude: $\sin\theta \sim \theta < \lambda^{-1}$

$$f(\theta) = \frac{1}{\mathrm{i}k}\left(\frac{\theta}{\sin\theta}\right)^{1/2} \sum_{m=-\infty}^{\infty} \mathrm{e}^{-\mathrm{i}m\pi} \int_0^\infty \lambda J_0(\lambda\theta) \times \left(\mathrm{e}^{2\mathrm{i}\eta(\lambda)} - 1\right)\mathrm{e}^{2\mathrm{i}m\pi\lambda}\,\mathrm{d}\lambda\,. \tag{45.218}$$

Stationary phase points: $\gamma'(\lambda_m) = 0$.

$$\chi(\lambda_m) = 2\left(\frac{\partial\eta}{\partial\lambda}\right) = -2m\pi\,. \tag{45.219}$$

Terms with $m < 0$ therefore make no SP contribution to $f(\theta)$ since $\chi \leq \pi$. The $m = 0$ term provides diffraction due to $\chi \to 0$, $\chi' \to 0$ at long range, and a forward glory due to $\chi \to 0$ at a finite λ_g and nonzero χ'_g.

C. Backward Amplitude: $\theta \sim \pi - \mathcal{O}(\lambda^{-1})$.

$$f(\theta) = \frac{1}{k}\left(\frac{\pi-\theta}{\sin\theta}\right)^{1/2} \times \sum_{m=-\infty}^{\infty} \mathrm{e}^{\mathrm{i}m\pi} \int_0^\infty \lambda J_0\left[\lambda(\pi-\theta)\right] \times \mathrm{e}^{\mathrm{i}[2\eta(\lambda)+(2m-1)\pi\lambda]}\,\mathrm{d}\lambda\,. \tag{45.220}$$

Stationary phase points:

$$\chi(\lambda_m) = 2\left(\frac{\partial\eta}{\partial\lambda}\right) = -(2m-1)\pi\,. \tag{45.221}$$

There are no SP for $m < 0$. The $m = 0$ term provides a normal backward amplitude due to repulsive collisions ($\chi = \pi$), and $m > 0$ terms are due to attractive half-windings.

D. Eikonal Amplitude

The $m = 0$ term of (45.218) gives

$$f_{\mathrm{E}}(\theta) = \frac{1}{\mathrm{i}k}\int_0^\infty \lambda\left(\mathrm{e}^{2\mathrm{i}\eta(\lambda)} - 1\right) J_0(\lambda\theta)\,\mathrm{d}\lambda \tag{45.222a}$$

$$= -\mathrm{i}k\int_0^\infty \left(\mathrm{e}^{2\mathrm{i}\eta(b)} - 1\right) J_0(kb\theta)b\,\mathrm{d}b\,. \tag{45.222b}$$

From the optical theorem,

$$\sigma_{\mathrm{E}}(E) = 8\pi\int_0^\infty \sin^2\eta(b,E)b\,\mathrm{d}b\,. \tag{45.223}$$

For potentials with cylindrical symmetry, $kb\theta$ can be replaced by $2kb\sin\frac{1}{2}\theta = \boldsymbol{K}\cdot\boldsymbol{b}$, and

$$f_{\mathrm{E}}(\theta) = -\frac{\mathrm{i}k}{2\pi}\int \left(\mathrm{e}^{2\mathrm{i}\eta(b)} - 1\right) J_0(\boldsymbol{K}\cdot\boldsymbol{b})\,\mathrm{d}\boldsymbol{b}\,. \tag{45.224}$$

45.3.4 Semiclassical Amplitudes and Cross Sections

Amplitude addition:

$$f(\theta) = \sum_{j=1}^{N} f_j(\theta)\,, \tag{45.225}$$

where each classical path $b_j = b_j(\theta)$ or SP-point $\lambda_j = \lambda_j(\theta)$ contributes $f_j(\theta)$ to the amplitude.

Primitive amplitudes:

$$f_j(\theta) = -\mathrm{i}\alpha_j\beta_j\sigma_j^{1/2}(\theta)\exp\left[\mathrm{i}S_j^{\mathrm{C}}(\theta)\right] \tag{45.226}$$

$$\chi'_j = (\mathrm{d}\chi/\mathrm{d}\lambda)_j \tag{45.227}$$

$$\alpha_j = \mathrm{e}^{\pm \mathrm{i}\pi/4}; \quad (+),\ \chi'_j > 0; \quad (-),\ \chi'_j < 0; \tag{45.228a}$$

$$\beta_j = \mathrm{e}^{\pm \mathrm{i}\pi/4}; \quad (+),\ \chi_j > 0; \quad (-),\ \chi_j < 0. \tag{45.228b}$$

Classical cross section:

$$\sigma_j(\theta) = \left| \frac{b\,\mathrm{d}b}{\mathrm{d}(\cos\chi)} \right|_{\chi_j} = \frac{1}{k^2} \frac{\lambda_j}{\sin\theta |\chi'_j|}. \tag{45.229}$$

N classical deflections χ_j provide the same θ:

$$\chi_j = \chi(\lambda_j) = \theta, -\theta, -2\pi \pm \theta, -4\pi \pm \theta, \ldots. \tag{45.230}$$

Classical collision action $S^{\mathrm{C}}(E, L; \chi)/\hbar$:

$$S^{\mathrm{C}}_j = 2\eta(\lambda_j) - \lambda_j \chi(\lambda_j) \tag{45.231a}$$
$$= 2\eta(\lambda_j) - \lambda_j \theta, \quad 0 \le \chi \le \pi \tag{45.231b}$$
$$= 2\eta(\lambda_j) + \lambda_j \theta - m\pi, \quad \chi < 0, \tag{45.231c}$$

where $m = 0, 1, 2, \ldots$ is the number of times the ray has traversed the backward direction during its attractive windings about the scattering center.

A. Amplitude Addition: For Three Well Separated Regions of Stationary Phase $\lambda_1 < \lambda_2 < \lambda_3$

A scattering angle θ in the range $0 < \theta < \theta_{\mathrm{r}}$ (rainbow angle) typically results from deflections χ_j at three impact parameters b (or λ): $\theta = \{\chi(b_1), -\chi(b_2), -\chi(b_3)\} \equiv \{\chi_j\}$. Scattering in the range $\theta_{\mathrm{r}} \le \theta < \pi$ results from one deflection at b_1. b_1 is in the positive branch (inner repulsion) and $b_{2,3}$ are in the negative branch (outer attraction) of the deflection function $\chi(b)$ such that $b_1 < b_2 < b_3$. $kb = (\ell + 1/2) = \lambda$. Thus

$$f(\theta) = \sum_{j=1}^{3} f_j(\theta) = \sum_{j=1}^{3} \left[\sigma_j(\theta)\right]^{1/2} \exp(\mathrm{i}S_j), \tag{45.232}$$

where the overall phases of each f_j are

$$S_1 = 2\eta(\lambda_1) - \lambda_1\theta - \pi/2, \tag{45.233a}$$
$$S_2 = 2\eta(\lambda_2) + \lambda_2\theta - \pi, \tag{45.233b}$$
$$S_3 = 2\eta(\lambda_3) + \lambda_3\theta - \pi/2, \tag{45.233c}$$

which are appropriate, respectively, to deflections $\chi_1 = \theta$ at λ_1, $\chi_2 = -\theta$ at λ_2 and $\chi_3 = -\theta$ at λ_3, within the range $-\pi \le \chi \le \pi$.

The elastic differential cross section

$$\sigma(\theta) = \sum_{j=1}^{3} \sigma_j(\theta) + 2\sum_{i<j}^{3} \left[\sigma_i(\theta)\sigma_j(\theta)\right]^{1/2} \cos(S_i - S_j) \equiv \sigma_{\mathrm{c}}(\theta) + \Delta\sigma(\theta) \tag{45.234}$$

exhibits interference effects. The first term σ_{c} is the classical background DCS with no oscillations. The second term $\Delta\sigma$ provides the oscillatory structure which originates from interference between classical actions associated with the different trajectories resulting in a given θ. The part of $S_{ij} = S_i - S_j$ most rapidly varying with θ are the angular action changes $(\lambda_1 + \lambda_2)\theta$, $(\lambda_1 + \lambda_3)\theta$ and $(\lambda_2 - \lambda_3)\theta$. Interference oscillations between the action phases S_1 and S_2 or between S_1 and S_3 then have angular separations

$$\Delta\theta_{1;(2,3)} = \frac{2\pi n}{(\lambda_1 + \lambda_{2,3})}, \tag{45.235}$$

which are much smaller than the separation

$$\Delta\theta_{2;3} = \frac{2\pi n}{(\lambda_2 - \lambda_3)} \tag{45.236}$$

for interference between phases S_2 and S_3. The oscillatory structure in $\Delta\sigma(\theta)$ is composed therefore of supernumerary rainbow oscillations with large angular separations $\Delta\theta_{2;3}$ from S_2 and S_3 interference, with superimposed rapid oscillations with smaller separation $\Delta\theta_{1,(2,3)}$ from interference between S_1 and S_2 or S_1 and S_3.

For deflections $\chi_j = \theta, -\theta, -2\pi \mp \theta, -4\pi \mp \theta, \cdots$, then the Δ^+-branch of (45.212) provides additional contributions to (45.232) with phases

$$S^{\pm}_{2m} = 2\eta(\lambda_{2m}) \pm \lambda_{2m}\theta - 2m\pi - \pi, \tag{45.237a}$$
$$S^{\pm}_{3m} = 2\eta(\lambda_{3m}) \pm \lambda_{3m}\theta - 2m\pi - \pi/2, \tag{45.237b}$$

for $m = 1, 2, 3, \ldots$.

B. Uniform Airy Result: For Two Regions of Stationary Phase Which can Coalesce

The combined contribution $f_{23}(\theta)$ from the λ_2 and λ_3 attractive regions in Δ^+ branch is

$$f_{23}(\theta) = \sigma_2^{1/2}\,\mathrm{e}^{\mathrm{i}S_2} F_{23} + \sigma_3^{1/2}\,\mathrm{e}^{\mathrm{i}S_3} F^*_{23}, \tag{45.238a}$$

$$F_{23} = (A + iB)\, e^{-i(S_{23}/2)}\,, \quad (45.238b)$$

$$S_{23} = S_2 - S_3 = 2(\eta_2 - \eta_3) + (\lambda_2 - \lambda_3)\theta - \frac{1}{2}\pi\,, \quad (45.238c)$$

$$A(z) = \pi^{1/2} z^{1/4} \mathrm{Ai}(-z)\,, \quad (45.238d)$$

$$B(z) = \pi^{1/2} z^{-1/4} \mathrm{Ai}'(-z)\,, \quad (45.238e)$$

$$\frac{4}{3}|z|^{3/2} = S_2^{\mathrm{C}} - S_3^{\mathrm{C}} = S_{23} + \frac{1}{2}\pi\,. \quad (45.238f)$$

The amplitude f_{23} tends to the primitive result $f_2(\theta) + f_3(\theta)$ in the limit of well-separated regions ($z \gg 1$) when $F_{23} \to 1$. An equivalent form of (45.238a) is

$$f_{23}(\theta) = \left[A\left(\sigma_2^{1/2} + \sigma_3^{1/2}\right) + iB\left(\sigma_2^{1/2} - \sigma_3^{1/2}\right)\right] \times \exp(i\bar{S})\,, \quad (45.239)$$

where the mean phase $\bar{S} = \frac{1}{2}(S_2 + S_3)$. This form is useful for analysis of caustic regions at $\theta \sim \theta_{\mathrm{r}}$ where $z \to 0$.

C. Transitional Result: Neighborhood of Caustic or Rainbow at $(\theta_{\mathrm{r}}, b_{\mathrm{r}}, \lambda_{\mathrm{r}})$

In the vicinity of rainbow angle $\theta \approx \theta_{\mathrm{r}}$,

$$\chi' = \frac{\partial \chi}{\partial \lambda} = \left[2(\theta_{\mathrm{r}} - \theta)\chi''(\lambda_{\mathrm{r}})\right]^{1/2} \quad (45.240)$$

$$z = (\theta_{\mathrm{r}} - \theta)\left[2/\chi''(\lambda_{\mathrm{r}})\right]^{1/3} > 0 \quad (45.241)$$

$$S_{\mathrm{r}} = \frac{1}{2}(S_1 + S_2) = 2\eta(\lambda_{\mathrm{r}}) + \lambda_{\mathrm{r}}\theta_{\mathrm{r}} - \frac{3}{4}\pi\,. \quad (45.242)$$

The scattering amplitude

$$f_{23}(\theta_r) = \left(\frac{2\pi\lambda_{\mathrm{r}}}{k^2 \sin\theta_{\mathrm{r}}}\right)^{1/2} \left(\frac{2}{\chi''(\lambda_{\mathrm{r}})}\right)^{1/3} \mathrm{Ai}(-z)\, e^{iS_{\mathrm{r}}}\,, \quad (45.243)$$

is finite at the rainbow angle θ_r. In (45.239), the $(\theta_{\mathrm{r}} - \theta)^{-1/4}$ divergence in $|\chi_\ell'|^{1/2}$ of (45.240) arising in the constructive interference term $\left(\sigma_2^{1/2} + \sigma_3^{1/2}\right)$ is exactly balanced by the $z^{1/4}$ term of $A(z)$. Also $\left(\sigma_2^{1/2} - \sigma_3^{1/2}\right) \to 0$ in (45.239) more rapidly than $z^{-1/4}$ in $B(z)$ so that (45.239) at θ_{r} is finite and reproduces (45.243).

The uniform semiclassical DCS

$$\frac{d\sigma}{d\Omega} = \left|\sigma_1^{1/2}(\theta)\, e^{iS_1} + \sigma_2^{1/2}(\theta) F_{23}\, e^{iS_2} + \sigma_3^{1/2} F_{23}^{*}\, e^{iS_3}\right|^2 \quad (45.244)$$

contains, in addition to the S_i/S_j interference oscillations in the primitive result (45.234), the θ-variation of the Airy Function $|\mathrm{Ai}(z)|^2$, which has a principal finite (rainbow) maximum at $\theta \le \theta_{\mathrm{r}}$, the classical rainbow angle, and subsidiary maxima (supernumary rainbows) at smaller angles. The DCS decreases exponentially as θ increases past θ_{r} into the classical forbidden region and tends to $\sigma_1(\theta)$ at larger angles. For $\theta_{\mathrm{r}} < \theta < \pi$,

$$f(\theta) = \sigma_1^{1/2}(\theta) \exp(iS_1)\,. \quad (45.245)$$

45.3.5 Diffraction and Glory Amplitudes

Diffraction. Diffraction arises from the outer (attractive) part of the potential. Many contributions arise from the attractive Δ^+ branch appropriate for negative χ at large b where η is small. Here χ, χ' both tend to zero.

Glory. The deflection function χ passes through zero at a finite λ_{g}. A confluence of the two maxima of each phase shift from the positive and negative branches of $\chi(b)$ occurs at $b_1 = b_2 = b_{\mathrm{g}} = \lambda_{\mathrm{g}}/k$. η_{g} is maximum for $\chi = 0$. In general $\chi(b_m) = -2m\pi$ (forward glory); $\chi(b_m) = -(2m-1)\pi$ (backward glory); $m = 0, 1, 2, \ldots$. There is only a forward glory at $\chi = 0$ when the deflection at the rainbow is $|\chi_{\mathrm{r}}| < 2\pi$. In contrast to diffraction, the glory contribution can be calculated by the *stationary phase approximation*.

Transitional Results for Forward and Backward Glories

Forward Glories. Contributions arise from $\chi = \pm\theta$, $-2\pi \pm \theta, \cdots, -2m\pi \pm \theta$ as $\theta \to 0$. The stationary phase points λ_m are located at

$$\chi(\lambda_m) = \chi_m = -2m\pi\,; \quad m \ge 0\,. \quad (45.246)$$

The phase function in the neighborhood of a glory is

$$\eta(\lambda) = \eta_m - m\pi(\lambda - \lambda_m) + \frac{1}{4}\chi_m'(\lambda - \lambda_m)^2\,. \quad (45.247)$$

The $m = 0$ term provides zero deflection χ due to a net balance of attractive and repulsive scattering for a finite impact parameter b_{g} or λ_{g} where $\eta(\lambda)$ attains its maximum value η_m. The glories at θ are due to a confluence of the two contributions from the deflections $\chi_m = -2m\pi \pm \theta$ at the stationary phase points $\lambda_{mn} = \lambda_{m1}$ and λ_{m2}. SP integration of (45.218) with

(45.247) yields the forward glory amplitude

$$f_{FG}=\frac{1}{k}\sum_{n=1}^{2}\sum_{m=0}^{\infty}\lambda_{mn}\left(\frac{2\pi}{|\chi'_m|}\right)^{1/2}\times J_0(\lambda_{mn}\theta)\,\mathrm{e}^{\mathrm{i}S^{(g)}_{mn}}\,, \quad (45.248)$$

$$S^{(g)}_{mn}=2\eta(\lambda_{mn})+m\pi(\lambda_{mn}-1)-\frac{3}{4}\pi\,. \quad (45.249)$$

Backward Glories. Contributions arising from $\chi=-\pi\pm\alpha,\ -3\pi\pm\alpha,\ \cdots,\ -(2m-1)\pi\pm\alpha$ coalesce as $\alpha\equiv\pi-\theta\to 0$. The phase function near a backward glory is

$$\eta(\lambda)=\eta(\lambda_m)+\frac{1}{2}\chi_m(\lambda-\lambda_m)+\frac{1}{4}\chi'_m(\lambda-\lambda_m)^2\,. \quad (45.250)$$

The $m=0$ term provides the normal backward amplitude due to head-on ($b=0$) repulsive collisions. $m>0$ terms provide contributions from attractive collisions for which there are two points λ_{mn} of stationary phase for each m in $\chi_m=-(2m-1)\pi\pm\alpha$.

The backward glory amplitude at $\theta=\pi-\alpha$ is

$$f_{BG}=\frac{1}{k}\sum_{n=1}^{2}\sum_{m=0}^{\infty}\lambda_{mn}\left(\frac{2\pi}{|\chi'_{mn}|}\right)^{1/2}\times J_0(\lambda_{mn}\alpha)\,\mathrm{e}^{\mathrm{i}S^{(g)}_{mn}}\,, \quad (45.251)$$

$$S^{(g)}_{mn}=2\eta(\lambda_{mn})+\pi(2m-1)\left(\lambda_{mn}-\frac{1}{2}\right)-\frac{3}{4}\pi\,. \quad (45.252)$$

In contrast to the Bessel amplitudes (below), these transitional formulae do not uniformly connect with the primitive semiclassical results for (f_1+f_2) away from the critical glory angles.

Uniform Bessel Amplitude for Glory Scattering

The combined contributions from $\chi_1=-N\pi+\theta$ and $\chi_2=-N\pi-\theta$, where $N=2m$, for forward and $N=2m-1$ for backward scattering, yield [45.9]

$$f_G(\theta)=\frac{\alpha_j}{2i}\,\mathrm{e}^{-\mathrm{i}N\pi/2}\left(\pi S^{(C)}_{21}\right)^{1/2}\exp\left[\mathrm{i}\bar{S}^{(C)}(\theta)\right]\times\left[(\sigma_1^{1/2}+\sigma_2^{1/2})J_0\left(\frac{1}{2}S^{(C)}_{21}\right)-\mathrm{i}(\sigma_1^{1/2}-\sigma_2^{1/2})J_1\left(\frac{1}{2}S^{(C)}_{21}\right)\right]\,, \quad (45.253)$$

where $S^{(C)}_{21}(\theta)=S^{(C)}_2-S^{(C)}_1$ is the difference of the collision actions (45.231a),

$$S^{(C)}_i(\theta)=2\eta(\lambda_i)-\lambda_i\chi_i\,,\quad i=1,2\,, \quad (45.254)$$

with mean

$$\bar{S}^{(C)}_{21}(\theta)=\frac{1}{2}\left(S^{(C)}_2+S^{(C)}_1\right)\,, \quad (45.255)$$

and phases

$$\alpha_j=\mathrm{e}^{\pm\mathrm{i}\pi/4};\quad (+),\ \chi'_j>0;\quad (-),\ \chi'_j<0\,, \quad (45.256)$$

and the ordinary Bessel functions $J_n(Z)$ satisfy the relationships $J_1(z)=-J'_0(z)$, $J_1(-z)=-J_1(z)$. This formula, valid for both forward ($\theta\sim 0$) and backward ($\theta\sim\pi$) glories, does uniformly connect the primitive result for (f_1+f_2), valid when $S^{(C)}_{21}\gg 1$ to the transitional results (45.248) and (45.251), valid only in the vicinity of the glories.

45.3.6 Small-Angle (Diffraction) Scattering

Diffraction originates from scattering in the forward direction by the long-range attractive tail of $V(R)$ where χ,χ' and $\eta\to 0$. The main contributions to (45.222a) arise from a large number of small $\eta(\lambda)$ at large λ. The Jeffrey–Born phase function (45.173) can therefore be used in (45.222b) for $f(\theta)$ and in (45.45) and (45.47), respectively, for $\sigma(E)$. A finite forward diffraction peak as $\theta\to 0$ is obtained for $f(\theta)$ in contrast to the classical infinite result.

Integral Cross Sections

For $V(R)=-C/R^n$, the Landau–Lifshitz (LL) and Massey–Mohr (MM) cross sections are [cf. (45.346)]

$$\sigma_{LL}(E)=\pi\left(\frac{2CF(n)}{(n-1)\hbar v}\right)^{2/(n-1)}\times\pi\left[\sin\left(\frac{\pi}{n-1}\right)\Gamma\left(\frac{2}{n-1}\right)\right]^{-1}\,, \quad (45.257)$$

$$\sigma_{MM}(E)=\pi\left(\frac{2CF(n)}{(n-1)\hbar v}\right)^{2/(n-1)}\left(\frac{2n-3}{n-2}\right)\,, \quad (45.258)$$

where $F(n)=\sqrt{\pi}\,\Gamma\left(\frac{1}{2}n+\frac{1}{2}\right)/\Gamma\left(\frac{1}{2}n\right)$ and v is the relative speed. For σ_{MM}, the phases are $\eta(\lambda)=\frac{1}{2}$ $(0<\lambda<\lambda_0)$ and $\eta(\lambda)=\eta_{JB}$ $(\lambda>\lambda_0)$. For σ_{LL}, phases are η_{JB}

for all λ. Both σ_{LL} and σ_{MM} have the general form

$$\sigma_D(E) = \gamma \left(\frac{C}{\hbar v}\right)^{2/(n-1)} . \quad (45.259)$$

Ion–Atom Collisions. For $n = 4$ attraction at low energy, $\sigma_D \sim v^{-2/3}$. $\gamma_{LL} = 11.373$, $\gamma_{MM} = 10.613$. For $n = 12$ repulsion at high energy, $\sigma_D \sim v^{-2/11}$. $\gamma_{LL} = 6.584$, $\gamma_{MM} = 6.296$.

Atom–Atom Collisions. For $n = 6$ (attraction), $\sigma_D \sim v^{-2/5}$, $\gamma_{LL} = 8.083$, $\gamma_{MM} = 7.547$ Fig. 45.1.

Exact numerical calculations favor σ_{LL} over σ_{MM} ([45.10], pp. 1325 for details).

Differential Cross Section

$$\frac{d\sigma}{d\Omega} = f_i^2(\theta) + f_r^2(\theta) , \quad (45.260a)$$

$$f_i = \frac{2}{k}\int_0^\infty \lambda \sin^2 \eta(\lambda)\left(1 - \frac{1}{4}\lambda^2\theta^2\right) d\lambda \quad (45.260b)$$

$$= \frac{k\sigma_D(E)}{4\pi}\left[1 - \left(\frac{k^2\sigma_D}{16\pi}\right) g_1(n)\theta^2\right] , \quad (45.260c)$$

$$f_r = \frac{1}{k}\int_0^\infty \lambda \sin 2\eta(\lambda)\left(1 - \frac{1}{4}\lambda^2\theta^2\right) d\lambda \quad (45.260d)$$

$$= \frac{k\sigma_D(E)}{4\pi}\left[1 - \left(\frac{k^2\sigma_D}{16\pi}\right) g_2(n)\theta^2\right] \times \tan\left(\frac{\pi}{n-1}\right) , \quad (45.260e)$$

where σ_D is given by (45.259), and

$$g_j(n) = \pi^{-1} \tan\left(\frac{j\pi}{n-1}\right) \frac{\{\Gamma[2/(n-1)]\}^2}{\Gamma[4/(n-1)]} . \quad (45.261)$$

The optical theorem (45.62) is satisfied, and

$$f_D(\theta \sim 0) = \sigma_D^{1/2}(E)\, e^{iS_D(n)} , \quad (45.262)$$

where the (energy-independent) phase is

$$S_D(n) = \frac{\pi(n-3)}{2(n-1)} . \quad (45.263)$$

45.3.7 Small-Angle (Glory) Scattering

Amplitude and Cross Section. The other contribution to forward scattering is the forward glory, which originates from the combined null effect of attraction and repulsion at a specified glory impact parameter $b_g = \lambda_g/k$, where $\eta(\lambda)$ attains a maximum value of η_g. The $m = 0$ term of (45.248) yields

$$f_G(\theta) = \sigma_G^{1/2}(\theta) \exp[iS_G(E)] , \quad (45.264a)$$

$$\sigma_G(\theta) = \frac{\lambda_g^2}{k^2}\left(\frac{2\pi}{|\chi_g'|}\right) J_0^2(\lambda_g\theta) , \quad (45.264b)$$

$$S_G(E) = 2\eta_g(E) - \frac{3}{4}\pi , \quad (45.264c)$$

where $J_0^2(x) \sim 1 - \frac{1}{4}x^2 + \cdots$. The classical result (45.30) is recovered by averaging (45.264b) over several oscillations with $\langle J_0^2(x)\rangle = (\pi x)^{-1}$.

Diffraction-Glory Oscillations.

$$\sigma(E) = \frac{4\pi}{k} \operatorname{Im}[f_D(0) + f_G(0)] \quad (45.265a)$$

$$= \sigma_D(E) + \Delta\sigma_G(E) , \quad (45.265b)$$

where the diffraction cross section is (45.259), and where

$$\Delta\sigma_G(E) = \frac{4\pi}{k^2}\lambda_g \left(\frac{2\pi}{|\chi_g'|}\right)^{1/2} \sin\left(2\eta_g(E) - \frac{3}{4}\pi\right) \quad (45.266)$$

oscillates with E.

For sufficiently deep attractive wells, the phase shift η_g successively decreases with increasing E

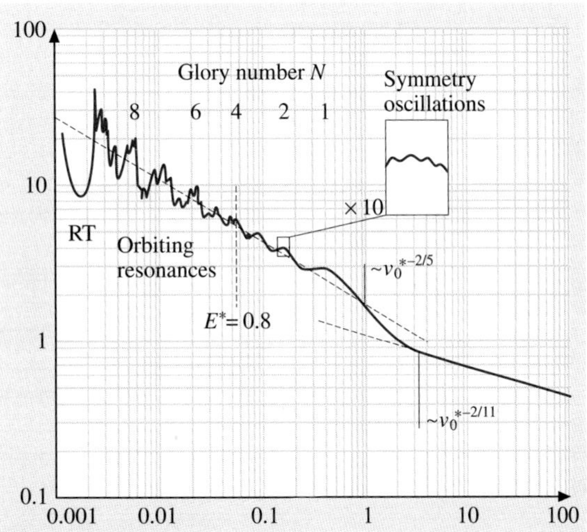

Fig. 45.1 Illustration of all the various oscillatory effects for elastic scattering by a Lennard–Jones (12,6) potential of well depth ϵ and equilibrium distance R_e. Ordinate $\sigma^* = \sigma/(2\pi R_e^2)$, abscissa $v^* = \hbar v/(\epsilon R_e)$

through a series of multiples of $\pi/2$. Writing $\eta_g(E) = \pi(N - 3/8)$, maxima appear at $N = 1, 2, \ldots$, and minima at $N = 3/2, 5/2, 7/2, \ldots$. The glories are indexed by N in order of appearance, starting at high energies. $\eta_g(E \to 0)$ is related to the number n of bound states by Levinson's theorem: $\eta_0(E \to 0) = (n + 1/2)\pi$. Diffraction-glory oscillations also occur in the DCS at a frequency governed entirely by the energy variation of $\eta_g(E)$ and n of (45.263).

JWKB Formulae for Shape Resonances and Tunneling Predissociation

For the three classical turning points $R_1 < R_2 < R_3$ at energies E below the orbiting threshold V_{max} at R_X, the JWKB phase shift

$$\eta_\ell = \left[\eta_\ell^{(0)} - \frac{1}{2}\phi(\gamma_\ell)\right] + \eta_\ell^{(r)} \tag{45.267}$$

is composed of (a) the phase shift

$$\eta_\ell^{(0)} = \lim_{R\to\infty}\left[\int_{R_3}^{\infty} k(R)\,\mathrm{d}R - kR + \frac{1}{2}(\ell + \frac{1}{2})\pi\right] \tag{45.268}$$

appropriate to one turning point at R_3, (b) a contribution $\eta^{(r)}$ arising from the region between the two inner turning points R_2 and R_3 due to penetration of the centrifugal barrier and given by

$$\tan\eta_\ell^{(r)}(E) = \left(\frac{[1+\exp(-2\gamma_\ell)]^{1/2} - 1}{[1+\exp(-2\gamma_\ell)]^{1/2} + 1}\right) \times \tan\left(\alpha_\ell - \frac{1}{2}\phi_\ell\right), \tag{45.269}$$

and (c) a phase correction factor

$$\phi_\ell(\gamma_\ell) = \arg\Gamma\left(\frac{1}{2} + \mathrm{i}\epsilon\right) - \epsilon\ln|\epsilon| + \epsilon, \tag{45.270}$$

where $\epsilon = -\gamma_\ell/\pi$. The radial action $J_R(E)$ is $2\hbar\alpha_\ell(E)$. For motion within the potential well α_ℓ is

$$\alpha_\ell(E) = \int_{R_1}^{R_2} k(R)\,\mathrm{d}R, \tag{45.271}$$

and is

$$\gamma_\ell(E < E_{max}) = \int_{R_2}^{R_3} |k(R)|\,\mathrm{d}R \tag{45.272}$$

in the classically forbidden region of the potential hump.

The above expressions also hold for energies $E > V_{max}$, except that (45.272) is replaced by

$$\gamma_\ell(E > E_{max}) = -\mathrm{i}\int_{R_-}^{R_+} k(R)\,\mathrm{d}R, \tag{45.273}$$

where $R_\pm$ are the complex roots of $k_\ell(R) = 0$. For the quadratic form

$$V(R) = V_{max} - \frac{1}{2}M\omega_*^2(R - R_{max})^2, \tag{45.274}$$

appropriate in the vicinity of the potential hump, γ for both cases reduces to

$$\gamma = \pi(V_{max} - E)/\hbar\omega_* . \tag{45.275}$$

The deflection function $\chi_\ell = 2(\partial\eta_\ell/\partial\ell)$ no longer diverges at the orbiting angular momentum ℓ_0 or impact parameter b_0. The singularities in η_ℓ of (45.51) are exactly canceled by $-\frac{1}{2}(\partial\phi/\partial\ell)$ in (45.270).

Limiting cases:

(a) $E \gg V_{max}$. Then $\gamma_\ell \to -\infty$ and $\phi \to -(\pi/24\gamma_\ell) \to 0$, so that $\eta_\ell^{(r)} \to \alpha_\ell$ and η_ℓ reduces to the single turning point result (45.268) with $R_3 = R_1$.

(b) $E \ll V_{max}$. Then $\gamma_\ell \gg 1$ and

$$\eta_\ell^{(r)}(E) = \tan^{-1}\left[\frac{1}{2}\mathrm{e}^{-2\gamma_\ell}\tan\left(\alpha_\ell - \frac{1}{2}\phi_\ell\right)\right], \tag{45.276}$$

which remains negligible except for those energies E close to quasibound energy levels $E_{n\ell}$ determined via the Bohr quantization condition

$$\alpha_\ell(E) - \frac{1}{2}\phi_\ell(E) = \left(n + \frac{1}{2}\right)\pi . \tag{45.277}$$

As E increases past each $E_{n\ell}$, $\eta_\ell^{(r)}$ increases rapidly by π. Since $(\partial J/\partial E)_{n\ell} = \nu_{n\ell}^{-1} = 2\pi/\omega_{n\ell}$, the time period for radial oscillation within the potential barrier, the level spacing is $\hbar\omega_{n\ell} = h\nu_{n\ell} = \pi(\partial E/\partial\alpha)_{n\ell}$.

Shape Resonance. In the neighborhood of $E_{n\ell} \sim E$,

$$\alpha_\ell(E) = \alpha_{n\ell}(E_{n\ell}) + \left(\frac{\pi}{\hbar\omega_{n\ell}}\right)(E - E_{n\ell}), \tag{45.278}$$

and, under the assumption that the energy variation of ϕ_ℓ can be neglected, (valid for E not close to V_{max}), then η_ℓ reduces to the Breit–Wigner form

$$\eta_\ell(E) = \eta_\ell^{(0)}(E) + \tan^{-1}\left(\frac{\Gamma_{n\ell}/2}{E_{n\ell} - E}\right), \tag{45.279}$$

with resonance width

$$\Gamma_{n\ell} = 2\left(\frac{\hbar\omega_{n\ell}}{\pi}\right)\frac{\left[1+\exp(-2\gamma_{n\ell})\right]^{1/2}-1}{\left[1+\exp(-2\gamma_{n\ell})\right]^{1/2}+1}, \tag{45.280}$$

where $\gamma_{n\ell} = \gamma(E_{n\ell})$. The partial cross sections are then determined by (45.166a,b) with $\eta_\ell^{(0)}$ replaced by $\eta_\ell^{(0)} - \frac{1}{2}\phi(\gamma)$ of (45.268).

Gamow's Result. For $E \ll V_{\max}$, $\gamma_{n\ell} \gg 1$, then

$$\Gamma_{n\ell} \xrightarrow{\gamma\gg1} \left(\frac{\hbar\omega_{n\ell}}{2\pi}\right)\exp(-2\gamma_{n\ell}) . \tag{45.281}$$

The probabilities of transmission through and reflection from a barrier for unit incident flux from the left are:

Transmission Probability:

$$T = \left(1+e^{2\gamma}\right)^{-1} \xrightarrow{\gamma\gg1} e^{-2\gamma} . \tag{45.282}$$

Reflection Probability:

$$R = \left(1+e^{-2\gamma}\right)^{-1} \xrightarrow{\gamma\gg1} 1 . \tag{45.283}$$

Frequency of leakage:

$$\nu_T = \Gamma_{n\ell}/\hbar = \left(\frac{\omega_{n\ell}}{2\pi}\right) e^{-2\gamma} . \tag{45.284}$$

45.3.8 Oscillations in Elastic Scattering

Figure 45.1 is an illustration [45.11] of all the various oscillatory structure and effects – Ramsauer–Townsend minimum (Sect. 45.2.4), orbiting resonances (45.340), diffraction-glory oscillations (45.265b) and symmetry oscillations (45.82) – for elastic scattering by a Lennard–Jones (12, 6) potential. Note the shift of velocity dependence from $v^{-2/5}$ at low v to $v^{-2/11}$ at high v. $\sigma = 2\pi R_e^2$ is the averaged cross section $2\pi b_0^2$ in (45.47) at $b_0 = R_e$. The region $\sigma^* > 1$ probes the attractive part of the potential at low speeds and $\sigma^* < 1$ probes the repulsive part at high speeds. The four distinct types of structure originate from nonrandom behavior of $\sin^2\eta$ in (45.45). Orbiting trajectories exist for $E < 0.8\epsilon$ (Sect. 45.5).

45.4 Elastic Scattering in Reactive Systems

All nonelastic processes (e.g. inelastic scattering and rearrangement collisions/chemical reactions) can be viewed as a net absorption from the incident beam current vector $\boldsymbol{J}$ and modeled by a complex optical potential

$$V(R) = V_r(R) + iV_i(R) . \tag{45.285}$$

The continuity equation is then

$$\nabla\cdot\boldsymbol{J} = -\frac{2}{\hbar}V_i(R)|\Psi(\boldsymbol{R})|^2 , \tag{45.286}$$

so that particle absorption implies $V_i > 0$ and particle creation $V_i < 0$. Since particle conservation implies $|S_\ell|^2 = 1$, the phase shift

$$\delta_\ell(k) = \eta_\ell(k) + i\gamma_\ell(k) \tag{45.287}$$

is also complex since then

$$S_\ell = A_\ell(k)\exp(2i\eta_\ell) , \tag{45.288}$$

where the absorption (inelasticity) factor is

$$A_\ell = \exp(-2\gamma_\ell) \le 1 . \tag{45.289}$$

45.4.1 Quantal Elastic, Absorption and Total Cross Sections

$$f_{el}(\theta) = \frac{1}{2ik}\sum_{\ell=0}^{\infty}(2\ell+1)\left(A_\ell e^{2i\eta_\ell}-1\right)P_\ell(\cos\theta) , \tag{45.290a}$$

$$\sigma_{el}(k) = \frac{\pi}{k^2}\sum_{\ell=0}^{\infty}(2\ell+1)|A_\ell e^{2i\eta_\ell}-1|^2 , \tag{45.290b}$$

$$\sigma_{abs}(k) = \frac{\pi}{k^2}\sum_{\ell=0}^{\infty}(2\ell+1)\left(1-A_\ell^2\right) , \tag{45.290c}$$

$$\sigma_{tot}(k) = \sigma_{el}(k)+\sigma_{abs}(k) = \frac{2\pi}{k^2}\sum_{\ell=0}^{\infty}(2\ell+1)(1-A_\ell\cos 2\eta_\ell) . \tag{45.290d}$$

Upper limits to the partial cross sections are

$$\sigma_\ell^{el} \le \frac{4\pi}{k^2}(2\ell+1), \quad \sigma_\ell^{abs} \le \frac{\pi}{k^2}(2\ell+1) , \tag{45.291a}$$

$$\sigma_\ell^{\text{tot}} \le \frac{4\pi}{k^2}(2\ell+1) = \frac{4\pi}{k}\,\text{Im}\left[f_\ell^{\text{el}}(\theta=0)\right]. \tag{45.291b}$$

For pure elastic scattering with no absorption, $A_\ell = 1$. All nonelastic processes ($0 \le A_\ell < 1$) are always accompanied by elastic scattering, even in the ($A_\ell = 0$) limit of full absorption.

Eikonal Formulae for Forward Reactive Scattering.

$$f_{\text{el}}(\theta) = -\text{i}k\int_0^\infty \left(\text{e}^{2\text{i}\delta} - 1\right) J_0(2kb\sin\frac{1}{2}\theta) b\,\text{d}b\,, \tag{45.292a}$$

$$\sigma_{\text{el}}(k) = 2\pi\int_0^\infty |\left(1 - \text{e}^{-2\gamma}\,\text{e}^{2\text{i}\eta}\right)|^2 b\,\text{d}b\,, \tag{45.292b}$$

$$\sigma_{\text{abs}}(k) = 2\pi\int_0^\infty \left(1 - e^{-4\gamma}\right) b\,\text{d}b\,, \tag{45.292c}$$

$$\sigma_{\text{tot}}(k) = 4\pi\int_0^\infty \left(1 - \text{e}^{-2\gamma}\cos 2\eta\right) b\,\text{d}b\,, \tag{45.292d}$$

where the phase shift function $\delta = \eta + \text{i}\gamma$ at impact parameter b can be either the Jeffrey–Born phase

$$\delta_{\text{JB}}(b) = -\frac{1}{2k}\int_b^\infty \frac{U(R)\,\text{d}R}{\left(1 - b^2/R^2\right)^{1/2}}\,, \tag{45.293}$$

where $kb = \lambda = (\ell + 1/2)$, or the eikonal phase

$$\delta_{\text{E}}(b) = -\frac{1}{4k}\int_{-\infty}^\infty U(b, Z)\,\text{d}Z\,, \tag{45.294}$$

where the reduced interaction is $U = \left(2m/\hbar^2\right)V$.

Fraunhofer Diffraction by a Black Sphere. For a complex spherical well U

$$U = \begin{cases} U_{\text{r}} + \text{i}U_i, & R < a \\ 0, & R > a\,. \end{cases} \tag{45.295}$$

The eikonal phase function (45.294) is

$$\delta(b) = \begin{cases} (U/2k)\left(a^2 - b^2\right)^{1/2}, & 0 \le b \le a \\ 0 & b > a\,. \end{cases} \tag{45.296}$$

The absorption factor is

$$A(b)^2 \equiv \text{e}^{-4\gamma} = \exp\left[-2\left(a^2 - b^2\right)^{1/2}/\lambda\right], \tag{45.297}$$

where $\lambda = k/U_i$ is the mean free path towards absorption. For strong absorption, $\lambda \ll a$, so that

$$f_{\text{el}}(\theta) = \text{i}k\int_0^a J_0(2kb\sin\frac{1}{2}\theta) b\,\text{d}b\,, \tag{45.298}$$

$$\frac{\text{d}\sigma_{\text{el}}}{\text{d}\Omega} = (ka)^2\left(\frac{J_1\left(2ka\sin\frac{1}{2}\theta\right)}{2ka\sin\frac{1}{2}\theta}\right)^2 a^2\,, \tag{45.299}$$

which has a diffraction shaped peak of width $\sim\theta \le (ka)^{-1}$ about the forward direction, and

$$\sigma_{\text{tot}} = \frac{4\pi}{k}\,\text{Im}\left[f_{\text{el}}(\theta = 0)\right] = 2\pi a^2 \tag{45.300}$$

is composed of πa^2 for classical absorption and πa^2 for edge diffraction or shadow (nonclassical) elastic scattering. This result also holds for the perfectly reflecting sphere (πa^2 for classical elastic and πa^2 for edge diffraction).

45.5 Results for Model Potentials

Exact results for various quantities in classical, quantal, and semiclassical elastic scattering are obtained for the model potentials (a)–(s) in Table 45.1.

Classical Deflection Functions for Model Potentials

(a) Hard Sphere.

$$\theta(b; E) = \chi = \begin{cases} \pi - 2\sin^{-1} b/a, & b \le a\,; \\ 0, & b > a\,. \end{cases} \tag{45.301}$$

$$b(\theta) = a\cos\frac{1}{2}\theta\,, \tag{45.302}$$

$$\sigma(\theta) = \frac{\text{d}\sigma}{\text{d}\Omega} = \frac{1}{4}a^2; \quad \text{isotropic}, \tag{45.303}$$

$$\sigma = \pi a^2 = \text{geometric cross section}; \tag{45.304}$$

θ, $\sigma(\theta)$ and σ are all independent of energy E.

(b) Potential Barrier. For $E < V_0$, classical scattering is the same as for hard sphere reflection as given by (45.301–45.304). For $E > V_0$ and $\theta = \chi$, define

$n^2 = 1 - V_0/E$, $b_0 = na$. Then

$$\theta(b) = \begin{cases} 2\left[\sin^{-1}(b/na) - \sin^{-1}(b/a)\right] & 0 \le b \le b_0 \\ \pi - 2\sin^{-1}(b/a)\,, & b_0 \le b \le a \end{cases} \quad (45.305)$$

and $\theta_{\max} = 2\cos^{-1} n$. For a given θ, the two impact parameters which contribute are

$$b_1(\theta) = \frac{an\sin\frac{1}{2}\theta}{\left(1 - 2n\cos\frac{1}{2}\theta + n^2\right)^{1/2}}\,, \quad 0 < b_1 \le b_0 \quad (45.306)$$

$$b_2(\theta) = a\cos\frac{1}{2}\theta\,, \quad b_0 < b_2 \le a \quad (45.307)$$

For $0 \le \theta \le \theta_{\max}$,

$$\frac{d\sigma}{d\Omega} = \frac{1}{4}a^2 + \frac{a^2n^2\left(n\cos\frac{1}{2}\theta - 1\right)\left(n - \cos\frac{1}{2}\theta\right)}{4\cos\frac{1}{2}\theta\left(1 + n^2 - 2\cos\frac{1}{2}\theta\right)}\,, \quad (45.308)$$

and $d\sigma/d\Omega = 0$ for $\theta_{\max} \le \theta \le \pi$. Finally,

$$\sigma = \int_{\theta=0}^{\theta_{\max}} \left(\frac{d\sigma}{d\Omega}\right) d\Omega = \pi a^2\,. \quad (45.309)$$

(c) Potential Well. Results are similar to the potential barrier case above, except that there is only a single scattering trajectory with $\theta = -\chi$, and $n = (1 + V_0/E)^{1/2}$ is the effective index of refraction for the equivalent problem in geometrical optics. Refraction occurs on entering

Table 45.1 Model interaction potentials

	Potential	$V(R)$
(a)	Hard sphere	$\infty,\ R \le a;\ 0,\ R > a$
(b)	Barrier	$V_0,\ R \le a;\ 0,\ R > a$
(c)	Well	$-V_0,\ R \le a;\ 0,\ R > a$
(d)	Coulomb ($\pm$)	$\pm k/R$
(e)	Finite-range Coulomb	$-k/R + k/R_s\ R \le R_s;\ 0,\ R > R_s$
(f)	Pure dipole	$\pm\alpha/R^2$
(g)	Finite-range dipole	$\pm\alpha\left(\frac{1}{R^2} - \frac{1}{a^2}\right),\ R \le a;\ 0,\ R > a$
(h)	Dipole + hard sphere	$\pm\alpha/R^2,\ R \le a;\ 0,\ R > a$
(i)	Power law attractive	$-C/R^n,\ (n > 2)$
(j)	Fixed dipole + polarization	$-\frac{De\cos\theta_d}{R^2} - \frac{\alpha_d e^2}{2R^4}$
(k)	Fixed dipole + Coulomb	$-\frac{De\cos\theta_d}{R^2} + \frac{e^2}{R}$
(l)	Lennard- Jones (n, 6)	$\frac{\epsilon n}{n-6}\left[\frac{6}{n}\left(\frac{R_e}{R}\right)^n - \left(\frac{R_e}{R}\right)^6\right]$
(m)	Polarization (n, 4)	$\frac{\epsilon n}{n-4}\left[\frac{4}{n}\left(\frac{R_e}{R}\right)^n - \left(\frac{R_e}{R}\right)^4\right]$
(n)	Multiple-term power law	$\frac{C_m}{R^m} - \frac{C_n}{R^n} = V_m(R) - V_n(R)$
(o)	Exponential	$V_0\exp(-\alpha R)$
(p)	Screened Coulomb	$V_0\exp(-\alpha R)/R$
(q)	Morse	$\epsilon\left[e^{2\beta(R_e-R)} - 2e^{\beta(R_e-R)}\right]$
(r)	Gaussian	$V_0\exp(-\alpha^2R^2)$
(s)	Polarization finite	$-V_0/(R^2 + R_0^2)^2$

and exiting the well. Then

$$\theta(b) = -2\left[\sin^{-1}(b/na) - \sin^{-1}(b/a)\right], \tag{45.310}$$

$$\theta(b=a) = \theta_{\max} = 2\cos^{-1}(1/n), \tag{45.311}$$

$$b(\theta) = \frac{-an\sin\frac{1}{2}\theta}{\left(1-2n\cos\frac{1}{2}\theta+n^2\right)^{1/2}}, \tag{45.312}$$

$$\frac{\mathrm{d}\sigma}{\mathrm{d}\Omega} = \frac{a^2n^2\left(n\cos\frac{1}{2}\theta-1\right)\left(n-\cos\frac{1}{2}\theta\right)}{4\cos\frac{1}{2}\theta\left(n^2+1-2n\cos\frac{1}{2}\theta\right)^2}, \tag{45.313}$$

$$\sigma = \pi a^2 \tag{45.314}$$

(d) Rutherford or Coulomb.

$$\theta(b,E) = |\chi| = 2\csc^{-1}\left[1+(2bE/k)^2\right]^{1/2}, \tag{45.315}$$

$$b(\theta,E) = (k/2E)\cot\frac{1}{2}\theta, \tag{45.316}$$

$$\sigma(\theta) = \frac{\mathrm{d}\sigma}{\mathrm{d}\Omega} = \left(\frac{k}{4E}\right)^2\csc^4\frac{1}{2}\theta. \tag{45.317}$$

(e) Finite Range Coulomb.

$$R_0(E) = \frac{k}{2E}, \quad \alpha(E) = R_0(E)/R_s,$$

$$\frac{\mathrm{d}\sigma}{\mathrm{d}\Omega} = \frac{R_0^2}{4}\left(\frac{1+\alpha}{\alpha^2+(1+2\alpha)\sin^2\frac{1}{2}\theta}\right)^2. \tag{45.318}$$

(f) Pure Dipole. $R_0^2(E) = \alpha/E$.

Repulsion (+): $\chi > 0$, $\chi = \theta$,

$$b^2(\chi) = R_0^2\left[\left(\frac{1}{\chi}+\frac{1}{2\pi-\chi}\right)\frac{\pi}{2}-1\right], \tag{45.319}$$

$$\frac{\mathrm{d}\sigma}{\mathrm{d}\Omega} = \frac{\pi R_0^2}{4\sin\theta}\left|\frac{1}{\theta^2}-\frac{1}{(2\pi-\theta)^2}\right|. \tag{45.320}$$

Attraction (−): $\chi < 0$.

$$b^2(\chi) = R_0^2\left[\left(\frac{1}{|\chi|}-\frac{1}{|\chi|+2\pi}\right)\frac{\pi}{2}+1\right]. \tag{45.321}$$

There is an infinite number of (negative) deflections $\chi = \chi_n^{\pm}$ associated with a given scattering angle θ:

$$|\chi_n^+| = 2\pi n+\theta, \quad n = 0,1,2,\ldots, \tag{45.322a}$$

$$|\chi_n^-| = 2\pi n-\theta, \quad n = 1,2,3,\ldots. \tag{45.322b}$$

The infinite sum over contributions from $b_n^{\pm} = b(\chi_n^{\pm})$ for the attractive dipole yields

$$\frac{\mathrm{d}\sigma}{\mathrm{d}\Omega} = \frac{\pi R_0^2}{4\sin\theta}\left|\frac{1}{\theta^2}+\frac{1}{(2\pi-\theta)^2}\right|. \tag{45.323}$$

(g) Finite Range Dipole Scattering. $R_0^2 = \alpha/E$, $\left(R_c^{\pm}\right)^2 = b^2 \pm R_0^2$, $\left(b_0^{\pm}\right)^2 = a^2 \pm R_0^2$.

Repulsion (+): for $b \le a$,

$$\chi(b) = \frac{\pi\left(R_c^+-b\right)}{R_c^+}+\frac{2b}{R_c^+}\sin^{-1}\left(\frac{R_c^+}{b_0^+}\right) - 2\sin^{-1}\left(\frac{b}{a}\right), \tag{45.324}$$

$$\chi(0) = \pi, \quad \chi(b\ge a) = 0, \quad \sigma = \pi a^2.$$

Attraction (−): for $b > R_0$,

$$\chi(b) = \frac{\pi\left(R_c^--b\right)}{R_c^-}+\frac{2b}{R_c^-}\sin^{-1}\left(\frac{R_c^-}{b_0^-}\right) - 2\sin^{-1}\left(\frac{b}{a}\right), \tag{45.325}$$

$$\chi(R_0)\to\infty, \quad \chi(b\ge a) = 0, \quad \sigma = \pi a^2.$$

(h) Dipole + Hard Sphere Scattering. $R_0^2 = \alpha/E$, $(R_c^{\pm})^2 = b^2 \pm R_0^2$, $(b_0^{\pm})^2 = a^2 \pm R_0^2$.

Repulsion (+): for $0 \le b \le b_0$,

$$\chi(b) = \frac{\pi\left(R_c^+-b\right)}{R_c^+}+\frac{2b}{R_c^+}\sin^{-1}\left(\frac{R_c^+}{a}\right) - 2\sin^{-1}\left(\frac{b}{a}\right), \tag{45.326}$$

$$b_0 \le b \le a; \quad \chi(b) = \pi - 2\sin^{-1}(b/a), \tag{45.327}$$

$$\chi(0) = \pi, \quad \chi(b\ge a) = 0, \quad \sigma = \pi a^2. \tag{45.328}$$

Attraction (−): for $b > R_0$,

$$\chi(b) = \frac{\pi(R_c^--b)}{R_c^-}+\frac{2b}{R_c^-}\sin^{-1}\left(\frac{R_c^-}{a}\right) - 2\sin^{-1}\left(\frac{b}{a}\right), \tag{45.329}$$

$\chi(b) = \chi_{\min}$ at $b = a$,

$$\chi(0) = \pi, \quad \chi(b\ge a) = 0, \quad \sigma = \pi a^2.$$

Orbiting or Spiraling Collisions

From Sect. 45.1.7, the parameters are

Orbiting radius: R_0.
Focusing factor: $F = [1 - V(R_0)/E]$.
Orbiting cross section: $\sigma_{\text{orb}} = \pi R_0^2 F$.

(i) Attractive Power Law Potentials.

$$V_{\text{eff}}(R_0) = (1 - \frac{1}{2}n)V(R_0), \quad n > 2,$$

$$R_0(E) = \left(\frac{(n-2)C}{2E}\right)^{1/n}, \quad F = \left(\frac{n}{(n-2)}\right), \tag{45.330}$$

$$\sigma_{\text{orb}}(E) = \pi\left(\frac{n}{(n-2)}\right)\left(\frac{(n-2)C}{2E}\right)^{2/n}. \tag{45.331}$$

For the case $n = 4$ with $V(R) = -\alpha_d e^2/2R^4$, this gives the Langevin cross section

$$\sigma_L(E) = 2\pi R_0^2 = 2\pi\left(\frac{\alpha_d e^2}{2E}\right)^{1/2} \tag{45.332}$$

for orbiting collisions, and the Langevin rate

$$k_L = v\sigma_L(E) = 2\pi\left(\alpha_d e^2/M\right)^{1/2}, \tag{45.333}$$

which is independent of E.

The case $n = 6$ with $V(R) = -C/R^6$ is the van der Waals potential for which

$$\sigma_{\text{orb}}(E) = \frac{3}{2}\pi R_0^2 = \frac{3}{2}\pi\,(2C/E)^{1/3}. \tag{45.334}$$

(j) Fixed Dipole plus Polarization Potential.

$$R_0^2(E) = \left(\frac{\alpha_d e^2}{2E}\right)^{1/2}, \tag{45.335}$$

$$\sigma_{\text{orb}}(E) = 2\pi\left(\frac{\alpha_d e^2}{2E}\right)^{1/2} + \left(\frac{\pi De\cos\theta_d}{E}\right). \tag{45.336}$$

For a locked-in dipole, the orientation angle is $\theta_d = 0$, and $\sigma_{\text{orb}}(E) > 0$ for all θ_d when $E > E_c = (D^2/2\alpha_d)$.

On averaging over all θ_d from 0 to $\theta_{\max} = \left[\frac{1}{2}\pi + \sin^{-1}\left(2ER_0^2/De\right)\right]$, which satisfies $\sigma_{\text{orb}}(E) > 0$ for all E, then

$$\langle\sigma_{\text{orb}}(E)\rangle_{\theta_d} = \pi\left[\left(\frac{\alpha_d e^2}{2E}\right)^{1/2} + \left(\frac{\alpha_d e^2}{2E_c}\right)^{1/2}\right] + \frac{\pi De}{4E}\left(1 - \frac{E}{E_c}\right) \tag{45.337a}$$

$$\to \sigma_L(E) \text{as} E \to E_c. \tag{45.337b}$$

(k) Fixed Dipole + Coulomb Repulsion.

$$R_0^2(E) = e^2/2E. \tag{45.338}$$

For all E and fixed rotations in the range $0 \le \theta_d \le \theta_{\max} = \cos^{-1}\left(e^2/2De\right)$,

$$\sigma_{\text{orb}}(E) = (\pi De\cos\theta_d)/E - \pi R_0^2(E). \tag{45.339}$$

(l) Lennard–Jones (n,6). For the following two interactions, there are two roots of $E = V_{\text{eff}}(R_0) = V(R_0) + \frac{1}{2}R_0 V'(R_0)$. They correspond to stable and unstable circular orbits [with different angular momenta associated with the minimum and maxima of the different $V_{\text{eff}}(R)$]. Analytical expressions can only be derived for the orbiting cross section at the critical energy $E_{\max}$ above which no orbiting can occur.

For the Lennard–Jones $(n, 6)$ potential, orbiting occurs for $E < E_{\max} = 2\epsilon[4/(n-2)]^{6/(n-6)}$. The orbiting radius at $E_{\max}$ is

$$R_0(E_{\max}) = R_e\,[(n-2)/4]^{1/(n-6)}.$$

The orbiting cross section at $E_{\max} = 2\epsilon(R_e/R_0)^6$ is

$$\sigma_{\text{orb}}(E_{\max}) = \pi b_0^2(E_{\max}) = \frac{3}{2}\pi R_0^2\left(\frac{n}{n-2}\right). \tag{45.340}$$

$$n = 12:\ E_{\max} = 4\epsilon/5,\ R_0 = 1.165R_e,$$
$$\sigma_{\text{orb}} = 2.4\pi R_e^2.$$

(m) Polarization (n,4). As discussed for case (l),

$$E_{\max} = \epsilon\left(\frac{2}{n-2}\right)^{4/(n-4)}, \tag{45.341}$$

$$R_0(E_{\max}) = R_e\left(\frac{n-2}{2}\right)^{1/(n-4)}, \tag{45.342}$$

$$\sigma_{\text{orb}}(E_{\max}) = 2\pi R_0^2\frac{n}{(n-2)}, \tag{45.343}$$

$$n = 12:\ E_{\max} = \epsilon/\sqrt{5};$$
$$R_0 = 1.22R_e;\ \sigma_{\text{orb}} = 3.6\pi R_e^2.$$

Small-Angle Scattering

For the power law potential $V(R) = -C/R^n$, (45.12) can be expanded in powers of $V(R)/E$ to obtain analytic expressions for χ and η_{JB}. The general form is

$$\chi(b) = \sum_{j=1}^{\infty}\left(\frac{V(b)}{E}\right)^j F_j(n) = \frac{2}{k}\frac{\partial\eta}{\partial b}, \tag{45.344}$$

$$F_j(n) = \frac{\pi^{1/2}\Gamma\left(\frac{1}{2}jn + \frac{1}{2}\right)}{\Gamma(j+1)\Gamma\left(\frac{1}{2}jn - j + 1\right)}. \tag{45.345}$$

The leading $j = 1$ terms equivalent to (45.16b) are $F_1(n) \equiv F(n)$, as defined following (45.257). Then to first order in V/E,

$$\eta_{JB} = -\left(\frac{k}{2E}\right)\left(\frac{CF(n)}{n-1}\right) b^{1-n} , \quad (45.346)$$

$$\frac{d\sigma}{d\Omega} = I_c(\theta) = \left(\frac{CF(n)}{E\theta}\right)^{2/n} \frac{1}{n\theta \sin\theta} . \quad (45.347)$$

From a log–log plot of $\sin\theta(d\sigma/d\Omega)$ versus E, C and n can both be determined.

The integral cross sections for scattering by $\theta \geq \theta_0$ is

$$\sigma(E) = 2\pi \int_{\theta_0}^{\pi} I_c(\theta)\, d(\cos\theta) = 2\pi \int_0^{b_{max}} b\, db$$

$$= \pi \left(\frac{CF(n)}{E\theta_0}\right)^{2/n} , \quad (45.348)$$

where θ_0 is the smallest measured scattering angle corresponding to a trajectory with impact parameter, $b_{max} = [CF(n)/E\theta_0]^{1/n}$. A plot of $\ln\sigma(E)$ versus $\ln E$ is a straight line with slope $(-2/n)$.

The Landau–Lifshitz cross section (45.257) and the Massey–Mohr cross section (45.258) follow from use of the JB phases (45.346).

The diffusion cross section in the Random Phase Approximation (45.49) is

$$\sigma_d(E) = 4\pi \int_0^{b_c} \langle \sin^2\theta/2 \rangle b\, db , \quad |\chi(b_c)| = \frac{2}{\pi}$$

$$= \pi(C/2E)^{2/n} [\pi F(n)]^{2/n} . \quad (45.349)$$

(n) Multiple-Term Power-Law Potentials.

$$\chi(E, b) = \frac{1}{E} [V_m(b)F(m) - V_n(b)F(n)] . \quad (45.350)$$

For example a Lennard–Jones $(n, 6)$ potential Table 45.1 has the following features:

Forward Glory: $\chi = 0$ when $b_g = \alpha_n^{1/(n-6)} R_e$, where $\alpha_n = 6F(n)/[nF(6)]$.

Rainbow: $d\chi/db = 0$ at $b_r = (n\alpha_n/6)^{1/(n-6)} R_e$.

$$\chi_r = -F(n)(\mathcal{E}/E)(R_e/b_r)^n , \quad (45.351)$$

$$\omega_r = \frac{1}{2}\left(d^2\chi_r/db^2\right)_r = \frac{3n}{b_r^2}|\chi(b_r)| . \quad (45.352)$$

(o) Exponential Potential.

$$\eta_{JB}(E, b) = -\frac{1}{2}kb\frac{V_0}{E}K_1(\alpha b) \underset{R\to\infty}{\longrightarrow} -\frac{1}{2}kb\frac{V(b)}{E}\left(\frac{\pi b}{2\alpha}\right)^{1/2} .$$

(p) Screened Coulomb Potential.

$$\chi(E, b) = \alpha (V_0/E) K_1(\alpha b) \xrightarrow{\text{large } b} \left(\frac{1}{2}\pi\alpha b\right)^{1/2} V(b)/E , \quad (45.353)$$

$$\eta_{JB}(E, b) = -\frac{k}{2E}V_0K_0(\alpha b) \xrightarrow{\text{large } b} -\frac{k}{2E}V(b)\left(\frac{\pi b}{2\alpha}\right)^{1/2} . \quad (45.354)$$

(q) Morse Potential.

$$\chi(E, b) = (2\beta b)\left(\frac{\epsilon}{E}\right)\left[e^{2\beta R_e}K_0(2\beta b) - e^{\beta R_e}K_0(\beta b)\right]$$

$$\xrightarrow{\text{large } b} (\pi\beta b)^{1/2}\left(\frac{\epsilon}{E}\right)\left[e^{2\beta(R_e-b)} - \sqrt{2}e^{\beta(R_e-b)}\right] ,$$

$$b_r = R_e + (2\beta)^{-1}\ln 2 ,$$

$$\chi_r = -(\pi\beta b_r)^{1/2}(\epsilon/2E) ,$$

$$\omega_r = \beta^2|\chi_r|R_e^2 .$$

Large-Angle Scattering

For power law potentials $V(R) = C/R^n$,

$$\chi(b) = \pi - \sum_{j=1}^{n}\left(\frac{E}{V(b)}\right)^{(2j-1)/n} G_j(n) , \quad (45.355)$$

$$G_j(n) = \frac{(-1)^{j-1}}{\Gamma(j)\Gamma(k)}\left(\frac{2\pi^{1/2}}{n}\right)\Gamma\left(\frac{2j-1}{n}\right) , \quad (45.356)$$

with $k = [(2j-1)/n] - j - \frac{1}{2}$. For the $j = 1$ term,

$$\chi(b) = \pi - \left(\frac{E}{C}\right)^{1/n} G_1(n)b , \quad (45.357)$$

$$I_c(\theta) = \frac{d\sigma}{d\Omega} = \left(\frac{C}{E}\right)^{2/n} G_1^{-2}(n) , \quad (45.358)$$

which is isotropic. Including both $j = 1$ and 2 terms provides a good approximation to the entire repulsive branch of the deflection function χ. Series (45.355) for large angles and (45.344) for small angles eventually diverge for impact parameters $b < b_c$ and $b > b_c$, respectively, where

$$b_c = n^{1/2}\left(\frac{C}{2E}\right)^{1/n} \frac{|n-2|^{1/n}}{|n-2|^{1/2}} . \tag{45.359}$$

45.5.1 Born Amplitudes and Cross Sections for Model Potentials

$$k^2 = 2ME/\hbar^2 , \quad K = 2k \sin\frac{1}{2}\theta ,$$
$$U_0 = 2MV_0/\hbar^2 , \quad U_0/k^2 = V_0/E .$$

(a) Exponential. $V(R) = V_0 \exp(-\alpha R)$

$$f_B(\theta) = -\frac{2\alpha U_0}{\left(\alpha^2+K^2\right)^2} , \tag{45.360}$$

$$\sigma_B(E) = \frac{16}{3}\pi U_0^2 \left(\frac{3\alpha^4 + 12\alpha^2 k^2 + 16k^4}{\alpha^4(\alpha^2+4k^2)^3}\right) ,$$
$$\xrightarrow{E\to\infty} \frac{4}{3}\pi \left(\frac{V_0}{E}\right)\left(\frac{U_0}{\alpha^4}\right) . \tag{45.361}$$

(b) Gaussian. $V(R) = V_0 \exp\left(-\alpha^2 R^2\right)$

$$f_B(\theta) = -\left(\frac{\pi^{1/2} U_0}{4\alpha^2}\right) \exp\left(-K^2/4\alpha^2\right) , \tag{45.362}$$

$$\sigma_B(E) = \left(\frac{\pi^2 U_0}{8\alpha^4}\right)\left(\frac{V_0}{E}\right)\left[1 - \exp\left(-2k^2/\alpha^2\right)\right] . \tag{45.363}$$

(c) Spherical Well. $V(R) = V_0$ for $R < a$, $V(R) = 0$ for $R > a$

$$f_B(\theta) = -\frac{U_0}{K^3}(\sin Ka - Ka\cos Ka) , \tag{45.364}$$

$$\sigma_B(E) = \frac{\pi}{2}\frac{V_0}{E}(U_0 a^4)\Big[1-(ka)^{-2}+(ka)^{-3}\sin 2ka - (ka)^{-4}\sin^2 2ka\Big] . \tag{45.365}$$

(d) Screened Coulomb. $V(R) = V_0 \exp(-\alpha R)/R$, $V_0 = Ze^2$, $U_0 = 2Z/a_0$

$$f_B(\theta) = -\frac{U_0}{\alpha^2+K^2} , \tag{45.366}$$

$$\sigma_B(E) = \frac{4\pi U_0^2}{\alpha^2\left(\alpha^2+4k^2\right)} \to \pi\left(\frac{V_0}{E}\right)\left(\frac{U_0}{\alpha^2}\right) . \tag{45.367}$$

When $\alpha \to 0$, then $f_B(\theta) = -U_0/K^2$.

(e) e^--Atom.

$$V(R) = -Ne^2\left[Z/a_0 + 1/R\right]\exp(-2ZR/a_0) , \tag{45.368}$$

H(1s): N= 1, Z= 1; He(1s^2): N= 2; Z= 27/16.

$$f_B(\theta) = \frac{2N}{a_0}\left(\frac{2\alpha^2+K^2}{\left(\alpha^2+K^2\right)^2}\right) , \quad \alpha = 2Z/a_0 , \tag{45.369}$$

$$\sigma_B(E) = \frac{\pi a_0^2 N^2\left(12Z^4 + 18Z^2k^2a_0^2 + 7k^4a_0^4\right)}{3Z^2\left(Z^2+k^2a_0^2\right)^3} . \tag{45.370}$$

Also, f_B decomposes (45.187) as

$$f_B(K) = f_B^{eZ}(K) + f_B^{ee}(K)F(K) , \tag{45.371}$$

where f_B^{ij} are two-body Coulomb amplitudes for (i, j) scattering, and where

$$F(K) = \int |\Psi_0(\boldsymbol{R})|^2 \exp(\mathrm{i}\boldsymbol{K}\cdot\boldsymbol{R})\,\mathrm{d}\boldsymbol{R} \tag{45.372}$$

is the elastic form factor.

(f) Dipole. $V(R) = V_0/R^2$.

$$f_B(\theta) = \pi U_0/2K . \tag{45.373}$$

(g) Polarization potential. $V(R) = V_0/\left(R^2+R_0^2\right)^2$

$$f_B(\theta) = -\frac{1}{4}\pi\left(\frac{U_0}{R_0}\right)\exp(-KR_0) , \tag{45.374}$$

$$\sigma_B(E) = \left(\frac{\pi^3 U_0}{32R_0^4}\right)\left(\frac{V_0}{E}\right)[1-(1 +4kR_0)\exp(-4kR_0)] . \tag{45.375}$$

References

45.1 T. F. O'Malley, L. Spruch, L. Rosenberg: J. Math. Phys. **2**, 491 (1961)

45.2 T. F. O'Malley: Phys. Rev. **130**, 1020 (1962)

45.3 L. M. Biberman, G. E. Norman: Soviet Phys. (JETP) **18**, 1353 (1964)

45.4 O. Hinckelmann, L. Spruch: Phys. Rev. A **3**, 642 (1971)

45.5 G. F. Gribakin, V. V. Flambaum: Phys. Rev. A **48**, 546 (1993)
45.6 Z. L. Petovi c, T. F. O'Malley, R. W. Crompton: J. Phys. B **28**, 3309 (1995)
45.7 K. W. Ford, J. A. Wheeler: Ann. Phys. (NY) **7**, 259 (1959)
45.8 K. W. Ford, J. A. Wheeler: Ann. Phys. (NY) **7**, 287 (1959)
45.9 M. V. Berry, K. E. Mount: Rep. Prog. Phys. **35**, 315 (1972)
45.10 H. S. W. Massey, E. H. S. Burhop, H. B. Gilbody (Eds.): *Electronic and Ionic Impact Phenomena*, Vol. 1–5 (Clarendon, Oxford 1969–1974)
45.11 E. W. McDaniel, J. B. A. Mitchell, M. E. Rudd: *Atomic Collisions: Heavy Particle Projectiles* (Wiley, New York 1993) Chap. 1, p. 47

(A) Textbooks on Scattering Theory

G.1 N. F. Mott, H. S. W. Massey: *The Theory of Atomic Collisions*, 3rd edn., (Clarendon, Oxford 1965) Chaps. 2–5
G.2 R. G. Newton: *Scattering Theory of Waves and Particles* (McGraw-Hill, New York 1966)
G.3 L. S. Rodberg, R. M. Thaler: *Introduction to the Quantum Theory of Scattering* (Academic, New York 1967)
G.4 H. S. W. Massey, E. H. S. Burhop, H. B. Gilbody (Eds.): *Electronic and Ionic Impact Phenomena* (Clarendon, Oxford 1969–1974) Vols. 1–5
G.5 M. R. C. McDowell, J. P. Coleman: *Introduction to the Theory of Ion-Atom Collisions* (North-Holland, Amsterdam 1970)
G.6 M. S. Child: *Molecular Collision Theory* (Academic, New York 1974)
G.7 C. J. Joachain: *Quantum Collision Theory* (North-Holland, Amsterdam 1975)
G.8 L. D. Landau, E. M. Lifshitz: *Course of Theoretical Physics*, 3rd edn. (Pergamon, Oxford 1976) Vol. 1, Chap. 4, p. 41
G.9 H. Eyring, S. H. Lin, S. M. Lin: *Basic Chemical Kinetics* (Wiley, New York 1980) Chap. 3, p. 81
G.10 B. H. Bransden: *Atomic Collision Theory*, 2nd edn. (Benjamin-Cummings, Menlo Park 1983)
G.11 B. H. Bransden, C. J. Joachain: *Physics of Atoms and Molecules* (Wiley, New York 1983)
G.12 E. E. Nikitin, S. Ya. Umanskiĭ: *Theory of Slow Atomic Collisions* (Springer, Berlin, Heidelberg 1984)
G.13 E. W. McDaniel: *Atomic Collisions: Electron and Photon Projectiles* (Wiley, New York 1989) Chaps. 3 and 4, p. 71
G.14 M. S. Child: *Semiclassical Mechanics with Molecular Applications* (Clarendon, Oxford 1991)
G.15 A. G. Sitenko: *Scattering Theory* (Springer, Berlin, Heidelberg 1991)
G.16 E. W. McDaniel, J. B. A. Mitchell, M. E. Rudd: *Atomic Collisions: Heavy Particle Projectiles* (Wiley, New York 1993) Chap. 1, p. 1

(B) Chapters in Edited Books

G.17 E. H. Burhop: *Quantum Theory I. Elements*, ed. by D. R. Bates (Academic, New York 1961), Chap. 9, p. 300
G.18 E. A. Mason, J. T. Vanderslice: *High-Energy Elastic Scattering of Atoms, Molecules and Ions*, in *Atomic and Molecular Processes*, ed. by D. R. Bates (Academic, New York 1962) Chap. 17, p. 663
G.19 B. L. Moiseiwitsch: *Atomic and Molecular Processes*, edited by D. R. Bates (Academic, New York 1962), Chap. 9, p. 281
G.20 A. Dalgarno: *Ion-Molecule Reactions*, ed. by E. W. McDaniel, V. Čermák, A. Dalgarno, E. E. Ferguson, L. Friedman (Wiley, New York 1970) Chap. 3, p. 159
G.21 K. W. Ford, J. A. Wheeler: *Semiclassical Description of Scattering*, Ann. Phys. (NY) **7**, 259 (1959)
G.22 K. W. Ford, J. A. Wheeler: *Applications of Semiclassical Scattering Analysis*, *ibid.* 287 (1959)
G.23 M. V. Berry, K. E. Mount: *Semiclassical Approximations in Wave Mechanics*, Rep. Prog. Phys. **35**, 315 (1972)
G.24 W. H. Miller: Adv. Chem. Phys. **25**, 69 (1974)
G.25 W. H. Miller: Adv. Chem. Phys. **30**, 77 (1975)
G.26 M. S. Child: *Dynamics of Molecular Collisions: Part B*, edited by W. H. Miller (Plenum Press, New York 1976) Chap. 4, p. 171
G.27 H. Pauly: *Atom-Molecule Collision Theory: A Guide for the Experimentalist* ed. by R. B. Bernstein (Plenum Press, New York 1979) Chap. 4, p. 111

G.28 S. Stolte, J. Reuss: *Atom-Molecule Collision Theory: A Guide for the Experime?ntalist* ed. by R. B. Bernstein (Plenum Press, New York 1979) Chap. 5, p. 231
G.29 J. N. L. Connor: *Semiclassical Methods in Molecular Scattering and Spectroscopy*, ed. by M. S. Child (D. Reidel, Boston 1980) p. 45
G.30 R. E. Johnson: *Encylopedia of Physical Science and Technology* (Academic, New York 1987) Vol. 2, p. 224
G.31 J. J. H. van Biesen: Vol. I, ed. by G. Scoles (Oxford Univ. Press, New York 1988) Chap. 19, p. 472
G.32 U. Buck: *Atomic and Molecular Beam Methods*, ed. by G. Scoles (Oxford University Press, New York, 1988) Chap. 20, p. 499

46. Orientation and Alignment in Atomic and Molecular Collisions

This chapter deals with the concepts of orientation and alignment in atomic and molecular physics. The terms refer to parameters related to the shape and dynamics of an excited atomic or molecular level, as it is manifested in a nonstatistical population of the magnetic sublevels. To take full advantages of the possibilities of this approach, one utilizes "third generation experiments", i. e., scattering experiments which exploit the planar scattering symmetry, contrary to an angular differential cross section determination (a "second generation experiment") having cylindrical symmetry, or a total cross section measurement (a "first generation experiment") integrating over all scattering angles. In this way one is often able to probe atomic collision theories at a more fundamental level, and in favorable cases approach a "perfect scattering experiment" in which the complex quantal scattering amplitudes are completely determined. This term was coined by *Bederson* [46.1] and has since served as an ideal towards which scattering experiments attempt to strive.

46.1 **Collisions Involving Unpolarized Beams** . 694
46.1.1 The Fully Coherent Case 694
46.1.2 The Incoherent Case with Conservation of Atomic Reflection Symmetry 697
46.1.3 The Incoherent Case without Conservation of Atomic Reflection Symmetry 697

46.2 **Collisions Involving Spin-Polarized Beams** 699
46.2.1 The Fully Coherent Case 699
46.2.2 The Incoherent Case with Conservation of Atomic Reflection Symmetry 699
46.2.3 The Incoherent Case without Conservation of Atomic Reflection Symmetry 700

46.3 **Example** 702
46.3.1 The First Born Approximation 702

46.4 **Recent Developments** 703
46.4.1 S → D Excitation 703
46.4.2 P → P Excitation 703
46.4.3 Relativistic Effects in S → P Excitation 703

46.5 **Summary** 703

References 703

The study of anisotropies has a long history in atomic physics, with light polarization, or Stokes parameter analysis, as a prominent example. A pioneering review by *Fano* and *Macek* [46.2] layed the mathematical and conceptual foundation for most of the later work. Advances in coincidence techniques, laser preparation methods, and development of efficient sources for polarized electrons [46.3] have boosted the field further. Parallel developments of powerful computational codes have enabled a matching theoretical effort. In this way, detailed insights into the collision dynamics have been obtained, such as the locking radius model for low energy atomic collisions, propensity rules for orientation in fast electronic and atomic collisions, and spin effects in polarized electron-heavy atom scattering. Presentations of these developments are contained in [46.4] and Chapts. 37, 50, 51, 63, 64, and 66. A comprehensive and critical review of the literature and the mathematical formalism was initiated by NIST [46.5–7].

Space limitations allow only presentation of the formalism for the simplest case of excitation. Recent developments in the description of processes involving polarized beams are included. The presentation is restricted to atomic outer-shell excitation. Other reviews describe excitation of inner shells [46.8] and molecular levels [46.9]. Related material on density matrices is contained in Chapt. 7 [46.10].

46.1 Collisions Involving Unpolarized Beams

46.1.1 The Fully Coherent Case

Consider first the simplest nontrivial case of S → P excitation. A general property of an atomic collision is that the total reflection symmetry with respect to the scattering plane of the total wave function describing the system is conserved. In simple cases, such as electron impact excitation of He singlet states, or atom excitation by fast, heavy particle impact, the projectile acts as a structureless particle, with only the target atom changing its quantum state. In this case, the reflection symmetry of the target wave function alone is conserved.

Figure 46.1 shows the angular parts of the simplest $|lm\rangle$ states with $l = 0, 1, 2$. The arrows indicate how the two families of states with positive or negative reflection symmetry may couple internally. Thus, in an S → P transition, an atom initially in an S-state may be excited to the (p_{+1}, p_{-1}) subspace, while the p_0 state is not accessible. A characteristic feature of the corresponding electron charge cloud is that it has "zero height", i.e., zero electron density in the direction perpendicular to the scattering plane. Furthermore, the expectation value of the orbital angular momentum has a nonvanishing component in this direction only.

The resulting electron charge cloud may thus have a shape as shown on Fig. 46.2a. We shall now discuss the parametrization of the wave function of this state and analyze the connection between the wave function and the corresponding photon radiation pattern emitted when the state decays back to the initial S-state, using the properties of electric dipole radiation.

Basic Definitions and Coordinate Frames

The coordinate frames of particular use in describing the wave function are as follows. The *collision frame*, (x^c, y^c, z^c), is defined by $\hat{z}^c \parallel \boldsymbol{k}_{\text{in}}$ and $\hat{y}^c \parallel \boldsymbol{k}_{\text{in}} \times \boldsymbol{k}_{\text{out}}$. The *natural frame*, $(x^{\text{n}}, y^{\text{n}}, z^{\text{n}})$, is defined by $\hat{x}^{\text{n}} \parallel \boldsymbol{k}_{\text{in}}$ and $\hat{z}^{\text{n}} \parallel \boldsymbol{k}_{\text{in}} \times \boldsymbol{k}_{\text{out}}$. Finally, the *atomic frame* is identical to the natural frame, except that the frame is rotated by an angle γ around $z^{\text{n}} = z^{\text{a}}$ such that x^{a} is parallel to the major symmetry axis of the P-state charge cloud, as shown in Fig. 46.2. Most scattering

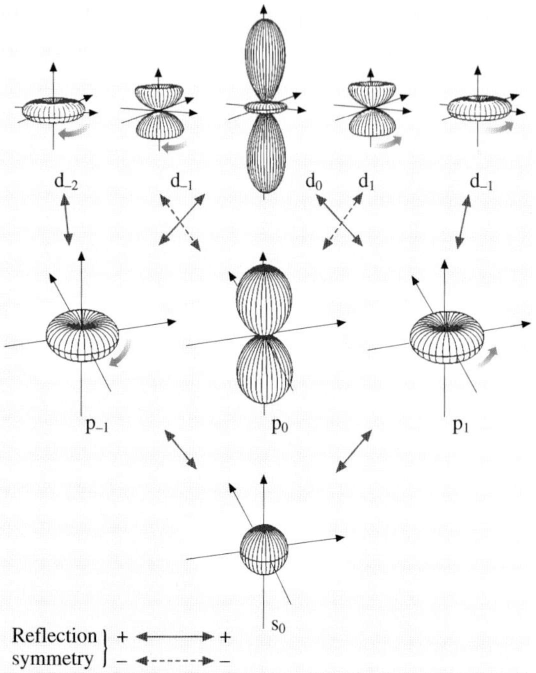

Fig. 46.1 Illustration of the reflection symmetry of the simplest spherical harmonics, corresponding to S, P, and D states

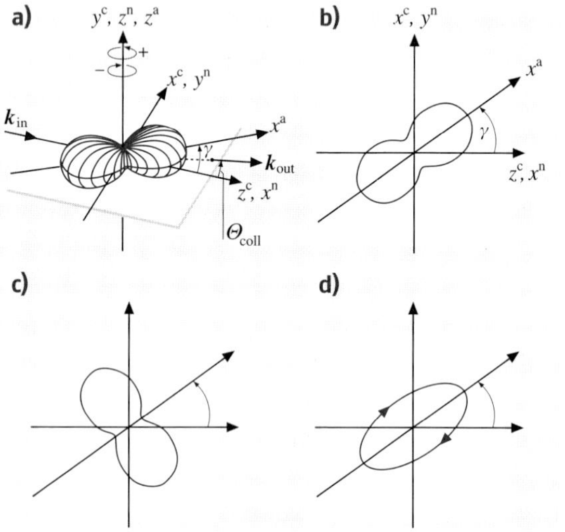

Fig. 46.2 (**a**) Shows the shape, or the angular part, of a P-state with positive reflection symmetry with respect to the scattering plane. Some relevant coordinate frames are also indicated; (**b**) shows a cut of this shape in the scattering plane. (**c**) is the angular distribution of photons emitted in the scattering plane; (**d**) is the polarization ellipse for light emitted in the direction perpendicular to the plane, as observed from above

calculations use the collision coordinate system as reference frame, while mathematical analysis is often most conveniently performed in the natural frame, where the algebra is particularly simple. Many expressions are even simpler in the atomic frame, but it has the disadvantage that the angle γ varies with collision parameters, such as impact velocity, impact parameter, etc.

The expansion of the P-state wave function in the three coordinate systems is

$$|\Psi\rangle = a^{c}_{+1}|p^{c}_{+1}\rangle + a^{c}_{0}|p^{c}_{0}\rangle + a^{c}_{-1}|p^{c}_{-1}\rangle \quad (46.1)$$
$$= a^{n}_{+1}|p^{n}_{+1}\rangle + a^{n}_{-1}|p^{n}_{-1}\rangle \quad (46.2)$$
$$= a^{a}_{+1}|p^{a}_{+1}\rangle + a^{a}_{-1}|p^{a}_{-1}\rangle . \quad (46.3)$$

Conservation of reflection symmetry in the scattering plane implies

$$a^{c}_{+1} = -a^{c}_{-1} , \quad (46.4)$$
$$a^{n}_{0} = 0 , \quad (46.5)$$
$$a^{a}_{0} = 0 . \quad (46.6)$$

The normalization condition implies

$$|a^{c}_{0}|^2 + 2|a^{c}_{+1}|^2 = |a^{n}_{+1}|^2 + |a^{n}_{-1}|^2 = |a^{a}_{+1}|^2 + |a^{a}_{-1}|^2 = 1 . \quad (46.7)$$

Finally, the a_m coefficients are related to the scattering amplitudes f_m and the differential cross section $\sigma(\theta)$ by

$$a_m = f_m (k_{out}/k_{in})^{\frac{1}{2}} \sigma(\theta)^{-\frac{1}{2}} \quad (46.8)$$
$$\sigma(\theta) = (k_{out}/k_{in}) \left(|f^{c}_{0}|^2 + 2|f^{c}_{+1}|^2\right) , \text{ etc.} \quad (46.9)$$

Except for an arbitrary phase factor, the pure state (46.1) may thus be characterized by two dimensionless parameters. Traditionally, they have been chosen as (λ, χ) defined as [46.11]

$$\lambda = \frac{|a^{c}_{0}|^2}{|a^{c}_{0}|^2 + 2|a^{c}_{+1}|^2} , \quad (46.10)$$
$$\chi = \arg\left(a^{c}_{+1} a^{c*}_{0}\right) . \quad (46.11)$$

An alternative parametrization $(L_{\perp}, \gamma)$ is given by [46.5]

$$L_{\perp} = \frac{|a^{n}_{+1}|^2 - |a^{n}_{-1}|^2}{|a^{n}_{+1}|^2 + |a^{n}_{-1}|^2} , \quad (46.12)$$
$$\gamma = \frac{1}{2}\arg\left(a^{n}_{-1} a^{n}_{+1}{}^{*}\right) \pm \frac{\pi}{2} ,$$
$$= -\frac{1}{2}(\delta \pm \pi) , \quad (46.13)$$

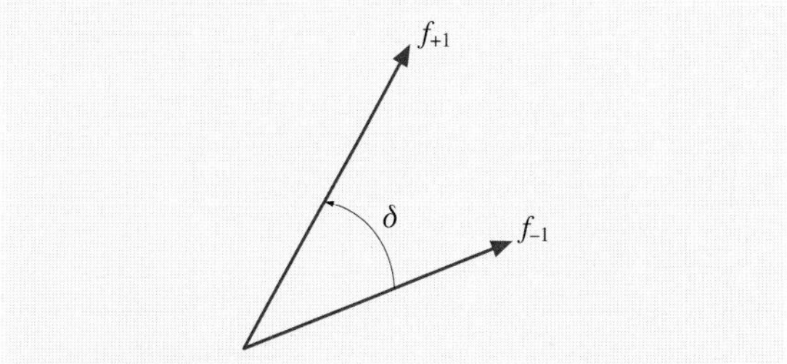

Fig. 46.3 Fully coherent S → P excitation may be described by two independent scattering amplitudes, characterized by their relative size and phase

with the notation of Fig. 46.3. Referring to the natural coordinate frame, the expectation value of the electronic orbital angular momentum is thus

$$\langle\Psi|\boldsymbol{L}|\Psi\rangle = (0, 0, L_{\perp}) . \quad (46.14)$$

Coherence Analysis: Stokes Parameters

We now discuss the information obtainable from a polarization analysis of the emitted light. In the notation of classical optics (see, e.g., *Born* and *Wolf* [46.12]) the components of the Stokes vector (P_1, P_2, P_3) measured in the direction $+y^{c}$ $(+z^{n})$ perpendicular to the scattering plane and defined by

$$IP_1 = I(0°) - I(90°) , \quad (46.15)$$
$$IP_2 = I(45°) - I(135°) , \quad (46.16)$$
$$IP_3 = I(\text{RHC}) - I(\text{LHC}) , \quad (46.17)$$

with

$$I = I(0°) + I(90°)$$
$$= I(45°) + I(135°)$$
$$= I(\text{RHC}) + I(\text{LHC}) , \quad (46.18)$$

are given by

$$P_1 = 2\lambda - 1 , \quad (46.19)$$
$$P_2 = -2\sqrt{\lambda(1-\lambda)} \cos\chi , \quad (46.20)$$
$$P_3 = 2\sqrt{\lambda(1-\lambda)} \sin\chi , \quad (46.21)$$

or, alternatively,

$$P_1 = P_l \cos 2\gamma , \quad (46.22)$$
$$P_2 = P_l \sin 2\gamma , \quad (46.23)$$
$$P_3 = -L_{\perp} , \quad (46.24)$$

with

$$P_l = \sqrt{P_1^2 + P_2^2} . \quad (46.25)$$

Here, $I(\theta)$ is the intensity transmitted through an ideal linear polarizer with transmission direction tilted by an angle θ with respect to the z^c or x^n direction. Similarly, RHC (LHC) refers to photons with negative (positive) helicity. Inspection of (46.19) to (46.25) shows that determination of the Stokes vector in the direction perpendicular to the collision plane determines the wave function (46.1) completely. A determination of the Stokes vector thus constitutes a "perfect scattering experiment".

The Stokes vector (P_1, P_2, P_3) is a unit vector characterizing the state $|\Psi\rangle$, and it may be represented by a point on the Poincaré sphere, see Fig. 46.4.

Correlation Analysis

Another experimental approach is to map the angular distribution of the photons emitted in the subsequent $S \to P$ decay. For this purpose we first note that the angular part $\Upsilon(\theta, \phi)$ of the electron probability density $\langle\Psi|\Psi\rangle$ corresponding to the wave function (46.1) may be written as

$$\Upsilon(\theta, \phi) = \frac{1}{2}[1 + P_l \cos 2(\phi - \gamma)] \sin^2 \theta \,. \qquad (46.26)$$

Figure 46.2(b) shows a cut of this charge cloud in the scattering plane,

$$\Upsilon\left(\frac{1}{2}\pi, \phi\right) = \frac{1}{2}[1 + P_l \cos 2(\phi - \gamma)] \,. \qquad (46.27)$$

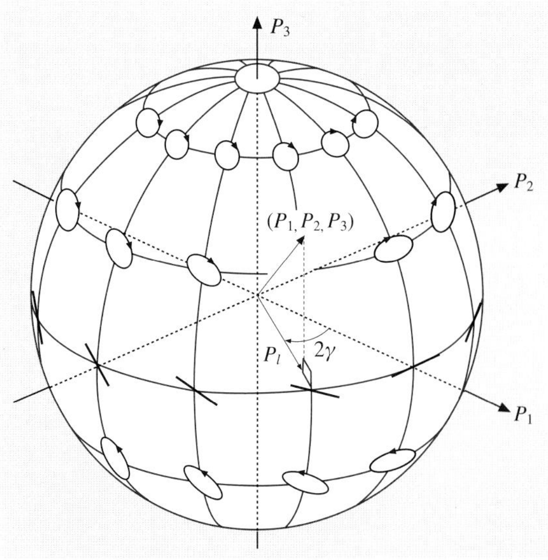

Fig. 46.4 The Poincaré sphere. The Stokes vector for a pure state $|\Psi\rangle$ corresponds to a characteristic polarization ellipse, represented by a point on the unit sphere

The relative length, l, and width, w of the charge cloud are given by

$$l = \frac{1}{2}(1 + P_l) \,, \qquad (46.28)$$

$$w = \frac{1}{2}(1 - P_l) \,. \qquad (46.29)$$

The degree of linear polarization P_l is thus a measure of the shape of the charge cloud, since

$$P_l = \frac{l - w}{l + w} \,. \qquad (46.30)$$

According to the properties of electric dipole radiation, the pattern $I(\phi)$ in the scattering plane is

$$I(\phi) = \frac{1}{2}[1 - P_l \cos 2(\phi - \gamma)] \,. \qquad (46.31)$$

A mapping of the radiation pattern in the collision plane determines γ and P_l, and thereby the absolute value of the angular momentum by

$$|L_\perp| = \sqrt{1 - P_l^2} \,. \qquad (46.32)$$

However, the sign of $L_\perp$ cannot be determined by correlation analysis, and this approach accordingly does not qualify for classification as "a perfect scattering experiment".

Density Matrix Representation

Recalling that $\rho_{mn} = a_m a_n^*$, the density matices in the various basis sets introduced above are

$$\rho^c = \frac{1}{\sqrt{8}} \begin{pmatrix} \frac{1}{\sqrt{2}}(1-P_1) & -P_2+\mathrm{i}P_3 & -\frac{1}{\sqrt{2}}(1-P_1) \\ -P_2-\mathrm{i}P_3 & \sqrt{2}(1+P_1) & P_2+\mathrm{i}P_3 \\ -\frac{1}{\sqrt{2}}(1-P_1) & P_2-\mathrm{i}P_3 & \frac{1}{\sqrt{2}}(1-P_1) \end{pmatrix} , \qquad (46.33)$$

$$\rho^n = \frac{1}{2} \begin{pmatrix} 1 + L_\perp & 0 & -P_l\, \mathrm{e}^{-2\mathrm{i}\gamma} \\ 0 & 0 & 0 \\ -P_l\, \mathrm{e}^{2\mathrm{i}\gamma} & 0 & 1 - L_\perp \end{pmatrix} , \qquad (46.34)$$

$$\rho^a = \frac{1}{2} \begin{pmatrix} 1 + L_\perp & 0 & -P_l \\ 0 & 0 & 0 \\ -P_l & 0 & 1 - L_\perp \end{pmatrix} . \qquad (46.35)$$

The density matrices illustrate the algebraic simplifications obtained using the natural coordinate frame. In the following, this frame will be used unless otherwise stated, and we therefore suppress the superscript n below.

Part D | 46.1

Postcollisional Depolarization due to Fine Structure and Hyperfine Structure Effects

After the collision, the isolated atom evolves under the influence of internal forces, such as fine structure and hyperfine structure, until the optical decay. While typical collision times are of the order 10^{-15} s, the light emission occurs after a time interval of 10^{-9} s. This is long compared with the Larmor precession time of the electron spin ($\sim 10^{-12}$ s) and one can thus assume that the atom has completely relaxed into, e.g., its $^2P_{1/2}$ and $^2P_{3/2}$ states before the photon emission. For the simplest nontrivial case of electron spin $S = \frac{1}{2}$, the Stokes vector $\boldsymbol{P}(S)$ of the subsequent $^2P \rightarrow {}^2S$ transition is modified to become

$$P_{1,2}\left(\frac{1}{2}\right) = \frac{3}{7} P_{1,2}(0) , \tag{46.36}$$

$$P_3\left(\frac{1}{2}\right) = P_3(0) , \tag{46.37}$$

and the Stokes vector is thus no longer a unit vector. In general, the Stokes vector components are reduced by depolarization factors c_i (a table of c_i coefficients for the most common values of electron and nuclear spin is given in Appendix B of [46.5])

$$P_i(S) = c_i P_i(0) . \tag{46.38}$$

This suggests introduction of *reduced polarizations*

$$\overline{P}_i = P_i(S)/c_i . \tag{46.39}$$

The *reduced* Stokes vector $(\overline{P}_1, \overline{P}_2, \overline{P}_3)$ is again a unit vector, and the formalism developed above for the spinless case can then be applied. We shall assume that this correction has been made, and the "bar" will be dropped.

46.1.2 The Incoherent Case with Conservation of Atomic Reflection Symmetry

The picture outlined above has to be modified in cases where the experiment sums over several, in principle distinguishable, channels, each of which, however, conserves reflection symmetry. Prototype examples are electron impact excitation of hydrogen or light alkali atoms. Here, the excitation process is described by singlet and triplet scattering amplitudes respectively, and we have the possibility of direct and exchange scattering. This doubles the number of scattering amplitudes from two to four. The experimental results are thus an incoherent sum over these channels (singlet and triplet), the unraveling of which would require application of spin-polarized beams, see Sect. 46.2. The atomic P-state can no longer be described as a single pure state, (46.1), but as a mixed state (Chapt. 7). The expressions (46.33) to (46.35) for the density matrix are unchanged, but the matrix elements are now sums of the contributions from the individual channels (We shall discuss its decomposition below). P_l and $L_\perp$ are now independent quantities, and the (reduced) Stokes vector $\boldsymbol{P}$ is generally no longer a unit vector. The degree of polarization P may thus be less than one, i. e.,

$$P^2 = P_1^2 + P_2^2 + P_3^2 = P_l^2 + L_\perp^2 \leq 1 . \tag{46.40}$$

For electron impact excitation, deviation of the parameter P from unity may thus serve as a measure of the effect of electron exchange. We have accordingly three independent observables, e.g., $(L_\perp, \gamma, P_l)$. This set of variables is frame-independent. A photon correlation experiment in the scattering plane may extract the (γ, P_l) pair, while coherence analysis in the z direction provides the complete set. An alternative set of (frame-dependent) parameters used in particular for hydrogen is (λ, R, I), with

$$\lambda = \frac{1}{2}(1 + P_1) , \tag{46.41}$$

$$R = -\frac{1}{\sqrt{8}} P_2 , \tag{46.42}$$

$$I = \frac{1}{\sqrt{8}} P_3 . \tag{46.43}$$

46.1.3 The Incoherent Case without Conservation of Atomic Reflection Symmetry

In the general case, the assumption of positive reflection symmetry with respect to the scattering plane for the wave function of the excited atom cannot be maintained. For example, for electron impact excitation of a P-state of a heavy atom, spin-orbit effects may be so strong that they flip the electron spin, thereby allowing population of the $|p_0^{\rm n}\rangle$ state. Thus, the total number of scattering amplitudes is now six. Typical examples are mercury, or the heavy noble gases. The total angular momentum of the excited state is $J = 1$, the fine structure being completely resolved in these cases. Strictly speaking, the density matrix elements no longer describe the electronic charge cloud, but rather the excited state ($J = 1$) distribution. Similarly, $L_\perp$ should be replaced by $J_\perp$, but for simplicity we keep the notation. This state radiates as a set of classical oscillators, completely analogous to the $^1P_1$ state described above, so we shall maintain our previous notation, simply replacing the term charge cloud density by oscillator density. The main difference from the cases considered above is that the density now displays a height, see Fig. 46.5.

Blum, da Paixão and collaborators, [46.13, 14], were the first to formulate a parametrization of the general case. Here, we shall use as starting point the expression for the density matrix in the natural frame and decompose it into the two terms with positive and negative reflection symmetry, respectively,

$$\rho^{\mathrm{n}} = (1-h)\,\frac{1}{2}\begin{pmatrix} 1-P_3 & 0 & -P_l^+\,\mathrm{e}^{-2\mathrm{i}\gamma} \\ 0 & 0 & 0 \\ -P_l^+\,\mathrm{e}^{2\mathrm{i}\gamma} & 0 & 1+P_3 \end{pmatrix} + h\begin{pmatrix} 0 & 0 & 0 \\ 0 & 1 & 0 \\ 0 & 0 & 0 \end{pmatrix}, \tag{46.44}$$

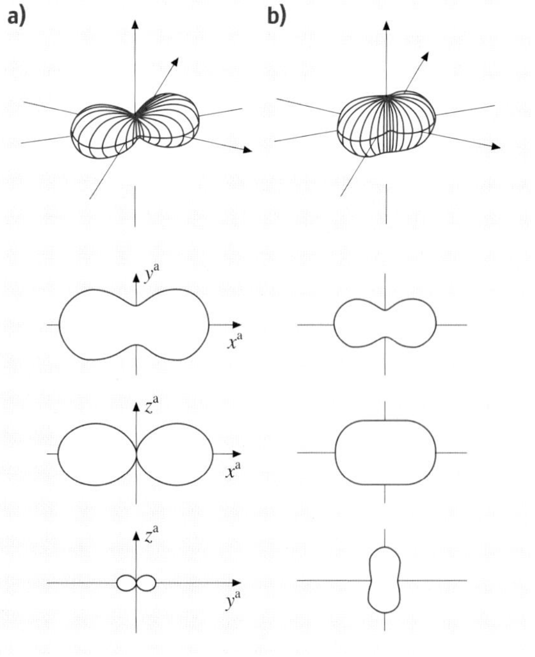

Fig. 46.5 The *top row* shows classical oscillator densities for height parameters $h=0$ and $h=1/3$, respectively. The alignment angle is $\gamma=35°$ and $P_l^+=0.6$ in both cases. *Below* are shown cuts along the symmetry axes

where the linear polarization P_l^+ is labeled with a "+" referring to the positive reflection symmetry. Similarly, we define

$$L_\perp^+ \equiv -P_3\,, \tag{46.45}$$

$$P^{+2} \equiv P_1^2 + P_2^2 + P_3^2 \le 1\,. \tag{46.46}$$

The shape of the density is now characterized by the three parameters

$$l = (1-h)\,\frac{1}{2}(1+P_l)\,, \tag{46.47}$$

$$w = (1-h)\,\frac{1}{2}(1-P_l)\,, \tag{46.48}$$

$$h = \rho_{00}\,, \tag{46.49}$$

with $l+w+h=1$. There are thus four independent observables, chosen as $(L_\perp^+, \gamma, P_l^+, h)$. Again, this set of variables is frame-independent. Determination of the height parameter evidently requires observation from a direction other than the z^{n}-direction. Traditionally, the Stokes parameter $(P_4, 0, 0)$ is measured from the y^{n}-direction to obtain

$$h = \frac{(1+P_1)(1-P_4)}{4-(1-P_1)(1-P_4)}\,. \tag{46.50}$$

Thus, all four parameters may be obtained from analysis of the light coherence, but two directions of observation are necessary. Similarly, photon correlation analysis in two planes are required to extract (γ, P_l^+, h), see, e.g., [46.5] for a discussion.

The various cases with unpolarized beams discussed in this Section are summarized in Table 46.1.

Table 46.1 Summary of cases of increasing complexity, and the orientation and alignment parameters necessary for unpolarized beams. N_p is the number of independent parameters, and N_d is the number of observation directions required

Variable	Sect. 46.1.1	Sect. 46.1.2	Sect. 46.1.3
Forces	Coulomb	+exchange	+spin–orbit
Representation	wave func.	ρ_{mn}	ρ_{mn}
Refl. symmetry	+	+	+, −
Ang. mom.	$L_\perp$	$L_\perp$	$L_\perp^+$
Align. angle	γ	γ	γ
Linear pol.	P_l	P_l	P_l^+
Degree of pol.	$P=1$	$P\le 1$	$P^+\le 1$
Height	$h=0$	$h=0$	$h\ge 0$
N_p	2	3	4
N_d	1	1	2

46.2 Collisions Involving Spin-Polarized Beams

In this section we discuss the additional information that can be gained by application of a polarized beam compared with an unpolarized beam. Collisions with polarized electron beams are discussed in particular, but most of the ideas presented are easily generalized to beams of atoms in polarized, or otherwise prepared, states. We shall keep the order of increasing complexity introduced in the previous section.

46.2.1 The Fully Coherent Case

As seen in Sect. 46.1.2, a simple example of a fully coherent excitation process is He $1\,^1\mathrm{S} \rightarrow n\,^1\mathrm{P}$ excitation, for which complete information can be obtained from experiments using unpolarized beams. Consequently, in this case application of polarized electrons will add nothing new. Also, the polarization of the scattered electron can be trivially predicted, since no change is possible in the scattering process.

46.2.2 The Incoherent Case with Conservation of Atomic Reflection Symmetry

The targets of interest here are hydrogen or light alkali atoms with an electron spin $S=\frac{1}{2}$. Since we now have the possibility of triplet (t) and singlet (s) scattering, this doubles the number of scattering amplitudes from two to four (recall that $f_0=0$) see Fig. 46.6.

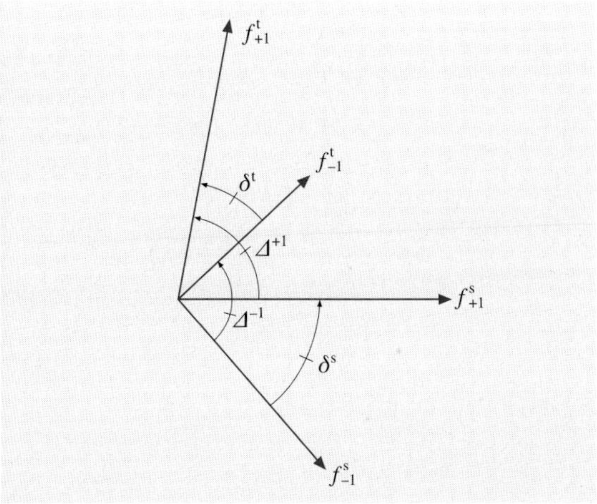

Fig. 46.6 For $^2\mathrm{S} \rightarrow {}^2\mathrm{P}$ electron impact excitation of hydrogen or light alkali atoms, four scattering amplitudes come into play

The amplitudes of interest are

$$f^{\mathrm{t}}_{+1} = \alpha_+ \mathrm{e}^{\mathrm{i}\phi_+}\,, \tag{46.51}$$

$$f^{\mathrm{t}}_{-1} = \alpha_- \mathrm{e}^{\mathrm{i}\phi_-}\,, \tag{46.52}$$

$$f^{\mathrm{s}}_{+1} = \beta_+ \mathrm{e}^{\mathrm{i}\psi_+}\,, \tag{46.53}$$

$$f^{\mathrm{s}}_{-1} = \beta_- \mathrm{e}^{\mathrm{i}\psi_-}\,. \tag{46.54}$$

Neglecting an overall phase, seven independent parameters are needed to characterize the amplitudes completely. Traditionally, one is chosen as the differential cross section σ_u corresponding to unpolarized particles. Six additional dimensionless parameters may be defined: three to characterize the relative lengths of the four vectors, and three to define their relative phase angles.

The density matrix is parametrized, in analogy to the unpolarized beam case, according to [46.15]

$$\rho^{\mathrm{t}} = \sigma^{\mathrm{t}}\,\frac{1}{2}\begin{pmatrix} 1+L^{\mathrm{t}}_{\perp} & 0 & -P^{\mathrm{t}}_{l}\,\mathrm{e}^{-2\mathrm{i}\gamma^{\mathrm{t}}} \\ 0 & 0 & 0 \\ -P^{\mathrm{t}}_{l}\,\mathrm{e}^{2\mathrm{i}\gamma^{\mathrm{t}}} & 0 & 1-L^{\mathrm{t}}_{\perp} \end{pmatrix}\,, \tag{46.55}$$

$$\rho^{\mathrm{s}} = \sigma^{\mathrm{s}}\,\frac{1}{2}\begin{pmatrix} 1+L^{\mathrm{s}}_{\perp} & 0 & -P^{\mathrm{s}}_{l}\,\mathrm{e}^{-2\mathrm{i}\gamma^{\mathrm{s}}} \\ 0 & 0 & 0 \\ -P^{\mathrm{s}}_{l}\,\mathrm{e}^{2\mathrm{i}\gamma^{\mathrm{s}}} & 0 & 1-L^{\mathrm{s}}_{\perp} \end{pmatrix}\,, \tag{46.56}$$

where

$$\sigma^{\mathrm{t}} = \alpha_+^2 + \alpha_-^2\,, \tag{46.57}$$

$$\sigma^{\mathrm{s}} = \beta_+^2 + \beta_-^2\,, \tag{46.58}$$

$$L^{\mathrm{t}}_{\perp} = \frac{1}{\sigma^{\mathrm{t}}}\left(\alpha_+^2 - \alpha_-^2\right)\,, \tag{46.59}$$

$$L^{\mathrm{s}}_{\perp} = \frac{1}{\sigma^{\mathrm{s}}}\left(\beta_+^2 - \beta_-^2\right)\,, \tag{46.60}$$

$$P^{\mathrm{t}}_{l}\,\mathrm{e}^{2\mathrm{i}\gamma^{\mathrm{t}}} = P^{\mathrm{t}}_{1} + \mathrm{i}\,P^{\mathrm{t}}_{2} = -\frac{2\alpha_+\alpha_-}{\sigma^{\mathrm{t}}}\,\mathrm{e}^{-\mathrm{i}\delta^{\mathrm{t}}}\,, \tag{46.61}$$

$$P^{\mathrm{s}}_{l}\,\mathrm{e}^{2\mathrm{i}\gamma^{\mathrm{s}}} = P^{\mathrm{s}}_{1} + \mathrm{i}\,P^{\mathrm{s}}_{2} = -\frac{2\beta_+\beta_-}{\sigma^{\mathrm{s}}}\,\mathrm{e}^{-\mathrm{i}\delta^{\mathrm{s}}}\,. \tag{46.62}$$

In the case of an unpolarized beam, the total density matrix becomes the weighted sum of the two matrices

ρ^{s} and ρ^{t}, i.e.,

$$\rho_{\mathrm{u}} = \sigma_{\mathrm{u}} \frac{1}{2} \begin{pmatrix} 1+L_{\perp} & 0 & -P_l \,\mathrm{e}^{-2\mathrm{i}\gamma} \\ 0 & 0 & 0 \\ -P_l \,\mathrm{e}^{2\mathrm{i}\gamma} & 0 & 1-L_{\perp} \end{pmatrix}$$
$$= 3w^{\mathrm{t}} \sigma_{\mathrm{u}} \frac{1}{2} \begin{pmatrix} 1+L_{\perp}^{\mathrm{t}} & 0 & -P_l^{\mathrm{t}} \,\mathrm{e}^{-2\mathrm{i}\gamma^{\mathrm{t}}} \\ 0 & 0 & 0 \\ -P^{\mathrm{t}_l} \,\mathrm{e}^{2\mathrm{i}\gamma^{\mathrm{t}}} & 0 & 1-L_{\perp}^{\mathrm{t}} \end{pmatrix}$$
$$+ w^{\mathrm{s}} \sigma_{\mathrm{u}} \frac{1}{2} \begin{pmatrix} 1+L_{\perp}^{\mathrm{s}} & 0 & -P_l^{\mathrm{s}} \,\mathrm{e}^{-2\mathrm{i}\gamma^{\mathrm{s}}} \\ 0 & 0 & 0 \\ -P_l^{\mathrm{s}} \,\mathrm{e}^{2\mathrm{i}\gamma^{\mathrm{s}}} & 0 & 1-L_{\perp}^{\mathrm{s}} \end{pmatrix} \tag{46.63}$$

where

$$w^{\mathrm{t}} = \frac{\sigma^{\mathrm{t}}}{\sigma^{\mathrm{s}}+3\sigma^{\mathrm{t}}} = \frac{\sigma^{\mathrm{t}}}{4\sigma_{\mathrm{u}}}\,, \tag{46.64}$$

$$w^{\mathrm{s}} = \frac{\sigma^{\mathrm{s}}}{\sigma^{\mathrm{s}}+3\sigma^{\mathrm{t}}} = \frac{\sigma^{\mathrm{s}}}{4\sigma_{\mathrm{u}}} = 1-3w^{\mathrm{t}}\,, \tag{46.65}$$

$$\sigma_{\mathrm{u}} = (3w^{\mathrm{t}}+w^{\mathrm{s}})\sigma_{\mathrm{u}} = \frac{3}{4}\sigma^{\mathrm{t}} + \frac{1}{4}\sigma^{\mathrm{s}}\,. \tag{46.66}$$

The six parameters σ_{u}, w^{t}, $L_{\perp}^{\mathrm{t}}$, $L_{\perp}^{\mathrm{s}}$, γ^{t}, and γ^{s} have now been introduced, leaving one parameter still to be chosen. Inspection of Fig. 46.6 suggests, for example, the angle Δ^{+}. The fourth angle, Δ^{-}, is then fixed through the relation

$$\Delta^{+} - \Delta^{-} = \delta^{\mathrm{t}} - \delta^{\mathrm{s}} = 2\left(\gamma^{\mathrm{s}} - \gamma^{\mathrm{t}}\right). \tag{46.67}$$

The following set of six dimensioness parameters is thus complete:

$$w^{\mathrm{t}},\, L_{\perp}^{\mathrm{t}},\, L_{\perp}^{\mathrm{s}},\, \gamma^{\mathrm{t}},\, \gamma^{\mathrm{s}},\, \Delta^{+}\,. \tag{46.68}$$

Detailed recipes for extraction of the parameters from coherence experiments are somewhat complicated, and we refer to discussions in the literature [46.7, 15, 16]. Using spin-polarized electrons and spin-polarized targets, all parameters may be determined, except for information about singlet-triplet phase differences, such as Δ^{+}.

The reduced Stokes vector $\boldsymbol{P}$ of the unpolarized beam experiment is given by the singlet and triplet (unit) Stokes vectors $\boldsymbol{P}^{\mathrm{s,t}}$ as

$$\boldsymbol{P} = 3\,w^{\mathrm{t}}\,\boldsymbol{P}^{\mathrm{t}} + w^{\mathrm{s}}\boldsymbol{P}^{\mathrm{s}}\,, \tag{46.69}$$

from which the set of parameters $(L_{\perp}, \gamma, P_l)$ for the unpolarized beam experiment may be evaluated from

$$L_{\perp} = 3\,w^{\mathrm{t}}\,L_{\perp}^{\mathrm{t}} + w^{\mathrm{s}}\,L_{\perp}^{\mathrm{s}}\,, \tag{46.70}$$

$$P_l\,\mathrm{e}^{2\mathrm{i}\gamma} = 3w^{\mathrm{t}}\,P_l^{\mathrm{t}}\,\mathrm{e}^{2\mathrm{i}\gamma^{\mathrm{t}}} + w^{\mathrm{s}}\,P_l^{\mathrm{s}}\,\mathrm{e}^{2\mathrm{i}\gamma^{\mathrm{s}}}\,. \tag{46.71}$$

Since in general, $L_{\perp}^{\mathrm{t}} \neq L_{\perp}^{\mathrm{s}}$ and $\gamma^{\mathrm{t}} \neq \gamma^{\mathrm{s}}$, this causes the (reduced) degree of polarization P to be smaller than unity.

To summarize this section, Stokes parameter analysis may provide five dimensionless parameters, the relative phase between the two f_{+1}^{t} and f_{-1}^{t} amplitudes and the relative phase between the two f_{+1}^{s} and f_{-1}^{s} amplitudes, as well as the relative sizes of all four amplitudes. However, none of the relative phases between any triplet and singlet amplitude can be determined, and coherence analysis alone is thus not able to provide a "perfect scattering experiment". The missing phase may be extracted from the STU parameters of the scattered electron, ([46.16, 17] and Chapt. 7).

46.2.3 The Incoherent Case without Conservation of Atomic Reflection Symmetry

The results of coherence analysis of a $J=0^{\mathrm{e}} \to J=1^{\mathrm{o}}$ transition will now be discussed. We only analyze the photon polarization in the exit channel, not the electron spin parameters. There are six independent scattering amplitudes for a $J=0 \to J=1$ transition (Fig. 46.7), thereby requiring the determination of one absolute differential cross section, five relative magnitudes, and another five relative phases of the scattering amplitudes.

This large number of independent parameters leads to considerable complications. Nevertheless, the

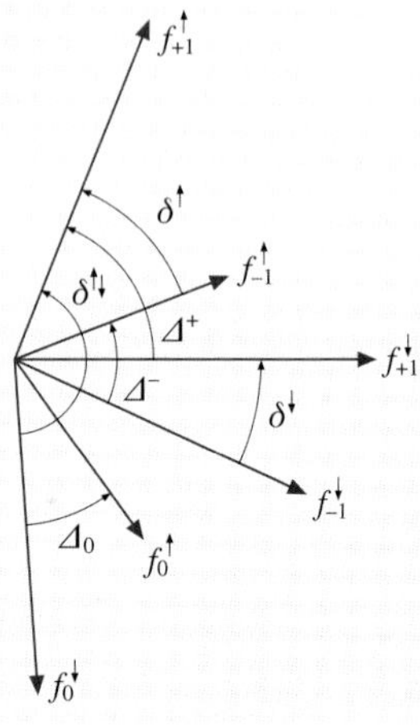

Fig. 46.7 For $J=0 \to J=1$ electron impact excitation of heavy atoms six scattering amplitudes come into play since spin-flip may occur

natural coordinate system enables disentangling of the scattering amplitudes and generalization of the parametrization of the density matrix for the case of unpolarized beams in a straightforward way [46.7, 18]. The nonvanishing amplitudes $f^{\mathrm{n}}(M_f, m_f, m_i)$ in the natural frame (Fig. 46.7) for a $J=0 \to J=1$ transition are

$$f^{\mathrm{n}}\left(1,\frac{1}{2},\frac{1}{2}\right) \equiv f_{+1}^{\uparrow} = \alpha_{+}\,\mathrm{e}^{\mathrm{i}\phi_{+}}\,, \tag{46.72}$$

$$f^{\mathrm{n}}\left(1,-\frac{1}{2},-\frac{1}{2}\right) \equiv f_{+1}^{\downarrow} = \beta_{+}\,\mathrm{e}^{\mathrm{i}\psi_{+}}\,, \tag{46.73}$$

$$f^{\mathrm{n}}\left(-1,\frac{1}{2},\frac{1}{2}\right) \equiv f_{-1}^{\uparrow} = \alpha_{-\mathrm{e}}^{\mathrm{i}\phi_{-}}\,, \tag{46.74}$$

$$f^{\mathrm{n}}\left(-1,-\frac{1}{2},-\frac{1}{2}\right) \equiv f_{-1}^{\downarrow} = \beta_{-}\,\mathrm{e}^{\mathrm{i}\psi_{-}}\,, \tag{46.75}$$

$$f^{\mathrm{n}}\left(0,\frac{1}{2},-\frac{1}{2}\right) \equiv f_{0}^{\downarrow} = \beta_{0}\,\mathrm{e}^{\mathrm{i}\psi_{0}}\,, \tag{46.76}$$

$$f^{\mathrm{n}}\left(0,-\frac{1}{2},\frac{1}{2}\right) \equiv f_{0}^{\uparrow} = \alpha_{0\mathrm{e}}^{\mathrm{i}\phi_{0}}\,, \tag{46.77}$$

where we have omitted $J_i = M_i = 0$ and $J_f = 1$. Equations (46.72–46.75) represent noflip amplitudes that leave the projectile spin unchanged while (46.76) and (46.77) describe the cases where the electron spin is flipped.

We first assume a polarization perpendicular to the scattering plane, i. e., along the z-direction. In (46.44), the density matrix for heavy atoms such as Xe or Hg was decomposed into a pair of matrices with one having positive reflection symmetry with respect to the scattering plane and the other one having negative reflection symmetry, respectively. The extension of this decomposition to the case of polarized electron beams is a pair of density matrices, one for spin-up electron impact excitation and one for spin-down excitation where "up" and "down" correspond to the initial spin component orientation with respect to the scattering plane. Hence,

$$\begin{aligned}
&\rho_{\mathrm{u}}\\
&= \sigma_{\mathrm{u}}\left[(1-h)\frac{1}{2}\begin{pmatrix} 1+L_{\perp}^{+} & 0 & -P_l^{+}\,\mathrm{e}^{-2\mathrm{i}\gamma} \\ 0 & 0 & 0 \\ -P_l^{+}\,\mathrm{e}^{2\mathrm{i}\gamma} & 0 & 1-L_{\perp}^{+}\end{pmatrix} + h\begin{pmatrix}0&0&0\\0&1&0\\0&0&0\end{pmatrix}\right]\\
&= w^{\uparrow}\rho^{\uparrow} + w^{\downarrow}\rho^{\downarrow}\\
&= w^{\uparrow}\sigma_{\mathrm{u}}\left[(1-h^{\uparrow})\frac{1}{2}\right.\\
&\quad\times\begin{pmatrix} 1+L_{\perp}^{+\uparrow} & 0 & -P_l^{+\uparrow}\,\mathrm{e}^{-2\mathrm{i}\gamma^{\uparrow}} \\ 0 & 0 & 0 \\ -P_l^{+\uparrow}\,\mathrm{e}^{2\mathrm{i}\gamma^{\uparrow}} & 0 & 1-L_{\perp}^{+\uparrow}\end{pmatrix}\\
&\quad\left.+h^{\uparrow}\begin{pmatrix}0&0&0\\0&1&0\\0&0&0\end{pmatrix}\right]\\
&\quad+w^{\downarrow}\sigma_{\mathrm{u}}\left[(1-h^{\downarrow})\frac{1}{2}\right.\\
&\quad\times\begin{pmatrix} 1+L_{\perp}^{+\downarrow} & 0 & -P_l^{+\downarrow}\,\mathrm{e}^{-2\mathrm{i}\gamma^{\downarrow}} \\ 0 & 0 & 0 \\ -P_l^{+\downarrow}\,\mathrm{e}^{2\mathrm{i}\gamma^{\downarrow}} & 0 & 1-L_{\perp}^{+\downarrow}\end{pmatrix}\\
&\quad\left.+h^{\downarrow}\begin{pmatrix}0&0&0\\0&1&0\\0&0&0\end{pmatrix}\right]\,.
\end{aligned} \tag{46.78}$$

Here we have defined

$$L_{\perp}^{+\uparrow} = \frac{\alpha_{+}^2-\alpha_{-}^2}{\alpha_{+}^2+\alpha_{-}^2} = -P_3^{\uparrow}\,, \tag{46.79}$$

$$L_{\perp}^{+\downarrow} = \frac{\beta_{+}^2-\beta_{-}^2}{\beta_{+}^2+\beta_{-}^2} = -P_3^{\downarrow}\,, \tag{46.80}$$

$$P_l^{+\uparrow}\,\mathrm{e}^{2\mathrm{i}\gamma^{\uparrow}} = P_1^{\uparrow}+\mathrm{i}P_2^{\uparrow} = -\frac{2\alpha_{+}\alpha_{-}\,\mathrm{e}^{\mathrm{i}(\phi_{-}-\phi_{+})}}{\alpha_{+}^2+\alpha_{-}^2}\,, \tag{46.81}$$

$$P_l^{+\downarrow}\,\mathrm{e}^{2\mathrm{i}\gamma^{\downarrow}} = P_1^{\downarrow}+\mathrm{i}P_2^{\downarrow} = -\frac{2\beta_{+}\beta_{-}\,\mathrm{e}^{\mathrm{i}(\psi_{-}-\psi_{+})}}{\beta_{+}^2+\beta_{-}^2}\,, \tag{46.82}$$

$$\sigma^{\uparrow} = \alpha_{+}^2+\alpha_{-}^2+\alpha_{0}^2\,, \tag{46.83}$$

$$\sigma^{\downarrow} = \beta_{+}^2+\beta_{-}^2+\beta_{0}^2\,, \tag{46.84}$$

$$\begin{aligned}\sigma_u &= \frac{1}{2}\left(\alpha_{+}^2+\alpha_{-}^2+\alpha_{0}^2+\beta_{+}^2+\beta_{-}^2+\beta_{0}^2\right)\\ &= \frac{1}{2}\left(\sigma^{\uparrow}+\sigma^{\downarrow}\right)\,,\end{aligned} \tag{46.85}$$

$$h^{\uparrow} = \alpha_0^2/\sigma^{\uparrow}\,, \tag{46.86}$$

$$h^{\downarrow} = \beta_0^2/\sigma^{\downarrow}\,, \tag{46.87}$$

$$w^{\uparrow} = \sigma^{\uparrow}/(2\,\sigma_{\mathrm{u}})\,, \tag{46.88}$$

$$w^{\downarrow} = \sigma^{\downarrow}/(2\,\sigma_{\mathrm{u}}) = 1-w^{\uparrow}\,. \tag{46.89}$$

From these definitions it follows that

$$(1-h)\,L_\perp^+ = w^\uparrow\left(1-h^\uparrow\right)L_\perp^{+\uparrow} + w^\downarrow\left(1-h^\downarrow\right)L_\perp^{+\downarrow}\,, \quad (46.90)$$

$$(1-h)\,P_l^+\,\mathrm{e}^{2\mathrm{i}\gamma} = w^\uparrow\left(1-h^\uparrow\right)P_l^{+\uparrow}\,\mathrm{e}^{2\mathrm{i}\gamma^\uparrow} + w^\downarrow\left(1-h^\downarrow\right)P_l^{+\downarrow}\,\mathrm{e}^{2\mathrm{i}\gamma^\downarrow}\,, \quad (46.91)$$

$$h = w^\uparrow h^\uparrow + w^\downarrow h^\downarrow = \left(\alpha_0^2+\beta_0^2\right)/(2\sigma_\mathrm{u})\,, \quad (46.92)$$

$$P_l^{+\uparrow 2} + L_\perp^{+\uparrow 2} = P_l^{+\uparrow 2} + P_3^{\uparrow 2} = 1\,, \quad (46.93)$$

$$P_l^{+\downarrow 2} + L_\perp^{+\downarrow 2} = P_l^{+\downarrow 2} + P_3^{\downarrow 2} = 1\,. \quad (46.94)$$

Extraction of these parameters is facilitated by introduction of "generalized Stoke parameters" [46.18]. In this way, $L_\perp^{+\uparrow}$, $L_\perp^{+\downarrow}$, $\gamma^\uparrow$, $\gamma^\downarrow$ may be determined. If, in addition, h is known, e.g., by polarization analysis in the y-direction, the following set of seven dimensionless independent parameters can be derived from the generalized Stokes parameters in the z-direction:

$$L_\perp^{+\uparrow},\ L_\perp^{+\downarrow},\ h^\uparrow,\ h^\downarrow,\ w^\uparrow;\gamma^\uparrow,\ \gamma^\downarrow. \quad (46.95)$$

This leaves three relative phases unknown. In the notation of Fig. 46.7, we see from inspection that

$$\Delta^+ - \Delta^- = \delta^\uparrow - \delta^\downarrow = 2\left(\gamma^\downarrow - \gamma^\uparrow\right), \quad (46.96)$$

in analogy to (46.67). A convenient choice for the remaining phase angles is $\left(\Delta^+,\ \Delta^0,\ \delta^{\uparrow\downarrow}\right)$, with

$$\delta^{\uparrow\downarrow} \equiv \phi_+ - \psi_0\,. \quad (46.97)$$

A complete set of dimensionless independent parameters is then given by

$$\left(w^\uparrow, L_\perp^{+\uparrow}, L_\perp^{+\downarrow}, h^\uparrow, h^\downarrow, \gamma^\uparrow, \gamma^\downarrow, \Delta^+, \Delta^0, \delta^{\uparrow\downarrow}\right). \quad (46.98)$$

Information about the remaining three phase angles may be obtained in experiments with *in-plane* spin polarization. Further analysis shows that the generalized Stokes parameters in the y (or x) direction with in-plane spin polarization P_y or P_x provides two additional phases. None of the relative phases Δ^+ between $f_{+1}^\uparrow$ and $f_{+1}^\downarrow$, etc. enter. Determination of the final remaining angle requires determination of generalized STU parameters, describing the electron spin in the exit channel.

Table 46.2 summarizes the various cases with polarized beams discussed in this section.

Table 46.2 Summary of cases of increasing complexity for spin-polarized beams. The number of independent dimensionless parameters N_p is listed, along with N_OA, the number determined from orientation and alignment only. N_d is the number of observation directions required

Variable	Sect. 46.2.1	Sect. 46.2.2	Sect. 46.2.3
Forces	Coulomb	+exchange	+spin–orbit
Representation	wave func.	$\rho_{mn}^{\mathrm{t,s}}$	$\rho_{mn}^{\uparrow,\downarrow}$
Refl. symmetry	+	+	+, −
N_p	2	6	10
N_OA	2	5	9
N_d	1	1	2

46.3 Example

46.3.1 The First Born Approximation

As a simple, illustrative example, consider the predictions of the first Born approximation (FBA) FBA. Here, S → P excitation by electron impact is described as creation of a pure p-orbital along the direction of the linear momentum transfer $\Delta\boldsymbol{k} = \boldsymbol{k}_\mathrm{in} - \boldsymbol{k}_\mathrm{out}$, along which there is axial symmetry. Evidently

$$L_\perp^\mathrm{FBA} = 0\,, \quad (46.99)$$

and consequently

$$P_l^\mathrm{FBA} = 1\,. \quad (46.100)$$

The alignment angle is found from simple geometrical considerations, see Fig. 46.8. Denoting the incident energy by E and the energy loss by ΔE, the relation between the projectile scattering angle Θ_col and the alignment angle γ is directly read from the

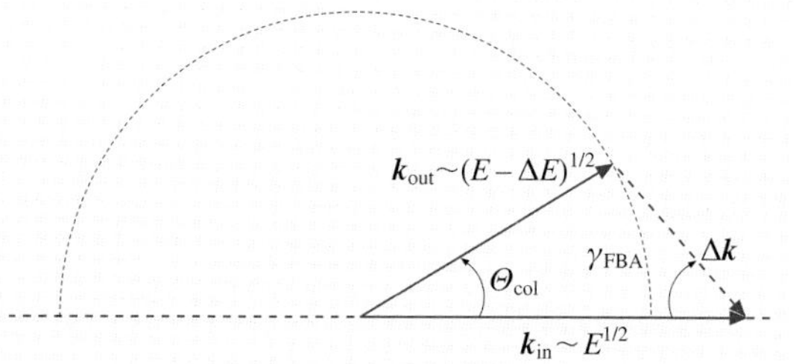

Fig. 46.8 Diagram for evaluation of the alignment angle γ in the first Born approximation

figure,

$$\tan \gamma^{\mathrm{FBA}} = \frac{\sin \Theta_{\mathrm{col}}}{\cos \Theta_{\mathrm{col}} - x}\,, \tag{46.101}$$

where $x = [E/(E - \Delta E)]^{1/2}$. For $\Delta E > 0$, γ^{FBA} is always negative, with its minimum value when $\Delta \boldsymbol{k} \perp \boldsymbol{k}_{\mathrm{out}}$. Any theoretical effort beyond the FBA involves serious computations.

46.4 Recent Developments

46.4.1 S → D Excitation

The generalization of the formalism of Sect. 46.2.2 to the case of S → D excitation involves the introduction of three scattering amplitudes, corresponding to a complete parameter set of one cross section, two relative amplitude sizes, and two relative phases. Analysis shows that a full coherence analysis of the light emitted in the subsequent D → P optical decay is not sufficient for a complete experiment, instead two solutions are obtained. A triple coincidence experiment may resolve the ambiguity [46.19].

46.4.2 P → P Excitation

By proper optical preparation of the atomic target, collision studies involving specific excited states may be performed as a function of scattering angle. For collision-induced P → P transitions, a systematic preparation of specific initial P states, combined with Stokes parameter analysis of the radiation pattern from the final P state, may lead to a complete scattering experiment. The corresponding complete set of nine parameters describes the process in terms of five independent scattering amplitudes. In addition to the charge cloud shape and orientation parameters, three Euler angles are needed to describe the atomic reference frame of the charge cloud with respect to the laboratory frame [46.20].

46.4.3 Relativistic Effects in S → P Excitation

It has been discussed to what extent relativistic effects can be studied for excitation of the two fine structure components of the resonance transitions of heavy alkali atoms, such as Rb or Cs. For electron-impact excitation, standard Stokes parameter analysis turns out to be extremely insensitive to the inclusion of relativistic effects in the numerical treatment, which explains the success of nonrelativistic theories. If spin-polarized electrons are used, either in the incident channel through measurement of spin asymmetries, or in the final channels by performing a time-reversed generalized Stokes parameter experiment with a laser-prepared target and a spin-polarized electron beam, distinct relativistic effects, typically at the 5% level, may be revealed [46.21].

46.5 Summary

A selection of fundamental formulas describing orientation and alignment in atomic collisions is given, with emphasis on the simplest case, S → P excitation. A tutorial introduction to the field with a series of examples and applications may be found in a recent textbook [46.22].

References

46.1 B. Bederson: Comments At. Mol. Phys. **1**, 41 and 65 (1969)

46.2 U. Fano, J. H. Macek: Rev. Mod. Phys. **45**, 553 (1973)

46.3 J. Kessler: *Polarized Electrons* (Springer, Berlin, Heidelberg 1985)

46.4 I. V. Hertel, H. Schmidt, A. Bähring, E. Meyer: Rep. Prog. Phys. **48**, 375 (1985)

46.5 N. Andersen, J. W. Gallagher, I. V. Hertel: Phys. Rep. **165**, 1 (1988)

46.6 N. Andersen, J. T. Broad, E. E. B. Campbell, J. W. Gallagher, I. V. Hertel: Phys. Rep. **278**, 107 (1997)

46.7 N. Andersen, K. Bartschat, J. T. Broad, I. V. Hertel: Phys. Rep. **279**, 251 (1997)

46.8 U. Wille, R. Hippler: Phys. Rep. **132**, 129 (1986)

46.9 C. H. Greene, R. N. Zare: Ann. Rev. Phys. Chem. **33**, 119 (1982)
46.10 K. Blum: *Density Matrix Theory and Applications* (Plenum, New York 1981)
46.11 J. Macek, D. H. Jaecks: Phys. Rev. A **4**, 2288 (1971)
46.12 M. Born, E. Wolf: *Principles of Optics* (Pergamon, New York 1970)
46.13 K. Blum, F. T. da Paixão, G. Csanak: J. Phys. B **13**, L257 (1980)
46.14 F. T. da Paixão, N. T. Padial, Gy. Csanak, K. Blum: Phys. Rev. Lett. **45**, 1164 (1980)
46.15 I. V. Hertel, M. H. Kelley, J. J. McClelland: Z. Phys. D **6**, 163 (1987)
46.16 N. Andersen, K. Bartschat: Comments At. Mol. Phys. **29**, 157 (1993)
46.17 K. Bartschat: Phys. Rep. **180**, 1 (1989)
46.18 N. Andersen, K. Bartschat: J. Phys. B **27**, 3189 (1994); corrigendum N. Andersen, K. Bartschat: J. Phys. B **29**, 1149 (1996), see [46.23]
46.19 N. Andersen, K. Bartschat: J. Phys. B **30**, 5071 (1997)
46.20 E. Y. Sidky, S. Grego, D. Dowek, N. Andersen: J. Phys. B **35**, 2005 (2002)
46.21 N. Andersen, K. Bartschat: J. Phys. **35**, 4507 (2002)
46.22 N. Andersen, K. Bartschat: *Polarization, Alignment, and Orientation in Atomic Collisions* (Springer, Berlin, Heidelberg 2001)
46.23 K. Muktavat, R. Srivastava, A. D. Stauffer: J. Phys. B **36**, 2341 (2003)

47. Electron–Atom, Electron–Ion, and Electron–Molecule Collisions

This chapter reviews the theory of electron collisions with atoms, ions and molecules. Section 47.1 discusses elastic, inelastic and ionizing collisions with atoms and atomic ions from close to threshold to high energies where the Born series becomes applicable. Section 47.2 extends the theory to treat electron collisions with molecules. Finally in Sect. 47.3 the theory of electron atom collisions in intense laser fields is discussed. This chapter will not present detailed comparisons of theoretical predictions with experiment. Such comparisons are given in recent review articles [47.1–3] and in Chapt. 63.

47.1 Electron–Atom and Electron–Ion Collisions 705
47.1.1 Low-Energy Elastic Scattering and Excitation 705
47.1.2 Relativistic Effects for Heavy Atoms and Ions......................... 708
47.1.3 Multichannel Resonance Theory.. 710
47.1.4 Multichannel Quantum Defect Theory 711
47.1.5 Solution of the Coupled Integrodifferential Equations...... 712
47.1.6 Intermediate and High Energy Elastic Scattering and Excitation.. 714
47.1.7 Ionization 717

47.2 Electron–Molecule Collisions................ 720
47.2.1 Laboratory Frame Representation 720
47.2.2 Molecular Frame Representation. 721
47.2.3 Inclusion of the Nuclear Motion .. 722
47.2.4 Electron Collisions with Polyatomic Molecules......... 723

47.3 Electron–Atom Collisions in a Laser Field................................... 723
47.3.1 Potential Scattering.................. 724
47.3.2 Scattering by Complex Atoms and Ions 725

References 727

47.1 Electron–Atom and Electron–Ion Collisions

47.1.1 Low-Energy Elastic Scattering and Excitation

In this section we consider the process

$$e^- + A_i \to e^- + A_j\,, \tag{47.1}$$

where A_i and A_j are bound states of the target atom or ion and where the velocity of the incident or scattered electron is of the same order or less than that of the target electrons actively involved in the collision.

Assume initially that all relativistic effects can be neglected, which restricts the treatment to low-Z atoms and ions. The Schrödinger equation describing the scattering of an electron by a target atom or ion containing N electrons and nuclear charge Z is then

$$H_{N+1}\Psi = E\Psi\,, \tag{47.2}$$

where E is the total energy of the system. The $(N+1)$-electron nonrelativistic Hamiltonian H_{N+1} is given in atomic units by

$$H_{N+1} = \sum_{i=1}^{N+1}\left(-\frac{1}{2}\nabla_i^2 - \frac{Z}{r_i}\right) + \sum_{i>j=1}^{N+1}\frac{1}{r_{ij}}\,, \tag{47.3}$$

where $r_{ij} = |\boldsymbol{r}_i - \boldsymbol{r}_j|$, and $\boldsymbol{r}_i$ and $\boldsymbol{r}_j$ are the vector coordinates of electrons i and j relative to the origin of coordinates taken to be the target nucleus, which is assumed to have infinite mass.

The target eigenstates Φ_i and the corresponding eigenenergies w_i satisfy the equation

$$\langle\Phi_i|H_N|\Phi_j\rangle = w_i\delta_{ij}\,, \tag{47.4}$$

where H_N is defined by (47.3) with $N+1$ replaced by N. The calculation of accurate target states is discussed in Chapt. 21. The solution of (47.2), corresponding to the process (47.1), then has the asymptotic form

$$\Psi_i \underset{r\to\infty}{\approx} \Phi_i \chi_{\frac{1}{2}m_i} \mathrm{e}^{\mathrm{i}k_i z} + \sum_j \Phi_j \chi_{\frac{1}{2}m_j} f_{ji}(\theta,\phi)\,\mathrm{e}^{\mathrm{i}k_j r} . \tag{47.5}$$

In (47.5), $\chi_{\frac{1}{2}m_i}$ and $\chi_{\frac{1}{2}m_j}$ are the spin eigenfunctions of the incident and scattered electrons, where the direction of spin quantization is usually taken to be the incident beam direction, and $f_{ji}(\theta,\phi)$ is the scattering amplitude, the spherical polar coordinates of the scattered electron being denoted by r, θ and ϕ. Also the wave numbers k_i and k_j are related to the total energy of the system by

$$E = w_i + \frac{1}{2}k_i^2 = w_j + \frac{1}{2}k_j^2 . \tag{47.6}$$

The outgoing wave term in (47.5) contains contributions from all target states that are energetically allowed; i. e., for which $k_j^2 \geq 0$. If the energy is above the ionization threshold, this includes target continuum states. For an atomic ion, a logarithmic phase factor is also needed as discussed below.

The differential cross section for a transition from an initial state $|i\rangle = \left|\boldsymbol{k}_i, \Phi_i, \chi_{\frac{1}{2}m_i}\right\rangle$ to a final state $|j\rangle = |\boldsymbol{k}_j, \Phi_j, \chi_{\frac{1}{2}m_j}\rangle$ is given by

$$\frac{\mathrm{d}\sigma_{ji}}{\mathrm{d}\Omega} = \frac{k_j}{k_i}|f_{ji}(\theta,\phi)|^2 , \tag{47.7}$$

and the total cross section is obtained by averaging over initial spin states, summing over final spin states and integrating over all scattering angles.

In order to solve the Schrödinger equation to obtain the scattering amplitude and cross section at low energies, we make a partial wave expansion of the total wave function

$$\begin{aligned}\Psi_j^\Gamma(\boldsymbol{X}_{N+1}) &= \mathcal{A}\sum_{i=1}^{n} \overline{\Phi}_i^\Gamma(\boldsymbol{x}_1,\ldots,\boldsymbol{x}_N;\hat{\boldsymbol{r}}_{N+1}\sigma_{N+1}) \\ &\quad\times r_{N+1}^{-1} F_{ij}^\Gamma(r_{N+1}) \\ &\quad+\sum_{i=1}^{m}\chi_i^\Gamma(\boldsymbol{x}_1,\ldots,\boldsymbol{x}_{N+1})\,b_{ij}^\Gamma ,\end{aligned} \tag{47.8}$$

where $\boldsymbol{X}_{N+1} \equiv \boldsymbol{x}_1, \boldsymbol{x}_2 \cdots \boldsymbol{x}_{N+1}$ represents the space and spin coordinates of all $N+1$ electrons, $\boldsymbol{x}_i \equiv \boldsymbol{r}_i\sigma_i$ represents the space and spin coordinates of the ith electron and $\mathcal{A}$ is the operator that antisymmetrizes the first summation with respect to exchange of all pairs of electrons in accordance with the Pauli exclusion principle. The channel functions $\overline{\Phi}_i^\Gamma$, assumed to be n in number, are obtained by coupling the orbital and spin angular momenta of the target states Φ_i with those of the scattered electron to form eigenstates of the total orbital and spin angular momenta, their z components and the parity π, where

$$\Gamma \equiv LM_L SM_S\pi \tag{47.9}$$

is conserved in the collision. The square integrable correlation functions χ_i^Γ allow for additional correlation effects not included in the first expansion in (47.8) that goes over a limited number of target eigenstates, and possibly pseudostates.

By substituting (47.8) into the Schrödinger equation (47.2), projecting onto the channel functions $\overline{\Phi}_i^\Gamma$ and onto the square integrable functions χ_i^Γ, and eliminating the coefficients b_{ij}^Γ, we obtain n coupled integrodifferential equations satisfied by the reduced radial functions F_{ij}^Γ representing the motion of the scattered electron of the form

$$\begin{aligned}&\left(\frac{\mathrm{d}^2}{\mathrm{d}r^2} - \frac{\ell_i(\ell_i+1)}{r^2} + \frac{2(Z-N)}{r} + k_i^2\right) F_{ij}^\Gamma(r) \\ &= 2\sum_\ell \Bigg\{ V_{i\ell}^\Gamma(r) F_{\ell j}^\Gamma(r) \\ &\quad + \int_0^\infty \left[K_{i\ell}^\Gamma(r,r') + X_{i\ell}^\Gamma(r,r')\right] F_{\ell j}^\Gamma(r')\,\mathrm{d}r' \Bigg\} .\end{aligned} \tag{47.10}$$

Here ℓ_i is the orbital angular momentum of the scattered electron, and $V_{i\ell}^\Gamma$, $W_{i\ell}^\Gamma$ and $X_{i\ell}^\Gamma$ are the local direct, nonlocal exchange and nonlocal correlation potentials respectively. If the correlation potential which arises from the χ_i^Γ terms in (47.8) is not included, then (47.10) are called the close coupling equations.

The direct potential can be written as

$$\begin{aligned}&V_{ij}^\Gamma(r_{N+1}) \\ &= \Bigg\langle \overline{\Phi}_i^\Gamma(\boldsymbol{x}_1,\ldots,\boldsymbol{x}_N;\hat{\boldsymbol{r}}_{N+1}\sigma_{N+1}) \\ &\quad \times \left|\sum_{i=1}^{N}\frac{1}{r_{iN+1}} - \frac{N}{r_{N+1}}\right| \\ &\quad \times \overline{\Phi}_j^\Gamma(\boldsymbol{x}_1,\ldots,\boldsymbol{x}_N;\hat{\boldsymbol{r}}_{N+1}\sigma_{N+1})\Bigg\rangle ,\end{aligned} \tag{47.11}$$

where the integral is taken over all electron space and spin coordinates, except the radial coordinate of the

$(N+1)$th electron. This potential has the asymptotic form

$$V_{ij}^{\Gamma}(r) = \sum_{\lambda=1}^{\lambda\,\max} a_{ij}^{\lambda} r^{-\lambda-1}\,, \quad r \geq a \tag{47.12}$$

where a is the range beyond which the orbitals in the target states Φ_i included in the first expansion in (47.8), are negligible. The $\lambda = 1$ term in (47.12) gives rise, in second-order, to the long-range attractive polarization potential

$$V(r) \underset{r\to\infty}{\to} -\frac{1}{2}\frac{\alpha}{r^4} \tag{47.13}$$

seen by an electron incident on an atom. For an s-state atom in a state Φ_0 the dipole polarizability α is given by

$$\alpha = 2\sum_j \frac{\left|\langle \Phi_0 | \left(\frac{4\pi}{3}\right)^{1/2} \sum_{i=1}^{N} r_i Y_{10}(\hat{\boldsymbol{r}}_i) | \Phi_j \rangle \right|^2}{w_j - w_0}\,. \tag{47.14}$$

These long-range potentials have a profound influence on low-energy scattering.

The exchange and correlation potentials, unlike the direct potential, are both nonlocal and the exchange potential vanishes exponentially for large r. Explicit expressions for these potentials are too complicated to write down, except in the case of e^-–H scattering where the direct and exchange potentials were first given by *Percival* and *Seaton* [47.4]. Instead they are determined by general computer programs.

The scattering amplitude and cross section can be obtained by solving (47.10) for all relevant conserved quantum numbers Γ subject to the following K-matrix asymptotic boundary conditions

$$\begin{aligned} F_{ij}^{\Gamma} &\underset{r\to\infty}{\approx} k_i^{-\frac{1}{2}} \left(\sin\theta_i \delta_{ij} + \cos\theta_i K_{ij}^{\Gamma}\right)\,, \\ &\qquad \text{open channels } k_i^2 \geq 0 \\ F_{ij}^{\Gamma} &\underset{r\to\infty}{\approx} 0\,, \\ &\qquad \text{closed channels } k_i^2 < 0 \end{aligned} \tag{47.15}$$

where

$$\theta_i = k_i r - \frac{1}{2}\ell_i \pi + \frac{z}{k_i} \ln(2k_i r) + \sigma_i \tag{47.16}$$

with $z = Z - N$, and $\sigma_i = \arg\Gamma(\ell_i + 1 - iz/k_i)$. The S-matrix and T-matrix are related to the K-matrix defined by (47.15) by the matrix equations

$$\boldsymbol{S}^{\Gamma} = \frac{\boldsymbol{I} + i\boldsymbol{K}^{\Gamma}}{\boldsymbol{I} - i\boldsymbol{K}^{\Gamma}}\,, \quad \boldsymbol{T}^{\Gamma} = \boldsymbol{S}^{\Gamma} - \boldsymbol{I} = \frac{2i\boldsymbol{K}^{\Gamma}}{\boldsymbol{I} - i\boldsymbol{K}^{\Gamma}}\,, \tag{47.17}$$

where the dimensions of the matrices in these equations are $n_a \times n_a$, where n_a is the number of open channels at the energy under consideration for the given Γ. The Hermiticity and time reversal invariance of the Hamiltonian ensures that $\boldsymbol{K}^{\Gamma}$ is real and symmetric, and $\boldsymbol{S}^{\Gamma}$ is unitary and symmetric.

The scattering amplitude defined by (47.5) can be expressed in terms of the T-matrix elements. For a neutral target,

$$\begin{aligned} f_{ji}(\theta,\phi) &= i\left(\frac{\pi}{k_i k_j}\right)^{\frac{1}{2}} \sum_{\substack{LS\pi \\ \ell_i \ell_j}} i^{\ell_i - \ell_j} (2\ell_i + 1)^{\frac{1}{2}} \\ &\times (L_i M_{L_i} \ell_i 0 | L_i \ell_i L M_L) \\ &\times \left(S_i M_{S_i} \frac{1}{2} m_i | S_i \frac{1}{2} S M_S\right) \\ &\times (L_j M_{L_j} \ell_j m_{\ell_j} | L_j \ell_j L M_L) \\ &\times \left(S_j M_{S_j} \frac{1}{2} m_j | S_j \frac{1}{2} S M_S\right) T_{ji}^{\Gamma} Y_{\ell_j m_{\ell_j}}(\theta,\phi)\,, \end{aligned} \tag{47.18}$$

which describes a transition from an initial state $\alpha_i L_i S_i M_{L_i} M_{S_i} m_i$ to a final state $\alpha_j L_j S_j M_{L_j} M_{S_j} m_j$, where α_i and α_j represent any additional quantum numbers required to completely define the initial and final states. The corresponding total cross section, obtained by averaging over the initial magnetic quantum numbers, summing over the final magnetic quantum numbers, and integrating over all scattering angles, is

$$\sigma_{\text{tot}}(i \to j) = \frac{\pi}{k_i^2} \sum_{\substack{LS\pi \\ \ell_i \ell_j}} \frac{(2L+1)(2S+1)}{2(2L_i+1)(2S_i+1)} \left|T_{ji}^{\Gamma}\right|^2\,, \tag{47.19}$$

which describes a transition from an initial target state $\alpha_i L_i S_i$ to a final target state $\alpha_j L_j S_j$. In applications, it is also useful to define a collision strength by

$$\Omega(i,j) = k_i^2 (2L_i + 1)(2S_i + 1)\sigma_{\text{tot}}(i \to j)\,, \tag{47.20}$$

which is dimensionless and symmetric with respect to interchange of the intial and final states denoted by i and j. For scattering by an ion, the above expression for $f_{ji}(\theta,\phi)$ is modified by the inclusion of the Coulomb scattering amplitude when the initial and final states are identical.

For incident electron energies insufficient to excite the atom or ion, only elastic scattering is possible and the above expressions simplify. Consider low energy elastic electron scattering by a neutral atom is a ^{1}S ground state. Then the expression for the scattering amplitude (47.18) reduces to

$$f(\theta) = \frac{1}{2ik} \sum_{\ell=0}^{\infty} (2\ell+1) \left(e^{2i\delta_\ell} - 1\right) P_\ell(\cos\theta) , \tag{47.21}$$

where $\ell \left(= L = \ell_i = \ell_j\right)$ is the angular momentum of the scattered electron, k is its wave number, and the phase shift δ_ℓ can be expressed in terms of the K-matrix, which now has only one element since $n_a = 1$, by

$$\tan\delta_\ell = K_{11}^{\Gamma} . \tag{47.22}$$

The corresponding expression for the total cross section is then

$$\sigma_{\text{tot}} = \frac{4\pi}{k^2} \sum_{\ell=0}^{\infty} (2\ell+1) \sin^2\delta_\ell . \tag{47.23}$$

A diffusion, or momentum transfer cross section, can also be defined as

$$\begin{aligned} \sigma_{\text{D}} &= 2\pi \int_0^{\pi} |f(\theta)|^2 (1-\cos\theta) \sin\theta \, d\theta \\ &= \frac{4\pi}{k^2} \sum_{\ell=0}^{\infty} (\ell+1) \sin^2(\delta_{\ell+1} - \delta_\ell) , \end{aligned} \tag{47.24}$$

which is important when considering the diffusion of electrons through gases.

At low incident electron energies, the behavior of the phase shift for an atom in a s-state is dominated by the long-range polarization potential (47.13). *O'Malley* et al. [47.5] showed that for s-wave scattering $k\cot\delta_0$ satisfies the effective range expansion

$$\begin{aligned} k\cot\delta_0 = &-\frac{1}{a_{\text{s}}} + \frac{\pi\alpha}{3a_{\text{s}}^2}k \\ &+ \frac{2\alpha}{3a_{\text{s}}} k^2 \ln\left(\frac{\alpha k^2}{16}\right) + O(k^2) , \end{aligned} \tag{47.25}$$

where a_{s} is the scattering length, while for $\ell \geq 1$

$$k^2\cot\delta_\ell = \frac{8\left(\ell+\frac{3}{2}\right)\left(\ell+\frac{1}{2}\right)\left(\ell-\frac{1}{2}\right)}{\pi\alpha} + \cdots . \tag{47.26}$$

It follows that close to threshold, the total elastic cross section has the form

$$\sigma_{\text{tot}} = 4\pi a_{\text{s}}^2 + \frac{8}{3}\pi^2\alpha a_{\text{s}} k + \cdots . \tag{47.27}$$

When an electron is elastically scattered by a positive or negative ion, then these formulae for the low-energy behavior of the phase shift are modified. For scattering by a positive ion, *Seaton* [47.6] has shown that

$$\frac{\cot\delta_\ell(k)}{1-e^{2\pi\eta}} = \cot\left[\pi\mu\left(k^2\right)\right] , \tag{47.28}$$

where $\eta = -z/k$, and where $\mu\left(k^2\right)$ is the analytic continuation of the quantum defects of the electron–ion bound states to positive energies. This quantum defect theory enables spectroscopic observations of bound state energies to be extrapolated to positive energies to yield electron–ion scattering phase shifts. For a negative ion, where the Coulomb potential is repulsive, the phase shift behaves as

$$\delta_\ell \underset{k\to 0}{\to} \exp(2\pi z/k) , \tag{47.29}$$

which vanishes rapidly as k tends to zero since z is now negative.

47.1.2 Relativistic Effects for Heavy Atoms and Ions

As the nuclear charge Z of the target increases, relativistic effects become important even for low energy scattering. There are two ways in which relativistic effects play a role. First, there is a direct effect corresponding to the relativistic distortion of the wave function describing the scattered electron by the strong nuclear Coulomb potential. Second, there is an indirect effect caused by the change in the charge distribution of the target due to the use of relativistic wave functions discussed in Chapt. 22. We will concentrate on the direct effect in this section.

For atoms and ions with small Z, the K-matrices can first be calculated in *LS* coupling, neglecting relativistic effects. The K-matrices are then recoupled to yield transitions between fine-structure levels. We introduce the pair-coupling scheme

$$\boldsymbol{L}_i + \boldsymbol{S}_i = \boldsymbol{J}_i , \quad \boldsymbol{J}_i + \boldsymbol{\ell}_i = \boldsymbol{K}_i , \quad \boldsymbol{K}_i + \boldsymbol{s} = \boldsymbol{J} , \tag{47.30}$$

where $\boldsymbol{J}_i$ is the total angular momentum of the target, $\boldsymbol{\ell}_i$ is the orbital angular momentum of the scattered electron, $\boldsymbol{s}$ is its spin, and $\boldsymbol{J}$ is the total angular momentum, which with the parity π is conserved in the collision.

The transition from *LS* coupling involves the recoupling coefficient

$$\left\langle [(L_i S_i) J_i, \ell_i] K_i, \frac{1}{2}; JM_J | (L_i \ell_i) L, \left(S_i \frac{1}{2}\right) S; JM_J \right\rangle$$
$$= [(2J_i+1)(2L+1)(2K_i+1)(2S+1)]^{\frac{1}{2}}$$
$$\times W(L\ell_i S_i J_i; L_i K_i)\, W\left(LJS_i\frac{1}{2}; SK_i\right) , \quad (47.31)$$

and the corresponding K-matrix transforms as

$$K_{ij}^{J\pi} = \sum_{LS} \left\langle ((L_i S_i) J_i, \ell_i) K_i, \frac{1}{2};\right.$$
$$\left. \times JM_J | (L_i \ell_i) L, \left(S_i \frac{1}{2}\right) S; JM_J \right\rangle \times K_{ij}^{\Gamma}$$
$$\times \left\langle (L_j \ell_j) L, \left(S_j \frac{1}{2}\right) S;\right.$$
$$\left. \times JM_J | ((L_j S_j) J_j, \ell_j) K_j, \frac{1}{2}; JM_J \right\rangle . \quad (47.32)$$

This transformation has been implemented in a computer program by *Saraph* [47.7, 8].

For intermediate-Z atoms and ions, relativistic effects can be included by adding terms from the Breit–Pauli Hamiltonian to the nonrelativistic Hamiltonian (*Jones* [47.9], *Scott* and *Burke* [47.10]). We write

$$H_{N+1}^{\mathrm{BP}} = H_{N+1}^{\mathrm{nr}} + H_{N+1}^{\mathrm{rel}} \quad (47.33)$$

where H_{N+1}^{nr} is defined by (47.3) and H_{N+1}^{rel} consists of one- and two-body relativistic terms. The one-body terms are (Sect. 21.1)

$$H_{N+1}^{\mathrm{mass}} = -\frac{1}{8}\alpha^2 \sum_{i=1}^{N+1} \nabla_i^4 \quad \text{mass-correction term} ,$$

$$H_{N+1}^{\mathrm{D}_1} = -\frac{1}{8}\alpha^2 Z \sum_{i=1}^{N+1} \nabla_i^2 \left(\frac{1}{r_i}\right)$$

one-body Darwin term ,

$$H_{N+1}^{\mathrm{so}} = \frac{1}{2}\alpha^2 \sum_{i=1}^{N+1} \frac{1}{r_i}\frac{\partial V}{\partial r_i} (\boldsymbol{\ell}_i \cdot \boldsymbol{s}_i) \quad \text{spin–orbit term} .$$

The two-body terms are less important and are usually not included in collision calculations.

The modified Schrödinger equation defined by (47.2), with H_{N+1} replaced by H_{N+1}^{BP}, is solved by adopting an expansion similar in form to (47.8), but now using the pair coupling scheme in the definition of the channel functions and quadratically integrable functions. We then obtain coupled integrodifferential equations similar in form to (47.10), from which the K-matrix, S-matrix and T-matrix can be obtained. The corresponding total cross section in the pair-coupling scheme analogous to (47.19) is

$$\sigma_{\mathrm{tot}}(i \to j) = \frac{\pi}{2k_i^2(2J_i+1)} \sum_{\substack{J\pi \\ K_i K_j \ell_i \ell_j}} (2J+1) \left| T_{ji}^{J\pi} \right|^2 , \quad (47.34)$$

which describes a transition from an initial target state $\alpha_i J_i$ to a final target state $\alpha_j J_j$. The corresponding collision strength is

$$\Omega(i, j) = k_i^2 (2J_i+1)\, \sigma_{\mathrm{tot}}(i \to j) . \quad (47.35)$$

For high-Z atoms and ions, the Dirac Hamiltonian [47.11, 12] (Sect. 47.2)

$$H_{N+1}^{\mathrm{D}} = \sum_{i=1}^{N+1} \left(c\boldsymbol{\alpha} \cdot \boldsymbol{p}_i + \beta' c^2 - \frac{Z}{r_i} \right) + \sum_{i>j=1}^{N+1} \frac{1}{r_{ij}} \quad (47.36)$$

must be used instead of (47.3), where $\beta' = \beta - 1$ and $\boldsymbol{\alpha}$ and β are the usual Dirac matrices. The expansion of the total wave functions for a particular $JM_J\pi$ takes the general form of (47.8). However, now both the bound orbitals in the target, and correlation functions and the orbitals representing the scattered electron are represented by Dirac orbitals. These are defined in terms of large and small components $P(r)$ and $Q(r)$ by

$$\phi(\boldsymbol{r}, \sigma) = \frac{1}{r} \begin{pmatrix} P_a(r)\, \chi_{\kappa m}(\hat{\boldsymbol{r}}, \sigma) \\ Q_a(r)\, \chi_{-\kappa m}(\hat{\boldsymbol{r}}, \sigma) \end{pmatrix} \quad (47.37)$$

for the bound orbitals, and

$$F(\boldsymbol{r}, \sigma) = \frac{1}{r} \begin{pmatrix} P_c(r)\, \chi_{\kappa m}(\hat{\boldsymbol{r}}, \sigma) \\ Q_c(r)\, \chi_{-\kappa m}(\hat{\boldsymbol{r}}, \sigma) \end{pmatrix} \quad (47.38)$$

for the continuum orbitals, where $a = n\kappa m$, $c = k\kappa m$ and the spherical spinor

$$\chi_{\kappa m}(\hat{\boldsymbol{r}}, \sigma) = \sum_{m_\ell m_i} \left(\ell m_\ell \frac{1}{2} m_i | \ell \frac{1}{2} jm \right) Y_{\ell m_\ell}(\theta, \phi)\, \chi_{\frac{1}{2} m_i}(\sigma) , \quad (47.39)$$

where $\kappa = j + \frac{1}{2}$ when $\ell = j + \frac{1}{2}$, and $\kappa = -j - \frac{1}{2}$ when $\ell = j - \frac{1}{2}$. We can now derive coupled integrodifferential equations for the functions $P_c(r)$ and $Q_c(r)$

in a similar way to the derivation of (47.10), except that these are now coupled first-order equations instead of coupled second-order equations. The K-matrix, and hence the S-matrix and T-matrix, can be obtained from the asymptotic form of these equations. The total cross section in the j–j coupling scheme, is then given by (47.34), and the corresponding collision strength is given by (47.35).

47.1.3 Multichannel Resonance Theory

General resonance theories have been developed by *Fano* [47.13, 14], *Feshbach* [47.15, 16], and *Brenig* and *Haag* [47.17]. They are also discussed in Chapt. 25. Here we will limit our discussion to the effect that resonances have on electron collision cross sections.

Following Feshbach, we introduce the projection operators P and Q, where P projects onto a finite set of low energy channels in (47.8) and Q projects onto the orthogonal space, where we restrict our consideration to the space corresponding to a particular set of conserved quantum numbers Γ. In this space we have

$$P^2 = P, \quad Q^2 = Q, \quad P + Q = 1 . \tag{47.40}$$

The Schrödinger equation (47.2) can then be written as

$$P(H - E)(P + Q)\Psi = 0 \tag{47.41}$$

and

$$Q(H - E)(P + Q)\Psi = 0 \tag{47.42}$$

where we have omitted the subscript $N+1$ on H and the superscript Γ on Ψ. After solving (47.42) for $Q\Psi$ and substituting into (47.41), we find that

$$P\left(H - PHQ\frac{1}{Q(H-E)Q}QHP - E\right)P\Psi = 0 , \tag{47.43}$$

where the term

$$V_{\mathrm{op}} = -PHQ\frac{1}{Q(H-E)Q}QHP , \tag{47.44}$$

called the *optical potential*, allows for propagation in the Q-space channels.

We now introduce the eigenfunctions ϕ_i and eigenvalues ε_i of the operator QHQ by

$$QHQ\phi_i = \varepsilon_i\phi_i . \tag{47.45}$$

It follows that the discrete eigenvalues ε_i each give rise to poles in V_{op} at ε_i. If the energy E is in the neighborhood of an isolated pole or bound state ε_i, we can rewrite (47.43) as

$$\begin{aligned}&\left(PHP - \sum_{j\neq i} PHQ\frac{|\phi_j\rangle\langle\phi_j|}{\varepsilon_j - E}QHP - E\right)P\Psi\\ &= PHQ\frac{|\phi_i\rangle\langle\phi_i|}{\varepsilon_i - E}QHP\Psi ,\end{aligned} \tag{47.46}$$

where the rapidly varying part of the optical potential has been separated and put on the right-hand-side of (47.46). This equation can be solved by introducing the Green's function G_0 and the solutions ψ_{0j} of the operator on the left-hand side of (47.46). We find that the pole term on the right-hand side of this equation gives rise to a Feshbach resonance whose position is

$$E_i = \varepsilon_i + \Delta_i - \frac{1}{2}i\Gamma_i = E_{i,\mathrm{r}} - \frac{1}{2}i\Gamma_i , \tag{47.47}$$

is where the resonance shift

$$\Delta_i = \langle\phi_i|QHP\,G_0PHQ|\phi_i\rangle , \tag{47.48}$$

and the resonance width is

$$\Gamma_i = 2\pi\sum_j |\langle\phi_i \mid QHP \mid \psi_{0j}\rangle|^2 , \tag{47.49}$$

where the summation in this equation is taken over all continuum states corresponding to the operator on the left-hand-side of (47.46) and these states are normalized to a delta function in energy.

In the neighborhood of the resonance energy $E_{i,\mathrm{r}}$, the S-matrix is rapidly varying with the form

$$\boldsymbol{S} = \boldsymbol{S}_0^{\frac{1}{2}}\left(\boldsymbol{I} - \mathrm{i}\Gamma\frac{\boldsymbol{\gamma}_i \times \boldsymbol{\gamma}_i}{E - E_{i,\mathrm{r}} + \frac{1}{2}\mathrm{i}\Gamma_i}\right)\boldsymbol{S}_0^{\frac{1}{2}} , \tag{47.50}$$

where $\boldsymbol{S}_0$ is the slowly varying nonresonant or background S-matrix corresponding to $\boldsymbol{\psi}_0$, and the partial widths $\boldsymbol{\gamma}_i$ are defined by

$$\langle\phi_i \mid QHP \mid \boldsymbol{\psi}_0\rangle = \Gamma_i^{\frac{1}{2}}\boldsymbol{\gamma}_i \cdot \boldsymbol{S}_0^{\frac{1}{2}} , \tag{47.51}$$

where $\boldsymbol{\gamma}_i \cdot \boldsymbol{\gamma}_i = 1$. A corresponding resonant expression can be derived for the K-matrix (*Burke* [47.18]).

Let us now diagonalize the S-matrix as follows:

$$\boldsymbol{S} = \boldsymbol{A}\exp(2\mathrm{i}\boldsymbol{\Delta})\boldsymbol{A}^{\mathrm{T}} , \tag{47.52}$$

where $\boldsymbol{A}$ is an orthogonal matrix and $\boldsymbol{\Delta}$ is a diagonal matrix whose diagonal elements, δ_i, $i = 1, \ldots, n_a$, are called the eigenphases. If we define the eigenphase sum δ_{sum} by

$$\delta_{\mathrm{sum}} = \sum_{i=1}^{n_a}\delta_i , \tag{47.53}$$

then we can show from (47.50) that in the neighborhood of the resonance

$$\delta_{\text{sum}}(E) = \delta_{0,\text{sum}}(E) + \tan^{-1}\left(\frac{\frac{1}{2}\Gamma_i}{E_{i,\text{r}} - E}\right), \tag{47.54}$$

where $\delta_{0,\text{sum}}$ is the slowly varying background eigenphase sum obtained by replacing $\boldsymbol{S}$ by $\boldsymbol{S}_0$ in (47.52) and (47.53). It follows from (47.54) that the eigenphase sum increases by π radians in the neighborhood of the resonance energy. If there are m resonances, that may be overlapping, (47.54) generalizes to

$$\delta_{\text{sum}}(E) = \delta_{0,\text{sum}}(E) + \sum_{i=1}^{m} \tan^{-1}\left(\frac{\frac{1}{2}\Gamma_i}{E_{i,\text{r}} - E}\right). \tag{47.55}$$

This result has proved to be very useful in analyzing closely spaced resonances to obtain the individual resonance positions and widths.

47.1.4 Multichannel Quantum Defect Theory

When an electron scatters from a positive ion, the operator QHQ in (47.45) has an infinite series of bound states, supported by the long-range attractive Coulomb potential, converging to the lowest energy threshold of QHQ. Multichannel quantum defect theory (MQDT), developed by *Seaton* [47.6,19,20], relates the S-matrices and K-matrices above and below thresholds using general analytic properties of the Coulomb wave function. In this way the cross sections above and below threshold can be related and whole Rydberg series of resonances rather than individual resonances can be predicted.

When all the channels are open, the S-matrix is related to the K-matrix by (47.17), rewritten here as

$$\boldsymbol{S} = (\mathrm{i}\boldsymbol{I} - \boldsymbol{K})(\mathrm{i}\boldsymbol{I} + \boldsymbol{K})^{-1}. \tag{47.56}$$

In the energy region where some channels are closed, Seaton showed that the S-matrix can be written as

$$\boldsymbol{S} = (\mathrm{i}\boldsymbol{I} - \boldsymbol{K})(\boldsymbol{t}\boldsymbol{I} + \boldsymbol{K})^{-1}, \tag{47.57}$$

where $\boldsymbol{t}$ is a diagonal matrix with nonzero matrix elements given by

$$\begin{aligned} t_{ii} &= i, && \text{open channels} \quad k_i^2 \geq 0 \\ t_{ii} &= \tan \pi\nu_i, && \text{closed channels} \quad k_i^2 < 0 \end{aligned} \tag{47.58}$$

where ν_i is defined in the closed channels by

$$k_i^2 = -z^2/\nu_i^2. \tag{47.59}$$

When some channels are closed, Rydberg series of resonances occur because of the terms $\tan \pi\nu_i$. We now introduce the matrix

$$\boldsymbol{\chi} = (\mathrm{i}\boldsymbol{I} - \boldsymbol{K})(\mathrm{i}\boldsymbol{I} + \boldsymbol{K})^{-1}, \tag{47.60}$$

which is analytic through the thresholds. When all the channels are open then clearly $\boldsymbol{S} = \boldsymbol{\chi}$. Let us partition the S-matrix and the χ-matrix into open (subscript o) and closed (subscript c) channel matrices as follows:

$$\boldsymbol{S} = \begin{pmatrix} \boldsymbol{S}_{\text{oo}} & \boldsymbol{S}_{\text{oc}} \\ \boldsymbol{S}_{\text{co}} & \boldsymbol{S}_{\text{cc}} \end{pmatrix}, \quad \boldsymbol{\chi} = \begin{pmatrix} \boldsymbol{\chi}_{\text{oo}} & \boldsymbol{\chi}_{\text{oc}} \\ \boldsymbol{\chi}_{\text{co}} & \boldsymbol{\chi}_{\text{cc}} \end{pmatrix}. \tag{47.61}$$

Then eliminating $\boldsymbol{K}$ between (47.57) and (47.60) and using (47.61), the open–open submatrix of $\boldsymbol{S}$ is

$$\boldsymbol{S}_{\text{oo}} = \boldsymbol{\chi}_{\text{oo}} - \boldsymbol{\chi}_{\text{oc}}\left(\boldsymbol{\chi}_{\text{cc}} - \mathrm{e}^{-2\pi\mathrm{i}\boldsymbol{\nu}_{\text{c}}}\boldsymbol{I}\right)^{-1}\boldsymbol{\chi}_{\text{co}}. \tag{47.62}$$

This is an expression for the open channel S-matrix in terms of quantities that can be analytically continued through the thresholds.

A similar expression can be obtained for the open channel, or contracted, K-matrix $\boldsymbol{K}_{\text{op}}$ defined in analogy with (47.56) by

$$\boldsymbol{S}_{\text{oo}} = \left(\mathrm{i}\boldsymbol{I} - \boldsymbol{K}_{\text{op}}\right)\left(\mathrm{i}\boldsymbol{I} + \boldsymbol{K}_{\text{op}}\right)^{-1}. \tag{47.63}$$

Partitioning the K-matrix into open and closed channels as

$$\boldsymbol{K} = \begin{pmatrix} \boldsymbol{K}_{\text{oo}} & \boldsymbol{K}_{\text{oc}} \\ \boldsymbol{K}_{\text{co}} & \boldsymbol{K}_{\text{cc}} \end{pmatrix}, \tag{47.64}$$

and substituting into (47.57) yields the expression

$$\boldsymbol{K}_{\text{op}} = \boldsymbol{K}_{\text{oo}} - \boldsymbol{K}_{\text{oc}}\left(\boldsymbol{K}_{\text{cc}} + \tan \pi\boldsymbol{\nu}_{\text{c}}\right)^{-1}\boldsymbol{K}_{\text{co}}. \tag{47.65}$$

This equation enables cross sections and the resonance parameters in an energy region where some channels are closed to be predicted by calculating $\boldsymbol{K}$ at a few energies where all channels are open and then extrapolating it to the region where some channels are closed.

A simple one channel example has already been discussed following (47.28). In this case, it follows from (47.22) that $K = \tan \delta_\ell$. Also, from (47.57), a pole in the S-matrix occurs below threshold when

$$\tan \pi\nu + K = 0. \tag{47.66}$$

Since from (47.59), the bound state energies correspond to $\nu = n - \mu$, where n is an integer and μ is the quantum defect, then (47.66) shows that δ_ℓ extrapolates below threshold to give $\pi\mu$. This result agrees with (47.28) since $1 - \exp(2\pi\eta) \approx 1$ close to threshold.

The above theory has been extended by *Gailitis* [47.21] who showed that the collision strengths defined by (47.20) can also be extrapolated through thresholds. Provided that the separation between the resonances is large compared with their widths, then the collision strength averaged over resonances below a threshold denoted by $\overline{\Omega}(i, f)$, can be written as

$$\overline{\Omega}(i, f) = \Omega^{>}(i, f) + \sum_{j} \frac{\Omega^{>}(i, j)\Omega^{>}(j, f)}{\sum_{k} \Omega^{>}(j, k)} , \tag{47.67}$$

where $\Omega^{>}$ are the collision strengths determined above the threshold and extrapolated to energies below this threshold. Also in (47.67), j is summed over the degenerate closed channels of the new threshold and k is summed over all open channels. This result is particularly useful in applications to plasmas where it is often only the collision strength averaged over a Maxwellian distribution of electron energies that is required.

Multichannel quantum defect theory has been extended to molecules by *Fano* [47.22], where it has proved to be important in the analysis of resonances occuring in partial cross sections involving rovibrational and dissociating channels as well as electronic channels [47.23, 24].

47.1.5 Solution of the Coupled Integrodifferential Equations

This section considers methods that have been developed for solving the coupled integrodifferential equations (47.10). These equations arise in low-energy electron–atom and electron–ion collisions. Section 47.2.2 shows that similar equations also arise in low energy electron–molecule collisions in the fixed-nuclei approximation. Thus the methods discussed in this section are also applicable for electron–molecule collisions.

R-Matrix Method

This method was first introduced in nuclear physics [47.25, 26] in a study of resonance reactions. It has now been applied to wide range of atomic, molecular and optical processes, as reviewed in [47.27].

This method starts by partitioning configuration space into two regions by a sphere of radius a, chosen so that the direct potential has achieved its asymptotic form given by (47.12) and the exchange and correlation potentials are negligible for $r \geq a$. The objective is then to calculate the R-matrix $R_{ij}^{\Gamma}(E)$, which is defined by

$$F_{ij}^{\Gamma}(a) = \sum_{\ell=1}^{n} R_{i\ell}^{\Gamma}(E) \left(a \frac{\mathrm{d}F_{\ell j}^{\Gamma}}{\mathrm{d}r} - b_{\ell} F_{\ell j}^{\Gamma} \right)_{r=a} , \tag{47.68}$$

by solving (47.10) in the internal region.

The collision problem is solved in the internal region by expanding the wave function, in analogy with (47.8), in the form

$$\begin{aligned} \Psi_k^{\Gamma}(\boldsymbol{X}_{N+1}) &= \mathcal{A} \sum_{ij} \overline{\Phi}_i^{\Gamma}(\boldsymbol{x}_1, \ldots, \boldsymbol{x}_N; \hat{\boldsymbol{r}}_{N+1}\sigma_{N+1}) \\ &\quad \times r_{N+1}^{-1} u_j(r_{N+1}) a_{ijk}^{\Gamma} \\ &\quad + \sum_{i} \chi_i^{\Gamma}(\boldsymbol{x}_1, \ldots, \boldsymbol{x}_{N+1}) b_{ik}^{\Gamma} , \end{aligned} \tag{47.69}$$

where the u_j are radial basis functions defined over the range $0 \leq r \leq a$. For radial basis functions u_j satisfying arbitrary boundary conditions at $r = a$, the Hamiltonian H_{N+1} defined by (47.3) is not Hermitian in the internal region due to the kinetic energy operators. It can however be made Hermitian by adding the Bloch operator [47.28]

$$L_{\mathrm{b}} = \sum_{i=1}^{n} \left| \overline{\Phi}_i^{\Gamma} \right\rangle \frac{1}{2} \delta(r-a) \left(\frac{\mathrm{d}}{\mathrm{d}r} - \frac{b_i - 1}{r} \right) \left\langle \overline{\Phi}_i^{\Gamma} \right| \tag{47.70}$$

to H_{N+1}, where b_i is an arbitrary parameter. The Schrödinger equation (47.2) then becomes

$$(H_{N+1} + L_{\mathrm{b}} - E)\Psi^{\Gamma} = L_{\mathrm{b}}\Psi^{\Gamma} , \tag{47.71}$$

which can be formally solved, giving

$$\Psi^{\Gamma} = (H_{N+1} + L_{\mathrm{b}} - E)^{-1} L_{\mathrm{b}}\Psi^{\Gamma} . \tag{47.72}$$

Next, expand the Green's function $(H_{N+1} + L_{\mathrm{b}} - E)^{-1}$ in terms of the basis Ψ_k^{Γ}, where the coefficients a_{ijk}^{Γ} and b_{ik}^{Γ} in (47.69) are chosen to diagonalize $H_{N+1} + L_{\mathrm{b}}$, giving

$$\left\langle \Psi_k^{\Gamma} \middle| H_{N+1} + L_{\mathrm{b}} \middle| \Psi_{k'}^{\Gamma} \right\rangle = E_k^{\Gamma} \delta_{kk'} . \tag{47.73}$$

Equation (47.72) can then be written as

$$\left| \Psi^{\Gamma} \right\rangle = \sum_{k} \frac{\left| \Psi_k^{\Gamma} \right\rangle \left\langle \Psi_k^{\Gamma} \right|}{E_k^{\Gamma} - E} L_{\mathrm{b}} \left| \Psi^{\Gamma} \right\rangle . \tag{47.74}$$

Finally, we project this equation onto the channel functions $\overline{\Phi}_i^{\Gamma}$, and evaluate it at $r = a$. Assuming Ψ^{Γ} is given by (47.8), we retrieve (47.68), where the R-matrix can be calculated from the expansion

$$R_{ij}^{\Gamma}(E) = \frac{1}{2a}\sum_k \frac{w_{ik}^{\Gamma} w_{jk}^{\Gamma}}{E_k^{\Gamma} - E}, \tag{47.75}$$

and where we have introduced the surface amplitudes

$$w_{ik}^{\Gamma} = \sum_j u_j(a)\, a_{ijk}^{\Gamma}. \tag{47.76}$$

The main part of the calculation involves setting up and diagonalizing the matrix given by (47.73). This has to be carried out once to determine the R-matrix for all energies E.

In the external region $r \geq a$, (47.10) reduces to coupled differential equations, coupled by the potential $V_{ij}^{\Gamma}(r)$ which has achieved its asymptotic form (47.12).These equations can be integrated outwards from $r = a$ for each energy of interest, subject to the boundary conditions (47.68), to yield the K-matrix given by (47.15), and hence the S-matrix and collision cross sections.

Kohn Variational Method

The application of variational methods in electron atom collision theory has been reviewed by *Nesbet* [47.29]. An S-matrix or complex Kohn version of this method has been developed [47.30], which has eliminated singularities present in earlier K-matrix versions of the theory. This new method has been particularly important in electron–molecule collision calculations [47.31].

This approach starts from the basic expansion given by (47.8), where the reduced radial functions are chosen to satisfy the T-matrix asymptotic boundary conditions

$$F_{ij}^{\Gamma} \underset{r\to\infty}{\sim} k_i^{-\frac{1}{2}}\left(\sin\theta_i \delta_{ij} + (2\mathrm{i})^{-1}\, \mathrm{e}^{\mathrm{i}\theta_i} T_{ij}^{\Gamma}\right), \tag{47.77}$$

which follows by taking linear combinations of the n_a solutions defined by (47.15) and using (47.17). We then define the integral

$$\boldsymbol{I}^{\Gamma} = -\frac{1}{2}\int_0^{\infty} \left(\boldsymbol{F}^{\Gamma}\right)^T \boldsymbol{L}\,\boldsymbol{F}^{\Gamma}\,\mathrm{d}r, \tag{47.78}$$

where $\boldsymbol{L}$ is the integrodifferential operator given by (47.10) with all terms taken onto the left-hand side of the equation.

Now consider variations of the integral $\boldsymbol{I}^{\Gamma}$ resulting from arbitrary variations $\delta\boldsymbol{F}^{\Gamma}$ of the functions $\boldsymbol{F}^{\Gamma}$ about the exact solution of (47.10), where these solutions satisfy the boundary conditions (47.77), and the variations satisfy the boundary condition

$$\delta\boldsymbol{F}^{\Gamma} \underset{r\to\infty}{\sim} (2\mathrm{i})^{-1}\boldsymbol{k}^{-\frac{1}{2}}\,\mathrm{e}^{\mathrm{i}\boldsymbol{\theta}}\,\delta\boldsymbol{T}^{\Gamma}. \tag{47.79}$$

The corresponding variation in $\boldsymbol{I}^{\Gamma}$ to first-order is

$$\delta\boldsymbol{I}^{\Gamma} = -\frac{1}{2}\int_0^{\infty}\left[\left(\delta\boldsymbol{F}^{\Gamma}\right)^T \boldsymbol{L}\,\boldsymbol{F}^{\Gamma} + \left(\boldsymbol{F}^{\Gamma}\right)^T \boldsymbol{L}\delta\boldsymbol{F}^{\Gamma}\right]\mathrm{d}r, \tag{47.80}$$

which after some manipulation yields

$$\delta\boldsymbol{I}^{\Gamma} = (4\mathrm{i})^{-1}\delta\boldsymbol{T}^{\Gamma}. \tag{47.81}$$

It follows that the functional

$$\left[\boldsymbol{T}^{\Gamma}\right] = \boldsymbol{T}^{\Gamma} - 4\mathrm{i}\,\boldsymbol{I}^{\Gamma} \tag{47.82}$$

is stationary for small variations about the exact solution. This is the complex Kohn variational principle.

This variational principle can be used to solve (47.10) by representing the reduced radial functions F_{ij}^{Γ} by the expansion

$$F_{ij}^{\Gamma}(r) = w_{1i}^{\Gamma}(r)\,\delta_{ij} + (2\mathrm{i})^{-1}\, w_{2i}^{\Gamma}(r)\, T_{ij}^{\Gamma} + \sum_k \phi_k^{\Gamma}(r)\, c_{ijk}^{\Gamma}, \tag{47.83}$$

where w_{1i}^{Γ} and w_{2i}^{Γ} are zero at the origin and have the asymptotic forms

$$w_{1i}^{\Gamma}(r) \underset{r\to\infty}{\sim} k_i^{-\frac{1}{2}} \sin\theta_i; \tag{47.84}$$

$$w_{2i}^{\Gamma}(r) \underset{r\to\infty}{\sim} k_i^{-\frac{1}{2}} \exp \mathrm{i}\theta_i, \tag{47.85}$$

while ϕ_k^{Γ} are square integrable basis functions. The coefficients c_{ijk}^{Γ} and the T-matrix elements T_{ij}^{Γ} can then be determined as variational parameters in the variational principle (47.82).

Schwinger Variational Method

This method has been used to calculate electron–molecule collision cross sections by *McKoy* and co-workers [47.32]. The method starts from the Lippmann–Schwinger integral equation, which corresponds to the integrodifferential equations (47.10), together with the K-matrix or T-matrix boundary condition defined by (47.15) or (47.77) respectively. The T-matrix form of the Lippmann–Schwinger integral equation is

$$F_{ij}^{\Gamma}(r) = w_{1i}^{\Gamma}(r)\,\delta_{ij} + \sum_{k\ell}\int_0^{\infty}\int_0^{\infty} G_{ik}^{\Gamma(+)}(r, r') \times U_{k\ell}^{\Gamma}\left(r', r''\right) F_{\ell j}^{\Gamma}\left(r''\right)\mathrm{d}r'\,\mathrm{d}r'' \tag{47.86}$$

where $U_{k\ell}^{\Gamma}$ represents the sum of the potential terms $2\left(V_{k\ell}^{\Gamma}+K_{k\ell}^{\Gamma}+X_{k\ell}^{\Gamma}\right)$ in (47.10), and the multichannel outgoing wave Green's function $G_{ik}^{\Gamma(+)}$ is defined assuming all channels are open by

$$G_{ij}^{\Gamma(+)}\left(r,r'\right)=\begin{cases}-w_{1i}^{\Gamma}(r)\,w_{2i}^{\Gamma}\left(r'\right)\delta_{ij}\,, & r<r'\\ -w_{2i}^{\Gamma}(r)\,w_{1i}^{\Gamma}\left(r'\right)\delta_{ij}\,, & r\geq r'\end{cases} \tag{47.87}$$

where w_{1i}^{Γ} and w_{2i}^{Γ} are solutions of the Coulomb equation

$$\left(\frac{\mathrm{d}^2}{\mathrm{d}r^2}-\frac{\ell_i\left(\ell_i+1\right)}{r^2}+\frac{2\left(Z-N\right)}{r}+k_i^2\right)w_i\left(r\right)=0 \tag{47.88}$$

that satisfy the asymptotic boundary conditions (47.84) and (47.85), respectively.

An integral expression for the T-matrix can be obtained by comparing (47.77) and (47.86). This gives

$$T_{ij}^{\Gamma}=-2\mathrm{i}\sum_k\int_0^{\infty}\int_0^{\infty}w_{1i}^{\Gamma}(r)\,U_{ik}^{\Gamma}\left(r,r'\right)F_{kj}^{\Gamma}\left(r'\right)\mathrm{d}r\,\mathrm{d}r' \tag{47.89}$$

or, rewriting this equation using Dirac bracket notation,

$$\boldsymbol{T}=-2\mathrm{i}\left\langle\boldsymbol{w}_1\middle|\boldsymbol{U}\middle|\boldsymbol{F}^{(+)}\right\rangle, \tag{47.90}$$

where the plus sign indicates that $\boldsymbol{F}^{(+)}$ satisfies the outgoing wave boundary condition (47.77), and where for convenience we have suppressed the superscript Γ. In a similar way, the Lippmann–Schwinger equation corresponding to the ingoing wave boundary condition

$$F_{ij}^{\Gamma}\underset{r\to\infty}{\sim}k_i^{-\frac{1}{2}}\left(\sin\theta_i\delta_{ij}-(2\mathrm{i})^{-1}\,\mathrm{e}^{-\mathrm{i}\theta_i}T_{ij}^{\Gamma*}\right), \tag{47.91}$$

can be introduced, from which follows the integral expression

$$\boldsymbol{T}=-2\mathrm{i}\left\langle\boldsymbol{F}^{(-)}\middle|\boldsymbol{U}\middle|\boldsymbol{w}_1\right\rangle, \tag{47.92}$$

where $\boldsymbol{F}^{(-)}$ satisfies the ingoing wave boundary condition (47.91). A further integral expression for the T-matrix is obtained by substituting for $\boldsymbol{w}_1$ in (47.92) from (47.86) giving

$$\boldsymbol{T}=-2\mathrm{i}\left\langle\boldsymbol{F}^{(-)}\middle|\boldsymbol{U}-\boldsymbol{U}\boldsymbol{G}^{(+)}\boldsymbol{U}\middle|\boldsymbol{F}^{(+)}\right\rangle. \tag{47.93}$$

Hence, a combination of (47.90), (47.92) and (47.93) yields the functional

$$\begin{aligned}[\boldsymbol{T}]=&-2\mathrm{i}\left\langle\boldsymbol{w}_1\middle|\boldsymbol{U}\middle|\boldsymbol{F}^{(+)}\right\rangle\\ &\times\left(\left\langle\boldsymbol{F}^{(-)}\middle|\boldsymbol{U}-\boldsymbol{U}\boldsymbol{G}^{(+)}\boldsymbol{U}\middle|\boldsymbol{F}^{(+)}\right\rangle\right)^{-1}\\ &\times\left\langle\boldsymbol{F}^{(-)}\middle|\boldsymbol{U}\middle|\boldsymbol{w}_1\right\rangle.\end{aligned} \tag{47.94}$$

This functional is stationary for small variations of $\boldsymbol{F}^{(+)}$ and $\boldsymbol{F}^{(-)}$ about the exact solution of (47.10) satisfying the boundary conditions (47.77) and (47.91) respectively, and forms the basis of numerical calculations.

Linear Algebraic Equations Method

In this method, the integrodifferential (47.10) are reduced directly to a set of linear algebraic (LA) equations [47.33], or alternatively, (47.10) are first converted to integral form, given by (47.86), which are then reduced to a set of linear algebraic equations [47.34].

A direct approach [47.33] has been widely used for electron–atom and electron–ion collisions. As in the R-matrix method, configuration space is first divided into two regions. A mesh of N points is then used to span the internal region $r\leq a$ where

$$r_1=0;\quad r_{k-1}<r_k,\quad k=1,\ldots,N;\quad r_N=a\,. \tag{47.95}$$

In addition, two further mesh points r_{N+1} and r_{N+2} are introduced to enable the solution in the internal region to be matched to the solution in the external region $r\geq a$. In the external region, (47.10) reduces to coupled differential equations that can be solved by one of the same methods as adopted in the R-matrix approach.

The n functions $F_{ij}^{\Gamma}(r)$, $i=1,\ldots,n$, in (47.10) are represented by their values at the mesh points r_k. Using a finite difference representation of the differential and integral operators, (47.10) then reduces to a set of linear algebraic equations for the unknown values $F_{ij}^{\Gamma}(r_k)$ in the internal region. These equations can be solved using standard methods for each linearly independent solution $j=1,\ldots,n_a$ defined by the asymptotic boundary conditions (47.15).

47.1.6 Intermediate and High Energy Elastic Scattering and Excitation

For electron–atom and electron–ion collisions at electron impact energies greater than the ionization threshold of the target, an infinite number of channels is open so that they cannot all be included explicitly in the expansion of the total wave function. Several approaches have been developed to treat collisions at these energies, such as extensions of low energy methods based on expansion (47.8) to intermediate energies, the development of optical potentials that take account of loss of flux into the infinity of open channels in some average way, and extensions of the Born approximation to lower energies by including higher-order terms in the Born Series.

Pseudostate Methods

In this approach, expansion (47.8) is extended by including a set of suitably chosen square-integrable pseudostates Φ_i^p that are orthogonal to the eigenstates Φ_i retained in expansion (47.8). These pseudostates are usually defined by diagonalizing the target Hamiltonian H_N in a square-integrable basis yielding an equation analogous to (47.4), viz

$$\langle \Phi_i^p | H_N | \Phi_j^p \rangle = \omega_i^p \delta_{ij} , \quad (47.96)$$

where the energies ω_i^p lie just below the ionization threshold or in the continuum. The pseudostates Φ_i^p thus represent in an average way the high lying Rydberg states and the continuum states of the target rather than just the low lying target bound states as in (47.4). This enables loss of flux into the continuum states to be represented enabling accurate excitation cross sections to be calculated at intermediate energies. In addition, by calculating the amplitudes for exciting these pseudostates accurate ionization cross sections can be determined, as discussed in Sect. 47.1.7.

A proposal to include pseudostates in expansion (47.8) for e–H scattering was first made by *Burke* and *Schey* [47.35]. Also a modification of this expansion to include polarized pseudostates which allowed for the long-range dipole and quadrupole polarizability in e–H scattering was proposed by *Damburg* and *Karule* [47.36]. Later detailed e–H scattering calculations which included pseudostates were carried out by a number of authors [47.37–40] with considerable success showing the validity of this approach.

More recently the R-matrix method, discussed in Sect. 47.1.5, has been extended to include pseudostates giving rise to the intermediate energy R-matrix (IERM) method [47.41–44] and the R-matrix with pseudostates (RMPS) method [47.45–48], the latter approach being applicable to multi-electron atoms and ions. Both of these methods have enabled excitation and ionization cross sections to be accurately calculated above the ionization threshold.

Methods have also been developed in which the pseudostates are represented by Sturmian functions. An expansion of this type was first proposed by *Rotenberg* [47.49] and in a more recent major development the convergent close coupling (CCC) method has been developed by *Bray*, *Stelbovics* and co-workers [47.50–56] which has been applied with considerable success for both electron impact excitation and ionization.

As an example of the CCC method, we consider e^-–H scattering. In this case the radial basis of the target states are expanded in the complete Laguerre basis

$$\xi_{k\ell}(r) = \left(\frac{\lambda_\ell (k-1)!}{(2\ell+1+k)!} \right)^{1/2} \times (\lambda_\ell r)^{\ell+1} \exp(\lambda_\ell r/2) L_{k-1}^{2\ell+2}(\lambda_\ell r) , \quad (47.97)$$

where $L_{k-1}^{2\ell+2}(\lambda_\ell r)$ are associated Laguerre polynomials and k ranges from 1 to the basis size N_ℓ which depends on the angular momentum ℓ. The target states $\Phi_{i\ell}$ are expanded for each angular momentum as follows

$$\Phi_{i\ell} = \sum_{k=1}^{N_\ell} \xi_{k\ell}(r) c_{ki\ell} , \quad i = 1, \ldots, N_\ell , \quad (47.98)$$

where the coefficients $c_{ki\ell}$ are obtained by diagonalizing the target Hamiltonian H_1 as follows

$$\langle \Phi_{i\ell} | H_1 | \Phi_{j\ell} \rangle = w_{i\ell} \delta_{ij} . \quad (47.99)$$

The states $\Phi_{i\ell}$ corresponding to the lowest energies $w_{i\ell}$ provide an accurate representation of the lowest bound eigenstates of the target for each ℓ, while the states corresponding to the higher energies $w_{i\ell}$ correspond to pseudostates representing the high lying Rydberg states and the continuum. As N_ℓ is increased for fixed range paramenter λ_ℓ in (47.98), more bound target eigenstates are accurately represented while at the same time the states corresponding to the higher energies provide a denser and more accurate representation of the continuum. The scattering amplitude is written in the close coupling approximation as

$$\langle \psi_f | H_2 - E | \Psi_i^{S+} \rangle \simeq \langle \psi_f | I(H_2 - E) \left[1 + (-1)^S P \right] I | \Psi_i^{S+} \rangle , \quad (47.100)$$

where P is the space exchange operator which ensures that the total wave function has the correct symmetry for each total spin S, I is the projection operator onto the target states $\Phi_{i\ell}$ retained in the calculation and $\langle \psi_f |$ is an eigenstate of the asymptotic Hamiltonian. The scattering amplitude is then obtained by solving the close coupling equations in momentum space [47.50].

Optical Potential Methods

An approach which is also capable in principle of including the effect of all excited and continuum states is the optical potential method [47.57]. Adopting Feshbach projection operators P and Q as in (47.40), an optical potential V_{op} is defined by (47.44), where now P projects onto those low-energy channels that can be treated exactly, for example by solving equations (47.10), and

Q allows for the remaining infinity of coupled channels, including the continuum.

At sufficiently high energies, it is appropriate to make a perturbation expansion of V_{op}, where the second-order term is given by

$$V^{(2)} = PVQ\frac{1}{E - \mathcal{T}_e - H_N + i\epsilon}QVP\,, \tag{47.101}$$

V being the electron–atom (ion) interaction potential, $\mathcal{T}_e$ the kinetic energy operator of the scattered electron, and H_N the target Hamiltonian. *Byron* and *Joachain* [47.58] converted the lowest-order terms of perturbation theory into an ab initio local complex potential for the elastic scattering of electrons and positrons from a number of atoms.

The optical potential calculated in second-order has also been used by *Bransden* et al. [47.59, 60] to describe e^-–H collisions, while *McCarthy* [47.61] has studied an optical-potential approximation that goes beyond second order, and also makes allowance for exchange. This method is called the coupled channels optical (CCO) model. Finally, *Callaway* and *Oza* [47.62] have constructed an optical potential using a set of pseudostates to evaluate the sum over intermediate states and have obtained encouraging results for elastic scattering and excitation of the $n = 2$ states in e^-–H collisions.

Born Series Methods

In the high energy domain, which can usually be assumed to extend from several times the ionization threshold of the target upwards, methods based on the Born series give reliable results. Ignoring electron exchange for the moment, the Born series for the direct scattering amplitude can be written as [47.63, 64]

$$f = \sum_{n=1}^{\infty} f_{n,B}\,, \tag{47.102}$$

where the nth Born term $f_{n,B}$ contains the interaction V between the scattered electron and the target atom or ion n times, and the Green's function

$$G_0^{(+)} = (H_N + \mathcal{T}_e - E - i\epsilon)^{-1} \tag{47.103}$$

$(n-1)$ times, where $\mathcal{T}_e$ and H_N are defined following (47.101).

It is important to retain consistently all terms in the Born series with similar energy and momentum-transfer $\Delta = |\boldsymbol{k}_i - \boldsymbol{k}_f|$ dependencies. For elastic scattering, the scattering amplitude converges to the first Born approximation for all Δ, but at lower energies it is necessary to include $\mathrm{Re}(f_{3,B})$, as well as the second Born terms, to obtain the cross section correct to k^{-2}. In the forward direction, convergence to the first Born approximation is slow because of the contribution from $\mathrm{Im}(f_{2,B})$ which, from the optical theorem, corresponds to loss of flux into all open channels.

For inelastic scattering, the first Born approximation does not give the correct high energy limit for large Δ. Instead this comes from $\mathrm{Im}(f_{2,B})$. Physically, this can be understood by noting that inelastic scattering at large angles involves two collisions: a collision of the incident electron with the nucleus to give the large scattering angle, followed or preceeded by an inelastic collision with the bound electrons to give excitation. Again, in the forward direction, it is necessary to retain $\mathrm{Re}(f_{3,B})$ to obtain the cross section correct to k^{-2}.

Since the third Born term is difficult to calculate, *Byron* and *Joachain* [47.65] suggested that to third-order, the scattering amplitude should be calculated using the eikonal Born series (EBS) approximation

$$f_{EBS} = f_{1,B} + f_{2,B} + f_{3,G} + g_{och}\,, \tag{47.104}$$

where $f_{3,G}$ is the third-order term in the expansion of the Glauber amplitude [47.66] in powers of V, which can be more easily calculated than $f_{3,B}$, and g_{och} is the Ochkur electron exchange amplitude [47.67]. The EBS method has been very successful when perturbation theory converges rapidly; namely, at high energies, at small and intermediate scattering angles, and for light atoms.

Distorted Wave Methods

Distorted wave methods are characterized by a separation of the interaction into two parts, one which is treated exactly and the other which is treated in first-order. The usual approach is based on the integral expressions for the T-matrix given by (47.90) or (47.92). The distorted wave approximation is obtained by replacing the exact solution of (47.10), denoted by $\boldsymbol{F}^{(+)}$ and $\boldsymbol{F}^{(-)}$ in (47.90) and (47.92) respectively, by approximate solutions, usually obtained by omitting all channels except the final or initial channels of interest. In addition, the potential interaction $\boldsymbol{U}$ is often approximated by just the direct term in (47.10).

This method becomes more accurate at intermediate energies as $Z - N$ increases, and as the angular momentum of the scattered electron becomes large. However, it gives poor results at low energies where the coupling between the channels in (47.10) is strong, and where resonances are often important. It has proved to be a very useful way of calculating electron–ion total and differential cross sections at intermediate and high energies well removed from threshold.

47.1.7 Ionization

This section discusses processes where electrons are ejected from the target during the collision, giving rise to ionization. The main focus is single ionization, or (e, 2e) processes given by

$$e^-(E_i)+A_i \to A_j^+ + e^-(E_A)+e^-(E_B)\,, \quad (47.105)$$

where E_i, E_A and E_B are the incident, scattered and ejected electron energies respectively. They are related through the energy conservation relation

$$E_i = E_A + E_B + \varepsilon\,, \quad (47.106)$$

where ε is the binding energy of the ejected atomic electron. The threshold behavior of the ionization cross section leading to the Wannier threshold law is treated in Chapt. 52.

The scattering amplitude for the direct ionization process is [47.68]

$$f_i(\boldsymbol{k}_A,\boldsymbol{k}_B) = -(2\pi)^{\frac{1}{2}}\, e^{i\chi(\boldsymbol{k}_A,\boldsymbol{k}_B)} \times \left\langle (H_{N+1}-E)\,\Phi \middle| \Psi_i^{(+)} \right\rangle \quad (47.107)$$

where $\Psi_i^{(+)}$ is an exact solution of (47.2) satisfying plane wave plus outgoing wave boundary conditions given by (47.5), and where Φ is a solution of the equation

$$(H_{N+1}-V-E)\,\Phi = 0 \quad (47.108)$$

satisfying ingoing wave boundary conditions. Also, $\boldsymbol{k}_A$ and $\boldsymbol{k}_B$ are the momenta of the two outgoing electrons defined in terms of E_A and E_B by

$$\frac{1}{2}k_A^2 = E_A, \qquad \frac{1}{2}k_B^2 = E_B\,. \quad (47.109)$$

In practice the potential V is chosen so that Φ has a simple form. Thus, for electron scattering by H-like ions with nuclear charge Z, V is often chosen as

$$V = -\frac{Z-Z_A}{r_1} - \frac{Z-Z_B}{r_2} + \frac{1}{r_{12}}\,, \quad (47.110)$$

so that

$$\Phi(\boldsymbol{r}_1,\boldsymbol{r}_2) = \Psi_C^{(-)}(Z_A,\boldsymbol{k}_A,\boldsymbol{r}_1)\,\Psi_C^{(-)}(Z_B,\boldsymbol{k}_B,\boldsymbol{r}_2)\,, \quad (47.111)$$

where the $\Psi_C^{(-)}$ are Coulomb wave functions satisfying ingoing-wave boundary conditions, and Z_A and Z_B are the effective charges seen by the two electrons. In order that the scattering amplitude does not contain a divergent phase factor, Z_A and Z_B must satisfy

$$\frac{Z_A}{k_A}+\frac{Z_B}{k_B} = \frac{Z}{k_A}+\frac{Z}{k_B}-\frac{1}{|\boldsymbol{k}_A-\boldsymbol{k}_B|}\,, \quad (47.112)$$

and the phase factor in (47.107) is then given by

$$\chi(\boldsymbol{k}_A,\boldsymbol{k}_B) = \frac{Z_A}{k_A}\ln\frac{k_A^2}{k_A^2+k_B^2} + \frac{Z_B}{k_B}\ln\frac{k_B^2}{k_A^2+k_B^2}\,. \quad (47.113)$$

The exchange ionization amplitude $g_i(\boldsymbol{k}_A,\boldsymbol{k}_B)$ can be obtained from the Peterkop theorem [47.69]

$$g_i(\boldsymbol{k}_A,\boldsymbol{k}_B) = f_i(\boldsymbol{k}_B,\boldsymbol{k}_A)\,. \quad (47.114)$$

This result follows from the fact that the amplitudes $f_i(\boldsymbol{k}_B,\boldsymbol{k}_A)$ and $g_i(\boldsymbol{k}_A,\boldsymbol{k}_B)$ describe the same physical process where the electron at $\boldsymbol{r}_{N+1}$ has momentum $\boldsymbol{k}_A$, and the electron at $\boldsymbol{r}_N$ has momentum $\boldsymbol{k}_B$. In practice, the Peterkop theorem is not valid if approximate wave functions are used to calculate the ionization amplitude. For this reason, a relative phase $\tau_i(\boldsymbol{k}_A,\boldsymbol{k}_B)$ between these two amplitudes is sometimes introduced by

$$g_i(\boldsymbol{k}_A,\boldsymbol{k}_B) = e^{i\tau(\boldsymbol{k}_A,\boldsymbol{k}_B)}\, f_i(\boldsymbol{k}_B,\boldsymbol{k}_A)\,. \quad (47.115)$$

This adds an element of arbitrariness into the calculation since different choices of this phase leads to different cross sections [47.70].

The ionization cross sections can be obtained directly from the scattering amplitudes. For random electron spin orientations, the triple differential cross section (TDCS) for ionization of a target with one active electron from an initial state denoted by $|i\rangle$ is given by

$$\frac{d^3\sigma_i}{d\Omega_A\, d\Omega_B\, dE} = \frac{k_A k_B}{k_i}\left(\frac{1}{4}|f_i+g_i|^2+\frac{3}{4}|f_i-g_i|^2\right)\,. \quad (47.116)$$

By integrating the TDCS with respect to $d\Omega_A$, $d\Omega_B$ or dE, we can form three different double differential cross sections, and three different single differential cross sections. The total ionization cross section obtained by integrating (47.116) over all outgoing electron scattering angles and energies is

$$\sigma_i = \frac{1}{k_i}\int_0^{E/2} dE\, k_A k_B \int d\Omega_A \times \int d\Omega_B \left[\frac{1}{4}|f_i+g_i|^2+\frac{3}{4}|f_i-g_i|^2\right]\,, \quad (47.117)$$

where the upper limit of integration over the energy variable is $E/2$ because the two outgoing electrons are indistinguishable.

We conclude this general discussion of the theory of electron impact ionization by mentioning recent work in which an integral representation for the ionization amplitude has been developed [47.71] which is free from the ambiguity and divergence problems associated with earlier work [47.68, 69, 72, 73]. An important aspect of this new development is that it has a form which can be used for practical calculations.

We now consider several approaches which have been used to obtain accurate ionization cross sections commencing, as in electron impact excitation with low energy methods and concluding with Born and distorted wave methods.

Pseudostate Methods

An important recent development is the realization that accurate ionization cross sections close to threshold can be obtained by representing the ionization continuum by suitably chosen square-integrable pseudostates. As already discussed when we considered pseudostate methods in Sect. 47.1.6 several methods including the intermediate energy R-matrix method [47.41–44], the R-matrix with pseudostates method [47.45–48] and the convergent close coupling method [47.50–56] have been used to obtain accurate ionization cross sections in this energy range.

Time-Dependent Close Coupling Method

Electron impact excitation and ionization amplitudes and cross sections can also be determined by solving the time-dependent Schrödinger equation directly [47.74–77]. In the case of electron scattering by atomic hydrogen or by an atom or atomic ion with one electron outside a closed inert shell, the total wave function can be expanded for each conserved $LS\pi$ symmetry as

$$\Psi^{\Gamma}(\boldsymbol{r}_1, \boldsymbol{r}_2, t) = \sum_{\ell_1\ell_2} (r_1 r_2)^{-1} P^{LS}_{\ell_1\ell_2}(r_1, r_2, t) \times \mathcal{Y}^{LM_L}_{\ell_1\ell_2}(\hat{\boldsymbol{r}}_1, \hat{\boldsymbol{r}}_2) , \quad (47.118)$$

where Γ is defined by (47.9) and the coupled spherical harmonics $\mathcal{Y}^{LM_L}_{\ell_1\ell_2}$ are defined by

$$\mathcal{Y}^{LM_L}_{\ell_1\ell_2}(\hat{\boldsymbol{r}}_1, \hat{\boldsymbol{r}}_2) = \sum_{m_1 m_2} (\ell_1 m_1 \ell_2 m_2 | \ell_1 \ell_2 L M_L) \times Y_{\ell_1 m_1}(\theta_1, \phi_1) Y_{\ell_2 m_2}(\theta_2, \phi_2) . \quad (47.119)$$

Substituting (47.118) into the time-dependent Schrödinger equation and projecting onto the coupled spherical harmonics $\mathcal{Y}^{LM_L}_{\ell_1\ell_2}$ then yields the following coupled differential equations

$$\mathrm{i}\frac{\partial P^{LS}_{\ell_1\ell_2}(r_1, r_2, t)}{\partial t} = T_{\ell_1\ell_2}(r_1, r_2) P^{LS}_{\ell_1\ell_2}(r_1, r_2, t) + \sum_{\ell_1'\ell_2'} V^L_{\ell_1\ell_2\ell_1'\ell_2'}(r_1, r_2, t) P^{LS}_{\ell_1'\ell_2'}(r_1, r_2, t) , \quad (47.120)$$

where

$$T_{\ell_1\ell_2}(r_1, r_2) = -\frac{1}{2}\frac{\partial^2}{\partial r_1^2} - \frac{1}{2}\frac{\partial^2}{\partial r_2^2} + V_{\ell_1}(r_1) + V_{\ell_2}(r_2) . \quad (47.121)$$

In this equation V_ℓ is an ℓ-dependent pseudopotential representing the interaction of the valence or scattered electrons with the closed shell core and $V^L_{\ell_1\ell_2\ell_1'\ell_2'}$ are the radial coupling potentials obtained by taking the matrix elements of the r_{12}^{-1} interaction between the valence and scattered electrons. Equations (47.120) are solved on a two-dimensional grid using an explicit time propagator which can be readily implemented on parallel computers. Commencing at time $t = 0$ with a wave function which is constructed as the appropriately symmetrized product of an incoming radial wave packet for one electron and the lowest energy bound stationary state for the other electron, the time-dependent equations (47.120) are integrated forward in time. Electron impact excitation and ionization amplitudes and cross sections are determined by projecting the time-evolved radial wave function $P^{LS}_{\ell_1\ell_2}(r_1, r_2, t)$ onto a complete set of target states.

Exterior Complex Scaling Method

In this approach, developed by *Rescigno* et al. [47.78–80], the time-independent Schrödinger equation for three charged particles is solve numerically on a two-dimensional grid without explicitly imposing asymptotic boundary conditions for three-body breakup. In the case of electron hydrogen atom scattering, the total wave function is partioned into the sum of an appropriately symmetrized unperturbed wave function $\psi^{\Gamma}_{k_i}$ describing a free electron with momentum k_i incident on the target ground state and a scattered wave function $\Psi^{\Gamma}_{\mathrm{sc}}$ which is expanded in the form

$$\Psi^{\Gamma}_{\mathrm{sc}}(\boldsymbol{r}_1, \boldsymbol{r}_2) = \sum_{L\ell_1\ell_2} r_1^{-1} r_2^{-1} \psi^{LS}_{\ell_1\ell_2}(r_1, r_2) \mathcal{Y}^{L0}_{\ell_1\ell_2}(\hat{\boldsymbol{r}}_1, \hat{\boldsymbol{r}}_2) . \quad (47.122)$$

Since the z-axis is defined to lie along the incident beam direction then M_L in the coupled spherical harmonics defined by (47.119) is zero. Also, since parity is conserved the summation in (47.122) is limited to terms for which $L+\ell_1+\ell_2$ is even.

Substituting the expression for the total wave function into the Schrödinger equation then yields the following set of coupled two-dimensional equations for the radial functions $\psi^{LS}_{\ell_1\ell_2}$ for each L and S:

$$\begin{aligned}&(E-H_{\ell_1}(r_1)-H_{\ell_2}(r_2))\psi^{LS}_{\ell_1\ell_2}(r_1,r_2)\\&-\sum_{\ell_1'\ell_2'}V^L_{\ell_1\ell_2\ell_1'\ell_2'}(r_1,r_2)\psi^{LS}_{\ell_1'\ell_2'}(r_1,r_2)\\&=\chi^{LS}_{\ell_1\ell_2}(r_1,r_2)\,.\end{aligned} \quad (47.123)$$

In this equation $H_\ell(r)$ is the radial hydrogenic Hamiltonian

$$H_\ell(r)=-\frac{1}{2}\frac{\mathrm{d}^2}{\mathrm{d}r^2}+\frac{\ell(\ell+1)}{2r^2}-\frac{1}{r}\,, \quad (47.124)$$

$V^L_{\ell_1\ell_2\ell_1'\ell_2'}$, as in (47.120), are the radial coupling potentials obtained by taking the matrix elements of the electron–electron interaction r_{12}^{-1} and the inhomogeneous term $\chi^{LS}_{\ell_1\ell_2}(r_1,r_2)$ arises from the partial wave expansion of the incident wave term.

The coupled differential equations (47.123) are solved on a two-dimensional grid using exterior complex scaling (ECS) boundary conditions which avoid imposing detailed asymptotic boundary conditions. The ECS transformation is taken as a mapping $r\to z(r)$ of all radial coordinates to a contour

$$z(r)\equiv\begin{cases}r, & r\le R_0\\ R_0+(r-R_0)\mathrm{e}^{\mathrm{i}\eta} & r>R_0\end{cases}\,, \quad (47.125)$$

that is real for $r\le R_0$ but is rotated into the upper half of the complex plane for $r>R_0$. This transformation has the desirable property that any outgoing wave evaluated on this contour dies exponentially as the coordinate becomes large. Thus the ECS procedure transforms any outgoing wave into a function that falls off exponentially outside R_0 but is equal to the infinite range wave over the finite region of space where the coordinates are real. Producing meaningful ionization cross sections at energies several eV above the ionization threshold requires R_0 to be at least 100 a_0. The grid must extend beyond R_0 far enough to allow the complex scaled radial function to decay effectively to zero at the edge of the grid requiring grids that extend an additional 25 a_0 beyond R_0. A detailed discussion of this method is given by *McCurdy* et al. [47.80].

Born and Distorted Wave Methods

The integral expression (47.107) for the direct ionization amplitude provides a starting point for the calculation of cross sections at higher energies. Both the Born series methods and distorted wave methods, which were described in Sect. 47.1.6 when we considered intermediate and high energy elastic scattering and excitation, have been used to obtain ionization amplitudes.

Recently an important development of the distorted wave method has been made by *Jones* and *Madison* [47.81–83]. This new approach entitled the continuum distorted wave with eikonal initial state (CDW-EIS), commences from the two-potential expression for the transition amplitude given by *Gell-Mann* and *Goldberger* [47.84]

$$T_{fi}=\langle\chi_f^-|W_f^+|\Psi_i^+\rangle+\langle\chi_f^-|V_i-W_f^+|\beta_i\rangle\,. \quad (47.126)$$

In the first term in this equation Ψ_i^+ is the exact scattering wave function developed from the initial state satisfying outgoing wave boundary conditions, χ_f^- is a distorted wave corresponding to the final state satisfying incoming wave boundary conditions and the corresponding perturbation W_f^+ is the adjoint of the operator W_f and operates to the left. In the second term, β_i is the unperturbed initial state which in the case of electron hydrogen atom scattering is

$$\beta_i=(2\pi)^{-3/2}\exp(\mathrm{i}\boldsymbol{k}_0\boldsymbol{r}_1)\psi_i(\boldsymbol{r}_2)\,, \quad (47.127)$$

and V_i is the initial state interaction potential given in this case by

$$V_\mathrm{i}=-\frac{1}{r_1}+\frac{1}{r_{12}}\,. \quad (47.128)$$

An eikonal approximation is made for the initial state wave function Ψ_i^+ in (47.126) and the final state wave function χ_f^- is represented by a CDW wave function [47.85]. In this way distortion effects are included in both initial and final state wave functions. Results using the CDC-EIS approximation have been compared with ECS calculations for electron hydrogen atom triple ionization cross sections at 54.4 eV [47.86]. The two calculations are generally in very good agreement for equal energy sharing between the outgoing electrons which was considered in this work.

A further development of the distorted wave method for ionization has been made when the incident electron is fast and interacts weakly with the target atom or ion and the ejected electron is slow and interacts strongly with the residual ion [47.87, 88]. In this case (47.107) is applicable where the ionizing electron is represented by plane waves or distorted waves while the initial target state and the ejected electron and residual ion state are both represented by the close coupling expansion (47.8). This approximation is particularly useful when the ejected electron can be captured into an autoionizing state of the target atom or ion which then decays giving rise to the following excitation-autoionization (EA) process

$$e^- + A^{q+} \to \left[A^{q+}\right] + e^- \to \left[A^{(q+1)+}\right] + 2e^- . \quad (47.129)$$

In this equation the bracket indicates a resonance state, while q is the charge on the atom A. This process together with related resonant excitation double autoionization (REDA) process

$$e^- + A^{q+} \to \left[A^{(q-1)+}\right] \to \left[A^{q+}\right] + e^- \to \left[A^{(q+1)+}\right] + 2e^- , \quad (47.130)$$

and the resonant excitation auto-double ionization (READI) process

$$e^- + A^{q+} \to \left[A^{(q-1)+}\right] \to \left[A^{(q+1)+}\right] + 2e^- , \quad (47.131)$$

have attracted considerable experimental and theoretical interest [47.89–93]. However, while the EA process can be accurately treated using a distorted wave method if the incident electron is fast, both the REDA and READI processes involve capture of the incident electron into a resonant state and a strong coupling approach is required.

An Example and Conclusions

We conclude this section by mentioning a recent comparison that has been made between theory and experiment for electron impact ionization of hydrogen at 17.6 eV [47.94]. At this energy, which is only 4 eV above the ionization threshold, strong coupling effects between the incident and ejected electrons and the residual proton are important and hence Born series and distorted wave methods are not applicable. This comparison thus provides a stringent test of theoretical calculations. In this work triple-differential cross section measurements with coplanar outgoing electrons both having 2 eV energy were compared with exterior complex scaling (ECS) and convergent close coupling (CCC) calculations. The two calculations show excellent overall agreement both with the shape and the magnitude of the experiment for a wide range of scattering angles.

It is clear that a detailed theoretical understanding of electron hydrogen atom ionization has now been obtained over a wide range of energies. Although further work is required to predict accurate cross sections involving highly excited states of interest in plasma physics and astrophysical applications, for example in astrophysical H II regions [47.95], methods have been developed which should enable these cross sections to be accurately determined.

Good progress has also been made in the study of electron impact ionization of multi-electron atoms and ions. However major problems still remain both due to the need to obtain accurate target states, which also applies to elastic scattering and excitation, and due to fundamental difficulties in carrying out accurate calculations for REDA and READI processes defined by (47.130) and (47.131) respectively. In the latter case major theoretical difficulties still arise in the accurate treatment of resonance states which decay with the emission of more than one electron.

47.2 Electron–Molecule Collisions

47.2.1 Laboratory Frame Representation

The processes that occur in electron collisions with molecules are more varied than those that occur in electron collisions with atoms and atomic ions because of the possibility of exciting degrees of freedom associated with the motion of the nuclei. In addition, the multicenter and nonspherical nature of the electron molecule interaction considerably complicates the solution of the collision problem by reducing its symmetry and by introducing multicenter integrals that are more difficult to calculate than those occurring for atoms and ions.

We first consider the derivation of the equations describing the collision in the laboratory frame of reference discussed by *Arthurs* and *Dalgarno* [47.96]. The Schrödinger equation describing the electron–molecule system is

$$(H_\mathrm{m} + \mathcal{T}_\mathrm{e} + V)\,\Psi = E\Psi , \quad (47.132)$$

where H_m is the molecular Hamiltonian, $\mathcal{T}_\mathrm{e}$ is the kinetic energy operator of the scattered electron and V is the electron–molecule interaction potential

$$V(\boldsymbol{R}, \boldsymbol{r}_\mathrm{m}, \boldsymbol{r}) = \sum_i \frac{1}{|\boldsymbol{r} - \boldsymbol{r}_i|} - \sum_i \frac{Z_i}{|\boldsymbol{r} - \boldsymbol{R}_i|} . \quad (47.133)$$

Here, $\boldsymbol{R}$ represents the coordinates $\boldsymbol{R}_i$ of all the nuclei, $\boldsymbol{r}_{\rm m}$ represents the coordinates $\boldsymbol{r}_i$ of the electrons in the target molecule, and $\boldsymbol{r}$ represents the coordinates of the scattered electron. The total energy E in (47.132) refers to the frame of reference where the c.m. of the whole system is at rest.

As in the case of electron–atom and electron–ion collisions, introduce target eigenstates, and possibly pseudostates Φ_i, by the equation

$$\langle \Phi_i | H_{\rm m} | \Phi_j \rangle = w_i \delta_{ij} \,, \tag{47.134}$$

and then expand the total wave function Ψ, in analogy with (47.8), in the form

$$\Psi_j = \mathcal{A} \sum_i \Phi_i (\boldsymbol{R}, \boldsymbol{r}_{\rm m}) \mathcal{F}_{ij} (\boldsymbol{r}) + \sum_i \chi_i (\boldsymbol{R}, \boldsymbol{r}_{\rm m}, \boldsymbol{r}) b_{ij} \,, \tag{47.135}$$

where the spin variables have been suppressed for notation simplicity, and we have not carried out a partial-wave decomposition of the wave function $\mathcal{F}_{ij}$ representing the scattered electron. The subscripts i and j now represent the rotational and vibrational states of the molecule as well as its electronic states.

Coupled equations for the functions $\mathcal{F}_{ij}$ can be obtained by substituting expansion (47.135) into (47.132), and projecting onto the target states Φ_i and onto the square integrable functions χ_j. After eliminating the coefficients b_{ij}, the coupled integrodifferential equations

$$\left(\nabla^2 + k_i^2\right) F_{ij} (r) = 2 \sum_\ell (V_{i\ell} + W_{i\ell} + X_{i\ell}) F_{\ell j} (r) \tag{47.136}$$

are obtained, where $k_i^2 = 2(E - w_i)$ and where $V_{i\ell}$, $W_{i\ell}$, and $X_{i\ell}$ are the direct, nonlocal exchange, and nonlocal correlation potentials. By expanding F_{ij} in partial waves, a set of coupled radial integrodifferential equations result, analagous to (47.10) for atoms and ions.

The scattering amplitude and cross section for a transition from an initial state $|i\rangle = |\boldsymbol{k}_i, \Phi_i, \chi_{\frac{1}{2}m_i}\rangle$ to a final state $|j\rangle = |\boldsymbol{k}_j, \Phi_j, \chi_{\frac{1}{2}m_j}\rangle$ is then given by (47.7), where now the subscripts i and j refer collectively to the ro-vibrational and electronic states of the molecule.

47.2.2 Molecular Frame Representation

The theory described in the previous section is completely general, and has been the basis of a number of early calculations for simple diatomic molecules such as H_2. However, major computational difficulties arise because of the very large number of rovibrational channels that need to be retained in expansion (47.135) for all but the simplest low-energy calculations.

This difficulty can be overcome by making a Born–Oppenheimer separation of the electronic and nuclear motion. The electronic motion is first determined with the nuclei held fixed. This is referred to as the fixed-nuclei approximation. The molecular rotational and vibrational motion is then included in a second step of the calculation. This procedure owes its validity to the large ratio of the nuclear mass to the electronic mass, and can be adopted when the collision time is much shorter than the periods of molecular rotation and vibration. Thus it is expected to be valid when the scattered electron energy is not close to a threshold, or when the energy does not coincide with that of a narrow resonance. In these cases, further developments described below are needed to obtain reliable cross sections.

In order to formulate the collision process in this representation, adopt a frame of reference that is rigidly attached to the molecule. The fixed-nuclei approximation then starts from the Schrödinger equation

$$(H_{\rm el} + \mathcal{T}_{\rm e} + V)\, \psi = E\psi \,, \tag{47.137}$$

where $H_{\rm el}$ is the electronic part of the target Hamiltonian obtained by assuming that the target nuclei have fixed coordinates denoted collectively by $\boldsymbol{R}$. It follows that $H_{\rm el}$ is related to $H_{\rm m}$ in (47.132) by

$$H_{\rm m} = H_{\rm el} + \mathcal{T}_{\rm R} \,, \tag{47.138}$$

where $\mathcal{T}_{\rm R}$ is the kinetic energy operator for the rotational and vibrational motion of the nuclei. The remaining quantities $\mathcal{T}_{\rm e}$ and V are the same as in (47.132).

The solution of (47.137) proceeds in an analogous way to the solution of (47.2) for electron collisions with atoms and ions. We adopt an expansion similar to (47.8), where we now expand the function representing the motion of the scattered electron in terms of symmetry-adapted angular functions that transform as an appropriate irreducible representation (IRR) of the molecular point group (*Burke* et al. [47.97]). Substituting this expansion into (47.137), and projecting onto the corresponding channel functions and onto the square integrable functions, yields a set of coupled integrodifferential equations with the form given by (47.10), where now the channel indices i, j and ℓ represent the component of the IRR, as well as the electronic state of the target, and where Γ represents the conserved quantum numbers that now include the IRR and the total spin.

The final step is to solve these coupled integrodifferential equations for each set of nuclear coordinates $\boldsymbol{R}$

of importance in the collision, using one of the methods discussed in Sect. 47.1.5. This yields the K-matrices, S-matrices and cross sections for fixed $\boldsymbol{R}$. For scattering calculations where only the ground electronic state has been included in the expansion, a number of approaches have been developed that replace the nonlocal exchange and correlation potentials by local potentials [47.2]. These approaches have proved particularly important in describing electronically elastic collisions of electrons with polyatomic molecules.

47.2.3 Inclusion of the Nuclear Motion

This section discusses how observables involving the nuclear motion, such as rotational and vibrational excitation cross sections, and dissociative attachment cross sections, can be obtained from the solutions of the fixed-nuclei equations.

The most widely used approach is the adiabatic-nuclei approximation [47.98–100]. In the case of diatomic molecules in a $^1\Sigma$ state, the scattering amplitude for a transition between electronic, vibrational and rotational states represented by $ivjm_j$ and $i'v'j'm'_j$ is given by

$$f_{i'v'j'm'_j,ivjm_j}(\hat{\boldsymbol{k}}\cdot\hat{\boldsymbol{r}}) = \left\langle \chi_{i'v'}(R)Y_{j'm'_j}(\hat{\boldsymbol{R}}) \middle| f_{i'i}(\hat{\boldsymbol{k}}\cdot\hat{\boldsymbol{r}};\boldsymbol{R}) \middle| \chi_{iv}(R)Y_{jm_j}(\hat{\boldsymbol{R}}) \right\rangle , \tag{47.139}$$

where $f_{i'i}(\hat{\boldsymbol{k}}\cdot\hat{\boldsymbol{r}};\boldsymbol{R})$ is the fixed-nuclei scattering amplitude, which depends parametrically on the nuclear coordinates $\boldsymbol{R}$, and χ_{iv} and Y_{jm_j} are the molecular vibrational and rotational eigenfunctions, respectively. This approximation is valid provided that the collision time is short compared with the vibration and/or rotation times, and is widely used in such situations.

The cross section is usually averaged over the degenerate sublevels m_j and summed over m'_j, giving the cross section for the transition ivj to $i'v'j'$. This leads to the relation

$$\frac{d\sigma_{i'v'j',ivj}}{d\Omega} = \sum_{j_t=|j-j'|}^{j+j'} \left[(j0j_t0 \mid jj_tj'0)\right]^2 \frac{d\sigma_{i'v'j_t,iv0}}{d\Omega} , \tag{47.140}$$

provided that the small differences in the wave numbers for the different rotational channels can be neglected. A similar relation holds for symmetric top molecules such as NH_3. For spherical top molecules such as CH_4, the equivalent relation is [47.101, 102]

$$\frac{d\sigma_{i'v'j',ivj}}{d\Omega} = \frac{2j'+1}{2j+1} \sum_{j_t=|j-j'|}^{j+j'} \frac{1}{2j_t+1} \frac{d\sigma_{i'v'j_t,iv0}}{d\Omega} . \tag{47.141}$$

The sum in (47.140) or (47.141) over the final rotational state j' is independent of the initial state j. Also, if the cross section is multiplied by the transition energy and then sum over j', the result, which is in the mean energy loss by the incident electron, is still independent of j.

The adiabatic-nuclei approximation breaks down close to threshold or in the neighborhood of narrow resonances [47.103]. A straightforward way of including nonadiabatic effects that arise in vibrational excitation is to retain the vibrational terms in the Hamiltonian, but still to treat the rotational motion adiabatically. Hence, instead of (47.137), the equation

$$(H_{el} + \mathcal{T}_{vib} + \mathcal{T}_e + V)\,\widetilde{\psi} = E\widetilde{\psi} \tag{47.142}$$

is solved, where $\mathcal{T}_{vib}$ is the kinetic energy operator for the nuclear vibrational motion, and where the other quantities have the same meaning as in (47.137). Adopting a frame of reference in which the molecule has fixed spacial orientation, and separating out the angular variables of the scattered electron, coupled integrodifferential equations coupling the target vibrational states as well the electronic states can be obtained. This approach has been adopted with success [47.104, 105], but is computationally demanding since the number of coupled channels can become very large.

Vibrational excitation and dissociative attachment are particularly important in resonance regions when the scattered electron spends an appreciable time in the neighborhood of the molecule. As a result, a number of approaches has been developed describing these processes based on electron molecule resonance theories (e.g., [47.106–110]). The basic idea is that a series of fixed-nuclei resonance states $\psi_n^{(r)}$ are introduced for a range of values of R, either by imposing Siegert outgoing wave boundary conditions [47.111], or by introducing Feshbach projection operators [47.15, 16]. The amplitude for a transition from an initial electronic–vibrational state iv to a final state $i'v'$ is then given by

$$T_{i'v',iv} = \sum_n \left\langle \chi_{i'v'}(R')\,\zeta_{ni'}(R') \right| G_n^{(r)}(R',R) \times \left| \zeta_{ni}(R)\,\chi_{iv}(R) \right\rangle , \tag{47.143}$$

where χ_{iv} are the vibrational eigenfunctions, ζ_{ni} are the "entry amplitudes" from the initial or final electronic states into the resonance states $\psi_n^{(\mathrm{r})}$, and $G_n^{(\mathrm{r})}$ are the Green's functions that describe the propagation in the intermediate resonance states $\psi_n^{(\mathrm{r})}$. Dissociative attachment can be described by a straightforward extension of this theory.

The fixed-nuclei R-matrix method has also been extended to treat vibrational excitation and dissociative attachment [47.112]. A generalized R-matrix is introduced by an equation which, in analogy with(47.143), can be written as

$$R_{i'v',iv} = \frac{1}{2a} \sum_k \left\langle \chi_{i'v'}\left(R'\right) w_{i'k}^{\Gamma}\left(R'\right) \middle| G_k^{RM}\left(R', R\right) \right. \\ \times \left| w_{ik}^{\Gamma}(R)\, \chi_{iv}(R) \right\rangle , \qquad (47.144)$$

where the surface amplitudes $w_{ik}^{\Gamma}(R)$ are defined by (47.76), and the Green's function G_k^{RM} now describes the propagation in the intermediate R-matrix states defined by (47.73). Once the generalized R-matrix has been determined, the final step in the calculation is to solve the collision problem in the external region which, for diatomic molecules, is defined by the condition that the scattered electron coordinate r is greater than some given radius a, and the internuclear coordinate R is greater than some given radius A.

Another approach that includes nonadiabatic effects is the energy-modified adiabatic approximation introduced by *Nesbet* [47.113]. In this approach, the S-matrix elements connecting the vibrational states are defined by

$$S_{i'v',iv} = \langle \chi_{i'v'} \mid S_{i'i}\left(E - H_i;\, R\right) \mid \chi_{iv} \rangle , \qquad (47.145)$$

where $S_{i'i}\left(E - H_i;\, R\right)$ is the S-matrix calculated in the fixed nuclei approximation at the internuclear separation R at an energy defined by the operator $H_i = E_i(R) + \mathcal{T}_{\mathrm{vib}}$. This has the effect of including the internal energy of the target into the S-matrix elements, giving the correct threshold energies.

Finally, we mention an off-shell T-matrix approach for including nonadiabatic effects discussed first by *Shugard* and *Hazi* [47.114]. This approach can also extend the range of validity of the adiabatic-nuclei approximation, while retaining much of its inherent simplicity. This has recently been applied with success to low-energy vibrational excitation of H_2 and CH_4 [47.115].

47.2.4 Electron Collisions with Polyatomic Molecules

In the last few years computer programs based on the ab initio methods described in Sect. 47.1.5 have been developed and used to calculate cross sections for electron collisions with polyatomic molecules of importance in many applications.

Recent work includes electron collisions with nitrous oxide which is an important species in the upper atmosphere where it plays a role in ozone destruction. N_2O lasers are also of importance. Fixed-nuclei total cross sections, calculated using independent Bonn [47.116] and UK [47.117] polyatomic R-matrix programs are in good agreement with experiment [47.118], showing a $^2\Pi$ shape resonance near 2 eV. The UK R-matrix program has also been used to study electron collisions with the open-shell radical OClO [47.119]. An important process in the stratospheric polar vortex involves the coupling of chlorine and bromine chemistry, in which OClO plays a key role. In this case OClO is formed by the reaction $\mathrm{BrO} + \mathrm{ClO} \rightarrow \mathrm{Br} + \mathrm{OClO}$. Fixed-nuclei total cross sections were calculated including eight electronic states in the R-matrix expansion, giving good agreement with experiment [47.120], reproducing a shoulder in the cross section between 2 and 6 eV.

Important advances have also been made in the theoretical treatment of vibrational excitation in polyatomic molecules. Electron collisions with CO_2 molecules were calculated in the fixed-nuclei approximation using an electron–polyatomic molecular scattering program based on the complex Kohn variational method [47.121]. At low energies the cross section is dominated by a virtual state at threshold and a $^2\Pi_{\mathrm{u}}$ shape resonance at 3.8 eV. As the molecule bends from its ground state linear configuration, the shape resonance splits into nondegenerate $^2A_1$ and $^2B_1$ configurations. The fixed-nuclei complex resonance energy surfaces are parametrized and motion on these surfaces is computed using a generalization of the Boomerang model [47.122]. The results reproduce the oscillations resulting from the interference between the nuclear and electronic motion seen experimentally by *Allen* [47.123, 124].

47.3 Electron–Atom Collisions in a Laser Field

Electron–atom collisions in the presence of an intense laser field have recently attracted considerable attention because of the importance of these processes in applications such as laser plasma interactions, and also because of their fundamental interest in atomic collision theory. This section summarizes the basic theory, commencing with the scattering of electrons by a potential in the presence of a laser field. The discussion is then generalized to the scattering of electrons by complex atoms and ions, where 'dressing' of the atomic eigenstates by the laser field must be considered, and where simultaneous electron photon excitation (SEPE) processes, defined by

$$nh\nu + e^- + A_i \rightarrow e^- + A_j\,, \tag{47.146}$$

can occur. Recent reviews of aspects of this subject have been given by *Mason* [47.3], *Newell* [47.125] and *Mittleman* [47.126].

47.3.1 Potential Scattering

We adopt a semiclassical description of the collision process in which the electrons and the target atom are described by the nonrelativistic Schrödinger equation, and the laser field is described classically. This is valid for most high intensity fields of current interest, where a typical coherence volume of the field contains a very large number of photons [47.126].

The time-dependent Schrödinger equation describing an electron scattered by a potential $V(\boldsymbol{r})$ in the presence of an external electromagnetic (laser) field is then

$$\mathrm{i}\frac{\partial}{\partial t}\Psi(\boldsymbol{r},t) = \left[-\frac{1}{2}\nabla^2 - \frac{\mathrm{i}}{c}\boldsymbol{A}(\boldsymbol{r},t)\cdot\nabla + \frac{1}{2c^2}\boldsymbol{A}^2(\boldsymbol{r},t) + V(\boldsymbol{r})\right]\Psi(\boldsymbol{r},t)\,, \tag{47.147}$$

in the Coulomb gauge such that the vector potential satisfies $\nabla\cdot\boldsymbol{A}=0$. We also assume that the laser field is monochromatic, monomode, linearly polarized, and spatially homogeneous (i. e., its wavelength is large compared with the range of the potential, or more generally with the size of the atom). Hence we can write

$$\boldsymbol{A}(\boldsymbol{r},t) = \boldsymbol{A}(t) = \hat{\boldsymbol{\epsilon}}A_0\cos\omega t\,, \tag{47.148}$$

where $\hat{\boldsymbol{\epsilon}}$ is a unit vector along the field polarization direction and ω is the angular frequency. The A^2 term in (47.147) can be removed by the unitary transformation

$$\Psi(\boldsymbol{r},t) = \exp\left(-\frac{\mathrm{i}}{2c^2}\int^t A^2(t')\,\mathrm{d}t'\right)\Psi_\mathrm{V}(\boldsymbol{r},t)\,, \tag{47.149}$$

where Ψ_V satisfies the Schrödinger equation in the velocity gauge given by

$$\mathrm{i}\frac{\partial}{\partial t}\Psi_\mathrm{V}(\boldsymbol{r},t) = \left[-\frac{1}{2}\nabla^2 - \frac{\mathrm{i}}{c}\boldsymbol{A}(t)\cdot\nabla + V(\boldsymbol{r})\right]\times\Psi_\mathrm{V}(\boldsymbol{r},t)\,. \tag{47.150}$$

The corresponding equation for the free electron with $V=0$ is readily solved to give the Volkov wave function [47.127]

$$\chi_k(\boldsymbol{r},t) = (2\pi)^{-\frac{3}{2}}\exp\left[\mathrm{i}\left(\boldsymbol{k}\cdot\boldsymbol{r} - \boldsymbol{k}\cdot\boldsymbol{\alpha}_0\sin\omega t - Et\right)\right]\,, \tag{47.151}$$

where $\boldsymbol{k}$ is the wave vector and $E=\frac{1}{2}k^2$ the kinetic energy. The quantity $\boldsymbol{\alpha}(t)$ is defined by

$$\boldsymbol{\alpha}(t) = \frac{1}{c}\int^t \boldsymbol{A}(t')\,\mathrm{d}t' = \boldsymbol{\alpha}_0\sin\omega t\,, \tag{47.152}$$

where $\boldsymbol{\alpha}_0 = \boldsymbol{\mathcal{E}}_0/\omega^2$, $\boldsymbol{\mathcal{E}}_0$ being the electric field strength.

To solve (47.150), introduce the causal Green's function $G_0^{(+)}(\boldsymbol{r},t;\boldsymbol{r}',t')$ satisfying the equation

$$\left(\mathrm{i}\frac{\partial}{\partial t} + \frac{1}{2}\nabla^2 + \frac{\mathrm{i}}{c}\boldsymbol{A}(t)\cdot\nabla\right)G_0^{(+)}(\boldsymbol{r},t;\boldsymbol{r}',t') = \delta(\boldsymbol{r}-\boldsymbol{r}')\,\delta(t-t')\,. \tag{47.153}$$

This Green's function is given by

$$G_0^{(+)}(\boldsymbol{r},t;\boldsymbol{r}',t') = -\mathrm{i}\theta(t-t')\int\chi_k(\boldsymbol{r},t)\,\chi_k^*(\boldsymbol{r}',t')\,\mathrm{d}\boldsymbol{k}\,, \tag{47.154}$$

where $\theta(x)=1$ for $x>0$ and $\theta(x)=0$ for $x<0$. The corresponding causal outgoing wave solution of (47.150) is

$$\Psi_k^{(+)}(\boldsymbol{r},t) = \chi_k(\boldsymbol{r},t) + \int_{-\infty}^{t}\mathrm{d}t'\int\mathrm{d}\boldsymbol{r}'\,G_0^{(+)}(\boldsymbol{r},t;\boldsymbol{r}',t')\times V(\boldsymbol{r}')\,\Psi_k^{(+)}(\boldsymbol{r}',t)\,, \tag{47.155}$$

and the S-matrix element for a transition $\boldsymbol{k}_i\rightarrow\boldsymbol{k}_f$ in the presence of the laser field is given by

$$S_{\boldsymbol{k}_f,\boldsymbol{k}_i} = -\mathrm{i}\left\langle\chi_{\boldsymbol{k}_f}\middle|V\middle|\Psi_{\boldsymbol{k}_i}^{(+)}\right\rangle\,, \tag{47.156}$$

where an integration is carried out over all space and time in this matrix element. The time integration in (47.156) can be performed using the relation

$$\exp(\mathrm{i}x\sin u) = \sum_{n=-\infty}^{\infty}J_n(x)\exp(\mathrm{i}nu)\,, \tag{47.157}$$

where $J_n(x)$ is an ordinary Bessel function of order n. Then

$$S_{\boldsymbol{k}_f,\boldsymbol{k}_i} = -2\pi \mathrm{i} \sum_{n=-\infty}^{\infty} \delta\left(E_{k_f} - E_{k_i} - n\omega\right) T^n_{\boldsymbol{k}_f,\boldsymbol{k}_i} \tag{47.158}$$

where the delta function ensures energy conservation, and n is the number of photons absorbed or emitted. The differential cross section for the scattering process $\boldsymbol{k}_i \to \boldsymbol{k}_f$ with exchange of n photons can then be defined in terms of the T-matrix elements $T^n_{\boldsymbol{k}_f,\boldsymbol{k}_i}$ by

$$\frac{\mathrm{d}\sigma_n}{\mathrm{d}\Omega} = (2\pi)^4 \frac{k_f}{k_i} \left|T^n_{\boldsymbol{k}_f,\boldsymbol{k}_i}\right|^2 . \tag{47.159}$$

There are two limiting cases in which considerable simplification in this expression occurs. First, at high energies or for weak potentials, the first Born approximation can be used to describe scattering by the potential $V(\boldsymbol{r})$. In this case (47.159) reduces to

$$\frac{\mathrm{d}\sigma_n^{1,\mathrm{B}}}{\mathrm{d}\Omega} = (2\pi)^4 \frac{k_f}{k_i} J_n^2\left(\boldsymbol{\Delta}\cdot\boldsymbol{\alpha}_0\right) |\mathcal{V}(\boldsymbol{\Delta})|^2 , \tag{47.160}$$

where $\boldsymbol{\Delta} = \boldsymbol{k}_i - \boldsymbol{k}_f$ is the momentum transfer vector and

$$\mathcal{V}(\boldsymbol{\Delta}) = (2\pi)^{-3} \int \exp(\mathrm{i}\boldsymbol{\Delta}\cdot\boldsymbol{r}) V(\boldsymbol{r})\, \mathrm{d}\boldsymbol{r} . \tag{47.161}$$

It follows immediately that

$$\frac{\mathrm{d}\sigma_n^{1,\mathrm{B}}}{\mathrm{d}\Omega} = J_n^2(\boldsymbol{\Delta}\cdot\boldsymbol{\alpha}_0) \frac{\mathrm{d}\sigma^{1,\mathrm{B}}}{\mathrm{d}\Omega} , \tag{47.162}$$

where $\mathrm{d}\sigma^{1,\mathrm{B}}/\mathrm{d}\Omega$ is the field-free first Born differential cross section. Using the sum rule

$$\sum_{n=-\infty}^{\infty} J_n^2(x) = 1 , \tag{47.163}$$

(47.162) immediately yields

$$\sum_{n=-\infty}^{\infty} \frac{\mathrm{d}\sigma_n^{1,\mathrm{B}}}{\mathrm{d}\Omega} = \frac{\mathrm{d}\sigma^{1,\mathrm{B}}}{\mathrm{d}\Omega} . \tag{47.164}$$

The second limiting case is the low frequency (soft photon) limit, where the laser photon energy ω is small compared with the electron energy E_{k_i}. In this limit, the T-matrix element is given by [47.128]

$$T^n_{\boldsymbol{k}_f,\boldsymbol{k}_i} = J_n(\boldsymbol{\Delta}\cdot\boldsymbol{\alpha}_0)\left\langle\chi_{\boldsymbol{k}'_f}\right|\mathcal{T}\left(E_{k'_i}\right)\left|\chi_{\boldsymbol{k}'_i}\right\rangle + O\left(w^2\right) , \tag{47.165}$$

where $\boldsymbol{k}'_i$ and $\boldsymbol{k}'_f$ are the shifted wave vectors

$$\boldsymbol{k}'_i = \boldsymbol{k}_i + \frac{n\omega}{\boldsymbol{\Delta}\cdot\boldsymbol{\alpha}_0}\boldsymbol{\alpha}_0 , \quad \boldsymbol{k}'_f = \boldsymbol{k}_f + \frac{n\omega}{\boldsymbol{\Delta}\cdot\boldsymbol{\alpha}_0}\boldsymbol{\alpha}_0 \tag{47.166}$$

and $\mathcal{T}(E_{k'_i})$ is the T-operator in the absence of the laser corresponding to the energy $E_{k'_i} = k_i'^2/2$. The differential cross section for the transfer of n photons is then

$$\frac{\mathrm{d}\sigma_n}{\mathrm{d}\Omega} = \frac{k_f}{k_i} J_n^2(\boldsymbol{\Delta}\cdot\boldsymbol{\alpha}_0) \frac{\mathrm{d}\sigma}{\mathrm{d}\Omega}(\boldsymbol{k}'_f, \boldsymbol{k}'_i) + O\left(\omega^2\right) , \tag{47.167}$$

where $\mathrm{d}\sigma(\boldsymbol{k}'_f, \boldsymbol{k}'_i)/\mathrm{d}\Omega$ refers to the transition $\boldsymbol{k}'_i \to \boldsymbol{k}'_f$ in the absence of the laser. If the frequency is small enough to neglect the n dependence of $\boldsymbol{k}'_i$ and $\boldsymbol{k}'_f$, then using the sum rule (47.163), (47.167) becomes

$$\sum_{n=-\infty}^{\infty} \frac{\mathrm{d}\sigma_n}{\mathrm{d}\Omega} = \frac{\mathrm{d}\sigma}{\mathrm{d}\Omega} , \tag{47.168}$$

where $\mathrm{d}\sigma/\mathrm{d}\Omega$ is the field-free differential cross section. The Kroll–Watson result (47.167) has been found to be surprisingly accurate, even for cases where $\omega/E_{k_i} \approx 0.5$ [47.129, 130]. A nonrigorous extension of the Kroll–Watson result to inelastic processes has been considered by a number of authors, and has been found to give qualitative agreement with experiments on helium [47.131].

47.3.2 Scattering by Complex Atoms and Ions

The time-dependent Schrödinger equation describing an electron scattered by an N-electron atom or ion in the presence of a laser field can be written in analogy with (47.147) as

$$\begin{aligned} &\mathrm{i}\frac{\partial}{\partial t}\Psi\left(\boldsymbol{X}_{N+1}, t\right) \\ &= \left[H_{N+1} - \frac{\mathrm{i}}{c}\sum_{i=1}^{N+1} \boldsymbol{A}\left(\boldsymbol{r}_i, t\right)\cdot\boldsymbol{\nabla}_i \right. \\ &\quad \left. + \frac{1}{2c^2}\sum_{i=1}^{N+1} \boldsymbol{A}^2\left(\boldsymbol{r}_i, t\right)\right]\Psi\left(\boldsymbol{X}_{N+1}, t\right) , \end{aligned} \tag{47.169}$$

where H_{N+1} is the $(N+1)$-electron Hamiltonian defined by (47.3) and $\boldsymbol{X}_{N+1}$ represents the space and spin coordinates of all $N+1$ electrons defined as in (47.8).

With the same assumptions made in the reduction of (47.147) to (47.150), (47.163) can be rewritten in the velocity gauge form

$$\begin{aligned} &\mathrm{i}\frac{\partial}{\partial t}\Psi_{\mathrm{V}}\left(\boldsymbol{X}_{N+1}, t\right) \\ &= \left(H_{N+1} + \frac{1}{c}\boldsymbol{A}(t)\cdot\boldsymbol{P}_{N+1}\right)\Psi_{\mathrm{V}}\left(\boldsymbol{X}_{N+1}, t\right) , \end{aligned} \tag{47.170}$$

where

$$P_{N+1} = \sum_{i=1}^{N+1} -\mathrm{i}\nabla_i \quad (47.171)$$

is the total momentum operator.

The solution of (47.170) for e$^-$–H scattering in a strong laser field at high electron impact energies has been considered by *Francken* and *Joachain* [47.132]. They discussed the use of the Born series and the eikonal Born series (EBS) approximations, given by (47.104), to describe the electron–atom collision, and included the dressed wave function of the atomic hydrogen target to first-order in the field strength $\mathcal{E}_0$. Important effects related to the dressing of the target are the appearance in the cross sections of asymmetries between the absorption and the emission of a given number of laser photons, as well as the appearance of new resonance structures in the cross sections.

So far, detailed studies of low-energy electron–atom and electron–ion collisions in laser fields have been very limited. Pioneering work on e$^-$–H$^+$ collisions by *Dimou* and *Faisal* [47.133], and by *Collins* and *Csanak* [47.134] have shown important resonance effects caused by the field coupling bound states to the continuum. In addition, multichannel quantum defect theory has been applied by *Zoller* and co-workers [47.135, 136] to study the behavior of Rydberg states in laser fields. We conclude this section by briefly describing the R-matrix–Floquet method [47.137] for treating electron collisions with complex atoms and ions in a laser field, based on the R-matrix method discussed in Sect. 47.1.5.

In this approach, configuration space is divided into internal and external regions, as in the field-free case. In the internal region, a further gauge transformation of the field is made to the length gauge, so that the time-dependent Schrödinger equation (47.170) now becomes

$$\mathrm{i}\frac{\partial}{\partial t}\Psi_{\mathrm{L}}(\boldsymbol{X}_{N+1}, t) = [H_{N+1} + \boldsymbol{\mathcal{E}}(t)\cdot \boldsymbol{R}_{N+1}]\Psi_{\mathrm{L}}(\boldsymbol{X}_{N+1}, t) , \quad (47.172)$$

where $\boldsymbol{R}_{N+1} = \sum_{i=1}^{N+1} \boldsymbol{r}_i$, and we assume that the electric field $\boldsymbol{\mathcal{E}}(t)$ is given by

$$\boldsymbol{\mathcal{E}}(t) = -\frac{1}{c}\frac{\mathrm{d}}{\mathrm{d}t}\boldsymbol{A}(t) = \hat{\boldsymbol{\epsilon}}\mathcal{E}_0 \cos\omega t . \quad (47.173)$$

In order to solve (47.172), we introduce the Floquet–Fourier expansion [47.138, 139]

$$\Psi_{\mathrm{L}}(\boldsymbol{X}_{N+1}, t) = \mathrm{e}^{-\mathrm{i}Et} \sum_{n=-\infty}^{\infty} \mathrm{e}^{-\mathrm{i}n\omega t}\Psi_n^{\mathrm{L}}(\boldsymbol{X}_{N+1}) . \quad (47.174)$$

Substituting this equation into (47.172), using (47.173) and equating the coefficients of $\exp[-\mathrm{i}(E+n\omega)t]$ to zero gives

$$(H_{N+1} - E - n\omega)\Psi_n^{\mathrm{L}} + D_{N+1}\left(\Psi_{n-1}^{\mathrm{L}} + \Psi_{n+1}^{\mathrm{L}}\right) = 0 , \quad (47.175)$$

where we have introduced the operator

$$D_{N+1} = \frac{1}{2}\mathcal{E}_0\hat{\boldsymbol{\epsilon}}\cdot\boldsymbol{R}_{N+1} . \quad (47.176)$$

The functions Ψ_n^{L} can be regarded as the components of a vector Ψ^{L} in photon space. Equation (47.175) can then be written in this space as

$$(\boldsymbol{H}_{\mathrm{F}} - E\boldsymbol{I})\Psi^{\mathrm{L}} = 0 , \quad (47.177)$$

where the Floquet Hamiltonian $\boldsymbol{H}_{\mathrm{F}}$ is an infinite tridiagonal matrix.

In order to solve (47.177) in the internal region, the components Ψ_n^{L} are expanded in a basis which, in analogy with (47.69), has the form

$$\begin{aligned}\Psi_{kn}^{\mathrm{L}}&(\boldsymbol{X}_{N+1})\\ &= \mathcal{A}\sum_{\Gamma ij}\overline{\Phi}_i^{\Gamma}\left(\boldsymbol{x}_1,\ldots,\boldsymbol{x}_N;\hat{\boldsymbol{r}}_{N+1}\sigma_{N+1}\right)r_{N+1}^{-1}\\ &\quad\times u_j(r_{N+1})\,a_{ijkn}^{\Gamma}\\ &\quad+\sum_{\Gamma i}\chi_i^{\Gamma}(\boldsymbol{x}_1,\ldots,\boldsymbol{x}_{N+1})\,b_{ikn}^{\Gamma} , \end{aligned} \quad (47.178)$$

where the summation over Γ is required since the total orbital angular momentum L and the total parity π in (47.9) are no longer conserved. The coefficients a_{ijkn}^{Γ} and b_{ikn}^{Γ} are then determined by diagonalizing $\boldsymbol{H}_{\mathrm{F}} + \boldsymbol{L}_{\mathrm{b}}$, where $\boldsymbol{L}_{\mathrm{b}}$ is an appropriate Bloch operator.

In the external region, the wave function describing the scattered electron is transformed to the velocity gauge, and the corresponding coupled equations integrated outwards from the internal region boundary for each energy of interest [47.140]. After a further transformation to the acceleration frame (or Kramers–Henneberger frame [47.141]) the wave function can be fitted to an asymptotic form to yield the K-matrix, S-matrix and collision cross sections. This approach has been used to calculate laser-assisted electron scattering by H and He atoms [47.142, 143] and a detailed discussion of the theory has been given [47.144].

References

47.1 P. G. Burke, C. J. Joachain (Eds.): *Photon and Electron Collisions with Atoms and Molecules* (Plenum, New York 1997)
47.2 M. J. Brunger, S. J. Buckman: Phys. Rep. **357**, 215 (2002)
47.3 N. J. Mason: Rep. Prog. Phys. **56**, 1275 (1993)
47.4 I. C. Percival, M. J. Seaton: Proc. Camb. Phil. Soc. **53**, 654 (1957)
47.5 T. F. O'Malley, L. Spruch, L. Rosenberg: J. Math. Phys. **2**, 491 (1961)
47.6 M. J. Seaton: Rep. Prog. Phys. **46**, 167 (1983)
47.7 H. E. Saraph: Comp. Phys. Commun. **3**, 256 (1972)
47.8 H. E. Saraph: Comp. Phys. Commun. **15**, 247 (1978)
47.9 M. Jones: Phil. Trans. R. Soc. London A **277**, 587 (1975)
47.10 N. S. Scott, P. G. Burke: J. Phys. B **13**, 4299 (1980)
47.11 J. J. Chang: J. Phys. B **8**, 2327 (1975)
47.12 P. H. Norrington, I. P. Grant: J. Phys. B **14**, L261 (1981)
47.13 U. Fano: Phys. Rev. **124**, 1866 (1961)
47.14 U. Fano: Rep. Prog. Phys. **46**, 97 (1983)
47.15 H. Feshbach: Ann. Phys. (N. Y.) **5**, 357 (1958)
47.16 H. Feshbach: Ann. Phys. (N. Y.) **19**, 287 (1962)
47.17 W. Brenig, R. Haag: Fortschr. Phys. **7**, 183 (1959)
47.18 P. G. Burke: Adv. At. Mol. Phys. **4**, 173 (1968)
47.19 M. J. Seaton: Proc. Phys. Soc. **88**, 801, 815 (1966)
47.20 M. J. Seaton: J. Phys. B **2**, 5 (1969)
47.21 M. Gailitis: Sov. Phys. JETP **17**, 1328 (1963)
47.22 U. Fano: Phys. Rev. A **2**, 353 (1970)
47.23 U. Fano: Comm. At. Mol. Phys. **10**, 223 (1981)
47.24 U. Fano: Comm. At. Mol. Phys. **13**, 157 (1983)
47.25 E. P. Wigner: Phys. Rev. **70**, 15, 606 (1946)
47.26 E. P. Wigner, L. Eisenbud: Phys. Rev. **72**, 29 (1947)
47.27 P. G. Burke, K. A. Berrington: *Atomic and Molecular Processes: An R-matrix Approach* (Institute of Physics, Bristol 1993)
47.28 C. Bloch: Nucl. Phys. **4**, 503 (1957)
47.29 R. K. Nesbet: *Variational Methods in Electron–Atom Scattering Theory* (Plenum, New York 1980)
47.30 W. H. Miller: Comm. At. Mol. Phys. **22**, 115 (1988)
47.31 B. I. Schneider, T. N. Rescigno: Phys. Rev. A **37**, 3749 (1988)
47.32 K. Takatsuka, V. McKoy: Phys. Rev. A **24**, 2743 (1981)
47.33 M. J. Seaton: J. Phys. B **7**, 1817 (1974)
47.34 B. I. Schneider, L. A. Collins: Comp. Phys. Rep. **10**, 49 (1989)
47.35 P. G. Burke, H. M. Schey: Phys. Rev. **126**, 147 (1962)
47.36 R. Damburg, E. Karule: Proc. Phys. Soc. **90**, 637 (1967)
47.37 P. G. Burke, T. G. Webb: J. Phys. B **3**, L131 (1970)
47.38 J. Callaway, J. W. Wooten: Phys. Lett. A **45**, 85 (1973)
47.39 J. Callaway, J. W. Wooten: Phys. Rev. A **9**, 1924 (1974)
47.40 J. Callaway, J. W. Wooten: Phys. Rev. A **11**, 1118 (1975)
47.41 P. G. Burke, C. J. Noble, M. P. Scott: Proc. Roy. Soc. A **410**, 289 (1987)
47.42 T. T. Scholz, H. R. J. Walters, P. G. Burke, M. P. Scott: J. Phys. B **24**, 2097 (1991)
47.43 K. M. Dunseath, M. LeDourneuf, M. Tereo-Dunseath, J.-M. Launay: Phys. Rev. A **54**, 561 (1996)
47.44 M. P. Scott, T. Stitt, N. S. Scott, P. G. Burke: J. Phys. B **35**, L323 (2002)
47.45 K. Bartschat, E. T. Hudson, M. P. Scott, P. G. Burke, V. M. Burke: J. Phys. B **29**, 115 (1996)
47.46 K. Bartschat, I. Bray: Phys. Rev. A **54**, R1002 (1996)
47.47 K. Bartschat: J. Phys. B **32**, L355 (1999)
47.48 D. M. Mitnik, D. C. Griffin, C. P. Balance, N. R. Badnell: J. Phys. B **36**, 717 (2003)
47.49 M. Rotenberg: Ann. Phys. (N. Y.) **19**, 262 (1962)
47.50 I. Bray, A. T. Stelbovics: Phys. Rev. A **46**, 6995 (1992)
47.51 I. Bray, A. T. Stelbovics: Phys. Rev. Lett. **69**, 53 (1992)
47.52 I. Bray, A. T. Stelbovics: Phys. Rev. Lett. **70**, 746 (1993)
47.53 I. Bray: Phys. Rev. A **49**, 1066 (1994)
47.54 I. Bray, D. V. Fursa: Phys. Rev. A **54**, 2991 (1996)
47.55 I. Bray, D. V. Fursa, A. S. Kheifets, A. T. Stelbovics: J. Phys. B **35**, R117 (2002)
47.56 I. Bray: Phys. Rev. Lett. **89**, 273201 (2002)
47.57 M. H. Mittleman, K. M. Watson: Phys. Rev. **113**, 198 (1959)
47.58 F. W. Byron Jr., C. J. Joachain: J. Phys. B **14**, 2429 (1981)
47.59 B. H. Bransden, J. P. Coleman: J. Phys. B **5**, 537 (1972)
47.60 B. H. Bransden, T. Scott, R. Shingal, R. K. Raychoudhurry: J. Phys. B **15**, 4605 (1982)
47.61 I. E. McCarthy: Comm. At. Mol. Phys. **24**, 343 (1990)
47.62 J. Callaway, D. H. Oza: Phys. Rev. A **32**, 2628 (1985)
47.63 H. R. J. Walters: Phys. Rep. **116**, 1 (1984)
47.64 C. J. Joachain: *Collision Theory for Atoms and Molecules*, ed. by F. A. Gianturco (Plenum, New York 1989) p. 59
47.65 F. W. Byron Jr., C. J. Joachain: Phys. Lett. C **44**, 233 (1977)
47.66 R. J. Glauber: *Lectures in Theoretical Physics*, Vol. 1, ed. by W. E. Brittin (Interscience, New York 1959) p. 315
47.67 V. I. Ochkur: Sov. Phys. JETP **18**, 503 (1964)
47.68 M. R. H. Rudge, M. J. Seaton: Proc. Roy. Soc. A **283**, 262 (1965)
47.69 R. K. Peterkop: Proc. Phys. Soc. **77**, 1220 (1961)
47.70 M. R. H. Rudge: Rev. Mod. Phys. **40**, 564 (1968)
47.71 A. S. Kadyrov, A. M. Mukhamedzhanov, A. T. Stelbovics, I. Bray: Phys. Rev. Lett. **91**, 253202 (2003)
47.72 R. K. Peterkop: Opt. Spektrosk. **1313**, 15387 (1962)
47.73 R. K. Peterkop: *Theory of Ionization of Atoms by Electron Impact* (Colorado Associated University Press, Boulder 1977)
47.74 M. S. Pindzola, D. R. Schultz: Phys. Rev. A **53**, 1525 (1996)

47.75 M.S. Pindzola, F.J. Robicheaux: Phys. Rev. A **54**, 2142 (1996)
47.76 M.S. Pindzola, F.J. Robicheaux: Phys. Rev. A **61**, 052707 (2000)
47.77 J. Colgan, M.S. Pindzola, F.J. Robicheaux, D.C. Griffin, M. Baertschy: Phys. Rev. A **65**, 042721 (2002)
47.78 T.N. Rescigno, M. Baertschy, W.A. Isaacs, C.W. McCurdy: Science **286**, 2474 (1999)
47.79 M. Baertschy, T.N. Rescigno, W.A. Isaacs, X. Lin, C.W. McCurdy: Phys. Rev. A **63**, 022712 (2001)
47.80 C.W. McCurdy, M. Baertschy, T.N. Rescigno: J. Phys B **37**, R137 (2004)
47.81 S. Jones, D.H. Madison: Phys. Rev. Lett. **81**, 2886 (1998)
47.82 S. Jones, D.H. Madison: Phys. Rev. A **62**, 042701 (2000)
47.83 S. Jones, D.H. Madison: Phys. Rev. A **65**, 052727 (2002)
47.84 M. Gell-Mann, M.L. Goldberger: Phys. Rev. **91**, 398 (1953)
47.85 D.S.F. Crothers, J.F. McCann: J. Phys. B **16**, 3229 (1983)
47.86 S. Jones, D.H. Madison, M. Baertschy: Phys. Rev. A **67**, 012703 (2003)
47.87 H. Jackubowicz, D.L. Moores: J. Phys. B **14**, 3733 (1981)
47.88 K. Bartschat, P.G. Burke: J. Phys. B **20**, 3191 (1987)
47.89 G. Müller: *The Physics of Electronic and Atomic Collisions*, ed. by A. Dalgarno, R.S. Freund, P.M. Koch, M.S. Lubell, T.B. Lucatarto (American Institute of Physics, New York 1990) p. 418
47.90 T. Rösel, J. Röder, L. Frost, K. Jung, H. Ehrhardt, S. Jones, D.H. Madison: Phys. Rev. A **46**, 2539 (1992)
47.91 K.J. LaGattata, Y. Hahn: Phys. Rev. A **24**, 2273 (1981)
47.92 D.L. Moores, K.J. Reed: Adv. At. Mol. Opt. Phys. **34**, 301 (1994)
47.93 M.P. Scott, H. Teng, P.G. Burke: J. Phys. B **33**, L63 (2000)
47.94 J. Röder, M. Baertschy, I. Bray: Phys. Rev. A **67**, 010702 (2003)
47.95 D.E. Osterbrock: *Astrophysics of Gaseous Nebulae and Active Galactic Nuclei* (University Science Books, Sausalito California 1989)
47.96 A.M. Arthurs, A. Dalgarno: Proc. Roy. Soc. A **256**, 540 (1960)
47.97 P.G. Burke, N. Chandra, F.A. Gianturco: J. Phys. B **5**, 2212 (1972)
47.98 S.I. Drozdov: Sov. Phys. JETP **1**, 591 (1955)
47.99 S.I. Drozdov: Sov. Phys. JETP **3**, 759 (1956)
47.100 D.M. Chase: Phys. Rev. **104**, 838 (1956)
47.101 I. Shimamura: Chem. Phys. Lett. **73**, 328 (1980)
47.102 I. Shimamura: J. Phys. B **15**, 93 (1982)
47.103 M.A. Morrison: Adv. At. Mol. Phys. **24**, 51 (1987)
47.104 N. Chandra, A. Temkin: Phys. Rev. A **13**, 188 (1976)
47.105 N. Chandra, A. Temkin: Phys. Rev. A **14**, 507 (1976)
47.106 J.N. Bardsley: J. Phys. B **1**, 365 (1968)
47.107 A. Herzenberg, F. Mandl: Proc. Roy. Soc. A **270**, 48 (1962)
47.108 A. Herzenberg: *Electron Molecule Collisions*, ed. by I. Shimamura, K. Takayanagi (Plenum, New York 1984) p. 351
47.109 J.N. Bardsley: J. Phys. B **1**, 349 (1968)
47.110 W. Domcke: Phys. Rep. **208**, 97 (1991)
47.111 A.F.J. Siegert: Phys. Rev. **56**, 750 (1939)
47.112 B.I. Schneider, M. Le Dourneuf, P.G. Burke: J. Phys. B **22**, L365 (1979)
47.113 R.K. Nesbet: Phys. Rev. A **19**, 551 (1979)
47.114 M. Shugard, A.U. Hazi: Phys. Rev. A **12**, 1895 (1975)
47.115 T.N. Rescigno: *Electron Collisions with Molecules, Clusters and Surfaces*, ed. by H. Ehrhardt, L.A. Morgan (Plenum, New York 1994) p. 1
47.116 B.K. Sarpal, K. Pfingst, B.M. Nestmann, S.D. Peyerimhoff: J. Phys. B **29**, 857 (1996)
47.117 L.A. Morgan, C.J. Gillan, J. Tennyson, X. Chen: J. Phys. B **30**, 4087 (1997)
47.118 Cz. Szmytkowski, K. Maciag, G. Karwasz: Chem. Phys. Lett. **107**, 481 (1984)
47.119 K.L. Baluja, N.J. Mason, L.A. Morgan, J. Tennyson: J. Phys. B **34**, 4014 (2001)
47.120 R.J. Gulley, T.A. Field, W.A. Steer, N.J. Mason, S.L. Lunt, J.-P. Zeisel, D. Field: J. Phys. B **31**, 5197 (1998)
47.121 C.W. McCurdy, W.A. Isaacs, H.-D. Meyer, T.N. Rescigno: Phys. Rev. A **67**, 042708 (2003)
47.122 D.T. Birtwistle, A. Herzenberg: J. Phys. B **4**, 53 (1971)
47.123 M. Allan: Phys. Rev. Lett. **87**, 033201 (2001)
47.124 M. Allan: J. Phys. B **35**, L387 (2002)
47.125 W.R. Newell: Comm. At. Mol. Phys. **28**, 59 (1992)
47.126 M.H. Mittleman: *Introduction to the Theory of Laser-Atom Interactions*, 2nd edn. (Plenum, New York 1993)
47.127 C.J. Joachain: *Fundamentals of Laser Interactions*, ed. by F. Ehlotzky (Springer, Berlin, Heidelberg 1985)
47.128 N.M. Kroll, K.M. Watson: Phys. Rev. **8**, 804 (1973)
47.129 R. Shakeshaft: Phys. Rev. A **28**, 667 (1983)
47.130 R. Shakeshaft: Phys. Rev. A **29**, 383 (1984)
47.131 S. Geltman, A. Maquet: J. Phys. B **22**, L419 (1989)
47.132 P. Francken, C.J. Joachain: J. Opt. Soc. Am. **7**, 554 (1990)
47.133 L. Dimou, F.H.M. Faisal: Phys. Rev. Lett. **59**, 872 (1987)
47.134 L.A. Collins, G. Csanak: Phys. Rev. **44**, 5343 (1991)
47.135 A. Giusti-Suzor, P. Zoller: Phys. Rev. A **36**, 5178 (1987)
47.136 P. Marte, P. Zoller: Phys. Rev. A **43**, 1512 (1991)
47.137 P.G. Burke, P. Francken, C.J. Joachain: J. Phys. B **24**, 761 (1991)
47.138 J.H. Shirley: Phys. Rev. **138**, 979 (1965)
47.139 R.M. Potvliege, R. Shakeshaft: *Atoms in Intense Laser Fields*, ed. by M. Gavrila (Academic, San Diego, London 1993) p. 373
47.140 M. Dörr, M. Terao-Dunseath, J. Purvis, C.J. Noble, P.G. Burke, C.J. Joachain: J. Phys. B **25**, 2809 (1992)
47.141 W.C. Henneberger: Phys. Rev. Lett. **21**, 838 (1968)

47.142 D. Charlo, M. Tereo-Dunseath, K. M. Dunseath, J.-M. Launay: J. Phys. B **31**, L539 (1998)
47.143 M. Tereo-Dunseath, K. M. Dunseath, D. Charlo, A. Hibbert, R. J. Allen: J. Phys. B. **34**, L263 (2001)
47.144 M. Tereo-Dunseath, K. M. Dunseath: J. Phys. B **35**, 125 (2002)
47.145 K. Takatsuka, V. McKoy: Phys. Rev. A **30**, 1734 (1984)

48. Positron Collisions

The positron is the antiparticle of the electron, having the same mass but opposite charge. Positrons undergo collisions with atomic and molecular systems in much the same way as electrons do. Thus, the standard scattering theory for electrons (see Chapt. 47) can also be applied to positron scattering. However, there are a number of important differences from electron scattering which we outline below.

Since the positron is a distinct particle from the atomic electrons, it cannot undergo an exchange process with the bound electrons during a collision, as is possible with electrons. Thus, the nonlocal exchange terms which arise in the description of electron scattering are not present for positrons. This leads to a simplification of the scattering equations from those for electrons. However, there are scattering channels available with positron scattering which do not exist with electrons. These are dealt with in Sect. 48.1.

Historically beams of low-energy positrons were difficult to obtain and consequently there is considerably less experimental data available for positrons than for electrons. This was particularly true for quantities which required large incident positron fluxes, such as differential scattering cross sections and coincidence parameters. However, the recent development of cold trap-based positron beams with high resolution and high brightness by the San Diego group [48.1] has the potential to revolutionize this field and put it on a par with electron scattering.

Throughout this chapter we will employ atomic units unless otherwise noted.

48.1 **Scattering Channels** 731
48.1.1 Postronium Formation 731
48.1.2 Annihilation 732

48.2 **Theoretical Methods** 733

48.3 **Particular Applications** 735
48.3.1 Atomic Hydrogen 735
48.3.2 Noble Gases 735
48.3.3 Other Atoms 736
48.3.4 Molecular Hydrogen 737
48.3.5 Other Molecules 737

48.4 **Binding of Positrons to Atoms** 737

48.5 **Reviews** 738

References 738

48.1 Scattering Channels

Positrons colliding with atomic and molecular systems have the same scattering channels available as for electrons, viz., elastic, inelastic, ionization, and for molecules, dissociation. However, two channels exist for positrons which do not exist for electrons, viz., positronium formation and annihilation.

48.1.1 Postronium Formation

Positronium, a bound state of an electron–positron pair (Chapt. 27), can be formed during the collision of a positron with an atomic or molecular target. The positronium 'atom' can escape to infinity leaving the target in a ionized state with a positive charge of one. Thus, this process can be difficult to distinguish experimentally from true ionization where both the incident positron and the ionized electron are asymptotically free particles. The positronium atom can exist in its ground state or in any one of an infinite number of excited states after the collision. The level structure of positronium is, to order α^2, where α is the fine-structure constant, identical to that of hydrogen but with each level having half the energy of the corresponding hydrogenic state.

Positronium formation is a rearrangement channel, and thus, is a two-centre problem. Because positronium is a light particle, having a reduced mass one-half of that of an electron, the semi-classical type of approximations used in ion–atom collisions (Chapt. 50) are not applicable here. We will discuss various theoretical approaches to this process in Sect. 48.2 and give references to experimental results in Sect. 48.3.

Positronium formation in the ground state has a threshold which is 1/4 of a Hartree (6.802 85 eV) below the ionization threshold of the target. This means that it is normally the lowest inelastic channel in positron scattering from neutral atoms. For atoms with a small ionization potential, such as the alkalis, this channel is always open. The energy range between the positronium threshold and the first excited state of the atom is known as the Ore gap. In this range, positronium formation is the only possible inelastic process.

48.1.2 Annihilation

Annihilation is a process in which an electron–positron pair is converted into two or more photons. It can occur either directly with a bound atomic electron or after positronium formation has taken place. The direct annihilation cross section for a positron of momentum $\boldsymbol{k}$ colliding with an atomic or molecular target can be written as [48.2]

$$\sigma_{\mathrm{a}} = \frac{\alpha^3 Z_{\mathrm{eff}}}{k}\left(\pi a_0^2\right), \tag{48.1}$$

where Z_{eff} can be thought of as the effective number of electrons in the target with which the positron can annihilate. If $\Psi(\boldsymbol{r}_1, \boldsymbol{r}_2, \ldots, \boldsymbol{r}_N, \boldsymbol{x})$ is the wave function for the system of a positron, with coordinate $\boldsymbol{x}$, colliding with an N-electron target, then

$$Z_{\mathrm{eff}} = \sum_{i=1}^{N} \int \mathrm{d}\boldsymbol{r}_1\, \mathrm{d}\boldsymbol{r}_2 \ldots \mathrm{d}\boldsymbol{r}_N \left|\Psi(\boldsymbol{r}_1, \boldsymbol{r}_2, \ldots, \boldsymbol{r}_N; \boldsymbol{r}_i)\right|^2 . \tag{48.2}$$

While this formula can be naively derived by assuming that the positron can only annihilate with an electron if it is at the identical location, it actually follows from a quantum electrodynamical treatment of the process [48.3]. If the wave function Ψ is approximated by the product of the undistorted target wave function times a positron scattering function $F(\boldsymbol{x})$, then

$$Z_{\mathrm{eff}} = \int \mathrm{d}\boldsymbol{r}\, \rho(\boldsymbol{r}) \left|F(\boldsymbol{r})\right|^2 , \tag{48.3}$$

where ρ is the electron number density of the target. Thus, in the Born approximation, where F is taken as a plane wave, Z_{eff} simply becomes the total number of electrons Z in the target. However, a pronounced enhancement of the annihilation rate in the vicinity of the Ps formation threshold due to virtual Ps formation was predicted [48.4, 5]. Subsequently, the Born approximation was shown to be grossly inadequate by the San Diego group, who found annihilation rates Z_{eff} at room temperature which are an order of magnitude larger for some atoms and even up to five orders of magnitude larger in large hydrocarbon molecules. Furthermore, there is evidence that only the outer shell of electrons takes part in the annihilation process. Two mechanisms have been proposed in order to explain these large values for Z_{eff}. One involves the enhancement of the direct annihilation process below the Ps formation threshold due to the attractive nature of the positron–electron interaction, which increases the overlap of positron and electron densities on the atom or molecule. The second mechanism is referred to as resonant annihilation, which occurs after the positron has been captured into a Feshbach resonance, where the positron is bound to a vibrationally excited molecule. A summary of the above results can be found in the recent article by *Barnes* et al. [48.6].

When a positron annihilates with an atomic electron, two 511 keV photons is the most likely result, if the positron–electron pair are in a singlet spin state (parapositronium). In the centre-of-mass frame of the pair, the photons are emitted in opposite directions to conserve momentum. However, in the laboratory frame the bound electron has a momentum distribution which is reflected in the photon directions not being exactly 180 degrees apart. This slight angular deviation, called the angular correlation, can be measured, and gives information about the momentum distribution of the bound electrons. This quantity is given by [48.3]

$$S(\boldsymbol{q}) = \sum_{i=1}^{N} \int \mathrm{d}\boldsymbol{r}_1 \ldots \mathrm{d}\boldsymbol{r}_{i-1}\, \mathrm{d}\boldsymbol{r}_{i+1} \ldots \mathrm{d}\boldsymbol{r}_N \times \int \mathrm{d}\boldsymbol{r}_i \mathrm{d}\boldsymbol{x}\, \mathrm{e}^{\mathrm{i}\boldsymbol{q}\cdot\boldsymbol{x}} \left|\Psi(\boldsymbol{r}_1, \boldsymbol{r}_2, \ldots, \boldsymbol{r}_N; \boldsymbol{x})\, \delta(\boldsymbol{r}_i - \boldsymbol{x})\right|^2 , \tag{48.4}$$

where $\boldsymbol{q}$ is the resultant momentum of the annihilating pair. In evaluating this quantity, the positron is assumed to be thermalized in the gas before undergoing annihilation. Experimentally, only one component of $\boldsymbol{q}$ is measured, so that $S(\boldsymbol{q})$ is integrated over the other two components of the momentum to obtain the measured quantity. The spin triplet component of an electron–positron pair (orthopositronium) can only decay with the emission of three or more photons which do not have well defined energies. This is a much less probable process than the two photon decay from the singlet component.

48.2 Theoretical Methods

The basic theoretical approaches to the calculation of positron scattering from atoms and molecules were originally developed for electron scattering and later applied to the positron case. Thus, we emphasize here only the differences that arise between the electron and positron cases, both in the theoretical formulations, and in later sections, in the nature of the results.

The lowest-order interaction between a free positron and an atomic or molecular target is the repulsive static potential of the target

$$V_s = \langle \psi_0 | V | \psi_0 \rangle , \quad (48.5)$$

where ψ_0 is the unperturbed target wave function and V is the electrostatic interaction potential between the positron and the target. Since this interaction has the opposite sign from that for electron scattering, the static potential also has the opposite sign in these two cases. On the other hand, the next higher-order of interaction is polarization, which arises from the distortion of the atom by the incident particle. If we represent this distortion of the target to first-order by the wave function ψ_1, as in the polarized-orbital approximation, for example [48.9], then the polarization interaction can be represented by the potential

$$V_p = \langle \psi_0 | V | \psi_1 \rangle . \quad (48.6)$$

This potential is attractive for both positron and electron scattering and has an asymptotic form with leading term $-\alpha_d/2r^4$, where α_d is the static dipole polarizability (Sect. 23.2.3) of the target. Thus, the static and polarization potentials for positron scattering from ground state systems are of opposite sign and tend to cancel one another. This leads to very different behaviour from the electron case where they are of the same sign. In particular, the elastic scattering cross sections for positron scattering from an atom are much smaller than for electron scattering, and the phase shifts (Sect. 47.1.1) have very different magnitudes and dependences on energy. This is illustrated for the case of scattering from helium in Figs. 48.1–48.3, where the results of the highly accurate variational calculations for scattering by electrons and positrons are shown. Note the difference in sign between the electron and positron s-wave phase shifts for very small values of the incident momentum. The fact that the positron phase shift goes through zero leads to the Ramsauer minimum in the positron total cross section, as shown in Fig. 48.3. The large difference in magnitudes between the electron and positron s- and p-wave phase shifts leads to the large difference in the total elastic cross sections as shown.

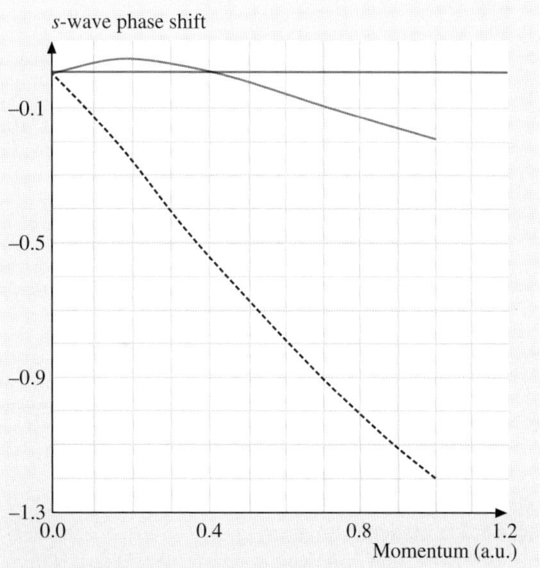

Fig. 48.1 Variational s-wave phase shifts for electron [48.7] (*dashed line*) and positron [48.8] (*solid line*) scattering from helium atoms

Higher-order terms in the interaction potential may also give important contributions to scattering cross sections. For a detailed discussion, see the article by *Drachman* and *Temkin* [48.10].

A simple potential scattering calculation using the sum of the static and polarization potentials, but without the exchange terms that are present for the electron case, can be applied to elastic scattering calculations for closed shell systems (see Sect. 48.3.3).

The potentials defined above can also be used in a distorted-wave approximation (Chapt. 47) which can be applied to excitation and ionization by positron impact. Once again, the complicated exchange terms which arise in electron scattering are absent here. *Sienkiwicz* and *Baylis* [48.11] have included such potentials in a Dirac–Fock formulation of positron scattering which treats the positron as an electron with negative energy.

For positrons with high enough incident energies ($\approx$ 1 keV), the first Born approximation will become valid (Chapt. 47). Since the first Born approximation is independent of the sign of the charge of the incident particles, this indicates that as the incident energy increases, the corresponding cross sections for electron

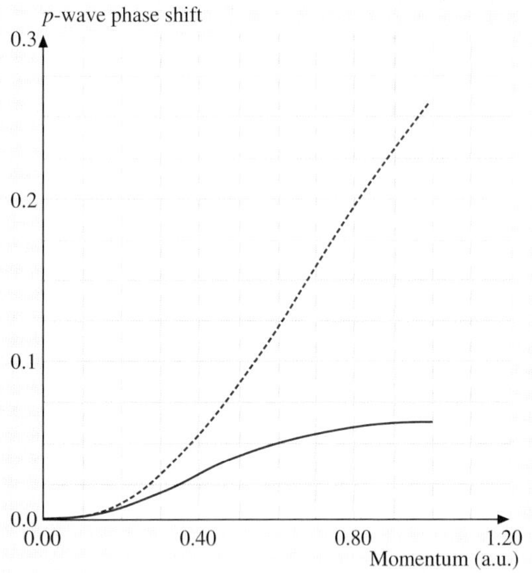

Fig. 48.2 Variational p-wave phase shifts for electron [48.7] (*dashed line*) and positron [48.15] (*solid line*) from helium atoms

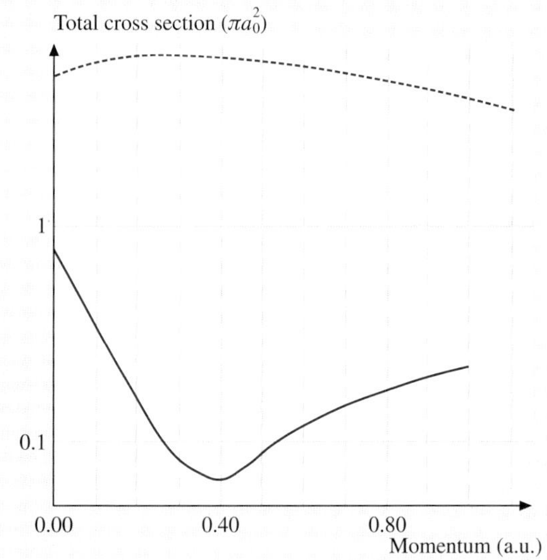

Fig. 48.3 Total elastic cross sections for electron (*dashed line*) and positron (*solid line*) from helium atoms calculated from the phase shifts shown in Figs. 48.1 and 48.2. The higher-order phase shifts were calculated from effective range theory (Sect. 47.1.1)

and positron scattering will eventually merge. From flux conservation arguments, this means that the positronium formation cross section will rapidly decrease as the incident energy increases. In fact, from experimental measurements, the total cross sections (summed over all possible channels) for electron and positron scattering appear to merge at a much lower energy than the cross sections for individual channels [48.12]. More elaborate calculations for high energy scattering have been carried out in the eikonal-Born series [48.13] (Chapt. 47). These approximations allow for both polarization and absorption (i. e., inelastic processes) and yield good agreement with elastic experimental measurements of differential cross sections at energies above 100 eV. A detailed analysis [48.13, 14] of the various contributions to the scattering indicates that absorption effects due to the various open inelastic channels plays a much more important role here than for electron scattering.

A more elaborate treatment of positron scattering is based on the close-coupling approximation (Chapt. 47), where the wave function for the total system of positron plus target is expanded using a basis set comprised of the wave functions of the target. Once again, there are no exchange terms involving the positron and, in principle, a complete expansion including the continuum states of the target would include the possibility of positronium formation. However, such an expansion is not practicable if one wants to calculate explicit cross sections for positronium formation. Even in cases where such cross sections are not required, the considerable effect that the positronium formation channels can have on the other scattering cross sections is best included by a close-coupling expansion that includes terms representing positronium states plus the residual target ion. There is a problem of double counting of states in such an expansion but, in practice, this does not appear to be a problem if the number of states in the expansion is not large. Also, in many cases, additional pseudostates have to be included in the expansion in order to correctly represent the long-range polarization interaction.

A close-coupling expansion including positronium states is a two-centre problem, i. e., it includes the centres of mass of both the target and the positronium states. Since positronium is a light system, the semi-classical approach often used to treat rearrangement collisions between heavy systems (Chapt. 50) is not applicable here. This means that one is faced with a problem of considerable computational complexity [48.16–21].

Another way to take into account the effects of open inelastic channels without the complications of a full close-coupling approach is to use optical potentials. These are often based on a close-coupling formalism [48.14, 22] and lead to a complex potential, the real part of which represents distortions of the target (such as polarization) while the complex part allows for absorption (i. e., flux into open channels not explicitly represented).

Bray and *Stelbovics* [48.23] have applied the convergent close-coupling method to the scattering of positrons from atoms. This method includes contributions from the continuum states of the target and sufficient terms in the expansion are included to ensure numerical convergence. At the present time, however, positronium states have not been explicitly included.

Finally, there is the variational method (Chapt. 47), which uses an analytic form of trial wave function to represent the total system. The parameters of this analytic function are determined as part of the method. Given a trial wave function with sufficient flexibility and a large enough number of parameters, essentially exact results can be obtained in the elastic energy range and the Ore gap. Because the complexity of the trial function increases as the square of the number of electrons in the target, only positron scattering from hydrogen, helium and lithium and the hydrogen molecule have been treated by this method, to date [48.8, 24].

In the case of ionization there appears to be quite distinct threshold behaviour of the cross sections for electron and positron collisions. For electrons, the Wannier threshold law (Sect. 52.2.1) has exponent 1.127, while a similar analysis for positrons [48.25] yields an exponent of 2.651. However, the existence of the positronium formation channel leaves in question whether this analysis will give the dominant term at threshold. For a fuller discussion see [48.26], and references therein.

There has been an investigation [48.27] of the behaviour of the elastic cross sections at the positronium formation threshold which predicts the occurrence of a Wigner cusp for the lighter noble gases.

48.3 Particular Applications

48.3.1 Atomic Hydrogen

Because of the difficulty of making measurements in atomic hydrogen, the available experimental data is restricted so far to total cross sections, as well as to total ionization and positronium formation cross sections. Essentially exact variational calculations have been carried out in the elastic energy regime and the Ore gap [48.8].

Ionization cross sections have been measured by both the Bielefeld and London groups [48.28, 29] (and references therein) and have been calculated in a number of approximations [48.23, 30–32]. However, disagreements between the experimental measurements mean that there is at present no reliable way of assessing the various approximations used. More elaborate calculations with asymptotically correct wave functions have been used to determine triple differential cross sections for ionization [48.33, 34]. However, the task of integrating these to produce total cross sections is a formidable one.

The total positron–hydrogen cross section has also been measured by the Detroit group [48.35], and is in quite good agreement with calculations based upon the coupled-pseudostate method [48.32] except at very low energies where the experimental uncertainties are the greatest.

In order to determine reliable positronium formation cross sections, the explicit positronium states have to be included. Several such calculations have been carried out [48.17–20]. These indicate the necessity of explicitly including positronium formation channels in the expansion of the total wave function in order to obtain accurate results, even for elastic scattering. The most recent calculations [48.32, 36] are in quite good agreement with various experiments [48.28, 37] over the majority of the energy range.

As is the case for electron scattering, positron cross sections exhibit resonances (Sect. 47.1.3). These have been extensively studied by *Ho* [48.38] using variational and complex rotation methods.

48.3.2 Noble Gases

Because the noble gases are convenient experimental targets, a good deal of effort has gone into calculations for these targets, particularly for elastic scattering, ionization, and Ps formation. In the purely elastic energy range, i. e., for energies below the positronium formation threshold, the simple potential scattering approach using the static and polarization potentials defined above yields quite good results. Since the long-range behaviour of the sum of the potentials is attractive, the scattering phase shifts for positrons must be positive for sufficiently low energies. However, as its incident energy increases, the positron probes the repulsive inner part of the potential and the phase shifts become negative. This behaviour leads to the well-known Ramsauer minimum in the integral elastic cross sections (Chapt. 47) for the lighter noble gases (helium, neon and possibly argon), but not for krypton and xenon [48.15, 39] (and references therein). This differs from electron scattering, where some low-energy phase shifts can be negative (modulo π) because of the existence of bound orbitals of the same symmetry.

Another difference between positron and electron scattering is exhibited by the differential cross sections (Chapt. 47). As a result of the low intensity and width of positron beams, most measurements of differential cross sections are relative, with the normalization often being made to theoretical calculations at specific angles. For electrons, the shape of the cross section is determined by a few dominant phase shifts, whereas for positrons, many phase shifts contribute to the final shape [48.40]. Because of this behaviour, the differential cross sections for positron scattering have much less overall structure than for electron scattering. However, the differential cross sections for positrons for many of the noble gases have a single minimum at relatively small angles, both below and above the first inelastic channel. These have been reviewed by *Kauppila* et al. [48.41]. At intermediate energies, the simple potential scattering approximation is no longer sufficient and the inelastic channels have to be taken into account via, for example, the use of an optical potential [48.13, 14, 42]. Furthermore, in the inelastic scattering regime, the existence of open channels has a much more marked effect on the shape of the differential cross sections for positron scattering than for electrons [48.43].

The first absolute differential cross sections were measured for argon and krypton at very low energies using a magnetized beam of cold positrons [48.44, 45]. These results are in excellent agreement with a variety of different theoretical predictions [48.46–48].

There is relatively very little experimental data for the excitation of the noble gases. Some experimental work has been carried out for the lighter noble gases, helium, neon and argon [48.49] (and references therein), and there is satisfactory agreement between these measurements and close-coupling [48.50], as well as distorted-wave [48.51] calculations.

The first state-resolved absolute excitation cross sections for the 4s $[1/2]_1^{\mathrm{o}}$ and 4s $[3/2]_1^{\mathrm{o}}$ states of argon and the 6s $[3/2]_1^{\mathrm{o}}$ state of xenon have been measured by the San Diego group [48.6, 52]. Relativistic distorted-wave calculations are in satisfactory agreement with the experiment for argon [48.53], but less so for xenon [48.54].

The total ionization and positronium cross sections have been measured extensively for all of the noble gases. In general, there is good agreement amongst the various experiments for the ionization cross section, but much less so for the positronium formation cross section. A summary of the experimental work on the ionization and positronium cross sections for neon, argon, krypton and xenon can be found in the article by *Laricchia* et al. [48.55] (and references therein), while a more detailed analysis of these cross sections in argon can be found in recent articles by the San Diego group [48.6, 56].

There has also been an extensive investigation of the energy dependence of the elastic and positronium formation cross section near the Ps formation threshold for all the inert gases [48.57] (and references therein).

There exist some measurements of Z_{eff} and angular correlation parameters for these gases [48.58], mainly at room temperature, and calculations for them have been made in the polarized-orbital approximation [48.59].

48.3.3 Other Atoms

In the case of positron scattering from the alkali atoms, the positronium formation channel is always open and the simple potential scattering approach does not yield reliable results. The Detroit group has measured the total cross section, as well as upper and lower bounds to the positronium formation cross section for sodium, potassium, and rubidium. Although early close-coupling calculations of the elastic and excitation cross sections [48.60–62] were in surprisingly good agreement with the experimental total cross section, these calculations did not include the positronium formation channel. Subsequently, much more sophisticated calculations were carried out by the Belfast group using the coupled-pseudostate method, which included both eigenstates of the target as well as positronium [48.32, 63–65]. These calculations also showed the increasing importance of positronium formation in excited states for the alkalis potassium, rubidium, and caesium. The overall agreement between experimental results and those from the coupled-pseudostate method are quite good for both potassium and rubidium, however, for sodium, the experimental positronium formation cross section is significantly above these theoretical calculations. A summary of the experimental work on the alkalis can be found in [48.35], while a corresponding summary of theoretical work is given in [48.32].

Substantial resonance features have been found in these positron–alkali atom cross sections [48.66].

The positronium formation threshold for magnesium is very low, only 0.844 eV, and hence, the elastic and positronium formation cross sections will dominate in the low energy region. Upper and lower bounds to the Ps formation cross section in magnesium have been determined [48.35, 67], and are in agreement with both close-coupling calculations [48.68] and the results of many-body theory [48.69].

48.3.4 Molecular Hydrogen

By its fundamental nature, molecular hydrogen has attracted considerable attention both experimentally and theoretically. The total elastic cross section has been measured by both the Detroit group [48.37] and the London group [48.70], with good agreement between both sets of data. There have been several theoretical calculations of this cross section by a variety of methods: Kohn variational [48.71], *R*-matrix [48.72], distributed positron model [48.73], and recently, a Schwinger multichannel method [48.74], with [48.72, 73] being in satisfactory agreement with experiment. Once again, the elastic cross section is strongly influenced by the positronium formation channel near threshold.

Quite recently the vibrational $(0 \to 1)$ excitation cross section of molecular hydrogen has been measured between 0.55 and 4 eV by *Sullivan* et al. [48.75]. Their data are in quite good agreement with theoretical calculations. The San Diego group has also measured the electronic excitation of the $B^1\Sigma$ state from threshold to 30 eV [48.52]. Their data are in reasonable agreement with the Schwinger multichannel calculation of *Lino* et al. [48.76]. Interestingly, the measured positron excitation cross section appears to be larger than that determined for electron excitation.

The ionization cross section has been determined over a wide range of energies by a number of different groups [48.77–81] (and references therein). Since all of the above measurements are relative, they must be normalized to one another at particular energies. Although there are some differences between these experiments, near threshold the positron cross section increases less rapidly, in general, than the corresponding electron cross section, in accordance with the Wannier law. Theoretical calculations are in satisfactory agreement with the measurements [48.82] (and references therein).

The positronium formation cross section has also been measured by a number of different groups with coupled-channels calculations being carried out for this process [48.83].

48.3.5 Other Molecules

For diatomic and triatomic molecules, most of the experimental and theoretical work has been carried out for CO, CO_2, O_2, and N_2. Total cross sections for O_2, N_2, and CO_2 have been measured from threshold to several hundred eV [48.70] (and references therein). Relative differential cross sections have been measured for CO, CO_2, O_2, N_2, as well as N_2O, on both sides of the positronium formation threshold [48.84]. Absolute differential cross sections have been measured for CO at 6.75 eV [48.45]. At low energies the gases N_2, O_2, and CO exhibit a minimum in the DCS at small angles, as per the heavier noble gases. This minimum gradually disappears as the energy increases.

Vibrational excitation cross sections for CO and CO_2 have been measured [48.75] and are in excellent agreement with the theoretical calculations of [48.85] for CO, and in satisfactory agreement with theory [48.86] for CO_2. Electron excitation of the $a^1\Pi$ and $a'^1\Sigma$ states of N_2 have been measured from threshold to 20 eV [48.52]. Interestingly, the positron cross section near threshold is approximately double that for electrons.

For polyatomic molecules, the majority of experimental and theoretical work has been carried out for CH_4. This includes the total cross section and quasi-elastic (summed over vibration–rotational levels) differential cross sections. At low energies there is a minimum in these DCS at small angles, as per the heavier noble gases which, in turn, also disappears at higher energies. The positronium formation cross section has also been measured.

48.4 Binding of Positrons to Atoms

There have been many recent investigations of the possible binding of positrons to a variety atoms. As was mentioned in Sect. 46.2.2, such binding could greatly enhance the annihilation cross section and help to explain the large measured values of Z_{eff} for both atoms and molecules. It has been shown theoretically that a positron will bind to a large number of one- and two-electron atomic systems [48.87] (and references therein). For one-electron systems, where the ionization potential is less than 6.802 85 eV, the dominant configuration is a polarized positronium (Ps) cluster moving in the field of the residual positive ion, while for two-electron systems, with an ionization potential greater than 6.802 85 eV, the dominant configuration involves a positron orbiting a polarized neutral atom [48.88]. So far, there is no experimental evidence for these positronic atoms. However, there is considerable evidence that positrons will bind to large hydrocarbon molecules [48.6] (and references therein).

48.5 Reviews

For a number of years a Positron Workshop has been held as a satellite of the International Conference on the Physics of Electronic and Atomic Collisions. Their proceedings [48.89–100] give an excellent summary of the state of positron scattering research, both experimental and theoretical, including such additional topics as positronium scattering from atoms, the formation of antihydrogen, inner shell ionization, and applications to astrophysics.

There are several review articles on positron scattering in gases, including the early historical development [48.101], more comprehensive articles [48.2, 102–104], as well as a more recent review [48.105], which also discusses the future for positron physics.

A recent book [48.106] discusses various aspects of both experimental and theoretical positron physics.

References

48.1 C. Kurz, S. J. Gilbert, R. G. Greaves, C. M. Surko: Nucl. Instrum. Methods B **143**, 188 (1998)
48.2 P. A. Fraser: Adv. At. Mol. Phys. **4**, 63 (1968)
48.3 R. Ferrell: Rev. Mod. Phys. **28**, 308 (1956)
48.4 J. W. Humberston, P. Van Reeth: Nucl. Instrum. Methods B **143**, 127 (1998)
48.5 G. Laricchia, C. Wilkin: Nucl. Instrum. Methods B **143**, 135 (1998)
48.6 L. D. Barnes, J. P. Marler, J. P. Sullivan, C. M. Surko: Phys. Scr. **T110**, 280 (2004)
48.7 R. K. Nesbet: Phys. Rev. A **20**, 58 (1979)
48.8 E. A. G. Armour, J. W. Humberston: Phys. Rep. **204**, 165 (1991)
48.9 R. P. McEachran, A. D. Stauffer: J. Phys. B **10**, 663 (1977)
48.10 R. J. Drachman, A. Temkin: *Case Studies in Atomic Collision Physics II*, ed. by E. W. McDaniel, M. R. C. McDowell (North-Holland, Amsterdam 1972) Chap. 6
48.11 J. E. Sienkiewicz, W. E. Baylis: Phys. Rev. A **43**, 1331 (1991)
48.12 W. E. Kauppila, T. S. Stein: Can. J. Phys. **60**, 471 (1982)
48.13 C. J. Joachain, R. M. Potvliege: Phys. Rev. A **35**, 4873 (1987)
48.14 K. Bartschat, R. P. McEachran, A. D. Stauffer: J. Phys. B. **21**, 2789 (1988)
48.15 J. W. Humberston, R. I. Campeanu: J. Phys. B **13**, 4907 (1980)
48.16 R. N. Hewitt, C. J. Noble, B. H. Bransden: J. Phys. B **26**, 3661 (1993)
48.17 K. Higgins, P. G. Burke: J. Phys. B **26**, 4269 (1993)
48.18 J. Mitroy: J. Phys. B **26**, 4861 (1993)
48.19 N. K. Sarkar, A. S. Ghosh: J. Phys. B **27**, 759 (1994)
48.20 G. Liu, T. T. Gien: Phys. Rev. A **46**, 3918 (1992)
48.21 C. P. Campbell, M. T. McAlinden, A. A. Kernoghan, H. R. J. Walters: Nucl. Instrum. Methods B **143**, 41 (1998)
48.22 I. E. McCarthy, K. Ratnavelu, Y. Zhou: J. Phys. B **26**, 2733 (1993)
48.23 I. Bray, A. Stelbovics: Phys. Rev. A **49**, R2224 (1994)
48.24 J. W. Humberston: J. Phys. B **25**, L491 (1992)
48.25 H. Klar: J. Phys. B **14**, 4165 (1981)
48.26 T.-Y. Kuo, H.-L. Sun, K.-N. Huang: Phys. Rev. A **67**, 012705 (2003)
48.27 G. Laricchia, J. Moxom, M. Charlton, Á. Kövér, W. E. Meyerhoff: Hyperfine Interact. **89**, 209 (1994)
48.28 M. Weber, A. Hofmann, W. Raith, W. Sperber: Hyperfine Interact. **89**, 221 (1994)
48.29 G. O. Jones, M. Charlton, J. Slevin, G. Laricchia, Á. Kövér, M. R. Poulsen, S. N. Chormaic: J. Phys. B **26**, L483 (1993)
48.30 P. Acacia, R. I. Campeanu, M. Horbatsch, R. P. McEachran, A. D. Stauffer: Phys. Lett. A **179**, 205 (1993)
48.31 S.-W. Hsu, T.-Y. Kuo, C. J. Chen, K.-N. Huan: Phys. Lett. **167A**, 277 (1992)

48.32 A. A. Kernoghan, D. J. R. Robinson, M. T. McAlinden, H. R. J. Walters: J. Phys. B **29**, 2089 (1996)

48.33 M. Brauner, J. S. Briggs, H. Klar: J. Phys. B **22**, 2265 (1989)

48.34 S. Jetzke, F. H. M. Faisal: J. Phys. B **25**, 1543 (1992)

48.35 T. S. Stein, J. Jiang, W. E. Kauppila, C. K. Kwan, H. Li, A. Surdutovich, S. Zhou: Can. J. Phys. **74**, 313 (1996)

48.36 J. Mitroy: J. Phys. B **29**, L263 (1996)

48.37 S. Zhou, H. Li, W. E. Kauppila, C. K. Kwan, T. S. Stein: Phys. Rev. A **55**, 361 (1997)

48.38 Y. K. Ho: Hyperfine Interact. **73**, 109 (1992)

48.39 G. Sinapius, W. Raith, W. G. Wilson: J. Phys. B **13**, 4079 (1980)

48.40 R. P. McEachran, A. D. Stauffer: *Positron (Electron)-Gas Scattering*, ed. by W. E. Kauppila, T. S. Stein, J. M. Wadehra (World Scientific, Singapore 1986) p. 122

48.41 W. E. Kauppila, C. K. Kwan, D. Przybyla, S. J. Smith, T. S. Stein: Can. J. Phys. **74**, 474 (1996)

48.42 A. Jain: Phys. Rev. A **41**, 2437 (1990)

48.43 W. E. Kauppila, T. S. Stein: Hyperfine Interact. **40**, 87 (1990)

48.44 S. J. Gilbert, R. G. Greaves, C. M. Surko: Phys. Rev. Lett. **82**, 5032 (1999)

48.45 J. P. Sullivan, S. J. Gilbert, J. P. Marler, R. G. Greaves, S. J. Buckman, C. M. Surko: Phys. Rev. A **66**, 042708 (2002)

48.46 R. P. McEachran, A. G. Ryman, A. D. Stauffer: J. Phys. B **12**, 1031 (1979)

48.47 R. P. McEachran, A. D. Stauffer, L. E. M. Campbell: J. Phys. B **13**, 1281 (1980)

48.48 V. A. Dzuba, V. V. Flambaum, G. F. Gribakin, W. A. King: J. Phys. B **29**, 3151 (1996)

48.49 S. Mori, O. Sueoka: J. Phys. B **27**, 4349 (1994)

48.50 R. N. Hewitt, C. J. Noble, B. H. Bransden: J. Phys. B **25**, 2683 (1992)

48.51 L. A. Parcell, R. P. McEachran, A. D. Stauffer: Nucl. Instrum. Methods B **177**, 113 (2000)

48.52 J. P. Sullivan, J. P. Marler, S. J. Gilbert, S. J. Buckman, C. M. Surko: Phys. Rev. Lett. **87**, 073201 (2001)

48.53 R. P. McEachran, A. D. Stauffer: Phys. Rev. A **65**, 034703 (2002)

48.54 L. A. Parcell, R. P. McEachran, A. D. Stauffer: Nucl. Instrum. Methods B **221**, 93 (2004)

48.55 G. Laricchia, P. Van Reeth, J. Moxom: J. Phys. B **35**, 2525 (2002)

48.56 J. P. Marler, L. D. Barnes, S. J. Gilbert, J. A. Young, J. P. Sullivan, C. M. Surko: Nucl. Instrum. Methods B **221**, 84 (2004)

48.57 W. E. Meyerhof, G. Laricchia: J. Phys. B **30**, 2221 (1997)

48.58 K. Iwata, G. F. Gribakin, R. G. Greaves, C. Kurz, C. M. Surko: Phys. Rev. A **61**, 022719 (2000)

48.59 D. M. Schrader, R. E. Svetic: Can. J. Phys. **60**, 517 (1982)

48.60 S. J. Ward, M. Horbatsch, R. P. McEachran, A. D. Stauffer: J. Phys. B **22**, 1845 (1989)

48.61 R. P. McEachran, M. Horbatsch, A. D. Stauffer: J. Phys. B **24**, 2853 (1991)

48.62 M. T. McAlinden, A. A. Kernoghan, H. R. J. Walters: Hyperfine Interact. **89**, 161 (1994)

48.63 M. T. McAlinden, A. A. Kernoghan, H. R. J. Walters: J. Phys. B **29**, 555 (1996)

48.64 A. A. Kernoghan, M. T. McAlinden, H. R. J. Walters: J. Phys. B **29**, 3971 (1996)

48.65 M. T. McAlinden, A. A. Kernoghan, H. R. J. Walters: J. Phys. B **30**, 1543 (1997)

48.66 S. J. Ward, M. Horbatsch, R. P. McEachran, A. D. Stauffer: J. Phys. B **22**, 3763 (1989)

48.67 E. Surdutovich, M. Harte, W. E. Kauppila, C. K. Kwan, S. Zhou: Phys. Rev. A **68**, 022709 (2003)

48.68 R. N. Hewitt, C. J. Noble, B. H. Bransden: Can. J. Phys. **74**, 559 (1996)

48.69 G. F. Gribakin, W. A. King: Can. J. Phys. **74**, 449 (1996)

48.70 M. Charlton, T. C. Griffith, G. R. Heyland, G. L. Wright: J. Phys. B **16**, 323 (1983)

48.71 E. A. G. Armour, D. J. Baker, M. Plummer: J. Phys. B **23**, 3057 (1990)

48.72 G. Danby, J. Tennyson: J. Phys. B **23**, 2471S (1990)

48.73 T. L. Gibson: J. Phys. B **25**, 1321 (1992)

48.74 J. S. E. Germano, M. A. P. Lima: Phys. Rev. A **47**, 3976 (1993)

48.75 J. P. Sullivan, S. J. Gilbert, C. M. Surko: Phys. Rev. Lett. **86**, 1494 (2001)

48.76 J. L. S. Lino, J. S. E. Germano, M. A. P. Lima: J. Phys. B **27**, 1881 (1994)

48.77 D. Fromme, G. Kruse, W. Raith, G. Sinapius: J. Phys. B **21**, L261 (1988)

48.78 H. Knudsen, L. Brun-Nielsen, M. Charlton, P. M. Poulsen: J. Phys. B **23**, 3955 (1990)

48.79 J. Moxom, P. Ashley, G. Laricchia: Can. J. Phys. **74**, 367 (1996)

48.80 P. Ashley, J. Moxom, G. Laricchia: Phys. Rev. Lett. **77**, 1250 (1996)

48.81 F. M. Jacobsen, N. P. Frandsen, H. Knudsen, U. Mikkelson, D. M. Schrader: J. Phys. B **28**, 4675 (1995)

48.82 R. I. Campeanu, V. Chis, L. Nagy, A. D. Stauffer: Phys. Lett. A **310**, 445 (2003)

48.83 P. K. Biswas, J. S. E. Germano, T. Frederico: J. Phys. B **35**, L409 (2002)

48.84 D. A. Przybyla, W. Addo-Asah, W. E. Kauppila, C. K. Kwan, T. S. Stein: Nucl. Instrum. Methods B **143**, 57 (1998)

48.85 F. A. Gianturco, T. Mukherjee, P. Paioletti: Phys. Rev. A **56**, 3638 (1997)

48.86 M. Kimura, M. Takekawa, Y. Itikawa, H. Takaki, O. Sueoka: Phys. Rev. Lett. **80**, 3936 (1998)

48.87 J. Mitroy, M. W. J. Bromley, G. G. Ryzhikh: J. Phys. B **35**, R81 (2002)

48.88 J. Mitroy: Phys. Rev. A **66**, 010501R (2002)

48.89 J. W. Darewych, J. W. Humberston, R. P. McEachran, D. A. L. Paul, A. D. Stauffer (Eds.): Can. J. Phys. **60**, 461–617 (1981)

48.90 J. W. Humberston, M. R. C. McDowell (Eds.): *Positron Scattering in Gases* (Plenum, New York 1984)

48.91 W. E. Kauppila, T. S. Stein, J. M. Wadehra (Eds.): *Positron (Electron)-Gas Scattering* (World Scientific, Singapore 1986)

48.92 J. W. Humberston, E. A. G. Armour (Eds.): *Atomic Physics with Positrons* (Plenum, New York 1987)

48.93 R. J. Drachman (Ed.): *Annihilation in Gases and Galaxies* (NASA Conference Publication, Washington 1990) p. 3058

48.94 L. A. Parcell (Ed.): Positron interaction with gases, Hyperfine Interact. **73**, 1–232 (1992)

48.95 W. Raith, R. P. McEachran (Eds.): Positron interactions with atoms, molecules and clusters, Hyperfine Interact. **89**, 1–496 (1994)

48.96 R. P. McEachran, A. D. Stauffer (Eds.): Proceedings of the 1995 positron workshop, Can. J. Phys. **74**, 313–563 (1996)

48.97 H. H. Andersen, E. A. G. Armour, J. W. Humberston, G. A. Laricchia (Eds.): Proceedings of the 1997 positron workshop, Nucl. Instrum. Methods B **143**, 1–232 (1998)

48.98 S. Hara, T. Hyodo, Y. Nagashima, L. Rehn (Ed.): Proceedings of the 1998 positron workshop, Nucl. Instrum. Methods B **171**, 1–250 (2000)

48.99 M. H. Holzscheiter (Ed.): Proceedings of the 2001 positron workshop, Nucl. Instrum. Methods B **192**, 1–237 (2002)

48.100 U. Uggerhøj, T. Ichioka, H. Knudsen (Eds.): Proceedings of the 2003 positron workshop, Nucl. Instrum. Methods B **221**, 1–242 (2004)

48.101 H. S. W. Massey: Can. J. Phys. **60**, 461 (1982)

48.102 W. E. Kauppila, T. S. Stein: Adv. At. Mol. Opt. Phys. **26**, 1 (1990)

48.103 B. H. Bransden: *Case Studies in Atomic Collision Physics I*, ed. by E. W. McDaniel, M. R. C. McDowell (Wiley, New York 1969) Chap. 4

48.104 T. S. Stein, W. E. Kauppila: Adv. At. Mol. Opt. Phys. **18**, 53 (1982)

48.105 S. J. Buckman: *New Directions in Antimatter in Chemistry and Physics*, ed. by C. M. Surko, F. A. Gianturco (Kluver, Amsterdam 2001) p. 391

48.106 M. Charlton, J. W. Humberston: *Positron Physics* (Cambridge Univ. Press, Cambridge 2000)

49. Adiabatic and Diabatic Collision Processes at Low Energies

Adiabatic and diabatic electronic states of a system of atoms are defined and their properties are described. Nonadiabatic interaction for slow quasiclassical motion of atoms is discussed within two-state common-trajectory approximation. Analytical formulae for nonadiabatic transition probabilities are presented for particular modles with reference to single and double passage of coupling regions (Landau–Zener–Stückelberg, Rosen–Zener–Demkov, Nikitin models). Generalization for multiple passage is described.

49.1 **Basic Definitions** ... 741
49.1.1 Slow Quasiclassical Collisions ... 741
49.1.2 Adiabatic and Diabatic Electronic States ... 742
49.1.3 Nonadiabatic Transitions: The Massey Parameter ... 742

49.2 **Two-State Approximation** ... 743
49.2.1 Relation Between Adiabatic and Diabatic Basis Functions ... 743
49.2.2 Coupled Equations and Transition Probabilities in the Common Trajectory Approximation ... 744
49.2.3 Selection Rules for Nonadiabatic Coupling ... 745

49.3 **Single-Passage Transition Probabilities: Analytical Models** ... 746
49.3.1 Crossing and Narrow Avoided Crossing of Potential Energy Curves: The Landau–Zener Model in the Common Trajectory Approximation ... 746
49.3.2 Arbitrary Avoided Crossing and Diverging Potential Energy Curves: The Nikitin Model in the Common Trajectory Approximation ... 747
49.3.3 Beyond the Common Trajectory Approximation ... 748

49.4 **Double-Passage Transition Probabilities and Cross Sections** ... 749
49.4.1 Mean Transition Probability and the Stückelberg Phase ... 749
49.4.2 Approximate Formulae for the Transition Probabilities ... 750
49.4.3 Integral Cross Sections for a Double-Passage Transition Probability ... 751

49.5 **Multiple-Passage Transition Probabilities** ... 751
49.5.1 Multiple Passage in Atomic Collisions ... 751
49.5.2 Multiple Passage in Molecular Collisions ... 751

References ... 752

49.1 Basic Definitions

49.1.1 Slow Quasiclassical Collisions

Slow collisions of atoms or molecules (neutral or charged) are defined as collisions for which the velocity of the relative motion of colliding particles v is substantially lower than the velocity of valence electrons v_e:

$$v/v_e \ll 1 . \tag{49.1}$$

If v_e is estimated as $v_e \approx 1$ a.u. $\approx 10^8$ cm/s, then (49.1) is fulfilled for medium mass nuclei (~ 10 amu) up to several keV.

Quasiclassical collisions are those for which the de Broglie wavelength λ_{dB} for the relative motion is substantially smaller than the range parameter a of the interaction potential

$$\lambda_{dB} \ll a . \tag{49.2}$$

The two conditions (49.1) and (49.2) define the energy range within which collisions are slow and quasiclassical. For medium mass nuclei, this energy range covers collision energies above room temperature and below hundreds of eV. The paramater a should not be confused with another important parameter L_0 which characterizes the extent of the interaction region. For instance, for the exchange interaction between two atoms, L_0 corresponds to the distance of closest approach of the colliding particles, while a is the range of the exponential decrease of the interaction. Typically, L_0 noticeably exceeds a.

49.1.2 Adiabatic and Diabatic Electronic States

Let r refer to a set of electronic coordinates in a body-fixed frame related to the nuclear framework of a colliding system, and let $\mathcal{R}$ refer to a set of nuclear coordinates determining the relative position of nuclei in this system. A configuration of electrons and nuclei in a frame fixed in space is completely determined by r, $\mathcal{R}$, and the set of Euler angles Ω, which relate the body-fixed frame to the space-fixed frame. If the total Hamiltonian of the system is $\mathcal{H}(r, \mathcal{R}, \Omega)$, the stationary state wave function satisfies the equation

$$\mathcal{H}(r, \mathcal{R}, \Omega)\Psi_E(r, \mathcal{R}, \Omega) = E\Psi_E(r, \mathcal{R}, \Omega) . \quad (49.3)$$

The electronic adiabatic Hamiltonian $H(r; \mathcal{R})$ is defined to be the part of $\mathcal{H}(r, \mathcal{R}, \Omega)$ in which the kinetic energy of the nuclei is ignored. The adiabatic electronic functions $\psi_n(r; \mathcal{R})$ are defined as eigenfunctions of $H(r; \mathcal{R})$ at a fixed nuclear configuration $\mathcal{R}$:

$$H(r; \mathcal{R})\psi_n(r; \mathcal{R}) = U_n(\mathcal{R})\psi_n(r; \mathcal{R}) . \quad (49.4)$$

The eigenvalues $U_n(\mathcal{R})$ are called adiabatic potential energy surfaces (adiabatic PES). In the case of a diatom, the set $\mathcal{R}$ collapses into a single coordinate, the internuclear distance R, and the PES become potential energy curves, $U_n(R)$. The functions $\psi_n(r; \mathcal{R})$ depend explicitly on $\mathcal{R}$ and implicitly on the Euler angles Ω. The significance of the adiabatic PES is related to the fact that in the limit of very low velocities, a system of nuclei will move across a single PES. In this approximation, called the adiabatic approximation, the function $U_n(\mathcal{R})$ plays the part of the potential energy which drives the motion of the nuclei.

An electronic diabatic Hamiltonian is defined formally as a part of H, i.e., $H_0 = H + \Delta H$. The partitioning of H into H_0 and ΔH is dictated by the requirement that the eigenfunctions of H_0, called diabatic electronic functions ϕ_n, depend weakly on the configuration $\mathcal{R}$. The physical meaning of this weak dependence is different for different problems. A perfect diabatic basis set $\phi_n(r)$ is $\mathcal{R}$-independent; for practical purposes one can use a diabatic set which is considered as $\mathcal{R}$-independent within a certain region of the configuration space $\mathcal{R}$.

Two basis sets ψ_n and ϕ_n generate the matrices

$$\begin{aligned}\langle\phi_m|H|\phi_n\rangle &= H_{mn} ,\\ \langle\phi_m|\mathcal{H}|\phi_n\rangle &= H_{mn} + D_{mn} ,\\ \langle\psi_m|\mathcal{H}|\psi_n\rangle &= U_n(\mathcal{R})\delta_{mn} + \mathcal{D}_{mn} . \end{aligned}\quad (49.5)$$

The eigenvalues of the matrix H_{mn} are U_n. D_{mn} is the matrix of dynamic coupling in the diabatic basis, and $\mathcal{D}_{mn}$ is the matrix of dynamic coupling in the adiabatic basis; the former matrix vanishes for a perfect diabatic basis. All the above matrices are, in principle, of infinite order. For low-energy collisions, the use of finite matrices of moderate dimension, will usually suffice.

Diabatic PES are defined as the diagonal elements H_{nn}. The significance of the diabatic PES is that for velocities which are high [but still satisfy (49.1)] the system moves preferentially across diabatic PES, provided that the additional conditions discussed in Sect. 49.3 are fulfilled.

For a given finite adiabatic basis $\psi_n(r; \mathcal{R})$, a perfect diabatic basis $\phi_{n'}(r)$ can be constructed by diagonalizing the matrix $\mathcal{D}_{nm}(\mathcal{R})$. The two basis sets are related by a unitary transformation

$$\psi_n\,(r; \mathcal{R}) = \sum_{n'} C_{nn'}\,(\mathcal{R})\,\phi_{n'}(r) . \quad (49.6)$$

49.1.3 Nonadiabatic Transitions: The Massey Parameter

Deviations from the adiabatic approximation manifest themselves in transitions between different PES which are induced by the dynamic coupling matrix $\mathcal{D}$. At low energies, the transitions usually occur in localized regions of nonadiabatic coupling (NAR). In these regions, the motion of nuclei in different electronic states is coupled, and in general it cannot be interpreted as being driven by a single potential.

An important simplifying feature of slow adiabatic collisions is that typically the distance between different NAR is substantially larger than the extents of each NAR. This makes it possible to formulate simple models for the coupling in isolated NAR, and subsequently to incorporate the solution for nonadiabatic coupling into the overall dynamics of the system.

For a system of s nuclear degrees of freedom, there are the following possibilities for the behavior of PES within NAR:

(i) If two s-dimensional PES correspond to electronic states of different symmetry, they can cross along an $(s-1)$-dimensional line. For a system of two atoms, $s=1$, and so two potential curves of different symmetry can cross at a point.
(ii) If two s-dimensional PES correspond to electronic states of the same symmetry, they can cross along an $(s-2)$-dimensional line. For a system of two atoms, $s=1$, and so two potential curves of different symmetry cannot cross. If they have a tendency to cross, they will exhibit a pattern which is called an avoided crossing or a pseudocrossing.
(iii) If two s-dimensional PES correspond to electronic states of the same symmetry in the presence of spin–orbit coupling, they can cross along an $(s-3)$-dimensional line.

Statement (ii) applied to a two-atom system is known as the Wigner–Witmer noncrossing rule.

The efficiency of the nonadiabatic coupling between two adiabatic electronic states is determined, according to the adiabatic principle of mechanics (both classical and quantum), by the value of the Massey parameter ζ, which represents the product of the electronic transition frequency ω_{el} and the time τ_{nuc} that characterizes the rate of change of electronic function due to nuclear motion. Putting $\omega_{\text{el}} \approx \Delta U(\mathcal{R})/\hbar$, ($\Delta U$ is the spacing between any two adiabatic PES), and $\tau_{\text{nuc}} = \Delta L/v(\mathcal{R})$, ($\Delta L$ is a certain range which depends on the type of coupling), we get

$$\zeta(\mathcal{R}) = \omega_{\text{el}}\tau_{\text{nuc}} = \Delta U(\mathcal{R})\Delta L/\hbar v(\mathcal{R}) . \tag{49.7}$$

The nonadiabatic coupling is inefficient at those configurations $\mathcal{R}$ where $\zeta(\mathcal{R}) \gg 1$. If $\zeta(\mathcal{R})$ is less than or of the order of unity, the nonadiabatic coupling is efficient, and a change in adiabatic dynamics of nuclear motion is very substantial.

The following relations usually hold for the parameters ΔL, a, L_0 for slow collisions:

$$\Delta L \ll a \ll L_0 . \tag{49.8}$$

When the nonadiabatic coupling is taken into account, the total (electronic and nuclear) wave function Ψ_{E} can be represented as a series expansion in ψ_n or ϕ_n (the Euler angles Ω are suppressed for brevity):

$$\begin{aligned} \Psi_{\text{E}}(r,\mathcal{R}) &= \sum_n \psi_n(r;\mathcal{R})\chi_{n\text{E}}(\mathcal{R}) \\ &= \sum_n \phi_n(r)\kappa_{n\text{E}}(\mathcal{R}) . \end{aligned} \tag{49.9}$$

Here $\chi_{n\text{E}}(\mathcal{R})$ and $\kappa_{n\text{E}}(\mathcal{R})$ are the functions which have to be found as solutions to the coupled equations formulated in the adiabatic or diabatic electronic basis, repectively [49.1, 2]. In general, different contributions to the first sum in (49.9) can be associated with nonadiabatic transition probabilities between different electronic states.

A practical means of calculating functions $\chi_{n\text{E}}(\mathcal{R})$ [or $\kappa_{n\text{E}}(\mathcal{R})$] consists of expanding them over certain basis functions $\Xi_{n\nu}(\mathcal{R}')$, where $\mathcal{R}'$ denotes all coordinates $\mathcal{R}$ except for the interparticle distance R. Writing

$$\chi_{n\text{E}}(\mathcal{R}) = \sum_\nu \Xi_{n\nu}(\mathcal{R}')\xi_{n\nu\text{E}}(R) , \tag{49.10}$$

one arrives at a set of coupled second-order equations for the unknown functions $\xi_{n\nu\text{E}}(R)$ (the scattering equations) [49.1]. In the semiclassical approximation, these equations become a set of first-order equations for the amplitudes of the WKB counterparts of $\xi_{n\nu\text{E}}(R)$. At the next step of simplification, in the common trajectory approximation, the variable R is changed into the time variable t, the latter being related to R via the classical trajectory $R=R(t)$ [49.2]. In the adiabatic approximation, the total wave function is represented by a single term in the first sum of (49.9):

$$\Psi_{\text{E}}(r,\mathcal{R}) = \psi_n(r;\mathcal{R})\chi_{n\text{E}}(\mathcal{R}) . \tag{49.11}$$

49.2 Two-State Approximation

49.2.1 Relation Between Adiabatic and Diabatic Basis Functions

In the two-state approximation, the basis of electronic functions consists of two states. In this case, the elements of the matrix C in (49.6) are expressed through a single parameter only, a mixing or rotation angle θ:

$$\begin{aligned} \psi_1(r;\mathcal{R}) &= \cos\theta(\mathcal{R})\phi_1(r) + \sin\theta(\mathcal{R})\phi_2(r) , \\ \psi_2(r;\mathcal{R}) &= -\sin\theta(\mathcal{R})\phi_1(r) + \cos\theta(\mathcal{R})\phi_2(r) . \end{aligned} \tag{49.12}$$

The rotation angle, $\theta(\mathcal{R})$, is expressed via the diagonal and off-diagonal matrix elements of the adiabatic Hamiltonian H in the diabatic basis ϕ_1, ϕ_2:

$$\tan 2\theta(\mathcal{R}) = \frac{2H_{12}(\mathcal{R})}{H_{11}(\mathcal{R}) - H_{22}(\mathcal{R})} . \tag{49.13}$$

The eigenvalues of H in terms of H_{ik} are

$$U_{1,2}(\mathcal{R}) = [H_{11}(\mathcal{R}) + H_{22}(\mathcal{R})]/2 \pm \Delta U(\mathcal{R})/2\,, \tag{49.14}$$

where

$$\Delta U(\mathcal{R}) = \left\{[H_{11}(\mathcal{R}) - H_{22}(\mathcal{R})]^2 + 4H_{12}^2(\mathcal{R})\right\}^{1/2}. \tag{49.15}$$

The matrix elements H_{ik} are expressed via the adiabatic potentials and the rotation angle by

$$\begin{aligned} H_{11}(\mathcal{R}) + H_{22}(\mathcal{R}) &= U_1(\mathcal{R}) + U_2(\mathcal{R})\,, \\ H_{11}(\mathcal{R}) - H_{22}(\mathcal{R}) &= \Delta U(\mathcal{R}) \cos 2\theta(\mathcal{R})\,, \\ H_{12}(\mathcal{R}) &= (1/2)\Delta U(\mathcal{R}) \sin 2\theta(\mathcal{R})\,. \end{aligned} \tag{49.16}$$

49.2.2 Coupled Equations and Transition Probabilities in the Common Trajectory Approximation

A two-state nonadiabatic wave function $\Psi(r, \mathcal{R})$ can be written as an expansion into either adiabatic or diabatic electronic wave functions:

$$\begin{aligned} \Psi(r;\mathcal{R}) &= \psi_1(r;\mathcal{R})\,\alpha_1(\mathcal{R}) + \psi_2(r;\mathcal{R})\,\alpha_2(\mathcal{R})\,, \\ \Psi(r;\mathcal{R}) &= \phi_1(r)\beta_1(\mathcal{R}) + \phi_2(r)\beta_2(\mathcal{R})\,, \end{aligned} \tag{49.17}$$

in which the nuclear wave functions satisfy two coupled s-dimensional Schrödinger equations [49.1].

In the common trajectory approximation, the motion of the nuclei is described by the classical trajectory, i. e., by a one-dimensional manifold $\mathcal{Q}(t)$ embedded in the s-dimensional manifold $\mathcal{R}$. A section of PES along this one-dimensional manifold determines a set of effective potential energy curves (PEC). In the case of atomic collisions, Q coincides with the interatomic distance R, and the effective PEC are just ordinary PEC.

A common trajectory counterpart of (49.17) is

$$\begin{aligned} \Psi(r,t) &= \psi_1\left[r;\mathcal{Q}(t)\right]a_1(t) + \psi_2\left[r;\mathcal{Q}(t)\right]a_2(t)\,, \\ \Psi(r;t) &= \phi_1(r)b_1(t) + \phi_2(r)b_2(t)\,. \end{aligned} \tag{49.18}$$

The adiabatic expansion coefficients $a_k(t)$ satisfy the set of equations

$$\begin{aligned} i\hbar\frac{da_1}{dt} &= U_1(\mathcal{Q})a_1 + i\dot{\mathcal{Q}}g(\mathcal{Q})a_2\,, \\ i\hbar\frac{da_2}{dt} &= -i\dot{\mathcal{Q}}g(\mathcal{Q})a_1 + U_2(\mathcal{Q})a_2\,, \end{aligned} \tag{49.19}$$

where $g(\mathcal{Q}) = \langle\psi_1|\partial/\partial\mathcal{Q}|\psi_2\rangle = d\theta/d\mathcal{Q}$, and $\mathcal{Q} = \mathcal{Q}(t)$. The diabatic expansion coefficients $b_k(t)$ satisfy the set of equations

$$\begin{aligned} i\hbar\frac{db_1}{dt} &= H_{11}(\mathcal{Q})b_1 + H_{12}(\mathcal{Q})b_2\,, \\ i\hbar\frac{db_2}{dt} &= H_{21}(\mathcal{Q})b_1 + H_{22}(\mathcal{Q})b_2\,. \end{aligned} \tag{49.20}$$

Clearly, for a system of two atoms, $\mathcal{Q} \equiv R$.

Solutions to (49.19) and (49.20) are equivalent, provided that the initial conditions are matched, and the transition probability is properly defined.

For a given trajectory, it is customary to identify the center of the NAR with a value of $\mathcal{Q} = \mathcal{Q}_p$ which corresponds to the real part of the complex-valued coordinate $\mathcal{Q}_c$ at which two adiabatic PES cross. The crossing conditions in the adiabatic and diabatic representations are

$$U_1(\mathcal{Q}_c) - U_2(\mathcal{Q}_c) = 0\,, \tag{49.21}$$

or

$$\left[H_{11}(\mathcal{Q}_c) - H_{22}(\mathcal{Q}_c)\right]^2 + 4H_{12}^2(\mathcal{Q}_c) = 0\,. \tag{49.22}$$

Since $\mathcal{Q}$ represents a one-dimensional manifold, the crossing condition (49.22) is satisfied for a complex value of $\mathcal{Q} = \mathcal{Q}_s$ unless $H_{12} = 0$. Then, by definition, the location of the NAR centeris identified with $\mathcal{Q}_p$ through

$$\mathcal{Q}_p = \mathrm{Re}(\mathcal{Q}_c)\,, \tag{49.23}$$

where $\mathcal{Q}_c$ is that value of $\mathcal{Q}_s$ which possesses the smallest imaginary part, and Re denotes the real part.

For the case when the regions of nonadiabatic coupling are well localized, the function $g(\mathcal{Q})$ possesses a pronounced maximum at (or close to) $\mathcal{Q}_p$, the width $\Delta\mathcal{Q}_p$ of which determines the range of the NAR; normally $\Delta\mathcal{Q}_p$ is about Im $(\mathcal{Q}_c)$, with Im denoting the imaginary part. The two Eqs. (49.19) decouple on both sides of this maximum. A solution of the equations for the nonadiabatic coupling across an isolated maximum of $g(\mathcal{Q})$ yields the so-called single-passage (or one-way) transition amplitude and transition probability. For this problem, the time $t = 0$ can be assigned to the maximum point of $g[\mathcal{Q}(t)]$. Assuming that away from $t = 0$ the decoupling occurs rapidly enough, the nonadiabatic transition probability

$$P_{12} = |a_2(\infty)|^2\,, \tag{49.24}$$

provided that a solution to (49.19) corresponds to the initial conditions,

$$a_1(-\infty) = 1\,, \quad a_2(-\infty) = 0\,. \tag{49.25}$$

In the limit of almost adiabatic conditions where P_{12} is very small, the following equation holds [49.3]:

$$P_{12} = \exp\left[-\frac{2}{\hbar}\left|\mathrm{Im}\left(\int_{t_r}^{t_c}\{U_1[\mathcal{Q}(t)]-U_2[\mathcal{Q}(t)]\}\,\mathrm{d}t\right)\right|\right], \tag{49.26}$$

where t_c is a root of

$$\mathcal{Q}(t_c) = \mathcal{Q}_c\,. \tag{49.27}$$

Here t_r is any real-valued time. Equation (49.26) is valid when the exponent is large, so that P_{12} is exponentially small.

The property of the function $g(\mathcal{Q})$ to pass through a single narrow maximum ensures that the rotation angle away from the maximum tends to constant values, and adiabatic functions in these regions are expressed by certain linear combinations of diabatic functions with constant mixing coefficients. These linear combinations should serve as the initial condition on one side of the coupling region, and as the proper final state on the other, when the problem of a nonadiabatic transition between adiabatic states is treated in the diabatic representation. The same property of the function $g(\mathcal{Q})$ implies that a common trajectory needs to be defined only locally, within a given NAR, and not globally, in the full configuration space.

49.2.3 Selection Rules for Nonadiabatic Coupling

In the general case, the coupling between adiabatic states or diabatic states is controlled by certain selection rules. The most detailed selection rules exist for a system of two colliding atoms, since this system possesses a high symmetry ($\mathcal{C}_{\infty v}$ or, for identical atoms, $D_{\infty h}$ point symmetry in the adiabatic approximation). In the adiabatic representation, the coupling is due to the elements of the matrix $\mathcal{D}$. They fall into two different categories: those proportional to the radial nuclear velocity (coupling by radial motion or radial coupling), and those proportional to the angular velocity of rotation of the molecular axis (coupling by rotational motion or Coriolis coupling).

In a diabatic representation, provided that the effect of the D matrix is neglected, the coupling is due to the parts of the interaction potential neglected in the definition of the diabatic Hamiltonian H_0. In typical cases, these parts are the electrostatic interaction between different electronic states constructed as certain electronic configurations (H_0 corresponds to a self-consistent field Hamiltonian); spin–orbit interaction (H_0 corresponds to a nonrelativistic Hamiltonian); hyperfine interaction (H_0 ignores the magnetic interaction of electronic and nuclear spins as well as the electrostatic interaction between electrons and nuclear quadrupole moments). The selection rules for the above interactions in the case of two atoms are listed in Table 49.1 for two conventional nomenclatures for molecular states: Hund's case (a), ${}^{2S+1}\Lambda_w^{(\sigma)}$ and Hund's case (c), $\Omega_w^{(\sigma)}$ [49.3].

For molecular systems with more than two nuclei, the selection rules cannot be put in a detailed form since, in general, the symmetry of the system is quite low. For the important case of three atoms, a general configuration is planar ($\mathcal{C}_s$ symmetry); particular configurations correspond to an isosceles triangle if two atoms are identical ($\mathcal{C}_{2v}$ symmetry), to an equilateral triangle for three identical atoms (D_{3h} symmetry) or to a linear configuration. For the last case, the selection rules are the same as for a system of two atoms.

Table 49.1 Selection rules for the coupling between diabatic and adiabatic states of a diatomic quasimolecule ($w = g, u$; $\sigma = +, -$)

Interaction	${}^{2S+1}\Lambda_w^{(\sigma)}$ nomenclature	$\Omega_w^{(\sigma)}$ nomenclature
Configuration interaction (electrostatic)	$\Delta\Lambda = 0,\ \Delta S = 0$ $g \not\rightleftharpoons u,\ + \not\rightleftharpoons -$	$\Delta\Omega = 0$ $g \not\rightleftharpoons u,\ + \not\rightleftharpoons -$
Spin–orbit interaction	$\Delta\Lambda = 0,\ \pm 1,\ \Delta S = 0,\ \pm 1$ $g \not\rightleftharpoons u,\ + \rightleftharpoons -$	$\Delta\Omega = 0$ $g \not\rightleftharpoons u,\ + \not\rightleftharpoons -$
Radial motion	$\Delta\Lambda = 0,\ \Delta S = 0$ $g \not\rightleftharpoons u,\ + \not\rightleftharpoons -$	$\Delta\Omega = 0$ $g \not\rightleftharpoons u,\ + \not\rightleftharpoons -$
Rotational motion	$\Delta\Lambda = \pm 1,\ \Delta S = 0$ $g \not\rightleftharpoons u,\ + \not\rightleftharpoons -$	$\Delta\Omega = \pm 1$ $g \not\rightleftharpoons u,\ + \not\rightleftharpoons -$
Hyperfine interaction	$\Delta\Lambda = 0,\ \pm 1,\ \Delta S = 0,\ \pm 1$ $g \rightleftharpoons u,\ + \rightleftharpoons -$	$\Delta\Omega = 0,\ \pm 1$ $g \rightleftharpoons u,\ + \rightleftharpoons -$

The selection rules for the dynamic coupling between adiabatic states classified according to the irreducible representations of the $\mathcal{C}_s$ and $\mathcal{C}_{2v}$ groups are listed in Table 49.2. In this table, z and y refer to two modes of the relative nuclear motion in the system plane, R_z and R_y refer to two rotations about principal axes of inertia lying in the system plane, and R_x refers to a rotation about the principal axis of inertia perpendicular to the system plane.

Table 49.2 Selection rules for dynamic coupling between adiabatic states of a system of three atoms

$\mathcal{C}_s$		A'		A''	
	$\mathcal{C}_{2v}$	A_1	B_1	A_2	B_2
A'	A_1	z	$y,\ R_x$	R_z	R_y
	B_1	$y,\ R_x$	z	R_y	R_z
A''	A_2	R_z	R_y	z	$y,\ R_x$
	B_2	R_y	R_z	$y,\ R_x$	z

49.3 Single-Passage Transition Probabilities: Analytical Models

49.3.1 Crossing and Narrow Avoided Crossing of Potential Energy Curves: The Landau–Zener Model in the Common Trajectory Approximation

The Landau–Zener model applies to a situation when the effective adiabatic PEC cross or show a narrow avoided crossing. The latter is defined by the condition that the spacing between adiabatic PEC within a NAR is much smaller than the spacing between adiabatic PEC away from the NAR. The cases of crossing and narrow avoided crossing of adiabatic PEC can be considered within a unified model since a narrow avoided crossing of adiabatic PEC corresponds to a crossing of diabatic PEC. Therefore, in both cases, one considers the crossing of zero-order PES (adiabatic or diabatic) and the interaction between them (dynamic or static). However, the definition of transition probability is different for crossing and avoided crossing.

In the framework of the Landau–Zener model [49.4–6], the two zero-order PEC which cross at a point $\mathcal{Q}_\mathrm{p}$ along a trajectory $\mathcal{Q}(t)$ are approximated by functions linear in $\Delta\mathcal{Q} = \mathcal{Q} - \mathcal{Q}_\mathrm{p}$, and the off-diagonal matrix element is assumed to be a constant.

For the avoided crossing adiabatic PEC (crossing diabatic PEC), the matrix H_{jk} within a NAR is approximated as

$$\begin{aligned} H_{11}(\mathcal{Q}) &= E_0 - F_1(\mathcal{Q} - \mathcal{Q}_\mathrm{p})\,,\\ H_{22}(\mathcal{Q}) &= E_0 - F_2(\mathcal{Q} - \mathcal{Q}_\mathrm{p})\,,\\ H_{12}(\mathcal{Q}) &= V = \text{constant}\,, \end{aligned} \tag{49.28}$$

from which the spacing between adiabatic PEC is

$$\Delta U = \left[(\Delta F)^2(\mathcal{Q} - \mathcal{Q}_\mathrm{p})^2 + 4|V|^2\right]^{1/2}\,, \tag{49.29}$$

with $\Delta F = |F_1 - F_2|$.

The common trajectory is assumed to be a linear function of t,

$$\mathcal{Q} = \mathcal{Q}_\mathrm{p} + v_\mathrm{p} t\,, \tag{49.30}$$

where v_p is the velocity of $\mathcal{Q}$-motion at point $\mathcal{Q}_\mathrm{p}$.

For this model, adiabatic wave functions on both sides of the nonadiabaticity region (in the limits $-\infty < t < +\infty$) coincide with diabatic functions, but their ordering is reversed. Explicitly,

$$\begin{aligned} &\psi_1 = \phi_1\,; \quad \psi_2 = \phi_2 \quad \text{for } t \to -\infty\,,\\ &\psi_1 = \phi_2\,; \quad \psi_2 = -\phi_1 \quad \text{for } t \to +\infty\,. \end{aligned} \tag{49.31}$$

The transition probability between pseudocrossing adiabatic curves for the Hamiltonian (49.28) and the trajectory (49.30) is given by the Landau–Zener formula

$$\begin{aligned} &P_{12}^{\mathrm{psc}} = P_{12}^{\mathrm{LZ}} = \exp\left(-2\pi\zeta^{\mathrm{LZ}}\right);\\ &\zeta^{\mathrm{LZ}} = V^2/(\hbar\Delta F v_\mathrm{p})\,, \end{aligned} \tag{49.32}$$

where ζ^{LZ} is the appropriate Massey parameter at the pseudocrossing point. Note that for the LZ model, the single-passage transition probability depends on one dimensionless parameter ζ^{LZ}. A remarkable property of the Landau–Zener model is that the probability P_{12}^{LZ} is given by (49.26) for an arbitrary value of the exponent, and not only for large ones when the probability is very low.

With a change of the velocity from very low to very high values, the transition probability varies from zero to unity. In the near-adiabatic limit, $\zeta^{\mathrm{LZ}} \gg 1$, the nuclei preferentially move across single adiabatic PEC, while in the sudden limit, $\zeta^{\mathrm{LZ}} \ll 1$, they move across single diabatic PEC. It is the latter property of the LZ transition probability that allows one to interpret diabatic energies H_{11} and H_{22} as the potentials which drive the nuclear motion at high velocities.

The region of applicability of the Landau–Zener formula is determined by the condition that the extension of the region of nonadiabatic interaction $\Delta\mathcal{Q}$ should be small compared with the range a over which potential curves deviate substantially from linear functions. The condition $\Delta\mathcal{Q} \ll a$ actually implies the two conditions [49.2]

$$2V/\Delta F \ll a\,, \tag{49.33}$$

and

$$2\left(\hbar v_{\rm p}/\Delta F\right)^{1/2} \ll a\,. \tag{49.34}$$

Clearly, the range parameter a does not enter the LZ formula since it controls the behavior of adiabatic curves away from the crossing point.

The constant velocity approximation (49.30) imposes yet another condition:

$$V \ll \mu v_{\rm p}^2/2\,. \tag{49.35}$$

The actual application of (49.32) requires the specification of V and $v_{\rm p}$ for each particular trajectory $\mathcal{Q}(t)$. For the case of avoided crossing between two potential curves of a diatom, V does not depend on the trajectory and represents, according to Table 49.2, the matrix element of the electrostatic interaction, spin–orbit interaction or hyperfine interaction.

For crossing adiabatic PEC, the Landau–Zener model assumes the following approximation for adiabatic potentials and the dynamic coupling:

$$\begin{aligned} U_1(\mathcal{Q}) &= E_0 - F_1(\mathcal{Q}-\mathcal{Q}_{\rm c})\,,\\ U_1(\mathcal{Q}) &= E_0 - F_2(\mathcal{Q}-\mathcal{Q}_{\rm c})\,,\\ \mathcal{D}_{12}(\mathcal{Q}) &= D = \text{constant}\,, \end{aligned} \tag{49.36}$$

with the trajectory parametrization given by (49.30) where $v_{\rm p}$ is replaced by $v_{\rm c}$. Since the ordering of adiabatic PEC for crossing and pseudocrossing is reversed on one side of a NAR, the following relation exists between transition probabilities for the crossing case $P_{12}^{\rm c}$, and the survival probability for the pseudocrossing case $1-P_{12}^{\rm psc}$:

$$P_{12}^{\rm c} = 1 - P_{12}^{\rm psc}\,, \tag{49.37}$$

provided that D and $v_{\rm c}$ in the crossing situation are replaced by V and $v_{\rm p}$ in the pseudocrossing situation. Conditions (49.33) and (49.34) applied to the case of the dynamic coupling often imply that this coupling is weak [49.2]. Therefore, (49.37) yields

$$P_{12}^{\rm c} = \frac{2\pi D^2}{\hbar\Delta F v_{\rm c}}\,. \tag{49.38}$$

Usually, the matrix element D is related to the Coriolis coupling, and it is proportional to the angular velocity of rotation of the molecular frame at the crossing point $\mathcal{Q}_{\rm c}$.

49.3.2 Arbitrary Avoided Crossing and Diverging Potential Energy Curves: The Nikitin Model in the Common Trajectory Approximation

The restrictions of narrow avoided crossing [(49.33) and (49.34) for the LZ model] are relaxed in a more general model suggested by *Nikitin* [49.7]. This model uses a more flexible exponential parametrization, instead of the linear parametrization for diabatic matrix elements (49.36).

In a diabatic basis, the model is formulated with the Hamiltonian

$$\begin{aligned} H_{11}(\mathcal{Q}) &= U_0(\mathcal{Q}) - \Delta E/2 + (A/2)\cos 2\vartheta \exp(-\alpha\mathcal{Q})\,,\\ H_{22}(\mathcal{Q}) &= U_0(\mathcal{Q}) + \Delta E/2 - (A/2)\cos 2\vartheta \exp(-\alpha\mathcal{Q})\,,\\ H_{12}(\mathcal{Q}) &= (A/2)\sin 2\vartheta \exp(-\alpha\mathcal{Q})\,. \end{aligned} \tag{49.39}$$

The spacing between adiabatic PEC is

$$\begin{aligned} \Delta U = \Delta E\big\{&1 - 2\cos 2\vartheta\,\exp\left[-\alpha\left(\mathcal{Q}-\mathcal{Q}_{\rm p}\right)\right]\\ &+\exp\left[-2\alpha\left(\mathcal{Q}-\mathcal{Q}_{\rm p}\right)\right]\big\}^{1/2}\,, \end{aligned} \tag{49.40}$$

where $\mathcal{Q}_{\rm p}$ is introduced instead of A via (49.21) and (49.23). At the center of an NAR, where $\mathcal{Q} = \mathcal{Q}_{\rm p}$, the spacing between adiabatic PEC, $\Delta U_{\rm p} = \Delta U(\mathcal{Q}_{\rm p})$, is

$$\Delta U_{\rm p} = 2\Delta E \sin\vartheta\,. \tag{49.41}$$

The common trajectory within the NAR is taken in a form identical to (49.30) in which $v_{\rm p}$ is now the velocity of $\mathcal{Q}$ motion at the center of the coupling region $\mathcal{Q}_{\rm p}$.

For this model, adiabatic wave functions coincide with diabatic functions before entering the coupling region (in the limit $\alpha(\mathcal{Q}-\mathcal{Q}_{\rm p}) \gg 1$), but after exiting the coupling region [in the limit $\alpha(\mathcal{Q}-\mathcal{Q}_{\rm p}) \ll -1$] they are linear combinations of the diabatic functions

$$\begin{aligned} &\psi_1(\boldsymbol{r};\mathcal{Q}) = \phi_1(\boldsymbol{r})\,, \quad \psi_2(\boldsymbol{r};\mathcal{Q}) = \phi_2(\boldsymbol{r})\,,\\ &\qquad \text{for}\,\alpha(\mathcal{Q}-\mathcal{Q}_{\rm p}) \gg 1\,;\\ &\psi_1 = \phi_1\cos\vartheta + \phi_2\sin\vartheta\,,\\ &\psi_2 = -\phi_1\sin\vartheta + \phi_2\cos\vartheta\,,\\ &\qquad \text{for}\,\alpha(\mathcal{Q}-\mathcal{Q}_{\rm p}) \ll -1\,. \end{aligned} \tag{49.42}$$

The latter equation identifies the parameter ϑ that enters into the definition of the diabatic Hamiltonian in (49.39) with the asymptotic value of the mixing angle θ (49.12) for $\alpha(\mathcal{Q}-\mathcal{Q}_p) \ll -1$.

The transition probability P_{12} between adiabatic PEC for the Hamiltonian (49.39) and the trajectory (49.30) is

$$P_{12}^{\mathrm{N}} = \exp(-\pi\zeta_p)\frac{\sinh(\pi\zeta - \pi\zeta_p)}{\sinh(\pi\zeta)} , \quad (49.43)$$

where $\zeta = \Delta E/(\hbar\alpha v_p)$ and $\zeta_p = \zeta \sin^2\vartheta$. With the change in velocity from very low to very high values, the transition probability varies from zero to $\cos^2\vartheta$. As ϑ changes from very small values to π, the pattern of adiabatic potential curves changes from narrow to wide pseudocrossing and ultimately to strong divergence.

Since the single-passage transition probability (49.43) depends on two parameters, the Nikitin model is more versatile than the Landau–Zener one.

In three limiting cases, $\vartheta \ll 1$, $\zeta \gg 1$, $\vartheta = \pi/4$, ζ arbitrary, and ϑ arbitrary, $\zeta = 0$, (49.43) may be simplified. In the first case, the diabatic Hamiltonian (49.39) becomes the Landau–Zener Hamiltonian (49.28). Also, (49.43) reduces to a single exponential which gives the LZ transition probability,

$$P_{12}^{\mathrm{N}} = P_{12}^{\mathrm{LZ}} = \exp(-2\pi\zeta_p) , \quad (49.44)$$

with ζ_p identical to ζ^{LZ}. The two conditions $\vartheta \ll 1$ and $\zeta \gg 1$ are equivalent to the two conditions (49.33) and (49.34). In the second case ($\vartheta = \pi/4$), the diabatic Hamitonian reads

$$\begin{aligned} H_{11}(R) &= E_0 - \Delta E/2 , \\ H_{22}(R) &= E_0 + \Delta E/2 , \\ H_{12}(R) &= (A/2)\exp(-\alpha\mathcal{Q}) . \end{aligned} \quad (49.45)$$

The transition probability in this case is given by the Rosen–Zener–Demkov formula [49.8, 9]

$$P_{12}^{\mathrm{RZD}} = \frac{\exp(-\pi\zeta)}{1+\exp(-\pi\zeta)} , \quad \zeta = \Delta E/(\hbar v_p \alpha) . \quad (49.46)$$

In the third case ($\zeta = 0$), also called the resonance case since $\Delta E = 0$, the transition probability reads

$$P_{12}^{\mathrm{Res}} = \cos^2\vartheta . \quad (49.47)$$

Equation (49.47) is a particular example of transitions between initially degenerate states. This kind of transition occurs in the recoupling of angular momenta in collisions of atoms possessing nonzero electronic angular momentum [49.10].

For the general case, the nuclei preferentially move across single adiabatic PEC in the near-adiabatic limit, $\zeta \gg 1$, while in the sudden limit, $\zeta \ll 1$, they move across both diabatic PEC, unless the condition of narrow avoided crossing, $\vartheta \ll 1$, is fulfilled.

49.3.3 Beyond the Common Trajectory Approximation

The common trajectory approximation is valid when the spacing between adiabatic PEC within an NAR is small compared to the local kinetic energy of the nuclei. The relaxation of this restriction is not unambiguous since one should pass from a one-dimensional manifold (time as a progress variable) to a multi-dimensional coordinate (configuration space manifold). Only if the latter is one-dimensional (a single coordinate as a progress variable, as is the case for atom-atom collisions), one can suggest a generalization of the common trajectory transition probability. We consider this case, taking R to be such a single coordinate, and assume that the quantum motion across the adiabatic PEC satisfies standard quasiclassical conditions [49.3]. For a two-state problem with adiabatic potentials $U_1(R)$ and $U_2(R)$, the general condition of the common trajectory apporximation reads

$$E - \frac{1}{2}\left[U_1(R_p) + U_2(R_p)\right] \gg \left|U_1(R_p) - U_2(R_p)\right| , \quad (49.48)$$

where E is the total (conserved) energy and R_p is the coordinate of the NAR center.

The quantum generalization of the expression for the transition probability in the near-adiabatic condition, (49.26), is given by the original Landau formula [49.4, 5]

$$P_{12} = \exp\left\{-\frac{2}{\hbar}\left|\mathrm{Im}\left(\int_R^{R_c}\sqrt{2\mu[E-U_1(R)]}\,\mathrm{d}R - \int_R^{R_c}\sqrt{2\mu[E-U_2(R)]}\,\mathrm{d}R\right)\right|\right\} , \quad (49.49)$$

where μ is the reduced mass of the colliding atoms, R_c is the complex-valued coordinate of the crossing of $U_1(R)$ and $U_2(R)$ and R is any value of the coordinate in the classically accessible region of motion of the nuclei. The nonadiabatic transition is localized in the region of width $\Delta R = \mathrm{Im}(R_c)$ centered at $R_p = \mathrm{Re}(R_c)$. Equation (49.49) becomes the common trajectory equation (49.26) under the condition (49.48).

The quantum generalization of the LZ transition probability can not be represented by an exact analytical expression though it is known that it depends on two parameters (and not on one, as in the case for the common trajectory approximation) [49.2]. A recommended approximate expression for $E > E_0$ reads [49.11]

$$P_{12}^{\mathrm{LZ}} = \exp\left[-2\pi\zeta^{\mathrm{LZ}}\left(\frac{2}{1+\sqrt{1+\epsilon_{\mathrm{p}}^{-2}\left[\frac{1}{160}\epsilon_{\mathrm{p}}(\zeta^{\mathrm{LZ}})^2+0.7\right]}}\right)^{1/2}\right], \tag{49.50}$$

where

$$\epsilon_{\mathrm{p}} = \frac{\mu v_{\mathrm{p}}^2}{2V}\frac{\Delta F}{2\sqrt{|F_1F_2|}} .$$

Equation (49.50) becomes the common trajectory (49.32) under the condition $\epsilon_{\mathrm{p}} \gg 1$; it turns out that the latter condition may be less restrictive than the general condition (49.48).

The quantum generalization of the Nikitin transition probability is possible provided $U_0(R)$ in (49.39) is given by an exponential function, $U_0(R) \approx \exp(-\alpha R)$. The transition probability reads [49.11]:

$$P_{12}^{\mathrm{N}} = \exp(-\pi\delta_{\mathrm{p}})\frac{\sinh(\pi\delta-\pi\delta_{\mathrm{p}})}{\sinh(\pi\delta)} \tag{49.51}$$

and depends on three parameters of the model (and not two as is the case for the common trajectory approximation). These parameters enter into δ_{p} and δ through complicated contour integrals. If the general condition of the common trajectory approximation, (49.48), is fulfilled, δ_{p} and δ reduce to ζ_{p} and ζ so that (49.51) becomes (49.43).

More discussions of two-state models within and beyond the common trajectory approximation can be found elsewhere [49.2, 11–14].

49.4 Double-Passage Transition Probabilities and Cross Sections

49.4.1 Mean Transition Probability and the Stückelberg Phase

In the case of an atomic collision, the set $\mathcal{R}$ shrinks into a single coordinate R. If there is only one NAR over the whole range of R, the colliding system traverses it twice, as the atoms approach and then recede. In this case, there are two different paths between the center of the NAR, R_{p}, and the turning points R_{t1} and R_{t2} on the adiabatic potential curves $U_1(R)$ and $U_2(R)$. The double-passage transition probability $\mathcal{P}_{12}$ is expressed via the single-passage transition probability P_{12}, the single-passage survival probability $1-P_{12}$, and the Stückelberg interference term $\cos\Delta\Phi_{12}$ [49.15],

$$\begin{aligned}\mathcal{P}_{12} &= 2P_{12}(1-P_{12})(1-\cos\Delta\Phi_{12})\\ &= 4P_{12}(1-P_{12})\sin^2(\Delta\Phi_{12}/2) .\end{aligned} \tag{49.52}$$

The Stückelberg phase $\Delta\Phi_{12}/2$ is expressed as the phase difference $\Delta\Phi_{12}^{(0)}/2$ which is accumulated during the motion of a diatom from the center of the NAR to the turning points, together with an additional phase ϕ_{12} by

$$\Delta\Phi_{12}/2 = \Delta\Phi_{12}^{(0)}/2 + \phi_{12} . \tag{49.53}$$

Generally, (49.53) is valid provided $\Delta\Phi_{12}/2 \gg 1$. For transitions between electronic states of the same axial symmetry, the relative angular momentum ℓ of the colliding atoms is conserved, and we have

$$\begin{aligned}\Delta\Phi_{12}^{(0)}/2 = &\int_{R_{\mathrm{p}}}^{R_{\mathrm{t1}}}\left[2\mu E-\hbar^2\ell(\ell+1)/R^2-2\mu U_1(r)\right]^{1/2}\mathrm{d}R/\hbar\\ &-\int_{R_{\mathrm{p}}}^{R_{\mathrm{t2}}}\left[2\mu E-\hbar^2\ell(\ell+1)/R^2-2\mu U_2(r)\right]^{1/2}\mathrm{d}R/\hbar ,\end{aligned} \tag{49.54}$$

where R_{t1} and R_{t2} are the turning points for adiabatic motion on potential curves U_1 and U_2, E is the total energy. Once R_{p} is chosen, $\Delta\Phi_{12}^{(0)}/2$ is well-defined and is independent of the dynamic details of a nonadiabatic transition. On the other hand, P_{12} and ϕ_{12} do depend on these details. In particular, for the models discussed in Sect. 49.3, the velocity v_{p} that enters into the Massey parameter, is

$$v_{\mathrm{p}} = \left(\frac{2}{\mu}\right)^{1/2}\left(E-U_{\mathrm{p}}-\frac{\hbar^2\ell(\ell+1)}{2\mu R_{\mathrm{p}}^2}\right)^{1/2} , \tag{49.55}$$

where $U_{\mathrm{p}} \approx U_1(R_{\mathrm{p}}) \approx U_2(R_{\mathrm{p}})$. As a function of E and ℓ, the double-passage transition probability is symmetric with respect to the initial and final states.

In many applications, one can use the mean transition probability $\langle \mathcal{P}_{12} \rangle$, which is obtained from $\mathcal{P}_{12}$ by averaging over several oscillations:

$$\langle \mathcal{P}_{12} \rangle = 2P_{12}(1 - P_{12}) . \quad (49.56)$$

The important limiting cases of the double-passage Nikitin model are:

(i) Double-passage Landau–Zener–Stückelberg equation,

$$\mathcal{P}_{12}^{\mathrm{LZS}} = 4\exp\left(-2\pi\zeta^{\mathrm{LZ}}\right)\left[1-\exp\left(-2\pi\zeta^{\mathrm{LZ}}\right)\right] \times \sin^2\left(\Delta\Phi_{12}^{(0)}/2 + \phi_{12}^{\mathrm{LZS}}\right) , \quad (49.57)$$

where ζ^{LZ} is given by (49.32), $\Delta\Phi_{12}^{(0)}/2$ by (49.54) and the expression for ϕ_{12}^{LZS} is available [49.2]. When ζ^{LZ} changes from zero to infinity, $\langle \mathcal{P}_{12}^{\mathrm{LZS}} \rangle$ passes through a maximum, $\langle \mathcal{P}_{12} \rangle_{\max} = 1/2$, and ϕ_{12}^{LZS} decreases from $\pi/4$ to zero.

(ii) Double-passage Rosen–Zener–Demkov equation:

$$\mathcal{P}_{12}^{\mathrm{RZD}} = \frac{\sin^2\left(\Delta\Phi_{12}^{(0)}/2 + \phi_{12}^{\mathrm{RZD}}\right)}{\cosh^2(\pi\zeta/2)} . \quad (49.58)$$

where ζ is given by (49.46) and the expression for ϕ_{12}^{RZD} is available [49.2]. Under certain conditions [49.2, 9], the Stückelberg phase in (49.58) can be identified with the phase accumulated during the motion of a diatom from infinitely large distance to the turning points.

(iii) Double-passage equation for a resonance process ($\Delta E = 0$):

$$\mathcal{P}_{12}^{\mathrm{res}} = \sin^2(2\vartheta)\sin^2(\Delta\Phi_{12}/2) . \quad (49.59)$$

The general resonance case (zero energy change, $\Delta E = 0$) is also called an accidental resonance. For the accidental resonance, the diagonal diabatic matrix elements are not equal to each other. A particular case of an accidental resonance is a symmetric resonance, for which the diabatic matrix elements are the same. For the Nikitin model, symmetric resonance corresponds to $\vartheta = \pi/4$, and (49.59) reads

$$\mathcal{P}_{12}^{\mathrm{symm}} = \sin^2(\Delta\Phi_{12}/2) . \quad (49.60)$$

Equation (49.60) also follows from (49.58) in the limit $\zeta \to 0$. Actually, (49.60) is valid for any symmetric resonance case [not necessarily for the model Hamiltonians (49.39) and (49.45)] and for the arbitrary values of the phase $\Delta\Phi_{12}/2$. This phase can be identified with $\Delta\Phi_{12}^{(0)}/2$ from (49.54) provided R_{p} is taken to be infinitely large.

49.4.2 Approximate Formulae for the Transition Probabilities

Several approximate formulae are available for $\langle \mathcal{P}_{12} \rangle$ in the case where H_{12} depends on time in a bell-shaped manner, and $\Delta H = H_{11} - H_{22}$ can be represented as $\hbar\omega + \Delta V$, with ΔV also having a bell-shaped form. Define

$$v = \left(\frac{1}{\hbar}\right)\int_{-\infty}^{+\infty} H_{12}(t)\,\mathrm{d}t ;$$

$$w_0 = \left(\frac{1}{\hbar}\right)\int_{-\infty}^{+\infty} H_{12}(t)\exp(\mathrm{i}\omega t)\,\mathrm{d}t ,$$

$$w = \left(\frac{1}{\hbar}\right)\int_{-\infty}^{+\infty} H_{12}(t)\exp\left[(\mathrm{i}/\hbar)\int_0^t \Delta H(t)\,\mathrm{d}t\right]\mathrm{d}t ,$$

$$u = \left(\frac{1}{2\hbar}\right)\int_{-\infty}^{+\infty} \Delta V_{12}(t)\,\mathrm{d}t , \quad (49.61)$$

and

$$S(t) = (1/\hbar)\int_0^t \left[\Delta H^2(t) + 4H_{12}^2(t)\right]^{1/2}\mathrm{d}t . \quad (49.62)$$

Then the various approximate formulae, as suggested by different authors [49.2], read

$$\mathcal{P}_{12} \cong (w_0/v)^2\sin^2 v , \quad (49.63)$$

$$\mathcal{P}_{12} \cong \sin^2 w , \quad (49.64)$$

$$\mathcal{P}_{12} \cong \frac{w_0^2}{u_0^2 + w_0^2}\sin^2\sqrt{(u_0^2 + w_0^2)} , \quad (49.65)$$

$$\mathcal{P}_{12} \cong \left|\int_{-\infty}^{+\infty} H_{12}(t)\exp[\mathrm{i}S(t)]\,\mathrm{d}t/\hbar\right|^2 , \quad (49.66)$$

$$\mathcal{P}_{12} \cong \frac{\sin^2 S_{\mathrm{c}}'}{\cosh^2 S_{\mathrm{c}}''} . \quad (49.67)$$

In (49.67), S_{c}' and S_{c}'' are the real and imaginary parts of the complex quantity $S_{\mathrm{c}} = S(t_{\mathrm{c}})$ from (49.62). The complex-valued time t_{c} is found from

$$\Delta H^2(t_{\mathrm{c}}) + 4H_{12}^2(t_{\mathrm{c}}) = 0 , \quad (49.68)$$

under the condition that t_{c} possesses the smallest imaginary part of all roots of this equation.

49.4.3 Integral Cross Sections for a Double-Passage Transition Probability

The quasiclassical inelastic integral cross section σ_{if} for the transition $i \to f$ is related to P_{if} by

$$\sigma_{if} = \frac{\pi}{\mu E_i} \int_0^\infty P_{if} \ell \, \mathrm{d}\ell \,, \tag{49.69}$$

where E_i is the initial collision energy, $E_i = E - U_i(\infty)$. The cross section defined by (49.69) typically shows the following qualitative dependence on the collision velocity: σ_{if} increases rapidly with E_i at low energies, reaches a maximum and then slowly falls off at high energies. The position of the maximum roughly corresponds to the energy $E_i = E_i^*$ at which the relevant Massey parameter at the NAR center, $\zeta(R_\mathrm{p})$, is of the order of unity. The conditions $E_i < E_i^*$ and $E_i > E_i^*$ correspond to the near-adiabatic and strongly nonadiabatic (also called diabatic) regimes, respectively.

In calculating σ_{if}, one usually neglects the Stückelberg oscillating term and sets the upper limit in the integral in (49.69) to a value $\ell = \ell_m$ beyond which the integrand begins to fall off quickly. Yet another simplification is possible, in the framework of the impact parameter approximation, when the relative motion of atoms is described by a rectilinear trajectory $R(t)$ with constant velocity v and impact parameter $b = \hbar\ell/\mu v$:

$$R(t) = \left(b^2 + v^2 t^2\right)^{1/2} . \tag{49.70}$$

For instance, for the Landau–Zener model $\ell_m = \mu v R_\mathrm{p}/\hbar$, and the cross section depends on one dimensionless parameter $\gamma = 2\pi V^2/(\Delta F \hbar v)$ according to

$$\sigma_{12}(\gamma) = 2\pi R_\mathrm{p}^2 \int_0^1 \exp(-\gamma/\sqrt{x}\,)\left[1 - \exp(-\gamma/\sqrt{x}\,)\right] \mathrm{d}x$$
$$= 4\pi R_\mathrm{p}^2 \left[E_3(\gamma) - E_3(2\gamma)\right] , \tag{49.71}$$

where $E_3(z)$ is the exponential integral.

For the symmetric resonance, the cross section reads

$$\sigma_{if}(v) = 2\pi \int_0^\infty \sin^2\left(\frac{\Delta\Phi_{if}^{\mathrm{Res}}(b, v)}{2}\right) b\,\mathrm{d}b \approx \frac{\pi}{2} b_m^2(v) \,, \tag{49.72}$$

where b_m is found from the Firsov criterion [49.2]:

$$\frac{\Delta\Phi_{if}^{\mathrm{Res}}(b_m, v)}{2} = \int_{b^*}^\infty \frac{\left[U_i(R) - U_f(R)\right]}{\hbar v \sqrt{R^2 - b_m^2}} \mathrm{d}R = \frac{2}{\pi} \,. \tag{49.73}$$

The cross section in (49.71) first increases and then decreases with the collision velocity v, while that in (49.72) slowly decreases with v.

49.5 Multiple-Passage Transition Probabilities

49.5.1 Multiple Passage in Atomic Collisions

In the case of atomic collisions, there is only one nuclear coordinate R. If there exist several NAR on the R-axis, those which are classically accessible (for given total energy E and total angular momentum J) can be traversed several times. In the semiclassical approximation [49.16], the multiple-passage transition amplitude A_{if} for a given transition between inital state i and final state f can be calculated as a sum of transition amplitudes $A_{if}^{\mathcal{L}}$, over all possible classical ways $\mathcal{L}$ which connect these states, and which run along a one-dimensional manifold R:

$$A_{if} = \sum_{\mathcal{L}} A_{if}^{\mathcal{L}} \,, \tag{49.74}$$

where each $A_{if}^{\mathcal{L}}$ can be expressed through the probability $\mathcal{P}_{if}^{\mathcal{L}}$ and the phase $\Phi_{if}^{\mathcal{L}}$ by [49.13]

$$A_{if}^{\mathcal{L}} = \left[\mathcal{P}_{if}^{\mathcal{L}}\right]^{1/2} \exp\left(\mathrm{i}\Phi_{if}^{\mathcal{L}}\right) . \tag{49.75}$$

The net transition probability is then

$$\mathcal{P}_{if} = |A_{if}|^2$$
$$= \sum_{\mathcal{L}} \mathcal{P}_{if}^{\mathcal{L}} + \sum_{\mathcal{L},\mathcal{L}'}{}' \left[\mathcal{P}_{if}^{\mathcal{L}} \mathcal{P}_{if}^{\mathcal{L}'}\right]^{1/2} \cos\left(\Phi_{if}^{\mathcal{L}} - \Phi_{if}^{\mathcal{L}'}\right) . \tag{49.76}$$

The first sum runs over all different paths, and the second (primed) over all different pairs of paths. The primed sum usually yields a contribution to the transition probability which oscillates rapidly with a change of the parameters entering into $\mathcal{P}_{if}$ (i. e., E and J) and represents a multiple-passage counterpart to the Stückelberg oscillations.

If the Stückelberg oscillations are neglected, $\mathcal{P}_{if}$ is equivalent to a mean transition probability $\langle\mathcal{P}_{if}\rangle$:

$$\langle\mathcal{P}_{if}\rangle = \sum_{\mathcal{L}} \mathcal{P}_{if}^{\mathcal{L}} \,. \tag{49.77}$$

For one NAR, there are two equivalent paths, and $\mathcal{P}_{if}^{(1)} = \mathcal{P}^{(2)}{}_{if} = P_{if}(1 - P_{if})$. Then (49.77) yields (49.56).

49.5.2 Multiple Passage in Molecular Collisions

For molecular collsions, (49.74) and (49.75) apply as well. However, the manifold of $\mathcal{R}$ to which a trajectory $\mathcal{Q}(t)$ belongs now comprises $3N-5$ (for a linear arrangement of nuclei) or $3N-6$ (for a nonlinear arrangement) degrees of freedom, where N is the number of atoms in the system. The approximation (49.77) is called, in the context of inelastic molecular collisions, the surface-hopping approximation [49.17, 18]. Each time a trajectory reaches an NAR, it bifurcates, and the system makes a hop from one PES to another with a certain probability. Keeping track of all the bifurcations and associated probabilities, one calculates $\mathcal{P}_{if}^{\mathcal{L}}$ along a path $\mathcal{L}$ made up of different portions of trajectories running across different PES. Because of the complicated sequence of nonadiabatic events leading from the initial state to the final state, each $\mathcal{P}_{if}^{\mathcal{L}}$ is a complicated function of different single passage transition probabilities P_{nm}, and survival probabilities $1-P_{nm}$. Even if all P_{nm} are known in analytical form, the calculation of $\langle \mathcal{P}_{if} \rangle$ requires numerical computations to keep track of individual nonadiabatic events [49.17].

The manifold $\mathcal{R}$ can be reduced in size if one treats other degrees of freedom, besides electronic ones, on the same footing. In this way one introduces adiabatic vibronic (vibrational + electronic) states and adiabatic vibronic PES, and considers nonadiabatic transitions between them [49.19]. In the vibronic representation, the formal theory remains the same; however its implementation is more difficult since there are many more possibilities for trajectory branching. Finally, under certain conditions, one can use a fully adiabatic description of all degrees of freedom save one – the intermolecular distance R. This approach provides a basis for the statistical adiabatic channel model (SACM) of unimolecular reactions [49.20] where the receding fragments are scattered adiabatically in the exit channels after leaving the region of a statistical complex.

For a latest review of the theory of molecular nonadiabatic dynamics, see [49.21] and papers in [49.22].

References

49.1 R. B. Bernstein (Ed.): *Atom-Molecule Collision Theory: A Guide for the Experimentalist* (Plenum, New York 1979)

49.2 E. E. Nikitin, S. Ya. Unamskii: *Theory of Slow Atomic Collisions* (Springer, Berlin, Heidelberg 1984)

49.3 L. D. Landau, E. M. Lifshitz: *Quantum Mechanics* (Pergamon, Oxford 1977)

49.4 L. D. Landau: Phys. Z. Sowjetunion **1**, 88 (1932)

49.5 L. D. Landau: Phys. Z. Sowjetunion **2**, 46 (1932)

49.6 C. Zener: Proc. Roy. Soc. **137**, 396 (1932)

49.7 E. E. Nikitin: Discuss. Faraday Soc. **33**, 14 (1962)

49.8 N. Rosen, C. Zener: Phys. Rev. **40**, 502 (1932)

49.9 Yu. N. Demkov: Sov. Phys. JETP **18**, 138 (1964)

49.10 E. I. Dashevskaya, E. E. Nikitin: Quasiclassical approximation in the theory of scattering of polarized atoms. In: *Atomic Physics Methods in Modern Research*, Lecture Notes in Physics, Vol. 499, ed. by K. Jungmann, J. Kowalski, I. Reinhard, F. Träger (Springer, Berlin, Heidelberg 1997) p. 185

49.11 H. Nakamura: *Nonadiabatic Transition: Concepts, Basic Theories and Applications* (World Scientific, Singapore 2002)

49.12 M. S. Child: *Semiclassical Mechanics with Molecular Applications* (Clarendon, Oxford 1994)

49.13 S. F. C. O'Rourke, B. S. Nesbitt, D. S. F. Crothers: Adv. Chem. Phys. **103**, 217 (1998)

49.14 E. S. Medvedev, V. I. Osherov: *Radiationless Transitions in Polyatomic Molecules* (Springer, Berlin, Heidelberg 1994)

49.15 E. C. G. Stückelberg: Helv. Phys. Acta **5**, 369 (1932)

49.16 W. H. Miller: Adv. Chem. Phys. **30**, 77 (1975)

49.17 S. Chapman: Adv. Chem. Phys. **82**, 423 (1992)

49.18 J. C. Tully: Nonadiabatic dynamics. In: *Modern Methods for Multidimensional Dynamics Computations in Chemistry*, ed. by D. L. Thompson (World Scientific, Singapore 1998) p. 34

49.19 V. Sidis: Adv. At. Opt. Phys. **26**, 161 (1990)

49.20 M. Quack, J. Troe: Statistical adiabatic channel models. In: *Encyclopedia of Computational Chemistry*, Vol. 4, ed. by P. v. R. Schleyer, N. L. Allinger, T. Clark, J. Gasteiger, P. A. Kollman, H. F. Schaefer III, P. R. Schreiner (Wiley, Chichester 1998) p. 2708

49.21 A. W. Jasper, B. K. Kendrick, C. A. Mead, D. G. Truhlar: Non-Born–Oppenheimer chemistry: Potential surfaces, couplings, and dynamics. In: *Modern Trends in Chemical Reaction Dynamics: Experiment and Theory (Part I)*, ed. by X. Yang, K. Lui (World Scientific, Singapore 2004) p. 329

49.22 A. Lagana, G. Lendvay (Eds.): *Theory of Chemical Reaction Dynamics* (Kluwer, Dordrecht 2004)

50. Ion–Atom and Atom–Atom Collisions

This chapter summarizes the principal features of theoretical treatments of ion–atom and atom–atom collisions. This is a broad topic and the goal here is a general overview that introduces the main concepts, terminology, and methods in the field. Attention will focus on intermediate and high collision velocities, for which the relative velocity between the projectile and target is on the order of, or larger than, the orbital speed of the electrons active in the transition.

50.1 **Treatment of Heavy Particle Motion** 754

50.2 **Independent-Particle Models Versus Many-Electron Treatments** 755

50.3 **Analytical Approximations Versus Numerical Calculations** 756

50.3.1 Single-Centered Expansion 757

50.3.2 Two-Centered Expansion 758

50.3.3 One-and-a-Half Centered Expansion 758

50.4 **Description of the Ionization Continuum** 758

References 759

Low energy collisions are treated in Chapt. 49. Charge transfer reactions are treated in Chapt. 51. We will emphasize therefore excitation and ionization transitions, and discuss charge transfer only as it is interrelated with these transitions. The description of these heavy particle collisions has many features in common with electron and positron collisions with atoms and ions (Chapts. 47 and 48) and this chapter should be studied in parallel with those. There are also chapters dealing with special phenomena in ion–atom collisions: excitation at high collision energies and the Thomas peak (Chapt. 57), electron emission in high energy ion–atom collisions (Chapt. 53), and alignment and orientation (Chapt. 46). Other chapters deal with certain specific theoretical methods: continuum distorted wave (CDW) approximations (Chapt. 52), the binary encounter approximation (Chapt. 56), and classical trajectory Monte Carlo (CTMC) techniques (Chapt. 58). The emphasis of the present chapter is coupled-states calculations of excitation and ionization. There are several review articles and monographs [50.1, 2].

The collisions considered here involve a projectile ion or atom and a target atom. The collision kinematics can be described in the lab frame, where the target atom is assumed to be initially at rest and the collision energy is the kinetic energy of the projectile when it is far from the target prior to the collision, or in the center of mass frame. The primary quantities of interest are the cross sections for producing various final states of the system for given initial states of the target and projectile. The total cross sections depend on the initial and final quantum state of the target and projectile and on the collision energy. Let A denote the projectile atom or ion with ionic charge q, and B denote the target atom. Let A^* and B^* denote excited states. Some examples of processes for which the cross section is of interest are

$$A^{q+} + B \to \begin{cases} (A^{q+})^* + B\,, & \text{projectile excitation} \\ A^{q+} + B^*\,, & \text{target excitation} \\ A^{q+} + B^+ + e^-\,, & \text{target single ionization} \\ A^{(q+1)+} + B + e^-\,, & \text{projectile single ionization} \\ A^{(q-1)+} + B^+, & \text{single e}^- \text{ charge transfer}\,. \end{cases}$$

For a multi-electron collision system, combinations of the above quantities are possible. A few representative examples are

$$A^{q+} + B \to \begin{cases} A^{q+} + B^{++} + 2e^-\,, & \text{target double ionization} \\ A^{q+} + (B^+)^* + e^-\,, & \text{target excitation-ionization} \\ A^{(q-1)+} + (B^+)^*\,, & \text{transfer-excitation} \\ A^{(q-1)+} + B^{++} + e^-\,, & \text{transfer-ionization}\,. \end{cases}$$

For each of these processes, the projectile and/or target can be initially in excited states. A class of ion–atom collisions that has received much theoretical attention because of the relative simplicity is the one where the projectile ion is initially bare, with charge $q = Z_P$, the projectile nuclear charge. One can also consider various differential cross sections: differential in the projectile scattering angle, in the energy and angle of emission of ionized electrons, in recoil momentum of the target, etc.

Theoretical calculations of these cross sections can be classified according to the approximations and/or methods used. We will discuss four such classifications: treatment of the heavy particle motion, independent particle model (IPM) versus inclusion of correlation in multi-electron systems, analytical approximations (PWBA, SCA, Glauber, etc.) versus numerical methods (such as coupled-states) for obtaining cross sections, and treatments of the ionization continuum.

50.1 Treatment of Heavy Particle Motion

Only at very low energies must the motion of the nuclei be described quantum mechanically. At the intermediate and high collision velocities considered here the semiclassical approximation, in which the motion of the nuclei can be described classically and only the electrons need be described by quantum mechanical wave functions, is accurate [50.3]. The projectile nucleus then moves on a predetermined classical path. At high collision energies this path can often be taken to be a straight line path with constant speed. At somewhat lower energies the deflection and change in speed of the projectile due to the projectile–target interaction is often incorporated [50.4–6]. The Coulomb trajectory due to the nucleus–nucleus interaction can be used, and the screening effects of the projectile and target electrons can be included. For a bare positive ion projectile, the Coulomb trajectory effects increase the distance of closest approach for a given impact parameter and reduce the projectile speed in the interaction region. Projectile trajectory effects are particularly strong for projectiles less massive than protons (positive or negative muons [50.7], for example). The semiclassical approach with trajectory effects has even been used for electron impact excitation and ionization [50.7]. Coulomb trajectories are also important when small impact parameter collisions are considered. The recoil of the target nucleus can also be treated classically and this recoil can affect the cross sections [50.8–10].

In the semiclassical approximation, the vector $\boldsymbol{R}(b, t)$ that locates the projectile relative to the target nucleus is a function of the impact parameter b and time t. The specific functional dependence is determined by the trajectory being used. For a straight line, constant velocity $\boldsymbol{v}$ path, $\boldsymbol{R} = \boldsymbol{b} + \boldsymbol{v}t$. For the case of a bare projectile ion and a one-electron target, the time-dependent projectile–target interaction is given by

$$V(b, t) = \frac{-Z_P e^2}{|\boldsymbol{r} - \boldsymbol{R}(b, t)|} + \frac{Z_P Z_T e^2}{R(b, t)}, \tag{50.1}$$

where Z_P and Z_T are the projectile and target nuclear charges and $\boldsymbol{r}$ is the position vector of the electron relative to the target nucleus. If the collision calculation starting from (50.1) is done exactly the nuclear repulsion term $Z_P Z_T e^2/R(b, t)$ does not involve the electronic coordinates and makes no contribution to total cross sections for anything other than elastic scattering. This term makes a nonzero contribution in an approximate calculation, such as a first-order perturbation theory calculation with nonorthogonal initial and final states, and this is a defect of such calculations.

A deficiency of the semiclassical approximation as described above is that, since a predetermined classical path is used, there is no coupling between the energy and momentum given to the target electron and that lost by the projectile. A simple improvement, in cases where the energy lost by the projectile when the target transition occurs is an appreciable fraction of its total energy, is to use some average of the projectile's initial and final speeds as the asymptotic projectile speed. This results in a projectile trajectory that depends on the cross section being calculated. This lack of coupling between the projectile motion and the states of the electrons can be a particular deficiency when cross sections differential in the scattering angle Θ of the projectile are computed. If the collision energy is large and the projectile is scattered primarily by the static potential of the target, a classical treatment of the scattering can be used to relate b to Θ. Even when straight line, constant speed projectile paths are used to calculate the transition probabilities, differential cross sections can be extracted by relating b to Θ. At lower energies, where the de Broglie wavelength of the projectile is not small compared with the range of the interaction, eikonal methods can be used to convert impact parameter dependent probabilities to differential cross sections [50.11–13]. A particularly striking example where there is strong coupling between the projectile motion and the final quantum state of the electron is proton impact single and double ionization of helium differential in the projectile scattering angle [50.14, 15]. The projectile can be deflected by interaction with the target nucleus. In this case the projectile scattering angle can be related to the impact parameter by the equations of classical mechanics, and the projectile can be scattered by up to 180° as b is decreased to zero. But the projectile can also be scattered by a close interaction with a target electron, and the kinematics then can be quite different. Such an interaction has a low probability for scattering the projectile to a large angle. Classically, from energy and momentum considerations, the maximum angle through which a proton can be scattered by an electron initially at rest is about 0.5 mrads; and if the proton is scattered through 0.5 mrads, the electron acquires large energy and momentum. The combined effects of both projectile scattering mechanisms are difficult to treat in a semiclassical model [50.16].

50.2 Independent-Particle Models Versus Many-Electron Treatments

In the semiclassical impact parameter method, as described in Sect. 50.1, for a single electron collision system one has to calculate a single particle wave function and from it single-electron transition amplitudes. A multi-electron collision problem can be treated as an effective single-electron problem if only one projectile–electron interaction is considered and the other electrons merely provide an effective single-particle potential in which the active electron moves.

In an independent particle model (IPM), the electron–electron interactions are replaced by effective single-electron potentials. Since the projectile–target interaction is a sum of single-electron interactions, the many-electron collision problem reduces to an uncoupled set of single electron problems. Their solution at each impact parameter gives single electron transition amplitudes a_{ij} and transition probabilities $\rho_{ij} = |a_{ij}|^2$. Cross sections for multi-electron transitions can still be calculated by combining the single-electron amplitudes in an appropriate way [50.17, 18]. In doing this, it is important to distinguish between inclusive and exclusive processes. For an inclusive cross section, one final orbital occupancy, or a few final orbital occupancies, is specified, but the final states of the remaining electrons are not specified and all possibilities are summed over. For a totally exclusive process, all final orbital occupancies are specified. For example, consider a $p+\mathrm{Ne}$ collision. For the inclusive cross section for K-shell vacancy production, at least one K-shell vacancy in the final state is specified. The other electrons could remain in their original orbitals, there could be two K-shell vacancies, the K-shell vacancy could be accompanied by any number of L-shell vacancies, etc., and all these possible final states are summed over. An example of an exclusive cross section is the cross section for producing a single K-shell vacancy with the specification that the final state of the target have no additional vacancies.

In calculating inclusive cross sections from single-electron transition amplitudes in an IPM it is important to take proper account of time-ordering and Pauli exclusion effects, as well as all multi-electron processes that lead to the specified final state. Again using the $p+\mathrm{Ne}$ collision example, a two-electron process that leads to a K-shell vacancy is for the projectile to first ionize an L-shell electron, and then in the same collision to excite a K-shell electron into the L-shell hole just produced. But the K-shell electron must have the same spin component as the L-shell electron that was removed, and the K to L excitation can occur only after the L-hole has been made. This can lead to correlations among the single-electron amplitudes that have been called Pauli correlations or Pauli blocking effects.

For a totally inclusive process where only one final occupancy is specified, all the multi-electron contributions cancel in any IPM and the probability for producing, for example, a hole in the initially occupied orbital labeled 1, without any other specification of final state orbital occupancies, is given by

$$\rho_1 = \sum_{k<N} |a_{k1}|^2 , \tag{50.2}$$

where k runs over the unoccupied bound orbitals of the target, the bound orbitals of the projectile, and the continuum orbitals. This expression says that the inclusive probability of producing a hole in the initially occupied orbital 1 is given by the sum of the probabilities of exciting an electron from this orbital to any initially unoccupied orbitals of the projectile–target system. The cross section is then exactly the same as if only a single target electron were interacting with the projectile. However, for an inclusive process in which two vacancies in orbitals of the same spin are specified in the final state, say orbitals 1 and 2, then the probability of producing this final state is not the product of the independent probabilities of producing a vacancy in orbital 1 and in orbital 2 [50.17, 18]. The correlation produced by antisymmetry destroys the independence of probabilities for this process. Specific expressions, and also a discussion of how to derive the correct expressions to use for any particular inclusive process, can be found in the literature [50.17–19]. If these Pauli correlations are neglected, and products of independent single-electron transition probabilities are used, then the distribution of multiple vacancies within a shell, such as the L-shell, are given in terms of a binomial distribution [50.20–22].

The IPM treatment is inadequate if electron correlation plays an important role in the collision. One notable set of examples is two-electron transitions in which the projectile–target interaction is weak, such as at high collision energies. It is useful to characterize multi-electron transitions as being either externally or internally induced [50.23]. In an externally induced two-electron process, the projectile interacts with both electrons and induces the transition of each. In the internal process, energy is transferred to just one electron; this energy is then shared with the other electron through the internal action of the correlating electron–electron force.

Consider double ionization. At sufficiently high energies, the external reaction has an amplitude proportional to Z_P^2. To lowest order, the reaction proceeds through two consecutive first-order Born collisions, perhaps more correctly described as a two-particle second-order Born collision. This is called a two step mechanism, TS2. The internal reaction amplitude, again at sufficiently high energies, is proportional to Z_P; one first-order Born reaction initiates the process. The sharing of energy with the second electron can be thought of in two ways. If the first electron leaves without interacting with the second, there is a change in the screening that can be thought of as being responsible for shaking the second electron off. The main role of the electron–electron force here is to introduce correlation into the initial ground state wave function. A second way for both electrons to be ionized is for the first electron to strike the second on the way out. This has been variously called interception or TS1 [50.21, 24–26]. The role of the electron–electron force here is to introduce correlation into the final two-electron wave function. At asymptotic energies the first Born term will eventually dominate, and the internal correlated mechanism is solely responsible for double ionization. At lower energies, the external and internal amplitudes are of the same order, and thus they interfere. The interference produces terms in the transition probability that are proportional to odd powers of Z_P, and therefore produces a cross section that changes when the sign of the projectile charge changes. One example is the double ionization of helium in the MeV/amu collision energy region, where antiprotons ($Z_\mathrm{P} = -1$) produce a double ionization cross section a factor of two larger than that for protons ($Z_\mathrm{P} = +1$) around 1 MeV/amu [50.25,26]. It is possible to account for electron correlation directly by putting the electron-electron interaction into the Hamiltonian and solving for the many-electron correlated wave function of the collision system. In coupled-states calculations (Sect. 50.3), this approach is computationally intractable if more than a few coupled channels are important, which is usually the case at intermediate collision energies; but some full electron calculations have been carried out [50.27–31]. In low-energy collisions where molecular basis expansions are appropriate, only a few channels can be important, i. e., those involved in potential curve crossings or near crossings, and full electron calculations are more feasible. Some alternatives to full electron quantum calculations have been developed, such as the independent event model [50.32], classical dynamics calculations (Chapt. 58), time-dependent Hartree–Fock (TDHF) [50.33, 34], and the forced impulse method (FIM) [50.25, 26]. The FIM divides the collision time into short sequential segments such that an impulse approximation is forced to be valid. The system is allowed to collapse back into a fully correlated eigenstate at the end of each segment, but the electrons propagate independently during each segment.

50.3 Analytical Approximations Versus Numerical Calculations

There is a wide variety of techniques used for computing cross sections for ion–atom and atom–atom collisions. One class uses approximate analytical methods. The expressions and the wave functions that enter into them may be sufficiently complicated that they must be evaluated on a computer, but they can be written down as closed form expressions. One important example is the first Born approximation, in which the projectile–target interaction is treated in first-order. Hence, the first Born approximation is most accurate for weak interactions, for example for small projectile charge and large collision energy. The first Born approximation is commonly expressed in either of two forms: the plane wave Born approximation (PWBA) [50.35], where the projectile nuclear motion wave function is taken to be a plane wave, or the semiclassical approximation (SCA) [50.3] which is an impact parameter approach that uses a classical path for the projectile motion. At the high collision velocities where the first Born approximation is accurate, and for projectiles that are at least as massive as a proton, the PWBA and SCA give identical results. Since it is based on first-order perturbation theory, the first Born approximation only requires computation of the matrix element of the projectile interaction between initial and final state wave functions. For example, the first Born transition amplitude for a bare ion (charge Z_P, mass M) incident on a one electron target is

$$f_{ji} = \frac{2MZ_Pe^2}{\hbar^2 K^2} \int \phi_j^*(\boldsymbol{r})\,\mathrm{e}^{\mathrm{i}\boldsymbol{K}\cdot\boldsymbol{r}}\phi_i(\boldsymbol{r})\,\mathrm{d}^3\boldsymbol{r}\,, \tag{50.3}$$

where $\hbar\boldsymbol{K}$ is the momentum transfer vector and ϕ_i, ϕ_j are the initial and final state wave functions. For a hydrogenic system, analytical expressions for the PWBA amplitudes for bound–bound transitions can be derived [50.36] and for bound–continuum transitions (ionization) the cross section can be expressed in terms of a two-dimensional integral over energy and momentum transfer, where the integrand is given by a simple analytical expression [50.37]. First Born calculations are also carried out with relativistic wave functions; relativistic target wave function effects are important for the inner shells of heavy atoms [50.38].

There is a number of other approximate methods that lead to simple analytic expressions, such as the Glauber approximation [50.39–41], continuum distorted wave (CDW) and eikonal methods (Chapt. 52), the binary encounter approximation (BEA, Chapt. 56), and second Born calculations [50.42]. A difficulty with all these methods is that they do not admit to a sequence of successive improvements; there is no procedure for systematically driving them to convergence.

One class of approximation methods very widely used for inner shell processes, such as inner shell vacancy production, is based on perturbed stationary state (PSS) methods. One variation is the energy loss and Coulomb deflection effects perturbed stationary state approximation with relativistic correction (ECPSSR) theory, developed originally by *Brandt* and co-workers and since extended and applied by others [50.43, 44]. This approximation involves procedures for correcting the first Born approximation. It is quite successful in fitting K-, L-, and M-shell vacancy production cross sections.

Another approach is to calculate the electronic wave function for the collision system by expanding it in some basis set [50.45, 46]. This procedure is called the coupled-channels, coupled-states, or close-coupling method. It is convenient to diagonalize the target and projectile systems in the chosen basis to produce matrix eigenfunctions for the target and projectile. Putting this expansion into the time-dependent Schrodinger equation leads to a set of coupled first-order equations for the time-dependent expansion coefficients. This set of equations is solved subject to the boundary condition that long before the collision, the expansion coefficient for the initial state wave function is unity and all other coefficients are zero. Then the expansion coefficients a long time after the collision give the transition amplitudes for transitions into those states.

Several types of expansions are used. The notation appropriate to a single-electron problem will be used for simplicity.

50.3.1 Single-Centered Expansion

In the single-centered expansion [50.47, 48], the wave function is expanded in a set of target centered basis functions. Diagonalization of the target Hamiltonian in this basis yields a set of target eigenfunctions $\phi_j^{\mathrm{T}}(\boldsymbol{r})$ and energies E_j. The expansion of the time-dependent wave function for the collision system is then

$$\psi_i(\boldsymbol{r}, b, t) = \sum_j a_{ji}(b, t)\,\mathrm{e}^{-\mathrm{i}(E_j - E_i)t/\hbar}\phi_j^{\mathrm{T}}(\boldsymbol{r})\,. \tag{50.4}$$

If the electron is initially in the target state i, the initial boundary condition is $a_{ji}(b, -\infty) = \delta_{ji}$. The expansion coefficients at times long after the collision, $a_{ji}(b, +\infty)$, give the transition amplitudes for transitions from target state i to j. Ionization probabilities and the description of the ionization continuum will be described in the next section. It will suffice to say here that some of the $\phi_j(\boldsymbol{r})$ represent ionized states and account for transitions that leave the electron no longer in a target bound state. This single centered expansion does not directly allow for calculation of charge transfer amplitudes. However, it has been shown that if the number of angular momenta included in the set of $\phi_j(\boldsymbol{r})$ is large enough, the basis can accurately describe electron loss from the target bound states even in cases where the charge transfer probability is large [50.47, 48]. If the charge transfer amplitude is small, it can be extracted from $\psi_i(\boldsymbol{r}, b, t)$ by use of a first-order T-matrix expression of the form

$$a^{\mathrm{P}}_{ki}(b, +\infty) = \left\langle \phi^{\mathrm{P}}_k(\boldsymbol{r} - \boldsymbol{R}) \middle| U(\boldsymbol{r}, \boldsymbol{R}) \middle| \psi_i(\boldsymbol{r}, b, t) \right\rangle , \tag{50.5}$$

where ϕ^{P}_k is a projectile state eigenfunction and $U(\boldsymbol{r}, \boldsymbol{R})$ is the projectile–target interaction [50.49, 50].

50.3.2 Two-Centered Expansion

In a two-centered expansion [50.45,46,51–54], the time-dependent wave function for the collision system is expanded in two sets of functions: a set of eigenfunctions centered on the target $\phi^{\mathrm{T}}_j(\boldsymbol{r})$ and a set of projectile centered eigenfunctions $\phi^{\mathrm{P}}_k(\boldsymbol{r} - \boldsymbol{R})$,

$$\begin{aligned}\psi_i(\boldsymbol{r}, b, t) = \sum_j a^{\mathrm{T}}_{ji}(b, t) & \\ \times \exp\left[-\mathrm{i}(E_j - E_i)t/\hbar\right] \phi^{\mathrm{T}}_j(\boldsymbol{r}) & \\ + \sum_k a^{\mathrm{P}}_{ki}(b, t) \exp\left[-\mathrm{i}(E_k - E_i)t/\hbar\right] & \\ \times \exp\left[\mathrm{i}\boldsymbol{p} \cdot \boldsymbol{r}/\hbar\right] \phi^{\mathrm{P}}_k(\boldsymbol{r} - \boldsymbol{R}) . &\end{aligned} \tag{50.6}$$

Here $\exp[\mathrm{i}\boldsymbol{p} \cdot \boldsymbol{r}/\hbar]$ is an electron translation factor that accounts for the momentum $\boldsymbol{p}$ that the electron has relative to the target when it moves with the projectile. The energies E_k also contain the kinetic energy in the target frame that the electron has when it moves with the projectile. The problem of electron translation factors when a two centered molecular basis is used instead of the atomic basis has been extensively addressed [50.45, 46].

The coefficients $a^{\mathrm{T}}_{ji}(b, +\infty)$ and $a^{\mathrm{P}}_{ki}(b, +\infty)$ are the transition amplitudes for transitions into target and projectile states. The two-centered expansion allows for charge transfer amplitudes to be calculated directly and is capable of an accurate description of the electron flux loss from the target due to charge transfer, but the computational difficulty is significantly greater. It is difficult in practice to include enough basis functions for convergence. Also, computational linear dependence can arise if the projectile and target are close because the projectile-centred basis set can become nearly identical to the target-centred basis set.

Calculations have also been done with a triple-centered expansion [50.55, 56]. Bound atomic states are centered on each nucleus and on a third center (the center of charge) in order to simulate the molecular character of slow collisions. The orbitals on this third center represent the united-atom character of the wave function.

50.3.3 One-and-a-Half Centered Expansion

To account for flux loss from the target region due to charge transfer, while retaining the computational efficiency of the single-centered expansion, a hybrid method called the one-and-a-half centered expansion (OHCE) was developed [50.57]. It is of the same form as the two-centered expansion except that the coefficients $a^{\mathrm{P}}_{ki}(b, t)$ of the projectile-centered states have predetermined rather than variable time dependence. The time dependence of the $a^{\mathrm{P}}_{ki}(b, t)$ must be such that the a^{P}_{ki} are all zero as $t \to -\infty$, and as $t \to +\infty$ they become constants that are identified as the charge transfer amplitudes. This ansatz greatly reduces the computational cost of propagating the coupled equations forward in time, but the projectile-centered functions are still present to allow explicitly for charge transfer channels at large times. At intermediate times, when the projectile and target separation is within the range of the target-centered basis, the target-centered part of the expansion, if enough basis functions are used, is sufficient to describe the time dependence of $\psi_i(\boldsymbol{r}, b, t)$ and the projectile-centered part of the expansion is redundant. The OHCE has been applied to several collision systems with good success [50.58].

50.4 Description of the Ionization Continuum

In ion–atom and atom–atom collision calculations description of the ionization continuum presents particular difficulties. For hydrogenic systems, the exact continuum wave functions are known analytically, and these may be used in analytic approximations such as the PWBA. For nonhydrogenic systems, screened hydrogenic wave functions or numerical wave functions computed from some local effective potential, such as the Hartree–Fock–Slater approximation, can be used [50.59]. But for coupled states calculations, that involve expansion of the system wave function in terms of a discrete set of functions, some discretization of the continuum must be performed.

A simple procedure for generating a discrete representation of the continuum is to diagonalize the target (or projectile) Hamiltonian in a finite basis of square integrable basis functions. Some commonly used basis functions are Sturmian [50.45, 46], Gaussian, or Slater-type orbitals. This diagonalization produces a set of discrete matrix eigenvalues and eigenvectors, called pseudostates. For a one-electron or effective one-electron system, the exact energy spectrum of the Hamiltonian consists of an infinite set of discrete bound states and an ionization continuum and the interpretation of the pseudostates is straightforward. The N matrix eigenvalues below the first ionization threshold represent the bound states. By the Hylleraas–Undheim Theorem, they are upper bounds to the first N exact energies (Sect. 11.2.1). The matrix eigenvectors with positive eigenvalues provide a discrete representation of the ionization continuum [50.60]. Each pseudostate is accurately proportional to the exact continuum wave function at the pseudostate energy for values of r out to the range of the basis. The proportionality constant gives the energy width of the pseudostate – the energy region of the continuum represented by that pseudostate [50.49,50]. These widths are approximately equal to the energy spacing between pseudostates. The pseudostates represent the continuum in the sense that the total ionization cross section is equal to the sum of the continuum pseudostate cross sections. Coupled-states calculations have also been carried out with wave packets constructed at discrete energies from the exact continuum wave functions [50.61]. What all these discretization methods do is to provide a quadrature rule for evaluating the integration over the energy of the ionized electron. Differential cross sections, differential in the ionized electron's energy, have also been extracted from discretized continuum calculations [50.62–65].

The situation is more complex for two-electron or multi-electron systems. The energy spectrum of a two-electron atom consists of an infinite set of bound states, an infinite set of overlapping single ionization continua that correspond to different residual ion states, and overlapping the higher energy part of these single ionization continua, a double ionization continuum where both electrons are unbound. The exact wave function for two electrons in the continuum of an ion is not known, but there are a number of approximate forms [50.66–68]. The interpretation of two electron pseudostates is also complicated. For energies below the second single ionization threshold the energy of a pseudostate clearly identifies whether it represents a bound state or an ionized state. But in the energy region where there are overlapping continua there are ambiguities. For example, a pseudostate with energy above the double ionization threshold can represent a doubly ionized state, a singly ionized state of the same total energy, or some admixture of the two. A method has been developed to solve this pseudostate interpretation problem [50.25, 26].

References

50.1 B. H. Bransden: *Atomic Collision Theory* (Benjamin-Cummings, Reading 1983)

50.2 J. S. Briggs, J. H. Macek: Adv. At. Mol. Opt. Phys. **28**, 1 (1991)

50.3 J. M. Hansteen: Phys. Scr. **42**, 299 (1990)

50.4 G. L. Swafford, J. F. Reading, A. L. Ford, E. Fitchard: Phys. Rev. A **16**, 1329 (1977)

50.5 A. Jakob, F. Rösel, D. Trautmann, G. Bauer: Z. Phys. A **309**, 13 (1982)

50.6 W. Fritsch: J. Phys. B **15**, L389 (1982)

50.7 M. H. Martir, A. L. Ford, J. F. Reading, R. L. Becker: J. Phys. B **15**, 1729 (1982)

50.8 P. A. Amudsen: J. Phys. B **11**, 3179 (1978)

50.9 M. Kleber, K. Unterseer: Zeit. Phys. A **292**, 311 (1979)

50.10 F. Rösel, D. Trautmann, G. Bauer: Nucl. Instrum. Methods **192**, 42 (1983)

50.11 B. H. Bransden, C. J. Noble: Phys. Lett. **70A**, 404 (1979)

50.12 J. M. Wadehra, R. Shakeshaft: Phys. Rev. A **26**, 1771 (1982)

50.13 R. Gayet, A. Salin: Nucl. Instrum. Methods **B56/57**, 82 (1991)

50.14 A. Salin: J. Phys. B **24**, 3211 (1991)

50.15 L. Meng, R. E. Olson, R. Dörner, J. Ullrich, H. Schmidt-Böcking: J. Phys. B **26**, 3387 (1991)

50.16 X. Fang, J. F. Reading: Nucl. Instrum. Meth. **B53**, 453 (1991)

50.17 J. Reinhardt, B. Muller, W. Greiner, G. Soff: Phys. Rev. Lett. **43**, 1307 (1979)

50.18 J. F. Reading, A. L. Ford: Phys. Rev. **21**, 124 (1980)

50.19 R. L. Becker, A. L. Ford, J. F. Reading: Phys. Rev. A **29**, 3111 (1984)

50.20 J. H. McGuire, P. Richard: Phys. Rev. A **8**, 1874 (1973)

50.21 J. H. McGuire: Adv. At. Mol. Opt. Phys. **29**, 217 (1991)

50.22 Yu S. Sayasov: J. Phys. B **26**, 1197 (1993)

50.23 J. F. Reading, A. L. Ford: Comments At. Mol. Phys. **23**, 301 (1990)

50.24 J. H. McGuire: Phys. Rev. Lett. **49**, 1153 (1982)

50.25 J. F. Reading, A. L. Ford: J. Phys. B **20**, 3747 (1987)

50.26 T. Bronk, J. F. Reading, A. L. Ford: J. Phys. B **31**, 2477 (1998)
50.27 W. Fritsch, C. D. Lin: Phys. Rev. A **41**, 4776 (1990)
50.28 L. F. Errea, L. Mendez, A. Riera: Phys. Rev. A **27**, 3357 (1983)
50.29 F. Martin, A. Riera, M. Yanez: Phys. Rev. A **34**, 4675 (1986)
50.30 H. A. Slim, B. H. Bransden, D. R. Flower: J. Phys. B **26**, 159 (1993)
50.31 K. Moribayashi, K. Hino, M. Matsuzawa, M. Kimura: Phys. Rev. A **44**, 7234 (1991)
50.32 D. P. Marshall, C. LeSech, D. S. F. Crothers: J. Phys. B **26**, L219 (1993)
50.33 J. D. Garcia: Nucl. Instrum. Methods **A240**, 552 (1985)
50.34 W. Stich, J. J. Lüdde, R. M. Driezler: J. Phys. B **18**, 1195 (1985)
50.35 E. Merzbacher, H. W. Lewis: *Encyclopedia of Physics*, Vol. 34 (Springer, Berlin, Heidelberg 1958) p. 166
50.36 D. R. Bates, G. Griffing: Proc. Phys. Soc. A **66**, 961 (1953)
50.37 O. Benka, A. Kropf: At. Data Nucl. Data Tables **22**, 219 (1978)
50.38 M. H. Chen, B. Crasemann: At. Data Nucl. Data Tables **33**, 217 (1985)
50.39 R. Glauber: *Lectures in Theoretical Physics*, Vol. 2 (Interscience, New York 1958)
50.40 V. Franco, B. K. Thomas: Phys. Rev. A **4**, 945 (1971)
50.41 J. Binstock, J. F. Reading: Phys. Rev. A **11**, 1205 (1975)
50.42 B. H. Bransden, D. P. Dewangen: J. Phys. B **12**, 1371 (1979)
50.43 G. Basbas, W. Brandt, R. H. Ritchie: Phys. Rev. A **7**, 1971 (1973)
50.44 W. Brandt, G. Lapicki: Phys. Rev. A **23**, 1717 (1981)
50.45 W. Fritsch, C. D. Lin: Phys. Rep. **202**, 1 (1991)
50.46 T. G. Winter: Phys. Rev. **56**, 2903 (1997)
50.47 C. O. Reinhold, R. E. Olson, W. Fritsch: Phys. Rev. A **41**, 4837 (1990)
50.48 A. L. Ford, J. F. Reading, K. A. Hall: J. Phys. B **26**, 4537, 4553 (1993)
50.49 J. F. Reading, A. L. Ford, G. L. Swafford, A. Fitchard: Phys. Rev. A **20**, 130 (1979)
50.50 R. L. Becker, A. L. Ford, J. F. Reading: J. Phys. B **13**, 4059 (1980)
50.51 R. Shakeshaft: Phys. Rev. A **18**, 1930 (1978)
50.52 W. Fritsch, C. D. Lin: Phys. Rev. A **27**, 3361 (1983)
50.53 A. M. Ermolaev: J. Phys. B **23**, L45 (1990)
50.54 H. A. Slim: J. Phys. B **26**, L743 (1993)
50.55 M. J. Antal, D. G. M. Anderson, M. B. McElroy: J. Phys. B **8**, 1513 (1975)
50.56 T. G. Winter, C. D. Lin: Phys. Rev. A **29**, 567 (1984)
50.57 J. F. Reading, A. L. Ford, R. L. Becker: J. Phys. B **14**, 1995 (1981)
50.58 J. F. Reading, A. L. Ford, R. L. Becker: J. Phys. B **15**, 625, 3257 (1982)
50.59 B. H. Choi: Phys. Rev. A **11**, 2004 (1975)
50.60 A. L. Ford, E. Fitchard, J. F. Reading: Phys. Rev. A **16**, 133 (1977)
50.61 G. Mehler, W. Greiner, G. Soff: J. Phys. B **20**, 2787 (1987)
50.62 T. Mukoyama, C. D. Lin, W. Fritsch: Phys. Rev. A **32**, 2490 (1985)
50.63 E. Y. Sidky, C. D. Lin: J. Phys. B **31**, 2949 (1998)
50.64 B. Pons: Phys. Rev. Lett. **84**, 4569 (2000)
50.65 J. Fu, M. J. Fitzpatrick, J. F. Reading, R. Garet: J. Phys. B **34**, 15 (2001)
50.66 M. Brauner, J. S. Briggs, H. Klar: J. Phys. B **22**, 2265 (1989)
50.67 J. S. Briggs: Phys. Rev. A **41**, 539 (1990)
50.68 C. Dal Cappello, H. Le Rouzo: Phys. Rev. A **43**, 1395 (1991)

51. Ion–Atom Charge Transfer Reactions at Low Energies

Ion–atom charge exchange reactions contribute significantly to the ionization balance of complex ions in many natural environments (astrophysical photoionized plasmas such as planetary nebulae, nova shells, etc.) [51.1]. For electron temperatures lower than 10^6 K, the abundance of neutral atomic H or He can be sufficient for the charge exchange rate to exceed the radiative (direct or dielectronic) recombination rate for ions with charge $q \geq 2$ [51.2]. In the case of singly charged ions, accidental resonance conditions are required for the charge exchange rate to be large at thermal energies. In this chapter, we shall limit discussion to the case of $q \geq 2$. We may distinguish two types of reaction: type I, for which electron capture takes place with no change in the configuration of the core electrons, and type II, for which electron capture is accompanied by a rearrangement of the core electron configuration.

We shall concentrate on the low energy range (less than a few hundred eV), where an adiabatic representation of the collision complex can successfully describe the collision dynamics [51.3]. Nonadiabatic transitions tend to occur in the vicinity of avoided potential energy crossings. For the transition probability to be appreciable, the energy separation Δ_X at the crossing radius R_X must be neither too large nor too small ($0.1 < \Delta_X < 3$ eV). Since the size of Δ_X is determined by electron exchange, whose effect varies exponentially with internuclear distance, only crossings in the range $3 \leq R_X \leq 15 a_0$ induce nonadiabatic transitions. Because of this constraint, the number of effective curve crossings is quite small and the electron capture process is very state selective [51.3].

51.1 **Molecular Structure Calculations** 762
51.1.1 Ab Initio Methods 762
51.1.2 Model Potential Methods 763
51.1.3 Empirical Estimates 764

51.2 **Dynamics of the Collision** 765

51.3 **Radial and Rotational Coupling Matrix Elements** 766

51.4 **Total Electron Capture Cross Sections** 767

51.5 **Landau–Zener Approximation** 769

51.6 **Differential Cross Sections** 769

51.7 **Orientation Effects** 770

51.8 **New Developments** 772

References 772

In general, electron capture is of type I for $q \geq 3$ and of type II for $q = 2$. In a few cases, type II processes can also occur for $q = 3$.

Although there are a few direct experimental measurements at eV energies [51.4], most measurements have been carried out at collision energies exceeding a few 100 eV [51.5]. Unfortunately, the processes which dominate in the higher energy range become negligible at eV energies (and vice versa). Most estimates of the cross sections in the thermal–eV range are based on theoretical models. Since at thermal energies, orbiting and tunneling effects may be considerable, a quantum mechanical description of the collision is preferable.

The dynamic coupling matrix elements between different molecular states depend on the origin of the electronic coordinates, and the corresponding scattering equations are not invariant with respect to a Galilean transformation. In a semiclassical formulation of the dynamics, this defect can be removed by the introduction of translation factors [51.6–12]. However, in a quantum mechanical formulation, it is more appropriate to introduce reaction coordinates of a more flexible form than those used traditionally in the Born–Oppenheimer approximation. The most convenient is the Eckart coordinate system [51.9], which leads to a simple modification of the nonadiabatic coupling matrix elements, and which is well adapted to cases where long range avoided crossings are dominant [51.10]. If a semiclassical approximation is introduced, the scattering equations become almost identical to those obtained directly using translation factors of the CTF (common translation factor) type [51.11, 12].

Three representative systems are chosen to illustrate the basic features of the problem: Al^{3+}/H and B^{3+}/He which are typical of type I, and O^{2+}/H which is typical of type II. In the Al^{3+}/H system, charge transfer takes place via the reaction [51.13]

$$Al^{3+}(3s^2)^2S + H \rightarrow Al^{2+}(3s^2 3p)^2P + H^+ \quad (51.1)$$

involving a network of two Σ and one Π adiabatic states. In the B^{3+}/He system, charge exchange occurs via a network of three Σ and one Π states involving two avoided crossings [51.14]. The two possible capture channels are

$$B^{3+}(1s^2) + He \rightarrow B^{2+}(1s^2 nl) + He^+ ,$$
$$nl = 2s, 2p \, . \quad (51.2)$$

The O^{2+}/H system is a good example of a type II reaction which plays an important role in astrophysical plasmas [51.15, 16]. The dominant reaction at low energies

$$O^{2+}(2s^2 2p^2)^3P + H \rightarrow O^+(2s2p^4)^4P + H^+ \quad (51.3)$$

involves only quartet states.

51.1 Molecular Structure Calculations

The construction of the network of adiabatic states of the molecular ion complex constitutes the first step in analyzing the dynamics of the ion–atom system. In principle, standard techniques of quantum chemistry can be employed. For doubly charged ions, where type II processes dominate and the effective avoided crossings take place at relatively short range, ab initio methods are required. On the other hand, for trebly and more highly charged ions, where type I processes dominate, the effective avoided crossings occur at long range where ab initio methods can be quite inaccurate. Model potential methods are then often more satisfactory.

51.1.1 Ab Initio Methods

Both molecular orbital and valence-bond methods have been extensively used and (provided a sufficient number of configurations are included) should be satisfactory in principle. A simple test is the location of the long range crossings, which is very sensitive to the energy difference of the initial and final channels. But this test is insufficient. For example, in the C^{3+}/H system, for which both the molecular orbital and valence bond methods yield comparable results in regions well away from avoided crossings, there are significant discrepancies in the important region of the avoided acrossings [51.17, 18]. In the present state of the art, the absolute accuracy of theoretical calculations in the crossing region is difficult to ascertain. Of course, since the computed electron capture cross sections depend sensitively on the minimum energy separation, experimental data at low collision energies makes some tests possible. Unfortunately, there is not yet much reliable data at low eV energies.

Figures 51.1 and 51.2 present the adiabatic network of the $^4\Sigma^-$, $^4\Pi$ and $^2\Sigma^+$, $^2\Pi$, $^2\Delta$ states of the O^{2+}/H system. These have been obtained by means of a configuration interaction molecular orbital method with several hundred configurations [51.15]. It is clear from these calculations that the only effective avoided crossings are those involving the $^4\Sigma^-$ and $^4\Pi$ at 3.7 and 4.8 a_0, for

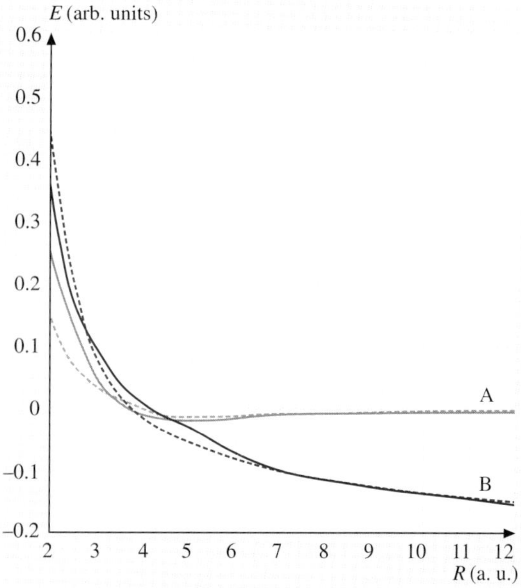

Fig. 51.1 Adiabatic potential energy curves of the $^4\Sigma^-$ and $^4\Pi$ states of the OH^{2+} molecular ion. The *solid curves* designate $^4\Pi$ states, the *dashed curves* $^4\Sigma^-$ states. The dissociation limits A and B correspond respectively to [O^{2+} (^3P) + H (^2S)] and [O^+ (^4P) + H^+]

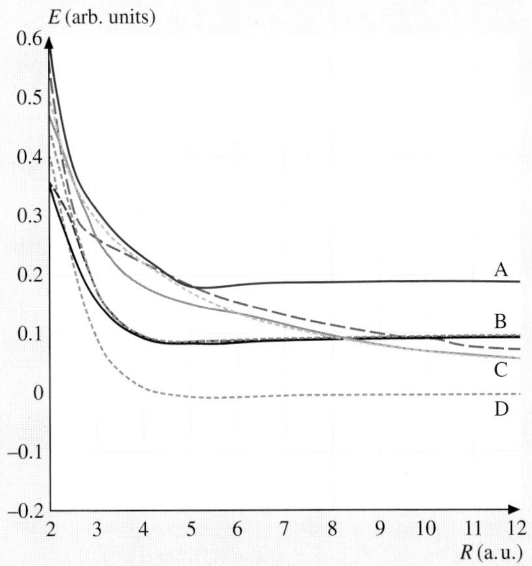

Fig. 51.2 Adiabatic potential energy curves of the $^2\Sigma^+$, $^2\Pi$ and $^2\Delta$ states of the OH^{2+} molecular ion. The *solid curves* designate $^2\Sigma^+$ states, the *dashed curves* $^2\Pi$ states and the *dotted curve* a $^2\Delta$ state. The dissociation limits A, B, C and D correspond respectively to [O^{2+} (^{1}S) + H (^{2}S)], [O^{2+} (^{1}D) + H (^{2}S)], [O^+ (^{2}D) + H+] and [O^{2+} (^{3}P) + H (^{2}S)]

which the energy separation is ≈ 0.1 eV. The avoided crossings involving the doublet states at 8.0, 8.5 and 10 a_0, for which the energy separation is ≈ 0.02 eV, are diabatic. The avoided crossings for type II transitions occur at much shorter distances than for type I transitions.

51.1.2 Model Potential Methods

Model potential methods offer an attractive alternative to ab initio methods in treating type I reactions. To illustrate the method [51.19, 20], we consider an effective one-electron system composed of a spherically symmetric ion X^{q+} and a hydrogen atom. The model Hamiltonian of the molecular ion XH^{q+} is written as

$$H = T + V_X(r_b) - \frac{1}{r_a} + V_X(R) , \quad (51.4)$$

where T is the electronic kinetic energy, $\boldsymbol{r}_a$ and $\boldsymbol{r}_b$ are respectively the position vectors of the Rydberg electron with respect to nuclei A and B, and $V_X(r)$ is the effective potential of the ion core. The latter is usually expressed in the parametric form

$$V_X(r) = -\left[q + (Z-q)\left(1 + \alpha r + \beta r^2 + \gamma r^3\right)\mathrm{e}^{-\delta r}\right] , \quad (51.5)$$

where Z is the ionic nuclear charge and the parameters α, β, γ, δ, ... are optimized to the spectroscopic data so that the asymptotic energies of those Rydberg states of $X^{(q-1)+}$ which govern the charge transfer process are essentially exact.

The eigenvalues of H for a given internuclear distance R are determined by standard variational techniques, using a basis set of Slater-type orbitals in prolate spheroidal coordinates [51.3]. As an example, we present in Fig. 51.3 the potential energy curves of the Al^{3+} system which presents one Σ–Σ avoided crossing around $R_X = 7.2\,a_0$. Model potential methods can also be used to treat multielectron systems such as $(XHe)^{q+}$. In long range collisions, the transition probabilities are primarily determined by the asymptotic form of the electron wave functions far from the nuclei. Thus even when multielectron targets are involved, the effect of dynamic correlation can be small. Since the asymptotic form of the wave functions may be easily generated by model potential techniques, the effect of static correlation can be taken into account in a rather simple way. The method proposed here is quite similar to that used by *Grice* and *Herschbach* [51.21] to treat the long range configuration interaction of ionic and covalent states in neutral atom–atom collisions.

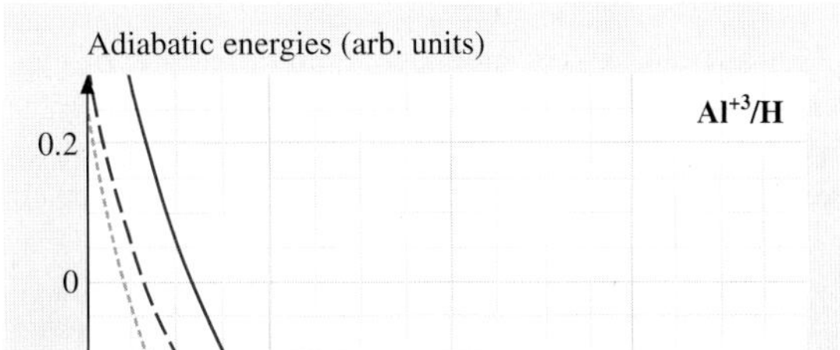

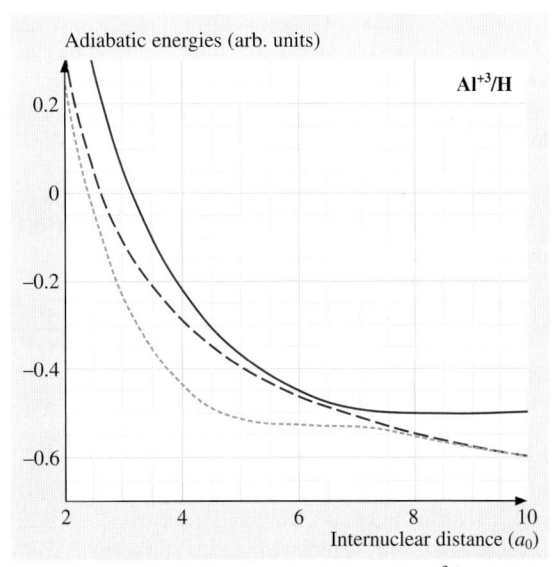

Fig. 51.3 Adiabatic potential energies of AlH^{3+}. The *full curve* designates the $^2\Sigma$ state correlated to the [Al^{3+}(3p)+ H (1s)] entry channel. The *dotted* and *dashed curves* designate respectively the $^2\Sigma$ and $^2\Pi$ states correlated to the [Al^{2+}(3p) + H^+] electron capture channel

For example, in the system X^{q+} He, where as previously X^{q+} is a closed shell ion, the energy separation at a long range avoided crossing is given by [51.22]

$$\Delta\varepsilon(R_c) \equiv 2\sqrt{2}\langle b_{1\sigma}|b'_{1s}\rangle\langle a_{nl\lambda}|V_X + q/R|b\rangle\,, \quad (51.6)$$

where the molecular orbitals $a_{nl\lambda}$ (dissociating to $X_{nl}^{(q-1)+}$) and $b_{1\sigma}$ (dissociating to a 1s orbital of He) are generated from a model Hamiltonian of the form

$$H = T + V_X(r_b) - V_{\mathrm{He}}(r_a) + V_{X,\mathrm{He}}(R)\,, \quad (51.7)$$

where V_{He} is an effective potential describing the 1s orbital of He. The orbital b'_{1s} is the 1s orbital of He^+. This expression is the optimal form of $\Delta\varepsilon(R_c)$ which can be achieved without the use of explicitly correlated wave functions. Since both the $a_{nl\lambda}$ and $b_{1\sigma}$ orbitals are generated from the same model Hamiltonian, the calculation of the interaction matrix element in (51.6) is simple. As an example, Fig. 51.4 presents the adiabatic energies of the B^{3+}/He system which has two avoided $\Sigma - \Sigma$ crossings, one at 4.6 a_0, the other at 7.4 a_0.

A similar extension to treat the case of capture by an ion with one electron in a p shell can be made along the same lines. See [51.23] for an application to the reaction

$$O^{3+}(2p)^2P + H \to O^{2+}(2p3p)^3P + H^+\,. \quad (51.8)$$

A combination of model potential methods to represent the core electrons and ab initio molecular orbital methods for the valence electrons has been successfully developed [51.24] to treat some complex systems such as C^{3+}/H where both type I and II transitions take place. The flexibility of the model potential parameters makes it possible to choose them so that the energies of the initial and final states are accurately reproduced.

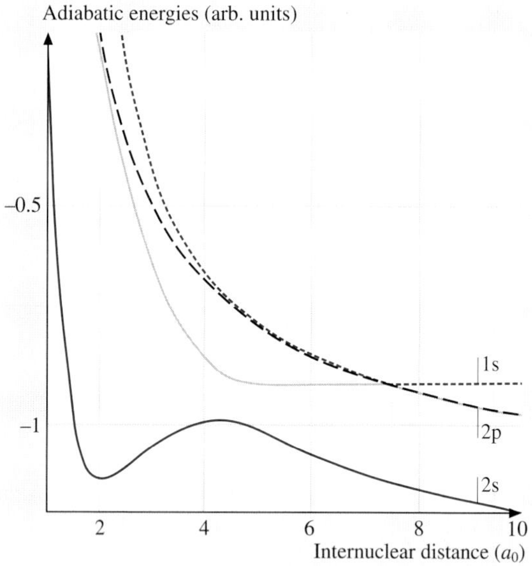

Fig. 51.4 Adiabatic potential energies of BHe^{3+}. The *dotted curve* designates the $^1\Sigma$ state correlated to the [B^{3+} + $He(1s^2)$] entry channel. The *short* and *long dashed curves* designate respectively the $^1\Sigma$ and $^1\Pi$ states correlated to the [B^{2+} (2p) + He^+(1s)] electron capture channel. The *full curve* designates the $^1\Sigma$ state correlated to the [B^{2+} (2s) + He^+(1s)] entry channel

51.1.3 Empirical Estimates

Electron capture cross sections at low energies are largely controlled by two parameters: the crossing radius R_X and the minimum energy separation Δ_X. These are the basic parameters required for the Landau–Zener model, and it is useful to have a simple way of estimating them without having recourse to a complex molecular structure calculation. Many empirical estimates [51.25–27] have been proposed. However, it is generally necessary to take into account the strong ℓ-dependence of the electron capture channel states. Defining $\alpha = \sqrt{I_t/13.6}$, where I_t is the ionization potential of the target (in eV), then Taulbjerg's formula is [51.27]

$$\Delta_X(R_c) = \frac{18.26}{\sqrt{q}} f_{nl} \exp(-1.324\alpha R_c\sqrt{q})\,, \quad (51.9)$$

where

$$f_{nl} = \frac{(-1)^{n+l-1}\sqrt{2l+1}\,\Gamma(n)}{[\Gamma(n+l+1)\Gamma(n-l)]^{1/2}}\,. \quad (51.10)$$

For type I systems, (51.9) is generally quite satisfactory when the energy levels of the capture states are well separated [51.28]. In that case, it can be used to predict the main electron capture reaction windows and obtain a reliable first estimate of the cross sections (in the energy range for which the Landau–Zener model is valid). On the other hand, in cases where there is a near degeneracy of the l states, serious errors can occur [51.28] and the predictions are less satisfactory.

For type II reactions, (51.9) must be modified by a corrective factor to take account of the simultaneous excitation of an electron from a 2s to 2p orbital. One such modification, proposed by *Butler* and *Dalgarno* [51.29], may give some useful idea of the main electron capture processes, but its precision is uncertain.

51.2 Dynamics of the Collision

The first step is to introduce a suitable set of scattering coordinates which can automatically describe both the excitation and rearrangement channels. The particular choice of coordinate system is conditioned by practical considerations. We have found Eckart coordinates to be convenient [51.10]. Their application is straightforward since it involves nonadiabatic matrix elements which can be calculated by the conventional techniques of quantum chemistry. The practical implementation of Eckart coordinates leads to the introduction of an adiabatic variable ξ defined as

$$\xi = \sqrt{\mu}\left(\boldsymbol{R} + \frac{1}{\mu}\boldsymbol{s}\right),$$

where

$$s = \frac{\boldsymbol{r}\cdot\boldsymbol{R}}{R^2}\left(\boldsymbol{r} - \frac{\boldsymbol{r}\cdot\boldsymbol{R}}{2R^2}\boldsymbol{R}\right), \tag{51.11}$$

μ is the reduced mass of the colliding system, $\boldsymbol{r}$ the coordinate of the active electron with respect to the c.m. of the colliding system and $\boldsymbol{R}$ the relative position vector of the nuclei.

The adiabatic Eckart and Born–Oppenheimer equations differ only by terms of the order of $1/\mu$. It may be assumed that the Eckart states are given to sufficient accuracy by the Born–Oppenheimer adiabatic states designated by $\chi_i(\boldsymbol{r}; R)$. We expand the total wave function of the system in the form

$$\Psi(\boldsymbol{r}, \xi) = \sum \chi_i(\boldsymbol{r}; \xi) F_i(\xi) . \tag{51.12}$$

(For a many-electron system, $\boldsymbol{r}$ represents the ensemble of electron coordinates.) Decomposing $F_i(\xi)$ on a basis set of symmetric top functions according to

$$F_i(\xi) = \sum_{K,M}(-1)^K\left(\frac{2K+1}{4\pi}\right)^{1/2} \times D^K_{\Lambda,M}(\theta,\phi)\frac{g_i^{(k)}(\xi)}{\xi}, \tag{51.13}$$

where $\{\theta, \phi\}$ are the spherical polar coordinates of ξ, then the radial functions $g_i^{(k)}(\xi)$ are solutions of the equation

$$\frac{d^2}{d\xi^2}\boldsymbol{g}^{(K)} + 2\boldsymbol{A}\frac{d}{d\xi}\boldsymbol{g}^{(K)} + \boldsymbol{W}\boldsymbol{g}^{(K)} = 0, \tag{51.14}$$

where

$$A_{mn}(\xi) = \left\langle \chi_m \left| \frac{\partial}{\partial\xi} + \frac{z}{\xi}\frac{\partial}{\partial z} \right| \chi_n \right\rangle \delta(\Lambda_m, \Lambda_n), \tag{51.15}$$

$$B_{mn} = \left\langle \chi_m \left| \frac{\partial^2}{\partial\xi^2} \right| \chi_n \right\rangle, \tag{51.16}$$

$$W_{mn} = \left[2\mu(E - \varepsilon_n) - \frac{\left[K(K+1) - \Lambda_n^2\right]}{R^2}\right]\delta_{mn} + B_{mn} + \frac{2}{R^2}\sqrt{K(K+1)}\,L_{mn}, \tag{51.17}$$

$$L_{mn} = \mp\left\langle \chi_m \left| -2x\frac{\partial}{\partial z} \right| \chi_n \right\rangle \delta(\Lambda_m, \Lambda_m \pm 1), \tag{51.18}$$

z, x are the components of $\boldsymbol{r}$ parallel to and perpendicular to the direction of ξ in the classical collision plane, respectively.

Since the ratio $1/\mu$ is small, it is legitimate to replace the matrix element $\langle\chi_m|\partial/\partial\xi|\chi_n\rangle$ by $\langle\chi_m|\partial/\partial R|\chi_n\rangle$ in (51.15) and (51.16). Furthermore, all the matrix elements A_{mn} and L_{mn} vanish asymptotically to first order in $1/\mu$. The modified radial and rotational matrix elements are identical to those obtained in the semiclassical formalism with common translation factors.

The simplest way to solve (51.14) is to eliminate the first-order derivative by transforming to a diabatic representation in which the radial matrix elements vanish [51.30]

$$\boldsymbol{g}^{(K)} = \boldsymbol{C}\boldsymbol{h}^{(K)}, \tag{51.19}$$

where

$$\frac{d}{d\xi}\boldsymbol{C} + \boldsymbol{A}\boldsymbol{C} = 0, \qquad \boldsymbol{C}(\infty) = \boldsymbol{I}. \tag{51.20}$$

Equation (51.14) then reduces to

$$\frac{d^2}{d\xi^2}\boldsymbol{h}^{(K)} - 2\mu\boldsymbol{V}^d\boldsymbol{h}^{(K)} + \left(2\mu E - \frac{K(K+1)}{\xi^2}\right)\boldsymbol{h}^{(K)} = 0 \tag{51.21}$$

with

$$\boldsymbol{V}^d = \boldsymbol{V}^d_E - \frac{\sqrt{K(K+1)}}{\mu R^2}\boldsymbol{V}^D_R, \tag{51.22}$$

$$\boldsymbol{V}^d_E = \boldsymbol{C}^{-1}\varepsilon\boldsymbol{C}, \quad \boldsymbol{V}^D_R = \boldsymbol{C}^{-1}\boldsymbol{L}\boldsymbol{C}. \tag{51.23}$$

The solution of (51.21) and the subsequent extraction of the scattering matrix S may be carried out using an extension of the log derivative method [51.31], adapted to the case of a repulsive Coulomb potential in one or more of the scattering channels [51.32].

This method is particularly stable and advantageous to use at low energies for problems of the type considered here. Of course, at energies exceeding a few hundred eV, rapid oscillations of the radial function render the method rather time consuming. But the method is usable up to keV energies. However, for higher energies exceeding 1 keV, semiclassical methods are preferable.

51.3 Radial and Rotational Coupling Matrix Elements

Two independent methods are employed for the determination of radial coupling matrix elements A_{mn}: the first based on a direct numerical differentiation of the expansion coefficients of the wave function, the second on a variant of the Hellman–Feynman (HF) theorem. In principle, the HF method is three times faster than the direct numerical method, since the eigenvectors need to be calculated only at the value of R concerned. However, the real gain in computing is not always appreciable, the HF theorem being more sensitive to errors in the wave function than the direct method. To achieve comparable accuracy, a larger basis set would be required in the molecular calculations, and this offsets much of the theoretical gain in computing time. The rotational coupling matrix elements L_{mn} are calculated numerically for each internuclear distance.

Typical results for the radial and rotational matrix elements are presented in Figs. 51.5 and 51.6 for the Al^{3+}/H system, illustrating the influence of translation effects. The origin dependence of the matrix element (without inclusion of translation) is weak in the vicinity of an avoided crossing. Away from the crossing, the origin dependence can be considerable. The corresponding diabatic energies and couplings are given in Figs. 51.7 and 51.8. For internuclear distances on the inward side of the crossing, the diabatic states correspond to a mixing of adiabatic states: it is clear that the mathematical definition (51.20) of the diabatic representation does not correspond to the empirical definition of the Landau–Zener model (Sect. 51.5).

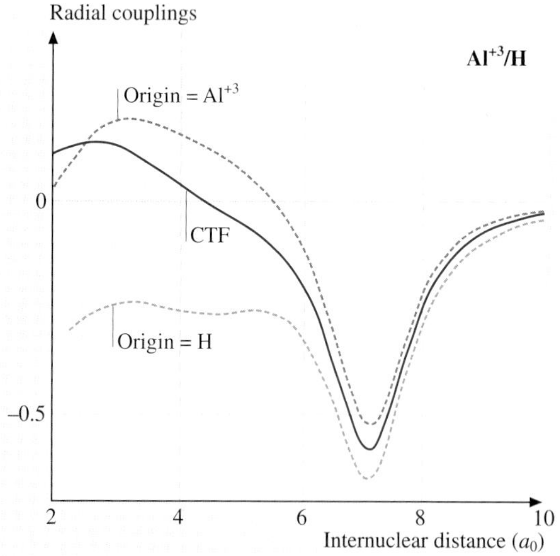

Fig. 51.5 Radial coupling matrix elements between the two $^2\Sigma$ states of AlH^{3+}. The *full curve* designates the matrix element of the CTF (or Eckart) type. The *short* and *long dashed curves* designate the radial derivative with the origin of electron coordinates respectively on the Al and H nuclei

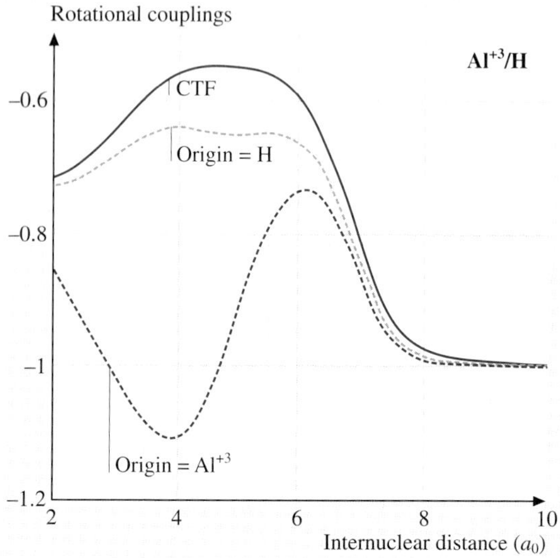

Fig. 51.6 Rotational coupling matrix elements between the $^2\Sigma$ and $^2\Pi$ states correlated to the electron capture channel. The designation of the *curves* is the same as for Fig. 51.5

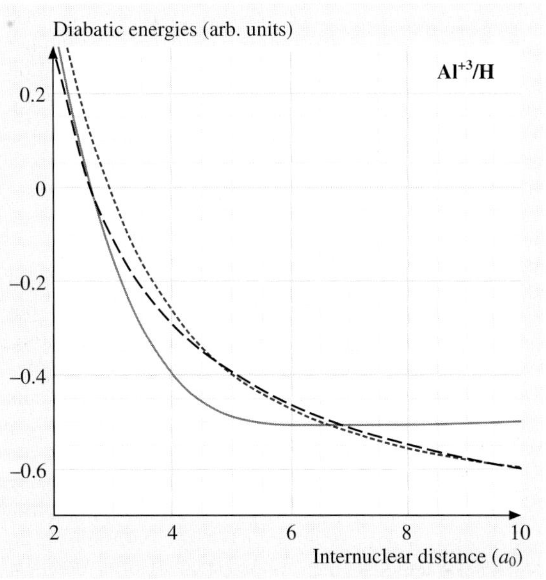

Fig. 51.7 Diagonal elements of the diabatic matrix of ALH^{3+}

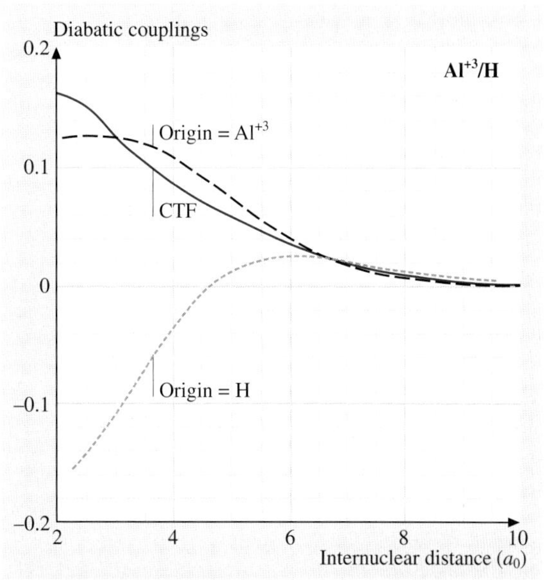

Fig. 51.8 Off-diagonal diabatic matrix element between the two $^2\Sigma$ states of AlH^{3+}

51.4 Total Electron Capture Cross Sections

Although the selection rules for electron capture at low energies are primarily governed by the avoided crossings between adiabatic states of the same symmetry, the combined effects of radial and rotational coupling can be quite complex. In general, rotational coupling effects are weak when capture occurs via an S state, but they can be strong for capture to P, D and higher L states.

In the case of Al^{3+}/H, where capture to the (3p) ^2P state of Al^{2+} takes place via a three-state network of two Σ and one Π states, rotational coupling between the Σ and Π states is strong in the vicinity of the avoided crossing and enhances the capture cross section considerably. This phenomenon is illustrated in Fig. 51.9, which shows the calculated total cross section at different energies as a function of Δ, the energy separation of the quasidegenerate diabatic $^2\Sigma$ and $^2\Pi$ exit channels at the $^2\Sigma$–$^2\Sigma$ crossing radius. The total cross section has a maximum for $\Delta = 0.08$ eV. The cause can be easily seen from the corresponding adiabatic potential energies (Fig. 51.10). For $\Delta = 0.8$ eV, the potential energies of the adiabatic $^2\Sigma$ entry channel and the $^2\Pi$ exit channel become tangential to one another, thereby inducing a resonant effect.

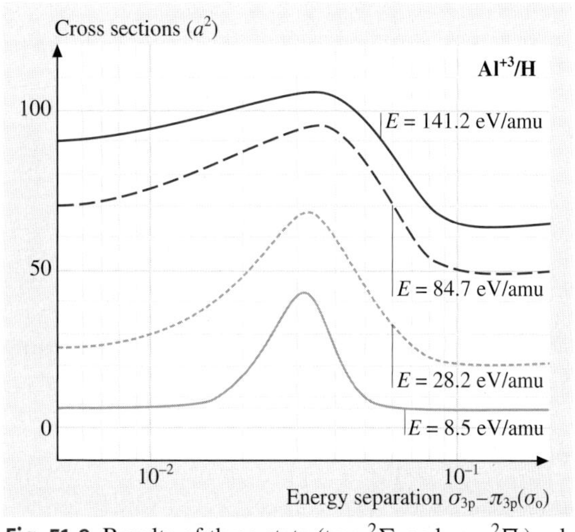

Fig. 51.9 Results of three state (two $^2\Sigma$ and one $^2\Pi$) calculations for electron capture in the Al^{3+}/H system. The cross sections are plotted as a function of the parameter Δ (see text). The extreme sensitivity to Δ illustrates the necessity of knowing the $^2\Sigma$ – $^2\Pi$ energy separation to high accuracy. The numerical values on the curves designate the collision energies in units of eV/amu

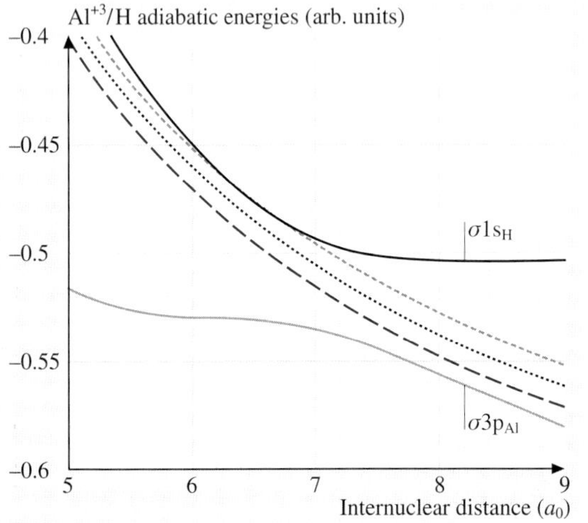

Fig. 51.10 Schematic diagram of the nondiagonal matrix elements of AlH^{3+} in the vicinity of the curve crossing. The three *broken curves* correspond to different $^2\Pi$ state potentials shifted from their calculated value by small amounts

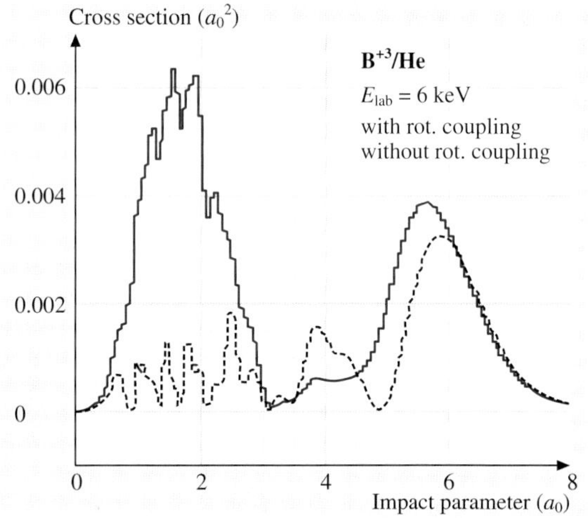

Fig. 51.12 Influence of rotational coupling on the ^{2}P electron capture cross section in B^{3+}/He collisions. The *solid curve* refers to the complete calculation, the *dotted curve* to the calculation with only radial coupling included

In the case of the B^{3+}/He system, four molecular states are implicated in the collision process: three Σ states (Σ_1 correlated to the ^{2}S exit channel, Σ_2 correlated to the ^{2}P exit channel and Σ_3 correlated to the entry channel) and one Π state (correlated to the ^{2}P exit channel). In order to understand better the rotational coupling mechanism, Figs. 51.11 and 51.12 show the ^{2}S and ^{2}P electron capture transition amplitudes as a function of angular momentum (impact parameter) for an energy of 6 keV, where rotational coupling is of major importance [51.14, 33]. The ^{2}P electron capture transition amplitude exhibits 2 maxima, one for an impact parameter of 1.6 a_0, the other for an impact parameter of 5.7 a_0, corresponding respectively to capture via Σ_2–Π crossing around $R = 2\,a_0$ and the outer Σ_2–Σ_3 crossing around $R = 7.4\,a_0$. It is clear from Fig. 51.12 that at very short internuclear distances, rotational coupling is much more important than radial coupling. At the outer crossing (which is nearly diabatic), radial coupling is dominant. This result may be generalized. When the avoided crossing has a largely diabatic character, as for 3p capture in Si^{2+}/H [51.34] or 3d capture in C^{4+}/H [51.10], the inclusion of rotational coupling is fairly weak, affecting principally the population of the sub-m levels and less appreciably the total capture cross section into a given l state. In this case, translation effects are of more importance than rotational coupling.

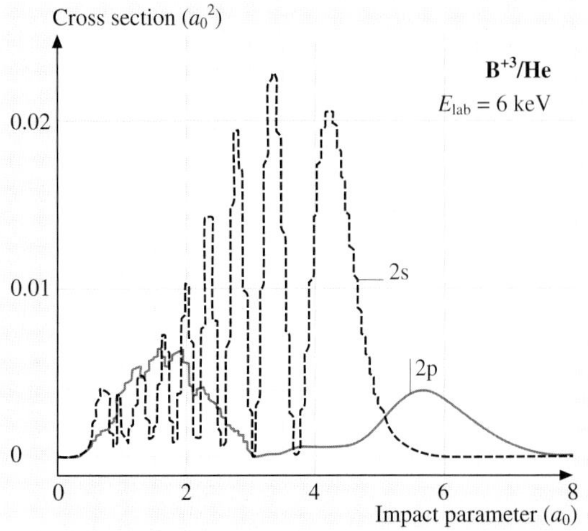

Fig. 51.11 Transition amplitudes for ^{2}S and ^{2}P electron capture as a function of impact parameter in B^{3+}/He collisions. The *solid curve* refers to the ^{2}P contribution, the *dotted curve* to the ^{2}S contribution

In the case of O^{2+}/H (a particularly important system in astrophysical plasmas), the existence of several adiabatic states correlated to the entry channel leads to many interesting features, typical of open p shell ions. The favored reaction channel via the $^4\Sigma^-$ and $^4\Pi$ states involves simultaneous electron capture into a 2p orbital and excitation of a 2s orbital. The avoided crossing is due to electron correlation. Rotational mixing of the $^4\Sigma^-$ and $^4\Pi$ states leads to a large enhancement of the cross section at energies exceeding a few tens of eV [51.16]. On the other hand, the avoided crossings involving the $^2\Sigma^+$ and $^2\Pi$ states around 8 a_0 are too diabatic to contribute to an electronic transition. As a consequence, the metastable ^{1}D ions can only react via a curve crossing at small distances (2.5 a_0). The cross section is much smaller than for ground ^{3}P state capture.

51.5 Landau–Zener Approximation

If a rapid estimation of the cross sections is all that is needed, the Landau–Zener method can be used with advantage, provided that the molecular structure parameters are known accurately. This method, based on an approximate solution of the dynamical equations in a semiclassical formalism (Chapt. 49), is satisfactory for the dominant channels. Aside from the transition being assumed to be localized at the crossing point, the method can easily take account of trajectory effects at low energies. On the other hand, it is unreliable for the weaker channels and it makes allowance neither for translation effects nor for rotational coupling.

The cross section for capture into a state n via a curve crossing located at R_X is given by

$$Q_n^{\mathrm{LZ}} = 2\pi \int_0^{\rho_{\max}} 2p(1-p)\rho\,\mathrm{d}\rho\,, \tag{51.24}$$

where p, the probability for a single passage trajectory (impact parameter) through the crossing, is given by

$$p = \exp\left[-2\pi\left(\frac{\Delta E_{nl}}{2}\right)^2 \frac{1}{v\Delta F}\right]\,, \tag{51.25}$$

ΔF being the difference in slope of the covalent and ionic diabatic curves at R_X, Δ_X the energy separation of the diabatic curves at R_X and v the radial velocity at R_X. It should be recalled that there is no very rigorous definition of the diabatic curves (which are required to obtain DF).For long distance crossings, the simplest (and probably the most satisfactory) estimate is that based on the asymptotic forms of the diabatic states

$$\Delta F = \frac{\partial}{\partial R}\left[V_1(R) - \frac{q-1}{R}\right]_{R=R_X}\,,$$

$$V_1(R) = -\frac{q^2\alpha_{\mathrm{d}}}{2R^4}\,, \tag{51.26}$$

where α_{d} is the polarizability of the target. The radial velocity is given by

$$v = \left[\frac{2E}{\mu}\left(1 - \frac{\rho^2}{R_X^2} - \frac{V_1(R_X)}{E}\right)\right]^{1/2}\,. \tag{51.27}$$

The inclusion of the attractive polarization potential V_1 considerably increases the cross section at low energies, since trajectories with impact parameters much greater than R_X contribute to the cross section. This effect can introduce a negative energy dependence of the cross section in the limit of low energies, of the same kind as the Langevin model for ion–molecule reactions.

51.6 Differential Cross Sections

Differential cross sections at large scattering angles (corresponding to small impact parameters) enable one to probe details of the collision dynamics not readily obtainable from total cross sections, which tend to be dominated by the contribution from small angle scattering (large impact parameters).

Their determination is staightforward once the S matrix elements have been extracted from the asymptotic solution of the coupled equations (51.21). The scattering amplitude $f_{fi}(\vartheta)$ for scattering through the c.m. angle ϑ is given by

$$f_{fi}(\vartheta) = \frac{1}{2\mathrm{i}(k_i k_f)^{1/2}} \sum_{l=0}^{\infty}(2l+1)\left[S_{fi}^l - \delta_{fi}\right] \times P_l(\cos\vartheta)\exp[\mathrm{i}\alpha_f(l)]\,, \tag{51.28}$$

where k_i and k_f are the wave numbers of the initial and final channels, S_{fi}^l the S-matrix element for the i to

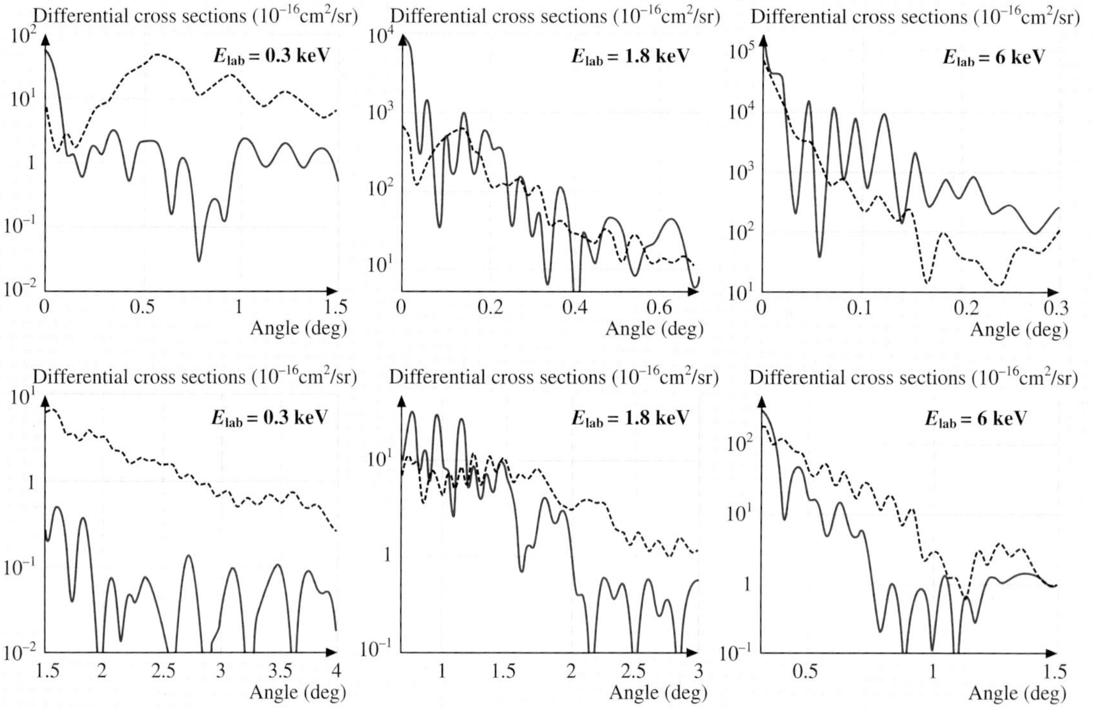

Fig. 51.13 Differential cross sections (10^{-16} cm^2/sr) for electron capture in B^{3+}/He collisions as a function of scattering angle for ion energies of 0.3, 1.8 and 6 keV. The *solid curves* refer to ^{2}S capture, the *dotted curves* to ^{2}P capture

f transition, and $\alpha_f(l)$ the partial wave Coulomb phase shift for the final channel:

$$\alpha_f(l) = \arg \Gamma(l+1+\mathrm{i}\gamma_f)\,, \quad \gamma_f = \frac{\mu q_1 q_2}{k_f}\,. \tag{51.29}$$

The charges of the two collision partners in the final state are designated q_1 and q_2, and μ is the nuclear reduced mass. The differential cross section is evaluated as

$$\frac{\mathrm{d}\sigma_{fi}}{\mathrm{d}\Omega}(\vartheta) = \frac{k_f}{k_i}|f_{fi}(\vartheta)|^2\,. \tag{51.30}$$

Figure 51.13 shows some typical differential cross sections for capture to the $(2s)^2$S and $(2p)^2$P states of B^{2+} in the B^{3+}/He system. The oscillations observed in the differential cross sections are of Stückelberg type. A knowledge of differential cross sections is essential for estimating acceptance angles in laboratory experiments.

51.7 Orientation Effects

Recent experiments on the B^{3+}/He system [51.35, 36] show that there is a strong tendency for electron capture to produce strongly oriented states at small scattering angles. This propensity for orientation is a direct measure of rotational coupling between the Σ and Π molecular states converging to the ^{2}P asymptotic state [51.37].

The orientation and alignment parameters, which characterize the polarization of the emitted photons, can be simply expressed in terms of the scattering amplitudes for electron capture to the magnetic sublevels. These scattering amplitudes can be directly obtained from the scattering matrix. However, care must be exercised in the definition of the scattering amplitudes. In most applications where an adiabatic representation of the collision dynamics is used, the quantization axis is taken to be in the direction of the internuclear axis. For most polarization measurements, it is more convenient to define the quantization axis with respect to an axis perpendicular to the collison plane. But it is straightforward to express the orientation and alignment parameters in terms of the scattering amplitudes obtained with respect to the molecular frame.

In accordance with customary conventions, in Fig. 51.14, the scattering plane contains the X and Z axes. The Y axis, perpendicular to the scattering plane, is taken to be the quantization axis. Let XYZ be the laboratory frame and xyz the body-fixed frame defined as above. The scattering amplitudes calculated in Sect. 51.3 are expressed with respect to the body-fixed frame xyz. The scattering amplitudes $f_{M_Y=\pm 1}$ in the laboratory frame are related to the amplitudes f_{Σ_Z}, $f_{\Pi_Z^+}$ in the body-fixed frame by

$$f_{M_Y=\pm 1} = \frac{1}{\sqrt{2}}\left(f_{\Sigma_Z} \mp f_{\Pi_Z^+}\right) . \tag{51.31}$$

The right-hand and left-hand circular polarizations (RHC and LHC respectively) are then defined as

$$\begin{aligned} \mathrm{RHC} &= |f_{M_Y=-1}|^2 \\ &= \frac{1}{2}\left(|f_{\Sigma_Z}|^2 + |f_{\Pi_Z^+}|^2\right) + \mathrm{Im}\left(f_{\Sigma_Z} f^*_{\Pi_Z^+}\right) , \end{aligned} \tag{51.32}$$

$$\begin{aligned} \mathrm{LHC} &= |f_{M_Y=+1}|^2 \\ &= \frac{1}{2}\left(|f_{\Sigma_Z}|^2 + |f_{\Pi_Z^+}|^2\right) - \mathrm{Im}\left(f_{\Sigma_Z} f^*_{\Pi_Z^+}\right) , \end{aligned} \tag{51.33}$$

and the circular polarization as

$$\mathrm{L} = \frac{\mathrm{RHC} - \mathrm{LHC}}{\mathrm{RHC} + \mathrm{LHC}} = \frac{2\,\mathrm{Im}\left(f_{\Sigma_Z} f^*_{\Pi_Z^+}\right)}{|f_{\Sigma_Z}|^2 + |f_{\Pi_Z^+}|^2} . \tag{51.34}$$

Figures 51.15 and 51.16 show the RHC, LHC and L quantities for an incident ion energy of 1.5 keV, where comparison with experiments [51.35, 36] can

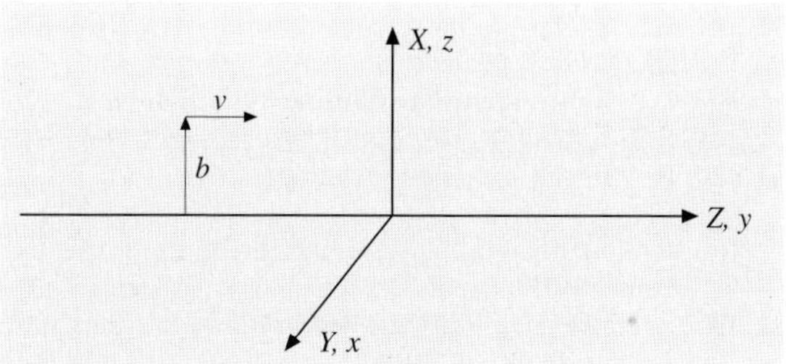

Fig. 51.14 coordinate systems for scattering

be made. The strong propensity for orientation of the ^{2}P state is clearly exhibited for small scattering angle ($\vartheta < 0.2°$). At larger angles (smaller impact parameters), the propensity decreases and even reverses ($\vartheta > 0.4°$). We have also plotted on Fig. 51.16 the experimental data of *Roncin* et al. [51.35]. The agreement with experiment is very satisfactory.

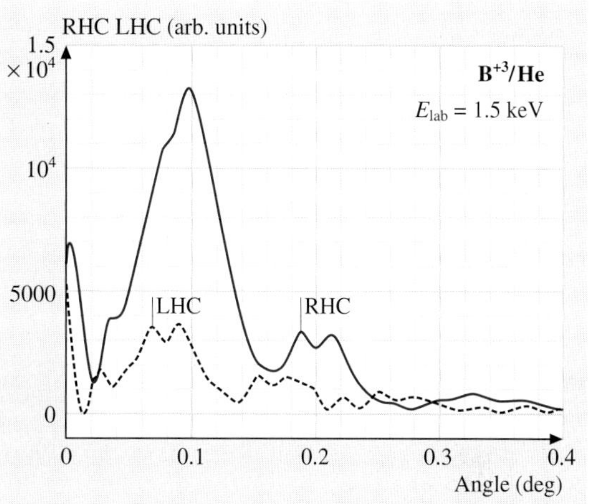

Fig. 51.15 Right-hand circular polarization (*full curve*) and left-hand circular polarization (*dashed curve*) for ^{2}P electron capture in B^{3+}/He collisions as a function of scattering angle for $E = 1.5$ keV

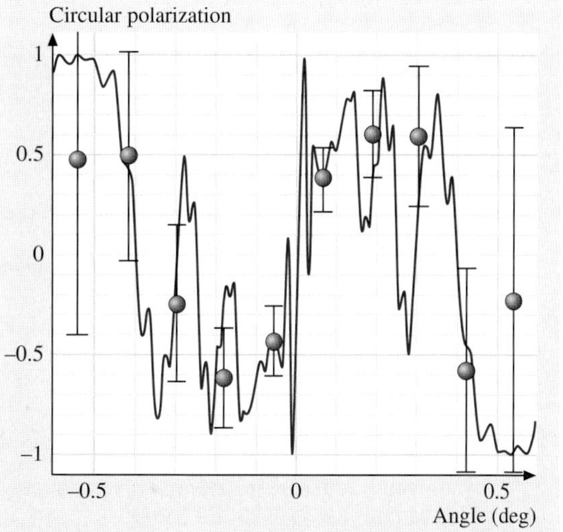

Fig. 51.16 Circular polarization for ^{2}P electron capture in B^{3+}/He collisions as a function of scattering angle for $E = 1.5$ keV. The *solid circles* (with the error bars) are taken from [51.35]

51.8 New Developments

During the last eight years there have been some interesting new developments [51.38–40] using hyperspherical coordinates to describe the dynamics of ion–atom collisions. Of particular interest for this chapter are the calculations of *Le* et al. [51.40] for charge transfer in the $Si^{4+}/H(D)$ and Be^{4+}/H systems. Their calculated cross sections are almost identical to those obtained by *Pieksma* et al. [51.41] using the Thorson–Delos-type [51.8] approximate Jacobi coordinates introduced in Sect. 49.3. This confirms the close connection between the hyerspherical and the Thorson–Delos-type coordinates which had already been observed by *Gargaud* et al. [51.10] for two-state systems.

Another aspect of the theoretical formulation which has been clarified recently concerns the adiabatic–diabatic transformation (51.20). The radial differential equations (51.21) only take this simple form if it can be assumed that

$$\boldsymbol{B} = \frac{\mathrm{d}}{\mathrm{d}\xi}\boldsymbol{A} - \boldsymbol{A}^2 . \tag{51.35}$$

However, (51.35) is only strictly satisfied if the basis set is complete. And indeed, it has been found [51.10] from direct calculations of the matrix elements B_{mn} that it is not well satisfied for any choice of Jacobi coordinates. On the other hand, the calculations show that (51.35) is well satisfied for a minimal basis set using the Thorson–Delos reaction coordinates. This result confirms that convergence of an adiabatic basis set can indeed be achieved using appropriate reaction coordinates and also explains why the calculations of *Le* et al. [51.40] and *Pieksma* et al. [51.41,42] agree so well.

References

51.1 A. Dalgarno, S. E. Butler: Comments At. Mol. Phys. **7**, 129 (1978)

51.2 D. Pequignot, S. M. V. Aldrovandi, G. Stasinska: Astron. Astrophys. **63**, 313 (1978)

51.3 M. Gargaud, J. Hanssen, R. McCarroll, P. Valiron: J. Phys. B **14**, 2259 (1981)

51.4 V. H. S. Kwong, Z. Fang: Phys. Rev. Lett. **71**, 4127 (1993)

51.5 T. K. McLaughlin, S. M. Wilson, R. W. McCullough, H. B. Gilbody: J. Phys. B **23**, 737 (1990)

51.6 D. R. Bates, R. McCarroll: Proc. R. Soc. London A **245**, 175 (1958)

51.7 S. B. Schneiderman, A. Russek: Phys. Rev. A **181**, 311 (1969)

51.8 W. R. Thorson, J. B. Delos: Phys. Rev. A **18**, 117 (1978)

51.9 R. McCarroll, D. S. F. Crothers: Adv. At. Mol. Opt. Phys. **32**, 253 (1994)

51.10 M. Gargaud, R. McCarroll, P. Valiron: J. Phys. B **20**, 1555 (1987)

51.11 L. F. Errea, L. Mendez, A. Riera: J. Phys. B **15**, 2255 (1982)

51.12 L. F. Errea, C. Harel, H. Jouin, L. Mendez, B. Pons, A. Riera: J. Phys. B **27**, 3603 (1994)

51.13 M. Gargaud, R. McCarroll, L. Opradolce: J. Phys. B **21**, 521 (1988)

51.14 M. Gargaud, F. Fraija, M. C. Bacchus-Montabonel, R. McCarroll: J. Phys. B **29**, 179 (1994)

51.15 P. Honvault, M. C. Bacchus-Montabonel, R. McCarroll: J. Phys. B **27**, 3115 (1994)

51.16 P. Honvault, M. Gargaud, M. C. Bacchusmontabonel, R. McCarroll: Astronom. Astrophys. **302**, 931 (1995)

51.17 D. L. Cooper, M. J. Ford, J. Gerratt, M. Raimondi: Phys. Rev. A **34**, 1752 (1986)

51.18 S. Bienstock, T. G. Heil, A. Dalgarno: Phys. Rev. A **25**, 2850 (1982)

51.19 C. Bottcher, A. Dalgarno: Proc. Soc. A **340**, 187 (1974)

51.20 P. Valiron, R. Gayet, R. McCarroll, F. Masnou-Seeuws, M. Philippe: J. Phys. B **12**, 53 (1979)

51.21 R. Grice, D. R. Herschbach: Mol. Phys. **27**, 159 (1974)

51.22 M. Gargaud, R. McCarroll: Phys. Scr. **51**, 752 (1995)

51.23 M. Gargaud, R. McCarroll, L. Opradolce: Astron. Astrophys. **208**, 251 (1989)

51.24 L. F. Errea, B. Herrero, L. Méndez, O. Mó, A. Riera: J. Phys. B **24**, 4049 (1991)

51.25 R. E. Olson, A. Salop: Phys. Rev. A **14**, 579 (1976)

51.26 M. Kimura, T. Iwai, Y. Kaneko, N. Kobayashi, A. Matumoto, S. Ohtani, K. Okuno, S. Takagi, H. Tawara, S. Tsurubuchi: J. Phys. Soc. (Japan) **53**, 2224 (1984)

51.27 K. Taulbjerg: J. Phys. B **19**, L367 (1986)

51.28 M. Gargaud: Transfert de charge entre ions multichargés et hydrogène atomique (et moléculaire) aux basses énergies. Ph.D. Thesis (Université de Bordeaux, France 1987)

51.29 S. E. Butler, A. Dalgarno: Astrophys. J. **241**, 838 (1980)

51.30 F. T. Smith: Phys. Rev. **179**, 111 (1969)

51.31 B. R. Johnson: J. Comput. Phys. **13**, 445 (1973)

51.32 P. Valiron: Echange de charge des ions C^{2+} et Si^{2+} avec l'hydrogène atomique dans le milieu interstellaire. Ph.D. Thesis (Université de Bordeaux, France 1976)

51.33 F. Fraija, M. C. Bacchus-Montabonel, M. Gargaud: Z. Phys. D **29**, 179 (1994)

51.34 M. Gargaud, R. McCarroll, P. Valiron: Astron. Astrophys. **106**, 197 (1982)

51.35 P. Roncin, C. Adjouri, N. Andersen, M. Barat, A. Dubois, M. N. Gaboriaud, J. P. Hansen, S. E. Nielsen, S. Z. Szilagyi: J. Phys. B **27**, 3079 (1994)

51.36 P. Roncin, C. Adjouri, M. N. Gaboriaud, L. Guillemot, M. Barat, N. Andersen: Phys. Rev. Lett. **65**, 3261 (1990)

51.37 M. Gargaud, M. C. Bacchus-Montabonel, T. Grozdanov, R. McCarroll: J. Phys. B **27**, 4675 (1994)

51.38 A. Igarashi, C. D. Lin: Phys. Rev. Lett. **83**, 4041 (1999)

51.39 C.-N. Liu, A.-T. Le, T. Morishita, B. D. Esry, C. D. Lin: Phys. Rev. A **67**, 52705 (2003)

51.40 A.-T. Le, M. Hesse, T. G. Lee, C. D. Lin: Phys. Rev. A **67**, 52705 (2003)

51.41 M. Pieksma, M. Gargaud, R. McCarroll, C. Havener: Phys. Rev. A **54**, 13 (1996)

51.42 D. Rabli: Extension de la méthode du potentiel modèle pour traiter la dynamique des systèmes diatomiques. Application au transfert de charge dans les collisions $Si^{3+}+He$ et $He^{2+}+He$ métastable. Ph.D. Thesis (Université Pierre et Marie Curie, Paris 2001)

52. Continuum Distorted Wave and Wannier Methods

The continuum distorted wave model has been extensively applied to charge transfer and ionization processes. We present both the perturbative and variational capture theories as well as highlighting the suitability of this model in describing the continuum final states in both heavy and light particle ionization. We then develop the Wannier theory for threshold ionization, and further theoretical work which led to the modern quantal semiclassical approximation. This very successful theory has provided the first absolute cross sections which are in good agreement with experiment.

52.1 **Continuum Distorted Wave Method** 775
52.1.1 Perturbation Theory.................. 775
52.1.2 Relativistic Continuum-Distorted Waves..................................... 778
52.1.3 Variational CDW 778
52.1.4 Ionization 779

52.2 **Wannier Method** 781
52.2.1 The Wannier Threshold Law........ 781
52.2.2 Peterkop's Semiclassical Theory 782
52.2.3 The Quantal Semiclassical Approximation.......................... 783

References 786

The most recent developments include third-order continuum distorted wave double-scattering 1s–1s transitions (Sect. 52.1.1), relativistic continuum distorted waves (Sect. 52.1.2), and new theory on magnetically quantized continuum distorted waves [52.1, 2] (Sect. 52.1.4). A novel ionization theory for low energies (below 80 keV) is also reported in which the target is considered as a one electron atom and the interactions between this active electron and the remaining target electrons are treated by a model potential including both short and long range effects. In the final channel the usual product of two continuum distorted wave functions, each associated with a distinct electron–nucleus interaction, is used [52.3].

In addition, Sect. 52.2 on the Wannier Method reports the major progress made over the last eight years. These major advances include (a) the development of below-threshold semiclassical theory for the study of doubly excited states [52.4–7] (b) a more accurate variant of the semi-classical quantum-mechanical treatment of *Crothers* [52.8].

52.1 Continuum Distorted Wave Method

52.1.1 Perturbation Theory

Continuum distorted wave theory (CDW) is one of the most advanced and complete perturbative theories of heavy particle collisions which has been formulated to date. It was originally introduced by *Cheshire* [52.9] to model the process of charge transfer during the collision of an atom/ion with an ion (specifically the resonant process of $p + H(1s) \rightarrow H(1s) + p$). These types of three-body collisions are made amenable to the perturbative approach when the ratio of the projectile impact velocity v to the electron initial bound state mean velocity v_b satisfies

$$\frac{v}{v_b} \gtrsim 3 \,. \quad (52.1)$$

The criterion for nonrelativistic collisions, in which electron capture is a dominant process, is that both v and v_b are small compared with the speed of light. The theoretical description of collisions which involve the disturbance of a bound electron of mass m_e by the field of a fast moving heavy particle of mass M can be greatly simplified by exploiting the fact that since the ratio m_e/M is so small, the heavy particle follows a straight-line trajectory throughout the collision.

This allows the parametrization of the internuclear vector $\boldsymbol{R}$ in terms of an impact parameter $\boldsymbol{b}$, such that

$$\boldsymbol{R}=\boldsymbol{b}+\boldsymbol{v}t\,. \tag{52.2}$$

This impact parameter picture (IPP) of the collision is equivalent to the full quantal or wave treatment when the eikonal criterion for small angle scattering is satisfied [52.10].

It has become standard to work in a generalized nonorthogonal coordinate system in which the vectors $\boldsymbol{r}_{\mathrm{T}}$ ($\boldsymbol{r}_{\mathrm{P}}$) from the target (projectile) to the electron are treated, along with $\boldsymbol{R}$, as independent variables [52.11]. Working in the frame centered on the target nucleus and using atomic units, the Lagrangian is given by

$$\begin{aligned}\mathrm{H}-\mathrm{i}\frac{\mathrm{d}}{\mathrm{d}t_{\boldsymbol{r}_{\mathrm{T}}}} &= -\frac{1}{2}\nabla^2_{\boldsymbol{r}_{\mathrm{T}}}+V_{\mathrm{T}}(\boldsymbol{r}_{\mathrm{T}})-\mathrm{i}\frac{\partial}{\partial t}-\frac{1}{2}\nabla^2_{\boldsymbol{r}_{\mathrm{P}}}\\ &\quad+V_{\mathrm{P}}(\boldsymbol{r}_{\mathrm{P}})+\mathrm{i}\boldsymbol{v}\cdot\boldsymbol{\nabla}_{\boldsymbol{r}_{\mathrm{P}}}\\ &\quad+V_{\mathrm{TP}}(\boldsymbol{R})-\boldsymbol{\nabla}_{\boldsymbol{r}_{\mathrm{P}}}\cdot\boldsymbol{\nabla}_{\boldsymbol{r}_{\mathrm{T}}}\,,\end{aligned} \tag{52.3}$$

where $\mathrm{d}/\mathrm{d}t_{\boldsymbol{r}_{\mathrm{T}}}$ refers to differentiation with respect to t, keeping $\boldsymbol{r}_{\mathrm{T}}$ fixed, and where

$$V_{\mathrm{T}}=\frac{-Z_{\mathrm{T}}}{r_{\mathrm{T}}},\quad V_{\mathrm{P}}=\frac{-Z_{\mathrm{P}}}{r_{\mathrm{P}}},\quad V_{\mathrm{TP}}=\frac{Z_{\mathrm{T}}Z_{\mathrm{P}}}{R}\,. \tag{52.4}$$

Since these potentials are pure Coulomb potentials, they continue to affect the relevant wave functions even at infinity. These long range Coulomb boundary conditions are defined in (52.10) and (52.11). The $+\mathrm{i}\boldsymbol{v}\cdot\boldsymbol{\nabla}_{\boldsymbol{r}_{\mathrm{P}}}$ term gives rise to the Bates–McCarroll electron translation factors which are required to satisfy Galilean invariance.

The Lagrangian above has been written in such a way as to highlight the three two-body decompositions exploited in CDW, with the $-\boldsymbol{\nabla}_{\boldsymbol{r}_{\mathrm{P}}}\cdot\boldsymbol{\nabla}_{\boldsymbol{r}_{\mathrm{T}}}$ term, the so-called nonorthogonal kinetic energy, coupling the systems. The essence of CDW is to treat the bound electron as simultaneously being in the continuum of the other heavy particle.

The initial wave function can be written as

$$\xi_i^{\pm}=D^{\pm}_{-\boldsymbol{v}}(\boldsymbol{r}_{\mathrm{P}})\Phi_i(\boldsymbol{r}_{\mathrm{T}},t)C(\boldsymbol{R},t)\,, \tag{52.5}$$

where $D^{\pm}_{-\boldsymbol{v}}$ is the distortion from the projectile, and C is due to the internuclear potential, V_{TP}. The bound state $\Phi_i=\phi_i(\boldsymbol{r}_{\mathrm{T}})\exp(-\mathrm{i}\epsilon_i t)$, where $\phi_i(\boldsymbol{r}_{\mathrm{T}})$ is the initial eigenstate, and ϵ_i is the initial eigenenergy.

In this form, the action of the Lagrangian can be split into three separate differential equations plus a residual interaction. This gives the following solutions: for the distortion D,

$$\begin{aligned}D^{+}_{-\boldsymbol{v}}&=N(\zeta_{\mathrm{P}})\,{}_1F_1(\mathrm{i}\zeta_{\mathrm{P}};1;\mathrm{i}\boldsymbol{v}\cdot\boldsymbol{r}_{\mathrm{P}}+\mathrm{i}vr_{\mathrm{P}})\,,\\ D^{-}_{\boldsymbol{v}}&=\left(D^{+}_{-\boldsymbol{v}}\right)^{*}\,,\end{aligned} \tag{52.6}$$

with

$$N(\zeta)=\exp(\pi\zeta/2)\Gamma(1-\mathrm{i}\zeta)\,,\quad \zeta_{T,P}=Z_{T,P}/v\,; \tag{52.7}$$

and for the internuclear function C,

$$C(\boldsymbol{R},t)=\exp\left[\mathrm{i}\frac{Z_{\mathrm{P}}Z_{\mathrm{T}}}{v}\ln(vR-v^2t)\right]. \tag{52.8}$$

Similarly it can be shown that

$$\begin{aligned}\xi_f^{\pm}&=D^{\pm}_{\boldsymbol{v}}(\boldsymbol{r}_{\mathrm{T}})\Phi_f(\boldsymbol{r}_{\mathrm{P}},t)C^*(\boldsymbol{R},-t)\\ &\quad\times\exp\left(\mathrm{i}\boldsymbol{v}\cdot\boldsymbol{r}_{\mathrm{T}}-\mathrm{i}\frac{v^2}{2}t\right)\,,\end{aligned} \tag{52.9}$$

where $\mathrm{i}\boldsymbol{v}\cdot\boldsymbol{r}_{\mathrm{T}}-\mathrm{i}\frac{v^2}{2}t$ results from the Galilean transformation to the target frame. The superscripts plus and minus refer to outgoing and incoming Coulomb boundary conditions respectively. These are determined by the asymptotic form of the wave functions

$$\begin{aligned}\lim_{t\to-\infty}\xi_i^{+}&\sim\Phi_i(\boldsymbol{r}_{\mathrm{T}},t)\\ &\quad\times\exp\left[\mathrm{i}\frac{Z_{\mathrm{P}}(Z_{\mathrm{T}}-1)}{v}\ln(vR-v^2t)\right]\,,\end{aligned} \tag{52.10}$$

and

$$\begin{aligned}\lim_{t\to+\infty}\xi_f^{-}&\sim\Phi_f(\boldsymbol{r}_{\mathrm{P}},t)\exp\left(\mathrm{i}\boldsymbol{v}\cdot\boldsymbol{r}_{\mathrm{T}}-\mathrm{i}\frac{v^2}{2}t\right)\\ &\quad\times\exp\left[-\mathrm{i}\frac{Z_{\mathrm{T}}(Z_{\mathrm{P}}-1)}{v}\ln(vR+v^2t)\right]\,.\end{aligned} \tag{52.11}$$

Of course, ξ_i^{+} and ξ_f^{-} are not exact solutions of the three-body Schrödinger equation; in fact,

$$\left(H-\mathrm{i}\frac{\mathrm{d}}{\mathrm{d}t_{\boldsymbol{r}_{\mathrm{T}}}}\right)\xi_i^{+}=W_i\xi_i^{+}=-\boldsymbol{\nabla}_{\boldsymbol{r}_{\mathrm{P}}}D^{+}_{-\boldsymbol{v}}\cdot\boldsymbol{\nabla}_{\boldsymbol{r}_{\mathrm{T}}}\Phi_i\,, \tag{52.12}$$

and

$$\begin{aligned}\left(H-\mathrm{i}\frac{\mathrm{d}}{\mathrm{d}t_{\boldsymbol{r}_{\mathrm{T}}}}\right)\xi_f^{-}&=W_f\xi_f^{-}\\ &=-\,\mathrm{e}^{\mathrm{i}\boldsymbol{v}\cdot\boldsymbol{r}_{\mathrm{T}}-\mathrm{i}\frac{v^2}{2}t}\nabla_{\boldsymbol{r}_{\mathrm{T}}}D^{-}_{\boldsymbol{v}}\cdot\nabla_{\boldsymbol{r}_{\mathrm{P}}}\phi_f\,.\end{aligned} \tag{52.13}$$

The CDW transition amplitude is written as

$$A_{if}=-\mathrm{i}\int_{-\infty}^{+\infty}\mathrm{d}t\left\langle\xi_f^{-}\middle|T_{\mathrm{CDW}}\middle|\xi_i^{+}\right\rangle\,. \tag{52.14}$$

A perturbative expansion via the distorted wave Lippmann–Schwinger equation can be made for T_{CDW}, either in the post form

$$T^{+}_{\text{CDW}} = W^{\dagger}_{f}(1+G_V W_i) + T^{+}_{\text{CDW}} G_i V G_V W_i \,, \tag{52.15}$$

or in the prior form

$$T^{-}_{\text{CDW}} = \left(1+W^{\dagger}_{f} G_V\right) W_i + W^{\dagger}_{f} G_V V G_f T^{-}_{\text{CDW}} \,, \tag{52.16}$$

where the Green functions are given by

$$G_{i,f} = \left(\mathrm{i}\frac{\mathrm{d}}{\mathrm{d}t_{r_{\mathrm{T}}}} - H + W_{i,f} + \mathrm{i}\epsilon\right)^{-1} \,, \tag{52.17}$$

$$G_V = \left(\mathrm{i}\frac{\mathrm{d}}{\mathrm{d}t_{r_{\mathrm{T}}}} - H + V + \mathrm{i}\epsilon\right)^{-1} \,, \tag{52.18}$$

and V is any potential which ensures that the kernels of the integral equations for T_{CDW} are continuous [52.13].

By taking the first term in the expansions (52.15) and (52.16), we get the post and prior forms of the CDW1 amplitude as used by Cheshire. When calculating these amplitudes, the separable nature of the CDW wave function is best exploited by using Fourier transforms to move to the time-independent wave picture. A similar transformation is not suitable in the coupled channel approach discussed in Sect. 52.1.3.

Crothers [52.13], working in the wave treatment, has calculated the second order CDW2 amplitude using various approximations for the Green functions, and has shown that the CDW perturbation series has converged very well to first-order in most parts of the differential cross section. This is in contrast to the standard Born or Brinkman–Kramers approximations which do not start to converge until expanded to second-order. CDW1 is the only first-order perturbation theory, apart from asymmetric hybrid models derived from it, which produces a Thomas peak. Unfortunately, due to the accidental cancellation of the leading order terms, CDW1 has an extreme dip at the Thomas angle, a defect removed in CDW2 [52.14]. This is illustrated in Fig. 52.1 which also includes both folded and unfolded versions of the asymmetric target CDW (TCDW) theory discussed below, as well as experimental data [52.12].

Further work in this area has included the development of the Thomas double-scattering electron capture at asymptotically high velocity within the third-order continuum distorted-wave perturbation theory for 1s–1s transitions in proton hydrogen collisions. It has been shown [52.15] that at the critical proton scattering angles, namely the forward peak, Thomas double encounter peak, small angles, and the interference minimum, the CDW series has converged at second order. Moreover, it is proven that the third-order correction makes no contribution to the velocity dependent v^{-11} and v^{-12} behavior of the Thomas double-scattering total cross section at the leading angles. In contrast, it may be seen in [52.15] that the Oppenheimer–Brinkman–Kramers (OBK) travelling atomic orbital theory (which in general suffers from a common phase factor which embraces intermediate elastic divergences [52.16] in the first and higher-order terms) has not converged at second order. It remains an open question as to whether fourth-order terms or higher in the OBK approximation contribute to various differential cross-sections. It is concluded that the CDW model gives a superior description of the Thomas double-scattering mechanism when compared with the OBK model.

Anomalously large cross sections are obtained at low energies if the CDW wave function is not normalized at all times throughout the collision [52.11]. This is best demonstrated by the presence of the $N(\zeta)$ terms in

$$\lim_{t\to+\infty} \xi^{+}_{i} = N(\zeta_{\mathrm{P}})\Phi_i(\boldsymbol{r}_{\mathrm{T}}, t) \times \exp\left[\mathrm{i}\frac{Z_{\mathrm{P}}Z_{\mathrm{T}}}{v}\ln(vR - v^2 t)\right] \tag{52.19}$$

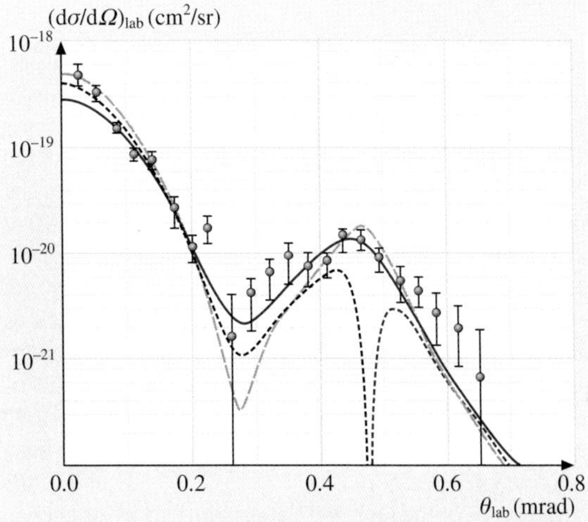

Fig. 52.1 Differential cross sections for electron capture in the collision $\mathrm{H}^+ + \mathrm{H(1s)} \to \mathrm{H(1s)} + \mathrm{H}^+$ as a function of laboratory scattering angle (θ_{lab}) for impact energy of 5 MeV: *solid line* TCDW, folded over the experimental resolution of *Vogt* et al. [52.12]. Unfolded theoretical results: *dashed line* TCDW; *dotted line* CDW1. Experimental data; *circles*, *Vogt* et al. [52.12]

and

$$\lim_{t\to-\infty} \xi_f^- = N^*(\zeta_T)\Phi_f(\boldsymbol{r}_P, t)\exp\left(\mathrm{i}\boldsymbol{v}\cdot\boldsymbol{r}_T - \mathrm{i}\frac{v^2t}{2}\right) \times\exp\left[-\mathrm{i}\frac{Z_T Z_P}{v}\ln(vR+v^2t)\right] . \tag{52.20}$$

Using

$$\lim_{v\to 0} \mid N(Z/v)\mid^2 \sim \frac{2\pi Z}{v} \tag{52.21}$$

it is clear that the problem gets worse as v decreases. It can be corrected by defining

$$\hat{\xi}_{i,f}^{\pm} = \xi_{i,f}^{\pm}\left\langle \xi_{i,f}^{\pm}\middle|\xi_{i,f}^{\pm}\right\rangle^{-1/2} . \tag{52.22}$$

Simpler distorted wave models can be generated through further approximations. Two of the most popular are Target CDW $[D_{-v}(\boldsymbol{r}_P) \to 1]$ and Projectile CDW $[D_v(\boldsymbol{r}_T) \to 1]$. These approximations are justified when $Z_T > Z_P$ and $Z_P > Z_T$, respectively, and are particularly simple to calculate when the simple Born-like residual interaction is used rather than the full CDW form.

The asymptotic forms of the CDW wave functions can be used throughout the collision, ensuring normalization, and this leads to the eikonal or symmetric eikonal models.

52.1.2 Relativistic Continuum-Distorted Waves

The CDW model can be naturally extended to the two-center time-dependent Dirac equation, so that a Lorentz invariant theory is obtained. When the electron orbital velocity αZc, or the collision velocity v, approaches the speed of light c, the kinematics are modified by time dilation. In addition, the particle interactions change because of retardation and the fact that spin-orbit effects are now important. Moreover, vacuum interactions such as radiative emissions and electron–positron pair production begin to play a role. A comprehensive account of atomic processes in relativistic heavy-particle collisions can be found in two recent books [52.17, 18]. At high collision energies, $\gamma \equiv (1 - v^2/c^2)^{-1/2} \gg 1$, and high-charge states of the ions, the Dirac sea of negative energy states becomes energetically accessible and strongly coupled. The process of electron capture, for example, may be mediated by spin-flip transitions [52.19], or spontaneous X-ray emission (radiative electron capture) [52.20] and even electron capture via pair production [52.21, 22]. Although the importance of vacuum processes diminishes with energy, these mechanisms dominate in the extreme relativistic regime. Indeed the last of these processes was used to produce antihydrogen in the laboratory [52.23] at GeV u^{-1} energies.

At relativistic energies, the principal inelastic process is collisional ionization [52.17, 18]. The extensions of the distorted-wave theory to accommodate Lorentz invariance has been developed by *Rivarola* and *Deco* [52.24, 25] and *Crothers* and coworkers [52.19] following work on the Born series [52.26] and impulse approximation [52.27]. In practical applications to electron capture cross sections, the symmetric semi-relativistic CDW theory of *Glass* et al. [52.19] was found to be in very good agreement with experiments in the GeV u^{-1} energy range with charges $Z_{P,T} \sim 6$–80. For non-radiative electron capture, it was found that second-order retardation dominates at extreme relativistic velocities so that $\sigma \sim \gamma^{-1}(\ln\gamma)^2$ [52.28]. However, the momentum transfer kinematics for this process are unfavorable, and a more efficient mechanism based on electron–positron pair production with capture of the created electron is more strongly coupled. Theoretical estimates of this process using relativistic CDW [52.29] compared with experiments [52.30] are in very good agreement.

The same model has been applied to estimate yields of antihydrogen following antiproton impact with neutral high-Z atoms [52.23] following experiments at CERN and Fermilab. The virtual photon model of *Baur* [52.30] gives cross sections that agree well with the limited data [52.23]. However, these estimates are roughly ten times larger than the relativistic CDW results [52.17] and one hundred times the first-Born estimate [52.31]. It appears that additional studies, both experimental and theoretical, would be worthwhile in order to understand this process more fully.

52.1.3 Variational CDW

As the ratio of v/v_b decreases, perturbation theory starts to fail. This is due to the effective interaction time between the projectile and target atoms being long enough for strong three-body coupling. In this environment variational methods have proved successful. This procedure ensures both gauge invariance and unitarity – two fundamental attributes perturbation theory usually cannot guarantee. Continuing in the IPP, we use the Sil variational principle, which gives

$$\delta\int_{-\infty}^{+\infty} \mathrm{d}t\langle\Psi\,|H - \mathrm{i}\frac{\mathrm{d}}{\mathrm{d}t_{r_T}}|\Psi\rangle = 0 . \tag{52.23}$$

In the two-state CDW approximation we may assume

$$\Psi_{\mathrm{CDW}} = c_0(t)\xi_i^+ + c_1(t)\xi_f^- \qquad (52.24)$$

subject to the boundary conditions $c_0(-\infty) = 1$ and $c_1(-\infty) = 0$. Variation of c_0^* and c_1^* gives the standard coupled equations

$$\mathrm{i}N_{00}\dot{c}_0 + \mathrm{i}N_{01}\dot{c}_1 = H_{00}c_0 + H_{01}c_1\,, \qquad (52.25)$$

$$\mathrm{i}N_{10}\dot{c}_0 + \mathrm{i}N_{11}\dot{c}_1 = H_{10}c_0 + H_{11}c_1\,, \qquad (52.26)$$

where

$$N_{00} = \left\langle \xi_i^+ \middle| \xi_i^+ \right\rangle,$$

$$N_{11} = \left\langle \xi_f^- \middle| \xi_f^- \right\rangle,$$

$$N_{01} = \left\langle \xi_i^+ \middle| \xi_f^- \right\rangle = N_{10}^*\,,$$

and

$$H_{00} = \langle \xi_i^+ | H - \mathrm{i}\frac{\mathrm{d}}{\mathrm{d}t_{r_\mathrm{T}}} | \xi_i^+ \rangle\,,$$

$$H_{01} = \langle \xi_i^+ | H - \mathrm{i}\frac{\mathrm{d}}{\mathrm{d}t_{r_\mathrm{T}}} | \xi_f^- \rangle\,,$$

$$H_{10} = \langle \xi_f^- | H - \mathrm{i}\frac{\mathrm{d}}{\mathrm{d}t_{r_\mathrm{T}}} | \xi_i^+ \rangle\,,$$

$$H_{11} = \langle \xi_f^- | H - \mathrm{i}\frac{\mathrm{d}}{\mathrm{d}t_{r_\mathrm{T}}} | \xi_f^- \rangle\,.$$

Equations (52.25) and (52.26) can clearly be written as a matrix equation,

$$\mathrm{i}\boldsymbol{N}\dot{\boldsymbol{c}} = \boldsymbol{H}\boldsymbol{c}\,, \qquad (52.27)$$

which is then easily generalized for larger expansions of Ψ. By using an orthogonalized basis set of normalized functions in the manner of *Löwdin* [52.32,33], the $\boldsymbol{N}$ matrix reduces to the unit matrix. This will be understood when considering expansions for Ψ from now on.

Another interesting, but potentially ruinous, result of the asymptotic forms (52.19) and (52.20) is their failure to obey the correct long-range Coulomb boundary conditions; compare this with their expressions at the opposite time extreme in (52.10) and (52.11). This has no consequence until second-order VCDW is calculated in which divergent integrals arise as a direct result of this feature of the wave functions. These terms are analogous to the well-known intermediate elastic divergences which occur in Born-type expansions which do not have the correct Coulomb phases.

A novel way to avoid this problem [52.34, 35] is to split the time plane into two parts, allowing the well-behaved set $\{\xi^+\}$ to be used exclusively for $t \le 0$, while the set $\{\xi^-\}$ forms the basis for $t \ge 0$. This phase integral halfway house VCDW is based on the factorization of the scattering matrix $\boldsymbol{S}$ into a product of two Møller matrices,

$$\boldsymbol{S} = \boldsymbol{\Omega}_-^\dagger \boldsymbol{\Omega}_+\,, \qquad (52.28)$$

where $\boldsymbol{\Omega}_-^\dagger$ represents the propagation of the initial state from $t = -\infty$ to $t = 0_-$, while $\boldsymbol{\Omega}_+$ represents the propagation of the final state from $t = +\infty$ to $t = 0_+$.

The total wave function is similarly split into two expansions over an orthogonal basis ψ, with

$$\boldsymbol{\Psi}^- = \boldsymbol{c}^-\psi^+\,, \quad t < 0\,, \qquad (52.29)$$

$$\boldsymbol{\Psi}^+ = \boldsymbol{c}^+\psi^-\,, \quad t > 0\,, \qquad (52.30)$$

where the superscripts on the $\boldsymbol{\Psi}$ correspond to the respective heavy-particle motion. This in turn divides the coupled equations into two sets

$$\mathrm{i}\dot{\boldsymbol{c}}^- = \boldsymbol{H}^{++}\boldsymbol{c}^-\,, \quad t < 0\,, \qquad (52.31)$$

$$\mathrm{i}\dot{\boldsymbol{c}}^+ = \boldsymbol{H}^{--}\boldsymbol{c}^+\,, \quad t > 0\,, \qquad (52.32)$$

where

$$\boldsymbol{H}^{\pm\pm} = \langle \boldsymbol{\Psi}^\pm \mid H - \mathrm{i}\frac{\mathrm{d}}{\mathrm{d}t_{r_\mathrm{T}}} \mid \boldsymbol{\Psi}^\pm \rangle\,. \qquad (52.33)$$

The coefficients $c^\pm$ then have to be matched over a local discontinuity in the total wave function at $t = 0$, such that $\boldsymbol{c}^+(0) = \boldsymbol{c}^-(0)$. Halfway house VCDW has all the appealing attributes of a variational theory but, by explicitly satisfying the long-range Coulomb boundary conditions, it is divergence free.

52.1.4 Ionization

CDW, by treating the Coulomb interactions to such a high degree, has obvious attractions for modelling the ionization process. Single ionization of an electron from an atom by a high-energy projectile is a perturbative process and the Born approximation will match experimental total cross sections rather well. However CDW-like representations of the initial and final states generate better results at lower energies, as well as producing features in the differential cross sections which are beyond the reach of the first Born approximation.

In full CDW ionization theory, the initial state is given by the usual charge transfer wave function ξ_i

(52.5), while the final state takes the form

$$\begin{aligned}\xi_f^- &= (2\pi)^{-2/3} \exp\left(\mathrm{i}\boldsymbol{k}\cdot\boldsymbol{r}_\mathrm{T} - \mathrm{i}\frac{k^2}{2}t\right)\\ &\times \exp\left[-\mathrm{i}\frac{Z_\mathrm{T}Z_\mathrm{P}}{v}\ln(vR+\boldsymbol{v}\cdot\boldsymbol{R})\right]\\ &\times N^*(Z_\mathrm{T}/k)\,{}_1F_1\\ &\times(-\mathrm{i}Z_\mathrm{T}/k;\ 1;\ -\mathrm{i}\boldsymbol{k}\cdot\boldsymbol{r}_\mathrm{T} - \mathrm{i}kr_\mathrm{T})\\ &\times N^*(Z_\mathrm{P}/p)\,{}_1F_1\\ &\times(-\mathrm{i}Z_\mathrm{P}/p;\ 1;\ -\mathrm{i}\boldsymbol{p}\cdot\boldsymbol{r}_\mathrm{P} - \mathrm{i}pr_\mathrm{P})\,,\end{aligned} \tag{52.34}$$

where $\boldsymbol{k}$ $(\boldsymbol{p}=\boldsymbol{k}-\boldsymbol{v})$ is the momentum of the electron relative to the target (projectile) nucleus.

CDW ionization theories presented prior to 1982 produced spuriously large results due to the unnormalized initial state. Since the matrix element $\langle\xi_i^+ \mid \xi_i^+\rangle$ is computationally expensive to calculate as a function of $\boldsymbol{b}$ and t, a very successful alternative is to take the initial state as an eikonal distorted state [52.36], thus ensuring normalization. The initial state in this CDW-EIS theory is taken to be

$$\xi_i^{\mathrm{EIS}} = \hat{D}^+_{-\boldsymbol{v}}(\boldsymbol{r}_\mathrm{P})\Phi_i(\boldsymbol{r}_\mathrm{T},t)C(\boldsymbol{R},t)\,, \tag{52.35}$$

where now

$$\hat{D}^+_{-\boldsymbol{v}}(\boldsymbol{r}_\mathrm{P}) = \exp\left[-\mathrm{i}\frac{Z_\mathrm{P}}{v}\ln(vr_\mathrm{P}+\boldsymbol{v}\cdot\boldsymbol{r}_\mathrm{P})\right]. \tag{52.36}$$

The final state remains as in (52.34).

The CDW final state is most effective when differential cross sections are studied. The most interesting features are in the forward $\theta=0$ ejection angle; i. e., the soft collision peak ($k\simeq 0$), the electron capture to the continuum peak ECC ($k\simeq v$) and the binary encounter peak ($k\simeq 2v$). CDW theories are especially suited to the description of the ECC peak which results theoretically from the presence of the $N(Z_\mathrm{P}/p)$ factor in the wave function. This peak can be analyzed in detail via a multipole expansion and much theoretical work has centered on the dipole parameter β. A negative β is strongly suggested by both experiment and physical intuition. CDW, CDW-EIS and halfway house VCDW all predict different values for β with the last theory being the only one which gives a high energy limit which remains negative [52.37].

Magnetically quantized continuum distorted wave theory also has been considered [52.1] in the description of ionization in ion–atom collisions. This generalizes the CDW-EIS theory of *Crothers* and *McCann* [52.2] to incorporate the azimuthal angle dependence of each CDW in the final state wave function. This is accomplished by the analytic continuation of hydrogenic-like wave functions from below to above threshold, using parabolic coordinates and quantum numbers, including magnetic quantum numbers, thus providing a more complete set of states. The continuation applies to excitation, charge transfer, ionization, and double and hybrid events for both light- and heavy-particle collisions. It has successfully been applied to the calculation of double differential cross sections for the single ionization of the hydrogen atom and for a hydrogen molecule by a proton for electrons ejected in the forward direction at a collision energy of 50 keV and 100 keV, respectively.

It is well known that the CDW-EIS models are the best suited to the intermediate and high energy regions. Recent results for proton-argon total ionization cross sections [52.38] highlight large discrepancies between CDW-EIS theory and experiment for energies below 80 keV. This problem has recently been addressed [52.3]. Here, following the theory of [52.39] the authors in [52.3] use a Born initial state wave function. In the final channel, the usual product of two continuum distorted wave functions each associated with a distinct electron–nucleus interaction is used. In their treatment the target is considered as a one-electron atom and the interactions between this active electron and remaining target electrons are treated by a model potential including both short- and long-range effects. The success of this new theory for low energies is shown in Fig. 52.2. Here it is clear that the calculation in [52.3] gives good agreement for the total cross sections in the energy range 10–300 keV with the measurements of *Rudd* et al. [52.40].

Double ionization in general remains an extremely difficult area for theoretical models based on perturbative expansions – even those with explicit distortions built in. In this process, the explicit correlation between the electrons in the target atom is vitally important. However, one example where CDW theory can overcome these problems is in the bound state wave function of *Pluvinage* [52.41]. This very successful treatment, which also includes a variationally determined parameter, is just the CDW analogy in bound state theory, although this appeared well before Cheshire's paper on scattering. The Pluvinage wave function for the ground state of a two-electron atom is given by

$$\begin{aligned}\phi &= c(\kappa)\left(\frac{Z_\mathrm{T}^3}{\pi}\right)\exp\left[-Z_\mathrm{T}(r_1+r_2)+\mathrm{i}\kappa r_{12}\right]\\ &\times {}_1F_1\left(1+\frac{1}{2\mathrm{i}\kappa};\ 2;\ 2\mathrm{i}\kappa r_{12}\right),\end{aligned} \tag{52.37}$$

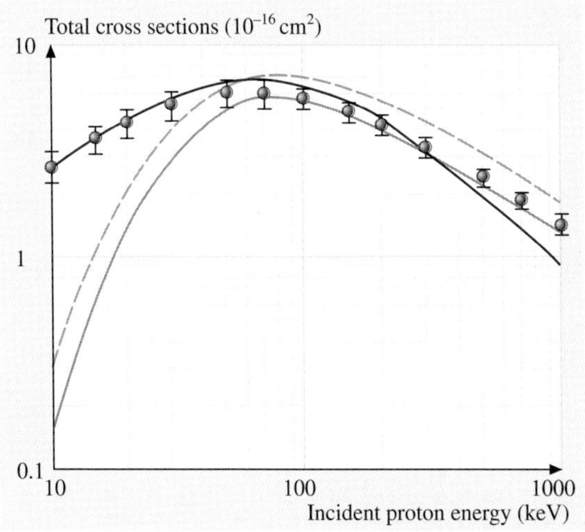

Fig. 52.2 Total ionization cross section for the proton-impact single ionization of Ar: *solid line* theoretical results of [52.3], *dashed line* theoretical results of OPM [52.38] and *dotted line* HFS [52.38]. Experimental data; *circles*, *Rudd* et al. [52.40]

where $r_{1,2}$ are the distances of the two electrons from the target nucleus, and r_{12} is the inter-electron separation.

The constant κ is variationally determined to be 0.41, giving the normalization constant $c = (0.364\,05)^{1/2}$.

An analogous wave function for two electrons in the target continuum was derived [52.11] and implemented later in a successful CDW treatment of ionization by electrons and positrons [52.42]. This BBK theory, so named in deference to the authors, demonstrates the high resolution CDW final states obtained, right down to triply differential cross sections. However, this model suffers from the low-energy normalization problems associated with CDW, and is also unable to describe the threshold effects which are in the domain of Wannier theory.

52.2 Wannier Method

52.2.1 The Wannier Threshold Law

In 1953, *Wannier* [52.43] deduced the relationship between the cross section at the threshold of a reaction, and the excess-of-threshold energy of the incident particle for a three-body ionization problem, where the final state consists of a residual unit positive charge and two electrons, with each body moving in the continuum of the other two. This extended to three bodies the earlier two-body threshold law derived by *Wigner* [52.44] (see Sect. 60.2.1).

For final states with $L = 0$, Wannier employed hyperspherical coordinates $(\rho, \alpha, \theta_{12})$, where

$$\rho^2 = r_1^2 + r_2^2 , \quad \alpha = \tan^{-1}\left(\frac{r_2}{r_1}\right) ,$$
$$\theta_{12} = \cos^{-1}(\hat{\boldsymbol{r}}_1 \cdot \hat{\boldsymbol{r}}_2) . \tag{52.38}$$

Here we assume that the residual ion is infinitely massive and at rest with respect to the two electrons. By converting the two-electron problem to the case of motion of a single point in six-dimensional space, we can take the hyperradius ρ as the 'size' of the hypersphere, α as the radial correlation of the electrons and θ_{12} as their angular correlation. This allows the Schrödinger equation for the final state to be written (in a.u.) as

$$\left(h_0 - \frac{l^2(\hat{\boldsymbol{r}}_1)}{\rho^2\cos^2\alpha} - \frac{l^2(\hat{\boldsymbol{r}}_2)}{\rho^2\sin^2\alpha} + 2E + \frac{2Z(\alpha,\theta_{12})}{\rho}\right)\Psi(\boldsymbol{r}_1,\boldsymbol{r}_2) = 0 , \tag{52.39}$$

where

$$h_0 = \frac{1}{\rho^5}\frac{\partial}{\partial\rho}\left(\rho^5\frac{\partial}{\partial\rho}\right) + \frac{1}{\rho^2\sin^2\alpha\cos^2\alpha}\frac{\partial}{\partial\alpha} \times\left(\sin^2\alpha\cos^2\alpha\frac{\partial}{\partial\alpha}\right) \tag{52.40}$$

and

$$Z(\alpha,\theta_{12}) = \frac{1}{\cos\alpha} + \frac{1}{\sin\alpha} - \frac{1}{(1-\cos\theta_{12}\sin 2\alpha)^{\frac{1}{2}}} \tag{52.41}$$

is the potential surface on which the particle is moving [52.45].

The most likely configuration of the electrons leading to double escape at threshold corresponds to the region $\boldsymbol{r}_1 = -\boldsymbol{r}_2$, i. e., the two electrons escape in opposite directions from the reaction zone, corresponding to the saddle point of the potential surface $Z(\alpha, \theta_{12})$, and defined as the Wannier ridge. Also, dynamic screening between the two electrons means there would be equal partitioning of the available energy for the two particles, and so they would have equal but opposite velocities on escape. In hyperspherical coordinates, the most important region for double escape is, therefore, $\alpha = \pi/4$, and $\theta_{12} = \pi$.

The main conclusion of Wannier's theory is that the total cross section for electron impact single ionization scales as

$$\sigma = kE^{m_{12}} , \tag{52.42}$$

where E is the excess-of-threshold energy,

$$m_{12} = -\frac{1}{4} + \frac{1}{4}\left(\frac{100Z-9}{4Z-1}\right)^{\frac{1}{2}} , \tag{52.43}$$

and Z the residual charge; m_{12} is 1.127 for unit residual charge, with $m_{12} \to 1$ as $Z \to \infty$.

52.2.2 Peterkop's Semiclassical Theory

The Wannier threshold law has been verified both semiclassically [52.46] and quantum mechanically [52.47]. *Peterkop* [52.46] adopted a semiclassical JWKB approach to the problem, by using the three-dimensional WKB ansatz

$$\Psi_0 = P^{\frac{1}{2}} \exp\left(\frac{\mathrm{i}S}{\hbar}\right) , \tag{52.44}$$

for the final-state wave function, where S and P are, respectively, the solutions of the Hamilton–Jacobi equation,

$$(\nabla_1 S)^2 + (\nabla_2 S)^2 = 2(E - V) , \tag{52.45}$$

and the continuity equation,

$$\nabla_1(P\nabla_1 S) + \nabla_2(P\nabla_2 S) = 0 . \tag{52.46}$$

In hyperspherical coordinates, these equations become

$$\left(\frac{\partial S}{\partial \rho}\right)^2 + \frac{1}{\rho^2}\left(\frac{\partial S}{\partial \alpha}\right)^2 + \frac{4}{\rho^2 \sin^2 2\alpha}\left(\frac{\partial S}{\partial \theta_{12}}\right)^2 = 2E + \frac{2Z}{\rho} \tag{52.47}$$

and

$$D_0\left(P\frac{\partial S}{\partial \rho}\right) + \frac{1}{\rho^2} \times\left[D_1\left(P\frac{\partial S}{\partial \alpha}\right) + D_2\left(\frac{\partial S}{\partial \theta_{12}}\right)\right] = 0 , \tag{52.48}$$

where

$$D_0 f = \frac{1}{\rho^5}\frac{\partial}{\partial \rho}(\rho^5 f) , \tag{52.49}$$

$$D_1 f = \frac{1}{\sin^2 2\alpha}\frac{\partial}{\partial \alpha}\left(f \sin^2 2\alpha\right) , \tag{52.50}$$

$$D_2 f = \frac{4}{\sin^2 2\alpha}\frac{1}{\sin\theta_{12}}\frac{\partial}{\partial \theta_{12}}(f \sin\theta_{12}) . \tag{52.51}$$

Following Wannier's hypothesis, solutions of these equations are found in the region $\alpha = \pi/4, \theta_{12} = \pi$. Taking the Taylor expansion for $Z(\alpha, \theta_{12})$ as

$$Z(\alpha, \theta_{12}) = Z_0 + \frac{1}{2}Z_1(\Delta\alpha)^2 + \frac{1}{8}Z_2(\Delta\theta_{12})^2 + \cdots , \tag{52.52}$$

where $\Delta\alpha = \alpha - \pi/4$ and $\Delta\theta_{12} = \theta_{12} - \pi$, it follows from (52.41) that

$$Z_0 = \frac{3}{\sqrt{2}},\ Z_1 = \frac{11}{\sqrt{2}},\ Z_2 = -\frac{1}{\sqrt{2}} . \tag{52.53}$$

Similarly, taking the solution of (52.47) in the form

$$S = S_0(\rho) + \frac{1}{2}S_1(\rho)(\Delta\alpha)^2 + \frac{1}{8}S_2(\rho)(\Delta\theta_{12})^2 + \cdots \tag{52.54}$$

gives

$$\frac{\mathrm{d}S_0}{\mathrm{d}\rho} = \omega , \tag{52.55}$$

$$\omega\frac{\mathrm{d}S_i}{\mathrm{d}\rho} + \frac{S_i^2}{\rho^2} = \frac{Z_i}{\rho} , \quad i = 1, 2 , \tag{52.56}$$

where $\omega = (2E + 2Z_0/\rho)^{\frac{1}{2}}$. The solutions are

$$S_0 = \rho\omega + \frac{Z_0}{\chi}\ln\frac{\rho(\chi+\omega)^2}{2Z_0} , \tag{52.57}$$

$$S_i = \rho^2\omega\frac{1}{u_i}\frac{\mathrm{d}u_i}{\mathrm{d}\rho} , \quad i = 1, 2 , \tag{52.58}$$

where $\chi = (2E)^{\frac{1}{2}}$ and

$$u_i = C_{i1}u_{i1} + C_{i2}u_{i2} , \tag{52.59}$$

$$u_{ij} = \rho^{m_{ij}}\,{}_2F_1\left(m_{ij}, m_{ij}+1; 2m_{ij}+\frac{3}{2}; \frac{-E\rho}{Z_0}\right) , \tag{52.60}$$

$$m_{i1} = -\frac{1}{4} - \frac{1}{2}\mu_i,\ m_{i2} = -\frac{1}{4} + \frac{1}{2}\mu_i , \tag{52.61}$$

$$\mu_i = \frac{1}{2}\left(1 + \frac{8Z_i}{Z_0}\right)^{\frac{1}{2}} . \tag{52.62}$$

where, for $i = 2$, the principal branch is understood.

Expanding P in the same form as S, and restricting the solution to finding P_0, gives

$$P_0 = \frac{C}{\rho^5 \omega u_1 u_2^2}\,, \tag{52.63}$$

where $C \sim C_{12} E^{1-m_{12}}$. By solving these equations, Peterkop extracted the Wannier cross section behavior by matching the exact wave function with an approximate one for which the energy dependence is known at some arbitrarily finite value r_0 of ρ, giving the total cross section as

$$\sigma_{\text{tot}} \sim E^{1.127}\,, \tag{52.64}$$

as required. However, neither this method of Peterkop nor the quantum mechanical approach of [52.47] were able to deduce the constant of proportionality.

52.2.3 The Quantal Semiclassical Approximation

As can be seen from the form of (52.63) for P_0, u_2 vanishes in the double limit $\rho \to +\infty$, $E \to 0$, and so, the semiclassical theory breaks down at the very configuration of importance. To avoid this problem, *Crothers* [52.8] adopted a change of the dependent variable to obtain a uniform JWKB approximation. Taking $\alpha = \pi/4$, (i. e., $\Delta\alpha = 0$) as the natural barrier, he set the final-state wave function as

$$\Psi^{-*} = \frac{x|\sin(\alpha - \pi/4)|^{1/2}}{\rho^{5/2}\sin\alpha\cos\alpha(\sin\theta_{12})^{1/2}}\,, \tag{52.65}$$

so that

$$\left[\frac{\partial^2}{\partial\rho^2} + \frac{1}{\rho^2 \sin|\Delta\alpha|}\frac{\partial}{\partial\alpha}\left(\sin|\Delta\alpha|\frac{\partial}{\partial\alpha}\right) + \frac{1}{\rho^2\sin^2\alpha\cos^2\alpha}\frac{\partial^2}{\partial\theta_{12}^2} + 2E + \frac{2Z}{\rho} + \frac{\left(\frac{1}{4}+\frac{1}{4}\csc^2\Delta\theta_{12}\right)}{\rho^2\sin^2\alpha\cos^2\alpha} - \frac{\csc^2|\Delta\alpha|}{4\rho^2}\right] x = 0\,, \tag{52.66}$$

where $|\Delta\alpha|$ and θ_{12} are, respectively, the polar and azimuthal angles. Near $\theta_{12} = \pi$ and $\alpha = \pi/4$, (i. e., at $\Delta\theta_{12} = \Delta\alpha = 0$), the term $(4\rho^2)^{-1}$ is negligible compared with $\csc^2\theta_{12}/(4\rho^2)$. Also the θ_{12} pseudopotential is clearly attractive while the α potential is repulsive, and both potentials are large just at the region of importance, i. e., at $\Delta\alpha = 0 = \Delta\theta_{12}$.

Again, following the method of Peterkop, the final state wave function is written in the form (52.44), (52.47) and (52.48), where now the action perturbation expression S must be generalized to

$$S = s_0 \ln|\Delta\alpha| + s_1 \ln(\Delta\theta_{12}) + S_0(\rho) + \frac{1}{2}S_1(\rho)(\Delta\alpha)^2 + \frac{1}{8}S_2(\rho)(\Delta\theta_{12})^2 + \ldots\,, \tag{52.67}$$

where the extra logarithmic phases indicate long-range Coulomb potentials. By applying the Kohn variational principle perturbatively, and invoking the Jeffreys' [52.48] connection formula on the Wannier ridge with $\rho = 0$ as the classical turning point, the final state wave function is [52.8]

$$\begin{aligned}\Psi_f^{-*} = {} & \frac{c^{1/2}E^{m_{12}/2}\rho^{m_{12}/2+1/4}r(2Z_0)^{1/4}(-\chi/2\pi)^{1/2}}{(2Z_0/\rho)^{1/4}\rho^{5/2}\sin\alpha\cos\alpha} \\ & \times\delta(\hat{\boldsymbol{k}}_1 - \hat{\boldsymbol{r}}_1)\delta(\hat{\boldsymbol{k}}_2 - \hat{\boldsymbol{r}}_2) \\ & \times\exp\left[4\mathrm{i}(8Z_0\rho)^{-1/2(\Delta\theta_{12})^{-2}}\right] \\ & \times\left\{\exp\left[-\mathrm{i}(8Z_0\rho)^{1/2} - \frac{1}{2}\mathrm{i}(\Delta\alpha)^2\right.\right. \\ & \times(2Z_0\rho)^{1/2}m_{12} - \frac{1}{8}\mathrm{i}(\Delta\theta_{12})^2 \\ & \left.\left.\times(2Z_0\rho)^{1/2}m_{21} - \frac{1}{4}\mathrm{i}\pi\right] - \mathrm{c.c}\right\}\,,\end{aligned} \tag{52.68}$$

where $\chi = 2\pi\,\mathrm{Im}(m_{21})$.

Taking the total cross section for distinguishable particles as

$$\begin{aligned}\sigma = {} & \frac{\pi^2 a_0^2}{k_0}\int\!\!\int \mathrm{d}\hat{\boldsymbol{k}}_1\mathrm{d}\hat{\boldsymbol{k}}_2 \frac{\pi Z_2\tanh\chi}{(2E)^{1/2}}\left|f(\hat{\boldsymbol{k}}_1,\hat{\boldsymbol{k}}_2)\right|^2 \\ & \times\exp\left[\frac{-Z_2}{4(2E)^{1/2}}(\Delta\Theta_{12})^2\pi\tanh\chi\right]\,,\end{aligned} \tag{52.69}$$

where f is the scattering amplitude, then the corresponding triple-differential cross section is

$$\begin{aligned}& \frac{\mathrm{d}^3\sigma}{\mathrm{d}\hat{\boldsymbol{k}}_1\,\mathrm{d}\hat{\boldsymbol{k}}_2\,\mathrm{d}\left(\frac{1}{2}k_1^2\right)} \\ & = \frac{2\pi^2 a_0^2}{k_0}\frac{\mathrm{d}}{\mathrm{d}E}\frac{\pi Z_2\tanh\chi}{(2E)^{1/2}} \\ & \times\exp\left[\frac{-Z_2}{4(2E)^{1/2}}(\Theta_{12} - R\pi)^2\pi\tanh\chi\right] \\ & \times\left|f\left(\hat{\boldsymbol{k}}_1,\hat{\boldsymbol{k}}_2\right)\right|^2.\end{aligned} \tag{52.70}$$

Assuming the contribution from triplet states is negligible, $|f|^2$ can be written as

$$\frac{1}{4}\left|f(\hat{\boldsymbol{k}}_1,\hat{\boldsymbol{k}}_2) + f(\hat{\boldsymbol{k}}_2,\hat{\boldsymbol{k}}_1)\right|^2\,, \tag{52.71}$$

where $f(\hat{\boldsymbol{k}}_1, \hat{\boldsymbol{k}}_2)$ (and by permutation $f(\hat{\boldsymbol{k}}_2, \hat{\boldsymbol{k}}_1)$, hereafter referred to as f and g respectively) is given by

$$f \simeq \frac{2\mathrm{i}}{\pi} \int \mathrm{d}\boldsymbol{r}_1 \mathrm{d}\boldsymbol{r}_2 \mathrm{d}\boldsymbol{r}_3 \Psi_f^{-*} \phi_f(2, \boldsymbol{r}_3) \times (H - E) \exp(\mathrm{i}\boldsymbol{k}_0 \cdot \boldsymbol{r}_1) \psi_i(\boldsymbol{r}_2, \boldsymbol{r}_3) . \quad (52.72)$$

As a test of the above formulation for the process $\mathrm{e}^- + \mathrm{He} \to \mathrm{He}^+ + 2\mathrm{e}^-$ near the ionization threshold, *Crothers* [52.8] used an independent-electron open-shell wave function for the initial bound state helium target, written as

$$\psi_i(\boldsymbol{r}_2, \boldsymbol{r}_3) = \frac{\phi(\boldsymbol{r}_2, z_0)\phi(\boldsymbol{r}_3, \beta) + \phi(\boldsymbol{r}_3, z_0)\phi(\boldsymbol{r}_2, \beta)}{[2(1+S)]^{1/2}} , \quad (52.73)$$

where

$$\phi(\boldsymbol{r}, z_0) = z_0^{3/2} \pi^{-1/2} \exp(-z_0 r) , \quad (52.74)$$

$$S = \left[\int \phi(\boldsymbol{r}, z_0)\phi(\boldsymbol{r}, \beta) \mathrm{d}\boldsymbol{r} \right]^2 = \left(\frac{4 z_0 \beta}{(z_0 + \beta)^2} \right)^3 , \quad (52.75)$$

and z_0 and β take the physical values $z_0 = 1.8072^{1/2}$ and $\beta = 2$. The total singlet cross section was found to be (in atomic units)

$$\sigma = 2.37 E^{m_{12}} a_0^2 , \quad (52.76)$$

in line with Wannier's threshold law, and with experiment [52.49], while the corresponding absolute triple differential cross sections (TDCS) were expressed as

$$\frac{\mathrm{d}^3\sigma}{\mathrm{d}\hat{\boldsymbol{k}}_1 \mathrm{d}\hat{\boldsymbol{k}}_2 \mathrm{d}\left(\frac{1}{2}k_1^2\right)} = \frac{70 c z_0^2 2^{1/2} \chi \tanh\chi}{\pi R_\infty Z_0^{1/2} m_{12}} |f + g|^2 \times \left[\frac{\mathrm{d}}{\mathrm{d}E} E^{m_{12} - 1/2} \exp\left(\frac{-Z_2(\Theta_{12} - \pi)^2 \pi \tanh\chi}{4(2E)^{1/2}} \right) \right] \quad (52.77)$$

in units of $10^{-19}\,\mathrm{cm}^2\,\mathrm{sr}^{-2}\,\mathrm{eV}^{-1}$, where

$$c = \frac{\Gamma(m_{12} + 3/2)\Gamma(m_{12} + 1)}{2\pi Z_0^{m_{12}} \Gamma(2m_{12} + 3/2)} , \quad (52.78)$$

and where f (and similarly g for Θ_2) is given by

$$f = \int_0^\infty \mathrm{d}\rho \rho^{3/2 + m_{21}/2 + 1/4} \sum_{L=0}^{L_{\max}} \mathrm{i}^L (2L+1) j_L\left(\frac{\rho z_0}{2^{1/2}}\right) \times P_L(\cos\Theta_1) \exp\left\{ \frac{1}{8} \times \mathrm{Im}\left[m_{21}(\Theta_{12} - \pi)^2 (2Z_0\rho)^{1/2} \right] \right\} \times 2\cos\left\{ (8Z_0\rho)^{1/2} + \frac{1}{8} \times \mathrm{Re}\left[m_{21}(\Theta_{12} - \pi)^2 (2Z_0\rho)^{1/2} \right] \right\} \times r(\rho, \Theta_{12}) \quad (52.79)$$

with

$$(1+S)^{1/2} r(\rho, \Theta_{12}) = \exp\left(-\frac{\rho z_0}{2^{1/2}}\right) \left(2^{1/2} z_0 - \frac{1}{(1 - \cos\Theta_{12})^{1/2}} \right) + \exp(-2^{1/2}\rho) \left[\frac{64(2)^{1/2}(z_0 - 1)}{(z_0 + 2)^3} + \frac{32}{(z_0 + 2)^3} [2(2)^{1/2} + (z_0 + 2)\rho] \exp\left(\frac{-\rho(z_0 + 2)}{2^{1/2}} \right) \right] . \quad (52.80)$$

These results have been found to compare favorably with both the relative experimental results of [52.50] and the absolute experimental results of [52.51]. The Crothers quantal semiclassical approximation has been successfully applied to other threshold (e, 2e) and (photon, 2e) collisions, namely two-electron photodetachment from H^- [52.52], $\mathrm{He}(^4\mathrm{P}^0_{5/2})$ [52.53] and K^- [52.54]. Further investigations of the TDCS for helium at threshold have since been carried out. The $^3\mathrm{P}^0$ triplet contribution to the TDCS was studied in [52.55], where small but notable improvements in the comparison with experimental results [52.50] and [52.51] were achieved for most configurations of the angles θ_1 and θ_2. The inclusion of contributions from $^3\mathrm{D}^{\mathrm{e},0}$ or $^3\mathrm{F}^0$ to the absolute TDCS were found to be negligible in comparison with the effect of the $^3\mathrm{P}^0$ [52.56], although the admittedly non-Wannier effective-charge investigation of $\theta_{12} = \pi$ by *Pan* and *Starace* [52.57] suggests that $^3\mathrm{F}^0$ may be important at $\theta_1 = \pi/6$ or $5\pi/6$, in line with the experiment of *Rösel* et al. [52.51]. Another aspect which is thought to contribute to the TDCS is explicit correlation in the initial bound state wave function for the helium target, in which the interelectronic distance r_{23} is explicitly contained. Absolute singlet triple-differential cross sections have been obtained [52.58], using a helium ground state wave function developed by *Le Sech* [52.59]. Again, excellent agreement with the singlet results of [52.8] has been achieved, and, in most configurations, notable improvements with the corresponding experimental data are obtained, as shown in Fig. 52.3 for scattering angle $\theta_1 = 60°$ (a) and $90°$ (b), indicating that electron correlation should also be considered if a full picture of threshold ionization is to be achieved.

The last eight years has seen significant new developments and contributions to the Wannier theory. One notable achievement has been the analytical continuation of the uniform semi-classical wave function [52.8] to below the energy threshold to calculate the complex eigenenergies for doubly excited states of helium using a complex Bohr–Sommerfeld quantization rule with at least one complex transition point [52.4, 5]. The real parts of the eigenvalues were found to be in good agreement with the experimental results of *Buckman* et al. [52.60, 61] for the resonance positions while the imaginary parts give the explicit widths of the resonances from which the intensities have been estimated. The theory in [52.4] was considered initially for the inaugural case of $L = 0$. Further investigation [52.5–7] has extended the theory to include resonant states for $L = 1$ and $L = 2$. In the case of $L = 1$, an irrational quantum number was obtained and attosecond lifetimes were obtained. Excellent results were obtained for the resonance positions, lifetimes, intensities, and scaling rules in comparison with the experimental data [52.60]. This success persuaded *Deb* and *Crothers* [52.62] to re-visit the problem of quantal near-threshold ionization of He by electron impact. In particular they re-examined the problem of above threshold ionization of He by electron impact by retaining the term $2L(L+1)/\rho^2$ in the hyperspherical equation

$$\left(\frac{1}{\rho^5}\frac{\partial}{\partial\rho}\rho^5\frac{\partial}{\partial\rho} + \frac{1}{\rho^2\sin^2 2\alpha}\frac{\partial}{\partial\alpha}\sin^2 2\alpha\frac{\partial}{\partial\alpha} + \frac{4}{\rho^2\sin\theta_{12}}\frac{\partial}{\partial\theta_{12}}\sin\theta_{12}\frac{\partial}{\partial\theta_{12}} + 2E + \frac{2\zeta(\alpha,\theta_{12})}{\rho}\frac{2L(L+1)}{\rho^2}\right)\Psi = 0\,. \tag{52.81}$$

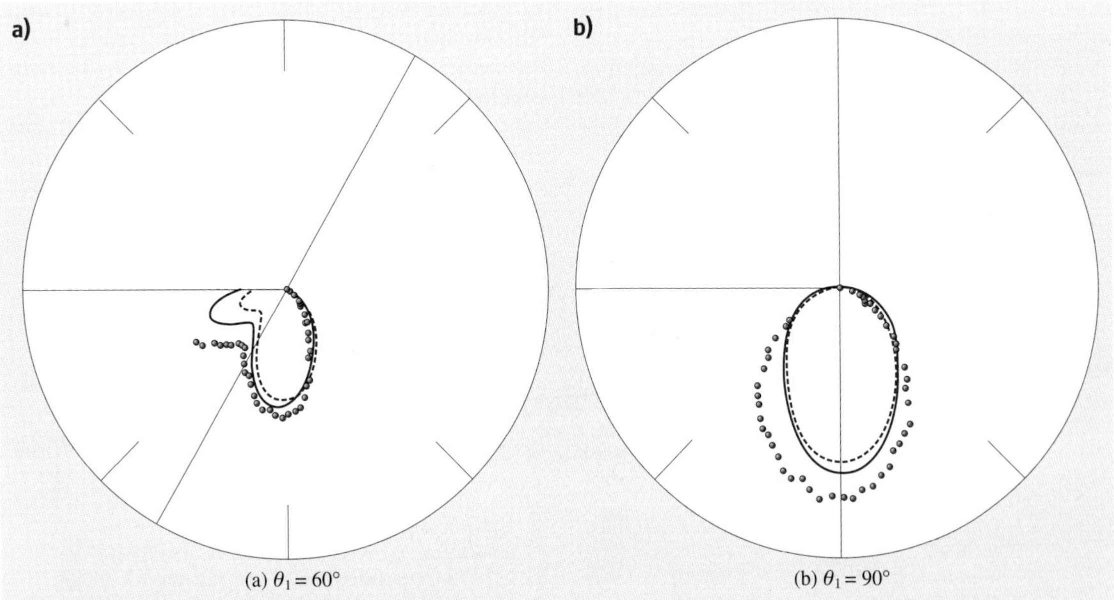

Fig. 52.3a,b Helium triply differential ionization cross section for coplanar geometry, $E_1 = E_2 = 1\,\text{eV}$ and $E = 2\,\text{eV}$, calculated (*full curve*) TDCS of *Copeland* and *Crothers* [52.58] in polar coordinates, with polar coordinates as θ_2, for scattering angles $\theta_1 = 60°$ **(a)** and $\theta_1 = 90°$ **(b)**, in comparison with absolute experimental (*circles*) data of *Rösel* et al. [52.44], and theoretical results (*broken curve*) of *Crothers* [52.40]. The radius of each circle is $1.0\times10^{-19}\,\text{cm}^2\text{sr}^{-2}\text{eV}^{-1}$

Following the procedure of *Crothers* [52.8], they obtained

$$\begin{aligned}\Psi_f^{-*} &= \frac{c^{1/2}E^{m_{12}/2}u_1^{1/2}}{\tilde{\omega}^{1/2}\rho^{5/2}\sin\alpha\cos\alpha}\delta(\hat{\boldsymbol{k}}_1-\hat{\boldsymbol{r}}_1)\\ &\times\delta(\hat{\boldsymbol{k}}_2-\hat{\boldsymbol{r}}2)\exp\left[4\mathrm{i}(8Z_0\rho)^{-1/2}(\Delta\theta_{12})^{-2}\right]\\ &\times\exp\left\{-\mathrm{i}\left[S_0+\frac{1}{2}S_1(\Delta\alpha)^2\right.\right.\\ &\left.\left.+\frac{1}{8}S_2(\Delta\theta_{12})^2+\frac{\pi}{4}\right]-\text{conjugate}\right\},\end{aligned} \tag{52.82}$$

where the classical action variables are given by

$$S_0=\int_0^\rho \mathrm{d}\tilde{\rho}\tilde{\omega}\,, \tag{52.83}$$

$$S_i=\rho^2\omega(\ln u_i)'\,,\quad i=1,2 \tag{52.84}$$

with

$$\tilde{\omega}=\left[\omega^2-\omega\left(\ln_{u2}-\mathrm{i}\ln u_1\right)'\right]^{1/2}, \tag{52.85}$$

$$\omega^2=2E+2Z_0/\rho-2L(L+1)/\rho^2\,. \tag{52.86}$$

The primes in (52.84) and (52.85) denote derivatives with respect to ρ and $\tilde{\rho}$ respectively. It is to be noted here that the original work used the approximated form of ω by dropping the L-dependent term in (52.86). The inclusion of this angular momentum term moves the classical turning point from the origin to ρ_+, where

$$\rho_+=\frac{-Z_0+\sqrt{Z_0^2+4EL(L+1)}}{2E}\,. \tag{52.87}$$

As a result, the lower limit of ρ integration will be replaced by ρ_+. The classical action variables S_1 and S_2 are now evaluated without introducing the limit $E\rho\to 0$. Using the final state wave function in (52.82) we have calculated first the direct ionization amplitude. The exchange ionization amplitudes for the two indistinguishable atomic electrons were then obtained by interchanging the angles θ_1 and θ_2 in the direct amplitude. Singlet and triplet contributions are then accounted for in the usual ratio of 1:3.

Equation (52.82) is a more accurate variant of (52.68) used in the original theory of *Crothers* [52.8]. Using this refinement of the wave function, all partial wave contributions for singlets and triplets are accounted for up to $L=6$ for the case of He by electron impact at an excess of 2 eV above threshold. It has been found that within the co-planar geometry, both the symmetric and asymmetric triple differential cross sections, peaking at and near the Wannier ridge, are greatly improved when compared with experiment [52.63]. However, far away from the Wannier ridge the triple differential cross sections tend to show qualitative differences from measurement [52.63].

The improved theory [52.62] has also successfully been applied to the calculation of total cross sections of positron impact ionization of helium for energies 0.5–10 eV above threshold [52.64]. Excellent agreement with available experimental data [52.65] was obtained for the absolute theoretical calculation of positron impact ionization [52.64].

Finally this recent work has answered some questions concerning near-threshold processes, a large and interesting area of study and we firmly believe that further development of this powerful theory will answer many more.

References

52.1 D. S. F. Crothers, D. M. McSherry, S. F. C. O'Rourke, M. B. Shah, C. McGrath, H. B. Gilbody: Phys. Rev. Lett **88**, 053201 (2002)

52.2 S. F. C. O'Rourke, D. M. McSherry, D. S. F. Crothers: J. Phys. B **36**, 314 (2003)

52.3 S. Bhattacharya, R. Das, N. C. Deb, K. Roy, D. S. F. Crothers: Phys. Rev. A **68**, 052702 (2003)

52.4 A. M. Loughan, D. S. F. Crothers: Phys. Rev. Lett. **79**, 4966 (1997)

52.5 A. M. Loughan: Adv. Chem. Phys. **114**, 312 (2000)

52.6 A. M. Loughan, D. S. F. Crothers: J. Phys. B **31**, 2153 (1998)

52.7 D. S. F. Crothers, A. M. Loughan: Philos. Trans. R. Soc. London A **357**, 1391 (1999)

52.8 D. S. F. Crothers: J. Phys. B **19**, 463 (1986)

52.9 I. M. Cheshire: Proc. Phys. Soc. **84**, 89 (1964)

52.10 B. H. Bransden, M. R. C. McDowell: *Charge Exchange and the Theory of Ion-Atom Collisions* (Clarendon, Oxford 1992)

52.11 D. S. F. Crothers: J. Phys. B **15**, 2061 (1982)

52.12 H. Vogt, R. Schuch, E. Justiniano, M. Schultz, W. Schwab: Phys. Rev. Lett. **57**, 2256 (1986)
52.13 D. S. F. Crothers: J. Phys. B **18**, 2893 (1985)
52.14 D. S. F. Crothers, J. F. McCann: J. Phys. B **17**, L177 (1984)
52.15 S. F. C. O'Rourke, D. S. F. Crothers: J. Phys. B **29**, 1969 (1996)
52.16 D. P. Dewangan, J. Eichler: J. Phys. B **18**, L65 (1985)
52.17 D. S. F. Crothers: *Relativistic Heavy-Particle Collision Theory* (Kluwer Academic Plenum, New York 2000)
52.18 J. Eichler, W. E. Meyerhof: *Relativistic Atomic Collisions* (Academic, San Diego 1995)
52.19 J. T. Glass, J. F. McCann, D. S. F. Crothers: J. Phys. B **27**, 3445 (1994)
52.20 J. T. Glass, J. F. McCann, D. S. F. Crothers: Proc. R. Soc. Lond. A **453**, 387 (1997)
52.21 U. Becker: J. Phys. B **20**, 6563 (1987)
52.22 C. A. Bertulani, G. Baur: Phys. Rev. D **58**, 4005 (1998)
52.23 G. Baur, G. Boero, A. Brauksiepe, A. Buzzo, W. Eyrich, R. Geyer, D. Grzonka, J. Hauffe, K. Kilian, M. LoVetere, M. Macri, M. Moosburger, R. Nellen, W. Oelert, S. Passaggio, A. Pozzo, K. Röhrich, K. Sachs, G. Schepers, T. Sefzick, R. S. Simon, R. Stratmann, F. Stinzing, M. Wolke: Phys. Lett. B **368**, 251 (1996)
52.24 G. R. Deco, R. D. Rivarola: J. Phys. B **19**, 1759 (1986)
52.25 G. R. Deco, O. Fojon, J. Maidagan, R. D. Rivarola: Phys. Rev. A **47**, 3769 (1993)
52.26 W. J. Humphries, B. L. Moiseiwitsch: J. Phys. B **17**, 2655 (1984)
52.27 D. H. Jakubassa-Amundsen, P. A. Amundsen: Z. Phys. A **298**, 13 (1980)
52.28 J. F. McCann, J. T. Glass, D. S. F. Crothers: J. Phys. B **29**, 6155 (1996)
52.29 R. J. S. Lee, J. V. Mullan, J. F. McCann, D. S. F. Crothers: Phys. Rev. A **63**, 062712 (2001)
52.30 G. Baur: Phys. Lett. B **311**, 343 (1993)
52.31 J. Eichler: Phys. Rev. Lett. **75**, 3653 (1995)
52.32 P. O. Löwdin: Ark. Nat. Astron. Phys. **35A**, 918 (1947)
52.33 P. O. Löwdin: J. Chem. Phys. **18**, 365 (1950)
52.34 D. S. F. Crothers: Nucl. Instrum. Methods B **27**, 555 (1987)
52.35 D. S. F. Crothers, L. J. Dubé: Adv. At. Mol. Phys. **30**, 287 (1993)
52.36 D. S. F. Crothers, J. F. McCann: J. Phys. B **16**, 3229 (1983)
52.37 D. S. F. Crothers, S. F. C. O'Rourke: J. Phys. B **25**, 2351 (1992)
52.38 T. Kirchner, L. Gulyas, H. J. Ludde, A. Henne, E. Engel, R. M. Dreizler: Phys. Rev. Lett **79**, 1658 (1997)
52.39 S. Sahoo, R. Das, N. C. Sil, S. C. Mukherjee, K. Roy: Phys. Rev. A **62**, 022716 (2000)
52.40 M. E. Rudd, Y. K. Kim, D. H. Madison, J. W. Gallagher: Rev. Mod. Phys. **57**, 965 (1985)
52.41 P. Pluvinage: Ann. Phys. N. Y. **5**, 145 (1950)
52.42 M. Brauner, J. S. Briggs, H. Klar: J. Phys. B **22**, 2265 (1989)
52.43 G. H. Wannier: Phys. Rev. **90**, 817 (1953)
52.44 E. P. Wigner: Phys. Rev. **73**, 1002 (1948)
52.45 M. R. H. Rudge, M. J. Seaton: Proc. R. Soc. A **283**, 262 (1965)
52.46 R. K. Peterkop: J. Phys. B **4**, 513 (1971)
52.47 A. R. P. Rau: Phys. Rev. A **4**, 207 (1971)
52.48 H. Jeffreys: Proc. Lond. Math. Soc. **23**, 428 (1923)
52.49 S. Cvejanović, F. H. Read: J. Phys. B **7**, 1841 (1974)
52.50 P. Selles, A. Huetz, J. Mazeau: J. Phys. B **20**, 5195 (1987)
52.51 T. Rösel, J. Röder, L. Frost, K. Jung, H. Ehrhardt: J. Phys. B **25**, 3859 (1992)
52.52 J. F. McCann, D. S. F. Crothers: J. Phys. B **19**, L399 (1986)
52.53 D. S. F. Crothers, D. J. Lennon: J. Phys. B **21**, L409 (1988)
52.54 D. R. J. Carruthers, D. S. F. Crothers: J. Phys. B **24**, L199 (1991)
52.55 D. R. J. Carruthers, D. S. F. Crothers: Z. Phys. D **23**, 365 (1992)
52.56 D. R. J. Carruthers: *Ph. D. Thesis* (Queen's University Belfast, Belfast 1993)
52.57 C. Pan, A. F. Starace: Phys. Rev. A **45**, 4588 (1992)
52.58 F. B. M. Copeland, D. S. F. Crothers: J. Phys. B **27**, 2039 (1994)
52.59 L. D. A. Siebbeles, D. P. Marshall, C. Le Sech: J. Phys. B **26**, L321 (1993)
52.60 S. J. Buckman, P. Hammond, F. H. Read, G. C. King: J. Phys B **16**, 4039 (1983)
52.61 S. J. Buckman, D. S. Newman: J. Phys. B **20**, L711 (1987)
52.62 N. C. Deb, D. S. F. Crothers: J. Phys. B **33**, L623 (2000)
52.63 T. Rösel, J. Röder, L. Frost, K. Jung, H. Ehrhardt, S. Jones, D. H. Madison: Phys. Rev. A **46**, 2539 (1992)
52.64 N. C. Deb, D. S. F. Crothers: J. Phys. B **35**, L85 (2002)
52.65 P. Ashley, J. Moxom, G. Laricchia: Phys. Rev. Lett. **77**, 1250 (1996)

53. Ionization in High Energy Ion–Atom Collisions

Atomic species moving at high velocities form an important component of ionizing radiation. When these species interact with matter, they effect chemical and biological changes which originate with primary collision processes, usually the ejection of electrons. This chapter gives an overview of this primary process that has emerged from studies of the energy and angular distribution of electrons ejected by the impact of atomic or ionic projectiles on atomic targets. It seeks to highlight those features which are most ubiquitous [53.1, 2]. Atomic units [53.3] are used, although e, m, and $\hbar$ are sometimes exhibited explicitly to show the connection with standard treatments of the first Born approximation [53.1].

53.1 **Born Approximation** 789
53.2 **Prominent Features** 792
53.2.1 Target Electrons 792
53.2.2 Projectile Electrons 796
53.3 **Recent Developments** 796
References 796

When high-velocity projectiles strike atomic targets, electrons are ejected from the target atoms, and, if the projectiles have some electrons, from the projectile ions also. For partially ionized projectiles, there are two groups of electrons, one from the target and one from the projectile. In principle, these two groups of electrons cannot be separately identified, but in practice each group often predominates in separate energy and angular regions, and can be identified with the help of computed distributions. These disstributions are expressed in terms of doubly differential cross sections (DDCS). Two somewhat different cross sections are used. The Galilean invariant cross sections are differential in the wave vectors $\boldsymbol{k}$ of the Schrödinger waves for the ejected electrons and are denoted by $\mathrm{d}^3\sigma/\mathrm{d}k^3$. The wave vectors $\boldsymbol{k}$ refer to the laboratory or target frame. The Galilean invariant cross sections take the same form in any reference frame, including the projectile frame. In this frame the electron wave vectors are denoted by primes, $\boldsymbol{k}'$. The DDCS is given in terms of an alternative expression in Sect. 53.1, (53.9).

Part D | 53

53.1 Born Approximation

The first Born approximation, often referred to as FBA or B1, provides an excellent framework to understand qualitatively most of the prominent features observed in fast ion–atom collisions. For a bare ion projectile of charge Z_P impinging upon a neutral atom target of nuclear charge Z_T, the Hamiltonian of the system is written as

$$H = H_0 + H_1 , \tag{53.1}$$

where

$$H_0 = \frac{P_\mathrm{P}^2}{2\mu} + \sum_{j=1}^{Z_\mathrm{T}} \left(\frac{p_j^2}{2m} - \frac{Z_\mathrm{T} e^2}{r_j} + \sum_{i>j}^{Z_\mathrm{T}} \frac{e^2}{|\boldsymbol{r}_i - \boldsymbol{r}_j|} \right) ,$$

$$H_1 = \frac{Z_\mathrm{P} Z_\mathrm{T} e^2}{R} - \sum_{j=1}^{Z_\mathrm{T}} \frac{Z_\mathrm{P}^2}{|\boldsymbol{R} - \boldsymbol{r}_j|} , \tag{53.2}$$

and where μ is the projectile reduced mass, m the electron mass, $\boldsymbol{R}$ the position vector of the projectile with respect to the target nucleus, $\boldsymbol{P}_\mathrm{P}$ the corresponding momentum operator, $\boldsymbol{r}_j$ the position vector of the ith target electron and $\boldsymbol{p}_j$ the corresponding momentum operator. The first Born approximation consists of breaking H as in (53.1): the wave functions employed are antisymmetric products of the target eigenfunctions of H_0 and a plane wave of relative motion for the target and projectile. Transitions are induced by the interaction term H_1. For an inelastic collision, the target wave function is orthogonal to the initial state wave function and the matrix element of the first term in H_1, $Z_\mathrm{P} Z_\mathrm{T} e^2/R$, vanishes. Using Bethe's integral

$$\int \frac{\mathrm{e}^{\mathrm{i}q\cdot R}}{|R-r|}\,\mathrm{d}^3r = \frac{4\pi}{q^2}\,\mathrm{e}^{\mathrm{i}q\cdot R}\,, \tag{53.3}$$

the inelastic cross section becomes [53.4]

$$\sigma_{if} = \frac{4\mu^2 e^4}{\hbar^4}\frac{K_f}{K_i}\int \mathrm{d}\Omega_{K_f}\frac{Z_\mathrm{P}^2}{q^4} \times |\langle\Phi_f|\sum_j \exp(\mathrm{i}\boldsymbol{q}\cdot\boldsymbol{r}_j)|\Phi_i\rangle|^2\,, \tag{53.4}$$

where $\hbar\boldsymbol{K}_i$ and $\hbar\boldsymbol{K}_f$ are the initial and final momenta of the projectile, $\hbar\boldsymbol{q}=\hbar(\boldsymbol{K}_f-\boldsymbol{K}_i)$ is the momentum transferred from the projectile to the target, $\Phi_{(f,i)}$ represents the wave functions for the final (f) and initial (i) states of the target electrons, Ω_{K_f} is the solid angle of the scattered projectile, j sums over all target electron coordinates $\boldsymbol{r}_j$, and an average over initial and sum over final states is assumed [53.4]. The integral over $\mathrm{d}\Omega_{K_f}$ can be performed by making the change of variables

$$\mathrm{d}\Omega_{K_f} \equiv \sin\theta_{K_f}\,\mathrm{d}\theta_{K_f}\,\mathrm{d}\phi_{K_f} = \frac{1}{K_i K_f}q\,\mathrm{d}q\,\mathrm{d}\phi_{K_f} \tag{53.5}$$

and substituting $\hbar K_i=\mu v_i$ to obtain

$$\sigma_{if} = \frac{8\pi a_0^2 Z_\mathrm{P}^2}{(v_i/v_\mathrm{B})^2} \times \int_{K_-}^{K_+}\frac{\mathrm{d}q}{q}\frac{|\langle\Phi_f|\sum_{j=1}^{Z_\mathrm{T}}\exp(\mathrm{i}\boldsymbol{q}\cdot\boldsymbol{r}_j)|\Phi_i\rangle|^2}{(qa_0)^2}\,, \tag{53.6}$$

where v_i is the initial relative velocity and $v_\mathrm{B}=\alpha c$ is the atomic unit of velocity. $(v_i/v_\mathrm{B})^2=T/R$ in the notation of [53.4]. From the definition of $\boldsymbol{q}$, the limits of integration are $K_\pm=|K_i\pm K_f|$. The cross section is independent of the mass of the incident projectile.

The dimensionless generalized oscillator strength (GOS) is defined by [53.5]

$$f_{if}(q) \equiv \frac{\Delta E}{R_\infty}\frac{|\langle\Phi_f|\sum_j \exp(\mathrm{i}\boldsymbol{q}\cdot\boldsymbol{r}_j)|\Phi_i\rangle|^2}{(qa_0)^2}\,, \tag{53.7}$$

where ΔE is the energy lost by the projectile in the collsion; again, an average over initial and sum over final magnetic substates is assumed. In terms of the GOS, the cross section becomes

$$\sigma_{if} = \frac{8\pi a_0^2 Z_\mathrm{P}^2}{(v_i/v_\mathrm{B})^2}\int_{K_-}^{K_+}\frac{f_{if}(q)}{\Delta E/R_\infty}\frac{\mathrm{d}q}{q}\,. \tag{53.8}$$

The DDCS, as defined, is related to the Galilean invariant cross section $\mathrm{d}^3\sigma/\mathrm{d}k^3$ by [53.6]

$$\frac{\mathrm{d}^3\sigma_{if}}{\mathrm{d}E_k\,\mathrm{d}\Omega_k} = \left(\frac{m}{\hbar^2}\right)k\frac{\mathrm{d}^3\sigma_{if}}{\mathrm{d}^3k}\,, \tag{53.9}$$

but it is more useful for describing the low energy ejected electrons since it connects smoothly with cross sections for discrete excitations of the target. The first Born approximation for both excitation and ionization by a bare projectile scale as Z_P^2. The advantage of the GOS is that, in the limit $q\to 0$, it approaches the optical oscillator strength – a relationship that connects ionization in fast ion–atom collisions with photoionization [53.7].

For an ionization process where an electron of momentum $\hbar\boldsymbol{k}$ is ejected from the target, the DDCS, differential in ejected electron energy E_k and angle Ω_k, denoted by $\mathrm{d}^3\sigma_{if}/\mathrm{d}E_k\,\mathrm{d}\Omega_k$, is also given by (53.4) with the proviso that the Φ_f becomes Φ_f^-, representing a target ion plus an ejected (continuum) electron with incoming wave boundary conditions [53.5], normalized on the energy scale [53.3], i. e.,

$$\langle\Phi_f^-(E)|\Phi_f^-(E')\rangle = \delta(E-E')\,. \tag{53.10}$$

If the energy is in Rydbergs, the asymptotic form of the radial part, $r\to\infty$, of the continuum wave function of angular momentum ℓ is given by [53.7]

$$\frac{1}{r}\left(\frac{1}{\pi k}\right)^{\frac{1}{2}}\sin\left(kr-\ell\pi/2+k^{-1}\ln 2kr+\sigma_\ell+\delta_\ell\right), \tag{53.11}$$

where $\sigma_\ell \equiv \arg\Gamma(\ell+1-\mathrm{i}/k)$ is the Coulomb phase shift, and δ_ℓ is the non-Coulomb phase shift due to the short range part of the potential. For normalization on the energy scale, in atomic units (1 a.u. $=2R_\infty$), this asymptotic form is multiplied by $\sqrt{2}$, and for normalization on the k-scale, (53.11) is multiplied by $\sqrt{2k}$ [53.3].

One can also obtain the single differential cross section (SDS), $\mathrm{d}\sigma_{if}/\mathrm{d}E_k$, by integrating over Ω_k. This cross section can be expressed in terms of the differential GOS density in the continuum [53.5],

$$\frac{\mathrm{d}f_{if}(q)}{\mathrm{d}E_k} = \frac{\Delta E}{R_\infty}\frac{1}{(qa_0)^2} \times \int |\langle\Phi_f^-|\sum_{j=1}^{Z_\mathrm{T}}\exp(\mathrm{i}\boldsymbol{q}\cdot\boldsymbol{r}_j)|\Phi_i\rangle|^2\,\mathrm{d}\Omega_k\,, \tag{53.12}$$

as

$$\frac{d\sigma_{if}}{dE_k} = \frac{8\pi a_0^2}{(v_i/v_B)^2} \int \frac{df_{if}(q)/dE_k}{\Delta E/R} \frac{dq}{q}, \quad (53.13)$$

with $\Delta E = E_k + I_T$, and I_T is the ionization energy of the target. Finally, the total ionization cross section is obtained by integration of $d\sigma_{if}/dE_k$ over E_k,

$$\sigma_{if} = \frac{8\pi a_0^2 Z_P^2}{(v_i/v_B)^2} \int Z_P^2 \frac{df_{if}(q)/dE_k}{\Delta E/R} \frac{dq}{q} dE_k . \quad (53.14)$$

Let us now consider a collision in which the projectile brings in its own N_P electrons. The Hamiltonian for the system becomes

$$H = H_0 + H_1$$

$$H_0 = \frac{p_P^2}{2\mu} + \sum_{j=1}^{Z_T} \left(\frac{p_j^2}{2m} - \frac{Z_T e^2}{r_j} + \sum_{i>j}^{Z_T} \frac{e^2}{|\boldsymbol{r}_i - \boldsymbol{r}_j|} \right) + \sum_{k=1}^{N_P} \left(\frac{p_k^2}{2m} - \frac{Z_P e^2}{r_k} + \sum_{i>k}^{N_P} \frac{e^2}{|\boldsymbol{r}_i - \boldsymbol{r}_k|} \right)$$

$$H_1 = \frac{Z_P Z_T e^2}{R} - \sum_{j=1}^{Z_T} \frac{Z_P e^2}{|\boldsymbol{R} - \boldsymbol{r}_j|} - \sum_{k=1}^{N_P} \frac{Z_T e^2}{|\boldsymbol{R} + \boldsymbol{r}_k|} + \sum_{j=1}^{Z_T} \sum_{k=1}^{N_P} \frac{e^2}{|\boldsymbol{R} + r_k - \boldsymbol{r}_j|} \quad (53.15)$$

where $\boldsymbol{p}_k$ and $\boldsymbol{r}_k$ refer to projectile electrons. Under these conditions, the solutions of H_0 include the wave function of the projectile electrons in the antisymmetric product. For an inelastic collision, the matrix element of the first term of H_1 vanishes as before. However, there are two extra terms in H_1: the interaction of the projectile electrons with the target nucleus, and the interaction of the target electrons with projectile electrons. These projectile electrons open physically distinct, alternative possibilities for the ionization process [53.2]: (a) target ionization, projectile remains in initial state; (b) target ionization, projectile excited (including ionization); (c) projectile ionization, target remains in initial state; (d) projectile ionization, target excited (including ionization).

For process (a), evaluation of the first Born cross section gives (53.4) with the projectile charge Z_P^2 replaced by $|Z_P - F_{ii}(q)|^2$, where the elastic form factor $F_{ii}(q)$ is given by

$$F_{ii}(q) = \langle \Phi_i^P | \sum_{k=1}^{N_P} \exp(i\boldsymbol{q} \cdot r_k) | \Phi_i^P \rangle , \quad (53.16)$$

and $\Phi_{(f,i)}^P$ represents the wave function for the final and initial electron state of the projectile. Thus, the effect of the projectile electrons on process (a) is to screen the projectile nucleus, and the dynamical screening depends upon the momentum transfer q. Clearly from (53.16), $F_{ii}(0) = N_P$ and $F_{ii}(\infty) = 0$. Thus, for small energy transfer, which implies small q (large impact parameter), $Z_P \to Z_P - N_P$, i. e., full screening of the projectile by its electrons. For large energy transfer, which implies q (small impact parameter), Z_P remains unmodified, i. e., no screening.

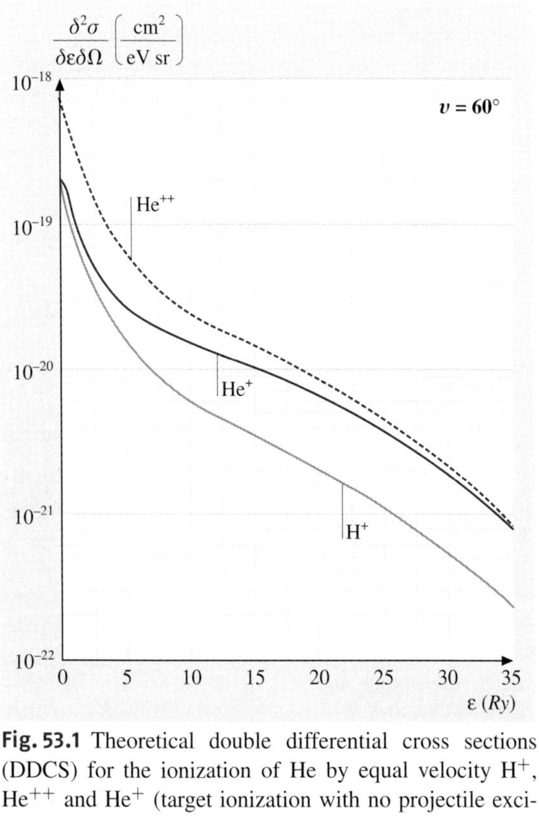

Fig. 53.1 Theoretical double differential cross sections (DDCS) for the ionization of He by equal velocity H^+, He^{++} and He^+ (target ionization with no projectile excitation only) as a function of ejected electron energy in Ry for an ejection angle of 60°. The incident velocity corresponds to 0.5 MeV H^+ and 2.0 MeV He^+ and He^{++} which all have the same velocity as a 20 Ry (272 eV) electron

To illustrate this effect, a calculation of process (a) is presented in Fig. 53.1 for the DDCS for 2 MeV $He^+ + He$ collisions, along with a DDCS calculation for equal velocity H^+ and He^{++} projectiles [53.8]. From these results it is seen that for small energy transfer, the He^+ projectile behaves almost like a heavy proton. With increasing ejected electron energy, however, it is seen to approach He^{++}-like behavior. Note that this result is process (a) alone.

Looking at process (b), the situation is rather different. Now two electrons change their state so that transition matrix elements of all terms except the electron–electron term in H_1 vanish. In this case, Z_P^2 in (53.4) is replaced by the square of the inelastic projectile form factor $|F_{fi}|^2$, where

$$F_{fi}(q) = \langle \Phi_f^P | \sum_{k=1}^{N_P} \exp(\mathrm{i}\boldsymbol{q} \cdot r_k) | \Phi_i^P \rangle . \tag{53.17}$$

Owing to the orthogonality of initial and final projectile states, this form factor vanishes in both the limits, $q \to 0$ and $q \to \infty$. Thus, the total target DDCS is process (a) plus the sum over all of the possible projectile excitations (including projectile ionization) of process (b). Since there is an infinite number of projectile excitations, it is useful to have a method for summing over all of them. Using the closure relation, the sum over all projectile excitations at fixed q is given by [53.2]

$$\sum_{f \neq i} |F_{fi}(q)|^2 = N_P - |F_{ii}(q)|^2 + \langle \Phi_i^P | \times \sum_{j,\, j',\, j \neq j'}^{N_P} \exp[\mathrm{i}\boldsymbol{q} \cdot (\boldsymbol{r}_j - \boldsymbol{r}_{j'})] | \Phi_i^P \rangle , \tag{53.18}$$

where only the initial state wave function of the projectile electrons appears. While this sum rule is exact, it cannot be substituted exactly into (53.4) because of the transformation between $\cos\theta_{K_f}$ and q, and hence the limits of integration over q vary with the excitation energy of the final state of the projectile. Various ways of choosing approximate integration limits have been suggested [53.9].

Projectile ionization alone, and with target excitation, process (c) and (d) above, are handled exactly as (a) and (b), but in the projectile reference frame. Thus, ionization of the projectile by the target is calculated using the methodology detailed above, after which the results are transformed into the laboratory frame using the invariance of $\mathrm{d}^3\sigma_{if}/\mathrm{d}^3k$ and $\boldsymbol{k}_L = \boldsymbol{k}_P + \boldsymbol{K}_i/\mu$. The subscripts L and P refer to the laboratory and projectile frame, respectively.

53.2 Prominent Features

A plot of $\mathrm{d}^3\sigma/\mathrm{d}k^3$ superimposes electrons from the target and electrons from the projectile. To sort out the main features of the two groups of electrons, first consider impact of bare ions on neutral targets, where the DDCS exhibits only electrons ejected from the target. Figure 53.2 shows a computed cross section for protons on H. The wave vectors are resolved into two components: a component $\boldsymbol{k}_\parallel$, which is parallel to the wave vector $\boldsymbol{K}_i$ of the projectile in the laboratory frame, and a component $\boldsymbol{k}_\perp$ perpendicular to $\boldsymbol{K}_i$. In most cases, integration is over the direction of $\boldsymbol{K}_f$ so that $\boldsymbol{K}_i$ is an axis of symmetry. Then the DDCS is independent of the direction of $\boldsymbol{k}_\perp$. Because Fig. 53.2 was computed using an approximate theory described in Sect. 53.2.1, the details are not necessarily accurate, but these computations are qualitatively reliable over an electron energy and angular range which excludes very low electron energies, i. e., the region around $k = 0$.

53.2.1 Target Electrons

The Bethe Ridge

Figure 53.2 provides a useful starting point to examine electron energy and angular distributions. The most prominent feature is the semicircular ridge, called the Bethe ridge, surrounding a valley. The center of the semicircle is at $\boldsymbol{k} = \boldsymbol{v}$, where $\boldsymbol{v} = \boldsymbol{K}_i/M_P$ is the velocity of the projectile in the lab frame. In the projectile frame, where $\boldsymbol{k}' = \boldsymbol{k} - \boldsymbol{v}$, the center of the circle is at $\boldsymbol{k}' = 0$ and the ridge is at $k' = v$. This region corresponds to ionization events where the momementum $\boldsymbol{J} = \boldsymbol{q} - \boldsymbol{k}$ of the recoiling target ion vanishes. Momentum conservation implies that

$$\boldsymbol{K}_i = \boldsymbol{K}_f + \boldsymbol{k} + \boldsymbol{J} , \tag{53.19}$$

while energy conservation is expressed as

$$\frac{K_i^2}{2M_P} + \varepsilon_i = \frac{K_f^2}{2M_P} + \frac{k^2}{2} + \frac{J^2}{2M_T} , \tag{53.20}$$

where $M_{\rm T}$ is the mass of the ionized target. At the Bethe ridge, $\boldsymbol{J} \approx 0$ so that (53.19) and (53.20) combine to determine the value of $\boldsymbol{k}$ at this point, called $\boldsymbol{k}_{\rm B}$;

$$[(1+m/M_{\rm P})\boldsymbol{k}_{\rm B} - \boldsymbol{v}]^2 = v^2 + (1+m/M_{\rm P})\varepsilon_i \tag{53.21}$$

where ε_i is the initial electron eigenenergy. In the approximation that $m/M_{\rm P}$ is set equal to zero and the inital binding energy is small compared with $1/2v^2$, (53.21) shows that the prominent ridge seen in Fig. 53.2 at $k' = v$ corresponds to collisions where all of the momentum lost by the projectile is transferred to the ejected electron. This ridge extends from $E_k = 0$ up to electron energies of the order of $2v^2$. Generally, ejected electrons move in the combined field of both the target and the projectile. Electrons with momenta $k_\parallel$ in the region $0 < k_\parallel < vZ_{\rm P}/(Z_{\rm P}+Z_{\rm P})$ will be referred to as low energy electrons since they move in regions where the target potential is stronger than that of the projectile. For very fast projectiles, many more electrons are in the low energy region than at higher energies, thus this region is of special interest. Since the target potential influences the motion significantly, the ionization process in this region represents a continuation of excitation across the ionization threshold. To analyze this region, cross sections differential in E_k and angle Ω_k in the target frame are preferable since they connect smoothly with cross sections for exciting target states (53.4). Because the DDCS is independent of the azimuthal angle φ_k, the DDCS integrated over φ_k is often employed.

As the variables E_k, θ_k suggest, the main features of the low energy electrons emerge in plots of energy and angular distributions. For fixed energy, the Bethe ridge appears at an angle $\theta_{\rm B}$ given by (53.21) and can be written as a relation between $\cos\theta_{\rm B}$ and E_k as

$$\cos\theta_{\rm B} = \left(\frac{E_k+\varepsilon_i}{\sqrt{2E_k}}\right)\frac{1}{v}\,. \tag{53.22}$$

For fixed $E_k \neq 0$, the DDCS maximizes at an angle that approaches 90° as $v \to \infty$; in particular this angle is obtained when $E_k = -\varepsilon_i$. Figure 53.3 shows the angular distribution of 13.6 eV electrons ejected from He by 5 MeV proton impact [53.10]. The cross section maximizes at, and is nearly symmetric about, 90°. This feature is understandable from the Born approximation.

For small deflection angles, (53.20) gives

$$q_\parallel \approx K_i - K_f \approx \frac{E_k+\varepsilon_i}{v}\,. \tag{53.23}$$

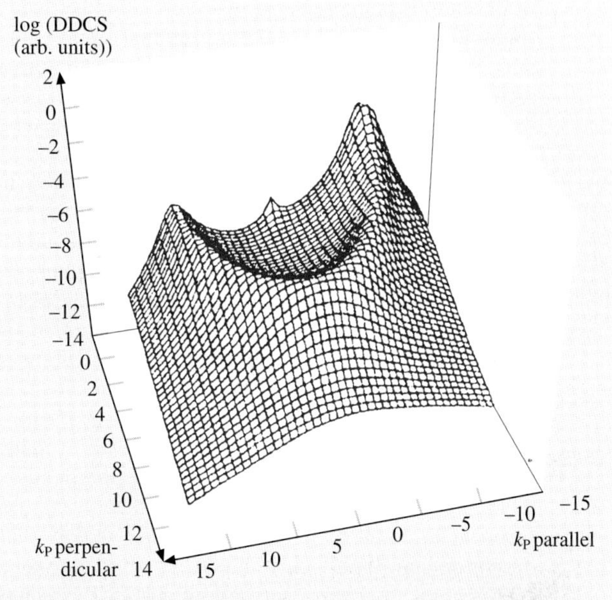

Fig. 53.2 Plot of the Galilean invariant DDCS for electrons ejected from atomic hydrogen by 1.5 keV proton impact computed in the distorted wave strong potential Born approximation (DSPB)

For electron energies E_k of the order of the initial binding energy $|\varepsilon_i|$ and large v, the parallel component of $\boldsymbol{q}$ is much smaller than the magnitude given by (53.19), so that $\boldsymbol{q}$ is predominantly perpendicular to the beam direction. From (53.4), electrons ejected

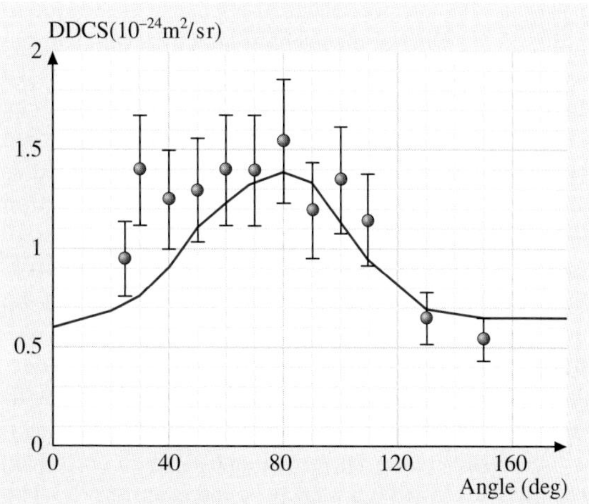

Fig. 53.3 Plot of the angular distribution of 13.6 eV electrons ejected from He by 5 keV proton impact [53.10]. The *solid curve* is the Born approximation and the *points* are the measured values

from isotropic inital states are distributed symmetrically about this axis, i.e., about 90°, in first approximation. Because $\boldsymbol{q}$ is averaged over in forming the cross section, all directions contribute to some extent, and the distribution is not completely symmetric. The solid curve in Fig. 53.3 represents a distribution computed in the Bethe–Born approximation, and it is seen to peak at 90° but is somewhat asymmeteric about this angle, reflecting effects of averaging over the direction of $\boldsymbol{q}$. Such distributions are well described by only a few partial waves for the outgoing electron. Indeed, the theoretical distribution is well described by a combination of s and p waves. The calculations fit the experimental data except in the forward direction where the experimental distribution increases sharply. Such sharp increases must reflect the presence of many partial waves not described by the Born approximation. Since the main disagreement between theory and experiment is confined to a small angular region near 0° which contributes only a small part to the total cross section, the Bethe–Born approximation represents a well-founded theoretical framework for interpreting features that appear in the low-electron-energy portion of the DDCS.

While the main contribution to total ionization cross sections comes from the low energy region, electrons in the high energy region ($v/2 < k$) carry more energy per electron, and therefore play a prominent role in processes initiated by fast electrons. The fast electrons on the Bethe ridge correspond to collisions where the momentum $\boldsymbol{q}$ lost by the incident projectile is transferred mainly to the ejected electron. Because the momentum $\boldsymbol{k}$ of the ejected electrons is much larger than the mean value of the electron momentum in the initial state, given by $\sqrt{2\varepsilon_i}$, this portion of the spectrum can be interpreted in terms of a binary collision between the incident projectile and the target electron treated as quasifree. By quasifree we understand that the electron has a nonzero binding energy ε_i, but is otherwise regarded as a free electron with an initial momentum $\boldsymbol{s}$, distributed over a range of values centered around the mean value. In the projectile reference frame, the electron has the momentum $\boldsymbol{s}-\boldsymbol{v}$. The electron scatters elastically from the projectile and emerges with a final momentum $\boldsymbol{k}' = \boldsymbol{k}-\boldsymbol{v}$. This simple picture is known as the *elastic scattering model*, and is often employed for processes involving weakly bound electrons [53.11]. There are several quantal versions of this picture, indeed any theory of ionization must reduce to the electron scattering model in the limit that the initial binding energy ε_i vanishes.

When the projectile P is a bare ion so that the binary scattering process is just Rutherford scattering, the Bethe–Born approximation is in accord with this picture. It works because the electrons are fast so that effects of the target potential in the final state are unimportant, and because a first-order computation of Rutherford scattering gives the same scattering cross section as the exact Rutherford amplitude. For that reason the domain of applicability of the first Born approximation includes the Bethe ridge, even in the high energy region.

The binary encounter peak shows new features when the projectile carries electrons. The target electrons scatter from a partially screened ion and the Bethe ridge reflects properties of the elastic scattering cross section for the complex projectile species. In contrast to the Rutherford cross section, elastic scattering cross sections for complex species may have maxima and minima as functions of the electron energy and angle. Such features have been identified in collisions of highly charged ions with neutral atomic targets [53.12].

The Continuum Electron Capture Cusp

Figure 53.2 shows a sharp cusp-like structure when the momentum of the ejected electron in the projectile frame k' vanishes. Then the ejected electron moves with a velocity exactly equal to the projectile velocity. For charged projectiles, it is virtually impossible to determine whether such electron states are Rydberg states of high principal quantum number n', or continuum states with $E_{k'} = \frac{1}{2}k'^2$ much less than ε_i. Indeed, the physical similarity of such atomic states implies that the cross section for electron capture to states of high n' differs from the ionization cross section near $k' = 0$ only by a density of states factor. This factor is just $\mathrm{d}E_{k'}$ for ionization processes and $\mathrm{d}E_{n'} = Z_\mathrm{P}^2 n'^{-3}$ for capture. It follows that the cross section $\mathrm{d}^3\sigma/\mathrm{d}E_{k'}\,\mathrm{d}\Omega_{k'}$ is a smooth function of $E_{k'}$ which is nonzero and finite at $E_{k'} = 0$. The Galilean invariant cross section of Fig. 53.2, however, is not smooth; rather it is given by

$$\frac{\mathrm{d}^3\sigma}{\mathrm{d}k^3} = \frac{\mathrm{d}^3\sigma}{\mathrm{d}k'^3} = \frac{1}{k'}\frac{\mathrm{d}^3\sigma}{\mathrm{d}E_{k'}\,\mathrm{d}\Omega_{k'}} . \tag{53.24}$$

Since $\mathrm{d}^3\sigma/\mathrm{d}E_{k'}\,\mathrm{d}\Omega_{k'}$ is nonzero at $k' = 0$, it follows that the DDCS has a k'^{-1} singularity at $k' = 0$. Experiments measure cross sections averaged over this singular factor. The averaging results in the cusp-shaped feature at $k' = 0$ seen in Fig. 53.2 [53.13–15].

The theoretical description of the cusp feature requires that the electron in the final state move mainly in the field of the projectile rather than that of the target. While the appropriate final state wave function can be introduced into (53.4), the result is quantitatively inaccurate. A more accurate amplitude emerges when the exact amplitude is expanded in powers the interaction potential V_T of the electron with the final state target ion [53.16], since V_T has a much smaller effect than the interaction V_P. Approximate evaluation of the ionization amplitude in an independent particle approximation gives [53.17]

$$T_{fi} = \int \mathrm{d}^3 s\, \tilde{V}_\text{T}(\boldsymbol{J}+\boldsymbol{s})\tilde{\varphi}_i(\boldsymbol{s}) \times \langle \psi^-_{\boldsymbol{k}'}(\boldsymbol{r})|\exp[\mathrm{i}(\boldsymbol{J}+\boldsymbol{s})]|\psi_{\boldsymbol{s}-\boldsymbol{v}}(\boldsymbol{r})\rangle\,, \quad (53.25)$$

and

$$\frac{\mathrm{d}^3\sigma}{\mathrm{d}E_k\,\mathrm{d}\Omega_k} = (2\pi)^4\mu^2 \int |T_{fi}|^2\,\mathrm{d}\Omega_{K_f} \quad (53.26)$$

where μ is the reduced mass given by

$$\mu = M_\text{P}(M_\text{P}+1)/(M_\text{P}+M_\text{T}+1)\,, \quad (53.27)$$

and $\tilde{\varphi}_i(\boldsymbol{s})$ is the momentum space independent-particle orbital wave function for the active electron in the intial state, $\tilde{V}_\text{T}$ is the corresponding effective interaction potential, and the $\psi^{\pm}_{\boldsymbol{k}'}$ are normalized on the momentum scale.

Figure 53.4 gives a pictorial interpretation of this amplitude. The initial momentum distribtution of the electron is given by $\tilde{\varphi}_i(\boldsymbol{s})$. In the projectile frame, this electron has momentum $\boldsymbol{s}-\boldsymbol{v}$ and moves in a projectile continuum state represented by $\psi^+_{\boldsymbol{s}-\boldsymbol{v}}(\boldsymbol{r})$. Owing to the interaction with the target represented by $\exp[\mathrm{i}(\boldsymbol{J}+\boldsymbol{s})]\,\tilde{V}_\text{T}(\boldsymbol{J}+\boldsymbol{s})$, a transition to the final projectile state $\psi^-_{\boldsymbol{k}'}$ occurs as illustrated in Fig. 53.4 (b). The final state $\psi^-_{\boldsymbol{k}'}$ has a $1/\sqrt{k'}$ normalization that gives rise to the cusp at $\boldsymbol{k}=\boldsymbol{v}$ seen in Fig. 53.2. This picture describes ionization in terms of the free–free transition, $\psi^-_{\boldsymbol{s}-\boldsymbol{v}} \to \psi^+_{\boldsymbol{k}'}$.

The amplitude in (53.25), known as the distorted wave strong potential Born amplitude (DSPB) [53.18], or the projectile impulse approximation amplitude [53.19], also describes binary encounter electrons. Thus the regions of applicability of (53.4) and (53.25) overlap considerably. In the binary encounter region, where $k_\parallel > v/2$ and $\boldsymbol{J} \approx 0$, the cross section given by (53.9) and (53.25) reduces to the elastic scattering model (ESM) [53.20], where the incident electron with projectile frame momentum $-\boldsymbol{J}-\boldsymbol{v}$ quasi-elastically scatters

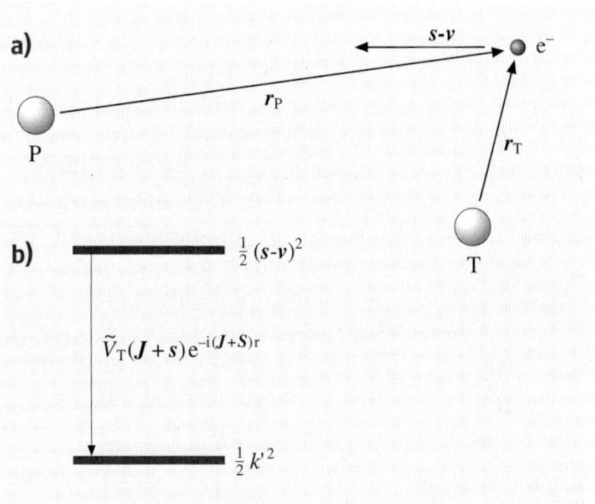

Fig. 53.4a,b Free–free transition picture of ionization. **(a)** Schematic representation of a target electron with momentum $\boldsymbol{s}-\boldsymbol{v}$ in the projectile frame. **(b)** A free–free transition from the projectile eigenstate $\psi^-_{\boldsymbol{s}-\boldsymbol{v}}$ to the final eigenstate $\psi^+_{\boldsymbol{k}'}$ occurs with the target interaction as the transition operator

from the projectile into a final state with projectile frame momentum $\boldsymbol{k}'$. This quasi-elastic cross section is averaged over the initial momentum distribution of the electron $|\tilde{\varphi}_i(-\boldsymbol{J})|^2\,\mathrm{d}^3 J$ so that

$$\mathrm{d}^5\sigma^\text{ESM} = \left[(2\pi)^4\frac{k'}{v}|T^\text{elas}(\boldsymbol{k}', -\boldsymbol{J}-\boldsymbol{v})|^2|\mathrm{d}\Omega_{k'}\right] \times |\tilde{\varphi}_i(-\boldsymbol{J})|^2\,\mathrm{d}^3 J\,. \quad (53.28)$$

The projectile frame cross section is recovered upon using the relation $\mathrm{d}^3 J = \mu_f\mu_i v\,\mathrm{d}\Omega_{K_f}k'^2\,\mathrm{d}k'$ in (53.25).

The two approximations (53.4) and (53.25) describe all of the major features that emerge in the ionization of one-electron targets by fully ionized projectiles. Both theories also incorporate some multi-electron effects that appear when either the target or projectile (or both) have several electrons. Equation (53.28) is useful for describing effects of target ionization by multi-electron projectiles when the projectile electrons remain in their ground state since T^elas incorporates, in principle, the exact amplitude for electrons to scatter from the projectile in its ground state. Of course, the multi-electron projectile can also become excited or ionized owing to its interaction with the target electrons. These processes are not incorporated in the ESM cross section (53.28); rather the Born amplitude of (53.4) represents a more tractable theory to analyze these features.

53.2.2 Projectile Electrons

When the incident atomic species carries some attached electrons, these electrons may be removed in the collision with a target. Essentially all of the features that appear for the target electrons also appear for the projectile electrons; however the DDCS is shifted by the Galilean transformation to the projectile frame. The transformation from the laboratory frame to the projectile frame is an essential step for interpreting features of the projectile electrons. This transformation has the effect of spreading out the electron energy distribution since at $\theta_k = 0°$ a small energy interval ΔE_k in the lab frame relates to a small energy interval $\Delta E'_k$ in the projectile frame according to

$$\Delta E_k = \Delta E'_k \left(1 + \frac{v}{\sqrt{2vE_{k'}}}\right) . \quad (53.29)$$

For example, when the projectile energy is 1.5 MeV/au, a projectile frame electron energy interval $0 < E'_k < 0.1$ eV maps into an interval of ± 8.3 eV centered at $E_k = 817$ eV. This amplification of both the electron energy and the electron energy interval proves useful for measuring features of projectile species with high resolution [53.21].

53.3 Recent Developments

Added by Mark M. Cassar. The theoretical study of ionization processes in atomic collisions remains an active research field. At present, there are three main quantum mechanical approaches to the simplest colliding systems with one active electron. The first approach is to solve directly the time-dependent Schrödinger equation by taking advantage of modern-day computing power. The accuracy of these calculations, however, is still insufficient, and further computational and technological advancements are necessary. The second approach expands the total wave function in atomic or molecular bases; satisfactory agreement with experimental results is obtained, but at the expense of having to include a large number of basis set members in the expansions. The third approach involves the use of Sturmian expansions, along with a specific scaling and transformation of the wave functions. This technique overcomes many of the difficulties that others suffer from, and at the same time provides the best agreement with experimental data over a broad range of energies.

Problems in our understanding of the ionization process remain, even for the simplest colliding system H^+-H, and extensions and refinements of the present theoretical and experimental methods are needed. The reader is referred to the in-depth review by *Macek* et al. [53.22].

References

53.1 U. Fano: Ann. Revs. Nucl. Sci. **13**, 1 (1963)

53.2 J. S. Briggs, K. Taulbjerg: Theory of inelastic atom–atom collisions. In: *Structure and Collisions in Ions and Atoms*, ed. by I. A. Sellin (Springer, Berlin, Heidelberg 1978) pp. 105–153

53.3 H. A. Bethe, E. E. Salpeter: *Quantum Mechanics of One- and Two-Electron Atoms* (Plenum, New York 1977)

53.4 M. Inokuti: Rev. Mod. Phys. **43**, 297 (1971)

53.5 S. Geltman: *Topics in Atomic Collision Theory* (Academic, New York 1969) pp. 41–44

53.6 F. Drepper, J. S. Briggs: J. Phys. B **9**, 2063 (1976)

53.7 A. F. Starace: Theory of atomic photoionization. In: *Handbuch der Physik*, Vol. 31, ed. by W. Mehlhorn (Springer, Berlin, Heidelberg 1982)

53.8 S. T. Manson, L. H. Toburen: Phys. Rev. Lett. **46**, 529 (1981)

53.9 H. M. Hartley, H. R. J. Walters: J. Phys. B **20**, 1983 (1987)

53.10 S. T. Manson, L. Toburen, D. Madison, N. Stolterfoht: Phys. Rev. A **12**, 60 (1975)

53.11 M. M. Duncan, M. G. Menendez: Phys. Rev. A **16**, 1799 (1977)

53.12 D. H. Lee, P. Richard, T. J. M. Zouros, J. M. Sanders, J. L. Shinpaugh, H. Hidmi: Phys. Rev. A **41**, 4816 (1990)

53.13 M. E. Rudd, J. H. Macek: Case Studies in Atomic Physics, **3**, 47 (1972)

53.14 G. G. Crooks, M. E. Rudd: Phys. Rev. Lett. **25**, 1599 (1970)

53.15 K. G. Harrison, M. W. Lucas: Phys. Lett. **33 A**, 142 (1970)
53.16 J. S. Briggs: J. Phys. B **10**, 3075 (1977)
53.17 M. Brauner, J. H. Macek: Phys. Rev. A **46**, 2519 (1992)
53.18 K. Taulbjerg, R. Barrachina, J. H. Macek: Phys. Rev. A **41**, 207 (1990)
53.19 D. Jakubassa-Amundsen: J. Phys. B **16**, 1767 (1983)
53.20 J. Macek, K. Taulbjerg: J. Phys. B **26**, 1353 (1993)
53.21 N. Stolterfoht: Phys. Rep. **146**, 315 (1987)
53.22 S. Yu. Ovchinnikov, G. N. Ogurtsov, J. H. Macek, Yu. S. Gordeev: Phys. Rep. **389**, 119 (2004)

54. Electron–Ion and Ion–Ion Recombination

Electron–ion and ion–ion recombination processes are of key importance in understanding the properties of plasmas, whether they are in the upper atmosphere, the solar corona, or industrial reactors on earth. This is a collection of formulae, expressions, and specific equations that cover the various aspects, approximations, and approaches to electron-ion and ion-ion recombination processes.

54.1 **Recombination Processes** 800
54.1.1 Electron–Ion Recombination 800
54.1.2 Positive–Ion Negative–Ion Recombination 800
54.1.3 Balances 800

54.2 **Collisional-Radiative Recombination**..... 801
54.2.1 Saha and Boltzmann Distributions............................ 801
54.2.2 Quasi-Steady State Distributions.................... 802
54.2.3 Ionization and Recombination Coefficients 802
54.2.4 Working Rate Formulae............. 802

54.3 **Macroscopic Methods** 803
54.3.1 Resonant Capture-Stabilization Model: Dissociative and Dielectronic Recombination . 803
54.3.2 Reactive Sphere Model: Three-Body Electron–Ion and Ion–Ion Recombination....... 804
54.3.3 Working Formulae for Three-Body Collisional Recombination at Low Density.... 805
54.3.4 Recombination Influenced by Diffusional Drift at High Gas Densities................................. 806

54.4 **Dissociative Recombination** 807
54.4.1 Curve-Crossing Mechanisms........ 807
54.4.2 Quantal Cross Section................ 808
54.4.3 Noncrossing Mechanism............ 810

54.5 **Mutual Neutralization** 810
54.5.1 Landau–Zener Probability for Single Crossing at R_X 811
54.5.2 Cross Section and Rate Coefficient for Mutual Neutralization 811

54.6 **One-Way Microscopic Equilibrium Current, Flux, and Pair-Distributions**..... 811

54.7 **Microscopic Methods for Termolecular Ion–Ion Recombination** 812
54.7.1 Time Dependent Method: Low Gas Density....................... 813
54.7.2 Time Independent Methods: Low Gas Density....................... 814
54.7.3 Recombination at Higher Gas Densities................................ 815
54.7.4 Master Equations 816
54.7.5 Recombination Rate 816

54.8 **Radiative Recombination** 817
54.8.1 Detailed Balance and Recombination-Ionization Cross Sections 817
54.8.2 Kramers Cross Sections, Rates, Electron Energy-Loss Rates and Radiated Power for Hydrogenic Systems 818
54.8.3 Basic Formulae for Quantal Cross Sections 819
54.8.4 Bound-Free Oscillator Strengths.. 822
54.8.5 Radiative Recombination Rate 822
54.8.6 Gaunt Factor, Cross Sections and Rates for Hydrogenic Systems 823
54.8.7 Exact Universal Rate Scaling Law and Results for Hydrogenic Systems 823

54.9 **Useful Quantities** 824

References 824

54.1 Recombination Processes

54.1.1 Electron–Ion Recombination

This proceeds via the following four processes:
(a) radiative recombination (RR)

$$e^- + A^+(i) \to A(n\ell) + h\nu , \quad (54.1)$$

(b) three-body collisional-radiative recombination

$$e^- + A^+ + e^- \to A + e^- , \quad (54.2a)$$

$$e^- + A^+ + M \to A + M , \quad (54.2b)$$

where the third body can be an electron or a neutral gas.
(c) dielectronic recombination (DLR)

$$e^- + A^{Z+}(i) \rightleftharpoons \left[A^{Z+}(k) - e^-\right]_{n\ell} \to A_{n'\ell'}^{(Z-1)+}(f) + h\nu , \quad (54.3)$$

(d) dissociative recombination (DR)

$$e^- + AB^+ \to A + B^* . \quad (54.4)$$

Electron recombination with bare ions can proceed only via (a) and (b), while (c) and (d) provide additional pathways for ions with at least one electron initially or for molecular ions AB^+. Electron radiative capture denotes the combined effect of RR and DLR.

54.1.2 Positive–Ion Negative–Ion Recombination

This proceeds via the following three processes:
(e) mutual neutralization

$$A^+ + B^- \to A + B^* , \quad (54.5)$$

(f) three-body (termolecular) recombination

$$A^+ + B^- + M \to AB + M , \quad (54.6)$$

(g) tidal recombination

$$AB^+ + C^- + M \to AC + B + M \quad (54.7a)$$

$$\to BC + A + M , \quad (54.7b)$$

where M is some third species (atomic, molecular or ionic). Although (e) always occurs when no gas M is present, it is greatly enhanced by coupling to (f). The dependence of the rate $\hat{\alpha}$ on density N of background gas M is different for all three cases, (e)–(g).

Processes (a), (c), (d) and (e) are elementary processes in that microscopic detailed balance (proper balance) exists with their true inverses, i. e., with photoionization (both with and without autoionization) as in (c) and (a), associative ionization and ion-pair formation as in (d) and (e), respectively. Processes (b), (f) and (g) in general involve a complex sequence of elementary energy-changing mechanisms as collisional and radiative cascade and their overall rates are determined by an input-output continuity equation involving microscopic continuum-bound and bound–bound collisional and radiative rates.

54.1.3 Balances

Proper Balances

Proper balances are detailed microscopic balances between forward and reverse mechanisms that are direct inverses of one another, as in

(a) Maxwellian: $e^-(v_1) + e^-(v_2) \rightleftharpoons e^-(v_1') + e^-(v_2')$, (54.8)

where the kinetic energy of the particles is redistributed;

(b) Saha: $e^- + H(n\ell) \rightleftharpoons e^- + H^+ + e^-$ (54.9)

between direct ionization from and direct recombination into a given level $n\ell$;

(c) Boltzmann: $e^- + H(n\ell) \rightleftharpoons e^- + H(n', \ell')$ (54.10)

between excitation and de-excitation among bound levels;

(d) Planck: $e^- + H^+ \rightleftharpoons H(n\ell) + h\nu$, (54.11)

which involves interaction between radiation and atoms in photoionization/recombination to a given level $n\ell$.

Improper Balances

Improper balances maintain constant densities via production and destruction mechanisms that are not pure inverses of each other. They are associated with flux activity on a macroscopic level as in the transport of particles into the system for recombination and net production and transport of particles (i. e. e^-, A^+) for ionization. Improper balances can then exist between dissimilar elementary production–depletion processes as in (a) coronal balance between electron-excitation into and radiative decay out of level n. (b) radiative balance between radiative capture into and radiative cascade out of level n. (c) excitation saturation balance between upward collisional excitations $n-1 \rightarrow n \rightarrow n+1$ between adjacent levels. (d) de-excitation saturation balance between downward collisional de-excitations $n+1 \rightarrow n \rightarrow n-1$ into and out of level n.

54.2 Collisional-Radiative Recombination

Radiative Recombination

Process (54.1) involves a free-bound electronic transition with radiation spread over the recombination continuum. It is the inverse of photoionization without autoionization and favors high energy gaps with transitions to low $n \approx 1, 2, 3$ and low angular momentum states $\ell \approx 0, 1, 2$ at higher electron energies.

Three-Body Electron–Ion Recombination

Processes (54.2a,b) favor free-bound collisional transitions to high levels n, within a few $k_B T$ of the ionization limit of $A(n)$ and collisional transitions across small energy gaps. Recombination becomes stabilized by collisional-radiative cascade through the lower bound levels of A. Collisions of the $e^- - A^+$ pair with third bodies becomes more important for higher levels n and radiative emission is important down to and among the lower levels n. In optically thin plasmas this radiation is lost, while in optically thick plasmas it may be re-absorbed. At low electron densities radiative recombination dominates with predominant transitions taking place to the ground level. For process (54.2a) at high electron densities, three-body collisions into high Rydberg levels dominate, followed by cascade which is collision dominated at low electron temperatures T_e and radiation dominated at high T_e. For process (54.2b) at low gas densities N, the recombination is collisionally-radiatively controlled while, at high N, it eventually becomes controlled by the rate of diffusional drift (54.61) through the gas M.

Collisional-Radiative Recombination

Here the cascade collisions and radiation are coupled via the continuity equation. The population n_i of an individual excited level i of energy E_i is determined by the rate equations

$$\frac{dn_i}{dt} = \frac{\partial n_i}{\partial t} + \nabla \cdot (n_i \boldsymbol{v}_i) \tag{54.12}$$

$$= \sum_{i \neq f} \left[n_f \nu_{fi} - n_i \nu_{if} \right] = P_i - n_i D_i \,, \tag{54.13}$$

which involve temporal and spatial relaxation in (54.12) and collisional-radiative production rates P_i and destruction frequencies D_i of the elementary processes included in (54.13). The total collisional and radiative transition frequency between levels i and f is ν_{if} and the f-sum is taken over all discrete and continuous (c) states of the recombining species. The transition frequency ν_{if} includes all contributing elementary processes that directly link states i and f, e.g., collisional excitation and de-excitation, ionization ($i \rightarrow c$) and recombination ($c \rightarrow i$) by electrons and heavy particles, radiative recombination ($c \rightarrow i$), radiative decay ($i \rightarrow f$), possibly radiative absorption for optically thick plasmas, autoionization and dielectronic recombination.

Production Rates and Processes

The production rate for a level i is

$$\begin{aligned} P_i = {} & \sum_{f \neq i} n_e n_f K^c_{fi} + n_e^2 N^+ k^R_{ci} \\ & + \sum_{f > i} n_f \left(A_{fi} + B_{fi} \rho_\nu \right) \\ & + n_e N^+ \left(\hat{\alpha}_i^{RR} + \beta_i \rho_\nu \right) \,, \end{aligned} \tag{54.14}$$

where the terms in the above order represent (1) collisional excitation and de-excitation by $e^- - A(f)$ collisions, (2) three-body $e^- - A^+$ collisional recombination into level i, (3) spontaneous and stimulated radiative cascade, and (4) spontaneous and stimulated radiative recombination.

Destruction Rates and Processes

The destruction rate for a level i is

$$\begin{aligned} n_i D_i = {} & n_e n_i \sum_{f \neq i} K^c_{if} + n_e n_i S_i \\ & + n_i \sum_{f < i} \left(A_{if} + B_{if} \rho_\nu \right) \\ & + n_i \sum_{f > i} B_{if} \rho_\nu + n_i B_{ic} \rho_\nu \,, \end{aligned} \tag{54.15}$$

where the terms in the above order represent (1) collisional destruction, (2) collisional ionization,

(3) spontantaneous and stimulated emission, (4) photoexcitation, and (5) photoionization.

54.2.1 Saha and Boltzmann Distributions

Collisions of $A(n)$ with third bodies such as e^- and M are more rapid than radiative decay above a certain excited level n^*. Since each collision process is accompanied by its exact inverse the principle of detailed balance determines the population of levels $i > n^*$.

Saha Distribution

This connects equilibrium densities $\tilde{n}_i$, $\tilde{n}_\mathrm{e}$ and $\tilde{N}^+$ of bound levels i, of free electrons at temperature T_e and of ions by

$$\frac{\tilde{n}_i}{\tilde{n}_\mathrm{e}\tilde{N}^+} = \left(\frac{g(i)}{g_\mathrm{e}g_A^+}\right)\frac{h^3}{(2\pi m_\mathrm{e}kT_\mathrm{e})^{3/2}}\exp(I_\mathrm{i}/k_\mathrm{B}T_\mathrm{e}) , \tag{54.16}$$

where the electronic statistical weights of the free electron, the ion of charge $Z+1$ and the recombined $e^- - A^+$ species of net charge Z and ionization potential I_i are $g_\mathrm{e} = 2$, g_A^+ and $g(i)$, respectively. Since $n_i \leq \tilde{n}_i$ for all i, then the Saha–Boltzmann distributions imply that $n_1 \gg n_i$ and $n_\mathrm{e} \gg n_i$ for $i \neq 1, 2$, where $i = 1$ is the ground state.

Boltzmann Distribution

This connects the equilibrium populations of bound levels i of energy E_i by

$$\tilde{n}_i/\tilde{n}_j = [g(i)/g(j)]\exp\left[-(E_i - E_j)/k_\mathrm{B}T_\mathrm{e}\right] . \tag{54.17}$$

54.2.2 Quasi-Steady State Distributions

The reciprocal lifetime of level i is the sum of radiative and collisional components and is therefore shorter than the pure radiative lifetime $\tau_\mathrm{R} \approx 10^{-7}Z^{-4}$ s. The lifetime τ_1 for the ground level is collisionally controlled, is dependent upon n_e, and generally is within the range of 10^2 and 10^4 s for most laboratory plasmas and the solar atmosphere. The excited level lifetimes τ_i are then much shorter than τ_1. The (spatial) diffusion or plasma decay (recombination) time is then much longer than τ_i and the total number of recombined species is much smaller than the ground-state population n_1. The recombination proceeds on a timescale much longer than the time for population/destruction of the excited levels. The condition for quasisteady state, or QSS-condition, $\mathrm{d}n_i/\mathrm{d}t = 0$ for the bound levels $i \neq 1$, therefore holds. The QSS distributions n_i therefore satisfy $P_i = n_i D_i$.

54.2.3 Ionization and Recombination Coefficients

Under QSS, the continuity equation (54.13) then reduces to a finite set of simultaneous equations $P_i = n_i D_i$. This gives a matrix equation which is solved numerically for $n_i (i \neq 1) \leq \tilde{n}_i$ in terms of n_1 and n_e. The net ground-state population frequency per unit volume ($\mathrm{cm}^{-3}\,\mathrm{s}^{-1}$) can then be expressed as

$$\frac{\mathrm{d}n_1}{\mathrm{d}t} = n_\mathrm{e}N^+\hat{\alpha}_\mathrm{CR} - n_\mathrm{e}n_1 S_\mathrm{CR} , \tag{54.18}$$

where $\hat{\alpha}_\mathrm{CR}$ and S_CR, respectively, are the overall rate coefficients for recombination and ionization via the collisional-radiative sequence. The determined $\hat{\alpha}_\mathrm{CR}$ equals the direct ($c \to 1$) recombination to the ground level supplemented by the net collisional-radiative cascade from that portion of bound-state population which originated from the continuum. The determined S_CR equals direct depletion (excitation and ionization) of the ground state reduced by the de-excitation collisional radiative cascade from that portion of the bound levels accessed originally from the ground level. At low n_e, $\hat{\alpha}_\mathrm{CR}$ and S_CR reduce, respectively, to the radiative recombination coefficient summed over all levels and to the collisional ionization coefficient for the ground level.

$\mathcal{C}$, $\mathcal{E}$ and $\mathcal{S}$ Blocks of Energy Levels

For the recombination processes (54.2a), (54.2b) and (54.6) which involve a sequence of elementary reactions, the $e^- - A^+$ or $A^+ - B^-$ continuum levels and the ground $A(n = 1)$ or the lowest vibrational levels of AB are therefore treated as two large particle reservoirs of reactants and products. These two reservoirs act as reactant and as sink blocks $\mathcal{C}$ and $\mathcal{S}$ which are, respectively, drained and filled at the same rate via a conduit of highly excited levels which comprise an intermediate block of levels $\mathcal{E}$. This $\mathcal{C}$ draining and $\mathcal{S}$ filling proceeds, via block $\mathcal{E}$, on a timescale large compared with the short time for a small amount from the reservoirs to be re-distributed within block $\mathcal{E}$. This forms the basis of QSS.

54.2.4 Working Rate Formulae

For electron–atomic–ion collisional-radiative recombination (54.2a), detailed QSS calculations can be fitted by the rate [54.1]

$$\hat{\alpha}_{\mathrm{CR}} = \left(3.8\times10^{-9}T_{\mathrm{e}}^{-4.5}n_{\mathrm{e}} + 1.55\times10^{-10}T_{\mathrm{e}}^{-0.63} + 6\times10^{-9}T_{\mathrm{e}}^{-2.18}n_{\mathrm{e}}^{0.37}\right)\ \mathrm{cm^3\,s^{-1}} \quad (54.19)$$

agrees with experiment for a Lyman α optically thick plasma with n_{e} and T_{e} in the range $10^9\,\mathrm{cm^{-3}} \le n_{\mathrm{e}} \le 10^{13}\,\mathrm{cm^{-3}}$ and $2.5\,\mathrm{K} \le T_{\mathrm{e}} \le 4000\,\mathrm{K}$. The first term is the pure collisional rate (54.49), the second term is the radiative cascade contribution, and the third term arises from collisional-radiative coupling.

For $(\mathrm{e}^- - \mathrm{He}_2^+)$ recombination in a high (5–100 Torr) pressure helium afterglow the rate for (54.2b) is [54.2]

$$\begin{aligned}\hat{\alpha}_{\mathrm{CR}} &= \left[(4\pm0.5)\times10^{-20}n_{\mathrm{e}}\right](T_{\mathrm{e}}/293)^{-(4\pm0.5)} \\ &\quad + \left[(5\pm1)\times10^{-27}n(\mathrm{He}) + (2.5\pm2.5)\times10^{-10}\right] \\ &\quad \times (T_{\mathrm{e}}/293)^{-(1\pm1)}\ \mathrm{cm^3/s}. \end{aligned} \quad (54.20)$$

The first two terms are in accord with the purely collisional rates (54.49) and (54.52b), respectively.

54.3 Macroscopic Methods

54.3.1 Resonant Capture-Stabilization Model: Dissociative and Dielectronic Recombination

The electron is captured dielectronically (54.41) into an energy-resonant long-lived intermediate collision complex of super-excited states d which can autoionize or be stabilized irreversibly into the final product channel f either by molecular fragmentation

$$\mathrm{e}^- + AB^+(i) \underset{\nu_a}{\overset{k_c}{\rightleftharpoons}} AB^{**} \overset{\nu_{\mathrm{S}}}{\rightarrow} A + B^*\,, \quad (54.21)$$

as in direct dissociative recombination (DR), or by emission of radiation as in dielectronic recombination (DLR)

$$\mathrm{e}^- + A^{Z+}(i) \underset{\nu_a}{\overset{k_c}{\rightleftharpoons}} \left[A^{Z+}(k) - \mathrm{e}^-\right]_{n\ell} \overset{\nu_{\mathrm{S}}}{\rightarrow} A_{n'\ell'}^{(Z-1)+}(f) + h\nu\,. \quad (54.22)$$

Production Rate of Super-Excited States d

$$\frac{\mathrm{d}n_d^*}{\mathrm{d}t} = n_{\mathrm{e}}N^+k_{\mathrm{c}}(d) - n_d^*\left[\nu_A(d) + \nu_{\mathrm{S}}(d)\right]\,; \quad (54.23)$$

$$\nu_A(d) = \sum_{i'} \nu_a(d\to i')\,, \quad (54.24a)$$

$$\nu_{\mathrm{S}}(d) = \sum_{f'} \nu_{\mathrm{s}}(d\to f')\,. \quad (54.24b)$$

Steady-State Distribution

For a steady-state distribution, the capture volume is

$$\frac{n_d^*}{n_{\mathrm{e}}N^+} = \frac{k_{\mathrm{c}}(d)}{\nu_A(d) + \nu_{\mathrm{S}}(d)}\,. \quad (54.25)$$

Recombination Rate and Stabilization Probability

The recombination rate to channel f is

$$\hat{\alpha}_f = \sum_d \left(\frac{k_{\mathrm{c}}(d)\nu_{\mathrm{s}}(\mathrm{d}\to f)}{\nu_A(d)+\nu_{\mathrm{S}}(d)}\right)\,, \quad (54.26a)$$

and the rate to all product channels is

$$\hat{\alpha} = \sum_d \frac{k_{\mathrm{c}}(d)\nu_{\mathrm{S}}(d)}{\nu_A(d)+\nu_{\mathrm{S}}(d)}\,. \quad (54.26b)$$

In the above, the quantities

$$P_f^{\mathrm{S}}(d) = \nu_{\mathrm{s}}(d\to f)/\left[\nu_A(d)+\nu_{\mathrm{S}}(d)\right]\,, \quad (54.27)$$

$$P^{\mathrm{S}}(d) = \nu_{\mathrm{S}}(d)/\left[\nu_A(d)+\nu_{\mathrm{S}}(d)\right]\,, \quad (54.28)$$

represent the corresponding stabilization probabilities.

Macroscopic Detailed Balance and Saha Distribution

$$K_{di}(T) = \frac{\tilde{n}_d^*}{\tilde{n}_{\mathrm{e}}\tilde{N}^+} = \frac{k_{\mathrm{c}}(d)}{\nu_a(d\to i)} = k_{\mathrm{c}}(d)\tau_a(d\to i) \quad (54.29a)$$

$$= \frac{h^3}{(2\pi m_{\mathrm{e}}k_{\mathrm{B}}T)^{3/2}}\left(\frac{\omega(d)}{2\omega^+}\right) \times\exp\left(-E_{di}^*/k_{\mathrm{B}}T\right)\,, \quad (54.29b)$$

where E_{di}^* is the energy of super-excited neutral levels AB^{**} above that for ion level $AB^+(i)$, and ω are the corresponding statistical weights.

Alternative Rate Formula

$$\hat{\alpha}_f = \sum_d K_{di}\left(\frac{\nu_a(d\to i)\nu_{\mathrm{s}}(d\to f)}{\nu_A(d)+\nu_{\mathrm{S}}(d)}\right)\,. \quad (54.30)$$

Normalized Excited State Distributions

$$\rho_d = n_d^*/\tilde{n}_d^* = \frac{\nu_a(d \to i)}{[\nu_A(d) + \nu_S(d)]}, \quad (54.31)$$

$$\hat{\alpha} = \sum_d k_c(d) P^S(d) = \sum_d K_{di} \rho_d \nu_S(d) \quad (54.32a)$$

$$= \sum_d k_c(d) [\rho_d \nu_S(d) \tau_a(d \to i)] . \quad (54.32b)$$

Although equivalent, (54.26a) and (54.30) are normally invoked for (54.21) and (54.22), respectively, since $P^S \leq 1$ for DR so that $\hat{\alpha}_{DR} \to k_c$; and $\nu_A \gg \nu_S$ for DLR with $n \ll 50$ so that $\hat{\alpha} \to K_{di} \nu_S$. For $n \gg 50$, $\nu_S \gg \nu_A$ and $\hat{\alpha} \to k_c$. The above results (54.26a) and (54.30) can also be derived from microscopic Breit–Wigner scattering theory for isolated (nonoverlapping) resonances.

54.3.2 Reactive Sphere Model: Three-Body Electron–Ion and Ion–Ion Recombination

Since the Coulomb attraction cannot support quasibound levels, three body electron–ion and ion–ion recombination do not in general proceed via time-delayed resonances, but rather by reactive (energy-reducing) collisions with the third body M. This is particularly effective for A–B separations $R \leq R_0$, as in the sequence

$$A + B \underset{\nu_d}{\overset{k_c}{\rightleftharpoons}} AB^*(R \leq R_0) , \quad (54.33a)$$

$$AB^*(R \leq R_0) + M \underset{\nu_{-s}}{\overset{\nu_s}{\rightleftharpoons}} AB + M . \quad (54.33b)$$

In contrast to (54.21) and (54.22) where the stabilization is irreversible, the forward step in (54.33b) is reversible. The sequence (54.33a) and (54.33b) represents a closed system where thermodynamic equilibrium is eventually established.

Steady State Distribution of AB^* Complex

$$n^* = \left(\frac{k_c}{\nu_s + \nu_d}\right) n_A(t) n_B(t) + \left(\frac{\nu_{-s}}{\nu_s + \nu_d}\right) n_s(t) . \quad (54.34)$$

Saha and Boltzmann balances:

$$\begin{aligned} \text{Saha:} \quad & \tilde{n}_A \tilde{n}_B k_c = \tilde{n}^* \nu_d , \\ \text{Boltzmann:} \quad & \tilde{n}_s \nu_{-s} = \tilde{n}^* \nu_s . \end{aligned} \quad (54.35)$$

$\tilde{n}^*$ is in Saha balance with reactant block $\mathcal{C}$ and in Boltzmann balance with product block $\mathcal{S}$.

Normalized Distributions

$$\rho^* = \frac{n^*}{\tilde{n}^*} = P^D \gamma_c(t) + P^S \gamma_s(t) , \quad (54.36a)$$

$$\gamma_c(t) = \frac{n_A(t) n_B(t)}{\tilde{n}_A \tilde{n}_B} , \quad \gamma_s(t) = \frac{n_s(t)}{\tilde{n}_s} . \quad (54.36b)$$

Stabilization and Dissociation Probabilities

$$P^S = \frac{\nu_s}{(\nu_s + \nu_d)} , \quad P^D = \frac{\nu_d}{(\nu_s + \nu_d)} . \quad (54.37)$$

Time Dependent Equations

$$\frac{dn_c}{dt} = -k_c P^S \tilde{n}_A \tilde{n}_B [\gamma_c(t) - \gamma_s(t)] , \quad (54.38a)$$

$$\frac{dn_s}{dt} = -\nu_{-s} P^D \tilde{n}_s [\gamma_s(t) - \gamma_c(t)] , \quad (54.38b)$$

$$\frac{dn_c}{dt} = -\hat{\alpha}_3 n_A(t) n_B(t) + k_d n_s(t) . \quad (54.39)$$

where the recombination rate coefficient $(\mathrm{cm^3/s})$ and dissociation frequency are, respectively,

$$\hat{\alpha}_3 = k_c P^S = \frac{k_c \nu_s}{(\nu_s + \nu_d)} , \quad (54.40)$$

$$k_d = \nu_{-s} P^D = \frac{\nu_{-s} \nu_d}{(\nu_s + \nu_d)} , \quad (54.41)$$

which also satisfy the macroscopic detailed balance relation

$$\hat{\alpha}_3 \tilde{n}_A \tilde{n}_B = k_d \tilde{n}_s . \quad (54.42)$$

Time Independent Treatment

The rate $\hat{\alpha}_3$ given by the time dependent treatment can also be deduced by viewing the recombination process as a source block $\mathcal{C}$ kept fully filled with dissociated species A and B maintained at equilibrium concentrations $\tilde{n}_A, \tilde{n}_B$ (i. e. $\gamma_c = 1$) and draining at the rate $\hat{\alpha}_3 \tilde{n}_A \tilde{n}_B$ through a steady-state intermediate block $\mathcal{E}$ of excited levels into a fully absorbing sink block $\mathcal{S}$ of fully associated species AB kept fully depleted with $\gamma_s = 0$ so that there is no backward re-dissociation from block $\mathcal{S}$. The frequency k_d is deduced as if the reverse scenario, $\gamma_s = 1$ and $\gamma_c = 0$, holds. This picture uncouples $\hat{\alpha}$ and k_d, and allows each coefficient to be calculated independently. Both dissociation (or ionization) and association (recombination) occur within block $\mathcal{E}$.

If $\gamma_c = 1$ and $\gamma_s = 0$, then

$$\rho^* = n^*/\tilde{n}^* = \nu_d/(\nu_s + \nu_d) , \quad (54.43a)$$

$$K = \tilde{n}^*/\tilde{n}_A \tilde{n}_B = k_c/\nu_d = k_c \tau_d , \quad (54.43b)$$

$$P^S = \nu_s/(\nu_s + \nu_d) = \rho^* \nu_s \tau_s , \quad (54.43c)$$

and the recombination coefficient is

$$\hat{\alpha} = k_c P^S = k_c \left(\rho^* \nu_s \tau_d\right) = K\rho^* \nu_s \,. \quad (54.44)$$

Microscopic Generalization

From (54.167), the microscopic generalizations of rate (54.40) and probability (54.43c) are, respectively,

$$\hat{\alpha} = \bar{v} \int_0^\infty \varepsilon \mathrm{e}^{-\varepsilon}\, \mathrm{d}\varepsilon \int_0^{b_0} 2\pi b\, \mathrm{d}b P^S(\varepsilon, b; R_0)\,, \quad (54.45a)$$

$$P^S(\varepsilon, b; R_0) = \oint_{R_i}^{R_0} \rho_i(R) \nu_i^b(R)\, \mathrm{d}t \equiv \langle \rho \nu_s \rangle\, \tau_d\,, \quad (54.45b)$$

where $\rho_i(R) = n(\varepsilon, b; R)/\tilde{n}(\varepsilon, b; R)$; $\nu_i^{(b)}$ is the frequency (54.164a) of $(A–B)–M$ continuum-bound collisional transitions at fixed $A–B$ separation R, R_i is the pericenter of the orbit, $|i\rangle \equiv |\varepsilon, b\rangle$, and

$$b_0^2 = R_0^2\,[1 - V(R)/E]\,, \quad \varepsilon = E/k_B T\,, \quad (54.45c)$$

$$\hat{\alpha} \equiv k_c \left\langle P^S \right\rangle_{\varepsilon,b}\,, \quad \bar{v} = (8k_B T/\pi M_{AB})^{1/2}\,, \quad (54.45d)$$

$$k_c = \left\{\pi R_0^2\,[1 - V(R_0)/k_B T]\, \bar{v}\right\}\,. \quad (54.45e)$$

where M_{AB} is the reduced mass of A and B.

Low Gas Densities

Here $\rho_i(R) = 1$ for $E > 0$,

$$P^S(\varepsilon, b; R_0) = \oint_{R_i}^{R_0} \nu(t)\, \mathrm{d}t = \oint_{R_i}^{R_0} \mathrm{d}s/\lambda_i\,. \quad (54.46)$$

$\lambda_i = (N\sigma)^{-1}$ is the microscopic path length towards the $(A–B)–M$ reactive collision with frequency $\nu = Nv\sigma$. For λ_i constant, the rate (54.45a) reduces at low N to

$$\hat{\alpha} = (v\sigma_0 N) \int_0^{R_0} \left(1 - \frac{V(R)}{k_B T}\right) 4\pi R^2\, \mathrm{d}R \quad (54.47)$$

which is linear in the gas density N.

54.3.3 Working Formulae for Three-Body Collisional Recombination at Low Density

For three-body ion–ion collisional recombination of the form $A^+ + B^- + M$ in a gas at low density N, set $V(R) = -\mathrm{e}^2/R$. Then (54.47) yields

$$\hat{\alpha}^c(T) = \left(\frac{8k_B T}{\pi M_{AB}}\right)^{1/2} \frac{4}{3}\pi R_0^3 \left(1 + \frac{3}{2}\frac{R_e}{R_0}\right) (\sigma_0 N)\,, \quad (54.48)$$

where $R_e = e^2/k_B T$, and the trapping radius R_0, determined by the classical variational method, is $0.41 R_e$, in agreement with detailed calculation. The special cases are:

(a) $e^- + A^+ + e^-$

Here, $\sigma_0 = \frac{1}{9}\pi R_e^2$ for $(e^- - e^-)$ collisions for scattering angles $\theta \geq \pi/2$ so that

$$\hat{\alpha}_{ee}^c(T) = 2.7 \times 10^{-20} \left(\frac{300}{T}\right)^{4.5} n_e\, \mathrm{cm}^3\, \mathrm{s}^{-1} \quad (54.49)$$

in agreement with *Mansbach* and *Keck* [54.3].

(b) $A^+ + B^- + M$

Here, $\sigma_0 \bar{v} \approx 10^{-9}\, \mathrm{cm}^3\, \mathrm{s}^{-1}$, which is independent of T for polarization attraction. Then

$$\hat{\alpha}_3(T) = 2 \times 10^{-25} \left(\frac{300}{T}\right)^{2.5} N\, \mathrm{cm}^3\, \mathrm{s}^{-1}\,. \quad (54.50)$$

(c) $e^- + A^+ + M$

Only a small fraction $\delta = 2m/M$ of the electron's energy is lost upon $(e^- - M)$ collision so that (54.45a) for constant λ is modified to

$$\hat{\alpha}_{eM} = \sigma_0 N \int_0^{R_0} 4\pi R^2\, \mathrm{d}R \int_0^{E_m} \tilde{n}(R, E) v\, \mathrm{d}E \quad (54.51a)$$

$$= \bar{v}_e \sigma_0 N \int_0^{R_0} 4\pi R^2\, \mathrm{d}R \int_0^{\varepsilon_m} \left(1 - \frac{V(R)}{E}\right) \varepsilon \mathrm{e}^{-\varepsilon}\, \mathrm{d}\varepsilon \quad (54.51b)$$

where $\varepsilon = E/k_B T$, and $E_m = \delta e^2/R = \varepsilon_m k_B T$ is the maximum energy for collisional trapping. Hence,

$$\hat{\alpha}_{eM}(T_e) = 4\pi\delta \left(\frac{8k_B T_e}{\pi m_e}\right)^{1/2} R_e^2 R_0\, [\sigma_0 N] \quad (54.52a)$$

$$\approx \frac{10^{-26}}{M} \left(\frac{300}{T}\right)^{2.5} N \mathrm{cm}^3\, \mathrm{s}^{-1}\,, \quad (54.52b)$$

where the mass M of the gas atom is now in u. This result agrees with the energy diffusion result of *Pitaevskiĭ* [54.4] when R_0 is taken as the Thomson radius $R_T = \frac{2}{3} R_e$.

54.3.4 Recombination Influenced by Diffusional Drift at High Gas Densities

Diffusional-Drift Current

The drift current of A^+ towards B^- in a gas under an A^+–B^- attractive potential $V(R)$ is

$$\boldsymbol{J}(R) = -D\nabla n(R) - \left[\frac{K}{e}\nabla V(R)\right] n(R) \quad (54.53a)$$

$$= -\left(D\tilde{N}_A\tilde{N}_B \mathrm{e}^{-V(R)/k_\mathrm{B}T}\frac{\partial\rho}{\partial R}\right)\hat{\boldsymbol{R}} . \quad (54.53b)$$

Relative Diffusion and Mobility Coefficients

$$D = D_A + D_B ,$$
$$K = K_A + K_B , \quad De = K(k_\mathrm{B}T) , \quad (54.54)$$

where the D_i and K_i are, respectively, the diffusion and mobility coefficients of species i in gas M.

Normalized Ion-Pair R-Distribution

$$\rho(R) = \frac{n(R)}{\tilde{N}_A\tilde{N}_B \exp[-V(R)/k_\mathrm{B}T]} . \quad (54.55)$$

Continuity Equations for Currents and Rates

$$\frac{\partial n}{\partial t} + \nabla\cdot\boldsymbol{J} = 0 , \quad R \ge R_0 \quad (54.56a)$$
$$\hat{\alpha}_\mathrm{RN}(R_0)\rho(R_0) = \hat{\alpha}\rho(\infty) \quad (54.56b)$$

The rate of reaction for ion-pairs with separations $R \le R_0$ is $\alpha_\mathrm{RN}(R_0)$. This is the recombination rate that would be obtained for a thermodynamic equilibrium distribution of ion pairs with $R \ge R_0$, i.e. for $\rho(R \ge R_0) = 1$.

Steady-State Rate of Recombination

$$\hat{\alpha}\tilde{N}_A\tilde{N}_B = \int_{R_0}^{\infty}\left(\frac{\partial n}{\partial t}\right)\mathrm{d}\boldsymbol{R} = -4\pi R_0^2 J(R_0) . \quad (54.57)$$

Steady-State Solution

$$\rho(R) = \rho(\infty)\left(1 - \frac{\hat{\alpha}}{\hat{\alpha}_\mathrm{TR}(R)}\right) , \quad R \ge R_0 \quad (54.58a)$$

$$\rho(R_0) = \rho(\infty)\left[\hat{\alpha}/\hat{\alpha}_\mathrm{RN}(R_0)\right] . \quad (54.58b)$$

Recombination Rate

$$\hat{\alpha} = \frac{\hat{\alpha}_\mathrm{RN}(R_0)\hat{\alpha}_\mathrm{TR}(R_0)}{\hat{\alpha}_\mathrm{RN}(R_0) + \hat{\alpha}_\mathrm{TR}(R_0)} \quad (54.59a)$$

$$\to \begin{cases} \hat{\alpha}_\mathrm{RN} , & N \to 0 \\ \hat{\alpha}_\mathrm{TR} , & N \to \infty . \end{cases} \quad (54.59b)$$

Diffusional-Drift Transport Rate

$$\hat{\alpha}_\mathrm{TR}(R_0) = 4\pi D\left(\int_{R_0}^{\infty}\frac{\mathrm{e}^{V(R)/k_\mathrm{B}T}}{R^2}\,\mathrm{d}R\right)^{-1} . \quad (54.60)$$

With $V(R) = -e^2/R$,

$$\hat{\alpha}_\mathrm{TR}(R_0) = 4\pi Ke\left[1 - \exp(-R_\mathrm{e}/R_0)\right]^{-1} , \quad (54.61)$$

where $R_\mathrm{e} = e^2/k_\mathrm{B}T$ provides a natural unit of length.

Langevin Rate

For $R_0 \ll R_\mathrm{e}$, the transport rate

$$\hat{\alpha}_\mathrm{TR} \to \hat{\alpha}_L = 4\pi Ke , \quad (54.62)$$

tends to the Langevin rate which varies as N^{-1}.

Reaction Rate

When R_0 is large enough that R_0-pairs are in (E, L^2) equilibrium (54.167),

$$\hat{\alpha}_\mathrm{RN}(R_0) = \bar{v}\int_0^{\infty}\varepsilon\mathrm{e}^{-\varepsilon}\,\mathrm{d}\varepsilon\int_0^{b_0}2\pi b\,\mathrm{d}b P^\mathrm{S}(\varepsilon, b; R_0) \quad (54.63a)$$

$$\equiv \bar{v}\int_0^{\infty}\varepsilon\mathrm{e}^{-\varepsilon}\,\mathrm{d}\varepsilon\left[\pi b_0^2 P^\mathrm{S}(\varepsilon; R_0)\right] \quad (54.63b)$$

$$\equiv \bar{v}\pi b_\mathrm{max}^2 P^\mathrm{S}(R_0) , \quad (54.63c)$$

where

$$b_0^2 = R_0^2\left[1 - V(R_0)/E\right] , \varepsilon = E/k_\mathrm{B}T , \quad (54.64a)$$
$$\bar{v} = (8kT/\pi M_{AB})^{1/2} , \quad (54.64b)$$
$$b_\mathrm{max}^2 = R_0^2\left(1 - \frac{V(R_0)}{k_\mathrm{B}T}\right) . \quad (54.64c)$$

The probability P^S and its averages over b and (b, E) for reaction between pairs with $R \le R_0$ is determined in (54.63a–c) from solutions of coupled master equations. P^S increases linearly with N initially and tends to unity at high N. The recombination rate (54.59a) with (54.63a) and (54.61) therefore increases linearly with N initially, reaches a maximum when $\hat{\alpha}_{TR} \approx \hat{\alpha}_{RN}$ and then decreases eventually as N^{-1}, in accord with (i).

Reaction Probability

The classical absorption solution of (54.157) is

$$P^S(E, b; R_0) = 1 - \exp\left(-\oint_{R_i}^{R_0} \frac{ds_i}{\lambda_i}\right) . \quad (54.65)$$

With the binary decomposition $\lambda_i^{-1} = \lambda_{iA}^{-1} + \lambda_{iB}^{-1}$,

$$P^S = P_A + P_B - P_A P_B . \quad (54.66)$$

Exact b^2-Averaged Probability

With $V_c = -e^2/R$ for the $A^+–B^-$ interaction in (54.63b), and at low gas densities N,

$$P_{A,B}(E, R_0) = \frac{\frac{4R_0}{3\lambda_{A,B}}\left(1 - \frac{3V_c(R_0)}{2E_i}\right)}{[1 - V_c(R_0)/E_i]} \quad (54.67)$$

appropriate for constant mean free path λ_i.

(E, b^2)-Averaged Probability

$P^S(R_0)$ in (54.63c) at low gas density is

$$P_{A,B}(R_0) = P_{A,B}(E = k_B T, R_0) . \quad (54.68)$$

Thomson Trapping Distance

When the kinetic energy gained from Coulomb attraction is assumed lost upon collision with third bodies, then bound $A - B$ pairs are formed with $R \le R_T$. Since $E = \frac{3}{2}k_B T - e^2/R$, then

$$R_T = \frac{2}{3}\left(\frac{e^2}{k_B T}\right) = \frac{2}{3}R_e . \quad (54.69)$$

Thomson Straight-Line Probability

The $E \to \infty$ limit of (54.65) is

$$P_{A,B}^T(b; R_T) = 1 - \exp\left[-2\left(R_T^2 - b^2\right)/\lambda_{A,B}\right] . \quad (54.70)$$

The b^2-average is the Thomson probability

$$P_{A,B}^T(R_T) = 1 - \frac{1}{2X^2}\left[1 - e^{-2X}(1 + 2X)\right] \quad (54.71a)$$

for reaction of $(A - B)$ pairs with $R \le R_T$. As $N \to 0$

$$P_{A,B}^T(R_T) \to \frac{4}{3}X\left(1 - \frac{3}{4}X + \frac{2}{5}X^2 - \frac{1}{6}X^3 + \cdots\right) \quad (54.71b)$$

and tends to unity at high N. $X = R_T/\lambda_{A,B} = N(\sigma_0 R_T)$. These probabilites have been generalized [54.5] to include hyperbolic and general trajectories.

Thomson Reaction Rate

$$\begin{aligned}\hat{\alpha}_T &= \pi R_T^2 \bar{v}\left(P_A^T + P_B^T - P_A^T P_B^T\right) \\ &\to \begin{cases} \frac{4}{3}\pi R_T^3\left(\lambda_A^{-1} + \lambda_B^{-1}\right), & N \to 0 \\ \pi R_T^2 \bar{v}, & N \to \infty . \end{cases}\end{aligned} \quad (54.72)$$

54.4 Dissociative Recombination

54.4.1 Curve-Crossing Mechanisms

Direct Process.

Dissociative recombination (DR) for diatomic ions can occur via a crossing at R_X between the bound and repulsive potential energy curves $V^+(R)$ and $V_d(R)$ for AB^+ and AB^{**}, respectively. Here, DR involves the two-stage sequence

$$e^- + AB^+(v_i) \underset{\nu_a}{\overset{k_c}{\rightleftharpoons}} (AB^{**})_R \overset{\nu_d}{\longrightarrow} A + B^* . \quad (54.73)$$

The first stage is dielectronic capture whereby the free electron of energy $\varepsilon = V_d(R) - V^+(R)$ excites an electron of the diatomic ion AB^+ with internal separation R and is then resonantly captured by the ion, at rate k_c, to form a repulsive state d of the doubly excited molecule AB^{**}, which in turn can either autoionize at probability frequency ν_a, or else in the second stage predissociate into various channels at probability frequency ν_d. This competition continues until the (electronically excited) neutral fragments accelerate past the crossing at R_X. Beyond R_X the increasing energy of relative separation reduces the total electronic energy to such an extent that autoionization is essentially precluded and the neutralization is then rendered permanent past the stabilization point R_X. This interpretation [54.6] has remained intact and robust in the current light of ab initio quantum chemistry and quantal scattering calculations for the simple diatomics $\left(O_2^+, N_2^+, Ne_2^+, \text{etc.}\right)$. Mechanism (54.73) is termed the direct process which, in terms of the macroscopic frequencies in (54.73), proceeds at the rate

$$\hat{\alpha} = k_c P_S = k_c [\nu_d/(\nu_a + \nu_d)] , \quad (54.74)$$

where P_S is probability for $A - B^*$ survival against autoionization from the initial capture at R_c to the crossing point R_X. Configuration mixing theories of this direct process are available in the quantal [54.7] and semiclassical-classical path formulations [54.8].

Indirect Process

In the three-stage sequence

$$\mathrm{e}^- + AB^+(v_i^+) \to \left[AB^+(v_f) - \mathrm{e}^-\right]_n \to (AB^{**})_d \to A + B^* \quad (54.75)$$

the so-called indirect process [54.7] might contribute. Here the accelerating electron loses energy by vibrational excitation $\left(v_i^+ \to v_f\right)$ of the ion and is then resonantly captured into a Rydberg orbital of the bound molecule AB^* in vibrational level v_f, which then interacts one way (via configuration mixing) with the doubly excited repulsive molecule AB^{**}. The capture initially proceeds via a small effect – vibronic coupling (the matrix element of the nuclear kinetic energy) induced by the breakdown of the Born-Oppenheimer approximation – at certain resonance energies $\varepsilon_n = E(v_f) - E\left(v_i^+\right)$ and, in the absence of the direct channel (54.73), would therefore be manifest by a series of characteristic very narrow Lorentz profiles in the cross section. Uncoupled from (54.73) the indirect process would augment the rate. Vibronic capture proceeds more easily when $v_f = v_i^+ + 1$ so that Rydberg states with $n \approx 7-9$ would be involved $\left[\text{for } H_2^+\left(v_i^+ = 0\right)\right]$ so that the resulting longer periods of the Rydberg electron would permit changes in nuclear motion to compete with the electronic dissociation. Recombination then proceeds as in the second stage of (54.73), i.e., by electronic coupling to the dissociative state d at the crossing point. A multichannel quantum defect theory [54.9] has combined the direct and indirect mechanisms

Interrupted Recombination

The process

$$\begin{array}{c} \mathrm{e}^- + AB^+(v_i) \underset{\nu_a}{\overset{k_c}{\rightleftharpoons}} (AB^{**})_d \overset{\nu_d}{\rightarrow} A + B^* \\ \nu_{nd} \upharpoonleft\downharpoonright \nu_{dn} \\ \left[AB^+(v) - \mathrm{e}^-\right]_n \end{array} \quad (54.76)$$

proceeds via the first (dielectronic capture) stage of (54.73) followed by a two-way electronic transition with frequency ν_{dn} and ν_{nd} between the d and n states. All (n, v) Rydberg states can be populated, particularly those in low n and high v since the electronic $d - n$ interaction varies as $n^{-1.5}$ with broad structure. Although the dissociation process proceeds here via a second order effect (ν_{dn} and ν_{nd}), the electronic coupling may dominate the indirect vibronic capture and interrupt the recombination, in contrast to (54.75) which, as written in the one-way direction, feeds the recombination. Both dip and spike structure has been observed [54.10].

54.4.2 Quantal Cross Section

The cross section for direct dissociative recombination

$$\mathrm{e}^- + AB^+\left(v_i^+\right) \rightleftharpoons \left(AB^{**}\right)_\mathrm{r} \longrightarrow A + B^* \quad (54.77)$$

of electrons of energy ε, wavenumber k_e and spin statistical weight 2, for a molecular ion $AB^+(v_i^+)$ of electronic statistical weight ω_{AB}^+ in vibrational level v_i^+ is

$$\begin{aligned} \sigma_\mathrm{DR}(\varepsilon) &= \frac{\pi}{k_\mathrm{e}^2}\left(\frac{\omega_{AB}^*}{2\omega^+}\right)|a_Q|^2 \\ &= \left(\frac{h^2}{8\pi m_\mathrm{e}\varepsilon}\right)\left(\frac{\omega_{AB}^*}{2\omega^+}\right)|a_Q|^2 . \end{aligned} \quad (54.78)$$

Here ω_{AB}^* is the electronic statistical weight of the dissociative neutral state of AB^* whose potential energy curve V_d crosses the corresponding potential energy curve V^+ of the ionic state. The transition T-matrix element for autoionization of AB^* embedded in the (moving) electronic continuum of $AB^+ + \mathrm{e}^-$ is the quantal probability amplitude

$$a_Q(v) = 2\pi \int_0^\infty V_{d\varepsilon}^*(R)\left[\psi_v^{+*}(R)\psi_d(R)\right] \mathrm{d}R \quad (54.79)$$

for autoionization. Here ψ_v^+ and ψ_d are the nuclear bound and continuum vibrational wave functions for AB^+ and AB^*, respectively, while

$$\begin{aligned} V_{d\varepsilon}(R) &= \langle\phi_d|\mathcal{H}_\mathrm{el}(\boldsymbol{r}, R(t))|\phi_\varepsilon(\boldsymbol{r}, \boldsymbol{R})\rangle_{\boldsymbol{r},\hat{\varepsilon}} \\ &= V_{\varepsilon d}^*(R) \end{aligned} \quad (54.80)$$

are the bound-continuum electronic matrix elements coupling the diabatic electronic bound state wave functions $\psi_d(\boldsymbol{r}, \boldsymbol{R})$ for AB^* with the electronic continuum state wave functions $\phi_\varepsilon(\boldsymbol{r}, \boldsymbol{R})$ for $AB^+ + \mathrm{e}^-$. The matrix element is an average over electronic coordinates $\boldsymbol{r}$ and all directions $\hat{\epsilon}$ of the continuum electron. Both continuum electronic and vibrational wave functions are energy normalized (Sect. 54.8.3), and

$$\Gamma(R) = 2\pi \left|V^*_{d\varepsilon}(R)\right|^2 \quad (54.81)$$

is the energy width for autoionization at a given nuclear separation R. Given $\Gamma(R)$ from quantum chemistry codes, the problem reduces to evaluation of continuum vibrational wave functions in the presence of autoionization. The rate associated with a Maxwellian distribution of electrons at temperature T is

$$\hat{\alpha} = \overline{v}_e \int \varepsilon\, \sigma_{DR}(\varepsilon)\, e^{-\varepsilon/k_B T}\, d\varepsilon/(k_B T)^2 \quad (54.82)$$

where $\overline{v}_e$ is the mean speed (Sect. 54.9).

Maximum Cross Section and Rate

Since the probability for recombination must remain less than unity, $\left|a_Q\right|^2 \leq 1$ so that the maximum cross section and rates are

$$\sigma_{DR}^{max}(\varepsilon) = \frac{\pi}{k_e^2}\left(\frac{\omega^*_{AB}}{2\omega^+}\right) = \left(\frac{h^2}{8\pi m_e \varepsilon}\right)(2\ell+1)\,, \quad (54.83)$$

where ω^*_{AB} has been replaced by $2(2\ell+1)\omega^+$ under the assumption that the captured electron is bound in a high level Rydberg state of angular momentum ℓ, and

$$\hat{\alpha}_{max}(T) = \overline{v}_e\, \sigma_{DR}^{max}(\varepsilon = k_B T) \quad (54.84a)$$

$$\approx 5\times 10^{-7}\left(\frac{300}{T}\right)^{1/2}(2\ell+1)\, cm^3/s\,. \quad (54.84b)$$

Cross section maxima of $5(2\ell+1)(300/T)\times 10^{-14}\, cm^2$ are therefore possible, being consistent with the rate (54.84b).

First-Order Quantal Approximation

When the effect of autoionization on the continuum vibrational wave function $\psi_d(R)$ for AB^* is ignored, then a first-order undistorted approximation to the quantal amplitude (54.79) is

$$T_B(v^+) = 2\pi \int_0^\infty V^*_{d\varepsilon}(R)\left[\psi_v^{+*}(R)\psi_d^{(0)}(R)\right] dR \quad (54.85)$$

where $\psi_d^{(0)}$ is ψ_d in the absence of the back reaction of autoionization. Under this assumption, (54.78) reduces to

$$\sigma_c(\varepsilon, v^+) = \frac{\pi}{k_e^2}\left(\frac{\omega^*_{AB}}{2\omega^+}\right)\left|T_B(v^+)\right|^2\,, \quad (54.86)$$

which is then the cross section for initial electron capture since autoionization has been precluded. Although the Born T-matrix (54.85) violates unitarity, the capture cross section (54.86) must remain less then the maximum value

$$\sigma_c^{max} = \frac{\pi}{k_e^2}\left(\frac{\omega^*_{AB}}{2\omega^+}\right) = \left(\frac{h^2}{8\pi m_e \varepsilon}\right)\left(\frac{\omega^*_{AB}}{2\omega^+}\right)\,, \quad (54.87)$$

since $\left|a_Q\right|^2 \leq 1$. So as to acknowledge after the fact the effect of autoionization, assumed small, and neglected by (54.85), the DR cross section can be approximated as

$$\sigma_{DR}(\varepsilon, v^+) = \sigma_c(\varepsilon, v^+) P_S\,, \quad (54.88)$$

where P_S is the probability of survival against autoionization on the V_d curve until stabilization takes place at some crossing point R_X.

Approximate Capture Cross Section

With the energy-normalized Winans–Stückelberg vibrational wave function

$$\psi_d^{(0)}(R) = \left|V_d'(R)\right|^{-1/2}\delta(R - R_c)\,, \quad (54.89)$$

where R_c is the classical turning point for $(A - B^*)$ relative motion, (54.86) reduces to

$$\sigma_c(\varepsilon, v^+) = \frac{\pi}{k_e^2}\left(\frac{\omega^*_{AB}}{2\omega^+}\right)[2\pi\Gamma(R_c)]\left\{\frac{\left|\psi_v^+(R_c)\right|^2}{\left|V_d'(R_c)\right|}\right\} \quad (54.90)$$

where the term inside the braces in (54.90) is the effective Franck–Condon factor.

Six Approximate Stabilization Probabilities

(1) A unitarized T-matrix is

$$T = \frac{T_B}{1+\left|\frac{1}{2}T_B\right|^2}\,, \quad (54.91)$$

so that $P_S = |T|^2 / |T_B|^2$ to give

$$P_S(\text{low } \varepsilon) = \left(1+\frac{1}{4}|T_B|^2\right)^{-2} = \left\{1+\pi^2\left|\int_0^\infty V^*_{d\varepsilon}(R)\left[\psi_v^{+*}(R)\psi_d^{(0)}(R)\right] dr\right|^2\right\}^{-2} \quad (54.92a)$$

which is valid at low ε when only one vibrational level v^+, i.e., the initial level of the ion is repopulated by autoionization.

(2) At higher ε, when population of many other ionic levels v_f^+ occurs, then

$$P_S(\varepsilon) = \left[1 + \frac{1}{4}\sum_f \left|T_B\left(v_f^+\right)\right|^2\right]^{-2}, \tag{54.92b}$$

where the summation is over all the open vibrational levels v_f^+ of the ion. When no intermediate Rydberg $AB^*(v)$ states are energy resonant with the initial $e^- + AB^+(v^+)$ state, i.e., coupling with the indirect mechanism is neglected, then (54.88) with (54.92b) is the direct DR cross section normally calculated.

(3) In the high-ε limit when an infinite number of v_f^+ levels are populated following autoionization, the survival probability, with the aid of closure, is then

$$P_S = \left[1 + \pi^2 \int_{R_c}^{R_X} \left|V_{d\varepsilon}^*(R)\right|^2 \left|\psi_d^{(0)}(R)\right|^2 \mathrm{d}R\right]^{-2}. \tag{54.93}$$

(4) On adopting in (54.93) the JWKB semiclassical wave function for $\psi_d^{(0)}$, then

$$P_S(\text{high } \varepsilon) = \left[1 + \frac{1}{2\hbar}\int_{R_c}^{R_X} \frac{\Gamma(R)}{v(R)} \mathrm{d}R\right]^{-2} = \left[1 + \frac{1}{2}\int_{t_c}^{t_X} \nu_a(t)\,\mathrm{d}t\right]^{-2}, \tag{54.94}$$

where $v(R)$ is the local radial speed of $A - B$ relative motion, and where the frequency $\nu_a(t)$ of autoionization is $\Gamma/\hbar$.

(5) A classical path local approximation for P_S yields

$$P_S = \exp\left(-\int_{t_c}^{t_X} \nu_a(t)\,\mathrm{d}t\right), \tag{54.95}$$

which agrees to first-order for small ν with the expansion of (54.94).

(6) A partitioning of (54.73) yields

$$P_S = \nu_d/(\nu_a + \nu_d) = (1 + \nu_a\tau_d)^{-1}, \tag{54.96}$$

on adopting macroscopic averaged frequencies ν_i and associated lifetimes $\tau_i = \nu_i^{-1}$. The six surivival probabilities in (54.92a,b), (54.93–54.96) are all suitable for use in the DR cross section (54.88).

54.4.3 Noncrossing Mechanism

The dissociative recombination (DR) processes

$$\begin{aligned} e^- + H_3^+ &\to H_2 + H \\ &\to H + H + H \end{aligned} \tag{54.97}$$

at low electron energy ε, and

$$e^- + HeH^+ \to He + H(n{=}2) \tag{54.98}$$

have spurred renewed theoretical interest because they both proceed at respective rates of $(2\times10^{-7}$ to $2\times10^{-8})$ $cm^3\,s^{-1}$ and $10^{-8}\,cm^3\,s^{-1}$ at 300 K. Such rates are generally associated with the direct DR, which involves favorable curve crossings between the potential energy surfaces, $V^+(R)$ and $V_d(R)$ for the ion AB^+ and neutral dissociative AB^{**} states. The difficulty with (54.97) and (54.98) is that there are no such curve crossings, except at $\varepsilon \geq 8$ eV for (54.97). In this instance, the previous standard theories would support only extremely small rates when electronic resonant conditions do not prevail at thermal energies. Theories [54.11, 12] are currently being developed for application to processes such as (54.97).

54.5 Mutual Neutralization

$$A^+ + B^- \to A + B\,. \tag{54.99}$$

Diabatic Potentials

$V_i^{(0)}(R)$ and $V_f^{(0)}(R)$ for initial (ionic) and final (covalent) states are diagonal elements of

$$V_{if}(R) = \langle\Psi_i(\boldsymbol{r}, \boldsymbol{R})|\mathcal{H}_{el}(\boldsymbol{r}, \boldsymbol{R})|\Psi_f(\boldsymbol{r}, \boldsymbol{R})\rangle_{\boldsymbol{r}}\,, \tag{54.100}$$

where $\Psi_{i,f}$ are diabatic states and $\mathcal{H}_{el}$ is the electronic Hamiltonian at fixed internuclear distance R.

Adiabatic Potentials for a Two-State System

$$V^{\pm}(R) = V_0(R) \pm \left[\Delta^2(R) + \left|V_{if}(R)\right|^2\right]^{1/2}, \tag{54.101a}$$

$$V_0(R) = \frac{1}{2}\left[V_i^{(0)}(R) + V_f^{(0)}(R)\right], \tag{54.101b}$$

$$\Delta(R) = \left[V_i^{(0)}(R) - V_f^{(0)}(R)\right]. \tag{54.101c}$$

For a single crossing of diabatic potentials at R_X then $V_i^{(0)}(R_X) = V_f^{(0)}(R_X)$ and the adiabatic potentials at R_X are,

$$V^{\pm}(R_X) = V_i^{(0)}(R_X) \pm V_{if}(R_X) \quad (54.102)$$

with energy separation $2V_{if}(R_X)$.

54.5.1 Landau–Zener Probability for Single Crossing at R_X

On assuming $\Delta(R) = (R - R_X)\Delta'(R_X)$, where $\Delta'(R) = \mathrm{d}\Delta(R)/\mathrm{d}R$, the probability for single crossing is

$$P_{if}(R_X) = \exp\left[\eta(R_X)/v_X(b)\right] \quad (54.103a)$$

$$\eta(R_X) = \left(\frac{2\pi}{\hbar}\right)\frac{|V_{if}(R_X)|^2}{\Delta'(R_X)} \quad (54.103b)$$

$$v_X(b) = \left[1 - V_i^{(0)}(R_X)/E - b^2/R_X^2\right]^{1/2} . \quad (54.103c)$$

Overall Charge-Transfer Probability

From the incoming and outgoing legs of the trajectory,

$$P^X(E) = 2P_{if}(1 - P_{if}) . \quad (54.104)$$

54.5.2 Cross Section and Rate Coefficient for Mutual Neutralization

$$\sigma_M(E) = 4\pi \int_0^{b_X} P_{if}(1 - P_{if}) b \, \mathrm{d}b$$
$$= \pi b_X^2 P_M , \quad (54.105a)$$

$$\pi b_X^2 = \pi \left(1 - \frac{V_i^{(0)}(R_X)}{E}\right) R_X^2$$
$$= \pi \left(1 + \frac{14.4}{R_X(\text{Å})E(\text{eV})}\right) R_X^2 . \quad (54.105b)$$

P_M is the b^2-averaged probability (54.104) for charge-transfer reaction within a sphere of radius R_X.

The rate is

$$\hat{\alpha}_M = (8k_B T/\pi M_{AB})^{1/2} \int_0^{\infty} \epsilon \sigma_M(\epsilon) \mathrm{e}^{-\epsilon} \, \mathrm{d}\epsilon \quad (54.106)$$

where $\epsilon = E/k_B T$.

54.6 One-Way Microscopic Equilibrium Current, Flux, and Pair-Distributions

Notation:

- M reduced mass $M_A M_B/(M_A + M_B)$
- R internal separation of $A - B$
- E orbital energy $\frac{1}{2}Mv^2 + V(R)$
- L orbital angular momentum
- L^2 $2MEb^2$ for $E > 0$
- v_R radial speed $|\dot{R}|$
- $\overline{v}$ mean relative speed $(8kT/\pi M_{AB})^{1/2}$
- ε normalized energy $E/k_B T$
- n_i pair distribution function $n_i^+ + n_i^-$
- $n_i^{\pm}$ component of n_i with $\dot{R} > 0$ (+) and $\dot{R} < 0$ (−).

All quantities on the RHS in the expressions (a)–(e) below are to be multiplied by $\tilde{N}_A \tilde{N}_B [\omega_{AB}/\omega_A \omega_B]$ where the ω_i denote the statistical weights of species i which are not included by the density of states associated with the E, L^2 orbital degrees of freedom.

Case (a). $|i\rangle \equiv |\boldsymbol{R}, E, L^2\rangle$.

Current: $j_i^{\pm}(R) = n^{\pm}(\boldsymbol{R}, E, L^2) v_R \equiv n_i^{\pm} v_R$

$$\text{Flux: } 4\pi R^2 j_i^{\pm}(R) \, \mathrm{d}E \, \mathrm{d}L^2 = \frac{4\pi^2 \mathrm{e}^{-E/k_B T}}{(2\pi M k_B T)^{3/2}} \mathrm{d}E \, \mathrm{d}L^2 . \quad (54.107)$$

This flux is independent of R. For dissociated pairs $E > 0$,

$$4\pi R^2 j_i^{\pm}(R) \, \mathrm{d}E \, \mathrm{d}L^2 = \left[\overline{v}\varepsilon \mathrm{e}^{-\varepsilon} \, \mathrm{d}\varepsilon\right] [2\pi b \, \mathrm{d}b] . \quad (54.108)$$

$(\boldsymbol{R}, E, L^2)$-Distribution:

$$n\left(\boldsymbol{R}, E, L^2\right) \mathrm{d}\boldsymbol{R} \, \mathrm{d}E \, \mathrm{d}L^2 = \frac{(8\pi^2/v_R)\mathrm{e}^{-E/k_B T}}{(2\pi M k_B T)^{3/2}} \left(\frac{\mathrm{d}\boldsymbol{R}}{4\pi R^2}\right) \mathrm{d}E \, \mathrm{d}L^2 . \quad (54.109)$$

Case (b). $|i\rangle \equiv |\boldsymbol{R}, E\rangle$; L^2-integrated quantities.

$$\text{Current:} \quad j_i^{\pm}(R) = \frac{1}{2} v n^{\pm}(\boldsymbol{R}, E) \equiv \frac{1}{2} v n_i^{\pm}, \tag{54.110}$$

$$\text{Flux:} \quad 4\pi R^2 j_i^{\pm}(R)\,\mathrm{d}E = \left[\bar{v}\varepsilon\,\mathrm{e}^{-\varepsilon}\,\mathrm{d}\varepsilon\right]\pi b_0^2, \tag{54.111a}$$

$$\pi b_0^2 = \pi R^2\,[1 - V(R)/E]\,, \tag{54.111b}$$

$(\boldsymbol{R}, E)$-Distribution:

$$\begin{aligned} &n(\boldsymbol{R}, E)\,\mathrm{d}\boldsymbol{R}\,\mathrm{d}E \\ &= \frac{2}{\sqrt{\pi}}\left[\frac{E - V(R)}{k_\mathrm{B}T}\right]^{1/2} \mathrm{e}^{-\varepsilon}\,\mathrm{d}\varepsilon\,\mathrm{d}\boldsymbol{R} \\ &\equiv G_{MB}(E, R)\,\mathrm{d}\boldsymbol{R}\,, \end{aligned} \tag{54.112}$$

which defines the Maxwell–Boltzmann velocity vistribution G_{MB} in the presence of the field $V(R)$.

Case (c). (E, L^2)-integrated quantities.

$$\text{Current:} \quad j^{\pm}(R) = \frac{1}{4}\bar{v}\,\mathrm{e}^{-V(R)/k_\mathrm{B}T}, \tag{54.113}$$

$$\text{Flux:} \quad 4\pi R^2 j^{\pm}(R) = \pi R^2 \bar{v}\,\mathrm{e}^{-V(R)/k_\mathrm{B}T}, \tag{54.114}$$

$$\text{Distribution:} \quad n(R) = \mathrm{e}^{-V(R)/k_\mathrm{B}T}. \tag{54.115}$$

When E-integration is only over dissociated states ($E > 0$), the above quantities are

$$j_d^{\pm}(R) = \frac{1}{4}\bar{v}\,[1 - V(R)/k_\mathrm{B}T]\,, \tag{54.116}$$

$$4\pi R^2 j_d^{\pm}(R) = \pi R^2\left(1 - \frac{V(R)}{k_\mathrm{B}T}\right)\bar{v} \equiv \pi b_{\max}^2\bar{v}\,, \tag{54.117}$$

$$n(R) = [1 - V(R)/k_\mathrm{B}T]\,. \tag{54.118}$$

Case (d). (E, L^2)-distribution. For Bound Levels

$$n(E, L^2)\,\mathrm{d}E\,\mathrm{d}L^2 = \frac{4\pi^2\tau_\mathrm{R}(E, L)}{(2\pi M k_\mathrm{B}T)^{3/2}}\,\mathrm{e}^{-E/k_\mathrm{B}T}\mathrm{d}E\,\mathrm{d}L^2, \tag{54.119}$$

where $\tau_\mathrm{R} = \oint \mathrm{d}t = (\partial J_\mathrm{R}/\partial E)$ is the period for bounded radial motion of energy E and radial action $J_\mathrm{R}(E, L) = M\oint v_\mathrm{R}\,\mathrm{d}R$.

Case (e). E-distribution. For bound levels

$$n(E)\,\mathrm{d}E = \frac{2\,\mathrm{e}^{-\varepsilon}}{\sqrt{\pi}}\,\mathrm{d}\varepsilon \int_0^{R_A}\left(\frac{E - V}{k_\mathrm{B}T}\right)^{1/2}\mathrm{d}\boldsymbol{R}\,, \tag{54.120}$$

where R_A is the turning point $E = V(R_A)$.

Example. For electron–ion bounded motion, $V(R) = -Ze^2/R$, $R_A = Ze^2/|E|$, $R_\mathrm{e} = Ze^2/k_\mathrm{B}T$, $\varepsilon = E/k_\mathrm{B}T$. Then $\tau_\mathrm{R} = 2\pi(m/Ze^2)^{1/2}(R_A/2)^{3/2}$,

$$\int_0^{R_A}\left(\frac{R_\mathrm{e}}{R} - |\varepsilon|\right)^{1/2}\mathrm{d}\boldsymbol{R} = \frac{\pi^2}{4}R_A^{5/2}R_\mathrm{e}^{1/2}\,, \tag{54.121}$$

and

$$n^s(E)\,\mathrm{d}E = \left(\frac{2\,\mathrm{e}^{-\varepsilon}}{\sqrt{\pi}}\,d\varepsilon\right)\frac{\pi^2}{4}R_A^{5/2}R_\mathrm{e}^{1/2} \tag{54.122}$$

$$= \left(\frac{2\,\mathrm{e}^{-\varepsilon}}{\sqrt{\pi}}\,\mathrm{d}\varepsilon\right)\left(\frac{\pi^2 R_\mathrm{e}^3}{4\,|\varepsilon|^{5/2}}\right). \tag{54.123}$$

For closely spaced levels in a hydrogenic $\mathrm{e}^- - A^{Z+}$ system,

$$n^s(p, \ell) = n(E, L^2)\left(\frac{\mathrm{d}E}{\mathrm{d}p}\right)\left(\frac{\mathrm{d}L^2}{\mathrm{d}\ell}\right) \tag{54.124a}$$

$$n^s(p) = n(E)\left(\frac{\mathrm{d}E}{\mathrm{d}p}\right). \tag{54.124b}$$

Using $E = -\left(2p^2\right)^{-1}\left(Z^2e^2/a_0\right)$ and $L^2 = (\ell + 1/2)^2\hbar^2$ for level (p, ℓ) then

$$\tau_\mathrm{R}(E, L)\frac{\mathrm{d}E}{\mathrm{d}p}\left(\frac{\mathrm{d}L^2}{\mathrm{d}\ell}\right) = \left(\frac{\mathrm{d}J_\mathrm{R}}{\mathrm{d}p}\right)\left(\frac{\mathrm{d}L^2}{\mathrm{d}\ell}\right) \tag{54.125}$$

$$= h\big((2\ell + 1)\hbar^2\big) \tag{54.126}$$

$$\frac{n^s(p, \ell)}{n_\mathrm{e}N^+} = \frac{2(2\ell + 1)}{2\omega_A^+}\,\frac{h^3}{(2\pi m_\mathrm{e}k_\mathrm{B}T)^{3/2}}\,\mathrm{e}^{I_p/k_\mathrm{B}T}, \tag{54.127a}$$

$$\frac{n^s(p)}{n_\mathrm{e}N^+} = \frac{2p^2}{2\omega_A^+}\,\frac{h^3}{(2\pi m_\mathrm{e}k_\mathrm{B}T)^{3/2}}\,\mathrm{e}^{I_p/k_\mathrm{B}T}, \tag{54.127b}$$

in agreement with the Saha ionization formula (54.16) where N^+ is the equilibrium concentration of A^{Z+} ions in their ground electronic states. The spin statistical weights are $\omega_{eA} = \omega_\mathrm{e} = 2$.

54.7 Microscopic Methods for Termolecular Ion–Ion Recombination

At low gas density, the basic process

$$A^+ + B^- + M \to AB + M \tag{54.128}$$

is characterized by nonequilibrium with respect to E. Dissociated and bound A^+–B^- ion pairs are in equilibrium with respect to their separation R, but bound pairs are not in E-equilibrium with each other. L^2-equilibrium can be assumed for ion–ion recombination but not for ion–atom association reactions.

At higher gas densities N, there is nonequilibrium in the ion-pair distributions with respect to R, E and L^2. In the limit of high N, there is only nonequilibrium with respect to R. See [54.13] and the appropriate reference list for full details of theory.

54.7.1 Time Dependent Method: Low Gas Density

Energy levels E_i of A^+–B^- pairs are so close that they form a quasicontinuum with a nonequilibrium distribution over E_i determined by the master equation

$$\frac{\mathrm{d}n_i(t)}{\mathrm{d}t} = \int_{-D}^{\infty} \left(n_i \nu_{if} - n_f \nu_{fi}\right) \mathrm{d}E_f \,, \tag{54.129}$$

where $n_i\,\mathrm{d}E_i$ is the number density of pairs in the interval $\mathrm{d}E_i$ about E_i, and $\nu_{if}\,\mathrm{d}E_f$ is the frequency of i-pair collisions with M that change the i-pair orbital energy from E_i to between E_f and $E_f + \mathrm{d}E_f$. The greatest binding energy of the A^+–B^- pair is D.

Association Rate

$$R^A(t) = \int_{-D}^{\infty} P_i^{\mathrm{S}} \left(\frac{\mathrm{d}n_i}{\mathrm{d}t}\right) \mathrm{d}E_i \tag{54.130a}$$

$$= \hat{\alpha} N_A(t) N_B(t) - k n_{\mathrm{s}}(t) \,, \tag{54.130b}$$

where P_i^{S} is the probability for collisional stabilization (recombination) of i-pairs via a sequence of energy changing collisions with M. The coefficients for $\mathcal{C} \to \mathcal{S}$ recombination out of the $\mathcal{C}$-block with ion concentrations $N_A(t)$, $N_B(t)$ (in cm^{-3}) into the $\mathcal{S}$ block of total ion-pair concentrations $n_{\mathrm{s}}(t)$ and for $\mathcal{S} \to \mathcal{C}$ dissociation are $\hat{\alpha}$ $\left(\mathrm{cm^3\,s^{-1}}\right)$ and $k\left(\mathrm{s^{-1}}\right)$, respectively.

One-Way Equilibrium Collisional Rate and Detailed Balance

$$C_{if} = \tilde{n}_i \nu_{if} = \tilde{n}_f \nu_{fi} = C_{fi} \,, \tag{54.131}$$

where the tilde denotes equilibrium (Saha) distributions.

Normalized Distribution Functions

$$\gamma_i(t) = n_i(t)/\tilde{n}_i^{\mathrm{S}} \,, \quad \gamma_{\mathrm{s}}(t) = n_s(t)/\tilde{n}_{\mathrm{s}}^{B}(t) \,, \tag{54.132}$$

$$\gamma_{\mathrm{c}}(t) = N_A(t) N_B(t)/\tilde{N}_A \tilde{N}_B \,, \tag{54.133}$$

where $\tilde{n}_i^{\mathrm{S}}$ and $\tilde{n}^B$ are the Saha and Boltzmann distributions.

Master Equation for $\gamma_i(t)$

$$\frac{\mathrm{d}\gamma_i(t)}{\mathrm{d}t} = -\int_{-D}^{\infty} \left[\gamma_i(t) - \gamma_f(t)\right] \nu_{if}\,\mathrm{d}E_f \,. \tag{54.134}$$

Quasi-Steady State (QSS) Reduction

Set

$$\gamma_i(t) = P_i^{\mathrm{D}} \gamma_{\mathrm{c}}(t) + P_i^{\mathrm{S}} \gamma_{\mathrm{s}}(t) \xrightarrow{t\to\infty} 1 \tag{54.135}$$

where P_i^{D} and P_i^{S} are the respective time-independent portions of the normalized distribution γ_i which originate, respectively, from blocks $\mathcal{C}$ and $\mathcal{S}$. The energy separation between the $\mathcal{C}$ and $\mathcal{S}$ blocks is so large that $P_i^{\mathrm{S}} = 0$ ($E_i \geq 0$, $\mathcal{C}$ block), $P_i^{\mathrm{S}} \leq 1$ ($0 > E_i \geq -S$, $\mathcal{E}$ block), $P_i^{\mathrm{S}} = 1$ ($-S \geq E_i \geq -D$, $\mathcal{S}$ block). Since $P_i^{\mathrm{S}} + P_i^{\mathrm{D}} = 1$, then

$$\frac{\mathrm{d}\gamma_i(t)}{\mathrm{d}t} = -\left[\gamma_{\mathrm{c}}(t) - \gamma_{\mathrm{s}}(t)\right] \int_{-D}^{\infty} \left(P_i^{\mathrm{D}} - P_f^{\mathrm{D}}\right) C_{if}\,\mathrm{d}E_f \,. \tag{54.136}$$

Recombination and Dissociation Coefficients

Equation (54.135) in (54.130a) enables the recombination rate in (54.130b) to be written as

$$\hat{\alpha} \tilde{N}_A \tilde{N}_B = \int_{-D}^{\infty} P_i^{\mathrm{D}}\,\mathrm{d}E_i \int_{-D}^{\infty} \left(P_i^{\mathrm{D}} - P_f^{\mathrm{D}}\right) C_{if}\,\mathrm{d}E_f \,. \tag{54.137}$$

The QSS condition ($\mathrm{d}n_i/\mathrm{d}t = 0$ in block $\mathcal{E}$) is then

$$P_i^{\mathrm{D}} \int_{-D}^{\infty} \nu_{if}\,\mathrm{d}E_f = \int_{-D}^{E} \nu_{if} P_f^{\mathrm{D}}\,\mathrm{d}E_f \,, \tag{54.138}$$

which involves only time independent quantities. Under QSS, (54.137) reduces to the net downward current across bound level $-E$,

$$\hat{\alpha} \tilde{N}_A \tilde{N}_B = \int_{-E}^{\infty} \mathrm{d}E_i \int_{-D}^{-E} \left(P_i^{\mathrm{D}} - P_f^{\mathrm{D}}\right) C_{if}\,\mathrm{d}E_f \,, \tag{54.139}$$

which is independent of the energy level $(-E)$ in the range $0 \geq -E \geq -S$ of block $\mathcal{E}$.

The dissociation frequency k in (54.130b) is

$$k \tilde{n}_{\mathrm{s}} = \int_{-D}^{-E} \mathrm{d}E_i \int_{-E}^{\infty} \left(P_i^{\mathrm{S}} - P_f^{\mathrm{S}}\right) C_{if}\,\mathrm{d}E_f \,, \tag{54.140}$$

and macroscopic detailed balance $\hat{\alpha}\tilde{N}_A\tilde{N}_B = k\tilde{n}_s$ is automatically satisfied. $\hat{\alpha}$ is the direct ($\mathcal{C} \to \mathcal{S}$) collisional contribution (small) plus the (much larger) net collisional cascade downward contribution from that fraction of bound levels which originated in the continuum $\mathcal{C}$. k_d is the direct dissociation frequency (small) plus the net collisional cascade upward contribution from that fraction of bound levels which originated in block $\mathcal{S}$.

54.7.2 Time Independent Methods: Low Gas Density

QSS-Rate. Since recombination and dissociation (ionization) involve only that fraction of the bound state population which originated from the $\mathcal{C}$ and $\mathcal{S}$ blocks, respectivel,y recombination can be viewed as time independent with

$$N_A\, N_B = \tilde{N}_A\tilde{N}_B\,,\quad n_s(t) = 0\,, \tag{54.141a}$$

$$\rho_i = n_i/\tilde{n}_i \equiv P_i^{\mathrm{D}} \tag{54.141b}$$

$$\hat{\alpha}\tilde{N}_A\tilde{N}_B = \int_{-E}^{\infty} \mathrm{d}E_i \int_{-D}^{-E} (\rho_i - \rho_f)\, C_{if}\, \mathrm{d}E_f\,. \tag{54.141c}$$

QSS Integral Equation.

$$\rho_i \int_{-D}^{\infty} \nu_{if}\, \mathrm{d}E_f = \int_{-S}^{\infty} \rho_f \nu_{if}\, \mathrm{d}E_f \tag{54.142}$$

is solved subject to the boundary condition

$$\rho_i = 1(E_i \geq 0)\,,\quad \rho_i = 0(-S \geq E_i \geq -D)\,. \tag{54.143}$$

Collisional Energy-Change Moments.

$$D^{(m)}(E_i) = \frac{1}{m!}\int_{-D}^{\infty} (E_f - E_i)^m C_{if}\, \mathrm{d}E_f\,, \tag{54.144}$$

$$D_i^{(m)} = \frac{1}{m!}\frac{\mathrm{d}}{\mathrm{d}t}\left\langle (\Delta E)^m \right\rangle\,. \tag{54.145}$$

Averaged Energy-Change Frequency. For an equilibrium distribution $\tilde{n}_i$ of E_i-pairs per unit interval $\mathrm{d}E_i$ per second,

$$D_i^{(1)} = \frac{\mathrm{d}}{\mathrm{d}t}\langle \Delta E\rangle\,.$$

Averaged Energy-Change per Collision.

$$\langle \Delta E\rangle = D_i^{(1)}/D_i^{(0)}\,.$$

Time Independent Dissociation. The time independent picture corresponds to

$$n_s(t) = \tilde{n}_s\,,\quad \gamma_c(t) = 0\,,\quad \rho_i = n_i/\tilde{n}_i \equiv P_i^{\mathrm{S}}\,, \tag{54.146}$$

in analogy to the macroscopic reduction of (54.38a,b).

Variational Principle

The QSS-condition (54.135) implies that the fraction P_i^{D} of bound levels i with precursor $\mathcal{C}$ are so distributed over i that (54.137) for $\hat{\alpha}$ is a minimum. Hence P_i^{D} or ρ_i are obtained either from the solution of (54.142) or from minimizing the variational functional

$$\hat{\alpha}\tilde{N}_A\tilde{N}_B = \int_{-D}^{\infty} n_i\, \mathrm{d}E_i \int_{-D}^{\infty} (\rho_i - \rho_f)\, \nu_{if}\, \mathrm{d}E_f \tag{54.147a}$$

$$= \frac{1}{2}\int_{-D}^{\infty} \mathrm{d}E_i \int_{-D}^{\infty} (\rho_i - \rho_f)^2\, C_{if}\, \mathrm{d}E_f \tag{54.147b}$$

with respect to variational parameters contained in a trial analytic expression for ρ_i. Minimization of the quadratic functional (54.147b) has an analogy with the principle of least dissipation in the theory of electrical networks.

Diffusion-in-Energy-Space Method

Integral Equation (54.142) can be expanded in terms of energy-change moments, via a Fokker–Planck analysis to yield the differential equation

$$\frac{\partial}{\partial E_i}\left(D_i^{(2)}\frac{\partial \rho_i}{\partial E_i}\right) = 0\,, \tag{54.148}$$

with the QSS analytical solution

$$\rho_i(E_i) = \left(\int_{E_i}^{0} \frac{\mathrm{d}E}{D^{(2)}(E)}\right)\left(\int_{-S}^{0} \frac{\mathrm{d}E}{D^{(2)}(E)}\right)^{-1} \tag{54.149}$$

of *Pitaevskiĭ* [54.4] for $(e^- + A^+ + M)$ recombination where collisional energy changes are small. This distribution does not satisfy the exact QSS condition (54.142). When inserted in the exact non-QSS rate (54.147b), highly accurate $\hat{\alpha}$ for heavy-particle recombination are obtained.

Bottleneck Method

The one-way equilibrium rate $(\mathrm{cm}^{-3}\,\mathrm{s}^{-1})$ across $-E$, i. e., (54.141c) with $\rho_i = 1$ and $\rho_f = 0$, is

$$\hat{\alpha}(-E)\tilde{N}_A\tilde{N}_B = \int_{-E}^{\infty} \mathrm{d}E_i \int_{-D}^{-E} C_{if}\,\mathrm{d}E_f\,. \tag{54.150}$$

This is an upper limit to (54.141c) and exhibits a minimum at $-E^*$, the bottleneck location. The least upper limit to $\hat{\alpha}$ is then $\hat{\alpha}(-E^*)$.

Trapping Radius Method

Assume that pairs with internal separation $R \le R_\mathrm{T}$ recombine with unit probability so that the one-way equilibrium rate across the dissociation limit at $E = 0$ for these pairs is

$$\hat{\alpha}(R_\mathrm{T})\tilde{N}_A\tilde{N}_B = \int_{0}^{R_\mathrm{T}} \mathrm{d}\boldsymbol{R} \int_{V(R)}^{0} C_{if}(R)\,\mathrm{d}E_f\,, \tag{54.151}$$

where $V(R) = -e^2/R$, and $C_{if}(R) = \tilde{n}_i(R)\nu_{if}(R)$ is the rate per unit interval $(\mathrm{d}\boldsymbol{R}\mathrm{d}E_i)\mathrm{d}E_f$ for the $E_i \to E_f$ collisional transitions at fixed R in

$$\left(A^+ - B^-\right)_{E_i,R} + M \to \left(A^+ - B^-\right)_{E_f,R} + M\,. \tag{54.152}$$

The concentration (cm^{-3}) of pairs with internal separation R and orbital energy E_i in the interval $\mathrm{d}\boldsymbol{R}\mathrm{d}E_i$ about $(\boldsymbol{R}, E_i)$ is $\tilde{n}_i(R)\,\mathrm{d}\boldsymbol{R}\mathrm{d}E_i$. Agreement with the exact treatment [54.13] is found by assigning $R_\mathrm{T} = (0.48 - 0.55)\left(\mathrm{e}^2/k_\mathrm{B}T\right)$ for the recombination of equal mass ions in an equal mass gas for various ion–neutral interactions. For further details on the above methods, see the appropriate references on termolecular recombination in the general references on page 825.

54.7.3 Recombination at Higher Gas Densities

As the density N of the gas M is raised, the recombination rate $\hat{\alpha}$ increases initially as N to such an extent that there are increasingly more pairs $n_i^-(R, E)$ in a state of contraction in R than there are those $n_i^+(R, E)$ in a state of expansion; i. e., the ion-pair distribution densities $n_i^\pm(R, E)$ per unit interval $\mathrm{d}E\mathrm{d}R$ are not in equilibrium with respect to R in blocks $\mathcal{C}$ and $\mathcal{E}$. Those in the highly excited block $\mathcal{E}$ in addition are not in equilibrium with respect to energy E. Basic sets of coupled master equations have been developed [54.13] for the microscopic nonequilibrium distributions $n^\pm\left(R, E, L^2\right)$ and $n^\pm(R, E)$ of expanding (+) and contracting (−) pairs with respect to A–B separation R, orbital energy E and orbital angular momentum L^2. With $n\left(\boldsymbol{R}, E_i, L_i^2\right) \equiv n_i(R)$, and using the notation defined at the beginning of Sect. 54.6, the distinct regimes for the master equations discussed in Sect. 54.7.4 are:

Low N	Equilibrium in R, but not in E, L^2 → master equation for $n\left(E, L^2\right)$.
Pure Coulomb attraction	Equilibrium in L^2 → master equation for $n(E)$.
High N	Nonequilibrium in R, E, L^2 → master equation for $n_i^\pm(R)$.
Highest N	Equilibrium in $\left(E, L^2\right)$ but not in R → macroscopic transport equation (54.56a) in $n(R)$.

Normalized Distributions

For a state $|i\rangle \equiv \left|E, L^2\right\rangle$,

$$\rho_i(R) = \frac{n_i(R)}{\tilde{n}_i(R)}, \quad \rho_i^\pm(R) = \frac{n_i^\pm(R)}{\tilde{n}_i^\pm(R)}\,,$$
$$\rho_i(R) = \frac{1}{2}\left(\rho_i^+ + \rho_i^-\right)\,. \tag{54.153}$$

Orbital Energy and Angular Momentum

$$E_i = \frac{1}{2}M_{AB}v^2 + V(R)\,, \tag{54.154a}$$

$$E_i = \frac{1}{2}M_{AB}v_\mathrm{R}^2 + V_i(R)\,, \tag{54.154b}$$

$$V_i(R) = V(R) + \frac{L_i^2}{2M_{AB}R^2}\,, \tag{54.154c}$$

$$L_i = |\boldsymbol{R} \times M_{AB}\boldsymbol{v}|\,,$$
$$L_i^2 = (2M_{AB}E_i)b^2,\ E_i > 0\,. \tag{54.154d}$$

Maximum Orbital Angular Momenta

(1) A specified separation R can be accessed by all orbits of energy E_i with L_i^2 between 0 and

$$L_{im}^2(E_i, R) = 2M_{AB}R^2\,[E_i - V(R)]\,. \tag{54.155a}$$

(2) Bounded orbits of energy $E_i < 0$ can have L_i^2 between 0 and

$$L_{ic}^2(E_i) = 2M_{AB}R_\mathrm{c}^2\,[E_i - V(R_\mathrm{c})]\,, \tag{54.155b}$$

where R_c is the radius of the circular orbit determined by $\partial V_i/\partial R = 0$, i. e., by $E_i = V(R_\mathrm{c}) + \frac{1}{2}R_\mathrm{c}(\partial V/\partial R)_{R_\mathrm{c}}$.

54.7.4 Master Equations

Master Equation for $n_i^{\pm}(R)^{\pm}(\boldsymbol{R}, E_i, L_i^2)$ [54.13]

$$\pm\frac{1}{R^2}\frac{\partial}{\partial R}\left[R^2 n_i^{\pm}(R)\,|v_\mathrm{R}|\right]_{E_i,L_i^2} = -\int_{V(R)}^{\infty}\mathrm{d}E_f\int_0^{L_{f\mathrm{m}}^2}\mathrm{d}L_f^2\Big[n_i^{\pm}(R)\nu_{if}(R) - n_f^{\pm}(R)\nu_{fi}(R)\Big]\,. \qquad (54.156)$$

The set of master equations [54.13] for n_i^+ is coupled to the n_i^- set by the boundary conditions $n_i^-\left(R_i^{\mp}\right) = n_i^+\left(R_i^{\mp}\right)$ at the pericenter R_i^- for all E_i and apocenter R_i^+ for $E_i < 0$ of the E_i, L_i^2-orbit.

Master Equations for Normalized Distributions [54.13]

$$\pm|v_\mathrm{R}|\frac{\partial\rho_i^{\pm}}{\partial R} = -\int_{V(R)}^{\infty}\mathrm{d}E_f\int_0^{L_{f\mathrm{m}}^2}\mathrm{d}L_f^2 \times\Big[\rho_i^{\pm}(R)-\rho_f^{\pm}(R)\Big]\nu_{if}(R)\,. \qquad (54.157)$$

Corresponding Master equations for the L^2 integrated distributions $n^{\pm}(R, E)$ and $\rho^{\pm}(R, E)$ have been derived [54.13].

Continuity Equations

$$J_i = \left[n_i^+(R) - n_i^-(R)\right]|v_\mathrm{R}| = \left(\rho_i^+ - \rho_i^-\right)\tilde{j}_i^{\pm} \qquad (54.158)$$

$$\frac{1}{R^2}\frac{\partial}{\partial R}(R^2 J_i) = -\int_{V(R)}^{\infty}\mathrm{d}E_f\int_0^{L_{f\mathrm{m}}^2}\mathrm{d}L_f^2 \times\left[n_i(R)\nu_{if}(R) - n_f(R)\nu_{fi}(R)\right]\,, \qquad (54.159)$$

$$\frac{1}{2}|v_\mathrm{R}|\frac{\partial\left[\rho_i^+(R)-\rho_i^-(R)\right]}{\partial R} = -\int_{V(R)}^{\infty}\mathrm{d}E_f\int_0^{L_{f\mathrm{m}}^2}\mathrm{d}L_f^2\left[\rho_i(R)-\rho_f(R)\right]\nu_{if}(R)\,. \qquad (54.160)$$

54.7.5 Recombination Rate

Flux Representation

The $R_0 \to \infty$ limit of

$$\hat{\alpha}\tilde{N}_A\tilde{N}_B = -4\pi R_0^2 J(R_0) \qquad (54.161)$$

has the microscopic generalization

$$\hat{\alpha}\tilde{N}_A\tilde{N}_B = \int_{V(R_0)}^{\infty}\mathrm{d}E_i\int_0^{L_{i\mathrm{c}}^2}\mathrm{d}L_i^2\left[4\pi R_0^2\tilde{j}_i^{\pm}(R_0)\right] \times\left[\rho_i^-(R_0)-\rho_i^+(R_0)\right]\,, \qquad (54.162)$$

where $L_{i\mathrm{c}}^2$ is given by (54.155b) with $R_\mathrm{c} = R_0$ for bound states and is infinite for dissociated states, and where

$$\rho_i^-(R_0)-\rho_i^+(R_0) = \oint_{R_i}^{R_0}\rho_i(R)\left[\nu_i^b(R)+\nu_i^\mathrm{c}(R)\right]\mathrm{d}t\,, \qquad (54.163)$$

with

$$\rho_i(R)\nu_i^b(R) = \int_{V(R)}^{V(R_0)}\mathrm{d}E_f\int_0^{L_{f\mathrm{m}}^2}\mathrm{d}L_f^2\left[\rho_i(R)-\rho_f(R)\right] \times\nu_{if}(R)\,, \qquad (54.164a)$$

$$\rho_i(R)\nu_i^\mathrm{c}(R) = \int_{V(R_0)}^{\infty}\mathrm{d}E_f\int_0^{L_{f\mathrm{m}}^2}\mathrm{d}L_f^2\left[\rho_i(R)-\rho_f(R)\right] \times\nu_{if}(R)\,. \qquad (54.164b)$$

Collisional Representation

$$\hat{\alpha}\tilde{N}_A\tilde{N}_B = \int_{V(R_0)}^{\infty}\mathrm{d}E_i\int_0^{L_{i\mathrm{c}}^2}\mathrm{d}L_i^2\int_{R_i}^{R_0}\tilde{n}_i(R)\,\mathrm{d}\boldsymbol{R} \times\left[\rho_i(R)\nu_i^b(R)\right]\,, \qquad (54.165)$$

which is the microscopic generalization of the macroscopic result $\hat{\alpha} = K\rho^*\nu_\mathrm{s} = \alpha_{RN}(R_0)\rho(R_0)$.

The flux for dissociated pairs $E_i > 0$ is

$$4\pi R^2|v_\mathrm{R}|\tilde{n}_i^{\pm}(R)\,\mathrm{d}E\,\mathrm{d}L^2 = \left[\bar{v}\varepsilon\,\mathrm{e}^{-\varepsilon}\,\mathrm{d}\varepsilon\right][2\pi b\,\mathrm{d}b]\,\tilde{N}_A\tilde{N}_B\,, \qquad (54.166)$$

so the rate (54.165) as $R_0 \to \infty$ is

$$\hat{\alpha} = \bar{v}\int_0^{\infty}\varepsilon\,\mathrm{e}^{-\varepsilon}\,\mathrm{d}\varepsilon\int_0^{b_0}2\pi b\,\mathrm{d}b\oint_{R_i}^{R_0}\rho_i(R)\nu_i^b(R)\,\mathrm{d}t\,, \qquad (54.167)$$

which is the microscopic generalization (54.45) of the macroscopic result $\hat{\alpha} = k_c P^S$ of (54.44).

Reaction Rate $\alpha_{RN}(R_0)$

On solving (54.157) subject to $\rho(R_0) = 1$, then according to (54.56b), $\hat{\alpha}$ determined by (54.162) is the rate $\hat{\alpha}_{RN}$ of recombination within the $(A-B)$ sphere of radius R_0. The overall rate of recombination $\hat{\alpha}$ is then given by the full diffusional-drift reaction rate (54.59b) where the rate of transport to R_0 is determined uniquely by (54.60).

For development of theory [54.13] and computer simulations, see the reference list on Termolecular Ion–Ion Recombination: Theory, and Simulations, respectively.

54.8 Radiative Recombination

In the radiative recombination (RR) process

$$e^-(E,\ell') + A^{Z+}(c) \to A^{(Z-1)+}(c, n\ell) + h\nu\,, \tag{54.168}$$

the accelerating electron e^- with energy and angular momentum (E,ℓ') is captured, via coupling with the weak quantum electrodynamical interaction $(e/m_e c)\boldsymbol{A}\cdot\boldsymbol{p}$ associated with the electromagnetic field of the moving ion, into an excited state $n\ell$ with binding energy $I_{n\ell}$ about the parent ion A^{Z+} (initially in an electronic state c). The simultaneously emitted photon carries away the excess energy $h\nu = E + I_{n\ell}$ and angular momentum difference between the initial and final electronic states. The cross section $\sigma_R^{n\ell}(E)$ for RR is calculated (a) from the Einstein A coefficient for free–bound transitions or (b) from the cross section $\sigma_I^{n\ell}(h\nu)$ for photoionization (PI) via the detailed balance (DB) relationship appropriate to (54.168).

The rates $\langle v_e\sigma_R\rangle$ and averaged cross sections $\langle\sigma_R\rangle$ for a Maxwellian distribution of electron speeds v_e are then determined from either

$$\hat{\alpha}_R^{n\ell}(T_e) = \bar{v}_e \int_0^\infty \varepsilon\sigma_R^{n\ell}(\varepsilon)\exp(-\varepsilon)\,d\varepsilon = \bar{v}_e\left\langle\sigma_R^{n\ell}(T_e)\right\rangle\,, \tag{54.169}$$

where $\varepsilon = E/k_B T_e$, or from the Milne DB relation (54.243) between the forward and reverse macroscopic rates of (54.168). Using the hydrogenic semiclassical σ_I^n of *Kramers* [54.5], together with an asymptotic expansion [54.14] for the g-factor of *Gaunt* [54.15], the quantal/semiclassical cross section ratio in (54.249), Seaton [54.16] calculated $\hat{\alpha}_R^{n\ell}$.

The rate of electron energy loss in RR is

$$\left\langle\frac{dE}{dt}\right\rangle_{n\ell} = n_e\bar{v}_e(k_B T_e)\int_0^\infty \varepsilon^2\sigma_R^{n\ell}(\varepsilon)e^{-\varepsilon}\,d\varepsilon\,, \tag{54.170}$$

and the radiated power produced in RR is

$$\left\langle\frac{d(h\nu)}{dt}\right\rangle_{n\ell} = n_e\bar{v}_e\int_0^\infty \varepsilon h\nu\sigma_R^{n\ell}(\varepsilon)e^{-\varepsilon}\,d\varepsilon\,. \tag{54.171}$$

Standard Conversions

$$E = p_e^2/2m_e = \hbar^2k_e^2/2m_e = k_e^2a_0^2(e^2/2a_0) \tag{54.172a}$$

$$= \kappa^2(Z^2e^2/2a_0) = \varepsilon(Z^2e^2/2a_0)\,, \tag{54.172b}$$

$$E_\nu = h\nu = \hbar\omega = \hbar k_\nu c = (I_n + E) \tag{54.172c}$$

$$\equiv (1+n^2\varepsilon)(Z^2e^2/2n^2a_0)\,, \tag{54.172d}$$

$$h\nu/I_n = 1 + n^2\varepsilon,\quad k_e^2a_0^2 = 2E/(e^2/a_0)\,, \tag{54.172e}$$

$$k_\nu a_0 = (h\nu)\alpha/(e^2/a_0)\,, \tag{54.172f}$$

$$k_\nu^2/k_e^2 = (h\nu)^2/(2Em_ec^2) \tag{54.172g}$$

$$= \alpha^2(h\nu)^2/\left[2E(e^2/a_0)\right]\,, \tag{54.172h}$$

$$I_H = e^2/2a_0,\quad \alpha = e^2/\hbar c = 1/137.035\,9895\,,$$
$$\alpha^{-2} = m_ec^2/(e^2/a_0)\,,\quad I_n = (Z^2/n^2)I_H\,. \tag{54.172i}$$

The electron and photon wavenumbers are k_e and k_ν, respectively.

54.8.1 Detailed Balance and Recombination-Ionization Cross Sections

Cross sections $\sigma_R^{n\ell}(E)$ and $\sigma_I^{n\ell}(h\nu)$ for radiative recombination (RR) into and photoionization (PI) out of level $n\ell$ of atom A are interrelated by the detailed balance relation

$$g_e g_A^+ k_e^2\sigma_R^{n\ell}(E) = g_\nu g_A\, k_\nu^2\sigma_I^{n\ell}(h\nu)\,, \tag{54.173}$$

where $g_e = g_\nu = 2$. Electronic statistical weights of A and A^+ are g_A and g_A^+, respectively. Thus, using

(54.172g) for k_ν^2/k_e^2,

$$\sigma_R^{n\ell}(E) = \left(\frac{g_A}{2g_A^+}\right)\left(\frac{(h\nu)^2}{Em_ec^2}\right)\sigma_I^{n\ell}(h\nu) . \tag{54.174}$$

The statistical factors are:

(a) For $(A^+ + e^-)$ state $c[S_c, L_c; \varepsilon, \ell', m']$:
$g_A^+ = (2S_c+1)(2L_c+1)$.

(b) For $A(n\ell)$ state $b[S_c, L_c; n, \ell]SL$:
$g_A = (2S+1)(2L+1)$.

(c) For $n\ell$ electron outside a closed shell:
$g_A^+ = 1,\ g_A = 2(2\ell+1)$.

Cross sections are averaged over initial and summed over final degenerate states. For case (c),

$$\sigma_I^n = \frac{1}{n^2}\sum_{\ell=0}^{n-1}(2\ell+1)\sigma_I^{n\ell} ; \tag{54.175a}$$

$$\sigma_R^n = \sum_{\ell=0}^{n-1} 2(2\ell+1)\sigma_R^{n\ell} . \tag{54.175b}$$

54.8.2 Kramers Cross Sections, Rates, Electron Energy-Loss Rates and Radiated Power for Hydrogenic Systems

These are all calculated from application of detailed balance (54.173) to the original $\sigma_I^n(h\nu)$ of *Kramers* [54.5].

Semiclassical (Kramers) Cross Sections

For hydrogenic systems,

$$I_n = \frac{Z^2e^2}{2n^2a_0},\quad h\nu = I_n + E . \tag{54.176}$$

The results below are expressed in terms of the quantities

$$b_n = \frac{I_n}{k_BT_e}, \tag{54.177}$$

$$\sigma_{I0}^n = \frac{64\pi a_0^2\alpha}{3\sqrt{3}}\left(\frac{n}{Z^2}\right) = 7.907\,071\times10^{-18}(n/Z^2)\,\text{cm}^2 , \tag{54.178}$$

$$\sigma_{R0}(E) = \left(\frac{8\pi a_0^2\alpha^3}{3\sqrt{3}}\right)\frac{(Z^2e^2/a_0)}{E}, \tag{54.179}$$

$$\hat{\alpha}_0(T_e) = \bar{v}_e\left(\frac{8\pi a_0^2\alpha^3}{3\sqrt{3}}\right)\frac{(Z^2e^2/a_0)}{k_BT_e} \tag{54.180}$$

PI and RR Cross Sections for Level n. In the Kramer (K) semiclassical approximation,

$$_K\sigma_I^n(h\nu) = \left(\frac{I_n}{h\nu}\right)^3\sigma_{I0}^n = {}_K\sigma_I^{n\ell}(h\nu) , \tag{54.181}$$

$$_K\sigma_R^n(E) = \sigma_{R0}(E)\left(\frac{2}{n}\right)\left(\frac{I_n}{I_n+E}\right) = 3.897\times10^{-20}\times\left[n\varepsilon(13.606+n^2\varepsilon^2)\right]^{-1}\text{cm}^2 , \tag{54.182}$$

where ε is in units of eV and is given by

$$\varepsilon = E/Z^2 \equiv (2.585\times10^{-2}/Z^2)(T_e/300) . \tag{54.183}$$

Equation (54.182) illustrates that RR into low n at low E is favored.

Cross Section for RR into Level $n\ell$.

$$_K\sigma_R^{n\ell} = \left[(2\ell+1)/n^2\right]{}_K\sigma_R^n . \tag{54.184}$$

Rate for RR into Level n.

$$\hat{\alpha}_R^n(T_e) = \hat{\alpha}_0(T_e)\,(2/n)\,b_n e^{b_n}E_1(b_n) , \tag{54.185a}$$

which tends for large b_n (i.e., $k_BT_e \ll I_n$) to

$$\hat{\alpha}_R^n(T_e\to0) = \hat{\alpha}_0(T_e)\,(2/n)\times\left(1-b_n^{-1}+2b_n^{-2}-6b_n^{-3}+\cdots\right) . \tag{54.185b}$$

The Kramers cross section for photoionization at threshold is σ_{I0}^n and

$$\sigma_{R0}^n = 2\sigma_{R0}/n;\quad \hat{\alpha}_0^n = 2\hat{\alpha}_0/n \tag{54.186}$$

provide the corresponding Kramers cross section and rate for recombination as $E\to0$ and $T_e\to0$, respectively.

RR Cross Sections and Rates into All Levels $n\ge n_f$.

$$\sigma_R^T(E) = \int_{n_f}^{\infty}\sigma_R^n(E)\,dn = \sigma_{R0}(E)\ln(1+I_f/E) , \tag{54.187a}$$

$$\hat{\alpha}_R^T(T_e) = \hat{\alpha}_0(T_e)\left[\gamma+\ln b_f+e^{b_f}E_1(b_f)\right] \tag{54.187b}$$

Useful Integrals.

$$\int_0^\infty e^{-x} \ln x \, dx = \gamma \,, \quad (54.188a)$$

$$\int_b^\infty x^{-1} e^{-x} \, dx = E_1(b) \,, \quad (54.188b)$$

$$\int_0^b e^x E_1(x) \, dx = \gamma + \ln b + e^b E_1(b) \,, \quad (54.188c)$$

$$\int_0^b \left[1 - x e^x E_1(x)\right] dx$$

$$= \gamma + \ln b + e^b (1-b) E_1(b) \,, \quad (54.188d)$$

where $\gamma = 0.577\,2157$ is Euler's constant, and $E_1(b)$ is the first exponential integral such that

$$b e^b E_1(b) \xrightarrow{b \gg 1} 1 - b^{-1} + 2b^{-2} - 6b^{-3} + 24b^{-4} + \cdots \,.$$

Electron Energy Loss Rate

Energy Loss Rate for RR into Level n.

$$\left\langle \frac{dE}{dt} \right\rangle_n = n_e \hat{\alpha}_R^n (T_e) k_B T_e \left(\frac{1 - b_n e^{b_n} E_1(b_n)}{e^{b_n} E_1(b_n)} \right) , \quad (54.189a)$$

which for large b_n (i. e. $(k_B T_e) \ll I_n$) tends to

$$n_e \hat{\alpha}_R^n (T_e) k_B T_e \left(1 - b_n^{-1} + 3b_n^{-2} - 13 b_n^{-3} + \cdots \right) \quad (54.189b)$$

with (54.185a) for $\hat{\alpha}_R^n$.

Energy Loss Rate for RR into All Levels $n \geq n_f$.

$$\left\langle \frac{dE}{dt} \right\rangle = n_e k_B T_e \hat{\alpha}_0(T_e) \left[\gamma + \ln b_f + e^{b_f} E_1(b_f)(1 - b_f)\right] \quad (54.190a)$$

$$= n_e (k_B T_e) \left[\hat{\alpha}_R^T (T_e) - \hat{\alpha}_0(T_e) b_f e^{b_f} E_1(b_f)\right] \quad (54.190b)$$

with (54.187b) and (54.180) for $\hat{\alpha}_R^T$ and $\hat{\alpha}_0$.

Radiated Power

Radiated Power for RR into Level n.

$$\left\langle \frac{d(h\nu)}{dt} \right\rangle_n = n_e \hat{\alpha}_R^n (T_e) I_n \left[b_n e^{b_n} E_1(b_n) \right]^{-1} , \quad (54.191a)$$

which for large b_n (i. e. $(k_B T_e) \ll I_n$) tends to

$$n_e \hat{\alpha}_R^n (T_e) I_n \left(1 + b_n^{-1} - b_n^{-2} + 3 b_n^{-3} + \cdots \right) . \quad (54.191b)$$

Radiated Power for RR into All Levels $n \geq n_f$.

$$\left\langle \frac{d(h\nu)}{dt} \right\rangle = n_e \hat{\alpha}_0 (T_e) I_f \,. \quad (54.192)$$

To allow n-summation, rather than integration as in (54.187a), to each of the above expressions is added $1/2\sigma_R^{n_f}$, $1/2\hat{\alpha}_R^{n_f}$, $1/2\langle dE/dt\rangle_{n_f}$ and $1/2\langle d(h\nu)/dt\rangle_{n_f}$, respectively. The expressions valid for bare nuclei of charge Z are also fairly accurate for recombination to a core of charge Z_c and atomic number Z_A, provided that Z is identified as $1/2(Z_A + Z_c)$.

Differential Cross Sections for Coulomb Elastic Scattering.

$$\sigma_c(E, \theta) = \frac{b_0^2}{4 \sin^4 \frac{1}{2}\theta} \,, \quad b_0^2 = (Ze^2/2E)^2 \,. \quad (54.193)$$

The integral cross section for Coulomb scattering by $\theta \geq \pi/2$ at energy $E = (3/2) k_B T$ is

$$\sigma_c(E) = \pi b_0^2 = \frac{1}{9} \pi R_e^2, \quad R_e = e^2 / k_B T \,. \quad (54.194)$$

Photon Emission Probability.

$$P_\nu = \sigma_R^n(E) / \sigma_c(E) \,. \quad (54.195a)$$

This is small and increases with decreasing n as

$$P_\nu(E) = \left(\frac{8\alpha^3}{3\sqrt{3}}\right) \frac{8}{n} \frac{E}{(e^2/a_0)} \left(\frac{I_n}{h\nu}\right) . \quad (54.195b)$$

54.8.3 Basic Formulae for Quantal Cross Sections

Radiative Recombination and Photoionization Cross Sections

The cross section $\sigma_R^{n\ell}$ for recombination follows from the continuum-bound transition probability P_{if} per unit time. It is also provided by the detailed balance relation (54.173) in terms of $\sigma_I^{n\ell}$ which follows from P_{fi}. The number of radiative transitions per second is

$$\left[g_e g_A^+ \rho(E)\,\mathrm{d}E\,\mathrm{d}\hat{\boldsymbol{k}}_e\right] P_{if}\left[\rho(E_\nu)\,\mathrm{d}E_\nu\,\mathrm{d}\hat{\boldsymbol{k}}_\nu\right]$$
$$= g_e g_A^+ v_e \frac{\mathrm{d}\boldsymbol{p}_e}{(2\pi\hbar)^3}\sigma_R(\boldsymbol{k}_e) = g_\nu g_A\, c\frac{\mathrm{d}\boldsymbol{k}_\nu}{(2\pi)^3}\sigma_I(\boldsymbol{k}_\nu)\,, \quad (54.196)$$

where the electron current $(\mathrm{cm}^{-2}\,\mathrm{s}^{-1})$ is

$$\frac{v_e\,\mathrm{d}\boldsymbol{p}_e}{(2\pi\hbar)^3} = \left(\frac{2mE}{h^3}\right)\mathrm{d}E\,\mathrm{d}\hat{\boldsymbol{k}}_e\,, \quad (54.197)$$

and the photon current $(\mathrm{cm}^{-2}\,\mathrm{s}^{-1})$ is

$$c\frac{\mathrm{d}\boldsymbol{k}_\nu}{(2\pi)^3} = c\frac{(h\nu)^2}{(2\pi\hbar c)^3}\,\mathrm{d}E_\nu\,\mathrm{d}\hat{\boldsymbol{k}}_\nu\,. \quad (54.198)$$

Time Dependent Quantum Electrodynamical Interaction.

$$V(\boldsymbol{r},t) = \frac{e}{mc}\boldsymbol{A}\cdot\boldsymbol{p} = \mathrm{i}e\left(\frac{2\pi\hbar\nu}{\mathcal{V}}\right)^{1/2}(\hat{\boldsymbol{\epsilon}}\cdot\boldsymbol{r})\,\mathrm{e}^{-\mathrm{i}(\boldsymbol{k}_\nu\cdot\boldsymbol{r}-\omega t)}$$
$$\equiv V(\boldsymbol{r})\,\mathrm{e}^{\mathrm{i}\omega t}\,. \quad (54.199)$$

In the dipole approximation, $\mathrm{e}^{-\mathrm{i}\boldsymbol{k}_\nu\cdot\boldsymbol{r}} \approx 1$.

Continuum-Bound State-to-State Probability.

$$P_{if} = \frac{2\pi}{\hbar}\left|V_{fi}\right|^2\delta\left[E_\nu-(E+I_n)\right]$$
$$V_{fi} = \langle\Psi_{n\ell m}(\boldsymbol{r})|\,V(\boldsymbol{r})\,|\Psi_i(\boldsymbol{r},\boldsymbol{k}_e)\rangle\,. \quad (54.200)$$

Number of Photon States in Volume $\mathcal{V}$.

$$\rho\left(E_\nu,\hat{k}_\nu\right)\mathrm{d}E_\nu\,\mathrm{d}\hat{\boldsymbol{k}}_\nu = \mathcal{V}(h\nu)^2/(2\pi\hbar c)^3\,\mathrm{d}E_\nu\,\mathrm{d}\hat{\boldsymbol{k}}_\nu \quad (54.201a)$$
$$= \mathcal{V}\left[\omega^2/(2\pi c)^3\right]\mathrm{d}\omega\,\mathrm{d}\hat{\boldsymbol{k}}_\nu\,. \quad (54.201b)$$

Continuum-Bound Transition Rate. On summing over the two directions ($g_\nu = 2$) of polarization, the rate for transitions into all final photon states is

$$A_{n\ell m}\left(E,\hat{\boldsymbol{k}}_e\right) = \int P_{if}\rho(E_\nu)\,\mathrm{d}E_\nu\,\mathrm{d}\hat{\boldsymbol{k}}_\nu$$
$$= \frac{4e^2}{3\hbar}\frac{(h\nu)^3}{(3\hbar c)^3}\left|\langle\Psi_{n\ell m}|\boldsymbol{r}|\Psi_i(\boldsymbol{k}_e)\rangle\right|^2\,. \quad (54.202)$$

Transition Frequency: Alternative Formula.

$$A_{n\ell m}\left(E,\hat{\boldsymbol{k}}_e\right) = (2\pi/\hbar)\left|D_{fi}\right|^2\,, \quad (54.203)$$

where the dipole atom-radiation interaction coupling is

$$D_{fi}(\boldsymbol{k}_e) = \left(\frac{2\omega^3}{3\pi c^3}\right)^{1/2}\langle\Psi_{n\ell m}|e\boldsymbol{r}|\Psi_i(\boldsymbol{k}_e)\rangle\,. \quad (54.204)$$

RR Cross Section into Level $(n\ell m)$.

$$\sigma_R^{n\ell m}(E) = \frac{1}{4\pi}\int\sigma_R^{n\ell m}(\boldsymbol{k}_e)\,\mathrm{d}\hat{\boldsymbol{k}}_e$$
$$= \frac{h^3\rho(E)}{8\pi m_e E}\int A_{n\ell m}\left(E,\hat{\boldsymbol{k}}_e\right)\mathrm{d}\hat{\boldsymbol{k}}_e\,. \quad (54.205)$$

RR Cross Section into Level $(n\ell)$.

$$\sigma_R^{n\ell}(E) = \frac{8\pi^2}{3}\left(\frac{(\alpha h\nu)^3}{2(e^2/a_0)E}\right)\rho(E)R_I^{n\ell}(E)$$
$$R_I^{n\ell}(E) = \int\mathrm{d}\hat{\boldsymbol{k}}_e\sum_m\left|\langle\Psi_{n\ell m}|\boldsymbol{r}|\Psi_i(\boldsymbol{k}_e)\rangle\right|^2\,. \quad (54.206)$$

Transition T-Matrix for RR.

$$\sigma_R^{n\ell}(E) = \frac{\pi a_0^2}{(ka_0)^2}\left|T_R\right|^2\rho(E)\,, \quad (54.207)$$
$$\left|T_R\right|^2 = 4\pi^2\int\sum_m\left|D_{fi}\right|^2\mathrm{d}\hat{\boldsymbol{k}}_e\,. \quad (54.208)$$

Photoionization Cross Section. From detailed balance in (54.196), $\sigma_I^{n\ell}$ is

$$\sigma_I^{n\ell}(h\nu) = \left(\frac{8\pi^2}{3}\right)\alpha h\nu\left(\frac{g_A^+}{g_A}\right)\rho(E)R_I^{n\ell}(E)\,. \quad (54.209)$$

Continuum Wave Function Expansion.

$$\Psi_i(\boldsymbol{k}_e,\boldsymbol{r}) = \sum_{\ell'm'}\mathrm{i}^{\ell'}\mathrm{e}^{\mathrm{i}\eta_{\ell'}}R_{E\ell'}(r)Y_{\ell'm'}\left(\hat{\boldsymbol{k}}_e\right)Y_{\ell'm'}(\hat{\boldsymbol{r}})\,. \quad (54.210)$$

Energy Normalization. With $\rho(E) = 1$,

$$\int\Psi_i(\boldsymbol{k}_e;\boldsymbol{r})\Psi_i^*(\boldsymbol{k}_e';\boldsymbol{r})\,\mathrm{d}\boldsymbol{r} = \delta(E-E')\delta\left(\hat{\boldsymbol{k}}_e-\hat{\boldsymbol{k}}_e'\right)\,. \quad (54.211)$$

Plane Wave Expansion.

$$\mathrm{e}^{\mathrm{i}\boldsymbol{k}\cdot\boldsymbol{r}} = 4\pi\sum_{\ell=0}^{\infty}\mathrm{i}^\ell j_\ell(kr)Y_{\ell m}^*\left(\hat{\boldsymbol{k}}\right)Y_{\ell m}(\hat{\boldsymbol{r}}) \quad (54.212)$$

$$j_\ell(kr) \sim \sin\left(kr-\frac{1}{2}\ell\pi\right)/(kr)\,. \quad (54.213)$$

For bound states,

$$\Psi_{n\ell m}(\boldsymbol{r}) = R_{n\ell}(r)Y_{\ell m}(\hat{\boldsymbol{r}})\,. \quad (54.214)$$

RR and PI Cross Sections and Radial Integrals.

$$\sigma_R^{n\ell}(E) = \frac{8\pi^2}{3}\left(\frac{(\alpha h\nu)^3}{2(e^2/a_0)E}\right)\rho(E)R_I(E;n\ell)\,. \quad (54.215)$$

For an electron outside a closed core,

$$g_A^+ = 1, \quad g_A = 2(2\ell+1)$$

$$\sigma_{\mathrm{I}}^{n\ell}(h\nu) = \frac{4\pi^2\alpha h\nu\rho(E)}{3(2\ell+1)} R_{\mathrm{I}}(E;n\ell), \qquad (54.216\mathrm{a})$$

$$R_{n\ell}^{\varepsilon,\ell'} = \int_0^\infty (R_{\varepsilon\ell'}\, r\, R_{n\ell})\, r^2\, \mathrm{d}r, \qquad (54.216\mathrm{b})$$

$$R_{\mathrm{I}}(E;n\ell) = \ell \left|R_{n\ell}^{\varepsilon,\ell-1}\right|^2 + (\ell+1)\left|R_{n\ell}^{\varepsilon,\ell+1}\right|^2. \qquad (54.216\mathrm{c})$$

For an electron outside an unfilled core (c) in the process $(A^+ + \mathrm{e}^-) \to A(n\ell)$, the weights are
State i: $[S_\mathrm{c}, L_\mathrm{c}; \varepsilon]$, $g_A^+ = (2S_\mathrm{c}+1)(2L_\mathrm{c}+1)$
State f: $[(S_\mathrm{c}, L_\mathrm{c}; n\ell)S, L]$, $g_A = (2S+1)(2L+1)$.

$$R_{\mathrm{I}}(E;n\ell) = \frac{(2L+1)}{(2L_\mathrm{c}+1)} \times \sum_{\ell'=\ell\pm1}\sum_{L'}(2L'+1)\begin{Bmatrix}\ell & L & L_\mathrm{c}\\ L' & \ell' & 1\end{Bmatrix}^2 \times \ell_{\max}\left|\int_0^\infty (R_{\varepsilon\ell'}\, r\, R_{n\ell})\, r^2\,\mathrm{d}r\right|^2. \qquad (54.217)$$

This reduces to (54.216c) when the radial functions $R_{i,f}$ do not depend on $(S_\mathrm{c}, L_\mathrm{c}, S, L)$.

Cross Section for Dielectronic Recombination

$$\sigma_{\mathrm{DLR}}^{n\ell}(E) = \frac{\pi a_0^2}{(ka_0)^2}\left|T_{\mathrm{DLR}}(E)\right|^2 \rho(E), \qquad (54.218)$$

$$\left|T_{\mathrm{DLR}}(E)\right|^2 = 4\pi^2 \int \mathrm{d}\hat{\boldsymbol{k}}_\mathrm{e} \times \sum_j \left|\frac{\langle \Psi_f|\, D\,|\Psi_j\rangle\langle\Psi_j|\, V\,|\Psi_i(\boldsymbol{k}_\mathrm{e})\rangle}{(E-\varepsilon_j + \mathrm{i}\Gamma_j/2)}\right|^2, \qquad (54.219)$$

which is the generalization of the T-matrix (54.208) to include the effect of intermediate doubly-excited autoionizing states $|\Psi_j\rangle$ in energy resonance to within width Γ_j of the initial continuum state Ψ_i. The electrostatic interaction $V = e^2\sum_{i=1}^{N}(\boldsymbol{r}_i - \boldsymbol{r}_{N+1})^{-1}$ initially produces dielectronic capture by coupling the initial state i with the resonant states j which become stabilized by coupling via the dipole radiation field interaction $\boldsymbol{D} = (2\omega^3/3\pi c^3)^{1/2}\sum_{i=1}^{N+1}(e\boldsymbol{r}_i)$ to the final stabilized state f. The above cross section for (54.3) is valid for isolated, nonoverlapping resonances.

Continuum Wave Normalization and Density of States

The basic formulae (54.206) for $\sigma_\mathrm{R}^{n\ell}$ depends on the density of states $\rho(E)$ which in turn varies according to the particular normalization constant N adopted for the continuum radial wave,

$$R_{E\ell}(r) \sim N\sin\left(kr - \frac{1}{2}\ell\pi + \eta_\ell\right)\Big/ r, \qquad (54.220)$$

in (54.210) where the phase is

$$\eta_\ell = \arg\Gamma(\ell+1+\mathrm{i}\beta) - \beta\ln 2kr + \delta_\ell. \qquad (54.221)$$

The phase corresponding to the Hartree–Fock short-range interaction is δ_ℓ. The Coulomb phase shift for electron motion under $(-Ze^2/r)$ is $(\eta_\ell - \delta_\ell)$ with $\beta = Z/(ka_0)$.

For a plane wave $\phi_{\boldsymbol{k}}(\boldsymbol{r}) = N'\exp(\mathrm{i}\boldsymbol{k}\cdot\boldsymbol{r})$,

$$\langle\phi_{\boldsymbol{k}}(\boldsymbol{r})|\phi_{\vec{k}'}(\boldsymbol{r})\rangle\,\mathrm{d}\boldsymbol{k} = (2\pi)^3\left|N'\right|^2\rho(\boldsymbol{k})\,\mathrm{d}\boldsymbol{k}\,\delta(\boldsymbol{k}-\mathbf{k}')$$
$$\equiv \left(\frac{h^3}{mp}\right)\left|N'\right|^2\rho(E,\hat{\boldsymbol{k}})\,\mathrm{d}E\,\mathrm{d}\hat{\boldsymbol{k}}\,\delta(E-E')\delta(\hat{\boldsymbol{k}}-\hat{\boldsymbol{k}}'). \qquad (54.222)$$

On integrating (54.222) over all E and $\hat{\boldsymbol{k}}$ for a single particle distributed over all $|E,\hat{\boldsymbol{k}}\rangle$ states, N' and ρ are then interrelated by

$$\left|N'\right|^2\rho(E,\hat{\boldsymbol{k}}) = mp/h^3. \qquad (54.223)$$

The incident current is

$$j\,\mathrm{d}E\,\mathrm{d}\hat{\boldsymbol{k}}_\mathrm{e} = v\left|N'\right|^2\rho(E,\hat{\boldsymbol{k}})\,\mathrm{d}E\,\mathrm{d}\hat{\boldsymbol{k}}_\mathrm{e} \qquad (54.224\mathrm{a})$$
$$= \left(2mE/h^3\right)\mathrm{d}E\,\mathrm{d}\hat{\boldsymbol{k}}_\mathrm{e} = v\,\mathrm{d}\boldsymbol{p}_\mathrm{e}/h^3. \qquad (54.224\mathrm{b})$$

Radial Wave Connection. From (54.210) and (54.212), $N = (4\pi N'/k)$, so that the connection between N of (54.220) and $\rho(E)$ is

$$|N|^2\rho(E,\hat{\boldsymbol{k}}) = \frac{(2m/\hbar^2)}{\pi k} = \frac{(2/\pi)}{ka_0e^2}. \qquad (54.225)$$

RR Cross Sections for Common Normalization Factors of Continuum Radial Functions

(a) $N = 1$; $\quad \rho(E) = \dfrac{(2m/\hbar^2)}{\pi k} = \dfrac{(2/\pi)}{(ka_0)e^2}$, $\qquad (54.226)$

$$\sigma_\mathrm{R}^{n\ell}(E) = \frac{8\pi^2a_0^2}{(ka_0)^3}\int\sum_m\left|D_{fi}\right|^2\mathrm{d}\hat{\boldsymbol{k}}_\mathrm{e}, \qquad (54.227)$$

where D_{fi} of (54.204) is dimensionless.

(b) $N = k^{-1}$; $\rho(E) = \left(2m/\hbar^2\right)(k/\pi)$, (54.228)

$$\sigma_{\mathrm{R}}^{n\ell}(E) = \frac{16\pi a_0^2}{3\sqrt{2}}\left(\frac{\alpha h\nu}{e^2/a_0}\right)^3 \sqrt{\frac{(e^2/a_0)}{E}}\left(\frac{R_{\mathrm{I}}}{a_0^5}\right), \tag{54.229}$$

where (54.216b) and (54.216c) for R_{I} has dimension $\left[L^5\right]$.

(c) $N = k^{-1/2}$; $\rho(E) = \dfrac{\left(2m/\hbar^2\right)}{\pi}$, (54.230)

$$\sigma_{\mathrm{R}}^{n\ell}(E) = \frac{8\pi a_0^2}{3}\left(\frac{\alpha^3(h\nu)^3}{\left(e^2/a_0\right)^2 E}\right)\left(\frac{R_{\mathrm{I}}}{a_0^4}\right), \tag{54.231}$$

where R_{I} has dimensions of $\left[L^4\right]$.

(d) $N = \left(2m/\hbar^2\pi^2 E\right)^{1/4}$; $\rho(E) = 1$, (54.232)

$$\sigma_{\mathrm{R}}^{n\ell}(E) = \frac{4(\pi a_0)^2}{3}\left(\frac{\alpha^3(h\nu)^3}{\left(e^2/a_0\right)^2 E}\right)\left(\frac{R_{\mathrm{I}}}{e^2 a_0}\right), \tag{54.233}$$

where R_{I} has dimensions of $\left[L^2E^{-1}\right]$.

54.8.4 Bound-Free Oscillator Strengths

For a transition $n\ell \to E$ to $E + \mathrm{d}E$,

$$\frac{\mathrm{d}f_{n\ell}}{\mathrm{d}E} = \frac{2}{3}\frac{(h\nu)}{(e^2/a_0)}\frac{1}{(2\ell+1)}\sum_m\sum_{\ell'm'}\left|\boldsymbol{r}_{n\ell m}^{\varepsilon\ell'm'}\right|^2, \tag{54.234}$$

$$R_{\mathrm{I}}(\varepsilon; n\ell) = \int \mathrm{d}\hat{\boldsymbol{k}}_e \sum_m |\langle\Psi_{n\ell m}|\boldsymbol{r}|\Psi_i(\boldsymbol{k}_{\mathrm{e}})\rangle\, E|^2$$

$$= \sum_{m,\ell',m'}\left|\boldsymbol{r}_{n\ell m}^{\varepsilon\ell'm'}\right|^2, \tag{54.235}$$

$$\sigma_{\mathrm{R}}^{n\ell}(E) = 2\pi^2\alpha a_0^2 g_A\left(\frac{k_\nu^2}{k_{\mathrm{e}}^2}\right)\left(\frac{e^2}{a_0}\right)\frac{\mathrm{d}f_{n\ell}}{\mathrm{d}E}, \tag{54.236a}$$

$$\sigma_{\mathrm{I}}^{n\ell}(h\nu) = 2\pi^2\alpha a_0^2 g_A^+\left(\frac{e^2}{a_0}\right)\frac{\mathrm{d}f_{n\ell}}{\mathrm{d}E}. \tag{54.236b}$$

Semiclassical Hydrogenic Systems

$$g_A = g_{n\ell} = 2(2\ell+1), \quad g_A^+ = 1,$$

$$\sigma_{\mathrm{R}}^{n}(E) = \sum_{\ell=0}^{n-1}\sigma_{\mathrm{R}}^{n\ell}(E) = 2\pi^2\alpha a_0^2\left(\frac{k_\nu^2}{k_{\mathrm{e}}^2}\right)\frac{\mathrm{d}F_n}{\mathrm{d}E}, \tag{54.237}$$

$$\frac{\mathrm{d}F_n}{\mathrm{d}E} = \sum_{\ell=0}^{n-1} g_{n\ell}\frac{\mathrm{d}f_{n\ell}}{\mathrm{d}E} = 2\sum_{\ell,m}\frac{\mathrm{d}f_{n\ell m}}{\mathrm{d}E}. \tag{54.238}$$

Bound-Bound Absorption Oscillator Strength.

For a transition $n \to n'$,

$$F_{nn'} = 2\sum_{\ell m}\sum_{\ell'm'} f_{n\ell m}^{n'\ell'm'} \tag{54.239a}$$

$$= \frac{2^6}{3\sqrt{3}\pi}\left[\left(\frac{1}{n^2}-\frac{1}{n'^2}\right)^{-3}\right]\frac{1}{n^3}\frac{1}{n'^3}, \tag{54.239b}$$

$$\frac{\mathrm{d}F_n}{\mathrm{d}E} = \frac{2^5}{3\sqrt{3}\pi}n\frac{I_n^2}{(h\nu)^3} = 2n^2\frac{\mathrm{d}f_{n\ell}}{\mathrm{d}E}, \tag{54.239c}$$

$$\sigma_{\mathrm{R}}^{n}(E) = \frac{2^5\alpha^3}{3\sqrt{3}}\left(\frac{nI_n^2}{E(h\nu)}\right)\pi a_0^2, \tag{54.239d}$$

$$\sigma_{\mathrm{I}}^{n\ell}(h\nu) = \frac{2^6\alpha}{3\sqrt{3}}\frac{n}{Z^2}\left(\frac{I_n}{h\nu}\right)^3\pi a_0^2, \tag{54.239e}$$

$$= 7.907\,071\left(\frac{n}{Z^2}\right)\left(\frac{I_n}{h\nu}\right)^3 \mathrm{Mb}. \tag{54.239f}$$

This semiclassical analysis yields exactly Kramers PI and associated RR cross sections in Sect. 54.8.2.

54.8.5 Radiative Recombination Rate

$$\hat{\alpha}_{\mathrm{R}}^{n\ell}(T_{\mathrm{e}}) = \bar{v}_{\mathrm{e}}\int_0^\infty \varepsilon\,\sigma_{\mathrm{R}}^{n\ell}(\varepsilon)\,\mathrm{e}^{-\varepsilon}\,\mathrm{d}\varepsilon \tag{54.240a}$$

$$\equiv \bar{v}_{\mathrm{e}}\left\langle\sigma_{\mathrm{R}}^{n\ell}(T_{\mathrm{e}})\right\rangle, \tag{54.240b}$$

where $\varepsilon = E/k_{\mathrm{B}}T$ and $\langle\sigma_{\mathrm{R}}^{n\ell}(T_{\mathrm{e}})\rangle$ is the Maxwellian-averaged cross section for radiative recombination.

In terms of the continuum-bound $A_{n\ell}(E)$,

$$\hat{\alpha}_{\mathrm{R}}^{n\ell}(T_{\mathrm{e}}) = \frac{h^3}{(2\pi m_{\mathrm{e}}k_{\mathrm{B}}T)^{3/2}}\int_0^\infty\left(\frac{\mathrm{d}A_{n\ell}}{\mathrm{d}\varepsilon}\right)\mathrm{e}^{-\varepsilon}\,\mathrm{d}\varepsilon, \tag{54.241}$$

$$\frac{\mathrm{d}A_{n\ell}}{\mathrm{d}E} = \rho(E)\sum_m\int A_{n\ell m}(E,\hat{\boldsymbol{k}}_{\mathrm{e}})\,\mathrm{d}\hat{\boldsymbol{k}}_{\mathrm{e}}. \tag{54.242}$$

Milne Detailed Balance Relation

In terms of $\sigma_{\mathrm{I}}^{n\ell}(h\nu)$,

$$\hat{\alpha}_{\mathrm{R}}^{n\ell}(T_{\mathrm{e}}) = \bar{v}_{\mathrm{e}}\left(\frac{g_A}{2g_A^+}\right)\left(\frac{k_{\mathrm{B}}T_{\mathrm{e}}}{mc^2}\right)\left(\frac{I_n}{k_{\mathrm{B}}T_{\mathrm{e}}}\right)^2\left\langle\sigma_{\mathrm{I}}^{n\ell}(T_{\mathrm{e}})\right\rangle, \tag{54.243}$$

where, in reduced units $\omega = h\nu/I_n$, $T = k_{\mathrm{B}}T_{\mathrm{e}}/I_n = b_n^{-1}$, the averaged PI cross section corresponding to (54.174) is

$$\left\langle \sigma_{\mathrm{I}}^{n\ell}(T)\right\rangle = \frac{\mathrm{e}^{1/T}}{T}\int_1^\infty \omega^2 \sigma_{\mathrm{I}}^{n\ell}(\omega)\,\mathrm{e}^{-\omega/T}\,\mathrm{d}\omega\,. \quad (54.244)$$

When $\sigma_{\mathrm{I}}^{n\ell}(\omega)$ is expressed in Mb $(10^{-18}\,\mathrm{cm}^2)$,

$$\hat{\alpha}_{\mathrm{R}}^{n\ell}(T_{\mathrm{e}}) = 1.508\times 10^{-13}\left(\frac{300}{T_{\mathrm{e}}}\right)^{1/2}\left(\frac{I_n}{I_{\mathrm{H}}}\right)^2\left(\frac{g_A}{2g_A^+}\right) \times\left\langle \sigma_{\mathrm{I}}^{n\ell}(T)\right\rangle \mathrm{cm}^3\,\mathrm{s}^{-1}\,. \quad (54.245)$$

When σ_{I} can be expressed in terms of the threshold cross section σ_0^n (54.178) as

$$\sigma_{\mathrm{I}}^{n\ell}(h\nu) = (I_n/h\nu)^p \sigma_0(n);\ (p=0,1,2,3)\,, \quad (54.246)$$

then $\left\langle \sigma_{\mathrm{I}}^{n\ell}(T)\right\rangle = S_p(T)\sigma_0(n)$, where

$$S_0(T) = 1+2T+2T^2\,,\quad S_1(T) = 1+T\,, \quad (54.247a)$$

$$S_2(T) = 1\,, \quad (54.247b)$$

$$S_3(T) = \left(\mathrm{e}^{1/T}/T\right)E_1(1/T) \quad (54.247c)$$

$$\stackrel{T\ll 1}{\sim} 1-T+2T^2-6T^3\,. \quad (54.247d)$$

The case $p=3$ corresponds to Kramers PI cross section (54.181) so that

$$_{\mathrm{K}}\hat{\alpha}_{\mathrm{R}}^{n\ell}(T_{\mathrm{e}}) = \frac{(2\ell+1)}{n^2}\frac{2}{n}\hat{\alpha}_0(T_{\mathrm{e}})S_3(T) \quad (54.248a)$$

$$\equiv {}_{\mathrm{K}}\hat{\alpha}_{\mathrm{R}}^{n\ell}(T_{\mathrm{e}}\to 0)S_3(T)\,, \quad (54.248b)$$

such that $_{\mathrm{K}}\hat{\alpha}_{\mathrm{R}}^{n\ell} \sim Z^2/(n^3 T_{\mathrm{e}}^{1/2})$ as $T=(k_{\mathrm{B}}T_{\mathrm{e}}/I_n)\to 0$.

54.8.6 Gaunt Factor, Cross Sections and Rates for Hydrogenic Systems

The Gaunt factor $G_{n\ell}$ is the ratio of the quantal to Kramers (K) semiclassical PI cross section such that

$$\sigma_{\mathrm{I}}^{n\ell}(h\nu) = {}_{\mathrm{K}}\sigma_{\mathrm{I}}^{n}(h\nu)G_{n\ell}(\omega)\,; \quad (54.249)$$

$$\omega = h\nu/I_n = 1+E/I_n\,.$$

(a) Radiative Recombination Cross Section

$$\sigma_{\mathrm{R}}^{n\ell}(E) = \left(\frac{g_A}{g_A^+}\right)\left(\frac{\alpha^2(h\nu)^2}{2E(e^2/a_0)}\right)G_{n\ell}(\omega)\,{}_{\mathrm{K}}\sigma_{\mathrm{I}}^{n}(h\nu) \quad (54.250a)$$

$$= G_{n\ell}(\omega)\,{}_{\mathrm{K}}\sigma_{\mathrm{R}}^{n\ell}(E) \quad (54.250b)$$

$$= \left[\frac{(2\ell+1)}{n^2}G_{n\ell}(\omega)\right]_{\mathrm{K}}\sigma_{\mathrm{R}}^{n}(E)\,, \quad (54.250c)$$

$$\sigma_{\mathrm{R}}^{n}(E) = G_n(\omega)\,{}_{\mathrm{K}}\sigma_{\mathrm{R}}^{n}(E) \quad (54.250d)$$

where the quantum mechanical correction, or Gaunt factor, to the semiclassical cross sections

$$G_{n\ell}(\omega) \to \begin{cases} 1, & \omega\to 1 \\ \omega^{-(\ell+1/2)}, & \omega\to\infty \end{cases} \quad (54.251)$$

favors low $n\ell$ states. The ℓ-averaged Gaunt factor is

$$G_n(\omega) = \left(1/n^2\right)\sum_{\ell=0}^{n-1}(2\ell+1)G_{n\ell}(\omega)\,. \quad (54.252)$$

Approximations for G_n: as ε increases from zero,

$$G_n(\varepsilon) = \left[1+\frac{4}{3}(a_n+b_n)+\frac{28}{18}a_n^2\right]^{-3/4} \quad (54.253a)$$

$$\simeq 1-(a_n+b_n)+\frac{7}{3}a_nb_n+\frac{7}{6}b_n^2 \quad (54.253b)$$

where $E=\varepsilon\left(Z^2e^2/2a_0\right)$, $\omega = 1+n^2\varepsilon$, and

$$a_n(\varepsilon) = 0.172\,825\left(1-n^2\varepsilon\right)c_n(\varepsilon)\,, \quad (54.254a)$$

$$b_n(\varepsilon) = 0.049\,59\left(1+\frac{4}{3}n^2\varepsilon+n^4\varepsilon^2\right)c_n^2(\varepsilon)\,, \quad (54.254b)$$

$$c_n(\varepsilon) = n^{-2/3}\left(1+n^2\varepsilon\right)^{-2/3}\,. \quad (54.254c)$$

Radiative Recombination Rate

$$\hat{\alpha}_{\mathrm{R}}^{n\ell}(T_{\mathrm{e}}) = {}_{\mathrm{K}}\hat{\alpha}_{\mathrm{R}}^{n\ell}(T_{\mathrm{e}}\to 0)F_{n\ell}(T)\,, \quad (54.255)$$

$$\hat{\alpha}_{\mathrm{R}}^{n\ell}(T_{\mathrm{e}}\to 0) = \frac{(2\ell+1)}{n^2}\left(\frac{2}{n}\right)\hat{\alpha}_0(T_{\mathrm{e}})\,, \quad (54.256)$$

in accordance with (54.185b).

$$F_{n\ell}(T) = \frac{\mathrm{e}^{1/T}}{T}\int_1^\infty \frac{G_{n\ell}(\omega)}{\omega}\mathrm{e}^{-\omega/T}\,\mathrm{d}\omega\,. \quad (54.257)$$

The multiplicative factors F and G convert the semiclassical (Kramers) $T_{\mathrm{e}}\to 0$ rate and cross section to their quantal values. Departures from the scaling rule $\left(Z^2/n^3T_{\mathrm{e}}^{1/2}\right)$ for RR rates is measured by $F_{n\ell}(T)$.

54.8.7 Exact Universal Rate Scaling Law and Results for Hydrogenic Systems

$$\hat{\alpha}_{\mathrm{R}}^{n\ell}(Z,T_{\mathrm{e}}) = Z\hat{\alpha}_{\mathrm{R}}^{n\ell}\left(1,T_{\mathrm{e}}/Z^2\right) \quad (54.258)$$

as exhibited by (54.243) with (54.239e) and (54.244).

Recombination rates are greatest into low n levels and the $\omega^{-\ell-1/2}$ variation of $G_{n\ell}$ preferentially populates states with low $\ell\approx 2$–5. Highly accurate analytical fits for $G_{n\ell}(\omega)$ have been obtained for $n \leq 20$ so that (54.249) can be expressed in terms of known functions of fit parameters [54.17]. This procedure (which does not violate the S_2 sum rule) has been extended to nonhydrogenic systems of neon-like Fe XVII, where $\sigma_{\mathrm{I}}^{n\ell}(\omega)$ is a monotonically decreasing function of ω.

The variation of the ℓ-averaged values

$$n^{-2} \sum_{\ell=0}^{n-1} (2\ell+1) F_{n\ell}(T)$$

is close in both shape and magnitude to the corresponding semiclassical function $S_3(T)$, given by (54.257) with $G_{n\ell}(\omega) = 1$. Hence the ℓ-averaged recombination rate is

$$\begin{aligned}\hat{\alpha}_{\mathrm{R}}^{n}(Z,T) &= (300/T)^{1/2}(Z^2/n)F_n(T) \\ &\quad \times 1.1932 \times 10^{-12}\,\mathrm{cm}^3\,\mathrm{s}^{-1}\,,\end{aligned}$$

where F_n can be calculated directly from (54.257) or be approximated as $G_n(1)S(T)$. A computer program based on a three-term expansion of G_n is also available [54.18]. From a three-term expansion for G, the rate of radiative recombination into all levels of a hydrogenic system is

$$\begin{aligned}\hat{\alpha}(Z,T) &= 5.2 \times 10^{-14} Z \lambda^{1/2} \\ &\quad \times \left(0.43 + \frac{1}{2}\ln\lambda + 0.47/\lambda^{1/2}\right)\,,\end{aligned} \tag{54.259}$$

where $\lambda = 1.58 \times 10^5 Z^2/T$ and $[\hat{\alpha}] = \mathrm{cm}^3/\mathrm{s}$. Tables [54.19] exist for the effective rate

$$\hat{\alpha}_E^{n\ell}(T) = \sum_{n'=n}^{\infty} \sum_{\ell'=0}^{n'-1} \hat{\alpha}_{\mathrm{R}}^{n'\ell'} C_{n'\ell',n\ell} \tag{54.260}$$

of populating a given level $n\ell$ of H via radiative recombination into all levels $n' \geq n$ with subsequent radiative cascade $(i \to f)$ with probability $C_{i,f}$ via all possible intermediate paths. Tables [54.19] also exist for the full rate

$$\hat{\alpha}_{\mathrm{F}}^{N}(T) = \sum_{n=N}^{\infty} \sum_{\ell=0}^{n-1} \hat{\alpha}_{\mathrm{R}}^{n\ell} \tag{54.261}$$

of recombination, into all levels above $N = 1, 2, 3, 4,$ of hydrogen. They are useful in deducing time scales of radiative recombination and rates for complex ions.

54.9 Useful Quantities

(a) Mean Speed

$$\begin{aligned}\bar{v}_{\mathrm{e}} &= \left(\frac{8k_{\mathrm{B}}T}{\pi m_{\mathrm{e}}}\right)^{1/2} = 1.076\,042 \times 10^7 \left(\frac{T}{300}\right)^{1/2} \mathrm{cm/s} \\ &= 6.692\,38 \times 10^7 T_{\mathrm{eV}}^{1/2}\,\mathrm{cm/s}\end{aligned}$$

$$\bar{v}_i = 2.511\,16 \times 10^5 \left(\frac{T}{300}\right)^{1/2} (m_{\mathrm{p}}/m_i)^{1/2}\,\mathrm{cm/s}$$

where $(m_{\mathrm{p}}/m_{\mathrm{e}})^{1/2} = 42.850\,352$, and $T = 11\,604.45\,T_{\mathrm{eV}}$ relates the temperature in K and in eV.

(b) Natural Radius

$|V(R_{\mathrm{e}})| = e^2/R_{\mathrm{e}} = k_{\mathrm{B}}T$.

$$R_{\mathrm{e}} = \frac{e^2}{k_{\mathrm{B}}T} = 557\left(\frac{300}{T}\right)\text{Å} = \left(\frac{14.4}{T_{\mathrm{eV}}}\right)\text{Å}\,.$$

(c) Boltzmann Average Momentum

$$\langle p \rangle = \int_{-\infty}^{\infty} \mathrm{e}^{-p^2/2mk_{\mathrm{B}}T}\,\mathrm{d}p = (2\pi m_{\mathrm{e}} k_{\mathrm{B}} T)^{1/2}\,.$$

(d) De Broglie Wavelength

$$\begin{aligned}\lambda_{\mathrm{dB}} &= \frac{h}{\langle p \rangle} = \frac{h}{(2\pi m_{\mathrm{e}} k_{\mathrm{B}} T)^{1/2}} \\ &= \frac{7.453\,818 \times 10^{-6}}{T_{\mathrm{e}}^{1/2}}\,\mathrm{cm} \\ &= 43.035 \left(\frac{300}{T_{\mathrm{e}}}\right)^{1/2} \text{Å} = \frac{6.9194}{T_{\mathrm{eV}}^{1/2}}\,\text{Å}\,.\end{aligned}$$

References

54.1 J. Stevefelt, J. Boulmer, J-F. Delpech: Phys. Rev. A **12**, 1246 (1975)

54.2 R. Deloche, P. Monchicourt, M. Cheret, F. Lambert: Phys. Rev. A **13**, 1140 (1976)

54.3 P. Mansbach, J. Keck: Phys. Rev. **181**, 275 (1965)

54.4 L. P. Pitaevskiĭ: Sov. Phys. JETP **15**, 919 (1962)

54.5 H. A. Kramers: Philos. Mag. **46**, 836 (1923)

54.6 D. R. Bates: Phys. Rev. **78**, 492 (1950)

54.7 J. N. Bardsley: J. Phys. A Proc. Phys. Soc. **1**, 365 (1968)

54.8 M. R. Flannery: *Atomic Collisions: A Symposium in Honor of Christopher Bottcher*, ed. by D. R. Schultz, M. R. Strayer, J. H. Macek (American Institute of Physics, New York 1995) p. 53

54.9 A. Giusti: J. Phys. B **13**, 3867 (1980)

54.10 P. van der Donk, F. B. Yousif, J. B. A. Mitchell, A. P. Hickman: Phys. Rev. Lett. **68**, 2252 (1992)

54.11 S. L. Guberman: Phys. Rev. A **49**, R4277 (1994)

54.12 M. R. Flannery: Int. J. Mass Spectrom. Ion Process **149/150**, 597 (1995)

54.13 M. R. Flannery: J. Chem. Phys. **95**, 8205 (1991)

54.14 A. Burgess: Mon. Not. R. Astron. Soc. **118**, 477 (1958)

54.15 J. A. Gaunt: Philos. Trans. Roy. Soc. A **229**, 163 (1930)

54.16 M. J. Seaton: Mon. Not. R. Astron. Soc. **119**, 81 (1959)

54.17 B. F. Rozsnyai, V. L. Jacobs: Astrophys. J. **327**, 485 (1988)

54.18 D. R. Flower, M. J. Seaton: Comp. Phys. Commun. **1**, 31 (1969)

54.19 P. G. Martin: Astrophys. J. Supp. Ser. **66**, 125 (1988)

General References (General Recombination)

G.1 M. R. Flannery: Adv. At. Mol. Phys. **32**, 117 (1994)

G.2 M. R. Flannery: *Recombination Processes*. In: *Molecular Processes in Space*, ed. by T. Watanabe, I. Shimamura, M. Shimiza, Y. Itikawa (Plenum, New York 1990), Chapt. 7, p. 145

G.3 D. R. Bates: *Recombination*. In: *Electronic and Atomic Collisions*, ed. by H. B. Gilbody, W. R. Newell, F. H. Read, A. C. H. Smith (North-Holland, Amsterdam 1988), p. 3

G.4 D. R. Bates: *Recombination*. In: *Case Studies in Atomic Physics*, ed. by E. W. McDaniel, M. R. C. McDowell (North-Holland, Amsterdam 1974), Vol. 4, p. 57

G.5 D. R. Bates, A. Dalgarno: *Electronic Recombination*. In: *Atomic and Molecular Processes*, ed. by D. R. Bates (Academic Press, New York 1962)

G.6 D. R. Bates: Adv. At. Mol. Phys. **15**, 235 (1979)

G.7 W. G. Graham (Ed.): *Recombination of Atomic Ions* (Plenum Press, New York 1992)

G.8 J. N. Bardsley: *Recombination Processes in Atomic and Molecular Physics*. In: *Atomic and Molecular Collision Theory*, ed. by F. A. Gianturco (Plenum Press, New York 1980), p. 123

G.9 J. Dubau, S. Volonte: Rep. Prog. Phys. **43**, 199 (1980)

G.10 Y. Hahn: Adv. At. Mol. Phys. **21**, 123 (1985)

G.11 Y. Hahn, K. J. LaGattuta: Phys. Rep. **116**, 195 (1988)

G.12 E. W. McDaniel, E. J. Mansky: Adv. At. Mol. Opt. Phys. **33**, 389 (1994)

(Three-Body Electron–Ion Collisional-Radiative Recombination: Theory)

G.13 D. R. Bates, A. E. Kingston, R. W. P. McWhirter: *Recombination between Electrons and Atomic Ions: I. Optically Thin Plasmas; II. Optically Thick Plasmas*; Proc. R. Soc. (London) Ser. A **267**, 297 (1962); **270**, 155 (1962)

G.14 D. R. Bates, S. P. Khare: Proc. Phys. Soc. **85**, 231 (1965)

G.15 D. R. Bates, V. Malaviya, N. A. Young: Proc. R. Soc. (London) Ser. A **320**, 437 (1971)

G.16 A. Burgess, H. P. Summers: Mon. Not. R. Astron. Soc. **174**, 345 (1976)

G.17 H. P. Summers: Mon. Not. R. Astron. Soc. **178**, 101 (1977)

G.18 N. N. Ljepojevic, R. J. Hutcheon, J. Payne: Comp. Phys. Commun. **44**, 157 (1987)

(Electron–Ion Recombination: Molecular Dynamics Simulations)

G.19 W. L. Morgan: J. Chem. Phys. **80**, 4564 (1984)

G.20 W. L. Morgan: Phys. Rev. A **30**, 979 (1984)

G.21 W. L. Morgan, J. N. Bardsley: Chem. Phys. Lett. **96**, 93 (1983).

G.22 W. L. Morgan: *Recent Studies in Atomic and Molecular Processes*, ed. by A. E. Kingston (Plenum Press, New York 1987), p. 149

(Ion–Ion Recombination: Review Articles)

G.23 M. R. Flannery: *Three-Body Recombination between Positive and Negative Ions*. In: *Case Studies in Atomic Collision Physics*, ed. by E.W. McDaniel, M. R. C. McDowell (North-Holland, Amsterdam 1972), Vol. 2, 1, p. 1
G.24 M. R. Flannery: *Ionic Recombination*. In: *Atomic Processes and Applications*, ed. by P. G. Burke, B. L. Moiseiwitsch (North-Holland, Amsterdam 1976), 12, p. 407
G.25 M. R. Flannery: *Ion–Ion Recombination in High Pressure Plasmas*. In: *Applied Atomic Collision Physics*, ed. by E.W. McDaniel and W. L. Nighan (Academic Press, New York 1983), Vol. 3 *Gas Lasers*, 5, p. 393
G.26 M. R. Flannery: *Microscopic and Macroscopic Perspectives of Termolecular Association of Atomic Reactants in a Gas*. In: *Recent Studies in Atomic and Molecular Processes*, ed. by A. E. Kingston (Plenum Press, London 1987), p. 167
G.27 D. R. Bates: Adv. At. Mol. Phys. **20**, 1 (1985)
G.28 B. H. Mahan: Adv. Chem. Phys. **23**, 1 (1973)
G.29 J. T. Moseley, R. E. Olson, J. R. Peterson: *Case Studies in Atomic Collision Physics*, ed. by M. R. C. McDowell, E. W. McDaniel (North-Holland, Amsterdam 1972), p. 1
G.30 D. Smith, N. G. Adams: *Studies of Ion–Ion Recombination using Flowing Afterglow Plasmas*. In: *Physics of Ion–Ion and Electron–Ion Collisions*, ed. by F. Brouillard, J. W. McGowan (Plenum Press, New York 1982), p. 501

(Termolecular Ion–Ion Recombination: Theory)(A) Low Gas Densities: Linear Region

G.31 M. R. Flannery, E. J. Mansky: J. Chem. Phys. **88**, 4228 (1988)
G.32 M. R. Flannery, E. J. Mansky: J. Chem. Phys. **89**, 4086 (1988)
G.33 M. R. Flannery: J. Chem. Phys. **89**, 214 (1988)
G.34 M. R. Flannery: J. Chem. Phys. **87**, 6947 (1987)
G.35 M. R. Flannery: J. Phys. B **13**, 3649 (1980)
G.36 M. R. Flannery: J. Phys. B **14**, 915 (1981)
G.37 D. R. Bates, I. Mendaš: J. Phys. B **15**, 1949 (1982)
G.38 D. R. Bates, P. B. Hays, D. Sprevak: J. Phys. B **4**, 962 (1971)
G.39 D. R. Bates, M. R. Flannery: Proc. R. Soc. (London) Ser. A **302**, 367 (1968)
G.40 D. R. Bates, R. J. Moffett: Proc. R. Soc. (London) Ser. A **291**, 1 (1966)

(B) All Gas Densities: Non-Linear Region

G.41 M. R. Flannery: *Microscopic and Macroscopic Theories of Termolecular Recombination between Atomic Ions*. In: *Dissociative Recombination*, ed. by B. R. Rowe, J. B. A. Mitchell, A. Canosa (Plenum Press, New York 1993), p. 205
G.42 M. R. Flannery: J. Chem. Phys. **95**, 8205 (1991)
G.43 M. R. Flannery: Phil. Trans. R. Soc. (London) Ser. A **304**, 447 (1982)
G.44 D. R. Bates, I. Mendaš: Proc. R. Soc. (London) Ser. A **359**, 275 (1978)
G.45 J. J. Thomson: Phil. Mag. **47**, 337 (1924)

Ion–Ion Recombination: Monte-Carlo Simulations

G.46 P. J. Feibelman: J. Chem. Phys. **42**, 2462 (1965)
G.47 A. Jones, J. L. J. Rosenfeld: Proc. R. Soc. (London) Ser. A **333**, 419 (1973)
G.48 D. R. Bates, I. Mendaš: Proc. R. Soc. (London) Ser. A **359**, 287 (1978)
G.49 D. R. Bates: Chem. Phys. Lett. **75**, 409 (1980)
G.50 J. N. Bardsley, J. M. Wadehra: Chem. Phys. Lett. **72**, 477 (1980)
G.51 D. R. Bates: J. Phys. B **14**, 4207 (1981)
G.52 D. R. Bates: J. Phys. B **14**, 2853 (1981)
G.53 D. R. Bates, I. Mendaš: Chem. Phys. Lett. **88**, 528 (1982)
G.54 W. L. Morgan, J. N. Bardsley, J. Lin, B. L. Whitten: Phys. Rev. A **26**, 1696 (1982)
G.55 B. L. Whitten, W. L. Morgan, J. N. Bardsley: J. Phys. B **15**, 319 (1982)

Ion–Ion Tidal Recombination: Molecular Dynamics Simulations

G.56 D. R. Bates, W. L. Morgan: Phys. Rev. Lett. **64**, 2258 (1990)

G.57 W. L. Morgan, D. R. Bates: J. Phys. B **25**, 5421 (1992)

Radiative Recombination: Theory

G.58 M. J. Seaton: Mon. Not. R. Astron. Soc. **119**, 81 (1959)

G.59 D. R. Flower, M. J. Seaton: Comp. Phys. Commun. **1**, 31 (1969)

G.60 A. Burgess, H. P. Summers: Mon. Not. R. Astron. Soc. **226**, 257 (1987)

G.61 Y. S. Kim, R. H. Pratt: Phys. Rev. A **27**, 2913 (1983)

G.62 D. J. McLaughlin, Y. Hahn: Phys. Rev. A **43**, 1313 (1991)

G.63 F. D. Aaron, A. Costescu, C. Dinu, J. Phys. II (Paris) **3**, 1227 (1993)

Dissociative Recombination: Theory and Experiment

G.64 D. Zajfman, J. B. A. Mitchell, B. R. Rowe, D. Schwalin (Eds.): *Dissociative Recombination: Theory, Experiment and Applications III* (World Scientific, Singapore 1996)

G.65 D. R. Bates: Adv. At. Mol. Phys. **34**, 427 (1994)

G.66 B. R. Rowe, J. B. A. Mitchell, A. Canosa (Eds.): *Dissociative Recombination: Theory, Experiment and Applications II* (Plenum, New York 1993)

G.67 J. B. A. Mitchell, S. L. Guberman (Eds.): *Dissociative Recombination: Theory, Experiment and Applications I* (World Scientific, Singapore 1989)

G.68 J. B. A. Mitchell: Phys. Rep. **186**, 215 (1990)

G.69 A. Giusti-Suzor: *Recent Developments in the Theory of Dissociative Recombination and Related Processes*. In: *Atomic Processes in Electron-Ion and Ion-Ion Collisions*, ed. by F. Brouillard (Plenum Press, New York 1986)

G.70 J. N. Bardsley, M. A. Biondi: Adv. At. Mol. Phys. **6**, 1 (1970)

55. Dielectronic Recombination

Dielectronic recombination (DR) is a two-step process that greatly increases the effciency for electrons and ions to recombine in a plasma. The process therefore plays an important role in the theoretical modeling of plasmas, whether in the laboratory or in astrophysical sources such as the solar corona. The purpose of this chapter is to present the theoretical formulation for DR, and the principal methods for calculating rate coefficients. The results are compared with experiment over a broad range of low-Z ions and high-Z ions where relativistic effects become important.

55.1 **Theoretical Formulation** 830

55.2 **Comparisons with Experiment** 831
55.2.1 Low-Z Ions 831
55.2.2 High-Z Ions and Relativistic Effects 831

55.3 **Radiative-Dielectronic Recombination Interference** 832

55.4 **Dielectronic Recombination in Plasmas** 833

References 833

Electron–ion recombination into a particular final recombined state may be schematically represented as

$$e^- + A_i^{q+} \rightarrow A_f^{(q-1)+} + \hbar\omega\,, \tag{55.1}$$

and

$$e^- + A_i^{q+} \rightarrow \left[A_j^{(q-1)+}\right] \rightarrow A_f^{(q-1)+} + \hbar\omega\,, \tag{55.2}$$

where q is the charge on the atomic ion A, ω is the frequency of the emitted light, and the brackets in (55.2) indicate a doubly-excited resonance state. The first process is called radiative recombination (RR), while the second is called dielectronic recombination (DR). Both recombination mechanisms are the inverse of photoionization. At sufficiently high electron density, three-body recombination becomes possible. The three-body mechanism is the inverse of electron impact ionization.

The review article by *Seaton* and *Storey* [55.1] includes an interesting history of the theoretical work on dielectronic recombination. The process was first referred to as dielectronic recombination by *Massey* and *Bates* [55.2], after a suggestion of its possible importance in the ionosphere by Sayers in 1939. However, estimates of the rate coefficient for this process indicated that DR is not an important process in the ionosphere, where the temperatures are too low to excite anything but the lower energy resonance states.

In 1961, Unsold, in a letter to Seaton, suggested that DR might account for a well-known temperature discrepancy in the solar corona. Seaton initially concluded that DR would not significantly increase recombination in the solar corona. However, he had only included the lower energy resonance states in his analysis; *Burgess* [55.3] showed that when one includes the higher members of the Rydberg series of resonance states that are populated at coronal temperatures, DR can indeed explain this discrepancy.

Dielectronic recombination has since received much theoretical attention due, in part, to its importance in modeling high temperature plasmas. Various approaches to the theory are discussed in a review by *Hahn* [55.4] and in [55.5]. Recently, there have been various projects aimed at the generation of large quantities of DR data for use in astrophysical and fusion plasma modeling. One such project is based on the results of the AUTOSTRUCTURE code, with both the total and partial (i.e., resolved by recombined level) DR rate coefficients being archived. The methodology is outlined in *Badnell* et al. [55.6]. Data are calculated for all members of an isoelectronic sequence from H to Ar, along with various ions relevant to astrophysics or fusion, namely Ca, Ti, Cr, Fe, Ni, Zn, Kr, Mo, and Xe. Work has been completed for the oxygen [55.7], beryllium [55.8], carbon [55.9], lithium [55.10], boron [55.11], neon [55.12] and nitrogen [55.13] isoelectronic sequences. There has also been a large quantity of data generated using a fully relativistic Dirac–Fock code [55.14]. This data includes calculations of Na-like ions [55.15] and H-like through to Ne-like [55.16] ions of certain astrophysically important elements.

Interest in dielectronic recombination has increased dramatically in the last twenty years. *Mitchell* et al. [55.17] published the DR cross section for C^+ using a merged electron–ion beams apparatus, and *Belic* et al. [55.18] reported on a crossed beams measurement of the DR cross section for Mg^+. Also, *Dittner* et al. [55.19] published merged beams measurements of the DR cross section for the multiply charged ions B^{2+} and C^{3+}. Since that time, atomic physics experiments carried out using heavy-ion traps, accelerators, and storage-cooler rings have produced high-resolution mappings of the resonance structures associated with electron-ion recombination. The experiments have been carried out using a wide range of facilities and technologies, such as the test storage ring (TSR) at Heidelberg, the experimental storage ring (ESR) at Darmstadt, the accelerator-cooler ring facility at Aarhus, the electron beam ion trap (EBIT) at Livermore, and the electron beam ion source (EBIS) at Kansas State. A good review of the dramatic experimental progress in DR measurements is again found in the NATO proceedings [55.5].

55.1 Theoretical Formulation

In the independent-processes approximation, the two paths for recombination are summed incoherently. The radiative recombination cross section for (55.1), in lowest-order of perturbation theory, is given by

$$\sigma_{\mathrm{RR}} = \frac{8\pi^2}{k^3} \times \sum_{\ell j} \sum_{JM} \sum_{M_0} \frac{1}{2g_1} \left| \langle \alpha_0 J_0 M_0 | D | \alpha_1 J_1 \ell j J M \rangle \right|^2 . \tag{55.3}$$

The set $(\alpha_1 J_1)$ represents the quantum numbers for the N-electron target ion state, $(\alpha_0 J_0 M_0)$ represents the quantum numbers for the $(N+1)$-electron recombined ion state, $(k\ell j)$ represents the quantum numbers for the continuum electron state, (JM) represents the quantum numbers for the $(N+1)$-electron system of target plus free electron state, and g_1 is the statistical weight of a J_1 level. The dipole radiation field operator is given by

$$D = \sqrt{\frac{2\omega^3}{3\pi c^3}} \sum_{s=1}^{N+1} \boldsymbol{r}_s . \tag{55.4}$$

Continuum normalization is chosen as one times a sine function, and atomic units ($e = \hbar = m = 1$) are used. In the isolated-resonance approximation, the dielectronic recombination cross section for (55.2), in lowest-order perturbation theory, is given by

$$\sigma_{\mathrm{DR}} = \frac{8\pi^2}{k^3} \sum_{\ell j} \sum_{JM} \sum_{M_0} \frac{1}{2g_1} \times \sum_{\alpha_i J_i M_i} \left| \frac{\langle \alpha_0 J_0 M_0 | D | \alpha_i J_i M_i \rangle \langle \alpha_i J_i M_i | V | \alpha_1 J_1 \ell j J M \rangle}{E_0 - E_i + \mathrm{i}\Gamma_i/2} \right|^2 , \tag{55.5}$$

where the set $(\alpha_i J_i M_i)$ represents the quantum numbers for a resonance state with energy E_i and total width Γ_i, and the electrostatic interaction between electrons is given by

$$V = \sum_{s=1}^{N} |\boldsymbol{r}_s - \boldsymbol{r}_{N+1}|^{-1} . \tag{55.6}$$

By the principle of detailed balance, σ_{RR} of (55.3) is proportional to the photoionization cross section from

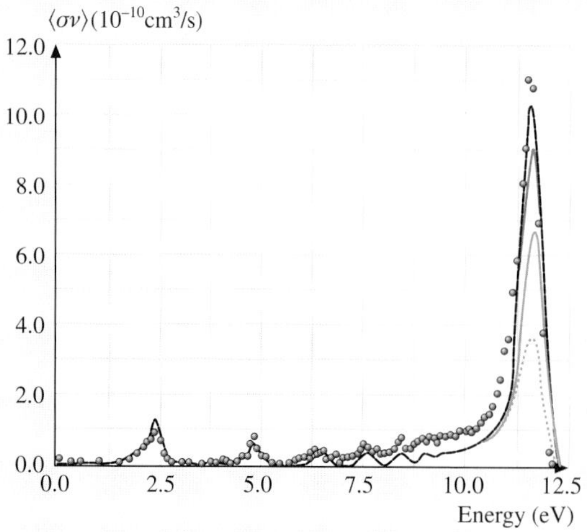

Fig. 55.1 Dielectronic Recombination for O^{5+}. Calculations were performed for fields of 0 V/cm (*dotted curve*), 3 V/cm (*dashed curve*), 5 V/cm (*chain curve*), and 7 V/cm (*solid curve*)

the bound state, while the energy-averaged σ_{DR} may be written as

$$\langle \sigma_{DR} \rangle = \frac{2\pi^2}{\Delta\epsilon k^2} \sum_{\alpha_i J_i M_i} \frac{1}{2g_1} \frac{A_a A_r}{\Gamma_i} , \quad (55.7)$$

where the autoionization decay rate A_a is given by

$$A_a = \frac{4}{k} \sum_{\ell j} \sum_{JM} |\langle \alpha_1 J_1 \ell j JM|V|\alpha_i J_i M_i \rangle|^2 , \quad (55.8)$$

the radiative decay rate A_r is given by

$$A_r = 2\pi \sum_{M_0} |\langle \alpha_0 J_0 M_0 |D| \alpha_i J_i M_i \rangle|^2 , \quad (55.9)$$

and $\Delta\epsilon$ is the energy bin width. Each resonance level in (55.7) makes a contribution at a fixed continuum energy $k^2/2$; thus $\langle \sigma_{DR} \rangle$ plotted as a function of energy is a histogram.

55.2 Comparisons with Experiment

55.2.1 Low-Z Ions

For the most part, the agreement between the recent high-resolution measurements and theoretical calculations based on the independent-processes and isolated-resonance (IPIR) approximations is quite good [55.20–38]. We illustrate the agreement with experiment obtained by calculations employing the IPIR approximations with three examples from the Li isoelectronic sequence. The pathways for dielectronic capture, in terms of specific levels, are given by

$$\begin{aligned} e^- + A^{q+}(2s_{1/2}) &\rightarrow \left[A^{(q-1)+}(2p_{1/2}nlj)\right] \\ &\searrow \left[A^{(q-1)+}(2p_{3/2}nlj)\right] , \end{aligned} \quad (55.10)$$

where a $1s^2$ core is assumed to be present. Both Rydberg series autoionize by the reverse of the paths in (55.10), while for sufficiently high n, the $2p_{3/2}\,nlj$ levels may autoionize to the $2p_{1/2}$ continuum. Both series radiatively stabilize by either a $2p \rightarrow 2s$ core orbital transition, or by a $nlj \rightarrow n'l'j'$ valence orbital transition, where $2p_j\,n'l'j'$ with $j = \frac{1}{2}, \frac{3}{2}$ is a bound level.

Dielectronic recombination cross section calculations [55.28] in the IPIR approximation for O^{5+} are compared with experiment in Fig. 55.1. Fine-structure splitting of the two series of (55.10) is minimal for this light ion, so there appears only one Rydberg series. The $2p\,6\ell$ resonances are located at 2.5 eV, the $2p\,7\ell$ at 5.0 eV, and so on; accumulating at the series limit around 11.3 eV. Electric field effects on the high-n resonances are strong in O^{5+}, so that calculations were done for fields of 0, 3, 5, and 7 V/cm. Since the precise electric field strength in the experiment is not known, the accuracy of an electric field dependent theory, in this case, has yet to be determined. However, the effects of state mixing by extrinsic fields in the collision region for the dielectronic recombination of Mg^+ have been investigated both experimentally [55.39] and theoretically [55.40].

There have been a significant number of recent experiments on low-Z ions. These experiments, in general, show good agreement with theory, see *Fogle* et al. [55.37] and *Schnell* et al. [55.38]. However, it is also clear that discrepancies remain between theory and experiment for certain low energy resonances, due to the difficulty in calculating the energy positions of such resonances. As has been pointed out in *Savin* et al. [55.41] and *Schippers* et al. [55.42], this can lead to significant uncertainties in low temperature DR rate coefficients. Calculating such low energy resonances to sufficient accuracy for low temperature DR rate coefficients remains a significant challenge for theory. Some success in this area has been achieved using relativistic many-body perturbation theory, obtaining very good agreement with low energy resonance positions for a range of systems, see, for example, *Lindroth* et al. [55.43], *Fogle* et al. [55.44], and *Tokman* et al. [55.45].

55.2.2 High-Z Ions and Relativistic Effects

Dielectronic recombination cross section calculations [55.31] in the IPIR approximation for Cu^{26+} are compared with experiment in Fig. 55.2. The two fine-structure Rydberg series are now clearly resolved; the fine-structure splitting is about 27 eV for Cu^{26+}. The $2p_{1/2}\,13\ell$ resonances are just above threshold, while the $2p_{3/2}\,11\ell$ resonances are found around 5.0 eV. Electric fields in the range 0–50 V/cm have little effect on the Cu^{26+} spectrum. Overall, the agreement between theory and experiment is excellent.

Electron-ion recombination cross section calculations [55.32] in the IPIR approximation for low-lying resonances in Au^{76+} are compared with experiment in Fig. 55.3. The $2p_{1/2}\,nlj$ series limit is at 217 eV, while the $2p_{3/2}\,nlj$ series limit is at 2.24 keV; yielding a fine

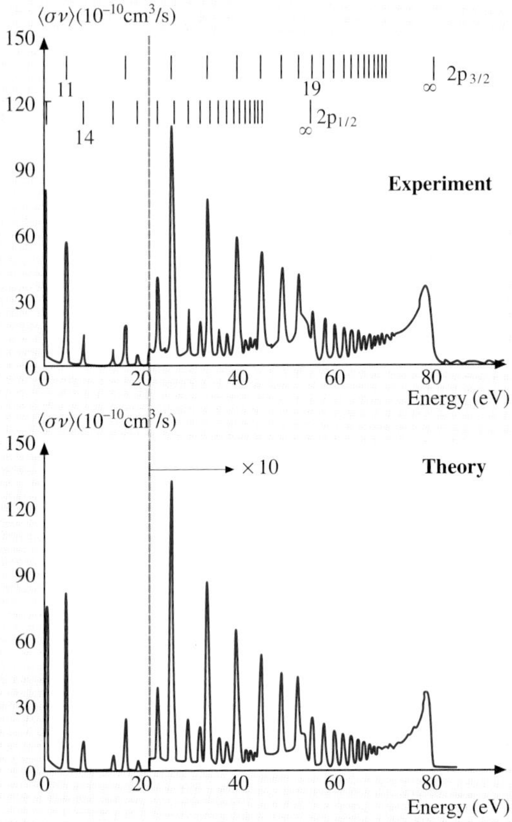

Fig. 55.2 Dielectronic Recombination for Cu^{26+}

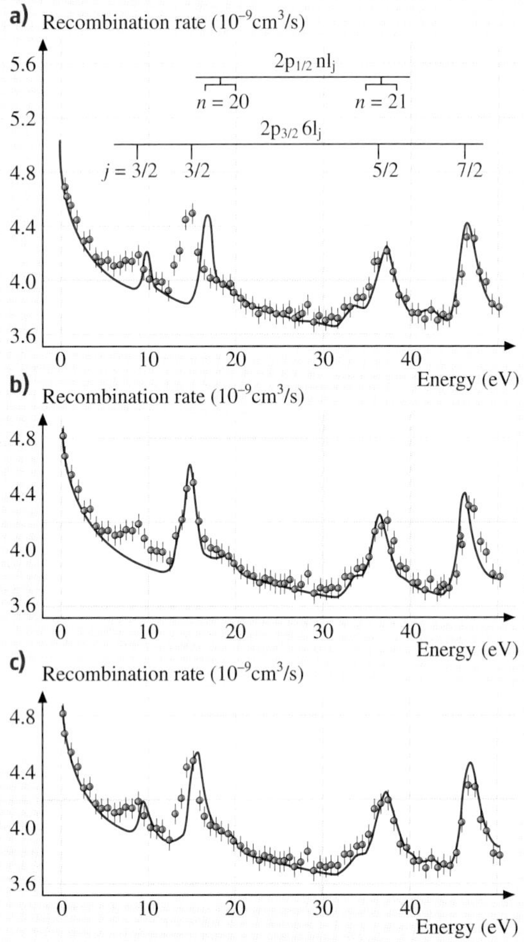

Fig. 55.3a–c Dielectronic recombination for Au^{76+}. The curves show (**a**) perturbative relativistic, (**b**) semirelativistic, and (**c**) fully relativistic calculations for the dielectronic recombination cross section

structure splitting of 2.03 keV. QED effects alone shift the $2p_{3/2}\,nlj$ series limit by 22.0 eV. Thus accurate atomic structure calculations must be made to locate the $2p_{3/2}\,6lj$ resonances in the 0–50 eV energy range of the experiment. The figure shows that the perturbative relativistic, semirelativistic, and fully relativistic calculations for the dielectronic recombination cross section ride on top of a strong radiative recombination background. In principle, the fully relativistic theory contains the most physics, and thus it is comforting that on the whole it is in good agreement with the experiment. It is instructive, however, to see how well the computationally simpler perturbative relativistic and semirelativistic theories do for such a highly charged ion.

There have been several recent experimental measurements on high-Z ions, in particular, for astrophysically abundant species. In general, there is good agreement between theory and experiment. Examples of high-Z element DR studies include those done on Fe XXI and Fe XXII by *Savin* et al. [55.41], and on Fe XX by *Savin* et al. [55.46].

55.3 Radiative–Dielectronic Recombination Interference

There has been a great deal of effort in recent years to develop a more general theory of electron-ion recombination which would go beyond the IPIR approximation to include radiative–dielectronic recombination interference and overlapping (and interacting) resonance structures [55.47–54]. In almost all cases, the interference between a dielectronic recombination resonance and the radiative recombination background is quite small and difficult to observe. The best possibility for observation of RR-DR interference appears to be in highly charged atomic ions. In the cases studied to date, the combination of electron and photon continuum coupling selection rules and the requirement of near energy degeneracy make the overlapping (and interacting) resonance effects small and difficult to observe. Heavy ions in relatively low stages of ionization are the best place to look, since there are resonance series attached to the large numbers of LS terms or fine structure levels.

The distorted wave approximation (Chapt. 52) has been so successful in describing dielectronic recombination cross sections for most atomic ions because, for low charged ions, the DR cross section is proportional to the radiative rate, while for highly charged ions the DR cross section is proportional to the autoionization rate. Thus the weakness of the distorted wave method in calculating accurate autoionization rates for low charged ions is masked by a DR cross section that is highly dependent on radiative atomic structure. As one moves to more highly charged ions, the DR cross section becomes more sensitive to the autoionization rates, but at the same time, the distorted-wave method becomes increasingly more accurate.

55.4 Dielectronic Recombination

Dielectronic recombination is an important atomic process that is included in the theoretical modeling of the ionization state and emission of radiating ions, which is fundamental to the interpretation of spectral emission from both fusion and astrophysical plasmas (Chapts. 82, 86, and 87). The dielectronic recombination rate coefficient, into a particular final recombined state, is given by

$$\alpha_{\mathrm{DR}} = \sqrt{\frac{2}{\pi T^3}} k^2 \Delta\epsilon \left\langle \sigma_{\mathrm{DR}}\left(i \rightarrow f\right)\right\rangle \exp\left(\frac{-k^2}{2T}\right), \tag{55.11}$$

where T is the electron temperature. Dielectronic recombination rate coefficients, from the ground and metastable states of a target ion into fully resolved low-lying states and bundled high-lying states of a recombined ion, are required for a generalized collisional radiative treatment [55.55, 56] of highly populated metastable states, the influence of finite plasma density on excited state populations, and of ionization in dynamic plasmas.

References

55.1 M. J. Seaton, P. J. Storey: *Atomic Processes and Applications*, ed. by P. G. Burke, B. L. Moiseiwitsch (North-Holland, Amsterdam 1976) p. 133

55.2 H. S. W. Massey, D. R. Bates: Rep. Prog. Phys. **9**, 62 (1942)

55.3 A. Burgess: Astrophys. J. **139**, 776 (1964)

55.4 Y. Hahn: Adv. At. Mol. Phys. **21**, 123 (1985)

55.5 W. G. Graham, W. Fritsch, Y. Hahn, J. A. Tanis (Eds.): *Recombination of Atomic Ions*, NATO ASI Ser. B, Vol. 296 (Plenum, New York 1992)

55.6 N. R. Badnell, M. G. O'Mullane, H. P. Summers, Z. Altun, M. A. Bautista, J. Colgan, T. W. Gorczyca, D. M. Mitnik, M. S. Pindzola, O. Zatzarinny: Astron., Astrophys. **406**, 1151 (2003)

55.7 O. Zatzarinny, T. W. Gorczyca, K. T. Korista, N. R. Badnell, D. W. Savin: Astron., Astrophys. **412**, 587 (2003)

55.8 J. Colgan, M. S. Pindzola, A. D. Whiteford, N. R. Badnell: Astron., Astrophys. **412**, 597 (2003)

55.9 O. Zatzarinny, T. W. Gorczyca, K. T. Korista, N. R. Badnell, D. W. Savin: Astron., Astrophys. **417**, 1173 (2003)

55.10 J. Colgan, M. S. Pindzola, N. R. Badnell: Astron., Astrophys. **417**, 1188 (2004)

55.11 Z. Altun, A. Yumak, N. R. Badnell, J. Colgan, M. S. Pindzola: Astron., Astrophys. **420**, 775 (2004)

55.12 O. Zatzarinny, T. W. Gorczyca, K. T. Korista, N. R. Badnell, D. W. Savin: Astron., Astrophys. **426**, 699 (2004)

55.13 D. M. Mitnik, N. R. Badnell: Astron., Astrophys. **425**, 1153 (2004)

55.14 M. F. Gu: Astrophys. J. **582**, 1241 (2003)

55.15 M. F. Gu: Astrophys. J. **153**, 389 (2004)

55.16 M. F. Gu: Astrophys. J. **590**, 1131 (2003)

55.17 J. B. A. Mitchell, C. T. Ng, J. L. Forand, D. P. Levac, R. E. Mitchell, A. Sen, D. B. Miko, J. Wm. McGowan: Phys. Rev. Lett. **50**, 335 (1983)

55.18 D. S. Belic, G. H. Dunn, T. J. Morgan, D. W. Mueller, C. Timmer: Phys. Rev. Lett. **50**, 339 (1983)
55.19 P. F. Dittner, S. Datz, P. D. Miller, C. D. Moak, P. H. Stelson, C. Bottcher, W. B. Dress, G. D. Alton, N. Neskovic: Phys. Rev. Lett. **51**, 31 (1983)
55.20 G. Kilgus, J. Berger, P. Blatt, M. Grieser, D. Habs, B. Hochadel, E. Jaeschke, D. Kramer, R. Neumann, G. Neureither, W. Ott, D. Schwalm, M. Steck, R. Stokstad, E. Szmola, A. Wolf, R. Schuch, A. Müller, M. Wagner: Phys. Rev. Lett. **64**, 737 (1990)
55.21 M. S. Pindzola, N. R. Badnell, D. C. Griffin: Phys. Rev. A **42**, 282 (1990)
55.22 D. R. DeWitt, D. Schneider, M. W. Clark, M. H. Chen, D. Church: Phys. Rev. A **44**, 7185 (1991)
55.23 L. H. Andersen, P. Hvelplund, H. Knudsen, P. Kvistgaard: Phys. Rev. Lett. **62**, 2656 (1989)
55.24 N. R. Badnell, M. S. Pindzola, D. C. Griffin: Phys. Rev. A **41**, 2422 (1990)
55.25 L. H. Andersen, G. Y. Pan, H. T. Schmidt, N. R. Badnell, M. S. Pindzola: Phys. Rev. A **45**, 7868 (1992)
55.26 D. A. Knapp, R. E. Marrs, M. B. Schneider, M. H. Chen, M. A. Levine, P. Lee: Phys. Rev. A **47**, 2039 (1993)
55.27 R. Ali, C. P. Bhalla, C. L. Cocke, M. Stockli: Phys. Rev. Lett. **64**, 633 (1990)
55.28 D. C. Griffin, M. S. Pindzola, P. Krylstedt: Phys. Rev. A **40**, 6699 (1989)
55.29 L. H. Andersen, J. Bolko, P. Kvistgaard: Phys. Rev. A **41**, 1293 (1990)
55.30 L. H. Andersen, G. Y. Pan, H. T. Schmidt, M. S. Pindzola, N. R. Badnell: Phys. Rev. A **45**, 6332 (1992)
55.31 G. Kilgus, D. Habs, D. Schwalm, A. Wolf, N. R. Badnell, A. Müller: Phys. Rev. A **46**, 5730 (1992)
55.32 W. Spies, A. Müller, J. Linkemann, A. Frank, M. Wagner, C. Kozhuharov, B. Franzke, K. Beckert, F. Bosch, H. Eickhoff, M. Jung, O. Klepper, W. König, P. H. Mokler, R. Moshammer, F. Nolden, U. Schaaf, P. Spädtke, M. Steck, P. Zimmerer, N. Grün, W. Scheid, M. S. Pindzola, N. R. Badnell: Phys. Rev. Lett. **69**, 2768 (1992)
55.33 N. R. Badnell, M. S. Pindzola, L. H. Andersen, J. Bolko, H. T. Schmidt: J. Phys. B **24**, 4441 (1991)
55.34 A. Lampert, D. Habs, G. Kilgus, D. Schwalm, A. Wolf, N. R. Badnell, M. S. Pindzola: *AIP Conference Proceedings # 274, The VIth International Conference on the Physics of Highly-Charged Ions, 1992*, ed. by M. Stockli, P. Richard (American Institute of Physics, New York 1992) p. 537
55.35 D. R. DeWitt, D. Schneider, M. H. Chen, M. W. Clark, J. W. McDonald, M. B. Schneider: Phys. Rev. Lett. **68**, 1694 (1992)
55.36 M. B. Schneider, D. A. Knapp, M. H. Chen, J. H. Scofield, P. Beiersdorfer, C. L. Bennett, J. R. Henderson, R. E. Marrs, M. A. Levine: Phys. Rev. A **45**, R1291 (1992)
55.37 M. Fogle, N. R. Badnell, N. Eklöw, T. Mohamed, R. Schuch: Astron., Astrophys. **409**, 781 (2003)
55.38 M. Schnell, G. Gwinner, N. R. Badnell, M. E. Bannister, R. Böhm, J. Colgan, S. Kieslich, S. D. Loch, D. Mitnik, A. Müller, M. S. Pindzola, S. Schippers, D. Schwalm, W. Shi, A. Wolf, S.-G. Zhou: Phys. Rev. Lett. **91**, 043001 (2003)
55.39 A. Müller, D. S. Belic, B. D. DePaola, N. Djuric, G. H. Dunn, D. W. Mueller, C. Timmer: Phys. Rev. Lett. **56**, 127 (1986)
55.40 C. Bottcher, D. C. Griffin, M. S. Pindzola: Phys. Rev. A **34**, 860 (1986)
55.41 D. W. Savin, S. M. Kahn, G. Gwinner, M. Grieser, R. Repnow, G. Saathoff, D. Schwalm, A. Wolf, A. Müller, S. Schippers, P. A. Zavodszky, M. H. Chen, T. W. Gorczyca, O. Zatzarinny, M. F. Gu: Astrophys. J. Suppl. Ser. **147**, 421 (2003)
55.42 S. Schippers, M. Schnell, C. Brandau, S. Kieslich, A. Müller, A. Wolf: Astron., Astrophys. **421**, 1185 (2004)
55.43 E. Lindroth, H. Danared, P. Glans, Z. Pesic, M. Tokman, G. Vikor, R. Schuch: Phys. Rev. Lett. **86**, 5027 (2001)
55.44 M. Fogle, N. Eklöw, E. Lindroth, T. Mohamed, R. Schuch, M. Tokman: J.Phys.B **36**, 2563 (2003)
55.45 M. Tokman, N. Eklöw, P. Glans, E. Lindroth, R. Schuch, G. Gwinner, D. Schwalm, A. Wolf, A. Hoffknecht, A. Müller, S. Schippers: Phys. Rev. A **66**, 012703 (2002)
55.46 D. W. Savin, E. Behar, S. M. Kahn, G. Gwinner, A. A. Saghiri, M. Schmitt, M. Grieser, R. Repnov, D. Schwalm, A. Wolf, T. Bartsch, A. Müller, S. Schippers, N. R. Badnell, M. H. Chen, T. W. Gorczyca: Astrophys. J. Suppl. Ser. **138**, 337 (2003)
55.47 P. C. W. Davies, M. J. Seaton: J. Phys. B **2**, 757 (1969)
55.48 E. Trefftz: J. Phys. B **3**, 763 (1970)
55.49 R. H. Bell, M. J. Seaton: J. Phys. B **18**, 1589 (1985)
55.50 G. Alber, J. Cooper, A. R. P. Rau: Phys. Rev. A **30**, 2845 (1984)
55.51 K. J. LaGattuta: Phys. Rev. A **40**, 558 (1989)
55.52 S. L. Haan, V. L. Jacobs: Phys. Rev. A **40**, 80 (1989)
55.53 N. R. Badnell, M. S. Pindzola: Phys. Rev. A **45**, 2820 (1992)
55.54 M. S. Pindzola, N. R. Badnell, D. C. Griffin: Phys. Rev. A **46**, 5725 (1992)
55.55 H. P. Summers, M. B. Hooper: Plasma Phys. **25**, 1311 (1983)
55.56 N. R. Badnell, M. S. Pindzola, W. J. Dickson, H. P. Summers, D. C. Griffin: Astrophys. J. **407**, L91 (1993)

56. Rydberg Collisions: Binary Encounter, Born and Impulse Approximations

Rydberg collisions are collisions of electrons, ions and neutral particles with atomic or molecular targets which are in highly excited Rydberg states characterized by large principal quantum numbers ($n \gg 1$). Rydberg collisions of atoms and molecules with neutral and charged particles include the study of collision-induced transitions both to and from Rydberg states and transitions among Rydberg levels. The basic quantum mechanical structural properties of Rydberg states are given in Chapt. 14. This Chapter collects together many of the equations used to study theoretically the collisional properties of both charged and neutral particles with atoms and molecules in Rydberg states or orbitals. The primary theoretical scattering approximations enumerated in this Chapter are the impulse approximation, binary encounter approximation and the Born approximation. The theoretical techniques used to study Rydberg collisions complement and supplement the eigenfunction expansion approximations used for collisions with target atoms and molecules in their ground ($n = 1$) or first few excited states ($n > 1$), as discussed in Chapt. 47. Direct application of eigenfunction expansion techniques to Rydberg collisions, wherein the target particle can be in a Rydberg orbital with principal quantum number in the range $n \geq 100$, is prohibitively difficult due to the need to compute numerically and store wave functions with n^3 nodes. For $n = 100$ this amounts to $\sim 10^6$ nodes for each of the wave functions represented in the eigenfunction expansion. Therefore, a variety of approximate scattering theories have been developed to deal specifically with the pecularities of Rydberg collisions.

56.1 **Rydberg Collision Processes** 836

56.2 **General Properties of Rydberg States** 836
- 56.2.1 Dipole Moments 836
- 56.2.2 Radial Integrals 836
- 56.2.3 Line Strengths 837
- 56.2.4 Form Factors 838
- 56.2.5 Impact Broadening 838

56.3 **Correspondence Principles** 839
- 56.3.1 Bohr–Sommerfeld Quantization 839
- 56.3.2 Bohr Correspondence Principle 839
- 56.3.3 Heisenberg Correspondence Principle 839
- 56.3.4 Strong Coupling Correspondence Principle 840
- 56.3.5 Equivalent Oscillator Theorem 840

56.4 **Distribution Functions** 840
- 56.4.1 Spatial Distributions 840
- 56.4.2 Momentum Distributions 840

56.5 **Classical Theory** 841

56.6 **Working Formulae for Rydberg Collisions** 842
- 56.6.1 Inelastic n,ℓ-Changing Transitions 842
- 56.6.2 Inelastic $n \to n'$ Transitions 843
- 56.6.3 Quasi-Elastic ℓ-Mixing Transitions 844
- 56.6.4 Elastic $n\,\ell \to n\,\ell'$ Transitions 844
- 56.6.5 Fine Structure $n\,\ell\,J \to n\,\ell\,J'$ Transitions 844

56.7 **Impulse Approximation** 845
- 56.7.1 Quantal Impulse Approximation 845
- 56.7.2 Classical Impulse Approximation 849
- 56.7.3 Semiquantal Impulse Approximation 851

56.8 **Binary Encounter Approximation** 852
- 56.8.1 Differential Cross Sections 852
- 56.8.2 Integral Cross Sections 853
- 56.8.3 Classical Ionization Cross Section 855
- 56.8.4 Classical Charge Transfer Cross Section 855

56.9 **Born Approximation** 856
- 56.9.1 Form Factors 856
- 56.9.2 Hydrogenic Form Factors 856
- 56.9.3 Excitation Cross Sections 858
- 56.9.4 Ionization Cross Sections 859
- 56.9.5 Capture Cross Sections 859

References 860

56.1 Rydberg Collision Processes

(A) State-Changing Collisions
Quasi-elastic ℓ-mixing collisions:

$$A^*(n\ell)+B \rightarrow A^*(n\ell')+B\,. \quad (56.1)$$

Quasi-elastic J-mixing collisions: Fine structure transitions with $J=|\ell \pm 1/2| \rightarrow J'=|\ell \pm 1/2|$ are

$$A^*(n\ell J)+B \rightarrow A^*(n\ell J')+B\,. \quad (56.2)$$

Energy transfer n-changing collisions:

$$A^*(n)+B(\beta) \rightarrow A^*(n')+B(\beta')\,, \quad (56.3)$$

where, if B is a molecule, the transition $\beta \rightarrow \beta'$ represents an inelastic energy transfer to the rotational-vibrational degrees of freedom of the molecule B from the Rydberg atom A^*.

Elastic scattering:

$$A^*(\gamma)+B \rightarrow A^*(\gamma)+B\,, \quad (56.4)$$

where the label γ denotes the set of quantum numbers n, ℓ or n, ℓ, J used.

Depolarization collisions:

$$A^*(n\ell m)+B \rightarrow A^*(n\ell m')+B\,, \quad (56.5a)$$
$$A^*(n\ell JM)+B \rightarrow A^*(n\ell JM')+B\,. \quad (56.5b)$$

(B) Ionizing Collisions
Direct and associative ionization:

$$A^*(\gamma)+B(\beta) \rightarrow \begin{cases} A^+ + B(\beta')+e^- \\ BA^+ + e^-\,. \end{cases} \quad (56.6)$$

Penning ionization:

$$A^*(\gamma)+B \rightarrow A+B^+ +e^-\,. \quad (56.7)$$

Ion pair formation:

$$A^*(\gamma)+B \rightarrow A^+ +B^-\,. \quad (56.8)$$

Dissociative attachment:

$$A^*(\gamma)+BC \rightarrow A^+B^- +C\,. \quad (56.9)$$

56.2 General Properties of Rydberg States

Table 56.1 displays the general n-dependence of a number of key properties of Rydberg states and some specific representative values for hydrogen.

56.2.1 Dipole Moments

Definition. $\boldsymbol{D}_{i\rightarrow f}=-e\boldsymbol{X}_{i\rightarrow f}$ where

$$\boldsymbol{X}_{i\rightarrow f}=\langle \phi_f| \sum_j \mathrm{e}^{\mathrm{i}\boldsymbol{k}\cdot\boldsymbol{r}_j}\boldsymbol{r}_j|\phi_i\rangle\,. \quad (56.10)$$

Hydrogenic Dipole Moments. See *Bethe* and *Salpeter* [56.1] and the references by *Khandelwal* and co-workers [56.2–5] for details and tables.

Exact Expressions. In the limit $|\boldsymbol{k}| \rightarrow 0$, the dipole allowed transitions summed over final states are

$$\left|X_{1s\rightarrow n}\right|^2=\frac{2^8}{3}n^7\frac{(n-1)^{2n-5}}{(n+1)^{2n+5}}\,, \quad (56.11a)$$

$$\left|X_{2s\rightarrow n}\right|^2=\frac{2^5}{3n^3}\frac{\left(\frac{1}{2}-1n\right)^{2n-7}}{\left(\frac{1}{2}+1n\right)^{2n+7}}\left(\frac{1}{4}-\frac{1}{n^2}\right)\left(1-\frac{1}{n^2}\right)\,, \quad (56.11b)$$

$$\left|X_{2p\rightarrow n}\right|^2=\frac{2^5}{144}\frac{1}{n^3}\frac{\left(\frac{1}{2}-\frac{1}{n}\right)^{2n-7}}{\left(\frac{1}{2}+\frac{1}{n}\right)^{2n+7}}\left(11-\frac{12}{n^2}\right)\,. \quad (56.11c)$$

Asymptotic Expressions. For $n \gg 1$,

$$n^3\left|X_{1s\rightarrow n}\right|^2 \approx 1.563+\frac{5.731}{n^2}+\frac{13.163}{n^4}+\frac{24.295}{n^6}+\frac{39.426}{n^8}+\frac{58.808}{n^{10}}\,, \quad (56.12a)$$

$$n^3\left|X_{2s\rightarrow n}\right|^2 \approx 14.658+\frac{180.785}{n^2}+\frac{1435.854}{n^4}+\frac{9341.634}{n^6}+\frac{54\,208.306}{n^8}+\frac{292\,202.232}{n^{10}}\,, \quad (56.12b)$$

$$n^3\left|X_{2p\rightarrow n}\right|^2 \approx 13.437+\frac{218.245}{n^2}+\frac{2172.891}{n^4}+\frac{17\,118.786}{n^6}+\frac{117\,251.682}{n^8}+\frac{731427.003}{n^{10}}\,. \quad (56.12c)$$

Table 56.1 General n-dependence of characteristic properties of Rydberg states. After [56.6]

Property	n−dependence	$n=10$	$n=100$	$n=500$	$n=1000$
Radius (cm)	$n^2 a_0/Z$	5.3×10^{-7}	5.3×10^{-5}	1.3×10^{-3}	5.3×10^{-3}
Velocity (cm/s)	$v_B Z/n$	2.18×10^{7}	2.18×10^{6}	4.4×10^{5}	2.18×10^{5}
Area (cm^2)	$\pi a_0^2 n^4/Z^2$	8.8×10^{-13}	8.8×10^{-9}	5.5×10^{-6}	8.8×10^{-5}
Ionization potential (eV)	$Z^2 R_\infty/n^2$	1.36×10^{-1}	1.36×10^{-3}	5.44×10^{-5}	1.36×10^{-6}
Radiative lifetime (s)[a]	$n^5(3\ln n-\frac{1}{4})/(A_0 Z^4)$	8.4×10^{-5}	17	7.3×10^{4}	7.22 hours
Period of classical motion (s)	$2\pi/\omega_{n,n\pm1}=hn^3/(2Z^2R_\infty)$	1.5×10^{-13}	1.5×10^{-10}	1.9×10^{-8}	1.5×10^{-7}
Transition frequency (s^{-1})	$\omega_{n,n\pm1}=2Z^2R_\infty/(\hbar n^3)$	4.1×10^{13}	4.1×10^{10}	3.3×10^{8}	4.1×10^{7}
Wavelength (cm)	$\lambda_{n,n\pm1}=2\pi c/\omega_{n,n\pm1}$	4.6×10^{-3}	4.6	570	4.5609×10^{3}

[a] $A_0=\left[8\alpha^3/(3\sqrt{3}\pi)\right](v_B/a_0)$

56.2.2 Radial Integrals

Definition.

$$R_{n\ell}^{n'\ell'} \equiv \int_0^\infty R_{n\ell}(r) r R_{n'\ell'}(r) r^2\,dr\,, \tag{56.13}$$

where $R_{n\ell}(r)$ are solutions to the radial Schrödinger equation. See Chapt. 9 for specific representations of $R_{n\ell}$ for hydrogen.

Exact Results for Hydrogen. For $\ell'=\ell-1$ and $n\neq n'$ [56.7],

$$\begin{aligned} R_{n\ell}^{n'\ell-1} &= \frac{a_0}{Z}\frac{(-1)^{n'-\ell}(4nn')^{\ell+1}(n-n')^{n+n'-2\ell-2}}{4(2\ell-1)!(n+n')^{n+n'}} \\ &\times\left[\frac{(n+\ell)!(n'+\ell-1)!}{(n'+\ell')!(n-\ell-1)!}\right]^{1/2} \\ &\times\Big\{{}_2F_1(-n+\ell+1,-n'+\ell;2\ell;Y) \\ &-\left(\frac{n-n'}{n+n'}\right)^2 {}_2F_1(-n+\ell-1,-n'+\ell;2\ell;Y)\Big\}\,, \end{aligned} \tag{56.14}$$

where $Y=-4nn'/(n-n')^2$. For $n=n'$,

$$R_{n\ell}^{n\ell-1} = \left(\frac{a_0}{Z}\right)\frac{3}{2}n\sqrt{n^2-\ell^2}\,. \tag{56.15}$$

Semiclassical Quantum Defect Representation [56.8].

$$\begin{aligned} \left|R_{n\ell}^{n'\ell'}\right|^2 &= \left(\frac{a_0}{Z}\right)^2 \left|\frac{n_c^2}{2\Delta}\left[\left(1-\frac{\Delta\ell\,\ell_>}{n_c}\right)\boldsymbol{J}_{\Delta-1}(-x)\right.\right. \\ &-\left(1+\frac{\Delta\ell\,\ell_>}{n_c}\right)\boldsymbol{J}_{\Delta+1}(-x) \\ &\left.\left.+\frac{2}{\pi}\sin(\pi\Delta)(1-e)\right]\right|^2\,, \end{aligned} \tag{56.16}$$

where

$$n_c = 2n^* n^{*\prime}/(n^*+n^{*\prime})\,, \tag{56.17a}$$

$$\Delta = n^{*\prime}-n^*\,, \tag{56.17b}$$

$$\Delta\ell=\ell'-\ell,\ \ell_>=\max(\ell,\ell')\,, \tag{56.17c}$$

$$x = e\Delta,\ e=\sqrt{1-(\ell_>/n_c)^2}\,, \tag{56.17d}$$

and $\boldsymbol{J}_n(y)$ is the Anger function.

The energies of the states $n\ell$ and $n'\ell'$ are given in terms of the quantum defects by

$$E_{n\ell} = -Z^2R_\infty/n^{*2},\ n^*=n-\delta_\ell\,, \tag{56.18a}$$

$$E_{n'\ell'} = -Z^2R_\infty/n^{*\prime 2},\ n^{*\prime}=n'-\delta_{\ell'}\,. \tag{56.18b}$$

Sum Rule. For hydrogen

$$\sum_{n'}\left|R_{n\ell}^{n'\ell-1}\right|^2 = \sum_{n'}\left|R_{n\ell}^{n'\ell+1}\right|^2 \tag{56.19a}$$

$$= \frac{n^2a_0^2}{2Z^2}\left[5n^2+1-3\ell(\ell+1)\right]. \tag{56.19b}$$

See §61 of [56.1] for additional sum rules.

56.2.3 Line Strengths

Definition.

$$S(n'\ell',n\ell) = e^2(2\ell+1)\left|\boldsymbol{r}_{n'\ell',n\ell}\right|^2 \tag{56.20a}$$

$$= e^2\max(\ell,\ell')\left|R_{n\ell}^{n'\ell'}\right|^2\,, \tag{56.20b}$$

where $\ell'=\ell\pm1$. For hydrogen

$$\begin{aligned} S(n',n) &= 32\left(\frac{ea_0}{Z}\right)^2 (nn')^6\frac{(n-n')^{2(n+n')-3}}{(n+n')^{2(n+n')+4}} \\ &\times\left\{\left[{}_2F_1(-n',-n+1;1;Y)\right]^2 \right. \\ &\left.-\left[{}_2F_1(-n'+1,-n;1;Y)\right]^2\right\}\,, \end{aligned} \tag{56.21}$$

where $Y=-4nn'/(n-n')^2$.

Semiclassical Representation [56.9].

$$S(n', n) = \frac{32}{\pi\sqrt{3}}\left(\frac{ea_0}{Z}\right)^2 \frac{(\varepsilon\varepsilon')^{3/2}}{(\varepsilon-\varepsilon')^4} G(\Delta n)\,, \quad (56.22)$$

where $\varepsilon = 1/n^2$, $\varepsilon' = 1/n'^2$, and the Gaunt factor $G(\Delta n)$ is given by

$$G(\Delta n) = \pi\sqrt{3}|\Delta n| \boldsymbol{J}_{\Delta n}(\Delta n)\boldsymbol{J}'_{\Delta n}(\Delta n)\,, \quad (56.23)$$

where the prime on the Anger function denotes differentiation with respect to the argument Δn. Equation (56.23) can be approximated to within 2% by the expression

$$1 - \frac{1}{4|\Delta n|}\,. \quad (56.24)$$

Relation to Oscillator Strength.

$$S(n', n) = \sum_{\ell,\ell'} S(n'\ell', n\ell) = 3e^2 a_0^2 \frac{R_\infty}{\hbar\omega} \sum_{\ell,\ell'} f_{n'\ell',n\ell}\,. \quad (56.25)$$

Connection with Radial Integral.

$$-f_{n'\ell',n\ell} = \frac{\hbar\omega}{3R_\infty} \frac{\max(\ell,\ell')}{(2\ell+1)} \left|R_{n\ell}^{n'\ell'}\right|\,. \quad (56.26)$$

Density of Line Strengths. For bound-free $n\ell \to E\ell'$ transitions in a Coulomb field, the semiclassical representation [56.6] is

$$\frac{\mathrm{d}}{\mathrm{d}E} S(n\ell, E) = 2n(2\ell+1)\left(\frac{R_\infty}{\hbar\omega}\right)^2 \times \left[\boldsymbol{J}'_\Delta(\mathrm{e}\Delta)^2 + \left(1 - \frac{1}{\mathrm{e}^2}\right)\boldsymbol{J}_\Delta(\mathrm{e}\Delta)^2\right] \frac{e^2 a_0^2}{R_\infty}\,, \quad (56.27)$$

where $\Delta = \hbar\omega n^3/2R_\infty$ and $\mathrm{e} = \sqrt{1 - \left(\ell + \frac{1}{2}\right)^2/n^2}$. Asymptotic expression for $\Delta \gg 1$:

$$\frac{\mathrm{d}}{\mathrm{d}E} S(n\ell, E) = \frac{2(2\ell+1)}{3\pi^2}\left(\frac{R_\infty}{\hbar\omega}\right)^2 \frac{\left(\ell+\frac{1}{2}\right)^4}{n^3} \times \left[K_{1/3}^2(\eta) + K_{2/3}^2(\eta)\right]\frac{e^2 a_0^2}{R_\infty}\,, \quad (56.28)$$

where $\eta = (E/R_\infty)(\ell + 1/2)^3/6$ and the $K_\nu(x)$ are Bessel functions of the third kind.

Line Strength of Line n.

$$S_n \equiv S(n) = \sum_{k\neq 0} S(n+k, n)\frac{1}{k^3}\,. \quad (56.29)$$

Born Approximation to Line Strength S_n [56.6].

$$S_n^B = \frac{Z^2 R_\infty}{E}\left[\frac{1}{2}\ln(1+\varepsilon_\mathrm{e}/\varepsilon)\sum_{k\neq 0}\left(1-\frac{1}{4k}\right)\frac{1}{k^4} + \frac{4}{3}\frac{\varepsilon_\mathrm{e}}{\varepsilon+\varepsilon_\mathrm{e}}\sum_{k\neq 0}\left(1-\frac{0.60}{k}\right)\frac{1}{k^3}\right] = \frac{Z^2 R_\infty}{E}\left[0.82\ln\left(1+\frac{\varepsilon_\mathrm{e}}{\varepsilon}\right) + \frac{1.47\varepsilon_\mathrm{e}}{\varepsilon+\varepsilon_\mathrm{e}}\right], \quad (56.30)$$

where $\varepsilon = |E_n|Z^2/R_\infty$ and $\varepsilon_\mathrm{e} = \varepsilon/Z^2 R_\infty$.

56.2.4 Form Factors

$$F_{n'n}(Q) = \sum_{\ell,m}\sum_{\ell',m'} \left|\langle n\ell m|\mathrm{e}^{\mathrm{i}\boldsymbol{Q}\cdot\boldsymbol{r}}|n'\ell'm'\rangle\right|^2\,. \quad (56.31)$$

Connection with Generalized Oscillator Strengths.

$$f_{n'n}(Q) = \frac{Z^2 \Delta E}{n^2 Q^2 a_0^2} F_{n'n}(Q)\,. \quad (56.32)$$

Semiclassical Limit.

$$\lim_{Q\to 0} f_{n'n}(Q) = \frac{32}{3n^2}\left(\frac{nn'}{\Delta n(n+n')}\right)^3 \times \Delta n J_{\Delta n}(\Delta n) J'_{\Delta n}(\Delta n)\,, \quad (56.33)$$

where $J_m(y)$ denotes the Bessel function.

Representation as Microcanonical Distribution.

$$F_{n',n\ell}(Q) = (2\ell+1)\frac{2Z^2 R_\infty}{n'^3}\int \mathrm{d}\boldsymbol{p}|g_{n\ell}(p)|^2 \times \delta\left(\frac{(\boldsymbol{p}-\hbar\boldsymbol{Q})^2}{2m} - \frac{p^2}{2m} - E_{n'} - E_{n\ell}\right)\,, \quad (56.34)$$

$$F_{n',n}(Q) = \frac{4Z^2 {R_\infty}^2}{(nn')^3}\int \frac{\mathrm{d}\boldsymbol{p}\,\mathrm{d}\boldsymbol{r}}{(2\pi\hbar)^3}\delta\left(\frac{p^2}{2m} - \frac{Ze^2}{r} - E_n\right) \times \delta\left(\frac{(\boldsymbol{p}-\hbar\boldsymbol{Q})^2}{2m} - \frac{Ze^2}{r} - E_{n'}\right)\,, \quad (56.35)$$

$$= \frac{2^9}{3\pi(nn')^3}\frac{\kappa^5}{(\kappa^2+\kappa_+^2)^3(\kappa^2+\kappa_-^2)^3}\,, \quad (56.36)$$

where $\kappa = Qa_0/Z$ and $\kappa_\pm = |1/n \pm 1/n'|$.

56.2.5 Impact Broadening

The total broadening cross section of a level n is

$$\sigma_n = \left(\pi a_0^2/Z^4\right) n^4 S_n \,. \tag{56.37}$$

The width of a line $n \to n+k$ is [56.10]

$$\gamma_{n,n+k} = n_e \left[\langle v\sigma_n\rangle + \langle v\sigma_{n+k}\rangle\right] , \tag{56.38}$$

where n_e is the number density of electrons, and

$$\langle v\sigma_n\rangle = \sum_{k\neq 0}\langle v\sigma_{n+k,n}\rangle = \frac{n^4}{Z^3}K_n \tag{56.39a}$$

$$= \frac{n^4\pi a_0^2 v_B}{Z^3\theta^{3/2}}\int_0^\infty \mathrm{e}^{-E/k_BT} S_n \frac{E\,\mathrm{d}E}{(Z^2R_\infty)^2} , \tag{56.39b}$$

where $\theta = k_BT/Z^2R_\infty$. See Chapt. 59 for collisional line broadening.

56.3 Correspondence Principles

Correspondence principles are used to connect quantum mechanical observables with the corresponding classical quantities in the limit of large n. See [56.11] for details on the equations in this section.

56.3.1 Bohr–Sommerfeld Quantization

$$A_i = J_i\Delta w_i \oint p_i\,\mathrm{d}q_i = 2\pi\hbar(n_i+\alpha_i) , \tag{56.40}$$

where $n_i = 0, 1, 2, \ldots$ and $\alpha_i = 0$ if the generalized coordinate q_i represents rotation, and $\alpha_i = 1/2$ if q_i represents a libration.

56.3.2 Bohr Correspondence Principle

$$E_{n+s} - E_n = h\nu_{n+s,n} \sim s\hbar\omega_n,\; s = 1, 2, \ldots \ll n , \tag{56.41}$$

where $\nu_{n+s,n}$ is the line emission frequency and ω_n is the angular frequency of classical orbital motion. The number of states with quantum numbers in the range Δn is

$$\Delta N = \prod_{i=1}^{D}\Delta n = \prod_{i=1}^{D}\left(\Delta J_i\Delta w_i\right)/(2\pi\hbar)^D$$

$$= \prod_{i=1}^{D}(\Delta p_i\Delta q_i)/(2\pi\hbar)^D , \tag{56.42}$$

for systems with D degrees of freedom, and the mean value $\bar{F}$ of a physical quantity $F(q)$ in the quantum state Ψ is

$$\bar{F} = \langle\Psi|F(q)|\Psi\rangle = \sum_{n,m} a_m^* a_n F_{mn}^{(q)}\mathrm{e}^{\mathrm{i}\omega_{mn}t} , \tag{56.43}$$

where the $F_{mn}^{(q)}$ are the quantal matrix elements between time independent states.

The first order S-matrix is

$$S_{fi} = -\frac{\mathrm{i}\omega}{2\pi\hbar}\int_{-\infty}^{\infty}\mathrm{d}t\int_0^{2\pi/\omega} V[\boldsymbol{R}(t), r(t_1)]\mathrm{e}^{\mathrm{i}s\omega(t_1-t)}\,\mathrm{d}t_1 , \tag{56.44}$$

where $\boldsymbol{R}$ denotes the classical path of the projectile and r the orbital of the Rydberg electron.

56.3.3 Heisenberg Correspondence Principle

For one degree of freedom [56.11],

$$F_{mn}^{(q)}(\boldsymbol{R}) = \int_0^\infty \phi_m^*(r)F(r,\boldsymbol{R})\phi_n(r)\,\mathrm{d}r \tag{56.45}$$

$$= \frac{\omega}{2\pi}\int_0^{2\pi/\omega} F^{(c)}\big[r(t)\big]\mathrm{e}^{\mathrm{i}s\omega t}\,\mathrm{d}t . \tag{56.46}$$

The three-dimensional generalization is [56.11]

$$F_{\boldsymbol{n},\boldsymbol{n}'}^{(q)} \sim F_{\boldsymbol{s}}^{(c)}(\boldsymbol{J}) = \frac{1}{8\pi^3}\int F^c\big[\boldsymbol{r}(\boldsymbol{J},\boldsymbol{w})\big]\mathrm{e}^{\mathrm{i}\boldsymbol{s}\cdot\boldsymbol{w}}\,\mathrm{d}\boldsymbol{w} , \tag{56.47}$$

where $\boldsymbol{n}$, $\boldsymbol{n}'$ denotes the triple of quantum numbers (n, ℓ, m), (n', ℓ', m'), respectively, and $\boldsymbol{s} = \boldsymbol{n} - \boldsymbol{n}'$.

The correspondence between the three dimensional quantal and classical matrix elements in (56.47) follows from the general Fourier expansion for any classical function $F^{(c)}(\boldsymbol{r})$ periodic in $\boldsymbol{r}$,

$$F^{(c)}\big[\boldsymbol{r}(t)\big] = \sum_{\boldsymbol{s}} F_{\boldsymbol{s}}^{(c)}(\boldsymbol{J})\exp(-\mathrm{i}\boldsymbol{s}\cdot\boldsymbol{w}) , \tag{56.48}$$

where $\boldsymbol{J}$, $\boldsymbol{w}$ denotes the action-angle conjugate variables for the motion. For the three dimensional Coulomb problem, the action-angle variables are

$$J_n = n\hbar\,, \qquad w_n = \left(\frac{\partial E}{\partial J_n}\right) t + \delta\,,$$

$$J_\ell = \left(\ell + \frac{1}{2}\right)\hbar\,, \qquad w_\ell = \psi_{\mathrm{E}}\,,$$

$$J_m = m\hbar\,, \qquad w_m = \phi_{\mathrm{E}}\,, \tag{56.49}$$

where ψ_{E} is the Euler angle between the line of nodes and a direction in the plane of the orbit (usually taken to be the direction of the perihelion or perigee), and is constant for a Coulomb potential. The Euler angle ϕ_{E} is the angle between the line of nodes and the fixed x-axis. See [56.11] for details.

The first order S-matrix is

$$S_{fi} = -\frac{\mathrm{i}\omega}{2\pi\hbar}\int_0^{2\pi/\omega} \mathrm{d}t_{\mathrm{e}} \int_{-\infty}^{\infty} \mathrm{d}t\, V\big[\boldsymbol{R}(t), r(t+t_{\mathrm{e}})\big]\mathrm{e}^{\mathrm{i}s\omega t_{\mathrm{e}}}\,, \tag{56.50}$$

with $s = i - f$, $\boldsymbol{R}$ is the classical path of the projectile, and $r(t_{\mathrm{e}})$ is the classical internal motion of the Rydberg electron.

56.3.4 Strong Coupling Correspondence Principle

The S-matrix is

$$S_{fi} = \frac{\omega}{2\pi}\int_0^{2\pi/\omega} \mathrm{d}t_{\mathrm{e}} \exp\left\{\mathrm{i}(s\omega t_{\mathrm{e}}) - \frac{\mathrm{i}}{\hbar}\int_{-\infty}^{\infty} V\big[\boldsymbol{R}(t), r(t+t_{\mathrm{e}})\big]\,\mathrm{d}t\right\}. \tag{56.51}$$

See [56.11–14] for additional details.

56.3.5 Equivalent Oscillator Theorem

$$\sum_n\!\!\!\!\!\!\int a_n(t) V_{fn}(t)\mathrm{e}^{\mathrm{i}\omega_{fn}t} = \sum_{d=-f}\!\!\!\!\!\!\int a_{d+f}(t) V_d(t)\mathrm{e}^{-\mathrm{i}d\omega t}\,. \tag{56.52}$$

The S-matrix is

$$S_{\boldsymbol{n}',\boldsymbol{n}} = a_{\boldsymbol{n}'}(t\to\infty) \tag{56.53}$$

$$= \int_0^{2\pi} \frac{\mathrm{d}\boldsymbol{w}}{8\pi^3} \exp\left[\mathrm{i}\boldsymbol{s}\cdot\boldsymbol{w} - \frac{\mathrm{i}}{\hbar}\int_{-\infty}^{\infty} V(\boldsymbol{w}+\omega t, t)\,\mathrm{d}t\right].$$

56.4 Distribution Functions

The function $W_\alpha(x)\,\mathrm{d}x$ characterizes the probability (distribution) of finding an electron in a Rydberg orbital α within a volume $\mathrm{d}x$ centered at the point x in phase space. Integration of the distribution function W_α over all phase space volumes $\mathrm{d}x$ yields, depending upon the normalization chosen, either unity or the density of states appropriate to the orbital α.

56.4.1 Spatial Distributions

Distribution over n, ℓ, m [56.6]:

$$W_{n\ell m}(r,\theta) r^2 \sin\theta\,\mathrm{d}r\,\mathrm{d}\theta \tag{56.54}$$

$$= \frac{r^2 \sin\theta\ \mathrm{d}r\ \mathrm{d}\theta}{\pi^2 a^2 r\left\{\left[\mathrm{e}^2 - (1-r/a)^2\right]\left[\sin^2\theta - (m/\ell)^2\right]\right\}^{1/2}}\,,$$

where $a = Ze^2/2|E| = n^2\hbar^2/mZe^2\hbar^2$ is the semimajor axis, and $\mathrm{e}^2 = 1-(\ell/n)^2$ is the eccentricity.

Distribution over n, ℓ:

$$W_{n\ell}(r,\theta) r^2 \sin\theta\ \mathrm{d}r\,\mathrm{d}\theta$$

$$= g(n\ell)\frac{r^2\sin\theta\ \mathrm{d}r\,\mathrm{d}\theta}{2\pi a^2 r\left[\mathrm{e}^2 - (1-r/a)^2\right]^{1/2}}\,, \tag{56.55}$$

where $g(n\ell) = 2\ell$.

Distribution over n:

$$W_n(r) r^2\,\mathrm{d}r = g(n)\frac{2}{\pi}\left[1 - \left(1-\frac{r}{a}\right)^2\right]^{1/2}\frac{r\,\mathrm{d}r}{a^2}\,, \tag{56.56}$$

with $g(n) = n^2$.

56.4.2 Momentum Distributions

Distribution over n, ℓ [56.6]:

$$W_{n\ell}(p) p^2\,\mathrm{d}p = g(n\ell)\frac{4}{\pi}\frac{\mathrm{d}x}{\left(1+x^2\right)^2}\,, \tag{56.57}$$

where $x = p/p_n$ and $p_n^2 = 2m|E|$.

Distribution over n:

$$W_n(p) p^2\,\mathrm{d}p = g(n)\frac{32}{\pi}\frac{x^2\,\mathrm{d}x}{\left(1+x^2\right)^4}\,. \tag{56.58}$$

Sum Rules.

$$\frac{1}{n^2}\sum_{\ell,m}\left|G_{n\ell m}(\boldsymbol{k})\right|^2 = \left(\frac{na_0}{Z}\right)^3 \frac{8}{\pi^2\left(x^2+1\right)^4}\,, \tag{56.59a}$$

$$\frac{1}{n^2}\sum_{\ell=0}^{n-1}(2\ell+1)\left|g_{n\ell}(k)\right|^2 k^2 = \frac{32na_0x^2}{\pi Z\left(x^2+1\right)^4}\,, \tag{56.59b}$$

where $x = nka_0/Z$, and

$$G_{n\ell m}(\boldsymbol{k}) = g_{n\ell}(k)Y_{\ell m}(\hat{\boldsymbol{k}})\,, \tag{56.60a}$$

$$\begin{aligned} g_{n\ell}(k) = {}&\left(\frac{2}{\pi}\frac{(n-\ell-1)!}{(n+\ell)!}\right)^{1/2}\left(\frac{a_0}{Z}\right)^{3/2}2^{2(\ell+1)}n^2\ell! \\ &\times\frac{(-\mathrm{i}x)^\ell}{\left(x^2+1\right)^{\ell+2}}C_{n-\ell-1}^{(\ell+1)}\left(\frac{x^2-1}{x^2+1}\right)\,, \end{aligned} \tag{56.60b}$$

where $C_i^{(j)}(y)$ is the associated Gegenbauer polynomial. See Chapt. 9 for additional details on hydrogenic wave functions.

Quantum Defect Representation [56.15].

$$\begin{aligned} g_{n\ell}(k) = {}&-\left(\frac{2}{\pi}\frac{\Gamma(n^*-1)}{\Gamma(n^*+\ell+1)}\right)^{1/2}n^*(a_0/Z)^{3/2}2^{2(\ell+1)} \\ &\times\frac{(\ell+1)!(-\mathrm{i}x)^\ell}{\left(x^2+1\right)^{\ell+2}}\mathcal{J}(n^*,\ell+1;X)\,, \end{aligned} \tag{56.61}$$

where $n^* = n-\delta$, δ being the quantum defect, and $x = n^*ka_0/Z$. The function $\mathcal{J}$ is given by the recurrence relation

$$\mathcal{J}(n^*,\ell+1;X) = -\frac{1}{2(2\ell+2)}\frac{\partial}{\partial X}\mathcal{J}(n^*,\ell;X)\,, \tag{56.62}$$

$$\begin{aligned} \mathcal{J}(n^*,0;X) = {}&-\frac{n^*\sin\left[n^*(\beta-\pi)\right]}{\sin(\beta-\pi)} \\ &-\frac{\sin n^*\pi}{\pi}\int_0^1\frac{\left(1-s^2\right)s^{n^*}}{\left(1-2Xs+s^2\right)}\,\mathrm{d}s\,, \end{aligned} \tag{56.63}$$

where $X = \left(x^2-1\right)/\left(x^2+1\right)$, and $\beta = \cos^{-1}X$. In the limit $\ell \ll n^*$, (56.61) becomes

$$|g_{n\ell}(k)|^2 = 4\left(\frac{n^*a_0}{Z}\right)^3\frac{1-(-1)^\ell\cos\left[2n^*(\beta-\pi)\right]}{\pi x^2(x^2+1)^2}\,. \tag{56.64}$$

Classical Density of States.

$$\rho(E) = \int\delta\left[E-H(\boldsymbol{p},\boldsymbol{r})\right]\frac{\mathrm{d}\boldsymbol{p}\,\mathrm{d}\boldsymbol{r}}{(2\pi\hbar)^3} = \frac{n^5\hbar^2}{mZ^2e^4}\,. \tag{56.65}$$

56.5 Classical Theory

The classical cross section for energy transfer ΔE between two particles, with arbitrary masses m_1, m_2 and charges Z_1, Z_2, is given by [56.16]

$$\begin{aligned} \sigma_{\Delta E}(\boldsymbol{v}_1,\boldsymbol{v}_2) = {}&\frac{2\pi(Z_1Z_2e^2V)^2}{v^2|\Delta E|^3} \\ &\times\left(1+\cos^2\bar{\theta}+\frac{\Delta E}{\mu vV}\cos\bar{\theta}\right)\,, \end{aligned} \tag{56.66}$$

valid for $-1 \le \cos\bar{\theta}-\Delta E/(\mu vV) \le 1$, and $\sigma_{\Delta E}(\boldsymbol{v}_1,\boldsymbol{v}_2) = 0$ otherwise, where

$$\boldsymbol{v} = \boldsymbol{v}_1-\boldsymbol{v}_2\,, \tag{56.67a}$$

$$\boldsymbol{V} = (m_1\boldsymbol{v}_1+m_2\boldsymbol{v}_2)/M\,, \tag{56.67b}$$

$$\cos\bar{\theta} = \frac{1}{vV}\boldsymbol{v}\cdot\boldsymbol{V}\,, \tag{56.67c}$$

and $\mu = m_1m_2/M$, $M = m_1+m_2$. If particle 2 has an isotropic velocity distribution in the lab frame, the effective cross section averaged over the direction $\hat{\boldsymbol{n}}_2$ of $\boldsymbol{v}_2$ is

$$v_1\sigma_{\Delta E}^{(\mathrm{eff})}(\boldsymbol{v}_1,\boldsymbol{v}_2) = \frac{1}{4\pi}\int\mathrm{d}\hat{\boldsymbol{n}}_2|\boldsymbol{v}_1-v_2\hat{\boldsymbol{n}}_2|\sigma_{\Delta E}(\boldsymbol{v}_1,\boldsymbol{v}_2)\,. \tag{56.68}$$

If $\boldsymbol{v}_1$ is also isotropic, then the average of (56.68), together with (56.66), gives for the special case of a Coulomb potential

$$\begin{aligned} &\sigma_{\Delta E}^{(\mathrm{eff})}(\boldsymbol{v}_1,\boldsymbol{v}_2) \\ &= \frac{\pi\left(Z_1Z_2e^2\right)^2}{4|\Delta E|3v_1^2v_2}\Big[\left(v_1^2-v_2^2\right)\left(v_2'^2-v_1'^2\right)\left(v_l^{-1}-v_u^{-1}\right) \\ &\quad+\left(v_1^2+v_2^2+v_1'^2+v_2'^2\right)(v_u-v_\ell)-\frac{1}{3}\left(v_u^3-v_l^3\right)\Big]\,, \end{aligned} \tag{56.69}$$

where

$$v_1' = \left(v_1^2-2\Delta E/m_1\right)^{1/2}\,, \tag{56.70}$$

$$v_2' = \left(v_2^2 + 2\Delta E/m_2\right)^{1/2}, \tag{56.71}$$

and v_u, v_l are defined below for cases 1.–4. With the definitions

$$\Delta\varepsilon_{12} = 4m_1m_2(E_1 - E_2)/M^2, \quad \Delta m_{12} = |m_1 - m_2|,$$
$$\Delta\tilde{\varepsilon}_{12} = \frac{4m_1m_2}{M^2}\left(E_1\frac{v_2}{v_1} - E_2\frac{v_1}{v_2}\right),$$

the four cases are

1. $\Delta E \geq \Delta\varepsilon_{12} + |\Delta\tilde{\varepsilon}_{12}| \geq 0$, and $2m_2v_2 \geq \Delta m_{12}v_1$:

$$v_l = v_2' - v_1', \quad v_u = v_1' + v_2', \quad \Delta E \geq 0; \tag{56.72a}$$

$$v_l = v_2 - v_1, \quad v_u = v_1 + v_2, \quad \Delta E \leq 0. \tag{56.72b}$$

If $2m_2v_2 < \Delta m_{12}v_1$, then $\sigma_{\Delta E}^{\text{eff}}(v_1, v_2) = 0$,

2. $\Delta\varepsilon_{12} - \Delta\tilde{\varepsilon}_{12} \leq \Delta E \leq \Delta\varepsilon_{12} + \Delta\tilde{\varepsilon}_{12}$, and $m_1 > m_2$:

$$v_l = v_2' - v_1', \quad v_u = v_1 + v_2, \quad \Delta E \geq 0; \tag{56.72c}$$

$$v_l = v_2 - v_1, \quad v_u = v_1' + v_2', \quad \Delta E \leq 0 \tag{56.72d}$$

3. $\Delta E \leq \Delta\varepsilon_{12} - |\Delta\tilde{\varepsilon}_{12}| \leq 0$, and $2m_1v_1 \geq \Delta m_{12}v_2$:

$$v_l = v_1 - v_2, \quad v_u = v_1 + v_2, \quad \Delta E \geq 0; \tag{56.72e}$$

$$v_l = v_1' - v_2', \quad v_u = v_1' + v_2', \quad \Delta E \leq 0. \tag{56.72f}$$

If $2m_1v_1 < \Delta m_{12}v_2$, then $\sigma_{\Delta E}^{\text{eff}}(v_1, v_2) = 0$,

4. $\Delta\varepsilon_{12} + \Delta\tilde{\varepsilon}_{12} \leq \Delta E \leq \Delta\varepsilon_{12} - \Delta\tilde{\varepsilon}_{12}$, and $m_1 < m_2$:

$$v_l = v_1 - v_2, \quad v_u = v_1' + v_2', \quad \Delta E \geq 0; \tag{56.72g}$$

$$v_l = v_1' - v_2', \quad v_u = v_1 + v_2, \quad \Delta E \leq 0. \tag{56.72h}$$

If $2m_1v_1 < \Delta m_{12}v_2$, then $\sigma_{\Delta E}^{\text{eff}}(v_1, v_2) = 0$.

Since v_1' and v_2', given by (56.70) and (56.71) respectively, must be real, $\sigma_{\Delta E}(\boldsymbol{v}_1, \boldsymbol{v}_2) = 0$ for ΔE outside the range

$$-\frac{1}{2}m_2v_2^2 \leq \Delta E \leq \frac{1}{2}m_1v_1^2, \tag{56.73}$$

which simply expresses the fact that the particle losing energy in the collision cannot lose more than its initial kinetic energy.

The cross section (56.69) must be integrated over the classically allowed range of energy transfer ΔE and averaged over a prescribed speed distribution $W(v_2)$ before comparison with experiment can be made. See [56.16, 17] for details.

Classical Removal Cross Section [56.18]. The cross section for removal of an electron from a shell is given by

$$\sigma_r(V) = \int_0^\infty f(v)\sigma_{\Delta E}(\boldsymbol{v}_1, \boldsymbol{v}_2)\,\mathrm{d}v. \tag{56.74}$$

Total Removal Cross Section [56.18]. In an independent electron model,

$$\sigma_r^{\text{total}}(V) = N_{\text{shell}}\sigma_r(V), \tag{56.75}$$

where N_{shell} is the number of equivalent electrons in a shell. In a shielding model,

$$\sigma_r^{\text{total}}(V) = \left[1 - \frac{(N_{\text{shell}} - 1)}{4\pi\bar{r}^2}\sigma_r(V)\right] N_{\text{shell}}\sigma_r(V), \tag{56.76}$$

where $\bar{r}^2$ is the root mean square distance between electrons within a shell. Experiment [56.19] favors (56.76) over (56.75). See Fig. 4a–e of [56.18] for details.

Classical trajectory and Monte-Carlo methods are covered in Chapt. 58.

56.6 Working Formulae for Rydberg Collisions

56.6.1 Inelastic n,ℓ-Changing Transitions

$$A^*(n\ell) + B \to A^*(n') + B + \Delta E_{n',n\ell}, \tag{56.77}$$

where $\Delta E_{n',n\ell} = E_{n'} - E_{n\ell}$ is the energy defect. The cross section for (56.77) in the quasifree electron model [56.20] is

$$\sigma_{n',n\ell}(V) = \frac{2\pi a_s^2}{(V/v_B)^2 n'^3} f_{n',n\ell}(\lambda), \quad \ell \ll n, \tag{56.78}$$

where a_s is the scattering length for $e^- + B$ scattering, $\lambda = n^* a_0 \omega_{n',n\ell}/V$, $\omega_{n',n\ell} = |\Delta E_{n',n\ell}|/\hbar$,

$E_{n'} = -R_\infty/n'^2$, and $E_{n\ell} = -R_\infty/n^{*2}$, with $n^* = n - \delta_\ell$. Also, v_B is the atomic unit of velocity (see Chapt. 1), and

$$f_{n',n\ell}(\lambda) = \frac{2}{\pi}\left[\tan^{-1}\left(\frac{2}{\lambda}\right) - \frac{\lambda}{2}\ln\left(1+\frac{4}{\lambda^2}\right)\right] . \quad (56.79)$$

Limiting cases: $f_{n',n\ell}(\lambda) \to 1$ as $\lambda \to 0$, and $f_{n',n\ell}(\lambda) \sim 8/\left(3\pi\lambda^3\right)$ for $\lambda \gg 1$. Then

$$\sigma_{n',n\ell} \sim \begin{cases} \dfrac{2\pi a_s^2}{(V/v_B)^2 n'^3} , & \lambda \to 0 , \\ \dfrac{16 a_s^2 V n^3}{3 v_B |\delta_\ell + \Delta n|^3} , & \lambda \gg 1 . \end{cases} \quad (56.80)$$

Rate Coefficients.

$$\langle\sigma_{n',n\ell}(V)\rangle \equiv \langle V\sigma_{n',n\ell}(V)\rangle/\langle V\rangle \quad (56.81a)$$

$$= \frac{2\pi a_s^2}{\left(V_T/v_B\right)^2 n'^3}\varphi_{n',n\ell}(\lambda_T) , \quad (56.81b)$$

where $V_T = \sqrt{2k_B T/\mu}$, $\lambda_T = n^* a_0 \omega_{n',n\ell}/V_T$, $\Delta n = n' - n$, and μ is the reduced mass of A–B. The function $\varphi_{n',n\ell}(\lambda_T)$ in (56.81b) is given by

$$\varphi_{n',n\ell}(\lambda_T) = e^{\lambda_T^2/4}\operatorname{erfc}\left(\frac{1}{2}\lambda_T\right) - \frac{\lambda_T}{\pi}\int_0^\infty \frac{du}{\sqrt{u}} e^{-u}\ln\left(1+\frac{4}{\lambda_T^2}\right) \quad (56.82a)$$

$$= \begin{cases} 1 - \dfrac{\lambda_T}{\sqrt{\pi}}\ln\left(1/\lambda_T^2\right) , & \lambda_T \to 0 \\ 2/\left(\sqrt{\pi}\lambda_T^3\right) , & \lambda_T \gg 1 \end{cases} \quad (56.82b)$$

and erfc (x) is the complementary error function.

56.6.2 Inelastic $n \to n'$ Transitions

$$A^*(n) + B \to A^*(n') + B + \Delta E_{n'n} . \quad (56.83)$$

(A) Cross Sections.

$$\sigma_{n'n} = \sum_{\ell\ell'} \frac{(2\ell+1)}{n^2}\sigma_{n'\ell',n\ell} , \quad (56.84)$$

$$\sigma_{n',n}(V) = \frac{2\pi a_s^2}{(V/v_B)^2 n'^3}F_{n'n}(\lambda) , \quad (56.85)$$

where $\lambda = n a_0 \omega_{n'n}/V = |\Delta n| v_B/(n^2 V)$, and

$$F_{n'n}(\lambda) = \frac{2}{\pi}\left[\tan^{-1}\left(\frac{2}{\lambda}\right) - \frac{2\lambda\left(3\lambda^2+20\right)}{3\left(4+\lambda^2\right)^2}\right] . \quad (56.86)$$

Limiting cases:

$$\sigma_{n'n} \sim \begin{cases} \dfrac{2\pi a_s^2}{\left(V/v_B^2 n'^3\right)} , & \lambda \ll 1 , \\ \dfrac{256\sigma_{e^--B}^{\text{elastic}}\left(V/v_B\right)^3 n^7}{15\pi|\Delta n|5} , & \lambda \gg 1 , \end{cases} \quad (56.87)$$

where $\sigma_e^- - B^{\text{elastic}}$ is the elastic cross section for $e^- + B$ scattering.

(B) Rate Coefficients.

$$K_{n'\ell',n\ell}(T) = \langle V\sigma_{n'\ell',n\ell}\rangle , \quad (56.88a)$$

$$K_{n'n}(T) = \sum_{\ell,\ell'} \frac{(2\ell+1)}{n^2}K_{n'\ell',n\ell} , \quad (56.88b)$$

$$K_{n'n}(T) = \frac{v_B\sigma_{e^--B}^{\text{elastic}}}{\sqrt{\pi}n^3\left(V_T/v_B\right)}\Phi_{n'n}(\lambda_T) , \quad (56.88c)$$

where

$$\Phi_{n'n}(\lambda_T) = e^{\lambda_T^2/8}\left[e^{\lambda_T^2/8}\operatorname{erfc}\left(\frac{1}{2}\lambda_T\right) - \frac{\lambda_T^2}{\sqrt{2\pi}}D_{-3}\left(\frac{\lambda_T}{2}\right) - \frac{5\lambda_T}{\sqrt{\pi}}D_{-4}\left(\frac{\lambda_T}{\sqrt{2}}\right)\right] \quad (56.89a)$$

$$\sim \begin{cases} 1 - 8\lambda_T/3\sqrt{\pi} , & \lambda_T \ll 1 \\ 2^6/\left(\sqrt{\pi}\lambda_T^5\right) , & \lambda_T \gg 1 \end{cases} , \quad (56.89b)$$

where $D_{-\nu}(y)$ denotes the parabolic cylinder function.

Limiting cases:

$$K_{n'n}(T) \sim \begin{cases} \left(\dfrac{\mu R_\infty}{\pi m_e k_B T}\right)^{1/2}\dfrac{v_B\sigma_{e^--B}^{\text{elastic}}}{n^3} , & \lambda_T \to 0 , \\ \dfrac{2^6 v_B\sigma_{e-B}^{\text{elastic}} n^7}{\pi|\Delta n|5}\left(\dfrac{2k_B T}{\mu v_B^2}\right)^2 , & \lambda_T \gg 1 . \end{cases} \quad (56.90)$$

Born Results:

$$\sigma_{n'n} = \frac{8\pi}{k^2}\frac{1}{n^2}\int_{|k-k'|}^{k+k'} F_{n'n}(Q)\frac{d(Qa_0)}{(Qa_0)^3} . \quad (56.91)$$

(A) Electron–Rydberg Atom Collision.

$$\sigma_{n'n} = \frac{8\pi a_0^2 R_\infty}{Z^2 E n^2}\left[\left(1-\frac{1}{4\Delta n}\right)\frac{(\varepsilon\varepsilon')^{3/2}}{(\Delta\varepsilon)^4}\ln\left(1+\varepsilon_e/\varepsilon\right) + \left(1-\frac{0.6}{\Delta n}\right)\frac{\varepsilon_e}{\varepsilon+\varepsilon_e}\frac{(\varepsilon')^{3/2}}{(\Delta\varepsilon)^2}\left(\frac{4}{3\Delta n}+\frac{1}{\varepsilon}\right)\right] \quad (56.92)$$

for $n' > n$, where $\varepsilon_e = E/(Z^2 R_\infty)$, $\varepsilon = 1/n^2$, $\varepsilon' = 1/n'^2$, and $\Delta\varepsilon = \varepsilon - \varepsilon'$.

(B) Heavy Particle–Rydberg Atom Collision.

$$\sigma_{n'n} = \frac{8\pi a_0^2 Z^2}{Z^4 n^2 \varepsilon_e}\left[\left(1 - \frac{1}{4\Delta n}\right)\frac{(\varepsilon\varepsilon')^{3/2}}{(\Delta\varepsilon)^4}\ln(1+\varepsilon_e/\varepsilon) + \left(1 - \frac{0.6}{\Delta n}\right)\varepsilon_{\frac{e}{\varepsilon+\varepsilon_e}}\frac{(\varepsilon')^{3/2}}{(\Delta\varepsilon)^2}\left(\frac{4}{3\Delta n} + \frac{1}{\varepsilon}\right)\right], \tag{56.93}$$

where $\varepsilon_e = m\varepsilon/MZ^2 R_\infty$ with heavy particle mass and charge denoted above by M and Z, respectively, and all other terms retain their meaning as in (56.92).

56.6.3 Quasi-Elastic ℓ-Mixing Transitions

$$\sigma_{n\ell}^{(\ell-\text{mixing})} \equiv \sum_{\ell' \neq \ell} \sigma_{n'\ell',n\ell} \tag{56.94a}$$

$$\sim \begin{cases} \sigma_{\text{geo}} = 4\pi a_0^2 n^4 , & n \ll n_{\max} , \\ 2\pi a_s^2 v_B^2 / V^2 n^3 , & n \gg n_{\max} . \end{cases} \tag{56.94b}$$

The two limits correspond to strong (close) coupling for $n \ll n_{\max}$, and weak coupling for $n \gg n_{\max}$, and expressions (56.94b) are valid when the quantum defect δ_ℓ of the initial Rydberg orbital $n\ell$ is small. $n_{\max}$ is the principal quantum number, where the ℓ-mixing cross section reaches a maximum [56.21],

$$n_{\max} \sim \left(\frac{v_B |a_s|}{V a_0}\right)^{2/7} . \tag{56.95}$$

For Rydberg atom–noble gas atom scattering, $n_{\max} = 8$ to 20, while for Rydberg atom–alkali atom scattering $n_{\max} = 15$ to 30.

56.6.4 Elastic $n\,\ell \rightarrow n\,\ell'$ Transitions

$$A^*(n\ell) + B \rightarrow A^*(n\ell') + B . \tag{56.96}$$

(A) Cross Sections.

$$\sigma_{ns}^{\text{elastic}}(V) = \frac{2\pi C_{ss} a_s^2}{(V/v_B)^2 n^{*4}} , \tag{56.97}$$

valid for $n^* \gg [v_B |a_s|/(4 V a_0)]^{1/4}$ with

$$C_{ss} = \frac{8}{\pi^2}\int_0^{1/\sqrt{2}} [K(k)]^2 \, dk , \tag{56.98}$$

where $K(k)$ denotes the complete elliptic integral of the first kind.

(B) Rate Coefficients [56.22] (Three Cases). With the definitions

$$\nu_B = v_B / v_{\text{rms}} , \quad v_{\text{rms}} = \sqrt{(8 k_B T)/\mu\pi} , \tag{56.99}$$

$$f(y) = y^{-1/2}\left(1 - (1-y)e^{-y}\right) + y^{3/2}\text{Ei}(y) , \tag{56.100}$$

$$y = (\nu_B a_s)^2 / \left(4\pi a_0^2 n^{*8}\right) , \tag{56.101}$$

$$n_1 = (|a_s| \nu_B / 4 a_0)^{1/4} , \tag{56.102}$$

$$n_2 = 0.7\left[|a_s| \nu_B^{5/6} / \left(\alpha_d a_0^3\right)^{1/6}\right]^{1/3} , \tag{56.103}$$

where α_d is the dipole polarizability of A^*, then

$$\left\langle \sigma_{ns}^{\text{el}} \right\rangle \sim \begin{cases} 8\pi a_0^2 n^{*4} , & n^* \leq n_1 , \\ 4\pi^{1/2} a_0 |a_s| \nu_B f(y) , & n_1 \leq n^* \leq n_2 , \\ 7(\alpha_d \nu_B)^{2/3} + \dfrac{4 a_s^2 \nu_B^2}{n^{*4}} - \dfrac{2.7 a_s^2 \nu_B^2 (\alpha_d \nu_B)^{1/3}}{a_0 n^{*6}} , & n^* \geq n_2 . \end{cases} \tag{56.104}$$

56.6.5 Fine Structure $n\,\ell\,J \rightarrow n\,\ell\,J'$ Transitions

$$A^*(n\ell J) + B \rightarrow A^*(n\ell J') + B + \Delta E_{J'J} . \tag{56.105}$$

(A) Cross Sections (Two Cases).

$$\sigma_{n\ell J}^{n\ell J'}(V) = \frac{2J'+1}{2(2\ell+1)} c_{\text{norm}} 4\pi a_0^2 n^{*4} , \tag{56.106}$$

valid for $n^* \leq n_0(V)$, and

$$\sigma_{n\ell J}^{n\ell J'}(V) = \frac{2\pi C_{J'J}^{(\ell)} a_s^2 v_B^2}{V^2 n^{*4}} \varphi_{J'J}^{(\ell)}(\nu_{J'J})\left(1 - \frac{n_0^8(V)}{2n^{*8}}\right) , \tag{56.107}$$

valid for $n^* \geq n_0(V)$, where the quantity $n_0(V)$ is the effective principal quantum number such that the impact parameter ρ_0 of B (moving with relative velocity V) equals the radius $2n^{*2}a_0$ of the Rydberg atom A^*. $n_0(V)$ is given by the solution to the following transcendental equation

$$n_0^8(V) = \frac{(2\ell+1) C_{J'J}^{(\ell)}}{2(2J'+1) c_{\text{norm}}}\left(\frac{v_B a_s}{V a_0}\right)^2 \varphi_{J'J}^{(\ell)}\left(\nu_{J'J}\left[n_0(V)\right]\right) . \tag{56.108}$$

The constant c_{norm} in (56.106) is equal to 5/8 if $\sigma_{\text{geo}} = \pi\langle r^2\rangle_{n\ell}$, or 1 if $\sigma_{\text{geo}} = 4\pi a_0^2 n^{*4}$. The function $\varphi_{J'J}^{(\ell)}(\nu_{J'J})$ in (56.107) is given in general by [56.23, 24]

$$\varphi_{J'J}^{(\ell)}(\nu_{J'J}) = \xi_{J'J}^{(\ell)}(\nu_{J'J})/\xi_{J'J}^{(\ell)}(0)\,, \tag{56.109a}$$

$$\xi_{J'J}^{(\ell)}(\nu_{J'J}) = \sum_{s=0}^{\ell} A_{\ell J',\ell J}^{(2s)} \int_{\nu_{J'J}}^{\infty} j_s^2(z) J_s^2(z) z\,\mathrm{d}z\,, \tag{56.109b}$$

$$\nu_{J'J} = |\delta_{\ell J'} - \delta_{\ell J}|\frac{v_{\mathrm{B}}}{Vn^*}\,, \tag{56.109c}$$

where $j_s(z)$ is the spherical Bessel function and the coefficients $C_{J'J}^{(\ell)}$ and $A_{\ell J',\ell J}^{(2s)}$ in (56.107) and (56.109b), respectively, are given in table 5.1 of *Beigman and Lebedev* [56.6]. The quantum defect of Rydberg state $n\ell J$ is $\delta_{\ell J}$. For elastic scattering, $\nu_{JJ} = 0$, and $\varphi_{JJ}^{(\ell)}(0) = 1$.

Symmetry relation:

$$\xi_{JJ'}^{(\ell)}(\nu_{J'J}) = \frac{2J+1}{2J'+1}\xi_{J'J}^{(\ell)}(\nu_{J'J})\,. \tag{56.110}$$

(B) Rate Coefficients.

$$\langle\sigma_{n\ell J}^{n\ell J'}\rangle = \left(\frac{c_{\text{norm}}(2J'+1)C_{J'J}^{(\ell)}}{2(2\ell+1)}\right)^{1/2} \times \pi a_0^2 F(\zeta)\left(\frac{v_{\mathrm{B}}|a_{\mathrm{s}}|}{V_{\mathrm{T}} a_0}\right)\,, \tag{56.111}$$

where $\zeta = n_0^8(V_{\mathrm{T}})/n^{*8}$, and

$$F(\zeta) \equiv \sqrt{\zeta}\left[\mathrm{E}_2(\zeta) + \frac{1}{\zeta}(1-\mathrm{e}^{-\zeta})\right]\,, \tag{56.112}$$

where $\mathrm{E}_2(x)$ is an exponential integral.

Limiting cases:

$$\left\langle\sigma_{n\ell J}^{n\ell J'}\right\rangle = \begin{cases} \dfrac{2J'+1}{2(2\ell+1)} c_{\text{norm}} 4\pi a_0^2 n^{*4}\,, & n^* \ll n_{\max}^*\,,\\[2ex] \dfrac{2\pi C_{J'J}^{(\ell)} a_{\mathrm{s}}^2 v_{\mathrm{B}}^2}{V_{\mathrm{T}}^2 n^{*4}}\,, & n^* \gg n_{\max}^*\,, \end{cases} \tag{56.113}$$

where $n_{\max}^* = (3/2)^{1/8} n_0(V)$ if $\nu_{J'J} \ll 1$.

56.7 Impulse Approximation

56.7.1 Quantal Impulse Approximation

Basic Formulation [56.25]

Consider a Rydberg collision between a projectile (1) of charge Z_1 and a target with a valence electron (3) in orbital ψ_i bound to a core (2). The full three-body wave function for the system of projectile + target is denoted by Ψ_i. The relative distance between 1 and the center-of-mass of 2 − 3 is denoted by $\boldsymbol{\sigma}$, while the separation of 2 from the center-of-mass of 1 − 3 is $\boldsymbol{\rho}$.

Formal Scattering Theory.

$$\Psi_i^{(+)} = \Omega^{(+)}\psi_i\,, \tag{56.114}$$

where the Möller scattering operator $\Omega^{(+)} = 1 + G^+ V_i$, and $V_i = V_{12} + V_{13}$.

Let χ_m be a complete set of free-particle wave functions satisfying

$$(H_0 - E_m)\chi_m = 0\,, \tag{56.115}$$

and define operators $\omega_{ij}^+(m)$ by

$$\omega_{ij}^+(m)\chi_m = \left(1 + \frac{1}{E_m - H_0 - V_{ij} + \mathrm{i}\epsilon}V_{ij}\right)\chi_m\,, \tag{56.116}$$

where V_{ij} denotes the pairwise interaction potential between particles i and j ($i, j = 1, 2, 3$). Then the action of the full Green's function G^+ on the two-body potential V_{ij} is

$$G^+ V_{ij} = \left[\omega_{ij}^+(m) - 1\right] + G^+\left[(E_m - E) + V_{12} + V_{13} + V_{23} - V_{ij}\right] \times \left[\omega_{ij}^+(m) - 1\right]\,. \tag{56.117}$$

Projection Operators.

$$b_{ij}^+(m) = \omega_{ij}^+(m) - 1\,, \tag{56.118a}$$

$$b_{ij}^+ = \sum_m b_{ij}^+(m)|\chi_m\rangle\langle\chi_m|\,, \quad \omega_{ij}^+ = b_{ij}^+ + 1\,. \tag{56.118b}$$

$$G^+ V_{ij}|\psi_i\rangle = \sum_m G^+ V_{ij}|\chi_m\rangle\langle\chi_m|\psi_i\rangle \tag{56.119a}$$

$$= \left[b_{ij}^+ + G^+\left[V_{23}, b_{ij}^+\right] + G^+\left[V_{12} + V_{13} - V_{ij}\right] b_{ij}^+(m)\right]|\psi_i\rangle\,. \tag{56.119b}$$

Möller Scattering Operator.

$$\Omega^+ = (\omega_{13}^+ + \omega_{12}^+ - 1) + G^+ [V_{23}, (b_{13}^+ + b_{12}^+)] + G^+ (V_{13} b_{12}^+ + V_{12} b_{13}^+) . \quad (56.120)$$

Exact T-Matrix.

$$T_{if} = \langle \psi_f | V_f | (\omega_{13}^+ + \omega_{12}^+ - 1) \psi_i \rangle + \langle \psi_f | V_f | G^+ [V_{23}, (b_{13}^+ + b_{12}^+)] \psi_i \rangle + \langle \psi_f | V_f | G^+ (V_{13} b_{12}^+ + V_{12} b_{13}^+) \psi_i \rangle . \quad (56.121)$$

The impulse approximation to the exact T-matrix element (56.121) is obtained by ignoring the second term involving the commutator of V_{23}.

$$\Psi_i \longrightarrow \Psi_i^{\mathrm{imp}} = (\omega_{13}^+ + \omega_{12}^+ - 1) \psi_i . \quad (56.122)$$

Impulse Approximation: Post Form.

$$T_{if}^{\mathrm{imp}} = \langle \psi_f | V_f | (\omega_{13}^+ + \omega_{12}^+ - 1) \psi_i \rangle . \quad (56.123)$$

The impulse approximation can also be expressed using incoming-wave boundary conditions by making use of the prior operators

$$\omega_{ij}^-(m) \chi_m = \left(1 + \frac{1}{E_m - H_0 - V_{ij} - i\epsilon} V_{ij} \right) \chi_m , \quad (56.124a)$$

$$\omega_{ij}^- = \sum_m \omega_{ij}^-(m) | \chi_m \rangle \langle \chi_m | . \quad (56.124b)$$

The impulse approximation (56.123) is exact if V_{23} is a constant since the commutator of V_{23} vanishes in that case.

Applications [56.25]

(1) Electron Capture. $X^+ + \mathrm{H}(i) \to X(f) + \mathrm{H}^+$.

$$T_{if}^{\mathrm{imp}} = \langle \psi_f | V_{12} + V_{23} (\omega_{12}^+ + \omega_{13}^+ - 1) \psi_i \rangle . \quad (56.125)$$

Wave functions: $\psi_i = \mathrm{e}^{\mathrm{i} \boldsymbol{k}_i \cdot \boldsymbol{\sigma}} \varphi_i(\boldsymbol{r})$, $\psi_f = e^{\mathrm{i} \boldsymbol{k}_f \cdot \boldsymbol{\rho}} \varphi_f(\boldsymbol{x})$, $\chi_m = (2\pi)^{-3} \exp[\mathrm{i}(\boldsymbol{K} \cdot \boldsymbol{x} + \boldsymbol{k} \cdot \boldsymbol{\rho})]$, where the φ_n are hydrogenic wave functions.

If X above is a heavy particle, the V_{12} term in (56.125) may be omitted due to the difference in mass between the projectile 1 and the bound Rydberg electron 3. See [56.25] and references therein for details.

With the definitions

$$z = \frac{4\alpha\delta^2}{(T - 2\delta)(T - 2\alpha\delta)} , \qquad T = \beta^2 + P^2 ,$$
$$\delta = \mathrm{i}\beta K - \boldsymbol{p} \cdot \boldsymbol{K} , \qquad \nu = a Z_1 / K ,$$
$$\boldsymbol{t}_1 = \boldsymbol{K}/a + \boldsymbol{v} , \qquad N(\nu) = \mathrm{e}^{\pi\nu/2} \Gamma(1 - \mathrm{i}\nu) ,$$
$$a = \frac{M_1}{M_1 + m_\mathrm{e}} , \qquad b = \frac{M_2}{M_2 + m_\mathrm{e}} ,$$
$$\boldsymbol{k} = a\boldsymbol{k}_2 - (1 - a)\boldsymbol{k}_f , \qquad \boldsymbol{K} = a\boldsymbol{k}_1 - (1 - a)\boldsymbol{k}_i ,$$
$$\boldsymbol{t} = (\boldsymbol{K} - \boldsymbol{p})/a , \qquad \boldsymbol{p} = a\boldsymbol{k}_f - \boldsymbol{k}_i ,$$
$$\beta = a Z_1 / n ,$$

and n is the principal quantum number, the impulse approximation to the T-matrix becomes, in this case,

$$T_{if}^{\mathrm{imp}} \sim \langle \psi_f | V_{23} | \omega_{13}^+ \psi_i \rangle \quad (56.126)$$

$$= \frac{-1}{2\pi^2 a^3} \int \frac{\mathrm{d}\boldsymbol{K}}{t^2} N(\nu) g_i(\boldsymbol{t}_1) \mathcal{F}(f, \boldsymbol{K}, \boldsymbol{p}) , \quad (56.127)$$

where, for a general final s-state,

$$\mathcal{F}(f, \boldsymbol{K}, \boldsymbol{p}) = \int \varphi_f^*(\boldsymbol{x}) \mathrm{e}^{\mathrm{i}\boldsymbol{p}\cdot\boldsymbol{x}} {}_1F_1[\mathrm{i}\nu, 1; \mathrm{i}(Kx - \boldsymbol{K} \cdot \boldsymbol{x})] \,\mathrm{d}\boldsymbol{x} , \quad (56.128)$$

and $g_i(\boldsymbol{t}_1)$ denotes the Fourier transform of the initial state. The normalization of the Fourier transform is chosen such that momentum and coordinate space hydrogenic wave functions are related by $\varphi_n(\boldsymbol{r}) = (2\pi)^{-3} \int \exp(\mathrm{i}\boldsymbol{t}_1 \cdot \boldsymbol{r}) g_n(\boldsymbol{t}_1) \,\mathrm{d}\boldsymbol{t}_1$. For the case $f = 1s$,

$$\mathcal{F}(1s, \boldsymbol{K}, \boldsymbol{p}) = -\frac{\beta^{3/2}}{\sqrt{\pi}} \frac{\partial}{\partial\beta} \mathcal{I}(\nu, 0, \beta, -\boldsymbol{K}, \boldsymbol{p}) = 8\sqrt{\pi} \beta^{3/2} \left[\frac{(1 - \mathrm{i}\nu)\beta}{T^2} + \frac{\mathrm{i}\nu(\beta - \mathrm{i}K)}{T(T - 2\delta)} \left(\frac{T}{T - 2\delta} \right)^{\mathrm{i}\nu} \right] \quad (56.129)$$

evaluated at $\beta = aZ_1$. For the case $f = 2s$,

$$\mathcal{F}(2s, \boldsymbol{K}, \boldsymbol{p}) = -\frac{\beta^{3/2}}{\sqrt{\pi}} \left[\left(\frac{\partial}{\partial\beta} + \beta \frac{\partial^2}{\partial\beta^2} \right) \times \mathcal{I}(\nu, 0, \beta, -\boldsymbol{K}, \boldsymbol{p}) \right] \quad (56.130)$$

evaluated at $\beta = aZ_1/2$. For a general final ns-state f,

$$\mathcal{I}(\nu, \alpha, \beta, \boldsymbol{K}, \boldsymbol{p}) = \frac{4\sqrt{\pi}}{T} \left(\frac{T - 2\alpha\delta}{T - 2\delta} \right)^{\mathrm{i}\nu} \times (U \cosh \pi\nu \pm \mathrm{i} V \sinh \pi\nu) , \quad (56.131)$$

where the complex quantity $U+\mathrm{i}V$ is

$$U+\mathrm{i}V=(4z)^{\mathrm{i}\nu}\frac{\Gamma\left(\frac{1}{2}+\mathrm{i}\nu\right)}{\Gamma(1+\mathrm{i}\nu)} \times {}_2F_1(-\mathrm{i}\nu,-\mathrm{i}\nu;1-2\mathrm{i}\nu;1/z)\,. \quad (56.132)$$

(2) Electron Impact Excitation.

$$\mathrm{e}^-+\mathrm{H}(i)\to \mathrm{e}^-+\mathrm{H}(f)\,. \quad (56.133)$$

Neglecting V_{12} and exchange yields the approximate T-matrix element

$$\begin{aligned}T_{if}^{\mathrm{imp}} &\sim \langle\psi_f|V_{13}|\omega_{13}^+\psi_i\rangle\\ &=\frac{-Z_1}{(2\pi a)^3}\int \mathrm{d}\boldsymbol{x}\int \mathrm{d}\boldsymbol{r}\;\mathrm{e}^{\mathrm{i}\boldsymbol{q}\cdot\boldsymbol{\sigma}}\varphi_f^*(\boldsymbol{r})\frac{1}{x}\\ &\quad\times\int \mathrm{d}\boldsymbol{K}\;N(\nu)g_i(\boldsymbol{t}_1)\mathrm{e}^{\mathrm{i}\boldsymbol{t}_1\cdot\boldsymbol{r}}{}_1F_1(\mathrm{i}\nu,1;\mathrm{i}Kx-\mathrm{i}\boldsymbol{K}\cdot\boldsymbol{x})\end{aligned} \quad (56.134)$$

$$=\frac{-Z_1}{(2\pi a)^3}\int \mathrm{d}\boldsymbol{K}\;N(\nu)g_i(\boldsymbol{t}_1)g_f^*(\boldsymbol{t}_2) \times\mathcal{I}(\nu,0,0,-\boldsymbol{K},-\boldsymbol{q})\,, \quad (56.135)$$

where

$$\begin{aligned}&\mathcal{I}(\nu,0,0,-\boldsymbol{K},-\boldsymbol{q})\\ &=\lim_{\beta\to 0}\frac{4\pi}{\beta^2+q^2}\left(\frac{\beta^2+q^2}{\beta^2+q^2+2\boldsymbol{q}\cdot\boldsymbol{K}-2\mathrm{i}\beta K}\right)^{\mathrm{i}\nu}\end{aligned} \quad (56.136)$$

$$=\frac{4\pi}{q^2}\left|1+\frac{2K}{q}\cos\theta\right|^{-\mathrm{i}\nu}A(\cos\theta)\,, \quad (56.137)$$

with

$$A(\cos\theta)=\begin{cases}1\,, & \cos\theta>-q/2K\,,\\ \mathrm{e}^{-\pi\nu}\,, & \cos\theta<-q/2K\,,\end{cases} \quad (56.138)$$

and $\cos\theta=\hat{\boldsymbol{K}}\cdot\hat{\boldsymbol{q}}$, $\boldsymbol{t}_2=\boldsymbol{t}_1+b\boldsymbol{q}$ and $\boldsymbol{q}=\boldsymbol{k}_i-\boldsymbol{k}_f$.

(3) Heavy Particle Excitation [56.26].

$$\mathrm{H}^++\mathrm{H}(1s)\to \mathrm{H}^++\mathrm{H}(2s)\,. \quad (56.139)$$

$$\begin{aligned}T_{if}^{\mathrm{imp}}&=-\frac{Z_1 2^{15/2}b^5}{\pi a^3q^2}\int_0^\infty \mathrm{d}K\;N(K)K^2\int_{-1}^{1}\mathrm{d}(\cos\theta)\\ &\quad\times\left|1+\frac{2K}{q}\cos\theta\right|^{-\mathrm{i}\nu}\widetilde{A}(\cos\theta)\,,\end{aligned} \quad (56.140)$$

where

$$\begin{aligned}\widetilde{A}(\cos\theta)=\frac{2\pi}{D^4}\Bigg(&\frac{\alpha D\left(D-2b^2\right)}{\left(\alpha^2-\beta^2\right)^{3/2}}+\frac{8\left(3b^2-D\right)}{\left(\alpha^2-\beta^2\right)^{1/2}}\\ &-\frac{48\gamma^2D^2b^2}{\left(\gamma^2-\delta^2\right)^{5/2}}+\frac{16D\left[\gamma D-(3\gamma+4\alpha)b^2\right]}{\left(\gamma^2-\delta^2\right)^{3/2}}\\ &+\frac{32\left(D-3b^2\right)}{\left(\gamma^2-\delta^2\right)^{1/2}}\Bigg)A(\cos\theta)\,,\end{aligned} \quad (56.141)$$

with $A(\cos\theta)$ given by (56.138), and

$$\alpha=b^2+v^2+\frac{K^2}{a^2}+\frac{K}{aq}\left(\frac{q^2}{\mu}+\Delta E\right)\cos\theta\,, \quad (56.142a)$$

$$\beta=\frac{K}{aq}\sin\theta\left[4v^2q^2-\left(\frac{q^2}{\mu}+\Delta E\right)^2\right]^{1/2}, \quad (56.142b)$$

$$\delta=4\beta,\quad \gamma=4\alpha+D\,, \quad (56.142c)$$

$$D=\frac{4bq}{a}(q+2K\cos\theta)\,, \quad (56.142d)$$

while ν and $N(\nu)$ retain their meaning from (56.127).

(4) Ionization. $\mathrm{e}^-+\mathrm{H}(i)\to \mathrm{e}^-+\mathrm{H}^++\mathrm{e}^-$.

$$T_{if}^{\mathrm{imp}}\sim-\frac{4\pi}{q^2}N(\nu)g_i(\boldsymbol{k}-b\boldsymbol{q})\left(\frac{q^2}{q^2-2\boldsymbol{q}\cdot\boldsymbol{K}}\right)^{\mathrm{i}\nu}, \quad (56.143)$$

where $\boldsymbol{K}=a(\boldsymbol{k}-b\boldsymbol{q}-\boldsymbol{v})$ and $\boldsymbol{q}=\boldsymbol{k}_i-\boldsymbol{k}_f$ and exchange and V_{12} are neglected.

(5) Rydberg Atom Collisions [56.11, 27].

$$A+B(n)\to A+B(n') \quad (56.144)$$

$$\to A+B^++\mathrm{e}^-\,. \quad (56.145)$$

Consider a Rydberg collision between a projectile (3) and a target with an electron (1) bound in a Rydberg orbital to a core (2) [56.11, 27].

Full T-matrix element:

$$\begin{aligned}&T_{fi}(\boldsymbol{k}_3,\boldsymbol{k}_3')\\ &=\left\langle\phi_f(\boldsymbol{r}_1)\mathrm{e}^{\mathrm{i}\boldsymbol{k}_3'\cdot\boldsymbol{r}_3}\middle|V(\boldsymbol{r}_1,\boldsymbol{r}_3)\middle|\Psi_i^{(+)}(\boldsymbol{r}_1,\boldsymbol{r}_3;\boldsymbol{k}_3)\right\rangle,\end{aligned} \quad (56.146)$$

with primes denoting quantities after the collision, and where the potential V is

$$V(\boldsymbol{r}_1,\boldsymbol{r}_3)=V_{13}(\boldsymbol{r})+V_{3\mathrm{C}}(\boldsymbol{r}_3+a\boldsymbol{r}_1)\,,\quad \boldsymbol{r}=\boldsymbol{r}_1-\boldsymbol{r}_3\,, \quad (56.147)$$

with $a=M_1/(M_1+M_2)$, while the subscript C labels the core. The impulse approximation to the full, outgoing wave function $\Psi_i^{(+)}$ is written

$$\Psi_i^{\text{imp}} = (2\pi)^{3/2} \int g_i(\boldsymbol{k}_1)\Phi(\boldsymbol{k}_1, \boldsymbol{k}_3; \boldsymbol{r}_1, \boldsymbol{r}_3)\,\mathrm{d}\boldsymbol{k}_1\,, \tag{56.148}$$

$$g_i(\boldsymbol{k}_1) = \frac{1}{(2\pi)^{3/2}} \int \phi_i(\boldsymbol{r}_1)\mathrm{e}^{-\mathrm{i}\boldsymbol{k}_1\cdot\boldsymbol{r}_1}\,\mathrm{d}\boldsymbol{r}_1\,. \tag{56.149}$$

Impulse approximation:

$$T_{fi}^{\text{imp}}(\boldsymbol{k}_3, \boldsymbol{k}_3') = \int \mathrm{d}\boldsymbol{k}_1 \int \mathrm{d}\boldsymbol{k}_1'\, g_f^*(\boldsymbol{k}_1')g_i(\boldsymbol{k}_1)T_{13}(\boldsymbol{k}, \boldsymbol{k}') \times\delta\left[\boldsymbol{P} - (\boldsymbol{k}_1' - \boldsymbol{k}_1)\right]\,, \tag{56.150}$$

where T_{13} is the exact off-shell T-matrix for 1–3 scattering,

$$T_{13}(\boldsymbol{k}, \boldsymbol{k}') = \left\langle \exp(\mathrm{i}\boldsymbol{k}'\cdot\boldsymbol{r})|V_{13}(\boldsymbol{r})|\psi(\boldsymbol{k}, \boldsymbol{r})\right\rangle\,. \tag{56.151}$$

The delta function in (56.150) ensures linear momentum, $\boldsymbol{K} = \boldsymbol{k}_1 + \boldsymbol{k}_3 = \boldsymbol{k}_1' + \boldsymbol{k}_3'$, is conserved in 1–3 collisions, with

$$\boldsymbol{k}_1' = \boldsymbol{k}_1 + (\boldsymbol{k}_3 - \boldsymbol{k}_3') \equiv \boldsymbol{k}_1 + \boldsymbol{P}\,, \tag{56.152a}$$

$$\boldsymbol{k}' = \frac{M_3}{M_1 + M_3}(\boldsymbol{k}_1 + \boldsymbol{k}_3) - \boldsymbol{k}_3' = \boldsymbol{k} + \boldsymbol{P}\,. \tag{56.152b}$$

Elastic scattering:

$$T_{ii}(\boldsymbol{k}_3, \boldsymbol{k}_3) = \int g_f^*(\boldsymbol{k}_1)g_i(\boldsymbol{k}_1)T_{13}(\boldsymbol{k}, \boldsymbol{k})\,\mathrm{d}\boldsymbol{k}_1\,, \tag{56.153}$$

where $\boldsymbol{k} = (M_3/M)\boldsymbol{k}_1 + (M_1/M)\boldsymbol{k}_3$ and $M = M_1 + M_3$.

Integral cross section: for 3–(1,2) scattering,

$$\sigma_{if}(k_3) = \left(\frac{M_{AB}}{M_{13}}\right)^2 \frac{k_3'}{k_3} \int \left|\langle g_f(\boldsymbol{k}_1 + \boldsymbol{P})\right| \times f_{13}(\boldsymbol{k}, \boldsymbol{k}')g_i(\boldsymbol{k}_1)\rangle\big|^2\,\mathrm{d}\hat{\boldsymbol{k}}_3'\,, \tag{56.154}$$

where M_{AB} is the reduced mass of the 3–(1,2) system, M_{13} the reduced mass of 1–3. The 1–3 scattering amplitude f_{13} is given by

$$f_{13}(\boldsymbol{k}, \boldsymbol{k}') = -\frac{1}{4\pi}\left(\frac{2M_{13}}{\hbar^2}\right)T_{13}(\boldsymbol{k}, \boldsymbol{k}')\,. \tag{56.155}$$

Six Approximations to (56.150)

(1) Optical Theorem.

$$\sigma_{\text{tot}}(k_3) = \frac{1}{k_3}\frac{2M_{AB}}{\hbar^2}T_{ii}(\boldsymbol{k}_3, \boldsymbol{k}_3') = \frac{1}{v_3}\int |g_i(\boldsymbol{k}_1)|^2\left[v_{13}\sigma_{13}^{\mathrm{T}}(v_{13})\right]\mathrm{d}\boldsymbol{k}_1\,, \tag{56.156}$$

where σ_{13}^{T} is the total cross section for 1–3 scattering at relative speed v_{13}. The resultant cross section (56.156) is an upper limit and contains no interference terms.

(2) Plane-Wave Final State.

$$\phi_f(\boldsymbol{r}_1) = (2\pi)^{-3/2}\exp(\mathrm{i}\boldsymbol{\kappa}_I'\cdot\boldsymbol{r}_1)\,, \tag{56.157}$$

$$g_f(\boldsymbol{k}_1') = \delta(\boldsymbol{k}_1' - \boldsymbol{\kappa}_I')\,, \tag{56.158}$$

$$T_{fi}(\boldsymbol{k}_3, \boldsymbol{k}_3') = g_i(\boldsymbol{k}_1)T_{13}(\boldsymbol{k}, \boldsymbol{k}')\,, \quad \boldsymbol{k}_1 = \boldsymbol{\kappa}_I - \boldsymbol{P}\,, \tag{56.159}$$

$$\frac{\mathrm{d}\sigma_{if}}{\mathrm{d}\hat{\boldsymbol{k}}_3'\,\mathrm{d}\boldsymbol{k}_1'} = \left(\frac{M_{AB}}{M_{13}}\right)^2\frac{k_3'}{k_3}|g_i(\boldsymbol{k}_1)|^2|f_{13}(\boldsymbol{k}, \boldsymbol{k}')|^2\,. \tag{56.160}$$

(3) Closure.

$$\sum\!\!\!\!\!\!\int_f g_f(\boldsymbol{k}_I')g_f^*(\boldsymbol{k}_I'') = \delta(\boldsymbol{k}_1' - \boldsymbol{k}_I'')\,, \tag{56.161}$$

$$\frac{\mathrm{d}\sigma_i^T}{\mathrm{d}\hat{\boldsymbol{k}}_3'} = \frac{\bar{k}_3'}{k_3}\left(\frac{M_{AB}}{M_{13}}\right)^2\int |g_i(\boldsymbol{k}_1)|^2|f_{13}(\boldsymbol{k}, \boldsymbol{k}')|^2\,\mathrm{d}\boldsymbol{k}_1\,, \tag{56.162}$$

where $\boldsymbol{k}_1' = (M_3/M)(\boldsymbol{k}_1 + \boldsymbol{k}_3) - \boldsymbol{k}_3'$.

Conditions for validity of (56.162): (a) k_3 is high enough to excite all atomic bound and continuum states, and (b) $k_3'^2 = (k_3^2 - 2\varepsilon_{fi}/M_{AB})$ can be approximated by k_3, or by an averaged wavenumber $\bar{k}_3' = (k_3^2 - 2\bar{\varepsilon}_{fi}/M_{AB})^{1/2}$, where the averaged excitation energy is

$$\bar{\varepsilon}_{fi} = \ln\langle\varepsilon_{fi}\rangle = \sum_j f_{ij}\ln\varepsilon_{ij}\left(\sum_j f_{ij}\right)^{-1}\,, \tag{56.163}$$

and the f_{ij} are the oscillator strengths.

(4) Peaking Approximation.

$$T_{fi}^{\text{peak}}(\boldsymbol{k}_3, \boldsymbol{k}_3') = F_{fi}(\boldsymbol{P})T_{13}(\boldsymbol{k}, \boldsymbol{k}')\,, \tag{56.164}$$

where F_{fi} is the inelastic form factor

$$F_{fi}(\boldsymbol{P}) = \int g_f^*(\boldsymbol{k}_1 + \boldsymbol{P})g_i(\boldsymbol{k}_1)\,\mathrm{d}\boldsymbol{k}_1 \tag{56.165}$$

$$= \langle\psi_f(\boldsymbol{r})\exp(\mathrm{i}\boldsymbol{P}\cdot\boldsymbol{r})|\psi_i(\boldsymbol{r})\rangle\,. \tag{56.166}$$

(5) $T_{13} = T_{13}(\boldsymbol{P})$.

$$T_{fi}(\boldsymbol{k}_3, \boldsymbol{k}_3') = T_{13}(\boldsymbol{P})F_{fi}(\boldsymbol{P})\,. \tag{56.167}$$

(6) $T_{13} =$ *constant.* $f_{13} \equiv a_s =$ constant scattering length.

$$\sigma_{if}(k_3) = \frac{2\pi a_s^2}{k_3^2}\left(\frac{M_{AB}}{M_{13}}\right)^2 \int\limits_{k_3-k_3'}^{k_3+k_3'} \left|F_{fi}(\boldsymbol{P})\right|^2 P\,\mathrm{d}P\,, \tag{56.168}$$

$$\sigma_{\text{tot}}(k_3) = \begin{cases} 4\pi a_s^2\,, & v_3 \gg v_1 \\ \langle v_1\rangle 4\pi a_s^2/v_3\,, & v_3 \ll v_1\,. \end{cases} \tag{56.169}$$

Validity Criteria

(A) Intuitive Formulation [56.27]. (i) Particle 3 scatters separately from 1 and 2, i.e., $r_{12} \gg A_{1,2}$; the relative separation of (1,2) $\gg$ the scattering lengths of 1 and 2. (ii) $\lambda\!\!\!^{-}_{13} \ll r_{12}$, i.e., the reduced wavelength for 1–3 relative motion $\ll r_{12}$. Interference effects of 1 and 2 can be ignored and 1, 2 treated as *independent* scattering centers.
(iii) 2–3 collisions do not contribute to inelastic 1–3 scattering.
(iv) Momentum impulsively transferred to 1 during collision (time τ_{coll}) with 3 $\gg$ momentum transferred to 1 due to V_{12}, i.e.,

$$P \gg \langle\psi_{n\ell}| - \nabla V_{12}|\psi_{n\ell}\rangle\tau_{\text{coll}}\,. \tag{56.170}$$

For a precise formulation of validity criteria based upon the two-potential formula see the Appendix of [56.27].

Two classes of interaction in A–$B(n)$ Rydberg collisions justify use of the impulse approximation for the T-matrix for 1–3 collisions: (i) quasiclassical binding with $V_{\text{core}} =$ const., and (ii) weak binding with

$$E_3 \gg \Delta E_c \sim \langle\psi_n(r)|V_{1C}(\boldsymbol{r})|\psi_n(\boldsymbol{r})\rangle\,, \tag{56.171a}$$

$$\langle\psi_n(\boldsymbol{r}_1)| - \frac{\hbar^2}{2M_{12}}\nabla_1^2|\psi_n(\boldsymbol{r}_1)\rangle \sim |\varepsilon_n|\,, \tag{56.171b}$$

where E_3 is the kinetic energy of relative motion of 3, and ΔE_c is the energy shift in the core. The fractional error is [56.28]

$$\frac{f_{13}}{\lambda\!\!\!^{-}}\frac{\Delta E_c}{\hbar}\left(\frac{\hbar}{E_3} + \tau_{\text{delay}}\right) \ll 1\,, \tag{56.172}$$

where $\lambda\!\!\!^{-} \sim k_3^{-1}$ is the reduced wavelength of 3, f_{13} is the scattering amplitude for 1–3 collisions and τ_{delay} is the time delay associated with 1–3 collisions.

Special Case: for nonresonant scattering with $\tau_{\text{delay}} = 0$

$$\frac{f_{13}}{\lambda\!\!\!^{-}}\frac{|\varepsilon_n|}{E_3} \ll 1\,, \tag{56.173}$$

which follows from (56.172) upon identifying the shift in the core energy ΔE_c with the binding energy $|\varepsilon|$.

Condition (56.173) is less restrictive than (56.171a) or (56.171b) since f_{13} can be either less than or greater than $\lambda\!\!\!^{-}$.

56.7.2 Classical Impulse Approximation

(A) Ionization. For electron impact on heavy particles [56.29], the cross section for ionization of a particle moving with velocity t by a projectile with velocity s is

$$Q(s,t) = \frac{1}{u^2}\frac{2s}{m_2}\int\limits_1^{s^2}\frac{\mathrm{d}z}{z^2}\left[\frac{A(z)}{z} + B(z)\right]\,, \tag{56.174}$$

where

$$\begin{aligned} A(z) = {}& \frac{1}{2st^3}\left[\frac{1}{3}\left(x_{02}^{3/2} - x_{01}^{3/2}\right)\right. \\ & \left. - 2\left(s^2+t^2\right)\left(x_{02}^{1/2} - x_{01}^{1/2}\right)\right] \\ & - \frac{\left(s^2-t^2\right)^2}{2ts^3}\left(x_{02}^{-1/2} - x_{01}^{-1/2}\right)\,, \end{aligned} \tag{56.175a}$$

$$\begin{aligned} B(z) = {}& \frac{1}{2m_2st^3}\left[(m_1+m_2)\left(s^2-t^2\right)\left(x_{02}^{-1/2} - x_{01}^{-1/2}\right)\right. \\ & \left. - (m_2-m_1)\left(x_{02}^{1/2} - x_{01}^{1/2}\right)\right]\,. \end{aligned} \tag{56.175b}$$

For electron impact, (56.175b) is evaluated at $m_1 = 1$. The remaining terms above are given by

$$s^2 = v_2^2/v_0^2\,,\quad t^2 = v_1^2/v_0^2 = E_1/u\,,$$
$$E_2 = m_2 v_2^2\,,$$
$$u = v_0^2 = \text{Ionization potential of target}\,,$$
$$x_{0i} = \left(s^2+t^2-2st\cos\theta_i\right)\,,\quad i = 1,2\,,$$

$$\cos\theta_i = \begin{cases} \kappa_0 \pm \kappa_1\,, & |\kappa_0 \pm \kappa_1| \le 1 \\ 1\,, & \kappa_0 \pm \kappa_1 > 1 \\ -1\,, & \kappa_0 \pm \kappa_1 < -1 \end{cases}\,,$$

$$\kappa_0 \pm \kappa_1 = -\frac{1}{2}\left(1 - \frac{m_1}{m_2}\right)\frac{z}{st} \pm \sqrt{\left(1+\frac{z}{t^2}\right)\left(1-\frac{z}{s^2}\right)}\,.$$

Equal Mass Case: ($m_1 = m_2$)

$$Q(s,t) = \begin{cases} \dfrac{4}{3s^2}\dfrac{1}{u^2}\dfrac{2(s^2-1)^{3/2}}{t}\,, & 1 \le s^2 \le t^2+1\,, \\ \dfrac{4}{3s^2}\dfrac{1}{u^2}\left(2t^2+3-\dfrac{3}{s^2-t^2}\right), & s^2 \ge t^2+1\,. \end{cases} \tag{56.176}$$

Integrating over the speed distribution (Sect. 56.4),

$$Q(s) = \frac{32}{\pi}\frac{1}{u^2}\int_0^\infty \frac{Q(s,t)t^2\,dt}{(t^2+1)^4}, \tag{56.177}$$

which is then numerically evaluated. For electrons, the integral can be done analytically with the result

$$\begin{aligned} Q(y) = {} & \frac{8}{3\pi y^2(y+1)^4} \\ & \times\Big[\left(5y^4+15y^3-3y^2-7y+6\right)(y-1)^{1/2} \\ & +\left(5y^5+17y^4+15y^3-25y^2+20y\right) \\ & \times\tan^{-1}(y-1)^{1/2} \\ & -24y^{3/2}\ln\left|\frac{\sqrt{y}+\sqrt{y-1}}{\sqrt{y}-\sqrt{y-1}}\right|\Big], \end{aligned} \tag{56.178}$$

with $y = s^2$.

Thomson's Result:

$$Q_T(y) = \frac{4}{y}\frac{1}{u^2}\left(1-\frac{1}{y}\right). \tag{56.179}$$

(B) Electron Loss Cross Section [56.18].

$$A(V) + B(u) \to A^+ + e^- + B(f), \tag{56.180}$$

where $B(f)$ denotes that the target B is left in any state (either bound or free) after the collision with the projectile A. The initial velocity of the projectile is V while the velocity of the Rydberg electron relative to the core is u, and the ionization potential of the target B is I.

$$\begin{aligned} \sigma_{\text{loss}} = {} & \frac{1}{3\pi\nu^2}\int_{\tau/4\nu}^\infty dx\,\sigma_T(x\bar{u})\left(\frac{8\nu x-1-(\nu-x)^2-2\tau}{\left[1+(\nu-x)^2\right]^3}\right. \\ & \left.+\frac{1}{\left[1+(\nu+x)^2-\tau\right]^2}\right), \end{aligned} \tag{56.181}$$

where $\nu = V/\bar{u}$, $\tau = I/\frac{1}{2}m_e\bar{u}^2$, $\bar{u} = \sqrt{2I/m_e}$, and σ_T is the total electron scattering cross section at speed $x\bar{u}$. The cross section (56.181) is valid only for particles being stripped (or lost from the projectile) which are not strongly bound. See [56.18, 30, 31] for details and numerous results.

(C) Capture Cross Section from Shell i [56.18].

$$\begin{aligned} & \sigma^i_{\text{capture}}(V) \\ & = \frac{2^{5/2}N_i\pi}{3V^7}\int_0^{r_i} dr \int_{\mathcal{C}(-1)}^{\mathcal{C}(+1)} d(\cos\eta')\,[P_i(r)]^2 \\ & \times\frac{\sqrt{1+y^2}\left[4\varepsilon^2-\left(\varepsilon^2-y^2\right)\left(1+\varepsilon^2+a^2-y^2\right)\right]}{r^{3/2}\varepsilon^{9/2}\left(1+a^2\right)^3\sqrt{y^2-a^2}}, \end{aligned} \tag{56.182}$$

where $\mathcal{C}$ denotes that the integration range $[-1,+1]$ is restricted such that the integrand is real and positive and that $|1-\varepsilon| < \sqrt{y^2-a^2}$. The dimensionless variables a and y above are defined as

$$y^2 = \frac{2}{m_e}|V(R)|V^2, \quad a^2 = \frac{2}{m_e}I_i V^2, \tag{56.183}$$

and with $P_i(r)/r$ representing the Hartree–Fock–Slater radial wave function for shell i, with normalization

$$\int_0^{r_i} [P_i(r)]^2\,dr = 1. \tag{56.184}$$

The ionization potential and number of electrons in shell i are denoted above, respectively, by I_i and N_i.

(D) Total Capture Cross Section [56.18].

$$\sigma^{\text{total}}_{\text{capture}}(V) = \sum_i \sigma^i_{\text{capture}}. \tag{56.185}$$

(E) Universal Capture Cross Section [56.18]. A universal curve independent of projectile mass M and charge Z is obtained from the above expressions for the capture cross section by plotting the scaled cross section

$$\tilde{\sigma}^{\text{total}}_{\text{capture}} = \frac{E^{11/4}}{M^{11/4}Z^{7/2}\lambda^{3/4}}\sigma^{\text{total}}_{\text{capture}} \tag{56.186}$$

versus the scaled energy

$$\widetilde{E} = \frac{m_e}{M}\frac{E}{I}, \tag{56.187}$$

where m_e is the mass of the particle captured, which is usually taken to be a single electron, and I is the ionization potential of the target. The term λ in (56.186) is the coupling constant in the target potential, $V(R) = m_e\lambda/R^2$, which the electron being captured experiences during the collision. See Fig. 11 of [56.18] for details.

56.7.3 Semiquantal Impulse Approximation

Basic Expression [56.27, 32, 33].

$$\frac{d\sigma}{d\varepsilon\, dP\, dk_1\, dk\, d\phi_1} = \frac{k_1'^2}{J_{55}}\frac{k_3'}{k_3}\left(\frac{M_3}{M_{13}}\right)^2 \left|g_i(\boldsymbol{k}_1)\right|^2 \times \left|f_{13}(\boldsymbol{k},\boldsymbol{k}')\right|^2 . \quad (56.188)$$

J_{55} is the 5-dimensional Jacobian of the transformation

$$(P,\varepsilon,k_1,k,\phi_1) \to \left(\hat{\boldsymbol{k}}_3', \boldsymbol{k}_1'\right), \quad (56.189a)$$

$$J_{55} = \frac{\partial\,(P,\varepsilon,k_1,k,\phi_1)}{\partial\left(\cos\theta_3'\phi_3', k_1', \cos\theta_1', \phi_1'\right)} . \quad (56.189b)$$

Expression for Elemental Cross Section [56.27]. In the $(P,\varepsilon,k_1,k,\phi_1)$ representation,

$$d\sigma = \frac{d\varepsilon\, dP}{M_{13}^2 v_3^2}\left[\frac{|g_i(\boldsymbol{k}_1)|^2 k_1^2\, dk_1\, d\phi_1}{v_1}\right] \times \frac{|f_{13}(\boldsymbol{k},\boldsymbol{k}')|^2\, dg^2}{\sqrt{\left(g_+^2-g^2\right)\left(g^2-g_-^2\right)}} , \quad (56.190)$$

where $g_\pm^2 = \frac{1}{2}B \pm \sqrt{\frac{1}{4}B^2 - C}$, and

$$\begin{aligned} B &= B(\varepsilon, P, v_1; v_3) \\ &= \frac{a}{(1+a)^2}\frac{P^2}{M_{13}^2} + \left(v_1^2 + v_1'^2 + v_3^2 + v_3'^2 + \frac{2\Delta_3}{M_{13}}\right) \\ &\quad - \frac{4\varepsilon(\varepsilon+\Delta_3)}{P^2} , \\ C &= C(\varepsilon, P, v_1; v_3) \\ &= \frac{v_1^2 + a v_3^2}{1+a}\frac{P^2}{M_{13}^2} + \left(v_1^2 - v_3^2\right)\left(v_1'^2 - v_3'^2\right) \\ &\quad + \frac{2\Delta_3}{M_{13}}\left(v_1^2+v_3^2\right) + \frac{4\Delta_3}{P^2}\left[v_1^2(\varepsilon+\Delta_3) - \varepsilon v_3^2\right] , \\ a &= \frac{M_2 M_3}{M_1(M_1+M_2+M_3)} , \\ \widetilde{M}_1 &= M_1(1+M_1/M_2) , \\ v_1'^2 &= v_1^2 + \frac{2\varepsilon}{\widetilde{M}_1} , \quad v_3'^2 = v_3^2 - \frac{2(\varepsilon+\Delta_3)}{M_{AB}} , \end{aligned}$$

and Δ_3 is the change in internal energy of particle 3, while ε denotes the energy change in the target $1-2$.

Hydrogenic Systems $g_i(\boldsymbol{k}_1) = g_{n'}(k_1)Y_{\ell m}(\theta_1,\phi_1)$. The $g_{n\ell}$ are the hydrogenic wave functions in momentum space. See Chapt. 9 for details on hydrogenic wave functions.

Elemental Cross Sections (m-Averaged and ϕ_1-Integrated).

$$d\sigma = \frac{d\varepsilon\, dP}{M_{13}^2 v_3^2}\frac{W_{n\ell}(v_1)\, dv_1}{2v_1}\frac{|f_{13}(P,g)|^2\, dg^2}{\sqrt{\left(g_+^2-g^2\right)\left(g^2-g_-^2\right)}} , \quad (56.191)$$

where the speed distribution $W_{n\ell}$ is given by (56.57).

Two Representations for 1–3 Scattering Amplitude [56.27]. (i) $f_{13} = f_{13}(P,g)$ is a function of momentum transferred and relative speed. Then

$$\sigma(v_3) = \frac{1}{M_{13}^2 v_3^2}\int_{\varepsilon_1}^{\varepsilon_2} d\varepsilon \int_{v_{10}}^{\infty}\frac{W_{n\ell}(v_1)\, dv_1}{v_1}\int_{P^-}^{P^+} dP \times \int_{g_-}^{g_+}\frac{|f_{13}(P,g)|^2\, dg^2}{\sqrt{\left(g_+^2-g^2\right)\left(g^2-g_-^2\right)}} , \quad (56.192)$$

where $v_{10}^2(\varepsilon) = \max\left[0, (2\varepsilon/M)\right]$, and the limits to the P integral are

$$\begin{aligned} P^+ &= P^+(\varepsilon, v_1; v_3) \\ &= \min\left[M(v_1'+v_1), M_{AB}(v_3'+v_3)\right] , \end{aligned} \quad (56.193a)$$

$$\begin{aligned} P^- &= P^-(\varepsilon, v_1; v_3) \\ &= \max\left[M\left|v_1'-v_1\right|, M_{AB}\left|v_3'-v_3\right|\right] , \end{aligned} \quad (56.193b)$$

and unless $P^+ > P^-$, the P integral is zero.
(ii) $f_{13} = f_{13}(g,\psi)$ is a function of relative speed and scattering angle. Then

$$\sigma(v_3) = \frac{1}{v_3^2}\int_{\varepsilon_1}^{\varepsilon_2} d\varepsilon \int_{v_{10}}^{\infty}\frac{W_{n\ell}(v_1)\, dv_1}{v_1}\int_{g_-}^{g_+}\frac{g\, dg}{S(v_1,g;v_3)} \times \int_{\psi^-}^{\psi^+}\frac{|f_{13}(g,\psi)|^2\, d(\cos\psi)}{\sqrt{(\cos\psi^+ - \cos\psi)(\cos\psi - \cos\psi^-)}} , \quad (56.194)$$

where

$$S(v_1,g;v_3) = \frac{M_{13}}{1+a}\left[(1+a)\left(v_1^2 + a v_3^2\right) - a g^2\right]^{1/2} .$$

Scattering angle ψ-limits,

$$\cos\psi^\pm = \cos\psi^\pm(\varepsilon, v_1, g; v_3) \quad (56.195)$$

$$= \frac{g}{g'}\frac{1}{\alpha^2+\beta^2}\left\{\alpha(\alpha+\tilde{\varepsilon}) \pm \beta\left[\omega^2\left(\alpha^2+\beta^2\right) - (\alpha+\tilde{\varepsilon})^2\right]^{1/2}\right\} , \quad (56.196)$$

where

$$\alpha = \alpha(v_1, g; v_3) = \frac{1}{2}M_{13}\left[v_1^2 - v_3^2 + \left(\frac{1-a}{1+a}\right)g^2\right],$$

$$\beta = \beta(v_1, g; v_3) = \frac{1}{2}M_{13}\left[\left(2v_1^2 + 2v_3^2 - g^2\right)g^2 - \left(v_1^2 - v_3^2\right)^2\right]^{1/2},$$

$$\omega = g'/g\,, \quad \tilde{\varepsilon} = \varepsilon + \frac{a}{1+a}\Delta_3\,.$$

Special Case: $f_{13} = f_{13}(P)$.

$$\sigma(v_3) = \frac{\pi}{M_{13}^2 v_3^2}\int_{\varepsilon_1}^{\varepsilon_2} d\varepsilon \int_{v_{10}}^{\infty} \frac{W_{n\ell}(v_1)\,dv_1}{v_1} \times \int_{P^-}^{P^+} |f_{13}(P)|^2\,dP\,. \quad (56.197)$$

56.8 Binary Encounter Approximation

The basic assumption of the binary encounter approximation is that an excitation or ionization process is caused solely by the interaction of the incoming charged or neutral projectile with the Rydberg electron bound to its parent ion. If, for example, the cross section depends only on the momentum transfer P to the Rydberg electron (as in the Born approximation), then the total cross section is obtained by integrating σ_P over the momentum distribution of the Rydberg electron. The basic cross sections required are given in the following section. For further details see [56.34] and references therein.

56.8.1 Differential Cross Sections

Cross Section per Unit Momentum Transfer

Let the masses, velocities and charges of the particles be $(m_1, \boldsymbol{v}_1, Z_1, e)$ and $(m_2, \boldsymbol{v}_2, Z_2, e)$, with $\boldsymbol{v} = |\boldsymbol{v}_1 - \boldsymbol{v}_2|$ denoting the relative velocity and quantities *after* the collision are denoted by primes. Then for distinguishable particles,

$$\sigma_P = \frac{8\pi Z_1^2 Z_2^2 e^4 P}{v^2}\left|\frac{\exp(i\eta_P)}{P^2}\right|^2, \quad (56.198)$$

where the phase shift η_P is

$$\eta_P = -2\gamma\ln(P/2\mu v) + 2\eta_0 + \pi\,, \quad (56.199)$$

and with

$$\mu = \frac{m_1 m_2}{m_1 + m_2}\,, \quad \gamma = \frac{Z_1 Z_2 e^2}{\hbar v}\,, \quad e^{2i\eta_0} = \frac{\Gamma(1+i\gamma)}{\Gamma(1-i\gamma)}\,. \quad (56.200)$$

For identical particles,

$$\sigma_P^{\pm} = \frac{8\pi Z_1^2 Z_2^2 e^4 P}{v^2}\left|\frac{e^{i\eta_P}}{P^2} \pm \frac{e^{i\eta_S}}{S^2}\right|^2, \quad (56.201)$$

where η_P is given by (56.199) and η_S is

$$\eta_S = -2\gamma\ln(S/2\mu v) + 2\eta_0 + \pi\,, \quad (56.202)$$

while η_0 is given by (56.200). The momenta $\boldsymbol{P}$ and $\boldsymbol{S}$ transferred by direct and exchange collisions, respectively, are given by

$$\boldsymbol{P} = m_1(\boldsymbol{v}_1 - \boldsymbol{v}_1') = m_2(\boldsymbol{v}_1 - \boldsymbol{v}_2')\,, \quad (56.203a)$$

$$\boldsymbol{S} = m_1(\boldsymbol{v}_1 - \boldsymbol{v}_2') = m_2(\boldsymbol{v}_1' - \boldsymbol{v}_2)\,. \quad (56.203b)$$

The collision rates (in cm^3/s) are

$$\hat{\alpha}_P = v_1\sigma_P\,, \quad \hat{\alpha}_P^{\pm} = v_1\sigma_P^{\pm}\,. \quad (56.204)$$

Cross Section per Unit Momentum Transferred per Unit Steradian

Differential relationships:

$$\alpha_{E,P} = \frac{d^2\alpha}{dP\,dE} = \frac{d^2\alpha}{dP\,d\varphi}\frac{d\varphi}{dE} = \alpha_{P,\varphi}\frac{d\varphi}{dE}\,. \quad (56.205)$$

For distinguishable particles,

$$\alpha_P = 2\pi v_1\sigma_{P,\varphi} = 2\pi\alpha_{P,\varphi}\,, \quad (56.206a)$$

$$\alpha_{P,\varphi} = \frac{4Z_1^2 Z_2^2 e^4 P}{v}\left|\frac{e^{i\eta_P}}{P^2}\right|^2, \quad (56.206b)$$

$$\alpha_{E,P} = v_1\sigma_{E,P} = \frac{8Z_1^2 Z_2^2 e^4}{v_1 v_2\sqrt{X}}\left|\frac{e^{i\eta_P}}{P^2}\right|^2. \quad (56.206c)$$

For identical particles,

$$\alpha_{P,\varphi}^{\pm} = \frac{4Z_1^2 Z_2^2 e^4 P}{v}\left|\frac{e^{i\eta_P}}{P^2} \pm \frac{e^{i\eta_S}}{S^2}\right|^2, \quad (56.207a)$$

$$\alpha_{E,P}^{\pm} = v_1\sigma_{E,P}^{\pm} = \frac{8Z_1^2 Z_2^2 e^4}{v_1 v_2\sqrt{X}}\left|\frac{e^{i\eta_P}}{P^2} \pm \frac{e^{i\eta_S}}{S^2}\right|^2, \quad (56.207b)$$

where

$$X = -\cos^2\phi + 2(\hat{v}_1\cdot\hat{P})(\hat{v}_2\cdot\hat{P})\cos\phi + 1 - (\hat{v}_1\cdot\hat{P})^2 - (\hat{v}_2\cdot\hat{P})^2 \quad (56.208a)$$

$$= (\cos\phi_{\min} - \cos\phi)(\cos\phi - \cos\phi_{\max}) \quad (56.208b)$$

$$= \left(\frac{v}{v_1 v_2 P}\right)^2 (E_{\max} - E)(E - E_{\min}), \quad (56.208c)$$

with ϕ being the angle between the velocity vectors $\boldsymbol{v}_1$ and $\boldsymbol{v}_2$.

For the special case of electron impact, $M_2 = m_e$, $Z_2 = -1$, and

$$\sigma_{E,P}(\phi) = \frac{8Z_1^2 e^4}{v_1^2 v_2 P^4 \sqrt{X}}, \quad (56.209)$$

$$\sigma^{\pm}_{E,P}(\phi) = 8e^4 v_1^2 v_2 \sqrt{X} \times \left(\frac{1}{P^4} + \frac{1}{S^4} \frac{2\cos(\eta_P - \eta_S)}{P^2 S^2}\right), \quad (56.210)$$

where $\eta_P - \eta_S = -2\gamma \ln(P/S) = (2e^2/\hbar v)\ln(S/P)$, and X is given by (56.208b).

Integrated Cross Sections

For incident heavy particles:

$$\sigma_{E,P} = \int_0^\pi \sigma_{E,P}(\phi)\frac{1}{2}\sin\phi\, d\phi = \frac{4\pi Z_1^2 e^4}{v_1^2 v_2 P^4}. \quad (56.211)$$

For incident electrons:

$$\sigma^{\pm}_{E,P} = \int_0^\pi \sigma^{\pm}_{E,P}(\phi)\frac{1}{2}\sin\phi\, d\phi \quad (56.212a)$$

$$= \frac{4\pi e^4}{v_1^2 v_2}\left(\frac{1}{P^4} + \frac{v_1^2 + v_2^2 - P^2/2m_e^2 - 2E^2/P^2}{m_e^4 |v_1^2 - v_2^2 - 2E/m_e|^3} \pm \frac{2\Phi}{m_e^2 P^2 |v_1^2 - v_2^2 - 2E/m_e|}\right), \quad (56.212b)$$

where Φ can be approximated [56.35] by

$$\Phi \sim \cos\left(\left|\frac{R_\infty}{E_3 - E_2}\right|^{1/2} \ln\left|\frac{E}{E_3 - E_2 - E}\right|\right). \quad (56.213)$$

and E_3 is defined in [56.35].

Cross Sections per Unit Energy

For incident heavy particles (three cases):

$$\sigma_E = \frac{2\pi Z_1^2 e^4}{m_e v_1^2}\left(\frac{1}{E^2} + \frac{2m_e v_2^2}{3E^2}\right), \quad (56.214)$$

which is valid for $2v_1 \geq v_2 + v_2'$, $E \leq 2m_e v_1(v_1 - v_2)$, or

$$\sigma_E = \frac{\pi Z_1^2 e^4}{3v_1^2 v_2 E^3}\left(4v_1^3 - \frac{1}{2}(v_2' - v_2)^2\right), \quad (56.215)$$

which is valid for $v_2' - v_2 \leq 2v_1 \leq v_2' + v_2$, $2m_e v_1 (v_1 - v_2) \leq E \leq 2m_e v_1(v_1 + v_2)$, or otherwise, $\sigma_E = 0$ for $E \geq m_e v_1(v_2' + v_2)$.

For incident electrons (two cases):

$$\sigma^{\pm}_E = \frac{2\pi e^4}{m_e v_1^2}\left(\frac{1}{E^2} + \frac{2m_e v_2^2}{3E^3} + \frac{1}{D^2} + \frac{2m_e v_2^2}{3D^3} \pm \frac{2\Phi}{ED}\right), \quad (56.216)$$

which is valid for $m_e(v_2 - v_2') \leq m_e(v_1' - v_1)$, $m_e(v_2' + v_2) \leq m_e(v_1 + v_1')$, $D \geq 0$, or

$$\sigma_E = \frac{2\pi e^4}{m_e v_1^2}\left(\frac{1}{E^2} + \frac{2m_e v_1'^2}{3E^3} + \frac{1}{D^2} + \frac{2m_e v_1'^2}{3|D|^3} \pm \frac{2\Phi}{E|D|}\right)\frac{v_1'}{v_1}, \quad (56.217)$$

which is valid for $m_e(v_2' - v_2) \leq m(v_1 - v_1')$, $m_e(v_1 + v_1') \leq m_e(v_2' + v_2)$, $D \leq 0$. In the expressions above, the exchange energy D transferred during the collision is

$$D = \frac{1}{2}m_e v_1'^2 - \frac{1}{2}m_e v_2'^2 = \frac{1}{2}m_e v_1^2 - \frac{1}{2}m_e v_2^2 - E. \quad (56.218)$$

56.8.2 Integral Cross Sections

$$e^-(T) + A(E_2) \to e^-(E) + A^+ + e^-, \quad (56.219)$$

where T is the initial kinetic energy of the projectile electron, while the Rydberg electron, initially bound in potential U_i to the core A^+, has kinetic energy E_2. The cross section per unit energy E is denoted below by σ_E. See the review by *Vriens* [56.34] for details.

For electron impact, there are two collision models: the *unsymmetrical collision model* of Thomson and Gryzinski assumes that the incident electron has zero potential energy, and the *symmetrical collision model* of Thomas and Burgess assumes that the incident electron is accelerated initially by the target (and thereby gains kinetic energy) while losing an equal amount of potential energy.

Unsymmetrical model (two cases):

$$\sigma_E = \frac{\pi e^4}{T}\left(\frac{1}{E^2} + \frac{4E_2}{3E^3} + \frac{1}{D^2} + \frac{4E_2}{3D^3} - \frac{\Phi}{ED}\right), \quad (56.220)$$

which is valid for $D = T - E_2 - E \geq 0$ or,

$$\sigma_E = \frac{\pi e^4}{T}\left(\frac{1}{E^2} + \frac{4T'}{3E^3} + \frac{1}{D^2} + \frac{4T'}{3|D|^3} - \frac{\Phi}{E|D|}\right) \times \left(\frac{T'}{E_2}\right)^{1/2} \quad (56.221)$$

which is valid for $D \leq 0$ and $T \geq E$; and where $T' \equiv T - E$.

Symmetrical model (two cases):

$$\sigma_E = \frac{\pi e^4}{T_i}\left(\frac{1}{E^2} + \frac{4E_2}{3E^3} + \frac{1}{X_i^2} + \frac{4E_2}{3X_i^3} - \frac{\Phi}{EX_i}\right), \quad (56.222)$$

which is valid for $X_i \equiv T + U_i - E \geq 0$, with $T_i \equiv T + U_i + E_2$, and

$$\sigma_E = \frac{\pi e^4}{T_i}\left(\frac{1}{E^2} + \frac{4T_i'}{3E^3} + \frac{1}{X_i^2} + \frac{4T_i'}{3|X_i|^3} - \frac{\Phi}{E|X_i|}\right) \times \left(\frac{T_i'}{E_2}\right)^{1/2} \quad (56.223)$$

which is valid for $0 \leq T_i' \leq E_2, T \geq 0$, with $T_i' \equiv T_i - E$, and where Φ is given by (56.213).

For incident heavy particles, the unsymmetrical model (56.220) should be used.

Single Particle Ionization

The total ionization cross section per atomic electron for incident heavy particles is

$$Q_i = \frac{2\pi Z_1^2 e^4}{m_e v_1^2}\left(\frac{1}{U_i} + \frac{m_e v_2^2}{3U_i^2} - \frac{1}{2m_e\left(v_1^2 - v_2^2\right)}\right), \quad (56.224)$$

which is valid for $U_i \leq 2m_e v_1(v_1 - v_2)$, or

$$Q_i = \frac{\pi Z_1^2 e^4}{m_e v_1^2}\left\{\frac{1}{2m_e v_2(v_1 + v_2)} + \frac{1}{U_i} + \frac{m_e}{3v_2U_i^2}\left[2v_1^3 + v_2^3 - \left(2U_i/m_e + v_2^2\right)^{3/2}\right]\right\}, \quad (56.225)$$

which is valid for $2m_e v_1(v_1 - v_2) \leq U_i \leq 2m_e v_1 (v_1 + v_2)$, or otherwise $Q_i = 0$ for $U_i \geq 2m_e v_1(v_1 + v_2)$.

For electron impact,

$$Q_i = \frac{1}{2}\left(Q_i^{\mathrm{dir}} + Q_i^{\mathrm{ex}} + Q_i^{\mathrm{int}}\right). \quad (56.226)$$

In the unsymmetrical model, Q_i^{ex} diverges, hence the exchange and interference terms above are omitted in the unsymmetrical model for electrons to obtain

$$Q_i^{\mathrm{dir}} = \frac{\pi e^4}{T}\left(\frac{1}{U_i} + \frac{2E_2}{3U_i^2} - \frac{1}{T - E_2}\right), \quad (56.227)$$

which is valid for $T \geq E_2 + U_i$, or

$$Q_i^{\mathrm{dir}} = \frac{2\pi e^4}{3T}\frac{(T - U_i)^{3/2}}{U_i^2\sqrt{E_2}}, \quad (56.228)$$

which is valid for $U_i \leq T \leq E_2 + U_i$.

In the symmetrical model,

$$Q_i^{\mathrm{dir}} = Q_i^{\mathrm{ex}} = \frac{\pi e^4}{T_i}\left[\frac{1}{U_i} - \frac{1}{T} + \frac{2}{3}\left(\frac{E_2}{U_i^2} - \frac{E_2}{T^2}\right)\right], \quad (56.229)$$

$$Q_i^{\mathrm{int}} = -\frac{\pi e^4}{T_i}\left(\frac{2\Phi'}{T + U_i}\ln\frac{T}{U_i}\right), \quad (56.230)$$

where Φ' can be approximated by [56.35]

$$\Phi' = \cos\left[\left(\frac{R_\infty}{E_1 + U_i}\right)^{1/2}\ln\frac{E_1}{U_i}\right]. \quad (56.231)$$

and E_1 is defined in [56.35].

The sum of (56.229) and (56.230) yields

$$Q_i = \frac{\pi e^4}{T_i}\left[\frac{1}{U_i} - \frac{1}{T} + \frac{2}{3}\left(\frac{E_2}{U_i} - \frac{E_2}{T^2}\right) - \frac{\Phi'}{T + U_i}\ln\frac{T}{U_i}\right], \quad (56.232)$$

which is also obtained by integrating the expression (56.223) for σ_E,

$$Q_i = \int_{U_i}^{\frac{1}{2}(T+U_i)} \sigma_E \, dE. \quad (56.233)$$

Ionization Rate Coefficients. For heavy particle impact [56.29],

$$\langle Q \rangle = \frac{a_0^2}{\kappa^2}\left\{\frac{128}{9}\left(\kappa^3 b^3 - b^{3/2}\right) + \frac{1}{3}\lambda b\left(35 - \frac{58}{3}b \frac{8}{3}b^2\right) + \frac{2}{3}\kappa ab\left[\left(5 - 4\kappa^2\right)\left(3a^2 + \frac{3}{2}ab + b^2\right) - \lambda\kappa\left(\frac{15}{2} + 9a + 5b\right)\right] - 16\kappa a^4 \ln\left(4\kappa^2 1\right) + \theta\left[\frac{35}{6} - \kappa^2 a\left(\frac{5}{2} + 3a + 4a^2 + 8a^3\right)\right]\right\}, \quad (56.234)$$

where

$$\kappa = v_1/v_0\,,$$
$$\lambda = \kappa - (4\kappa)^{-1}\,,$$
$$\theta = \pi + 2\tan^{-1}\lambda\,,$$
$$a = (1+\kappa^2)^{-1}\,,$$
$$b = (1+\lambda^2)^{-1}\,.$$

$$\langle \sigma_{\mathrm{E,P}}\,\mathrm{d}P\,\mathrm{d}E\rangle = \frac{64e^4 v_0^5}{3v_1^2 P^4} \times \left(\left| \frac{E}{P} - \frac{P}{2m_\mathrm{e}} \right|^2 v_0^2 \right)^{-3} \mathrm{d}P\,\mathrm{d}E\,, \tag{56.235}$$

where $\frac{1}{2}m_\mathrm{e}v_0^2$ is the ionization energy of H(1s).

Scaling Laws. Given the binary encounter cross section for ionization by protons of energy E_1 of an atom with binding energy u_a, the cross section for ionization of an atom with different binding energy u_b and scaled proton energy E_1' can be determined to be [56.18]

$$\sigma_{\mathrm{ion}}\left(E_1', u_b\right) = \left(\frac{u_a^2}{u_b^2}\right)\sigma_{\mathrm{ion}}(E_1, u_a)\,, \tag{56.236}$$

$$E_1' = (u_b/u_a)E_1\,, \tag{56.237}$$

where $\sigma_{\mathrm{ion}}(E, u)$ is the ionization cross section for removal of a single electron from an atom with binding energy u by impact with a proton with initial energy E.

Double Ionization. See [56.36] for binary encounter cross section formulae for the direct double ionization of two-electron atoms by electron impact.

Excitation. Excitation is generally less violent than ionization and hence binary encounter theory is less applicable. Binary encounter theory can be applied to exchange excitation transitions, e.g., $\mathrm{e}^- + \mathrm{He}(n^1L) \to \mathrm{e}^- + \mathrm{He}(n'^3L)$, with the restriction of large incident electron velocities. The cross section is

$$Q_\mathrm{e}^{\mathrm{ex}} = \int_{U_n}^{U_{n+1}} \sigma_{\mathrm{E,ex}}\,\mathrm{d}E = \frac{\pi e^4}{T_i}\left[\frac{1}{\mathcal{T}_{n+1}} - \frac{1}{\mathcal{T}_n} + \frac{2}{3}\left(\frac{E_2}{\mathcal{T}_{n+1}^2} - \frac{E_2}{\mathcal{T}_n^2}\right)\right], \tag{56.238}$$

valid for $T \geq U_{n+1}$, with $\mathcal{T}_n \equiv T + U_i - U_n$ and $\mathcal{T}_{n+1} \equiv T + U_i + U_{n+1}$, or

$$Q_\mathrm{e}^{\mathrm{ex}} = \int_{U_n}^{T} \sigma_{\mathrm{E,ex}}\,\mathrm{d}E \tag{56.239}$$
$$= \frac{\pi e^4}{T_i}\left[\frac{1}{U_i} - \frac{1}{\mathcal{T}_n} + \frac{2}{3}\left(\frac{E_2}{U_i^2} - \frac{E_2}{\mathcal{T}_n^2}\right)\right],$$

valid for $U_n \leq T \leq U_{n+1}$. U_n and U_{n+1} denote the excitation energies for levels n and $n+1$, respectively.

56.8.3 Classical Ionization Cross Section

Applying the classical energy-change cross section result (56.69) of *Gerjuoy* [56.16] to the case of electron-impact ionization yields the four cases [56.17]

$$\sigma_{\mathrm{ion}}(v_1, v_2) \sim \int_{\Delta E_\ell}^{\Delta E_u} \sigma_{\Delta E}^{\mathrm{eff}}(v_1, v_2; m_1/m_2)\, d(\Delta E) = \frac{\pi\left(Z_1Z_2e^2\right)^2}{3v_1^2 v_2}\left(\frac{-2v_2^3}{(\Delta E)^2} - \frac{6v_2}{m_2\Delta E}\right), \tag{56.240}$$

which is valid for $0 < \Delta E < b$, or

$$\sigma_{\mathrm{ion}}(v_1, v_2) = \frac{\pi\left(Z_1Z_2e^2\right)^2}{3v_1^2 v_2} \times \left(\frac{4\left(v_1 - 2v_1'\right)}{m_1^2\left(v_1 - v_1'\right)^2} + \frac{4\left(v_2 - 2v_2'\right)}{m_2^2\left(v_2 - v_2'\right)^2}\right), \tag{56.241}$$

which is valid for $b < \Delta E < a$, or

$$\sigma_{\mathrm{ion}}(v_1, v_2) = \frac{\pi(Z_1Z_2e^2)^2}{3v_1^2 v_2}\left(\frac{-2v_1'^3}{(\Delta E)^2}\right), \tag{56.242}$$

which is valid for $\Delta E > a$, $2m_2v_2 > |m_1 - m_2|v_1$, or otherwise is zero for $\Delta E > a$, $2m_2v_2 < |m_1 - m_2|v_1$.

The limits $\Delta E_{\ell,u}$ to the ΔE integration in each of the four cases is indicated in the appropriate validity conditions. The constants a and b above are given by

$$a = \frac{4m_1m_2}{(m_1+m_2)^2}\left[E_1 - E_2 + \frac{1}{2}v_1v_2(m_1 - m_2)\right],$$
$$b = \frac{4m_1m_2}{(m_1+m_2)^2}\left[E_1 - E_2 - \frac{1}{2}v_1v_2(m_1 - m_2)\right].$$

The expressions above for $\sigma_{\mathrm{ion}}(v_1, v_2)$ must be averaged over the speed distribution of v_2 before comparison with experiment. See [56.17] for explicit formulae for the case of a delta function speed distribution.

56.8.4 Classical Charge Transfer Cross Section

Applying the classical energy-change cross section result (56.69) of *Gerjuoy* [56.16] to the case of charge-transfer yields the four cases [56.17]

$$\sigma_{CX}(v_1, v_2) \sim \int_{\Delta E_\ell}^{\Delta E_u} \sigma_{\Delta E}^{(\text{eff})}(\boldsymbol{v}_1, \boldsymbol{v}_2)\,\mathrm{d}\Delta E = \frac{\pi e^4}{3v_1^2 v_2}\left(-\frac{2v_2^3}{(\Delta E)^2} - \frac{6v_2/m_2}{\Delta E}\right), \tag{56.243}$$

which is valid for $0 < \Delta E < b$, or

$$\sigma_{CX}(v_1, v_2) = \frac{\pi e^4}{3v_1^2 v_2}\left(3\frac{v_1/m_1 - v_2/m_2}{\Delta E} + \frac{\left(v_2'^3 - v_2^3\right) - \left(v_1'^3 + v_1^3\right)}{(\Delta E)^2}\right), \tag{56.244}$$

which is valid for $b < \Delta E < a$, or

$$\sigma_{CX}(v_1, v_2) = \frac{\pi e^4}{3v_1^2 v_2}\left(-\frac{2v_1'^2}{(\Delta E)^2}\right), \tag{56.245}$$

which is valid for $\Delta E > a$, $m_e v_2 > (m_1 - m_e)v_1$, or otherwise $\sigma_{CX} = 0$ when $\Delta E > a$, $m_e v_2 < (m_1 - m_e)v_1$. The above expressions for $\sigma_{CX}(v_1, v_2)$ must be averaged over the speed distribution $W(v_2)$ before comparison with experiment. See [56.17] for details. The constants a and b above are as defined in Sect. 56.8.3, and the limits $\Delta E_{\ell,u}$ are given by

$$\Delta E_\ell = \frac{1}{2}m_e v_1^2 + U_A - U_B, \tag{56.246a}$$

$$\Delta E_u = \frac{1}{2}m_e v_1^2 + U_A + U_B, \tag{56.246b}$$

$$v_2 = \sqrt{2U_A/m_e}, \tag{56.246c}$$

where $U_{A,B}$ are the binding energies of atoms A and B. The expressions above for σ_{CX} diverge for some $v_1 > 0$ if $U_A < U_B$. If $U_A = U_B$ then σ_{CX} diverges at $v_1 = 0$. To avoid the divergence, employ Gerjuoy's modification, $\Delta E_\ell = \frac{1}{2}m_e v_1^2 + U_A$.

56.9 Born Approximation

See reviews [56.37,38], as well as any standard textbook on scattering theory, for background details on the Born approximation, and [56.39–42] for extensive tables of Born cross sections.

56.9.1 Form Factors

The basic formulation of the first Born approximation for high energy heavy particle scattering is discussed in Sect. 53.1. For the general atom–atom or ion–atom scattering process

$$A(i) + B(i') \to A(f) + B(f'), \tag{56.247}$$

with nuclear charges Z_A and Z_B respectively, let $\hbar\boldsymbol{K}_i$ and $\hbar\boldsymbol{K}_f$ be the initial and final momenta of the projectile A, and $\hbar\boldsymbol{q} = \hbar\boldsymbol{K}_f - \hbar\boldsymbol{K}_i$ be the momentum transferred to the target. Then (53.6) can be written in the generalized form

$$\sigma_{if}^{i'f'} = \frac{8\pi a_0^2}{s^2}\int_{t_-}^{t_+}\frac{\mathrm{d}t}{t^3}\left|Z_A\delta_{if} - F_{if}^A(t)\right|^2 \times\left|Z_B\delta_{i'f'} - F_{i'f'}^B(t)\right|^2, \tag{56.248}$$

where the momentum transfer is $t = qa_0$, and $s = v/v_B$ is the initial relative velocity in units of v_B. The form factors are

$$F_{if}^A(t) = \left\langle \Phi_f^A \,\middle|\, \sum_{k=1}^{N_A}\exp(\mathrm{i}\boldsymbol{t}\cdot\boldsymbol{r}_a/a_0)\,\middle|\, \Phi_i^A\right\rangle, \tag{56.249}$$

where N_A is the number of electrons associated with atom A, and similarly for $F_{if}^B(t)$. The limits of integration are $t_\pm = |K_f \pm K_i|a_0$. For heavy particle collisions, $t_+ \sim \infty$ and

$$t_- = (K_i - K_f)a_0 \simeq \frac{\Delta E_{if}}{2s}\left(1 + \frac{m_e\Delta E_{if}}{4Ms^2}\right), \tag{56.250}$$

where $M = M_A M_B/(M_A + M_B)$.

Limiting Cases. As discussed in Sect. 53.1, for the case $i = f$, $F_{if}^A(t) \to N_A$ as $t \to 0$, so that $Z_A - F_{if}^A(t) \to Z_A - N_A$. For the case $i \neq f$, $F_{if}^A(t) \to 0$ as $t \to 0$ and $t \to \infty$.

56.9.2 Hydrogenic Form Factors

Bound–Bound Transitions. In terms of $\tau = t/Z$,

$$|F_{1s,1s}| = \frac{16}{\left(4+\tau^2\right)^2}, \tag{56.251a}$$

$$|F_{1s,2s}| = 2^{17/2}\frac{\tau^2}{(4\tau^2+9)^3}\,, \quad (56.251b)$$

$$|F_{1s,2p}| = 2^{15/2}\frac{3\tau}{(4\tau^2+9)^3}\,, \quad (56.251c)$$

$$|F_{1s,3s}| = 2^4 3^{7/2}\frac{(27\tau^2+16)\tau^2}{(9\tau^2+16)^4}\,, \quad (56.251d)$$

$$|F_{1s,3p}| = 2^{11/2} 3^3\frac{(27\tau^2+16)\tau}{(9\tau^2+16)^4}\,, \quad (56.251e)$$

$$|F_{1s,3d}| = 2^{17/2} 3^{7/2}\frac{\tau^2}{(9\tau^2+16)^4}\,. \quad (56.251f)$$

Bound–Continuum Transitions. In terms of the scaled wave vector $\kappa = \boldsymbol{k}a_0/Z$ for the ejected electron,

$$|F_{1s,\kappa}|^2 = \frac{2^8\kappa\tau^2\left(1+3\tau^2+\kappa^2\right)\exp\left(-2\theta/\kappa\right)}{3\left[1+(\tau-\kappa)^2\right]^3\left[1+(\tau+\kappa)^2\right]^3\left(1-\mathrm{e}^{-2\pi/\kappa}\right)} \quad (56.252)$$

where $\theta = \tan^{-1}\left[2\kappa/\left(1+\tau^2-\kappa^2\right)\right]$. Expressions for the bound–continuum Form Factors for the L-shell $(2\ell \to \kappa)$ and M-shell $(3\ell \to \kappa)$ transitions can be found in [56.43] and [56.44], respectively. See also §4 of [56.45] for further details.

General Expressions and Trends

For final ns states

$$|F_{1s,ns}|^2 = \frac{2^4 n\left[(n-1)^2+n^2\tau^2\right]^{n-1}}{\tau^2\left[(n+1)^2+n^2\tau^2\right]^{n+1}} \times \sin^2\left(n\tan^{-1}x+\tan^{-1}y\right)\,, \quad (56.253)$$

where

$$x = \frac{2\tau}{n(\tau^2+1-n^{-2})}\,, \quad y = \frac{2\tau}{\tau^2-1+n^{-2}}\,. \quad (56.254)$$

For final $n\ell$ states [56.46]

$$\begin{aligned} &F_{1s,n\ell}(\tau)\\ &= (i\tau)^\ell 2^{3(\ell+1)}\sqrt{2\ell+1}(\ell+1)!\left(\frac{(n-\ell-1)!}{(n+\ell)!}\right)^{1/2}\\ &\times n^{\ell+1}\frac{\left[(n-1)^2+n^2\tau^2\right]^{(n-\ell-3)/2}}{\left[(n+1)^2+n^2\tau^2\right]^{(n+\ell+3)/2}}\left[a_{n\ell}C^{(\ell+2)}_{n-\ell-1}(x)\right.\\ &\left.- b_{n\ell}C^{(\ell+2)}_{n-\ell-2}(x)+c_{n\ell}C^{(\ell+2)}_{n-\ell-3}(x)\right]\,, \end{aligned} \quad (56.255)$$

with coefficients $a_{n\ell}$, $b_{n\ell}$ and $c_{n\ell}$ given by

$$\begin{aligned} a_{n\ell} &= (n+1)\left[(n-1)^2+n^2\tau^2\right]\,,\\ b_{n\ell} &= 2n\sqrt{\left[(n-1)^2+n^2\tau^2\right]\left[(n+1)^2+n^2\tau^2\right]}\,,\\ c_{n\ell} &= (n-1)\left[(n+1)^2+n^2\tau^2\right]\,, \end{aligned}$$

and argument

$$x = \frac{n^2-1+n^2\tau^2}{\sqrt{\left[(n+1)^2+n^2\tau^2\right]\left[(n-1)^2+n^2\tau^2\right]}}\,.$$

Summation over final ℓ states:

$$\begin{aligned} |F_{1s,n}|^2 &= \sum_\ell |F_{1s,n\ell}|^2\\ &= 2^8 n^7\tau^2\left[\frac{1}{3}\left(n^2-1\right)+n^2\tau^2\right]\\ &\times \frac{\left[(n-1)^2+n^2\tau^2\right]^{n-3}}{\left[(n+1)^2+n^2\tau^2\right]^{n+3}}\,. \end{aligned} \quad (56.256)$$

which becomes for large n,

$$|F_{1s,n}|^2 \sim \frac{2^8\tau^2\left(3\tau^2+1\right)}{3n^3\left(\tau^2+1\right)^6}\exp\left(\frac{-4}{(\tau^2+1)}\right)\,. \quad (56.257)$$

For initial $2s$ and $2p$ states,

$$\begin{aligned} |F_{2s,n}|^2 &= 2^4 n^7\tau^2\left[-\frac{1}{3}+\frac{1}{2}n^2-\frac{3}{16}n^4+\frac{1}{48}n^6\right.\\ &+n^2\tau^2\left(\frac{1}{3}-\frac{2}{3}n^2+\frac{19}{48}n^4\right)\\ &\left.+n^4\tau^4\left(\frac{5}{3}-\frac{7}{6}n^2\right)+n^6\tau^6\right]\\ &\times\frac{\left[\left(\frac{1}{2}n-1\right)^2+n^2\tau^2\right]^{n-4}}{\left[\left(\frac{1}{2}n+1\right)^2+n^2\tau^2\right]^{n+4}}\,, \end{aligned} \quad (56.258)$$

$$\begin{aligned} |F_{2p,n}|^2 &= \frac{2^4 n^9\tau^2}{3}\left[\frac{1}{4}-\frac{7}{24}n^2+\frac{11}{192}n^4\right.\\ &\left.-n^2\tau^2\left(\frac{5}{6}-\frac{23}{24}n^2\right)+\frac{1}{4}n^4\tau^4\right]\\ &\times\frac{\left[\left(\frac{1}{2}n-1\right)^2+n^2\tau^2\right]^{n-4}}{\left[\left(\frac{1}{2}n+1\right)^2+n^2\tau^2\right]^{n+4}}\,. \end{aligned} \quad (56.259)$$

Power Series Expansion. $\tau^2 \ll 1$ [56.3]

$$|F_{1s,ns}(\tau)|^2 = A(n)\tau^4 + B(n)\tau^6 + C(n)\tau^8 + \cdots, \tag{56.260}$$

where

$$A(n) = \frac{2^8 n^9 (n-1)^{2n-6}}{3^2(n+1)^{2n+6}},$$
$$B(n) = -\frac{2^9 n^{11}\left(n^2+11\right)(n-1)^{2n-8}}{3^2 5(n+1)^{2n+8}},$$
$$C(n) = -\frac{2^8 n^{13}\left(313n^4 - 1654n^2 - 2067\right)}{3^2 5^2 7(n+1)^{2n+10}} \times (n-1)^{2n-10}.$$

For analytical expressions for $A(n)$, $B(n)$ and $C(n)$ for final np and nd states see [56.47, 48].

General Trends in Hydrogenic Form Factors [56.49]. The inelastic form factor $|F_{n\ell\to n'\ell'}|$ oscillates with ℓ' on an increasing background until the value

$$\ell'_{\max} = \min\left((n'-1),\, n\left(\frac{2(n+3)}{(n+1)}\right)^{1/2} - \frac{1}{2}\right) \tag{56.261}$$

is reached, after which a rapid decline for $\ell > \ell'_{\max}$ occurs. See [56.49] for illustrative graphs.

56.9.3 Excitation Cross Sections

Atom–Atom Collisions [56.50, 51]

Single Excitation. For the process

$$A(i) + B \to A(f) + B, \tag{56.262}$$

(56.248) reduces to

$$\sigma_{if} = \frac{8\pi a_0^2}{s^2}\int_{t_-}^{\infty}\frac{\mathrm{d}t}{t^3}\left|F_{if}^A\right|^2\left|Z_B - F_{i'i'}^B\right|^2. \tag{56.263}$$

Double Excitation. For the process

$$\mathrm{H}(1s) + \mathrm{H}(1s) \to \mathrm{H}(n\ell) + \mathrm{H}(n'\ell'), \tag{56.264}$$

$$\sigma_{1s,n\ell}^{1s,n'l'} = \frac{8\pi a_0^2}{s^2}\int_{t_-}^{\infty}\frac{\mathrm{d}t}{t^3}\left|F_{1s,n\ell}\right|^2\left|F_{1s,n'\ell'}\right|^2. \tag{56.265}$$

Special cases are [56.52]

$$\sigma_{1s,2s}^{1s,2s} = \frac{2^{30}\pi a_0^2\left(880t_-^4 + 396t_-^2 + 81\right)}{495s^2\left(4t_-^2+9\right)^{11}}, \tag{56.266a}$$

$$\sigma_{1s,2p}^{1s,2p} = \frac{2^{30}3^4\pi a_0^2}{11s^2\left(4t_-^2+9\right)^{11}}, \tag{56.266b}$$

$$\sigma_{1s,2s}^{1s,2p} = \frac{2^{29}3^2\left(44t_-^2+9\right)}{55s^2\left(4t_-^2+9\right)^{11}}, \tag{56.266c}$$

with $t_-^2 = \left[9/\left(16s^2\right)\right]\left[1 + 3m_e/\left(4Ms^2\right) + \cdots\right]$.

Ion–Atom Collisions. For the proton impact process

$$\mathrm{H}^+ + \mathrm{H}(1s) \to \mathrm{H}^+ + \mathrm{H}(n\ell), \tag{56.267}$$

(56.248) reduces to

$$\sigma_{1s,n\ell} = \frac{8\pi a_0^2}{s^2}\int_{t_-}^{\infty}\frac{\mathrm{d}t}{t^3}\left|F_{1s,n\ell}(t)\right|^2, \tag{56.268}$$

with $t_- = \left(1-n^{-2}\right)/(2s)$.

Asymptotic Expansions

$$\sigma_{1s,ns} = \frac{4\pi a_0^2\left(n^2-1\right)|X_{1s\to ns}|^2}{24s^2n^2}\left[C_s(n) - \frac{1}{s^2} + \frac{n^2+11}{20n^2s^4} + \frac{313n^4-1654n^2-2067}{8400n^4s^6}\right], \tag{56.269}$$

$$\sigma_{1s,np} = \frac{2^{10}\pi a_0^2 n^7(n-1)^{2n-5}}{3s^2(n+1)^{2n+5}}\left[C_p(n) + \ln s^2 + \frac{n^2+11}{10n^2s^2} + \frac{313n^4-1654n^2-2067}{5600n^4s^4}\right], \tag{56.270}$$

$$\sigma_{1s,nd} = \frac{2^{11}\pi a_0^2\left(n^2-4\right)n^5\left(n^2-1\right)^2(n-1)^{2n-7}}{3^2 5s^2(n+1)^{2n+7}} \times\left[C_d(n) - \frac{1}{s^2} + \frac{11n^2+13}{28n^2s^4}\right], \tag{56.271}$$

where $C_s(2) = 16/5$, $C_s(3) = 117/32$, $C_s(4) \approx 3.386$, and

$$\gamma_n C_p(n) = \frac{1.3026}{n^3} + \frac{1.7433}{n^5} + \frac{16.918}{n^7}, \tag{56.272}$$
$$\gamma_n C_d(n) = \frac{2.0502}{n^3} + \frac{7.6125}{n^5}, \tag{56.273}$$

with

$$\gamma_n \equiv \frac{2^8 n^7 (n-1)^{(2n-5)}}{3(n+1)^{(2n+5)}}. \tag{56.274}$$

Further asymptotic expansion results can be found in [56.2–5].

A general expression for Born excitation and ionization cross sections for hydrogenic systems in terms of a parabolic coordinate representation (Chapt. 9) is given in [56.53].

Number of Independent Transitions $\mathcal{N}_i$ Between Levels n and n′. [56.53]

$$\mathcal{N}_i = n^2\left[n' - \binom{n}{3}\right] + \binom{n}{3} . \tag{56.275}$$

Validity Criterion. The Born approximation is valid provided that [56.54]

$$nE \gg \ln\left(\frac{4E}{J_n}\right) \tag{56.276}$$

for transitions $n \to n'$ when $n, n' \gg 1$ and $|n - n'| \sim 1$. The constant J_n is undetermined (see [56.54] for details) but is generally taken to be the ionization potential of level n.

56.9.4 Ionization Cross Sections

$$\mathrm{e}^-(k) + \mathrm{H} \to \mathrm{e}^-(k') + \mathrm{H}^+ + \mathrm{e}^-(\kappa) . \tag{56.277}$$

The general expression for the Born differential ionization cross section can be evaluated in closed form using screened hydrogenic wave functions. The differential cross section per incident electron scattered into solid angle $\mathrm{d}\Omega_{k'}$, integrated over directions κ for the ejected electron (treated as distinguishable) is [56.55–57]

$$I(\theta, \phi)\,\mathrm{d}\Omega_{k'}\,\mathrm{d}\kappa' = \frac{4k'}{kq^4 a_0 \tilde{Z}_\mathrm{B}^4}\left|F_{n\ell,\kappa'}\right|^2(q)\,\mathrm{d}\Omega_{k'}\,\mathrm{d}\kappa' , \tag{56.278}$$

where the form factor is given by (56.253)) for the case $n\ell = 1s$, with the ejected electron wavenumber κ and momentum transferred q in the collision, $\kappa' = \kappa a_0/\tilde{Z}_B$, $\boldsymbol{q} = (\boldsymbol{k}' - \boldsymbol{k})a_0/\tilde{Z}_B$, being scaled by the screened nuclear charge $\tilde{Z}_B$ appropriate to the $n\ell$-shell from which the electron is ejected. The total Born ionization cross section per electron is

$$\sigma_\mathrm{ion}^\mathrm{B} = \int_0^{\kappa_\mathrm{max}} \mathrm{d}\kappa' \int_{k-k'}^{k+k'} I(q, \kappa')\,\mathrm{d}q , \tag{56.279}$$

which is generally evaluated numerically.

56.9.5 Capture Cross Sections

Electron Capture.

$$A^+ + B(n\ell) \to A(n'\ell') + B^+ . \tag{56.280}$$

In the Oppenheimer–Brinkman–Kramers (OBK) approximation [56.58],

$$\sigma_{n\ell,n'\ell'} = \frac{M^2}{2\pi\hbar^3}\frac{v_f}{v_i}\int_{-1}^{1}\mathrm{d}(\cos\theta)\left|F_{n\ell\to n'\ell'}\right|^2 , \tag{56.281}$$

where $\boldsymbol{v}_i = v_i\hat{\boldsymbol{n}}_i$, $\boldsymbol{v}_f = v_f\hat{\boldsymbol{n}}_f$, θ is the angle between $\hat{\boldsymbol{n}}_i$ and $\hat{\boldsymbol{n}}_f$, $M = M_A M_B/(M_A + M_B)$, and

$$\left|F_{n\ell,n'\ell'}\right| = \iint \mathrm{d}\boldsymbol{r}\,\mathrm{d}\boldsymbol{s}\,\varphi_i(\boldsymbol{r})\varphi_f^*(\boldsymbol{s})\left(\frac{Z_A e^2}{r}\right) \times \mathrm{e}^{\mathrm{i}(\boldsymbol{\alpha}\cdot\boldsymbol{r}+\boldsymbol{\beta}\cdot\boldsymbol{s})} , \tag{56.282}$$

with

$$\boldsymbol{\alpha} = k_f\hat{\boldsymbol{n}}_f + k_i\hat{\boldsymbol{n}}_i\frac{M_A}{M_A + m_\mathrm{e}} ,$$
$$\boldsymbol{\beta} = -k_i\hat{\boldsymbol{n}}_i - k_f\hat{\boldsymbol{n}}_f\frac{M_B}{M_B + m_\mathrm{e}} ,$$
$$k_i = \frac{v_f}{\hbar}\frac{M_B(M_A + m_\mathrm{e})}{(M_A + M_B + m_\mathrm{e})} ,$$
$$k_f = \frac{v_f}{\hbar}\frac{(M_B + m_\mathrm{e})M_A}{(M_A + M_B + m_\mathrm{e})} .$$

The Jackson–Schiff correction factor [56.59] is

$$\gamma_\mathrm{JS} = \frac{1}{192}\left(127 + \frac{56}{p^2} + \frac{32}{p^4}\right) - \frac{\tan^{-1}\frac{1}{2}p}{48p}\left(83 + \frac{60}{p^2} + \frac{32}{p^4}\right) + \frac{\left(\tan^{-1}\frac{1}{2}p\right)^2}{24p^2}\left(31 + \frac{32}{p^2} + \frac{16}{p^4}\right) , \tag{56.283}$$

and the capture cross section is

$$\sigma(n_i\ell_i, n_f\ell_f) = \frac{\gamma_\mathrm{JS}\pi a_0^2}{p^2}C(n_i\ell_i, n_f\ell_f) \times \int_x^\infty F(n_i\ell_i, n_f\ell_f; x)\,\mathrm{d}x , \tag{56.284}$$

with

$$p = \frac{mv_i}{\hbar} , \quad a = \frac{Z_A}{n_i} , \quad b = \frac{Z_B}{n_f} ,$$
$$x = \left[p^2 + (a+b)^2\right]\left[p^2 + (a-b)^2\right]/4p^2 .$$

Table 56.2 Coefficients $C(n_i\ell_i \to n_f\ell_f)$ in the Born capture cross section formula (56.284)

$n_f\ell_f$	$C(1s \to n_f\ell_f)$
$1s$	$2^8 Z_A^5 Z_B^5$
$2s$	$2^5 Z_A^5 Z_B^5$
$2p$	$2^5 Z_A^5 Z_B^7$
$3s$	$2^8 Z_A^5 Z_B^5/3^3$
$3p$	$2^{13} Z_A^5 Z_B^7/3^6$
$3d$	$2^{15} Z_A^5 Z_B^9/3^9$
$4s$	$2^2 Z_A^5 Z_B^5$
$4p$	$5 Z_A^5 Z_B^7$
$4d$	$Z_A^5 Z_B^9$
$4f$	$Z_A^5 Z_B^{11}/20$
$C(2s \to n_f\ell_f) = C(1s \to n_f\ell_f)/8$	
$C(2p \to n_f\ell_f) = C(1s \to n_f\ell_f)/24$	

Table 56.3 Functions $F(n_i\ell_i \to n_f\ell_f; x)$ in the Born capture cross section formula (56.284)

$n_f\ell_f$	$F(n_i\ell_i, n_f\ell_f; x)$
$1s$	x^{-6}
$2s$	$(x-2b^2)^2x^{-8}$
$2p$	$(x-b^2)x^{-8}$
$3s$	$(x^2 - \frac{16}{3}b^2x + \frac{16}{3}b^4)^2x^{-10}$
$3p$	$(x-b^2)(x-2b^2)^2x^{-10}$
$3d$	$(x-b^2)^2x^{-10}$
$4s$	$(x-2b^2)^2(x^2-8b^2x+8b^4)^2x^{-12}$
$4p$	$(x-b^2)(x^2 - \frac{24}{5}b^2x + \frac{24}{5}b^4)^2x^{-12}$
$4d$	$(x-b^2)^2(x-2b^2)^2x^{-12}$
$4f$	$(x-b^2)^3x^{-12}$
$F(2s, n_f\ell_f; x) = (x-2a^2)^2x^{-2}F(1s, n_f\ell_f; x)$	
$F(2p, n_f\ell_f; x) = (x-a^2)x^{-2}F(1s, n_f\ell_f; x)$	

The coefficients C in (56.284) are given in Table 56.2, while the functions F are given in Table 56.3 [56.58]. In Table 56.3, the appropriate value of a and b is indicated by the quantum numbers n_i, ℓ_i and n_f, ℓ_f.

References

56.1 H. A. Bethe, E. E. Salpeter: *Quantum Mechanics of One- and Two-Electron Atoms* (Springer, Berlin, Heidelberg 1957)
56.2 G. S. Khandelwal, B. H. Choi: J. B. Phys. **1**, 1220 (1968)
56.3 G. S. Khandelwal, E. E. Fitchard: J. Phys. B **2**, 1118 (1969)
56.4 G. S. Khandelwal, J. E. Shelton: J. Phys. B **4**, 109 (1971)
56.5 G. S. Khandelwal, B. H. Choi: J. Phys. B **2**, 308 (1969)
56.6 I. L. Beigman, V. S. Lebedev: Phys. Rep. **250**, 95 (1995)
56.7 W. Gordon: Ann. Phys. (Leipzig) **2**, 1031 (1929)
56.8 V. A. Davidkin, B. A. Zon: Opt. Spectrosk. **51**, 25 (1981)
56.9 L. A. Bureeva: Astron. Zh. **45**, 1215 (1968)
56.10 H. Griem: Astrophys. J. **148**, 547 (1967)
56.11 M. R. Flannery: *Rydberg States of Atoms and Molecules*, ed. by R. F. Stebbings, F. B. Dunning (Cambridge Univ. Press, Cambridge 1983) Chap. 11
56.12 A. Burgess, I. C. Percival: Adv. At. Mol. Phys. **4**, 109 (1968)
56.13 I. C. Percival: *Atoms and Molecules in Astrophysics*, ed. by T. R. Carson, M. J. Roberts (Academic, New York 1972) p. 65
56.14 I. C. Percival, D. Richards: Adv. At. Mol. Phys. **11**, 1 (1975)
56.15 M. Matsuzawa: J. Phys. B **8**, 2114 (1975)
56.16 E. Gerjuoy: Phys. Rev. **148**, 54 (1966)
56.17 J. D. Garcia, E. Gerjuoy, J. E. Welker: Phys. Rev. **165**, 66 (1968)
56.18 D. R. Bates, A. E. Kingston: Adv. At. Mol. Phys. **6**, 269 (1970)
56.19 J. M. Khan, D. L. Potter: Phys. Rev. **133**, A890 (1964)
56.20 V. S. Lebedev, V. S. Marchenko: Sov. Phys. JETP **61**, 443 (1985)
56.21 A. Omont: J. de Phys. **38**, 1343 (1977)
56.22 B. Kaulakys: J. Phys. B **17**, 4485 (1984)
56.23 V. S. Lebedev: J. Phys. B: At. Mol. Opt. Phys. **25**, L131 (1992)
56.24 V. S. Lebedev: Soviet Phys. (JETP) **76**, 27 (1993)
56.25 J. P. Coleman: *Case Studies in Atomic Collision Physics I*, ed. by E. W. McDaniel, M. R. C. McDowell (North Holland, Amsterdam 1969) Chap. 3
56.26 J. P. Coleman: J. Phys. B **1**, 567 (1968)
56.27 M. R. Flannery: Phys. Rev. A **22**, 2408 (1980)
56.28 M. L. Goldberger, K. M. Watson: *Collision Theory* (Wiley, New York 1964) Chap. 11
56.29 M. R. C. McDowell: Proc. Phys. Soc. **89**, 23 (1966)
56.30 D. R. Bates, J. C. G. Walker: Planetary Spac. Sci. **14**, 1367 (1966)
56.31 D. R. Bates, J. C. G. Walker: Proc. Phys. Soc. **90**, 333 (1967)
56.32 M. R. Flannery: Ann. Phys. **61**, 465 (1970)
56.33 M. R. Flannery: Ann. Phys. **79**, 480 (1973)
56.34 L. Vriens: *Case Studies in Atomic Collision Physics I*, ed. by E. W. McDaniel, M. R. C. McDowell (North Holland, Amsterdam 1969) Chap. 6
56.35 L. Vriens: Proc. Phys. Soc. **89**, 13 (1966)

56.36 B. N. Roy, D. K. Rai: J. Phys. B **5**, 816 (1973)
56.37 A. R. Holt, B. Moiseiwitsch: Adv. At. Mol. Phys. **4**, 143 (1968)
56.38 K. L. Bell, A. E. Kingston: Adv. At. Mol. Phys. **10**, 53 (1974)
56.39 L. C. Green, P. P. Rush, C. D. Chandler: Astrophys. J. Suppl. Ser. **3**, 37 (1957)
56.40 W. B. Sommerville: Proc. Phys. Soc. **82**, 446 (1963)
56.41 A. Burgess, D. G. Hummer, J. A. Tully: Phil. Trans. Roy. Soc. A **266**, 255 (1970)
56.42 C. T. Whelan: J. Phys. B **19**, 2343, 2355 (1986)
56.43 M. C. Walske: Phys. Rev. **101**, 940 (1956)
56.44 G. S. Khandelwal, E. Merzbacher: Phys. Rev. **144**, 349 (1966)
56.45 E. Merzbacher, H. W. Lewis: X-ray Production by Heavy Charged Particles. In: *Handbuch der Physik*, Vol. 34/2, ed. by E. Flügge (Springer, Berlin, Heidelberg 1958) p. 166
56.46 H. Bethe: Quantenmechanik der Ein- und Zwei-Elektronenprobleme. In: *Handbuch der Physik*, Vol. 24/1, ed. by E. Flügge (Springer, Berlin, Heidelberg 1933) p. 502
56.47 M. Inokuti: Argonne Nat. Lab. Radio Phys. Div. A Report **No. ANL-7220**, 1–10 (1965)
56.48 M. Inokuti: Rev. Mod. Phys. **43**, 297 (1971)
56.49 M. R. Flannery, K. J. McCann: Astrophys. J. **236**, 300 (1980)
56.50 D. R. Bates, G. Griffing: Proc. Phys. Soc. **66A**, 961 (1953)
56.51 D. R. Bates, G. Griffing: Proc. Phys. Soc. **67A**, 663 (1954)
56.52 D. R. Bates, A. Dalgarno: Proc. Phys. Soc. **65A**, 919 (1952)
56.53 K. Omidvar: Phys. Rev. **140**, A26 and A38 (1965)
56.54 A. N. Starostin: Sov. Phys. JETP **25**, 80 (1967)
56.55 E. H. S. Burhop: Proc. Camb. Phil. Soc. **36**, 43 (1940)
56.56 E. H. S. Burhop: J. Phys. B **5**, L241 (1972)
56.57 N. F. Mott, H. S. W. Massey: *The Theory of Atomic Collisions* (Clarendon Press, Oxford 1965) pp. 489–490
56.58 D. R. Bates, A. Dalgarno: Proc. Phys. Soc. **66A**, 972 (1953)
56.59 J. D. Jackson, H. Schiff: Phys. Rev. **89**, 359 (1953)

57. Mass Transfer at High Energies: Thomas Peak

Thomas peaks correspond to singular second-order quantum effects whose location may be determined by classical two step kinematics. The widths of these peaks (or ridges) may be estimated using the uncertainty principle. A second-order quantum calculation is required to obtain the magnitude of these peaks. Thomas peaks and ridges have been observed in various reactions in atomic and molecular collisions involving mass transfer and also ionization.

57.1 **The Classical Thomas Process** 863

57.2 **Quantum Description** 864
57.2.1 Uncertainty Effects 864
57.2.2 Conservation of Overall Energy and Momentum 864
57.2.3 Conservation of Intermediate Energy 865
57.2.4 Example: Proton–Helium Scattering 865

57.3 **Off-Energy-Shell Effects** 866

57.4 **Dispersion Relations** 866

57.5 **Destructive Interference of Amplitudes** 867

57.6 **Recent Developments**........................... 867

References 868

Transfer of mass is a quasiforbidden process. Simple transfer of a stationary mass M_2 to a moving mass M_1 is forbidden by conservation of energy and momentum. If $M_1 < M_2$ then M_1 rebounds, if $M_1 = M_2$ then M_1 stops and M_2 continues on, and if $M_1 > M_2$ then M_2 leaves faster than M_1. In none of these cases do M_1 and M_2 leave together. *Thomas* [57.1] understood this in 1927 and further realized that transfer of mass occurs only when a third mass is present and all three masses interact. The simplest allowable process is a two-step process now called a Thomas process [57.2, 3].

Quantum mechanically [57.4], the second Born term at high energies is the largest Born term and corresponds to the simplest allowed classical process, namely the Thomas process. While the classically forbidden first Born term is not zero (saved by the uncertainty principle), the first Born cross section varies at high v as v^{-12}, in contrast to the second Born cross section which varies as v^{-11}, thus dominating. Higher Born terms correspond to multistep processes that are unlikely in fast collisions where there is not enough time for complicated processes. The higher Born terms ($n > 2$) are also smaller than the second Born term.

57.1 The Classical Thomas Process

The basic Thomas process is shown in Fig. 57.1. Here, the entire collision is coplanar since particles 1 and 2 go off together (that is what is meant by mass transfer). We assume that all the masses and the incident velocities $\boldsymbol{v}$ are known. Thus, there are six unknowns, $\boldsymbol{v}'$, $\boldsymbol{v}_f$ and $\boldsymbol{v}_3$, each with two components. Conservation of momentum gives two equations of constraint for each collision. Conservation of overall energy gives a fifth constraint, and conservation of energy in the intermediate state (which only holds classically) gives a sixth constraint. With six equations of constraint, all six unknowns may be completely determined. The allowed values of $\boldsymbol{v}'$, $\boldsymbol{v}_f$ and $\boldsymbol{v}_3$ depend on the masses M_1, M_2 and M_3. For example, in the case of the transfer of an electron from atomic hydrogen to a proton, i. e., $p^+ = H \to H + p^+$, it is easily verified that the angles are $\alpha = (M_2/M_1)\sin 60°$, $\beta = 60°$ and $\gamma = 120°$, where $m' = M_2 = m_e$ and $M_1 = M_3 = M_p = 1836\,m_e$. For the case $e^+ + H \to Ps + p^+$, it may be verified that $\alpha = 45°$, $\beta = 45°$ and $\gamma = 90°$.

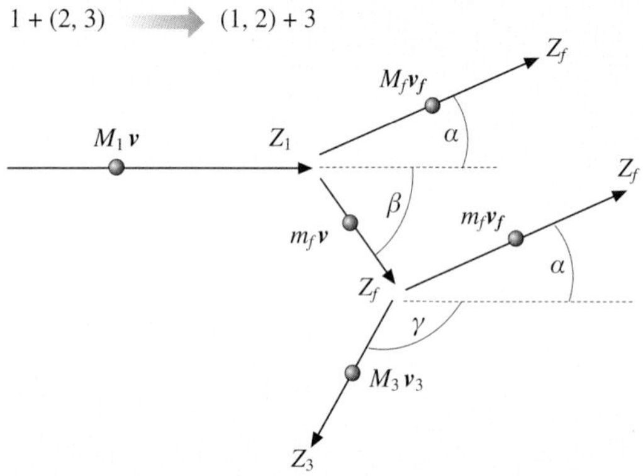

Fig. 57.1 Diagram for mass transfer via a Thomas two-step process

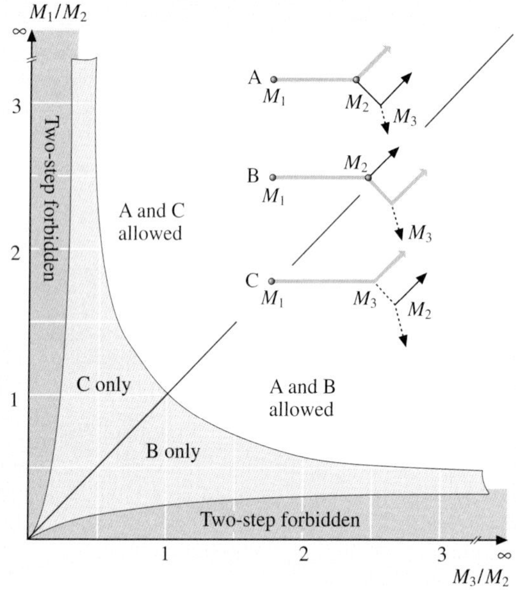

Fig. 57.2 Lieber diagram for mass transfer

In general, the intermediate mass m' may be equal to M_1, M_2 or M_3. We shall regard these as different Thomas processes, and label them B, A, and C respectively. The standard Thomas process (the one actually considered by Thomas in 1927) is case A, and corresponds to the first example given above. Lieber diagrams for the Thomas processes A, B, and C [57.5] are illustrated in Fig. 57.2. In the Lieber diagram, mass regions in which solutions exist for processes A, B, and C are shown. (An equivalent diagram was given earlier by *Detmann* and *Liebfried* [57.3].) There are some regions in which two-step processes are forbidden. In these regions the theory of mass transfer is not fully understood at present.

57.2 Quantum Description

57.2.1 Uncertainty Effects

In quantum mechanics, energy conservation in the intermediate states may be violated within the limits of the uncertainty principle, $\Delta E \geq \hbar/\Delta t$, where Δt is the uncertainty in time of mass transfer. It is not possible to determine if mass transfer actually occurs at the beginning, in the middle, or at the end of the collision. Thus, we choose $\Delta t = \bar{r}/\bar{v}$, where $\bar{r}$ is the size of the collision region and $\bar{v}$ is the mean collision velocity. Taking $\bar{r} \approx a_0/Z_{\text{target}}$ and $\bar{v} \approx v$, the projectile velocity, we have

$$\Delta E \approx \frac{\hbar}{\Delta t} = \frac{\hbar \bar{v}}{\bar{r}} = \frac{\hbar v}{a_0/Z_{\text{target}}} = \frac{\hbar Z_{\text{target}}}{a_0} . \tag{57.1}$$

Here, a_0 is the Bohr radius and Z_{target} is the nuclear charge of the target in units of the electron charge. Within this range of energy ΔE, the constraint of energy conservation in the intermediate state does not apply.

57.2.2 Conservation of Overall Energy and Momentum

Conservation of overall energy and momentum then gives three equations of constraint on the four unknowns $\boldsymbol{v}_f$ and $\boldsymbol{v}_3$ [57.6], namely,

$$M_1 \boldsymbol{v} = \left(M_f + \tilde{M}_f\right) \boldsymbol{v}_f + M_3 \boldsymbol{v}_3 , \tag{57.2}$$

$$M_1 v^2 = \left(M_f + \tilde{M}_f\right) v_f^2 + M_3 v_3^2 , \tag{57.3}$$

where M_f $\left(\tilde{M}_f\right)$ is the mass of the upper (lower) particle in the final state of the bound system shown in Fig. 57.1, in which m' is the mass of the intermediate particle M_1, M_2 or M_3. From (57.2) and (57.3) it may be shown that the velocity of the recoil particle is constrained by the condition

$$2\boldsymbol{v}_3 \cdot \hat{\boldsymbol{v}} = 2\cos\gamma = \frac{M_1 + M_2 + M_3}{M_1} \frac{v_3}{v} - \frac{M_2}{M_3} \frac{v}{v_3} . \tag{57.4}$$

Thus, the magnitude and the direction of $\boldsymbol{v}_3$ are not independent. Specifying either v_3 or $\hat{\boldsymbol{v}}_3$ is sufficient, together with the equations of constraint, to determine the energies and directions of all particles in the final state. Similarly, one may express the equations of constraint in terms of v_f and $\hat{\boldsymbol{v}}_f$.

57.2.3 Conservation of Intermediate Energy

In a classical two-step process, the projectile hits a particle in the target, and the intermediate mass m' then propagates and subsequently undergoes a second collision. Quantum mechanically, this corresponds to a second-Born term represented by $V_1 G_0 V_2$, where V represents an interaction and G_0 is the propagator of the intermediate state, namely,

$$G_0 = (E - H_0 + \mathrm{i}\epsilon)^{-1} = -\mathrm{i}\pi\delta(E - H_0) + \wp\frac{1}{E - H_0}\,, \tag{57.5}$$

where $\wp$ is the Cauchy principal value of G_0 which excludes the singularity at $E = H_0$. This singularity corresponds to conservation of energy in the intermediate state. It is this singularity which gives rise to the weaker secondary ridge in Fig. 57.3 at $v_3 = v$. The width of the secondary ridge is given approximately by $\Delta E = \hbar/\Delta t$, as discussed above. The intersection of the ridges is the Thomas peak. At very high collision velocities, the Thomas peak dominates the total cross section for mass transfer.

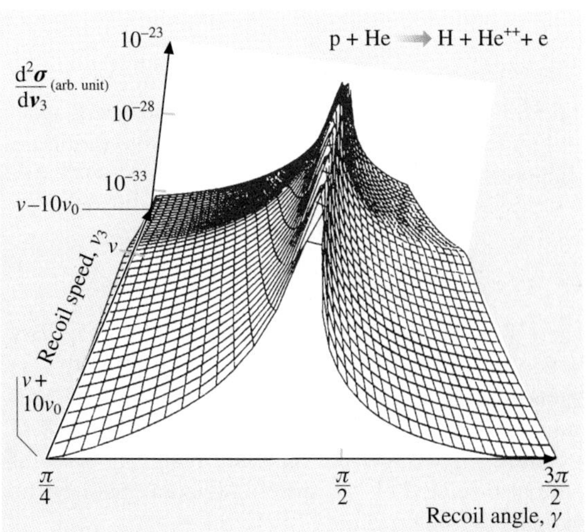

Fig. 57.3 Counting rate (or cross section) on the *vertical axis* versus recoil speed v_3 and recoil angle γ for a Thomas process in which a proton (projectile) picks up an electron from helium (target). The captured electron bounces off the second target electron

The constraint imposed by conservation of intermediate energy may be expressed by replacing the speed of the recoil particle v_3 by the scaled variable $K = M_3 v_3 / m'v$. Then it may be shown that the conservation of intermediate energy may be expressed in the form [57.6]:

$$\left(\frac{M_3 v_3}{m'v}\right)^2 \equiv K^2 = 1\,. \tag{57.6}$$

The constraints of conservation of overall energy and momentum, i. e., (57.4), may be easily written in terms of K as

$$2\cos\gamma = r\frac{m'}{M_2}K - \frac{M_2}{m'}\frac{1}{K}\,, \tag{57.7}$$

where $r = (M_1 + M_2 + M_3)M_2/(M_1 M_3)$.

57.2.4 Example: Proton–Helium Scattering

The effect of the constraints of conservation of overall energy and momentum may be seen in Fig. 57.3, where a sharp ridge is evident in the reaction

$$\mathrm{p}^+ + \mathrm{He} \rightarrow \mathrm{H} + \mathrm{He}^{++} + \mathrm{e}^-\,, \tag{57.8}$$

and $M_1 = M_\mathrm{p}$, $M_2 = M_3 = m_\mathrm{e}$. Here v_3 is the speed of the recoiling ionized target electron, and the target nucleus is not directly involved in the reaction. The width

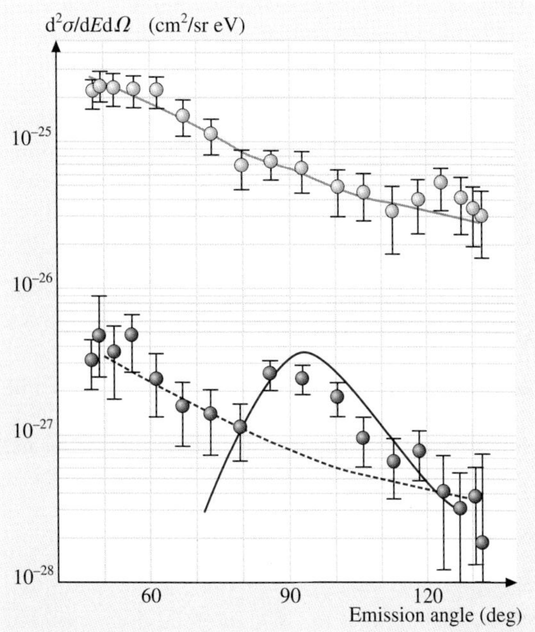

Fig. 57.4 Observation of a slice of the Thomas ridge structure in $\mathrm{p}^+ + \mathrm{He} \rightarrow \mathrm{H} + \mathrm{He}^{++} + \mathrm{e}^-$ at 1 MeV by *Palinkas* et al. [57.7]

of the sharp ridge is due to the momentum spread of the electrons in helium and may be regarded as being caused by the uncertainty principle since this momentum (or velocity) spread corresponds to $\Delta p = \hbar/\Delta r$ where Δr is taken as the radius of the helium atom.

The locus of the sharp ridge in Fig. 57.3, corresponding to conservation of overall energy and momentum, is given by (57.8). The locus of the weaker ridge, corresponding to the conservation of energy in the intermediate state, is given by (57.7). The intersection of these two loci gives the unique classical result suggested by Thomas. The width of these ridges may be estimated from the uncertainty principle as described above.

Experimental evidence for the double ridge structure has been reported by *Palinkas* et al. [57.7] corresponding to the calculations given in Fig. 57.3, but at a collision energy of 1 MeV, as shown below.

The data in Fig. 57.4 corresponds to a slice across the sharp ridge of Fig. 57.3 at $v = v_3$. The solid line is a second Born calculation [57.8,9] at 1 MeV. The bump of data above a smooth backgroud is the indication of the ridge structure.

57.3 Off-Energy-Shell Effects

In (57.6), the Green's function G_0 contains an energy-conserving term $\mathrm{i}\pi\delta(E - H_0)$ which is imaginary, and a real energy-nonconserving term $\wp[1/(E - H_0)]$. The latter does not occur classically; it is permitted by the uncertainty principle and represents the contribution of virtual (off-the-energy-shell or energy-nonconserving) states within $\pm\Delta E = \hbar/\Delta t$ about the classical value $E = H_0$. This quantum term also represents the effect of time-ordering in the second Born amplitude [57.10]. In plane wave second-Born calculations, the off-energy-shell term gives the real part of the scattering amplitude f_2, while the on-shell (energy conserving) term gives the imaginary part of f_2. These two contributions are shown in Fig. 57.5.

Half of the Thomas peak comes from energy-nonconserving contributions which are not included in a classical description. Also, the energy-nonconserving contribution plays a significant role in determining the shape of the standard Thomas peak, which has been observed [57.11].

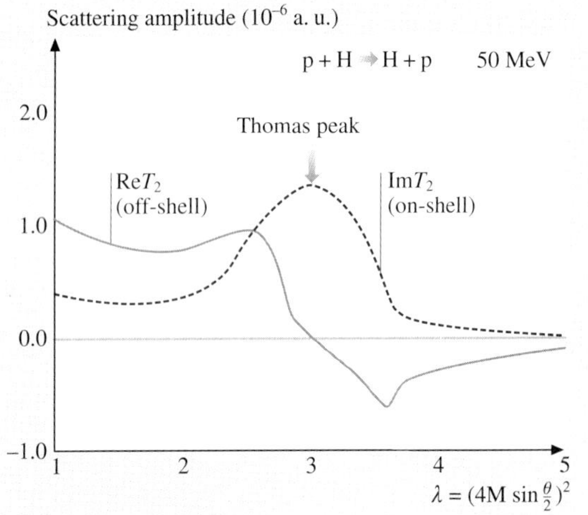

Fig. 57.5 Energy-conserving (on-shell) and energy-nonconserving contributions to the second Born scattering amplitude

57.4 Dispersion Relations

Because of the form of the Green's function of (57.6), the second Born contribution f_2 to the scattering amplitude has a single pole in the lower half of the complex plane. Consequently it obeys the dispersion relation

$$\mathrm{Re}[f_2(\lambda)] = -\frac{1}{\pi}\wp\int_{-\infty}^{+\infty}\frac{\mathrm{Im}[f_2(\lambda')]}{\lambda-\lambda'}\,\mathrm{d}\lambda' ,$$

$$\mathrm{Im}[f_2(\lambda)] = \frac{1}{\pi}\wp\int_{-\infty}^{+\infty}\frac{\mathrm{Re}[f_2(\lambda')]}{\lambda-\lambda'}\,\mathrm{d}\lambda' . \qquad (57.9)$$

where Re f_2 and Im f_2 denote the real and imaginary parts of f_2. Thus the energy-nonconserving part of f_2 is related to an integral over the energy-conserving part and vice versa. In the case of the dielectric constant it is well known that the real and imaginary parts of ϵ are also related by a dispersion relation, namely the Kramers–Kronig relatio [57.12].

Resonances are usually a function of energy E. The width of a resonance gives the lifetime τ of the resonance. Classically, τ is how long the projectile orbits the target before it leaves, corresponding to a delay or shift in time of the projectile during the interaction. If the width of the resonance is ΔE, then the lifetime is $\tau = \hbar/\Delta E$. E and τ are conjugate variables. The Thomas peak is an overdamped resonance in scattering angle, corresponding to a shift in the impact parameter of the scattering event [57.13]. However, unlike energy resonances, our Thomas resonance in the scattering angle seems to have no classical analog [57.14].

57.5 Destructive Interference of Amplitudes

It has already been noted that the location of the Thomas peaks depends on the mass of the collision partners. For the process

$$p^+ + \text{Atom} \rightarrow \text{Atom} + \text{H}\,, \tag{57.10}$$

there are two separate Thomas peaks [57.2, 15] corresponding to cases A and B in the Lieber diagram (Fig. 57.2). Experimental evidence exists for both peaks. The standard Thomas peak occurs at small forward angles [57.11], while the second peak [57.16] occurs at about 60°. If the mass of the projectile is reduced, the positions of these Thomas peaks move toward one another [57.17] as illustrated in Fig. 57.6.

When $M_1 = M_2$, then both Thomas peaks occur at 45°. This occurs in positronium formation where $M_1 = M_2 = m_e$, i. e.,

$$e^+ + \text{He} \rightarrow \text{Ps} + \text{He}^{++} + e^-\,. \tag{57.11}$$

In cases A and B of Fig. 57.2, the two $V_1 G_0 V_2$ second Born terms are of opposite sign because V_2 is of opposite sign in diagrams A and B. This leads to destructive interference for 1s − 1s electron capture (which is dominant at high velocities) as was first discussed by *Shakeshaft* and *Wadehra* [57.17]. Consequently, the observed Thomas peak structure is expected [57.18, 19] to be quite different for e^+ impact than for impact of p^+ or other projectiles heavier than an electron. The double ridge structure for transfer ionization of helium by e^+ is expected to differ significantly from the structure shown in Fig. 57.3. Understanding such destructive interference between resonant amplitudes may give deeper insight into the physical nature of the intermediate states in this special few-body collision system.

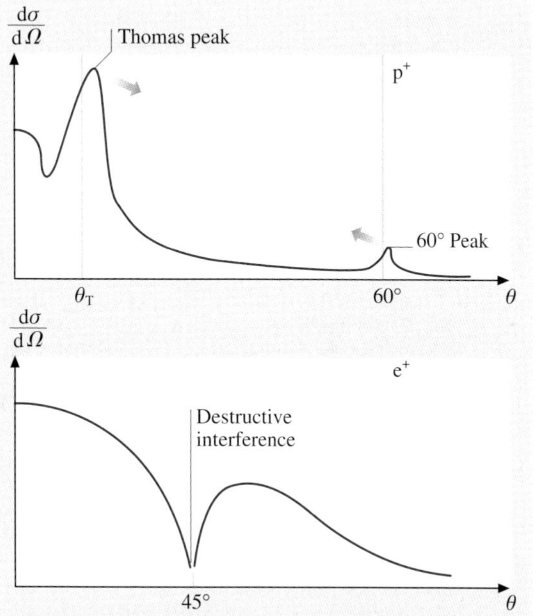

Fig. 57.6 Change of position and nature of the Thomas peaks with decreasing projectile mass

57.6 Recent Developments

In the late 1990's observations [57.20] of the Thomas peak in the case of transfer ionization (where one electron is ionized and another is transferred) differential in the momentum of the ejected electron provided new specific detail on the kinematics of the two step process [57.21]. In 2001 the Thomas peak was discussed [57.22] in the context of quantum time ordering. In this case time ordering surprisingly is not significant at the center of the peak, in contradiction to the classical picture that there is a definite order in the two step process for transfer ionization. However, time ordering does contribute to the shape of the Thomas peak. In 2002 the Stockholm group [57.23] reported that at very high velocities, the ratio of transfer ionization to total transfer approaches the same asymptotic limit as in double to single ionization in (non-Compton) photoionization, namely 1.66%. This was interpreted in terms of a common shake process occurring when the wavefunction collapses after a sudden collision.

References

57.1 L. H. Thomas: Proc. Soc. A **114**, 561 (1927)

57.2 R. Shakeshaft, L. Spruch: Rev. Mod. Phys. **51**, 369 (1979)

57.3 K. Detmann, G. Liebfried: Z. Phys. **218**, 1 (1968)

57.4 P. R. Simony: A second order calculation for charge transfer. Ph.D. Thesis (Kansas State University, Manhattan, KS, USA 1981)

57.5 M. Lieber: private communication (1987)

57.6 J. H. McGuire, J. C. Straton, T. Ishihara: *The Application of Many-Body Theory to Atomic Physics*, ed. by M. S. Pindzola, J. J. Boyle (Cambridge Univ. Press, Cambridge 1994)

57.7 J. Palinkas, R. Schuch, H. Cederquist, O. Gustafsson: Phys. Rev. Lett. **22**, 2464 (1989)

57.8 J. H. McGuire, J. C. Straton, W. C. Axmann, T. Ishihara, E. Horsdal: Phys. Rev. Lett. **62**, 2933 (1989)

57.9 J. S. Briggs, K. Taulbjerg: J. Phys. B **12**, 2565 (1979)

57.10 J. H. McGuire: Adv. At. Mol. Opt. Phys. **29**, 217 (1991)

57.11 E. Horsdal-Pederson, C. L. Cocke, M. Stöckli: Phys. Rev. Lett. **57**, 2256 (1986)

57.12 J. D. Jackson: *Classical Electrodynamics* (Wiley, New York 1975) p. 286

57.13 O. L. Weaver, J. H. McGuire: Phys. Rev. A **32**, 1435 (1985)

57.14 J. H. McGuire, O. L. Weaver: J. Phys. **17**, L583 (1984)

57.15 J. H. McGuire: Indian J. Phys. **62B**, 261 (1988)

57.16 E. Horsdal-Pederson, P. Loftager, J. L. Rasmussen: J. Phys. B **15**, 7461 (1982)

57.17 R. Shakeshaft, J. Wadehra: Phys. Rev. A **22**, 968 (1980)

57.18 A. Igarashi, N. Toshima: Phys. Rev. A **46**, R1159 (1992)

57.19 J. H. McGuire, N. C. Sil, N. C. Deb: Phys. Rev. A **34**, 685 (1986)

57.20 V. Mergel, R. Dörner, M. Achler, Kh. Khayyat, S. Lencinas, J. Euler, O. Jagutski, S. Nüttgens, M. Unverzagt, L. Spielberger, W. Ru, R. Ali, J. Ullrich, H. Cederquist, A. Salin, C. J. Wood, R. E. Olson, Dz. Belkic, C. L. Cocke, H. Schmidt-Böcking: Phys. Rev. Lett. **79**, 387 (1977)

57.21 S. G. Tolmanov, J. H. McGuire: Phys. Rev. A **62**, 032771 (2000)

57.22 A. L. Godunov, A. L. Godunov, J. H. McGuire, P. B. Ivanov, V. A. Shipakov, H. Merabet, R. Bruch, J. Hanni, Kh. Shakov: J. Phys. B **34**, 5055 (2001)

57.23 H. T. Schmidt, H. Cederquist, A. Fardi, R. Schuch, H. Zettergren, L. Bagge, A. Kallberg, J. Jensen, K. G. Resfelt, V. Mergel, L. Schmidt, H. Schmidt-Boecking, C. L. Cocke: *Photonic, Electronic and Atomic Collisions*, ed. by J. Burgdoerfer, J. S. Cohen, S. Datz, C. R. Vane (Rinton, Princeton, NJ 2002) p. 720

58. Classical Trajectory and Monte Carlo Techniques

The classical trajectory Monte Carlo (CTMC) method originated with Hirschfelder, who studied the H + D_2 exchange reaction using a mechanical calculator [58.1]. With the availability of computers, the CTMC method was actively applied to a large number of chemical systems to determine reaction rates, and final state vibrational and rotational populations (see, e.g., *Karplus* et al. [58.2]). For atomic physics problems, a major step was introduced by *Abrines* and *Percival* [58.3] who employed Kepler's equations and the Bohr–Sommerfield model for atomic hydrogen to investigate electron capture and ionization for intermediate velocity collisions of H^+ + H. An excellent description is given by *Percival* and *Richards* [58.4]. The CTMC method has a wide range of applicability to strongly-coupled systems, such as collisions by multiply-charged ions [58.5]. In such systems, perturbation methods fail, and basis set limitations of coupled-channel molecular- and atomic-orbital techniques have difficulty in representing the multitude of active excitation, electron capture, and ionization channels. Vector- and parallel-processors now allow increasingly detailed study of the dynamics of the heavy projectile and target, along with the active electrons.

58.1 **Theoretical Background** 869
58.1.1 Hydrogenic Targets 869
58.1.2 Nonhydrogenic One-Electron Models 870
58.1.3 Multiply-Charged Projectiles and Many-Electron Targets......... 870

58.2 **Region of Validity** 871

58.3 **Applications** 871
58.3.1 Hydrogenic Atom Targets............ 871
58.3.2 Pseudo One-Electron Targets 872
58.3.3 State-Selective Electron Capture.. 872
58.3.4 Exotic Projectiles 873
58.3.5 Heavy Particle Dynamics 873

58.4 **Conclusions** 874

References 874

58.1 Theoretical Background

58.1.1 Hydrogenic Targets

For a simple three-body collision system comprised of a fully-stripped projectile (a), a bare target nucleus (b), and an active electron (c), one begins with the classical Hamiltonian for the system,

$$\begin{aligned} H = {} & p_a^2/2m_a + p_b^2/2m_b + p_c^2/2m_c \\ & + Z_aZ_b/r_{ab} + Z_aZ_c/r_{ac} + Z_bZ_c/r_{bc}\,, \end{aligned} \tag{58.1}$$

where p_i are the momenta and Z_iZ_f/r_{if} are the Coulomb potentials between the individual particles. From (58.1), one obtains a set of 18 coupled, first-order differential equations arising from the necessity to determine the time evolution of the Cartesian coordinates of each particle,

$$\mathrm{d}q_i/\mathrm{d}t = \partial H/\partial p_i\,, \tag{58.2}$$

and their corresponding momenta,

$$\mathrm{d}p_i/\mathrm{d}t = -\partial H/\partial q_i\,. \tag{58.3}$$

Five random numbers, constrained by Kepler's equation, are then used to initialize the plane and eccentricity of the electron's orbit, and another is used to determine the impact parameter within the range of interaction [58.4, 5]. A fourth-order Runge–Kutta integration method is suitable because of its ease of use and its ability to vary the time step-size. This latter requirement is essential since it is not uncommon for the time step to vary by three orders of magnitude during a single trajectory.

In essence, the CTMC method is a computer experiment. Total cross sections for a particular process are determined by

$$\sigma_\mathrm{R} = (N_\mathrm{R}/N)\pi b_\mathrm{max}^2\,, \tag{58.4}$$

where N is the total number of trajectories run within a given maximum impact parameter $b_{\max}$, and N_R is the number of positive tests for a reaction, such as electron capture or ionization. Angle and energy differential cross sections are easily generalized from the above. As in an experiment, the cross section given by (58.4) has a standard deviation of

$$\Delta\sigma_R = \sigma[(N - N_R)/NN_R]^{1/2} , \quad (58.5)$$

which for large N is proportional to $1/N_R^{1/2}$. Here lies one of the major difficulties associated with the CTMC method: it takes considerable computation time to determine minor or highly differential cross sections. Present day desktop workstations can provide a partial remedy of this statistics problem. To decrease the statistical error of a cross section by a factor of two, four times as many trajectories must be evaluated.

58.1.2 Nonhydrogenic One-Electron Models

For many-electron target atoms, it is sometimes adequate to treat the problem within a one-electron model and employ the independent electron approximation to approximate atomic shell structure [58.6]. For an accurate calculation, it is necessary to use an interaction potential that simulates the screening of the target nucleus by the electrons. One can simply apply a Coulomb potential with an effective charge Z_{eff} obtained from, for example, Slater's rules. Then, the computational procedure is the same as for the hydrogenic case. However, the boundary conditions for the long- and short-range interactions are poorly satisfied.

To improve the electronic representation of the target, potentials derived from quantum mechanical calculations are now routinely used. Here, the simple solution of Kepler's equation cannot be applied. However, *Peach* et al. [58.7] and *Reinhold* and *Falcón* [58.8] have provided the appropriate methods that yield a target representation that is correct under the microcanonical distribution. The method of Reinhold and Falcón is popular because of its ease of use and generalizability. For the effective interaction potential, *Garvey* et al. [58.9] have performed a large set of Hartree–Fock calculations, and have parametrized their results in the form

$$V(R) = -[Z - NS(R)]/R , \quad (58.6)$$

with the screening of the core given by

$$S(R) = 1 - \{(\eta/\xi)[\exp(\xi R) - 1] + 1\}^{-1/2} , \quad (58.7)$$

where Z and N denote the nuclear charge and number of nonactive electrons in the target core, and η and ξ are screening parameters. Screening parameters are given in [58.9] for all ions and atoms with $Z \le 54$. This potential can also be used for the representation of partially-stripped projectile ions.

58.1.3 Multiply-Charged Projectiles and Many-Electron Targets

Multiple ionization and electron capture mechanisms in energetic collisions between multiply-charged ions and many-electron atoms is poorly understood because major approximations must be made to solve a many-electron problem associated with transitions between two centers. For a representative collision system such as

$$A^{q+} + B \to A^{(q-j)+} + B^{i+} + (i - j)e^- , \quad (58.8)$$

it is essential that the theoretical method be able to predict simultaneously the various charge states of the projectile and recoil ions, and also the energy and angular spectra of the ejected electrons. Theoretical methods based on the independent electron model fail because the varying ionization energies of the electrons are not well represented by a constant value, especially for outer shells. To present, only the nCTMC method, which is a direct extension of the hydrogenic CTMC method to an n-electron system, has been able to make reasonable predictions of the cross sections and scattering dynamics of such strongly-coupled systems [58.10]. As such, the number of coupled equations rises to $6n + 12$, where n is the number of electrons included in the calculation. However, computing time does not increase linearly, since modern vector-processors become very efficient with coupled equations for multiples of 64.

In the nCTMC technique, all interactions of the projectile and target nuclei with each other and the electrons are implicitly included in the calculations. The inclusion of all the particles then allows a direct determination of their angular scattering, along with an estimate of the energy deposition to electrons and heavy particles. Post collision interactions are included between projectile and recoil ions with the electrons; however, electron–electron interactions are introduced only in the bound initial state via a screening factor in a central-field approximation. This theoretical model has been very successful in predicting the single and double differential cross sections for the ionized electron spectra. Moreover, since a fixed target nucleus approximation is not used, this method has been the only one available to help understand and predict the results for the new field of recoil-ion momentum spectroscopy [58.11].

Only a modest amount of work has been completed on molecular targets. At present, only an H_2 target has been formulated for application to the CTMC method [58.12]. For H_2, a fixed internuclear axis is assumed which is then randomly orientated for each trajectory. The electrons are initialized in terms of two one-electron microcanonical distributions constructed from the quantum mechanical wave function for the ground state of H_2. Total cross sections for a variety of projectiles are in reasonable accord with experiment. The effect of the orientation of the molecular axis on the cross sections was also investigated, along with tests as to the validity of assuming that the cross sections for H_2 are simply the product of twice the H values.

58.2 Region of Validity

The CTMC method has a demonstrated region of applicability for ion–atom collisions in the intermediate velocity regime, particularly in the elucidation of both heavy-particle and electron collision dynamics. The method can be termed a semiclassical method in that the initial conditions for the electron orbits are determined by quantum mechanically determined interaction potentials with the parent nucleus. Since the method is most applicable to strongly-coupled systems, it has been applied successfully to a variety of intermediate energy multiply-charged ion collisions. Figure 58.1 describes pictorially the regions of validity of theoretical models. Both the atomic orbital (AO) and molecular orbital (MO) basis set expansion methods (Chapt. 50) work well until ionization strongly mediates the collision, since the theoretical description of the ionization continuum is not well-founded and relies on pseudostates to span all ejected electron energies and angles. We have arbitrarily limited these methods to a projectile charge to target charge ratio of, Z_P/Z_T, $\simeq$ 8, since above this value the number of terms in the basis set becomes prohibitively large. The CTMC method does not include molecular effects, and thus it is restricted from low velocities, except in the case of high-charge-state projectiles that capture electrons into high-lying Rydberg states which are well-described classically. Likewise, at high velocities the method is inapplicable in the perturbation regime where quantum tunneling is important, and thus is restricted to strongly coupled systems. The continuum distorted wave (CDW) method (Chapt. 52) greatly extends the region of applicability of first-order perturbation methods and has demonstrated validity in high-charge state ionization collisions.

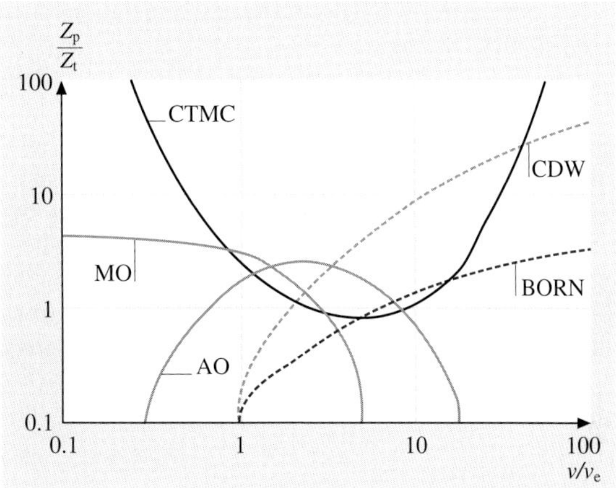

Fig. 58.1 Approximate regions of validity of various theoretical methods. Z_P/Z_T is the ratio of the projectile charge to the target charge, and v/v_e is the ratio of the collision velocity to the velocity of the active target electron. Theoretical methods: molecular orbital (MO), atomic orbital (AO), classical trajectory Monte Carlo (CTMC), continuum distorted wave (CDW), and first-order perturbation theory (BORN)

58.3 Applications

58.3.1 Hydrogenic Atom Targets

The original application of the CTMC method to atomic physics collisions were done on the $H^+ + H$ system [58.3]. Here, the electron capture and ionization total cross sections were found to be in very good accord with experiment. The Abrines and Percival procedure casts the coupled equations into the c.m. coordinate system to reduce the three-body problem to 12 coupled equations. However, this reduction complicates extensions of the code to laser processes and collisions in electric fields or with many electrons.

An ideal application for the CTMC method is for collisions involving excited targets. Such processes are well-described classically, and basis set expansion methods show limited applicability due to computer memory constraints. Considerable early work has been done on Rydberg atom collisions which includes state-selective electron capture, ionization, and electric fields [58.13–15]. Presently, there is a resurgence of work on Rydberg atom collisions because new crossed-field experimental techniques allow the production of these atoms with specific spatial orientations and eccentricities [58.16].

For hydrogenic ion–ion collision processes, one must be careful to apply the CTMC method only for projectile charges $Z_P \geq Z_T$ because after the initialization of the active electron's orbit and energy, there is no classical constraint on the orbital energy of a captured electron. For a low-charge-state ion colliding with a ground state highly charged ion, one will obtain unphysical results because a captured electron will tend to preserve its original binding energy. Thus, excess probability will be calculated for electron orbits that lead to unrealistic deeply bound states of the projectile.

58.3.2 Pseudo One-Electron Targets

Collisions involving alkali atoms are of interest because of their relevance to applied programs, such as plasma diagnostics in tokamak nuclear fusion reactors. They are also a testing ground for theoretical methods since experimental benchmarks are difficult to realize with hydrogenic targets, but are amenable for such cases as alkali atoms. In such collisions, it is essential that a theoretical formalism be used that correctly simulates the screening of the nucleus by the core electrons (Sect. 58.1.2), since a simple $-Z_{\text{eff}}/R$ Coulomb potential is inadequate for both large and small R.

One can also apply the methods of Sect. 58.1.2 to partially or completely filled atomic shells. This works reasonably well for collisions with a low charge state ion such as a proton, but fails for strong collisions involving multiply-charged ions. The reason is that the independent electron approximation [58.6] must be applied to the calculated transition probabilities in order to simulate the shell structure. This latter method can only maintain its validity if the transition probability is low. Otherwise, it will greatly overestimate the multiple electron removal processes since the first ionization potential is inherently assumed for each subsequent electron that is removed from the shell, leading to an underestimate of the energy deposition.

58.3.3 State-Selective Electron Capture

One of the powers of the CTMC method is that it can be applied to electron capture and excitation of high-lying states that are not accessible with basis set expansion techniques [58.17]. For the C^{4+} + Li system, where AO calculations and experimental data exist, the CTMC method agrees quite favorably with both [58.18].

The procedure is first to define a classical number n_c related to the calculated binding energy E of the active electron to either the projectile (electron capture) or target nucleus (excitation) as

$$E = -Z^2/\left(2n_c^2\right) . \tag{58.9}$$

Then, n_c is related to the principal quantum number n of the final state by the condition [58.19]

$$\left[n\left(n-\frac{1}{2}\right)(n-1)\right]^{1/3} < n_c \leq \left[n\left(n+\frac{1}{2}\right)(n+1)\right]^{1/3} . \tag{58.10}$$

From the electron's normalized classical angular momentum $l_c = (n/n_c)(\boldsymbol{r} \times \boldsymbol{k})$, l_c is related to the orbital quantum number l of the final state by

$$l < l_c \leq l+1 . \tag{58.11}$$

The magnetic quantum number m_l is then obtained from

$$\frac{2m_l - 1}{2l+1} \leq \frac{l_z}{l_c} < \frac{2m_l+1}{2l+1} , \tag{58.12}$$

where l_z is the z-projection of the angular momentum obtained from calculations [58.20]. In principle, it is also possible to analyze the final-state distributions from the effective quantum number

$$n^* = n - \delta_l , \tag{58.13}$$

where δ_l is the quantum defect. In this latter case, it is necessary to sort the angular momentum quantum numbers first, and then sort the principal quantum numbers.

The CTMC method has been widely applied to collisions of multiply-charged ions and hydrogen targets in the nuclear fusion program. Here, the calculated nl

charge exchange cross sections are used to predict the resulting visible and UV line emissions arising after electron capture to high principle quantum numbers. These line emission cross sections are routinely used as a diagnostic for tokamak fusion plasmas [58.21]. Likewise, for low energy collisions, CTMC results have been used to provide an explanation for the X-ray emission discovered from comets as they orbit through our solar system [58.22, 23].

58.3.4 Exotic Projectiles

The study of collisions involving antimatter projectiles, such as positrons and antiprotons, is a rapidly growing field which is being spurred on by recent experimental advances. Such scattering processes are of basic interest, and they also contribute to a better understanding of normal matter-atom collisions. Antimatter-atom studies highlight the underlying differences in the dynamics of the collision, as well as on the partitioning of the overall scattering. In the Born approximation, ionization cross sections depend on the square of the projectile's charge and are independent of its mass. Thus, the comparison of the cross sections for electron, positron, proton and antiproton scattering from a specific target gives a direct indication of higher-order corrections to scattering theories.

Early work using the CTMC method concentrated on the spectra of ionized electrons for antimatter projectiles [58.24]. Later work focused on the angular scattering of the projectiles during electron removal collisions, such as positronium formation, on ratios of the electron removal cross sections, and on ejected electron 'cusp' and 'anticusp' formation. A recent review that compares various theoretical results and available experiments is given in [58.25].

58.3.5 Heavy Particle Dynamics

A major attribute of the CTMC method is that it inherently includes the motion of the heavy particles after the collision. A straight-line trajectory for the projectile is not assumed, nor is the target nucleus constrained to be fixed. This allows one to compute easily the differential cross sections for projectile scattering or the recoil momenta of the target nucleus. As a computational note, the angular scattering of the projectile should be computed from the momentum components, not the position coordinates after the collision, since faster convergence of the cross sections using the projectile momenta is obtained. For recoil ion momentum transfer studies, one must initialize the target atom such that the c.m. of the nucleus plus its electrons has zero momentum so that there is no initial momentum associated with the target. A common error is to initialize only the target nucleus momenta to zero. Then the target atom after a collision has an artificial residual momentum that is associated with the Compton profile of the electrons because target–electron interactions are included in the calculations. Examples of recoil and projectile scattering cross sections are given in [58.10, 11].

The field of recoil ion momentum spectroscopy is rapidly expanding and the CTMC theoretical method has impacted the interpretation and understanding of experimental results because the method inherently provides a kinematically complete description of the collision products. For the studied systems, primarily He targets because of experimental constraints, it is necessary that a theoretical method be able to follow all ejected electrons and the heavy particles after a collision. As an example, it has been possible to observe the backward recoil of the target nucleus in electron capture reactions, which is due to conservation of momentum when the active electron is transferred from the target's to the projectile's frame of [58.26]. Theoretical methods are being tested further with the recent development of magneto-optical-taps (MOT) that provide frozen alkali metal atomic targets ($T \ll 1$ mK) from which to perform recoil ion studies [58.27].

For three- and four-body systems, it is now possible to measure the momenta of all collision products. These observations provide a severe test of theory, since all projectile and target interactions must be included in calculations. The CTMC method includes all projectile interactions with the target nucleus and electrons. Thus, it is possible to calculate fully differential cross sections. It is of interest that recent triply differential cross sections calculated using the CTMC method compare very favorably with sophisticated continuum distorted wave methods [58.28].

The CTMC technique allows one to incorporate electrons on both the projectile and target nuclear centers. All interactions between centers are included. The only interaction that needs to be approximated is the electron–electron interactions on a given center. Here, simple screening parameters derived from Hartree–Fock calculations are employed to eliminate nonphysical autoionization. Within this many-electron model, the signatures of the electron–electron and electron–nuclear interactions on the dynamics of the collisions have been observed [58.29, 30]. Further, projectile ionization studies can be undertaken [58.31].

58.4 Conclusions

In many ways it is surprising that a classical model can be successful in a quantum mechanical world, especially since the classical radial distribution for the hydrogen atom is described so poorly. However, hydrogen's classical momentum distribution is exactly equivalent to the quantum one, and since collision processes are primarily determined by velocity matching between projectile and electron, reasonable results can be expected. Moreover, the CTMC method preserves conservation of flux, energy, and momentum; and Coulomb scattering is the same in both quantal and classical frameworks.

Of significant importance is that the CTMC method is not restricted to one-electron systems and can easily be extended to more complicated systems involving electrons on both projectile and target. For these latter cases, multiple electron capture and ionization reactions can be investigated.

References

58.1 J. Hirschfelder, H. Eyring, B. Topley: J. Chem. Phys. **4**, 170 (1936)

58.2 M. Karplus, R. N. Porter, R. D. Sharma: J. Chem. Phys. **43**, 3259 (1965)

58.3 R. Abrines, I. C. Percival: Proc. Phys. Soc. **88**, 861 (1966)

58.4 I. C. Percival, D. Richards: Adv. At. Mol. Phys. **11**, 1 (1975)

58.5 R. E. Olson, A. Salop: Phys. Rev. A **16**, 531 (1977)

58.6 J. H. McGuire, L. Weaver: Phys. Rev. A **16**, 41 (1977)

58.7 G. Peach, S. L. Willis, M. R. C. McDowell: J. Phys. B **18**, 3921 (1985)

58.8 C. O. Reinhold: Phys. Rev. A **33**, 3859 (1986)

58.9 R. H. Garvey, C. H. Jackman, A. E. S. Green: Phys. Rev. A **12**, 1144 (1975)

58.10 R. E. Olson, J. Ullrich, H. Schmidt-Böcking: Phys. Rev. A **39**, 5572 (1989)

58.11 C. L. Cocke, R. E. Olson: Phys. Rep. **205**, 153 (1991)

58.12 L. Meng, C. O. Reinhold, R. E. Olson: Phys. Rev. A **40**, 3637 (1989)

58.13 R. E. Olson: J. Phys. B **13**, 483 (1980)

58.14 R. E. Olson: Phys. Rev. Lett. **43**, 126 (1979)

58.15 R. E. Olson, A. D. MacKellar: Phys. Rev. Lett. **46**, 1451 (1981)

58.16 D. Delande, J. C. Gay: Europhys. Lett. **5**, 303 (1988)

58.17 R. E. Olson: Phys. Rev. A **24**, 1726 (1981)

58.18 R. Hoekstra, R. E. Olson, H. O. Folkerts, W. Wolfrum, J. Pascale, F. J. de Heer, R. Morgenstern, H. Winter: J. Phys. B **26**, 2029 (1993)

58.19 R. C. Becker, A. D. MacKellar: J. Phys. B **17**, 3923 (1984)

58.20 S. Schippers, P. Boduch, J. van Buchem, F. W. Bliek, R. Hoekstra, R. Morgenstern, R. E. Olson: J. Phys. B **28**, 3271 (1995)

58.21 H. Anderson, M. G. von Hellermann, R. Hoekstra, L. D. Horton, A. C. Howman, R. W. T. Konig, R. Martin, R. E. Olson, H. P. Summers: Plasma Phys. Control. Fusion **42**, 781 (2000)

58.22 P. Beiersdorfer, R. E. Olson, G. V. Brown, H. Chen, C. L. Harris, P. A. Neill, L. Schweikhard, S. B. Utter, K. Widmann: Phys. Rev. Lett. **85**, 5090 (2000)

58.23 P. Beiersdorfer, C. M. Lisse, R. E. Olson, G. V. Brown, H. Chen: Astrophys. J. **549**, 147 (2001)

58.24 R. E. Olson, T. J. Gay: Phys. Rev. Lett. **61**, 302 (1988)

58.25 D. R. Schultz, R. E. Olson, C. O. Reinhold: J. Phys. B **24**, 521 (1991)

58.26 V. Frohne, S. Cheng, R. Ali, M. Raphaelian, C. L. Cocke, R. E. Olson: Phys. Rev. Lett. **71**, 696 (1993)

58.27 J. W. Turkstra, R. Hoekstra, S. Knoop, D. Meyer, R. Morgenstern, R. E. Olson: Phys. Rev. Lett. **87**, 123202 (2001)

58.28 J. Fiol, R. E. Olson: J. Phys. B **35**, 1759 (2002)

58.29 H. Kollmus, R. Moshammer, R. E. Olson, S. Hagmann, M. Schulz, J. Ullrich: Phys. Rev. Lett. **88**, 103202 (2002)

58.30 J. Fiol, R. E. Olson, A. C. F. Santos, G. M. Sigaud, E. C. Montenegro: J. Phys. B **34**, 503 (2001)

58.31 R. E. Olson, R. L. Watson, V. Horat, K. E. Zaharakis: J. Phys. B **35**, 1893 (2002)

59. Collisional Broadening of Spectral Lines

One-photon processes only are discussed and aspects of line broadening directly related to collisions between the emitting (or absorbing) atom and one perturber are considered. Molecular lines and bands are not considered here. Pointers to other aspects are included and a comprehensive bibliography of work on atomic line shapes, widths, and shifts already exists [59.1–7]. The perturber may be an electron, a neutral atom or an atomic ion and can interact weakly or strongly with the emitter. The emitter is either a hydrogenic or nonhydrogenic atom that is either neutral or ionized. In general, transitions in nonhydrogenic atoms can be treated as isolated, that is the separation between neighboring lines is much greater than the width of an individual line. When the emitter is hydrogen or a hydrogenic ion, the additional degeneracy of the energy levels with respect to orbital angular momentum quantum number means that lines overlap and are coupled.

Pressure broadening is a general term that describes any broadening and shift of a spectral line produced by fields generated by a background gas or plasma. The term Stark broadening implies that the perturbers are atomic ions and/or electrons, and collisional broadening implies that the 'collision' model is appropriate; this term is often used to describe an isolated line perturbed by electrons. Neutral atom broadening indicates neutral atomic perturbers; this implies short-range emitter-perturber interactions which in turn influence the approximations made.

59.1 **Impact Approximation** 875

59.2 **Isolated Lines** 876
59.2.1 Semiclassical Theory 876
59.2.2 Simple Formulae 877
59.2.3 Perturbation Theory.................. 878
59.2.4 Broadening by Charged Particles 879
59.2.5 Empirical Formulae 879

59.3 **Overlapping Lines** 880
59.3.1 Transitions in Hydrogen and Hydrogenic Ions 880
59.3.2 Infrared and Radio Lines........... 882

59.4 **Quantum-Mechanical Theory** 882
59.4.1 Impact Approximation 882
59.4.2 Broadening by Electrons 883
59.4.3 Broadening by Atoms 884

59.5 **One-Perturber Approximation** 885
59.5.1 General Approach and Utility...... 885
59.5.2 Broadening by Electrons 885
59.5.3 Broadening by Atoms 886

59.6 **Unified Theories and Conclusions** 888

References 888

General reviews of the theory of pressure broadening have been given [59.8–10], and Chapt. 2, Chapt. 10, Chapt. 14, Chapt. 19, Chapt. 45, Chapt. 47, and Chapt. 86 discuss topics relevant to the theory of collisional broadening of spectral lines. The International Conference on Spectral Line Shapes (ICSLS) is devoted exclusively to this subject.

59.1 Impact Approximation

If the perturbers are rapidly moving, the broadening and shift of the line arise from a series of binary collisions between the atom and one of the perturbers. The theory assumes that although weak collisions may occur simultaneously, strong collisions are relatively rare and only occur one at a time. The impact approximation is valid if

$$w\tau \ll 1\,, \quad \bar{V}\tau/\hbar \ll 1\,, \tag{59.1}$$

where w is the half width at half maximum, τ is the average time of collision and $\bar{V}$ is the average emitter-perturber interaction. It is not only widely applicable to electron and neutral atom broadening, but also, for certain plasma conditions, to broadening by atomic ions. The power radiated per unit time and per unit interval in circular frequency ω, in terms of the line profile $I(\omega)$, is

$$P(\omega) = \frac{4}{3}\frac{\omega^4}{c^3} I(\omega) \ . \tag{59.2}$$

For an isolated line produced by a transition from an upper energy level i to a lower level f, the line profile is Lorentzian with a shift d:

$$I(\omega) = \frac{1}{\pi} \langle\!\langle if^* | \boldsymbol{\Delta} | if^* \rangle\!\rangle \frac{w}{(\omega - \omega_{if} - d)^2 + w^2} \ , \tag{59.3}$$

and if the profile is for a transition between an upper set of levels i, i' and a lower set f, f' with quantum numbers $J_i M_i$, $J_{i'} M_{i'}$, $J_f M_f$, and $J_{f'} M_{f'}$, then

$$I(\omega) = \frac{1}{\pi} \mathrm{Re}\left(\sum_{ii'ff'} \langle\!\langle if^* | \boldsymbol{\Delta} | i' f'^* \rangle\!\rangle \times \left\langle\!\left\langle i' f'^* \left| \left[\boldsymbol{w} + \mathrm{i}\boldsymbol{d} - \mathrm{i}(\omega - \omega_{if}) \boldsymbol{I} \right]^{-1} \right| if^* \right\rangle\!\right\rangle \right) , \tag{59.4}$$

where $\boldsymbol{I}$ is the unit operator and $\boldsymbol{w}$ and $\boldsymbol{d}$ are width and shift operators. In (59.4) $\boldsymbol{\Delta}$ is an operator corresponding to the dipole line strength defined by

$$\langle\!\langle if^* | \boldsymbol{\Delta} | i' f'^* \rangle\!\rangle \equiv q^2 \langle i | \boldsymbol{r} | f \rangle \cdot \langle i' | \boldsymbol{r} | f' \rangle \ , \tag{59.5}$$

where $\boldsymbol{r}$ represents the internal emitter coordinates, $q^2 = e^2/(4\pi\epsilon_0)$ is the square of the electronic charge, and ϵ_0 is the permittivity of vacuum in SI units of J m. On taking the average over all degenerate magnetic sublevels,

$$I(\omega) = \frac{1}{\pi} \mathrm{Re}\left(\sum_{ii'ff'} \langle\!\langle if^* \| \boldsymbol{\Delta} \| i' f'^* \rangle\!\rangle \times \left\langle\!\left\langle i' f'^* \left\| \left[\boldsymbol{w} + \mathrm{i}\boldsymbol{d} - \mathrm{i}(\omega - \omega_{if}) \boldsymbol{I} \right]^{-1} \right\| if^* \right\rangle\!\right\rangle \right) \tag{59.6}$$

in terms of reduced matrix elements that are independent of magnetic quantum numbers. They are defined by

$$\langle\!\langle if^* | \boldsymbol{\Delta} | i' f'^* \rangle\!\rangle = D_{ifi'f'} \langle\!\langle if^* \| \boldsymbol{\Delta} \| i' f'^* \rangle\!\rangle \ , \tag{59.7}$$

where

$$D_{ifi'f'} = \sum_{\mu} (-1)^{J_i + J_{i'} - M_i - M_{i'}} \times \begin{pmatrix} J_i & 1 & J_f \\ -M_i & \mu & M_f \end{pmatrix} \begin{pmatrix} J_{i'} & 1 & J_{f'} \\ -M_{i'} & \mu & M_{f'} \end{pmatrix} , \tag{59.8}$$

$$\begin{aligned} &\langle\!\langle i' f'^* \| [\boldsymbol{w} + \mathrm{i}\boldsymbol{d} - \mathrm{i}(\omega - \omega_{if})\boldsymbol{I}]^s \| if^* \rangle\!\rangle \\ &= \sum_{M_i, M_{i'}, M_f, M_{f'}} D_{ifi'f'} \\ &\quad \times \langle\!\langle i' f'^* | [\boldsymbol{w} + \mathrm{i}\boldsymbol{d} - \mathrm{i}(\omega - \omega_{if})\boldsymbol{I}]^s | if^* \rangle\!\rangle \end{aligned} \tag{59.9}$$

with $s = -1, 1$. For the line profile (59.3), the width and each have a single matrix element:

$$\gamma = 2w = \langle\!\langle if^* \| 2\boldsymbol{w} \| if^* \rangle\!\rangle, \quad d = \langle\!\langle if^* \| \boldsymbol{d} \| if^* \rangle\!\rangle \ , \tag{59.10}$$

where γ is the full width at half maximum.

Throughout the rest of this article it will be assumed that collisions only connect the set of upper levels i, i' or the set of lower levels f, f' which is valid when $w \ll \omega$; pressure broadening of spectral lines is also assumed to be independent of Doppler broadening. However, for microwave spectra of molecules, w can be of the order of ω and collisions connecting the upper to the lower levels become important; for further details see *Ben-Reuven* [59.11, 12]. Also for microwave spectra, pressure broadening and Doppler broadening cannot be considered to be independent effects and a generalized theory has been developed by *Ciuryło* and *Pine* [59.13].

59.2 Isolated Lines

59.2.1 Semiclassical Theory

The motion of the perturber relative to the emitter is treated classically and is assumed to be independent of the internal states of the emitter and perturber. This common trajectory is specified by an emitter–perturber separation

$$\boldsymbol{R} \equiv \boldsymbol{R}(\boldsymbol{b}, \boldsymbol{v}, t) \ , \quad \boldsymbol{b} \cdot \boldsymbol{v} = 0 \ , \tag{59.11}$$

where $\boldsymbol{v}$ is the relative velocity and $\boldsymbol{b}$ is the impact parameter. The time-dependent wave equation for the emitter-perturber system is

$$\mathrm{i}\hbar \frac{\mathrm{d}\Psi}{\mathrm{d}t} = H\Psi \tag{59.12}$$

and the eigenfunctions ψ_i for the unperturbed emitter obey

$$H_0 \psi_i = E_i \psi_i \ , \quad i = 0, 1, 2, \ldots \ . \tag{59.13}$$

If $\Psi(\boldsymbol{r},\boldsymbol{R})$ is expanded in the form

$$\Psi(\boldsymbol{r},\boldsymbol{R})=\sum_j a_{ji}(t)\psi_j(\boldsymbol{r})\exp(-\mathrm{i}E_jt/\hbar)\,, \quad (59.14)$$

where initially at time $t=-\infty$

$$a_{ji}(-\infty)=\delta_{ji}\,, \quad (59.15)$$

and (59.12)–(59.14) give

$$\mathrm{i}\hbar\frac{\mathrm{d}a_{ji}}{\mathrm{d}t}=\sum_k a_{ki}V_{jk}\exp\left(\mathrm{i}\omega_{jk}t\right)\,,$$
$$i,j,k=0,1,2,\dots\,, \quad (59.16)$$

where

$$V_{jk}(\boldsymbol{R})=\int\psi_j^*(\boldsymbol{r})\,V(\boldsymbol{r},\boldsymbol{R})\,\psi_k(\boldsymbol{r})\;\mathrm{d}\boldsymbol{r} \quad (59.17)$$

in which $V(\boldsymbol{r},\boldsymbol{R})$ is the emitter-perturber interaction and

$$\hbar\omega_{jk}=E_j-E_k\,. \quad (59.18)$$

Integration of equations (59.16) for $-\infty\le t\le\infty$ gives the unitary scattering matrix $\boldsymbol{S}$, with elements

$$S_{ji}(\boldsymbol{b},\boldsymbol{v})\equiv a_{ji}(\infty)\,,\quad i,j=0,1,2,\dots\,. \quad (59.19)$$

Then

$$w+\mathrm{i}d=2\pi N\int_0^\infty vf(v)\;\mathrm{d}v \times\int_0^\infty\left[\delta_{i'i}\delta_{f'f}-S_{i'i}(\boldsymbol{b},\boldsymbol{v})\,S_{f'f}^*(\boldsymbol{b},\boldsymbol{v})\right]_{\mathrm{av}}b\,\mathrm{d}b\,, \quad (59.20)$$

where $J_{i'}=J_i$ and $J_{f'}=J_f$, N is the perturber density and $[\cdots]_{\mathrm{av}}$ denotes an average over all orientations of the collision and over the magnetic sublevels [see Eq. (59.9)]. In (59.20), $f(v)$ is the Maxwell velocity distribution at temperature T for an emitter-perturber system of reduced mass μ:

$$f(v)=4\pi v^2\left(\frac{\mu}{2\pi k_{\mathrm{B}}T}\right)^{3/2}\exp\left(-\frac{\mu v^2}{2k_{\mathrm{B}}T}\right)\,,$$
$$\int_0^\infty f(v)\;\mathrm{d}v=1\,. \quad (59.21)$$

59.2.2 Simple Formulae

These are useful for making quick estimates, but in individual cases may give results in error by a factor of two or more. If it is assumed in (59.17) that

$$V_{ij}(\boldsymbol{R})=V_{jj}(\boldsymbol{R})\,\delta_{ij}\,, \quad (59.22)$$

where $V_{jj}(\boldsymbol{R})$ is a simple central potential

$$V_{jj}(\boldsymbol{R})=\hbar C_jR^{-p}\,,\quad j=i,f\,, \quad (59.23)$$

and C_j depends only on the state j of the emitter, and if the relative motion is along the straight line

$$\boldsymbol{R}=\boldsymbol{b}+\boldsymbol{v}t\,, \quad (59.24)$$

then

$$\left[S_{i'i}(\boldsymbol{b},\boldsymbol{v})\,S_{f'f}^*(\boldsymbol{b},\boldsymbol{v})\right]_{\mathrm{av}}=\exp\left[2\mathrm{i}\left(\eta_i-\eta_f\right)\right]\,, \quad (59.25)$$

where the phase shifts are

$$\eta_j(b,v)=-\frac{1}{2\hbar}\int_{-\infty}^{\infty}V_{jj}(R)\;\mathrm{d}t\,,\quad j=i,f\,. \quad (59.26)$$

Equations (59.20)–(59.26) give

$$w+\mathrm{i}d=\pi N\bar{v}\left(\frac{\beta_p\left|C_p\right|}{\bar{v}}\right)^{2/(p-1)} \times\Gamma\left(\frac{p-3}{p-1}\right)\exp\left(\pm\frac{\mathrm{i}\pi}{p-1}\right)\alpha_p\,, \quad (59.27)$$

where

$$\alpha_p=\Gamma\left(\frac{2p-3}{p-1}\right)\left(\frac{\pi}{4}\right)^{-1/(p-1)}\,,$$
$$\beta_p=\sqrt{\pi}\frac{\Gamma\left(\frac{1}{2}(p-1)\right)}{\Gamma\left(\frac{1}{2}p\right)}\,,$$
$$C_p=C_i-C_f\,,$$
$$\bar{v}=\int_0^\infty vf(v)\;\mathrm{d}v=\left(\frac{8k_{\mathrm{B}}T}{\pi\mu}\right)^{1/2}\,. \quad (59.28)$$

In (59.27) the $\pm$ sign indicates the sign of C_p, $\alpha_p\simeq 1$ for $p\ge 3$, and $\Gamma(\cdots)$ is the gamma function.

The cases $p=3,4$, and 6 correspond to resonance, quadratic Stark and van der Waals broadening, respectively. This approximation is invalid for the dipole case ($p=2$) for which (59.27) is not finite. The dipole–dipole interaction ($p=3$) occurs when emitter and perturber are identical atoms (apart from isotopic differences). If the level i is connected to the ground state by a strong allowed transition with absorption oscillator strength f_{gi}, and the perturbation of the level f can be neglected by comparison, then

$$C_3 = c_d \frac{q^2 f_{gi}}{2m_e \left|\omega_{gi}\right|} ,$$
$$c_d = 1 + \frac{1}{2\sqrt{3}} \ln\left(2+\sqrt{3}\right) = 1.380\,173 . \tag{59.29}$$

Also, if g_j is the statistical weight of level j, the constant c_d may be replaced by an empirical value

$$c_d = \frac{4}{\pi}\left(\frac{g_g}{g_i}\right)^{1/2} , \tag{59.30}$$

and this gives a width correct to about 10% [59.14]. Equation (59.27) does not predict a finite shift. Quadratic Stark broadening occurs when a nonhydrogenic emitter is polarized by electron perturbers. Then

$$C_4 = -\frac{q^2}{2\hbar}\left(\alpha_i - \alpha_f\right) , \tag{59.31}$$

where α_i and α_f are the dipole polarizabilities of states i and f, respectively. Van der Waals broadening occurs when the emitter and perturber are nonidentical neutral atoms. If energy level separations of importance in the perturbing atom are much greater than those of the emitter (e.g., alkali spectra broadened by noble gases), C_6 is given by

$$C_6 = -\frac{q^2}{\hbar}\alpha_d\left(\overline{r_i^2} - \overline{r_f^2}\right) , \tag{59.32}$$

where α_d is the dipole polarizability of the perturber. The mean square radii can be calculated from the normalized radial wave functions $\frac{1}{r}P_{n_j^* l_j}(r)$ or estimated from

$$\overline{r_j^2} = \int_0^\infty P^2_{n_j^* l_j}(r)\, r^2 \, dr$$
$$\simeq \frac{n_j^{*2} a_0^2}{2z^2}\left[5n_j^{*2} + 1 - 3l_j\left(l_j+1\right)\right] ,$$
$$j = i, f \tag{59.33}$$

in which the effective principal quantum numbers n_j^* are given by

$$E_j \equiv -\frac{z^2}{n_j^{*2}} I_h , \quad z = Z_e + 1 , \tag{59.34}$$

where $I_h = hcR_\infty$ is the Rydberg energy, Z_e is the charge on the emitter and $z = 1$ in this case.

59.2.3 Perturbation Theory

An approximate solution of (59.16) is given by [59.8–10, 15, 16]

$$S_{ji}(\boldsymbol{b}, \boldsymbol{v}) = \delta_{ji} - \frac{i}{\hbar}\int_{-\infty}^{\infty} V_{ji}(t) \exp\left(i\omega_{ji} t\right) dt$$
$$- \frac{1}{\hbar^2}\sum_k \left[\int_{-\infty}^{\infty} V_{jk}(t) \exp\left(i\omega_{jk} t\right) dt \right.$$
$$\left. \times \int_{-\infty}^{t} V_{ki}(t') \exp\left(i\omega_{ki} t'\right) dt'\right] . \tag{59.35}$$

This gives a cross section for the collisional transition $i \to j$:

$$\sigma_{ij}(v) = 2\pi \int_0^\infty \left[P_{ij}(\boldsymbol{b}, \boldsymbol{v})\right]_{av} b \, db ,$$
$$i, j = 0, 1, 2, \ldots , \tag{59.36}$$

where

$$P_{ij}(\boldsymbol{b}, \boldsymbol{v}) = \left|\delta_{ji} - S_{ji}(\boldsymbol{b}, \boldsymbol{v})\right|^2 , \tag{59.37}$$
$$2\,\mathrm{Re}\left[1 - S_{ii}(\boldsymbol{b}, \boldsymbol{v})\right] = \sum_j P_{ij}(\boldsymbol{b}, \boldsymbol{v}) , \tag{59.38}$$

correct to second-order in $V(\boldsymbol{r}, \boldsymbol{R})$ on both sides. Using (59.10), (59.20) and (59.36)–(59.38), the full width is

$$\gamma = N \int_0^\infty v f(v)\, dv$$
$$\times \left[\sum_{j \neq i} \sigma_{ij}(v) + \sum_{j \neq f} \sigma_{fj}(v) + \widetilde{\sigma}_{if}(v)\right] , \tag{59.39}$$

where the sums are taken over all energy-changing transitions, the tilde indicates an interference term, and

$$\widetilde{\sigma}_{if}(v) = 2\pi \int_0^\infty \left[\widetilde{P}_{if}(\boldsymbol{b}, \boldsymbol{v})\right]_{av} b \, db , \tag{59.40}$$

in which

$$\widetilde{P}_{if}(\boldsymbol{b}, \boldsymbol{v}) = \left|\frac{1}{\hbar}\int_{-\infty}^{\infty}\left[V_{ii}(t) - V_{ff}(t)\right] dt\right|^2 . \tag{59.41}$$

59.2.4 Broadening by Charged Particles

The total emitter-perturber interaction is

$$\frac{zZ_pq^2}{R} - \frac{Z_pq^2}{|\boldsymbol{R}-\boldsymbol{r}|} \simeq V_0(R) + V(\boldsymbol{r}, \boldsymbol{R}) , \quad (59.42)$$

where Z_p is the charge on the perturber and

$$V_0(R) = \frac{Z_eZ_pq^2}{R} , \quad V(\boldsymbol{r}, \boldsymbol{R}) = -Z_pq^2\frac{\boldsymbol{r}\cdot\boldsymbol{R}}{R^3} . \quad (59.43)$$

If $Z_e = 0$, the relative motion is described by (59.24), but if $Z_e \neq 0$, the trajectory is hyperbolic and is given by

$$\mu\frac{d^2\boldsymbol{R}}{dt^2} = -\nabla V_0 = \frac{Z_eZ_pq^2}{R^3}\boldsymbol{R} , \quad (59.44)$$

with the resulting hyperbola characterized by a semi-major axis a and an eccentricity ϵ, where

$$b^2 = a^2\left(\epsilon^2 - 1\right) , \quad a = \frac{|Z_eZ_p|\, q^2}{\mu v^2} . \quad (59.45)$$

On using (59.17), (59.20), (59.36)–(59.41) and (59.43),

$$V_{ii}(t) = 0 , \quad \widetilde{P}_{if}(\boldsymbol{b}, \boldsymbol{v}) = 0 , \quad \widetilde{\sigma}_{if}(v) = 0 , \quad (59.46)$$

and

$$w + \mathrm{i}d = 2\pi N \int_0^\infty v f(v)\, dv \times\left[\int_0^\infty \sum_{j\neq i} Q_{ij}(b, v) + \sum_{j\neq f} Q_{fj}(b, v)\right] b\, db , \quad (59.47)$$

where

$$Q_{ij}(v) = \frac{4Z_p^2 I_h^2 a_0^2}{\hbar m_e |\omega_{ij}|}\frac{f_{ij}}{b^2v^2}\left[A(\beta, \xi) + \mathrm{i}B(\beta, \xi)\right] ,$$
$$2\,\mathrm{Re}[Q_{ij}(v)] = \left[P_{ij}(\boldsymbol{b}, \boldsymbol{v})\right]_{\mathrm{av}} . \quad (59.48)$$

If

$$\xi \equiv \frac{a|\omega_{ij}|}{v} , \quad \beta \equiv \xi\epsilon , \quad \delta \equiv \frac{\epsilon^2 - 1}{\epsilon^2} , \quad (59.49)$$

the functions $A(\beta, \xi)$ and $B(\beta, \xi)$ in (59.48) are given by

$$A(\beta, \xi) = \delta\exp(\mp\pi\xi)\,\beta^2 \times\left[\left|K'_{\mathrm{i}\xi}(\beta)\right|^2 + \delta\left|K_{\mathrm{i}\xi}(\beta)\right|^2\right] \quad (59.50)$$

$$B(\beta, \xi) = \frac{2\beta}{\pi}\wp\int_0^\infty \frac{A(\beta', \xi)\, d\beta'}{\left(\beta^2 - \beta'^2\right)} , \quad (59.51)$$

where $K_{\mathrm{i}\xi}(\beta)$ is a modified Bessel function. In (59.50), the $\mp$ sign corresponds to $Z_eZ_p = \pm|Z_eZ_p|$, and in (59.51), $\wp$ indicates the Cauchy principal value. If $Z_e = 0$, then

$$\xi = 0 , \quad \beta = b\frac{|\omega_{ij}|}{v} , \quad \delta = 1 \quad (59.52)$$

in (59.50) and (59.51).

Approximation (59.48) breaks down at small values of b because of assumption (59.43) and the lack of unitarity of $\boldsymbol{S}$ as given by (59.35). This problem is discussed elsewhere [59.8, 15, 16], and all methods used involve choosing a cutoff at $b = b_0$, where $b_0^2 \gtrsim \overline{r_f^2}$, and using (59.48) only for $b > b_0$. For $b \leq b_0$, an effective constant probability is introduced and the method works well as long as the contribution from $b \leq b_0$ is small. For $b > b_0$ (or $\beta > \beta_0$), the contribution to $\sigma_{ij}(v)$ in (59.39) is evaluated using (59.47)–(59.50) and (59.52), where

$$\int_{b_0}^\infty A(\beta, \xi)\frac{db}{b} = -e^{\mp\pi\xi}\beta_0 K'_{i\xi}(\beta_0) K_{i\xi}(\beta_0) . \quad (59.53)$$

A similar treatment exists for the quadrupole contribution to $V(\boldsymbol{r}, \boldsymbol{R})$ in (59.43) [59.8, 15, 16].

59.2.5 Empirical Formulae

An empirical formula based on the theory of Sect. 59.2.4 for the width of an atomic line Stark broadened by electrons has been developed [59.17]. *Konjević* [59.18] has reviewed the data available for nonhydrogenic lines and has provided simple analytical representations of the experimental results for widths and shifts.

The full half-width is given by (59.39), where $\widetilde{\sigma}_{if} = 0$ and

$$\int_0^\infty v f_e(v)\sum_{k\neq j}\sigma_{jk}(v)\, dv = \frac{8\pi^2}{3\sqrt{3}}\left(\frac{\hbar}{m_e}\right)^2\overline{v^{-1}} \times\left[T_j g(x_j) + \sum_{l_{j'}=l_j\pm 1}\widetilde{T}_{jj'}\widetilde{g}(x_{jj'})\right] , j = i, f , \quad (59.54)$$

where

$$\overline{v^{-1}} = \int_0^\infty v^{-1} f_e(v)\, dv = \left(\frac{2m_e}{\pi k_B T}\right)^{1/2} , \quad (59.55)$$

and $f_e(v) = f(v)$ with $\mu = m_e$. In (59.54),

$$T_j = \left(\frac{3n_j^*}{2z}\right)^2 \frac{1}{9}\left[n_j^{*2} + 3l_j(l_j+1) + 11\right] , \quad (59.56)$$

$$\widetilde{T}_{jj'} = \frac{l_>}{(2l_j+1)} R^2_{jj'}\left(n^*_{l_>}, n^*_{l_<}, l_>\right) , \quad (59.57)$$

$$l_< = \min(l_j, l_{j'}) , \quad l_> = \max(l_j, l_{j'}) ,$$

with

$$R_{jj'}\left(n^*_{l_>}, n^*_{l_<}, l_>\right) \equiv a_0^{-1} \int_0^\infty P_{n^*_{l_>} l_>}(r)\, r P_{n^*_{l_<} l_<}(r)\, dr . \quad (59.58)$$

The radial matrix element (59.58) can be written as

$$\begin{aligned} &R_{jj'}\left(n^*_{l_>}, n^*_{l_<}, l_>\right) \\ &= \widetilde{R}_{jj'}\left(n^*_{l_>}, l_>\right) \phi\left(n^*_{l_>}, n^*_{l_<}, l_>\right) \\ &\equiv \frac{3n^*_{l_>}}{2z}\left(n^{*2}_{l_>} - l^2_>\right)^{1/2} \phi\left(n^*_{l_>}, n^*_{l_<}, l_>\right) \quad (59.59) \end{aligned}$$

and $\phi\left(n^*_{l_>}, n^*_{l_<}, l_>\right)$ is tabulated elsewhere [59.19]. The effective principal quantum numbers $n^*_{l_>}$ and $n^*_{l_<}$ of the states j and j' in (59.57)–(59.59) both correspond to principal quantum number n_j and $\phi \simeq 1$ for $\left(n^*_{l_>} - n^*_{l_<}\right) \ll 1$. The effective Gaunt factors $g(x)$ and $\widetilde{g}(x)$ are given by

$$\widetilde{g}(x) = 0.7 - 1.1/z + g(x) , \quad x = \frac{3k_B T}{2\Delta E} , \quad (59.60)$$

where

x	≤ 2	3	5	10	30	100
$g(x)$	0.20	0.24	0.33	0.56	0.98	1.33

is used for $x < 50$, and for $x > 50$

$$\widetilde{g}(x) = g(x) = \frac{\sqrt{3}}{\pi}\left[\frac{1}{2} + \ln\left(\frac{4}{3z}\frac{|E|}{I_h}x\right)\right] , \quad (59.61)$$

with (59.60) and (59.61) joined smoothly near $x = 50$. The energy $E = E_j$ is given by (59.34) and x_j and $x_{jj'}$ in (59.54) are evaluated using

$$\Delta E_j = \frac{2z^2}{n_j^{*3}} I_h , \quad \Delta E_{jj'} = \left|E_j - E_{j'}\right| . \quad (59.62)$$

For $Z_e = 2$ and 3, (59.54) is generally accurate to within ±30% and ±50%. For $Z_e \ge 4$, (59.54) is less accurate, as relativistic effects and resonances become more important. Accuracy increases for transitions to higher Rydberg levels as long as the line remains isolated.

Tables in appendices IV and V of [59.8] give widths for atoms with $Z_e = 0, 1$ and other semi-empirical formulas based on detailed calculations have been developed by Seaton [59.20, 21] for use in the Opacity Project where simple estimates of many thousands of line widths are required.

59.3 Overlapping Lines

59.3.1 Transitions in Hydrogen and Hydrogenic Ions

The most important case is that of lines of hydrogenic systems emitted by a plasma with overall electrical neutrality, broadened by perturbing electrons and atomic ions. The line profile is given by (59.4)–(59.9) in which (59.20) is generalized to give

$$\begin{aligned} &\left\langle\!\left\langle i'f'^* \| \boldsymbol{w} + i\boldsymbol{d} \| if^* \right\rangle\!\right\rangle = 2\pi N \int_0^\infty v f(v)\, dv \\ &\times \int_0^\infty \left[\delta_{i'i}\delta_{f'f} - S_{i'i}(\boldsymbol{b}, \boldsymbol{v}) S^*_{f'f}(\boldsymbol{b}, \boldsymbol{v})\right]_{av} b\, db . \end{aligned} \quad (59.63)$$

The superscripts and suffices e and i will be used to denote electron and ion quantities, and e*m* indicates that averaging over magnetic quantum numbers has not been carried out. In the impact approximation, electron and ion contributions evaluated using (59.63) are additive and the matrix to be inverted is of order $n_i n_f$. However, under typical conditions in a laboratory plasma, e.g., a hydrogen plasma with $N_e = N_i = 10^{22}\ m^{-3}$, perturbing atomic ions cannot be treated using the impact approximation. The ions collectively generate a static field at the emitter which produces first-order Stark splitting of the upper and lower levels. The ions are randomly distributed around the emitter and the field distribution $W(\boldsymbol{F})$ used assumes that each ion is Debye screened by electrons; allowance is made for these heavy composite perturbers interacting with each other as well as with the emitter. If the ion field has a slow time variation, ion dynamic effects on the line are produced [59.8, 9, 22].

The shift produced by electron perturbers is very small and the usual model adopted is to assume that the ions split the line into its Stark components, and that each component is broadened by electron impact. In both cases, only the dipole interactions in (59.43) are included and the profile is symmetric. Then (59.4) takes the form

$$I(\omega)=\frac{1}{\pi}\operatorname{Re}\int W(\boldsymbol{F})\,\mathrm{d}\boldsymbol{F}\sum_{ii'ff'}\langle\!\langle if^*|\boldsymbol{\Delta}|i'f'^*\rangle\!\rangle \times\left\langle\!\!\left\langle i'f'^*\left|\left[\boldsymbol{w}+\mathrm{i}\boldsymbol{d}-\mathrm{i}(\omega-\omega_{if})\boldsymbol{I}\right]^{-1}\right|if^*\right\rangle\!\!\right\rangle, \tag{59.64}$$

and the destruction of the degeneracy by the ion field means that the matrix to be inverted in (59.64) is of order $(n_i n_f)^2$. Inclusion of higher multipoles in $V(\boldsymbol{r},\boldsymbol{R})$ in (59.43) introduces small asymmetries. The Stark representation for the hydrogenic wave functions is often used because it diagonalizes the shift matrix in (59.64). The transformation is given by (see Sect. 9.1.2 and 13.4.2)

$$|n_jK_jm_j\rangle=\sum_{l_j=|m_j|}^{n_j-1}(-1)^K(2l_j+1)^{1/2}\times\begin{pmatrix}N & N & l_j\\ M_1 & M_2 & -m_j\end{pmatrix}|n_jl_jm_j\rangle\,,\quad j=i,i',f,f'\,, \tag{59.65}$$

where quantum number K_j replaces l_j and

$$\begin{aligned}&n_j=K_j+K_j'+|m_j|+1\,,\quad N=\frac{1}{2}(n_j-1)\,,\\&K=\frac{1}{2}\left(2K_j'+|m_j|+m_j\right)+1\,,\\&0\le K_j\le(n_j-1)\,,\\&M_1=\frac{1}{2}\left(m_j+K_j'-K_j\right)\,,\\&M_2=\frac{1}{2}\left(m_j+K_j-K_j'\right)\,.\end{aligned} \tag{59.66}$$

For the electron impact broadening, it is convenient to separate the energy-changing and the zero energy-change transitions, so that (59.39) is generalized to give

$$\gamma_{em}=\gamma_{em}^0+\tilde{\gamma}_{em}\equiv\langle\!\langle i'f'^*|2\boldsymbol{w}_\mathrm{e}|if^*\rangle\!\rangle\,, \tag{59.67}$$

where

$$\gamma_{em}^0\equiv\left\langle\!\!\left\langle i'f'^*\left|2\boldsymbol{w}_\mathrm{e}^0\right|if^*\right\rangle\!\!\right\rangle=N_\mathrm{e}\int_0^\infty vf_\mathrm{e}(v)\,\mathrm{d}v\times\left[\sum_{j\neq i}\sigma_{ij}^{em}(v)+\sum_{j\neq f}\sigma_{fj}^{em}(v)\right]\delta_{ii'}\delta_{ff'}\,, \tag{59.68}$$

$$\tilde{\gamma}_{em}\equiv\langle\!\langle i'f'^*|2\tilde{\boldsymbol{w}}_\mathrm{e}|if^*\rangle\!\rangle=N_\mathrm{e}\int_0^\infty vf_\mathrm{e}(v)\,\mathrm{d}v\,\tilde{\sigma}_{i'f'if}^{em}(v)\,. \tag{59.69}$$

In (59.68), $|n_i-n_j|\neq0$ and $|n_f-n_j|\neq0$ in the first and second terms, respectively, and in (59.69) $(n_{i'}-n_i)=(n_{f'}-n_f)=0$. The matrix element (59.68) can be evaluated using (59.36), (59.48), (59.50), and (59.52) as before. In (59.68),

$$\sigma_{ij}^{em}(v)=2\pi\int_0^\infty\left[P_{ij}^\mathrm{e}(\boldsymbol{b},\boldsymbol{v})\right]_{\mathrm{av}_0}b\,\mathrm{d}b\,,\quad i,j=0,1,2,\ldots\,, \tag{59.70}$$

and in (59.69)

$$\tilde{\sigma}_{i'f'if}^{em}(v)=2\pi\int_0^\infty\left[\tilde{P}_{i'f'if}^\mathrm{e}(\boldsymbol{b},\boldsymbol{v})\right]_{\mathrm{av}_0}b\,\mathrm{d}b \tag{59.71}$$

by analogy with (59.36) and (59.40), where av_0 indicates an average over all orientations of the collision only. On using (59.48)–(59.50), (59.59) and (59.8) with $j_i=l_i$, $j_{i'}=l_{i'}$, $j_f=l_f$, and $j_{f'}=l_{f'}$,

$$\left[\tilde{P}_{i'f'if}^\mathrm{e}(\boldsymbol{b},\boldsymbol{v})\right]_{\mathrm{av}_0}=\frac{8I_ha_0^2}{3m_\mathrm{e}b^2v^2}D_{ifi'f'}\tilde{R}_{i'f'if}A(0,0)\,, \tag{59.72}$$

where

$$\begin{aligned}\tilde{R}_{i'f'if}&\equiv\left[\sum_{l_j=l_i\pm1}\tilde{R}_{ij}^2(n_i,l_{i>})+\sum_{l_j=l_f\pm1}\tilde{R}_{fj}^2(n_f,l_{f>})\right]\\&\quad\times\delta_{l_{i'}l_i}\delta_{l_{f'}l_f}\\&\quad-2\tilde{R}_{i'i}(n_i,l_{i>})\tilde{R}_{f'f}(n_f,l_{f>})\,\delta_{l_{i'}l_i\pm1}\delta_{l_{f'}l_f\pm1}\,.\end{aligned} \tag{59.73}$$

From (59.45) and (59.49)–(59.53)

$$A(0,0)=\delta\,,$$

$$\int_{b_0}^{b_1} A(0,0)\,\frac{db}{b} = \begin{cases} \ln(\epsilon_1/\epsilon_0)\ , & Z_e \neq 0\ ,\\ \ln(b_1/b_0)\ , & Z_e = 0\ . \end{cases} \qquad (59.74)$$

(59.53). The impact approximation neglects electron–electron correlations and the finite duration of collisions, so the long-range dipole interaction leads to a logarithmic divergence at large impact parameters in (59.74). Therefore, a second cutoff parameter is introduced which is chosen to be the smaller of the Debye length b_D and $v\tau$:

$$b_1 = \min\left[b_D \equiv \left(\frac{k_B T}{4\pi q^2 N_e}\right)^{1/2}, \ v\tau \right], \qquad (59.75)$$

but estimating τ in this case is not straightforward; it depends on the splitting of the Stark components [59.8].

59.3.2 Infrared and Radio Lines

If the density N_i is low enough, the impact approximation becomes valid for the perturbing atomic ions, and since impact shifts are unimportant, (59.6) gives

$$I(\omega) = \frac{1}{\pi}\,\mathrm{Re}\sum_{ii'ff'} \langle\!\langle if^* \| \boldsymbol{\Delta} \| i'f'^* \rangle\!\rangle \times \left\langle\!\!\left\langle i'f'^* \left\| \left[\boldsymbol{w}_e + \boldsymbol{w}_i - \mathrm{i}(\omega - \omega_{if})\,\boldsymbol{I} \right]^{-1} \right\| if^* \right\rangle\!\!\right\rangle \qquad (59.76)$$

where in (59.76)

$$\gamma_{e,i} = \gamma_{e,i}^0 + \tilde{\gamma}_{e,i} \equiv \langle\!\langle i'f'^* \| 2\boldsymbol{w}_{e,i} \| if^* \rangle\!\rangle\ , \qquad (59.77)$$

$$\gamma_{e,i}^0 \equiv \left\langle\!\!\left\langle i'f'^* \left\| 2\boldsymbol{w}_{e,i}^0 \right\| if^* \right\rangle\!\!\right\rangle = N_{e,i} \int_0^\infty v f_{e,i}(v)\ \mathrm{d}v \times \left[\sum_{j\neq i} \sigma_{ij}^{e,i}(v) + \sum_{j\neq f} \sigma_{fj}^{e,i}(v) \right] \delta_{i'i}\delta_{f'f}\ , \qquad (59.78)$$

$$\tilde{\gamma}_{e,i} \equiv \langle\!\langle i'f'^* \| 2\tilde{\boldsymbol{w}}_{e,i} \| if^* \rangle\!\rangle = N_{e,i} \int_0^\infty v f_{e,i}(v)\ \mathrm{d}v\, \tilde{\sigma}_{i'f'if}^{e,i}(v)\ . \qquad (59.79)$$

In general, cross sections for electron and heavy-particle impact are roughly comparable for the same velocity and hence different impact energies. Therefore, using (59.78) and (59.79),

$$\gamma_e^0 \gg \gamma_i^0\ , \quad \tilde{\gamma}_e \ll \tilde{\gamma}_i\ , \qquad (59.80)$$

and this result is consistent with approximation (59.64) for high density plasmas. If $(n_i - n_f) = 1, 2$, say, as n_f increases, the relative contributions from (59.78) and (59.79) decrease because there is increasing coherence, and hence cancellation in $\tilde{\sigma}_{i'f'if}^{e}(v)$ and $\tilde{\sigma}_{i'f'if}^{i}(v)$ between the effects of levels i, i' and f, f'.

Radio lines of hydrogen are observed in galactic HII regions where principal quantum numbers are of the order of $n_f \simeq 100$, temperatures are $T_e = T_i \simeq 10^4$ K, and densities are $N_e = N_i \simeq 10^9\ \mathrm{m}^{-3}$. If γ is the full-half width and $\tilde{\gamma}$ is the full-half width when only contributions (59.79) are retained, the effect of cancellation is illustrated by

$n_i - n_f = 1$	electrons	protons + electrons
n_f	$\tilde{\gamma}/\gamma$	$\tilde{\gamma}/\gamma$
5	0.81	0.99
10	0.44	0.95
15	0.21	0.87
20	0.11	0.75
25	0.06	0.61
50	0.00	0.32
100	0.00	0.16

59.4 Quantum-Mechanical Theory

59.4.1 Impact Approximation

The scattering amplitude for a collisional transition $i \to j$ is given in terms of elements of the transition matrix $\boldsymbol{T} = \boldsymbol{1} - \boldsymbol{S}$ by

$$f(\boldsymbol{k}_j, \boldsymbol{k}_i) \equiv f(\chi_j M_j \boldsymbol{k}_j, \chi_i M_i \boldsymbol{k}_i) = \frac{2\pi\mathrm{i}}{(k_i k_j)^{1/2}} \sum_{lml'm'} \mathrm{i}^{l-l'}\, Y_{lm}^*(\hat{\boldsymbol{k}}_i)\, Y_{l'm'}(\hat{\boldsymbol{k}}_j) \times T(\chi_j M_j l'm'; \chi_i M_i lm)\ , \qquad (59.81)$$

where the quantities $\boldsymbol{k}_i lm$ and $\boldsymbol{k}_j l'm'$ refer to the motion of the perturber relative to the emitter before and after the collision, and χ_i and χ_j represent all nonmagnetic quantum numbers associated with the unperturbed states i and j of the emitter. The total energy of the emitter-perturber system is given by

$$E_J = E_j + \varepsilon_j\ , \quad \varepsilon_j = \frac{\hbar^2}{2\mu} k_j^2\ , \quad (J, j) = (I, i)\ ,\ (F, f)\ , \qquad (59.82)$$

and for an isolated line, γ is given by (59.39), where

$$\sigma_{ij}(v) = \frac{k_j}{k_i}\frac{1}{4\pi g_i} \times \sum_{M_i M_j} \int \left| f\left(\chi_j M_j \boldsymbol{k}_j, \chi_i M_i \boldsymbol{k}_i\right)\right|^2 \mathrm{d}\hat{\boldsymbol{k}}_i \, \mathrm{d}\hat{\boldsymbol{k}}_j \,, \quad k_i = \mu v/\hbar \,, \tag{59.83}$$

and the interference term $\tilde{\sigma}_{if}(v) \equiv \tilde{\sigma}_{ifif}(v)$ is given by

$$\tilde{\sigma}_{if}(v) = \frac{1}{4\pi} \sum_{\substack{M_i M_{i'} \\ M_f M_{f'}}} D_{ifi'f'} \int \left| f(\chi_i M_{i'} \mathbf{k}', \chi_i M_i \boldsymbol{k}) - f(\chi_f M_{f'} \boldsymbol{k}', \chi_f M_f \boldsymbol{k})\right|^2 \mathrm{d}\hat{\boldsymbol{k}} \, \mathrm{d}\hat{\boldsymbol{k}}' \,, \tag{59.84}$$

with $k = k' = \mu v/\hbar$. From (59.8) and (59.20),

$$w + \mathrm{i}d = \pi \left(\frac{\hbar}{\mu}\right)^2 N \int_0^\infty \frac{1}{v} f(v) \, \mathrm{d}v \times \sum_{l=0}^\infty (2l+1) \left[1 - S_{ii}(l, v) S_{ff}^*(l, v)\right] , \tag{59.85}$$

where

$$(\mu v b)^2 \Longrightarrow \hbar^2 l\,(l+1) \tag{59.86}$$

and the integral over b has been replaced by a summation over l . In (59.85)), $S_{ii}(l, v)\, S_{ff}^*(l, v)$ is given by

$$S_{ii}(l, v)\, S_{ff}^*(l, v) = \frac{1}{(2l+1)} \sum_{\substack{M_i M_i' M_f M_f' \\ mm'}} D_{ifi'f'} S_I\left(\chi_{i'} M_{i'} l m'; \chi_i M_i l m\right) \times S_F^*\left(\chi_{f'} M_{f'} l m'; \chi_f M_f l m\right) \tag{59.87}$$

and subscripts I and F are introduced to emphasize that the S-matrix elements correspond to different total energies E_I and E_F defined by (59.82). If scattering by the emitter in a state j is treated using a central potential, the amplitude for elastic scattering is

$$f\left(k', k\right) = \frac{\mathrm{i}}{2k} \sum_{l=0}^\infty (2l+1)\, T_{jj}(l, v)\, P_l\left(\hat{\boldsymbol{k}}' \cdot \hat{\boldsymbol{k}}\right) , \tag{59.88}$$

where

$$1 - T_{jj}(l, v) = S_{jj}(l, v) = \exp\left[2\mathrm{i}\eta_j(l, k)\right] , \quad j = i, f \,, \tag{59.89}$$

[*cf.* (59.23), (59.25), and (59.26)]. For the case of overlapping lines, (59.63) becomes

$$\left\langle\!\left\langle i' f'^* \left\| \boldsymbol{w} + \mathrm{i}\boldsymbol{d} \right\| i f^* \right\rangle\!\right\rangle = \pi \left(\frac{\hbar}{\mu}\right)^2 N \int_0^\infty \frac{1}{v} f(v) \, \mathrm{d}v \times \sum_{l=0}^\infty (2l+1) \left[\delta_{i'i}\delta_{f'f} - S_{i'i}(l, v)\, S_{f'f}^*(l, v)\right] \tag{59.90}$$

on generalizing (59.85) and using (59.87). Formulae (59.85) and (59.90) have been obtained by assuming that a collision produces no change in the angular momentum of the relative emitter-perturber motion. This corresponds to the assumption of a common trajectory in semiclassical theory, and means that the total angular momentum of the emitter-perturber system is not conserved. This assumption is removed in the derivation of the more general expressions given in the following sections.

59.4.2 Broadening by Electrons

Different coupling schemes can be used to describe the emitter-perturber collision. For LS coupling,

$$\chi_j M_j \Rightarrow \chi_j^0 L_j M_j S M_S \,, \quad j = i, i', f, f' \,, \tag{59.91}$$

in (59.81), where χ_j^0 denotes all other quantum numbers required to describe state j that do not change during the collision. Then

$$\left| L_j M_j S M_S l m \tfrac{1}{2} m_s \right\rangle = \sum_{L_j^T M_j^T S^T M_S} C^{L_j l L_j^T}_{M_j m M_S^T} C^{S \frac{1}{2} S^T}_{M_S m_s M_S^T} \times \left| L_j S l \tfrac{1}{2} L_j^T M_j^T S^T M_S^T \right\rangle , \tag{59.92}$$

where $C^{j_1 j_2 j_3}_{m_1 m_2 m_3}$ is a vector coupling coefficient, the superscript T denotes quantum numbers of the emitter-perturber system, and $\frac{1}{2}$, m_s are the spin quantum numbers of the scattered electron. On using (59.92),

(59.90) is replaced by

$$\begin{aligned}&\langle\!\langle i' f'^{*} \| \boldsymbol{w}+\mathrm{i}\boldsymbol{d} \| i f^{*} \rangle\!\rangle = \pi\,(\hbar/m_{\mathrm{e}})^2 N\\ &\times \sum_{L_i^T L_f^T S^T l l'} (-1)^{L_i+L_{i'}+l+l'} \left(2L_i^T+1\right)\left(2L_f^T+1\right)\\ &\times \frac{\left(2S^T+1\right)}{2\,(2S+1)} \begin{Bmatrix} L_f^T & L_i^T & 1 \\ L_i & L_f & l \end{Bmatrix} \begin{Bmatrix} L_f^T & L_i^T & 1 \\ L_{i'} & L_{f'} & l' \end{Bmatrix}\\ &\times \int_0^\infty \frac{1}{v} f_{\mathrm{e}}(v)\,\mathrm{d}v \Big[\delta_{l'l}\delta_{L_{i'}L_i}\delta_{L_{f'}L_f}\\ &- S_I\left(L_{i'}Sl'\frac{1}{2}L_i^T S^T; L_i S l \frac{1}{2} L_i^T S^T\right)\\ &\times S_F^*\left(L_{f'}Sl'\frac{1}{2}L_f^T S^T; L_f S l \frac{1}{2} L_f^T S^T\right)\Big], \end{aligned} \quad (59.93)$$

where, for an isolated line, the width and shift are given by (59.93) with $L_{i'} = L_i$ and $L_{f'} = L_f$. For hydrogenic systems, where states i, i' and f, f' with different angular momenta are degenerate, a logarithmic divergence occurs for large values of l and l' [*cf.* Eq. (59.74)], and must be removed by using (59.75) and (59.86).

If a jj coupling scheme is used $\chi_j M_j \Rightarrow \chi_j^0 J_j M_j$, $j = i, i', f, f'$ in (59.81), and

$$\begin{aligned}&\left| J_j M_j l m \frac{1}{2} m_s \right\rangle\\ &= \sum_{J_j^T M_j^T jm'} C^{J_j j J_j^T}_{M_j m' M_j^T} C^{l \frac{1}{2} j}_{m m_s m'} \left| J_j l j J_j^T M_j^T \right\rangle, \end{aligned} \quad (59.94)$$

then (59.93) becomes

$$\begin{aligned}&\langle\!\langle i' f'^{*} \| \boldsymbol{w}+\mathrm{i}\boldsymbol{d} \| i f^{*} \rangle\!\rangle = \pi\,(\hbar/m_{\mathrm{e}})^2 N\\ &\times \sum_{J_i^T J_f^T jj'll'} (-1)^{J_i+J_{i'}+2J_f^T+j+j'} \frac{1}{2}(2J_i^T+1)(2J_f^T+1)\\ &\times \begin{Bmatrix} J_f^T & J_i^T & 1 \\ J_i & J_f & j \end{Bmatrix} \begin{Bmatrix} J_f^T & J_i^T & 1 \\ J_{i'} & J_{f'} & j' \end{Bmatrix}\\ &\times \int_0^\infty \frac{1}{v} f_{\mathrm{e}}(v)\,\mathrm{d}v \Big[\delta_{l'l}\delta_{j'j}\delta_{J_{i'}J_i}\delta_{J_{f'}J_f}\\ &- S_I(J_{i'}l'j'J_i^T; J_i l j J_i^T)\, S_F^*(J_{f'}l'j'J_f^T; J_f l j J_f^T)\Big], \end{aligned} \quad (59.95)$$

where $J_{i'} = J_i$ and $J_{f'} = J_f$ for an isolated line. If the spectrum of the emitter is classified using LS coupling, it is often sufficient to use energies defined by

$$E_{L_j S} = \sum_{J_j} \frac{(2J_j+1)}{(2L_j+1)(2S_j+1)} E_{L_j S J_j}, \quad (59.96)$$

and obtain the $\boldsymbol{S}$-matrix elements in an LS coupling scheme. They are then transformed to jj coupling by using the algebraic transformation

$$\begin{aligned}&S\left(J_{j'}l'j'J_j^T; J_j l j J_j^T\right)\\ &= \left[(2J_j+1)(2J_{j'}+1)(2j+1)(2j'+1)\right]^{1/2}\\ &\times \sum_{L_j^T S^T} (2L_j^T+1)(2S^T+1)\\ &\times \begin{Bmatrix} L_j & l & L_j^T \\ S & \frac{1}{2} & S^T \\ J_j & j & J_j^T \end{Bmatrix} \begin{Bmatrix} L_{j'} & l' & L_j^T \\ S & \frac{1}{2} & S^T \\ J_{j'} & j' & J_j^T \end{Bmatrix}\\ &\times S\left(L_{j'}Sl'\frac{1}{2}L_j^T S^T; L_j S l \frac{1}{2} L_j^T S^T\right), \end{aligned} \quad (59.97)$$

and introducing the splitting of the fine structure components in (59.4) or (59.6). If in LS coupling the line is isolated, but nevertheless the broadened fine structure components overlap significantly, then the interference terms in (59.6) must be included.

59.4.3 Broadening by Atoms

The formal result is very similar to (59.95), but in this case, the relative motion only gives rise to orbital angular momentum. Thus

$$\begin{aligned}&\langle\!\langle i' f'^{*} \| \boldsymbol{w}+\mathrm{i}\boldsymbol{d} \| i f^{*} \rangle\!\rangle = \pi\,(\hbar/\mu)^2 N\\ &\times \sum_{J_i^T J_f^T ll'} (-1)^{J_i+J_{i'}+2J_f^T+l+l'} (2J_i^T+1)(2J_f^T+1)\\ &\times \begin{Bmatrix} J_f^T & J_i^T & 1 \\ J_i & J_f & l \end{Bmatrix} \begin{Bmatrix} J_f^T & J_i^T & 1 \\ J_{i'} & J_{f'} & l' \end{Bmatrix}\\ &\times \int_0^\infty \frac{1}{v} f(v)\,dv \big[\delta_{l'l}\delta_{J_{i'}J_i}\delta_{J_{f'}J_f}\\ &- S_I\left(J_{i'}l'J_i^T; J_i l J_i^T\right) S_F^*\left(J_{f'}l'J_f^T; J_f l J_f^T\right)\big]. \end{aligned} \quad (59.98)$$

For many cases of practical interest, transitions of type $J_i \to J_f$ are isolated and so have line profiles given by (59.3), (59.10), and (59.98), where $J_i = J_{i'}$ and $J_f = J_{f'}$ [59.14]. In order to obtain the $\boldsymbol{S}$-matrix elements in

(59.98), it is usually sufficient to use adiabatic potentials for the emitter-perturber system that have been calculated neglecting fine structure. Since T is typically a few hundred degrees, only coupling between adiabatic states that tend to the appropriate separated-atom limit are retained in the scattering problem. The coupled scattering equations are then solved with fine structure introduced by applying an algebraic transformation to the adiabatic potentials, and using the observed splittings of the energy levels. The Born–Oppenheimer approximation is valid, and details are given in [59.23].

59.5 One-Perturber Approximation

59.5.1 General Approach and Utility

If only one perturber is effective in producing broadening, $I(\omega)$ can be obtained by considering a dipole transition between initial and final states I and F of the emitter-perturber system. Then $P(\omega)$ is given by (59.2), where

$$I(\omega) = \left[\delta(\omega - \omega_{IF}) \langle\!\langle IF^* \,|\boldsymbol{\Delta}|\, IF^* \rangle\!\rangle\right]_{\text{av}} ,$$
$$\hbar\omega_{IF} = E_I - E_F , \tag{59.99}$$

and av denotes an average over states I and a sum over states F [59.9]. Wave functions Ψ_J are given by

$$\Psi_J(\boldsymbol{r}, \boldsymbol{R}) = \mathcal{O} \sum_j \psi_j(\boldsymbol{r})\, \phi(\boldsymbol{k}_j, \boldsymbol{k}_{j0}; \boldsymbol{R}) , \tag{59.100}$$

where $J = I, F$, and $\mathcal{O}$ is an operator that takes account of any symmetry properties of the emitter-perturber system. The energies E_I and E_F are given by (59.82). The perturber wave functions for initial state j_0 and final state j are expanded in the form

$$\phi(\boldsymbol{k}_j, \boldsymbol{k}_{j0}; \boldsymbol{R}) = 2\pi\mathrm{i} \sum_{l_{j0} m_{j0} l_j m_j} \mathrm{i}^{l_{j0}} k_{j0}^{-1/2} Y^*_{l_{j0} m_{j0}}(\hat{\boldsymbol{k}}_{j0}) \times Y_{l_j m_j}(\hat{\mathbf{R}}) \frac{1}{R} F(\Gamma_j, \Gamma_{j0}; R) \tag{59.101}$$

where Γ_j denotes a channel characterized by

$$\Gamma_j = \chi_j M_j l_j m_j , \quad j = 0, 1, 2, \ldots , \tag{59.102}$$

[see Eq. (59.81)]. In (59.101), the radial perturber wave function has the limiting forms

$$F(\Gamma_j, \Gamma_{j0}; R) \overset{R\to 0}{\sim} R^{l_j+1}$$
$$\overset{R\to\infty}{\sim} k_j^{-1/2} \left[\delta_{\Gamma_j \Gamma_{j0}} \exp(-i\theta_j) - S_J(\Gamma_j; \Gamma_{j0}) \exp(i\theta_j)\right] , \tag{59.103}$$

with

$$\theta_j = k_j R - \frac{1}{2} l_j \pi - \frac{z}{k_i} \ln(2k_j R) + \arg\Gamma\left(l_j + 1 + \mathrm{i}\frac{z}{k_j}\right) ,$$
$$z = \frac{\mu q^2}{\hbar^2} Z_\mathrm{e} Z_\mathrm{p} . \tag{59.104}$$

The coupled equations obtained by using (59.12), (59.13), (59.100), and (59.101), where

$$\Psi(\boldsymbol{r}, \boldsymbol{R}, t) = \Psi_J(\boldsymbol{r}, \boldsymbol{R}) \exp(-\mathrm{i}E_J t/\hbar) , \tag{59.105}$$

are integrated to give functions $F(\Gamma_j, \Gamma_{j0}; R)$. Using (59.100) and (59.101), (59.99) becomes

$$I(\omega) = \frac{1}{2} N \sum_{\Gamma_{i0}\Gamma_i\Gamma_{i'}\Gamma_{f0}\Gamma_f\Gamma_{f'}} u_{i0} \langle\!\langle \Gamma_i \Gamma_f{}^* \,|\boldsymbol{\Delta}|\, \Gamma_{i'} \Gamma_{f'}{}^* \rangle\!\rangle \times \int_0^\infty \frac{1}{v} f(v)\, \mathrm{d}v\, \mathcal{F}(\Gamma, v) , \tag{59.106}$$

where

$$\mathcal{F}(\Gamma, v) = \int_0^\infty F^*(\Gamma_i, \Gamma_{i0}; R)\, F(\Gamma_f, \Gamma_{f0}; R)\, \mathrm{d}R \times \int_0^\infty F(\Gamma_{i'}, \Gamma_{i0}; R)\, F^*(\Gamma_{f'}, \Gamma_{f0}; R)\, \mathrm{d}R , \tag{59.107}$$

$$u_{i0} = g_{i0} \Big/ \sum_{i'_0} g_{i'_0} , \quad v = \frac{\hbar k_{i0}}{\mu} . \tag{59.108}$$

The one-perturber approximation is valid when

$$\Delta\omega \equiv |\omega - \omega_{if}| \gg w ; \quad V \gg \bar{V} , \tag{59.109}$$

where V is the effective interaction potential required to produce a shift $\Delta\omega$. In the center of the line, many-body effects are always important and the one-perturber approximation diverges as $\Delta\omega \to 0$. In many cases, there is a region of overlap where criteria (59.1) and (59.109) are all valid, but when $\Delta\omega\,\tau \gg 1$, (59.99) is a static approximation, since the average time between collisions is $\Delta\omega^{-1}$.

59.5.2 Broadening by Electrons

If LS coupling is used, definition of channel Γ_j in (59.104) is replaced by

$$\Gamma_j = L_j S l_j \frac{1}{2} L_j^T S^T\,, \quad j = i_0, i, i', f_0, f, f'\,, \tag{59.110}$$

[*cf.* (59.91) and (59.92)]. Then assuming that the weights u_{i_0} of all the levels i_0 that effectively contribute to the line are the same, (59.106) becomes

$$\begin{aligned} I(\omega) &= \frac{1}{2} N_e \sum_{\substack{\Gamma_{i_0}\Gamma_i\Gamma_{i'} \\ \Gamma_{f_0}\Gamma_f\Gamma_{f'}}} \langle\!\langle L_i S (L_f S)^* \|\boldsymbol{\Delta}\| L_{i'} S (L_{f'} S)^* \rangle\!\rangle \\ &\times \delta_{l_i l_f} \delta_{l_{i'} l_{f'}} \delta_{L_{i_0}^T L_i^T} \delta_{L_{f_0}^T L_f^T} \delta_{L_{i_0}^T L_{i'}^T} \delta_{L_{f_0}^T L_{f'}^T} \\ &\times (-1)^{L_i + L_{i'} + l_i + l_{i'}} \frac{(2S^T+1)}{2(2S+1)} \\ &\times \left(2L_i^T+1\right)\left(2L_f^T+1\right) \\ &\times \begin{Bmatrix} L_f^T & L_i^T & 1 \\ L_i & L_f & l \end{Bmatrix} \begin{Bmatrix} L_f^T & L_i^T & 1 \\ L_{i'} & L_{f'} & l_{i'} \end{Bmatrix} \\ &\times \int_0^\infty \frac{1}{v} f_e(v)\, dv\, \mathcal{F}(\Gamma, v)\,, \end{aligned} \tag{59.111}$$

where $\mathcal{F}(\Gamma, v)$ is defined by (59.107) and (59.111). If the functions $F\left(\Gamma_{j'}, \Gamma_{j_0}; R\right)$ in (59.107) are replaced by their asymptotic forms (59.103), then

$$\begin{aligned} \mathcal{F}(\Gamma, v) &\simeq \Delta\omega^{-2} (\hbar/m_e)^2 \\ &\times \left[\delta_{\Gamma_{i_0}\Gamma_i} \delta_{\Gamma_{f_0}\Gamma_f} - S_I\left(\Gamma_i; \Gamma_{i_0}\right) S_F^*\left(\Gamma_f; \Gamma_{f_0}\right)\right] \\ &\times \left[\delta_{\Gamma_{i_0}\Gamma_{i'}} \delta_{\Gamma_{f_0}\Gamma_{f'}} - S_I^*\left(\Gamma_{i'}; \Gamma_{i_0}\right) S_F\left(\Gamma_{f'}; \Gamma_{f_0}\right)\right]. \end{aligned} \tag{59.112}$$

On substituting (59.112) into (59.111), summing over Γ_{i_0} and Γ_{f_0} and using the unitary property of the S-matrix,

$$\begin{aligned} &I(\omega) \\ &= \frac{1}{\pi\Delta\omega^2} \sum_{L_i^T L_f^T S^T l l'} \langle\!\langle L_i S (L_f S)^* \|\boldsymbol{\Delta}\| L_{i'} S (L_{f'} S)^* \rangle\!\rangle \\ &\times \langle\!\langle i' f'^* \|\mathbf{w}\| i f^* \rangle\!\rangle\,, \end{aligned} \tag{59.113}$$

where $\langle\!\langle i' f'^* \|\mathbf{w}\| i f^* \rangle\!\rangle$ is given by (59.93). Line shape (59.113) is identical to that obtained from (59.6) when $\Delta\omega \gg w$. If the jj coupling scheme specified by (59.94) is used, and channel Γ_j is defined by

$$\Gamma_j = J_j l_j j_j J_j^T\,, \quad j = i_0, i, i', f_0, f, f'\,, \tag{59.114}$$

Equation (59.106) becomes [*cf.* (59.95)]

$$\begin{aligned} &I(\omega) \\ &= \frac{1}{2} N_e \sum_{\Gamma_{i_0}\Gamma_i\Gamma_{i'}\Gamma_{f_0}\Gamma_f\Gamma_{f'}} \langle\!\langle J_i (J_f)^* \|\boldsymbol{\Delta}\| J_{i'} (J_{f'})^* \rangle\!\rangle \\ &\times \delta_{l_i l_f} \delta_{l_{i'} l_{f'}} \delta_{j_i j_f} \delta_{j_{i'} j_{f'}} \delta_{J_{i_0}^T J_i^T} \delta_{J_{f_0}^T J_f^T} \delta_{J_{i_0}^T J_{i'}^T} \delta_{J_{f_0}^T J_{f'}^T} \\ &\times (-1)^{J_i + J_{i'} + 2J_f^T + j_i + j_{i'}} \\ &\times \frac{1}{2}\left(2J_i^T+1\right)\left(2J_f^T+1\right) \\ &\times \begin{Bmatrix} J_f^T & J_i^T & 1 \\ J_i & J_f & j_i \end{Bmatrix} \begin{Bmatrix} J_f^T & J_i^T & 1 \\ J_{i'} & J_{f'} & j_{i'} \end{Bmatrix} \\ &\times \int_0^\infty \frac{1}{v} f_e(v)\, dv\, \mathcal{F}(\Gamma, v)\,, \end{aligned} \tag{59.115}$$

where $\mathcal{F}(\Gamma, v)$ is given by (59.107) and (59.114).

59.5.3 Broadening by Atoms

In the wings of a line where $\hbar\,|\Delta\omega| \simeq \left|E_{J_i} - E_{J_{i'}}\right|$, $j = i_0, i, i', f_0, f, f'$, coupling between the fine structure levels is important. If channel Γ_j is defined by

$$\Gamma_j = J_j l_j J_j^T\,, \quad j = i_0, i, i', f_0, f, f'\,, \tag{59.116}$$

Equation (59.106) becomes [*cf.* (59.115)]

$$\begin{aligned} I(\omega) &= \frac{1}{2} N \sum_{\Gamma_{i_0}\Gamma_i\Gamma_{i'}\Gamma_{f_0}\Gamma_f\Gamma_{f'}} \langle\!\langle J_i (J_f)^* \|\boldsymbol{\Delta}\| J_{i'} (J_{f'})^* \rangle\!\rangle \\ &\times \delta_{l_i l_f} \delta_{l_{i'} l_{f'}} \delta_{J_{i_0}^T J_i^T} \delta_{J_{f_0}^T J_f^T} \delta_{J_{i_0}^T J_{i'}^T} \delta_{J_{f_0}^T J_{f'}^T} \\ &\times (-1)^{J_i + J_{i'} + 2J_f^T + l_i + l_{i'}} \left(2J_i^T+1\right)\left(2J_f^T+1\right) \\ &\times \begin{Bmatrix} J_f^T & J_i^T & 1 \\ J_i & J_f & l_i \end{Bmatrix} \begin{Bmatrix} J_f^T & J_i^T & 1 \\ J_{i'} & J_{f'} & l_{i'} \end{Bmatrix} \\ &\times \int_0^\infty \frac{1}{v} f(v)\, dv\, \mathcal{F}(\Gamma, v)\,, \end{aligned} \tag{59.117}$$

where $\mathcal{F}(\Gamma, v)$ is given by (59.107) and (59.116). In the far wings, where $\hbar\,|\Delta\omega| \gg \left|E_{J_i} - E_{J_{i'}}\right|$, $j = i_0, i, i', f_0, f, f'$, an adiabatic approximation is valid.

Adiabatic states of the diatomic molecule formed by the emitter-perturber system are considered in which the total spin is assumed to be decoupled from the total orbital angular momentum of the electrons. The coupling between rotational and electronic angular momentum can also be neglected, because typically, contributions to the line profile come from $0 \leq l_j \lesssim 400$, whereas $\Lambda_j \lesssim 2$. Therefore, transitions take place between channels defined by

$$\Gamma_j = \Lambda_j L_j S L_j \, , \quad j = i, f \, , \tag{59.118}$$

where the unperturbed emitter in state j has quantum numbers $L_j S$, the quantum number Λ_j represents the projection of the orbital angular momentum on the internuclear axis, and (59.100) is replaced by

$$\Psi_J(\boldsymbol{r}, \boldsymbol{R}) = \mathcal{O} \sum_j \psi_j(\boldsymbol{r}; \boldsymbol{R}) \, \phi\left(\boldsymbol{k}_j, \boldsymbol{k}_{j0}; \boldsymbol{R}\right) \, , \tag{59.119}$$

where $k_j = k_{j0}$ and $\psi_j(\boldsymbol{r}; R)$ is the wave function for molecular state Λ_j. In (59.119), the only molecular states retained are those that correlate with emitter states i and f. The scattering is described by

$$\left[\frac{d^2}{dR^2} - \frac{l_j(l_j+1)}{R^2} - \frac{2z}{R} - \frac{2\mu}{\hbar^2} V_{\Lambda_j}(R) + k_j^2\right] \times F_j(R) = 0 \, , \tag{59.120}$$

where $k_j = k_{j0}$ and

$$F_j(R) = P_{v_j l_j}(R) \text{ or}$$
$$F_j(R) \equiv F\left(\Gamma_j, \Gamma_j; R\right) = F_{k_i l_i}(R) \tag{59.121}$$

for vibrational or free states, respectively, and $V_{\Lambda_j}(R)$ is the potential energy of state Λ_j. Free–free transitions always contribute to the line profile, but bound-free and free-bound transitions only contribute on the red and blue wings, respectively. On using (59.82), (59.99), and (59.109), $\hbar\Delta\omega = \varepsilon_i - \varepsilon_f$, and ε_j becomes the energy of bound state j with vibrational quantum number v_j when $\varepsilon_j < 0$. If

$$\mathcal{G}\left(\Gamma, \varepsilon_i, \varepsilon_f\right) \equiv \left| \int_0^\infty F_i^*(R) \, \bar{\Delta}(R) \, F_f(R) \, dR \right|^2 \tag{59.122}$$

[*cf.* (59.107)], where

$$\bar{\Delta}(R) = -q \int \psi_i^*(\boldsymbol{r}; \boldsymbol{R}) \boldsymbol{r} \, \psi_f(\boldsymbol{r}; \boldsymbol{R}) \, d\boldsymbol{r} \tag{59.123}$$

is the dipole moment, then using (59.119)), the free–free contribution is given by

$$I_0(\omega) = \frac{N}{2} \sum_{\Gamma_i \Gamma_f} u_{\Lambda_i} \delta_{l_i l_f} (2l_i + 1) \times \int_0^\infty \frac{f(v)}{v} \, dv \, \mathcal{G}(\Gamma, \varepsilon_i, \varepsilon_f) \, , \tag{59.124}$$

where $\varepsilon_i = \frac{1}{2}\mu v^2$, and u_{Λ_i} is the relative weight of state Λ_i [*cf.* Eqs. (59.106)–(59.108)]. The bound-free contribution is

$$I_1(\omega) = \frac{N}{2} \sum_{\Gamma_i \Gamma_f} u_{\Lambda_i} \delta_{l_i l_f} (2l_i + 1) \times \sum_i g(\varepsilon_i) \, \mathcal{G}\left(\Gamma, \varepsilon_i, \varepsilon_f\right) \tag{59.125}$$

and the free–bound contribution is

$$I_2(\omega) = \frac{1}{2} N \sum_{\Gamma_i \Gamma_f} u_{\Lambda_i} \delta_{l_i l_f} (2l_i + 1) \exp\left(-\frac{\hbar\Delta\omega}{k_B T}\right) \times \sum_f g(\varepsilon_f) \, \mathcal{G}\left(\Gamma, \varepsilon_i, \varepsilon_f\right) \, , \tag{59.126}$$

where

$$g(\varepsilon) = 2\pi \left(\frac{\hbar}{\mu}\right)^2 \frac{f(v)}{v^2} = 8\pi^2 \left(\frac{\hbar}{\mu}\right)^2 \left(\frac{\mu}{2\pi k_B T}\right)^{3/2} \exp\left(-\frac{\varepsilon}{k_B T}\right) \, , \tag{59.127}$$

on using (59.21) with $\varepsilon = \frac{1}{2}\mu v^2$. The full line profile is then given by (59.124)–(59.126), so that

$$I(\omega) = \sum_{j=0}^{2} I_j(\omega) \, . \tag{59.128}$$

The satellite features that are often seen in line wings arise because turning points in the difference potential $\left[V_{\Lambda_i}(R) - V_{\Lambda_f}(R)\right]$ produce a phenomenon analogous to the formation of rainbows in scattering theory. The JWKB approximation is often used for the functions $F_{k_j l_j}(R)$, and can be shown to lead to the correct static limit in which transitions take place at fixed values of R called 'Condon points', i.e., the Franck–Condon principle is valid. Further details are given in [59.9, 10, 24–26].

59.6 Unified Theories and Conclusions

The pressure broadening of spectral lines is in general a time-dependent many-body problem and as such cannot be solved exactly. After all, even the problem of two free electrons scattered by a proton is still a subject of active research. There is no practical theory that leads to the full static profile in the limit of high density (or low temperature) and to the full impact profile in the limit of low density (or high temperature). As with so many problems in physics, it is the intermediate problem that is intractable because no particular feature can be singled out as providing a weak perturbation on a known physical situation. However, much progress has been made over the last thirty years in developing theories that take into account many of the key features of the intermediate problem and they are often successful in predicting line profiles for practical applications [59.8–10, 24–26].

More recently, time-dependent many-body problems have been tackled using computer-oriented approaches that invoke Monte Carlo and other simulation methods to study line broadening in dense, high-temperature plasmas, see for example [59.27]. In this chapter, the emphasis has been on aspects of the subject that relate directly to electron-atom and low-energy atom–atom scattering. Many experts in the fields of electron–atom and atom–atom collisions are still not exploiting the direct applicability of their work to line broadening. It is hoped that this contribution will encourage more research workers to study these fascinating problems that not only provide links with plasma physics and in particular with the physics of fusion plasmas, but also with a quite distinct body of laboratory-based experimental data.

References

59.1 J. R. Fuhr, W. L. Wiese, L. J. Roszman: *Bibliography on Atomic Line Shapes and Shifts*, Special Publication 366 (National Bureau of Standards, Washington, DC. 1972)

59.2 J. R. Fuhr, L. J. Roszman, W. L. Wiese: *Bibliography on Atomic Line Shapes and Shifts*, Special Publication 366 (National Bureau of Standards, Washington, DC. 1974), Supplement 1

59.3 J. R. Fuhr, G. A. Martin, B. J. Specht: *Bibliography on Atomic Line Shapes and Shifts*, Special Publication 366 (National Bureau of Standards, Washington, DC. 1974), Supplement 2

59.4 J. R. Fuhr, B. M. Miller, G. A. Martin: *Bibliography on Atomic Line Shapes and Shifts*, Special Publication 366 (National Bureau of Standards, Washington, DC. 1978), Supplement 3

59.5 J. R. Fuhr, A. Lesage: *Bibliography on Atomic Line Shapes and Shifts*, Special Publication 366 (National Institute of Standards and Technology, Washington, DC. 1992), Supplement 4

59.6 A. Lesage, J. R. Fuhr: *Bibliography on Atomic Line Shapes and Shifts* (Observatoire de Paris, Meudon 1998), Supplement 5

59.7 http://www.physics.nist.gov/PhysRefData </Redirect/www.physics.nist.gov/PhysRefData>

59.8 H. R. Griem: *Spectral Line Broadening in Plasmas* (Academic, New York 1974)

59.9 G. Peach: Adv. Phys. **30**, 367 (1981)

59.10 N. Allard, J. R. Kielkopf: Rev. Mod. Phys. **54**, 1103 (1982)

59.11 A. Ben-Reuven: Phys. Rev. **145**, 7 (1966)

59.12 A. Ben-Reuven: Adv. Atom. Molec. Phys. **5**, 201 (1969)

59.13 R. Ciuryto, A. S. Pine: J. Quant. Spectrosc. Radiat. Transfer **67**, 375 (2000)

59.14 E. L. Lewis: Phys. Rep. **58**, 1 (1980)

59.15 S. Sahal-Bréchot: Astron. & Astrophys. **1**, 91 (1969)

59.16 S. Sahal-Bréchot: Astron. & Astrophys. **2**, 322 (1969)

59.17 M. S. Dimitrijević, N. Konjević: J. Quant. Spectrosc. Radiat. Transfer **24**, 451 (1980)

59.18 N. Konjević: Phys. Rep. **316**, 339 (1999)

59.19 G. K. Oertel, L. P. Shomo: Astrophys. J. Suppl. **16**, 175 (1969)

59.20 M. J. Seaton: J. Phys. B **21**, 3033 (1988)

59.21 M. J. Seaton: J. Phys. B **22**, 3603 (1989)

59.22 V. I. Kogan, V. S. Lisitsa, G. V. Sholin: Rev. Plasma Phys. **13**, 261 (1987)

59.23 P. J. Leo, G. Peach, I. B. Whittingham: J. Phys. B **28**, 591 (1995)

59.24 J. Szudy, W. E. Baylis: J. Quant. Spectrosc. Radiat. Transfer **15**, 641 (1975)

59.25 J. Szudy, W. E. Baylis: J. Quant. Spectrosc. Radiat. Transfer **17**, 269 (1977)

59.26 J. Szudy, W. E. Baylis: Phys. Rep. **266**, 127 (1996)

59.27 A. Calisti, L. Godbert, R. Stamm, B. Talin: J. Quant. Spectrosc. Radiat. Transfer **51**, 59 (1994)

Part E

Scattering

Part E Scattering Experiments

60 Photodetachment
David J. Pegg, Knoxville, USA

61 Photon–Atom Interactions: Low Energy
Denise Caldwell, Arlington, USA
Manfred O. Krause, Oak Ridge, USA

62 Photon–Atom Interactions: Intermediate Energies
Bernd Crasemann, Eugene, USA

63 Electron–Atom and Electron–Molecule Collisions
Sandor Trajmar, Redwood City, USA
William J. McConkey, Windsor, Canada
Isik Kanik, Pasadena, USA

64 Ion–Atom Scattering Experiments: Low Energy
Ronald Phaneuf, Reno, USA

65 Ion–Atom Collisions – High Energy
Lew Cocke, Manhattan, USA
Michael Schulz, Rolla, USA

66 Reactive Scattering
Arthur G. Suits, Stony Brook, USA
Yuan T. Lee, Taipei, Taiwan

67 Ion–Molecule Reactions
James M. Farrar, Rochester, USA

60. Photodetachment

Investigations of photon-ion interactions have grown rapidly over the past few decades due primarily to the increased availability of laser and synchrotron light sources. At photon energies below about 1 keV the dominant radiative process is the electric dipole induced photoelectric effect. In the gaseous phase the photoelectric effect is referred to as either photoionization (atoms and positive ions) or photodetachment (negative ions). This chapter reviews developments in the field of photodetachment that have taken place over the past decade. The focus will be on accelerator-based investigations of the photodetachment of atomic negative ions. The monographs of *Massey* [60.1] and *Smirnov* [60.2] offer a good introduction to the subject of negative ions. Recent reviews of negative ions and photodetachment include those of *Bates* [60.3], *Buckman* and *Clark* [60.4], *Blondel* [60.5], *Andersen* [60.6], *Andersen* et al. [60.7] and *Bilodeau* and *Haugen* [60.8].

60.1 **Negative Ions** 891
60.2 **Photodetachment** 892
60.2.1 Threshold Behavior 892
60.2.2 Resonance Structure 892
60.2.3 Higher Order Processes 893
60.3 **Experimental Procedures** 893
60.3.1 Production of Negative Ions 893
60.3.2 Interacting Beams 893
60.3.3 Light Sources 894
60.3.4 Detection Schemes 895
60.4 **Results** 895
60.4.1 Threshold Measurements 895
60.4.2 Resonance Parameters 896
60.4.3 Lifetimes of Metastable Negative Ions 897
60.4.4 Multielectron Detachment 898
References 898

60.1 Negative Ions

Interest in negative ions stems from the fact that their structure and dynamics are qualitatively different from those of isoelectronic atoms and positive ions. This can be traced to the nature of the force that binds the outermost electron. In the case of atoms and positive ions, the outermost electron moves asymptotically in the long range Coulomb field of the positively charged core. The relatively strong $1/r$ potential is able to support an infinite spectrum of bound states that converge on the ionization limit. In contrast, the outermost electron in a negative ion experiences the short-range induced-dipole field of the atomic core. The relatively weak $1/r^4$ polarization potential is shallow and typically can only support a single bound state. The weakness of the binding is reflected in the magnitudes of electron affinities of atoms, which are numerically equal to the binding energies of the outermost electron in the corresponding negative ion. Electron affinities are typically an order of magnitude smaller than the ionization energies of atoms. Excited bound states of negative ions are rare. With the possible exception of Os^-, all such states that exist have the same configuration, and therefore parity, as the ground state. A rich spectrum of unbound excited states, however, are associated with most ions. These discrete states are embedded in the continua lying above the first detachment limit.

Electron correlation plays an important role in determining the structure and dynamics of many-electron systems [60.9]. Weakly bound systems such as negative ions are ideally suited for investigations of the effects of correlation. As a result of the more efficient shielding of the nucleus by the atomic core, the electron–electron interactions become relatively more important than the electron-nucleus interaction in negative ions. The goal of photodetachment experiments is to measure, in high resolution, correlation-sensitive quantities such as electron affinities and the energies and widths of resonant states. These quantities provide sensitive tests of the ability of theorists to incorporate electron correlation into their calculations. The stimulating interplay between experiment and theory continues to help elucidate the role of many-electron effects in the structure and dynamics of atomic systems.

60.2 Photodetachment

Essentially all information about the structure and dynamics of negative ions comes from controlled experiments in which electrons are detached from the ions when they interact with photons or other particles. Photodetachment is the preferred method of studying negative ion structure and dynamics since the energy resolution associated with such measurements is typically much higher than that attainable in any particle-induced detachment process. Generally, one or more electrons are detached from a negative ion following the absorption of one or more photons in the photodetachment process. Most measurements to date, however, involve the simplest process of single electron detachment following single photon absorption. Cross sections for this process start at zero at threshold, rise to a maximum a few eV above threshold and then decrease monotonically. Photodetachment cross sections at their maximum have a typical magnitude of ≈ 10–100 Mb. Threshold behavior and resonance structure in detachment cross sections are of particular interest since they both involve a high degree of correlation between the electrons.

60.2.1 Threshold Behavior

Cross sections for photodetachment are zero at threshold, in contrast to the finite value characteristic of photoionization cross sections. The threshold behavior is determined by the dynamics of two particles in the final continuum state. The Wigner law [60.10] governs the energy dependence of the near-threshold cross section for the photodetachment of a single electron from an atomic negative ion. The Wigner law can be written as

$$\sigma = Ak^{2l+1} = B(E-E_t)^{l+1/2}\,, \tag{60.1}$$

where k represents the wavenumber of the detached electron, $(E-E_t)$ is the excess energy of the electron above threshold and l is the smallest value of the orbital angular momentum quantum number. As a result of the electric dipole selection rules, the detached electron is represented, in general, by two partial waves with $l = l_0+1$ and l_0-1, where l_0 is the angular momentum of the bound electron in the negative ion prior to detachment. Wigner demonstrated that for a two-body final state the near-threshold cross section depends only on the dominant long-range interaction between the two product particles. In the case of photodetachment involving electrons with $l > 0$, this contribution arises from the centrifugal force. Shorter-range interactions, such as the polarization force, will not change the form of the threshold behavior but they will limit the range of validity of the Wigner law. There is no a priori way of determining the range of validity of the Wigner law in any particular experiment. It depends on the strengths of short-range interactions. Measured threshold data is usually fit to the Wigner law in order to determine the threshold energy. In principle, it is possible to extend the range of the fit beyond that of the Wigner law. *O'Malley* [60.11], for example, considered the effects of multipole forces on threshold behavior. O'Malley's formalism, however, does not treat polarization explicitly. This is, however, accounted for in the modified effective range theory of *Watanabe* and *Greene* [60.12]. Recently, *Sandstroem* et al. [60.13] have used a modified effective range theory to fit photodetachment data taken at excited state thresholds of the alkali-metal atoms, Li and K. In these cases the dipole polarizability is very high and consequently the range of validity of the Wigner law is correspondingly small.

60.2.2 Resonance Structure

Negative ion resonances correspond to states in which an electron and an atom are transiently associated. Such states are the subject of a review by *Buckman* and *Clark* [60.4]. In photodetachment they arise when more than one electron, or a core electron, is excited. These unbound discrete states are embedded in the continua above the first detachment limit and are therefore subject to decay via the spontaneous process of autodetachment. The allowed autodetachment process is induced by the relatively strong electrostatic interaction between the outermost electrons. This process causes discrete continuum states to be very short lived. If the selection rules on the allowed Coulomb-induced autodetachment process are violated, however, the state may live much longer. Metastable states eventually decay via autodetachment processes induced by the weaker magnetic interactions. The He^- ion is the prototypical metastable negative ion. It is formed in the spin-aligned $1s2s2p\ ^4P^0$ state when an electron attaches itself to a He atom in the metastable $1s2s\ ^3S$. It is bound by 77.516 meV [60.14]. The decay of a discrete state in the continuum by autodetachment is manifested as a resonance structure in the detachment cross section. The shape of a resonance is determined by the interference between the two pathways for reaching the same final continuum state: direct detachment and detachment via the discrete state embedded in the continuum. A resonance can be parametrized by fitting it to a *Fano* [60.15] or *Shore* profile [60.16]. The energy and width of the discrete continuum state are extracted from the fit.

60.2.3 Higher Order Processes

With the advent of high power, pulsed lasers it became possible to observe multiphoton detachment. In this process a single electron is ejected following the absorption of two or more photons. Early work in this area has been reviewed by *Crance* [60.17], *Davidson* [60.18] and *Blondel* [60.5]. More recently, Haugen and coworkers have used two photon E1 transitions to determine fine structure splittings in the ground state of negative ions and to measure the binding energies of excited states of negative ions that have the same parity as the ground state. *Bilodeau* and *Haugen* [60.8] have reviewed these measurements.

Multielectron detachment involves the detachment of two or more electrons following the absorption of a single photon. This process, which appears to be initiated by the detachment of an inner shell electron, requires photons with energies higher than can be generated by lasers. Such measurements can be performed at synchrotron radiation sites.

60.3 Experimental Procedures

60.3.1 Production of Negative Ions

Negative ions are created in exoergic attachment processes when an electron is captured by an atom or molecule. These quantum systems are weakly bound with diffuse outer orbitals. As a consequence, they are easily destroyed in collisions with other particles. Due to their fragility they are rarely observed in bulk matter. The production of negative ions with a density sufficiently high for spectroscopic studies poses a challenge to the experimentalist since processes involved in their creation must compete with more probable destruction processes. The most versatile source of production of negative ions for accelerator-based experiments is the Cs sputter ion source [60.19]. This source has been used to generate a wide variety of atomic, molecular, and cluster negative ions.

Negative ions can be produced and maintained in ion traps [60.20]. In this case, the ions are produced inside the trap by electron-induced dissociative attachment collisions and photodetachment is investigated by monitoring the depletion of the negative ions. The most commonly used source for spectroscopic studies of negative ions is, however, a beam produced by an accelerator. In an accelerator-based apparatus the ions are extracted from the ion source and focused to form a collimated beam that is accelerated to a desired energy, typically 1–10 keV. Mass analysis of the ions is used to produce an elementally and isotopically pure beam that is essentially mono–energetic and unidirectional. The directed particles then drift to the interaction region through a beam line that is maintained at low pressure to minimize destructive collisions between the ions and the residual gas. Recently, negative ions have been injected into storage rings. In this case the ions make repeated passes through the interaction region. The enhanced luminosity associated with multiple-pass experiments makes it possible to investigate relatively rare processes that would be impossible in single-pass experiments.

60.3.2 Interacting Beams

The well-defined spatial dimensions of an ion beam readily permit an efficient overlap with a beam of photons. The two interacting beams are most often mated in either crossed or collinear beam geometries. The choice of geometry is typically determined by the types of particles to be detected, the detection geometry to be used, and the level of sensitivity and resolution required in the experiment. The crossed beam arrangement is best suited for spectroscopic studies of the photoelectrons ejected following photodetachment, since the electrons can most easily be collected from a spatially well defined interaction region. In a collinear beam arrangement it is better to detect the residual heavy particles produced in the photodetachment process since they all travel in the same direction, the direction of motion of the ion beam, and can be collected with high efficiency. Both the sensitivity and energy resolution attainable using a collinear beam apparatus are typically much higher than for a crossed beam apparatus. Nowadays, most experiments employ an apparatus in which the photon and ions are collinearly merged and the present chapter will focus on this arrangement. Figure 60.1 shows a typical collinear beam apparatus that was designed by *Hanstorp* [60.21]. The signal is enhanced when the photon and ion beams are collinearly merged due to the extended interaction region and the high collection and detection efficiencies of the heavy residual particles. The major source of background noise in collinear beam experiments is associated with the production of atoms or positive ions by collisions of the beam ions with the atoms or molecules of the residual gas in the vacuum chamber. By maintaining a high vacuum, typically 10^{-9} mTorr or better, one can keep the background contribution to a tolerable level. In interacting beam experiments involving the detection of the heavy residual particles, the energy resolution that is attainable is usually limited by kinematic broadening. The amount of broadening is determined by the properties of the ion and photon beams and how they are overlapped. The longitudinal velocity distribution of the ions in a beam is compressed when the ions undergo acceleration after leaving the ion source [60.22]. If the photon beam is merged collinearly with the "cooled" ion beam, the photons sample the narrowed velocity distribution, thus significantly reducing the contribution from Doppler broadening. If kinematic broadening is rendered negligible, the energy resolution is usually determined by the bandwidth of the light source.

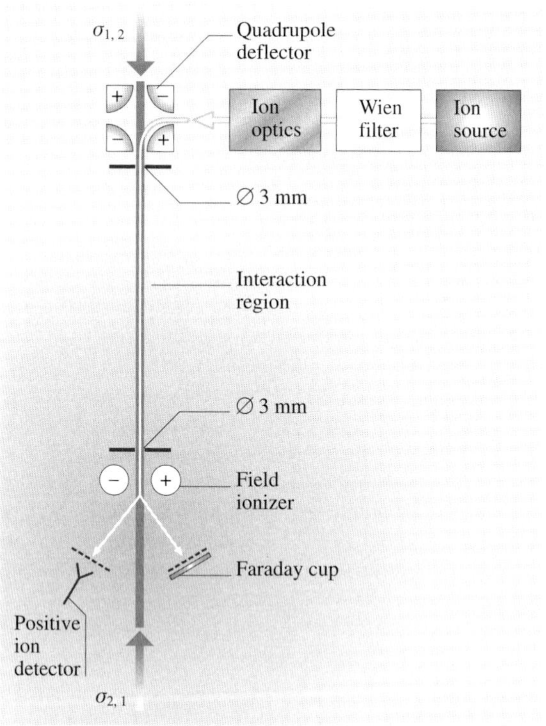

Fig. 60.1 A schematic of a collinear laser-negative ion beam apparatus. The quadrupole deflector is used to merge the laser and ion beam in the interaction region. The first laser is used to photodetach electrons from the ions. A second laser beam is directed along the common path of the first laser beam and the ion beam. This laser is used in the state-selective detection scheme based on resonance ionization. The positive ions produced in the sequential interaction of the negative ions with both laser beams and the external electric field are detected in a channel electron multiplier. The directions of the laser beams can be reversed

60.3.3 Light Sources

Since the particle density in the ion beam is typically low, it is important to have a light source that generates an intense beam of photons. In addition, the output of the light source must be tunable. Pulsed lasers are most often used in photodetachment experiments. Their time structure is often used to advantage in time-of-flight schemes to enhance the signal-to-background ratio. The large peak powers characteristic of pulsed lasers are required in multiphoton experiments. Lasers or laser-based sources used in photodetachment experiments span the wavelength range from the ultraviolet to the infrared. Second harmonic generation in a nonlinear crystal is the conventional method of producing UV radiation. The generation of tunable infrared radiation with wavelengths of a few μm has proven to be more difficult. Recently, however, Haugen and coworkers have performed experiments using infrared radiation produced in a laser-pumped Raman conversion cell [60.8]. Commercial optical parametric oscillators are also becoming more readily available. In order to investigate inner shell excitation and detachment processes it is necessary to access the VUV or X-ray region. These regions are currently outside the limits of lasers and can only be accessed at synchrotron radiation facilities.

60.3.4 Detection Schemes

Photodetachment events can be monitored by either measuring the attenuation of the negative ions or by detecting the particles (electrons, atoms or positive ions) produced in the breakup of the ion. In accelerator-based measurements the ion beam is too tenuous to be able to monitor attenuation and particle detection must be employed. The heavy residual particles, atoms or positive ions, are usually detected in experiments that employ collinearly merged beams of photons and negative ions. The selectivity and sensitivity of a measurement is improved significantly if the residual particles are state-selectively detected. In the case of residual excited atoms, the method most often employed is based on the use of a second laser to excite the atoms to a state near the ionization limit. This resonance step is followed by electric field ionization. The resulting positive ions constitute the signal.

60.4 Results

There have been several new developments in accelerator-based photodetachment measurements during the past decade. Tunable infrared radiation has been used in single photon and multiphoton experiments. State-selective detection schemes based on resonance ionization have been successfully employed in measurements of thresholds and resonances. The lifetimes of long-lived negative ions have been determined by the use of magnetic storage rings. Synchrotron radiation sources have been instrumental in the pioneering studies of inner shell processes in negative ions.

60.4.1 Threshold Measurements

A measurement of a threshold energy using photodetachment allows one to determine the binding energy of the extra electron in the negative ion or, equivalently, the electron affinity of the parent atom. *Andersen* et al. [60.7] have recently published a review of the methods currently used to measure the binding energies of atomic negative ions. The article includes an up-to-date compilation of recommended electron affinities. The simplest, and potentially the most accurate, method of determining binding energies is the laser photodetachment threshold (LPT) method. In this technique, the normalized yield of residual atoms is recorded as a function of the photon energy in the near-threshold region of the cross section. In most cases, the Wigner law can be fitted to the data and the threshold energy is determined by extrapolation. The Wigner law demonstrates that not all thresholds have the same energy dependence. The most accurate measurements to date involve detachment into an s-wave continuum. In the case of $l = 0$, the threshold energy dependence of $E^{1/2}$ is more pronounced than for cases with $l > 0$. S-wave photodetachment requires that a p-orbital electron be ejected. Haugen and coworkers have used tunable infrared spectroscopy to measure the binding energies of negative ions with open p-shells [60.23–25]. In these experiments the detachment process left the residual atom in its ground state so that state-selective detection was not needed.

Considerable experimental and theoretical effort have gone into investigating the negative ions of the alkaline earth elements since the experimental discovery [60.26] and subsequent theoretical confirmation [60.27] of the existence of a stable Ca^- ion in 1987. Prior to this time it was generally accepted that the closed s-shell configurations of the alkaline earth atoms would inhibit the production of stable negative ions. *Andersen* et al. [60.28] have reviewed progress in this field. *Andersen* and coworkers used the LPT method combined with state-selective detection to determine the binding energies of the negative ions of the heavier alkaline earths Ca^-, Sr^- and Ba^- [60.29–31]. No stable negative ions of Be and Mg have been found, but the Be^- ion is known to be metastable. These heavier ions are weakly bound but tunable infrared sources were not available at the time to detach them into the ground state of the parent atom. Instead, UV radiation was used to access an excited state threshold. In the case of Ca^- the 4s5s ^3S threshold was used since it allowed access to an s-wave continuum. In order to suppress the background noise in the experiment, the Ca atoms left in this excited state following detachment were selectively detected by a method based on resonance ionization. Before the excited Ca atom could radiatively decay, a second laser was used to induce a transition from the excited state to a high lying Rydberg state. The Rydberg atoms were efficiently ionized in an electrostatic field applied to the beam. The Ca^+ ion thus produced were used as the signal that, once normalized, was proportional to the photodetachment cross section. The structures of the heavy alkaline earths are very difficult to calculate. Three relatively loosely bound electrons move in the field of a highly polarizable core. Electron correlation and relativistic effects must be included in a theoretical description of their structure. As calculations became more sophisticated it became clear that correlations between the core electrons and between the valence and core electrons had to be taken into account in addition to the correlations between the valence electrons [60.32].

The negative ions of the alkali-metal elements have a closed s-shell configuration. In this case it is necessary to access an excited state threshold in order to detach into an s-wave continuum. Hanstorp and coworkers have used the LPT method combined with state-selective detection to measure the electron affinities of Li [60.33] and K [60.34]. Since accelerator-based measurements involve the use of fast and unidirectional beams of ions, one must take into account Doppler shifts in accurate measurements of threshold energies. In the K^- experiment [60.34], two separate sets of data were accumulated, one with the laser and ion beams co-propagating and the other with them counter-propagating. The Doppler shift can be eliminated to all orders by taking the geometric mean of the measured red-shifted and blue-shifted threshold energies [60.35].

LPT measurements can be used to selectively suppress one isotope relative to other isotopes of the same element, thereby changing the relative abundances from their natural values. This technique could be applied, for example, to the problem of sensitivity enhancement in mass spectrometry by suppressing unwanted isotopic interferences. *Sandstroem* et al. [60.36] recently performed a proof-of-principle experiment using the ^{34}S and ^{32}S isotopes. The goal of the experiment was to enrich the ^{34}S isotope relative to the more abundant ^{32}S isotope. Due to the large differential Doppler shifts associated with the fast moving ions of the two isotopes of different masses, it was possible to selectively photodetach one isotope and leave the other untouched. In this feasibility experiment, the $^{34}S/^{32}S$ ration was enhanced by a factor of > 50 over its natural value. With a better vacuum and the selection of a more suitable laser, it is predicted that the enhancement ratio could be significantly improved. The application of LPT to mass spectrometry clearly has the potential for enhancing the sensitivity in measurements of the abundances of rare and ultra-rare isotopes.

60.4.2 Resonance Parameters

The simplest negative ion is the two-electron H^- ion. This three-body Coulomb system is fundamentally important in our understanding of the role played by electron correlation in atomic structure. The pioneering measurements of the photodetachment of one and two electrons from the H^- ion were performed by *Bryant* and coworkers [60.37–39] several decades ago. The ASTRID (Aarhus storage ring Denmark) heavy ion storage ring has been used in two new measurements of the resonance structure in the vicinity of the $H(n=2)$ threshold [60.40, 41]. The energy resolution of these storage ring experiments was much higher than that attained in previous experiments. As a consequence, *Andersen* et al. [60.41] were able to observe, for the first time, a second resonance below the $H(n=2)$ threshold. In principle, the $1/r^2$ dipolar potential should support an infinite series of resonances below each excited state of the H atom [60.42]. Calculations, however, indicate that the series will be truncated after the third member by relativistic and radiative interactions [60.43].

Detachment continua contain a wealth of structure and many measurements of Feshbach resonances in non-hydrogenic negative ions have been reported during the past decade. The dipole polarizability of an atom increases with the degree of excitation, making it easier for electrons to attach to the excited parent atom. Series of Feshbach resonances containing several members are often found below excited state thresholds. Resonances in the photodetachment spectra of the metastable He^- ion [60.44] and the alkali-metal negative ions [60.45–49] have been studied extensively by Hanstorp and coworkers using the collinear beam apparatus shown in Fig. 60.1. *R*-matrix calculations [60.50–53] have generally been successful in predicting the energies and widths of most of the resonances observed in the experiments. There has been keen interest in the similarities and differences between the photodetachment spectra of Li^- and H^-.

The He^- ion is a metastable negative ion but it is sufficiently long lived to pass from the ion source to the interaction region with relatively little attenuation via autodetachment. Electric dipole selection rules limit photon-induced transitions from the 1s2s2p $^4P^0$ ground state to excited states with ^4S, ^4P and ^4D symmetry. The spectra of Feshbach resonances that lie below the $He(n = 3, 4, 5)$ thresholds have been investigated using the collinear beam apparatus shown in Fig. 60.1 [60.44, 54, 55]. Resonance ionization was used to state selectively detect the residual excited He atoms.

Figure 60.2 shows a high resolution spectrum of the resonance structure in the range 3.7–4.0 eV, a range that encompasses the H($n = 4$) thresholds. In this relatively small energy range *Kiyan* et al. [60.44] found many resonances exhibiting a variety of different shapes. The resonances labeled a,c,e are members of the ^{4}P series with dominant configurations of 1s4pnp ($n = 4, 5, 6$). The resonances labeled b,d appear to be the $n = 5, 6$ members of the 1s4sns ^{4}S series.

Recent studies using synchrotron radiation have revealed resonances in photodetachment cross sections in the X-ray and VUV regions that can be associated with the excitation of inner shell electrons. Resonances arising from K-shell excitation in the Li^- ion have been reported by *Kjeldsen* et al. [60.56] and *Berrah* et al. [60.57]. Similarly, resonances were found in a study of He^- [60.58] and C^- [60.59]. Resonances associated with L-shell excitation of the Na^- ion were observed by *Covington* et al. [60.60]. Figure 60.3 shows

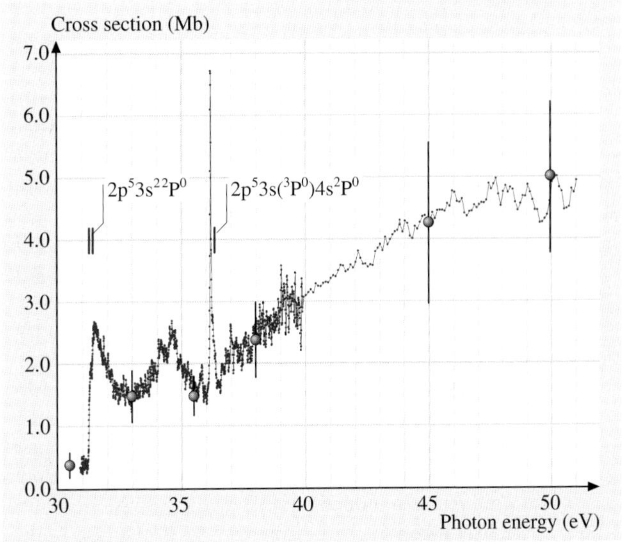

Fig. 60.3 Total cross section for the photodetachment of Na^- over the range 30–51 eV. Thresholds are indicated by vertical lines. The peaks are resonances associated with the excitation of a 2p core electron accompanied, in most cases, by the excitation of a 3s valence electron

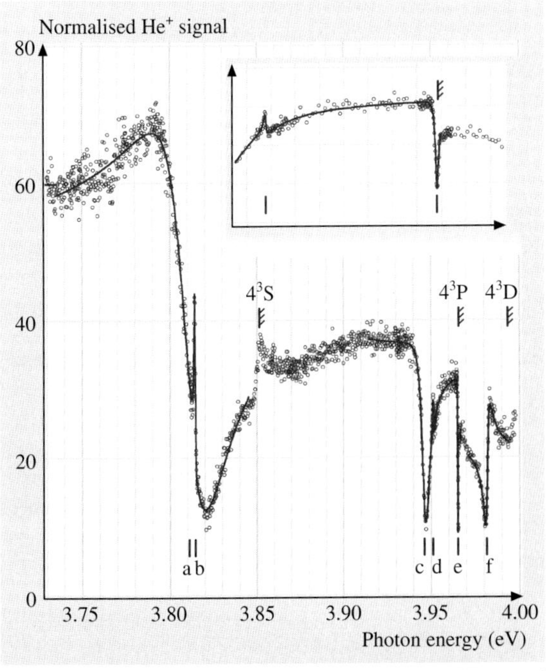

Fig. 60.2 Partial cross section for the photodetachment of He^- via the He(1s3p ^{3}P)+e(kp) continuum channel in the energy range 3.73–4.00 eV. The *open circles* represent the measured data. The fits to the sum of Shore profiles are shown by the *solid lines*. The energies of the resonances obtained from the fits are shown as *short vertical lines*. The *inset* shows the region near the He(1s4p ^{3}P) threshold in finer detail

part of the spectrum in which the dominant feature is a resonance at ≈ 36 eV that arises from the excitation of a pair of electrons – a 2p core electron and a 3s valence electron. Absolute cross sections were measured in most of the experiments. *R*-matrix calculations of the cross sections at energies corresponding to K-shell excitation have successfully accounted for most of the observations [60.61–63].

60.4.3 Lifetimes of Metastable Negative Ions

Heavy ion storage rings are well suited for studies of the radiative or autodetaching decay of long lived excited states of negative ions. They have also been used to investigate the effect of blackbody radiation on weakly bound stable negative ions [60.64]. Andersen and coworkers have used the ASTRID facility to measure the lifetimes of the metastable negative ions Be^- [60.65] and He^- [60.66] against autodetaching decay. The decay rate was measured by simply detecting the neutral atoms produced in the ring as a function of time after injection. The range of autodetaching lifetimes that can be measured in a storage ring depends on the size of the ring and on the destruction rate of the ions by collisional detachment with the residual gas in the ring. More recently, *Ellman* et al. [60.67] have used the CRYRING facility (at the Manne Siegbahn Laboratory in Stockholm) to measure the radiative lifetime of a bound excited state of a negative ion. In this proof-of-principle experiment, the lifetime of the $5p^5\ ^2P_{1/2}$ level of Te^- was measured to be 0.42(5) s. This value is in excellent agreement with the result of a multi-configuration Dirac Hartree-Fock (MCDHF) calculation. The $J = 1/2$ level radiatively decays to the $J = 3/2$ ground level, primarily via M1 transitions. The idea of the experiment was to monitor the population of the $J = 1/2$ level as a function of time after injection of the Te^- into the ring. This was accomplished by selectively photodetaching ions in the $J = 1/2$ level as the Te^- ions repeatedly passed through the field of a laser beam situated along one arm of the ring. The neutral Te atoms thus produced were used as the signal. Corrections were made for collisionally-induced detachment and repopulation. Data was taken at four different ring pressures. A linear fit to this data yielded the zero-pressure radiative lifetime of the excited $J = 1/2$ level.

60.4.4 Multielectron Detachment

Synchrotron radiation has been used over the past few years in order to study how negative ions respond to the absorption of high-energy photons. Photons in the VUV and X-ray regions will excite and/or detach inner shell electrons. Multiple electron detachment appears to be initiated by the detachment of a core electron. This process triggers the ejection of one or more valence electrons either by shake off or by interactions of the detached core electron with the valence electrons as it leaves the atom. The ALS (Advanced Light Source) has been used to investigate multiple electron detachment. Measurements of the absolute cross sections for the detachment of two electrons from the closed shell ions Cl^- [60.68] and F^- [60.69] have been reported.

References

60.1 H. S. W. Massey: *Negatives Ions* (Cambridge Univ. Press, London 1976)

60.2 B. M. Smirnov: *Negative Ions* (Mcgraw-Hill, New York 1972)

60.3 D. R. Bates: Adv. At. Mol. and Opt. Phys. **27**, 1 (1991)

60.4 S. J. Buckman, C. W. Clark: Rev. Mod. Phys. **66**, 539 (1994)

60.5 C. Blondel: Physica Scripta **T58**, 31 (1995)

60.6 T. Andersen: Physica Scripta **T43**, 23 (1991)

60.7 T. Andersen, H. K. Haugen, H. Hotop: J. Phys. Chem. Ref. Data **28**, 1511 (1999)

60.8 R. C. Bilodeau, H. K. Haugen: *Photonic, Electronic and Atomic Collisions* (Rinto Press, New York 2002)

60.9 U. Fano: Rep. Prog. Phys. **46**, 96 (1983)

60.10 E. P. Wigner: Phys. Rev. **73**, 1002 (1948)

60.11 T. F. O'Malley: Phys. Rev. **137**, 1668 (1965)

60.12 S. Watanabe, C. H. Greene: Phys. Rev. A **22**, 158 (1980)

60.13 J. Sandstroem, G. Haeffler, I. Kiyan, U. Berzinsh, D. Hanstorp, D. J. Pegg, J C. Hunnell, S. J. Ward: Phys. Rev. A **70**, 052707 (2004)

60.14 P. Kristensen, U. V. Pedersen, V. V. Petrunin, T. Andersen, K. T. Chung: Phys. Rev. A **55**, 978 (1997)

60.15 U. Fano: Phys. Rev.A **124**, 1866 (1961)

60.16 B. W. Shore: Phys. Rev. **171**, 43 (1968)

60.17 M. Crance: Comments At. Mol. Phys. **2**, 95 (1990)

60.18 M. D. Davidson, H. G. Muller, H. B. van Linden van den Heuvell: Comments At. Mol. Phys. **29**, 65 (1993)

60.19 R. Middleton: Nucl. Instrum. Methods **214**, 139 (1983)

60.20 D. J. Larson C. J. Edge, R. E. Elmquist, N. B. Mansour, R. Trainham: Physica Scripta **T22**, 183 (1988)

60.21 D. Hanstorp, M. Gustafsson: J. Phys. B **25**, 1773 (1992)

60.22 S. L. Kauffman: Opt. Comm. **17**, 309 (1976)

60.23 M. Scheer, R. C. Bilodeau, H. K. Haugen: Phys. Rev. Lett **80**, 2562 (1998)

60.24 M. Scheer, R. Bilodeau, J. Thägersen, H. K. Haugen: Phys. Rev. A **57**, 1493 (1998)

60.25 M. Scheer, R. Bilodeau, C. A. Brodie, H. K. Haugen: Phys. Rev. A **58**, 2844 (1998)

60.26 D. J. Pegg, J. S. Thompson, R. N. Compton, G. D. Alton: Phys. Rev. Lett. **59**, 2267 (1987)

60.27 C. Froese Fischer, J. B. Lagowski, S. H. Vosko: Phys. Rev. Lett. **59**, 2263 (1987)

60.28 T. Andersen, H. H. Andersen, P. Balling, P. Kristensen, V. V. Petrunin: J. Phys. B **30**, 3317 (1997)

60.29 V. V. Petrunin, H. H. Andersen, P. Balling, T. Andersen: Phys. Rev. Lett. **76**, 744 (1996)

60.30 P. Kristensen, C. A. Brodie, U. V. Pedersen, V. V. Petrunin, T. Andersen: Phys. Rev. Lett. **78**, 2329 (1997)

60.31 V. V. Petrunin, J. D. Voldstad, P. Balling, P. Kristensen, T. Andersen, H. K. Haugen: Phys. Rev. Lett. **75**, 3317 (1995)

60.32 S. Salmonson, H. Warston, I. Lindgren: Phys. Rev. Lett. **76**, 3092 (1996)

60.33 G. Haeffler, D. Hanstorp, I. Yu. Kiyan, A. E. Klinkmüller, U. Ljungblad, D. J. Pegg: Phys. Rev. A **53**, 4127 (1996)

60.34 K.T. Andersen, J. Sandström, I.Yu. Kiyan, D. Hanstorp, D.J. Pegg: Phys. Rev. A **62**, 022503 (2000)
60.35 P. Juncar, C.R. Bingham, J.A. Bounds, D.J. Pegg, H.K. Carter, R.L. Mlekodaj, J.D. Cole: Phys. Rev. Lett. **54**, 11 (1985)
60.36 J. Sandstroem P. Andersson, K. Fritioff, D. Hanstorp, R. Thomas, D.J. Pegg, K. Wendt: Nucl. Instrum. Methods B **217**, 513 (2004)
60.37 H.C. Bryant, B.D. Dieterle, J. Donahue, H. Sarifian, H. Tootoonchi, D.M. Wolfe, P.A.M. Gram, M.A. Yates-Williams: Phys. Rev. Lett. **38**, 228 (1977)
60.38 D.W. MacArthur, K.B. Butterfield, D.A. Clark, J.B. Donahue, P.A.M. Gram, H.C. Bryant, C.J. Harvey, W.W. Smith, G. Comtet: Phys. Rev. A **32**, 1921 (1985)
60.39 P.G. Harris, H.C. Bryant, A.H. Mohagheghi, R.A. Reeder, C.Y. Tang, J.B. Donahue, C.R. Quick: Phys. Rev. A **42**, 6443 (1990)
60.40 P. Balling, P. Kristensen, H.H. Andersen, U.V. Pedersen, V.V. Petrunin, L. Præstegaard, H.K. Haugen, T. Andersen: Phys. Rev. Lett. **77**, 2905 (1996)
60.41 H.H. Andersen, P. Balling, P. Kristensen, U.V. Pedersen, S.A. Aseyev, V.V. Petrunin, T. Andersen: Phys. Rev. Lett. **79**, 4770 (1997)
60.42 M. Gailitis, R. Damburg: Sov. Phys. JETP **17**, 1107 (1963)
60.43 E. Lindroth, A. Burgers, N. Brandefelt: Phys. Rev. **57**, 685 (1998)
60.44 I.Yu. Kiyan, U. Berzinsh, D. Hanstorp, D.J. Pegg: Phys. Rev. Lett. **81**, 2874 (1998)
60.45 U. Berzinsh, G. Haeffler, D. Hanstorp, A. Klinkmüller, E. Lindroth, U. Ljungblad, D.J. Pegg: Phys. Rev. Lett. **74**, 4795 (1995)
60.46 U. Ljungblad, D. Hanstorp, U. Berzinsh, D.J. Pegg: Phys. Rev. Lett. **77**, 3751 (1996)
60.47 G. Haeffler, I.Yu. Kiyan, U. Berzinsh, D. Hanstorp, N. Brandefelt, E. Lindroth, D.J. Pegg: Phys. Rev. A **63**, 053409 (2001)
60.48 G. Haeffler, I.Yu. Kiyan, D. Hanstorp, B.J. Davies, D.J. Pegg: Phys Rev. A **59**, 3655 (1999)
60.49 I. Yu Kiyan, U. Berzinsh, J. Sandström, D. Hanstorp, D.J. Pegg: Phys. Rev. Lett. **84**, 5979 (2000)
60.50 C. Pan, A.F. Starace, C.H. Greene: J Phys. B **27**, 137 (1994)
60.51 C. Pan, A.F. Starace, C.H. Greene: Phys. Rev. A **53**, 840 (1996)
60.52 C-N. Liu, A.F. Starace: Phys Rev. A **58**, 4997 (1998)
60.53 C-N. Liu, A.F. Starace: Phys Rev. A **59**, 3643 (1999)
60.54 A.E. Klinkmüller, G. Haeffler, D. Hanstorp, I.Yu. Kiyan, U. Berzinsh, C.W. Ingram, D.J. Pegg, J. Peterson: Phys. Rev. A **56**, 2788 (1997)
60.55 A.E. Klinkmüller, G. Haeffler, D. Hanstorp, I.Yu. Kiyan, U. Berzinsh, D.J. Pegg: J. Phys. B **31**, 2549 (1998)
60.56 H.K. Kjeldsen, P. Andersen, F. Folkmann, B. Kristensen, T. Andersen: J. Phys. B **34**, L353 (2001)
60.57 N. Berrah: Phys. Rev. Lett. **87**, 253002 (2001)
60.58 N. Berrah, J.D. Bozek, G. Turi, G. Akerman, B. Rude, H.-L. Zhou, S.T. Manson: Phys. Rev. Lett. **88**, 093001 (2002)
60.59 N.D. Gibson et al.: Phys. Rev. A **67**, 03070 (2003)
60.60 A.M. Covington, A. Aguilar, V.T. Davis, I. Alvarez, H.C. Bryant, C. Cisneros, M. Halka, D. Hanstorp, G. Hinojosa, A.S. Schlacter, J.S. Thompson, D.J. Pegg: J. Phys. B **34**, L735 (2001)
60.61 H.-L. Zhou, S.T. Manson, L. Voky, N. Feautrier, A. Hibbert: Phys. Rev. Lett. **87**, 02301 (2001)
60.62 H.-L. Zhou, S.T. Manson, L. Voky, A. Hibbert, N. Feautrier: Phys, Rev. A **64**, 012714 (2001)
60.63 O. Satsarinny, T.W. Gorczyca, C. Froese Fischer: J. Phys. B **35**, 4161 (2002)
60.64 H.K. Haugen, L.H. Andersen, T. Andersen, P. Balling, N. Hertel, P. Hvelplund, S.D. M'ller: Phys. Rev. A **46**, R1 (1992)
60.65 P. Balling, L.H. Andersen, T. Andersen, H.K. Haugen, P. Hvelplund, K. Taulbjerg: Phys. Rev. Lett. **69**, 1042 (1992)
60.66 T. Andersen, L.H. Andersen, P. Balling, H.K. Haugen, P. Hvelplund, W.W. Smith, K. Taulbjerg: Phys. Rev. A **47**, 890 (1993)
60.67 A. Ellmann, P. Schef, P. Lundin, P. Royen, S. Mannervik, K. Fritioff, P. Andersson, D. Hanstorp, C. Froese Fischer, F. Österdahl, D.J. Pegg, N.D. Gibson, H. Danared, A. Källberg: Phys. Rev. Lett. **92**, 253002 (2004)
60.68 A. Aguillar, J.S. Thompson, D. Calabrese, A.M. Covington, C. Cisneros, V.T. Davis, M.S. Gulley, M. Halka, D. Hanstorp, J. Sandström, B.M. McLaughlin, D.J. Pegg: Phys. Rev.A **69**, 022711 (2004)
60.69 V.T. Davis, A. Aguilar, J.S. Thompson, D. Calabrese, A.M. Covington, C. Cisneros, M.S. Gulley, M. Halka, D. Hanstorp, J. Sandström, B.M. Mclaughlin, G.F. Gribakin, D.J. Pegg: J. Phys. B (to be published)

61. Photon–Atom Interactions: Low Energy

Theoretical and experimental aspects of the atomic photoelectric effect at photon energies up to about 1 keV are presented. Relevant formulae and interpretations are given for the various excitation and decay processes. Techniques and results of photoelectron spectrometry in conjunction with synchrotron radiation are emphasized.

61.1 **Theoretical Concepts** 901
61.1.1 Differential Analysis 901
61.1.2 Electron Correlation Effects 904
61.2 **Experimental Methods** 907
61.2.1 Synchrotron Radiation Source 907
61.2.2 Photoelectron Spectrometry 908
61.2.3 Resolution and Natural Width 910
61.3 **Additional Considerations** 911
References 912

61.1 Theoretical Concepts

Scattering of low-energy photons proceeds predominantly through the photoelectric effect. In this process a photon γ of energy $h\nu$, angular momentum $j_\gamma = 1$, and parity $\pi_\gamma = -1$ interacts with a free atom or molecule A, having total energy E_i, angular momentum J_i, and parity π_i to produce an electron of energy ε, spin $s = 1/2$, orbital angular momentum ℓ, total angular momentum j, and parity $\pi_\mathrm{e} = (-1)^\ell$ and an ion A^+ with final total energy E_f, angular momentum J_f, and parity π_f. This process can be written as the reaction

$$\begin{aligned}&\gamma(h\nu,\, j_\gamma{=}1,\, \pi_\gamma{=}-1) + A(E_i,\, J_i,\, \pi_i)\\ &\quad \to A^+(E_f,\, J_f,\, \pi_f) + e^-[\varepsilon,\, \ell s j,\, \pi_\mathrm{e}{=}(-1)^\ell]\,.\end{aligned} \tag{61.1}$$

Conservation laws require that

$$\begin{aligned}&h\nu + E_i = \varepsilon + E_f\,,\\ &\boldsymbol{J}_i + \boldsymbol{j}_\gamma = \boldsymbol{J}_f + \boldsymbol{s} + \boldsymbol{\ell}\,,\\ &\pi_i \cdot \pi_\gamma = \pi_f \cdot \pi_\mathrm{e} = (-1)^\ell \cdot \pi_f\,.\end{aligned} \tag{61.2}$$

Since $E_f - E_i$ becomes quite large for inner shells or deep core levels, scattering of low-energy photons involves the removal of an electron from a valence or shallow core level. In the low-energy regime, from the first ionization threshold to $h\nu \approx 1$ keV, the photoelectric effect accounts for more than 99.6% of the photon interactions in the elements, with elastic scattering contributing the remainder [61.1, 2]. Ionization by inelastic scattering, the Compton effect, assumes increasing importance with the higher photon energies and the lower Z elements. Above the first ionization potential, the total photoabsorption cross section and the photoionization cross section are essentially equivalent at the lower photon energies.

The cross section σ_{if} for producing a given final ionic state in the photoionization process is given by

$$\sigma_{if} = \frac{4\pi^2\alpha^2}{k} \sum |\langle \Psi_f \mid \hat{T} \mid \Psi_i \rangle|^2\,, \tag{61.3}$$

where k is the photon momentum, $\hat{T}$ is the transition operator, Ψ_i and Ψ_f are the wave functions of the initial and final states, and the summation includes an average over all initial states and a summation over all the unobserved variables in the final state. A detailed derivation of this expression, including the different forms for $\hat{T}$, is given in the articles by *Fano* and *Cooper* [61.3] and *Starace* [61.4] and in Chapt. 24. The total cross section is given by the sum of all these different partial cross sections, σ_{if}.

61.1.1 Differential Analysis

Detailed information about the photoionization process can be obtained most directly in emission measurements, especially those involving the photoelectron. The resulting photoelectron spectrum yields the energy and intensity for a given interaction. Further differentiation is obtained by varying the angle of observation and by a spin analysis of the photoelectron. Hence, electron emission analysis can reveal all energetically allowed photoprocesses connecting an initial atomic state i to a final ionic state f and yield their dynamic properties. When averaged over the spin, the differential cross section $\mathrm{d}\sigma_{if}/\mathrm{d}\Omega$ is given in terms of the partial cross section σ_{if} and an expression involving an expansion in Legendre polynomials of order n with the coefficients B_n:

$$\frac{\mathrm{d}\sigma_{if}}{\mathrm{d}\Omega} = \left(\frac{\sigma_{if}}{4\pi}\right) \sum_n B_n P_n(\cos\theta) , \tag{61.4}$$

where the angle θ is measured between the direction of the emitted electron and the unpolarized incoming photon beam. In the *dipole* approximation, which describes the dominant process at low energy, only the terms containing P_0 and P_2 contribute. Then (61.4) reduces, for a photon beam with linear polarization p, to

$$\frac{\mathrm{d}\sigma_{if}}{\mathrm{d}\Omega} = \left(\frac{\sigma_{if}}{4\pi}\right) \left[1 + \frac{\beta_{if}}{4}(1+3p\cos 2\theta)\right] , \tag{61.5}$$

where the angle θ lies in the plane perpendicular to the direction of propagation and is measured with respect to the major axis of the polarization ellipse [61.5]. Then, the differential cross section, or photoelectron angular distribution, is characterized by the single angular distribution or anisotropy parameter β_{if} for a particular process $i \to f$. For observation at the so-called pseudomagic angle θ_{m}, defined as

$$\theta_{\mathrm{m}} = \frac{1}{2}\cos^{-1}\left(\frac{-1}{3p}\right) , \tag{61.6}$$

the differential cross section $\mathrm{d}\sigma_{if}/\mathrm{d}\Omega$ becomes proportional to the angle-integrated, or partial, cross section σ_{if}.

In the absence of correlation effects, the partial cross section σ_{if} for the production of an individual final state and the corresponding anisotropy parameter β_{if} are given by simple expressions derived from a single-particle model [61.1, 4]. For the central field potential,

$$\sigma_{if} = \frac{4\pi^2\alpha}{3} a_0^2 N_{n\ell} h\nu \times \left[\left(\frac{\ell}{2\ell+1}\right) R_-^2 + \left(\frac{\ell+1}{2\ell+1}\right) R_+^2\right] , \tag{61.7}$$

where $N_{n\ell}$ is the occupation number of the subshell, and

$$\beta_{if} = \frac{\ell(\ell-1)R_-^2 + (\ell+1)(\ell+2)R_+^2}{(2\ell+1)\left[\ell R_-^2 + (\ell+1)R_+^2\right]} - \frac{6\ell(\ell+1)R_+R_-\cos\Delta}{(2\ell+1)\left[\ell R_-^2 + (\ell+1)R_+^2\right]} . \tag{61.8}$$

The subscripts + and − refer to the $(\ell+1)$ and $(\ell-1)$ channels respectively, and $\Delta = \delta_+ - \delta_-$ is the difference in phase shift between these two allowed outgoing waves. The parameter $R_\pm$ is the radial dipole matrix element connecting the electron in the bound orbital with orbital angular momentum ℓ with the outgoing wave having orbital angular momentum $\ell \pm 1$.

Effects of the electron correlation on the direct photoionization process can result in values for β which are not reproduced by the Cooper–Zare expression (61.8) [61.4]. The contribution of the different partial waves to the outgoing wave function can, however, be ascertained through the angular momentum transfer formalism developed by *Fano* and *Dill* [61.6]. In this approach, one defines the angular momentum transferred from the photon to the unobserved variables j_{t} as

$$\boldsymbol{j}_{\mathrm{t}} = \boldsymbol{j}_\gamma - \boldsymbol{\ell} = \boldsymbol{J}_f + \boldsymbol{s} - \boldsymbol{J}_i , \tag{61.9}$$

where the second portion of the equality results from the conservation of angular momentum. For each allowed value of j_{t} the associated transfer can be defined as either parity favored or parity unfavored according to whether the product $\pi_i\pi_f$ is equal to $+(-1)^{j_{\mathrm{t}}}$ or $-(-1)^{j_{\mathrm{t}}}$ respectively. (All symbols have the same definition as in (61.1).) Calculation of the partial cross section for the production of a given final state characterized by the values $\boldsymbol{J}_f$ and s is then determined from the cross section corresponding to each angular momentum transfer according to

$$\sigma_{if} = \sum_{j_{\mathrm{t}}} \sigma(j_{\mathrm{t}}) . \tag{61.10}$$

The associated anisotropy parameter β_{if} is derived from a similar sum:

$$\sigma_{if}\beta_{if} = \sum_{j_{\mathrm{t}}=\mathrm{fav}} \sigma(j_{\mathrm{t}})_{\mathrm{fav}}\, \beta(j_t)_{\mathrm{fav}} - \sum_{j_{\mathrm{t}}=\mathrm{unfav}} \sigma(j_{\mathrm{t}})_{\mathrm{unfav}} . \tag{61.11}$$

The second equation derives from the fact that $\beta(j_{\mathrm{t}})$ for each parity-favored value must be calculated separately, whereas for the parity-unfavored case $\beta(j_{\mathrm{t}}) = -1$ always.

The physical effect described by the angular momentum transfer approach is the interaction between an electron and the anisotropic distribution of the other electrons in the atom. Thus, it becomes most useful in the case of ionization from an open-shell atom having an extra electron or a hole in a shell with $\ell \neq 0$. An illustrative example is the 3s ionization of chlorine. Here $\beta_{if} = 2$ identically in (61.8) because only the single value $\ell = 1$ is allowed in the single-particle model. However, the three possible values, $j_t = 0,\ 1,\ 2$ are allowed, of which only the first corresponds to the Cooper–Zare or single-particle, central field result [61.4]. That $\beta \neq 2$ for 3s ionization of atomic chlorine has been demonstrated experimentally [61.7].

It is generally the case for ionization of elements which are found naturally in the atomic state that there is an equal population in all the fine-structure components of the initial state. This is because of the relatively small energies associated with the fine-structure splitting. (This does not necessarily apply to atomic species generated through a process of molecular dissociation or high-temperature metal vaporization.) Thus, the determination of all cross sections and angular distributions involves an average over these fine-structure components. However, it is possible to generate atoms in which one of the fine-structure components is preferentially populated. In this case there can also be a preferential ionization to a particular J-component of the final ionic state even in the limit in which the electron correlation is neglected, i.e., the geometrical limit. The partial intensities for the production of a given ionic state characterized by the angular momenta $L_f S_f J_f$ by removal of an electron from an orbital ℓ of a state characterized by $L_i S_i J_i$ are given by

$$R_\ell^{L_f S_f J_f} = \frac{[J_f][L_f][S_f]}{[1/2][\ell]} g(\ell, L_i, S_i, L_f, S_f) \times \sum_{j=\ell-1/2}^{j=\ell+1/2} [j] \begin{Bmatrix} \ell & 1/2 & j \\ L_i & S_i & J_i \\ L_f & S_f & J_f \end{Bmatrix}^2 . \tag{61.12}$$

Here the term in curly brackets is a 9–j symbol, and the notation $[J] = 2J+1$ is used. The quantities $g(\ell, L_i, S_i, L_f, S_f)$ are weighting factors determined solely by the initial-state wave function. For the case in which ℓ represents a closed shell, these factors are equal to unity [61.8].

In situations in which the target atoms possess an initial orientation, i.e., have an average value $\langle J_z \rangle \neq 0$, or if the ionization is performed with circularly polarized radiation, the electrons which are produced have a net spin [61.9, 10]. It is also possible that unpolarized atoms which are ionized by unpolarized photons can have a net spin, provided that the detection is carried out at a specific angle, and the ionization is from a given fine-structure component of the initial state to a given fine-structure component of the final state. In the latter case, the transverse spin polarization is given by

$$P = \frac{-2\xi \sin\theta \cos\theta}{1 + \beta P_2(\cos\theta)} , \tag{61.13}$$

for linearly polarized radiation, and by

$$P = \frac{2\xi \sin\theta \cos\theta}{2 - \beta P_2(\cos\theta)} , \tag{61.14}$$

for unpolarized radiation [61.11, 12]. The angle θ is the same as in the angular distribution measurement; the parameter ξ is the spin parameter analogous to β; and $P_2(\cos\theta)$ is the Legendre polynomial of order 2.

Yet another parameter which describes the differentiation inherent in the photoionization process is the alignment A, which reflects an anisotropy in the quadrupole distribution of the angular momentum J_f of the ion [61.12]. For the cylindrically symmetric coordinate system appropriate to dipole photon excitation, only one moment A_0 of the distribution is nonzero. This is defined by

$$A_0 = \frac{\sum_{m_j} \left[3m_j^2 - J_f(J_f+1)\right] \sigma(m_j)}{J_f(J_f+1) \sum_{m_j} \sigma(m_j)} , \tag{61.15}$$

where $\sigma(m_j)$ is the partial cross section for production of a given m_j component of J_f. A very useful approach to the interpretation of the alignment can be obtained through the angular-momentum-transfer formalism [61.13]. In this approach the angular momentum transfer j_t is defined as

$$\boldsymbol{j}_t = \boldsymbol{j}_\gamma - \boldsymbol{J}_f . \tag{61.16}$$

In contrast to the case of the electron angular distribution, it is possible to derive an alignment for each value of j_t as a function of the angular momentum J_f of the ion. The net alignment is then the incoherent sum of the contributions corresponding to each:

$$A_0 = \sum_{j_t} A_0(j_t)\sigma(j_t) \Big/ \sum_{j_t} \sigma(j_t) \tag{61.17}$$

If the photoionization produces an ion in an excited state which decays by photoemission, the parameter A_0 is reflected either in the angular distribution of the fluorescence photons $I(\theta)$ or the linear polarization P measured at one angle, typically 90°, according to

$$I(\theta) = I_0 \left[1 - \frac{1}{2} h^{(2)} A_0 P_2(\cos\theta) + \frac{3}{4} h^{(2)} A_0 \sin^2\theta \cos(2\chi)\cos(2\eta)\right] \tag{61.18}$$

Part E | 61.1

or, for $\theta = \pi/2$ and $\chi = 0$,

$$P = \frac{I(\eta=0) - I(\eta=\pi/2)}{I(\eta=0) + I(\eta=\pi/2)} = \frac{3h^{(2)}A_0}{4+h^{(2)}A_0} , \tag{61.19}$$

respectively. The angle θ is the angle at which the fluorescence is determined, and the angle χ is measured between the axis of the polarization selected by the detector and the quantization axis. The polarization of the fluorescence is given by $\zeta = (\cos\eta,\ \mathrm{i}\sin\eta,\ 0)$. The quantity $h^{(2)}$ is a ratio of 6–j symbols depending on the angular momenta J_f of the intermediate ion and the final state J'_f:

$$h^{(2)} = (-1)^{J_f - J'_f} \times \left\{ \begin{matrix} J_f & J_f & 2 \\ 1 & 1 & J'_f \end{matrix} \right\} \Big/ \left\{ \begin{matrix} J_f & J_f & 2 \\ 1 & 1 & J_f \end{matrix} \right\} . \tag{61.20}$$

When it is energetically allowed, a hole in a shallow inner-shell will preferentially undergo Auger decay, emitting an electron with an energy ε_A determined by the energy difference between the energy E_f of the ion and E'_f of the state of the doubly-charged ion to which the decay occurs. Angular analysis of the Auger electrons reflects the alignment of the intermediate ionic state, which is different from, and does not bear a one-to-one relationship to, the angular distribution parameter β of the photoelectrons. Normally, Auger decay is regarded as a two-step process in which the first step is the production of the hole and the release of the primary photoelectron, followed by the decay and the release of the second electron. Within this approximation [61.14], the angular distribution of the Auger electrons takes on the simple form

$$I(\theta) = \left(\frac{I_0}{4\pi}\right) [1 + \alpha_2 A_0 P_2(\cos\theta)] . \tag{61.21}$$

Here $P_2(\cos\theta)$ is the second-order Legendre polynomial, and α_2 is the matrix element corresponding to the Auger decay. For the specific case in which the Auger decay is to a final ionic state of $^1S_0$ symmetry, α_2 is purely geometric, and a measurement of the angular distribution leads directly to a determination of the alignment. Correspondingly, if the alignment of a specific state can be determined through such a decay, then analysis of the angular distribution of the decay to other states provides a value for α_2.

61.1.2 Electron Correlation Effects

The primary focus of advanced studies in photoionization is to determine the role played by electron correlation in the structure and dynamics of electron motion above the lowest ionization threshold. Because the form of the interaction potential for the Coulomb interaction is very well known, theory can focus on the many-body aspects of the process (Chapt. 23). Electron correlation manifests itself in many ways. Most prominent are the appearance of autoionization structure due to the excitation of one or two electrons, the production of correlation satellites due primarily to the ionization of one electron accompanied by the excitation of another, and the creation of two continuum electrons in a double ionization process.

Autoionization resonances are perhaps the oldest known features associated with electron correlation (Chapt. 25). These features arise when the absorption of a photon creates a localized state which lies in energy above at least one ionization limit. This state is then degenerate in energy with a state of an electron in the continuum, and the interaction between these states results in the decay of the quasi-localized state into the continuum. Such resonance states appear in an absorption spectrum in the form of strong, localized variations over an energy range characteristic of the width Γ of the feature, which is in turn related to the lifetime τ of the state by

$$\Gamma = \hbar/\tau . \tag{61.22}$$

In contrast to absorption features between bound states, autoionization resonances are characterized by having an asymmetric line shape. When only one localized state and one continuum are involved, these line shapes can be derived analytically, as first shown by *Fano* [61.15] and later by *Shore* [61.16], resulting in simple parametrized forms which are suitable for numerical calculation of overlap integrals for determining widths. For the Fano profile

$$\sigma(\epsilon) = \sigma_a \frac{(\epsilon+q)^2}{\epsilon^2+1} + \sigma_b , \tag{61.23}$$

with

$$\epsilon = \frac{E - E_r}{(\Gamma/2)} , \tag{61.24}$$

the parameter q describes the asymmetry of the line, E_r is the resonance position, and Γ is the width of the line. The parameters σ_a and σ_b reflect the relative contributions to states in the continuum which do and do not interact with the autoionizing state respectively. The energy E_r does not correspond to the peak energy E_m of the resonance feature but is related to the maximum through

$$E_m = E_r + \frac{\Gamma}{2q} . \quad (61.25)$$

The Shore profile,

$$\sigma(\epsilon) = C(\epsilon) + \frac{A\epsilon + B}{\epsilon^2 + 1} , \quad (61.26)$$

describes the same phenomenon except that the interpretation of the parameters A and B is different. In this case, they represent products of dipole and Coulomb matrix elements. $C(\epsilon)$ is the continuum contribution.

From an experimental point of view, the parametrized forms for the Fano and Shore profiles are very useful as a basis for fitting autoionization spectra. However, they both have the limitation that they only describe the interaction of an isolated state with the continuum. While they can be extended to include several continuua [61.4], they do not allow for an interaction among two or more localized states [61.17, 18]. Nevertheless, it is possible to use these functions to achieve often good fits of states which do interact with each other, as these functions are mathematical representations of localized resonances in a continuous spectral distribution. If this is done, the parameters no longer have the physical meaning which they have for the noninteracting case.

Mixing of discrete ionization channels with competing continuum channels adds complexity to the photoionization process, not just in the classical autoioinization regime but also in the vicinity of inner shells [61.19–21]. In a rigorous application of the Mies formalism [61.17], feasible with modern computer power, even complex experimental spectra can now be satisfactorily interpreted and reproduced. A case in point is the excitation spectrum from the 2p level of the open-shell chlorine atom [61.19].

The process of autoionization is discussed in more detail in Chapt. 25. In Fig. 61.1 an example is shown of the set of $2s^2p^3(^4S) \to 2s2p^3np$, $n \geq 3$ autoionization resonances which decay into the ^3P ground state of the N^+ ion [61.22]. The energies E_n of these resonances are related to the ionization limit E_∞ of the series by the Rydberg formula

$$E_n = E_\infty - R_\infty/(n - \mu_s)^2 \quad (61.27)$$

where n is the principal quantum number and μ_s is the quantum defect characteristic of a given series and reflecting the short-range electrostatic interactions of the electron with the ion core. Values of μ_s for s, p, d, and f electrons have been calculated for atoms and ions up to $Z = 50$ [61.23]. For high precision work, the reduced Rydberg constant

$$R_M = \frac{R_\infty}{1 + 5.485\,799 \times 10^{-4}/(M_A - m_e)} \quad (61.28)$$

should be used instead of the value R_∞ for infinite nuclear mass (Chapt. 1). The atomic mass M_A and the electron mass m_e are in a.u.

A process closely related to the autoionization phenomenon is resonant Auger decay. This process differs from the ordinary Auger process [61.24] in that an electron from an inner shell is not ionized but excited to either a partially filled or an empty subshell. It may be viewed either as an Auger process or as autoionization. Such an inner-shell excited state of a neutral atom (molecule) lies above one or more of the ionization limits of the singly ionized species and consequently must decay by electron emission unless the decay is forbidden by selection rules. As a result, resonance structure will appear superimposed on the continua of direct photoionization from the various subshells. From a most general point of view, the resonant Auger process can be considered as resonances in the continua of single photoionization, while the ordinary Auger process can be regarded as resonances in the continua of double photoionization. If excitation proceeds to a partially filled subshell *within* a principal shell, as, for example, Mn 3p → 3d [61.20], interference between the direct photoionization channels and the indirect resonance channel may be strong, and the lineshapes are

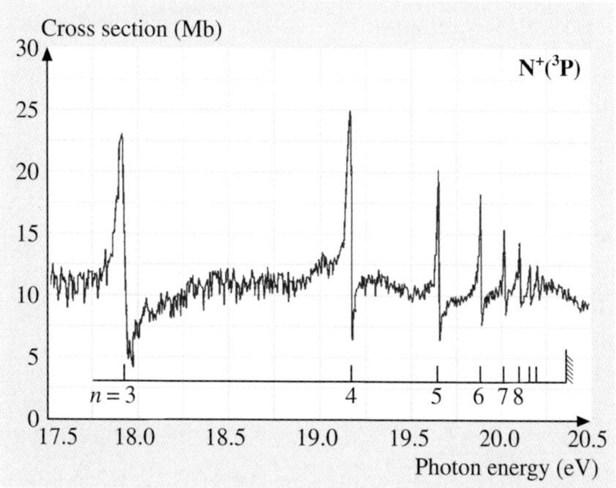

Fig. 61.1 Autoionization resonances $2s^22p^3(^4S) \to 2s2p^3np$ in atomic nitrogen

given by (61.23) with arbitrary q values and σ_a/σ_b ratios. If, however, the excitation proceeds to an empty shell, as, for example, Mg 2p $\rightarrow$ ns or nd, $n \geq 4$, interference with the direct channels is likely to be negligible, and the resulting resonances are distinguished by essentially Lorentzian line shapes (as for normal Auger lines) with $q \gg 1$ and $\sigma_a/\sigma_b \gg 1$ in (61.23). For a given excitation state a number of resonance peaks may arise because more than one ionization channel is usually available and, in addition, the excited electron can change its orbital from n to $n' = n \pm 1, 2, \ldots$ in a shakeup or shakedown process [61.25].

As a consequence of the electron–electron interactions which occur simultaneously with the electron–photon interaction, ions are produced in states which do not correspond to those which would be expected based on an interpretation using an independent particle model, which allows for only a single-electron transition (Chapt. 24). Evidence for these states appears as correlation satellites in the photoelectron spectrum, the Auger electron spectrum or the X-ray spectrum [61.26]. Figure 61.2 [61.27] presents as an example the photoelectron spectrum of argon produced by photons with $h\nu$ =60.6 eV. In addition to the 3s main line of single electron photoionization (and the 3p main lines not shown) numerous satellite lines are seen as the manifestation of two-electron transitions involving ionization-with-excitation correlations. It is convenient to categorize the satellites in a photoelectron spectrum according to various electron correlations, as, for example, initial state interactions which mix different configurations into the initial state, and final state interactions, which include core relaxation and electron–electron interactions in the final ionic state, electron–continuum, and continuum–continuum interactions. While initial-state correlations are essentially independent of the photon energy, final-state correlations depend on the energy of the photon through the interactions with the continuum channels. However, the heuristic value of placing correlation effects into a strict classification scheme is limited by the fact that their relative strengths depend on the basis set used in a particular theoretical model and its expansion into a "fully correlated" system within a given gauge [61.3, 4, 28, 29] (see also Chapt. 24).

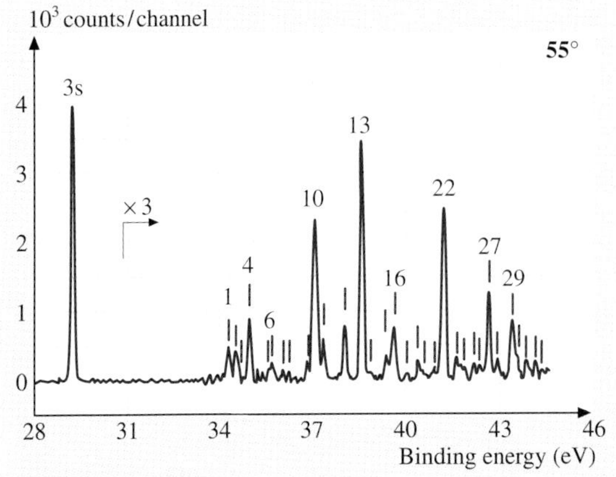

Fig. 61.2 Photoelectron spectrum (PES) of 3s, 3p satellites in argon at a photon energy of 60.6 eV. Note the reduced intensity of the satellites compared to the 3s main line

Another manifestation of double-electron processes is the simultaneous excitation of two electrons to bound states. These states may decay by electron or photon emission and are seen as resonance structures above the thresholds of inner-shell ionization or near autoionizing members of Rydberg series. As single or double ionization continua are usually strong in the spectral range of the double excitations, interference occurs, and the lineshapes can display dispersion forms. Typically, the cross section for the sum of all correlation processes is between 10 and 30% of that for single photoionization, but may exceed this range considerably in special cases.

Photon scattering near thresholds is complex because of the possibility of strong interactions between the various particles created and the different modes of deexcitation (Chapt. 62). In the case of ionization-with-excitation processes, the threshold cross section is finite, as it is for single electron photoionization, in accord with Wigner's theorem [61.30] (Sect. 60.2.1). In the case of double photoionization, the cross section is zero at threshold and then rises according to Wannier's law [61.31] (Chapt. 52). For the motion of two electrons with essentially zero kinetic energies in the field of the ionic core,

$$\sigma \propto E^{(2\mu-1)/4}\,, \tag{61.29}$$

where μ depends on the value of the nuclear charge Z through

$$\mu = \frac{1}{2}\left(\frac{(100Z-9)}{(4Z-1)}\right)^{\frac{1}{2}}\,. \tag{61.30}$$

For $Z = 1$ the Wannier exponent has the value 1.127.

In the case of Auger decay following ionization at threshold, interaction between the two electrons results in a shift in the energy of the Auger electron and a corresponding shift in that of the photoelectron (to conserve energy), as well as an asymmetry in the shape of the Auger electron peak and a corresponding asymmetry in the photoelectron peak shape. In this so-called post-collision interaction (Chapt. 62), the lineshape, averaged over angles, has the form [61.32]

$$K(\varepsilon)=\frac{(\Gamma/2\pi)}{(\varepsilon-\varepsilon_{\mathrm{A}})^2+(\Gamma/2)^2}f(\varepsilon) \tag{61.31}$$

with

$$f(\varepsilon)=\frac{\pi\psi}{\sinh(\pi\psi)}\exp\left[2\psi\tan^{-1}\left(\frac{\varepsilon-\varepsilon_{\mathrm{A}}}{(\Gamma/2)}\right)\right]. \tag{61.32}$$

In the above equations, ε is the energy of the Auger line, ε_{A} is the nominal Auger energy, Γ is the initial hole-state width, and the parameter $\psi=1/\sqrt{2\varepsilon_{\mathrm{e}}}-1/\sqrt{2\varepsilon_{\mathrm{A}}}$, with ε_{e} being the energy of the photoelectron, and $\varepsilon_{\mathrm{A}}\geq\varepsilon_{\mathrm{e}}$.

61.2 Experimental Methods

An overview of the experimental approaches to the study of photon interactions at low energies is given in Fig. 61.3. The sketch emphasizes the interaction of a polarized photon beam with a small static or particle-beam target of atoms, ions, molecules, or clusters, and the detection of the reaction products at various angles in a plane perpendicular to the direction of propagation of the photon beam, where the general equation (61.5) is valid. Emission products, such as electrons, ions, or photons, may be studied by way of the total yields, which can be related to the total photoionization cross section, or by differential analysis in a spectrometer according to energy, intensity, emission angle, and polarization. The various particles may be measured independently, simultaneously, or in coincidence. The photon monitor provides the information for normalization of the data with regard to flux and polarization. The photon monitor can also be used for a measurement of the total photoionization cross section, equivalent to the photoabsorption or photoattenuation coefficient at low photon energies. For this purpose, the size of the target source is advantageously increased in the direction of the photon beam. While experimental apparatus differs, sometimes drastically, for the photon sources as well as for the spectrometry of electrons, ions, and fluorescence photons, many features are common, and the relationships of the measured quantities to basic properties of the atoms and the photon–atom interaction are similar. Thus, the following will place emphasis only on the roles of the synchrotron radiation source and photoelectron spectrometry, whereas specific references to other methods can be found elsewhere [61.5, 33–38].

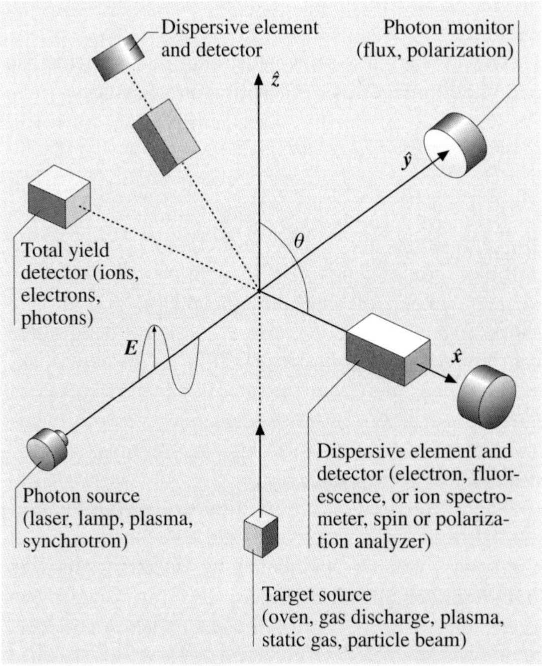

Fig. 61.3 Generic arrangement for detection of particles in an emission measurement. The incoming radiation is assumed to be linearly polarized along the z-axis

61.2.1 Synchrotron Radiation Source

The primary source of photons over a broad energy range for experiments in the VUV and soft X-ray region of the spectrum is the synchrotron radiation source [61.39, 40]. In a synchrotron or electron storage ring, radiation is produced as the electrons are bent to maintain the closed orbit. Such bending magnet radiation is emitted in a broad continuous spectrum which begins in the infra-red and ends sharply at a critical photon energy given by $h\nu_{\mathrm{c}}=\kappa E_{\mathrm{e}}^2$, where E_{e} is the energy of the electrons in the ring and κ is a constant characteristic of the ring. Synchrotron radiation can also be generated by introducing additional magnetic field structures [61.41] into the ring, such as undulators or wigglers, which produce a deviation of the electron motion from a straight path in a well-defined manner. Wiggler radiation has the same spectrum as a bending magnet, except that the critical energy is generally much higher because the effective magnetic fields can be larger than those of the bending magnet. Undulator radiation is very different in that it consists of a sharp spiked profile of about a 1% bandwidth at energies determined by the magnetic field within the undulator and by the electron beam energy in the storage ring.

Synchrotron radiation, no matter what the magnetic field structure of the source, requires monochromatization before it can be used for experiments. For the wavelengths of interest in low-energy photon scattering, this can be achieved by using grating instruments with a metallic coating on the grating surface. The highest resolution possible is obtained through the use of a normal incidence monochromator (NIM) with a plane grating set at normal incidence. However, because the reflectivity of the metallic coating at normal beam incidence decreases drastically as the photon energy increases, use of a NIM has an upper limit of about 40 eV. At energies above this, up to about 1 keV, gratings can still be used but must be mounted at grazing incidence. There is a number of functional designs for these grazing-incidence instruments which vary in the shape of the grating – spherical, toroidal, or plane surfaces (SGM, TGM, PGM) – and the associated optics. Above 1 keV, gratings are no longer suitable, and crystal diffraction must be used. While the radiation emerging from a beamline which couples the monochromator to a bending magnet, wiggler, or undulator has a high degree of linear polarization, varying from 80 to 99% in the plane of the electron orbit, a useful flux of circularly polarized radiation can be derived from out-of-plane radiation [61.33], by the use of multiple reflection optics, or from a helical undulator.

61.2.2 Photoelectron Spectrometry

The primary particle emitted in photoionization is the photoelectron. Hence, a photoelectron spectrum provides a detailed view of the photon interaction by (a) specifying the individual processes from an initial state i to a final state f by way of the electron energy, (b) determining their differential and partial cross sections by recording the number of electrons as a function of emission angle, and (c) measuring the polarization of the electrons by a spin analysis (spin polarimetry). The experimental approach is governed largely by the relations (61.3), (61.5), and (61.6). The number of electrons $N_{if}(\mathrm{e})$ detected per unit time at an angle θ within an energy interval $\mathrm{d}\varepsilon$ and within a solid angle $\mathrm{d}\Omega$ is given by

$$N_{if}(\mathrm{e}) = GN(h\nu)N(A)f(h\nu)f(\varepsilon)\frac{\mathrm{d}\sigma_{if}}{\mathrm{d}\Omega}\,\mathrm{d}\varepsilon \quad (61.33)$$

where G is a geometry factor, which includes the source dimensions, $N(h\nu)$ the number of photons, $N(A)$ the number of atoms in the source, $f(h\nu)$ and $f(\varepsilon)$ efficiency factors depending, respectively, on photon and electron kinetic energies, and $\mathrm{d}\sigma_{if}/\mathrm{d}\Omega$ is the differential cross section for a particular transition $i \to f$. Equation (61.33) assumes that $\mathrm{d}\Omega$ and $\mathrm{d}\varepsilon$ are sufficiently small that integration over the pertinent parameters is not needed. Since $N_{if} \propto \mathrm{d}\sigma_{if}/\mathrm{d}\Omega$, a measurement at two angles, e.g., $\theta = 0°$ and $90°$, yields the electron angular distribution parameter β_{if} according to (61.5), and a measurement at θ_m (61.6) yields the partial cross section σ_{if}. In the case of closed-shell atoms, $n\ell j$ notation is sufficient to designate single ionization to an $\varepsilon\ell$ continuum, e.g., $3\mathrm{p}_{1/2,3/2} \to \varepsilon\mathrm{s}$ or $\varepsilon\mathrm{d}$ in argon, but for open-shell atoms LSJ notation is required, e.g., $3\mathrm{p}^5(^2\mathrm{P}^\mathrm{o}_{3/2}) \to 3\mathrm{p}^4(^3\mathrm{P}^\mathrm{e}_{2,1,0}, {}^1\mathrm{D}^\mathrm{e}_2, {}^1\mathrm{S}^\mathrm{e}_0)\varepsilon\ell(^2\mathrm{D}^\mathrm{e}, {}^2\mathrm{P}^\mathrm{e}, {}^2\mathrm{S}^\mathrm{e})$ in chlorine. Similarly, for ionization-with-excitation transitions, the final state requires an open-shell designation.

The sum of the partial cross sections is equal to the total photoionization or absorption cross section

$$\sigma_\mathrm{tot} = \sum_{i,f} \sigma_{if} \quad (61.34)$$

where the σ_{if} encompass (a) single ionization events in all energetically accessible subshells $n\ell j$, or the LSJ multiplet components, (b) ionization-with-excitation events (shakeup or shakedown), and (c) double ionization events (shakeoff). All σ_{if} can be determined from a photoelectron spectrum from its discrete peaks (cf. Fig. 61.2) and from the continuum distribution of multiple ionization. However, the latter process is measured more readily by observing the multiply charged ions in a mass spectrometer.

The differentiation afforded by measuring the various partial cross sections, and the associated β parameters, can be augmented by differentiating the continuum channels according to the spin using a spin polarimeter [61.33]. In closed-shell atoms, this allows for an experimental determination of the relevant matrix elements and phase shifts, and hence for a direct comparison with theory at the most basic level [61.34, 35]. In a more global measurement, the cross section σ_tot is obtained by ion or mass spectrometry from

$$\sigma_{tot} = \sigma(A^+) + \sigma(A^{2+}) + \sigma(A^{3+}) + \cdots . \quad (61.35)$$

Generally, the charge states can be correlated with the various initial photoionization processes if allowance is made for Auger transitions and the fluorescence yield upon exceeding the binding energies of core levels.

If the charge states are not distinguished, as in a total ion yield measurement, σ_{tot} is obtained directly. Similarly, a direct measurement of the global quantity σ_{tot} is obtained by the total electron yield, although care must be exercised to avoid discrimination by angular distribution effects. At photon energies below about 1 keV, ionization, absorption, and attenuation are virtually equivalent, and σ_{tot} can also be determined in an ion chamber setup [61.36] or in a photoabsorption measurement in which the number of photons ΔN absorbed in a source of length d and having an atom density n is given by

$$\Delta N = N(\text{ph})[1 - \exp(-\sigma_{tot} n d)] \quad (61.36)$$

with N(ph) being the flux of incident photons. As a rule, in all experiments employing the relation (61.5), the total, partial, or differential cross sections are determined on a relative rather than an absolute scale because it is very difficult to know accurately such factors as the geometry of the source volume and the number density. However, once a single absolute value of σ_{tot} or any σ_{if} is available, all relative values of the other quantities can be converted to absolute values.

An electron spectrometric experiment can be carried out in three different operational modes, as defined in Fig. 61.4. In the most conventional mode, PES, the photon energy is fixed, and a scan of the electron kinetic energy reveals all the electron-emission processes possible and yields their properties. The CIS (constant ionic state) mode is especially suited to follow continuously a selected process as a function of photon energy by locking onto a given state $E_f - E_i$ (denoted by E_B) which requires a strict synchronization of the photon energy ($h\nu$) and electron kinetic energy (ε) during a scan. This mode is particularly advantageous to elucidate resonance features, such as autoionization resonances. Finally, a CKE (constant kinetic energy) scan allows one to access various processes sequentially or, most importantly, follow a process of fixed energy, such as an Auger transition, as a function of photon energy. This description also includes the technique of zero-

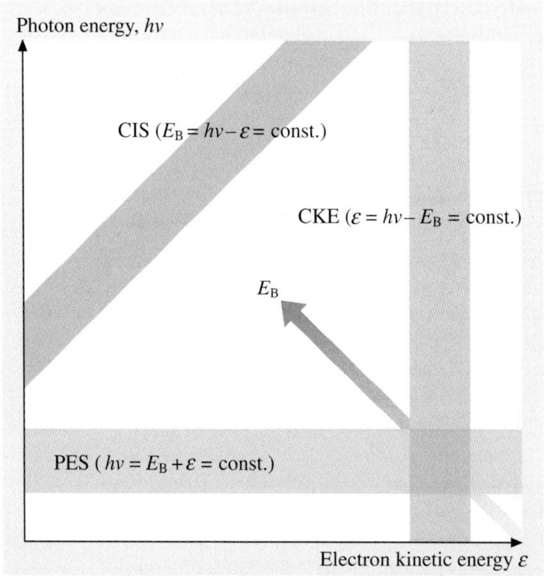

Fig. 61.4 Energy relationship among the three different operational modes of the technique of photoelectron spectrometry. E_B is the binding energy of the level

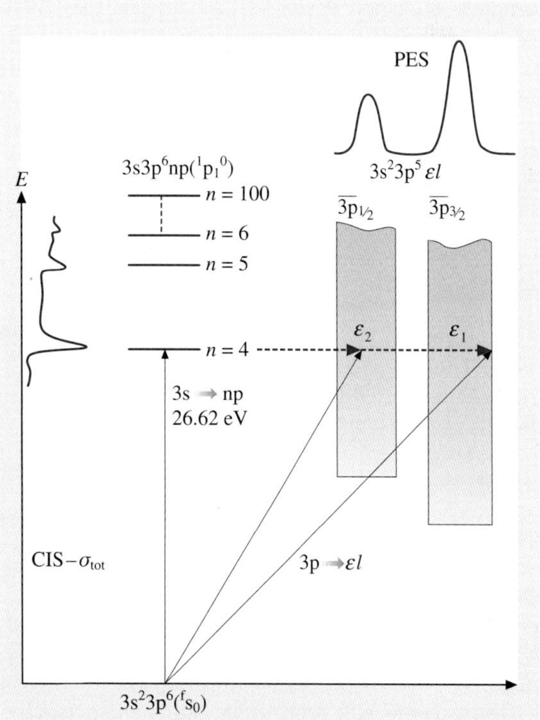

Fig. 61.5 Connection between the PES and CIS techniques as illustrated by the $3s \rightarrow np$ autoionization resonances in argon

kinetic-energy measurements. Most frequently, the PES and CIS modes are employed, and Fig. 61.5 gives a self-explanatory example of an actual experiment directed at the characterization of the argon $3s \rightarrow np$ autoionizing resonances. It should be stressed that the cross section σ_{tot} can be partitioned into its components by CIS scans that differentiate between the $3p_{1/2}$ and $3p_{3/2}$ doublet states.

Energy analysis can be performed either by electrostatic energy analyzers or by time-of-flight techniques. The latter is well-suited to those electrons which have very low kinetic energy, including threshold electrons with $\varepsilon \approx 0$ eV. Of the electrostatic energy analyzers now in use, two designs are prevalent, the cylindrical mirror analyzer (CMA) and the hemispherical analyzer, where the latter readily lends itself to the application of multichannel detectors.

61.2.3 Resolution and Natural Width

The details that can be gleaned from an experiment using photons depend on the resolution achievable with the particular photon source and spectrometry used, the particular excitation or analysis modes and the target conditions chosen, and, ultimately, on the natural width of either the levels or transitions examined as well as any fine structure present. Generally, the instrumental and operational resolution should approach, but need not exceed by much, the natural width inherent in the photoprocess under scrutiny. The demands are most severe for processes involving outer levels because of their typically very narrow widths, and are relatively mild for processes involving inner levels [61.5,35]. It is desirable that in the former case the resolving power (the inverse of the resolution) of the instrument exceed 10^5, while in the latter case 10^4 may suffice. If the target atoms move randomly, a resolution limit is set by the thermal motion, namely

$$\Delta\varepsilon = 0.723\,(\varepsilon T/M)^{1/2}\ (\text{meV})\,, \qquad (61.37)$$

where, in an experiment involving photoelectron spectrometry, ε is the kinetic energy of the photoelectron in eV, T the temperature in K, and M the mass in a.u. of the target atom. This contribution can be limited in first order by employing a suitably directed atomic beam.

The experimental peak-width and shape generally contain the natural width; the extent to which instrumental factors enter depends on the specific experiment. In a measurement of the total or partial photoionization cross section in which either the fluorescence photons, the electrons, or the ions are monitored, the resolution of the photon source (often called the bandpass) is the only instrumental contributor. This applies specifically to the CIS mode of electron spectrometry, in which features are scanned in photon energy and the electron serves solely as a monitor. However, in such a CIS study, or a corresponding fluorescence study, the resolution of the electron or fluorescence spectrometer must be adequate to be able to distinguish adjacent processes. For the example of Fig. 61.5, the $3p_{3/2}$ and $3p_{1/2}$ levels of Ar need to be separated if the partial photoionization cross sections are to be determined across the resonances. In such a case of more than one open ionization channel, the natural widths of the features will be identical in all channels [61.4], but the shapes may be different. In emission processes subsequent to initial photoionization, namely electron (Auger) decay or photon (X-ray) emission, the resolution of only the spectrometer performing the detection counts on the instrumental side. In the PES mode (Fig. 61.4) the observed lines contain contributions from all sources, the photon source or photon monochromator, the electron analyzer, thermal broadening, and the natural level width. In the special case of photoprocesses near inner thresholds, the post-collision interaction influences the position and shape of photoelectron and Auger lines (Chapt. 62).

Excluding threshold regions and the resonant Raman effect, the line profile observed in the various experiments is given by the Voigt function

$$V(\omega, \omega_0) = \int L(\omega' - \omega) G(\omega' - \omega_0)\, d\omega'\,. \qquad (61.38)$$

In this the Lorentzian function L represents the natural level or transition profile, and the Gaussian function G is representative of the window functions of the dispersive apparatus. Although the integral representing the Voigt profile has no analytic form, it can be represented for practical purposes by the analytic Pearson-7 function [61.42]

$$P_7(\varepsilon) = A\left(1 + \frac{(\varepsilon - \varepsilon_0)^2}{B^2 C}\right)^{-C}\,, \qquad (61.39)$$

where A is the peak height, ε_0 the peak position, B the nominal half-width-half-maximum of the peak, and C the shape of the peak. In the limit in which $C = 1$, this function is identically a Lorentzian; in the limit $C \rightarrow \infty$, the function is essentially Gaussian. Use of this function allows one to fit the resulting photoelectron spectrum using standard numerical techniques. If the width of the feature is the only quantity of interest, the simple approximate expression

$$\frac{\Gamma_L}{\Gamma_V} = 1 - \left(\frac{\Gamma_G}{\Gamma_V}\right)^2 \tag{61.40}$$

which relates the Voigt width Γ_V with the Lorentzian width Γ_L and the Gaussian width Γ_G can be used to determine either Γ_L or Γ_G from the measured Γ_V [61.5]. Often the observed feature exhibits a dispersive shape given by the Fano or Shore profile. In this instance the instrumental function must be convoluted with the resonance profiles given by (61.23) or (61.26) in order to fit the data and to extract the parameters [61.43].

For the special case of resonant Auger decay in which the bandpass of the exciting radiation is very narrow compared with the natural width of the excited state, the experimental linewidth is governed by the width of the exciting radiation, and will be more narrow than the natural width of the line. The resulting lineshape in this resonant Raman effect is then the simple product [61.24]

$$L'(\omega, \omega_0) = L(\omega)G(\omega - \omega_0) , \tag{61.41}$$

where $L(\omega)$ is the line profile as determined by the natural width, typically Lorentzian for resonant Auger decay, and $G(\omega - \omega_0)$ is the, usually, Gaussian function representing the bandpass of the exciting radiation.

61.3 Additional Considerations

Although low-energy photon interactions are well described nonrelativistically in the dipole approximation, relativistic and higher multipole effects which become increasingly important at higher energies cannot be ignored even below 1 keV. Spin-orbit effects [61.44] and relativistic effects [61.45] are of special significance even at low energies in Cooper minima, where one of the transition matrix elements becomes zero. Moreover, the use of intermediate coupling, which includes both the spin-orbit and electrostatic interactions [61.46], is required in open-shell systems, as exemplified for the halogen atoms and atomic oxygen [61.47]. Level energies of heavy elements also require a relativistic treatment [61.48] (Chapt. 22), and it is natural to employ relativistic formulations for calculating the spin parameters appearing in photoionization [61.49].

Although low energy photon scattering is dominated by the dipole contribution, experiments and theory have shown higher multipole effects to be present at $h\nu < 1$ keV [61.50–57]. As a result, measurements that take the dipole formulations as a basis (61.5) and (61.6) can incur a discernible error in both the differential cross sections $\mathrm{d}\sigma_{if}/\mathrm{d}\Omega$ and the partial cross sections σ_{if}. A more accurate determination of $\mathrm{d}\sigma_{if}/\mathrm{d}\Omega$ can be made [61.51,55,56] on the basis of the equation

$$\frac{\mathrm{d}\sigma_{if}}{\mathrm{d}\Omega} = \left(\frac{\sigma_{if}}{4\pi}\right)\Big[1 + \beta P_2(\cos\theta) + (\delta + \gamma\cos^2\theta)\,\sin\theta\,\cos\phi\Big] , \tag{61.42}$$

using linearly polarized radiation, where $P_2(\cos\theta)$ is the second-order Legendre polynomial, θ is the angle between the electron emission direction and the electric vector, ϕ is the angle between the electron and photon directions, β is the angular distribution parameter related to B_2, and the parameters δ and γ are related to B_1 and B_3 in (61.4). Figure 61.6 shows the geometry for the relationship between the photoelectron momentum vector, the polarization vector, and the photon propagation vector as used in (61.42). It serves as the template for the arrangement and motion of the electron detector, or, for added efficiency and accuracy, several detectors [61.51–53, 56]. The parameters β, δ, and γ have been calculated for most subshells of the noble gases for photoelectron energies between 100 eV and 5 keV [61.55]

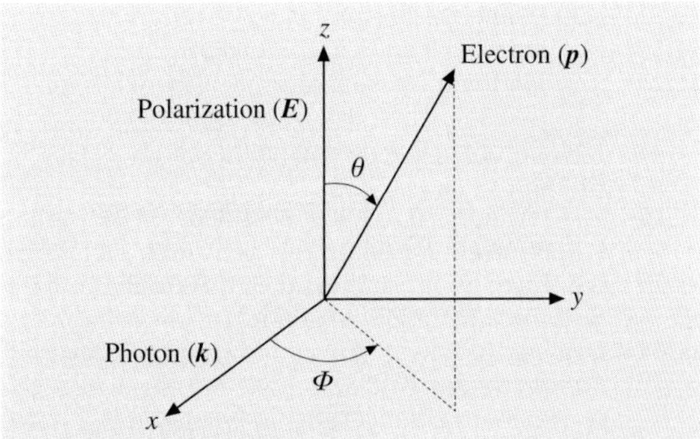

Fig. 61.6 Geometry of the relationship between the photoelectron momentum $\boldsymbol{p}$, the polarization vector $\boldsymbol{E}$, and the photon momentum $\boldsymbol{k}$. (Courtesy of O. Hemmers)

and 20 eV to 5 keV [61.57]. Generally, the electric quadrupole (E2) and magnetic dipole (M1) photoionization channels are the most important beyond the dipole (E1) photoionization channel. Similar to the dipole angular distribution parameter β, the parameter γ may also be subject to interchannel coupling and relativistic effects, as demonstrated for xenon 5s photoionization [61.51].

References

61.1 W. J. Veigele: At. Data Nucl. Data Tables **5**, 51 (1973)
61.2 J. H. Hubbell, M. J. Berger: Photon cross section, attenuation coefficients, and energy absorption coefficients. In: *Engineering Compendium on Radiation Shielding*, Vol. 1, ed. by R. G. Jaeger (Springer, Berlin 1968) p. 167
61.3 U. Fano, J. W. Cooper: Rev. Mod. Phys. **40**, 441 (1963)
61.4 A. F. Starace: Theory of photoionization. In: *Handbuch der Physik*, Vol. XXI, ed. by S. Flügge, W. Mehlhorn (Springer, Berlin 1982) p. 1
61.5 M. O. Krause: Electron spectrometry of atoms and molecules. In: *Synchrotron Radiation Research*, ed. by H. Winick, S. Doniach (Plenum, New York 1980) Chap. 5
61.6 U. Fano, D. Dill: Phys. Rev. A **6**, 185 (1972)
61.7 S. B. Whitfield, K. Kehoe, M. O. Krause, C. D. Caldwell: Phys. Rev. Lett. **84**, 4818 (2000)
61.8 J. Schirmer, L. S. Cederbaum, J. Kiessling: Phys. Rev. A **22**, 2696 (1990)
61.9 V. W. Hughes, R. L. Long Jr., M. S. Lubell, M. Posner, W. Raith: Phys. Rev. A **5**, 195 (1972)
61.10 U. Fano: Phys. Rev. **178**, 131 (1969)
61.11 N. A. Cherepkov: Adv. At. Mol. Phys. **19**, 395 (1983)
61.12 U. Fano, J. H. Macek: Rev. Mod. Phys. **45**, 553 (1973)
61.13 C. H. Greene, R. N. Zare: Phys. Rev. A **25**, 2031 (1982)
61.14 W. Mehlhorn: Auger-electron spectrometry of core levels in atoms. In: *Atomic Inner-Shell Physics*, ed. by B. Crasemann (Plenum, New York 1985) Chap. 4
61.15 U. Fano: Phys. Rev. **124**, 1866 (1961)
61.16 B. Shore: J. Opt. Soc. Am. **57**, 881 (1967)
61.17 F. H. Mies: Phys. Rev. **175**, 164 (1968)
61.18 J. P. Connerade, A. M. Lane: Rep. Prog. Phys. **51**, 1439 (1988)
61.19 M. Martins: J. Phys. B **34**, 1321 (2001)
61.20 S. B. Whitfield, M. O. Krause, P. van der Meulen, C. D. Caldwell: Phys. Rev. A **50**, 1269 (1994)
61.21 T. W. Gorczyca, B. M. McLaughlin: J. Phys. B **33**, 859 (2000)
61.22 S. J. Schaphorst, S. B. Whitfield, H. P. Saha, C. D. Caldwell, Y. Azuma: Phys. Rev. A **47**, 3007 (1993)
61.23 C. E. Theodosiou, M. Inokuti, S. T. Manson: At. Data Nucl. Data Tables **35**, 473 (1986)
61.24 T. Åberg, G. Howat: Theory of the Auger effect. In: *Handbuch der Physik*, Vol. XXI, ed. by S. Flügge, W. Mehlhorn (Springer, Berlin 1982) p. 469
61.25 S. B. Whitfield, J. Tulkki, T. Åberg: Phys. Rev. A **44**, R6985 (1994)
61.26 M. O. Krause: J. Phys. (Paris) **32**, C4-67 (1971)
61.27 M. O. Krause, S. B. Whitfield, C. D. Caldwell, J.-Z. Wu, P. van der Meulen, C. A. de Lange, R. W. C. Hansen: J. Elec. Spectros. Rel. Phen. **58**, 79 (1992)
61.28 R. L. Martin, D. A. Shirley: Phys. Rev. A **13**, 1475 (1976)
61.29 T. Åberg: Nucl. Instrum. Methods **B87**, 5 (1994)
61.30 E. P. Wigner: Phys. Rev. **73**, 1002 (1948)
61.31 G. H. Wannier: Phys. Rev. **90**, 817 (1953)
61.32 M. Y. Kuchiev, S. A. Sheinermann: Sov. Phys. JETP **63**, 986 (1986)
61.33 U. Heinzmann: J. Phys. B **13**, 4353 (1980)
61.34 V. Schmidt: Rep. Prog. Phys. **55**, 1483 (1992)
61.35 U. Becker, D. A. Shirley (Eds.): *VUV and Soft X-ray Radiation Studies* (Plenum, New York 1996)
61.36 J. A. R. Samson: Atomic photoionization. In: *Handbuch der Physik*, Vol. XXI, ed. by S. Flügge, W. Mehlhorn (Springer, Berlin 1982) p. 123
61.37 B. Sonntag, P. Zimmermann: Rep. Prog. Phys. **55**, 911 (1992)
61.38 C. A. de Lange: Adv. Chem. Phys. **117**, 1 (2001)
61.39 H. Winick, G. Brown: Nucl. Instrum. Methods **195**, 1 (1987)
61.40 E. E. Koch: *Handbook of Synchrotron Radiation* (North-Holland, Amsterdam 1983)
61.41 J. A. Clarke: *The Science, Technology of Undulators, Wigglers* (Oxford Univ. Press, Oxford 2004)
61.42 S. B. Whitfield, C. D. Caldwell, D. X. Huang, M. O. Krause: J. Phys. B **25**, 4755 (1992)
61.43 J. Jiménez-Mier: J. Quant. Spectros. Radiat. Transfer **51**, 741 (1994)
61.44 U. Fano: Phys. Rev. **178**, 131 (1969)
61.45 S. T. Manson, Z. J. Lee, R. H. Pratt, I. B. Goldberg, B. R. Tambe, A. Ron: Phys. Rev. A **28**, 2885 (1983)
61.46 R. D. Cowan: *The Theory of Atomic Structure and Spectra* (Univ. of California Press, Berkeley 1981)
61.47 C. D. Caldwell, M. O. Krause: Rad. Phys. Chem. **70**, 43 (2004)
61.48 I. P. Grant: Aust. J. Phys. **39**, 649 (1986)
61.49 K. N. Huang, W. R. Johnson, K. T. Cheng: At. Data Nucl. Data Tables **26**, 33 (1981)
61.50 H. K. Tseng, R. H. Pratt, S. Yu, A. Ron: Phys. Rev. A **17**, 1060 (1978)
61.51 O. Hemmers, R. Guillemin, D. W. Lindle: Rad. Phys. Chem. **70**, 123 (2004)
61.52 N. L. S. Martin, D. B. Thompson, R. P. Bauman, C. D. Caldwell, M. O. Krause, S. P. Frigo, M. Wilson: Rev. Lett. **81**, 1199 (1998)

61.53 R. W. Dunford, E. P. Kanter, B. Krässig, S. H. Southworth, L. Young: Rad. Phys. Chem. **70**, 149 (2004)

61.54 M. O. Krause: Phys. Rev. **177**, 151 (1969)

61.55 J. W. Cooper: Phys. Rev. A **47**, 1841 (1993)

61.56 P. S. Shaw, U. Arp, S. H. Southworth: Phys. Rev. A **54**, 1463 (1996)

61.57 A. Derevianko, W. R. Johnson, K. T. Cheng: At. Data Nucl. Data Tables **73**, 153 (1999)

62. Photon–Atom Interactions: Intermediate Energies

The main photon-atom interaction processes in the energy range from ≈1 keV to ≈1 MeV are outlined. The atomic response to inelastic photon scattering is discussed; essential aspects of radiative and radiationless transitions are described, including hole-state widths and fluorescence yields, in the two-step approximation which applies well above threshold where atomic de-excitation is nearly independent of the excitation process. Multi-electron photoexcitation that transcends the independent-electron model is briefly considered.

Threshold phenomena that arise when the two-step model breaks down, so that excitation and de-excitation must be treated as a single second-order quantum process are described. The time-independent resonant scattering approach that leads to a unified description of photoexcitation, ionization, and Auger electron or X-ray emission is outlined. In terms of this theory, Raman processes and post-collision interaction are described.

62.1 **Overview** 915
62.1.1 Photon–Atom Processes 915

62.2 **Elastic Photon–Atom Scattering** 916
62.2.1 Rayleigh Scattering 916
62.2.2 Nuclear Scattering 917

62.3 **Inelastic Photon–Atom Interactions** 918
62.3.1 Photoionization 918
62.3.2 Compton Scattering 919

62.4 **Atomic Response to Inelastic Photon–Atom Interactions** 919
62.4.1 Auger Transitions 919
62.4.2 X-Ray Emission 921
62.4.3 Widths and Fluorescence Yields 921
62.4.4 Multi-Electron Excitations 921
62.4.5 Momentum Spectroscopy 922
62.4.6 Ultrashort Light Pulses 922
62.4.7 Nondipolar Interactions 923

62.5 **Threshold Phenomena** 923
62.5.1 Raman Processes 924
62.5.2 Post-Collision Interaction 925

References 925

62.1 Overview

62.1.1 Photon-Atom Processes

It is convenient to divide photon-atom interactions into elastic and inelastic processes, depending on whether or not the photon energy changes in the c.m. frame. This separation, while useful, is somewhat arbitrary [62.1]: the radiative corrections of quantum electrodynamics (Chapt. 27) and the possibility of emitting very soft photons, as well as target recoil, make all processes in fact inelastic, while experimental comparison of incident and scattered photon energies is limited by source bandwidth and detector resolution.

Elastic photon-atom interactions (Chapt. 45) include scattering from bound electrons (Rayleigh scattering) with transition amplitude A^{R} and nuclear Thomson scattering, Delbrück scattering, and nuclear resonance scattering, with amplitudes A^{NT}, A^{D}, and A^{NR}, respectively. These amplitudes are coherent, i. e., not physically distinguishable: they add, retaining relative phase information, in the total atom elastic scattering amplitude A [62.1]. In the commonly used approximation, that is however "neither unique nor exact" [62.2], we have

$$A = A^{\mathrm{R}} + A^{\mathrm{NT}} + A^{\mathrm{D}} + A^{\mathrm{NR}} . \quad (62.1)$$

The total atom elastic cross section is proportional to A^2.

Inelastic photon-atom interactions are photoexcitation (including ionization) (Chapt. 24), Compton scattering, and pair production. The pair production threshold at $2m_{\mathrm{e}}c^2$ lies above the energy region considered here (Chapt. 27). Inelastic scattering processes are generally incoherent, e.g., ejected Compton electrons originating from different orbitals can be distinguished [62.1]. However, coherence can enter even between photoionization and Compton scattering in the infrared-divergence limit [62.1, 3–5].

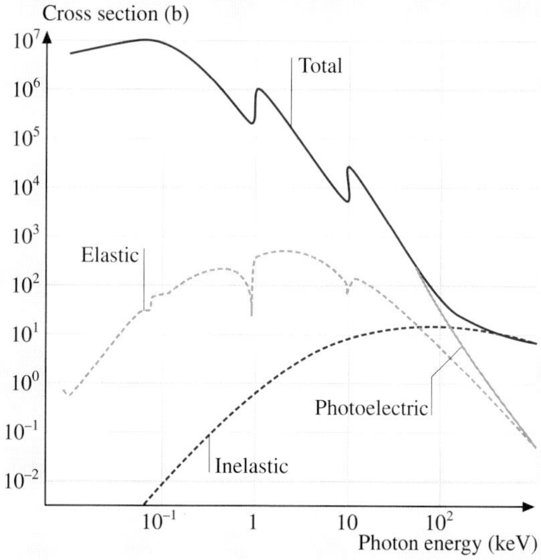

Fig. 62.1 Contributions of the photoeffect, elastic, and inelastic scattering to the total photon interaction cross section of copper. (After [62.6], by permission, University of California, Lawrence Livermore National Laboratory; work performed under the auspices of the U.S. Department of Energy)

The cross sections for the main elastic and inelastic photon-atom processes comprising the photoeffect are illustrated in Fig. 62.1 for Cu [62.6] in which, however, resonant pieces of the inelastic (Compton) cross section are omitted. Elastic exceeds inelastic scattering up to well above the K edge; this is true for all atomic numbers. The total cross section is dominated by the photoeffect up to photon energies that increase with atomic number, from ≈ 6 keV for He [62.5] to ≈ 25 keV for C and ≈ 700 keV for U [62.6–8]. (See also Sect. 24.2.2 and Sect. 92.3.2). For a summary of some current developments in the field see [62.9]

62.2 Elastic Photon–Atom Scattering

62.2.1 Rayleigh Scattering

The term Rayleigh scattering is applied to elastic photon scattering from bound atomic electrons (Sect. 68.6.1). The Rayleigh scattering amplitude for neutral atoms dominates the total atom elastic scattering amplitude, at all angles, for photon energies below ≈ 1 MeV and dominates forward scattering at higher energies [62.1]. At angles other than 0°, Rayleigh scattering occurs increasingly from inner atomic shells as the photon energy is raised. Depending on the energy, the process can thus sensitively probe atomic structure in distinct regions [62.1].

Rayleigh scattering has been reviewed by *Henke* [62.10], *Gavrila* [62.11], *Kissel* and *Pratt* [62.1], and *Kane* et al. [62.12]. The process has commonly been estimated in a simple *form-factor approximation*, which represents scattering from a free charge distribution; this underlies comprehensive tabulations of elastic scattering cross sections and attenuation coefficients [62.7, 8]. The form factor relates cross sections to classical theory, and has been derived from both nonrelativistic and relativistic quantum mechanics. It is attractive because of calculational simplicity [62.2, 13].

The form factor $f(q)$ for a spherically symmetric charge number density $\rho(r)$ is the Fourier transform of the charge density:

$$f(q) \equiv \int \rho(\boldsymbol{r})\,\mathrm{e}^{\mathrm{i}\boldsymbol{q}\cdot\boldsymbol{r}}\,\mathrm{d}\boldsymbol{r} = \int_0^\infty \rho(r)\frac{\sin(qr)}{qr}r^2\,\mathrm{d}r\,, \tag{62.2}$$

where

$$\hbar q = (2\hbar\omega/c)\sin(\theta/2) \tag{62.3}$$

is the momentum transfer, and θ the scattering angle. In elastic photon scattering, $f(q)$ corrects the Thomson point charge scattering cross section for scattering from an extended charge distribution. In the form factor approximation, the differential Rayleigh scattering cross section (neglecting other coherent processes) for unpolarized photons, averaged over final polarization, is

$$\frac{\mathrm{d}\sigma}{\mathrm{d}\Omega} = \frac{1}{2}r_\mathrm{e}^2\left(1+\cos^2\theta\right)|f(q)|^2\,, \tag{62.4}$$

where $r_\mathrm{e} = e^2/\left(m_\mathrm{e}c^2\right)$ is the classical electron radius and the terms multiplying $|f(q)|^2$ are the Thomson cross section.

Even with modifications and corrections [62.1, 2], the range of validity for the form factor approximation for elastic photon scattering is limited [62.13]. Predictions based on this approach are often wrong in quantitative detail, although they lend insight into qualitative features of the Rayleigh amplitude [62.2] Fig. 62.2.

The shortcomings of the form factor approximation arise from the absence of binding effects in the Thomson cross section, which pertains to photon scattering from a free charged particle. Specifically, accurate Rayleigh amplitudes must include the interaction of the electron with the atomic potential in the initial, intermediate, and final states [62.1]. The form factor approximation fails, especially for photons of low energy compared with the electron binding energy, and for large scattering angles (high momentum transfer), but $\mathcal{O}(Z\alpha^2)$ discrepancies remain at all energies and angles [62.2].

Substantial progress in the understanding of elastic photon-atom scattering was made with the advent of numerical calculations of the amplitudes based on the second-order S-matrix of quantum electrodynamics and relativistic wave functions [62.2, 14]. These numerical partial wave calculations include binding effects in the intermediate state exactly to all orders. They constitute an approach that leads to amplitudes with errors of $\mathcal{O}(1\%)$. However, some limitations do exist: (1) great computer time requirements make it necessary to employ simpler (e.g., modified form factor) methods for outer shell contributions, (2) only lowest-order terms of the S-matrix expansion in powers of e are included, and (3) most electron-electron correlation effects are not taken into account [62.1].

Measurements of Rayleigh scattering in the energy range here considered are rather scarce [62.12]. Gamma rays from radioactive sources have been used in most experiments, but discrimination from inelastic contributions has been a problem [62.15]. Comparison of measurements on medium- and high-Z targets with theory, albeit incomplete, has however been encouraging [62.1, 12, 15].

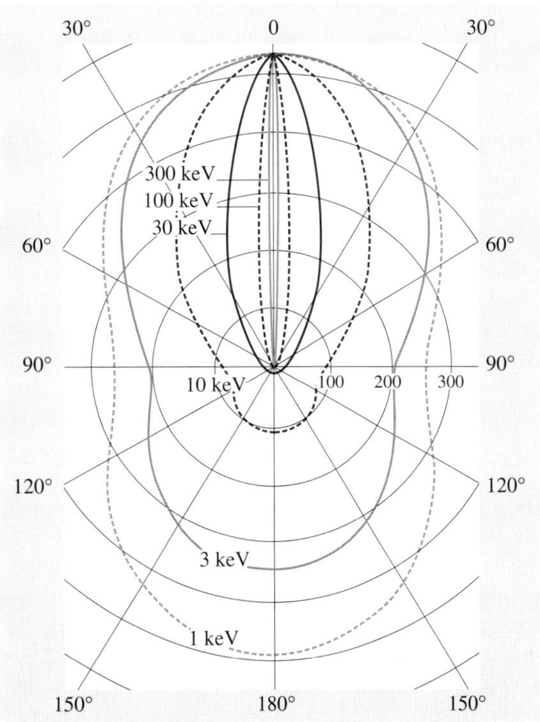

Fig. 62.2 Differential elastic photon-atom scattering cross section (in b/atom) for neutral lead in form factor approximation. The contribution of nuclear Thomson scattering (see Sect. 62.2.2) to the amplitude is included (After [62.1])

62.2.2 Nuclear Scattering

The nucleus contributes to the total atom elastic scattering amplitude through nuclear Thomson, Delbrück, and nuclear resonance scattering. Only the first two of these processes are important in the energy range considered here.

Nuclear Thomson Scattering

Photon scattering by the nuclear charge can be approximated by treating the nucleus as a rigid, spinless sphere of charge Ze, mass M, and radius r_N. In the simplest point-charge form factor approximation, nuclear Thomson scattering is included by replacing $|f(q)|$ in (62.4) by $|f(q) + Z^2(m_e/M)|$. If the finite nuclear radius $\boldsymbol{r}_N$ is taken into account, the nuclear form factor is approximately [62.1]

$$f_N \approx Z^2 \left(1 - \frac{\omega^2}{3c^2} \langle r_N^2 \rangle \right) . \tag{62.5}$$

The nuclear Thomson scattering cross section is

$$\frac{d\sigma^{NT}}{d\Omega} = \frac{r_e^2}{2} \left| \frac{m_e}{M} f_N \right|^2 (1 + \cos^2 \theta) \tag{62.6}$$

if other coherent processes can be neglected. For some energies, scattering angles, and target atoms, the nuclear Thomson scattering contribution to elastic scattering can dominate. A few measurements exist that confirm the preceding predictions with $f_N \cong Z^2$ [62.1].

Delbrück Scattering

Scattering of photons by virtual electron-positron pairs created in the screened nuclear Coulomb potential is called Delbrück scattering [62.1, 16]. The imaginary part of the amplitude describes photon absorption due to pair production; it vanishes for $\hbar\omega \leq 2m_ec^2$. The real part of the amplitude, which adds coherently to the Rayleigh and nuclear Thomson amplitudes, arises from vacuum polarization. Together with photon-photon scattering, photon splitting and coalescence in an external field, Delbrück scattering belongs to the nonlinear effects of quantum electrodynamics that have no classical analog (Sect. 27.2) [62.17]. In the photon energy range 0.1–1 MeV, the Delbrück amplitude can be significant at intermediate scattering angles for high-Z targets [62.1]. Theory and experiment have been reviewed by *Papatzacos* and *Mork* [62.16] and more recently by *Kissel* and *Pratt* [62.1] and by *Milstein* and *Schumacher* [62.18]. Lowest-order Born approximation calculations of the Delbrück amplitude scale as $\alpha(Z\alpha^2)$. Experiments to verify the process have been performed since 1933, but only recently has the necessary accuracy been attained [62.18]; a major obstacle in the interpretation was the lack of accurate predictions of the Rayleigh amplitude, a problem that is now being solved to a considerable extent (Sect. 62.2.1). Recent measurements agree with the Born approximation predictions to a few percent below $Z = 60$; but for heavier targets, Coulomb corections appear to play a significant role that have only recently been calculated [62.19].

62.3 Inelastic Photon–Atom Interactions

62.3.1 Photoionization

Photoexcitation and ionization are important processes for the investigation of atomic structure and dynamics (Chapt. 61). Early reviews of the theory of the atomic photoelectric effect are by *Fano* and *Cooper* [62.20] (low photon energy) and by *Pratt* et al. [62.21] ($\hbar\omega > 10$ keV). See also recent comprehensive monographs by *Starace* [62.22] and *Amusia* [62.23]. In the present volume, the theory of atomic photoionization for photon energies below 1 keV is summarized in Chapt. 24. Experimental aspects have been reviewed, e.g., by *Samson* [62.22, 24] and by *Siegbahn* and *Karlsson* [62.25].

An extensive compilation of measured X-ray attenuation coefficients or total absorption cross sections, as well as calculated photoelectric cross sections (computed with relativistic Hartree-Slater wave functions), has been published by *Saloman* et al. [62.26]. Based on the National Bureau of Standards (now the National Institute of Standards and Technology) data base, this work covers the photon energy range from 0.1–100 keV and includes all elements with atomic numbers $1 \leq Z \leq 92$; an extensive bibliography is included. Very useful tables of theoretical subshell photoionization cross sections covering all elements and the photon energy range 1–1500 keV have been computed by *Scofield* [62.27]; the atomic electrons are treated relativistically as moving in a Hartree-Slater central potential, and all relevant multipoles as well as retardation effects are included. Accuracy of the results is borne out by systematic comparisons with experiment and detailed tests [62.28]. A useful energy range for applications in electron spectrometry is covered by a tabulation of subshell cross sections calculated by *Yeh* and *Lindau* [62.29] in the dipole approximation with nonrelativistic Hartree-Fock-Slater wave functions. Because they were obtained in a frozen-core model, these cross sections like those of *Scofield* [62.27] automatically include all multiple excitations [62.30]. Such multi-electron processes (Sect. 62.4.4) can make substantial contributions, e.g., $\approx 40\%$ of the $1s$ photoionization in the threshold region of Kr [62.31]. A very useful family of codes being developed by I Grant's Oxford group is "GRASP – A General Purpose Relativistic Atomic Structure Program", several versions of which are being distributed through the *Computer Physics Communications Program Library* (Elsevier Science).

The usefulness of photoexcitation and ionization for experimental studies of atomic structure and processes has been greatly enhanced by the advent of *synchrotron radiation sources* beginning in the 1960's [62.1, 32, 33]. At first, experiments were performed mostly "parasitically". Later, they were done with dedicated sources, taking advantage of the high brightness, narrow bandwidth and wide tunability of the radiation (with suitable monochromators, from the visible to the hard X-ray regime), its high degree of polarization and sharp time structure [62.34]. In the 1990's, "third-generation" sources are being commissioned in which radiation is primarily derived from insertion devices, wigglers and undulators [62.35], entailing further increase of the power of this experimental tool. There is a growing literature on research applications of synchrotron radiation in atomic and molecular physics, including handbooks [62.36–38], proceedings of the series of conferences on X-ray and atomic inner-shell physics [62.39], and monographs [62.40–43]. Much of the physics described in the remainder of this chapter is being explored with synchrotron radiation.

62.3.2 Compton Scattering

Compton scattering denotes the scattering of photons from free electrons. The term is also used for inelastic photon scattering from bound electrons, which approaches the free electron case when the photon energy greatly exceeds the electron binding energy [62.1]. The theory of the process is discussed in standard textbooks [62.17, 44, 45]. Photon scattering from a free electron at rest is expressed, within lowest-order relativistic quantum electrodynamics, by the Klein–Nishina formula. A low energy approximation to the Klein–Nishina cross section is

$$\sigma^{\mathrm{KN}} = \sigma^{\mathrm{T}}\left(1+\frac{2\hbar\omega}{m_{\mathrm{e}}c^2}+\cdots\right), \quad \hbar\omega \ll m_{\mathrm{e}}c^2, \tag{62.7}$$

where σ^{T} is the classical total Thomson cross section,

$$\sigma^{\mathrm{T}} = (8\pi/3)r_{\mathrm{e}}^2 . \tag{62.8}$$

The energy $\hbar\omega'$ of a photon scattered through an angle θ is related to the incident photon energy $\hbar\omega$ by the Compton relation

$$\hbar\omega' = \frac{\hbar\omega}{1+(\hbar\omega/m_{\mathrm{e}}c^2)(1-\cos\theta)} . \tag{62.9}$$

The differential cross section for Compton scattering from free electrons, averaged over initial and summed over final photon and electron polarizations, is

$$\frac{\mathrm{d}\sigma^{\mathrm{KN}}}{\mathrm{d}\Omega} = \frac{r_{\mathrm{e}}^2}{2}\left(\frac{\omega'}{\omega}\right)^2\left(\frac{\omega}{\omega'}+\frac{\omega'}{\omega}-\sin^2\theta\right) . \tag{62.10}$$

In the limit $\omega' \to \omega$, this reduces to the Thomson differential cross section,

$$\frac{\mathrm{d}\sigma^{\mathrm{T}}}{\mathrm{d}\Omega} = \frac{r_{\mathrm{e}}^2}{2}(1+\cos^2\theta) \tag{62.11}$$

contained in (62.4).

Compton scattering from bound electrons has extensive applications in the study of electron momentum distributions in atoms and solids [62.46]. An exact second-order S-matrix code for the relativistic numerical calculation of cross sections for Compton scattering of photons by bound electrons within the independent particle approximation has been developed [62.47].

A systematic treatment of elastic and inelastic photon scattering, starting from many-body formalism, has been described by *Pratt*, *Kissel*, and *Bergstrom* [62.48].

62.4 Atomic Response to Inelastic Photon–Atom Interactions

62.4.1 Auger Transitions

Atomic inner shell vacancy states tend to decay predominantly through radiationless or Auger transitions, which are autoionization processes (see also Chapt. 25 and Chapt. 61) that arise from the Coulomb interaction between electrons [62.1,1,22,49–55]. Radiationless transitions dominate over radiative ones, except for 1s vacancies in atoms with atomic numbers $Z < 30$, primarily because of the magnitude of the matrix elements and because far fewer channels are allowed by electric dipole X-ray emission selection rules (Sect. 62.4.2, 62.4.3) than by the selection rules for Auger transitions, which in pure LS coupling are

$$\begin{aligned} &\Delta L = \Delta S = \Delta M_L = \Delta M_S = 0 , \\ &\Delta J = \Delta M = 0 , \quad \Pi_i = \Pi_f . \end{aligned} \tag{62.12}$$

Thus, for example, 2784 electron-electron interaction matrix elements can contribute to the radiationless decay of a $2p_{3/2}$ hole state in Hg.

In the simplest approach, ionization and subsequent Auger decay are treated as two distinct steps. This approximation is valid if the electron which is ejected in the ionization process is sufficiently energetic so that it does not interact significantly with the Auger electron, and the interaction of the core hole state with the Auger continuum is weak. Then the hole state can be considered to be quasi-stationary and the decay rate can be expressed according to Wentzel's ansatz, formulated in 1927 and later known as Fermi's Golden Rule No. 2 of time dependent perturbation theory [62.49]. In the independent-electron central-field approximation, this leads to the following nonrelativistic matrix element for a direct Auger transition:

$$\begin{aligned} D = &\iint \psi^*_{n''\ell''f''}(1)\psi^*_{\infty\ell_A j_A}(2)\left|\frac{e^2}{\boldsymbol{r}_1-\boldsymbol{r}_2}\right| \\ &\times \psi_{n\ell j}(1)\psi_{n'\ell'j'}(2)\,\mathrm{d}\tau_1\,\mathrm{d}\tau_2 , \end{aligned} \tag{62.13}$$

where the quantum numbers n, ℓ, j characterize electrons that are identified schematically in Fig. 62.3. The

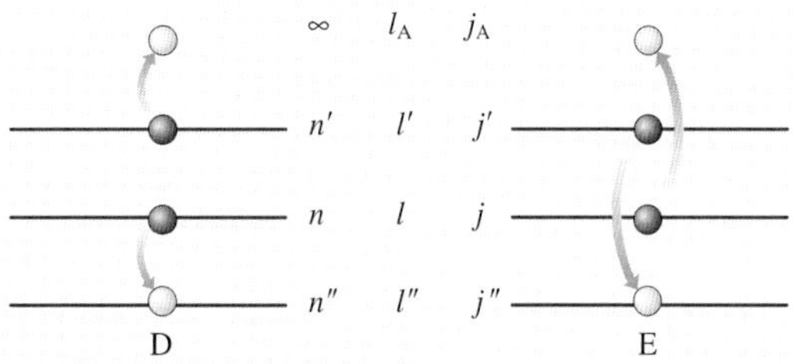

Fig. 62.3 Energy levels involved in the direct (D) and exchange (E) Auger processes, and notation for the principal, orbital angular momentum, and total angular momentum quantum numbers that characterize the pertinent electron states

state of the continuum (Auger) electron is labeled by $\infty\ell_A j_A$. In the physically indistinguishable exchange process, described by a matrix element E, the roles of electrons $n\ell j$ and $n'\ell' j'$ are interchanged (Fig. 62.3). The radiationless transition probability per unit time is

$$w_{fi} = \frac{2\pi}{\hbar}\left|D - E\right|^2 \rho(E_f)\,, \tag{62.14}$$

where $\rho(E_f)$ is the density of final states for the energy E_f that satisfies energy conservation. With the continuum wave function normalized to one electron ejected per unit time, we have $\rho(E_f) = (2\pi\hbar)^{-1}$ ([62.22, 52], Appendix E).

The matrix elements D and E in (62.14) can be separated into radial and angular factors. Evaluation of the angular factors depends upon the choice of an appropriate angular momentum coupling scheme [62.1, 51, 53], ranging from the $(LSJM)$ representation of Russell-Saunders coupling for the lightest atoms in which the spin-orbit interaction can be neglected, through intermediate coupling, to j–j coupling in the high-Z limit. The coupling scheme particularly affects intensity ratios among multiplet components of Auger spectra, which can vary by more than an order of magnitude depending upon its choice. Furthermore, realistic calculations of Auger energies and rates call for the use of relativistic wave functions and inclusion of correlation and relaxation effects [62.1, 54]. It is this extreme sensitivity of Auger transitions to the details of the atomic model that makes them an exceedingly useful probe of atomic structure (see also Chapt. 23 and Chapt. 61).

Classification of Auger transitions in the central field model is conventionally based on the nomenclature summarized in Table 62.1.

The spectroscopic notation $s, p, d, f, \ldots$ is employed for orbital angular momentum quantum numbers

Table 62.1 Nomenclature for vacancy states. The subscripts in the column headings are the values of $j = l \pm s$, and n is the principal quantum number

n	$s_{1/2}$	$p_{1/2}$	$p_{3/2}$	$d_{3/2}$	$d_{5/2}$	$f_{5/2}$	$f_{7/2}$
1	K						
2	L_1	L_2	L_3				
3	M_1	M_2	M_3	M_4	M_5		
4	N_1	N_2	N_3	N_4	N_5	N_6	N_7

$\ell = 0, 1, 2, 3, \ldots$, and shells with principal quantum numbers $n = 1, 2, 3, \ldots$ are denoted by $K, L, M, \ldots$. Let an inner shell vacancy be created initially in a subshell X_ν and consider a radiationless transition in which an electron from subshell Y_μ fills that vacancy and a Z_ξ-electron is ejected, in the direct process. This transition is denoted by X_ν-$Y_\mu Z_\xi$, which stands for $(n''\ell'' j'')$-$(n\ell j)(n'\ell' j')$ in terms of the vacancies indicated in Fig. 62.3

Coster–Kronig transitions are a subclass of Auger transitions in which a vacancy "bubbles up" among subshells of the same shell, i. e., X_ν-$X_\mu Y_\xi$. These are exceptionally fast because the low energy of the ejected electron tends to prevent cancellations of its wave function overlap with the other factors in the matrix element (62.13) and the overlap of the bound state wave functions can be large owing to their similarity. Coster–Kronig transitions can therefore lead to hole-state widths in excess of 10 eV (lifetimes $< 10^{-16}$ s) for the $L_{1,2}$ subshells of the heaviest elements, for example. McGuire has coined the name super-Coster–Kronig transitions for the (even faster) type X_ν-$X_\mu X_\xi$ [62.50]. Here the initial state can no longer be considered quasi-stationary, the width becomes energy-dependent and the two-step model breaks down.

The Auger transition energy or kinetic energy of the ejected electron is, within the central field model,

$$E_A = E_{n''\ell'' j''} - E_{n\ell j, n'\ell' j'}\,, \tag{62.15}$$

where, in the notation of Fig. 62.3, $E_{n''\ell'' j''}$ is the absolute value of the energy of the atom with an $n''\ell'' j''$ vacancy, with reference to the neutral atom energy, and $E_{n\ell j, n'\ell' j'}$ is the energy of the atom with *two* vacancies in the states indicated by the subscripts. This latter energy can be approximated by a mean of measured single-ionization energies of the atom under consideration and that with the next higher atomic number (the "$Z+1$ rule" [62.49]); with present-day computers it is readily calculated through one of the nonrelativistic [62.56, 57] or relativistic [62.58] self-consistent-field codes.

Photoionization and, even more so, ionization by charged particle impact, often produces more than one vacancy in the target atom (see Sect. 61.1.2 and Sect. 62.4.4). The ensuing Auger spectrum can then exhibit satellite lines shifted in energy, due to the presence of spectator vacancies and electrons, with respect to the diagram lines that arise from the decay of a singly ionized atom [62.1, 50, 53].

62.4.2 X-Ray Emission

The emission of X-rays by atoms ionized in inner shells was studied long before the discovery of the much more probable radiationless transitions, and has played a crucial role in the investigation of atomic structure and interactions with the environment [62.59]. At present, X-ray spectrometry is providing important insights into the structure of highly ionized species produced, e.g., in electron beam ion traps [62.60] or plasmas [62.61], and for the very precise experimental evaluation of inner-vacancy level energies [62.1, 62].

A classical treatment of multipole radiation by atomic and nuclear systems is given by Jackson [62.44], who shows by a simple argument that for atoms the electric dipole transition rate is $\approx (137/Z_{\text{eff}})^2$ times more intense than electric quadrupole or magnetic dipole emission, where Z_{eff} is the effective (screened) nuclear charge. The quantum theory of atomic radiation is presented in the treatises by *Sakurai* [62.63], *Mizushima* [62.64], *Berestetskii* et al. [62.17], and *Sobel'man* [62.65], for example. X-ray emission has been reviewed by *Scofield* [62.50, 66] and, with emphasis on relativistic effects, by *Chen* [62.1, 54, 60, 67]. In the present volume, radiative transitions are treated in Chapt. 33 to which the reader is referred for details.

62.4.3 Widths and Fluorescence Yields

The energy profile of an atomic hole state that decays exponentially with time is found to be the same from quantum as from classical theory [62.68]; it has the resonance or Lorentz shape

$$I(\omega)\,\mathrm{d}\omega = \frac{I_0(\Gamma/2\pi)\,\mathrm{d}(\hbar\omega)}{(\hbar\omega-\hbar\omega_0)^2+\frac{1}{4}\Gamma^2}\,. \tag{62.16}$$

The full width Γ at half maximum of this energy distribution is proportional to the total decay rate τ^{-1} of the state, in accordance with Heisenberg's uncertainty principle $\Gamma\tau = \hbar$.

The decay rate $\tau^{-1} = \Gamma/\hbar$ of an atomic hole state is commonly given in units of $\text{eV}/\hbar$ or in atomic units (a.u.) of inverse time, with the corresponding level width Γ in eV or a.u. Thus, for the decay rate,

$$\frac{1}{\tau} = 1\,\text{a.u.} = 4.134\,14\times10^{16}\,\text{s}^{-1} = 27.2116\,\frac{\text{eV}}{\hbar}\,. \tag{62.17}$$

K-level (1s hole-state) widths increase monotonically with atomic number, from 0.24 eV for Ne to 96 eV for U, closely following the approximate relation [62.49]

$$\Gamma(1\,\text{s}) = 1.73\,Z^{3.93}\times10^{-6}\,\text{eV}\,. \tag{62.18}$$

L-subshell widths, as functions of atomic number, exhibit sharp discontinuities that correspond to energetic cutoffs and onsets of intense Coster–Kronig channels (Sect. 62.4.1) [62.69].

If there are several decay channels, their partial widths Γ_i add:

$$\Gamma = \sum_i \Gamma_i\,. \tag{62.19}$$

The main decay channels for inner shell hole states are radiative, with a partial width Γ_R, and radiationless or Auger, with a partial width Γ_A.

The fluorescence yield ω_i of a hole state i is defined as the relative probability that the state decays radiatively:

$$\omega_i \equiv \Gamma_{\text{R}}(i)/\Gamma(i)\,. \tag{62.20}$$

Macroscopically, for example, the K-shell fluorescence yield ω_K is the number of characteristic K X-ray photons emitted from a sample divided by the number of primary K-shell vacancies created in the sample. This ratio rises from 0.04 for Al to 0.97 for U [62.49]. The definition of fluorescence yields of shells above the K-shell is more complicated because the higher shells consist of several subshells and Coster–Kronig transitions (Sect. 62.4.1) can shift primary vacancies to higher subshells before they are filled radiatively.

The concept of fluorescence yields is useful in many applications, particularly where the transport of radiant energy through matter is an issue, as in space sciences, medical radiology, and some fields of engineering as well as in physics. Tables of fluorescence yields can be found in the review by *Bambynek* et al. [62.49] and in the compilations by *Krause* [62.70] and *Hubbell* et al. [62.71]. Considerable uncertainty persists in the fluorescence yields of higher shells.

62.4.4 Multi-Electron Excitations

In atomic inner-shell photoionization, more than one electron can be excited with significant probability. Final states are thus produced that can be described approximately by removal of a core electron and excitation of additional electrons to higher bound states (shakeup) or to the continuum (shakeoff). These multiple excitation processes exhibit themselves through satellites in photoelectron spectra (Sect. 61.1.2) and in the Auger and X-ray spectra emitted when the excited states decay, as well as in some cases by features in absorption spectra [62.72]. As the photon-electron interaction is described by a one-electron operator, the frozen-core, central field model does not predict changes of state under photon impact by more than one electron. Direct multiple excitation processes are thus a result of electron-electron correlation (Chapt. 23, Sect. 61.1.2).

Cross sections for photoexcitation and ionization of two inner shell electrons have been evaluated [62.72] by the multichannel multiconfiguration Dirac–Fock (MMCDF) method [62.73]. Within this model, the correlation effects that cause direct two-electron processes are (i) relaxation or core rearrangement, (ii) initial state configuration interaction, (iii) final ionic state configuration interaction, and (iv) final continuum state configuration, or final state channel interaction.

Mechanisms that contribute to two-electron photoionization of the outermost shells of noble gas atoms have been separated within many-body perturbation theory [62.74]. In first order of the combined perturbations by the photon field and electron correlations, the important contributions are core rearrangement, initial state correlations, virtual Auger transitions, and "direct collisions" by the photoelectron with another orbital electron. If treated nonperturbatively, these mechanisms belong within the MMCDF scheme to categories (i)–(iv) described above. The different mechanisms have been found to be of varying importance, depending on the photon energy relative to the double ionization threshold energy, the orbitals which are ionized, and the relative final state energies of the active electrons. Interest in the theory [62.75, 76] has been enhanced by experiments with synchrotron radiation on the double ionization of He [62.77], a process that epitomizes the problem. Above $\approx 5\,\mathrm{keV}$ photon energy, the double ionization of He has been shown to be dominated by Compton scattering, not photoeffect [62.78].

62.4.5 Momentum Spectroscopy

We close with a few examples of fruitful recent advances in approaches to the study of electron-atom interactions. A step-function advance in viewing atomic collision dynamics, including photon-electron interactions, arose from the development (first in the University of Frankfurt by Schmidt-Böcking and his group) of Cold-Target Recoil-Ion Momentum Spectroscopy (COLTRIMS), a "momentum microscope" to view the dynamics of photon-atom collisions [62.79–81]. This novel momentum space imaging technique allows the investigation of the dynamics of ion, electron, or photon impact reactions with atoms or molecules. Studies performed with this technique yield kinematically complete pictures of the fragments of atomic and molecular breakup processes, unprecedented in resolution, detail and completeness. The multiple-dimensional momentum-space images often directly unveil the physical mechanism underlying the many-particle transitions that are being investigated [62.82]. Important applications are being developed [62.83, 84].

62.4.6 Ultrashort Light Pulses

Advances in laser technology have made it possible to obtain high-intensity (up to $10^{14}\,\mathrm{W/cm^2}$), short (5 fs) pulses, making laser-atom [62.85] and laser-molecule interactions [62.86–89] amenable to experimental study. In the latter category, nonlinear phenomena such as above-threshold ionization and laser-induced molecular potentials have already been explained theoretically [62.90–92].

Ultrashort light pulses make it possible to follow ultrafast relaxation processes on never-before-accessed time scales and to study light-matter interactions at unprecedented intensity levels [62.85, 93].

Ultrafast optics has permitted the generation of light wave packets comprising only a few oscillation cycles of the electric and magnetic fields. The spatial extension of these wave packets along the direction of their propagation is limited to a few times the wavelength of the radiation. The pulses can be focused to a spot size comparable to the wavelength. Radiation can thus be temporarily confined to a few cubic micrometers, forming a "light bullet". The extreme temporal and spatial confinement allows moderate pulse energies of the order of one microjoule to result in peak intensities higher than $10^{15}\,\mathrm{W/cm^2}$. These field strengths exceed those of the static field seen by outer-shell electrons in atoms. The laser field consequently is strong enough to suppress the binding Coulomb potential in atoms and triggers optical field ionization [62.94, 95].

Attosecond $(10^{-18}\,\mathrm{s})$ trains from high-harmonic generation have been discovered [62.90–92]. High harmonics, generated in a gas jet by a femtosecond laser pulse, have a well-defined phase relation with respect to each other. The relative phases result in strong amplitude beating between the various harmonics. The emerging radiation thus consists of attosecond pulses.

The process of harmonic generation can be thought of as consisting of three steps [62.92]: a (rate-limiting) field ionization of the atom at times near the electric field maxima, followed by acceleration of the free electron which, due to the ac character of the driving light, eventually makes the electron recollide at high energy with its parent ion. Radiative recapture of this fast electron into the ground state, strongly favored due to the remaining coherence between the two parts of what was initially the same wave function, completes the harmonic generation process without any change to the atom.

In summary, intense few-cycle light pulses open up never yet accessed parameters in high-field physics [62.85]. Strong-field processes induced by few-cycle laser fields may permit access to the phase of the carrier wave and hence to the light fields for the first time. Phase-controlled light pulses may allow control of high-intensity light-matter interaction on a subcycle time scale. The search for laser-driven ultrafast X-ray sources is also drawing benefits from this research area. Substantial major progress can be anticipated in this new field.

62.4.7 Nondipolar Interactions

A powerful tool for exploring photon-atom interactions has been photoionization of atoms and molecules [62.96]. Analysis of photoelectron angular distribution measurements was routinely performed in the electric-dipole approximation, in which all higher-order multipoles are neglected [62.97]. While breakdown of the dipole approximation at higher photon energies (above 5 keV) was known, so that a proper description requires inclusion of many multipoles [62.98], experimental limitations appeared to make such extension unnecessary. A notable exception was a hint found by Krause, who noticed deviations from the dipole approximation in measurements on rare gases using unpolarized X-rays [62.99].

Only when more sophisticated radiation sources became available, notably synchrotron radiation, and advances in detector technology took place, did it become possible to gain data on nondipolar photoionization [62.98]. Intense experimental and theoretical work ensued.

Photoelectron angular distributions depend upon dynamical parameters of the target atoms and upon the polarization of the photon beam. Peshkin developed a systematic analysis of the restrictions imposed by symmetry, through a geometric description of polarization that is easily visualized and has been found useful in planning experiments [62.100].

Earlier theories predicted nondipolar asymmetries of 1s photoelectron distributions in the point-Coulomb potential. More recently, photoionization amplitudes are calculated in screened potentials, predicting richer Z-dependent and subshell-dependent energy variations of nondipole asymmetries [62.97]. A full relativistic multipole expansion, incuding screening effects, in the independent-particle approximation was discussed by *Tseng* et al. [62.97]. In addition, leading correction terms to the dipole approximation can be included by adopting the "retardation expansion" of the exponential factor $\exp(\mathrm{i}\boldsymbol{k}\cdot\boldsymbol{r})$ [62.101, 102] which, unlike the multipole expansion, includes a finite number of multipoles within the long-wavelength limit [62.103]. A lucid summary of the development of the subject can be found in the paper by *Krässig* et al.on nondipole asymmetries of Kr 1s photoelectrons [62.104]. Extending the subject, experiments and theory on electric-octupole and pure-electric-quadruple effects on soft-X-ray photoemission have been described by *Derevianko* et al. [62.105].

62.5 Threshold Phenomena

During the creation of a core vacancy by photoexcitation or ionization, the escape of the photoelectron from an inner shell involves complex dynamics of electron excitations with multiple correlational aspects. The entire process of inner shell photoionization and subsequent de-excitation has drastically different characteristics near threshold from that in the high energy limit. In the latter case, discussed in the preceding sections, if an X-ray photon promotes an inner shell electron to an energetic continuum state, the atom first relaxes in its excited (hole) state. In a practically distinct second step, the hole is then filled—most often by a radiationless transition, under emission of an Auger electron with *diagram* energy that can readily be calculated from the wave function of the stationary, real intermediate state [62.1, 53].

On the other hand, in the vicinity of core-level energy thresholds, atomic photoexcitation and the ensuing X-ray or Auger electron emission occur in a single second-order quantum process, the resonant Raman effect. Here, the intermediate states are virtual and there is no relaxation phase. The resonant Raman regime, comprising the energy region just below threshold, is linked by the post-collision interaction regime (Sect. 62.5.2) above threshold to the (asymptotically) two-step regime.

The primary process of photoionization and Auger electron emission can be viewed as resonant double photoionization mediated by a complete set of intermediate states; these correspond to an intermediate virtual inner shell hole and an electron in excited bound or continuum states. A unified treatment of the process, as well as of resonant X-ray Raman scattering, has been developed within the framework of relativistic time independent resonant scattering theory [62.48, 106–108]. As the energy of the incoming photons is increased,

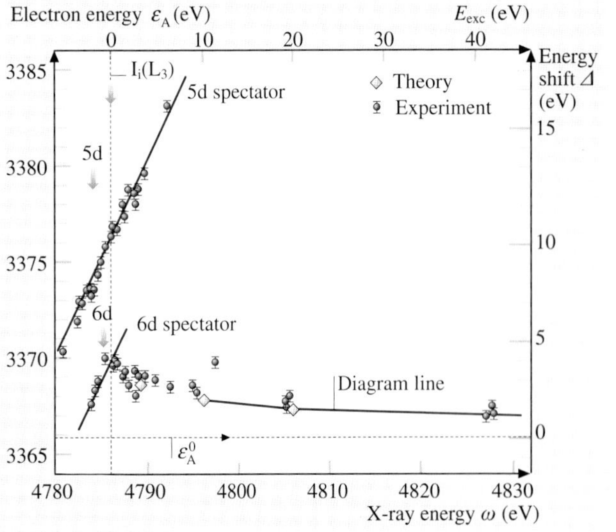

Fig. 62.4 Energy of the Xe L_3–M_4M_5 ($^1G_4$) Auger electron peak, as a function of exciting photon energy. Near threshold Auger satellites, caused by the spectator photoelectron in bound orbits, exhibit linear Raman dispersion. The post-collision interaction shift (*right-hand scale*) vanishes only in the asymptotic limit (After [62.83])

sweeping through an inner shell threshold, continuous asymmetric electron distributions associated with different two-hole multiplets evolve into Auger-electron lines with characteristic energies. The shapes and energies of the characteristic lines continue to change until the energy of the photoelectron exceeds that of the Auger electron. The nonresonant two-electron emission amplitude is usually negligible in this region, compared with the resonant amplitude.

62.5.1 Raman Processes

Resonant Raman transitions to bound final states are the inelastic analog of resonance fluorescence [62.68], with which they share two characteristic features: (1) the emitted radiation can have a narrower bandwidth than the natural lifetime width of the corresponding diagram line, and (2) the energy of the emitted radiation exhibits linear dispersion with excitation energy. Radiative resonant Raman scattering was first observed by *Sparks* in 1974 [62.109] and subsequently explored with synchrotron radiation by *Eisenberger* et al.(see [62.48, 48, 108, 110]). Radiationless resonant Raman scattering was identified by *Brown* et al. [62.111] and in subsequent studies [62.112] of the deformation of Auger-electron lines near threshold (Fig. 62.4). The connection between resonant Raman and Compton scattering has been pointed out by *Bergstrom* et al. [62.47].

According to a generalization of time independent resonant scattering theory [62.106, 107], a transition matrix element can be constructed that accounts for photoexcitation and either radiative or radiationless de-excitation, including resonance structure [62.48, 108]. The wave functions represent both the atomic and photon fields and include all electrons throughout the process. The conventional two-step model of X-ray fluorescence and Auger electron emission is but a special case of the transition amplitude

$$T_{\beta\alpha} = \left\langle \psi^-_{\beta\varepsilon\varepsilon'} \middle| V_\gamma + \sum_\nu \int \mathrm{d}\tau \frac{(H-E)|\psi_{\tau\nu}\rangle\langle\psi_{\tau\nu}|V_\gamma}{E - E_{\tau\nu}(E)} \middle| \psi_\alpha \right\rangle \tag{62.21}$$

where H is the total Hamiltonian, V_γ the photon-electron interaction operator, and $E = \hbar\omega_1 + E_\alpha$ is the total energy. The first term in $T_{\beta\alpha}$ represents direct nonresonant Raman scattering; the resonance behavior is embedded in the second amplitude. The complex eigenenergies $E_{\tau\nu}(E)$ satisfy a complicated secular equation that involves diagonal and nondiagonal matrix elements of the level shifts and widths [62.1, 3, 48, 108]. The initial state wave function ψ_α is a direct product of the initial atomic state wave function and a one-photon state $|\boldsymbol{k}_1,\boldsymbol{e}_1\rangle$, where $\boldsymbol{k}_1$ is the wave vector and $\boldsymbol{e}_1$, the polarization vector. The final state scattering wave function $\psi^-_{\beta\varepsilon\varepsilon'}$ corresponds, in the radiative case, to a scattered photon in a definite state $|\boldsymbol{k}_2,\boldsymbol{e}_2\rangle$ and an electron plus ion in a given quantum state, which may involve the linear momentum $\hbar\boldsymbol{k}_e$ of the electron if the latter is in the continuum. In the radiationless case, there are an ion and two electrons, both of which may be in a continuum state $|\boldsymbol{k}_e, \boldsymbol{k}'_e\rangle$. The wave function $\psi^-_{\beta\varepsilon\varepsilon'}$ fulfills the ingoing-wave boundary condition with respect to the electron(s) and accounts for final state channel interaction effects.

Work on inner shell Raman processes is as yet in its infancy. Synchrotron radiation experiments, interpreted in light of the theory outlined above, promise to become a useful tool in atomic physics and materials science [62.113, 114].

62.5.2 Post-Collision Interaction

In the radiationless decay of an atom that has been photoionized near an inner shell threshold, the Coulomb field of the receding photoelectron perturbs the Auger electron energy and line shape. This post-collision interaction provides continuity in the energy evolution of inner shell dynamics between the Raman and asymptotic two-step regimes. In Auger decay, following photoionization, the Auger electron initially screens the ionic Coulomb field seen by the receding photoelectron. This screening subsides when the (usually fast) Auger electron passes the slow photoelectron. Distortion of the Auger line shape results, and the Auger energy is raised at the expense of the photoelectron energy (Fig. 62.4).

A semiclassical potential curve model of post-collision interaction leads to intuitive insight and surprisingly accurate predictions [62.115, 116]. In simple terms, the attractive potential well in which the slow photoelectron moves is deepened suddenly when the residual singly ionized atom undergoes Auger decay and its net charge changes from $+e$ to $+2e$. The photoelectron sinks down into this deeper well, there being no time for it to speed up, and the energy lost by the photoelectron is transferred to the Auger electron. The photoelectron slows down and may even be recaptured by the atom from which it was emitted [62.117].

Quantum theoretically, post-collision interaction can be treated from the point of view of resonant scattering theory (Sect. 62.5.1). A lowest-order line shape formula, corresponding to a shakedown mechanism, can be shown to emerge from approximations of the general multichannel transition matrix element [62.118]; the phenomenon thus arises as a consequence of a resonant rearrangement collision in which a photon and an atom in the initial channel turn into an ion and two electrons (one of which has nearly characteristic energy) in the final channel. In this lowest order, the results can be interpreted in terms of an analytic line shape formula based on asymptotic Coulomb wave functions [62.118]. The line shape depends only on the excess photon energy, the lifetime of the initial state of the Auger process, and the change of the ionic charge during Auger electron emission.

The interaction between the photoelectron and the Auger electron in the final state can be included by reinterpreting the ionic charge seen by the photoelectron on the basis of asymptotic properties of the continuum wave function pertaining to the two outgoing electrons [62.119, 120]. This results in a procedure which is consistent with semiclassical models that account for the time required for the Auger electron to overtake the photoelectron. Calculations based on this approach agree very well with measurements performed with synchrotron radiation [62.120].

References

62.1 L. Kissel, R. H. Pratt: In: *Atomic Inner-Shell Physics*, ed. by B. Crasemann (Plenum, New York 1985) Chap. 11

62.2 L. Kissel, R. H. Pratt, S. C. Roy: Phys. Rev. A **22**, 1970 (1980)

62.3 J. Tulkki, T. Åberg: [62.1, Chap. 10]

62.4 V. A. Bushuev, R. N. Kuz'min: Sov. Phys. Usp. **20**, 406 (1977)

62.5 K.-I. Hino, P. M. Bergstrom Jr., J. H. Macek: Phys. Rev. Lett. **72**, 1620 (1994)

62.6 For up-to-date atomic data see *Photon Interaction Data (EPDL97); Electron Interaction Data (EEDL), Atomic Relaxation Data (EADL)*. Contact for users within the U.S.A.: National Nuclear Data Center (NNDC), Brookhaven National Laboratory, Vicki McLane, services@bnlnd2.dne.bnl.gov. Contact for users outside the U.S.A.: Nuclear Data Section (NDS), International Atomic Energy Agency (IAEA), Vienna, Austria, Vladimir Pronyaev, v.pronyaev@iaea.org.

62.7 J. H. Hubbell: Radiation physics. In: *Encyclopedia of Physical Sciences and Technology*, 3rd edn., ed. by R. Meyers (Academic, San Diego 2001)

62.8 U. W. Arndt, D. C. Creagh, R. D. Deslattes, J. H. Hubbell, P. Indelicato, E. G. Kessler Jr., E. Lindroth: *X-rays: International Tables for Crystallography*, Vol. C, 2nd edn., ed. by A. J. C. Wilson, E. Prince (Kluwer Academic, Dordrecht 1999)
62.9 R. H. Pratt: Some frontiers of X-ray/atom interactions. In: *X-Ray and Inner-Shell Processes*, ed. by R. W. Dunford et al. (American Institute of Physics, Melville 2000)
62.10 B. L. Henke: Low energy X-ray interactions: photoionization, scattering, specular and Bragg reflection. In: *Am. Inst. Phys. Conf. Proc. 75*, ed. by D. T. Atwood, B. L. Henke (American Institute of Physics, New York 1981) p. 146
62.11 M. Gavrila: Photon–atom elastic scattering. In: *Am. Inst. Phys. Conf. Proc. 94*, ed. by B. Crasemann (American Institute of Physics, New York 1982)
62.12 P. P. Kane, L. Kissel, R. H. Pratt, S. C. Roy: Phys. Rep. **140**, 75 (1986)
62.13 L. Kissel, B. Zhou, S. C. Roy, S. K. SenGupta, R. H. Pratt: Acta Crystallogr. **A51, No. 3** (1985)
62.14 W. R. Johnson, K.-T. Cheng: Phys. Rev. A **13**, 692 (1976)
62.15 W. Mückenheim, M. Schumacher: J. Phys. G **6**, 1237 (1980)
62.16 P. Papatzacos, K. Mork: Phys. Rep. **21**, 81 (1975)
62.17 V. B. Berestetskii, E. M. Lifshitz, L. P. Pitaevskii: *Quantum Electrodynamics*, 2nd edn. (Pergamon, New York 1982)
62.18 A. I. Milstein, M. Schumacher: Phys. Rep. **243**, 183 (1994)
62.19 R. Solberg, K. Mork, I. Øverbø: Phys. Rev. A **51**, 359 (1995)
62.20 U. Fano, J. W. Cooper: Rev. Mod. Phys. **40**, 441 (1968)
62.21 R. H. Pratt, A. Ron, H. K. Tseng: Rev. Mod. Phys. **45**, 273 (1973)
62.22 A. F. Starace: *Handbuch der Physik*, Vol. XXI, ed. by S. Flügge, W. Mehlhorn (Springer, Berlin 1982) p. 1
62.23 M. Ya. Amusia: *Atomic Photoeffect* (Plenum, New York 1990)
62.24 J. A. R. Samson: In: *[62.22]* p. 123
62.25 H. Siegbahn, L. Karlsson: In: *[62.22]* p. 215
62.26 E. B. Saloman, J. H. Hubbell, J. H. Scofield: At. Data Nucl. Data Tables **38**, 1 (1988)
62.27 J. H. Scofield: Lawrence Livermore Laboratory Report No. UCRL-51326, 1973 (unpublished)
62.28 J. H. Hubbell, W. J. Veigele: *National Bureau of Standards (U. S.) Technical Report*, Vol. 901 (U.S. GPO, Washington, DC 1976)
62.29 J. J. Yeh, I. Lindau: At. Data Nucl. Data Tables **32**, 1 (1985)
62.30 T. Åberg: Shake theory of multiple photoexcitation processes. In: *Photoionization and Other Probes of Many-Electron Interactions*, ed. by J. Wuilleumier F. (Plenum, New York 1976) p. 49
62.31 D. L. Wark, R. Bartlett, T. J. Bowles, R. G. H. Robertson, D. S. Sivia, W. Trela, J. F. Wilkerson, G. S. Brown, B. Crasemann, S. L. Sorensen, S. J. Schaphorst, D. A. Knapp, J. Henderson, J. Tulkki, T. Åberg: Phys. Rev. Lett. **67**, 2291 (1991)
62.32 B. Crasemann, F. Wuilleumierin: [62.1, Chap. 7]
62.33 B. Crasemann: Atomic and molecular physics with synchrotron radiation. In: *Electronic and Atomic Collisions*, ed. by W. R. MacGillivray, I. E. McCarthy, M. C. Standage (Adam Hilger, Bristol 1992) p. 69
62.34 H. Winick, S. Doniach (Eds.): *Synchrotron Radiation Research* (Plenum, New York 1980)
62.35 Nucl. Instrum. Methods A **246**, 1 (1986)
62.36 C. Kunz (Ed.): *Synchrotron Radiation–Techniques and Applications* (Springer, Berlin, Heidelberg 1979)
62.37 E. E. Koch (Ed.): *Handbook on Synchrotron Radiation* (North-Holland, Amsterdam 1983)
62.38 J. F. Moulder et al.: *Handbook of X-Ray Photoelectron Spectroscopy* (Perkin-Elmer, Eden Prairie 1992)
62.39 Most recent in the series are *X-Ray and Inner-Shell Processes*, edited by P. Lagarde, F. J. Wuilleumier, and J. P. Briand, J. Phys. (Paris) C **9**, 48 (1987); *idem*, *American Institute of Physics Conference Proceedings*, No. 215, edited by T. A. Carlson, M. O. Krause, and S. T. Manson, (American Institute of Physics, New York, 1990). For a recent example, see *X-Ray and Inner-Shell Processes*, edited by A. Bianconi, A. Marcelli, and N. L. Saini (American Institute of Physics, Melville, 2002)
62.40 T. A. Carlson: *Photoelectron and Auger Spectroscopy* (Plenum, New York 1975)
62.41 T. A. Carlson: *X-Ray Photoelectron Spectroscopy* (Dowden, Hutchinson, Ross, Stroudsburg 1978)
62.42 C. R. Brundle, A. D. Baker (Eds.): *Electron Spectroscopy: Theory, Techniques, and Applications* (Academic, New York 1977) Vol. 1 & Vol. 2.
62.43 J. Berkowitz: *Photoabsorption, Photoionization, and Photoelectron Spectroscopy* (Academic, New York 1979)
62.44 J. D. Jackson: *Classical Electrodynamics*, 2nd edn. (Wiley, New York 1975)
62.45 J. D. Bjorken, S. D. Drell: *Relativistic Quantum Mechanics* (McGraw-Hill, New York 1964)
62.46 B. Williams (Ed.): *Compton Scattering* (McGraw-Hill, New York 1977)
62.47 P. M. Bergstrom Jr., T. Surić, K. Pisk, R. H. Pratt: Phys. Rev. A **48**, 1134 (1993)
62.48 R. H. Pratt, L. Kissel, P. M. Bergstrom Jr.: In: *X-Ray Anomalous (Resonant) Scattering: Theory and Experiment*, ed. by G. Materlik, C. J. Sparks, K. Fischer (North Holland, Amsterdam 1994)
62.49 W. Bambynek et al.: Rev. Mod. Phys. **44**, 716 (1972)
62.50 E. J. McGuire: *Atomic Inner-Shell Processes*, ed. by B. Crasemann (Academic, New York 1975), Vol. 1, Chap. 7
62.51 D. Chattarji: *The Theory of Auger Transitions*, Vol. 1 (Academic, New York 1976) Chap. 7
62.52 T. Åberg, G. Howat: [62.22[p. 469]]

62.53 W. Mehlhorn: [62.1[Chap. 4]]
62.54 M. H. Chen: [62.1[Chap. 2]]
62.55 W. Mehlhorn: Atomic auger spectroscopy: historical perspective and recent highlights. In: *X-Ray and Inner-Shell Processes* (American Institute of Physics, Melville 2000)
62.56 C. Froese Fischer: Comput. Phys. Commun. **14**, 145 (1978)
62.57 C. Froese Fischer: Comput. Phys. Commun. **64**, 369 (1991)
62.58 I. P. Grant: Comput. Phys. Commun. **21**, 207 (1980)
62.59 A. Meisel, G. Leonhardt, R. Szargan: *X-Ray Spectra and Chemical Binding* (Springer, Berlin, Heidelberg 1989)
62.60 P. Beiersdorfer: X-ray spectroscopy of high-A ions. In: *American Institute of Physics Conference Proceedings*, ed. by T. A. Carlson, M. O. Krause, S. T. Manson (American Institute of Physics, New York 1990) p. 648
62.61 E. Källne, J. Källne: Phys. Scr. **T17**, 152 (1987)
62.62 R. D. Deslattes, E. G. Kessler Jr.: [62.1[Chap. 5]]
62.63 J. J. Sakurai: *Advanced Quantum Mechanics* (Addison-Wesley, Reading 1967)
62.64 M. Mizushima: *Quantum Mechanics of Atomic Spectra and Atomic Structure* (Benjamin, New York 1970)
62.65 I. I. Sobel'man: *An Introduction fo the Theory of Atomic Spectra* (Pergamon Press, Oxford 1972)
62.66 J. H. Scofield: In: *[62.50] Vol. 1, Chap. 6.*
62.67 M. H. Chen: In: *[62.60]* p. 391
62.68 W. Heitler: *The Quantum Theory of Radiation*, 3rd edn. (Clarendon, Oxfordrd 1954), Chap. V.
62.69 M. H. Chen, B. Crasemann, K.-N. Huang, M. Aoyagi, H. Mark: At. Data Nucl. Data Tables **19**, 97 (1977)
62.70 M. O. Krause: J. Phys. Chem. Ref. Data **8**, 307 (1979)
62.71 J. H. Hubbell, P. N. Trehan, N. Singh, B. Chand, D. Mehta, M. I. Garg, R. R. Garg, S. Singh, S. Puri: J. Phys. Chem. Ref. Data **23**, 339 (1994)
62.72 S. J. Schaphorst, A. F. Kodre, J. Ruscheinski, B. Crasemann, T. Åberg, J. Tulkki, M. H. Chen, Y. Azuma, G. S. Brown: Phys. Rev. A **47**, 1953 (1993)
62.73 J. Tulkki, T. Åberg, A. Mäntykenttä, H. Aksela: Phys. Rev. A **46**, 1357 (1992)
62.74 T. N. Chang, T. Ishihara, R. T. Poe: Phys. Rev. Lett. **27**, 838 (1971)
62.75 C. Pan, H. P. Kelly: Phys. Rev. A **41**, 3624 (1990)
62.76 K.-I. Hino, T. Ishihara, F. Shimizu, N. Toshima, J. H. McGuire: Phys. Rev. A **48**, 1271 (1993)
62.77 J. C. Levin, I. A. Sellin, B. M. Johnson, D. W. Lindle, R. D. Miller, N. Berrah, Y. Azuma, H. G. Berry, D.-H. Lee: Phys. Rev. A **47**, R16 (1993)
62.78 J. A. R. Samson, C. H. Greene, R. J. Bartlett: Phys. Rev. Lett. **71**, 201 (1993)
62.79 R. Dörner, V. Mergel, O. Jagutzki, L. Spielberger, J. Ullrich, R. Moshammer, H. Schmidt-Böcking: Phys. Rep. **330**, 96 (2000)
62.80 Th. Weber, M. Weckenbrock, A. Staudte, L. Spielberger, O. Jagutzki, V. Mergel, F. Afaneh, G. Urbasch, M. Vollmer, H. Giessen, R. Dörner: Phys. Rev. Lett. **84**, 443 (2000)
62.81 R. Mosshammer: Phys. Rev. A **65**, 035401 (2002)
62.82 L. Cocke: Momentum Imaging in Atomic Collisions. In: *ICPEAC XXIII* (2003), *Physica Scripta* Vol. T110, 9 (2004)
62.83 G. B. Armen, T. Åberg, J. C. Levin, B. Crasemann, M. H. Chen, G. E. Ice, G. S. Brown: Phys. Rev. Lett. **54**, 1142 (1985)
62.84 J. Ullrich, R. Dörner, R. Moshammer, H. Rottke, W. Sander: In: *Proceedings of the XVIII International Conference on Atomic Physics*, ed. by H. R. Sadgehpour, E. J. Heller, D. E. Pritchard (World Scientific, New Jersey 2003) p. 219
62.85 T. Brabec, F. Krausz: Rev. Mod. Phys. **72**, 545 (2000)
62.86 M. Gavrila: *Atoms in Intense Laser Fields* (Academic, New York 1992)
62.87 A. D. Bandrauk: *Molecules in Laser Fields* (Dekker, New York 1994)
62.88 A. D. Bandrauk, E. Constant: J. Phys. **1**, 1033 (1991)
62.89 A. D. Bandrauk: Phys. Rev. A **67**, 013407 (2003)
62.90 H. G. Muller: Characterization of attosecond pulse trains from high-harmonic generation. In: *Proceedings of the XVIII International Conference on Atomic Physics*, ed. by H. R. Sadgehpour, E. J. Heller, D. E. Pritchard (World Scientific, New Jersey 2003) p. 209
62.91 M. Ferray, A. L'Huillier, X. F. Li, L. A. Lompre, G. Mainfray, C. Manus: J. Phys. B **21**, L31 (1988)
62.92 P. B. Corkum: Phys. Rev. Lett. **71**, 1993 (1994)
62.93 C. Wunderlich, E. Kobler, H. Figger, Th. W. Hänsch: Phys. Rev. Lett. **78**, 2333 (1997)
62.94 L. V. Keldysh: Sov. Phys. JETP **20** (1965)
62.95 L. V. Keldysh: *Electron Spectrometry of Atoms Using Synchrotron Radiation* (Cambridge Univ. Press, Cambridge 1997)
62.96 V. Schmidt: *Electron Spectrometry of Atoms Using Synchrotron Radiation* (Cambridge Univ. Press, Cambridge 1997)
62.97 H. K. Tseng, R. H. Pratt, S. Yu, A. Ron: Phys. Rev. A **17**, 1061 (1978)
62.98 B. Krässig: Phys. Rev. Lett. **75**, 4736 (1995)
62.99 M. O. Krause: Phys. Rev. **177**, 151 (1969)
62.100 M. Peshkin: Photon beam polariziation and nondipolar angular distributions. In: *Atomic Physics with Hard X-Rays from High-Brilliance Synchrotron Light Sources*, ed. by S. Southworth, D. Gemmell (Argonne National Laboratory, Argonne 1996)
62.101 A. Bechler, R. H. Pratt: Phys. Rev. A **39**, 1774 (1989)
62.102 A. Bechler, R. H. Pratt: Phys. Rev. A **42**, 6400 (1990)
62.103 M. Peshkin: Adv. Chem. Phys. **18**, 1 (1970)
62.104 B. Krässig, J. C. Levin, I. A. Sellin, B. M. Johnson, D. W. Lindle, R. D. Miller, N. Berrah, Y. Azuma, H. G. Berry, D.-H. Lee: Phys. Rev. A **67**, 022707 (2003)
62.105 A. Derevianko: Phys. Rev. Lett. **84**, 2116 (2000)
62.106 T. Åberg: Phys. Scr. **21**, 495 (1980)
62.107 T. Åberg: Phys. Scr. **T41**, 71 (1992)

62.108 B. Crasemann, T. Åberg: [62.48]
62.109 C. J. Sparks Jr.: Phys. Rev. Lett. **33**, 262 (1974)
62.110 P. L. Cowan: [62.48]
62.111 G. S. Brown, M. H. Chen, B. Crasemann, G. E. Ice: Phys. Rev. Lett. **45**, 1937 (1980)
62.112 G. B. Armen, T. Åberg, J. C. Levin, B. Crasemann, M. H. Chen, G. E. Ice, G. S. Brown: Phys. Rev. Lett. **54**, 1142 (1985)
62.113 H. Wang, J. C. Woicik, T. Åberg, M. H. Chen, A. Herrera-Gomez, T. Kendelewicz, A. Mänttykenttä, K. E. Miyano, S. Southworth, B. Crasemann: Phys. Rev. A **50**, 1359 (1994)
62.114 B. Craseman: Can. J. Phys. **76**, 251 (1998)
62.115 A. Niehaus: J. Phys. B **10**, 1845 (1977)
62.116 A. Niehaus, C. J. Zwakhals: J. Phys. BL **16**, L135 (1983)
62.117 W. Eberhardt, S. Bernstorff, H. W. Jochims, S. B. Whitfield, B. Crasemann: Phys. Rev. A **38**, 3808 (1988)
62.118 J. Tulkki, G. B. Armen, T. Åberg, B. Crasemann, M. H. Chen: Z. Phys. D **5**, 241 (1987)
62.119 W. Eberhardt, G. Kalkoffen, C. Kunz: Phys. Rev. Lett. **41**, 156 (1978)
62.120 G. B. Armen, J. Tulkki, T. Åberg, B. Crasemann: Phys. Rev. A **36**, 5606 (1987)
62.121 Nucl. Instrum. Methods A **261**, 1 (1987)
62.122 Nucl. Instrum. Methods A **266**, 1 (1988)
62.123 Nucl. Instrum. Methods A **282**, 369 (1989)
62.124 Nucl. Instrum. Methods A **291**, 1 (1990)
62.125 Nucl. Instrum. Methods A **303**, 397 (1991)
62.126 Nucl. Instrum. Methods A **319**, 1 (1991)

63. Electron–Atom and Electron–Molecule Collisions

Electron–atom and electron–molecule collision processes play a prominent role in a variety of systems ranging from discharge or electron-beam lasers and plasma processing devices to aurorae and solar plasmas. Early studies of these interactions contributed significantly to the understanding of the quantum nature of matter. Experimental activities in this field, initiated by *Franck* and *Hertz* [63.1], flourished in the 1930s and, after a dormant period of about a quarter of a century, have had a renaissance in recent years.

When electrons collide with atomic or molecular targets, a large variety of reactions can take place (see Sect. 63.1.1). We limit our discussion to electron collisions with gaseous targets, where single collision conditions prevail. Furthermore, we discuss only low-energy (threshold to few hundred eV) impact processes where the interaction between the valence-shell electrons of the target and the free electron dominates.

Comprehensive discussions on electron–atom (molecule) collision physics can be found in the books of *Massey* et al. [63.2], *McDaniel* [63.3] and the volumes of Advances in Atomic and Molecular (since 1990 Atomic, Molecular and Optical) Physics. The latest developments are usually published in Physical Review A, Journal of Physics B, Journal of Chemical Physics and Journal of Physical and Chemical Reference Data, and are presented at the biannual International Conference on Photonic, Electronic and Atomic Collisions (ICPEAC).

63.1 **Basic Concepts** ... 929
63.1.1 Electron Impact Processes ... 929
63.1.2 Definition of Cross Sections ... 929
63.1.3 Scattering Measurements ... 930

63.2 **Collision Processes** ... 933
63.2.1 Total Scattering Cross Sections ... 933
63.2.2 Elastic Scattering Cross Sections ... 933
63.2.3 Momentum Transfer Cross Sections ... 933
63.2.4 Excitation Cross Sections ... 933
63.2.5 Dissociation Cross Sections ... 935
63.2.6 Ionization Cross Sections ... 935

63.3 **Coincidence and Superelastic Measurements** ... 936

63.4 **Experiments with Polarized Electrons** ... 938

63.5 **Electron Collisions with Excited Species** ... 939

63.6 **Electron Collisions in Traps** ... 939

63.7 **Future Developments** ... 940

References ... 940

63.1 Basic Concepts

63.1.1 Electron Impact Processes

Lox energy electrons can very effectively interact with the valence-shell electrons of atoms and molecules, in part because they have similar speeds. In elastic scattering, the continuum electron changes direction and transfers momentum to the target. Inelastic collisions include also a transfer of kinetic energy to the target, such as excitation of valence electrons to discrete energy levels, to the ionization continuum, and, in the case of molecules, excitation of nuclear motion (rotational, vibrational) and excitation to states which dissociate into neutral or ionic fragments. Various combinations of these processes are also possible, e.g., dissociative attachment, excitation or ionization. Excitation of more than one valence electron at the same time, or excitation of electrons from intermediate and inner shells, may also occur but these processes are more likely at impact energies of a few keV. These excitations lie above the first ionization limit and lead, therefore, with high probability, to autoionization, except for heavy elements where X-ray emission is an important competing process.

63.1.2 Definition of Cross Sections

The parameters which characterize collision processes are the cross sections. Electron collision cross sections depend on impact energy E_0 and scattering polar angles θ and ϕ. The differential cross section, for a specific well-defined excitation process indicated by the index n is defined as

$$\frac{\mathrm{d}\sigma_n(E_0,\Omega)}{\mathrm{d}\Omega}=\frac{k_f}{k_i}|f_n(E_0,\Omega)|^2\,, \tag{63.1}$$

where Ω is the polar angle of detection, k_i and k_f are the initial and final electron momenta, and f_n is the complex scattering amplitude ($n=0$ refers to elastic scattering). Integration over the energy-loss profile is assumed. If the energy-loss spectrum is broad, differentiation with respect to energy loss also has to be included. For certain processes it may be necessary to define differential cross sections with respect to angle and energy for both primary and secondary particles.

Integration over all scattering angles yields the integral cross sections

$$\sigma_n(E_0)=\int_0^{2\pi}\int_0^{\pi}\frac{\mathrm{d}\sigma_n(E_0,\Omega)}{\mathrm{d}\Omega}\sin\theta\,\mathrm{d}\theta\,\mathrm{d}\phi\,. \tag{63.2}$$

In the case of elastic scattering, the momentum transfer cross section is defined as

$$\sigma_0^{\mathrm{M}}(E_0)=\int_0^{2\pi}\int_0^{\pi}\frac{\mathrm{d}\sigma_0(E_0,\Omega)}{\mathrm{d}\Omega}(1-\cos\theta)\sin\theta\,\mathrm{d}\theta\,\mathrm{d}\phi\,. \tag{63.3}$$

The total electron scattering cross section is obtained by summing all integral cross sections:

$$\sigma_{\mathrm{tot}}(E_0)=\sum_n\sigma_n(E_0)+\sum_m\sigma_m(E_0)\,, \tag{63.4}$$

where σ_m are the cross sections for other possible channels. Experimental cross sections typically represent averages over indistinguishable processes (e.g., magnetic sublevels, hyperfine states, rotational states etc.). The cross section obtained this way corresponds to an average over initial and sum over final indistinguishable states with equal weight given to the initial states. (This may not always be true, as discussed later.) If the target molecules are randomly oriented, the cross section averaged over these orientations is independent of ϕ. In addition, there is an averaging over the finite energy and angular resolution of the apparatus. It is important, therefore, to specify clearly the nature of the measured cross section, otherwise their utilization and comparison with other experimental and theoretical cross sections become meaningless. We denote the conventionally measured differential and integral cross sections by $D_n(E_0,\theta)$ and $Q_n(E_0)$, with the various averagings implied. Similarly, $Q^{\mathrm{M}}(E_0)$ and $Q_{\mathrm{tot}}(E_0)$ are the corresponding momentum transfer and total scattering cross sections.

The collision strength for a process $i\to j$, which is the particle equivalent of the oscillator strength, is defined by

$$\Omega_{ij}(E_0)=q_iE_0\sigma_{ij}(E_0)\,, \tag{63.5}$$

where q_i is the statistical weight of the initial state [$q_i=(2L_i+1)(2J_i+1)$], E_0 is in Rydbergs and σ_{ij} is in units of πa_0^2.

The rate for a specific collision process (e.g., excitation) for electrons of energy E_0 is given as

$$R_{ij}(E_0)=NI(E_0)\sigma_{ij}(E_0)\,, \tag{63.6}$$

where N is the density of the target particles (m^{-3}), and $I(E_0)$ is the electron flux $(\mathrm{m}^{-2}\,\mathrm{s}^{-1})$; σ_{ij} is in m^2, yielding R_{ij} in units of $\mathrm{m}^{-3}\,\mathrm{s}^{-1}$. For nonmonoenergetic electron beams, (63.6) must be integrated over E_0 to get the average rate.

63.1.3 Scattering Measurements

Most scattering experiments, are carried out in a beam–beam arrangement (Fig. 63.1). A beam of nearly

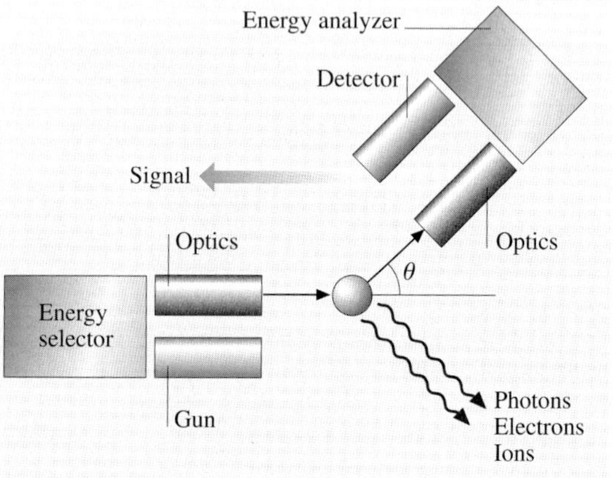

Fig. 63.1 A schematic diagram for electron scattering measurements

monoenergetic electrons is formed by extracting electrons from a hot filament and selecting a narrow segment of the thermal energy distribution. For the formation and control of the electron beam, electrostatic lenses are used and the energy selection is achieved with electrostatic energy selectors. A magnetic field may also be applied to obtain a magnetically collimated electron beam. The target beam is formed by letting the sample gas effuse from an orifice, tube or capillary array with various degrees of collimation. Target species which are in the condensed phase at room temperature need to be placed in a crucible and evaporated by heating. Extensive discussion of this technique has been given by *Scoles* [63.4]. The electron beam intersects the target beam at a 90° angle and electrons scattered into a specific direction, over a small solid angle $\left(\approx 10^{-3}\,\text{sr}\right)$, are detected. However the scattered electron is not necessarily the same as the incoming electron. Exchange with the target electrons may occur, and is required for spin-forbidden transitions in light elements. The detector system consists of electron lenses and energy analyzers similar to those used in the electron gun. The actual detector is an electron multiplier which generates a pulse for each electron. In the scattering process, secondary species (electrons, photons, ions, neutral fragments) may also be generated and can be detected individually or in various coincidence schemes. The experiments are carried out in a vacuum chamber and it is important to minimize stray electric and magnetic fields. More details about the apparatus and procedures can be found in a review by *Trajmar* and *Register* [63.5]. The primary information gained in these experiments is the energy and angular distribution of the scattered electrons.

There are several methods used to carry out scattering measurements. In the most commonly used energy-loss mode, the impact energy and scattering angle are fixed, and the scattering signal as a function of energy lost by the electron is measured by applying pulse counting and multichannel scaling techniques. The result of such an experiment is an energy-loss spectrum. The elastic scattering feature appears at zero energy loss; the other features correspond to various excitation processes and to ionization. Energy-loss spectra can also be generated in the constant residual energy mode. In this case, the detector is set to detect only electrons with a specific residual energy $E_R = E_0 - \Delta E$ at a fixed scattering angle, and E_0 and ΔE are simultaneously varied. Each feature in the energy-loss spectrum is obtained then at the same energy above its own threshold. An example, taken from the work of *Allan* [63.6], ist shown in Fig. 63.2.

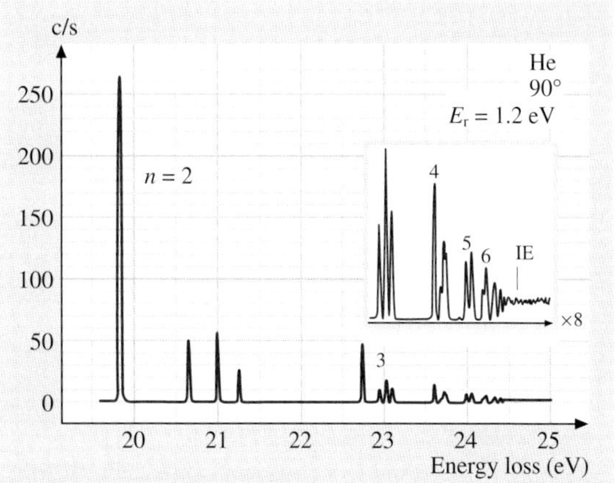

Fig. 63.2 Energy-loss spectrum of He at a constant residual energy of 1.2 eV and scattering angle of 90°. The inelastic features with the corresponding principal quantum numbers are shown. IE is the ionization potential of He (24.58 eV). No background is subtracted and true signal zero is indicated by a dotted line under the expand portion of the spectrum m. (Taken from *Allen* [63.6])

The energy loss spectrum becomes equivalent to the photoabsorption spectrum in the limit of small momentum transfer $\boldsymbol{K}$, where $\boldsymbol{K} = \boldsymbol{k}_f - \boldsymbol{k}_i$ (i. e., high impact energy, small scattering angle). The equivalence of electrons and photons in this limit follows from the Born approximation, and it can be used to obtain optical absorption and ionization cross sections. The correspondence is defined through the Limit Theorem

$$\lim_{\boldsymbol{K}\to 0} f_n^{\mathrm{G}}(\boldsymbol{K}) = f_n^{\mathrm{opt}}\,, \tag{63.7}$$

$$f_n^{\mathrm{G}}(\boldsymbol{K}) = \frac{\Delta E}{2}\frac{\boldsymbol{k}_i}{\boldsymbol{k}_f}\boldsymbol{K}^2\frac{\mathrm{d}\sigma_n}{\mathrm{d}\Omega}(\boldsymbol{k})\,, \tag{63.8}$$

where f_n^{G} is the generalized oscillator strength for excitation process n, and f_n^{opt} is the corresponding optical f-value. Equation (63.7) was originally derived by *Bethe* [63.7] from the Born cross section. It was extended later by *Lassettre* et al. [63.8] to cases where the Born approximation does not hold. In this case, $f_n^{\mathrm{G}}(\boldsymbol{K})$ is replaced by $f_n^{\mathrm{app}}(\boldsymbol{K})$, the apparent generalized oscillator strength. The Limit Theorem implies that, in the limit of small K, optical selection rules apply to electron impact excitation. The practical problem is that the limit is nonphysical and the extrapolation to zero K involves some arbitrariness. When K is significantly different from zero,

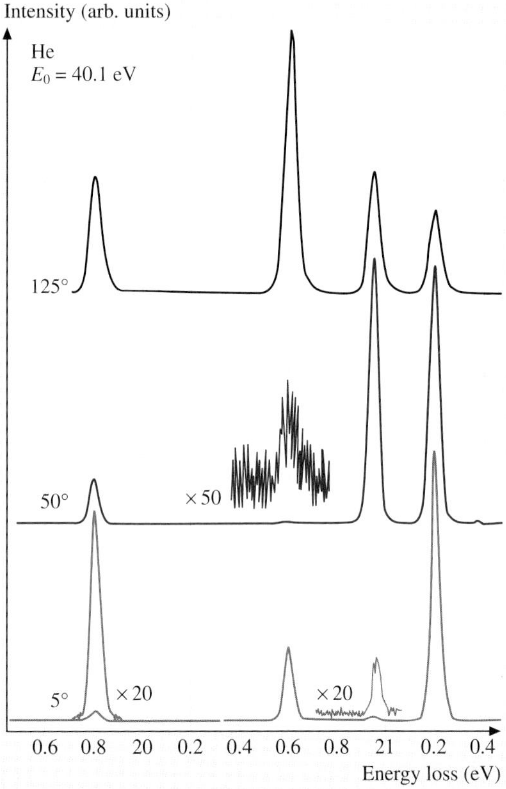

Fig. 63.3 Variation of energy-loss spectra (and DCSs) with scattering angle for He at 40 eV impact energy. Spectra are shown at 5°, 50° and 125° scattering angles

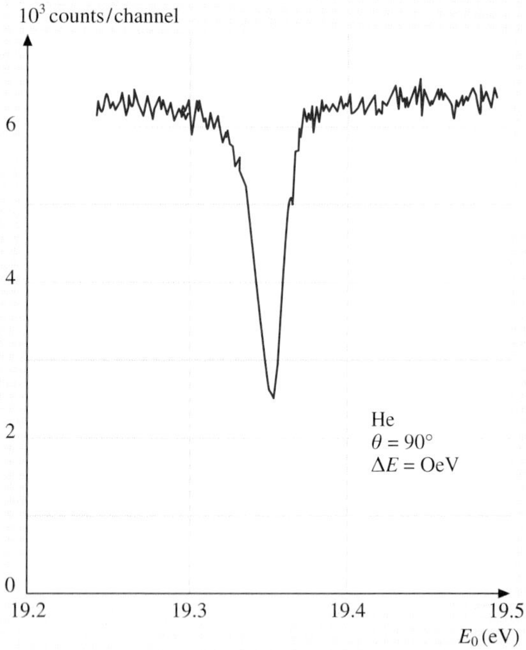

Fig. 63.4 The 19.37 eV He resonance observed in the elastic channel at 90° scattering angle

optical-type selection rules do not apply to electron-impact excitation. As can be seen from Fig. 63.3, spin and/or symmetry forbidden transitions then readily occur and can be an efficient way of producing metastable species.

Selection rules for electron impact excitations can be derived from group theoretical arguments [63.9, 10]. For atoms, the selection rule $S_g \leftrightarrow S_u$ applies in general and scattering to 0° and 180° is forbidden if $(L_i + \Pi_i + L_f + \Pi_f)$ is odd. Here, L_i and L_f are the angular momenta and Π_i and Π_f are the parities. For molecules, selection rules can be derived under two special conditions: (a) rules concerning 0° and 180° scattering for arbitrary orientation of the molecule, and (b) rules concerning scattering to any angle but for specific orientation of the molecule. An important example of the first case is the $\Sigma^- \nleftrightarrow \Sigma^+$ selection rule for linear molecules at 0° and 180° scattering angles.

The energy dependence of cross sections is obtained by fixing the energy-loss value (scattering channel) and studying the variation of scattering signal with impact energy at a given angle or integrated over all scattering angles. In general, cross sections associated with spin forbidden and optically allowed transitions peak near and at several times the threshold impact energy respectively, and they usually vary smoothly with impact energy. However, resonances may appear at certain specific impact energies. These sudden changes are associated with temporary electron capture and are the result of quantum mechanical interference between two indistinguishable paths. An example is shown in Fig. 63.4 for He in the elastic channel at 19.37 eV impact energy.

Integral cross sections can be obtained from extrapolation of the measured $D_n(E_0, \theta)$ to 0° and 180° scattering angles and integration over all angles. Recently, incorporation of an "angle-changing" device has enabled measurements to be extended over the whole range of scatering angles [63.11, 12]. In certain cases it is possible to measure integral cross sections directly by detecting secondary products such as photons and ions. These procedures and the resulting cross sections will be discussed in some detail in Sect. 63.2.

63.2 Collision Processes

In addition to the basic elastic and inelastic processes defined in Sect. 63.1.2, we now also explicitly include dissociation (to neutral and charged fragments) cross sections $Q_D(E_0)$; and ionization cross sections $Q_I(E_0)$. Each of these is now considered separately.

63.2.1 Total Scattering Cross Sections

Total electron scattering cross sections represent the sum of all integral cross sections:

$$Q_{tot}(E_0) = \sum_n Q_n(E_0) + Q_I(E_0) + Q_D(E_0) , \quad (63.9)$$

$Q_{tot}(E_0)$ are useful for checking the validity of scattering theories, and the consistency of available data, for normalization of integral and differential cross sections, and as input to the Boltzmann equation. At low impact energies, elastic scattering dominates, while at intermediate and high impact energies, electronic excitations and ionization become major contributors to Q_{tot}. Figure 63.5 shows the various cross sections for electron–helium collisions. The data are from the recommended values of *Trajmar* and *Kanik* [63.13].

Two methods are commonly used for measuring $Q_{tot}(E_0)$: the transmission method and the target recoil method (for details see [63.5, 14]). Total scattering cross sections measured by these techniques are, in general, accurate to within a few percent. The extensive reviews by *Zecca* and co-workers [63.15–17] should be noted.

63.2.2 Elastic Scattering Cross Sections

Elastic scattering cross sections $Q_0(E_0)$ are not as readily available as $Q_{tot}(E_0)$. They are obtained from differential scattering experiments over limited angular ranges by extrapolation and integration of the measured values. Typical error limits are 5 to 20%. For molecular species rotational excitation is usually not resolved but is included in the $D_0(E_0, \theta)$ and $Q_0(E_0)$ values. In order to obtain the absolute $D_0(E_0, \theta)$ directly from the scattering signal, one has to know the electron flux, the number of scattering species, the scattering geometry and the overall response function of the apparatus. A direct measurement of these parameters can be made at high energies (> 100 eV). However, at low electron energies, this approach is not feasible. A number of methods have been devised to derive relative $D_0(E_0, \theta)$ from the measured scattering intensities and then to normalize the $D_0(E_0, \theta)$ to the absolute scale. We briefly outline here only the most commonly used procedure.

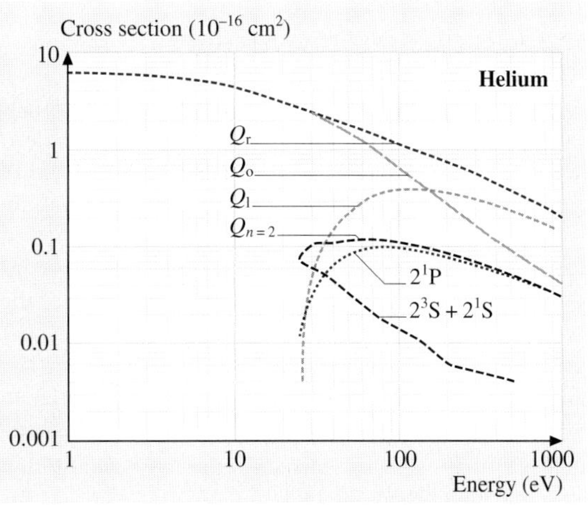

Fig. 63.5 Cross sections for various processes in the electron–helium collision (see text for data sources)

The most practical and reliable method of obtaining the absolute $D_0(E_0, \theta)$ is the relative flow technique in which scattering signals for a known standard gas and an unknown test gas are compared at each energy and angle [63.5, 18–20]. The He elastic cross section is the natural choice of standard since it is known accurately over a wide energy and angular range, and He is experimentally easy to handle. Only the relative electron beam flux and molecular beam densities (and their distributions) need be known in the two measurements. The flow rate of the test gas is adjusted so that the flux and density distribution patterns of the two gases are identical, and all geometrical factors cancel in the scattering intensity ratios. The absolute $D_0(E_0, \theta)$ for the sample gas is obtained from the measured scattering intensity, target density, and electron beam intensity ratios and the standard $D_0(E_0, \theta)$ value. See [63.5, 20, 21] for a detailed discussion of this technique.

63.2.3 Momentum Transfer Cross Sections

$Q^M(E_0)$ can be obtained both from the elastic DCSs and from swarm measurements. At low electron-impact energies (from 0.05 to a few eV), where only a few collision channels are open, the electron swarm technique is the most accurate (≈ 3%) way to determine the momentum transfer cross sections. Beam–beam experiments are mandatory at higher energies. A detailed discussion of these techniques is given by *Trajmar* and *Register* [63.5].

63.2.4 Excitation Cross Sections

$D_n(E_0, \theta)$ and $Q_n(E_0)$ can be derived from energy-loss spectra obtained in beam–beam scattering experiments. The relative $D_n(E_0, \theta)$ is usually normalized to $D_0(E_0, \theta)$ which in turn is normalized to the helium $D_0(E_0, \theta)$ by the relative flow technique described in Sect. 63.2.2. There are, however, complications and uncertainties associated with this technique because of the sensitivity of the instrument response function to the residual energy of the scattered electrons. For more details, see *Trajmar* and *McConkey* [63.21]. Data obtained by this procedure are rather limited, partly due to experimental difficulties and partly due to the time required to carry out such measurements.

For cases where an excited state j is formed which can radiatively decay by means of a short-lived (dipole-allowed) transition to a lower lying state i, the intensity of the resultant radiation is directly related to the cross section for production of the excited state in the original collision process. An optical emission cross section, $Q_{ji}(E_0)$, is defined by

$$Q_{ji}(E_0) = \frac{N_j \Gamma_{ji}}{I n_0 \tau_j}, \tag{63.10}$$

where N_j and n_0 are the densities in the excited and ground states, respectively, Γ_{ji} is the branching ratio for radiative decay from state j to state i, I is the electron beam flux, and τ_j is the natural radiative lifetime of state j. Since the excited state may be produced either by direct electron impact or by cascade from higher-lying states k, also formed in the collision process, we may define the direct excitation cross section $Q_j^{\mathrm{d}}(E_0)$ by

$$Q_j^{\mathrm{d}}(E_0) = \sum_i Q_{ji}(E_0) - \sum_k Q_{kj}(E_0). \tag{63.11}$$

The last term subtracts the cascade contribution from higher lying states. The quantity $Q_j^{\mathrm{a}}(E_0) = \sum_i Q_{ji}(E_0)$ is known as the apparent excitation cross section for level j. Clearly, to obtain $Q_j^{\mathrm{d}}(E_0)$ from $Q_j^{\mathrm{a}}(E_0)$, the cascade contribution must be known.

In (63.10), $N_j \Gamma_{ji}/\tau_j$ gives the steady state number of $j \to i$ photons per unit time per unit volume emitted from the interaction region. Since observation is made in a particular direction care must be taken to correct for any anisotropy in the radiation pattern. Alternatively, if observation is made at the so-called "magic" angle ($54° 44'$) to the electron beam direction, the emission intensity per unit solid angle is equal to the average intensity per unit solid angle irrespective of the polarization of the emitted radiation. However, even at this magic angle, care must be taken to avoid problems with polarization sensitivity of the detection equipment [63.22, 23].

The phenomenon of radiation trapping is often a problem if the radiative decay channels of the excited state include a dipole allowed channel to the ground state. Repeated absorption and re-emission of the radiation can occur and can lead to a diffuse emitting region much larger than the original interaction region, and the polarization of the emitted light can also be altered. Often a study of the variation of the emitted intensity or polarization with the target gas pressure is sufficient to reveal the presence of radiation trapping or other secondary effects.

The emission cross sections of certain lines have been measured with great care and now serve as bench-marks for other work. Examples of these are the measurements of *van Zyl* et al. [63.24] on the n^1S levels of He in the visible spectral region or the measurements of the cross section for production of Lyman α from H_2 in the VUV region (see [63.25] for a full discussion of this including many references). Use of secondary standards is particularly important when crossed-beam measurements are being carried out because of the cancellation of geometrical and other effects which occur.

The Bethe–Born theory [63.26] provides a convenient calibration of the detection system for optically allowed transitions of known oscillator strength. At sufficiently high energies, the excitation cross section, Q_n, of level n is given by

$$Q_n = \frac{4\pi a_0^2}{E_0/R_\infty} \frac{f_n^{\mathrm{opt}}}{\Delta E_n/R_\infty} \ln(4c_n E_0/R_\infty). \tag{63.12}$$

Here ΔE_n is the excitation energy, and c_n is a constant dependent on the transition. A plot of $Q_n E_0$ versus $\ln E_0$ is a straight line with a slope proportional to f_n^{opt} and the intercept with the $\ln E_0$ axis yields an experimental value for c_n independent of the normalization. For example, the He n^1P–1^1S optical oscillator strengths are very accurately known, as are cascade contributions. Thus accurate normalization of the slope of the Bethe plot can be made, yielding accurate excitation cross sections.

As mentioned above, the excitation cross sections display characteristic shapes as a function of energy. For optically allowed transitions, the cross section rises relatively slowly from threshold to a broad maximum approximately five times the threshold energy. At higher energies the $(\ln E_0)/E_0$ dependence of the cross section predicted by (63.12) is observed. For exchange processes, e.g., a triplet excited state from a singlet ground state, the cross section peaks sharply close to threshold and falls off at high energy as E^{-3}. If the excitation is spin allowed but optically forbidden, e.g., He n^1D from 1^1S, then the Bethe theory predicts an E^{-1} dependence of the cross section at high energies.

When excitation occurs to a long-lived (metastable or Rydberg) state following electron impact, it is often possible to detect the excited particle directly. Time-of-flight (T.O.F.) techniques are used to distinguish the long lived species from other products, e.g., photons, produced in the collisions.

63.2.5 Dissociation Cross Sections

Dissociation of a molecular target can result in fragments which may be excited or ionized. Such processes may be studied using the techniques discussed in the previous section or in the following section, where charged particle detection is considered. Because a repulsive state of the molecule is accessed, the fragments can leave the interaction region with considerable kinetic energy (several eV). If the fragment is in a long-lived metastable or Rydberg state, T.O.F. techniques may be used to distinguish the long-lived species from other products such as photons, and also to measure the energies of the excited fragments, and thus provide information on the repulsive states responsible for the dissociation. For further discussion see the reviews by *Compton* and *Bardsley* [63.27], *Freund* [63.28], and *Zipf* [63.29]. If the detector can be made sensitive to a particular excited species, its excitation can be isolated and studied. Examples are the work of *McConkey* and co-workers [63.30,31] on O(^{1}S) and S(^{1}S$_0$) production from various molecules. The detection of unexcited neutral fragments is more challenging. One early method was to trap selectively the dissociation products using a getter and measure the resulting pressure decrease. In a more sophisticated approach, *Cosby* [63.32] produced a fast (≈ 1 keV) target molecular beam by resonant charge exchange and subjected it to electron impact dissociation. The fast dissociation products were detected by conventional particle detectors in a time correlated measurement. Laser techniques, such as laser-induced fluorescence or multiphoton ionization, have also been used recently to detect the dissociation products.

The Franck–Condon principle largely governs molecular dissociation. The principle states that if the excitation takes place on a time scale which is short compared with vibrational motion of the atomic nuclei the transition occurs vertically between potential energy curves. Since dissociation rapidly follows a vertical transition to the repulsive part of a potential energy curve, compared with the period of molecular rotation, the dissociation products tend to move in the direction of vibrational motion. Since the excitation probability depends on the relative orientation of the electron beam and the molecule, dissociation products often demonstrate pronounced anisotropic angular distributions. The angular distributions have been analyzed by *Dunn* [63.33] using symmetry considerations.

63.2.6 Ionization Cross Sections

Tate and *Smith* [63.34] some 60 years ago developed the basic techniques for measuring total ionization cross sections. These were later improved by *Rapp* and *Englander-Golden* [63.35]. Full details of the experimental methods are given in the reviews and books already cited. *Märk* and *Dunn* [63.36] reviewed the situation as it existed in the mid 1980s. In the basic "parallel plate" method, the electron beam is directed through a beam or a static target gas between collector plates which detect the resultant ions. Unstable species can be studied by the "fast neutral beam" technique [63.37], in which the neutral target species is formed by charge neutralization from a fast ion beam, and is subsequently ionized by a crossed electron beam. For the determination of partial ionization cross sections specific to a given ion species in a given ionization stage, mass spectroscopic (quadrupole mass spectrometer, electrostatic or magnetic charged particle analyzer or time of flight) methods are used. Fourier Transform Mass Spectrometry (FTMS) has also been used effectively to study fragmentation with formation of both positive and negative ions. Reference [63.38] is a recent example of this. Absolute total ionization cross sections have been measured for a large number of species with an accuracy of better than 10%. *Christophorou* and colleagues have presented helpful compilations of ionization and other data of particular relevance to the plasma processing industry, [63.39, and earlier references in this journal].

A large number of mechanisms can contribute to the ionization of atoms and molecules by electron impact. For targets with only a few atomic electrons, the dominant process is single ionization of the outer shell, with the resultant ion being left in its ground state. The process is direct and is characterized by large impact parameters b and small momentum transfers. The cross section varies with incident electron energy in a way very similar to the optically allowed excitation processes discussed in Sect. 63.2.4. Processes involving ionization of more than one outer shell electron become more important as the size of the target increases. These events are associated with small b and electron–electron correlations are usually strong. Autoionization increases in significance for heavier targets. Here also, collisions with small b dominate and electron–electron correlations are strong. For heavier targets, inner shell effects, such as Auger electron or X-ray emission, become progressively more important. For molecular targets, dissociative ionization (either directly or through a highly excited intermediate state) and ion pair formation also play a significant role.

In addition to measurement of gross ion production, it is also possible to study the ionization process by monitoring the electron(s) ejected or scattered inelastically. Conventional electron spectroscopic techniques are used for this purpose. The addition of coincidence techniques (e–2e measurements) in which the momenta of all the electrons involved are completely specified has allowed many of the fine details of the ionization process to be extracted [63.40].

63.3 Coincidence and Superelastic Measurements

The cross section measurements described so far do not yield complete information on electron scattering processes. As mentioned in Sect. 63.1.2, these cross sections do not distinguish for magnetic sublevels, electron spin etc.and represent summation of cross sections over these experimentally indistinguishable processes (summation of the square moduli of the corresponding scattering amplitudes). The quantum mechanical description of a scattering process is given in terms of scattering amplitudes and under certain conditions requires summing up amplitudes and squaring the sum. This leads to interference terms which arise from the coherent nature of the scattering process. A complete characterization of a scattering process, therefore, requires knowledge of the complex scattering amplitudes.

Sophisticated experimental techniques have been developed in recent years, which go beyond the conventional scattering cross section measurements and yield information on magnetic sublevel specific scattering amplitudes and the polarization (alignment and orientation) of the excited atomic ensemble. The experimental techniques fall into two main categories: a) electron–photon coincidence measurements, and b) superelastic scattering measurements involving coherently excited species. (We still consider unpolarized electron beams in the description of these two techniques here and will address the question of spin polarization in the following section.)

In electron–photon coincidence measurements, the radiation pattern emitted by the excited atom is determined for a given direction of the scattered electron. A scattering plane is defined by $\boldsymbol{k}_i$ and $\boldsymbol{k}_f$, and hence the symmetry is lowered from cylindrical (around the incident beam direction) to planar, (Fig. 63.6).

It is now possible to determine, at least in principle, both the atomic alignment (i. e., the shape of the excited state charge cloud and its alignment in space) and its orientation (i. e., the angular momentum transferred to the atom during the course of the collision). Complete sets of excitation amplitudes for the coherently excited atomic states and their relative phases have been measured in some cases. A comparison with theory can then be made at the most fundamental level. See [63.41] for full discussion and analysis.

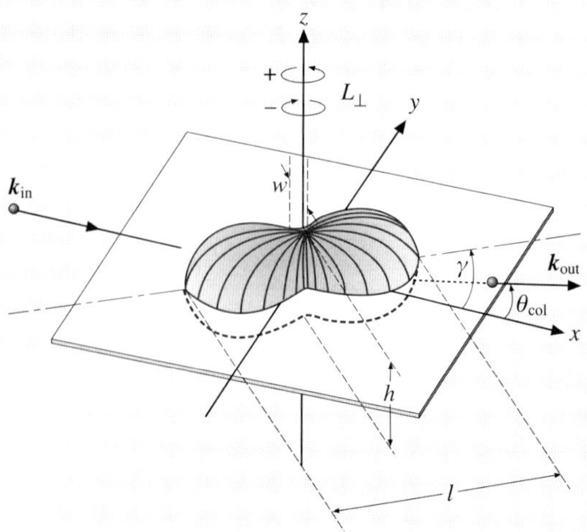

Fig. 63.6 Schematic illustration of a collisionally induced charge cloud in a p-state atom. The scattering plane is fixed by the direction of incoming $\boldsymbol{k}_{in}$ and outgoing $\boldsymbol{k}_{out}$ momentum vectors of the electrons. The atom is characterized by the relative length (l), width (w), and height (h) of the charge cloud, by its alignment angle γ, and by its inherent angular momentum $L_\perp$. The coordinate frame is the natural frame with the z-axis perpendicular to the scattering plane and with the x- and y-axes defined as shown in the figure relative to $\boldsymbol{k}_{in}$ and $\boldsymbol{k}_{out}$

The electron–photon coincidence measurements can be carried out in two ways: (1) measuring polarization correlations, and (2) measuring angular correlations. In (1), polarization analysis of the emitted photon in a given direction occurs, while in (2), the angular distribution of the emitted photons is determined without polarization analysis. We will describe here only method (1) in some detail.

Method (1) has the advantage that it measures directly the angular momentum (perpendicular to the scattering plane) transferred in the collision. For P-state excitation from a $^1S_0$ ground state, four parameters plus a cross section are needed to describe fully the collisionally excited P-state. The natural parameters introduced by *Andersen* et al. [63.41] are defined as follows (Fig. 63.6): γ is the alignment angle of the excited state charge cloud relative to the electron beam axis, P_ℓ^+ is the linear polarization in the scattering plane, $L_\perp^+$ is the orbital angular momentum perpendicular to the scattering plane that is transferred to the atom in the collision, and ρ_{00} is the relative height of the charge cloud perpendicular to the scattering plane at the point of origin. The $+$ superscript indicates positive reflection symmetry with respect to the scattering plane.

In polarization correlation experiments, one typically measures two linear (P_1, P_2) and one circular (P_3) polarization correlation parameters perpendicular to the scattering plane. One additional linear polarization correlation parameter P_4 is measured in the scattering plane. Each parameter is the result of two intensity measurements for different orientations of the polarization analyzer:

$$\begin{aligned}
P_1 &= \frac{I(0°) - I(90°)}{I(0°) + I(90°)}\beta^{-1} ,\\
P_2 &= \frac{I(45°) - I(135°)}{I(45°) + I(135°)}\beta^{-1} ,\\
P_3 &= \frac{I^R - I^L}{I^R + I^L}\beta^{-1} ,\\
P_4 &= \frac{I(0°) - I(90°)}{I(0°) + I(90°)}\beta^{-1} .
\end{aligned} \tag{63.13}$$

Here $I(\alpha)$ denotes the photon intensity measured for a polarizer orientation α with respect to the electron beam axis, I^R and I^L refer to right- and left-handed circularly polarized light and β denotes the polarization sensitivity of the polarization analyzer. The relationships between the experimentally determined polarization correlation parameters and the natural parameters are given by

$$\begin{aligned}
\gamma &= \frac{1}{2}\tan^{-1}(P_2/P_1) ,\\
P_\ell^+ &= \left(P_1^2 + P_2^2\right)^{1/2} ,\\
L_\perp^+ &= -P_3 ,\\
\rho_{00} &= \frac{(1+P_1)(1-P_4)}{4-(1-P_1)(1-P_4)} .
\end{aligned} \tag{63.14}$$

The total polarization P_{tot}^+, which is defined as

$$\begin{aligned}
P_{\text{tot}}^+ &= \left[\left(P_\ell^+\right)^2 + \left(L_\perp^+\right)^2\right]^{1/2}\\
&= \left(P_1^2 + P_2^2 + P_3^2\right)^{1/2} ,
\end{aligned} \tag{63.15}$$

is a measure for the degree of coherence in the excitation process. In the absence of atomic depolarizing effects due to, for example, fine and/or hyperfine interactions, a value of $P_{\text{tot}}^+ = +1$ for the emitted radiation indicates total coherence of the excitation process.

Much of the earlier work involved excitation of helium n^1P state. Here the situation is simplified as L–S coupling applies strictly: $P_4 = 1$ and ρ_{00} is zero. Excitation of the 2^1P state, for example, is fully coherent and hence the excitation is completely specified by just two parameters, γ and $L_\perp$ (or P_ℓ since $P_\ell = \left(1 - L_\perp^2\right)^{1/2}$). More recently, the techniques have been applied to heavier targets and more complicated excitation processes [63.42–46].

The superelastic scattering experiments could be looked at as time inverse electron–photon coincidence experiments (although this is not exactly the case). In these experiments, a laser beam is utilized to prepare a coherently excited, polarized ensemble of target atoms for the electron scattering measurement. The superelastic scattering intensity is then measured as a function of laser-beam polarization and/or angle with respect to a reference direction. Linearly polarized laser light produces an aligned target (uneven population in magnetic sublevels for quantum numbers of different $|M_J|$ value). Circularly polarized laser light produces oriented targets (uneven population in $M_J = +m$ and $M_J = -m$ magnetic sublevels). From these measurements the same electron impact coherence parameters can be deduced as from the coincidence experiments. This approach has been applied to atomic species (mainly metal atoms) which are conveniently excited with available lasers. Detailed descriptions of the experimental techniques, the underlying theoretical background, and the interpretation of the experimental data are given in [63.47–56].

It should be noted that electron scattering by coherently excited atoms can be utilized not only for obtaining electron impact coherence parameters for inelastic processes originating from ground state but for elastic, inelastic, and superelastic transitions involving excited states. These measurements yield information on creation, destruction, and transfer of alignment and orientation in electron collision processes which is needed, e.g., in the application of plasma polarization spectroscopy [63.57].

63.4 Experiments with Polarized Electrons

So far we have considered the utilization of unpolarized electron beams which yield spin averaged cross sections. Little information on spin dependent interactions is gained from these experiments. However, these interactions can be studied using polarized electron beam techniques. Developments on both the production and detection of spin-polarized electron beams have resulted in a wide range of elegant experiments probing these effects. The theory is also highly developed. For a detailed discussion see the works of *Kessler* [63.58, 59], *Blum* and *Kleinpoppen* [63.60], *Hanne* [63.61–63] and *Andersen* et al. [63.44, 45] and the references therein. Some basic concepts are presented here.

The degree of polarization P of an electron beam is given by

$$P = \frac{N(\uparrow) - N(\downarrow)}{N(\uparrow) + N(\downarrow)} , \tag{63.16}$$

where $N(\uparrow)$ and $N(\downarrow)$ are the numbers of electrons with spins respectively parallel and antiparallel to a particular quantization direction. Measurements of P both before and after the collision enable one to probe directly for specific spin dependent processes. For example, in elastic scattering from heavy spinless atoms any changes in the polarization of the electrons must be caused by spin–orbit interactions alone since, in this instance, it is not possible to alter the polarization of the electron beam by electron exchange. Measurements have been carried out for Hg and Xe and both direct (f) and spin-flip (g) scattering amplitudes, as well as their phase differences, have been determined [63.59].

The spin–orbit interaction for the continuum electron caused by the target nucleus leads to different scattering potentials and consequently to different cross sections for spin-up and spin-down electrons (called Mott scattering). Consequently, an initially unpolarized electron beam can become spin polarized after scattering by a specific angle according to

$$\boldsymbol{P}' = S_p(\theta)\hat{\boldsymbol{n}} , \tag{63.17}$$

where $\boldsymbol{n}$ is the unit vector normal to the scattering plane, $S_p(\theta)$ is the polarization function and $\boldsymbol{P}'$ is the polarization of the scattered beam. For the same reasons, when a spin-polarized electron beam is scattered by an angle θ to the left and to the right, an asymmetry is found in the scattering cross sections. Furthermore, an existing polarization $\boldsymbol{P}'$ can be detected through the left–right asymmetry A in the differential cross section, which is given by

$$A \equiv \frac{\sigma_l(\theta) - \sigma_r(\theta)}{\sigma_l(\theta) + \sigma_r(\theta)} = S_{\mathrm{A}}(\theta)\boldsymbol{P}' \cdot \hat{\boldsymbol{n}} , \tag{63.18}$$

where $\sigma_{\mathrm{l}}(\theta)$ and $\sigma_{\mathrm{r}}(\theta)$ are the differential cross sections for scattering at an angle θ relative to the incident beam axis to the left and to the right, respectively. For elastic scattering, the polarization function S_{P}, and the asymmetry function S_{A} are identical and are called the Sherman function.

When electron exchange is studied under conditions where other explicitly spin-dependent forces can be neglected, the cross sections for scattering of polarized electrons from polarized atoms depend on the relative orientation of the polarization vectors. According to [63.59]

$$\sigma(\theta) = \sigma_u(\theta)[1 - A_{\mathrm{ex}}(\theta)\boldsymbol{P}_e \cdot \boldsymbol{P}_A] , \tag{63.19}$$

where $\boldsymbol{P}_{\mathrm{e}}$ and $\boldsymbol{P}_{\mathrm{A}}$ are the electron and atom polarization vectors, and $\sigma_{\mathrm{u}}(\theta)$ is the cross section for unpolarized electrons. Hence, an "exchange asymmetry" $A_{\mathrm{ex}}(\theta)$ can be defined by

$$A_{\mathrm{ex}}(\theta)\, \boldsymbol{P}_e \cdot \boldsymbol{P}_A = \frac{\sigma_{\uparrow\downarrow}(\theta) - \sigma_{\uparrow\uparrow}(\theta)}{\sigma_{\uparrow\downarrow}(\theta) + \sigma_{\uparrow\uparrow}(\theta)} , \tag{63.20}$$

where $\sigma_{\uparrow\uparrow}(\theta)$ and $\sigma_{\uparrow\downarrow}(\theta)$ denote the cross sections for parallel and antiparallel polarization vectors respectively. As *Bartschat* [63.64] points out, an asymmetry can occur even if the scattering angle is not defined. In this case the function $A_{\mathrm{ex}}(\theta)$ is averaged over all angles. Differential and integral measurements of this kind have been performed for elastic scattering, excitation and ionization.

For heavy target systems it is necessary to consider a combination of effects together with a description of the target states in the intermediate or fully coupled scheme. Consequently, the number of independent parameters can become very large and the "complete" experiments, which disentangle the various contributions to any observed asymmetry in the scattering, are rarely possible.

Even for very light target atoms, where conventional Mott scattering is negligible, *Hanne* [63.61] has shown that the "fine structure" effect, in which electron scattering from individual fine structure levels of a multiplet occurs, can lead to polarization effects. In fact it can be a dominating effect for inelastic collision processes.

For full details of these various mechanisms and how density matrix theory and other theoretical techniques have been applied to scattering involving polarization effects, the reader is referred to the review articles cited, particularly *Andersen* et al. [63.44].

In certain cases, experiments involving spin polarized electron beams coupled with coincidence (or superelastic) measurements allow one to extract the maximum possible information for a given process, and are termed as complete or perfect in the sense defined by *Bederson* [63.65–67].

63.5 Electron Collisions with Excited Species

There are many plasma systems where electron collisions with excited atoms and molecules play a prominent role, e.g., electron beam and discharge pumped lasers, planetary and astrophysical plasmas. Especially important are electron collisions with metastable species because of the long lifetime, large cross section and large amount of excitation energy associated with them. Electron collision studies and cross section data in this area are scarce mainly due to experimental difficulties associated with the production of target beams with sufficiently high densities of excited species. With the application of lasers for the preparation of the excited species this problem can be overcome. However, this approach has not yet been extensively exploited. Reviews of this field are given by *Lin* and *Anderson* [63.68], *Trajmar* and *Nickel* [63.69] and *Christophorou* and *Olthoff* [63.70].

Since electron collisions with excited species necessarily involve a method of preparation, they are two step processes. Excitation and ionization in these cases are frequently referred to as stepwise excitation and ionization. The target preparation leads to mixed beams containing both ground and excited atoms or molecules. Preparation of excited atoms is achieved by electron impact or photoabsorption. Fast metastable beams can be produced by near-resonance charge exchange. For more details see [63.69, 71, 72]. Electron impact excitation is simple and effective but highly nonspecific, and characterization of the composition of the mixture is difficult. Laser excitation is more involved but very well defined. Specific fine and hyperfine levels of individual isotopes can be excited. When laser excitation is used in conjunction with superelastic electron scattering, an energy resolution of 10^{-8} eV is easily achieved, compared with the 10^{-2} eV resolution possible in conventional electron scattering.

Depending on the method of preparation, the population distribution in the magnetic sublevels of the target atoms may be uneven and some degree of polarization (alignment or orientation) may be present. The scattering will then be ϕ-dependent. For polarized target atoms the measured electron collision cross sections do not correspond to the conventional cross sections (which are summed over final and averaged over initial experimentally indistinguishable states, with equal populations in the initial states). One, therefore, has to characterize precisely the state of the target beam in order to be able to deduce a well defined, meaningful cross section. Polarization of atoms can be conveniently controlled, in the case of excitation with laser light, through the control of the laser light polarization (as discussed in Sect. 63.3). Since the atomic ensemble is coherently excited in this case, the scattering cross sections will depend on the azimuthal scattering angle ϕ. These considerations also come into play when one tries to relate measured inelastic and corresponding inverse superelastic cross sections by the principle of detailed balancing.

63.6 Electron Collisions in Traps

A technique which has recently begun to be exploited involves collisions with trapped atoms. Pioneered by *Lin* and colleagues [63.73, 74], using Rb targets the technique has many advantages over more conventional techniques, not least of which is the fact that the absolute number density of the target need not be known. Cross section data are obtained from measurements of trap loss and electron beam current density. Because up to half of the atoms in the trap can be in the excited state, it is possible to make measurements of cross sections involving excited states as well [63.75]. Measurements involving Cs targets have also been reported [63.76].

63.7 Future Developments

Electron-driven processes have been identified as being of fundamental importance in a wide range of environmentally and technologically significant areas [63.77]. *Boudaiffa* et al. [63.78] have shown that electron attachment is a significant process in bond-breaking in DNA. Electron-initiated dissociation of large molecules can act as a catalyst for reactive chemistry in environmentally sensitive situations. Developments in large scale computing have opened the door to calculations involving large molecules which could not even have been contemplated a few years ago. Electron collisions in intense laser field situations is an exciting new field which is rapidly expanding [63.79, 80]. Electron–cluster interactions allow one to probe how interactions change as one progresses from the gaseous to the solid phase.

References

63.1 J. Franck, G. Hertz: Verh. dtsch. Phys. Ges. **16**, 457 (1914)

63.2 H. S. N. Massey, E. H. S. Burhop, H. B. Gilbody: *Electronic and Ionic Impact Phenomena*, Vol. 1 & 2 (Clarendon Press, Oxford 1969)

63.3 E. W. McDaniel: *Atomic Collisions, Electron and Photon Projectiles* (John Wiley and Sons, New York 1989)

63.4 G. Scoles: *Atomic and Molecular Beam Methods*, Vol. 1 (Oxford Univ. Press, New York 1988)

63.5 S. Trajmar, D. F. Register: Experimental techniques for cross section measurements. In: *Electron Molecule Collisions*, ed. by I. Shimamura, K. Takayanagi (Plenum, New York 1984)

63.6 M. Allen: J. Phys. B **25**, 1559 (1992)

63.7 H. A. Bethe: Ann. Phys. **5**, 325 (1930)

63.8 E. N. Lassettre, A. Skerbele, M. A. Dillon: J. Chem. Phys. **50**, 1829 (1969)

63.9 D. C. Cartwright, S. Trajmar, W. Williams, D. L. Huestis: Phys. Rev. Lett. **27**, 704 (1971)

63.10 W. A. Goddard, D. L. Huestis, D. C. Cartwright, S. Trajmar: Chem. Phys. Lett. **11**, 329 (1971)

63.11 M. Zubek, B. Mielewska, J. Channing, G. C. King, F. H. Read: J. Phys. B **32**, 1351 (1996)

63.12 J. Channing, F. H. Read: Rev. Sci. Instr. **67**, 2372 (1996)

63.13 S. Trajmar, I. Kanik: Elastic and excitation electron collisions with atoms. In: *Atomic and Molecular Processes in Fusion Edge Plasmas*, ed. by R. K. Janev, H. P. Winter, W. Fritsch (Plenum, New York 1995)

63.14 B. Bederson, L. J. Kieffer: Rev. Mod. Phys. **43**, 601 (1971)

63.15 A. Zecca, G. P Carwasz, R. S. Brusa: La Rivista del Nuovo Cimento **19**(3), 1 (1996)

63.16 G. P Carwasz, R. S. Brusa, A. Zecca: La Rivista del Nuovo Cimento **24**(1), 1 (2001)

63.17 G. P. Carwasz, R. S. Brusa, A. Zecca: La Rivista del Nuovo Cimento **24**(4), 1 (2001)

63.18 S. Srivastava, A. Chutjian, S. Trajmar: J. Chem. Phys. **63**, 2659 (1975)

63.19 J. C. Nickel, C. Mott, I. Kanik, D. C. McCollum: J. Phys. B **21**, 1867 (1988)

63.20 J. C. Nickel, P. Zetner, G. Shen, S. Trajmar: J. Phys. E **22**, 730 (1989)

63.21 S. Trajmar, J. W. McConkey: Adv. At. Mol. Opt. Phys. **33**, 63 (1994)

63.22 P. N. Clout, D. W. O. Heddle: J. Opt. Soc. Am. **59**, 715 (1969)

63.23 F. G. Donaldson, M. A. Hender, J. W. McConkey: J. Phys. B **5**, 1192 (1972)

63.24 B. van Zyl, G. H. Dunn, G. Chamberlain, D. W. O. Heddle: Phys. Rev. A **22**, 1916 (1980)

63.25 A. R. Filippelli, C. C. Lin, L. W. Anderson, J. W. McConkey: Adv. At. Mol. Opt. Phys. **33**, 1 (1994)

63.26 M. Inokuti: Rev. Mod. Phys. **43**, 297 (1971)

63.27 R. N. Compton, J. N. Bardsley: Dissociation of molecules by slow electrons. In: *Electron Molecule Collisions*, ed. by I. Shimamura, K. Takayanagi (Plenum, New York 1984)

63.28 R. S. Freund: *Rydberg States of Atoms and Molecules*, ed. by R. F. Stebbings, F. B. Dunning (Cambridge Univ. Press, Cambridge 1983)

63.29 E. C. Zipf: Dissociation of molecules by electron impact. In: *Electron-Molecule Interactions and Their Applications*, Vol. 1, ed. by L. G. Christophorou (Academic, New York 1984)

63.30 L. R. LeClair, J. W. McConkey: J. Chem. Phys. **99**, 4566 (1993)

63.31 W. Kedzierski, J. Borbely, J. W. McConkey: J. Phys. B **34**, 4027 (2001)

63.32 P. C. Cosby: J. Chem. Phys. **98**, 7804, 9544 and 9560 (1993)

63.33 G. H. Dunn: Phys. Rev. Lett. **8**, 62 (1962)

63.34 J. T. Tate, P. T. Smith: Phys. Rev. **39**, 270 (1932)

63.35 D. Rapp, P. Englander-Golden: J. Chem. Phys. **43**, 1464 (1965)

63.36 T. D. Märk, G. H. Dunn: *Electron Impact Ionization* (Springer, Vienna, New York 1985)

63.37 A. J. Dixon, M. F. A. Harrison, A. C. H. Smith: J. Phys. B **9**, 2617 (1986)

63.38 C. Q. Jaio, C. A. DeJoseph, A. Garscadden: J. Phys. Chem **107**, 9040 (2003)
63.39 L. G. Christophorou, J. K. Olthoff: J. Phys. Chem. Ref. Data **31**, 971 (2002)
63.40 H. Ehrhardt, K. Jung, G. Knoth, P. Schlemmer: Z. Phys. D **1**, 3 (1986)
63.41 N. Andersen, J. W. Gallagher, I. V. Hertel: Phys. Rep. **165**, 1 (1988)
63.42 K. Becker, A. Crowe, J. W. McConkey: J. Phys. B **25**, 3885 (1992)
63.43 M. Sohn, G. F. Hanne: J. Phys. B **25**, 4627 (1992)
63.44 N. Andersen, K. Bartschat, J. T. Broad, I. V. Hertel: Phys. Rep **279**, 251 (1997)
63.45 N. Andersen, K. Bartschat: *Polarization, Alignment, and Orientation in Atomic Collisions* (Springer, Berlin 2001)
63.46 C. Herting, G. F. Hanne: Correlation and polarization in photonic, electronic and atomic collisions, AIP Conf. Proc. **697**, 181 (2004)
63.47 I. V. Hertel, W. Stoll: Adv. At. Mol. Phys. **13**, 113 (1977)
63.48 P. W. Zetner, S. Trajmar, G. Csanak: Phys. Rev. A **41**, 5980 (1990)
63.49 U. Fano: Rev. Mod. Phys. **29**, 74 (1957)
63.50 K. Blum: *Density Matrix Theory and Applications* (Plenum Press, New York 1981)
63.51 U. Fano, J. H. Macek: Rev. Mod. Phys. **45**, 553 (1973)
63.52 J. H. Macek, I. V. Hertel: J. Phys. B **7**, 2173 (1974)
63.53 W. R. MacGillivray, M. C. Standage: Phys. Rep. **168**, 1 (1988)
63.54 V. Karaganov, I. Bray, P. J. O. Teubner: Phys. Rev. A **57**, 208 (1998)
63.55 K. A. Stockman, V. Karaganov, I. Bray, P. J. O. Teubner: J. Phys. B **34**, 1105 (2001)
63.56 P. W. Zetner, P. V. Johnson, Y. Li, G. Csanak, R. E. H. Clark, J. Abdallah: J. Phys. B **34**, 1619 (2001)
63.57 S. A. Kazantsev, J.-C. Hénoux: *Polarization Spectroscopy of Ionized Gases* (Kluwer, Dordrecht 1995)
63.58 J. Kessler: *Polarized Electrons*, 2nd edn. (Springer, Berlin, Heidelberg 1985) p. 2
63.59 J. Kessler: Adv. At. Mol. Opt. Phys. **27**, 81 (1991)
63.60 K. Blum, H. Kleinpoppen: Adv. At. Mol. Phys. **19**, 187 (1983)
63.61 G. F. Hanne: Phys. Rep. **95**, 95 (1983)
63.62 G. F. Hanne: *Coherence in Atomic Collision Physics*, ed. by H. J. Beyer, K. Blum, R. Hippler (Plenum, New York 1988) p. 41
63.63 G. F. Hanne: Collisions of polarized electrons with atoms and molecules. In: *Proceedings of the 17th ICPEAC, Electronic and Atomic Collisions*, ed. by W. R. MacGillivray, I. E. McCarthy, M. C. Standage (Hilger, Bristol 1992) p. 199
63.64 K. Bartschat: Comments At. Mol. Phys. **27**, 239 (1992)
63.65 B. Bederson: Comments At. Mol. Phys. **1**, 41 (1969)
63.66 D. H. Yu, J. F. Williams, X. J. Chen, P. A. Hayes, K. Bartschat, V. Zeman: Phys. Rev. A **67**, 032707 (2003)
63.67 H. M. Al-Khateeb, B. G. Birdsey, T. J. Gay: Phys. Rev. Lett **85**, 4040 (2000)
63.68 C. C. Lin, L. W. Anderson: Adv. At. Mol. Opt. Phys. **29**, 1 (1992)
63.69 S. Trajmar, J. C. Nickel: Adv. At. Mol. Opt. Phys. **30**, 45 (1993)
63.70 L. G. Christopherou, J. K. Olthoff: Adv. At. Mol. Opt. Phys. **44**, 156 (2001)
63.71 J. B. Boffard, M. E. Lagus, L. W. Anderson, C. C. Lin: Rev. Sci. Inst **67**, 2738 (1996)
63.72 J. B. Boffard, M. F. Gehrke, M. E. Lagus, L. W. Anderson, C. C. Lin: Europhys. J. D **8**, 193 (2000)
63.73 R. S. Schappe, T. Walker, L. W. Anderson, C. C. Lin: Europhys. Lett **29**, 439 (1995)
63.74 R. S. Schappe, T. Walker, L. W. Anderson, C. C. Lin: Phys. Rev. Lett **76**, 4328 (1996)
63.75 M. L. Keeler, L. W. Anderson, C. C. Lin: Phys. Rev. Lett **85**, 3353 (2000)
63.76 J. A. MacAskill, W. Kedzierski, J. W. McConkey, J. Domyslawska, I. Bray: J. Elect. Spect. Rel. Phen **123**, 173 (2002)
63.77 K H. Becker, C. W. McCurdy, T. M. Orlando, T. N. Resigno: *Electron-Driven Processes: Scientific Challenges and Technological Opportunities* (US DOE Report, 2000)
63.78 B. Boudaiffa, P. Cloutier, D. Hunting, M. A. Huels, L. Sanche: Science **287**, 1658 (2000)
63.79 H. Niikura, F. Légaré, R. Hasbani, M. Y. Ivanov, D. M. Villeneuve, P. B. Corkum: Nature **421**, 826 (2003)
63.80 M. Weckenbrock, A. Becker, A. Staudte, S. Kammer, M. Smolarski, V. R. Bhardwaj, D. M. Rayner, D. M. Villeneuve, P. B. Corkum, R. Dorner: Phys. Rev. Lett **91**, 123004 (2003)

64. Ion–Atom Scattering Experiments: Low Energy

This chapter outlines the physical principles and experimental methods used to investigate low energy ion–atom collisions. A low energy collision is here defined as one in which the initial ion–atom relative velocity is less than the mean orbital velocity $\langle v_e \rangle$ of the electrons affected by the collision. For outer or valence electrons, $\langle v_e \rangle \simeq v_B$, where $v_B = 2.1877 \times 10^8$ cm/s is the Bohr velocity. In terms of the energy of a projectile ion, v_B corresponds to 24.8 keV/N, where N is the projectile nuclear number (e.g., 16 for O^+).

The theory and results of ion–atom scattering studies are further discussed in Chapts. 37, 38, and 47 to 51. The focus here is on the experimental techniques. Since several of these depend on the characteristics of a specific process, the following section presents a summary of the physics of low-energy ion–atom collisions. See Chapts. 50 and 51 for more detailed information.

64.1 **Low Energy Ion–Atom Collision Processes** 943

64.2 **Experimental Methods for Total Cross Section Measurements** 945
- 64.2.1 Gas Target Beam Attenuation Method 945
- 64.2.2 Gas Target Product Growth Method 945
- 64.2.3 Crossed Ion and Thermal Beams Method 945
- 64.2.4 Fast Merged Beams Method 946
- 64.2.5 Trapped Ion Method 946
- 64.2.6 Swarm Method 947

64.3 **Methods for State and Angular Selective Measurements** 947
- 64.3.1 Photon Emission Spectroscopy 947
- 64.3.2 Translational Energy Spectroscopy 947
- 64.3.3 Electron Emission Spectroscopy ... 948
- 64.3.4 Angular Differential Measurements 948
- 64.3.5 Recoil Ion Momentum Spectroscopy 948

References 948

64.1 Low Energy Ion–Atom Collision Processes

The most important and widely studied inelastic ion–atom collision process in the low energy region is electron capture (also referred to as charge exchange, charge transfer or electron transfer) represented by

$$A^{+q} + B \to A^{+q-k} + B^{k+} + Q \,, \tag{64.1}$$

where Q is the potential energy difference between the initial and final states. For an exoergic process, $Q > 0$ and this energy appears as excess kinetic energy of the products after the collision. For an endoergic process, $Q < 0$ and must be provided by the initial kinetic energy of the reactants, so that the corresponding cross section is usually small at low collision energies. Cross sections for electron capture are appreciable even at very low energies if Q is zero or very small (resonant or near-resonant process). Electron capture by multiply charged ions from atoms is predominantly an exoergic process, for which cross sections may also be large at low energies. In this case, electrons are preferentially captured into excited levels of A^{+q-k}. The typical cross section behavior for single electron capture ($k = 1$) by a multiply charged ion from atomic hydrogen is shown in Fig. 64.1. The initial ionic charge is the major determinant of the cross section at intermediate and high collision energies, whereas the cross section at low energies depends strongly on the structure of the transient quasimolecule formed during the collision.

Multiple electron capture ($k > 1$) from multielectron atoms occurs predominantly into multiply excited levels, which stabilize either radiatively, leading to stabilized or "true" double capture, or via autoionization. The latter process is usually referred to as transfer ionization,

$$A^{+q} + B \to A^{+q-k} + B^{m+} + (m-k)e^- + Q \,. \tag{64.2}$$

At low energies, transfer ionization is particularly important in collisions of highly charged ions with multi-electron atoms.

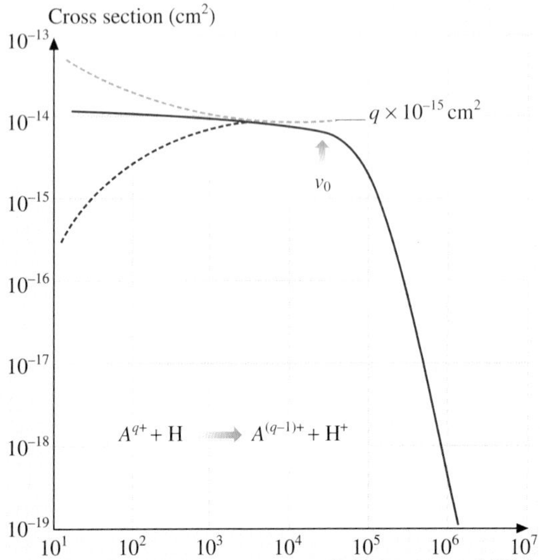

Fig. 64.1 Typical cross section variation with collision energy for electron capture by a multiply charged ion from hydrogen. The low energy behavior depends on the structure of the quasimolecule formed during the collision

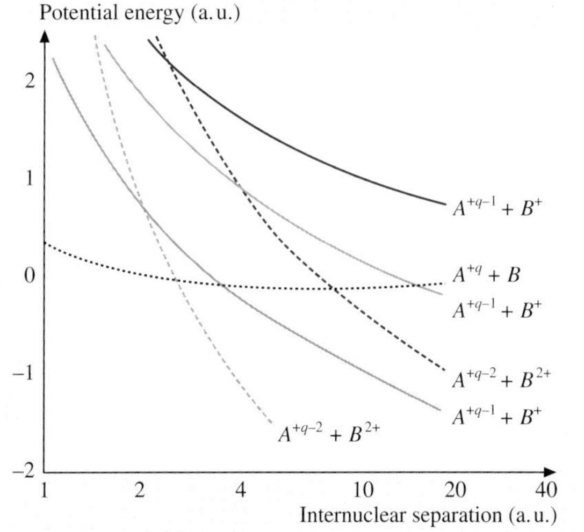

Fig. 64.2 Schematic representation of potential energy curves for the collision of a multiply charged ion A^{q+} with a multielectron atom B

Adiabatic potential energy curves representing the collision of a multiply charged ion with a neutral atom are presented in Fig. 64.2. For such collisions, the Coulomb repulsion in the final state produces avoided crossings of the initial and final state potential curves, the positions of which are determined by the binding energy of the atomic electron and electronic energy level structure of the product ion. Generally, reaction channels that are moderately exoergic produce curve crossings at internuclear separations where there is sufficient overlap of the electron clouds for electron capture to be a likely process. The Landau–Zener curve-crossing model [64.1] (Chapt. 49), the classical over-barrier model for single capture [64.2] and the extended classical over-barrier model for multiple capture [64.3], are useful in predicting the important final product states, and in providing a semiquantitative interpretation. In the case of single electron capture by a bare multiply charged ion (of charge $q = Z$) from a hydrogen atom, the principal quantum number n_p of the most probable final ionic state is given in this model [64.2] by

$$n_\text{p} = \sqrt{\frac{2Z^{1/2}+1}{Z+2Z^{1/2}}} \,. \tag{64.3}$$

The internuclear separation R_p at which the potential curves cross in this case is given by

$$R_\text{p} = \frac{2(Z-1)}{Z^2/n_\text{p}^2 - 1} \,. \tag{64.4}$$

There has been much discussion of the role of electron correlation in the multiple electron capture process. At issue is the relative importance of the mechanism whereby several electrons are transferred (in a correlated manner) at a single curve crossing compared with that whereby single electrons are transferred successively at different curve crossings. Experimental evidence exists for both mechanisms, with the relative importance depending on the electronic structure of the transient quasimolecule that is formed during the collision. Measurements of the distribution of final ion product electronic states provides the major insight into such collision mechanisms [64.4].

Other inelastic ion–atom collision processes, such as direct electronic excitation and ionization, are endothermic, with relatively small cross sections that fall off with decreasing energy below a few tens of keV/N. Exceptions are collisions involving Rydberg atoms and collisional excitation of fine structure transitions, for which the required energy transfer is relatively small. Relatively little experimental data are available for direct excitation and ionization processes at low collision energies.

64.2 Experimental Methods for Total Cross Section Measurements

In the present context, a total cross section measurement refers to an integration or summation over scattering angles, product kinetic energies and (frequently) electronic states. The total cross section is usually measured as a function of relative collision energy or velocity.

64.2.1 Gas Target Beam Attenuation Method

The attenuation of a collimated ion beam of incident intensity I_0 in a differentially pumped gas target cell or gas jet is related to the collision cross section σ by

$$I = I_0 \mathrm{e}^{-\sigma NL} , \qquad (64.5)$$

where I is the intensity after traversing an effective length L of the target gas, and N is the number density of target atoms. For a gas target cell with entrance and exit apertures of diameter d_1 and d_2 which are much less than the physical length z of the gas cell, L is given to a good approximation by $z+(d_1+d_2)/2$. This is valid under molecular flow conditions, for which the mean free path between collisions of target atoms is much larger than the dimensions of the gas cell.

In designing the gas cell for measurements of total cross sections, d_2 and the beam detector must be large enough that elastic scattering may be eliminated as a contributor to the measured attenuation. Usually d_2 is made larger than d_1. Measurement of the gas pressure in a target cell is usually made using a capacitance manometer connected to the cell via a tube whose conductance is much larger than that of the gas cell apertures so that, to a good approximation, the pressure will be the same in both the manometer and the gas cell. For gas jet targets, the effective target thickness NL is usually determined by in situ normalization to some well-known cross section.

The quantity σ in (64.5) refers to an effective cross section for removing projectile ions from the incident beam, which is the sum of cross sections for all such processes. In many cases, a single process (e.g., electron capture) is dominant, and σ primarily describes that process. Whether a collision process removes a projectile ion from the reactant beam or not depends on the configuration of the experiment. For example, the projectile particle may remain physically in the beam after passing through the target, but with a changed charge due to a collision. This would be registered as an attenuation of the primary ion beam if the beam is charge analyzed after the reaction.

64.2.2 Gas Target Product Growth Method

The product growth method is similar to the beam attenuation method; the major difference that the growth of reaction products is measured rather than the loss of reactant projectiles. The products may be derived from either the projectile beam or the gas target, or both. The main advantage of this method is its higher degree of selectivity of a specific collision process. In addition, the reactants and products can usually be registered simultaneously, or in some cases in coincidence, eliminating the sensitivity of the measurement to temporal variation of ion beam intensity.

An important criterion is that the target gas density be low enough that single collision conditions prevail (i. e. that the likelihood of an ion passing through the gas target and interacting with more than one target atom is negligibly small). This must in general be satisfied in order for (64.5) to relate correctly the measured attenuation to the collision cross section of interest, and is critical to the product growth method [64.5]. In this case, under single collision conditions, one may set the number of products $I_\mathrm{p} = I_0 - I$, and (64.5) then may be written as

$$\frac{I_\mathrm{p}}{I_0} = 1 - \mathrm{e}^{-\sigma NL} \approx \sigma NL . \qquad (64.6)$$

The approximate expression is useful for $\sigma NL \ll 1$, which is a requirement for single collision conditions. It is also important that the products not be lost in a subsequent collision in the gas cell, so the magnitudes of cross sections for loss of products in the target gas must also be considered. The products in such an experiment may be derived from either the projectile beam or the target gas (e.g., collection of slow product ions in a gas cell), or from both in coincidence to enhance the specificity of the method. The method may in principle be used to determine either total or differential cross sections, depending on the degree of selectivity of collision products.

64.2.3 Crossed Ion and Thermal Beams Method

Replacement of the gas target cell by an effusive thermal beam is advantageous for studying collisions of ions with reactive species such as atomic hydrogen, as well as for collecting slow ion products. The use of accelerating electrodes or grids for slow charged products allows coincident detection of fast and slow products, permitting measurement of ionization as well as electron-capture cross sections. Use of an effusive source or gas jet precludes accurate measurement of the effective target thickness, and in situ normalization to the cross section for some well known process is usually employed. A comprehensive discussion of such methods as applied to collisions of multiply charged ions with atomic hydrogen is given by *Gilbody* [64.6].

64.2.4 Fast Merged Beams Method

Cross sections for ion–atom collision processes at very low energies have been measured by merging fast beams of ions and neutral atoms [64.7], as in Fig. 64.3. In this case, σ is determined from experimental parameters by

$$\sigma = \frac{R}{\epsilon}\frac{e}{I_+ I_0}\frac{v_+ v_0}{|v_+ - v_0|}F\,, \tag{64.7}$$

where R is the number of products detected per second, ϵ is the product detection efficiency, e is the electronic charge, I_+ is the ion current, I_0 is the flux of atoms, v_+ and v_0 are the laboratory velocities of the ion and atom beams, $v_{\rm rel}$ is their relative velocity, and F is the form factor that describes the spatial overlap of the two beams. If the z-axis is chosen to be the direction of propagation of the beams, the form factor has units of length and is given by

$$F = \frac{\iint I_+(x,y,z)\,\mathrm{d}x\,\mathrm{d}y \iint I_0(x,y,z)\,\mathrm{d}x\,\mathrm{d}y}{\iiint I_+(x,y,z) I_0(x,y,z)\,\mathrm{d}x\,\mathrm{d}y\,\mathrm{d}z}\,. \tag{64.8}$$

The two-dimensional integrals in the numerator represent the total intensities of the two beams, which are independent of z, so that F is also independent of z.

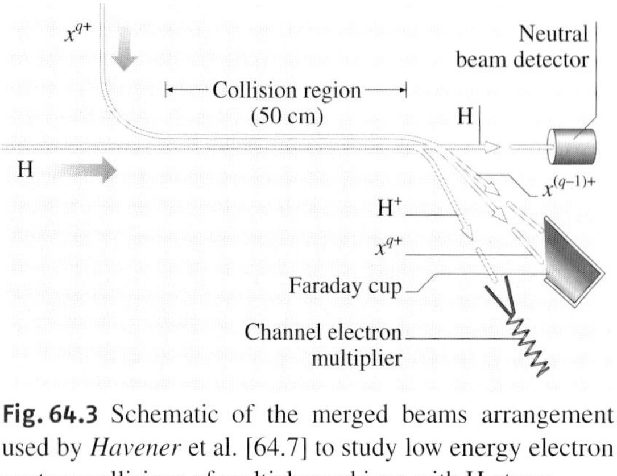

Fig. 64.3 Schematic of the merged beams arrangement used by *Havener* et al. [64.7] to study low energy electron capture collisions of multicharged ions with H atoms

The relative collision energy $E_{\rm rel}$ in eV/u is given by

$$\frac{E_{\rm rel}}{\mu} = \frac{E_+}{m_+} + \frac{E_0}{m_0} - 2\sqrt{\frac{E_+}{m_+}\frac{E_0}{m_0}}\cos\theta\,, \tag{64.9}$$

where E_+, m_+ and E_0, m_0 are the energies and masses of the ion and atom, respectively, and μ is their reduced mass. For collinear merged beams, $\theta = 0$, and $E_{\rm rel}$ can be reduced to zero by making the two beam velocities the same. In practice, the finite divergences of the beams place a lower limit on the energy and the energy resolution. The fast neutral atom beam is created by neutralizing an accelerated ion beam by electron capture by a positive ion beam in a gas, by electron detachment of a negative ion beam either in a gas, or using a laser beam. In gas collisions, a small fraction of the neutral beam is produced in excited n-levels (with an n^{-3} distribution), which may influence the measurements.

With fast colliding beams, the maximum effective beam densities are invariably much smaller than the background gas density, even under ultrahigh vacuum conditions. For example, a typical 10 keV proton beam with a circular cross section of diameter 3 mm and a current $I = 10\,\mu$A has an average effective density $n = I/(eAv) = 1.6\times10^6\,\mathrm{cm}^{-3}$ (A is the beam cross sectional area and v is its velocity). It is therefore necessary either to modulate the beams or to use coincidence techniques to separate signal events due to beam–beam collisions from background events produced by collisions of either beam with background gas. A typical two-beam modulation scheme is shown in Fig. 64.4. To eliminate the production of spurious signals, the detector gates are delayed for a short time after the beams are switched, and the beam modulation period is made much shorter than the pressure time constant of the vacuum system.

Absolute electron-capture cross sections have been measured for O^{5+} + H collisions to energies below 1 eV/N, where the attractive ion-induced-dipole (polarization) interaction is expected to play a role [64.7]. The inverse velocity dependence of the cross section in this region is suggestive of the classical Langevin orbiting model for ion–neutral collisions [64.8].

64.2.5 Trapped Ion Method

The trapped ion method is used to determine rate coefficients and effective cross sections for ion–atom collisions at near thermal energies. The technique involves storing ions in an electrostatic or electromagnetic

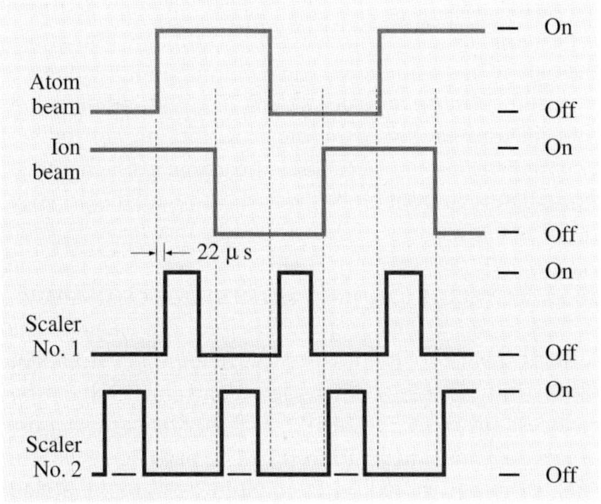

Fig. 64.4 Fast two-beam modulation scheme to separate the signal due to beam–beam collisions from events due to beam collisions with residual gas or surfaces [64.7]

trap, and measuring the rate of loss of ions from the trap after a small quantity of neutral gas is admitted [64.9]. Like the beam attenuation method, the trap technique cannot distinguish different processes that cause ions to be lost from the trap. The mean collision energy is estimated from an analysis of the ion dynamics in the trap.

64.2.6 Swarm Method

The swarm method, using the flowing afterglow, drift tube or selected-ion flow tube, has been used successfully to study ion–atom collisions at very low energies [64.10]. Ions are injected into a homogeneous electric field and drift through a suitable low-density buffer gas such as helium. The ions move as a swarm whose mean energy depends on the applied electric field and on collisions with the buffer gas, and can be varied from the near-thermal region to tens of eV. The method involves measuring the additional attenuation of the directed ion swarm by a known quantity of added reactant gas, and is the major technique that has been used for the study of ion–atom collisions at near-thermal energies [64.11]. As with ion beam attenuation and ion trap methods, the drift tube is not selective of the process that leads to attenuation of the ion swarm.

64.3 Methods for State and Angular Selective Measurements

Three principal methods have been developed and applied to the measurement of partial cross sections for population of specific product electronic states in ion–atom collisions involving electron capture [64.12]. These are based on spectroscopic measurements of photon emission, translational energy spectroscopy and electron emission in collisions of ion beams with gas targets.

64.3.1 Photon Emission Spectroscopy

Since electron capture from atoms by multiply charged ions populates excited levels, photon emission spectroscopy may be employed to determine state-selective partial cross sections [64.13]. An ion beam is directed through a gas cell or jet, and a photomultiplier or suitable detector registers photons analyzed by an optical filter, a grating or a crystal spectrometer. The measured photon signal at a given angle depends on the detection solid angle, the polarization of the emitted radiation, the absolute efficiencies of the dispersive device and detector, and (for emission by fast ions) the lifetime of the radiating state. Depending on the spectral region, the photon detection system may be absolutely calibrated using a standard photon source or detector, or by using a reference ion or electron beam and well established cross section data for photon emission [64.14]. Cascading from higher populated levels must also be considered whenever measured spectral line emission intensities are used to infer cross sections for populating specific energy levels. Successful state-selective cross section measurements have been made for transitions in the visible, VUV and X-ray spectral regions.

64.3.2 Translational Energy Spectroscopy

Translational energy loss or energy gain spectroscopy provides a convenient method to determine the distribution of final states in low energy ion–atom collisions that are either endoergic or exoergic ($Q \neq 0$). For example, this method has been used extensively by the Belfast group [64.15] to study the predominantly exoergic process of electron capture by multiply charged ions from hydrogen atoms. An ion beam with a well defined energy is directed through a gas target, and the energy of the product ion beam is energy analyzed after the collision. Since the energy gain or loss to be measured is only a very small fraction of the initial kinetic energy, it is usually necessary to reduce the initial energy spread to a few tenths of an eV by decelerating the reactant ion beam prior to energy selection by an electrostatic analyzer. If the scattering angle of the ion is very small, its energy change is approximately equal to Q. Since the ion beam is attenuated by deceleration and energy analysis, cross sections for collisionally populating specific states are determined by normalizing the measured product-state distributions to total cross section data. The attainable state resolution is not as good as for photon emission spectroscopy.

64.3.3 Electron Emission Spectroscopy

As noted in Sect. 64.1, multiple electron capture by multiply charged ions from atoms at low energies occurs primarily into multiply excited states, which decay either radiatively or via autoionization (with a branching ratio) [64.4]. The latter decay pathway (transfer ionization) provides an experimental method to determine the product ionic states by ejected-electron spectroscopy. Analysis of electrons emitted into the forward (ion beam) direction (zero degree spectroscopy) offers significant advantages for analysis of low energy electrons with high resolution [64.16]. Since a gas jet is often employed and absolute electron collection and spectrometer efficiencies are difficult to determine, some normalization procedure is usually employed to determine state-selective cross sections by this method. Electrons and product projectile or recoil ions have also been detected in coincidence to increase the specificity of the method.

64.3.4 Angular Differential Measurements

The measurement of angular distributions of scattered ions in low energy ion–atom collisions has been facilitated by the availability of position-sensitive particle detectors consisting of a microchannel plate and a resistive or segmented anode [64.17]. The method for processes that have forward-peaked angular distributions involves directing a highly collimated ion beam through a gas target cell, and counting the scattered projectile ions on a position-sensitive particle detector. Product ions produced by electron capture can be selected by the use of electrostatic retarding grids mounted immediately in front of the detector, to reject the primary ion beam.

64.3.5 Recoil Ion Momentum Spectroscopy

Perhaps the most significant experimental development of the last decade, cold target recoil-ion momentum spectroscopy (COLTRIMS) [64.18] has been made possible by advances in position-sensitive particle detection. This technique, based on momentum imaging and also called a "reaction microscope", has yielded important new insights into the dynamics of ion–atom collision processes [64.19] as well as other types of interactions involving cold atoms and molecules. The method is particularly suited to studies of charge-changing and molecular fragmentation processes.

In this method, an ion beam intersects a cold supersonic atomic beam in an interaction volume within which small electric and magnetic fields are imposed in order to guide slow ions and ejected electrons to fast position-sensitive detectors. The charge and the transverse and longitudinal components of the momentum of the slow (recoil) ion are measured by a combination of position and time-of-flight measurements, permitting the Q-value of the collision and the final electronic states of the projectile and recoil ions to be uniquely determined. The position measurement additionally provides information about the angular scattering during the collision. Coincident measurement of the scattered projectile ion and determination of its charge state by electrostatic deflection, and/or time-of-flight measurements of ejected electrons have provided new insights into complex multielectron processes occurring in low-energy ion–atom collisions.

References

64.1 F. W. Meyer, A. M. Howald, C. C. Havener, R. A. Phaneuf: Phys. Rev. Lett. **54**, 2663 (1985)
64.2 H. Ryufuku, K. Sasaki, T. Watanabe: Phys. Rev. A **23**, 745 (1980)
64.3 A. Niehaus: J. Phys. B **19**, 2925 (1986)
64.4 M. Barat, P. Roncin: J. Phys. B **25**, 2205 (1992)
64.5 H. B. Gilbody: Adv. At. Mol. Phys. **22**, 143 (1986)
64.6 F. W. Meyer, A. M. Howald, C. C. Havener, R. A. Phaneuf: Phys. Rev. A **32**, 3310 (1985)
64.7 C. C. Havener, M. S. Huq, H. F. Krause, P. A. Schulz, R. A. Phaneuf: Phys. Rev. A **39**, 1725 (1989)
64.8 G. Gioumoussis, D. P. Stevenson: J. Chem. Phys. **29**, 294 (1958)
64.9 D. A. Church: Phys. Rep. **228**, 254 (1993)

64.10 Y. Kaneko: Comm. At. Mol. Phys. **10**, 145 (1981)
64.11 W. Lindinger: Phys. Scrip. T **3**, 115 (1983)
64.12 R. K. Janev, H. Winter: Phys. Rep. **117**, 266–387 (1985)
64.13 D. Ćirić, A. Brazuk, D. Dijkkamp, F. J. de Heer, H. Winter: J. Phys. B **18**, 3639 (1985)
64.14 B. Van Zyl, G. H. Dunn, G. Chamberlain, D. W. O. Heddle: Phys. Rev. A **22**, 1916 (1980)
64.15 H. B. Gilbody: Adv. At. Mol. Opt. Phys. **32**, 149 (1994)
64.16 N. Stolterfoht: Phys. Rep. **146**, 316 (1987)
64.17 L. N. Tunnell, C. L. Cocke, J. P. Giese, E. Y. Kamber, S. L. Varghese, W. Waggonner: Phys. Rev. A **35**, 3299 (1987)
64.18 R. Dörner, V. Mergel, O. Jagutzki, L. Spielberger, J. Ullrich, R. Moshammer, H. Schmidt-Böcking: Phys. Rep. **330**, 95 (2000)
64.19 J. Ullrich, R. Moshammer, A. Dorn, R. Dörner, L. Ph. H. Schmidt, H. Schmidt-Böcking: Rep. Prog. Phys. **66**, 1463 (2003)

65. Ion–Atom Collisions – High Energy

This chapter deals with inelastic processes which occur in collisions between fast, often highly charged, ions and atoms. Fast collisions are here defined to be those for which $V/v_e \geq 1$, where V is the projectile velocity and v_e the orbital velocity of this electron. For processes involving outer shell target electrons, this implies $V \gtrsim 1$ a.u., or the projectile energy $\gtrsim 25$ keV/a.m.u. For inner shell electrons, typically, $V \gtrsim Z_2/n$ a.u., where Z_2 is the target nuclear charge and n the principal quantum number of the active electron. A useful relationship is $V = 6.35\sqrt{E/M}$, where V is in a.u., E is in MeV, and M is in a.m.u. Fast collisions involving outer shell processes can be studied using relatively small accelerators, while those involving inner shell processes require larger van de Graaffs, LINACs, etc. Because the motion of the inner shell electrons is dominated by the nuclear Coulomb field of the target, and because transitions involving these electrons take place rather independently of what transpires with the outer shell electrons, it has proven somewhat easier to understand one electron processes involving inner shell electrons. Thus, for a long time, a great deal of the work on fast ion–atom collisions has concentrated on inner shell processes involving heavy target atoms. However, more recently, new experimental techniques have led to a shift of this focus to inelastic processes involving light target atoms. Furthermore, present investigations go beyond the one-electron picture to include the influence of the electron–electron interaction. The present chapter outlines some of the developments in this area over a very active past few decades. The literature is vast, and only a small sampling of references is given. Emphasis is on experimental results (for the theory see Chapts. 45–57)

65.1 **Basic One-Electron Processes** 951
65.1.1 Perturbative Processes 951
65.1.2 Nonperturbative Processes 955

65.2 **Multi-Electron Processes** 957

65.3 **Electron Spectra in Ion–Atom Collisions**. 959
65.3.1 General Characteristics............... 959
65.3.2 High Resolution Measurements... 960

65.4 **Quasi-Free Electron Processes in Ion–Atom Collisions** 961
65.4.1 Radiative Electron Capture 961
65.4.2 Resonant Transfer and Excitation 961
65.4.3 Excitation and Ionization 961

65.5 **Some Exotic Processes**.......................... 962
65.5.1 Molecular Orbital X-Rays 962
65.5.2 Positron Production from Atomic Processes 962

References 963

65.1 Basic One-Electron Processes

65.1.1 Perturbative Processes

Inner Shell Ionization of Heavy Targets

For ion–atom collisions involving projectile and target nuclear charges Z_1 and Z_2 respectively, the parameters $\eta_1 = \left[\hbar V/(Z_1 e^2)\right]^2$ and $\eta_2 = \left[\hbar v_e/(Z_2 e^2)\right]^2$ are useful in characterizing the strength of the interaction between Z_1, Z_2, and the target electron. If $\eta_2 \ll \eta_1$, (i.e., $Z_1/Z_2 \ll V/v_e$), the effect of the projectile on the target wave function can be treated perturbatively. Perturbation treatments of inner shell ionization by lighter projectiles have been extensively studied and reviewed [65.1–8]. Two well-known formulations have been used: the plane wave Born approximation (PWBA) [65.1–4], and the semiclassical approximation (SCA) [65.9, 10]. The former represents the nuclear motion with plane waves, while the latter is formulated in terms of the impact parameter b with the nuclear motion treated classically. For straight line motion of the nuclei, the results are equivalent [65.11]. The total cross section for ionizing the K-shell of a target of charge Z_2 by a projectile of charge Z_1 is given within the PWBA by

$$\sigma_i = \left(8\pi Z_1^2/Z_2^4 \eta_2\right) f(\theta_2, \eta_2)\, a_0^2\,, \tag{65.1}$$

where $\theta_2 = 2u_k n^2/Z_2^2$ and u_k is the target binding energy. The function f rises rapidly for $V < v_e$, reaching a value near unity near $V = v_e$ and falling very slowly thereafter. Tables of f for K- and L-shell ionization are given in [65.3, 4]. Figure 65.1 shows a comparison of experimental data for K vacancy production by protons with PWBA calculations, and with a classical binary encounter approximation [65.12] for a large range of proton data [65.6]. For larger Z_1, corrections to the PWBA and SCA must be made for the effective increase of u_k due to the presence of the projectile during the ionization, for nuclear projectile deflection, for relativistic corrections, and for the polarization of the electron cloud, as reviewed in [65.13–17]. Total cross section measurements for inner shell vacancy production in the perturbative region are reviewed in [65.15, 16].

In the SCA treatment, the heavy particle motion is taken to be classical, and the evolution of the electronic wave function under the influence of the projectile field is calculated by time-dependent perturbation theory. The assumption of classical motion is valid if the Bohr parameter $K = 2Z_1Z_2e^2/(\hbar V)$ is much larger than unity [65.18]. If this condition is satisfied, the projectile scattering angle can be associated with a particular b through a classical deflection function. For K-shell ionization, the action occurs typically at sufficiently small b that a screened Coulomb potential is sufficient for calculating the deflection. In the absence of screening, $\theta = r_0/b$, where $r_0 = Z_1Z_2e^2/E$ with θ and E expressed in either the laboratory or c.m. system. Calculations for K- and L-shell ionization have been carried out [65.10]. The typical ionization probability $P(b)$ for $V \sim v_e$ and $b = 0$ is $P(0) \sim (Z_1/Z_2)^2$. For $V < v_e$, $P(b)$ decreases with increasing b with a characteristic scale length of $r_{ad} = V/\omega$, the *adiabatic radius*, where ω is the transition energy. For $V > v_e$, $P(b)$ cuts off near the K-shell radius of the target. A more sophisticated relativistic SCA program has been written [65.19], and is widely used for calculating $P(b)$, cross sections, and probabilities differential in final electron energy and angle. Experimentally, the probability $P(b)$ for inner shell ionization can be determined from

$$P(b) = \frac{1}{\epsilon\omega\Delta\Omega}\left(\frac{Y}{N[\theta(b)]}\right), \tag{65.2}$$

where Y is the coincidence yield for the scattering of $N(\theta)$ ions into a well-defined angle $\theta(b)$ accompanied by X-ray (or Auger electron) emission with fluorescence yield ω into a detector of efficiency ϵ and solid angle $\Delta\Omega$ [65.20]. The necessary ω can be obtained from calculations for neutral targets [65.21] (Chapt. 62). However, they must be corrected for changes due to extensive outer shell ionization during the collision. Such corrections are particularly important for targets with low fluorescence yields, for Z_2 below 30, and for collisions in which the L-shell is nearly depleted in the collision [65.15,16]. Values of $P(b)$ have been measured for many systems and generally show good agreement (better than 10%) with the SCA for fast light projectiles such as protons, with increasing deviation as higher Z_1 or slower V are used [65.22]. Examples of $P(b)$ for K vacancy production for several systems are shown in Fig. 65.2, showing the evolution away from the SCA as the collision becomes less perturbative.

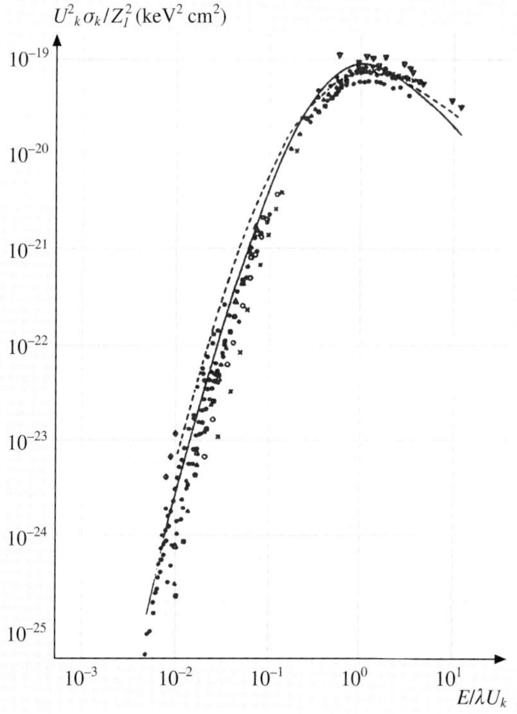

Fig. 65.1 Comparison of experimental cross sections for K-shell vacancy production with PWBA (*dashed*) and binary encounter (*solid*) theories. U_k is the target binding energy in keV and λ the projectile/electron mass ratio [65.6]

Ionization of Light Target Atoms

Ionization of light target atoms by bare ion impact is a particularly suitable process to study the atomic few-body problem. In the case of an atomic hydrogen target the collision represents a three-body system, i. e., the simplest system for which the Schrödinger equation is not analytically solvable. However, because of the ex-

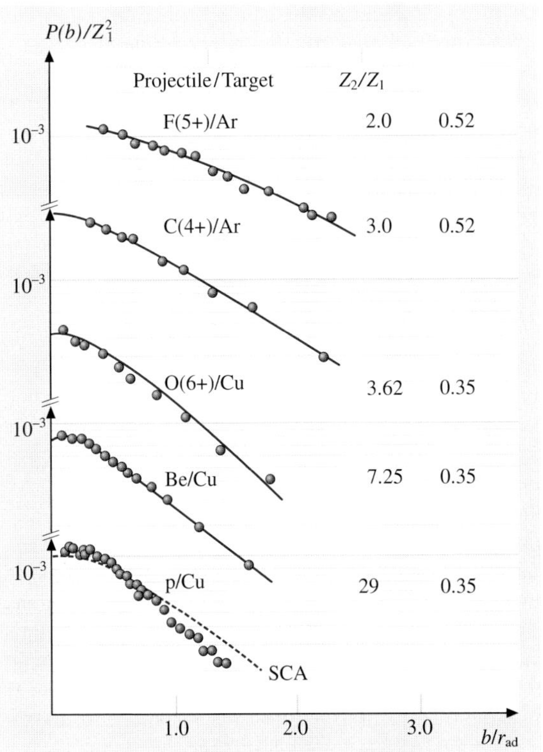

Fig. 65.2 $P(b)$ for K-shell vacancy production versus b/r_{ad} for several systems (see text). The ratio V/v_{e} is designated as "V" in this figure. For protons p, agreement with the SCA theory is found [65.10], while for higher Z_1/Z_2, $P(b)$ moves to larger impact parameters as one leaves the perturbative region [65.22]

perimental difficulties associated with atomic hydrogen, measurements with this target species are rare [65.23] and experimental studies have focused on helium targets. Here, the collision still constitutes a relatively simple four-body system. With regard to the few-body problem, studies of ionization processes have the important advantage that, in contrast to pure excitation and capture processes, the final state involves at least three independently moving particles.

Detailed information about the few-body dynamics in a collision can be extracted from multiply differential measurements. This can be accomplished by measuring the kinematic properties (e.g., energy, momentum, ejection angle) of one or more of the collision fragments. The first experimental multiply differential single ionization cross sections were obtained by studying the ionized electron spectra as a function of energy and ejection angle. Such studies were reviewed by *Rudd* et al. [65.24] and are discussed in more detail in Sect. 65.3. More recently, complementary multiply differential data were obtained by measuring projectile energy-loss spectra as a function of scattering angle in p + He collisions [65.25, 26].

A comprehensive picture of ionizing collisions can be obtained from kinematically complete experiments. In such a study the momentum vectors of all collision fragments need to be determined. However, in the case of single ionization it is sufficient to directly measure the momentum vectors of any two particles in the final state; the third one is then readily determined by momentum conservation. For ionization by electron impact, this has been accomplished by momentum-analyzing the scattered and the ionized electrons (for a review see [65.27]). For ion impact, this approach is difficult because of the very small scattering angles and energy losses (relative to the initial collision energy) resulting from the large projectile mass. Consequently, the only kinematically complete experiments involving a direct projectile-momentum analysis were reported for light ions at relatively low projectile energies [65.28]. For heavy-ion impact at high projectile energies, in contrast, the complete determination of the final space state is only possible through a direct measurement of the ionized electron and recoil-ion momenta [65.29].

The technology to measure recoil-ion momenta with sufficient resolution, and therefore to perform kinematically complete experiments for heavy-ion impact, has only become available over the last decade (for reviews, see [65.30–32]). Figure 65.3 shows measured (top) and calculated (bottom) three-dimensional angular distributions of electrons ionized in 100 MeV/a.m.u. C^{6+} + He collisions for fully determined kinematic conditions [65.33]. The arrows labeled p_o and q indicate the direction of the initial projectile momentum and the momentum transfer defined as the difference between p_0 and the final projectile momentum p_f. This plot is rich in information about the dynamics of the ionization process. The main feature is a pronounced peak in the direction of q. It can be explained in terms of a binary interaction between the projectile and the electron, i. e., a first-order process, and is thus dubbed the "Binary Peak". A second, significantly smaller, structure is a contribution centered on the direction of $-q$ (called the "Recoil Peak"). This has been interpreted as a two-step mechanism where the electron is initially kicked by the projectile in the direction of q and then backscattered by the residual target ion by 180°. Although this process involves two interactions of the electron, it is nevertheless a first-order process in the projectile–target atom inter-

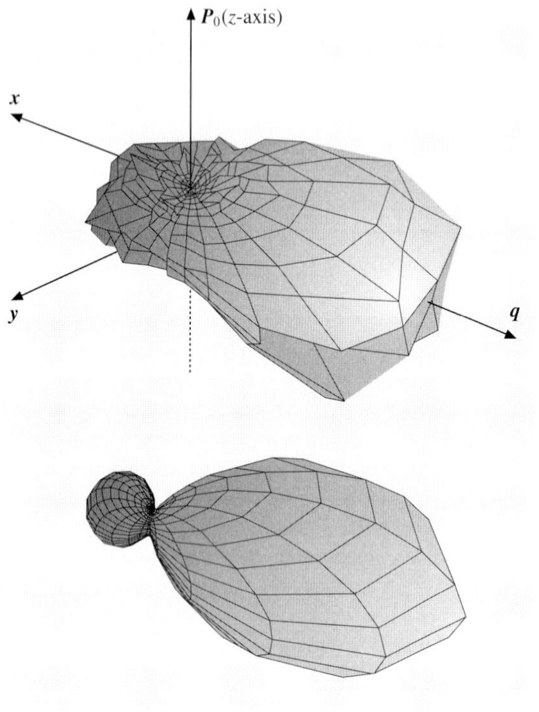

Fig. 65.3 Three-dimensional angular distribution for fully determined kinematic conditions of electrons ionized in 100 MeV/a.m.u. C^{6+} + He collisions. *Top*, experimental data; *bottom*, CDW calculation (see text)

action. Therefore, as expected for this very large value of $\eta_1 = 100$ (in a.u.), the ionization cross sections are dominated by first-order contributions.

The basic features of the data in Fig. 65.3 are well reproduced even by the relatively simple first Born approximation (FBA). Furthermore, the calculation shown in the bottom of Fig. 65.3, which is based on the more sophisticated continuum distorted wave approach (CDW)([65.35–37] see also Chapt. 52), yields practically identical results to the FBA. In the CDW method, higher-order contributions are accounted for in the final-state scattering wavefunction. Apart from this good overall agreement, a closer inspection of the comparison between experiment and theory also reveals some significant discrepancies. While in the calculation the Binary and Recoil peaks are sharply separated by a minimum near the origin, in the data this minimum is almost completely filled up giving rise to a "ring-like" shape of the recoil peak. This was explained by a higher-order ionization mechanism involving an interaction between the projectile and the residual target ion [65.33,37,38]. Although the contribution of this process to the total cross section is negligible, it is a very surprising result that for selected kinematic conditions higher-order processes can be important even at large projectile energies. A sobering conclusion of recent research on ionization of light target atoms is that even well inside the perturbative regime the atomic few-body problem is not nearly as well understood as was previously assumed based on studies for restricted collision geometries. At large perturbation, the lack of understanding is dramatic [65.39].

Excitation

Inner shell excitation can be treated within the same perturbative framework, which leads to a cross section given in terms of the generalized oscillator strength for the transition [65.40–42]. For inner shell vacancy production by light projectiles, the excitation is generally much smaller than the ionization, since the strongest oscillator strengths are to low-lying occupied orbitals, as reviewed by *Inokuti* [65.41]. Excitation cross sections can be deduced from photon production cross sections and from inelastic energy loss experiments. An example of the cross section for excitation of the $n = 2$ level of H by protons, measured by the latter technique, is shown in Fig. 65.4 [65.34].

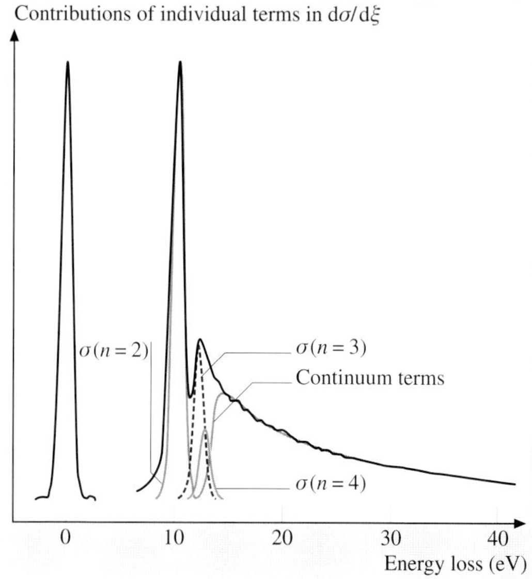

Fig. 65.4 Energy loss spectrum for 50 keV protons in atomic hydrogen, showing excitation to discrete states in H proceeding smoothly into ionization at the continuum limit [65.34]

Capture

As Z_1/Z_2 is raised, the probability for direct transfer of inner shell electrons from projectile to target becomes competitive with, and can even exceed, that for ionization of the target electron into the continuum. The first-order perturbation treatment for electron capture, given by *Oppenheimer* [65.43] and by *Brinkman* and *Kramers* [65.44] (OBK) ([65.11], p. 379) results in a cross section per atom

$$\sigma_{\text{OBK}} = 2^9\pi(Z_1Z_2)^5/5\,V^2\nu^5n^3\beta^5\,a_0^2\,, \tag{65.3}$$

from a filled shell ν to all final states n, where

$$\beta = \frac{1}{4}V^2\Big[V^4 + 2V^2\big(Z_2^2/\nu^2 + Z_1^2/n^2\big) + \big(Z_2^2/\nu^2 - Z_1^2/n^2\big)^2\Big]\,. \tag{65.4}$$

Both the PWBA/SCA and the OBK cross sections maximize near the matching velocity, but the OBK falls off much more strongly with increasing V beyond this, eventually falling as V^{-12}, while the ionization cross section only falls as $V^{-2}\ln V$. The OBK amplitude for capture is simply the momentum space overlap of the initial wave function with the final state wave function, where the latter is simply a bound state on the projectile but moving at a velocity V relative to the initial bound state. The integral is done only over the transverse momentum, since the longitudinal momentum transfer is fixed by energy conservation [65.11]. This capture amplitude thus depends heavily on there being enough momentum present in the initial and/or final wave function to enable the transfer, and the loss of this match is what leads to the steep decrease in the OBK cross section above velocity matching. Cross sections for K-shell capture have been measured by detection of K Auger electrons and K X-rays in coincidence with charge capture by the projectiles [65.22, 45, 46]. On the basis of these and many other data on electron capture, the OBK is a factor of approximately three too large [65.45–48]. This factor comes from a fundamental failure of first-order perturbation theory for electron capture. As pointed out already in 1927 by *Thomas* [65.49], who proposed a classical two-collision mechanism for capture, it is essential that the electron interacts with both nuclei during the collision in order to be captured (Chapt. 57). In quantum theories, this corresponds to the fundamental need to include second-order terms (and higher) in the capture amplitude. In the limit of large V, the second-order cross section decreases more slowly than the OBK term, as V^{-10}, and thus is asymptotically larger than the first-order term [65.50]. At large V, the coefficient of the V^{-12} term, the dominant one at most experimentally reachable V, is 0.29 times the OBK cross section when the theory is carried out to second-order in the projectile potential [65.50, 51]. Roughly speaking, this provides an explanation for the factor of three. Much more sophisticated treatments of high velocity capture are now available [65.52–60]. The underlying role of the second-order scattering process was confirmed experimentally by the detection of the Thomas peak in the angular distribution of protons capturing electrons from He and H [65.61, 62] (Chapt. 57).

In spite of the basic importance of second-order amplitudes in perturbative capture, the OBK gives an excellent account of the relative contributions from and to different final shells over a large range of V above v_e, and is thus, when appropriately reduced, still useful as an estimate for perturbative capture cross sections between well defined ν and n for large V.

For electron capture, as in the case of ionization (see previous section), the development of recoil-ion momentum spectroscopy (RIMS) has enabled much more detailed studies of the collision dynamics. The transverse (perpendicular to the beam direction) recoil-ion momentum component $p_\perp$ reflects the closeness of the collision both relative to the target nucleus and the electrons. The longitudinal (parallel to the beam direction) component p_z, on the other hand, is related to the internal energy transfer Q in the collision by (in a.u.)

$$p_z = -Q/V - nV/2\,, \tag{65.5}$$

where n is the number of captured electrons. A measurement of p_z is therefore equivalent to a measurement of Q. The advantage over measuring Q from the projectile energy loss is that at large collision energies a much better energy resolution is achievable. A sample Q measurement with RIMS is shown in Fig. 65.5 [65.63]. Very recently, RIMS was applied to study capture processes in collisions with an atomic hydrogen target [65.64]. This could be an important breakthrough in advancing our understanding of the atomic few-body problem as it opens the possibility to perform kinematically complete experiments on the true three-body system $X^{Z+} + H$, where X can be any bare projectile.

65.1.2 Nonperturbative Processes

Fano–Lichten Model

When the collision becomes increasingly perturbative, either due to a decreased V or increased Z_1/Z_2, higher-order effects become generally more important. One approach to account for such contributions

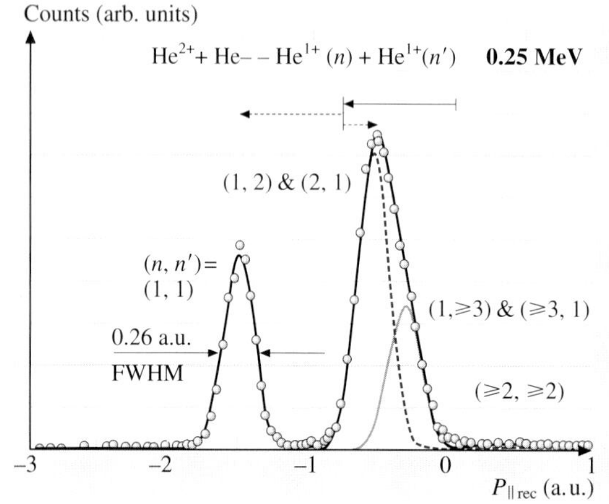

Fig. 65.5 Longitudinal momentum spectrum of recoil ions from 0.25 MeV He^{2+} capturing a single electron from a cold He target, showing clear resolution of capture to $n = 1$ from that which leaves target or projectile excited [65.63]

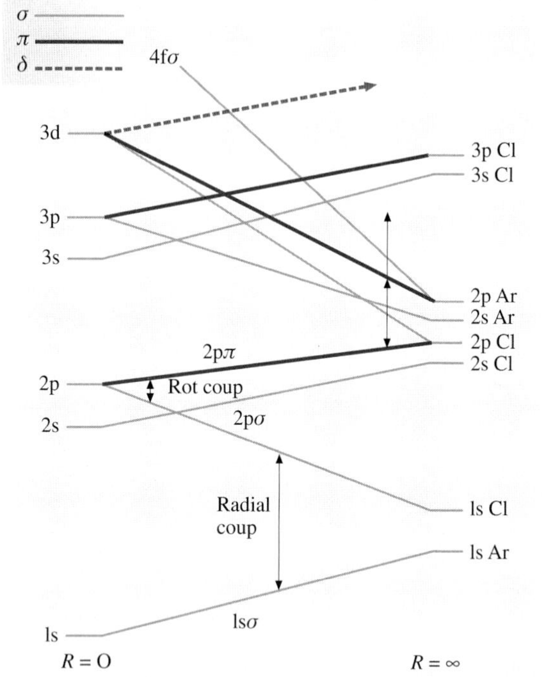

Fig. 65.6 Schematic correlation diagram for the Cl–Ar system, indicating the rotational coupling and radial couplings important for K vacancy production and the 4fσ orbital whose promotion leads to L vacancy production [65.66]

is the continuum distorted wave–eikonal initial state (CDW-EIS)model ([65.35, 36, 65], see also Chapt. 52 and Sect. 65.1.1). The range of validity of CDW-EIS is roughly given by $Z_1/V^2 \ll 1$ [65.35]. Therefore, if the perturbation is large due to the projectile charge, the collision may still be treated perturbatively provided that the collision energy is sufficiently large. Otherwise, the perturbation treatment is replaced by a molecular orbital treatment.

Fano and *Lichten* [65.67] pointed out that the ratio V/v_e can be small for inner orbitals even for V of several a.u., and thus an adiabatic picture of the collision holds. K vacancy production cross sections become much larger than the perturbation treatments above predict and extend to much larger b. In the molecular orbital picture, the collision system is described in terms of time-dependent molecular orbitals (MO) formed when the inner shells of the systems overlap. Vacancy production occurs due to rotational, radial, and potential coupling terms between these orbitals during the collision. The independent electron model is used, but the results in any specific collision are quite sensitive to the occupation numbers (or vacancies) in the initial orbitals. These are very difficult to control in ion–atom collisions in solids and even problematic in gases, since outer shell couplings can produce vacancies at large internuclear distances which then enable transfers at smaller distances. Numerous reviews of the subject are available, including [65.68–72]. The most famous MO ionization mechanism involves the promotion of the 4fσ orbital in a symmetric collision (Fig. 65.6), which promotes both target and projectile L electrons to higher energies where they are easily lost to the continuum during the collision. There are now many treatments of inner shell vacancy production mechanisms based on MO expansions (Chapts. 50, 51). For the case of K vacancy production in quasisymmetric collisions, an important MO mechanism is the transfer of L vacancies in the projectile to the K-shell of the target through the rotational coupling between 2pπ and 2pσ orbitals which correlate to the L- and K-shells respectively of the separated systems (Fig. 65.6) [65.73, 74]. The process can be dynamically altered by the sharing at large b between L vacancies of target and projectile through a radial coupling mechanism [65.75, 76]. This sharing mechanism can also give rise to the direct transfer of K vacancies from projectile to target (KK sharing). In symmetric systems, the KK sharing results in an oscillation of the K vacancy back and forth between target and projectile during the collision, and leads to an oscillatory behavior of the transfer probability with V and b [65.77, 78]. Both of the above vacancy production mechanisms are electron transfer processes rather than direct ionization processes, in that no inner shell electron need be liberated into the continuum. Between the perturbation region and the full MO region the importance of transfer increases relative to ionization. While the MO correlation diagrams and mechanisms are qualitatively useful, actual close coupling calculations for both inner and outer shell processes are often carried out using atomic orbitals instead of molecular orbitals, as well as other basis sets (Chapts. 50, 51).

65.2 Multi-Electron Processes

In a single collision between multi-electron partners, two or more electrons may be simultaneously excited or ionized. The electric fields created during a violent ion–atom collision are so large that the probability of such multi-electron processes can be of order unity. While there are many similarities between ion–atom collisions and the interaction of atoms with photons (X-rays or short laser pulses) or electrons, the dominance of multi-electron processes is very much less common in the photon and electron cases. As an example, when a K-shell electron is removed from a target atom by the passage of a fast highly charged ion through its heart, the probability that L-shell electrons will be removed at the same time can be large. This gives rise to target X-ray and Auger-electron spectra which are dominated by satellite structure [65.20]. For example, the spectrum of X-rays from Ti bombarded by 30 MeV oxygen shows that the production of the K vacancy is accompanied by multiple L vacancy production, and that the dominant K X-rays are those of systems which are missing several L electrons [65.79] (Fig. 65.7).

When a gas target is used, the recoil target ion is heavily ionized and/or excited electronically without receiving much translational kinetic energy. In the impulse approximation the transverse momentum $\Delta p_{\perp}$ received by the target from a projectile passing at impact parameter b is given in a.u. by $\Delta p_{\perp} = 2Z_1 Z_2/bV$. This expression ignores the exchange of electronic translational momentum but gives a good estimate. The resulting recoil energies are typically quite small, ranging from thermal to a few eV. This subject has been reviewed in [65.63, 80]. These slow moving recoils have

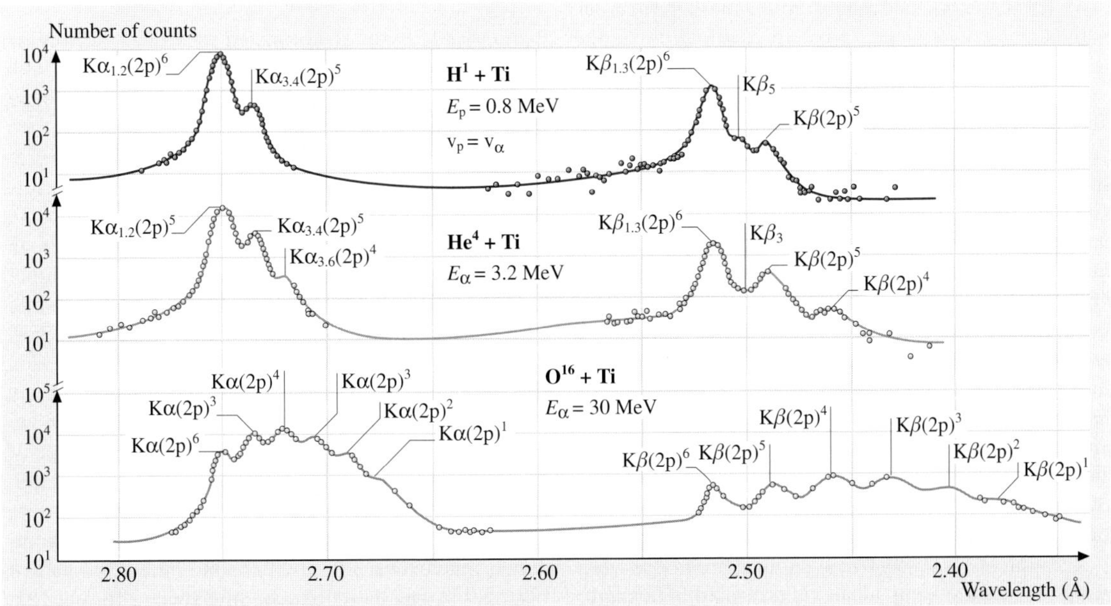

Fig. 65.7 K X-ray spectrum of Ti for various projectiles, showing dominance of multi-electron transitions when K vacancies are collisionally produced by heavily ionizing projectiles (^{16}O beam) [65.79]

been used to provide information about the primary collision dynamics, and as secondary highly charged ions from a fast-beam-pumped ion source. Such an ion source has, for moderately charged ions, a high brightness and has been used extensively for energy-gain measurements. The primary recoil production process is difficult to treat without the independent electron model, and even in this model the nonperturbative nature of the collision makes the theory difficult. The most successful treatments have been the CTMC (see Chapt. 58) and a solution of the Vlasov equation [65.64].

Studies of many-electron transitions in collisions of bare projectiles with a He target are particularly suitable to investigate the role of electron–electron correlation effects because such collisions represent the simplest systems where the electron–electron interaction is present. Such studies have been performed extensively for a variety of processes, such as double ionization, transfer-ionization, double excitation, transfer-excitation, or double capture (for reviews see [65.82–84]). It is common to distinguish (somewhat artificially) between such correlations in the initial state, the final state, and during the transition (dynamic correlation). From a theoretical point of view, the biggest challenge is to describe electron–electron correlation effects and the dynamics of the two-center potential generated by the projectile and the target nucleus simultaneously with sufficient accuracy.

In the case of double ionization, an experimental method, based on the so-called correlation function [65.81], was developed to analyze electron–electron correlations independently of the collision dynamics. Here, a measured two-electron spectrum (for example the momentum difference spectrum of both ionized electrons) is normalized to the corresponding spectrum one would obtain for two independent electrons. An example of such a correlation function R is shown in Fig. 65.8 for three very different collision systems (η_1 ranging from 0.05 to 100 and η_2 from 0.01 to 0.5 in a.u.). The similarity in these three data sets illustrates that R is remarkably insensitive to the collision dynamics. Rather, the shape of R is determined predominantly by correlations in the final state [65.81, 85]. However, for selected kinematic conditions, R can also be sensitive to initial-state correlations [65.86]. Clear signatures of initial-state correlations were found in the recoil-ion momentum spectra for transfer-ionization [65.87].

Early attempts to identify dynamic correlations were based on measurements of the ratio of double to single ionization cross sections [65.88, 89]. From such studies, it was found that at small V double ionization is dominated by an uncorrelated mechanism involving two independent interactions of the projectile with both electrons. In contrast, at large V the double to single ionization ratio asymptotically approaches a common value for all collision systems [65.90]. This is indicative of the dominance of first-order double ionization mechanisms, where the projectile interacts with only one electron and the second electron is ionized through an electron–electron correlation effect. This may either be a rearrangement process of the target atom adjusting to a new Hamiltonian (shake-off, an initial-state correlation), or a direct interaction with the first electron (i. e., dynamic correlation). However, a recent nearly kinematically complete experiment on double ionization in p + He collisions revealed that even at large V higher-order contributions are not negligible [65.91]. In Fig. 65.9 the ejection angles of both electrons are plotted against each other for almost completely determined kinematics. For comparison, the bottom part of Fig. 65.9 shows the corresponding spectra for electron impact at the same V [65.92]. For both projectiles, the basic features of these spectra are determined by the

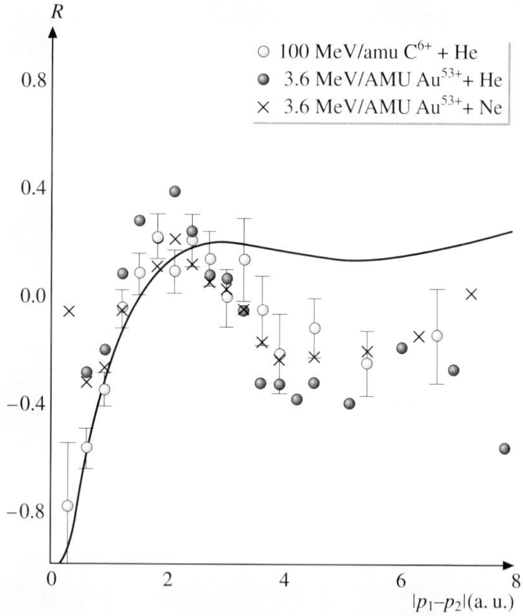

Fig. 65.8 Correlation function R for double ionization in the collisions indicated in the legend as a function of the momentum difference between the two electrons [65.81]. R is defined as $R = I_{\text{exp}}/I_{\text{IEM}} - 1$, where I_{exp} is the directly measured momentum spectrum and I_{IEM} the one obtained for independent electrons

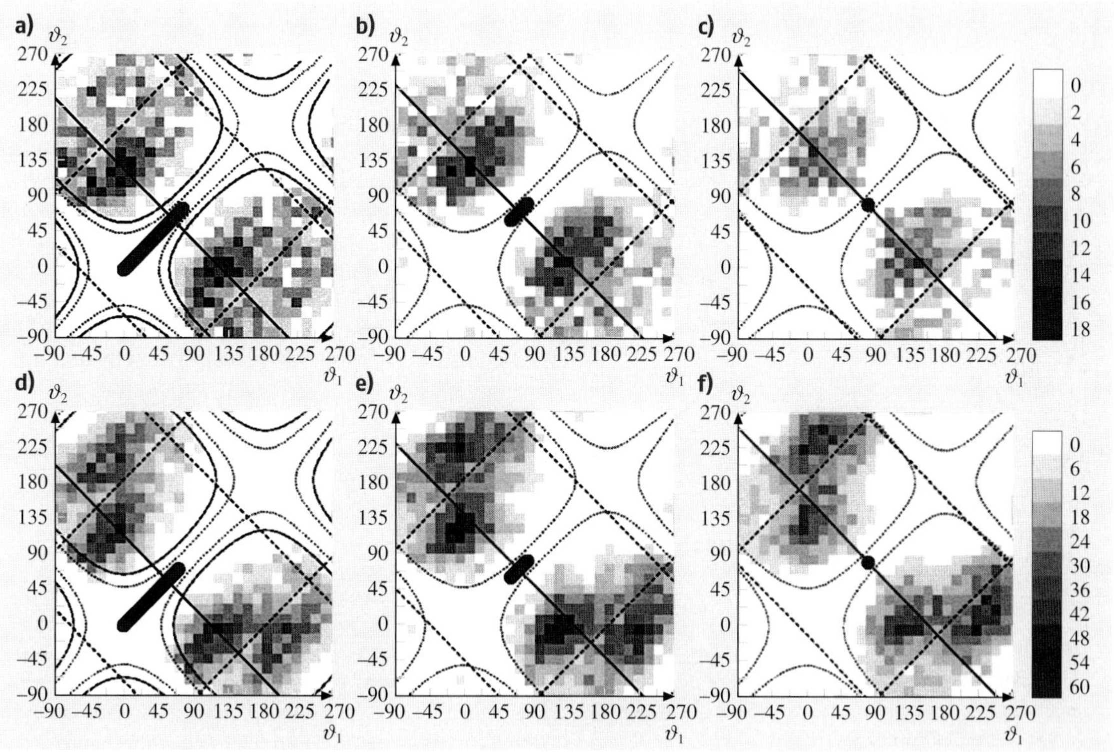

Fig. 65.9a–f Differential double ionization cross sections in 6 MeV p + He (*top*) and 2 keV e^- + He (*bottom*) collisions as a function of the polar emission angle of both electrons, which are emitted into the scattering plane. The electrons have equal energy and data are shown for small (*left*), medium (*center*), and large momentum transfers (*right*) [65.91]

electric dipole selection rules, which again is indicative of dominating first-order contributions. However, a closer inspection of the comparison between the proton and electron impact data shows some non-negligible differences. Since in a first-order treatment the cross sections should be identical for both projectile species, this demonstrates that higher-order contributions cannot be ignored.

65.3 Electron Spectra in Ion–Atom Collisions

65.3.1 General Characteristics

An ionizing collision between a single ion and a neutral atom ejects electrons into the continuum via two major processes. Electrons ejected during the collision form broad features or continua, and are traditionally referred to as delta rays; electrons ejected after the collisions from the Auger decay of vacancies created during the collision form sharp lines in the spectra. The distributions of energy and angle of all electrons determine the electronic stopping power and characteristics of track formation of ions in matter (Chapt. 91), and the study of these distributions in the binary encounter of one ion with one atom form the basis of any detailed understanding of these averaged quantities. Figure 65.10 shows a typical electron spectrum from the collision of a fast O ion with O_2 [65.93, 94]. Electrons from the projectile can be identified in the cusp peak (electron loss, P) or ELC, and the O-K-Auger (P) peak. Electrons from the target include the soft (large b) collision electrons (T) which are ejected directly by Coulomb ionization by the projectile, the binary collision (or encounter) electrons coming from hard collisions between projectile and quasifree target electrons, and the target O-K-Auger (T)

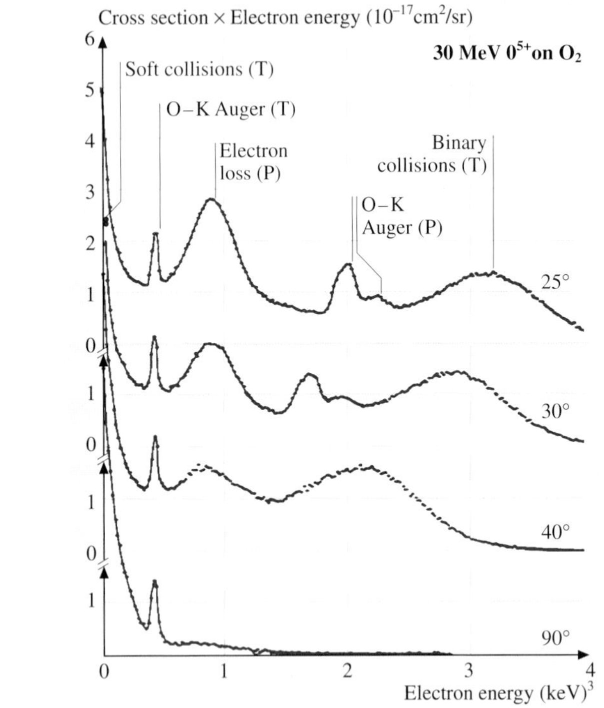

Fig. 65.10 Electron spectrum from 30 MeV O on molecular oxygen. See text for explanation of features [65.93, 94]

electrons. The electron loss peak is widely called the *cusp* peak because the doubly differential cross section in the laboratory $d^2\sigma/dE\,d\Omega$ becomes infinite, in principle, if it is finite in the projectile frame. In general, this peak may also contain capture to the continuum. All of these features have been heavily studied; some reviews are [65.95, 96]. Capture to the continuum [65.96] is an extension of normal capture into the continuum of the projectile, and is not a weak process. Both it and ELC produce a heavy density of events in the electron momentum space centered on the projectile velocity vector, and thus appear strongly only at or near zero degrees in the laboratory and at $v_e \simeq V$.

The binary encounter electrons at forward angles occur at $v_e \cong 2V$. For relatively slow collisions it was found that the ELC and binary peaks are just part of a more general and complex structure of the electron spectra [65.98]. Additional peaks in the forward direction were found for $v_e \cong nV$, where n in principle can be any integer number. These structures reflect a "bouncing back and forth" (known as Fermi shuttle) between the projectile and the target core before the electron eventually gets ejected from the collision system.

In electron spectra for molecular targets, additional structures were found that were not observed for atomic targets [65.99]. These were initially interpreted as an interference effect. The electronic wavefunction has maxima at the atomic centers of the molecule. Since in the experiment it cannot be distinguished from which center the electron is ionized, both possibilities have to be treated coherently. However, more recent studies showed that at small electron energies the structures in the electron spectra reflect vibrational excitation of the molecule [65.100].

65.3.2 High Resolution Measurements

The Auger electron spectra provide detailed information about inner shell vacancy production mechanisms.

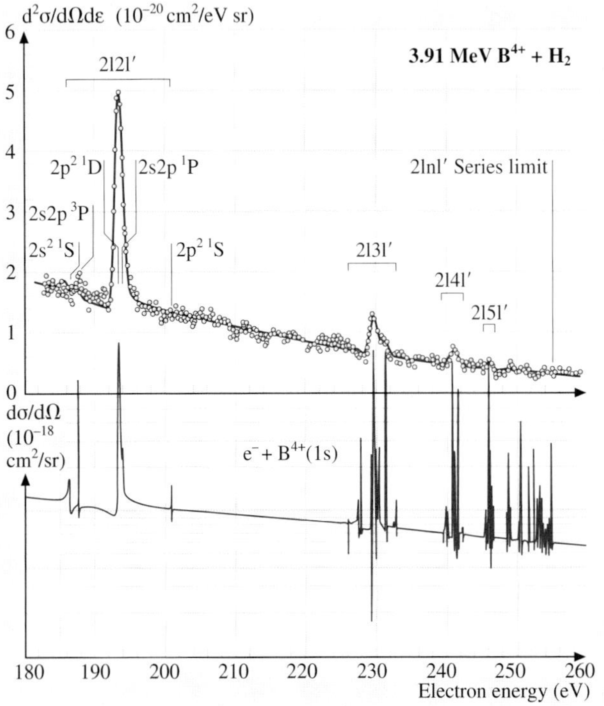

Fig. 65.11 High resolution Auger electron spectrum from H-like B on H_2, showing resolved lines from doubly excited projectile states lying on top of a continuum due to electron elastic scattering [65.97]. The *bottom part* shows an R-matrix calculation which does not account for the experimental resolution. The smooth line in the upper figure is the R-matrix calculation convoluted with the experimental spectrometer resolution

When coupled with fluorescence yields, Auger electron production probabilities and cross sections can be converted into the corresponding quantities for vacancy production [65.15, 95]. This is best done when sufficient resolution can be obtained to isolate individual Auger lines. The Auger spectra in ion–atom collisions are often completely different from those obtained from electron or photon bombardment because of the multiple outer shell ionization which attends the inner shell vacancy producing event, in close analogy to X-ray spectra (see previous section). Projectile Auger electron spectra suffer from kinematic broadening due to the finite solid angle of the spectrometer and velocity of the emitter, but at 0° to the beam this problem vanishes, and the resolution in the emitter frame is actually enhanced by the projectile motion, such that for electrons with eV energies in the projectile frame, resolutions in the meV region are possible [65.101, 102]. The highest resolution Auger lines from ion–atom collisions has been done on the projectiles. A sample spectrum is shown in Fig. 65.11. From such high resolution spectra, one-electron processes in which one electron is excited, captured or ionized can be distinguished from the configuration of the emitting state.

65.4 Quasi-Free Electron Processes in Ion–Atom Collisions

At sufficiently low V, those electrons not actively involved in a transition play only a passive role in screening the Coulomb potential between the nuclei, and thereby create a coherent effective potential for their motion. However, at high V the colliding electrons begin behaving as incoherent quasifree particles capable of inducing transitions directly via the electron–electron interaction. Such processes signal their presence through their free-particle kinematics, as if the parent nucleus were not present. For example, a projectile ionization process requiring energy U has a threshold at $\frac{1}{2}m_e V^2 \simeq U$, in collisions with light targets where the quasifree picture is meaningful. The threshold is not sharp, due to the momentum distribution or Compton profile of the target electrons. Within the impulse approximation, the cross section for any free electron process can be related to the corresponding cross section for the ion–atom process by folding the free electron cross section into the Compton profile [65.97, 103, 104].

65.4.1 Radiative Electron Capture

The first quasifree electron process to be observed was radiative electron capture (REC), the radiative capture of a free electron by an ion. Conservation of energy and momentum is achieved by the emission of a photon which carries away the binding energy. The cross section exceeds that for bound state capture at high V. Radiative electron capture was observed through the X-ray spectra from fast heavy projectiles for which the electrons of light targets appear to be 'quasi-free'. The corresponding free electron process was seen [65.105] and has recently been heavily studied in EBIT [65.106], cooler [65.107], and storage rings [65.108]. Total cross sections for REC have also been deduced from measured total capture cross sections at large V where REC dominates bound state capture [65.109]. At high velocities, the cross section for radiative capture to the K-shell of a bare projectile is given approximately by

$$\sigma_n = \left[n/\left(\kappa^{-2}+\kappa^{-4}\right)\right]\times 2.1\times 10^{-22}\,\mathrm{cm}^2\,, \qquad (65.6)$$

where [65.110] $\kappa = \sqrt{E_B/E_0}$, E_B is the binding energy of the captured electron, E_0 the energy of the initial electron in the ion frame and n the principal quantum number of the captured electron. The theory seems to be in good agreement with experiment for capture to all shells of fast bare projectiles, although a small unexplained discrepancy between theory and experiment exists for capture to the K-shell [65.108].

65.4.2 Resonant Transfer and Excitation

Dielectronic recombination in electron–ion collisions is the process whereby an incident electron excites one target electron and, having suffered a corresponding energy loss, drops into a bound state on the projectile (Chapt. 55). If the doubly excited state so populated decays radiatively, resonant radiative recombination is achieved (DR); if it Auger decays, resonant elastic scattering has occurred. The process has long been known to be important as a recombination process in hot plasmas [65.111], but was not observed in the laboratory until 1983 [65.112–114]. The corresponding ion–atom process, known as resonant transfer and excitation (RTE) was seen a bit earlier by *Tanis* et al. [65.115]. (See [65.116, 117] for reviews of both DR and RTE.)

65.4.3 Excitation and Ionization

Excitation and ionization of inner shells of fast projectile ions by the quasifree electrons of light targets (usually He or H_2) have been identified and studied. This process competes with excitation and ionization by the target nucleus [65.118–120], and special signatures must be sought to distinguish the processes. In the case of excitation, the e–e excitation populates states through the exchange part of the interaction which is excluded for the nuclear excitation, and this has been used to separate this mechanism [65.121]. For ionization, enhancements of the ionization cross section above the Born result for nuclear ionization [65.122], coincident charge exchange measurements [65.123], and projectile [65.124] and recoil ion momentum spectroscopy [65.125, 126] have been used to distinguish the two processes. Rapid development in the production of good sources of beams of highly charged ions (EBIS/T, ECR; see [65.127]) have made these studies possible. Continued study of this field in heavy ion storage rings is now achieving resolutions of meV and opening broad new opportunities for data of unprecedented high quality for electron-ion collisions [65.128].

Recently, the first kinematically complete experiment on projectile ionization by quasifree electrons was reported [65.129]. The observed features are qualitatively similar to those found for ionization of neutral target atoms by free electron impact.

65.5 Some Exotic Processes

65.5.1 Molecular Orbital X-Rays

A typical time duration for an ion–atom encounter is $\sim 10^{-17}$ s, which is much shorter than Auger and X-ray lifetimes, so that hard characteristic radiation is emitted by the products long after the collision. There remains, however, a small but finite probabiltiy that X-rays or Auger electrons can be emitted during the collision, in which case the radiation proceeds between the time-dependent molecular orbitals formed in the collision and reflects the time evolution of the energies and transition strengths between the orbitals. Such molecular orbital X-radiation (MOX) has now been observed in many collision systems [65.130] and is reviewed in [65.131–133]. MOX spectra have been studied in total cross sections as well as a function of impact parameter. In the latter case, oscillating structures in the MOX spectra are seen [65.132, 134], due to the interference between amplitudes for the emission of X-rays with the same energy on the incoming and outgoing parts of the trajectory.

The formation of transient molecular orbitals in close collisions between highly charged ions provides opportunities for studying the electrodymamics of very highly charged systems [65.135]. For example, two uranium nuclei passing within one K-shell radius of each other form a transient molecule whose energy levels resemble those of an atom of charge 184.

65.5.2 Positron Production from Atomic Processes

Investigating the MOX interference patterns in such exotic systems offers the interesting prospect of performing spectroscopy on superheavy ions.

Reinhart et al. [65.136] predicted that, for such highly charged species, the binding energy of the united atom K-shell exceeds twice the rest mass energy of the electron, and that if a K vacancy is either brought into the collision or created during it, spontaneous electron-positron pair production occurs (the decay of the charged vacuum) with the electron filling the K-hole. However, further analysis showed that the dominant mechanism for positron production (other than those resulting from the decay of nuclear excitations) likely results from the dynamic time dependence of the fields during the collision [65.137]. Experiments showed evidence for such positron production in collisions at 6 MeV/u [65.138], but reported sharp lines in the positron spectra were later attributed to an error in the data analysis.

Present theories for the production of lepton pairs in the close collision of two highly charged systems predict that the cross section grows rapidly with collision energy. Electrons produced in such a process may end up in bound states on either collision partner, and thus represent a new charge changing mechanism. At highly relativistic velocities, the cross section for this process exceeds that for any other charge changing process. In a heavy ion collider, such as RHIC, this process could limit the ultimate storage time for the counter-propagating beams, since charge-exchanged ions are lost. The cross section has been measured recently by *Vane* et al. [65.139], for 6.4 TeV S on several targets (the highest energy ion–atom collision experiment performed to date), and are in good agreement with theory [65.140]. The bound state capture has been measured at lower energy by *Belkacem* et al. [65.141], with similarly good agreement. The extension of ion–atom collisions to such extreme velocities has opened the field for the study of processes not even imagined a short time ago.

References

65.1 E. Merzbacher, H. W. Lewis: *Handbuch der Physik*, Vol. 34th (Springer, Berlin 1958) p. 166

65.2 D. Madison, E. Merzbacher: Theory of charge-particle excitation. In: *Atomic Inner Shell Processes*, ed. by B. Craseman (Academic, New York 1975)

65.3 B. S. Khandelwal, B. H. Choi, E. Merzbacher: At. Data **15**, 103 (1969)

65.4 B. S. Khandelwal, B. H. Choi, E. Merzbacher: At. Data **5**, 291 (1973)

65.5 B. Crasemann (Ed.): *Atomic Inner Shell Processes* (Academic, New York 1975)

65.6 J. D. Garcia, R. J. Fortner, T. M. Kavanagh: Rev. Mod. Phys. **45**, 111 (1973)

65.7 P. Richard: Atomic physics: Accelerators. In: *Meth. Expt. Physics*, Vol. 17, ed. by L. Williams (Academic, New York 1980)

65.8 D. Williams: Spectroscopy. In: *Meth. Expt. Physics*, Vol. 13 (Academic, New York 1976)

65.9 J. Bang, J. M. Hansteen: Kgl. Dan. Viden. Mat.-Fys. Medd. **31**, 13 (1973)

65.10 J. M. Hansteen, O. M. Johnsen, L. Kocbach: At. Data Nucl. Data Tables **15**, 305 (1975)

65.11 M. R. C. McDowell, J. P. Coleman: *Introduction to the Theory of Ion–Atom Collisions* (North Holland, Amsterdam 1970)

65.12 J. D. Garcia: Phys. Rev. A **4**, 955 (1971)

65.13 R. Anholt: Phys. Rev. A **17**, 983 (1978)

65.14 G. Basbas, W. Brandt, R. Laubert: Phys. Rev. A **7**, 983 (1973)

65.15 J. Gray T. *Meth. Expt. Physics*, Vol. 17 (Academic, New York 1980) p. 193

65.16 H. Paul, J. Muhr: Phys. Rep. **135**, 47 (1986)

65.17 W. Brandt, G. Lapicki: Phys. Rev. A **10**, 474 (1974)

65.18 N. Bohr: K. Dan. Vidensk Selsk. Mat. Fys. Medd. **18** (1948) No. 8

65.19 M. Pauli, D. Trautmann: J. Phys. B **11**, 667, 2511 (1978)

65.20 R. L. Kauffman, P. Richard: *Meth. Expt. Physics*, Vol. 17 (Academic, New York 1980) p. 148

65.21 W. Bambynek, C. Crasemann, R. W. Fink, H.-U. Freund, H. Mark, C. D. Swift, R. E. Price, P. V. Rao: Rev. Mod. Phys. **44**, 716 (1972)

65.22 C. L. Cocke: *Meth. Expt. Physics*, Vol. 17 (Academic, New York 1980) p. 303

65.23 J. T. Park, J. E. Aldag, J. M. George, J. L. Peacher, J. H. McGuire: Phys. Rev. **A15**, 508 (1977)

65.24 M. E. Rudd, Y.-K. Kim, D. H. Madison, T. J. Gay: Rev. Mod. Phys. **64**, 441 (1992)

65.25 M. Schulz, T. Vajnai, A. D. Gaus, W. Htwe, R. E. Olson: Phys. Rev. **54**, 2951 (1996)

65.26 T. Vajnai, A. D. Gaus, J. A. Brandt, W. Htwe, D. H. Madison, R. E. Olson, J. L. Peacher, M. Schulz: Phys. Rev. Lett. **74**, 3588 (1995)

65.27 H. Ehrhardt, K. Jung, G. Knoth, P. Schlemmer: Z. Phys. **D1**, 3 (1986)

65.28 L. An, Kh. Khayyat, M. Schulz: Phys. Rev. **A63**, 030703(R) (2001)

65.29 R. Moshammer, J. Ullrich, M. Unverzagt, W. Schmidt, P. Jardin, R. E. Olson, R. Mann, R. Dörner, V. Mergel, U. Buck, H. Schmidt-Böcking: Phys. Rev. Lett. **73**, 3371 (1994)

65.30 R. Dörner, V. Mergel, O. Jagutzki, L. Spielberger, J. Ullrich, R. Moshammer, H. Schmidt-Böcking: Physics Reports **330**, 95 (2000)

65.31 J. Ullrich, R. Moshammer, A. Dorn, R. Dörner, L. P. H. Schmidt, H. Schmidt-Böcking: Rep. Prog. Phys. **66**, 1463 (2003)

65.32 C. L. Cocke: Physics Scripta **T110**, 9 (2004)

65.33 M. Schulz, R. Moshammer, D. Fischer, H. Kollmus, D. H. Madison, S. Jones, J. Ullrich: Nature **422**, 48 (2003)

65.34 J. T. Park, J. E. Aldag, J. M. Geroge, J. L. Peacher: Phys. Rev. A **14**, 608 (1976)

65.35 D. S. F. Crothers, J. F. McCann: J. Phys. **B16**, 3229 (1983)

65.36 D. H. Madison, M. Schulz, S. Jones, M. Foster, R. Moshammer, J. Ullrich: J. Phys. **B35**, 3297 (2002)

65.37 D. H. Madison, D. Fischer, M. Foster, M. Schulz, R. Moshammer, S. Jones, J. Ullrich: Phys. Rev. Lett. **91**, 253201 (2003)

65.38 M. Schulz, R. Moshammer, D. Fischer, J. Ullrich: J. Phys. **B36**, L311 (2003)

65.39 M. Schulz, R. Moshammer, A. N. Perumal, J. Ullrich: J. Phys. **B35**, 161L (2002)

65.40 N. F. Mott, S. N. Massey: *The Theory of Atomic Collisions* (Oxford Univ. Press, Oxford 1965)

65.41 M. Inokuti: Rev. Mod. Phys. **43**, 297 (1971)

65.42 H. Bethe: Ann. Phys. **5**, 325 (1930)

65.43 J. R. Oppenheimer: Phys. Rev. **31**, 349 (1928)

65.44 H. S. Brinkman, H. A. Kramers: Proc. Acad. Sci. Amsterdam **33**, 973 (1930)

65.45 M. Rødbro, E. Horsdal Pedersen, C. L. Cocke, J. R. Macdonald: Phys. Rev. A **19**, 1936 (1979)
65.46 J. R. Macdonald: *Meth. Expt. Physics*, Vol. 17 (Academic, New York 1980) p. 303
65.47 Nikolaev: Zh. Eksp. Teor. Fiz. **51**, 1263 (1966)
65.48 Nikolaev: Sov. Phys. JETP **24**, 847 (1967)
65.49 L. H. Thomas: Proc. R. Soc. London A **114**, 561 (1927)
65.50 R. Shakeshaft, L. Spruch: Rev. Mod. Phys. **51**, 369 (1979)
65.51 B. H. Bransden, I. M. Cheshire: Proc. Phys. Soc. **81**, 820 (1963)
65.52 P. R. Simony, J. H. McGuire: J. Phys. B **14**, L737 (1981)
65.53 J. Macek, S. Alston: Phys. Rev. A **26**, 250 (1982)
65.54 J. Macek, K. Taulbjerg: Phys. Rev. Lett. **46**, 170 (1981)
65.55 A. L. Ford, J. R. Reading, R. L. Becker: Phys. Rev. A **23**, 510 (1981)
65.56 Dz. Belkic, A. Salin: J. Phys. B **11**, 3905 (1978)
65.57 J. Eichler, T. Chan: Phys. Rev. A **20**, 104 (1978)
65.58 D. P. Dewangan, J. Eichler: Comm. At. Mol. Phys. **27**, 317 (1992)
65.59 H. Marxer, J. S. Briggs: J. Phys. B **18**, 3823 (1992)
65.60 J. S. Briggs, J. H. Macek, K. Taulbjerg: Comments At. Mol. Phys. **12**, 1 (1983)
65.61 E. Horsdal-Pedersen, C. L. Cocke, M. Stöckli: Phys. Rev. Lett. **50**, 1910 (1983)
65.62 H. Vogt, R. Schuch, E. Justiniano, M. Schultz, W. Schwab: Phys. Rev. Lett. **57**, 2256 (1986)
65.63 V. Mergel, R. Dörner, J. Ullrich, O. Jagutzki, S. Lencinas, S. Nüttgens, L. Spielberger, M. Unverzagt, C. L. Cocke, R. E. Olson, M. Schulz, U. Buck, E. Zanger, W. Theisinger, M. Isser, S. Geis, , H. Schmidt-Böcking: Phys. Rev. Lett. **74**, 220 (1995)
65.64 E. Edgu-Fry, C. L. Cocke, J. Stuhlman: Bull. Am. Phys. Soc. **48**, 12 (2003)
65.65 P. D. Fainstein, L. Gulyas, A. Salin: J. Phys. **B27**, L259 (1994)
65.66 C. L. Cocke, R. R. Randall, S. L. Varghese, B. Curnutte: Phys. Rev. A **14**, 2026 (1976)
65.67 U. Fano, W. Lichten: Phys. Rev. Lett. **14**, 627 (1965)
65.68 U. Wille, R. Hippler: Phys. Rep. **132**, 131 (1986)
65.69 Q. Kessel, B. Fastrup: Case Stud. At. Phys. **3**, 137 (1973)
65.70 J. Briggs: Rep. Prog. Phys. **39**, 217 (1976)
65.71 W. E. Meyerhof, K. Taulbjerg: Ann. Rev. Nucl. Sci. **27**, 279 (1977)
65.72 I. A. Sellin (Ed.): *Topics in Current Physics 5: Structure and Collisions of Ions and Atoms* (Springer, Berlin 1975)
65.73 K. Taulbjerg, J. S. Briggs: J. Phys. B **8**, 1895 (1975)
65.74 J. S. Briggs, J. Macek: J. Phys. B **5**, 579 (1972)
65.75 Y. Demkov: Sov. Phys. JETP **18**, 138 (1964)
65.76 W. E. Meyerhof: Phys. Rev. Lett. **31**, 1973 (1341)
65.77 G. T. Lockwood, E. Everhart: Phys. Rev. **125**, 567 (1962)
65.78 S. Hagmann, C. L. Cocke, J. R. Macdonald, P. Richard, H. Schmidt-Böcking, R. Schuch: Phys. Rev. A **25**, 1918 (1982)
65.79 C. F. Moore, M. Senglaub, B. Johnson, P. Richard: Phys. Rev. Lett. **40**, 107 (1972)
65.80 I. A. Sellin: Extensions of beam-foil spectoscopy. In: *Topics in Current Physics 5: Structure and Collisions of Ions and Atoms*, ed. by I. A. Sellin (Springer, Berlin 1978) Chap. 7
65.81 M. Schulz, R. Moshammer, W. Schmitt, H. Kollmus, B. Feuerstein, R. Mann, S. Hagmann, J. Ullrich: Phys. Rev. Lett. **84**, 863 (2000)
65.82 J. H. McGuire: Adv. At. Mol. Phys. **29**, 217 (1992)
65.83 M. Schulz: Int. J. Mod. Phys. **B9**, 3269 (1995)
65.84 J. H. McGuire: Electron correlation dynamics in atomic collisions. In: *Cambridge Monographs on Atomic, Molecular, and Chemical Physics*, Vol. 8, ed. by J. H. McGuire (Cambridge Univ. Press, Cambridge 1997)
65.85 L. G. Gerchikov, S. A. Sheinerman: J. Phys. **B34**, 647 (2001)
65.86 M. Schulz, M. Moshammer, L. G. Gerchikov, S. A. Sheinerman, J. Ullrich: J. Phys. **B34**, L795 (2001)
65.87 V. Mergel, R. Dörner, Kh. Khayyat, M. Achler, T. Weber, O. Jagutzki, H. J. Lüdde, C. L. Cocke, H. Schmidt-Böcking: Phys. Rev. Lett. **86**, 2257 (2001)
65.88 L. H. Andersen, P. Hvelplund, H. Knudsen, S. P. Møller, K. Elsener, K.-G. Rensfelt, E. Uggerhøj: Phys. Rev. Lett. **57**, 2147 (1986)
65.89 J. P. Giese, E. Horsdal: Phys. Rev. Lett. **60**, 2018 (1988)
65.90 J. Ullrich, R. Moshammer, H. Berg, R. Mann, H. Tawara, R. Dörner, H. Schmidt-Böcking, S. Hagmann, C. L. Cocke, M. Unverzagt, S. Lencincas, V. Mergel: Phys. Rev. Lett. **71**, 1697 (1993)
65.91 D. Fischer, R. Moshammer, A. Dorn, J. R. Crespo López-Urrutia, B. Feuerstein, C. Höhr, C. D. Schröter, S. Hagmann, H. Kollmus, R. Mann, B. Bapat, J. Ullrich: Phys. Rev. Lett. **90**, 243201–1 (2003)
65.92 A. Dorn, A. Kheifets, C. D. Schröter, B. Najjari, C. Höhr, R. Moshammer, J. Ullrich: Phys. Rev. Lett. **86**, 3755 (2001)
65.93 N. Stolterfoht, D. Schneider, D. Burch, H. Wieman, J. S. Risely: Phys. Rev. Lett. **33**, 59 (1974)
65.94 N. Stolterfoht, D. Schneider, D. Burch, H. Wieman, J. S. Risely: Phys. Rev. Lett. **64**, 441 (1992)
65.95 N. Stolterfoht: Electronic screening in heavy-ion-atom collisions. In: *Topics in Current Physics*, Vol. 5, ed. by I. A. Sellin (Springer, Berlin 1978)
65.96 G. B. Crooks, M. E. Rudd: Phys. Rev. Lett. **25**, 1599 (1970)
65.97 P. Richard: *Proceedings of 15th International Conference on X-Ray and Inner Shell Processes, 1990*, ed. by T. Carlson (American Institute of Physics, New York 1990)
65.98 B. Sulik, Cs. Koncz, K. Tokési, A. Orbán, D. Berényi: Phys. Rev. Lett. **88**, 073201 (2002)
65.99 N. Stolterfoht, B. Sulik, V. Hoffmann, B. Skogvall, J. Y. Chesnel, J. Rangama, F. Frémont, D. Hennecart, A. Cassimi, X. Husson, A. L. Landers,

J. A. Tanis, M. E. Galassi, R. D. Rivarola: Phys. Rev. Lett. **87**, 023201 (2001)
65.100 C. Dimopoulou, R. Moshammer, D. Fischer, C. Höhr, A. Dorn, P. D. Fainstein, J. R. C. L. Urrutia, C. D. Schröter, H. Kollmus, R. Mann, S. Hagmann, J. Ullrich: Phys. Rev. Lett. **93**, 123203 (2004)
65.101 A. Itoh, T. Schneider, G. Schiwietz, Z. Roller, H. Platten, G. Nolte, D. Schneider, N. Stolterfoht: J. Phys. B. **16**, 3965 (1983)
65.102 D. H. Lee, P. Richard, T. J. M. Zouros, J. M. Sanders, J. L. Shinpaugh, H. Hidmi: Phys. Rev. A. **41**, 4816 (1990)
65.103 C. L. Cocke: Recent trends in ion–atom collisions. In: *Electronic and Atomic Collisions, Invited Papers, XVII ICPEAC, Book of Abstracts*, ed. by I. E. MacCarty, W. R. MacGillivray, M. C. Standage (Adam Hilger, Bristol, Philadelphia and New York 1992) p. 49
65.104 I. A. Sellin: , Vol. 376, ed. by D. Berenyi, D. Hock (Springer, Berlin 1990)
65.105 H. W. Schnopper, H. D. Betz, J. P. Delvaille, K. Kalata, A. R. Sohval, K. W. Jones, H. E. Wegner: Phys. Rev. Lett. **29**, 898 (1972)
65.106 R. W. Marrs: Phys. Rev. Lett. **60**, 1757 (1988)
65.107 L. H. Andersen, J. Bolko: J. Phys. B **23**, 3167 (1990)
65.108 T. Stöhlker, C. Kozhuharov, A. E. Livongston, P. H. Mokler, Z. Stachura, A. Warczak: Z. Phys. D. **23**, 121 (1992)
65.109 H. Gould, D. Greiner, P. Lindstrom, T. J. M. Symons, H. Crawford: Phys. Rev. Lett. **52**, 180 (1984)
65.110 H. A. Bethe, E. E. Salpeter: *Encyclopidia of Physics*, Vol. 35, ed. by S. Fluegge (Springer, Berlin 1957) p. 408
65.111 A. Burgess: Astrophy. J. **139**, 776 (1964)
65.112 J. B. A. Mitchell, C. T. Ng, J. L. Forand, D. P. Levac, R. E. Mitchell, A. Sen, D. B. Miko, J. Wm. McGowan: Phys. Rev. Lett. **50**, 335 (1983)
65.113 D. S. Belic, G. H. Dunn, T. J. Morgan, D. W. Mueller, C. Timmer: Phys. Rev. Lett. **50**, 339 (1983)
65.114 P. F. Dittner, S. Datz, P. D. Miller, C. D. Moak, P. H. Stelson, C. Bottcher, W. B. Dress, G. D. Alton, N. Neskovic, C. M. Fou: Phys. Rev. Lett. **51**, 31 (1983)
65.115 J. A. Tanis, S. M. Shafroth, J. E. Willis, M. Clark, J. Swenson, E. N. Strait, J. R. Mowat: Phys. Rev. Lett. **47**, 828 (1981)
65.116 W. G. Graham, W. Fritsch, Y. Hahn, J. H. Tanis (Eds.): *Recombination of Atomic Ions*, NATO ASI Ser. B (Plenum, New York 1992) p. 296
65.117 J. A. Tanis: Resonant transfer excitation (RTE) associated with single X-ray emission. In: *Recombination of Atomic Ions*, NATO ASI Series B, Vol. 296, ed. by W. G. Graham, W. Fritsch, Y. Hahn, J. H. Tanis (Plenum, New York 1992) p. 241
65.118 D. R. Bates, G. Griffing: Proc. Phys. Soc. London A **66**, 961 (1955)
65.119 D. R. Bates, G. Griffing: Proc. Phys. Soc. London A **67**, 663 (1954)
65.120 D. R. Bates, G. Griffing: Proc. Phys. Soc. London A **68**, 90 (1955)
65.121 T. Zouros, D. H. Lee, P. Richard: *Proceedings of the XVI International Conference on the Physics of Electronic and Atomic Collisions, New York, 1989*, AIP Conference Proceedings No. 205, ed. by A. Dalgarno, R. S. Freund, M. S. Lubell, T. B. Lucatorto (AIP, New York 1990) p. 568
65.122 W. E. Meyerhof, H.-P. Hülskötter, Q. Dai, J. H. McGuire, Y. D. Wang: Phys. Rev. A. **43**, 5907 (1991)
65.123 E. C. Montenegro, W. S. Melo, W. E. Meyerhof, A. G. de Pinho: Phys. Rev. Lett. **69**, 3033 (1992)
65.124 E. C. Montenegro, A. Belkacem, D. W. Spooner, W. E. Meyerhof, M. B. Shah: Phys. Rev. A **47**, 1045 (1993)
65.125 W. Wu, K. L. Wong, R. Ali, C. Y. Chen, C. L. Cocke, V. Frohne, J. P. Giese, M. Raphaelian, B. Walch, R. Dörner, V. Mergel, H. Schmidt-Böcking, W. E. Meyerhof: Phys. Rev. Lett. **72**, 3170 (1994)
65.126 R. Dörner, V. Mergel, R. Ali, U. Buck, C. L. Cocke, K. Froschauer, O. Jagutzki, S. Lencinas, W. E. Meyerhof, S. Nüttgens, R. E. Olson, H. Schmidt-Böcking, L. Spielberger, K. Tökesi, J. Ullrich, M. Unverzagt, W. Wu: Phys. Rev. Lett. **72**, 3166 (1994)
65.127 C. L. Cocke: Progress in atomic collisions with multiply charged ions. In: *Review of Fundemental Processes and Application of Ions and Atoms*, ed. by C. D. Lin (World Scientific, Singapore 1993) p. 138
65.128 R. Schuch: Cooler storage rings: New tools for atomic physics. In: *Review of Fundemental Processes and Application of Ions and Atoms*, ed. by C. D. Lin (World Scientific, Singapore 1993) p. 169
65.129 H. Kollmus, R. Moshammer, R. E. Olson, S. Hagmann, M. Schulz, J. Ullrich: Phys. Rev. Lett. **88**, 103202–1 (2002)
65.130 F. W. Saris, W. F. van der Weg, H. Tawara, R. Laubert: Phys. Rev. Lett. **28**, 717 (1972)
65.131 P. O. Mokler: Quasi molecular radiation. In: *Topics in Current Physics*, Vol. 5, ed. by I. A. Sellin (Springer, Berlin 1978) p. 245
65.132 R. Schuch, M. Meron, B. M. Johnson, K. W. Jones, R. Hoffmann, H. Schmidt-Böcking, I. Tserruya: Phys. Rev. A. **37**, 3313 (1988)
65.133 R. Anholt: Rev. Mod. Phys. **57**, 995 (1985)
65.134 I. Tserruya, R. Schuch, H. Schmidt-Böcking, J. Barrette, W. Da-Hai, B. M. Johnson, M. Meron, K. W. Jones: Phys. Rev. Lett. **50**, 30 (1983)
65.135 W. Pieper, W. Geiner: Z. Phys. **218**, 126 (1969)
65.136 J. Reinhardt, U. Müller, B. Müller, W. Greiner: Z. Phys. A **303**, 173 (1981)
65.137 G. Soff, J. Reinhard, B. Müller, W. Greiner: Z. Phys. A **294**, 137 (1980)
65.138 U. Müller-Nehrer, G. Soff: Electron excitations in superheavy quasi-molecules, Phys. Rep. **246** (1994)
65.139 C. R. Vane, S. Datz, P. F. Dittner, H. F. Krause, C. Bottcher, M. Strayer, R. Schuch, H. Gao, R. Hutton: Phys. Rev. Lett. **69**, 1911 (1992)
65.140 C. Bottcher, M. Strayer: Phys. Rev. A **39**, 4 (1989)

65.141 A. Belkacem, H. Gould, B. Feinberg, R. Bossingham, W.E. Meyerhof: Phys. Rev. Lett. **71**, 1514 (1993)

65.142 W. König, F. Bosch, P. Kienle, C. Kozhukarov, H. Tsertos, E. Berdermann, S. Muckler, W. Wagner: Z. Phys. A. **328**, 129 (1987)

66. Reactive Scattering

This chapter presents a résumé of the methods commonly employed in scattering experiments involving neutral molecules at chemical energies, i. e., less than about 10 eV. These experiments include the study of intermolecular potentials, the transfer of energy in molecular collisions, and elementary chemical reaction dynamics. Closely related material is presented in Chapts. 35, 37, and 38 as well as in other chapters on quantum optics.

66.1 **Experimental Methods** ... 967
66.1.1 Molecular Beam Sources ... 967
66.1.2 Reagent Preparation ... 968
66.1.3 Detection of Neutral Products ... 969
66.1.4 A Typical Signal Calculation ... 971

66.2 **Experimental Configurations** ... 971
66.2.1 Crossed-Beam Rotatable Detector 971
66.2.2 Doppler Techniques ... 973
66.2.3 Product Imaging ... 973
66.2.4 Laboratory to Center-of-Mass Transformation ... 975

66.3 **Elastic and Inelastic Scattering** ... 976
66.3.1 The Differential Cross Section ... 976
66.3.2 Rotationally Inelastic Scattering ... 977
66.3.3 Vibrationally Inelastic Scattering ... 977
66.3.4 Electronically Inelastic Scattering ... 978

66.4 **Reactive Scattering** ... 978
66.4.1 Harpoon and Stripping Reactions ... 978
66.4.2 Rebound Reactions ... 979
66.4.3 Long-lived Complexes ... 979

66.5 **Recent Developments** ... 980

References ... 980

66.1 Experimental Methods

66.1.1 Molecular Beam Sources

The development of molecular beam methods in the past two decades has transformed the study of chemical physics [66.1]. Supersonic molecular beam sources allow one to prepare reagents possessing a very narrow velocity distribution with very low internal energies, ideal for use in detailed studies of intermolecular interactions. Early experiments generally employed continuous beam sources, but in recent years intense pulsed beam sources have come into common usage [66.2]. The advantages of pulsed beams primarily arise from the lower gas loads associated with their use, hence reduced demands on the pumping system. If any component of the experiment is pulsed (pulsed laser detection, for example) then considerable advantage may be obtained by also pulsing the beam. Although the theoretical descriptions of pulsed and continuous expansions are essentially equivalent, in practice some care is required in employing pulsed beams because the temperature and velocity distributions may change dramatically through the course of the pulse. Free jet expansions are supersonic because the dramatic drop in the local temperature in the beam is associated with a drop in the local speed of sound. A detailed description of the supersonic expansion may be found in [66.3–5]. In practice, many of the detailed features associated with a supersonic expansion may be ignored and one may assume an isentropic expansion into the vacuum. For an isentropic nozzle expansion of an ideal gas, the maximum terminal velocity is given by

$$v_{\max} = \sqrt{2\hat{C}_p T_0}\,, \tag{66.1}$$

where, for an ideal gas, the heat capacity is

$$\hat{C}_p = \left(\frac{\gamma}{\gamma - 1}\right)\frac{R}{m}\,, \tag{66.2}$$

R is the gas constant, m is the molar molecular mass, T_0 the temperature in the stagnation region, and γ the heat capacity ratio. For ideal gas mixtures, and assuming C_p independent of temperature for the range encountered in the expansion, one may use

$$\bar{C}_p = \sum_i X_i C_{p_i} = \sum_i X_i \left(\frac{\gamma_i}{\gamma_i - 1} \right) R , \quad (66.3)$$

and the average molar mass

$$\bar{m} = \sum_i X_i m_i , \quad (66.4)$$

where X_i is the mole fraction of component i, to obtain an estimate of the maximum velocity for a mixture:

$$v_{\max} = \sqrt{2\bar{C}_p T_0 / \bar{m}} . \quad (66.5)$$

By seeding heavy species in light gases one may accelerate them to superthermal energies. Supersonic beams are characterized by the speed ratio, i. e., the mean velocity divided by the velocity spread:

$$S \equiv \frac{v}{\sqrt{2kT/m}} , \quad (66.6)$$

where T is the local translational temperature, or by the Mach number

$$M \equiv \frac{v}{\sqrt{\gamma kT/m}} . \quad (66.7)$$

For the purpose of order-of-magnitude calculations, the number density on axis far from the nozzle may be estimated as

$$n \approx n_0 \, (d/x)^2 , \quad (66.8)$$

where n_0 is the number density in the stagnation region, d is the nozzle diameter, and x is the distance from the nozzle. The number density versus speed distribution of a nozzle beam is well described as a Gaussian characterized by the speed ratio S and a parameter $\alpha = v_0/S$, where v_0 is the most probable velocity:

$$n(v) = v^2 \exp\left[-(v/\alpha - S)^2 \right] . \quad (66.9)$$

Cooling efficiencies for the various internal degrees of freedom correlate with the efficiency of coupling of these modes with translation, hence they vary widely. Coupling of modes A and B is expressed by the collision number Z_{A-B}:

$$Z_{A-B} \equiv Z\tau_{A-B} , \quad (66.10)$$

where τ_{A-B} is the bulk relaxation time, and Z the collision frequency. This represents the number of collisions between effective inelastic events. Typical values are summarized in Table 66.1. R–T coupling is relatively efficient, while V–T coupling is quite inefficient, so that vibrational excitation may not be effectively cooled in the expansion.

Table 66.1 Collision numbers for coupling between different modes. V, R, T refer to vibrational, rotational, and translational energy, respectively. Each entry is the typical range of Z_{A-B}

	V	R	T
V	$10^{0.5-3}$	10^{3-4}	10^{5-6}
R		10^{0-1}	10^{2-3}

66.1.2 Reagent Preparation

Molecular beam methods may be used in conjunction with a variety of other techniques to prepare atoms or molecules in excited or polarized initial states (Chapt. 46), to generate unstable molecules or radicals [66.6,7] or to produce beams of refractory materials such as transition metals or carbon [66.8,9]. Some of the common techniques are outlined below. Optical pumping of atoms to excited electronic states is a useful means of reagent preparation, and this topic is presented in detail in Chapt. 10. This technique further allows one, using polarized lasers, to explore the influence of angular momentum polarization in the reagents on the collision dynamics [66.10]. Most of these studies have been performed using alkali and alkaline earth metals since there exist strong electronic transitions and convenient narrow-band visible lasers suitable for use with these systems. Laser excitation may also be used to generate vibrationally excited molecules in their ground electronic states. The techniques employed include direct IR excitation using an HF chemical laser [66.11], population depletion methods [66.12] and various Raman techniques [66.13].

Metastable atoms may also be prepared by laser photolysis of a suitable precursor. $O(^1D)$ preparation is readily prepared by photolysis of ozone or N_2O, for example [66.14]. Alternatively, rf or microwave discharges may be used to produce metastable species or reactive atoms or radicals [66.15]. These techniques may also be used to prepare ground state atoms; for example, hot H atom beams are frequently produced by photolysis of HI or H_2S [66.16]. Such atomic or molecular radical beams may also be generated by pyrolysis in the nozzle. In this case care must be taken to minimize recombination through careful choice of the temperature, nozzle geometry, and transit time through the heated region.

Beams of refractory materials are now commonly generated using laser ablation sources [66.8, 9]. Typically these employ a rod or disk of the substrate of interest which is simultaneously rotated and translated to provide a fresh surface for ablation at each laser pulse. A laser beam is focused on the substrate and timed to fire just as a carrier gas pulse passes over. Laser power and wavelength must be optimized for a given substrate. Lasers operating in the IR, visible, and UV have all been employed. Aligned or oriented molecules have been prepared using multipole focussing [66.17, 18], and more recently using strong electric fields ("brute force") [66.19]. In the former case, specific quantum states are focused by the field. In the latter case, so called pendular states are prepared from the low rotational levels of molecules possessing large dipole moments and small rotational constants. The ability to orient these molecules can be estimated on the basis of the Stark parameter $\omega = \mu E/B$, where μ is the dipole moment, E the electric field strength, and B the rotational constant. Orientation is feasible for low rotational levels of molecules when the Stark parameter is on the order of 10 or higher [66.19].

66.1.3 Detection of Neutral Products

Broadly speaking, detection of neutral molecules is accomplished either by optical (spectroscopic) or nonoptical techniques. Nonoptical methods usually involve nonspecific ionization of neutral particles, most commonly by electron impact, followed by mass selection and ion counting. Thermal detectors such as cryogenic bolometers are also finding widespread application in molecular beam experiments owing to their remarkable sensitivity [66.20]. In general, optical methods may rely on resonant or nonresonant processes, hence they may or may not enjoy quantum state selectivity. Both photoionization and laser-induced fluorescence methods are now in common usage, usually in applications where quantum state resolved information is desired. The advantage of nonoptical methods is primarily one of generality: all neutral molecules may be detected, and branching into different channels readily measured. Quantum state resolution is more difficult to achieve using nonoptical detection methods, but both vibrationally- and rotationally-resolved measurements have been obtained by these means [66.21, 22].

The primary advantage of spectroscopic detection is the aforementioned possibility of quantum state specificity. Another unique opportunity afforded by spectroscopic probes is the measurement of product aligment and orientation. In addition, in some cases background interference may be reduced or eliminated using state-specific probes, thereby affording enhanced signal-to-noise ratios.

Nonoptical Techniques

Detectors based on nonspecific ionization remain the most commonly used in molecular beam experiments, owing to the ease of subsequent mass selection, and the convenience and sensitivity of ion detection. Surface ionization is a sensitive means of detecting alkali atoms and other species exhibiting low ionization potentials [66.23]. Surface ionization occurs when a neutral atom or molecule with a low ionization potential sticks on a surface with a high work function and is subsequently desorbed. Typically these detectors employ a hot platinum or oxidized tungsten wire or ribbon for formation and subsequent desorption of the ions, which is surrounded by an ion collector. They are very efficient for the detection of alkali atoms and molecules whose ionization potentials are $\lesssim 6$ eV.

All neutral gas molecules may be ionized by collision with energetic electrons, and electron beam ionizers may be produced that couple conveniently to quadrupole mass spectrometers [66.24]. Collision of a molecule with a 100–200 eV electron leads predominantly to formation of the positive ion and a secondary electron. Other processes also occur and can be very significant: doubly or triply charged ions may be formed and, importantly, molecules can fragment yielding many daughter ions in addition to the parent ion. These fragmentation patterns vary with different molecules, and may further show a strong dependence on molecular internal energy, so particular care must be taken to determine the role of these phenomena in each particular application. It is often necessary to record data for the parent ion and daughter ions for a given product channel and compare them to eliminate contributions arising from cracking of the parent molecule or other species [66.25]. Electron impact ionization probabilities for most species exhibit a similar dependence on electron energy, rising rapidly from the ionization potential to a peak at 80–100 eV, then falling more slowly with increasing collision energy. The ionization cross section for different species scales with molecular polarizability according to a well-established empirical relation [66.26]:

$$\sigma_{\text{ion}} = 36\sqrt{\alpha} - 18\,, \tag{66.11}$$

where σ_{ion} is in Å^2 and α, the molecular polarizability, is in Å^3. This relation can be used to estimate branching ratios in the absence of any other means of calibrating the relative contributions of two different channels. The ionization rate is given by

$$\frac{d[M^+]}{dt} = I_e\sigma[M] , \tag{66.12}$$

where I_e is the electron beam intensity, typically $10\,mA/cm^2$ or 6×10^{16} electrons/cm^2 s, and $[M]$ is the number density of molecules M in the ionizer. If one assumes an ionization cross section σ_{ion} of $10^{-16}\,cm^2$ for collision with 150 eV electrons (a typical value for a small molecule), the ionization probability for molecules residing in the ionizer is then

$$\frac{d[M^+]}{dt}\frac{1}{[M]} = I_e\sigma = 6\times10^{16}\times10^{-16} = 6\,s^{-1} . \tag{66.13}$$

However, product molecules arriving in the detector are not stationary. Typically, product velocities are on the order of 500 m/s. If the ionization region has a length of 1 cm, the residence time τ of a product molecule is on the order of 2×10^{-5} s. Consequently, the ionization probability of product molecules passing through the ionizer is

$$\frac{d[M^+]}{dt}\frac{\tau}{[M]} = 2\times10^{-5}\times6 = 1.2\times10^{-4} . \tag{66.14}$$

Although this does not appear very efficient (indeed, it is 4 orders of magnitude less so than surface ionization), nevetheless, if the background count rate is sufficiently low, then good statistics may be obtained with signal levels as low as 1 Hz. Thus, for detection based on electron impact ionization, a key factor determining the sensitivity of the experiment is the background count rate at the masses of interest.

Spectroscopic Detection

Spectroscopic detection methods usually involve either laser-induced fluorescence (LIF) or resonant photoionization (REMPI) (Chapt. 44). Alternative techniques such as laser-induced grating methods and nonresonant VUV photoionization are also being applied to scattering experiments. Essential to the use of spectroscopic methods for reactive scattering studies is an understanding of the spectrum of the species of interest. This may be challenging for many reactive systems because the products may be produced in highly excited vibrational or electronic states that may not be well characterized. Additional spectroscopic data may be required. Franck–Condon factors are necessary to compare the intensities of different product vibrational states, while a calibration of the relative intensities of different electronic bands requires a measure of the electronic transition moments. In some cases, one must include the specific dependence of the electronic transition moment on the internuclear distance by integrating over the vibrational wave function. Populations corresponding to different rotational lines may be compared after the appropriate correction, which is represented by the Hönl–London factors, only for isotropic irradiation and detection. This is certainly not the case for most laser-based experiments. Generally, the detailed dependence of the excitation and detection on the relevant magnetic sublevels must be considered [66.27–29]. Caution is required in using any spectroscopic method involving a level that is predissociated. This may lead to a dramatic decrease in the associated fluorescence or photoionization yield if the predissociation rate approaches or exceeds the rate of fluorescence or subsequent photoionization. An important question in any experiment based on spectroscopic detection is whether product flux or number density is probed. This question is considered in detail in several articles [66.13,30]. It depends on the lifetime of the state that is probed, the relative time that the molecule is exposed to the probe laser field, and its residence time in the interaction region. Saturation phenomena are also important, yet not necessarily easily anticipated. Complete saturation does not readily occur because excitation in the wings of the laser beam profile becomes more significant as the region in the center of the beam becomes saturated [66.31].

LIF is currently the most widely used spectroscopic technique in inelastic and reactive scattering experiments [66.27, 32, 33]. It has been used to measure state-resolved total cross sections [66.34] and differential cross sections in electronic [66.35], vibrational and rotationally inelastic scattering [66.12] as well as reactive scattering [66.36].

With the development of high-power tunablebreak lasers and the discovery of useful photoionization schemes, REMPI is becoming a more general technique [66.37,38]. REMPI has the advantages associated with ion detection, namely considerable convenience in mass selection and efficient detection, in addition to the capability for quantum state selectivity. Disadvantages associated with REMPI arise primarily from higher laser power employed compared with LIF. Caution is required in attempting to extract quantitative information from REMPI spectra if one or several of the steps involved in the ionization process are saturated. This is of particular concern at the high laser powers necessary for multiple photon transitions. An alternative to direct photoionization involves excitation of products to metastable Rydberg states, followed by field ionization some distance from the interaction region. This technique has the advantage of very low background and is capable of extraordinary time-of-flight resolution. Remarkable results have recently been obtained for the reaction $D + H_2$ using this method [66.39]. Photoionization techniques are becoming more widely used in scattering experiments as the basis for product imaging detection schemes discussed below.

66.1.4 A Typical Signal Calculation

For a crossed-beam system in which a beam of atoms A collides with a beam of molecules B yielding products C and D, the rate of formation of C is given by

$$\frac{dN_C}{dt} = n_A n_B \sigma_r g \Delta V \,, \qquad (66.15)$$

where n_A and n_B are the number densities of the respective reagents at the interaction region, σ_r is the reaction cross section, g the magnitude of the relative velocity between the reactants, and ΔV the volume of intersection of the beams. For a typical experiment employing continuous supersonic beams, the number densities of the atomic and molecular reactants are $\sim 10^{11}$–$10^{12}\,\text{cm}^{-3}$ and the scattering volume $10^{-2}\,\text{cm}^3$. For $g = 10^5\,\text{cm/s}$ and $\sigma_r = 10^{-15}\,\text{cm}^2$, the rate of product formation $dN_C/dt = 10^{11}$ molecules/s. The kinematics and energetics of the reaction then determines the range of laboratory angles into which the products scatter, and the magnitude of the scattered signal.

If the products scatter into 1 sr of solid angle, and the detector aperture is 3×10^{-3} sr (roughly 1 degree in both directions perpendicular to the detector axis), then the detector receives 3×10^7 product molecules/s. Given the detection probability obtained above, 3600 product ions/s are detected. This is adequate to obtain very good statistics in a short time as long as the background count rate is not considerably higher.

For a nonspecific detection technique, such as electron bombardment ionization coupled with mass filtering, it is necessary to use ultrahigh vacuum (10^{-10} torr) in the detector region to minimize interference from background gases. The residual gases are then primarily H_2 and CO, with number densities on the order of $10^6\,\text{cm}^3$. Differential pumping stages, each of which may reduce the background by 2 orders of magnitude, are generally used to lower the background from gases whose partial pressures are lower than the ultrahigh vacuum limit of the detector chamber. However, this differential pumping helps only for those molecules that do not follow a straight trajectory through the detector. The contribution from the latter is given by

$$n' = \frac{nA}{4\pi x^2} \,, \qquad (66.16)$$

where n is the number density of molecules effusing from an orifice of area A, and n' is their number density at a distance x on axis downstream. For a distance of 30 cm and a main chamber pressure of 3×10^{-7} torr, this corresponds to a steady state density of 10^5 molecules/cm^3 at the ionizer, a reduction of 6 orders of magnitude. Three stages of differential pumping are thus the maximum useful under these conditions, since the primary source of background is then molecules following a straight trajectory from the main chamber. A liquid helium cooled surface opposite the detector entrance may then be useful to minimize scattering of background molecules into the ionizer.

66.2 Experimental Configurations

66.2.1 Crossed-Beam Rotatable Detector

The configuration illustrated in Fig. 66.1 represents a standard now widely used [66.24], usually with two continuous beams fixed at 90 degrees. The molecular beam sources are differentially pumped and collimated to yield an angular divergence of about 2 degrees. The beams cross as close as possible to the nozzles, with a typical interaction volume of 3 mm^3. Scattered products pass through an aperture on the front of the detector, thence through several stages of differential pumping before reaching the ionizer. Ions formed by electron impact on the neutral products are then extracted into a quadrupole mass spectrometer with associated ion counter. A chopper wheel is generally used at the entrance to the detector to provide a time origin for recording time-of-flight spectra. Pseudorandom sequence chopper disks provide optimal counting statistics while maintaining a high duty cycle (50%) [66.40]. The detector may be rotated about the interaction region, typically through a range of 120° or so, allowing one to examine products scattered at a range of laboratory angles. In addition to time-of-flight detection, one of the beams may be gated on and off for background sub-

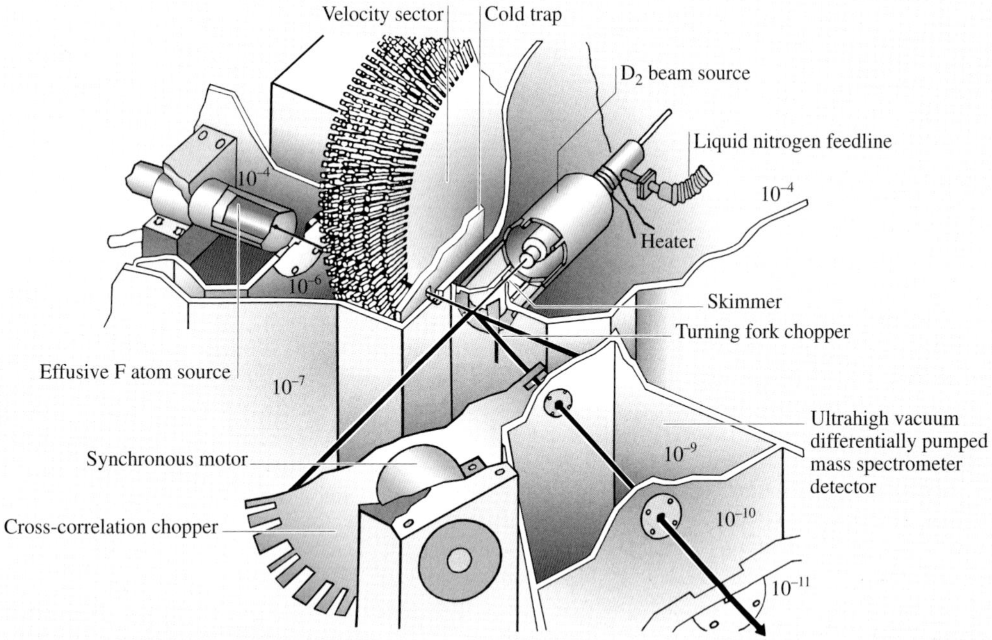

Fig. 66.1 Experimental arrangement for $F + D_2 \rightarrow DF + F$ reactive scattering. Pressure (in torr) indicated in each region. Components are (1) effusive F atom source; (2) velocity selector; (3) cold trap; (4) D_2 beam source; (5) heater; (6) liquid nitrogen feedline; (7) skimmer; (8) tuning fork chopper; (9) synchronous motor; (10) cross-correlation chopper; (11) ultrahigh vacuum differentially pumped mass spectrometer detector

traction and the detector moved to record the integrated signal at each laboratory angle.

Two kinds of measurements are typically made in these experiments: time-of-flight spectra and angular distributions. Usually one is interested in obtaining the complete product-flux vs. velocity contour map, since this contains full details of the scattering process. This is obtained by measuring a full angular distribution as well as time-of-flight data at many laboratory angles. The results are then simulated using a forward convolution fitting procedure to obtain the underlying contour map [66.41–43]. Because scattering of isotropic reagents exhibits cylindrical symmetry about the relative velocity vector, it is sufficient to measure products scattered in any plane containing this vector to determine the product distribution. This is not true for structured particles (e.g. involving atoms in P states); however, this azimuthal anisotropy has been used to explore the impact parameter dependence of the reaction dynamics [66.44].

In a typical reactive scattering experiment, $A + BC \rightarrow AB + C$, either of the two products may be detected. Conservation of linear momentum requires that the cm frame momenta of the two products must sum to zero. It is thus only necessary to obtain the contour map for one of the products. The choice of detected product is usually dictated by kinematic considerations, although one may choose to detect a product that is kinematically disfavored if its partner happens to have a mass with a large natural background in the detector. Kinematic considerations can be critical in assessing the suitability of a given system for study. It is very important that one of the products be scattered entirely within the viewing range of the detector in order to obtain a complete picture of the reaction dynamics.

The advantages of crossed-beams employed in conjunction with an electron impact ionizer-mass spectrometer detector derive primarily from the universality of the detector. No spectroscopic information is required and there are no invisible channels, such as may occur with spectroscopic detection methods. In addition, the resolution of these machines may be increased almost arbitrarily; indeed, even rotationally inelastic scattering has been studied [66.45]. The disadvantages are complementary to the advantages: the universal detector implies that quantum state resolution is not achieved directly, although in favorable cases the product vibrational states may be resolved in the translational energy distributions [66.21, 46]. In addition, if the product of interest represents a mass that receives interference from one of the beam masses, background interference may be problematic. Kinematic considerations mentioned above may also preclude study of certain systems. However, the kinematic requirements for the Doppler and imaging approach discussed below are complementary to those of the rotatable-detector configuration.

66.2.2 Doppler Techniques

Spectroscopic detection methods in crossed-beam experiments allows the measurement of state-resolved differential cross sections, and thus the ultimate level of insight into the reaction dynamics. A method developed by *Kinsey* and others [66.47, 48] determines differential cross sections by measurement of product Doppler profiles using LIF (called ADDS for Angular Distribution by Doppler Spectroscopy). For a laser directed parallel to the relative velocity vector, a particle scattered with a cm velocity of $\boldsymbol{u}$ perceives the photon as having the Doppler shifted frequency

$$\nu' = \nu\left[1 - (\boldsymbol{u} + \boldsymbol{V}_{\mathrm{cm}}) \cdot \hat{\boldsymbol{n}}/c\right] , \qquad (66.17)$$

where ν is the laser frequency and $\boldsymbol{V}_{\mathrm{cm}}$ is the velocity of the center of mass, both in the laboratory frame of refernce, and $\hat{\boldsymbol{n}}$ is the unit vector in the probe laser direction. For the case of a single possible recoil speed, one may obtain the full differential cross section directly in the cm frame by reconstruction of a single Doppler profile [66.47]. In this case, the angular resolution is a maximum for the sideways scattered products, and a minimum at the poles. An alternative approach is to measure the Doppler profile with a laser perpendicular to the relative velocity vector. This approach (PADDS for Perpendicular ADDS) affords complementary angular resolution, but folds the forward and backward scattered products into a single symmetric component [66.48]. For the case in which the detected product does not possess a known recoil speed (for example if the thermodynamics of the process is not known, or if one probes the atomic fragment in an $A + BC \to AB + C$ reaction), a single Doppler profile is insufficient to reconstruct the double differential cross sections. Nevertheless, Kinsey's earliest experimental results were for one such example: the reaction $H + NO_2 \to OH + NO$ [66.49].

More recently *Mestdagh* et al. [66.50] have studied electronically inelastic collision processes using this approach by measuring the Doppler profiles over a range of probe laser angles, as illustrated in Fig. 66.2. A beam of barium atoms is crossed at 90 degrees by a beam of some molecular perturber. At the interaction region, the barium atoms are electronically excited using a narrow band dye laser. Scattered barium atoms that have undergone a specific electronic transition as a result of the collision are probed at the interaction region using a second dye laser, which is scanned across the Doppler profile. The product-flux vs. velocity contour map is then reconstructed by means of a forward convolution simulation procedure analogous to that described in the preceding section.

In addition to state-resolved detection, another difference between the Doppler methods and the traditional crossed-beam configuration is that the kinematic considerations favor detection of fast particles, and almost any system that is spectroscopically suitable may be considered. The primary disadvantage of Doppler methods is the limited angular and translational energy resolution possible. Often, however, modest angular resolution is sufficient to achieve a global picture of the reaction dynamics. Much current work involves the study of photoinitiated reactions in cells, relying on the short excited state lifetimes to guarantee single collision conditions, and using iterative fitting procedures to probe product velocity distributions and angular momentum polarization [66.14]. Angular momentum polarization can have a profound effect on the measured distributions and can afford a powerful additional means of exploring the collision dynamics. Examples of 3- and 4-vector correlation experiments approach a "complete description" of the scattering process (Chapt. 46) [66.50–52].

66.2.3 Product Imaging

Another spectroscopic technique is based on direct imaging of the scattered product distribution. The technique was first used to record state-resolved angular distributions of methyl radicals from the photodissociation of methyl iodide [66.53]. The method has since been widely employed to study photodissociation, and more recently to record state-resolved inelastic scattering in a crossed-beam experiment [66.54]. Recently it has been applied to a crossed-beam reactive scattering system [66.55]. The crossed-beam configuration used by Houston and coworkers is shown schematically in Fig. 66.3. The two skimmed supersonic beams cross at right angles, and scattered products are state-selectively ionized on the axis of a time-of-flight mass spectrometer using resonant photoionization. The ion cloud thus formed continues to expand with its nascent recoil vel-

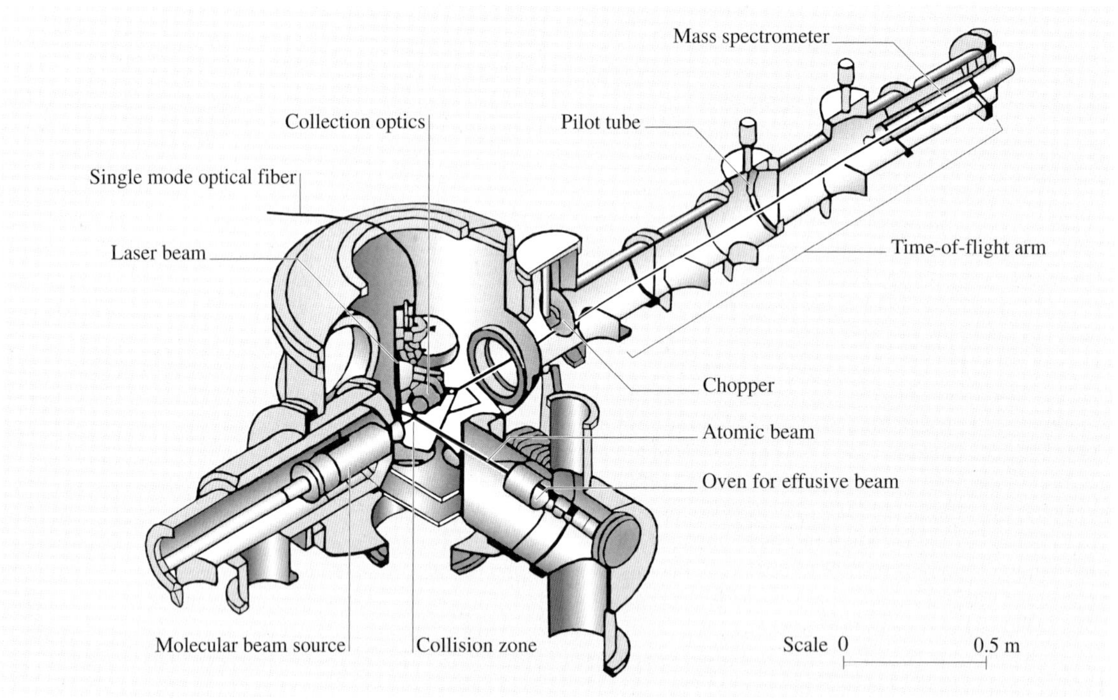

Fig. 66.2 Schematic view of crossed molecular beam apparatus with LIF-Doppler detection

ocity as it drifts through the flight tube. The ions then strike a microchannel plate coupled to a phosphor screen. The latter is viewed by a video camera gated to record the signal at the mass of interest. The images are thus two-dimensional projections of the nascent three dimensional product distributions.

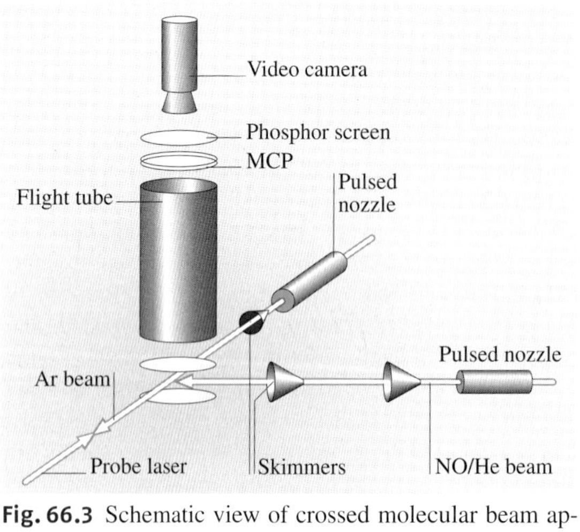

Fig. 66.3 Schematic view of crossed molecular beam apparatus with product imaging detection

There now exist two alternatives for regenerating the three dimensional distribution from the projection. The first, a tomographic reconstruction using an inverse Abel transform, is widely used in photodissciation studies [66.56, 57]. It is a direct inversion procedure feasible for cases in which the image is the projection of a cylindrically symmetric object, with its axis of symmetry parallel to the image plane. This analysis yields a unique product contour map directly from the image, but it is difficult to incorporate apparatus functions, and is sensitive to noise in the data. The second alternative is a forward convolution fitting method. A Monte Carlo based simulation has the advantage that one may treat the averaging over experimental parameters quite rigorously.

The advantages of the imaging method again derive from its reliance on a spectroscopic probe, so that quantum state resolution is possible and background interference may be avoided. In addition, it possesses a multiplexing advantage since the velocity distribution is recorded for all angles simultaneously. Imaging relies exclusively on photoionization, unlike the Doppler methods which may use either photoionization or LIF. This is somewhat disadvantageous since the available photoionization schemes are limited and often high laser power is necessary to achieve adequate signal intensity. As a result, background ions can be a problem. In general, resonantly enhanced two-photon ionization, i. e. $[1+1]$, detection schemes are thus preferable.

66.2.4 Laboratory to Center-of-Mass Transformation

Angular and velocity distributions measured in the laboratory frame must be transformed to the cm frame for theoretical interpretation. Accounts of this transformation and details concerning the material presented below may be found in [66.58–61], among others. The Newton diagram is useful to aid in visualizing the transformation, and in understanding the kinematics of a given collision system. For the scattering of $F+D_2$ for example, shown in Fig. 66.4, a beam of fluorine atoms with a velocity $\boldsymbol{v}_F$ is crossed by a beam of D_2, velocity $\boldsymbol{v}_{D_2}$, at 90 degrees. The relative velocity between the two reactants is $\boldsymbol{g}=\boldsymbol{v}_F-\boldsymbol{v}_{D_2}$, and the velocity of the cm of the entire system is

$$\boldsymbol{V}_{cm}=\frac{M_F\boldsymbol{v}_F+M_{D_2}\boldsymbol{v}_{D_2}}{M_F+M_{D_2}}\,. \tag{66.18}$$

$\boldsymbol{V}_{cm}$ divides the $\boldsymbol{g}$ into two segments corresponding to the cm velocities of the two reagents. The magnitude of these vectors, u_F and u_{D_2} are inversely proportional to the respective masses. If scattered DF products are formed with a laboratory scattering angle Θ and a laboratory velocity $\boldsymbol{v}_{DF}$ as shown in Fig. 66.4, this corresponds to DF *backscattered* with respect to the incident F atom, in the cm system. It is common to refer the scattering frame direction to the atomic reagent in an $A+BC\to AB+C$ reaction, for example, to make clear the dynamics of the process. In this case the backscattered DF arises as a result of a direct rebound collision. Some useful kinematic quantities are summarized here. For beams A and BC intersecting at 90 degrees, the angle of the cm velocity vector with respect to A is given by

$$\Theta_{cm}=\arctan\frac{M_{BC}v_{BC}}{M_Av_A}\,. \tag{66.19}$$

For an arbitrary Newton diagram with angle α between the two beams the magnitude of the relative velocity is

$$g^2=v_A^2+v_{BC}^2-2v_Av_{BC}\cos\alpha\,, \tag{66.20}$$

the relative velocity vector is

$$\boldsymbol{g}=\boldsymbol{v}_A-\boldsymbol{v}_{BC}\,, \tag{66.21}$$

and the collision energy is

$$E_{coll}=\frac{1}{2}\mu_i g^2\,, \tag{66.22}$$

where μ_i is the reduced mass of the initial collision system. The magnitude of the cm frame velocity of particle A before collision is

$$\boldsymbol{u}_A=\frac{m_{BC}}{m_A+m_{BC}}\boldsymbol{g}\,. \tag{66.23}$$

The final relative velocity is

$$\boldsymbol{g}'=\boldsymbol{v}_{AB}-\boldsymbol{v}_C\,, \tag{66.24}$$

with magnitude

$$g'=\sqrt{2E_{avail}/\mu_F}\,, \tag{66.25}$$

where the available energy E_{avail} is

$$E_{avail}=E_{coll}+E_{int,reac}+E_{exo}-E_{int,prod}\,, \tag{66.26}$$

in which $E_{int,reac}$ is the internal energy of the reactants, E_{exo} is the exoergicity of the reaction, and $E_{int,prod}$ is the internal energy of the products.

One must transform the laboratory intensity $I(\Omega)\equiv d^2\sigma/d^2\Omega$ into $I(\omega)\left(\equiv d^2\sigma/d^2\omega\right)$, the corresponding cm quantity. For the crossed-beam configuration described in Sect. 66.2.1, the laboratory distributions are distorted by a transformation Jacobian that arises because the laboratory detector views different cm frame solid angles depending on the scattering angle and recoil velocity. For the spectroscopic experiments described in Sects. 66.2.2 and 66.2.3, the Jacobian is unity (the cm velocity represents a simple frequency offset of the Doppler profiles, for example); however, the transformation of the scattering distributions from the recorded quantities (2-dimensional projections or intensity vs. wavelength) to recoil velocity distributions may be complex. Two cases must be considered for the

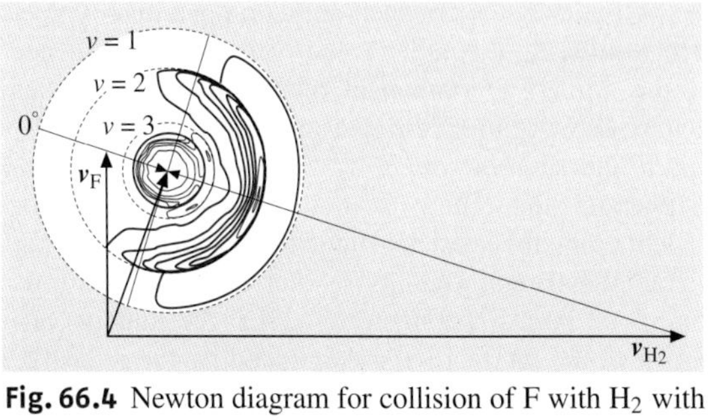

Fig. 66.4 Newton diagram for collision of F with H_2 with superimposed c.m. flux vs. velocity contour map

configuration discussed in Sect. 66.2.1: one in which discrete velocities result (such as elastic or state-resolved scattering experiments), and one in which continuous final velocities are measured. For the first case, the laboratory and cm differential cross sections are independent of the respective product velocities v and u and these quantities are related by

$$\frac{\mathrm{d}^2\sigma}{\mathrm{d}^2\Omega} = J\frac{\mathrm{d}^2\sigma}{\mathrm{d}^2\omega}\,, \tag{66.27}$$

so that the transformation Jacobian is given by

$$J = \frac{\mathrm{d}^2\omega}{\mathrm{d}^2\Omega}\,. \tag{66.28}$$

For discrete recoil velocities, the cm solid angle is

$$\mathrm{d}^2\omega = \frac{\mathrm{d}A}{u^2}\,, \tag{66.29}$$

where $\mathrm{d}A$ is a surface element of the product Newton sphere. The laboratory solid angle corresponding to this quantity is

$$\mathrm{d}^2\Omega = \frac{\cos(\boldsymbol{u},\boldsymbol{v})}{v^2}\,\mathrm{d}A\,, \tag{66.30}$$

so that the Jacobian for the first case is given by

$$J = \frac{v^2}{u^2\cos(\boldsymbol{u},\boldsymbol{v})}\,. \tag{66.31}$$

For the case of continuous final velocities, the σ are velocity-dependent and are related by

$$\frac{\mathrm{d}^3\sigma}{\mathrm{d}^2\Omega\;\mathrm{d}v} = J\frac{\mathrm{d}^3\sigma}{\mathrm{d}^2\omega\;\mathrm{d}u}\,, \tag{66.32}$$

so that here the Jacobian is given by

$$J = \frac{\mathrm{d}^2\omega\;\mathrm{d}u}{\mathrm{d}^2\Omega\;\mathrm{d}v}\,. \tag{66.33}$$

In this case we consider a recoil volume element $\mathrm{d}\tau$ (in velocity space), which must be the same in both coordinate frames:

$$\mathrm{d}\tau_{\mathrm{cm}} = u^2\;\mathrm{d}u\;\mathrm{d}^2\omega = \mathrm{d}\tau_{\mathrm{lab}} = v^2\;\mathrm{d}v\;\mathrm{d}^2\Omega\,, \tag{66.34}$$

so that the Jacobian is

$$J = v^2/u^2\,. \tag{66.35}$$

The laboratory intensity is then related to that in the cm frame by

$$I_{\mathrm{lab}}(v,\Theta) = (v^2/u^2)\,I_{\mathrm{cm}}(\theta,u)\,. \tag{66.36}$$

For a mass spectrometer detector with electron bombardment ionizer, one measures number density of particles rather than flux, so that the recorded signal is given by

$$N_{\mathrm{lab}}(v,\Theta) = \frac{I_{\mathrm{lab}}(v,\Theta)}{v} = \frac{v}{u^2}\,I_{\mathrm{cm}}(u,\theta)\,. \tag{66.37}$$

The usual flux vs. velocity contour map is a polar plot of the quantity $I_{\mathrm{cm}}(u,\theta)$. The product velocity distributions are then

$$\begin{aligned} I(u) &= \iint I(\theta,u)\,\sin\theta\;\mathrm{d}\theta\;\mathrm{d}\phi \\ &= 2\pi\int_0^{\pi} I(u,\theta)\sin\theta\;\mathrm{d}\theta\,, \end{aligned} \tag{66.38}$$

and the translational energy distributions are

$$I(E_{\mathrm{T}}) = I(u)\left|\frac{\mathrm{d}u}{\mathrm{d}E_{\mathrm{T}}}\right|\,. \tag{66.39}$$

66.3 Elastic and Inelastic Scattering

When particles collide, they may exchange energy or recouple it into different modes, they may change their direction of motion, and they may even change their identity. The study of these processes reveals a great deal of information about the forces acting between the particles and their internal structure. It is useful to begin with a summary of the dominant features of elastic and inelastic scattering.

66.3.1 The Differential Cross Section

Figure 66.5 illustrates the relation between the deflection function χ and the impact parameter b for a realistic potential containing an attractive well and a repulsive core. For large b there is no interaction, hence no deflection. At smaller values of b, the attractive part of the potential is experienced and some positive deflection results. At a smaller value of b, b_{r}, the influence of the attractive component of the potential reaches a maximum, giving the greatest positive deflection: this is the *rainbow angle* by analogy with the optical phenomenon. There is another value of the b for which point the attractive and repulsive parts of the potential balance, yielding no net deflection. This is the *glory* impact parameter b_{g}. For yet smaller values of b, the interaction is dominated by the repulsive core and rebound scattering gives a negative deflection function.

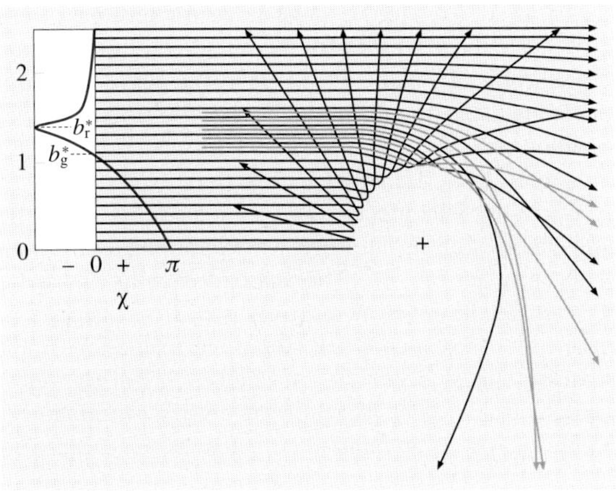

Fig. 66.5 Schematic diagram showing the relation between impact parameter b and deflection function χ

The important expressions related to the differential cross section are summarized here [66.62]. For scattering involving an isotropic potential, the deflection angle is $\Theta = |\chi|$. The differential cross section $d\sigma$ gives the rate of all collisions leading to deflection angles in the solid angle element $d\omega$:

$$\frac{dN(\theta)}{dt} \propto I(\theta)\,d\omega = I(\theta)2\pi\sin(\theta)\,d\theta\,. \quad (66.40)$$

The incremental cross section is $d\sigma = I(\theta)\,d\omega = 2\pi b\,db$, so

$$I(\theta) = \frac{b}{\sin\theta\,(d\theta\,/\,db)}\,. \quad (66.41)$$

For classical particles, the relation between the deflection function and the potential is

$$\chi = \pi - 2b\int_{R_0}^{\infty}\frac{dR}{R^2}\left(1 - \frac{V(R)}{E_T} - \frac{b^2}{R^2}\right)^{-1/2}, \quad (66.42)$$

where $V(R)$ is the potential as a function of interparticle distance R, R_0 is the turning point of the collision, and E_T the collision energy.

In the high energy limit, for large $b \approx R_0$,

$$\chi(b, E_T) \propto V(b)/E_T\,. \quad (66.43)$$

For a long-range potential $V(R)$ proportional to R^{-s},

$$E_T^{2/s}\theta^{2(1+1/s)}I(\theta) = \text{const}\,. \quad (66.44)$$

For a potential exhibiting a minimum, the rainbow angle θ_r is proportional to the collision energy, and clearly resolved when the collision energy is 3 to 5 times the well depth. In addition, supernumery rainbows and quantum mechanical "fast osillations" occur in the $d\sigma$, and these provide a sensitive probe of the interaction. Accurate interatomic potentials are routinely obtained from elastic scattering experiments [66.60, 63].

66.3.2 Rotationally Inelastic Scattering

Classical scattering involving an anisotropic potential results in another rainbow phenomenon, distinct from that seen in pure elastic scattering and notable in that it does not require an attractive component in the potential. These rotational rainbows are equivalently seen in a plot of integral cross section against change in rotational angular momentum Δj, or in the differential cross section for a particular value of Δj. The rotational rainbow peaks arise from the range of possible orientation angles γ in a collision involving an anisotropic potential. When there is a minimum in $d\gamma/\,d\theta$ for a given Δj, the differential cross section reaches a maximum [66.64]. The rotational rainbow peak occurs at the most forward classically allowed value of the scattering angle, and $d\sigma$ drops rapidly at smaller angles. The rainbow moves to more backward angles with increasing Δj because the larger j-changing collisions require greater momentum transfer, hence must arise from lower impact parameter collisions. For heteronuclear molecules, two rainbow peaks may be observed, corresponding to scattering off either side of the molecule. One can relate the location of the rainbow peak to the shape of the potential using a classical hard ellipsoid model [66.65]:

$$A - B = \frac{j}{p_0}\left[2\sin\left(\frac{\theta_{r,cl}}{2}\right)\right]^{-1}, \quad (66.45)$$

where j is the rotational angular momentum, p_0 is the inital linear momentum, $\theta_{r,cl}$ is the classical rainbow position, and A and B are the semimajor and semiminor axes of a hard ellipse potential. The classical rainbow positions occur somewhat behind the quantum mechanical and experimental rainbow positions, so the classical rainbow may be estimated as the point at which the peak has fallen to 44% of the experimental value. Real molecular potentials may be far from ellipsoids, however, so detailed quantitative insight into the potential requires a comparison of scattering data with trajectory calculations.

66.3.3 Vibrationally Inelastic Scattering

There has been no direct observation of the differential cross section of T–V or V–T energy transfer involving neutral molecules owing to the small cross sections for these processes. Integral cross section data are available, however. Above threshold, the latter has shown a linear dependence of σ on collision energy for $\Delta v = 1$, quadratic for $\Delta v = 2$ and cubic for $\Delta v = 3$ [66.66]. In addition, a great deal of information on vibrational relaxation processes has been obtained in cell experiments [66.67].

66.3.4 Electronically Inelastic Scattering

A wealth of information is available on electronically inelastic scattering systems, since these in general exhibit much larger cross sections than V–T processes [66.68, 69]. In addition, spectroscopic methods may be used to overcome some of the background problems that hamper the study of the latter. Often, quenching of electronically excited states involves curve crossing mechanisms, so that very effective coupling of electronic to vibrational energy may occur. Spin-orbit changing collisions of $Ba(^1P)$ with O_2 or NO, for example, occur by a near-resonant process and result in almost complete conversion of electronic energy to vibrational excitation of the product [66.70]. The analogous collisions with N_2 and H_2, however, reveal very repulsive energy release with little concomitant vibrational excitation. Both processes likely occur via curve crossings of the relevant electronic states, but the near-resonant mechanism occurs by way of an ionic intermediate.

66.4 Reactive Scattering

Reactive differential cross sections reveal several distinct aspects of the chemical encounter. The angular distributions themselves may be used to infer the lifetime of the collision intermediate: long-lived complexes exhibit forward-backward symmetry along the relative velocity vector. In this case "long-lived" means on the order of several rotational periods. The rotational period of the complex may thus be used as a clock to study the energy dependence of the intermediate's lifetime. The angular distributions further reveal the relation of initial and final orbital angular momentum. Sharply peaked angular distributions generally indicate strongly correlated initial and final orbital angular momentum vectors. Finally, the product translational energy release contains the details of the energy disposal, and reveals a wealth of information about the thermodynamics of the process, the existence of barriers, and sometimes even the geometry of the transition state. Together, the angular and tranlational energy distributions reveal many of the details of the potential energy surface.

The dynamics of reactive collisions fall broadly into three main categories characterized by distinct angular and energy distributions. The three categories are harpoon/stripping reactions, rebound reactions, and long-lived complex formation. Some reactions may exhibit more than one of these mechanisms at once, or the dynamics may change from one to another as the collision energy is varied.

66.4.1 Harpoon and Stripping Reactions

It was known in the 1930s that collisions of alkali atoms with halogen molecules exhibit very large cross sections and yield highly excited alkali halide products. These observations were accounted for by the *harpoon* mechanism proposed by M. Polanyi. Because alkali atoms have low ionization potentials and halogen molecules large electron affinities, as the alkali atom approaches the molecule, electron transfer may occur at long range. These processes are considered in detail in Chapt. 49 [66.71, 72]. The harpooning distance R_c at which this curve crossing takes place may be estimated simply as the distance at which the Coulomb attraction of the ion pair is sufficient to compensate for the endoergicity of the electron transfer:

$$R_c = e^2/(I_P - A_e)\,, \tag{66.46}$$

where I_P and A_e represent the ionization potential and electron affinity of the electron donor and recipient, respectively. For R in Å and E in eV, this relation is

$$R_c = 14.4/(I_P - A_e)\,. \tag{66.47}$$

Owing to the large Coulombic attraction between the ion pair, reaction proceeds immediately following electron transfer. The crossing distance may then be used to estimate the effective reaction cross section. The vertical A_e is not necessarily the appropriate value to use in estimating these crossing distances; stretching of the halogen bond may occur during approach, so the effective A_e is generally somewhere between the vertical and adiabatic values. Often there exists some repulsion between the atoms in the resulting halogen molecular ion, so that electron transfer is accompanied by dissociation of the molecule in the strong field of the ion pair. The alkali ion, having sent out the electron as the "harpoon", then reels in the negative ion, leaving the neutral halogen atom nearly undisturbed as a spectator. Because these events occur at long range, there is no momentum transfer to the spectator atom, and it is a simple matter to estimate the anticipated angular and translational energy distributions in this *spectator stripping* limit. The product molecule is scattered forward (relative to the direction of the incident atom) and for the reaction $A+BC \rightarrow AB+C$, the final cm velocity for the product AB is given by

$$\boldsymbol{u}'_{AB} = -M_C \boldsymbol{u}_{BC}/M_{AB} , \qquad (66.48)$$

where $\boldsymbol{u}_{BC}$ is the initial cm velocity of the BC molecule. This spectator stripping mechanism may occur in systems other than harpoon reactions, and is useful to remember as a limiting case.

The likelihood of electron transfer at these crossings may be estimated using a simple Landau–Zener model [66.72] (Chapt. 49). For relative velocity g, impact parameter b and crossing distance R_c, the probability for undergoing a transition from one adiabatic curve to another (that is, the probability for remaining on the diabatic curve) is given by

$$p = 1 - e^{-\delta} , \qquad (66.49)$$

where

$$\delta = \frac{2\pi H_{12}^2 R_c^2}{g} \left(1 - \frac{b^2}{R_c^2}\right)^{-\frac{1}{2}} , \qquad (66.50)$$

and H_{12} is the coupling matrix element between the two curves. H_{12} may be estimated from an empirical relation which is accurate within a factor of three over a range of 10 orders of magnitude [66.73]. In atomic units

$$H_{12} = \sqrt{I_1 I_2} R_c^* e^{-0.86 R_c^*} , \qquad (66.51)$$

where

$$R_c^* = \left(\sqrt{I_1} + \sqrt{I_2}\right) \qquad (66.52)$$

is the reduced crossing distance, and I_1 and I_2 are the initial and final ionization potentials of the transferred electron. One finds electron transfer probabilities near unity for curve crossing distances below about 5 Å, dropping to zero for crossing at distances greater than about 8 Å. These estimates are based on electron transfer in atom–atom collisions, and it is important to remember that atom–molecule collisions occur on surfaces rather than curves, so the crossing seam may cover a broad range of internuclear distances.

66.4.2 Rebound Reactions

Another common direct reaction mechanism is the *rebound* reaction exemplified by $F+D_2 \rightarrow DF+D$ [66.21]. The cm product flux vs. velocity contour map obtained for this reaction is shown in Fig. 66.4. Owing to the favorable kinematics and energetics in this case, the FD product vibrational distribution is clearly resolved, and peaks at $v = 2$. The dominant $v = 2$ product peaks at a cm angle of 180 degrees (referred to the direction of the incident F atom). This rebound scattering is characteristic of reactions exhibiting a barrier in the entrance channel. Rebound scattering implies small b collisions, and this serves to couple the translational energy efficiently into overcoming the barrier. Small b collisions have necessarily smaller cross sections however, since cross section scales quadratically with b.

66.4.3 Long-lived Complexes

A third important reaction mechanism involves the formation of an intermediate that persists for some time before dissociating to give products. If the collision complex survives for many rotational periods $(\sim 10^{-11}\,s)$, then the cm angular distribution exhibits a characteristic forward–backward symmetry, usually with peaking along the poles. The latter occurs because the initial and final orbital angular momenta tend to be parallel (and perpendicular to the initial relative velocity vector). When there exist dynamical constraints enforcing some other relation, as in the case $F+C_2H_4$, then sideways scattering may be observed, despite a lifetime of several rotational periods [66.74–76]. For some systems exhibiting this long-lived behavior, the rotational period may be used as a *molecular clock* to monitor the lifetime of the complex. By increasing the collision energy until the distribution begins to lose its forward–backward symmetry, one can investigate the internal energy of the system just when its lifetime is on the order of a rotational period.

Systems that have an inherent symmetry may exhibit this forward–backward symmetry in the scattering distributions despite lifetimes that are considerably shorter than a rotational period. This is the case for $O(^1D)$ reacting with H_2, for example [66.77]. This reaction involves insertion of the O atom into the H_2 bond resulting in an intermediate that accesses the deep H_2O well and contains considerable vibrational excitation. Trajectory calculations show that the complex dissociates after a few vibrational periods, but the distribution exhibits forward–backward symmetry because the O atom is equally likely to depart with either H atom.

66.5 Recent Developments

Astonishing progress in reactive scattering methods has continued in the past decade, and a few highlights are summarized here. These advances have taken the form of improvements in detection methods or, in some cases, entirely new experimental approaches.

One of the most important of these is the H atom Rydberg time-of-flight (HRTOF) method [66.78, 79] pioneered by the late Karl Welge and coworkers for the hydrogen exchange reaction. This approach employs a conventional scattering geometry, and is suitable only for experiments yielding product H or D atoms. Despite this narrow focus, owing to the general importance of hydrogen elimination reactions and the remarkable resolution of the technique, this has been an important development. The H or D atom products are excited to long-lived high-n Rydberg states in a $1+1'$ excitation scheme in the interaction region. The atoms fly through a field free region and impinge upon a rotatable field-ionization detector. The result is very high velocity resolution, largely because the spreads in the beam velocities make a negligible contribution to the product velocity spread since the H atoms are moving so fast. In addition, the dimensions of the scattering volume and ionization region may easily be made small relative to the flight length. Using this technique, Welge and coworkers achieved fully rotationally-resolved differential cross sections for the hydrogen exchange reaction.

A second, widely-used approach is a variation of the state-resolved Doppler probe in a bulb configuration. This strategy, pioneered by *Hall* at Brookhaven [66.80] and *Brouard* et al. in Oxford [66.81], has been applied most notably to study excited oxygen atom reactions. Although not a true crossed-beam approach, through appropriate exploitation of kinematic constraints, energy conservation, and careful analysis, state-resolved doubly differential cross sections may be obtained, sometimes with additional vector properties as well [66.82].

Another significant new direction in detection strategies is the use of near-threshold VUV product ionization. This is a universal approach, in that little advance spectroscopic information is required, but it is selective in that dissociative ionization is minimized and sometimes isomer-selective detection may be achieved. This approach has been used in synchrotron-based studies of Cl atom reactions [66.83], in transition metal reactions [66.84], and in product imaging studies of oxygen and chlorine atom reactions using the F_2 excimer at 157 nm [66.85]. Inspired by the threshold VUV detection methods, *Casavecchia* and coworkers have recently advanced the use of near-threshold electron impact ionization in a conventional universal crossed-beam configuration [66.86]. Their recent results show the great promise of this technique to deliver higher signal-to-noise and to minimize fragmentation processes in the detection step that may obscure the underlying dynamics.

A final note concerns advances in imaging techniques applied to reactive scattering. Mention has been made of the successful application of the VUV excimer probe for imaging radical products of reaction of Cl and $O(^3P)$ reactions with alkanes. Two significant advances in imaging strategies have made it a very powerful technique. The first of these is "velocity mapping", developed by *Eppink* and *Parker* at Nijmegen [66.87], a simple but important strategy that eliminates spatial blurring in the images. The second is "slicing", or 3-D, methods that allow the velocity-flux contour map to be recorded directly [66.88–90]. This has seen its most beautiful illustration in recent work by *Liu* and coworkers at IAMS in state-resolved detection of methyl radicals following reaction of F atoms with methane [66.91]. Their images provide quantum state correlated differential cross sections for this reaction directly, allowing comparison to theory at an unprecedented level of detail.

References

66.1 Y. T. Lee: Science **236**, 793 (1987)

66.2 W. R. Gentry: *Atomic and Molecular Beam Methods*, ed. by G. Scoles (Oxford Univ. Press, Oxford 1988) Chap. 3, p. 54

66.3 J. B. Anderson, R. P. Andres, J. B. Fenn: Adv. Chem. Phys. **10**, 275 (1966)

66.4 D. R. Miller: *Atomic and Molecular Beam Methods*, ed. by G. Scoles (Oxford Univ. Press, Oxford 1988) Chap. 2, p. 14

66.5 R. Campargue: J. Phys. Chem. **88**, 275 (1984)

66.6 H. F. Davis, B. Kim, H. S. Johnston, Y. T. Lee: J. Phys. Chem. **97**, 2172 (1993)

66.7 P. Chen: *Unimolecular and Bimolecular Reaction Dynamics*, ed. by C.Y. Ng, T. Baer, I. Powis (John Wiley and Sons, New York 1994) Chap. 8, p. 371
66.8 R. E. Smalley: Laser Chem. **2**, 167 (1983)
66.9 M. D. Morse, J. B. Hopkins, P. R. R. Langridge-Smith, R. E. Smalley: J. Chem. Phys. **79**, 5316 (1983)
66.10 I. V. Hertel, W. Stoll: Adv. At. Mol. Phys. **13**, 113 (1978)
66.11 J. G. Pruett, R. N. Zare: J. Chem. Phys. **64**, 1774 (1976)
66.12 K. Bergmann, U. Hefter, J. Witt: J. Chem. Phys. **72**, 4777 (1980)
66.13 K. Bergmann: *Atomic and Molecular Beam Methods*, ed. by G. Scoles (Oxford Univ. Press, Oxford 1988) Chap. 12
66.14 M. Brouard, S. Duxon, P. A. Enriquez, J. P. Simons: J. Chem. Soc. Faraday Trans. **89**, 1435 (1991)
66.15 S. J. Sibener, R. J. Buss, P. Cassavechia, T. Hirooka, Y. T. Lee: J. Chem. Phys. **72**, 4341 (1980)
66.16 R. E. Continetti, B. A. Balko, Y. T. Lee: J. Chem. Phys. **93**, 5719 (1990)
66.17 P. R. Brooks: Science **193**, 11 (1976)
66.18 S. Stolte: Ber. Bunsenges. Phys. Chem. **86**, 413 (1982)
66.19 B. Friedrich, D. R. Herschbach: Nature **353**, 412 (1991)
66.20 D. Bassi, A. Boschetti, M. Scotoni, M. Zen: Appl. Phys. B **26**, 99 (1981)
66.21 D. M. Neumark, A. M. Wodtke, G. N. Robinson, C. C. Hayden, K. Shobatake, R. K. Sparks, T. P. Schaefer, Y. T. Lee: J. Chem. Phys. **82**, 3067 (1985)
66.22 M. Faubel: Adv. At. Mol. Phys. **19**, 345 (1983)
66.23 T. R. Touw, J. W. Trischka: J. Appl. Phys. **34**, 3635 (1963)
66.24 Y. T. Lee, J. D. McDonald, P. R. LeBreton, D. Herschbach: Rev. Sci. Intstrum. **40**, 1402 (1969)
66.25 Y. T. Lee: *Atomic and Molecular Beam Methods*, ed. by G. Scoles (Oxford Univ. Press, Oxford 1988) Chap. 22, p. 563
66.26 R. E. Center, A. Mandl: J. Chem. Phys. **57**, 4104 (1972)
66.27 C. H. Greene, R. N. Zare: J. Chem. Phys. **78**, 6741 (1983)
66.28 D. A. Case, G. M. McLelland, D. R. Herschbach: Mol. Phys. **35**, 541 (1978)
66.29 G. Hall, P. Houston: Ann. Rev. Phys. Chem. **40**, 375 (1989)
66.30 D. M. Sonnenfroh, K. Liu: Chem. Phys. Lett. **176**, 183 (1991)
66.31 N. Billy, B. Girard, G. Gouédard, J. Vigué: Mol. Phys. **61**, 65 (1987)
66.32 R. N. Zare, P. J. Dagdigian: Science **185**, 739 (1974)
66.33 R. Altkorn, R. N. Zare: Ann. Rev. Phys. Chem. **35**, 265 (1984)
66.34 H. Joswig, P. Andresen, R. Schinke: J. Chem. Phys. **85**, 1904 (1986)
66.35 J. M. Mestdagh, J. P. Visticot, A. G. Suits: *The Chemical Dynamics and Kinetics of Small Radicals*, ed. by K. Liu, W. Wagner (World Scientific, Singapore 1994)
66.36 S. D. Jons, J. E. Shirley, M. T. Vonk, C. F. Giese, W. R. Gentry: J. Chem. Phys. **92**, 7831 (1992)
66.37 D. L. Feldman, R. K. Lengel, R. N. Zare: Chem. Phys. Lett. **52**, 413 (1977)
66.38 E. E. Marinero, C. T. Rettner, R. N. Zare: J. Chem. Phys. **80**, 4142 (1984)
66.39 L. Schneider, K. Seekamp-Rahn, F. Liedeker, H. Stewe, K. H. Welge: Farad. Discuss. Chem. Soc. **91**, 259 (1991)
66.40 G. Comsa, R. David, B. J. Schumacher: Rev. Sci. Instrum. **52**, 789 (1981)
66.41 E. Entemann, D. R. Herschbach: J. Chem. Phys. **55**, 4872 (1971)
66.42 R. Buss: *Ph. D. thesis* (University of California, Berkeley 1972)
66.43 R. T. Pack: J. Chem. Phys. **81**, 1841 (1984)
66.44 A. G. Suits, H. Hou, H. F. Davis, Y. T. Lee: J. Chem. Phys. **95**, 8207 (1991)
66.45 M. Faubel, K.-H. Kohl, J. P. Toennies, K. T. Tang, Y. Y. Yung: Faraday Discuss. Chem. Soc. **73**, 205 (1982)
66.46 R. E. Continetti, B. A. Balko, Y. T. Lee: J. Chem. Phys. **93**, 5719 (1990)
66.47 J. L. Kinsey: J. Chem. Phys. **66**, 2560 (1977)
66.48 J. A. Serri, J. L. Kinsey, D. E. Pritchard: J. Chem. Phys. **75**, 663 (1981)
66.49 E. L. Murphy, J. H. Brophy, G. S. Arnol, W. L. Dimpfl, J. L. Kinsey: J. Chem. Phys. **70**, 5910 (1979)
66.50 J. M. Mestdagh, J. P. Visticot, P. Meynadier, O. Sublemontier, A. G. Suits: J. Chem. Soc. Faraday Trans. **89**, 1413 (1993)
66.51 C. J. Smith, E. M. Spain, M. J. Dalberth, S. R. Leone, J. P. J. Driessen: J. Chem. Soc. Faraday Trans. **89**, 1401 (1993)
66.52 H. A. J. Meijer, T. J. C. Pelgrim, H. G. M. Heideman, R. Morgenstern, N. Andersen: J. Chem. Phys. **90**, 738 (1989)
66.53 D. W. Chandler, P. L. Houston: J. Chem. Phys. **87**, 1445 (1987)
66.54 L. S. Bontuyan, A. G. Suits, P. L. Houston, B. J. Whitaker: J. Phys. Chem. **97**, 6342 (1993)
66.55 T. A. Kitsopoulos, D. P. Baldwin, M. A. Buntine, R. N. Zare, D. W. Chandler: Science **260**, 1605 (1993)
66.56 R. N. Strickland, D. W. Chandler: Appl. Opt. **30**, 1811 (1990)
66.57 K. R. Castleman: *Digital Image Processing* (Prentice Hall, Englewood Cliffs 1979)
66.58 T. T. Warnock, R. B. Bernstein: J. Chem. Phys. **49**, 1878 (1968)
66.59 R. K. B. Helbing: J. Chem. Phys. **48**, 472 (1968)
66.60 J. P. Toennies: *in Physical Chemistry, an Advanced Treatise*, Vol. VIA, ed. by W. Jost (Academic, New York 1974) Chap. 5
66.61 M. Faubel, J. P. Toennies: Adv. At. Mol. Phys. **13**, 229 (1978)
66.62 R. D. Levine, R. B. Bernstein: *Molecular Reaction Dynamics and Chemical Reactivity* (Oxford, New York 1987) Chap. 3

66.63 J. M. Farrar, T. P. Schafer, Y. T. Lee: *AIP Conference Proceedings No. 11, Transport Phenomena*, ed. by J. Kestin (American Institute of Physics, New York 1973)

66.64 R. Schinke, J. Bowman: *Molecular Collision Dynamics*, ed. by J. Bowman (Springer, Berlin, Heidelberg 1983)

66.65 S. Bosanac: Phys. Rev. A **22**, 2617 (1980)

66.66 G. Hall, K. Liu, M. J. McAuliffe, C. F. Giese, W. R. Gentry: J. Chem. Phys. **81**, 5577 (1984)

66.67 X. Yang, A. Wodtke: Int. Rev. Phys. Chem. **12**, 123 (1993)

66.68 I. V. Hertel: Adv. Chem. Phys. **50**, 475 (1982)

66.69 W. H. Breckenridge, H. Umemoto: Adv. Chem. Phys. **50**, 325 (1982)

66.70 A. G. Suits, P. de Pujo, O. Sublemontier, J. P. Visticot, J. Berlande, J. Cuvellier, T. Gustavsson, J. M. Mestdagh, P. Meynadier, Y. T. Lee: J. Chem. Phys. **97**, 4094 (1992)

66.71 J. Los, A. W. Kleyn: *Alkali Halide Vapors*, ed. by P. Davidovits, D. L. McFadden (Academic, New York 1979) Chap. 8

66.72 E. A. Gislason: *Alkali Halide Vapors*, ed. by P. Davidovits, D. L. McFadden (Academic, New York 1979) Chap. 13

66.73 R. E. Olson, F. T. Smith, E. Bauer: Appl. Optics **10**, 1848 (1971)

66.74 W. B. Miller, S. A. Safron, D. R. Herschbach: Discuss. Faraday Soc. **44**, 108 (1967)

66.75 W. B. Miller: . Ph.D. Thesis (Harvard Univ., Harvard 1969)

66.76 J. M. Farrar, Y. T. Lee: J. Chem. Phys. **65**, 1414 (1976)

66.77 R. J. Buss, P. Casavecchia, T. Hirooka, Y. T. Lee: Chem. Phys. Lett. **82**, 386 (1981)

66.78 L. Schnieder, K. Seekamp-Rahn, E. Wrede, K. H. Welge: J. Chem. Phys. **107**, 6175 (1997)

66.79 B. Strazisar, C. Lin, H. F. Davis: Science **290**, 958 (2000)

66.80 H. L. Kim, M. A. Wickramaaratchi, X. Zheng, G. E. Hall: J. Chem. Phys. **101**, 2033 (1994)

66.81 M. Brouard, S. P. Duxon, P. A. Enriquez, J. P. Simons: J. Chem. Phys. **97**, 7414 (1992)

66.82 T. P. Rakitzis, S. A. Kandel, R. N. Zare: J. Chem. Phys. **107**, 9382 (1997)

66.83 D. A. Blank, N. Hemmi, A. G. Suits, Y. T. Lee: Chem. Phys. **231**, 261 (1998)

66.84 P. A. Willis, H. U. Stauffer, P. Z. Hinrichs, H. F. Davis: J. Chem. Phys. **108**, 2665 (1998)

66.85 M. Ahmed, D. S. Peterka, A. G. Suits: Phys. Chem. Chem. Phys. **2**, 861 (2000)

66.86 G. Capozza, E. Segoloni, F. Lenori, G. G. Volpi, P. Casavecchia: J. Chem. Phys. **120**, 4557 (2004)

66.87 A. T. J. B. Eppink, D. H. Parker: Rev. Sci. Instrum. **68**, 3457 (1997)

66.88 T. N. Kitsopoulos, C. R. Gebhardt, T. P. Rakitzis: Rev. Sci. Instrum. **72**, 3848 (2001)

66.89 D. Townsend, M. P. Minitti, A. G. Suits: Rev. Sci. Instrum. **74**, 2530 (2003)

66.90 J. J. Lin, J. Zhou, W. Shiu, K. Liu: Rev. Sci. Instrum. **74**, 2495 (2003)

66.91 J. Zhou, J. J. Lin, K. Liu: J. Chem. Phys. **119**, 8289 (2003)

67. Ion–Molecule Reactions

The observation of ion–molecule reactions has a history that goes back to the beginning of the twentieth century, when J. J. Thomson discovered that operating his positive ray parabola apparatus in a hydrogen atmosphere produced signals at a mass to charge ratio of 3, which he correctly attributed to the species H_3 [67.1]. Later studies showed that this species was produced by a reaction between the primary ionization product H_2^+ and molecular hydrogen. Most ion-molecule reactions proceed without an activation barrier and their cross sections are governed by the long range attractive potential of the approaching reactants (Sect. 64.2.4). Reaction rates based on long range potential capture models [67.2] predict rates in excess of 10^{-9} cm^{3}molecule^{-1}s^{-1}, corresponding to thermal energy cross sections (Sect. 47.1.7) of 10^{-16}–10^{-15} cm^2. The importance of ion-molecule reactions in such widely diverse areas as planetary atmospheres, (Sect. 84.1), electrical discharges and plasmas (Sect. 87.1.4), particularly in semiconductor processing, in the formation of molecules in interstellar space (Chapt. 82), and in flames and combustion systems (Sect. 88.1), has borne out that prediction. This chapter discusses applications of single-collision scattering methods to the study of reactive collision dynamics of ionic species with neutral partners.

67.1 **Instrumentation** ... 985
67.2 **Kinematic Analysis** ... 985
67.3 **Scattering Cross Sections** ... 987
67.3.1 State-to-State Differential Cross Sections ... 987
67.3.2 Velocity-Angle Differential Cross Sections ... 988
67.3.3 Total Cross Sections with State-Selected Reactants ... 989
67.3.4 Product-State Resolved Total Cross Sections ... 989
67.3.5 State-to-State Total Cross Sections ... 990
67.3.6 Energy Dependent Total Cross Sections ... 990
67.4 **New Directions: Complexity and Imaging** ... 991
References ... 992

A number of different physical processes can be categorized as ion-molecule reactions, with examples such as

$A^{\pm} + BC \rightarrow A + BC^{\pm}\,,$	charge transfer
$A^{\pm} + BC \rightarrow A + (B + C)^{\pm}\,,$	dissociative charge transfer
$A^{\pm} + BC \rightarrow AB^{\pm} + C\,.$	particle transfer

The ± superscript indicates charge appropriate to anions and cations. The parentheses indicate that the charge can reside on either the B or C fragment. Particle transfer reactions often involve the transfer of a hydrogen atom or a proton, but heavy particle <transfer processes are often important ones as well. An interesting example in which new carbon-carbon bonds are formed, termed a condensation reaction, is the following:

$$C^+ + CH_4 \rightarrow C_2H_3^+ + H\,.$$

An additional process unique to anionic systems is detachment, occurring when the intermediate collision complex is internally excited above its autodetachment threshold:

$$A^- + BC \rightarrow [ABC^-]^* \rightarrow A + BC + e^-\,,$$

detachment

$$A^- + BC \rightarrow [ABC^-]^* \rightarrow ABC + e^-\,.$$

associative detachment

For exothermic reactions at low collision energies where the long range attraction dominates the interaction potential, cross sections are consistent with Langevin orbiting, generally having $E^{-1/2}$ energy dependence.

At higher energies, cross sections drop below this limit, as surface crossings and short-range repulsive features become important. Endothermic reactions exhibit cross section thresholds, as illustrated in Sect. 67.3.4 on the $N^+ + D_2$ system.

Scattering measurements probe the potential energy surface, or surfaces, governing the collision dynamics with techniques that measure the fluxes from specific reactant quantum states into product quantum states, scattering angles, and product translational energies. Scattering experiments define more precisely than in a bulb the initial and final conditions in a collision [67.3]. A scattering experiment measures a cross section (Sect. 91.1.1) rather than a rate constant. The measured cross section represents an average over initial conditions and a sum over final states inherent in the technique used.

Cross sections of various forms can be defined with respect to the rate of formation of product states under single-collision conditions. The total rate of product formation $\mathrm{d}N_\pm/\mathrm{d}t$ for a beam of ions with number density n_1 intersecting a gas of number density n_2 within a scattering volume ΔV, defined by the overlap of the ion beam with the target gas, is given by

$$\mathrm{d}N_\pm/\mathrm{d}t = \sigma v_{\mathrm{rel}} n_1 n_2 \Delta V \,, \tag{67.1}$$

where σ is the total reaction cross section (Sect. 47.1.7) and v_{rel} is the relative speed of the collision partners. Using the experimental ion beam current $I_\pm$, cross section σ, neutral number density n_2, and attenuation length L, this expression converts to the particularly useful form

$$\mathrm{d}N_\pm/\mathrm{d}t = 6.25 \times E - 7\sigma I_\pm n_2 L \tag{67.2}$$

for computing signal levels, where σ is expressed in Å^2, $I_\pm$ in nA, n_2 in cm^{-3}, and L in cm. An ion beam of current 1 nA, intersecting a target of length 1 cm at a pressure of 10^{-3} Torr, corresponding to a number density of 3.5×10^{13} cm^{-3} at STP, and reacting with a cross section of 1 Å^2 yields a total rate of product formation of 2×10^7 s^{-1}.

The detector observes only a fraction of this total rate. If reaction products are scattered isotropically over 4π steradians in the laboratory, and the detector entrance slit of area $\mathrm{d}S$ is located a distance r from the collision center, subtending a solid angle $\mathrm{d}S/r^2$, the fraction of the total signal scattered into the slit is $\mathrm{d}S/(4\pi r^2)$ [67.4]. The fraction of the collision volume, or of the product state distribution accessible to a particular experimental method, determines the tradeoffs between resolution and signal level, and therefore the feasibility of a given experiment.

In an "ideal" experiment, one collides reactants, with well specified quantum numbers collectively denoted $\boldsymbol{n}$, at a precisely defined relative velocity $\boldsymbol{v}_{\mathrm{rel}}$, resolving products in quantum states $\boldsymbol{n}'$ scattered through center-of-mass scattering angle θ. The resulting detailed differential cross section (DCS) (Sect. 47.1.1) is denoted by $\sigma(\boldsymbol{n}', \theta | \, n, \boldsymbol{v}_{\mathrm{rel}})$. However, most experiments involve at least partial averages over initial states and/or summations over final states. Figure 67.1 shows the result of averaging $\sigma(\boldsymbol{n}', \theta | \, n, \boldsymbol{v}_{\mathrm{rel}})$ over θ to yield the state-to-state cross section at fixed $\boldsymbol{v}_{\mathrm{rel}}$, denoted by $\sigma(\boldsymbol{n}' | \boldsymbol{n}, \boldsymbol{v}_{\mathrm{rel}})$. Averaging this cross section over a Maxwell–Boltzmann distribution of molecular speeds at a specified temperature T yields the detailed state-to-state rate constant $k(\boldsymbol{n}' | \boldsymbol{n}, T)$, while summation over the final states $\boldsymbol{n}'$ and averaging over the initial states $\boldsymbol{n}$ yields the thermal rate constant $k(T)$. These latter two quantities are thermally averaged, multiple collision properties and not the subject of this chapter, although they play an important role in practical applications. Another pathway for averaging the detailed DCS $\sigma(\boldsymbol{n}', \theta | \boldsymbol{n}, \boldsymbol{v}_{\mathrm{rel}})$ arises from averaging over $\boldsymbol{n}$ and summing over $\boldsymbol{n}'$ to yield cross sections differential in product velocity $\boldsymbol{v}'_{\mathrm{rel}}$ and θ at fixed $\boldsymbol{v}_{\mathrm{rel}}$, denoted by $\sigma(\boldsymbol{v}'_{\mathrm{rel}}, \theta | \boldsymbol{v}_{\mathrm{rel}})$. An average over θ and $\boldsymbol{v}'_{\mathrm{rel}}$ yields the velocity dependent total cross section $\sigma(\boldsymbol{v}_{\mathrm{rel}})$, and its Maxwell-Boltzmann average produces the thermal rate constant $k(T)$ once again. The subject of this chapter is a discussion of the various cross sections σ shown in Fig. 67.1. As one moves from

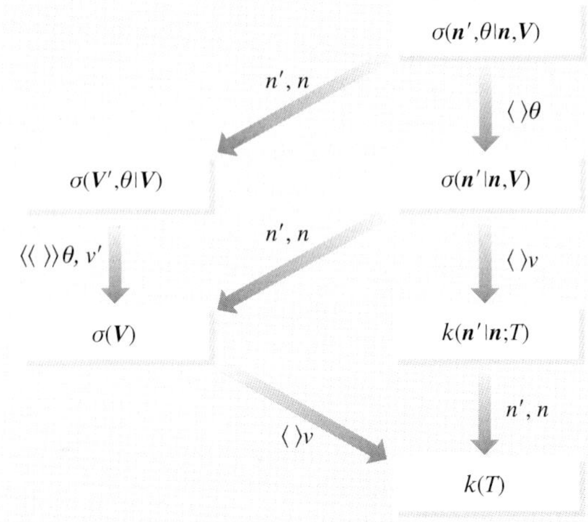

Fig. 67.1 Relationships among differential and total cross sections, and rate constants. *Brackets* denote averages over indicated variables, or averages over initial states and summations over final states

more highly averaged quantities to the detailed cross sections, more sophisticated reactant preparation and product detection schemes are required to extract the desired information, at the expense of decreased signal levels. Technological advances, particularly in laser preparation of quantum state-selected reactants and in state-specific product detection, have made the "ideal" experiment a near reality in favorable circumstances, particularly in neutral-neutral interactions, e.g., in experiments on H + H_2 and its isotopic variants. This chapter discusses the cross sections $\sigma(\boldsymbol{n}', \theta | n, \boldsymbol{v}_{\text{rel}})$, $\sigma(\boldsymbol{v}'_{\text{rel}}, \theta | \boldsymbol{v}_{\text{rel}})$, $\sigma(\boldsymbol{n}' | n, \boldsymbol{v}_{\text{rel}})$, and $\sigma(\boldsymbol{v}_{\text{rel}})$, emphasizing the dynamical information that can be extracted from each kind of measurement.

67.1 Instrumentation

Instrumentation for studying ion-molecule reactions is quite diverse, and numerous literature sources are available for further discussion [67.5]. A typical instrument has an ionization source, a primary mass selector, a collision region, and detector, consisting of a mass spectrometer or employing a spectroscopic technique allowing molecular identification.

67.2 Kinematic Analysis

The transformation of laboratory measured speeds, angles and intensities to their center of mass (cm) counterparts can be accomplished with appropriate geometric constructions [67.6, 7]. The geometric relationships can be understood by considering a kinematic Newton diagram for the collision process $A + BC$, as shown in Fig. 67.2. The diagram is constructed for the special case in which the reactant beams intersect at 90°. The laboratory scattering angle and velocity are Θ and v respectively, while the corresponding center of mass quantities are θ and u. The beam velocity vectors $\boldsymbol{v}_A$ and $\boldsymbol{v}_{BC}$ define the initial conditions, and the relative *velocity vector* $\boldsymbol{v}_{\text{rel}}$ is defined by their vector difference. The velocity $\boldsymbol{c}$ of the center of mass, or centroid, of the collision system is determined by conservation of linear momentum; the vector $\boldsymbol{c}$ divides the $\boldsymbol{v}_{\text{rel}}$ in inverse proportion to the masses of A and BC:

$$\boldsymbol{c} = \frac{m_A \boldsymbol{v}_A + m_{BC} \boldsymbol{v}_{BC}}{M} , \tag{67.3}$$

$$\boldsymbol{u}_A = -\frac{m_{BC}}{M} \boldsymbol{v}_{\text{rel}} , \tag{67.4}$$

$$\boldsymbol{u}_{BC} = \frac{m_A}{M} \boldsymbol{v}_{\text{rel}} , \tag{67.5}$$

where M is the total mass of the reactants with masses m_A and m_{BC}. An observer moving away from the laboratory origin in the direction of the center of mass vector $\boldsymbol{c}$ would see the reactants A and BC approach along the direction of $\boldsymbol{v}_{\text{rel}}$, with the products retreating along the direction of $\boldsymbol{v}'_{\text{rel}}$. The angle θ between these vectors defines the center of mass scattering angle. For a single Newton diagram, detection of products at recoil speed u and scattering angle θ requires that measurements be made at laboratory coordinates (v, Θ). For monoenergetic incident beams, the measurement of both laboratory scattering angle and speed results in a unique lab to cm coordinate transformation. Linear momentum conservation also relates the center of mass speeds of products AB and C to the relative velocity of the separating products according

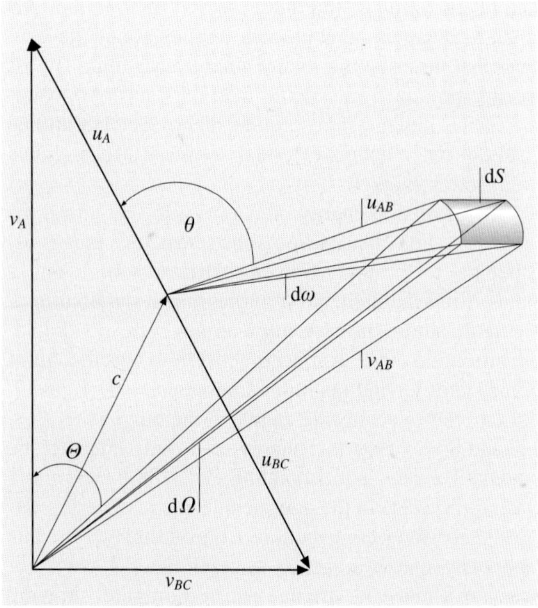

Fig. 67.2 Kinematic Newton diagram showing laboratory and center of mass velocities and scattering angles

to

$$u'_{AB} = \frac{m_C}{M} v'_{rel} , \quad (67.6)$$

$$u'_C = -\frac{m_{AB}}{M} v'_{rel} . \quad (67.7)$$

The final relative kinetic energy of the separating products is

$$\mathcal{T}'_{rel} = \frac{1}{2}\mu' v'_{rel} \cdot v'_{rel} , \quad (67.8)$$

where μ' is the reduced mass of the products.

The total energy of a collision system is computed from the energies of the incident reactants and the energy accessible to the products:

$$E_{tot} = \mathcal{T}_{rel} + E_{int} - \Delta D_0^o = \mathcal{T}'_{rel} + E'_{int} ; \quad (67.9)$$

$\mathcal{T}_{rel}$ and E_{int} refer to the incident kinetic and internal excitation energies of the reactants and the primed quantities correspond to the products. ΔD_0^o is the zero point energy difference of reactants and products. Specification of the vibrational and rotational energy of a product determines the relative velocity v'_{rel} with which the products separate, according to

$$v'_{rel} = \left[\left(\frac{2}{\mu'}\right)(\mathcal{T}_{rel} + E_{int} - \Delta D_0^o - E'_{int})\right]^{1/2} . \quad (67.10)$$

The quantization of internal energy of the product AB leads to a series of concentric circles about the centroid that describe the loci of final translational speeds for AB produced in specific internal states. An example of the kinematic resolution of product vibrational states in the reaction $O^- + HF \rightarrow F^- + OH$ is shown in Fig. 67.3 [67.8,9].

At a collision energy of 0.47 eV (45.0 kJ/mol), the laboratory flux distribution shows structure that is attributable to the formation of OH in $v' = 0$, 1, and 2 states. The kinematic Newton diagram at the bottom of the figure shows the concentric circles corresponding to the formation of F^- in concert with OH in specific vibrational states having quantum numbers $v' = 0$, 1, and 2. For laboratory scattering angles in the range $0° \leq \Theta \leq 11°$ and $80° \leq \Theta \leq 90°$ energy scans only intersect the kinematic circles corresponding to OH in $v' = 0$ and 1 states, while data in the intermediate range of angles $18° \leq \Theta \leq 70°$ show contributions from all three states. The laboratory speeds at which these vibrational states appear, for a given Θ, are marked in the figure, showing clear correspondence with the structure in the experimental data. Rotational excitation in the products will broaden the contributions from individual vibrational states in experiments lacking rotational resolution. The best resolution achieved in electrostatic energy analyzers is approximately 5 meV (40 cm^{-1}); and consequently, scattering measurements of product fluxes based on kinetic energy analysis generally do not have rotational state resolution. However, photoelectron spectroscopy measurements on H_2 [67.10] have yielded spectra of H_2^+ energy levels with rotational resolution.

The transformation of intensities and cross sections between laboratory and cm coordinate systems relies on conservation of flux in a transformation between two coordinate systems moving with respect to one another. Figure 67.2 shows the nature of this conservation: in laboratory coordinates, the flux into solid angle $d\Omega$ is given by $I_{lab}(\Omega)\,d\Omega$, where the laboratory intensity I_{lab} is the flux per unit solid angle, while the corresponding flux in the cm frame is $I_{cm}(\omega)\,d\omega$, where I_{cm} is the cm flux per unit solid angle. The solid angles subtended by the detector in the laboratory and cm frames expressed in velocity space may be computed from the surface element dS subtended by the detector:

$$d\omega = \frac{dS}{u^2} , \quad d\Omega = \frac{dS}{v^2} . \quad (67.11)$$

The flux equality $I_{lab}(\Omega)\,d\Omega = I_{cm}(\omega)\,d\omega$ leads to the intensity transformation

$$I_{lab}(v, \Theta) = \frac{v^2}{u^2} I_{cm}(u, \theta) . \quad (67.12)$$

The widths of the beam velocity distributions must be accounted for in extracting accurate cm cross sections from laboratory data. In the case where the laboratory flux at a given Θ is comprised of contributions from individual quantum states of the products, the laboratory flux is recovered from the relation

$$I_{lab}(v, \Theta) = v^2 \int_0^\infty dv_2 f_2(v_2) \int_0^\infty dv_1 f_1(v_1) \frac{v_{rel}}{u^2} \times \left[\sum_{n'} \sigma(n', \theta | n, v_{rel}) \delta(u - u_{n'})\right] . \quad (67.13)$$

In this expression, the velocity distributions for the primary ion beam and the secondary neutral beam are given by $f_1(v_1)$ and $f_2(v_2)$ respectively. The final product internal states are labeled by the index n'; the cm cross section for producing collective quantum state n' is given by $\sigma(n', \theta | n, v_{rel})$ and the cm speed corresponding to the formation of this state is given by

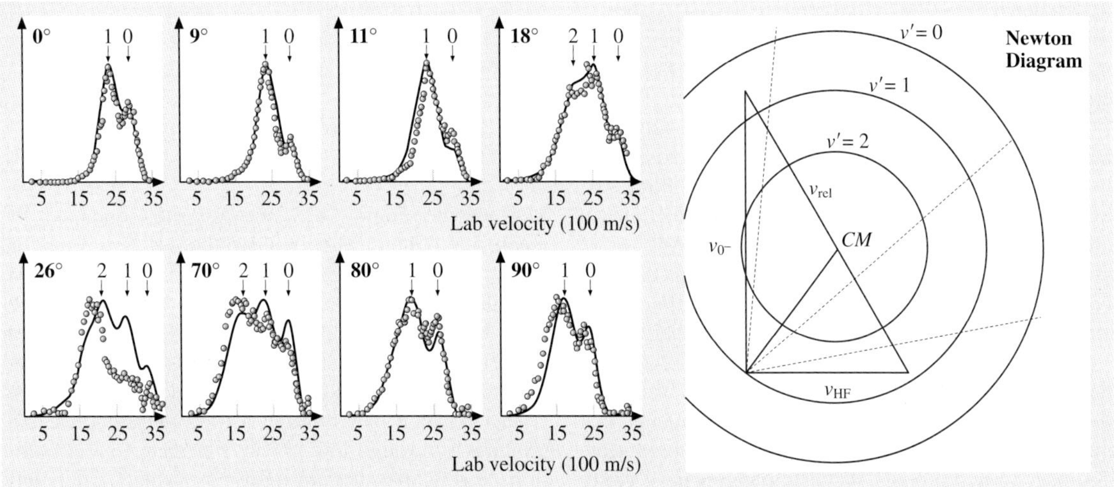

Fig. 67.3 F^- reactive fluxes in the O^- + HF system at a collision energy of 45 kJ/mol. After [67.8, 9] by permission

$u_{n'}$. The solution of this equation for $\sigma(\boldsymbol{n}', \theta|\boldsymbol{n}, \boldsymbol{v}_{rel})$ can be accomplished both by forward convolution integration fitting procedures [67.11] and by iterative unfolding [67.12]. Forward convolution methods assume parametric forms for $\sigma(\boldsymbol{n}', \theta|\boldsymbol{n}, \boldsymbol{v}_{rel})$ that are substituted into (67.13). The parameters are then varied until the calculated fluxes agree with the data within experimental error. Iterative deconvolution methods generally extract the cm cross section summed over final states,

$$\sum_{n'} \sigma(\boldsymbol{n}', \theta|\boldsymbol{n}, \boldsymbol{v}_{rel}) , \tag{67.14}$$

but the finite energy of the detection scheme also determines whether quantum states are completely resolved or if the data represent a summation over product energy levels.

67.3 Scattering Cross Sections

Measurements of the single-collision cross sections shown schematically in Fig. 67.1 require a variety of sophisticated techniques. Each of these are discussed in the context of measurement techniques and information content, with illustrative examples.

67.3.1 State-to-State Differential Cross Sections

Information most diagnostic of the potential energy surface for chemical reaction comes from $\sigma(\boldsymbol{n}', \theta|\boldsymbol{n}, \boldsymbol{v}_{rel})$. The experimental data of Fig. 67.3 on the O^- + HF system provide an example of a case in which $\boldsymbol{n}'$ refers to product OH vibrations in the ground electronic state, but without resolution of product rotations. Iterative deconvolution of the laboratory fluxes results in a flux distribution that is a sum of cross sections, as described in (67.14). Figure 67.4 shows this distribution in cm velocity space as a function of u and θ. In this representation, the relative velocity vector lies along the 0° - 180° line, and the symmetric peaks of the data near 0° (forward scattering) and 180° (backward scattering) indicate that the reaction proceeds through a transient collision complex living for at least several rotational periods. The symmetry of the flux distribution with respect to 90° is the most important diagnostic for the participation of a long-lived transient complex. Whether the distribution is forward-backward peaked, as in the present example, or more isotropic reflects the geometry of the complex (i. e., oblate or prolate symmetric top) and the manner in which angular momentum is partitioned in the products [67.13]. In Fig. 67.4, the forward-backward scattering is indicative of the fact that the orbital angular momentum $\boldsymbol{L}$ of the approaching reactants is partitioned preferentially into orbital angular momentum of the products, $\boldsymbol{L}'$, through a near-linear $[O \cdots H \cdots F]^-$ intermediate. The various peaks in the flux distribution of Fig. 67.4 can be assigned to OH vibrations $v' = 0$, 1, and 2. Integration of the cross sections over the appropriate angular and velocity ranges yields the result $P(v' = 0) : P(v' = 1) : P(v' = 2) = 0.38 : 0.43 : 0.18$.

Crossed–beam experiments (Sect. 60.2.2) with sufficient angular and kinetic energy resolution yield angular distributions for individual vibrational states. Similarly, laser-induced fluorescence experiments [67.15] yield rotational distributions. Recent work on the $Ar^+ + N_2 \rightarrow Ar + N_2^+(v' = 0, J')$ system illustrates this capability [67.16] (Sects. 64.2.1 and 38.4.1). However, these experiments probe N_2^+ in the collision volume, thereby averaging over all possible scattering angles, and are more properly examples of product-state resolved cross sections, discussed in Sect. 67.3.4. Experiments that provide complete product vibrational-rotational state specification and product scattering angles have yet to be carried out.

In the above examples the neutral beam is produced by supersonic expansion, producing reactants in their ground vibrational states, with rotational temperatures of only a few kelvins. Experiments with reactants prepared in excited states, either by laser absorption [67.17] or photoionization, and with full final state selection and angular resolution, remain a major goal of ion-molecule chemistry.

The production of reagent ions in selected vibrational and vibrational-rotational states by resonance enhanced multiphoton ionization (REMPI) (Sect. 74.1.2) has been accomplished in a number of systems [67.17, 18]. The use of photoelectron spectroscopy to assess the purity of the state-selection process is essential [67.19].

67.3.2 Velocity–Angle Differential Cross Sections

The velocity-angle DCS $\sigma(\boldsymbol{v}'_{\mathrm{rel}}, \theta|\boldsymbol{v}_{\mathrm{rel}})$ obtained by measuring kinetic energy distributions as a function of Θ can be plotted as a distribution in u and θ. Figure 67.5 shows $\sigma(\boldsymbol{v}'_{\mathrm{rel}}, \theta|\boldsymbol{v}_{\mathrm{rel}})$ for the reaction $C^+ + H_2O \rightarrow [COH]^+ + H$ at a collision energy of 2.14 eV (206.5 kJ/mol) [67.14]. The scattering is predominately backward, and the data indicate a reaction taking place in part through a transient complex, but principally through low impact parameter direct collisions leading to backward scattered products. Collisions leading to such backward scattered products are called rebound collisions and are dominated by the repulsive part of the potential surface. Although the reaction of C^+ with H_2O is exothermic, high kinetic energy release in the rebound component suggests that a potential energy barrier is in the exit channel, i. e., it acts as the products separate.

At lower collision energies, the forward peak in the flux distribution becomes more pronounced for the $C^+ + H_2O$ reaction, suggesting that the reaction is mediated by a collision complex living for a fraction of a rotational period. Under these conditions, it is possible

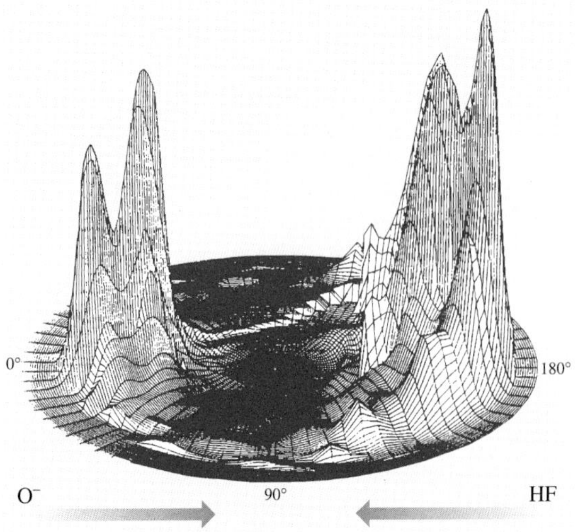

Fig. 67.4 Axonometric plot of F^- fluxes in velocity space in the reaction $O^- + HF \rightarrow OH + F^-$ at 45 kJ/mol. Data from [67.8, 9]

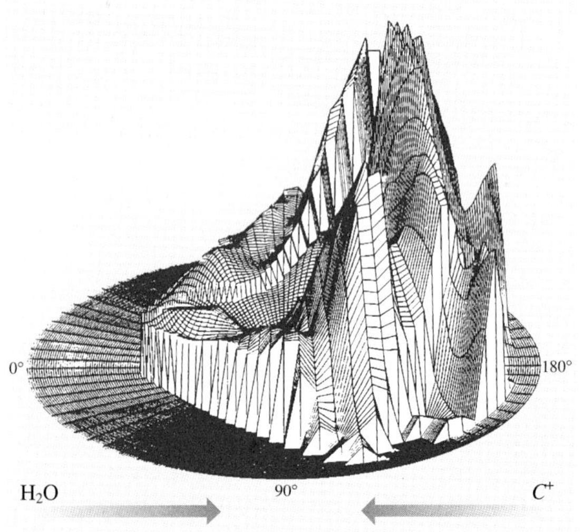

Fig. 67.5 Axonometric plot of $[COH]^+$ fluxes in velocity space in the reaction $C^+ + H_2O \rightarrow [COH]^+ + H$ at 2.14 eV (206.5 kJ/mol). Data from [67.14]

to extract collision complex lifetimes from the angular distribution asymmetry through the osculating model for chemical reactions [67.20]. In this model, the lifetime is a parametric function of the complex's rotational period, evidenced by the forward-backward asymmetry

$$I(180°)/I(0°) = 1/\cosh(\tau_R/2\tau)\,. \tag{67.15}$$

In this expression, τ is the lifetime of the complex and τ_R is the rotational period, estimated from the moment of inertia I and the maximum orbital angular momentum L_{max} of the complex. The total reaction cross section σ can be used to estimate this latter quantity from

$$\sigma = \left(\frac{\pi\hbar^2}{2\mu\mathcal{T}_{rel}}\right)(L_{max}+1)^2\,. \tag{67.16}$$

In the present case, the lifetime estimate for the $[CHOH]^+$ complex is less than 1×10^{-13} s (Chapt. 35).

At high energies, rebound collisions dominate the dynamics of the $C^+ + H_2O$ reaction. Another limit often encountered in high energy ion-molecule reactions is the stripping process in which the incoming ionic projectile removes a particle from its molecular collision partner, and the new ionic product travels in the same direction as the incident ionic reactant. The spectator stripping limit [67.22] for a reaction of the type $A + BC \to AB + C$ corresponds to the case in which the cm speed of product C is unchanged from the cm speed of reactant BC, i.e., C "spectates" as the reaction occurs.

67.3.3 Total Cross Sections with State-Selected Reactants

State-selected cross sections $\sigma(\boldsymbol{n}'|\boldsymbol{n}, \boldsymbol{v}_{rel})$ were first measured by *Chupka* and collaborators [67.23, 24] in the endothermic reaction of $H_2^+ + He \to HeH^+ + H$, with H_2^+ prepared in states $v \geq 2$ by photoionization, and HeH^+ detected without product state analysis. Thus, $\boldsymbol{n}'$ represents a summation over all accessible product states in state-selected reactant experiments. At a given vuv ionization wavelength, H_2^+ is prepared, subject to Franck-Condon factor limitations (Sect. 33.6.1), in all possible vibrational states up to the maximum allowed by the photon energy. Therefore, extraction of cross sections for individual vibrational states requires VUV wavelength dependent studies in which the number of reactant vibrational states is increased smoothly from the ground state up to the limit allowed by instrumental resolution or photon sources. A number of reactive and charge transfer systems have been studied with photoionization techniques at the reactant state-selected level with unresolved product states. Data are typically in the form of $P(v')$ vs. v', as a function of collision energy. A comprehensive review by *Ng* is available [67.25].

67.3.4 Product-State Resolved Total Cross Sections

The resolution of product states in cross section measurements can be accomplished in a variety of ways. We will first discuss the case in which the reactants have not been state-selected. In principle, spectroscopic methods such as laser-induced fluorescence can be used for complete product quantum state specification. Earlier reference to N_2^+ vibration-rotation state-resolved charge transfer experiments [67.16] provides an excellent example of the state of the art of such methods, and their extension to chemical reactions represents an important future development. A beautiful example is mass spectrometric detection of N_2^+ in the $Ar^+ + N_2$ charge transfer system [67.21]. Guided ion-beam production and detection methods [67.26] in conjunction with a supersonic beam of N_2 allow 4π detection of the charge transfer products as a function of collision energy, yielding accurate total cross sections. The results, plotted in Fig. 67.6, show thresholds for the production of excited vibrational states of N_2^+ X^2Sigma$_g$, $v' = 1, 2$, and 4) as well as the formation of the excited $B\,^2\Sigma_g$ state. The detection of chemiluminescence from electronically ex-

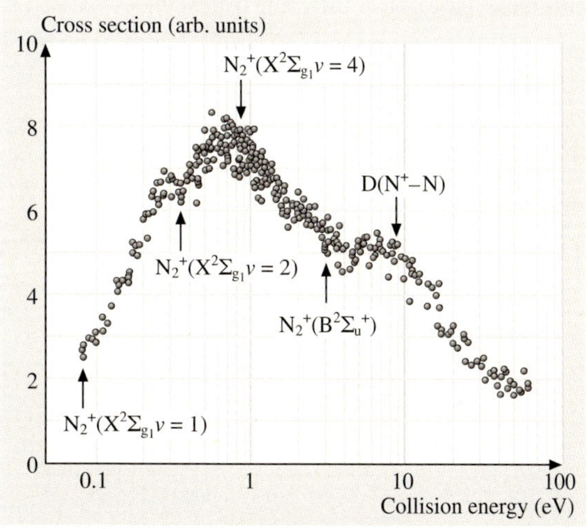

Fig. 67.6 Energy dependent cross section for production of N_2^+ in states as indicated in the $Ar^+ + N_2$ charge transfer system. After [67.21] by permission

cited states of ion-molecule reactions is also a powerful method for determining such cross sections; methods and results are reviewed in [67.27].

67.3.5 State-to-State Total Cross Sections

The most detailed probe of an ion-molecule reaction is the state-to-state cross section. The measurement of angularly-resolved state-to-state cross sections has only been achieved for reactants in their ground states, but true state-to-state total cross sections have now been measured in a few favorable cases. A particularly novel method for product-state determination with reactant-state selection has been developed [67.25]. In the differential reactivity method, product vibrational states are distinguished by their differing cross sections for charge transfer with selected molecules. Although the method is limited at present because state-selected charge transfer cross sections have been measured in only a few systems, the technique has great potential. The charge transfer reaction H_2^+ (v)+ $H_2 \rightarrow H_2 + H_2^+$ (v'), in which H_2^+ is prepared in $v = 0$ and 1 by VUV photoionization, provides a particularly good example of its power. The H_2^+ charge transfer products, formed in states $v' = 0-3$, are first mass-analyzed, accelerated to 10 eV, and then passed through a collision cell containing N_2, Ar, or CO, where charge transfer occurs once again. The N_2^+, Ar^+, or CO^+ products are mass analyzed and the cross sections σ_m for forming these ions by charge transfer from the H_2^+ reaction products are measured. The charge transfer cross sections for H_2^+ with these gases have different dependences on v', as shown in Fig. 67.7, and therefore, the gases can be used to probe the product states of H_2^+ formed in the symmetric charge exchange reaction. Letting $X_{v'}$ denote the fraction of H_2^+ formed in the vibrational state v', n denotethe number density of the neutral collision gas, and l denote the attenuation length, the following set of simultaneous equations can be solved for $X_{v'}$, since thecross sections $\sigma_{v'}$ are known and the σ_m are measured:

$$X_0 + X_1 + X_2 + X_3 = 1 \qquad (67.17)$$

and, for a particular ion (e.g., N_2^+, Ar^+, CO^+),

$$X_0 nl\sigma_0 + X_1 nl\sigma_1 + X_2 nl\sigma_2 + X_3 nl\sigma_3 = nl\sigma_m \,. \qquad (67.18)$$

[An equation like (67.18) will exist for each collision gas.] An extensive set of data at collision energies from 2 eV up to 16 eV has been obtained [67.29]. When the reactant ions are in the vibrational ground state, at low kinetic energies the charge transfer product is also in the ground state, indicating that resonant charge transfer is the dominant process. At increasing collision energies, X_1 increases from 0.0 to 0.17. For vibrationally excited H_2^+ reactants, inelastic relaxation to form H_2^+ in $v' = 0$ is important at all collision energies, increasing in magnitude with increasing collision energy. Of particular interest is the fact that inelastic relaxation forming $v' = 0$ is substantially more important than inelastic excitation producing $v' = 1$ or 2. This trend is predicted by theory, but underestimated at lower collision energies.

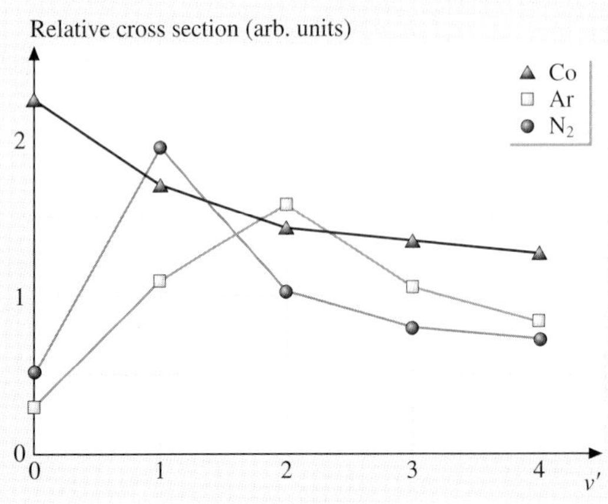

Fig. 67.7 Cross sections for charge transfer of $H_2^+(v')$ vs. v' with N_2, Ar, and CO at a collision energy of 10 eV. After [67.25] by permission of John Wiley & Sons, Inc

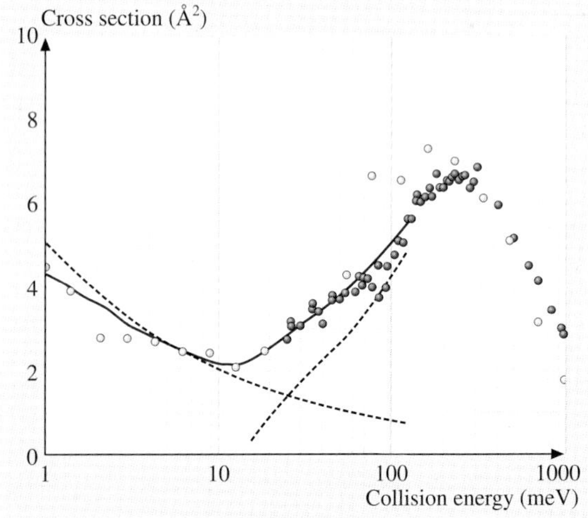

Fig. 67.8 Total cross section for ND^+ formation in the N^+ + D_2 system, showing a low energy threshold at 15 meV. After [67.28] by permission

67.3.6 Energy Dependent Total Cross Sections

Although non-state-selected cross sections, denoted $\sigma(v)$, lack information about product energy disposal, important features of the potential energy surface can be obtained from their measurement. Guided ion beam methods have been especially important in the determination of accurate cross sections, particularly those that employ a supersonic beam rather than a gas as a neutral target. Energy-dependent cross sections provide crucial information on ion and neutral thermochemistry through accurate measurements of reaction thresholds, and help to elucidate important potential surface features such as thresholds, barriers, and crossings [67.30]. A particularly illustrative example of threshold formation concerns the reaction $N^+(^3P) + H_2 \to NH^+ + H$ and its isotopic variants. This reaction is important at low (< 0.1 eV) collision energies as the first step in the chain reaction that leads to formation of ammonia in dense interstellar clouds, but the bond energy of NH^+ is uncertain enough to prevent knowing whether the above reaction is endothermic or exothermic. Recent total cross section measurements on the isotopic variant $N^+(^3P) + D_2 \to ND^+ + D$ at very low collision energies now appear to answer this question [67.28]. Figure 67.8 shows experimental results performed on two different instruments, one a guided ion beam-crossed neutral beam machine and the other a guided merged-beam apparatus. The data show a very clear threshold at 15 meV, demonstrating that the reaction is endothermic and allowing for a more accurate estimate of the ND^+ bond energy. This experiment shows that the high energy resolution afforded in total cross section measurements employing crossed and merged beams, rather than thermal collision cells, can answer thermochemical questions of importance in a wide variety of applications, including astrophysics, combustion, electrical discharges, and atmospheric processes.

67.4 New Directions: Complexity and Imaging

The crossed beam technique provides the precise kinematic definition required to extract the most intimate details of reactive collisions. Although the highest resolution examples of the technique involve systems with three or four atoms, several recent examples of systems of greater complexity have appeared in the literature. The multiply-charged $CF_2^{2+} + D_2$ system [67.31] is appreciably more complex experimentally than its singly-charged analog, owing to the possibility of forming two charged fragments with correlated product distributions. The nine-atom $C_2H_2^+ + CH_4$ system has been studied previously with guided ion beam-gas cell methods under conditions where the reactant ions are produced by multiphoton ionization [67.32]. Those early experiments, which yielded total reaction cross sections for vibrational state-selected ions, gave significant indications of mode-selectivity in complex reactions. A more recent study [67.33], conducted without reactant state-selection or product state analysis, maps out product angular distributions and disposal of energy in relative translation with ion trapping techniques that measure longitudinal velocity components directly and provide upper bounds on the transverse velocity components.

The majority of extant crossed beam experimental studies construct three-dimensional velocity space distributions with a series of one-dimensional sections through the full distribution such that, in any interval of time, only a single detection element in velocity space can be measured. Recent advances in imaging methods [67.34–36], in which many velocity space elements can be observed in a single time interval, promise to increase the sensitivity of the crossed beams method by orders of magnitude. The signal gains that can be achieved with the implementation of such multiplex advantage methods will allow the lower reactant and product signal levels associated with increasing state-specification and/or system complexity to be tolerated without undue increases in data acquisition times.

Figure 67.9 shows a schematic of an instrument that illustrates the imaging principle.

As indicated in the diagram, the locus of points of reaction products with a constant center of mass speed is a sphere whose radius increases with time. Individual quantum states of a given reaction product form a set of nested spheres in velocity space. Projecting this set of product spheres onto a detection plane allows all product velocity elements to be observed in a single time window. Application of the inverse Abel transform [67.37] allows the full three-dimensional velocity distribution to be extracted from the two-dimensional projection on the detection plane. Velocity focusing methods [67.38] allow reaction products originating from spatially distinct regions of the collision volume

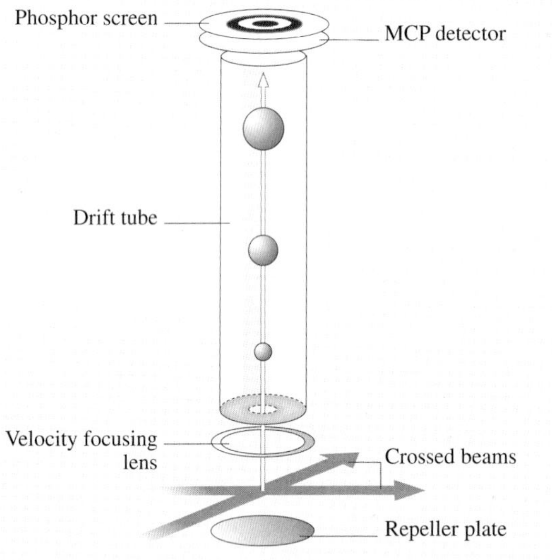

Fig. 67.9 Schematic of crossed beam imaging instrument. Product flux extracted from the collision volume with a single center of mass speed describes a sphere in velocity space. This flux is projected onto a multichannel plate/phosphor screen detector and recorded with a CCD camera

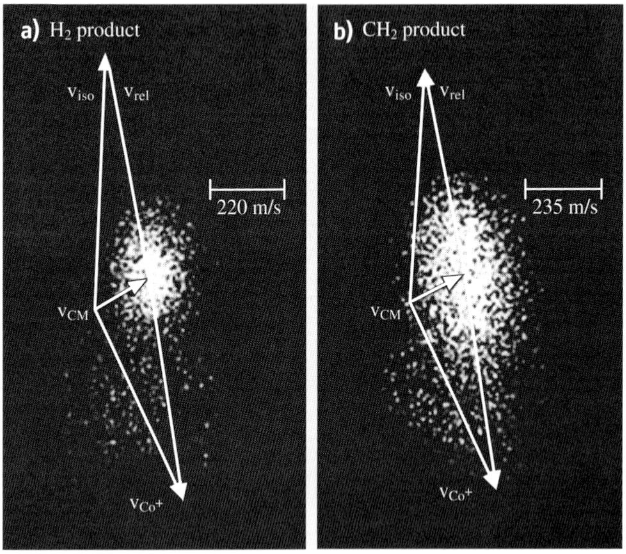

Fig. 67.10a,b *Panel* (**a**): image of ionic products formed by H_2 ejection from Co^+ C_4H_10 collision complex. *Panel* (**b**): image of ionic products formed by CH_4 ejection from Co^+ C_4H_10 collision complex

that have the same lab velocity components in the plane of the beams to be focused to a single point. Figure 67.10 shows velocity space images of the H_2 and CH_4 elimination products arising from decay of the transient collision complex formed in collisions of Co^+ with iso-butane C_4H_{10} [67.39].

This velocity space image is equivalent to the c.m. cross section expressed in equation (67.14), and can be appropriately averaged over initial states or summed over final states to yield product angular or energy distributions, total cross sections, or rate constants according to the hierarchy of Fig. 67.1. The time window that imaging requires is ideally mated with pulsed molecular beam sources, and thus with pulsed lasers. This last connection opens up a number of possibilities for reactant preparation, including electron impact ionization via pulsed electron beams formed by laser-induced photoemission from metals, pulsed IR laser excitation of ro-vibrationally state-selected molecules, formation of free radicals by UV laser photolysis, and multiphoton ionization preparation of state-selected ions.

As laser-based photolysis, state-selection, and ionization methods continue to advance, additional routes for preparing atomic and molecular ions, free radicals, and clusters with quantum state-specificity will appear. These schemes will enhance all crossed beam methodologies, but will have an especially important impact on methods that exploit time-based snapshots of the velocity space product distribution in coordinate space. Point-by-point methods for reconstruction of the three-dimensional flux distribution in velocity space will continue to play a role in the development of the crossed beam technique, but technological advances in lasers and in imaging systems virtually assure that the most impressive gains in crossed beam technology in the next several years will come from multichannel product detection.

References

67.1 J. J. Thomson: *Rays of Positive Electricity* (Longmans Green, New York 1913)

67.2 T. Su, M. T. Bowers: *Gas Phase Ion Chemistry*, Vol. 1, ed. by M. T. Bowers (Academic, New York 1979) p. 84

67.3 R. D. Levine, R. B. Bernstein: *Molecular Reaction Dynamics and Chemical Reactivity* (Oxford, New York 1987)

67.4 Y. T. Lee: *Atomic and Molecular Beam Methods*, ed. by G. Scoles, U. Buck (Oxford, New York 1988) p. 553

67.5 J. M. Farrar, W. H. Saunders, Jr.: *Techniques for the Study of Ion–Molecule Reactions* (Wiley-Interscience, New York 1988)

67.6 T. T. Warnock, R. B. Bernstein: J. Chem. Phys. **49**, 1878 (1968)

67.7 G. L. Catchen, J. Husain, R. N. Zare: J. Chem. Phys. **69**, 1737 (1978)

67.8 D. J. Levandier, D. F. Varley, M. A. Carpenter, J. M. Farrar: J. Chem. Phys. **99**, 148 (1993)

67.9 M. A. Carpenter, M. T. Zanni, D. J. Levandier, D. F. Varley, J. M. Farrar: J. Chem. Phys. **72**, 828 (1994)

67.10 J. E. Pollard, D. J. Trevor, J. E. Reutt, Y. T. Lee, D. A. Shirley: J. Chem. Phys. **77**, 34 (1982)

67.11 P. E. Siska: Ph. D. Thesis, Harvard University (1969), Available from University Microfilms, Ann Arbor, MI

67.12 P. E. Siska: J. Chem. Phys. **59**, 6052 (1973)

67.13 S. A. Safron, W. B. Miller, D. R. Herschbach: Discuss. Faraday Soc. **44**, 108 (1967)

67.14 D. M. Sonnenfroh, R. A. Curtis, J. M. Farrar: J. Chem. Phys. **83**, 3958 (1985)

67.15 J. L. Kinsey: Ann. Rev. Phys. Chem. **28**, 349 (1977)

67.16 L. Hüwel, D. R. Guyer, G.-H. Lin, S. R. Leone: J. Chem. Phys. **81**, 3520 (1984)

67.17 S. L. Anderson: Adv. Chem. Phys. **82**, 177 (1992)

67.18 L. A. Posey, R. D. Guettler, N. J. Kirchner, R. N. Zare: J. Chem. Phys. **101**, 3772 (1994)

67.19 R. N. Compton, J. C. Miller: *Laser Applications in Physical Chemistry*, ed. by K. Evans D. (Marcel Dekker, New York 1989) p. 221

67.20 M. K. Bullitt, C. H. Fisher, J. L. Kinsey: J. Chem. Phys. **64**, 1914 (1974)

67.21 P. Tosi, O. Dmitrijev, D. Bassi: Chem. Phys. Lett. **200**, 483 (1992)

67.22 A. Henglein, K. Lacmann, G. Jacobs: Ber. Bunsenges. Phys. Chem. **69**, 279 (1965)

67.23 W. A. Chupka, M. E. Russell: J. Chem. Phys. **4849**, 1527 (1968)

67.24 W. A. Chupka, M. E. Russell: J. Chem. Phys. **49**, 5426 (1968)

67.25 C. Y. Ng: Adv. Chem. Phys. **82**, 401 (1992)

67.26 D. Gerlich: Adv. Chem. Phys. **82**, 1 (1992)

67.27 M. Tsuji: *Techniques for the Study of Ion-Molecule Reactions*, ed. by J. M. Farrar, W. H. Saunders, Jr. (Wiley-Interscience, New York 1988) p. 489

67.28 P. Tosi, O. Dmitriev, D. Bassi, O. Wick, D. Gerlich: J. Chem. Phys. **100**, 4300 (1994)

67.29 C.-L. Liao, C. Y. Ng: J. Chem. Phys. **84**, 197 (1986)

67.30 P. B. Armentrout: Ann. Rev. Phys. Chem. **41**, 313 (1990)

67.31 Z. Dolejsek, M. Farnik, Z. Herman: Chem. Phys. Lett. **235**, 99 (1995)

67.32 Y. H. Chiu, H. S. Fu, J. T. Huang, S. L. Anderson: J. Chem. Phys. **101**, 5410 (1994)

67.33 J. Zabka, O. Dutuit, Z. Dolejsek, J. Polach, Z. Herman: PCCP Phys. Chem. Chem. Phys. **2**, 781 (2000)

67.34 D. W. Chandler, P. L. Houston: J. Chem. Phys. **87**, 1445 (1987)

67.35 P. L. Houston: Acct. Chem. Res. **28**, 453 (1995)

67.36 A. G. Suits, R. E. Continetti (Eds.): *Imaging in Chemical Dynamics*, Vol. 770 (American Chemical Society, Washington DC 2001)

67.37 V. Dribinski, A. Ossadtchi, V. A. Mandelshtam, H. Reisler: Rev. Sci. Instrum. **73**, 2634 (2002)

67.38 A. Eppink, D. H. Parker: Rev. Sci. Instrum. **68**, 3477 (1997)

67.39 E. L. Reichert, G. Thurau, J. C. Weisshaar: J. Chem. Phys. **117**, 653 (2002)

Part F

Part F Quantum Optics

68 Light–Matter Interaction
Pierre Meystre, Tucson, USA

69 Absorption and Gain Spectra
Stig Stenholm, Stockholm, Sweden

70 Laser Principles
Peter W. Milonni, Los Alamos, USA

71 Types of Lasers
Richard C. Powell, Tuscon, USA

72 Nonlinear Optics
Alexander L. Gaeta, Ithaca, USA
Robert W. Boyd, Rochester, USA

73 Coherent Transients
Joseph H. Eberly, Rochester, USA
Carlos R. Stroud Jr., Rochester, USA

74 Multiphoton and Strong-Field Processes
Kenneth C. Kulander, Livermore, USA
Maciej Lewenstein, Barcelona, Spain

75 Cooling and Trapping
Juha Javanainen, Storrs, USA

76 Quantum Degenerate Gases
Juha Javanainen, Storrs, USA

77 De Broglie Optics
Carsten Henkel, Potsdam, Germany
Martin Wilkens, Potsdam, Germany

78 Quantized Field Effects
Matthias Freyberger, Ulm, Germany
Karl Vogel, Ulm, Germany
Wolfgang P. Schleich, Ulm, Germany
Robert F. O'Connell, Baton Rouge, USA

79 Entangled Atoms and Fields: Cavity QED
Dieter Meschede, Bonn, Germany
Axel Schenzle, München, Germany

80 Quantum Optical Tests of the Foundations of Physics
Aephraim M. Steinberg, Toronto, Canada
Paul G. Kwiat, Urbana, USA
Raymond Y. Chiao, Berkeley, USA

81 Quantum Information
Sir Peter L. Knight, London, UK
Stefan Scheel, London, UK

68. Light–Matter Interaction

Optical physics is concerned with the dynamical interactions of atoms and molecules with electromagnetic to fields. Semiclassical theories, which study the interaction of atoms with classical fields, are often said to comprise optical physics, while quantum optics treats the interaction of atoms or molecules with quantized electromagnetic fields. A significant part of optical physics and quantum optics is the study of near-resonant atom-field interactions, and concentrates on nonperturbative dynamics, where the effects of the optical fields have to be kept to all orders. The atomic properties themselves are assumed to be known.

The vast majority of problems in light-matter interactions can be treated quite accurately within semiclassical theories. However, an important class of problems where this is not the case are presented in Chapt. 78. While much of optical physics and quantum optics ignores the effects of the electromagnetic fields on the center-of-mass motion of the atoms, important topics such as atomic trapping and cooling (Chapt. 75) and de Broglie optics (Chapt. 77) rely in an essential way on such mechanical effects of light. The present chapter deals with more "traditional" aspects of optical physics, where these effects are ignored.

68.1 **Multipole Expansion** ... 997
68.1.1 Electric Dipole (E1) Interaction ... 998
68.1.2 Electric Quadrupole (E2) Interaction ... 998
68.1.3 Magnetic Dipole (M1) Interaction ... 999
68.2 **Lorentz Atom** ... 999
68.2.1 Complex Notation ... 999
68.2.2 Index of Refraction ... 999
68.2.3 Beer's Law ... 1000
68.2.4 Slowly-Varying Envelope Approximation ... 1000
68.3 **Two-Level Atoms** ... 1000
68.3.1 Hamiltonian ... 1001
68.3.2 Rotating Wave Approximation ... 1001
68.3.3 Rabi Frequency ... 1001
68.3.4 Dressed States ... 1002
68.3.5 Optical Bloch Equations ... 1003
68.4 **Relaxation Mechanisms** ... 1003
68.4.1 Relaxation Toward Unobserved Levels ... 1003
68.4.2 Relaxation Toward Levels of Interest ... 1004
68.4.3 Optical Bloch Equations with Decay ... 1004
68.4.4 Density Matrix Equations ... 1004
68.5 **Rate Equation Approximation** ... 1005
68.5.1 Steady State ... 1005
68.5.2 Saturation ... 1005
68.5.3 Einstein A and B Coefficients ... 1005
68.6 **Light Scattering** ... 1006
68.6.1 Rayleigh Scattering ... 1006
68.6.2 Thomson Scattering ... 1006
68.6.3 Resonant Scattering ... 1006
References ... 1007

68.1 Multipole Expansion

Consider a test charge q of mass m localized within an atom and acted upon by an external electromagnetic field with electric field $\boldsymbol{E}(\boldsymbol{r}, t)$ and magnetic field $\boldsymbol{B}(\boldsymbol{r}, t)$. In the multipole expansion formalism [68.1, 2], the electric and magnetic interaction energies between the charge and the electromagnetic field are

$$V_{\mathrm{e}} = V_{\mathrm{E0}}(t) - q \int_0^{\boldsymbol{r}} \mathrm{d}\boldsymbol{s} \cdot \boldsymbol{E}(\boldsymbol{R}+\boldsymbol{s}, t) , \tag{68.1}$$

$$V_{\mathrm{m}} = -q \int_0^{\boldsymbol{r}} \mathrm{d}\boldsymbol{s} \cdot \boldsymbol{v} \times \boldsymbol{B}(\boldsymbol{R}+\boldsymbol{s}, t) , \tag{68.2}$$

respectively. The material of this chapter is discussed in detail in a number of texts and review articles. We cite such references rather than the original sources whenever possible. These energies correspond to the work done by the electric and magnetic components of the Lorentz force in first moving the charge to a stationary origin of coordinates at a point $\boldsymbol{R}$ and then to a location $\boldsymbol{r}$ relative to $\boldsymbol{R}$. Here, $V_{\mathrm{E0}}(t)$ represents the energy of the charge when located at the reference point $\boldsymbol{R}$. It may be expressed in terms of the electrostatic potential $\phi(\boldsymbol{R},t)$ as $V_{\mathrm{E0}}(t)=+q\phi(\boldsymbol{R},t)$.

A Taylor series expansion of $V_{\mathrm{e}}(t)$ and $V_{\mathrm{m}}(t)$ about $\boldsymbol{r}=0$ yields

$$V_{\mathrm{e}}(t)=V_{\mathrm{E0}}(t)-q\sum_{n=1}^{\infty}\frac{1}{n!}\left(\boldsymbol{r}\cdot\frac{\partial}{\partial\boldsymbol{R}}\right)^{n-1}\boldsymbol{r}\cdot\boldsymbol{E}(\boldsymbol{R},t)\,, \tag{68.3}$$

$$V_{\mathrm{m}}(t)=-\frac{q\hbar}{m}\sum_{n=1}^{\infty}\frac{n}{(n+1)!}\left(\boldsymbol{r}\cdot\frac{\partial}{\partial\boldsymbol{R}}\right)^{n-1}\boldsymbol{\ell}\cdot\boldsymbol{B}(\boldsymbol{R},t)\,, \tag{68.4}$$

where $\hbar\boldsymbol{\ell}=\boldsymbol{r}\times\boldsymbol{p}$ is the angular momentum of the test charge relative to the coordinate origin $\boldsymbol{R}$. Here, use of the mechanical momentum $\boldsymbol{p}=m\dot{\boldsymbol{r}}$, instead of the canonical momentum, neglects the electromagnetic component of the momentum responsible for diamagnetic effects.

In addition to the electromagnetic interaction, electrons and nuclei are characterized by a spin magnetic moment $\boldsymbol{m}_s=(q\hbar/2m)g_s\boldsymbol{s}$, where $\boldsymbol{s}$ is the spin of the test charge and g_s its gyromagnetic factor, equal to 2.002... for electrons. The factor $q\hbar/2m$ is the particle's magneton. The spin magnetic moment yields an additional term to the magnetic energy V_{m}, which becomes

$$V_{\mathrm{m}}(t)=-\frac{e\hbar}{2m}\sum_{n=1}^{\infty}\frac{1}{n!}\left(\boldsymbol{r}\cdot\frac{\partial}{\partial\boldsymbol{R}}\right)^{n-1}\times\left(\frac{2}{n+1}g_\ell\boldsymbol{\ell}+g_s\boldsymbol{s}\right)\cdot\boldsymbol{B}(\boldsymbol{R},t)\,, \tag{68.5}$$

where the orbital g-factor is $g_\ell=q/e$ ($g_\ell=-1$ for an electron). For an ensemble $\{\alpha\}$ of charged particles q_α in an atom, these expressions are to be summed over all particles. Thus, the electric energy becomes

$$\begin{aligned}V_{\mathrm{e}}(t)&=\sum_\alpha q_\alpha\phi(\boldsymbol{R},t)-\sum_{i=1}^{3}\sum_\alpha q_\alpha r_i(\alpha)E_i(\boldsymbol{R},t)\\&\quad-\frac{1}{2}\sum_{i,j=1}^{3}\left[\sum_\alpha q_\alpha r_i(\alpha)r_j(\alpha)\right]\\&\quad\times\frac{\partial}{\partial R_j}E_i(\boldsymbol{R},t)+\cdots\\&\equiv V_{\mathrm{E0}}(t)+V_{\mathrm{E1}}(t)+V_{\mathrm{E2}}(t)+\cdots\,,\end{aligned} \tag{68.6}$$

and the magnetic energy becomes

$$\begin{aligned}V_{\mathrm{m}}(t)&=\sum_{i=1}^{3}B_i(\boldsymbol{R},t)\\&\quad\times\sum_\alpha\frac{e\hbar}{2m_\alpha}\left[g_\ell(\alpha)\ell_i(\alpha)+g_s(\alpha)s_i(\alpha)\right]\\&\quad-\sum_{i,j=1}^{3}\frac{\partial B_i(\boldsymbol{R},t)}{\partial R_j}\sum_\alpha\frac{e\hbar}{2m_\alpha}\left[\frac{2}{3}g_\ell(\alpha)\ell_i(\alpha)r_j(\alpha)\right.\\&\quad\left.+g_s(\alpha)s_i(\alpha)\right]+\cdots\\&\equiv V_{\mathrm{M1}}(t)+V_{\mathrm{M2}}(t)+\cdots\,.\end{aligned} \tag{68.7}$$

68.1.1 Electric Dipole (E1) Interaction

For optical fields whose wavelength is large compared with the interacting atom, only the first few terms in the Taylor expansions of $V_{\mathrm{e}}(t)$ and $V_{\mathrm{m}}(t)$ need be retained. The first term, $V_{\mathrm{E0}}(t)$, of $V_{\mathrm{e}}(t)$ is the net charge of the atom, which vanishes for neutral atoms. The second term, $V_{\mathrm{E1}}(t)$, is the electric dipole interaction energy. Introducing the electric dipole (E1) moment,

$$\boldsymbol{d}=\sum_\alpha q_\alpha\boldsymbol{r}(\alpha)\,, \tag{68.8}$$

or

$$\boldsymbol{d}=\int\mathrm{d}^3r\rho(\boldsymbol{r})\boldsymbol{r}\,, \tag{68.9}$$

for a charge distribution, this contribution to the interaction energy may be re-expressed as

$$V_{\mathrm{E1}}(t)=-\boldsymbol{d}\cdot\boldsymbol{E}(\boldsymbol{R},t)\,. \tag{68.10}$$

The E1 term dominates most optical phenomena.

68.1.2 Electric Quadrupole (E2) Interaction

The $V_{\mathrm{E2}}(t)$ contribution to $v_{\mathrm{e}}(t)$, describes electric quadrupole (E2) interactions. In terms of the quadrupole tensor

$$Q=3\int\mathrm{d}^3r\rho(\boldsymbol{r})r_ir_j\,, \tag{68.11}$$

$V_{\mathrm{E2}}(t)$ becomes

$$V_{\mathrm{E2}}(t)=-\frac{1}{6}\sum_{i,j=1}^{3}Q_{ij}\frac{\partial}{\partial R_i}E_j(\boldsymbol{R},t)\,. \tag{68.12}$$

Alternatively, E2 interactions can be expressed in terms of the traceless quadrupole tensor $Q_{ij}^{(2)} = \int \mathrm{d}^3 r \rho(\boldsymbol{r}) \times(3r_i r_j - \delta_{ij} r^2)$. E2 interactions are typically weaker than E1 interactions by a factor a_0/λ, where a_0 is the Bohr radius and λ is the wavelength of the transition. Since a_0/λ is very small for optical transitions, E2 interactions are typically neglected in quantum optics.

68.1.3 Magnetic Dipole (M1) Interaction

The first term in the multipole expansion the magnetic interaction (Sect. 68.6) is the magnetic dipole M1 interaction,

$$V_{\mathrm{M1}} = -\boldsymbol{m} \cdot \boldsymbol{B}(\boldsymbol{R}, t) , \tag{68.13}$$

of a magnetic moment $\boldsymbol{m}$ in a magnetic field, where

$$\begin{aligned} \boldsymbol{m} &= \sum_\alpha \left(\frac{q_\alpha \hbar}{2m_\alpha} \right) [g_\ell(\alpha)\boldsymbol{\ell}(\alpha) + g_s(\alpha)\boldsymbol{s}(\alpha)] \\ &= -\mu_{\mathrm{B}}(\boldsymbol{L}+2\boldsymbol{S}) , \end{aligned} \tag{68.14}$$

and we have used the fact that for electrons $g_\ell = -1$ and $g_s \simeq -2$. The Bohr magneton μ_{B} is

$$\mu_{\mathrm{B}} = \frac{e\hbar}{2mc} = \frac{\alpha e a_0}{2} , \tag{68.15}$$

where α is the fine structure constant. Thus M1 interactions tend to be smaller than E1 interactions by a factor of order $\alpha/2$. The connection between $\boldsymbol{m}$ and angular momentum $\boldsymbol{J}$ is $\boldsymbol{m} = \gamma \boldsymbol{J}$, where γ is the gyromagnetic ratio.

68.2 Lorentz Atom

The Lorentz atom consists of a classical electron harmonically bound to a proton. It provides a framework for understanding a number of elementary aspects of the electric dipole interaction between a single atom and light [68.3–7]. Assuming that the c.m. motion of the atom is unaffected by the field, and neglecting magnetic effects, the equation of motion of the electron is

$$\left(\frac{\mathrm{d}^2}{\mathrm{d}t^2} + \frac{\Gamma_0}{2}\frac{\mathrm{d}}{\mathrm{d}t} + \omega_0^2 \right) \boldsymbol{r} = -\frac{e}{m}\boldsymbol{E}(\boldsymbol{R}, t) , \tag{68.16}$$

where ω_0 is the electron's natural oscillation frequency, and Γ_0 represents a frictional decay rate that accounts for the effects of radiative damping. For the classical Lorentz atom

$$\Gamma_0 = 2\omega_0^2 r_0/3c , \tag{68.17}$$

where $r_0 = e^2/4\pi\epsilon_0 mc^2$ is the classical electron radius. This damping arises physically from the radiation reaction of the field radiated by the atom on itself. In the E1 approximation, the electric field is evaluated at the location $\boldsymbol{R}$ of the atomic c.m.

68.2.1 Complex Notation

The study of light-matter interactions is simplified by the introduction of complex variables [68.8–11]. For example, an electric field

$$\boldsymbol{E}(\boldsymbol{R}, t) = \sum_{n,\mu} \boldsymbol{\epsilon}_\mu \mathcal{E}_n \cos(\omega_n t) , \tag{68.18}$$

where $\boldsymbol{\epsilon}_\mu$ is the polarization vector of the field Fourier component at frequency ω_n is expressed as

$$\boldsymbol{E}(\boldsymbol{R}, t) = \boldsymbol{E}^+(\boldsymbol{R}, t) + \boldsymbol{E}^-(\boldsymbol{R}, t) , \tag{68.19}$$

where the *positive frequency part* of the field is

$$\boldsymbol{E}^+(\boldsymbol{R}, t) = \frac{1}{2}\sum_{n,\mu} \boldsymbol{\epsilon}_\mu \mathcal{E}_n \exp[\mathrm{i}(\boldsymbol{k}_n \cdot \boldsymbol{R} - \omega_n t)] . \tag{68.20}$$

Due to the linearity of (68.16), it is sufficient to study the response of the Lorentz atom to a plane monochromatic electric field of frequency ω, complex amplitude $\mathcal{E}$ and polarization $\boldsymbol{\epsilon}$. Introducing the *complex dipole moment*

$$\boldsymbol{d} = -e\boldsymbol{r} = \boldsymbol{\epsilon}\mathcal{P} \exp[\mathrm{i}(\boldsymbol{k} \cdot \boldsymbol{R} - \omega t)] + \text{c.c.} , \tag{68.21}$$

where $\mathcal{P}$ is in general complex for $\mathcal{E}$ real, and the complex polarizability $\alpha(\omega)$ via $\mathcal{P} = \alpha(\omega)\mathcal{E}$, then

$$\alpha(\omega) = \frac{e^2/m}{\omega_0^2 - \omega^2 - \mathrm{i}\gamma\omega} . \tag{68.22}$$

68.2.2 Index of Refraction

From the Maxwell wave equation

$$\left(\nabla^2 - \frac{1}{c^2}\frac{\partial^2}{\partial t^2} \right) \boldsymbol{E}(\boldsymbol{R}, t) = \frac{1}{\epsilon_0 c^2}\frac{\partial^2 \boldsymbol{P}(\boldsymbol{R}, t)}{\partial t^2} , \tag{68.23}$$

where $\boldsymbol{P}(\boldsymbol{R}, t)$ is the electric polarization, given by the electric dipole density of the medium $\boldsymbol{P} = N\boldsymbol{d}$, N being the atomic density, the plane wave dispersion relation is

$$k^2 = \frac{\omega^2}{c^2} n^2(\omega) , \quad (68.24)$$

where the index of refraction $n(\omega)$ is

$$n(\omega) = \sqrt{1 + \frac{N\alpha(\omega)}{\epsilon_0}} . \quad (68.25)$$

68.2.3 Beer's Law

Since the polarizability $\alpha(\omega)$ is normally complex, so is the index of refraction. Its real part leads to dispersive effects, while its imaginary part leads to absorption. Specifically, $\mathrm{Re}[n(\omega)] - 1$ has the form of a standard dispersion curve, positive for $\omega - \omega_0 < 0$ and negative for $\omega - \omega_0 > 0$, while $\mathrm{Im}[n(\omega)]$ is a Lorentzian curve peaked at $\omega = \omega_0$. The intensity absorption coefficient $a(\omega)$ is

$$a(\omega) = 2\,\mathrm{Im}\,[n(\omega)]\,\omega/c \quad (68.26)$$

$$= \frac{2\omega}{c}\,\mathrm{Im}\left[1 + \left(\frac{Ne^2}{m\epsilon_0}\right)\frac{\mathrm{i}\gamma\omega + (\omega_0^2 - \omega^2)}{(\omega_0^2 - \omega^2)^2 + \gamma^2\omega^2}\right]^{1/2} .$$

For atomic vapors, the corrections to the vacuum index of refraction are normally small, so that the square root in (68.26) can be expanded to first order, giving

$$a(\omega) = \left(\frac{Ne^2}{\epsilon_0 mc}\right)\frac{\gamma\omega^2}{(\omega_0^2 - \omega^2)^2 + \gamma^2\omega^2} . \quad (68.27)$$

The intensity of a monochromatic field propagating along the z-direction through a gas of Lorentz atoms is therefore attenuated according to *Beer's law* given by

$$I(\omega, z) = I(\omega, 0)\,\mathrm{e}^{-a(\omega)z} . \quad (68.28)$$

If the index of refraction at a given frequency becomes purely imaginary, no electromagnetic wave can propagate inside the medium. This is the case for field frequencies smaller than the plasma frequency

$$\omega_\mathrm{p} = \sqrt{\frac{Ne^2}{m\epsilon_0}} . \quad (68.29)$$

While the Lorentz atom model gives an adequate description of absorption and dispersion in a weakly excited absorbing medium, it fails to predict the occurence of important phenomena such as saturation and light amplification. This is because, in this model, the phase of the induced atomic dipoles with respect to the incident field is always such that the polarization field adds destructively to the incident field. The description of light amplification requires a quantum treatment of the medium, which gives a greater flexibility to the possible relative phases between the incident and polarization fields.

68.2.4 Slowly-Varying Envelope Approximation

Light–matter interactions often involve quasi-monochromatic fields for which the electric field (taken to propagate along the z-axis) can be expressed in the form

$$\boldsymbol{E}(\boldsymbol{R}, t) = \frac{1}{2}\boldsymbol{\epsilon}\mathcal{E}^+(\boldsymbol{R}, t)\,\mathrm{e}^{\mathrm{i}(kz-\omega t)} + \mathrm{c.c.} , \quad (68.30)$$

such that

$$\left|\frac{\partial \mathcal{E}^+}{\partial t}\right| \ll \omega\left|\mathcal{E}^+\right| , \quad \left|\frac{\partial \mathcal{E}^+}{\partial z}\right| \ll k\left|\mathcal{E}^+\right| . \quad (68.31)$$

It is further consistent within this approximation to assume that the polarization takes the form

$$\boldsymbol{P}(\boldsymbol{R}, t) = \frac{1}{2}\boldsymbol{\epsilon}\mathcal{P}^+(\boldsymbol{R}, t)\,\mathrm{e}^{\mathrm{i}(kz-\omega t)} + \mathrm{c.c.} , \quad (68.32)$$

with

$$\left|\frac{\partial \mathcal{P}^+}{\partial t}\right| \ll \omega\left|\mathcal{P}^+\right| . \quad (68.33)$$

Under these conditions, known as the *slowly-varying envelope approximation* ([68.5]), Maxwell's wave equation reduces to

$$\left(\frac{\partial}{\partial z} + \frac{1}{c}\frac{\partial}{\partial t}\right)\mathcal{E}^+(z, t) = -\frac{k}{2i\epsilon_0}\mathcal{P}^+(z, t) . \quad (68.34)$$

Hence, in the slowly-varying envelopes approximation we ignore the backward propagation of the field [68.12]. The slowly-varying amplitude and phase approximation is essentially the same, except that it expresses the electric field envelope in terms of a real amplitude and phase.

68.3 Two-Level Atoms

A large number of optical phenomena can be understood by considering the interaction between a quasi-monochromatic field of central frequency ω and a two-level atom, which simulates a (dipole-allowed) atomic transition [68.5, 7, 10, 11, 13–17]. This approximation is well justified for near-resonant interactions;

i. e., $\omega \simeq \omega_0$. The next three sections discuss the model Hamiltonian for this system in the *semiclassical* approximation where the electromagnetic field can be described classically. The formal results are then extended to the case of a quantized field, where the electric field is treated as an operator.

68.3.1 Hamiltonian

In the absence of dissipation mechanisms, the dipole interaction between a quasi-monochromatic classical field and a two-level atom is

$$H = \hbar\omega_e|e\rangle\langle e| + \hbar\omega_g|g\rangle\langle g| - \boldsymbol{d}\cdot\boldsymbol{E}(\boldsymbol{R}, t)\,, \quad (68.35)$$

where $|e\rangle$ and $|g\rangle$ label the upper and lower atomic levels, of frequencies ω_e and ω_g, respectively, with $\omega_e - \omega_g = \omega_0$, and $\boldsymbol{R}$ is the location of the center of mass of the atom. The electric dipole operator (68.8), couples the excited and ground levels, and may be expressed as

$$\boldsymbol{d} = \boldsymbol{\epsilon}_\mathrm{d} d\,(|e\rangle\langle g| + |g\rangle\langle e|)\,, \quad (68.36)$$

where $\boldsymbol{\epsilon}_\mathrm{d}$ is a unit vector in the direction of the dipole and d the matrix element of the electric dipole operator between the ground and excited state, which we take to be real for simplicity. We also neglect the vector character of $\boldsymbol{d}$ and $\boldsymbol{E}(\boldsymbol{R}, t)$ in the following, assuming, for example, that both $\boldsymbol{\epsilon}_\mathrm{d}$ and $\boldsymbol{\epsilon}$ are parallel to x-axis. The Hamiltonian (68.35) may then be expressed as

$$\begin{aligned} H = {} & \hbar\omega_e|e\rangle\langle e| + \hbar\omega_g|g\rangle\langle g| - d\,(|e\rangle\langle g| + |g\rangle\langle e|) \\ & \times\left[E^+(\boldsymbol{R}, t) + E^-(\boldsymbol{R}, t)\right]\,, \end{aligned} \quad (68.37)$$

where we have generalized the notation of (68.20) in an obvious way. One can introduce the *pseudo-spin operators*

$$\begin{aligned} s_z &= (|e\rangle\langle e| - |g\rangle\langle g|)/2\,, \\ s_+ &= s_-^\dagger = |e\rangle\langle g|\,, \end{aligned} \quad (68.38)$$

and redefine the zero of atomic energy to introduce the commonly used form

$$H = \hbar\omega_0 s_z - d\,(s_+ + s_-)\left[E^+(\boldsymbol{R}, t) + E^-(\boldsymbol{R}, t)\right]\,. \quad (68.39)$$

68.3.2 Rotating Wave Approximation

Under the influence of a monochromatic electromagnetic field of frequency ω, atoms undergo transitions between their lower and upper states by interacting with either the positive or the negative frequency part of the field. The corresponding contributions to the atomic dynamics oscillate at frequencies $\omega_0 - \omega$ and $\omega_0 + \omega$, respectively, and their contributions to the probability amplitudes involve denominators containing this same frequency dependence. For near-resonant atom-field interactions, the rapidly oscillating contributions lead to small corrections, the first-order one being the Bloch–Siegert shift, whose value near resonance, is $\omega \simeq \omega_0$ [68.17].

$$\delta\omega_{eg} = -\frac{\left(d|E^+|/\hbar\right)^2}{4\omega} \quad (68.40)$$

to lowest-order in $d\mathcal{E}/\hbar\omega$. The neglect of these terms is the Rotating Wave Approximation (RWA). Note that it is normally inconsistent to regard an atom as a two-level system and not to perform the RWA. In the RWA, the atomic system is described by the Hamiltonian

$$H = \hbar\omega_0 s_z - d\left[s_+ E^+(\boldsymbol{R}, t) + s_- E^-(\boldsymbol{R}, t)\right]\,, \quad (68.41)$$

or, in a frame rotating at the frequency ω of the field,

$$H = \hbar\Delta s_z - \frac{1}{2}d\left(s_+\mathcal{E}\,\mathrm{e}^{\mathrm{i}\boldsymbol{k}\cdot\boldsymbol{R}} + \mathrm{h.c.}\right)\,, \quad (68.42)$$

where $\Delta = \omega_0 - \omega$ is the atom-light detuning. (Note that the alternate definition $\delta = \omega - \omega_0$ is frequently used in the literature.) In the rest of this chapter, we consider atoms placed at $\boldsymbol{R} = 0$.

68.3.3 Rabi Frequency

The dynamics of the two-level atom is conveniently expressed in terms of its density operator ρ, whose evolution is given by the Schrödinger equation

$$\frac{\mathrm{d}\rho}{\mathrm{d}t} = -\frac{\mathrm{i}}{\hbar}[H, \rho]\,, \quad (68.43)$$

where $\rho_{ee} = \langle e|\rho|e\rangle$ and $\rho_{gg} = \langle g|\rho|g\rangle$ are the upper and lower state populations P_e and P_g, respectively, while the off-diagonal matrix elements $\rho_{eg} = \langle e|\rho|g\rangle = \rho_{ge}^\star$ are called the atomic coherences, or simply coherences, between levels $|e\rangle$ and $|g\rangle$. These coherences play an essential role in optical physics and quantum optics, since they are proportional to the expectation value of the electric dipole operator.

The evolution of $P_g(t)$ and $P_e(t) = 1 - P_g(t)$, is characterized by oscillations at the generalized Rabi frequency

$$\Omega = \left(\Omega_1^2 + \Delta^2\right)^{1/2}\,, \quad (68.44)$$

where the *Rabi frequency* Ω_1 is $\Omega_1 = d\mathcal{E}/\hbar$, (or $\Omega_1 = d\mathcal{E}(\boldsymbol{\epsilon}_{\rm d} \cdot \boldsymbol{\epsilon})/\hbar$ when the vector character of the electric field and dipole moment are included). Specifically, assuming that the atom is initially in its ground state $|g\rangle$, the probability that it is in the excited state $|e\rangle$ at a subsequent time t is given by Rabi's formula

$$P_e(t) = (\Omega_1/\Omega)^2 \sin^2(\Omega t/2) . \quad (68.45)$$

At resonance ($\Delta = 0$), the generalized Rabi frequency Ω reduces to the Rabi frequency Ω_1. (In addition to the texts on quantum optics already cited, see also [68.18].)

68.3.4 Dressed States

Semiclassical Case

The atomic dynamics can alternatively be described in terms of a *dressed states* basis instead of the *bare states* $|e\rangle$ and $|g\rangle$ (see especially [68.17]). The dressed states $|1\rangle$ and $|2\rangle$ are eigenstates of the Hamiltonian (68.42), and, by convention, the state $|1\rangle$ is the one with the greatest energy. They are conveniently expressed in terms of the bare states via the Stückelberg angle $\theta/2$ as

$$\begin{aligned} |1\rangle &= \sin\theta|g\rangle + \cos\theta|e\rangle , \\ |2\rangle &= \cos\theta|g\rangle - \sin\theta|e\rangle , \end{aligned} \quad (68.46)$$

where $\sin(2\theta) = -\Omega_1/\Omega$, $\cos(2\theta) = \Delta/\Omega$. The corresponding eigenenergies are

$$\begin{aligned} E_1 &= +\frac{1}{2}\hbar\Omega , \\ E_2 &= -\frac{1}{2}\hbar\Omega . \end{aligned} \quad (68.47)$$

These energies are illustrated in Fig. 68.1 as a function of the field frequency ω. The dressed levels repel each other and form an anticrossing at resonance $\omega = \omega_0$. As the detuning Δ varies from positive to negative values, state $|1\rangle$ passes continuously from the excited state $|e\rangle$ to the bare ground state $|g\rangle$, with both bare states having equal weights at resonance. The distances between the perturbed levels and their asymptotes for $|\Delta| \gg \Omega_1$ represent the ac Stark shifts, or light shifts, of the atomic states when coupled to the laser. From Fig. 68.1, the ac Stark shift of $|g\rangle$ is positive for $\Delta < 0$ and negative for $\Delta > 0$, while the $|e\rangle$ state shift is negative for $\Delta < 0$ and positive for $\Delta > 0$.

Quantized Field

The concept of dressed states can readily be generalized to a two-level atoms interacting with a single-mode quantized field in the dipole and rotating wave approximations. The atom and its dipole interaction with the field are still described by the Hamiltonian (68.41), except that the positive and negative frequency components of the field are now operators, and the free field Hamiltonian must be included. The Hamiltonian of the total atom-field system becomes

$$H = \hbar\omega_0 s_z + \hbar\omega\left(a^\dagger a + \frac{1}{2}\right) + \hbar g\left(s_+ a + a^\dagger s_-\right) , \quad (68.48)$$

where the creation and annihilation operators $a^\dagger$ and a obey the boson commutation relation $\left[a, a^\dagger\right] = 1$ (Chapt. 6), and the coupling constant

$$g = d\sqrt{\frac{\omega}{2\epsilon_0\hbar V}} \quad (68.49)$$

is the vacuum Rabi frequency, with V being a photon normalization volume. This Hamiltonian defines the Jaynes–Cummings model, [68.7, 19] which is discussed in more detail in Chapt. 78.

The dressed states of the atom-field system are the eigenstates of the Jaynes–Cummings model. Since, in the RWA, the dipole interaction only couples states of same "excitation number", e.g., $|e, n\rangle$ and $|g, n+1\rangle$, where $|n\rangle$ is an eigenstate of the photon number operator, $a^\dagger a|n\rangle = n|n\rangle$, with n an integer, the diagonalization of the Jaynes–Cummings model reduces to that of the semiclassical driven two-level atom in each of these manifolds. Hence, the dressed states are

$$\begin{aligned} |1, n\rangle &= \sin\theta_n|g, n+1\rangle + \cos\theta_n|e, n\rangle , \\ |2, n\rangle &= \cos\theta_n|g, n+1\rangle - \sin\theta_n|e, n\rangle , \end{aligned} \quad (68.50)$$

with

$$\tan(2\theta_n) = -2g\sqrt{n+1}/\Delta . \quad (68.51)$$

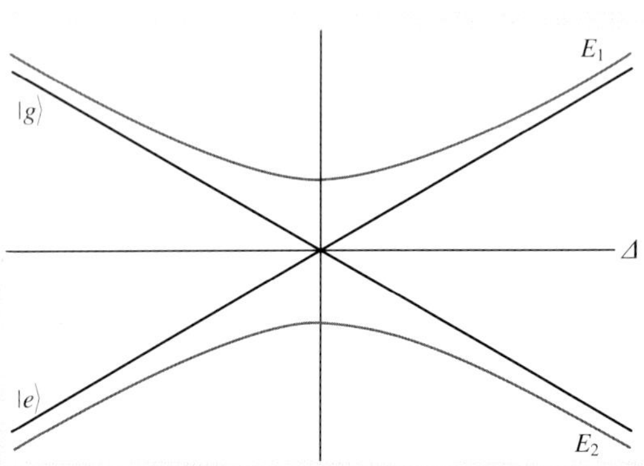

Fig. 68.1 Dressed levels of a two-level atom driven by a classical monochromatic field as a function of the detuning $\Delta = \omega_0 - \omega$

(The factor of 2 difference between this and the semiclassical case is due to the use of a running waves quantization scheme, while the semiclassical discussion was for standing waves.) The corresponding eigenenergies are

$$\begin{aligned} E_{1n} &= \hbar(n+1)\omega - \hbar R_n \,, \\ E_{2n} &= \hbar(n+1)\omega + \hbar R_n \,, \end{aligned} \tag{68.52}$$

where

$$R_n = \frac{1}{2}\sqrt{\Delta^2 + 4g^2(n+1)} \,. \tag{68.53}$$

Chapter 78 shows that by including the effects of spontaneous emission, this picture yields a straightforward interpretation of a number of effects, including the Burshtein–Mollow resonance fluorescence spectrum. Dressed states also help to elucidate the interaction between two-level atoms and quantized single-mode fields, as occur for example in cavity QED ([68.19] and Chapt. 79). Their generalization to the case of moving atoms offers simple physical interpretations of several aspects of laser cooling, see Chapt. 75 and [68.20].

68.3.5 Optical Bloch Equations

Introducing the density operator matrix elements $\rho_{ab} = \langle a|\rho|b\rangle$, where a, b can be either e or g, as well as the real quantities

$$\begin{aligned} U &= \rho_{eg}\,\mathrm{e}^{\mathrm{i}\omega t} + \text{c.c.} \,, \\ V &= \mathrm{i}\rho_{eg}\,\mathrm{e}^{\mathrm{i}\omega t} + \text{c.c.} \,, \\ W &= \rho_{ee} - \rho_{gg} \,, \end{aligned} \tag{68.54}$$

the equations of motion for the density matrix elements $\rho_{ij} = \langle i|\rho|j\rangle$ may be expressed, with (68.43), as

$$\begin{aligned} \frac{\mathrm{d}U}{\mathrm{d}t} &= -\Delta V \,, \\ \frac{\mathrm{d}V}{\mathrm{d}t} &= \Delta U + \Omega_1 W \,, \\ \frac{\mathrm{d}W}{\mathrm{d}t} &= -\Omega_1 V \,. \end{aligned} \tag{68.55}$$

These are the optical Bloch equations, as discussed extensively in [68.5, 17]. Physically, U describes the component of the atomic coherence in phase with the driving field, V the component in quadrature with the field, and W the atomic inversion. The optical Bloch equations have a simple geometrical interpretation offered by thinking of U, V and W as the three components of a vector called the Bloch vector $\boldsymbol{U}$, whose equation of motion is

$$\frac{\mathrm{d}\boldsymbol{U}}{\mathrm{d}t} = \boldsymbol{\Omega} \times \boldsymbol{U} \,, \tag{68.56}$$

where $\boldsymbol{\Omega} = (-\Omega_1, 0, \Delta)$. Thus $\boldsymbol{U}$ precesses about $\boldsymbol{\Omega}$, of length Ω, while conserving its length. The evolution of a two-level atom driven by a monochromatic field is thus mathematically equivalent to that of a spin-$\frac{1}{2}$ system in two magnetic fields $\boldsymbol{B}_0$ and $2\boldsymbol{B}_1 \cos\omega t$ which are parallel to the z- and x-axis, respectively, and whose amplitudes are such that the Larmor spin precession frequencies around them are ω and $2\Omega_1 \cos\omega t$, respectively. In optics, this vectorial picture is often referred to as the Feynman–Vernon–Hellwarth picture [68.21]. It is very useful in discussing the coherent transient phenomena discussed in Chapt. 73.

68.4 Relaxation Mechanisms

In addition to their coherent interaction with light fields, atoms suffer incoherent relaxation mechanisms, whose origin can be as diverse as elastic and inelastic collisions and spontaneous emission. Collisional broadening is discussed in Chapt. 59, while a QED microscopic discussion of spontaneous emission is described in Chapt. 78 in terms of reservoir theory. One advantage of describing the atomic state in terms of a density operator ρ is that the physical interpretation of its elements allows us phenomenologically to add various relaxation terms directly to its elements.

68.4.1 Relaxation Toward Unobserved Levels

If the relaxation mechanisms transfer populations or atomic coherences toward uninteresting or unobserved levels, their description can normally be given in terms of a Schrödinger equation, but with a complex Hamiltonian. In contrast, if all levels involved in the relaxation mechanism are observed, a more careful description, e.g. in terms of a master equation, is required. Specifically, in the case of relaxation to unobserved levels, the evolution of the atomic density operator, restricted to the levels of interest, is of the general form [68.16]

$$\frac{\mathrm{d}\rho}{\mathrm{d}t} = -\frac{\mathrm{i}}{\hbar}\left(H_{\text{eff}}\rho - \rho H_{\text{eff}}^{\dagger}\right) , \quad (68.57)$$

where

$$H_{\text{eff}} = H + \hat{\Gamma} , \quad (68.58)$$

H being the atom-field Hamiltonian and $\hat{\Gamma}$ the non-Hermitian relaxation operator, defined by its matrix elements

$$\langle n|\hat{\Gamma}|m\rangle = \frac{\hbar}{2i}\gamma_n \delta_{nm} . \quad (68.59)$$

Both inelastic collisions and spontaneous emission to unobserved levels can be described by this form of evolution. In the framework of this chapter, inelastic, or strong, collisions are defined as collisions that can induce atomic transitions into other energy levels.

68.4.2 Relaxation Toward Levels of Interest

A master equation description is necessary when all involved levels are observed [68.7, 17]. This master equation can rapidly take a complicated form if more than two levels are involved. We give results only for the case of a two-level atom and upper to lower-level spontaneous decay and elastic or soft collisions; i. e., collisions that change the separation of energy levels during the collision, but leave the level populations unchanged. In that case, the atomic master equation takes the form

$$\begin{aligned}\frac{\mathrm{d}\rho}{\mathrm{d}t} = &-\frac{\mathrm{i}}{\hbar}[H,\rho] - \frac{\Gamma}{2}\left(s_+s_-\rho + \rho s_+s_- - 2s_-\rho s_+\right)\\ &-\frac{1}{2}\gamma_{\text{ph}}\rho + 2\gamma_{\text{ph}}s_z\rho s_z ,\end{aligned} \quad (68.60)$$

where the free-space spontaneous decay rate Γ is found from QED to be

$$\Gamma = \frac{1}{4\pi\epsilon_0}\frac{4d^2\omega_0^3}{3\hbar c^3} , \quad (68.61)$$

and γ_{ph} is the decay rate due to elastic collisions.

It is possible to express the classical decay rate (68.17) in terms of the quantum spontaneous emission rate (68.61) as

$$\Gamma = \Gamma_{\text{cl}} f_{ge} , \quad (68.62)$$

where f_{ge} is the oscillator strength of the transition. The various oscillator strengths characterizing the dipole-allowed transitions from a ground state $|e\rangle$ to excited levels $|e\rangle$ obey the Thomas–Reiche–Kuhn sum rule

$$\sum_e f_{ge} = 1 , \quad (68.63)$$

where the sum is on all levels dipole-coupled to $|g\rangle$. Assuming that $\boldsymbol{d}$ and the polarization of the field are both parallel to the x-axis, this gives

$$f_{ge} = \frac{2m\omega_0}{\hbar}|\langle g|x|e\rangle|^2 . \quad (68.64)$$

68.4.3 Optical Bloch Equations with Decay

In general, the optical Bloch equations cannot be generalized to cases where relaxation mechanisms are present. There are, however, two notable exceptions corresponding to situations where

1. the upper level spontaneously decays to the lower level only, while the atom undergoes only elastic collisions;
2. spontaneous emission between the upper and lower levels can be ignored in comparison with decay to unobserved levels, which occur at equal rates $\gamma_e = \gamma_g = 1/T_1$.

Under these conditions, (68.55) generalizes to

$$\begin{aligned}\frac{\mathrm{d}U}{\mathrm{d}t} &= -U/T_2 - \Delta V ,\\ \frac{\mathrm{d}V}{\mathrm{d}t} &= -V/T_2 + \Delta U + \Omega_1 W ,\\ \frac{\mathrm{d}W}{\mathrm{d}t} &= -(W - W_{\text{eq}})/T_1 - \Omega_1 V ,\end{aligned} \quad (68.65)$$

where we have introduced the longitudinal and transverse relaxation times T_1 and T_2, with $T_1 = 1/\Gamma$ and $T_2 = (1/2T_1 + \gamma_{\text{ph}})^{-1}$ in the first case, and $T_2 = (1/T_1 + \gamma_{\text{ph}})^{-1}$ in the second case. The equilibrium inversion W_{eq} is equal to zero in the second case since the decay is to unobserved levels.

68.4.4 Density Matrix Equations

In the general case, it is necessary to consider the density operator equation (68.43) instead of the optical Bloch equations. The equations of motion for the components of ρ become, for the general case of complex Ω_1,

$$\begin{aligned}\frac{\mathrm{d}\rho_{ee}}{\mathrm{d}t} &= -\gamma_e\rho_{ee} - \frac{1}{2}\left(\mathrm{i}\Omega_1^*\tilde{\rho}_{eg} + \text{c.c.}\right) ,\\ \frac{\mathrm{d}\rho_{gg}}{\mathrm{d}t} &= -\gamma_g\rho_{gg} + \frac{1}{2}\left(\mathrm{i}\Omega_1^*\tilde{\rho}_{eg} + \text{c.c.}\right) ,\\ \frac{\mathrm{d}\tilde{\rho}_{eg}}{\mathrm{d}t} &= -(\gamma + \mathrm{i}\Delta)\tilde{\rho}_{eg} - \mathrm{i}\frac{\Omega_1}{2}\left(\rho_{ee} - \rho_{gg}\right) ,\end{aligned} \quad (68.66)$$

where $\gamma = (\gamma_e + \gamma_g)/2 + \gamma_{ph}$, and $\tilde{\rho}_{eg} = \rho_{eg}\, e^{i\omega t}$. In the case of spontaneous decay from the upper to the lower level, these equations become

$$\frac{d\rho_{ee}}{dt} = -\Gamma\rho_{ee} - \frac{1}{2}\left(i\Omega_1^*\tilde{\rho}_{eg} + \text{c.c.}\right),$$
$$\frac{d\rho_{gg}}{dt} = +\Gamma\rho_{ee} + \frac{1}{2}\left(i\Omega_1^*\tilde{\rho}_{eg} + \text{c.c.}\right),$$
$$\frac{d\tilde{\rho}_{eg}}{dt} = -(\gamma + i\Delta)\tilde{\rho}_{eg} - i\frac{\Omega_1}{2}\left(\rho_{ee} - \rho_{gg}\right), \quad (68.67)$$

where $\gamma = \Gamma/2 + \gamma_{ph}$. Equations (68.67) are completely equivalent to the optical Bloch equations (68.65).

68.5 Rate Equation Approximation

If the coherence decay rate γ is dominated by elastic collisions, and hence is much larger than the population decay rates γ_e and γ_g, $\tilde{\rho}_{eg}$ can be adiabatically eliminated from the equations of motion (68.66) and (68.67) to obtain the rate equations (Sect. 68.2)

$$\frac{d\rho_{ee}}{dt} = -\gamma_e\rho_{ee} - R\left(\rho_{ee} - \rho_{gg}\right),$$
$$\frac{d\rho_{gg}}{dt} = -\gamma_g\rho_{gg} + R\left(\rho_{ee} - \rho_{gg}\right), \quad (68.68)$$

and

$$\frac{d\rho_{ee}}{dt} = -\Gamma\rho_{ee} - R\left(\rho_{ee} - \rho_{gg}\right),$$
$$\frac{d\rho_{gg}}{dt} = +\Gamma\rho_{gg} + R\left(\rho_{ee} - \rho_{gg}\right), \quad (68.69)$$

respectively, where the transition rate is

$$R = |\Omega_1|^2 \mathcal{L}(\Delta)/(2\gamma), \quad (68.70)$$

and we have introduced the dimensionless Lorentzian

$$\mathcal{L}(\Delta) = \gamma^2/\left(\gamma^2 + \Delta^2\right). \quad (68.71)$$

The transitions between the upper and lower state are thus described in terms of simple rate equations.

Adding phenomenological pumping rates Λ_e and Λ_g on the right-hand side of (68.68) provides a description of the excitation of the upper and lower levels from some distant levels, as would be the case in a laser. The equations then form the basis of conventional, single-mode laser theory. In the absence of such mechanisms, the atomic populations eventually decay away.

68.5.1 Steady State

In the case of upper to lower-level decay, the state populations reach a steady state with inversion [68.5, 17]

$$W_{st} = -\frac{\Gamma}{\Gamma + 2R} = -\frac{1}{1+s}, \quad (68.72)$$

where s is the saturation parameter. In the case of pure radiative decay, $\gamma_{ph} = 0$, s is given by

$$s = \frac{\Omega_1^2/2}{\Gamma^2/4 + \Delta^2}. \quad (68.73)$$

In steady state, the other two components of the Bloch vector $\boldsymbol{U}$ are given by

$$U_{st} = -\frac{2\Delta}{\Omega_1}\left(\frac{s}{1+s}\right) \quad (68.74)$$

and

$$V_{st} = \frac{\Gamma}{\Omega_1}\left(\frac{s}{1+s}\right). \quad (68.75)$$

U_{st} varies as a dispersion curve as a function of the detuning Δ, while V_{st} is a Lorentzian of power-broadened half-width at half maximum $\left(\Gamma^2/4 + \Omega_1^2\right)^{1/2}$.

68.5.2 Saturation

As the intensity of the driving field, or Ω_1^2, increases, U_{st} and V_{st} first increase linearly with Ω_1, reach a maximum, and finally tend to zero as $\Omega_1 \to \infty$. The inversion W_{st}, which equals -1 for $\Omega_1 = 0$, first increases quadratically, and asymptotically approaches $W_{st} = 0$ as $\Omega_1 \to \infty$. At this point, where the upper and lower state populations are equal, the transition is said to be saturated, and the medium becomes effectively transparent, or bleached. (This should not be confused with self-induced transparency discussed in Chapt. 73.) The inversion is always negative, which means in particular that no steady-sate light amplification can be achieved in this system. This is one reason why external pump mechanisms are required in lasers.

68.5.3 Einstein A and B Coefficients

When atoms interact with broadband radiation instead of the monochromatic fields considered so far, (68.69) still apply, but the rate R becomes

$$R \to B_{eg}\varrho(\omega)\,, \tag{68.76}$$

where $\varrho(\omega)$ is the spectral energy density of the inducing radiation. Einstein's A and B coefficients apply to an atom in thermal equilibrium with the field, which is described by Planck's black-body radiation law

$$\varrho(\omega) = \frac{\hbar\omega^3}{\pi^2 c^3}\frac{1}{e^{\hbar\omega/k_B T}-1}\,, \tag{68.77}$$

where T is the temperature of the source and k_B is Boltzmann's constant. Invoking the principle of detailed balance, which states that at thermal equilibrium, the average number of transitions between arbitrary states $|i\rangle$ and $|k\rangle$ must be equal to the number of transitions between $|k\rangle$ and $|i\rangle$, one finds

$$\frac{A_{ki}}{B_{ki}} = \frac{\hbar\omega^3}{\pi^2 c^3}\,, \tag{68.78}$$

where A_{ki} is the rate of spontaneous emission from $|k\rangle$ to $|i\rangle$, and $\Gamma_k = \sum_i A_{ki}$ is the level width.

68.6 Light Scattering

Far from resonance, the approximation of a two- or few-level atom is no longer adequate. Two limiting cases, which are always far from resonance, are Rayleigh scattering for low frequencies, and Thomson scattering for high frequencies.

68.6.1 Rayleigh Scattering

Rayleigh scattering is the elastic scattering of a monochromatic electromagnetic field of frequency ω, wave vector $\boldsymbol{k}$ and polarization $\boldsymbol{\epsilon}$ by an atomic system in the limit where ω is very small compared with its excitation energies [68.4, 17]. To second-order in perturbation theory, the Rayleigh scattering differential cross section into the solid angle Ω' about the wave vector $\boldsymbol{k}'$ with $k' = k$, and polarization ϵ' is

$$\frac{d\sigma}{d\Omega'} = r_0^2\omega^4(\boldsymbol{\epsilon}\cdot\boldsymbol{\epsilon}')^2\left(\sum_e \frac{f_{ge}}{\omega_{eg}^2}\right)^2\,, \tag{68.79}$$

where r_0 is the classical electron radius, the sum is over all states $|e\rangle$, f_{ge} is the E1 oscillator strength (68.64) and ω_{ge} is the transition frequency. The corresponding total cross section is

$$\sigma = \frac{8\pi r_0^2\omega^4}{3}\left(\sum_e \frac{f_{ge}}{\omega_{eg}^2}\right)^2\,. \tag{68.80}$$

68.6.2 Thomson Scattering

Thomson scattering is the corresponding elastic photon scattering by an atom in the limit where ω is very large compared with the atomic ionization energy, yet small enough compared to $\alpha mc^2/\hbar$, that the dipole approximation can be applied. The differential cross section for this process is [68.4, 17]

$$\frac{d\sigma}{d\Omega'} = r_0^2\left(\boldsymbol{\epsilon}\cdot\boldsymbol{\epsilon}'\right)^2\,, \tag{68.81}$$

and the total cross section is

$$\sigma = \frac{8}{3}\pi r_0^2\,. \tag{68.82}$$

This is a completely classical result, which exhibits no frequency dependence.

68.6.3 Resonant Scattering

We finally consider elastic scattering in the limit where ω is close to the transition frequency ω_0 between $|g\rangle$ and $|e\rangle$. Provided that no other level is near-resonant with the ground state, the resonant scattering differential cross section is [68.6, 17]

$$\frac{d\sigma}{d\Omega'} = \frac{9}{16\pi^2}\lambda_0^2\left(\boldsymbol{\epsilon}\cdot\boldsymbol{\epsilon}'\right)^2\frac{(\Gamma/2)^2}{\Delta^2+(\Gamma/2)^2}\,, \tag{68.83}$$

where $\lambda_0 = 2\pi c/\omega_0$ is the wavelength of the transition and Γ is the spontaneous decay rate (68.61). The total elastic scattering cross section is

$$\sigma = \frac{3}{2\pi}\lambda_0^2\,. \tag{68.84}$$

In contrast to the nonresonant Rayleigh and Thomson scattering cross sections, which scale as the square of the classical electron radius, the resonant scattering cross section scales as the square of the wavelength. For optical to frequencies, $\lambda_0/r_0 \simeq 10^4$, giving a *resonant enhancement* of about eight orders of magnitude. This illustrates why near resonant phenomena, which form the bulk of the following chapters, are so important in optical physics and quantum optics.

References

68.1 B. W. Shore: *The Theory of Coherent Atomic Excitation* (Wiley-Interscience, New York 1990)

68.2 C. Cohen-Tannoudji, J. Dupont-Roc, G. Grynberg: *Photons and Atoms: Introduction to Quantum Electrodynamics* (Wiley-Interscience, New York 1989)

68.3 A. Sommerfeld: *Optics* (Academic, New York 1967)

68.4 J. D. Jackson: *Classical Electrodynamics* (Wiley, New York 1975)

68.5 L. Allen, J. H. Eberly: *Optical Resonance and Two-Level Atoms* (Dover, New York 1987)

68.6 P. W. Milonni, J. H. Eberly: *Lasers* (Wiley-Interscience, New York 1988)

68.7 P. Meystre, M. Sargent III: *Elements of Quantum Optics*, 3rd edn. (Springer, Berlin, Heidelberg 1999)

68.8 M. Born, E. Wolf: *Principles of Optics*, 4th edn. (Pergamon, Oxford 1970)

68.9 R. J. Glauber: Optical coherence and photon statistics. In: *Quantum Optics and Electronics*, ed. by C. DeWitt, A. Blandin, C. Cohen-Tannoudji (Gordon and Breach, New York 1965)

68.10 L. Mandel, E. Wolf: *Optical Coherence and Quantum Optics* (Cambridge Univ. Press, Cambridge 1995)

68.11 M. S. Zubairy, M. O. Scully: *Quantum Optics* (Cambridge Univ. Press, Cambridge 1997)

68.12 Y. R. Shen: *The Principles of Nonlinear Optics* (Wiley, New York 1984)

68.13 M. Orszag: *Quantum Optics* (Springer, Berlin, Heidelberg 2000)

68.14 W. P. Schleich: *Quantum Optics in Phase Space* (Wiley-VCH, Weinheim 2001)

68.15 M. Sargent III, M. O. Scully Jr., W. E. Lamb: *Laser Physics* (Wiley, Reading 1977)

68.16 S. Stenholm: *Foundations of Laser Spectroscopy* (Wiley-Interscience, New York 1984)

68.17 C. Cohen-Tannoudji, J. Dupont-Roc, G. Grynberg: *Atom-Photon Interactions: Basic Processes and Applications* (Wiley-Interscience, New York 1992)

68.18 P. L. Knight, P. W. Milonni: Phys. Rep. **66**, 21 (1980)

68.19 P. P. Berman (Ed.): *Cavity QED* (Academic, Boston 1994)

68.20 C. Cohen-Tannoudji: Atomic motion in laser light. In: *Fundamental Systems in Quantum Optics*, ed. by J. Dalibard, J. M. Raimond, J. Zinn-Justin (North-Holland, Amsterdam 1992)

68.21 R. P. Feynman, F. L. Vernon, R. W. Hellwarth: J. Appl. Phys. **28**, 49 (1957)

69. Absorption and Gain Spectra

This chapter develops theoretical techniques to describe absorption and emission spectra, using concepts introduced in Chapt. 68, and density matrix methods from Chapt. 7. The simplest cases are treated, compatible with the physics involved, and more realistic applications are referred to in other chapters. Vector notation is not used, but it can be inserted as required. Only steady-state spectroscopy is covered; for time-resolved transient techniques see Chapt. 73.

Laser technology has greatly expanded the potential of atomic and molecular spectroscopy, but the same techniques for describing the interaction of light with matter also apply to the traditional arc lamps and flash discharges, and the more recent synchrotron radiation sources.

69.1 Index of Refraction 1009
69.2 Density Matrix Treatment of the Two-Level Atom 1010
69.3 Line Broadening 1011
69.4 The Rate Equation Limit 1013
69.5 Two-Level Doppler-Free Spectroscopy ... 1015
69.6 Three-Level Spectroscopy 1016
69.7 Special Effects in Three-Level Systems ... 1018
69.8 Summary of the Literature 1020
References 1020

In many cases, departures from the thin sample limit, such as beam attenuation, light scattering and radiation trapping (Sect. 69.2) may be important. However, the properties of laser devices themselves depend in an essential way on these effects, making a self-consistent treatment of their properties necessary.

At the other extreme, the spectroscopy of dilute gases is well characterized by ensemble averages over the properties of the individual particles, interrupted by occasional brief collisions. Ensemble averages, however, may no longer apply to recent experiments probing a single atomic particle in a trap, as discussed in Chapt. 75.

69.1 Index of Refraction

As discussed in Sect. 68.2.2, the complex index of refraction for a medium containing harmonically bound charges (electrons) [69.1] with natural frequency ω_0 is

$$\begin{aligned} n(\omega) &= \sqrt{1+\frac{N\alpha(\omega)}{\varepsilon_0}} \approx 1+\frac{N\alpha(\omega)}{2\varepsilon_0} \\ &= 1+\frac{Ne^2}{2m\varepsilon_0}\left(\frac{\mathrm{i}\gamma\omega+(\omega_0^2-\omega^2)}{\left(\omega_0^2-\omega^2\right)^2+\gamma^2\omega^2}\right) \\ &= n'+\mathrm{i}n'' . \end{aligned} \tag{69.1}$$

The expansion is valid when the density of atoms N is low.

A plane wave can be written in the form

$$\mathcal{E} \propto \mathrm{e}^{\mathrm{i}kz} = \mathrm{e}^{\mathrm{i}\omega nz/c} = \mathrm{e}^{\mathrm{i}\omega n'z/c}\,\mathrm{e}^{-\omega n''z/c} . \tag{69.2}$$

The absorption of light through the medium then shows a resonant behavior near $\omega \approx \omega_0$ determined by

$$\begin{aligned} n'' &= \frac{Ne^2}{2m\varepsilon_0}\left(\frac{\gamma\omega}{\left(\omega_0^2-\omega^2\right)^2+\gamma^2\omega^2}\right) \\ &\approx \frac{\pi Ne^2}{4m\varepsilon_0\omega_0}\left(\frac{\gamma/2\pi}{(\omega-\omega_0)^2+\gamma^2/4}\right) . \end{aligned} \tag{69.3}$$

This is called an absorptive lineshape. When the single electron is harmonically bound, its interaction with radiation is found in this response. For a real atom, the response of the electron is divided among the various transitions to other states. The fraction assigned to one single transition is characterized by the oscillator strength f_n as discussed in Sect. 68.4.2.

In ordinary linear spectroscopy, the laser is tuned through the resonance $\omega \approx \omega_0$, and the value of ω_0 is determined from the lineshape (69.3). Several closely spaced resonances can be *resolved* if their spacing is larger than their widths

$$\left|\omega_0^{(1)} - \omega_0^{(2)}\right| > \gamma \,. \tag{69.4}$$

This defines the *spectral resolution* (Chapt. 10).

The velocity of light in the medium is seen to be given by the expression

$$c_{\text{eff}} = \frac{c}{n'} \approx c\left[1 - \frac{Ne^2}{2m\varepsilon_0}\left(\frac{\omega_0^2 - \omega^2}{\left(\omega_0^2 - \omega^2\right)^2 + \gamma^2\omega^2}\right)\right] . \tag{69.5}$$

This expression shows a *dispersive behavior* around the position $\omega = \omega_0$, where the modification of the velocity disappears. Below resonance $\omega < \omega_0$, the velocity of light is lower than in vacuum. This derives from the fact that the polarization is in phase with the driving field. Thus, by storing the incoming energy, the driving field retards the propagation of the radiation.

For a harmonically bound charge, the refractive index (69.1) always stays absorptive and it is independent of the intensity of the laser radiation. This no longer holds for discrete level atomic systems. In order to see this, we consider the *two-level atom* in Sect. 69.2 (Sect. 68.3).

69.2 Density Matrix Treatment of the Two-Level Atom

The response of atoms to light is conveniently expressed in terms of the density matrix ρ. In addition to the direct physical meaning of the density matrix elements discussed in Sect. 68.3.3, the density matrix formalism is advantageous because the various relaxation mechanisms effecting the atomic resonances can be introduced phenomenologically into its equations of motion (Sect. 68.4), and theoretical derivations often provide *master equations* for the density matrix (Chapt. 7).

The two-level Hamiltonian (68.35) can be written as

$$\boldsymbol{H} = \begin{bmatrix} \hbar\omega_0/2 & -d\mathcal{E}(R,t) \\ -d\mathcal{E}(R,t) & -\hbar\omega_0/2 \end{bmatrix} . \tag{69.6}$$

The equation of motion for the density matrix (68.67) is then

$$\begin{aligned}
\frac{\mathrm{d}}{\mathrm{d}t}\rho_{ee} &= -\Gamma\rho_{ee} + \frac{\mathrm{i}d\mathcal{E}}{\hbar}\cos\omega t\left(\rho_{ge} - \rho_{eg}\right) ,\\
\frac{\mathrm{d}}{\mathrm{d}t}\rho_{gg} &= \Gamma\rho_{ee} - \frac{\mathrm{i}d\mathcal{E}}{\hbar}\cos\omega t\left(\rho_{ge} - \rho_{eg}\right) ,\\
\frac{\mathrm{d}}{\mathrm{d}t}\rho_{eg} &= -(\gamma + \mathrm{i}\omega_0)\rho_{eg} + \frac{\mathrm{i}d\mathcal{E}}{\hbar}\cos\omega t\left(\rho_{gg} - \rho_{ee}\right) .
\end{aligned} \tag{69.7}$$

Here Γ is the spontaneous decay rate given by (68.61) and

$$\gamma = \Gamma/2 + \gamma_{\text{ph}} \,, \tag{69.8}$$

where γ_{ph} derives from all processes that tend to randomize the phase between the quantum states $|e\rangle$ and $|g\rangle$, such as collisions (Chapts. 7 and 19), noise in the laser fields and thermal excitation of the environment in solid state spectroscopy. The Greek letters Γ and γ correspond to the longitudinal relaxation rate (T_1-process) and the transverse relaxation rate (T_2-process), respectively (Sect. 68.4.3).

In the rotating wave approximation (RWA), discussed in Sect. 68.3.2, the density matrix equations (69.7) become identical to (68.67) with

$$\rho_{eg} = \tilde{\rho}_{eg}\,\mathrm{e}^{-\mathrm{i}\omega t} \,. \tag{69.9}$$

Using the condition of conservation of probability

$$\rho_{ee} + \rho_{gg} = 1 \,, \tag{69.10}$$

the steady state solutions to (68.67) are [69.2]

$$\rho_{ee} = \frac{\Omega_1^2\gamma}{2\Gamma}\left(\frac{1}{\Delta^2 + \gamma^2 + \Omega_1^2\gamma/\Gamma}\right) , \tag{69.11}$$

$$\begin{aligned}
\tilde{\rho}_{eg} &= \frac{\mathrm{i}\Omega_1}{2}\frac{\left(\rho_{gg} - \rho_{ee}\right)}{\gamma + \mathrm{i}\Delta}\\
&= \frac{\mathrm{i}\Omega_1}{2}\left(\frac{\gamma - \mathrm{i}\Delta}{\Delta^2 + \gamma^2 + \Omega_1^2\gamma/\Gamma}\right) ,
\end{aligned} \tag{69.12}$$

where Ω_1 is the Rabi frequency from (68.44), and $\Delta = \omega_0 - \omega$ is the detuning. The induced polarization is then

$$\begin{aligned}
\mathcal{P} &= N\mathrm{Tr}\left(\hat{d}\rho\right) = N\mathrm{Tr}\left[\begin{pmatrix} 0 & d \\ d & 0 \end{pmatrix}\begin{pmatrix} \rho_{ee} & \rho_{eg} \\ \rho_{ge} & \rho_{gg} \end{pmatrix}\right]\\
&= Nd\left(\mathrm{e}^{-\mathrm{i}\omega t}\tilde{\rho}_{eg} + \mathrm{e}^{\mathrm{i}\omega t}\tilde{\rho}_{ge}\right)\\
&= N\left(\alpha\mathcal{E}^{(+)} + \alpha^*\mathcal{E}^{(-)}\right) ,
\end{aligned} \tag{69.13}$$

where N is the density of active two-level atoms. Setting

$$\mathcal{E}^{(+)} = \frac{1}{2}\mathcal{E}\,\mathrm{e}^{-\mathrm{i}\omega t}\,, \tag{69.14}$$

the complex polarization is

$$\alpha(\omega) = \frac{d^2}{\hbar}\left(\frac{\mathrm{i}\gamma+\Delta}{\Delta^2+\gamma^2+\Omega_1^2\gamma/\Gamma}\right), \tag{69.15}$$

and from (69.1), the complex index of refraction is

$$\begin{aligned} n(\omega) &= 1+\frac{N\alpha(\omega)}{2\varepsilon_0} \\ &= 1+\frac{\pi N e^2}{4\varepsilon_0 m\omega_0}\left(\frac{f_0}{\pi}\right)\frac{\mathrm{i}\gamma+\Delta}{\Delta^2+\gamma^2+\Omega_1^2\gamma/\Gamma}\,, \end{aligned} \tag{69.16}$$

where $f_0 = 2d^2 m\omega_0/\hbar e^2$ is the oscillator strength. Summing over all possible transitions yields the f-sum rule (68.63) (Chapt. 21).

The imaginary part of (69.16) shows exactly the same absorptive behavior as in the harmonic oscillator model of Sect. 69.1 [see (69.3)]. However, the additional factor of $\left(\Omega_1^2\gamma/\Gamma\right)$ in the denominator makes the line appear broader than in the harmonic case; the line is power broadened. Physically, this derives from a saturation of the two-level system in which the population of the upper level becomes an appreciable fraction of that of the lower level. In the limit $\Omega_1 \to \infty$, (69.16) shows that $n(\omega) \to 1$ and the atom-field interaction effectively vanishes. In this limit, $\rho_{ee} \to \frac{1}{2}$ (69.11), and the field induces as many upward transitions as downward transitions.

When radiation at the frequency ω propagates in a medium of two-level atoms, the energy density is

$$I(z) \propto \mathcal{E}^*\mathcal{E} \propto \exp\left[-\left(\frac{Nd^2\omega}{\hbar\varepsilon_0 c}\right)\frac{\gamma z}{(\omega-\omega_0)^2+\gamma^2}\right]. \tag{69.17}$$

Far from resonance ($|\omega-\omega_0| \gg \gamma$), the medium is transparent; but near resonance, damping is observed. The impinging radiation energy is deposited in the medium and propagation is impeded. This is called radiation trapping. In spectroscopy, the phenomenon is seen as a prolongation of the radiative decay time; the spontaneously emitted energy is seen to emerge from the sample more slowly than the single atom lifetime implies.

69.3 Line Broadening

The effective width of a spectral line from (69.16) is

$$\gamma_{\text{eff}} = \sqrt{\gamma^2+\Omega_1^2\frac{\gamma}{\Gamma}} \simeq \frac{1}{2}\Gamma+\gamma_{\text{ph}}+\frac{\Omega_1^2}{2\Gamma}+O\left(\Omega_1^4\right). \tag{69.18}$$

The various contributions are as follows [69.3]. The term γ contains all the transverse relaxation mechanisms. If decay to additional levels occurs, these must be included (Sect. 69.4). The term γ_{ph} contains all perturbing effects effecting each single atom. For low enough pressures, collisional perturbations are proportional to the density of perturbing atoms so that

$$\gamma_{\text{ph}} = \eta_{\text{coll}}\,p\,, \tag{69.19}$$

where p is the pressure of the perturbing gas and η_{coll} is a constant of proportionality. This is called *collision* or *pressure broadening* (Chapt. 59), whose order of magnitude can be estimated to be the inverse of the average free time between collisions. For high pressure (usually of the order of torrs), the linearity in (69.19) breaks down. When identical atoms collide, *resonant exchange* of energy may also take place. The third term in (69.18) is the *power broadening* term. It derives from the effect of the laser field on each individual atom. All such relaxation processes that are active on each and every individual atom separately are called *homogeneous broadening* processes. For a detailed discussion of *line broadening*, consult Chapt. 19.

In contrast to homogeneous broadening, Doppler broadening is characterized by an atomic velocity parameter v which varies over the observed assembly. In a thermal assembly at temperature T (e.g., a gas cell), the velocity distribution is

$$P(v) = \frac{1}{\sqrt{2\pi u^2}}\exp\left(-\frac{v^2}{2u^2}\right), \tag{69.20}$$

where $u^2 = k_{\text{B}}T/M$. A particular atom with velocity v in the direction of the optical beam with wave vector k then experiences the Doppler-shifted frequency $\omega - kv$ relative to a stationary atom, and the effective detuning becomes

$$\Delta' = \Delta + kv\,, \tag{69.21}$$

replacing Δ. The population in the lower level from (69.11) is then

$$\rho_{gg} = 1 - \frac{\Omega_1^2 \gamma}{2\Gamma} \left(\frac{1}{(\Delta + kv)^2 + \gamma^2 + \Omega_1^2 \gamma / \Gamma} \right) . \tag{69.22}$$

The atoms in the lower level, originally distributed according to (69.20), are now depleted from the velocity group around

$$v = (\omega - \omega_0)/k . \tag{69.23}$$

The width of the depleted region is given by γ_{eff} of (69.18). This region is called a *Bennett hole*. When the laser frequency ω is tuned, the hole sweeps over the velocity distribution of the atoms. The atomic response is saturated at the velocity group of the hole, indicating that *spectral hole burning* has occurred.

The observed spectrum is obtained by averaging the single atom response (69.15) over the velocity distribution. From the imaginary part, the absorption response is

$$\alpha''(\omega) = \frac{d^2}{\hbar \sqrt{2\pi u^2}} \int_{-\infty}^{+\infty} \frac{\gamma \, \mathrm{e}^{-v^2/2u^2}}{(\Delta + kv)^2 + \gamma^2 + \Omega_1^2 \gamma / \Gamma} \, \mathrm{d}v . \tag{69.24}$$

In the limits $\Omega_1 \to 0$ (no saturation), and $\gamma \ll ku$ (the Doppler limit), the Lorentzian line shape sweeps over the entire velocity profile, finding a resonant velocity group according to (69.23) as long as $v \le u$. Thus, the linear spectroscopy sees a *Doppler broadened line* of width ku. This is called *inhomogeneous broadening*.

In the unsaturated regime, the atomic response function (69.24) is proportional to the imaginary part of the function

$$V(z) = \frac{1}{\sqrt{2\pi\sigma^2}} \int_{-\infty}^{+\infty} \frac{\exp\left(-x^2/2\sigma^2\right)}{z - x} \, \mathrm{d}x , \tag{69.25}$$

at $z = -\Delta - \mathrm{i}\gamma$ and $\sigma = ku$. This is the Hilbert transform of the Gaussian, and its shape is called a *Voigt profile* [69.4]. For $\gamma \ll ku$, it traces over the Gaussian, but for large detunings it always goes to zero as slowly as the Lorentzian, i. e., as Δ^{-1}. The profile has been widely used to interpret the data of *linear spectroscopy*. The function is tabulated [69.5] and its expansion is

$$V(z) = \sqrt{-\frac{\pi}{2\sigma^2}} \sum_{n=0}^{\infty} \frac{\left(\mathrm{i}z/\sqrt{2}\sigma\right)^n}{\Gamma(1 + n/2)} , \tag{69.26}$$

and it has the continued fraction representation,

$$V(z) = \cfrac{1}{z - \cfrac{\sigma^2}{z - \cfrac{2\sigma^2}{z - \cfrac{3\sigma^2}{z - \cfrac{4\sigma^2}{z - \dots}}}}} . \tag{69.27}$$

In addition to velocity, any other parameter shifting the individual atomic resonance frequencies ω_0 by different amounts for the different individuals leads to *inhomogeneous broadening*. The detuning Δ is then different for different members of the observed assembly, and a line shape similar to (69.24) applies. The distribution function must be replaced by the one relevant for the problem. In practice a Gaussian is almost always assumed.

An example of inhomogeneous broadening is the influence of the lattice environment on impurity spectroscopy in solids. The resonant light selectively excites atoms at those particular positions which make the atoms resonant. Thus only these spatial locations are saturated, and the phenomenon of *spatial hole burning* occurs. This has been investigated as a method for storing information, signal processing, and volume holography.

It is, however, possible that the effects of collisions can counteract the inhomogeneous broadening. In order to see this we observe that the induced atomic dipole is proportional to the induced density matrix element, from (69.9),

$$\rho_{eg} = \tilde{\rho}_{eg} \, \mathrm{e}^{\mathrm{i}(kz(t) - \omega t)} \propto \mathrm{e}^{\mathrm{i}kvt} . \tag{69.28}$$

If the atoms now experience collisions characterized by an average free time of flight τ, the phase kvt cannot build up coherently for times longer than this duration, the atomic velocity is quenched on the average, and the full Doppler profile cannot be observed. To see how this comes about, consider a time $t \gg \tau$. During this period the atom experiences on the average $\bar{n} = t/\tau$ collisions. Assuming a Poisson distribution, the probability of n collisions in time t is

$$p_n = \frac{\mathrm{e}^{-\bar{n}}}{n!} \bar{n}^n . \tag{69.29}$$

Taking the average of (69.28) over the time $t = n\tau$ with the distribution p_n yields

$$\begin{aligned} \bar{\rho}_{eg} &\propto \sum_{n=0}^{\infty} \mathrm{e}^{\mathrm{i}kv\tau n} \left(\frac{\mathrm{e}^{-\bar{n}}}{n!} \bar{n}^n \right) \\ &= \mathrm{e}^{-\bar{n}} \exp\left(\bar{n} \, \mathrm{e}^{\mathrm{i}kv\tau} \right) \\ &\approx \mathrm{e}^{\mathrm{i}kvt} \exp\left(-\frac{1}{2} t k^2 v^2 \tau \right) . \end{aligned} \tag{69.30}$$

This heuristic derivation suggests that for long enough interaction times ($t \gg \tau$), the large velocity components are suppressed. This tends to prevent the tails of the velocity distribution from contributing to the observed spectral profile. The effect is called *collisional narrowing* or *Dicke narrowing*. It is an observable effect, but the narrowing cannot be very large. To overcome the Doppler broadening one has to turn to nonlinear laser methods (Sect. 69.5).

69.4 The Rate Equation Limit

Consider now a generalized theory for the case of several incoming electromagnetic fields of the form

$$\mathcal{E}(R,t) = \sum_i \frac{1}{2}\mathcal{E}_i(R)\,\mathrm{e}^{-\mathrm{i}\omega_i t + \mathrm{i}\varphi_i} + \text{c.c.}\,. \tag{69.31}$$

The index i may range over several laser sources, the output of a multimode laser or the multitude of components of a flashlight or a thermal source. Each component carries its own amplitude $\mathcal{E}_i$.

In steady state, the generalization of (69.12) becomes

$$\rho_{eg} = \frac{\mathrm{i}}{2}\sum_i \frac{d\mathcal{E}_i}{\hbar}\frac{\left(\rho_{gg}-\rho_{ee}\right)}{\gamma + i\Delta_i}\mathrm{e}^{-\mathrm{i}\omega_i t + \mathrm{i}\varphi_i}\,, \tag{69.32}$$

where the detuning is

$$\Delta_i = \omega_0 - \omega_i\,. \tag{69.33}$$

The response of the atom now separates into individual contributions oscillating at the various frequencies ω_i according to

$$\rho_{eg} = \sum_i \rho_{eg}^{(i)}\,\mathrm{e}^{-\mathrm{i}\omega_i t + \mathrm{i}\varphi_i}\,. \tag{69.34}$$

This resolution is of key importance to the theory.

With the multimode field, (68.67) for the level occupation probabilities become

$$\begin{aligned}\frac{\mathrm{d}}{\mathrm{d}t}\rho_{ee} =& -\frac{\mathrm{d}}{\mathrm{d}t}\rho_{gg}\\ =& -\Gamma\rho_{ee} + \frac{\mathrm{i}}{2}\sum_i\\ &\times\left(\frac{d\mathcal{E}_i}{\hbar}\mathrm{e}^{-\mathrm{i}\omega_i t + \mathrm{i}\varphi_i}\rho_{ge} - \text{c.c.}\right)\,. \end{aligned}\tag{69.35}$$

Insertion of the steady state result (69.32) into this equation yields a closed set of equations for the level occupation probabilities. These are called *rate equations.*

To justify the above steps, consider the single frequency case again. The off-diagonal time derivatives can be neglected when

$$\left|\frac{\mathrm{d}}{\mathrm{d}t}\rho_{eg}\right| \ll |\gamma + \mathrm{i}\Delta||\rho_{eg}|\,. \tag{69.36}$$

This can be surmised to hold when the phase relaxation contributions to γ are large (69.8) or the detuning $|\Delta|$ is large. Insertion of (69.32) into (69.35) for the single mode case gives

$$\frac{\mathrm{d}}{\mathrm{d}t}\rho_{ee} = -\Gamma\rho_{ee} - W\left(\rho_{ee}-\rho_{gg}\right)\,, \tag{69.37}$$

where the rate coefficient is given by

$$W = 2\pi\left(\frac{d\mathcal{E}}{2\hbar}\right)^2\frac{\gamma/\pi}{\Delta^2+\gamma^2}\,. \tag{69.38}$$

Multiplying (69.37) by the density of active atoms N then produces the conventional rate equations for the populations $N_{ee} = N\rho_{ee}$ and $N_{gg} = N\rho_{gg}$.

Two physical effects can be discerned in (69.37): *induced* and *spontaneous emission*. The term with Γ gives the spontaneous emission which forces the entire population to the lower level. The terms proportional to W describe induced emission, with upward transitions proportional to $W\rho_{gg}$ and downward transitions proportional to $W\rho_{ee}$. In the absence of spontaneous emission, they strive to equalize the population of the two levels. Using (69.10), the steady-state solution is

$$\begin{aligned}\rho_{ee} &= \frac{W}{\Gamma + 2W}\\ &= \frac{1}{2}\left(\frac{d\mathcal{E}}{\hbar}\right)^2\frac{\gamma}{\Gamma}\frac{1}{\Delta^2+\gamma^2+(d\mathcal{E}/\hbar)^2\gamma/\Gamma}\,.\end{aligned}\tag{69.39}$$

This is clearly seen to agree with the solution (69.11) as is expected in steady state.

Although the rate equations were derived in the limit (69.36), the rate coefficient (69.38) has a special significance in the limit $\gamma \to 0$. In this limit, the factor

$$\begin{aligned}\frac{1}{\hbar}\lim_{\gamma\to 0}\left(\frac{\gamma/\pi}{\Delta^2+\gamma^2}\right) &= \frac{1}{\hbar}\delta(\Delta)\\ &= \delta(\hbar\omega - \hbar\omega_0)\,,\end{aligned}\tag{69.40}$$

enforces energy conservation in the transition. Using the field in (69.14), and the interaction from (69.6), the off-diagonal matrix element is

$$|\langle e|H|g\rangle| = \frac{1}{2} d\mathcal{E} \,. \tag{69.41}$$

With these results, (69.38) can be written in the form of *Fermi's Golden Rule*

$$W = \frac{2\pi}{\hbar} |\langle e|H|g\rangle|^2 \delta(\hbar\omega - \hbar\omega_0) \,, \tag{69.42}$$

usually derived from time dependent perturbation theory.

Returning to the multimode rate equations, an incoherent *broad band light source* has many components that contribute to the sum over field frequencies. In the case of flash pulses, thermal light sources, or free-running multimode lasers the spectral components are uncorrelated. Inserting (69.32) into (69.35) yields

$$\begin{aligned}\frac{\mathrm{d}}{\mathrm{d}t}\rho_{ee} &= -\frac{\mathrm{d}}{\mathrm{d}t}\rho_{gg} \\ &= -\Gamma\rho_{ee} + \frac{d^2}{2\hbar^2}\left(\rho_{gg} - \rho_{ee}\right) \\ &\quad \times \sum_{i,j} \mathcal{E}_i \mathcal{E}_j \, \mathrm{e}^{\mathrm{i}(\omega_j - \omega_i)t} \, \mathrm{e}^{\mathrm{i}(\varphi_i - \varphi_j)} \frac{\gamma}{\Delta_j^2 + \gamma^2} \,.\end{aligned} \tag{69.43}$$

The contributions from the different terms $i \neq j$ average to zero either by beating at the frequencies $|\omega_i - \omega_j|$ or by incoherent effects from the random phases φ_j. Thus, only the coherent sum survives to give

$$\begin{aligned}\frac{\mathrm{d}}{\mathrm{d}t}\rho_{ee} &= -\frac{\mathrm{d}}{\mathrm{d}t}\rho_{gg} \\ &= -\Gamma\rho_{ee} \times \sum_i W^{(i)}\left(\rho_{gg} - \rho_{ee}\right) \,,\end{aligned} \tag{69.44}$$

where the rate coefficients $W^{(i)}$ are given by (69.38) with the appropriate detunings $\Delta_j = \omega_0 - \omega_j$. This is a *rate equation* in the limit of many uncorrelated components of light, i. e., for a *broad band light source*. In this case the incoherence between the different components justifies the use of a rate approach, and no assumption like (69.36) is needed. Thus, the limit $\gamma \to 0$ (69.40) is legitimate, and the $W^{(i)}$ can be calculated in time dependent perturbation theory from Fermi's Golden Rule.

In the limit of an incoherent broad band light source, the sum in (69.44) can be replaced by an integral. In particular, this is allowed for incandescent light sources as used in *optical pumping* experiments [69.6]. Pumping of lasers by strong lamps or flashes are also describable by the same rate equations.

In amplifiers and lasers, the atoms must be brought into states far from equilibrium by incoherent optical excitation or resonant transfer of excitation energy in collisions (Chapt. 70). In the two-level description, the atomic levels are constantly replenished. The normalization condition (69.10) is then no longer appropriate; often the density matrix is normalized so that $\mathrm{Tr}(\rho)$ directly gives the density of active atoms.

With pumping into the levels, one must allow for decay out of the two-level system in order to prevent the atomic density from growing in an unlimited way. This decay takes the atom to unobserved levels. In the rate equation approximation, the pumping and decay processes can be described by terms added to the equations of the form (68.68)

$$\begin{aligned}\frac{\mathrm{d}}{\mathrm{d}t}\rho_{ee} &= \lambda_e - \gamma_e\rho_{ee} \,, \\ \frac{\mathrm{d}}{\mathrm{d}t}\rho_{gg} &= \lambda_g - \gamma_g\rho_{gg} \,.\end{aligned} \tag{69.45}$$

In a laser, the level $|g\rangle$ is usually not the ground state of the system.

From (69.45), the steady state population is

$$\rho_{gg}^{(0)} - \rho_{ee}^{(0)} = \lambda_g/\gamma_g - \lambda_e/\gamma_e \,. \tag{69.46}$$

A *population inversion* exists when this is negative. The population difference (69.46) is modified when the effects of spontaneous and induced processes are added, as in (69.37). Then the transitions saturate because of the *induced processes*.

Using (69.46) in (69.12), the calculated polarizability without saturation is

$$\alpha(\omega) = \left(\frac{d^2}{\hbar}\right)\left(\frac{\mathrm{i}\gamma + \Delta}{\gamma^2 + \Delta^2}\right)\left(\rho_{gg}^{(0)} - \rho_{ee}^{(0)}\right) \,. \tag{69.47}$$

According to (69.1), the index of refraction is

$$\begin{aligned}n'(\omega) &= 1 + \frac{N\alpha'(\omega)}{2\varepsilon_0} \\ &= 1 - \left(\frac{Nd^2}{2\varepsilon_0\hbar}\right)\frac{(\omega - \omega_0)}{(\omega - \omega_0)^2 + \gamma^2}\left(\rho_{gg}^{(0)} - \rho_{ee}^{(0)}\right) \,.\end{aligned} \tag{69.48}$$

For weakly excited atoms, $\rho_{gg} \approx 1$, and (69.47) agrees with the unsaturated limit of (69.16). The dispersion of a light signal behaving according to (69.48) is called *normal*, i. e., according to the harmonic model in Sect. 69.1. Below resonance ($\omega < \omega_0$), n' is larger than unity, implying a reduction of the velocity of light. As a function of ω, the curve (69.48) starts above unity, and passes below unity for $\omega > \omega_0$. This is *normal dispersion*.

However, for an *inverted medium* $\left(\rho_{gg}^{(0)} < \rho_{ee}^{(0)}\right)$, n' is less than unity for low frequencies and goes through unity with a positive slope. This is called *anomalous dispersion* and signifies the presence of a *gain profile*. In such a medium, $\alpha'' = \mathrm{Im}[\alpha(\omega)]$ is of opposite sign, as seen from (69.47); in the inverted medium, α'' becomes negative near $\Delta = 0$. From (69.2), this indicates a growing electromagnetic field, i. e., an *amplifying medium*. The amplitude grows, and the assumption of a small signal becomes invalid. Then saturation has to be included, either at the rate equation level or by performing a full density matrix calculation. This regime describes a laser with *saturated gain*. In steady state, the two levels become nearly equally populated ($\rho_{ee} \approx \rho_{gg}$), and the operation is stable. The theory of the laser is discussed in detail in Chapt. 70.

69.5 Two-Level Doppler-Free Spectroscopy

The *linear absorption* of a scanned laser signal defines *linear spectroscopy* and gives information characterizing the sample. However, Sect. 69.3 shows that inhomogeneous broadening masks the desired information by dominating the line shape. The availability of laser sources has made it possible to overcome this limitation, and to use the *saturation* properties of the medium to perform *nonlinear spectroscopy*. This section discusses how Doppler broadening can be eliminated to achieve *Doppler-free spectroscopy*. Similar techniques may be used to overcome other types of *inhomogeneous line broadening*; a general name is then *hole-burning spectroscopy* (see the discussion in Sect. 69.3). Other aspects of nonlinear matter-light interaction are found in Chapt. 72.

Equation (69.24) shows that a single laser cannot resolve beyond the Doppler width. However, if a strong laser is used to pump the transition, a weak *probe signal* can see the hole burned into the spectral profile by the pump. This technique is called *pump-probe spectroscopy*. Because the probe is taken to be weak, perturbation theory may be used to calculate the induced polarization to lowest order in the probe amplitude only.

In the field expansion (69.31), define the strong pump amplitude to be $\mathcal{E}_1$ and the weak probe $\mathcal{E}_2$ at frequency ω_2. From the resolution (69.34), the component $\rho_{eg}^{(2)}$ carries the information about the linear response at frequency ω_2. If the field $\mathcal{E}_2$ propagates in a direction opposite to that of $\mathcal{E}_1$, its detuning is

$$\Delta_2' = \Delta_2 - kv \tag{69.49}$$

(as compared with $\Delta_1' = \Delta_1 + kv$ for $\mathcal{E}_1$). Since $\omega_1 \simeq \omega_2$, the two k-vectors are nearly equal in magnitude.

The linear response now becomes

$$\rho_{eg}^{(2)} = \left(\frac{\mathrm{i}d\mathcal{E}_2}{2\hbar}\right)\frac{1}{\gamma + \mathrm{i}(\Delta_2 - kv)}\left(\rho_{gg} - \rho_{ee}\right). \tag{69.50}$$

With only the signal $\mathcal{E}_1$ present, the population difference follows directly from (69.22)). The linear response at frequency ω_2 is then

$$\rho_{eg}^{(2)} = \left(\frac{d\mathcal{E}_2}{2\hbar}\right)\frac{\mathrm{i}\gamma + (\Delta_2 - kv)}{\gamma^2 + (\Delta_2 - kv)^2} \times \left[1 - \frac{\Omega_1^2\gamma}{\Gamma} \times \left(\frac{1}{(\Delta_1 + kv)^2 + \gamma^2 + \Omega_1^2\gamma/\Gamma}\right)\right]. \tag{69.51}$$

This is the linear response of atoms moving with velocity v. To obtain the polarization of the whole sample, we must average over the velocity distribution using the Gaussian weight (69.20). The first term in (69.51) gives the linear response in the form of a *Voigt profile*, as discussed in Sect. 69.3. This part of the response carries no Doppler-free information. The second terms contain the nonlinear response. This shows the details of the homogeneous features under the Doppler line shape. For simplicity we assume the Doppler limit, $\gamma \ll ku$, and neglect the variation of the Gaussian over the atomic line shape. We also neglect the power broadening due to the field $\mathcal{E}_1$ and obtain, using (69.13), (69.20), and (69.51),

$$\begin{aligned}\alpha''(\omega) &= -\left(\frac{d^2}{\hbar}\right)\frac{\Omega_1^2\gamma^2}{\sqrt{2\pi}\,\Gamma u} \times \int_{-\infty}^{+\infty}\frac{\mathrm{d}v}{\left[(\Delta_2 - kv)^2 + \gamma^2\right]\left[(\Delta_1 + kv)^2 + \gamma^2\right]} \\ &= -\left(\frac{d^2}{\hbar}\right)\frac{\sqrt{2\pi}\,\Omega_1^2}{4\Gamma ku}\frac{\gamma}{(\omega - \omega_0)^2 + \gamma^2}.\end{aligned} \tag{69.52}$$

This denotes the energy absorbed from the field $\mathcal{E}_1$, as induced nonlinearly by the intensity $\mathcal{E}_1^2$. The resonance is still at $\omega = \omega_0$, but with a homogeneous atomic line shape. In the Doppler limit, the Doppler broadening is only seen in the prefactor

$$\frac{\Omega_1^2}{\Gamma ku} = \left(\frac{\Omega_1^2}{\Gamma\gamma}\right)\frac{\gamma}{ku}\,, \tag{69.53}$$

which shows that only the fraction (γ/ku) of all atoms can contribute to the resonant response. The first factor on the right-hand side of (69.53) is the dimensionless saturation factor.

The Doppler-free character of this spectroscopy derives from the fact that the two fields burn their two separate Bennett holes at the velocities

$$kv_1 = -\Delta_1\,, \quad \text{and } kv_2 = \Delta_2\,. \tag{69.54}$$

When these two groups coincide, i. e., when $v_1 = v_2$, the probe $\mathcal{E}_2$ sees the absorption saturated by the pump field $\mathcal{E}_1$ and a decreased absorption is observed. For $\omega_1 = \omega_2$, the two holes meet in the middle at zero velocity. With two different frequencies, one can make the holes meet at a nonzero velocity to one side of the Doppler profile. The decreased absorption seen in these experiments is called an *inverted Lamb dip* Chapt. 70.

The results derived here are based on a simplified view of the pump-probe response which in turn is based on the rate equation approach. Certain coherent effects are neglected, which would considerably complicate the treatment [69.2]. Section 69.6 discusses these effects in the three-level system where they are more important.

In addition to measuring the probe absorption induced by the pump field, it is also possible to observe the dispersion of the probe signal caused by the saturation induced by the pump. Assuming that the pump introduces the velocity dependent population difference

$$\rho_{ee} - \rho_{gg} = \Delta\rho(v)\,, \tag{69.55}$$

then from (69.50), Re(n) is

$$n' = 1 + \frac{Nd^2}{2\hbar\varepsilon_0}\int \frac{\Delta_2 - kv}{(\Delta_2 - kv)^2 + \gamma^2}\Delta\rho(v)\,\mathrm{d}v\,. \tag{69.56}$$

From (69.2), the phase of the electromagnetic signal feels the value of n' through the factor

$$\mathcal{E} \propto \exp(\mathrm{i}\omega n' z/c)\,. \tag{69.57}$$

Thus, by modifying $\Delta\rho(v)$, a pump laser can control the phase acquired by light traversing the sample, and thereby control the optical length of the sample.

A real atom has magnetic sublevels, which are coupled to light in accordance with *dipole selection rules* (see the discussion in Chapt. 33). If a pump laser is used to affect populations in the various sub-levels differently, the optical paths experienced by the differently circularly polarized components of a linearly polarized probe signal are different. Thus, its plane of polarization will turn, corresponding to the *Faraday effect*. By tuning the pump and the probe over the spectral lines of the sample, the turning of the probe polarization provides a signal to investigate the atomic level structure. This method of *polarization spectroscopy* can be used both with two-level and three-level systems [69.7].

69.6 Three-Level Spectroscopy

New nonlinear phenomena appear when one of the levels in the two-state configuration, $|e\rangle$ say, is coupled to a final level $|f\rangle$ by a weak probe. This may be above the level $|e\rangle$ (the cascade configuration denoted by Ξ) or below $|e\rangle$ (the lambda configuration Λ) (if the third level were coupled to $|g\rangle$ we would talk about the inverted lambda or V configuration) [69.8]. For simplicity, only the Ξ configuration is discussed here.

Assume that the level pair $|g\rangle \leftrightarrow |e\rangle$ is pumped by the field $\mathcal{E}_1$ and its effect probed by the field $\mathcal{E}_2$ coupling $|e\rangle \leftrightarrow |f\rangle$. The dipole matrix element is $d_2 = \langle f|H|e\rangle$. The RWA is now achieved by introducing slowly varying quantities through the definitions

$$\rho_{fe} = \mathrm{e}^{\mathrm{i}(k_2 z - \omega_2 t)}\tilde{\rho}_{fe}\,,$$
$$\rho_{fg} = \mathrm{e}^{\mathrm{i}[(k_1+k_2)z-(\omega_1+\omega_2)t]}\tilde{\rho}_{fg}\,, \tag{69.58}$$

and omitting all components oscillating at multiples of the optical frequencies. From the equation of motion for the density matrix, the steady state equations are [69.2]

$$(\Delta_2 + k_2 v - \mathrm{i}\gamma_{fe})\tilde{\rho}_{fe} = \frac{d_2\mathcal{E}_2}{2\hbar}\rho_{ee} - \frac{d_1\mathcal{E}_1}{2\hbar}\tilde{\rho}_{fg}\,,$$
$$[(\Delta_1+\Delta_2)+(k_1+k_2)v - \mathrm{i}\gamma_{fg}]\tilde{\rho}_{fg} = \frac{d_2\mathcal{E}_2}{2\hbar}\tilde{\rho}_{eg} - \frac{d_1\mathcal{E}_1}{2\hbar}\tilde{\rho}_{fe}\,, \tag{69.59}$$

where the second step detuning is now

$$\Delta_2 = \omega_{fe} - \omega_2\,. \tag{69.60}$$

These equations contain the lowest order response proportional to $\mathcal{E}_2$, including some coherence effects. The coherence effects remain to lowest order, even when the last term in (69.59) is neglected; in that case, the solution is

$$\begin{aligned}\tilde{\rho}_{fe} = &\left(\frac{d_2\mathcal{E}_2}{2\hbar}\right)\frac{\rho_{ee}}{\Delta_2+k_2v-\mathrm{i}\gamma_{fe}}\\ &-\left(\frac{d_1\mathcal{E}_1}{2\hbar}\right)\left(\frac{d_2\mathcal{E}_2}{2\hbar}\right)\\ &\times\frac{\tilde{\rho}_{eg}}{(\Delta_2+k_2v-\mathrm{i}\gamma_{fe})}\\ &\times\frac{1}{\left[(\Delta_1+\Delta_2)+(k_1+k_2)\,v-\mathrm{i}\gamma_{fg}\right]}\ . \end{aligned} \quad (69.61)$$

The result to lowest order in $\mathcal{E}_2$ follows by replacing the density matrix elements for the two-level system $|e\rangle \leftrightarrow |g\rangle$ by their results calculated without this field. This shows that the polarization induced at the frequency ω_2 consists of two parts: one induced by the population excited to level $|e\rangle$ by the field $\mathcal{E}_1$ proportional to ρ_{ee}, and the other one is induced by the coherence $\tilde{\rho}_{eg}$ created by the pump. Retaining only the former produces a rate equation approximation, called a *two-step process*. This misses important physical features which are included in the second term called a *two-photon* or *coherent process*.

In order to see the effects of the two terms most clearly, consider the two-level matrix elements from (69.11) and (69.12) in lowest perturbative order with respect to the pump field $\mathcal{E}_1$, i. e.,

$$\begin{aligned}\rho_{ee} &= \left(\frac{d_1\mathcal{E}_1}{\hbar}\right)^2\frac{\gamma_{eg}}{2\gamma_{ee}}\frac{1}{(\Delta_1+k_1v)^2+\gamma_{eg}^2}\ ,\\ \tilde{\rho}_{eg} &= \frac{d_1\mathcal{E}_1}{2\hbar}\,\frac{1}{\Delta_1+k_1v-\mathrm{i}\gamma_{eg}}\end{aligned} \quad (69.62)$$

From (69.61), these matrix elements give

$$\begin{aligned}\tilde{\rho}_{fe} = &-\left(\frac{A}{\Delta_1+k_1v-\mathrm{i}\gamma_{eg}}\right)\left[\left(\frac{2\gamma_{eg}}{\gamma_{ee}}\right)\right.\\ &\times\frac{1}{(\Delta_1+k_1v+\mathrm{i}\gamma_{eg})(\Delta_2+k_2v-\mathrm{i}\gamma_{fe})}\\ &-\frac{1}{(\Delta_2+k_2v-\mathrm{i}\gamma_{fe})}\\ &\left.\times\frac{1}{\left[(\Delta_1+\Delta_2)+(k_1+k_2)\,v-\mathrm{i}\gamma_{fg}\right]}\right]\ ,\end{aligned} \quad (69.63)$$

where

$$A = -\left(\frac{d_2\mathcal{E}_2}{2\hbar}\right)\left(\frac{d_1\mathcal{E}_1}{2\hbar}\right)^2\ . \quad (69.64)$$

The imaginary part of this yields the absorptive part of the polarization at the frequency ω_2. The first term becomes the product of two Lorentzians, and is an incoherent rate contribution. It dominates when the induced population ρ_{ee} decays much more slowly than the induced coherence $\tilde{\rho}_{eg}$, i. e., when $\gamma_{ee} \ll \gamma_{eg}$.

The significance of the second term in (69.63) is evident in the limit when no phase perturbing processes intervene, so that

$$\begin{aligned}\gamma_{ee} &= \Gamma\ , & \gamma_{eg} &= \frac{1}{2}\Gamma\ ,\\ \gamma_{fg} &= \frac{1}{2}\gamma_{ff}\ , & \gamma_{fe} &= \frac{1}{2}(\gamma_{ff}+\Gamma)\ .\end{aligned} \quad (69.65)$$

For $v=0$, $\tilde{\rho}_{fe}$ becomes

$$\tilde{\rho}_{fe} = -\frac{A}{\Delta_1^2+(\Gamma/2)^2}\left(\frac{1}{\Delta_1+\Delta_2-\mathrm{i}\frac{1}{2}\gamma_{ff}}\right)\ . \quad (69.66)$$

Neglecting the decay of the final level $|f\rangle$, $\gamma_{ff}\to 0$, the absorption becomes proportional to

$$\operatorname{Im}\left(\tilde{\rho}_{fe}\right) = \frac{\pi|A|}{\Delta_1^2+(\Gamma/2)^2}\delta\left(\Delta_1+\Delta_2\right)\ . \quad (69.67)$$

In this limit, strict energy conservation between the ground state and the final state must prevail; the final state must be reached by the absorption of exactly two quanta. The delta function in (69.67) indicates precisely this:

$$\begin{aligned}\Delta_1+\Delta_2 &= \omega_{fe}+\omega_{eg}-(\omega_1+\omega_2)\\ &= \omega_{fg}-\omega_1-\omega_2\\ &= 0\ .\end{aligned} \quad (69.68)$$

The detuning and the width of the intermediate state affect the total transition rate, but not the condition of energy conservation. The presence of the second term in (69.63) makes the resonance contributions at

$$\Delta_2 = \omega_{fg}-\omega_2 = 0 \quad (69.69)$$

cancel approximately. Only a *two-photon transition* remains; in the Λ configuration this would be a *Raman process* (Chapts 62 and 72).

If the velocity dependence in (69.63) is retained, the nonlinear response of an atomic sample must be averaged over the velocity distribution given by the Gaussian (69.20). The computations become involved, but the results show more or less well resolved resonances around the two positions (69.68) and (69.69), i. e., the *coherent two-photon process* and the single step rate process $|e\rangle \to |f\rangle$ appearing due to a previous single step process $|g\rangle \to |e\rangle$; this is the *two-step process*.

A special situation arises when the intermediate step is detuned so much that no velocity group is in resonance, i. e., $|\Delta_1| \approx |\Delta_2| \gg kv$ for all velocities contributing significantly to the spectrum. Then the second coherent term of (69.63) can be written in the form

$$\tilde{\rho}_{fe} = -\frac{A}{\Delta_1^2}\left(\frac{1}{(\Delta_1+\Delta_2)+(k_1+k_2)v - \mathrm{i}\gamma_{fg}}\right) . \tag{69.70}$$

If the two fields $\mathcal{E}_1$ and $\mathcal{E}_2$ have the same frequency but are counterpropagating, then $k_1+k_2=0$, and no velocity dependence occurs in (69.70) for the two-photon resonance. All atoms in the sample contribute to the strength of the resonance, and then polarization is obtained directly by multiplication with the total atomic density. Equation (69.13) then gives

$$\alpha_2'' = \frac{d_2^2}{\hbar}\left(\frac{d_1\mathcal{E}_1}{2\hbar\Delta_1}\right)^2 \frac{\gamma_{fg}}{\left(2\omega-\omega_{fg}\right)^2+\gamma_{fg}^2} . \tag{69.71}$$

This is a sharp *Doppler-free resonance* on the two-photon transition ω_{fg}. The advantage is that all atoms contribute with the sharp line width γ_{fg}, which is not easily affected by phase perturbations because of the two-photon nature of the transition. The disadvantage is the weakness of the transition, which is caused by the large detuning, making $d_1\mathcal{E}_1/|\Delta_1| \ll 1$ in most cases. With tunable lasers, however, this Doppler-free spectroscopy method has been used successfully in many cases.

69.7 Special Effects in Three-Level Systems

We continue our considerations of a three-level system but this time in the V-configuration. Thus, we have a ground state $|g\rangle$ coupled to a doublet of excited states $\{|e\rangle, |f\rangle\}$ through the interaction

$$V = \Omega_1|g\rangle\langle e| + \Omega_2|g\rangle\langle f| + \text{h.c.} , \tag{69.72}$$

where the couplings are due to radiation fields and given by

$$\Omega_i = \frac{d_i\mathcal{E}_i}{2\hbar} . \tag{69.73}$$

We use the rotating wave approximation and set the detunings to

$$\Delta\omega_e = \omega_e - \omega_1 \quad \text{and } \Delta\omega_f = \omega_f - \omega_2 , \tag{69.74}$$

where the frequency ω_i derives from the coupling field $\mathcal{E}_i$. The energy $E_g = 0$.

The time-dependent Schrödinger equation for this system is then written as

$$\mathrm{i}\dot{c}_g = \Omega_1 c_e + \Omega_2 c_f , \tag{69.75}$$

$$\mathrm{i}\dot{c}_e = \Delta\omega_e c_e + \Omega_1 c_g , \tag{69.76}$$

$$\mathrm{i}\dot{c}_f = \Delta\omega_f c_f + \Omega_2 c_g . \tag{69.77}$$

We now introduce the two new variables

$$c_\mathrm{C} = \bar{\Omega}^{-1}\left(\Omega_1 c_e + \Omega_2 c_f\right) \tag{69.78}$$

$$c_\mathrm{NC} = \bar{\Omega}^{-1}\left(\Omega_2 c_e - \Omega_1 c_f\right) , \tag{69.79}$$

where

$$\bar{\Omega}^2 = \Omega_2^2 + \Omega_1^2 . \tag{69.80}$$

The equations of motion follow:

$$\begin{aligned}\mathrm{i}\dot{c}_\mathrm{NC} &= \left(\frac{\Omega_2}{\bar{\Omega}}\Delta\omega_e c_e - \frac{\Omega_1}{\bar{\Omega}}\Delta\omega_f c_f\right)\\ &= \frac{1}{\bar{\Omega}^2}\Big[\Omega_1\Omega_2\left(\Delta\omega_e - \Delta\omega_f\right)c_\mathrm{C} .\\ &\quad + \left(\Delta\omega_e\Omega_2^2 + \Delta\omega_f\Omega_1^2\right)c_\mathrm{NC}\Big] ,\end{aligned} \tag{69.81}$$

$$\begin{aligned}\mathrm{i}\dot{c}_\mathrm{C} &= \left(\frac{\Omega_1}{\bar{\Omega}}\Delta\omega_e c_e + \frac{\Omega_2}{\bar{\Omega}}\Delta\omega_f c_f\right) + \bar{\Omega}c_g\\ &= \frac{1}{\bar{\Omega}^2}\Big[\left(\Delta\omega_e\Omega_1^2 + \Delta\omega_f\Omega_2^2\right)c_\mathrm{C}\\ &\quad + \Omega_1\Omega_2\left(\Delta\omega_e - \Delta\omega_f\right)c_\mathrm{NC}\Big] + \bar{\Omega}c_g .\end{aligned} \tag{69.82}$$

We notice that c_NC is not coupled to the ground state, but, in general, its coupling to the state c_C provides an indirect coupling. This indirect coupling, however, can be made to disappear, if we set $\Delta\omega_e = \Delta\omega_f \equiv \Delta\omega$. Then we find the equations

$$\begin{aligned}\mathrm{i}\dot{c}_\mathrm{NC} &= \Delta\omega c_\mathrm{NC} ,\\ \mathrm{i}\dot{c}_\mathrm{C} &= \Delta\omega c_\mathrm{C} + \bar{\Omega}c_g ,\\ \mathrm{i}\dot{c}_g &= \bar{\Omega}c_\mathrm{C} .\end{aligned} \tag{69.83}$$

This describes a pair of coupled quantum levels and a single uncoupled level. Thus, if we start in the ground state, this latter level can never be populated. It remains unpopulated and is called a dark state. This state corresponds to the superposition

$$|\mathrm{NC}\rangle = \frac{1}{\bar{\Omega}}\left(\Omega_2|e\rangle - \Omega_1|f\rangle\right) . \quad (69.84)$$

Using the coupling operator (69.72), we find the matrix element

$$\langle g|V|\mathrm{NC}\rangle = 0 . \quad (69.85)$$

We also notice that the dark state can be found if the states $|e\rangle$ and $|f\rangle$ are the lower ones, i.e., we have the Λ-configuration.

Alternatively, if we start the system off in the dark state, it will never be able to get out of this state. This is also taken to hold true if we let the couplings depend slowly on time. In this case, we may let Ω_1 come on later than Ω_2. Then we may have

$$\lim_{t\to-\infty}\left(\frac{\Omega_1}{\sqrt{\Omega_2^2+\Omega_1^2}}\right) = 0 ,$$
$$\lim_{t\to-\infty}\left(\frac{\Omega_2}{\sqrt{\Omega_2^2+\Omega_1^2}}\right) = 1 \quad (69.86)$$

and

$$\lim_{t\to\infty}\left(\frac{\Omega_1}{\sqrt{\Omega_2^2+\Omega_1^2}}\right) = 1 ,$$
$$\lim_{t\to\infty}\left(\frac{\Omega_2}{\sqrt{\Omega_2^2+\Omega_1^2}}\right) = 0 . \quad (69.87)$$

Both couplings are thus pulses, but they occur with a slight time delay. If we now start the system in the state $|e\rangle$, we find from (69.84) that

$$\lim_{t\to-\infty}|\mathrm{NC}\rangle = |e\rangle , \quad (69.88)$$

and

$$\lim_{t\to+\infty}|\mathrm{NC}\rangle = -|f\rangle . \quad (69.89)$$

Thus, by keeping the system in this uncoupled state we can adiabatically transfer its population beteen the states without involving any population of the intermediate state $|g\rangle$. Especially if this is an upper state, which may decay and dephase rapidly, the proposed population transfer may be greatly advantageous. Because it is usually applied in the Λ-configuration, it is termed Stimulated Raman Adiabatic Passage (STIRAP).

The dark state has found a wide range of applications in laser physics. As we may use the method to pass radiation through a medium without any absorption, it has led to the phenomenon of light-induced transparency. It can also be used to affect the index of refraction without having the accompanying absorption. The absorptive part of a resonance normally manifests itself as a quantum noise; utilizing the dark state idea one may reduce the noise in quantum devices. The dressing of the levels due to the special features of the interaction has also made it possible to achieve lasing without an inversion of the bare levels. These topics, however, will not be treated here.

A special application of the method to affect the refractive index deserves a more detailed consideration. We look at the relationship between the electric field vectors in the medium:

$$D(\omega) = \varepsilon_0 E + N\alpha(\omega)E(\omega) = \chi(\omega)\varepsilon_0 E(\omega) , \quad (69.90)$$

where $\chi(\omega)$ is the susceptibility of the medium. From Maxwell's equations we find the relation (68.24)

$$k^2 = n^2(\omega)\frac{\omega^2}{c^2} = \frac{\omega^2}{c^2}\left[1+\chi(\omega)\right] . \quad (69.91)$$

If we take the real parts of the quantities, this describes the propagation of waves in the medium.

Now we may use the relations (69.16) to estimate the function $\alpha(\omega)$ in the case when we have two weak fields exciting the three-level system in the V-configuration. We assume that both couplings have the same frequency, $\omega_1 = \omega_2 = \omega$. We write

$$\chi(\omega) \approx \Lambda\left(\frac{(\omega-\omega_e)}{(\omega-\omega_e)^2+\gamma^2} + \frac{(\omega-\omega_f)}{(\omega-\omega_f)^2+\gamma^2}\right) ; \quad (69.92)$$

in the weak field limit we may assume the two processes to add independently. The parameter

$$\Lambda = \frac{Nd^2}{2\hbar\varepsilon_0} \quad (69.93)$$

has the dimension of a frequency and indicates the strength of the interaction. It is clear that tuning the frequency between the levels may give rise to a value of zero for the susceptibility. For a large enough γ, we write, in the neighbourhood of the zero,

$$\chi(\omega) \approx \frac{2\Lambda}{\gamma^2}\left(\omega - \bar{\omega}\right) , \tag{69.94}$$

where $\bar{\omega} = \frac{1}{2}(\omega_e + \omega_f)$.

We now have the relation (69.91) to determine the dispersion relation, and assuming the effect of the medium to be substantial, we may derive an expression for the group velocity in the medium

$$v_g^{-1} = \frac{\partial k}{\partial \omega} . \tag{69.95}$$

We find

$$\frac{2k}{v_g} = \frac{2\omega}{c^2}(1+\chi) + \frac{\omega^2}{c^2}\left(\frac{2\Lambda}{\gamma^2}\right) . \tag{69.96}$$

Even though $\chi = 0$ near the point $\bar{\omega}$, we still have $\omega \sim ck$, so that

$$v_g = \frac{c}{1 + \frac{\Lambda\omega}{\gamma^2}} \approx c\left(\frac{\gamma^2}{\Lambda\omega}\right) \ll c . \tag{69.97}$$

The last inequality follows from the fact that in all cases $\gamma \ll \omega$.

We have thus found that utilizing the interference of two near quantum levels, the refractive index may acquire a very strong dependence on frequency. This may manifest itself in an exceedingly slow propagation of light pulses. Such slow light has been shown to travel at only a few kilometers per hour, which is a most remarkable result. The drawback is, however, that this can only occur over a very narrow frequency range, as follows from the assumption of a strong dependence on frequency.

69.8 Summary of the Literature

Much of the material needed to formulate the basic theory of interaction between light and matter can be found in the text book [69.2]. A comparison between the harmonic model and the two-level model is given by *Feld* [69.1]. The density matrix formulation is presented in detail in [69.2]. The influence of various line broadening mechanisms on laser spectroscopy is discussed in the book [69.3]. The Voigt profile is related to the error function, which is treated in the compilation [69.5]. The numerical evaluation of the Voigt profile is discussed in [69.4]. Rate equations are commonly used in laser theory and they are derived for optical pumping and laser-induced processes in the lectures [69.6]. The Doppler-free spectroscopy was developed in the 1960s and 1970s by many authors following the initial discovery of the Benett hole by Bill R. Bennett Jr. and the Lamb dip by Willis E. Lamb Jr. Much of the pioneering work can be found in the book [69.3]. The three-level work has been reviewed by *Chebotaev* [69.8]. Various applications of lasers in spectroscopy are treated by *Levenson* and *Kano* [69.7]. For references to other topics, we refer to the specialized chapters of the present book.

Many features of the quantum dynamics of a few-level system are found in [69.9, 10]. The theoretical methods to treat such systems are presented in detail in [69.11]. The ensuing physical processes are presented in [69.12] with much additional material on quantum optics phenomena. The basic theory of the dark state and many of its applications in spectroscopy and laser physics are found in [69.13]. A rather complete review of adiabatic processes induced by delayed pulses is the article [69.14]. The slowing down and stopping of light is reviewed in [69.15]. A very recent article with earlier references is [69.16].

References

69.1 M. S. Feld: *Frontiers in Laser Spectroscopy*, ed. by R. Balian, S. Haroche, S. Liberman (North-Holland, Amsterdam 1977) p. 203

69.2 S. Stenholm: *Foundations of Laser Spectroscopy* (Wiley, New York 1984)

69.3 V. S. Letokhov, V. P. Chebotaev: *Nonlinear Laser Spectroscopy* (Springer, Berlin, Heidelberg 1977)

69.4 W. J. Thompson: Comp. Phys. **7**, 627 (1993)

69.5 M. Abramowitz, I. E. Stegun: *Handbook of Mathematical Functions* (Dover, New York 1970)

69.6 C. Cohen-Tanoudji: *Frontiers in Laser Spectroscopy*, ed. by R. Balian, S. Haroche, S. Liberman (North-Holland, Amsterdam 1977) p. 1

69.7 M. D. Levenson, S. S. Kano: *Introduction to Nonlinear Laser Spectroscopy* (Academic, New York 1988)

69.8 V. P. Chebotaev: *High-Resolution Spectroscopy*, ed. by K. Shimoda (Springer-Verlag, Berlin, Heidelberg 1976)

69.9 B. W. Shore: *The Theory of Coherent Atomic Excitation: Vol. 1. Simple Atoms and Fields* (Wiley, New York 1990)

69.10 B. W. Shore: *The Theory of Coherent Atomic Excitation: Vol. 2. Multilevel Atoms and Incoherence* (Wiley, New York 1990)

69.11 C. Cohen-Tannoudji, J. Dupont-Roc, G. Grynberg: *Atom-Photon Interactions, Basic Processes and Applications* (Wiley, New York 1992)

69.12 L. Mandel, E. Wolf: *Optical Coherence and Quantum Optics* (Cambridge Univ. Press, Cambridge 1995)

69.13 M. O. Scully, M. S. Zubairy: *Quantum Optics* (Cambridge Univ. Press, Cambridge 1997)

69.14 N. V. Vitanov, M. Fleischauer, B. W. Shove, K. Bergmann: In: *Advances of Atomic, Molecular and Optical Physics*, Vol. 46, ed. by B. Bederson, H. Walther (Academic, New York 2001) p. 55

69.15 A. B. Matsko, O. Kocharovskaya, Y. Restoutsev, G. R. Welch, A. S. Zibrov, M. O. Scully: *Advances in Atomic, molecular, and Optical Physics*, Vol. 46, ed. by B. Bederson, H. Walther (Academic, New York 2001) p. 191

69.16 M. Bajcsy, A. S. Zibrov, M. D. Lukin: Nature **26**, 368 (2003)

70. Laser Principles

Despite their great variety and range of power, wavelength, and temporal characteristics, all lasers involve certain basic concepts, such as gain, threshold, and electromagnetic modes of oscillation [70.1–3]. In addition to these universal characteristics are features, such as Gaussian beam modes, that are important to such a wide class of devices that they must be included in any reasonable compendium of important laser concepts and formulas. We have therefore included here both generally applicable results as well as some more specific but widely applicable ones.

70.1 Gain, Threshold, and Matter–Field Coupling 1023

70.2 Continuous Wave, Single-Mode Operation 1025

70.3 Laser Resonators 1028

70.4 Photon Statistics 1030

70.5 Multi-Mode and Pulsed Operation 1031

70.6 Instabilities and Chaos 1033

70.7 Recent Developments 1033

References 1034

70.1 Gain, Threshold, and Matter–Field Coupling

All lasers involve some medium that amplifies an electromagnetic field within some band of frequencies. At the simplest level of description, the amplifying medium changes the intensity I of a field according to the equation

$$\frac{\mathrm{d}I}{\mathrm{d}z} = gI\,, \tag{70.1}$$

where z is the coordinate along the direction of propagation and g is the gain coefficient, typically expressed in cm^{-1}. Amplification occurs as a consequence of stimulated emission of radiation from the upper state (or band of states) of a transition for which a population inversion exists; i. e., for which an upper state has greater likelihood of occupation than a lower state. Different types of lasers may be classified by the pump mechanisms used to achieve population inversion (Chapt. 71). In the case that the amplifying transition involves two discrete energy levels, E_1 and $E_2 > E_1$, the gain coefficient at the frequency ν is given by

$$g(\nu) = \frac{\lambda^2 A}{8\pi n^2}\left(N_2 - \frac{g_2}{g_1}N_1\right) S(\nu)\,. \tag{70.2}$$

Here $\lambda = c/\nu$ is the transition wavelength, A $\left(\mathrm{s}^{-1}\right)$ is the Einstein A coefficient for spontaneous emission for the transition, and g_2, g_1 are the degeneracies of the upper and lower energy levels. These quantities in nearly every case are fixed characteristics of the medium, independent of the laser intensity or the pump mechanism. N_2 and N_1 are the population densities $\left(\mathrm{cm}^{-3}\right)$ of the upper and lower levels, respectively, and $S(\nu)$ is the normalized transition lineshape function (Chapts. 19 and 69). n is the refractive index at frequency ν of the "background" host medium and in general has contributions from all nonlasing transitions. Equation (70.2) describes either amplification or absorption, depending on whether $N_2 - (g_2/g_1)N_1$ is positive (amplification) or negative (absorption).

By far the most common configuration is that in which the gain medium is contained in a cavity bounded on two sides by reflecting surfaces. The mirrors allow feedback; i. e., the redirection of the field back into the gain medium for multipass amplification and sustained laser action. The two mirrors allow the field to build up along the directions parallel to the "optical axis" and to form a pencil-like beam of light. In order to sustain laser action, the gain in intensity due to stimulated emission must equal or exceed the loss due to imperfect mirror reflectivities, scattering, absorption in the host medium, and diffraction.

Typically the imperfect mirror reflectivities dominate the other sources of loss. If the mirror reflectivities are r_1 and r_2, then the intensity I is reduced by the factor $r_1 r_2$ in a round trip pass through the cavity, while according to (70.1) the gain medium causes the intensity to increase by a factor $\exp(2g\ell)$ in the two passes through the gain cell of length ℓ. Equating of the gain and loss factors leads to the threshold condition for laser oscillation: $g \geq g_t$, where the threshold gain is

$$g_t = -\frac{1}{2\ell}\ln(r_1 r_2) + \alpha\,, \tag{70.3}$$

α being an attenuation coefficient associated with any loss mechanisms that may exist in addition to reflection losses at the mirrors.

Suppose, for example, that a laser has a 50 cm gain cell and mirrors with reflectivities 0.99 and 0.97, and that absorption within the host medium of the gain cell is negligible. Then the threshold gain is $g_t = 4\times10^{-4}\,\text{cm}^{-1}$. If the lasing transition is the 6328 Å Ne transition of the He–Ne laser, we have $A \cong 1.4\times10^6\,\text{s}^{-1}$, $n \cong 1.0$ and, assuming a pure Doppler lineshape,

$$S(\nu) = \left(\frac{4\ln 2}{\pi}\right)^{1/2}\frac{1}{\delta\nu_D} \tag{70.4}$$

at line center, where $\delta\nu_D$ is the width (FWHM) of the Doppler lineshape (Sect. 69.3). For $T = 400\,\text{K}$ and the Ne atomic weight, $\delta\nu_D \cong 1500\,\text{MHz}$ and $S(\nu) \cong 6.3\times10^{-10}\,\text{s}$. Then the threshold population difference required for laser oscillation is

$$\left(N_2 - \frac{g_2}{g_1}N_1\right)_t = \frac{8\pi n^2 g_t}{\lambda^2 A S(\nu)} \cong 2.8\times10^9\,\text{cm}^{-3}\,. \tag{70.5}$$

This is a typical result: the population inversion required for laser oscillation is small compared with the total number of active atoms.

Calculations of population inversions and other properties of the gain medium are based on rate equations, or more generally, the density matrix ρ. In many instances, the medium is fairly well described in terms of two energy eigenstates, other states appearing only indirectly through pumping and decay channels. In this case, ρ is a 2×2 matrix whose elements satisfy [70.3] (68.66) and (68.67)

$$\begin{aligned}\dot\rho_{22} &= -(\Gamma_2+\Gamma)\rho_{22} - \frac{1}{2}\mathrm{i}\left(\Omega_1^*\tilde\rho_{21} - \Omega_1\tilde\rho_{12}\right),\\ \dot\rho_{11} &= -\Gamma_1\rho_{11} + \Gamma\rho_{22} + \frac{1}{2}\mathrm{i}\left(\Omega_1^*\tilde\rho_{21} - \Omega_1\tilde\rho_{12}\right),\\ \dot{\tilde\rho}_{21} &= -(\gamma+\mathrm{i}\Delta)\tilde\rho_{21} - \frac{1}{2}\mathrm{i}\Omega_1(\rho_{22}-\rho_{11})\,,\end{aligned} \tag{70.6}$$

with $\tilde\rho_{12} = \tilde\rho_{21}^*$. Here, Γ_2 and Γ_1 are, respectively, the rates of decay of the upper and lower states due to all processes other than the spontaneous decay from state 2 to state 1 described by the rate $\Gamma = A$. γ, which is 2π times the homogeneous linewidth (HWHM) of the transition, is the rate of decay of off-diagonal coherence due to both elastic and inelastic processes; in general, $\gamma \geq (\Gamma_1+\Gamma_2+\Gamma)/2$. $\Omega_1 = \mathbf{d}_{21}\cdot\boldsymbol{\mathcal{E}}/\hbar$ is the Rabi frequency (The Rabi frequency is often defined as $2\boldsymbol{d}_{21}\cdot\boldsymbol{\mathcal{E}}/\hbar$.) (Sect. 68.3.3 and Chapt. 73), with $\boldsymbol{\mathcal{E}}$ the complex amplitude of the electric field; i. e., the electric field is

$$\begin{aligned}\boldsymbol{E}(\boldsymbol{r},t) &= \mathrm{Re}\left[\boldsymbol{\mathcal{E}}(\boldsymbol{r},t)\,\mathrm{e}^{\mathrm{i}(\boldsymbol{k}\cdot\boldsymbol{r}-\omega t)}\right]\\ &\cong \mathrm{Re}\left[\boldsymbol{\epsilon}\mathcal{E}(\boldsymbol{r},t)\,\mathrm{e}^{\mathrm{i}kz}\,\mathrm{e}^{-\mathrm{i}\omega t}\right].\end{aligned} \tag{70.7}$$

It is assumed that $\mathcal{E}$ is slowly varying in time compared with the oscillations at frequency ω $(=2\pi\nu)$, and that the wave vector $\boldsymbol{k}$ is approximately $k\hat{z} = (n\omega/c)\hat{z}$, where $\hat{z}$ is a unit vector pointing in the direction of propagation. Finally, $\Delta = \omega_0 - \omega$ in (70.6) is the detuning of ω from the central transition frequency $\omega_0 = (E_2 - E_1)/\hbar$ of the lasing transition. Rapidly oscillating terms involving $\omega_0 + \omega$ are ignored in the rotating wave approximation that pervades nearly all of laser theory (Sect. 68.3.2).

In most lasers, γ is so large compared with the diagonal decay rates that the off-diagonal elements of ρ may be assumed to relax quickly to the quasisteady values obtained by setting $\dot\rho_{12} = 0$ in (70.6). Then the diagonal density matrix elements satisfy the rate equations (69.37)

$$\begin{aligned}\dot\rho_{22} &= -(\Gamma_2+\Gamma)\rho_{22} - \frac{|\Omega_1|^2\gamma/2}{\Delta^2+\gamma^2}(\rho_{22}-\rho_{11})\,,\\ \dot\rho_{11} &= -\Gamma_1\rho_{11} + \Gamma\rho_{22} + \frac{|\Omega_1|^2\gamma/2}{\Delta^2+\gamma^2}(\rho_{22}-\rho_{11})\,.\end{aligned} \tag{70.8}$$

Such rate equations, usually expressed equivalently in terms of population densities N_2, N_1 rather than occupation probabilities ρ_{22}, ρ_{11}, are the basis of most practical computer models of laser oscillation. These equations, or, more generally, the density matrix equations, must also include terms accounting for the pump mechanism. In the simplest model of pumping, one adds a constant pump rate Λ_2 to the right-hand side of the equation for ρ_{22} to obtain (69.45)

$$\dot\rho_{22} = \Lambda_2 - (\Gamma_2 + A)\rho_{22} - \frac{|\Omega_1|^2\gamma/2}{\Delta^2+\gamma^2}(\rho_{22}-\rho_{11})\,. \tag{70.9}$$

In the case of an inhomogeneously broadened laser transition (Sect. 69.3), equations of the type (70.6) and (70.8) apply separately to each detuning Δ arising from the distribution of atomic or molecular transition frequencies. In writing these equations, we have assumed a nondegenerate electric dipole transition. The generalization to magnetic or multiphoton transitions, or to a case where the amplification is due, for instance, to a Raman process, is straightforward but of less general interest.

A more realistic treatment of the electromagnetic field than that based on (70.1) proceeds from the Maxwell equations, which, for a homogeneous and nonmagnetic medium, lead to the equation

$$\frac{1}{2ik}\nabla_T^2\mathcal{E}+\left(\frac{\partial}{\partial z}+\frac{1}{c}\frac{\partial}{\partial t}\right)\mathcal{E}=\frac{4\pi i\omega}{nc}N\mu^*\rho_{21}\,. \tag{70.10}$$

Here N is the density of active atoms, $\mu\equiv(\boldsymbol{d}_{12}\cdot\boldsymbol{\epsilon})^*$, and $\nabla_T^2\equiv\partial^2/\partial x^2+\partial^2/\partial y^2$. The result (70.10) assumes the validity of the rotating wave approximation as well as the assumption that $\mathcal{E}$ is slowly varying compared with $\exp(ikz)$ and $\exp(-i\omega t)$. In the plane wave approximation, (70.10) becomes (More generally, the velocity c on the left sides of (70.10) and (70.11) should be replaced by the group velocity v_g associated with nonresonant transitions. If there is substantial group velocity dispersion, it is sometimes necessary to include a term involving the second derivative of $\mathcal{E}$ with respect to t)

$$\left(\frac{\partial}{\partial z}+\frac{1}{c}\frac{\partial}{\partial t}\right)\mathcal{E}=\frac{4\pi i\omega}{nc}N\mu^*\rho_{21}\,. \tag{70.11}$$

Equations (70.6) or (70.8) and (70.10) or (70.11) are coupled matter–field equations whose self consistent solutions determine the operating characteristics of the laser. The density matrix or rate equations must be modified to include pumping, as in (70.9), and the field equations must be supplemented by boundary conditions and loss terms. With these modifications, the equations are the basis of semiclassical laser theory, wherein the particles constituting the gain medium are treated quantum mechanically whereas the field is treated according to classical electromagnetic theory [70.4]. Aside from fundamental linewidth considerations and photon statistics (see Sects. 70.2 and 70.4), very few aspects of lasers require the quantum theory of radiation.

70.2 Continuous Wave, Single-Mode Operation

In the case of steady state, continuous wave (cw) operation, the appropriate matter–field equations are those obtained by setting all time derivatives equal to zero. Equation (70.11), for instance, becomes

$$\frac{d\mathcal{E}}{dz}=\frac{2\pi\omega N|d|^2}{3n\hbar c}\frac{1}{\gamma+i\Delta}(\rho_{22}-\rho_{11})\mathcal{E}\,, \tag{70.12}$$

or, in terms of the intensity I,

$$\begin{aligned}\frac{dI}{dz}&=\frac{4\pi\omega N|d|^2}{3n\hbar c}\frac{\gamma}{\Delta^2+\gamma^2}(\rho_{22}-\rho_{11})I\\&=\frac{\lambda^2A}{8\pi n^2}(N_2-N_1)S(\nu)I=g(\nu)I\end{aligned} \tag{70.13}$$

for the nondegenerate case under consideration. Here, $|d|^2=3|\boldsymbol{d}_{12}\cdot\boldsymbol{\epsilon}|^2$, $N_j=N\rho_{jj}$, $S(\nu)=\gamma/(\Delta^2+\gamma^2)$ is the Lorentzian lineshape function for homogeneous broadening, and $A=4\omega^3|d|^2n/3\hbar c^3$ is the spontaneous emission rate in the host medium of (real) refractive index n. Local (Lorentz–Lorenz) field corrections will in general modify these results, but such corrections are ignored here [70.5].

The steady state solution of the density matrix or rate equations gives, similarly,

$$g(\nu)=\frac{g_0(\nu)}{1+I/I_{sat}}\,, \tag{70.14}$$

where the saturation intensity I_{sat}, like the small signal gain coefficient $g_0(\nu)$, depends on decay rates and other characteristics of the lasing species. Thus, in the plane wave approximation, the growth of intensity in a homogeneously broadened laser medium is typically described by the equation

$$\frac{dI}{dz}=\frac{g_0(\nu)I}{1+I/I_{sat}}\,. \tag{70.15}$$

This equation, supplemented by boundary conditions at the mirrors, and possibly other terms on the right side to account for any distributed losses within the medium, determines the intensity in cw, single-mode operation.

The simplest model for calculating output intensity assumes that the intensity is uniform throughout the laser cavity. In steady state, the gain exactly compensates for the loss; i. e., $g(\nu)=g_t$, the gain clamping condition for cw lasing. Equation (70.14) then implies that the steady state intracavity intensity is

$$I=I_{sat}\,[g_0(\nu)/g_t-1]\,. \tag{70.16}$$

If I is assumed to be the sum of the intensities of waves propagating in the $+z$ and $-z$ directions, i. e. $I=I_++I_-$, then the output intensity from the laser is

$$I_{out}=t_2I_++t_1I_-\,, \tag{70.17}$$

where t_2, t_1 are the mirror transmissivities at the right and left mirrors, respectively. The uniform intensity approximation implies that $I_+ = I_- = I/2$ and

$$I_{\text{out}} = \frac{1}{2}(t_2 + t_1) I_{\text{sat}} \left[g_0(\nu)/g_{\text{t}} - 1 \right] . \tag{70.18}$$

Suppose one of the mirrors is perfectly transmitting, so that $t_1 = 0$ and $t_2 = t > 0$. Furthermore, if the reflectivity r of the transmitting mirror is close to unity, then $g_{\text{t}} \cong (1/2\ell)(1-r) = (1/2\ell)(t+s)$, where s is the fraction of the incident beam power that is scattered or absorbed at the output mirror. Then

$$I_{\text{out}} \cong \frac{1}{2} I_{\text{sat}} t \left(\frac{2g_0(\nu)\ell}{t+s} - 1 \right) , \tag{70.19}$$

and it follows that the optimal output coupling, i. e., the transmissivity that maximizes the output intensity, is

$$t_{\text{opt}} = \sqrt{2g_0(\nu)\ell s} - s . \tag{70.20}$$

This output coupling gives the output intensity

$$I_{\text{out}}^{\max} = I_{\text{sat}} \left[\sqrt{g_0(\nu)\ell} - \sqrt{s/2} \right]^2 . \tag{70.21}$$

$g_0(\nu) I_{\text{sat}}$ is the largest possible power per unit volume extractable as output laser radiation at the frequency ν.

More generally, when mirror reflectivities are not necessarily close to unity, $I_+ \neq I_-$ and both I_+ and I_- vary with the axial coordinate z. In this more general case, (70.14) and (70.15) are replaced by

$$g(\nu, z) = \frac{g_0(\nu)}{1 + [I_+(z) + I_-(z)]/I_{\text{sat}}} \tag{70.22}$$

and

$$\frac{\mathrm{d}I_+}{\mathrm{d}z} = g(\nu, z) I_+ , \qquad \frac{\mathrm{d}I_-}{\mathrm{d}z} = -g(\nu, z) I_- , \tag{70.23}$$

where it is assumed that $g_0(\nu)$ is independent of z, and that all cavity loss processes occur at the mirrors. The solution of these equations with the boundary conditions $I_-(L) = r_2 I_+(L)$, $I_+(0) = r_1 I_-(0)$ for mirrors at $z = 0$ and $z = L$ gives, for $I_{\text{out}} = t_1 I_-(0) + t_2 I_+(L)$, the formula [70.3, 6]

$$I_{\text{out}} = I_{\text{sat}} \left(t_2 + \sqrt{\frac{r_2}{r_1}} t_1 \right) \frac{\sqrt{r_1}}{(\sqrt{r_1} + \sqrt{r_2})(1 - \sqrt{r_1 r_2})} \times \left[g_0(\nu)\ell + \ln \sqrt{r_1 r_2} \right] . \tag{70.24}$$

Analysis of this result gives an optimal output coupling that reduces to (70.19) in the limit $t + s \ll 1$. Curves for optimal output coupling and I_{out} as a function of g_0 and s are given by *Rigrod* [70.6]. These results are based on several assumptions and approximations: the gain medium is assumed to be homogeneously broadened and to saturate according to the formula (70.14); g_0 and I_{sat} are taken to be constant throughout the medium; the field is approximated as a plane wave; field loss processes occur only at the mirrors; and interference between the left- and right-going waves is ignored.

Interference of the counterpropagating waves in a standing wave, single-mode laser modifies the gain saturation formula (70.14) as follows:

$$g(\nu, z) = \frac{g_0(\nu)}{1 + (2I_+/I_{\text{sat}}) \sin^2 kz} \tag{70.25}$$

in the case of small output coupling, where $I_+ \cong I_-$ as assumed in (70.18). (The general case of arbitrary output coupling with spatial interference of counterpropagating waves is somewhat complicated and is not considered here.) The $\sin^2 kz$ term is responsible for spatial hole burning: "holes" are "burned" in the curve of $g(\nu, z)$ versus z at points where $\sin^2 kz$ is largest. This spatially dependent saturation acts to reduce the output intensity, typically by as much as about 30%, compared with the case where interference of counterpropagating waves is absent or ignored. Spatial hole burning tends to be washed out by atomic motion in gas lasers, and is absent entirely in pure traveling wave ring lasers. If complete spatial hole burning based on (70.25) is assumed, the output intensity is

$$I_{\text{out}} = \frac{t}{2} I_{\text{sat}} \left(\frac{g_0(\nu)}{g_{\text{t}}} - \frac{1}{4} - \sqrt{\frac{g_0(\nu)}{2g_{\text{t}}} + \frac{1}{16}} \right) \tag{70.26}$$

in the case where one mirror is perfectly reflecting.

In inhomogeneously broadened media, the gain coefficient is obtained by integrating the contributions from all possible values of ν_0. The different contributions saturate differently, depending on the detuning of ν_0 from the cavity mode frequency ν. If spatial hole burning and power broadening (see Sect. 69.3) are ignored, then, to a good approximation, the gain saturates as

$$g(\nu) = \frac{g_0(\nu)}{\sqrt{1 + I/I_{\text{sat}}}} \tag{70.27}$$

in typical inhomogeneously broadened media.

Oscillation on a single longitudinal mode (see Sect. 70.3) may be realized simply by making the cavity length L small enough that the mode spacing $c/2nL$ (70.28) exceeds the spectral width of the gain curve. This is possible in many gas lasers where the spectral width is small, and in semiconductor lasers, where L is very small. More generally the gain clamping condition $g(\nu)=g_t$ implies that the cavity mode frequency having the largest small-signal gain $g_0(\nu)$ saturates the gain $g(\nu)$ down to the threshold value g_t, while the gain at all other mode frequencies then lies below g_t. In other words, the gain clamping condition implies single-mode oscillation. However, this conclusion assumes homogeneous broadening and also that spatial hole burning is unimportant, so that the gain saturates uniformly throughout the cavity. High pressure gas lasers, where the line broadening is due primarily to collisions and therefore is homogeneous, tend to oscillate on a single mode because spatial hole burning is largely washed out by atomic motion. On the other hand, homogeneously broadened solid state lasers can be multi-mode as a consequence of spatial hole burning.

Single-mode oscillation in inhomogeneously broadened media is generally more difficult to achieve because of spectral hole burning (see Sect. 69.3), which makes the simple gain clamping argument inapplicable. However, single-mode oscillation can be enforced in any case by introducing, in effect, an additional loss mechanism for all mode frequencies except one. This is commonly done with a Fabry–Perot etalon having a free spectral range that is large compared with the spectral width of the gain curve. By choosing the tilt angle appropriately, a particular resonant frequency of the etalon can be brought close to the center of the gain curve while all other resonance frequencies lie outside the gain bandwidth.

Laser oscillation at a fixed polarization can likewise be achieved by discriminating against the orthogonal polarization, as is done when Brewster windows are employed.

Laser oscillation, in general, does not occur precisely at one of the allowed cavity mode frequencies. Associated with the $\sin kz$ dependence of the intracavity field is the condition $kL=N\pi$, or

$$\nu = N\frac{c}{2nL} \quad (N \text{ an integer}) , \tag{70.28}$$

for the cavity mode frequencies in the plane wave approximation. If the gain medium of refractive index n does not fill the entire length L between the mirrors, then (70.28) must be replaced by

$$\nu = \frac{Nc/2}{n\ell + (L-\ell)} , \tag{70.29}$$

or

$$\frac{\ell}{L}[n(\nu)-1]\nu = \nu_N - \nu , \tag{70.30}$$

where ℓ ($\leq L$) is the length of the gain medium and $\nu_N = Nc/2L$ is an empty cavity mode frequency. The laser oscillation frequency will therefore be different, in general, from any of the allowed empty-cavity mode frequencies. If the refractive index $n(\nu)$ is attributable primarily to the lasing transition, as opposed to the host material, or other nonlasing transitions, then the following relation between $n(\nu)$ and $g(\nu)$ may be used in the case of homogeneous broadening [70.3]:

$$n(\nu) - 1 = -\frac{\lambda_0}{4\pi}\frac{\nu_0-\nu}{\delta\nu_0}g(\nu) , \tag{70.31}$$

where λ_0, ν_0, and $\delta\nu_0$ are the wavelength, frequency, and homogeneous linewidth (HWHM), respectively, of the lasing transition. This implies that

$$\nu = \frac{\nu_0\delta\nu_c + \nu_N\delta\nu_0}{\delta\nu_c + \delta\nu_0} \tag{70.32}$$

for the laser oscillation frequency ν, where

$$\delta\nu_c \equiv \frac{cg(\nu)\ell}{4\pi L} \tag{70.33}$$

is the cavity bandwidth. Thus, the actual lasing frequency is not simply one of the allowed empty-cavity frequencies ν_N, but rather is "pulled" away from ν_N toward the center of the gain profile. This is called frequency pulling.

If spatial hole burning is absent or ignored, then $g(\nu)=g_t$ in steady state oscillation, and the cavity bandwidth $\delta\nu_c = cg_t\ell/4\pi L$ is largest for lossy cavities. Most lasers fall into the "good cavity" category, that is $\delta\nu_c \ll \delta\nu_0$, so that (70.32) may be approximated by

$$\nu \cong \nu_N + (\nu_0 - \nu_N)\delta\nu_c/\delta\nu_0 . \tag{70.34}$$

Similar results apply to inhomogeneously broadened lasers. For a Doppler broadened medium, for instance, the frequency pulling formula is

$$\nu \cong \nu_N + (\nu_0 - \nu_N)(\delta\nu_c/\delta\nu_D)\sqrt{4\ln 2/\pi} \tag{70.35}$$

for good cavities. These results show that frequency pulling is most pronounced in lasers with large peak gain coefficients and narrow gain profiles, as observed experimentally.

Spectral hole burning leads to especially interesting consequences in Doppler broadened gas lasers. Since two traveling waves propagating in opposite ($\pm z$) directions will strongly saturate spectral packets of atoms with oppositely Doppler shifted frequencies, a standing wave field will burn two holes on opposite sides of the peak of the Doppler profile. When the mode frequency is exactly at the center of the Doppler profile, however, the two holes merge, the field now being able to saturate strongly only those atoms with zero velocity along the z-direction. In this case, since the field "feeds" off a single spectral packet of more strongly saturated atoms, there is a dip in the output power compared with the case when the mode frequency is detuned from line center. This dip in the output power at line center is called the Lamb dip. It can be used to determine whether a gas laser is Doppler broadened, and more importantly, to stabilize the laser frequency to the center of the dip. Lamb dip frequency stabilization employs a feedback circuit to control the bias voltage across a piezoelectric element used to vary the cavity length, and thereby sweep the laser frequency.

The above semiclassical laser theory suggests that cw laser radiation should be perfectly monochromatic, since the amplitude and phase of the field given by (70.7) are time independent. However, when quantum electrodynamical considerations are built into laser theory, it is found that spontaneous emission, which adds to the number of photons put in the lasing mode by stimulated emission, causes a phase diffusion that results in a Lorentzian linewidth (HWHM)

$$\Delta\nu = \frac{N_2}{N_2 - N_1} \frac{8\pi h\nu}{P_{\text{out}}} (\delta\nu_c)^2 , \tag{70.36}$$

where P_{out} is the output laser power. This is the Schawlow–Townes linewidth, which implies a fundamentally quantum mechanical, finite linewidth that persists no matter how small various sources of "technical noise", such as mirror jitter, are made. Although the Schawlow–Townes linewidth has been observed in highly stabilized gas lasers, it is negligible compared with technical noise in conventional lasers. But in semiconductor lasers, L is very small and consequently $\delta\nu_c$ is large, and quantum noise associated with the Schawlow–Townes formula can be the dominant contribution to the laser linewidth.

However, the 10–100 MHz linewidths typically observed in semiconductor lasers are too large to be explained by the Schawlow–Townes formula (70.36), and two modifications to this formula are necessary, each of them involving a multiplication of the Schawlow–Townes linewidth $\Delta\nu$ by a certain factor:

$$\Delta\nu \to \Delta\nu' = \left(1+\alpha^2\right) K \Delta\nu , \tag{70.37}$$

where α is called the "Henry α parameter" and is associated with a coupling between phase and intensity fluctuations above the laser threshold [70.7]. Values of α between about 4 and 6 are typical in semiconductor injection lasers, and consequently, the correction to the Schawlow-Townes linewidth due to the Henry α parameter is substantial. The K factor [70.8–10] arises as a consequence of the deviation from the spatially uniform intracavity intensity assumed in the derivation of the Schawlow–Townes formula [70.11, 12]. (Intracavity intensities along the optical axis are approximately uniform only in the case of low output couplings.) The fundamental quantum mechanical linewidth under consideration can be associated with vacuum field fluctuations, which, according to general fluctuation–dissipation ideas, will increase as the cavity loss increases. This explains why the "Petermann K factor" deviates increasingly from unity as the output coupling (cavity loss) increases. Values of K between 1 and 2 appear to be typical for lossy, stable resonators [70.13, 14], but much larger values are possible for unstable resonators [70.15] (see also *Siegman* [70.2]).

70.3 Laser Resonators

The assumption that the complex field amplitude $\mathcal{E}(\boldsymbol{r}, t)$ is slowly varying in z compared with $\exp(\mathrm{i}kz)$ leads to the paraxial wave equation (70.10)

$$\nabla_T^2 \mathcal{E} + 2\mathrm{i}k \frac{\partial \mathcal{E}}{\partial z} = 0 \tag{70.38}$$

for a monochromatic field in vacuum. If $\mathcal{E}(x, y, z)$ satisfies the paraxial wave equation and is specified in the plane $(x, y, z = 0)$, it follows that

$$\mathcal{E}(x, y, z) = -\frac{\mathrm{i}}{\lambda z} \iint \mathrm{d}x' \mathrm{d}y' \mathcal{E}(x', y', 0) \times \mathrm{e}^{\mathrm{i}k[(x-x')^2+(y-y')^2]/2z} . \tag{70.39}$$

Thus, in the case of a laser resonator, the field $\mathcal{E}(x, y, L)$ at the mirror at $z = L$ is related to the field $\mathcal{E}(x, y, 0)$ at the mirror at $z = 0$ by

$$\mathcal{E}(x, y, L) = -\frac{\mathrm{i}}{\lambda L} \iint \mathrm{d}x' \mathrm{d}y' \mathcal{E}(x', y', 0) \times \mathrm{e}^{\mathrm{i}k[(x-x')^2+(y-y')^2]/2L} \equiv \iint \mathrm{d}x' \mathrm{d}y' K(x, y; x', y') \mathcal{E}(x', y', 0) . \quad (70.40)$$

Similarly, the field at $z = 0$ after one round trip pass through the resonator is

$$\begin{aligned} &\mathcal{E}(x, y, 0) \\ &= \iint \mathrm{d}x' \mathrm{d}y' K(x, y; x', y') \mathcal{E}(x', y', L) \\ &= \iint \mathrm{d}x' \mathrm{d}y' K(x, y; x', y') \iint \mathrm{d}x'' \mathrm{d}y'' \\ &\quad \times K(x', y'; x'', y'') \mathcal{E}(x'', y'', 0) \\ &\equiv \iint \mathrm{d}x'' \mathrm{d}y'' \tilde{K}(x, y; x'', y'') \mathcal{E}(x'', y'', 0) . \end{aligned} \quad (70.41)$$

By definition, a mode of the resonator is a field distribution that does not change on successive round-trip passes through the resonator. More precisely, since an empty cavity is assumed, a mode will be such that the field spatial pattern remains the same except for a constant decrease in amplitude per pass. A longitudinal mode is defined by the value of k in $\exp(\mathrm{i}kz)$, whereas a transverse mode is defined by the corresponding (x, y) dependence and satisfies the integral equation

$$\gamma \mathcal{E}(x, y, z) = \int \mathrm{d}x' \mathrm{d}y' \tilde{K}(x, y; x', y') \mathcal{E}(x', y', z) , \quad (70.42)$$

where z defines any plane between the mirrors and $|\gamma| < 1$. Iterative numerical solutions of this equation for laser resonator modes were first discussed by *Fox* and *Li* [70.16].

Laser resonators may be classified as stable or unstable according to whether a paraxial ray traced back and forth through the resonator remains confined in the resonator or escapes. This leads to the condition

$$0 \leq g_1 g_2 \leq 1 \quad (70.43)$$

for stability, where the g parameters are defined in terms of the (spherical) mirror curvatures R_i and the mirror separation L by $g_i \equiv 1 - L/R_i$. R_i is defined as positive or negative depending on whether the mirror i is concave or convex, respectively, with respect to the interior of the resonator. Plane–parallel mirrors ($R_i \to \infty$) can be used, but they are difficult to keep aligned and have much larger diffractive losses.

Among stable configurations, the symmetric confocal resonator with $R_1 = R_2 = L$ has the smallest mode spot sizes at the mirrors, while the concentric resonator with $R_1 = R_2$ slightly greater than $L/2$, has the smallest beam waist (Fig. 70.1). The widely used hemispherical resonator ($R_1 = \infty$, R_2 slightly greater than L) is relatively easy to keep aligned and allows the spot size at mirror 2 to be adjusted by slight changes in L.

The fundamental Gaussian beam modes of stable resonators may be constructed from the free space solutions of the paraxial wave equation. The most important (lowest-order) solution for this purpose is

$$\mathcal{E}(x, y, z) = \frac{A\,\mathrm{e}^{-\mathrm{i}\phi(z)}\,\mathrm{e}^{\mathrm{i}k(x^2+y^2)/2R(z)}\,\mathrm{e}^{-(x^2+y^2)/w^2(z)}}{\sqrt{1+z^2/z_0^2}} , \quad (70.44)$$

where A is a constant, $\phi(z) = \tan^{-1}(z/z_0)$, and $R(z)$, $w(z)$, and z_0 are the radius of curvature of surfaces of constant phase, the spot size, and the Rayleigh range, respectively. The confocal parameter, $2z_0$, is also used to characterize Gaussian beams. Here $z_0 \equiv \pi w_0^2/\lambda$, where w_0 is the spot size at the beam waist at $z = 0$ (Fig. 70.1), and $R(z)$ and $w(z)$ vary with the distance z as follows:

$$R(z) = z + z_0^2/z , \quad w(z) = w_0\sqrt{1+z^2/z_0^2} . \quad (70.45)$$

The divergence angle of a Gaussian beam (Fig. 70.1) is given by $\theta = \lambda/\pi w_0$.

The ABCD law for Gaussian beams allows the effects of various optical elements on Gaussian beam propagation to be calculated in a relatively simple fashion [70.1–3]. For instance, a Gaussian beam incident on a lens of focal length f at its waist is focused to a new waist at a distance $d = f/(1 + f^2/z_0^2)$ behind the lens,

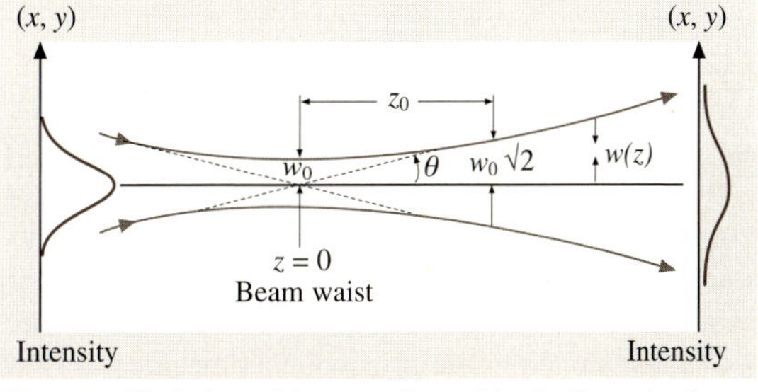

Fig. 70.1 Variation of the spot size $w(z)$ of a Gaussian beam with the propagation distance z from the beam waist

Part F | 70.3

and the spot size at the new waist is

$$w_0' = \frac{f\lambda}{\pi w_0}\frac{1}{\sqrt{1+f^2/z_0^2}}\,, \tag{70.46}$$

which is approximately $f\lambda/\pi w_0 = f\theta$ for tight focusing. This shows that a Gaussian beam can be focused to a very small spot. On the other hand, beam expanders consisting of two appropriately spaced lenses may be used to increase the spot size by the ratio of the focal lengths.

Gaussian beam modes of laser resonators have radii of curvature that match in magnitude those of the mirrors. The spot sizes at the mirrors and the location of the beam waist with respect to the mirrors may be expressed in terms of λ, L, and g_1g_2. The empty-cavity mode frequencies are given by

$$\nu_N = \frac{c}{2L}\left(N + \frac{1}{\pi}\cos^{-1}\sqrt{g_1g_2}\right)\,, \tag{70.47}$$

where N is an integer. For a host medium of refractive index n, $c/2L$ is replaced by $c/2nL$, and (70.47) then generalizes the plane wave result (70.28) to account for both longitudinal and transverse effects in the determination of the cavity mode frequencies.

The assumption of Gaussian modes presupposes that the resonator mirrors are large enough to intercept the entire beam without any "spillover"; i. e., that $a \gg w_1, w_2$, where w_1, w_2 are the spot sizes at the mirrors and a is an effective mirror cross sectional radius. This implies that the Fresnel number $N_\text{F} \equiv a^2/\lambda L \gg 1$. Diffraction losses generally increase with decreasing Fresnel numbers.

Higher-order Gaussian modes, where the Gaussian functions of x and y in (70.44) are replaced by higher-order Hermite polynomials, are often more difficult to realize than the fundamental lowest-order mode because their larger spot sizes imply higher diffractive losses, and beyond a certain mode order the spot sizes are too large to satisfy the no-spillover condition for a pure Gaussian mode.

It is not possible in general to write closed form expressions for laser modes. There are at least two reasons for this, the first being that the gain medium cannot in general be regarded as a simple amplifying element that preserves the basic empty-cavity mode structure. In low power gas lasers, the spatial variations of the gain and refractive index are sufficiently mild that the lasing modes can be accurately described as Gaussians, but more generally, there can be strong gain and index variations that themselves play an important role in determining the modes of the laser, as in "index-guided" and "gain-guided" semiconductor lasers.

Secondly, the resonator structure itself may introduce complications that preclude closed form solutions even for the empty cavity. This is generally true, for instance, of unstable resonators, where iterative numerical solutions of the integral equation (70.42) are necessary for accurate predictions of modes. In such computer simulations, the mode structure must be determined self consistently with the numerical solutions of density matrix or, more commonly, rate equations for the gain medium.

Unstable resonators, though inherently lossy, offer some important advantages for high-gain lasers. Thus, whereas the Gaussian modes of stable resonators typically involve very small beam spot sizes, the distinctly non-Gaussian modes of unstable resonators can have large mode volumes and make more efficient use of the available gain volume. Unstable resonators also tend to yield higher output powers when they oscillate on the lowest loss transverse mode, whereas in stable resonators higher output powers are generally associated with multitransverse mode oscillation. For very high power lasers, the fact that unstable resonators involve all-reflective optics, as opposed to transmissive output coupling, can be important in avoiding optical damage.

70.4 Photon Statistics

Optical fields may be characterized and distinguished by their photon statistical properties (Chapt. 78 and [70.17–21]). In a photon counting experiment, the number of photons registered at a photodetector during a time interval T is measured and used to infer the probability $P_n(T)$ that n photons are counted in a time interval T. If the probability of counting a photon at the time t in the interval $\mathrm{d}t$ is denoted $\alpha I(t)\,\mathrm{d}t$, where $I(t)$ is the intensity of the field and α is a factor depending on the microscopic details of the photoelectric process and on the phototube geometry, then it may be shown from largely classical considerations that [70.3, 17–21]

$$P_n(T) = \left\langle \frac{1}{n!}\left[\alpha\int_0^T \mathrm{d}t'\, I(t')\right]^n \mathrm{e}^{-\alpha\int_0^T \mathrm{d}t'\, I(t')}\right\rangle\,, \tag{70.48}$$

where the average $\langle \cdots \rangle$ is over the intensity variations during the counting interval. This is Mandel's formula. In the simplest case of constant intensity, P_n follows the Poisson distribution:

$$P_n = \bar{n}^{\,n}\, \mathrm{e}^{-\bar{n}}/n!\,, \tag{70.49}$$

where $\bar{n} = \alpha I T$. Since a single-mode laser field can be thought of approximately as a "classical stable wave" [70.20,21], it is not surprising that its photocount distribution is found both theoretically and experimentally to satisfy (70.49), which is characteristic of a coherent state of the field (Chapt. 78). A thermal light source, by contrast, follows the Bose–Einstein distribution,

$$P_n = \bar{n}^n/(1+\bar{n})^{n+1}\,, \tag{70.50}$$

if the time interval T is short compared with the coherence time of the light; i. e., if $T\Delta\nu \ll 1$, where $\Delta\nu$ is the bandwidth. If $T\Delta\nu \gg 1$, P_n is again Poissonian.

Thus, if a quasi-monochromatic beam of light is made from a natural source by spatial and spectral filtering, it has measurably different photon counting statistics from a single-mode laser beam of exactly the same bandwidth and average intensity. Laser radiation approaches the ideal coherent state that, of all possible quantum mechanical states of the field, most closely resembles the "classical stable wave". These differences are exhibited in other experiments, such as the measurement of intensity correlations of the Brown–Twiss type [70.3, 17–21]. In such experiments, thermal photons have a statistical tendency to arrive in pairs (photon bunching), whereas the photons from a laser arrive independently.

70.5 Multi-Mode and Pulsed Operation

Multi-mode laser theory is generally much more complicated than single-mode theory, particularly in the case of inhomogeneous broadening with both spectral and spatial hole burning. In certain situations, however, considerable simplification is possible. For instance, when the cavity-mode frequency spacing is small compared with the homogeneous linewidth $\delta\nu_0$, the gain tends to saturate homogeneously, and the total output power on all modes is well described by the Rigrod analysis outlined in Sect. 70.2.

Pulsed laser operation adds the further complication of temporal variations to the cw theory outlined in the preceding sections. It is possible, nevertheless, to understand some of the most important types of multi-mode and pulsed operation using relatively simple models.

One method of obtaining short, high power laser pulses is Q-switching. In a very lossy cavity, the gain can be pumped to large values before the threshold condition is met and gain saturation occurs. If the cavity loss is suddenly decreased, there will be a rapid build-up of intensity because the small-signal gain is far above the (now reduced) threshold value. The switching of the cavity loss is called Q-switching, the "quality factor" Q being defined as $\nu/2\delta\nu_c$. This type of Q-switching produces intense pulses of duration typically in the range 10–100 ns, as dictated by the fact that a light pulse must make several passes through the gain cell for amplification. Q-switching requires that the gain medium be capable of retaining a population inversion over a time much larger than the Q-switched pulse duration, and in particular that the spontaneous emission lifetime be relatively long. The pulse duration can be reduced to approximately a single cavity round trip time by Q-switching from low reflectivity mirrors to 100% reflectivities, and then switching the reflectivity of the outcoupling mirror from 100% to 0% at the peak of the amplified pulse. In addition to this pulsed transmission mode is cavity dumping, where both mirrors have nominally 100% reflectivity and the intracavity power is "dumped" by an acousto-optic or electro-optic intracavity element that deflects the light out of the cavity. The pulse duration achieved in this way is again roughly a round trip time. Cavity dumping can be employed with cw lasers and does not require the long energy storage times necessary for ordinary Q-switching.

Shorter pulses can be realized by mode-locking, where the phases of N longitudinal cavity modes are locked together. In the simplest model, assuming equal amplitudes and phases of the individual modes, the net field amplitude is proportional to

$$\begin{aligned} X(t) &= \sum_{n=-(N-1)/2}^{(N-1)/2} X_0 \sin[(\omega_0 + n\Delta)t + \phi_0] \\ &= X_0 \sin(\omega_0 t + \phi_0) \frac{\sin(N\Delta t/2)}{\sin(\Delta t/2)}\,. \end{aligned} \tag{70.51}$$

The temporal variation described by this function for large N is a train of "spikes" of amplitude NX_0 at times $t_m = 2\pi m/\Delta$, $m = 0, \pm 1, \pm 2, \cdots$, the width of each spike being $2\pi/(N\Delta)$. In the case of a mode-locked laser,

$\Delta = 2\pi(c/2L)$, and the output field consists of a train of pulses separated in time by $T = 2\pi/\Delta = 2L/c$. The peak amplitude of the spikes is proportional to N, and the duration of each spike is approximately $2L/cN$.

The maximum number $N_{\max}$ of modes that can actually be phase-locked is limited by the spectral width $\Delta\nu_g$ of the gain curve:

$$N_{\max} = \frac{\Delta\nu_g}{c/2L} = \frac{2L}{c}\Delta\nu_g \,. \quad (70.52)$$

Similarly, the shortest pulse duration is

$$\tau_{\min} = \frac{2L}{cN_{\max}} = \frac{1}{\Delta\nu_g} \,. \quad (70.53)$$

Mode-locking thus requires a gain bandwidth large compared with the cavity mode spacing, and the shortest and most intense mode-locked pulse trains are obtained in gain media having the largest gain bandwidths. Trains of picosecond pulses are routinely obtained with liquid dye and solid gain media having gain bandwidths $\Delta\nu_g \approx 10^{12}\,\mathrm{s}^{-1}$ or more.

Various techniques, employing acoustic or electro-optic modulation or saturable absorbers, are used to achieve mode-locking [70.1–3]. The different methods all rely basically on the fact that a modulation of the gain or loss at the mode separation frequency $c/2L$ tends to cause the different modes to oscillate in phase. Such a modulation is achieved "passively" when a saturable absorber is placed in the laser cavity: the multi-mode intensity oscillates with a beat frequency that is impressed on the saturated loss coefficient. Dye lasers, with their large gains across a broad range of optical frequencies, are often employed to generate mode-locked picosecond pulses.

Ultrashort pulse generation is possible with additional nonlinear or frequency chirping techniques [70.22]. The colliding-pulse laser [70.23] is a three-mirror ring laser in which two mode-locked pulse trains propagate in opposite senses and overlap in a very thin ($\approx 10\,\mu\mathrm{m}$) saturable absorber placed in the ring in addition to the gain cell. The cavity loss is least when the two pulses synchronize to produce the highest intensity, and therefore the lowest loss coefficient in the saturable absorber. The short length of the absorbing cell forces the pulses to overlap within a very small distance and therefore to produce very short pulses ($10\,\mu\mathrm{m}/\mathrm{c} \approx 30\,\mathrm{fs}$ pulse duration with $c/L \approx 100\,\mathrm{MHz}$ repetition rate).

Another method of ultrashort pulse generation relies on frequency chirping; i. e., a time dependent shift of the frequency of an optical pulse [70.24–26]. In a medium (e.g., a glass fiber) with linear and nonlinear refractive index coefficients n_0 and n_2, the refractive index is

$$n = n_0 + n_2 I \,, \quad (70.54)$$

so that there is an instantaneous phase shift $\phi(t)$ that depends on the instantaneous intensity. Therefore, as $I(t)$ increases toward the peak intensity of the pulse, $\phi(t)$ increases (assuming $n_2 > 0$), whereas $\phi(t)$ decreases as $I(t)$ decreases from its peak value. The frequency shift $\dot{\phi}(t)$ is such that the instantaneous frequency of the pulse is smaller at the leading edge and larger at the trailing edge of the pulse (Fig. 70.2), resulting in a stretching of the pulse bandwidth.

Following this spectral broadening by the nonlinear medium, the pulse can be compressed in time by means of a frequency dependent delay line such that the smaller frequencies, say, are delayed more than the higher frequencies. The trailing edge of the pulse can therefore "catch up" to the leading edge, resulting in a shorter pulse whose duration is given by the inverse of the chirp bandwidth. The delay line can be realized with a pair of diffraction gratings (see Fig. 70.2). Using this pulse compression technique, 40 fs amplified pulses from a colliding pulse laser have been compressed to 8 fs, corresponding to about four optical cycles [70.27].

In chirped-pulsed-amplification (CPA) lasers [70.28] a laser pulse is chirped, temporally stretched, and then passed through an amplifier. The lengthening in time of the pulse prior to amplification allows greater energy extraction from the amplifier. After amplification,

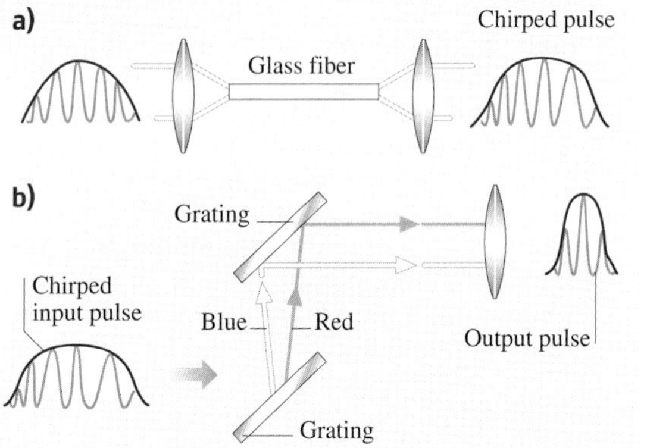

Fig. 70.2a,b Nonlinear pulse compression by frequency chirping. In (**a**), the nonlinear refractive index of a glass fiber results in a time dependent frequency of the transmitted pulse, and in (**b**) a pair of diffraction gratings is used to produce frequency dependent path delays such as to temporally compress the pulse

pulse compression is performed with a grating pair. The Ti :sapphire (Ti : Al_2O_3) amplifier is particularly attractive for femtosecond CPA because of its very large spectral width and high saturation fluence.

The wavelength dependence of the linear refractive index n_0 in (70.54) results in a group velocity

$$v_g = c[n_0 - \lambda \, dn_0/d\lambda]^{-1} \quad (70.55)$$

that, if $dn_0/d\lambda < 0$, is such that higher frequencies propagate more rapidly than lower frequencies. Assuming $n_2 > 0$, on the other hand, the nonlinear part of the index causes a delay of higher frequencies with respect to lower ones, as discussed above. This leads to soliton solutions of the wave equation, such that the opposing effects of the linear and nonlinear dispersion are balanced and the pulse propagates without distortion. Soliton lasers, with pulse durations ranging from picoseconds down to ≈ 100 fs, depending on the fiber length, have been made with solid state lasers and intracavity optical fibers [70.29].

70.6 Instabilities and Chaos

Mode-locked pulses and solitons exemplify an ordered dynamics, as opposed to the erratic and seemingly random intensity fluctuations that are sometimes observed in the output of a laser. In fact, it is possible, under certain circumstances, for laser oscillation to exhibit deterministic chaos; i. e., an effectively random behavior that can nevertheless be described by purely deterministic equations of motion [70.30–34].

Lasers are nonlinear and dissipative systems, and as such exhibit essentially all the modes of behavior characteristic of such systems. It was shown by *Haken* [70.35] that (70.6) and (70.11), with $\Delta = 0$ and with pumping and field loss terms included, can be put into the form of the Lorenz model for chaos. For a single-mode, homogeneously broadened ring laser, the Lorenz model instability requires a "bad cavity" in the sense that the field loss rate is larger than the sum of the homogeneous linewidth of the lasing transition and the population decay rate. It also requires the gain medium to be pumped at least nine times above the threshold gain value, a condition sometimes referred to as a "second laser threshold."

When the single-mode laser equations corresponding to the Lorenz model are generalized to the case of inhomogeneous broadening, an instability occurs at small–signal gain values much less than nine times threshold. In particular, the cw output of the laser gives way to an oscillatory intensity, even though the pumping and loss terms in the equations are assumed to be time independent. As the pumping and loss parameters are varied, this self-pulsing instability can give way to chaotic behavior, and numerical studies of the set of coupled matter–field equations reveal period doubling, two-frequency, and intermittency routes to chaos in different regimes [70.30]. This unstable behavior of single-mode inhomogeneously broadened lasers was first discovered experimentally and analyzed by *Casperson* [70.36–38] for low pressure, 3.51 μm He−Xe lasers. *Arecchi* et al. [70.39] reported the first observation and characterization of chaotic behavior in a laser system; by modulating the cavity loss of a CO_2 laser they observed a period doubling route to chaos as the modulation frequency was varied.

Extensive experimental and theoretical work on unstable and chaotic behavior in a wide variety of other laser systems has been reported [70.30–34, 40], including work aimed at the control of chaotic laser oscillation by the so-called occasional proportional feedback technique [70.41, 42]. Instabilities of single transverse mode dynamics have also been studied, especially in connection with spontaneous spatial pattern formation [70.40].

70.7 Recent Developments

Recent developments in the basic physics of lasers include the application of cavity QED techniques to produce a single-atom laser [70.43] that emits $< 10^5$ photons/s, and a two-photon laser [70.44] that operates on the basis of amplification on a two-photon transition. Recent progress in the development of ultrashort pulses includes the generation of attosecond pulses by high-order harmonic generation [70.45, 46], and the application of such pulses to a measurement of the photoionization time of Auger electrons [70.47].

References

70.1 A. Yariv: *Quantum Electronics*, 2nd edn. (Wiley, New York 1989)
70.2 A. E. Siegman: *Lasers* (University Science Books, Mill Valley. 1986)
70.3 P. W. Milonni, J. H. Eberly: *Lasers* (Wiley, New York 1988)
70.4 P. W. Milonni: Phys. Rep. **25**, 1 (1976)
70.5 B. DiBartolo: *Optical Interactions in Solids* (Wiley, New York 1968) p. 405
70.6 W. W. Rigrod: J. Appl. Phys. **36**, 27 (1965)
70.7 C. H. Henry: IEEE J. Quantum Electron. **18**, 259 (1982)
70.8 K. Petermann: IEEE J. Quantum Electron. **19**, 1391 (1979)
70.9 A. E. Siegman: Phys. Rev. A **39**, 1253 (1989)
70.10 A. E. Siegman: Phys. Rev. A **39**, 1264 (1989)
70.11 P. Goldberg, P. W. Milonni, B. Sundaram: Phys. Rev. A **44**, 1969 (1991)
70.12 P. Goldberg, P. W. Milonni, B. Sundaram: Phys. Rev. A **44**, 4556 (1991)
70.13 W. A. Hamel, J. P. Woerdman: Phys. Rev. Lett **64**, 1506 (1990)
70.14 M. A. Eijkelenborg, A. M. Lindberg, M. S. Thijssen, J. P. Woerdman: Phys. Rev. A. **55**, 4556 (1997)
70.15 K.-J. Cheng, P. Mussche, A. E. Siegman: IEEE J. Quantum Electron. **30**, 1498 (1994)
70.16 A. G. Fox, T. Li: Bell System Tech. J. **40**, 453 (1961)
70.17 R. J. Glauber: Phys. Rev. **130**, 2529 (1963)
70.18 R. J. Glauber: Phys. Rev. **131**, 2766 (1963)
70.19 L. Mandel, E. Wolf: Rev. Mod. Phys. **37**, 231 (1965)
70.20 R. Loudon: *The Quantum Theory of Light*, 3rd edn. (Clarendon Press, Oxford 2000) p. 32
70.21 P. Meystre, M. Sargent III: *Elements of Quantum Optics*, 2nd edn. (Springer, Berlin, Heidelberg 1991) p. 32
70.22 E. B. Treacy: IEEE J. Quantum Electron. **5**, 454 (1969)
70.23 R. L. Fork, B. I. Greene, C. V. Shank: Appl. Phys. Lett. **38**, 671 (1981)
70.24 D. Grischkowsky, A. C. Balant: Appl. Phys. Lett **41**, 1 (1982)
70.25 L. F. Mollenauer, R. H. Stolen, J. P. Gordon, W. J. Tomlinson: Opt. Lett. **8**, 289 (1983)
70.26 J. G. Fujimoto, A. M. Weiner, E. P. Ippen: Appl. Phys. Lett. **44**, 832 (1984)
70.27 C. V. Shank: Science **233**, 1276 (1986) and references therein
70.28 P. Maine, D. Strickland, P. Bado, M. Pessot, G. Mourou: IEEE J. Quantum Electron. **24**, 398 (1988)
70.29 L. F. Mollenauer, R. H. Stolen: Opt. Lett. **9**, 13 (1984)
70.30 P. W. Milonni, M.-L. Shih, J. R. Ackerhalt: *Chaos in Laser–Matter Interactions* (World Scientific, Singapore 1987)
70.31 D. K. Bandy, A. N. Oraevsky, J. R. Treddice: Special issues of J. Opt. Soc. Am. **5**, 5 (1988)
70.32 N. B. Abraham, W. J. Firth: Special issues of J. Opt. Soc. Am. **7**, 6 (1990)
70.33 N. B. Abraham, W. J. Firth: Special issues of J. Opt. Soc. Am. **7**, 7 (1990)
70.34 N. B. Abraham, F. T. Arecchi, L. A. Lugiato (Eds.): *Instabilities and Chaos in Quantum Optics II* (Plenum, New York 1988)
70.35 H. Haken: Phys. Lett. A **53**, 77 (1975)
70.36 L. W. Casperson: *IEEE J. Quantum Electron*, Vol. 14 (Springer, Berlin, Heidelberg 1978) p. 756
70.37 L. W. Casperson: Spontaneous pulsations in lasers. In: *Laser Physics*, ed. by J. D. Harvey, D. F. Walls (Springer, Berlin, Heidelberg 1983)
70.38 P. Chenkosol, L. W. Casperson: J. Opt. Soc. Am. B **20**, 2539 (2003)
70.39 F. T. Arecchi, R. Meucci, G. Puccione, J. Tredicce: Phys. Rev. Lett. **49**, 1217 (1982)
70.40 L. A. Lugiato, W. Kaige, N. B. Abraham: Phys. Rev. A **49**, 2049 (1994) and references therein.
70.41 R. Roy, T. W. Murphy Jr., T. D. Maier, Z. Gills: Phys. Rev. Lett. **68**, 1259 (1992)
70.42 Z. Gills, C. Iwata, R. Roy, I. B. Schwartz, I. Triandaf: Phys. Rev. Lett. **69**, 3169 (1992) see also Optics and Photonics News (May,1994)
70.43 J. McKeever, A. Boca, A. D. Boozer, J. R. Buck, H. J. Kimble: Nature **425**, 268 (2003)
70.44 D. J. Gauthier: Progress in Optics **45**, 205 (2003)
70.45 M. Hentschel, R. Kienberger, C. Spielmann, G. A. Reider, N. Milosevic, T. Brabec, P. Corkum, U. Heinzmann, M. Drescher, F. Krausz: Nature **414**, 509 (2001)
70.46 P. M. Paul, E. S. Toma, P. Breger, G. Mullot, F. Auge, P. Balcou, H. G. Muller, P. Agostini: Science **292**, 1689 (2001)
70.47 M. Drescher, M. Hentschel, R. Kienberger, M. Uiberacker, V. Yakovlev, A. Scrinizi, T. Westerwalbesloh, U. Kleineberg, U. Heinzmann, F. Krausz: Nature **419**, 803 (2002)

71. Types of Lasers

The availability of coherent light sources (i. e., lasers) has revolutionized atomic, molecular, and optical science. Since its invention in 1960, the laser has become the basic tool for atomic and molecular spectroscopy and for elucidating fundamental properties of optics and optical interactions with matter. The unique properties of laser light have spawned new types of spectroscopy, as discussed in Chapt. 72 to Chapt. 80.

There are now literally hundreds of different types of lasers. However, only a few of these are commercially available, and lasers tailor made with operational properties optimized for specific applications are often needed. Chapter 70 describes the principles of laser operation leading to specific output characteristics. This chapter summarizes the current status of the development of different types of lasers, emphasizing those that are commercially available. There are several ways to categorize types of lasers; for example, in terms of spectral range, temporal characteristics, pumping mechanism, or lasing media.

71.1 **Gas Lasers** ... 1036
71.1.1 Neutral Atom Lasers ... 1036
71.1.2 Ion Lasers ... 1036
71.1.3 Metal Vapor Lasers ... 1037
71.1.4 Molecular Lasers ... 1037
71.1.5 Excimer Lasers ... 1038
71.1.6 Nonlinear Mixing ... 1038
71.1.7 Chemical Lasers ... 1039

71.2 **Solid State Lasers** ... 1039
71.2.1 Transition Metal Ion Lasers ... 1040
71.2.2 Rare Earth Ion Lasers ... 1040
71.2.3 Color Center Lasers ... 1042
71.2.4 New Types of Solid State Laser Systems ... 1043
71.2.5 Frequency Shifters ... 1043

71.3 **Semiconductor Lasers** ... 1043

71.4 **Liquid Lasers** ... 1044
71.4.1 Organic Dye Lasers ... 1044
71.4.2 Rare Earth Chelate Lasers ... 1045
71.4.3 Inorganic Rare Earth Liquid Lasers ... 1045

71.5 **Other Types of Lasers** ... 1045
71.5.1 X-Ray and Extreme UV Lasers ... 1045
71.5.2 Nuclear Pumped Lasers ... 1046
71.5.3 Free Electron Lasers ... 1046

71.6 **Recent Developments** ... 1046

References ... 1048

For this chapter, the types of lasers are categorized in terms of the lasing media, as shown in Table 71.1.

The variety of different laser types offers a wide range of beam parameters. The spectral output of lasers covers the X-ray to far IR regions as shown in Fig. 71.1. Spectral linewidths can be as narrow as 20 Hz and diffraction limited beam quality with coherence lengths up to 10 m can be obtained. Temporal pulse widths of a few femtoseconds have been generated. Some lasers can produce peak powers of over 10^{13} W and average powers of 10^5 W with pulse energies greater than 10^4 J. The important operational characteristics, such

Table 71.1 Categories of lasers

Gas lasers	Special lasers	Solid State lasers	Liquid lasers	Miscellaneous
Atomic	Metal vapor	Transition metal ion	Organic dye	X-ray
Ionic	Chemical	Rare earth ion	Rare earth chelate	Nuclear pumped
Molecular		Color center	Inorganic solvents with rare earth ions	Free electron
Excimer		Semiconductor		

as frequency range and output power, are given for each of the types of lasers described in the following sections. Extensive tables of laser properties have been published in several different handbooks [71.1–5] and all of these details for every laser are not repeated here due to space limitations. The data that are quoted come from these reference books and from recent proceedings of conferences such as CLEO (Conference on Lasers and Electro-Optics) and ASSL (Advanced Solid State Laser Conference). The reader is referred to these sources for further details of specific laser operation parameters.

71.1 Gas Lasers

Gas lasers can be separated into subclasses based on the lasing media: neutral atoms; ions; and molecules. In addition, molecular lasers contain the special classes of excimer lasers and chemical lasers. Except for chemical lasers, the pumping mechanism is generally an electrical discharge in a gas filled tube. This discharge causes the acceleration of electrons that transfer their kinetic energy to the lasing species through collisions, leaving them in a variety of excited states. These relax back to the ground state with different rates, resulting in the possibility of a population inversion for some transitions. These can be electronic, vibrational, or rotational transitions with wavelengths ranging from the near ultraviolet (UV) through the far infrared (IR) spectral regions. Systems lasing at short wavelengths generally operate only in the pulsed mode because of the short radiative lifetimes of the transitions involved. Because of their low gain, gas lasers usually have relatively long linear cavity designs. Gas lasers generally have narrow spectral emission lines with the possibility of lasing at several different wavelengths. They are broadened both by collisions (homogeneous broadening) and the Doppler effect of the motion of the atoms or molecules (inhomogeneous broadening). To date, 51 elements in the periodic table have shown either ionic or neutral ion gas laser emission.

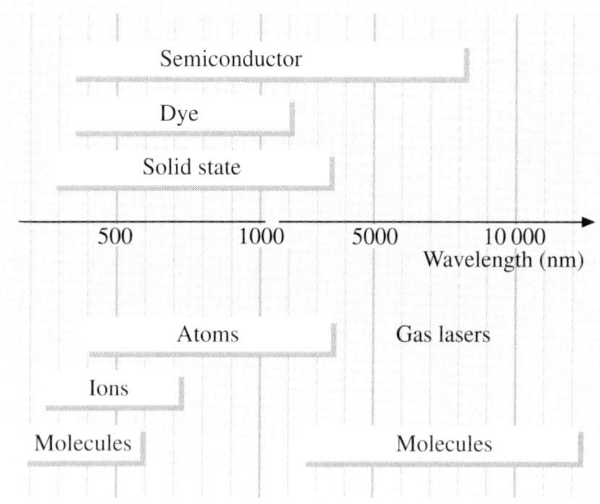

Fig. 71.1 Spectral range of laser emission

71.1.1 Neutral Atom Lasers

These lasers generally emit in the visible and near IR spectral range. The first and most common laser of this class is the He–Ne laser. Its most prominent emission line is at 632.8 nm. It is usually operated in the continuous wave (cw) configuration with typical power outputs between 0.5 to 50 mW, although powers of over 100 mW have been achieved. Excitation occurs through electrical discharge which pumps both the He and Ne atoms to excited states. The more abundant He atoms transfer their energy to several excited states of Ne atoms. Several radiative transitions of Ne are available for laser transitions. These provide emission at 543.36, 632.8, 1152.27, and 3391.32 nm. Final relaxation back to the ground state occurs through collisions with the walls of the gas tube. The desired laser line can be selected by adjusting the reflectivity of the cavity mirrors to discriminate against unwanted transitions. In addition, it is possible to adjust the discharge current, gas ratio, and pressure to optimize a specific emission transition. Using an external magnetic field to produce a Zeeman effect can also be helpful in tuning the laser emission. In addition to the normal red He–Ne lasers, green lasers are now available. Typical He–Ne lasers operate with a coherence length of 0.1–0.3 m, a beam divergence of 0.5–2 mrad, and a stability of 5%/h. The gain at the red line is 0.5 dB/m. This line can be operated as a single mode with a linewidth of 0.0019 nm and a coherence length between 20 and 30 cm.

71.1.2 Ion Lasers

The most important ion lasers are based on noble gas ions such as Ar, Kr, Ne, or Xe in various states of ionization. These operate in either a pulsed or cw mode, and their emission covers the wavelength range from the near UV through the visible part of the spectrum. In the electrical-discharge excitation process, electrons collide with neutral atoms in their ground states, transferring enough energy to ionize them and leave the ions in several possible excited states. For example, low discharge currents produce Ar^+ giving rise to visible emission lines, while high discharge currents produce Ar^{2+} giving rise to UV emission lines. Radiative emission then occurs to lower excited levels of the ions, followed by subsequent spontaneous emission to the ground state of the ion, and then radiationless relaxation back to the neutral atom ground state. This transition scheme limits the wall plug efficiency to about 0.1% for visible and 0.01% for UV operation. Heat management is accomplished through either water cooling or air cooling.

In the visible region of the spectrum, argon lasers have blue and green emission lines with the strongest ones at 488 and 515 nm, respectively. Krypton lasers have several strong emission lines in the green and red, with the most prominent ones at 521, 568, and 647 nm. Mixed gas lasers can produce all of these lines. In the near UV, argon has a strong laser line at 351 nm as well as several other emission lines down to 275 nm. The same methods for selecting specific laser lines on neutral atom lasers, discussed in Sect. 71.1.1, are used for ion lasers. These lasers can operate at powers of over 20 W of cw emission in the visible and at powers of several watts cw in the UV. The Doppler-limited linewidth of noble gas ion lasers is generally $\approx 5-10$ GHz. By using special techniques for stabilization of the cavity, linewidths of ≈ 500 MHz with drifts of 100 MHz/hr can be obtained. By mode-locking argon or krypton lasers with 10 GHz linewidths, it is possible to produce trains of pulses with pulse lengths of 100 ps and peak powers of 1 kW, at a pulse repetition frequency of 150 MHz. Cavity-dumping produces narrow pulses at pulse repetition frequencies of ≈ 1 MHz with peak powers over 100 times the cw power. Lower power cw lasers of this type typically have beam divergences of 1.5 mrad and a stability of 5%, while high power lasers have beam divergences of ≈ 0.4 mrad with a stability of 0.5%.

71.1.3 Metal Vapor Lasers

These lasers can operate with either neutral atoms or ions. Their excitation process begins with vaporizing a solid or liquid to produce the gas for lasing, followed by normal electrical discharge pumping. Either cw or pulsed operation can be obtained, with laser emission lines in the near UV and visible spectral regions.

One important ion laser of this class is the Helium–Cadmium laser. For the excitation processes, metallic Cd is evaporated and mixed with He. Then a d.c. electric discharge excites the He ions and ionizes the Cd. The excited He atoms transfer their energy to the Cd atoms and the laser transitions take place between electronic levels of the Cd atom. The main emission line of a He–Cd laser is the blue line at 441.6 nm. This typically has a cw output from 130 mW for single-mode operation up to 150 mW for multimode operation. The laser linewidth can be as narrow as 0.003 nm. This system also has an important laser emission at 325.029 nm, which typically has cw powers between 5 and 10 mW single-mode and 100 mW for multimode emission. The wall plug efficiency is between 0.002% and 0.02%.

The most important neutral atom metal vapor laser of this type is the copper vapor laser. This has important emission lines in the green at 510.55 nm and in the yellow at 578.21 nm. These lasers operate in the pulsed mode with temporal pulse widths between 10 and 20 ns at pulse repetition rates of up to 20 000 pps. Typical pulse energies are ≈ 1 mJ, yielding average powers of 20 W. It is possible to increase the repetition rate significantly to achieve average powers of 120 W or even higher. However, this laser is self-terminating since the lower levels of the laser transitions are metastable. This restricts the pulse sequencing of the laser and requires fast discharge risetimes. Copper vapor lasers have high gain (10% to 30%/cm), and very high wall plug efficiency ($\approx 1.0\%$). Gold vapor lasers have similar properties but operate in the red at 624 nm at several watts of power.

71.1.4 Molecular Lasers

There are several types of molecular lasers that can be classified with respect to their spectral emission range, their mode of excitation, or the energy levels involved in the lasing transition. In the far IR, molecular lasers operate on transitions between rotational energy levels. These include water vapor lasers that emit between 17 and 200 μm, cyanide lasers at 337 μm, methyl fluoride lasers emitting between 450 and 550 μm, and ammonia lasers that operate at 81 μm. These are generally excited through optical pumping by a CO_2 laser. They are built with a metal or dielectric wave guide cavity. The former design results in lower thresholds but gives multimode, mixed polarization output, while the latter design results in propagation losses, giving higher thresholds but linearly polarized outputs.

CO_2 lasers are some of the most widely used, with a variety of medical and industrial applications. They emit in the mid-IR range at 10.6 and 9.6 μm. The CO_2 molecules are excited by electrical discharge, and it is common to mix CO_2 with other gases, such as nitrogen, to enhance the efficiency of the excitation process through energy transfer, or with helium, to keep the average electron energy high and to depopulate the lower levels of the laser transition. Both the initial and final states of the lasing transition are vibrational levels that have many rotational sublevels. This allows discrete tuning of the output within the 9.4 and 10.4 μm bands. Using high pressures of the gas broadens the laser line into a continuum so that the emission can be continuously tuned over several microns near 10 μm. Without any frequency selective element in the cavity, the system oscillates on the transition with highest gain, which is near 10.6 μm. Under pulsed operating conditions, CO_2 lasers can also emit at bands near 4.3 μm and at several bands between 11 and 18 μm.

These laser systems typically operate at a few milliwatts of power to over 100 kW cw. Waveguide and slab cavity designs have been developed for heat removal. In the pulsed or Q-switched mode of operation, pulse widths are between a few microseconds and a few milliseconds, with energies as high as 10 000 J/pulse. This leads to peak powers more than 100 times higher than cw powers. It is also possible to mode-lock these systems to get a train of nanosecond pulses. In the transverse-excitation-atmospheric-pressure (TEA) configuration, CO_2 lasers are pumped very rapidly compared with the lifetime of the metastable state, resulting in a large population inversion in the gain medium when the electromagnetic field builds up in the cavity. This results in an intense "gain-switched pulse" of between 100 and 200 ns in duration followed by a lower intensity emission due to continued pumping of the upper state. These systems provide up to 3 J/pulse at pulse repetition frequencies of up to 50 Hz. At much lower pulse repetition frequencies, energies as high as 1000 J/pulse have been obtained. The typical beam divergence of these lasers is less than 3 mrad.

Carbon monoxide lasers operate on the vibrational levels of CO. They can be excited either through electrical discharge or chemical reaction (as discussed in Sect. 71.1.7). CO lasers have a tunable emission between 5 and 7 μm, operating with powers as high as 1 kW cw. In the pulsed mode, the typical energy emission is 10 mJ/pulse with 1 μs pulses at a pulse repetition frequency of 10 Hz. N_2O lasers extend the molecular laser wavelength range to beyond 10 μm.

Nitrogen lasers operate in the near UV at 337.1 nm. These are based on transitions between electronic energy levels of the N_2 molecule excited by electrical discharge. The typical emission from an N_2 laser is a single pulse of 10 ns in duration. The peak pulse power can be as high as 1 MW with 10 mJ/pulse at a pulse repetition frequency of less than 100 Hz. It is also possible to design these systems to obtain picosecond pulses. H_2 operates in a similar way in the 120 to 160 nm region of the UV.

71.1.5 Excimer Lasers

The most important lasers in the near UV to VUV for industrial and medical applications are based on rare-gas halide excimers such as XeF at 351 nm, XeCl at 308 nm, KrF at 248 nm, ArF at 193 nm, and F_2 at 153 nm. The major problem with these systems is the corrosive nature of the gases. They can be pumped by electric discharges, electron or proton beams, or optical excitation. Using electrical discharge excitation, the electrons ionize the noble gas molecules, and these react by pulling an atom off the halide molecule to create an excited state dimer molecule (excimer) that radiates to an unstable lower state where dissociation occurs. The short radiative lifetimes of excimers result in laser pulses of 10 to 50 ns duration. Systems can be configured to have pulse durations ranging from picoseconds to microseconds. These lasers typically operate at high pulse repetition frequencies between 200 and 1000 Hz. The energy per pulse of discharge-pumped excimer lasers ranges from several mJ to 0.1 J, with typical average powers up to 200 W. Single shot, electron-beam-pumped excimer lasers can have as high as 10^4 J/pulse for an average power of almost 1 kW.

Typical emission bands for excimer lasers are approximately 100 cm^{-1} wide due to vibrational sublevels. With the use of etalons and gratings, the laser line can be narrowed to 0.3 cm^{-1}. These frequency-selective elements can be used to tune the emission over several nanometers. Special configurations of excimer lasers have been developed to obtain specific operational parameters. For example, pulses as short as 45 fs have been obtained by mode-locking or by using excimer amplifiers to amplify frequency shifted dye laser pulses [71.3, 4]. Stimulated Brillouin scattering has been used as a phase conjugate mirror to minimize phase-front distortion in excimer lasers. Both master oscillator power amplifiers and injection locked resonator techniques have been used to achieve low spatial divergence with narrow bandwidths [71.3,4]. Input pulses for the latter system can be from either frequency shifted Nd:YAG or dye lasers.

One interesting recent development has been the demonstration of laser operation of solid state excimer systems [71.6–8] . This is discussed in Sect. 71.2.4.

71.1.6 Nonlinear Mixing

Another way to produce tunable VUV laser transitions is to use nonlinear four-wave sum mixing in atomic vapors and molecular gases. This requires optical pumping with two sources and can be achieved with any of a variety of laser combinations such as excimer lasers with dye lasers, or Nd:YAG lasers with dye lasers. This technique has been used to obtain laser emission in the 57 to 195 nm spectral range with visible to VUV conversion efficiencies as high as 10^{-3}.

71.1.7 Chemical Lasers

It is possible for some chemical reactions between molecules to leave the final molecule in an excited state. This type of "pumping" can result in a population inversion with respect to one of the lower states and laser transitions can occur between vibrational or rotational states of the molecule. Using mixtures of different types of molecules, pumping can be enhanced through energy transfer. Both pulsed and cw laser operation have been obtained with chemical lasers. Typical emission is of the order of a few hundred watts in the IR with a tunable output wavelength. Visible emission from chemical lasers has been demonstrated [71.9–11]. This type of laser provides the possibility of a system with a self contained chemical power supply for use in remote environments. The problems associated with handling hazardous chemicals have restricted the applications of chemical lasers.

The HF chemical laser is associated with an exothermic chain reaction between H_2 and F_2 molecules to yield vibrationally excited HF [71.4]. The fluoride atoms are generated in an electrical discharge tube from the dissociation of SF_6. These are injected into the optical resonator along with the H_2 or D_2 gas which flows perpendicular to the lasing direction. Controlling the gas flow for mixing the reactants is critical to the laser design. The chemical reaction for direct excitation is $F + H_2 \rightarrow HF^* + H$. For this reaction, $\Delta H = -32\,\text{kcal/mole}$, resulting in laser emission energies of between 100 and 400 kJ/kg, at wavelengths between 2.5 and 3.7 μm. This corresponds to vibrational-rotational transitions in the HF molecule. It is possible to select a single line for the laser output and then tune the laser output wavelength by selecting different lines. Single line output for cw operation can produce up to 100 W of power. If D_2 replaces H_2, the emission shifts to between 3.6 and 4.2 μm, and the single line cw output power drops to about 50 W. The output power for a cw laser of this type, operating in the multiple line mode, can be as high as 2.2 MW, while in the pulsed mode of operation pulse energies of 5 kJ can be obtained with multiple line emission.

An example of energy transfer pumping of chemical lasers is the $DFCO_2$ system. Pumping of vibrational-rotational transitions of DF occurs through multiple chemical reactions of fluorine and deuterium, followed by energy transfer to excited states of the CO_2 molecules. This exhibits laser emission with kilowatts of power at 10.6 μm, as described above. Another important laser system using energy transfer pumping is the chemical-oxygen-iodine laser (COIL). This is based on transitions between electronic levels in which singlet oxygen is excited and transfers its energy to a metastable state of iodine. Emission occurs at 1.3 μm and cw powers of up to 25 kW have been obtained.

One example of chemical laser action in the visible region is excited GeO transferring its energy to atomic Tl, which lases at 535 nm. Only a few systems of this type have been demonstrated, and none are developed to the level of commercial availability [71.9–11].

71.2 Solid State Lasers

Solid state lasers are based on luminescence centers randomly distributed in a crystalline or glass host material. These can be classified in terms of the type of their laser active centers: transition metal ion lasers; rare earth ion lasers; and color center lasers. The use of dye molecules in plastic host media is a new type of solid state laser that is discussed in Sect. 71.4.1. The active ions are substitutionally "doped" into the host during crystal growth or glass melting, whereas the lattice defects that produce color centers are generally produced by post-growth radiation or heat treatments. The excitation mechanism is through optical pumping by either another laser or lamps. The spectral range covered by solid state lasers spans the visible and near IR. The variety of combinations of active centers and hosts provides the ability for both pulsed and cw operation with either narrow band or broad band emission. The latter type can provide frequency tunable output with the appropriate frequency-selective element in the cavity. Temperature tuning of narrow emission lines is also possible over a limited range. The spectral lines are broadened by internal strains in the host lattice (inhomogeneous broadening) and by radiationless relaxation and scattering processes involving thermal vibrations of the host (homogeneous broadening). Standard Q-switching and mode-locking techniques can be used with most solid state lasers.

71.2.1 Transition Metal Ion Lasers

The laser ions of this type have optically active electrons in unfilled $3d^n$ electron configurations. They include the positively charged ions Cr^{3+}, Cr^{4+}, Co^{2+}, V^{2+}, Ni^{2+}, and Ti^{3+}. The transitions involved in pumping and lasing are associated with the optically active electrons. Because these are outer shell electrons, they are sensitive to their local crystal field environment. Typical host materials are ionic crystals formed from oxides such as sapphire, emerald, chrysoberyl, forsterite, and garnets, or fluorides such as MgF_2, $KMgF_3$, $LiCaAlF_6$, and $LiSrAlF_6$. The host determines the bulk optical, mechanical, and thermal properties of the laser material, and it influences the spectroscopic properties of the active ions.

The most successful transition metal laser ion is Cr^{3+}. It has been made to lase in many different types of host crystals with both strong and weak crystal field environments. The most common strong field host is sapphire, and Al_2O_3:Cr^{3+}, commonly known as ruby, which was the first laser invented. The typical characteristic of a strong field laser material is a sharp laser line associated with a spin-flip electronic transition between states of the same crystal field configuration. In ruby this occurs at 699.7 nm. Because of its strong, broad absorption bands, ruby can be efficiently pumped by lamps and operates in either a pulsed or cw mode. Typical cw power output is a few watts with an efficiency of $\approx 0.1\%$. Ruby can be Q-switched to produce 10 ns pulses with several joules of energy per pulse, and mode-locked to produce pulses that are 5 ps in duration.

The typical characteristic of a weak field laser material is a broad gain curve associated with vibronic transitions between states of different crystal field configurations. Although chrysoberyl is a host with intermediate crystal field, $BeAl_2O_4$:Cr^{3+}, commonly known as *alexandrite*, is sufficiently close to a weak field case to operate as a laser in this regime. Using frequency selective elements in the cavity, alexandrite lasers can be tuned in the 700 to 820 nm range. The gain of alexandrite increases with temperature, with the cross section at the peak of the gain curve near 10^{-20} cm^2. Typical laser outputs are 4.5 J/pulse at 20 Hz pulse repetition rate and 90 W average power with 2% overall efficiency. In the Q-switched mode, 40 ns pulses with 2 J/pulse are obtained and the pulse width can be stretched to much longer values. Alexandrite lasers can also be mode-locked to obtain 28 ps pulses with 0.5 mJ/pulse.

Fluoride crystals, such as $LiCaAlF_6$ and $LiSrAlF_6$, are also weak field hosts for tunable Cr^{3+} lasers in the near IR. The latter is termed Cr:LiSAF and has a peak stimulated emission cross section of 0.4×10^{-19} cm^2 with a tuning range from 780 to 1020 nm. This system can be either flashlamp pumped or diode laser pumped. Single pulses with energies of 75 J have been generated by these lasers. Kerr lens mode-locking has produced pulses shorter than 100 fs.

One of the most interesting tunable solid state laser systems is Ti-sapphire (Al_2O_3:Ti^{3+}) because it has the broadest tuning range of any ion, extending from about 660 nm to about 1180 nm. Pumping can be provided by either lasers or flashlamps, resulting in either cw or pulsed operation. This results in a versatile source of excitation in this spectral region and sub-picosecond pulses can be obtained through mode-locking. The single 3d electron of Ti^{3+} gives a simplified energy level scheme which minimizes losses due to excited state absorption, which can be a problem in Cr^{3+} lasers. However, the metastable state lifetime of Ti^{3+} is significantly shorter than that of Cr^{3+}, and therefore Cr^{3+} lasers have much greater energy storage capability. Thus, Ti-sapphire lasers are difficult to pump by flashlamp, and are therefore generally pumped by argon lasers or frequency-doubled Nd-YAG lasers. Ti-sapphire has a very high peak gain cross section of about 4×10^{-19} cm^2. On the other hand, it is difficult to grow Ti-sapphire crystals with high concentrations of Ti^{3+} ions because of valance state stability.

71.2.2 Rare Earth Ion Lasers

All of the trivalent lanthanide ions (Ce^{3+}, Pr^{3+}, Nd^{3+}, Pm^{3+}, Sm^{3+}, Eu^{3+}, Gd^{3+}, Tb^{3+}, Dy^{3+}, Ho^{3+}, Er^{3+}, Tm^{3+}, Yb^{3+}) and the divalent ions Sm^{2+}, Dy^{2+}, Tm^{2+} have been used as active ions in solid state lasers. These ions are characterized by unfilled $4f^n$ electron configurations and the most common source of laser emission comes from electronic transitions among their energy levels. Because the inner shell 4f electrons are shielded by outer shell electrons, the energy levels are not strongly affected by the environment of the local host material. This leads to sharp lines in both absorption and emission.

One exception to this is vibronic emission from Ho^{3+} which can produce tunable laser emission in the near IR. In some cases, transitions between 5d and 4f levels are involved in the laser emission. Since the 5d levels are broadened by the environment, broadly tunable laser emission can be obtained. Examples are Ce^{3+} in the UV and Sm^{2+} in the near IR The only actinide ion that has been made into a laser is U^{3+}.

Both crystals and glasses can be used as host materials for rare earth ion lasers. Common oxide crystal hosts include the garnets such as $Y_3Al_5O_{12}$ (commonly referred to as YAG) and a typical fluoride crystal host is $YLiF_3$. A wide variety of glass hosts has been used, including silicates, phosphates, heavy metal fluorides, and mixtures of these. The major difference between crystal and glass hosts is that crystals provide similar crystal field sites for every dopant ion, leading to a minimum of inhomogeneous broadening, while the disorder associated with glass structure gives many different types of local crystal field sites for the dopant ions and thus significant inhomogeneous broadening.

Because of the abundance of their energy levels, many trivalent rare earth ions have more than one metastable state, and laser emission is possible from several transitions. This results in over 100 possible laser emission lines ranging from the near UV through the visible and near IR. Both pulsed and cw operation can be obtained. The standard configuration for a rare earth solid state laser is a rod of laser material pumped by a lamp. Other configurations are used for special situations, such as a slab of laser material for high power glass lasers where heat management is a problem, and microchip lasers for photonics applications. Glass *fiber lasers* and amplifiers are becoming important configurations for some applications.

A major problem is the inefficiency of coupling the excitation energy of a lamp source with a broad spectral output into the spectrally sharp absorption bands of the trivalent rare earth ions. This can be overcome by using a laser as a pump source. One of the major advances in solid state laser technology has been the development of bars of high power diode laser arrays as pump sources. This has significantly increased the efficiency and decreased the thermal problems in these lasers. Currently available diode laser pump sources cover a limited range of wavelengths and thus can only be used to excite a limited set of metastable states. Several schemes have been adapted to excite other metastable states. One of these is *up-conversion pumping* [71.12–15] in which an ion is excited to a low-lying metastable state by a photon from the pump source and then re-excited to a higher metastable state either by another photon from the pump source or by energy transfer interaction with a neighboring ion that has also been excited to the low-lying metastable state. *Avalanche pumping* [71.16] relies on thermal fluctuations to populate a low-lying energy level of an ion which can then be re-excited by a pump photon to a high energy metastable state. Another method involves the addition of a second dopant ion to the host. This "sensitizer" ion is one with broad pump bands (such as Cr^{3+}) that can efficiently absorb the energy from the pump lamp. The excited sensitizer then interacts with the "activator" (lasing) ion through a radiationless, resonant energy transfer process. This deactivates the sensitizer and excites the activator ion.

Nd^{3+} has been made to lase in a greater number of host materials than any other active ion. $Y_3Al_5O_{12}$:Nd^{3+}, commonly referred to as Nd-YAG, is one of the most successful commercially available lasers. Although it is possible for Nd-YAG lasers to operate at several different wavelengths around 1 μm, the standard lasers emit at 1.06 μm. Continuous wave powers of 250 W are available and pulsed performance of several megawatts at 10 Hz and 1 J/pulse can be obtained. To obtain visible (532 nm) and near UV (354 and 266 nm) emission, nonlinear optical crystals are used to modify the near IR output through second, third, and fourth harmonic generation. The optimum concentration of Nd^{3+} in YAG is a few percent, and above this amount, concentration quenching of the emission occurs through energy transfer and cross-relaxation processes. Attempts to co-dope Nd-YAG with Cr^{3+} to enhance pumping through energy transfer have not been successful [71.17, 18]. However, in other garnet crystal hosts such as $Gd_3Sc_2Ga_3O_{12}$ (GSGG), the Cr–Nd energy transfer is strong enough to enhance pumping efficiency. Diode laser pumping of Nd-YAG has significantly increased pumping efficiency.

Other crystal hosts such as the pentaphosphates do not exhibit concentration quenching, so materials with 100% Nd^{3+} can be used as "stoichiometric laser materials". These can be important for some mini-laser applications. In glass hosts, Nd^{3+} ions can produce pulses with energies of $\approx$ 100 kJ/pulse, but these systems must operate at very low pulse repetition rates of one pulse every few minutes to allow for heat dissipation. New athermal glass compositions decrease the problems with thermal lensing for high power laser operation. Nd:YVO lasers have been operated in a microchip configuration pumped by diode lasers.

A major area of recent development involves rare earth doped crystal lasers operating at specific wavelengths in the near IR [71.19–21]. This includes both the "eye safe" region between 1.35 and 2.2 μm for applications involving atmospheric transmission, and the 2 to 3 μm range to match water overtone absorption bands for medical applications. The ions of most interest are Er^{3+}, Tm^{3+}, and Ho^{3+}. Er^{3+} can lase on 13 transitions at wavelengths ranging from 0.56 to 4.8 μm. Both Tm^{3+} and Ho^{3+} exhibit laser transitions near 2 μm, but they are difficult to pump efficiently. One successful laser in this spectral region is triply doped Cr;Tm;Ho:YAG. The flashlamp energy is absorbed by the Cr^{3+} ions and transferred to Tm^{3+} ions. Cross-relaxation between two thulium ions doubles the quantum efficiency by leaving two Tm^{3+} ions in the excited state. Energy transfer among the thulium ions with transfer to Ho^{3+} then occurs and the lasing transition occurs on the holmium ions.

Fiber lasers have been developed that consist of trivalent rare earth ions doped in either oxide or fluoride glass fibers and pumped by other lasers [71.22]. A useful application of active fibers is Er^{3+} and Pr^{3+} optical amplifiers for fiber communication systems. The extended length of the gain media and the nonlinear dispersion effects in fiber transmission allow precise tailoring of laser emission properties. Heavy metal fluoride fibers give improved IR transmission in the 2 to 3 μm spectral region. An important recent development involves writing laser-induced gratings in fibers to produce distributed feedback lasers with stable, single-mode operation [71.23]. Nd-doped fiber lasers pumped by diode lasers have produced 5 W of power. Mode-locked fiber ring lasers have produced solitons. Efficient up-conversion laser operation has been achieved in fibers. In addition, the efficient nonlinear optical properties of fibers has led to fiber Raman lasers.

71.2.3 Color Center Lasers

In color center lasers, the optically active center is a point defect in the lattice. For example, in alkali halide host crystals, such as NaCl, a typical color center consists of an electron trapped at a halide ion vacancy. Similar color centers occur in oxide host crystals such as diamond and sapphire. Color centers can be produced by thermal treatment or exposure to radiation. Many times, these centers are stable only at low temperatures due to ion and electron mobility. In some cases, impurity ions act to stabilize the defect center. A neutral Tl atom at a cation site next to a anion vacancy in KCl is an example of this. A major recent advance in color center lasers involves the development of room temperature stable pulsed laser systems. Systems based on the vibrational transitions of molecular defects such as CN^- have been demonstrated to operate as lasers in the 5 μm spectral region.

Color center absorption generally occurs in the visible, and they are optically pumped by Ar, Kr, or Nd:YAG lasers. Typical color center emission occurs as a broad band in the near IR between 0.8 and 4.0 μm. The emission is based on allowed transitions with high oscillator strengths leading to high gain cross sections.

The homogeneously broadened emission band of color centers allows for efficient, tunable laser emission and single mode operation. Optically pumped cw output powers of ≈ 2 W have been obtained, and mode-locked pulses of less than 100 fs and 1 MW peak power at repetition rates of 100 MHz and hundreds of milliwatts average power have been generated. Laser linewidths of less than 4 kHz have been obtained.

One problem with a high gain medium such as a color center crystal is a tendency for multimode laser operation. In a linear standing wave cavity, a primary oscillating mode reaches gain saturation and burns spatially periodic holes in the population inversion of the gain medium. The high gain allows secondary modes to oscillate with peaks at the nodes of the primary mode. A grating/etalon combination can be used to select and tune the laser output frequency [71.4]. The etalon selects one cavity mode and the grating selects one order of the etalon. This results in single mode tunable output. The single mode power output is 70% of the multimode laser power due to energy loss in the hole burning mode. A ring laser configuration in a traveling wave operation can be used to give uniform saturation of the gain medium, and thus no hole burning. A ring laser cavity needs additional optics such as a Faraday rotator and an optically active plate to force oscillation in only one direction. Active frequency stabilization circuits are necessary to obtain linewidths of less than 4 kHz.

Synchronously pumping a color center laser with a mode-locked Nd:YAG laser gives mode-locked output with pulses typically between 5 and 15 ps. If passive mode-locking is obtained through use of a saturable absorber, pulses of ≈ 200 fs can be obtained. Additive pulse mode-locking consists of two coupled cavities, one with a color center gain medium and the other with a single mode optical fiber. Self phase modulation in the fiber gives a broader frequency spectrum to the pulses and thus shorter time widths. This scheme has generated pulses of about 75 fs [71.24]. Soliton lasers are also obtained by coupling a color center laser with a fiber laser.

71.2.4 New Types of Solid State Laser Systems

One type of new solid state laser system being studied involves the fourth and fifth row transition metal ions. So far, stimulated emission and gain have been reported for Rh^{3+} [71.25] under strong pumping conditions, but no laser operation has been achieved.

Solid state dye lasers [71.26] consist of organic dye molecules doped in crystal or glass host materials. The first systems of this type used the same class of organic dyes as in liquid lasers (such as rhodamine 6G) and host materials such as sol-gels or polymethylmethacrylate. These systems have traditionally had a problem with photo-degradation of the material after a limited number of shots. However, recent combinations of new dyes and new host materials have produced outputs of tens of millijoules per pulse, over 50% slope efficiency, and a degradation to 60% of the initial output after 30 000 pulses. One example of these new materials is pyrromethene-BF_2 complex dye doped in an acrylic plastic host [71.26]. Another material system that has exhibited good laser performance involves xerogel hosts doped with perylene or pyrromethene dyes [71.27]. The performance of some of these new systems has reached the point that they may be useful for tunable laser applications in the visible spectral region.

Another new type of system can be described as a *solid state excimer laser* [71.6–8]. These materials consist of noble gas solids as host crystals, such as Ar and Ne, doped with excimer molecules, such as XeF. The emission lines occur in the UV and visible, with major lines at 286, 411, and 540 nm. These systems have very high stimulated emission cross sections and large gain coefficients.

71.2.5 Frequency Shifters

Since solid state lasers are commercially available at only a limited number of wavelengths, it is sometimes easier to use nonlinear optical techniques to shift the frequency of an available laser than to develop a new primary laser system. The techniques for this include harmonic frequency generation, frequency mixing, optical parametric oscillators and amplifiers, and Raman shifting. Significant advances have been made recently [71.28] in developing new types of materials for these applications. For frequency mixing, harmonic generation and OPOs in the visible and near UV, important new crystals include $KTiOPO_4$ (KTP), BaB_2O_4 (BBO), and LiB_3O_5 (LBO). In the 3 to 5 μm region, new materials for frequency mixing and OPOs include KTA (the arsinate analog of KTP) and $ZnGeP_2$. Gas phase Raman cells have been commercially available for solid state laser systems for many years. Recently, it has been demonstrated that crystals such as $Ba(NO_3)_2$ can be used as efficient solid state Raman shifters. New waveguide configurations with periodic poling for quasiphase matching have greatly enhanced the efficiency of harmonic generation [71.29]. Optical damage threshold is still the limiting parameter for nonlinear optical materials.

71.3 Semiconductor Lasers

The light emission from semiconductor diode lasers is generally associated with the radiative recombination of electrons and holes. This occurs at the junction of an n-type material with excess electrons and a p-type material with excess holes. The excitation is provided by an external electric field applied across the p-n junction that causes the two types of charges to come together. The most common semiconductor laser emission lines occur in the near or mid-IR. These are generally made of III-V compounds such as gallium arsenide in the red and near IR, and lead salts in the mid-IR region. Wide bandgap II-VI materials are currently being explored for use in the green and blue spectral regions. The spectral lines are generally narrow with broadening due to lattice defects (inhomogeneous broadening) and radiationless relaxation and scattering processes associated with the thermal vibration of the host (homogeneous broadening). Temperature tuning can be used to change the output wavelength over a narrow spectral range. Direct modulation of the laser output can be achieved by modulating the external current. This is an important feature of electric current pumped semiconductor lasers, and leads to applications where high frequency modulation is required. Modulation bandwidths in excess of 11 GHz have been obtained.

The ability to design and grow specialized structures one atomic layer at a time using techniques such as molecular beam epitaxy (MBE) has led to the design of quantum well lasers with enhanced properties, as well as more esoteric designs of quantum wires and quantum boxes [71.30, 31]. Quantum well microlasers can be as small as 100 μm and operate with milliamps of current at a few volts. These are generally based on material systems such as GaAs/GaAlAs or InP/InGaAsP. Heterostructure lasers have layers of aluminum, indium, and phosphorus on the sides of the junction to confine the electronic current to the junction region to minimize the amount of current required, and thus minimize heat dissipation compared with homostructures of the same materials. Special device structures can be fabricated to produce gain-guiding and index-guiding to enhance the operating characteristics of the lasers. Grating structures can be fabricated to give distributed feedback lasers that narrow the laser linewidths to ≈ 1 MHz. External cavity lasers with gratings have achieved linewidths as low as 1 kHz. cw operating powers of up to 10 mW have been achieved from a single p-n junction, while phased arrays have reached combined powers of well over 10 W.

Gallium arsenide (GaAs) was the first compound semiconductor diode laser. It can produce laser emission at wavelengths between 750 and 870 nm. The development of strained-layer technology has allowed the use of mixed compounds of gallium aluminum arsenide ($Ga_{1x}Al_xAs$) to fabricate different types of laser structures, and the concentration of aluminum determines the laser emission wavelength. Wavelengths from 620 nm to 905 nm have been obtained. The most common diode laser structures are simple double-heterostructure lasers, and monolithic arrays of laser stripes can be fabricated for higher power. In a typical GaAlAs laser, the active layer is sandwiched between two layers having larger bandgaps and lower refractive indices. The former characteristic produces electrical confinement and the latter produces optical confinement. This improves the efficiency and allows cw laser operation. In the conventional horizontal-cavity structure, cleaved end facets of the chip produce the optical feedback required for laser oscillation. Fabricating quantum well structures in the active layer produces improved confinement, and thus higher efficiency operation. For low power lasers, high beam quality is achieved through an index-guiding structure that concentrates the optical beam in the laser stripe. In high power lasers, the current is concentrated in the laser stripe to achieve gain-guiding. Laser arrays can generate cw powers of the order 20 W and in a quasicw mode they can produce peak powers of 100 W. Stacking diode bars in planar arrays can generate kilowatts of power. Another approach to obtaining high powers is a master oscillator power amplifier (MOPA) configuration. This has the advantage of maintaining high beam quality, and gallium arsenide MOPAs have produced single frequency operation at 1 W of power.

Long wavelength IR diode lasers are made of IV–VI compounds such as PbS. These lasers operate at cryogenic temperatures and provide tunable emission from 4 to 32 μm. The tunability is achieved by changing temperature or current.

One of the most important areas of research in semiconducting lasers is the development of new device configurations such as vertical cavity surface emitting lasers (VCSELs) [71.32]. These have lower round trip gain but significantly reduced divergence of the output beam. This configuration allows for the fabrication of two dimensional arrays of independently modulated lasers.

Another major research area is generating new laser wavelengths. Using strained layer technology, a variety of different combinations of direct bandgap materials can be made into semiconductor lasers [71.33]. The range of available bandgaps can conceivably result in lasers with emission wavelengths spanning the visible and near IR spectral regions. There is currently special emphasis on the development of lasers in the blue and green spectral regions using wide bandgap II–VI materials such as ZnSe [71.34].

71.4 Liquid Lasers

There are three classes of liquid lasers. The most widely used are based on organic solvents with organic dye molecules as the active laser species. The other two types are based on rare earth ions for the lasing entity. In one case, the lasing system involves rare earth chelates in organic solvents, while in the other, the rare earth ions are in inorganic solvents. These systems are optically pumped with either flashlamps or other lasers.

71.4.1 Organic Dye Lasers

Dye lasers are generally based on fluorescent dyes in liquid solvents, using optical pumping by either flashlamps or other lasers as the mechanism for excitation. They can operate in either a pulsed or cw mode at wavelengths as short as 310 nm out to about 1.5 μm. Dye lasers provide the versatility of varying wavelength, bandwidth, and pulse length as desired.

The fluorescence emission of dye molecules appears as broad spectral bands due to coupling of the electronic energy levels with molecular vibrations. This gives a broadband gain curve for lasing, and thus with a dispersive element such as a grating, prism, filter, or etalon in the cavity, dye laser outputs can be tuned over a range of several hundred angstroms (30–60 nm). There are now over 200 organic laser dyes. One of the most successful dyes is rhodamine 6G which covers the spectral range 570–630 nm. Alcohol is a typical solvent. The various types of cavity designs include folded cavities and ring cavities. Flowing dye configurations are useful for heat management. Oscillator/amplifier configurations are used to suppress amplified spontaneous emission. Dispersive elements plus frequency stabilization have been used for spectral line narrowing to single mode cw operation. Frequencies as narrow as 10 GHz or less have been obtained. For cw operation, output powers of a few watts can be obtained, while in the pulsed mode the energy per pulse can be up to 100 mJ in 10 ns pulses. With mode-locking, trains of femtosecond pulses with intervals of 20 ns can be produced.

Mode-locking takes advantage of the broad emission spectrum of the dye molecules to get short pulses. The standard techniques of synchronous pumping, active, and passive mode-locking have been used. In addition, colliding pulse mode-locking has been used in a ring configuration. In this case, pump beams going the opposite direction in the cavity collide in an absorber jet dye to produce interference fringes and saturation. The gain dye is located half way around the ring from the absorber dye. Fiber compression techniques have also been used with dye laser systems. The shortest pulses obtained so far are 6 fs [71.4]. Hybrid mode-locking utilizing synchronous pumping plus a saturable absorber and prism dispersion compensators has been employed to achieve powers of 350 mW and greater tunability than colliding pulse systems [71.4].

Several new technological advances have increased the spectral coverage of dye lasers. These include better pump sources, such as the increased power of argon pump lasers and the availability of UV pump lines, and the use of Ti-sapphire pump lasers. Combining these pump sources with new dyes has provided extended dye laser output in both the blue and near IR [71.4]. Also, improved nonlinear crystals have allowed coverage of the near UV from 260 to 960 nm through harmonic frequency generation and frequency mixing the dye laser output with pump laser wavelengths. Another developing technology is solid state dye lasers. As mentioned in Sect. 71.2.4, significant progress has been made recently in decreasing the photodegradation problems associated with dye molecules doped in solid host materials [71.26, 27].

71.4.2 Rare Earth Chelate Lasers

In these systems, the active lasing center is a rare earth complex with organic molecules in an organic solvent. The chelate ligands are organic phosphates, carboxylate ions, or β-diketonate. The optical pump energy is absorbed by the ligand and efficiently transferred to the rare earth ion. Energy transfer quenching from the rare earth ion to the organic molecule vibrational levels decreases the efficiency of these lasers. This is especially true for Nd^{3+}. The three ions that have been most effective in these systems are Eu^{3+}, Tb^{3+}, and Nd^{3+}.

71.4.3 Inorganic Rare Earth Liquid Lasers

In these systems, the active lasing center is a rare earth ion inorganic complex of heavy metal halides or oxyhalides (the optical pumping is directly into the rare earth ion). Nd^{3+} lasers can produce several hundred joules of energy per pulse in the long pulse mode and peak powers of 180 MW in a Q-switched mode. These lasers have also been mode-locked to obtain 3 ps long pulses having 1 GW of peak power. Self mode-locking and self Q-switching is also observed in these systems.

71.5 Other Types of Lasers

Several more complex laser systems have been developed that have significant interest for scientific studies, but so far have had limited applications outside the laboratory. These include X-ray lasers, particle-beam-pumped lasers, and free electron lasers.

71.5.1 X-Ray and Extreme UV Lasers

These systems are based on highly ionized ions produced by powerful laser sources. Examples are krypton at 93 nm, molybdenum at 13 nm, and carbon at 18 nm.

There is significant interest in developing X-ray lasers for applications in lithography and medical imaging, but so far the lack of reliable X-ray optics for the laser cavities has limited the technology.

71.5.2 Nuclear Pumped Lasers

These are gas lasers excited by high energy charged particles or gamma rays resulting from nuclear reactions. Either nuclear reactors or nuclear explosives are used as pump sources. These can operate in either a pulsed or cw mode and produce emission covering the spectral range from the UV through the IR [71.4]. Typical gases range from Xe_2^* at 170 nm to CO at 5.4 μm. Other gases that have been used include argon and nitrogen. Typical pulsed outputs produce 10 ns pulses with energies from 2×10^{-7} J to 3 J. There are significant problems with radiation damage of the laser components in these systems.

71.5.3 Free Electron Lasers

These lasers are based on a high-energy beam of electrons in a spatially varying magnetic field. The varying field causes the electrons to oscillate, and thus to emit radiation at the oscillation frequency. The stimulated emission produced under these conditions provides the laser output. Since the electrons are making transitions between continuum states instead of discrete states, these systems can give high power output over the entire spectral range from the VUV to the far IR [71.35, 36]. Powers as high as 1 GW and efficiencies as high as 35% have been obtained. Beam spread is controlled by the use of tapered instead of uniform wigglers. Both cw and picosecond pulsed emission can be obtained.

There are three types of free electron laser configurations [71.35, 36]. The first is a master oscillator power amplifier (MOPA), in which an electron beam is injected into a wiggler in synchronism with the signal to be amplified. The external radiation source to drive the amplifier is a master oscillator such as a conventional laser system. This is a single pass, high gain system. The second configuration is an oscillator. This is designed with reflection at the ends of the wiggler so that the signal makes multiple passes in the cavity. This can operate with low gain. Since it amplifies spontaneous noise, no injected signal is necessary. The third configuration is a super-radiant amplifier. In this configuration, shot noise is amplified over a single pass through the wiggler. These systems require high current accelerators to drive them. Because their operation is based on broad band shot noise, super-radiant amplifier radiation has a broader band than radiation from a master oscillator power amplifier.

Different accelerator configurations can be used to produce the electron beam, storage rings, induction linacs, pulse line accelerators, etc. [71.35, 36]. These give different beam properties such as quality, current, and energy. They also each give a limited range of wavelengths and temporal structure of the output.

Advances have been made recently in designing smaller and less complex free electron lasers [71.35, 36]. As this trend continues, these systems will find important applications in medicine and industry.

71.6 Recent Developments

Over the past eight years, the designs of all types of lasers have continued to evolve, driven by applications requirements for lasers with specific operating parameters. Some of the major advances in this time period are summarized here.

The requirements for laser outputs of several kilowatts with near-diffraction-limited beam quality in the infrared wavelength region has lead to improved designs of CO_2 lasers. This includes a diffusion cooled, annular discharge design with free-space propagation instead of waveguiding [71.37].

The push for solid state lasers in the ultraviolet has led to recent progress in cerium-doped fluoride crystal lasers that provides direct laser emission tunable in the 280 to 330 nm spectral region. Using host crystals such as $LiCaAlF_6$, $LiSrAlF_6$ or $LiLuF_4$ has helped to overcome the problems with excited state absorption and color center formation that has been a major problem with more common hosts such as YAG and YLF crystals [71.38]. These lasers are pumped by either excimer lasers or frequency-quadrupled Nd:YAG lasers and have produced up to 60 mJ per pulse and 60% slope efficiency. The output can be pulse compressed to 115 fs.

Thin-disk solid state laser configurations using rare earth doped crystals and semiconductor saturable absorber mirrors for passive mode-locking have produced ultrashort (femtosecond) pulse trains with high average powers [71.39]. Yb:YAG thin disks reduce the problem of thermal lensing and have achieved output powers of up to 100 W in cw mode locked operation with near-diffraction-limited beam quality [71.40]. To achieve shorter pulses, a thin-disk Yb:KYW laser has obtained 22 W with 240 fs pulses [71.41].

Advances continue to be made in the materials for solid state organic dye lasers. The use of polymer materials, organic-inorganic matrices, and nanoparticles has provided advances in this type of laser [71.42–46]. Use of semiconductor excitation and electrical excitation may play important roles in the future of solid state organic dye lasers.

Commercially available solid state Raman lasers have been developed for frequency shifting to a wide variety of wavelengths and for pulse compression to the 0.1–1 ns region [71.47–49]. Both internal and external cavity designs have been demonstrated. New materials, such as $KGd(WO_4)_2$ and $KY(WO_4)_2$, have been used for Raman lasers along with $Ba(NO_3)_2$, which has excellent properties for this application. An external-resonator Raman laser using $Ba(NO_3)_2$ has reached 1.3 W of power [71.50].

The major advancement in fiber lasers has been in cladding geometry to allow for higher powers. Double clad fiber lasers operating in single transverse modes have exceeded 100 W of output for four-level systems and over 1 W for three-level systems [71.51]. These involve a variety of geometric shapes of cladding. A new breakthrough in fiber delivery systems that will impact the future of fiber lasers is the use of holey fibers (photonic crystal fibers) [71.52]. These are glass fibers that have a periodic array of air holes running their entire length. These fibers can be engineered to produce a photonic bandgap and allow for dispersion control and minimized nonlinear effects compared to standard fibers. This is useful for short pulse delivery.

In the field of semiconductor lasers, vertical-cavity surface-emitting lasers (VCSELs) have developed as a competitive alternative to the conventional edge-emitting semiconductor lasers. The most recent advance involves designs that have a horizontal laser cavity but emits from the surface [71.53]. This combines the ease of packaging of VCSELs with the high power and good stability properties of edge-emitting lasers.

Several new designs of high power semiconductor lasers that can be frequency shifted to the blue and green spectral regions have been developed as rugged, efficient sources [71.54, 55]. One of these is a GaAs-based vertical external cavity surface-emitting laser (VECSEL) optically pumped with an 808 nm semiconductor laser. This emits at 976 nm with a cavity that includes a wavelength selector and a doubling crystal. The second configuration uses a semiconductor material as a gain medium in an external cavity with a wavelength selector. In this case the frequency doubling crystal is outside the cavity. These configurations have reached 20 mW cw operation with high beam quality. One enabling technology is the use of microelectromechanical systems (MEMS) for mirrors to tune VCELs [71.56].

New material configurations offer some advantages. Photonic crystals can be used to produce nanocavity lasers [71.57] while band-structure engineering can be used to design quantum-cascade lasers [71.58]. The latter are multiple-quantum-well heterostructures based on intersub-band transitions. They operate in the infrared to terahertz spectral region.

The broad spectral band available for laser gain is the major distinguishing feature of liquid organic dye lasers. This provides the ability to have tunable output over a broad range of wavelength, the ability to generate ultrashort pulses, narrow linewidth cw operation, and high average power operation. This variety of operating parameters keeps this class of lasers competitive in a variety of applications. They have been especially useful for laser spectroscopy in the visible region of the spectrum and in the area of laser cooling. They have achieved energies of up to 800 J per pulse and average powers greater than 1 kW. The discovery of highly stable water-soluble dyes is an important advancement in this field [71.59].

Free-electron lasers (FELs) can produce coherent emission at a wide range of wavelengths. Output formats include ultrashort pulses and high powers [71.60]. Using a photocathode electron gun in a single pass, self-amplified spontaneous emission mode, FELs can operate at wavelengths where there are no mirrors with high reflectivity. This type of laser emission has been extended to the vacuum ultraviolet and the hard X-ray regions. The technique of using a subharmonic seed laser has been developed to improve the spectral purity of FEL emission. The use of an energy-recovering accelerator produces FELs with high average power (over 1 MW). In the spectral region around 1 mm, FELs have produced over 300 W of picosecond pulses.

Laser systems using various nonlinear optics techniques continue to be developed. There has been significant research on high intensity, ultrashort pulse lasers because of their special characteristics for atmospheric propagation. Using femtosecond pulses above a critical power level produces "light strings" that propagate without dispersion for many kilometers due to the balance between Kerr self-focusing and air ionization [71.61]. Stimulated Brillouin scattering in water has been shown to be effective in pulse compression and the production of nondiffracting laser beams [71.62].

References

71.1 M.J. Weber (Ed.): *Handbook of Laser Science and Technology*, Vol. 1 (CRC, Boca Raton 1982)

71.2 M.J. Weber (Ed.): *Handbook of Laser Science and Technology* (CRC, Boca Raton 1991)

71.3 M. Bass, M.L. Stitch (Eds.): *Laser Handbook*, Vol. 5 (North Holland, Amsterdam 1985)

71.4 R.A. Meyers (Ed.): *Encyclopedia of Lasers and Optical Technology* (Academic Press, San Diego 1991)

71.5 P.K. Cheo (Ed.): *Handbook of Solid State Lasers* (Marcel Dekker, New York 1989)

71.6 N. Schwentner, V.A. Apkarian: Chem. Phys. Lett **154**, 413 (1989)

71.7 G. Zerza, G. Sliwinski, N. Schwentner: Appl. Phys. **B55**, 331 (1992)

71.8 G. Zerza, G. Sliwinski, N. Schwentner: Appl. Phys. **A56**, 156 (1993)

71.9 W.H. Crumly, J.L. Gole, D.A. Dixon: J. Chem. Phys. **76**, 6439 (1982)

71.10 S.H. Cobb, J.R. Woodward, J.L. Gole: Chem. Phys. Lett. **143**, 205 (1988)

71.11 S.H. Cobb, J.R. Woodward, J.L. Gole: Chem. Phys. Lett. **157**, 197 (1989)

71.12 R.M. Macfarlane, F. Tong, A.J. Silversmith, W. Lenth: Appl. Phys. Lett. **52**, 1300 (1988)

71.13 R.A. Macfarlane: Appl. Phys. Lett. **54**, 2301 (1989)

71.14 T. Hebert, R. Wannemacher, W. Lenth, R.M. Macfarlane: Appl. Phys. Lett. **57**, 1727 (1990)

71.15 R.A. Macfarlane: Opt. Lett. **16**, 1397 (1991)

71.16 M.E. Koch, A.W. Kueny, W.E. Case: J. Appl. Phys. **56**, 1083 (1990)

71.17 N. Karayianis, D.E. Wortman, C.A. Morrison: Solid State Comm. **18**, 1299 (1976)

71.18 W.F. Krupke, M.D. Shinn, J.E. Marion, J.A. Caird, S.E. Stokowski: J. Opt. Soc. Am. **B3**, 102 (1986)

71.19 M.J. Weber, M. Bass, G.A. deMars: J. Appl. Phys. **42**, 301 (1971)

71.20 G.J. Quarles, A. Rosenbaum, C.L. Marquardt, L. Esterowitz: Opt. Lett. **15**, 42 (1990)

71.21 L. Esterowitz: Opt. Eng. **29**, 676 (1990)

71.22 P. Urquhart: IEE Proc. **J135**, 385 (1988)

71.23 I.M. Jauncey, L. Reekie, R.J. Mears, D.N. Payne, C.J. Rowe, D.C.J. Reid, I. Bennion, C. Edge: Electron. Lett. **22**, 987 (1986)

71.24 L.F. Mollenhauer, R.H. Stolen: Opt. Lett. **9**, 12 (1984)

71.25 R.C. Powell, G.J. Quarles, J.J. Martin, C.A. Hunt, W.A. Sibley: Opt. Lett. **10**, 212 (1985)

71.26 R.E. Hermes, T.H. Allik, S. Chandra, J.A. Hutchinson: Appl. Phys. Lett. **63**, 877 (1993)

71.27 B. Dunn, F. Nishida, R. Toda, J.I. Zink, T.H. Allik, S. Chandra, J.A. Hutchinson: Mat. Res. Soc. Symposium, Proc. **329**, 267 (1994)

71.28 V.G. Dmitriev, G.G. Girzodyan, D.N. Nikogosyan: *Handbook of Nonlinear Optical Crystals* (Springer, Berlin, Heidelberg 1991)

71.29 E.J. Lim, M.M. Fejer, R.L. Byer, W.J. Kozlovsky: Electron. Lett. **25**, 731 (1989)

71.30 Y. Arakawa, K. Vahala, A. Yariv: Surf. Sci. **174**, 155 (1986)

71.31 K. Vahala: IEEE J. Quantum Electron. **24**, 523 (1988)

71.32 R.E. Slusher: Opt., Photon. News **4**, 8 (1993)

71.33 W.W. Chow, S.W. Koch, M.II.I. Sargent: *Semiconductor Laser Physics* (Springer, Berlin, Heidelberg 1994)

71.34 M.A. Hasse, J. Qui, J.M. DePuydt, H. Cheng: Appl. Phys. Lett. **58**, 1272 (1991)

71.35 C.A. Brau: *Free Electron Lasers* (Academic Press, San Diego 1990)

71.36 H.P. Freund, G.R. Neil: Proc. IEEE. **87**, 782 (1999)

71.37 A. Lapucci, F. Rossetti, P. Burlamacchi: Opt. Com **111**, 290 (1994)

71.38 A.J.S. McGonigle, D.W. Coutts: Laser Focus World **39**, 127 (2003)

71.39 E. Innerhofer, T. Südmeyer, F. Brunner, R. Häring, A. Aschwanden, R. Paschotta, U. Keller, C. Hönninger, M. Kumkar: Opt. Lett. **28**, 376 (2003)

71.40 A. Giesen, H. Hügel, A. Voss, K. Wittig, U. Brauch, H. Popwer: Appl. Phys. B **58**, 363 (1994)

71.41 F. Brunner, T. Südmeyer, E. Innerhofer, R. Paschotta, F. Morier-Genoud, J. Gao, K. Contag, A. Giesen, V.E. Kisel, V.G. Shcherbitsky, N.V. Kuleshov, U. Keller: Opt. Lett. **27**, 1162 (2002)

71.42 A. Costela, I. Garcia-Moreno, J.M. Figuera, F. Amat-Guerri, R. Sastre: Laser Chem. **18**, 63 (1998)

71.43 I. Braun, G. Ihlein, J.U. Nöckel, G. Schulz-Ekloff, F. Schüth, U. Vietze, D. Wöhrle: Appl. Phys. B **70**, 335 (2000)

71.44 F.J. Duarte: Appl. Opt. **38**, 6347 (1999)

71.45 W.J. Wadsworth, I.T. McKinnie, A.D. Woolhouse, T.G. Haskell: Appl. Phys. B **69**, 163 (1999)

71.46 X. Zhu, S.K. Lam, D. Lo: Appl. Opt. **39**, 3104 (2000)

71.47 J.T. Murray, R.C. Powell, N. Peyghambarian, D. Smith, W. Austin, R.A. Stolzenberger: Opt. Lett. **20**, 1017 (1995)

71.48 A.A. Kaminskii, H.G. Eichler, K. Ueda, N.V. Klassen, B.S. Redkin, L.E. Li, J. Findeisen, D. Jaque, J. Garcia-Sole, J. Fernandez, R. Balda: Appl. Opt. **38**, 4533 (1999)

71.49 P.G. Zverev, T.T. Basiev, A.M. Prokhorov: Opt. Materials **11**, 335 (1999)

71.50 H.M. Pask, S. Myers, J.A. Piper, J. Richards, T. McKay: Opt. Lett. **28**, 435 (2003)

71.51 L.A. Zenteno, J.D. Minelly, A. Liu, A.J.G. Ellison, S.G. Crigler, D.T. Walton, D.V. Kuksenkov, M.J. Dejneka: Electron. Lett. **37**, 819 (2001)

71.52 H. Sabert, J. Knight: Photonics Spectra **37**, 92 (2003)

71.53 N. Anscombe: Photonics Spectra 37 **60** (2003)

71.54 E.H. Wahl, B.A. Richman, C.W. Rella, G.M.H. Knippels, B.A. Paldus: Opt., Photonics News **14**, 36 (2003)

71.55 S. Lutgen, T. Albrecht, P. Brick, W. Reill, J. Luft, J. Späth: Appl. Phys. Lett. **82**, 3620 (2003)
71.56 J. Hecht: Laser Focus World **37**, 121 (2001)
71.57 G. G. Park, J. K. Hwang, J. Huh, H. Y. Ryu, Y. H. Lee: Appl. Phys. Lett. **79**, 3032 (2001)
71.58 R. Köhler, A. Tredicucci, F. Beltram, H. E. Beere, E. H. Linfield, A. G. Davies, D. A. Ritchie, S. S. Dhillon, C. Sirtori: Appl. Phys. Lett. **82**, 1518 (2003)
71.59 F. J. Duarte: Opt., Photonics News **14**, 20 (2003)
71.60 H. P. Freund, P. O'Shea: Science **292**, 1853 (2001)
71.61 F. Courvoisier, V. Boutou, J. Kasparian, E. Salmon, G. Mejean, J. Yu, J.-P. Wolf: Appl. Phys. Lett. **83**, 213 (2003)
71.62 F. Brandi, I. Velchev, D. Neshev, W. Hogervorst, W. Ubachs: Rev. Sci. Inst. **74**, 32 (2003)
71.63 A. Costela, I. Garcia-Moreno, C. Gomez, O. Garcia, R. Sastre: Phys. Lett. **369**, 656 (2003)

72. Nonlinear Optics

Nonlinear optics is concerned with the propagation of intense beams of light through a material system. The optical properties of the medium can be modified by the intense light beam, leading to new processes that would not occur in a material that responded linearly to an applied optical field. These processes can lead to the modification of the spectral, spatial, or polarization properties of the light beam, or the creation of new frequency components. More complete accounts of nonlinear optics including the origin of optical nonlinearities can be found in references [72.1–4].

Both the Gaussian and MKS system of units are commonly used in nonlinear optics. Thus, we have chosen to express the equations in this chapter in both the Gaussian and MKS systems. Each equation can be interpreted in the MKS system as written or in the Gaussian system by omitting the prefactors (e.g., $1/4\pi\varepsilon_0$) that appear in square brackets at the beginning of the expression on the right-hand-side of the equation.

72.1 **Nonlinear Susceptibility** 1051
- 72.1.1 Tensor Properties 1052
- 72.1.2 Nonlinear Refractive Index 1052
- 72.1.3 Quantum Mechanical Expression for $\chi^{(n)}$ 1052
- 72.1.4 The Hyperpolarizability 1053

72.2 **Wave Equation in Nonlinear Optics** 1054
- 72.2.1 Coupled-Amplitude Equations 1054
- 72.2.2 Phase Matching 1054
- 72.2.3 Manley–Rowe Relations 1055
- 72.2.4 Pulse Propagation 1055

72.3 **Second-Order Processes** 1056
- 72.3.1 Sum Frequency Generation 1056
- 72.3.2 Second Harmonic Generation 1056
- 72.3.3 Difference Frequency Generation 1056
- 72.3.4 Parametric Amplification and Oscillation 1056
- 72.3.5 Focused Beams 1056

72.4 **Third-Order Processes** 1057
- 72.4.1 Third-Harmonic Generation 1057
- 72.4.2 Self-Phase and Cross-Phase Modulation 1057
- 72.4.3 Four-Wave Mixing 1058
- 72.4.4 Self-Focusing and Self-Trapping 1058
- 72.4.5 Saturable Absorption 1058
- 72.4.6 Two-Photon Absorption 1058
- 72.4.7 Nonlinear Ellipse Rotation 1059

72.5 **Stimulated Light Scattering** 1059
- 72.5.1 Stimulated Raman Scattering 1059
- 72.5.2 Stimulated Brillouin Scattering 1060

72.6 **Other Nonlinear Optical Processes** 1061
- 72.6.1 High-Order Harmonic Generation 1061
- 72.6.2 Electro-Optic Effect 1061
- 72.6.3 Photorefractive Effect 1061
- 72.6.4 Ultrafast and Intense-Field Nonlinear Optics 1062

References 1062

72.1 Nonlinear Susceptibility

In linear optics it is customary to describe the response of a material in terms of a macroscopic polarization $\tilde{\boldsymbol{P}}$ (i. e., dipole moment per unit volume) which is linearly related to the applied electric field $\tilde{\boldsymbol{E}}$ through the linear susceptibility $\chi^{(1)}$. In order to extend the relationship between $\tilde{\boldsymbol{P}}$ and $\tilde{\boldsymbol{E}}$ into the nonlinear regime, the polarization is expanded in a power series of the electric field strength. We express this relationship mathematically by first decomposing the field and the polarization into their frequency components such that

$$\tilde{\boldsymbol{E}}(\boldsymbol{r},t)=\sum_l \boldsymbol{E}(\boldsymbol{r},\omega_l)\,\mathrm{e}^{-\mathrm{i}\omega_l t}\,, \tag{72.1}$$

$$\tilde{\boldsymbol{P}}(\boldsymbol{r},t)=\sum_l \boldsymbol{P}(\boldsymbol{r},\omega_l)\,\mathrm{e}^{-\mathrm{i}\omega_l t}\,, \tag{72.2}$$

where the summations are performed over both positive and negative frequencies. The reality of $\tilde{E}$ and $\tilde{P}$ is then assured by requiring that $\boldsymbol{E}(\boldsymbol{r},\omega_l)=\boldsymbol{E}^*(\boldsymbol{r},-\omega_l)$ and $\boldsymbol{P}(\boldsymbol{r},\omega_l)=\boldsymbol{P}^*(\boldsymbol{r},-\omega_l)$. In this case the general expression for the Cartesian component i of the polarization at frequency ω_σ is given by

$$\begin{aligned}P_i(\omega_\sigma) = [\varepsilon_0]\Big[&\sum_j \chi_{ij}^{(1)}(\omega_\sigma)E_j(\omega_\sigma) \\ &+\sum_{jk}\sum_{(mn)} \chi_{ijk}^{(2)}(\omega_\sigma;\omega_m,\omega_n) \\ &\times E_j(\omega_m)E_k(\omega_n) \\ &+\sum_{jkl}\sum_{(mno)} \chi_{ijkl}^{(3)}(\omega_\sigma;\omega_m,\omega_n,\omega_o)E_j(\omega_m) \\ &\times E_k(\omega_n)E_l(\omega_o) \\ &+\cdots\Big]\,, \end{aligned} \quad (72.3)$$

where $ijkl$ refer to field components, and the notation (mn), for example, indicates that the summation over n and m should be performed such that $\omega_\sigma=\omega_m+\omega_n$ is held constant. Inspection of (72.3) shows that the $\chi^{(n)}$ can be required to satisfy intrinsic permutation symmetry, i.e., the Cartesian components and the corresponding frequency components [e.g., (j,ω_j) but not (i,ω_σ)] associated with the applied fields may be permuted without changing the value of the susceptibility. For example, for the second-order susceptibility,

$$\chi_{ijk}^{(2)}(\omega_\sigma;\omega_m,\omega_n)=\chi_{ikj}^{(2)}(\omega_\sigma;\omega_n,\omega_m)\,. \quad (72.4)$$

If the medium is lossless at all the field frequencies taking part in the nonlinear interaction, then the condition of full permutation symmetry is necessarily valid. This condition states that the pair of indices associated with the Cartesian component and the frequency of the nonlinear polarization [i.e., (i,ω_σ)] may be permuted along with the pairs associated with the applied field components. For example, for the second-order susceptibility, this condition implies that

$$\chi_{ijk}^{(2)}(\omega_\sigma;\omega_m,\omega_n)=\chi_{kji}^{(2)}(-\omega_n;\omega_m,-\omega_\sigma)\,. \quad (72.5)$$

If full permutation symmetry holds, and in addition all the frequencies of interest are well below any of the transition frequencies of the medium, the $\chi^{(n)}$ are invariant upon free permutation of all the Cartesian indices. This condition is known as the Kleinman symmetry condition.

72.1.1 Tensor Properties

The spatial symmetry properties of a material can be used to predict the tensor nature of the nonlinear susceptibility. For example, for a material that possesses inversion symmetry, all the elements of the even-ordered susceptibilities must vanish (i.e., $\chi^{(n)}=0$ for n even). The number of independent elements of the nonlinear susceptibility for many materials can be substantially fewer than than the total number of elements. For example, in general $\chi^{(3)}$ consists of 81 elements, but for the case of isotropic media such as gases, liquids, and glasses, only 21 elements are nonvanishing and only three of these are independent. The nonvanishing elements consist of the following types: $\chi_{iijj}^{(3)}$, $\chi_{ijij}^{(3)}$, and $\chi_{ijji}^{(3)}$, where $i\neq j$. In addition, it can be shown that

$$\chi_{iiii}^{(3)}=\chi_{iijj}^{(3)}+\chi_{ijij}^{(3)}+\chi_{ijji}^{(3)}\,. \quad (72.6)$$

72.1.2 Nonlinear Refractive Index

For many materials, the refractive index n is intensity-dependent such that

$$n=n_0+n_2I\,, \quad (72.7)$$

where n_0 is the linear refractive index, n_2 is the nonlinear refractive index coefficient , and $I=[4\pi\varepsilon_0]n_0c|E|^2/2\pi$ is the intensity of the optical field. For the case of a single, linearly polarized light beam traveling in an isotropic medium or along a crystal axis of a cubic material, n_2 is related to $\chi^{(3)}$ by

$$n_2=\left(\frac{1}{16\pi^2\varepsilon_0}\right)\frac{12\pi^2}{n_0^2c}\chi_{iiii}^{(3)}(\omega;\omega,\omega,\omega,-\omega)\,. \quad (72.8)$$

For the common situation in which n_2 is measured in units of $\mathrm{cm^2/W}$ and $\chi^{(3)}$ is measured in Gaussian units, the relation becomes

$$n_2\left(\frac{\mathrm{cm^2}}{\mathrm{W}}\right)=\frac{12\pi^2\times10^7}{n_0^2c}\chi_{iiii}^{(3)}(\omega;\omega,\omega,\omega,-\omega)\,. \quad (72.9)$$

There are various physical mechanisms that can give rise to a nonlinear refractive index. For the case of induced molecular orientation in CS_2, $n_2=3\times10^{-14}\,\mathrm{cm^2/W}$. If the contribution to the nonlinear refractive index is electronic in nature (e.g., glass), then $n_2\approx2\times10^{-16}\,\mathrm{cm^2/W}$.

72.1.3 Quantum Mechanical Expression for $\chi^{(n)}$

The general quantum mechanical perturbation expression for the $\chi^{(n)}$ in the nonresonant limit is (Under conditions of resonant excitation, relaxation phenomena must be included in the treatment, and the density matrix formalism must be used [72.4]. The resulting equation for the nonlinear susceptibility is then more complicated)

$$\begin{aligned}&\chi^{(n)}_{i_0\cdots i_n}(\omega_\sigma;\omega_1,\dots,\omega_n)\\&=\left[\frac{1}{\varepsilon_0}\right]\frac{N}{\hbar^n}\mathcal{P}_F\sum_{ga_1\cdots a_n}\rho_0(g)\\&\times\frac{1}{(\omega_{a_1g}-\omega_1-\dots-\omega_n)}\\&\times\frac{\mu^{i_0}_{ga_1}\mu^{i_1}_{a_1a_2}\cdots\mu^{i_{n-1}}_{a_{n-1}a_n}\mu^{i_n}_{a_ng}}{(\omega_{a_2g}-\omega_2-\dots-\omega_n)\cdots(\omega_{a_ng}-\omega_n)}\end{aligned}\quad(72.10)$$

where $\omega_\sigma=\omega_1+\dots+\omega_n$, N is the density of atoms or molecules that compose the material, $\rho_0(g)$ is the probability that the atomic or molecular population is initially in the state g in thermal equilibrium, $\mu^{i_1}_{a_1a_2}$ is the i_1th Cartesian component of the (a_1a_2) dipole matrix element, ω_{a_1g} is the transition frequency between the states a_1 and g, and $\mathcal{P}_F$ is the full permutation operator which is defined such that the expression that follows it is to be summed over all permutations of the pairs (i_0,ω_σ), $(i_1,\omega_1)\cdots(i_n,\omega_n)$ and divided by the number of permutations of the input frequencies. Thus the full expression for $\chi^{(2)}$ consists of six terms and that for $\chi^{(3)}$ consists of 24 terms.

In the limit in which the frequencies of all the fields are much smaller than any resonance frequency of the medium, the value of $\chi^{(n)}$ can be estimated to be

$$\chi^{(n)}\approx\left[\frac{1}{\varepsilon_0}\right]\left(\frac{2\mu}{\hbar\omega_0}\right)^nN\mu,\quad(72.11)$$

where μ is a typical value for the dipole moment and ω_0 is a typical value of the transition frequency between the ground state and the lowest-lying excited state. For the case of $\chi^{(3)}$ in Gaussian units, the predicted value is $\chi^{(3)}=3\times10^{-14}$, which is consistent with the measured values of many materials (e.g., glass) in which the nonresonant electronic nonlinearity is the dominant contribution.

72.1.4 The Hyperpolarizability

The nonlinear susceptibility relates the macroscopic polarization $\boldsymbol{P}$ to the electric field strength $\boldsymbol{E}$. A related microscopic quantity is the hyperpolarizability, which relates the dipole moment $\boldsymbol{p}$ induced in a given atom or molecule to the electric field $\boldsymbol{E}^{\text{loc}}$ (the Lorentz local field) that acts on that atom or molecule. The relationship between $\boldsymbol{p}$ and $\boldsymbol{E}^{\text{loc}}$ is

$$\begin{aligned}&p_i(\omega_\sigma)\\&=[\varepsilon_0]\Big[\sum_j\alpha_{ij}(\omega_\sigma)E^{\text{loc}}_j(\omega_\sigma)\\&\quad+\sum_{jk}\sum_{(mn)}\beta_{ijk}(\omega_\sigma;\omega_m,\omega_n)E^{\text{loc}}_j(\omega_m)E^{\text{loc}}_k(\omega_n)\\&\quad+\sum_{jkl}\sum_{(mno)}\gamma_{ijkl}(\omega_\sigma;\omega_m,\omega_n,\omega_o)\\&\quad\times E^{\text{loc}}_j(\omega_m)E^{\text{loc}}_k(\omega_n)E^{\text{loc}}_l(\omega_o)+\cdots\Big],\end{aligned}\quad(72.12)$$

where α_{ij} is the linear polarizability, β_{ijk} is the first hyperpolarizability, and γ_{ijkl} is the second hyperpolarizability. The nonlinear susceptibilities and hyperpolarizabilities are related by the number density of molecules N and by local-field factors, which account for the fact that the field $\boldsymbol{E}^{\text{loc}}$ that acts on a typical molecule is not in general equal to the macroscopic field $\boldsymbol{E}$. Under many circumstances, it is adequate to relate $\boldsymbol{E}^{\text{loc}}$ to $\boldsymbol{E}$ through use of the Lorentz approximation

$$\boldsymbol{E}^{\text{loc}}(\omega)=\boldsymbol{E}(\omega)+\left[\frac{1}{4\pi\varepsilon_0}\right]\frac{4\pi}{3}\boldsymbol{P}(\omega).\quad(72.13)$$

To a good approximation, one often needs to include only the linear contribution to $\boldsymbol{P}(\omega)$, and thus the local electric field becomes

$$\boldsymbol{E}^{\text{loc}}(\omega)=\mathcal{L}(\omega)\boldsymbol{E}(\omega),\quad(72.14)$$

where $\mathcal{L}(\omega)=\{[\varepsilon_0^{-1}]\varepsilon(\omega)+2\}/3$ is the local field correction factor and $\varepsilon(\omega)$ is the linear dielectric constant. Since $\boldsymbol{P}(\omega)=N\boldsymbol{p}(\omega)$, (72.3) and (72.12) through (72.14) relate the $\chi^{(n)}$ to the hyperpolarizabilities through

$$\chi^{(1)}_{ij}(\omega_\sigma)=\mathcal{L}(\omega_\sigma)N\alpha_{ij}(\omega_\sigma),\quad(72.15)$$

$$\begin{aligned}\chi^{(2)}_{ijk}(\omega_\sigma;\omega_m,\omega_n)&=\mathcal{L}(\omega_\sigma)\mathcal{L}(\omega_m)\mathcal{L}(\omega_n)\\&\quad\times N\beta_{ijk}(\omega_\sigma;\omega_m,\omega_n),\end{aligned}\quad(72.16)$$

$$\begin{aligned}\chi^{(3)}_{ijk}(\omega_\sigma;\omega_m,\omega_n,\omega_o)&=\mathcal{L}(\omega_\sigma)\mathcal{L}(\omega_m)\mathcal{L}(\omega_n)\mathcal{L}(\omega_o)\\&\quad\times N\gamma_{ijkl}(\omega_\sigma;\omega_m,\omega_n,\omega_o).\end{aligned}\quad(72.17)$$

For simplicity, the analysis above ignores the vector character of the interacting fields in calculating $\mathcal{L}(\omega)$. A generalization that does include these effects is given in [72.5].

72.2 Wave Equation in Nonlinear Optics

72.2.1 Coupled-Amplitude Equations

The propagation of light waves through a nonlinear medium is described by the wave equation

$$\nabla^2 \tilde{\boldsymbol{E}} - \frac{1}{c^2}\frac{\partial^2}{\partial t^2}\tilde{\boldsymbol{E}} = \left[\frac{1}{4\pi\varepsilon_0}\right]\frac{4\pi}{c^2}\frac{\partial^2}{\partial t^2}\tilde{\boldsymbol{P}} . \quad (72.18)$$

For the case in which $\tilde{\boldsymbol{E}}$ and $\tilde{\boldsymbol{P}}$ are given by (72.1), the field amplitudes associated with each frequency component can be decomposed into their plane wave components such that

$$\boldsymbol{E}(\boldsymbol{r},\omega_l) = \sum_l \boldsymbol{A}_n(\boldsymbol{r},\omega_l)\,\mathrm{e}^{\mathrm{i}\boldsymbol{k}_n\cdot\boldsymbol{r}} ,$$

$$\boldsymbol{P}(\boldsymbol{r},\omega_l) = \sum_l \boldsymbol{P}_n(\boldsymbol{r},\omega_l)\,\mathrm{e}^{\mathrm{i}\boldsymbol{k}_n\cdot\boldsymbol{r}} , \quad (72.19)$$

where $k_n = n(\omega_l)\omega_l/c$ is the magnitude of the wavevector $\boldsymbol{k}_n$. The amplitudes $\boldsymbol{A}_n$ and $\boldsymbol{P}_n$ are next decomposed into vector components whose linear optical properties are such that the polarization associated with them does not change as the field propagates through the material. For example, for a uniaxial crystal these eigenpolarizations could correspond to the ordinary and extraordinary components. In order to describe the propagation and the nonlinear coupling of these eigenpolarizations, the vector field amplitudes are expressed as

$$\boldsymbol{A}_n(\boldsymbol{r},\omega_l) = \hat{\boldsymbol{u}}_{ln} A_n(\boldsymbol{r},\omega_l) ,$$

$$\boldsymbol{P}_n(\boldsymbol{r},\omega_l) = \hat{\boldsymbol{u}}_{ln} \mathcal{P}_n(\boldsymbol{r},\omega_l) , \quad (72.20)$$

where $\hat{\boldsymbol{u}}_{ln}$ is the unit vector associated with the eigenpolarization of the spatial mode n at frequency ω_l. If the fields are assumed to travel along the z-direction, and the slowly-varying amplitude approximation $\partial^2 A_n/\partial z^2 \ll 2k_n \partial A_n/\partial z$ is made, the change in the amplitude of the field as it propagates through the nonlinear medium with no linear absorption is described by the differential equation

$$\frac{\mathrm{d}A_n(\omega_l)}{\mathrm{d}z} = \pm\left[\frac{1}{4\pi\varepsilon_0}\right]\frac{\mathrm{i}2\pi\omega_l}{n(\omega_l)c}\mathcal{P}_n^{\mathrm{NL}}(\omega_l) , \quad (72.21)$$

where $\mathcal{P}_n^{\mathrm{NL}}$ is the nonlinear contribution to the polarization amplitude $\mathcal{P}_n$, $n(\omega_l)$ is the linear refractive index at frequency ω_l, and the plus (minus) sign indicates propagation in the positive (negative) z-direction. Sections 72.3 and 72.4 give expressions for the $\mathcal{P}_n^{\mathrm{NL}}$ for various second- and third-order nonlinear optical processes. Equation (72.21)) is used to determine the set of coupled-amplitude equations describing a particular nonlinear process. For example, for the case of sum-frequency generation , the two fields of frequency ω_1 and ω_2 are combined through second-order nonlinear interaction to create a third wave at frequency $\omega_3 = \omega_1 + \omega_2$. Assuming full permutation symmetry, the amplitudes of the nonlinear polarization for each of the waves are

$$\mathcal{P}^{\mathrm{NL}}(z,\omega_1) = [\varepsilon_0]2\chi_{\mathrm{eff}}^{(2)} A(z,\omega_3)A^*(z,\omega_2)\,\mathrm{e}^{-\mathrm{i}\Delta kz} , \quad (72.22)$$

$$\mathcal{P}^{\mathrm{NL}}(z,\omega_2) = [\varepsilon_0]2\chi_{\mathrm{eff}}^{(2)} A(z,\omega_3)A^*(z,\omega_1)\,\mathrm{e}^{-\mathrm{i}\Delta kz} , \quad (72.23)$$

$$\mathcal{P}^{\mathrm{NL}}(z,\omega_3) = [\varepsilon_0]2\chi_{\mathrm{eff}}^{(2)} A(z,\omega_1)A(z,\omega_2)\,\mathrm{e}^{\mathrm{i}\Delta kz} , \quad (72.24)$$

where $\Delta k = k_1 + k_2 - k_3$ is the wavevector mismatch (see Sect. 72.2.2) and $\chi_{\mathrm{eff}}^{(2)}$ is given by

$$\chi_{\mathrm{eff}}^{(2)} = \sum_{ijk} \chi_{ijk}^{(2)} (\hat{\boldsymbol{u}}_1^*)_i (\hat{\boldsymbol{u}}_2)_j (\hat{\boldsymbol{u}}_3)_k , \quad (72.25)$$

where $(\hat{\boldsymbol{u}}_l)_i = \hat{\boldsymbol{u}}_l \cdot \hat{\boldsymbol{\imath}}$. For simplicity, the subscripts on each of the field amplitudes have been dropped, since only one spatial mode at each frequency contributed. The resulting coupled amplitude equations are

$$\frac{\mathrm{d}A(\omega_1)}{\mathrm{d}z} = \left[\frac{1}{4\pi}\right]\frac{\mathrm{i}4\pi\omega_1\chi_{\mathrm{eff}}^{(2)}}{n(\omega_1)c} A(\omega_3)A^*(\omega_2)\,\mathrm{e}^{-\mathrm{i}\Delta kz} , \quad (72.26)$$

$$\frac{\mathrm{d}A(\omega_2)}{\mathrm{d}z} = \left[\frac{1}{4\pi}\right]\frac{\mathrm{i}4\pi\omega_2\chi_{\mathrm{eff}}^{(2)}}{n(\omega_2)c} A(\omega_3)A^*(\omega_1)\,\mathrm{e}^{-\mathrm{i}\Delta kz} , \quad (72.27)$$

$$\frac{\mathrm{d}A(\omega_3)}{\mathrm{d}z} = \left[\frac{1}{4\pi}\right]\frac{\mathrm{i}4\pi\omega_3\chi_{\mathrm{eff}}^{(2)}}{n(\omega_3)c} A(\omega_1)A(\omega_2)\,\mathrm{e}^{\mathrm{i}\Delta kz} . \quad (72.28)$$

72.2.2 Phase Matching

For many nonlinear optical processes (e.g., harmonic generation) it is important to minimize the wave vector mismatch in order to maximize the efficiency. For example, if the field amplitudes $A(\omega_1)$ and $A(\omega_2)$ are constant, the solution to (72.28) yields for the output intensity

$$I(L,\omega_3)=\left[\frac{1}{64\pi^3\varepsilon_0}\right]\times\frac{32\pi^3\left[\chi^{(2)}_{\text{eff}}\right]^2\omega_3^2 I(\omega_1)I(\omega_2)L^2}{n(\omega_1)n(\omega_2)n(\omega_3)c^3}\times\text{sinc}^2(\Delta kL/2)\,, \quad (72.29)$$

in terms of sinc $x=(\sin x)/x$, where $I(L,\omega_3)=(4\pi\varepsilon_0)n(\omega_3)c|A(L,\omega_3)|^2/2\pi$, and $I(\omega_1)$ and $I(\omega_2)$ are the corresponding input intensities. Clearly, the effect of the wavevector mismatch is to reduce the efficiency of the generation of the sum frequency wave. The maximum propagation distance over which efficient nonlinear coupling can occur is given by the coherence length

$$L_c=\frac{2}{\Delta k}\,. \quad (72.30)$$

As a result of the dispersion in the linear refractive index that occurs in all materials, achieving phase matching over typical interaction lengths (e.g., 5 mm) is nontrivial. For the case in which the nonlinear material is birefringent, it is sometimes possible to achieve phase matching by insuring that the interacting waves possess some suitable combination of ordinary and extraordinary polarization. Other techniques for achieving phase matching include quasiphase matching [72.5] and the use of the mode dispersion in waveguides [72.6].

However, the phase matching condition is automatically satisfied for certain nonlinear optical processes, such as two-photon absorption (see Sect. 72.4.6) and Stokes amplification in stimulated Raman scattering (see Sect. 72.5.1). One can tell when the phase matching condition is automatically satisfied by examining the frequencies that appear in the expression for the nonlinear susceptibility. For a nonlinear susceptibility of the sort $\chi^{(3)}(\omega_1;\omega_2,\omega_3,\omega_4)$ the wave vector mismatch is given in general by $\Delta\boldsymbol{k}=\boldsymbol{k}_2+\boldsymbol{k}_3+\boldsymbol{k}_4-\boldsymbol{k}_1$. Thus, for the example of Stokes amplification in stimulated Raman scattering, the nonlinear susceptibility is given by $\chi^{(3)}(\omega_1;\omega_1,\omega_0,-\omega_0)$ where $\omega_0(\omega_1)$ is the frequency of the pump (Stokes) wave, and consequently the wave vector mismatch vanishes identically.

72.2.3 Manley–Rowe Relations

Under conditions of full permutation symmetry, there is no flow of power from the electromagnetic fields to the medium, and thus the total power flow of the fields is conserved. The flow of energy among the fields can be described by the Manley–Rowe relations. For example, for the case of sum-frequency generation, one can deduce from (72.26, 27, 28) that

$$\frac{\mathrm{d}}{\mathrm{d}z}\left[\frac{I(\omega_1)}{\omega_1}\right]=\frac{\mathrm{d}}{\mathrm{d}z}\left[\frac{I(\omega_2)}{\omega_2}\right]=-\frac{\mathrm{d}}{\mathrm{d}z}\left[\frac{I(\omega_3)}{\omega_3}\right]. \quad (72.31)$$

The expressions in square brackets are proportional to the flux of photons per unit area per unit time, and imply that the creation of a photon at ω_3 must be accompanied by the annihilation of photons at both ω_1 and ω_2. Similar relations can be formulated for other nonlinear optical processes that are governed by a nonlinear susceptibility that satisfies full permutation symmetry. Since this behavior occurs at the photon level, nonlinear optical processes can lead to the generation of light fields that have esoteric quantum statistical properties (Chapt. 78 and Chapt. 80).

A nonlinear optical process that satisfies the Manley–Rowe relations is called a parametric process. Conversely, a process for which field energy is not conserved, and thus Manley–Rowe relations cannot be formulated, is said to be nonparametric. Thus, parametric processes are described by purely real $\chi^{(n)}$, whereas nonparametric proceses are described by complex $\chi^{(n)}$.

72.2.4 Pulse Propagation

If the optical field consists of ultrashort (<100 ps) pulses, it is more convenient to work with the temporally varying amplitude, rather than with the individual frequency components. Thus, for a linearly polarized plane wave pulse propagating along the z-axis, the field is decomposed into the product of a slowly varying amplitude $A(z,t)$ and a rapidly varying oscillatory term such that

$$\tilde{E}(\boldsymbol{r},t)=A(z,t)\,\mathrm{e}^{\mathrm{i}(k_0z-\omega_0t)}+\text{c.c.}\,, \quad (72.32)$$

where $k_0=n_0\omega_0/c$. For a pulse propagating in a material with an intensity-dependent refractive index, the propagation can be described by the nonlinear Schrödinger equation

$$\frac{\partial A}{\partial z}+\frac{\mathrm{i}\beta_2}{2}\frac{\partial^2 A}{\partial\tau^2}=\mathrm{i}\gamma|A|^2A\,, \quad (72.33)$$

where $\beta_2=\left(\mathrm{d}^2k/\mathrm{d}\omega^2\right)|_{\omega=\omega_0}$ is the group velocity dispersion parameter, $\tau=t-z/v_g$ is the local time for the pulse, $v_g=[(\mathrm{d}k/\mathrm{d}\omega)|_{\omega=\omega_0}]^{-1}$ is the group velocity, and $\gamma=[4\pi\varepsilon_0]n_2n_0\omega_0/2\pi$ is the nonlinear refractive index parameter.

72.3 Second-Order Processes

Second-order nonlinear optical processes occur as a consequence of the second term in expression (72.3), i. e., processes whose strength is described by $\chi^{(2)}(\omega_\sigma;\omega_m,\omega_n)$. These processes entail the generation of a field at frequency $\omega_\sigma=\omega_m+\omega_n$ in response to applied fields at (positive and/or negative) frequencies ω_m and ω_n. Several examples of such processes are described in this Section.

72.3.1 Sum Frequency Generation

Sum frequency generation produces an output field at frequency $\omega_3=\omega_1+\omega_2$ for ω_1 and ω_2 both positive. It is useful, for example, for the generation of tunable radiation in the uv if ω_1 and/or ω_2 are obtained from tunable lasers in the visible range. Sum frequency generation is described in detail in Sects. 72.2.1–72.2.3.

72.3.2 Second Harmonic Generation

Second harmonic generation is routinely used to convert the output of a laser to a higher frequency. It is described by $\chi^{(2)}(2\omega;\omega,\omega)$. Let η be the power conversion efficiency from frequency ω to 2ω. Assuming that phase matching is perfect, and the pump wave at frequency ω is undepleted by the interaction, a derivation analogous to that for (72.29) yields

$$\eta=\tanh^2\left(z/l\right), \tag{72.34}$$

where the characteristic conversion length l is given by

$$l=\left[4\pi\right]\frac{c\sqrt{n(\omega)n(2\omega)}}{4\pi\omega\chi^{(2)}|A_1(0)|}. \tag{72.35}$$

Note that the conversion efficiency asymptotically approaches unity. In practice, conversion efficiencies exceeding 80% can be achieved.

72.3.3 Difference Frequency Generation

Difference frequency generation can be used to create light in the infrared and far infrared by generating the difference frequency $\omega_2=\omega_3-\omega_1$ (where ω_3 and ω_1 are positive and $\omega_3>\omega_1$) of two incident lasers. Consider the case in which a strong (undepleted) pump wave at frequency ω_3 and a weak (signal) wave at ω_1 are incident on a nonlinear medium described by $\chi^{(2)}(\omega_2;\omega_3,-\omega_1)=\chi^{(2)}(\omega_1;\omega_3,-\omega_2)$. The amplitude $A(\omega_3)$ of the strong wave can be taken as a constant, and thus the interaction can be described by finding simultaneous solutions to (72.26) and (72.27) for $A(\omega_1)$ and $A(\omega_2)$. In the limit of perfect phase matching (i. e., $\Delta k=0$), the solutions are

$$A(z,\omega_1)=A(0,\omega_1)\cosh\kappa z\,, \tag{72.36}$$

$$A(z,\omega_2)=\mathrm{i}\sqrt{\frac{n_1\omega_2}{n_2\omega_1}}\frac{A(\omega_3)}{|A(\omega_3)|}A^*(0,\omega_1)\sinh\kappa z\,, \tag{72.37}$$

where

$$\kappa^2=\left[\frac{1}{16\pi^2}\right]\frac{16\pi^2\left[\chi^{(2)}\right]^2\omega_1^2\omega_2^2}{k_1k_2c^4}|A(\omega_3)|^2\,. \tag{72.38}$$

Equation (72.37) describes the spatial growth of the difference frequency signal.

72.3.4 Parametric Amplification and Oscillation

For the foregoing case of a strong wave at frequency ω_3 and a weak wave with $\omega_1<\omega_3$ incident on a second-order nonlinear optical material, the lower frequency input wave is amplified by the nonlinear interaction; this process is known as parametric amplification. Difference frequency generation is a consequence of the Manley–Rowe relations, as described above in Sect. 72.2.3. Since $\omega_3=\omega_1+\omega_2$, the annihilation of an ω_3 photon must be accompanied by the simultaneous creation of photons ω_1 and ω_2.

An optical parametric oscillator can be constructed by placing the nonlinear optical material inside an optical resonator that provides feedback at ω_1 and/or ω_2. When such a device is excited by a wave at ω_3, it can produce output frequencies ω_1 and ω_2 that satisfy $\omega_1+\omega_2=\omega_3$. Optical parametric oscillators are of considerable interest as sources of broadly tunable radiation [72.7].

72.3.5 Focused Beams

For conceptual clarity, much of the discussion so far has assumed that the interacting beams are plane waves. In practice, the incident laser beams are often focused into the nonlinear material to increase the field strength within the interaction region and consequently to increase the nonlinear response. However, it is undesirable to focus too tightly, because doing so leads to a decrease in the effective length of the interaction region. In particular, if w_0 is the radius of the laser beam at the beam waist, the beam remains focused only over a distance of the order $b = 2\pi w_0^2/\lambda$ where λ is the laser wavelength measured in the nonlinear material. For many types of nonlinear optical processes, the optimal nonlinear response occurs if the degree of focusing is adjusted so that b is several times smaller than the length L of the nonlinear optical material.

72.4 Third-Order Processes

A wide variety of nonlinear optical processes are possible as a result of the nonlinear contributions to the polarization that are third-order in the applied field. These processes are described by $\chi^{(3)}(\omega_\sigma; \omega_m, \omega_n, \omega_o)$ (72.3) and can lead not only to the generation of new field components (e.g., third-harmonic generation) but can also result in a field affecting itself as it propagates (e.g., self-phase modulation). Several examples are described in this section.

72.4.1 Third-Harmonic Generation

Assuming full-permutation symmetry, the nonlinear polarization amplitudes for the fundamental and third-harmonic beams are

$$\begin{aligned}\mathcal{P}^{NL}(z,\omega) &= [\varepsilon_0] 3\chi^{(3)}_{eff} A(z,3\omega)[A^*(z,\omega)]^2 e^{-i\Delta kz} ,\\ \mathcal{P}^{NL}(z,3\omega) &= [\varepsilon_0]\chi^{(3)}_{eff}[A(z,\omega)]^3 e^{i\Delta kz} , \qquad (72.39)\end{aligned}$$

where $\Delta k = 3k(\omega) - k(3\omega)$ and $\chi^{(3)}_{eff}$ is the effective third-order susceptibility for third-harmonic generation and is defined in a manner analogous to the $\chi^{(2)}_{eff}$ in (72.25). If the intensity of the fundamental wave is not depleted by the nonlinear interaction, the solution for the output intensity $I(L, 3\omega)$ of the third-harmonic field for a crystal of length L is

$$\begin{aligned}I(L,3\omega) = &\left[\frac{1}{256\pi^4\varepsilon_0^2}\right]\frac{48\pi^2\omega^2\left[\chi^{(3)}_{eff}\right]^2}{n(3\omega)n(\omega)^3c^4}\\ &\times I(\omega)^3L^2\sinh^2[\Delta kL/2] , \qquad (72.40)\end{aligned}$$

where $I(\omega)$ is the input intensity of the fundamental field. As a result of the typically small value of $\chi^{(3)}_{eff}$ in crystals, it is generally more efficient to generate the third harmonic by using two $\chi^{(2)}$ crystals in which the first crystal produces second harmonic light and the second crystal combines the second harmonic and the fundamental beams via sum-frequency generation. It is also possible to use resonant enhancement of $|\chi^{(3)}|$ in gases to increase the efficiency of third-harmonic generation [72.8].

72.4.2 Self-Phase and Cross-Phase Modulation

The nonlinear refractive index leads to an intensity-dependent change in the phase of the beam as it propagates through the material. If the medium is lossless, the amplitude of a single beam at frequency ω propagating in the positive z-direction can be expressed as

$$A(z,\omega) = A(0,\omega)\, e^{i\phi^{NL}(z)} , \qquad (72.41)$$

where the nonlinear phase shift $\phi^{NL}(z)$ is given by

$$\phi^{NL}(z) = \frac{\omega}{c} n_2 I z , \qquad (72.42)$$

and $I = [4\pi\varepsilon_0]n_0c|A(0,\omega)|^2/2\pi$ is the intensity of the laser beam. If two fields at different frequencies ω_1 and ω_2 are traveling along the z-axis, the two fields can affect each other's phase; this effect is known as cross-phase modulation. The nonlinear phase shift $\phi^{NL}_{1,2}(z)$ for each of the waves is given by

$$\phi^{NL}_{1,2}(z) = \frac{\omega_{1,2}}{c} n_2(I_{1,2} + 2I_{2,1})z . \qquad (72.43)$$

For the case of a light pulse, the change in the phase of the pulse inside the medium becomes a function of time. In this case the solution to (72.33) for the time-varying amplitude $A(z,\tau)$ shows that in the absence of group-velocity dispersion (GVD) (i.e., $\beta_2 = 0$) that the solution for $A(z,\tau)$ is of the form of (72.41), except that the temporal intensity profile $I(\tau)$ replaces the steady-state intensity I in (72.42). As the pulse propagates through the medium, its frequency becomes time dependent, and the instantaneous frequency shift from the central frequency ω_0 is given by

$$\delta\omega(\tau) = -\frac{\partial\phi^{NL}(\tau)}{\partial\tau} = -\frac{\omega n_2 z}{c}\frac{\partial I}{\partial t} . \qquad (72.44)$$

This time-dependent self-phase modulation leads to a broadening of the pulse spectrum and to a frequency chirp across the pulse.

If the group velocity dispersion parameter β_2 and the nonlinear refractive index coefficient n_2 are of opposite sign, the nonlinear frequency chirp can be compensated by the chirp due to group velocity dispersion, and (72.33) admits soliton solutions . For example, the fundamental soliton solution is

$$A(z,t) = \sqrt{\frac{1}{L_D}} \operatorname{sech}\left(\frac{\tau}{\tau_p}\right) e^{iz/2L_D} , \qquad (72.45)$$

where τ_p is the pulse duration and $L_D = \tau_p^2/|\beta_2|$ is the dispersion length. As a result of their ability to propagate in dispersive media without changing shape, optical solitons show a great deal of promise in applications such as optical communications and optical switching. For further discussion of optical solitons see [72.9].

72.4.3 Four-Wave Mixing

Various types of four-wave mixing processes can occur among different beams. One of the most common geometries is backward four-wave mixing used in nonlinear spectroscopy and optical phase conjugation. In this interaction, two strong counterpropagating pump waves with amplitudes A_1 and A_2 and with equal frequencies $\omega_{1,2} = \omega$ are injected into a nonlinear medium. A weak wave, termed the probe wave, (with frequency ω_3 and amplitude A_3) is also incident on the medium. As a result of the nonlinear interaction among the three waves, a fourth wave with an amplitude A_4 is generated which is counterpropagating with respect to the probe wave and with frequency $\omega_4 = 2\omega - \omega_3$. For this case, the third-order nonlinear susceptibilities for the probe and conjugate waves are given by $\chi^{(3)}(\omega_{3,4}; \omega, \omega, -\omega_{4,3})$. For constant pump wave intensities and full permutation symmetry, the amplitudes of the nonlinear polarization for the probe and conjugate waves are given by

$$\mathcal{P}^{NL}(z,\omega_{3,4}) = \pm[\varepsilon_0]6\chi^{(3)}\big[\big(|A_1|^2+|A_2|^2\big)A_{3,4} + A_1A_2A_{4,3}^* e^{i\Delta kz}\big] , \qquad (72.46)$$

where $\Delta k = k_1 + k_2 - k_3 - k_4$ is the phase mismatch, which is nonvanishing when $\omega_3 \neq \omega_4$. For the case of optical phase conjugation by degenerate four-wave mixing (i. e., $\omega_3 = \omega_4 = \omega$ and $A_4(L) = 0$), the phase conjugate reflectivity R_{PC} is

$$R_{PC} = \frac{|A_4(0)|^2}{|A_3(0)|^2} = \tan^2(\kappa L) , \qquad (72.47)$$

where $\kappa = \big[1/16\pi^2\varepsilon_0\big]\big[24\pi^2\omega\chi^{(3)}/(n_0c)^2\big]\sqrt{I_1I_2}$ and $I_{1,2}$ are the intensities of the pump waves. Phase-conjugate reflectivities greater than unity can be routinely achieved by performing four-wave mixing in atomic vapors or photorefractive media.

72.4.4 Self-Focusing and Self-Trapping

Typically a laser beam has a transverse intensity profile that is approximately Gaussian. In a medium with an intensity-dependent refractive index, the index change at the center of the beam is different from the index change at the edges of the beam. The gradient in the refractive index created by the beam can allow it to self-focus for $n_2 > 0$. For this condition to be met, the total input power of the beam must exceed the critical power P_{cr} for self-focusing which is given by

$$P_{cr} = \frac{\pi(0.61\lambda)^2}{8n_0n_2} , \qquad (72.48)$$

where λ is the vacuum wavelength of the beam. For powers much greater than the critical power, the beam can break up into various filaments, each with a power approximately equal to the critical power. For a more extensive discussion of self-focusing and self-trapping see [72.10, 11].

72.4.5 Saturable Absorption

When the frequency ω of an applied laser field is sufficiently close to a resonance frequency ω_0 of the medium, an appreciable fraction of the atomic population can be placed in the excited state. This loss of population from the ground state leads to an intensity-dependent saturation of the absorption and the refractive index of the medium (see Sect. 69.2 for more detailed discussion) [72.4]. The third-order susceptibility as a result of this saturation is given by

$$\chi^{(3)} = \left[\frac{1}{\varepsilon_0}\right]\frac{|\mu|^2T_1T_2\alpha_0c}{3\pi\omega_0\hbar^2}\frac{\delta T_2 - i}{\left[1+(\delta T_2)^2\right]^2} , \qquad (72.49)$$

where μ is the transition dipole moment, T_1 and T_2 are the longitudinal and transverse relaxation times, respectively (see Sect. 68.4.3), α_0 is the line-center weak-field intensity absorption coefficient, and $\delta = \omega - \omega_0$ is the detuning. For the 3s $\leftrightarrow$ 3p transition in atomic sodium vapor at 300 °C, the nonlinear refractive index $n_2 \approx 10^{-7}\,\mathrm{cm^2/W}$ for a detuning $\delta T_2 = 300$.

72.4.6 Two-Photon Absorption

When the frequency ω of a laser field is such that 2ω is close to a transition frequency of the material, it is possible for two-photon absorption (TPA) to occur. This process leads to a contribution to the imaginary part of $\chi^{(3)}(\omega;\,\omega,\,\omega,\,-\omega)$. In the presence of TPA, the intensity $I(z)$ of a single, linearly polarized beam as a function of propagation distance is

$$I(z) = \frac{I(0)}{1+\beta I(0)z}\,, \tag{72.50}$$

where $\beta = \left[1/16\pi^2\varepsilon_0\right]24\pi^2\omega\,\mathrm{Im}\left[\chi^{(3)}\right]/(n_0c)^2$ is the TPA coefficient. For wide-gap semiconductors such as ZnSe at 800 nm, $\beta \approx 10^{-8}$ cm/W.

72.4.7 Nonlinear Ellipse Rotation

The polarization ellipse of an elliptically polarized laser beam rotates but retains its ellipticity as the beam propagates through an isotropic nonlinear medium. Ellipse rotation occurs as a result of the difference in the nonlinear index changes experienced by the left- and right-circular components of the beam, and the angle θ of rotation is

$$\begin{aligned}\theta &= \frac{1}{2}\Delta n\omega z/c \\ &= \left[\frac{1}{16\pi^2\varepsilon_0}\right]\frac{12\pi^2}{n_0^2c}\chi^{(3)}_{xyyx} \\ &\quad\times(\omega;\,\omega,\,\omega,\,-\omega)(I_+ - I_-)z\,,\end{aligned} \tag{72.51}$$

where $I_\pm$ are the intensities of the circularly polarized components of the beam with unit vectors $\hat{\boldsymbol{\sigma}}_\pm = (\hat{\boldsymbol{x}} \pm i\hat{\boldsymbol{y}})/\sqrt{2}$. Nonlinear ellipse rotation is a sensitive technique for determining the nonlinear susceptibility element $\chi^{(3)}_{xyyx}$ for isotropic media and can be used in applications such as optical switching.

72.5 Stimulated Light Scattering

Stimulated light scattering occurs as a result of changes in the optical properties of the material that are induced by the optical field. The resulting nonlinear coupling between different field components is mediated by some excitation (e.g., acoustic phonon) of the material that results in changes in its optical properties. The nonlinearity can be described by a complex susceptibility and a nonlinear polarization that is of third order in the interacting fields. Various types of stimulated scattering can occur. Discussed below are the two processes that are most commonly observed.

72.5.1 Stimulated Raman Scattering

In stimulated Raman scattering (SRS), the light field interacts with a vibrational mode of a molecule. The coupling between the two optical waves can become strong if the frequency difference between them is close to the frequency ω_v of the molecular vibrational mode. If the pump field at ω_0 and another field component at ω_1 are propagating in the same direction along the z-axis, the steady-state nonlinear polarization amplitudes for the two field components are given by

$$\begin{aligned}\mathcal{P}^{\mathrm{NL}}(z,\omega_{0,1}) &= [\varepsilon_0]6\chi_{\mathrm{R}}(\omega_{0,1}) \\ &\quad\times|A(z,\omega_{1,0})|^2A(z,\omega_{0,1})\,,\end{aligned} \tag{72.52}$$

where $\chi_{\mathrm{R}}(\omega_{0,1}) \equiv \chi^{(3)}(\omega_{0,1};\,\omega_{0,1},\omega_{1,0},-\omega_{1,0})$, the Raman susceptibility, actually depends only on the frequency difference $\Omega = \omega_0 - \omega_1$ and is given by

$$\chi_{\mathrm{R}}(\omega_{0,1}) = \left[\frac{1}{\varepsilon_0}\right]\frac{N(\partial\alpha/\partial q)_0^2}{6\mu_M}\frac{1}{\omega_v^2 - \Omega^2 \mp 2\mathrm{i}\gamma\Omega}\,, \tag{72.53}$$

where the minus (plus) sign is taken for the ω_0 (ω_1) susceptibility, μ_M is the reduced nuclear mass, and $(\partial\alpha/\partial q)_0$ is a measure of the change of the polarizability of the molecule with respect to a change in the intermolecular distance q at equilibrium. If the intensity of the pump field is undepleted by the interaction with the ω_1 field and is assumed to be constant, the solution for the intensity of the ω_1 field at $z = L$ is given by

$$I(L,\omega_1) = I(0,\omega_1)\mathrm{e}^{G_{\mathrm{R}}}\,, \tag{72.54}$$

where the SRS gain parameter G_{R} is

$$\begin{aligned}G_{\mathrm{R}} &= \left[\frac{1}{16\pi^2\varepsilon_0}\right]48\pi^2\frac{\omega_1}{(n_1c)^2}\,\mathrm{Im}[\chi_{\mathrm{R}}(\omega_1)]I_0L \\ &= g_{\mathrm{R}}I_0L\,,\end{aligned} \tag{72.55}$$

g_{R} is the SRS gain factor, and I_0 is the input intensity of the pump field. For $\omega_1 < \omega_0$ ($\omega_1 > \omega_0$), the ω_1 field is termed the Stokes (anti-Stokes) field, and it experiences exponential amplification (attenuation). For sufficiently large gains (typically $G_{\mathrm{R}} \gtrsim 25$), the Stokes wave can be seeded by spontaneous Raman scattering and can grow to an appreciable fraction of the pump field. For a complete discussion of the spontaneous initiation of SRS see [72.12]. For the case of CS_2, $g_R = 0.024\,\mathrm{cm/MW}$.

Four-wave mixing processes that couple a Stokes wave having $\omega_1 < \omega_0$ and an anti-Stokes wave having $\omega_2 > \omega_0$, where $\omega_1 + \omega_2 = 2\omega_0$, can also occur [72.4]. In this case, additional contributions to the nonlinear polarization are present and are characterized by a Raman susceptibility of the form $\chi^{(3)}(\omega_{1,2}; \omega_0, \omega_0, -\omega_{2,1})$. The technique of coherent anti-Stokes Raman spectroscopy is based on this four-wave mixing process [72.13].

72.5.2 Stimulated Brillouin Scattering

In stimulated Brillouin scattering (SBS), the light field induces and interacts with an acoustic wave inside the medium. The resulting interaction can lead to extremely high amplification for certain field components (i. e., Stokes wave). For many optical media, SBS is the dominant nonlinear optical proccess for laser pulses of duration > 1 ns. The primary applications for SBS are self-pumped phase conjugation and pulse compression of high-energy laser pulses.

If an incident light wave with wave vector $\boldsymbol{k}_0$ and frequency ω_0 is scattered from an acoustic wave with wave vector $\boldsymbol{q}$ and frequency Ω, the wave vector and frequency of the scattered wave are determined by conservation of momentum and energy to be $\boldsymbol{k}_1 = \boldsymbol{k}_0 \pm \boldsymbol{q}$ and $\omega_1 = \omega_0 \pm \Omega$, where the (+) sign applies if $\boldsymbol{k}_0 \cdot \boldsymbol{q} > 0$ and the (−) applies if $\boldsymbol{k}_0 \cdot \boldsymbol{q} < 0$. Here, Ω and $\boldsymbol{q}$ are related by the dispersion relation $\Omega = v|\boldsymbol{q}|$ where v is the velocity of sound in the material. These Bragg scattering conditions lead to the result that the Brillouin frequency shift $\Omega_B = \omega_1 - \omega_0$ is zero for scattering in the forward direction (i. e., in the $\boldsymbol{k}_0$ direction) and reaches its maximum for scattering in the backward direction given by

$$\Omega_B = 2\omega_0 v n_0/c\,, \tag{72.56}$$

where n_0 is the refractive index of the material.

The interaction between the incident wave and the scattered wave in the Brillouin-active medium can become nonlinear if the interference between the two optical fields can coherently drive an acoustic wave, either through electrostriction or through local density fluctuations resulting from the absorption of light and consequent temperature changes. The following discussion treats the more common electrostriction mechanism.

Typically, SBS occurs in the backward direction (i. e., $\boldsymbol{k}_0 = k_0\hat{z}$ and $\boldsymbol{k}_1 = -k_1\hat{z}$), since the spatial overlap between the Stokes beam and the laser beam is maximized under these conditions and, as mentioned above, no SBS occurs in the forward direction. The steady-state nonlinear polarization amplitudes for backward SBS are

$$\begin{aligned}\mathcal{P}^{\mathrm{NL}}(z, \omega_{0,1}) &= [\varepsilon_0] 6\chi_B(\omega_{0,1}) \\ &\quad \times |A(z, \omega_{1,0})|^2 A(z, \omega_{0,1})\,,\end{aligned} \tag{72.57}$$

where $\chi_B(\omega_{0,1}) \equiv \chi^{(3)}(\omega_{0,1}; \omega_{0,1}, \omega_{1,0}, -\omega_{1,0})$, the Brillouin susceptibility, depends only on $\Omega = \omega_0 - \omega_1$ and is given by

$$\chi_B(\omega_{0,1}) = \left[\frac{1}{\varepsilon_0}\right] \frac{\omega_0^2 \gamma_e^2}{24\pi^2 c^2 \rho_0} \frac{1}{\Omega_B^2 - \Omega^2 \mp \mathrm{i}\Gamma_B \Omega}\,, \tag{72.58}$$

where the minus (plus) sign is taken for the ω_0 (ω_1) susceptibility, γ_e is the electrostrictive constant, ρ_0 is the mean density of the material, and Γ_B is the Brillouin linewidth given by the inverse of the phonon lifetime. If the pump field is undepleted by the interaction with the ω_1 field and is assumed to be constant, the solution for the output intensity of the ω_1 field at $z = 0$ is given by

$$I(0, \omega_1) = I(L, \omega_1)\,\mathrm{e}^{G_B}\,, \tag{72.59}$$

where the Brillouin gain coefficient G_B is given by

$$\begin{aligned}G_B &= \left[\frac{1}{16\pi^2\varepsilon_0}\right] 48\pi^2 \frac{\omega_1}{(n_0 c)^2} \mathrm{Im}[\chi_B(\omega_1)] I_0 L\,,\\ &= g_0 \frac{\Omega\Omega_B \Gamma_B^2}{\left[\Omega_B^2 - \Omega^2\right]^2 + (\Omega\Gamma_B)^2} I_0 L \\ &= g_B I_0 L\,,\end{aligned} \tag{72.60}$$

g_B is the SBS gain factor, I_0 is the input intensity of the pump field, and

$$g_0 = \left[\frac{1}{\varepsilon_0^2}\right] \frac{\omega_0^2 \gamma_e^2}{n_0 c^3 \rho_0 v \Gamma_B} \tag{72.61}$$

is the line-center (i. e., $\Omega = \pm\Omega_B$) SBS gain factor. For $\Omega > 0$ ($\Omega < 0$), the ω_1 field is termed the Stokes (anti-Stokes) field, and it experiences exponential amplification (attenuation). For sufficiently large gains (typically $G_B \gtrsim 25$), the Stokes wave can be seeded by spontaneous Brillouin scattering and can grow to an appreciable fraction of the pump field. For a complete discussion of the spontaneous initiation of SBS see [72.14]. For CS_2, $g_0 = 0.15\,\mathrm{cm/MW}$.

72.6 Other Nonlinear Optical Processes

72.6.1 High-Order Harmonic Generation

If full permutation symmetry applies and the fundamental field ω is not depleted by nonlinear interactions, then the intensity of the qth harmonic is given by

$$I(z,q\omega)=\left[\frac{1}{4\pi(4\pi\varepsilon_0)^{(q-1)/2}}\right]\times\frac{2\pi q^2\omega^2}{n^2(q\omega)c}\left[\frac{2\pi I(\omega)}{n(\omega)c}\right]^q\times\left|\chi^{(q)}(q\omega;\omega,\ldots,\omega)J_q(\Delta k,z_0,z)\right|^2,\qquad(72.62)$$

where $\Delta k=[n(\omega)-n(q\omega)]\omega/c$,

$$J_q(\Delta k,z_0,z)=\int_{z_0}^{z}\frac{\mathrm{e}^{\mathrm{i}\Delta k z'}\,\mathrm{d}z'}{(1+2\mathrm{i}z'/b)^{q-1}},\qquad(72.63)$$

$z=z_0$ at the input face of the nonlinear medium, and b is the confocal parameter Sect. 72.3.5 of the fundamental beam. Defining $L=z-z_0$, the integral J_q can be easily evaluated in the limits $L\ll b$ and $L\gg b$. The limit $L\ll b$ corresponds to the plane-wave limit in which case

$$|J_q(\Delta k,z_0,z)|^2=L^2\mathrm{sinc}^2\left(\frac{\Delta kL}{2}\right).\qquad(72.64)$$

The limit $L\gg b$ corresponds to the tight-focusing configuration in which case

$$J_q(\Delta k,z_0,z)=\begin{cases}0, & \Delta k\le 0,\\ \dfrac{\pi b}{(q-2)!}\left(\dfrac{b\Delta k}{2}\right)^{q-2}\mathrm{e}^{-b\Delta k/2}, & \Delta k>0.\end{cases}\qquad(72.65)$$

Note that in this limit, the qth harmonic light is only generated for positive phase mismatch. *Reintjes* et al. [72.15, 16] observed both the fifth and seventh harmonics in helium gas which exhibited a dependence on $I(\omega)$ which is consistent with the $I^q(\omega)$ dependence predicted by (72.62). However, more recent experiments in gas jets have demonstrated the generation of extremely high-order harmonics which do not depend on the intensity in this simple manner (see Chapt. 74 for further discussion of this nonperturbative behavior).

72.6.2 Electro-Optic Effect

The electro-optic effect corresponds to the limit in which the frequency of one of the applied fields approaches zero. The linear electro-optic effect (or Pockels effect) can be described by a second-order susceptibility of the form $\chi^{(2)}(\omega;\omega,0)$. This effect produces a change in the refractive index for light of certain polarizations which depends linearly on the strength of the applied low-frequency field. More generally, the linear electro-optic effect induces a change in the amount of birefringence present in an optical material. This electrically controllable change in birefringence can be used to construct amplitude modulators, frequency shifters, optical shutters, and other optoelectronic devices. Materials commonly used in such devices include KDP and lithium niobate [72.17]. If the laser beam is propagating along the optic axis (i. e., z-axis) of the material of length L and the low-frequency field E_z is also applied along the optic axis, the nonlinear index change $\Delta n=n_y-n_x$ between the components of the electric field polarized along the principal axes of the crystal is given by

$$\Delta n=\left[\frac{1}{4\pi}\right]n_0^3r_{63}E_z\qquad(72.66)$$

where r_{63} is one of the electro-optic coefficients.

The quadratic electro-optic effect produces a change in refractive index that scales quadratically with the applied dc electric field. This effect can be described by a third-order susceptibility of the form $\chi^{(3)}(\omega;\omega,0,0)$.

72.6.3 Photorefractive Effect

The photorefractive effect leads to an optically induced change in the refractive index of a material. In certain ways this effect mimics that of the nonlinear refractive index described in Sect. 72.1.2, but it differs from the nonlinear refractive index in that the change in refractive index is independent of the overall intensity of the incident light field, and depends only on the degree of spatial modulation of the light field within the nonlinear material. In addition, the photorefractive effect can occur only in materials that exhibit a linear electro-optic effect, and contain an appreciable density of trapped electrons and/or holes that can be liberated by the application of a light field. Typical photorefractive materials include lithium niobate, barium titanate, and strontium barium niobate.

A typical photorefractive configuration might be as follows: two beams interfere within a photorefractive crystal to produce a spatially modulated intensity distribution. Bound charges are ionized with greater probability at the maxima than at the minima of the distribution and, as a result of the diffusion process, carriers tend to migrate away from regions of large light intensity. The resulting modulation of the charge distribution leads to the creation of a spatially modulated electric field that produces a spatially modulated change in refractive index as a consequence of the linear electro-optic effect. For a more extensive discussion see [72.18].

72.6.4 Ultrafast and Intense-Field Nonlinear Optics

Additional nonlinear optical processes are enabled by the use of ultrashort (< 1 ps) or ultra-intense laser pulses. For reasons of basic laser physics, ultra-intense pulses are necessarily of short duration, and thus these effects normally occur together. Ultrashort laser pulses possess a broad frequency spectrum, and therefore the dispersive properties of the optical medium play a key role in the propagation of such pulses. The three-dimensional nonlinear Schrödinger equation must be modified when treating the propagation of these ultrashort pulses by including contributions that can be ignored under other circumstances [72.19,20]. These additional terms lead to processes such as space-time coupling, self-steepening, and shock wave formation [72.21, 22]. The process of self-focusing is significantly modified under short-pulse (pulse duration shorter than approximately 1 ps) excitation. For example, temporal splitting of a pulse into two components can occur; this pulse splitting lowers the peak intensity, and can lead to the arrest of the usual collapse of a pulse undergoing self-focusing [72.23]. Moreover, optical shock formation, the creation of a discontinuity in the intensity evolution of a propagating pulse, leads to supercontinuum generation, the creation of a light pulse with an extremely broad frequency spectrum [72.24]. Shock effects and the generation of supercontinuum light can also occur in one-dimensional systems, such as a microstructure optical fiber. The relatively high peak power of the ultrashort pulses from a mode-locked laser oscillator and the tight confinement of the optical field in the small ($\approx 2\,\mu$m) core of the fiber yield high intensities and strong self-phase modulation, which results in a spectral bandwidth that spans more than an octave of the central frequency of the pulse [72.25]. Such a coherent octave-spanning spectrum allows for the stabilization of the underlying frequency comb of the mode-locked oscillator, and has led to a revolution in the field of frequency metrology [72.26]. Multiphoton absorption [72.27] constitutes an important loss process that becomes important for intensities in excess of $\approx 10^{13}\,\mathrm{W/cm^2}$. In addition to introducing loss, the electrons released by this process can produce additional nonlinear effects associated with their relativistic motion in the resulting plasma [72.28, 29]. For very large laser intensities (greater than approximately $10^{16}\,\mathrm{W/cm^2}$), the electric field strength of the laser pulse can exceed the strength of the Coulomb field that binds the electron to the atomic core, and nonperturbative effects can occur. A dramatic example is that of high-harmonic generation [72.30–32]. Harmonic orders as large as the 341-st have been observed, and simple conceptual models have been developed to explain this effect [72.33]. Under suitable conditions the harmonic orders can be suitably phased so that attosecond pulses are generated [72.34].

References

72.1 N. Bloembergen: *Nonlinear Optics* (Benjamin, New York 1964)

72.2 Y. R. Shen: *Nonlinear Optics* (Wiley, New York 1984)

72.3 P. N. Butcher, D. Cotter: *The Elements of Nonlinear Optics* (Cambridge Univ. Press, Cambridge 1990)

72.4 R. W. Boyd: *Nonlinear Optics* (Academic, Boston 1992)

72.5 J. A. Armstrong, N. Bloembergen, J. Ducuing, P. S. Pershan: Phys. Rev. **127**, 1918 (1962)

72.6 G. I. Stegeman: *Contemporary Nonlinear Optics*, ed. by G. P. Agrawal, R. W. Boyd (Academic, Boston 1992) Chap. 1

72.7 See for example the Special Issue on: Optical Parametric Oscillation and Amplification, J. Opt. Soc. Am. B **10** (1993) No. 11

72.8 R. B. Miles, S. E. Harris: IEEE J. Quant. Electron. **9**, 470 (1973)

72.9 G. P. Agrawal: *Nonlinear Fiber Optics* (Academic, Boston 1989)

72.10 S. A. Akhmanov, R. V. Khokhlov, A. P. Sukhorukov: *Laser Handbook*, ed. by F. T. Arecchi, E. O. Schulz-Dubois (North-Holland, Amsterdam 1972)

72.11 J. H. Marburger: Prog. Quant. Electr. **4**, 35 (1975)

72.12 M. G. Raymer, I. A. Walmsley: *Prog. Opt.*, Vol. 28, ed. by E. Wolf (North-Holland, Amsterdam 1990)

72.13 M. D. Levenson, S. Kano: *Introduction to Nonlinear Spectroscopy* (Academic, Boston 1988)
72.14 R. W. Boyd, K. Rzazewski, P. Narum: Phys. Rev. A **42**, 5514 (1990)
72.15 J. Reintjes, C. Y. She, R. C. Eckardt, N. E. Karangelen, R. C. Elton, R. A. Andrews: Phys. Rev. Lett. **37**, 1540 (1976)
72.16 J. Reintjes, C. Y. She, R. C. Eckardt, N. E. Karangelen, R. C. Elton, R. A. Andrews: Appl. Phys. Lett. **30**, 480 (1977)
72.17 I. P. Kaminow: *An Introduction to Electro-optic Devices* (Academic, New York 1974)
72.18 P. Günter, J.-P. Huignard (Eds.): *Photorefractive Materials and Their Applications* (Springer, Berlin, Heidelberg, Part I (1988), Part II (1989))
72.19 T. Brabec, F. Krausz: Phys. Rev. Lett. **78**, 3283 (1997)
72.20 J. K. Ranka, A. L. Gaeta: Opt. Lett. **23**, 534 (1998)
72.21 J. E. Rothenberg: Opt. Lett. **17**, 1340 (1992)
72.22 G. Yang, Y. R. Shen: Opt. Lett. **9**, 510 (1984)
72.23 J. K. Ranka, R. Schirmer, A. L. Gaeta: Phys. Rev. Lett. **77**, 3783 (1996)
72.24 A. L. Gaeta: Phys. Rev. Lett. **84**, 3582 (2000)
72.25 J. K. Ranka, R. S. Windeler, A. J. Stentz: Opt. Lett. **25**, 25 (2000)
72.26 D. J. Jones, S. A. Diddams, J. K. Ranka, A. Stentz, R. S. Windeler, J. L. Hall, S. T. Cundiff: Science **288**, 635 (2000)
72.27 W. Kaiser, C. G. B. Garrett: Phys. Rev. Lett. **7**, 229 (1961)
72.28 P. Sprangle, C.-M. Tang, E. Esarez: IEEE Transactions on Plasma Science **15**, 145 (1987)
72.29 R. Wagner, S.-Y. Chen, A. Maksemchak, D. Umstadter: Phys. Rev. Lett. **78**, 3125 (1997)
72.30 P. Agostini, F. Fabre, G. Mainfray, G. Petite, N. K. Rahman: Phys. Rev. Lett. **42**, 1127 (1979)
72.31 Z. Chang: Phys. Rev. Lett. **79**, 2967 (1997)
72.32 Z. Chang: Phys. Rev. Lett. **82**, 2006 (1999)
72.33 P. B. Corkum: Phys. Rev. Lett. **71**, 1994 (1993)
72.34 H. R. Kienberger, Ch. Spielmann, G. A. Reider, N. Milosevic, T. Brabec, P. Corkum, U. Heinzmann, M. Drescher, F. Krausz: Nature **414**, 509 (2001)

73. Coherent Transients

Coherent optical transients are excited in atomic and molecular systems when a stable phase relation persists between an exciting light field and the system's electronic response. The extreme sensitivity of phase-dependent effects is responsible for the many applications of optical transient techniques in atomic and molecular physics [73.1–7].

The theory of coherent transients distinguishes carefully between two types of relaxation: homogeneous and inhomogeneous. Relaxation occurs whenever the environment of a physical system fluctuates randomly. By random environment one means the combination of all interactions that are too complex to be treated fundamentally, and that can be seen to lead to degradation of the degree of coherence of a particular interaction of main interest. The time scale of environmental fluctuations then provides the division between the two types.

When environmental fluctuations are sufficiently rapid that all dynamical systems in a macroscopic sample experience the whole range of fluctuations in a time short compared with the time of an experiment, the resultant relaxation is called homogeneous. If environmental fluctuations exist randomly over a macroscopic sample, but change relatively slowly in time, then the relaxation is called inhomogeneous. For example, weak distant collisions are experienced constantly by all atoms at thermal equilibrium in a vapor cell, and give rise to homogeneous relaxation. If the vapor is sufficiently dilute, the same atoms may nevertheless retain for long times their own individual velocities. These velocities are relatively fixed in time, but they are random over the Maxwellian distribution of velocities and so give rise to inhomogeneous relaxation. Fundamentally, the distinction between homogeneous and inhomogeneous relaxation is artificial, depending on a separation of time scales that may not always exist. Nevertheless, when it exists, the distinction provides an extremely useful way to classify coherent transients. It is one of the foundations of the subject.

The presence of quantum entanglement leads to nonintuitive effects in coherent transients.

73.1 **Optical Bloch Equations** 1065
73.2 **Numerical Estimates of Parameters** 1066
73.3 **Homogeneous Relaxation** 1066
73.3.1 Rabi Oscillations 1067
73.3.2 Bloch Vector and Bloch Sphere 1067
73.3.3 Pi Pulses and Pulse Area 1067
73.3.4 Adiabatic Following 1068
73.4 **Inhomogeneous Relaxation** 1068
73.4.1 Free Induction Decay 1068
73.4.2 Photon Echoes 1069
73.5 **Resonant Pulse Propagation** 1069
73.5.1 Maxwell–Bloch Equations 1069
73.5.2 Index of Refraction and Beers Law 1070
73.5.3 The Area Theorem and Self-Induced Transparency 1070
73.6 **Multi-Level Generalizations** 1071
73.6.1 Rydberg Packets and Intrinsic Relaxation 1071
73.6.2 Multiphoton Resonance and Two-Photon Bloch Equations 1072
73.6.3 Pump–Probe Resonance and Dark States 1073
73.6.4 Three-Level Transparency 1074
73.7 **Disentanglement and "Sudden Death" of Coherent Transients** 1074
References 1076

73.1 Optical Bloch Equations

A very weakly excited dipole transition in an atom responds linearly to an applied time-dependent electric field. This is the basis of classical Lorentzian dielectric theory, but because any transition can be inverted, an atom is more than a classical linear oscillator [73.8]. The three atomic variables that describe the primary coherent optical transients in a dipole-allowed transition include the intrinsically quantum mechanical inversion variable, as well as the components of the expectation value of the atomic dipole moment that are in-phase and in-quadrature with the field.

We write the time-dependent atomic dipole moment of a transition excited by light near exact resonance in the form

$$\begin{aligned} -\langle ex(t)\rangle &= -\langle \Psi(t)|ex|\Psi(t)\rangle \\ &= \mathrm{Re}\left[d(U-\mathrm{i}V)\,\mathrm{e}^{-\mathrm{i}\omega t}\right] , \end{aligned} \tag{73.1}$$

where d is the transition dipole matrix element, ω is the frequency of the impressed optical field, and U and V are the time-dependent amplitudes of the in-phase and in-quadrature dipole components. The impressed field is taken in the quasi-monochromatic form

$$E(t) = \frac{1}{2}\left(\mathcal{E}\,\mathrm{e}^{-\mathrm{i}\omega t} + \mathrm{c.c.}\right) \tag{73.2}$$

Both dipole moment and field will be taken to be real scalars because the complications of vector notation add little to a first discussion of the principles of coherent transients.

Section 68.3.5 shows that U and V are dynamically coupled to each other and to the inversion W through the optical Bloch equations (OBE). When relaxation terms are included, the OBEs are given by (68.55). These are

$$\begin{aligned} \frac{\mathrm{d}U}{\mathrm{d}t} &= -\Delta V - U/T_2 , \\ \frac{\mathrm{d}V}{\mathrm{d}t} &= \Delta U + \Omega_1 W - V/T_2 , \\ \frac{\mathrm{d}W}{\mathrm{d}t} &= -\Omega_1 V - (W - W_{\mathrm{eq}})/T_1 , \end{aligned} \tag{73.3}$$

the key equations of the theory of optical transients [73.1, 2].

As in Chapt. 68, $\Delta = (E_e - E_g)/\hbar - \omega$ is the detuning and $\Omega_1 = d\mathcal{E}/\hbar$ is the Rabi frequency. It is the dipole interaction energy in frequency units, but has a significance beyond this, as discussed in Sect. 73.3.1.

73.2 Numerical Estimates of Parameters

The nature of the coherent interaction between an atom or molecule and an optical field is controlled by the relative size of a number of frequencies or rates. In the case of single photon transitions they include: Δ and Ω_1, the detuning and Rabi frequency defined above, $1/T_2$ the transverse, and $1/T_1$ the longitudinal damping rates, $1/T^*$ the inhomogenenous linewidth, and $2\pi/\tau_{\mathrm{p}}$ the transform bandwidth of the optical pulse. All of these frequencies with the exception of the last are defined in Chapt. 68. In the case of multiphoton transitions and simultaneous excitation by a number of resonant laser fields, the appropriately generalized versions of these same parameters apply.

A laser pulse with $\tau_{\mathrm{p}} \geq 1$ ns and with an intensity less than about 1 GW/cm^2 can be tuned to an isolated atomic resonance and the interaction can be described in terms of a simple two-level theory. Laser pulses as short as a few fs in duration, or with intensities as high as 10^{22} W/cm^2, have been produced and such extreme pulses quasi-resonantly excite more than one upper level. A 1 ps pulse has a bandwidth of approximately 20 cm^{-1}, while a 1 fs pulse has a bandwidth of about 20 000 cm^{-1}. If a 1 ps pulse were tuned so that it resonantly excited the $n = 95$ Rydberg state of an atom, it would simultaneously and coherently excite all the levels from $n = 67$ to the continuum limit, while a 10 fs pulse could excite all the levels from $n = 4$ to the continuum.

Similarly, when laser pulses are intense enough so that the electric field amplitude approaches that of the Coulomb field holding the atom together – in hydrogen this occurs at an intensity of 3.6×10^{16} W/cm^2 – a Rabi frequency on the order of 200 000 cm^{-1} is generated, again much more than enough to excite a coherent superposition of all atomic bound states [73.9].

73.3 Homogeneous Relaxation

Homogeneous relaxation is dominant in well-collimated atomic and molecular beams as well as in high-pressure vapor cells. In the absence of a laser field ($\Omega_1 = 0$) the solutions of the OBEs are

$$(U - \mathrm{i}V) = (U - \mathrm{i}V)_0\,\mathrm{e}^{-(1/T_2 + \mathrm{i}\Delta)t} ,$$

$$W = -1 + (W_0 + 1)\mathrm{e}^{-t/T_1} , \quad (73.4)$$

where the subscript denotes values at $t = 0$. The roles of T_1 and T_2 as relaxation times are clear. They are homogeneous because they apply to each atom individually.

73.3.1 Rabi Oscillations

The OBEs predict coherent damped oscillations of the inversion with the angular Rabi frequency Ω_1 if Ω_1 is large enough, such that $\Omega_1 T_1 \gg 1$ and $\Omega_1 T_2 \gg 1$. These oscillations were originally called optical nutations following the terminology of nuclear magnetic resonance, however they are now usually called Rabi oscillations. Figure 73.1 shows the behavior of the atomic variables undergoing Rabi oscillations in a representative case.

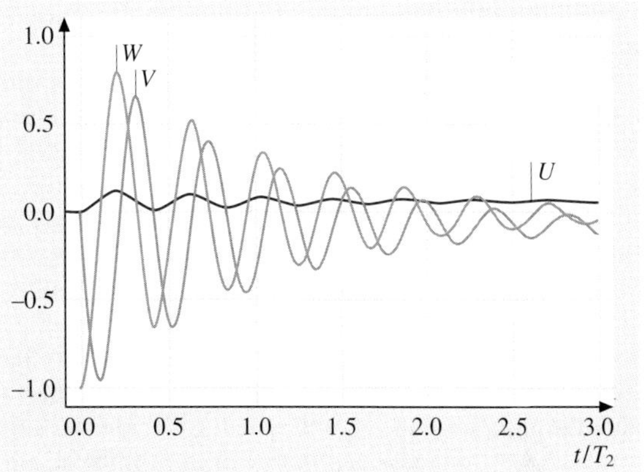

Fig. 73.1 Damped Rabi oscillations of the atomic variables after sudden turn-on of the field. In this example $T_1 = T_2$, $\Delta T_2 = 1$, and $\Omega_1 T_2 = 15$

73.3.2 Bloch Vector and Bloch Sphere

Coherent dynamical behavior is simplest for times much shorter than the relaxation times T_1 and T_2. In this case, the damping terms can be dropped from the OBEs and the resulting equations written in the form (Sect. 68.3.5)

$$\frac{\mathrm{d}\boldsymbol{U}}{\mathrm{d}t} = \boldsymbol{\Omega} \times \boldsymbol{U} , \quad (73.5)$$

where $\boldsymbol{U} = (U, V, W)$ is the Bloch vector, and $\boldsymbol{\Omega} = (-\Omega_1, 0, \Delta)$ acts as a torque vector defining the axis and rate of precession. By conservation of probability, $\boldsymbol{U} \cdot \boldsymbol{U} = 1$.

All possible quantum states of the two-level atom are mapped onto a unit sphere in U–V–W space. Conventionally, W defines the polar axis with the atomic ground state the south pole, and the excited state the north pole. Points on the sphere between the poles are coherent superpositions of the two states. The azimuthal angle ϕ represents the phase between the expectation value of the dipole moment and the optical field. In Fig. 73.2 the solutions to (73.5) are shown for the case of a square pulse applied to an atom in its ground state at $t = 0$. The solutions in this case are

$$\begin{aligned} U(t, \Delta) &= \frac{\Omega_1}{\Omega} \sin \Omega t , \\ V(t, \Delta) &= -\frac{\Delta \Omega_1}{\Omega^2}(1 - \cos \Omega t) , \\ W(t, \Delta) &= -1 + \frac{\Omega_1^2}{\Omega^2}(1 - \cos \Omega t) . \end{aligned} \quad (73.6)$$

For any Δ the solution orbit is a circle on the surface of the sphere with the orbit passing through the south pole. The rate at which the system precesses about the circle is given by the generalized Rabi frequency $\Omega \equiv \sqrt{\Delta^2 + \Omega_1^2}$.

73.3.3 Pi Pulses and Pulse Area

The exactly resonant ($\Delta = 0$) undamped OBEs can be solved analytically even for arbitrarily time dependent laser pulse envelopes. The solutions are

$$\begin{aligned} U(t, 0) &= 0 , \\ V(t, 0) &= -\sin \theta(t) , \\ W(t, 0) &= -\cos \theta(t) , \end{aligned} \quad (73.7)$$

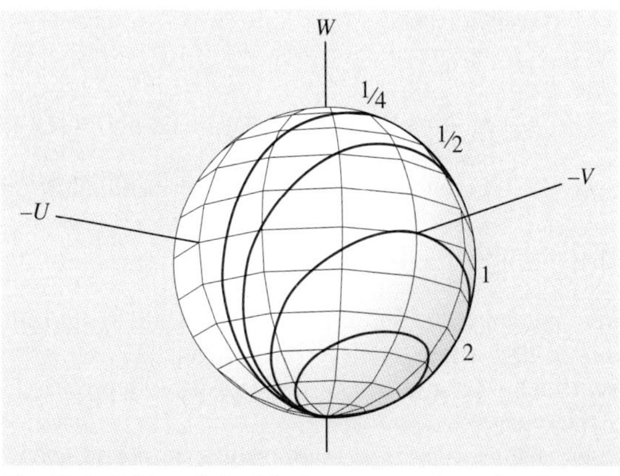

Fig. 73.2 Orbits of the Bloch vector on the unit sphere for various ratios of the detuning Δ to the Rabi frequency Ω_1

a rotation of the Bloch vector in the V–W plane. The angle θ is called the *pulse area,* defined by

$$\theta(t) \equiv \int_{-\infty}^{t} \mathrm{d}t'\,\Omega_1(t') = \frac{d}{\hbar}\int_{-\infty}^{t} \mathrm{d}t'\,\mathcal{E}(t')\,. \tag{73.8}$$

The area under the envelope of the Rabi frequency is thus the same as the angle through which an exactly-resonant Bloch vector turns due to the pulse. If $\theta = \pi$, the atom is driven from the ground state exactly to the excited state. This is "π-pulse" inversion. A "2π-pulse" takes the atom from the ground state through the excited state and back to the ground state. For $\Delta = 0$, the Bloch vector rotation angle does not depend upon the shape of the field pulse, only on the area of the pulse.

73.3.4 Adiabatic Following

The Bloch vector picture is used for semiquantitative predictions. These are reliable even if Δ and Ω_1 are time-dependent, if the parameters change slowly (adiabatically). For example, if $\boldsymbol{\Omega}$ is moved slowly the Bloch vector follows closely. It is possible to achieve complete inversion smoothly in this way. If the field is initially tuned far below resonance so $\Delta \ll \Omega_1$ then $\boldsymbol{\Omega}$ points approximately toward the south pole of the Bloch sphere. As shown in Fig. 73.3, the Bloch vector then precesses in a very small circle about the torque vector. If the field frequency is now slowly changed (chirped) so that Δ goes from a large negative value to a large positive value then every atomic Bloch vector will continue to precess rapidly around the torque vector, and follow it as it proceeds from pointing straight down to pointing straight up. In this way the population is transferred between the two levels.

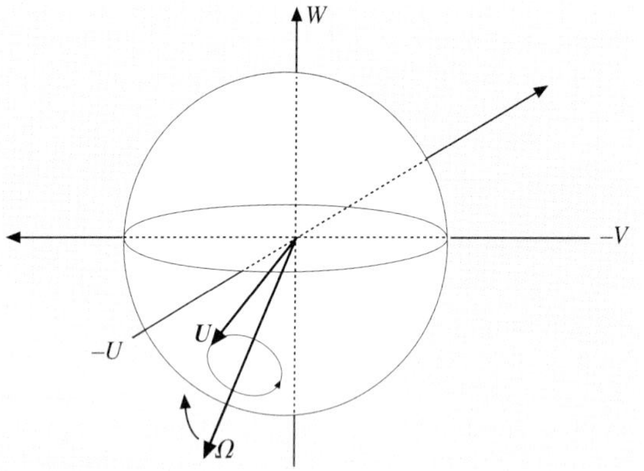

Fig. 73.3 In adiabatic inversion the Bloch vector of each atom precesses in a small cone about the torque vector as the torque vector goes from straight down to straight up

73.4 Inhomogeneous Relaxation

The fact that the various atoms in a sample may have different resonance frequencies produces a number of novel phenomena. Given a distribution $g(\Delta)$ of detunings in a dilute gas of density N, the macroscopic polarization can be written

$$\begin{aligned} P(t) &= -N\langle ex(t)\rangle \\ &= Nd\int g(\Delta)\,\mathrm{Re}\left[(U - \mathrm{i}V)\mathrm{e}^{-\mathrm{i}\omega t}\right]\mathrm{d}\Delta\,, \end{aligned} \tag{73.9}$$

where $U - \mathrm{i}V$ generally depends on both t and Δ.

73.4.1 Free Induction Decay

Free induction refers to evolution of the polarization in the absence of a laser field. For $\Omega_1 = 0$, the Bloch vector of an atom with $\Delta < 0$ precesses counterclockwise in the U–V plane. In a macroscopic sample, there are many values of Δ and about as many are positive as negative. Thus an oriented collection of Bloch vectors, all pointing in the V direction at $t = 0$, will rapidly fan out in the U–V plane due to differing precession rates, and after a short time the net V value will be zero, as will the net U value. This is free induction decay (FID) of polarization.

More precisely, if all atoms are first exposed to a θ_0 pulse, so that at $t = 0$

$$\begin{aligned} U(0,\Delta) &= 0\,, \\ V(0,\Delta) &= -\sin\theta_0\,, \\ W(0,\Delta) &= -\cos\theta_0\,, \end{aligned} \tag{73.10}$$

then if $\mathcal{E} = 0$ for $t > 0$, an individual atom with detuning Δ evolves according to (73.4):

$$U - \mathrm{i}V = \mathrm{i}\sin\theta_0\,\mathrm{e}^{-(1/T_2 - \mathrm{i}\Delta)t}\,. \tag{73.11}$$

The macroscopic polarization is found by summing the individual $(U - \mathrm{i}V)$ values over the detuning distribution $g(\Delta)$.

For simplicity, in this subsection we will ignore competition from homogeneous decay (take $1/T_2 \approx 0$)

and assume the most common inhomogeneous lineshape (i.e., Doppler-Maxwellian):

$$g(\Delta) = \frac{T^*}{\sqrt{2\pi}} e^{-(\Delta-\Delta_0)^2 T^{*2}/2}, \qquad (73.12)$$

where $1/T^*$ is here defined as the width (standard deviation) of the Doppler distribution and Δ_0 is the detuning of the zero-velocity atoms. The collective result is

$$P(t) = Nd \sin\theta_0 \sin\omega t \, e^{i\Delta_0 t} \exp\left(-\frac{t^2}{2T^{*2}}\right). \qquad (73.13)$$

The detuning "inhomogeneity" in the sample leads to dephasing of the collective dipole moment, and the inhomogeneous relaxation time is obviously T^*. This is illustrated by the decrease of collective alignment of Bloch vectors in the top row of Fig. 73.4. For a typical room temperature gas a visible transition has a width given by $1/(2\pi T^*) \approx 1.5$ GHz so that $T^* \approx 10^{-10}$ s.

73.4.2 Photon Echoes

A photon echo is generated by pulse-induced recovery of a nonzero $P(t)$ after $P(t) \to 0$ due to free induction decay (FID). This analog of the spin echo effect is possible because each atom retains its own detuning for a relatively long time, usually up to an average collision time T_2.

During FID, the U–V projection of every atom's Bloch vector precesses steadily clockwise or counterclockwise depending on the sign of its Δ. Thus the Bloch vectors could be rephased if they could all be forced at the same moment to reverse their relative sense of precession. The prototypical echo scenario has FID beginning at $t = 0$, with $P(t) \to 0$ for $t \gg T^*$, followed by a π–pulse at the time t', where $t' \gg T^*$. The effect of the π-pulse is to reverse the sign of V and W [recall (73.10)], in effect flipping the equatorial plane of the Bloch sphere upside down. Thus for $t \gg t'$ we have Bloch vectors fanning back together. The macroscopic

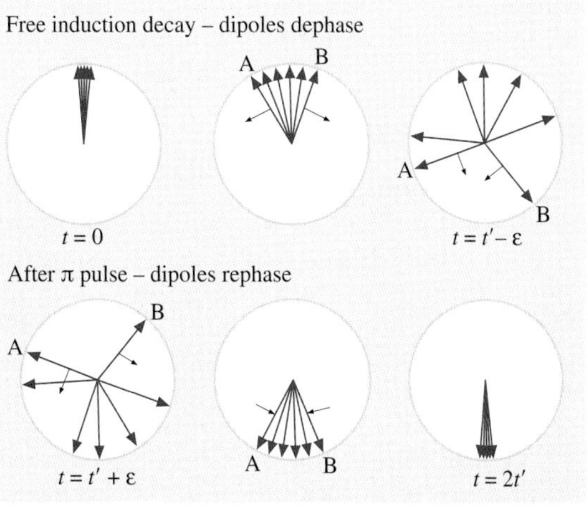

Fig. 73.4 The ensemble of dipole moments spreads due to the distribution of resonance frequencies. The distribution of Bloch vectors in the U–V plane is shown at various times after the initial short pulse excitation. By the time t', the dipoles have spread uniformly around the unit circle. A π-pulse then flips the relative orientation of the dipoles so that they subsequently rephase

polarization obeys

$$P(t) = Nd \sin\theta_0 \sin\omega t \exp\left[-\frac{(t-2t')^2}{2T^{*2}}\right] e^{-t/T_2}, \qquad (73.14)$$

where the last factor recovers the effect of homogeneous dipole damping. We require $T_2 \gg t' \gg T^*$ for a strong echo signal in the neighborhood of $t = t'$.

The result is illustrated in Fig. 73.4. An "echo" of the initial excitation at $t = 0$ appears at the time $t = 2t'$. After this, FID occurs again, and this second decay can also be reversed by applying another π–pulse, and so on, until $t \approx T_2$, at which time the inevitable and irreversible homogeneous relaxation cannot be avoided. The scenario of π–pulse reversal is only the most ideal, leading to the most complete echo, and other pulse areas will also lead to echos; a more important factor is that the reversing pulse must be short enough that negligible dephasing takes place during its application.

73.5 Resonant Pulse Propagation

73.5.1 Maxwell–Bloch Equations

Time-dependent atomic dipole moments created by applied fields are themselves a source of fields, another form of coherent transient. We limit discussion to plane-wave propagation in the z–direction. Note that we use z rather than Z for convenience, although in the dipole approximation the coordinate entering our equations is the coordinate of the center of mass of the atom rather than the internal electron coordinate. The field is generalized from (73.2) to

$$E(t,z)=\frac{1}{2}\left[\mathcal{E}(t,z)\,\mathrm{e}^{\mathrm{i}(Kz-\omega t)}+\mathrm{c.c.}\right], \tag{73.15}$$

and the macroscopic polarization is the correspondingly generalized form of (73.9):

$$P(t,z)=Nd\int g(\Delta)\,\mathrm{Re}\left[(U-\mathrm{i}V)\,\mathrm{e}^{\mathrm{i}(Kz-\omega t)}\right]\mathrm{d}\Delta\,. \tag{73.16}$$

The difference between K and $k=\omega/c$ indicates that the refractive index is nonzero. The traveling-pulse rotating frame is also obtained by replacing ωt by $\omega t-Kz$.

If the field envelope $\mathcal{E}$ is slowly varying, its second derivatives can be dropped when $E(t,z)$ is substituted into Maxwell's wave equation. The resulting dispersive and absorptive reduced wave equations are

$$\left(K^2-k^2\right)\mathcal{E}=4\pi k^2 Nd \times\int g(\Delta)U(t,z,\Delta)\,\mathrm{d}\Delta\,, \tag{73.17}$$

$$\left(K\frac{\partial}{\partial z}+k\frac{\partial}{\partial(ct)}\right)\mathcal{E}=2\pi k^2 Nd \times\int g(\Delta)V(t,z,\Delta)\,\mathrm{d}\Delta\,.$$

The Bloch equations along with these reduced wave equations form the self-consistent Maxwell–Bloch equations that are used to treat most resonant propagation problems in quantum optics and laser theory [73.1, 2, 5, 8].

73.5.2 Index of Refraction and Beers Law

If a weak pulse of duration τ propagates in a medium of ground-state atoms ($W\approx -1$), the Bloch equations have simple quasisteady-state solutions ($\tau \gg T_2$)

$$U=\frac{\Omega_1\Delta}{\Delta^2+1/T_2^2}\,,\quad V=\frac{-\Omega_1/T_2}{\Delta^2+1/T_2^2}\,. \tag{73.18}$$

When U and V are substituted back into the reduced wave equations (73.17), the dispersive equation gives the index of refraction $n=K/k$ due to the ground-state atoms:

$$n^2-1=\frac{4\pi Nd^2}{\hbar}\int\frac{\Delta g(\Delta)}{\Delta^2+1/T_2^2}\,\mathrm{d}\Delta\,, \tag{73.19}$$

and the absorptive equation predicts steady state field attenuation during propagation:

$$\frac{\partial}{\partial z}\mathcal{E}=-\frac{1}{2}\alpha_{\mathrm{B}}\mathcal{E}\,. \tag{73.20}$$

The constant α_{B} given by

$$\alpha_{\mathrm{B}}=\frac{4\pi Nd^2\omega}{\hbar cT_2}\int\frac{g(\Delta)}{\Delta^2+1/T_2^2}\,\mathrm{d}\Delta \tag{73.21}$$

is called the extinction coefficient, or the reciprocal Beers length.

Since field intensity I is proportional to $\mathcal{E}^2$, the solution to the absorptive equation is

$$I(z,t)=I(0,t)\,\mathrm{e}^{-\alpha_{\mathrm{B}}z}\,. \tag{73.22}$$

This relation is called Beers Law. Both the dispersive and absorptive results are familiar from classical physics [73.10], with the important distinction that here the $\hbar$-dependent oscillator strength enters naturally rather than as an empirical parameter from Lorentzian dielectric theory [73.8, 10].

73.5.3 The Area Theorem and Self-Induced Transparency

A form of pulse propagation with no classical analog arises in the short-pulse limit ($\tau \ll T_2, T_1$, but $\tau \gg T_2^*$). By integration over the entire pulse, the absorptive Maxwell equation becomes an equation for $\partial\theta/\partial z$, where θ is the pulse area defined in (73.8). In the short-pulse limit, the relaxation terms in the OBEs can be ignored and when substituting from them we obtain the McCall-Hahn Area Theorem [73.1]:

$$\frac{\partial}{\partial z}\theta(z)=-\frac{1}{2}\alpha_B\sin\theta(z)\,. \tag{73.23}$$

This predicts the same exponential attenuation as (73.20) in the case of a small area pulse, $\theta(z)\ll\pi$, but in the case of larger area pulses the behavior is quite different. In general, the area decreases during propagation for areas in the range $0<\theta(z)<\pi$, but it increases for areas $\pi<\theta(z)<2\pi$. As seen from Fig. 73.5, this change of area with propagation causes the pulse area to evolve to one of the stable values $0, 2\pi, 4\pi, \ldots$.

There is one special pulse, a soliton solution with area exactly 2π, which propagates without shape change in the short pulse limit, given by

$$\mathcal{E}(t,z)=\frac{2\hbar}{\tau d}\,\mathrm{sech}\left(\frac{t-z/v}{\tau}\right), \tag{73.24}$$

where τ is the pulse duration, which is arbitrary but must satisfy the short-pulse inequality $\tau \ll T_1, T_2$. The soliton's group velocity is determined by the corresponding

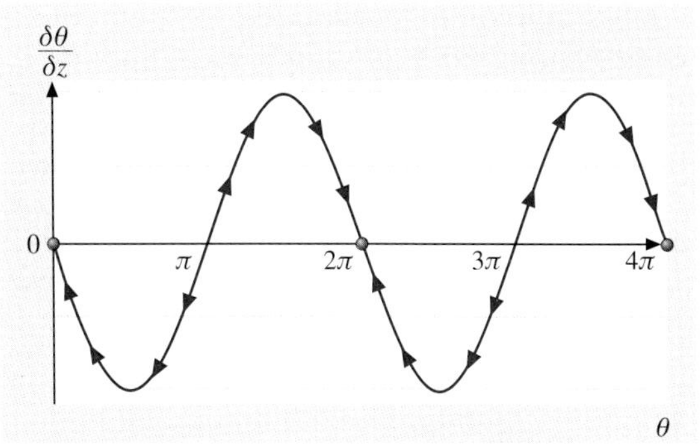

Fig. 73.5 McCall-Hahn area theorem for an absorbing medium. On propagation the area of the pulse will follow the arrows toward one of the stable values, $0, 2\pi, 4\pi, \ldots$

soliton solutions to the OBE's and the dispersive

$$v = \frac{c}{1+\frac{1}{2}\alpha_B c\tau}, \tag{73.25}$$

where α_B is to be taken in the limit $T_2^* \ll T_2$. The group velocity can be slower than the speed of light by orders of magnitude if $\alpha_B c\tau \gg 1$.

73.6 Multi-Level Generalizations

73.6.1 Rydberg Packets and Intrinsic Relaxation

A short laser pulse can populate a band of excited states whose probability amplitudes will exhibit mutual coherence. This single-atom coherence is transient, even without collisions or other external perturbations to disrupt it, and its decay can be called intrinsic relaxation. The decay is basically a dephasing. The dipole moments associated with the excited band interfere due to the wide variety of resonance frequencies of the states in the superposition. Because of the discreteness of the energy levels of any bounded quantum system, this relaxation has its own unique characteristics, including similarities with both homogeneous and inhomogeneous decay.

The wave function for a coherently excited atom can be expressed in the interaction picture in the form [73.9]

$$\Psi(r,t) = a(t)\psi_g(r) + \sum_n b_n(t)\,\mathrm{e}^{-\mathrm{i}\omega_n t}\psi_n(r), \tag{73.26}$$

where $\psi_g(r)$ is the ground state wave function, and n labels the states in a band with excitation frequencies $\omega_n \approx \omega$. If $|\omega_n - \omega| \ll \omega_n$, the transition frequency ω_n can be expanded about the principal quantum number $\overline{n}$ of the resonant excited state $E_{\overline{n}} = \hbar\omega$ to obtain

$$\begin{aligned}\omega_n &= \omega_{\overline{n}} + (n-\overline{n})\frac{\partial\omega_n}{\partial n} + \frac{1}{2!}(n-\overline{n})^2\frac{\partial^2\omega_n}{\partial n^2} + \cdots \\ &= \omega_{\overline{n}} + (n-\overline{n})\frac{2\pi}{T_K} + (n-\overline{n})^2\frac{2\pi}{T_R} + \cdots .\end{aligned} \tag{73.27}$$

Thus $2\pi/T_K$ is the mean frequency separating neighboring levels, i. e., $T_K/2\pi = \hbar\rho(E)$, where $\rho(E)$ is the density of excited states, and $2\pi/T_R$ is the mean change in this frequency separation. T_K is the same as the Kepler period a classical orbit, and T_R is the revival time.

Substituting the Bohr frequencies into the definitions for T_K and T_R yields $T_R = \overline{n}T_K/3$.

For times $t \le T_R$ the expansion can be truncated after the third term. Then the wave function is given approximately by

$$\begin{aligned}\Psi(r,t) \approx{}& a(t)\psi_g(r) + \mathrm{e}^{-\mathrm{i}\omega_{\overline{n}}t}\sum_m b_{\overline{n}+m}(t)\,\mathrm{e}^{-\mathrm{i}2\pi m t/T_K} \\ &\times \mathrm{e}^{-\mathrm{i}2\pi m^2 t/T_R}\psi_{\overline{n}+m}(r),\end{aligned} \tag{73.28}$$

where $m = n - \overline{n}$. For high Rydberg states, $\overline{n} \gg 1$, the time scales associated with the two exponentials inside the sum are quite different. For times $t \ll T_K$ the individual levels are not resolved, thus the laser excites what is effectively a continuum with a density of states $\rho(E)$. In that case the ground state population simply decays exponentially at the rate given by first-order perturbation theory, $\Gamma = (2\pi/\hbar)d^2\mathcal{E}^2\rho(E)$. At longer times $t \approx T_K$, the first exponential contributes, but the second does not, giving a simple Fourier series time dependence. In this regime the evolution of the wave function is just periodic motion of a wave packet around a Kepler orbit, as is illustrated in Fig. 73.6a.

The coherent quantum wave packet behaves like a classical particle for many Kepler periods, gradually spreading out as the second exponential in the sum (73.28) begins to contribute. This spreading of the wave packet produces the intrinsic relaxation of the collective dipole moments from the various transitions. However, because the levels are discrete, this decay is not permanent, but is reversed and leads to a spontaneous "revival" of the original wave packet [73.6], without the need for a π-pulse to produce an "echo."

In its evolution toward the revival, the wave packet passes through a number of fractional revivals in which miniature replicas of the original wave packet are equally spaced around the orbit, each traveling at the velocity of a particle traveling in a corresponding classical Kepler orbit. This complex time evolution arises from the spreading of the wave packet all the way around the orbit so that the head and tail of the packet interfere with each other, producing interference fringes. The further evolution of this fringe pattern produces the various revival phenomena shown in Fig. 73.6.

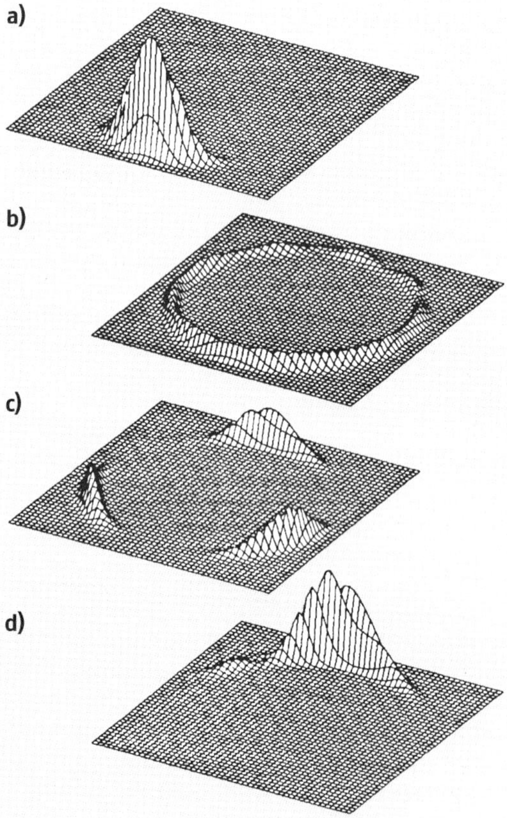

Fig. 73.6a–d The free evolution of a Rydberg wave packet made up of a superposition of circular-orbit states centered about $n = 360$. **(a)** Initially the wave packet is to good approximation a minimum uncertainty wave packet in all three dimensions, but after 12 orbits **(b)** the packet has spread all the way around the orbit so that the head and tail of the wave packet overlap, producing interference fringes. **(c)** After 40 orbits, $t = T_R/3$, the fringes have produced the one-third fractional revival in which three miniature replicas of the original wave packet are equally spaced around the orbit. **(d)** After 120 orbits, $t = T_R$, the complete wave packet revival occurs

73.6.2 Multiphoton Resonance and Two-Photon Bloch Equations

Multiphoton transitions Fig. 73.7a introduce new coherent transient phenomena. If levels $|g\rangle$ and $|e\rangle$ have the same parity, two photons from the same laser field are sufficient to excite level $|e\rangle$. For simplicity we regard $|e\rangle$ as a single state, but any number of intermediate levels $|j\rangle$ of opposite parity may be present.

Substituting the state vector

$$|\Psi(t)\rangle = a_g(t)|g\rangle + \sum_j b_j(t)\,\mathrm{e}^{-\mathrm{i}\omega t}|j\rangle + a_e(t)\,\mathrm{e}^{-\mathrm{i}2\omega t}|e\rangle\,, \tag{73.29}$$

into the Schrödinger equation yields

$$\mathrm{i}\frac{da_g}{dt} = -\frac{1}{2}\sum_j \Omega_{gj} b_j\,, \tag{73.30}$$

$$\mathrm{i}\frac{db_j}{dt} = \Delta_{jg} b_j - \frac{1}{2}\left(\Omega_{jg} a_g + \Omega_{je} a_e\right)\,, \tag{73.31}$$

$$\mathrm{i}\frac{da_e}{dt} = \Delta_{eg} a_e - \frac{1}{2}\sum_j \Omega_{ej} b_j\,, \tag{73.32}$$

where the Δs are the detunings and the Ωs are the Rabi frequencies for the dipole-allowed transitions. For example, $\Omega_{gj} = d_{gj}\mathcal{E}/\hbar$, and $\Delta_{jg} = (E_j - E_g)/\hbar - \omega$ and $\Delta_{eg} = (E_e - E_g)/\hbar - 2\omega$.

If the states $|j\rangle$ are not too close to resonance, the b_j oscillate rapidly and to a first approximation average to zero. A better approximation is to retain the small nonzero solution for b_j obtained by setting $\mathrm{d}b_j/\mathrm{d}t = 0$ in (73.31) to obtain

$$b_j = -\frac{\left[\Omega_{jg} a_g + \Omega_{je} a_e\right]}{2\Delta_{jg}}\,, \tag{73.33}$$

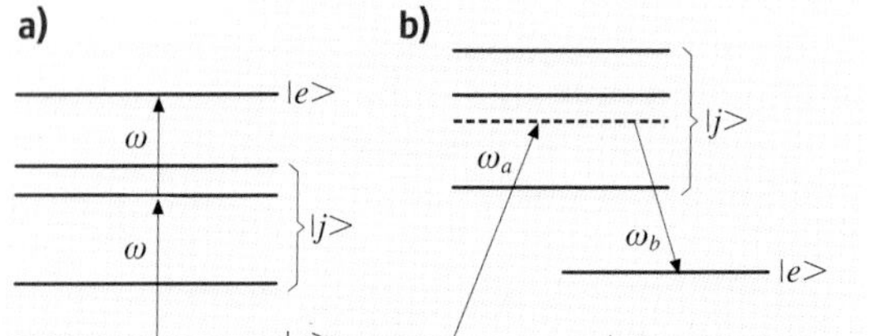

Fig. 73.7a,b Model two-photon resonances. Two photons couple the ground state $|g\rangle$ with an excited state $|e\rangle$. Many intermediate nonresonant levels $|j\rangle$ are present. In **(a)** we have the cascade system, and in **(b)** the Λ pump–probe system

which can be used to eliminate b_j from the equations for a_g and a_e. This is called adiabatic elimination of dipole coherence. In this approximation, levels $|g\rangle$ and $|e\rangle$ are directly coupled to each other and two-photon coherence arises. The coupling of levels $|g\rangle$ and $|e\rangle$ is similar to the two-level coupling described in Sect. 73.1 and two-photon Bloch equations analogous to (73.5) are the result:

$$\frac{\mathrm{d}U^{(2)}}{\mathrm{d}t} = -\Delta^{(2)} V^{(2)} ,$$
$$\frac{\mathrm{d}V^{(2)}}{\mathrm{d}t} = \Delta^{(2)} U^{(2)} + \Omega^{(2)} W^{(2)} ,$$
$$\frac{\mathrm{d}W^{(2)}}{\mathrm{d}t} = -\Omega^{(2)} V^{(2)} , \tag{73.34}$$

Here the superscript (2) indicates that the variables are identified with the two-photon $|g\rangle \to |e\rangle$ transition.

The various coefficients are similarly generalized [73.11]. For example, the two-photon Rabi frequency is given by

$$\Omega^{(2)} \equiv \frac{1}{2} \sum_j \frac{d_{gj} d_{je}}{\hbar^2 \Delta_{jg}} \mathcal{E}^2 . \tag{73.35}$$

and the two-photon detuning $\Delta^{(2)}$ incorporates the laser-induced level shifts

$$\Delta^{(2)} \equiv \Delta_{ej} + \Delta_{jg} + \frac{1}{4} \sum_j \frac{|d_{ej}|^2 \mathcal{E}^2(t)}{\hbar^2 \Delta_{ej}} + \frac{1}{4} \sum_j \frac{|d_{jg}|^2 \mathcal{E}^2(t)}{\hbar^2 \Delta_{jg}} . \tag{73.36}$$

The last two terms give the difference in the ac Stark shifts of the upper and lower levels produced by the laser field.

$W^{(2)}$ is the inversion as before, but $U^{(2)}$ and $V^{(2)}$ are somewhat different. They cannot be directly tied to the expectation value of a dipole moment because levels $|g\rangle$ and $|e\rangle$ have the same parity. Thus, while the quantity $U^{(2)} - \mathrm{i}V^{(2)}$ is the two-photon analog of $U - \mathrm{i}V$ in the original OBE's, it cannot serve as a source term in the Maxwell equation.

In the case of cw applied fields, the solutions to the two-photon OBEs are formally identical to those for a two-level atom. In the case of pulsed fields, however, the detuning $\Delta^{(2)}(t)$ is automatically "chirped" in frequency by the Stark shifts. This chirping may significantly modify the dynamics. Multiphoton generalizations of the Bloch equations can be made for other arrangements and numbers of levels. The two-photon version applies as well to three-level Λ and V configurations as to the cascade system shown in Fig. 73.7a, for which they were derived.

73.6.3 Pump–Probe Resonance and Dark States

Dark states or trapping states occur whenever a field-dependent linear combination of active levels is dynamically disconnected from the other levels. This occurs, e.g., in a pump-probe interaction, which fits the scenario of Fig. 73.7 if two lasers instead of one are used to excite level $|e\rangle$ from the ground level via two-photon resonance. A strong steady laser a is applied for the $|g\rangle \to |j\rangle$ transitions and a weak tunable "probe" laser b for the $|j\rangle \to |e\rangle$ transitions. In the simplest format, $\Delta_{eg} = 0$, and all the $|j\rangle$ levels can be combined into a single level labeled $|2\rangle$.

The three-level state vector can be written in terms of field-free states

$$|\Psi(t)\rangle = a_g|g\rangle + b_2|2\rangle + a_e|e\rangle , \tag{73.37}$$

or, in terms of field-dependent dressed states

$$|\Psi(t)\rangle = A_\mathrm{T}|T\rangle + b_2|2\rangle + A_\mathrm{S}|S\rangle , \tag{73.38}$$

where $\Omega|T\rangle \equiv \Omega_a|e\rangle - \Omega_b|g\rangle$, $\Omega|S\rangle \equiv \Omega_a|g\rangle + \Omega_b|e\rangle$ and

$$A_\mathrm{S}(t) \equiv \Omega^{-1}[\Omega_a a_g + \Omega_b a_e] ,$$
$$A_\mathrm{T}(t) \equiv \Omega^{-1}[\Omega_a a_e - \Omega_b a_g] . \tag{73.39}$$

The normalizing factor is $\Omega \equiv \sqrt{\Omega_a{}^2 + \Omega_b{}^2}$.

The state $|T\rangle$ is an eigenvector of the three-level RWA Hamiltonian, with eigenvalue zero, and the amplitude $A_\mathrm{T}(t)$ is a constant of motion. Thus $|T\rangle$ is termed a trapping state, and population in $|T\rangle$ is inaccessible to the (possibly very strong) laser fields. At two-photon resonance this conclusion is robust, not depending strongly on the idealized conditions assumed here. In fact, A_T, the trapping state amplitude, is an adiabatic invariant, remaining constant to first order even under slow changes in Ω_a and Ω_b. In a pump-probe experiment, this trapping state is observed as an abrupt drop in probe absorption as the probe frequency is tuned through two-photon resonance. Since only two-photon resonance is required (both transitions can be equally detuned) this coherent transient effect has no analog in two-level physics.

The ideal method for exciting the trapping state from the ground state uses another coherent transient process called counter-intuitive pulse sequencing in which pulse b is turned on first. The trapping state $|T\rangle$ is essentially the ground state $|g\rangle$ if $\Omega_a = 0$. Thus if Ω_b is turned on first, and then Ω_a turned on later, the ground state adiabatically becomes the trapping state and all initial probability flows smoothly with it.

An essential point is the ease with which pump-probe adiabaticity is maintained, particularly for strong fields on resonance, in contrast to one-photon adiabaticity, which is never achieved at strong-field resonance. In the pump-probe case one must only satisfy the inequality

$$\left|\frac{\mathrm{d}\Omega_a}{\mathrm{d}t}\Omega_b - \frac{\mathrm{d}\Omega_b}{\mathrm{d}t}\Omega_a\right| \ll \left(\Omega_a^2+\Omega_b^2\right)^{3/2} , \quad (73.40)$$

which is automatically accomplished by counter-intuitive pulse sequencing. The inequality allows one to tolerate rapid change of Ω_b while $\Omega_a = 0$. Then after Ω_b has reached a high value, Ω_a can also be turned on very rapidly because the right side of (73.40) is already very large. This is "counter-intuitive" excitation because if the population is in level $|g\rangle$ it is "natural" to turn on pulse Ω_a first, not Ω_b. It can be dramatically beneficial to use counter-intuitive excitation when it is important to avoid relaxation associated with level $|2\rangle$.

73.6.4 Three-Level Transparency

The foregoing results for three-level excitation can be extended to resonant pulse propagation in three-level media. The equations governing simultaneous two-pulse evolution in the local-time coordinates $cT \equiv ct - z$ and $\zeta \equiv z$ are

$$\begin{aligned}\frac{\partial \mathcal{E}_a}{\partial \zeta} &= \mathrm{i}\frac{\hbar}{d_a}\mu_a a_g^* b_2 ,\\ \frac{\partial \mathcal{E}_b}{\partial \zeta} &= \mathrm{i}\frac{\hbar}{d_b}\mu_b a_e^* b_2 ,\end{aligned} \quad (73.41)$$

where

$$\mu_a = \frac{4\pi d_a^2 N\omega_a}{\hbar c} , \quad \text{etc.} \quad (73.42)$$

Note that the bilinear combination $2a_g^* b_2$ corresponds to $U - \mathrm{i}V$ in (73.16).

Soliton-like pulses can propagate in three-level media. Both pulses must compete for interaction with level $|2\rangle$. They depend only on a single variable $Z \equiv \zeta - uT$ where u is the pulse's constant velocity in the moving frame. Soliton solutions are given (for $\mu_a = \mu_b$) for Λ media by

$$\begin{aligned}\mathcal{E}_a &= \hbar/d_a = A \operatorname{sech} KZ ,\\ \mathcal{E}_b &= \hbar/d_b = B \tanh KZ ,\end{aligned} \quad (73.43)$$

and

$$\begin{aligned}a_g &= -\tanh KZ\\ b_2 &= \frac{-2\mathrm{i}Ku}{A}\operatorname{sech} KZ ,\\ a_e &= \frac{B}{A}\operatorname{sech} KZ ,\end{aligned} \quad (73.44)$$

where the parameters A, B, Ku are nonlinearly related to the pulse length τ:

$$Ku \equiv 1/\tau \quad \text{and} \quad (2/\tau)^2 = A^2 - B^2 . \quad (73.45)$$

The moving frame velocity is given by $1/u = 2\mu/A^2$ and the expression for the lab frame velocity V is $1/V = 1/c + 2\mu/A^2$.

If $B \to 0$, then $\mathcal{E}_a \to (2\hbar/\tau d_a)\operatorname{sech} KZ$, which is the exact McCall–Hahn formula for the two-level one-pulse soliton amplitude [73.1]. No adiabatic condition was invoked in obtaining the soliton solutions. The physical measure of adiabaticity comes from the pulse duration τ. If τ is short, an appreciable population appears transiently in level $|2\rangle$, but if τ is long (an adiabatic pulse), the population skips level $|2\rangle$ and goes directly from $|g\rangle$ to $|e\rangle$ and back again during the pulse.

Note that the sech and tanh functions are ideally counter-intuitive, with pulse b starting infinitely far ahead of pulse a. In practice, the infinite leading edge of the tanh function plays no role and can be truncated to several times τ without appreciable change in the character of the pulse pair.

73.7 Disentanglement and "Sudden Death" of Coherent Transients

The existence of entanglement (non-separability) of states is the most prominent evidence that quantum mechanics is a truly nonlocal theory. This has consequences for coherent transients, allowing them to exhibit non-intuitive effects unlike any discussed up to this point. We will choose an example that directly illustrates this point by showing that two entangled atoms whose inversion and coherence decay exponentially can have an entanglement that not only does not decay exponentially but which reaches its steady state long before the atoms reach their final states. In this sense the nonlocal transients of the pair of atoms are not at all intuitively related to the local transients affecting the atoms separately.

We imagine the situation sketched in Fig. 73.8, two two-level atoms A and B of the type discussed at length already, prepared in a partially excited entangled mixed state and located remotely from each other, without direct or indirect interaction. They each must eventually, because of spontaneous emission, come to their ground states, creating the final joint state $|-_A\rangle \otimes |-_B\rangle$, which is clearly in factored form (disentangled). The question is, what is the manner of evolution by which the quantum entanglement feature evolves toward zero.

The standard Master Equation methods [73.12] for investigating spontaneous emission [73.13] can be applied to each atom separately since they are not interacting with each other. For any initial state $\rho^{AB}(0)$, the density operator at t can be expressed as

$$\rho^{AB}(t) = \sum_{\mu=1}^{4} K_\mu(t)\rho^{AB}(0)K_\mu^\dagger(t), \tag{73.46}$$

where the so-called Kraus operators [73.14] $K_\mu(t)$ are available in closed form in this case [73.15].

For illustration we will choose a partially coherent initial state, expressed by a two-atom density matrix with a single free parameter a:

$$\rho^{AB}(0) = \frac{1}{3}\begin{pmatrix} a & 0 & 0 & 0 \\ 0 & 1 & 1 & 0 \\ 0 & 1 & 1 & 0 \\ 0 & 0 & 0 & 1-a \end{pmatrix}. \tag{73.47}$$

Here the convention is to label the rows and columns in the order $++$, $+-$, $-+$, $--$. The transient decay of either atom's excitation can be calculated separately from their reduced density matrices $\rho^A \equiv \mathrm{Tr_B}\{\rho^{AB}\}$, etc. For example:

$$\begin{aligned}\rho^A &= \langle +_B|\rho^{AB}|+_B\rangle + \langle -_B|\rho^{AB}|-_B\rangle \\ &= \frac{1}{3}\begin{pmatrix} 1+a & 0 \\ 0 & 2-a \end{pmatrix},\end{aligned} \tag{73.48}$$

and the upper level excitation of atom A can be found to behave exactly as expected: $\rho^A_{++}(t) = \frac{a+1}{3}\,\mathrm{e}^{-t/\tau_0}$, where τ_0 is the usual spontaneous emission lifetime.

For nonlocal transients we will need a time-dependent measure of entanglement, and there are several options [73.16], all related to the joint entropy of the two-atom system. We will use the concurrence $C(t)$ of *Wootters* [73.17], which has the convenient normalization $1 \geq C(t) \geq 0$, where $C = 1$ represents completely entangled atoms (such as in a pure Bell state, for example) and $C = 0$ denotes the complete absence of entanglement.

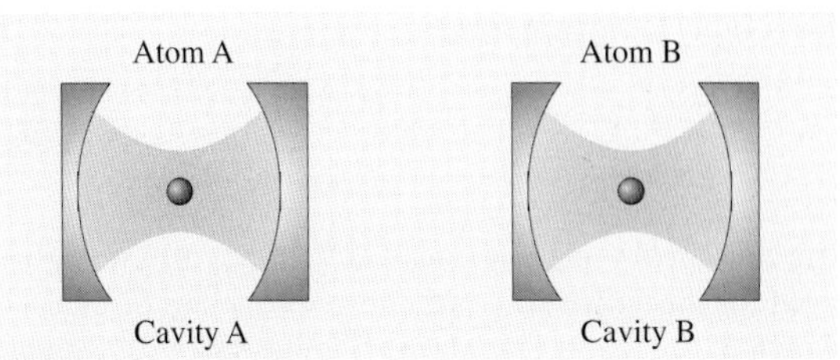

Fig. 73.8 Illustration of a set-up in which two partially excited atoms A and B are located inside two spatially separated cavities that are possibly very remote from each other. The two atoms are assumed initially entangled, but they have no interaction

For the specific case $a = 1$ one finds for the state (73.47) the initial concurrence $C(0) = \frac{2}{3}$, indicating a state with partial two-party coherence (incomplete entanglement). At time t one then finds:

$$C^{AB}(t) = \frac{2}{3}\max\left\{0,\, \mathrm{e}^{-t/\tau_0} f(t)\right\}, \tag{73.49}$$

where $f(t) = 1 - \sqrt{2\omega^2 + \omega^4}$, and $\omega \equiv 1 - \mathrm{e}^{-t/\tau_0}$. The strikingly non-intuitive consequence of this expression is that $C(t) \to 0$ abruptly after a short time (disentanglement suffers a "sudden death") if $2\omega^2 + \omega^4 \geq 1$. In fact, only a minor algebraic rearrangement shows that this strange condition must occur, and the finite sudden

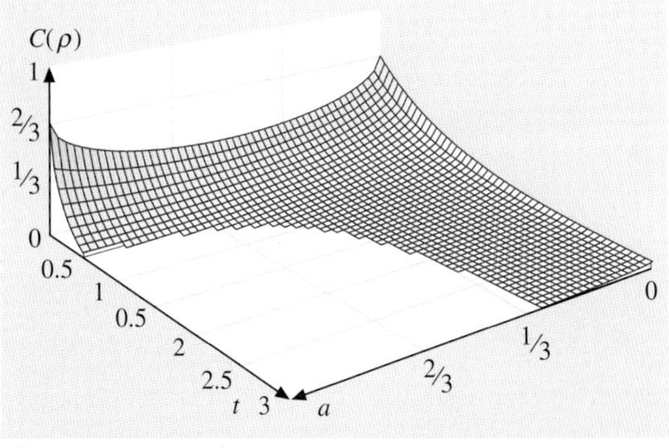

Fig. 73.9 Effect of spontaneous emission on concurrence of two two-level atoms given the initially entangled mixed state (73.47) depending on the single parameter $1 \geq a \geq 0$

death time t_0 is given by

$$\frac{t_0}{\tau_0} \equiv \ln\left(\frac{2+\sqrt{2}}{2}\right) \approx 0.53\,. \qquad (73.50)$$

The non-local coherent transient behavior of entanglement for the entire range of allowed a values [73.15] is shown in Fig. 73.9. This shows that the concurrence undergoes familiar smooth and infinitely long decay only for a values in the limited range $\frac{1}{3} > a \geq 0$. Otherwise sudden death occurs sooner or later. In our example, non-local coherence becomes zero most abruptly after a finite time for $a = 1$, the case that was calculated in (73.49). Although not yet observed experimentally, it appears that these results are not exceptional. Sudden termination of entanglement has also been predicted for two-party continuum states as well as for qubit pairs experiencing only T_2 decay, in contrast to the combined T_1 and T_2 decay appropriate to spontaneous emission as treated here.

References

73.1 L. Allen, J. H. Eberly: *Optical Resonance and Two-Level Atoms* (Dover, New York 1987)
73.2 B. W. Shore: *Theory of Coherent Atomic Excitation*, Vol. 1,2 (Wiley, New York 1990)
73.3 P. Meystre, M. Sargent III: *Elements of Quantum Optics* (Springer, Berlin, Heidelberg 1990)
73.4 C. Cohen-Tannoudji, J. Dupont-Roc, G. Grynberg: *Atom–Photon Interactions* (Wiley, New York 1992)
73.5 M. O. Scully, M. S. Zubairy: *Quantum Optics* (Cambridge Univ. Press, Cambridge 1997)
73.6 W. P. Schleich: *Quantum Optics in Phase Space* (Wiley-VCH, New York 2001)
73.7 C. C. Gerry, P. L. Knight: *Introductory Quantum Optics* (Cambridge Univ. Press, Cambridge 2005)
73.8 P. W. Milonni, J. H. Eberly: *Lasers* (Wiley, New York 1988)
73.9 M. V. Fedorov: *Atomic and Free Electrons in a Strong Light Field* (World Scientific, Singapore 1997)
73.10 J. D. Jackson: *Classical Electrodynamics*, 2nd edn. (Wiley, New York 1975)
73.11 F. H. M. Faisal: *Theory of Multiphoton Processes* (Plenum, New York 1987)
73.12 See Sect. 78.7 in this book
73.13 See Sect. 78.12 in this book
73.14 A complete discussion of Kraus operators appropriate to Bloch vector evolution is given in S. Daffer, K. Wódkiewicz, J. K. McIver: J. Mod. Optics **51**, 1843 (2004)
73.15 Ting Yu, J. H. Eberly: Phys. Rev. Lett. **93**, 140404 (2004)
73.16 See Chapt. 81 in this book
73.17 W. K. Wootters: Phys. Rev. Lett. **80**, 2245 (1998)

74. Multiphoton and Strong-Field Processes

Multiphoton

The excitation of atoms by intense laser pulses can be divided into two broad regimes: the first regime involves relatively weak optical laser fields of long duration, and the second involves strong fields of short duration. In the first case, the intensity is presumed to be high enough for multiphoton transitions to occur. The resulting spectroscopy is not limited by the single-photon selection rules for radiative transitions. However, the intensity is still low enough for a theoretical description based on perturbations of field-free atomic states to be valid, and the time dependence of the field amplitude does not play an essential role. In the second case, the field intensities are too large to be treated by perturbation theory, and the time dependence of the pulse must be taken into account. A discussion on the generation of sub-femtosecond pulses is included.

74.1 **Weak Field Multiphoton Processes** 1078
74.1.1 Perturbation Theory 1078
74.1.2 Resonant Enhanced Multiphoton Ionization 1078
74.1.3 Multi-Electron Effects 1079
74.1.4 Autoionization 1079
74.1.5 Coherence and Statistics 1079
74.1.6 Effects of Field Fluctuations 1079
74.1.7 Excitation with Multiple Laser Fields 1080

74.2 **Strong-Field Multiphoton Processes** 1080
74.2.1 Nonperturbative Multiphoton Ionization 1081
74.2.2 Tunneling Ionization 1081
74.2.3 Multiple Ionization 1081
74.2.4 Above Threshold Ionization 1081
74.2.5 High Harmonic Generation 1082
74.2.6 Stabilization of Atoms in Intense Laser Fields 1083
74.2.7 Molecules in Intense Laser Fields 1084
74.2.8 Microwave Ionization of Rydberg Atoms 1084

74.3 **Strong-Field Calculational Techniques** ... 1086
74.3.1 Floquet Theory 1086
74.3.2 Direct Integration of the TDSE 1086
74.3.3 Volkov States 1086
74.3.4 Strong Field Approximations 1087
74.3.5 Phase Space Averaging Method ... 1087

References 1088

The excitation of atoms by intense laser pulses can be divided into two broad regimes determined by the characteristics of the laser pulse relative to the atomic response. The first regime involves relatively weak optical laser fields of long duration (>1 ns), and the second involves strong fields of short duration (<10 ps). These will be referred to as the weak-long (WL) and strong-short (SS) cases respectively.

In the case of atomic excitation by WL pulses, the intensity is presumed to be high enough for multiphoton transitions to occur. The resulting spectroscopy of absorption to excited states is potentially much richer than single-photon excitation because it is not limited by the single-photon selection rules for radiative transitions. However, the intensity is still low enough for a theoretical description based on perturbations of field-free atomic states to be valid, and the time dependence of the field amplitude does not play an essential role.

The SS case is fundamentally different in that the atomic electrons are strongly driven by fields too large to be treated by perturbation theory, and the time dependence of the pulse as it switches on and off must be taken into account. Atoms may absorb hundreds of photons, leading to the emission of one or more electrons, as well as photons of both lower and higher energy. Because the flux of incident photons is high, a classical description of the laser field is adequate, but the time dependent Schrödinger equation (TDSE) must be solved directly to obtain an accurate representation of the atom–field interaction.

For SS pulses of optical wavelength, it is sufficient in most cases to consider only the electric dipole (E1) interaction term defined in Chapt. 68. The atom–field interaction can then be expressed in either the length gauge or the velocity gauge [74.1] (Chapt. 21). In the length gauge, the TDSE is

$$\mathrm{i}\hbar\frac{\partial\Psi(\boldsymbol{r},t)}{\partial t}=\left[H_0+e\boldsymbol{r}\cdot\boldsymbol{E}(t)\right]\Psi(\boldsymbol{r},t)\,, \tag{74.1}$$

where H_0 is the field-free atomic Hamiltonian, $\boldsymbol{r}$ the collective coordinate of the electrons, and $\boldsymbol{E}(t)$ the electric field of the laser given by

$$\boldsymbol{E}(t)=\mathcal{E}(t)\left[\hat{\boldsymbol{x}}\cos(\omega t+\varphi)+\epsilon\hat{\boldsymbol{y}}\sin(\omega t+\varphi)\right]. \tag{74.2}$$

Here φ is the phase and ϵ defines the polarization: linear if $\epsilon=0$ and circular if $|\epsilon|=1$. In the velocity gauge, the TDSE is

$$\mathrm{i}\hbar\frac{\partial\psi(\boldsymbol{r},t)}{\partial t}=\left[H_0-\frac{\mathrm{i}e\hbar}{mc}\boldsymbol{A}(t)\cdot\nabla+\frac{e^2}{2mc^2}A^2(t)\right]\times\psi(\boldsymbol{r},t)\,. \tag{74.3}$$

Here $\boldsymbol{A}(t)=-c\int^t\boldsymbol{E}(t')\,\mathrm{d}t'$ is the vector potential of the laser field. The solutions of (74.1) and (74.3) are related by the phase transformation

$$\Psi(\boldsymbol{r},t)=\exp\left[\frac{\mathrm{i}e}{\hbar c}\boldsymbol{r}\cdot\boldsymbol{A}(t)\right]\psi(\boldsymbol{r},t)\,. \tag{74.4}$$

Since lasers usually must be focussed to reach the strong field regime, measured electron and ion yields include contributions from a distribution of field strengths. The photoemission spectrum, on the other hand, contains a coherent component due to the macroscopic polarization of all the atoms and therefore is sensitive also to the laser phase variations within the focal volume. In this chapter methods for solving (74.1) and (74.3) are discussed along with details of the atomic emission processes.

References [74.1–3] are three recent books which provide excellent introductions to this subject. Further developments are well described in the proceedings of the International Conferences on Multiphoton Physics [74.4–7] and the NATO workshop on Super-Intense Laser-Atom Physics [74.8, 9].

74.1 Weak Field Multiphoton Processes

74.1.1 Perturbation Theory

Since atomic ionization energies are generally $\gtrsim 10\,\mathrm{eV}$, while optical photons have energies of only a few eV, several photons must be absorbed to produce ionization, or even electronic excitation in the case of the noble gases. For WL pulses, the electronic states are only weakly perturbed by the electromagnetic field. The rate of an n-photon transition can then be calculated using nth order perturbation theory for the atom–field interaction. For an incident photon number flux ϕ of frequency ω, the rate is

$$W^{(n)}_{i\to f}=2\pi\left(\frac{2\pi\alpha\phi\omega}{e^2}\right)^n\left|T^{(n)}_{i\to f}\right|^2\delta(\omega_i+n\omega-\omega_f)\,, \tag{74.5}$$

where

$$T^{(n)}_{i\to f}=\langle f|d\,G[\omega_i+(n-1)\omega]\times d\,G[\omega_i+(n-2)\omega]\cdots d\,G[\omega_i+\omega]\,d|i\rangle\,, \tag{74.6}$$

$|i\rangle$ is the ith eigenstate of the field-free atomic Hamiltonian, $d=e\hat{\boldsymbol{\epsilon}}\cdot r$, with $\hat{\boldsymbol{\epsilon}}$ the polarization direction and

$$G(\omega)=\sum_j\frac{|j\rangle\langle j|}{(\omega-\omega_j+\mathrm{i}\Gamma_j/2)}\,. \tag{74.7}$$

The sum over j includes an integration over the continuum for all sequences of E1 transitions allowed by angular momentum and parity selection rules. Methods for calculating cross sections and rates in the weak-field regime are described in [74.1, 10] and in Chapt. 24.

74.1.2 Resonant Enhanced Multiphoton Ionization

For multiphoton ionization, ω can be continuously varied because the final state in (74.5) lies in the continuum. If ω is tuned so that $\omega_i+m\omega\simeq\omega_j$ for some contributing intermediate state $|j\rangle$ in (74.7), then that state lies an integer m photons above the initial state, and the corresponding denominator vanishes (to within the level width Γ_j), producing a strongly peaked resonance. Since it takes $k=n-m$ additional photons for ionization, the process is called m,k resonant enhanced multiphoton ionization (REMPI). Measurements of the photoelectron angular distribution are useful in characterizing the resonant intermediate state.

Calculations using the semi-empirical multichannel quantum defect theory to provide the needed matrix elements have been very successful in describing experimental results. This technique is discussed in more detail in Chapt. 24.

The perturbation equation (74.5) indicates that the rate for nonresonant multiphoton ionization scales as ϕ^n for an n-photon process [74.11]. However, this is not the case for REMPI since the resonant transition saturates and (74.5) no longer applies. Then the rate can be controlled either by the m-photon resonant excitation step, or by the number of photons k needed for the ionization step.

74.1.3 Multi-Electron Effects

Multiply excited states can play a role in multiphoton excitation dynamics. These states are particularly important if their energies are below or not too far above the first ionization potential. Configuration expansions including these states have been used successfully in studies, for example, of the alkaline earth atoms, which have many low-lying doubly excited states. The presence of these states also can enhance the direct double ionization of an atom [74.12].

74.1.4 Autoionization

The configuration interaction between a bound state and an adjacent continuum leads to an absorption profile in the single photon ionization spectrum with a Fano lineshape. The actual lineshape reflects the interference between the two pathways to the continuum. Autoionizing states can also be probed via multiphoton excitation [74.13, 14]. Because, in the strong field regime, coupling strengths and phases change with intensity, the lineshapes can be strongly distorted by changing the incident intensity. At particular intensities, the phases of the excited levels can be manipulated to prevent autoionization completely. Then a trapped population with energy above the ionization limit can be created [74.15].

74.1.5 Coherence and Statistics

Real laser fields exhibit various kinds of fluctuations, and so are never perfectly coherent. The effects of such fluctuations on the complex electric field amplitude

$$E(t) = \mathcal{E}(t)\exp\left[-\mathrm{i}\omega t + \varphi(t)\right] \tag{74.8}$$

can be modeled by a variety of stochastic processes [74.16], depending on the conditions [74.10, 17–20], as follows.

For cw lasers, a phase diffusion model (PDM) is often used for which $\mathcal{E}(t) = \mathcal{E} = \text{const.}$ and

$$\dot\varphi(t) = \sqrt{2b}F(t)\,, \tag{74.9}$$

where $F(t)$ describes white noise by a real Gaussian function [74.16] characterized by the averaged values $\langle F(t)\rangle = 0$, $\langle F(t)F(t')\rangle = 2b\delta(t-t')$. The stochastic electric field then has an exponential autocorrelation function

$$\langle E(t)E^*(t')\rangle = \mathcal{E}^2\exp\left[-b|t-t'| - \mathrm{i}\omega(t-t')\right]\,, \tag{74.10}$$

and a Lorentzian spectrum of width b. Far off resonance, such a Lorentzian spectrum often gives unrealistic results, and the model (74.9) is then replaced by an Ornstein–Uhlenbeck process,

$$\ddot\varphi(t) = -\beta\dot\varphi(t) + \sqrt{2b\beta}F(t)\,, \tag{74.11}$$

where the parameter β for $\beta \ll b$ plays the role of a cut-off of the Lorentzian spectrum.

A multimode laser with a large number M of independent modes has a field of the form $E(t) = \sum_{j=1}^{M}\mathcal{E}_j \exp[-\mathrm{i}\omega_j t + \mathrm{i}\varphi_j(t)]$, and according to the central limit theorem [74.16], can be described for large M as a complex Gaussian process defined to be a chaotic field,

$$\dot E(t) = -(b+\mathrm{i}\omega)E(t) + \sqrt{2b\langle|E(t)|^2\rangle}F(t)\,, \tag{74.12}$$

where $F(t)$ is now a complex white noise, and ω is the central frequency of the field. The field, (74.12) has an exponential autocorrelation function, and a Lorentzian spectrum of width b.

Various other stochastic models have been discussed in the literature. These include Gaussian fluctuations of the real amplitude of the field $\mathcal{E}(t)$; Gaussian chaotic fields with non-Lorentzian spectra; non-Gaussian, nonlinear diffusion processes (that describe for instance a laser close to threshold [74.16]); multiplicative stochastic processes (that describe a laser with pump fluctuations [74.18]) and jump-like Markov processes [74.21–23]. Statistical properties of laser fields can sometimes be controlled experimentally to a great extent [74.19, 20].

74.1.6 Effects of Field Fluctuations

Since the response of systems undergoing multiphoton processes is in general a nonlinear function of the field intensity (and, in particular, of the field amplitude), it depends in a complex manner on the statistics of the field. The enhancement of the nonresonant multiphoton ionization rate illustrates the point. According to the perturbation equation (74.5), the rate of an n-photon process is proportional to ϕ^n; i. e., to I^n, where I is the field intensity. For fluctuating fields, the average response is thus

$$W_{i\to f}^{(n)} \propto \left\langle I^n \right\rangle \propto \left\langle |E(t)|^{2n} \right\rangle . \qquad (74.13)$$

Phase fluctuations (as described by PDM) do not affect the average. On the other hand, for complex chaotic fields, the average is

$$\langle I^n \rangle \simeq n! \langle I \rangle^n , \qquad (74.14)$$

i. e., significant enhancement of the rate for $n > 1$.

Field fluctuations lead to more complex effects in resonant processes. Two well-studied examples are the enhancement of the ac Stark shift in resonant multiphoton ionization [74.24], and the spectrum of double optical resonance – a process in which the ac Stark splitting of the resonant line is probed by a slightly detuned fluctuating laser field [74.18]. Double optical resonance is very sensitive not only to the bandwidth of the probing field, but also to the shape of its frequency spectrum.

74.1.7 Excitation with Multiple Laser Fields

The simultaneous application of more than one laser field produces interesting and novel effects. If a laser and its second (2ω) or third (3ω) harmonic are combined and the relative phase between the fields controlled, product state distributions and yields can be altered dramatically. The effects include reducing the excitation or ionization rates in the $\omega - 3\omega$ case [74.25] or altering the photoelectron angular distributions and the harmonic emission parity selection rules using $\omega - 2\omega$ [74.26].

A laser field can dress or strongly mix the field-free excited states, including the continuum, of an atom. This can produce a number of effects depending on how the dressed system is probed. By coupling a bound, excited state with the continuum, ionization strengths and dynamics are altered, resulting in new resonance-like structures where none existed before. This effect is called laser-induced continuum structure, or LICS [74.15,27]. This general idea has been exploited to design schemes for lasers without inversion [74.28] in which the dressed atom can have an inverted population, allowing gain even though in terms of the undressed states the lower level has the largest population. A laser can produce dramatic changes in the index of refraction of an atomic medium [74.29], creating, at specific frequencies, laser-induced transparency for a second, probe laser field.Multistep ionization, where each step is driven by a laser at its resonant frequency, has resulted in two useful applications. These are: efficient atomic isotope separation [74.23]; and the detection of small numbers of atoms in a sample, called single-atom detection. This technique is extremely sensitive because the use of exact resonance for each step yields very large cross sections for ionization, and the efficiency of collecting ions is high [74.30].

74.2 Strong-Field Multiphoton Processes

Recently developed laser systems can produce very short pulses, some as short as a few to tens of femtoseconds, while at the same time maintaining the pulse energy so that the peak power becomes very high. Focused intensities up to $10^{19}\,\mathrm{W/cm^2}$ have been achieved. Because the pulses are short, atoms survive to much higher intensities before ionizing, making possible studies of laser-atom interactions in an entirely new regime. A discussion of the status of short pulse laser development is given in Chapt. 71 and in [74.31].

With increasing intensity, higher-order corrections to (74.6) contribute to the transition rate. The next order correction comes from transitions involving two additional photons, one absorbed and one emitted, leading to the same final state. One effect of these terms is to shift the energies of the excited states in response to the oscillating field. This is called the dynamic or ac Stark shift. The ac Stark shift of the ground state tends to be small because of the large detuning from the excited states for long wavelength photons. On the other hand, in strong fields the shift of the higher states and the continuum can become appreciable. Electrons in highly excited states respond to the oscillating field in the same manner as a free electron. Their energies shift with the continuum by the amount

$$U_\mathrm{p} = \frac{(1+\epsilon)e^2\mathcal{E}^2}{4m\omega^2} , \qquad (74.15)$$

where U_p is the cycle-averaged kinetic energy of a free electron in the field and ϵ defines the polarization of the field in (74.2). U_p is called the ponderomotive or quiver energy of the electron. For strong laser fields, U_p can be several eV or more, meaning that during a pulse, many states shift through resonance as their energies change by an amount larger than the incident photon energy. The resulting intensity-induced resonances can dominate the ionization dynamics.

Electrons promoted into the continuum acquire the ponderomotive energy, oscillating in phase with the field. In a linearly polarized field, the amplitude of the quiver motion of a free electron, given by $e\mathcal{E}/4m\omega^2$, can become many times larger than the bound state orbitals. If the initial velocity of an electron is small after ionization, it can be accelerated by the field back into the ion core. The subsequent rescattering changes the photoelectron energy and angular distributions, and allows the emission of high energy photons [74.32, 33]. This simple dynamical picture forms the basis of the current understanding of many strong-field multiphoton processes.

74.2.1 Nonperturbative Multiphoton Ionization

The breakdown of perturbation theory for nth-order multiphoton processes occurs when the higher-order correction terms become comparable to the nth-order term. Assuming that the dipole strength is $\propto ea_0$, where a_0 is the Bohr radius, and the detuning is $\delta \propto \omega$, the ratio of an $(n+2)$-order contribution to the nth-order term from (74.6) is [74.1]

$$R_{n+2,n} \simeq \left(\frac{2\pi\alpha\phi\omega}{e^2}\right)\left(\frac{ea_0}{\omega}\right)^2 = \left(\frac{I}{I_\gamma}\right)\left(\frac{\omega_a}{2\omega}\right)^2 , \tag{74.16}$$

where $\hbar\omega_a \simeq 27.2114\,\text{eV}$ is the atomic unit of energy e^2/a_0, and $I_\gamma \simeq 3.509\,45\times10^{16}\,\text{W/cm}^2$ is the intensity corresponding to an atomic unit of field strength, given by $E_a = \alpha c\left(m/a_0^3\right)^{1/2} \simeq 5.1422\times10^9\,\text{V/cm}$. The atomic unit of intensity itself is defined by

$$I_a = \phi_a\hbar\omega_a = \alpha c E_a^2 , \tag{74.17}$$

which is $6.436\,414\,(4)\times10^{15}\,\text{W/cm}^2$. Thus $I_\gamma = I_a/(8\pi\alpha)$. For photon energies of 1 eV, $R_{n+2,n}$ becomes unity for $I \sim 10^{14}\,\text{W/cm}^2$. Because of the large number of $(n+2)$-order terms, perturbation theory actually fails for $I > 10^{13}\,\text{W/cm}^2$. Above this critical intensity, nonresonant n-photon ionization ceases to scale with the ϕ^n dependence predicted by perturbation theory.

74.2.2 Tunneling Ionization

At sufficiently high intensity and low frequency, a tunneling mechanism changes the character of the ionization process. For lasers in the ir or optical range, a strongly bound electron can respond to the instantaneous laser field since the oscillating electric field varies slowly on the time scale of the electron. The Coulomb attraction of the ion core combines with the laser electric field to form an oscillating barrier through which the electron can escape by tunneling, if the amplitude of the laser field is large enough. The dc rate for this process is $e\mathcal{E}/\sqrt{2mI_P}$, where I_P is the ionization potential of the electron. When this rate is comparable to the laser frequency, tunneling becomes the most probable ionization mechanism [74.34–36]. The ratio of the incident laser frequency to the tunneling rate is called the Keldysh parameter, and is given by

$$\gamma = \sqrt{I_P/2U_p} , \tag{74.18}$$

which is less than unity when tunneling dominates and larger than unity when multiphoton ionization dominates.

74.2.3 Multiple Ionization

Excitation and ionization dynamics are dominated by single electron transitions in the strong field regime. Although atoms can lose several electrons during a single pulse, the electrons are released sequentially. There is no convincing evidence of significant collective excitation in atoms in strong fields, even though it has been extensively sought. Once one electron is excited in an atom, the remaining electrons have much higher binding energies. As a result, the laser field is unable to affect them significantly until it reaches much higher intensity. By that time the first electron has been emitted.

Simultaneous ejection of two electrons occurs as a minor channel ($< 1\%$) in strong field multiple ionization. Although it is possible that doubly excited states of atoms could assist in the double ionization, in the helium and neon cases studied, these states are unlikely to be contributors [74.37].

74.2.4 Above Threshold Ionization

In strong optical and ir laser fields, electrons can gain more than the minimum amount of energy required for ionization. Rather than forming a single peak, the emitted electron energy spectrum contains a series of peaks separated by the photon energy. This is called above threshold ionization, or ATI [74.38–40]. The peaks appear at the energies

$$E_s = (n+s)\hbar\omega - I_P , \tag{74.19}$$

where n is the minimum number of photons needed to exceed I_P, and $s = 0, 1, \ldots$ is called the number of excess photons or above threshold photons carried by the electron. Calculations in the perturbative regime for ATI are given for hydrogen in [74.11].

Peak Shifting

As the intensity approaches the nonperturbative regime, the ac Stark shift of the atomic states begins to play a significant role in the structure of the ATI spectrum. The first effect is a shift of the ionization potential, given roughly by the ponderomotive energy U_p. Additional photons may then be required in order to free the electron from the atom; i. e., enough to exceed $I_P + U_p$. If the emitted electron escapes from the focal volume while the laser is still on, it is accelerated by the gradient of the field. The quiver motion is converted into radial motion, increasing the kinetic energy by U_p and exactly canceling the shift of the continuum. The electron energies are still given by (74.19). However, when U_p exceeds the photon energy, the lowest ATI peaks disappear from the spectrum. In this long pulse limit, no electron is observed with energy less than U_p. This is called peak shifting in that the dominant peak in the ATI spectrum moves to higher order as the intensity increases.

ATI Resonance Substructure

If the laser pulse is short enough (< 1 ps for the typical laser focus), the field turns off before the electron can escape from the focal volume. Then the quiver energy is returned to the field and the ATI spectrum becomes much more complicated. The observed electron energy corresponds to the energy

$$E_s(\text{shortpulse}) = (n+s)\hbar\omega - (I_P + U_p)\,. \qquad (74.20)$$

relative to the shifted ionization potential. Electrons from different regions of the focal volume are thus emitted with different ponderomotive shifts, introducing substructure in the spectrum which can be directly associated with ac Stark-shifted resonances [74.38, 39].

ATI in Circular Polarization

The above discussion is appropriate for the case of linear polarization where the excited states of the atom can play a significant role in the excitation. In a circularly polarized field, the orbital angular momentum L must increase one unit with each photon absorbed so that multiphoton ionization is allowed only to states which have high L, and hence a large centrifugal barrier. The lower energy scattering states then cannot penetrate into the vicinity of the initial state. Thus the ATI spectrum in circular polarization peaks at high energy and is very small near threshold.

74.2.5 High Harmonic Generation

High-order harmonic generation (HG) in noble gases is a rapidly developing field of laser physics [74.41, 42]. When an SS pulse interacts with an atomic gas, the atoms respond in a nonlinear way, emitting coherent radiation at frequencies that are integral multiples of the laser frequency. Due to the inversion symmetry of the atom, only odd harmonics of the fundamental are emitted. In the high intensity ($> 10^{13}\,\text{W/cm}^2$), low frequency regime, the harmonic strengths fall off for the first few orders, followed by a broad plateau of nearly constant conversion efficiency, and then a rather sharp cut-off [74.41, 42]. The plateaux extend to well beyond the hundredth order of the 800–1000 nm incident wavelengths, using the light noble gases as the active medium. There has also been experimental evidence of HG from ions. Harmonic generation provides a source of very bright, short-pulse, coherent XUV radiation which can have several advantages over the other known sources, such as the synchrotron.

Plateau and Cut-off

A recently developed two-step model [74.32, 33], which combines quantum and classical aspects of laser-atom physics, accounts for many strong field phenomena. In this model, the electron first tunnels [74.43] from the ground state of the atom through the barrier formed by the Coulomb potential and the laser field. Its subsequent motion can be treated classically, and primarily consists of oscillatory motion in phase with the laser field. If the electron returns to the vicinity of the nucleus with kinetic energy $\mathcal{T}$, it may recombine into the ground state with the emission of a photon of energy $(2n+1)\hbar\omega \le \mathcal{T} + I_P$, where n is an integer. The maximum kinetic energy of the returning electron turns out to be $\mathcal{T} \simeq 3.2\,U_p$, resulting in a cut-off in the harmonic spectrum at the harmonic of order

$$N_{\max} \simeq (I_P + 3.2U_p)/\hbar\omega\,. \qquad (74.21)$$

Theoretical Methods

Calculation of harmonic strengths requires the evaluation of the time-dependent dipole moment of the atom,

$$\boldsymbol{d}(t) = \langle \Psi(t)|e\boldsymbol{r}|\Psi(t)\rangle\,. \qquad (74.22)$$

The strength of harmonics emitted by a single atom are then related to Fourier components of $\boldsymbol{d}(t)$, or more precisely, its second time derivative, $\ddot{\boldsymbol{d}}(t)$.

The induced dipole moment $\boldsymbol{d}(t)$ can be directly evaluated from the numerical [74.44] or *Floquet* [74.26] solutions of the TDSE. Good agreement with numerical and experimental data is also obtained using a strong field approximation discussed below and a Landau–Dyhne formula. This approach can be considered to be a quantum mechanical implementation of the two-step model [74.45].

Propagation and Phase Matching Effects

A single atom response is not sufficient to determine the macroscopic response of the atomic medium. Because different atoms interact with different parts of the focused laser beam, they feel different peak field intensities and phases (which actually undergo a rapid π shift close to the focus). The total harmonic signal results from coherently adding contributions from single atoms, accounting for propagation and interference effects. The latter effects can wipe out the signal completely a constructive phase matching takes place.

The propagation and phase matching effects in the strong field regime [74.46] can be studied by solving the Maxwell's equations for a given harmonic component of the electric field $\mathcal{E}_M(\boldsymbol{r})$ (68.23),

$$\nabla^2 \mathcal{E}_{\mathrm{M}}(\boldsymbol{r}) + n_{\mathrm{M}}(\boldsymbol{r})\mathcal{E}_{\mathrm{M}}(\boldsymbol{r}) = -\frac{1}{\epsilon_0}\left(\frac{\mathrm{M}\omega}{c}\right)^2 \mathcal{P}_{\mathrm{M}}(\boldsymbol{r}) , \tag{74.23}$$

where $n_{\mathrm{M}}(\boldsymbol{r})$ is the refractive index of the medium (which depends on atomic, electronic and ionic dipole polarizabilities), while $\mathcal{P}_{\mathrm{M}}(\boldsymbol{r})$ is the polarization induced by the fundamental field only. It can be expressed as

$$\mathcal{P}_{\mathrm{M}}(\boldsymbol{r}) \propto N(\boldsymbol{r}) d_{\mathrm{M}} M(\boldsymbol{r}) \exp\left[-i\mathrm{M}\Delta(\boldsymbol{r})\right] , \tag{74.24}$$

where $N(\boldsymbol{r})$ is the atomic density, $d_{\mathrm{M}}(\boldsymbol{r})$ is the Mth Fourier component of the induced dipole moment, and $\Delta(\boldsymbol{r})$ is a phase shift coming from the phase dependence of the fundamental beam due to focusing. All of these quantities may have a slow time dependence, reflecting the temporal envelope of the laser pulse.

Phase matching is most efficient in the forward direction. In general, the strength and spatial properties of an harmonic depend in a very complex way on the focal parameters, the medium length and the coherence length of a given harmonic. Propagation and phase matching effects can lead to a shift of the location of the cut-off in the harmonic spectrum [74.47].

Harmonic Generation by Elliptically Polarized Fields

The two-step model implies that for harmonic emission it is necessary that the tunneling electrons return to the nucleus and recombine into their initial state. According to classical mechanics, there are many trajectories in a linearly polarized field that involve one or more returns to the origin. However, there are practically no such trajectories in elliptically polarized fields. As a result, the two-step model predicts a strong decrease of the harmonic strengths as a function of the laser ellipticity. This prediction has been confirmed experimentally [74.48].

The Generation of Sub-Femtosecond XUV Pulses

Manipulation of generated harmonics by allowing the temporal beating of superposed high-order harmonics can produce a train of very short intensity spikes, on the order of ~ 100 attoseconds and shorter, where $1\,\mathrm{as} = 10^{-18}\,\mathrm{s}$ [74.49]. The structural characteristics of the generated pulse-trains depend on the relative phases of the harmonics combined. Employing driving pulses that were themselves only a few femtoseconds long, experimental groups in Vienna [74.50] and Paris [74.51] reported the first observations and measurements of such sub-femtosecond UV/XUV light pulse-trains. The scientific importance of breaking the femtosecond barrier is obvious: the time-scale necessary for probing the motion of an electron in a typical bound, valence state is measured in attoseconds (atomic unit of time $\equiv 24$ as). Attosecond pulses will allow the study of the time-dependent dynamics of correlated electron systems by freezing the electronic motion, in essence exploring the structure with ultra-fast snapshots. A crucial aspect for all attosecond pulse generation is the control of spectral phases. Measurements of the timing of the attosecond peaks relative to the absolute phase of the ir driving field have been accomplished [74.52]. This provides insight into the recollision, harmonic generation process. Also, the control of the group velocity phase relative to the envelope of the few cycle driving pulses allows the production of reproducible pulse-trains [74.53]. Thus, the highly non-perturbative, nonlinear multiphoton interactions of very short ir or visible light pulses with atoms or molecules is becoming a novel, powerful, and unique source for studies of very rapid quantum-electronic processes.

74.2.6 Stabilization of Atoms in Intense Laser Fields

It has been argued [74.54] that in very intense laser fields of high frequency, atoms undergo dynamical stabilization and do not ionize. The stabilization effect can be explained by gauge transforming the TDSE (74.1) to the Kramers–Henneberger (K–H) frame; i. e., a noninertial oscillating frame which follows the motion of the free electron in the laser field. The K–H transformation consists of replacing $\boldsymbol{r} \to \boldsymbol{r} + \boldsymbol{\alpha}(t)$, where, for the linearly polarized monochromatic laser field, $\boldsymbol{\alpha}(t) = \hat{\boldsymbol{x}}\alpha_0 \cos(\omega t - \varphi)$; $\alpha_0 = e\mathcal{E}/(m\omega^2)$ is the excursion amplitude of a free electron, and $\hat{\boldsymbol{x}}$ is the polarization direction. The TDSE in the K–H frame is

$$\mathrm{i}\hbar \frac{\partial \Psi(\boldsymbol{r}, t)}{\partial t} = \left\{ -\frac{\hbar^2 \nabla^2}{2m} + V\big[\boldsymbol{r} + \boldsymbol{\alpha}(t)\big] \right\} \Psi(\boldsymbol{r}, t) , \tag{74.25}$$

i. e., it describes the motion of the electron in an oscillatory potential. In the high frequency limit, this potential may be replaced by its time average

$$V_{\text{K–H}}(\boldsymbol{r}) = \frac{\omega}{2\pi} \int_0^{2\pi/\omega} \mathrm{d}t\, V\big[\boldsymbol{r} + \boldsymbol{\alpha}(t)\big] , \tag{74.26}$$

and the remaining Fourier components of $V[\boldsymbol{r} + \boldsymbol{\alpha}(t)]$ treated as a perturbation. When α_0 is large, the effective potential (74.26) has two minima close to $\boldsymbol{r} = \pm\hat{\boldsymbol{x}}\alpha_0$. The corresponding wave functions of the bound states are centered near these minima, thus exhibiting a dichotomy. The ionization rates from the K–H bound states are induced by the higher Fourier components of $V(\boldsymbol{r} + \boldsymbol{\alpha}(t))$. For large enough α_0, the rates decrease if either the laser intensity increases or the frequency decreases.

Numerical solutions of the TDSE [74.55, 56] show that stabilization indeed occurs for laser field strengths and frequencies of the order of one atomic unit. More importantly, stabilization is possible even when the laser excitation is not monochromatic, but rather, is produced by a short laser pulse. Physically, free electrons in a monochromatic laser field cannot absorb photons due to the constraints imposed by energy and momentum conservation. Absorption is possible only in the vicinity of a potential, such as the Coulombic attraction of the nucleus. In the case of strong excitation, i. e., when α_0 is much larger than the Bohr radius, the electron spends most of the time very far from the nucleus, and therefore does not absorb energy from the laser beam. Thus stabilization, as viewed from the K–H frame, has a classical analog. Other mechanisms of stabilization based on the quantum mechanical effects of destructive interference between various ionization paths have also been proposed [74.57].

Due to classical scaling (Sect. 74.2.8) stabilization is predicted to occur for much lower laser frequencies if the atoms are initially prepared in highly excited states. If additionally, the initial state has a large orbital angular momentum corresponding to classical trajectories that do not approach the nucleus, stabilization is even more easily accomplished. The stabilization of a 5g Rydberg state of neon has recently been reported [74.58].

74.2.7 Molecules in Intense Laser Fields

Molecular systems are more complex than atoms because of the additional degrees of freedom resulting from nuclear motion. Even in the presence of a laser field, the electron and nuclear degrees of freedom can be separated by the Born–Oppenheimer approximation, and the dynamics of the system can be described in terms of motions on potential energy surfaces. In strong fields, the Born–Oppenheimer states become dressed, or mixed by the field, creating new molecular potentials. Because of avoided crossings between the dressed molecular states, the field induces new potential wells in which the molecules become trapped. These states, known as laser-induced bound states, are stable against dissociation, but exist only while the laser field is present [74.59]. Their existence affects the spectra of photoelectrons, photons, and the fragmentation dynamics. If the field is strong enough, many electrons can be ejected from a molecule before dissociation, producing highly charged, energetic fragments [74.60]. Such experiments are similar to beam-foil Coulomb explosion studies of molecular structure. However, because of changes from the field-free equilibrium geometries in laser dissociation, the energies of the fragments lie systematically below the corresponding values from Coulomb explosion studies.

74.2.8 Microwave Ionization of Rydberg Atoms

Similar phenomena appear in the ionization of highly excited hydrogen-like (Rydberg) atoms by microwave fields [74.61, 62], but the dynamical range of the parameters involved is different from the case of tightly bound electrons. Recent developments have greatly extended techniques for the preparation and detection of Rydberg states. Since, according to the equivalence principle, highly excited Rydberg states exhibit many classical properties, a classical perspective of ionization yields useful insights (Sect. 74.3.5).

Classical Scaling

The classical equations of motion for an electron in both a Coulomb field and a monochromatic laser field polarized along the z-axis are invariant with respect to the following scaling transformations:

$$\begin{aligned} \boldsymbol{p} &\propto n_0^{-1}\tilde{\boldsymbol{p}}\,, & \boldsymbol{r} &\propto n_0^2\tilde{\boldsymbol{r}}\,,\\ t &\propto n_0^3\tilde{t}\,, & \varphi &\propto \tilde{\varphi}\,,\\ \omega &\propto n_0^{-3}\tilde{\omega}\,, & \mathcal{E} &\propto n_0^{-4}\tilde{\mathcal{E}}\,. \end{aligned} \tag{74.27}$$

In the scaled units, the Hamiltonian $\tilde{H} = n_0^{-2} H$ of the system becomes

$$\tilde{H} = \frac{\tilde{p}^2}{2m} - \frac{1}{\tilde{r}} + \tilde{z}e\tilde{\mathcal{E}}\cos\left(\tilde{\omega}\tilde{t} + \tilde{\varphi}\right), \tag{74.28}$$

i. e., it depends only on $\tilde{\omega}$ and $\tilde{\mathcal{E}}$. In experiments, the principal quantum number n_0 of the prepared initial state typically ranges from 1 to 100.

Classical scaling extends to the fields of other polarization and to pulsed excitation, provided that the number of cycles in the rise, top and fall of the pulse is kept fixed. This scaling does not hold for a quantum Hamiltonian, unless one also rescales Planck's constant, $\tilde{\hbar} = \hbar/n_0$. In practice, increasing n_0, keeping $\tilde{\mathcal{E}}$ and $\tilde{\omega}$ constant, corresponds to a decrease in the effective $\hbar$ toward the classical limit. In view of this classical scaling, experimental and theoretical results are usually analyzed in terms of the scaled variables. Since the classical dynamics generated by the Hamiltonian (74.28) exhibits chaotic behavior in some regimes, the dynamics of the corresponding quantum system is frequently referred to as an example of quantum chaos [74.63–65].

Regimes of Response

By varying the initial n_0, several regimes of the scaled parameters can be covered. The experimentally measured response of Rydberg atoms in microwave fields can be divided into six categories:

The Tunneling Regime. For $\tilde{\omega} \le 0.07$, the response of the system is accurately represented as tunneling through the slowly oscillating potential barrier composed of the Coulomb and microwave potentials.

The Low Frequency Regime. For $0.05 \le \tilde{\omega} \le 0.3$, the ionization probability exhibits structure (bumps, steps, changes of slope) as a function of the field strength. The quantum probability curves might be lower or higher than the corresponding classical counterparts calculated with the aid of the phase averaging method (Sect. 74.3.5).

The Semiclassical Regime. For $0.1 \le \tilde{\omega} \le 1.2$, the ionization probabilities agree well for mostfrequencies with the results obtained from the classical theory. In particular, the onset of ionization and appearance intensities (i. e., the intensities at which a given degree of ionization is achieved) coincide with the onset of chaos in the classical dynamics. Resonances in the ionization probabilities appear that correspond to the classical trapping resonances [74.63–66].

The Transition Region. For $1 \le \tilde{\omega} \le 2$, the differences between the quantum and classical results are visible. Quantum ionization probabilities are frequently lower and appearance intensities higher than their classical counterparts.

The High Frequency Regime. For $\tilde{\omega} \ge 2$, quantum results for ionization probabilities are systematically lower and appearance intensities higher than their classical counterparts. This apparent stability of the quantum system has been attributed to three kinds of effects: quantum localization [74.66], quantum scars [74.67], and perhaps to the stabilization of atoms in intense laser fields (Sect. 74.2.6).

The Photoeffect Regime. When the scaled frequency becomes greater than the single photon ionization threshold, the system undergoes single photon ionization (the photoeffect).

Quantum Localization

The classical dynamics changes as the field increases. Chaotic trajectories start to fill phase space and, as the KAM tori (describing periodic orbits) [74.63–65] break down, the motion becomes stochastic, resembling a random walk. This process, in which the mean energy grows linearly in time, is termed diffusive ionization. In the quantum theory, diffusion corresponds to a random walk over a ladder of suitably chosen quantum levels. However, both diagonal and off-diagonal elements of the evolution operator, which describe quantum mechanical amplitudes for transitions between the levels, depend in a quasiperiodic manner on the quantum numbers of the levels involved. Such quasiperiodic behavior is quite analogous to a random one. Electronic wave packets that initially spread in accordance with the classical laws tend to remain localized for longer times due to destructive quantum interference effects. Quantum localization is an analog of the Anderson localization of electronic wave functions propagating in random media [74.66].

Quantum Scars

Even in the fully chaotic regime, classical phase space contains periodic, though unstable, orbits. Nevertheless, quantum mechanical wave function amplitudes can become localized around these unstable orbits, resulting in what are called quantum scars. The increased stability of the hydrogen atom at $\tilde{\omega} \simeq 1.3$ has been in fact attributed [74.67] to the effects of quantum scars. These effects are very sensitive to fluctuations in the driving laser field. Control of the laser noise therefore provides a powerful spectroscopic tool to study such quantal phenomena [74.68]. Using this tool, it has recently become possible to demonstrate the effects of quantum scars in the intermediate regime of the scaled frequencies (less than but close to 1).

74.3 Strong-Field Calculational Techniques

The SS pulse regime requires a nonperturbative solution of the TDSE. Two approaches have been developed: the explicit numerical solution of the TDSE and the Floquet expansion technique. In addition to these, several approximate methods have been proposed.

74.3.1 Floquet Theory

The excitation and ionization dynamics of an atom in a strong laser field can be determined by turning the problem into a time-independent eigenvalue problem [74.26,69]. From Floquet's theorem, the eigenfunctions for a perfectly periodic Hamiltonian of the form

$$H = H_0 + \sum_{N \neq 0} H_N \,\mathrm{e}^{-\mathrm{i}N\omega t} \tag{74.29}$$

can be expressed in the form

$$\Psi(t) = \mathrm{e}^{-\mathrm{i}Xt/\hbar} \sum_{N} \mathrm{e}^{-\mathrm{i}N\omega t} \psi_N \,. \tag{74.30}$$

Putting this into the time-dependent Schrödinger equation results in an infinite set of coupled Floquet equations for the harmonic components ψ_N. In the velocity gauge, the Floquet equations are

$$(X + N\hbar\omega - H_0)\psi_N = V_+\psi_{N-1} + V_-\psi_{N+1} \,, \tag{74.31}$$

where, for a vector potential of amplitude $\mathcal{A}$,

$$V_+ = -\frac{e}{2mc}\mathcal{A} \cdot \boldsymbol{p} \,, \tag{74.32}$$

and $V_- = V_+^\dagger$. The equations (74.31) have been solved, after truncation to a manageable number of terms, using many techniques to provide what are called the quasi-energy states of the laser-atom system. The eigenvalues X of these equations are complex, with $\mathrm{Im}(X)$ giving the decay or ionization rate for the system. The generated rates are found to be very accurate as long as the pulse length of the laser field is not too short, at least hundreds of cycles. The eigenfunctions provide the amplitudes for the photoelectron energy spectra, and the time-dependent dipole of the state can be related to the photo-emission spectrum of the system. Yields for slowly varying pulses can be constructed by combining the results from the individual, fixed-intensity calculations [74.26].

The Floquet method can be applied for any periodic Hamiltonian. In strong enough fields of high frequency, the Floquet equations can be truncated to a very small set in the K–H frame [74.54].

74.3.2 Direct Integration of the TDSE

Methods for the direct solution of the time-dependent Schrödinger equation are described in general in Chapt. 8 and in [74.70,71] for multiphoton processes. The wave functions are defined on spatial grids or in terms of an expansion in basis functions. The time evolution is obtained by either explicit or implicit time propagators. All these methods are capable of generating numerically exact results for an atom with a single electron in a short pulsed field for a wide range of pulse shapes, wavelengths and intensities. The solutions are time-dependent wave functions for the electrons which can be analyzed to obtain excitation and ionization rates, photoelectron energies, angular distributions, and photoemission yields. The ability to generate an explicit solution of the TDSE allows the study of arbitrary pulse shapes and provides insight into the excitation dynamics.

For multi-electron atoms, one generally has to limit the calculations to that for a single electron in effective potentials which represent, as well as possible, the influences of the remaining atomic electrons. This approach is called the single active electron approximation, and it gives generally accurate results for systems with no low-lying doubly excited states, for example, the noble gases [74.70]. In these cases, the excitation dynamics are dominated by the sequential promotion of a single electron at a time.

74.3.3 Volkov States

A laser interacting with free a electron superimposes an oscillatory motion on its drift motion in response to the field. The wave function for an electron with drift velocity $\boldsymbol{v} = \hbar\boldsymbol{k}/m$ is given by

$$\Psi_V(\boldsymbol{r}, t) = \exp\left(-\frac{\mathrm{i}}{2m\hbar}\int^t \left[\hbar\boldsymbol{k} - \frac{e}{c}\boldsymbol{A}(t')\right]^2 \mathrm{d}t'\right) \times \mathrm{e}^{\mathrm{i}[\boldsymbol{k} - e\boldsymbol{A}(t)/\hbar c]\cdot\boldsymbol{r}} , \quad (74.33)$$

where $\boldsymbol{A}(t) = -c\int^t \boldsymbol{E}(t')\,\mathrm{d}t'$ is the vector potential of the field. Ψ_V is called a Volkov state. In a linearly polarized field, the electron oscillates along the direction of polarization with an amplitude $\alpha_0 = e\hbar\mathcal{A}/(mc\omega)$. In the strong field regime, this amplitude can greatly exceed the size of a bound state orbital. Volkov states provide a useful tool that can be applied in various strong field approximations discussed in the next Section.

74.3.4 Strong Field Approximations

There have been several attempts to solve the TDSE in the strong field limit using approximate, but analytic methods. Such strong field approximations (SFA) typically neglect all the bound states of the atom except for the initial state. In the tunneling regime ($\gamma < 1$), and a quasistatic limit ($\omega \to 0$), one can use a theory [74.43] in which the ionization occurs due to the tunneling through the Coulomb barrier distorted by the electric field of the laser. The wave function is constructed as a combination of a bare initial wave function of the electron (close to the nucleus) and a wave function describing a motion of the electron in a quasistatic electric field (far from the nucleus). In an second approach [74.34–36], the elements of the scattering matrix $\hat{S}$ are calculated assuming that initially the electronic wave function corresponds to a bare bound state. On the other hand, the final, continuum states of the electron are described by dressed wave functions that account for the free motion of the electron in the laser field. In the simplest case, such dressed states are Volkov states (74.33). Alternatively, the time-reversed $\hat{S}$-Matrix is obtained by dressing the initial state and using field-free scattering states for the final state.

Yet another method consists of expanding the electronic continuum–continuum dipole matrix elements in terms corresponding to matrix elements for free electrons plus corrections due to the potential [74.45]. In the latter version of SFA, the amplitude of the electronic wave function $b(\boldsymbol{p})$ corresponding to outgoing momentum $\boldsymbol{p}$ is given by

$$b(\boldsymbol{p}) = \mathrm{i}\int_0^{t_F} \mathrm{d}t\, \boldsymbol{d}[\boldsymbol{p} - e\boldsymbol{A}(t)/c]\cdot\boldsymbol{E}(t) \times \exp\left[-\mathrm{i}S(t_F, t)/\hbar\right] . \quad (74.34)$$

Here $\boldsymbol{d}[\boldsymbol{p} - e\boldsymbol{A}(t)/c]$ denotes the dipole matrix element for the transition from the initial bound state to the continuum state in which the electron has the kinetic momentum $\boldsymbol{p} - e\boldsymbol{A}(t)/c$, t_F is the switch-off time of the laser pulse, and

$$S(t_F, t) = \int_t^{t_F} \mathrm{d}t' \left[\frac{\left[\boldsymbol{p} - e\boldsymbol{A}(t')/c\right]^2}{2m} + I_P\right] \quad (74.35)$$

is a quasiclassical action for an electron which is born in the continuum at t and propagates freely in the laser field. The form of the expression (74.34) is generic to the SFA.

74.3.5 Phase Space Averaging Method

The methods of classical mechanics are particularly useful in describing the microwave excitation of highly excited (Rydberg) atoms [74.61, 62] (Sect. 74.2.8), but have also been applied to describe high harmonic generation, stabilization of atoms in super intense fields and two electron ionization [74.72–74].

The classical phase space averaging method [74.75] solves Newton's equations of motion

$$\dot{\boldsymbol{r}} = \boldsymbol{p}/m , \quad (74.36)$$

$$\dot{\boldsymbol{p}} = -\nabla V(\boldsymbol{r}) - e\boldsymbol{E}(t) , \quad (74.37)$$

for the electron interacting with the ion core and the laser field. A distribution of initial conditions in phase space is chosen to mimic the initial quantum mechanical state of the system, and a sample of classical trajectories generated. Quantum mechanical averages of physical observables are then identified with ensemble averages of those observables over the initial distribution. Since the dynamics of multiphoton processes is very complex, the neglected phases in this approach generally cause negligible errors and the results can be in quite good agreement with quantum calculations. Additionally, an examination of the trajectories provides details of the excitation dynamics which are often difficult to extract from a complex time-dependent wave function.

References

74.1 F. H. M. Faisal: *Theory of Multiphoton Processes* (Plenum, New York 1987)

74.2 M. Gavrila (Ed.): *Atoms in Intense Laser Fields* (Academic Press, San Diego 1992)

74.3 M. H. Mittleman: *Theory of Laser–Atom Interactions*, 2nd edn. (Plenum, New York 1993)

74.4 P. Lambropoulos, S. J. Smith (Eds.): *Proceedings of the International Conference of Multiphoton Processes III, 1984*, Vol. 2 (Springer, Berlin, Heidelberg 1984)

74.5 S. J. Smith, P. L. Knight (Eds.): *Proceedings of the International Conference on Multiphoton Processes IV, 1988*, Vol. 8 (Cambridge Univ. Press, Cambridge 1988)

74.6 G. Mainfray, P. Agostini (Eds.): *Proceedings of the International Conference on Multiphoton Processes V, 1991* (Centre d'Etudes de Saclay, Saclay 1991)

74.7 L. F. DiMauro, R. R. Freeman, K. C. Kulander (Eds.): *Proceedings of the International Conference of Multiphoton Processes VIII*, AIP Conference Proceedings, Vol. 525 (American Institute of Physics, Melville 2000)

74.8 B. Piraux, A. L'Huillier, K . Rzążewski (Eds.): *Super-Intense Laser–Atom Physics*, Vol. 316 (Plenum, New York 1993)

74.9 B. Piraux, K. Rzążewski (Eds.): *Super-Intense Laser-Atom Physics*, NATO ASI Series Ii, Vol. 12 (Kluwer Academic, The Netherlands 2001)

74.10 P. Lambropoulos: Adv. At. Mol. Phys. **12**, 87–158 (1976)

74.11 Y. Gontier, M. Trahin: J. Phys. B **13**, 4383 (1980)

74.12 X. Tang, P. Lambropoulos: Phys. Rev. Lett. **58**, 108 (1987)

74.13 J. H. Eberly: Phys. Rev. Lett. **47**, 408 (1981)

74.14 P. Lambropoulos, P. Zoller: Phys. Rev. A. **24**, 379 (1981)

74.15 P. L. Knight, M. A. Lauder, B. J. Dalton: Phys. Rep. **190**, 1 (1990)

74.16 H. Risken: *The Fokker–Planck Equation: Methods of Solution and Applications*, ed. by H. Haken (Springer, Berlin, Heidelberg 1984)

74.17 J. H. Eberly: *Laser Spectroscopy*, ed. by H. Walther, K. W. Rothe (Springer, Berlin, Heidelberg 1979) p. 80

74.18 P. Zoller: *Proceedings of the International Conference of Multiphoton Processes III, 1984*, Vol. 2, ed. by P. Lambropoulos, S. J. Smith (Springer, Berlin, Heidelberg 1984) pp. 68–75

74.19 D. S. Elliot: *Proceedings of the International Conference of Multiphoton Processes III, 1984*, Vol. 2, ed. by P. Lambropoulos, S. J. Smith (Springer, Berlin, Heidelberg 1984) pp. 76–81

74.20 D. S. Elliot et al.: Phys. Rev. A. **32**, 887 (1985)

74.21 A. I. Burshtein et al.: Sov. Phys. JETP. **21**, 597 (1965)

74.22 A. I. Burshtein et al.: Sov. Phys. JETP. **22**, 939 (1996)

74.23 B. W. Shore: *The Theory of Coherent Atomic Excitation* (Wiley, New York 1990)

74.24 L. A. Lompré, G. Mainfray, C. Manus, J. P. Marinier: J. Phys. B **14**, 4307 (1981)

74.25 C. Chen, D. S. Elliot: Phys. Rev. Lett. **65**, 1737 (1990)

74.26 R. M. Potvliege, R. Shakeshaft: *Atoms in Intense Laser Fields*, ed. by M. Gavrila (Academic Press, San Diego 1992) pp. 373–434

74.27 O. Faucher et al.: Phys. Rev. Lett. **70**, 3004 (1993)

74.28 J. E. Field, S. E. Harris: Phys. Rev. Lett. **66**, 1154 (1991)

74.29 S. E. Harris: Phys. Rev. Lett. **70**, 552 (1993)

74.30 G. S. Hurst, M. G. Payne, S. D. Kramer, J. P. Young: Rev. Mod. Phys. **52**, 767 (1979)

74.31 M. D. Perry, G. Mourou: Science **264**, 917 (1994)

74.32 K. C. Kulander, K. J. Schafer, J. L. Krause: *Super-Intense Laser–Atom Physics*, NATO ASI Series Ii, Vol. 12, ed. by B. Piraux, K. Rzążewski (Kluwer Academic, The Netherlands 2001) pp. 95–110.

74.33 P. B. Corkum: Phys. Rev. Lett. **73**, 1995 (1993)

74.34 L. V. Keldysh: Sov. Phys. JETP **20**, 1307 (1965)

74.35 H. R Reiss: Phys. Rev. A **22**, 1786 (1980)

74.36 F. Faisal: J. Phys. B **6**, 312 (1973)

74.37 D. Fittinghoff, P. R. Bolton, B. Chang, K. C. Kulander: Phys. Rev. A **49**, 2174 (1994)

74.38 H. G. Muller, P. Agostini, G. Petite: *Atoms in Intense Laser Fields*, ed. by M. Gavrila (Academic Press, San Diego 1992) pp. 1–42

74.39 R. R. Freeman et al.: *Atoms in Intense Laser Fields*, ed. by M. Gavrila (Academic Press, San Diego 1992) pp. 43–65

74.40 J. H. Eberly, J. Javanainen, K. Rzążewski: Phys. Rep. **204**, 331 (1991)

74.41 A. L'Huillier, L.-A. Lompré, G. Mainfray, C. Manus: *Atoms in Intense Laser Fields*, ed. by M. Gavrila (Academic Press, San Diego 1992) pp. 139–205

74.42 Y. Liang, M. V. Ammosov, S. L. Chin: J. Phys. B **27**, 1269 (1994)

74.43 M. V. Ammosov, N. B. Delone, V. P. Krainov: Sov. Phys. JETP **64**, 1191 (1986)

74.44 J. L. Krause, K. J. Schafer, K. C. Kulander: Phys. Rev. A **45**, 4998 (1992)

74.45 M. Lewenstein, Ph. Balcou, M. Yu. Ivanov, A. L'Huillier, P. Corkum: Phys. Rev. A **49**, 2117 (1994)

74.46 A. L'Huillier, K. J. Schafer, K. C. Kulander: J. Phys. B **24**, 3315 (1991)

74.47 A. L'Huillier, M. Lewenstein, P. Salières, Ph. Balcou, J. Larsson, C. G. Wahlström: Phys. Rev. A **48**, 4091 (1993)

74.48 K. S. Budil, P. Salières, A. L'Huillier, T. Ditmire, M. D. Perry: Phys. Rev. A **48**, 3437 (1993)

74.49 S. E. Harris, J. L. Macklin, T. W. Hänsch: Opt. Comm. **100**, 487 (1993)

74.50 M. Drescher et al.: Science **291**, 1923 (2001)
74.51 P. M. Paul et al.: Scienc **292**, 1689 (2001)
74.52 L. C. Dinu et al.: Phys. Rev. Lett. **91**, 063901 (2003)
74.53 A. Baltuška et al.: Nature **421**, 611 (2003)
74.54 M. Gavrila: *Atoms in Intense Laser Fields*, ed. by M. Gavrila (Academic Press, San Diego 1992) pp. 435–510 and references therein
74.55 Q. Su, J. H. Eberly, J. Javanainen: Phys. Rev. Lett. **64**, 862 (1990)
74.56 K. C Kulander, K. J. Schafer, J. L. Krause: Phys. Rev. Lett. **66**, 2601 (1991)
74.57 M. V. Fedorov: *Super-Intense Laser-Atom Physics*, NATO ASI Series Ii, Vol. 12, ed. by B. Piraux, K. Rzążewski (Kluwer Academic, The Netherlands 2001) pp. 245–259
74.58 M. P. de Boer, J. H. Hoogenraad, R. B. Vrijen, L. D. Noordam, H. Muller: Phys. Rev. Lett. **71**, 3263 (1993)
74.59 A. D. Bandrauk (Ed.): *Molecules in Laser Fields* (Dekker, New York 1994)
74.60 D. Normand, C. Cornaggia: *Super-Intense Laser-Atom Physics*, NATO ASI Series Ii, Vol. 12, ed. by B. Piraux, K. Rzążewski (Kluwer Academic, The Netherlands 2001) pp. 351–362
74.61 P. M. Koch: *Super-Intense Laser-Atom Physics*, NATO ASI Series Ii, Vol. 12, ed. by B. Piraux, K. Rzążewski (Kluwer Academic, The Netherlands 2001) pp. 305–316
74.62 P. M. Koch: *Proceedings of the Eigth South African Summer School in Physics, 1993* (Springer, Berlin, Heidelberg 1993)
74.63 M. C. Gutzwiller: *Chaos in Classical and Quantum Mechanics* (Springer, Berlin, Heidelberg 1990)
74.64 F. Haake: *Quantum Signatures of Chaos* (Springer, Berlin, Heidelberg 1991)
74.65 G. Casati, B. Chirikov, D. L. Shepelyansky, I. Guarnieri: Phys. Rep. **154**, 77 (1987)
74.66 G. Casati, I. Guarneri, D. L. Shepelyansky: Physica A **163**, 205 (1990) and references therein
74.67 R. V. Jensen, M. M. Sanders, M. Saraceno, B. Sundaram: Phys. Rev. Lett. **63**, 2771 (1989)
74.68 L. Sirko, M. R. W. Bellermann, A. Haffmans, P. M. Koch, D. Richards: Phys. Rev. Lett. **71**, 2895 (1993)
74.69 S. I. Chu: Adv. Chem. Phys. **73**, 739 (1989)
74.70 K. C. Kulander, K. J. Schafer, J. L. Krause: *Atoms in Intense Laser Fields*, ed. by M. Gavrila (Academic Press, San Diego 1992) pp. 247–300
74.71 K. Burnett, V. C. Reed, P. L. Knight: J. Phys. B **26**, 561 (1993)
74.72 M. Lewenstein, K. Rzążewski, P. Salières: *Super-Intense Laser-Atom Physics*, NATO ASI Series Ii, Vol. 12, ed. by B. Piraux, K. Rzążewski (Kluwer Academic, The Netherlands 2001) pp. 425–434
74.73 P. B. Lerner, K. LaGattuta, J. S. Cohen: *Super-Intense Laser-Atom Physics*, NATO ASI Series Ii, Vol. 12, ed. by B. Piraux, K. Rzążewski (Kluwer Academic, The Netherlands 2001) pp. 413–424
74.74 V. Vèniard, A. Maquet, T. Mènis: *Super-Intense Laser-Atom Physics*, NATO ASI Series Ii, Vol. 12, ed. by B. Piraux, K. Rzążewski (Kluwer Academic, The Netherlands 2001) pp. 225–232
74.75 J. G. Leopold, I. C. Percival: J. Phys. B **12**, 709 (1979)

75. Cooling and Trapping

Interactions of light with an atomic particle are accompanied by exchange of momentum between the electromagnetic field and the atom. Narrow-band resonance radiation from tunable lasers enhances the ensuing mechanical effects of light to the extent that it is literally possible to stop atoms emanating from a thermal gas, and to trap atoms with light.

References [75.1] and [75.2] are two early sources on laser cooling and trapping of the traditional two-state model atom. A number of articles on cooling and trapping of atoms with the inclusion of angular momentum degeneracy are contained in [75.3]. While the development based on the atom-field dressed states is followed sparingly in the present Chapter, an authoritative survey of this approach is given in [75.4]. Reviews on traps for charged particles include [75.5] and [75.6]. These articles, as well as [75.2], also discuss cooling of trapped particles.

Cooling and trapping of atomic particles are now standard technologies, but development and extension of the methods to new applications continues. The special issues [75.7] and [75.8] give a current snapshot. Additional references are occasionally listed here to accentuate specific points. These citations are to either particularly representative papers or to the most recent articles on the subject, and are intended as entries to the literature. No assignment of credit or priority is implied.

75.1 **Notation** 1091

75.2 **Control of Atomic Motion by Light** 1092
- 75.2.1 General Theory 1092
- 75.2.2 Two-State Atoms 1094
- 75.2.3 Multistate Atoms 1097

75.3 **Magnetic Trap for Atoms** 1099

75.4 **Trapping and Cooling of Charged Particles** 1099
- 75.4.1 Paul Trap 1099
- 75.4.2 Penning Trap 1101
- 75.4.3 Collective Effects in Ion Clouds 1102

75.5 **Applications of Cooling and Trapping** 1103
- 75.5.1 Neutral Atoms 1103
- 75.5.2 Trapped Particles 1104

References 1105

75.1 Notation

In this Chapter, the lower and upper states of an optical transition are denoted by the respective labels g and e, for "ground" and "excited". The notation $J_g \to J_e$ stands for a transition in which the lower and upper levels have the angular momentum degeneracies $2J_g+1$ and $2J_e+1$. The resonance frequency of the transition is ω_0.

The detuning of the driving monochromatic light of frequency ω from the atomic resonance is $\Delta = \omega_0 - \omega$. Γ is the spontaneous decay rate. Spontaneous emission is taken to be the sole mechanism of line broadening, so that the HWHM linewidth of the transition is $\gamma = \Gamma/2$. The Rabi frequency is $\Omega = \mathcal{D}\mathcal{E}/\hbar$, where $\mathcal{D}$ is the reduced dipole moment matrix element that would apply to a transition with unit Clebsch–Gordan coefficient, and $\mathcal{E}$ is the electric field amplitude of the laser. The corresponding intensity scale is the saturation intensity

$$I_s = \frac{4\pi^2 \hbar c \Gamma}{3\lambda^3}\,, \tag{75.1}$$

defined in such a way that the light intensity I satisfies

$$\Omega = \Gamma \Rightarrow I = I_s\,. \tag{75.2}$$

If multiple laser beams are explicitly mentioned, laser intensity and Rabi frequency are quoted for each of the equally intense beams.

The momentum of a photon with the wave vector $\boldsymbol{k}$ is $\hbar\boldsymbol{k}$. The recoil velocity

$$v_r = \frac{\hbar k}{m} \tag{75.3}$$

Table 75.1 Laser cooling parameters for the lowest $S_{1/2}$–$P_{3/2}$ transition of hydrogen and most alkalis (the D_2 line). Also shown are the nuclear spin I and the ground state hyperfine splitting $\Delta\nu_{hfs}$. Γ is typically known to within a few per cent, so these values of Γ, T_D and I_s may not all be accurate to the full displayed precision

Parameter	^{1}H	^{6}Li	^{7}Li	^{23}Na	^{39}K	^{40}K	^{85}Rb	^{87}Rb	^{133}Cs	Units
m	1.67	9.99	11.7	38.2	64.7	66.4	141	144	221	10^{-27} kg
λ	121.6	670.8	670.8	589.0	766.5	766.5	780.0	780.0	852.1	nm
v_r	326	9.89	8.48	2.95	1.34	1.30	0.602	0.589	0.352	cm/s
Γ	98.9	5.92	5.92	9.90	6.16	6.16	5.89	5.89	5.22	2π MHz
T_D	2390	142	142	238	148	148	141	141	125	μK
ε_r	13 396	73.7	63.2	25.0	8.72	8.50	3.86	3.77	2.07	2πkHz
T_r	643	3.54	3.03	1.20	0.418	0.408	0.185	0.181	0.0992	μK
I_s	14 509	5.13	5.13	12.7	3.58	3.58	3.24	3.24	2.21	mW/cm^2
I	1/2	1	3/2	3/2	3/2	4	5/2	3/2	7/2	
$\Delta\nu_{hfs}$	1420	228.2	803.5	1772	461.7	1286	3036	6835	9193	2πMHz

equals the change of the velocity of an atom of mass m when it absorbs a photon with wave number $k = 2\pi/\lambda$. The kinetic energy of an atom with velocity v_r and the corresponding frequency,

$$R = \frac{1}{2} m v_r^2 ,\ \varepsilon_r = \frac{R}{\hbar} \tag{75.4}$$

are referred to as recoil energy and recoil frequency. Two temperatures, the Doppler limit T_D and the recoil limit T_r are often cited in laser cooling. They are

$$T_D = \frac{\hbar\gamma}{k_B} , \quad T_r = \frac{R}{k_B} , \tag{75.5}$$

where k_B is the Boltzmann constant.

Table 75.1 lists numerical values of pertinent parameters for laser cooling and trapping using the D_2 line for most stable and long-lived alkali isotopes and hydrogen.

75.2 Control of Atomic Motion by Light

75.2.1 General Theory

Hamiltonian

The mechanical effects of light may be derived from the Hamiltonian

$$\hat{H} = \hat{H}_A + \hat{H}_{cm} + \hat{H}_F - \hat{\boldsymbol{d}} \cdot \hat{\boldsymbol{E}}(\hat{\boldsymbol{r}}) , \tag{75.6}$$

where $\hat{H}_A$, $\hat{H}_{cm}$ and $\hat{H}_F$ are the Hamiltonians for the internal degrees of freedom of the atom, center-of-mass (cm) motion of the atom, and free electromagnetic field. The quantized electric field is $\hat{\boldsymbol{E}}(\hat{\boldsymbol{r}})$, where $\hat{\boldsymbol{r}}$ is the cm position operator. The dipole operator $\hat{\boldsymbol{d}}$ acts on the internal degrees of freedom of the atom. Since the $\hat{\boldsymbol{d}} \cdot \hat{\boldsymbol{E}}(\hat{\boldsymbol{r}})$ term couples all degrees of freedom, the possibility of influencing cm motion by light immediately follows. The inclusion of the quantized cm motion is the essential ingredient not contained in traditional theories of light-matter interactions. For an atom with mass m trapped in a possibly anisotropic harmonic oscillator potential with frequencies ν_i ($i = x, y, z$), the cm Hamiltonian is

$$H_{cm} = \frac{\hat{\boldsymbol{p}}^2}{2m} + \sum_{i=x,y,z} \frac{m\nu_i^2 \hat{r}_i^2}{2} , \tag{75.7}$$

where $\hat{\boldsymbol{p}}$ is the cm momentum operator. For a free atom, $\nu_i = 0$.

Master Equation

With the aid of Markov and Born approximations, the vacuum modes of the electromagnetic field may be eliminated as described in Sect. 78.7. This gives a master equation for the reduced density operator $\hat{\rho}$ that contains the internal and cm degrees of freedom of the atom. Relaxation terms proportional to Γ and γ are all that is left of the quantized fields.

Consider as an example a two-state atom in a traveling wave of light with the electric field strength

$$E(\boldsymbol{r}, t) = \frac{1}{2} \mathcal{E}\, e^{i(\boldsymbol{k}\cdot\boldsymbol{r} - \omega t)} + \text{c.c.} . \tag{75.8}$$

Master equations are conveniently written using *Wigner functions* to represent the cm motion. Given the internal-state labels i and $j = g$ or e, and the three-dimensional variables $\boldsymbol{r}$, $\boldsymbol{p}$, the Wigner functions are defined as

$$\rho_{ij}(\boldsymbol{r},\boldsymbol{p}) = \frac{1}{(2\pi\hbar)^3}\int \mathrm{d}^3u\, \mathrm{e}^{\mathrm{i}\boldsymbol{u}\cdot\boldsymbol{p}/\hbar} \times \left\langle \boldsymbol{r} - \frac{1}{2}\boldsymbol{u}|\langle i|\hat{\rho}|j\rangle|\boldsymbol{r} + \frac{1}{2}\boldsymbol{u}\right\rangle . \tag{75.9}$$

The Wigner function is one of the quantum mechanical quasiprobability distributions, Sect. 78.5, with the special property that the marginal distribution of $\boldsymbol{r}$ obtained by integrating over $\boldsymbol{p}$ coincides with the correct quantum probability distribution for position, and vice versa with $\boldsymbol{r}$ and $\boldsymbol{p}$ interchanged. In the rotating wave approximation, Sect. 68.3.2, the master equations are

$$\frac{\mathrm{d}}{\mathrm{d}t}\rho_{ee}(\boldsymbol{p}) = -\Gamma\rho_{ee}(\boldsymbol{p}) + \mathrm{i}\frac{\Omega}{2}\left[\mathrm{e}^{\mathrm{i}\boldsymbol{k}\cdot\boldsymbol{r}}\hat{\rho}_{ge}\left(\boldsymbol{p} - \frac{1}{2}\hbar\boldsymbol{k}\right) - \mathrm{e}^{-\mathrm{i}\boldsymbol{k}\cdot\boldsymbol{r}}\hat{\rho}_{eg}\left(\boldsymbol{p} - \frac{1}{2}\hbar\boldsymbol{k}\right)\right] , \tag{75.10}$$

$$\frac{\mathrm{d}}{\mathrm{d}t}\rho_{gg}(\boldsymbol{p}) = \Gamma\int \mathrm{d}^2n\, W(\hat{\boldsymbol{n}})\rho_{ee}(\boldsymbol{p} + \hbar k\hat{\boldsymbol{n}}) - \frac{\mathrm{i}\Omega}{2} \times\left[\mathrm{e}^{\mathrm{i}\boldsymbol{k}\cdot\boldsymbol{r}}\hat{\rho}_{ge}\left(\boldsymbol{p} + \frac{1}{2}\hbar\boldsymbol{k}\right) - \mathrm{e}^{-\mathrm{i}\boldsymbol{k}\cdot\boldsymbol{r}}\hat{\rho}_{eg}(\boldsymbol{p} + 12\hbar\boldsymbol{k})\right] , \tag{75.11}$$

$$\frac{\mathrm{d}}{\mathrm{d}t}\hat{\rho}_{ge}(\boldsymbol{p}) = -(\gamma - \mathrm{i}\Delta)\hat{\rho}_{ge}(\boldsymbol{p}) - \frac{\mathrm{i}\Omega}{2} \times\left[\mathrm{e}^{-\mathrm{i}\boldsymbol{k}\cdot\boldsymbol{r}}\rho_{gg}(\boldsymbol{p} - \frac{1}{2}\hbar\boldsymbol{k}) - \mathrm{e}^{\mathrm{i}\boldsymbol{k}\cdot\boldsymbol{r}}\rho_{ee}\left(\boldsymbol{p} + \frac{1}{2}\hbar\boldsymbol{k}\right)\right] , \tag{75.12}$$

$$\frac{\mathrm{d}}{\mathrm{d}t}\hat{\rho}_{eg}(\boldsymbol{p}) = -(\gamma + \mathrm{i}\Delta)\rho_{eg}(\boldsymbol{p}) + \frac{\mathrm{i}\Omega}{2} \times\left[\mathrm{e}^{\mathrm{i}\boldsymbol{k}\cdot\boldsymbol{r}}\rho_{gg}\left(\boldsymbol{p} - \frac{1}{2}\hbar\boldsymbol{k}\right) - \mathrm{e}^{-\mathrm{i}\boldsymbol{k}\cdot\boldsymbol{r}}\rho_{ee}\left(\boldsymbol{p} + \frac{1}{2}\hbar\boldsymbol{k}\right)\right] . \tag{75.13}$$

Here the convective derivative that describes the motion of the atom in the absence of light is

$$\frac{\mathrm{d}}{\mathrm{d}t} = \frac{\partial}{\partial t} + \frac{\boldsymbol{p}}{m}\cdot\frac{\partial}{\partial \boldsymbol{r}} - \sum_i m v_i r_i \frac{\partial}{\partial p_i} ; \tag{75.14}$$

cf. H_{cm} in (75.7). $W(\hat{\boldsymbol{n}})$ is the angular distribution of spontaneous photons, and the integral runs over the unit sphere. Representative expressions for $W(\hat{\boldsymbol{n}})$ are

$$W(\hat{\boldsymbol{n}}) = \frac{1}{4\pi} , \quad \frac{3}{8\pi}\left[1 - (\hat{\boldsymbol{e}}\cdot\hat{\boldsymbol{n}})^2\right] , \quad \frac{3}{16\pi}\left[1 + (\hat{\boldsymbol{e}}\cdot\hat{\boldsymbol{n}})^2\right] . \tag{75.15}$$

These apply, respectively, for isotropic spontaneous emission, for spontaneous emission in a $\Delta m = 0$ transition, and in $\Delta m = \pm 1$ transitions; $\hat{\boldsymbol{e}}$ stands for the unit vector in the direction of the quantization axis for angular momentum. Only the $\boldsymbol{p}$ dependence has been denoted explicitly in the Wigner functions, as the recoil effects displayed on the right-hand sides of (75.10–75.13) take place at a fixed position $\boldsymbol{r}$.

Semiclassical Theory

Suppose that $v_{\mathrm{cm}} \gg v_{\mathrm{r}}$ and $\tau \ll \tau_{\mathrm{cm}}$, where τ and τ_{cm} are the time scales for light-driven changes of the internal state and cm motion of the atom. Then the internal degrees of freedom may be eliminated adiabatically from the master equations in favor of the position-momentum distribution for the cm motion,

$$f(\boldsymbol{r},\boldsymbol{p},t) = \sum_i \rho_{ii}(\boldsymbol{r},\boldsymbol{p},t) , \tag{75.16}$$

where the sum runs over the internal states of the atom. Technically, the recoil velocity v_{r} is treated as an asymptotically small expansion parameter. The result is the Fokker–Planck equation for the cm motion

$$\frac{\mathrm{d}}{\mathrm{d}t}f = -\frac{\partial}{\partial \boldsymbol{p}}\cdot(\boldsymbol{F}f) + \sum_{i,j}\frac{\partial^2}{\partial p_i \partial p_j}(D_{ij}f) . \tag{75.17}$$

In this semiclassical theory the cm motion of the atom is regarded as classical. The atom moves under the optical force $\boldsymbol{F}(\boldsymbol{r},\boldsymbol{p},t)$, which models the coarse-grained flow of momentum between the electromagnetic field and the atom. $D_{ij}(\boldsymbol{r},\boldsymbol{p},t)$, with $i, j = x, y, z$, is the diffusion tensor. Diffusion is an attempt to model quantum mechanics with a classical stochastic process, including discreteness of recoil kicks, random directions of spontaneous photons, and random timing of optical absorption and emission processes.

A general prescription exists for calculating the force and the diffusion tensor for an arbitrary atomic level scheme and light field [75.9]. However, the study of diffusion amounts to an involved analysis of photon statistics of the scattered light, and here only the force is considered explicitly. Let $\hat{V}(\boldsymbol{r})$ be the dipole interaction operator coupling the driving field and the internal state for an atom at position $\boldsymbol{r}$. By assumption, $\hat{V}(\boldsymbol{r})$ has been rendered slowly varying in time with the aid of a suitable rotating wave approximation. To compute the force, one takes an atom that travels along a hypothetical trajectory unperturbed by light in such a way that at time t it arrives at the phase space point $(\boldsymbol{r}, \boldsymbol{p})$, whereupon the density operator of the internal degrees of freedom is $\hat{\varrho}$. The force is then

$$F_i(\boldsymbol{r}, \boldsymbol{p}, t) = -\mathrm{Tr}\left(\hat{\varrho}\,\frac{\partial \hat{V}}{\partial r_i}\right) . \tag{75.18}$$

Quantum Theory

When either $v_{\mathrm{cm}} \lesssim v_{\mathrm{r}}$ or $\tau \gtrsim \tau_{\mathrm{cm}}$, the full quantum theory of cooling and trapping is needed. Master equations such as (75.10–75.13) must then be solved without the assumption that v_{r} is small. Most practical calculations have been numerical case studies [75.10–12]. A truncated basis, e.g., of plane waves, is used to expand the cm state. Density matrix equations are solved numerically, either directly, or by resorting to quantum trajectory simulations, Sect. 78.11.

Qualitative Origin of Laser Cooling

Velocity dependent dissipative forces are needed for cooling. They arise because the evolution of the internal state of a moving atom has a finite response time τ. The atom conveys the memory of the field it has sampled over the length $\ell = v\tau$ on its past trajectory. If $\ell \ll \lambda$, a nonequilibrium component proportional to ℓ is present in the density operator of the internal state of the atom. Further interactions with light convert this component into a velocity dependent force of the form

$$F = -m\beta v\,, \quad \beta \propto I\tau\,. \tag{75.19}$$

If the damping constant β is positive, (75.19) describes exponential damping of the velocity on the time scale β^{-1}. In the contrary case $\ell \gg \lambda$, when the atom travels many wavelengths during the memory time, linear dependence of force on velocity breaks down. The watershed is the critical velocity or velocity capture range

$$v_{\mathrm{c}} \approx \frac{\lambda}{\tau}\,. \tag{75.20}$$

One-Dimensional Considerations

Most specific results cited here are one-dimensional. By default, the propagation direction of light and the direction of vector quantities other than light polarization is $\hat{\boldsymbol{e}}_x$. The relevant components of position, velocity and momentum are denoted by x, v, and p.

The general one-dimensional Fokker–Planck equation for a particle trapped in a harmonic oscillator potential with a cm oscillation frequency ν is

$$\left(\frac{\partial}{\partial t} + \frac{p}{m}\frac{\partial}{\partial x} - m\nu^2 x\frac{\partial}{\partial p}\right) f = -\frac{\partial}{\partial p}(Ff) + \frac{\partial^2}{\partial p^2}(Df)\,. \tag{75.21}$$

For the force (75.19) with constant $\beta = \beta_0$ and $D(z, p) = D_0$, the steady state of the Fokker–Planck equation is a thermal distribution of the form

$$f(x, p) = K \exp\left[-\frac{\beta_0 m}{D_0}\left(\frac{p^2}{2m} + \frac{m\nu^2 x^2}{2}\right)\right]\,, \tag{75.22}$$

where K is a normalization coefficient.

Since Wigner functions give correct quantum mechanical marginal distributions for $\boldsymbol{r}$ and $\boldsymbol{p}$, expectation values of kinetic and potential energy may be calculated from the distribution function (75.22) as if it were a classical phase space density. For a free atom with $\nu = 0$, the temperature is directly proportional to the kinetic energy,

$$T = \frac{D_0}{\beta_0 m k_{\mathrm{B}}}\,. \tag{75.23}$$

However, for a trapped particle with $\nu \neq 0$ the Fokker–Planck equation may be valid all the way to the quantum mechanical zero-point energy. Then temperature and energy are no longer directly proportional to one another. For a trapped particle, the safe interpretation of (75.22) is that the total cm energy of the particle is

$$E = \frac{D_0}{\beta_0 m}\,. \tag{75.24}$$

75.2.2 Two-State Atoms

A two-state or two-level atom, discussed in detail in Sect. 68.3, stands for a closed (recycling) transition with one lower state and one excited state. In practice, a two-state system is often realized by driving a $J \to J+1$ transition with circularly polarized light. This leads to optical pumping to the states with maximal (or minimal) component of angular momentum along the quantization axis, say, to the transition $m = J \to J+1$.

Two types of force are generally distinguished: light pressure, or scattering force, or spontaneous force, and dipole, or gradient, or induced force. However, the distinction is neither exclusive, nor exhaustive. Here the two types of force are approached by way of examples.

Traveling Waves

Light Pressure. Consider a cycle of absorption and spontaneous emission. In an absorption, the atom receives a photon recoil kick in the direction of the laser beam, while in spontaneous emission the recoil kick has a random direction and zero average. The atom is on the average left with a velocity change equal to v_r. The corresponding force is along $\boldsymbol{k}$, and is given by

$$F = F_m \frac{\Omega^2/2}{\gamma^2 + \Delta^2(v) + \Omega^2/2} . \tag{75.25}$$

Here the maximum of light pressure force, a convenient scale for optical forces, is

$$F_m = \frac{1}{2} M v_r \Gamma , \tag{75.26}$$

and

$$\Delta(v) = \Delta + kv \tag{75.27}$$

is the effective detuning, which includes the Doppler shift experienced by the moving atom.

Diffusion. For a traveling wave, the diffusion coefficient accompanying light pressure is

$$\frac{D}{\hbar^2 k^2 \Gamma} = \frac{(1+\alpha)\Omega^2}{4\left[\Delta^2(v) + \gamma^2 + \Omega^2/2\right]} - \frac{\left[\Delta^2(v) - 3\gamma^2\right]\Omega^4}{4\left[\Delta^2(v) + \gamma^2 + \Omega^2/2\right]^3} , \tag{75.28}$$

where

$$\alpha = \int d^2 n W(\hat{\boldsymbol{n}}) (\hat{\boldsymbol{e}}_x \cdot \hat{\boldsymbol{n}})^2 \tag{75.29}$$

depends on $W(\hat{\boldsymbol{n}})$, see (75.15). Representative values are $\alpha = 1/3$ for isotropic spontaneous emission, and $\alpha = 2/5$ ($\alpha = 3/10$) for spontaneous emission with $\Delta m = 0$ ($\Delta m = \pm 1$) with respect to a quantization axis that is perpendicular (parallel) to the direction $\hat{\boldsymbol{e}}_x$.

Spontaneously emitted photons cover all of the 4π solid angle, and so do the directions of photon recoil kicks on the atom. Absorption from a light wave traveling in a particular direction leads to transverse diffusion also in the orthogonal directions, which is not accounted for by the one-dimensional (75.28).

Phenomenology in Multimode Fields

Doppler Cooling in Standing Waves. Next take an atom in two counterpropagating plane waves of light. At low intensity, $\Omega \ll \Gamma$, forces of the form (75.25) for the two beams may be added when averaged over a wavelength. For velocities well below the critical velocity

$$v_{c,D} = \frac{\Gamma}{k} , \tag{75.30}$$

the wavelength-averaged force is of the form of (75.19),

$$F = -m\bar{\beta} v , \bar{\beta} = \frac{4\Omega^2 \gamma \Delta}{\left(\Delta^2 + \gamma^2\right)^2} \varepsilon_r . \tag{75.31}$$

When light is tuned below the atomic resonance ("red detuning" with $\Delta > 0$), exponential damping of the atomic velocity with the time constant $\bar{\beta}^{-1}$ ensues. No matter which way the atom moves, it is always Doppler tuned toward resonance with the light wave that propagates opposite to its velocity, and away from resonance with the light wave that propagates along its velocity. Net momentum transfer therefore opposes the motion of the atom. This is known as Doppler cooling.

Optical Molasses. For three pairs of counterpropagating waves in three orthogonal directions, (75.31) is valid in all coordinate directions, and hence as a vector equation between the force $\boldsymbol{F}$ and velocity $\boldsymbol{v}$. For $\Delta > 0$ an atom experiences an isotropic viscous damping force, as if it were moving in a thick liquid. Such a field configuration is dubbed optical molasses. Two counterpropagating beams make a one-dimensional optical molasses.

Limit of Doppler Cooling. Under the conditions of (75.31), the diffusion coefficients for the two counterpropagating beams averaged over a wavelength may be added, and the $v = 0$ form suffices for slow atoms. This yields

$$\frac{\bar{D}(v=0)}{\hbar^2 k^2 \Gamma} = \frac{(1+\alpha)\,\Omega^2}{2\left(\Delta^2 + \gamma^2\right)} . \tag{75.32}$$

The random diffusive motion of the atom corresponds to diffusive heating that competes with Doppler cooling. In equilibrium, the temperature is

$$T = \frac{\bar{D}(v=0)}{m\bar{\beta} k_B} = \frac{\hbar\gamma}{4k_B}(1+\alpha)\left(\frac{\Delta}{\gamma} + \frac{\gamma}{\Delta}\right) . \tag{75.33}$$

Equation (75.33) also applies to three-dimensional Doppler cooled molasses, provided one uses $\alpha = 1$ corresponding to added transverse diffusion. The minimum temperature is reached at

$$\Delta = \gamma = \frac{\Gamma}{2} . \tag{75.34}$$

For three-dimensional molasses, the Doppler limit T_D of (75.5) is obtained.

For $\Omega > \Gamma$, the performance of Doppler cooling deteriorates. Qualitatively, power broadening increases the effective linewidth γ.

Dipole Forces. Dipole forces are the resonant analog of ponderomotive forces discussed in Sect. 74.2. They arise from successions of absorption and induced emission driven by photons with different momenta. Such processes occur only if there is more than one wave vector present in the field, i. e., if there is an intensity gradient. For a zero-velocity two-state atom, the gradient force in a monochromatic field with the local total intensity $I(\boldsymbol{r})$ is

$$\boldsymbol{F}_g(\boldsymbol{r}) = \frac{4\hbar\Delta\gamma^2}{\Delta^2 + \gamma^2[1 + 2I(\boldsymbol{r})/I_s]} \frac{\nabla I(\boldsymbol{r})}{I_s} . \tag{75.35}$$

The dipole force may be derived from the potential energy

$$V_g(\boldsymbol{r}) = -2\hbar\Delta \ln\left(1 + \frac{2\gamma^2 I(\boldsymbol{r})/I_s}{\Delta^2 + \gamma^2}\right) . \tag{75.36}$$

The atoms are strong field seekers for $\Delta > 0$, and weak field seekers when $\Delta < 0$.

Optical Trap and Optical Lattice. Dipole forces are utilized in the optical trap for atoms, and even molecules. A common configuration consists of a focused laser beam tuned below resonance. The focus becomes the trap. The detuning from resonance may be substantial; lasers such as CO_2 and Nd–YAG have been used.

A standing wave of light makes a periodic array of optical traps called an optical lattice. Optical lattices may be set up in 1D, 2D, and 3D configurations.

Induced Diffusion. Random motion of atoms in velocity space owing to absorptions and induced emissions of photons with different momenta leads to induced diffusion. Contrary to diffusion in a traveling wave as in (75.28), induced diffusion does not saturate at high intensity. Instead, the diffusion coefficient continues to grow linearly with I. Induced diffusion is another reason why the lowest Doppler cooling temperatures are generally reached at low ($I < I_s$) light intensities.

Sisyphus Effect. In a standing wave at high intensity and large detuning, another kind of optical force becomes important that cannot be categorized either as light pressure or gradient force.

As explained in Sect. 68.3.4, one may diagonalize the Hamiltonian to obtain the dressed atom-field states. Because the light field depends on position, so do the energies of the dressed states and their decompositions into plain atomic states. In Fig. 75.1 drawn for $\Delta < 0$, the dressed state with a minimum (maximum) at the field nodes coincides with the bare ground state (excited state) at the nodes. At the antinodes the admixtures of ground and excited states are evened out to some extent.

The energy of a dressed state acts as potential energy for the cm motion of an atom residing in that particular state. In fact, the gradient force is the force derived from these potential energies, averaged over the occupation probabilities of the dressed states. The occupation probability is larger for the dressed state with a larger ground state admixture. From Fig. 75.1 one therefore sees that the atom predominantly resides in the dressed state that has a minimum of energy at the nodes. The atom is a weak-field seeker, as it should for $\Delta < 0$.

Spontaneous emission remains to be considered. It gives rise to transitions between the dressed states. These transitions may go both ways between the dressed states, because the states are in general superpositions of the bare ground state and the excited state. The rate of spontaneous transitions from one dressed state to another

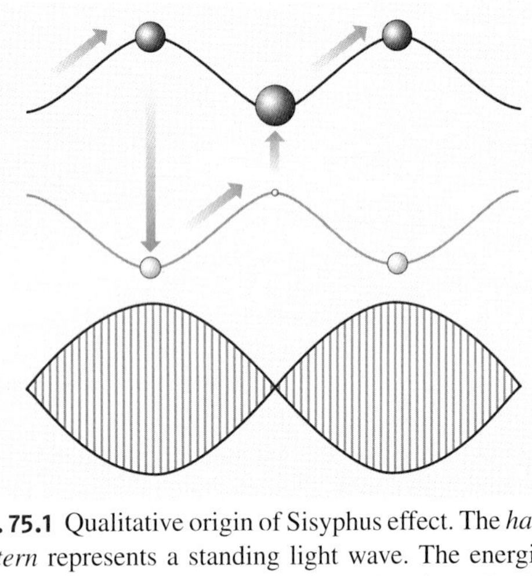

Fig. 75.1 Qualitative origin of Sisyphus effect. The *hatched pattern* represents a standing light wave. The energies of the two dressed states are drawn as a function of position, along with a few *filled circles* representing the admixture of the ground state in each dressed state at selected field positions. *Larger circles* correspond to larger ground state admixtures, and hence, larger equilibrium populations of the dressed state. This figure applies for tuning of the laser above the atomic resonance ("blue detuning")

increases (decreases) with the excited (ground) state admixture of the initial state.

In reference to Fig. 75.1, suppose that the atom is coming from the left in the upper dressed state. The probability that the atom makes a transition to the lower dressed state, as marked by the downward vertical arrow, is largest at the node. If this transition takes place, near the next antinode the most probable transition is as shown by the upward vertical arrow. In this manner the atom spends most of its time at an uphill climb against the potential, and is therefore slowed down. In reference to Greek mythology, this is called the Sisyphus effect. In the two-state model atom, cooling takes place when the laser frequency is higher than the atomic resonance frequency.

Exact Results for Standing Waves

Force. The force on a two-state atom in a one-dimensional standing light wave may be expanded analytically to first order in velocity. With the field phase chosen so that the antinode is at $x = 0$, the force is

$$F(x, v) = F_g(x) - m\beta(x)v \,, \tag{75.37}$$

where the gradient force is

$$F_g(x) = -\frac{\hbar k \Delta\, \Omega^2 \sin 2kx}{d} \,, \tag{75.38}$$

and the damping coefficient is

$$\begin{aligned}\beta(x) &= 8\Delta\Omega^2\varepsilon_r\gamma^{-1}d^{-3}\left(1-\cos^2 kx\right)\\ &\quad\times\Big(\Delta^2\gamma^2+\gamma^4-2\gamma^2\Omega^2\cos^2 kx\\ &\quad -2\Omega^4\cos^4 kx\Big)\,,\end{aligned} \tag{75.39}$$

with $d = \Delta^2 + \gamma^2 + 2\Omega^2\cos^2 kx$.

Diffusion. For a standing wave, the $v = 0$ form of the diffusion coefficient is

$$\begin{aligned}&\frac{D(v=0)}{\hbar^2k^2\Gamma}\\ &= \frac{\Omega^2\left(\alpha\gamma^2\cos^2 kx+\gamma^2\sin^2 kx+2\Omega^2\sin^2 kx\cos^2 kx\right)}{2\gamma^2 d}\\ &\quad -\frac{\Delta^2\Omega^4\sin^2 kx\cos^2 kx\left(\Delta^2+5\gamma^2+4\Omega^2\cos^2 kx\right)}{\gamma^2 d^3}\,.\end{aligned} \tag{75.40}$$

Semiclassical versus Quantum Theory

When $\gamma \gg \varepsilon_r$, the r.m.s. velocity of a cooled two-state atom is always $\gg v_r$, and semiclassical theory is valid. Under the same condition $\gamma \gg \varepsilon_r$, the Doppler-limit r.m.s. velocity also is less than the critical velocity $v_{c,D}$ from (75.30). Velocity expansions such as in (75.37) and (75.40) are then justified.

In the contrary case, $\gamma \lesssim \varepsilon_r$, the full quantum theory of trapping and cooling must be employed. The cooled velocity distribution is not thermal, and temperature is ill-defined. The lowest expectation value of kinetic energy for a two-state atom in a linearly polarized standing wave occurs at low I for $\Delta = 4.4\varepsilon_r$, and is equal to 0.53 R.

75.2.3 Multistate Atoms

Energy levels of atomic systems usually have angular momentum degeneracy. In addition, the polarization of light in general depends on position. A combination of these aspects leads to phenomena beyond the two-state atomic model.

Polarization Gradient Cooling

As explained in connection with (75.19), a finite memory time of the internal atomic state may lead to damping of the cm motion. For a two-state atom, internal equilibration arises from spontaneous emission. The time scale is $\tau_D \sim \Gamma^{-1}$, and Doppler cooling ensues. However, an atom whose ground state has angular momentum degeneracy is also subject to optical pumping. If the polarization of light varies as a function of position, optical pumping is needed to reach local equilibrium. The pumping time scale $\tau_p \propto I^{-1}$ then becomes relevant for a moving atom. The associated cooling is known as polarization gradient cooling. Its hallmark is that, for low I, the damping coefficient $\beta \propto I\tau_p$ is independent of intensity.

Two detailed mechanisms of polarization gradient cooling have been described [75.13], although in three-dimensional light fields they are intertwined. The Sisyphus effect works like the Sisyphus effect for a two-state atom, except that it relies on light shifts and optical pumping within the ground state manifold. Induced orientation cooling is analogous to Doppler cooling. Velocity dependence of optical pumping in counterpropagating waves leads to pumping to a state for which the force due to the wave propagating opposite to the atom exceeds the force due to the wave propagating along with the atom.

Lin $\perp$ Lin Molasses. One-dimensional lin $\perp$ lin molasses consists of two counterpropagating waves with orthogonal linear polarizations. The net polarization varies from linear – via elliptical – to circular over a distance of $\lambda/8$. With the phases chosen such that light is linearly polarized at $x = 0$, in the limit of nonsaturating intensity, and at low velocity, the semiclassical force on an atom with a $J = 1/2 \rightarrow 3/2$ transition is

$$F = -\hbar k \Delta \left[\frac{1}{3} s \sin 4kx - \frac{2}{3} q v \gamma^{-1} (1 + \cos 4kx) \right] . \tag{75.41}$$

Here the saturation parameter s is

$$s = \frac{\Omega^2/2}{\Delta^2 + \gamma^2} . \tag{75.42}$$

In this configuration, only the Sisyphus effect contributes to cooling. Cooling takes place for $\Delta > 0$, and the resulting temperature obtained from the position-averaged quantities is

$$T = \frac{135\Delta^2 + 296\gamma^2}{1080(\Delta^2 + \gamma^2)} \left(\frac{2I\Gamma}{I_s |\Delta|} \right) T_D . \tag{75.43}$$

σ^+–σ^- Molasses. One-dimensional σ^+–σ^- molasses consists of two counterpropagating waves with opposite circular polarizations. The net polarization is linear everywhere, but the direction of polarization rotates as the point of observation is displaced along the propagation axis; hence, the alternative name corkscrew molasses. At low intensity and low velocity, the force on an atom with a $J = 1 \rightarrow 2$ transition is

$$F = -\frac{60}{17} \frac{\Delta \gamma}{5\gamma^2 + \Delta^2} \hbar k^2 v . \tag{75.44}$$

In this configuration only induced orientation cooling contributes. Cooling again takes place for $\Delta > 0$, and the resulting temperature is

$$T = \frac{29\Delta^2 + 1045\gamma^2}{300\,(\Delta^2 + \gamma^2)} \left(\frac{2I\Gamma}{I_s |\Delta|} \right) T_D . \tag{75.45}$$

Lin $\parallel$ Lin Molasses. The designation lin $\parallel$ lin denotes a standing wave with the same linear polarization for both counterpropagating beams. In one dimension there is no polarization gradient, but three lin $\parallel$ lin pairs in orthogonal directions (often with mutually orthogonal polarizations) make a three-dimensional optical molasses with potential polarization gradient cooling.

Experimental Molasses. For $I\Gamma/I_s|\Delta| < 1$ and $|\Delta| > \Gamma$, the temperatures (75.43) and (75.45) both reduce to the form

$$T = C \frac{\hbar \Omega^2}{k_B |\Delta|} = C \frac{2I\Gamma}{I_s |\Delta|} T_D . \tag{75.46}$$

Under these conditions the same scaling is approximately observed also in three-dimensional six-beam optical molasses operating with atoms that have a degenerate ground state. The constant C depends on the degeneracy of the transitions and on the polarizations of the molasses beams. Measured values are mostly in the range $0.25 < C < 0.5$ [75.14].

Limit of Cooling. While the expressions (75.43), (75.45) and (75.46) suggest that T goes all the way to zero as $I \rightarrow 0$ or $\Delta \rightarrow \infty$, there is a lower limit of T reached in polarization gradient cooling. T eventually starts to rise abruptly when $|\Delta|$ is increased or I is decreased. The empirical rule of thumb is that $T \sim 10\, T_r$ is the lowest temperature one can expect.

Semiclassical versus Quantum Theory. According to the semiclassical theory, $T \propto I$, so the r.m.s. velocity of cooled atoms is proportional to $I^{1/2}$. Now, the critical velocity of polarization gradient cooling, estimated roughly as

$$v_{c,p} = \frac{\lambda}{\tau_p} \approx \frac{\lambda \gamma \Omega^2}{\Delta^2 + \gamma^2} , \tag{75.47}$$

is proportional to I. At low enough I, $v_{c,p}$ is therefore smaller than the velocity width of the cooled atoms. Expansions of force and diffusion in velocity are no longer useful, and temperature predictions of the type (75.46) fail. This occurs, at the latest, when the r.m.s. velocity equals a few recoil velocities.

Semiclassical theory does not lead to predictions that grossly violate its key premise that the ensuing velocity distribution is much broader than v_r. However, reliable theoretical limits of temperature for polarization gradient cooling may only be obtained from the full quantum treatment.

Magneto-Optical Trap

Since spontaneous forces may be strong already at modest light intensities $\sim I_{mathrm S}$ use of light pressure to trap neutral atoms appears desirable. However, the optical Earnshaw theorem states that (in the limit of low I) the spontaneous force on a two-state atom is sourceless. While confinement may be possible in some directions, escape routes for atoms remain open in others. Three-dimensional trapping of a two-state atom with light pressure is not possible.

A Magneto-optical trap (MOT) defeats the Earnshaw theorem by relying on angular momentum degeneracy. Consider an atom with a $J = 0 \rightarrow 1$ transition in a magnetic field $\boldsymbol{B}$ that depends linearly on position around the zero at $x = 0$. Suppose that the gradient of $\boldsymbol{B}$ is chosen in such a way that the $m = 1$ ($m = -1$) magnetic substate of the excited state has the higher (lower) energy for $x > 0$, and that the atom is illuminated by $\sigma^{\pm}$ polarized beams propagating in the $\pm x$-directions, tuned below resonance. When the atom is displaced from $x = 0$ in either direction, it is closer to resonance with the beam that pushes it back toward $x = 0$. This makes the restoring force responsible for trapping.

A magneto-optical trap can be set up also in three dimensions. A quadrupole magnetic field of the form

$$\boldsymbol{B}(\boldsymbol{r}) \simeq \left.\frac{\partial B_z}{\partial z}\right|_{r=0} \left(z\hat{\boldsymbol{e}}_z - \frac{1}{2}x\hat{\boldsymbol{e}}_x - \frac{1}{2}y\hat{\boldsymbol{e}}_y \right) \tag{75.48}$$

is produced by reversing the direction of current in one of the two Helmholtz coils. Three orthogonal pairs of light beams, each in the σ^+–σ^- configuration, complete the trap. The magnetic field is sourceless. To compensate for the ensuing signs of the field gradients, one of the σ^+–σ^- corkscrews has the opposite handedness from the other two.

The mechanism of the magneto-optical trap for the $J = 0 \to 1$ configuration is the same as the mechanism for Doppler cooling, except that position dependent level shifts of the excited states take the place of velocity dependent Doppler shifts. The restoring force and the damping coefficient of Doppler cooling are closely related. For the coordinate directions $u = x, y, z$ the relation is

$$F_u = -\kappa_u u\,, \; \kappa_u = \beta \frac{m g_{\mathrm{e}} \mu_B}{\hbar k} \left| \frac{\partial B_u}{\partial u} \right|\,, \tag{75.49}$$

where g_{e} is the Landé factor for the excited state.

A magneto-optical trap may similarly be based on the induced orientation mechanism of polarization gradient cooling. In that case, the Landé factor of the ground state, g_{g}, should probably be used in (75.49). This may be the true mechanism of most magneto-optical traps, but insufficient quantitative understanding precludes firm conclusions.

Atoms in a well-aligned magneto-optical trap reside near the zero of $\boldsymbol{B}$, so that the magnetic field has little effect on polarization gradient cooling. Trapping and cooling are achieved simultaneously.

75.3 Magnetic Trap for Atoms

The magnitude $B(\boldsymbol{r})$ of a magnetic field may have a minimum in free space, as in (75.48). A particle with a magnetic dipole moment $\boldsymbol{\mu}$ then experiences a trapping potential $U(\boldsymbol{r}) = \mu B(\boldsymbol{r})$ if $\boldsymbol{\mu}$ and $\boldsymbol{B}$ are antiparallel. $\boldsymbol{\mu}$ remains locked antiparallel to $\boldsymbol{B}$ if the field seen by the moving dipole satisfies the adiabatic condition

$$\frac{1}{B}\left|\frac{\mathrm{d}B}{\mathrm{d}t}\right| \ll \frac{\mu B}{\hbar} \tag{75.50}$$

(Section 73.3.4). However, if $B(\boldsymbol{r}) = 0$ at the minimum, the adiabatic condition is violated and the dipole may flip (Majorana transition). The particle may end up in a repulsive potential, and get expelled from the trap. This becomes a problem at low temperatures, when the particles accumulate near the minimum of the potential. Trap configurations are therefore designed in which $B(\boldsymbol{r}) \neq 0$ at the minimum.

Evaporative Cooling.

A magnetic trap is often combined with evaporative cooling. The most energetic atoms from the tail of the thermal distribution escape from the trap, whereupon the average energy of the remaining atoms decreases. Successful operation of evaporative cooling requires a high enough rate of elastic collisions so that the atoms thermalize in a time short compared with the lifetime of the sample. In order to sustain the rate of evaporation, the effective depth of the trap is lowered as the atoms cool.

75.4 Trapping and Cooling of Charged Particles

Since the potential $\Phi(\boldsymbol{r})$ of a static electric field satisfies Laplace's equation, $\Phi(\boldsymbol{r})$ cannot have an extremum in free space. A static electric field therefore cannot serve as an ion trap (Earnshaw's theorem). Paul and Penning traps circumvent this limitation by making use of an alternating electric field and a magnetic field, respectively. Cooling is often employed to assist trapping.

75.4.1 Paul Trap

Trapping

Configuration. Consider an ideal trap whose surfaces are hyperboloids of revolution; see Fig. 75.2. The two "endcaps" and the intervening "ring" are equipotential surfaces of the quasistatic electric

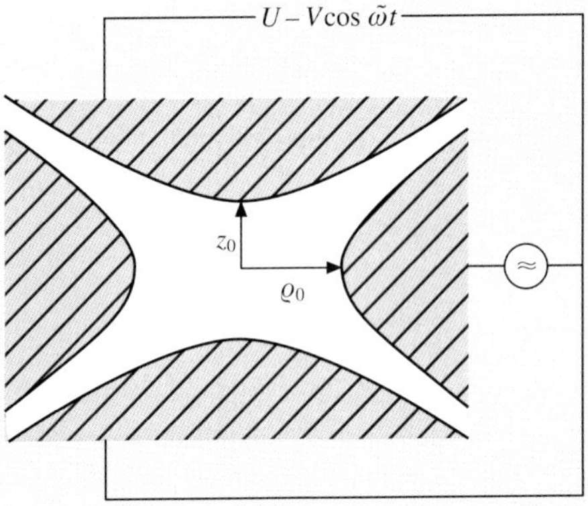

Fig. 75.2 Electrode configuration and voltages of an ideal hyperboloid Paul trap

potential

$$\Phi(x, y, z) = \frac{\Phi_0(t)\left(2z^2 - x^2 - y^2\right)}{2\varrho_0^2} , \quad (75.51)$$

where ϱ_0 is the distance from the center to the ring, $z_0 = \varrho_0/\sqrt{2}$ is the distance to the endcaps, and $\Phi_0(t)$ is a voltage applied between the endcaps and the ring,

$$\Phi_0(t) = U - V \cos \tilde{\omega} t . \quad (75.52)$$

Motion of an Ion. In the ideal three-dimensional Paul trap, Newton's equations of motion for the coordinates $u = x$, y or z may be recast as Mathieu's equations,

$$\frac{d^2u}{d\tau^2} + (a_u - 2q_u \cos 2\tau)u = 0 , \quad (75.53)$$

where $\tau = \tilde{\omega} t/2$ is a dimensionless quantity proportional to time, the parameters are

$$a_z = -2a_{x,y} = -\frac{8qU}{m\tilde{\omega}^2\varrho_0^2} , \quad (75.54)$$

$$q_z = -2q_{x,y} = -\frac{4qV}{m\tilde{\omega}^2\varrho_0^2} , \quad (75.55)$$

and m and q are the mass and charge of the particle. Stable trapping ensues when a_u and q_u are such that the motion of the ion is stable in all directions. A Paul trap normally operates in the first stability region of (75.53).

Stable motion may be qualitatively divided into forced micromotion at the frequency $\tilde{\omega}$ of the external drive, and into slower secular motion of the center of the micromotion. If $U = 0$, the secular motion takes place in an effective ponderomotive potential U_P, Sect. 74.2, equal to the cycle-averaged kinetic energy in the micromotion. Explicitly,

$$U_P(\boldsymbol{r}) = \frac{q^2\mathcal{E}^2(\boldsymbol{r})}{4m\tilde{\omega}^2} = \frac{q^2V^2\left(x^2 + y^2 + 4z^2\right)}{4m\tilde{\omega}^2\varrho_0^4} , \quad (75.56)$$

where $\mathcal{E}(\boldsymbol{r})$ is the ac field amplitude. This is an anisotropic harmonic oscillator potential characterized by the oscillation frequencies

$$\nu_z = 2\nu_{x,y} = \frac{\sqrt{2}qV}{m\tilde{\omega}\varrho_0^2} . \quad (75.57)$$

Quantization of C.M. Motion. The separation of micromotion and secular motion is excellent, and the trap is stable, when $\nu_{x,y,z} \ll \tilde{\omega}$. Ignoring the micromotion, the cm motion of the ions in the potential $U_P(\boldsymbol{r})$ may be quantized readily. The energy of a state with $n_{x,y,z}$ quanta in the coordinate directions x, y, z is

$$E = \sum_{i=x,y,z} \hbar\nu_i \left(n_i + \frac{1}{2}\right) . \quad (75.58)$$

Variations of Paul Trap. Little practical advantage usually arises from a realization of the ideal shape. Even a single electrode with an applied ac voltage may work as a Paul trap. A linear trap is basically a two-dimensional Paul trap with an added static longitudinal potential to prevent escape of the ions from the ends of the trap. A closed race track Paul trap is obtained by bending a linear trap into a ring.

Cooling

Laser Cooling in One Dimension. The secular motion of an ion may be cooled using lasers. Consider the motion of the ion in one of the principal-axis directions x, y, z, with ν denoting the corresponding cm frequency. In the common case where $\gamma \gg \nu$, Doppler cooling works basically as with a free atom. In the contrary case, $\nu \gg \gamma$, cooling may be achieved by tuning the laser to $\omega = \omega_0 - \nu$. Resonant photoabsorption starting with n cm quanta decreases the quantum number from n to $n - 1$, and subsequent spontaneous emission on the average leaves the cm energy nearly untouched. The net effect is reduction of the cm energy by $\hbar\nu$ in such a Raman process. Since the oscillating ion sees a frequency-modulated laser with sidebands, this method of ion cooling is called sideband cooling.

For one-dimensional motion of a two-state ion in a traveling light wave at low I, the velocity damping rate is

$$\beta = \frac{2\Omega^2\gamma\Delta}{\left[(\Delta+\nu)^2+\gamma^2\right]\left[(\Delta-\nu)^2+\gamma^2\right]}\varepsilon_r\,, \tag{75.59}$$

and the expectation value of the cm energy is

$$E = \frac{\hbar}{4\Delta}\left(\Delta^2+\gamma^2+\nu^2 + \alpha\frac{\left[(\Delta-\nu)^2+\gamma^2\right]\left[(\Delta+\nu)^2+\gamma^2\right]}{\Delta^2+\gamma^2}\right), \tag{75.60}$$

where α characterizes the angular distribution of spontaneous emission, see (75.29). The result (75.60) is useful when either $\varepsilon_r \ll \gamma$ or $\varepsilon_r \ll \nu$. The limit $\nu \ll \gamma$ is for Doppler cooling; the temperature from (75.60) coincides with (75.33) for a free atom. The case with both $\nu \gg \gamma$ and $\varepsilon_r \ll \nu$ corresponds to sideband cooling in the Lamb–Dicke regime, in which the cooled ion is confined to a region much smaller than λ.

In connection with sideband cooling, it is convenient to cite the expectation number of harmonic oscillator quanta $\langle n\rangle$ instead of energy or temperature; the latter are

$$E = \hbar\nu\left(\langle n\rangle + \frac{1}{2}\right), T = \frac{\hbar\nu}{k_B}\left[\ln\left(1+\langle n\rangle^{-1}\right)\right]^{-1}. \tag{75.61}$$

For optimal sideband cooling, $\nu \gg \gamma$ and $\Delta = \nu$, the result is

$$\langle n\rangle = \frac{1}{4}(1+4\alpha)\left(\frac{\gamma}{\nu}\right)^2 + O\left[\left(\frac{\gamma}{\nu}\right)^4\right]. \tag{75.62}$$

In principle, by decreasing the linewidth γ, the ion may be put arbitrarily close to the ground state of the cm harmonic oscillator. Such a decrease is not practical in real two-state systems, but is routinely achieved by using the two-photon resonance in a three-state Λ configuration as an effective two-state system; see Sect. 73.6.2.

Laser Cooling in Three Dimensions. Either by design or chance, no two of the ν_i are precisely degenerate. If the damping rate β and the trap frequencies ν_i satisfy

$$\beta \lesssim |\nu_i - \nu_j|, i \neq j\,, \tag{75.63}$$

the motion of the ion in each principal axis direction of the trap is cooled independently of the other directions. For $\gamma \gg \nu_{x,y,z}$, a single laser beam propagating approximately in the direction $\left(1/\sqrt{3}\right)(\hat{\boldsymbol{e}}_x+\hat{\boldsymbol{e}}_y+\hat{\boldsymbol{e}}_z)$ suffices to cool all components of the secular motion.

Energy in Micromotion. Possibly with the aid of compensating static electric fields, one cooled ion may be confined near the zero of the trapping ac electric fields. Then the energy in the micromotion is comparable to the energy in the secular motion.

75.4.2 Penning Trap

Trapping

Configuration. In the Penning trap, a dc voltage U is applied between the endcaps and the ring, and a constant magnetic field $\boldsymbol{B}$ in the direction of the trap axis z is added. The magnetic field forces an ion escaping toward the ring to turn back.

Motion of an Ion. The motion of an ion is a superposition of three periodic components. For the same ideal hyperboloid shape that was discussed with the Paul trap, (75.51) and Fig. 75.2, the three components are completely decoupled. Firstly, in the axial direction, the ion executes oscillations at the axial frequency

$$\nu_z = \left(\frac{2qU}{m\varrho_0^2}\right)^{1/2}. \tag{75.64}$$

Secondly, the ion undergoes cyclotron motion in the plane perpendicular to the trap axis. As a result of the electric field, the frequency of the cyclotron motion

$$\nu_c' = \frac{1}{2}\nu_c + \left(\frac{1}{4}\nu_c^2 - \frac{1}{2}\nu_z^2\right)^{1/2} \tag{75.65}$$

is displaced from the cyclotron frequency $\nu_c = qB/m$ of a free ion. Thirdly, the guiding center of cyclotron motion rotates about the trap axis at the magnetron frequency

$$\nu_m = \frac{1}{2}\nu_c - \left(\frac{1}{4}\nu_c^2 - \frac{1}{2}\nu_z^2\right)^{1/2}. \tag{75.66}$$

The frequencies typically satisfy

$$\nu_m \ll \nu_z \ll \nu_c'\,. \tag{75.67}$$

Magnetron motion has unusual properties. It takes up the majority of the electrostatic energy in the transverse directions, which in the absence of the magnetic field would lead to expulsion of the ion. Relative to a stationary ion at the trap center, the energy of the magnetron motion is bounded from above by zero. The radius, as well as velocity and kinetic energy of magnetron motion, decreases with increasing total energy. The energy for a state with n_c, n_z and n_m quanta in the cyclotron, axial and magnetron motions is therefore

$$E = \hbar\nu_c'\left(n_c + \frac{1}{2}\right) + \hbar\nu_z\left(n_z + \frac{1}{2}\right) - \hbar\nu_m\left(n_m + \frac{1}{2}\right) . \quad (75.68)$$

Cooling

Laser Cooling. For ions in a practical Penning trap, the frequencies ν_c', ν_z and ν_m are $< \gamma$. If $\boldsymbol{k}$ is not orthogonal to either the cyclotron motion or the axial motion, Doppler cooling proceeds essentially as for a free atom. However, energy should be added to the magnetron motion in order to reduce the magnetron radius and velocity. The solution is to aim a finite-size laser beam off the center of the trap in such a way that an ion experiences a higher (lower) intensity over the part of its magnetron orbit in which it travels in the direction of (opposite to) the laser beam. With a proper choice of the parameters, the ensuing addition of energy overcomes Doppler cooling of the magnetron motion.

Other Cooling Methods. Precision measurements are carried out in Penning traps with objects that do not have an internal level structure suitable for laser cooling; and thus, other cooling methods are used.

For light particles such as electrons, characteristic times of radiative damping of the cyclotron motion are in the subsecond regime, and hence, so are the equilibration times with blackbody radiation. Cooling is accomplished by enclosing the trap in a low-temperature (e.g. liquid helium) environment. For protons and heavier particles, the equilibration times of the cyclotron motion with the environment are impracticably long, and the same applies to the axial and magnetron motions even for electrons.

A workable cooling scheme for the axial motion is based on the charges that the oscillating particle induces on the endcaps. The charges generate currents in an external circuit connecting the endcaps. The endcaps are thus coupled to a cooled resonant circuit tuned to the axial frequency, and axial motion relaxes to thermal equilibrium with the resonant circuit. A variant of this resistive cooling , in which the ring is split into electrically insulated segments, is used to cool the cyclotron motion of protons and heavier ions.

Magnetron motion of an electron or proton is cooled by sideband cooling. An electric field with components in both the z-direction and xy-plane, and tuned to $\omega = \nu_z + \nu_m$, drives transitions which may either increase or decrease the number of quanta in each mode. However, the matrix elements favor transitions with $\Delta n_z = 1$ and $\Delta n_m = -1$. Pumping of the axial motion is canceled by axial cooling, while an equilibrium with low kinetic energy ensues for the magnetron motion. Ideally, the ratio of kinetic energies becomes

$$\frac{\mathcal{T}_{kin,m}}{\mathcal{T}_{kin,z}} = \frac{\nu_m}{\nu_z} . \quad (75.69)$$

75.4.3 Collective Effects in Ion Clouds

As soon as there is more than one ion in the trap, Coulomb interactions between the ions profoundly shape the physics [75.15, 16].

Ion Crystal

In the standard Paul trap radio frequency heating due (presumably) to transfer of energy from micromotion to secular motion limits the number of ions that can be cooled efficiently by a laser. Nevertheless, at a low temperature, the ions settle to equilibrium positions corresponding to a minimum of the joint trapping and Coulomb potentials, and form a "crystal". Depending on the trap parameters, the ions may also execute quasiperiodic or chaotic collective motion, or move nearly independently of one another. Changes between crystalline and liquid forms of the ion cloud resembling phase transitions are observed.

Strongly Coupled Plasma

Cooling of a large number of ions is possible in a Penning trap. However, magnetron motion becomes uniform rotation of the entire cloud, and Coulomb interactions set a lower limit on the attainable radius of the cloud. This leads to a lower limit on the kinetic energy and second-order Doppler shift.

In a co-rotating frame, the ions behave like a one-component plasma on a neutralizing background. The characteristic parameter for a one-component plasma with charge per particle q and density n is

$$\Gamma_P = \left(\frac{4\pi n}{3}\right)^{1/3} \frac{q^2}{4\pi\epsilon_0 k_B T} , \quad (75.70)$$

essentially the ratio of the Coulomb energy between two nearest-neighbor ions divided by the thermal kinetic energy. $\Gamma_P > 1$ indicates a strongly coupled plasma; for $\Gamma_P > 2$ and $\Gamma_P > 170$ solid and liquid phases are expected in an infinite plasma. Experiments with a Penning trap have produced $\Gamma_P \gtrsim 300$. Concentric shells of ions or various more or less crystalline arrangements are seen depending on the experimental conditions.

Sympathetic Cooling

In a trap that holds two or more species of charged particles, cooling of the motion of one species is transferred by Coulomb interactions to the other species. This sympathetic cooling broadens the scope of ion cooling methods.

75.5 Applications of Cooling and Trapping

Trapping and cooling offer increased interaction times between the atoms/ions and the light. This leads to reduced transit time broadening, and indeed to macroscopic (> 1 s) interaction times. Laser cooling in a magneto-optical trap routinely gives temperatures so low that the Doppler width is below the natural linewidth of the cooling transition. A homogeneously broadened atomic sample is thus prepared. Cooling also enables reduction of the second-order Doppler effect.

Various frequency measurements are the prime beneficiary of cooling and trapping. Potential applications range from detection of the change of natural constants in time Chapt. 30 to such feats of technology as the Global Positioning System.

75.5.1 Neutral Atoms

Experimental Considerations

Originally the experiments often started with a longitudinal deceleration and cooling of an atomic beam by a counterpropagating laser beam. To compensate for the change of the Doppler shift of the atoms while they slowed down, the position dependent magnetic field of a tapered solenoid shifted the transition frequency of the atoms to keep them near resonance while they moved down the solenoid. A magneto-optical trap then scooped some atoms, cooled them further, and captured them. Nowadays this Zeeman slower ist mostly supplanted by various schemes in which a MOT directly captures atoms from the low-velocity wing of the thermal distribution. Depths of neutral-atom traps are below 1 K. Storage times are typically of the order of 1 s, limited at high densities by exothermal binary collisions and at low densities by collisions with the atoms in the background gas.

The temperature of cooled atoms may be measured by the time-of-flight method. All cooling and trapping fields are turned off, whereupon the atoms fall freely under gravity. The distribution of arrival times of atoms at a probe laser beam underneath the initial molasses is compared with a numerical model for the disintegration of the molasses. A fit gives the temperature.

The focus is on Li, Rb and Cs, to a large extent because the required laser frequencies can be generated using inexpensive diode lasers. Hyperfine structure of the ground state of the alkalis complicates experiments because the atoms may end up in an inert hyperfine level outside the active cooling/trapping transitions. To counteract this, a second appropriately tuned repump laser is added to return such atoms to circulation. A few experiments use lanthanide atoms or metastable states of rare-gas atoms, some isotopes of which do not have hyperfine structure.

Cold Collisions

In the molecular picture of a collision involving a laser, optical excitation takes place at the interatomic distance for which the difference between the potential curves of the incoming and excited states equals the energy of a laser photon [75.17]. The end products of an inelastic collision (fine or hyperfine structure changing collision, associative ionization, radiative escape, etc.) are normally determined at shorter interatomic distances, when the potential curve of the excited state has an anti-crossing with the potential curve of the product channel. The novel feature of ultracold collisions is the long duration due to the low velocity of the collision partners. Spontaneous emissions and other phenomena irrelevant at room temperature may take place during the collision.

On the scale of typical resonance widths, at very low temperatures the collision partners are in effect in a single continuum state with zero energy. This facilitates photoassociation spectroscopy. Laser-induced transitions from the initial continuum state to bound vibrational states of the molecule are observed.

Collisions are of practical importance in that they limit the achievable atom density in a trap: an inelastic collision may release more kinetic energy than the trap can contain, which results in a loss of an atom (or two atoms) from the trap. Collision rates are actually measured by monitoring the loss rate of atoms as a function of density. Precise energies of the molecular vibrational states from photoassociation spectroscopy are used as input to determine [75.18] s-wave scattering lengths for atoms and to measure molecular parameters to an accuracy that far exceeds the capabilities of ab-initio calculations.

Frequency Standards

An atomic fountain starts with an optical molasses or a magneto-optical trap. The laser beams are then manipulated to give an upward push to the atoms. The atoms fly up against gravity for a few tens of centimeters, then turn back and, because of the initial transverse velocities, fan out to a "fountain". In a fountain clock [75.19], the fountain erupts through a microwave cavity that drives a hyperfine transition in the atoms. The clock is in effect an accurate measurement of the transition frequency. The fountain is beneficial because the interrogation times ≈ 1 s are longer and the atomic velocities ≈ 1 m/s slower than in traditional beam clocks.

Bose–Einstein Condensate

At present, the most prominent basic-physics applications of cooling and trapping of atoms undoubtedly are in studies of Bose–Einstein condensation and quantum degenerate Fermi gases in dilute atomic vapors. This topic is covered in Chapt. 76.

75.5.2 Trapped Particles

Experimental Considerations

Both Paul and Penning traps behave like a conservative potential, and scatter rather than confine a particle coming from the outside. One method to load a trap is to generate the ions in situ, e.g., by letting a beam of atoms and electrons collide inside the trap. Time dependent electric potentials are another loading method. The trapped species is injected thorough a hole in the endcap, and the opposing endcap is raised to an electric potential that makes the entering particles stop. The potential is then lowered before it ejects the particles. A single electron, positron, proton, antiproton or ion may be loaded. Typical depths of ion traps are ≈ 1 eV or $\approx 10^4$ K. With the aid of cooling, the storage time may be made infinite for all practical purposes.

Trap frequencies are measured by observing the resonance excited by added ac fields. For instance, an electric field near the axial cm resonance frequency may be coupled between the ring and one endcap. A resonance circuit coupled between the ring and the other endcap is used to detect the resonance. Alternatively, ejection of the driven ions is monitored.

For an electron in a Penning trap, the cyclotron frequency is in the extreme microwave region. Detection of the resonance is achieved indirectly. The uniform magnetic field is perturbed with a piece of a ferromagnet to make a magnetic bottle. The axial motion and the cyclotron motion are then coupled. A resonant microwave drive adds energy to the cyclotron motion, which detectably alters the axial frequency.

The three trap frequencies satisfy

$$\nu_{\mathrm{c}}^2 = \nu_{\mathrm{c}}'^2 + \nu_{\mathrm{m}}^2 + \nu_z^2 \,. \tag{75.71}$$

This relation remains valid even if the magnetic field is misaligned with respect to the trap axis, and is also insensitive to small imperfections in the cylindrical symmetry of the electrodes. The bare cyclotron frequency may therefore be deduced accurately.

For an ion with a dipole-allowed resonance transition, fluorescence of a single ion is readily detected. Even absorption of a single ion may be measurable. Various methods of finding the temperature have been devised. At temperatures of 1 K and higher, Doppler broadening of a dipole-allowed optical transition is observable. The size of the single-ion cloud is a measure of temperature. Finally, motional sidebands in the absorption of a narrow transition ($\gamma \ll \nu$), not necessarily the same transition as the one used for cooling, may be measured to find $\langle n \rangle$. In the Lamb–Dicke regime only the carrier absorption at $\Delta = 0$ and sidebands at $\Delta = \pm\nu$ are significant, and the ratios of the peak absorptions are

$$\alpha_- : \alpha_0 : \alpha_+ = \langle n \rangle \frac{\varepsilon_{\mathrm{r}}}{\nu} : 1 : (1 + \langle n \rangle) \frac{\varepsilon_{\mathrm{r}}}{\nu} \,. \tag{75.72}$$

In an ion crystal, the ions have collective vibration modes akin to phonons, instead of the three vibration modes along the principal axes of the trap of a single ion. Doppler cooling and sideband cooling work for such collective modes much like they work for the vibration modes of a single ion.

Quantum Jumps

Ion traps make it possible to isolate an individual atomic scale particle for studies for a practically indefinite time, which enables clean experiments on various fundamental aspects of quantum mechanics and quantum electrodynamics. Quantum jumps are a case in point. Suppose that, in addition to an optically driven two-level system, a single ion has a third shelving state. The ion infrequently makes a transition to the shelving state, stays there for a long time compared with the time scale of spontaneous emission of the active system, and then returns to the two-level system. When the ion makes a transition to the shelving state, fluorescence from the two-level system suddenly ceases; and the fluorescence reappears, equally abruptly, when the ion returns to the two-level system. The jumps in light scattering are the quantum jumps [75.20]. They are a method to detect a weak transition with an enormous amplification: a single transition to or from the shelving state may mean the difference between the presence or absence of billions of fluorescence photons.

$g-2$ Measurements

For an electron, the cyclotron frequency ν_c and the spin-flip frequency ν_s are related by

$$\nu_s = \frac{1}{2} g \nu_c \,. \tag{75.73}$$

Due to quantum electrodynamic corrections $g \neq 2$, and so the anomaly frequency $\nu_a = \nu_c - \nu_s \sim 10^{-3} \nu_c$ is nonzero. A magnetic field at the anomaly frequency causes a simultaneous flip of the spin and a loss or gain of one quantum of energy in the cyclotron motion. In a magnetic bottle, the change in the cyclotron motion causes a change in the axial resonance frequency, which is detected. The anomaly frequency can thus be measured accurately. Together with a measurement of the cyclotron frequency, this yields a measurement of the g-factor of the electron, or positron [75.21].

Measurements of Mass Ratios

As the cyclotron frequency is inversely proportional to the mass of the ion (or electron, positron, proton, antiproton, etc.), an accurate measurement of the cyclotron frequencies of two species in the same Penning trap amounts to an accurate measurement of the ratio of the masses [75.22]. A sufficient resolution to weigh molecular bonds is conceivable.

Quantum System of Motional States

A vibrational mode in a trapped ion and an effective two-state system for the internal degrees of freedom make a realization of the Jaynes–Cummings model (discussed in detail in Sect. 79.5.1). Moreover, sideband cooling enables an experimenter to put this mode cleanly in its lowest quantum state. These observations have inspired quantum-state engineering with the objective of generating an arbitrary state of the vibrational motion of the ion [75.23]. In many-ion crystals the collective vibration modes may be used to couple and entangle the internal degrees of freedom of two or more ions. As discussed in Chapt. 81, prototype quantum gates have been demonstrated in this manner.

More generally, experiments have come to the point when it is possible to address joint quantum states for the internal and cm degrees of freedom almost at will. This facilitates new cooling schemes. Time evolution derived from a Hamiltonian can never lead to cooling; an irreversible mechanism such as spontaneous emission is always needed. The idea of many cooling schemes therefore is to pump atoms optically around the quantum states in such a way there is no pathway out of the target state, so that the atoms eventually accumulate there. Velocity selective coherent population trapping [75.24], Raman sideband cooling of an ion, and Raman cooling of an atom in an optical lattice [75.25] work in this way.

References

75.1 The Mechanical Effects of Light, J. Opt. Soc. Am. B **2**(11) (1985), special issue

75.2 S. Stenholm: Rev. Mod. Phys. **58**, 699 (1986)

75.3 Laser Cooling and Trapping of Atoms, J. Opt. Soc. Am. B **6**(11) (1989), special issue

75.4 C. Cohen-Tannoudji: *Fundamental Systems in Quantum Optics*, ed. by J. Dalibard, J. M. Raymond, J. Zinn-Justin (Elsevier, Amsterdam 1991) p. 1

75.5 D. J. Wineland, W. M. Itano, R. S. Van Dyck Jr.: Adv. At. Mol. Phys. **19**, 135 (1984)

75.6 L. S. Brown, G. Gabrielse: Rev. Mod. Phys. **58**, 233 (1986)

75.7 Physics of Trapped Ions, J. Opt. Soc. Am. B **20**(5) (2003), special issue

75.8 Laser Cooling of Matter, J. Phys. B **36**(3) (2003), special issue

75.9 J. Javanainen: Phys. Rev. A **46**, 5819 (1992)

75.10 Y. Castin, K. Mølmer: Phys. Rev. Lett. **74**, 3772 (1995)

75.11 M. R. Doery, E. J. D. Vredenbregt, T. Bergeman: Phys. Rev. A **51**, 4881 (1995)

75.12 A. C. Doherty, T. W. Lynn, C. J. Hood, H. J. Kimble: Phys. Rev. A **63**, 013401 (2001)

75.13 J. Dalibard, C. Cohen-Tannoudji: J. Opt. Soc. Am. B **6**, 2023 (1989)

75.14 C. Gerz, T. W. Hodapp, P. Jessen, K. M. Jones, C. I. Westbrook, K. Mølmer: Europhys. Lett. **21**, 661 (1993)

75.15 H. Walther: Adv. At. Mol. Opt. Phys. **31**, 137 (1993)

75.16 J. J. Bollinger, D. J. Wineland, D. H. E. Dubin: Phys. Plasmas **1**, 1403 (1994)

75.17 P. S. Julienne, A. M. Smith, K. Burnett: Adv. At. Mol. Opt. Phys. **30**, 141 (1993)

75.18 E. G. M. van Kempen, S. J. J. M. F. Kokkelmans, D. J. Heinzen, B. J. Verhaar: Phys. Rev. Lett. **88**, 093201 (2002)

75.19 Y. Sortais, S. Bize, M. Abgrall, S. Zhang, C. Nicolas, C. Mandache, R. Lemonde, P. Laurent, G. Santarelli, P. Petit, A. Clairon, A. Mann, S. Chang, C. Salomon: Phys. Scr. **T95**, 50 (2001)

75.20 R. Blatt, P. Zoller: Eur. J. Phys. **9**, 250 (1988)

75.21 R. S. Van Dyck Jr., P. B. Schwinberg, H. G. Dehmelt: Phys. Rev. Lett. **59**, 26 (1987)

75.22 M. P. Bradley, J. V. Porto, S. Rainville, J. K. Thompson, D. E. Pritchard: Phys. Rev. Lett. **83**, 4510 (1999)

75.23 D. Leibfried, R. Blatt, C. Monroe, D. Wineland: Rev. Mod. Phys. **75**, 281 (2003)

75.24 J. Lawall, F. Bardou, B. Saubamea, K. Shimizu, M. Leduc, A. Aspect, C. Cohen-Tannoudji: Phys. Rev. Lett. **73**, 1915 (1994)

75.25 S. P. Hamann, D. L. Haycock, G. Klose, P. H. Pax, I. H. Deutsch, P. S. Jessen: Phys. Rev. Lett. **80**, 4149 (1998)

76. Quantum Degenerate Gases

The purpose of this Chapter is to summarize the basic physics of dilute quantum degenerate gases. Given the broad activity in the field, many choices have to be made regarding the topics to include and the style of the discussion. Emphasis is placed on AMO physics, as opposed to condensed matter physics. One related choice is that virtually nothing is said about temperature dependence. Inside AMO physics the approach is in the vein of quantum optics, as opposed to atomic/molecular structure and collisions. For the most part, the coverage is on elementary concepts and basic material. The exception to this is Sect. 76.5, where a few topical issues are addressed.

The review article [76.1] has become the standard reference on the basic properties of a Bose–Einstein condensate (BEC), [76.2] is its contemporary with more of a quantum optics slant, [76.3] concentrates on conceptual issues, [76.4] makes connections between the present theories and traditional condensed matter physics, and [76.5] is particularly explicit about the structure and excitations of a BEC. Here, references are usually not given for topics that are discussed in these reviews, or where a full discussion is easily traced from them. Otherwise, references are meant to be entries to the literature only. Assignment of credit or priority is never implied.

76.1 **Elements of Quantum Field Theory** 1107
76.1.1 Bosons 1108
76.1.2 Fermions 1109
76.1.3 Bosons versus Fermions 1109

76.2 **Basic Properties of Degenerate Gases** 1110
76.2.1 Bosons 1110
76.2.2 Meaning of Macroscopic Wave Function 1114
76.2.3 Fermions 1115

76.3 **Experimental** 1115
76.3.1 Preparing a BEC 1115
76.3.2 Preparing a Degenerate Fermi Gas 1117
76.3.3 Monitoring Degenerate Gases 1117

76.4 **BEC Superfluid** 1117
76.4.1 Vortices 1117
76.4.2 Superfluidity 1118

76.5 **Current Active Topics** 1119
76.5.1 Atom–Molecule Systems 1119
76.5.2 Optical Lattice with a BEC 1121

References 1123

Bose–Einstein condensation in dilute alkali metal vapors has realized a source of atoms with properties analogous to the properties of laser light, and more recently, ultralow-temperature Fermi gases have come under study. The field of quantum degenerate gases has become a main theme in AMO physics. Dilute-vapor systems are weakly interacting, and subject to a degree of experimental control not seen before in traditional low-temperature condensed matter systems. Ultralow-temperature gases have thereby also given a new lease on life to investigations of superfluid systems in condensed matter physics. The result is a broad interdisciplinary effort that is still expanding at the time of writing.

76.1 Elements of Quantum Field Theory

A Bose–Einstein condensate and a degenerate Fermi gas are both consequences of particle statistics, exchange symmetries of the many-particle wave function. It is possible, in principle, to deal directly with the wave functions, but in practice analyses of many-body systems are usually carried out using the methods of second quantization and field theory. In first quantization, the particles are labeled as if each one had a unique tag on it, and the wave function for more than one indistinguishable particle must be symmetrized explicitly. In second quantization, the question is how many particles are in a given state without a distinction between identical particles. The exchange symmetries are then taken care of automatically. Here we briefly summarize [76.6] elementary features of quantum field theories for both bosons and fermions.

76.1.1 Bosons

Particles with an integer value of the angular momentum obey the Bose–Einstein statistics. The characteristic property is that a one-particle state can accommodate an arbitrary number of bosons.

State Space for Bosons. Specifically, first consider one particle whose states are completely specified by a set of quantum numbers k. As a notational device for the purposes of the present Chapter, all of the quantum numbers are assumedly mapped in a one-to-one fashion to nonnegative integers, and correspondingly the quantum numbers are written $k=0,1,2,\ldots$. The quantum numbers written here always incorporate a description of the state of the c.m. motion of the particle. We therefore have an orthonormal basis of wave functions to represent any state of a particle, $\{u_k(\boldsymbol{x})\}_k$, where $\boldsymbol{x}$ stands for the c.m. coordinate.

Given the one-particle states, the postulate is that the Fock states $|n_0, n_1, \ldots, n_\infty\rangle$ with $n_k = 0, 1, 2, \ldots$ particles in the states $k = 0, 1, 2, \ldots$ form an orthonormal basis for the many-body system.

Second-Quantized Operators for Bosons. The annihilation operator for the state k, a_k, is defined by

$$\begin{aligned} &a_k|n_0, n_1, \ldots, n_k, \ldots, n_\infty\rangle \\ &= \sqrt{n_k}\,|n_0, n_1, \ldots, n_k-1, \ldots, n_\infty\rangle . \end{aligned} \tag{76.1}$$

Its Hermitian conjugate, the creation operator, behaves as

$$a_k^\dagger|\ldots, n_k, \ldots\rangle = \sqrt{n_k+1}\,|\ldots, n_k+1, \ldots\rangle . \tag{76.2}$$

It follows that

$$a_k^\dagger a_k|\ldots, n_k, \ldots\rangle = n_k\,|\ldots, n_k, \ldots\rangle , \tag{76.3}$$

and so $\hat{n}_k = a_k^\dagger a_k$ is called the number operator for the state k. Correspondingly,

$$\hat{N} = \sum_k a_k^\dagger a_k \tag{76.4}$$

is the operator for the total number of particles in the system. The annihilation and the creation operators have the usual boson commutators,

$$[a_k, a_{k'}] = \left[a_k^\dagger, a_{k'}^\dagger\right] = 0, \quad \left[a_k, a_{k'}^\dagger\right] = \delta_{kk'} . \tag{76.5}$$

The boson field operator is defined as

$$\hat{\psi}(\boldsymbol{x}) = \sum_k u_k(\boldsymbol{x}) a_k . \tag{76.6}$$

The commutator for the field operator,

$$\left[\hat{\psi}(\boldsymbol{x}), \hat{\psi}^\dagger(\boldsymbol{x}')\right] = \delta(\boldsymbol{x}-\boldsymbol{x}') , \tag{76.7}$$

follows from boson commutators and the completeness of the wave functions $\{u_k(\boldsymbol{x})\}_k$. The orthogonality of the wave functions gives the expression

$$\hat{N} = \int d^3x\, \hat{\psi}^\dagger(\boldsymbol{x})\, \hat{\psi}(\boldsymbol{x}) \tag{76.8}$$

for the particle number operator. The positive operator

$$\hat{n}(\boldsymbol{x}) = \hat{\psi}^\dagger(\boldsymbol{x})\, \hat{\psi}(\boldsymbol{x}) \tag{76.9}$$

evidently represents the density of the particles at the position $\boldsymbol{x}$.

The second-quantized operators introduced thus far can be used to express all observables acting on indistinguishable bosons. The most relevant here are the one- and two-particle operators. One-particle operators, such as the kinetic energy, act on one particle at a time, while two-particle operators, such as atom–atom interactions, refer to two particles. In first quantization, these are of the form

$$O_1 = \sum_n V(\boldsymbol{x}_n), \quad O_2 = \frac{1}{2}\sum_{n,n'} u(\boldsymbol{x}_n, \boldsymbol{x}_{n'}) , \tag{76.10}$$

where the sums run over the labels of the particles. The corresponding second-quantized operators are

$$\hat{O}_1 = \int d^3x\, \hat{\psi}^\dagger(\boldsymbol{x}) V(\boldsymbol{x}) \hat{\psi}(\boldsymbol{x}), \tag{76.11}$$

$$\begin{aligned} \hat{O}_2 = \frac{1}{2} \int & d^3x\, d^3x'\, \hat{\psi}^\dagger(\boldsymbol{x}) \hat{\psi}^\dagger(\boldsymbol{x}') \\ & \times u(\boldsymbol{x}, \boldsymbol{x}') \hat{\psi}(\boldsymbol{x}') \hat{\psi}(\boldsymbol{x}) . \end{aligned} \tag{76.12}$$

When the particles have internal degrees of freedom in addition to the c.m. motion, such as hyperfine and Zeeman states, it is convenient for the present purposes to regard particles in each internal state as a separate species. Thus, if the quantum number breaks up into

$k \equiv \{p, \alpha\}$, where p stands for quantum numbers of the center of the mass and α for the quantum numbers of the internal state, it is expedient to define a quantum field for each species α as

$$\hat{\psi}_\alpha(\boldsymbol{x}) = \sum_p u_{p\alpha}(\boldsymbol{x}) a_{p\alpha} \,. \tag{76.13}$$

Mechanisms that cause transitions between the internal states couple the fields $\hat{\psi}_\alpha(\boldsymbol{x})$ for different α.

States of Bosons. To complete the transformation from wave function quantum mechanics to second quantization, the state of the system must be specified in second quantization. For instance, take the Hamiltonian $\hat{H}$ and the particle number operator $\hat{N}$. According to statistical mechanics a system characterized by the temperature T and chemical potential μ is in the state with the density operator $\hat{\rho} = \mathrm{e}^{-(\hat{H}-\mu\hat{N})/k_\mathrm{B}T}/\mathcal{Z}$, where the grand partition function is $\mathcal{Z} = \mathrm{Tr}\, \mathrm{e}^{-(\hat{H}-\mu\hat{N})/k_\mathrm{B}T}$.

Bose–Einstein Condensate. The state of a boson system of particular interest here is the BEC. In an ideal gas condensation entails a macroscopic fraction of the particles occupying the ground state of the c.m. motion. Condensation is a phase transition that occurs when either the density of the gas is increased or the temperature is lowered. In a homogeneous ideal Bose gas the governing parameter is the phase space density ζ defined as

$$\zeta = \left(\frac{2\pi\hbar^2}{mk_\mathrm{B}T}\right)^{3/2} n \,, \tag{76.14}$$

where m is the mass of the condensing atoms, T is the temperature, and n is the density of the condensing species. For the purposes of quantum degeneracy, each internal state of an atom behaves as a separate species. Bose–Einstein condensation takes place when the phase space density satisfies $\zeta = 2.612$. Depending on whether density or temperature is regarded as a constant, (76.14) may be regarded as an equation for the critical temperature T_c or the critical density n_c for Bose–Einstein condensation.

76.1.2 Fermions

Particles with a half-integer angular momentum obey the Fermi–Dirac statistics. Each Fock state may then only have the occupation numbers $n_k = 0, 1$. The conventional definition of the annihilation operator contains a phase factor,

$$\begin{aligned} &a_k|n_0, n_1, \ldots, n_k, \ldots, n_\infty\rangle \\ &= n_k\,(-1)^{\sum_{p=0}^{k-1} n_p}\,|n_0, n_1, \ldots, n_k - 1, \ldots, n_\infty\rangle \,, \end{aligned} \tag{76.15}$$

and fermion operators are governed by the anticommutator

$$[A, B]_+ \equiv AB + BA \tag{76.16}$$

rather than the commutator. For instance,

$$[a_k, a_{k'}]_+ = 0, \quad \left[a_k, a_{k'}^\dagger\right]_+ = \delta_{kk'} \,. \tag{76.17}$$

Except for the use of anticommutators in lieu of commutators, all formal expressions for field operators and one- and two-particle operators written down for bosons in Sect. 76.1.1 remain valid as stated.

Degenerate Fermi Gas. A degenerate Fermi gas realized in a dilute atom vapor is the fermion counterpart of a BEC. The basic parameter of a free noninteracting Fermi gas is the Fermi energy, the chemical potential at zero temperature. It is given by

$$\epsilon_\mathrm{F} = \frac{\hbar^2 k_\mathrm{F}^2}{2m}; \quad k_\mathrm{F} = \left(6\pi^2 n\right)^{1/3} , \tag{76.18}$$

where n once more is the density for the relevant fermion species. In the limit of zero temperature the Fermi gas makes a Fermi sea; the states below the Fermi energy are filled with one particle each, the states above the Fermi energy are empty. The gas begins to show substantial deviations from the Maxwell-Boltzmann statistics and may be regarded as degenerate when the temperature is below the Fermi temperature, $T \leq T_\mathrm{F} = \epsilon_\mathrm{F}/k_\mathrm{B}$. Except for a numerical factor, in terms of density and temperature the condition $T \leq T_\mathrm{F}$ is the same as the condition for Bose–Einstein condensation.

76.1.3 Bosons versus Fermions

Isotopes of alkali metals with an odd mass number (^{7}Li, ^{23}Na, ^{39}K, ^{85}Rb, ^{87}Rb, ^{133}Cs) make Bose–Einstein gases, while isotopes with an even mass number (^{6}Li, ^{40}K) make Fermi–Dirac gases. Atoms are composite particles consisting of fermions, and how they may act as bosons is a legitimate question. Whether a satisfactory formal answer to this question exists may be debatable, but in practice atoms seem to obey the correct statistics in processes that do not expose their individual constituents.

When two bosonic atoms with integer angular momenta combine into a molecule, the molecule has an integer angular momentum and behaves as a boson. On the other hand, two fermionic atoms also make a bosonic molecule. Models for this latter type of system are basically ad hoc since, at this point in time, no microscopic theory for such a reorganization of the statistics exists. Nonetheless, empirically, diatomic molecules formed by combining two fermionic atoms indeed appear to be bosons.

76.2 Basic Properties of Degenerate Gases

Atoms Are Trapped. Quantum degenerate alkali vapors are typically prepared in an atom trap. Close to the bottom almost every trap is a three-dimensional harmonic oscillator potential completely characterized by the principal-axis directions and the corresponding trap frequencies ω_i, the (angular) frequencies at which a single atom would oscillate back and forth in the given principal-axis direction. In the principal-axis coordinate system the trapping potential reads

$$V(\boldsymbol{x}) = \frac{1}{2}\sum_{i=1}^{3} m\omega_i^2 x_i^2 . \tag{76.19}$$

It is convenient to introduce the characteristic harmonic-oscillator frequency scale and the corresponding harmonic-oscillator length scale as

$$\bar{\omega} = (\omega_1\omega_2\omega_3)^{1/3} , \quad \ell = \sqrt{\frac{\hbar}{m\bar{\omega}}} . \tag{76.20}$$

Atom–Atom Interactions. At low temperatures/energies only s-wave collisions are significant. In the theory of quantum degenerate gases these are frequently represented by a pseudopotential tailored to give the correct s-wave phase shift. For two atoms the atom–atom interaction is

$$u(\boldsymbol{x}_1, \boldsymbol{x}_2) = \frac{4\pi\hbar^2 a}{m}\delta(\boldsymbol{x}_1 - \boldsymbol{x}_2) , \tag{76.21}$$

where a is the s-wave scattering length. Qualitatively, the scattering length is positive (negative) if the interaction is repulsive (attractive).

Model Hamiltonian. Quantum field theory for a single-component Bose gas usually starts with the Hamiltonian

$$\hat{H} = \int \mathrm{d}^3x\, \hat{\mathcal{H}}(\boldsymbol{x}) , \tag{76.22}$$

where the Hamiltonian density is

$$\hat{\mathcal{H}}(\boldsymbol{x}) = \hat{\psi}^\dagger(\boldsymbol{x})\left[-\frac{\hbar^2}{2m}\nabla^2 + V(\boldsymbol{x})\right]\hat{\psi}(\boldsymbol{x}) + \frac{2\pi\hbar^2 a}{m}\hat{\psi}^\dagger(\boldsymbol{x})\hat{\psi}^\dagger(\boldsymbol{x})\hat{\psi}(\boldsymbol{x})\hat{\psi}(\boldsymbol{x}) . \tag{76.23}$$

As in Sect. 76.1.1, $\hat{\psi}(\boldsymbol{r})$ is the boson field operator, the kinetic energy $-\hbar^2\nabla^2/2m$ and trapping potential $V(\boldsymbol{x})$ are one-particle operators, and atom–atom interactions are governed by the two-body operator u of (76.21). Analogous models can be written down for multicomponent boson and fermion fields, for coupling between atoms and molecules, and so on.

76.2.1 Bosons

Gross–Pitaevskii Equation

Mean-Field approximation. Conventionally, the next step for bosons is to go over to the corresponding classical field theory. The result is referred to as mean-field theory or semiclassical theory.

Formally, one first writes down explicitly the Heisenberg equation of motion for the boson field $\hat{\psi}$,

$$\mathrm{i}\hbar\frac{\partial}{\partial t}\hat{\psi}(\boldsymbol{x}, t) = \left[\hat{\psi}(\boldsymbol{x}, t), \hat{H}\right] , \tag{76.24}$$

and then declares that in the equations of motion $\hat{\psi} \to \psi$ is a classical field not a quantum field anymore. We call ψ the macroscopic wave function of the condensate. This approximation is precisely analogous to using the classical instead of the quantum description for the electric and magnetic fields of the light coming out of a laser.

Time-Dependent Gross–Pitaevskii Equation. The time-dependent Gross–Pitaevskii equation (GPE) is

$$\mathrm{i}\hbar\frac{\partial}{\partial t}\psi(\boldsymbol{r}, t) = \left[-\frac{\hbar^2}{2m}\nabla^2 + V(\boldsymbol{r})\right]\psi(\boldsymbol{r}, t) + \frac{4\pi\hbar^2 a}{m}|\psi(\boldsymbol{r}, t)|^2\psi(\boldsymbol{r}, t) . \tag{76.25}$$

This equation is nonlinear, and normalization of the macroscopic wave function ψ is important. Quantum mechanically, the particle number operator is given by (76.8), so that the normalization for a system with N particles naturally reads

$$\int \mathrm{d}^3x\, |\psi(\boldsymbol{x}, t)|^2 = N . \tag{76.26}$$

Time evolution under (76.25) preserves the normalization. Obviously, and in accordance with (76.9),

$$n(\boldsymbol{x}) = |\psi(\boldsymbol{x})|^2 \tag{76.27}$$

is the local density of the gas.

Time-Independent Gross–Pitaevskii Equation. Solutions to the time-dependent GPE of the form $\psi(\boldsymbol{x},t) = \phi(\boldsymbol{x})\,\mathrm{e}^{-\mathrm{i}\mu t/\hbar}$ are stationary states with no time evolution in the physics. The analog of the energy of a stationary state is called the chemical potential μ. The corresponding wave function ϕ satisfies the time-independent GPE

$$\mu\phi = \left(-\frac{\hbar^2}{2m}\nabla^2 + V\right)\phi + \frac{4\pi\hbar^2 a}{m}\,|\phi|^2\phi\,. \tag{76.28}$$

Both GPEs are nonlinear variants of the Schrödinger equation, and in other contexts they are often referred to as nonlinear Schrödinger equations. The nonlinear term approximates the interaction energy of an atom with the other atoms in an averaged way by relying on the local density of the atoms, hence the term mean-field theory.

Sign of Scattering Length. The qualitative properties of a condensate, as per the GPE, depend on the sign of the scattering length. For repulsive atom–atom interactions or no atom–atom interactions, $a \geq 0$, both the time-dependent and the time-independent forms are mathematically well behaved. Unless otherwise noted, the scattering length is always assumed non-negative.

In the case of a negative scattering length a BEC may, in principle, decrease its energy without a bound by collapsing to a point. Mechanisms such as three-body recombination or molecule formation would eventually set in as the density increases and the collapse would stop, but the condensate must then be presumed lost. In the absence of an external potential, a condensate with a negative scattering length is unconditionally unstable against collapse. For a bounded condensate the increase in the kinetic energy coming with the decreasing size may hold off the collapse, provided the number of atoms in the condensate is sufficiently small. Simple dimensional-analysis arguments give the condition of stability in a harmonic trap as $N|a| \lesssim \ell$.

Behavior attributed to a collapse has been observed in ^{7}Li for trapped states with a negative scattering length. By using a Feshbach resonance it is also possible to adjust the (apparent) scattering length (Sect. 76.5.1) which has led to further demonstrations of collapse-like physics.

Healing Length. Consider the time-independent GPE (76.28) without an external potential, and scale the various quantities as follows:

$$\phi = \sqrt{n}\bar{\phi},\quad \mu = \bar{\mu}\,\frac{4\pi\hbar^2 na}{m},$$
$$\boldsymbol{x} = \xi\bar{\boldsymbol{x}};\quad \xi = \frac{1}{\sqrt{8\pi na}}\,. \tag{76.29}$$

Here n is the density scale for the gas, and the length scale is ξ. In terms of these new variables the time-independent GPE reads

$$\bar{\mu}\,\bar{\phi} = -\bar{\nabla}^2\bar{\phi} + |\bar{\phi}|^2\bar{\phi}\,. \tag{76.30}$$

There is a solution in all of space with $\bar{\mu} = 1$, $\bar{\phi} = 1$. If for some reason, such as at an edge of the sample, the condensate wave function must vanish, the length scale over which the wave function grows back to one (in the scaled units) is of the order of unity. In fact, (76.30) has the solution $\bar{\phi}(\bar{x}) = \tanh\left(\bar{z}/\sqrt{2}\right)$ in the half-space $\bar{z} \geq 0$. The quantity ξ is the minimum length scale over which a condensate wave function can build up to the density n. It is called the healing length.

Thomas–Fermi Approximation. Without atom–atom interactions, the ground state of the trapping potential $V(\boldsymbol{x})$ would be the lowest-energy (lowest μ) solution to (76.28). However, experience has shown that even modest repulsive atom–atom interactions ($a > 0$) spread out the macroscopic wave function of the condensate a great deal. With increasing size comes decreasing kinetic energy, according to the Heisenberg uncertainty principle. This suggests the Thomas–Fermi approximation, in which the kinetic energy term in (76.28) is simply ignored. The density of the gas is then easily solved to be

$$n(\boldsymbol{x}) = \begin{cases} \dfrac{m\left[\mu - V(\boldsymbol{x})\right]}{4\pi\hbar^2 a}, & \mu > V(\boldsymbol{x}) \\ 0, & \text{otherwise,} \end{cases} \tag{76.31}$$

an inverted image of the trapping potential. The normalization (76.26) can be used to find the relation between chemical potential and particle number, and all of the unknown quantities may, in principle, be found.

For a harmonic potential the Thomas–Fermi approximation can be worked out explicitly with the results

$$\mu = \frac{1}{2}\hbar\bar{\omega}\left(\frac{15Na}{\ell}\right)^{2/5},\qquad R = \ell\left(\frac{15Na}{\ell}\right)^{1/5},$$
$$n(0) = \frac{1}{8\pi\ell^3}\,\frac{\ell}{a}\left(\frac{15Na}{\ell}\right)^{2/5}\,. \tag{76.32}$$

The quantity R represents the size of the condensate. In particular, it equals the radius of the spherical condensate if the trap is isotropic with $\omega_1 = \omega_2 = \omega_3$. Finally, $n(0)$ is the central, maximum, density of the atoms.

The relevant dimensionless parameter is Na/ℓ, which can easily be much larger than unity in the experiments. When the Thomas–Fermi approximation is accurate, the chemical potential exceeds the typical level spacing of the harmonic-oscillator trap, and the condensate is larger than the ground-state wave function of the harmonic oscillator would be. Ordinarily, the condensate is also much larger than the healing length.

Small Excitations in a BEC

Linearizing the GPE. The time-dependent GPE is nonlinear, but may be linearized around a stationary solution. Consider the special case without a trapping potential, $V(\boldsymbol{x}) \equiv 0$. The stationary solutions are plane waves,

$$\phi(\boldsymbol{x}) = \sqrt{n}\, \mathrm{e}^{\mathrm{i}\boldsymbol{p}\cdot\boldsymbol{x}} . \tag{76.33}$$

This corresponds to a flow of the gas at the velocity $\boldsymbol{v} = \hbar\boldsymbol{p}/m$ and with a momentum $\hbar\boldsymbol{p}$ per atom. The chemical potential of such a mode is

$$\frac{\mu}{\hbar} = \epsilon_p + \frac{1}{2}\epsilon_0 , \tag{76.34}$$

where

$$\epsilon_p = \frac{\hbar p^2}{2m}, \quad \epsilon_0 = \frac{mc^2}{\hbar}, \quad c = \frac{\hbar\sqrt{8\pi na}}{m} = \frac{\hbar}{m\xi} \tag{76.35}$$

are the dispersion relation of free atoms, a peculiar analog of the rest energy, and the speed of sound in the BEC.

The ansatz for small deviations from the stationary solution is written as

$$\psi(\boldsymbol{x}, t) = \sqrt{n}\, \mathrm{e}^{\mathrm{i}(\boldsymbol{p}\cdot\boldsymbol{x} - \frac{\mu}{\hbar}t)} \big[1 + u\, \mathrm{e}^{\mathrm{i}(\boldsymbol{q}\cdot\boldsymbol{x} - \nu t)} + v^* \mathrm{e}^{-\mathrm{i}(\boldsymbol{q}\cdot\boldsymbol{x} - \nu^* t)}\big] , \tag{76.36}$$

where $\hbar\boldsymbol{q}$ and $\hbar\nu$ are the momentum and energy associated with the excitation relative to the momentum and energy of the original flow. The GPE mixes the field ψ and its complex conjugate ψ^*, so that two small amplitudes u and v are needed for the excitations. The ansatz (76.36) is a solution to the time-dependent GPE to the lowest nontrivial order in u and v if these amplitudes and the frequency of the excitation satisfy the eigenvalue equation

$$\begin{pmatrix} \epsilon_q + \epsilon_0 + \boldsymbol{q}\cdot\boldsymbol{v} & \epsilon_0 \\ \epsilon_0 & \epsilon_q + \epsilon_0 - \boldsymbol{q}\cdot\boldsymbol{v} \end{pmatrix} \begin{pmatrix} u \\ v \end{pmatrix} = \nu \begin{pmatrix} u \\ -v \end{pmatrix} . \tag{76.37}$$

The remaining problem is that the eigenvalue equation has two solutions for each $\boldsymbol{q}$, which gives twice as many small-excitation modes as there are degrees of freedom. The extra modes are the penalty one pays for the linearization of the GPE. The criterion $|u|^2 - |v|^2 > 0$ picks out the correct small-excitation modes. The corresponding dispersion relation for the excitations is

$$\nu(\boldsymbol{q}) = \boldsymbol{q}\cdot\boldsymbol{v} + \sqrt{\epsilon_q(\epsilon_q + 2\epsilon_0)} . \tag{76.38}$$

For a stationary BEC with $\boldsymbol{v} = 0$ and in the limit $q \to 0$, (76.38) gives $\nu \simeq cq$. This confirms the identification of c as the speed of sound.

In the BEC experiments the condensates are trapped, but in principle the same analysis of small-excitation modes may be carried out both numerically and in a myriad of analytical approximations. The generic result is that the trap frequencies lend their frequency scale to small excitations. At low enough temperatures, excitation frequencies calculated in this way agree well with the experiments.

Within the mean-field approximation small excitations may be analyzed similarly in all boson systems, for instance, in a multi-component Bose–Einstein condensate or a joint atom–molecule condensate. The evolution frequencies may be complex, which signals a dynamical instability of the stationary configuration; there are small-excitation modes that grow exponentially. The instability of a free gas with a negative scattering length, which is apparent in (76.38) for $\epsilon_0 < 0$, is a simple example.

Bogoliubov Theory. Bogoliubov theory is the many-body quantum version of the analysis of small excitations. The idea is to treat the condensate mode ψ_0, containing n_0 atoms, separately in the field operator

$$\hat{\psi} = \sqrt{n_0}\, \psi_0 + \delta\hat{\psi} , \tag{76.39}$$

expand the Hamiltonian in the lowest nontrivial (second) order in the remnant quantum field $\delta\hat{\psi}$, and diagonalize. The result is small-excitation modes with the annihilation operators A_k, where k stands for the appropriate quantum numbers. It turns out that the core mathematics of Bogoliubov theory is the same as the mathematics of small excitations, but two features are added.

First, Bogoliubov theory explicitly shows that the coefficients u and v in the analog of (76.37) need to satisfy $|u|^2 - |v|^2 = 1$ to ensure boson commutators for the operators A_k. Second, with quantum fluctuations, atom–atom interactions force atoms out of the condensate even at zero temperature. In a homogeneous (untrapped) condensate, in the limit $na^3 \ll 1$, at $T = 0$, the fraction of noncondensate atoms is

$$\frac{N - n_0}{N} = \frac{8}{3}\sqrt{\frac{na^3}{\pi}} . \tag{76.40}$$

When the gas parameter na^3 is much smaller than unity, at low enough temperatures most of the atoms are in the condensate. Mean-field theory and the GPE are expected to apply, and empirically, they do.

Numerical Methods for GPE

Mathematical Properties of the GPE. Let us momentarily assume that by separation of variables, or by some fiat, the problem of solving the time-independent GPE has been rendered one-dimensional. The Schrödinger equation is linear and any constant multiple of a solution is also a solution. One parameter, e.g., the logarithmic derivate of the wave function at a given point in space, determines a stationary state completely. This does not hold for the corresponding GPE, for which the values of the wave function and its derivative can be specified independently at (almost) every fixed point in space. As a result of the added flexibility, and unlike the Schrödinger equation, the GPE has bounded solutions for continuous ranges of the values of the chemical potential μ.

However, the time-independent GPE (76.28) comes with the added normalization condition (76.26). Normalization quantizes the values of μ for the bound states. In practice one might, for instance, find a solution that satisfies the boundary conditions for a given μ with the shooting method, then adjust the value of μ until normalization holds. Techniques used in the first numerical analyses of the time-independent GPE in the context of atom vapor condensates were variations of this theme. Such schemes are not feasible in spatial dimensions greater than one.

In general there is one solution to the time-independent GPE that can be chosen to be positive everywhere, the ground state with the lowest chemical potential. Excited steady states exist, but only a few, such as the flowing states of (76.33) and vortices discussed in Sect. 76.4.1, have obvious physical meanings. As the GPE is nonlinear, excited states are not the same as small excitations.

Split-Step Fourier Method. The superposition principle does not hold for the solutions of the time-dependent GPE, and the excited states are usually not orthogonal to one another in any useful sense. Methods based on eigenstate expansions for solving the time-dependent GPE are cumbersome at best. Instead, one often simply integrates the GPE as a partial differential equation in time. A number of different methods are used, but here we only discuss an elementary split-step Fourier method [76.7]. This is an exceedingly popular algorithm for parabolic equations, easy to implement, and with minor modifications also solves the time-independent GPE in any number of dimensions.

Thus, consider integration of (76.25) forward in time over a step from t to $t + \Delta t$. For this purpose assume first that $|\psi|^2$ in the GPE were a constant equal to its value at time t, then the evolution over the time step Δt would be given by

$$\psi(\boldsymbol{x}, t + \Delta t) = \exp\left[-\mathrm{i}\Delta t\left(-\frac{\hbar\nabla^2}{2m} + U(\boldsymbol{x})\right)\right] \times \psi(\boldsymbol{x}, t) , \tag{76.41}$$

where $U(\boldsymbol{x})$ is a given function of position. In the algorithm the exponential is first split approximately, for instance, as

$$\begin{aligned} &\exp\left[-\mathrm{i}\,\Delta t\left(-\frac{\hbar\nabla^2}{2m} + U(\boldsymbol{x})\right)\right] \\ &\simeq \exp\left(\mathrm{i}\frac{\Delta t}{2}\frac{\hbar\nabla^2}{2m}\right)\exp\left[-\mathrm{i}\,\Delta t\, U(\boldsymbol{x})\right] \\ &\quad\times \exp\left(\mathrm{i}\frac{\Delta t}{2}\frac{\hbar\nabla^2}{2m}\right) \\ &\equiv \tilde{T}\tilde{U}\tilde{T} . \end{aligned} \tag{76.42}$$

The exponential of the kinetic-energy operator is diagonal in the Fourier representation. Consequently, carrying out the Fourier transform $\mathcal{F}$ and its inverse with the aid of the Fast Fourier Transformation gives the split-step algorithm

$$\psi(t + \Delta t) = \mathcal{F}^{-1}\tilde{T}\mathcal{F}\tilde{U}\mathcal{F}^{-1}\tilde{T}\mathcal{F}\ \psi(t) \tag{76.43}$$

with obvious efficient implementations. The inaccurate constant $|\psi|^2$ may be improved upon in a corrector step in which the average of the initial wave function $\psi(t)$ and the $\psi(t + \Delta t)$ obtained in the first pass is used as $|\psi|^2$, and step (76.43) is taken again. This split-step algorithm preserves the normalization of the macroscopic wave function, and features a high-order approximation to the exponential operator of the kinetic energy.

Integration in Imaginary Time. The split-step algorithm also provides a global method to find the ground state. To this end the time-dependent GPE is integrated in imaginary time, i. e., replacing $\Delta t \to -\mathrm{i}\Delta t$, starting from a more or less random initial wave function and normalizing after every step. If the GPE were linear, this procedure would emphasize the lowest-energy component of the wave function until it is the only one that remains to within a prescribed accuracy. It is not clear that the same should apply to the nonlinear GPE, but often this is the case.

In nonlinear problems, split-operator methods, in spite of their seeming simplicity, often exhibit spurious behavior. Successful applications of these techniques require skill and experience in the art of numerical methods.

Local-Density Approximation

While an experimental BEC is usually trapped, it is often much easier to study the theory for a formally infinite homogeneous condensate. As long as the phenomena under investigation involve length scales much smaller than the size of the condensate and time scales much shorter than the inverse trap frequencies, trapping cannot affect the behavior of the gas locally. Under such conditions one may analyze the gas at each position $\boldsymbol{x}$ as if it were homogeneous, and at the end of the calculations average over the density distribution. The unit-normalized distribution of the density of the gas used in the averaging is

$$P(\varrho) = \frac{\int \delta[\varrho - n(\boldsymbol{x})]n(\boldsymbol{x})\,\mathrm{d}^3x}{\int n(\boldsymbol{x})\,\mathrm{d}^3x} . \qquad (76.44)$$

For instance,

$$P(\varrho) = \frac{15\sqrt{n_0 - \varrho}\,\varrho}{4n_0^{5/2}} H(\varrho)H(n_0 - \varrho) \qquad (76.45)$$

holds for the Thomas–Fermi approximation with the maximum density $n_0 \equiv n(0)$. The Heaviside step functions H restrict the density to the correct range $0 \le \varrho \le n_0$. As an example, the average density in the Thomas–Fermi model is

$$\int \varrho P(\varrho)\,\mathrm{d}\varrho = \frac{4}{7} n_0 . \qquad (76.46)$$

76.2.2 Meaning of Macroscopic Wave Function

Here the macroscopic wave function ψ has been introduced by replacing a boson field theory with a classical field theory.

The intuitive interpretation is that, for interacting particles, the atoms condense not to the ground state of the confining potential, but to the one-body state whose wave function is the macroscopic wave function. This notion may be criticized on various grounds, but in practice it makes a useful picture.

A precise formal meaning of the macroscopic wave function, and of Bose–Einstein condensation for interacting systems, is found by considering the one-particle density matrix

$$\rho(\boldsymbol{x}, \boldsymbol{x}') = \langle \hat{\psi}^\dagger(\boldsymbol{x})\hat{\psi}(\boldsymbol{x}')\rangle , \qquad (76.47)$$

which is sufficient to determine the expectation value of any one-particle operator. This is the position representation of a positive Hermitian operator with the trace equal to particle number (or its expectation value) N. In this way, with an orthonormal set of functions $\{\psi_k(\boldsymbol{x})\}_k$ and nonnegative eigenvalues n_k, such that

$$\sum_k n_k = N , \qquad (76.48)$$

an expansion of the form

$$\rho(\boldsymbol{x}, \boldsymbol{x}') = \sum_k n_k\, \psi_k(\boldsymbol{x})\psi_k(\boldsymbol{x}') \qquad (76.49)$$

exists. The system is a BEC if at least one eigenvalue n_k is of the order of the number of particles and does not formally go to zero in the thermodynamic limit (if the limit exists and is sensible). The usual case is that only one eigenstate, call it $k = 0$, has such a large eigenvalue. The macroscopic wave function is the corresponding eigenfunction $\psi \equiv \psi_0$, and n_0 gives the number of condensate atoms. If there is more than one macroscopic eigenvalue, the condensate is called fragmented.

Another interpretation of the macroscopic wave function comes from statistical mechanics. In a continuous (second-order) phase transition typically a symmetry of the system is spontaneously broken. For example, below the Curie temperature a single-domain ferromagnet magnetizes in some specific direction, and the state has a lower symmetry than the rotationally invariant Hamiltonian of an isotropic ferromagnet. Any quantity that appears in a continuous phase transition and characterizes the breaking of the symmetry may be called an order parameter. The macroscopic wave function can be viewed as the order parameter associated with spontaneous breaking of the global phase or "gauge" symmetry of quantum mechanics. Specifically, in quantum mechanics the state of the system is unchanged if the wave function is multiplied by an arbitrary complex phase factor $e^{i\varphi}$. But to write the wave function as $\psi(\boldsymbol{x})$ already implies a preferred phase, and likewise even if the wave function is adorned with random but, for any given condensate, fixed phase, as in $e^{i\varphi}\psi(\boldsymbol{x})$.

For one condensate the random phase is inconsequential. Suppose, however, that two BECs with the wave functions $e^{i\varphi_1}\psi_1(\boldsymbol{x})$ and $e^{i\varphi_2}\psi_2(\boldsymbol{x})$ are combined. If the macroscopic wave functions behave as wave functions should, the combination of two condensates displays the density $n(\boldsymbol{x}) = |e^{i\varphi_1}\psi_1(\boldsymbol{x}) + e^{i\varphi_2}\psi_2(\boldsymbol{x})|^2$. There should be an interference pattern between the condensates. Two BECs indeed produce an interference pattern when they are combined, although the randomness or the absence thereof of the phases $\varphi_{1,2}$ is difficult to verify experimentally.

From the quantum optics viewpoint, a condensate is a given number of atoms in a given one-particle state, a number state, and cannot possess any phase at all. This seems to contradict the observations of an interference pattern. The resolution is that the process of measurement in itself produces a phase difference between the condensates even if there initially is none.

76.2.3 Fermions

Static Fermi Gas

Thomas–Fermi Approximation. Consider an ideal single-species Fermi gas of trapped atoms. The original Thomas–Fermi approximation (see Chapt. 20) was formulated for fermions, namely, electrons, and in the present case it is modified as follows. For the atom density $n(\boldsymbol{x})$ at position $\boldsymbol{x}$, at a low temperature, the corresponding local internal chemical potential is approximated according to (76.18) as

$$\epsilon_F(\boldsymbol{x}) = \frac{\hbar^2\left[6\pi^2 n(\boldsymbol{x})\right]^{2/3}}{2m} . \quad (76.50)$$

Given the trapping potential $V(\boldsymbol{x})$, the density of the gas adjusts in such a way that the sum of the external potential energy and the local internal chemical potential, the Fermi energy, is a constant across the gas,

$$V(\boldsymbol{x}) + \frac{\hbar^2\left[6\pi^2 n(\boldsymbol{x})\right]^{2/3}}{2m} = \mu , \quad (76.51)$$

the global chemical potential. One may solve the density for a given chemical potential as

$$n(\boldsymbol{x}) = \begin{cases} \dfrac{\sqrt{2}\, m^{\frac{3}{2}} \left[\mu - V(\boldsymbol{x})\right]^{\frac{3}{2}}}{3\pi^2\hbar^3}, & V(\boldsymbol{x}) < \mu ; \\ 0, & \text{otherwise.} \end{cases} \quad (76.52)$$

Finally, the integral of the density over all space should equal the atom number, which gives an equation to determine the chemical potential μ.

For a harmonic trap this program can be carried out in an exact, analytical manner with the result that

$$\mu = 6^{1/3} N^{1/3}\hbar\bar{\omega} , \quad R = 2^{2/3} 3^{1/3} N^{1/6} \ell ,$$
$$n(0) = \frac{2\sqrt{N}}{\sqrt{3}\pi^2 \ell^3} . \quad (76.53)$$

The quantities $\bar{\omega}$, ℓ, and R have the same meaning as in the BEC case. The Thomas–Fermi approximation for fermions should be applicable whenever $N \gg 1$.

In a one-component Fermi gas at low temperature atom–atom interactions are typically negligible for a multitude of reasons. There is no s-wave scattering, and the presence of the Fermi sea tends to suppress repulsive interactions. However, in the case of attractive interactions between two species, the Fermi sea may be thermodynamically unstable; the energy may be lowered by pairing fermions into Cooper pairs. This is the mechanism behind the BCS theory of superconductivity [76.8].

Excitations in a Fermi Gas

If the interactions do not render a fermion system into a superfluid, see Sect. 76.5.1, the elementary excitations of a degenerate Fermi gas with short-range interactions are basically atom–hole pairs. What happens in the contrary case for trapped and strongly interacting atoms is presently an active area of research.

76.3 Experimental

76.3.1 Preparing a BEC

In a trapped gas the density $n(\boldsymbol{x})$ is self-determined from the atom number N, and the condition for a BEC in an ideal gas is most readily expressed in terms of the total number of atoms as

$$k_B T_c = 0.94\, \hbar\bar{\omega} N^{1/3} . \quad (76.54)$$

In practice, at the bottom of the trap the conditions on temperature and density for a BEC are similar to the conditions for a BEC in a free gas. In the thermodynamic limit, such that $\bar{\omega} \to 0$, $N \to \infty$ with $\bar{\omega} N^{1/3}$ held constant, below the critical temperature T_c the fraction of condensate atoms behaves as a function of temperature T as

$$\frac{n_0}{N} = 1 - \left(\frac{T}{T_c}\right)^3 . \tag{76.55}$$

The experimental realizations of alkali vapor condensates are based on techniques of laser cooling and trapping of atoms. The following discussion relies heavily on material from Chapt. 75.

A BEC in a dilute atomic gas is usually prepared using a two-stage process. First, a magneto-optical trap is used to capture a sample of cold atoms and to cool it to a temperature of the order of a few tens of microkelvin. The atoms are then transferred to a magnetic trap for evaporative cooling that leads to condensation.

A magnetic trap is based on a combination of two ideas. First, if an atom that starts out with its magnetic moment antiparallel to the magnetic field moves slowly enough in a position dependent magnetic field, its magnetic moment remains adiabatically locked antiparallel to the magnetic field. The energy of the atom is then a minimum where the magnetic field is a minimum. Second, the absolute value of the magnetic field may have a minimum in free space. The minimum is then a trap for atoms whose magnetic moments are suitably oriented. The downside is that only atoms in the right magnetic (Zeeman) states are trapped. While the atoms cool down, they accumulate at the center of the trap. The center should not be a zero of the magnetic field, because at zero field an atom would lose the lock between the directions of the magnetic moment and the magnetic field necessary for trapping.

A time orbiting potential (TOP) trap starts with the same kind of magnetic field that is used in a magneto-optical trap. A time-dependent magnetic field is then added in such a way that the zero of the magnetic field orbits around the center of the trap. If the frequency at which the zero orbits is high enough so that the atoms cannot follow, they see an effective potential with a minimum at the center of the trap and do not sample the zero. Alternatively, it is possible to wind a coil in such a way that it makes a magnetic field whose absolute value has a minimum that is not zero. In this type of a Ioffe–Pritchard trap the winding of the wire resembles the seams on a US baseball.

The basic idea of evaporative cooling is that the most energetic atoms escape from the trap, then the remaining atoms thermalize to a lower temperature. Some atoms are lost in the process, but with the decreasing temperature the density at the trap center nonetheless tends to increase and the phase space density increases even more due to the cooling.

The cooling is usually forced by an rf drive. The transition frequency between the Zeeman states depends on the magnetic field, and increases toward the edges of the trap. Atoms are removed where the rf frequency is on resonance and drives transitions to untrapped Zeeman states. Thus, while the atoms cool and concentrate at the center, the radio frequency is swept down in such a way that the "rf knife" removing the atoms slides in from the edge of the trap. At some radio frequency a condensate abruptly emerges. The temperature can be further lowered by continuing evaporative cooling, albeit at the expense of loss of atoms. As a rule of thumb, an atom needs to experience a hundred collisions before condensation occurs, and a typical time needed to prepare a condensate is a few seconds. In a good vacuum a condensate may live for tens of seconds.

It is also possible to condense atoms trapped in a far-off resonant optical trap based on the dipole forces of light, instead of the magnetic trap [76.9]. For tuning below the resonance, atoms are strong-field seekers. A focused laser beam is a three-dimensional trap for atoms, as is an arrangement with two crossed beams focused to the same spot. Furthermore, with extreme off-resonant light from a CO_2 or a Nd-YAG laser, absorption of photons and the associated photon recoil kicks and heating may be negligible.

An optical trap may also be added after a BEC is prepared in a magnetic trap. The advantage is that an optical trap will hold the atoms regardless of their magnetic state, so that multicomponent "spinor" condensates may be studied. Moreover, while an adiabatic change of the strength of a trap cannot change the phase space density, the phase space density may be altered by changing the shape of the trap by adding a tight optical trap to the bottom of a much wider magnetic trap. Reversible condensation inside an added optical subtrap based on such an increase in the phase space density has been demonstrated.

Methods to condense atoms that might be suited for future technological applications are being pursued. For instance, by lithographic techniques it is possible to put conducting wires on a substrate to make an atom chip. With currents flowing, the wires produce magnetic fields that guide the atoms. Condensation in such a configuration has been reported [76.10]. Two-dimensional condensation in what is known as a gravito-optical surface trap has also been achieved experimentally [76.11].

There is an analogy between a condensate and a beam of light from a laser that we rely on extensively. However, the analogy is only partial. By dropping a condensate under gravity one makes a pulsed atom laser, and by coupling a trapped Zeeman state to an untrapped state by rf excitation it is possible to make a condensate leak slowly out of the trap. Nonetheless, at this time a method to produce a continuous beam of condensate atoms, a continuous-wave atom laser, is yet to be demonstrated.

76.3.2 Preparing a Degenerate Fermi Gas

A single-species, very-low temperature Fermi gas is an uninteresting system, as the Fermi–Dirac statistics forbids s-wave interactions between the atoms and the gas is nearly ideal. In experiments the gas usually has two species, different states of the same atom. The interactions between the species are comparable in strength to the interactions between bosonic atoms.

Evaporative cooling works in a two-species gas, and can be used to prepare a degenerate Fermi gas either in a magnetic trap [76.12] or in an optical trap [76.13]. Second, one can use a gas of bosons, and indeed a BEC, as a refrigerator [76.14]. At this writing the lowest temperatures are of the order of $0.1\,T_F$. Reaching lower temperatures is complicated by various factors, such as the very low heat capacity of a BEC and collisions becoming inefficient at low temperature because the inert Fermi sea reduces the available phase space. Nonetheless, lower temperatures seem to be mainly a matter of advances in technology.

76.3.3 Monitoring Degenerate Gases

Orders of Magnitude. As a rule of thumb, the trapping frequencies in a magnetic trap are $\bar{\omega} \approx 2\pi \times 10$ Hz, while the frequencies in an optical trap may reach into the kHz regime. A typical oscillator length is $\ell \approx 1\,\mu$m. A usual number of atoms is $N \approx 10^6$. Scattering lengths are of the order of $a \approx 10$ nm. The size of a degenerate gas is in the neighborhood of $R \approx 0.1$ mm, the maximum density is about $n_0 \approx 10^{15}\,\text{cm}^{-3}$, and the BEC transition temperature and the Fermi temperature are of the order of $T_c \approx T_F \approx 1\,\mu$K. However, much lower temperatures are readily reached in a BEC.

Phase Contrast Imaging. It is possible to monitor condensate features substantially larger than the wavelength of the light used in the measurements nondestructively, in situ, by using phase contrast imaging. In this method the light is detuned far off resonance so that absorptions with the accompanying photon recoil kicks on the atoms are rare, but the phase of the light nonetheless changes upon propagation through the sample. The phase change may be detected by interfering the transmitted light with the original light, with the phase of the latter suitably shifted.

Time-of-Flight Imaging. Usually, though, the observation of a degenerate gas at the end of an experiment is by time-of-flight imaging. The trap is suddenly removed, whereupon the gas expands freely. After the atom cloud has grown to a size large enough compared to the wavelength of the resonant light used to monitor the gas, an absorption image of the cloud is taken. This gives the projection of the density of the gas onto a plane perpendicular to the direction of propagation of the light. Except for the effects of atom–atom interactions, after a sufficiently long time of free flight the density reflects the initial momentum distribution of the atoms. Time-of-flight images bear the signs of both condensation in a Bose gas and quantum degeneracy in a Fermi gas. Nontrivially, other features of interest such as vortex cores are also preserved and can be detected after the free expansion. The downside is that the time-of-flight method is destructive. After each snapshot the sample will have to be prepared again.

76.4 BEC Superfluid

76.4.1 Vortices

Flow Velocity in a Superfluid. By manipulating the Heisenberg equations of motion for a Bose field under the Hamiltonian (76.22) it is easy to derive the equation of continuity for the atoms,

$$\frac{\partial}{\partial t}\hat{n} + \nabla \cdot \hat{\mathrm{J}} = 0\,, \tag{76.56}$$

$$\hat{n} = \hat{\psi}^\dagger \hat{\psi}, \quad \hat{\mathrm{J}} = \mathrm{i}\frac{\hbar}{2m}\left(\psi \nabla \psi^\dagger - \psi^\dagger \nabla \psi\right), \tag{76.57}$$

which identifies $\hat{n}$ and $\hat{J}$ as the operators for atom density and atom current density. The corresponding mean-field quantities are obtained when again the boson fields are replaced with the corresponding classical fields. Writing the classical field in terms of the density $n(\boldsymbol{x}, t)$ and phase $\varphi(\boldsymbol{x}, t)$ in the form

$$\psi = \sqrt{n}\, e^{i\varphi}\,, \tag{76.58}$$

the local flow velocity $\boldsymbol{v} = \boldsymbol{j}/n$ becomes

$$\boldsymbol{v} = \frac{\hbar}{m} \nabla\varphi\,. \tag{76.59}$$

The velocity field is irrotational.

Quantization of Circulation. Integration of the flow velocity around an arbitrary loop gives

$$\oint d\boldsymbol{l} \cdot \boldsymbol{v} = \frac{\hbar}{m} \Delta\varphi = p\,\frac{2\pi\hbar}{m}\,, \quad p = 0, \pm 1, \pm 2, \ldots\,, \tag{76.60}$$

since the change of the phase $\Delta\varphi$ around a closed loop must be an integer multiple of 2π. Equation (76.60) expresses the quantization of circulation in a superfluid. A medium described by a macroscopic wave function, such as a BEC, cannot sustain arbitrary flow velocities.

Vortices. As an example, in bulk rotation at an angular velocity Ω the line integral around a loop at the distance r from the axis of rotation would be $2\pi r^2 \Omega$, which is not permitted for an arbitrary r. Instead, upon an attempt to make a BEC rotate, the angular velocity will be carried by vortex lines. These are lines through the condensate, entering and exiting at the surface, such that each vortex carries one quantum of circulation. At the core of a vortex the flow velocity should be infinite to sustain a finite circulation, which is physically impossible. Nature solves this problem by making the vortex core normal (not BEC), so that the macroscopic wave function does not apply. The diameter of the vortex core is of the order of the healing length ξ, given by (76.29).

When the trapping potential on the atoms is rotated, it is convenient to study the physics in the co-rotating frame. Given a frame rotating at the angular velocity $\boldsymbol{\Omega}$ and the angular momentum operator $\boldsymbol{L} = \boldsymbol{x} \times \boldsymbol{p}$ per particle, transformation to the rotating frame adds the one-particle term

$$H_r = -\boldsymbol{\Omega} \cdot \boldsymbol{L} \tag{76.61}$$

to the Hamiltonian. Any particular configuration of vortices is a thermodynamically stable equilibrium if it is the minimum of energy in the co-rotating frame. For a trapped condensate, at zero rotation velocity the state without vortices is the energy minimum, and increasing the rotation speed makes states with an increasing number of vortices the stable configuration. However, a vortex configuration may be metastable and live for a long time even if it is not the minimum of energy. Conversely, even the energy-minimum configuration of vortices must first be nucleated. Since the circulation can only have quantized values, it cannot change in a continuous process. It takes a zero condensate density somewhere to create or destroy a vortex.

These alternatives provide a large number of experimental scenarios involving rotation of the trap or stirring of the condensate, condensation of a rotating normal gas by taking it across the transition temperature, and so forth. For instance, when a trap containing a BEC is rotated, vortices are generated at the surface where they start their lives as dynamical instabilities. The vortices then drift in and form a regular vortex array [76.15]. When the rotation is halted, the vortices drift out to the surface and disappear.

76.4.2 Superfluidity

A BEC also has the remarkable property that it may sustain persistent currents that are completely immune to viscosity. The qualitative reason may be seen from the dispersion relation of small excitations (76.38). As long as the flow speed $|\boldsymbol{v}|$ is less than the speed of sound c, all excitation energies are positive, so that the flowing state is the state of lowest energy and is thermodynamically stable. On the other hand, when the flow velocity exceeds the speed of sound, the system has excitations that lower the energy, $\nu(\boldsymbol{q}) < 0$ for some $\boldsymbol{q}$. The flowing state is then not a minimum of energy. The flow is not thermodynamically stable, and it decays when it interacts with an environment by sending off small excitations. The speed of sound gives the Landau critical velocity for superfluidity.

The critical velocity c tends to zero when the atom–atom interactions vanish with $a \to 0$. While the condensate wave function may be written down whether the atoms interact or not, superfluidity and persistent flows rely on the interactions. The same applies to vortices, as in the limit of a noninteracting gas the healing length and the radius of the vortex core tend to infinity.

The conventional picture is that superfluid flow in an inhomogeneous medium is unstable if the local flow velocity exceeds the local (density dependent) speed of sound. In practice, in liquid He experiments and numerical simulations of dilute condensates the current often dissipates by shedding vortices when it flows too fast past an obstacle. It is not clear if the conventional picture is, or should be expected to be, quantitatively accurate. In fact, at this time there are no experiments with alkali vapor condensates in toroidal geometries that would offer natural conduits for persistent currents.

76.5 Current Active Topics

76.5.1 Atom–Molecule Systems

Diatomic molecules and conversion between atoms and molecules at temperatures low enough to render the system quantum degenerate are at present probably the most active frontier in the studies of ultracold gases. Experimental achievements include a condensate of molecules, and coherent transitions between two chemically different species. More broadly, unforeseen new angles open up into long-standing issues in superfluid systems in condensed matter physics, such as strongly interacting superfluids and BEC-BCS crossover. A snapshot of the field at this time is given in the present section.

Two colliding asymptotically free atoms cannot in general combine into a diatomic molecule, as energy and momentum would not be conserved in the process. However, there are two mathematically equivalent methods to adjust energy conservation, photoassociation and Feshbach resonance [76.16], both of which may lead to molecule formation.

The underlying idea is that two seemingly free atoms may be regarded as a dissociated state of a corresponding diatomic molecule. In photoassociation a laser drives transitions from a dissociated two-atom state to a bounded state of the molecule. Energy conservation is adjusted by tuning the laser. In a Feshbach resonance hyperfine interactions drive transitions from a two-atom state in a particular manifold of electronic states to a molecular state in another manifold of electronic states. The magnetic moments of the two-atom state and the bounded molecular state are different, so that the resonance may be tuned by varying the magnetic field applied on the atom–molecule gas.

Basic Atom–Molecule Model for Bosons. Consider a minimal model for conversion of bosonic atoms (boson field $\hat{\phi}$) into bosonic molecules ($\hat{\psi}$). The Hamiltonian density reads

$$\begin{aligned}\frac{\hat{\mathcal{H}}}{\hbar} &= \hat{\phi}^\dagger\left(-\frac{\hbar}{2m}\nabla^2\right)\hat{\phi} + \hat{\psi}^\dagger\left(-\frac{\hbar}{4m}\nabla^2+\delta\right)\hat{\psi} \\ &\quad + g(\hat{\psi}^\dagger\hat{\phi}\hat{\phi} + \hat{\psi}\hat{\phi}^\dagger\hat{\phi}^\dagger)\,. \end{aligned} \tag{76.62}$$

This model is for free atoms. Atom–molecule coupling is described as a contact interaction characterized by the coupling coefficient g. The detuning δ may be adjusted in photoassociation by tuning the laser, and in a Feshbach resonance by varying the magnetic field.

Effective Scattering Length. Heisenberg equations of motion for the atomic and molecular fields read

$$\mathrm{i}\frac{\mathrm{d}}{\mathrm{d}t}\hat{\phi} = -\frac{\hbar}{2m}\nabla^2\hat{\phi} + 2g\hat{\phi}^\dagger\hat{\psi}\,, \tag{76.63}$$

$$\mathrm{i}\frac{\mathrm{d}}{\mathrm{d}t}\hat{\psi} = \left(-\frac{\hbar}{4m}\nabla^2+\delta\right)\hat{\psi} + g\hat{\phi}\hat{\phi}\,. \tag{76.64}$$

Suppose now that the detuning from resonance, $|\delta|$, is the largest frequency parameter in the problem, then one may solve the molecular field adiabatically from (76.64) as $\hat{\psi} = -g\hat{\phi}\hat{\phi}/\delta$. Inserting this into (76.63) gives

$$\mathrm{i}\frac{\mathrm{d}}{\mathrm{d}t}\hat{\phi} = -\frac{\hbar}{2m}\nabla^2\hat{\phi} - \frac{2g^2}{\delta}\hat{\phi}^\dagger\hat{\phi}\hat{\phi}\,, \tag{76.65}$$

which is the Heisenberg equation of motion for the atomic field that ensues from an effective Hamiltonian density

$$\mathcal{H}_\mathrm{E} = \hat{\phi}^\dagger\left(-\frac{\hbar^2}{2m}\nabla^2\right)\hat{\phi} + \frac{2\pi\hbar^2 a_\mathrm{E}}{m}\hat{\phi}^\dagger\hat{\phi}^\dagger\hat{\phi}\hat{\phi} \tag{76.66}$$

with the effective scattering length

$$a_\mathrm{E} = -\frac{mg^2}{2\pi\hbar\delta}\,. \tag{76.67}$$

Experimentally, the modification of the scattering length is in addition to a constant "background" scattering length a_0, and the sum of the two scattering lengths

$$a = a_0 + a_\mathrm{E}\,. \tag{76.68}$$

is usually reported.

As long as one stays sufficiently far away from an atom–molecule resonance, tuning the resonance condition is tantamount to tuning the atom–atom scattering length. When the detuning is negative (positive), the energy of a molecule is lower (higher) than the energy of an on-threshold pair of atoms, the corresponding induced scattering length is positive (negative), and the induced atom–atom interactions are in effect repulsive (attractive). It should be noted, though, that the scattering length does not simply become very large. Vernacular of this style, and associated attempts to study the theory of an interacting Bose gas in the limit when the gas parameter na_E^3 is not small, are occasionally misguided and misleading. The physics rather is in resonant, energy-conserving conversion between atoms and molecules.

Atom–Molecule Coupling Strength. Given the difference in magnetic moments $\Delta\mu$ between a molecule and two atoms and the position of the Feshbach resonance B_0, the detuning is

$$\delta = \frac{\Delta\mu(B - B_0)}{\hbar} . \tag{76.69}$$

The variation of the scattering length is conventionally parametrized in terms of the magnetic field width of the Feshbach resonance ΔB as

$$a = a_0 \left(1 - \frac{\Delta B}{B - B_0}\right) . \tag{76.70}$$

A combination of (76.67–76.70) gives the relation between the field width and the contact interaction parameter characterizing the Feshbach resonance,

$$g = \sqrt{\frac{2\pi |a_0 \, \Delta\mu \, \Delta B|}{m}} . \tag{76.71}$$

Unfortunately, the difference in magnetic moments $\Delta\mu$ is not always publicized and one may be forced to estimate, say, $\Delta\mu \approx \mu_B$.

Two-Mode Model. Consider next, in the mean-field approximation, the case when only uniform atomic and molecular condensates are present. The condensates are represented by complex amplitudes α and β such that $\hat\phi \to \sqrt{n}\,\alpha$ and $\hat\psi \to \sqrt{n/2}\,\beta$, where n now is the invariant density equal to atom density plus twice the density of the molecules. It follows from (76.63) and (76.64) that the probability amplitudes for atoms and molecules, normalized as $|\alpha|^2 + |\beta|^2 = 1$, satisfy

$$\mathrm{i}\dot\alpha = \frac{\Omega}{\sqrt{2}}\alpha^*\beta, \quad \mathrm{i}\dot\beta = \delta\beta + \frac{\Omega}{\sqrt{2}}\alpha^2 . \tag{76.72}$$

These are nonlinear variations of the usual two-level equations (see Chapt. 73) of quantum optics, with the quantity $\Omega = \sqrt{n}\,g$ playing the role of the Rabi frequency. This system displays analogs of coherent optical transients (see Chapt. 73), such as Rabi oscillations between atomic and molecular condensates, and adiabatic following from an atomic condensate to a molecular condensate when the detuning is swept through the resonance [76.17].

The two-mode model is simplistic in that it ignores processes in which molecules dissociate into correlated pairs of noncondensate atoms [76.18, 19]. There are also secondary complications. The molecules created in a Feshbach resonance are highly vibrationally excited, and tend to get quenched in collisions with atoms and other molecules. Typical lifetimes are in the millisecond regime. Usual one-color photoassociation from two atoms to a molecule with the absorption of a photon, on the other hand, creates an electronically excited molecule, which decays spontaneously on a time scale far shorter than the typical photoassociation time scales. To mitigate spontaneous emission, one usually resorts to two-color photoassociation, in which a second laser takes the photoassociated molecules to another more stable level [76.20].

At present, the two-mode system with just an atomic and a molecular condensate has never been realized cleanly in an experiment. Nonetheless, experiments using a Feshbach resonance have demonstrated Ramsey fringes in transitions between atomic and molecular condensates [76.21], and formation of what probably is a (short-lived) molecular condensate from the bosonic isotope ^{23}Na [76.22].

Fermion Systems. Combining two fermionic atoms gives a bosonic molecule, and Feshbach resonance in a Fermi gas is currently a popular topic. The main interest is in the BEC-BCS crossover. Basically, if the magnetic field is tuned so that the detuning is negative, molecules have a lower energy than atoms. Thermal equilibration then leads to molecules that will condense if the temperature is low enough. On the other hand, if the detuning is positive, atom–atom interactions are attractive. The atoms may undergo a phase transition into a fermion superfluid that is analogous to the BCS phase transition in a superconductor. What happens in between has been a question in theory for a while [76.23], and is finally accessible to experiments.

Collisionless adiabatic transfer from atoms to molecules by sweeping the magnetic field across a Feshbach resonance works with fermions much like with bosons [76.24]. Moreover, molecules formed in the 834 G Feshbach resonance in the fermionic isotope ^{6}Li may live for seconds. It is now possible to study thermal equilibrium and long-lived excited states in the neighborhood of the resonance [76.25–27]. In particular, the observation of a vortex lattice over a wide range of magnetic fields on both sides of the resonance [76.24] indicates that close to the resonance the gas is a strongly interacting superfluid.

76.5.2 Optical Lattice with a BEC

A BEC confined to a periodic potential enjoys a long-standing popularity for reasons that have varied in time. In the early days of BEC the Josephson effect and phase behavior of a BEC were topical. The possibility of a quantum phase transition in an optical-lattice system was the next broad topic to emerge, and nowadays speculations about using condensates in a lattice either as supporting technology or as the active element in quantum information processing abound. The deceptively simple theoretical models of these systems add to their staying power. In this section, a brief discussion of an optical lattice holding a BEC is presented. The quantum information view is not pursued here, as so far no specific experimental progress has appeared in print.

Optical Lattice. Dipole forces of standing-wave fields of light generate a periodic potential, an optical lattice, on the atoms. If the light is detuned far enough from atomic resonances, absorption and spontaneous emission are negligible and the potential is conservative. In one dimension, the potential energy for the motion of the atoms is typically of the form

$$V(x) = V_0 \sin^2 kx \,, \tag{76.73}$$

where k is the wave number of the lattice light, and the depth of the lattice V_0 can be inferred from the known parameters of the atoms and the light as explained in Sect. 75.2.2.

Double-Well Potential. One can integrate the GPE numerically for an arbitrary potential, but many more insights have been gained from restricted models. The simplest one is a double-well potential in the two-state approximation, in which only the ground state of the atoms in both wells is taken into account. The Hamiltonian is

$$\frac{\hat{H}}{\hbar} = -\frac{\Delta}{2}\left(a_\mathrm{l}^\dagger a_\mathrm{r} + a_\mathrm{r}^\dagger a_\mathrm{l}\right) + 2\kappa\left[\left(a_\mathrm{l}^\dagger a_\mathrm{l}\right)^2 + \left(a_\mathrm{r}^\dagger a_\mathrm{r}\right)^2\right] . \tag{76.74}$$

Here Δ is a parameter characterizing tunneling between the "left" and the "right" potential well, $a_{\mathrm{l,r}}$ are the annihilation operators for ground-state bosons in each well, and κ is a measure of atom–atom interactions. The one-particle states in a symmetric double-well potential with weak tunneling come in doublets, one even state ϕ_+ and one odd state ϕ_-, with respect to the center of the double-well trap. The choices of signs of the wave functions are assumed to work out in such a way that the left and right states are $\phi_{\mathrm{l,r}} = \sqrt{1/2}\,(\phi_+ \pm \phi_-)$. Equation (76.74) could be the version of the Hamiltonian (76.22) restricted to the basis of the two states $\phi_{\mathrm{l,r}}$.

There are many forms of the Hamiltonian (76.74) that differ by a polynomial of the conserved particle number $\hat{N} = \left(a_\mathrm{l}^\dagger a_\mathrm{l} + a_\mathrm{r}^\dagger a_\mathrm{r}\right)$. Inasmuch as particle number is fixed, adding any function of the conserved quantity $\hat{N}$ to the Hamiltonian has no effect on the dynamics, and so such forms are functionally equivalent. Here polynomials of $\hat{N}$ are added to the Hamiltonian without further notice to produce the simplest-looking results.

Semiclassical Approximation. The usual method of going to the classical field theory gives the equations for the semiclassical amplitudes defined by $a_{\mathrm{l,r}}/\sqrt{N} \to \alpha_{\mathrm{l,r}}$,

$$\begin{aligned} \mathrm{i}\dot{\alpha}_\mathrm{l} &= -\frac{1}{2}\Delta\,\alpha_\mathrm{r} + \chi\,|\alpha_\mathrm{l}|^2\alpha_\mathrm{l}, \\ \mathrm{i}\dot{\alpha}_\mathrm{r} &= -\frac{1}{2}\Delta\,\alpha_\mathrm{l} + \chi\,|\alpha_\mathrm{r}|^2\alpha_\mathrm{r}\,, \end{aligned} \tag{76.75}$$

with $\chi = 4\kappa N$. Without atom–atom interactions these are the equations of a resonant two-level system and describe Josephson oscillations of the atoms between the sides of the double-well potential. Interactions temper the oscillations, or stop them completely [76.28].

Phase Diffusion in the Double Well. The model (76.74) is unusual in that one can easily go beyond the semiclassical approximation [76.29]. The entire state space for N atoms is spanned by the vectors $|n_\mathrm{l}, n_\mathrm{r}\rangle = |n_\mathrm{l}, N - n_\mathrm{l}\rangle$ with $n_\mathrm{l} = 0, \ldots, N$. The Hamiltonian can be diagonalized and the time dependence of the system solved numerically even for large N. Moreover, the common case in which the atom number fluctuations at the sites are at least of the order unity, but small in a relative sense, is amenable to a simple analytical approximation.

The phase difference of the condensates between the two traps, $\hat{\varphi}$, is a case in point. One can measure it by releasing the atoms from the trap and letting them interfere. Although there are serious in-principle problems with this interpretation, phase difference can be viewed qualitatively as the canonical conjugate of the difference between the number of atoms on the sides of the double well, $\hat{n} \equiv \hat{n}_\mathrm{l} - \hat{n}_\mathrm{r}$, with $[\hat{n}, \hat{\varphi}] = -i$. The minimum uncertainty product of phase difference and atom number difference is then 1/2. On this basis, results for various experiments can be qualitatively and quantitatively predicted.

For instance, suppose the system is prepared in the ground state of the Hamiltonian (76.74). The ground state is an even split of the atoms between the two traps, but atom number fluctuations depend on the ratio between tunneling and atom–atom interactions. For $|\Delta|/\chi \ll 1$, atom–atom interactions dominate, and since moving an atom from one side of the trap to the other costs much interaction energy, the ground state is close to a number state with half of the atoms in each potential well. In the contrary case, the ground state is essentially the many-body state with all N atoms in the symmetric state ϕ_+, which breaks up into a Poissonian distribution of the atoms between the states ϕ_l and ϕ_r. As long as the standard deviation of the atom number difference is at least of the order of unity, it is given by

$$\Delta n = \sqrt{N}\left(\frac{\Delta}{\Delta + 4N\kappa}\right)^{1/4}, \tag{76.76}$$

and the phase fluctuations are

$$\Delta\varphi = \frac{1}{2\,\Delta n}. \tag{76.77}$$

As discussed above in Sect. 76.3.2, a measurement of the phase difference will produce a definite result. In the interaction-dominated case the result should in effect be random, while in the tunneling-dominated case the phase difference should come out the same every time, save for fluctuations of the order $1/\sqrt{N}$. Similarly, if one were to start from the ground state in the case when tunneling dominates, and then suddenly turn off tunneling by adjusting the potential well, atom–atom interactions would lead to diffusion (or rather, dispersion) of the phase difference. The result of a phase measurement becomes increasingly random with time according to

$$\Delta\varphi(t) = \frac{1}{2}\sqrt{\frac{1}{N} + 16N\kappa^2 t^2}. \tag{76.78}$$

Bose–Hubbard Model. The corresponding multi-well problem goes under the rubrics fo the Bose–Hubbard model and tight-binding approximation. The Hamiltonian is

$$\frac{\hat{H}}{\hbar} = \sum_n \left[-\frac{\Delta}{2}\left(a^\dagger_{n+1}a_n + a^\dagger_{n-1}a_n\right) + 2\kappa\left(a^\dagger_n a_n\right)^2\right]. \tag{76.79}$$

The sum runs over the sites of the optical lattice, which is taken to be one-dimensional in this example, and Δ and κ again characterize tunneling and atom–atom interactions. At the ends of the lattice there are some boundary conditions, but for a long enough lattice they do not influence the physics.

One-particle eigenstates in a periodic lattice are organized in energy bands, but there is a well-known transformation in condensed matter physics that makes orthonormal Wannier states, more or less localized in the lattice sites, out of the states in each band. The most rigorous interpretation of the Hamiltonian (76.79) is that it is the representation of the Hamiltonian (76.22) in the Wannier states belonging to the lowest energy band of the lattice, ignoring tunneling between non-adjacent sites. Viewed in this way, the model is only valid in the limit when the energy per atom for atom–atom interactions is small compared to the energy spacing between the bands.

Nonetheless, direct integrations of the GPE with atom–atom interactions also show energy bands, and for suitably picked parameters the Hamiltonian (76.79) should be generic for the case when interband transitions are negligible. Now, without atom–atom interactions the width of the energy band from the Bose–Hubbard model would be $\hbar\Delta$. On the other hand, the band structure for the potential energy (76.73) of an optical lattice comes from the Mathieu equation [76.30]. A comparison gives the estimate

$$\Delta = \frac{8}{\sqrt{\pi}}\frac{E_\text{r}}{\hbar}\left(\frac{V_0}{E_\text{r}}\right)^{3/4}\exp\left(-2\sqrt{\frac{V_0}{E_\text{r}}}\right). \tag{76.80}$$

This is an asymptotic expression for the limit when the recoil energy $E_\text{r} = \hbar^2 k^2/2m$ and the lattice depth V_0 satisfy $V_0 \gg E_\text{r}$. Next suppose that one ascribes to each potential well the unit-normalized wave function $\phi(\boldsymbol{x})$, then an estimate for the atom–atom interaction parameter κ comes from (76.23) in the form

$$\kappa = \frac{\pi\hbar a}{m}\int \mathrm{d}^3x\,|\phi(\boldsymbol{x})|^4. \tag{76.81}$$

Phase Diffusion in the Bose–Hubbard Model. The dominant feature of the Bose–Hubbard model again is competition between tunneling and atom–atom interactions. Also, phase and atom number fluctuations may again be studied analytically under the assumptions that atom number fluctuations at each site are at least of the order of unity, but small in a relative sense. For the same numbers of atoms per site, the results are basically the same in the two- and multi-well cases.

In fact, multi-well counterparts [76.31] of the experiments on the phase relations are well ahead of the experiments on two-well systems [76.32]. The multi-well experiments are in satisfactory agreement with the theory.

Superfluid–Mott Insulator Transition. Because the size of the state space tends to grow as L^L with the number of lattice sites L, direct numerical solutions to the multi-well problem are computationally intractable. This is somewhat unfortunate, as a long lattice also presents behaviors with no obvious analogs in the two-well case, and for which analytical approximations have proven hard to come by. When tunneling dominates, the system is in what is referred to as the superfluid phase. Fluctuations of atom number between the states are relatively large. On the other hand, if atom–atom interactions dominate, it becomes costly in energy to put anything but an exact number state of atoms at each lattice site. This is the Mott insulator phase. According to calculations carried out using the so-called Gutzwiller ansatz, the ground state of an optical lattice inserted in an atom trap consists of regions with the same integer number of atoms at the lattice sites within each region [76.33]. When the parameters of the system are varied, in what is known as a quantum phase transition, the system should abruptly switch between these phases.

The superfluid-Mott insulator transition has been observed experimentally [76.34]. Lattice parameters, especially tunneling, can be varied easily by changing the intensity of the lattice light. The observation of the transition is by means of phase coherence. In the superfluid state the system is characterized by a global macroscopic wave function. When the atoms are released from the lattice, atoms originating from different sites are capable of interference, and the interference pattern reflects the lattice structure. On the other hand, in the Mott insulator phase the lattice sites are in number states with little phase coherence between them, and there is no interference pattern.

References

76.1 F. Dalfovo, S. Giorgini, L. P. Pitaevskii, S. Stringari: Rev. Mod. Phys. **71**, 463 (1999)
76.2 A. S. Parkins, D. F. Walls: Phys. Rep. **C303**, 1 (1998)
76.3 A. J. Leggett: Rev. Mod. Phys. **73**, 307 (2001)
76.4 S. Stenholm: Phys. Rep. **C363**, 173 (2002)
76.5 A. L. Fetter: J. Low. Temp. Phys. **129**, 263 (2002)
76.6 A. L. Fetter, J. D. Walecka: *Quantum Theory of Many-Particle Systems* (McGraw-Hill, New York 1971)
76.7 M. D. Feit, J. A. Fleck Jr., A. Steiger: Solution of the Schrödinger equation by a spectral method, J. Comput. Phys. **47**, 412 (1982)
76.8 J. R. Schrieffer: *Theory of Superconductivity* (Addison-Wesley, Redwood City 1988)
76.9 M. D. Barrett, J. A. Sauer, M. S. Chapman: All-optical formation of an atomic Bose-Einstein condensate, Phys. Rev. Lett. **87**, 010404 (2001)
76.10 S. Schneider, A. Kasper, Ch. vom Hagen, M. Bartenstein, B. Engeser, T. Schumm, I. Bar-Joseph, R. Folman, L. Feenstra, J. Schmiedmayer: Bose-Einstein condensation in a simple microtrap, Phys. Rev. A **67**, 023612 (2003)
76.11 D. Rychtarik, B. Engeser, H.-C. Nägerl, R. Grimm: Two-dimensional Bose–Einstein condensate in an optical surface trap, Phys. Rev. Lett. **92**, 173003 (2004)
76.12 B. DeMarco, D. S. Jin: Onset of Fermi degeneracy in a trapped atomic gas, Science **285**, 1703 (1999)
76.13 K. M. O'Hara, S. L. Hemmer, M. E. Gehm, S. R. Granade, J. E. Thomas: Observation of a strongly interacting degenerate Fermi gas of atoms, Science **298**, 2179 (2002)
76.14 A. G. Truscott, K. E. Strecker, W. I. McAlexander, G. B. Partridge, R. G. Hulet: Observation of Fermi pressure in a gas of trapped atoms, Science **291**, 2570 (2001)
76.15 J. R. Abo-Shaeer, C. Raman, J. M. Vogels, W. Ketterle: Observation of Vortex Lattices in Bose-Einstein Condensates, Science **292**, 476 (2001)
76.16 E. Timmermans, P. Tommasini, M. Hussein, A. Kerman: Feshbach resonances in atomic Bose-Einstein condensates, Phys. Rep. **C315**, 199 (1999)
76.17 J. Javanainen, M. Mackie: Coherent photoassociation of a Bose-Einstein condensate, Phys. Rev. A **59**, R3186 (1999)
76.18 S. J. J. M. F. Kokkelmans, M. J. Holland: Ramsey fringes in a Bose–Einstein condensate between atoms and molecules, Phys. Rev. Lett. **89**, 180401 (2002)
76.19 M. Mackie, K.-A. Suominen, J. Javanainen: Mean-field theory of Feshbach-resonant interactions in ^{85}Rb condensates, Phys. Rev. Lett. **89**, 180403 (2002)
76.20 R. Wynar, R. S. Freeland, D. J. Han, C. Ryu, D. J. Heinzen: Molecules in a Bose-Einstein condensate, Science **287**, 1016 (2000)
76.21 E. A. Donley, N. R. Claussen, S. T. Thompson, C. E. Wieman: Atom-molecule coherence in a Bose-Einstein condensate, Nature **417**, 529 (2002)
76.22 K. Xu, T. Mukaiyama, J. R. Abo-Shaeer, J. K. Chin, D. E. Miller, W. Ketterle: Formation of quantum-degenerate sodium molecules, Phys. Rev. Lett. **91**, 210402 (2003)
76.23 M. Holland, S. J. J. M. F. Kokkelmans, M. L. Chiofalo, R. Walser: Resonance superfluidity in a quantum

degenerate Fermi gas, Phys. Rev. Lett. **87**, 120406 (2001)
76.24 C. A. Regal, C. Ticknor, J. L. Bohn, D. S. Jin: Creation of ultracold molecules from a Fermi gas of atoms, Nature **424**, 47 (2003)
76.25 S. Jochim, M. Bartenstein, A. Altmeyer, G. Hendl, S. Riedl, C. Chin, J. Hecker Denschlag, R. Grimm: Bose–Einstein condensation of molecules, Science **302**, 2101 (2003)
76.26 M.W. Zwierlein, C.A. Stan, C.H. Schunck, S.M.F. Raupach, A.J. Kerman, W. Ketterle: Condensation of pairs of fermionic atoms near a Feshbach resonance, Phys. Rev. Lett. **92**, 120403 (2004)
76.27 M.W. Zwierlein, J.R. Abo-Shaeer, A. Schirotzek, C.H. Schunck, W. Ketterle: Vortices and superfluidity in a strongly interacting Fermi gas, Nature **435**, 1047 (2005)
76.28 S. Raghavan, A. Smerzi, S. Fantoni, S.R. Shenoy: Coherent oscillations between two weakly coupled Bose-Einstein condensates: Josephson effects, π oscillations, and macroscopic quantum self-trapping, Phys. Rev. A **59**, 620 (1999)
76.29 J. Javanainen, M. Yu. Ivanov: Splitting a trap containing a Bose–Einstein condensate: Atom number fluctuations, Phys. Rev. A **60**, 2351 (1999)
76.30 M. Abramowitz, I. A. Stegun: *Handbook of Mathematical Functions* (Dover, New York 1970)
76.31 C. Orzel, A. K. Tuchman, M. L. Fenselau, M. Yasuda, M. A. Kasevich: Squeezed states in a Bose–Einstein condensate, Science **291**, 2386 (2001)
76.32 Y. Shin, M. Saba, T.A. Pasquini, W. Ketterle, D.E. Pritchard, A.E. Leanhardt: Atom interferometry with Bose-Einstein condensates in a double-well potential, Phys. Rev. Lett. **92**, 050405 (2004)
76.33 D. Jaksch, C. Bruder, J.I. Cirac, C.W. Gardiner, P. Zoller: Cold bosonic atoms in optical lattices, Phys. Rev. Lett. **81**, 3108 (1998)
76.34 M. Greiner, O. Mandel, T. Esslinger, T.W. Hänsch, I. Bloch: Quantum phase transition from a superfluid to a Mott insulator in a gas of ultracold atoms, Nature **415**, 39 (2002)

77. De Broglie Optics

De Broglie optics concerns the propagation of matter waves, their reflection, refraction, diffraction and interference. The main subfields of de Broglie optics are electron optics [77.1], neutron optics [77.2], and atom optics [77.3]. Well-established applications are found in electron diffraction and microscopy [77.4], electron holography [77.5], neutron diffraction and interferometry [77.6]. The subject of atom optics is relatively new, and applications are currently being developed in precision spectroscopy, precision measurement, atom lithography, atom microscopy, and atom interferometry [77.7–9]. This chapter concentrates on the principles of de Broglie optics. Illustrations of these principles will be presented mainly in the framework of atom optics.

A typical de Broglie optical experiment involves a source, a beam of particles produced by that source, an array of optical elements, possibly a probe, other optical elements placed behind the probe, and finally a detector. Optical elements are collimators, apertures, lenses, mirrors, and beam-splitters. A probe could be a crystal, a biological sample, or just another unknown optical element whose properties are to be investigated.

To avoid proliferation of notation, we refer to any object placed in the beam path simply as an "optical element".

77.1 **Overview** ... 1125

77.2 **Hamiltonian of de Broglie Optics** ... 1126
- 77.2.1 Gravitation and Rotation ... 1127
- 77.2.2 Charged Particles ... 1127
- 77.2.3 Neutrons ... 1127
- 77.2.4 Spins ... 1127
- 77.2.5 Atoms ... 1127

77.3 **Principles of de Broglie Optics** ... 1129
- 77.3.1 Light Optics Analogy ... 1129
- 77.3.2 WKB Approximation ... 1130
- 77.3.3 Phase and Group Velocity ... 1130
- 77.3.4 Paraxial Approximation ... 1130
- 77.3.5 Raman–Nath Approximation ... 1131

77.4 **Refraction and Reflection** ... 1131
- 77.4.1 Atomic Mirrors ... 1131
- 77.4.2 Atomic Cavities ... 1132
- 77.4.3 Atomic Lenses ... 1132
- 77.4.4 Atomic Waveguides ... 1132

77.5 **Diffraction** ... 1133
- 77.5.1 Fraunhofer Diffraction ... 1133
- 77.5.2 Fresnel Diffraction ... 1133
- 77.5.3 Near-Resonant Kapitza–Dirac Effect ... 1133
- 77.5.4 Atom Beam Splitters ... 1134

77.6 **Interference** ... 1135
- 77.6.1 Interference Phase Shift ... 1135
- 77.6.2 Internal State Interferometry ... 1136
- 77.6.3 Manipulation of Cavity Fields by Atom Interferometry ... 1137

77.7 **Coherence of Scalar Matter Waves** ... 1137
- 77.7.1 Atomic Sources ... 1137
- 77.7.2 Atom Decoherence ... 1138

References ... 1139

77.1 Overview

Any collection of particles of mass M and momentum p has a de Broglie wavelength $\lambda_{\mathrm{dB}} = 2\pi\hbar/p$. Originally meant to explain the orbits of a single Coulomb-bound electron, the ultimate wave character of matter has been confirmed for all fundamental particles, and also for composite particles such as ions, atoms and molecules.

In terms of the Bohr radius a_0, the fine-structure constant α, and the electron mass m_e

$$\lambda_{\mathrm{dB}} = \frac{2\pi\alpha a_0}{(v/c)(M/m_\mathrm{e})} = 2\pi\left(\frac{E_\mathrm{h}}{2\mathcal{T}}\frac{m_\mathrm{e}}{M}\right)^{1/2} a_0\,, \tag{77.1}$$

where $v = p/M$ is the velocity, $\mathcal{T} = \frac{1}{2}Mv^2$ is the kinetic energy, and $E_{\rm h}$ is the atomic unit of energy (Table 1.4 of Sect. 1.2). For electrons, $\lambda_{\rm dB} \approx 1.226\,426\,{\rm nm}/(\mathcal{T}/{\rm eV})^{1/2}$. The thermal de Broglie wavelength is defined in (77.80).

Regardless of the particle species, the theory of de Broglie optics divides into two distinct parts: the theory of dispersion and the theory of the optical phenomena. In the theory of dispersion an effective Hamiltonian is derived which describes the interaction of the particles with the optical elements. In the theory of optical phenomena, the ensuing Schrödinger equation is solved and put into the context of geometrical and Fourier optics (e.g., by relating distributions of intensity on the detector screen to properties of a given optical element).

Dispersion of de Broglie Waves

The theory of dispersion is highly particle specific, since it depends on any internal degrees of freedom which may give rise to permanent magnetic moments or to induced electromagnetic moments either in static or in optical light fields. Interaction with the latter fields also may give rise to spontaneous emission. Spontaneous emission has important consequences for the coherence properties of atomic matter waves because of the random recoil associated with it (Chapt. 75).

Dispersion of matter waves differs from that of light waves in a number of regards. First, the dispersion relation of free particles is quadratic in the wavenumber giving rise to spatial spreading of wavepackets even in one-dimensional configurations. Second, particles may be brought to rest, which is impossible for photons. And finally, particles in beams may show self-interaction giving rise to nonlinear optical phenomena even for freely traveling particle waves. Electrons and ions, for example, experience a particle density dependent Coulomb broadening in the focus of a lens (Boersch effect [77.5]).

In atom optics, nonlinear phenomena occur due to atom–atom interactions in the ensemble, the nature of which may be significantly influenced by laser light [77.10]. In the presence of laser light, atom–atom interactions mainly result from photon exchange which, in most cases, leads to a repulsive interaction. Details of the microscopic basis of atom–atom interactions, in particular, those concerning cold collisions, are presented in Sect. 75.5.1. Characteristic effects of nonlinear atom optics like four-wave mixing [77.11] and parametric amplification [77.12] have been observed with the highly dense samples provided by Bose–Einstein degenerate gases (Chapt. 76).

Optics of de Broglie Waves

In contrast to the theory of dispersion, the theory of optical phenomena of matter waves is quite universal and bears strong resemblance to ordinary light optics. This resemblance is closest if the particles can be described by a scalar wave function of a structureless point particle. However, in the presence of resonant laser fields, electronic levels, their Zeeman sublevels and possible hyperfine structure may play an important role, in which case a multicomponent spinor wave function must be used to describe the atom optical phenomena properly.

77.2 Hamiltonian of de Broglie Optics

A large class of phenomena of particle optics is well-described by an effectively one-particle Hamiltonian model in which the particles are not assumed to mutually interact. In such a model the ensemble of particles is described by a wave function $\psi(\boldsymbol{x}, t) = \langle \boldsymbol{x} | \psi(t) \rangle$ whose time evolution is governed by the Schrödinger equation

$$\mathrm{i}\hbar \frac{\partial}{\partial t} |\psi(t)\rangle = H(t) |\psi(t)\rangle \tag{77.2}$$

with a one-particle Hamiltonian of the generic form

$$H(t) = \frac{\left[\hat{\boldsymbol{p}} - \mathcal{A}(\hat{\boldsymbol{x}}, t)\right]^2}{2M} + \mathcal{U}(\hat{\boldsymbol{x}}, t) \,. \tag{77.3}$$

Here, $\hat{\boldsymbol{p}}$ and $\hat{\boldsymbol{x}}$ denote the canonically conjugate operators of the center of mass momentum and position, respectively. The cartesian components of $\hat{\boldsymbol{p}}$ and $\hat{\boldsymbol{x}}$ obey the fundamental commutation relation

$$\left[\hat{x}_i, \hat{p}_j\right] = \mathrm{i}\hbar\delta_{ij} \,. \tag{77.4}$$

The vector potential $\mathcal{A}(\boldsymbol{x}, t)$ and scalar potential $\mathcal{U}(\boldsymbol{x}, t)$ account for the interaction of the particle with all the optical elements, probes, and samples which are placed between the source and the detector, including the effects of gravitation and rotation.

77.2.1 Gravitation and Rotation

All particles are subject to the influence of gravitation and rotation, which both may be viewed as being special cases of an accelerated frame of reference. Effects of uniform gravitation are described by

$$\mathcal{U}(\boldsymbol{x}) = -M\boldsymbol{g}\cdot\boldsymbol{x}\,, \quad \mathcal{A}(\boldsymbol{x}) = 0\,, \tag{77.5}$$

where $\boldsymbol{g}$ describes the direction and magnitude of gravitational acceleration. Effects of uniform rotation are described by

$$\mathcal{U}(\boldsymbol{x}) = -\frac{M}{2}\,(\boldsymbol{\Omega}\times\boldsymbol{x})^2\,, \quad \mathcal{A}(\boldsymbol{x}) = M\,(\boldsymbol{\Omega}\times\boldsymbol{x})\,, \tag{77.6}$$

where the direction and magnitude of $\boldsymbol{\Omega}$ refer to the orientation of the axis of rotation and the angular velocity, respectively. Here, $\mathcal{U}(\boldsymbol{x})$ is the potential of the centrifugal force while $\mathcal{A}(\boldsymbol{x})$ is the potential of the Coriolis force.

77.2.2 Charged Particles

The interaction of charged particles (electrons, ions) with the electromagnetic field is described by

$$\mathcal{U}(\boldsymbol{x},t) = q\Phi(\boldsymbol{x},t)\,, \quad \mathcal{A}(\boldsymbol{x},t) = (q/c)\boldsymbol{A}(\boldsymbol{x},t)\,, \tag{77.7}$$

where q is the particle charge ($q=-e$ for electrons), and $\boldsymbol{A}$ and Φ denote the gauge potentials of the electromagnetic fields $\boldsymbol{E}$ and $\boldsymbol{B}$, respectively. The fields are given by

$$\boldsymbol{E}(\boldsymbol{x},t) = -\frac{1}{c}\frac{\partial}{\partial t}\boldsymbol{A}(\boldsymbol{x},t) - \nabla\Phi(\boldsymbol{x},t)\,, \tag{77.8}$$

$$\boldsymbol{B}(\boldsymbol{x},t) = \nabla\times\boldsymbol{A}(\boldsymbol{x},t)\,. \tag{77.9}$$

77.2.3 Neutrons

For neutrons interacting with a spatially homogeneous gas, liquid, or amorphous solid contained in a volume V, a common model is

$$\mathcal{U}(\boldsymbol{x}) = \begin{cases} \mathcal{U}_0 & \text{inside } V\,, \\ 0 & \text{outside } V\,, \end{cases} \tag{77.10}$$

where $\mathcal{U}_0 = 2\pi\hbar^2\varrho b/M$, ϱ being the number density of scatterers in the volume, and b the bound coherent scattering length.

For neutrons interacting with perfect crystals, $\mathcal{U}(\boldsymbol{x})$ is a smooth periodic function with the same periodicity as the lattice and an average value $\mathcal{U}_0$ within a single unit cell (see [77.2] for details).

77.2.4 Spins

The interaction of the spin related magnetic moment $\boldsymbol{\mu}$ with the electromagnetic field is described by

$$\hat{\mathcal{U}}(\boldsymbol{x},t) = -\hat{\boldsymbol{\mu}}\cdot\boldsymbol{B}(\boldsymbol{x},t)\,,$$
$$\hat{\mathcal{A}}(\boldsymbol{x},t) = \left[\hat{\boldsymbol{\mu}}\times\boldsymbol{E}(\boldsymbol{x},t)\right]/c \tag{77.11}$$

Here, the vector potential is due to the motional correction of the magnetic dipole interaction. Usually, it is neglected. However, it does play an important role for the Aharonov–Casher effect (Sect. 77.7).

For fundamental particles with spin $\frac{1}{2}$, one has

$$\hat{\boldsymbol{\mu}} = g\frac{\hbar e}{4Mc}\boldsymbol{\sigma}\,, \tag{77.12}$$

where $\boldsymbol{\sigma}$ is the vector of Pauli matrices, and g is the g-factor of the particle (Sect. 75.5.2).

Here and in what follows, a hat on $\mathcal{U}$ and $\mathcal{A}$ indicates the matrix character of the hatted quantity, the matrix indices referring to the internal degrees of freedom, like spin. Similarly, $\Psi(\boldsymbol{x},t)$ denotes a spinor-valued wave function. The wave function of a spin-$\frac{1}{2}$ particle, for example, is displayed in the form

$$\Psi(\boldsymbol{x},t) = \begin{pmatrix} \psi_\uparrow(\boldsymbol{x},t) \\ \psi_\downarrow(\boldsymbol{x},t) \end{pmatrix}\,, \tag{77.13}$$

where $\psi_\uparrow$ ($\psi_\downarrow$) is the scalar wave function of the $\sigma_3 = +1(-1)$ component of the state $|\Psi(t)\rangle$.

77.2.5 Atoms

Many optical elements in atom optics are based on the mechanical effects of the radiation interaction. In the electric dipole approximation, the interaction of a single atom with the electromagnetic field is described by

$$\hat{\mathcal{U}}(\boldsymbol{x},t) = -\hat{\boldsymbol{d}}\cdot\boldsymbol{E}(\boldsymbol{x},t)\,,$$
$$\hat{\mathcal{A}}(\boldsymbol{x},t) = \boldsymbol{B}(\boldsymbol{x},t)\times\hat{\boldsymbol{d}}/c\,, \tag{77.14}$$

where $\hat{\boldsymbol{d}}$ is the operator of electric dipole transition. The vector potential $\hat{\mathcal{A}}$ is due to the motional correction of the electric dipole interaction. Usually it is neglected; see however, the paragraph Electric Dipole Phase in Sect. 77.7.

For a monochromatic field of frequency ω

$$\boldsymbol{E}(\boldsymbol{x},t) = \boldsymbol{E}^{(+)}(\boldsymbol{x})\,\mathrm{e}^{-\mathrm{i}\omega t} + \text{h.c.}\,, \tag{77.15}$$

where $\boldsymbol{E}^{(+)}(\boldsymbol{x})$ defines both polarization and spatial characteristics of the field. A standing wave laser field with optical axis in the x-direction and linear polarization $\boldsymbol{\epsilon}$, for example, is described by

$$\boldsymbol{E}^{(+)}(\boldsymbol{x}) = \boldsymbol{\epsilon}\, \mathcal{E}_0 f(x, y, z) \cos(kx) , \tag{77.16}$$

where $\mathcal{E}_0$ is the electric field amplitude, $k = \omega/c$ is the wavenumber, and the slowly varying function $f(x, y, z)$ accounts for the transverse profile of the laser field.

The electric dipole operator $\hat{\boldsymbol{d}}$ acts in the Hilbert space of electronic states of the atom. In the particular case that the polarization of the laser field is spatially uniform and that spontaneous emission does not play a role, two states are generally sufficient and the atom may be modeled by a two-level atom with electronic levels $|e\rangle$ and $|g\rangle$ of energy E_e and E_g, respectively ($E_e > E_g$). With the spinor representation

$$|e\rangle = \begin{pmatrix} 1 \\ 0 \end{pmatrix} , \quad |g\rangle = \begin{pmatrix} 0 \\ 1 \end{pmatrix} , \tag{77.17}$$

the electric dipole operator assumes the form

$$\hat{\boldsymbol{d}} = d\, \boldsymbol{\epsilon}_{\mathrm{d}} \begin{pmatrix} 0 & 1 \\ 1 & 0 \end{pmatrix} , \tag{77.18}$$

where d is the matrix element of the dipole transition and $\boldsymbol{\epsilon}_{\mathrm{d}}$ denotes its polarization.

The laser field is assumed to be near resonant with the $e \leftrightarrow g$ transition at $\omega_0 = (E_e - E_g)/\hbar$, and we denote by $\Delta \equiv \omega_0 - \omega$ the atom-laser detuning. Using the rotating wave approximation (Sect. 68.3.2) in an interaction picture with respect to the laser frequency, the Hamiltonian describing the atomic dynamics – both internal and center-of-mass – is given by

$$H = \frac{\hat{\boldsymbol{p}}^2}{2M} - \frac{\hbar}{2} \begin{pmatrix} -\Delta & \Omega_1(\hat{\boldsymbol{x}}) \\ \Omega_1{}^{\dagger}(\hat{\boldsymbol{x}}) & \Delta \end{pmatrix} , \tag{77.19}$$

where

$$\Omega_1(\boldsymbol{x}) = 2d\, \boldsymbol{\epsilon}_{\mathrm{d}} \cdot \boldsymbol{E}^{(+)}(\boldsymbol{x})/\hbar \tag{77.20}$$

is the spatially dependent bare Rabi frequency. In the field (77.16), $\Omega_1(\boldsymbol{x})$ is cosinusoidal with peak value $\Omega_0 = 2d\mathcal{E}_0/\hbar$.

Atom Optical Stern–Gerlach Effect

The Hamiltonian (77.19) may be written in the form

$$H = \frac{\hat{\boldsymbol{p}}^2}{2M} + \frac{\hbar}{2} \boldsymbol{\sigma} \cdot \boldsymbol{B}_{\mathrm{eff}}(\hat{\boldsymbol{x}}) , \tag{77.21}$$

where

$$\boldsymbol{B}_{\mathrm{eff}}(\boldsymbol{x}) = \big(-\mathrm{Re}[\Omega_1(\boldsymbol{x})],\ \mathrm{Im}[\Omega_1(\boldsymbol{x})],\ \Delta \big) . \tag{77.22}$$

As it stands, the Hamiltonian (77.21) describes the precession and center of mass motion of a fictitious spin in an external “magnetic field” $\boldsymbol{B}_{\mathrm{eff}}(\boldsymbol{x})$. Spatial variations of this field give rise to the Stern–Gerlach effect, i. e., the splitting of the atomic center of mass wave function (see also Sect. 77.5.3).

Adiabatic Approximation

In the position representation, the Hamiltonian (77.19) acts on state vectors of the form of a bispinor

$$\Psi(\boldsymbol{x}, t) = \begin{pmatrix} \psi_e(\boldsymbol{x}, t) \\ \psi_g(\boldsymbol{x}, t) \end{pmatrix} . \tag{77.23}$$

Alternatively, this state vector may be expanded in terms of the local eigenvectors $\alpha_\pm(\boldsymbol{x})$, $\beta_\pm(\boldsymbol{x})$ of the interaction matrix, also called dressed states,

$$\Psi(\boldsymbol{x}, t) = \psi_+(\boldsymbol{x}, t) \begin{pmatrix} \alpha_+(\boldsymbol{x}) \\ \beta_+(\boldsymbol{x}) \end{pmatrix} + \psi_-(\boldsymbol{x}, t) \begin{pmatrix} \alpha_-(\boldsymbol{x}) \\ \beta_-(\boldsymbol{x}) \end{pmatrix} , \tag{77.24}$$

where

$$\hat{\mathcal{U}}(\boldsymbol{x}) \begin{pmatrix} \alpha_\pm(\boldsymbol{x}) \\ \beta_\pm(\boldsymbol{x}) \end{pmatrix} = \pm \frac{\hbar\Omega(\boldsymbol{x})}{2} \begin{pmatrix} \alpha_\pm(\boldsymbol{x}) \\ \beta_\pm(\boldsymbol{x}) \end{pmatrix} , \tag{77.25}$$

with eigenvalues determined by the dressed Rabi frequency

$$\Omega(\boldsymbol{x}) = \sqrt{|\Omega_1(\boldsymbol{x})|^2 + \Delta^2} , \tag{77.26}$$

and coefficients (we suppress the $\boldsymbol{x}$-dependence for notational clarity)

$$\begin{pmatrix} \alpha_+ & \alpha_- \\ \beta_+ & \beta_- \end{pmatrix} = \begin{pmatrix} \cos\frac{\vartheta}{2} & -\mathrm{e}^{-\mathrm{i}\varphi}\sin\frac{\vartheta}{2} \\ \mathrm{e}^{\mathrm{i}\varphi}\sin\frac{\vartheta}{2} & \cos\frac{\vartheta}{2} \end{pmatrix} \equiv \hat{S} . \tag{77.27}$$

Here, the Stückelberg angle $\vartheta \equiv \vartheta(\boldsymbol{x})$ and phase angle $\varphi \equiv \varphi(\boldsymbol{x})$ are defined in terms of the polar representation of the effective magnetic field

$$\boldsymbol{B}_{\mathrm{eff}} = \Omega\, (\cos\varphi \sin\vartheta,\ \sin\varphi \sin\vartheta,\ \cos\vartheta) . \tag{77.28}$$

In the dressed state basis, the transformed Hamiltonian assumes the form

$$\tilde{\hat{H}} \equiv \hat{S}^{\dagger} \hat{H} \hat{S} = \frac{\left[\hat{\boldsymbol{p}} - \hat{\boldsymbol{\mathcal{A}}}(\hat{\boldsymbol{x}})\right]^2}{2M} + \frac{\hbar}{2} \Omega(\hat{\boldsymbol{x}}) \sigma_3 , \tag{77.29}$$

with matrix-valued vector potential

$$\hat{\mathcal{A}}(\boldsymbol{x}) = \mathrm{i}\hbar S^\dagger(\boldsymbol{x})\,[\nabla S(\boldsymbol{x})]\ . \tag{77.30}$$

This matrix is not diagonal; its off-diagonal elements describe nonadiabatic transitions between the dressed states. If the detuning is sufficiently large, and the atom moves sufficiently slowly, these nonadiabatic transitions may be neglected in a first approximation. In such an approximation, which is akin to the Born–Oppenheimer approximation in molecular physics, the dynamics of the atom is described by two decoupled Hamiltonians of the generic form (77.3), with scalar potentials $\mathcal{U}$ given by

$$\mathcal{U}(\boldsymbol{x}) = \pm\frac{\hbar}{2}\Omega(\boldsymbol{x}) \tag{77.31}$$

and vector potentials $\mathcal{A}(\boldsymbol{x})$ given by the diagonal elements of (77.30). The vector potential is usually neglected. If included, it describes the Berry phase of the mechanical effects of the radiation interaction of a two-level atom.

The idea behind the adiabatic approximation is that the internal state of the atom, which is described by any of the locally varying dressed states, has enough time to adjust smoothly to the motion of the atom. For the important case of strong detuning $|\Delta| \gg |\Omega_1(\boldsymbol{x})|$, this assumption is usually well justified. In this case, an atom in the dressed ground state experiences a potential which is approximately given by

$$\mathcal{U}(\boldsymbol{x}) = -\frac{\hbar|\Omega_1(\boldsymbol{x})|^2}{4\Delta}\ . \tag{77.32}$$

For red detuning we have $\Delta > 0$, in which case the atom is attracted towards regions of high intensity (high field seeker). For blue detuning, the atom is repelled from such regions (low field seeker). A potential similar to (77.32) also applies for complex particles like molecules whose transitions are far detuned from the light frequency. The potential is proportional to the dynamic polarizability $\alpha(\omega)$ and the field intensity.

Atom Optical Nonlinearity

In very dense atom ensembles at low temperatures, collisions between particles can be described by a contact interaction whose strength depends, in the simplest case, on a single parameter, the s-wave scattering length a (Sect. 75.5.1). In the mean field approximation, these interactions translate into a density-dependent potential

$$\mathcal{U}(\boldsymbol{x}) = \frac{4\pi\hbar^2 a(N-1)}{M}|\psi(\boldsymbol{x})|^2\ , \tag{77.33}$$

where N is the number of particles. This nonlinearity leads to the occurrence of atom solitons [77.13], four-wave mixing [77.11], and parametric amplification [77.12].

77.3 Principles of de Broglie Optics

Since the Schrödinger equation is a linear partial differential equation, de Broglie optics shares most of its principles with principles of other wave phenomena, and in particular with the optical principles of electromagnetic waves.

77.3.1 Light Optics Analogy

The analogy of de Broglie optics and light optics becomes particularly transparent for monoenergetic beams of scalar particles. Such beams are described by a time harmonic wave function $\psi(\boldsymbol{x}, t) = \mathrm{e}^{-\mathrm{i}Et/\hbar}\psi_\mathrm{E}(\boldsymbol{x})$, where $\psi_\mathrm{E}(\boldsymbol{x})$ obeys the stationary Schrödinger equation

$$H\psi_\mathrm{E} = E\psi_\mathrm{E}\ . \tag{77.34}$$

Setting $\mathcal{A}(\boldsymbol{x}) = 0$ in (77.3) for simplicity, this equation assumes the form

$$\left[\nabla^2 + \tilde{k}_0^2\left(1 - \frac{\mathcal{U}(\boldsymbol{x})}{E}\right)\right]\psi_\mathrm{E}(\boldsymbol{x}) = 0\ , \tag{77.35}$$

where the wavenumber $\tilde{k}_0$ is related to the energy E via the dispersion relation

$$E \equiv \frac{\hbar^2\tilde{k}_0^2}{2M}\ . \tag{77.36}$$

If $\mathcal{U} = 0$ at the entrance to the interaction region, E is the kinetic energy of the freely traveling de Broglie wave and $\tilde{k}_0$ is the related wavenumber.

Comparing (77.35) with the scalar Helmholtz equation of electromagnetic theory, and identifying

$$\tilde{n}_\mathrm{E}(\boldsymbol{x}) \equiv \left[1 - \mathcal{U}(\boldsymbol{x})/E\right]^{1/2} \tag{77.37}$$

as an index of refraction for matter waves, one observes the complete analogy of scalar optics of stationary matter waves and monochromatic light waves. This analogy can be generalized for spinor valued wave functions which would correspond to vector wave optics in anisotropic index media. However, in contrast to light, spinor-valued wave functions do not obey a transversality condition.

In (77.35–77.37), the parameter E describes the kinetic energy of the incoming beam. Thus, E is positive, and therefore $\tilde{n}_E < 1$ for positive values of the potential, while $\tilde{n}_E > 1$ for negative values of the potential. For neutrons, one generally has $\tilde{n}_E < 1$. This contrasts to the index of refraction for light waves which is generally larger than one. For electrons, ions, and atoms both $\tilde{n}_E < 1$ and $\tilde{n}_E > 1$ may be realized.

77.3.2 WKB Approximation

Waves are described by amplitude and phase. Particles are described by position and momentum. The link between these concepts is provided by Hamilton's ray optics. For scalar matter waves, a ray follows a classical trajectory. The optical signature of the ray is the phase associated with it. The quantum mechanical version of Hamilton's ray optics is obtained in the WKB approximation of the stationary Schrödinger equation (77.34).

Any solution of (77.34) may be written in the form

$$\psi_E(\boldsymbol{x}) = A(\boldsymbol{x})\,e^{iW(\boldsymbol{x})} \tag{77.38}$$

with real-valued $A(\boldsymbol{x})$ and $W(\boldsymbol{x})$. In the WKB approximation $A(\boldsymbol{x}) \approx 1$, and

$$W(\boldsymbol{x}) = \tilde{k}_0 \int_{P_i}^{P} \tilde{n}_E(s)\,ds + \hbar^{-1} \int_{\boldsymbol{x}_i}^{\boldsymbol{x}} \mathcal{A}(\boldsymbol{x}') \cdot d\boldsymbol{x}' , \tag{77.39}$$

which is called eikonal in Hamilton's ray-optics. In this expression, $\tilde{n}_E(s) \equiv \tilde{n}_E(\boldsymbol{x}(s))$, where $\boldsymbol{x}(s)$ denotes the classical trajectory of energy E connecting the point P_i with the point P, $\boldsymbol{x}_i$ and $\boldsymbol{x}$ are the coordinates of P_i and P, respectively, and $ds \equiv |d\boldsymbol{x}|$ is the element of arc length measured along the classical trajectory. Note that the second contribution in (77.39) is generally gauge-dependent. However, for closed loops which are frequently encountered in interferometry, the gauge-dependence disappears by virtue of Stokes theorem which transforms the path integral into an area integral over the rotor of $\mathcal{A}$.

The eikonal (77.39) may also be written in the form of a reduced action

$$W(\boldsymbol{x}) = \frac{1}{\hbar} \int_{\boldsymbol{x}_i}^{\boldsymbol{x}} \boldsymbol{p}(\boldsymbol{x}') \cdot d\boldsymbol{x}' = \int_{\boldsymbol{x}_i}^{\boldsymbol{x}} \tilde{\boldsymbol{k}}(\boldsymbol{x}') \cdot d\boldsymbol{x}' , \tag{77.40}$$

where $\boldsymbol{p}(\boldsymbol{x})$ is the local value of the canonical momentum of the particle, $\tilde{\boldsymbol{k}}(\boldsymbol{x}) \equiv \boldsymbol{p}(\boldsymbol{x})/\hbar$ is the corresponding wave vector, and the integral is evaluated along the classical trajectory of the particle. Note that in the presence of a vector potential, $\boldsymbol{p}(\boldsymbol{x})$ and $d\boldsymbol{x}$ are no longer parallel as a result of the difference between canonical momentum $\boldsymbol{p}$ and kinetic momentum $M(d/dt)\,\boldsymbol{x} \equiv \boldsymbol{p} - \mathcal{A}$.

The WKB approximation becomes invalid in the vicinity of caustics where neighboring rays intersect. There, connection formulae are used to find the proper phase factors picked up by the ray in traversing the caustics. Depending on the topology of the intersecting rays, different classes of diffraction integrals provide uniform approximations for the wave amplitude near caustics. For further details see [77.14] and [77.15].

77.3.3 Phase and Group Velocity

The velocity of a particle which traverses a region of negative potential increases so that $p(\boldsymbol{x}) > p_0$, and the phase advances: $\delta W = \int [\boldsymbol{p}(\boldsymbol{x}) - \boldsymbol{p}_0] \cdot d\boldsymbol{x} > 0$. In quantum mechanics, the classical velocity corresponds to the group velocity, while the evolution of the phase is determined by the phase velocity. The phase and group velocities of de Broglie waves are given by

$$v_p(\boldsymbol{x}) \equiv \frac{E}{p(\boldsymbol{x})} = \frac{1}{\tilde{n}_E(\boldsymbol{x})} \sqrt{\frac{E}{2M}} , \tag{77.41}$$

$$v_g(\boldsymbol{x}) \equiv \frac{\partial E}{\partial p(\boldsymbol{x})} = \tilde{n}_E(\boldsymbol{x}) \sqrt{\frac{2E}{M}} , \tag{77.42}$$

respectively. Note that the product $v_p v_g = E/M$ is independent of $\tilde{n}_E(\boldsymbol{x})$.

77.3.4 Paraxial Approximation

The paraxial approximation is useful in describing the evolution of wave-like properties and/or distortion of wavefronts in the immediate neighborhood of an optical ray.

Let the z-axis be the central optical axis of symmetry along which the optical elements are aligned. Using the Ansatz $\psi_E(\boldsymbol{x}) = e^{i\tilde{k}_0 z}\phi(x, y; z)$, and dropping $\partial^2\phi/\partial z^2$ in a slowly varying envelope approximation, one obtains

$$i\hbar v_0 \frac{\partial}{\partial z}\phi(x, y; z) = \left[-\frac{\hbar^2}{2M}\nabla_\perp^2 + \mathcal{U}(x, y; z)\right]\phi(x, y; z) , \tag{77.43}$$

where $\nabla_\perp^2 = \partial^2/\partial x^2 + \partial^2/\partial y^2$, and $v_0 = \hbar\tilde{k}_0/M$ is the longitudinal velocity of incoming particles. This equation has exactly the form of a time-dependent

Schrödinger equation in two dimensions, with z/v_0 playing the role of a fictitious time t. With this interpretation, the spatial evolution of phase fronts along z can be analyzed in dynamical terms of particles moving in the xy-plane.

77.3.5 Raman–Nath Approximation

In the Raman–Nath approximation (RNA) (also called the short-time, thin-hologram, or thin-lens approximation), the $\nabla_\perp^2$ term in (77.43) is neglected. The potential $\mathcal{U}(x, y; z)$ then acts as a pure phase structure, and the solution of (77.43) becomes

$$\phi(x, y; z) = \exp\left[-\frac{\mathrm{i}}{\hbar v_0}\int_{z_i}^{z} \mathrm{d}z' \mathcal{U}(x, y; z')\right]\phi(x, y; z_i). \tag{77.44}$$

In terms of a classical particle moving under the influence of $\mathcal{U}$, the approximation loses validity for $\frac{1}{2}Mv_\perp^2 \gtrsim \mathcal{U}$, which is just a quarter cycle for a harmonic oscillator.

77.4 Refraction and Reflection

Consider a particle beam of energy E incident on a medium with constant index of refraction $\tilde{n}_\mathrm{E}$. The boundary plane at $z = 0$ in Fig. 77.1 divides the vacuum from the medium. At the boundary, the beam is partially reflected and partially transmitted, with the angles determined by Snell's law of refraction

$$\sin\alpha = \tilde{n}_\mathrm{E} \sin\beta , \tag{77.45}$$

and the law of reflection

$$\alpha = \alpha' . \tag{77.46}$$

The coefficients of reflectivity $\tilde{R}$, and transmittivity $\tilde{T} = 1 - \tilde{R}$ are given by the Fresnel formula

$$\tilde{R} = \left|\frac{\cos\alpha - \tilde{n}_\mathrm{E}\cos\beta}{\cos\alpha + \tilde{n}_\mathrm{E}\cos\beta}\right|^2 . \tag{77.47}$$

For $\tilde{n}_\mathrm{E} > 1$, the interface is "attractive" and $\tilde{R} \ll 1$, with $\tilde{R} \to 1$ only for glancing incidence $\alpha \to \pi/2$. For $\tilde{n}_\mathrm{E} < 1$, the interface is "repulsive" and total reflection $(\tilde{R} = 1)$ occurs for $\alpha \geq \tilde{\alpha}_\mathrm{c}$, where $\tilde{\alpha}_\mathrm{c} = \sin^{-1}\tilde{n}_\mathrm{E}$ is the critical angle. For $\alpha > \tilde{\alpha}_\mathrm{c}$, the de Broglie wave becomes evanescent, with $\psi_\mathrm{E}(z > 0) \sim \mathrm{e}^{-\tilde{\kappa}z}$ inside the medium $(z > 0)$, where

$$\tilde{\kappa} = \tilde{k}_0\left(\sin^2\alpha - \sin^2\tilde{\alpha}_\mathrm{c}\right)^{1/2} . \tag{77.48}$$

For thermal neutrons, $\pi/2 - \tilde{\alpha}_\mathrm{c} \sim 3\times 10^{-3}$ radians.

If $E < \mathcal{U}$, then $\tilde{n}_\mathrm{E}$ is imaginary and total reflection occurs for all α. In neutron optics, this total mirror reflection requires ultracold neutrons ($T \approx 0.5$ mK). It has important applications for storage of ultracold neutrons in material cavities, and neutron microscopy using spherical mirrors. For details see [77.2].

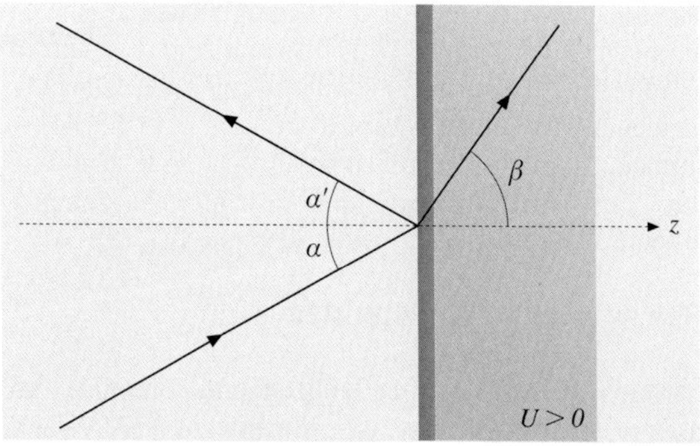

Fig. 77.1 Reflection geometry

77.4.1 Atomic Mirrors

Inelastic processes, such as diffuse scattering and absorption, inhibit coherent reflection of atoms from bare surfaces. The surface must therefore be coated either with material of low adsorptivity (noble gas, see [77.16]) or electromagnetic fields (evanescent light or magnetic fields, see below).

Reflection of Atoms by Evanescent Laser Light

Evanescent light fields are produced by total internal reflection of a light beam at a dielectric–vacuum interface [77.17]. In the vacuum, the field decays exponentially away from the interface on a characteristic length κ^{-1} where

$$\kappa = k\left(n^2\sin^2\theta_i - 1\right)^{1/2} . \tag{77.49}$$

Here, n is the light index of refraction of the dielectric, k is the wave number of the light beam in vacuo, and θ_i is its angle of incidence.

If the light is blue-detuned from the atomic resonance, an incident beam of ground state atoms experiences the repulsive potential

$$\mathcal{U}(\boldsymbol{x}) = \frac{\hbar}{2}\left[\left({\Omega_0}^2 e^{-2\kappa|z|} + \Delta^2\right)^{1/2} - |\Delta|\right] . \tag{77.50}$$

For $\alpha > \tilde{\alpha}_c$, the evanescent field acts as a nearly perfect mirror, the imperfections being due to nonadiabatic transitions into the "wrong" dressed state, and possible spontaneous emission. Reflection of atoms by evanescent laser light was demonstrated by *Balykin* et al. [77.18] at grazing incidence and by *Kasevich* et al. [77.19] at normal incidence.

Reflection of Atoms by Magnetic Near Fields

Magnetic near fields are produced above substrates with a spatially modulated permanent magnetization or close to arrays of stationary currents. In the vacuum above the substrate, the field decays approximately exponentially over a length comparable to the scale of the magnetic modulation. The motion of atoms that cross such inhomogeneous magnetic fields sufficiently slowly is governed by the analog of the adiabatic potential described in Sect. 77.2.5:

$$\mathcal{U}(\boldsymbol{x}) = -\mu\left(m_s/s\right)\left|\boldsymbol{B}(\boldsymbol{x})\right| , \tag{77.51}$$

where μ is the magnetic moment and m_s is the (conserved) projection of the atomic spin s onto the local magnetic field direction. A repulsive mirror potential is achieved for spin states with $\mu m_s < 0$; these weak field seekers are repelled from the strong fields close to the substrate.

Experiments have used magnetic recording media like magnetic tapes or hard disks [77.20], arrays of current-carrying wires [77.21], or amorphous magnetic substrates [77.22].

77.4.2 Atomic Cavities

Atomic reflections are used in the two kinds of cavities proposed so far: the trampoline cavity and the Fabry–Perot resonator.

The trampoline cavity, also called the gravito-optical cavity, consists of a single evanescent mirror facing upwards, the second mirror being provided by gravitation. A stable cavity is realized with the evanescent laser field of a parabolically shaped dielectric–vacuum interface, see [77.23] and the experiment by *Aminoff* et al. [77.24]. A cavity with transverse confinement provided by a hollow blue-detuned laser beam has been demonstrated by *Hammes* et al. [77.25].

In the atomic Fabry–Perot resonator both mirrors are realized by laser light [77.26].

77.4.3 Atomic Lenses

De Broglie waves may be focused by refraction from a parabolic potential or by diffraction, e.g., by a Fresnel zone plate (Sect. 77.5.2). Consider focusing by the parabolic potential

$$\mathcal{U}(x, y; z) = \begin{cases} \left(\begin{matrix} \frac{1}{2} M\omega_f^2 (x^2 + y^2) & , \; -w \le z \le 0 \\ 0 & , \; \text{otherwise} . \end{matrix}\right) \end{cases} \tag{77.52}$$

For ground state atoms, such a potential is realized in the vicinity of the node of a blue detuned standing wave laser field of transverse width w. In this case

$$\omega_f = \Omega_0\left(\omega_{\text{rec}}/|\Delta|\right)^{1/2} , \tag{77.53}$$

where $\omega_{\text{rec}} = \hbar k^2 / 2M$ is the recoil frequency.

Comparison with the Raman–Nath approximation (77.44) at $z = 0$, with the phase fronts of a spherical wave converging towards a point $\boldsymbol{x}_f = (0, 0, f)$, shows that $\mathcal{U}$ describes a lens of focal length

$$f = \frac{v_0^2}{w\omega_f^2} . \tag{77.54}$$

The Raman–Nath approximation is only valid for a thin lens $w \ll f$, and breaks down for $w > w_{\text{RN}} = \pi v_0 / 2\omega_f$. In the latter case, oscillations of the particles in the harmonic potential become relevant, a phenomenon sometimes called channeling. Channeling may be used to realize thick lenses with focal length $f = w_{\text{RN}}$ corresponding to a quarter oscillation period.

Focusing of a metastable helium beam using the anti-node of a large period standing wave laser field has been demonstrated [77.27]. Such a field is produced by reflecting a laser beam from a mirror under glancing incidence. The standing wave forms normal to the mirror surface. Similar interference patterns provide arrays of thick lenses that have been exploited in atom lithography to focus an atomic beam onto a substrate [77.28].

77.4.4 Atomic Waveguides

Atomic waveguides can be realized with potentials that confine atoms in one or two dimensions [77.29–31]. These devices are key elements for integrated atom optics, a field that has seen a rapid evolution recently [77.32, 33]. A planar waveguide is provided by the one-dimensional confinement in an optical standing wave [77.34] or an atomic mirror combined with gravity [77.31]. The discrete nature of the waveguide modes in that case could be demonstrated by lowering the amplitude of the confining potential. Linear waveguides can be modeled by the parabolic transverse potential (77.52) that now extends along the waveguide axis (the z-axis). Physical realizations include hollow, blue-detuned laser beams [77.30], hollow fibers whose inner wall is coated with blue-detuned evanescent light [77.35], elongated foci of red-detuned light created by cylindrical lenses [77.36], and magnetic field minima along current-carrying wires, possibly combined with homogeneous bias fields [77.37]. With typical thermal atomic ensembles, these waveguides operate in a multimode regime, and coherent operation has been demonstrated only with Bose–Einstein condensates. A strong transverse confinement that facilitates monomode operation, can be achieved with miniaturized wire networks deposited on a solid substrate [77.38, 39]. This approach may lead to the fabrication of atom chips [77.33]. Even multimode waveguides, however, can yield robust atomic interferometers, as suggested theoretically in [77.40] and [77.41].

77.5 Diffraction

The diffraction of matter waves is described by the solution of the Schrödinger equation (77.34) subject to the boundary conditions imposed by the diffracting object. For a plane screen Σ made of opaque portions and apertures, the solution in the source-free region behind the screen is given by the Rayleigh–Sommerfeld formulation of the Huygens principle

$$\psi_{\mathrm{E}}(\boldsymbol{x}) = \frac{\tilde{k}_0}{\mathrm{i}} \int_{\Sigma} \frac{\mathrm{d}\xi\,\mathrm{d}\eta}{2\pi} \frac{\mathrm{e}^{\mathrm{i}\tilde{k}_0 R}}{R} \times \left(1 + \frac{\mathrm{i}}{\tilde{k}_0 R}\right) \frac{\boldsymbol{n}\cdot\boldsymbol{R}}{R} \psi(\xi)\,, \tag{77.55}$$

where $\xi = (\xi, \eta, \zeta)$ denotes coordinates of points on Σ, $\boldsymbol{n}$ is an inwardly directed normal to Σ at a point $\boldsymbol{\xi}$, and $\boldsymbol{R} = \boldsymbol{x} - \boldsymbol{\xi}$.

A diffraction pattern only becomes manifest in the diffraction limit $r \gg d$, where r is the distance to the observation point, and d is the length scale of the diffracting system. The two diffraction regimes are then the Fraunhofer limit $r \gg d^2/\lambda_{\mathrm{dB}}$ and the Fresnel regime $r \approx d^2/\lambda_{\mathrm{dB}}$, also called near-field optics.

77.5.1 Fraunhofer Diffraction

In the Fraunhofer limit, the field at position (x, y) on a screen at a distance L downstream from the diffracting object is given by

$$\psi(x, y; L) = \tilde{k}_0 \frac{\mathrm{e}^{\mathrm{i}k_0 L}}{\mathrm{i}L} \int \frac{\mathrm{d}\xi\,\mathrm{d}\eta}{2\pi} \times \mathrm{e}^{-\mathrm{i}(\tilde{k}_x \xi + \tilde{k}_y \eta)} \psi(\xi, \eta; 0)\,, \tag{77.56}$$

where $\tilde{k}_x = \tilde{k}_0 x/L$, $\tilde{k}_y = \tilde{k}_0 y/L$. The field at the observation screen is thus given by the Fourier transform of the field in the object plane; i.e. the momentum representation of the diffracted state. Since most diffraction experiments in atom optics are performed in the Fraunhofer limit, most calculations are done in the momentum representation.

Atomic diffractions from microfabricated transmission gratings [77.42] and double slits [77.43] have been observed. Recent experiments have extended de Broglie wave diffraction to heavier, complex particles like fullerence molecules (C_{60}) [77.44].

77.5.2 Fresnel Diffraction

Typical applications of Fresnel diffraction are Fresnel zone plates and the effects of Talbot and Lau. Fresnel zone plates are microfabricated concentric amplitude structures which act like lenses. They are frequently employed in optics of α-particles and neutrons. In atom optics, focusing with a Fresnel zone plate was first demonstrated by *Carnal* et al. [77.45].

The Talbot effect and the related Lau effect refer to the self-imaging of a grating of period d, which appears downstream at distances that are integral multiples of the Talbot length $L = 2d^2/\lambda_{\mathrm{dB}}$. For a discrete set of smaller distances than the Talbot length, images of the grating appear with smaller periods d/n, $n = 2, 3 \cdots$. For applications in matter wave interferometry, [77.46]; for applications in atom lithography see [77.47].

77.5.3 Near-Resonant Kapitza–Dirac Effect

The near resonant Kapitza–Dirac effect refers to the diffraction of two-level atoms from a standing wave laser field with a spatially uniform polarization. The dynamics of the effect is described by the Hamiltonian (77.19) with the mode function of the laser field given by (77.16). Consider atoms traveling predominantly in the z-direction, i. e., orthogonal to the axis of the laser field, with energy $\frac{1}{2}Mv_0^2 \gg \hbar\Omega_0$ (Fig. 77.2). Kapitza–Dirac diffraction is then observed in transmission, which in the theory of diffraction is called Laue geometry. In the paraxial approximation (Sect. 77.3.4) for motion in the z-direction, and assuming that the laser profile is homogeneous in the y-direction, the description reduces to an effectively one-dimensional model for the quantum mechanical motion along the x-axis of the laser field.

Due to the periodicity of the standing wave light field, the transverse momenta of the transmitted atom waves differ by multiples of $\hbar k$ from the transverse momentum of the undiffracted wave. For the important case of strong detuning, and assuming that the incoming atoms are in their electronic ground state moving with transverse momentum $p_x = 0$, the p_x-distribution of the outgoing wave is given in the Raman–Nath approximation (Sect. 77.3.5) by

$$\text{Prob}(p_x = 2n\hbar k) \propto \left| J_n\left(\frac{{\Omega_0}^2}{8|\Delta|}\tau\right)\right|^2 , \qquad (77.57)$$

where J_n is a Bessel function of order n, $\tau = w/v_0$ is an effective interaction time, w being the width of the laser field, and v_0 the longitudinal velocity of the atoms. The distribution (77.57) was observed by *Gould* et al. [77.48].

For $\tau \gtrsim \tau_{\text{RN}}$, where $\tau_{\text{RN}} = \left(\omega_{\text{rec}}{\Omega_0}^2/|\Delta|\right)^{-1/2}$, the Raman–Nath approximation becomes invalid. As a result of Doppler related phase-mismatch, the momentum spread saturates and shows a sequence of collapse and revival as a function of τ [77.49].

If the detuning is too small to allow for a scalar description, the two-level character of the atoms must be taken into account. For the particular case of $\Delta = 0$, the ground state evolves into an equal superposition of the two diabatic states $1/\sqrt{2}(|e\rangle \pm |g\rangle)$ while entering the interaction region. Inside the interaction region, these states experience potentials which differ only by their sign. For atomic beams with a small spatial spread $\delta x \ll 2\pi/k$, the diabatic states experience opposite forces, leading to a splitting of the atomic beam called the atom-optical Stern–Gerlach effect [77.50].

In the general case of arbitrary Δ, the Kapitza–Dirac Hamiltonian (77.21) is most conveniently analyzed using band theoretical methods of solid state theory [77.51].

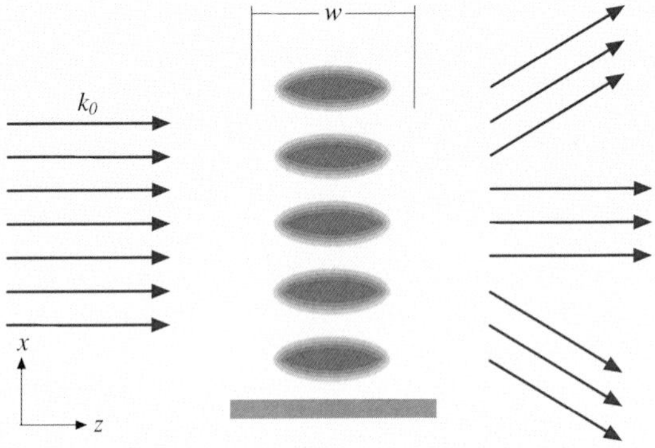

Fig. 77.2 Geometry of the Kapitza–Dirac effect

77.5.4 Atom Beam Splitters

Beam splitters are optical devices which divide an incoming beam into two outgoing beams traveling in different directions. For thermal neutrons, beam splitters may be realized by diffraction from perfect crystals in Laue geometry. For atoms, they can be realized using diffraction from crystalline surfaces, microfabricated structures (see Sect. 77.5.1), or by using diffraction from an optical standing wave.

The Kapitza–Dirac effect, for example, may be exploited to split an atomic beam coherently using Bragg reflection at the "lattice planes" provided by the periodic intensity variations of a standing wave laser field [77.52]. This process is resonant for an incoming atomic beam traveling with transverse momentum $p_x = \hbar k$ because it is energetically degenerate with the diffracted beam traveling with transverse momentum $\bar{p}_x = -\hbar k$. This level degeneracy is lifted while the atoms enter the interaction region. In the Bragg regime, the lifting happens slowly enough that only the momentum states $|\pm\hbar k\rangle$ participate in the diffraction (two-beam resonance), and their populations show Pendellösung type oscillations as a function of the transit time. The frequency of the oscillations is given by $\delta E/\hbar$, where δE is the energy splitting of the two beams inside the interaction region. For a transit time given by a quarter period of the Pendellösung, a 50% beam splitting is observed [77.52]. In principle, Bragg resonances may also be realized for higher diffraction order $p_x = n\hbar k \leftrightarrow \bar{p}_x = -n\hbar k$. However, in this case, intermediate momentum states become populated (multibeam resonance), which makes the higher-order Bragg resonances less suitable for beam splitting purposes.

More promising for the realization of an atomic beam splitter is the magneto-optical diffraction which refers to the diffraction of three-level atoms from a laser field with a periodic polarization gradient (lin ⊥ lin configuration) (Chapt. 75), and a magnetic field aligned parallel to the optical axis of the laser field. This configuration realizes an interaction potential in the form of a blazed grating, i. e., a phase grating with an approximately triangular variation of phase. In an experiment by *Pfau* et al. [77.53], transverse splitting of a beam of metastable helium by $42\hbar k$ was observed [77.49].

Diffraction from an evanescent standing wave involves Bragg reflection of atoms under glancing incidence from the periodic grating of a blue detuned evanescent standing wave laser field [77.54–56]. Diffraction at normal incidence has been demonstrated with sufficiently slow atoms and can be described by a generalization of the RNA (Sect. 77.3.5) [77.57,58].

77.6 Interference

While the overall phase of a wave function ψ is not observable, interferometry makes detectable the relative phases of two components ψ_1, ψ_2 in a superposition $\psi = \psi_1 + \psi_2$. Two types of interferometers are most common: the Young double slit as a paradigm for interferometers based on division of wavefront, and the Mach–Zehnder interferometer as a paradigm for interferometers based on division of amplitude. In de Broglie optics, the latter type is realized in the form of a three-grating interferometer, since division of amplitude is achieved by diffraction at gratings rather than by semi-transparent mirrors. Experiments with this geometry have been reported for atoms [77.59] and recently for more massive, complex molecules (fullerenes) [77.60].

77.6.1 Interference Phase Shift

From a fundamental point of view, any interferometer is a ring. At the entrance port of a three-grating interferometer displayed in Fig. 77.3, for example, the wave function is split into two coherent parts which spatially evolve along different paths and subsequently come together at the exit port where they are superimposed to produce two outgoing waves

$$\psi_\pm = \frac{1}{\sqrt{2}}(\psi_1 \pm \psi_2) , \tag{77.58}$$

where the components from path 1 and 2 are given by

$$\begin{aligned} \psi_1 &= A_1 \exp(iW_1) , \\ \psi_2 &= A_2 \exp(iW_2) . \end{aligned} \tag{77.59}$$

For simplicity, assume $A_1 = A_2 = A_0/\sqrt{2}$. The relative flux of the outgoing waves is then

$$I_\pm = \frac{1}{2}(1 \pm \cos\chi) , \tag{77.60}$$

where $\chi = W_1 - W_2$ is the relative phase of the two components. The sinusoidal variations of $I_\pm$ with varying χ are called interference fringes, and $\chi/2\pi$ is called the fringe order number.

In the WKB approximation, the phases W_1 and W_2 are given by

$$W_i = W_0 + \int_i \tilde{\boldsymbol{k}}(\boldsymbol{x}) \cdot \mathrm{d}\boldsymbol{x}, \quad i = 1, 2 , \tag{77.61}$$

where the integrals extend over the classical paths 1 and 2, respectively, and $\mathrm{d}\boldsymbol{x}$ is an element of displacement along the paths. Using (77.61), the relative phase is

$$\chi = \int_C \tilde{\boldsymbol{k}}(\boldsymbol{x}) \cdot \mathrm{d}\boldsymbol{x} , \tag{77.62}$$

where C is the closed interferometer loop. Note, that on path 2, the path element $\mathrm{d}\boldsymbol{x}$ and $\tilde{\boldsymbol{k}}$ are antiparallel.

Usually, the absolute value of χ is not measured, but only variations, called phase shifts, which result from displacements of the diffraction gratings or placement of optical elements into the beam path. Phase shifts

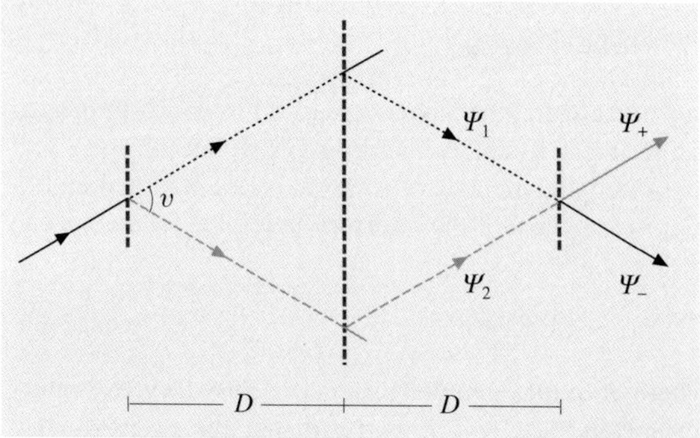

Fig. 77.3 Geometry of the three-grating interferometer

come in two categories: dispersive and geometric. If a phase shift χ depends on v_0, it is called dispersive, otherwise it is called geometric. Geometric phases depend only on the geometry of the interferometer loop. The Sagnac effect (see below), for example, may be geometric. A phase which depends neither on v_0 nor on the geometry of the interferometer loop is called topological. The Aharonov–Bohm effect (see below) is topological.

Using (77.39), χ becomes

$$\chi = \chi_0 + \chi\{\mathcal{U}\} + \chi\{\mathcal{A}\} , \tag{77.63}$$

where for weak potentials $\mathcal{U}$ and $\mathcal{A}$,

$$\chi_0 = \frac{M v_0}{\hbar}(L_1 - L_2) , \tag{77.64}$$

$$\chi\{\mathcal{U}\} = -\frac{1}{\hbar v_0}\oint \mathcal{U}[\boldsymbol{x}(s)]\,\mathrm{d}s , \tag{77.65}$$

$$\chi\{\mathcal{A}\} = \frac{1}{\hbar}\oint \mathcal{A}(\boldsymbol{x})\cdot \mathrm{d}\boldsymbol{x} , \tag{77.66}$$

and L_i is the geometric length of the path i. For a constant potential $\mathcal{U}_0$ intersecting the interferometer on a length w, $\chi\{\mathcal{U}\}$ is given by

$$\chi\{\mathcal{U}\} = -\frac{\mathcal{U}_0 w}{\hbar v_0} . \tag{77.67}$$

Using Stokes theorem, $\chi\{\mathcal{A}\}$ may be written in the manifestly gauge-invariant form

$$\chi\{\mathcal{A}\} = \frac{1}{\hbar}\int [\nabla \times \mathcal{A}(\boldsymbol{x})]\cdot \mathrm{d}\boldsymbol{a} , \tag{77.68}$$

where the integral extends over the area enclosed by the interferometer, and $\mathrm{d}\boldsymbol{a}$ is an infinitesimal area element.

Dispersive Phase Shifts. Atom interferometers have been able to measure phase shifts of the form (77.65) due to, for example, the atomic level shift in an electric field (Stark effect) [77.59] or to coherent forward scattering by background gas atoms, see (77.10) and [77.61].

Sagnac Effect. The Sagnac effect refers to $\chi\{\mathcal{A}\}$ in a rotating interferometer. Inserting (77.6) into (77.68), and assuming that the axis of rotation is oriented perpendicular to the plane of the interferometer, the Sagnac phase shift is given by

$$\chi_{\mathrm{Sa}} = 4M\Omega A/\hbar , \tag{77.69}$$

where A is the geometric area enclosed by the interferometer loop. χ_{Sa} may be dispersive or geometric depending on the type of interferometer. In a Young double slit, A is independent of energy and χ_{Sa} is geometric. In a three-grating interferometer, the area is $A \approx \vartheta D^2$, where ϑ is the splitting angle; see Fig. 77.3. In this case, χ_{Sa} is dispersive because of the velocity dependence of $\vartheta \approx 2\hbar k/M v_0$. The Sagnac effect for de Broglie waves was first observed by *Werner* et al. [77.62] using a neutron interferometer.

Aharonov–Bohm Effect. The Aharonov–Bohm effect refers to the $\chi\{\mathcal{A}\}$ of charged particles encircling a magnetic flux line [77.63]. Inserting $\mathcal{A}$ from (77.7) into (77.68), and assuming particles of charge q encircling a line of flux Φ once, one finds

$$\chi_{\mathrm{AB}} = q\Phi/\hbar . \tag{77.70}$$

A characteristic feature of the Aharonov–Bohm effect is that the particles actually never "see" the magnetic field of the flux line which is confined to some region inaccessible to the particles. χ_{AB} is strictly topological, and only depends on the linking number of the interferometer loop and the flux line. Its appearance is characteristic for all gauge theories. For further details and a summary on its experimental verification see [77.5].

Aharonov–Casher Effect. The Aharonov–Casher effect refers to $\chi\{\mathcal{A}\}$ of a magnetic spin encircling an electric line charge [77.64]. Inserting $\mathcal{A}$ from (77.11) into (77.68), one obtains for proper alignment of $\boldsymbol{\mu}$ and $\boldsymbol{E}$

$$\chi_{\mathrm{AC}} = 2\pi\frac{|\boldsymbol{\mu}|}{\mu_{\mathrm{B}}}\frac{r_0}{\xi} , \tag{77.71}$$

where r_0 is the classical electron radius, μ_{B} is the Bohr magneton, and $\xi = e/\rho_{\mathrm{el}}$, ρ_{el} being the electric line charge density. χ_{AC} is topological only if the spin is aligned parallel to the electric line charge and both are oriented perpendicular to the plane of the interferometer. χ_{AC} for atoms has been observed by *Sangster* et al. [77.65].

Electric Dipole Phase. Electric dipole phase refers to the $\chi\{\mathcal{A}\}$ of an electric dipole moment encircling a magnetic line charge [77.66]. Inserting $\mathcal{A}$ from (77.14) into (77.68), one obtains for proper alignment of $\boldsymbol{d}$ and $\boldsymbol{B}$

$$\chi_{\mathrm{dE}} = 2\pi\frac{|\boldsymbol{d}|}{e a_0}\frac{a_0}{\xi} , \tag{77.72}$$

where a_0 is the Bohr radius, and $\xi = \Phi_0/\varrho_{\mathrm{mg}}$, Φ_0 being the magnetic flux unit, and ϱ_{mg} being the magnetic line charge density. In analogy to the Aharonov–Casher effect, χ_{dE} is topological, provided that $\boldsymbol{d}$ is aligned parallel to the magnetic line charge, and both are oriented perpendicular to the interferometer plane.

77.6.2 Internal State Interferometry

Manipulation of the internal state of atoms by means of electromagnetic fields makes it possible to realize interferometric setups which involve separation of paths in internal space rather than in real space. Examples of such interferometers are the Optical Ramsey interferometer [77.67], the stimulated Raman interferometer [77.68], and the interferometers using static electric and magnetic fields [77.69, 70].

77.6.3 Manipulation of Cavity Fields by Atom Interferometry

The entanglement of atomic states and quantized field states opens novel possibilities for manipulating and/or measuring nonclassical field states in a cavity. In the adiabatic limit, for example, and assuming sufficient detuning between the atom and the cavity field, the interaction and c.m. motion of an atom traversing a cavity is well described by the potential (77.32)

$$\mathcal{U}(\boldsymbol{x}) = -\hbar\frac{g^2}{\Delta} f(\boldsymbol{x})^2 a^\dagger a\,, \tag{77.73}$$

where g is the vacuum Rabi frequency, $f(\boldsymbol{x})$ is a cavity mode function, and a, $a^\dagger$ denote cavity photon annihilation and creation operators, respectively.

Because of the presence of the photon-number operator $a^\dagger a$ in (77.73), the deflection and phase shift of an atom traversing the cavity is quantized, displaying essentially the photon number statistics in the cavity. The quantized deflection is sometimes called the inverse Stern–Gerlach effect.

Due to the entanglement of atom and cavity states, and the position dependence of the interaction strength, the phase shift induced by $\mathcal{U}(\boldsymbol{x})$ in a standing wave cavity may be used to measure either the atomic position via homodyne detection of the cavity field [77.71, 72], or the photon statistics via atom interferometry [77.73, 74]. In a ring cavity, the entanglement of c.m. motion and cavity field may be used to measure the atomic momentum [77.75] via homodyne detection of the cavity field. For further details see [77.76] and Chapt. 78.

77.7 Coherence of Scalar Matter Waves

The general solution of the free Schrödinger equation (77.2) may be written in the form

$$\psi(\boldsymbol{x}, t) = \int \mathrm{d}^3\tilde{\boldsymbol{k}}\, a(\tilde{\boldsymbol{k}})\,\mathrm{e}^{\mathrm{i}(\tilde{\boldsymbol{k}}\cdot\boldsymbol{x} - \omega(\tilde{\boldsymbol{k}})t)}\,, \tag{77.74}$$

where $\omega(\tilde{\boldsymbol{k}}) \equiv E/\hbar = \hbar\tilde{\boldsymbol{k}}^2/2M$. If the coefficients $a(\tilde{\boldsymbol{k}})$ are known, the state represented by $\psi(\boldsymbol{x}, t)$ is called a pure state. Otherwise it is called a mixed state, and physical quantities are obtained by an ensemble average over the possible realizations of $a(\tilde{\boldsymbol{k}})$.

The degree of coherence of matter waves is described by the autocorrelation function of $\Psi(\boldsymbol{x}, t)$:

$$\Gamma(\boldsymbol{x}, t; \boldsymbol{x}', t') \equiv \overline{\Psi(\boldsymbol{x}', t')^* \Psi(\boldsymbol{x}, t)}\,, \tag{77.75}$$

where the overline $\overline{(\cdots)}$ denotes the ensemble average over the possible realizations of $a(\tilde{\boldsymbol{k}})$. In light optics, $\Gamma(\boldsymbol{x}, t; \boldsymbol{x}', t')$ is called the mutual coherence function. In particular, for equal times, $\Gamma(\boldsymbol{x}, t; \boldsymbol{x}', t)$ describes the spatial coherence, and for equal positions, $\Gamma(\boldsymbol{x}, t; \boldsymbol{x}, t')$ describes the temporal coherence.

For a beam of particles, coherence may be either longitudinal (measured along the beam) or transverse (measured across the beam). In contrast to light optics, there is no simple relation between longitudinal coherence and temporal coherence because the dispersion relation of matter waves is quadratic in the wavenumber.

The spatial coherence function is intimately related to the quantum mechanical density operator of the particles (Chapt. 7)

$$\rho(t) = \overline{|\Psi(t)\rangle\langle\Psi(t)|}\,. \tag{77.76}$$

In the position representation, one has

$$\langle \boldsymbol{x}|\rho(t)|\boldsymbol{x}'\rangle \equiv \rho(\boldsymbol{x}, \boldsymbol{x}'; t) = \Gamma(\boldsymbol{x}, t; \boldsymbol{x}', t)\,. \tag{77.77}$$

Longitudinal and temporal coherence of a particle beam is determined mainly by the source of the beam. The thermal fission reactors used in neutron optics and the ovens used in atom optics are analogous to black-body sources in light optics. In contrast, the transverse coherence is mainly determined by the way the particles are extracted from the oven to form a beam.

77.7.1 Atomic Sources

To describe thermal sources, consider a single particle in an oven of temperature T and volume V, assuming that $a(\tilde{\boldsymbol{k}})$ and $a(\tilde{\boldsymbol{k}}')$ are statistically independent:

$$\overline{a(\tilde{\boldsymbol{k}})^* a(\tilde{\boldsymbol{k}}')} = \rho(\tilde{k})\delta(\tilde{\boldsymbol{k}} - \tilde{\boldsymbol{k}}') , \tag{77.78}$$

where

$$\rho(\tilde{k}) = \frac{\lambda_{\text{th}}^3}{V} \exp\left(-\frac{\hbar^2 \tilde{k}^2}{2Mk_\text{B}T}\right) \tag{77.79}$$

accounts for the thermal distribution of wavenumbers, and

$$\lambda_{\text{th}} = \left(\frac{2\pi\hbar^2}{Mk_\text{B}T}\right)^{1/2} \tag{77.80}$$

denotes the thermal de Broglie wavelength.

Using (77.78)–(77.80) in (77.75), the mutual coherence function becomes

$$\begin{aligned} &\Gamma(\boldsymbol{x}, t; \boldsymbol{x}', t') = \\ &\frac{1}{V} \frac{1}{\{1 + [\mathrm{i}(t-t')/\tau_{\text{th}}]\}^{\frac{3}{2}}} \\ &\times \exp\left\{-\pi \frac{(\boldsymbol{x}-\boldsymbol{x}')^2}{\lambda_{\text{th}}^2 (1 + [\mathrm{i}(t-t')/\tau_{\text{th}}])}\right\} , \end{aligned} \tag{77.81}$$

where

$$\tau_{\text{th}} = \frac{\hbar}{k_\text{B}T} , \tag{77.82}$$

is the thermal coherence time.

According to (77.81), the spatial coherence of a thermal state falls off in a Gaussian manner on a scale given by λ_{th}. The temporal coherence, in contrast, falls off algebraically on a time scale given by τ_{th}. Expressed in physical units, one has

$$\begin{aligned} \lambda_{\text{th}} &= \frac{1.74(5)\times 10^{-9}\,\text{m}}{\sqrt{(M/\text{u})(T/\text{K})}} , \\ \tau_{\text{th}} &= \frac{7.63\times 10^{-12}\,\text{s}}{(T/\text{K})} , \end{aligned} \tag{77.83}$$

where u is the atomic mass unit.

Atomic Beams

Effusive Beams. Effusive beams are produced from thermal sources by a suitable set of collimators placed in front of the opening of the oven. This produces a Maxwell–Boltzmann distribution of atomic velocities in the longitudinal direction. The coherence properties in the transverse direction are described by the van Cittert–Zernike theorem [77.77]; for details see any textbook on classical optics.

Supersonic Beams. Supersonic beams are produced by supersonic expansion of a high pressure gas which is forced through an appropriately designed nozzle. The expansion produces a velocity distribution in the longitudinal direction which is approximately Gaussian with a velocity ratio $v/\delta v \approx 10$–20.

Pulsed Beams. Pulsed beams are produced by chopping any of the beams described above. Important applications for pulsed beams are the resolution of temporal coherence and the mapping of the relative phases of the $a(\tilde{\boldsymbol{k}})$ in matter wave interferometry.

Laser-like Source of Atoms. In these sources, many atoms with integral spin (Bosons) occupy one and the same quantum state of motion [77.78]. Their operational principle is rooted in the quantum statistical effects of indistinguishability. It may be viewed in close analogy to the operational principle of an ordinary laser (Chapt. 70) and the mechanism underlying Bose–Einstein Condensation (Sect. 76.1.1). Laser-like sources have indeed been achieved by letting a small current of atoms leak out of a trapped Bose–Einstein condensate [77.79].

77.7.2 Atom Decoherence

In any interferometer, the contrast of the interference fringes quantitatively measures the coherence of the wave involved. Partially coherent beams show an output flux given by

$$I_{\pm} = \frac{1}{2}\left[1 \pm \text{Re}\left(C\,\mathrm{e}^{\mathrm{i}\chi}\right)\right] , \tag{77.84}$$

instead of (77.60), with a complex number C. One has $|C| \leq 1$, with the maximum achieved for a pure state; the phase of C is measured by scanning the interferometer phase shift χ.

In de Broglie interferometry, coherence can be lost when the interfering matter wave gets entangled with other systems. This happens for atoms, for example, due to the emission or scattering of photons, as soon as the detection of these photons permits, in principle, the resolution of spatially separated paths in the interferometer. In fact, the width of $\Gamma(\boldsymbol{x}, t; \boldsymbol{x}', t)$ as a function of $\boldsymbol{x} - \boldsymbol{x}'$, also called the spatial coherence length, is reduced to the photon wavelength after a single scattering event, see [77.80, 81]. Interference can be restored when the emitted photons are detected and correlated with the atom output [77.82, 83]. Collisions with background gas atoms between the optical elements of a three-grating interferometer also reduce coherence, as has been shown with fullerene molecules [77.84]. Finally, coherence is lost when atoms interact with random electromagnetic fields. This has become relevant for atom reflection from evanescent light because of the roughness of the dielectric surface used [77.85] (see also [77.86]). The coherent operation of integrated atom optics near metallic surfaces is limited by thermally excited electromagnetic near fields as shown in experiments by *Harber* et al. [77.87] (see also [77.88]).

References

77.1 P. W. Hawkes, E. Kasper: *Principles of Electron Optics* (Academic Press, London 1989)

77.2 V. F. Sears: *Neutron Optics*, Oxford Series on Neutron Scattering in Condensed Matter (Oxford Univ. Press, Oxford 1989)

77.3 P. Meystre: *Atom Optics* (Springer, Berlin, Heidelberg 2001)

77.4 L. Reimer: *Transmission Electron Microscopy*, Springer Series in Optical Sciences, Vol. 36, 2nd edn. (Springer, Berlin, Heidelberg 1989)

77.5 A. Tonomura: Electron Holography. In: *Springer Series in Optical Sciences*, Vol. 70, ed. by K. Shimoda (Springer, Berlin, Heidelberg 1993)

77.6 D. M. Greenberger, A. W. Overhauser: Rev. Mod. Phys. **51**, 43 (1979)

77.7 C. S. Adams, M. Sigel, J. Mlynek: Phys. Rep. **240**, 143 (1994)

77.8 H. Wallis: Phys. Rep. **255**, 203 (1995)

77.9 H. J. Metcalf, P. van der Straten: *Laser Cooling and Trapping*, 2nd edn. (Springer, Berlin, Heidelberg 2001)

77.10 B. P. Anderson, P. Meystre: Contemp. Phys. **44**, 473 (2003)

77.11 L. Deng et al.: Nature **398**, 218 (1999)

77.12 S. Inouye et al.: Nature **402**, 641 (1999)

77.13 G. Lenz, P. Meystre, E. M. Wright: Phys. Rev. Lett. **71**, 3271 (1993)

77.14 A. B. Migdal, V. Krainov: Approximation Methods in Quantum Mechanics. In: *Frontiers in Physics Series*, ed. by D. Pines (W. A. Benjamin, New York 1969)

77.15 M. V. Berry, C. Upstill: Catastrophe optics: morphology of caustics and their diffraction patterns. In: *Progress in Optics*, Vol. XVIII, ed. by E. Wolf (North-Holland, Amsterdam 1980) pp. 259–346

77.16 J. J. Berkhout et al.: Phys. Rev. Lett. **63**, 1689 (1989)

77.17 J. P. Dowling, J. Gea-Banacloche: Evanescent light-wave atom mirrors, resonators, waveguides, and traps. In: *Adv. At. Mol. Opt. Phys.*, Vol. 37, ed. by P. R. Berman (Academic, New York 1997) pp. 1–94

77.18 V. I. Balykin, V. S. Letokhov, Y. B. Ovchinnikov, A. I. Sidorov: Phys. Rev. Lett. **60**, 2137 (1988)

77.19 M. A. Kasevich, D. S. Weiss, S. Chu: Opt. Lett. **15**, 607 (1990)

77.20 T. M. Roach et al.: Phys. Rev. Lett. **75**, 629 (1995)

77.21 D. C. Lau et al.: Eur. Phys. J. **D5**, 193 (1999)

77.22 P. Rosenbusch et al.: Appl. Phys. B **70**, 661 (2000)

77.23 H. Wallis, J. Dalibard, C. Cohen-Tannoudji: Appl. Phys. B **54**, 407 (1992)

77.24 C. G. Aminoff et al.: Phys. Rev. Lett. **71**, 3083 (1993)

77.25 M. Hammes et al.: J. mod. Optics. **47**, 2755 (2000)

77.26 M. Wilkens, E. Goldstein, B. Taylor, P. Meystre: Phys. Rev. A **47**, 2366 (1993)

77.27 T. Sleator, T. Pfau, V. Balykin, J. Mlynek: Appl. Phys. B **54**, 375 (1992)

77.28 G. Timp et al.: Phys. Rev. Lett. **69**, 1636 (1992)

77.29 J. Schmiedmayer: Eur. Phys. J. D. **4**, 57 (1998)

77.30 V. I. Balykin: Atom waveguides. In: *Adv. At. Mol. Opt. Phys.*, Vol. 41, ed. by B. Bederson, H. Walther (Academic, San Diego 1999) pp. 181–260

77.31 E. A. Hinds, I. G. Hughes: J. Phys. D: Appl. Phys. **32**, R199 (1999)

77.32 J. Reichel, W. Hänsel, P. Hommelhoff, T. W. Hänsch: Appl. Phys. B **72**, 81 (2001)

77.33 R. Folman et al.: Microscopic atom optics: from wires to an atom chip. In: *Adv. At. Mol. Opt. Phys.*, Vol. 48, ed. by B. Bederson (Academic, San Diego 2002) pp. 263–356

77.34 H. Gauck et al.: Phys. Rev. Lett. **81**, 5298 (1998)

77.35 H. Ito et al.: Phys. Rev. Lett. **76**, 4500 (1996)

77.36 O. Houde, D. Kadio, L. Pruvost: Phys. Rev. Lett. **85**, 5543 (2000)

77.37 J. Schmiedmayer: Phys. Rev. A **52**, R13 (1995)

77.38 J. Fortágh, A. Grossmann, C. Zimmermann, T. W. Hänsch: Phys. Rev. Lett. **81**, 5310 (1998)

77.39 N. H. Dekker et al.: Phys. Rev. Lett. **84**, 1124 (2000)

77.40 E. Andersson et al.: Phys. Rev. Lett. **88**, 100401 (2002)

77.41 H. Kreutzmann et al.: Phys. Rev. Lett. **92**, 163201 (2004)

77.42 D. W. Keith, M. L. Schattenburg, H. I. Smith, D. E. Pritchard: Phys. Rev. Lett. **61**, 1580 (1988)

77.43 O. Carnal, J. Mlynek: Phys. Rev. Lett. **66**, 2689 (1991)

77.44 M. Arndt et al.: Nature **401**, 680 (1999)

77.45 O. Carnal et al.: Phys. Rev. Lett. **67**, 3231 (1991)

77.46 J. F. Clauser, M. Reinsch: Appl. Phys. B **54**, 380 (1992)

77.47 U. Janicke, M. Wilkens: J. Physique (France) II **4**, 1975 (1994)

77.48 P. L. Gould, G. A. Ruff, D. E. Pritchard: Phys. Rev. Lett. **56**, 827 (1986)

77.49 U. Janicke, M. Wilkens: Phys. Rev. A **50**, 3265 (1994)

77.50 T. Sleator et al.: Phys. Rev. Lett. **68**, 1996 (1992)

77.51 M. Wilkens, E. Schumacher, P. Meystre: Phys. Rev. A **44**, 3130 (1991)
77.52 P. J. Martin, B. G. Oldaker, A. H. Miklich, D. E. Pritchard: Phys. Rev. Lett. **60**, 515 (1988)
77.53 T. Pfau et al.: Phys. Rev. Lett. **71**, 3427 (1993)
77.54 R. Deutschmann, W. Ertmer, H. Wallis: Phys. Rev. A **47**, 2169 (1993)
77.55 M. Christ et al.: Opt. Commun. **107**, 211 (1994)
77.56 R. Brouri et al.: Opt. Commun. **124**, 448 (1996)
77.57 A. Landragin et al.: Europhys. Lett. **39**, 485 (1997)
77.58 C. Henkel et al.: Appl. Phys. B. **69**, 277 (1999)
77.59 D. W. Keith, C. R. Ekstrom, Q. A. Turchette, D. E. Pritchard: Phys. Rev. Lett. **66**, 2693 (1991)
77.60 B. Brezger et al.: Phys. Rev. Lett. **88**, 100404 (2002)
77.61 J. Schmiedmayer et al.: Phys. Rev. Lett. **74**, 1043 (1995)
77.62 S. A. Werner, J. Staudenmann, R. Colella: Phys. Rev. Lett. **42**, 1102 (1979)
77.63 Y. Aharonov, D. Bohm: Phys. Rev. **115**, 485 (1959)
77.64 Y. Aharonov, A. Casher: Phys. Rev. Lett. **53**, 319 (1984)
77.65 K. Sangster, E. A. Hinds, S. M. Barnett, E. Riis: Phys. Rev. Lett. **71**, 3641 (1993)
77.66 M. Wilkens: Phys. Rev. Lett. **72**, 5 (1994)
77.67 C. J. Bordé: Phys. Lett. A. **140**, 10 (1989)
77.68 M. Kasevich, S. Chu: Phys. Rev. Lett. **67**, 181 (1991)
77.69 Y. L. Sokolov, V. P. Yakovlov: Sov. Phys. JETP **56**, 7 (1982)
77.70 J. Robert et al.: Europhys. Lett. **16**, 29 (1991)
77.71 M. A. M. Marte, P. Zoller: Appl. Phys. B **54**, 477 (1992)
77.72 P. Storey, M. Collett, D. Walls: Phys. Rev. Lett. **68**, 472 (1992)
77.73 P. Meystre, E. Schumacher, S. Stenholm: Opt. Commun. **73**, 443 (1989)
77.74 A. M. Herkommer, V. M. Akulin, W. P. Schleich: Phys. Rev. Lett. **69**, 3298 (1992)
77.75 T. Sleator, M. Wilkens: Phys. Rev. A **48**, 3286 (1993)
77.76 S. Haroche, J. M. Raimond: Manipulation of non classical field states by atom interferometry. In: *Cavity Quantum Electrodynamics, Adv. At. Mol. Opt. Phys., suppl. 2*, ed. by P. R. Berman (Academic Press, Boston 1994) pp. 123–170
77.77 B. Taylor, K. J. Schernthanner, G. Lenz, P. Meystre: Opt. Commun. **110**, 569 (1994)
77.78 R. J. C. Spreeuw, T. Pfau, U. Janicke, M. Wilkens: Europhys. Lett. **32**, 469 (1995)
77.79 M. Köhl, T. W. Hänsch, T. Esslinger: Phys. Rev. Lett. **87**, 160404 (2001)
77.80 T. Pfau et al.: Phys. Rev. Lett. **73**, 1223 (1994)
77.81 O. Steuernagel, H. Paul: Phys. Rev. A **52**, 905 (1995)
77.82 M. S. Chapman et al.: Phys. Rev. Lett. **75**, 3783 (1995)
77.83 C. Kurtsiefer et al.: Phys. Rev. A **55**, R2539 (1996)
77.84 L. Hackermüller et al.: Appl. Phys. B **77**, 781 (2003)
77.85 A. Landragin et al.: Opt. Lett. **21**, 1581 (1996)
77.86 C. Henkel et al.: Phys. Rev. A **55**, 1160 (1997)
77.87 D. M. Harber, J. M. McGuirk, J. M. Obrecht, E. A. Cornell: J. Low Temp. Phys. **133**, 229 (2003)
77.88 C. Henkel, M. Wilkens: Europhys. Lett. **47**, 414 (1999)

78. Quantized Field Effects

The electromagnetic field appears almost everywhere in physics. Following the introduction of Maxwell's equations in 1864, Max Planck initiated quantum theory when he discovered $h = 2\pi\hbar$ in the laws of black-body radiation. In 1905 Albert Einstein explained the photoelectric effect on the hypothesis of a corpuscular nature of radiation and in 1917 this paradigm led to a description of the interaction between atoms and electromagnetic radiation.

The study of quantized field effects requires an understanding of the quantization of the field which leads to the concept of a quantum of radiation, the photon. Specific nonclassical features arise when the field is prepared in particular quantum states, such as squeezed states. When the radiation field interacts with an atom, there is an important difference between a classical field and a quantized field. A classical field can have zero amplitude, in which case it does not interact with the atom. On the other hand a quantized field always interacts with the atom, even if all the field modes are in their ground states, due to vacuum fluctuations. These lead to various effects such as spontaneous emission and the Lamb shift.

The interaction of an atom with the many modes of the radiation field can conveniently be described in an approximate manner by a master equation where the radiation field is treated as a reservoir. Such a treatment gives a microscopic and quantum mechanically consistent description of damping.

78.1 **Field Quantization** 1142

78.2 **Field States** 1142
78.2.1 Number States 1143
78.2.2 Coherent States 1143
78.2.3 Squeezed States 1144
78.2.4 Phase States 1145

78.3 **Quantum Coherence Theory** 1146
78.3.1 Correlation Functions 1146
78.3.2 Photon Correlations 1146
78.3.3 Photon Bunching and Antibunching 1147

78.4 **Photodetection Theory** 1147
78.4.1 Homodyne and Heterodyne Detection 1147

78.5 **Quasi-Probability Distributions** 1148
78.5.1 s-Ordered Operators 1148
78.5.2 The P Function 1149
78.5.3 The Wigner Function 1149
78.5.4 The Q Function 1151
78.5.5 Relations Between Quasi-Probabilities 1151

78.6 **Reservoir Theory** 1151
78.6.1 Thermal Reservoir 1152
78.6.2 Squeezed Reservoir 1152

78.7 **Master Equation** 1152
78.7.1 Damped Harmonic Oscillator 1153
78.7.2 Damped Two-Level Atom 1153

78.8 **Solution of the Master Equation** 1154
78.8.1 Damped Harmonic Oscillator 1154
78.8.2 Damped Two-Level Atom 1155

78.9 **Quantum Regression Hypothesis** 1156
78.9.1 Two-Time Correlation Functions and Master Equation 1156
78.9.2 Two-Time Correlation Functions and Expectation Values 1156

78.10 **Quantum Noise Operators** 1157
78.10.1 Quantum Langevin Equations 1157
78.10.2 Stochastic Differential Equations 1158

78.11 **Quantum Monte Carlo Formalism** 1159

78.12 **Spontaneous Emission in Free Space** 1159

78.13 **Resonance Fluorescence** 1160
78.13.1 Equations of Motion 1160
78.13.2 Intensity of Emitted Light 1160
78.13.3 Spectrum of the Fluorescence Light 1161
78.13.4 Photon Correlations 1161

78.14 **Recent Developments** 1162
78.14.1 Literature 1162
78.14.2 Field States 1162
78.14.3 Reservoir Theory 1162

References 1163

Part F | 78

78.1 Field Quantization

This section provides the basis for the quantized field effects discussed in this Chapter [78.1]. We expand the field in a complete set of normal modes which reduces the problem of field quantization to the quantization of a one dimensional harmonic oscillator corresponding to each normal mode.

The classical free electromagnetic field, i. e., the field in a region without charge and current densities, obeys the Maxwell equations

$$\nabla\cdot\boldsymbol{B}=0\,, \tag{78.1}$$

$$\nabla\cdot\boldsymbol{D}=0\,, \tag{78.2}$$

$$\nabla\times\boldsymbol{E}+\frac{\partial\boldsymbol{B}}{\partial t}=0\,, \tag{78.3}$$

$$\nabla\times\boldsymbol{H}-\frac{\partial\boldsymbol{D}}{\partial t}=0\,, \tag{78.4}$$

where $\boldsymbol{B}=\mu_0\boldsymbol{H}$, $\boldsymbol{D}=\epsilon_0\boldsymbol{E}$. The magnetic permeability μ_0 connects the magnetic induction $\boldsymbol{B}$ with the magnetic field $\boldsymbol{H}$ and the electric permittivity ϵ_0 of free space connects the displacement $\boldsymbol{D}$ with the electric field $\boldsymbol{E}$. In the case of a free field, $\boldsymbol{E}$ and $\boldsymbol{B}$ may be obtained from

$$\boldsymbol{B}=\nabla\times\boldsymbol{A}\,, \tag{78.5}$$

$$\boldsymbol{E}=-\frac{\partial\boldsymbol{A}}{\partial t}\,, \tag{78.6}$$

where the vector potential $\boldsymbol{A}$ obeys the Coulomb gauge condition $\nabla\cdot\boldsymbol{A}=0$ and satisfies a wave equation. In order to solve this wave equation we expand the vector potential

$$\boldsymbol{A}(\boldsymbol{x},t)=\sum_{\boldsymbol{k}}\sum_{\sigma=1}^{2}\left(\frac{\hbar}{2\omega_{\boldsymbol{k}}\epsilon_0 V}\right)^{1/2}\times\left[\alpha_{\boldsymbol{k}\sigma}\boldsymbol{\epsilon}_{\boldsymbol{k}\sigma}\,\mathrm{e}^{\mathrm{i}(\boldsymbol{k}\cdot\boldsymbol{x}-\omega_{\boldsymbol{k}}t)}+\mathrm{c.c.}\right] \tag{78.7}$$

in a set of normal modes $V^{-1/2}\exp(\mathrm{i}\boldsymbol{k}\cdot\boldsymbol{x})\boldsymbol{\epsilon}_{\boldsymbol{k}\sigma}$ which are orthonormal in the volume V. Due to the gauge condition $\nabla\cdot\boldsymbol{A}=0$, we obtain two orthogonal polarization vectors $\boldsymbol{\epsilon}_{\boldsymbol{k}1}$ and $\boldsymbol{\epsilon}_{\boldsymbol{k}2}$ with $\boldsymbol{\epsilon}_{\boldsymbol{k}\sigma}\cdot\boldsymbol{k}=0$ for each wave vector $\boldsymbol{k}$. The dispersion relation is $\omega_{\boldsymbol{k}}=c|\boldsymbol{k}|$. The Fourier amplitudes $\alpha_{\boldsymbol{k}\sigma}$ are complex numbers in the classical theory.

The field is quantized by replacing the classical amplitude $\alpha_{\boldsymbol{k}\sigma}$ by the mode annihilation operator $a_{\boldsymbol{k}\sigma}$. The complex conjugate $\alpha^*_{\boldsymbol{k}\sigma}$ is replaced by the mode creation operator $a^\dagger_{\boldsymbol{k}\sigma}$. They obey the commutation relation

$$\left[a_{\boldsymbol{k}\sigma},a^\dagger_{\boldsymbol{k}'\sigma'}\right]=\delta_{\boldsymbol{k}\boldsymbol{k}'}\delta_{\sigma\sigma'}\,. \tag{78.8}$$

The representation of the electric field operator

$$\begin{aligned}\boldsymbol{E}(\boldsymbol{x},t)&=\mathrm{i}\sum_{\boldsymbol{k},\sigma}\left(\frac{\hbar\omega_{\boldsymbol{k}}}{2\epsilon_0 V}\right)^{1/2}\times\left[a_{\boldsymbol{k}\sigma}\boldsymbol{\epsilon}_{\boldsymbol{k}\sigma}\,\mathrm{e}^{\mathrm{i}(\boldsymbol{k}\cdot\boldsymbol{x}-\omega_{\boldsymbol{k}}t)}-\mathrm{h.c.}\right]\\&\equiv\boldsymbol{E}^+(\boldsymbol{x},t)+\boldsymbol{E}^-(\boldsymbol{x},t)\end{aligned} \tag{78.9}$$

in terms of these operators follows from (78.6) and the operator for the vector potential. Note also the often used decomposition of the electric field operator into the positive and negative frequency parts $\boldsymbol{E}^+$ and $\boldsymbol{E}^-$ respectively. A similar relation holds for the operator describing the magnetic induction $\boldsymbol{B}$.

Using the operators for the electric and magnetic field, one can transform the field energy

$$H=\frac{1}{2}\int\mathrm{d}V\left(\epsilon_0\boldsymbol{E}^2+\boldsymbol{B}^2/\mu_0\right) \tag{78.10}$$

into the form

$$H=\sum_{\boldsymbol{k},\sigma}\hbar\omega_{\boldsymbol{k}}\left(a^\dagger_{\boldsymbol{k}\sigma}a_{\boldsymbol{k}\sigma}+1/2\right)\,, \tag{78.11}$$

which is a sum of independent harmonic oscillator Hamiltonians corresponding to each mode $(\boldsymbol{k},\sigma)$. The number operator $N_{\boldsymbol{k}\sigma}=a^\dagger_{\boldsymbol{k}\sigma}a_{\boldsymbol{k}\sigma}$ represents the number of photons in the mode $(\boldsymbol{k},\sigma)$, while $\hbar\omega_{\boldsymbol{k}}/2$ is the energy of the vacuum fluctuations.

Hence each mode of the electromagnetic field is equivalent to a harmonic oscillator. In the next section we discuss specific states of a single mode. The general quantum state of the electromagnetic field consisting of many modes is given by a superposition of product states that are composed out of these single mode states.

78.2 Field States

This section summarizes the properties of several important states of the electromagnetic field. From the independence of the normal modes, the discussion may be restricted to a single normal mode. With the mode index $(\boldsymbol{k},\sigma)$ suppressed, a single mode Hamiltonian is

$$H=\hbar\omega\left(a^\dagger a+1/2\right)=\frac{\hbar\omega}{2}p^2+\frac{\hbar\omega}{2}x^2\,. \tag{78.12}$$

In the second step, the quadrature operators

$$x = \frac{1}{\sqrt{2}}\left(a + a^{\dagger}\right), \quad (78.13)$$

$$p = \frac{1}{\mathrm{i}\sqrt{2}}\left(a - a^{\dagger}\right) \quad (78.14)$$

are introduced, which are equivalent to scaled position and momentum operators of a massive particle in a harmonic potential. The quadratures of a quantized field are measurable with the help of homodyne detection as discussed in Sect. 78.4.1

We shall now describe several states of this quantized field mode: number states, coherent states, squeezed states, Schrödinger cats, and phase states. A quantized field in a coherent state shows the most classical behavior. A superposition of two coherent states, which is a Schrödinger cat, already shows nonclassical features. Number states and squeezed states are further typical examples of nonclassical states.

78.2.1 Number States

The eigenstates of the Hamiltonian (78.12) are the eigenstates of the number operator $N = a^{\dagger}a$,

$$N|n\rangle = n|n\rangle, \quad (78.15)$$

where $n = 0, 1, 2, \ldots$ denotes the excitations or the number of photons in the mode. The vacuum state of the mode $|0\rangle$, is defined by

$$a|0\rangle = 0. \quad (78.16)$$

The ladder of excitations can be climbed up and down via the application of creation and annihilation operators

$$a^{\dagger}|n\rangle = \sqrt{n+1}|n+1\rangle, \quad (78.17)$$

$$a|n\rangle = \sqrt{n}|n-1\rangle, \quad (78.18)$$

on a Fock state $|n\rangle$. These number or Fock states form a complete and orthonormal set of states so that

$$\sum_{n=0}^{\infty}|n\rangle\langle n| = 1, \quad \langle n|k\rangle = \delta_{nk}. \quad (78.19)$$

Their quadrature representations are

$$\langle x|n\rangle = \left(\sqrt{\pi}2^n n!\right)^{-1/2} H_n(x)\,\mathrm{e}^{-x^2/2}, \quad (78.20)$$

$$\langle p|n\rangle = \left(\sqrt{\pi}2^n n!\right)^{-1/2}(-\mathrm{i})^n H_n(p)\,\mathrm{e}^{-p^2/2}. \quad (78.21)$$

The states $|x\rangle$ and $|p\rangle$ are eigenstates of the quadrature operators x and p, (78.13) and (78.14).

Number states provide a frequently used representation of a pure quantum state

$$|\psi\rangle = \sum_{n=0}^{\infty} c_n|n\rangle, \quad (78.22)$$

or a mixed quantum state given by the density operator

$$\rho = \sum_{n,k}\rho_{nk}|n\rangle\langle k| \quad (78.23)$$

(Chapt. 7).

78.2.2 Coherent States

The coherent state is a specific superposition of number states. In contrast to a number state, a coherent state does not possess a definite number of photons: the photon distribution is Poissonian. For a large average photon number, the electric and magnetic fields have rather well defined amplitudes and phases with vanishing relative quantum fluctuations. Hence the Poissonian photon distribution frequently serves as a borderline between classical and nonclassical field states. Nonclassical states show a sub-Poissonian behavior. An extreme example is a field prepared in a number state. A parameter which quantifies the deviations from Poissonian behavior is the Q parameter introduced by *Mandel* [78.2].

We define the coherent state $|\alpha\rangle$ as an eigenstate of the annihilation operator

$$a|\alpha\rangle = \alpha|\alpha\rangle \quad (78.24)$$

with the complex amplitude $\alpha = |\alpha|e^{\mathrm{i}\theta}$. The coherent state can be represented by

$$|\alpha\rangle = \mathrm{e}^{\alpha a^{\dagger} - \alpha^* a}|0\rangle = D(\alpha)|0\rangle, \quad (78.25)$$

that is, by the action of the displacement operator $D(\alpha)$ on the vacuum.

The number state representation of $|\alpha\rangle$ reads

$$|\alpha\rangle = \mathrm{e}^{-|\alpha|^2/2}\sum_{n=0}^{\infty}\frac{\alpha^n}{\sqrt{n!}}|n\rangle. \quad (78.26)$$

A coherent state $|\alpha_0\rangle$ that evolves in time according to the free field Hamiltonian (78.12) stays coherent, i. e.,

$$|\psi(t)\rangle = \exp\left(-\mathrm{i}Ht/\hbar\right)|\alpha_0\rangle = \mathrm{e}^{-\mathrm{i}\omega t/2}|\alpha(t)\rangle, \quad (78.27)$$

with amplitude $\alpha(t) = \alpha_0\exp[-\mathrm{i}\omega t]$.

Another important representation of a coherent state is the x representation

$$\langle x|\alpha\rangle = \pi^{-1/4}\exp\left\{-[\mathrm{Re}(\alpha)]^2\right\} \times \exp\left(-x^2/2+\sqrt{2}\alpha x\right) , \quad (78.28)$$

where $|x\rangle$ denotes again the x quadrature eigenstate.

The photon distribution in a coherent state

$$|\langle n|\alpha\rangle|^2 = \frac{|\alpha|^{2n}\,\mathrm{e}^{-|\alpha|^2}}{n!} \quad (78.29)$$

is a Poisson distribution with average photon number $\langle N\rangle = |\alpha|^2$ and variance $(\Delta N)^2 = \langle N^2\rangle - \langle N\rangle^2 = |\alpha|^2$. Hence the relative fluctuations $(\Delta N)/\langle N\rangle = \langle N\rangle^{-1/2}$ vanish for a large average photon number.

The Mandel Q parameter

$$Q \equiv \frac{(\Delta N)^2 - \langle N\rangle}{\langle N\rangle} \quad (78.30)$$

vanishes for a field in a coherent state. A nonclassical field may show sub-Poissonian behavior with $Q < 0$. As an example, the Schrödinger cat state is a *macroscopic* superposition of two coherent states

$$|\mathrm{cat}\rangle = \left[2+2\cos\left(\alpha^2\sin\phi\right)\mathrm{e}^{-2\alpha^2\sin^2(\phi/2)}\right]^{-1/2} \times\left[\left|\alpha\,\mathrm{e}^{\mathrm{i}\phi/2}\right\rangle + \left|\alpha\,\mathrm{e}^{-\mathrm{i}\phi/2}\right\rangle\right] , \quad (78.31)$$

where α is assumed to be real. The Q parameter for this superposition state, shown in Fig. 78.1, takes on negative values for specific angles ϕ. The nonclassical behavior of such a $|\mathrm{cat}\rangle$-state can be explained [78.3] as a result of quantum interference between the two coherent states present in (78.31). The incoherent superposition described by the density operator

$$\rho = \frac{1}{2}\left(\left|\alpha\,\mathrm{e}^{\mathrm{i}\phi/2}\right\rangle\!\left\langle\alpha\,\mathrm{e}^{\mathrm{i}\phi/2}\right| + \left|\alpha\,\mathrm{e}^{-\mathrm{i}\phi/2}\right\rangle\!\left\langle\alpha\,\mathrm{e}^{-\mathrm{i}\phi/2}\right|\right) \quad (78.32)$$

does not have this nonclassical character: its Q parameter vanishes. Coherent states have a direct physical significance: the quantum state of a stabilized laser operating well above threshold can be approximated by a coherent state.

78.2.3 Squeezed States

Squeezed states [78.4–6] minimize the uncertainty product of the quadrature components of the electromagnetic field. The quadrature components x and p of the single mode field are defined in (78.13) and (78.14). They obey the commutation relation $[x, p] = \mathrm{i}$. Their uncertainties $(\Delta x)^2 \equiv \langle x^2\rangle - \langle x\rangle^2$ and $(\Delta p)^2 \equiv \langle p^2\rangle - \langle p\rangle^2$ fulfill the Heisenberg inequality

$$\Delta x \Delta p \geq 1/2 . \quad (78.33)$$

The coherent state is a special minimum uncertainty state with equal uncertainties $\Delta x = \Delta p = 1/\sqrt{2}$. Squeezed states comprise a more general class of minimum uncertainty states with reduced uncertainty in one quadrature at the expense of increased uncertainty in the other. These states $|\alpha, \epsilon\rangle$ are obtained by applying the displacement operator $D(\alpha)$ and the unitary squeeze operator

$$S(\epsilon) = \mathrm{e}^{\frac{1}{2}\epsilon^* a^2 - \frac{1}{2}\epsilon a^{\dagger 2}} \quad (78.34)$$

to the vacuum

$$|\alpha, \epsilon\rangle = D(\alpha)S(\epsilon)|0\rangle . \quad (78.35)$$

The squeeze operator $S(\epsilon)$ transforms a and $a^\dagger$ according to

$$S^\dagger(\epsilon)a\,S(\epsilon) = a\cosh r - a^\dagger\,\mathrm{e}^{-2\mathrm{i}\phi}\sinh r , \quad (78.36)$$

$$S^\dagger(\epsilon)a^\dagger S(\epsilon) = a^\dagger\cosh r - a\,\mathrm{e}^{2\mathrm{i}\phi}\sinh r , \quad (78.37)$$

where $\epsilon = r\,\mathrm{e}^{-2\mathrm{i}\phi}$. The rotated quadratures

$$X_1 = x\cos\phi - p\sin\phi , \quad (78.38)$$

$$X_2 = p\cos\phi + x\sin\phi , \quad (78.39)$$

transform according to

$$S^\dagger(\epsilon)(X_1 + \mathrm{i}X_2)S(\epsilon) = X_1\,\mathrm{e}^{-r} + iX_2\,\mathrm{e}^{r} , \quad (78.40)$$

which yields the uncertainties

$$\Delta X_1 = \mathrm{e}^{-r}/\sqrt{2} , \quad (78.41)$$

$$\Delta X_2 = \mathrm{e}^{r}/\sqrt{2} . \quad (78.42)$$

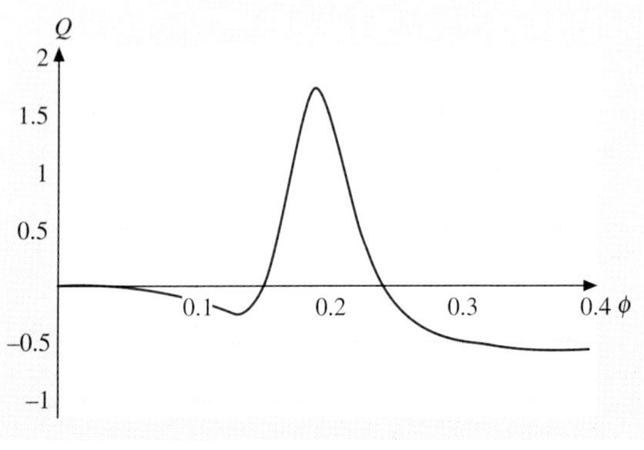

Fig. 78.1 The Q parameter for a Schrödinger cat state (78.31) with amplitude $\alpha = 4$

In particular, for $\phi = 0$ the squeezed state $|\alpha, r\rangle$ is a minimum uncertainty state for the quadratures x and p with $\Delta x = \mathrm{e}^{-r}/\sqrt{2}$ and $\Delta p = \mathrm{e}^{r}/\sqrt{2}$. The degree of squeezing in the quadrature x is determined by the squeeze factor r.

The average photon number

$$\langle N \rangle = |\alpha|^2 + \sinh^2 r \tag{78.43}$$

of a squeezed state and its photon number variance

$$(\Delta N)^2 = \left|\alpha \cosh r - \alpha^* \mathrm{e}^{-2\mathrm{i}\phi} \sinh r\right|^2 + 2\cosh^2 r \sinh^2 r \tag{78.44}$$

contain the coherent contribution α as well as squeezing contributions expressed by r and ϕ. In particular, for $\phi = 0$, the Q parameter becomes negative for a large enough amplitude α and $r > 0$. The photon number distribution $W_n = |\langle \alpha, r|n\rangle|^2$ becomes narrower than the one for the corresponding coherent state with the same α. This sub-Poissonian behavior is one of the nonclassical features of a squeezed state. Furthermore, W_n shows oscillations [78.7] for larger squeezing. The two regimes with sub-Poissonian and oscillating photon statistics W_n are shown in Fig. 78.2.

A second representation of squeezed states has been introduced by Yuen [78.8]. In his notation, a squeezed state is an eigenstate of the operator

$$b = \mu a + \nu a^\dagger , \tag{78.45}$$

with $|\mu|^2 - |\nu|^2 = 1$ and eigenvalue β. This eigenstate can be written in the form

$$|\epsilon, \beta\rangle = S(\epsilon) D(\beta) |0\rangle , \tag{78.46}$$

which connects the squeezing operator $S(r\,\mathrm{e}^{-2\mathrm{i}\phi})$ with the parameters $\mu = \cosh r$ and $\nu = \mathrm{e}^{-2\mathrm{i}\phi} \sinh r$. In contrast to the definition (78.35), the displacement operator $D(\beta)$ and the squeezing operator $S(\epsilon)$ are applied now in reversed order. Nevertheless, the two equations (78.35) and (78.46) define the same state if the relation $\alpha = \beta\mu + \beta^*\nu$ is fulfilled.

Several experiments have demonstrated the generation of squeezed light. *Slusher* et al. [78.9] obtained squeezing in the sidemodes of a four-wave mixing process. An optical parametric oscillator below threshold has been used by *Wu* et al. [78.10] in order to generate squeezed light. Nonclassical features can also be found in a down conversion process. This second-order process creates so-called signal and idler photons from one pump photon. Signal and idler beam are distinguished by frequency or polarization. *Heidmann* et al. [78.11] have shown that the difference intensity of these twin beams may exhibit reduced quantum fluctuations. Pulsed twin beams also contain reduced noise in the difference of their intensities.

78.2.4 Phase States

The problem of a correct quantum mechanical description of phase has a long history in quantum mechanics [78.12]. First attempts to define a quantum phase are due to London and Dirac. The London phase state is

$$|\phi\rangle = \frac{1}{\sqrt{2\pi}} \sum_{n=0}^{\infty} \mathrm{e}^{\mathrm{i}n\phi} |n\rangle , \tag{78.47}$$

which is an eigenstate of the exponential phase operator

$$\widehat{\mathrm{e}^{\mathrm{i}\phi}} = \sum_{n=0}^{\infty} |n\rangle\langle n+1| . \tag{78.48}$$

Since this operator is not unitary, it does not define a Hermitian operator $\hat{\phi}$ for the phase. Nevertheless many treatments of the phase of a quantum state $|\psi\rangle = \sum c_n |n\rangle$ are based on the London phase distribution

$$Pr(\phi) = |\langle\phi|\psi\rangle|^2 = \frac{1}{2\pi} \left|\sum_n c_n \mathrm{e}^{-\mathrm{i}n\phi}\right|^2 . \tag{78.49}$$

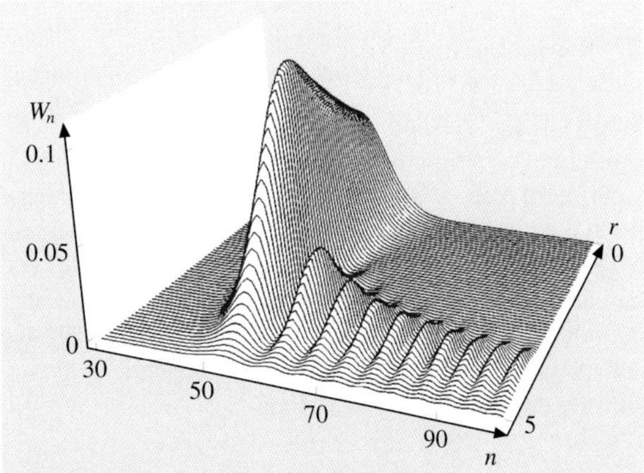

Fig. 78.2 The photon number distribution W_n of a squeezed state $|\alpha, r\rangle$ with the coherent amplitude $\alpha = 7$. For a squeezing parameter $r = 0$, the Poisson distribution of a coherent state $|\alpha = 7\rangle$ is just visible. When r increases the photon distribution first becomes sub-Poissonian and then oscillatory

Later treatments [78.13] rely on the Hermitian operators

$$\widehat{\sin\phi} = \frac{1}{2i}\left(\widehat{e^{i\phi}} - \widehat{e^{i\phi}}^{\dagger}\right), \tag{78.50}$$

$$\widehat{\cos\phi} = \frac{1}{2}\left(\widehat{e^{i\phi}} + \widehat{e^{i\phi}}^{\dagger}\right), \tag{78.51}$$

for the sine and cosine function of the phase.

Recently [78.14] a Hermitian phase operator was constructed starting from the phase state (78.47), restricted to a finite Hilbert space. An operational phase description has been proposed [78.15] in which a classical phase measurement is translated to the quantum realm by using an eight-port homodyne detector.

78.3 Quantum Coherence Theory

This section introduces the correlation functions of the electromagnetic field. Ideal photon correlation measurements can bring out the phenomenon of photon bunching and antibunching.

78.3.1 Correlation Functions

Correlation functions were originally introduced to describe an ideal photodetection process. Glauber [78.16] has presented a treatment based on an absorption mechanism in the detector which is sensitive to the positive frequency part E^+ of the electric field evaluated at the detector's space-time position $x \equiv (\boldsymbol{x}, t)$. This leads to an average field intensity

$$I(x) = \mathrm{Tr}\left[\rho E^-(x) E^+(x)\right] \tag{78.52}$$

at point x. Here the density operator ρ describes the state of the field. The ordering of the operators, i. e., $E^- E^+ \sim a^\dagger a$, is known as normal ordering with all annihilation operators to the right of all creation operators.

The expression (78.52) now immediately generalizes to the correlation function of first order

$$G^{(1)}(x_1, x_2) = \mathrm{Tr}\left[\rho E^-(x_1) E^+(x_2)\right], \tag{78.53}$$

with $x_1 = (\boldsymbol{x}_1, t_1)$ and $x_2 = (\boldsymbol{x}_2, t_2)$. The classical interference experiments, such as Young's double slit experiment, can be described in terms of $G^{(1)}$. Furthermore, the correlation function of first order is connected to the power spectrum $S(\omega)$ of a quantized field via the Wiener–Khintchine theorem [78.17]. Under the assumption of a stationary process, i. e., when the autocorrelation function $\langle E^-(t)E^+(t')\rangle$ depends only on the time difference $\tau = t - t'$, then

$$S(\omega) = \frac{1}{2\pi}\int_0^\infty d\tau \langle E^-(\tau)E^+(0)\rangle e^{-i\omega\tau} + \text{c.c.} \tag{78.54}$$

This relation between the spectrum and the first-order correlation function is known as Wiener–Khintchine theorem.

In order to analyze the Hanbury–Brown and Twiss experiment [78.18] it is necessary to define higher order correlation functions. The general nth order correlation function is defined by

$$\begin{aligned} &G^{(n)}(x_1, \dots, x_n, x_{n+1}, \dots, x_{2n}) \\ &= \mathrm{Tr}\big[\rho E^-(x_1)\cdots E^-(x_n) \\ &\quad \times E^+(x_{n+1})\cdots E^+(x_{2n})\big], \end{aligned} \tag{78.55}$$

where the field operators are again normal ordered.

These correlation functions fulfill a generalized Schwartz inequality

$$G^{(1)}(x_1, x_1)G^{(1)}(x_2, x_2) \geq \left|G^{(1)}(x_1, x_2)\right|^2, \tag{78.56}$$

which becomes, for the nth order functions,

$$\begin{aligned} &G^{(n)}(x_1, \dots, x_n, x_n, \dots, x_1) \\ &\times G^{(n)}(x_{n+1}, \dots, x_{2n}, x_{2n}, \dots, x_{n+1}) \\ &\geq \left|G^{(n)}(x_1, \dots, x_n, x_{n+1}, \dots, x_{2n})\right|^2. \end{aligned} \tag{78.57}$$

A field is said to be first-order coherent when its normalized correlation function

$$g^{(1)}(x_1, x_2) = \frac{G^{(1)}(x_1, x_2)}{\left[G^{(1)}(x_1, x_1)G^{(1)}(x_2, x_2)\right]^{1/2}} \tag{78.58}$$

satisfies $\left|g^{(1)}(x_1, x_2)\right| = 1$. In a Young type experiment, this case gives maximum fringe visibility. A more general definition of first order coherence is the condition that $G^{(1)}(x_1, x_2)$ factorizes

$$G^{(1)}(x_1, x_2) = \mathcal{G}^*(x_1)\mathcal{G}(x_2), \tag{78.59}$$

where $\mathcal{G}$ denotes some complex function. This definition can be readily generalized to the nth order case. The nth order coherence applies when the relation

$$\begin{aligned} &G^{(n)}(x_1, \dots, x_{2n}) \\ &= \mathcal{G}^*(x_1)\cdots\mathcal{G}^*(x_n)\mathcal{G}(x_{n+1})\cdots\mathcal{G}(x_{2n}) \end{aligned} \tag{78.60}$$

holds. A field in a coherent state possesses nth order coherence.

78.3.2 Photon Correlations

The Young experiment demonstrates the appearance of first-order correlations. However, experiments that can distinguish between the classical and quantum domains have to be based on measurements of second-order correlations. These experiments are of the Hanbury–Brown and Twiss type, and determine the arrival of a photon at detector position $\boldsymbol{x}$ and time t and another photon at time $t+\tau$. Following the theory of Glauber, the second-order correlation function

$$G^{(2)}(\tau) = \langle E^{-}(t)E^{-}(t+\tau)E^{+}(t+\tau)E^{+}(t)\rangle \tag{78.61}$$

is measured. In this formula we have omitted the variable for the position $\boldsymbol{x}$. Usually the normalized correlation function

$$g^{(2)}(\tau) = \frac{G^{(2)}(\tau)}{\left[G^{(1)}(0)\right]^2} \tag{78.62}$$

is introduced. The function $g^{(2)}$ is always positive, which is true for classical as well as for quantum fields; but there exists a purely quantum domain given by

$$0 \le g^{(2)}(0) < 1 \,. \tag{78.63}$$

For example, for a number state $|n\rangle$,

$$g^{(2)}_{|n\rangle}(0) = 1 - 1/n \,, \tag{78.64}$$

with $n \ge 1$. In contrast, a coherent state $|\alpha\rangle$ yields $g^{(2)}_{|\alpha\rangle}(0) = 1$.

78.3.3 Photon Bunching and Antibunching

In a realistic theory (but not in an oversimplified one-mode model), the correlation function $G^{(2)}(\tau)$ always factorizes on a sufficiently long time scale, and $g^{(2)}(\tau) \to 1$. The photons are then no longer correlated, and they arrive randomly as in the case of coherent light; see for example (78.195).

If $g^{(2)}(0) > 1$, the photons show a tendency to arrive in bunches, an effect known as photon bunching. This effect has been observed for chaotic light. The opposite situation with $0 \le g^{(2)}(0) < 1$ demonstrates the reverse effect, namely photon antibunching. As seen from (78.63), this is a regime only accessible to nonclassical light. An example is given by the resonance fluorescence of a two-level atom, treated in Sect. 78.13. Note that we can rewrite $g^{(2)}(0)$ with the help of Mandel's Q parameter

$$g^{(2)}(0) = \frac{\langle a^\dagger a^\dagger a\, a\rangle}{\langle a^\dagger a\rangle^2} = 1 + Q \,. \tag{78.65}$$

Hence, a field state with $Q < 0$ shows the effect of photon antibunching.

78.4 Photodetection Theory

So far we have used a very simple theory of photodetection: any absorbed photon leads to a photoelectric emission which can be observed. But in any real experiment, these photons are counted over some time interval T and the observed photoelectric emissions are dominated by two statistics: (i) the statistics of photoelectric emission which is also present for a classical field and (ii) the specific quantum statistics of a quantized field. A detailed discussion of the quantum theory of photoelectric detection has been given by *Kelley* and *Kleiner* [78.19]. A central result is the formula

$$p(n,t,T) = \left\langle : \frac{I^n}{n!} \exp(-I) : \right\rangle \tag{78.66}$$

for the probability of counting n photoelectrons in the time interval from t to $t+T$. This photocounting distribution contains the integrated intensity operator

$$I = \eta \int_t^{t+T} \mathrm{d}t'\, E^{-}(t')E^{+}(t') \tag{78.67}$$

containing the quantum efficiency η of the detector. The notation $\langle : \cdots : \rangle$ indicates a quantum average where the operators have to be normally ordered and time ordered. This operator ordering reflects the process on which a photodetector is based. It annihilates or absorbs photons, one after the other. A good treatment of photoelectric detection can be found in [78.20].

78.4.1 Homodyne and Heterodyne Detection

These detection methods allow the extraction of specific quantum features of a single mode quantum field, the signal field. Figure 78.3 summarizes the principle of optical homodyning. Two quantum fields described by the annihilation operators a and b are mixed at a 50/50 beam splitter BS. Both fields have the same frequency. The mode a represents the signal mode whose quantum state is given by the density operator ρ. Mode b serves as a reference field, the local oscillator. The coherent state $|\alpha\rangle = ||\alpha| e^{i\theta}\rangle$ determines the quantum state of the local oscillator. Two ideal photodetectors 1 and 2 measure the number of photons in the output modes of the beam splitter. For a highly excited coherent state, i. e., a classical local oscillator, the statistics of the photocurrent difference ΔI can be described by the moments of the signal mode operator

$$X_\theta = \frac{1}{\sqrt{2}}\left(a\, e^{-i\theta} + a^\dagger e^{i\theta}\right). \tag{78.68}$$

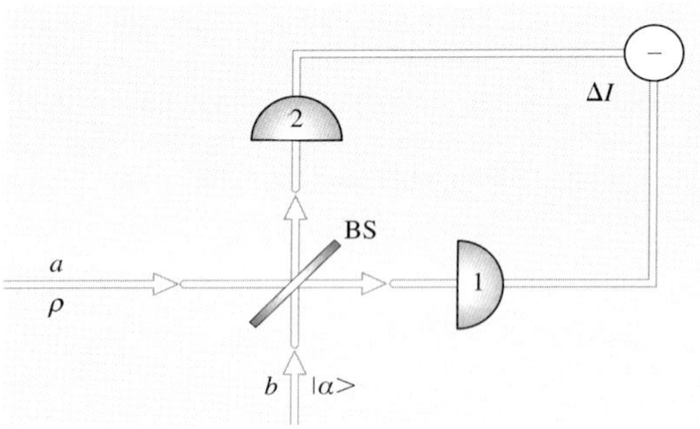

Fig. 78.3 The principle of optical homodyning

For example, the photocurrent difference

$$\langle \Delta I\rangle \sim \langle X_\theta\rangle = \mathrm{Tr}(\rho X_\theta) \tag{78.69}$$

is proportional to the expectation value of X_θ. In particular, for $\theta = 0$ and $\theta = \frac{\pi}{2}$ one is able to measure all the moments of the two quadratures x and p (78.13) and (78.14) of the signal mode. In general, the statistics of the photocurrent ΔI reveal the probability distribution

$$Pr(X_\theta) = \langle X_\theta|\rho|X_\theta\rangle \tag{78.70}$$

of the observable X_θ when the signal mode is in the state ρ. The states

$$|X_\theta\rangle = \pi^{-1/4}\exp\left[-X_\theta^2/2\right] \times \sum_{n=0}^{\infty}\frac{1}{\sqrt{2^n n!}} H_n(X_\theta)\, e^{in\theta}|n\rangle \tag{78.71}$$

are eigenstates of the operator X_θ, and are known as rotated quadrature states.

The heterodyne technique [78.21,22] relies on a similar mixing of a signal field with a local oscillator at a beam splitter, but this time the local oscillator frequency is offset by the intermediate frequency $\Delta\omega$ with respect to the frequency ω_0 of the signal mode. Filters select the beat frequency components in the photocurrent of the detectors. This photocurrent contains the quantum statistics of the two quadratures of the signal field [78.21,22].

78.5 Quasi-Probability Distributions

Quasi-probability distributions play an important role in quantum optics for three reasons. First, they are a complete representation of the density operator of a quantum field. Second, they allow one to calculate expectation values in the spirit of classical statistical physics. Third, they offer the possibility of converting a master equation for the density operator into an equivalent c-number partial differential equation. In this section, we relate a specific quasi-probability function to a specific operator ordering.

78.5.1 s-Ordered Operators

A normally ordered product of a and $a^\dagger$ is a product of the form $\left(a^\dagger\right)^m a^n$: the annihilation operators a stand to the right of the creation operators $a^\dagger$. In an antinormally ordered product like $a^n\left(a^\dagger\right)^m$, the order of a and $a^\dagger$ has changed. A generalized s-ordered product can be defined as

$$\left\{a^m\left(a^\dagger\right)^n\right\}_s \equiv \left(\frac{\partial}{\partial\xi}\right)^n\left(-\frac{\partial}{\partial\xi^*}\right)^m \times D(\xi,\xi^*,s)\Big|_{\xi=\xi^*=0} \tag{78.72}$$

with the generalized displacement operator

$$D(\xi,\xi^*,s) \equiv \exp\left(\xi a^\dagger - \xi^* a + \frac{1}{2}s\xi\xi^*\right). \tag{78.73}$$

For $s = 1$ we find again normal ordering. The values $s = 0$ and $s = -1$ produce symmetric and antinormal ordered products. As an example we note $\{a^\dagger a\}_s = a^\dagger a - (s-1)/2$.

Expectation values of those s-ordered products are easily derived from the characteristic function

$$\chi(\xi, \xi^*, s) = \mathrm{Tr}\big[\rho D(\xi, \xi^*, s)\big] \qquad (78.74)$$

via differentiation

$$\left(\frac{\partial}{\partial \xi}\right)^n \left(-\frac{\partial}{\partial \xi^*}\right)^m \chi(\xi, \xi^*, s)\Big|_{\xi=\xi^*=0} = \langle\{(a^\dagger)^n a^m\}_s\rangle \,. \qquad (78.75)$$

The Fourier transform of χ yields the quasi-probability distribution of Cahill and Glauber [78.23]

$$W(\alpha, s) = \frac{1}{\pi^2}\int \mathrm{d}^2\xi \chi(\xi, \xi^*, s)\, \mathrm{e}^{\alpha\xi^* - \alpha^*\xi} \,, \qquad (78.76)$$

where $\mathrm{d}^2\xi = d\,\mathrm{Re}(\xi)\,\mathrm{d}\,\mathrm{Im}(\xi)$. With this distribution one is able to calculate the expectation value of any s-ordered operator product

$$\left\langle\left\{\left(a^\dagger\right)^n a^m\right\}_s\right\rangle = \int (\alpha^*)^n \alpha^m W(\alpha, s)\, \mathrm{d}^2\alpha \,. \qquad (78.77)$$

We concentrate now on three important quasi-probability distributions, namely the cases $s = 1$, $s = 0$, and $s = -1$, corresponding to the Glauber–Sudarshan distribution, the Wigner function, and the Q function respectively. A detailed discussion of these three functions can be found in [78.24].

78.5.2 The *P* Function

The quasi-probability function $P(\alpha)$ was introduced [78.25, 26] as a diagonal representation of the density operator

$$\rho = \int P(\alpha)|\alpha\rangle\langle\alpha|\, \mathrm{d}^2\alpha \qquad (78.78)$$

in terms of coherent states $|\alpha\rangle$, (78.26). It is related to the Cahill–Glauber function $W(\alpha, s)$ via

$$P(\alpha) = W(\alpha, s = 1) \,. \qquad (78.79)$$

The expectation value of any normally ordered operator product

$$\left\langle\left(a^\dagger\right)^m a^n\right\rangle = \int (\alpha^*)^m \alpha^n P(\alpha)\, \mathrm{d}^2\alpha \qquad (78.80)$$

has a particularly simple form in terms of $P(\alpha)$. From (78.78) the P function of a coherent state $|\alpha_0\rangle$ becomes

$$P(\alpha) = \delta^{(2)}(\alpha - \alpha_0) = \delta\big[\mathrm{Re}(\alpha) - \mathrm{Re}(\alpha_0)\big]\delta\big[\mathrm{Im}(\alpha) - \mathrm{Im}(\alpha_0)\big] \,. \qquad (78.81)$$

Quantized fields for which the P function is positive do not show nonclassical effects such as squeezing and antibunching. For nonclassical states, such as number states or squeezed states, the P function only exists in terms of generalized functions, such as delta functions and their derivatives, which have a highly singular character.

The positive P-representation $P(\alpha, \beta)$ for a nondiagonal decomposition of a density operator is [78.27, 28]

$$\rho = \iint \mathrm{d}^2\alpha\, \mathrm{d}^2\beta \frac{|\alpha\rangle\langle\beta^*|}{\langle\beta^*|\alpha\rangle} P(\alpha, \beta) \,. \qquad (78.82)$$

The function $P(\alpha, \beta)$ is a direct generalization of the Glauber–Sudarshan function $P(\alpha)$, (78.78). $P(\alpha, \beta)$ exists for any physical density operator ρ [78.27] and is given by

$$P(\alpha, \beta) = \frac{1}{4\pi^2} \exp\left(-\frac{1}{4}|\alpha - \beta^*|^2\right) \times\left\langle\frac{1}{2}(\alpha+\beta^*)\middle|\rho\middle|\frac{1}{2}(\alpha+\beta^*)\right\rangle \,. \qquad (78.83)$$

Here the state $|1/2(\alpha+\beta^*)\rangle$ denotes a coherent state with the complex amplitude $1/2(\alpha+\beta^*)$. Note that $P(\alpha, \beta)$ is always positive.

78.5.3 The Wigner Function

This quasi-probability was first introduced by Wigner [78.29] and may be defined as the distribution function for a symmetrically ordered operator product which is obtained in the case $s = 0$. The Wigner function plays an important role in other branches of physics, such as quantum chaology, and in particular in any semiclassical phenomenon when one considers the transition from quantum mechanics to classical mechanics.

Consider the Wigner function of a quantum mechanical particle of position

$$x = \sqrt{\frac{\hbar}{2m\omega}}\left(a + a^\dagger\right) , \qquad (78.84)$$

and momentum

$$p = \mathrm{i}\sqrt{\frac{m\hbar\omega}{2}}\left(a^\dagger - a\right) . \qquad (78.85)$$

The Wigner function may be written in terms of position and momentum variables

$$W(x, p) = \frac{1}{\pi\hbar}\int_{-\infty}^{\infty} \mathrm{d}y\, \mathrm{e}^{-2\mathrm{i}py/\hbar}\langle x + y|\rho|x - y\rangle \,, \qquad (78.86)$$

where $|x \pm y\rangle$ denotes position eigenstates. The $s = 0$ Cahill–Glauber definition and the above definition of the Wigner function are related by

$$W(x, p) = \frac{1}{2} W\left(\alpha = \frac{m\omega x + \mathrm{i}p}{\sqrt{2m\hbar\omega}},\ s = 0\right) . \qquad (78.87)$$

The position and momentum distributions of a particle, or equivalently the quadruture distributions in the case of a quantized field mode, are

$$Pr(x) = \int_{-\infty}^{\infty} W(x, p)\, \mathrm{d}p\,, \tag{78.88}$$

$$Pr(p) = \int_{-\infty}^{\infty} W(x, p)\, \mathrm{d}x\,. \tag{78.89}$$

Furthermore, the scalar product

$$|\langle\psi_1|\psi_2\rangle|^2 = 2\pi\hbar \iint \mathrm{d}x\,\mathrm{d}p\, W_{|\psi_1\rangle}(x, p) \times W_{|\psi_2\rangle}(x, p) \tag{78.90}$$

of two quantum states is expressed by the phase space overlap of the two corresponding Wigner functions. Consequently, any Wigner function $W(x, p)$ has to obey the necessary condition

$$\iint \mathrm{d}x\,\mathrm{d}p W(x, p) W_{|\psi\rangle}(x, p) \geq 0 \tag{78.91}$$

for all $W_{|\psi\rangle}$ representing a pure state. For a normalized state $|\psi\rangle$,

$$\iint \mathrm{d}x\,\mathrm{d}p\big[W_{|\psi\rangle}(x, p)\big]^2 = \frac{1}{2\pi\hbar}\,. \tag{78.92}$$

Instead of solving the Schrödinger equation for the dynamics of a massive particle in a potential $V(x)$, we can try to solve the equation

$$\frac{\partial W}{\partial t} = -\frac{p}{m}\frac{\partial W}{\partial x} + \sum_{r=1,3,..} \frac{1}{r!}\left(\frac{\mathrm{i}\hbar}{2}\right)^{r-1} \frac{\partial^r V}{\partial x^r}\frac{\partial^r W}{\partial p^r} \tag{78.93}$$

for its Wigner function $W(x, p, t)$. Note that here only the odd derivatives of the potential V enter. This equation is the quantum analogue of the classical Liouville equation, to which it reduces in the limit of $\hbar \to 0$. However, the initial distribution $W(x, p, t=0)$ has to be a Wigner function in the sense of (78.86).

Furthermore the Wigner function of an energy eigenfunction in the potential $V(x)$ may be obtained from the equations

$$\left[\frac{p^2}{2m} + V(x) - \frac{\hbar^2}{8m}\frac{\partial^2}{\partial x^2} - \frac{\hbar^2}{8}\frac{\partial^2 V}{\partial x^2}\frac{\partial^2}{\partial p^2}\right] W(x, p) + \sum_{r=4,6,..} \frac{1}{r!}\left(\frac{\mathrm{i}\hbar}{2}\right)^{r} \frac{\partial^r V}{\partial x^r}\frac{\partial^r W}{\partial p^r} = EW(x, p)\,, \tag{78.94}$$

and

$$\left(-\frac{p}{m}\frac{\partial}{\partial x} + \frac{\partial V}{\partial x}\frac{\partial}{\partial p}\right) W(x, p) + \sum_{r=3,5,..} \frac{1}{r!}\left(\frac{\mathrm{i}\hbar}{2}\right)^{r-1} \frac{\partial^r V}{\partial x^r}\frac{\partial^r W}{\partial p^r} = 0\,. \tag{78.95}$$

The Wigner function has negative parts for most quantum states. For example, the Wigner function of a Fock state $|n\rangle$,

$$W_{|n\rangle}(\bar{x}, \bar{p}) = \frac{(-1)^n}{\pi}\mathrm{e}^{-\bar{x}^2-\bar{p}^2} L_n\big(2\bar{x}^2 + 2\bar{p}^2\big)\,, \tag{78.96}$$

clearly becomes negative due to the oscillating Laguerre polynomial L_n as shown in Fig. 78.4. Note that we have introduced the dimensionless position $\bar{x} = \sqrt{m\omega/\hbar}\, x$ and momentum $\bar{p} = 1/\sqrt{m\omega\hbar}\, p$.

On the other hand, the Wigner function

$$W_{|\alpha,r\rangle}(\bar{x}, \bar{p}) = \frac{1}{\pi}\exp\Big\{-\mathrm{e}^{2r}\big[\bar{x} - \sqrt{2}\,\mathrm{Re}(\alpha)\big]^2 - \mathrm{e}^{-2r}\big[\bar{p} - \sqrt{2}\,\mathrm{Im}(\alpha)\big]^2\Big\} \tag{78.97}$$

of a squeezed state (78.35) is always positive as shown in Fig. 78.5. It is a long thin ellipse in phase space (i. e. a Gaussian cigar). Concerning the negative parts of the Wigner function, the Hudson theorem [78.30] states that a necessary and sufficient condition for the Wigner function of a pure state $|\psi\rangle$ to be nonnegative is that it can be described by a wave function of the

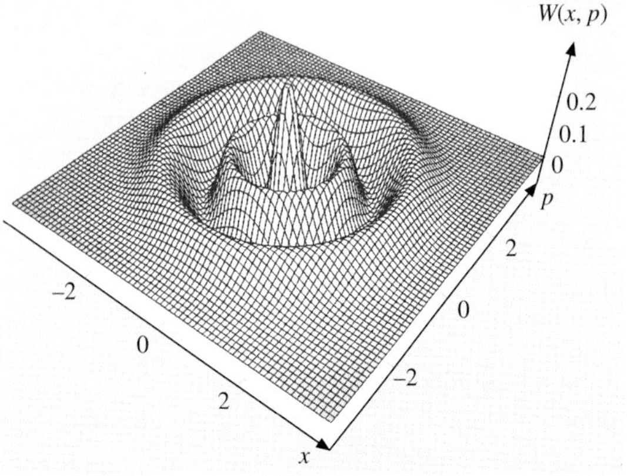

Fig. 78.4 The Wigner function (78.96) of a Fock state $|n=4\rangle$. The negative parts can be seen clearly

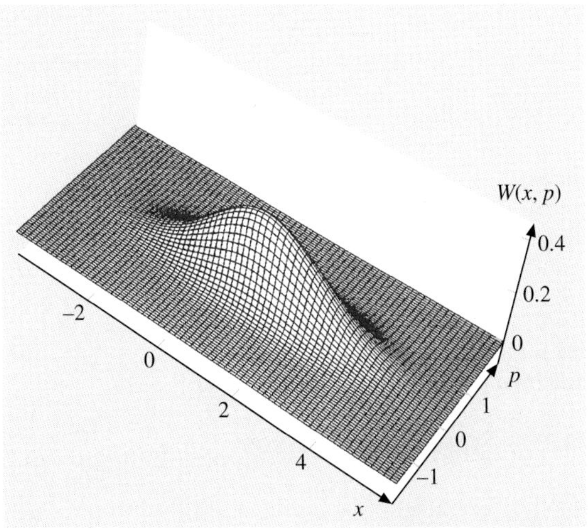

Fig. 78.5 The Wigner function of a squeezed state $|\alpha, r\rangle$ with coherent amplitude $\alpha = 1$ and squeezing $e^{2r} = 0.25$. For these values the phase space ellipse is oriented along the x-axis and squeezed in the p-direction. Note that this function is positive everywhere

form

$$\langle x|\psi\rangle = \exp\left[-\frac{1}{2}(ax^2 + bx + c)\right] . \tag{78.98}$$

Here a, b and c denote some constants with $\mathrm{Re}(a) > 0$.

Finally, the Kirkwood distribution function

$$K(x, p) = \frac{1}{\pi\hbar}\int_{-\infty}^{\infty} \mathrm{d}y\, \mathrm{e}^{-2\mathrm{i}py/\hbar}\langle x|\rho|x - 2y\rangle \tag{78.99}$$

is a phase space function that resembles the Wigner function. In the case of a pure state $\rho = |\psi\rangle\langle\psi|$ this function reduces to

$$K(x, p) = \psi(x)\tilde{\psi}(p)\,\mathrm{e}^{-\mathrm{i}xp/\hbar} , \tag{78.100}$$

where $\tilde{\psi}(p)$ denotes the Fourier transform of $\psi(x)$.

78.5.4 The Q Function

The Q function is defined by the diagonal matrix elements

$$Q(\alpha) = \langle\alpha|\rho|\alpha\rangle/\pi \tag{78.101}$$

of the density operator ρ, where $|\alpha\rangle$ denotes a coherent state. The $Q(\alpha)$ function is always a positive and bounded function, which exists for any density operator ρ. The Q function is also known as Husimi's function. It allows one to calculate expectation values of antinormally ordered operator products of the form

$$\left\langle a^n \left(a^\dagger\right)^m\right\rangle = \int \mathrm{d}^2\alpha\, \alpha^n (\alpha^*)^m Q(\alpha) . \tag{78.102}$$

Moreover, since the Q function corresponds to the case $s = -1$ of the Cahill–Glauber distribution,

$$Q(\alpha) = W(\alpha, -1) . \tag{78.103}$$

78.5.5 Relations Between Quasi-Probabilities

In general, the relation

$$W(\alpha, s) = \frac{2}{\pi(s' - s)}\int \mathrm{d}^2\beta \exp\left(-\frac{2|\alpha - \beta|^2}{s' - s}\right) \times W(\alpha, s') \tag{78.104}$$

holds between two Cahill–Glauber distributions with the parameters $s' > s$. In particular, the non-negative Q function

$$Q(\alpha) = \frac{2}{\pi}\int \mathrm{d}^2\beta \exp[-2|\alpha - \beta|^2] W(\beta, s = 0) \tag{78.105}$$

turns out to be a smoothed Wigner function $W(\beta, s = 0)$. It is this smoothing process that washes out possible negative parts in the Wigner function.

78.6 Reservoir Theory

Reservoir theory treats the interaction of one system with a few degrees of freedom, called the system, with another system with many degrees of freedom, called the reservoir. A typical application of reservoir theory is a microscopic theory of damping: the system interacts with a reservoir, called the heat bath. The system dissipates energy into the heat bath whereas the heat bath introduces additional fluctuations to the system. Since the present chapter focuses on quantized field effects, the reservoir consists of the many modes of the radiation field in free space. Such a reservoir is modeled by a large number of independent harmonic oscillators

$$H_r = \sum_i \hbar\omega_i \left(b_i^\dagger b_i + \frac{1}{2} \right) , \quad (78.106)$$

where b_i and $b_i^\dagger$ are the annihilation and creation operators for the ith harmonic oscillator of the reservoir. For convenience the interaction with the system is frequently approximated by a Hamiltonian of the form

$$H_{\text{int}} = \hbar \sum_i \left(g_i A\, b_i^\dagger + g_i^* A^\dagger b_i \right) , \quad (78.107)$$

where A is an operator of the small system and g_i is the coupling strength of this system to the ith oscillator of the reservoir. For example, A may be an annihilation operator if the system is a harmonic oscillator or a Pauli spin matrix in the case of a two-level atom coupled to the free space radiation field.

Reservoir theory has important applications, and a detailed discussion can be found in various books, for example [78.17, 20, 27, 28, 31–33].

78.6.1 Thermal Reservoir

The most commonly used reservoir is the thermal reservoir or thermal heat bath. Its characteristic properties are

$$\langle b_i \rangle = \langle b_i^\dagger \rangle = \langle b_i b_j \rangle = \langle b_i^\dagger b_j^\dagger \rangle = 0 , \quad (78.108)$$

$$\langle b_i^\dagger b_j \rangle = \bar{n}_i \delta_{ij} . \quad (78.109)$$

Here

$$\bar{n}_i = \frac{1}{\exp(\hbar\omega_i / k_{\text{B}} T) - 1} \quad (78.110)$$

is the average number of photons at frequency ω_i, T is the temperature of the reservoir, and k_{B} denotes the Boltzmann constant.

78.6.2 Squeezed Reservoir

Another example of a reservoir is a squeezed vacuum or squeezed reservoir. If, for example, multiwave mixing is used to squeeze the radiation field, conjugate pairs of the reservoir operators b are correlated. Therefore, the expectation values $\langle b_i b_j \rangle$ and $\langle b_i^\dagger b_j^\dagger \rangle$ may be nonvanishing. Apart from the average number $\bar{n}_i$ of photons at frequency ω_i, which take into account nonvanishing expectation values $\langle b_i^\dagger b_i \rangle$, additional complex squeezing parameters are needed to describe the reservoir [78.28, 33, 34]. The characterization of a squeezed reservoir based on noise operators is discussed in Sect. 78.10.

78.7 Master Equation

In quantum mechanics, density operators are used to describe mixed states, and are discussed in Chapt. 7. Here we introduce the concept of the reduced density operator

$$\rho_{\text{s}} = \text{Tr}_{\text{r}} (\rho_{\text{sr}}) , \quad (78.111)$$

which is the density operator ρ_{sr} of the complete system traced over the degrees of freedom of the reservoir. The equation of motion for ρ_{s} in the Schrödinger picture is

$$\dot{\rho}_{\text{s}}(t) = -\frac{\text{i}}{\hbar} \text{Tr}_{\text{r}} \{ [H_{\text{sr}}, \rho_{\text{sr}}(t)] \} . \quad (78.112)$$

In the Born–Markov approximation the trace over the reservoir can be evaluated and leads to an equation of motion for ρ_{s} which no longer contains reservoir operators. This equation of motion is usually called the master equation. The Born–Markov approximation consists of two different parts:

1. Born approximation: The coupling to the reservoir is assumed to be sufficiently weak to allow a perturbative treatment of the interaction between the reservoir and the system.
2. Markov approximation: The correlations of the reservoir are assumed to decay very rapidly on a typical time scale of the system, or equivalently, the reservoir has a very broad spectrum. This approximation involves the assumption that the modes of the reservoir are spaced closely together, so that the frequency ω_i is a smooth function of i.

Since a general treatment is rather technical, we consider two typical examples. A more general discussion can be found in [78.17, 20, 27, 28, 31–33]

78.7.1 Damped Harmonic Oscillator

The universally accepted Hamiltonian in nonrelativistic QED for a harmonic oscillator of frequency ω coupled to a reservoir consisting of a large number of harmonic oscillators is given by the total Hamiltonian [78.35, 36]

$$H_{\mathrm{sr}} = \hbar\omega\left(a^\dagger a + \frac{1}{2}\right) + \sum_i \hbar\omega_i\left(b_i^\dagger b_i + \frac{1}{2}\right) + H_{\mathrm{lc}} + H_{\mathrm{si}}\,, \tag{78.113}$$

with the linear coupling term

$$H_{\mathrm{lc}} = \hbar\sum_i g_i\left(a + a^\dagger\right)\left(b_i + b_i^\dagger\right)\,, \tag{78.114}$$

and the self-interaction term

$$H_{\mathrm{si}} = \sum_i \frac{\hbar g_i^2}{\omega_i}\left(a + a^\dagger\right)^2\,. \tag{78.115}$$

The approach used in quantum optics is to drop the term H_{si} and to make the rotating-wave approximation, that is, to drop the terms $a\,b_i$ and $a^\dagger b_i^\dagger$, see also Chapt. 68. Then the approximate total Hamiltonian reads

$$H_{\mathrm{sr}} = \hbar\omega\left(a^\dagger a + \frac{1}{2}\right) + \sum_i \hbar\omega_i\left(b_i^\dagger b_i + \frac{1}{2}\right) + \hbar\sum_i g_i\left(a\,b_i^\dagger + a^\dagger b_i\right)\,. \tag{78.116}$$

Despite the problems with this approximate Hamiltonian (see Sect. V.D of [78.35, 36] for a discussion) we adopt it in the present context because it leads to the widely used master equation for the damped harmonic oscillator. We consider two reservoirs: a thermal bath and a squeezed bath.

Harmonic Oscillator in a Thermal Bath

Within the Born–Markov approximation the master equation is

$$\dot\rho = \frac{1}{2}\gamma\left(\bar n + 1\right)\left(2a\rho a^\dagger - a^\dagger a\rho - \rho a^\dagger a\right) + \frac{1}{2}\gamma\bar n\left(2a^\dagger\rho a - a\,a^\dagger\rho - \rho a\,a^\dagger\right)\,, \tag{78.117}$$

where

$$\rho(t) = \mathrm{e}^{\mathrm{i}\omega a^\dagger a(t-t_0)}\rho_{\mathrm{s}}(t)\,\mathrm{e}^{-\mathrm{i}\omega a^\dagger a(t-t_0)} \tag{78.118}$$

is the reduced density operator in the interaction picture. The damping constant γ is given by

$$\gamma = 2\pi\mathcal{D}(\omega)|g(\omega)|^2\,, \tag{78.119}$$

where $g(\omega)$ denotes the coupling strength at frequency ω. The number of thermal photons at frequency ω is

$$\bar n = \frac{1}{\exp\left(\hbar\omega/k_{\mathrm{B}}T\right) - 1}\,. \tag{78.120}$$

Thus the Born–Markov approximation replaces the discrete reservoir modes by a continuum of modes with a density $\mathcal{D}(\omega)$.

Harmonic Oscillator in a Squeezed Bath

Within the Born–Markov approximation, the reduced density operator (78.118) in the interaction picture satisfies the master equation

$$\begin{aligned}\dot\rho &= \frac{1}{2}\gamma\left(\bar n + 1\right)\left(2a\rho a^\dagger - a^\dagger a\rho - \rho a^\dagger a\right)\\ &\quad + \frac{1}{2}\gamma\bar n\left(2a^\dagger\rho a - a\,a^\dagger\rho - \rho a\,a^\dagger\right)\\ &\quad - \frac{1}{2}\gamma\bar m\left(2a^\dagger\rho a^\dagger - a^\dagger a^\dagger\rho - \rho a^\dagger a^\dagger\right)\\ &\quad - \frac{1}{2}\gamma\bar m^*\left(2a\rho a - a\,a\rho - \rho a\,a\right)\,.\end{aligned} \tag{78.121}$$

Here γ is again given by (78.119). The squeezed reservoir is characterized by a real number $\bar n$ and a complex number $\bar m$. Physically $\bar n$ is the number of photons at frequency ω, i.e., similar to the thermal reservoir, it measures the average energy at frequency ω. The complex number $\bar m$ determines the amount of squeezing. In general, the positivity of the density operator requires

$$|\bar m|^2 \le \bar n\left(\bar n + 1\right)\,. \tag{78.122}$$

A more quantitatively definition of $\bar n$ and $\bar m$ in terms of noise operators is given in Sect. 78.10.

78.7.2 Damped Two-Level Atom

The interaction of a two-level atom with a classical electromagnetic field is already discussed in Chapt. 68. For a quantum mechanical treatment of the field we only have to replace the classical field by its quantum mechanical counterpart (78.9). We then find in the rotating-wave approximation (Chapt. 68), that the dynamics of a two-level atom with a transition frequency ω_0 coupled to a reservoir consisting of a large number of harmonic oscillators is approximately described by the total Hamiltonian

$$H_{\mathrm{sr}} = \frac{1}{2}\hbar\omega_0\sigma_z + \sum_i \hbar\omega_i\left(b_i^\dagger b_i + \frac{1}{2}\right) + \hbar\sum_i\left(g_i\sigma_- b_i^\dagger + g_i^*\sigma_+ b_i\right)\,, \tag{78.123}$$

where

$$\sigma_+ = \begin{pmatrix} 0 & 1 \\ 0 & 0 \end{pmatrix}, \quad \sigma_- = \begin{pmatrix} 0 & 0 \\ 1 & 0 \end{pmatrix}, \tag{78.124}$$

$$\sigma_z = \begin{pmatrix} 1 & 0 \\ 0 & -1 \end{pmatrix}. \tag{78.125}$$

Again, two reservoirs are considered: a thermal bath and a squeezed bath.

Two-Level Atom in a Thermal Bath

Within the Born–Markov approximation, the master equation is

$$\dot{\rho} = \frac{1}{2}\gamma\,(\bar{n}+1)\,(2\sigma_-\rho\sigma_+ - \sigma_+\sigma_-\rho - \rho\sigma_+\sigma_-) + \frac{1}{2}\gamma\bar{n}\,(2\sigma_+\rho\sigma_- - \sigma_-\sigma_+\rho - \rho\sigma_-\sigma_+)\,, \tag{78.126}$$

where

$$\rho(t) = \mathrm{e}^{\mathrm{i}\omega_0\sigma_z(t-t_0)/2}\rho_\mathrm{s}(t)\,\mathrm{e}^{-\mathrm{i}\omega_0\sigma_z(t-t_0)/2} \tag{78.127}$$

is the reduced density operator in the interaction picture, and γ and $\bar{n}$ are given by (78.119) and (78.120).

Two-Level Atom in a Squeezed Bath

Within the Born–Markov approximation, the reduced density operator in the interaction picture, (78.127), satisfies the master equation

$$\begin{aligned}\dot{\rho} = {} & \frac{1}{2}\gamma\,(\bar{n}+1)\,(2\sigma_-\rho\sigma_+ - \sigma_+\sigma_-\rho - \rho\sigma_+\sigma_-) \\ & + \frac{1}{2}\gamma\bar{n}\,(2\sigma_+\rho\sigma_- - \sigma_-\sigma_+\rho - \rho\sigma_-\sigma_+) \\ & - \gamma\bar{m}\sigma_+\rho\sigma_+ - \gamma\bar{m}^*\sigma_-\rho\sigma_-\,,\end{aligned} \tag{78.128}$$

where γ, $\bar{n}$ and $\bar{m}$ have the same meaning as in (78.121).

78.8 Solution of the Master Equation

78.8.1 Damped Harmonic Oscillator

We consider only a thermal reservoir and present the solution of the master equation (78.117). For $\bar{n}=0$ it can be solved in terms of coherent states, see (78.26). For $\bar{n}\neq 0$ we give solutions in terms of quasi-probability distributions.

Coherent States

For $\bar{n}=0$, which is a good approximation for optical frequencies, if the system is initially in a coherent state $|\alpha_0\rangle$ with a density operator

$$\rho(t_0) = |\alpha_0\rangle\langle\alpha_0|\,, \tag{78.129}$$

then there exists a simple analytical solution of the master equation (78.117)

$$\rho(t) = |\alpha_0\,\mathrm{e}^{-\gamma(t-t_0)/2}\rangle\langle\alpha_0\,\mathrm{e}^{-\gamma(t-t_0)/2}|\,. \tag{78.130}$$

A coherent state thus remains a coherent state with an exponentially decaying amplitude $\alpha_0 e^{-\gamma(t-t_0)/2}$. According to (78.78) a general solution

$$\rho(t) = \int \mathrm{d}^2\alpha_0\, P(\alpha_0)\big|\alpha_0\,\mathrm{e}^{-\gamma(t-t_0)/2}\big\rangle\big\langle\alpha_0\,\mathrm{e}^{-\gamma(t-t_0)/2}\big| \tag{78.131}$$

can be constructed for an initial density operator

$$\rho(t_0) = \int \mathrm{d}^2\alpha_0\, P(\alpha_0)|\alpha_0\rangle\langle\alpha_0|\,. \tag{78.132}$$

If the system is initially in a superposition

$$|\psi(t_0)\rangle = \sum_i c_i|\alpha_i\rangle \tag{78.133}$$

of coherent states, the time evolution is given by

$$\begin{aligned}\rho(t) = {} & \sum_{i,k} c_i c_k^* \exp\left[-\frac{1}{2}\left(1-\mathrm{e}^{-\gamma(t-t_0)}\right)|\alpha_i-\alpha_k|^2\right] \\ & \times\exp\left[\mathrm{i}\left(1-\mathrm{e}^{-\gamma(t-t_0)}\right)\mathrm{Im}\left(\alpha_i\alpha_k^*\right)\right] \\ & \times\big|\alpha_i\,\mathrm{e}^{-\gamma(t-t_0)/2}\big\rangle\big\langle\alpha_k\,\mathrm{e}^{-\gamma(t-t_0)/2}\big|\,.\end{aligned} \tag{78.134}$$

For $\gamma(t-t_0)\ll 1$, the interference terms $|\alpha_i\rangle\langle\alpha_k|$, $i\neq k$ decay with an effective decay constant $\gamma|\alpha_i-\alpha_k|^2/2$. Thus the damping constant is modified by the separation of the two coherent states in phase space.

Fokker–Planck Equation

A widely used procedure for solving the master equation for a damped harmonic oscillator, (78.117), or for similar problems, is to derive an equation of motion for the quasi-probability distributions $W(\alpha,\alpha^*;s)$ defined in (78.76) from the master equation. The operators a and $a^\dagger$ are replaced by appropriate differential operators. The substitution rules can be derived from (78.73), (78.74) and (78.76) and are

$$
\begin{aligned}
\left(a^\dagger\right)^k a^\ell \rho &\to \left(\alpha^* - \frac{s+1}{2}\frac{\partial}{\partial\alpha}\right)^k \\
&\quad\times \left(\alpha - \frac{s-1}{2}\frac{\partial}{\partial\alpha^*}\right)^\ell W\,,\\
\rho\left(a^\dagger\right)^k a^\ell &\to \left(\alpha - \frac{s+1}{2}\frac{\partial}{\partial\alpha^*}\right)^\ell \\
&\quad\times \left(\alpha^* - \frac{s-1}{2}\frac{\partial}{\partial\alpha}\right)^k W\,,\\
a\rho a^\dagger &\to \left(\alpha - \frac{s-1}{2}\frac{\partial}{\partial\alpha^*}\right) \\
&\quad\times \left(\alpha^* - \frac{s-1}{2}\frac{\partial}{\partial\alpha}\right) W\,,\\
a^\dagger \rho a &\to \left(\alpha^* - \frac{s+1}{2}\frac{\partial}{\partial\alpha}\right) \\
&\quad\times \left(\alpha - \frac{s+1}{2}\frac{\partial}{\partial\alpha^*}\right) W\,,\\
a^\dagger \rho a^\dagger &\to \left(\alpha^* - \frac{s+1}{2}\frac{\partial}{\partial\alpha}\right) \\
&\quad\times \left(\alpha^* - \frac{s-1}{2}\frac{\partial}{\partial\alpha}\right) W\,,\\
a\rho a &\to \left(\alpha - \frac{s-1}{2}\frac{\partial}{\partial\alpha^*}\right) \\
&\quad\times \left(\alpha - \frac{s+1}{2}\frac{\partial}{\partial\alpha^*}\right) W\,. \qquad (78.135)
\end{aligned}
$$

In general, this procedure leads to equations of motion which involve higher derivatives of W as exemplified by the quantum mechanical Liouville equation (78.93) for the Wigner function. For simple Hamiltonians, however, this equation has the form of a Fokker–Planck equation which is well known in classical stochastic problems [78.27, 37] (Sect. 78.10.2). In particular, for a damped harmonic oscillator described by the master equation (78.117), one obtains

$$
\frac{\partial W}{\partial t} = \frac{\gamma}{2}\left[\frac{\partial}{\partial\alpha}(\alpha W) + \frac{\partial}{\partial\alpha^*}\left(\alpha^* W\right)\right] + \gamma\bar{n}_s \frac{\partial^2 W}{\partial\alpha\partial\alpha^*}\,, \qquad (78.136)
$$

where

$$
\bar{n}_s = \bar{n} + \frac{1-s}{2} = \frac{1}{\exp(\hbar\omega/k_B T) - 1} + \frac{1-s}{2}\,. \qquad (78.137)
$$

The time-dependent solution of this Fokker–Planck equation has the form

$$
W(\alpha,\alpha^*,t;s) = \int G(\alpha,\alpha^*,t|\alpha',\alpha'^*,t';s) \times W(\alpha',\alpha'^*,t';s)\,\mathrm{d}^2\alpha'\,, \qquad (78.138)
$$

where

$$
G(\alpha,\alpha^*,t|\,\alpha',\alpha'^*,t';s) = \frac{\exp\left(-\dfrac{\left|\alpha-\alpha' e^{-\gamma(t-t')/2}\right|^2}{\bar{n}_s\left[1-e^{-\gamma(t-t')}\right]}\right)}{\pi\bar{n}_s\left(1-e^{-\gamma(t-t')}\right)} \qquad (78.139)
$$

is the Green's function of the Fokker–Planck equation (78.136). The steady-state solution is

$$
W(\alpha,\alpha^*,t\to\infty;s) = \frac{1}{\pi\bar{n}_s}\,\mathrm{e}^{-|\alpha|^2/\bar{n}_s}\,, \qquad (78.140)
$$

which is the distribution function of a harmonic oscillator in thermal equilibrium with a reservoir of temperature T.

78.8.2 Damped Two-Level Atom

The density operator

$$
\rho = \begin{pmatrix} \rho_{ee} & \rho_{eg} \\ \rho_{ge} & \rho_{gg} \end{pmatrix} \qquad (78.141)
$$

for a two-level atom can be written as

$$
\rho = \begin{pmatrix} \frac{1}{2}[1+\langle\sigma_z\rangle] & \langle\sigma_-\rangle \\ \langle\sigma_+\rangle & \frac{1}{2}[1-\langle\sigma_z\rangle] \end{pmatrix}\,. \qquad (78.142)
$$

Thus, a two-level atom is completely described by the expectation values

$$
\begin{aligned}
\langle\sigma_z\rangle &= \rho_{ee} - \rho_{gg}\,,\\
\langle\sigma_+\rangle &= \rho_{ge}\,,\\
\langle\sigma_-\rangle &= \rho_{eg}\,. \qquad (78.143)
\end{aligned}
$$

Hence the master equation (78.128) can be cast into the equations of motions for these expectation values

$$
\frac{\mathrm{d}}{\mathrm{d}t}\langle\sigma_+\rangle = -\gamma\left(\bar{n}+\frac{1}{2}\right)\langle\sigma_+\rangle - \gamma\bar{m}^*\langle\sigma_-\rangle\,,
$$

$$\frac{d}{dt}\langle\sigma_-\rangle = -\gamma\left(\overline{n}+\frac{1}{2}\right)\langle\sigma_-\rangle - \gamma\overline{m}\langle\sigma_+\rangle ,$$
$$\frac{d}{dt}\langle\sigma_z\rangle = -2\gamma\left(\overline{n}+\frac{1}{2}\right)\langle\sigma_z\rangle - \gamma , \quad (78.144)$$

which can easily be solved for arbitrary initial conditions. In contrast to a thermal reservoir ($\overline{m}=0$), a squeezed reservoir results in two different transverse decay constants $\gamma\left(\overline{n}+\frac{1}{2}+|\overline{m}|\right)$ and $\gamma\left(\overline{n}+\frac{1}{2}-|\overline{m}|\right)$ [78.34].

78.9 Quantum Regression Hypothesis

In the Schrödinger picture, time-dependent expectation values for system operators A_j can be calculated from the reduced density operator $\rho_s(t)$ via

$$\langle A_j\rangle = \mathrm{Tr}_s\left\{A_j\rho_s(t)\right\} . \quad (78.145)$$

The reduced density operator, however, is not sufficient to calculate two-time correlation functions such as $\langle A_j(t+\tau)A_k(t)\rangle$. For a definition of two-time correlation functions, the Heisenberg picture is more appropriate. Here, expectation values follow from

$$\langle A_j\rangle = \mathrm{Tr}_{sr}\left[U^\dagger_{sr}(t,t_0)A_j(t_0)U_{sr}(t,t_0)\rho_{sr}(t_0)\right] , \quad (78.146)$$

where $U_{sr}(t,t_0)$ describes the unitary time evolution of the complete system and $\rho_{sr}(t_0)$ is the density operator in the Heisenberg picture. Similarly, two-time correlation functions such as $\langle A_j(t+\tau)A_k(t)\rangle$ can be defined as

$$\begin{aligned}&\langle A_j(t+\tau)A_k(t)\rangle\\ &= \mathrm{Tr}_{sr}\Big[U^\dagger_{sr}(t+\tau,t_0)A_j(t_0)U_{sr}(t+\tau,t_0)\\ &\quad\times U^\dagger_{sr}(t,t_0)A_k(t_0)U_{sr}(t,t_0)\rho_{sr}(t_0)\Big] . \end{aligned} \quad (78.147)$$

The quantum regression hypothesis avoids the calculation of $U_{sr}(t,t_0)$. Two equivalent formulations exist, one based on the master equation for ρ_s and another based on the equation of motion for the expectation values $\langle A_j\rangle$, see for example [78.20, 27, 28, 31, 33].

78.9.1 Two-Time Correlation Functions and Master Equation

It follows from their definition (78.147) in the Heisenberg picture that two-time correlation functions $\langle A_j(t+\tau)A_k(t)\rangle$ for system operators A_j and A_k can be calculated with the help of the operator

$$\begin{aligned}R_s(t+\tau,t) = \mathrm{Tr}_r\Big[&U_{sr}(t+\tau,t)\\ &\times A_k\rho_{sr}(t)U^\dagger_{sr}(t+\tau,t)\Big] ,\end{aligned} \quad (78.148)$$

where $U_{sr}(t+\tau,t)$ describes the unitary time evolution of the complete system between t and $t+\tau$. We find

$$\langle A_j(t+\tau)A_k(t)\rangle = \mathrm{Tr}_s\left[A_jR_s(t+\tau,t)\right] . \quad (78.149)$$

Note, that in (78.148) and (78.149) we interpret A_j and A_k as operators in the Schrödinger picture and have omitted the argument t_0. Because the reduced density operator

$$\rho_s(t) = \mathrm{Tr}_r\left[U_{sr}(t,t_0)\rho_{sr}(t_0)U^\dagger_{sr}(t,t_0)\right] \quad (78.150)$$

satisfies the master equation, it is plausible to assume, that when the time derivative is taken with respect to τ, the operator $R_s(t+\tau,t)$ also satisfies the master equation for ρ_s, subject to the initial condition $R_s(t,t)=A_k\rho_s(t)$. However, this requires the additional assumption that the approximations made in the derivation of the master equation for $\rho_s(t)$ are also valid for $R_s(t+\tau,t)$.

78.9.2 Two-Time Correlation Functions and Expectation Values

A second formulation of the quantum regression hypothesis asserts that two-time correlation functions $\langle A_j(t+\tau)A_k(t)\rangle$ obey

$$\begin{aligned}&\frac{\partial}{\partial\tau}\langle A_j(t+\tau)A_k(t)\rangle\\ &= \sum_\ell G_{j\ell}(\tau)\langle A_\ell(t+\tau)A_k(t)\rangle ,\end{aligned} \quad (78.151)$$

provided that the expectation values of a set of system operators A_j satisfy

$$\frac{\partial}{\partial t}\langle A_j(t)\rangle = \sum_\ell G_{j\ell}(t)\langle A_\ell(t)\rangle . \quad (78.152)$$

This is the form of the quantum regression hypothesis that was first formulated by *Lax* [78.38].

The equivalence of the two formulations follows from the interpretation of $R_s(t+\tau,t)$ on the right side of (78.149) as a "density operator". Then $\mathrm{Tr}_s\left\{A_jR_s(t+\tau,t)\right\}$ is an "expectation value" for which we assume that (78.152) is valid; i.e.,

$$\begin{aligned}&\frac{\partial}{\partial\tau}\mathrm{Tr}_s\left\{A_jR_s(t+\tau,t)\right\}\\ &= \sum_\ell G_{j\ell}(\tau)\mathrm{Tr}_s\left\{A_\ell R_s(t+\tau,t)\right\} .\end{aligned} \quad (78.153)$$

According to (78.149), this is identical to (78.151).

78.10 Quantum Noise Operators

The master equation is based on the Schrödinger picture in quantum mechanics: the state of the system described by a density operator is time-dependent, whereas operators corresponding to observables are time independent. If we use the Heisenberg picture instead and make similar approximations as in the derivation of the master equation, we arrive at equations of motion for the Heisenberg operators, see for example [78.17,28,31,32]. Due to the interaction with a reservoir these equations have additional noise terms and damping terms.

78.10.1 Quantum Langevin Equations

Again consider a damped harmonic oscillator. The equation of motion for the annihilation operator

$$\tilde{a}(t) = \mathrm{e}^{\mathrm{i}\omega(t-t_0)} a(t) \tag{78.154}$$

in the interaction picture follows from the Heisenberg equations for the operators a, $a^\dagger$, b_i and $b_i^\dagger$ and reads

$$\begin{aligned}\frac{\mathrm{d}\tilde{a}}{\mathrm{d}t} = &-\sum_i |g_i|^2 \int_{t_0}^{t} \mathrm{e}^{-\mathrm{i}(\omega_i-\omega)(t-t')}\tilde{a}(t')\,\mathrm{d}t' \\ &-\mathrm{i}\sum_i g_i^* \,\mathrm{e}^{-\mathrm{i}(\omega_i-\omega)(t-t_0)} b_i(t_0)\,.\end{aligned} \tag{78.155}$$

In general, the noise operator

$$F(t) = -\mathrm{i}\sum_i g_i^* \,\mathrm{e}^{-\mathrm{i}(\omega_i-\omega)(t-t_0)} b_i(t_0) \tag{78.156}$$

is not delta-correlated, and there are also memory effects in (78.155). The noise operator $F(t)$ can be used to classify the reservoir: if it is delta-correlated, that is, if the reservoir has a very broad spectrum, one speaks of white noise, see below. If the correlation time is finite so that there are memory effects, one speaks of colored noise.

If the spectrum of the noise is very broad (as in the derivation of the master equation for the reduced density operator), the operator $\tilde{a}(t)$ satisfies the quantum Langevin equation

$$\frac{\mathrm{d}\tilde{a}}{\mathrm{d}t} = -\frac{\gamma}{2}\tilde{a}(t) + F(t)\,, \tag{78.157}$$

with a damping term $-\gamma\tilde{a}(t)/2$ and a noise term $F(t)$. Note that a simple damping equation such as

$$\dot{\tilde{a}}(t) = -\frac{\gamma}{2}\,\tilde{a}(t) \tag{78.158}$$

is unphysical since it does not preserve the commutation relation $\left[\tilde{a}, \tilde{a}^\dagger\right] = 1$. It is the noise term which saves the commutation relation.

For a thermal reservoir with a sufficiently small correlation time, the standard derivations [78.32] give

$$\begin{aligned}&\langle F(t)\rangle = \left\langle F^\dagger(t)\right\rangle = \left\langle F(t)F(t')\right\rangle = \left\langle F^\dagger(t)F^\dagger(t')\right\rangle = 0\,,\\ &\left\langle F^\dagger(t)F(t')\right\rangle = \gamma\bar{n}\delta(t-t')\,,\\ &\langle F(t)F^\dagger(t')\rangle = \gamma(\bar{n}+1)\delta(t-t')\,,\end{aligned} \tag{78.159}$$

where the averages are taken over the reservoir. The damping constant γ and the number of thermal photons $\bar{n}$ are given in (78.119) and (78.120). For more general relations, see [78.35, 36], where it is shown explicitly that correlation functions involving the fluctuation force do not in fact depend on the oscillator frequency. The condition of a sufficiently small reservoir correlation time requires that $\tau_c \approx \hbar/(k_B T)$ is small compared with the time scales of the systems. The only time scale in (78.157) is γ^{-1}. The relevant condition is therefore $\tau_c \ll \gamma^{-1}$. For typical applications in quantum optics, a is the annihilation operator and $a^\dagger$ is the creation operator of a single-mode cavity field. Here one can have quality factors of the cavity on the order of $Q = \omega/\gamma \approx 10^6$. In terms of the quality factor, the condition of sufficiently small reservoir correlation times requires $\hbar\omega/(k_B T) \ll Q$. For optical frequencies $(\omega \approx 3\times 10^{15}\,\mathrm{Hz})$ and $T \approx 300\,\mathrm{K}$ one has $\hbar\omega/(k_B T) \approx 75$. In the microwave regime ($\omega \approx 30\,\mathrm{GHz}$) one can have temperatures as low as $T \approx 3\,\mathrm{mK}$ and still have $\hbar\omega/(k_B T) \approx 75$. Therefore the assumption of delta-correlated noise is a good approximation for typical applications in quantum optics.

Similarly, for a squeezed reservoir one has

$$\begin{aligned}&\langle F(t)\rangle = \left\langle F^\dagger(t)\right\rangle = 0\,,\\ &\langle F(t)F(t')\rangle = \gamma\bar{m}\delta(t-t')\,,\\ &\left\langle F^\dagger(t)F^\dagger(t')\right\rangle = \gamma\bar{m}^*\delta(t-t')\,,\\ &\left\langle F^\dagger(t)F(t')\right\rangle = \gamma\bar{n}\delta(t-t')\,,\\ &\left\langle F(t)F^\dagger(t')\right\rangle = \gamma(\bar{n}+1)\delta(t-t')\,,\end{aligned} \tag{78.160}$$

which gives a quantitative definition of the parameters $\bar{n}$ and $\bar{m}$ in the master equations (78.121) and (78.128). Again, a detailed discussion in [78.35, 36] shows that correlation functions involving the fluctuation forces do not depend on the oscillator frequency.

The Langevin equation (78.157) is based on the use of the approximate Hamiltonian given in (78.116),

i. e., it is based on the rotating-wave approximation and the neglect of self-interaction terms. The corresponding Langevin equation for $x = \sqrt{\hbar/(2m\omega)}\,(a + a^\dagger)$ may be calculated and, not unexpectedly, it disagrees with the Abraham–Lorentz equation which *Ford* and *O'Connell* [78.39] showed could be derived systematically using the exact Hamiltonian (78.113). In fact, Ford and O'Connell showed that an improved equation for the radiating electron (improved in the sense that it is second-order and is not subject to the analyticity problems and the problems with runaway solutions associated with the Abraham–Lorentz equation) may be obtained by generalizing the Hamiltonian (78.113) to include electron structure. The implications following from these different equations are presently under study.

78.10.2 Stochastic Differential Equations

In Sect. 78.10.1 we discussed one of the simplest quantum systems with dissipation, the damped harmonic oscillator. For more complicated systems the noise term can also contain system operators. In such cases there are two different ways to interpret the Langevin equation. In order to give a feeling for the two possible interpretations, we discuss the one dimensional classical Langevin equation

$$\frac{\mathrm{d}x}{\mathrm{d}t} = g(x,t) + h(x,t)F(t) \tag{78.161}$$

for the stochastic variable $x(t)$ with delta-correlated noise $\langle F(t)F(t')\rangle = \delta(t-t')$. Due to the singular nature of delta-correlated noise, such a Langevin equation does not exist from a strictly mathematical point of view. A mathematically more rigorous treatment is based on stochastic differential equations [78.27, 28, 37]. The variable $x(t)$ is said to obey a stochastic differential equation

$$\begin{aligned}\mathrm{d}x(t) &= g(x,t)\,\mathrm{d}t + h(x,t)F(t)\,\mathrm{d}t\\ &= g(x,t)\,\mathrm{d}t + h(x,t)\,\mathrm{d}W(t)\,,\end{aligned} \tag{78.162}$$

if, for all times t and t_0, $x(t)$ is given by

$$\begin{aligned}x(t) = x(t_0) &+ \int_{t_0}^{t} g(x(t'),t')\,\mathrm{d}t'\\ &+ \int_{t_0}^{t} h[x(t'),t']\,\mathrm{d}W(t')\,.\end{aligned} \tag{78.163}$$

Here the last term is a Riemann–Stieltjes integral defined by

$$\begin{aligned}&\int_{t_0}^{t} h\big[x(t'),t'\big]\,\mathrm{d}W(t')\\ &= \lim_{n\to\infty}\sum_{i=0}^{n-1} h[x(\tau_i),\tau_i]\big[W(t_{i+1}) - W(t_i)\big]\,,\end{aligned} \tag{78.164}$$

where τ_i is in the interval (t_i, t_{i+1}).

There are two different approaches to such problems: the Ito approach and the Stratonovich approach. They differ in the definition of stochastic integrals.

In the Stratonovich approach, one evaluates $h\big[x(\tau_i),\tau_i\big]$ at $\tau_i = (t_i + t_{i+1})/2$, whereas in the Ito approach one evaluates $h\big[(x(\tau_i),\tau_i)\big]$ at $\tau_i = t_i$. This slightly different definition of τ_i leads to different results because, as a consequence of the delta-correlated noise term, $x(t)$ is not a continuous path. However, there is a relation between the solution of a Stratonovich stochastic differential equation and an Ito stochastic differential equation. Suppose $x(t)$ is a solution of the Stratonovich stochastic differential equation

$$\mathrm{d}x(t) = g(x,t)\,\mathrm{d}t + h(x,t)\,\mathrm{d}W(t)\,. \tag{78.165}$$

Then $x(t)$ satisfies the Ito stochastic differential equation

$$\begin{aligned}\mathrm{d}x(t) = &\left[g(x,t) + \frac{1}{2}h(x,t)\frac{\partial h(x,t)}{\partial x}\right]\mathrm{d}t\\ &+ h(x,t)\,\mathrm{d}W(t)\,.\end{aligned} \tag{78.166}$$

Instead of dealing with stochastic differential equations, one can derive a Fokker–Planck equation for the conditional probability $P(x,t|x_0,t_0)$. For the Stratonovich stochastic differential equation (78.165), the Fokker–Planck equation is

$$\begin{aligned}\frac{\partial P}{\partial t} = &-\frac{\partial}{\partial x}\left[g(x,t) + \frac{1}{2}h(x,t)\frac{\partial h(x,t)}{\partial x}\right]P\\ &+ \frac{1}{2}\frac{\partial^2}{\partial x^2}h^2(x,t)P\,,\end{aligned} \tag{78.167}$$

which takes the form

$$\frac{\partial P}{\partial t} = -\frac{\partial}{\partial x}g(x,t)P + \frac{1}{2}\frac{\partial^2}{\partial x^2}h^2(x,t)P\,, \tag{78.168}$$

if (78.165) is interpreted as a stochastic differential equation in the Ito sense.

The two approaches have the following properties: (i) in most of the models used in physics the Stratonovich definition of a stochastic integral is needed to give correct results, (ii) rules from ordinary calculus are applicable only in the Stratonovich approach, and (iii) for Langevin equations with $h(x,t) = \text{const}$, as in (78.157), the Stratonovich interpretation and the Ito interpretation of stochastic integrals are equivalent.

78.11 Quantum Monte Carlo Formalism

The quantum Monte Carlo formalism was developed to solve numerically master equations of the Lindblad type [78.20, 40–44]

$$\dot{\rho}_s = -\frac{i}{\hbar}[H_s, \rho_s] + \frac{1}{2}\sum_j \left(2C_j\rho_s C_j^\dagger - C_j^\dagger C_j \rho_s - \rho_s C_j^\dagger C_j\right) . \quad (78.169)$$

Here C_j are arbitrary system operators.

As an illustrative example, consider

$$\dot{\rho}_s = -\frac{i}{\hbar}[H_s, \rho_s] + \frac{1}{2}\left(2C\rho_s C^\dagger - C^\dagger C \rho_s - \rho_s C^\dagger C\right) , \quad (78.170)$$

where C is an arbitrary system operator. Instead of solving the master equation, one defines quantum trajectories or stochastic wave functions as follows. Starting from $|\psi(t)\rangle$, there are two possibilities for the time evolution during the interval dt:

1. The system evolves according to the non-Hermitian Hamiltonian $H_s - \frac{i\hbar}{2}C^\dagger C$; i. e.

$$|\psi(t+dt)\rangle = \frac{\left[1 - i\,dt\left(H_s - \frac{i\hbar}{2}C^\dagger C\right)/\hbar\right]|\psi(t)\rangle}{\sqrt{1 - \langle\psi(t)|C^\dagger C|\psi(t)\rangle\, dt}} . \quad (78.171)$$

2. The system makes a jump; i. e.

$$|\psi(t+dt)\rangle = \frac{C|\psi(t)\rangle}{\sqrt{\langle\psi(t)|C^\dagger C|\psi(t)\rangle}} . \quad (78.172)$$

Since both possibilities describe a nonunitary time evolution, $|\psi\rangle$ must be normalized after each step. For each time interval dt one of these two possibilities is randomly chosen according to the probability

$$P(t)dt = \langle\psi(t)|C^\dagger C|\psi(t)\rangle\, dt \quad (78.173)$$

to make a jump between t and $t + dt$. We can now define a density operator

$$\rho_s(t) = |\psi(t)\rangle\langle\psi(t)| \quad (78.174)$$

for a specific quantum trajectory $|\psi(t)\rangle$. The density operator

$$\overline{\rho_s(t)} = \overline{|\psi(t)\rangle\langle\psi(t)|} \quad (78.175)$$

averaged over all trajectories (indicated by the bar) is then a solution of the master equation (78.170).

This method can easily be generalized to master equations of the form (78.169).

78.12 Spontaneous Emission in Free Space

Consider an atom which is initially in one of its excited states and which interacts with the quantized electromagnetic field of free space. Even if none of the modes of the electromagnetic field is excited, there are still the vacuum fluctuations which "interact" with the atom and give rise to important effects:

1. Spontaneous emission: the atom spontaneously emits a photon and decays from the excited state.
2. Natural linewidth: due to the finite lifetime of the atomic levels, the radiation from an atomic transition has a finite linewidth, called the natural linewidth.
3. Lamb shift: the energy levels of the atom are shifted.

The standard theory of spontaneous emission is the Wigner–Weisskopf theory [78.17, 32]. Here an initially excited atomic state $|\ell\rangle$ decays exponentially according to

$$|c_\ell(t)|^2 = e^{-\Gamma_\ell t} , \quad (78.176)$$

where the decay constant Γ_ℓ is given by

$$\Gamma_\ell = \sum_i \frac{\omega_{\ell i}^3 |\boldsymbol{d}_{\ell i}|^2}{3\pi\epsilon_0\hbar c^3} , \quad (78.177)$$

and the sum is over all atomic states with an energy E_i lower than the energy E_ℓ of the state $|\ell\rangle$. $\omega_{\ell i} = (E_\ell - E_i)/\hbar$ is the transition frequency for the transition $|\ell\rangle \to |i\rangle$, and $\boldsymbol{d}_{\ell i} = e\langle\ell|\boldsymbol{r}|i\rangle$ is the corresponding dipole moment.

The same decay constant Γ_ℓ is also observed as a linewidth in the spectrum of the radiation scattered by an atom when the incoming photon excites the atom to the level $|\ell\rangle$.

The energy level shift is more troublesome and needs the concept of mass renormalization, a standard problem in quantum electrodynamics. The theory and results are discussed in Chapt. 27 and Chapt. 28.

Recent calculations of *Pachucki* [78.45] based on fully relativistic quantum electrodynamics and including two-loop corrections predict 1 057 838(6) kHz for the energy difference between the $2s_{1/2}$-state and the $2p_{1/2}$-state which is in excellent agreement with the experimental result of 1 057 839(12) kHz [78.46].

For a discussion of energy levels and transition frequencies in hydrogen and deuterium atoms see also Sect. 28.3.

78.13 Resonance Fluorescence

Consider a two-level atom driven by a continuous monochromatic wave which is treated classically. The excited state of the atom can decay by spontaneous emission into vacuum modes of the electromagnetic field. This emission is called resonance fluorescence. Of particular interest are the properties of the emitted light. For a detailed discussion of resonance fluorescence, see for example [78.20, 31–33].

The far field at position $\boldsymbol{R}$ emitted by an atom at the origin is proportional to its dipole moment and can be expressed in terms of the dipole operators σ_+ and σ_- according to the relation [78.20]

$$\begin{aligned} E^+(\boldsymbol{R},t) &= -\frac{\omega_0^2\,(\boldsymbol{d}\times\boldsymbol{R})\times\boldsymbol{R}}{4\pi\epsilon_0 c^2 R^3}\,\sigma_-(t-r/c)\,,\\ E^-(\boldsymbol{R},t) &= -\frac{\omega_0^2\,(\boldsymbol{d}^*\times\boldsymbol{R})\times\boldsymbol{R}}{4\pi\epsilon_0 c^2 R^3}\,\sigma_+(t-r/c)\,, \end{aligned} \tag{78.178}$$

where

$$\boldsymbol{d} = e\langle g|\boldsymbol{r}|e\rangle \tag{78.179}$$

is the atomic dipole matrix element and the field operators $\boldsymbol{E}^+(\boldsymbol{R},t)$ and $\boldsymbol{E}^-(\boldsymbol{R},t)$ as well as the dipole operators $\sigma_+(t)$ and $\sigma_-(t)$ are in the Heisenberg picture. Knowledge of the operators $\sigma_+(t)$ and $\sigma_-(t)$ is therefore sufficient to study the properties of the emitted light in the far field.

78.13.1 Equations of Motion

The total Hamiltonian for the system reads

$$\begin{aligned} H_{\text{sr}} &= \frac{1}{2}\hbar\omega_0\sigma_z + \sum_i \hbar\omega_i\left(b_i^\dagger b_i + \frac{1}{2}\right)\\ &\quad + \hbar\sum_i\left(g_i\sigma_- b_i^\dagger + g_i^*\sigma_+ b_i\right)\\ &\quad - \frac{1}{2}\left(\mathrm{d}\,\mathcal{E}^*\sigma_-\,\mathrm{e}^{i\omega_0 t} + d^*\,\mathcal{E}\,\sigma_+\,\mathrm{e}^{-i\omega_0 t}\right)\,, \end{aligned} \tag{78.180}$$

where a resonant driving term has been added to the Hamiltonian (78.123). Here d is the projection of the dipole matrix element $e\langle g|\boldsymbol{r}|e\rangle$ onto the polarization vector of the driving field with an amplitude $\mathcal{E}$. The corresponding master equation in the interaction picture is

$$\begin{aligned} \dot{\rho} &= -\mathrm{i}\frac{1}{2}\Omega_1\left[\sigma_+ + \sigma_-, \rho\right]\\ &\quad + \frac{1}{2}\gamma\left(2\sigma_-\rho\sigma_+ - \sigma_+\sigma_-\rho - \rho\sigma_+\sigma_-\right)\,, \end{aligned} \tag{78.181}$$

where $\Omega_1 = -\mathcal{E}\,d^*/\hbar$ is the Rabi frequency associated with the driving field. The vacuum modes of the field are described by a thermal reservoir at zero temperature.

The equations of motion for the expectation values $\langle\sigma_+\rangle$, $\langle\sigma_-\rangle$, and $\langle\sigma_z\rangle$ are the optical Bloch equations with radiative damping (Chapt. 68) and are

$$\begin{aligned} \frac{\mathrm{d}}{\mathrm{d}t}\langle\sigma_+\rangle &= -\frac{\gamma}{2}\langle\sigma_+\rangle - i\frac{\Omega_1}{2}\langle\sigma_z\rangle\,,\\ \frac{\mathrm{d}}{\mathrm{d}t}\langle\sigma_-\rangle &= -\frac{\gamma}{2}\langle\sigma_-\rangle + i\frac{\Omega_1}{2}\langle\sigma_z\rangle\,,\\ \frac{\mathrm{d}}{\mathrm{d}t}\langle\sigma_z\rangle &= -\gamma\left(\langle\sigma_z\rangle + 1\right) - i\Omega_1\left(\langle\sigma_+\rangle - \langle\sigma_-\rangle\right)\,. \end{aligned} \tag{78.182}$$

These expectation values determine the density operator (78.142) of the two-level atom. Because (78.182) are a system of linear differential equations for $\langle\sigma_+\rangle$, $\langle\sigma_-\rangle$, and $\langle\sigma_z\rangle$, they can be solved analytically. Furthermore, the quantum regression hypothesis allows one to calculate two-time correlation functions as shown in Sect. 78.9.

78.13.2 Intensity of Emitted Light

According to (78.52)) and (78.178), the intensity of the fluorescence light at position $\boldsymbol{R}$ is given by

$$I = \langle E^-(\boldsymbol{R},t)E^+(\boldsymbol{R},t)\rangle \propto \langle\sigma_+\sigma_-\rangle\,, \tag{78.183}$$

and can be decomposed into two parts: the coherent intensity

$$I_{\text{coh}} \propto \langle\sigma_+\rangle\langle\sigma_-\rangle \tag{78.184}$$

originating from the mean motion of the dipole, and the incoherent intensity

$$I_{\text{inc}} \propto \langle\sigma_+\sigma_-\rangle - \langle\sigma_+\rangle\langle\sigma_-\rangle\,, \tag{78.185}$$

which is due to fluctuations of the dipole motion around its average value. The steady state intensities are

$$I_{\text{coh}} \propto \frac{\Omega_1^2\gamma^2}{\left(\gamma^2+2\Omega_1^2\right)^2}\,, \tag{78.186}$$

and

$$I_{\text{inc}} \propto \frac{2\Omega_1^4}{\left(\gamma^2+2\Omega_1^2\right)^2}\,. \tag{78.187}$$

For weak laser intensities (Ω_1 small) the intensity of the fluorescence light is dominated by the coherent part whereas for high intensities (Ω_1 large) it is dominated by the incoherent part.

78.13.3 Spectrum of the Fluorescence Light

The Wiener–Khintchine theorem (78.54) allows one to express the steady state spectrum of the fluorescence light as the Fourier transform of the correlation function $\langle\sigma_+(\tau)\sigma_-(0)\rangle_{\text{ss}}$ in the form

$$\begin{aligned}S(\omega) &= \frac{1}{2\pi}\int_0^\infty \mathrm{e}^{-\mathrm{i}\omega\tau}\langle E^-(\tau)E^+(0)\rangle_{\text{ss}}\,\mathrm{d}\tau + \text{c.c.}\\ &\propto \frac{1}{2\pi}\int_0^\infty \mathrm{e}^{-\mathrm{i}(\omega-\omega_0)\tau}\langle\sigma_+(\tau)\sigma_-(0)\rangle_{\text{ss}}\,\mathrm{d}\tau + \text{c.c.}\end{aligned} \tag{78.188}$$

Again it consists of two contributions: a coherent part $S_{\text{coh}}(\omega)$, and an incoherent part $S_{\text{inc}}(\omega)$. The coherent part is

$$S_{\text{coh}}(\omega) \propto \frac{\Omega_1^2\gamma^2}{\left(\gamma^2+2\Omega_1^2\right)^2}\delta(\omega-\omega_0)\,. \tag{78.189}$$

The incoherent part of the fluorescence light has two qualitatively different spectra. For $\Omega_1 < \gamma/4$, it has a single peak at ω_0, whereas it consists of three peaks for $\Omega_1 > \gamma/4$. For $\Omega_1 \gg \gamma/4$ it is given by

$$\begin{aligned}S_{\text{inc}}(\omega) \propto \frac{1}{2\pi\gamma}\Bigg(&\frac{(\gamma/2)^2}{(\omega-\omega_0)^2+(\gamma/2)^2}\\ &+\frac{1}{3}\frac{(3\gamma/4)^2}{(\omega-\omega_0+\Omega_1)^2+(3\gamma/4)^2}\\ &+\frac{1}{3}\frac{(3\gamma/4)^2}{(\omega-\omega_0-\Omega_1)^2+(3\gamma/4)^2}\Bigg)\,.\end{aligned} \tag{78.190}$$

The central peak at $\omega = \omega_0$ has a width of $\gamma/2$ whereas the width of the two side peaks at $\omega = \omega_0 \pm \Omega_1$ is $3\gamma/4$. Their heights are one third of the height of the central peak. This spectrum was predicted by *Burshtein* [78.47] and *Mollow* [78.48] and experimentally confirmed by *Schuda* et al. [78.49], *Wu* et al. [78.50], and *Hartig* et al. [78.51].

This triplet can be explained in terms of the dressed states $|1,n\rangle$ and $|2,n\rangle$ introduced in Chapt. 68 (78.50). If the driving field is resonant with the atomic transition, these states have the energies

$$\begin{aligned}E_{1,n} &= \hbar\left(n+\frac{1}{2}\right)\omega_0 - \hbar R_n\,,\\ E_{2,n} &= \hbar\left(n+\frac{1}{2}\right)\omega_0 - \hbar R_n\,,\end{aligned} \tag{78.191}$$

(78.52). The energy differences between the allowed transitions are

$$\begin{aligned}E_{2,n}-E_{2,n-1} &= \hbar\omega_0+\hbar R_n-\hbar R_{n-1} \approx \hbar\omega_0\,,\\ E_{2,n}-E_{1,n-1} &= \hbar\omega_0+\hbar R_n+\hbar R_{n-1} \approx \hbar\omega_0+\hbar\Omega_1\,,\\ E_{1,n}-E_{2,n-1} &= \hbar\omega_0-\hbar R_n-\hbar R_{n-1} \approx \hbar\omega_0-\hbar\Omega_1\,,\\ E_{1,n}-E_{1,n-1} &= \hbar\omega_0+\hbar R_n-\hbar R_{n-1} \approx \hbar\omega_0\,,\end{aligned} \tag{78.192}$$

where we have made the approximations

$$\begin{aligned}R_n - R_{n-1} &\approx 0\,,\\ R_n + R_{n+1} &\approx 2g\sqrt{\bar{n}+1} \approx \Omega_1\,.\end{aligned} \tag{78.193}$$

This is a good approximation for an intense driving field which can approximated by a highly excited coherent state with an average photon number $\bar{n}$.

Figure 78.6 shows these energy levels and the allowed transition. Obviously, the transitions correspond to frequencies ω_0, $\omega_0-\Omega_1$ and $\omega_0+\Omega_1$. The dressed

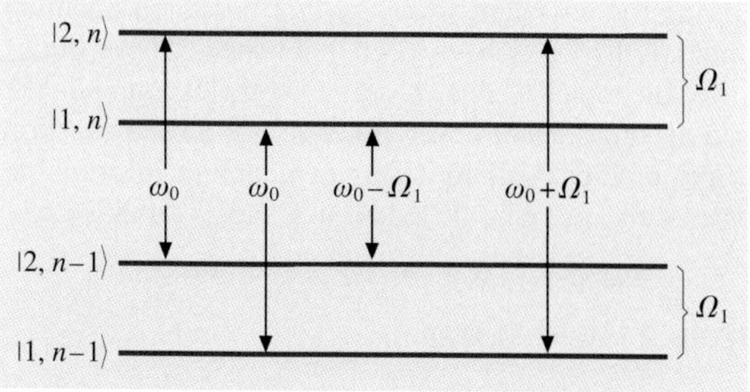

Fig. 78.6 Energy level diagram of dressed states. The transition frequencies are ω_0, $\omega_0-\Omega_1$ and $\omega_0+\Omega_1$

Part F | 78.13

state picture also explains the 2:1 ratio for the integrated intensities of the central peak and the side peak in (78.190).

78.13.4 Photon Correlations

In addition to the spectrum which is based on the correlation function $\langle E^-(\tau)E^+(0)\rangle_{ss}$ in Sect. 78.13.3, the second-order correlation function

$$G^{(2)}_{ss}(\tau) = \langle E^-(0)E^-(\tau)E^+(\tau)E^+(0)\rangle_{ss} \propto \langle \sigma_+(0)\sigma_+(\tau)\sigma_-(\tau)\sigma_-(0)\rangle_{ss} \quad (78.194)$$

can be measured to gain more insight into the fluorescence light, see also Sect. 78.3.2. Experimentally this is done by measuring the joint probability for detecting a photon at time $t = 0$ and a subsequent photon at time $t = \tau$. Again the result can be obtained from the quantum regression hypothesis and reads

$$g^{(2)}(\tau) = \frac{G^{(2)}_{ss}(\tau)}{|G^{(1)}_{ss}(0)|^2} = 1 - e^{-3\gamma\tau/4}\left(\cos\delta\tau + \frac{3\gamma}{4\delta}\sin\delta\tau\right), \quad (78.195)$$

where δ is given by

$$\delta = \sqrt{\Omega_1^2 - \gamma^2/4}\,. \quad (78.196)$$

For $\tau = 0$, $g^{(2)}(0) = 0$, indicating a tendency of photons to be separated. This tendency is known as photon antibunching and was first predicted by *Carmichael* and *Walls* [78.52, 53] and experimentally verified by *Kimble* et al. [78.54, 55]. Photon antibunching of radiation emitted from a two-level atom has a simple explanation: After the atom has emitted a photon it is in the ground state and must first be excited again before it can emit another photon.

78.14 Recent Developments

This chapter has discussed the fundamentals of the quantized electromagnetic field and applications to the broad area of quantum optics. However, in the last eight years, quantum optics has blossomed in several new directions particularly in the key role it is playing in recent investigations of the fundamentals of quantum theory and related applications. In particular, the superposition principle (the bedrock of quantum mechanics), entanglement, the quantum-classical interface, and precision measurements have become very topical research areas, especially in respect to their relevance to quantum information processing.

78.14.1 Literature

During the last eight years, several books on quantum optics [78.56–62] have been published. These books cover the topics of this chapter to some extent and take into account recent developments. For an introduction to the rapidly evolving fields of quantum information processing, we refer the reader to Chapt. 81 and [78.63–65].

78.14.2 Field States

Recently, number states of the radiation field were observed in a cavity-QED experiment [78.66].

78.14.3 Reservoir Theory

New research topics, such as quantum information processing, rely on the superposition principle and entangled quantum states. Since these states are very sensitive to decoherence, reservoir theory has attracted a lot of interest in recent years. Furthermore, as discussed in [78.67–72], decoherence is the physical process by which the classical world emerges from its quantum underpinning.

Many investigations in this area involve the presence of a reservoir (heat-bath/environment) and master equations are a ubiquitous tool. The familiar master equations of quantum optics are in Lindblad form [78.73], which guarantees that the density matrix is always positive definite during time evolution. In the derivation of this equation [78.74, 75], rapidly oscillating terms are omitted by the method of coarse-graining in time; the high frequencies correspond to the oscillator frequency ω_0 and, in the usual weak coupling limit, $\omega_0 \gg \gamma$, where γ is a typical decay constant. This is the rotating wave approximation (Sect. 66.3.2).

We have referred to the equations obtained prior to coarse-graining in time as pre-master (or pre-Lindblad) equations [78.74, 76], and such equations have been used extensively in other areas of physics [78.77, 78]; other authors have simply referred to them as master equations

but, to avoid confusion, we reserve the latter term for equations in Lindblad form. Pre-master equations, like the master equations, describe an approach to the equilibrium state. This equilibrium state is the same in either case [78.76], but with pre-master (non-Lindblad) equations the approach can be through non-physical states of negative probability. However, as recently demonstrated, pre-master equations have other advantages vis à vis master equations:
(a) they lead to the exact expression for the mean value of $x(t)$ (as obtained from the exact Langevin equation for the problem);
(b) they lead, in the classical limit ($\hbar \to 0$), to the familiar Fokker–Planck equation of classical probability; and
(c) the exact master equation [78.79–83] is for long times of pre-master form. However, the general expectation (based on the time dependence of the coefficients) that the exact master equation preserves positivity for all times has not been realized since Ford and O'Connell have recently shown that, even in high temperature regime, the density matrix is not necessarily positive [78.84].

In traditional quantum optics, the emphasis has been on long-time $\left(t \gg \gamma^{-1}\right)$ phenomena, for which the use of either master or pre-master equations is justified. However, they are both inadequate for dealing with short-time $\left(t \ll \gamma^{-1}\right)$ phenomena (as can be shown most simply by calculating the mean-square displacement, a key ingredient in decoherence calculations), which are of much recent interest. Thus, it is desirable to use exact master equations. In that respect, the exact master equation of *Hu* et al. [78.79, 80] for an oscillator is an arbitrary dissipative environment has proved to be a popular and useful tool for which an exact solution has now been obtained [78.83]. However, it should also be mentioned that the solution of the initial value quantum Langevin equation gives all the same information as the exact master equation, and in fact, the solutions of the former were used to obtain the solutions of the latter [78.83].

The familiar two-level atom master equation is, of course, similar in form to the usual Lindblad-type master equation for the oscillator. However, motivated in particular by the desire to study decoherence and other short time phenomena, an exact master equation was derived [78.85] to study the non-Markovian dynamics of a two-level atom interacting with the electromagnetic field.

In addition, motivated by the desire to study a driven oscillator, the usual two-level atom master equation was generalized to include the case of an external force field [78.86, 87]. This generalized equation was then used not only to obtain the familiar zero-temperature Burshtein–Mollow spectrum, but also the corresponding high temperature results. For strong resonant driving at high temperature, the same three-peaked structure was observed in the zero temperature case, but a much larger width was found. The analysis, following other investigations, used the Lax formula for calculating two-time correlation functions. This formula is not a "quantum regression theorem" as it is often designated (see also Sect. 78.9), but simply an approximation (which more resembles an Onsager classical regression theorem [78.88]) which works very well in the case of weak coupling and for frequencies near a resonant frequency, but not otherwise [78.86, 87].

In Sect. 78.6, we stressed the usefulness of quasi-probability distributions instead of the density matrix, with particular attention to the Wigner distribution. In particular, for simple Hamiltonians, we pointed out that the equation for the corresponding Wigner function has the form of a Fokker–Planck equation and we considered the explicit form describing the usual master equation. The more general equations associated with an exact master equation and their solution was the subject of [78.83] and interesting limits of that equation, including the pre-master equation for both momentum coupling and coordinate coupling were discussed at length in [78.89, 90]. In the case of two-level systems, it is not convenient to use quasi-probability distributions; instead, it is found that the preferred tool is the polarization vector [78.91]. Surprisingly, it has not been generally adopted by the quantum optics community although its usefulness in that context has been demonstrated recently in [78.86, 87].

References

78.1 C. Cohen-Tannoudji, J. Dupont-Roc, G. Grynberg: *Photons, and Atoms. An Introduction to Quantum Electrodynamics* (Wiley, New York 1989)
78.2 L. Mandel: Opt. Lett. **4**, 205 (1979)
78.3 W. Schleich, M. Pernigo, Fam Le Kien: Phys. Rev. A **44**, 2172 (1991)
78.4 H.J. Kimble, D.F. Walls (Eds.): J. Opt. Soc. Am. B **4**(10), 1453–1737 (1987)

78.5 P. Knight, R. London (Eds.): J. Mod. Opt. **34**(6) (1987)
78.6 E. Giacobino, C. Fabre (Eds.): Appl. Phys. B **55**(3) (1992)
78.7 W. Schleich, J. A. Wheeler: J. Opt. Soc. Am. B **4**, 1715 (1987)
78.8 H. P. Yuen: Phys. Rev. A **13**, 2226 (1976)
78.9 R. E. Slusher, L. W. Hollberg, B. Yurke, J. C. Mertz, J. F. Valley: Phys. Rev. Lett. **55**, 2409 (1985)
78.10 L. A. Wu, M. Xiao, H. J. Kimble: J. Opt. Soc. Am. B **4**, 1465 (1987)
78.11 A. Heidmann, R. J. Horowicz, S. Reynaud, E. Giacobino, C. Fabre: Phys. Rev. Lett. **59**, 2555 (1987)
78.12 W. P. Schleich, S. M. Burnett (Eds.): Special issue on Quantum Phase and Phase Dependent Measurements, Phys. Scr.T. **T48** (1993)
78.13 P. Carruthers, M. M. Nieto: Rev. Mod. Phys. **40**, 411 (1968)
78.14 D.T. Pegg, S.M. Barnett: Phys. Rev. A **39**, 1665 (1989)
78.15 J. W. Noh, A. Fougères, L. Mandel: Phys. Rev. A **45**, 424 (1992)
78.16 R. J. Glauber: Phys. Rev. **130**, 2529 (1963)
78.17 W. H. Louisell: *Quantum Statistical Properties of Radiation* (Wiley, New York 1973)
78.18 R. Hanbury Brown, R. Q. Twiss: Nature **177**, 27 (1956)
78.19 P. L. Kelley, W. H. Kleiner: Phys. Rev. **136**, 316 (1964)
78.20 H. Carmichael: *An Open Systems Approach to Quantum Optics* (Springer, Berlin, Heidelberg 1993)
78.21 H. Yuen, H. P. Shapiro: IEEE Trans. Inf. Theory **26**, 78 (1980)
78.22 J. H. Shapiro, S. S. Wagner: IEEE J. Quantum Electron. **20**, 803 (1984)
78.23 K. E. Cahill, R. J. Glauber: Phys. Rev. A **177**, 1882 (1969)
78.24 M. Hillery, R. F. O'Connell, M. O. Scully, E. P. Wigner: Phys. Rep. **106**, 121 (1984)
78.25 R. J. Glauber: Phys. Rev. **131**, 2766 (1963)
78.26 E. C. G. Sudarshan: Phys. Rev. Lett. **10**, 277 (1963)
78.27 C. W. Gardiner: *Handbook of Stochastic Methods* (Springer, Berlin, Heidelberg 1985)
78.28 C. W. Gardiner: *Quantum Noise* (Springer, Berlin, Heidelberg 1991)
78.29 E. Wigner: Phys. Rev. **40**, 749 (1932)
78.30 R. L. Hudson: Rep. Math. Phys. **6**, 249 (1974)
78.31 P. Meystre, M. Sargent III: *Elements of Quantum Optics* (Springer, Berlin, Heidelberg 1991)
78.32 C. Cohen-Tannoudji, J. Dupont-Roc, G. Grynberg: *Atom-Photon Interactions* (Wiley, New York 1992)
78.33 D. F. Walls, G. J. Milburn: *Quantum Optics* (Springer, Berlin, Heidelberg 1994)
78.34 C. W. Gardiner: Phys. Rev. Lett. **56**, 1917 (1986)
78.35 G. W. Ford, J. T. Lewis, R. F. O'Connell: Phys. Rev. A **37**, 4419 (1988)
78.36 R. F. O'Connell: Dissipation in a squeezed-state environment, Proceedings of the Second International Workshop on Squeezed States and Uncertainty Relations, Moscow, Russia 1992, ed. by D. Han, Y.-S. Kim, V. I. Man'ko (NASA, Maryland, USA 1993)
78.37 H. Risken: *The Fokker-Planck Equation* (Springer, Berlin, Heidelberg 1989)
78.38 M. Lax: Fluctuations and Coherence Phenomena in Classical and Quantum Physics. In: *Brandeis University Summer Institute Lectures*, Vol. 2, ed. by M. Chretin, E. P. Gross, S. Deser (Gordon and Breach, New York 1966)
78.39 G. W. Ford, R. F. O'Connell: Phys. Lett. A **157**, 217 (1991)
78.40 J. Dalibard, Y. Castin, K. Mølmer: Phys. Rev. Lett. **68**, 580 (1992)
78.41 K. Mølmer, Y. Castin, J. Dalibard: J. Opt. Soc. Am B **10**, 524 (1993)
78.42 R. Dum, P. Zoller, H. Ritsch: Phys. Rev. A **45**, 4879 (1992)
78.43 C. W. Gardiner, A. S. Parkins, P. Zoller: Phys. Rev. A **46**, 4363 (1992)
78.44 R. Dum, A. S. Parkins, P. Zoller, C. W. Gardiner: Phys. Rev. A **46**, 4382 (1992)
78.45 K. Pachucki: Phys. Rev. Lett. **72**, 3154 (1994)
78.46 E. W. Hagley, F. M. Pipkin: Phys. Rev. Lett. **72**, 1172 (1994)
78.47 A. I. Burshtein: Sov. Phys. JETP **22**, 939 (1966)
78.48 B. R. Mollow: Phys. Rev. **188**, 1969 (1969)
78.49 F. Schuda, C. R. Straud, Jr., M. Hercher: J. Phys. B **7**, L198 (1974)
78.50 F. Y. Wu, R. E. Grove, S. Ezekiel: Phys. Rev. Lett. **35**, 1426 (1975)
78.51 W. Hartig, W. Rasmussen, R. Schieder, H. Walther: Z. Phys. A **278**, 205 (1976)
78.52 H. J. Carmichael, D. F. Walls: J. Phys. B **9**, L43 (1976)
78.53 H. J Carmichael, D. F. Walls: J. B. Phys **9**, 1199 (1976)
78.54 H. J. Kimble, M. Dagenais, L. Mandel: Phys. Rev. Lett. **39**, 691 (1977)
78.55 M. Dagenais, L. Mandel: Phys. Rev. A **18**, 201 (1978)
78.56 M. O. Scully, M. S. Zubairy: *Quantum Optics* (Cambridge Univ. Press, Cambridge 1996)
78.57 H.-A. Bachor: *A Guide to Experiments in Quantum Optics* (Wiley-VCH, Weinheim 1998)
78.58 H. J. Carmichael: *Statistical Methods in Quantum Optics 1: Master Equations and Fokker-Planck Equations* (Springer, Berlin, Heidelberg 1999)
78.59 P. Meystre, M. Sargent III: *Elements of Quantum Optics* (Springer, Berlin, Heidelberg 1999)
78.60 C. W. Gardiner, P. Zoller: *Quantum Noise* (Springer, Berlin, Heidelberg 2000)
78.61 M. Orszag: *Quantum Optics* (Springer, Berlin, Heidelberg 2000)
78.62 W. P. Schleich: *Quantum Optics in Phase Space* (VCH-Wiley, Weinheim 2001)
78.63 D. Bouwmeester, A. Ekert, A. Zeilinger (Eds.): *The Physics of Quantum Information: Quantum Cryptography, Quantum Teleportation, Quantum Information* (Springer, Berlin, Heidelberg 2000)

78.64 M. A. Nielsen, I. L. Chuang: *Quantum Computation and Quantum Information* (Cambridge Univ. Press, Cambridge 2000)

78.65 G. Alber, T. Beth, M. Horodecki, P. Horodecki, R. Horodecki, M. Rötteler, H. Weinfurter, R. Werner, A. Zeilinger: *Quantum Information: An Introduction to Basic Theoretical Concepts and Experiments*, Springer Tracts in Modern Physics, Vol. 173 (Springer, Berlin, Heidelberg 2001)

78.66 B. T. H. Varcoe, S. Brattke, M. Weidinger, H. Walther: Nature **403**, 743 (2000)

78.67 H. D. Zeh: Found. Phys. **1**, 69 (1970)

78.68 A. J. Leggett: Suppl. Prog. Theor. Phys. **69**, 80 (1980)

78.69 E. P. Wigner: In: *Quantum Optics, Experimental Gravity and Measurement Theory*, ed. by P. Meystre, M. O. Scully (Plenum, New York 1983) p. 43

78.70 D. F. Walls, G. J. Milburn: Phys. Rev. A **31**, 2403 (1985)

78.71 E. Joos, H. D. Zeh, C. Kiefer, D. Giulini, J. Kupsch, I.-O. Stamatescu: *Decoherence and the Appearence of a Classical World in Quantum Theory*, 2nd edn. (Springer, Berlin, Heidelberg 2003)

78.72 W. H. Zurek: Rev. Mod. Phys. **75**, 715 (2003)

78.73 G. Lindblad: Commun. Math. Phys. **48**, 119 (1976)

78.74 G. W. Ford, J. T. Lewis, R. F. O'Connell: Ann. Phys. (NY) **252**, 362 (1996)

78.75 J. T. Lewis, R. F. O'Connell: Ann. Phys. (NY) **269**, 51 (1998)

78.76 G. W. Ford, R. F. O'Connell: Phys. Rev. Lett. **82**, 3376 (1999)

78.77 A. O. Caldeira, A. J. Leggett: Physica A **121**, 587 (1983)

78.78 R. Dekker: Phys. Rep. **80**, 1 (1981)

78.79 B. L. Hu, J. P. Paz, Y. Z. Zhang: Phys. Rev. D **45**, 2843 (1992)

78.80 J. J. Halliwell, T. Yu: Phys. Rev. D **53**, 2012 (1996)

78.81 F. Haake, R. Reibold: Phys. Rev. A **32**, 2462 (1985)

78.82 R. Karrlein, H. Grabert: Phys. Rev. E **55**, 153 (1997)

78.83 G. W. Ford, R. F. O'Connell: Phys. Rev. D **64**, 105020 (2001)

78.84 G. W. Ford, R. F. O'Connell: Limitations on the utility of exact master equations, Ann. Phys (NY) **319**, in press

78.85 C. Anastopoulos, B. L. Hu: Phys. Rev. A **62**, 033821 (2000)

78.86 R. F. O'Connell: Optics Comm. **179**, 451 (2000) Reprinted in Ode to a Quantum Physicist edited by W. Schleich, H. Walther, W. E. Lamb (Elsevier,Amsterdam, 2000)

78.87 R. F. O'Connell: Optics Comm. **179**, 477 (2000)

78.88 G. W. Ford, R. F. O'Connell: Phys. Rev. Lett. **77**, 798 (1996)

78.89 G. W. Ford, R. F. O'Connell: Acta Phys. Hung. B **20**, 91 (2004)

78.90 R. F. O'Connell: J. Optics B **5**, S349 (2003)

78.91 E. Merzbacher: *Quantum Mechanics*, 3rd edn. (Wiley, New York 1998) p. 394

79. Entangled Atoms and Fields: Cavity QED

Although the concept of a "free atom" is of use as a first approximation, a full quantum description of the interaction of atoms with an omnipresent electromagnetic radiation field is necessary for a proper account of spontaneous emission and radiative level shifts such as the Lamb shift (Chapt. 27). This chapter is concerned with the changes in the atom–field interaction that take place when the radiation field is modified by the presence of a cavity. An atom in the vicinity of a plane perfect mirror serves as an example of cavity quantum electrodynamics [79.1–5].

The primary focus in this chapter is the two extreme cases of weak coupling and strong coupling, as exemplified by spontaneous emission.

79.1 **Atoms and Fields** 1167
79.1.1 Atoms 1167
79.1.2 Electromagnetic Fields 1168

79.2 **Weak Coupling in Cavity QED** 1169
79.2.1 Radiating Atoms in Waveguides .. 1169
79.2.2 Trapped Radiating Atoms and Their Mirror Images 1170
79.2.3 Radiating Atoms in Resonators ... 1170
79.2.4 Radiative Shifts and Forces 1171
79.2.5 Experiments on Weak Coupling... 1172
79.2.6 Cavity QED and Dielectrics 1173

79.3 **Strong Coupling in Cavity QED** 1173

79.4 **Strong Coupling in Experiments** 1174
79.4.1 Rydberg Atoms and Microwave Cavities 1174
79.4.2 Strong Coupling in Open Optical Cavities 1174

79.5 **Microscopic Masers and Lasers** 1175
79.5.1 The Jaynes–Cummings Model 1175
79.5.2 Fock States, Coherent States and Thermal States 1175
79.5.3 Vacuum Splitting 1177

79.6 **Micromasers** 1178
79.6.1 Maser Threshold 1178
79.6.2 Nonclassical Features of the Field 1179
79.6.3 Trapping States 1179
79.6.4 Atom Counting Statistics 1180

79.7 **Quantum Theory of Measurement** 1180

79.8 **Applications of Cavity QED** 1181
79.8.1 Detecting and Trapping Atoms through Strong Coupling 1181
79.8.2 Generation of Entanglement 1181
79.8.3 Single Photon Sources 1182

References 1182

In the weak coupling regime, the coupling of an excited atom to a broad continuum of radiation modes leads to exponential decay (Fig. 79.1a), as first described by *Weisskopf* and *Wigner* [79.6]. Spontaneous emission may be enhanced or suppressed in structures such as waveguides or "bad" cavities. Cavities also introduce van der Waals forces and the subtle Casimir level shifts [79.7].

In the strong coupling regime, the excited atom is strongly coupled to an isolated resonant cavity mode. In the absence of damping, an oscillatory exchange of energy between the atom and the field replaces exponential decay (Fig. 79.1b) with a coherent evolution in time. Experimental investigations of these effects began [79.8] with the development of suitable resonators and techniques for producing atoms with long lived excited states and strong dipole transition moments.

79.1 Atoms and Fields

79.1.1 Atoms

The essential features of cavity QED are elucidated by the two-level model atom discussed in Chapts. 68, 69, 70, and 77 (see also [79.9]). A ground state $|g\rangle$ and an excited state $|e\rangle$ are coupled to the radiation field by a dipole interaction. Using the formal equivalence to a spin-1/2 system, the Pauli spin operators are

$$\begin{aligned}\sigma_x &= \sigma^\dagger + \sigma\,,\\ \sigma_y &= -\mathrm{i}\left(\sigma^\dagger - \sigma\right)\,,\\ \sigma_z &= \sigma^\dagger\sigma - \sigma\sigma^\dagger = \left[\sigma^\dagger, \sigma\right]\,,\end{aligned} \tag{79.1}$$

with $\sigma^\dagger = |e\rangle\langle g|$ and $\sigma = |g\rangle\langle e|$. The quadratures (out of phase components) of the atomic polarization are given by σ_x and σ_y, while σ_z is the occupation number difference. The free atom Hamiltonian is

$$\mathcal{H}_{\text{atom}} = \frac{1}{2}\hbar\omega_0\sigma_z\,, \tag{79.2}$$

where $\hbar\omega_0 = E_e - E_g$ is the transition energy.

79.1.2 Electromagnetic Fields

Classical Fields

Classical electromagnetic fields have longitudinal and transverse components:

$$\boldsymbol{E}(\boldsymbol{r}, t) = \boldsymbol{E}^{\mathrm{l}}(\boldsymbol{r}, t) + \boldsymbol{E}^{\mathrm{t}}(\boldsymbol{r}, t)\,. \tag{79.3}$$

In the Coulomb gauge, the longitudinal part is the instantaneous electric field. The transverse part is the radiation field which obeys the wave equation

$$\left(\nabla^2 - \frac{1}{c^2}\frac{\partial^2}{\partial t^2}\right)\boldsymbol{E}^{\mathrm{t}}(\boldsymbol{r}, t) = \frac{1}{\epsilon_0 c^2}\frac{\partial}{\partial t}\boldsymbol{j}(\boldsymbol{r}, t)\,. \tag{79.4}$$

In empty space, the driving current density $\boldsymbol{j}(\boldsymbol{r}, t)$ vanishes, and the field may be expanded in a set of orthogonal modes as

$$\boldsymbol{E}^{\mathrm{t}}(\boldsymbol{r}, t) = \sum_\mu E_\mu(t)\,\mathrm{e}^{-\mathrm{i}\omega_\mu r}\boldsymbol{u}_\mu(\boldsymbol{r}) + \text{c.c.}\,, \tag{79.5}$$

with slowly varying amplitudes $E_\mu(t)$. The spatial distributions $\boldsymbol{u}_\mu(\boldsymbol{r})$ obey the vector Helmholtz equation

$$\left[\nabla^2 + \left(\frac{\omega_\mu}{c}\right)^2\right]\boldsymbol{u}_\mu(\boldsymbol{r}, t) = 0\,, \tag{79.6}$$

depending on geometric boundary conditions as imposed by conductive or dielectric mirrors, waveguides, and resonators. In free space, plane wave solutions $\boldsymbol{u}_\mu(\boldsymbol{r}, t) = \boldsymbol{u}_\epsilon\,\mathrm{e}^{\mathrm{i}\boldsymbol{k}\cdot\boldsymbol{r}}$ have a continuous index $\mu = (\boldsymbol{k}, \epsilon)$ with wave vector $\boldsymbol{k}$ and an index ϵ for the two independent polarizations. The orthogonality relation

$$\frac{1}{V}\int_V \boldsymbol{u}_\mu \cdot \boldsymbol{u}_\nu^*\,\mathrm{d}^3\boldsymbol{r} = \delta_{\mu\nu} \tag{79.7}$$

applies. For a closed cavity, V is the resonator volume. In waveguides and free space, an artificial boundary is introduced and then increased to infinity at the end of a calculation, such that the final results do not depend on V.

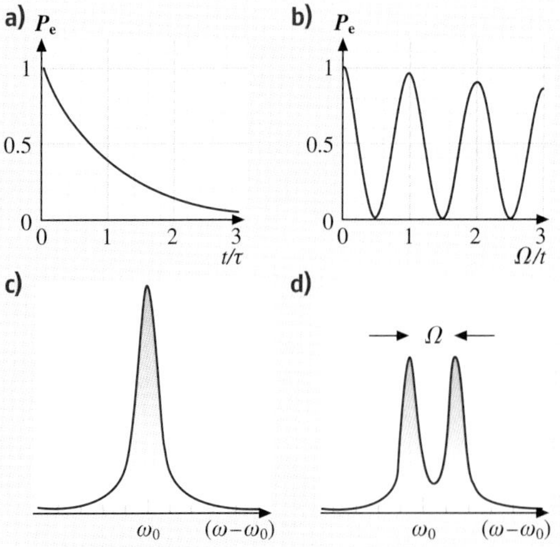

Fig. 79.1a–d *Upper row*: Excitation probability of an excited atom. **(a)** Exponential decay in free space or bad cavities in the weak coupling limit. **(b)** Oscillatory evolution in good cavities or in the strong coupling case. *Lower row:* The spectral signature of exponential decay is a Lorentzian line shape **(c)** while the so-called vacuum Rabi splitting **(d)** is observed in the strong coupling case

Quantum Fields

The quantum analog of the classical transverse field in (79.4) is obtained through a quantization of its harmonic modes leading to a number state expansion. Field operators obey standard commutation relations $[a_\mu, a_\nu^\dagger] = \delta_{\mu,\nu}$, and for a single mode with index μ, the amplitude E_μ in (79.5) is replaced by the corresponding operator

$$E_\mu(t) = \mathcal{E}_\mu a_\mu\,\mathrm{e}^{-\mathrm{i}\omega t}\,, \quad E_\mu^\dagger(t) = \mathcal{E}_\mu{}^* a_\mu^\dagger\,\mathrm{e}^{\mathrm{i}\omega t}\,. \tag{79.8}$$

The normalization factor $\mathcal{E}_\mu$ is chosen such that the energy difference between number states $|n\rangle_\mu$ and $|n+1\rangle_\mu$ in the volume V is $\hbar\omega_\mu$, giving

$$\mathcal{E}_\mu\mathcal{E}_\mu{}^\dagger = \frac{\hbar\omega_\mu}{2\epsilon_0 V}\,. \tag{79.9}$$

The Hamiltonian of the free field is

$$\mathcal{H}_{\text{Field}} = \sum_\mu \hbar\omega_\mu\left(a_\mu^\dagger a_\mu + \frac{1}{2}\right)\,. \tag{79.10}$$

In the Coulomb gauge, the vector potential $\boldsymbol{A}(\boldsymbol{r})$ is related to the electric field $\boldsymbol{E} = -\partial \boldsymbol{A}/\partial t$ by

$$\boldsymbol{A}_\mu(\boldsymbol{r}, t) = -\frac{\mathcal{E}_\mu}{\omega_\mu}\left(a_\mu \, \mathrm{e}^{-\mathrm{i}\omega t} + a_\mu^\dagger \, \mathrm{e}^{\mathrm{i}\omega t}\right) \boldsymbol{u}_\mu(\boldsymbol{r}) \,. \tag{79.11}$$

The ground state $|0\rangle_\mu$ is called the vacuum state. While the expectation value $\langle n|\boldsymbol{E}|n\rangle = 0$ for a number state, the variance is not zero, since $\langle n|\boldsymbol{E}\boldsymbol{E}^*|n\rangle > 0$, giving rise to nonvanishing "fluctuations" of the free electromagnetic field.

Dipole Coupling of Fields and Atoms

The combined system of atoms and fields can be described by the product quantum states $|a, n\rangle$ of atom states $|a\rangle$ and field states $|n\rangle$. The interaction Hamiltonian $\mathcal{H}_I$ of the atom and the radiation field is given by (The $\boldsymbol{A}^2$-term plays an important role in energy shifts and can only be neglected when radiative processes involving energy exchange are considered.)

$$\mathcal{H}_\mathrm{I} = -\frac{q}{m}\boldsymbol{p} \cdot \boldsymbol{A}(\boldsymbol{r}) + \frac{q^2}{2m}\boldsymbol{A}^2(\boldsymbol{r}) \,. \tag{79.12}$$

This interaction causes the atom to exchange energy with the radiation field. In the dipole approximation, the coupling strength is proportional to the component of the atomic dipole moment $\boldsymbol{d}_{eg} = q\langle e|\boldsymbol{r}_{eg}|g\rangle$ along the electric field, with coupling constant

$$g_\mu(\boldsymbol{r}) = |\boldsymbol{d}_{eg} \cdot \boldsymbol{u}_\mu(\boldsymbol{r})\mathcal{E}_\mu|/\hbar \,. \tag{79.13}$$

In the rotating wave approximation (RWA) (Chapts. 68, 69, and 70),

$$\mathcal{H}_\mathrm{RWA} = \sum_\mu \hbar\left(g_\mu \sigma^\dagger a_\mu + g_\mu^* a_\mu^\dagger \sigma\right) \,, \tag{79.14}$$

where we have used the atomic operators of (79.1).

In a continuous electromagnetic spectrum, the atom interacts with a large number of modes having quantum numbers μ, yielding exponential decay of an excited atomic level at the rate [79.6]

$$\Gamma_{eg} = \frac{2\pi}{\hbar^2}\sum_{\tilde{\mu}}\sum_{\boldsymbol{k}} |g_\mu|^2 \, \delta\left(\omega_\mu - \omega_0\right) \,. \tag{79.15}$$

Here we have separated the discrete ($\tilde{\mu}$) and the continuous part (wave vector $\boldsymbol{k}$) of the mode index μ. If g_μ (79.13) does not vary much across a narrow resonance, then

$$\Gamma_{eg} \simeq 2\pi \left|g_\mu(\omega_0)\right|^2 \sum_{\tilde{\mu}} \rho_{\tilde{\mu}}(\omega_0) \,. \tag{79.16}$$

The density of states corresponding to the continuous mode index $\boldsymbol{k}$ of dimension ν can be evaluated on a ν-dimensional fictitious volume $V^{(\nu)}$ as

$$\rho_{\tilde{\mu}} = \sum_{\boldsymbol{k}} \delta(\omega_{\tilde{\mu},\boldsymbol{k}} - \omega) \to \frac{V^{(\nu)}}{(2\pi)^\nu}\int_0^\infty \mathrm{d}^\nu k \delta(\omega_{\tilde{\mu},\boldsymbol{k}} - \omega) \,, \tag{79.17}$$

provided $\omega(\boldsymbol{k})$ is known, and by converting the sum (This is formally accomplished by taking the limit of $\Delta k = 2\pi/l$ for large l, where l is a linear dimension of an artificial resonator, and the resonator volume is $V = l^3$. If the relation between mode spacing Δk and geometric dimension is nonlinear in a more complex geometry, this analysis can be very complicated.) over plane wave vectors $\boldsymbol{k}$ into an integral.

The Rate of Spontaneous Emission

In free space $[\omega(\boldsymbol{k})^2 = (c\boldsymbol{k})^2]$, the sum in (79.16) contributes a factor of two, due to polarization, to the total density of states in free space, $\rho_\mathrm{free}(\omega) = V\omega^2/\pi^2 c^3$. When the vector coupling of atom and field (79.13) is replaced by its average in isotropic free space, that is, by 1/3, the result

$$\Gamma_{eg} = A_{eg} = \frac{e^2 r_{eg}^2 \omega^3}{3\pi\epsilon_0 \hbar c^3} \tag{79.18}$$

is obtained for the decay rate A_{eg} as measured by the natural linewidth Γ_{eg}.

79.2 Weak Coupling in Cavity QED

The regime of weak cavity QED generally applies when an atom is coupled to a continuum of radiation modes. This is always the case with mirrors, waveguides, or bad cavities. The signatures of weak cavity QED are modifications of the rate of spontaneous emission, as well as the existence of van der Waals and Casimir forces. Formally, this regime is well described by perturbation theory.

79.2.1 Radiating Atoms in Waveguides

Within the continuous spectrum of a waveguide, radiative decay of an excited atomic level remains exponential, and Γ_{eg} may be determined as in the preceding section. We now consider the modifications of spontaneous decay in a parallel plate waveguide. According to (79.16), the theoretical problem is reduced to a geometric evaluation of mode densities. Between a pair of mirrors it is convenient to distinguish TE_{nk} and TM_{nk} modes, where n is the number of half waves across the gap of width d. The dispersion relation $\omega(\boldsymbol{k})$ reflects the discrete standing wave part $(n\pi/d)$ and a running wave part as in free space,

$$\omega_{n,\boldsymbol{k}}^2 = c^2\left(|\boldsymbol{k}|^2 + n\pi/d\right)^2 \quad \begin{array}{ll} n = 0, 1, 2, \ldots & \mathrm{TM} \\ n = 1, 2, \ldots & \mathrm{TE}\,. \end{array} \tag{79.19}$$

The average mode density [$\boldsymbol{du} = 1$, (79.13)] is evaluated [(79.17), $\nu = 2$] with an appropriate quantization volume containing the area of the plates, $V = Ad$, giving

$$\rho^{\mathrm{TE}}(\omega) = \frac{\omega_\mathrm{c}[\omega]}{2\omega_\mathrm{c}^2}\rho_\mathrm{free}(\omega_\mathrm{c})\,,$$
$$\rho^{\mathrm{TM}}(\omega) = \frac{\omega_\mathrm{c}[\omega+1]}{2\omega_\mathrm{c}^2}\rho_\mathrm{free}(\omega_\mathrm{c})\,, \tag{79.20}$$

where $[x]$ is the largest integer in x, and $\omega_\mathrm{c} = \pi c/d$ gives the waveguide cutoff frequency. Below ω_c, the TE-mode density clearly vanishes and, with the pictorial notion of turning off the vacuum introduced by *Kleppner* [79.10], inhibition of radiative decay is obvious. Figure 79.2 shows the calculated mode density for a parallel plate waveguide. The decay rate can be calculated from (79.16), with the spatial variation of g_μ included. This configuration was used for the first experiments which showed the suppression of spontaneous emission in both the microwave and the near optical frequency domain [79.11, 12] with atomic beams.

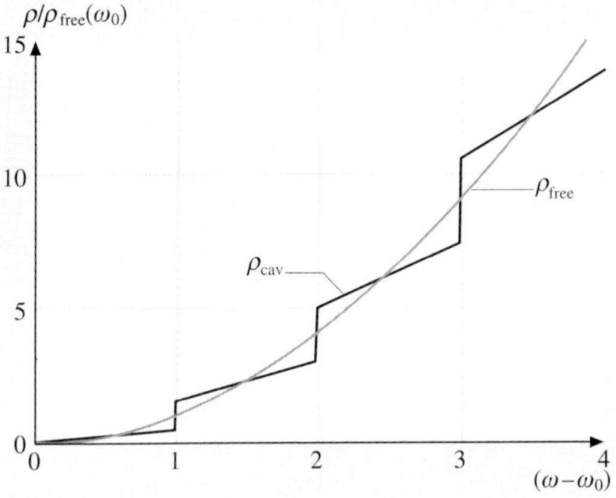

Fig. 79.2 Modification of the average vacuum spectral density $(\rho^{\mathrm{TE}} + \rho^{\mathrm{TM}})$ in a parallel plate cavity (*thick line*) compared with free space (*thin line*)

79.2.2 Trapped Radiating Atoms and Their Mirror Images

Boundary conditions imposed by conductive surfaces may also be simulated by appropriately positioned image charges. Inspired by classical electrodynamics, this image charge model can be successfully used to determine the modifications of radiative properties in confined spaces. In the simplest case, an atom is interacting with its image produced by a plane mirror. Trapped atoms and ions allow one to control their relative position with respect to a mirror to distances below the wavelength of light. Hence they are ideal objects for studying the spatial dependence of the mirror induced modifications of their radiative properties. In an experiment with a single trapped ion (see Fig. 79.3), its radiation field was superposed onto its mirror image [79.13, 14], yielding a sinusoidal variation of both the spontaneous decay rate and the mirror induced level shift with excellent contrast.

79.2.3 Radiating Atoms in Resonators

Resonators

In a resonator, the electromagnetic spectrum is no longer continuous and the discrete mode structure can also be resolved experimentally. While a resonator is only weakly coupled to external electromagnetic fields, it still interacts with a large thermal reservoir through currents induced in its walls. The total damping rate is due to resistive losses in the walls (κ_wall) and also due to transmission at the radiation ports, $1/\tau_\mu = \kappa_\mu = \kappa_\mathrm{wall} + \kappa_\mathrm{out}$. An empty resonator stores energy for times

$$\tau_\mu = Q/\omega_\mu\,, \tag{79.21}$$

and the power transmission spectrum is a Lorentzian with width $\Delta\omega_\mu = \omega_\mu/Q_\mu$. The index μ, for instance, represents the TE_{lm} and TM_{lm} modes of a "pillbox" microwave cavity, or the TEM_{klm} modes of a Fabry–Perot interferometer (Fig. 79.4).

When cavity damping remains strong, $\Gamma_\mu \gg \Gamma_{eg}$, the atomic radiation field is "immediately" absorbed and Weisskopf–Wigner perturbation theory remains valid. In this so-called bad cavity limit, resonator damping can be accounted for by an effective mode density of Lorentzian width $\Delta\omega_\mu$ for a single isolated mode,

$$\rho_\mu(\omega) = \frac{1}{\pi}\frac{\omega_\mu/2Q_\mu}{(\omega-\omega_\mu)^2 + (\omega_\mu/2Q_\mu)^2}\,. \tag{79.22}$$

Bad and Good Cavities

The modification of spontaneous decay is again calculated from (79.16). For an atomic dipole aligned parallel to the mode polarization, and right at resonance, $\omega_\mu = \omega_0$, the enhancement of spontaneous emission is found to be proportional to the Q-value of a selected resonator mode:

$$\frac{\Gamma_{eg}^{\text{cav}}}{\Gamma_{eg}^{\text{free}}} = \frac{\rho_\mu |\boldsymbol{u}(\boldsymbol{r})|}{\rho_{\text{free}}} = \frac{3Q\lambda^3}{4\pi^2 V}|\boldsymbol{u}(r)|^2 = \frac{3Q\lambda^3}{4\pi^2 V_{\text{eff}}}, \tag{79.23}$$

where the effective mode volume is $V_{\text{eff}} = V/|u(\boldsymbol{r})|^2$. The lowest possible value $V_{\text{eff}} \simeq \lambda^3$ is obtained for ground modes of a closed resonator. For an atom located at the waist of an open Fabry–Perot cavity with length L, it is much larger. Special limiting cases for concentric and confocal cavities are $V_{\text{eff}}^{\text{conc}} = \lambda^2 L(R/D)$ and $V_{\text{eff}}^{\text{conf}} = \lambda L^2/2\pi$, respectively, where (R/D) gives the ratio of mirror radius to cavity diameter.

At resonance, the atomic decay rate Γ_μ grows with Q_μ, whereas the resonator damping time constant κ_μ is reduced. Eventually, the energy of the atomic radiation field is stored for such a long time that reabsorption becomes possible. Perturbative Weisskopf–Wigner theory is no longer valid in this good cavity limit, which is separated from the regime of bad cavities by the more formal condition

$$\Gamma_{eg}^{\text{cav}} > \kappa_\mu \,. \tag{79.24}$$

The strong coupling case is considered explicitly in Sect. 79.3.

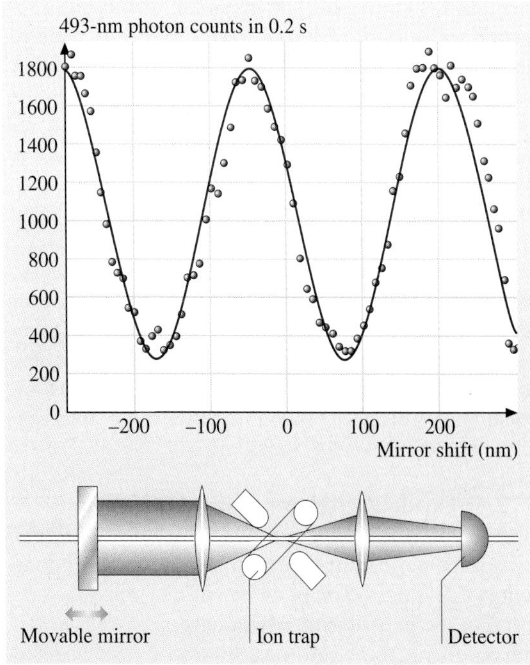

Fig. 79.3 Sinusoidal variation of the $\lambda = 493$ nm spontaneous emission rate of a single trapped Ba ion caused by self-interference from a retroreflecting mirror. The experimental arrangement is sketched at the bottom [79.13]

Antenna Patterns

Since the reflected radiation field of an atomic radiator is perfectly coherent with the source field, the combined radiation pattern modifies the usual dipole distribution of a radiating atom. The new radiation pattern can be understood in terms of antenna arrays [79.15]. For a single atomic dipole in front of a reflecting mirror for example, one finds a quadrupole type pattern due to the superposition of a second, coherent image antenna. In some of the earliest experimental investigations on radiating molecules in cavities, modifications of the radiation pattern were observed [79.16].

79.2.4 Radiative Shifts and Forces

When the radiation field of an atom is reflected back onto its source, an energy or radiative shift is caused by the corresponding self polarization energy. An atom in

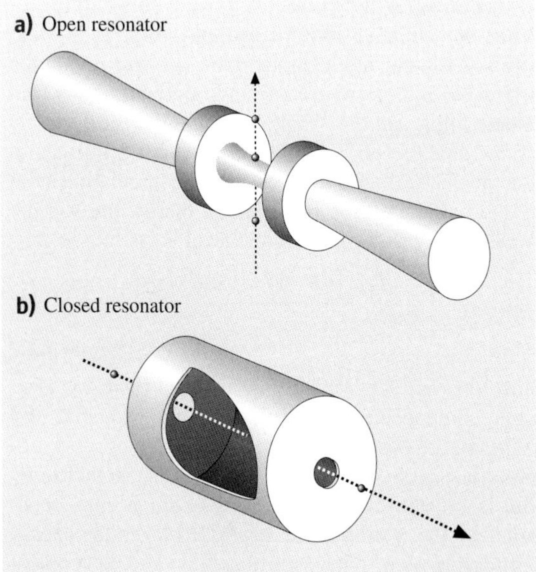

Fig. 79.4a,b Two frequently used resonator types for cavity QED: **(a)** Open Fabry–Perot optical cavity. **(b)** Closed "pillbox" microwave cavity

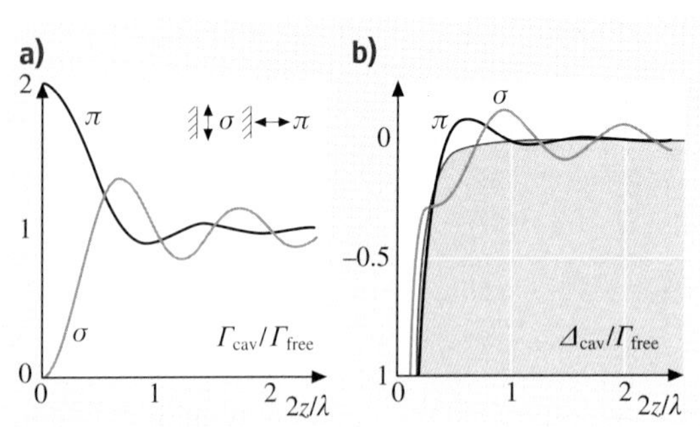

Fig. 79.5 (**a**) Normalized rate of modified spontaneous emission in the vicinity of a perfectly reflecting wall for σ and π orientation of the radiating dipole. (**b**) Corresponding energy shift of the resonance frequency. *Shaded area* indicates contribution of static van der Waals interaction

the vicinity of a plane mirror (Fig. 79.5) again makes a simple model system. Since the energy shift depends on the atom wall separation z, it is equivalent to a dipole force $\boldsymbol{F}_{dip}$ whose details depend on the role of retardation. Here we distinguish between the two cases where no radiation energy is exchanged between the atom and the field (van der Waals, Casimir forces) and where the atomic radiation causes forces by self-interference.

The Unretarded Limit: van der Waals Forces

When the radiative round trip time $t_r = 2z/c$ is short compared with the characteristic atomic revolution period $2\pi/\omega_{eg}$, retardation is not important. In this quasistatic limit, van der Waals energy shifts for decaying atomic dipoles vary as z^{-3} with the atom–wall separation. Such a shift is also present for a nonradiating atom in its ground state. In perturbation theory, the van der Waals energy shift of an atomic level $|a\rangle$ is

$$\Delta_{vdW} = -\frac{\langle a|q^2\left[(\boldsymbol{d}^2\cdot\widehat{\boldsymbol{x}}_t)^2 + 2(\boldsymbol{d}^2\cdot\widehat{\boldsymbol{z}})^2\right]|a\rangle}{64\pi\epsilon_0 z^3} . \tag{79.25}$$

Since the van der Waals force is anisotropic for electronic components parallel ($\widehat{\boldsymbol{z}}$) and perpendicular ($\widehat{\boldsymbol{x}}_t$) to the mirror normal, the degeneracy of magnetic sublevels in an atom is lifted near a surface. The total energy shift is ≈ 1 kHz for a ground state atom at 1 μm separation, and very difficult to detect. However, the energy shifts grow as n^4 since the transition dipole moment scales as n^2. With Rydberg atoms, van der Waals energy shifts have been successfully observed in spectroscopic experiments [79.17].

The Retarded Limit: Casimir Forces

At large separation, retardation becomes relevant, since the contributions of individual atomic oscilllation frequencies in (79.25) cancel by dephasing, thus reducing the Δ_{vdW}. A residual Casimir–Polder [79.18] shift may be interpreted as the polarization energy of a slowly fluctuating field with squared amplitude $\langle\mathcal{E}^2\rangle = 3\hbar c/64\epsilon_0 z^4$ originating from the vacuum field noise

$$\Delta_{CP} = -\frac{1}{4\pi\epsilon_0}\frac{3\hbar c\alpha_{st}}{8\pi z^4} , \tag{79.26}$$

where α_{st} is the static electric polarizability. The vacuum field noise Δ_{CP} replaces Δ_{vdW} at distances larger than characteristic wavelengths, and is even smaller. Only indirect observations have been possible to date, relying on a deflection of polarizable atoms by this force [79.19, 20]. The Casimir–Polder force can also be regarded as an ultimate, cavity induced consequence of the mechanical action of light on atoms [79.21]. It is an example of the conservative and dispersive dipole force which is even capable of binding a polarizable atom to a cavity [79.22].

Radiative Self-Interference Forces

Spontaneous emission of atoms in the vicinity of a reflecting wall also provides an example of cavity induced modification of the dissipative type of light forces, or radiation pressure. If the returning field is reabsorbed, the spontaneous emission rate is reduced and a recoil force directed away from the mirror is exerted. If the returning radiation field causes enhanced decay, a recoil towards the mirror occurs due to stimulated emission.

If the photon is detected at some angle with respect to the normal vector connecting the atom with the mirror surface, two paths for the photon are possible: It can reach a detector directly, or following a reflection off the wall. At small atom–mirror separation these paths are indistinguishable, the atom is thus left in a superposition of two recoil momentum states.

79.2.5 Experiments on Weak Coupling

Perhaps the most dramatic experiment in weak coupling cavity QED is the total suppression of spontaneous emission. For the experiments which have been carried out with Rydberg atoms and for a low-lying near infrared atomic transition [79.11, 12], it is essential to prepare atoms in a single decay channel. In addition, the atoms must be oriented in such a way that they are only coupled to a single decay mode (see the model waveguide in Fig. 79.4). This may be interpreted as an anisotropy of the electromagnetic vacuum, or as a specific antenna pattern.

An important problem in detecting the modification of radiative properties – changes in emission rates as well as radiative shifts – arises from their inhomogeneity due to the dependence on atom–wall separation. This difficulty has been overcome by controlling the atom–wall separation at microscopic distances through light forces [79.17], or by using well localized trapped ions [79.13, 14]. Furthermore, spectroscopic techniques that are only sensitive to a thin layer of surface atoms [79.23] have been used to clearly detect van der Waals shifts.

An atom emitting a radiation field in the vicinity of a reflecting wall will experience an additional dipole optical force caused by its radiation field. This force has been observed as a modification of the trapping force holding an ion at a fixed position with respect to the reflector [79.24].

Conceptually most attractive and experimentally most difficult to detect is the elusive Casimir interaction. Only for atomic ground states is this effect observable, free from other much larger shifts. The influence of the corresponding Casimir force on atomic motion has been observed in a variant of a scattering experiment, confirming the existence of this force in neutral atoms [79.19, 20].

The success of this experiment shows that spectroscopic techniques involving the exchange of photons are not suitable for the Casimir problem. A notable exception could be Raman spectroscopy of the magnetic substructure in the vicinity of a surface. In general, scattering or atomic interferometry experiments are more promising methods. The experiment by *Brune* et al. [79.25] may be interpreted in this way.

79.2.6 Cavity QED and Dielectrics

There are two variants of dielectric materials employed to study light-matter interaction in confined space: Conventional materials such as glass or sapphire, and artificial materials called photonic materials or metamaterials.

While dielectric materials are theoretically more difficult to treat than perfect mirrors, since the radiation at least partially enters the medium, they have a similar influence on radiative decay processes. One new aspect is, however, the coupling of atomic excitations to excitations of the medium, which was observed for the case of a surface-polariton in [79.26].

Cavities with dimensions comparable to the wavelength promise the most dramatic modification of radiative atomic properties, but micrometer sized cavities for optical frequencies with highly reflecting walls are difficult to manufacture. So-called whispering gallery modes of spherical microcavities [79.27] have been intensely studied, but no simple way of coupling atoms to these resonator modes has been found yet.

On the other hand, dielectric materials with a periodic modulation of the index of refraction may exhibit photonic bandgaps in analogy with electronic bandgaps in periodic crystals [79.28, 29]. Electronic phenomena of solid state physics can then be transferred to photons. For example, excited states of a crystal dopant or a quantum dot cannot radiate into a photonic bandgap, the radiation field cannot propagate, and the excitation energy remains localized. The bandgap behaves like an empty resonator, and if a resonator structure is integrated into the device, the regime of strong coupling [79.30, 31] can be achieved with such photonic structures. An overview of suitable systems can be found in [79.32].

79.3 Strong Coupling in Cavity QED

Strong coupling of atoms and fields is realized in a good cavity when $\Gamma_\mu < \Gamma_{eg}$ (79.24). The Hilbert space of the combined system is then the product space of a single two-level atom and the countable set of Fock-states of the field,

$$\mathcal{H} = \mathcal{H}_{\text{atom}} \otimes \mathcal{H}_{\text{field}} \,, \tag{79.27}$$

which is spanned by the states

$$|n; a\rangle = |n\rangle|a\rangle \,. \tag{79.28}$$

The interaction of a single cavity mode with an isolated atomic resonance is now characterized by the Rabi nutation frequency, which gives the exchange frequency of the energy between atom and field. For an amplitude $\mathcal{E}$ corresponding to n photons,

$$\Omega(n) = g_\mu \sqrt{n+1} \,. \tag{79.29}$$

This is the simplest possible situation of a strongly coupled atom–field system. The new energy eigenvectors are conveniently expressed in the dressed atom model [79.33]:

$$
\begin{aligned}
|+,n\rangle &= \cos\theta|g,n+1\rangle + \sin\theta|e,n\rangle\,,\\
|-,n\rangle &= -\sin\theta|g,n+1\rangle + \cos\theta|e,n\rangle\,,
\end{aligned} \tag{79.30}
$$

with $\tan 2\theta = 2g_\mu\sqrt{n+1}/(\omega_0-\omega_\mu)$. The separate energy structures of free atom and empty resonator are now replaced by the combined system of Fig. 79.6. At resonance, the new eigenstates are separated by $2\hbar\Omega_R$, where $\Omega_R = g_\mu$ is the vacuum Rabi frequency.

79.4 Strong Coupling in Experiments

In order to achieve strong coupling experimentally, it is necessary to use a high-Q resonator in combination with a small effective mode volume. This condition was first realized for ground modes of a closed microwave cavity [79.8], and later also for open cavity optical resonators (Fig. 79.6) [79.34]. It is interesting to control the interaction time of the atoms with the cavity field. In earlier experiments, this was typically achieved by selecting the passage time for an atom transiting the cavity. The advancement of atom trapping methods has also led to the observation of a truly one-atom laser at optical frequencies [79.35].

More recently, this situation has also been realized for artificial atoms including superconducting systems [79.36, 37] and quantum dots [79.30, 31].

79.4.1 Rydberg Atoms and Microwave Cavities

At microwave frequencies, very low loss superconducting niobium cavities are available with $Q \approx 10^{10}$. Resonator frequencies are typically several tens of GHz and can be matched by atomic dipole transitions between two highly excited Rydberg states. By selective field ionization, the excitation level of Rydberg atoms can be detected, and hence it is possible to measure whether a transition between the levels involved has occurred. The efficiency of this method approaches unity, so that experiments can be performed at the single atom level. The interaction or transit time T is usually much shorter than the lifetime τ_{Ry} of the Rydberg states involved. For this reason, circular Rydberg states with quantum numbers $l = m = n-1$ are particularly suitable.

Rydberg atoms [79.38] are prepared in an atomic beam, selectively excited to an upper level, and then sent through a microwave cavity where the upper and lower levels are coupled by the electromagnetic field. If the atom is detected in the lower of the coupled levels as it leaves the resonator, the excitation energy has been stored in the resonator field. Thus the evolution of the resonator field is recorded as a function of the atomic interaction.

A microwave cavity in interaction with a single or a few Rydberg atoms is called a micromaser (formerly a one atom maser) [79.8]. The experimental conditions may be summarized as

$$
g_\mu > 1/T > 1/\tau_{Ry} > \kappa_\mu\,. \tag{79.31}
$$

79.4.2 Strong Coupling in Open Optical Cavities

At optical wavelengths, a cavity with small V_{eff} in (79.23) is clearly more difficult to construct than at centimeter wavelengths. However, dielectric coatings are now available which allow very low damping rates ω_μ/Q_μ for optical cavities. Very high finesse $\mathcal{F} \simeq 10^7$ (which is a more convenient measure for the damping rate of an optical Fabry–Perot interferometer) has been achieved. By reducing the volume of such a high-Q cavity mode, strong coupling of

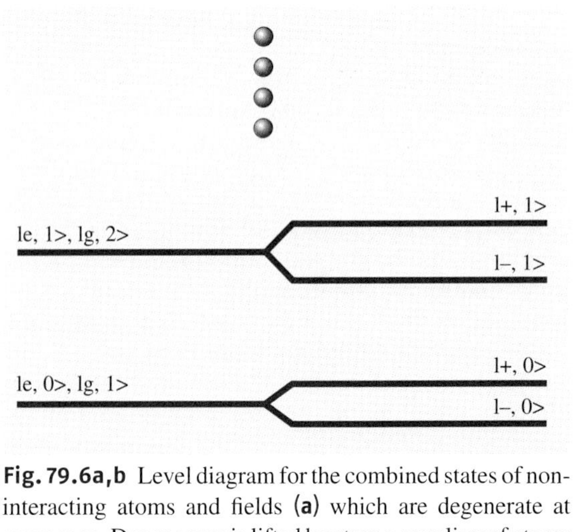

Fig. 79.6a,b Level diagram for the combined states of non-interacting atoms and fields **(a)** which are degenerate at resonance. Degeneracy is lifted by strong coupling of atoms and fields **(b)** yielding new "dressed" eigenstates

atoms and fields at optical frequencies has been demonstrated [79.34].

In open structures, the atoms can still decay into the continuum states with a rate γ. Therefore the condition for strong coupling in such systems is usually given as

$$\frac{g_\mu^2}{\kappa_\mu \gamma} > 1 . \tag{79.32}$$

79.5 Microscopic Masers and Lasers

In a microscopic laser, simple atoms are strongly coupled to a single mode of a resonant or near resonant radiation field. Collecting atomic and field operators from (79.2), (79.10), and (79.14), this situation is described by the Jaynes–Cummings model Hamiltonian [79.39, 40]

$$\begin{aligned}\mathcal{H}_{\text{JC}} &= \mathcal{H}_{\text{atom}} + \mathcal{H}_{\text{field}} + \mathcal{H}_{\text{RWA}} \\ &= \frac{1}{2}\hbar\omega_0\sigma_z + \hbar\omega_\mu\left(a_\mu^\dagger a_\mu + \frac{1}{2}\right) \\ &\quad + \hbar g_\mu(\sigma^\dagger a_\mu + a_\mu^\dagger \sigma) .\end{aligned} \tag{79.33}$$

79.5.1 The Jaynes–Cummings Model

The Jaynes–Cummings model (79.33) represents the most basic and, at the same time, the most informative model of strong coupling in quantum optics. It consists of a single two-level atom interacting with a single mode of the quantized cavity field. The time evolution of the system is determined by

$$\mathrm{i}\hbar\frac{\partial\psi}{\partial t} = H\psi . \tag{79.34}$$

This model can be solved exactly due to the existence of the additional constant of motion

$$N = a^\dagger a + \sigma_z + 1 , \tag{79.35}$$

i. e., conservation of the "number of excitations". Its eigenvalues are the integers N which are twofold degenerate except for $N = 0$. The simultaneous eigenstates of H and N are the pairs of dressed states defined in (79.30) which are not degenerate with respect to the energy H. The initial state problem corresponding to (79.34) is solved by elementary methods in terms of the expansion

$$|\Psi(t)\rangle = \sum_{n=0}^{\infty}\sum_{j=1}^{2} C_n^j(t)|n, j\rangle , \tag{79.36}$$

where the expansion coefficients are

$$\begin{aligned}C_n^1(t) = &\Bigg(C_n^1(0)\left\{\cos[\Omega(n)t] - \mathrm{i}\frac{\delta}{2\Omega(n)}\sin[\Omega(n)t]\right\} \\ &- \mathrm{i}\frac{\sqrt{n}g_\mu}{\Omega(n)}C_{n-1}^2(0)\sin[\Omega(n)t]\Bigg) \\ &\times \exp\left[-\mathrm{i}\omega_\mu\left(n - \frac{1}{2}\right)t\right]\end{aligned} \tag{79.37}$$

and

$$\begin{aligned}C_n^2(t) = &\Bigg(C_n^2(0)\Bigg\{\cos[\Omega(n+1)t] \\ &+ \mathrm{i}\frac{\delta}{2\Omega(n+1)}\sin[\Omega(n+1)t]\Bigg\} \\ &- \mathrm{i}\frac{g_\mu\sqrt{n+1}}{\Omega(n+1)}C_{n+1}^1(0)\sin[\Omega(n+1)t]\Bigg) \\ &\times \exp\left[-\mathrm{i}\omega_\mu\left(n + \frac{1}{2}\right)t\right] ,\end{aligned} \tag{79.38}$$

with $\delta = \omega_\mu - \omega_0$ the detuning between the atom and cavity and $\Omega(n) = \frac{1}{2}(\delta^2 + 4g_\mu^2 n)^{1/2}$ is the generalized Rabi frequency. The coefficients $C_n^j(0)$ are determined by the initial preparation of atom and cavity mode. The result simplifies considerably for $\delta = 0$ to

$$\begin{aligned}|\Psi(t)\rangle = \sum_{m=0}^{\infty} &\Big\{C_m^1(0)\,\mathrm{e}^{-\mathrm{i}\omega_\mu(m-1/2)t} \\ &\times\big[\cos\left(g_\mu\sqrt{m}t\right)|m;\,1\rangle \\ &- \mathrm{i}\sin\left(g_\mu\sqrt{m}\,t\right)|m-1;\,2\rangle\big] \\ &+ C_m^2(0)\,\mathrm{e}^{-\mathrm{i}\omega_m u(m+1/2)t} \\ &\times\big[\cos\left(g_\mu\sqrt{m+1}\,t\right)|m;\,2\rangle \\ &- i\sin\left(g_\mu\sqrt{m+1}t\right)|m+1;\,1\rangle\big]\Big\} .\end{aligned} \tag{79.39}$$

The coefficients $C_n^j(0)$ represent any initial state of the system, from uncorrelated product states to entangled states of atom and field. There exist numerous generalizations of this model which include more atomic levels and several coherent fields.

79.5.2 Fock States, Coherent States and Thermal States

We now illustrate the properties of the Jaynes–Cummings model by specifying the initial state. Assume that the atom and field are brought into contact at time $t=0$ and that all correlations that might exist due to previous interactions are suppressed.

Rabi Oscillations

If the atom is initially in the excited state and the field contains precisely m quanta, then

$$C_n^j(t=0)=\delta_{n,m}\delta_{j,2}\,. \tag{79.40}$$

The solution of (79.34) assumes the form

$$\begin{aligned}|\Psi(t)\rangle &= \mathrm{e}^{-\mathrm{i}\omega_\mu(m+1/2)t}\big[\cos\left(g_\mu\sqrt{m+1}\,t\right)|m;2\rangle\\ &\quad -i\sin\left(g_\mu\sqrt{m+1}\,t\right)|m+1;1\rangle\big]\,.\end{aligned} \tag{79.41}$$

The occupation probabilities of the atomic states evolve in time according to

$$n_2(t)=\langle\Psi(t)|2\rangle\langle 2|\Psi(t)\rangle=\cos^2\left(g_\mu\sqrt{m+1}\,t\right), \tag{79.42}$$

$$n_1(t)=\langle\Psi(t)|1\rangle\langle 1|\Psi(t)\rangle=\sin^2\left(g_\mu\sqrt{m+1}\,t\right). \tag{79.43}$$

The photon number and its variance are

$$\langle n(t)\rangle=\left\langle\Psi(t)a^\dagger a\Psi(t)\right\rangle=m+\sin^2\left(g_\mu\sqrt{m+1}\,t\right), \tag{79.44}$$

$$\begin{aligned}\langle\Delta^2 n\rangle &= \left\langle\Psi(t)\left(a^\dagger a-\left\langle a^\dagger a\right\rangle\right)^2\Psi(t)\right\rangle\\ &=\frac{\sin^2\left(2g_\mu\sqrt{m+1}\,t\right)}{4}\,.\end{aligned} \tag{79.45}$$

In the limit of large m, $g_\mu\sqrt{m+1}$ is proportional to the field amplitude and the classical Rabi oscillations in a resonant field are recovered. The nonclassical features of the states are characterized by Mandel's parameter

$$Q_\mathrm{M}=\frac{\left\langle\Delta^2 n\right\rangle-\langle n\rangle}{\langle n\rangle}\geq -1\,. \tag{79.46}$$

For the present example,

$$Q_\mathrm{M}=-1+\frac{1}{4}\frac{\sin^2\left(2g_\mu\sqrt{m+1}\,t\right)}{m+\sin^2\left(g_\mu\sqrt{m+1}\,t\right)}\,. \tag{79.47}$$

$Q_\mathrm{M}\geq 0$ indicates the classical regime, while $Q\leq 0$ can only be reached by a quantum process.

The Coherent State

Consider the case where the field is initially prepared in a coherent state

$$|\alpha\rangle=\exp\left(\alpha a^\dagger-\alpha^* a\right)|0\rangle=\mathrm{e}^{-|\alpha|^2/2}\sum_{n=0}^{\infty}\frac{\alpha^n}{\sqrt{n!}}|n\rangle\,, \tag{79.48}$$

while the atom starts from the excited state

$$C_n^j(0)=\mathrm{e}^{-|\alpha|^2/2}\frac{|\alpha|^n}{\sqrt{n!}}\delta_{j,2}\,. \tag{79.49}$$

In this case, the general solution specializes to

$$\begin{aligned}|\Psi(t)\rangle &= \sum_{n=0}^{\infty}\frac{\alpha^n}{\sqrt{n!}}\mathrm{e}^{-\mathrm{i}\omega(n+1/2)t}\,\mathrm{e}^{-|\alpha|^2/2}\\ &\quad\times\left[\cos\left(g_\mu\sqrt{n+1}\,t\right)|n;2\rangle\right.\\ &\quad\left.-i\sin\left(g_\mu\sqrt{n+1}\,t\right)|n+1;1\rangle\right],\end{aligned} \tag{79.50}$$

and the occupation probability of the excited state is

$$n_2(t)=\frac{1}{2}\left[1+\sum_{n=0}^{\infty}\frac{|\alpha|^{2n}}{n!}\mathrm{e}^{-|\alpha|^2}\cos\left(2g_\mu\sqrt{n+1}\,t\right)\right]. \tag{79.51}$$

From here, detailed quantitative results can only be obtained by numerical methods [79.41]. However, if the coherent state contains a large number of photons $|\alpha|^2\gg 1$, the essential dynamics can be determined by elementary methods. Initially, the population oscillates with the Rabi frequency $\Omega_1\approx g_\mu|\alpha|$, which is proportional to the average amplitude of the field, as expected from its classical counterpart. With increasing time, the coherent oscillations tend to cancel due to the destructive interference of the different Rabi frequencies in the sum:

$$n_2(t)=\frac{1}{2}\left[1+\cos(2g_\mu\,|\alpha|\,t)\,\mathrm{e}^{-(gt)^2/2}\right]. \tag{79.52}$$

However, strictly aperiodic relaxation of $n_2(t)$ is impossible since the exact expressions, (79.36) and (79.37), represent a quasiperiodic function which, given enough time, approaches its initial value with arbitrary accuracy.

For short times, the oscillating terms in the sum cancel each other due to the slow evolution of their frequency with n. However, consecutive terms interfere constructively for larger times t_r, such that the phases satisfy

$$\phi_{n+1}(t_\mathrm{r})-\phi_n(t_\mathrm{r})=2\pi\,. \tag{79.53}$$

For $|\alpha|^2\gg 1$, the increment of the arguments is

$$\phi_{n+1}-\phi_n=g_\mu t_\mathrm{r}/|\alpha|\,, \tag{79.54}$$

and therefore the first revival of the Rabi oscillations occurs approximately at $t_r = \pi|\alpha|/g_\mu$. A clear distinction of Rabi oscillation, collapses, and revivals requires a clear separation of the three time scales

$$t_1 \ll t_2 \ll t_3 \,, \tag{79.55}$$

where $t_1 \approx (g_\mu|\alpha|)^{-1}$ for Rabi oscillation, $t_2 \approx g_\mu^{-1}$ for collapse, and $t_3 \approx |\alpha|/g_\mu$ for revival.

The typical features of the transient evolution starting from a coherent state are shown in Fig. 79.7. With time increasing even further, revivals of higher order occur which spread in time, and finally can no longer be separated order by order.

The Thermal State

Consider a microwave resonator brought into thermal contact with a reservoir, inducing loss on a time scale κ^{-1} and thermal excitation. The dissipative time evolution is described by the master equation

$$\begin{aligned}\dot{\rho} &= (L_0 + L)\rho \\ &\equiv \mathrm{i}[H,\rho]/\hbar + \kappa(n_{\text{th}}+1)\left\{\left[a,\rho a^\dagger\right] + \left[a\rho, a^\dagger\right]\right\} \\ &\quad + \kappa n_{\text{th}}\left\{\left[a^\dagger,\rho a\right] + \left[a^\dagger\rho, a\right]\right\} \,,\end{aligned} \tag{79.56}$$

where $n_{\text{th}} = [\exp(\beta\hbar\omega) - 1]^{-1}$, at $T = k_B\beta^{-1}$, is the equilibrium population of the cavity mode, L_0 symbolizes the unitary evolution according to the Jaynes–Cummings dynamics and L is a dissipation term.

The solution of this model can be expressed in terms of an eigenoperator expansion of the equation

$$L\rho = -\lambda\rho \,. \tag{79.57}$$

The eigenvalues λ that determine the relaxation rates, as well as the eigenoperators, are known in closed form for

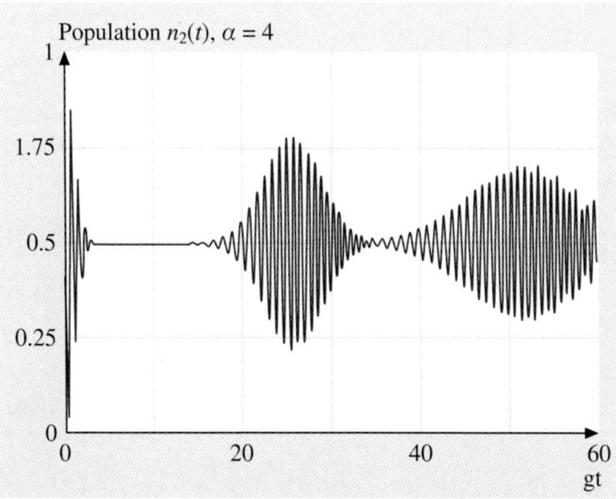

Fig. 79.7 Rabi oscillations, dephasing, and quantum revival

the case of vanishing temperature [79.42]. Since energy is exchanged between the nondecaying atom and the decaying cavity mode, cavity damping is modified in a characteristic way due to the presence of the atom. The technical details can be found in [79.43].

79.5.3 Vacuum Splitting

In the classical case, the eigenvalues of the interaction free Hamiltonian are degenerate at resonance. The atom–field interaction splits the eigenvalues and determines the Rabi frequency of oscillation between the two states. One consequence is the existence of side bands in the resonance fluorescence spectrum [79.44]. In the quantum case, the field itself is treated as a quantized dynamical variable determined from a self-consistent solution for the complete system of atom plus field. The vacuum Rabi frequency $\Omega_{\text{vac}} = g_\mu$ remains finite, and accounts for the spontaneous emission of radiation from an excited atom placed in a vacuum. In the limiting case of a single atom interacting with the quantized field, the photon number n can only change by ± 1, and the population oscillates with the frequency $\Omega(n)$ given by (79.29). For an ensemble of N atoms, n can in principle change by up to $\pm N$. However, if the field and atoms are only weakly excited, the collective frequency of the ensemble is determined by the linearized Maxwell–Bloch equations. The eigenfrequencies are given by

$$\lambda^{\pm} = \frac{1}{2}\left[\mathrm{i}(\gamma_\perp + \kappa) \pm \sqrt{4g_\mu^2 N - (\gamma_\perp - \kappa)^2}\,\right] , \tag{79.58}$$

where $\gamma_\perp^{-1}$ is the phase relaxation time of the atom and κ^{-1} the decay time of the resonator. This is the polariton dispersion relation in the neighborhood of the polariton gap. The spectral transmission

$$T(\omega) = T_0 \left| \frac{\kappa[\gamma + \mathrm{i}(\omega_0 - \omega)]}{(\omega - \lambda^+)(\omega - \lambda^-)} \right|^2 \tag{79.59}$$

of an optical cavity containing a resonant atomic ensemble of N atoms reveals the internal dynamics of the coupled system and a splitting of the resonance line occurs. T_0 is the peak transmission of the empty cavity. The splitting increases either with the number of photons, approaching $\sqrt{n+1}$ in the presence of a single atom, or with the number of atoms, approaching $\sqrt{N}$ in the resonator when the field is weak. The latter case is demonstrated in Fig. 79.8 [79.34] for an optical resonator with 1–10 atoms interacting with a field that contains, on average, much less than a single photon.

79.6 Micromasers

Sustained oscillations of a cavity mode in a microwave resonator can be achieved by a weak beam of Rydberg atoms excited to the upper level of a resonant transition. For a cavity with a $Q \approx 10^{10}$, much less than a single atom at a time, on average, suffices to balance the cavity losses. Operation of a single atom maser has been demonstrated [79.8]. The atoms enter the cavity at random times, according to the Poisson statistics of a thermal beam, and interact with the field only for a limited time. In order to restrict the fluctuations of the atomic transit time, the velocity spread is reduced. This is achieved either by Fizeau chopping techniques, or by making use of Doppler velocity selection in the initial laser excitation process. Since most of the time no atom is present, it is natural to separate the dynamics into two parts [79.45]:

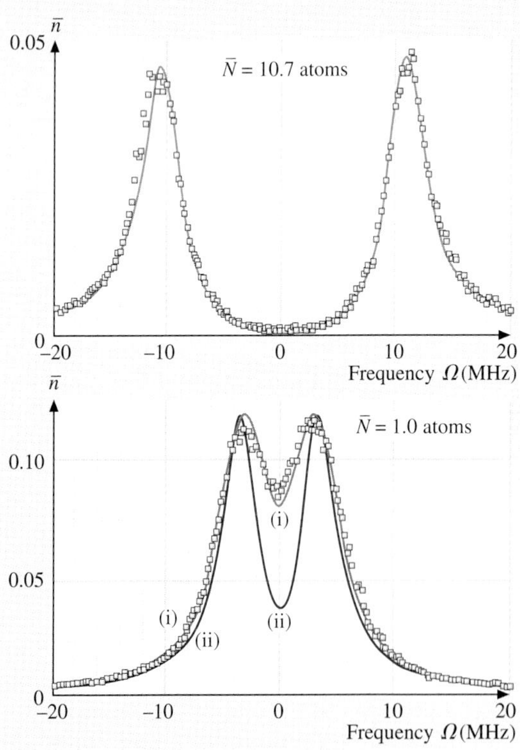

Fig. 79.8 Intracavity photon number (measured from a transmission experiment, [79.34]) as a function of probe frequency detuning, and for two values of N, the average number of atoms in the mode. *Thin lines* give theoretical fits to the data, including atomic number and position fluctuations. *Curve (ii)* in the *lower graph* is for a single intracavity atom with optimal coupling g_μ

1. For the short time while an atom is present, the state evolves according to the Jaynes–Cummings dynamics, where H is defined in (79.33),
$$\dot{\rho}(t) = \mathrm{i}[H, \rho]/\hbar\,, \tag{79.60}$$
and damping can safely be neglected. The formal solution is abbreviated by $\rho(t) = F(t - t_0)\rho(t_0)$.
2. During the time interval between successive atoms, the cavity field relaxes freely toward the thermal equilibrium according to (79.56) with $L_0 = 0$:
$$\dot{\rho}(t) = L\rho\,, \tag{79.61}$$
with the formal solution $\rho(t) = \exp[L(t - t_0)]\rho(t_0)$.

The time development of the micromaser therefore consists of an alternating sequence of unitary $F(t)$ and dissipative $\mathrm{e}^{(Lt)}$ evolutions. Atoms enter the cavity one by one at random times t_i. Until the next atom enters at time t_{i+1}, the evolution t_i is given by

$$\rho(t_{i+1}) = \exp(Lt_p)F(\tau)\rho(t_i)\,, \tag{79.62}$$

where $t_p = t_{i+1} - t_i - \tau$, and τ is the transit time. If $\tau \ll t_{i+1} - t_i$ on average, then $t_p \approx t_{i+1} - t_i$. After averaging (79.62) over the Poisson distribution $P(t) = R\exp(-Rt_p)$ for t_p, where R is the injection rate, the mean propagator from atom to atom is

$$\langle\rho(t_{i+1})\rangle = \frac{R}{R - L}F(\tau)\langle\rho(t_i)\rangle\,. \tag{79.63}$$

After excitation, the reduced density matrix of the field alone becomes diagonal after several relaxation times κ^{-1}:

$$\langle n|\mathrm{Tr}_{\mathrm{atom}}(\rho)|m\rangle = P_n\delta_{n,m}\,. \tag{79.64}$$

Due to the continuous injection of atoms, the field never becomes time independent, but may relax toward a stroboscopic state defined by

$$\langle\rho(t_{i+1})\rangle = \langle\rho(t_i)\rangle\,. \tag{79.65}$$

The state of the cavity field can be determined in closed form by iteration:

$$P_n = N\prod_{k=1}^{n}\frac{n_{\mathrm{th}}\kappa + A_k}{(n_{\mathrm{th}} + 1)\kappa}\,, \tag{79.66}$$

where N guarantees normalization of the trace and $A_k = (R/n)\sin^2(g_\mu\tau\sqrt{n})$, and exact resonance between cavity mode and atom is assumed. Since all off-diagonal elements vanish in steady state, (79.66) provides a complete description for the photon statistics of the field.

79.6.1 Maser Threshold

The steady state distribution determines the mean photon number of the resonator as a function of the operating conditions:

$$\langle n\rangle = \sum_{n=0}^{\infty} n P_n \ . \tag{79.67}$$

A suitable dimensionless control parameter is

$$\Theta = \frac{1}{2} g_\mu \tau \sqrt{R/\kappa} \ . \tag{79.68}$$

For $\Theta \ll 1$, the energy input is insufficient to counterbalance the loss of the cavity, effectively resulting in a negligible photon number. With increasing pump rate R, a threshold is reached at $\Theta \simeq 1$, where $\langle n\rangle$ increases rapidly with R. In contrast to the behavior of the usual laser, the single atom maser displays multiple thresholds with a sequence of minima and maxima of $\langle n\rangle$ as a function of Θ [79.46]. This can be related to the rotation of the atomic Bloch vector. When the atom undergoes a rotation of about π during the transit time τ, a maximum of energy is transferred to the cavity and $\langle n\rangle$ is maximized. The converse applies if the average rotation is a multiple of 2π. This behavior is shown in Fig. 79.9. The minima in $\langle n\rangle$ are at $\Theta \simeq 2n\pi$.

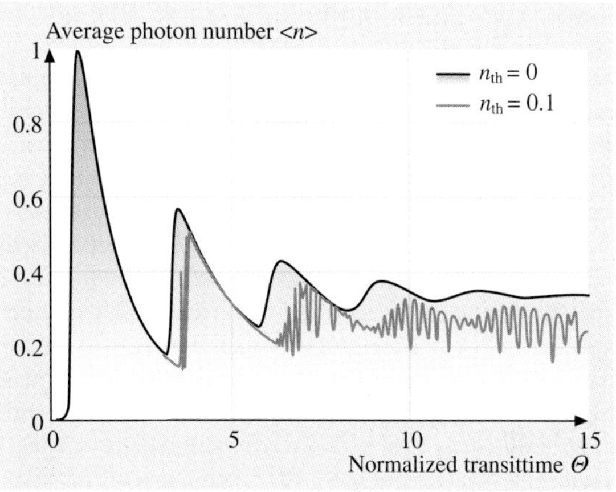

Fig. 79.9 Average photon number as a function of the normalized transit time defined by (79.68)

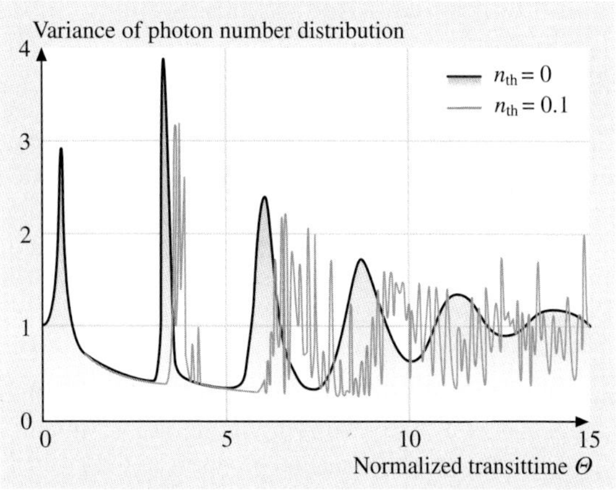

Fig. 79.10 Variance normalized on the average photon number $\sigma^{2<n>}/\langle\sigma\rangle$. Values below unity indicate regions of nonclassical behavior

79.6.2 Nonclassical Features of the Field

Fluctuations can be of classical or of quantum origin. The variance of the photon number

$$\sigma^2 = \left(\langle n^2\rangle - \langle n\rangle^2\right) \tag{79.69}$$

is a measure of the randomness of the field intensity. Classical Poisson statistics require that $\sigma^2 \geq \langle n\rangle$. A value below unity indicates quantum behavior, which has no classical analog. In Fig. 79.10, the variance is plotted as a function of Θ. Regions of enhanced fluctuations $\sigma^2 > \langle n\rangle$ alternate with regions with sub-Poissonian character $\sigma^2 < \langle n\rangle$ [79.47]. When $\langle n\rangle$ approaches a local maximum it is accompanied by large fluctuations, while at points of minimum field strength the fluctuations are reduced below the classical limit. This feature is repeated with a period of $\Theta \simeq 2\pi$, but finally washes out at large values of Θ.

The large variance of n is caused by a splitting of the photon distribution P_n into two peaks, which gives rise to bistability in the transient response [79.48]. The sub-Poissonian behavior of the field is reflected in an increased regularity of the atoms leaving the cavity in the ground state.

79.6.3 Trapping States

If cavity losses are neglected, operating conditions exist which lead directly to nonclassical, i. e., Fock states. If the cavity contains precisely n_q photons, an atom that enters the resonator in the excited state leaves it again in the same state provided the condition [79.49]

$$g_\mu \tau \sqrt{n_q + 1} = 2q\pi \tag{79.70}$$

is satisfied, i. e., the Bloch vector of the atom undergoes q complete rotations. If the maser happens to reach such a trapping state $|n_q\rangle$, the photon number n_q can no longer increase irrespective of the flux of pump atoms. With the inclusion of cavity damping at zero temperature, n_q still represents an upper barrier that cannot be overcome, since damping only causes downward transitions. Even in the presence of dissipation, generalized trapping states exist with a photon distribution that vanishes for $n > n_q$ and has a tail towards smaller photon numbers $n \leq n_q$. However, thermal fluctuations at finite temperatures destabilize the trapping states since they can momentarily increase the photon number and allow the distribution to jump over the barrier $n = n_q$. Nevertheless, even for $n_{\text{th}} < 10^{-7}$, remnants of the trapping behavior persist, and can be seen in the transient response of the micromaser (Sect. 79.6.4).

79.6.4 Atom Counting Statistics

Direct measurements of the field in a single atom maser resonator are not possible because detector absorption would drastically degrade its quality. However, the field can be deduced from the statistical signature of the atoms leaving the resonator.

The probability $P(n)$ of finding n atoms in a beam during an observation interval t is given by the classical Poisson distribution

$$P(n) = (Rt)^n \, \mathrm{e}^{-Rt}/n! \,. \tag{79.71}$$

Information on the field inside is then revealed by the conditional probability $W(n, |g\rangle, m, |e\rangle, T)$ of finding n atoms in the ground state and m atoms in the excited state during a time t. Since there are only two states, it is sufficient to determine the probability

$$W(n, |g\rangle, t) = \sum_{m=0}^{\infty} W(n, |g\rangle, m, |e\rangle, t) \tag{79.72}$$

for being in the ground state [79.50]. For $n = 0$, the probability of observing no atom in the ground state during the period t is

$$W(0, |g\rangle, t) = \quad \mathrm{Tr}(\rho_{\text{stst}}) \exp\{L + R[O_{|g\rangle} + (1-\eta) O_{|e\rangle} - 1]t\} \,, \tag{79.73}$$

where $O_{|j\rangle} = \langle j|F(\tau)|j\rangle$ (79.60) and ρ_{stst} is the steady state of the maser field. This probability is closely related to the waiting time statistic $P_2(0, |g\rangle, t)$ between two successive ground state atoms, a property which is easily determined in a start-stop experiment. For an atom detector with finite quantum efficiency η for state selective detection, the waiting time probability is

$$\begin{aligned} &P_2(0, |g\rangle, t) \\ &= \{\mathrm{Tr}(\rho_{\text{stst}}) O_{|g\rangle} \\ &\quad \times \exp[L + R[O_{|g\rangle} + (1-\eta) O_{|e\rangle} - 1]T] O_{|g\rangle}\} \\ &\quad / [\mathrm{Tr}(\rho_{\text{stst}}) O_{|g\rangle}]^2 \end{aligned} \tag{79.74}$$

How a specific field state is reflected in the atom counting statistics will be illustrated for two situations: the region of sub-Poisson statistics and the region where the trapping condition is satisfied. Increased regularity of the cavity field $Q_{\text{M}} \leq 0$ manifests itself in increased regularity of ground state atoms in the beam. The statistical behavior exhibits "anti-bunching", i. e., $P_2(0, |g\rangle, t)$ has a maximum at finite t, indicating "repulsion" between successive atoms in comparison with a Poissonian beam. If the transit time τ is chosen in such a way that $g\tau \simeq 2\pi$, the chance of observing an initially excited atom in the ground state is negligible. At some point, however, an unlikely thermal fluctuation occurs, adding a photon. The rotation angle of the Bloch vector suddenly increases to $2\pi\sqrt{2} \simeq 3\pi$, and the atoms tend to leave the cavity in the ground state. After a typical cavity lifetime, the field decays and the trapping condition is restored again. Under this operation condition, the statistics of ground state atoms is governed by two time constants:

1. a short interval, in which successive atoms leave the cavity in the ground state after a thermal fluctuation;
2. a long time interval, in which the trapping condition is maintained and all atoms leave the resonator in their excited state until the next fluctuation occurs.

The probability $P_2(0, |g\rangle, t)$ is plotted in Fig. 79.11. The plot clearly shows the two time regimes that govern the imperfect trapping situation.

79.7 Quantum Theory of Measurement

When the object of interest consists of only a few atoms and a few photons, the puzzling consequences of quantum mechanical measurement become visible. In the case of the micromaser, the information on the state of the field is imprinted in a subtle way on the atomic beam. While photon counting is normally a destructive operation, the dispersive part of the photon-atom interaction may be used to determine the photon number inside

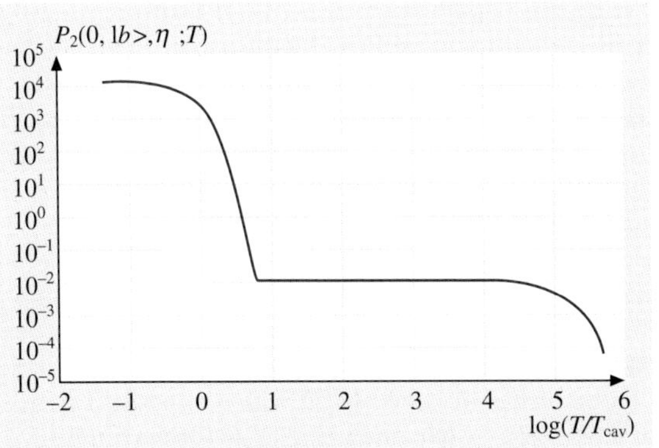

Fig. 79.11 Waiting time probability for atoms in the ground state while cavity is operated at vacuum trapping-state condition

a resonator without altering it, on average. Dispersive effects shift the phase of an oscillating atomic dipole without changing its state.

The phase shift due to the field in the resonator can be measured in a Ramsey-type experiment [79.51]. Consider an atom with two transitions $|g\rangle \rightarrow |e\rangle$ and $|e\rangle \rightarrow |i\rangle$. The first is far from resonance with the cavity and the second is close to resonance, but with a detuning $\delta_{ie} = \omega - \omega_{ie}$ large enough so as not to change the cavity photon number as the atom passes through. The dynamic Stark effect of the $|g\rangle \rightarrow |e\rangle$ transition frequency due to state $|i\rangle$ is then

$$\Delta\omega_{eg} = \left[g_{ie}\sqrt{n+1}\right]^2 / \delta_{ie} . \qquad (79.75)$$

If the resonator is now placed between the two Ramsey cavities, which are tuned to $\omega_R \approx \omega_{eg}$, such that the polarization of the $|e\rangle \rightarrow |g\rangle$ transition is rotated by $\approx \pi/2$, then the additional phase shift $\Delta\omega_{eg}\tau$, where τ is the transit time through the optical resonator, can be measured, and hence the photon number n. Since Rydberg states have a large coupling constant g_μ, the phase shift due to a single atom is detectable [79.51].

A complete measurement of n requires a sequence of N atoms because a single Ramsey measurement only determines whether the atom is in state $|e\rangle$ or $|g\rangle$, and hence $\Delta\omega_{eg}\tau$ to within $\pm\pi/2$. Since each measurement provides one binary bit of information, a sequence of N measurements can in principle distinguish 2^N possible Fock states for the photon field. However, with a monoenergetic beam, integral multiples of 2π remain undetermined. A distribution of velocities, and hence transit times, is therefore desirable. An entropy reduction strategy for selecting an optimal velocity distribution, based on the outcome of previous measurements, is described in [79.52].

As a consequence of the uncertainty principle, a measurement of the photon number destroys all information about the phase of the field. In the present case, the noise in the conjugate variable (the phase) is prevented from coupling back on the measured one, and hence the measurement is called a quantum nondemolition experiment. Many other aspects of phase diffusion, entangled states, and quantum measurements in the micromaser are discussed in [79.53].

79.8 Applications of Cavity QED

79.8.1 Detecting and Trapping Atoms through Strong Coupling

From Fig. 79.8 it is obvious that an atom travelling through the cavity will modify the transmission properties of this cavity. Strong coupling thus enables the experimenter to detect the presence of a single atom dispersively by monitoring cavity transmission or reflection. Laser cooled atoms have low velocities and spend sufficient time in the cavity even in free flight to generate the transmission signal shown in Fig. 79.12. The signals correspond to individual atom transits, and the shape depends on the detuning of the probe laser from the resonantly interacting cavity-atom system.

If an atom absorbs a photon inside the cavity, a strong dipole force can be exerted due to the inhomogeneous field distribution of the cavity mode. Trapping of atoms with a single photon was achieved [79.54], and from the time variation of the cavity transmission a reconstruction of atomic trajectories became possible.

79.8.2 Generation of Entanglement

In the middle of the 1990s, it was realized that fully controlled quantum systems could be used to implement a revolutionary type of information processing now called quantum computing [79.55]. From the beginning, cavity QED has conceptually played an important role for experimental realizations, since it offers a route to manipulate, in principle, all physical parameters of a coherently interacting system. With the well established microwave-cavity–Rydberg-atom system, it was proven

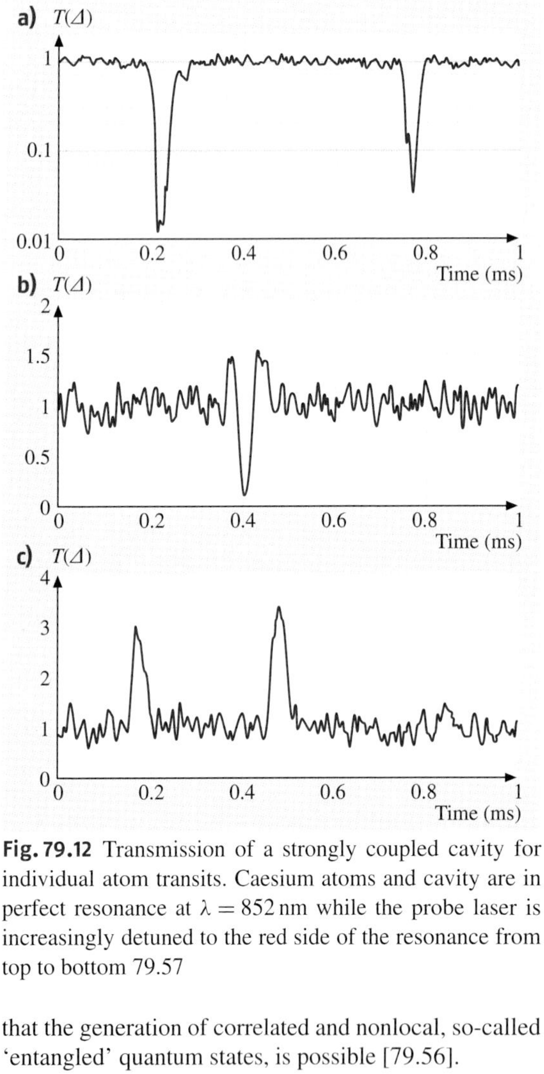

Fig. 79.12 Transmission of a strongly coupled cavity for individual atom transits. Caesium atoms and cavity are in perfect resonance at $\lambda = 852\,\text{nm}$ while the probe laser is increasingly detuned to the red side of the resonance from top to bottom 79.57

that the generation of correlated and nonlocal, so-called 'entangled' quantum states, is possible [79.56].

The first 'application' of cavity QED was the transfer of the strong coupling idea to the combined internal and motional quantum states of trapped ions [79.58]. Here the harmonic oscillation of the ion replaces the electric field of the conventional cavity-QED system. This quantum gate was realized with a system of two trapped ions coupled to each other by Coulomb forces [79.59].

Ideas about how to use the strong coupling of atoms and photons [79.60–62] for the generation of atom–photon, or atom–atom (by insertion of more than one atom) entanglement abound, but entanglement generation by means of cavity-QED with a controlled source of atoms or ions remains a challenge for the future.

79.8.3 Single Photon Sources

Coherent laser fields are considered the ultimate source of classical radiation fields, and they are characterized by the random arrival time of photons. Nonclassical light sources with, for instance, a regularized stream of photons offer interesting properties for low-noise measurement applications.

Cavity-QED systems offer an attractive light-matter process for the generation of such 'photon-bit-streams', or single photon sources [79.63]. In such devices, a single photon state can, for instance, be created by Raman processes involving a classical field, which serves as the control parameter for the process, and the vacuum field of the optical resonator. The Raman process leaves a single photon in the cavity, which only weakly interacts with the atom. If the resonator has suitable transmission properties, this photon will then escape with predetermined frequency, shape, and propagation direction. Deterministic single photon sources have been realized with quantum dots [79.64,65], single molecules [79.66], and also with slow [79.67] or trapped [79.68] cold atoms and ions [79.69] inside optical cavities.

References

79.1 Cavity QED is reviewed in detail in S. Haroche: *Fundamental Systems in Quantum Optics*, ed. by J. Dalibard et al. (Elsevier, Amsterdam 1992)
79.2 D. Meschede: Phys. Rep. **211**(5), 201–250 (1992)
79.3 P. Berman (Ed.): *Cavity Quantum Electrodynamics* (Academic, Amsterdam 1994)
79.4 An introduction into the more general framework of low energy Quantum Electrodynamics may be found in the recent textbook P. Milonni (Ed.): *The Quantum Vacuum* (Academic, Boston 1994)
79.5 P. Berman (Ed.): *Cavity Quantum Electrodynmics* (Academic, Boston 1994)
79.6 V. Weisskopf, E. Wigner: Z. Phys. **63**, 54 (1930)
79.7 F. Levin, D. Micha (Eds.): *Long-Range Casimir Forces* (Plenum, New York 1993)
79.8 D. Meschede, H. Walther, G. Müller: Phys. Rev. Lett. **54**, 551 (1985)
79.9 L. Allen, J. Eberly: *Optical Resonance and Two-Level-Atoms* (Dover, New York 1987), reprint of the original 1975 edn.
79.10 D. Kleppner: Phys. Rev. Lett. **47**, 233 (1981)

79.11 R. G. Hulet, E. Hilfer, D. Kleppner: Phys. Rev. Lett. **55**, 2137 (1985)
79.12 W. Jhe, A. Anderson, E. Hinds, D. Meschede, L. Moi, S. Haroche: Phys. Rev. Lett. **58**, 666 (1987)
79.13 J. Eschner, Ch. Raab, F. Schmidt-Kaler, R. Blatt: Nature **413**, 495 (2001)
79.14 M. A. Wilson, P. Bushev, J. Eschner, F. Schmidt-Kaler, C. Becher, R. Blatt, U. Dorner: Phys. Rev. Lett. **91**, 213602 (2003)
79.15 J. Dowling: *Quantum Measurements in Optics*, ed. by P. Tombesi, D. Walls (Plenum, New York 1992)
79.16 K. H. Drexhage: *Progress in Optics*, Vol. 12, ed. by E. Wolf (North-Holland, Amsterdam 1974)
79.17 V. Sandoghar, C. Sukenik, E. Hinds, S. Haroche: Phys. Rev. Lett. **68**, 3432 (1992)
79.18 H. Casimir, D. Polder: Phys. Rev. **73**, 360 (1948)
79.19 C. Sukenik, M. Boshier, D. Cho, V. Sandoghar, E. Hinds: Phys. Rev. Lett. **70**, 560 (1993)
79.20 A. Shih, D. Raskin, P. Kusch: Phys. Rev. A **9**, 652 (1974)
79.21 C. Cohen-Tannoudji: *Fundamental Systems in Quantum Optics*, ed. by J. Dalibard et al. (Elsevier, Amsterdam 1992)
79.22 S. Haroche, M. Brune, J. M. Raimond: Europhys. Lett. **14**, 19 (1991)
79.23 M. Chevrollier, M. Fichet, M. Oria, G. Rahmat, D. Bloch, M. Ducloy: J. Phys. (Paris) **2**, 631 (1992)
79.24 P. Bushev, A. Wilson, J. Eschner, C. Raab, F. Schmidt-Kaler, C. Becher, R. Blatt: Phys. Rev. Lett. **92**, 223602 (2004)
79.25 M. Brune, P. Nussenzveig, F. Schmidt-Kaler, R. Bernadot, A. Maali, J. M. Raimond, S. Haroche: Phys. Rev. Lett. **72**, 3339 (1994)
79.26 H. Failache, S. Saltiel, A. Fischer, D. Bloch, M. Ducloy: Phys. Rev. Lett. **88**, 243603 (2002)
79.27 W. von Klitzing, R. Long, V. S. Ilchenko, J. Hare, V. Lefèvre-Seguin: Opt. Lett. **26**, 166 (2001)
79.28 E. Yablonovitch: Phys. Rev. Lett. **58**, 2059 (1987)
79.29 K. Inoue, K. Ohtaka (Eds.): *Photonic Crystals* (Springer, Berlin, Heidelberg 2004)
79.30 J. P. Reithmaier, G. Sek, A. Löffler, C. Hofmann, S. Kuhn, S. Reitzenstein, L. V. Keldysh, V. D. Kulakovskii, T. L. Reinecke, A. Forchel: Nature **432**, 197 (2004)
79.31 T. Yoshie, A. Scherer, J. Hendrickson, G. Khitrova, H. M. Gibbs, G. Rupper, C. Ell, O. B. Shchekin, D. G. Deppe: Nature **432**, 200 (2004)
79.32 K. Vahala: Nature **424**, 839 (2003)
79.33 C. Cohen-Tannoudji: *Frontiers in Laser Spectroscopy*, ed. by R. Balian et al. (North-Holland, Amsterdam 1977)
79.34 R. Thompson, G. Rempe, H. Kimble: Phys. Rev. Lett. **68**, 1132 (1992)
79.35 J. McKeever, A. Boca, A. D. Boozer, J. R. Buck, H. J. Kimble: Nature **425**, 268 (2003)
79.36 I. Chiorescu, P. Bertet, K. Semba, Y. Nakamura, C. J. P. M. Harmans, J. E. Mooij: Nature **431**, 159 (2004)
79.37 A. Wallraff, D. I. Schuster, A. Blais, L. Frunzio, R.-S. Huang, J. Majer, S. Kumar, S. M. Girvin, R. J. Schoelkopf: Nature **431**, 164 (2004)
79.38 R. F. Stebbings, F. B. Dunning (Eds.): *Rydberg States of Atoms and Molecules* (Cambridge Univ. Press, Cambridge 1983)
79.39 E. T. Jaynes, F. W. Cummings: Proc. IEEE **51**, 89 (1963)
79.40 H. Paul: Ann. Phys. **11**, 411 (1963)
79.41 J. H. Eberly, N. B. Narozhny, J. J. Sanchez-Mondragon: Phys. Rev. Lett. **44**, 1323 (1980)
79.42 H. J. Briegel, B. G. Englert: Phys. Rev. A **47**, 3311 (1993)
79.43 C. Ginzel, H. J. Briegel, U. Martini, B. G. Englert, A. Schenzle: Phys. Rev. A **48**, 732 (1993)
79.44 B. R. Mollow: Phys. Rev. **188**, 1969 (1969)
79.45 P. Filipowicz, J. Javanainen, P. Meystre: Phys. Rev. A **34**, 3077 (1986)
79.46 G. Rempe, H. Walther: Phys. Rev. A **42**, 1650 (1990)
79.47 G. Rempe, F. Schmidt-Kaler, H. Walther: Phys. Rev. Lett. **64**, 2783 (1990)
79.48 U. Benson, G. Raithel, H. Walther: Phys. Rev. Lett. **72**, 3506 (1994)
79.49 P. Meystre, G. Rempe, H. Walther: Opt. Lett. **13**, 1078 (1988)
79.50 C. Wagner, A. Schenzle, H. Walther: Opt. Commun. **107**, 318 (1994)
79.51 M. Brune, S. Haroche, V. Lefevre, J. M. Raimond, N. Zagury: Phys. Rev. Lett. **65**, 976 (1990)
79.52 R. Schack, A. Breitenbach, A. Schenzle: Phys. Rev. A **45**, 3260 (1992)
79.53 C. Wagner, A. Schenzle, H. Walther: Phys. Rev. A **47**, 5068 (1993)
79.54 A. C. Doherty, T. W. Lynn, C. J. Hood, H. J. Kimble: Phys. Rev. A **63**, 013401 (2001) and references therein
79.55 D. P. DiVincenzo: Fortschr. Phys. **48**, 771–783 (2000)
79.56 J. M. Raimond, M. Brune, S. Haroche: Rev. Mod. Phys. **73**, 565 (2001)
79.57 C. J. Hood, M. S. Chapman, T. W. Lynn, H. J. Kimble: Phys. Rev. Lett. **80**, 4157 (1998)
79.58 J. I. Cirac, P. Zoller: Phys. Rev. Lett. **74**, 4091 (1995)
79.59 F. Schmidt-Kaler, H. Häffner, M. Riebe, S. Gulde, G. P. T. Lancaster, T. Deuschle, C. Becher, C. F. Roos, J. Eschner, R. Blatt: Nature **422**, 408 (2003)
79.60 T. Pellizari, S. A. Gardiner, J. I. Cirac, P. Zoller: Phys. Rev. Lett. **75**, 3788 (1995)
79.61 A. S. Soerensen, K. Moelmer: Phys. Rev. Lett. **91**, 097905 (2003)
79.62 L. You, X. X. Yi, X. H. Su: Phys. Rev. A **67**, 032308 (2003)
79.63 C. K. Law, H. J. Kimble: J. Mod. Opt. **44**, 2067 (1997)
79.64 P. Michler, A. Kiraz, C. Becher, W. V. Schoenfeld, P. M. Petroff, E. Hu. Lidong Zhang, A. Imamoglu: Science **290**, 2282 (2000)

79.65 C. Santori, M. Pelton, G. Solomon, Y. Dale, Y. Yamamoto: Phys. Rev. Lett. **86**, 1502 (2001)
79.66 B. Lounis, W. E. Moerner: Nature **407**, 491 (2000)
79.67 A. Kuhn, M. Hennrich, G. Rempe: Phys. Rev. Lett. **89**, 067901 (2002)
79.68 J. McKeever, A. Boca, A. D. Boozer, R. Miller, J. R. Buck, A. Kuzmich, H. J. Kimble: Science **303**, 1992 (2004)
79.69 M. Keller, B. Lange, K. Hayasaka, W. Lange, H. Walther: Nature **431**, 1075 (2004)

80. Quantum Optical Tests of the Foundations of Physics

Quantum mechanics began with the solution of the problem of blackbody radiation by Planck's quantum hypothesis: in the interaction of light with matter, energy can only be exchanged between the light in a cavity and the atoms in the walls of the cavity by the discrete amount $E = h\nu$, where h is Planck's constant and ν is the frequency of the light. Einstein, in his treatment of the photoelectric effect, reinterpreted this equation to mean that a beam of light consists of particles ("light quanta") with energy $h\nu$. The Compton effect supported this particle viewpoint of light by demonstrating that photons carried momentum, as well as energy. In this way, the wave–particle duality of quanta made its first appearance in connection with the properties of light.

It might seem that the introduction of the concept of the photon as a particle would necessarily also introduce the concept of locality into the quantum world. However, in view of observed violations of Bell's inequalities, exactly the opposite seems to be true. Here we review some recent results in quantum optics which elucidate nonlocality and other fundamental issues in physics.

In spite of the successes of quantum electrodynamics, and of the standard model in particle physics, there is still considerable resistance to the concept of the photon as a particle. Many papers have been written trying to explain all optical phenomena semiclassically, i. e., with the light viewed as a classical wave, and the atoms treated quantum mechanically [80.1–4]. We first present some quantum optics phenomena which exclude this semiclassical viewpoint.

80.1 **The Photon Hypothesis** 1186

80.2 **Quantum Properties of Light** 1186
- 80.2.1 Vacuum Fluctuations: Cavity QED 1186
- 80.2.2 The Down-Conversion Two-Photon Light Source 1187
- 80.2.3 Squeezed States of Light 1187

80.3 **Nonclassical Interference** 1188
- 80.3.1 Single-Photon and Matter–Wave Interference ... 1188
- 80.3.2 "Nonlocal" Interference Effects and Energy–Time Uncertainty 1189
- 80.3.3 Two-Photon Interference 1190

80.4 **Complementarity and Coherence** 1191
- 80.4.1 Wave–Particle Duality 1191
- 80.4.2 Quantum Eraser 1191
- 80.4.3 Vacuum-Induced Coherence 1192
- 80.4.4 Suppression of Spontaneous Down-Conversion 1192

80.5 **Measurements in Quantum Mechanics** ... 1193
- 80.5.1 Quantum (Anti-)Zeno Effect 1193
- 80.5.2 Quantum Nondemolition 1193
- 80.5.3 Quantum Interrogation 1194
- 80.5.4 Weak and "Protected" Measurements 1195

80.6 **The EPR Paradox and Bell's Inequalities** 1195
- 80.6.1 Generalities 1195
- 80.6.2 Polarization-Based Tests 1196
- 80.6.3 Nonpolarization Tests 1196
- 80.6.4 Bell Inequality Loopholes 1198
- 80.6.5 Nonlocality Without Inequalities . 1199

80.7 **Quantum Information** 1200
- 80.7.1 Information Content of a Quantum: (No) Cloning 1200
- 80.7.2 Super-Dense Coding 1200
- 80.7.3 Teleportation 1200
- 80.7.4 Quantum Cryptography 1201
- 80.7.5 Issues in Causality 1202

80.8 **The Single-Photon Tunneling Time** 1202
- 80.8.1 An Application of EPR Correlations to Time Measurements 1202
- 80.8.2 Superluminal Tunneling Times 1203
- 80.8.3 Tunneling Delay in a Multilayer Dielectric Mirror ... 1203
- 80.8.4 Interpretation of the Tunneling Time 1204
- 80.8.5 Other Fast and Slow Light Schemes 1205

80.9 **Gravity and Quantum Optics** 1206

References 1207

80.1 The Photon Hypothesis

In an early experiment, Taylor reduced the intensity of a thermal light source in Young's two-slit experiment, until, on the average, there was only a single photon passing through the two slits at a time. He then observed a two-slit interference pattern which was identical to that for a more intense classical beam of light. In Dirac's words, the apparent conclusion is that "each photon then interferes only with itself" [80.6]. However, a coherent state, no matter how strongly attenuated, always remains a coherent state (Sect. 78.2.2); since a thermal light source can be modeled as a statistical ensemble of coherent states, a stochastic classical wave model yields complete agreement with Taylor's observations. The one-by-one darkening of grains of film can be explained by treating the matter alone quantum mechanically [80.2]; consequently, the concept of the photon need not be invoked, and the claim that this experiment demonstrates quantum interference of individual photons is unwarranted [80.7].

This weakness in Taylor's experiment can be removed by the use of nonclassical light sources; as discussed by *Glauber* [80.8], classical predictions diverge from quantum ones only when one considers counting statistics, or photon correlations. In particular, two-photon light sources, combined with coincidence detection, allow the production of single-photon ($n = 1$ Fock) states with near certainty. In the first such experiment [80.9], two photons, produced in an atomic cascade within nanoseconds of each other, impinged on two beam splitters, and were then detected in coincidence by means of four photomultipliers placed at all possible exit ports. In a simplified version of this experiment [80.5], one of the beam splitters and its two detectors are replaced with a single detector D_1 (Fig. 80.1). We define

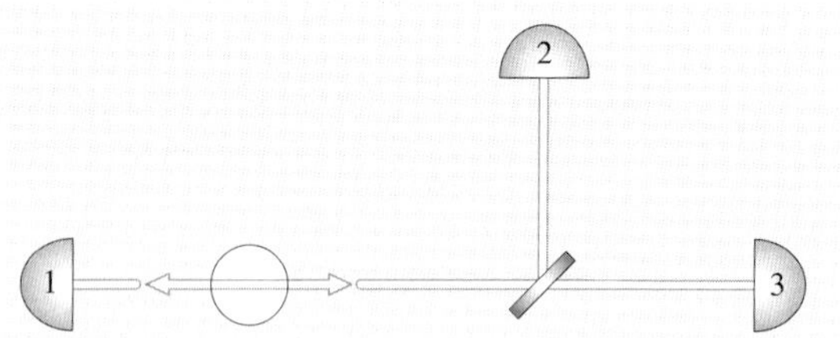

Fig. 80.1 Triple-coincidence setup of *Grangier* et al. [80.5]

the anticorrelation parameter

$$\alpha \equiv N_{123}N_1/N_{12}N_{13}\,, \tag{80.1}$$

where N_{123} is the rate of triple-coincidences between detectors D_1, D_2 and D_3; N_1 is the singles rate at D_1; and N_{12} and N_{13} are double-coincidence rates. Then from Schwarz's inequality [80.5, 7, 10], $\alpha \geq 1$ for any classical wave. In essence, since the wave divides smoothly, the coincidence rate between D_2 and D_3 is never smaller than the "accidental" coincidence rate, even when measurements are conditioned on an event at D_1. (The Hanbury–Brown and Twiss experiment [80.11] can be explained classically, because the thermal fluctuations lead to "bunching," or a mean coincidence rate which is greater than the mean accidental rate; Sect. 78.3.3.) By contrast, the indivisibility of the photon leads to strong anticorrelations between D_2 and D_3, making α arbitrarily small. In agreement with this quantum mechanical picture, *Grangier* et al. observed a 13-standard-deviation violation of the inequality [80.5], corroborating the notion of the "collapse of the wave packet" as proposed by *Heisenberg* [80.12].

80.2 Quantum Properties of Light

80.2.1 Vacuum Fluctuations: Cavity QED

The above considerations necessitate the quantization of the electromagnetic field, which in turn leads to the concept of vacuum fluctuations [80.4] (Sect. 78.1). Difficulties with this idea, such as the implied infinite zero-point energy of the universe, have led some reseachers to attempt to dispense with this concept altogether, along with that of the photon, in every explanation of electromagnetic interactions with matter. Of course, it is impossible to explain all phenomena, such as spontaneous emission and the Lamb shift, without some kind of fluctuating electromagnetic fields (Chapt. 78 and Sect. 79.2.4), but one can go a long way with an ad hoc ambient classical electromagnetic noise-field filling all of space, in conjunction with the radiation reaction [80.1, 4].

In particular, even the Casimir attraction between two conducting plates (Sect. 79.2.4), which has now been verified with high precision [80.13–19] can be explained semiclassically in terms of dipole forces between electrons in each plate as they undergo zero-point motion that induce image charges in the other plate. Nevertheless, the effects of cavity QED (Chapt. 79) [80.20–22], including the influence of cavity-induced boundary conditions on energy levels and spontaneous emission rates, are most easily unified via quantization of the electromagnetic field.

By coupling highly excited Rydberg atoms to photons in a high-finesse superconducting microwave cavity, *Haroche* et al. have observed single-photon driven Rabi oscillations [80.23], and have used these to study decoherence effects [80.24], atom–photon entanglement [80.25], and quantum nondemolition measurements [80.26] (Sect. 80.5.2). *Kimble* et al. [80.27] and *Rempe* et al. [80.28] have performed similar experiments, coupling atoms to small optical cavities, and even trapping the atoms with light fields at the single-photon level [80.29, 30]. By monitoring the amplitude of the light transmitted through the cavity (which depends on the precise location of the atom inside the cavity volume), the trajectory of the atom can be determined with ultrahigh resolution, much smaller than an optical wavelength [80.31, 32].

80.2.2 The Down-Conversion Two-Photon Light Source

The quantum aspects of electromagnetism are made more striking with a two-photon light source, in which two highly correlated photons are produced in spontaneous parametric down-conversion, or parametric fluorescence [80.33–36]. In this process, an ultraviolet "pump" photon produced in a laser spontaneously decays inside a crystal with a $\chi^{(2)}$ nonlinearity into two highly correlated red photons, conventionally called the "signal" and the "idler" (Sect. 72.3.4). (The quantum state of the light is more correctly written as $|\psi\rangle \propto |vacuum\rangle + \epsilon|1\rangle_s|1\rangle_i + \epsilon^2|2\rangle_s|2\rangle_i + \ldots$, but since the amplitude of the down-conversion process itself is very weak (ϵ is of order 10^{-6}), one often neglects the terms containing 2 or more pairs. However, recent experiments have begun to exploit these higher-order terms, e.g., to investigate 3-, 4-, 5-photon quantum effects [80.37–40]. Very recently, a stimulated down-conversion process [80.41] has indicated the presence of 12-photon entanglement [80.42].) As shown in Fig. 80.2, a rainbow of colored cones is produced around an axis defined by the direction of the uv beam (for the case of type-I phase-matching), with the correlated down-conversion photons always emitted on opposite sides of the UV beam (Sect. 72.2.2). Their emission times

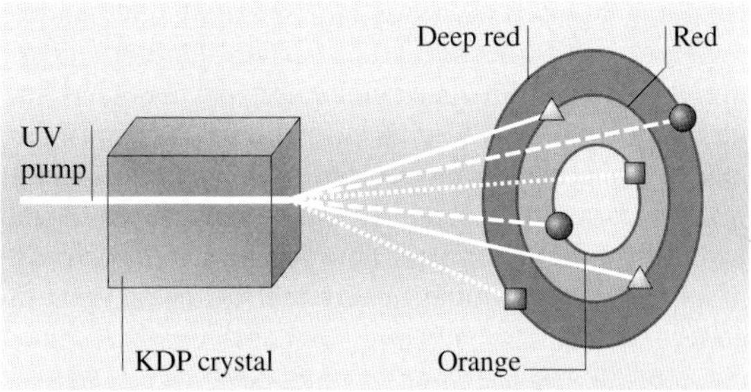

Fig. 80.2 Conical emissions of down-conversion from a nonlinear crystal (for type-I phase-matching). Photon energy depends on the cone opening angle, and conjugate photons lie on opposite sides of the axis, e.g., the inner "circle" orange photon is conjugate to the outer "circle" deep-red photon, etc.

are within femtoseconds of each other, so that detection of one photon implies with near certainty that there is exactly one quantum present in the conjugate mode [80.43]. In type-I phase-matching the correlated photons share the same polarization, while in type-II phase-matching they have orthogonal polarizations . We will see below (Sect. 80.6.2) how both of these can enable the production of photons that are entangled in polarization, as well as in other degrees of freedom.

This production technique allowed for the first reconstruction of the Wigner distribution for a single photon [80.44], which is manifestly non-classical in that the quasiprobability for both quadratures of the field to vanish is negative. In contrast to an earlier demonstration of a negative Wigner function [80.45] (using atoms, not photons), this measurement was possible using essentially classical measurement techniques with no quantum assumptions, and is in this sense a direct demonstration of the non-classical nature of the electromagnetic field. Later work [80.46] used these techniques to demonstrate that a single photon forced to choose between two output ports of a beam splitter exhibits quantum correlations.

80.2.3 Squeezed States of Light

The creation of correlated photon pairs is closely related to the process of quadrature-squeezed light production (Sect. 78.2.2, and the review in [80.47]). For example, when the gain arising from parametric amplification in a down-conversion crystal becomes large, there is a transition from spontaneous to stimulated emission of pairs. This gain is dependent on the phase of amplified light relative to the phase of the pump light. As a result, the vacuum fluctuations are reduced ("squeezed") below the standard quantum limit (SQL) in one quadrature, but increased in the other, in such a way as to preserve the minimum uncertainty-principle product [80.48]. This periodicity of the fluctuations at 2ω is a direct consequence of the fact that the light is a superposition of states differing in energy by $2\hbar\omega$ – the quadrature-squeezed vacuum state

$$|\xi\rangle = \exp\left(\frac{1}{2}\xi^* aa - \frac{1}{2}\xi a^\dagger a^\dagger\right)|0\rangle \quad (80.2)$$

represents a vacuum state transformed by the creation $(a^\dagger a^\dagger)$ and destruction (aa) of photons two at a time. Essentially any optical processes operating on photon pairs (e.g., four-wave mixing [80.49, 50]) can also produce such squeezing.

Amplitude squeezing involves preparation of states with well-defined photon number, i. e., states lacking the Poisson fluctuations of the coherent state. The possibility of producing such states (e.g., via a constant-current-driven semiconductor laser [80.51]) demonstrates that "shot noise" in photodetectionshould not be thought of as merely the result of the probabilistic (à la Fermi's Golden Rule, Sect. 69.4) excitation of quantum mechanical atoms in a classical field, but as representing real properties of the electromagnetic field, accessible to experimental control.

The highest level of number-squeezing reported (-5.7 ± 0.1) dB, corresponding to a noise reduction 73% below the SQL) employed an asymmetric fiber loop to squeeze solitons [80.52]. The highest level of quadrature-squeezing reported has been (-7.0 ± 0.2) dB (80% below the SQL), using a $\chi^{(2)}$-crystal optical parametric oscillator [80.53]. This was a continuous-wave vacuum-squeezed beam – the same technique was used to produce -5.0 dB of bright continuous-wave squeezing locked for several hours.

Squeezed light has begun to have impact in metrology. A few examples suffice: the generation of audio-band squeezed light [80.54] and the demonstration of a squeezing-enhanced power-recycled Michelson interferometer [80.55], both suitable for gravity-wave interferometry (Sect. 80.9); the use of squeezing to measure displacement of a light beam below the standard quantum limit [80.56], suitable for atomic force microscopy; and demonstrations of squeezed light spectroscopy [80.57–59].

80.3 Nonclassical Interference

80.3.1 Single-Photon and Matter–Wave Interference

The first truly one-photon interference experiment [80.5] used the cascade source discussed in Sect. 80.1. One of the photons was directed to a "trigger" detector, while the other, thus prepared in an $n = 1$ Fock state, was sent through a Mach–Zehnder interferometer. The output photon, detected in coincidence with the trigger photon, showed fringes with a visibility $> 98\%$. Dirac's statement that a single photon interferes with itself is thus verified.

Of course, matter can also display interference, determined by the deBroglie wavelength (Chapt. 77). There have been significant recent advances in atomic matter–wave interferometry and its applications [80.60], ever since the early experiments of Pritchard, which used standing-light gratings and nanofabricated diffraction gratings to construct Mach–Zehnder-type interferometers for sodium atoms [80.61] and molecules [80.62] from a supersonic source. *Chu* and *Kasevich* introduced the use of STIRAP (Stimulated Raman Adiabatic fast Passage; Sect. 69.7) to produce coherent beam splitters for cold atomic beams, also in a Mach–Zehnder-type interferometer, but whose source were cesium atoms cooled in and launched from a magneto-optical trap (MOT) [80.63]. Matter–wave interferometry has now been applied to precision measurements of the acceleration g due to Earth's gravity [80.64], gravity gradiometry [80.65], and Sagnac matter–wave gyroscopes [80.66, 67].

To date some of the largest systems to display quantum interference are large molecules like carbon 60 ("Buckyballs") and carbon 70 [80.68]. These are significant in that the average deBroglie wavelength of the molecules, emitted from an oven, was 2.8 pm, actually about 350 times smaller than the molecule itself. *Arndt* et al. have also demonstrated multislit diffraction with the biological molecule porphyrin, and with fluorofullerenes ($C_{60}F_{48}$) [80.69]. With a mass of 1632 amu, the latter are currently the largest single objects to display interference. Looking ahead, others have suggested that it may be possible to put a micron-scale mirror (with $\approx 10^{14}$ atoms) into a superposition of resolvable spatial locations [80.70] – the mirror, part of a high-finesse optical cavity forming one arm of a Michelson interferometer, could be mounted on a high-quality mechanical oscillator, whereby the interaction with a single photon would change the frequency of the oscillator.

Two other systems, demonstrating Bose–Einstein condensation (BEC) (see Chapt. 76), have also produced evidence of macroscopic quantum coherence. In the experiments of *Ketterle* et al., atoms from two different atomic vapor BEC clouds were allowed to fall onto the same detection region, and display interference fringes [80.71] (more recently interference from an array of 30 independent BECs has been observed [80.72]). In some ways this is the matter–wave equivalent of the famous *Pfleegor–Mandel* experiment [80.73], in which light from two separate lasers displays interference, even when attenuated to the single-photon level. The explanation in terms of the indistinguishability of the underlying processes is that one cannot ascertain from which laser source a given photon originated. However, this explanation must be applied carefully to the situation of the two atomic BECs: Unlike the lasers, the BEC clouds can – at least in principle – be prepared with a definite number of atoms, and it would therefore seem that one could in principle determine which cloud emitted a given detected atom. However, this determinacy is rapidly lost after a few atoms are detected [80.74]. Once the number becomes uncertain, a well-defined relative phase of the two BECs is established, according to the number-phase uncertainty relation $\Delta(N_2 - N_1)\Delta(\phi_2 - \phi_1) \gtrsim 1/2$. In fact, discussions have recently arisen over whether or not lasers should not also be viewed as incoherent number-state combinations, instead of the usual coherent state $|\alpha\rangle$ (the issue is that in principle there is nothing in a laser to break the symmetry and select a particular phase) [80.75–77].

Finally, quantum coherence (though not explicitly spatial interference as in the previous examples), has been detected in the operation of a Josephson-junction linked superconducting loop – the group of *Mooij* was able to prepare a superposition of clockwise and counterclockwise circulating electrical currents [80.78]. Since the $\approx 0.5\,\mu$A currents corresponded to the motion of millions of Cooper pairs, this is arguably the largest system thus far to have displayed quantum coherence. This superconducting system also holds promise for quantum computing (see Chapt. 81), as Rabi oscillations between the different flux states have been observed [80.79, 80].

80.3.2 "Nonlocal" Interference Effects and Energy–Time Uncertainty

The energy–time uncertainty principle, $\Delta E \Delta t \geq \hbar/2$ has been tested in a down-conversion interference experiment [80.81]. The down-conversion process conserves energy and momentum:

$$\hbar\omega_0 = \hbar\omega_1 + \hbar\omega_2 \,, \qquad (80.3)$$

$$\hbar\mathbf{k}_0 \approx \hbar\mathbf{k}_1 + \hbar\mathbf{k}_2 \,, \qquad (80.4)$$

where $\hbar\omega_0$ ($\hbar\mathbf{k}_0$) is the energy (momentum) of the parent photon, and $\hbar\omega_1$ ($\hbar\mathbf{k}_1$) and $\hbar\omega_2$ ($\hbar\mathbf{k}_2$) are the energies (momenta) of the daughter photons; $\mathbf{k}_1$ and $\mathbf{k}_2$ sum to $\mathbf{k}_0$ to within an uncertainty given by the reciprocal of the crystal length [80.82]. Since there are many ways of partitioning the parent photon's energy, each daughter photon may have a broad spectrum, and hence a wave packet narrow in time. However, $\omega_1 + \omega_2 = \omega_0$ is extremely well-defined, so that the difference in the daughter photons' arrival times, and the sum of their energies can be simultaneously known to high precision. Thus, the daughter photons of a parent photon of sharp energy E_0 are in an energy-"entangled" state, a nonfactorizable sum of product states [80.83]:

$$|\Psi\rangle = \int_0^{E_0} \mathrm{d}E\, A(E)\, |E\rangle\, |E_0 - E\rangle \,, \qquad (80.5)$$

where $A(E)$ is the probability amplitude for the production of two photons of energies E and $E_0 - E$. A measurement of the energy of one of the photons to be E_1 can be interpreted as causing an instantaneous "collapse" of the system to the state $|E_1\rangle|E_0 - E_1\rangle$, implying an instantaneous increase of the width of the other photon's wave packet. (Of course, the notion of collapse need not be invoked to explain such results. One can view the detection of the trigger photon as conditionally selecting a particular subensemble of the pairs. However, as discussed in Sect. 80.6.3, it is not correct to interpret the down-conversion photons as possessing a well-defined energy prior to measurement.) In the experiment, one photon was used as a trigger, while the other was sent into an adjustable Michelson interferometer (Fig. 80.3), used to measure its coherence length. (The same apparatus was also used to demonstrate that Berry's phase in optics has a quantum origin [80.84].) If the trigger photon passed through an interference filter F1 of narrow width ΔE and was detected, then the conjugate photon occupied a broad wavepacket of duration $\Delta t \approx \hbar/\Delta E$, and displayed interference. When

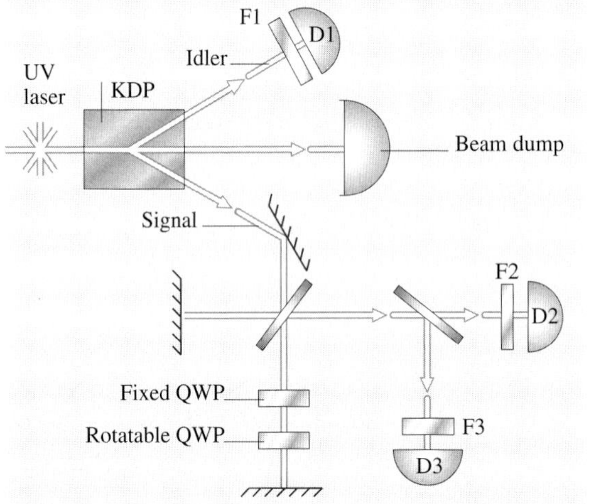

Fig. 80.3 The energy–time uncertainty relation and wave function collapse were studied by investigating the effect of various filters before the detectors in a single-photon interference experiment [80.5, 81]

there was no trigger, no fringes were observed, implying a much shorter wave packet. This is a nonlocal effect in that the photons could in principle be arbitrarily far away from each other when the collapse occurs.

80.3.3 Two-Photon Interference

In the above experiments, interference occurs between two paths taken by a single photon. An early experiment to demonstrate two-photon interference using the down-conversion light source was performed by *Ghosh* and *Mandel* [80.85]. They looked at the counting rate of a detector illuminated by both of the twin beams. No interference was observed at the detector, because although the sum of the phases of the two beams emitted in parametric fluorescence is well defined (by the phase of the pump), their difference is not, due to the number-phase uncertainty principle. However, the rate of coincidence detections between two such detectors whose separation was varied did display high-visibility interference fringes. Whereas in the standard two-slit experiment, interference occurs between the two paths a single photon could have taken to reach a given point on a screen, in this case it occurs between the possibility that the signal photon reached detector 1 and the idler photon detector 2, and the possibility that the reverse happened. This experiment provides a manifestation of quantum nonlocality; interference occurs between alternate global histories of a system, not between local fields. At a null of the coincidence fringes, the detection of one photon at detector 1 excludes the possibility of finding the conjugate photon at detector 2.

Such interference becomes clearer in the related interferometer of *Hong* et al. [80.86] (Fig. 80.4). The identically polarized conjugate photons from a down-conversion crystal are directed to opposite sides of a 50–50 beam splitter, such that the transmitted and reflected modes overlap. If the difference in the path lengths ΔL prior to the beam splitter is larger than the two-photon correlation length (of the order of the coherence length of the down-converted light), the photons behave independently at the beam splitter, and coincidence counts between detectors in the two output ports are observed half of the time – the other half of the time both photons travel to the same detector. However, when $\Delta L \approx 0$, such that the photon wave packets overlap at the beam splitter, the probability of coincidences is reduced, in principle to zero if $\Delta L = 0$. One can explain the coincidence null at zero path-length difference using the Feynman rules for calculating probabilities: add the probability amplitudes of indistinguishable processes which lead to the same final outcome, and then take the absolute square. The two indistinguishable processes here are both photons being reflected at the beam splitter (with Feynman amplitude $\mathrm{r}\cdot\mathrm{r}$) and both photons being transmitted (with Feynman amplitude $\mathrm{t}\cdot\mathrm{t}$). The probability of a coincidence detection is then

$$P_\mathrm{c} = |\mathrm{r}\cdot\mathrm{r}+\mathrm{t}\cdot\mathrm{t}|^2 = \left|\frac{\mathrm{i}}{\sqrt{2}}\cdot\frac{\mathrm{i}}{\sqrt{2}}+\frac{1}{\sqrt{2}}\cdot\frac{1}{\sqrt{2}}\right|^2 = 0\,, \tag{80.6}$$

assuming a real transmission amplitude, and where the factors of i come from the phase shift upon reflection at a beam splitter [80.87, 88].

The possibility of a perfect null at the center of the dip is indicative of a nonclassical effect. Indeed, classical field predictions allow a maximum coincidence-fringe visibility of only 50% [80.89]. The tendency of the photons to travel off together at the beam splitter can be thought of as a manifestation of the Bose–Einstein statistics for the photons [80.90]. In practice, the bandwidth of the photons, and hence the width of the null, is determined by filters and/or irises before the detectors [80.82]. Widths as small as $5\,\mu$m have been observed, corresponding to time delays of only 15 fs [80.91]. Consequently, one application is the determination of single-photon propagation times with extremely high time resolution (Sect. 80.8).

80.4 Complementarity and Coherence

80.4.1 Wave–Particle Duality

The complementary nature of wave-like and particle-like behavior is frequently interpreted as follows: due to the uncertainty principle, any attempt to measure the position (particle aspect) of a quantum leads to an uncontrollable, irreversible disturbance in its momentum, thereby washing out any interference pattern (wave aspect) [80.93, 94]. This picture is incomplete though; no "state reduction," or "collapse," is necessary to destroy interference, and measurements which do not involve reduction can be reversible. One must view the loss of coherence as arising from an entanglement of the system wave function with that of the measuring apparatus (MA) [80.95]. Previously interfering paths can thereby become distinguishable, such that no interference is observed. Consider the simplest experiment, a Mach–Zehnder interferometer with a 90° polarization rotator in arm 1. If horizontally polarized light is input, the state before the recombining beam splitter is $(|1\rangle|V\rangle + |2\rangle|H\rangle)/\sqrt{2}$, where $|1\rangle$ and $|2\rangle$ indicate the path of the photon. Because the polarization – playing the role of the MA – labels the path, no interference is observed at the output. *Englert* [80.96] has introduced a generalized relation quantifying the interplay between the wave-like attributes of a system (as measured by the fringe visibility V) and the particle-like character (as measured by the distinguishability D of the underlying quantum processes):

$$V^2 + D^2 \le 1 \,. \qquad (80.7)$$

The equality holds for pure input states. This relation has now been well verified in optical systems like that described above [80.97, 98], as well as in atom interferometry (Sect. 77.6) [80.99]. In the latter, the role of the polarization was played by internal energy states of an atom diffracted off a standing light wave.

80.4.2 Quantum Eraser

The interference lost to entanglement may be regained if one manages to "erase" the distinguishing information. This is the physical content of quantum erasure [80.100, 101]. The primary lesson is that one must consider the total physical state, including any MA with which the interfering quantum has become entangled, even if that MA does not allow accessible which-path information [80.102]. If the coherence of the MA is maintained, then interference may be recovered.

The first demonstration of a quantum eraser was based on the interferometer in Fig. 80.4 [80.92]. A half waveplate inserted into one of the paths before the beam splitter serves to rotate the polarization of light in that path. In the extreme case, the polarization is made orthogonal to that in the other arm, and the r·r and t·t processes become distinguishable; hence, the destructive interference which led to a coincidence null does not occur. The distinguishability can be erased, however, by using polarizers just before the detectors. In particular, if the initial polarization of the photons is horizontal, and the waveplate rotates one of the photon polarizations to vertical, then polarizers at 45° before both detectors restore the original interference dip. If one polarizer is at 45° and the other at −45°, interference is once again seen, but now in the form of a peak instead of a dip (Fig. 80.5). There are four basic measurements possible on the MA (here the polarization) – two of which yield which-path information, one of which recovers the initial interference fringes (here the coincidence dip), and one of which yields interference anti-fringes (the peak instead of the dip). In some implementations, the decision to measure wave-like or particle-like behavior may even be delayed until after detection of the original quantum, an irreversible process [80.103–105]. (This is an extension of the original delayed-choice discussion by

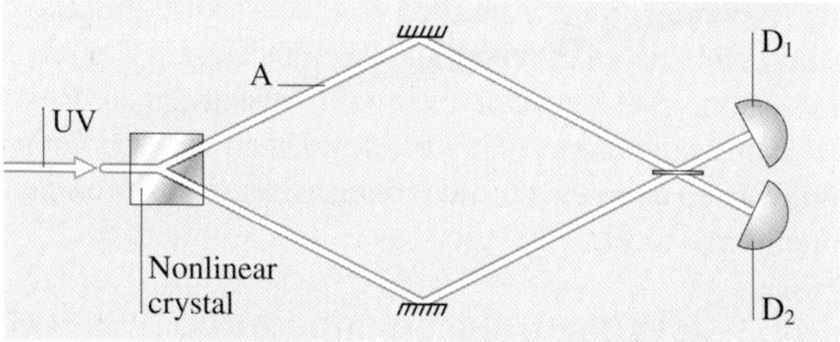

Fig. 80.4 Simplified setup for a Hong–Ou–Mandel (HOM) interferometer [80.86]. Coincidences may result from both photons being reflected, or both being transmitted. When the path lengths to the beam splitter are equal, these processes destructively interfere, causing a null in the coincidence rate. In a modified scheme, a half waveplate in one arm of the interferometer (at "A") serves to distinguish these otherwise interfering processes, so that no null in coincidences is observed. Using polarizers before the detectors, one can "erase" the distinguishability, thereby restoring interference [80.92] (Sects. 80.4.2 and 80.6.2)

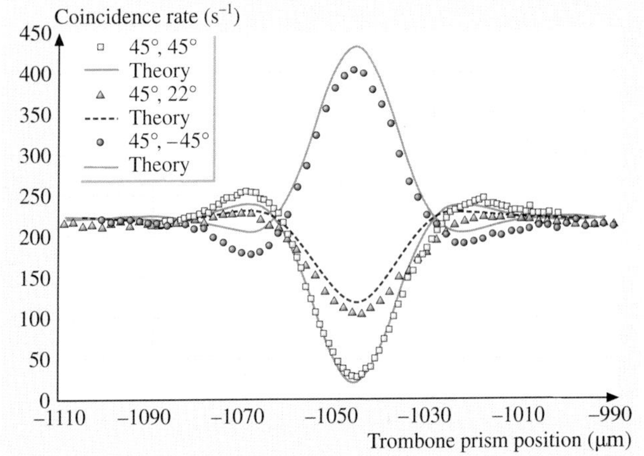

Fig. 80.5 Experimental data and scaled theoretical curves (adjusted to fit observed visibility of 91%) with polarizer 1 at 45° and polarizer 2 at various angles. Far from the dip, there is no interference and the angle is irrelevant [80.92]

Wheeler [80.106], and the experiments by *Hellmuth* et al. and *Alley* et al. [80.107, 108], in which the decision to display wave-like or particle-like aspects in a light beam may be delayed until after the beam has been split by the appropriate optics.) But in all cases, one must correlate the results of measurements on the MA with the detection of the originally interfering system. This requirement precludes any possibility for superluminal signaling.

In a related atom-optics experiment, researchers observed contrast loss in an atom interferometer when single photons were scattered off the atoms (yielding which-path information) [80.110]. They further demonstrated that the lost coherence could be recovered by observing only atoms that were correlated with photons emitted into a limited angular range, in essence realizing a quantum eraser.

80.4.3 Vacuum-Induced Coherence

A somewhat different demonstration [80.109, 111] of complementarity involves two down-conversion crystals, NL1 and NL2, aligned such that the trajectories of the idler photons from each crystal overlap (Fig. 80.6). A beam splitter acts to mix the signal modes. If the path lengths are adjusted correctly, and the idler beams overlap precisely, there is no way to tell, even in principle, from which crystal a photon detected at D_S originated. Interference appears in the signal *singles* rate at D_S, as any of the path lengths is varied. If the idler beam from crystal NL1 is prevented from entering crystal NL2, then the interference vanishes, because the presence or absence of an idler photon at D_i then "labels" the parent crystal. One explanation for the effect of blocking this path is that coherence is established by the idler-mode vacuum field seen by both crystals.

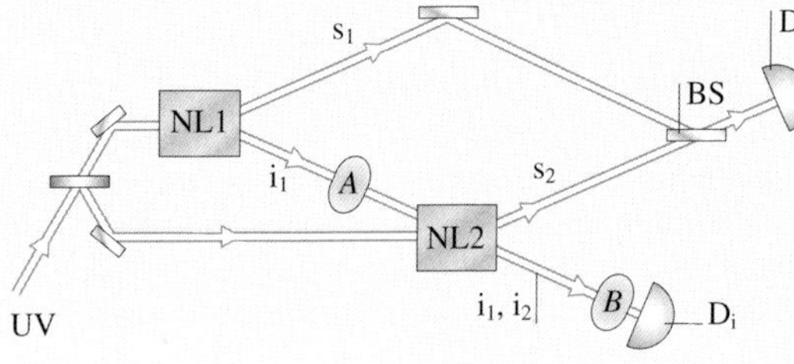

Fig. 80.6 Schematic of setup used in [80.109]. The idler photons from the two crystals are indistinguishable; consequently, interference fringes may be observed in the signal singles rate at detector D_S. Additional elements at *A* and *B* can be used to make a quantum eraser

Experiments have also been performed in which a time-dependent gate is introduced in the idler arm between the two crystals [80.112]. As one expects, the presence or absence of interference depends on the earlier state of the gate, at the time when the idler photon amplitude was passing through it.

80.4.4 Suppression of Spontaneous Down-Conversion

A modification [80.113] of this two-crystal experiment uses only a single nonlinear crystal (Fig. 80.7). A given pump photon may down-convert in its initial right-ward passage through the crystal, or in its left-going return trip (or not at all, the most likely outcome). As in the previous experiment, the idler modes from these two processes are made to overlap; moreover, the signal modes are also aligned to overlap. Thus, the left-going and right-going production processes are indistinguishable and interfere. The result is that fringes are observed in all of the counting rates (i. e., the coincidence rate and both singles rates) as any of the mirrors is translated. A different interpretation is as a change in the spontaneous emission of the down-converted photons, akin to the suppression of spontaneous emission in cavity QED demonstrations, discussed in Sects 80.2.1 and 79.2. Subsequent theoretical and experimental work has shown that, in a sense, there are always photons between the down-conversion crystal and the mirrors, even in the case of complete suppression of the spontaneous emission process [80.114, 115]. This same conclusion should also apply in the atom case [80.116], though in contrast to that system, for the down-conversion experiment the distances to the mirrors are much longer than the coherence lengths of the spontaneously emitted photons.

One recent application of this phenomenon is to study effective nonlinearities at the single-photon level, which has enabled the construction of a 2-photon switch [80.117]. Finally, with the inclusion of wave-plates to label the photons' paths, and polarizers to erase this information, an improved quantum eraser experiment was also completed, in which the which-path information for one photon was carried by the other photon [80.104].

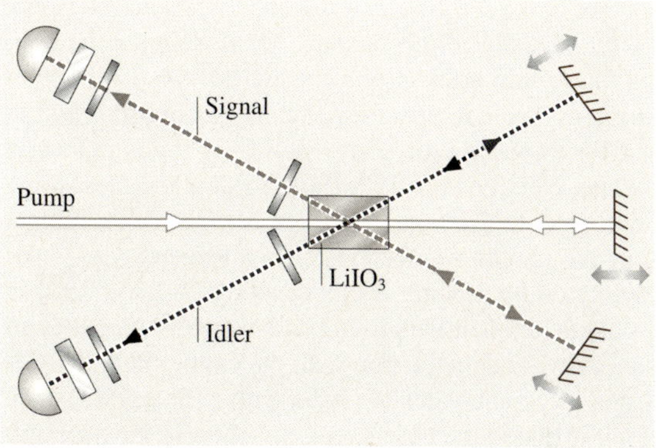

Fig. 80.7 Schematic of the experiment to demonstrate enhancement and suppression of spontaneous down-conversion [80.113]

80.5 Measurements in Quantum Mechanics

80.5.1 Quantum (Anti-)Zeno Effect

A strong measurement of a quantum system will project it into one of its eigenstates [80.95]. If the system evolves slowly out of its initial state: $|\Psi(t=0)\rangle = |\Psi_0\rangle \rightarrow (1-t^2/\tau^2)^{1/2}|\Psi_0\rangle + t/\tau|\Psi_1\rangle$, then repeated measurements with an interval much less than τ can inhibit this evolution. If there are N total measurements within τ, then the probability for the system to still be in the initial state is $P(\tau) = \left(1-(\tau/N)^2/\tau^2\right)^N \rightarrow 1$ as $N \rightarrow \infty$. This phenomenon, known as the quantum Zeno effect [80.118], has been experimentally observed using 3 levels in $^9Be^+$ ions [80.119]. The ions were prepared in state $|i\rangle$, and weakly coupled to state $|f\rangle$ via RF radiation that induced a slow Rabi oscillation between the two states. Thus, in the absence of any intervening measurements, the ions evolved sinusoidally into state $|f\rangle$. When rapid measurements were made (by a laser strongly coupling state $|f\rangle$ to readout state $|r\rangle$, hence leading to strong fluorescence only if the atom was in state $|f\rangle$), the effect was to inhibit the $|i\rangle \rightarrow |f\rangle$ transition. Note that here it was the absence of fluorescent photons which projected the state at each measurement back into the state $|i\rangle$ (Sect. 80.5.3). Also, although we have explained the effect in terms of a repeated "collapse" of the wave function back into its initial state, equally valid explanations without such reductions are also possible [80.120, 121].

Koffman and *Kurizki* have pointed out that the above inhibition phenomenon depends on there being a bounded number of final states (the ion example had only one). If instead the measurement process actually increases the number of accessible final states $|f\rangle$, then one obtains the "anti-Zeno" effect, in which the $|i\rangle \rightarrow |f\rangle$ rate is enhanced rather than suppressed by frequent measurements [80.122]. For example, this would be the case in the ion example if the $|i\rangle \rightarrow |f\rangle$ transition were spontaneous (allowing all frequencies) instead of driven (proceeding only at the driving Rabi frequency). The anti-Zeno effect has been observed by monitoring the survival time (against tunneling escape) of atoms trapped in an accelerating far-detuned standing wave of light [80.123].

80.5.2 Quantum Nondemolition

The uncertainty principle between the number of quanta N and phase ϕ of a beam of light,

$$\Delta N \Delta\phi \geq 1/2\,, \tag{80.8}$$

implies that to know the number of photons exactly, one must give up all knowledge of the phase of the wave. In theory, a quantum nondemolition (QND) process is possible [80.124]: without annihilating any of the light quanta, one can count them. It might seem that this would make possible successive measurements on noncommuting observables of a single photon, in violation of the uncertainty principle; it is the unavoidable introduction of phase uncertainty by any number measurement which prevents this.

QND schemes [80.125] often employ the intensity-dependent index of refraction arising from the optical

Kerr effect (Sect. 72.4.2) – the change in the index due to the intensity of the "signal" beam changes the optical phase shift on a "probe" beam [80.126–128]. Other proposals include using the Aharonov–Bohm effect to sense photons via the phase shift their fields induce in passing electrons [80.129, 130]. To date the closest experimental realization [80.26] of a QND measurement – of the photon number in a microwave cavity – was performed by passing Rydberg atoms [80.131, 132] in a superposition of the ground and excited states through the cavity. The interaction with the cavity photon is adjusted to be equivalent to a 2 π-pulse (Sect. 79.5). The result is that in the absence of any photon, the quantum state of the atoms after the cavity was unchanged; with a photon in the cavity, the ground state acquired an extra relative phase of π, which was then detected by measuring the atom's quantum state. An efficiency approaching 50% was achieved. Recently, it was suggested that optical QND measurements could enable scalable quantum computing [80.133].

80.5.3 Quantum Interrogation

The previous section described techniques to measure the presence of a photon without absorbing it. Now we discuss a method – quantum interrogation – to optically detect the presence of an object without absorbing or scattering a photon. The possibility that the absence of a detection event – a "negative-result" measurement – can lead to wavepacket reduction was first discussed by *Renninger* [80.134] and later by *Dicke* [80.135]. Here we consider the Gedankenexperiment proposed by *Elitzur* and *Vaidman* (EV), a simple single-particle interferometer, with particles injected one at a time [80.136]. The path lengths are adjusted so that all the particles leave a given output port (A), and never the other (B). Now suppose that a nontransmitting object is inserted into one of the interferometer's two arms – to emphasize the result, EV considered an infinitely sensitive "bomb", such that interaction with even a single photon would cause it to explode. By classical intuition, any attempt to check for the presence of the bomb involves interacting with it in some way, and, by hypothesis, inevitably setting it off.

Quantum mechanics, however, allows one to be certain some fraction of the time that the bomb is in place, without setting it off. After the first beam splitter of the interferometer, a photon has a 50% chance of heading towards the bomb, and thus exploding it. On the other hand, if the photon takes the path without the bomb, there is no more interference, since the nonexplosion of the bomb provides welcher Weg ("which way") information (see Sect. 80.4.1). Thus the photon reaches the final beam splitter and chooses randomly between the two exit ports. Some of the time (25%), it leaves by output port B, something which never happened in the absence of the bomb. This immediately implies that the bomb (or some nontransmitting object) is in place – even though (since the bomb is unexploded) it has not interacted with any photon; EV termed this an "interaction-free measurement". (We prefer the more general description "quantum interrogation", which then includes cases – e.g., detecting a semi-transparent or quantum object – where it may not be possible to logically exclude the possibility of an interaction.) It is the mere possibility that the bomb could have interacted with a photon which destroys interference. An initial experimental implementation of these ideas [80.137] used down-conversion to prepare the single photon states (Sect. 80.1), and a single-photon detector as the "bomb". Subsequently the technique was implemented incorporating focusing lenses, which would enable the image (more correctly, the silhouette) of an object to be determined with less than one photon per "pixel" being absorbed [80.138].

By adjusting the beamsplitter reflectivities in the above example, one can achieve at most a 50% fraction of measurements that are interaction-free. An improved method, relying on the quantum Zeno effect [80.118] (Sect. 80.5.1), was discovered with which one can in principle make this fraction arbitrarily close to 1 [80.137]. For example, consider a photon initially in cavity #1 of two identical cavities coupled by a lossless beam splitter whose reflectivity $R = \cos^2(\pi/2N)$. If the photon's coherence length is shorter than the cavity length, after N cycles the photon will with certainty be located in cavity #2, due to an interference effect (the equivalent of a π-pulse interaction). However, if cavity #2 instead contains an absorbing object (e.g., the ultra-sensitive bomb), at each cycle there is only a small chance ($= 1 - R$) that the photon will be absorbed; otherwise, the *non* absorption projects the photon wave packet entirely back onto cavity #1. After all N cycles, the total probability for the photon to be absorbed by the object is $1 - R^N$, which goes to 0 as N becomes large. (In practice, unavoidable losses in the system limit the maximum number of cycles and hence the achievable performance [80.139].) The photon effectively becomes trapped in cavity #1, thus indicating umambiguously the presence of the object in cavity #2.

This quantum-Zeno version of interrogation was first implemented using the inhibited rotation of a photon's polarization (the object to be detected blocked one arm of a polarizing interferometer through which the photon was repeatedly cycled), achieving an efficiency of 75% [80.139]. A cavity-based implementation, in which the presence of the absorbing object inside a high-finesse cavity vastly increased the reflection off the cavity [80.140], detected the presence of the object with only 0.15 photons on average being absorbed or scattered [80.141].

80.5.4 Weak and "Protected" Measurements

Aharonov, *Albert*, and *Vaidman* extended quantum measurement theory by introducing "weak" measurement, a procedure that determines a physical property of a quantum system belonging to an ensemble that is both preselected and postselected [80.142, 143]. In the standard theory, a quantum system is measured by entangling its eigenstates with distinguishable pointer states of a measurement device, completely resolving the observable eigenvalue spectrum. The measurement can be weakened by increasing the overlap of the pointer states, consequently reducing the resolution of the eigenstates. When performed between two measurements this can give surprising results – in contrast to ordinary expectation values, the pointer can lie outside the range of the eigenvalue spectrum of the measured observable $\mathbf{O}$. The "weak value" $\mathbf{O}_w \equiv \langle \Psi_{\text{ini}}|\mathbf{O}|\Psi_{\text{fin}}\rangle / \langle \Psi_{\text{fin}}|\Psi_{\text{ini}}\rangle$ between preselected ($|\Psi_{\text{ini}}\rangle$) and postselected ($|\Psi_{\text{fin}}\rangle$) states completely characterizes the outcome of the weak measurement. Weak measurements do not disturb each other, so that the weak values of non-commuting observables can be measured simultaneously [80.144, 145]. Furthermore, they have proved to yield meaningful results in many different circumstances, e.g., Hardy's paradox of Sect. 80.6.5 [80.146], measurement of negative kinetic energies [80.147], and the "observation" of a single particle in two locations [80.148]. One other potentially powerful application of weak measurements is the amplification of weak signals, which was first demonstrated by amplifying the birefringence-induced small displacement of optical fields [80.149, 150]. It was recently shown that weak measurements of this kind in fact arise naturally in fiber optics telecom networks, due to polarization-mode dispersion and polarization-dependent losses [80.151]. One recent proposal [80.152] suggests that the controversy over whether or not "welcher Weg" information may be obtained (and interference consequently destroyed) without disturbing a particle's momentum (Sect. 80.4.2) may be resolved by making weak measurements of momentum inside an interferometer.

The above results push us to reexamine our interpretation of wave functions. We customarily use the wave function only as a calculational tool, but we have also learned that it is in some sense physical, and should not be regarded merely as some distribution from classical statistics. One proposal [80.153, 154] suggests that the wave function of a single particle should be regarded as a real entity. When a state is "protected" from change, e.g., by an energy gap, and measurements are performed sufficiently "gently", one should be able to determine not just the expectation value of position, but the wave function at many different positions, without altering the state of the particle. (This idea of measuring the entire wave function of a single particle should not be conflated with *Raymer* et al.'s fascinating work on the reconstruction of the quantum state of a light field by repeated sampling of a large ensemble; see [80.155] and Sect. 78.4.) No violation of the uncertainty principle or the no-cloning theorem (Sect. 80.7.1) arises from this, as the ability to "protect" a state relies on some preexisting knowledge about the state; but it assigns a deeper significance to the wave function, one *Aharonov* terms "ontological," as opposed to merely epistemological (but see also [80.156]).

80.6 The EPR Paradox and Bell's Inequalities

80.6.1 Generalities

Nowhere is the nonlocal character of the quantum mechanical entangled state as evident as in the "paradox" of *Einstein*, *Podolsky*, and *Rosen* (EPR) [80.157], the version of *Bohm* [80.158], and the related inequalities by *Bell* [80.159, 160]. Consider two photons traveling off back-to-back, described by the entangled state

$$|\psi^-\rangle = (|\mathrm{H}_1, \mathrm{V}_2\rangle - |\mathrm{V}_1, \mathrm{H}_2\rangle)/\sqrt{2}\,, \quad (80.9)$$

where the letters denote horizontal (H) or vertical (V) polarization, and the subscripts denote photon propagation direction. This state, analogous to the singlet state of a pair of spin-1/2 particles, is isotropic – it has the same form regardless of what basis is used to describe it. Measurement of any polarization component for one of the particles will yield a count with 50% probability; individually, each particle is unpolarized. Nevertheless, if one measures the polarization component of particle 1 in any basis, one can predict with certainty the polarization of particle 2 in the same basis, seemingly without disturbing it, since it may be arbitrarily remote. Therefore, according to EPR, to avoid any nonlocal influences one should ascribe an "element of reality" to every component of polarization. A quantum mechanical state cannot specify that much information, and is consequently an incomplete description, according to the EPR argument. The intuitive explanation implied by EPR is that the particles leave the source with definite, correlated properties, determined by some local "hidden variables" not present in quantum mechanics (QM).

For two entangled particles, a local hidden variable (LHV) theory can be made which correctly describes perfect correlations or anti-correlations (i. e., measurements made in the same polarization basis). The choice of an LHV theory versus QM is then a philosophical decision, not a physical one. However, in 1964 *John Bell* [80.159] discovered that QM gives different statistical predictions than does any LHV theory, for situations of nonperfect correlations (i. e., analyzers at intermediate angles). Bell's inequality (BI) constrains various joint probabilities given by any local realistic theory, and was later generalized to include any model incorporating locality [80.161, 162], and also extended to apply to real experimental situations [80.163, 164]. With the caveat of supplementary assumptions (Sect. 80.6.4), Bell's inequalities have now been tested many times, and the vast majority of experiments have violated them, in support of QM. One general interpretation is that the predictions of QM cannot be reproduced by any completely local theory. It must be that the results of measurements on one of the particles depend on the results for the other, and these correlations are not merely due to a common cause at their creation [80.165, 166].

80.6.2 Polarization-Based Tests

The first BI tests were performed with pairs of photons produced via an atomic cascade, and a later version incorporated rapid (albeit periodic) switching of the analyzers [80.167, 168]. Unfortunately, the angular correlation of the cascade photons is not very strong. In contrast, the strong correlations of the down-converted photons make them ideal for such tests, the first of which were performed using setups essentially identical to that already discussed in connection with quantum erasure (Fig. 80.4). Orthogonally polarized (e.g., horizontal and vertical) but otherwise identical photons are combined on a nonpolarizing 50–50 beam splitter. If one considers only the events with a single photon in each output (i. e., ignoring the cases for which both photons use the same beam splitter output port), one obtains the (post-selected) entangled state (80.9) [80.169, 170]. (In fact, this technique is now used as a method for characterizing the indistinguishability of photons from independent sources, e.g., quantum dots [80.171] or independent down-conversion crystals [80.172, 173].)

Down-conversion schemes have also been developed to produce entangled states without the need to post-select out half of the photons. For example, consider a type-I phase-matched crystal (Sect. 72.2.2) that down-converts H-polarized pump photons into V-polarized pairs; and an adjacent, identical crystal that is rotated by 90°, thus down-converting V-polarized pump photons into H-polarized pairs. By coherently pumping the two crystals with light polarized at $|45\rangle \equiv (|V\rangle + |H\rangle)/\sqrt{2}$, one obtains the entangled state $(|HH\rangle + |VV\rangle/\sqrt{2}$. (More generally, pumping $\alpha|V\rangle + \mathrm{e}^{i\varphi}\beta|H\rangle$ produces arbitrary nonmaximally entangled states of the form $\alpha|HH\rangle + \mathrm{e}^{i\varphi}\beta|VV\rangle$[80.174].) Such a source has produced the largest and fastest violations of Bell inequalities to date (over 200-σ violation in less than 1 second) [80.175, 176].

Using type-II phase-matching one also can produce polarization entanglement from a single crystal [80.177]. One member of each down-conversion pair is emitted along an ordinary polarized cone while the other is emitted along an extraordinary polarized cone. If the photons happen to be emitted along the intersection of the two cones, neither photon will have a definite polarization – they will be in the state $(|HV\rangle + |VH\rangle/\sqrt{2}$. This entanglement source has now been used in a variety of quantum investigations, including Bell inequality tests [80.177, 178], quantum cryptography [80.179, 180] (Sect. 80.7.4 and Chapt. 81) and teleportation [80.181, 182], and as a resource for studying entanglement of more than 2 photons [80.37–42].

80.6.3 Nonpolarization Tests

The advent of parametric down-conversion has also led to the appearance of several nonpolarization-based BI tests, using, for example, an entanglement of the photon momenta (Fig. 80.8) [80.183]. By use of small irises (labeled 'A' in the figure), *Rarity* and *Tapster* examined four down-conversion modes: 1s, 1i, 2s, and 2i. Beams

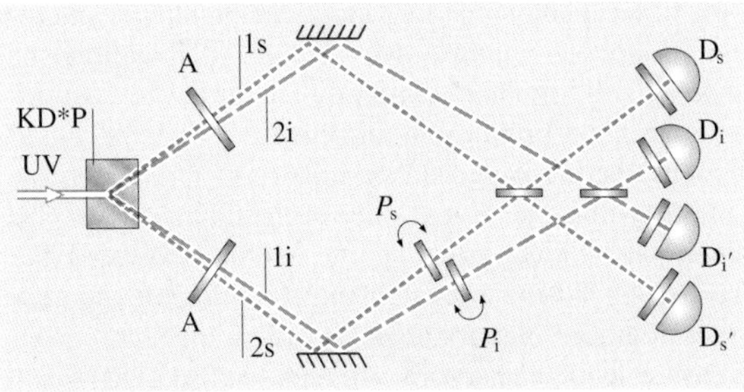

Fig. 80.8 Outline of *Rarity* and *Tapster* apparatus used to demonstrate a violation of a Bell's inequality based on momentum entanglement [80.183]

1s and 1i correspond to one pair of conjugate photons; beams 2s and 2i correspond to a different pair. Photons in beams 1s and 2s have the same wavelength, as do photons in beams 1i and 2i. With proper alignment, after the beam splitters there is no way to tell whether a pair of photons came from the 1s–1i or the 2s–2i paths. Consequently, the coincidence rates display interference, although the singles rates at the four detectors indicated in Fig. 80.8 remain constant. This interference depends on the difference of phase shifts induced by rotatable glass plates P_i and P_s in paths 1i and 2s, respectively, and is formally equivalent to the polarization case considered above, in which it is the difference of polarization-analyzer angles that is relevant. By measuring the coincidence rates for two values for each of the phase shifters – a total of four combinations – the experimenters were able to violate an appropriate BI. One interpretation is that the emission directions of a given pair of photons are not elements of reality.

Momentum conservation in the down-conversion process (80.4) also leads to entanglement directly in the spatial modes in the correlated photons. For example, *Zeilinger* et al. [80.184, 185] and *White* et al. [80.186] have demonstrated entanglement between the orbital angular momentum of the photons, of the form $(|+1,-1\rangle + \epsilon|0,0\rangle + |-1,+1\rangle$, where 0 and ± 1 respectively denote modes with no orbital angular momentum (gaussian spatial profiles) and $\pm\hbar$ (Laguerre–Gauss-Vortex modes). Note that this enables one to investigate correlations for degrees of freedom that reside in larger Hilbert spaces than do the 2-level systems (e.g., polarization) discussed above. The nonlocal spatial correlations of the down-conversion photons have also given rise to many interesting experiments in the area of quantum imaging [80.187–190], where one is able to obtain spatial resolution beyond that predicted by the usual $\sqrt{N}$ shot-noise limitations.

Several groups [80.191, 192] have violated a BI based on energy–time entanglement of the photons [80.193]. In the method due to *Franson*, one member of each down-converted pair is directed into an unbalanced Mach–Zehnder-like interferometer, allowing both a short and long path to the final beam splitter; the other photon is directed into a separate but similar interferometer. There arises interference between the indistinguishable processes ("short–short" and "long–long") which could lead to coincidence detection. Using fast detectors to select out only these processes, the reduced state (80.9) is

$$|\psi\rangle = \frac{1}{2}\left(|S_1, S_2\rangle - e^{i\phi}|L_1, L_2\rangle\right) , \qquad (80.10)$$

where the letters indicate the short or long path, and the phase is the sum of the relative phases in each interferometer. Although no fringes are seen in any of the singles count rates, the high-visibility coincidence fringes (Fig. 80.9) lead to a violation of an appropriate BI. One conclusion is that it is incorrect to ascribe to the photons a definite time of emission from the crystal, or even a definite energy, unless these observables are explicitly measured.

This same sort of arrangement, modified to work with a pulsed pump, has been used to demonstrate the longest violation of local realism, in which *Gisin's* group has observed a 16-σ BI violation (modulo the detection and timing loopholes discussed in Sect. 80.6.4) with photons separated by 10.9 km [80.194]. In a related experiment, they have used a similar system to place limits on the "speed of collapse" of the 2-photon wave function, i. e., how fast a nonlocal "signal" would need to propagate from one side of the experiment to the other to account for the measured nonclassical correlations. Depending on some assumptions about the detection process and which inertial frame of reference is considered, the nonlocal-influence speed was constrained to be at least $10^4\,c$ to $10^7\,c$ [80.195]. In one interesting variant, the researchers arranged to have moving detectors, such that in the local reference frame of each detector, it was the other detector which initiated the collapse. (Due to the experimental difficulty of accelerating actual detectors to high velocities, a rapidly rotating absorbing disk was placed close to one output port of a polarizing beamsplitter; following ideas discussed in Sect. 80.5.3, the non-absorbance of the photon by the absorber was deemed sufficient to cause a reduction of the wave function.) As expected, the measured correlations were in no way reduced, but this experiment did rule out one alternative theory of nonlocal collapse [80.196, 197]. Finally,

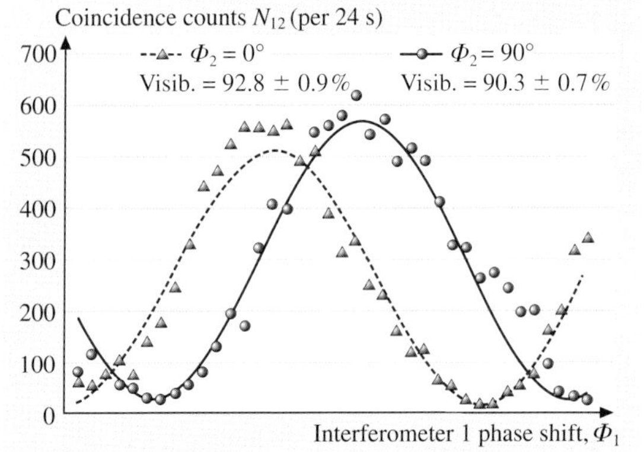

Fig. 80.9 High-visibility coincidence fringes in a Franson dual-interferometer experiment [80.192] for two values of the phase in interferometer 2 as the phase in interferometer 1 is slowly varied. The curves are sinusoidal fits

by using more than two possible creation times, e.g., with a mode-locked pulsed laser, Gisin's group demonstrated entanglement for a two-photon state in a Hilbert space of at least dimension 11 [80.198].

EPR-like correlations have also been observed directly between two correlated field modes via homodyne tomography [80.199], though the joint Wigner function is positive-definite here. In contrast, a homodyne measurement on the state of a *single* photon split between two paths (Sect. 80.4.4) took advantage of the nonclassical nature of the initial state to violate a Bell inequality [80.46]. Like all existing Bell-inequality experiments, this work suffered from several loopholes (Sect. 80.6.4), but a new proposal suggests that the high efficiency of homodyne detection may provide a unique opportunity for a loophole-free test of Bell's inequalities [80.200].

80.6.4 Bell Inequality Loopholes

In fact, to date no single experiment has unambiguously violated a Bell inequality, due to the existence of two experimental challenges, the detection and locality "loopholes". All of the experiments discussed thus far have required supplementary assumptions, e.g., the "fair-sampling" assumption that the fraction of pairs detected is representative of the entire ensemble emitted by the source. [In fact, for the entangled photons emitted in the atomic cascade experiments, this assumption is manifestly false, because the strong polarization correlations only exist for those photons emitted nearly in opposite directions. If one were to collect all of the emitted photon pairs, they would not lead to a violation (see [80.201] for a fuller discussion).] To close the locality (or "timing") loophole requires that the analyzer settings be switched rapidly and randomly, in order to guarantee that no (sub)luminal information transfer could account for the observed correlations. The necessary conditions have been met only in the down-conversion experiment by *Zeilinger* et al., which separated the photons by 400 m and used ultrafast random number generation and electronic polarization-analysis choice to ensure space-like separated observers [80.178]. However, in that experiment the detection efficiency was less than 5%.

In order to understand the detection loophole, consider the Clauser–Horne (CH) form of the Bell inequality [80.164], which relates the directly observable singles rates S_1 and S_2 and the coincidence rate C_{12}, rather than "inferred" probabilities, by

$$C_{12}(a,b)+C_{12}(a,b')+C_{12}(a',b)-C_{12}(a',b') \le S_1(a)+S_2(b)\,, \tag{80.11}$$

where a and a' (b and b') are any pair of analyzer (e.g., polarizer) settings at detector 1 (2). For certain choices of a, a', b, and b', quantum mechanics predicts the left hand side of the CH inequality can exceed the right hand side. However, in practice this is very difficult to observe, since the coincidence rates fall as η^2 (η is the detection efficiency), compared to the singles rates on the right hand side, which fall only as η. In order to close the detection loophole, one requires $\eta \ge 83\%$ [80.201] (for maximally entangled photons. (*Eberhard* has shown that the required detection efficiency may be reduced to 67% by using nonmaximally entangled quantum systems [80.202, 203]. The idea is that one can choose the analysis settings a and b to reduce the value of the RHS of the CH inequality.) In essence, one requires high detection efficiencies to ensure that the contributions from undetected events are not sufficient to cause the total ensemble to satisfy the Bell inequality even while the detected events violate it.

To date, only the entangled-ion experiment of *Wineland* et al. [80.204] has had sufficiently high efficiencies to close the detection loophole. In this experiment the entangled variables were the hyperfine energy levels of $^9Be^+$ ions. By employing a cycling transition that leads to the emission of many photons if the atom is in one of the states, a detection efficiency in excess of 98% was achieved, allowing an 8-σ Bell inequality violation. However, because the ions were separated by only 3 μm in the same linear Paul trap, and in fact were measured using the same laser pulse, there was no possibility of closing the locality loophole. More recently, *Monroe* et al. have demonstrated the entanglement of a trapped ion and a photon [80.205], and have used this to violate a Bell inequality [80.206]. Similar experiments have also enabled the production of up to four entangled ions [80.207]. Though neither of the experimental loopholes was closed in these experiments, they are noteworthy as the first controlled demonstrations of entanglement in massive particles. Efforts are now underway to attempt a direct violation of (80.11) with no auxiliary assumptions using down-conversion photons and high-efficiency ($> 85\%$) single-photon detectors [80.208, 209], in addition to atomic schemes [80.210] and homodyne schemes [80.200] (Sect. 80.6.3), which enjoy even higher intrinsic efficiency.

80.6.5 Nonlocality Without Inequalities

In the above experiments for testing nonlocality, the disagreement between quantum predictions and Bell's constraints on local realistic theories are only statistical. *Greenberger*, *Horne*, and *Zeilinger* (GHZ) pointed out that in some systems involving three or more entangled particles, a contradiction could arise even at the level of perfect correlations [80.211, 212]. A schematic of one version of the GHZ Gedankenexperiment is shown in Fig. 80.10. The source at the center is posited to emit

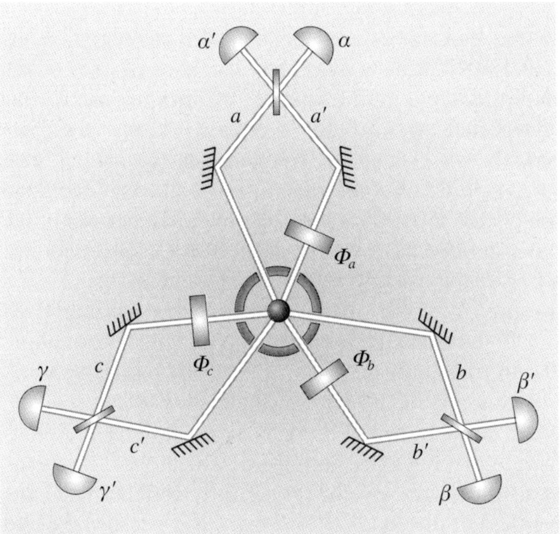

Fig. 80.10 A three-particle Gedankenexperiment to demonstrate the inconsistency of quantum mechanics and any local realistic theory. All beam splitters are 50–50 [80.111]

trios of correlated particles. Just as the *Rarity–Tapster* experiment selected two pairs of photons (Fig. 80.8), the GHZ source selects two trios of photons; these are denoted by abc and $a'b'c'$. Hence, the state coming from the source may be written

$$|\psi\rangle = \left(|abc\rangle + |a'b'c'\rangle\right)/\sqrt{2}\,. \tag{80.12}$$

After passing through a variable phase shifter (e.g., ϕ_a), each primed beam is recombined with the corresponding unprimed beam at a 50–50 beam splitter. Detectors (denoted by Greek letters) at the output ports signal the occurrence of triple coincidences. The following simplified argument conveys the spirit of the GHZ result.

Given the state (80.12), one can calculate from standard QM the probability of a triple coincidence as a function of the three phase shifts:

$$P(\phi_1, \phi_2, \phi_3) = \frac{1}{8}\left[1 \pm \sin(\phi_a + \phi_b + \phi_c)\right]\,, \tag{80.13}$$

where the plus sign applies for coincidences between all unprimed detectors, and the minus sign for coincidences between all primed detectors. For the case in which all phases are 0, it will occasionally happen (1/8th of the time) that there will be a triple coincidence of all primed detectors. Using a "contrafactual" approach, we ask what would have happened if ϕ_a had been $\pi/2$ instead. By the locality assumption, this would not change the state from the source, nor the fact that detectors β' and γ' went off. But from (80.13) the probability of a triple coincidence for primed detectors is zero in this case; therefore, we can conclude that detector α would have "clicked" if ϕ_a had been $\pi/2$. Similarly, if ϕ_b or ϕ_c had been $\pi/2$, then detectors β or γ would have clicked. Consequently, if all the phases had been equal to $\pi/2$, we would have seen a triple coincidence between unprimed detectors. But according to (80.13) this is impossible: the probability of triple coincidences between unprimed detectors when all three phases are equal to $\pi/2$ is strictly zero! Hence, if one believes the quantum mechanical predictions for these cases of perfect correlations, it is not possible to have a consistent local realistic model.

Down-conversion experiments have enabled the production of 3- and 4-photon GHZ states, with results in good agreement with theory (the all-or-nothing arguments given above become inequalities in any real experiment) [80.37–39].

By similar arguments, *Hardy* has shown the inconsistency of quantum mechanics and local realism in a Gedankenexperiment with just two particles [80.213, 214]. When the arguments are suitably modified to deal with real experiments, inequalities once again result; these have also been experimentally violated, using non-maximally entangled states from down-conversion [80.215, 216], further underscoring the inconsistency between quantum theory and locality.

80.7 Quantum Information

80.7.1 Information Content of a Quantum: (No) Cloning

The inherent nonlocality of particles in an entangled state cannot be used to transmit superluminal messages. For example, if A and B receive a polarization-entangled pair of photons, which A then collapses in a certain basis by performing a polarization measurement, B can only extract one bit of information from a measurement on his photon – this bit corresponds not to A's choice of basis, but to the (random) outcome of A's measurement. However, instantaneous communication would be possible if one could make copies ("clones") of a single photon in an unknown polarization state: by performing measurements on n copies of his photon, B could determine its polarization to a resolution of n bits, thereby accurately determining A's choice of basis. Taking into account quantum fluctuations [80.217], one finds that no physically allowed amplifier can make a sufficiently faithful copy for such a scheme to work – it is impossible to clone an unknown quantum state [80.218]. A simple proof is as follows. Consider an ideal cloner, initially in the state $|i\rangle_c$, which would take $|0\rangle|i\rangle_c \rightarrow |0\rangle|0\rangle_c$ and $|1\rangle|i\rangle_c \rightarrow |1\rangle|1\rangle_c$. For an accurate copier, one would expect $(|0\rangle+|1\rangle)|i\rangle_c \rightarrow (|0\rangle+|1\rangle)(|0\rangle_c+|1\rangle_c)$, but the linearity of quantum transformations instead yields $|0\rangle|0\rangle_c+|1\rangle|1\rangle_c$, i.e., an entangled state.

Although perfect cloning is impossible, it is nevertheless possible to create copies which are "pretty good". Specifically, using an optimal cloning strategy, one can in principle create a copy with a fidelity of 5/6 with the original state [80.219, 220]. Such a cloning procedure has been experimentally realized by several groups, e.g., relying on stimulated emission, or sending the photon to be cloned through a low noise optical amplifier, with results matching the theoretical predictions [80.221–223].

80.7.2 Super-Dense Coding

The previous considerations make the work by *Bennett* et al. on "quantum teleportation" and related effects all the more remarkable. In the quantum dense-coding protocol, a single photon can be used to transfer two bits of information, when it is part of an entangled EPR pair [80.224]. Again consider A and B, each possessing one photon of such a pair. By manipulating only her photon (via a polarization rotator and a phase shifter), A can convert the initial joint state $|\psi^-\rangle$ (80.9) into any of the four two-particle "Bell states" $[|\psi^{\pm}_{AB}\rangle = (|\mathrm{H}_A, \mathrm{V}_B\rangle \pm |\mathrm{V}_A, \mathrm{H}_B\rangle)/\sqrt{2}$, $|\phi^{\pm}_{AB}\rangle = (|\mathrm{H}_A, \mathrm{H}_B\rangle \pm |\mathrm{V}_A, \mathrm{V}_B\rangle)/\sqrt{2}]$, and then send her photon to B. By making a suitable measurement on both photons, B can then in principle determine which of the four states A produced [80.225]: A's single photon carried two bits. This protocol has been experimentally realized using down-conversion photons [80.226], though only two of the four Bell states could be reliably distinguished. (Standard polarization Bell state analysis is implemented by combining the two photons on a nonpolarizing 50–50 beam splitter. For any of the triplet states $|\psi^+\rangle$, $|\phi^{\pm}\rangle$, the photons will both travel to the same output port due to the Hong–Ou–Mandel interference discussed in Sect. 80.3.3; only for the state $|\psi^-\rangle$ will the photons travel to different outputs, resulting in a coincidence detection.)

80.7.3 Teleportation

In the even more striking quantum teleportation effect [80.227], an unknown polarization state f (with its in-principle infinite amount of information) can be "teleported" from A to B, if each already possesses one photon of an EPR pair (e.g., in the singlet state $|\psi^-_{AB}\rangle$). First, A jointly measures her EPR photon and the photon F (whose state f is to be teleported) in the basis defined by the Bell states of these two photons. Via a mere two bits of classical information, A then informs B which of the four (equally probable) Bell states she actually measured. With this information, B can transform the state of his EPR particle into f. For example, if A found the singlet state $|\psi^-_{BF}\rangle$, then the polarization of her EPR photon must have been orthogonal to that of F (because the polarizations of particles in a singlet state are always perfectly anticorrelated, regardless of the quantization basis). But because the two EPR photons were initially also in a singlet state, their polarizations must also be orthogonal, so B's EPR photon is already in state f. If instead A found $|\phi^-_{BF}\rangle$, for example, then B simply makes the same transformation (with a polarization rotator) that would have changed $|\psi^-_{AB}\rangle$ into $|\phi^-_{AB}\rangle$, again leaving his photon in the state f. Thus, although one may only extract one bit of (normally useless) information from an EPR particle, the perfect correlations may be used to transfer an infinite amount of information, i. e., precise specification of a point in the state space of the particle (the Poincaré sphere for a photon or the Bloch sphere for an electron). The "no-cloning" theorem (Sect. 80.7.1) is not violated, since A irrevocably alters the state of F by the measurement she performs, leaving only one particle in f.

A number of experiments have now experimentally realized quantum teleportation. The first of these used polarization-entangled down-conversion photons [80.181], but was limited by the impossibility to resolve all four Bell states using only linear optics. Teleportation of continous-variable states has also been observed [80.228], using twin-beam squeezed states (Sect. 80.2.3). Recently, the first teleportation in a matter system has been achieved: the groups of *Blatt* et al. [80.229] and *Wineland* et al. [80.230] have successfully teleported the (energy) quantum state of an ion to a separate ion. Although the overall distance was less than 1 mm, these challenging experiments are significant because they incorporate most of the techniques necessary for scalable quantum information processing in an ion-trap system (see Sect. 81.7.2).

In an interesting extension of the original teleportation protocol, one can ask what happens if the photon to be teleported is itself entangled to a 4th photon G. In this case, known as "entanglement swapping", a successful teleportation will lead to the entanglement of photons G and B, even though these have never directly interacted. Entangled down-conversion photons have been used to demonstrate such "entanglement swapping" achieving a violation of Bell's inequalities between the two noninteracting photons [80.231]; similar results have been achieved with continous variables as well [80.232]. Such procedures may one day enable construction of a quantum "repeater" which could enable the transmission of quantum information over long distances [80.233, 234].

A fundamental connection between quantum information and black holes has recently been suggested by Lloyd and Ng [80.235] and others. The basic idea is that every physical object, including a black hole, can be thought of as a quantum computer that unitarily transforms input states to output states; i. e., "in" quantum bits ("qubits") can always be reversibly interconverted into "out" qubits, thus obeying time-reversal symmetry. To resolve the paradox of the apparent loss of information of matter falling into a black hole, Lloyd and Ng propose that pairs of entangled photons can materialize at the event horizon of a black hole. One member of the photon pair flies outward to become the Hawking blackbody radiation; the other falls into the black hole and hits the singularity together with the matter that formed the hole. The annihilation of the infalling photon acts as a measurement on the infalling matter in a quantum teleportation-like process, transporting the information contained in the infalling matter to the outgoing Hawking radiation, using the Horowitz–Maldacena mechanism [80.236].

80.7.4 Quantum Cryptography

Although EPR schemes cannot send signals superluminally, they have other potential applications in cryptography. In the "one-time pad" of classical cryptography [80.237], two collaborators share a secret "key" (a random string of binary digits) in order to encode and decode a message. Such a key may provide an absolutely unbreakable code, provided that it is unknown to an eavesdropper. The problem arises in key distribution: any classical distribution scheme is subject to noninvasive eavesdropping, e.g., using a fiber-coupler to tap the line, without disturbing the transmitted classical signal. In quantum cryptography proposals, security is guaranteed by using single-photon states [80.238, 239], some of the schemes employing particles prepared in an EPR-entangled state. Each collaborator receives one member of each correlated pair, and measures the polarization (or whatever degree of freedom is carrying the information) in a random basis. After repeating the process many times, the two then discuss publicly which bases were used for each measurement, but not the actual measurement results. The cases where different bases were chosen are not used for conveying the key, and may be discarded, along with instances where one party detected no photon. In cases where the same bases were used, however, the participants will now have correlated information, from which a random, shared key can be generated. As long as single photons are used, any attempt at eavesdropping, even one relying on QND, will necessarily introduce errors due to the uncertainty principle. If the eavesdropper uses the wrong basis to study a photon before sending it on to the real recipient, the very act of measuring will disturb the original state.

Although one can perform quantum cryptography with single-photon states or even weak coherent states (Sect. 81.2.1), there are a number of advantages to using entanglement: there is inherent randomness; photons from different pairs show no correlations (unlike multiple photons in an attenuated coherent state, which may be used by an eavesdropper to gain information), and any information leaked to other degrees of freedom automatically shows up as increased error rate [80.240]. This last feature means that the sources are "self-checking" in a sense, and one could even let an adversary have control over the source, secure in the knowledge that any tampering would become evident in the nonlocal correlations. A number of quantum cryptography implementations using down-conversion photon pairs have been realized, either using polarization entanglement [80.179, 180, 241, 242], or energy–time entanglement [80.243, 244]. One main experimental challenge will be to increase the rates of entangled pair production (the typical source emission rate for these cryptography experiments is only 10 000 /s), to make them competitive with current weak coherent pulse schemes (which can easily operate at over 10 MHz).

80.7.5 Issues in Causality

Outstanding causal paradoxes in optics include the paradox of *Barton* and *Scharnhorst* [80.245], closely related to the Casimir effect (Sect. 79.2.4). In this case, the amplitude for light-by-light scattering is modified by the presence of closely-spaced parallel conducting plates. It appears that this can lead to propagation of light in vacuum (albeit a vacuum "colored" by the presence of the Casimir plates) faster than c. *Hegerfeldt* has pointed out similar paradoxes in connection with localization of any particle in a quantum field theory [80.246] and with the *Glauber* theory of photodetection (Sect. 78.4) [80.247]. However, at least for the simplest example – the interaction between two widely separated atoms – as long as one considers only probabilities that depend on the separation r, the second atom cannot be excited by light from the first until after a time r/c [80.248].

80.8 The Single-Photon Tunneling Time

80.8.1 An Application of EPR Correlations to Time Measurements

In this section we discuss experiments involving the quantum propagation of light in matter. Due to the sharp time correlations of the paired photons from spontaneous down-conversion, one can use the HOM interferometer (Sect. 80.3.3, and Fig. 80.4) to measure very short relative propagation delay times for the signal and idler photons. One early application was therefore to confirm that single photons in glass travel at the group velocity [80.91]. At least until recently, the only quantum theory of light in dispersive media was an ad hoc one [80.249–256]. The shift of the interference dip resulting from a medium introduced into one of the interferometer arms can be accurately measured by determining how much the path lengths must be changed to compensate the shift and recover the dip. This result suggests that when looking for a microscopic description of dielectrics, it is unnecessary to consider the medium as being polarized by an essentially classical electric field due to the collective action of all photons present, and reradiating accordingly. Linear dielectric response is not a collective effect in this sense – each photon interferes only with itself (as per Dirac's dictum) as it is partially scattered from the atoms in the medium. The single-photon group velocity thus demonstrates "wave–particle unity."

The standard limitation for measurements of short-time phenomena is that to have high time-resolution, one needs short pulses (or at least short coherence lengths), but these in turn require broad bandwidths and are therefore very susceptible to dispersive broadening. It is a remarkable consequence of the EPR energy correlations of the down-conversion photons (Sect. 80.6.3) that time measurements made with the HOM interferometer are essentially immune to such broadening [80.91, 257, 258]. In effect, the measurement is sensitive to the difference in emission times while the broadening is sensitive to the sum of the frequencies. While frequency and time cannot both be specified for a given pulse, the crucial feature of EPR correlations such as those exhibited by down-converted photons is that this difference and this sum correspond to commuting observables, and both may be arbitrarily well defined. The photon which reaches detector 1 could either have traversed the dispersive medium and been transmitted, or traversed the empty path and been reflected, leaving its twin to traverse the medium. The medium thus samples both of the (anticorrelated) frequencies, leading to an automatic cancellation of any first-order (and in fact, all odd-order) dispersive broadening. Measurements can be more than 5 orders of magnitude more precise than would be possible via electronic timing of direct detection events, and in principle better than those performed with nonlinear autocorrelators (which rely on the same nonlinear physics as down-conversion, but do not benefit from a cancellation of dispersive broadening).

80.8.2 Superluminal Tunneling Times

Another well-known problem in the theory of quantum propagation is the delay experienced by a particle as it tunnels. There are difficulties associated with calculating the "duration" of the tunneling process, since evanescent waves do not accumulate any phase [80.259–261]. First, the kinetic energy in the barrier region is negative, so the momentum is imaginary. Second, the transit time of a wavepacket peak through the barrier, defined in the stationary phase approximation by

$$\tau^{(\phi)} \equiv \partial\left[\arg t(\omega)\,\mathrm{e}^{\mathrm{i}kd}\right]/\partial\omega\,, \qquad (80.14)$$

tends to a constant as the barrier thickness diverges, in seeming violation of relativistic causality. (Actually, it is shown in [80.88] that such saturation of the delay time is a natural consequence of time-reversal symmetry, and in [80.262] that one can deduce from the principle of causality itself that *every* system possesses a superluminal group delay, at least at the frequency where its transmission is a minimum.) For example, the transmission function for a rectangular barrier,

$$t(k,\kappa) = \frac{\mathrm{e}^{-\mathrm{i}kd}}{\cosh\kappa d + \mathrm{i}\frac{\kappa^2-k^2}{2k\kappa}\sinh\kappa d}\,, \qquad (80.15)$$

leads in the opaque limit ($\kappa d \gg 1$) to a traversal time of $2m/\hbar k\kappa$, independent of the barrier width d. The same result applies to photons undergoing frustrated total internal reflection [80.88], when the mass m is replaced by $n^2\hbar\omega/c^2$, and similar results apply to other forms of tunneling.

Some researchers have therefore searched for some more meaningful "interaction time" for tunneling, which might accord better with relativistic intuitions and perhaps have implications for the ultimate speed of devices relying on tunneling [80.263]. The "semiclassical time" corresponds to treating the magnitude of the (imaginary) momentum as a real momentum. This time is of interest mainly because it also arises in *Büttiker* and *Landauer's* calculation of the critical timescale in problems involving oscillating barriers, which they take to imply that it is a better measure of the duration of the interaction than is the group delay [80.259]. The Larmor time [80.264] is one of the early efforts to attach a "clock" to a tunneling particle, in the form of a spin aligned perpendicular to a small magnetic field confined to the barrier region. The basic idea is that the amount of Larmor precession experienced by a transmitted particle is a measure of the time spent by that particle in the barrier. This clock turns out to contain components corresponding both to the distance-independent "dwell time" and the linear-in-distance semiclassical time. Curiously, the most common theories for tunneling times become superluminal in certain cases anyway, whether or not they deal with the motion of wave packets.

Here, we will restrict ourselves to discussing the time of appearance of a peak of a single-photon wave packet. While other tunneling-time experiments have been performed in the past [80.265], optical tests offer certain unique advantages [80.266], including the ease of construction of a barrier with no dissipation, very little energy-dependence, and a superluminal group delay. The transmitted wave packets suffer little distortion, and are essentially indistinguishable from the incident wave packets. At a theoretical level, the fact that photons are described by Maxwell's (fully relativistic) equations is an important argument against interpreting superluminal tunneling predictions as a mere artifact of the nonrelativistic Schrödinger equation. Also, one is denied the recourse suggested by some workers [80.267] of interpreting the superluminal appearance of transmitted peaks to mean that only the high-energy components (which, for matter waves, traveled faster even before reaching the barrier [80.268]) were transmitted.

80.8.3 Tunneling Delay in a Multilayer Dielectric Mirror

A suitable optical tunnel barrier can be a standard multilayer dielectic mirror. The alternating layers of low and high index material, each one quarter-wave thick at the design frequency of the mirror, lead to a photonic bandgap [80.269] analogous to that in the Kronig–Penney model of solid state physics (Sect. 79.2.6). The gap represents a forbidden range of energies, in which the multiple reflections will interfere constructively so as to exponentially damp any incident wave. The analogy with tunneling in nonrelativistic quantum mechanics arises because of the exponential decay of the field envelope within the periodic structure, i. e., the imaginary value of the quasimomentum. The same qualitative features arise for the transmission time: for thick barriers, it should saturate at a constant value, as was verified in a recent experiment employing short classical pulses [80.270]. (A more direct analogy, that of waveguides beyond cutoff, yielded similar results in a classical microwave experiment [80.271], while another paper has reported superluminal effects related to the penetration of diffracted or "leaky" microwaves into a shadow region [80.272]. All these experiments involve very small detection probabilities, just as in *Chu* and *Wong's* pioneering experiment on propagation within an absorption line [80.273]. However, it has been predicted that superluminal propagation could occur without high loss or reflection [80.274–276], by operating outside the resonance line of an inverted medium (Sect. 70.1). One can understand the effect as off-resonance "virtual amplification" of the leading edge of a pulse.)

The phenomenon was investigated at the single-photon level by using the high time-resolution techniques discussed in Sect. 80.8.1 to measure the relative delay experienced by down-conversion photons [80.277] when such a tunnel barrier (consisting of 11 layers) was introduced into one arm of a HOM interferometer. The transmissivity of the barrier was relatively flat throughout the bandgap (extending from 600 nm to 800 nm; Fig. 80.11), with a value of 1% at the gap center (700 nm), where the experiment was performed. The HOM coincidence dip was measured both with the barrier (and its substrate) and with the substrate alone (Fig. 80.12). Each dip was subsequently fitted to a Gaussian, and the difference between their centers was calculated. When several such runs were combined, it was found that the tunneling peak arrived 1.47 ± 0.21 fs earlier than the one traveling through air, in reasonable agreement with the theoretical prediction of approximately 1.9 fs. Taking into account the 1.1 μm thickness of the barrier, this implies an effective photon tunneling velocity of $1.7\,c$. The results exclude the "semiclassical" time but are consistent with the group delay. An investigation of the energy-dependence of the tunneling time was also performed, by angling the dielectric mirror, thus shifting its bandgap [80.278]. The data confirm the group delay in this limit as well, and rule out identification of Büttiker's Larmor time with a peak propagation time, at least in this optical system.

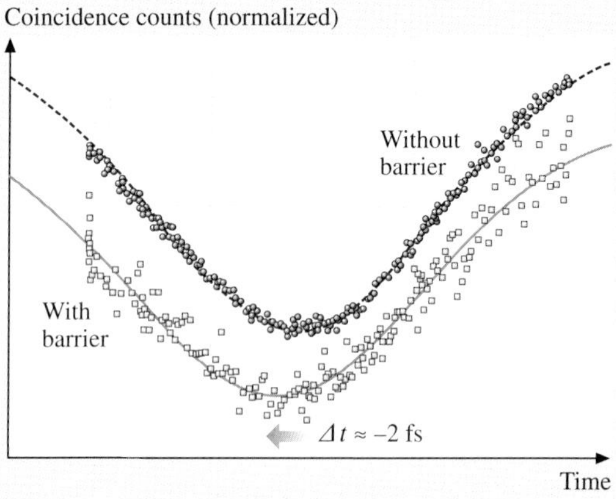

Fig. 80.12 Coincidence profiles with and without the tunnel barrier map out the single-photon wave packets. The *lower profile* shows the coincidences with the barrier; this profile is shifted by ≈ 2 fs to negative times relative to the one with no barrier (*upper curve*): the average particle which tunnels arrives earlier than the one which travels the same distance in air

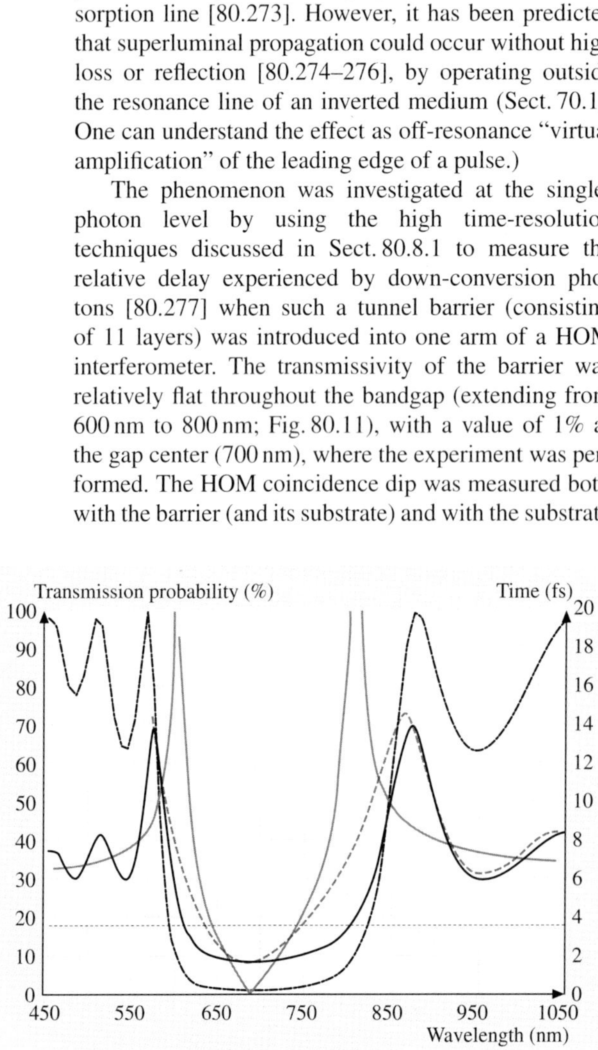

Fig. 80.11 Transmission probability for the tunnel barrier used in [80.277] (*heavy dotted curve*); *heavy black*, *dashed brown*, and *solid grey* curves show group delay, Larmor time, and semiclassical time. Also shown for comparison is the "causality limit" $d/c = 3.6$ fs (horizontal line)

80.8.4 Interpretation of the Tunneling Time

Even though a wave packet peak may appear on the far side of a barrier sooner than it would under allowed propagation, it is important to stress that no information is transmitted faster than c, nor on average is any energy. These effects occur in the limit of low transmission, where the transmitted wavepacket can be considered as a "reshaped" version of the leading edge of the incident pulse [80.273, 279, 280]. At a physical level, the reflection from a multilayer dielectric is due to destructive interference among coherent multiple reflections between the different layers. At times before the field inside the structure reaches its steady-state value, there is little interference, and a non-negligible fraction of the wave is transmitted. This preferential treatment of the leading ramp-up engenders a sort of "optical illusion," shifting the transmitted peak earlier in time. A signal, such as a front with a sharp onset, relies on arbitrarily high-frequency components, which would not benefit from this illusion, but instead travel arbitrarily close to the vacuum speed of light c. Even for a smooth wave packet, no energy travels faster than light; most is simply reflected by the barrier. Only if one considers the Copenhagen interpretation of quantum mechanics, with its instantaneous collapse, does one find superluminal propagation of those particles which happen to be transmitted. This leads to the question of whether it is possible to ask which part of a wave packet a given particle comes from.

One paper argued that transmitted particles do in fact stem only from the leading ramp-up of the wave packet [80.281]. While it is true that the transmission only depends on causally connected portions of the incident wave packet, further analysis revealed that simultaneous discussion of such particle-like questions and the wave nature of tunneling ran afoul of the complementarity principle [80.282]. In essence, labeling the initial positions of a tunneling particle destroys the careful interference by which the reshaping occurs (as in the quantum eraser, Sect. 80.4.1). However, one picture in which the transmitted particles really *do* originate earlier is the Bohm–de Broglie model of quantum mechanics [80.283, 284]. This theory considers Ψ to be a real field (residing however in configuration space, thus incorporating nonlocality) which guides pointlike particles in a deterministic manner. It reproduces all the predictions of quantum mechanics without incorporating any randomness; the probabilistic predictions of QM arise from a range of initial conditions. Bohm's equation of motion has the form of a fluid-flow equation, $\mathbf{v}(\mathbf{x}) = \hbar\nabla \arg \Psi(\mathbf{x})/m$, implying that particle trajectories may never cross, as velocity is a single-valued function of position. Consequently, all transmitted particles originate earlier in the ensemble than all reflected particles [80.282, 285]. This approach yields trajectories with well-defined (and generally subluminal) dwell times in the barrier region. However, the fact that the mean tunneling delay of Bohm particles diverges as the incident bandwidth becomes small, along with other interpretational issues [80.286, 287], leaves open the question of whether time scales as defined by the Bohm model have any physical meaning.

The "weak measurement" approach of *Aharonov* et al. [80.142, 143] or equivalently, complex conditional probabilitiy amplitudes obeying Bayes's theorem [80.288, 289], can be used to address the question of tunneling interaction times in an experimentally unambiguous way. The real part of the resulting complex times determine the effect a tunneling particle would have on a "clock" to which it coupled, while the imaginary part indicates the clock's back-action on the particle. They unify various approaches such as the Larmor and Büttiker–Landauer times, as well as Feynman-path methods. In addition, they allow one to discuss separately the histories of particles which have been transmitted or reflected by a barrier, rather than discussing only the wave function as a whole. Interestingly, these calculations do not support the assertion that transmitted particles originate in the leading ramp-up of a wave packet.

80.8.5 Other Fast and Slow Light Schemes

In addition to the case of tunneling described already, apparently superluminal propagation was observed in a Bessel-beam geometry [80.290], and in the case of an inverted medium [80.291], as described in [80.274–276]. While the former case may be explained geometrically, the latter – in which superluminal group velocities occur without significant gain, loss, or distortion – raises difficult questions about the speeds of propagation of both energy and information. For two contrasting perspectives, see [80.292] and [80.293]. Much more work has followed, including theoretical treatments of the role of quantum noise in preventing superluminal information transfer [80.294], and attempts to experimentally compare the velocity of information transfer with the group velocity [80.295]. The latter work seemed to verify the claim, previously tested only in an electronic analog [80.296], that even in the regime of superluminal propagation, new information was limited to causal speeds. Some dispute has persisted [80.297], and it seems clear that a more rigorous definition of information velocity is required; somewhere between the idealized extremes of infinitely sharp signal fronts and strictly finite signal bandwidths lies the real world, and neither the front velocity nor the group velocity should be expected to completely describe the behavior of actual information-carrying pulses. At the same time, the definition of the energy velocity in active media cannot be resolved without reviving long-standing conundra about how to apportion energy between the propagating field and energy stored in the medium [80.298].

Since 1999, there has actually been much more excitement over so-called "slow light" than over "fast light" [80.299]. Building on the concepts of electromagnetically-induced transparency [80.300] (Sect. 69.7), two groups succeeded in that year in utilizing the steep dispersion curves which can be generated by extremely narrow holes in absorption lines to slow light by a remarkable factor, with group velocities as low as 17 m/s [80.301, 302]. Two years later, two experiments succeeded in bringing light to a standstill [80.303, 304]. This can be understood in terms of a "dark-state polariton" model, in which the propagating photon is adiabatically converted into a (stationary) metastable atomic excitation. In addition to the obvious possibilities for storage of optical pulses, particularly tantalizing at the quantum level [80.305], there have been several proposals for generating extremely strong optical nonlinearities in this system, perhaps directly applicable to quantum information processing [80.306, 307]. Both slow light and fast light have now been observed in solid-state systems [80.308, 309], as well. In both cases, the extension to the single-photon level remains a major goal, for both practical and theoretical reasons. For a recent review on anomalous optical propagation velocities, see [80.310].

80.9 Gravity and Quantum Optics

According to general relativity, gravitational radiation can be produced and detected by moving mass distributions [80.311–313]. However, gravitational radiation is coupled only to time-varying mass quadrupole moments in lowest order, since the mass dipole moment is $\sum_j m_j \boldsymbol{r}_j = M \boldsymbol{R}_{\text{cm}}$ and $\boldsymbol{R}_{\text{cm}}$ for a closed system can only exhibit uniform rectilinear motion. Current efforts focus on detecting gravitational waves (typically at 100 Hz to 1 kHz) from astrophysical sources, such as supernovae or collapsing binary stellar systems. For example, it is expected that in the nearby Virgo cluster of galaxies, several such events should occur per year, each yielding a fractional strain ($\Delta L/L$) of 10^{-21} on Earth. However, there are large uncertainties in this estimate.

Two main efforts have been pursued toward gravitational wave detection. The first type of detector, the resonant-mass detector (sometimes known as a "Weber bar," after its inventor), utilizes a large cylindrical mass whose fundamental mode of acoustical oscillation is resonantly excited by time-varying tidal forces produced by the passage of a gravitational wave. The induced motions are typically detected by piezoelectric crystals, or by SQUIDs (superconducting quantum interference devices) [80.314], yielding strain sensitivities better than $10^{-18}/\sqrt{\text{Hz}}$. Such detectors were first constructed in the 1960's [80.313], and are still in use (e.g., such as the 2.3-ton bar at Louisiana State University, ALLEGRO [80.315]), but to date no incontrovertible detections have been reported. (Attempts to improve the signal-to-noise ratio in resonant-mass GW detectors led to the consideration of back-action-evading sensors (a special case of QND measurements, Sect. 80.5.2) to circumvent the standard quantum noise limit [80.316].)

More recently, a large amount of research has been devoted to using optical interferometry to detect gravitational radiation. A passing gravitational wave alters the relative path length in the arms of a Michelson interferometer, thereby slightly shifting the output fringes. Although the effective gravitational mass of the light is much smaller than that of the Weber bar, very long interferometer arms (2–4 km, with a Fabry–Perot cavity in each arm to increase the effective length) more than make up for this disadvantage. The signal-to-noise ratio for the detection of a fringe shift depends on the power of the light. The US initiative, called LIGO (Laser Interferometer Gravitational-Wave Observatory [80.317]) uses 10 W from a Nd:YAG laser, and an additional external mirror to recirculate the unmeasured light, thus increasing the stored light power up to 10 kW.

There are three LIGO interferometers, located respectively in Hanford, Washington (with both a 2 km and a 4 km version) and Livingston, Louisiana (4 km version). The registration of coincident events at the separated interferometers allows one to rule out terrestial artifacts, but many problems involving seismic and thermal isolation, absorption and heating, intrinsic thermal noise and optical quality had to be addressed. The LIGO interferometers have had several preliminary science runs since Fall 2002; the first true "search run" is scheduled for 2005 [80.318]. The present sensitivities, which range from 10^{-21} to $10^{-22}/\sqrt{\text{Hz}}$ (at ≈ 200–300 Hz) are approaching the initial design goal of $3\times10^{-23}/\sqrt{\text{Hz}}$.

Other similar detectors nearing, or currently in, operation are VIRGO (in Pisa, Italy), GEO 600 (near Hannover, Germany), and TAMA (in Tokyo, Japan); planning for an instrument (ACIGA) in Australia is underway, as is planning for the NASA/ESA collaborative Laser Interferometer Space Antenna (LISA), currently aiming for launch in 2012 [80.319]); this space-based interferometer, with arms up to 5×10^6 km long, would probe frequencies from 10^{-4} to 10^{-1} Hz, inaccessible to terrestial experiments due to seismic and atmospheric disturbances.

Because the standard quantum noise limit of these detectors is ultimately determined by the vacuum fluctuations incident on the unused input ports of the interferometers, it is in principle possible to achieve reduced noise levels by using squeezed vacuum instead [80.313, 320] (Sect. 80.2.3). This has been demonstrated experimentally in table-top experiments [80.55, 321], though there are no plans to incorporate it into the current version of LIGO.

It has also been suggested that matter waves which interact with gravity waves inside a matter–wave interferometer (Sect. 80.3.1) could lead to a sensitive method to detect gravitation radiation [80.322, 323]. Such a "Matter–wave Interferometric Gravitational-wave Observatory" (MIGO) may allow the detection of primordial gigahertz gravity waves arising from the Big Bang [80.324]. Moreover, quantum mechanical detectors based on the use of macroscopically coherent entangled states may enable quantum transducers which can interconvert between electromagnetic and gravitational radiations, based on time-reversal symmetry [80.325, 326].

References

80.1 E. T. Jaynes: Phys. Rev. A **2**, 260 (1970)

80.2 J. F. Clauser: Phys. Rev. A **6**, 49 (1972)

80.3 A. O. Barut, J. P. Dowling: Phys. Rev. A **41**, 2284 (1990)

80.4 P. W. Milonni: *The Quantum Vacuum: An Introduction to Quantum Electrodynamics* (Academic, Boston 1994)

80.5 P. Grangier, G. Roger, A. Aspect: Europhys. Lett. **1**, 173 (1986)

80.6 P. A. M. Dirac: *The Principles of Quantum Mechanics*, 4th edn. (Oxford Univ. Press, London 1958)

80.7 A. Aspect, P. Grangier, G. Roger: J. Opt. (Paris) **20**, 119 (1989)

80.8 F. J. Glauber: Optical coherence and photon statistics. In: *Quantum Optics and Electronics*, ed. by DeWitt, Blandin, Cohen-Tannoudji (Les Houches, Haute-Savoie 1964)

80.9 J. F. Clauser: Phys. Rev. D **9**, 853 (1974)

80.10 R. Y. Chiao, P. G. Kwiat, I. H. Deutsch, A. M. Steinberg: Observation of a nonclassical Berry's phase in quantum optics. In: *Recent Developments in Quantum Optics*, ed. by R. Inguva (Plenum, New York 1993) p. 145

80.11 R. Hanbury-Brown, R. Q. Twiss: Proc. R. Soc. London A **248**, 199 (1958)

80.12 W. Heisenberg: *The Physical Principles of the Quantum Theory* (Dover, New York 1930) p. 39

80.13 S. K. Lamoreaux: Phys. Rev. Lett. **78**, 5 (1997)

80.14 S. K. Lamoreaux: Phys. Rev. Lett. **81**, 5475 (1998)

80.15 A. Lambrecht, S. Reynaud: Phys. Rev. Lett. **84**, 5672 (2000)

80.16 U. Mohideen, A. Roy: Phys. Rev. Lett. **81**, 4549 (1998)

80.17 S. K. Lamoreaux: Phys. Rev. Lett. **83**, 3340 (1999)

80.18 G. Bressi, G. Carugno, R. Onofrio, G. Ruoso: Phys. Rev. Lett. **88**, 041804 (2002)

80.19 R. S. Decca, D. López, E. Fischbach, D. E. Krause: Phys. Rev. Lett. **91**, 050402 (2003)

80.20 P. R. Berman (Ed.): *Cavity Quantum Electrodynamics*, Adv. At. Mol. Opt. Phys., Suppl. 2 (Academic, San Diego 1994)

80.21 C. I. Sukenik, M. G. Boshier, D. Cho, V. Sandoghdar, E. A. Hinds: Phys. Rev. Lett. **70**, 560 (1993)

80.22 G. Rempe, R. J. Thompson, H. J. Kimble: Phys. Scr. T **51**, 67 (1994)

80.23 M. Brune, F. Schmidt-Kaler, A. Maali, J. Dreyer, E. Hagley, J. M. Raimond, S. Haroche: Phys. Rev. Lett. **76**, 1800 (1996)

80.24 X. Maître, E. Hagley, J. Dreyer, A. Maali, C. Wunderlich, M. Brune, J. M. Raimond, S. Haroche: J. Mod. Optics. **44**, 2023 (1997)

80.25 J. M. Raimond, M. Brune, S. Haroche: Rev. Mod. Phys. **73**, 565 (2001)

80.26 G. Nogues: Nature **400**, 239 (1999)

80.27 H. Mabuchi, J. Ye, H. J. Kimble: Appl. Phys. B **68**, 1095 (1999)

80.28 P. Münstermann, T. Fischer, P. Maunz, P. W. H. Pinkse, G. Rempe: Phys. Rev. Lett. **82**, 3791 (1999)

80.29 A. C. Doherty, T. W. Lynn, C. J. Hood, H. J. Kimble: Phys. Rev. A **63**, 013401 (2000)

80.30 P. W. H. Pinkse, T. Fischer, P. Maunz, G. Rempe: Nature **404**, 365 (2000)

80.31 C. J. Hood, T. W. Lynn, A. C. Doherty, A. S. Parkins, H. J. Kimble: Science **287**, 1457 (2000)

80.32 T. Fischer, P. Maunz, P. W. H. Pinkse, T. Puppe, G. Rempe: Phys. Rev. Lett. **88**, 163002 (2002)

80.33 D. C. Burnham, D. L. Weinberg: Phys. Rev. Lett. **25**, 84 (1970)

80.34 S. E. Harris, M. K. Oshman, R. L. Byer: Phys. Rev. Lett. **18**, 732 (1967)
80.35 D. N. Klyshko: Pis'ma Zh. Eksp. Teor. Fiz. Sov. Phys. JETP **6**, 490 (1967)
80.36 D. N. Klyshko: JETP Lett. **6**, 23 (1967)
80.37 D. Bouwmeester, J.-W. Pan, M. Daniell, H. Weinfurter, A. Zeilinger: Phys. Rev. Lett. **82**, 1345 (1999)
80.38 J.-W. Pan, D. Bouwmeester, M. Daniell, H. Weinfurter, A. Zeilinger: Nature **403**, 515 (2000)
80.39 J.-W. Pan, M. Daniell, S. Gasparoni, G. Weihs, A. Zeilinger: Phys. Rev. Lett. **86**, 4435 (2001)
80.40 Z. Zhao, YA. Chen, A.-N. Zhang, T. Yang, H. J. Briegel, J.-W. Pan: Nature **430**(54), 02463 (2004)
80.41 A. Lamas-Linares, J. C. Howell, D. Bouwmeester: Nature **412**, 887 (2001)
80.42 H. S. Eisenberg, G. Khoury, G. A. Durkin, C. Simon, D. Bouwmeester: Phys. Rev. Lett. **93**, 193901 (2004)
80.43 C. K. Hong, L. Mandel: Phys. Rev. Lett. **56**, 58 (1986)
80.44 A. I. Lvovsky, H. Hansen, T. Aichele, O. Benson, J. Mlynek, S. Schiller: Phys. Rev. Lett. **87**, 050402 (2001)
80.45 D. Liebfried, D. M. Meekhof, B. E. King, C. Monroe, W. M. Itano, D. J. Wineland: Phys. Rev. Lett. **77**, 4281 (1996)
80.46 S. A. Babichev, J. Appel, A. I. Lvovsky: Phys. Rev. Lett. **92**, 193601 (2004)
80.47 H. J. Kimble, D. F. Walls: J. Opt. Soc. Am. B **4**, 1450 (1987)
80.48 J. G. Rarity: Appl. Phys. B **55**, 250 (1992)
80.49 R. E. Slusher, L. W. Hollberg, B. Yurke, J. C. Mertz, J. F. Valley: Phys. Rev. Lett. **55**, 2409 (1985)
80.50 A. Lambrecht, T. Coudreau, A. M. Steinberg, E. Giacobino: Squeezing with laser cooled atoms. In: *Coherence and Quantum Optics VII*, ed. by J. H. Eberly, L. M. Mandel, E. Wolf (Plenum, New York 1996)
80.51 S. Machida, Y. Yamamoto, Y. Itaya: Phys. Rev. Lett. **58**, 1000 (1987)
80.52 D. Kylov, K. Bergman: Opt. Lett. **23**, 1390 (1998)
80.53 P. K. Lam, T. C. Ralph, B. C. Buchler, D. E. McClelland, H. A. Bachor, J. Gao: J. Opt. B **1**, 469 (1999)
80.54 K. McKenzie, N. Grosse, W. P. Bowen, S. E. Whitcomb, M. B. Gray, D. E. McClelland, P. K. Lam: Phys. Rev. Lett. **93**, 161105 (2004)
80.55 K. McKenzie, D. A. Shaddock, D. E. McClelland, B. C. Buchler, P. K. Lam: Phys. Rev. Lett. **88**, 231102 (2002)
80.56 N. Treps, U. L. Andersen, B. C. Buchler, P. K. Lam, A. Maître, H.-A. Bachor, C. Fabre: Phys. Rev. Lett. **88**, 203601 (2002)
80.57 E. S. Polzik, J. Carri, H. J. Kimble: Appl. Phys. B **55**, 279 (1992)
80.58 Z. Y. Ou, S. F. Pereira, H. J. Kimble: Appl. Phys. B **55**, 265 (1992)
80.59 Y.-Q. Li, P. Lynam, M. Xiao, P. J. Edwards: Phys. Rev. Lett. **78**, 3105 (1997)
80.60 P. R. Berman (Ed.): *Atom Interferometry* (Academic, San Diego 1997)
80.61 D. E. Keith, C. R. Ekstrom, Q. A. Turchette, D. E. Pritchar: Phys. Rev. Lett. **66**, 2693 (1991)
80.62 M. S. Chapman, C. R. Ekstrom, T. D. Hammond, R. A. Rubenstein, J. Schmiedmayer, S. Wehinger, D. E. Pritchard: Phys. Rev. Lett. **74**, 4783 (1995)
80.63 M. A. Kasevich, S. Chu: Phys. Rev. Lett. **67**, 181 (1991)
80.64 A. Peters, K. Y. Chung, S. Chu: Metrologia **38**, 25 (2001)
80.65 J. M. McGuirk, G. T. Foster, J. B. Fixler, M. J. Snadden, M. A. Kasevich: Phys. Rev. A **65**, 033608 (2002)
80.66 F. Riehle, T. Kisters, A. Witte, J. Helmcke, C. Bordé: Phys. Rev. Lett. **67**, 177 (1991)
80.67 T. L. Gustavson, A. Landragin, M. A. Kasevich: Quantum Grav., **17**, 2385 (2000)
80.68 M. Arndt, O. Nairz, J. Voss-Andreae, C. Keller, G. van der Zouw, A. Zeilinger: Nature **401**, 680 (1999)
80.69 L. Hackermüller, S. Uttenthaler, K. Hornberger, E. Reiger, B. Brezger, A. Zeilinger, M. Arndt: Phys. Rev. Lett. **91**, 090408 (2003)
80.70 W. Marshall, C. Simon, R. Penrose, D. Bouwmeester: Phys. Rev. Lett. **91**, 130401 (2003)
80.71 M. R. Andrews, C. G. Townsend, H.-J. Miesner, D. S. Durfee, D. M. Kurn, W. Ketterle: Science **275**, 637 (1997)
80.72 Z. Hadzibabic: Phys. Rev. Lett. **93**, 180403 (2004)
80.73 R. L. Pfleegor, L. Mandel: Phys. Rev. **159**, 1084 (1967)
80.74 J. Javanainen, S. M. Yoo: Phys. Rev. Lett. **76**, 161 (1996)
80.75 K. Molmer: Phys. Rev. A **55**, 3195 (1997)
80.76 K. Molmer: Phys. Rev. A **58**, 4247 (1998)
80.77 J. Gea-Banacloche: Phys. Rev. A **58**, 4244 (1998)
80.78 C. H. van der Wal, A. C. J. ter Haar, F. K. Wilhelm, R. N. Schouten, C. J. P. M. Harmans, T. P. Orlando, S. Lloyd, J. E. Mooij: Science **290**, 773 (2000)
80.79 I. Chiorescu, Y. Nakamura, C. J. P. M. Harmans, J. E. Mooij: Science **299**, 1869 (2003)
80.80 I. Siddiqi, R. Vijay, F. Pierre, C. M. Rigetti, L. Frunzio, M. H. Devoret: Phys. Rev. Lett. **93**, 207002 (2004)
80.81 R. Y. Chiao, P. G. Kwiat, A. M. Steinberg: The energy–time uncertainty principle and the EPR paradox: Experiments involving correlated two-photon emission in parametric down-conversion. In: *Proceedings Workshop on Squeezed States and Uncertainty Relations*, ed. by D. Han, Y. S. Kim, W. W. Zachary (NASA Conference Publication 3135, Washington DC 1992) p. 61
80.82 A. Joobeur, B. Saleh, M. Teich: Phys. Rev. A **50**, 3349 (1994)
80.83 E. Schrödinger: Proc. Am. Phil. Soc. **124**, 323 (1980)
80.84 P. G. Kwiat, R. Y. Chiao: Phys. Rev. Lett. **66**, 588 (1991)
80.85 R. Ghosh, L. Mandel: Phys. Rev. Lett. **59**, 1903 (1987)
80.86 C. K. Hong, Z. Y. Ou, L. Mandel: Phys. Rev. Lett. **59**, 2044 (1987)
80.87 Z. Y. Ou, L. Mandel: Phys. Rev. Lett. **57**, 66 (1989)

80.88 A. M. Steinberg, R. Y. Chiao: Phys. Rev. A **49**, 3283 (1994)
80.89 L. Mandel: Phys. Rev. A **28**, 929 (1983)
80.90 H. Fearn, R. Loudon: J. Opt. Soc. Am. **B6**, 917 (1989)
80.91 A. M. Steinberg, P. G. Kwiat, R. Y. Chiao: Phys. Rev. Lett. **68**, 2421 (1992)
80.92 P. G. Kwiat, A. M. Steinberg, R. Y. Chiao: Phys. Rev. A **45**, 7729 (1992)
80.93 N. Bohr: Discussion with Einstein on epistemological problems in atomic physics. In: *Quantum Theory and Measurement*, ed. by J. A. Wheeler, W. H. Zurek (Princeton University Press, Princeton 1983) p. 9
80.94 R. P. Feynman, R. B. Leighton, M. Sands: *The Feynman Lectures on Physics* (Addison-Wesley, Reading 1965)
80.95 J. von Neumann: *Mathematical Foundations of Quantum Mechanics* (Princeton Univ. Press, Princeton 1955)
80.96 B.-G. Englert: Phys. Rev. Lett. **77**, 2154 (1996)
80.97 P. D. D. Schwindt, P. G. Kwiat, B.-G. Englert: Phys. Rev. A **60**, 4285 (1999)
80.98 A. Trifonov, G. Bjork, J. Soderholm, T. Tsegaye: Euro. Phys. J. D **18**, 251 (2002)
80.99 S. Dürr, T. Nonn, G. Rempe: Phys. Rev. Lett. **81**, 5705 (1998)
80.100 M. Hillery, M. O. Scully: On state reduction and observation in quantum optics: Wigner's friends and their amnesia. In: *Quantum Optics, Experimental Gravitation, and Measurement Theory*, ed. by P. Meystre, M. O. Scully (Plenum, New York 1983) p. 65
80.101 M. O. Scully, B.-G. Englert, H. Walther: Nature **351**, 111 (1991)
80.102 P. G. Kwiat, B.-G. Englert: Quantum-erasing the nature of reality or, perhaps, the reality of nature? In: *Science and Ultimate Reality: Quantum Theory, Cosmology, and Complexity*, ed. by J. D. Barrow, P. C. W. Davies, C. L. Harper Jr. (Cambridge Univ. Press, Cambridge 2004) p. 306
80.103 P. G. Kwiat, A. M. Steinberg, R. Y. Chiao: Phys. Rev. A **49**, 61 (1994)
80.104 T. Herzog, P. G. Kwiat, H. Weinfurter, A. Zeilinger: Phys. Rev. Lett. **75**, 3034 (1995)
80.105 Y.-H. Kim, R. Yu, S. P. Kulik, Y. Shih, M. O. Scully: Phys. Rev. Lett. **84**, 1 (2000)
80.106 J. A. Wheeler: Frontiers of Time. In: *Proceedings in the formulation of physics*, ed. by G. T. diFrancia (North-Holland, Amsterdam 1979)
80.107 T. Hellmuth, H. Walther, A. Zajonc, W. Schleich: Phys. Rev. A **35**, 2532 (1987)
80.108 W. C. Wickes, C. O. Alley, O. G. Jakubowicz: A 'delayed-choice' quantum mechanics experiment. In: *Quantum Theory and Measurement*, ed. by J. A. Wheeler, W. H. Zurek (Princeton University Press, Princeton 1983) p. 457
80.109 X. Y. Zou, L. J. Wang, L. Mandel: Phys. Rev. Lett. **67**, 318 (1991)
80.110 M. S. Chapman, T. D. Hammond, A. Lenef, J. Schmiedmayer, R. A. Rubenstein, E. Smith, D. E. Pritchard: Phys. Rev. Lett. **75**, 3783 (1995)
80.111 D. M. Greenberger, M. A. Horne, A. Zeilinger: Phys. Today **22** (1993)
80.112 L. J. Wang, J.-K. Rhee: Phys. Rev. A **59**, 1654 (1999)
80.113 T. J. Herzog, J. G. Rarity, H. Weinfurter, A. Zeilinger: Phys. Rev. Lett. **72**, 629 (1994)
80.114 M. Zukowski, A. Zeilinger, H. Weinfurter: Entangling photons radiated by independent pulsed sources. In: *Proceedings 'Fundamental Problems on Quantum Theory'*, Ann. N. Y. Acad. Sci., Vol. 755, ed. by D. M. Greenberger, A. Zeilinger (New York Academy of Sciences, New York 1995) p. 91
80.115 D. Branning, A. L. Migdall: Observation of time-dependent inhibited spontaneous emission. In: *Coherence and Quantum Optics VIII*, ed. by N. P. Bigelow, J. H. Eberly, C. L. Stroud, I. A. Walmsley (Kluwer Academic/Plenum Publishing, New York 2003) p. 507
80.116 P. W. Milonni, H. Fearn, A. Zeilinger: Phys. Rev. A **53**, 4556 (1996)
80.117 K. J. Resch, J. S. Lundeen, A. M. Steinberg: Phys. Rev. Lett. **87**, 123603 (2001)
80.118 B. Misra, E. C. G. Sudarshan: J. Math. Phys. **18**, 756 (1977)
80.119 W. M. Itano, D. J. Heinzen, J. J. Bollinger, D. J. Wineland: Phys. Rev. A **41**, 2295 (1990)
80.120 V. Frerichs, A. Schenzle: Phys. Rev. A **44**, 1962 (1991)
80.121 S. Pascazio, M. Namiki: Phys. Rev. A **50**, 4582 (1994)
80.122 A. G. Kofman, G. Kurizki: Nature **405**, 546 (2000)
80.123 M. C. Fischer, B. Gutiérrez-Medina, M. G. Raizen: Phys. Rev. Lett. **87**, 040402 (2001)
80.124 V. B. Braginsky, Y. I. Vorontsov, K. S. Thorne: Science **209**, 547 (1980)
80.125 J.-F. Roch, G. Roger, P. Grangier, J. Courty, S. Reynaud: Appl. Phys. B **55**, 291 (1992)
80.126 N. Imoto, H. A. Haus, Y. Yamamoto: Phys. Rev. A **32**, 2287 (1985)
80.127 M. Kitagawa, Y. Yamamoto: Phys. Rev. A **34**, 3974 (1986)
80.128 P. Grangier, J.-P. Poizat, J.-F. Roch: Phys. Scr. T **51**, 51 (1994)
80.129 R. Y. Chiao: Phys. Lett. A **33**, 177 (1970)
80.130 B. Lee, E. Yin, T. K. Gustafson, R. Y. Chiao: Phys. Rev. A **45**, 4319 (1992)
80.131 S. Haroche, M. Brune, J. M. Raimond: Atomic motion in the field of a few photons stored in a high Q cavity and matter-wave interferometry. In: *Atomic Physics 13*, ed. by H. Walther, T. W. Hänsch, D. Niezart (American Institute of Physics, New York 1993) p. 261
80.132 D. Meschede, H. Walther, G. Müller: Phys. Rev. Lett. **54**, 551 (1985)
80.133 W. J. Munro, K. Nemoto, T. P. Spiller: New J. Phys. **7**, 137 (2005)
80.134 M. Renninger: Z. Phys. **158**, 417 (1960)

80.135 R. H. Dicke: Am. J. Phys. **49**, 925 (1981)
80.136 A. C. Elitzur, L. Vaidman: Found. Phys. **23**, 987 (1993)
80.137 P. G. Kwiat, H. Weinfurter, T. Herzog, A. Zeilinger, M. A. Kasevich: Phys. Rev. Lett. **74**, 4763 (1995)
80.138 A. G. White, J. R. Mitchell, O. Nairz, P. G. Kwiat: Phys. Rev. A **58**, 605 (1998)
80.139 P. G. Kwiat, A. G. White, J. R. Mitchell, O. Nairz, G. Weihs, H. Weinfurter, A. Zeilinger: Phys. Rev. Lett. **83**, 4725 (1999)
80.140 H. Paul, M. Pavicic: Int. J. Theor. Phys. **35**, 2085 (1996)
80.141 T. Tsegaye, E. Goobar, A. Karlsson, G. Björk, M. Y. Loh, K. H. Lim: Phys. Rev. A **57**, 3987 (1998)
80.142 Y. Aharonov, D. Z. Albert, L. Vaidman: Phys. Rev. Lett. **60**, 1351 (1988)
80.143 Y. Aharonov, L. Vaidman: Phys. Rev. A **41**, 11 (1990)
80.144 O. Schulz, R. Steinhübl, M. Weber, B.-G. Englert, C. Kurtsiefer, H. Weinfurter: Phys. Rev. Lett. **90**, 177901 (2003)
80.145 K. J. Resch, A. M. Steinberg: Phys. Rev. Lett. **92**, 130402 (2004)
80.146 Y. Aharonov, A. Botero, S. Popescu, B. Reznik, J. Tollaksen: Phys. Rev. A **301**, 130 (2002)
80.147 Y. Aharonov, S. Popescu, D. Rohrlich, L. Vaidman: Phys. Rev. A **48**, 4084 (1993)
80.148 K. J. Resch, J. S. Lundeen, A. M. Steinberg: Phys. Lett. A **324**, 125 (2004)
80.149 N. W. M. Ritchie, J. G. Story, R. G. Hulet: Phys. Rev. Lett. **66**, 1107 (1991)
80.150 D. R. Solli, C. F. McCormick, R. Y. Chiao, S. Popescu, J. M. Hickmann: Phys. Rev. Lett. **92**, 043602 (2004)
80.151 N. Brunner, A. Acin, D. Collins, N. Gisin, V. Scarani: Phys. Rev. Lett. **91**, 180402 (2003)
80.152 H. M. Wiseman: Phys. Rev. A **311**, 285 (2003)
80.153 Y. Aharonov, J. Anandan: Phys. Rev. A **47**, 4616 (1993)
80.154 Y. Aharonov, L. Vaidman: Phys. Rev. A **178**, 38 (1993)
80.155 D. T. Smithey, M. Beck, M. G. Raymer, A. Faridani: Phys. Rev. Lett. **70**, 1244 (1993)
80.156 W. G. Unruh: Phys. Rev. A **50**, 882 (1994)
80.157 A. Einstein, B. Podolsky, N. Rosen: Phys. Rev. **47**, 777 (1935)
80.158 D. Bohm: A suggested interpretation of the quantum theory in terms of "hidden" variables, Phys. Rev. **85**, 166 (1982) Reprinted in *Quantum Theory and Measurement*, ed. by J. A. Wheeler, W. H. Zurek (Princeton Univ. Press, Princeton 1983)
80.159 J. S. Bell: Physics **1**, 195 (1964)
80.160 J. S. Bell: *Speakable and Unspeakable in Quantum Mechanics* (Cambridge Univ. Press, Cambridge 1987)
80.161 H. P. Stapp: Phys. Rev. D **3**, 1303 (1971)
80.162 P. H. Eberhard: Nuovo Cimento B **38**, 75 (1977)
80.163 J. F. Clauser, M. A. Horne, A. Shimony, R. A. Holt: Phys. Rev. Lett. **23**, 880 (1969)
80.164 J. F. Clauser, M. A. Horne: Phys. Rev. D **10**, 526 (1974)
80.165 J. Jarrett: Noûs **18**, 569 (1984)
80.166 A. Shimony: *An Exposition of Bell's Theorem* (Plenum, New York 1990)
80.167 S. J. Freedman, J. F. Clauser: Phys. Rev. Lett. **28**, 938 (1972)
80.168 A. Aspect, J. Dalibard, G. Roger: Phys. Rev. Lett. **49**, 1804 (1982)
80.169 Y. H. Shih, C. O. Alley: Phys. Rev. Lett. **61**, 2921 (1988)
80.170 Z. Y. Ou, L. Mandel: Phys. Rev. Lett. **61**, 50 (1988)
80.171 C. Santori, D. Fattal, J. Vuckovic, G. S. Solomon, Y. Yamamoto: Nature **419**, 594 (2002)
80.172 T. B. Pittman, J. D. Franson: Phys. Rev. Lett. **90**, 240401 (2003)
80.173 H. d. Riedmatten, I. Marcikic, W. Tittel, H. Zbinden, N. Gisin: Phys. Rev. A **67**, 022301 (2003)
80.174 A. G. White, D. F. V. James, P. H. Eberhard, P. G. Kwiat: Phys. Rev. Lett. **83**, 3103 (1999)
80.175 P. G. Kwiat, E. Waks, A. G. White, I. Appelbaum, P. H. Eberhard: Phys. Rev. A **60**, 773 (1999)
80.176 P. G. Kwiat, J. B. Altepeter, J. T. Barreiro, M. E. Goggin, E. Jeffrey, N. A. Peters, A. VanDevender: The conversion revolution: down, up and sideways. In: *Quantum Communication, Measurement and Computing*, Vol. 734, ed. by S. M. Barnett, E. Andersson, J. Jeffers, P. Öhberg, O. Hirota (AIP Conf. Proc., Glasgow 2004) p. 337
80.177 P. G. Kwiat, K. Mattle, H. Weinfurter, A. Zeilinger: Phys. Rev. Lett. **75**, 4337 (1995)
80.178 G. Weihs, T. Jennewein, C. Simon, H. Weinfurter, A. Zeilinger: Phys. Rev. Lett. **81**, 5039 (1998)
80.179 T. Jennewein, C. Simon, G. Weihs, H. Weinfurter, A. Zeilinger: Phys. Rev. Lett. **84**, 4729 (2000)
80.180 A. Poppe et al. http://arxiv.org/abs/quant-ph/0404115 (2004)
80.181 D. Bouwmeester, J.-W. Pan, K. Mattle, M. Eibl, H. Weinfurther, A. Zeilinger: Nature **390**, 575 (1997)
80.182 D. Boschi, S. Branca, F. De Martini, L. Hardy, S. Popescu: Phys. Rev. Lett. **80**, 1121 (1998)
80.183 J. G. Rarity, P. R. Tapster: Phys. Rev. Lett. **64**, 2495 (1990)
80.184 A. Mair, A. Vaziri, G. Weihs, A. Zeilinger: Nature **412**, 313 (2001)
80.185 A. Vaziri, G. Weihs, A. Zeilinger: Phys. Rev. Lett. **89**, 240401 (2002)
80.186 N. K. Langford: Phys. Rev. Lett. **93**, 053601 (2004)
80.187 A. F. Abouraddy, B. E. A. Saleh, A. V. Sergienko, M. C. Teich: Phys. Rev. Lett. **87**, 123602 (2001)
80.188 A. F. Abouraddy, B. E. A. Saleh, A. V. Sergienko, M. C. Teich: Opt. Expr. **9**, 498 (2001)
80.189 L. A. Lugiato, A. Gatti, E. Brambilla: J. B. Opt Quantum Semicl. Opt. **4**, 176 (2002)
80.190 P. Navez, E. Brambilla, A. Gatti, L. A. Lugiato: Phys. Rev. A **65**, 013813 (2002)
80.191 J. Brendel, E. Mohler, W. Martienssen: Europhys. Lett. **20**, 575 (1992)
80.192 P. G. Kwiat, A. M. Steinberg, R. Y. Chiao: **47**, R2472 (1993)
80.193 J. D. Franson: Phys. Rev. Lett. **62**, 2205 (1989)

80.194 W. Tittel, J. Brendel, H. Zbinden, N. Gisin: Phys. Rev. Lett. **81**, 3563 (1998)
80.195 H. Zbinden, J. Brendel, N. Gisin, W. Tittel: Phys. Rev. A **63**, 022111 (2001)
80.196 A. Suarez, V. Scarani: Phys. Lett. A **232**, 9 (1997)
80.197 A. Suarez: Phys. Lett. A **236**, 383 (1997)
80.198 H. d. Riedmatten, I. Marcikic, H. Zbinden, N. Gisin: Quant. Inf. Comp. **2**, 425 (2002)
80.199 Z. Y. Ou, S. F. Pereira, H. J. Kimble, K. C. Peng: Phys. Rev. Lett. **68**, 3663 (1992)
80.200 R. García-Patrón, J. Fiurášek, N. J. Cerf, J. Wenger, R. Tualle-Brouri, P. Grangier: Phys. Rev. Lett. **93**, 130409 (2004)
80.201 J. F. Clauser, A. Shimony: Rep. Prog. Phys. **41**, 1881 (1978)
80.202 P. H. Eberhard: Phys. Rev. A **47**, R747 (1993)
80.203 P. G. Kwiat, P. H. Eberhard, A. M. Steinberg, R. Y. Chiao: Phys. Rev. A **49**, 3209 (1994)
80.204 M. A. Rowe, D. Kielpinski, V. Meyer, C. A. Sacket, W. M. Itano, C. Monroe, D. J. Wineland: Nature **409**, 791 (2001)
80.205 B. B. Blinov, D. L. Moehring, L.-M. Duan, C. Monroe: Nature **428**, 153 (2004)
80.206 D. L. Moehring, M. J. Madsen, B. B. Blinov, C. Monroe: Phys. Rev. Lett. **93**, 090410 (2004)
80.207 C. A. Sackett, D. Kielpinski, B. E. King, C. Langer, V. Meyer, C. J. Myatt, M. Rowe, Q. A. Turchette, W. M. Itano, D. J. Wineland, C. Monroe: Nature **3**, 256, 404 (2000)
80.208 P. G. Kwiat, A. M. Steinberg, R. Y. Chiao, P. H. Eberhard, M. D. Petroff: Appl. Opt. **33**, 1844 (1994)
80.209 E. Waks, K. Inoue, E. Diamanti, Y. Yamamoto: http://arxiv.org/abs/quant-ph/0308054 (2003)
80.210 E. S. Fry, T. Walther, S. Li: Phys. Rev. A **52**, 4381 (1995)
80.211 N. D. Mermin: Phys. Rev. Lett. **65**, 1838 (1990)
80.212 D. M. Greenberger, M. A. Horne, A. Shimony, A. Zeilinger: Am. J. Phys. **58**, 1131 (1990)
80.213 L. Hardy: Phys. Rev. Lett. **68**, 2981 (1992)
80.214 L. Hardy: Phys. Rev. A **167**, 17 (1992)
80.215 J. Torgerson, D. Branning, L. Mandel: Appl. Phys. B **60**, 267 (1995)
80.216 A. G. White, D. F. V. James, P. H. Eberhard, P. G. Kwiat: Phys. Rev. Lett. **83**, 3103 (1999)
80.217 R. Glauber: Ann. N. Y. Acad. Sci. **480**, 336 (1986)
80.218 W. K. Wootters, W. H. Zurek: Nature **299**, 802 (1982)
80.219 N. Gisin, S. Massar: Phys. Rev. Lett. **79**, 2153 (1997)
80.220 D. Bruss, A. Ekert, C. Macchiavello: Phys. Rev. Lett. **81**, 2598 (1998)
80.221 A. Lamas-Linares, C. Simon, J. C. Howell, D. Bouwmeester: Science **296**, 712 (2002)
80.222 S. Fasel, N. Gisin, G. Ribordy, V. Scarani, H. Zbinden: Phys. Rev. Lett. **89**, 107901 (2002)
80.223 W. T. M. Irvine, A. L. Linares, M. J. A. de Dood, D. Bouwmeester: Phys. Rev. Lett. **92**, 047902 (2004)
80.224 C. Bennett, S. J. Wiesner: Phys. Rev. Lett. **69**, 2881 (1992)
80.225 S. L. Braunstein, A. Mann, M. Revzen: Phys. Rev. Lett. **68**, 3259 (1992)
80.226 K. Mattle, H. Weinfurter, P. G. Kwiat, A. Zeilinger: Phys. Rev. Lett. **76**, 4656 (1996)
80.227 C. H. Bennett, G. Brassard, C. Crepeau, R. Jozsa, A. Peres, W. K. Wootters: Phys. Rev. Lett. **70**, 1895 (1993)
80.228 A. Furusawa, J. L. Sørensen, S. L. Braunstein, C. A. Fuchs, H. J. Kimble, E. S. Polzik: Science **282**, 706 (1998)
80.229 M. Riebe, H. Häffner, C. F. Roos, W. Hänsel, J. Benhelm, G. P. T. Lancaster, T. W. Körber, C. Becher, F. Schmidt-Kaler, D. F. V. James: Nature **429**, 734 (2004)
80.230 M. D. Barrett, J. Chiaverini, T. Schaetz, J. Britton, W. M. Itano, J. D. Jost, E. Knill, C. Langer, D. Leibfried, R. Ozeri: Nature **429**, 737 (2004)
80.231 T. Jennewein, G. Weihs, J.-W. Pan, A. Zeilinger: Phys. Rev. Lett. **88**, 017903 (2002)
80.232 X. Jia, X. Su, Q. Pan, J. Gao, C. Xie, K. Peng: Phys. Rev. Lett. **93**, 250503 (2004)
80.233 H.-J. Briegel, W. Dürr, J. I. Cirac, P. Zoller: Phys. Rev. Lett. **81**, 5932 (1998)
80.234 Z. Zhao, T. Yang, Y.-A. Chen, A.-N. Zhang, J.-W. Pan: Phys. Rev. Lett. **90**, 207901 (2003)
80.235 S. Lloyd, Y. J. Ng: Sci. Amer. **291**, 52 (2004)
80.236 G. T. Horowitz, J. Maldacena: High Energy Physics – Theory, abstract hep-th/0310281 Journal-ref: JHEP **0402**, 008 (2004)
80.237 G. Brassard (Ed.): *Modern Cryptology: A Tutorial* (Springer, New York 1988)
80.238 A. K. Ekert: Phys. Rev. Lett. **67**, 661 (1991)
80.239 C. H. Bennett: Phys. Rev. Lett. **68**, 3121 (1992)
80.240 N. Gisin, G. Ribordy, W. Tittel, H. Zbinden: Rev. Mod. Phys. **74**, 145 (2002)
80.241 D. S. Naik, C. G. Peterson, A. G. White, A. J. Berglund, P. G. Kwiat: Phys. Rev. Lett. **84**, 4733 (2000)
80.242 D. G. Enzer, R. J. Hughes, C. G. Peterson, P. G. Kwiat: Focus Issue on Quantum Cryptographyin, New J. Phys. **4**, 45 (2002)
80.243 W. Tittel, J. Brendel, H. Zbinden, N. Gisin: Phys. Rev. Lett. **84**, 4737 (2000)
80.244 I. Marcikic, H. de. Riedmatten, W. Tittel, H. Zbinden, M. Legré, N. Gisin: Phys. Rev. Lett. **93**, 180502 (2004)
80.245 G. Barton, K. Scharnhorst: J. Phys. A **26**, 2037 (1993)
80.246 G. C. Hegerfeldt: Phys. Rev. Lett. **54**, 2395 (1985)
80.247 V. P. Bykov, V. I. Tatarskii: Phys. Rev. A **136**, 77 (1989)
80.248 P. W. Milonni, D. F. V. James, H. Fearn: Phys. Rev. A **52**, 1525 (1995)
80.249 P. D. Drummond: Phys. Rev. A **42**, 6845 (1990)
80.250 R. J. Glauber, M. Lewenstein: Phys. Rev. A **43**, 467 (1991)
80.251 B. Huttner, S. M. Barnett: Phys. Rev. A **46**, 4306 (1992)
80.252 R. Matloob, R. Loudon, S. M. Barnett: Phys. Rev. A **52**, 4823 (1995)
80.253 T. Gruner, D.-G. Welsch: Phys. Rev. A **53**, 1818 (1996)

80.254 P.W. Milonni: J. Mod. Opt. **42**, 1991 (1995)
80.255 R. Loudon: J. Mod. Opt. **49**, 821 (2002)
80.256 J.C. Garrison, R.Y. Chiao: Phys. Rev. A **70**, 053826 (2004)
80.257 J.D. Franson: Phys. Rev. A **45**, 3126 (1992)
80.258 A.M. Steinberg, P.G. Kwiat, R.Y. Chiao: Phys. Rev. A **45**, 6659 (1992)
80.259 M. Büttiker, R. Landauer: Phys. Rev. Lett. **49**, 1739 (1982)
80.260 E.H. Hauge, J.A. Støvneng: Rev. Mod. Phys. **61**, 917 (1989)
80.261 E.P. Wigner: Phys. Rev. **98**, 145 (1955)
80.262 E.L. Bolda, R.Y. Chiao, J.C. Garrison: Phys. Rev. A **48**, 3890 (1993)
80.263 M. Büttiker, R. Landauer: Phys. Scr. **32**, 429 (1985)
80.264 M. Büttiker: Phys. Rev. B **27**, 6178 (1983)
80.265 R. Landauer: Nature **341**, 567 (1989)
80.266 T. Martin, R. Landauer: Phys. Rev. A **45**, 2611 (1992)
80.267 R.S. Dumont, T.L. Marchioro II: Phys. Rev. A **47**, 85 (1993)
80.268 E.H. Hauge, J.P. Falck, T.A. Fjeldly: Phys. Rev. B **36**, 4203 (1987)
80.269 E. Yablonovitch: J. Opt. Soc. Am. B **10**(2), 283 (1993), and other articles in this special issue on photonic bandgaps
80.270 Ch. Spielmann, R. Szipöcs, A. Stingl, F. Krausz: Phys. Rev. Lett. **73**, 2308 (1994)
80.271 A. Enders, G. Nimtz: J. Phys. (Paris) **3**, 1089 (1993)
80.272 A. Ranfagni, P. Fabeni, G.P. Pazzi, D. Mugnai: Phys. Rev. E **48**, 1453 (1993)
80.273 S. Chu, S. Wong: Phys. Rev. Lett. **48**, 738 (1982)
80.274 R.Y. Chiao: Phys. Rev. A **48**, R34 (1993)
80.275 A.M. Steinberg, R.Y. Chiao: Phys. Rev. A **49**, 2071 (1994)
80.276 E.L. Bolda, J.C. Garrison, R.Y. Chiao: Phys. Rev. A **49**, 2938 (1994)
80.277 A.M. Steinberg, P.G. Kwiat, R.Y. Chiao: Phys. Rev. Lett. **71**, 708 (1993)
80.278 A.M. Steinberg, R.Y. Chiao: Phys. Rev. A **51**, 3525 (1995)
80.279 C.G.B. Garrett, D.E. McCumber: Phys. Rev. A **1**, 305 (1970)
80.280 A.E. Siegman: *Lasers* (University Science Books, Mill Valley 1986)
80.281 J.M. Deutch, F.E. Low: Ann. Phys. **228**, 184 (1993)
80.282 A.M. Steinberg, P.G. Kwiat, R.Y. Chiao: Found. Phys. Lett. **7**, 223 (1994)
80.283 D. Bohm, B.J. Hiley: *The Undivided Universe: An Ontological Interpretation of Quantum Mechanics* (Routledge, London 1993)
80.284 P.R. Holland: *The Quantum Theory of Motion* (Cambridge Univ. Press, Cambridge 1993)
80.285 C.R. Leavens, G.C. Aers: Bohm trajectories and the tunneling time problems. In: *Scanning Tunneling Microscopy III*, Springer Series in Surface Sciences, Vol. 29, ed. by R. Wiesendanger, H.-J. Güntherodt (Springer, Berlin, Heidelberg 1993) p. 105
80.286 B.-G. Englert, M.O. Scully, G. Süssmann, H. Walther: Z. Naturforsch. A **47**, 1175 (1992)
80.287 C. Dewdney, L. Hardy, E.J. Squires: Phys. Rev. A **184**, 6 (1993)
80.288 A.M. Steinberg: Phys. Rev. A **52**, 32 (1995)
80.289 A.M. Steinberg: Phys. Rev. Lett. **74**, 2405 (1995)
80.290 D. Mugnai, A. Ranfagni, R. Ruggeri: Phys. Rev. Lett. **84**, 4830 (2000)
80.291 L.J. Wang, A. Kuzmich, A. Dogariu: Nature **406**, 277 (2000)
80.292 G. Nimtz, W. Heitmann: Prog. Quant. Electr. **21**, 81 (1997)
80.293 R.Y. Chiao, A.M. Steinberg: Prog. Opt. **37**, 345 (1997)
80.294 A. Kuzmich, A. Dogariu, L.J. Wang, P.W. Milonni, R.Y. Chiao: Phys. Rev. Lett. **86**, 3925 (2001)
80.295 M.D. Stenner, D.J. Gauthier, M.A. Neifeld: Nature **425**, 695 (2003)
80.296 M.W. Mitchell, R.Y. Chiao: Phys. Rev. A **230**, 133 (1997)
80.297 G. Nimtz: Nature **429**, 695–698 (2004), Brief Communications Arising (6 May 2004)
80.298 G. Diener: Ann. Phys. (Leipzig) **7**, 639 (1998)
80.299 P.W. Milonni: *Fast Light, Slow Light, and Left-Handed Light* (Institute of Physics, Bristol 2004)
80.300 J.P. Marangos: J. Mod. Opt **45**, 471 (1998)
80.301 L.V. Hau, S.E. Harris, Z. Dutton, C.H. Behroozi: Nature **397**, 594 (1999)
80.302 M.M. Kash, V.A. Sautenkov, A.S. Zibrov, L. Hollberg, G.R. Welch, M.D. Lukin, Y. Rostovtsev, E.S. Fry, M.O. Scully: Phys. Rev. Lett. **82**, 5229 (1999)
80.303 C. Liu, Z. Dutton, C.H. Behroozi, L.V. Hau: Nature **409**, 490 (2001)
80.304 D.F. Phillips, A. Fleischhauer, A. Mair, R.L. Walsworth, M.D. Lukin: Phys. Rev. Lett **86**, 783 (2001)
80.305 C.H. van der Wal, M.D. Eisaman, A. André, R.L. Walsworth, D.F. Phillips, A.S. Zibrov, M.D. Lukin: Science **301**, 196 (2003)
80.306 S.E. Harris, L.V. Hau: Phys. Rev. Lett. **82**, 4611 (1999)
80.307 M.D. Lukin, M. Fleischhauer, R. Cote: Phys. Rev. Lett. **87**, 037901 (2001)
80.308 A.V. Turukhin, V.S. Sudarshanam, M.S. Shahriar: Phys. Rev. Lett. **88**, 023602 (2002)
80.309 N.S. Bigelow, N.N. Lepeshkin, R.W. Boyd: Science **301**, 200 (2003)
80.310 R.W. Boyd, D.J. Gauthier: Prog. Opt. **43**, 497 (2002)
80.311 S. Weinberg: *Gravitation and Cosmology* (Wiley, New York 1972)
80.312 C.W. Misner, K.S. Thorne, J.A. Wheeler: *Gravitation* (Freeman, San Francisco 1987)
80.313 K.S. Thorne: Gravitational radiation. In: *Three Hundred Years of Gravitation*, ed. by S.W. Hawking, W. Israel (Cambridge Univ. Press, Cambridge 1987) p. 400
80.314 P.F. Michelson, J.C. Price, R.C. Taber: Science **237**, 150 (1987)
80.315 http://sam.phys.lsu.edu/

80.316 C. M. Caves, K. S. Thorne, W. P. Drever, V. D. Sandberg, M. Zimmerman: Rev. Mod. Phys. **52**, 341 (1980)
80.317 B. C. Barish, R. Weiss: Phys. Today **52**, 44 (1999)
80.318 http://www.ligo.caltech.edu/
80.319 http://lisa.jpl.nasa.gov/; http://sci.esa.int/science-e/www/area/index.cfm?fareaid=27
80.320 C. M. Caves: Phys. Rev. D **23**, 1693 (1981)
80.321 M. Xiao, L.-A. Wu, H. J. Kimble: Phys. Rev. Lett. **59**, 278 (1987)
80.322 R. Y. Chiao, A. D. Speliotopoulos: J. Mod. Opt. **51**, 861 (2004)
80.323 A. D. Speliotopoulos, R. Y. Chiao: Phys. Rev. D. **69**, 084013 (2004)
80.324 S. Foffa, A. Gasparini, M. Papucci and R. Sturani: http://arxiv.org/abs/gr-qc/0407039 (2004)
80.325 R. Y. Chiao, J. Boyce, J. C. Garrison: Ann. N. Y. Acad. of Sci. **755**, 400 (1995)
80.326 R. Y. Chiao: Conceptual tensions between quantum mechanics and general relativity: Are there experimental consequences? In: *Science and Ultimate Reality*, ed. by J. D. Barrow, P. C. W. Davies, C. L. Harper Jr. (Cambridge Univ. Press, Cambridge 2004) p. 254

81. Quantum Information

For many years atomic physicists had used quantum mechanics very successfully to calculate energy levels, cross sections and other practical quantities, and for the most part, left the philosophical issues of interpretation to others. But after the work of Bell in the 1960's showed that the peculiarly nonlocal nature of quantum correlations could be tested in the lab, a number of atomic physicists turned to the experimental study of entanglement and quantum measurement. A second phase began at the start of the 1990's when it was realized that correlations and quantum superpositions could be exploited in quantum information processing and secure communication. This has led to an explosive growth of the subject over the past 10 years, fuelled by the long-term prospects of quantum computing and the nearer goal of quantum cryptography. We review some of these developments in this chapter.

81.1 **Quantifying Information** 1216
81.1.1 Separability Criterion 1216
81.1.2 Entanglement Measures 1217

81.2 **Simple Quantum Protocols** 1218
81.2.1 Quantum Key Distribution 1219
81.2.2 Quantum Teleportation 1219
81.2.3 Dense Coding 1220

81.3 **Unitary Transformations** 1221
81.3.1 Single-Qubit Operations 1221
81.3.2 Two-Qubit Operations 1221
81.3.3 Multi-Qubit Gates and Networks . 1222

81.4 **Quantum Algorithms** 1222
81.4.1 Deutsch–Jozsa Algorithm 1222
81.4.2 Grover's Search Algorithm 1223

81.5 **Error Correction** 1223

81.6 **The DiVincenzo Checklist** 1224
81.6.1 Qubit Characterization, Scalability 1224
81.6.2 Initialization 1224
81.6.3 Long Decoherence Times 1224
81.6.4 Universal Set of Quantum Gates .. 1225
81.6.5 Qubit-Specific Measurement 1225

81.7 **Physical Implementations** 1225
81.7.1 Linear Optics 1225
81.7.2 Trapped Ions 1226
81.7.3 Cavity QED 1226
81.7.4 Optical Lattices, Mott Insulator 1227

81.8 **Outlook** 1227

References 1228

Quantum information theory is regarded as a mainly mathematics-based subject area which straddles the fields of theoretical physics (quantum mechanics and statistics), mathematics, and theoretical computer science. Its success stems from the introduction of novel methods into both physics and mathematics.

The fundamental quantity and resource in many applications in quantum information processing is quantum-mechanical entanglement between spatially separated subsystems. Entanglement is a purely quantum-mechanical effect and has led to numerous speculations about the validity of quantum mechanics itself for its apparent paradoxical implications. Most, if not all, of these difficulties have been resolved and can be mostly attributed to the simple fact that paradoxical behaviour is incompatible with common sense or everyday experience. This initial upsetting seems to be common to all revolutionary theories and has occurred most notably in Einstein's theory of relativity [81.1].

These quantum-mechanical correlations have numerous applications in quantum cryptography [or rather quantum key distribution (QKD)], quantum communication, dense coding, and act as the main resource in quantum computing. We will briefly touch upon some mathematical issues concerning separability, quantification of entanglement and channel capacities before describing how quantum key distribution, teleportation and dense coding work. After that, a brief discussion of single-qubit and two-qubit quantum gates follows before we describe the simplest quantum algorithms. The issues of error correction and fault tolerant computation as well as DiVincenzo's checklist (which any realization should satisfy) provide the background for the discussion of some physical implementations.

We are acutely aware of the fact that we can give only a brief introduction into what has become a major field of investigation over the last decade. There are already a number of review articles and textbooks on the market that cover the vast literature on this emerging subject. Amongst those are the first quantum computing compendium by *Gruska* [81.2] and the quantum information textbooks by *Nielsen* and *Chuang* [81.3] and *Stolze* and *Suter* [81.4]. A regularly updated annotated bibliography on this subject, compiled by *Cabello*, forms an invaluable resource for those interested in the subject of this chapter [81.5, 6].

81.1 Quantifying Information

As already noted, entanglement comes about if a quantum-mechanical system can be divided into several parts. As an example, consider a two-photon emission process from a spin-zero particle by which two photons escape in opposite directions. Given that the photons are spin-one particles, their spin projections onto some axis must be mutually opposite. As there is no prior information about the actual orientation of the spin, the part of the photon wave function associated with the spin degree is therefore

$$|\psi\rangle = \frac{1}{\sqrt{2}}\left(|\uparrow\downarrow\rangle - |\downarrow\uparrow\rangle\right) . \tag{81.1}$$

The striking feature of this type of quantum state is that it describes correlations of two spatially separated particles. If the polarization state of one photon is measured, the state of the other particle, which can be far away, is then instantly predetermined. These (nonlocal) correlations that exist between the particles are of purely quantum origin and are called entanglement. Note, however, that no information can be transmitted faster than the speed of light with this type of set up because the (classical) information concerning the measurement result on one particle needs to be transmitted via a necessarily causal classical channel.

The issue of nonlocality has been seen as a vital part in understanding the foundations of quantum mechanics itself (Chapt. 80). In 1935, *Einstein*, *Podolsky*, and *Rosen* argued on the basis of entangled states that quantum mechanics is incomplete [81.7]. They were most concerned about the existence of elements of reality within strongly correlated quantum systems and initiated the debate on quantum nonlocality. The non-existence of so-called local hidden variable theories for the description of states like (81.1) was finally demonstrated by *Bell* [81.8, 9]. He showed that maximally entangled states violate certain inequalities (now called Bell's inequalities) which local hidden variable models would have obeyed. Later experiments showed the correctness of Bell's demonstration [81.10–14].

In classical information theory, the unit of information is called a bit, which can be defined as the amount of information contained in a yes–no question. As a matter of fact, 'bit' is the abbreviation for 'binary digit' and refers to Boolean algebra in which the allowed states of a system are the logical 0 and the logical 1. Therefore, by abuse of language, one bit (as a unit) is the information carried by one bit (as a binary digit) [81.15].

In quantum mechanics, however, due to its inherent linearity, two 'quantum bits' (qubits for short) can be in superpositions of the logical states $|01\rangle$ and $|10\rangle$, or $|\uparrow\downarrow\rangle$ and $|\downarrow\uparrow\rangle$, as in the example above. This typical example of an entangled state shows that quantifying the amount of information contained in a quantum state is different from what is known in classical information theory because of the superposition property. The very same linearity prohibits us from copying an arbitrary quantum state. This effect is known as the no-cloning theorem [81.16]. However, universal copying machines can be constructed within the constraints of quantum mechanics [81.17].

81.1.1 Separability Criterion

From the above it is clear that entangled states play a major role in defining the differences between classical and quantum information. Let us begin by asking under which circumstances a particular given quantum state is entangled or not. For this, we need to give a criterion which allows one to decide this crucial question. Consider a bipartite quantum state, i. e., a state which is decomposed into two distinct, albeit possibly correlated, subsystems A and B. Note that these subsystems themselves might consist of ensembles of particles, in which case we are looking at a bipartite cut through the whole system. Then we say that a bipartite state is not entangled and hence separable if its density operator can be written as a convex combination of tensor product states, viz.

$$\hat{\varrho} = \sum_i p_i \hat{\varrho}^i_A \otimes \hat{\varrho}^i_B \,, \quad \sum_i p_i = 1 \,. \tag{81.2}$$

The range of summation in (81.2) is limited by a theorem due to Caratheodory which states that every point in a convex set can be reached by suitable convex combinations of its extreme points. All states that cannot be written in the form of (81.2) are said to be entangled. Note that the set of separable states form a convex subset of the convex set of all possible states.

We will now give a simple criterion which decides whether a given state is actually separable or not. For this, one notes that by transposing the part of the density operator associated with the subsystem B, an operation which is called partial transposition, the resulting operator will not necessarily stay positive. However, if the density operator is separable, then its partial transpose is again a positive operator, and hence is a valid density operator. This condition of a state possessing a positive partial transpose is a necessary separability criterion [81.18] but sufficient only in the case of density matrices having Hilbert space dimensions 2×2 or 2×3 [81.19]. In higher-dimensional Hilbert spaces there exist states with positive partial transposes (PPT) which are nevertheless inseparable. This phenomenon is called bound entanglement [81.20, 21].

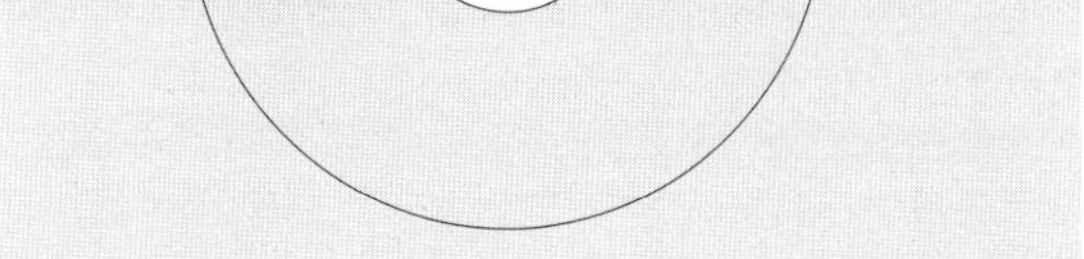

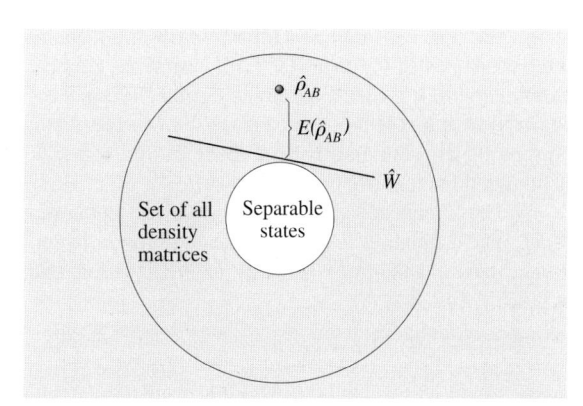

Fig. 81.1 Convex set of bipartite density matrices; the inner convex set represents the separable states. The witness operator $\hat{W}$ forms a hyperplane that separates $\hat{\varrho}_{AB}$ from the set of separable states

Because of the convexity of the set of separable states, one can construct an operator (a hyperplane) $\hat{W}$ that separates an entangled state from the disentangled states,

$$\begin{aligned} &\mathrm{tr}\left(\hat{W}\hat{\varrho}_{AB}\right) < 0 \Leftrightarrow \hat{\varrho}_{AB} \text{ inseparable} \,, \\ &\mathrm{tr}\left(\hat{W}\hat{\varrho}_{AB}\right) \geq 0 \Leftrightarrow \hat{\varrho}_{AB} \text{ separable} \,. \end{aligned} \tag{81.3}$$

Such an operator is called an entanglement witness [81.22, 23], and its existence is ensured by a consequence of the Hahn–Banach theorem [81.24].

A similar separability criterion can be found for a particularly interesting class of quantum states in infinite-dimensional Hilbert spaces, the Gaussian states. Gaussian states are most frequently encountered in quantum optics. They comprise all coherent, squeezed and thermal states, and combinations of them. Although being infinite-dimensional, these states permit a complete description in terms of their first and second moments. The characteristic function of a single-mode Gaussian state with $\boldsymbol{\lambda}^{\mathrm{T}} = (x, p)$ is given by [81.25]

$$\chi(\boldsymbol{\lambda}) = \exp\left\{ i\boldsymbol{m}^{\mathrm{T}}\boldsymbol{\lambda} - \frac{1}{4}\boldsymbol{\lambda}^{\mathrm{T}}\boldsymbol{V}\boldsymbol{\lambda} \right\} \,, \tag{81.4}$$

where $\boldsymbol{m}$ is a vector containing the first moments and $\boldsymbol{V}$ is the covariance matrix containing the second moments. A necessary and sufficient criterion for separability of a bipartite Gaussian state is that the partially transposed covariance matrix still possesses positive symplectic eigenvalues [81.26, 27].

81.1.2 Entanglement Measures

Once one has checked for inseparability, the obvious question to ask concerns the amount of entanglement, hence the amount of nonclassical correlations in the given state. For bipartite pure states the answer is unique and given by the von Neumann entropy of one subsystem, viz.,

$$E(|\psi_{AB}\rangle) = S_A(\hat{\varrho}_B) = S_B(\hat{\varrho}_A) \,, \tag{81.5}$$

with $S_A(\hat{\varrho}_B) = -\,\mathrm{tr}\,\hat{\varrho}_B \ln \hat{\varrho}_B$ where

$$\hat{\varrho}_{A(B)} = \underset{B(A)}{\mathrm{tr}} |\psi_{AB}\rangle\langle\psi_{AB}| \,.$$

The second equality in (81.5) follows from the left-hand side of the Araki–Lieb inequality [81.28]

$$|S_A - S_B| \leq S_{AB} \leq S_A + S_B \tag{81.6}$$

when noting that the entropy of a pure state vanishes. Obviously, since $S_{AB} = 0$, no information can be extracted from the total state, all information is contained in the correlations between the subsystems A and B which are revealed by performing a measurement on one of the subsystems.

For bipartite quantum systems prepared in mixed states the answer is not so obvious. However, some insight can already be gained by looking at the Schmidt decomposition of the state (which, for bipartite states, always exists) [81.29, 30], in particular, the number of elements in the decomposition, named the Schmidt number [81.31].

For a more precise definition of mixed bipartite entanglement, something more is needed. Recall that the set of separable density matrices forms a convex subset of all feasible density matrices. It therefore makes sense to look for a distance-type measure between the given state and the convex hull of product states. Note that the possibility of defining such a measure is provided by the convexity of the separable states and a consequence of the Hahn–Banach theorem [81.24]. Generally, agreement has been reached on what properties any feasible entanglement measure must fulfil [81.32–34]. Let $E(\hat{\varrho}_{AB})$ be a real-valued functional over the tensor-product Hilbert space of bipartite density matrices. If in addition $E(\hat{\varrho}_{AB})$ has the following properties:

1. $E(\hat{\varrho}_{AB}) = 0$ for all separable states;
2. $E(\hat{\varrho}_{AB})$ is invariant under local unitary transformations, viz., $E\big[(\hat{U}_A \otimes \hat{U}_B)\hat{\varrho}_{AB}(\hat{U}_A^\dagger \otimes \hat{U}_B^\dagger)\big] = E(\hat{\varrho}_{AB})$;
3. $E(\hat{\varrho}_{AB})$ is non-increasing under general local operations assisted by classical communication, viz., $E\big(\sum_i \hat{V}_A^i \otimes \hat{W}_B^i \hat{\varrho}_{AB} \hat{V}_A^{i\dagger} \hat{W}_B^{i\dagger}\big) \le E(\hat{\varrho}_{AB})$;
4. $E(\hat{\varrho}_{AB})$ reduces to the reduced von Neumann entropy for pure states,

then $E(\hat{\varrho}_{AB})$ is called an entanglement measure.

Important examples of widely used entanglement measures are the entanglement of formation [81.35]

$$E_{\mathrm{F}}(\varrho_{AB}) = \min_{\hat{\varrho}_{AB}=\sum_i p_i |\psi_i\rangle\langle\psi_i|} \sum_i p_i E(|\psi_i\rangle) \,, \quad (81.7)$$

and the relative entropy of entanglement [81.32, 33]

$$E_{\mathrm{R}}(\hat{\varrho}_{AB}) = \min_{\hat{\sigma}=\sum_i p_i \hat{\sigma}_i^A \otimes \hat{\sigma}_i^B} \mathrm{tr}\left[\hat{\varrho}_{AB}(\ln \hat{\varrho}_{AB} - \ln \hat{\sigma})\right] \,. \quad (81.8)$$

In general, both of these measures are hard to evaluate. Analytical formulas are known only in special cases. For qubits, the entanglement of formation is also a monotonic function of the concurrence [81.36,37]. The definition of the entanglement of formation, (81.7), can also be extended to cover Gaussian states [81.38].

The number of singlets, i. e., states of the form (81.1), that can be distilled from an ensemble of non-maximally entangled states is called the entanglement of distillation [81.39]. The entanglement of formation and the entanglement of distillation differ by the amount of bound entanglement (Sect. 81.1.1).

In some instances, when it is not necessary to comply with all of the above properties of entanglement measures, other quantities can be used to assess the entanglement content of a bipartite state. Particularly useful is the logarithmic negativity [81.40, 41]

$$E_N(\hat{\varrho}_{AB}) = \log_2 \left\|\hat{\varrho}_{AB}^{\mathrm{P.T.}}\right\|_1 \,, \quad (81.9)$$

where $\|\cdot\|_1$ denotes the trace norm and $\hat{\varrho}_{AB}^{\mathrm{P.T.}}$ the partial transpose of $\hat{\varrho}_{AB}$. This measure is often used in connection with Gaussian states.

In close analogy to classical information theory, the amount of nonclassical correlations is measured in ebits when one computes entropies with the dual logarithm ($\log_2$). For example, a pure state with state vector

$$|\psi\rangle = \frac{1}{\sqrt{2}}(|01\rangle + |10\rangle) \quad (81.10)$$

in an abstract two-particle Hilbert space spanned by the basis states $\{|00\rangle, |01\rangle, |10\rangle, |11\rangle\}$ contains 1 ebit of entanglement. It is also a maximally entangled state associated with this Hilbert space since the von Neumann entropy of any state in a Hilbert space of dimension N is bounded from above by $\log_2 \dim N$.

We have concentrated here on bipartite entanglement. The extension to multipartite systems is by no means trivial and much remains to be done on this subject [81.42–44].

81.2 Simple Quantum Protocols

In this section we describe the historically first and simplest quantum protocols – quantum key distribution, quantum teleportation, and super-dense coding – that make use of inherently 'quantum' properties of quantum-mechanical systems. These are either entanglement or, in the case of the simplest version of quantum key distribution, properties of the quantum-mechanical measurement process. We should mention here the pioneering work of *Holevo* [81.45] who showed that there are fundamental limits on the amount of information that can be extracted by measurements. The application of his ideas to channel capacity and communication [81.46,47] are well described in [81.3] and space limitations prevent us from elaborating on it in this chapter.

81.2.1 Quantum Key Distribution

Historically, the earliest protocol that used quantum-mechanical features in order to realize some specific task that could not have been performed classically was a protocol for secure distribution of a key in cryptography, known as the BB84-protocol after its inventors *Bennett* and *Brassard* and the year of its invention [81.48]. Although it is commonly referred to as the first example of quantum information processing, it does not make use of entanglement which was only done some years later, by *Ekert* [81.49].

The BB84 protocol works in the following way. The sender A prepares a random sequence (or string) of single photons in a polarization state which is chosen out of a set of four basis states, horizontally and vertically (H and V) polarized, and 45° and 135° (L and R) polarized. In each of the two basis sets $\{H, V\}$, $\{L, R\}$ one of the states is used to encode the logical value 0 (say in H and L) and the other states encode the logical value 1 (V and R). The random sequence is sent to the receiver B who performs measurements on the sequence of signals by randomly choosing analogous basis states. The result will be another string of 0's and 1's that generically does not coincide completely with the original string. To rectify this problem, sender and receiver communicate over a classical public channel where the sender announces the sequence of basis sets in which the photon states were prepared. The receiver compares its sequence of randomly chosen basis states with the announced string and keeps all measurement results for which the choice of basis had been the same. In that way a common secret key is established (Table 81.1).

Table 81.1 BB84 protocol for secret key distribution. The sender A sends information encoded in either of two basis sets. The receiver B randomly chooses a measurement basis which is publicly communicated. For those cases when sender and receiver chose the same basis, the receiver's measurement yields a secure bit

Sender A	↗	↑	↘	→	↗	↑	→
Receiver B	↑→	↗↘	↗↘	↑→	↗↘	↗↘	↑→
Key			1	1	0		1

The security against eavesdropping of this simple protocol comes from the fact that even by knowing the measurement basis (say $\{H, V\}$) no information has been revealed about the choice of the actual bit value (H or V). Hence, it is the quantum-mechanical measurement process itself that provides security of the protocol. The first quantum key distribution experiments were reported in [81.50–52]. However, imperfections in the generation and detection of photons, transmission losses and polarization drift causes an actual experimental realization to be far from ideal. In practise, encodings other than polarization may be used (for example a time-binned interferometric basis [81.53]). Despite these error sources, unconditionally secure quantum key distribution can be [81.54–57] and has been achieved [81.58]. Some fiber-based systems have reached distances of more than 100 km [81.59, 60], but discussions of their security continue. For a review of theoretical and experimental aspects of quantum cryptography, see [81.61].

81.2.2 Quantum Teleportation

An important utilization of entanglement as a necessary resource can be found in what is commonly known as quantum teleportation. The task of teleportation is to transmit the complete information of an arbitrary unknown quantum state to a spatially different location with the aim of re-creating it. The simplest and obvious way to perform this task would be to take the quantum object which is prepared in the original state and physically transport it to a different location. But sometimes this is not possible because for example an ion needs to be stored in a trap and cannot be moved. The next obvious thing to do would be to measure the quantum state and to re-create it at a different position using the classical information obtained during the measurement. However, single measurements on a quantum system yield only partial information and multiple measurements on many identically prepared copies would have to be performed.

The protocol, which was originally proposed in [81.62] for qubits and later generalized to states in infinite-dimensional Hilbert spaces in [81.63], makes use of the existence of maximally entangled states. Let the unknown quantum state which is to be teleported be a qubit superposition state of the form

$$|\psi\rangle = \alpha|0\rangle + \beta|1\rangle\,, \quad |\alpha|^2 + |\beta|^2 = 1\,. \tag{81.11}$$

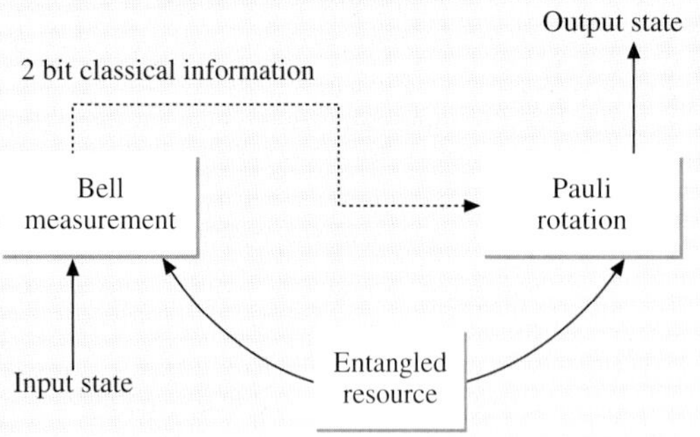

Fig. 81.2 Schematic outline of an ideal teleportation protocol

Then one prepares a maximally entangled state of the form (81.10) which is one of the four so-called Bell states defined by

$$|\Psi^{\pm}\rangle = \frac{1}{\sqrt{2}}\left(|01\rangle \pm |10\rangle\right) ,$$
$$|\Phi^{\pm}\rangle = \frac{1}{\sqrt{2}}\left(|00\rangle \pm |11\rangle\right) . \qquad (81.12)$$

We then form the tensor product state $|\psi\rangle|\Psi^+\rangle$ as

$$\begin{aligned}|\psi_A\rangle|\Psi^+_{BC}\rangle &= \frac{\alpha}{\sqrt{2}}(|0_A0_B1_C\rangle + |0_A1_B0_C\rangle)\\ &\quad+\frac{\beta}{\sqrt{2}}(|1_A0_B1_C\rangle + |1_A1_B0_C\rangle)\\ &= \frac{1}{2}\Big[(\alpha|0_C\rangle + \beta|1_C\rangle)|\Psi^+_{AB}\rangle\\ &\quad+(\alpha|0_C\rangle - \beta|1_C\rangle)|\Psi^-_{AB}\rangle\\ &\quad+(\alpha|1_C\rangle + \beta|0_C\rangle)|\Phi^+_{AB}\rangle\\ &\quad+(\alpha|1_C\rangle - \beta|1_C\rangle)|\Phi^-_{AB}\rangle\Big] , \qquad (81.13)\end{aligned}$$

where we have explicitly indexed the relevant subsystems. After performing a joint measurement on subsystems A and B in the Bell basis (this is called a Bell-state measurement [81.64, 65]) one obtains one of four possible results. If the measurement result was $|\Psi^+\rangle$, then the subsystem C is indeed prepared in the original unknown quantum state $|\psi\rangle$, hence the state has been 'teleported' from subsystem A to C. For all other measurement results the outcome is not exactly the same quantum state as intended, but the difference is just a unitary transformation which is uniquely determined by the outcome of the Bell measurement. For example, measuring $|\Psi^-\rangle$ means one has to perform a $\hat{\sigma}_z$-operation that flips the sign of the state $|1\rangle$, whereas on obtaining $|\Phi^+\rangle$ or $|\Phi^-\rangle$ the operations to be applied have to be $\hat{\sigma}_x$ or $\hat{\sigma}_z\hat{\sigma}_x$, respectively.

Note that this quantum teleportation protocol works with perfect fidelity only if a maximally entangled state has been used, i.e., a state containing 1 ebit of quantum information. In the course of the Bell measurement, the quantum information is used up, and two classical bits of information (the measurement result) have to be communicated to C in order to restore the original quantum state. In this sense, entanglement can be regarded as a resource or 'fuel' for certain tasks in quantum information processing. The first experimental demonstrations of teleportation of qubits were performed in [81.66–68] and of continuous variables in [81.69–71]. Recently, a teleportation experiment over 2 km standard telecommunication fibre has been reported [81.72]. A generalization of teleportation is entanglement swapping, in which EPR correlations are established between previously uncorrelated particles by Bell-state measurements [81.73, 74].

81.2.3 Dense Coding

The complementary protocol to teleportation is characterized by the name of (super) dense coding [81.75]. The idea here is to transmit two classical bits of information at the expense of consuming 1 ebit of quantum information. The similarity to teleportation is best seen by noting that if the experimental apparatus of sender and receiver are interchanged and the protocol reversed (Fig. 81.3), then one reduces to the other. The mathematical equivalence of the teleportation and dense coding schemes has been beautifully shown in [81.76]. As in teleportation, sender and receiver initially share a two-particle maximally entangled state, i.e., one of the Bell states defined in (81.12). By acting with one of the four operations $\hat{I}$, $\hat{\sigma}_x$, $\hat{\sigma}_z$, or $\hat{\sigma}_z\hat{\sigma}_x$ on the qubit on the sender's side, the total

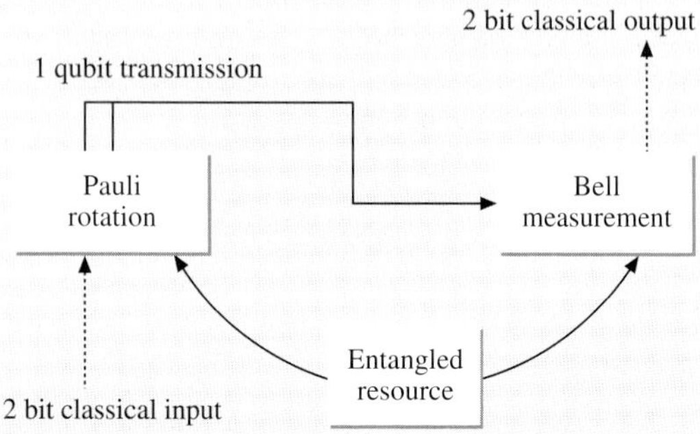

Fig. 81.3 Schematic outline of an ideal superdense coding protocol

two-qubit state is again in one of the four Bell states (81.12). Since they are mutually orthogonal to each other, the receiver can tell them apart by measuring in the Bell basis. In that way, two classical bits of information (the information about the single-qubit unitaries) can be transmitted using only a single qubit at a time.

An experiment using entangled photon pairs was reported in [81.77], which demonstrated dense coding in practise.

81.3 Unitary Transformations

As in classical information theory, one has to define a certain set of allowed operations or maps between states of an information-carrying system. Classical information processing allows operations such as the NAND (Not-AND), which is defined by the Boolean operation $\overline{X_1 \wedge X_2 \wedge \cdots \wedge X_n}$ on the Boolean variables $X_1, X_2, \ldots, X_n$. This operation is not reversible in the sense that, given the outcome of the operation, there is no unique way of determining what the input was. Hence, such types of classical gates destroy information during the course of their operation.

Loss of information or irreversibility of an operation is accompanied by an increase in entropy of the state that has been operated on. For an initially pure quantum state having zero entropy, this means mixing the state and destroying its superposition nature and hence its quantum-mechanical entanglement. Therefore, the advantages of parallelism inherent in the superposition is lost. Thus, valid quantum operations in this sense can only be those that preserve the purity of states, hence unitary operations and partial projective measurements. Of course, in order to reset a quantum register, information has to be erased [81.78, 79]. This erasure procedure is described by a completely positive map (see, e.g. [81.3]). Note that any allowed quantum operation is completely positive; unitary operations are a special class of these. An example of how this constrains operations in quantum rather than classical information theory is the absence of a NOT operation in the former, as a NOT operation cannot be described in terms of completely positive maps [81.80].

81.3.1 Single-Qubit Operations

It is instructive to give an example of how to classify all possible unitary operations that can act on a single qubit. A unitary operation acting upon the basis states $\{|0\rangle, |1\rangle\}$ can be represented by a unitary (2×2)-matrix, hence a matrix that represents an element of the unitary group U(2). This group has four generators, the identity matrix and the three Pauli matrices. Hence, all unitary (2×2)-matrices are linear combinations of those four matrices. Given the way they act upon basis states, they can be written as

$$\hat{\mathrm{I}} = |0\rangle\langle 0| + |1\rangle\langle 1| \,, \tag{81.14}$$

$$\hat{\sigma}_x = |0\rangle\langle 1| + |1\rangle\langle 0| \,, \tag{81.15}$$

$$\hat{\sigma}_z = |0\rangle\langle 0| - |1\rangle\langle 1| \,, \tag{81.16}$$

and, by virtue of the commutation rules for U(2)-generators, $\hat{\sigma}_y = \mathrm{i}\hat{\sigma}_x\hat{\sigma}_z$. Sometimes, the short-hand notation $X \equiv \hat{\sigma}_x$, etc., is used.

A particularly useful single-qubit gate which is not just one of the Pauli operators is the Hadamard gate H. In terms of Pauli operators it is defined as $H = (X+Z)/\sqrt{2}$. Its purpose is to transform each basis state into equal superpositions of basis states, i. e., $|0\rangle \mapsto (|0\rangle + |1\rangle)/\sqrt{2}$ and $|1\rangle \mapsto (|0\rangle - |1\rangle)/\sqrt{2}$. The Hadamard gate is used to initialize an equal superposition of all possible N-qubit basis states from the state $|0\rangle^{\otimes N}$. Hence,

$$\bigotimes_{i=1}^{N} H_i |0\rangle^{\otimes N} = \frac{1}{\sqrt{N!}} \sum_k |x_k\rangle \,, \tag{81.17}$$

where the $|x_k\rangle$ are all $N!$ possible words of length N containing 0's and 1's.

81.3.2 Two-Qubit Operations

Similarly to the single-qubit case, one can write down all possible unitary operations on two qubits by noting that they constitute a representation of U(4). We will not give an exhaustive list of all 16 generators of this group since they can be found in the literature. Instead, we give examples of particularly useful two-qubit gates. Trivially, the group U(4) contains an 8-parameter subgroup U(2)×U(2) which consists of operations such as $X_1 \otimes X_2$, etc.

Particular examples of nontrivial two-qubit gates are the controlled-NOT and the controlled-phase gate,

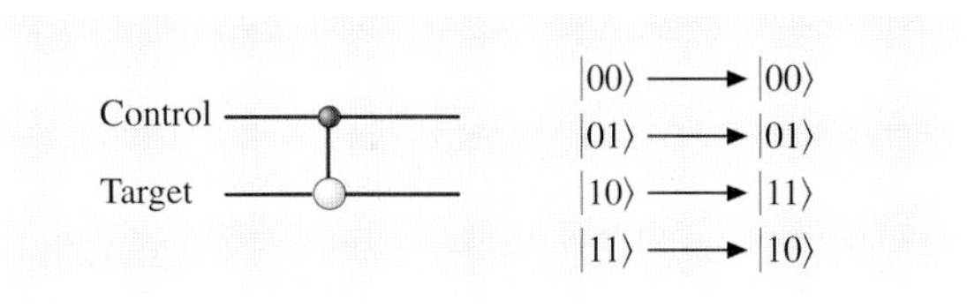

Fig. 81.4 Symbol and truth table of the controlled-NOT gate. The target qubit is flipped depending on the state of the control qubit

defined in terms of Pauli operators as

$$\overline{C}_{12} = |0_1\rangle\langle 0_1| \otimes I_2 + |1_1\rangle\langle 1_1| \otimes X_2 \,, \tag{81.18}$$

$$\Pi_{12} = |0_1\rangle\langle 0_1| \otimes I_2 + |1_1\rangle\langle 1_1| \otimes Z_2 \,, \tag{81.19}$$

where in both cases qubit 1 acts as the 'control' and qubit 2 acts as the 'target' (Fig. 81.4). The net effect of the controlled-NOT gate is to interchange the states $|10\rangle \leftrightarrow |11\rangle$, whereas the controlled-phase gate changes the phase of the basis state $|11\rangle$ by π and leaves all other states unchanged. The controlled-NOT gate has an interpretation as a sum gate in that it performs a mapping $|x, y\rangle \mapsto |x, x \oplus y\rangle$ where the addition has to be taken modulo 2. Moreover, it acts as an entangling gate when acting on tensor products of superpositions.

81.3.3 Multi-Qubit Gates and Networks

To realize a unitary operation on many qubits for a particular algorithm one would need a network of single-particle and multi-particle quantum gates. Quantum networks enable a prepared input state to be transformed by the appropriate unitary operator to a final state which is then measured. Deutsch's model of quantum networks enables us to decompose the network into component gates in diagrammatic form [81.81]. The task is then to optimize the sequence of gates. One can treat quantum gates acting on N qubits as being elements of the group $\mathrm{U}(N^2)$ which has N^4 generators. This, however, is not a particularly transparent or useful way of looking at these gates. Much more useful, and of immense practical importance, is a result essentially from linear algebra which states that every N-qubit gate can be decomposed into a network of single-qubit and two-qubit operations [81.82]. As a matter of fact, there is an even deeper result which says that every N-qubit gate can be generated by a network that consists only of very few elementary building blocks, the so-called universal set of quantum gates [81.83–85]. This set contains all possible single-qubit rotations and one nontrivial two-qubit gate, such as the above-mentioned controlled-NOT or controlled-phase gate.

81.4 Quantum Algorithms

The search for algorithms that would run faster on a quantum computer than on any classical computer is a formidable task. When we say faster, we actually mean that the temporal complexity in performing a given task should be drastically reduced. The hope is that eventually one will find algorithms that provably run exponentially faster on a quantum computer compared to a classical computer.

81.4.1 Deutsch–Jozsa Algorithm

Let us give a particularly instructive example known as the Deutsch–Jozsa algorithm [81.86]. Let us suppose one is given a string of N bits and a Boolean N-bit function $f(x)$ such that $|x\rangle|y\rangle \mapsto |x\rangle|y \oplus f(x)\rangle$ for $x \in \{0, \ldots, 2^N - 1\}$. From $f(x)$ is known that it either returns always 0 or 1 (in which case one calls it 'constant') or returns 0 and 1 with equal probability (in which case it is 'balanced'). The task is to find out whether $f(x)$ is constant or balanced. Classically, one needs at least $2^{N-1} + 1$ strings to find the answer. Quantum-mechanically, one prepares the trial input in a superposition of all possible computational basis states using the Hadamard gate from (81.17) and uses one function evaluation on all basis states simultaneously (Fig. 81.5). A measurement outcome other than 0 on any of the N query qubits then tells that the function $f(x)$ is balanced. If it were constant, the measurement outcome would be 0 in all query qubits. This quantum parallelism is at the heart of the increase in speed that occurs in quantum computation. Versions of the Deutsch–Jozsa

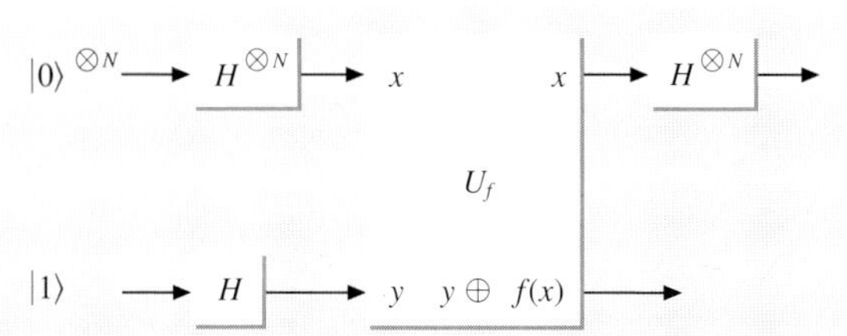

Fig. 81.5 Gate network for implementing the Deutsch–Jozsa algorithm

algorithm involving a few qubits have been implemented in nuclear magnetic resonance systems [81.87, 88] as well as ion traps [81.89].

81.4.2 Grover's Search Algorithm

In contrast to the preceding example, which always gives the desired answer after exactly one trial, the quantum search algorithm by *Grover* [81.90] uses a procedure that amplifies the sought after result by a method called 'inversion about average'. The goal of Grover's algorithm is to search an unsorted database with 2^N entries out of which only one fulfils a given criterion. As in the Deutsch–Jozsa algorithm described above, the query is simultaneously run on all 2^N possible N-qubit basis states, being prepared in an equal superposition. It is assumed that the state that satisfies the search criterion will acquire a phase shift of π. After this step, the inversion about average is carried out. It is represented by a diffusion operator

$$\hat{D} = 2\hat{P} - \hat{I}\,, \tag{81.20}$$

where $\hat{I}$ is the identity operator and $\hat{P}$ a projection operator that averages each input vector with respect to its components. Compared to the previous average value of probability amplitudes, after each of these steps the magnitude of the desired state increases by $\mathcal{O}\left(2^{-N/2}\right)$. This procedure is repeated, and after only $\mathcal{O}\left(2^{N/2}\right)$ steps a projective measurement yields the desired result with probability of $\mathcal{O}(1)$, or more precisely, of more than a half. This is a quadratic increase in speed compared to classical search algorithms, which need $\mathcal{O}\left(2^N\right)$ steps.

We have seen in these two examples that the use of quantum-mechanical superpositions can lead to a speed increase compared to the best classical algorithms. The most prominent example of such increases in speed is found in Shor's algorithm for factoring large numbers [81.91,92]. Its core element is essentially a quantum Fourier transform to find the period of a Boolean string. This algorithm provides an exponential speed increase over any known classical algorithm. It should be noted, however, that the fastest known classical algorithm has not yet been proven to be optimal. Implementations in nuclear magnetic resonance systems with a few qubits have been reported for the quantum Fourier transform [81.93] as well as Shor's factoring algorithm [81.94].

It turns out that there exists a whole class of algorithmic problems, the hidden subgroup problems [81.95–97], whose quantum-mechanical analogues can lead to exponential increases in speed over their classical counterparts, a particular example of which is Shor's factoring algorithm. Another instance of quantum-mechanically exponentially faster processes is found in quantum random walks on hypercubes where hitting times, i. e., the traversal time of an excitation across the cube's diagonal, can be exponentially faster than for classical random walks [81.98].

81.5 Error Correction

As we have discussed, the essence of quantum information processing is the use of quantum superpositions, interference, and entanglement. But quantum interference is fragile. It appears in practice that it is very difficult to maintain a superposition of states of many particles in which each particle is physically separated from all the others. Entanglement turns out to be incredibly delicate. The reason for this is that all systems, quantum or classical, are not isolated; they interact with everything around them: local fluctuating electromagnetic fields, the presence of impurity ions, coupling to unobserved degrees of freedom of the system containing the qubit, etc. These fluctuations destroy quantum interference. A simple analogy is the interference of optical waves in Young's double slit experiment. In that apparatus waves from two spatially separated portions of a beam are brought together. If the two parts of the beam have the same phase, then the fringe pattern remains stable. But if the phase of one part of the beam is drifting with respect to the other, then the fringe pattern will be washed out. And the more slits there are in the screen, the lower the visibility for the same amount of phase randomization per pair of slits. The sensitivity of an N-qubit register to decoherence is even worse, as a maximally entangled N-particle state decoheres at a rate N times faster than a single particle [81.99], one of the reasons why the world around us appears so classical. A single bit of information lost to an unobserved degree of freedom will result in the reduction of the quantum superposition to a mixed state. Yet correcting errors due to environmental interactions is essential if a quantum computer is to be constructed: to do 'fault tolerant' computing we need to be able to execute many gate operations coherently within the decoherence time if we are to have a chance of building a scalable quantum register [81.100].

It might appear that the problem of stabilizing a register of qubits is hopeless, like trying to balance several pencils on their tips while on the deck of a ship in a storm. But, amazingly, quantum mechanics provides a way to solve this problem, through the use of even higher levels of entanglement. *Shor* and coworkers, and independently *Steane* showed in the mid-1990's that encoding information in entangled sets of qubits offered the opportunity to execute quantum error correction [81.101, 102]. That one can do this is a remarkable consequence of entanglement. In classical information processing, inevitable environmental noise is dealt with by error correction. In its simplest form, this involves repeating the message transmission or calculation until a majority result is obtained.

But there are more efficient ways, for example, use of a parity check on a block of bits. It turns out that a similar notion can be applied to a quantum register. However, the application is not straight forward because the contents of a register cannot be measured without destroying the superposition state encoded in it. The problem then is to determine what errors might be present in a quantum register without looking at the qubits. The elegant solution is to entangle the qubits in question with those in an ancillary register and measure the ancillary register. Because the two registers are correlated, the results of the measurement of the ancilla reveal any errors present in the processing register without destroying any coherent superpositions in the processing register itself.

The first experimental demonstration of quantum error correction used NMR techniques [81.103–105], but given the inherently mixed nature of NMR quantum computing [81.106], this has had limited impact on quantum information processing (QIP). However, very recently, Wineland's group in Boulder has succeeded in implementing quantum error correction using laser-cooled trapped ions [81.107].

Another way to prevent the register coherence from falling apart is to know a little about the sort of noise that is acting on it. If the noise has some very slow components (or those with very long wavelengths), then it is sometimes possible to find certain combinations of qubit states for which the noise on one qubit exactly cancels the noise on another. These qubit states live in a 'decoherence-free subspace' (DFS) [81.108–111]. A computer will then be immune to environmental perturbations if all the computational states lie in this DFS. The connection between DFS and quantum error correction codes has been shown in [81.112]. *Kwiat* [81.113], using photonic qubits, and Wineland's group [81.114], using trapped ions, have demonstrated the use of DFS experimentally.

Although these results are encouraging, we are still a long way from the figure of merit for gate time to decoherence time needed for fault tolerance.

81.6 The DiVincenzo Checklist

DiVincenzo gave a list of requirements that a physical implementation must fulfil in order to qualify as a sensible candidate for an implementation of quantum information processing [81.115].

81.6.1 Qubit Characterization, Scalability

Each physical implementation must be tested upon how qubits should be encoded. For a qubit being essentially a two-level system, this task is generally not too difficult. Several candidates, such as electronic or nuclear spin, photon polarization, choice of path in an interferometer, degenerate ground states of an atom or ion, charge or flux states in superconducting quantum interference devices (SQUIDs) or exciton population, have all been recently explored. Much more challenging will be the question as to whether there are fundamental or technological limitations of having many of those qubits being operated upon seperately, hence whether the system can be scaled up to contain potentially many qubits.

81.6.2 Initialization

Once the qubits have been specified, each quantum information processing or quantum computation task needs to be able to start from a well-defined state. This can be basically any quantum state of the many-qubit system as long as it is a product state and can be prepared error-free. Commonly, this state is then called the ground state and denoted by $|0\rangle^{\otimes N}$.

81.6.3 Long Decoherence Times

In order to ensure error-free computation without loss of purity of quantum superpositions, the decoherence times that are relevant for the quantum operation should be much longer than the gate operation time itself. In most situations, decoherence limits the number of qubits that can be worked on simultaneously, thus affecting the scalability of the system. Typical examples of decoherence processes are heating mechanisms in ultracold systems, such as ion trap or atom chip experiments, spin relaxation in NMR-type experiments, or absorption in linear optical elements.

Generally, decoherence is unavoidable due to the basic principles upon which quantum information processing is supposed to work. Avoiding decoherence means isolating the system from the outside world, the environment. But controlling the interaction between subsystems always has the negative effect of bringing the system in contact with the environment and therefore necessarily introduces decoherence. Once one has accepted that decoherence is unavoidable, ways have to be found to guard against it. Several error-correction schemes have been proposed that can correct for certain small amounts of decoherence as described in Sect. 81.5.

81.6.4 Universal Set of Quantum Gates

A necessary prerequisite for quantum information processing and quantum computing is the ability to generate a set of quantum gates that can be considered universal. With such a set it will then be possible to generate all other quantum gates by concatenating them to form suitably arranged networks. The choice of which set out of the many possible is taken, depending on the physical implementation itself. Basically, it is determined by the operations that are intrinsically simple for the given interaction Hamiltonian. In some applications, such as the ion trap experiments, the controlled-NOT gate is preferred as the nontrivial two-qubit gate, whereas in linear optical networks one rather works with the controlled-phase gate.

81.6.5 Qubit-Specific Measurement

The last requirement is to be able to read out the result of the computation. That is, there has to be a way of providing a selective projective measurement. This proves to be a major challenge in most proposals for implementing quantum computing. Examples of the challenges involved are the necessity to provide photon-number resolving photodetectors, single-electron charge measurement devices, or single-spin measurements.

81.7 Physical Implementations

Quantum information theory regards relevant objects as abstract quantities in a Hilbert space of a certain dimension. The different strands in its development can be roughly divided into generalized spin systems (qubits, qudits, the d-dimensional generalization of qubits) living in finite-dimensional Hilbert spaces, and harmonic oscillator systems which naturally live in infinite-dimensional Hilbert space. The latter are hence called continuous-variable (cv) systems. Examples for generalized spin systems are polarization states of a photon, magnetic sublevels of atomic hyperfine states or, to a good approximation, electronic levels of atoms and ions. Harmonic oscillator systems can be found in atomic populations in optical lattices, electromagnetic-field modes or indeed any excitation of a bosonic quantized field.

Both strands have their own virtues and disadvantages. Harmonic oscillator systems are naturally abundant and, in their materialization as photons, rather easily accessible and manipulable. However, due to their Hilbert space dimensionality, the nonlinear operations that are required for quantum gates are generally hard to achieve. In contrast, spin-like systems (apart from photon polarization) require more experimental effort in preparing them but, unlike harmonic oscillator systems, they can show effective nonlinearities due to the finite dimensionality of their Hilbert space (the nonlinearities appear when coupled to another physical system which can then be traced out).

81.7.1 Linear Optics

The use of photons as carriers of quantum information seems to be a straightforward matter for several reasons. First, they are easy to produce and to manipulate, and second, they both show spin-like behaviour (polarization) and can be treated as continuous-variable systems (Gaussian states). There exists, however, yet another possibility to store and manipulate quantum information in photons, namely when encoding information in number states or Fock states. But as noted before, for photons being bosonic systems, there are no natural nonlinearities (at least none which is strong enough) on the level of single or few photons. The trick here is to use conditional measurements or measurement-induced nonlinearities. The idea was first

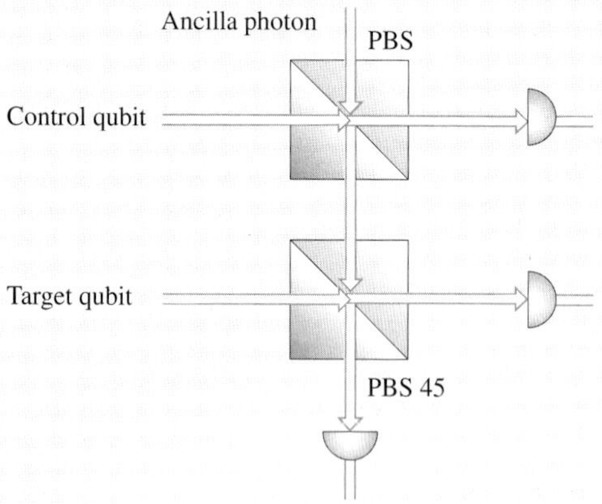

Fig. 81.6 Schematic setup of an all-optical controlled-NOT gate. Control and target qubits are encoded in the polarization of single photons. These are fed into polarizing beam splitters (PBS), one of which is rotated by 45 degrees

put forward in [81.116, 117] and realized experimentally in [81.118, 119], where a polarization-encoding was used. Figure 81.6 shows the schematic setup of a simplified version of a controlled-NOT gate with one single ancilla photon (after [81.120]).

Measurement-induced nonlinearities make use of the fact that unitary transformations in a larger Hilbert space, e.g., with added auxiliary photon modes combined with photodetection, can yield effective nonlinearities [81.121]. The drawback is, however, that the wanted nonlinearity is conditioned on the appearance of a certain measurement pattern which means that these schemes work only with a certain probability. Bounds for certain classes of gates have been reported in [81.122–125].

The set of quantum gates that can be considered fundamental differs slightly from most other physical implementations. Within the qubit encoding in photon-number states the gates that can actually be implemented efficiently are those that act within Fock layers (subspaces of fixed total photon number), such as Z, the controlled-phase gate, or the swap gate. Other gates that do not fall into this class require excessively more resources unless other types of qubit encodings are used simultaneously.

Gate operation times can be very fast and are only limited by the gating times of the photodetectors. However, a major experimental challenge is mode-matching in larger networks and interferometric set-ups.

A complementary approach to the gate model is based on so-called cluster states which were originally introduced to describe the properties in 3D optical lattices [81.126] (Sect. 81.7.4). In the cluster-state model the computational process is not described by a succession of elementary gates that act upon an (in principal arbitrary) input state, but by a well-defined sequence of single-qubit measurements performed on a maximally entangled 'cluster' of qubits. It has been realized that the cluster-state model represents an alternative model for quantum computing, the so-called 'one-way quantum computer' [81.127]. It was later found that there also exists a linear optical realization of the cluster-based approach [81.128, 129] which has been experimentally verified [81.130].

81.7.2 Trapped Ions

So far, the most advanced method in terms of the number of qubits and the number of gates that have been generated is by using ultracold ions stored in linear Paul traps (Chapt. 75) in which radio-frequency fields are used to generate confining potentials. The ions are trapped in the radial direction by electric quadrupole fields and in the axial direction by a static repulsive Coulomb force [81.131, 132]. The ions are cooled into their motional ground states by Doppler cooling [81.133, 134] and further cooled by resolved sideband cooling [81.135]. The qubits are encoded into two metastable electronic states. Various groups have used either transitions to metastable states or Raman coupling to avoid the decoherence that an upper-state lifetime would generate. The coupling between qubits is provided by the common vibrational motion in the Paul trap [81.136, 137]. For reviews of the dynamics of laser-cooled ions [81.138–141]. To date, a few qubits have been entangled and coherently manipulated. Simple quantum algorithms have been demonstrated [81.142] and teleportation achieved [81.143, 144].

81.7.3 Cavity QED

Cavity QED provided the very first examples of atom-field entanglement. Single atoms interacting with single cavity-field modes are well described by the Jaynes–Cummings model [81.145]. In this model, excitation is transferred periodically between atoms and field provided the Q-factor of the cavity is high enough (Chapt. 79). The Rabi flopping can be used to generate controlled superpositions, and the cavity field used as a catalyst to entangle atoms [81.146]. Although it is possible to coherently manipulate single or few qubits in cavity QED, scaling to large numbers of qubits would

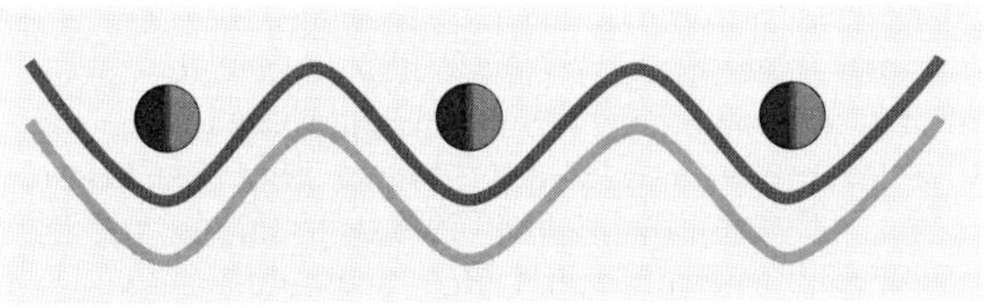

Fig. 81.7 Counter-propagating laser beams induce a periodic spatially varying trapping potential through the AC Stark shift

seem very difficult. Nevertheless, trapped atoms within cavity QED environments offer great potential as local processors linked by quantum communication channels [81.147]. Progress towards this has been reported by several groups [81.148–151].

81.7.4 Optical Lattices, Mott Insulator

Another possible way of implementing quantum computation is with cold atom technology. This includes the application of optical lattices in a sufficiently cold cloud of atoms showing Bose–Einstein condensation (BEC).

In recent years it has been realized that Bose–Einstein condensates can undergo a phase transition if loaded into a three-dimensional periodic potential which, for example, can be realized by standing-wave optical fields [81.152, 153]. That is, one starts off with a BEC in its superfluid phase, in which the relative phases (or rather correlations) between the atoms are well-defined such that the whole ensemble of atoms can be described by a single macroscopic wave function (in first approximation). By loading this condensate into the optical lattice (Fig. 81.7) the number of atoms per lattice site is undetermined and can vary widely. However, when increasing the strength of the potential by increasing the power of the laser beams that create the standing-wave potential, eventually there will be a phase-transition to a state of the condensate in which each lattice site is occupied by a fixed and well-defined number of atoms (ideally we would like to have exactly one atom per site). In this so-called Mott-insulator phase, the relative phases (or correlations) between neighboring lattice sites are undetermined. Experimental evidence of this phase-transition has been obtained in the beautiful experiments described in [81.154–156]. Although a Bose–Einstein condensate really exists only in three dimensions (since only there one finds a phase transition from a thermal cloud to a condensate), there are analogous systems, such as the quasi-condensate [81.157] and the Tonks–Girardeau gas [81.158] (for recent experiments, see [81.159, 160]), in one dimension that have similar properties.

Such a system is well described by the Bose–Hubbard Hamiltonian [81.152, 153], in which the collisional interaction between atoms at the same lattice site provides the necessary nonlinearity. Atoms trapped in a one-dimensional optical lattice could serve as an atomic register that promises well controlled single-qubit and two-qubit manipulability. A universal set of quantum gates can be realised by manipulations of the lattice potential with additional laser fields [81.161]. The different types of quantum gates could, for example, be realized if the atoms possess two degenerate ground states that are used for the qubit encoding (as sketched in Fig. 81.7). A Raman transition between the ground states would result in single-qubit operations, whereas controlled collisions between atoms in neighboring lattice sites would produce two-qubit gates. Recently, three-qubit gates [81.162] and global adressing of strings of qubits have been proposed [81.163].

The estimated gate evolution times in the adiabatic regime are roughly $\mathcal{O}(100\,\text{ms})$, which is just one to two orders of magnitude below the trapping lifetime measured in recent experiments [81.164–166]. The dominant loss effect is thereby a thermally induced spin flip mechanism that causes the atoms to leave the trapping region [81.167–169]. The gate evolution times could be reduced by several orders of magnitude in non-adiabatic regimes. The price to pay is that the temporal evolution of the laser pulse envelope has to be controlled much more precisely. Although quantum information processing using cold atom technology is still in its infancy, it promises to provide relatively long decoherence times. Moreover, scalability seems possible as rather long one-dimensional strings of atoms could be formed. Experimental evidence for this has of course yet to be shown.

81.8 Outlook

We have discussed the very basic ideas behind quantum information and described a few possible applications. However, the immense wealth of ideas and possible routes have barely been touched upon. Quantum key distribution is already at a stage where private companies are selling component parts to set up commercial

QKD systems. Only a few qubits have been maximally entangled and manipulated experimentally, so far. But a great number of qubits has been partially entangled within optical lattices [81.170] or in atomic vapours [81.171]. The dynamics of qubit interactions in such systems is closely related to systems studied in phase-transition theory, pointing to yet another application of the subject. The simulation of many-particle quantum systems is of course intrinsically difficult and could well require a quantum computer for its analysis [81.172–175]. If and when a quantum computer can be built remains shrouded in mist. However, the ideas and methods that have already come out of quantum information theory provide useful tools for tackling other, seemingly unrelated, problems. One direction of current research regards many-body problems in condensed matter systems and quantum field theory (see, for example, [81.176]).

We have touched upon only a small part of a rapidly developing subject – one in which quantum effects are the enablers of new technology [81.177]. We are confident that much more remains to be discovered.

References

81.1 B. Russell: *The ABC of Relativity* (Allen Unwin, London 1969)

81.2 J. Gruska: *Quantum Computing* (McGraw-Hill, London 1999)

81.3 M. A. Nielsen, I. L. Chuang: *Quantum Computation and Quantum Information* (Cambridge Univ. Press, Cambridge 2000)

81.4 J. Stolze, D. Suter: *Quantum Computing: A Short Course from Theory to Experiment* (Wiley-VCH, Weinheim 2004)

81.5 A. Cabello: *quant-ph/0012089*

81.6 A. Cabello: URL www.adancabello.com

81.7 A. Einstein, B. Podolsky, N. Rosen: Phys. Rev. **47**, 777 (1935)

81.8 J. S. Bell: Physics **1**, 195 (1964)

81.9 J. S. Bell: *Speakable and unspeakable in quantum mechanics* (Cambridge Univ. Press, Cambridge 1987)

81.10 S. J. Freedman, J. F. Clauser: Phys. Rev. Lett. **28**, 938 (1972)

81.11 A. Aspect, P. Grangier, G. Roger: Phys. Rev. Lett. **47**, 460 (1981)

81.12 A. Aspect, J. Dalibard, G. Roger: Phys. Rev. Lett. **49**, 1804 (1982)

81.13 J. G. Rarity, P. R. Tapster: Phys. Rev. Lett. **64**, 2495 (1990)

81.14 W. Tittel, J. Brendel, H. Zbinden, N. Gisin: Phys. Rev. Lett. **81**, 3563 (1998)

81.15 T. M. Cover, J. A. Thomas: *Elements of Information Theory* (Wiley, New York 1991)

81.16 W. H. Zurek, W. K. Wootters: Nature (London) **299**, 802 (1981)

81.17 M. Hillery: Phys. Rev. A **54**, 1844 (1996)

81.18 A. Peres: Phys. Rev. Lett. **77**, 1413 (1996)

81.19 M. Horodecki, P. Horodecki, R. Horodecki: Phys. Lett. A **223**, 1 (1996)

81.20 P. Horodecki: Phys. Lett. A **232**, 333 (1997)

81.21 M. Horodecki, P. Horodecki, R. Horodecki: Phys. Rev. Lett. **80**, 5239 (1998)

81.22 B. Terhal: Phys. Lett. A **271**, 391 (2000)

81.23 M. Lewenstein, B. Kraus, J. I. Cirac, P. Horodecki: Phys. Rev. A **62**, 052310 (2000)

81.24 E. Zeidler: *Applied functional analysis: main principles and their applications*, AMS, Vol. 109 (Springer, Berlin, Heidelberg 1995)

81.25 C. C. Gerry, P. L. Knight: *Introductory Quantum Optics* (Cambridge Univ. Press, Cambridge 2005)

81.26 L. M. Duan, G. Giedke, J. I. Cirac, P. Zoller: Phys. Rev. Lett. **84**, 2722 (2000)

81.27 R. Simon: ibid. **84**, 2726 (2000)

81.28 H. Araki, E. Lieb: Commun. Math. Phys. **18**, 160 (1970)

81.29 E. Schmidt: Math. Annalen **63**, 433 (1906)

81.30 A. Ekert, P. L. Knight: Am. J. Phys. **63**, 415 (1995)

81.31 K. W. Chan, C. K. Law, J. H. Eberly: Phys. Rev. A **68**, 022110 (2003)

81.32 V. Vedral, M. B. Plenio: Phys. Rev. A **57**, 1619 (1998)

81.33 M. B. Plenio, V. Vedral: Contemp. Phys. **39**, 431 (1998)

81.34 M. Horodecki, P. Horodecki, R. Horodecki: Phys. Rev. Lett. **84**, 2014 (2000)

81.35 C. H. Bennett, D. P. DiVincenzo, J. A. Smolin, W. K. Wootters: Phys. Rev. A **54**, 3824 (1996)

81.36 S. Hill, W. K. Wootters: Phys. Rev. Lett. **78**, 5022 (1997)

81.37 W. K. Wootters: Phys. Rev. Lett. **80**, 2245 (1998)

81.38 M. M. Wolf, G. Giedke, O. Krüger, R. F. Werner, J. I. Cirac: Phys. Rev. A **69**, 052320 (2004)

81.39 C. H. Bennett, H. Bernstein, S. Popescu, B. Schumacher: Phys. Rev. A **53**, 2046 (1996)

81.40 G. Vidal, R. F. Werner: Phys. Rev. A **65**, 032314 (2002)

81.41 J. Eisert, M. B. Plenio: Int. J. Quant. Inf. **1**, 479 (2003)

81.42 N. Linden, S. Popescu: Fortschr. Phys. **46**, 567 (1998)

81.43 P. van Loock, S. L. Braunstein: Phys. Rev. Lett. **84**, 3482 (2000)

81.44 S. Wu, Y. Zhang: Phys. Rev. A **63**, 012308 (2001)

81.45 A. S. Holevo: *Proceedings of the Second Japan–USSR Symposium on Probability Theory* (Springer, Berlin, Heidelberg 1973) pp. 104–119

81.46 A. S. Holevo: IEEE Trans. Inf. Theory **44**, 269 (1998)

81.47 B. Schumacher, M. D. Westmoreland: Phys. Rev. A **56**, 131 (1997)

81.48 C. H. Bennett, G. Brassard: *Proceedings of IEEE Int. Conf. on Computers Systems, Signal Processing* (IEEE, Bangalore, India 1984) pp. 175–179

81.49 A. K. Ekert: Phys. Rev. Lett. **67**, 661 (1991)

81.50 C. H. Bennett, F. Bessette, G. Brassard, L. Salvail, J. Smolin: J. Crypt. **5**, 3 (1992)

81.51 P. D. Townsend, J. G. Rarity, P. R. Tapster: Electron. Lett. **29**, 1291 (1993)

81.52 A. Muller, J. Breguet, N. Gisin: Europhys. Lett. **23**, 383 (1993)

81.53 W. Tittel, J. Brendel, H. Zbinden, N. Gisin: Phys. Rev. Lett. **84**, 4737 (2000)

81.54 N. Lütkenhaus: Phys. Rev. A **54**, 97 (1996)

81.55 N. Lütkenhaus: Phys. Rev. A **59**, 3301 (1999)

81.56 N. Lütkenhaus: Phys. Rev. A **61**, 052304 (2000)

81.57 N. Lütkenhaus: Photonics News Optics **3**, 24 (2004)

81.58 M. Bourennane, F. Gibson, A. Karlsson, A. Hening, P. Jonsson, T. Tsegaye, D. Ljunggren, E. Sundberg: Opt. Express **4**, 383 (1999)

81.59 H. Kosaka, A. Tomita, Y. Nambu, T. Kimura, K. Nakamura: Electr. Lett. **39**, 1199 (2003)

81.60 C. Gobby, Z. L. Yuan, A. J. Shields: Appl. Phys. Lett. **84**, 3762 (2004)

81.61 N. Gisin, G. Ribordy, W. Tittel, H. Zbinden: Rev. Mod. Phys. **74**, 145 (2002)

81.62 C. H. Bennett, G. Brassard, C. Crepeau, R. Josza, A. Peres, W. K. Wootters: Phys. Rev. Lett. **70**, 1895 (1993)

81.63 L. Vaidman: Phys. Rev. A **49**, 1473 (1994)

81.64 S. L. Braunstein, A. Mann: Phys. Rev. A **51**, R1727 (1995)

81.65 S. L. Braunstein, A. Mann: Phys. Rev. A **53**, 630 (1996)

81.66 D. Bouwmeester, J. W. Pan, K. Mattle, M. Eibl, H. Weinfurter, A. Zeilinger: Nature (London) **390**, 575 (1997)

81.67 D. Boschi, S. Branca, F. DeMartini, L. Hardy, S. Popescu: Phys. Rev. Lett. **80**, 1121 (1998)

81.68 D. Bouwmeester, K. Mattle, J. W. Pan, H. Weinfurter, A. Zeilinger, M. Zukowski: Appl. Phys. B **67**, 749 (1998)

81.69 S. L. Braunstein, H. J. Kimble: Phys. Rev. Lett. **80**, 869 (1998)

81.70 A. Furusawa, J. L. Sorensen, S. L. Braunstein, C. A. Fuchs, H. J. Kimble, E. S. Polzik: Science **282**, 706 (1998)

81.71 W. P. Bowen, N. Treps, B. C. Buchler, R. Schnabel, T. C. Ralph, H. A. Bachor, T. Symul, P. K. Lam: Phys. Rev. A **67**, 032302 (2003)

81.72 I. Marcikic, H. deRiedmatten, W. Tittel, H. Zbinden, N. Gisin: Nature (London) **421**, 509 (2003)

81.73 M. Zukowski, A. Zeilinger, M. A. Horne, A. K. Ekert: Phys. Rev. Lett. **71**, 4287 (1993)

81.74 J.-W. Pan, D. Bouwmeester, H. Weinfurter, A. Zeilinger: Phys. Rev. Lett. **80**, 3891 (1998)

81.75 C. H. Bennett, S. J. Wiesner: Phys. Rev. Lett. **69**, 2881 (1992)

81.76 R. F. Werner: J. Phys. A **34**, 7081 (2001)

81.77 K. Mattle, H. Weinfurter, P. G. Kwiat, A. Zeilinger: Phys. Rev. Lett. **76**, 4656 (1996)

81.78 R. Landauer: IBM J. Res. Develop. **5**, 183 (1961)

81.79 M. B. Plenio, V. Vitelli: Contemp. Phys. **42**, 25 (2001)

81.80 M. Hillery, R. F. Werner: Phys. Rev. A **60**, R2626 (1999)

81.81 D. Deutsch: Proc. R. Soc. London A **425**, 73 (1989)

81.82 M. Reck, A. Zeilinger, H. J. Bernstein, P. Bertani: Phys. Rev. Lett. **73**, 58 (1994)

81.83 A. Barenco: Proc. R. Soc. London A **449**, 679 (1995)

81.84 T. Sleator, H. Weinfurter: Phys. Rev. Lett. **74**, 4087 (1995)

81.85 D. P. DiVincenzo: Phys. Rev. A **51**, 1015 (1995)

81.86 D. Deutsch, R. Jozsa: Proc. R. Soc. London A **439**, 553 (1992)

81.87 J. A. Jones, M. Mosca: J. Chem. Phys. **109**, 1648 (1998)

81.88 I. L. Chuang, L. M. K. Vandersypen, X. L. Zhou, D. W. Leung, S. Lloyd: Nature (London) **393**, 143 (1998)

81.89 S. Gulde, M. Riebe, G. P. T. Lancaster, C. Becher, J. Eschner, H. Häffner, F. Schmidt-Kaler, I. L. Chuang, R. Blatt: Nature (London) **421**, 48 (2003)

81.90 L. Grover: Phys. Rev. Lett. **79**, 325 (1997)

81.91 P. W. Shor: *Proceedings, 35th Annual Symposium on Foundations of Computer Science, 1994, Los Alamitos* (IEEE Computer Society Press, New York 1994) pp. 124–134

81.92 P. W. Shor: SIAM J. Comput. **26**, 1484 (1997)

81.93 Y. S. Weinstein, M. A. Pravia, E. M. Fortunato, S. Lloyd, D. G. Cory: Phys. Rev. Lett. **86**, 1889 (2001)

81.94 L. M. K. Vandersypen, M. Steffen, G. Breyta, C. S. Yannoni, M. H. Sherwood, I. L. Chuang: Nature (London) **414**, 883 (2001)

81.95 D. R. Simon: *Proceedings, 35th Annual Symposium on Foundations of Computer Science, 1994, Los Alamitos* (IEEE Computer Society Press, New York 1994) pp. 116–123

81.96 D. R. Simon: SIAM J. Comput. **26**, 1474 (1997)

81.97 R. Jozsa: *quant-ph/9707033*

81.98 J. Kempe: *Proc. 7th Intern. Workshop Randomization, Approximation Techniques Comput. Sci. (RANDOM'03)* (Springer, Berlin, Heidelberg 2003) pp. 354–369

81.99 W. H. Zurek: Rev. Mod. Phys. **75**, 715 (2003)

81.100 J. Preskill: Proc. R. Soc. London A **454**, 385 (1998)

81.101 A. R. Calderbank, P. W. Shor: Phys. Rev. A **54**, 1098 (1996)

81.102 A. Steane: Proc. R. Soc. London A **452**, 2551 (1996)

81.103 D. G. Cory, M. D. Price, W. Maas, E. Knill, R. Laflamme, W. H. Zurek, T. F. Havel, S. S. Somaroo: Phys. Rev. Lett. **81**, 2152 (1998)

81.104 D. Leung, L. Vandersypen, X. Zhou, M. Sherwood, C. Yannoni, M. Kubinec, I. Chuang: Phys. Rev. A **60**, 1924 (1999)

81.105 E. Knill, R. Laflamme, R. Martinez, C. Negrevergne: Phys. Rev. Lett. **86**, 5811 (2001)

81.106 S. L. Braunstein, C. M. Caves, R. Jozsa, N. Linden, S. Popescu, R. Schack: Phys. Rev. Lett. **83**, 1054 (1999)
81.107 J. Chiaverini, D. Leibfried, T. Schaetz, M. D. Barrett, R. B. Blakestad, J. Britton, W. M. Itano, J. D. Jost, E. Knill, C. Langer, R. Ozeri, D. J. Wineland: Nature (London) **432**, 602 (2004)
81.108 G. A. Palma, K.-A. Suominen, A. K. Ekert: Proc. R. Soc. London A **452**, 567 (1996)
81.109 L.-M. Duan, G.-C. Guo: Phys. Rev. Lett. **79**, 1953 (1997)
81.110 P. Zanardi, M. Rasetti: Phys. Rev. Lett. **79**, 3306 (1997)
81.111 D. A. Lidar, I. L. Chuang, K. B. Whaley: Phys. Rev. Lett. **81**, 2594 (1998)
81.112 D. A. Lidar, D. Bacon, K. B. Whaley: Phys. Rev. Lett. **82**, 4556 (1999)
81.113 P. G. Kwiat, A. J. Berglund, J. B. Altepeter, A. G. White: Science **290**, 498 (2000)
81.114 D. Kielpinski, C. Monroe, D. J. Wineland: Nature (London) **417**, 7097 (2002)
81.115 D. P. DiVincenzo: Fortschr. Phys. **48**, 7718 (2000)
81.116 E. Knill, R. Laflamme, G. J. Milburn: Nature (London) **409**, 46 (2001)
81.117 M. Koashi, T. Yamamoto, N. Imoto: Phys. Rev. A **63**, 030301 (2001)
81.118 T. B. Pittman, B. C. Jacobs, J. D. Franson: Phys. Rev. A **64**, 062311 (2001)
81.119 T. B. Pittman, B. C. Jacobs, J. D. Franson: Phys. Rev. Lett. **88**, 257902 (2002)
81.120 T. B. Pittman, M. J. Fitch, B. C. Jacobs, J. D. Franson: Phys. Rev. A **68**, 032316 (2003)
81.121 S. Scheel, K. Nemoto, W. J. Munro, P. L. Knight: Phys. Rev. A **68**, 032310 (2003)
81.122 E. Knill: Phys. Rev. A **68**, 064303 (2003)
81.123 S. Scheel, N. Lütkenhaus: New J. Phys. **6**, 51 (2004)
81.124 J. Eisert: *quant-ph/0409156*
81.125 S. Scheel, K. M. R. Audenaert: New J. Phys. **7**, 149 (2005)
81.126 H. J. Briegel, R. Raussendorf: Phys. Rev. Lett. **86**, 910 (2001)
81.127 R. Raussendorf, H. J. Briegel: Phys. Rev. Lett. **86**, 5188 (2001)
81.128 M. A. Nielsen: Phys. Rev. Lett. **93**, 040503 (2004)
81.129 D. E. Browne and T. Rudolph: *quant-ph/0405157*
81.130 A. N. Zhang, C. Y. Lu, X. Q. Zhou, Y. A. Chen, Z. Zhao, T. Yang, J. W. Pan: *quant-ph/0501036*
81.131 W. Paul, O. Osberghaus, E. Fischer: *Forschungsberichte des Wirtschafts- und Verkehrsministeriums Nordrhein-Westfalen Nr. 415* (Westdeutscher Verlag, Köln 1958)
81.132 W. Neuhauser, M. Hohenstatt, P. E. Toschek, H. G. Dehmelt: Phys. Rev. A **22**, 1137 (1980)
81.133 D. J. Wineland, R. Drullinger, D. F. Walls: Phys. Rev. Lett. **40**, 1639 (1978)
81.134 W. Neuhauser, M. Hohenstatt, P. E. Toschek, H. G. Dehmelt: Phys. Rev. Lett. **41**, 233 (1978)
81.135 D. J. Wineland, H. G. Dehmelt: Bull. Am. Phys. Soc. **20**, 637 (1975)
81.136 J. I. Cirac, P. Zoller: Phys. Rev. Lett. **74**, 4094 (1995)
81.137 C. Monroe, D. M. Meekhof, B. E. King, W. M. Itano, D. J. Wineland: Phys. Rev. Lett. **75**, 4714 (1995)
81.138 J. F. Poyatos, J. I. Cirac, P. Zoller: Fortschr. Phys. **48**, 785 (2000)
81.139 J. F. Poyatos, J. I. Cirac, P. Zoller: J. Mod. Opt. **49**, 1593 (2002)
81.140 D. Leibfried, R. Blatt, C. Monroe, D. J. Wineland: Rev. Mod. Phys. **75**, 281 (2003)
81.141 D. J. Wineland, M. Barrett, J. Britton, J. Chiaverini, B. DeMarco, W. M. Itano, C. Langer, D. Leibfried, V. Meyer, T. Rosenband, T. Schätz: Philos. Trans. R. Soc. London A **361**, 1349 (2003)
81.142 S. Gulde, M. Riebe, G. P. T. Lancaster, C. Becher, J. Eschner, H. Häffner, F. Schmidt-Kaler, I. L. Chuang, R. Blatt: Nature (London) **421**, 48 (2003)
81.143 M. Riebe, H. Häffner, C. F. Roos, W. Hänsel, J. Benhelm, G. P. T. Lancaster, T. W. Körber, C. Becher, F. Schmidt-Kaler, D. F. V. James, R. Blatt: Nature (London) **429**, 734 (2004)
81.144 M. Barrett, J. Chiaverini, T. Schaetz, J. Britton, W. M. Itano, J. D. Jost, E. Knill, C. Langer, D. Leibfried, R. Ozeri, D. J. Wineland: Nature (London) **429**, 737 (2004)
81.145 E. T. Jaynes, F. W. Cummings: Proc. IEEE **51**, 89 (1963)
81.146 S. Haroche: Philos. Trans. R. Soc. London A **361**, 1339 (2003) and references therein
81.147 C. Monroe: Nature (London) **416**, 238 (2002)
81.148 T. Pellizzari, S. A. Gardiner, J. I. Cirac, P. Zoller: Phys. Rev. Lett. **75**, 3788 (1995)
81.149 J. I. Cirac, P. Zoller, H. J. Kimble, H. Mabuchi: Phys. Rev. Lett. **78**, 3221 (1997)
81.150 A. Kuhn, M. Hennrich, G. Rempe: Phys. Rev. Lett. **89**, 067901 (2002)
81.151 J. McKeever, A. Boca, A. D. Boozer, R. Miller, J. R. Buck, A. Kuzmich, H. J. Kimble: Science **303**, 1992 (2004)
81.152 D. Jaksch, C. Bruder, J. I. Cirac, C. W. Gardiner, P. Zoller: Phys. Rev. Lett. **81**, 3108 (1998)
81.153 S. R. Clark, D. Jaksch: Phys. Rev. A **70**, 043612 (2004)
81.154 M. Greiner, I. Bloch, O. Mandel, T. W. Hänsch, T. Esslinger: Phys. Rev. Lett. **87**, 160405 (2001)
81.155 M. Greiner, O. Mandel, T. Esslinger, T. W. Hänsch, I. Bloch: Nature (London) **415**, 39 (2002)
81.156 C. Orzel, A. K. Tuchman, M. L. Fenselau, M. Yasuda, M. A. Kasevich: Science **291**, 2386 (2001)
81.157 D. S. Petrov, G. V. Shlyapnikov, J. T. M. Walraven: Phys. Rev. Lett. **85**, 3745 (2000)
81.158 V. Dunjko, V. Lorent, M. Olshanii: Phys. Rev. Lett. **86**, 5413 (2001)
81.159 B. Paredes, A. Widera, V. Murg, O. Mandel, S. Folling, J. I. Cirac, G. V. Shlyapnikov, T. W. Hänsch, I. Bloch: Nature (London) **429**, 277 (2004)
81.160 T. Kinoshita, T. Wenger, D. S. Weiss: Science **305**, 1125 (2004)

81.161 J. K. Pachos, P. L. Knight: Phys. Rev. Lett. **91**, 107902 (2003)

81.162 J. K. Pachos, E. Rico: Phys. Rev. A **70**, 053620 (2004)

81.163 A. Kay, J. K. Pachos: New J. Phys. **6**, 126 (2004)

81.164 M. P. A. Jones, C. J. Vale, D. Sahagun, B. V. Hall, E. A. Hinds: Phys. Rev. Lett. **91**, 080401 (2003)

81.165 D. M. Harber, J. M. McGuirk, J. M. Obrecht, E. A. Cornell: J. Low Temp. Phys. **133**, 229 (2003)

81.166 Y. J. Lin, I. Teper, C. Chin, V. Vuletić: Phys. Rev. Lett. **92**, 050404 (2004)

81.167 C. Henkel, S. Pötting, M. Wilkens: Appl. Phys. B **69**, 379 (1999)

81.168 P. K. Rekdal, S. Scheel, P. L. Knight, E. A. Hinds: Phys. Rev. A **70**, 013811 (2004)

81.169 S. Scheel, P.K. Rekdal, P.L. Knight, and E.A. Hinds: *quant-ph/0501149*

81.170 A. Widera, O. Mandel, M. Greiner, S. Kreim, T. W. Hänsch, I. Bloch: Phys. Rev. Lett. **92**, 160406 (2004)

81.171 B. Julsgaard, A. Kozhekin, E. S. Polzik: Nature (London) **413**, 400 (2001)

81.172 R. P. Feynman: Int. J. Theor. Phys. **21**, 467 (1982)

81.173 S. Lloyd: Science **273**, 1073 (1996)

81.174 H. Rabitz, R. deVivie-Riedle, M. Motzkus, K. Kompa: Science **288**, 824 (2000)

81.175 C. H. Bennett, J. I. Cirac, M. S. Leifer, D. W. Leung, N. Linden, S. Popescu, G. Vidal: Phys. Rev. A **66**, 012305 (2002)

81.176 M. B. Plenio, J. Eisert, M. Cramer: quant-ph/0405142, Phys. Rev. Lett. (2005)

81.177 J. P. Dowling, G. J. Milburn: Philos. Trans. R. Soc. London A **361**, 1655 (2003)

Part G

Part G Applications

82 Applications of Atomic and Molecular Physics to Astrophysics
Alexander Dalgarno, Cambridge, USA
Stephen Lepp, Las Vegas, USA

83 Comets
Paul D. Feldman, Baltimore, USA

84 Aeronomy
Jane L. Fox, Dayton, USA

85 Applications of Atomic and Molecular Physics to Global Change
Kate P. Kirby, Cambridge, USA
Kelly Chance, Cambridge, USA

86 Atoms in Dense Plasmas
Jon C. Weisheit, Pullman, USA
Michael S. Murillo, Los Alamos, USA

87 Conduction of Electricity in Gases
Alan Garscadden, Wright Patterson Air Force Base, USA

88 Applications to Combustion
David R. Crosley, Menlo Park, USA

89 Surface Physics
Erik T. Jensen, Prince George, Canada

90 Interface with Nuclear Physics
John D. Morgan III, Newark, USA
James S. Cohen, Los Alamos, USA

91 Charged-Particle–Matter Interactions
Hans Bichsel, Seattle, USA

92 Radiation Physics
Mitio Inokuti, Argonne, USA

82. Applications of Atomic and Molecular Physics to Astrophysics

The range of physical conditions of density, temperature, and radiation fields encountered in astrophysical environments is extreme and can rarely be reproduced in a laboratory setting. It is not only reliable data on known processes that are needed but also a deep understanding so that the relevant processes can be identified and the influence of the conditions in which they occur fully taken into account.

We present here a summary of the processes that take place in photoionized gas, collisionally ionized gas, the diffuse interstellar medium, molecular clouds, circumstellar shells, supernova ejecta, shocked regions and the early Universe.

Almost all our information about the Universe reaches us in the form of photons. Observational astronomy is based on measurements of the distribution in frequency and intensity of the photons that are emitted by astronomical objects and detected by instrumentation on ground-based and space-borne telescopes. Information about the earliest stages in the evolution of the Universe before galaxies and stars had formed is carried to us by blackbody background photons that attended the beginning of the Universe. The photons that are the signatures of astronomical phenomena are the result of many processes of nuclear physics, plasma physics and atomic, molecular and optical physics. The processes that modify the photons on their journey from distant origins through intergalactic and interstellar space to the Earth belong mostly to the domain of atomic, molecular, and optical physics, as do the instruments that detect and measure the arriving photons and their spectral distribution. The spectra are used to classify galaxies and stars and to identify the astronomical entities and phenomena such as quasars, active galactic nuclei, gravitational lensing, jets and outflows, pulsars, supernovae, novae, supernova remnants, nebulae, masers, protostars, shocks, molecular clouds, circumstellar shells, accretion disks and black holes.

Quantitative analyses of the spectra of astronomical sources of photons and of the atomic, molecular, and optical processes that populate the atomic and molecular energy levels and give rise to the observed absorption and emission require accurate data on transition frequencies and wavelengths, oscillator strengths, cross sections for electron impact, rate coefficients for radiative, dielectronic and dissociative recombination, and cross sections for heavy particle collisions involving charge transfer, excitation, ionization, dissociation, fine structure, and hyperfine structure transitions, collision-induced absorption and line broadening. Data on radiative association and ion–molecule and neutral particle reaction rate coefficients are central to the interpretation of measurements of chemical composition in molecular clouds, circumstellar shells and supernova ejecta.

82.1 Photoionized Gas

The Universe contains copious sources of energetic photons most often in the form of hot stars, and much of the material of the Universe exists as photoionized gas. Photoionized gas produces the visible emission from emission nebulae, planetary nebulae, nova shells, starburst galaxies and probably active galactic nuclei [82.1].

Emission nebulae are extended regions of luminosity in the sky. They arise from the absorption of stellar radiation by the gas surrounding one or more hot stars. The gas is ionized by the photons and excited and heated by the electrons released in the photoionizing events. A succession of ionization zones is created in which highly ionized regions give way to less ionized gas with increasing distance from the central star as the photon flux is diminished by geometrical dilution and by absorption. The outer edge of a nebula is a front of ionization pushing out into the neutral interstellar gas. The densities are typically between 100 and 10 000 cm^{-3} and the temperatures between 5000 and 15 000 K. Nebulae are also called HII regions. At low densities, the luminosity is low, but the ionized regions can still be detected by radio observations.

Planetary nebulae are smaller in extent and more dense. They have a passing similarity in appearance to planets. Planetary nebulae are produced by the photoionization of shells of gas that have been ejected from the parent star as it evolved to its final white dwarf stage. Because the core of the parent star is very hot the irradiated gas is more highly ionized than are emisssion nebulae and has a distinctive spectrum.

Photoionized gas is also found around novae. Novae are stars that have undergone spasmodic outbursts and they are surrounded by faint shells of ejected gas, photoionized and excited by the stellar radiation. Some supernova remnants, which are what remains after a massive star has exploded, have spectra that also appear to be emanating from photoionized gas. The source of ionization may be synchrotron radiation. Figure 82.1 shows the X-ray emission spectrum of a supernova remnant.

The nuclei of starburst galaxies have spectra like those of emission nebulae. They result from gas photoionized by radiation from hot stars created in a period of rapid star formation. Active galactic nuclei, such as quasars, have a different spectrum characterized by broad lines indicating a large range of velocities. Photoionized gas is the most likely interpretation. The ionizing source may be an accretion disk around a compact object such as a black hole.

The ionization structure in a photoionized gas is determined by a balance of photoionization

$$X^{(m-1)+} + h\nu \to X^{m+} + \mathrm{e}^- \tag{82.1}$$

and radiative

$$X^{m+} + \mathrm{e}^- \to X^{(m-1)+} + h\nu \tag{82.2}$$

and dielectronic

$$X^{m+} + \mathrm{e}^- \to \left(X^{(m-1)+}\right)^* \to X^{(m-1)+} + h\nu \tag{82.3}$$

recombination, and in plasmas with a significant population of neutral hydrogen and helium, by charge transfer recombination

$$X^{m+} + \mathrm{H} \to X^{(m-1)+} + \mathrm{H}^+ \tag{82.4}$$

$$X^{m+} + \mathrm{He} \to X^{(m-1)+} + \mathrm{He}^+ \,. \tag{82.5}$$

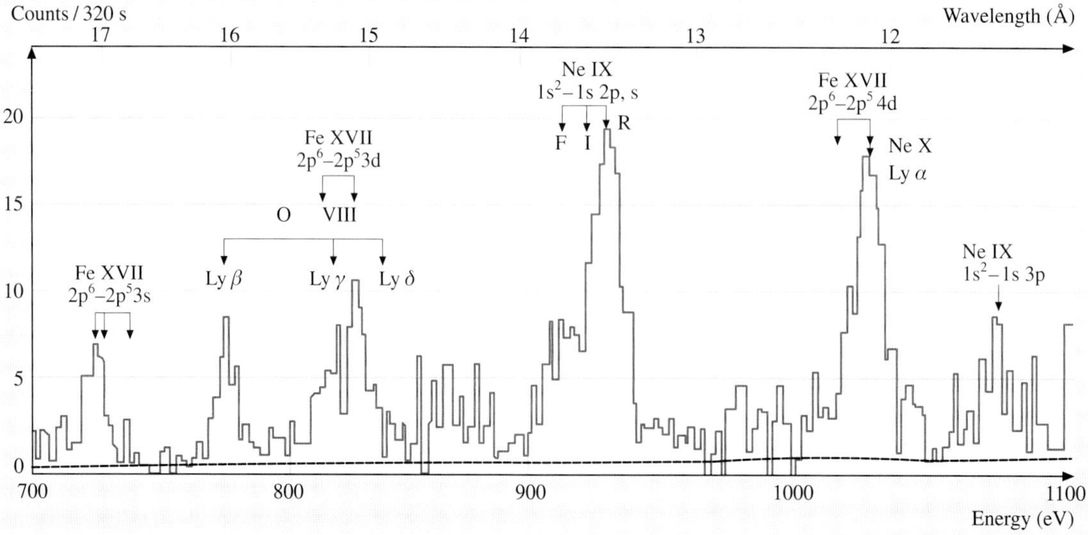

Fig. 82.1 X-ray spectrum of the supernova remnant Puppis A as observed by the Einstein satellite. Note the high level of ionization with hydrogen-like ions of oxygen and neon, suggesting a high temperature. After [82.2]

Many detailed calculations of the ionization structure of photoionized regions have been carried out [82.1].

The ionizing source spectra of hot stars can be obtained from calculations of stellar atmospheres Sect. 82.5. Approximate values of cross sections for photoionization for a wide range of atomic and ionic systems in many stages of ionization are available [82.3–6]. Calculations of higher precision and reliability that incorporate the contributions from autoionizing resonance structures exist for specific systems [82.7]. They are undergoing continual improvements as increasingly powerful computational techniques are brought to bear on the calculations.

The cross sections for radiative recombination are obtained by summing the cross sections for capture into the ground and excited states of the recombining system. Because of the contribution from highly excited states which are nearly hydrogenic, the rate coefficients are similar for different ions of the same excess nuclear charge. They vary slowly with temperature. In contrast, dielectronic recombination is a specific process whose efficiency depends on the energy level positions of the resonant states. For nebular temperatures, the rate coefficients vary exponentially with temperature. Explicit calculations have been carried out for many ionic systems [82.8–10]. Because the photoionization cross sections of the major cosmic gases hydrogen and helium diminish rapidly at high frequencies, multiply charged ions and neutral gas coexist in cosmic plasmas produced by energetic photons and charge transfer recombination may control the ionization structure. For multiply charged ions with excess charge greater than two, charge transfer is rapid. For doubly charged and singly charged ions, the cross sections are sensitive to the details of the potential energy curves of the quasimolecule formed in the approach of the ion and the neutral particle. Few reliable data exist. Some recent calculations may be found in the papers [82.11–13].

Photoionized gas is heated by collisions of the energetic photoelectrons and cooled by electron impact excitation of metastable levels, principally of O^+ and O^{++}, N^+ and N^{++}, and S^+ and S^{++}, followed by emission of photons which escape from the nebula. Considerable attention has been given to the determination of the rate coefficients [82.14]. The resulting cooling rates increase exponentially with temperature and keep the temperature of the gas between narrow limits. Some contribution to cooling occurs from recombination and from free-free emission by electrons moving in the field of the positive ions.

The luminosity of the photoionized gas comes from the photons emitted in the cooling processes and from radiative and dielectronic recombination. The radiative recombination spectrum of hydrogen extends from the Ly α line at 121.6 nm to radio lines at meter wavelengths.

The recombination spectrum can be predicted to high accuracy, and calculations for a wide range of temperature, density, and radiation environments have been carried out for diagnostic purposes [82.15–18]. Electron impact and proton impact induced transitions are important in determining the energy level populations and the resulting spectrum. Stimulated emission often affects the intensities of the radio lines, especially those from extragalactic sources. Comparisons of the predicted intensities in the visible and infrared with theoretical predictions yield information on interstellar extinction in the nebula and along the line of sight.

The relative intensities of the lines emitted by different metastable levels depend exponentially on the temperature. The relative intensities of the lines at 500.7 nm and 436.3 nm originating in the $^1D_2$ and $^1S_0$ levels of O^{++} vary as $\exp(33170/T)$, and are commonly used to derive the temperature T.

The electron density can be inferred from the lines emitted from neighboring levels with different radiative lifetimes for which there occurs a competition between spontaneous emission and quenching by electron impact. There are many possible combinations of lines. The lines at 372.89 nm and 372.62 nm emitted by the $^2D_{3/2}$ and $^2D_{5/2}$ levels of N^+ are readily observable and their relative intensity yields the electron density.

Radiative and dielectronic recombination lines are often seen in the spectra, as are a few lines due to charge transfer recombination. Fluorescence of starlight and resonance fluorescence of lines emitted in the nebula (called Bowen fluorescence by astronomers) also contribute to the spectra of photoionized gases. Many data are needed to adequately interpret the observations.

82.2 Collisionally Ionized Gas

Hot gas is found in the coronae of stars and particularly the Sun, and in young supernova remnants, in the hot phase of the interstellar medium, and in intergalactic space. In a hot gas the ionization is produced by the impact ionization of the fast thermal electrons and recombination is radiative and dielectronic [82.19].

The rate coefficients for electron impact ionization and for recombination for any given ionization stage are functions only of temperature, and hence so is the resulting ionization distribution. When ionization and recombination balance, coronal equilibrium is attained in which the ionization structure is specified by the temperature.

Recombination at high temperatures is dominated by dielectronic recombination. At high temperatures, dielectronic recombination is stabilized by transitions in which the core electrons are the active electrons. The associated emission lines lie close in frequency to that of the resonant transition of the parent ion. They are called satellite lines. Together with lines generated by electron impact excitation, they provide a powerful diagnostic probe of density and temperature. In many circumstances such as in supernova remnants, coronal equilibrium does not hold, and the ionization and recombination must be followed as functions of time. The temperature also evolves as the hot plasma is cooled by electron impact excitation and ionization.

The recombining gas produces X-rays and extreme UV radiation which modify the ionization structure. There is a particular need for more reliable data on high energy photoionization cross sections, on collision cross sections for electrons and positive ions, and on the energy levels and transition probabilities of highly stripped complex ions. Figure 82.2 shows the emissivity of coronal gas.

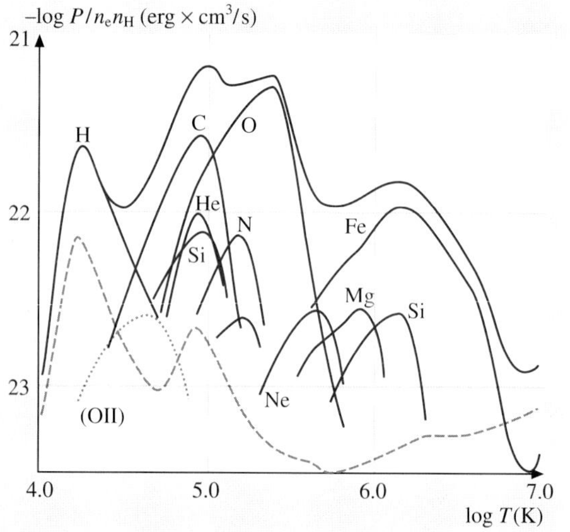

Fig. 82.2 Total emissivity and emissivity by element as a function of temperature in coronal equilibrium. The *heavy solid curve* is total emissivity and the *lighter lines* are contributions from individual elements. After [82.20]

82.3 Diffuse Molecular Clouds

Diffuse molecular clouds are intermediate between the hot phase of the galaxy and the giant molecular clouds where much of the gas resides. They are called diffuse because they have optical depths of order unity, so photons can penetrate from outside the cloud and affect the chemical composition. The atoms and molecules are observed in absorption against background stars. Translucent clouds with optical depths between about 2 and 5 are intermediate between diffuse and dark clouds where photons from the outside still affect the chemistry. They can be observed both in absorption against a background source or in emission in the radio.

The temperature is 100–200 K at the edges of a diffuse cloud with a density of about $100\,\mathrm{cm}^{-3}$. In a typical diffuse cloud the temperature decreases to about 30 K at the center while the density increases to about $300–800\,\mathrm{cm}^{-3}$. The chemistry is driven by ionization from interstellar UV photons and from cosmic rays.

Interstellar UV photons ionize species which have ionization potentials less then that of atomic hydrogen. Atomic hydrogen is so pervasive in the galaxy that UV photons with energies higher than 13.6 eV are absorbed near the source. The UV flux is a very important parameter in determining the composition of a diffuse cloud. Photodissociation provides destruction which limits the buildup of more complex species and so diffuse clouds are dominated by simpler diatomic species.

Species with ionization potentials greater then hydrogen are mainly ionized by cosmic rays. Cosmic rays are high energy nuclei which stream through the galaxy. The cosmic ray ionization rate, the number of cosmic ray ionizations per second per particle, is an important parameter in interstellar chemistry. A lower limit to the cosmic ray ionization rate may be set by measured high energy cosmic rays reaching earth, giving an ionization rate of $\approx 10^{-17}\,\mathrm{s}^{-1}$. More realistic estimates of the cosmic ray ionization rate from looking at recombination lines suggest values of a few $\times 10^{-17}\,\mathrm{s}^{-1}$.

The hydroxyl radical OH is produced in a manner similar to that discussed below in Sect. 82.5, and removed by photodissociation. Thus in diffuse clouds,

OH may be used to measure the cosmic ray ionization rate, subject to uncertainties in the OH photodissociation rate and the H_3^+ recombination rate. The OH abundances give rates of several $\times 10^{-17}\,s^{-1}$ for many diffuse clouds.

The carbon chemistry begins with the ionization of C by UV photons:

$$C + h\nu \rightarrow C^+ + e^- . \quad (82.6)$$

The carbon ion cannot react directly with H_2 by

$$C^+ + H_2 \rightarrow CH^+ + H , \quad (82.7)$$

as this reaction is exothermic by 0.4 eV. Instead, the chemistry proceeds by the slow radiative association process

$$C^+ + H_2 \rightarrow CH_2^+ + h\nu . \quad (82.8)$$

The CH_2^+ ion may either dissociatively recombine

$$CH_2^+ + e^- \rightarrow CH + H , \quad (82.9)$$

or react with molecular hydrogen

$$CH_2^+ + H_2 \rightarrow CH_3^+ + H . \quad (82.10)$$

The CH_3^+ then undergoes dissociative recombination

$$CH_3^+ + e \rightarrow C + H_2 + H \quad (82.11)$$
$$\rightarrow CH + H + H \quad (82.12)$$
$$\rightarrow CH_2 + H \quad (82.13)$$
$$\rightarrow CH + H_2 , \quad (82.14)$$

where the products are listed in order of decreasing likelihood. The CH is removed by photodissociation

$$CH + h\nu \rightarrow C + H \quad (82.15)$$

and by photoionization

$$CH + h\nu \rightarrow CH^+ + e^- . \quad (82.16)$$

CH may also be removed by reactions with oxygen or nitrogen atoms to form CO and CN respectively.

One of the outstanding problems in diffuse clouds is to understand the large abundance of CH^+ relative to CH. The problem is producing the CH^+ without producing additional CH. Since most reaction paths go through reaction (82.7), this is the most likely candidate. What is needed is some extra energy to overcome the endothermicity. This energy must come from either hot C^+ or from hot or vibrationally excited H_2. The most popular model is gas heated by a shock, possibly a magnetic shock in which ions stream relative to the neutrals, giving a high effective energy. Unfortunately, though these shock models can reproduce the CH^+ abundances, they also predict relative velocities between the CH^+ and CH which are not often observed. Recently there have been suggestions that turbulence in the cloud could account for the CH^+ abundance.

The most comprehensive models of diffuse clouds are by *van Dishoeck* and *Black* [82.21]. A collection of photodissociation rates and photoionization rates is given in *Roberge* et al. [82.22].

The UV flux is predominantly from stars and may be as much as 10^5 times larger near an HII region Sect. 82.1 than it is in the general interstellar medium. Regions in which the chemistry is dominated by photons are referred to as *photon dominated regions* or *photodissociation regions* (PDR's). In the presence of high UV flux the cloud is much warmer than in a typical diffuse cloud. Temperatures may reach 1000 K near the edge of the cloud and 100 K far into the cloud. The chemistry differs from traditional diffuse cloud chemistry in that the high temperatures allow endothermic reactions to proceed. *Sternberg* and *Dalgarno* [82.23] have published a comprehensive model of photodominated regions.

82.4 Dark Molecular Clouds

Much of the mass of the galaxy is in the form of dark molecular clouds. The molecular clouds are sites of forming new stars. They are composed primarily of hydrogen, with about 10% helium and trace amounts of heavier elements. They have densities of approximately 10^3 or $10^4\,cm^{-3}$ and temperatures between 10 and 20 K, and often contain denser clumps. The clouds are optically thick and so photons from the outside are absorbed on the surface of the clouds. The interiors are heated and ionized by cosmic rays which penetrate deep into the cloud.

The temperatures are too low to sustain much neutral chemical activity in the clouds, and cosmic ray ionization is important in driving the chemistry. In dense clouds, the cosmic rays both initiate the chemistry and limit it through the production of He^+ and through cosmic ray induced photons.

Table 82.1 is a list of molecules that have been observed in the interstellar medium, many of which have also been found in other galaxies. It is likely that all but H_2 are formed in the gas phase by ion–molecule re-

Table 82.1 Molecules observed in interstellar clouds

H_2	Hydrogen	CH	Methylidyne
CH^+	Methylidyne ion	OH	Hydroxyl
C_2	Carbon	CN	Cyanogen
CO	Carbon monoxide	CO^+	Carbon monoxide ion
NH	Amidogen	NO	Nitric oxide
CS	Carbon monosulphide	SiO	Silicon monoxide
SO	Sulphur monoxide	SO^+	Sulphur monoxide ion
NS	Nitrogen sulphide	SiS	Silicon sulphide
PN	Phosphorus nitride	HCl	Hydrogen chloride
SiN	Silicon nitride	NH_2	Amino radical
H_2O	Water	C_2H	Ethynyl
HCN	Hydrogen cyanide	HNC	Hydrogen isocyanide
HCO	Formyl	HCO^+	Formyl ion
N_2H^+	Protonated nitrogen	H_2S	Hydrogen sulphide
HNO	Nitroxyl	OCS	Carbonyl sulphide
SO_2	Sulphur dioxide	HCS^+	Thioformyl ion
C_2O	Carbon suboxide	C_2S	Dicarbon sulphide
N_2O	Nitrous oxide	H_2CN	Methylene amidogen
H_2CO	Formaldehyde	H_2CS	Thioformaldehyde
NH_3	Ammonia	HCNS	Isothiocyanic acid
HNCO	Isocyanic acid	$HOCO^+$	Protonated carbon dioxide
C_3H	Propynylidyne	C_3N	Cyanoethynyl
C_3S	Tricarbon sulphide	C_3O	Tricarbon monoxide
C_2H_2	Acetylene	H_3O^+	Hydronium ion
$HCNH^+$	Protonated hydrogen cyanide	C_3H_2	Cyclopropenylidene
CH_4	Methane	H_2CCC	Propadienylidene
HCOOH	Formic acid	CH_2CO	Ketene
HC_3N	Cyanoacetylene	HNCCC	Cyanoacetylene isomer
HCCNC	Ethynyl isocyanide	C_4H	Butadinyl
NH_2CH	Cyanamide	CH_2CN	Cyanomethyl radical
CH_2NH	Methanimine	CH_3CH	Methyl cyanide
H_2CCCC	Butatrienylidene	CH_3SH	Methyl mercaptan
C_5H	Pentynylidyne	HCC_2HO	Propynal
CH_3OH	Methyl alcohol	HC_3NH^+	Protonated cyanoacetylene
NH_2CHO	Formamide	CH_3C_2H	Methyl acetylene
CH_2CHCN	Vinyl cyanide	HC_5N	Cyanodiacetylene
C_6H	Hexatrinyl	CH_3NH_2	Methylamine
CH_3CHO	Acetaldehyde	$HCOOCH_3$	Methyl formate
CH_3C_3N	Methyl cyanoacetylene	CH_3C_4H	Methyl diacetylene
$CH_{32}O$	Dimethyl ether	CH_3CH_2CN	Ethyl cyanide
HC_7N	Cyanohexatriyne	CH_3CH_2OH	Ethyl alcohol
HC_9N	Cyano-octatetra-yne	$HC_1 1N$	Cyano-decapenta-yne

action sequences initiated by cosmic ray ionization. The fact that isomers such as HCN and HNC are seen in approximately equal abundances suggests a low density gas phase environment. Reactions on surfaces and the formation of grains are not well understood, but are surely important.

The chemistry of molecular clouds is dominated by ion–molecule reactions driven by cosmic ray ionization.

The cosmic rays primarily ionize H_2:

$$H_2 \to H_2^+ + e^- , \quad (82.17)$$

producing both H_2^+ and fast electrons. The fast electrons produce additional ionizations. The H_2^+ quickly reacts with H_2 to form H_3^+

$$H_2^+ + H_2 \to H_3^+ + H . \quad (82.18)$$

The H_3^+ reacts with other species by proton transfer, which then drives much of the interstellar chemistry.

As an example of the production of more complex molecules in interstellar chemistry, we examine the reaction networks leading to the production of water H_2O and the hydroxl radical OH. The H_3^+ ions formed by cosmic ray ionization react with atomic oxygen to form OH^+

$$O + H_3^+ \to OH^+ + H_2 , \quad (82.19)$$

which quickly reacts with H_2 to form H_3O^+ in an abstraction sequence

$$OH^+ + H_2 \to H_2O^+ + H , \quad (82.20)$$

$$H_2O^+ + H_2 \to H_3O^+ + H . \quad (82.21)$$

The H_3O^+ then undergoes dissociative recombination to form water and OH

$$H_3O+ + e^- \to H_2O + H , \quad (82.22)$$

$$\to OH + H_2 . \quad (82.23)$$

The water is removed by reactions with neutral or ionized carbon, which eventually lead to the production of CO. OH is primarily removed by reactions with atomic oxygen leading to O_2. The CO and O_2 are removed by reactions with He^+. The He^+, generated by cosmic ray ionization of helium, does not react with H_2 and so is available to remove species by reactions such as

$$CO + He^+ \to C^+ + O + He . \quad (82.24)$$

Water and OH are also removed by UV photons generated within the cloud. The clouds are too thick for external UV photons to penetrate, but cosmic rays excite H_2 into electronically excited states which decay through emission of UV photons. These internally generated photons play an important role in determining the composition of the cloud. *Gredel* et al. [82.24] have compiled a list of the photodissociation and photoionization rates for cosmic ray induced photons.

Modern chemical networks for molecular clouds include several hundred species and several thousand reactions. A standard set of reaction rates is provided by the UMIST (University of Manchester Institute of Science and Technology) dataset [82.25, 26]. The dataset may be obtained from the UMIST Astrophysics Group homepage (http://saturn.ma.umist.ac.uk:8000/).

The clouds also contain dust particles as evidenced by the extinction curves for clouds and the observed depletions of heavier elements. The importance of surface chemistry on these dust particles to interstellar clouds is still uncertain. Dust particles are the best candidate to be the site of formation of molecular hydrogen, because known gas phase reactions fail to produce H_2 in the quantities observed.

82.5 Circumstellar Shells and Stellar Atmospheres

The continuum emission from a star is very nearly that of a blackbody. This emission is then absorbed and redistributed by the atmosphere of the star. The spectrum of the star is thus determined by its atmosphere. In the hottest stars, most of the material is ionized and the absorption lines are predominantly those of ions, while in the coldest stars, molecular lines are prominent. *Kurucz* has calculated models with continuum spectra and the inclusion of a large number of absorption lines [82.27]. There are two major projects for calculating the required atomic data. In 1984 an international collaboration named the *Opacity Project* was set up to calculate accurate atomic data needed for opacity calculations [82.28, 29]. The other earlier project is called OPAL. The two sources of data are compared in [82.30, 31].

Low and intermediate mass stars eject circumstellar envelopes in their red giant phase near the end of their evolution. Circumstellar envelopes are an important part of astronomy and they are a likely location for dust formation. They provide an interesting environment for studying molecules because they represent a transition between very high density stellar atmosphere environments to low density interstellar environments. These objects evolve to become planetary nebulae Sect. 82.1.

We are fortunate in having one example, IRC 10216, which is very close to the Sun. The brightest 10 μm source beyond the solar system, IRC 10216 is a carbon-rich star surrounded by dust and gas it ejected in a strong stellar wind. The central star is so shielded that it is almost undetectable at optical wavelengths, and was not discovered until the 2 μm survey. IRC 10216 is where most of the circumstellar molecules are detected, and has greatly increased our understanding of circumstellar envelopes.

The envelopes are ejected by the red giant in its final phase of evolution. The mass loss rates increase to $\sim 10^{-4}$ solar masses per year and temperatures in the envelope are of order 1000 K. Close to the star, the density is high and the chemistry is characteristic of thermal equilibrium. The situation is quite unlike any interstellar environments. For example in IRC 10216, the HNC is over one hundred times less abundant than HCN, whereas in molecular clouds, they have about the same abundance.

If the star is oxygen-rich, large amounts of H_2O are formed and if it is carbon-rich, C_2H_2. This high temperature environment forms both molecules and grains. The high obscuration of the central source indicates that grains are formed in these envelopes. Polycyclic aromatic hydrocarbons (PAH's) or some similar species are observed in carbon rich planetary nebulae. These large molecules must have been produced when the object was a carbon rich circumstellar envelope.

As the material flows out from the star the density and temperature decrease. As the density becomes lower, three body reactions become less important and at some point the products of these reactions are frozen out in a similar manner to the evolution of molecules in the early universe (Sect. 82.10). In the outermost portions of the circumstellar envelope, molecules are dissociated by interstellar UV photons. The penetrating UV radiation is shielded by dust, H_2, and CO. The relative abundances can vary rapidly with radius, and observations provide abundance and radial distribution, a wealth of data for modelers. Circumstellar chemistry is reviewed by Omont [82.32] and recent chemical models are given in [82.33–35].

82.6 Supernova Ejecta

A supernova, the explosion of a massive star following core collapse, is one of the most spectacular displays in the Universe. The explosion occurs when the iron core of a massive star collapses to form a neutron star and the rebound shock and neutrino flux eject the outer portion of the star. The ejected portions of the star are rich in heavy elements produced in the interior of the progenitor star.

We are fortunate to have had in our lifetime a supernova which was close and in an unobscured line of sight. Supernova 1987A, the first supernova observed in 1987, went off in the Large Magellanic Cloud, a small satellite galaxy to our own. It was the first supernova visible to the naked eye in nearly 400 years (since the Kepler supernova in 1604). Using the full range of modern astronomical instruments has allowed us to get detailed spectra of the evolving ejecta which has greatly increased our knowledge of supernovae. We will use SN1987A as an example of supernova ejecta.

Initially the temperature of the ejecta of SN1987A was high, $\approx 10^6$ K, but it quickly cooled through adiabatic expansion and radiation from the photosphere. The temperature leveled off at several thousand degrees because of heating by radioactive nuclei, first ^{56}Ni and then ^{56}Co, formed in the explosion. The dynamics is homologous free expansion: the velocity scales linearly with the radius $r(t) = vt$ where v the velocity and t the time since the explosion.

The ejecta at first were optically thick and the spectrum resembled that of a hot star continuum with absorption lines from the surface. After a few days, the temperature dropped, but the ejecta remained optically thick and continued to show strong continuum emission. As the ejecta expand, the temperature drops, the ejecta become optically thin, and the spectrum is dominated by strong emission lines, superficially resembling an emission nebula Sect. 82.2. The emission is dominated by neutral atoms and singly ionized species.

The gas is heated and ionized by the gamma rays from radioactive decay. The gamma rays Compton scatter, producing X-rays and fast electrons. The X-rays further ionize the gas and produce multiply charged ions through the Auger process. These multiply charged ions recombine through charge transfer with neutral atoms. Further charge transfer determines the relative ionization of different species, with the lowest ionization potential species more ionized than the higher ionization potential species. The development of the infrared and optical spectrum of Supernova 1987A has been recently reviewed by *McCray* [82.36].

One of the great surprises in the spectrum of Supernova 1987A was the discovery of molecules in the infrared region. CO, SiO and possibly H_3^+ have been identified. In the absence of grains, molecules must be formed through either three-body or radiative processes. In the supernova ejecta, the densities are too low for three-body processes to be effective and molecules are formed through radiative association reactions. The molecules are removed by reactions with He^+ and the molecular abundances put a constraint on how much helium can be mixed back into the region with carbon and oxygen [82.37].

82.7 Shocked Gas

Shock waves occur in compressible fluids when the pressure gradients are large enough to generate supersonic motion, or when a disturbance is propagating through the fluid at supersonic velocities. Because information about the disturbance cannot propagate upstream in the fluid faster than the speed of sound, the fluid cannot respond dynamically until the shock arrives. The shock then compresses, heats, and accelerates the fluid. The boundary separating the hot compressed gas and the upstream gas is the shock front in which the energy of directed motion of the shock is converted to random thermal energy.

Shocks are ubiquitous in the interstellar medium where they are driven by the ionization fronts of expanding HII regions or nebulae, by outflowing gas accompanying stellar birth and evolution, and by supernova explosions. If the shock velocity is above 50 km/s, the shock gas is excited, dissociated, and ionized. The subsequent recombination and cooling radiation produces photons that may ionize and dissociate the gas components ahead of and behind the shock. This precursor radiation modifies the effects of the shock and influences its dynamical and thermal evolution. Fast shocks destroy all molecules by dissociating H_2 by collisions with H, H_2 and He and with electrons. Exchange reactions with H atoms destroy the other molecular species. At low densities, radiative stabilization occurs and dissociation is less efficient. Molecules reform in the cooling postshock gas. Slower shocks do not cause ionization or dissociation, but the chemical composition and the ion composition are modified by reactions taking place in the warm gas. The response of the interstellar gas to slow shocks is significantly affected by the presence of a magnetic field. In some ionization conditions, a magnetic precursor may occur in which a magnetosonic wave carries information about the shock, and the ionized and neutral components of the gas react differenly to the shock. Many different kinds of shock have been identified [82.38].

A very fast shock with a velocity of hundreds of km/s such as are driven by supernova explosions, creates a hot dilute cavity in the interstellar medium with a temperature of millions of degrees. The density is low and the gas cools and recombines slowly. Overlapping supernova-induced cavities may be responsible for the hot gas that occupies a considerable volume of the interstellar medium in the Galaxy and in some external galaxies. The conditions are far from coronal equilibrium as the gas cools more rapidly than it recombines. The cooling radiation appears as soft X-rays and UV emission lines with a characteristic spectrum.

As the gas cools below 10 000 K, molecular formation occurs. Molecular hydrogen is formed on the surfaces of grains as in molecular clouds and by the negative ion sequence that is effective in the early Universe Sect. 82.8. With the formation of H_2 in a still warm gas, the chemistry is driven by exothermic and endothermic reactions with H_2. Thus OH is produced by the reaction of O atoms, and H_2O by the further reaction of OH with H_2. Enhanced abundances of other neutral and ionic molecules are the products of subsequent reactions with OH. The reactions of S^+ and S with OH lead to SO^+ and SO, and their simultaneous presence may be an indicator of a dissociative shock. There are in addition physical indicators of shocks, such as asymmetric line profiles indicating high velocities.

In a nondissociative shock in a molecular gas, reactions with warm H_2 dominate the chemistry as it does in the cooling zone of a dissociative shock. The composition is controlled by the post shock temperature and the H/H_2 ratio. The warm H_2 changes the ionic composition by converting C^+ into CH^+. Evidence for a nondissociative shock is the infrared emission from H_2. The thermal emission from collisionally excited vibrational levels in shock-heated gas is readily distinguished from that discussed in Sect. 82.3 arising from UV pumping in a PDR. Emission from H_2 has been detected in numerous objects in the Galaxy and in many distant external galaxies. In external galaxies, X-rays may contribute to the H_2 infrared sprectrum through heating the gas and through excitation by photoelectron pumping to excited states followed by a downward cascade [82.39, 40].

82.8 The Early Universe

Molecules appeared first in the Universe after the adiabatic expansion had reduced the matter and radiation temperature to a few thousand degrees and recombination occurred, creating a nearly neutral Universe. The small fractional ionization that remained was essential to the formation of molecules. Molecular hydrogen formed through the sequences

$$H^+ + H \rightarrow H_2^+ + h\nu \quad (82.25)$$

$$H_2^+ + H \rightarrow H_2 + H^+ \quad (82.26)$$

and

$$H + e^- \rightarrow H^- + h\nu \quad (82.27)$$

$$H^- + H \rightarrow H_2 + e^- , \quad (82.28)$$

the protons and electrons acting as catalysts. Many other atomic and molecularprocesses occurred Fig. 82.3, some involving excited hydrogen atoms. Thus

$$H^* + H_2 \rightarrow H_3^+ + e^- \quad (82.29)$$

was a source of H_3^+.

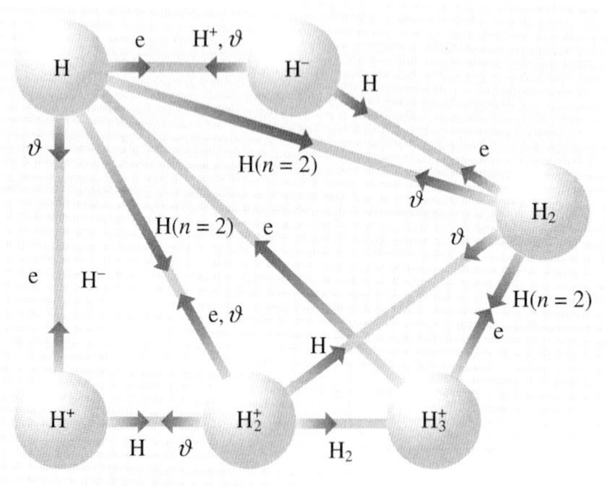

Fig. 82.3 Diagram showing the important reactions in the production of hydrogen molecules in the early Universe

The Universe contained trace amounts of deuterium and ^{7}Li nuclei with which heteronuclear molecules could be made. Molecules with dipole moments may leave an imprint on the cosmic blackbody background radiation that occupies the Universe. The deuterated molecules HD form from

$$D^+ + H_2 \rightarrow HD + H^+ , \quad (82.30)$$

and H_2D^+ from

$$D + H_3^+ \rightarrow H_2D^+ + H , \quad (82.31)$$

$$HD^+ + H_2 \rightarrow H_2D^+ + H , \quad (82.32)$$

and

$$HD + H_2^+ \rightarrow H_2D^+ + H . \quad (82.33)$$

Lithium hydride is formed through

$$Li + H \rightarrow LiH + h\nu \quad (82.34)$$

$$Li + H^- \rightarrow LiH + e^- \quad (82.35)$$

$$Li^- + H \rightarrow LiH + e^- . \quad (82.36)$$

There are many destruction processes, of which

$$LiH + H \rightarrow Li + H_2 \quad (82.37)$$

may be the most severe, though its rate coefficient is uncertain. The chemistry of the early Universe is summarized in [82.41].

The formation of molecules was a crucial step in the fragmentation of the first gravitationally collapsing objects which separated out of the cosmic flow. Three-body recombination

$$H + H + H \rightarrow H_2 + H \quad (82.38)$$

$$Li + H + H \rightarrow LiH + H \quad (82.39)$$

may be a major source of molecules as the density increases.

82.9 Recent Developments

While the core atomic and molecular process outlined are still unchanged, our understanding of the astrophysical environment has been greatly enhanced by a number of recent satellites. The Wilkinson Microwave Anisotropy Probe (WMAP) has given us the best map of the universe at the time of recombination and given us general confirmation of the Big Bang model. Perhaps the most surprising result is that stars seem to have formed much sooner than would have been expected, about 180 million years after the big bang [82.42]. This makes it even more difficult to understand how the universe goes from the relative uniformity at the time of recombination to the collapse and formation of the first objects so quickly, a problem which is certainly controlled by atomic and molecular processes. The relevant atomic and molecular processes have been recently reviewed by *Lepp*, *Stancil* and *Dalgarno* [82.43]. Two recent X-ray satellites, Chandra and XMM-Newton, both launched in 1999, have greatly increased our ability to detect hot ionized gas in stars, supernova remnants, active galaxies and other regions [82.44]. In particular, Chandra has allowed us for the first time to directly observe the hot gas between galaxies [82.45]. The Infrared Space Observatory (ISO) has provided a tremendous amount of data on cold regions in our own galaxy and allowed us to directly observe the icy mantles of dust grains [82.46].

Since the detection of molecules in SN 1987A, there have been many more observations of CO molecules in Type II supernova and they may even occur in every Type II [82.47]. CO has also been observed in a Type Ic supernova [82.48]. It remains a puzzle as to why the molecules are not rapidly removed by helium ions. A recent calculation of the $O + He^+$ system [82.49] finds that radiative charge transfer is much faster then direct charge transfer for temperatures below 10^6 K, but still too slow to significantly reduce the helium ion abundance in supernova ejecta. The most likely explanation remains that mixing is not complete in supernova ejecta, and the molecules survive in regions of relatively low helium abundance.

The state of modeling photoionized clouds has been recently reviewed by Ferland [82.50]. He also highlights the great need for atomic and molecular data for analyzing these clouds. New satellite data along with continued ground observations continually raise new astrophysical puzzles, puzzles which are controlled and probed by atomic and molecular processes. The astrophysical community owes a great debt to both atomic and molecular laboratory measurements and theoretical models of energy levels, reaction rates, and transition probabilities. In order to continue to progress in our understanding of the universe we will need to continue to fund the understanding of the atomic and molecular processes which control it.

82.10 Other Reading

Astronomy is one of the oldest sciences and one of the fastest evolving. Advances in technology are rapidly increasing the sensitivity and resolution of our instruments and so new observations and more sophisticated models lead to an ever greater understanding of the Universe. This means that books will often be somewhat dated when they appear. However, the series *Annual Review of Astronomy and Astrophysics* is a good source of recent review articles.

In addition, good introductions or overviews of a particular field are given in [82.1, 51–56]. Many sources of atomic and molecular data are listed and discussed in [82.7]. For details on atomic spectroscopy, see [82.57, 58]. For details on molecular spectroscopy, see [82.59, 60].

References

82.1 D. E. Osterbrock: *Astrophysics of Gaseous Nebulae and Active Galactic Nuclei* (Univ. Science Books, Mill Valley 1989)

82.2 P. F. Winkler, G. W. Clark, T. H. Markert, K. Kalata, H. W. Schnopper, C. R. Canizares: Astrophys. J. Lett. **246**, 27L (1981)

82.3 R. F. Reilman, S. T. Manson: Astrophys. J. Suppl. Ser. **40**, 815 (1979)

82.4 B. L. Henke, P. Lee, T. J. Tanaka, R. L. Shimabukoro, B. K. Fujikawa: Atom. Nucl. Data Tables **27**, 1 (1982)

82.5 M. Balucinska-Church, D. McCammon: **400**, 699 (1992)

82.6 D. A. Verner, D. G. Yakovlav, J. M. Band, A. B. Trzhaskavskaya: Atom. Nucl. Data Tables **55**, 233 (1993)

82.7 Special issue of Revista Mexicana de Astronomia y Astrofisica, March 23 (1992)

82.8 H. Nussbaumer, P. J. Storey: Astron. Astrophys. **178**, 324 (1978)

82.9 H. R. Ramadan, Y. Hahn: Phys. Rev. A **39**, 3350 (1989)

82.10 N. R. Badnell: Phys. Scr. T **28**, 33 (1989)

82.11 M. C. Bacchus-Montabonel, K. Amezian: Z. Phys. D **25**, 323 (1993)

82.12 P. Honvault, M. C. Bacchus-Montabonel, R. McCarroll: J. Phys. B **27**, 3115 (1994)

82.13 B. Herrero, I. L. Cooper, A. S. Dickinson, D. R. Flower: J. Phys. B **28**, 711 (1995)

82.14 C. Mendoza: *Planetary Nebulae IAU Symp. 103*, ed. by D. R. Flower (Reidel, Dordrecht 1983) p. 143

82.15 M. Brocklehurst, M. Salem: Comp. Phys. Commun. **13**, 39 (1977)

82.16 M. Salem, M. Brocklehurst: Astrophys. J. Suppl. Ser. **39**, 633 (1979)

82.17 P. G. Martin: Astrophys. J. Suppl. Ser. **66**, 125 (1988)

82.18 P. J. Storey, D. G. Hummer: Mon. Not. R. Astron. Soc. **272**, 41 (1995)

82.19 M. Arnaud, R. Rothenflug: Astron. Astrophys. Suppl. **60**, 425 (1985)
82.20 T. J. Gaetz, E. E. Salpeter: Astrophys. J. Suppl. Ser. **52**, 155 (1983)
82.21 E. van Dishoeck, J. Black: Astrophys. J. Suppl. Ser. **62**, 109 (1987)
82.22 W. G. Roberge, D. Jones, S. Lepp, A. Dalgarno: Astrophys. J. Suppl. Ser. **77**, 287 (1991)
82.23 A. Sternberg, A. Dalgarno: Astrophys. J. Suppl. Ser. **99**, 565 (1995)
82.24 R. Gredel, S. Lepp, A. Dalgarno, E. Herbst: Astrophys. J. **347**, 289 (1989)
82.25 T. J. Millar, A. Bennett, J. M. C. Rawlings, P. D. Brown, S. B. Charnley: Astron. Astrophys. Suppl. **87**, 585 (1991)
82.26 P. R. A. Farquhar, T. J. Millar: CCP7 Newsletter **18**, 6 (1993)
82.27 R. L. Kurucz: Astrophys. J. Suppl. Ser. **40**, 1 (1979)
82.28 M. J. Seaton: J. Phys. B **20**, 6363 (1987)
82.29 A. E. Lynas-Gray, M. J. Seaton, P. J. Storey: J. Phys. B **28**, 2817 (1995)
82.30 M. J. Seaton, Y. Yan, B. Mihalas, A. K. Pradhan: Mon. Not. R. Astron. Soc. **266**, 805 (1994)
82.31 C. A. Iglesias, F. J. Rogers: Astrophys. J. **443**, 460 (1995)
82.32 A. Omont: Circumstellar Chemistry. In: *Chemistry in Space*, ed. by J. M. Greenberg, V. Pirronello (Kluwer Academic, Dordrecht 1991) p. 171
82.33 G. A. Mamon, A. E. Glassgold, A. Omont: Astrophys. J. **323**, 306 (1987)
82.34 L. A. M. Nejad, T. J. Millar: Astron. Astrophys. **183**, 279 (1987)
82.35 T. J. Millar, E. Herbst: Astron. Astrophys. **288**, 561 (1994)
82.36 R. McCray: Ann. Rev. Astron. Astrophys. **31**, 175 (1993)
82.37 W. Liu, S. Lepp, A. Dalgarno: Astrophys. J. **396**, 679 (1992)
82.38 B. T. Draine, C. F. McKee: Ann. Rev. Astron. Astrophys. **31**, 373 (1993)
82.39 S. Lepp, R. McCray: Astrophys. J. **269**, 560 (1983)
82.40 R. Gredel, A. Dalgarno: Astrophys. J. **446**, 852 (1995)
82.41 A. Dalgarno, J. Fox: Ion Chemistry in Atmospheric and Astrophysical Plasmas. In: *Unimolecular and Bimolecular Reaction Dynamics*, ed. by C.-Y. Ng, T. Baer, I. Powis (Wiley, New York 1994)
82.42 C. L. Bennett et al.: Astrophys. J. Suppl. **148**, 1 (2003)
82.43 S. Lepp, P. Stancil, A. Dalgarno: J. Phys. B **35**, R57 (2002)
82.44 F. Paerels, S. Kahn: Ann. Rev. Ast. Appl. **41**, 291 (2003)
82.45 F. Nicasto et al.: Nature **433**, 495 (2005)
82.46 C. Cesarsky, A. Salama: *ISO Science Legacy: A Compact Review of ISO Major Achievements* (Springer, 2005)
82.47 J. Spyrimilo, B. Leibundgut, R. Gilmozzi: Ast. Ap. **376**, 188 (2001)
82.48 C. Gerardy et al.: PASJ **54**, 905 (2002)
82.49 L. B. Zhao et al.: Astrophys. J. **615**, 1063 (2004)
82.50 G. Ferland: Ann. Rev. Ast. Appl. **41**, 517 (2003)
82.51 C. W. Allen: *Astrophysical Quantities* (Athlone, London 1973)
82.52 K. R. Lang: *Astrophysical Formula* (Springer, Berlin, Heidelberg 1980)
82.53 W. W. Duley, D. A. Williams: *Interstellar Chemistry* (Academic, London 1984)
82.54 L. Spitzer: *Physical Processes in the Interstellar Medium* (Wiley, New York 1978)
82.55 A. Dalgarno, D. R. Layzer: *Spectroscopy of Astrophysical Plasmas* (Cambridge Univ. Press, Cambridge 1987)
82.56 T. Hartquist: *Molecular Astrophysics* (Cambridge Univ. Press, Cambridge 1990)
82.57 R. D. Cowan: *The Theory of Atomic Structure and Spectra* (Univ. California Press, Berkeley 1981)
82.58 I. I. Sobelman: *Atomic Spectra and Radiative Transitions* (Springer, Berlin, Heidelberg 1979)
82.59 G. Herzberg: *Molecular Spectra and Molecular Structure* (Prentice-Hall, New York 1939)
82.60 P. F. Bernath: *Spectra of Atoms and Molecules* (Oxford Univ. Press, Oxford 1995)

83. Comets

With the exception of the in situ measurements made by the *Giotto* and *Vega* spacecraft at comet 1P/Halley (the P/ signifies a periodic comet) during March 1986, all determinations of the volatile composition of the coma are derived from spectroscopic analyses. Detailed modeling is then used to infer the volatile composition of the cometary nucleus. This chapter focuses on the principal atomic and molecular processes that lead to the observed spectrum as well as the needs for basic atomic and molecular data in the interpretation of these spectra. The largely collisionless and low density coma, with no gravity or magnetic field, is a unique spectroscopic laboratory, as evidenced by the discovery of C_3 before its identification in terrestrial laboratories [83.1]. Many key discrepancies remain to be resolved concerning the basic molecular composition and the elemental abundances of both the volatile and refractory components of the cometary nucleus, as well as the comet-to-comet variation (particularly between "new" and evolved periodic comets) of these quantities. These issues (and many others) are discussed in the recent analytical review of *Festou* et al. [83.2, 3] or in the compendia of Halley results [83.4]. The former also contains a comprehensive bibliography. Other sources concentrating largely on the physics and chemistry of comets include the volumes edited by *Wilkening* [83.5] and *Huebner* [83.6] and the pre-Halley review of *Mendis* et al. [83.7].

Comets are small bodies of the solar system believed to be remnants of the primordial solar disk. Formed near the orbits of Uranus and Neptune and subsequently ejected into an "Oort cloud" of some 40 000 AU in extent, these objects likely preserve a record of the volatile composition of the early outer solar system, and so are of great interest for the physical and chemical modeling of solar system formation. The comets arrive in the inner solar system as a result of galactic perturbations. The cometary volatiles are vaporized as their orbits bring them closer to the sun and it is solar radiation that initiates all of the processes that lead to the extended coma. Gas vaporization also leads to the release of dust into the coma, and the scattering of sunlight by dust is the major source of the visible coma and dust tail of comets. Somewhat fainter, and much more extended, is the plasma tail, resulting from photoionization by solar extreme UV radiation of the neutral volatiles and their subsequent interaction with the solar wind.

83.1 Observations

In a review in 1965, *Arpigny* [83.8] summarized the known molecular and atomic emissions detected in the visible region of the spectrum (here defined as 3000 to 11 000 Å) as follows:

radicals:	OH, NH, CN, CH, C_3, C_2, and NH_2
ions:	OH^+, CH^+, CO_2^+, CO^+, and N_2^+;
metals:	Na, Fe .

The only known atomic feature was the O I forbidden red doublet at 6300 and 6364 Å. From the radicals and ions one could infer the presence of their progenitor "parent" molecules such as H_2O, NH_3, HCN, CO and CO_2, directly vaporizing from the comet's nucleus. The metals, seen only in comets passing close to the sun, were assumed to come from the vaporization of refractory grains. The inventory of metals was soon expanded to include K, Ca^+, Ca, V, Cr, Mn, Ni and Cu, from observations of the sun-grazing comet Ikeya-Seki (C/1965 S1) [83.9, 10] and H_2O^+ was identified in comet Kohoutek (C/1973 E1). This latter comet was also the first to be extensively studied at wavelengths both shortward and longward of the visible spectral range.

The first parent molecule to be directly identified was CO, which fluoresces in the Fourth Positive system $(A\,^1\Pi_u - X\,^1\Sigma^+)$ in the VUV [83.11], although the in situ neutral mass spectrometer measurements made of Halley disclosed the presence of an extended, dominant source of CO [83.12] whose origin is still being debated [83.13]. Ideally, the molecular species should be detectable through their radio and sub-mm rotational transitions or through the detection of vibrational bands or individual ro-vibrational lines in the near IR. Water was first directly detected through ro-vibrational lines near 2.7 μm in comet Halley and again in comet Wilson (C/1986 P1) [83.14, 15]. However, due to the low column densities of the other expected species, typically ≈1 or less than that of H_2O, the direct detection of species such as H_2CO, H_2S and CH_3OH has only recently been made possible by the development of more sensitive instrumental techniques together with the fortuitous apparitions of two bright comets, C/1996 B2 (Hyakutake) and C/1995 O1 (Hale-Bopp) in 1996 and 1997. To date, more than two dozen parent molecules have been identified [83.16]. Isotope ratios, particularly the D/H ratio, in molecules such as HDO have been determined from sub-millimeter observations [83.17].

The ultimate result of solar photolysis (and to a lesser degree, the interaction with the solar wind) is the reduction of all of the cometary volatiles to their atomic constituents. The atomic inventory is somewhat easier to derive as the resonance transitions of the cosmically abundant elements H, O, C, N and S all lie in the VUV and, in principle, the total content of these species in the coma can be determined by an instrument with a suitably large field of view. Of course, a fraction of the atomic species of each element will be produced directly in ionic form, and will not be counted using this approach. In addition, another fraction exists in the coma in the solid grains, and this component will also not be included, except for a small amount volatilized by evaporation or sputtering by energetic particles. The composition of the grains, though not the absolute abundance, has been determined from in situ measurements made by the Halley encounter spacecraft [83.18], and can be inferred, though not unambiguously, from reflection spectroscopy of cometary dust in the 3–5 μm range.

The advent of space-borne platforms for observations in the VUV has produced a wealth of new information about the volatile constituents of the coma. The $A\,^2\Sigma^+ - X\,^2\Pi$ (0, 0) band of the OH radical at ≈ 3085 Å was well known from ground-based spectroscopic observations, but as this wavelength lies very close to the edge of the atmospheric transparency window, the strength of this feature (relative to that of other species) was not appreciated until 1970 when comet Bennett (C/1969 Y1) was observed from space by the Orbiting Astronomical Observatory (OAO-2). The OAO-2 spectrum also showed a very strong, broadened H I Ly-α emission from H, the other principal dissociation product of H_2O. The broad shape of Ly-α seen in the OAO-2 spectrum is due to the large spatial extent of the atomic H envelope, the result of a high velocity acquired in the photodissociation process and a long lifetime against ionization. Later, at the apparition of comet Kohoutek (C/1973 E1), atomic O and C were identified in the spectra and direct UV images of the H coma, as well as of the O I and C I emissions, were obtained from sounding rocket experiments. These experiments were repeated for comet West (C/1975 V1) and led to the first detection of CO [83.11].

Between 1978 and 1996, over 50 comets were observed spectroscopically over the wavelength range 1200–3400 Å by the International Ultraviolet Explorer (IUE) satellite observatory [83.19, 20]. Most of the spectra were obtained at moderate resolution ($\Delta\lambda$ = 6–10 Å), although high dispersion echelle spectra ($\Delta\lambda$ = 0.2–0.3 Å) are useful for some studies, particularly those of fluorescence equilibrium (Sect. 83.2.2). For Halley alone, over 200 UV spectra were obtained from September 1985 to July 1986. The launch of the *Hubble Space Telescope* (HST) in 1990, together with subsequent enhancements to the spectroscopic instrumentation that were made on-orbit, marked another advance in sensitivity as well as the ability to observe in a small field-of-view very close to the nucleus. This yielded the first detection of CO Cameron band emission, a direct measure of CO_2 being vaporized from the nucleus [83.21]. For an overview of a cometary spectrum, a composite spectrum of 103P/Hartley 2 spanning the region from H I Ly-α to 7000 Å taken with the Faint Object Spectrograph of HST, using five separate gratings, is shown in Fig. 83.1. The launch of the *Far Ultraviolet Spectroscopic Explorer* (FUSE) in 1999 provides access to the spectral region between 900 and 1200 Å at very high spectral resolution, and has led to the detection of H_2 (Sect. 83.2.4), upper limits on Ar and N_2, and some three dozen unidentified emission lines [83.22, 23].

Several other satellite observatories have contributed unique cometary observations in the UV and sub-mm spectral windows. The *Solar and Heliospheric Observatory* (SOHO) has two valuable instruments: The SWAN (Solar Wind Anisotropies) instrument provides sky maps in H I Ly-α at 1° resolution and has observed over 20 comets since 1996 [83.25]. The UVCS (Ultraviolet Coronograph Spectrometer) provides far-uv spectra and images of comets close to the Sun, where HST and FUSE are prohibited from observing, and has recently detected C^{++} in the tail of comet C/2002 X5 (Kudo-Fujikawa) [83.26]. The direct detection of H_2O in the fundamental rotational line at 557 GHz in several comets was made by the *Submillimeter Wave Astronomy Satellite* (SWAS) [83.27] and the *Odin satellite* [83.28]. This line cannot be observed from ground-based telescopes because of the strong absorption by water vapor in the terrestrial atmosphere.

Prior to 1996, X-rays had not been detected in comets and the conventional wisdom was that they were unlikely to be produced in the cold, rather thin cometary atmosphere. The discovery of soft X-ray emission ($E < 2$ keV) from comet Hyakutake (C/1996 B2) by the *Röntgen Satellite* (ROSAT) thus came as a surprise [83.29]. Since then, X-ray emission has been detected from over a dozen comets using ROSAT and four other space observatories, the *Extreme Ultraviolet Explorer*, *BeppoSAX*, the *Chandra X-ray Observatory* (CXO), and *Newton-XMM* [83.30]. The earliest observations were at very low spectral resolution, making it difficult to select amongst the possible excitation mechanisms: charge exchange, scattering of solar photons by attogram dust particles, energetic electron impact and bremsstrahlung, collisions between cometary and interplanetary dust, and solar X-ray scattering and fluorescence. The more recent CXO observations, at higher spectral resolution, favor the charge exchange of energetic minor solar wind ions such as O^{6+}, O^{7+}, C^{5+}, C^{6+}, and others, with cometary gas, principally H_2O, CO, and CO_2, as the primary mechanism. This mechanism would explain why the X-ray intensity appears to be independent of the gas

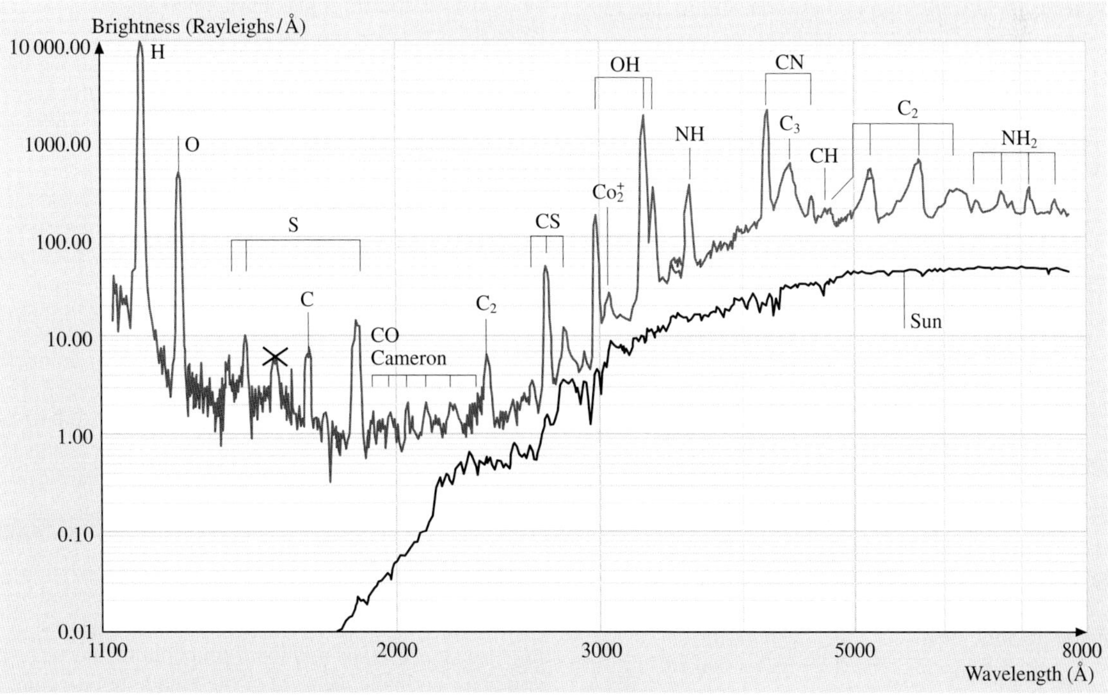

Fig. 83.1 Composite FOS spectrum of comet 103P/Hartley 2. After [83.24]

production rate of the comet, and that the peak emission is offset from the location of the comet's nucleus. This conclusion is also supported by recent laboratory work on the charge transfer of highly ionized species with cometary molecules [83.31] and by theoretical calculations of state specific cascades [83.32]. The X-ray emission thus tells us more about the solar wind than about the gaseous composition of comets.

83.2 Excitation Mechanisms

83.2.1 Basic Phenomenology

Coma abundances may be derived from spectrophotometric measurements of either the total flux or the surface brightness in a given spectral feature. The uncertainty in the derived abundances includes not only the measurement uncertainty, but also uncertainties in the atomic and molecular parameters and, in the case of surface brightness measurements, uncertainties in the model parameters used. Thus relative abundances, derived from observations of different comets with the same instrument and under similar geometrical conditions, are often more reliable.

Atoms and ions in the cometary coma emit radiation primarily by means of resonance re-radiation of solar photons. For the cosmically abundant elements H, C, N, O, and S, their strongest resonance transitions are in the VUV. The few exceptions are noted below. Assume that the coma is optically thin in these transitions. The total number of species i in the coma is

$$M_i = Q_i \tau_i(r) , \tag{83.1}$$

where Q_i is the production rate (atoms or molecules s^{-1}) of species i and $\tau_i(r)$ is its lifetime at heliocentric distance r, $\tau_i(r) = \tau_i(1\,\mathrm{AU})r^2$. The r dependence arises from photolytic destruction processes induced by solar UV radiation, and to a lesser degree by the solar wind, as described in Sect. 83.3.1.

The luminosity, in photons s^{-1}, in a given transition at wavelength λ, is then

$$L_{i\lambda} = M_i g_{i\lambda}(r) , \tag{83.2}$$

where the fluorescence efficiency, or "g-factor", $g_{i\lambda}(r) = g_{i\lambda}(1\,\mathrm{AU})r^{-2}$, is

$$g_{i\lambda}(1\,\mathrm{AU}) = \frac{\pi e^2}{mc^2}\lambda^2 f_\lambda \pi F_\odot \tilde{\omega} \quad \text{photons } s^{-1}\text{atom}^{-1} , \tag{83.3}$$

where f_λ is the absorption oscillator strength, $\pi F_\odot$ is the solar flux per unit wavelength interval at 1 AU and $\tilde{\omega}$ is the albedo for single scattering, defined for a line in an atomic multiplet as

$$\tilde{\omega} = \frac{A_j}{\sum_j A_j} , \tag{83.4}$$

where A_j is the decay rate. If a given multiplet is not resolved, then $\tilde{\omega} = 1$. For diatomic molecules, fluorescence to other vibrational levels becomes important and the evaluation of $\tilde{\omega}$ depends on the physical conditions in the coma, as discussed in Sect. 83.2.2. Thus, for a comet at a geocentric distance Δ, the total flux from the coma for the transition is

$$F_{i\lambda} = \frac{L_{i\lambda}}{4\pi\Delta^2} = \frac{Q_i g_{i\lambda}(r)\tau_i(r)}{4\pi\Delta^2} , \tag{83.5}$$

and the product $g_{i\lambda}(r)\tau_i(r)$ is independent of r.

Unfortunately, the scale lengths (the product of lifetime and outflow velocity) of almost all of the species of interest in the UV are $\approx 10^5 - 10^6$ km at 1 AU. Thus, total flux measurements require fields of view ranging from several arc-minutes to a few degrees. This has been done only in the case of a few isolated sounding rocket experiments. Most information about the UV spectra comes from observations made by orbiting satellite observatories whose spectrographs have small apertures (e.g., $10'' \times 20''$ for IUE) and thus sample only a very small part of the total coma. In this case, again assuming an optically thin coma, the measured flux $F'_{i\lambda}$ in the aperture can be converted to an average surface brightness $B_{i\lambda}$ (in units of Rayleighs):

$$B_{i\lambda} = 4\pi 10^{-6} F'_{i\lambda}\Omega^{-1} , \tag{83.6}$$

where Ω is the solid angle subtended by the aperture. The brightness, in turn, is related to $\overline{N}_i$, the average column density of species i within the field of view by

$$B_{i\lambda} = 10^{-6} g_{i\lambda}(r)\overline{N}_i . \tag{83.7}$$

The evaluation of Q_i from $\overline{N}_i$ requires the use of a model of the density distribution of the species i (Sect. 83.3.2).

A similar treatment can be applied to the excitation of the near infrared vibrational transitions of cometary parent molecules since the direct pumping by solar IR radiation far exceeds the indirect pumping of ground state vibrational levels through electronic transitions excited by the solar UV flux [83.16]. However, this does not apply to the rotational transitions which are controlled by collisional excitation, primarily collisions with H_2O. In this case, the observed rotational temperature may be regarded as a reliable measure of the kinetic temperature of the coma gas.

83.2.2 Fluorescence Equilibrium

In the case of low resolution spectroscopy, where an atomic multiplet or molecular band is unresolved, the evaluation of the g-factor (83.3) does not require knowledge of the population of either atomic fine structure levels or molecular rotational levels in the ground state of the transition. Furthermore, the assumption that the solar flux does not vary over the multiplet or band allows us to use the total transition oscillator strength. This assumption is more often than not invalid because of the Fraunhofer structure of the solar spectrum in the near uv and visible region and the emission line nature of the spectrum below 2000 Å. For high resolution spectra, the g-factors for each individual line must be calculated separately and the relative populations of the ground state levels must be included. There are three cases to be considered:

1. The g-factor, or probability of absorption of a solar photon, is less than the probability that the species will be dissociated or ionized, i. e., $g_{i\lambda} < (\tau_i)^{-1}$. In this case, the ground state population is not affected by fluorescence and a Boltzmann distribution at a suitable temperature (typically $\approx$200 K at 1 AU) corresponding to the production of the species may be used. This is often the case in the far UV, where the solar flux is low, such as for the Fourth Positive band system of CO. For atomic transitions from triplet ground states, such as is found with O, C and S, downward fine structure transitions are fast enough to effectively depopulate all but the lowest fine structure level.
2. The species undergoes many photon absorption and emission cycles in its lifetime, and the ground state population is determined (usually after 5 or 6 cycles) by the fluorescence branching ratios. This is the condition of fluorescence equilibrium, which applies for almost all radicals observed in the visible and near UV regions. The general procedure is to solve a set of coupled equations of the form
$$\frac{\mathrm{d}n_a}{\mathrm{d}t} = -n_a \sum_{b=1}^{N} p_{ab} + \sum_{b=1}^{N} n_b p_{ba} , \qquad (83.8)$$
where n_a is the relative population of level a and p_{ab} and p_{ba} are transition rates out of and into this level, respectively. The n_a are normalized to unity. The steady state, obtained after many cycles, is given by $\mathrm{d}n_a/\mathrm{d}t = 0$. Since the downward transition rates are determined only by quantum mechanics, while the absorption rates depend on the magnitude of the solar flux, the steady state population varies with distance from the sun with higher rotational levels (as for the case of CN [83.8]) populated closer to the sun. In some cases, where only a few cycles occur, the equations are integrated numerically. As the g-factor varies with time, it also effectively varies with the position of a species in the coma. Care must also be taken when spectra taken with small apertures are analyzed as the transit time for an atom or molecule to cross the aperture may be $\lesssim g_{i\lambda}(r)^{-1}$. In practice, these considerations are often not important.
3. The same as 2. except that the density is sufficiently high that collisional transitions must be included in addition to the radiative transitions between levels. However, as the collisional rates are poorly known, in practice a "collision sphere" is defined such that a molecule traveling radially outward from this sphere suffers only one collision with other molecules or atoms. Outside this sphere, fluorescence equilibrium is assumed to hold, while inside a thermal distribution of ground state levels is used. A rough estimate of the radius R_c of the collision sphere, based on the radial outflow model of Sect. 83.3.2 is given by [83.2, 3]
$$R_\mathrm{c} = 10^3 \frac{Q}{10^{29}}\ \mathrm{km} , \qquad (83.9)$$
where Q is the total production rate in molecules s^{-1}.

83.2.3 Swings and Greenstein Effects

Swings [83.33] first pointed out that because of the Fraunhofer absorption lines in the visible region of the solar spectrum, the absorption of solar photons in a molecular band would vary with the comet's heliocentric velocity $\dot{r}$, leading to differences in the structure of a band at different values of $\dot{r}$ when observed at high resolution. In (83.8), this corresponds to evaluating the $p_{ba} = p_{ba}(\dot{r})$. For typical comets observed near 1 AU, $\dot{r}$ can range from –30 to +30 km/s, while in certain cases of comets with small perihelia the range can be twice as large. This effect of the Doppler shift between the sun and the comet is commonly referred to as the Swings effect. Even for observations at low resolution, the Swings effect must be taken into account in the calculation of the total band g-factor, and this has been done for a number of important species such as OH, CN and NH. A particularly important case, that of the OH $A\,^2\Sigma^+ - X\,^2\Pi$ (0,0) band at $\approx$ 3085 Å, which is often used to derive the water production rate of a comet, is illustrated in Fig. 83.2, which also shows the dependence of fluorescence equilibrium on heliocentric distance [83.34].

While this effect was first recognized in the spectra of radicals in the visible range, a similar phenomenon occurs in the excitation of atomic multiplets below 2000 Å, where the solar spectrum makes a transition to

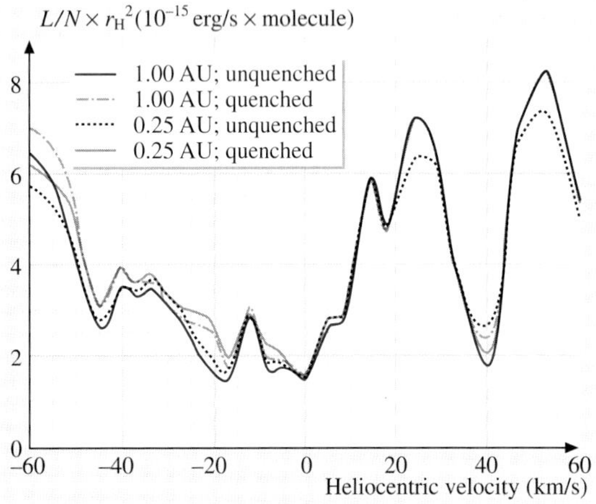

Fig. 83.2 OH (0,0) band g-factor as a function of heliocentric velocity. After [83.34]

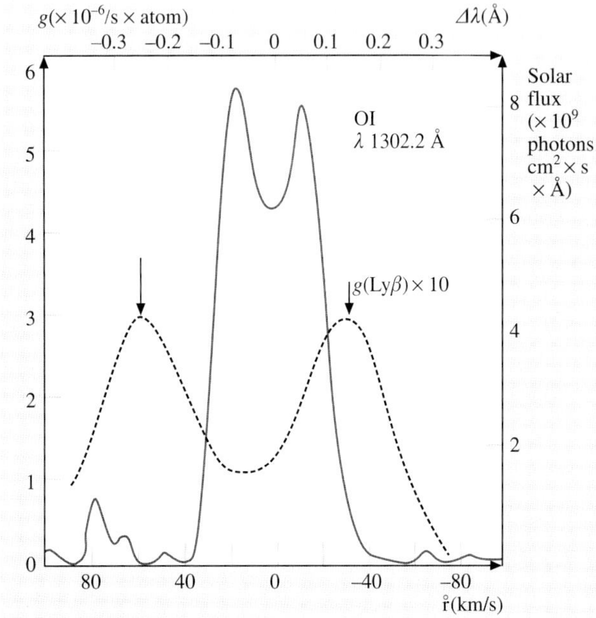

Fig. 83.3 Solar flux and fluorescence efficiency for O I λ1302 as a function of heliocentric velocity. After [83.35]

an emission line spectrum. For example, the three lines of O I λ1302 have widths of ≈ 0.1 Å, corresponding to a velocity of ≈25 km/s, so that knowledge of exact solar line shapes is essential for a reliable evaluation of the g-factor for this transition [83.35], as illustrated in Fig. 83.3.

A differential Swings effect occurs in the coma since atoms and molecules on the sunward side of the coma, flowing outward towards the sun, have a net velocity that is different from those on the tailward side, and so, if the absorption of solar photons takes place on the edge of a line, the g-factors will be different in the two directions. Differences of this type appear in long-slit spectra in which the slit is placed along the sun-comet line (the *Greenstein* effect [83.36]). Again, an analogue in the far UV has been observed in the case of O I λ1302 [83.37], as can be seen in Fig. 83.3. Although it is also possible to explain the observation by nonuniform outgassing, this was considered unlikely as all of the other observed emissions had symmetric spatial distributions. The measurement of the Greenstein effect leads immediately to a determination of the mean outflow velocity of the given species.

83.2.4 Bowen Fluorescence

Figure 83.3 also demonstrates that for heliocentric velocities > 30 km/s, the Doppler shift reduces the solar flux at the center of the absorption line to a very small value, so that the O I λ1302 line is expected to appear weakly, if at all, in the observed spectrum. Thus, it was a surprise that this line appeared fairly strongly in two comets, Kohoutek (C/1973 E1) and West (C/1975 V1), whose values of $\dot{r}$ were both > 45 km/s at the times of observation. The explanation invoked the accidental coincidence of the solar H I Ly-β line at 1025.72 Å with the O I ^{3}D – ^{3}P transition at 1025.76 Å, cascading through the intermediate ^{3}P state as shown in the simplified energy level diagram of Fig. 83.4 [83.35]. This mechanism, well known in the study of planetary nebulae, is referred to as *Bowen* fluorescence [83.38]. The g-factor due to Ly-β pumping is an order of magnitude smaller than that for resonance scattering, as shown in Fig. 83.3, but sufficient to explain the observations and to confirm that H_2O is the dominant source of the observed oxygen in the coma.

Ly-β is also coincident with the P1 line of the (6,0) band of the H_2 Lyman system $(B\,^1\Sigma_u^+ - X\,^1\Sigma_g^+)$ leading to fluorescence in the same line of several $(6,v'')$ bands, the strongest being that of the (6,13) band at 1608 Å [83.39]. This line is, however, difficult to ob-

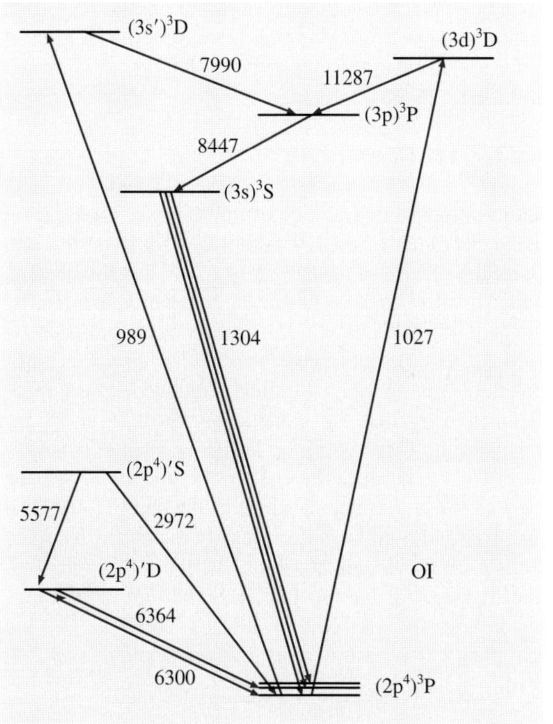

Fig. 83.4 Simplified O I energy level diagram showing transitions of interest in cometary spectra

serve because of the nearby strong CO Fourth Positive bands. Recently, the shorter wavelength (6,1), (6,2), and (6,3) bands have been detected in three comets using FUSE and the derived H_2 column abundance was found to be consistent with a water photodissociation source [83.22, 23]. Another interesting example occurs for Ne I, where the second resonance transition at 629.74 Å coincides with the strong solar O V line at 629.73 Å. This line was used to set a sensitive upper limit on the Ne abundance in the coma of comet Hale-Bopp (C/1995 O1) [83.40].

83.2.5 Electron Impact Excitation

The photoionization of the parent molecules and their dissociation products leads to the formation of a cometary ionosphere whose characteristics are only poorly known. Planetary ionospheres serve only as a poor model since the cometary atmosphere is gravitationally unbound and there is no constraining magnetic field. The in situ measurements of Halley provided order of magnitude confirmation of the theoretical modeling. In principle, electron impact excitation, which is often the dominant source of airglow in the atmospheres of the terrestrial planets Chapt. 84, also contributes to the observed emissions, particularly in the UV, and so must be accounted for in deriving column densities from the observed emission brightnesses. However, one can use a very simple argument, based on the known energy distribution of solar UV photons, to demonstrate that electron impact excitation is only a minor source for the principal emissions. Since the photoionization rate of water (and of the important minor species such as CO and CO_2) is $\approx 10^{-6}\,s^{-1}$ at 1 AU, and the efficiency for converting the excess electron energy into excitation of a single emission is of the order of a few percent, the effective excitation rate for any emission will be $\approx 10^{-8}\,s^{-1}$ or less at 1 AU [83.41]. Since the efficiencies for resonance scattering or fluorescence for almost all the known cometary emissions are much larger, electron impact may be safely neglected except in a few specific cases.

The cases of interest are those of forbidden transitions, where the oscillator strength, and consequently the *g*-factor, is very small. Examples include the O I $^5S_2 - {}^3P_{2,1}$ doublet at 1356 Å, which was observed in comets West and Halley by rocket-borne spectrographs and more recently in comets Hyakutake and Hale-Bopp, the O I $^1D - {}^3P$ red lines at 6300 and 6364 Å, observed in many comets, and the CO Cameron bands [83.21]. However, the excitation of these latter two is dominated by prompt emission in the inner coma, the same region of the coma where electron excitation is important, as described in the next section.

83.2.6 Prompt Emission

In cases where the dissociation or ionization of a molecule leaves the product atom or molecule in an excited state, the decay of this state with the prompt emission of a photon provides a useful means for tracing the spatial distribution of the parent molecule in the inner coma. The products of interest for the water molecule are described in Sect. 83.3.1. Prompt emission includes both allowed radiative decays (such as from the $A\,^2\Sigma^+$ state of OH) as well as those from metastable states such as O(^{1}D), since the latter will move ~150 km (for a comet at a geocentric distance of 1 AU, 1 arcsecond corresponds to a projected distance of 725 km) in its lifetime. The O I $^1D - {}^3P$ transition at 6300 and 6364 Å Fig. 83.4 has been used extensively as a ground-based monitor of the water production rate with the caveat that other species such as OH, CO and CO_2 may also contribute to the observed red line emission.

In addition, when the density of H_2O is sufficient to produce observable red line emission, it is also sufficient to produce collisional quenching of the ^{1}D state, and this must also be considered in the interpretation of the observations. The analogous $^1D - {}^3P$ transitions in carbon occur at 9823 and 9849 Å and provide similar information about the production rate of CO. Carbon atoms in the ^{1}D state, whose lifetime is ≈4000 s, are known to be present from the observation of the resonantly scattered $^1P^o - {}^1D$ transition at 1931 Å [83.42], and the 9849 Å line has been detected in comet Hale-Bopp [83.43].

The OH $A\,^2\Sigma^+ - X\,^2\Pi$ prompt emission competes with that produced by the resonance fluorescence of OH and is difficult to detect, except close to the nucleus (inside 100 km) where the density of water molecules exceeds that of OH by a few orders of magnitude [83.44]. Again, the reason is that at the wavelengths below the threshold for simultaneous dissociation and excitation, the sun has much less flux than at the resonance wavelength. On the other hand, only a few rotational lines are excited in fluorescence equilibrium [83.34], while the prompt emission is characterized by a very "hot" rotational distribution, so in principle the two components may be separated although observations at very high spatial and spectral resolution are required. OH prompt emission has also recently been detected in the infrared at 3.28 μm [83.45].

83.2.7 OH Level Inversion

An important consequence of fluorescence equilibrium in the OH radical is the UV pumping of the hyperfine and Λ-doublet levels of the $X\,^2\Pi_{3/2}(J=3/2)$ ground state, which results in a deviation of the population from statistical equilibrium [83.34]. Depending on the heliocentric velocity, this departure may be either "inverted" or "anti-inverted" giving rise to either stimulated emission or absorption against the galactic background at 18 cm wavelength. This technique has been used extensively since 1974 to monitor the OH production rate in comets, even of those that appear close to the sun [83.46]. The resulting radio emissions are easily quenched by collisions with molecules and ions, the latter giving rise to a fairly large R_c that must be accounted for in interpreting the derived OH column density. Nevertheless, the radio and UV measurements give reasonably consistent results [83.46].

83.3 Cometary Models

83.3.1 Photolytic Processes

As an example of the photolytic destruction processes occurring in the coma, consider the dominant molecular species, water. Water vapor is assumed to leave the surface of the nucleus with some initial velocity v_0 and flow radially outward, expanding into the vacuum, and increasing its velocity according to thermodynamics [83.7]. Even though collisions are important at distances typically up to 10^4 km (depending on the density and consequently, on the total gas production rate), the net flow of H_2O molecules is radially outward, such that the density varies as R^{-2} near the nucleus, where R is the cometocentric distance. This is the basis for spherically symmetric coma models (the number of particles flowing through a spherical surface is conserved), which assume isotropic gas production, but appears to hold equally well for the case of Halley, which clearly was not outgassing uniformly over its surface [83.4].

The photolysis of H_2O can proceed by:

a	$H_2O + h\nu \rightarrow OH + H$	2424.6 Å
a'	$\rightarrow OH(A\,^2\Sigma^+) + H$	1357.1 Å
b	$H_2O + h\nu \rightarrow H_2 + O(^1D)$	1770 Å
b'	$\rightarrow H_2 + O(^1S)$	1450 Å
b''	$\rightarrow H + H + O(^3P)$	1304 Å
c	$\rightarrow H_2O^+ + e^-$	984 Å
d	$\rightarrow H + OH^+ + e^-$	684.4 Å
e	$\rightarrow H_2 + O^+ + e^-$	664.4 Å
f	$\rightarrow OH + H^+ + e^-$	662.3 Å

The right-hand column gives the energy threshold for each reaction, in wavelength units. The products are subsequently removed by:

g	$OH + h\nu \rightarrow O + H$	2823.0 Å
h	$\rightarrow OH^+ + e^-$	928 Å
i	$H_2 + h\nu \rightarrow H + H$	844.79 Å
j	$\rightarrow H_2^+ + e^-$	803.67 Å
k	$\rightarrow H + H^+ + e^-$	685.8 Å
l	$O + h\nu \rightarrow O^+ + e^-$	910.44 Å
m	$H + h\nu \rightarrow H^+ + e^-$	911.75 Å

Reactions l and m can also occur by resonant charge exchange with solar wind protons. Reactions a', b and b' correspond to the production of prompt emission, as discussed in Sect. 83.2.6. The determination of column densities of H, O and OH simultaneously was convincing evidence that the dominant volatile species in the cometary nucleus was H_2O, long before the direct infrared detection of this species in the coma.

Detailed cross sections for the absorption of UV photons by each of the reactants, including proper identification of the final product states, is necessary for the evaluation of the photodestruction rate in the solar radiation field of each of the above reactions. These rates are evaluated at 1 AU using whole disk measurements of the solar flux by integrating the cross section

$$J_\mathrm{d} = \int_0^{\lambda_\mathrm{th}} \pi F_\odot \sigma_\mathrm{d} \, \mathrm{d}\lambda \,. \tag{83.10}$$

Another quantity of interest in coma modeling is the excess velocity (or energy) of the dissociation or ionization products, and this requires knowledge of the partitioning of energy between internal and translational modes for each reaction [83.47].

Qualitatively, the photodissociation and photoionization rates can be estimated from the threshold energies given in the table above, since the solar flux is decreasing very rapidly to shorter wavelengths, as can be seen in Fig. 83.1. It is customary to specify the lifetime against photodestruction τ_i, of species i, which is just

$$(\tau_i)^{-1} = \sum_j (J_\mathrm{d})_j \,, \tag{83.11}$$

where the sum is over all possible reaction channels, as well as the lifetimes into specific channels. Processes with thresholds near 3000 Å have lifetimes $\approx 10^4$ s, those with thresholds near 2000 Å an order of magnitude longer, while those with thresholds below Ly-α, such as most photoionization channels, have lifetimes $\approx 10^6$ s, all at 1 AU. In addition to uncertainties in the details of the absorption cross sections, further uncertainty is introduced into the calculation of J_d by the lack of knowledge of the solar flux at the time of a given observation due to the variability of the solar radiation below 2000 Å, and most importantly, below Ly-α, where there have not been continuous space observations for more than a decade. The solar UV flux is known to vary considerably both with the 27-day solar rotation period and with the 11-year solar activity cycle. Also, at any given point in its orbit, a comet sees a different hemisphere of the sun than what is seen from Earth. *Huebner* et al. [83.48] have compiled an extensive list of useful photodestruction rates using mean solar fluxes to represent the extreme conditions of solar minimum and solar maximum. They also include the excess energies of the dissociation products. A detailed analysis of the rates for H_2O and OH, using surrogate solar indices such as the 10.7 cm solar radio flux, or the equivalent width of the He I line at 1.083 μm, is in good agreement with observations [83.49]. Similar analyses still remain to be carried out for other important species, such as CO and NH_3.

83.3.2 Density Models

For parent molecules produced directly by sublimation from the surface of the comet, a spherically symmetric radial outflow model is often adopted. Such a model assumes a steady-state gas production rate Q_i and a constant outflow velocity v, and gives rise to a density distribution as a function of cometocentric distance R given by

$$n_i(R) = \frac{Q_i}{4\pi v R^2} \mathrm{e}^{-R/\beta_i} \,, \tag{83.12}$$

where $\beta_i = v\tau_i$ is the scale length of species i. The basic validity of this model was demonstrated by the *Giotto* neutral mass spectrometer measurements of H_2O and CO_2 [83.4], although detailed analysis revealed that the velocity of the water molecules increased from 0.8 km/s at about 1000 km from the nucleus to 1.1 km/s at a radial distance of 10 000 km. The dependence of outflow velocity on heliocentric distance remains uncertain, although *Delsemme* [83.50] has suggested an $r^{-1/2}$ dependence based on thermodynamic arguments. Sub-millimeter observations of H_2O have sufficient spectral resolution to permit the mapping of outflow velocities along various lines-of-sight to the comet [83.28].

For the dissociation products, the modeling is more complex. The simplest model assumes continued radial outflow, although at a different velocity, such as to maintain a constant flux of the initial particle across an arbitrary spherical surface surrounding the nucleus [83.51]. This model, which is valid only at distances equal to a few β_i, is widely used as the densities can be easily expressed in analytical form. However, as surface brightness measurements are often made with small fields of view close to the nucleus, this model can lead to a factor of two error from the neglect of the dissociation kinematics. Since the solar photodissociation often leaves the product fragments with 1–2 eV of kinetic energy [83.47, 48], the resultant motion (which is assumed to be isotropic in the parent molecule's rest frame), will contain a large nonradial component. Several approaches have been developed to account properly for the kinematics; notably, the vectorial model of *Festou* [83.52], and the average random walk model (a Monte Carlo method) of *Combi* and *Delsemme* [83.53]. The latter model has been extended to include time-dependent gas kinetics so as to properly account for regions of the coma where the gas is not in local thermodynamic equilibrium [83.54].

In addition to the photodestruction chains, chemical reactions, particularly ion-molecule reactions, can alter the composition within the collision zone defined in Sect. 83.2.2. While such reactions may produce numerous minor species, they do not erase the signatures of the original parent molecules. In fact, detailed chemical models have clearly demonstrated the need for complex molecules to serve as the parents of the observed C_2 and C_3 radicals in the coma, strengthening the connection between comet formation and molecular cloud abundances. Thus, the photochemical chains provide a valid means of relating the coma composition to that of the nucleus.

83.3.3 Radiative Transfer Effects

The results of a model calculation for the density of a species must be integrated over the line of sight to obtain the column density at a given projected distance from the nucleus, and then integrated over the instrumental field of view for comparison with the observed average surface brightness or derived average column density $\overline{N_i}$. This assumes that the coma is optically thin, and that all atoms or molecules have an equal probability of absorbing a solar photon. In practice, this is true for all molecular emissions except perhaps within 1000 km of the nucleus (i. e., for observations made at better than 1″ resolution). Since the cross sections at line center can be very large for an atomic resonance transition, the optical depth for the abundant species can exceed unity and radiative transfer along both the line of sight to the sun and that to the Earth must be considered. This is not a trivial problem as the velocity distribution of the atoms, particularly the component due to the excess energy of the photodestruction process, must be well known, as must be the shape of the exciting solar line. The most thoroughly studied case to date is that of H I Ly-α, whose angular extent, in direct images, can exceed several degrees on the sky [83.55].

An interesting case arises for resonance transitions between an excited $^3S_1$ state and the ground $^3P_{2,1,0}$ state, as for O and S, particularly the latter, as its concentration near the nucleus can be quite large due to the rapid decay of one of its parents, CS_2. For S I the three lines at 1807, 1820 and 1826 Å are not observed to have their statistical intensity ratio of 5:3:1, except at large distances from the nucleus. This is explained by noting that fine structure transitions will lead to all of the S atoms reaching the $J = 2$ ground state in a time short compared with that for absorbing a solar photon, and that the emitted 1807 Å photons will be re-absorbed and can then branch into the other two lines. The detailed solution to this problem has led to the conclusion that H_2S was the primary source of sulfur rather than CS_2, whose other product, CS, was simultaneously observed in the UV [83.56]. Millimeter and sub-millimeter observations of comet Hale-Bopp (C/1995 O1) subsequently showed that SO, SO_2, and OCS were also minor sources of atomic sulfur, comparable in abundance to CS_2 [83.57]. Another minor source is S_2, initially observed in only one comet, IRAS-Araki-Alcock (C/1983 H1) [83.58], but recently seen in three additional comets by HST. The origin of S_2 in the cometary nucleus remains a puzzle.

83.4 Summary

This brief chapter can only hint at the wealth of observational data spanning the entire electromagnetic spectrum now routinely acquired at almost every comet apparition allowing for a statistically significant assessment of comet diversity and formation scenarios. Reference [83.59] will bring the interested reader up to date on all aspects of comet science. The next few years will see several spacecraft missions to comets, *Stardust*, *Deep Impact* and *Rosetta*, whose primary objective is the study of the cometary nucleus whose properties can only be inferred from remote observations. Nevertheless, Earth-based observations of comets will continue to play an important role in understanding the physical and chemical environments of these objects left over from the formation of the Solar System. There are still significant challenges in understanding the atomic and molecular physics of the cometary atmosphere, an example being the identification of the large number of unidentified lines seen in high resolution spectra in both the far UV [83.22, 23] and visible [83.60] regions of the spectrum.

References

83.1 G. Herzberg: *The Spectra and Structures of Simple Free Radicals: An Introduction to Molecular Spectroscopy* (Dover, New York 1971)

83.2 M. C. Festou, H. Rickman, R. M. West: Astron. Astrophys. Rev. **4**, 363 (1993)

83.3 M. C. Festou, H. Rickman, R. M. West: Astron. Astrophys. Rev. **5**, 37 (1993)

83.4 D. Krankowsky: *Comets in the Post-Halley Era*, ed. by R. L. Newburn, Jr. M. Neugebauer, J. Rahe (Kluwer, Dordrecht 1989) p. 855

83.5 L. Wilkening (Ed.) L. *Comets* (Univ. Arizona Press, Tucson 1982)

83.6 M. F. A Hearn, M. C. Festou: *Physics and Chemistry of Comets*, ed. by W. F. Huebner (Springer, New York 1990) p. 69

83.7 D. A. Mendis, H. L. F. Houpis, M. L. Marconi: Fund. Cosmic Phys. **10**, 1 (1985)

83.8 C. Arpigny: Ann. Rev. Astron. Astrophys. **3**, 351 (1965)

83.9 G. W. Preston: Astrophys. J. **147**, 718 (1967)

83.10 C. D. Slaughter: Astron. J. **74**, 929 (1969)

83.11 P. D. Feldman, W. H. Brune: Astrophys. J. Lett. **209**, L45 (1976)

83.12 P. Eberhardt, D. Krankowsky, W. Schulte, U. Dolder, P. Lämmerzahl, J. J. Berthelier, J. Woweries, U. Stubbemann, R. R. Hodges, J. H. Hoffman, J. M. Illiano: Astron. Astrophys. **187**, 484 (1987)

83.13 M. A. DiSanti, M. J. Mumma, N. D. Russo, K. Magee-Sauer: Icarus **153**, 361 (2001)

83.14 M. J. Mumma, H. A. Weaver, H. P. Larson, D. S. Davis, M. Williams: Science **232**, 1523 (1986)

83.15 H. P. Larson, H. A. Weaver, M. J. Mumma, S. Drapatz: Astrophys. J. **338**, 1106 (1989)

83.16 D. Bockelée-Morvan, J. Crovisier, M. J. Mumma, H. A. Weaver: *Comets II*, ed. by M. C. Festou, H. A. Weaver, H. U. Keller (Univ. Arizona Press, Tucson 2004)

83.17 R. Meier, T. C. Owen, H. E. Matthews, D. C. Jewitt, D. Bockelée-Morvan, N. Biver, J. Crovisier, D. Gautier: Science **279**, 842 (1998)

83.18 E. K. Jessberger, J. Kissel: *Comets in the Post-Halley Era*, ed. by R. L. Newburn, Jr. M. Neugebauer, J. Rahe (Kluwer, Dordrecht 1989) p. 1075

83.19 M. C. Festou, P. D. Feldman: *Exploring the Universe with the IUE Satellite*, ed. by Y. Kondo (Reidel, Dordrecht 1987) p. 101

83.20 M. C. Festou: *International Ultraviolet Explorer – Uniform Low Dispersion Data Archive Access Guide: Comets*, ESA SP-1134 (ESA, Noordwijk 1990)

83.21 H. A. Weaver, P. D. Feldman, J. B. McPhate, M. F. A'-Hearn, C. Arpigny, T. E. Smith: Astrophys. J. **422**, 374 (1994)

83.22 P. D. Feldman, H. A. Weaver, E. B. Burgh: Astrophys. J. Lett. **576**, L91 (2002)

83.23 H. A. Weaver, P. D. Feldman, M. R. Combi, V. Krasnopolsky, C. M. Lisse, D. E. Shemansky: Astrophys. J. Lett. **576**, L95 (2002)

83.24 H. A. Weaver, P. D. Feldman: *Science with the Hubble Space Telescope*, ed. by P. Benvenuti, E. Schreier (ESO, Garching 1992) p. 475

83.25 J. T. T. Mäkinen, J.-L. Bertaux, T. I. Pulkkinen, W. Schmidt, E. Kyrölä, T. Summanen, E. Quémerais, R. Lallement: Astron. Astrophys. **368**, 292 (2001)

83.26 M. S. Povich, J. C. Raymond, G. H. Jones, M. Uzzo, Y.-K. Ko, P. D. Feldman, P. L. Smith, B. G. Marsden, T. N. Woods: Science **302**, 1949 (2003)

83.27 D. A. Neufeld et al.: Astrophys. J. Lett. **539**, L151 (2000)

83.28 A. Lecacheux et al.: Astron. Astrophys. **402**, L55 (2003)

83.29 C. M. Lisse, and 11 colleagues: Science **274**, 205 (1996)

83.30 T. E. Cravens: Science **296**, 1042 (2002)

83.31 J. B. Greenwood, I. D. Williams, S. J. Smith, A. Chutjian: Phys. Rev. A **63**, 62707 (2001)

83.32 V. Kharchenko, A. Dalgarno: Astrophys. J. Lett. **554**, L99 (2001)

83.33 P. Swings: Lick. Obs. Bull. **XIX**, 131 (1941)

83.34 D. G. Schleicher, M. F. A'Hearn: Astrophys. J. **331**, 1058 (1988)

83.35 P. D. Feldman, C. B. Opal, R. R. Meier, K. R. Nicolas: *The Study of Comets*, NASA SP-393, ed. by B. Donn, M. Mumma, W. Jackson, M. A'Hearn, R. Harrington (NASA, Washington 1976) p. 773

83.36 J. L. Greenstein: Astrophys. J. **128**, 106 (1958)

83.37 K. F. Dymond, P. D. Feldman, T. N. Woods: Astrophys. J. **338**, 1115 (1989)

83.38 I. S. Bowen: Publ. Astron. Soc. Pacific **59**, 196 (1947)

83.39 P. D. Feldman, W. G. Fastie: Astrophys. J. Lett. **185**, L101 (1973)

83.40 V. A. Krasnopolsky, M. J. Mumma, M. Abbott, B. C. Flynn, K. J. Meech, D. K. Yeomans, P. D. Feldman, C. B. Cosmovici: Science **277**, 1488 (1997)

83.41 T. E. Cravens, A. E. S. Green: Icarus **33**, 612 (1978)

83.42 G. P. Tozzi, P. D. Feldman, M. C. Festou: Astron. Astrophys. **330**, 753 (1998)

83.43 R. J. Oliversen, N. Doane, F. Scherb, W. M. Harris, J. P. Morgenthaler: Astrophys. J. **581**, 770 (2002)

83.44 J.-L. Bertaux: Astron. Astrophys. **160**, L7 (1986)

83.45 M. J. Mumma, and 18 colleagues: Astrophys. J. **546**, 1183 (2001)

83.46 E. Gérard: Astron. Astrophys. **230**, 489 (1990)

83.47 H. Okabe: *The Photochemistry of Small Molecules* (Wiley, New York 1978)

83.48 W. F. Huebner, J. J. Keady, S. P. Lyon: *Solar Photo Rates for Planetary Atmospheres and Atmospheric Pollutants* (Springer, New York 1992)

83.49 S. A. Budzien, M. C. Festou, P. D. Feldman: Icarus **107**, 164 (1994)

83.50 A. H. Delsemme: *Comets*, ed. by L. L. Wilkening (Univ. Arizona Press, Tucson 1982) p. 85
83.51 L. Haser: Bull. Acad. R. Sci. Liège **43**, 740 (1957)
83.52 M. C. Festou: Astron. Astrophys. **95**, 69 (1981)
83.53 M. R. Combi, A. H. Delsemme: Astrophys. J. **237**, 633 (1980)
83.54 M. R. Combi: Icarus **123**, 207 (1996)
83.55 K. Richter, M. R. Combi, H. U. Keller, R. R. Meier: Astrophys. J. **531**, 599 (2000)
83.56 R. Meier, M. F. A'Hearn: Icarus **125**, 164 (1997)
83.57 D. Bockelée-Morvan, D. C. Lis, J. E. Wink, D. Despois, J. Crovisier, R. Bachiller, D. J. Benford, N. Biver, P. Colom, J. K. Davies, E. Gérard, B. Germain, M. Houde, D. Mehringer, R. Moreno, G. Paubert, T. G. Phillips, H. Rauer: Astron. Astrophys. **353**, 1101 (2000)
83.58 M. F. A'Hearn, D. G. Schleicher, P. D. Feldman: Astrophys. J. Lett. **274**, L99 (1983)
83.59 M. C. Festou, H. A. Weaver, H. U. Keller (Eds.): *Comets II* (Univ. Arizona Press, Tucson 2004)
83.60 A. L. Cochran, W. D. Cochran: Icarus **157**, 297 (2002)

84. Aeronomy

We describe here the neutral and ionic structures of atmospheres, including the processes that determine the atmospheric layers, the distribution of the species, and the temperature profiles. We focus on the upper atmosphere, which comprises the thermosphere and the ionosphere, two regions which overlay and interact with each other. We describe the interaction of near and extreme ultraviolet solar photons and energetic electrons with the atmosphere and their role in ionization and dissociation of atmospheric species. We also review the production and loss processes that are important in the formation of the different layers of the dayside and nightside ionospheres, including ion and neutral diffusion. The processes that determine the neutral, ion and electron temperatures are discussed. We review the processes that are important in production of the luminosity of the upper atmospheres, including dayglow, nightglow and auroras. Finally, we describe atmospheric escape processes, including thermal and non-thermal mechanisms.

84.1 Basic Structure of Atmospheres

84.1.1 Introduction

In a stationary atmosphere, the force of gravity is balanced by the plasma pressure gradient force in the vertical direction, and the variation of pressure $P(z)$ with altitude above the surface z is governed by the hydrostatic relation

$$\frac{\mathrm{d}P(z)}{\mathrm{d}z} = -\rho(z)g(z)\,, \tag{84.1}$$

where $\rho(z) = n(z)m(z)$ is the mass density, $n(z)$ is the number density, and $m(z)$ is the average mass of the atmospheric constituents. In general, variables such as P, ρ, g, n and even m are functions of altitude, although it will often not be shown explicitly in the equations that follow for the sake of compactness. The acceleration of gravity g is usually taken to be the vector sum of the gravitational attraction per unit mass and the centrifugal acceleration due to the rotation of the planet:

$$g(r) = GM/r^2 - \omega^2 r \cos^2\phi\,, \tag{84.2}$$

where $r = r_0 + z$ is the distance from the center of the planet, r_0 is the planetary radius, M is the planetary mass, $G = 6.670\times10^{-8}\,\mathrm{dyn\,cm^2\,g^{-2}}$ is the gravitational constant, ϕ is the latitude, and ω is the angular velocity of the planet.

When the hydrostatic relation (84.1) is combined with the ideal gas law in the form

$$P = nk_\mathrm{B}T\,, \tag{84.3}$$

where k_B is Boltzmann's constant and T is the temperature, and integrated, the the barometric formula

$$P(z) = P_0 \exp\left(-\int_{z_0}^{z} \frac{1}{H(z')}\,\mathrm{d}z'\right)\,, \tag{84.4}$$

for the pressure $P(z)$ above a reference level (denoted by the subscript 0) as a function of altitude results. The pressure scale height $H(z)$ is defined as

$$H(z) = \frac{k_B T}{mg} . \quad (84.5)$$

In the lower and middle atmosphere, the mass m in (84.5) is the weighted average mass of the atmospheric constituents.

When the ideal gas law (84.3) is substituted into the barometric formula (84.4), the altitude distribution

$$n(z) = n_0 \frac{T_0}{T(z)} \exp\left(- \int_{z_0}^{z} \frac{1}{H(z')} \mathrm{d}z' \right) , \quad (84.6)$$

for the number density $n(z)$ above a reference altitude is obtained. Integration of (84.1) or (84.6) from z to infinity shows that the column density above that altitude is approximately $N(z) = n(z)H(z)$. Thus the scale height can be thought of as the effective thickness of the atmosphere.

In its lower and middle regions, the homosphere, the atmosphere is well-mixed by convection and/or turbulence. The upper boundary of this region is called the homopause (or turbopause), and above this level, the major transport process is diffusion. The homopause is defined as the level at which the time constants for mixing and diffusion are equal, and usually occurs at $n(z) \sim 10^{11}–10^{13}\,\mathrm{cm}^{-3}$, depending on the strength of vertical mixing for a given planet. Since molecular diffusion coefficients vary from one species to another, the exact altitude of the homopause is species-dependent, with smaller species having lower homopause altitudes. In the terrestrial atmosphere, the homopause is near 100 km at $n(z) \sim 10^{13}\,\mathrm{cm}^{-3}$. Below the homopause, the mixing ratios (or fractions by number) of the constituent gases, apart from those minor or trace species whose density profiles are determined by photochemistry or physical loss processes, are fairly constant with altitude.

Throughout the atmosphere, gravity exerts a force on each particle that is proportional to its mass. Below the homopause, however, the tendency of the species to separate out under the force of gravity is overpowered by large scale mixing processes, such as convection and turbulence. Above the homopause, each species is distributed according to its own scale height. Characteristics of the homopauses of the planets are presented in Table 84.1.

Table 84.1 Homopause characteristics of planets and satellites

Planet	Altitude (km)	K (cm^2/s)	T (K)	n_t (cm^{-3})	P (μbar)	Composition (Fraction by number)
Venus[b]	135	4(7)[a]	199	1.4(11)	4.7(-3)	CO_2(76), N_2(7.6), O(9.3), CO(6.7), N(0.16), C(0.01)
Earth[c]	100	1(6)	185	1.3(13)	0.3	N_2(77%), O_2(18%), O(3–4%), Ar(0.7%), He(9.5 ppm), H(1.3 ppm)
Mars[d]	120–125	5(7)	154	1.3(11)	2.8(-3)	CO_2(95%), N_2(2.5%), Ar(1.5%), O(0.9%), CO(0.42%) O_2(0.12%), NO(0.007%)
Jupiter[e]	500[f]	2(6)	600	1.4(13)	0.4	H_2(95%), He(4.1%), H(0.055%), CH_4(200 ppb), C_2H_2(1.2 ppb),C_2H_4(2.5 ppb), C_2H_6(0.12 ppb)
Saturn[g]	1100[f]	1.3(8)	200	1.2(11)	3(-3)	H_2(94%), He(6%), CH_4(178 ppm)
Uranus[h]	300[f]	1(4)	130	1(15)	20	H_2(85%), He(15%), CH_4(20 ppm), C_2H_2(10 ppb), C_2H_6(0.1 ppb), C_4H_2(0.05 ppt)
Neptune[i]	750[f]	2(7)	280	1.2(12)	3.9(-2)	H_2(83%), He(16%), CH_4(3 ppm)
Titan[j]	1100	1(9)	155	3(9)	6.4(-5)	N_2(96%), CH_4(2.4%), C_2H_2(0.07%), C_2H_4(0.7%), C_2H_6(404 ppm), C_4H_2(105 ppm)
Triton[k]	30	3(3)	40	3.3(14)	1.8	N_2(99.9%), H_2(190 ppm), CH_4(37 ppm), H(3.1 ppm), N(4.1 ppb)

[a] Read as 4×10^7
[b] K from *von Zahn* et al. [84.1] and model atmosphere from *Hedin* et al. [84.3], for 1500 h, 15° N latitude, $F_{10.7} = 150$
[c] From The US Standard Atmosphere [84.6]
[d] Viking model from [84.8]
[e] From [84.13] and photochemical model of *Kim* [84.14].
[f] Altitude above the 1 bar level.
[g] From [84.2]
[h] From [84.4, 5]
[i] From [84.7]
[j] From [84.9–12]
[k] From [84.15, 16]

84.1.2 Atmospheric Regions

The division of atmospheres into regions is based on the temperature structure of the terrestrial atmosphere, which is shown in Fig. 84.1. In the troposphere of a planet, above the boundary layer, T decreases at close to the adiabatic lapse rate (Γ) for the constituent gases from the surface to the tropopause. For an atmosphere that is a mixture of ideal gases, $\Gamma = g/c_\mathrm{p}$, where c_p is the specific heat of the gas mixture at constant pressure. The presence of a condensible constituent, such as water vapor in the terrestrial troposphere, and ammonia or methane in the atmospheres of the outer planets and satellites, decreases Γ because upward motion leads to cooling and condensation, which releases latent heat. On Earth, the dry adiabatic lapse rate is about 10 K/km and the moist adiabatic lapse rate is about 4–6 K/km in the lower to middle troposphere. The average lapse rate is about 6.5 K/km, and the altitude of the tropopause varies from about 9 to 16 km from the poles to the equator. The composition of the lower atmosphere of the Earth is given in Table 84.2.

Above the terrestrial tropopause lies the stratosphere, a region of increasing T that is terminated at the stratopause, near 50 km. This increase in T is caused by absorption of solar near UV radiation by ozone in the Hartley bands and continuum (200–310 nm). In the terrestrial mesosphere, which lies above the stratosphere, T decreases again to an absolute minimum at the mesopause, where $t \approx 180$ K and $n(z) \approx 10^{14}\,\mathrm{cm}^{-3}$. Above the mesopause, in the thermosphere, T increases rapidly to a constant value, the exospheric temperature, T_∞. The value of T_∞ in the terrestrial atmosphere depends on solar activity and is usually between about 700 and 1500 K. Fig. 84.1 also shows altitude profiles of the noon and midnight thermospheric temperature for four values of the $F_{10.7}$ index, (the 2800 MHz flux in units of $10^{-22}\,\mathrm{Wm^{-2}Hz^{-1}}$ at 1 AU), which represent different levels of solar activity.

The exosphere is a nearly collisionless region of the thermosphere that is bounded from below by the exobase. A particle traveling upward at or above the exobase will, with high probability, escape from the gravitational field of the planet. The exobase on Earth is located at about 450–500 km, depending upon solar activity. The surface P and T on Mars are about 6 mbar

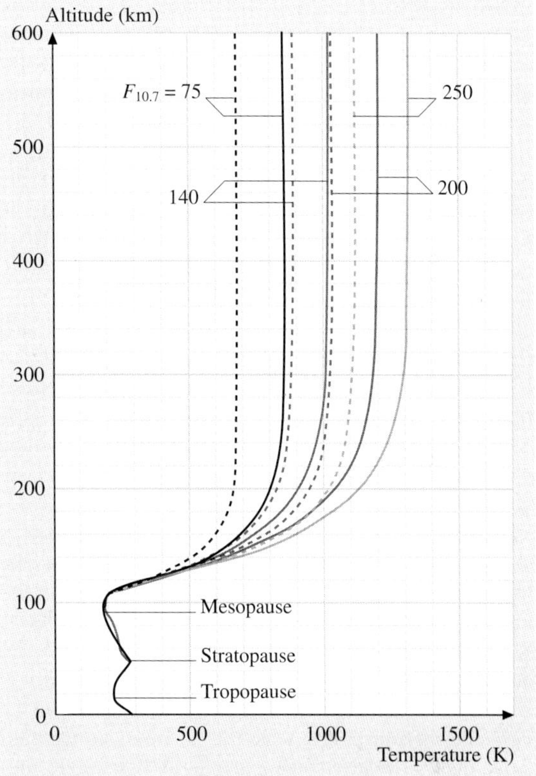

Fig. 84.1 Vertical distribution of temperature in the terrestrial atmosphere. The altitudes of the tropopause, stratopause and mesopause are indicated. The thermospheric temperatures depend on solar activity and profiles are shown for four values of the $F_{10.7}$ index, from 75 (low solar activity) to 250 (high solar activity). The *solid* and *dashed curves* are for noon and midnight, respectively. After the MSIS model of *Hedin* [84.17]

Table 84.2 Molecular weights and fractional composition of dry air in the terrestrial atmosphere[a]

Species	Molecular Weight (g/mole)	Fraction by volume
N_2	28.0134	0.780 84
O_2	31.9988	0.209 476
Ar	39.948	0.009 34
CO_2	44.009 95	0.000 3756[b]
Ne	20.183	0.000 018 18
He	4.0026	0.000 005 24
Kr	83.80	0.000 001 14
Xe	131.30	0.000 000 087
CH_4	16.043 03	0.000 002
H_2	2.015 94	0.000 0005

[a] Taken from The US Standard Atmosphere [84.6], except as noted

[b] 2003 annual average value. The CO_2 mixing ratio is increasing at an annual rate of about 0.45%. Value is from [84.18]

Part G | 84.1

and 230 K, respectively. Due to the effect of dust storms, the extent of the Martian troposphere is highly variable, with a lapse rate that is 2–3 K/km compared with the adiabatic lapse rate of 4.5 K/km and a variable thickness of 20–50 km [84.19]. The atmosphere of Mars, like many planetary atmospheres, does not have a stratosphere. A roughly isothermal mesosphere extends from the tropopause to the base of the thermosphere at about 90 km. The thermospheric T is sensitive to solar activity and, since Mars has a very eccentric orbit, to heliocentric distance; T_∞ varies from about 180 to 350 K.

Near the surface of Venus, $T \gtrsim 700$ K and $P \sim 95$ bar. T decreases with a mean lapse rate of 7.7 K/km, compared with the adiabatic lapse rate of 8.9 K/km, from the surface to about 50 km. The region from 50 to 60 km contains the major cloud layer, and the tropopause is usually considered to be at about 60 km. In the mesosphere, between about 60 and 85 km, T decreases slowly from about 250 K to 180 K, and is nearly constant from 85 km to the mesopause at 100 km. The daytime exospheric temperature is only weakly dependent on solar activity, varying from about 230 to 300 K from low to high solar activity. The slow retrograde rotation of the planet, which results in a period of darkness that lasts 58 days, leads to the relative isolation of the nightside thermosphere, where T is found to decrease above the mesopause to an exospheric $T \approx 100$ K. Because of this, the nightside Venus thermosphere has been called the "cryosphere". The compositions of the lower atmospheres of Mars and Venus are given in Table 84.3.

The giant planets, Jupiter, Saturn, Uranus and Neptune do not have solid surfaces, so their atmospheric regions are defined either in terms of pressure, altitude above the 1 bar level, or altitude above the cloud tops. The temperature structures of all but Uranus are influenced by internal heat sources that the terrestrial planets do not possess. The temperature structures near the tropopause can be determined from IR observations and radio occultation data, and at thermospheric altitudes from ultraviolet solar and stellar occultations performed by the Voyager spacecraft. In between these regions, there is a substantial gap in which only average temperatures can be inferred. Thus the location and temperature of the mesopauses are largely unknown.

Below 300 mbar on Jupiter, the lapse rate is close to adiabatic (1.9 K/km). T at 1 bar is about 165 K, and the tropopause occurs near 140 mbar, where $T \approx 110$ K. At 1 mbar, T again reaches 160–170 K. Temperature inversions have been reported in the stratosphere, and are probably due to absorption of solar radiation by dust or aerosols. Temperatures derived from the Voyager UV stellar and solar occultations show that T increases from about 200 K near 1 μbar to an exospheric value of about 1100 K [84.2].

For $P > 500$ mbar on Saturn, the lapse rate approaches the adiabatic value of 0.9 K/km, and the tropopause, near the 100 mbar level, is characterized by $T \approx 80$ K. Above the tropopause the temperature increases to about 140 K near a $P \approx 1$ mbar, and above that altitude there are no measurements of T up to a pressure of about 10^{-8} bar, about 1000 km above the 1 bar level, where T is again about 140 K. Application of the hydrostatic equation to the altitude range 300–1000 km yields an average temperature near 140 K for the region. Above 1000 km, T increases to a $T_\infty \approx 800$ K [84.2]. The mixing ratios of the species in the lower atmospheres of Jupiter and Saturn are given in Table 84.4.

The tropopauses on both Uranus and Neptune occur near 100 mbar, where $T \approx 50$ K. The lapse rates in the troposphere are 0.7 and 0.85 K/km for Uranus and Neptune, respectively. The temperatures in the Uranus thermosphere range from 500 K near 10^{-7} bar (about 1000 km above the 1 bar level) to an exospheric value of about 800 K. At 300 km on Neptune, T attains a nearly constant value in the range 150 to 180 K. At 600 km where $P \approx 1$ μbar, T increases again to a value that is

Table 84.3 Composition of the lower atmospheres of Mars and Venus[a]

Species	Mixing Ratio Mars	Venus
CO_2	0.953	0.96
N_2	0.027	0.04
^{40}Ar	0.016	50–120 ppm[b]
O_2	0.0013	20–40 ppm
CO	0.0008	20–30 ppm
H_2O	0.0003[c]	30 ppm[d]
He	4 ppm[e]	10 ppm
Ne	2.5 ppm	5–13 ppm
Kr	0.3 ppm	0.02–0.4 ppm
Xe	0.08 ppm	–
SO_2	–	150 ppm
H_2S	–	1–3 ppm
H_2	15 ppm[f]	0.1 ppm[g]

[a] From [84.20], except as noted.
[b] Includes all isotopes of Ar.
[c] Variable
[d] From [84.21].
[e] From *Krasnopolsky* and *Gladstone*. [84.22]
[f] *Krasnopolsky* and *Feldman* [84.23]
[g] *Yung* and *DeMore* [84.24]

Table 84.4 Composition of the lower atmospheres of Jupiter and Saturn

Species	Mixing Ratio Jupiter	Saturn
H_2	0.864[b]	0.94[a]
He	0.136[b]	0.06[a]
CH_4	0.001 81[b]	0.0045[a]
NH_3	< 0.002[b]	(0.5–0.2 ppm[c])
H_2O	520 ppm[b]	
C_2H_6	5 ppm[a]	7.0 ppm[c]
PH_3	0.6 ppm[a]	1.4 ppm[c]
C_2H_2	0.02 ppm[a]	0.3 ppm[c]
^{20}Ne	≤ 26 ppm[b]	
^{36}Ar	≤ 9 ppm[b]	
^{84}Kr	≤ 3.2 ppb[b]	
^{132}Xe	≤ 0.38 ppb[b]	

[a] After *Strobel* [84.25]
[b] After *Niemann* et al. [84.26]
[c] After *Lodders* and *Fegley* [84.27]

probably about 600 K. Because the Voyager data have not been fully analyzed, the value of T_∞ is uncertain [84.12]. The compositions of the lower atmospheres of Uranus and Neptune are given in Table 84.5.

Titan, which is a satellite of Saturn, has an N_2/CH_4 atmosphere of intermediate oxidation state. The mixing ratios of components of the lower atmosphere are given in Table 84.6. The surface P and T are 1.496 bar and 94 K, respectively. T decreases above the surface to about 71 K at the tropopause, which occurs at an altitude of 42 km and a pressure of 128 mbar. A reanalysis of the Voyager 1 solar occultation experiment showed that, above the tropopause, the temperature increases to a peak value of about 176 K at an altitude of about 300 km. The temperature then decreases to a T_∞ of 153–158 K [84.9]

Table 84.5 Composition of the lower atmospheres of Uranus and Neptune[a]

Species	Mixing Ratio Uranus	Neptune
H_2	≈ 0.825	≈ 0.80
He	≈ 0.152	≈ 0.19
CH_4	≈ 0.023	≈ 0.01–0.02
HD	≈ 148 ppm	≈ 192 ppm
CH_3D	≈ 8.3 ppm	≈ 12 ppm
C_2H_6	≈ 1–20 ppb	≈ 1.5 ppm
C_2H_2	≈ 10 ppb	≈ 60 ppb
CO	< 40 ppb	2.7 ± 1.8 ppm[b]
NH_3	< 100 ppb	< 600 ppb
H_2O	5–12 ppb	1.5–3.5 ppb

[a] After *Lodders* and *Fegley* [84.27], except as noted
[b] *Courtin* et al. [84.28]

Table 84.6 Composition of the lower atmosphere of Titan[a]

Species	Mixing Ratio
N_2	0.90–0.98
CH_4	0.01–0.03[b]
H_2	2.0×10^{-3}
CO	60–150 ppm
C_2H_6	20 ppm
C_3H_8	4 ppm
C_2H_2	2 ppm
C_2H_4	0.4 ppm
HCN	0.2 ppm

Titan's atmosphere may also contain up to 14% Ar [84.10]
[a] From [84.29], except as noted
[b] From [84.9]

Triton is a satellite of Neptune. It also has an N_2 atmosphere with small amounts of methane, CO, H_2, and other species. The mixing ratios at 10 km are given in Table 84.7. The surface P is about 14–19 μbar. Methane in the troposphere is in equilibrium with a surface methane frost at about 38–50 K. The tropopause temperature is about 36 K, and occurs in the 8 to 12 km region. The middle atmosphere is isothermal with a temperture of about 52 K from 25 to 50 km, increasing to 78 K near 150 km [84.30]. T rises to an T_∞ of about 100 K.

Io and Europa are satellites of Jupiter. Both have transient atmospheres, with mean lifetimes of 2–3 days. The radius of Io is about 1821 km, and its atmosphere is

Table 84.7 Composition of the atmosphere of Triton[a]

Species	Mixing Ratio	Comments
N_2	0.99 ± 0.01	Below ≈ 200 km
CO	0.0001–0.01	Uncertain
CH_4	113 ppm	
H_2	75 ppm	
N	3.8×10^{-5} ppm	
N	290 ppm	100 km (near peak)
H	0.092 ppm	
H	1 ppm	30 km (near peak)
C_2H_4	3.9×10^{-4} ppm	
C_2H_4	2.9×10^{-2} ppm	26 km (near peak)

[a] From [84.15]. Values are at 10 km, except as noted

Part G | 84.1

mostly SO_2, which is produced by volcanic plumes. One model predicts that the average column density of SO_2 is about $10^{16}\,cm^{-3}$, and is larger at the equator than at the poles [84.31]. The atmospheric temperatures range from 100 to 2000 K, and the exospheric temperature is about 1800 K. The altitude of the exobase is about 1400 km. A plot of the number density and temperature as a function of altitude is shown in Figure 80–9 k. Europa is characterized by a radius of 1596 km. The atmosphere is mostly O_2 with column density of $5\times10^{14}\,cm^{-2}$ and a scale height of 145 km. The O_2 is produced by sputtering of the ice-covered surface, and is removed in sputtering by torus thermal ions [84.32]. The ionosphere is produced by impact of electrons in Jupiter's magnetosphere, and the maximum density of electrons is about $4\times10^4\,cm^{-3}$ [84.33].

Mercury does not have a troposphere, mesosphere, or stratosphere; The pressure at the surface is on the order of a picobar; thus the surface of the planet is the exobase. Nevertheless, several atomic species have been identified in fluorescence. They are listed in Table 84.8. Among the possible sources of atmospheric species are evaporation, ion sputtering, meteoroid bombardment, and photon-stimulated desorption. Ions produced by photoionization of neutrals may be picked up by the solar wind and lost from the atmosphere.

Pluto and its satellite Charon form what is sometimes referred to as a double planet system. The radius of Pluto is 1150–1200 km, and that of Charon is about 600 km. The atmosphere of Pluto is mostly N_2, with small amounts of methane, CO, H_2, and H. Only upper limits are available for the mixing ratio of CO. The surface pressure and temperature are in the ranges 1 to 10 μbar and 35 to 57 K, respectively. Although the pressure at the surface is approximately the same as that of the base of the thermosphere on most planets, the thermal structure of the atmosphere is influenced by the large thermal escape flux at the top of the atmosphere and by adiabatic cooling. T maximizes near 1200–1260 km radius at about 100 K due to absorption of solar UV radiation. Above that radius, T decreases asymptotically to a value of about 80 K [84.35, 36].

Table 84.8 Number densities of species at the surface of Mercury[a]

Species	Number density (cm^{-3})
H	230
hot H	23
He	6×10^3
O	$<4.4\times10^4$
Na[b]	$(1.7–3.8)\times10^4$
K[b]	5×10^2

[a] from *Hunten* et al. [84.34]
[b] Variable spatially and temporally

84.2 Density Distributions of Neutral Species

84.2.1 The Continuity Equation

The density distribution of a minor neutral species j in an atmosphere is determined by the continuity equation:

$$\frac{\partial n_j}{\partial t} + \nabla\cdot\boldsymbol{\Phi}_j = P_j - L_j\,, \tag{84.7}$$

where $\boldsymbol{\Phi}_j$ is the flux of species j, and P_j and L_j are the chemical production and loss rates, respectively. If only the vertical direction is considered, the divergence of the flux becomes $\partial\Phi_j/\partial z$, and $\Phi_j = n_j w_j$, where w_j is the vertical velocity of the species and n_j is its number density. In one-dimensional models, transport due to turbulence and other macroscopic motions of air masses is often parametrized like molecular diffusion, using an eddy diffusion coefficient K in place of the molecular diffusion coefficient D_j. The total transport velocity w_j is then the sum of the diffusion velocity w_j^{D} and the eddy diffusion velocity w_j^{K}:

$$w_j = w_j^{\mathrm{D}} + w_j^{\mathrm{K}}\,. \tag{84.8}$$

If there are no net flows of major constituents, w_j^{D} and w_j^{K} satisfy the equations

$$w_j^{\mathrm{D}} = -D_j\left(\frac{1}{n_j}\frac{\mathrm{d}n_j}{\mathrm{d}z} + \frac{1}{H_j} + \frac{(1+\alpha_j^T)}{T}\frac{\mathrm{d}T}{\mathrm{d}z}\right), \tag{84.9}$$

$$w_j^{\mathrm{K}} = -K\left(\frac{1}{n_j}\frac{\mathrm{d}n_j}{\mathrm{d}z} + \frac{1}{H_{\mathrm{avg}}} + \frac{1}{T}\frac{\mathrm{d}T}{\mathrm{d}z}\right). \tag{84.10}$$

In these expressions, α_j^{T} is the thermal diffusion factor (the ratio of the thermal diffusion coefficient to the molecular diffusion coefficient), and the pressure scale height H_{avg} for a mixed atmosphere is given by (84.5) with $m = m_{\mathrm{avg}}$, the average molecular mass.

For a stationary atmosphere, if molecular diffusion greatly exceeds eddy diffusion and if photochemistry can be neglected, then $w_j^{\mathrm{D}} = 0$. The resulting number density distribution is called diffusive equilibrium, and is given by

$$n_j(z) = n_j(z_0)\left(\frac{T_0}{T}\right)^{\left(1+\alpha_j^{\mathrm{T}}\right)} \exp\left(-\int_{z_0}^{z} \frac{\mathrm{d}z'}{H_i}\right) . \tag{84.11}$$

When mixing processes dominate and $w_j^{\mathrm{K}} = 0$, the distribution is given by (84.6), with $H = H_{\text{avg}}$.

84.2.2 Diffusion Coefficients

In the thermosphere of a planet, above the homopause, the major transport mechanism is diffusion, or transport by random molecular motions. The characteristic time τ_{D} for molecular diffusion is approximately H_j^2/D_j. The diffusion coefficient for a species j in a multicomponent mixture is usually taken as a weighted mean of binary diffusion coefficients D_{jk}

$$\frac{1}{D_j} = \sum_{k \neq j} \frac{f_k}{D_{jk}} , \tag{84.12}$$

where f_k is the mixing ratio of species k. The binary diffusion coefficient can be expressed as

$$D_{jk} = \frac{3k_{\mathrm{B}}T}{16 n_{\mathrm{t}} \mu_{jk} \Omega_{jk}} , \tag{84.13}$$

where μ_{jk} is the reduced mass

$$\mu_{jk} = \frac{m_j m_k}{m_j + m_k} \tag{84.14}$$

and $n_{\mathrm{t}} = n_j + n_k$ is the total number density. The collision integral Ω_{jk} is given by

$$\Omega_{jk} = \frac{1}{2\pi^{1/2}}\left(\frac{\mu}{2k_{\mathrm{B}}T}\right)^{5/2} \times \int_0^{\infty} Q^{\mathrm{D}}(v) v^5 \exp\left(-\mu v^2/2k_{\mathrm{B}}T\right) \mathrm{d}v \tag{84.15}$$

where v is the relative velocity of the particles, $Q^{\mathrm{D}}(v)$ is the diffusion or momentum transfer cross section

$$Q^{\mathrm{D}}(v) = 2\pi \int_0^{\pi} \sigma_{jk}^{\mathrm{el}}(\theta, v)(1-\cos\theta)\sin\theta\, \mathrm{d}\theta , \tag{84.16}$$

and $\sigma_{jk}^{\mathrm{el}}(\theta, v)$ is the differential cross section for elastic scattering of species j and k through angle θ. In practice, D_{jk} is often expressed as b_{jk}/n_{t} where n_{t} is the total number density and b_{jk} is the binary collision parameter, which is usually given in tabulations in the semi-empirical form $b = AT^s$. Here A and s $(0.5 \leq s \leq 1.0)$ are parameters that are fit to the data. The binary collision parameter appears, for example, in the expression for the diffusion limited flux of a light species to the exobase of a planet (Sect. 84.7).

84.3 Interaction of Solar Radiation with the Atmosphere

84.3.1 Introduction

The source for all atmospheric processes is ultimately the interaction of solar radiation, either photons or particles, with atmospheric gases. Since visible photons arise from the photosphere of the sun, which is characterized by $T \approx 6000$ K, the solar spectrum in the visible and IR is similar to that of a black body at 6000 K. At longer (radio) and shorter (UV and X-ray) wavelengths, the photons arise from parts of the chromosphere and corona where the temperatures are higher (10^4 to 10^6 K). Thus the photon fluxes differ substantially from those which would be predicted for a 6000 K black body. Photons in the extreme and far UV regions of the spectrum are absorbed in the terrestrial thermosphere and X-rays in the lower thermosphere and mesosphere. The solar Lyman α line at 1216 Å penetrates through a window in the O_2 absorption cross sections to about 75 km. Near UV photons are absorbed by ozone in the stratosphere, and visible radiation is not appreciably attenuated by the atmosphere.

The wavelength ranges that are most important for aeronomy are the UV and X-ray regions. A solar spectrum in the UV and soft X-ray regions at low solar activity is presented in Fig. 84.2a, and the ratio of a high solar activity photon fluxes to those at low solar activity is shown in Fig. 84.2b. The ratio is near unity at wavelengths longward of 2000 Å, but increases to factors that range between 2 and 3 over much of the extreme UV. At wavelengths between about 100 and

550 Å, the ratio of high to low solar activity fluxes reaches values as high as 100. The fluxes at X-ray wavelengths arise principally from solar flares and can increase by orders of magnitude from low to high solar activity.

The sun also emits a stream of charged particles, the solar wind, which flows radially outward in all directions, and consists mostly of protons, electrons, and alpha particles. The average number density of solar wind protons is about $5\,\mathrm{cm}^{-3}$, and the average speed is about 400–450 km/s at Earth orbit (1 AU). The interaction of these particles with the magnetic field (either induced or intrinsic) of a planet, and ultimately with the atmosphere, is the source of auroral activity. Terrestrial auroras arise mostly from precipitation of electrons with energies in the kilovolt range, although measured spectra vary widely. An example of a primary electron auroral spectrum is shown in Fig. 84.3. Terrestrial auroral emissions maximize in the midnight sector, but dayside cusp auroras are produced by lower energy electrons, and diffuse proton auroras are also observed.

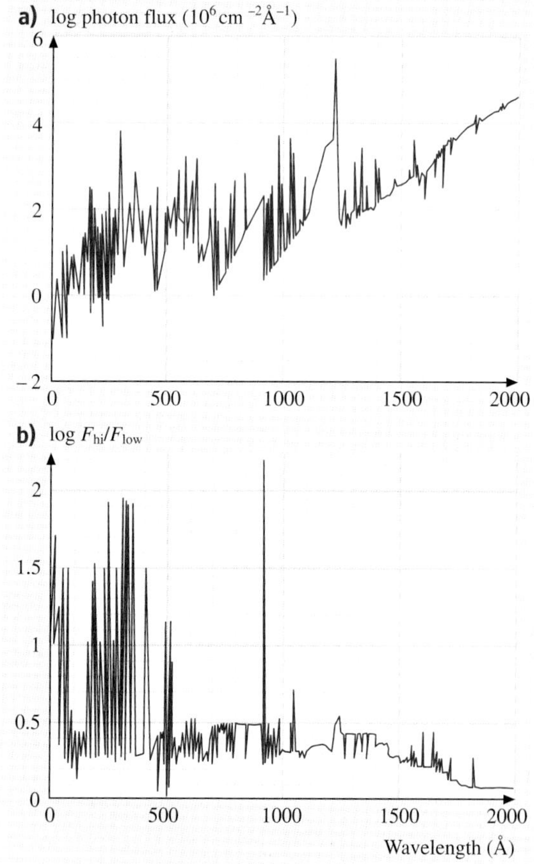

Fig. 84.2 **(a)** Solar spectrum at 1 AU for 18–2000 Å. **(b)** Ratio of the flux at high solar activity to low solar activity. Plotted with data from *Tobiska* [84.37]

Since charged particles are constrained to move along magnetic field lines, for planets with intrinsic magnetic fields, auroras usually occur in an oval near the magnetic poles, where the dipole field lines enter the atmosphere. For Venus, which has no intrinsic magnetic field, auroras are seen as diffuse and variable emissions on the nightside of the planet. On Earth, low latitude auroras, which arise from heavy particle precipitation, have also been observed. The primary particles that are responsible for Jovian aurora may be heavy ions originating from its satellite Io, protons, or electrons. Due to charge transfer, heavy particles spend part of their lifetime as neutral species, and their paths may then diverge from magnetic field lines. In any case, a large fraction of the effects of auroral precipitation is due to secondary electrons, regardless of the identity of the primary particles. In addition to producing emissions of atmospheric species in the visible, UV and IR portions of the spectrum, auroral particles ionize and dissociate atmospheric species and contribute to heating the neutrals, ions and electrons.

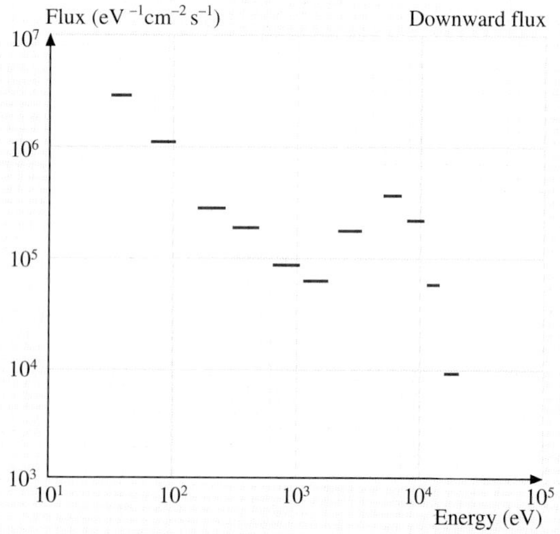

Fig. 84.3 Downward electron flux as a function of energy measured by electron spectrometers on board a rocket traversing an auroral arc near Poker Flat, Alaska. After [84.38] with kind permission from Elsevier Science Ltd., UK

84.3.2 The Interaction of Solar Photons with Atmospheric Gases

The number flux of solar photons in a small wavelength interval around λ at an altitude z can, for the most part, be computed from the Beer–Lambert absorption law

$$F_\lambda(z) = F_\lambda^\infty \exp[-\tau(\lambda, z)] , \quad (84.17)$$

where F_λ^∞ is the solar photon flux outside the atmosphere, and $\tau(\lambda, z)$ is the optical depth which, in the plane parallel approximation, is given by

$$\tau(\lambda, z) = \sum_j \int_z^\infty n_j(z')\sigma_j^{\mathrm{a}}(\lambda) \sec\chi \, \mathrm{d}z' . \quad (84.18)$$

Here, $\sigma_j^{\mathrm{a}}(\lambda)$ is the absorption cross section of species j at wavelength λ, and the solar zenith angle χ is the angle of the sun with respect to the local vertical.

For χ greater than about 75°, the variation of the solar zenith angle along the path of the radiation cannot be neglected; the optical depth must be computed by numerical integration along this path in spherical geometry. For $\chi \leq 90°$ the optical depth is

$$\tau(\lambda, z) = \sum_j \int_z^\infty n_j(z')\sigma_j^{\mathrm{a}}(\lambda) \times \left[1 - \left(\frac{r_\mathrm{o}+z}{r_\mathrm{o}+z'}\right)^2 \sin^2\chi\right]^{-0.5} \mathrm{d}z' . \quad (84.19)$$

For χ larger than 90°, the optical depth is given by

$$\tau(\lambda, z) = \sum_j \left\{ 2\int_{z_s}^\infty n_j(z')\sigma_j^{\mathrm{a}}(\lambda) \times \left[1 - \left(\frac{r_\mathrm{o}+z_s}{r_\mathrm{o}+z'}\right)^2 \sin^2 90°\right]^{-0.5} \mathrm{d}z' - \int_z^\infty n_j(z')\sigma_j^{\mathrm{a}}(\lambda) \times \left[1 - \left(\frac{r_\mathrm{o}+z}{r_\mathrm{o}+z'}\right)^2 \sin^2\chi\right]^{-0.5} \mathrm{d}z' \right\} , \quad (84.20)$$

where z_s is the tangent altitude, the point at which the solar zenith angle is 90° for the path of solar radiation through the atmosphere.

In a one-species atmosphere, the rate of absorption of solar photons of wavelength λ is

$$q^{\mathrm{a}}(\lambda) = F_\lambda \sigma^{\mathrm{a}}(\lambda) n . \quad (84.21)$$

For an isothermal atmosphere in which $H(z) \approx$ const., the absorption maximizes where $\tau(\lambda, z) = 1$. This is a fairly good approximation even for regions of the atmosphere where the $H(z)$ is not constant. The altitude of unit optical depth is shown for wavelengths from X-rays to the near UV for overhead sun in the terrestrial atmosphere in Fig. 84.4a. Similar plots for Venus, Mars and Jupiter are shown in Fig. 84.4b, Fig. 84.4c, and Fig. 84.4d, respectively. N_2 does not absorb longward of about 100 nm, so in the terrestrial atmosphere, O_2 and O_3 are the primary absorbers between 100 and 220 nm, while ozone dominates the absorption for wavelengths in the range 220–320 nm. On Venus and Mars, CO_2 is the main absorber of FUV and EUV radiation, although at wavelengths less than about 100 nm, N_2, CO, and O also contribute. On Titan, methane is the primary absorber of UV radiation between 1400 Å and the absorption threshold of N_2, near 1000 Å.

The interaction of UV photons with atmospheric gases produces ions and photoelectrons through photoionization, which may be represented as

$$X + h\nu \to X^+ + \mathrm{e}^- , \quad (84.22)$$

and photodissociative ionization

$$AB + h\nu \to A^+ + B + \mathrm{e}^- . \quad (84.23)$$

In these equations, X represents any atmospheric species; A is either an atom or a molecular fragment and AB a molecule. The energy of the photoelectron in reaction (84.22) is given by

$$E_{\mathrm{pe}} = h\nu - I_X - E_{\mathrm{ex}} \quad (84.24)$$

and in reaction (84.23) is

$$E_{\mathrm{pe}} = h\nu - E_\mathrm{d} - I_A - E_{\mathrm{ex}} , \quad (84.25)$$

where I_j is the ionization potential of species j, E_d is the dissociation energy of molecule AB, and E_{ex} is the internal excitation energy of the products. Neutral fragments, which may be reactive radicals, are also produced in photodissociation

$$AB + h\nu \to A + B . \quad (84.26)$$

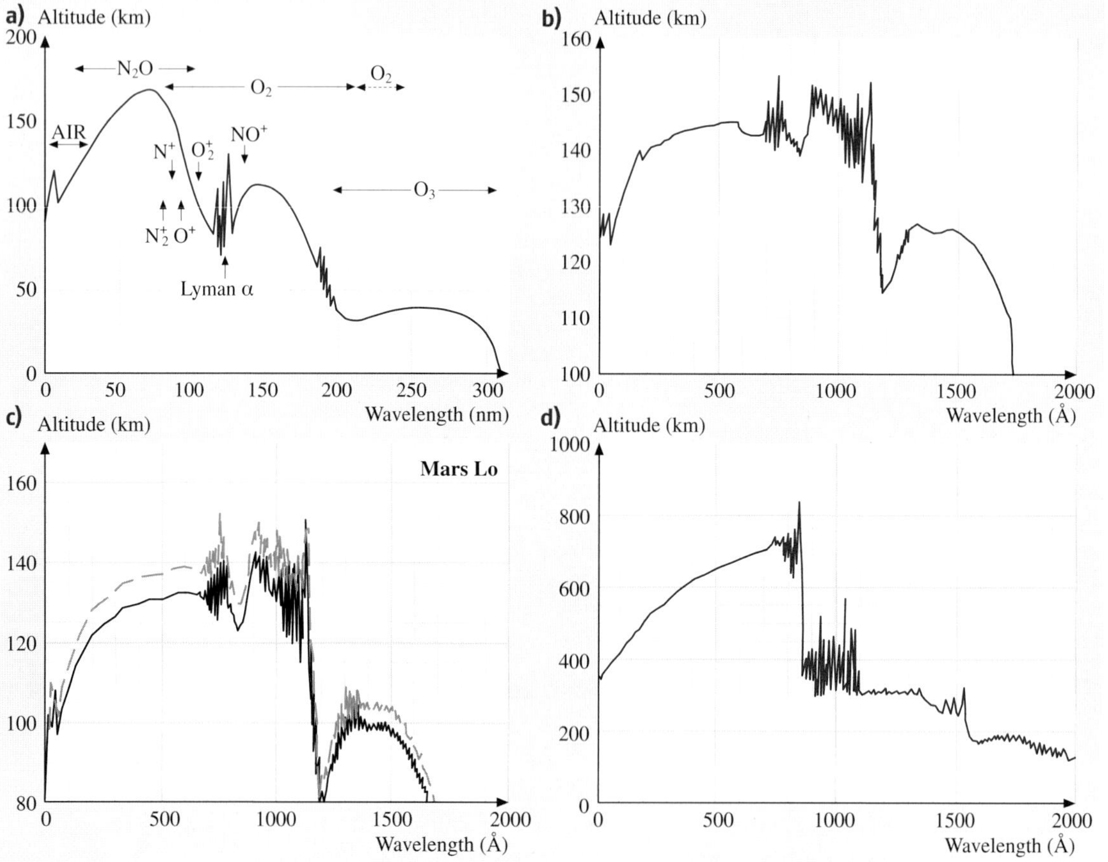

Fig. 84.4a–d The altitude where $\tau = 1$ versus wavelength **(a)** Earth [84.39] **(b)** Venus **(c)** Mars [84.40] **(d)** Jupiter (Y. H. Kim, unpublished)

The rate of ionization of a species j by a photon of wavelength λ at an altitude z is given by

$$q_j^{\mathrm{i}}(\lambda, z) = F_\lambda(z)\sigma_j^{\mathrm{i}}(\lambda)n_j(z) , \tag{84.27}$$

where $\sigma_j^{\mathrm{i}}(\lambda)$ is the photoionization cross section. The rate for photodissociation is given by a similar expression, with the photoionization cross section replaced by the photodissociation cross section. The expression above must be integrated over the solar spectrum to give the total rate. In addition, it is often necessary to take into account ionization and/or dissociation to different final internal states of the products, so the partial cross sections or yields are needed.

In the atmospheres of magnetic planets, photoelectrons may travel upward along the magnetic field lines to the conjugate point, where the field line re-enters the atmosphere. In order to model this effect, the differential (with respect to angle) cross sections for photoionization $\sigma_j^{\mathrm{i}}(\lambda, \theta)$ are necessary. The differential cross section is sometimes expressed as

$$\sigma_j^{\mathrm{i}}(\lambda, \theta) = \frac{\sigma_j^{\mathrm{i}}(\lambda)}{4\pi}[1 - \frac{1}{2}\beta(\lambda)P_2(\cos\theta)] , \tag{84.28}$$

where θ is the angle between the incident photon beam and the ejected electron, P_2 is a Legendre polynomial, and β is an asymmetry parameter.

84.3.3 Interaction of Energetic Electrons with Atmospheric Gases

Suprathermal electrons, which are denoted here e^{*-}, and include both photoelectrons and auroral primary electrons, can also ionize species through electron-impact ionization

$$X + \mathrm{e}^{*-} \rightarrow X^+ + \mathrm{e}^- + \mathrm{e}'^- \tag{84.29}$$

and electron-impact dissociative ionization

$$AB + e^{*-} \rightarrow A^{+} + B + e^{-} + e'^{-} . \quad (84.30)$$

In these reactions, e^- represents the energy degraded photoelectron or primary electron, and e'^- the secondary electron. The energy of the secondary electron $E_{e'}$ in an electron-impact ionization process (84.29) is given by

$$E_{e'} = E_{e^*} - I_X - E_{ex} - E_e , \quad (84.31)$$

where E_{e^*} is the energy of the primary or photoelectron, E_e is the energy of the degraded primary or photoelectron, and E_{ex} is the internal excitation energy of the product ions and/or neutral fragments. For the dissociative ionization process (84.30) the dissociation energy of the molecule must also be subtracted as well.

Energetic electrons can also dissociate atmospheric species. In this process

$$AB + e^{*-} \rightarrow A + B + e^{-} \quad (84.32)$$

the energy of the degraded electron is

$$E_e = E_{e^*} - D_{AB} - E_{ex} , \quad (84.33)$$

where D_{AB} is the dissociation energy of molecule AB. Collisions with suprathermal electrons can also promote species to excited electronic, vibrational or rotational states:

$$AB + e^{*-} \rightarrow AB^{\dagger} + e^{-} , \quad (84.34)$$

where the dagger denotes internal excitation. The energy lost by the electron is thus the excitation energy of the species.

In determining the rate of ionization, dissociation and excitation by photoelectrons, the local energy loss approximation, that is, the assumption that the electrons lose their energy at the same altitude where they are produced, is fairly good near the altitude of peak photoelectron production. The mean free path of an electron near 150 km is about 30 m. Substantially above the altitude of peak production of photoelectrons, transport of electrons from below is important, and use of the local energy loss assumption causes the excitation, ionization, and dissociation rates to be underestimated. For keV auroral electrons, the computation of the energy deposition of the electrons must consider their transport through the atmosphere. Thus the elastic total and differential cross sections for electrons colliding with neutral species must be employed, as well as the inelastic cross sections, and the angles through which the electrons scatter must be taken into account.

In general, the excitation rate $q_j^k(z)$ of a species j to an excited level k with a threshold energy E_k at an altitude z by electron impact is given by:

$$q_j^k(z) = n_j(z) \int_{E_k}^{\infty} \sigma_j^k(E) \frac{dF(z, E)}{dE} \, dE , \quad (84.35)$$

where $\sigma_j^k(E)$ is the excitation cross section at electron energy E, and $dF(z, E)/dE$ is the differential flux of electrons (between energies E and $E + dE$). The ionization rate $q_j^i(z)$ of a species with ionization potential I_j due to electron impact is given by

$$q_j^i(z) = n_j(z) \int_{I_j}^{\infty} \int_{0}^{(E-I_j)/2} \frac{d\sigma_j^i(E)}{dW_s} \frac{dF(z, E)}{dE} \, dW_s \, dE , \quad (84.36)$$

where $d\sigma_j^i(E)/dW_s$ is the differential cross section for production of a secondary electron with energy W_s by a primary electron with energy E. The integral over secondary energies W_s terminates at $(E - I_j)/2$ because the secondary electron is by convention considered to be the one with the smaller energy. Since the average energy of photoelectrons is less than 20 eV, the error incurred in cutting off the integrals in equations (84.35) and (84.36) at 200 eV or so, rather than $(E - I_j)/2$ is not serious, although for high energy auroral electrons a larger upper limit may be required.

An estimate of the number of ionizations in a gas produced by a primary electron with energy E_p is E_p/W_{ip}, where W_{ip} is the energy loss per ion pair produced, which approaches a constant value as the energy of the electron increases. Empirical values are available for W_{ip} for many gases, and usually fall in the range 30–40 eV [84.41].

The total loss function or stopping cross section for an electron with incident energy E in a gas j is given by the expression

$$L_j(E) = \sum_k \sigma_j^k(E) W_j^k + \int_{0}^{(E-I_j)/2} (I_j + W_s) \frac{d\sigma^{i_j}(E)}{dW_s} \, dW_s , \quad (84.37)$$

where W_j^k is the energy loss associated with excitation of species j to excited state k. The differential cross section is usually adopted from an empirical formula that is normalized so that

$$\sigma_j^{\mathrm{i}}(E) = \int_0^{(E-I_j)/2} \frac{\mathrm{d}\sigma_j^{\mathrm{i}}(E)}{\mathrm{d}W_{\mathrm{s}}} \mathrm{d}W_{\mathrm{s}} , \quad (84.38)$$

where $\sigma_j^{\mathrm{i}}(E)$ is the total ionization cross section at primary electron energy E. One formula in common use is that employed by *Opal* et al. [84.42] to fit to their data:

$$\frac{\mathrm{d}\sigma_j^{\mathrm{i}}(E)}{\mathrm{d}W_{\mathrm{s}}} = \frac{A(E)}{1+\left(W_{\mathrm{s}}/\overline{W}\right)^{2.1}} , \quad (84.39)$$

where $A(E)$ is a normalization factor and $\overline{W}$ is an empirically determined constant, which has been found to be equal to within a factor of about 50% to the ionization potential for a number of species.

For energy loss due to elastic scattering by thermal electrons, an analytic form of the loss function such as that proposed by *Swartz* et al. [84.43] may be used:

$$L_{\mathrm{e}}(E) = \frac{3.37\times 10^{-12}}{E^{0.94} n_{\mathrm{e}}^{0.03}} \left(\frac{E - k_{\mathrm{B}}T_{\mathrm{e}}}{E - 0.53 k_{\mathrm{B}} T_{\mathrm{e}}}\right)^{2.36} , \quad (84.40)$$

where T_{e} is the electron temperature and n_{e} is the number density of ambient thermal electrons.

For high energy auroral electrons, the rate of energy loss per electron per unit distance over the path s of the electrons in the atmosphere can be estimated using the continuous slowing down approximation (CSDA) as

$$-\frac{\mathrm{d}E}{\mathrm{d}s} = \sum_j n_j(z) L_j(E) \sec\theta + n_{\mathrm{e}}(z) L_{\mathrm{e}}(E) \sec\theta , \quad (84.41)$$

where θ is the angle between the path of the primary electron s and the local vertical. In the CSDA, all the electrons of a given energy are assumed to lose their energy continuously and at the same rate. The rate of energy loss $(-\mathrm{d}E/\mathrm{d}s)$ is integrated numerically over the path of the electron, which degrades in energy until it is thermalized. In this approximation, inelastic processes are assumed always to scatter the electrons forward, so cross sections that are differential in angle are not required. Because electrons actually lose energy at different rates, however, and because elastic and inelastic scattering processes do change the direction of the electrons, the CSDA gives an estimate for the rates of electron energy loss processes that is increasingly inaccurate as the energy of the electron decreases.

In practice, discrete energy loss of electrons can be easily treated numerically if the local energy loss approximation is valid. The spectrum of electrons is divided into energy bins that are smaller than the energy losses for the processes, and the integrals in (84.35, 36) are replaced by sums over energy bins. Since elastic scattering of electrons by neutrals changes mostly the direction of the incident electron, and not its energy, only inelastic processes need be considered. In order to compute excitation and dissociation rates, only integral cross sections are required; the scattering angle is unimportant. For ionization, of course, the energy distribution of the secondary electrons must be considered, but not the scattering angles of either the primary or secondary electrons. Below the lowest thresholds for excitations, energetic electrons lose their energy in elastic collisions with thermal electrons. The process of energy loss to thermal electrons is often approximated as continuous, rather than discrete.

The collision frequency ν_j^k for a discrete electron-impact excitation process k of a species j is given by

$$\nu_j^k(E) = n_j(z) v_{\mathrm{e}}(E) \sigma_j^k(E) . \quad (84.42)$$

For energy loss due to elastic scattering from thermal electrons, a pseudo-collision frequency ν_{e} may be defined as

$$\nu_{\mathrm{e}}(E) = \frac{1}{\Delta E}\left(-\frac{\mathrm{d}E}{\mathrm{d}t}\right) , \quad (84.43)$$

where ΔE is the grid spacing in the calculation, and the energy loss rate is

$$-\frac{\mathrm{d}E}{\mathrm{d}t} = v_{\mathrm{e}}(E) n_{\mathrm{e}} L_{\mathrm{e}}(E) , \quad (84.44)$$

where L_{e} is taken from (84.40).

Since the energy bins should be smaller than the typical energy loss in order to obtain accurate rates for the excitation processes, it is often convenient to treat rotational excitation also as a continuous process, with a pseudo-collision frequency similar to that for elastic scattering from ambient electrons (84.43) with

$$-\frac{\mathrm{d}E}{\mathrm{d}t} = v_{\mathrm{e}}(E) n_j L_j^{\mathrm{rot}}(E) , \quad (84.45)$$

where the loss function for rotational excitation is given by

$$L_j^{\mathrm{rot}}(E) = \sum_J \eta_j^J \sum_{J'} \sigma_j^{J,J'}(E) W_j^{J,J'} . \quad (84.46)$$

In this expression, η_j^J is the fraction of molecules $j$measured or computed cross section for electron-impact excitation of species j from rotational state J to rotational state J', and $W_j^{J,J'}$ is the associated energy loss.

The slowing down of high energy auroral primary electrons or photoelectrons arises from both elastic and inelastic scattering processes, and cannot be treated using the local energy loss approximation. In solving the equations for electron transport, theangle through which the primary electron is scattered, as well as the change in energy of the primary electron and the production of any secondaries, must be taken into account. Thus differential cross sections for the elastic and inelastic scattering of electrons by neutral species are required. The detailed equations for electron transport have been presented by, for example, *Rees* [84.44].

Several methods for approximating the energy deposition of auroral electrons are currently in use. The CSDA has already been discussed, but it provides only a rough approximation to the depth of penetration of the electrons, and the rates of excitation, ion production, and other energy loss processes. In the two-stream approximation, the electrons are assumed to be scattered in either the forward or backward direction [84.45]. Implementation of this method requires only the backscattering probabilities, rather than complete differential cross sections. The method has been generalized to multi-stream models, in which the solid angle range of the electrons is divided into 20 or more intervals, so more or less complete differential cross sections are required [84.46,47]. Monte Carlo methods have also been used to model auroral precipitation [84.48].

84.4 Ionospheres

84.4.1 Ionospheric Regions

The division of the ionosphere into regions is based on the structure of the terrestrial ionosphere, which consists of overlapping layers of ions. These layers are the result of changes both in the composition of the thermosphere and in the sources of the ionization, and are shown schematically in Fig. 84.5. The major molecular ion layer is the F_1 layer, which is produced by absorption of EUV (100–1000 Å) photons by the major thermospheric species, and occurs where the ion production maximizes. The E layer is below the F_1 layer and is produced by shorter and longer wavelength photons that are absorbed deeper in the atmosphere: soft X-rays and Lyman β, which can ionize O_2 and NO (Fig. 84.4a). In the D region, the densities of negative ions ecome appreciable and large densities of positive cluster ions appear. These ions are produced by harder X-rays, with $\lambda \lesssim 10$ Å, and Lyman α, which penetrates to about 75 km, where it ionizes NO. The highest altitude peak in the terrestrial ionosphere is the F_2 peak, which occurs near or slightly below 300 km, where the major ion is O^+. The peak density occurs where the chemical lifetime of the ion is equal to the characteristic time for transport by diffusion ($\sim H^2/D$).

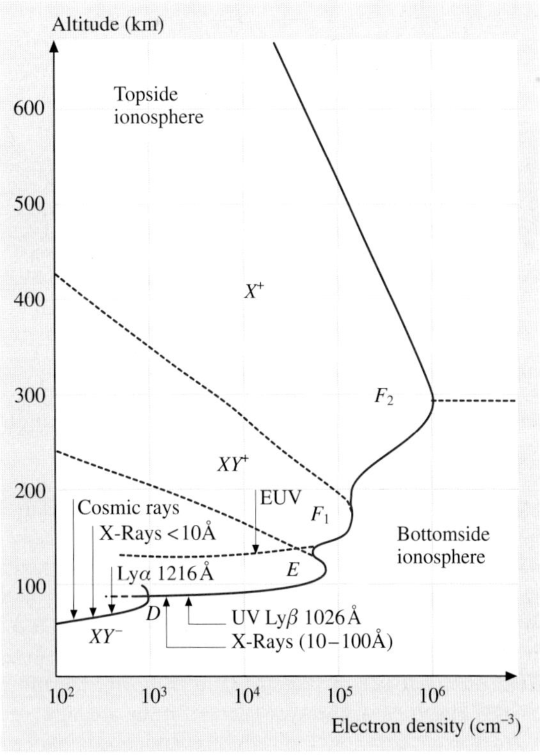

Fig. 84.5 Ionospheric regions and primary ionization sources. After *Bauer* [84.49]

84.4.2 Sources of Ionization

As discussed in Sect. 84.3, ionization can be produced either by solar photons and photoelectrons during the daytime or by energetic particles and secondary electrons during auroral events. Photoelectrons have sufficient energy to carry out further ionization if they are produced by photons with $\lambda \lesssim 500$ Å. These photons penetrate further and exhibit larger solar activity variations than longer-wavelength ionizing photons. Thus the ionization rate due to photoelectrons peaks below the main photoionization peak. Primary flux spectra of photoelectrons produced near the F_1 peak (172 km) and below the ion peak (100 km) are shown in Fig. 84.6. The primary spectrum at the ion peak consists mostly of low energy electrons, whereas at 100 km, the low energy primaries are depleted, and there are relatively larger fluxes of electrons with $E \gtrsim 50$ eV. Figure 84.7 shows the primary and steady-state photoelectron spectra near the ion peaks on Venus and Titan.

The major ions produced in the ionospheres of the earth and planets are usually those from the major thermospheric species: N_2^+, O_2^+, and O^+ on Earth; CO_2^+, O^+, N_2^+, and CO^+ on Venus and Mars; and H_2^+, H^+, and He^+ on the outer planets; N_2^+, N^+, and CH_4^+ in the ionosphere of Titan, and N_2^+, N^+, and C^+ in the ionosphere of Triton. In the presence of sufficient neutral densities, however, ion–molecule reactions transform ions whose parent neutrals have high ionization potentials to ions whose parent neutrals have low ionization potentials. This is a rigorous rule only for charge transfer reactions, but it applies more often than not in other ion–molecule reactions as well.

Because of transformations by ion–molecule reactions, the major ions in the F_1 regions of the ionospheres of Earth, Venus and Mars are O_2^+ and NO^+, in spite of the large differences in composition between the thermosphere of the earth and the thermospheres of Venus and Mars. A diagram illustrating the ion chemistry in the ionospheres of the terrestrial planets is shown in Fig. 84.8. The vertical positions of the ions in this figure represent the relative ionization potentials of the parent neutrals. In regions where there are sufficient neutral densities the ionization flows downward.

Table 84.9 shows ionization potentials (I_P) for several major and minor species present in planetary thermospheres. Major atmospheric species generally have $I_P \gtrsim 12$–13 eV ($\lambda < 900$–1000 Å). Only a few species can be ionized by the strong solar Lyman alpha line (1216 Å, 10.2 eV), including NO, and a few small hydrocarbons and radicals, such as CH_3 and C_2H_5. Metal atoms, which are produced in the lower thermospheres and mesospheres of planets from ablation of meteors, have very low ionization potentials, and some can be ionized by photons with wavelengths longer than 2000 Å.

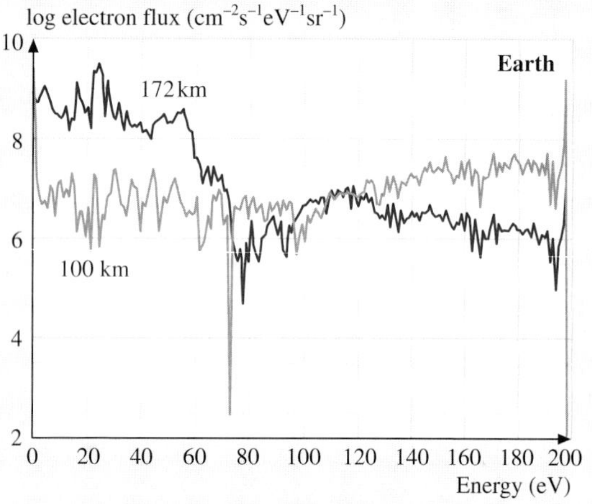

Fig. 84.6 Primary photoelectron spectrum for the terrestrial atmosphere at 172 km (near the F_1 peak) and at 100 km. The spectrum at 100 km is significantly harder than that at 172 km

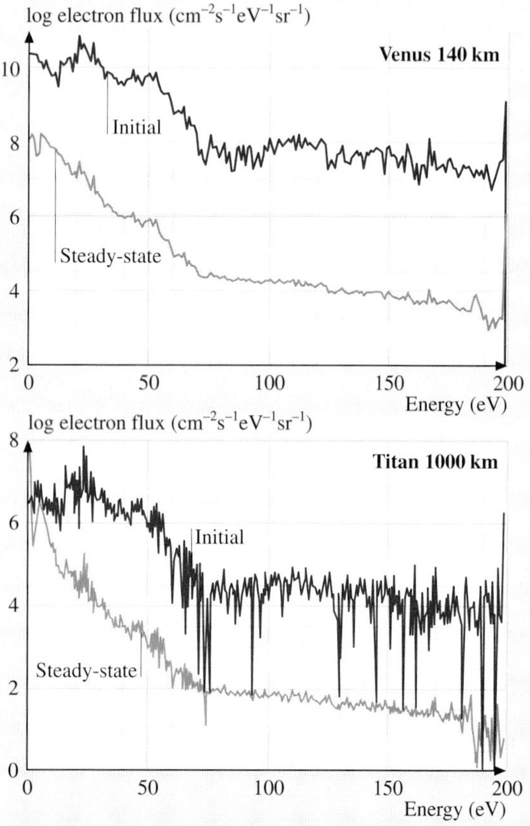

Fig. 84.7 Computed primary and steady-state spectra for photoelectrons near the F_1 peak on Venus at 1 eV resolution (*top*), and Titan at 0.5 eV resolution (*bottom*). The steady-state spectra are averaged over three intervals in both plots

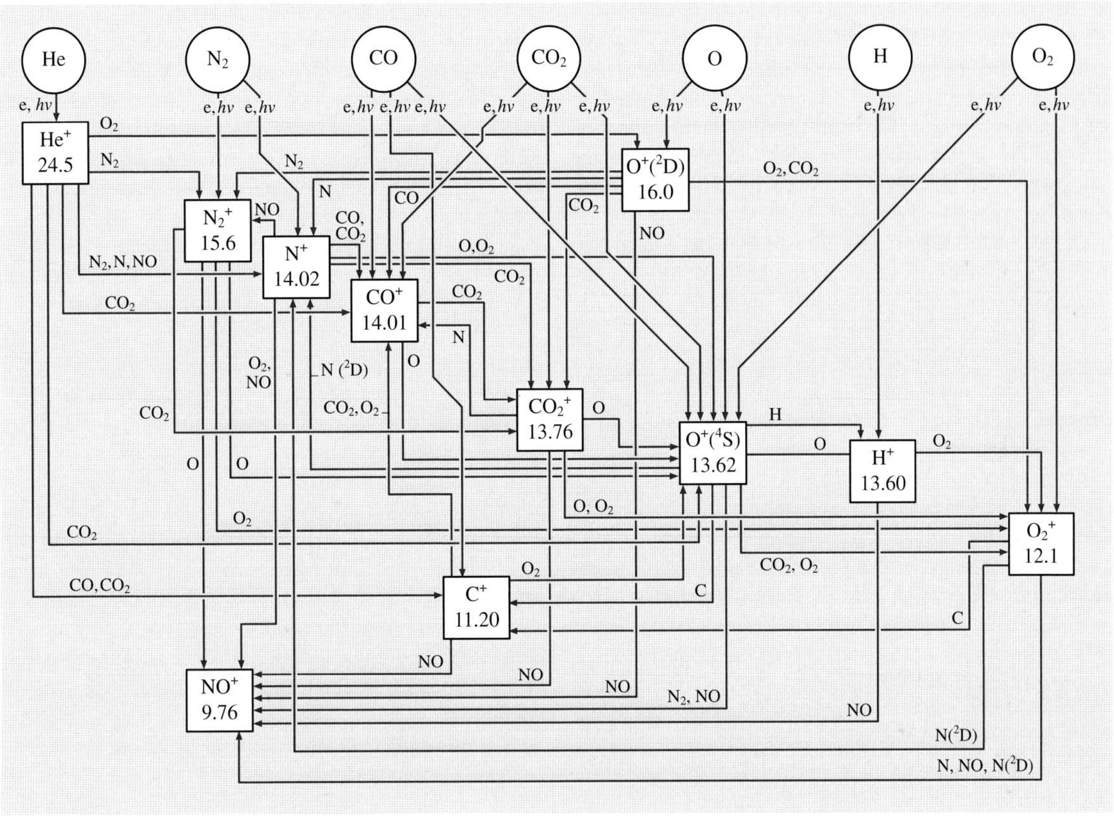

Fig. 84.8 Diagram illustrating the ion chemistry in the ionospheres of the terrestrial planets. The numbers under the names of the ions indicate the ionization potentials of the parent neutral. In the presence of sufficient neutral densities, the ionization flows downward, the importance of dissociative recombination for the molecular ions increases as the ionization potentials of the parent neutrals decrease

Table 84.9 Ionization potentials (I_P) of common atmospheric species[a]

High I_P Species	I_P (eV)	Medium I_P Species	I_P (eV)	Ionized by Ly α Species	I_P (eV)
He	24.59	CH_4	12.61	C_4H_2	10.18
Ne	21.56	CH_4	12.51	CH_3	9.84
Ar	15.76	O_2	12.32	C_3H_6	9.73
N_2	15.58	O_2	12.07	NO	9.264
H_2	15.43	C_2H_6	11.52	C_2H_5	8.13
N	14.53	C_2H_2	11.40	HCO	8.10
CO	14.01	C	11.26	C_3H_7	8.09
CO_2	13.77	C_3H_8	10.95	Mg	7.65
O	13.62	CH	10.64	trans-HCNH	7.0[b]
H	13.60	C_2H_4	10.51	cis-HCNH	6.8[b]
HCN	13.60	CH_2	10.40	Ca	6.11
OH	13.00	S	10.35	Na	5.14

[a] Computed with data taken from [84.50], except as noted; [b] From [84.51]

In ionospheres where hydrogen is abundant and sufficient neutral densities are present, ionization flows to species formed by protonation of neutrals that have large proton affinities. There are no *in situ* measurements of the ion composition of the outer planets, but models predict that H_3^+ and hydrocarbon ions dominate the lower ionospheres. In regions where meteor ablation occurs, metal ions may also be found.

Many ion–molecule reactions proceed at or near gas kinetic (or collision) rates. The interaction of an ion with a nonpolar molecule is dominated by the ion-induced-dipole interaction, for which the interaction potential is $-\frac{1}{2}\alpha_d q^2/r^4$, where α_d is the polarizability of the neutral, q is the charge on the ion, and r is the distance between the particles. The Langevin rate coefficient is then given by

$$k_L = 2\pi q(\alpha_d/\mu)^{1/2} , \quad (84.47)$$

where μ is the reduced mass of the two species. For a singly charged ion, with α_d in Å^3 and μ in atomic mass units, this formula reduces to $2.34\times 10^{-9}(\alpha_d/\mu)^{1/2}$. The rate coefficient for an ion with a polar molecule is

$$k_d = \frac{2\pi q}{\mu^{1/2}}\left[\alpha_d^{1/2} + c\mu_d\left(\frac{2}{\pi k_B T}\right)^{1/2}\right] , \quad (84.48)$$

where μ_d is the dipole moment and c is a constant that is unity in the locked dipole approximation and is about 0.1 in the average dipole orientation (ADO) theory. Theories for ion–quadrupole interactions have also been developed, and the resulting formulas can be found in, for example, the review by *Su* and *Bowers* [84.59]. Measured rate coefficients for ion–molecule reactions have been compiled by *Anicich* et al. [84.60] and *Ikezoe* et al. [84.61].

Loss of ionization in planetary atmospheres proceeds mainly by dissociative recombination of molecular ions, which may be represented by

$$AB^+ + e^- \rightarrow A + B . \quad (84.49)$$

Dissociative recombination coefficients are characteristically large, about $10^{-7}\,cm^3/s$, at the electron temperatures T_e typical of planetary ionospheres, which are usually within a factor of two or so of the neutral $T \approx 200–2000\,K$ near the molecular ion density peak. Daytime peak electron densities are usually in the range $10^4–10^6\,cm^{-3}$, and fractional ionizations are small, about 10^{-5}, near the F_1 peak.

The relative importance of ion–molecule reactions and dissociative recombination in the destruction of a particular ion is determined by the relative densities of electrons and neutrals with which the ions can react. In general, molecular ions whose parent neutrals have high I_P are transformed by ion–molecule reactions preferentially to loss by dissociative recombination, and their peak densities occur higher in the atmosphere. Ions for which dissociative recombination is an important loss mechanism near the ion peaks of the terrestrial planets include NO^+ and O_2^+, and in the atmospheres of the outer planets, H_3^+ and hydrocarbon ions. For ions with very high I_P, such as N_2^+ and H_2^+, dissociative recombination is rarely important as a loss process, except at very high altitudes. It may, however, be important as a source of vibrationally or electronically excited fragments or hot atoms.

Atomic ions may be destroyed by radiative recombination:

$$X^+ + e^- \rightarrow X + h\nu , \quad (84.50)$$

but the rate coefficients are small, about $10^{-12}\,cm^3/s$ at the typical T_e of planetary ionospheres [84.62]. Atomic ions may dominate at high altitudes, where neutral densities are low, but in such regions, loss by downward diffusion is more important than chemical recombination. The ions diffuse downward to altitudes where the neutral densities are higher and are then destroyed in ion–molecule reactions. The major ions in the topside ionospheres of the planets tend to be atomic ions: O^+ in the ionospheres of Earth and Venus, and H^+ in the iono-

Fig. 84.9a–k Model thermospheres for the Earth and planets. The curves are number density profiles and are labeled by the species they represent. **(a)** Earth, based on the MSIS model of [84.17] for a latitude of 45°, a local time of noon for low (*top*) and high (*bottom*) solar activities; **(b)** High solar activity model of Venus, based on the model of [84.3] for 15° N latitude 15 h local time; **(c)** Low solar activity model of Mars, based on Viking 1 measurements [84.8]. Adapted after [84.52]; **(d)** Jupiter, After [84.53]. **(e)** Saturn, After [84.54]. **(f)** Uranus, after [84.55]; **(g)** Neptune [84.56]; **(h)** Pluto [84.57]. The abscissa is given in units of Pluto radii, and the ordinate in the *top plot* is number density in cm^{-3}; the bottom is temperature in K. **(i)** Titan, After [84.10, 12]; and **(j)** Triton model, after [84.15]; **(k)** Io model, temperature (*upper scale*), and number density (*lower scale*). The major constituent is SO_2, and transient with a lifetime of 2–3 days. From *Strobel* and *Wolven* [84.58]. The *short dashed curves* are for solar ionization only, and the *solid* and *long-dashed curves* are for different assumptions about the interaction of Triton's thermosphere with electrons from Neptune's magnetosphere. The *solid curves* are the recommended model **(j)** ▶

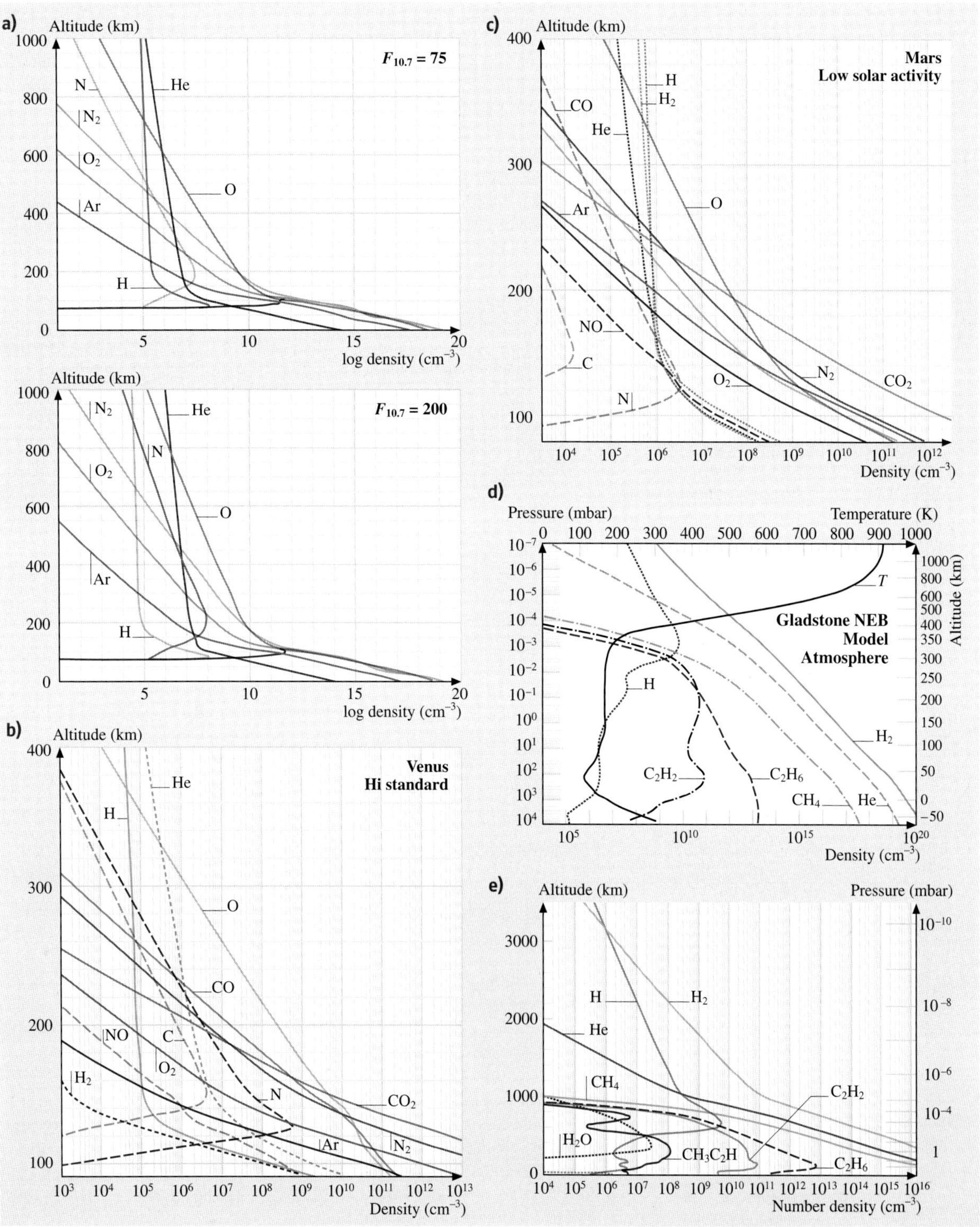
a)
Altitude (km)
$F_{10.7}$ = 75
N
He
N_2
O_2
O
Ar
H
log density (cm^{-3})
Altitude (km)
$F_{10.7}$ = 200
N_2
He
N
O_2
O
Ar
H
log density (cm^{-3})
b)
Altitude (km)
Venus
Hi standard
He
H
O
CO
NO
C
O_2
H_2
N
CO_2
Ar
N_2
Density (cm^{-3})
c)
Altitude (km)
Mars
Low solar activity
H
H_2
CO
He
O
Ar
NO
C
N
O_2
N_2
CO_2
Density (cm^{-3})
d)
Pressure (mbar)
Temperature (K)
T
Gladstone NEB
Model
Atmosphere
H
H_2
C_2H_2
C_2H_6
CH_4
He
Altitude (km)
Density (cm^{-3})
e)
Altitude (km)
Pressure (mbar)
H
H_2
He
CH_4
C_2H_2
H_2O
CH_3C_2H
C_2H_6
Number density (cm^{-3})

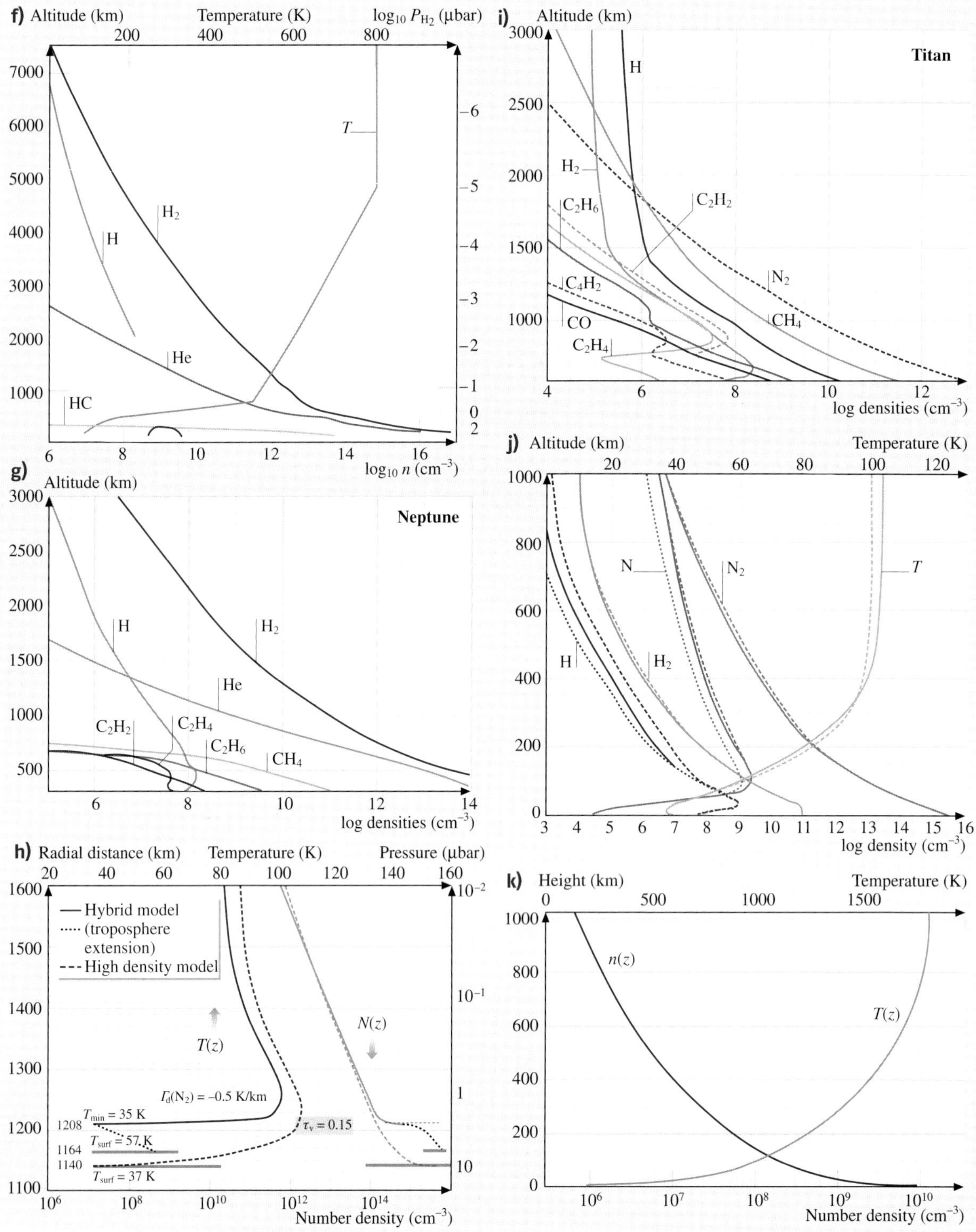
f) Altitude (km)
Temperature (K)
log10 PH2 (µbar)
200
400
600
800
7000
6000
5000
4000
3000
2000
1000
T
H2
H
He
HC
−6
−5
−4
−3
−2
−1
0
2
6
8
10
12
14
16
log10 n (cm−3)
g) Altitude (km)
Neptune
3000
2500
2000
1500
1000
500
H
H2
He
C2H2
C2H4
C2H6
CH4
6
8
10
12
14
log densities (cm−3)
h) Radial distance (km)
Temperature (K)
Pressure (µbar)
20
40
60
80
100
120
140
160
1600
1500
1400
1300
1200
1100
Hybrid model
(troposphere extension)
High density model
T(z)
N(z)
Γd(N2) = −0.5 K/km
Tmin = 35 K
Tsurf = 57 K
Tsurf = 37 K
τv = 0.15
1208
1164
1140
10−2
10−1
1
10
10^6
10^8
10^10
10^12
10^14
Number density (cm−3)
i) Altitude (km)
Titan
3000
2500
2000
1500
1000
H
H2
C2H6
C2H2
N2
C4H2
CH4
CO
C2H4
4
6
8
10
12
log densities (cm−3)
j) Altitude (km)
Temperature (K)
20
40
60
80
100
120
1000
800
600
400
200
0
N
N2
T
H
H2
3
4
5
6
7
8
9
10
11
12
13
14
15
16
log density (cm−3)
k) Height (km)
Temperature (K)
0
500
1000
1500
1000
800
600
400
200
0
n(z)
T(z)
10^6
10^7
10^8
10^9
10^10
Number density (cm−3)

spheres of the outer planets. On Mars, however, the O^+ peak density does not exceed that of O_2^+ even at high altitudes.

Model thermospheres for Earth and selected planets and satellites are shown in Fig. 84.9a–k. Measured or computed ion density profiles for the Earth and selected planets and satellites are shown in Fig. 84.10a–h.

84.4.3 Nightside Ionospheres

Nightside ionospheres can result from several sources including remnant ionization from dayside, like O^+ in the terrestrial ionosphere. While the lower molecular ion layers recombine, the F_2 peak persists through the night, although it rises and the peak density is reduced by a factor of 10. Electron density profiles for day and night at high and low solar activities are shown in Fig. 84.11. In the auroral regions of the Earth, the precipitating electrons may also produce significant ionization, which maximizes in the midnight sector of the auroral oval.

The nightside ionosphere of Venus is highly variable, but has been shown to contain the same ions as the dayside ionosphere. The densities are, however, lower by factors of 10 or more than those of the dayside ionosphere, and the average peak in the electron density profile is about $(1\text{ to }2)\times 10^4\,\text{cm}^{-3}$. It is produced by a combination of precipitation of suprathermal electrons that have been observed at high altitudes in the umbra, and transport of atomic ions (mostly O^+) at high altitudes from the dayside. For Mars, only a narrow range of solar zenith angles near the terminator at low solar activity has been measured by the radio occultation experiments on the Viking spacecraft [84.63], the Mariner 9 spacecraft, and more recently by the Mars Global Surveyor (MGS) spacecraft radio sciences (RS) experiment [84.64]. The electron densities are apparently low, and no composition information is available. At this time, there is no information available about the nightside ionospheres of the other planets.

84.4.4 Ionospheric Density Profiles

Density profiles of molecular ions can often be approximated as idealized Chapman layers. A Chapman layer of ions is one in which the ions are produced by photoionization and lost locally by dissociative recombination. The ionization rate q^{i} in a one-species Chapman layer for monochromatic radiation is given by

$$q^{\mathrm{i}} = F\sigma^{\mathrm{i}} n\,, \tag{84.51}$$

where σ^{i} is the ionization cross section, and $F = F^{\infty}\exp[-\tau(z)]$ is the local solar flux. For an isothermal atmosphere, the scale height is approximately constant and therefore $n = n_0 \exp(-z/H)$. Sometimes an ionization efficiency η^{i} is defined such that

$$\sigma^{\mathrm{i}} = \eta^{\mathrm{i}}\sigma^{\mathrm{a}}\,. \tag{84.52}$$

Near threshold, the ionization efficiency for molecules is usually about 0.3–0.7 but it increases rapidly to 1.0 at shorter wavelengths.

Since the maximum ionization rate in an isothermal atmosphere occurs where the optical depth ($\tau = nH\sigma^{\mathrm{a}}\sec\chi$) is unity and therefore $n = 1/(\sigma^{\mathrm{a}} H\sec\chi)$, the maximum ionization rate in a Chapman layer is

$$q^{\mathrm{i}}_{\max,\chi} = \frac{F^{\infty}}{\mathrm{e}}\frac{\sigma^{\mathrm{i}}}{\sigma^{\mathrm{a}} H\sec\chi} = \frac{q^{\mathrm{i}}_{\max,0}}{\sec\chi}\,. \tag{84.53}$$

If the altitude of maximum ionization for overhead sun is defined as $z = 0$, then $n_0 = (\sigma^{\mathrm{a}} H)^{-1}$, and, expressing F^{∞} in terms of $q^{\mathrm{i}}_{\max,0}$, the ionization rate is

$$q^{\mathrm{i}}(z) = q^{\mathrm{i}}_{\max,0}\exp\left(1 - \frac{z}{H} - \sec\chi\,\mathrm{e}^{-z/H}\right)\,. \tag{84.54}$$

It is apparent that at high altitudes ($z \to \infty$) the ionization profile follows that of the neutral density, and below the peak ($z \to -\infty$), the ionization rate rapidly approaches zero. As the solar zenith angle increases, the peak rises and the magnitude of the density maximum decreases. Figure 84.12 shows a production profile for an idealized Chapman layer on both linear and semilog plots. The asymmetry with respect to the maximum is more obvious for the semilog plot.

If photochemical equilibrium prevails, the production rate of the major ion is equal to the loss rate due to dissociative recombination

$$q^{\mathrm{i}}(z) = \alpha_{\mathrm{dr}} n_{\mathrm{i}} n_{\mathrm{e}} = \alpha_{\mathrm{dr}} n_{\mathrm{i}}^2\,, \tag{84.55}$$

where α_{dr} is the dissociative recombination coefficient, n_{i} is the ion density, and n_{e} is the electron density. Therefore the density of an ion in a Chapman layer (in the photochemical equilibrium region) is given by

$$\begin{aligned} n_{\mathrm{i}}(z) &= \left(\frac{q^{\mathrm{i}}(z)}{\alpha_{\mathrm{dr}}}\right)^{1/2} \\ &= \left(\frac{q^{\mathrm{i}}_{\max,0}}{\alpha_{\mathrm{dr}}}\right)^{1/2} \exp\left(\frac{1}{2} - \frac{z}{2H} - \frac{1}{2}\sec\chi\,\mathrm{e}^{-z/H}\right)\,. \end{aligned} \tag{84.56}$$

Actual ionization profiles differ from the idealized Chapman profile for several reasons. First, ionization is

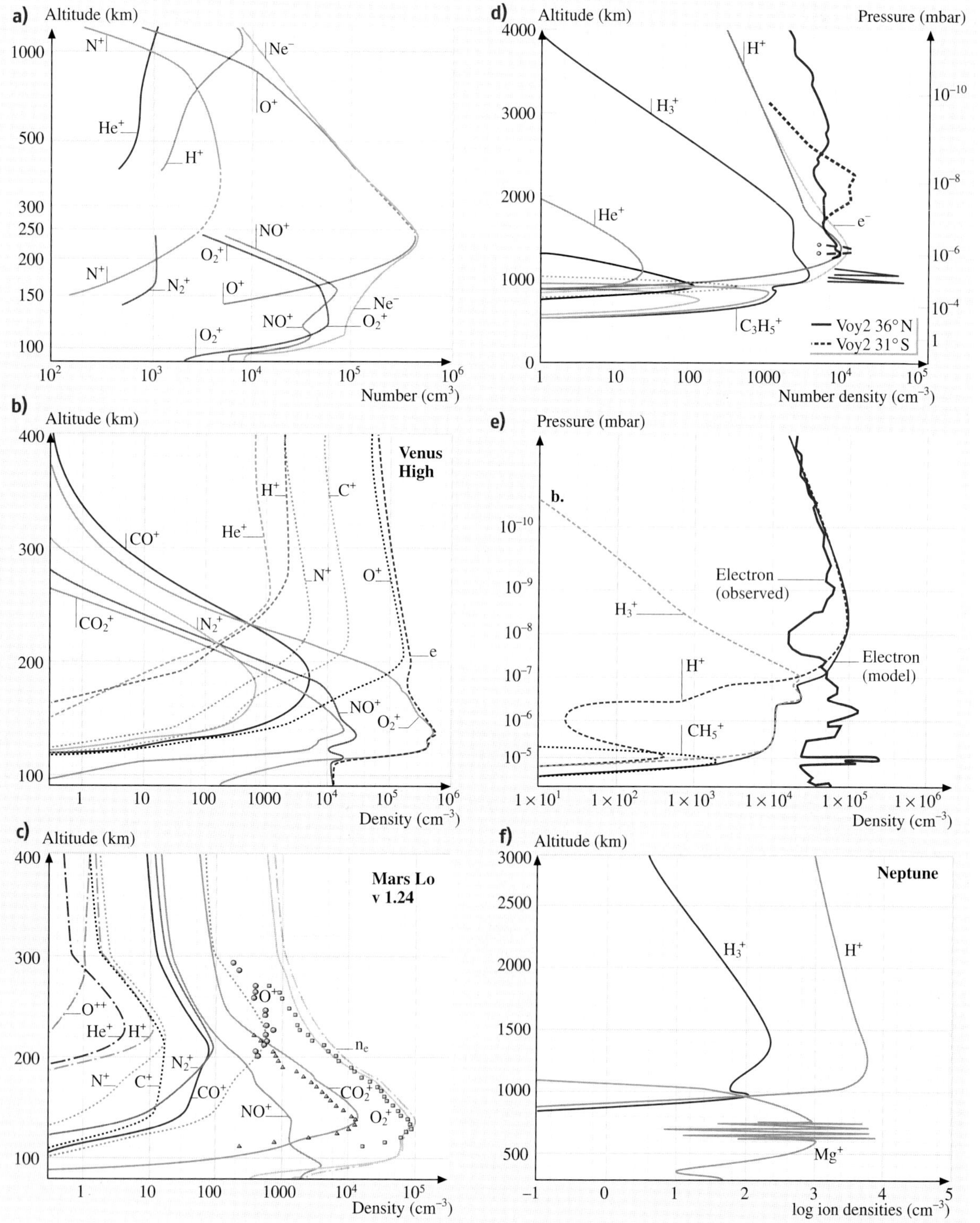
a) Altitude (km)
1000
500
300
250
200
150
100
N⁺
Ne⁻
O⁺
He⁺
H⁺
NO⁺
O₂⁺
N⁺
N₂⁺
O⁺
Ne⁻
NO⁺
O₂⁺
O₂⁺
10² 10³ 10⁴ 10⁵ 10⁶
Number (cm⁻³)
b) Altitude (km)
400
300
200
100
Venus High
H⁺
C⁺
He⁺
CO⁺
N⁺
O⁺
CO₂⁺
N₂⁺
e
NO⁺
O₂⁺
1 10 100 1000 10⁴ 10⁵ 10⁶
Density (cm⁻³)
c) Altitude (km)
400
300
200
100
Mars Lo v 1.24
O⁺⁺
He⁺
H⁺
O⁺
n_e
N₂⁺
N⁺
C⁺
CO⁺
CO₂⁺
NO⁺
O₂⁺
1 10 100 1000 10⁴ 10⁵
Density (cm⁻³)
d) Altitude (km)
Pressure (mbar)
4000
3000
2000
1000
0
H⁺
H₃⁺
He⁺
e⁻
C₃H₅⁺
Voy2 36° N
Voy2 31° S
10⁻¹⁰
10⁻⁸
10⁻⁶
10⁻⁴
1
1 10 100 1000 10⁴ 10⁵
Number density (cm⁻³)
e) Pressure (mbar)
b.
10⁻¹⁰
10⁻⁹
10⁻⁸
10⁻⁷
10⁻⁶
10⁻⁵
H₃⁺
Electron (observed)
Electron (model)
H⁺
CH₅⁺
1 × 10¹ 1 × 10² 1 × 10³ 1 × 10⁴ 1 × 10⁵ 1 × 10⁶
Density (cm⁻³)
f) Altitude (km)
3000
2500
2000
1500
1000
500
Neptune
H₃⁺
H⁺
Mg⁺
−1 0 1 2 3 4 5
log ion densities (cm⁻³)

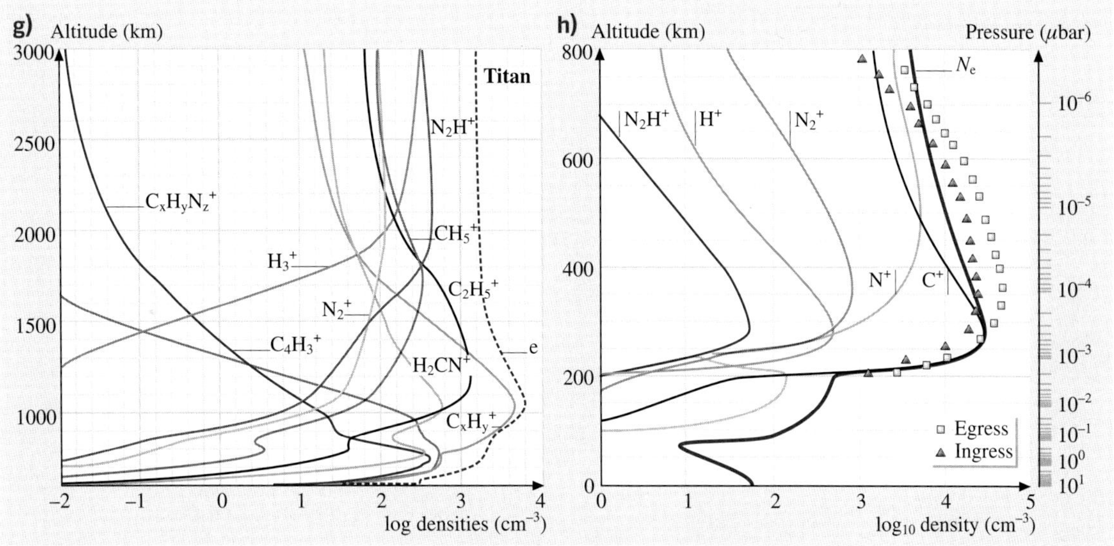

Fig. 84.10a–h Ionospheric density profiles. (**a**) Measured profiles for Earth [84.65]; Computed profiles for (**b**) Venus, for high solar activity and a solar zenith angle of 45°; (**c**) Mars, for a low solar activity model for Viking 1 conditions and a solar zenith angle of 45°; the *dashed curves* are measured profiles from Viking [84.66]; (**d**) Jupiter, After [84.67]; (**e**) Saturn, After [84.54]; (**f**) Neptune [84.56]. See also [84.68] (**g**) Titan, After *Fox* and *Yelle* (unpublished, 1995); and (**h**) Triton [84.16]

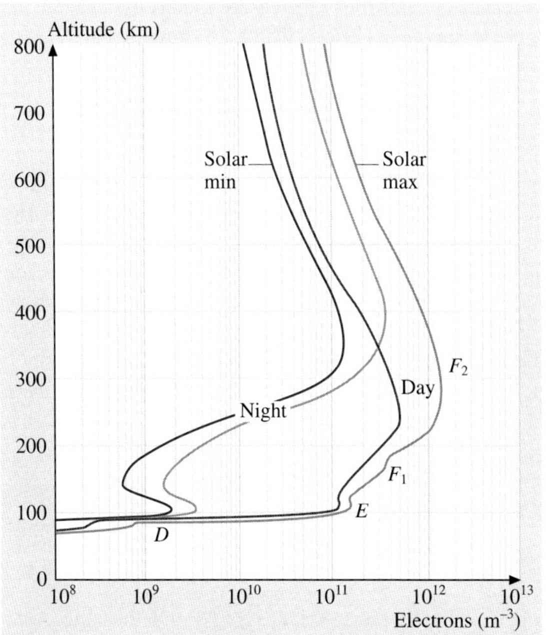

Fig. 84.11 Typical midlatitude ionospheric electron density profiles for sunspot maximum and minimum, day and night. After *Richmond* [84.69]

produced by photons over a range of wavelengths, which do not all reach unit optical depth at the same altitude. Second, thermospheres are often not isothermal near the altitude of peak ion production. Third, photoionization is supplemented by photoelectron-impact ionization, which peaks lower in the atmosphere; and finally, the major ion produced is often transformed by ion–molecule reactions before it can recombine dissociatively. Nonetheless, the idealized concept of the Chapman profile is useful in understanding the general shape of ion profiles and their behavior as the solar zenith angle changes. In addition, ion layers produced by auroral precipitation may take on a similar appearance to a Chapman-type layer, although energetic electrons are not always extinguished, as are photons, in ion production.

84.4.5 Ion Diffusion

Above the photochemical equilibrium layer of the ionosphere, upward and downward transport of ions must be considered. The motions of ions, neutrals, and electrons are coupled, and the momentum equation, which determines the fluxes or velocities of the ions must take into account these interactions. The interaction of an ion, denoted by a subscript i, with a neutral species de-

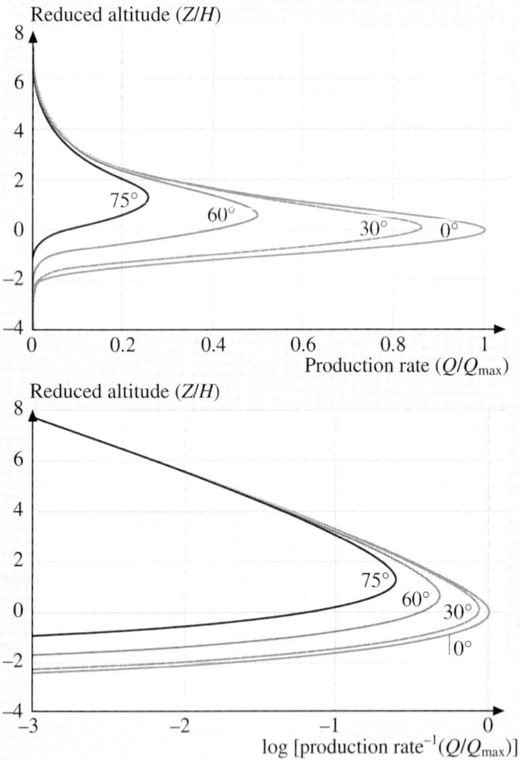

Fig. 84.12 Chapman production profile as a function of altitude on linear (*top*) and semilog (*bottom*) plots. The production rate is divided by the maximum production rate and the altitude by the scale height H, which is assumed to be constant. The origin on the altitude scale corresponds to the point of maximum absorption for overhead sun

noted by a subscript n, is through the ion-induced-dipole attraction or, for the diffusion of an ion through its parent neutral, by resonant charge transfer. For the former process, the ion–neutral diffusion coefficient is given by

$$D_{\mathrm{in}} = \frac{k_{\mathrm{B}}T}{m_{\mathrm{i}}\overline{\nu}_{\mathrm{in}}} , \tag{84.57}$$

where m_{i} is the mass of the ion, and the ion–neutral momentum transfer collision frequency

$$\overline{\nu}_{\mathrm{in}} = 2.21\pi \frac{n_{\mathrm{n}}m_{\mathrm{n}}}{m_{\mathrm{i}+m_{\mathrm{n}}}} \left(\frac{\alpha_{\mathrm{n}}e^2}{\mu_{\mathrm{in}}}\right)^{1/2} , \tag{84.58}$$

where α_{n} is the polarizability of the neutral species (*Dalgarno* et al. [84.70]; *Schunk* and *Nagy* [84.71]). For the resonant charge transfer interaction, the diffusion coefficient is

$$D_{\mathrm{in}}^{\mathrm{ct}} = \frac{3(\pi/2)^{1/2}}{8n_{\mathrm{n}}\overline{Q_{\mathrm{in}}^{\mathrm{ct}}}} \left(\frac{k_{\mathrm{B}}T_{\mathrm{i}}}{m_{\mathrm{i}}}\right)^{1/2} \frac{1}{(1+T_{\mathrm{n}}/T_{\mathrm{i}})^{1/2}} , \tag{84.59}$$

where $\overline{Q_{\mathrm{in}}^{\mathrm{ct}}}$ is the average charge transfer cross section [84.72].

The momentum transfer collision frequency for Coulomb interactions between an ion i and another ion or electron denoted by the subscript s is

$$\overline{\nu}_{\mathrm{i}s} = \frac{16\pi^{1/2}}{3} \frac{n_s m_s}{m_{\mathrm{i}}+m_s} \left(\frac{\mu_{\mathrm{i}s}}{2k_{\mathrm{B}}T_{\mathrm{i}s}}\right)^{3/2} \frac{e_{\mathrm{i}}^2 e_s^2}{\mu_{\mathrm{i}s}^2} \ln\Lambda , \tag{84.60}$$

where e_s is the charge on species s, $\ln\Lambda$ is related to the Debye shielding length, and $T_{\mathrm{is}} = (m_s T_{\mathrm{i}} + m_{\mathrm{i}} T_s)/(m_{\mathrm{i}}+m_s)$ is a reduced temperature. Numerically, $\ln\Lambda$ is about 15, and the collision frequency is approximately [84.71]

$$\overline{\nu}_{is} = 1.27 \frac{Z_{\mathrm{i}}^2 Z_s^2 \mu_{\mathrm{is}}^{1/2} n_s}{m_{\mathrm{i}} T_{\mathrm{is}}^{3/2}} \,\mathrm{s}^{-1} , \tag{84.61}$$

where Z is the species charge number, μ and m are in amu, and the number density is in units of cm^{-3}.

The ion densities can be computed by solving the ion continuity equation, which is similar to (84.7) for neutral species, and in one dimension is

$$\frac{\partial n_{\mathrm{i}}}{\partial t} + \frac{\partial \Phi_{\mathrm{i}}}{\partial z} = P_{\mathrm{i}} - L_{\mathrm{i}} , \tag{84.62}$$

where the ion flux is given by $\Phi_{\mathrm{i}} = n_{\mathrm{i}} w_{\mathrm{i}}$. In general it is impossible to solve the momentum equation for the ion diffusion velocity w_{i} in closed form, except for the special cases of a single major ion and of a minor ion moving through a dominant ion species. If motion of the ions only parallel to magnetic field lines is considered, the vertical velocity of a dominant ion (for which $n_{\mathrm{i}} \approx n_{\mathrm{e}}$) moving through a stationary neutral atmosphere is

$$\begin{aligned} w_{\mathrm{i}} &= -D_{\mathrm{a}} \sin^2 I \\ &\times \left(\frac{1}{n_{\mathrm{i}}}\frac{\mathrm{d}n_{\mathrm{i}}}{\mathrm{d}z} + \frac{m_{\mathrm{i}}g}{k(T_{\mathrm{e}}+T_{\mathrm{i}})} + \frac{1}{T_{\mathrm{e}}+T_{\mathrm{i}}}\frac{\mathrm{d}(T_{\mathrm{e}}+T_{\mathrm{i}})}{\mathrm{d}z}\right) , \end{aligned} \tag{84.63}$$

where I is the magnetic dip angle and the ambipolar diffusion coefficient defined as

$$D_{\mathrm{a}} = \frac{k_{\mathrm{B}}(T_{\mathrm{e}}+T_{\mathrm{i}})}{m_{\mathrm{i}}\overline{\nu}_{\mathrm{in}}} . \tag{84.64}$$

For a minor ion i diffusing through a major ion species j, its velocity is given by

$$w_\mathrm{i} = -\frac{k_\mathrm{B}T_\mathrm{i}/m_\mathrm{i}}{\overline{\nu}_{\mathrm{i}j}+\overline{\nu}_\mathrm{in}}\left(\frac{1}{n_\mathrm{i}}\frac{\mathrm{d}n_\mathrm{i}}{\mathrm{d}s}+\frac{T_\mathrm{e}/T_\mathrm{i}}{n_\mathrm{e}}\frac{\mathrm{d}n_\mathrm{e}}{\mathrm{d}s} + \frac{m_\mathrm{i}g}{k_\mathrm{B}T_\mathrm{i}}+\frac{1}{T_\mathrm{i}}\frac{\mathrm{d}(T_\mathrm{e}+T_\mathrm{i})}{\mathrm{d}s}\right)-\frac{\overline{\nu}_{\mathrm{i}j}w_j}{\overline{\nu}_{\mathrm{i}j}+\overline{\nu}_\mathrm{in}}\,. \quad (84.65)$$

For regions in which there are large gradients in the ion or electron temperatures, thermal diffusion may also be important in determining the ion density profiles, especially those of light ions such as H^+ and He^+. Equations for ion distributions in which thermal diffusion is included have been presented by, for example, *Schunk* and *Nagy* [84.71] and references therein.

84.5 Neutral, Ion and Electron Temperatures

The temperature distribution in planetary thermospheres/ionospheres can be modeled by solving the equation for conservation of energy, which, in simplified form in the vertical direction is

$$n_m\frac{\mathcal{N}}{2}k_\mathrm{B}\frac{\partial T_m}{\partial t}-\frac{\partial}{\partial z}\left(\kappa_m\frac{\partial T_m}{\partial z}\right)=Q_m-L_m \quad (84.66)$$

where $\mathcal{N}$ is the number of degrees of freedom (3 for an atom, and 5 for a diatomic molecule), the subscript m refers to the neutrals, ions or electrons, κ_m is the thermal conductivity, Q_m is the volume heating rate, and L_m is the volume cooling rate. If horizontal variations are considered, the model becomes multidimensional and advective terms must be added to the equations. The T_n are also affected by compression or expansion due to subsidence or upwelling, respectively. Viscous heating may be a factor where there are local regions of intense energy input, such as in auroral arcs. These terms are not shown in the energy equation above, but may be found in standard aeronomy texts, such as *Banks* and *Kockarts* [84.72] or *Rees* [84.44] and *Schunk* and *Nagy* [84.73]. For planets with intrinsic magnetic fields, the electrons and ions are constrained to move along magnetic field lines, and the second term on the left-hand side of (84.66) must be multiplied by a factor $\sin^2 I$.

The neutral thermospheres of planets are mostly heated by absorption of solar radiation in the 10 to 2000 Å range, although on planets with powerful auroras, electron precipitation may be an important source of heat. Absorption of EUV radiation (100–1000 Å) largely results in ionization of the major thermospheric species, in which most of the excess energy is carried away by the photoelectron. The photoelectron, however, may produce further dissociation or excitation of neutral species along the path to thermalization, and these processes may result in neutral heating. Photons near and longward of ionization thresholds in the FUV may lose their energy in photodissociation, in which the excess energy of the photon appears as kinetic or internal energy of the fragments.

Chemical reactions that follow ionization or dissociation release much of the absorbed solar energy as heat. Although the partitioning of kinetic energy released between the product species can be determined easily by conservation of energy and momentum, the fraction of energy that appears as internal or kinetic energy must be determined by measurements or theoretical calculations. If vibrationally or electronically excited states are produced in these interactions, however, the energy may be radiated to space, thus producing cooling. This may occur promptly if the radiative lifetime is short, or subsequent to an energy transfer process from a long-lived metastable species to a species for which radiation to a lower state is allowed. If the metastable species is quenched, however, its energy can also appear as heat. Thus the energy partitioning in chemical reactions and in the interactions of photons and photoelectrons with atmospheric species is important in understanding the temperature structure of thermospheres.

A heating efficiency ϵ is often defined as the fraction of energy absorbed at a given altitude that appears locally as heat. The heating efficiencies are in the range 30–40% in the terrestrial lower thermosphere. Above 200 km, the heating efficiency decreases because the energy of the important metastable species $O(^1D)$ is lost as radiation rather than by quenching [84.74]. The heating efficiencies in the thermospheres of Venus and Mars are about 20% from 100 to 200 km [84.75, 76], and on Titan, they range from 20 to 30% from 800 to 2000 km. A column averaged heating efficiency for the Jovian thermosphere has been computed as 53% [84.77].

On Venus and Mars, CO_2 is the major absorber of far UV radiation, whereas on the Earth, O_2 plays that role. In the F_1 regions of the ionospheres of the terrestrial planets, dissociative recombination of molecular ions tends to be the major source of heating. Below the F_1 peak, photodissociation and neutral-neutral reactions, including quenching of metastable species, dominate. Since CH_4 is a very strong absorber, the major heating mechanisms in the thermosphere of Titan are photodissociation and neutral-neutral reactions, both above and below the F_1 peak. The few data that exist suggest that electron-impact dissociation is unimportant as a source of neutral heating, although further measurements would certainly be of benefit.

Profiles of the heat sources in the terrestrial thermosphere and that of Mars are also shown in Fig. 84.13.

Important cooling processes in planetary thermospheres include downward transport of heat by molecular and eddy conduction and infrared cooling from rotational and vibrational excitation of IR active species such as NO, CO and CO_2. Excitation of the fine structure levels of atomic oxygen and subsequent emission at 63 and 147 μm also plays a role in cooling the neutral species in the thermospheres of the terrestrial planets. In the outer planets and their satellites, hydrocarbon molecules such as CH_4 and C_2H_2 are the primary thermospheric IR radiators. The global circulation may play a role in redistributing the heat that is deposited in the dayside or auroral thermosphere [84.78, 79].

In order to model heating rates, cross sections for processes in which solar photons or photoelectrons interact with neutral species, and rate coefficients and product yields for chemical reactions of ions and neutral atmospheric species are necessary. In addition, it is necessary to know, for example, how much of the energy released appears as internal energy of the products in chemical reactions and how much appears as kinetic energy of the products. Knowledge of energy transfer processes, including vibration–vibration (V–V) and translation–vibration (T–V) transfer between atmospheric species is also important. For example, a particularly important cooling process for the thermospheres of the terrestrial planets is excitation of the CO_2 15 μm bending mode in collisions with energetic O, and subsequent radiation [84.80]. The de-excitation rate is several percent of gas kinetic, which is anomalously large for a V–T process [84.81].

In the lower ionosphere, the electrons and ions are in thermal equilibrium with the neutral species, but at higher altitudes the plasma temperatures deviate from the neutral temperatures. Near the F_1 peak, T_e is usu-

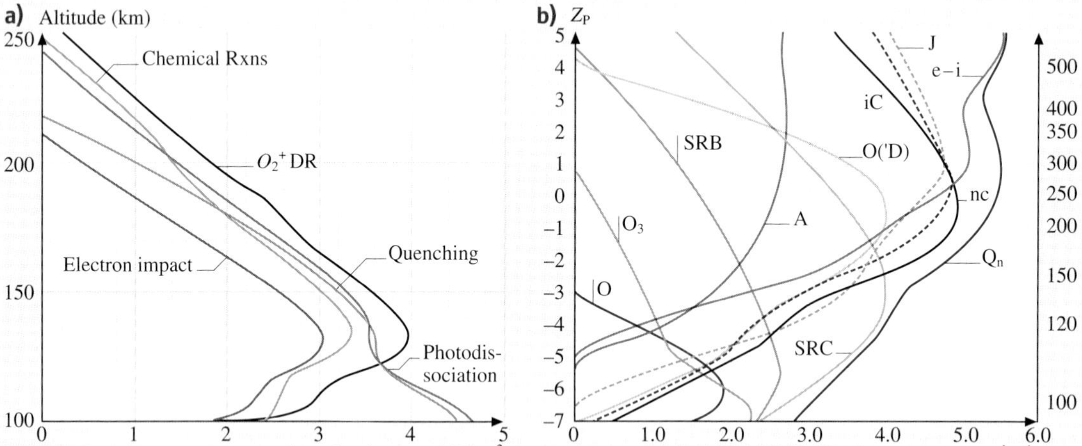

Fig. 84.13a,b Heating rates for the thermospheres of **(a)** Mars and **(b)** Earth. In **(b)**, the curve labeled Q_n is the total heating rate; e–i is the heating rate due to collisions between the neutrals and electron and ions, iC and nc are the heating rates due to exothermic ion–neutral and neutral–neutral chemical reactions, respectively; J is that due to Joule heating for a superimposed electric field of 3.6 mV m^{-1}; A is that from auroral particle precipitation; O(^{1}D) is the heating due to quenching of O(^{1}D); SRC and SRB are the heating rates due to absorption in the Schumann Runge continuum and bands, respectively; O is the heating from recombination of atomic oxygen; O_3 is the heating rate due to absorption of photons by O_3 in the Hartley bands. After [84.74]. In **(a)** the curve labeled "O_2^+ DR" is the heating rate due to dissociative recombination of O_2^+; that labeled "photodissociation" is heating due to the production of energetic neutrals in photodissociation; the curve labeled "quenching" is that due to quenching of metastable species, such as O(^{1}D); and "chemical reactions" denotes heating due to exothermic chemical reactions other than quenching of metastable species. After [84.76]

ally larger than T_i, but T_i begin to diverge from T_e at slightly higher altitudes. The energy source for the electrons on the dayside is largely photoionization, which, as discussed above, produces electrons with average energies in the 15 to 20 eV range. In slowing down, these energetic electrons lose their energy in inelastic processes with neutrals until $E \approx 1{-}2\,\text{eV}$. At this point, elastic scattering by the thermal electron population becomes the dominant energy loss process for the suprathermal electrons and the major source of heat for the thermal electrons. Other electron heating mechanisms include deactivation of electronically or vibrationally excited species, and, for the terrestrial planets, quenching of the fine structure levels of O.

As for the neutrals, heat in the electron gas is redistributed by conduction at a rate that depends on the electron thermal conductivity. This quantity is inversely proportional to the sum of the momentum transfer collision frequencies of electrons with ions, neutrals, and ambient electrons. Cooling mechanisms for thermal electrons include Coulomb collisions with ions, rotational excitation of molecules, and, for the terrestrial planets, excitation of the fine structure levels of O. Because of the large mass difference, elastic collisions between neutrals and electrons are not effective in transferring kinetic energy.

T_i in the ionosphere is elevated above T_n at high altitudes principally because of Coulomb collisions with energetic electrons. Another potentially important source of heat input to the ions near the ion peak is exothermic ion–neutral reactions, including quenching of metastable ions, such as $O^+(^2D)$, by neutrals. In the presence of electric fields, joule heating may be important and can cause T_i to exceed T_e.

The ions cool in elastic collisions and resonant charge transfer with neutral species, which are characterized by lower temperatures than the ions. The cooling rate for elastic collisions is

$$L_{in} = -2n_i \frac{m_i}{m_i + m_n} \overline{\nu}_{in} \frac{3}{2} k_B (T_i - T_n) \,. \tag{84.67}$$

Collisions between ions and neutrals (other than their parents) are dominated by the ion-induced-dipole interaction. The momentum transfer collision frequency is thus given by (84.58). Resonant charge transfer between an ion and its parent neutral, such as

$$O^{+*} + O \rightarrow O^* + O^+ \,, \tag{84.68}$$

leads to very effective ion cooling, which dominates at sufficiently high temperatures.

Examples of T_i and T_e profiles are shown in Fig. 84.14a–c for Earth, Venus and Titan. The International Reference Ionosphere (IRI) temperatures profiles

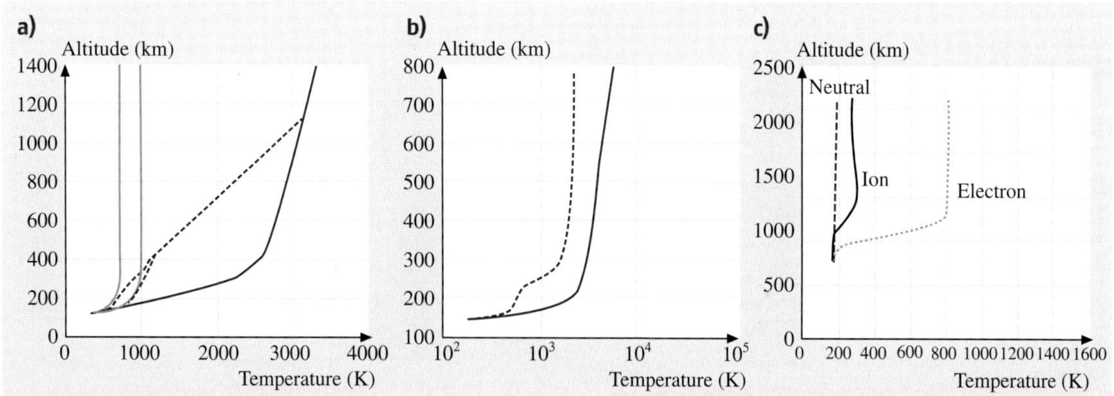

Fig. 84.14a–c Ion and electron temperature profiles. **(a)** Neutral (*dot-dashedcurve*), ion (*dashed curve*) and electron (*solid curve*) temperatures for the terrestrial ionosphere from the International Reference Ionosphere for equinox, noon and low and high solar activities. The electron temperature is found not to vary substantially with solar activity. After *Bilitza* and *Hoegy* [84.82] with kind permission from Elsevier Science Ltd., Kidlington UK. **(b)** Smoothed median ion (*dashed curve*) and electron (*solid curve*) temperatures in the Venus ionosphere as measured by the PV retarding potential analyzer and the Langmuir probe, respectively. The electron temperature profile is essentially constant with solar zenith angle. The ion temperature profile applies to solar zenith angles between 0 and 90°. After [84.83]. **(c)** Computed electron and ion temperatures for the ionosphere of Titan, including only solar photoionization as the source of electron heating. After [84.84]

for the terrestrial thermosphere show the close coupling between the electrons, ions and neutrals at low altitudes and the ions and electrons at high altitudes. T_i increases with increasing solar activity at low altitudes and approaches T_e at high altitudes. The values of T_e and T_i are about 3000 K at 1000 km [84.82]. Electron and ion temperatures in the terrestrial ionosphere are discussed in *Rees* [84.44], *Banks* and *Kockarts* [84.72], and *Whitten* and *Popoff* [84.85].

T_e and T_i in the Venus atmosphere were measured by instruments on the Pioneer Venus spacecraft. T_i, which approaches values of 2000–2500 K at high altitudes, has been found to be insensitive to solar zenith angle, except near the antisolar point, where it increases to values of 5000–6000 K [84.83]. T_e also does not vary appreciably with solar zenith angle; the high altitude values are in the range 4000–6000 K [84.86].

The retarding potential analyzer (RPA) on the Viking spacecraft found that T_i on Mars decouples from the T_n near 180 km, and approach values of about 3000 K at high altitudes [84.66]. T_e is predicted to diverge from the T_n in the lower ionosphere, and to approach values of 3000–4000 K at high altitudes [84.87].

There are no measurements of plasma temperatures in the ionospheres of the outer planets. The plasma temperatures on Titan have been predicted by a model [84.84]. The computed T_i are grater than T_n near the n_e maximum near 1000 km, but approach values of about 300 K at high altitude. For the solar source only, T_e increases rapidly to a constant value of about 800 K near 1200 km. Electrons from Saturn's magnetosphere may interact with the Titan ionosphere during the part of its orbit that is within the magnetosphere, and in this case T_e up to about 5000 K near 2000 km are predicted.

84.6 Luminosity

The luminosity that originates in the atmospheres of the planets is generally classified as dayglow, nightglow, or aurora. Dayglow is the luminosity of the dayside atmosphere that occurs as a more or less direct result of the interaction of solar radiation with atmospheric gases. Among the sources of dayglow are photodissociative excitation and simultaneous photoionization and excitation. Dayglow may also include scattering of solar radiation by processes that are selective, such as resonance scattering by atoms and fluorescent scattering by molecules, but the term generally excludes nonselective scattering processes, such as Rayleigh scattering.

In resonance scattering, the absorption of a photon by an atom in the ground state, causes a (usually dipole allowed) transition to a higher electronic state:

$$A + h\nu \rightarrow A^* , \tag{84.69}$$

followed by the emission of a photon as the state decays back to the ground state:

$$A^* \rightarrow A + h\nu . \tag{84.70}$$

The wavelength of the emitted radiation is very nearly the same as the wavelength of the radiation absorbed. The cross section for absorption of a line in the solar spectrum is

$$\sigma_{12}^{a}(\nu) = \frac{\varpi_2}{\varpi_1}\frac{c^2}{8\pi\nu^2}A_{21}\phi(\nu) , \tag{84.71}$$

where the subscript 1 indicates the lower state and 2 the upper state, ϖ is the statistical weight of the state and $\nu = c/\lambda$ is the frequency of the transition. A_{21} is the Einstein A coefficient for the transition, and $\phi(\nu)$ is the lineshape function, which in this equation is normalized so that the integral over all frequencies is unity.

If the linewidth is determined by the spread of velocities of the species, the lineshape $\phi(\nu)$ is a Doppler (Gaussian) profile

$$\phi_D(\nu) = \frac{c}{u\nu_0\sqrt{\pi}} \exp\left[-\left(\frac{\nu-\nu_0}{\nu_0}\right)^2 \frac{c^2}{u^2}\right] , \tag{84.72}$$

where ν_0 is the frequency at line center. The variable $u = (2k_BT/m)^{1/2}$ is the modal velocity of a gas in thermal equilibrium at temperature T. The width of the line at half maximum, $\Delta\nu_D$ is

$$\Delta\nu_D = \frac{2\nu_0 u}{c}(\ln 2)^{1/2} . \tag{84.73}$$

If the linewidth is determined by the natural lifetime, the profile is a Lorentzian

$$\phi_L(\nu) = \frac{\Delta\nu_L/2\pi}{(\nu-\nu_0)^2 + (\Delta\nu_L/2)^2} , \tag{84.74}$$

where $\Delta\nu_L = \Gamma_R/2\pi$ is the line width at half maximum, and

$$\Gamma_R = \Gamma_2 + \Gamma_1 , \tag{84.75}$$

where Γ_2 and Γ_1 are the inverse radiative lifetimes of the levels 2 and 1, respectively. Collisional broadening also results in a Lorentzian lineshape. If both Doppler and natural broadening mechanisms are important, the lineshape is a convolution of the two profiles, called a Voigt profile:

$$\int_{-\infty}^{\infty} \phi_{\mathrm{D}}(\nu')\phi_{\mathrm{L}}(\nu-\nu')\,\mathrm{d}\nu' . \tag{84.76}$$

The absorption cross section σ_{12}^{a} integrated over all frequencies is proportional to the absorption oscillator strength f_{12}:

$$\int_{0}^{\infty} \sigma_{12}^{\mathrm{a}}(\nu)\,\mathrm{d}\nu = \frac{\pi e^2}{m_{\mathrm{e}}c} f_{12} , \tag{84.77}$$

where m_{e} is the mass of the electron. A_{21} is related to the oscillator strength through

$$A_{21} = \frac{\varpi_1}{\varpi_2}\frac{8\pi^2 e^2 \nu^2}{m_{\mathrm{e}}c^3} f_{12} = \frac{\varpi_1}{\varpi_2}\frac{8\pi^2 e^2}{m_{\mathrm{e}}c\lambda^2} f_{12} . \tag{84.78}$$

The excitation rate q_2 of an upper level 2 by resonance scattering is given by

$$q_2 = F(\nu)\frac{\pi e^2}{m_{\mathrm{e}}c} f_{12} = F(\lambda)\frac{\varpi_2}{\varpi_1}\frac{\lambda^4}{8\pi c} A_{21} , \tag{84.79}$$

where $F(\nu)$ is the solar flux in units of photons $\mathrm{cm^{-2}\,s^{-1}\,Hz^{-1}}$ and $F(\lambda)$ is the flux in units of photons $\mathrm{cm^{-2}\,s^{-1}}$ per unit wavelength interval. It should be noted that for radiative transfer purposes the photon flux that we have called $F(\nu)$ is sometimes denoted $\pi F(\nu)$. It is customary in aeronomy to define a "g-factor," which is the probability per atom that a photon will be resonantly scattered in a particular transition:

$$g_{21} = q_2 A_{21} / \sum_i A_{2i} , \tag{84.80}$$

where the sum in the denominator is over all the lower states i that are accessible from the upper state 2. The g-factor for unattenuated solar radiation is often quoted at the mean sun-earth distance or at a particular planet. The volume emission rate $\varepsilon_{21}(z)$ for resonance scattering of a solar photon is then given by

$$\varepsilon_{21}(z) = g_{21} n_1(z) , \tag{84.81}$$

where $n_1(z)$ is the number density of atoms in level 1.

In fluorescent scattering, a photon is absorbed by a molecule in a vibrational state v producing an excited electronic state with a vibrational quantum number v'

$$AB(v) + h\nu \rightarrow AB^*(v') . \tag{84.82}$$

This is followed by emission, at wavelengths that are usually the same as or longer than that of the absorbed photon, to a range of vibrational levels v'' of a lower state

$$AB^*(v') \rightarrow AB(v'') + h\nu' . \tag{84.83}$$

The volume emission rate of a transition from a level v' of the upper electronic state to a vibrational level v'' of a lower electronic state at an altitude z is given by

$$\varepsilon_{v'v''}(z) = n(z) g_{v'v''} = n(z) q_{v'} \frac{A_{v'v''}}{\sum_{v''} A_{v'v''}} , \tag{84.84}$$

where $A_{v'v''}$ is the transition probability, $n(z)$ is the number density of the molecular species at altitude z, and $q_{v'}$ is the excitation rate of vibrational level v' of the upper electronic state from a range of lower states v. The latter quantity is

$$q_{v'} = \sum_v \eta_v F(\lambda) \frac{\pi e^2}{m_{\mathrm{e}}c^2} \lambda^2 f_{vv'} , \tag{84.85}$$

where η_v is the fraction of molecules in the v vibrational level.

Dayglow also includes emissions that are the result of the interaction of atmospheric species with the photoelectrons produced in solar photoionization, either by direct excitation

$$A + \mathrm{e}^{*-} \rightarrow A^* + \mathrm{e}^- , \tag{84.86}$$

or by simultaneous dissociation and excitation

$$AB + \mathrm{e}^{*-} \rightarrow A^* + B + \mathrm{e}^- , \tag{84.87}$$

or ionization and excitation

$$X + \mathrm{e}^{*-} \rightarrow X^{+*} + \mathrm{e}^- . \tag{84.88}$$

Electron impact processes are particularly important in producing excited states that are connected to the ground state by dipole forbidden transitions, whereas resonance and fluorescent scattering are largely limited to transitions that are dipole allowed. Dayglow emissions may also result from prompt chemiluminescent reactions, which occur when fragments or ions produced by dissociation or ionization recombine with the emission of a photon.

As an example, the dayglow spectrum of the earth from 1200 to 9000 Å is shown in Fig. 84.15. An ultraviolet spectrum of Mars as measured by the Mariner 9 spectrometer is shown in Fig. 84.16a, and the UV dayglow of Saturn and Uranus, which were measured by the Voyager spacecraft, are compared in Fig. 84.16b.

Nightglow arises from chemiluminescent reactions of species whose origin can be traced to species produced during the daytime or which have been transported

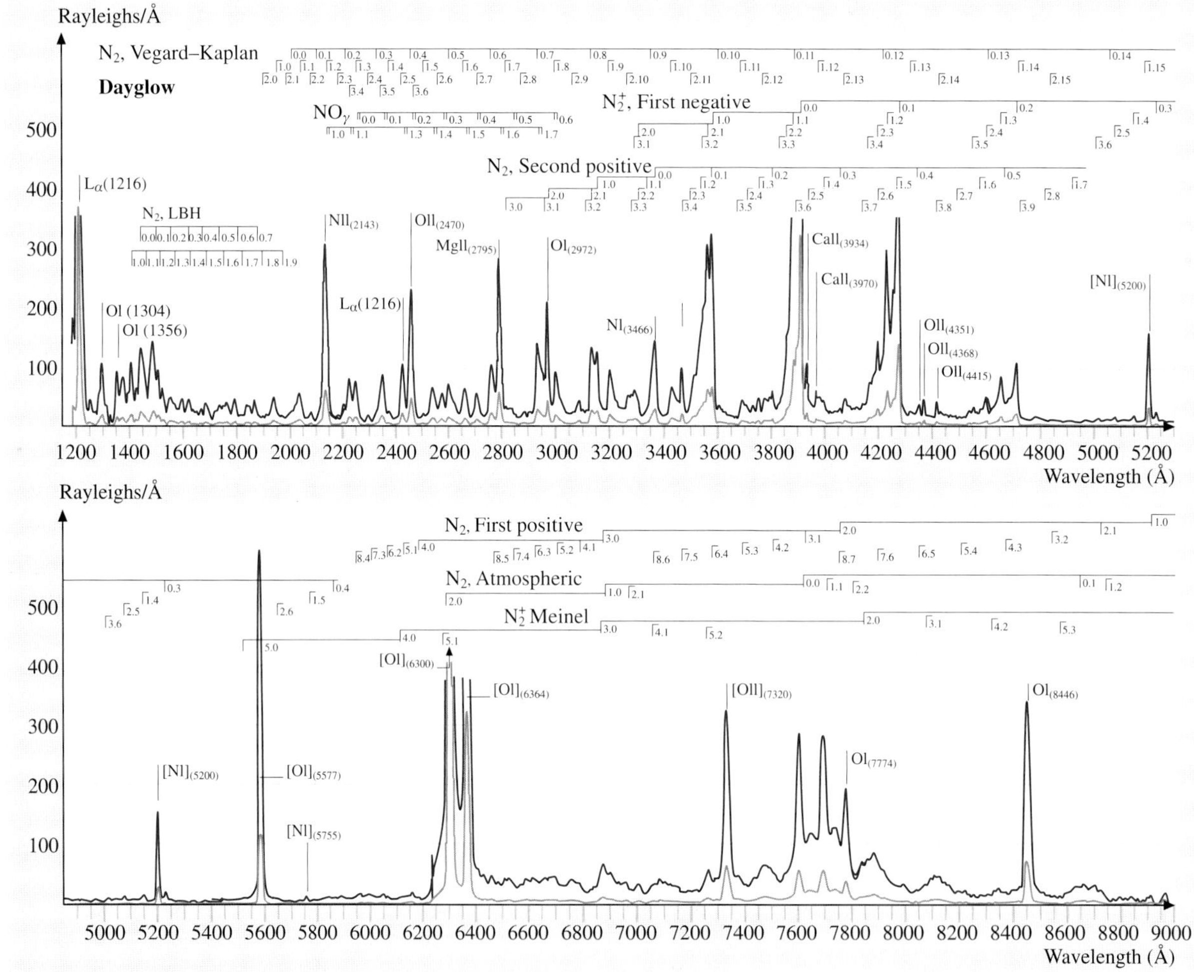

Fig. 84.15 Terrestrial dayglow spectrum measured in a single 32-second exposure by the Arizona Imager/Spectrograph on board the Space Shuttle [84.88]

from the dayside. For example, on Venus, O and N produced on the dayside are transported by the subsolar to antisolar circulation to the nightside, where they subside and radiatively associate:

$$N + O \rightarrow NO + h\nu\,, \tag{84.89}$$

producing emission in the δ and γ bands of NO (e.g., *Stewart* et al. [84.89]). The Venus UV nightglow spectrum is shown in Fig. 84.17. Similar phenomena have recently been observed by the ultraviolet spectrometer on the Mars Express spacecraft [84.90].

Auroral emissions are defined here as those produced by impact of particles other than photoelectrons. Although aurorae are usually thought of as confined to the polar regions of the Earth and the outer planets, Venus, which does not have an intrinsic magnetic field, exhibits UV emissions on the nightside that are highly variable and cannot be explained as nightglow. It has been proposed that the emissions are produced by precipitation of soft electrons into the nightside thermosphere. Mars has recently been observed to exhibit auroral emissions, which are concentrated over magnetic field anomalies in the martian crust [84.91].

The intensities of airglow and aurora are usually measured in units of brightness called Rayleighs. One Rayleigh is an apparent column emission rate at the source of 10^6 photons $cm^{-2}\,s^{-1}$ integrated over all angles, or $10^6/4\pi$ photons $cm^{-2}\,s^{-1}\,sr^{-1}$. A comparison

Fig. 84.16a,b Dayglow spectra of selected planets. (**a**) Martian airglow spectrum recorded by Mariner 9 at 15 Å resolution. After [84.92]. (**b**) Comparison of dayglow spectra from Uranus (*heavy line*) and Saturn (*thin line*) recorded by Voyager 2 [84.4] ▶

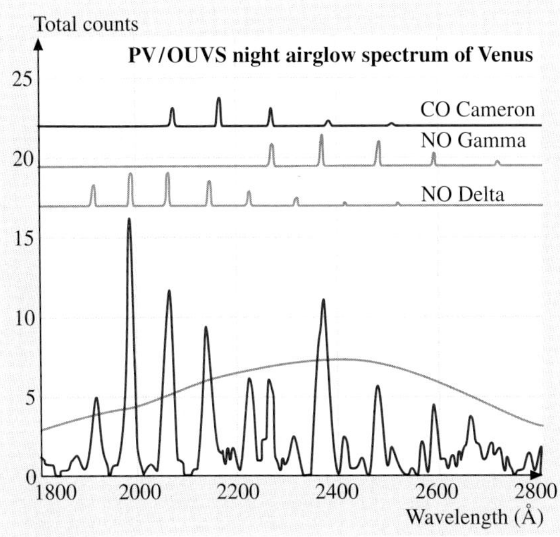

Fig. 84.17 Far ultraviolet nightglow spectrum of Venus obtained with the Pioneer Venus orbiter ultraviolet spectrometer. The predicted responses for three different band systems are also shown. After [84.93]

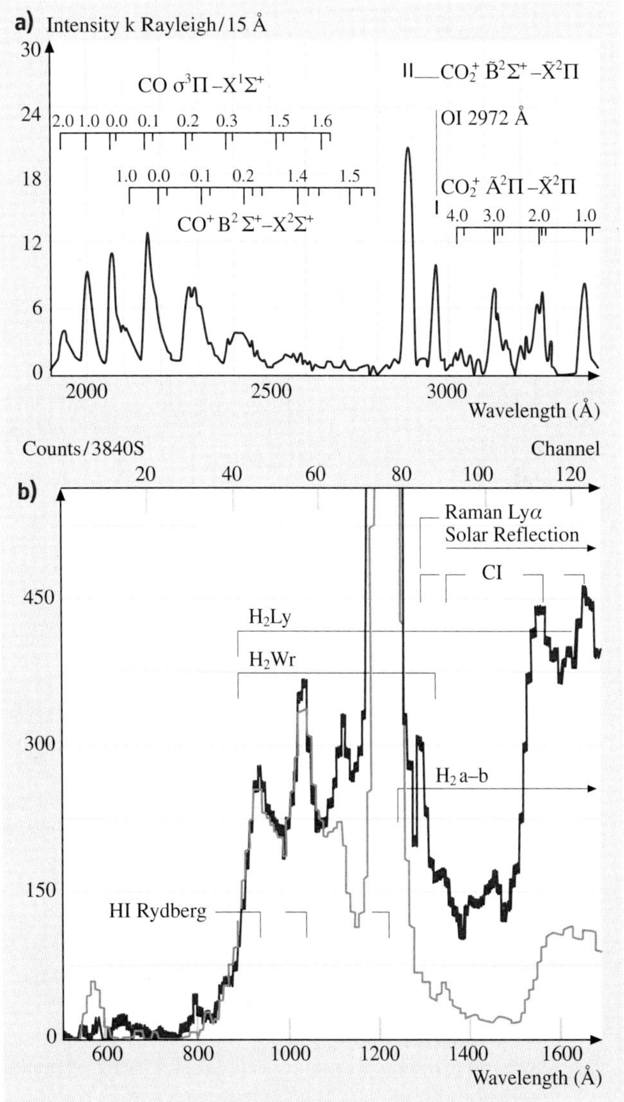

of auroral and dayglow emissions as measured by the Cassini UVIS as the spacecraft flew by Jupiter is shown in Fig. 84.18. (A. I. F. Stewart, private communication, 2004).

A discussion of terrestrial airglow and auroral emissions can be found in *Rees* [84.44]. *Meier* [84.94] has reviewed spectroscopy and remote sensing of the terrestrial ultraviolet emissions. The airglows of Mars and Venus have been discussed by *Barth* [84.19], by *Fox* [84.95] and by *Paxton* and *Anderson* [84.40]. Airglow and auroral emissions on the outer planets have been reviewed by *Atreya* et al. [84.96], *Atreya* [84.2], and *Strobel* [84.4]. Airglow in the atmospheres of the planets has been reviewed by by *Slanger* and *Wolven* [84.97].

84.7 Planetary Escape

Escape of species from atmospheres can occur by thermal and nonthermal mechanisms. Thermal processes include Jeans escape and hydrodynamic escape. Jeans escape is essentially evaporation of the energetic tail of the Maxwell–Boltzmann distribution, while hydrodynamic escape is a large-scale "blow-off" of the atmosphere that occurs when the average molecular velocity is near or above the escape velocity. Although Jeans escape still occurs for light species in the thermospheres of small planets, hydrodynamic escape is thought to have occurred only in the early history of the terrestrial planets when the solar flux in the UV was

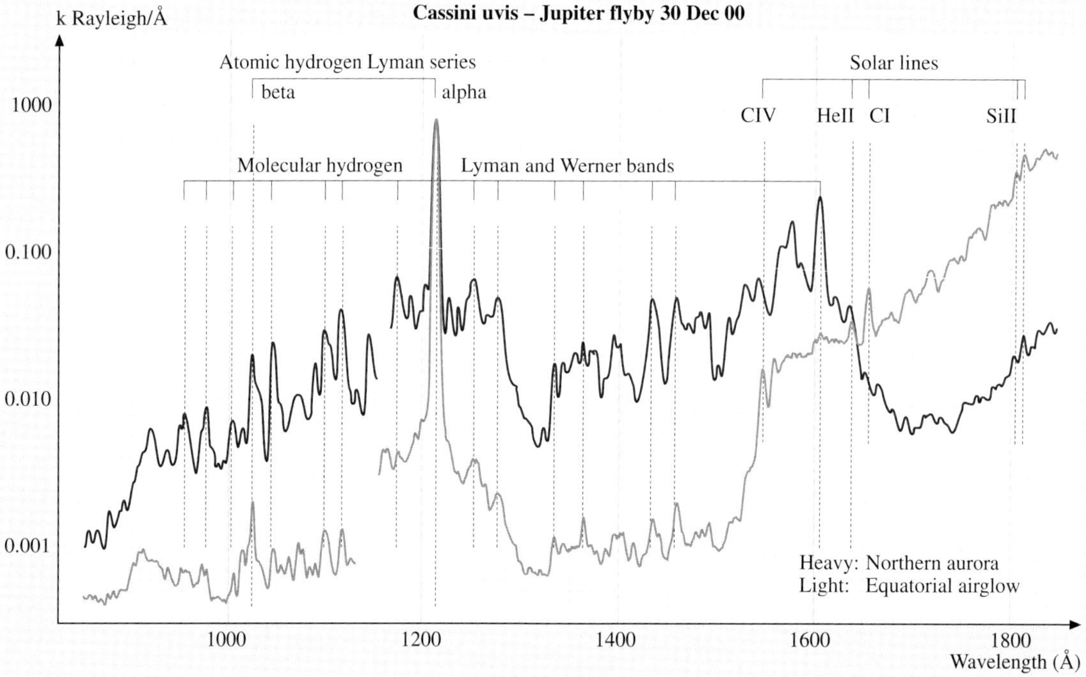

Fig. 84.18 Spectra of the northern aurora (*heavy line*) and the equatorial airglow (*light line*) taken by the UVIS spectrometer the Cassini spacecraft on closest approach on 30 December 2000. In both spectra, the Lyman and Werner bands of molecular hydrogen are prominent. They are excited by photoelectron impact in the airglow, and by primary and secondary auroral electrons. Locations of some of the brighter bands are indicated by *dotted lines*. Also shown are the Lyman-alpha and -beta lines of atomic hydrogen. At wavelengths longer than 1550 Å, the airglow spectrum is swamped by reflected UV sunlight; some solar lines are indicated. At shorter wavelengths, UV sunlight is absorbed by methane and acetylene, rather than reflected. The solar signal is much weaker in the auroral spectrum than in the equatorial, due to the higher angles of incidence and emission in the former [84.99]

higher. Nonthermal escape mechanisms, which dominate for heavy species on smaller bodies such as Mars and Titan, and for all species on Venus and Earth, include both photochemical and mechanical processes. A pedagogical discussion of escape processes can be found in *Chamberlain* and *Hunten* [84.98]

Because of the exponential rate of change of density with altitude in atmospheres, escape is sometimes assumed to occur only at and above the exobase. The exobase is mathematically defined as the altitude where the mean free path $l = (n\sigma^{c})^{-1}$ (where σ^{c} is the collision cross section), is equal to the atmospheric scale height. The probability that a particle, moving upward from the exobase with sufficient velocity will actually escape without suffering another collision is $1/\mathrm{e}$. The condition $l = H$ therefore reduces to $nH\sigma = 1$ or, equivalently, to $N = (\sigma^{c})^{-1}$, where N is the column density. Since a typical collision cross section is about $3\times10^{-15}\,\mathrm{cm}^2$, the exobase is located near the altitude above which the column density is about $3.3\times10^{14}\,\mathrm{cm}^{-3}$, although the collision cross section and thus the location of the exobase is different for different escaping species.

Whether the trajectory of a particle moving upward at the exobase is ballistic (bound) or escaping (free) is determined by its total energy E, which is the sum of its kinetic and potential energies:

$$E = \frac{1}{2}mv_{\mathrm{c}}^2 + \int_{\infty}^{r_{\mathrm{c}}} \frac{mGM}{r^2}\,\mathrm{d}r\,, \qquad (84.90)$$

where the symbols have the same meaning as in equation (84.2), and the subscript c refers to the critical level or exobase. If $E < 0$, the particle is bound. Expression (84.90) reduces to

$$E = \frac{1}{2}mv_{\mathrm{c}}^2 - mg_{\mathrm{c}}r_{\mathrm{c}} \qquad (84.91)$$

where g_c is the gravitational acceleration at the exobase. The escape velocity at the exobase, v_{esc}, is then defined by the condition $v_{esc} = (2g_c r_c)^{1/2}$. Particles with velocities greater than the escape velocity are assumed to escape if their velocity vector is oriented in the upward hemisphere and if they undergo no further collisions. The radius, gravitational acceleration, escape velocities and scale height at the equatorial exobases of the planets are given in Table 84.10.

In the Jeans process, escape occurs when particles in the high energy tail of the Maxwellian distribution attain the escape velocity. The escape flux, Φ_J is given by

$$\Phi_J = \frac{n_c u}{2\sqrt{\pi}}(1+\lambda_c)\exp(-\lambda_c)\,, \tag{84.92}$$

where $u = (2k_B T/m)^{1/2}$ is the modal velocity and λ is the gravitational potential energy in units of $k_B T$

$$\lambda = \frac{GMm}{rk_B T} = \frac{mgr}{k_B T} = \frac{r}{H}\,. \tag{84.93}$$

Table 84.10 Exobase properties of the planets

Planet	r_c (km)	g_c^a ($cm\,s^{-2}$)	$v_{esc,c}^a$ ($km\,s^{-1}$)	$H_{avg}^{a,b}$ (km)
Mercury[c]	2439	378	4.29	–
Venus[d]	6250	831	10.2	17
Earth[e]	6878	842	10.8	71
Mars[f]	3593	333	4.89	17
Jupiter[g]	73 000	2236	57.2	250
Saturn[h]	67 000	731	31.3	910
Uranus[i]	31 800	561	18.9	66
Neptune[j]	27 300	919	22.4	250
Titan[k]	4175	51.4	2.07	116
Triton[l]	2222	28.9	1.13	140

[a] Values given are those at the equatorial exobase, and assume that the thermosphere co-rotates with the planet.
[b] Average value computed from $H_{avg} = k_B T/m_{avg}g$, but does not represent the local pressure scale height, except for cases where there is one major constituent.
[c] Exobase is at the surface.
[d] Model from *Hedin* et al. [84.3] for $F_{10.7}$=150, 45° solar zenith angle.
[e] MSIS model for $F_{10.7}$=150, equator, 45° Solar Zenith Angle [84.17].
[f] Model from *Nier* and *McElroy* [84.8], and pertains to low solar activity conditions.
[g] Model from [84.2, 14].
[h] Model from *Atreya* [84.2].
[i] Model from [84.55].
[j] Model from [84.56].
[k] Model from *Strobel* et al. [84.10].
[l] Model from *Krasnopolsky* et al. [84.15]

Sometimes a correction factor is applied to the expression for the escape flux to account for the suppression of the tail of the distribution due to the escape of the energetic particles [84.100, 101].

Photochemical processes that produce energetic fragments include photodissociation and photodissociative ionization, photoelectron impact dissociation and dissociative ionization, as well as exothermic chemical reactions. The most important example of the latter are dissociative recombination reactions, which are very exothermic and tend to produce neutral fragments with large kinetic energies. Charge transfer processes such as

$$H^{+*} + O \rightarrow O^+ + H^* \tag{84.94}$$

can produce fast neutrals if the ion temperature is larger than the neutral temperature, as is usually the case near the exobase of a planet. In modeling these processes, the kinetic energy distribution of the product species is important, as well as the cross sections or reaction rates.

Physical or collisional escape mechanisms include sputtering and "knock-on." Sputtering can occur when a heavy ion picked up by the solar wind collides with an atmospheric species near or above the exobase, and in the process produces a "back-splash" in which the accelerated neutral may be ejected from the atmosphere. In knock-on, hot atmospheric neutral species, such as O atoms produced in exothermic chemical reactions near the exobase can collide with a lighter species, such as H, imparting sufficient kinetic energy to allow it to escape. Modeling these processes requires knowledge of the ion–neutral or neutral–neutral collision cross sections.

The escape rate of a light species from a planetary atmosphere may be controlled by diffusion of the species from the lower atmosphere to the exobase, rather than by the escape process itself [84.102]. The limiting upward flux, ϕ_l of a species i with mixing ratio f_i can be estimated as

$$\phi_l \approx b_i f_i / H_a\,, \tag{84.95}$$

where H_a is the average scale height of the atmosphere and b_i is the binary collision parameter introduced in Sect. 84.2.2. Equation (84.95) above is usually evaluated at the homopause, with the mixing ratio taken from a suitable altitude in the middle atmosphere, but above the cold trap (where the species condenses), if one exists. The limiting flux obtains if and only if the mixing ratio is constant with altitude. The effect of photochemistry can be accounted for if all chemical forms of the species are considered in the calculation of f_i.

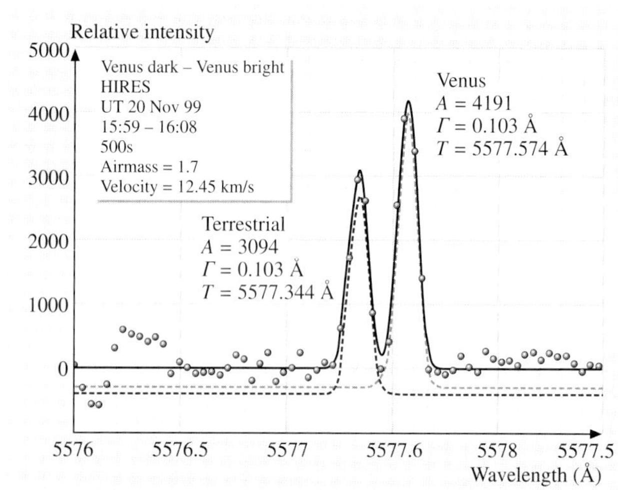

Fig. 84.19 Spectrum of the oxygen green line taken on the nightside of Venus taken by the Keck/HIRES on 20 November 1999. The individual terrestrial and Doppler-shifted Venusian components are shown by the *dashed lines* (from *Slanger* et al. 84.103)

Even if energetic particles released at the exobase of a planet do not have enough energy to escape, they may travel to great heights along ballistic orbits before falling back to the atmosphere. These particles are said to form a hot atom "corona". Hot H and O coronas have been found to surround the Earth and Venus, and have been predicted for Mars (Fig. 84.19). Reviews of the H and O coronas of Venus have been presented by *Fox* and *Bougher* [84.79] and by *Nagy* et al. [84.104]. See *Chamberlain* and *Hunten* [84.98] for a detailed discussion of planetary coronal population processes.

References

84.1 U. von Zahn, S. Kumar, J. Niemann, R. Prinn: Composition of the Venus atmosphere. In: *Venus*, ed. by D. M. Hunten, L. Colin, T. M. Donahue, V. I. Moroz (Univ. Arizona Press, Tucson 1983)

84.2 S. K. Atreya: *Atmospheres and Ionospheres of the Outer Planets and Their Satellites* (Springer, New York 1986)

84.3 A. E. Hedin, H. B. Neimann, W. T. Kasprzak, A. Seiff: J. Geophys. Res. **88**, 73 (1983)

84.4 D. F. Strobel, R. V. Yelle, D. E. Shemansky, S. K. Atreya: The upper atmosphere of Uranus. In: *Uranus*, ed. by J. T. Bergstralh, E. D. Miner, M. S. Matthews (Univ. Arizona Press, Tucson 1991)

84.5 S. K. Atreya, B. R. Sandel, P. N. Romani: Photochemical and vertical mixing. In: *Uranus*, ed. by J. T. Bergstralh, E. D. Miner, M. S. Matthews (Univ. Arizona Press, Tucson 1991)

84.6 *The U. S. Standard Atmosphere* (National Oceanic and Atmospheric Administration, U.S. Government Printing Office 1976)

84.7 R. V. Yelle, F. Herbert, B. R. Sandel, R. J. Vervack, T. M. Wentzel: Icarus **104**, 38 (1993)

84.8 A. O. Nier, M. B. McElroy: J. Geophys. Res. **82**, 4341 (1977)

84.9 R. J. Vervack Jr., B. R. Sandel, D. F. Strobel: Icarus **170**, 91 (2004)

84.10 D. F. Strobel, M. E. Summers, X. Zhu: Icarus **100**, 512 (1992)

84.11 Y. L. Yung, M. Allen, J. P. Pinto: Astrophys. J. Suppl. **55**, 465 (1984)

84.12 R. V. Yelle: private communication

84.13 R. V. Yelle, L. A. Young, R. J. Vervack, R. Young, L. Pfister, B. R. Sandel: J. Geophys. Res. **101**, 2149 (1996)

84.14 Y. H. Kim: *The Jovian Ionosphere.* Ph.D. Thesis (State University of New York at Stony Brook, New York 1991)

84.15 V. A. Krasnopolsky, B. R. Sandel, F. Herbert, R. J. Vervack: J. Geophys. Res. **98**, 3065 (1993)

84.16 D. F. Strobel, M. E. Summers: Triton's upper atmosphere and ionosphere. In: *Neptune*, ed. by D. Cruikshank, M. S. Matthews (Univ. Arizona Press, Tucson 1996)

84.17 A. E. Hedin: J. Geophys. Res. **96**, 1159–1172 (1991)

84.18 C. D. Keeling, T. P. Whorf: Atmospheric CO_2 records from sites in the SIO sampling network. In: *Trends: A Compendium of Data on Global Change. Carbon Dioxide Information Analysis Center*, ed. by T. A. Boden, D. P. Kaiser, R. J. Sepanski, W. F. Stoss (Oak Ridge National Laboratory, U.S. Department of Energy, Oak Ridge, Tenn., U.S.A. 2004)

84.19 C. A. Barth, A. I. F. Stewart, S. W. Bougher, D. M. Hunten, S. J. Bauer, A. F. Nagy: Aeronomy of the current Martian atmosphere. In: *Mars*, ed. by H. Kiefer, B. M. Jakosky, C. W. Snyder, M. S. Matthews (Univ. Arizona Press, Tucson 1992)

84.20 J. S. Lewis, R. G. Prinn: *Planets and their atmospheres: Origin and Evolution* (Academic Press, Orlando 1984)

84.21 T. M. Donahue, R. R. Hodges Jr.: Geophys. Res. Lett. **20**, 591 (1993)

84.22 V. A. Krasnopolsky, G. R. Gladstone: J. Geophys. Res. **101**, 15,765 (1996)

84.23 V. A. Krasnopolsky, P. D. Feldman: Science **294**, 1914 (2001)
84.24 W. B. DeMore: Icarus **51**, 199 (1982)
84.25 D. F. Strobel: *The Photochemistry of Atmospheres: Earth, the Other Planets, and Comets*, ed. by J. S. Levine (Academic, New York 1985)
84.26 H. B. Niemann: J. Geophys. Res. **103**, 22,831 (1998)
84.27 K. Lodders, B. Fegley: *The Planetary Scientist's Companion* (Oxford, New York 1998)
84.28 R. Courtin, D. Gautier, D. Strobel: Icarus **123**, 37 (1996)
84.29 D. M. Hunten, M. G. Tomasko, F. M. Flasar, J. R. E. Samuelson, D. F. Strobel, D. J. Stevenson: Titan. In: *Saturn*, ed. by T. Gehrels, M. S. Matthews (Univ. Arizona Press, Tuscon 1984)
84.30 J. L. Elliot, M. J. Person, S. Qu: The Astronomical Journal **126**, 1041 (2003)
84.31 D. F. Strobel, B. C. Wolven: Astrophysics and Space Science **277**, 271 (2001)
84.32 J. Saur, D. F. Strobel, F. M. Neubauer: J. Geophys. Res. **103**, 19,947 (1998)
84.33 A. J. Kliore, D. P. Hinson. F. M. Flasar, A. F. Nagy, T. E. Cravens: Science **277**, 355 (1997)
84.34 D. M. Hunten, T. H. Morgan, D. E. Shemansky: The Mercury atmosphere. In: *Mercury*, ed. by F. Vilas, C. R. Chapman, M. S. Matthews (Univ. Arizona Press, Tucson 1988)
84.35 X. Zhu, M. E. Summers, M. H. Stevens: Icarus **120**, 266 (1996)
84.36 L. M. Lara, W. H. Ip, R. Rodrigo: Icarus **130**, 16 (1997)
84.37 W. K. Tobiska: J. Atmos. Terr. Phys. **53**, 1005 (1991)
84.38 D. Lummerzheim, M. H. Rees, H. R. Anderson: Planet. Space Sci. **37**, 109 (1989)
84.39 L. Herzberg: Solar optical radiation and its role in upper atmospheric processes. In: *Physics of the Earth's upper atmosphere*, ed. by C. Hines, I. Paghis, T. R. Hartz, J. A. Fejer (Prentice-Hall, Englewood Cliffs, NJ 1965)
84.40 L. J. Paxton, D. E. Anderson: Far ultraviolet remote sensing of Venus and Mars. In: *Venus and Mars: Atmospheres, Ionospheres and Solar Wind Interactions*, Geophysical Monograph, Vol. 66, ed. by J. G. Luhmann, M. Tatrallay, R. O. Pepin (American Geophysical Union, Washington, D.C. 1992) pp. 113–190
84.41 J. M. Valentine, S. C. Curran: Rep. Prog. Phys. **21**, 1 (1958)
84.42 C. B. Opal, W. K. Peterson, E. C. Beaty: J. Chem. Phys. **55**, 4100 (1971)
84.43 W. E. Swartz, J. S. Nisbet, A. E. S. Green: J. Geophys. Res. **74**, 6415 (1971)
84.44 M. H. Rees: *Physics and Chemistry of the Upper Atmosphere* (Cambridge Univ. Press, Cambridge 1989)
84.45 A. F. Nagy, P. M. Banks: J. Geophys. Res. **75**, 6260 (1970)
84.46 H. S. Porter, F. Varosi, H. G. Mayr: J. Geophys. Res. **92**, 5933 (1987)
84.47 D. J. Strickland, R. E. Daniell, B. Basu, J. R. Jasperse: J. Geophys. Res. **98**, 21533 (1993)
84.48 S. C. Solomon: Geophys. Res. Lett. **20**, 185 (1993)
84.49 S. J. Bauer: *Physics of Planetary Ionospheres* (Springer, New York 1973)
84.50 S. G. Lias, J. E. Bartmess, J. F. Liebman, J. L. Holmes, R. D. Levin, W. G. Mallard: Gas-phase ion, and neutral thermochemistry, J. Phys. Chem. Ref. Data **17**, Suppl. 1 (1988)
84.51 F. L. Nesbitt, G. Marston, L. J. Stief, M. A. Wickramaaratchi, W. Tao, R. B. Klemm: J. Phys. Chem. **95**, 7613 (1991)
84.52 J. L. Fox, A. Dalgarno: J. Geophys. Res. **84**, 7315 (1979)
84.53 G. R. Gladstone: private communication (2004)
84.54 J. I. Moses, S. F. Bass: J. Geophys. Res. **105**, 7013 (2000)
84.55 F. Herbert, B. R. Sandel, R. V. Yelle, J. B. Holberg, A. L. Broadfoot, D. E. Shemansky: J. Geophys. Res. **92**, 15093 (1987)
84.56 J. R. Lyons, private communication
84.57 M. Summers, D. F. Strobel: private communication (2005)
84.58 D. F. Strobel, B. C. Wolven: Astrophys. Space Sci. **277**, 271 (2001)
84.59 T. Su, M. T. Bowers: Classical ion-molecule collision theory. In: *Gas Phase Ion Chemistry*, Vol. 1, ed. by M. T Bowers (Academic, New York 1979)
84.60 V. G. Anicich: Astrophys. J. Suppl. Ser. **84**, 215–315 (1993)
84.61 Y. Ikezoe, S. Matsuoka, M. Takabe, A. Viggiano: *Gas Phase Ion-Molecule Rate Constants Through 1986* (Maruzen Co., Tokyo 1986)
84.62 D. R. Bates, A. Dalgarno: Electronic recombination. In: *Atomic and Molecular Processes*, ed. by D. R. Bates (Academic, New York 1962) pp. 245–271
84.63 M. H. G. Zhang, J. G. Luhmann, A. J. Kliore: J. Geophys. Res. **95**, 17095 (1990)
84.64 D. P. Hinson and the Mars Radio Science team: Public access to MGS RS Standard electron density profiles, http://nova.stanford.edu/projects/mgs/eds-public.html (2003)
84.65 C. Y. Johnson: J. Geophys. Res. **71**, 330 (1966)
84.66 W. B. Hanson, S. Sanatani, D. R. Zuccaro: J. Geophys. Res. **82**, 4351 (1977)
84.67 A. N. Maurelis, T. E. Craveus: Icarus **154**, 350 (2001)
84.68 J. R. Lyons: Nature **267**, 648 (1995)
84.69 A. D. Richmond: The ionosphere. In: *The Solar Wind and the Earth*, ed. by S.-I. Akasofu, Y. Kamide (Terra Scientific Publishing Co., Tokyo 1987)
84.70 A. Dalgarno, M. R. C. McDowell, A. Williams: Phil. Trans. Roy. Soc. London, Ser. A **250**, 411 (1958)
84.71 R. W. Schunk, A. F. Nagy: Rev. Geophys. Space Phys. **18**, 813 (1980)
84.72 P. M. Banks, G. Kockarts: *Aeronomy* (Academic, New York 1973)

84.73 R.W. Schunk, A.F. Nagy: *Ionospheres: Physics, Plasma Physics and Chemistry* (Cambridge Univ. Press, Cambridge 2000)
84.74 R.G. Roble, E.C. Ridley, R.E. Dickinson: J. Geophys. Res. **92**, 8745–8758 (1987)
84.75 J.L. Fox: Planet. Space Sci. **37**, 36 (1988)
84.76 J.L. Fox, P. Zhou, S.W. Bougher: Adv. Space Res. **17**, (11)203–(11)218 (1995)
84.77 J.H. Waite, T.E. Cravens, J. Kozyra, A.F. Nagy, S.K. Atreya, R.H. Chen: J. Geophys. Res. **88**, 6143 (1983)
84.78 R.G. Roble: Rev. Geophys. **21**, 217–233 (1983)
84.79 J.L. Fox, S.W. Bougher: Space Sci. Rev. **55**, 357 (1991)
84.80 S.W. Bougher, D.M. Hunten, R.G. Roble: J. Geophys. Res. **99**, 14609 (1994)
84.81 R.P. Wintersteiner, R.H. Picard, R.D. Sharma, J.R. Winick, R.A. Joseph: J. Geophys. Res. **97**, 18083 (1992)
84.82 D. Bilitza, W.R. Hoegy: Adv. Space Res. **10**, (8)81–(8)90 (1990)
84.83 K.L. Miller, W.C. Knudsen, K. Spenner, R.C. Whitten, V. Novak: J. Geophys. Res. **85**, 7759 (1980)
84.84 A. Roboz, A.F. Nagy: J. Geophys. Res. **99**, 2087 (1994)
84.85 R.C. Whitten, I.G. Poppoff: *Fundamentals of Aeronomy* (Wiley, New York 1971)
84.86 R.F. Theis, L.H. Brace, H.G. Mayr: J. Geophys. Res. **85**, 7787 (1980)
84.87 W.B. Hanson, G.P. Mantas: J. Geophys. Res. **93**, 7538 (1988)
84.88 A.L. Broadfoot, B.R. Sandel, D. Knecht, R. Viereck, E. Murad: Appl. Opt. **31**, 3083 (1992)
84.89 A.I.F. Stewart, J.-C. Gerard, D.W. Rusch, S.W. Bougher: J. Geophys. Res. **85**, 7861 (1980)
84.90 J.L. Bertaux, F. Leblanc, S.V. Perrier, E. Quemerais, O. Korablev, E. Dimarellis, A. Reberac, F. Forget, P.C. Simon, B. Sandel: Nightglow in the upper atmosphere of Mars and implication for atmospheric transport, Science **307**(5709), 566–569 (2005)
84.91 J.L. Bertaux, F. Leblanc, O. Witasse, E. Quemerais, J. Lilenstein, S.A. Stern, B. Sandel, O. Korablev: Nature, **435**(7043), 790–794 (2005)
84.92 C.A. Barth, C.W. Hord, A.I. Stewart, A.L. Lane: Science **175**, 309 (1972)
84.93 A.I.F. Stewart, C.A. Barth: Science **205**, 59 (1979)
84.94 R.R. Meier: Space Sci. Rev. **58**, 1–187 (1991)
84.95 J.L. Fox: Airglow and aurora in the atmospheres of Venus and Mars. In: *Venus and Mars: Atmospheres, Ionospheres and Solar Wind Interactions*, Geophysical Monograph, Vol. 66, ed. by J.G. Luhmann, M. Tatrallay, R.O. Pepin (American Geophysical Union, Washington, D.C. 1992) pp. 191–222
84.96 S.K. Atreya, J.H. Waite Jr., T.M. Donahue, A.F. Nagy, J.C. McConnell: Theory, measurements, and models of the upper atmosphere and ionosphere of Saturn. In: *Saturn*, ed. by T. Gehrels, M.S. Matthews (Univ. Arizona Press, Tucson 1984)
84.97 T.G. Slanger, B.C. Wolven: Airglow processes in planetary atmospheres. In: *Atmospheres in the Solar System: Comparative Aeronomy*, Geophysical Monograph 130 (AGU, Washington, D.C. 2002)
84.98 J.W. Chamberlain, D.M. Hunten: *Theory of Planetary Atmospheres* (Academic, New York 1987)
84.99 A.I.F. Stewart: private communication (2004)
84.100 J.W. Chamberlain: Planet. Space Sci. **11**, 901 (1963)
84.101 D.M. Hunten: Planet. Space Sci. **30**, 773 (1982)
84.102 D.M. Hunten: J. Atmos. Sci. **30**, 1481 (1973)
84.103 T.G. Slanger, D.L. Huestis, P.C. Cosby, N. Chanover: The Venus nightglow: Ground based observations and chemical mechanisms, Icarus (2004) in press
84.104 A.F. Nagy, J. Kim, T.E. Cravens: Ann. Geophys. **8**, 251 (1990)

85. Applications of Atomic and Molecular Physics to Global Change

While there has been a general understanding and appreciation of the science involved in both global warming and stratospheric ozone depletion by atmospheric scientists for some time, detailed understanding and rigorous proof has often been lacking. Over the last ten years, there have been many advances made in filling in the details and there will continue to be rapid advances in the future. This means that any article or book discussing this topic becomes out of date as soon as it is written. Nevertheless several recent references on these topics are recommended [85.1–3].

Atomic and molecular structure and spectroscopy, as well as collision processes involving atoms, molecules, ions and electrons, are important to the study of all planetary atmospheres. For additional information on this topic, see Chapt. 84 in this volume.

85.1 Overview

85.1.1 Global Change Issues

Over the last several decades there has been increasing concern about the global environment and the effect of human perturbations on it. This whole area, which involves a wide range of scientific disciplines, has become known as *Global Change*. Knowledge of processes taking place in the atmosphere, oceans, land masses, and plant and animal populations, as well as the interactions between these various earth-system components is essential to an overall understanding of global change – both natural and human-induced.

The processes of atomic and molecular physics find greatest application in the area of atmospheric global change. The two major issues which have received significant attention in both the media and the scientific literature are: (1) global warming, due to the buildup of infrared-active gases; and (2) stratospheric ozone depletion due to an enhancement of destructive catalytic cycles. Although both of these problems are thought to be caused by atmospheric pollutants due to industrialized human society, the general problem of air pollution and its direct effects on plant and animal populations will not be addressed here.

85.1.2 Structure of the Earth's Atmosphere

The vertical temperature structure of the earth's atmosphere shown in Fig. 85.1 provides an important nomenclature that is widely used [85.4]. The atmosphere is divided into regions called "-spheres", in which the sign of the temperature gradient with respect to altitude, dT/dz, is constant. The regions in which the temperature gradient changes sign are called "-pauses". The precise altitude pertaining to each of these regions can vary depending upon latitude and the time of year.

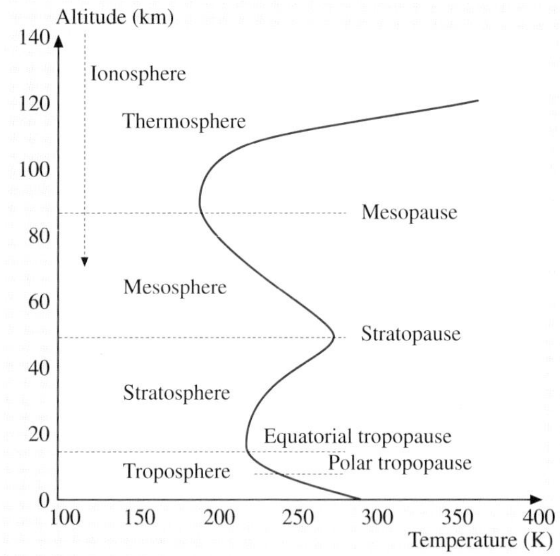

Fig. 85.1 Vertical temperature profile of the atmosphere

In the troposphere, covering the range from 0 to ≈ 15 km above the earth, the temperature steadily decreases with altitude. This is the most complex region of the atmosphere, as it interacts directly with plant and animal life, land masses and the oceans. It is the region in which weather occurs.

The change in sign of the temperature gradient at the tropopause to a positive dT/dz in the stratosphere is due to heating by absorption of solar ultraviolet radiation which photodissociates O_2 and O_3. The stratosphere, extending from ≈ 15 to 50 km above the earth, contains the ozone layer which shields the earth's surface from harmful ultraviolet radiation in the range of 280–320 nm.

At the stratopause, the heating processes have become too weak to compete with the cooling processes, and throughout the mesosphere, approximately 50–85 km, the temperature again decreases with increasing altitude. Cooling processes, which will be discussed in Sect. 85.2.2, involve collisional excitation of molecular vibrational modes which decay by radiating to space. The coldest temperatures in the atmosphere are found at the mesopause, where the temperature gradient once again becomes positive. From approximately 70 km upward, a very diffuse plasma called the ionosphere exists due to photoionization of atoms and molecules by short wavelength (UV and EUV) solar radiation.

Throughout the thermosphere, which extends from approximately 90 km upward, heating occurs because the atmosphere has become so thin that there are very few collisions and thus inefficient equilibration of the highly translationally excited atoms and ions with the molecular species which can radiate in the infrared. This "bottleneck" for energy loss causes increased heating. In the thermosphere and ionosphere, the thermal inertia is very small and there are huge temperature variations, both diurnally, and with respect to solar activity.

The densities are low enough in the thermosphere, ionosphere, and mesosphere, that the primary processes determining the chemical and physical characteristics of these regions are two-body processes and "half-collision events" discussed elsewhere in this volume: dissociative recombination, photoionization, photodissociation, charge transfer, and collisional excitation of molecular rotation and vibration. As the altitude decreases, the density increases. Then three-body interactions, interactions on surfaces (of aerosols and ices), and complex chemical cycles together with dynamical effects such as winds determine the chemical and physical characteristics of the stratosphere and troposphere.

85.2 Atmospheric Models and Data Needs

While models are absolutely essential to the study of any system as complex as the earth's atmosphere, they play a particularly fundamental role in exploring global change issues. Models not only provide predictions of future changes, but also allow exploration of sensitivities to particular parameters. Comparing the results of a model with observations ultimately tests and challenges scientific understanding. Of critical importance is the atomic and molecular data which goes into the models.

Generally, atmospheric models become increasingly complex as altitude decreases. General Circulation Models (GCMs), incorporating thousands of chemical reactions, global wind patterns, and abundances of large numbers of trace species, require supercomputers in order to model aspects of the troposphere and stratosphere. Tropospheric chemistry and transport models, such as GEOS-CHEM [85.5], model the sources, evolution, transport, and sinks of pollution, as well as the oxidative capacity of the troposphere.

85.2.1 Modeling the Thermosphere and Ionosphere

The data necessary for modeling of the thermosphere and ionosphere are described here. The primary components of such a model include: a solar spectrum of photon fluxes as a function of wavelength, concentrations of neutral species, photoabsorption, photodissociation and photoionization cross sections as a function of wavelength. The computer code brings all these elements together, calculating opacity as the solar radiation propagates downward through the atmosphere, keeping track of ion production and electron production. Additional steps are required to calculate the abundances of trace species such as NO^+, necessitating the inclusion of all relevant ion-neutral reactions. Electron energy degradation can be tracked by including inelastic collisions of electrons with ions, atoms, and molecules.

The primary neutral species are N_2, O_2, O, and He. Below about 100 km, the atmosphere is fully mixed by turbulence in the ratio 78% N_2, 21% O_2 and 1% trace species such as O_3 and CO_2. Above 100 km, turbulence dies out and the atmospheric species are in diffusive equilibrium, distributed by their molecular weight, with atomic oxygen dominating above ≈ 150 km and He dominating much higher. The major ions are N_2^+, O_2^+, O^+ and NO^+.

85.2.2 Heating and Cooling Processes

In the upper atmosphere, heating occurs through absorption of short wavelength solar radiation to produce ionization and dissociation, and is mediated by collisions between electrons, ions, and neutrals. Ions and electrons are created during the daytime and to a great extent disappear during the night with the absence of solar radiation. Processes such as dissociative recombination, the primary electron loss mechanism, heat the gas:

$$e + O_2^+ \rightarrow O(^3P) + O(^3P) + 7\,\text{eV}\,, \qquad (85.1)$$

$$e + N_2^+ \rightarrow N(^4S) + N(^4S) + 6\,\text{eV}\,. \qquad (85.2)$$

Cooling takes place when the kinetic energy of the gas is transformed through collisions into internal energy which can then be radiated away. The primary coolant above ≈ 200 km is the fine structure transition of atomic oxygen, $O(^3P_1) \xrightarrow{h\nu} O(^3P_2)$, which is excited by thermal collisions and radiates at 63 μm. From approximately 120 km to 200 km, the fundamental band of NO, $v = 1 \rightarrow v = 0$, which is excited by collisions with atomic oxygen and radiates at 5.3 μm, dominates the cooling. Below 120 km and throughout the mesosphere and stratosphere, the primary coolant is the ν_2 band of CO_2 radiating at 15 μm. This transition is excited by collisions of CO_2 and atomic oxygen. Cooling throughout most of the atmosphere is accomplished through trace species because the major molecular species, N_2 and O_2, are not infrared active.

85.2.3 Atomic and Molecular Data Needs

Knowledge of rate coefficients for ion-neutral and neutral-neutral reactions as a function of vibrational and rotational excitation of the reactants is becoming increasingly important, as there is recent evidence of more internal excitation of molecular species than had previously been thought [85.6]. Accurate photoabsorption, photodissociation and photoionization cross sections as a function of wavelength for all the relevant species are important parameters determining the reliability and ultimate accuracy of an atmospheric model. Compilations of data, such as that by *Conway* [85.7] and *Kirby* et al. [85.8] are very useful, but can become rapidly outdated. The Smithsonian Astrophysical Observatory maintains the world standard database, HITRAN, for molecular line parameters and absorption cross sections from the microwave through the ultraviolet for analysis of atmospheric spectra [85.9]. Discussions of the needs for atomic and molecular data in the context of space astronomy, but including applications to atmospheric physics, can be found in a book edited by *Smith* and *Wiese* [85.10].

85.3 Tropospheric Warming/Upper Atmosphere Cooling

85.3.1 Incoming and Outgoing Energy Fluxes

The overall temperature of a planet is determined by a balance between incoming and outgoing energy fluxes. In a steady state, the planet must radiate as much energy as it absorbs from the sun. The Earth, radiating as a black-body at an effective temperature T_E, obeys the Stefan–Boltzmann law in which the energy emitted is expressed as $\sigma T_E^4 4\pi R_E^2$, with σ the Stefan–Boltzmann constant, and R_E the radius of the earth. An equation expressing the equality of energy absorbed and energy emitted can be written [85.11] as

$$F_S \pi R_E^2 (1 - A) = \sigma T_E^4 4\pi R_E^2 , \quad (85.3)$$

where A is the albedo of the earth (the fraction of solar radiation reflected from, rather than absorbed by, the Earth), F_S is the solar flux at the edge of the earth's atmosphere, and πR_E^2 is the Earth's area normal to the solar flux. Solving this equation for T_E, one obtains $T_E \approxeq 255$ K (−18 °C).

The sun, which has a surface temperature of approximately 6000 K, emits most of its radiation in the 0.2–4.0 μm region of the spectrum (200–4000 nm). The upper atmosphere of the Earth (thermosphere, ionosphere, mesosphere, and stratosphere) absorbs all the solar radiation shortward of 320 nm. The atmosphere of the earth absorbs only weakly in the visible region of the spectrum where the solar flux peaks.

The Earth, with an effective radiating temperature of 255 K, emits mainly long-wavelength radiation in the 4–100 μm region. Molecules naturally present in the atmosphere in trace amounts, such as carbon dioxide, water and methane, absorb strongly in this wavelength region [85.12]. Radiation coming from the Earth is thus absorbed, reradiated back to the surface, and thermalized through collisions with the ambient gas. This trapping of the radiation produces an additional warming of 33 K. Thus the mean surface temperature of the Earth is 288 K, not 255 K as found for T_E above. This effect of the Earth's atmosphere is known as the greenhouse effect. The greenhouse effect is what makes Earth habitable for life as we know it. Gases, both natural and man-made, which absorb strongly in the 4–100 μm region, are known collectively as greenhouse gases.

85.3.2 Tropospheric "Global" Warming

According to a 2000 National Research Council Report, "the global-mean temperature at the earth's surface is estimated to have risen by 0.25 to 0.4 °C during the past 20 years" [85.13]. The Intergovernmental Panel on Climate Change (IPCC) has also concluded that global surface temperatures have increased and that "there is new and stronger evidence that most of the warming over the last 50 years is attributable to human activities" [85.14]. The Arctic region has warmed by an estimated 1 °C in the past two decades, leading to substantial changes in the cryosphere [85.15]. Antarctic sea ice was stable from 1840 to 1950, but has since declined sharply. Sea ice extent shows a 20% decline since about 1950 [85.16, 17].

From air bubbles trapped at different depths in polar ice, it is possible to determine carbon dioxide and methane concentrations several thousand years ago. Over the last two hundred years, CO_2 levels have increased by 20%, from 280 to 330 ppm. Over the next century the total amount of CO_2 in the atmosphere since 1900 is expected to double to as much as 600 ppm [85.18]. This increase is due primarily to the burning of fossil fuels.

Although methane is present at levels several orders of magnitude less than CO_2, it is increasing much more rapidly. Methane concentrations have more than doubled over the last two hundred years due to industrial processes, fuels, and agriculture [85.18].

The man-made chlorofluorocarbons (CFCs), which have been widely used as refrigerants and in industry, have been increasing in the atmosphere at a rate of over 5% per year since the 1970s. Only recently has there been an indication that this trend is slowing down [85.19].

Ozone, which is a primary component of chemical smog, is a pollutant when it occurs in the troposphere and an effective greenhouse gas. It has been increasing worldwide also.

This buildup of CO_2, CH_4, CFCs and tropospheric O_3 causes a problem. In much of the spectral region from 5–100 μm, there is 100% absorption of radiation by the atmosphere – due mainly to naturally occurring water vapor. There is, however, a region of rather weak absorption, from ≈ 7–15 μm, known as the "atmospheric window". Increased concentrations of the greenhouse gases strengthen the absorption in this region, tending to "close" this window, thus increasing the infrared opacity of the atmosphere. The increased opacity causes an immediate decrease in the thermal radiation from the planet-atmosphere system, forcing the temperature to rise until the energy balance is restored [85.20].

It is difficult to prove that the buildup of greenhouse gases is the cause of the observed temperature rise. Other possible causes include slight changes in solar activity and irradiance, and changes in ocean currents, which may have a profound effect on global temperature and climate. These are areas of active research.

Given the increase in concentrations of greenhouse gases that has occurred and is predicted to continue, the change in radiative heating of the troposphere can be calculated. Models generally predict an increase in tropospheric temperatures ranging from 1.5 to 4.5 °C, upon doubling the CO_2 concentrations over the next century. The 3 °C range in temperature is due to the ways that different models incorporate climate feedbacks. Climate feedbacks include water vapor, snow and sea ice, and clouds. Rising temperatures increase the concentration of water vapor, which is itself a greenhouse gas, producing further warming. Rising temperatures reduce the extent of reflective snow and ice, thus reducing the Earth's albedo. This leads to increased absorption of solar radiation, further increasing temperatures. Clouds both contribute to the albedo, thereby reducing the solar flux reaching the Earth, and absorb infrared radiation causing temperatures to rise. The modeling of clouds and their radiative properties is very difficult, and is one of the largest sources of uncertainty in the climate models. Understanding the role that the ocean, with its giant heat capacity, plays in global warming, and identifying and quantifying the various interactions occurring at the ocean-atmosphere interface, are vital areas of research which will affect the size of the predicted temperature increase. At present, there are few obvious opportunities for traditional atomic and molecular physics to play a significant part in global-warming research.

85.3.3 Upper Atmosphere Cooling

The buildup of CO_2 has an even greater effect on the temperature in the upper atmosphere than on that in the troposphere [85.21]. As discussed in the Sect. 85.2.2, CO_2 is a coolant in the stratosphere, mesosphere and thermosphere, but as a greenhouse gas is involved in heating the troposphere. The explanation for this revolves around the collision physics issue of quenching versus radiating.

In the troposphere, CO_2 absorbs infrared radiation coming from the Earth, exciting the ν_2 vibrational bending mode at 15 μm. The excited molecule can either reradiate or collisionally de-excite. In the lower atmosphere where densities are large, the lifetime against collisions is very short and the excited molecule is rapidly quenched. This transfer of energy from radiation through collisions into the kinetic energies of the colliding partners results in a net heating.

In the stratosphere and above, atomic oxygen collisions with CO_2 excite this same bending mode. But at these higher altitudes, densities are lower and quenching is greatly reduced. The excited molecule radiates and the radiation escapes to space. A net cooling results because the opacity is low at these altitudes.

Roble and co-workers [85.22, 23] have investigated the doubling of CO_2 and CH_4 concentrations (as predicted for the next century) in the mesosphere and at the lower boundary of the thermosphere. Using sophisticated atmospheric general circulation models, they predict that the stratosphere, mesosphere and thermosphere will show significant cooling — the largest cooling of 40–50 °C occurring in the thermosphere.

The extent of this cooling very much depends on the rate coefficient for the O + CO_2 excitation of the ν_2 bending mode. *Rishbeth* and *Roble* [85.22] assumed a value for this rate coefficient of 1×10^{-12} cm^3/s, intermediate between the value of *Sharma* and *Wintersteiner* [85.24] (6×10^{-12} cm^3/s) and an earlier value of 2×10^{-13} cm^3/s used by *Dickenson* [85.25]. The Sharma and Wintersteiner value, based on observations of 15 μm emission in the atmosphere around 100–150 km, was recently confirmed by *Rodgers* et al. [85.26], but *Pollock* et al. [85.27] obtain a value of 1.2×10^{-12} cm^3/s in laboratory experiments. Using the larger rate coefficient would result in even greater cooling [85.28].

The overall consequences of such a large temperature decrease in the upper atmosphere have not been fully explored — particularly the question as to how the dynamics of the atmosphere will be affected. Since many chemical reactions depend on temperature, there may be considerable readjustments in the vertical distribution of minor species in the atmosphere. Cooler temperatures cause the atmosphere to contract, reducing densities and, consequently, satellite drag. Cooler temperatures may also increase the occurrence of polar stratospheric clouds, thereby affecting ozone depletion (Sect. 85.4).

Most significantly, tropospheric warming and upper atmosphere cooling both result from a buildup of CO_2. The size of the predicted cooling is greater by an order of magnitude than the amount of the predicted heating. Thus it may be possible to monitor the global warming trend by observing the predicted cooling in the upper atmosphere.

There is evidence in the mesosphere that this cooling has already begun. Temperatures appear to have decreased by 3–4 °C over the last decade [85.29, 30]. *Gadsden* [85.31] has also found that the frequency of occurrence of noctilucent clouds, the highest-lying clouds in the atmosphere, has more than doubled over the last twenty-five years. He has calculated that this change could result from a decrease in the mean temperature at the mesopause of 6.4 °C during this time period. However, increased concentrations of water produced by oxidation of increased amounts of methane may be responsible for the more frequent appearance of the clouds. This is an ongoing area of research.

85.4 Stratospheric Ozone

85.4.1 Production and Destruction

Ozone production takes place continually in the stratosphere during daylight hours, as molecular oxygen is photodissociated and the resulting oxygen atoms undergo three-body recombination with O_2:

$$O_2 + h\nu \rightarrow 2O\,, \quad (85.4)$$

$$O + O_2 + M \rightarrow O_3 + M\,. \quad (85.5)$$

Ozone can be destroyed through photodissociation:

$$O_3 + h\nu \rightarrow O_2 + O\,, \quad (85.6)$$

but because an oxygen atom is produced which immediately recombines with another O_2 to form O_3, no net loss of O_3 results. The photodissociation of O_2 and O_3 are important heating processes in the stratosphere.

The amount of ozone in the stratosphere is quite variable, changing significantly with the seasons and with latitude. In the lower stratosphere, much of the ozone is created over the equatorial regions and then transported toward the poles.

Besides being photodissociated, O_3 is destroyed by reactions with radicals that are involved in catalytic cycles. The short-hand notation for the major cycles, NO_x, HO_x, ClO_x (BrO_x) refers to the catalytically active forms involved in the cycles. Our knowledge about the relative importance of these catalytic cycles in ozone destruction has increased dramatically over the last decade. A number of these cycles are given below, with the ozone-destroying step listed first, and the rate-limiting step closing the catalytic cycle and regenerating the ozone-destroying radical, listed last. The net effect in each of these cases is to convert ozone and atomic oxygen (otherwise known as odd-oxygen) into molecular oxygen:

$$\begin{aligned} NO + O_3 &\rightarrow NO_2 + O_2 \\ NO_2 + O &\rightarrow NO + O_2 \\ \text{NET: } O_3 + O &\rightarrow 2O_2\,, \end{aligned} \quad (85.7)$$

and

$$\begin{aligned} NO + O_3 &\rightarrow NO_2 + O_2 \\ NO_2 + O_3 &\rightarrow NO_3 + O_2 \\ NO_3 + h\nu &\rightarrow NO + O_2 \\ \text{NET: } 2O_3 &\rightarrow 3O_2\,. \end{aligned} \quad (85.8)$$

$$\begin{aligned} OH + O_3 &\rightarrow HO_2 + O_2 \\ HO_2 + O_3 &\rightarrow OH + 2O_2 \\ \text{NET: } 2O_3 &\rightarrow 3O_2\,, \end{aligned} \quad (85.9)$$

and the halogen cycle, in which $Z = Cl$ or Br:

$$\begin{aligned} Z + O_3 &\rightarrow ZO + O_2 \\ ZO + O &\rightarrow Z + O_2 \\ \text{NET: } O_3 + O &\rightarrow 2O_2\,. \end{aligned} \quad (85.10)$$

The following series of reactions couples the HO_x and halogen cycles:

$$\begin{aligned} HO_2 + ZO &\rightarrow HOZ + O_2 \\ HOZ + h\nu &\rightarrow OH + Z \\ Z + O_3 &\rightarrow ZO + O_2 \\ OH + O_3 &\rightarrow HO_2 + O_2 \\ \text{NET: } 2O_3 &\rightarrow 3O_2\,. \end{aligned} \quad (85.11)$$

Finally the reaction set

$$\begin{aligned} BrO + ClO &\rightarrow Br + Cl + O_2 \\ Br + O_3 &\rightarrow BrO + O_2 \\ Cl + O_3 &\rightarrow ClO + O_2 \\ \text{NET: } 2O_3 &\rightarrow 3O_2 \end{aligned} \quad (85.12)$$

is also important in the halogen destruction cycle. The coupling between these different cycles by reactions such as

$$HO_2 + NO \rightarrow OH + NO_2 \quad (85.13)$$

turns out to be very important in understanding the details of ozone destruction, such as how much each mechanism contributes to the destruction as a function of altitude and in the presence of aerosols. *Wennberg* et al. [85.32] have recently shown that catalytic destruction by NO_2, which for two decades was considered to be the predominant loss process, accounted for less than 20% of the O_3 removal in the lower stratosphere during May 1993. They further show that the cycle involving the hydroxyl radical accounted for nearly 50% of the total O_3 removal and the halogen-radical chemistry was responsible for the remaining 33%.

The NO_x and HO_x cycles are naturally occurring, whereas the ClO_x and BrO_x cycles are due mainly to man-made chemicals – the CFCs and halons. The amplification that takes place through a catalytic cycle is the reason that these chemicals, which are only present at the level of parts per trillion, can have such a destructive effect.

It is useful to think in terms of a total chemical budget for a radical such as Cl which enters into a cycle. Chlorine is put into the stratosphere when chemicals such as CF_2Cl_2 are released into the atmosphere. Such compounds are chemically inert and insoluble in water, and therefore are not easily cleansed out of the lower atmosphere. In the stratosphere, however, the CF_2Cl_2 is subjected to solar UV radiation and is photodissociated, producing the Cl radical. Chlorine exists in the upper atmosphere in catalytically active forms, Cl and ClO, as well as in stable, reservoir species, HCl and $ClONO_2$. The total chlorine budget consists of both the catalytically active plus reservoir species. Reactions which reduce the formation of reservoir species, or convert reservoir species to catalytically active forms, contribute to the ozone destruction. Photolysis of stable reservoir species, such as $ClONO_2$, can produce catalytically active forms. Bromine has an identical cycle to that of chlorine, but is 50 to 100 times more destructive than Cl because it does not react readily to go into its reservoir form, HBr. A knowledge of the photodestruction rates of all such species is important to an understanding of the overall ozone photochemical depletion problem.

Studies of the Antarctic ozone hole show that gas phase photochemical cycles, as given above, are not the whole story with respect to ozone depletion. Heterogeneous chemistries taking place on the surfaces of ice crystals and sulfate aerosols play an important role also. These are discussed briefly in Sect. 85.4.2 and Sect. 85.4.4.

85.4.2 The Antarctic Ozone Hole

The ozone depletion problem was largely theoretical until the discovery of the ozone hole over Antarctica. Following the 1985 announcement by *Farman* et al. [85.33] of ground-based observations of significant decline in O_3 concentrations during springtime in the Southern Hemisphere, it was possible to map this event using archived satellite data beginning in 1979. The data depict a worsening event throughout the early 1980s. In 1987, 70% of the total O_3 column over Antarctica was lost during the month of September and early October, and the areal extent of the hole was $\approx 10\%$ of the Southern Hemisphere. The ozone hole has continued to grow in depth and width [85.34]. Recent data shows that this phenomenon continues, with the 2003 ozone hole the second largest observed to date (the largest yet observed was on September 10, 2000) [85.35].

The causal link between the release and buildup of man-made CFCs and the ozone hole over Antarctica has been quite convincingly established by *Anderson* et al. [85.36] through in situ observations from high altitude aircraft flights into the polar vortex during the end of polar night and the beginning of Antarctic spring in 1987.

The polar vortex is a stream of air circling Antarctica in the winter, creating an isolated region which becomes very cold during the polar night. Flights into the vortex were able to document a heightened, increasing level of ClO and a monotonically decreasing O_3 concentration over a 3–4 week time period during late September and early October.

The mechanism which appears to be repartitioning the chlorine from its reservoir form into its catalytically active form is a heterogeneous process occurring on the surfaces of polar stratospheric clouds. At the cold temperatures during the polar night, polar stratospheric clouds form, consisting of ice and nitric acid trihydrate. Gaseous $ClONO_2$ collides with HCl that has been adsorbed onto the surface of the cloud crystals. Chlorine gas is liberated and the nitric acid formed in the reaction remains in the ice [85.36]:

$$\mathrm{HCl} + \mathrm{ClONO_2} \rightarrow \mathrm{Cl_2}(g) + \mathrm{HNO_3}\,. \tag{85.14}$$

As solar radiation starts to penetrate the region at the beginning of spring, the Cl_2 molecules are rapidly photodissociated, producing Cl atoms which initiate the catalytic destruction of O_3.

As there are no oxygen atoms around to complete the catalytic cycles, several mechanisms for regenerating the Cl and Br radicals have been proposed which involve only the ClO and BrO molecules themselves.

Mechanism I [85.37]

$$\begin{aligned}
\mathrm{ClO} + \mathrm{ClO} + M &\rightarrow (\mathrm{ClO})_2 + M\\
(\mathrm{ClO})_2 + h\nu &\rightarrow \mathrm{Cl} + \mathrm{ClOO}\\
\mathrm{ClOO} + M &\rightarrow \mathrm{Cl} + \mathrm{O_2} + M\\
2\times(\mathrm{Cl} + \mathrm{O_3} &\rightarrow \mathrm{ClO} + \mathrm{O_2})\\
\text{NET: } 2\mathrm{O_3} &\rightarrow 3\mathrm{O_2}\,;
\end{aligned} \tag{85.15}$$

Mechanism II [85.38]

$$\begin{aligned}
\mathrm{ClO} + \mathrm{BrO} &\rightarrow \mathrm{Cl} + \mathrm{Br} + \mathrm{O_2}\\
\mathrm{Cl} + \mathrm{O_3} &\rightarrow \mathrm{ClO} + \mathrm{O_2}\\
\mathrm{Br} + \mathrm{O_3} &\rightarrow \mathrm{BrO} + \mathrm{O_2}\\
\text{NET: } 2\mathrm{O_3} &\rightarrow 3\mathrm{O_2}\,.
\end{aligned} \tag{85.16}$$

While Mechanism I accounts for 75% of the observed ozone loss, the sum of I and II yields a destruction rate in harmony with the observed O_3 loss rates [85.36].

85.4.3 Arctic Ozone Loss

The region around the North Pole does not appear to exhibit an ozone hole as severe as that found in the Antarctic. Several factors lessen the probability of a significant ozone hole developing in the Arctic. First, a stable polar vortex does not get well established due to increased atmospheric turbulence from the greater land surface area in the Northern Hemisphere. Second, temperatures during the Arctic winter do not get as cold as during the Antarctic winter, so that polar stratospheric clouds (PSCs) do not form as easily. As seen in the preceding section, the surfaces of PSCs play an essential role in the O_3 destruction mechanisms in the Antarctic.

However, ozone levels are showing 20–25% reductions during February and March [85.39] over a much larger area around the North Pole than in the South. Thus ozone destruction is taking place during the transition from polar winter to spring in the Arctic but the phenomenon is more widespread, diffuse, and not as well-contained as in the Antarctic.

85.4.4 Global Ozone Depletion

Over the last twenty-five years, satellite instruments have measured the total ozone column in the atmosphere. During this time ozone levels have been steadily decreasing globally, especially at mid- to high-latitudes. Recent analysis indicates the first evidence of recovery of stratospheric ozone levels, with diminished rates of ozone loss at altitudes of 35–45 km, coupled with a slowdown in the increase in stratospheric loading of chlorine [85.40].

Heterogeneous reactions on aerosol surfaces, as well as the homogeneous gas phase chemical cycles mentioned earlier, must be invoked to explain the global decline in ozone levels. A particularly important reaction appears to be the hydrolysis of N_2O_5 on sulfate aerosols. This occurs very rapidly, converting reactive nitrogen, NO_2, into its reservoir species HNO_3:

$$N_2O_5 + H_2O \xrightarrow{\text{sulfate aerosol}} 2HNO_3 \,. \tag{85.17}$$

The N_2O_5 is formed at night by reaction of NO_2 and NO_3. Following the hydrolysis of N_2O_5, there is less reactive NO_2 around to convert ClO into its reservoir species, $ClONO_2$, and less NO_2 around to convert OH into the reservoir species, HNO_3. A heightened sensitivity of the ozone to increasing levels of CFCs develops [85.41]. It has been shown that certain regions of significantly depleted ozone also show high concentrations of sulfate aerosols. In addition, measurements of the ratio of catalytically active nitrogen to total nitrogen can be reproduced using the above heterogeneous reaction, and not by using gas phase processes alone. Study of further mechanisms at varying altitudes and latitudes is an active area of research.

Record low global ozone measurements, 2% to 3% lower than any previous year, were reported beginning in 1992 [85.42] and continuing well into 1993. The increase in naturally occurring aerosols due to the eruption of Mount Pinatubo in June 1991 appears to explain this decline. During the winter of 1993–1994, total ozone levels returned to levels slightly above normal [85.43], presumably because the excess aerosols had been removed from the stratosphere by natural sedimentation processes.

The continuing buildup of CO_2 is predicted to contribute to increased cooling of the stratosphere. Declining temperatures in the stratosphere may increase the frequency of formation of polar stratospheric clouds which drive the destructive heterogeneous chemistry creating the Antarctic ozone hole. An increased occurrence of these clouds outside of the polar regions could affect ozone levels globally. There are also indications that certain ozone depletion chemistries taking place on the surface of sulfate aerosols may also be enhanced by lower temperatures [85.41].

Ozone itself is the dominant heat source in the lower stratosphere. Decreasing the amount of ozone drives temperatures still lower [85.44]. It is unfortunate that the two most significant atmospheric global change effects — the buildup of CO_2 and the enhanced ozone destruction due to man-made CFCs — both cause decreasing temperatures in the stratosphere which may further enhance the destructiveness of the ozone photochemical cycles.

85.5 Atmospheric Measurements

Ground-based observations, as well as measurements made by instruments carried aloft in satellites, balloons, and high-flying aircraft, allow one to explore the atmosphere.

Measurements may be made either in situ or by remote-sensing techniques. The region of the atmosphere from ≈ 60 km to 120 km, encompassing the mesosphere and lower thermosphere and ionosphere, cannot be studied in situ as it is too high for balloons and aircraft, and too low (i. e., too much drag) for satellites. For this region, remote sensing experiments are essential and a comprehensive book on the subject is recommended [85.45].

Most of the instruments used to make atmospheric measurements have been developed in molecular physics and spectroscopy laboratories. Even experiments utilizing sophisticated techniques, such as laser induced fluorescence, and instruments, such as Fourier transform spectrometers, are being flown on payloads. An excellent compendium of ozone-measuring instruments for stratospheric research has been assembled by *Grant* [85.46].

A combination of good laboratory experiments, theoretical calculations, and ingenuity are necessary to extract accurate information from measurements made in the atmosphere. For instance, in order to understand the complicated interactions of the different photochemical cycles involved in ozone chemistry, spectroscopic emissions and absorptions of the many trace species are used to measure concentration profiles. An accurate knowledge of the emission spectroscopy of species such as OH, HO_2, H_2O_2, H_2O, O_3, HNO_3, NO_2, N_2O, N_2O_5, HNO_3, $ClNO_3$, BrO, HCl, HOCl, and ClO is essential. Such measurements provide a rigorous test of atmospheric models. The recently-launched NASA EOS Aura satellite carries instruments that will make global measurements of a number of these species [85.47].

In order to analyze the data and deconvolve some of the line profiles to give information on concentrations as a function of altitude, molecular data such as line strengths and pressure broadening coefficients are needed [85.9].

Until recently, it has been impossible for remote-sensing experiments to distinguish between ozone occurring in the stratosphere (where it is formed naturally) and ozone occurring in the troposphere (where it is a pollutant). Satellite instruments such as the ESA Global Ozone Monitoring Experiment (GOME), the Scanning Imaging Absorption Spectrometer for Atmospheric Chartography (SCIAMACHY) and the Ozone Monitoring Instrument (OMI) have broad enough spectral coverage and high enough resolution that the temperature dependence of the ozone absorption features from 300–340 nm, known as the Huggins bands, can be used to separate out the ozone concentrations in the middle and lower atmospheres [85.48, 49].

References

85.1 T. E. Graedel, P. J. Crutzen: *Atmospheric Change: An Earth System Perspective* (Freeman, New York 1993)

85.2 J. Houghton: *Global Warming: The Complete Briefing*, 2nd edn. (Cambridge Univ. Press, Cambridge 1997)

85.3 J. Staehelin, N. R. P. Harris, C. Appenzeller, J. Eberhard: Rev. Geophys. **39**, 231 (2001)

85.4 R. G. Roble: *Encyclopedia of Applied Physics*, Vol. 2 (VCH, Weinheim 1991) pp. 201–224

85.5 I. Bey, D. J. Jacob, R. M. Yantosca, J. A. Logan, B. D. Field, A. M. Fiore, Q. Li, H. Y. Liu, L. J. Mickley, J. Loretta, M.G. Schultz: J. Geophys. Res. **106**(D19), 23073 (2001)

85.6 P. S. Armstrong, J. J. Lipson, J. A. Dodd, J. R. Lowell, W. A. M. Blumberg, R. M. Nadile: Geophys. Res. Lett. **21**, 2425 (1994)

85.7 R. R. Conway: NRL Memorandum Report, MR-6155, 79 pp. (1988)

85.8 K. Kirby, E. R. Constantinides, S. Babeu, M. Oppenheimer, G. A. Victor: At. Data Nucl. Data Tables **23**, 63 (1979)

85.9 L. S. Rothman, A. Barbe, D. C. Benner, L. R. Brown, C. Camy-Peyret, M. R. Carleer, K. Chance, C. Clerbaux, V. Dana, V. M. Devi, A. Fayt J.-M. Flaud, R. R. Gamache, A. Goldman, D. Jacquemart, K. W. Jucks, W. J. Lafferty, J.-Y. Mandin, S. T. Massie, V. Nemtchinov, D. A. Newnham, A. Perrin, C. P. Rinsland, J. Schroeder, K. M. Smith, M. A. H. Smith, K. Tang, R. A. Toth, J. Vander Auwera, P. Varanasi, K. Yoshino: The HITRAN Molecular Spectroscopic Database: Edition of 2000 including Updates Through 2001, J. Quant. Spectrosc. Radiat. Transfer **82**, 5–44 (2003)

85.10 P. L. Smith, W. L. Wiese: *Atomic and Molecular Data for Space Astronomy* (Springer, Berlin, Heidelberg 1992)

85.11 R. P. Wayne: *Chemistry of Atmospheres* (Oxford Univ. Press, New York 1991) p. 41

85.12 J. F. B. Mitchell: Rev. Geophys. **27**, 115 (1989)

85.13 National Research Council: *Reconciling Observations of Global Temperature Change* (National Academy Press, Washington, DC 2000)

85.14 IPCC Third Assessment Report Climate Change (The Intergovernmental Panel on Climate Change 2001)

85.15 J. C. Comiso, C. L. Parkinson: Phys. Today **57**, 38 (2004)

85.16 E. W. Wolff: Science **302**, 1164 (2003)
85.17 M. A. J. Curran, T. D. van Ommen, V. I. Morgan, K. L. Phillips, A. S. Palmer: Science **302**, 1203 (2003)
85.18 J. Firor: *The Changing Atmosphere* (Yale University Press, New Haven 1990)
85.19 J. W. Elkins, T. M. Thompson, T. H. Swanson, J. H. Butler, B. D. Hall, S. O. Cummings, D. A. Fishers, A. G. Raffo: Nature **364**, 780–783 (1993)
85.20 J. Hansen, D. Johnson, A. Lacis, S. Lebedeff, P. Lee, D. Rind, G. Russell: Science **213**, 957 (1981)
85.21 R. J. Cicerone: Nature **344**, 104 (1990)
85.22 H. Rishbeth, R. G. Roble: Planet. Space Sci. **40**, 1011 (1992)
85.23 R. G. Roble, R. E. Dickinson: Geophys. Res. Lett. **16**, 1441 (1989)
85.24 R. D. Sharma, P. P. Wintersteiner: Geophys. Res. Lett. **17**, 2201 (1990)
85.25 R. E. Dickenson: J. Atmos. Terr. Phys. **46**, 995 (1984)
85.26 C. D. Rodgers, F. W. Taylor, A. H. Muggeridge, M. Lopez-Puertas, M. A. Lopez-Valverde: Geophys. Res. Lett. **19**, 589 (1992)
85.27 D. S. Pollock, G. B. I. Scott, L. F. Phillips: Geophys. Res. Lett. **20**, 727 (1993)
85.28 S. W. Bougher, D. M. Hunten, R. G. Roble: J. Geophys. Res. **99**, No. E7, 14,609 (1994)
85.29 A. C. Aiken, M. L. Charin, J. Nash, D. J. Kendig: Geophys. Res. Lett. **18**, 416 (1991)
85.30 A. Hauchecorne, M.-L. Chanin, R. Keckhut: J. Geophys. Res. **96**(D8), 15297 (1991)
85.31 M. Gadsden: J. Atm. Terr. Phys. **52**, 247 (1990)
85.32 P. O. Wennberg, R. C. Cohen, R. M. Stimpfle, J. P. Koplow, J. G. Anderson, R. J. Salawitch, D. W. Fahey, E. L. Woodbridge, E. R. Keim, R. S. Gao, C. R. Webster, R. D. May, D. W. Toohey, L. M. Avallone, M. H. Proffitt, M. Loewenstein, J. R. Podolske, K. R. Chan, S. C. Wofsy: Science **266**, 398–404 (1994)
85.33 J. C. Farman, B. G. Gardiner, J. D. Shankin: Nature **315**, 207 (1985)
85.34 R. A. Kerr: Science **266**, 217 (1994)
85.35 http://jwocky.gsfc.nasa.gov/multi/multi.html
85.36 J. G. Anderson, D. H. Toohey, W. H. Brune: Science **251**, 39 (1991)
85.37 L. T. Molina, M. J. Molina: J. Phys. Chem. **91**, 433 (1987)
85.38 M. B. McElroy, R. J. Salawitch, S. C. Wofsy, J. A. Logan: Nature **321**, 759 (1986)
85.39 G. L. Manney, L. Froidevaux, J. W. Waters, R. W. Zurek, W. G. Read, L. S. Elson, J. B. Kumer, J. L. Mergenthaler, A. E. Roche, A. O'Neill, R. S. Harwood, I. MacKenzie, R. Swinbank: Nature **370**, 429–434 (1994)
85.40 M. J. Newchurch, E.-S. Yang, D. M. Cunnold, G. C. Reinsel, J. M. Zawodny, J. M. Russell III: J. Geophys. Res. **108**(D16), 4507 (2003)
85.41 D. W. Fahey, S. R. Kawa, E. L. Woodbridge, P. Tin, J. C. Wilson, H. H. Jonsson, J. E. Dye, D. Baumgardner, S. Borrmann, D. W. Toohey: Nature **363**, 509–514 (1993)
85.42 J. F. Gleason, P. K. Bhartia, J. R. Herman, R. McPeters, P. Newman, R. S. Stolarski, L. Flynn, G. Labow, D. Larko, C. Seftor, C. Wellemeyer, W. D. Komhyr, A. J. Miller, W. Planet: Science **260**, 523–526 (1993)
85.43 D. J. Hofmann, S. J. Oltmans, J. M. Harris, J. A. Lathrop, G. L. Koenig, W. D. Komhyr, R. D. Evans, D. M. Quincy, T. Deshler, B. J. Johnson: Geophys. Res. Lett. **21**, 1779–1782 (1994)
85.44 V. Ramaswamy, M.-L. Chanin, J. Angell, J. Barnett, D. Gaffen, M. Gelman, P. Keckhut, Y. Koshelkov, K. Labitzke, J.-J. R. Lin, A. O'Neill, J. Nash, W. Randel, R. Rood, K. Shine, M. Shiotani, R. Swinbank: Stratospheric Temperature Trends: Observations and Model Simulations, Rev. Geophys. **39**, 71 (2001)
85.45 J. T. Houghton, F. W. Taylor, C. D. Rodgers: *Remote Sounding of Atmospheres* (Cambridge Univ. Press, Cambridge 1984)
85.46 W. B. Grant (Ed.): *Ozone Measuring Instruments for the Stratosphere* (Optical Society of America, Washington, DC 1989)
85.47 Further information is available at http://aura.gsfc.nasa.gov/
85.48 K. V. Chance, J. P. Burrows, D. Perner, W. Schneider: J. Quant. Spectrosc. Radiat. Transfer **57**, 467 (1997)
85.49 R. Munro, R. Siddans, W. J. Reburn, B. Kerridge: Nature **392**, 168 (1998)

86. Atoms in Dense Plasmas

When plasma densities are high enough that interparticle separations are comparable to atomic dimensions, there are important "environmental" consequences for atomic structure and atomic processes. Such conditions are found not only within stars and giant planets but, nowadays, also in the laboratory – especially in experiments related to the quest for inertial confinement fusion. After introducing important plasma concepts, we examine these consequences in regard to several issues: modification of atomic bound states, ionization balance, equation of state, and radiative and collisional processes that regulate transport coefficients and the spectral emission of non-equilibrium plasmas. Finally, we describe modern simulation methods that are being used to tackle various many-body problems in this subject. For nearly every issue we raise there is a need for better understanding and for more, and more precise, data.

Ionized gases, or plasmas, are the predominant form of matter throughout the universe, and physical conditions in laboratory and cosmic plasmas vary greatly. No single experimental methodology or theoretical construct suffices to explore all aspects of the plasma state.

Systematic study of plasmas began early in the 20th century, but until recently the physics of atoms in plasmas has been largely synonymous with the physics of isolated ions. This perspective is valid as long as the interparticle spacing is very much larger than the relevant atomic dimensions, typically a few to a few tens of Bohr radii. For example, ions are isolated in this sense in interstellar space where the electron density $n_e \approx 1\,\text{cm}^{-3}$, or even in a tokamak, where $n_e \approx 10^{14}\,\text{cm}^{-3}$. For neutral and moderately charged atoms, data such as energy levels, oscillator strengths, and collision cross sections have long been obtained from traditional kinds of experiments and quantal calculations, as discussed elsewhere in this book. Progress in these areas continues to be made, with the X-ray spectrum of highly-charged Fe [86.1] and the Lamb shift in U^{+91} [86.2] being noteworthy examples of the kinds of accurate measurements that can now be made using electron beam ion traps and storage rings.

The focus of the present chapter is partially ionized matter in which important atomic phenomena are influenced by a dense plasma environment. As Fig. 86.1 (which is discussed in detail below) reveals, the densities in many laboratory and astrophysical plasmas are high enough to invalidate the presumption of isolated systems. The interaction of intense laser or particle beams with solid matter produces rapidly evolving, hot, and dense plasmas [86.3] that mimic some of the most extreme conditions in nature, including the thermonuclear environment of stellar interiors; these plasmas, as well as some produced in z-pinch implosions [86.4], are the basis of world-wide inertial confinement fusion (ICF) efforts. Dense plasmas also can be transient gain media for amplified spontaneous emission at X-ray wavelengths [86.5]. Additional impetus for the study of atoms in dense plasmas now comes from experiments involving irradiation of solids by ultra-short (sub-picosecond) laser pulses [86.6]. The moderate temperatures (tens of eV) but high (near-solid) densities typical of this so-called warm dense matter (WDM) regime [86.7] produce severely perturbed bound ionic states.

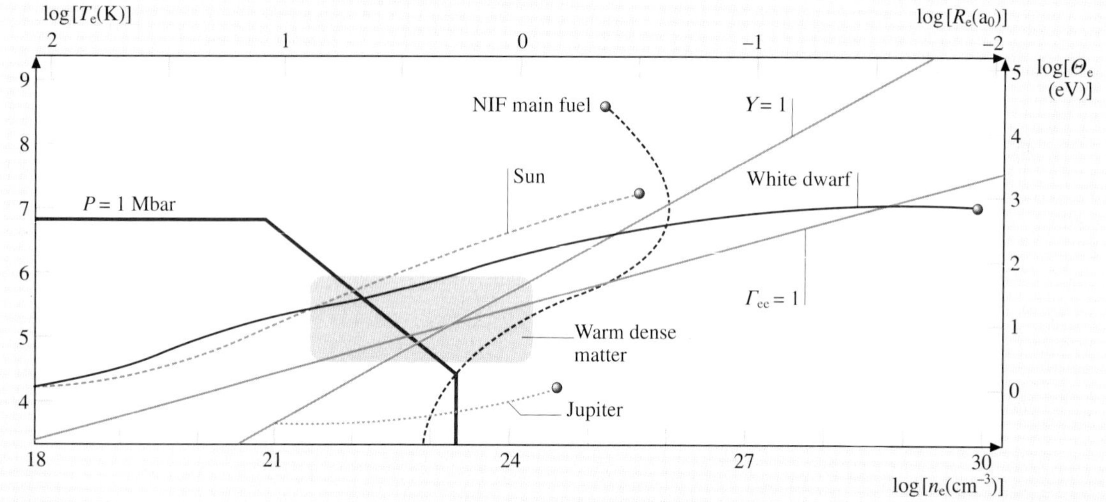

Fig. 86.1 Plasma conditions discussed in the text are identified on this temperature-density plane. Plasmas below the line $\Gamma_{ee} = 1$ are strongly coupled, and those below the line $\Upsilon = 1$ contain degenerate electrons. The HEDP regime, which lies above and to the right of the line marked $P = 1$ Mbar, includes some conditions characteristic of warm dense matter. Also plotted are tracks representing conditions (as a function of radius) within the Sun, Jupiter and a typical white dwarf star; time-dependent conditions are also shown – from early compression through ignition – within the main (DT) fuel of a prototype target capsule for the National Ignition Facility

Dense cosmic plasmas – specifically, the interiors of stars and giant planets – are very large and have very long lifetimes. Their thermodynamic variables change only slowly with position or time and, hence, macroscopic regions can be considered as statistical systems evolving through a succession of states in local thermodynamic equilibrium (LTE). The equation of state (EOS), and the radiation and heat transport coefficients, viz. the opacity and thermal conductivity, are key to understanding the behavior of LTE plasmas. A recent monograph [86.8], plus comprehensive articles by *More* et al. [86.9], by *Rogers* and *Iglesias* [86.10], and by *Saumon* et al. [86.11] discuss many of the high density consequences for the opacity and for EOS.

In contrast, the short lifetimes of dense plasmas created by intense beam irradiation or by explosive pinch devices often preclude the establishment of a thermal distribution of atomic level populations; in extreme cases, there is not even enough time to establish a Maxwellian distribution of particle velocities. Populations in highly nonequilibrium (non-LTE) laboratory plasmas, which must be found by solving rate equations [86.12, 13], are essential information for using X-ray line emission to diagnose conditions in ICF targets, or for identifying likely gain media for X-ray lasing. And, as we discuss in Sect. 86.3, the dense plasma environment modifies transition rates themselves, further complicating the interplay of numerous collisional and radiative processes in such atomic kinetics calculations.

The topics addressed here are needed for understanding non-LTE situations, as well as LTE ones. After characterizing the perturbing plasma environment in Sect. 86.1, we summarize well-known prescriptions for atomic structure and ionization balance in Sect. 86.2, and then discuss modified transition rates in Sect. 86.3 for ions in dense plasmas. Finally, we review in Sect. 86.4 how simulations are now being used to address a wide array of issues needed to accurately describe atoms in dense plasmas.

There are several periodic meetings devoted to various aspects of this subject, and especially relevant ones include: Atomic Processes in Plasmas; Radiative Properties of Hot, Dense Matter; Spectral Line Shapes; and Strongly Coupled Coulomb Systems. Printed proceedings of these conferences are an excellent guide to recent developments in the topics discussed here, as well as numerous other, related ones. Additionally, three recent textbooks [86.14–16] provide detailed treatments of many of the subjects surveyed here.

The present topic is an important part of what is now being termed "high energy-density physics"

(HEDP). Conventionally this interdisciplinary subject, which involves collective and/or non-linear phenomena in many-body systems, is defined as the study of matter in regimes where the total (matter plus electromagnetic field) pressure exceeds one megabar; this boundary is also marked in Fig. 86.1. Reference [86.17] is a recent National Research Council report on key issues and opportunities in HEDP.

86.1 The Dense Plasma Environment

Most plasmas are charge neutral, so the mean number densities n_i and n_e of constituent ions (charge $Z_\text{i}e$, mass m_i) and electrons (charge $-e$, mass m_e) satisfy

$$\sum_a Z_a n_a = 0\,, \tag{86.1}$$

where the sum ranges over all particle species. Moreover, plasma conditions usually change slowly enough that each of the species is able to establish a thermal distribution of velocities, fixed by its temperature $\Theta_a = k_\text{B} T_a$ (in energy units). Here, we assume these conditions hold.

86.1.1 Plasma Parameters

A few key quantities characterize the plasmas under consideration. Derivations, and discussions of the roles of these and other auxiliary quantities can be found in standard plasma physics texts [86.18, 19], as well as a recent tutorial article [86.20].

1. The plasma frequency
$$\omega_\text{p} = \left[\sum_a 4\pi n_a (Z_a e)^2/m_a\right]^{1/2} = \left(\sum_a \omega_a^2\right)^{1/2} \tag{86.2}$$
defines a timescale $\left(\sim \omega_\text{e}^{-1}\right)$ for the particle oscillations in response to a non-equilibrium charge density in the plasma.
2. The Debye length
$$\lambda_\text{D} = \left[\sum_a 4\pi n_a (Z_a e)^2/\Theta_a\right]^{-1/2} = \left(\sum_a D_a^{-2}\right)^{-1/2} \tag{86.3}$$
is the distance beyond which plasma particles effectively screen any localized charge imbalance.
3. The ion-sphere (or electron-sphere) radius
$$R_a = \left(\frac{3}{4\pi n_a}\right)^{1/3} \tag{86.4}$$
defines a spherical volume associated with a single particle and is a measure of interparticle spacing (among particles of species "a").
4. The Fermi energy
$$\Theta_\text{F} = \frac{\hbar^2 \left(3\pi^2 n_\text{e}\right)^{2/3}}{2m_\text{e}} \tag{86.5}$$
characterizes the highest occupied energy level in a zero temperature system of electrons. A dimensionless measure of degeneracy is $\Upsilon = \Theta_\text{F}/\Theta_\text{e}$. Velocity distributions are either Maxwellian or Fermi–Dirac in the limits $\Upsilon \ll 1$ or $\Upsilon \gg 1$, respectively.
5. The Coulomb coupling parameters
$$\Gamma_{ab} = \frac{Z_a Z_b e^2}{R_{ab}\Theta_{ab}} \tag{86.6}$$
give the average ratio of potential to kinetic energies between species a and b. The reduced ion-sphere radius and temperature are $R_{ab} = \frac{1}{2}(R_a + R_b)$ and $\Theta_{ab} = (m_a\Theta_b + m_b\Theta_a)/(m_a + m_b)$, respectively. When Γ_{ab} is greater (less) than 1, that species is said to be strongly (weakly) coupled. And, when the number of particles a in a sphere of radius D_a (a Debye sphere), $\left[4\pi n_a D_a^3/3\right] = 1/(3\Gamma_{aa})^{3/2}$, is small, discreteness of the charge density can be important in describing certain plasma phenomena.

Figure 86.1 shows that dense plasmas can be strongly or weakly coupled; further, some of these plasma conditions involve degenerate electrons while others do not. And, in WDM, one encounters the situation where $\Upsilon \sim 1$ and $\Gamma_\text{ee} \sim 1$, which is particularly difficult to treat theoretically because several effects are competing amongst each other. In this figure, we plot the run of (n_e, Θ_e) values for the sun, for Jupiter, and for a typical white dwarf star (0.6 solar mass, pure C/O core, H/He outer layers). Also plotted is the track of DT fuel conditions in an imploding ICF capsule designed for the National Ignition Facility. Note that all of these systems sample wide portions of parameter space, and therefore an accurate description of each requires some very different plasma models.

86.1.2 Quasi-Static Fields in Plasmas

A simple, yet useful description of the plasma environment is given by the one-component plasma (OCP) model [86.22], in which particles of a single kind (Z_a,m_a) move against a smooth background of matter having on average the opposite charge density, $\rho(\boldsymbol{r}) = -Z_a e n_a$. Most plasma phenomena can be described within the context of either this or the two-component model (electrons and their parent ions, with $n_i = n_e/Z_i$). In the more realistic, two-component picture, electron screening of the (slower moving) ions is established on a timescale $\omega_e^{-1} \ll \omega_i^{-1}$, so it makes sense to speak of screened, quasi-static ionic fields.

When there are many electrons in each Debye sphere, the electrostatic potential $\Phi(r)$ near a test charge Ze placed in an otherwise uniform, neutral plasma is exponentially reduced from the Coulomb expression, viz.,

$$\Phi_D(r) = \frac{Ze}{r}\exp(-r/D_e)\ . \quad (86.7)$$

(A more elaborate version of this formula, for partially ionized atoms, has recently been proposed [86.23].) This "Debye screening" obtains only in weakly coupled plasmas and applies only to a test charge at rest. The faster the charge Ze moves, the less effective is the plasma at screening it, since only plasma particles with higher velocities can form the shielding cloud [86.18, 24].

In the opposite limit of large Γ-values, quasi-static screening is better described by the ion-sphere (IS) picture [86.19], in which each stationary ion of charge $Z_i e$ is surrounded by Z_i electrons, uniformly distributed throughout a sphere of radius R_i, to produce the potential

$$\Phi_{IS}(r) = Z_i e\left[1/r - (1/2R_i)\left(3 - r^2/R_i^2\right)\right] \quad (86.8)$$

inside the sphere, and zero potential outside.

Consideration of a plasma's electric microfield illustrates these concepts. Moreover, microfields are a key ingredient in calculations of spectral line broadening in plasmas – an important subject discussed in Chapts. 59, 19 and in [86.14–16].

Local fluctuations in the density of any species about its mean value n_a create a microscopic electric field $\tilde{\boldsymbol{E}}_a(\boldsymbol{r}, t)$. There is a probability distribution $P(\tilde{E}_a)$ that characterizes the strengths of these microfields, which are quasi-static within time intervals short compared with the fluctuation timescale, $1/\omega_i$. Holtsmark first calculated this distribution at an arbitrary position in an infinite, isotropic gas of noninteracting particles, and *Chandrasekhar* [86.25] gives a thorough account of this famous stochastic problem. Holtsmark's formula is

$$P_H(\varepsilon) = (2\varepsilon/\pi)\int_0^\infty x\sin(\varepsilon x)\exp\left[-x^{3/2}\right]dx\ , \quad (86.9)$$

where $\varepsilon = \tilde{E}_a/\left[(8\pi/25)^{1/3}E_a\right]$ is the scaled field, and $E_a = |Z_a|e/R_a^2$. The mean Holtsmark field is $\langle\varepsilon\rangle \simeq 2.99$, and for $\varepsilon \gg 1$, $P_H(\varepsilon)$ is well approximated by the distribution of fields due to a single nearest-neighbor in the gas,

$$P_{nn}(\varepsilon) \simeq 3/\left(2\varepsilon^{5/2}\right)\ . \quad (86.10)$$

Both of these distributions ignore the interactions among charged particles that become increasingly important as Γ_{ii} grows, because particle positions then tend to be correlated. Quasi-static ion microfields – at the position of an ion – therefore become weaker, on average, as the coupling increases. Figure 86.2 illustrates this point and shows distributions computed with the

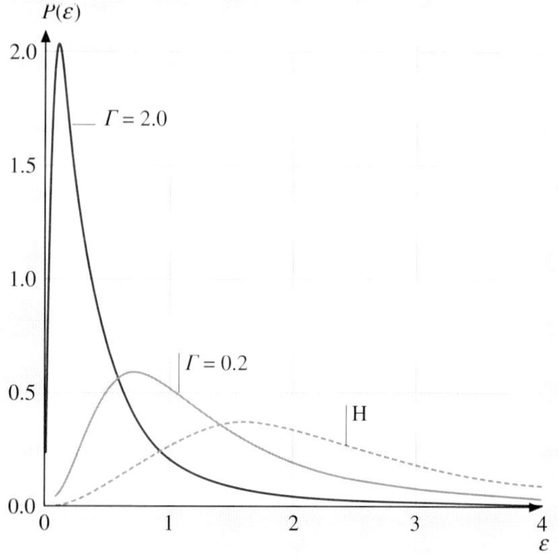

Fig. 86.2 The probability distribution $P(\varepsilon)$ of scaled microfield strengths ε for different plasma conditions. The curve marked H represents the Holtsmark distribution, which applies to an idealized case of non-interacting particles (i.e., $\Gamma = 0$). The other two curves, with $\Gamma = 0.2$ and $\Gamma = 2.0$, represent distributions determined by the APEX model for interacting ions (charge $Z = 1$) that are Debye-screened by plasma electrons. In these latter two cases, the ion density and temperature are, respectively, $1.0\times10^{18}\ \mathrm{cm^{-3}}$ and 1.15 eV, and $2.6\times10^{24}\ \mathrm{cm^{-3}}$ and 16 eV. After [86.21]

APEX method [86.26, 27], which uses a parameterized, two-particle distribution function "tuned" to yield the exact second moment $\langle \varepsilon^2 \rangle$ for the distribution function of field strengths experienced by any one ion in a plasma whose ions are all Debye-screened by electrons only.

The potentials and microfields discussed above are based on classical statistical mechanics, even though dense plasmas are inherently quantum many-body systems. Reference [86.28] provides a careful discussion of the merits and limitations of this approach.

86.1.3 Coulomb Logarithms and Collision Frequencies

It is well known that the total cross section for elastic scattering of two charged particles, $Z_a e$ and $Z_b e$, diverges at any collision energy $\mathcal{E}$, as a consequence of the infinite range of the Coulomb interaction. Plasma transport coefficients (e.g., electrical conductivity and thermal conductivity), however, which depend ultimately upon momentum transfer in this elementary process, are finite. This comes about because scattering at small center-of-mass angles ϕ, or, correspondingly, at large impact parameters b, is diminished by plasma screening, as in (86.7) and (86.8).

Following *Spitzer* [86.29], we argue that there is some minimum effective scattering angle ϕ_{min}. And, since the analysis will be based on classical formulae, there is some maximum scattering angle ϕ_{max} beyond which quantum effects are important. Between these limits the Coulomb interaction is taken to be unscreened and Rutherford's differential cross section $\sigma_{\text{R}}(\phi)$ is applicable. By assuming that ϕ_{max} also is small, it follows that the momentum transfer cross section can be approximated as

$$\sigma_m(\mathcal{E}) = 2\pi \int_{\phi_{\text{min}}}^{\phi_{\text{max}}} \mathrm{d}\phi \, [\sin\phi\,(1-\cos\phi)\,\sigma_{\text{R}}(\phi)]$$

$$\approx \pi a^2 \ln\left(\frac{\phi_{\text{max}}}{\phi_{\text{min}}}\right) , \tag{86.11}$$

where $a = \left|Z_a Z_b e^2\right| / \mathcal{E}$ is a characteristic length. The familiar result $(a/2b) = \tan(\phi/2) \approx (\phi/2)$ gives σ_m in terms of minimum and maximum impact parameters, $b_{\text{min}} = a$ (from $\phi_{\text{max}} = 1$) and $b_{\text{max}} = \lambda_{\text{D}}$.

Finally, if the actual collision energy in the argument of the logarithm, (λ_{D}/a), is replaced by its mean value, $\mathcal{E} = \frac{3}{2}\Theta$, the momentum transfer cross section takes the simple form

$$\sigma_m(\mathcal{E}) = \pi \left(\frac{Z_a Z_b e^2}{\mathcal{E}}\right)^2 \ln \Lambda \, ; \tag{86.12}$$

for a two-component, electron-ion plasma, the argument of this Coulomb logarithm is

$$\Lambda \approx 1/2\Gamma_{\text{e}Z}^{3/2} \approx (\text{\# particles in Debye sphere}) \, . \tag{86.13}$$

Spitzer's result for σ_m yields the simplest expression for the frequency $\nu_{\text{e}Z}$ of electron-ion collisions in a two-component plasma, defined [86.18] as the mean value of the reciprocal of the time between collisions,

$$\nu_{\text{e}Z} = \frac{n_Z Z^2 e^4}{\sqrt{m_{\text{e}}}} \left(\frac{\pi}{2\Theta}\right)^{3/2} \ln \Lambda \, . \tag{86.14}$$

Equation (86.12) is commonly used to determine transport coefficients in weakly coupled plasmas, where $\ln \Lambda \gg 1$. In the dense plasma regime, however, the Coulomb logarithm can be small or even negative at high enough density, which yields meaningless results for the cross section. Physically, a small $\ln \Lambda$ arises from a small λ_{D} (high density and/or low temperature), which means that collisions can only occur at very small separations where the Coulomb potential is largest. Strong collisions can be included in the above analysis simply by not making a small-angle approximation in the evaluation of the cross section; the result is [86.30]

$$\sigma_m(\mathcal{E}) = \frac{\pi}{2} \left(\frac{Z_a Z_b e^2}{\mathcal{E}}\right)^2 \ln\left[1 + \Lambda^2\right] . \tag{86.15}$$

In obtaining this result – which no longer yields a negative cross section – no assumption need be made about the value of ϕ_{max}. One must still choose, however, b_{max}, which will not generally be given by λ_{D}, since Debye screening is invalid in the dense plasma regime. Strong collisions at small separations also bring in the effects of quantal scattering. These issues are best circumvented by obtaining the cross section directly from a quantal calculation involving the chosen screened potential [86.30, 31]. (Note that the formulae in [86.31] actually describe the scattering of one unscreened charge by another charge that is screened, so they are most relevant to the scattering of fast electrons by slow ions.)

Eventually, even this formulation will fail when plasma kinetic processes affect the collision. For example, hard collisions can alter the velocity distribution function and collective modes can modify the static screened interaction potential. Furthermore, strong coupling introduces ionic structure that correlates the collisions. Accurate calculations must consider the collision process in the context of an appropriate kinetic equation [86.30, 32–34]. When such a process is carried out, the result can be inverted to yield "effective" or "generalized" Coulomb logarithms that typically are process dependent [86.18]. *Li* and *Petrasso* [86.32] obtained high-order correction terms for the Fokker–Planck equation, for example, and *Boercker* et al. [86.33] generalized the collision term in the Lenard–Balescu quantum kinetic equation to include strong coupling effects. Additionally, *Berkovsky* and *Kurilenkov* [86.34] extended the strong coupling description to also include "strong" collisions (those poorly treated within the Born approximation).

The physics of collisions is directly measured by resistivity experiments, which employ some method to heat a solid and attempt to tamp the high-pressure plasma that is formed. The resistivity is then measured either by the reflectivity or directly through current and voltage probes. A review of these methods has recently been given by *Benage* [86.35].

86.2 Atomic Models and Ionization Balance

A pervasive issue in the study of both laboratory and astrophysical plasmas is ionization balance: What is the distribution of charge states Z_i of atomic ions in a particular plasma? Answers impact subjects as diverse as cosmic abundances deduced from astrophysical spectra, and the temporal behavior of laser-heated foils. Table 86.1 lists the charge-state dependence of several plasma quantities [86.36]. Here, $\langle Z\rangle$ and $\langle Z^2\rangle$ denote the mean and mean square ionic charge, respectively, viz.,

$$\langle Z^n\rangle = \sum_{\text{ions}} Z_\mathrm{i}^n n_\mathrm{i} \Big/ \sum_{\text{ions}} n_\mathrm{i} \tag{86.16}$$

For a dense plasma, experimental determination of actual charge-state distributions, or even $\langle Z\rangle$, has proven difficult. Traditional spectroscopic methods (as described in the next section) require large atomic data bases and sophisticated kinetics models, which typically are run several times to find the best match to measured line shapes and intensities. Recently, however, an X-ray scattering method (based on the Compton effect) has been developed to determine $\langle Z\rangle$ in rapidly evolving plasmas [86.37]. Instead of detailed atomic data, this method requires accurate knowledge of the plasma's dynamic structure factor [86.18], which in general must be obtained from a molecular dynamics simulation (as discussed in Sect. 86.4).

Table 86.1 Some plasma quantities that depend on its ionization balance

Quantity	Z-scaling (fixed nucleon density)
(Ideal) gas pressure	$\sim (\langle Z\rangle + 1)$
Electrical resistivity	$\sim \langle Z^2\rangle / \langle Z\rangle$
Thermal conductivity	$\sim \langle Z\rangle / \langle Z^2\rangle$
Ionic viscosity	$\sim 1/\langle Z^2\rangle^2$
Bremsstrahlung	$\sim \langle Z\rangle \langle Z^2\rangle$

86.2.1 Dilute Plasma Models

Consider a nondegenerate plasma in thermal equilibrium at a temperature Θ (for instance, some region of a star's interior). The time independent ionization balance for each element is given by the Saha–Boltzmann formula [86.38]

$$\frac{n_{Z+1}}{n_Z} = \frac{1}{n_\mathrm{e}}\left(\frac{2G_{Z+1}}{G_Z}\right)\left(\frac{m_\mathrm{e}\Theta}{2\pi\hbar^2}\right)^{3/2} \exp(-I_Z/\Theta)\,, \tag{86.17}$$

for the density ratio of successive charge states, where G_Z and I_Z are, respectively, the partition function and ionization potential for the Z-times ionized atom.

The solution of (86.17) is shown in Fig. 86.3 (top panel) for the case of solid density aluminum over a wide range of temperatures. The partition functions were determined from atomic states of the ground configuration only. The aluminum plasma is predominantly neutral at temperatures in the few electron volt range and ionizes stage by stage until it is nearly fully ionized just above one kilovolt. Of course, real aluminum is not an insulator at solid density and low temperatures, as Fig. 86.3 would suggest. Major corrections to (86.17) are evidently needed to incorporate the physics of WDM ($\Gamma \sim 1$, $\Upsilon \sim 1$), especially corrections for partial electron degeneracy. We will return to this problem in later subsections.

When conditions change too rapidly for LTE to be established, the plasma may evolve through a succession of "steady states" in which the relative abundances of different ion stages are determined by a balance of certain ionization and recombination rates. Then, in order to answer the straightforward question of what are the relative abundances of atomic ionization stages, one needs a vast data base of atomic energy levels, plus collisional and radiative rates. A set of rate equations must be solved to determine the populations $n_Z(\alpha)$ for each quantum state α of each ion stage Z. Each equation involves transitions to and from all other states [86.12, 13].

The balance of photoionizations and dielectronic plus radiative recombinations in low-density, steady-state plasmas is termed nebular equilibrium, because these are the conditions appropriate to astrophysical nebulae – regions of ionized gas surrounding hot stars [86.39]. The balance of collisional ionizations and dielectronic plus radiative recombinations in low-density, steady-state plasmas is termed coronal equilibrium, because these are the conditions appropriate to the solar corona. Most tokamak plasmas also are in coronal equilibrium. Under coronal conditions, essentially all ions are in their respective ground states and the ionization balance is a function only of the plasma temperature [86.40]. As the plasma density increases, three-body collisions become important, and the resulting steady-state ionization balance, which depends on density as well as temperature, is termed collisional-radiative equilibrium [86.12]. This situation exists in most ICF experiments.

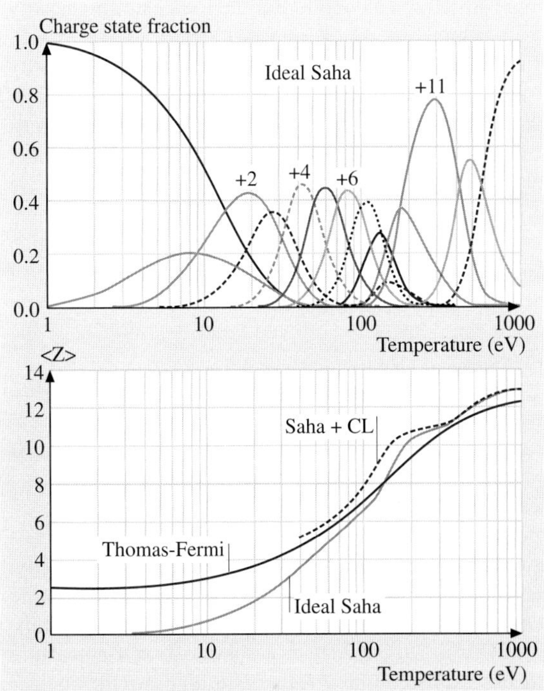

Fig. 86.3 The *top panel* shows fractional abundances of different charge states of aluminum in thermal equilibrium at solid density over a range of temperatures, as computed with the ideal Saha–Boltzmann equation (86.17). The corresponding mean charge $\langle Z\rangle$ is shown in the *lower panel*, in addition to the result from Thomas–Fermi theory, as given by (86.18). Also shown is a modified Saha formulation that includes continuum lowering shifts

Finally, conditions in sub-picosecond laser-plasma experiments can change so rapidly that none of the above simplifications apply. Ionization is strongly time-dependent, and may involve multiphoton processes.

As we describe below, a dense plasma environment vastly complicates the determination of ionization balance. In LTE cases, energy levels are changed and partition functions are truncated by the phenomenon of continuum lowering. In non-LTE cases these effects still occur, but, in addition, radiative and collisional rates themselves are altered.

86.2.2 Dense Plasma "Chemical" Models

There are two distinct strategies taken to extend the results of the previous section. One strategy, known as the "chemical picture", formulates the Saha–Boltzmann equation in terms of a free energy $F(T, V, \{n_a\})$, where the species populations $\{n_a\}$ are to be determined, and various corrections due to couplings and degeneracy can be added to yield a thermodynamically consistent equation of state that includes atomic physics [86.41].

Because plasma screening attenuates the Coulomb interaction at long range, atoms and ions no longer have an infinite number of bound (Rydberg) states, and atomic partition functions are truncated in a natural way. The simplest chemical picture is that the onset of continuum energies has been "lowered" by some amount ΔI (relative to the atom's ground state). When ΔI is a fixed quantity, continuum lowering eliminates bound states whose (unscreened) ionization potentials were less than ΔI, and moves all remaining states closer to the continuum by this same amount. Thus, levels get shifted but spectral lines do not. Schemes that use an effective, single-particle potential to determine a spectrum of modified eigenstates produce distinct plasma shifts for different levels and, hence, spectral line shifts. Experiments show, however, that almost all such predictions have been inaccurate: actual plasma-induced shifts are very small and, for most applications, ignorable ([86.14, Sect. 4.10], [86.15, Sect. 3.5]).

A variety of arguments has been put forth to quantify continuum lowering, including in particular:

1. determine ΔI from the last distinct spectral line near a series limit (the Inglis–Teller formula [86.42]);
2. determine ΔI from the atom's dipolar interaction with the plasma's microfield [86.41];
3. determine ΔI from the binding energy of the ground state in some specified, screened Coulomb potential [86.43, 44];
4. determine ΔI from a rigorous, statistical mechanical treatment of the atomic partition functions [86.45].

Figure 86.3 (lower panel) illustrates the effect of continuum lowering on the average ionization state $\langle Z \rangle$; here, we solve (86.17) for solid density aluminum with the ionization potentials shifted by an amount determined by electron screening in the Debye approximation, $\Delta I = \langle Z \rangle e^2 / D_e$. Although we do not plot $\langle Z \rangle$ for this case when the number of particles in a Debye sphere is less than ten, it is obvious that a somewhat higher degree of ionization exists when continuum lowering is accounted for.

A recent experiment [86.46] suggests that the Inglis–Teller prescription for line merging accurately describes the disappearance of the uppermost members of a spectral series. But simply truncating the number of bound states and, hence, the internal partition function, does not yield a self-consistent thermodynamic description of the plasma [86.41, 47]. In this regard, the true situation in dense plasmas is far more complicated for two reasons, and both give rise to a gradual disappearance of high-lying bound states.

First, excited ionic states can be strongly perturbed by one or more nearby ions, which means that (as the density increases) bound, quasi-molecular states form and eventually evolve to a conduction band. Models with names such as "incipient Rydberg states" [86.48], "quasi-localized states" [86.45], "cluster states" [86.49], "negative-energy continuum states" [86.50], and "collectivized states" [86.51] have been developed to capture the complicated physics of this intermediate regime. Second, space- and time-dependent density fluctuations give rise to different perturbing configurations, which means that the plasma is more accurately described by the average of an ensemble of perturbed ionic states than by the individual states of an ion experiencing the mean (usually spherical) perturbation. In the chemical picture, the most common approach [86.41, 51, 52] reduces the effective statistical weight of each (unperturbed) ionic state by a factor representing the probability that the plasma's microfield is sufficiently strong to Stark ionize it.

The actual inclusion of dynamical plasma screening effects on ionic bound states requires a much more elaborate model [86.53] that, as yet, has not been incorporated into computer codes simulating high energy-density plasma experiments. Other computational studies, involving simple continuum lowering prescriptions [86.54, 55], indicate that an accurate treatment of this phenomenon is essential for understanding non-LTE, as well as LTE, situations.

86.2.3 Dense Plasma "Physical" Models

An alternative strategy abandons the distinction between atomic and plasma electrons; this is known as the "physical picture" [86.47]. The simplest model that accomplishes this is that of a nucleus centered in a charge-neutral, spherical cell of radius R_s. An electronic structure calculation for the total electron density $n_e(r)$ at temperature Θ, subject to the boundary condition $dn_e(R_s)/dr = 0$, is carried out and, once the density is known, various physical quantities can be obtained. The advantage of this approach is that effects such as continuum lowering and degeneracy are naturally and self-consistently incorporated. Models of this kind are referred to as either "statistical" models or as "average atom" (AA) models depending on the manner in which the electronic structure is determined. The accuracy of the approach depends on both the sophistication with which the density is computed and the validity of the spherical cell boundary condition.

The simplest way to obtain the electronic density is with a statistical model, such as the finite-temperature Thomas–Fermi approximation and its various extensions to include exchange ("Thomas–Fermi–Dirac") and gradient corrections ("Thomas–Fermi–Dirac–Weizsacker"); these models are covered in detail in Chapt. 20 for free atoms at $\Theta = 0$. Briefly, the Thomas–Fermi (TF) model describes atomic charge densities by treating all electrons as a partially degenerate Fermi gas subject to a spherical, self-consistent electrostatic potential $\Phi_{TF}(r)$ resulting from the nuclear charge $Z_n e$ and the electrons themselves. Given the simplicity of the TF model, agreement with experiment (for binding energies) is surprisingly good, usually well within a factor of two for the thousands of ions in the periodic table.

Feynman and coworkers [86.56] were the first to use such models to describe hot, compressed atoms and their thermodynamic properties. Quantities such as the internal energy, free energy, and pressure are readily computable. Extended Thomas–Fermi models are useful for describing properties of matter in dense, cold stars [86.57], for example. As we have emphasized, a quantity of particular interest is the average ionization state $\langle Z\rangle$ of the plasma, which can be determined from the electronic density that extends from cell to cell, viz.,

$$\langle Z\rangle = \frac{4\pi}{3} R_s^3 n_e(R_s)\,. \qquad (86.18)$$

The definition (86.18) is not unique, however, and some authors prefer to define bound electrons as all those in negative energy states [86.58, 59].

It is interesting to note that this intuitive definition of $\langle Z\rangle$, (86.18), is considerably different from that used in the Saha formulation. This $\langle Z\rangle$ generally will have a nonzero value even at $\Theta = 0$, a phenomenon known as "pressure ionization," because the ionization occurs solely due to the finite value of R_s. More [86.59] published a convenient prescription for finding ionization potentials and total energies, as predicted by the TF model, of ions with net charge $Z_i e$ between $0.1\,Z_n e$ and $0.9\,Z_n e$. In the lower panel of Fig. 86.3 we show $\langle Z\rangle$ based on the More/TF result. There is general agreement with Saha at high temperatures, where ionic bound states are much smaller than the interparticle spacing, but important differences occur at low temperatures.

Average atom models extend the statistical models by directly employing the Schrödinger equation for the electron structure. Typically, a self-consistent electronic structure calculation is carried out such that the single-particle levels are thermally populated according to a Fermi–Dirac distribution. These models describe atomic shell structure, which is absent in the statistical models. Modern versions of AA are detailed quantum mechanical calculations based on, usually, finite-temperature density functional theory (DFT), with some approximation for the exchange-correlation potential. A good review of the finite-temperature DFT approach has been given by *Gupta* and *Rajagopol* [86.60], and [86.61] contains several numerical comparisons. This approach was pioneered by *Rozsnyai* [86.62, 63], who employed a TF approximation for the free electrons, and by *Liberman* [86.64] who constructed an AA based on a self-consistent field model with a thermal population of Dirac orbitals for all states.

There are two major weaknesses of the AA method. First, the spherical cell neglects asymmetrical ionic configurations in the plasma and assumes that no ion can penetrate within the radius R_s. And, the AA does not straightfowardly yield the distribution of ionic stages, which is important for opacity and transport calculations. *Ying* and *Kalman* [86.65] have introduced a model that addresses the $\langle Z\rangle$ issue while also incorporating strong ionic correlations from neighboring ions. A DFT-based model that describes both strong coupling and the distribution of ionic stages also has been published [86.66].

86.3 Elementary Processes

In a truly equilibrium plasma, atomic transitions do not modify the plasma's physical state. However, the evolution of LTE and nonequilibrium plasmas is regulated by the time rate of change of quantities such as Θ and n_e, and these in turn depend on transport coefficients such as the radiative opacity and the thermal conductivity.

The processes controlling these coefficients are induced by various radiative and collisional interactions. Indeed, so many processes can occur that a major task is the identification of those which are most important in a particular situation. The plasma environment may also alter rates applicable to isolated atoms, through the perturbation of the atomic states involved and/or the screening of long-range Coulomb forces. Further complicating the usual two-body collision picture is the close proximity of many scattering centers in a dense plasma. Presently, analysis of any of these many-body problems requires considerable simplification.

86.3.1 Radiative Transitions and Opacity

For a radiative transition between atomic states α and β, the absorption and emission rates are proportional to quantities of the form $(\Delta E)^n |\langle\alpha|\boldsymbol{d}|\beta\rangle|^2$ summed over degenerate substates, where $\boldsymbol{d}$ is the electric dipole operator Chapt. 10, and $n = 1$ or 3 for the Einstein B and A coefficients, respectively. In a dense plasma, changes in these radiative quantities are due primarily to changes in the atomic wave functions. Theory predicts that plasma screening reduces line strengths, and that the reduction factor increases toward the series limit [86.67, 68].

Also, the cross section for photoejection of an electron bound by any screened Coulomb potential must vanish at threshold – in marked contrast to the nonzero photoionization cross sections of isolated atoms and ions. Since the oscillator strength sum rule still holds, any diminution of the total bound-bound oscillator strength must be offset by an increase in the bound-free contribution; continuum lowering partly accounts for this latter increase.

Spectroscopic observations of the reduction of decay rates by plasma screening have been reported [86.69]. But, the plasma densities evidently were too low for this effect to clearly manifest itself, and alternative explanations have since been given [86.70, 71]. In addition, there is the more fundamental question of whether static screening models are even appropriate for the description of radiative processes. As discussed below, this issue also arises in connection with inelastic collision processes in dense plasmas.

Several large-scale computer codes are in wide use to calculate the opacity of hot, dense matter (in LTE). Among these, we note the code HOPE [86.62], which is based on the average atom model; the code LEDCOP [86.72], which uses accurate (Hartree–Fock) atomic term data; and the code OPAL [86.73], which uses detailed configuration accounting and parametric, (static) screened potentials to compute wave functions and energy levels. Also, there are some published results from the new code IDEFIX [86.74], which is based on a non-spherical (di-center) screened potential arising from the radiating ion and its nearest neighbor. When making comparisons among these models, it should be realized that the codes use quite different line-broadening and continuum lowering prescriptions.

86.3.2 Collisional Transitions

Various screened Coulomb interactions also can be used to study plasma effects on inelastic scattering. References [86.31, 75] and citations therein use either Debye or ion-sphere potentials, and Born, distorted-wave, or close coupling approximations, to investigate excitation processes in plasmas; however, bound states were left unperturbed. More elaborate static potentials and perturbed bound states were treated by *Davis* and *Blaha* [86.76, 77], but they did not self-consistently screen the interaction between projectile and target. For excitations involving a small transition energy ΔE, *Kitamura* [86.78] has recently published a self-consistent treatment of both (1) the quasi-static perturbations of the target ion by the microfield, and (2) the dynamically screened electron-target interaction.

The use of static screening models is invalid when $\Delta E \gg \hbar\omega_{\mathrm{e}}$ because the collision duration is too short for any average description of the plasma's screening to apply. In such cases, ionization being a particular example, one must consider the response of the target to electrodynamic disturbances [86.79]. Reference [86.80] gives a thorough discussion of this issue, and presents numerical examples of the effects of projectile and target screening in ionizing collisions.

Bremsstrahlung is another important plasma collision process for which static screening models have been extensively used [86.81–83]. Unfortunately, most bremsstrahlung radiation emanating from hot plasmas represents free–free transitions in which $\Delta E \gg \hbar\omega_{\mathrm{e}}$ (lower frequency emission being attenuated), and in these situations static screening models are suspect. In contrast, the formation of laser plasmas occurs mainly through inverse bremsstrahlung (free–free absorption) under conditions such that $\Delta E = \hbar\omega_{\mathrm{laser}} \gtrsim \hbar\omega_{\mathrm{e}}$, making static screening models relevant here.

More sophisticated treatments of bremsstrahlung in dense plasmas [86.84–86] include one or more of the following: strong coupling effects among the plasma ions (introduced via radial distribution functions [86.22]), dynamic screening effects involving the electrons (introduced via frequency-dependent dielectric response functions [86.18]), possible degeneracy effects (introduced via Fermi–Dirac distribution functions for occupation probabilities of initial and final states), and only partial screening of the nuclear charge by the target ion's bound electrons (introduced via a form factor for the target (Chapt. 56)). Calculations for plasmas with moderate coupling parameters ($\Gamma \leq$ few) reveal that the first three of these effects tend to reduce free-free emission and absorption rates, while the last effect tends to enhance the rates. At larger Γ-values, strong ion-ion coupling tends to drive these rates back up [86.87].

Advances in simulation capability (which we discuss next) are yielding ever more realistic descriptions of the dense plasma environment, but what is proving difficult to improve upon is the ubiquitous use of the Born approximation to treat all electron-ion scattering events (see, however, *Berkovsky* and *Kurilenkov* [86.34]). Strong collisions, i. e., those in which the photon energy $\hbar\omega$ is comparable with the relative kinetic energy of the collision, are particulary important for radiative losses from plasmas, but these also are just the collisions most likely to be poorly described by the Born approximation.

86.4 Simulations

Most of the challenges pertaining to atomic phenomena in dense plasmas arise from the many-body nature of the atom-plasma interaction. Simple models are therefore subject to inaccuracies that may arise from inconsistencies or severe approximations which, in turn, degrade our understanding of experimental data. Because of this complexity, simulations are playing an increasingly important role in this field.

Historically, simulations of dense plasmas have used simplified plasma models to address issues ranging from equations of state [86.88, 89] to plasma microfields [86.27, 28, 90]. Atomic physics is typically ignored within the simulation by assuming that an average charge $\langle Z \rangle$, somehow known, can be assigned to each ion. For these simulation methods, there are many excellent textbooks that introduce the basic ideas [86.91–93]. Here, we focus instead on simulations that attempt to describe both the dense plasma and the atomic physics self-consistently. These simulation methods are categorized in terms of the underlying algorithm used and fall either into Monte Carlo or molecular dynamics methods; we discuss each in turn.

86.4.1 Monte Carlo

Monte Carlo is a method to evaluate average quantities in thermodynamic equilibrium using random numbers. For example, to obtain a property of a classical system we might write

$$\langle \mathcal{O} \rangle = \frac{\int \mathrm{d}^{3N} r \, \mathrm{d}^{3N} p \, \mathcal{O}\left(r^{3N}, p^{3N}\right) \exp\left(-\beta \mathcal{H}\right)}{\int \mathrm{d}^{3N} r \, \mathrm{d}^{3N} p \, \exp\left(-\beta \mathcal{H}\right)} , \tag{86.19}$$

and, in principle, we could sample the multidimensional integral randomly to obtain a good estimate of the average value of $\mathcal{O}$ given a many-body Hamiltonian $\mathcal{H}$. In practice, however, there are large portions of phase space that give very little contribution to the average and some method of "importance sampling" must be carried out. This problem was originally solved by *Metropolis* and coworkers [86.94] who introduced the Metropolis method, which uses a Markov chain of states in phase space that preferentially migrates towards states of higher probability [86.92].

The computation of properties of atomic systems in dense plasmas requires a quantum Monte Carlo (QMC) method because atomic systems are inherently quantum systems and the plasma itself can be degenerate. Although there are a variety of QMC methods [86.93, 95], the most important for our purposes is the path-integral Monte Carlo (PIMC) method, which is formulated at finite temperatures and, in fact, exploits this condition by constructing an equivalent system of many more particles at a higher effective temperature. This is achieved by writing spatial matrix elements of the quantal version of the Boltzmann factor of (86.19) as

$$\begin{aligned} &\langle \boldsymbol{r} | \exp\left(-\beta \hat{\mathcal{H}}\right) | \boldsymbol{r}' \rangle \\ &= \int \mathrm{d}^3 r'' \, \langle \boldsymbol{r} | \exp\left(-\beta \hat{\mathcal{H}}/2\right) | \boldsymbol{r}'' \rangle \\ &\quad \times \langle \boldsymbol{r}'' | \exp\left(-\beta \hat{\mathcal{H}}/2\right) | \boldsymbol{r}' \rangle . \end{aligned} \tag{86.20}$$

This expression can be used recursively to obtain matrix elements evaluated at higher and higher temperatures; this allows a high-temperature approximation to be made, albeit at the expense of having many more matrix elements to evaluate. It can be shown that in the simplest approximation each quantum particle can be replaced by a polymer of M classical particles linked by springs; this picture is referred to as the "classical isomorphism" [86.96]. All electrons in the system (bound and free) are treated on an equal footing.

Although the PIMC method is, in principle, simple to implement and can be quite accurate, there are several issues that arise when it is applied to dense plasmas. There are difficulties with the deep attractive Coulomb well that is crucial for describing atomic physics; this leads to the need for enormous numbers of fictitious classical particles. A partially analytical or numerical solution can greatly mitigate this problem [86.93, 97]. PIMC also suffers from difficulty when treating fermion systems because of the so-called fermion sign problem in which many terms of opposite sign arise from the antisymmetric form of the N-electron wave function. Progress has been made in this direction as well [86.95]. Finally, the long-ranged nature of the Coulomb potential causes additional difficulties for describing bulk systems with periodic boundary conditions [86.98]. When these additional considerations are taken into account, good agreement with other methods is found, and results have led to important conclusions about experiments [86.99], which are detailed below.

86.4.2 Molecular Dynamics

Molecular dynamics is a simulation method based on the time evolution of a many-body system [86.91–93].

The vast majority of MD simulations are based on the solution of Newton's equations for N classical particles in a main cell with periodic boundary conditions. Simulations can be carried out in various ensembles and one assumes that long simulations in the canonical ensemble will correctly sample canonical averages in the same manner as (86.19). Atomic properties in dense plasmas are usually obtained in the Born–Oppenheimer approximation, which freezes the ionic dynamics between time-steps and performs an electronic structure calculation to obtain the electronic density for that particular ionic configuration. The ions are then advanced using the forces from the resulting electronic density, and the procedure is repeated. Many methods are available, including basic TF [86.100], Hartree–Fock [86.101], and tight-binding [86.102] to compute the electronic structure. A commonly-used DFT code for warm matter is the Vienna Ab-Initio Simulation Package (VASP), which is widely distributed [86.103]. In contrast to most PIMC implementations, the deep Coulomb potential is often treated in VASP by softening the electron-ion interaction with some form of pseudopotential [86.103, 104]. These methods trade between accuracy and range of validity; for example, TF theory is less accurate than DFT-MD, but is useful at higher temperatures ($\Theta > 50\,\text{eV}$) and for very dense plasmas.

There are three advantages to the MD method. First, the MD approach greatly extends the AA model by including many nuclear centers in the main simulation cell, and these centers may arise from different elements. (This can also be done with PIMC.) Futhermore, DFT-MD, which can be very accurate for cold systems ([86.105], p. 117), has the additional advantage that single-particle (Kohn–Sham) orbitals can be used in formulae for linear, frequency-dependent response properties, such as the electrical conductivity. (It may seem paradoxical that electron dynamical information can be obtained from a static calculation. Strictly speaking this is not possible, although reasonable results can be obtained for some quantities [86.105, p. 49]). And finally, MD has the additional advantage that dynamical ionic properties can be obtained from the time evolution that is simulated. For example, DFT-MD simulations for the computation of the self-diffusion coefficient of dense hydrogen [86.106] have recently been performed; in principle, a wide range of other dynamical ionic properties are available, such as viscosity, thermal conductivity, and collective behavior.

Quantum simulations beyond the Born–Oppenheimer approximation that treat electrons and ions on an equal, dynamical footing are much more difficult since they involve a numerical solution of the N-body Schrödinger equation. Such simulations are necessary, however, for obtaining dynamical electron properties, especially under nonequilibrium conditions – those with time-dependent temperatures or nonthermal momentum distributions. Furthermore, these simulations can describe electronic properties, such as atomic physics, strong scattering, and degeneracy.

Very simple models have been developed by *Deutsch* and coworkers, who constructed effective interactions between the electrons that yield some known property, such as the high-temperature pair correlation function [86.107]. These interactions can then be used directly in a simulation, but with modified equations of motion – an approach pioneered for dense plasmas by *Hansen* and coworkers [86.108]. Diffractive and symmetry effects can be accounted for in the high-temperature limit. For example, one model of the diffractive potential for the electron-ion interaction is

$$v_{\text{ei}}(r) = -\frac{Ze^2}{r}\left[1 - \exp\left(-r/\lambda\right)\right] , \qquad (86.21)$$

where λ is on the order of the electron deBroglie wavelength. This potential is finite at the origin, which prevents the classical collapse of a neutral system during the simulation. Unfortunately, such a method suffers from several weaknesses, including being limited to high temperatures, only describing Pauli exclusion by pair-interactions, and having incorrect atomic binding energies.

There have been several attempts to improve upon simple potentials of the form (86.21). Since one of the main features of quantum mechanics is that conjugate space- and momentum-dependent quantities do not commute, it is natural to construct potentials $v(r, p)$ that depend on both r and p. Such momentum-dependent potentials have been formulated in the context of nuclear physics and have recently been applied to dense plasmas by *Ebeling* and coworkers [86.109]. Although quite good atomic properties can be obtained, these potentials suffer from the fact that they are ad hoc, and one does not know how to choose adjustable parameters for unexplored conditions.

A direct approach is to solve the time-dependent, many-particle Schrödinger equation, albeit approximately. This can be done by reducing the (infinite) degrees of freedom to a smaller, more manageable set. For example, *Heller* [86.110] first suggested using a Gaussian wavepacket to describe electron semiclassical dynamics. This has been applied to ionization in a dense plasma by *Ebeling* [86.111]. In general, a many-body, antisymmetric wave function can be parametrized in terms of a few parameters for each particle and a time-dependent variational principle can be used to obtain the equations of motion of the parameters. For example, each particle in a Slater determinant can be chosen to be a Gaussian with a width $w(t)$ with its conjugate momentum $p_w(t)$. This method is referred to as "Fermion molecular dynamics" or "wavepacket molecular dynamics" (WPMD). A review of this approach has recently been given by *Feldmeir* and *Schnank* [86.112]. Wavepacket shapes other than Gaussians, which can better reproduce atomic properties, have been proposed by *Murillo* and *Timmermans* [86.113].

86.4.3 The Deuterium EOS Problem

Experiments can be notoriously difficult to perform in the dense plasma regime because of the enormous pressures produced. Recently, experiments on compressed deuterium have been performed that pass from the molecular fluid phase into the dense plasma phase and therefore probe the physics of atomic and molecular states in a dense environment. The first of these was conducted at Livermore [86.114] using the Nova laser to shock compress liquid deuterium to 2 Mbar. The experiments indicated a higher compressibility compared with commonly used equation of state properties. These interesting results led to new experiments, again at Livermore [86.115], with pressures exceeding 3.0 Mbar, and at Sandia using a magnetically driven flyer plate on the Z machine [86.116] to achieve 0.7 Mbar. The Sandia results did not show the unexpected higher compressibility. Later, additional laser-based experiments were carried out at the Naval Research Laboratory [86.117] to 6 Mbar, which agreed with the orignial laser-based experiments but had significant error bars. The Livermore and Sandia results are shown in Fig. 86.4. (Error bars are not shown.) Interpreting these experiments has, in turn, led to increased activity in the use and development of various simulation methods.

Also shown in Fig. 86.4 are results from PIMC and WPMD simulations. Early PIMC results, which treated the Fermion sign problem using properties of free particles, did not agree well with either experiment. A later calculation [86.99] improved upon that treatment and

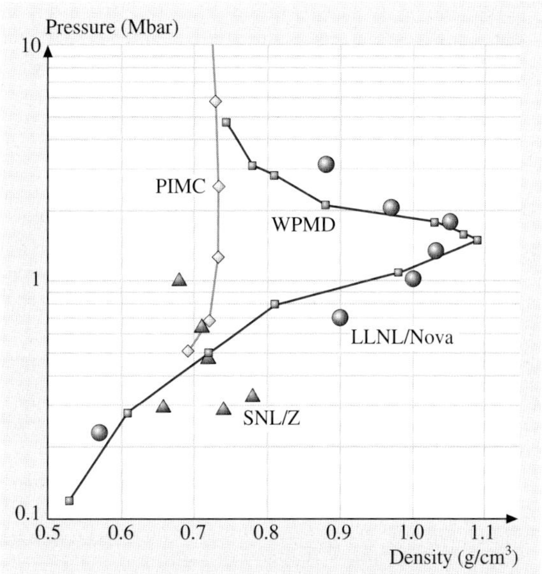

Fig. 86.4 The equation of state of shocked deuterium; here pressure (in Mbars) versus density (in g/cm^3) is plotted along the Hugoniot. Results from two experiments are shown: Sandia (*triangles*) and Livermore (*circles*). Also shown are the results from two simulations: path integral monte carlo (*diamonds*) and wavepacket molecular dynamics (*squares*)

showed that details of antisymmetric wave functions become important below 2 Mbar. These results are shown in the figure, and better agreement with the Sandia result is found. Also shown in the figure are results from a WPMD calculation [86.118], which tend to agree with the Livermore data. The WPMD calculations, however, did not include full antisymmetrization of the electron-electron interaction. The temperatures predicted by the simulations tend to be in the fraction of an eV range at the lower part of the figure and up to tens of eV toward the top of the figure; thus the experiments and the simulations are probing the very interesting WDM regime in which molecular and atomic species are heated into a cool plasma state.

Together with experiments on resistivity and $\langle Z \rangle$, a more complete picture of the physics of atoms in dense plasmas is emerging. But, more theoretical and experimental developments are needed before we can tackle with confidence the wide range of dense plasmas that occur in the HEDP regime.

References

86.1 V. Decaux, V. L. Jacobs, P. Beiersdorfer, D. A. Liedahl, S. M. Kahn: Phys. Rev. A **68**, 012509 (2003)
86.2 Th. Stöhlker et al.: Phys. Rev. Lett. **71**, 2184 (1993)
86.3 J. D. Lindl, B. A. Hammel, B. G. Logan, D. D. Meyerhofer, S. A. Payne, J. D. Sethian: Plasma Phys. Control. Fusion **45**, A217 (2003)
86.4 S. A. Slutz, J. E. Bailey, G. A. Chandler, G. R. Bennett, G. Cooper, J. S. Lash, S. Lazier, P. Lake, R. W. Lemke, T. A. Mehlhorn, T. J. Nash, D. S. Nielson, J. McGurn, T. C. Moore, C. L. Ruiz, D. G. Schroen, J. Torres, W. Varnum, R. A. Vesey: Phys. Plasmas **10**, 1875 (2003)
86.5 H. Daido: Rept. Prog. Phys. **65**, 1513 (2002)
86.6 P. Audebert, R. Shephard, K. B. Fournier, O. Peyrusse, D. Price, R. Lee, P. Springer, J.-C. Gauthier, L. Klein: Phys. Rev. Lett. **89**, 265001 (2002)
86.7 R. W. Lee, S. J. Moon, H. K. Chung, W. Rozmus, H. A. Baldis, G. Gregori, R. C. Cauble, O. L. Landen, J. S. Wark, A. Ng, S. J. Rose, C. L. Lewis, D. Riley, J. C. Gauthier, P. Audebert: J. Opt. Soc. Am. B **20**, 770 (2003)
86.8 S. Eliezer, A. Ghatak, H. Hora: *Fundamentals of Equations of State* (World Scientific, Singapore 2002)
86.9 R. M. More, K. H. Warren, D. A. Young, G. B. Zimmerman: Phys. Fluids **31**, 3059 (1988)
86.10 F. J. Rogers, C. A. Iglesias: Astrophys. J. Suppl. **79**, 507 (1992)
86.11 D. Saumon, G. Chabrier, H. M. Van Horn: Astrophys. J. Suppl. **99**, 713 (1995)
86.12 D. R. Bates, A. E. Kingston, R. W. P. McWhirter: Proc. R. Soc. London **A267**, 297 (1962)
86.13 A. Sasaki, T. Utsumo, K. Moribayashi, T. Tajima, H. Takuma: J. Quant. Spectrosc. Radiat. Transfer **65**, 501 (2000)
86.14 H. R. Griem: *Principles of Plasma Spectroscopy* (Cambridge, New York 1997)
86.15 D. Salzmann: *Atomic Physics in Hot Plasmas* (Oxford Univ. Press, Oxford 1998)
86.16 T. Fujimoto: *Plasma Spectroscopy* (Oxford Univ. Press, Oxford 2004)
86.17 *Frontiers in High Energy Density Physics, The X-Games of Contemporary Science* (The National Academies Press, Washington, D.C. 2003)
86.18 S. Ichimaru: *Statistical Plasma Physics*, Vol. 1 (Addison-Wesley, Redwood City 1992)
86.19 S. Ichimaru: *Statistical Plasma Physics*, Vol. 2 (Addison-Wesley, Redwood City 1994)
86.20 M. S. Murillo: Phys. Plasmas **11**, 2964 (2004)
86.21 C. A. Iglesias: private communication (2004)
86.22 N. H. March, M. P. Tosi: *Atomic Dynamics in Liquids* (Dover, New York 1991) Chap. 7
86.23 J. M. Gil, P. Martel, E. Minguez, J. G. Rubiano, R. Rodrigeuz, F. H. Ruano: J. Quant. Spectrosc., Radiat. Transfer **75**, 539 (2002)
86.24 L. Chen, A. B. Langdon, M. A. Liebman: J. Plasma Phys. **9**, 311 (1973)
86.25 S. Chandrasekhar: Rev. Mod. Phys. **15**, 1 (1943)
86.26 C. A. Iglesias, F. J. Rogers, R. Shepherd, A. Bar-Shalom, M. S. Murillo, D. P. Kilcrease, A. Calisti, R. W. Lee: J. Quant. Spectrosc. Radiat. Transfer **65**, 303 (2000)
86.27 A. Y. Potekhin, G. Chabrier, D. Gilles: Phys. Rev. E **65**, 036412 (2002)
86.28 B. Talin, A. Calisti, J. Dufty: Phys. Rev. E **65**, 056406 (2002)
86.29 L. Spitzer, Jr.: *Physics of Fully Ionized Gases*, 2nd edn. (Wiley-Interscience, New York 1962)
86.30 D. O. Gericke, M. S. Murillo, M. Schlanges: Phys. Rev. E **65**, 036418 (2002)
86.31 J. C. Weisheit: *Applied Atomic Collision Physics*, Vol. 2 (Academic Press, Orlando 1984)
86.32 C.-K. Li, R. D. Petrasso: Phys. Rev. Lett. **70**, 3063 (1993)
86.33 D. B. Boercker, F. J. Rogers, H. E. DeWitt: Phys. Rev. A **25**, 1623 (1982)
86.34 M. A. Berkovsky, Yu. K. Kurilenkov: Physica A **197**, 676 (1993)
86.35 J. F. Benage, Jr.: Phys. Plasmas **7**, 2040 (2000)
86.36 J. C. Weisheit: *Physics of Strongly Coupled Plasmas* (World Scientific, Singapore 1996)
86.37 G. Gregori et al.: Phys. Plasmas, **11**, 2754 (2004)
86.38 Ya. B. Zeldovich, Yu. P. Raizer: *Physics of Shock Waves and High-Temperature Hydrodynamic Phenomena* (Academic, New York 1966) Chap. 3
86.39 D. E. Osterbrock: *Astrophysics of Gaseous Nebulae and Active Galactic Nuclei* (University Science Books, Mill Valley 1989)
86.40 R. A. Hulse: Nucl. Tech. Fusion **3**, 259 (1983)
86.41 D. G. Hummer, D. Mihalas: Astrophys. J. **331**, 794 (1988)
86.42 D. R. Inglis, E. Teller: Astrophys. J. **90**, 439 (1939)
86.43 F. J. Rogers, H. C. Graboske, D. J. Harwood: Phys. Rev. A **1**, 1577 (1970)
86.44 J. C. Stewart, K. D. Pyatt: Astrophys. J. **144**, 1203 (1966)
86.45 M. W. C. Dharma-Wardana, F. Perrot: Phys. Rev. A **45**, 5883 (1992)
86.46 A. Maksimchuk, M. Nantel, G. Ma, S. Gu, C. Y. Côte, D. Umstadter, S. A. Pikuz, I. Yu. Skobelev, A. Ya. Faenov: J. Quant. Spectrosc. Radiat. Transfer **65**, 367 (2000)
86.47 F. J. Rogers: Astrophys. J. **310**, 723 (1986)
86.48 S. Tanaka, X.-Z. Yan, S. Ichimaru: Phys. Rev. A **41**, 5616 (1990)
86.49 J. Stein, D. Salzmann: Phys. Rev. A **45**, 3943 (1992)
86.50 E. Oks: Phys. Rev. E **63**, 057401 (2001)
86.51 D. V. Fisher, Y. Maron: J. Quant. Spectrosc. Radiat. Transfer **81**, 147 (2003)

86.52 J. Al-Kuzee, T. A. Hall, H.-D. Frey: Phys. Rev. E **57**, 7060 (1998)
86.53 J. Seidel, S. Arndt, W. D. Kraeft: Phys. Rev. E **52**, 5387 (1995)
86.54 S. R. Stone, J. C. Weisheit: J. Quant. Spectrosc. Radiat. Transfer **35**, 67 (1986)
86.55 C. A. Iglesias, R. W. Lee: J. Quant. Spectrosc. Radiat. Transfer **58**, 637 (1997)
86.56 R. P. Feynman, N. Metropolis, E. Teller: Phys. Rev. **73**, 1561 (1949)
86.57 D. Lai, A. M. Abrahams, A. L. Shapiro: Astrophys. J. **377**, 612 (1991)
86.58 W. Zakowicz, I. J. Feng, R. H. Pratt: J. Quant. Spectrosc. Radiat. Transfer **27**, 329 (1982)
86.59 R. M. More: Adv. Am. Mol. Phys. **21**, 305 (1985)
86.60 U. Gupta, A. K. Rajagopal: Phys. Rep. **87**, 259 (1982)
86.61 P. Fromy, C. Deutsch, G. Maynard: Phys. Plasmas **3**, 714 (1996)
86.62 B. F. Rozsnyai: Phys. Rev. A **5**, 1137 (1972)
86.63 B. F. Rozsnyai: Phys. Rev. E **55**, 7507 (1997)
86.64 D. A. Liberman: J. Quant. Spectrosc. Radiat. Transfer **27**, 335 (1982)
86.65 R. Ying, G. Kalman: Phys. Rev. A **40**, 3927 (1989)
86.66 F. Perrot, M. W. C. Dharma-Wardana: Phys. Rev. E **52**, 5352 (1995)
86.67 J. C. Weisheit, B. W. Shore: Astrophys. J. **194**, 519 (1974)
86.68 L. G. D'yachkov, G. A. Kobzev, P. M. Pankratov: J. Quant. Spectrosc. Radiat. Transfer **44**, 123 (1990)
86.69 Y. Chung, H. Hirose, S. Suckewer: Phys. Rev. A **40**, 7142 (1989)
86.70 Y.-C. Chen, J. L. Libowitz: Phys. Rev. A **41**, 2127 (1990)
86.71 F. Aumayr, W. Lee, C. H. Skinner, S. Suckewer: J. Phys. B **24**, 4489 (1991)
86.72 N. H. Magee, A. L. Merts, J. J. Keady, D. P. Kilcrease: Los Alamos Report LA-UR-97-1038 (1997)
86.73 C. A. Iglesias, F. J. Rogers: Astrophys. J. **464**, 943 (1996)
86.74 E. Minguez, P. Sauvan, J. M. Gil, R. Rodriguez, J. G. Rubiano, R. Florido, P. Martel, P. Angelo, R. Schott, F. Philippe, R. Leboucher-Dalimier, R. Mancini: J. Quant. Spectrosc. Radiat. Transfer **81**, 301 (2003)
86.75 B. L. Whitten, N. F. Lane, J. C. Weisheit: Phys. Rev. A **29**, 945 (1984)
86.76 J. Davis, M. Blaha: J. Quant. Spectrosc. Radiat. Transfer **27**, 307 (1982)
86.77 J. Davis, M. Blaha: *Physics of Electronic and Atomic Collisions* (North-Holland, Amsterdam 1982)
86.78 H. Kitamura: Phys. Plasmas **11**, 771 (2004)
86.79 J. C. Weisheit: Adv. At. Mol. Phys. **25**, 101 (1988)
86.80 M. S. Murillo, J. C. Weisheit: Physics Reports **302**, 1 (1998)
86.81 I. P. Grant: Mon. Not. Roy. Astron. Soc. **118**, 241 (1958)
86.82 B. F. Rozsnyai, J. Quant: Spectrosc. Radiat. Transfer **22**, 337 (1979)
86.83 L. Kim, R. H. Pratt, H. K. Tseng: Phys. Rev. A **32**, 1693 (1985)
86.84 R. Kawakami, K. Mima, H. Totsuji, Y. Yokoyama: Phys. Rev. A **38**, 3618 (1988)
86.85 F. Perrot: Laser and Particles Beams **14**, 731 (1996)
86.86 C. A. Iglesias: J. Plasma Phys. **58**, 381 (1997)
86.87 Th. Bornath, M. Schlanges, P. Hilse, D. Kremp: J. Physics A **36**, 5941 (2003)
86.88 S. Hamaguchi, R. T. Farouki, D. H. E. Dubin: Phys. Rev. E **56**, 4671 (1997)
86.89 G. S. Stringfellow, H. E. Dewitt, W. L. Slattery: Phys. Rev. A **41**, 1105 (1990)
86.90 M. S. Murillo, D. P. Kilcrease, L. A. Collins: Phys. Rev. E **55**, 6289 (1997)
86.91 M. P. Allen, D. J. Tildesley: *Computer Simulation of Liquids* (Oxford Science Publications, Oxford 1994)
86.92 D. Frenkel, B. Smit: *Understanding Molecular Simulation, From Algorithms to Applications* (Academic Press, San Diego 1996)
86.93 J. M. Thijssen: *Computational Physics* (Cambridge Univ. Press, New York 1999)
86.94 N. Metropolis, A. W. Rosenbluth, M. N. Rosenbluth, A. H. Teller, E. Teller: J. Chem. Phys. **21**, 1087 (1953)
86.95 D. Ceperley, B. Alder: Science **231**, 555 (1986)
86.96 D. Chandler, P. G. Wolynes: J. Chem. Phys. **74**, 4078 (1981)
86.97 J. Theilhaber, B. J. Alder: Phys. Rev. A **43**, 4143 (1991)
86.98 V. Natoli, D. M. Ceperley: J. Comp. Phys. **117**, 171 (1995)
86.99 B. Militzer, D. M. Ceperley: Phys. Rev. Lett. **85**, 1890 (2000)
86.100 G. Zérah, J. Clérouin, E. L. Pollock: Phys. Rev. Lett. **69**, 446 (1992)
86.101 S. M. Younger, A. K. Harrison, G. Sugiyama: Phys. Rev. A **40**, 5256 (1989)
86.102 T. J. Lenosky, J. D. Kress, L. A. Collins, R. Redmer, H. Juranek: Phys. Rev. E **60**, 1665 (1999)
86.103 G. Kresse, J. Furthmüller: Phys. Rev. B **54**, 11169 (1996)
86.104 P. L. Silvestrelli, A. Alavi, M. Parinello: Phys. Rev. B **55**, 15515 (1997)
86.105 W. Koch, M. C. Holthausen: *A Chemist's Guide to Density Functional Theory* (Wiley-VCH, New York 2000)
86.106 J. Clérouin, J. F. Dufreche: Phys. Rev. E **64**, 066406 (2001)
86.107 M.-M. Gombert, H. Minoo, C. Deutsch: Phys. Rev. A **29**, 940 (1984)
86.108 J. P. Hansen, I. R. McDonald: Phys. Rev. A **23**, 2041 (1981)
86.109 W. Ebeling, F. Shautz, J. Ortner: Phys. Rev. E **56**, 3498, 4665 (1997)
86.110 E. J. Heller: J. Chem. Phys. **62**, 1544 (1975)
86.111 W. Ebeling, A. Förster, V.,Yu. Podlipchuk: Phys. Lett. A **218**, 297 (1996)
86.112 H. Feldmeier, J. Schnack: Rev. Mod. Phys. **72**, 655 (2000)

86.113 M. S. Murillo, E. Timmermans: Contr. Plasma Phys. **43**, 333 (2003)

86.114 L. B. Da Silva, P. Celliers, G. W. Collins, K. S. Bundil, N. C. Holmes, T. W. Barbee Jr., B. A. Hammel, J. D. Kilkenny, R. J. Wallace, M. Ross, R. Cauble, A. Ng, G. Chiu: Phys. Rev. Lett. **78**, 483 (1997)

86.115 G. W. Collins, L. B. Da Silva, P. Celliers, D. M. Gold, M. E. Foord, R. J. Wallace, A. Ng, S. V. Weber, K. S. Bundil, R. Cauble: Science **281**, 1178 (1998)

86.116 M. D. Knudsen, D. L. Hansen, J. E. Bailey, C. A. Hall, J. R. Asay, W. W. Anderson: Phys. Rev. Lett. **87**, 225501 (2001)

86.117 A. N. Mostovych, Y. Chan, T. Lehecha, A. Schmitt, J. D. Sethian: Phys. Rev. Lett. **85**, 3870 (2000)

86.118 M. Knaup, P.-G. Reinhard, C. Toepffer, G. Zwicknagel: J. Phys. A **36**, 6165 (2003)

87. Conduction of Electricity in Gases

The conduction of electricity through gases has played ubiquitous roles in science and technology. It was responsible for many of the fundamental discoveries in atomic and molecular physics; gas discharge lighting is essential to every night operations; gas discharge lasers are still important in research and manufacturing; and all of advanced microelectronics depends on plasma enhanced processing. To a large extent, the efficiencies of the above cited applications of gaseous electronics depend on the maintenance of the distinct non-equilibrium between the electrons and the gas or vapor. This non-equilibrium can be achieved by operating at low pressures or under pulsed excitation, where the duration of the energy input is less than the energy equilibration time between the electrons and the heavy particles. The term gas discharge originally described a transient or spark condition, but has been extended to mean the continuous conduction of electricity through gases.

Section 87.1 treats the electron-velocity distribution and its effect on various measurements involving a swarm or distribution of electron velocities. In this section, low fractional ionization ($<10^{-6}$) is assumed; electron–ion collisions are negligible relative to electron–atom (and electron–molecule) collisions in so far as they affect electron mobility, diffusion or energy loss.

Section 87.2 introduces the glow discharge and considers the cold cathode and hot cathode discharge phenomena. Section 87.3 discusses ionization by electron collision, electron attachment, ion mobility, ion–ion and electron–ion recombination, and other important processes that affect the conduction of electricity in gases. Section 87.4 illustrates the importance of gaseous electronics with several important phenomena and technical applications.

The physics and chemistry of the conduction of electricity in ionized gases involves the interactions of the electrons and ions in the gas among themselves, with ground-state and excited-state gas atoms or molecules, with any surfaces that may be present, and with any electric or magnetic fields that are externally applied or generated by movement of the charged species. The electron collision-induced excitation and dissociation can result in new compounds being formed at rates which are orders of magnitude larger than those without a plasma present. For a gas pressure of 1 torr, the electron mean free path λ_{mfp} between collisions is ~ 1 mm ($\sim 10^9$ collisions/second). The assumption that λ_{mfp} is small relative to the plasma physical dimensions is usually valid, however it is often comparable to, or larger than the space charge sheaths that form near electrodes and surfaces.

Since the electron–atom interactions, in most cases of interest, have ranges shorter than the average gas-atom separation, they may be treated as binary collisions and the effects of neighboring gas atoms on a given electron–atom encounter can be neglected. Quantities involving such interactions in a partially ionized gas, where the electrons are distributed both in velocity and position, can be directly related to the corresponding quantities in beam experiments, where the collisions are binary and the electrons have a defined velocity. The measurements and calculations of the electron velocity distributions, or of the electron energy distribution functions (EEDF), are therefore of fundamental importance in the description of partially ionized gases.

87.1 Electron Scattering and Transport Phenomena

87.1.1 Electron Scattering Experiments

In 1903, *Lenard* [87.1] determined the attenuation of a beam of mono-energetic electrons by several gases. He measured the fraction of an electron beam that was transmitted without scattering through a field-free region containing the gas. Let an electron beam of density n_e electrons/cm^3, traveling with velocity v, pass a distance $\mathrm{d}x$ through the gas. Let N be the number of atoms/cm^3. The loss of electrons by scattering per unit time due to collisions with atoms is then given by

$$\frac{\mathrm{d}n_e}{\mathrm{d}t} = -Nv\sigma n_e\,, \qquad (87.1)$$

where σ is the collision cross section for electrons of velocity v. Writing $v\,\mathrm{d}t = \mathrm{d}x$, we obtain

$$n_e = n_{e0}\,\mathrm{e}^{-N\sigma x}\,. \qquad (87.2)$$

In relation to the conduction of electricity in gases, a more useful concept than σ is the momentum transfer cross section σ_m. In general, σ_m is not equal to σ. Experimental determinations of σ_m, generally from electron swarm experiments, are summarized in [87.2]. Results for the noble gases are presented in Fig. 87.1 and for representative diatomic gases in Fig. 87.2. Above energies of about 10 eV, σ_m decreases as the electron energy increases. Towards higher electron energies in the monatomic gases, σ_m is inversely proportional to the ionization potential and directly proportional to the gas polarizability. The fact that σ_m has extremely low values in certain gases at low energies is known as the Ramsauer–Townsend (RT) effect. This can be explained by partial wave scattering theory as the point where the radial wave function has exactly one more oscillation inside the interaction potential well than does the corresponding free function. Outside the well the two functions are indistinguishable and there is no $\ell = 0$ (s-wave) scattering [87.3] (Sect. 45.2.7). For the scattering potentials of the heavier noble gases, this occurs at

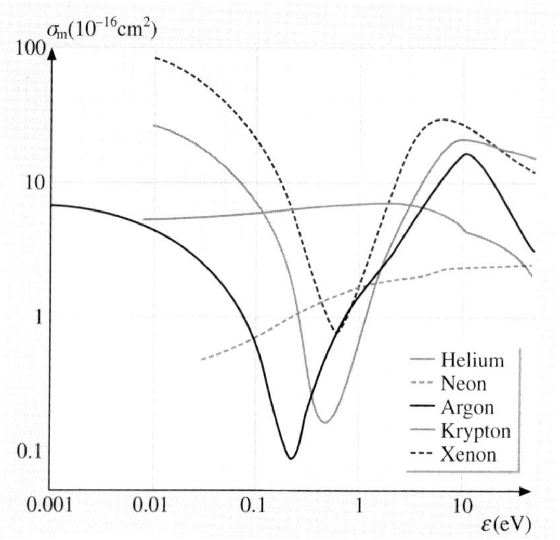

Fig. 87.1 Momentum transfer cross section for noble gases

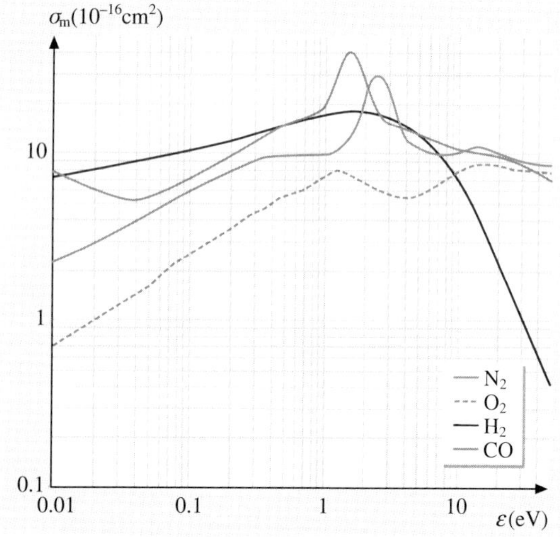

Fig. 87.2 Momentum transfer cross section for some diatomic gases

low energies where the partial cross sections for the higher angular momenta ($\ell = 1, 2, \dots$) are still negligible. Methane and silane (Fig. 87.3, [87.4, 5]) and possibly other nonpolar polyatomic gases, also display Ramsauer–Townsend (RT) minima. There is great similarity between the σ_m curves for atoms and molecules with similar external electron arrangements. The σ_m curves for H_2, O_2, N_2 and CO (Fig. 87.2) show that these cross sections are large over the range 0.1 to 10 eV. Around 1.5 eV, N_2 and CO display resonances which are associated with the formation of temporary negative ions followed by efficient vibrational excitation of these molecules [87.6].

87.1.2 Electron Transport Phenomena

Electron transport phenomena in an ionized gas, such as diffusion under the influence of a density gradient and drift under the influence of an electric field, are directly related to the electron current density Γ_e which is caused by these influences. In order to calculate Γ_e, it is necessary to know the electron spatial and velocity density distribution $f(\boldsymbol{v}, \boldsymbol{r}, t)$. The electron density n_e and current density Γ_e are given by

$$n_e(\boldsymbol{r}, t) = \int f(\boldsymbol{v}, \boldsymbol{r}, t)\, d^3v \,, \tag{87.3}$$

$$\Gamma_e(\boldsymbol{r}, t) = \int v f(\boldsymbol{v}, \boldsymbol{r}, t)\, d^3v \,. \tag{87.4}$$

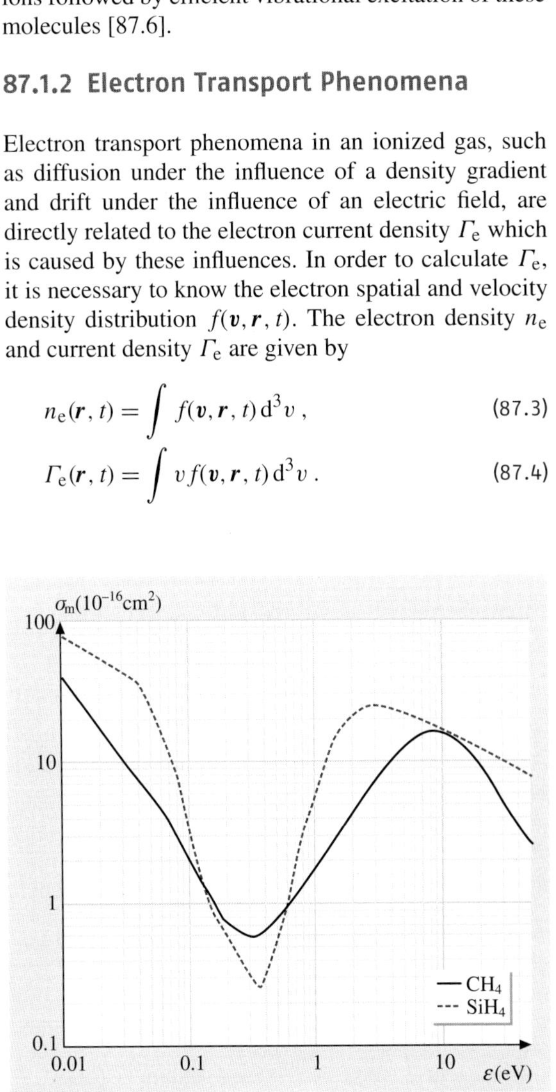

Fig. 87.3 Momentum transfer cross section for CH_4 and SiH_4

The electron drift velocity w_d is related to Γ_e and n_e by

$$w_d = \Gamma_e / n_e \,. \tag{87.5}$$

The electron drift velocity is usually small compared with the electron random velocity. The other limiting case is important in connection with ionic mobilities. It can also occur in electron transport in Ramsauer gases or gas mixtures that are used in fast particle counters [87.7]. The theoretical analysis of electron motion in gases uses primarily the Boltzmann transport equation approach [87.8]. Mean free path methods [87.9] and Monte-Carlo methods [87.10] have also been applied.

87.1.3 The Boltzmann Equation

The electron density distribution $f(\boldsymbol{v}, \boldsymbol{r}, t)$ of a given kind of particle in phase space (configuration and velocity space) is determined by the combined effect of all the interactions to which the particle is subjected. If there are no sources or sinks, the number of particles in a volume element of the six-dimensional space which moves with the particles does not change with time. In configuration space, the continuity equation for $f(\boldsymbol{v}, \boldsymbol{r}, t)$ is

$$\frac{\partial f}{\partial t} + \nabla_r \cdot \boldsymbol{v} f = 0\,; \tag{87.6}$$

in phase space, it has the form

$$\frac{\partial f}{\partial t} + \nabla_r \cdot \boldsymbol{v} f + \nabla_v \cdot \boldsymbol{a} f = 0\,, \tag{87.7}$$

where the subscripts on the divergence operators denote the independent variable, $\boldsymbol{v}$ is the particle flow velocity in configuration space, and $\boldsymbol{a}$ is the particle acceleration (i. e., the flow velocity velocity space). This may be written as $q(\boldsymbol{E} + \boldsymbol{v} \times \boldsymbol{B})/m$ for a particle of charge q and mass m in the presence of electric and magnetic fields, $\boldsymbol{E}$ and $\boldsymbol{B}$ respectively. If there are sources or sinks of particles, then terms representing these appear on the right-hand side of (87.7).

In principle, the effect of elastic and inelastic collisions may be included in the expressions for $\boldsymbol{E}$ and $\boldsymbol{B}$, but they are usually treated as source terms which transport the colliding particles instantaneously from one volume element of velocity space to another. This is a good approximation, for example, in the case of electron–atom collisions, but it is not valid for the longer range electron–ion and electron–electron interactions in the presence of high-frequency electric fields. Other collisions, such as ionizing collisions, act to change the total n_e as well as transporting electrons from one element of velocity space to another.

The electron–electron collisions will produce a Maxwellian velocity distribution when the fractional ionization is sufficiently large. For molecular gases, this generally requires $n_e/N_0 > 10^{-5}$. Since the electron–electron Coulomb collision frequency varies as v^{-3}, these collisions are especially effective in the low energy elastic regime where there are no inelastic energy loss processes, and they may have an influence for n_e/N_0 as low as 10^{-7} for the noble gases [87.11]. *Spitzer* [87.12], *Dreicer* [87.13], and *Butler* and *Buckingham* [87.14] treat cases where the electron–electron term is included explicitly. A full treatment of this subject is given by *Shkarofsky* [87.15].

Using $\nabla_v \cdot \boldsymbol{a} = 0$, and including collisions as a source term, (87.7) becomes

$$\frac{\partial f}{\partial t} + \boldsymbol{v} \cdot \nabla_r f + \boldsymbol{a} \cdot \nabla_v f = \left(\frac{\partial f}{\partial t}\right)_{\text{collisions}} , \qquad (87.8)$$

where the subscript 'collisions' identifies the source term due to all types of collisions.

87.1.4 Electron-Atom Elastic Collisions

For electrons in a weakly ionized gas, where only electron–atom elastic collisions need be explicitly considered, the collision term is given by the Boltzmann collision integral [87.16]. This integral describes the rate at which electrons are brought into and removed from an element of volume in velocity space in terms of the differential elastic scattering cross section .

The Boltzmann equation may be solved by a perturbation method [87.8] in which $\boldsymbol{E}$, $\boldsymbol{B}$, and the density gradients are the perturbations, starting from

$$f(\boldsymbol{v}, \boldsymbol{r}, t) \approx f_0(v, \boldsymbol{r}, t) + \frac{\boldsymbol{v}}{v} \cdot \boldsymbol{f}_1(v, \boldsymbol{r}, t) , \qquad (87.9)$$

where f_0 is the unperturbed isotropic term and $\boldsymbol{f}_1$ is the anisotropy due to $\boldsymbol{E}$ and $\boldsymbol{B}$. Substituting into (87.8) and equating terms of the same angular type then yields

$$\frac{\partial f_0}{\partial t} + \frac{v}{3}\nabla_r \cdot \boldsymbol{f}_1 - \frac{e}{3mv^2}\frac{\partial}{\partial v}\left(v^2 \boldsymbol{E} \cdot \boldsymbol{f}_1\right) = \frac{g}{2v^2}\frac{\partial}{\partial v}\left[v^3 \nu_m \left(f_0 + \frac{k_B T_g}{mv}\frac{\partial f_0}{\partial v}\right)\right] \qquad (87.10)$$

and

$$\frac{\partial \boldsymbol{f}_1}{\partial t} + v\nabla_r f_0 - \frac{e\boldsymbol{E}}{m}\frac{\partial f_0}{\partial v} + \boldsymbol{\omega}_B \times \boldsymbol{f}_1 = -\nu_m \boldsymbol{f}_1 , \qquad (87.11)$$

where $\nu_m = e\boldsymbol{E}/(w_d m)$ is the momentum transfer collision frequency from the Boltzmann collisions integral, $\boldsymbol{\omega}_B = e\boldsymbol{B}/m$ is a vector whose magnitude is the cyclotron frequency, and $g = 2m/(M+m)$ controls the partition of kinetic energy in an elastic collision between particles with masses M and m.

The right-hand term of (87.10) is the first-order correction due to elastic kinetic energy exchange between electrons and atoms. The $\boldsymbol{E} \cdot \boldsymbol{f}_1$ term arises from work done by $\boldsymbol{E}$ on the electrons, and the $\nabla_r \cdot \boldsymbol{f}_1$ term accounts for particle loss from the volume element in phase space due to electric field drift and density gradients. Similarly, in (87.11), the $\boldsymbol{E}$ term accounts for loss due to electric field acceleration, and the $v\nabla_r f_0$ term accounts for changes due to electron drift across a density gradient. The $-\nu_m \boldsymbol{f}_1$ term results from randomization of the electron velocity direction in electron–atom collisions.

The neglect of higher order terms in (87.9) is justified if $\delta\mathcal{T}_e/\mathcal{T}_e \ll 1$, where $\mathcal{T}_e$ is the gain in electron kinetic energy between collisions, and $\delta f_0/f_0 \ll 1$, where δf_0 is the change in f_0 over λ_{mfp}. For the case of static $\boldsymbol{E}$ and $\boldsymbol{B}$ with $\boldsymbol{E} \perp \boldsymbol{B}$, the orbital time $1/\omega_B$ replaces the time between collisions if it is shorter, making the above conditions less stringent. However, for an ac field with the resonant frequency ω_B, the original conditions apply. The off-resonant case in between. See [87.8, 17] for a further discussion of validity.

87.1.5 The Electron Drift Current

Transport properties of a partially ionized gas are calculated from the moments of $f(\boldsymbol{v}, \boldsymbol{r}, t)$. To obtain Γ_e, (87.11) must be solved for $\boldsymbol{f}_1$. Assume that $\omega_B = 0$ and that $\boldsymbol{E}$ is constant in time. Then, since f_0 usually changes slowly compared with ν_m, $\partial \boldsymbol{f}_1/\partial t$ may be neglected to a first approximation, and

$$\boldsymbol{f}_1 \simeq -\frac{1}{\nu_m}\left(v\nabla_r\left(n_e f_v^0\right) - \frac{e\boldsymbol{E}}{m} n_e \frac{\partial f_v^0}{\partial v}\right) , \qquad (87.12)$$

where $n_e(\boldsymbol{r}, t) f_v^0(v, \boldsymbol{r}, t) = f_0(v, \boldsymbol{r}, t)$. Use of (87.3) gives

$$\Gamma_e = -\nabla_r(D_e n_e) - n_e \mu_e \boldsymbol{E} , \qquad (87.13)$$

which defines the electron free diffusion coefficient

$$D_e = \int \frac{v^2}{3\nu_m} f_v^0 4\pi v^2 \, dv = \frac{\langle v^2 \rangle}{3\nu_m} , \qquad (87.14)$$

and the static electron mobility

$$\mu_e = -\frac{e}{m}\int \frac{v}{3\nu_m}\frac{\partial f_v^0}{\partial v} 4\pi v^2 \, dv . \qquad (87.15)$$

In (87.13), ∇_r should be interpreted as acting only on that part of D_e whose spatial dependence enters through an energy change, such as a variation of $\mathcal{T}_e$ with position (as in the theory of striations [87.18], but not on variations due to pressure gradients which would affect ν_m [87.8]).

Electron Free Diffusion

When the ionization density is low enough, the electrons diffuse freely to the walls, being unaffected by the space charge field caused by the heavier ions, which diffuse more slowly than the electrons. Setting $\boldsymbol{E}=0$ in (87.13), the electron current due to diffusion is

$$\Gamma_e = -\nabla_r(D_e n_e) . \quad (87.16)$$

Since the electron heating and elastic recoil terms in (87.10) do not contribute to a net gain or loss of electrons, integration of (87.10) over $\boldsymbol{v}$ gives

$$\frac{\partial n_e}{\partial t} + \nabla_r \cdot \Gamma_e = \frac{\partial n_e}{\partial t} - \nabla_r^2(D_e n_e) = 0 , \quad (87.17)$$

for a decaying electron density. This equation is identical in form to the thermal conductivity equation, which can be solved analytically for certain simple geometries. For the case of a long cylindrical container of radius R, the equation becomes

$$\frac{\partial n_e}{\partial t} = D_e \frac{1}{r}\frac{\partial}{\partial r}\left(r\frac{\partial n_e}{\partial r}\right) , \quad (87.18)$$

if D_0 is constant and n_e depends only on r. The general solution for n_e satisfying $n_e(R,t)=0$ is

$$n_e = \sum_{m=0} b_m \exp\left(-\frac{t}{t_m}\right) J_0\left(\frac{a_m r}{R}\right) , \quad (87.19)$$

where $t_m = [D_e(a_m/R)^2]^{-1}$, and J_0 is the zeroth-order Bessel function with roots $J_0(a_m)=0$, $a_{m+1} > a_m$. The b_m are determined by the radial dependence of n_e at $t=0$. (If the coaxial wall of radius $r_0 < R$ confines the plasma to an annular region $r_0 < r < R$, the general solution must also include the zeroth-order Bessel function of the second kind, Y_0, to satisfy the boundary conditions). The time constants t_m decrease rapidly with increasing m so that for sufficiently long t, the density distribution relaxes (e.g., after the microwave or static excitation is switched off) to

$$n_e \approx b_0 \exp\left(-\frac{t}{t_0}\right) J_0\left(\frac{a_0 r}{R}\right) . \quad (87.20)$$

This is called the fundamental diffusion mode. The above treatment gives a good approximation to $n_e(r)$ provided that $\lambda_{mfp} \ll R$.

When ν_m is velocity-independent and f_0 is the Maxwellian of temperature T_e, (87.14) becomes

$$D_e = \frac{k_B T_e}{m\nu_m} . \quad (87.21)$$

For the case where a magnetic field is applied to the ionized gas, parallel to the cylinder axis so as to impede the electron diffusion to the walls, the free-diffusion coefficient in equation (87.18) is replaced by

$$D_{eB} = \int \frac{\nu_m v^2 f_v^0}{3(\nu_m^2 + w_B^2)} 4\pi v^2 \, dv = \left\langle \frac{\nu_m v^2}{3(\nu_m^2 + w_B^2)} \right\rangle . \quad (87.22)$$

In this case, the electrons cannot diffuse to the walls unless they make collisions that interrupt their spiraling motion.

Electron DC Mobility in Static Fields

From (87.13), the electron particle current induced by an electric field $\boldsymbol{E}$ is $\Gamma_e = -n_e\mu_e E$, and the electron drift velocity is

$$w_d = \frac{\Gamma_e}{n_e} = -\mu_e E . \quad (87.23)$$

Measurement of w_d by a time-of-flight measurement can in principle, through the use of (87.14), provide some information about ν_m, provided that f_v^0 is known. Since a static electric field may perturb considerably the distribution function, and since n_e is low in such measurements, f_v^0 should be determined using (87.10) after having eliminated f_1 using (87.11). In the case where ν_m is independent of v, (87.15) after integration by parts, reduces to

$$\mu_e = \frac{e}{m\nu_m} \int f_v^0 4\pi v^2 \, dv = \frac{e}{m\nu_m} , \quad (87.24)$$

which is independent of f_v^0.

In general, w_d exhibits a complex dependence on E/N, as shown for some example gases in Fig. 87.4 [87.19]. The high mobility of electrons in methane is of technical interest because of potential applications in plasma switches used for current interruption [87.20]. The local maximum of w_d in methane for E/N = 3 Td (1 Townsend $= 10^{-17}$ Vcm^2) and its decrease with further increase of E/N arise because of the onset of inelastic collisions due to vibrational excitation at an energy close to the energy of the RT minimum [87.21]. As shown in Fig. 87.4, the drift velocities are generally in the range of 10^6–10^7 cm/s.

A detailed summary of electron drift velocities measured in many pure gases and gas mixtures of technical interest is given in [87.22, 23], and the techniques are elaborated in [87.24, 25].

The classic article on the drift velocity of ions in electric fields measured by a transit time technique is that

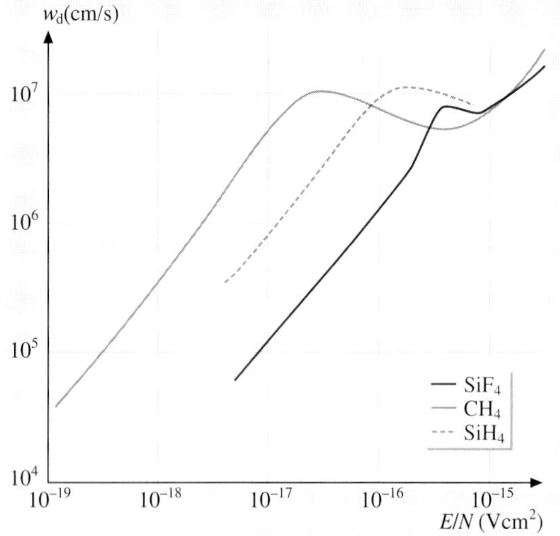

Fig. 87.4 Electron drift velocities in polyatomic gases

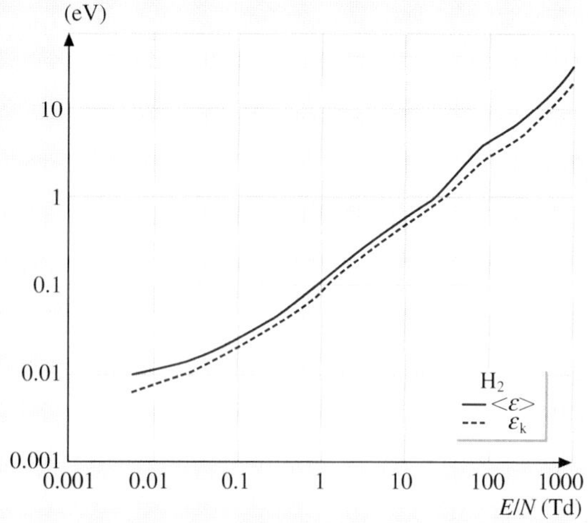

Fig. 87.5 Calculated u_{av} and D_e/μ_e coefficients in H_2

of *Hornbeck* [87.26]. He showed that in some instances earlier workers had identified molecular ions as atomic ions. Contemporary measurements often use the guided ion beam (GIB) technique to measure such quantities as the symmetric charge transfer cross sections [87.27]. A review of measurement data has been given [87.28].

The Ratio of D_e/μ_e for Electrons

From (87.14) and (87.15),

$$\frac{D_e}{\mu_e} = -\frac{m}{e}\frac{\int_0^\infty (v^4/\nu_m) f_v^0 \, dv}{\int_0^\infty (v^3/\nu_m)\partial f_v^0/\partial v \, dv} . \tag{87.25}$$

The above ratio of the free-diffusion coefficient to the mobility is termed the characteristic energy and it is a measure of the average electron energy in certain cases. When f_v^0 is Maxwellian, then

$$D_e/\mu_e = k_B T_e/e \tag{87.26}$$

becomes a direct measure of the electron energy, independent of ν_m. Since the average electron energy $u_{av} = \frac{3}{2} k_B T_e$, (87.26) becomes

$$D_e/\mu_e = \frac{2}{3} u_{av}/e . \tag{87.27}$$

In this form it is known as the Einstein relation. Furthermore, if ν_m is independent of velocity, (87.27) applies regardless of the velocity distribution.

Values of D_e/μ_e can be determined from experiments where a static electric field E_z is applied to electrons drifting under steady state conditions of diffusion. If D_e is taken to be independent of position, then the generalization of (87.17) is

$$\nabla^2 n_e = -\frac{\mu_e}{D_e} E_z \frac{\partial n_e}{\partial z} . \tag{87.28}$$

Solutions to this equation for n_e were for many years used to deduce D_e/μ_e from the measurements [87.29]. However, *Wagner* et al. [87.30] showed that there are two effective diffusion coefficients D_L and D_T in the longitudinal and transverse directions with $D_L \ll D_T$. For H_2, N_2 and He, $D_L \approx D_T/2$, while for Ar, $D_L \approx D_T/8$. See [87.31, 32] for a quantitative discussion. The different coefficients arise because n_e and w_d are functions of $\partial n_e/\partial z$.

For H_2, f_v^0 is fairly well approximated by a Maxwellian distribution. A comparison of the average and characteristic energies u_{av} and D_e/μ_e calculated from the Boltzmann transport equation for H_2 is shown in Fig. 87.5.

Ambipolar Diffusion

In the case of electron free diffusion discussed above, diffusion of the electrons (and ions) is unaffected by space-charge fields caused by an imbalance of positive and negative charges. When the charge density is high, this no longer holds. The resulting electric field caused by space-charge separation retards the electron diffusion and enhances the positive ion motion. In the steady state the flows of positive ions and of electrons are equal, i. e., $\boldsymbol{\Gamma}_e = \boldsymbol{\Gamma}_+$. For electrons in the absence of temperature gradients,

$$\boldsymbol{\Gamma}_e = -D_e \nabla n_e - n_e \mu_e \boldsymbol{E} , \tag{87.29}$$

where D_e and μ_e are given by (87.14) and (87.15) respectively. The analogous equation for the positive ions is

$$\boldsymbol{\Gamma}_+ = -D_+\nabla n_+ - n_+\mu_+\boldsymbol{E} . \tag{87.30}$$

Eliminating $\boldsymbol{E}$ between the two equations and setting $n_e \approx n_+ = n$, and $\boldsymbol{\Gamma}_e \approx \boldsymbol{\Gamma}_+ = \boldsymbol{\Gamma}$ gives

$$\boldsymbol{\Gamma} = -\left(\frac{D_+\mu_e + D_e\mu_+}{\mu_e + \mu_+}\right)\nabla n . \tag{87.31}$$

The quantity in parentheses is a diffusion coefficient that is applicable both to the electrons and to the ions. Since they are interacting, they diffuse together. This quantity is termed the ambipolar diffusion coefficient:

$$D_a = \frac{D_+\mu_e + D_e\mu_+}{\mu_e + \mu_+} . \tag{87.32}$$

Franklin, however, has proposed that the term ambipolar flow is a more correct physical description [87.33]. In the case where the electrons have a temperature T_e and the ions a temperature T_+,

$$\frac{D_+}{\mu_+} = \frac{k_B T_+}{e} , \quad \frac{D_e}{\mu_e} = \frac{k_B T_e}{e} , \tag{87.33}$$

and since $\mu_e \gg \mu_+$,

$$D_a \approx D_+\left(1 + \frac{T_e}{T_+}\right) . \tag{87.34}$$

If $T_e \gg T_+$, then $D_a \approx \mu - k_B T_e/e$, i.e., the ambipolar diffusion is determined by the ion mobility and the electron energy. *Allis* and *Rose* [87.34] have studied the transition from free to ambipolar diffusion as the electron density is increased.

For the case in which the electrons and ions are diffusing together in the radial direction across a static B_z, and the electron–atom and ion collision rates may be assumed to be independent of T_e, the magneto-ambipolar diffusion coefficient is given by

$$D_{aB} = \frac{D_a}{1 + \mu_+\mu_e B^2} . \tag{87.35}$$

References [87.35, 36] review theoretical and experimental work on diffusion in a magnetic field.

Debye Shielding

Consider an ensemble of positive ions and electrons in thermal equilibrium. Each ion repels other ions in its neighborhood, but attracts electrons. The space charge around each ion (assumed stationary) is $e(n_+ - n_e)$, which by Poisson's equation creates a potential ϕ satisfying

$$\nabla^2\phi = -\frac{e}{\epsilon_0}(n_+ - n_e) , \tag{87.36}$$

where n_+ is the average ion density. For time independent fields in the absence of collisions, (87.8) becomes

$$\boldsymbol{v}\cdot\nabla_r f + e\boldsymbol{E}\cdot\nabla_v f = \boldsymbol{v}\cdot\nabla_r f - e\nabla_r\phi\cdot\nabla_v f = 0 . \tag{87.37}$$

Assuming a solution of the form

$$f = f_0 \exp\left(-\frac{e\phi}{k_B T_e}\right) , \tag{87.38}$$

where f_0 is Maxwellian, and using $\int f_0\,d^3v = n_e$ gives

$$\left(-\frac{e}{k_B T_e}\boldsymbol{v}\cdot\nabla_r\phi + e\nabla_r\phi\cdot\frac{\boldsymbol{v}}{k_B T}\right) f \equiv 0 . \tag{87.39}$$

Hence,

$$\int f\,d^3v = n_e \exp\left(-\frac{e\phi}{k_B T_e}\right) . \tag{87.40}$$

Since the assembly is macroscopically neutral, $n_e = n_+$, and

$$\nabla^2\phi = -\frac{e}{\epsilon_0}n_+\left[1 - \exp\left(-\frac{e\phi}{k_B T_e}\right)\right] . \tag{87.41}$$

If $e\phi \ll k_B T_e$, then

$$\nabla^2\phi \approx \frac{e^2}{\epsilon_0}n_+\frac{\phi}{k_B T_e} . \tag{87.42}$$

If spherical symmetry is assumed,

$$\phi = \frac{e}{4\pi\epsilon_0 r}\exp\left(-\frac{r}{\lambda_D}\right) , \tag{87.43}$$

where

$$\lambda_D = \left(\frac{\epsilon_0 k_B T_e}{n_+ e^2}\right)^{1/2} \tag{87.44}$$

is the Debye shielding length for the Coulomb potential. λ_D arises from the screening effect of the electron cloud about each ion. The dimensions of a plasma must be much larger than λ_D for the electrons to act collectively, i. e., to undergo plasma oscillations.

87.1.6 Cross Sections Derived from Swarm Data

The use of transport theory to unfold low energy electron collision cross sections from experimentally measured electron transport data was originally introduced by *Townsend* [87.29]. Beginning with trial input values for the cross sections as a function of energy, the electron transport coefficients are calculated and compared with experiments (such as w_d, D_e, ϵ_k, and ionization/attachment coefficients, measured as a function of E/N). The input cross sections are then adjusted in the appropriate energy range, and the comparison procedure is iterated until theory and experiment come into agreement. The above trial and error procedure was greatly advanced by the use of computers to solve the Boltzmann equation [87.37,38]. A large body of fairly complete sets of low energy cross sections has been obtained in this way [87.23].

Assuming that the electron kinetics are accurately described by the solution of the Boltzmann equation, the above unfolding procedure is severely limited by the lack of uniqueness in the derived cross sections (especially in the molecular gases where many inelastic processes dominate in determining the electron energy distribution). For the noble gases, the method works well for He and Ne at low energies, where only elastic scattering occurs [87.24]. For Ar, Kr, and Xe, the method becomes questionable again for electron energies near the RT minimum. The slow redistribution of electron energies leads to lack of sensitivity of the calculated transport coefficients to the σ_m.

It has been shown that the swarm analyses of He-Ramsauer noble gas mixtures lead to unique σ_m of the Ramsauer noble gases (Ar, Kr, Xe) [87.39]. The measured electron drift velocity data in He-Xe mixtures shown in Fig. 87.6 satisfy the two important requirements (1) high sensitivity of the drift velocity data in gas mixtures over that in either pure He and Xe, and (2) the σ_m of He is very accurately known, to warrant the uniqueness of the derived σ_m of Xe.

Once the cross sections of the Ramsauer noble gases have been more accurately defined by the above approach, the RT minimum can be used to advantage in addressing the uniqueness problem for molecular gases. The electron transport data in molecular-gas–rare-mixtures exhibit large sensitivity due to the low

Part G | 87.1

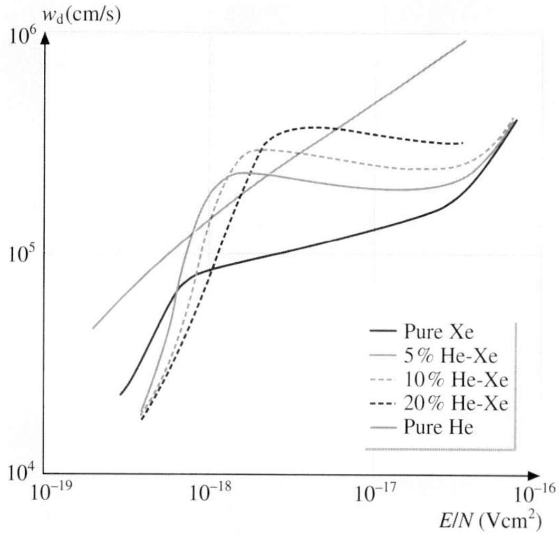

Fig. 87.6 Measured electron drift velocities in He–Xe mixtures

energy inelastic collisions in the molecular gas (which increase the f_1 component of the electron energy distribution) [87.21,40]. Since the cross sections of the buffer noble gas are now accurately defined, the demanding fits to the mixture transport data can be conveniently exploited to enhance the accuracy of the cross sections of the molecular gases.

The application of the two-term approximation in many papers treating molecular gases has been criticized [87.41]. Fortunately, methods are now available for the use of multi-term and Monte Carlo approaches to solve the collisional Boltzmann equation. These approaches do require differential cross sections. *Schmidt* et al. [87.42] report that their experiments using both electric fields and crossed electric and magnetic fields allow the extraction of the drift velocity, the Lorentz deflection angle, the longitudinal and transverse diffusion coefficients, and the ionization coefficient. This approach has lead to a fully automated procedure for extraction of cross sections from accurate experimental transport data.

When approximate answers (20% or so) are needed for comparisons with measurements on complex discharges, rather than measurements on transport properties, the program Bolsig and its database provided by *Morgan* [87.43] and the Paul Sabatier University plasma group [87.44] are justifiably popular.

Under very high E/n conditions, approximate answers can be obtained quickly using the completely anisotropic beam assumption.

87.2 Glow Discharge Phenomena

87.2.1 Cold Cathode Discharges

The cold cathode discharge operates because of feedback by the electrons that are generated by ion bombardment of the cathode (secondary electron emission) in turn generating sufficient ions that flow to the cathode to keep the process going. There can be contributions to secondary electron emission from photons and metastable atoms which are created by electron collisions with the gas. Since the secondary emission coefficients are functions of the cathode material, ion species, and ion energy, the magnitude of the needed accelerating potential between the plasma and the cathode varies. This potential creates a space charge sheath, and various excitation features occur near the cold cathode. The potential required is typically 200–600 V. The electrons from the cathode (called primaries) are accelerated through the sheath and acquire sufficient energy to excite the gas efficiently, resulting in a region termed the negative glow. As one moves further from the cathode, the light from the negative glow tapers off due to the degradation of the primaries in creating slow electrons (termed secondaries and ultimates). A sufficiently high density of electrons is usually created to satisfy the external circuit continuity at very low fields (or even slight field reversal). Since the low energy electrons do not create visible excitation, the region to the anode side of the negative glow forms the Faraday dark space. At even greater distance from the cathode, where the slow electron density has decayed by drift, diffusion, recombination, and in some gases by attachment, the electric field increases and the discharge develops long diffuse bright regions which occupy the remainder of the inter-electrode space. This is termed the positive column. If one increases the inter-electrode spacing, the cathode regions remain essentially constant and the positive column extends with average uniform properties. Temporal variations called moving striations are common in all gases, and spatial variations called standing striations, or ionization waves are frequently observed in molecular gases. The different cathodic regions occur because the abrupt boundary conditions create local discharge conditions of excitation and ionization which are not in equilibrium with the local electric field. Even in the so-called uniform positive column, the ionization waves correspond to nonlocal equilibrium solutions of the discharge equations that permit more efficient discharge ionization and conductivity than would be given by local equilibrium

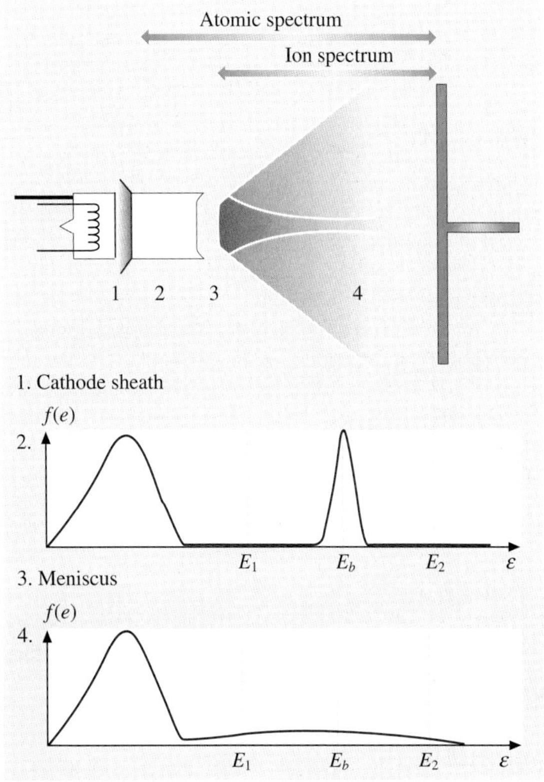

Fig. 87.7 Illustration of the appearance of a hot cathode discharge

with the electric field. The ionization waves can be absolutely or convectively stable, and have dispersion properties of forward or backward waves (the latter corresponding to opposite directions of phase and group velocities).

At any plasma boundary a positive space charge sheath is usually formed in order to equalize the fluxes of ions and electrons to the boundary. *Bohm* [87.45], and much earlier *Langmuir* [87.46], showed that in a collisionless sheath, to avoid oscillatory solutions, the ions must enter the sheath with velocities at least equal to the ion sound speed ($\sqrt{kT_e/m_i}$). The Bohm criterion continues to attract research activity. The review paper by *Riemann* [87.47] treats the issues addressed up to that time. There are important recent studies on satisfying the neutrality boundary conditions in electro-negative gases and in gas discharges with multiple positively- and negatively-charged species [87.48].

87.2.2 Hot Cathode Discharges

Low pressure, high current discharges are important in thyratrons and other high current switches. The hot cathode permits much higher current densities than are usually obtained from a cold cathode. The voltage drop of the cathode sheath is much less, and a discharge will operate if the applied potential exceeds I_P for the gas. Hot cathode discharges with low P and Γ_e show three groups of electrons: the primaries, which are beam electrons that essentially have retained the energy acquired in the cathode sheath; secondaries which are randomized electrons with T_e approximately proportional to the energy of the primaries (≈ 5 eV for 30 eV primaries) and approximately independent of the gas; and the ultimates which are the bulk of the electrons with $T - E < 1$ eV. The primaries degrade slowly with distance from the cathode, provided that $\Gamma_e \leq 40$ mA for argon at 10 μm (or more generally when the density of the beam electrons is still less than n_e). Above this critical current (a function of the gas and its pressure), the discharge adopts a definite new structure. The beam is no longer quasi-homogeneous but displays a meniscus about 7 mm from the cathode (Fig. 87.7). Langmuir probe measurements show that the primary electrons are abruptly scattered in energy at the meniscus. The energy scattering of the fast electrons by the electrostatic waves far exceeds the electron–atom collisional scattering. Spectroscopic measurements in argon, for example, with a cathode fall of 26 V, show the Ar spectrum throughout the discharge, whereas the Ar^+ lines appear only to the anode side of the meniscus. The Ar^+ lines appear even though the minimum energy required to excite them is 34.8 to 39.9 eV and the total voltage across the discharge can be 5 to 10 V lower. Coincident with onset of the meniscus, distinct GHz and MHz frequency oscillations are detected. These frequencies have been interpreted as electron- and ion-plasma oscillations, respectively.

87.3 Atomic and Molecular Processes

If the cross section for a given process is σ_j, then the rate of excitation per electron for that process is

$$Z_j = N \int_0^\infty \sigma_j(v) v f_v(v, \boldsymbol{r}, t) 4\pi v^2 \, dv \, . \qquad (87.45)$$

For inelastic processes, $\sigma_i(v) = 0$ for velocities below the threshold energy. The ionization threshold is the ionization potential I_P. The ionization cross sections typically increase monotonically with energy from threshold to about 100 eV. An exception to this is the alkali metal group for which the ionization cross section maximum occurs below 15 eV.

Assuming that the discharge excitation conditions are spatially uniform, the electron conservation equation is

$$\frac{dn_e}{dt} = (Z_i - Z_a - Z_d) n_e - Z_r n_e^2 \, , \qquad (87.46)$$

where $Z_i n_e$ is the total rate of ionization, $Z_a n_e$ is the rate of attachment, $Z_d n_e$ is the net diffusion rate out of the region considered, and $Z_r n_e^2$ is the rate of recombination, all per unit volume. The coefficients are usually functions of both P and E, either directly or indirectly. We consider these processes separately in Sect. 87.3.1 to Sect. 87.3.3.

87.3.1 Ionization

The ionization frequency (in s^{-1}) for a Maxwellian distribution of electron velocities is given by [87.49]

$$Z_i \approx 9 \times 10^7 a \, P \exp\left(-\frac{I_P}{\Theta}\right) \Theta^{1/2} I_P \, , \qquad (87.47)$$

where $\Theta = (k_B T_e / e)$ and a is the initial slope of the ionization efficiency curve in electron energy (ion pairs/Torr/V/electron). (The ionization efficiency is defined as the number of ion pairs produced per electron per cm of path at 1 Torr and 0°C. The ionization efficiency is proportional to the ionization cross section.) For a Maxwellian distribution, the quantity

$$\frac{dZ_i}{d\Theta} = Z_i \frac{(V_i + \Theta/2)}{\Theta^2} \, , \qquad (87.48)$$

together with Z_i, is useful in estimating the electron density and electric field for steady state discharges, or the onset of ionization instabilities [87.18].

If more accuracy is desired in estimating the ionization rate, a numerical solution of the Boltzmann equation is needed to calculate f_v^0. An energy balance calculation can then be performed by integrating the electron kinetic equation over all electron energies. The electron energy balance and the average electron energy are independent of n_e in a weakly ionized gas. Therefore the energy transfer into different excitations can be treated on a single electron basis. The total power density in the discharge as a function of E/N is then given by

$$\boldsymbol{J} \cdot \boldsymbol{E} = e w_d \frac{E}{N} N n_e \,. \qquad (87.49)$$

The relative contributions of the various electron loss processes are then obtained. Figure 87.8 shows the fractional input power deposition into the principal electron loss channels of hydrogen [87.50] as a function of the normalized electric field. Over a wide range of E/N, a large fraction of the discharge energy goes into vibrational excitation. Significant energy deposition into dissociation occurs only above 40 Townsends. This explains why it is necessary to use a capillary tube discharge (or Wood's tube) to obtain the atomic spectrum, or to provide a source of atomic hydrogen. The higher diffusion losses of the narrow bore cause the discharge to run at high E/N in order to maintain the electron density. The high E/N also provides large dissociation rates.

To obtain the total rate of excitation at high Γ_e, it is necessary that an assessment of the excited state densities also be made. This is most important for excited states that are known to be metastable so that the effective lifetimes are determined by collisional quenching or diffusion. The metastables can be very important in the excitation of higher states and in ionization. First, the threshold to ionize a metastable is usually much less than the I_P of the ground state. Second, the cross section for ionization of an excited state is often much larger than that of the ground state. Evaluations [87.51] of the low pressure helium discharge above 1 torr show that at current densities above a few mA cm^{-2} the two-step ionization through metastable states (proportional to n_e^2) exceeds the single step ionization (proportional to n_e) from the ground state. A multi-temperature approximation introduced by *Vriens* [87.52] has proven to be very useful in analyzing noble gas-alkali discharges and also noble gas discharges. The decreasing electric field required at higher discharge currents (higher n_e) is thus due to a combination of more efficient ionization through the metastables and increased gas heating.

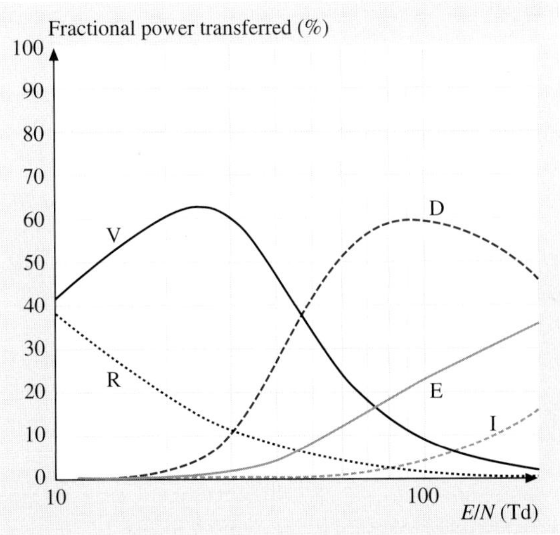

Fig. 87.8 Fractional power deposited into the different inelastic modes of hydrogen as a function of E/N. The labels V, D, E, R, and I correspond to vibrational, dissociation, electronic, rotational, and ionization processes respectively. The remainder is transferred into elastic collisions

Also, molecular dissociation changes the ionization rate through a combination of changes in f_v^0. At high energy density depositions (≈ 0.1 eV/molecule), the ionization rate and the dissociation rate are influenced by the degree of vibrational excitation, and by the changed Franck-Condon probabilities. However, the dissociation rate is changed primarily by two processes: collisions of the second kind [87.49]

$$e_{(\text{slow})} + N_2(v) \rightarrow e_{(\text{fast})} + N_2(v = 0) \,, \qquad (87.50)$$

which increase the number of fast electrons, and by increased anharmonic pumping of vibrational states to the dissociation limit [87.53]. Anharmonic pumping [87.54] occurs when large populations of vibrationally excited states are created in a molecular gas (usually diatomic) with a small amount of anharmonicity between adjacent levels, and the translational energy of the molecules is low. Because of the anharmonicity, the vibrational energy exchange favors the transfer from low vibrational levels to higher vibrational levels. This process is the basis of infrared laser action in carbon monoxide [87.54].

For discharges in gas mixtures, the component with the larger inelastic cross section and lower threshold tends to control the mean energy of the electron energy distribution. This situation is of common occurrence in discharge lasers. Numerical solutions of the Boltzmann equation have been obtained for a large number of gas mixtures of technical interest [87.55, 56]. Usually the two-term approximation in (87.9) is adequate, although there are circumstances where higher order terms are needed [87.57].

87.3.2 Electron Attachment

The phenomenon of electron attachment to a neutral atom or molecule to form a negative ion is a common occurrence for gases whose outer electronic shells are nearly filled. The energy of formation of a negative ion is called the electron affinity (E_A). This varies from about 3.5 eV (for the halogens) to nearly zero among the gases that exhibit electron attachment. Atoms having closed electronic shells do not form negative ions. These atoms, which have $^1S_0$ ground states, include the noble gases. Molecules in the $^1\Sigma$ ground state also do not form permanent negative ions.

A general experimental technique for measurement of attachment coefficients involves passing electrons through a gas target and measuring the attenuation of the electrons due to attachment, and usually also the negative ions produced [87.6]. Mass spectrometers are often included to identify the ions. A comprehensive survey of negative ions is available in the revised edition of *Massey*'s text [87.58], and an update with emphasis on discharge lasers has been given by *Chantry* [87.55, 59]. There is not as yet any experiment [87.41] that can measure separately the flux and reactive components of swarm transport when nonconservative interactions are present. The problem becomes more complex in inhomogeneous fields when the negative ions are weakly bound and experience collisional detachment in higher field regions. Plasma etching almost always involves electro-negative gases. The effects of negative ions in positive column discharges have been described by *Franklin* [87.48]. The discharge radial profile is sensitive to the relative values of ionization, attachment, detachment, and recombination. Additional discharge instability modes are possible, including an ionization-attachment-detachment mode that gives rise to distinct high and low field regions that have been observed in oxygen discharges and in other electro-negative gas discharges.

87.3.3 Recombination

One of the most common loss mechanisms for ions is the recombination of negative ions and electrons with positive ions. The loss of ions due to recombination is proportional to the product of their concentrations:

$$\frac{dn_+}{dt} = \frac{dn_-}{dt} = -\alpha n_- n_+ . \quad (87.51)$$

Here α is called the recombination coefficient and n_- is the negative ion or electron density. The value of α is quite different for negative ions and electrons. If there is only one negative ion species, and the positive and negative ion species have equal concentrations, then

$$\frac{dn}{dt} = -\alpha n^2 , \quad (87.52)$$

which on integration gives

$$n^{-1} = n_0^{-1} + \alpha t , \quad (87.53)$$

where n_0 is the initial ion concentration at $t = 0$. Recombination phenomena therefore often exhibit a linear relation between $1/n$ and t after switch-off of a discharge.

The three primary processes for positive ion–negative ion recombination are: three body recombination

$$A^+ + B^- + M \rightarrow AB + M ; \quad (87.54)$$

radiative recombination

$$A^+ + B^- \rightarrow AB + \text{photon} ; \quad (87.55)$$

mutual neutralization

$$A^+ + B^- \rightarrow A^* + B^* . \quad (87.56)$$

There are two additional processes for positive ion–electron recombination: dissociative recombination

$$e + AB^+ \rightarrow [AB]^* \rightarrow A + B ; \quad (87.57)$$

dielectronic recombination

$$e + A^+ \rightarrow A^{**} \rightarrow A + \text{photon} . \quad (87.58)$$

87.4 Electrical Discharge in Gases: Applications

Some of the many different plasma sources have been reviewed by *Conrads* and *Schmidt* [87.60]. The various discharge types are well described by *Raizer* [87.61]. In the examples of rf discharges, many of the applications employ control of power input to the plasma by means of automatic impedance matching networks. We present some notes on the more important technical plasma sources.

87.4.1 High Frequency Breakdown

In the theory of high frequency breakdown, the electron energy distribution function is calculated as a function of the applied ac electric field $\boldsymbol{E}$, using (87.12) and (87.13). From this the ionization frequency is derived, so that the rate of ion production can be expressed in terms of $\boldsymbol{E}$. At breakdown, the ionization rate equals the sum of the losses due to diffusion, attachment and recombination. The μ_e, ν_e and attachment coefficients, as well as the various rate constants producing new species, depend on $\boldsymbol{E}$. The electron continuity equation is solved to obtain the breakdown field. (In the absence of appreciable pre-ionization, the electron loss is governed by free diffusion. When the electron loss is controlled by ambipolar diffusion, the operating field of the discharge is obtained. This effect, in addition to cumulative ionization and gas heating, gives an operating field lower than the breakdown field)

Rewriting (87.13) for a constant ν_m, and assuming a Maxwellian velocity distribution, the electron current density becomes

$$\boldsymbol{J} = -e\Gamma_e = \frac{n_e e^2 E_0 \,\mathrm{e}^{-i\omega t}}{m(\nu_m - i\omega)}\,, \tag{87.59}$$

where the time dependence of $\boldsymbol{E}$ is written explicitly. The power P gained per unit volume is the time average

$$P = \langle \boldsymbol{J} \cdot \boldsymbol{E} \rangle = \frac{n_e e^2 \nu_m}{m(\nu_m^2 + \omega^2)} \frac{E_0^2}{2}\,. \tag{87.60}$$

For a given $\boldsymbol{E}$, P has a maximum at $\nu_m = \omega$. Equation (87.60) can also be written as

$$P = n_e \frac{e^2 E_{eff}^2}{m\nu_m}\,, \tag{87.61}$$

where the effective electric field is defined by

$$E_{eff}^2 = \frac{E_0^2}{2}\left(\frac{\nu_m^2}{\nu_m^2 + \omega^2}\right)\,. \tag{87.62}$$

E_{eff} is useful for comparing the relative heating effect of alternating and static fields. These considerations are carried further by *MacDonald* [87.62], and are important in determining microwave antenna breakdown fields at high altitudes.

87.4.2 Parallel Plate Reactors and RF Discharges

A situation of interest for plasma deposition and etching is the discharge between two parallel plates driven by a rf power supply. The industrial standard excitation frequency is 13.56 MHz. There have been several quite different approaches. One of the first models due to *Bell* [87.63] treats the discharge as a circuit element, and applies the boundary condition that the sum of the conduction and displacement currents remain constant:

$$\boldsymbol{J} = \sigma \boldsymbol{E} + \epsilon_0 \frac{\partial \boldsymbol{E}}{\partial t}\,. \tag{87.63}$$

Substituting for σ, $\boldsymbol{E}$ and $\partial \boldsymbol{E}/t$ in (87.63) for the case where electrons are the major current carriers, and assuming a velocity-independent ν_m, the amplitude of the electric field as a function of position becomes

$$E = \frac{J}{\epsilon_0 \omega}\left(\frac{(1+q^2)}{(\delta - 1)^2 + q^2}\right)^{\frac{1}{2}}\,, \tag{87.64}$$

where $\delta = \omega_p^2(x)/\omega^2$ (86.2) and $q = \nu_m/\omega$. At high electron densities or low frequency fields, $\delta \gg 1$ and

$$E = E_0\left(\frac{(\delta_0 - 1)^2 + q^2}{(\delta - 1)^2 + q^2}\right)^{\frac{1}{2}}\,, \tag{87.65}$$

where E_0, δ_0 denote values at the centre of the discharge. At low densities or high frequency fields, $\delta \ll 1$ and $E = J/(\epsilon_0\omega)$, i.e. the system acts like a capacitor. It is also necessary to consider the self-bias that the electrodes acquire under the influence of an applied electric field. The bias arises because of the requirement that, averaged over time, no net charge can collect at the electrode. Since the anode current-voltage characteristic is very nonlinear, application of a sinusoidal voltage to the electrode creates at each electrode a negative dc bias approximately equal to the peak amplitude of the rf voltage. The system therefore acts like a hollow cathode discharge with modulated sheath voltages. The ions from the plasma are accelerated across these sheaths and, over most of the rf cycle, acquire sufficient energy to cause secondary electron emission at the electrodes.

The ion flux and energy distribution hitting the substrate depend on the plasma conditions: the values of the dc and rf bias voltages, and the ratio of the transit time of the ions across the sheath to the period of the rf excitation frequency. The anisotropy and selectivity of the etching processes are determined by the ion energy and its directionality, the etching radical, and the substrate materials. Above energies of several hundred volts, ion bombardment of the substrate causes sputtering, which is relatively unselective. Pure chemical etching by radicals alone may be selective, but it is relatively isotropic. The aim in discharge enhanced processing is to obtain plasma conditions that generate selected radicals with high efficiency, concurrent with bias voltages that provide highly directional ions to prepare the substrate for etching. The combination of these processes gives etch rates that are typically an order of magnitude higher than the sum of the individual processes. The microstructures that can be fabricated using processing are critical to all of high density integrated circuit manufacturing. The achievements are illustrated by 0.2 μm wide trenches in silicon that are 4 microns deep with vertical sides [87.64].

On the other hand, the electrons form an energy-modulated fast electron beam that is injected into the plasma region. These fast electrons maintain much of the ionization in a symmetrically excited, equal electrode area reactor. In principle, it is possible to develop a complete self-consistent model of the rf discharge, consisting of the Boltzmann equation for the energy distribution function of the electrons, positive ions, and negative ions coupled to the Poisson's equation for the electric field. Simpler models based on moments of the Boltzmann equation have been developed for higher pressures [87.65] (where a fluid approximation is more appropriate) and particle-in-cell Monte-Carlo simulations for low pressures [87.64, 66]. The above studies have led to the identification of four principal mechanisms affecting energy deposition. These involve:

1. the impedance of the bulk plasma;
2. the energy deposition by fast electrons created by secondary emission due to ion bombardment of the electrodes;
3. collisionless absorption due to the asymmetrical sheath boundary-plasma electron interaction; and
4. wave-riding, or collisional sheath interactions causing electron heating in the sheath modulated electric field. The electron is regarded as surfing on the expanding sheath field.

Processes (3) and (4) are related, with the difference being that in (4) electron collisions occur during the sheath expansion. There is also an additional interaction due to the changes in the complex impedance (capacitive sheaths and resistive bulk plasma) of the discharge and the consequent changes in the power transferred from the rf generator. The mechanism of collisionless heating is an active research area [87.67]. It is proposed that low pressure rf plasmas can be maintained mainly by collisionless heating in the rf modulated sheaths, and that electron inertia plays a dominant role.

Other less general plasma modes occur due to resonance when the electric field and period are such that one electron transit requires one half period of the rf cycle. Under these conditions at very low pressures, when λ_{mfp} is larger than the gap spacing, and secondary electron emission can occur, one has the multipactor discharge mode. Otherwise, when λ_{mfp} corresponds approximately to small integral fractions of the gap spacing, plasmoid modes are excited.

Dielectric Barrier Discharge

In many applications, such as ozone generation (for water treatment), distributed uv sources, and the creation of radicals for surface treatments, there are economic incentives to operate at high pressures, especially at atmospheric pressure. The dielectric barrier discharge (DBD, also called the silent discharge in earlier literature) is a high voltage of very short pulse duration (1–10 ns) or an ac discharge between two electrodes where at least one of the electrodes is covered with a dielectric [87.68]. The dielectric acts like a high value impedance which prevents discharge current runaway and tends to distribute the average current fairly uniformly across the surface area. The charge transferred by the streamer to the dielectric essentially compensates for the external electric field and limits the discharge duration. Close examination shows that the current terminations on the cathode dielectric are many microdischarges; the termination at the dielectric anode is usually diffuse. The cathode spot diameters are typically only 200 μm so that with a peak current of 0.1 A, the current density is hundreds of A/cm^2. However, because of the usually very low duty cycle, the neutral gas temperature is close to ambient. The electron temperatures during the discharge pulse are estimated in the range of 1 – 10 eV, so the DBD is a very non-equilibrium plasma. The charge transferred to the dielectric on the previous pulse strongly affects the next pulse. The high field discharge is efficient at producing excitation, dissociation, and ionization. Miniature DBDs are used in plasma display panels; these are discharges between electrode arrays coated with dielectric and separated by typically 100 μm in a Penning gas mixture at about 500 Torr [87.69].

87.5 Conclusions

The study of the conduction of electricity in gases has had several renaissances because of application incentives. Discharge lighting technology science and development are described by *Waymouth* [87.70]. Atmospheric plasma physics and cosmic plasma physics are covered by *Rees* [87.71] and *Alfvén* [87.72], respectively. The role of plasma physics in gas discharge laser development was recently reviewed [87.73], and the fundamental interactions in laser generated plasmas are detailed by *Hughes* [87.74]. The scaling of discharge volumes to many liters is again of interest because of applications to large area surface treatments and thin films. The discharge enhanced chemistry of complex gas mixtures is of particular relevance to etch processing used for most microelectronics fabrication. Work on discharge-generated and -trapped particulates in microelectronics processing plasmas has re-established links with cluster physics and space plasmas [87.75]. The physics of many of the effects was formulated earlier by *Emeleus* and *Breslin* [87.76]. Future prospects for further exploitation of non-equilibrium gas discharge physics include waste/toxic hazard decontamination, fabrication of defined complex molecules, some with strain energy, selected cluster morphologies, and the further development of large area plasma flat panel displays.

References

87.1 P. Lenard: Ann. Physik **12**, 714 (1903)
87.2 M. Hayashi: Tech. Rep. No.1PPJ-AM-19. Ph.D. Thesis (Institute of Plasma Physics, Nagoya 1981) (unpublished)
87.3 H. S. W. Massey: *Handbuch der Physik*, Vol. 36 (Springer, Berlin, Heidelberg 1956) p. 232
87.4 W. J. Pollock: Trans. Farad. Soc. **64**, 2919 (1968)
87.5 R. Nagpal, A. Garscadden: J. Appl. Phys. **75**, 703 (1994)
87.6 G. J. Schulz: Rev. Mod. Phys. **45**, 423 (1973)
87.7 P. Kleban, H. T. Davis: Phys. Rev. Lett. **39**, 456 (1977)
87.8 W. P. Allis: *Handbuch der Physik*, Vol. 21 (Springer, Berlin, Heidelberg 1956) p. 383
87.9 G. Cavaleri, G. Sesta: Phys. Rev. **170**, 286 (1968)
87.10 G. L. Braglia: Physica **92C**, 91 (1979)
87.11 W. L. Nighan: Phys. Fluids **12**, 162 (1969)
87.12 L. Spitzer: *Physics of Fully Ionized Gases* (Wiley–Interscience, New York 1956)
87.13 H. Dreicer: Phys. Rev. **117**, 343 (1960)
87.14 S. T. Butler, M. J. Buckingham: Phys. Rev. **126**, 1 (1962)
87.15 I. P. Shkarofsky, T. W. Johnston, M. P. Bachynski: *The Particle Kinetics of Plasmas* (Addison-Wesley, Reading 1966)
87.16 A. M. Weinberg, E. P. Wigner: *The Physical Theory of Neutron Chain Reactors* (Univ. Chicago Press, Chicago 1958)
87.17 V. L. Ginzburg: *The Propagation of Electromagnetic Waves in Plasmas* (Addison-Wesley, Reading 1964)
87.18 L. Pekarek: Uspeki Fiz. Nauka **94**, 463 (1968)
87.19 T. L. Cottrell, I. C. Walker: Trans. Farad. Soc. **61**, 1585 (1965)
87.20 M. R. Hallada, P. Bletzinger, W. F. Bailey: IEEE Trans. Plasma Sc. **PS-10**, 218 (1982)
87.21 W. H. Long Jr.W. F. Bailey, A. Garscadden: Phys. Rev. A **13**, 471 (1976)
87.22 J. Dutton: J. Phys. Chem. Ref. Data **4**, 577 (1975)
87.23 S. R. Hunter, L. G. Christophorou: *Electron-Molecule Interactions and Their Applications*, Vol. II, ed. by L. G. Christophorou (Academic, New York 1984) p. 89
87.24 L. G. H. Huxley, R. W. Crompton: *The Diffusion and Drift of Electrons in Gases* (Wiley–Interscience, New York 1974)
87.25 A. L. Gilardini: *Low Energy Electron Collisions in Gases* (Wiley– Interscience, New York 1972)
87.26 J. A. Hornbeck: Phys. Rev. **84**, 615 (1951)
87.27 S. H. Pullins, R. A. Dressler, R. Torrents, D. Gerlich: Z. Phys. Chem. **214**, 1279 (2000)
87.28 S. Sakabe, I. Izawa: At. Data Nucl. Data **49**, 257 (1991)
87.29 J. S. E. Townsend: *Electricity in Gases* (Clarendon, Oxford 1915)
87.30 E. B. Wagner, F. J. Davis, G. S. Hurst: J. Chem. Phys. **47**, 3138 (1967)
87.31 J. H. Parker, J. J. Lowke: Phys. Rev. **181**, 290 (1969)
87.32 H. R. Skullerud: J. Phys. B: At. Mol. Opt. Phys. **2**, 696 (1969)
87.33 R. N. Franklin: J. Phys. D. **36**, 828 (2003)
87.34 W. P. Allis, D. J. Rose: Phys. Rev. **93**, 84 (1954)
87.35 F. C. Hoh: Rev. Mod. Phys. **34**, 267 (1962)
87.36 F. F. Chen: *Introduction to Plasma Physics* (Plenum, New York 1974)
87.37 L. S. Frost, A. V. Phelps: Phys. Rev. **127**, 1621 (1962)
87.38 A. G. Englehardt, A. V. Phelps, C. G. Risk: Phys. Rev. **135**, 1566 (1964)
87.39 R. Nagpal, A. Garscadden: Phys. Rev. Lett. **73**, 1598 (1994)
87.40 M. Kurachi, Y. Nakamura: J. Phys. D: Appl. Phys. **21**, 602 (1988)
87.41 R. D. White, R. E. Robson, B. Schmidt, M. A. Morrison: J. Phys. D **36**, 3125 (2003)

87.42 B. Schmidt, K. Berkan, B. Goetz, M. Mueller: Phys. Scr. **53**, 30 (1994)
87.43 W. L. Morgan, B. M. Penetrante: Comput. Phys. Commun. **58**, 127 (1990)
87.44 http://www.sni.net/siglo/database/x-sect/siglo.sec
87.45 D. Bohm: *Characteristics of Electrical Discharges in Magnetic Fields*, ed. by A. Guthrie, R. K. Wakerling (McGraw-Hill, New York 1949)
87.46 I. Langmuir: Gen. Elect. Rev. **26**, 731 (1923)
87.47 K-U. Riemann: J. Phys. D **24**, 493 (1991)
87.48 R. N. Franklin: J. Phys. D **36**, 823 (2003)
87.49 A. Von Engel: *Ionized Gases* (Clarendon Press, Oxford 1965)
87.50 A. Garscadden, W. F. Bailey: *Progress in Astronautics and Aeronautics*, Vol. 74, ed. by S. S. Fisher (AIAA, New York 1981) p. 74
87.51 B. E. Cherrington: IEEE Trans. Elec. Devices ED **26**, 148 (1979)
87.52 L. Vriens: J. Appl. Phys. **45**, 1191 (1974)
87.53 M. Cacciatore, M. Capitelli, C. Gorse: J. Phys. D: Appl. Phys. **13**, 575 (1980)
87.54 C. E. Treanor, J. W. Rich, R. G. Rehm: J. Chem. Phys. **48**, 1798 (1968)
87.55 E. W. McDaniel, W. L. Nighan: *Gas Lasers*, Vol. 3 (Academic, New York 1982)
87.56 A. Garscadden, R. Nagpal: Plasma Sources Sci. Technol. **4**, 268 (1995)
87.57 L. C. Pitchford, S. V. O. NeilJ. R. Rumble Jr.: Phys. Rev. A **23**, 294 (1981)
87.58 H. S. W. Massey: *Negative Ions*, 3rd edn. (Cambridge Univ. Press, Cambridge 1976)
87.59 P. J. Chantry: *Gas Lasers*, Vol. 3 (Academic, New York 1982)
87.60 H. Conrads, M. Schmidt: Plasma Sources Sci. Technol. **9**, 441 (2000)
87.61 Y. Raizer: *Gas Discharge Physics* (Springer, Berlin, Heidelberg 1997)
87.62 A. D. MacDonald: *Microwave Breakdown in Gases* (Wiley Interscience, New York 1969)
87.63 A. T. Bell: In: *Engineering, Chemistry and Use of Plasma Reactors*, ed. by J. E. Flinn (American Institute of Chemical Engineering, New York 1971)
87.64 M. A. Lieberman, A. J. Lichtenberg: *Principles of Plasma Discharges and Materials Processing* (Wiley Interscience, New York 1994)
87.65 F. F. Young, C. J. Wu: IEEE Trans. Plasma Sci. **21**, 312 (1993)
87.66 R. Krimke, H. M. Urbassek, D. Korzec, J. Engemann: J. Phys. D: Appl. Phys. **27**, 1653 (1994)
87.67 G. Gozadinos, D. Vender, M. M. Turner, M. A. Lieberman: Plasma Sources Sci. Technol. **10**, 117 (2001)
87.68 U. Kogelschatz: Plasma Chem. Plasma Process **23**, 1 (2003)
87.69 J. P. Boeuf: J. Phys. D **36**, R53 (2003)
87.70 J. F. Waymouth: *Electric Discharge Lamps* (MIT Press, Cambridge 1971)
87.71 M. H. Rees: *Physics and Chemistry of the Upper Atmosphere* (Cambridge Univ. Press, Cambridge 1989)
87.72 H. Alfvén: *Cosmical Electrodynamics* (Clarendon, Oxford 1950)
87.73 A. Garscadden, M. J. Kushner, J. G. Eden: IEEE Trans. Plasma Sci. **19**, 1013 (1991)
87.74 T. P. Hughes: *Plasmas and Laser Light* (Wiley-Interscience, New York 1975)
87.75 A. Evans: *The Dusty Universe* (Horwood, New York 1993)
87.76 K. G. Emeleus, A. C. Breslin: Int. J. Elect. **29**, 1 (1970)

88. Applications to Combustion

However, there have been two major advances in combustion research which contain new ideas and directions in detailed chemistry and physics. The first is the use of powerful computers for the numerical solutions to combustion problems, so that a predictive description can be built beginning with fundamental physical principles. The second is the use of laser diagnostic techniques for the determination of the detailed properties of the combustion system, especially temperature, velocity, and composition, including both major components and trace chemical intermediates. These have not only provided tests of the computational models, but have also furnished new insights and approaches to an understanding of combustion. This has occurred by measuring properties of the system not previously available, or via improved scales of spatial and temporal resolution.

This chapter concentrates on these two areas of physical models and laser diagnostics, and the outstanding physical questions that remain. Each is greatly influencing the development of the two important combustion science issues: describing turbulent flows and incorporating realistic chemistry.

Combustion processes are vital to the operation of present-day society. Although there exist alternative energy sources, combustion of a variety of fuels is worldwide a major mode of energy production, and will remain so for many years to come. Despite a well-established technology, combustion is not without need of improvement. The pollutant emissions associated with burning pose major barriers to more widespread use. Increasingly stringent environmental regulations make this a current focus of combustion research and development.

Both scientific inquiry and empirical solutions will be important in advancing combustion in the future. Most of the progress in combustion technology has been made without benefit of an understanding of the science involved; rather, it has been accomplished through an approach of trial and error, which has led to many ingenious, innovative solutions to problems. In fact, as witnessed by the early development of thermodynamics, the technology has often driven the science rather than the other way around. Nonetheless, future advances in combustion are expected to rely more and more on a planned implementation of fundamental knowledge of physics and chemistry.

Combustion may be thought of as self-sustaining reactive flow, in which chemical energy is converted into extractable, useful heat. This is usually accompanied by an abrupt change in properties of the system in space and/or time, particularly the chemical composition and the temperature. The description of combustion processes involves the subdisciplines of thermochemistry, chemical kinetics, fluid mechanics, and transport. The major challenge is to apply the known principles from these to a description and understanding of the entire combustion process. This endeavor involves three approaches: experiment, computation, and theory, the latter two differentiated via the characteristic of numerical versus analytical solutions. Each part now plays an important role in combustion science.

The efficiency of combustion processes, be they steady state such as a burner flame or rocket motor, or transitory such as an explosion or internal combustion engine, is generally governed by the overall thermochemistry, fluid dynamics, and heat transfer characteristics of the system. Details of the chemical reactions and the role of trace intermediates are usually unimportant for this question. The latter do play a key part, however, when considering problems of pollutant formation (emission of NO_x, SO_x, soot, toxic organic compounds), and flame ignition and inhibition. The current challenges in combustion science are solutions to these problems of efficiency and pollution, with advancement likely to occur through two major thrusts: a better understanding of turbulent reactive flow, and the incorporation of detailed chemical kinetics into combustion models (as contrasted to simplified versions such as one- or two-step reaction schemes). Additionally, new areas of science involve combustion under unusual conditions (e.g., microgravity combustion aboard the space shuttle) and new applications (e.g., flame synthesis of exotic materials such as diamond films).

From the viewpoint of the basic science, most of the general, qualitative understanding of behavior at the molecular scale needed to describe combustion is at hand. In many engineering applications, an empirical approach is taken in which a quantitative description of that behavior is condensed and parametrized, using previously acquired experimental data acquired directly on the system of interest. As models become more detailed (often through computational advances), there is a need for further quantitative supporting molecular data, such as improved values of transport properties of molecules.

88.1 Combustion Chemistry

The addition of detailed chemical kinetics to flame models will greatly advance our understanding of combustion, at least for nonturbulent flows. Detailed chemistry is becoming a part of codes describing systems of greater and greater complexity. Numerical integration packages, including full chemistry for a one-dimensional flow, are routinely available [88.1], and progress is being made for the two-dimensional case. A sensitivity analysis enables the modeler to examine the value predicted for some variable (e.g., concentration of a transient species) at a given point in the flame as a function of the rate coefficients and other input variables. This facilitates interpreting the output in terms of physically meaningful variables, and extrapolating the model predictions to other applications.

The quality of the predictions of a combustion model is a strong function of the quality of the reaction rate coefficients for important elementary reaction steps, and the thermochemical properties of reactive species, especially free radicals. A flame chemical mechanism can be quite complex: a current methane/air model [88.2] containing reactions involving only hydrogen, oxygen and carbon (i. e., no nitrogen chemistry) contains 30 species and 177 reactions (plus their reverses). Although it is necessary to include most of these steps in the reaction scheme, a much smaller number play dominant roles in determining any given parameter such as flame speed. The determination of sufficiently accurate rate coefficients for those important reactions is made difficult by the complexity of the reactions, and the difficult conditions (high temperature) and reactants (usually free radicals) that are involved.

There are many unimolecular reactions and their reverses that play important roles in combustion chemistry, including the recombination of fuel-derived radicals such as $CH_3 + H + M \rightarrow CH_4 + M$, and low temperature decomposition/recombination reactions important in ignition, e.g., $H + O_2 + M \rightarrow HO_2 + M$, compared with chain-branching steps. Under most combustion conditions, the important reactions in this class are either in the low pressure limit, where the rate constant is proportional to total pressure P of collider M, or in the fall-off region between that limit and the P-independent high pressure limit. Knowledge of the collisional energy transfer characteristics for highly excited vibrational levels of polyatomic molecules is apropos to a fundamental understanding of these reactions. For purposes of combustion models, these rate constants can be formulated in a modified Arrhenius mode, requiring as many as nine parameters to describe correctly the combined P, T dependence needed for a realistic range of combustion conditions.

A further complication is posed by the multiple potential energy surfaces that can be accessed by many reactions. For example, two methyl radicals can combine to form not only C_2H_6, but also C_2H_5 + H. Measurements are often made of the overall rate coefficient (in this case, disappearance of CH_3) but seldom are determinations made of branching ratios into these multiple paths. Especially, the temperature dependence of that branching is rarely known accurately enough. Since reaction rate coefficients can change orders of magnitude over the range of temperatures encountered in combustion, simple extrapolations are seldom sufficient.

Theoretical chemistry has begun addressing these problems. Electronic structure calculations provide thermochemical properties of transient, reactive species. Using theories which include semi-empirical adjustments, enthalpies of formation of free radicals and potential barriers on reactive surfaces can be calculated with accuracies of about 2 kcal/mole, sufficient for modeling purposes. Multidimensional potential surfaces provide starting points for conventional transition state theory calculation of rate coefficients, as well as variational transition state approaches. Theoretical calculations may be useful for understanding the energy transfer questions of unimolecular and recombination rate theory. Also useful are recent laser experiments which examine the foundations of rate theory. These include laser photolysis investigations of the reaction rate coefficient as a function of energy just above threshold [88.3], and the formation of molecules from dissociating complexes on subpicosecond scales [88.4], as well as stimulated emission pumping experiments which probe the density of states and energy transfer in high vibrational levels of diatomic and polyatomic molecules [88.5].

We conclude this section with brief comments on some individual combustion systems. Flames of H_2 and CO burning in (moist) O_2 are relatively simple, and their mechanisms are important subsets of reaction networks for more complex systems. The chemistry of these flames is relatively well understood on a fundamental basis. So, generally, is that of methane, although there remain crucial questions, for example, the rates for CH_3 + OH, a multisurface reaction producing singlet and triplet CH_2 + H_2O, as well as CH_2OH. For hydrocarbon fuels more complex than methane, the complexity of the mechanism rapidly escalates, due to the many possible ways to combine radicals containing multiple carbon-carbon bonds. For a fuel such as isooctane (where few rate coefficients are known) it is necessary to use a computer algorithm simply to write down the mechanism without making mistakes or omissions [88.6].

Many pollutant emissions are produced by reactions of trace intermediate species. Nitric oxide has received much recent attention due to increasingly stringent environmental standards. NO is formed in combustion processes in three ways: (i) so-called thermal (or Zeldovich) NO, formed by reaction of O + N_2, N + O_2, and OH + N_2, where the atoms and OH are usually present at nonequilibrium, kinetically controlled concentrations; (ii) prompt NO, which begins with the CH + N_2 reaction breaking the strong N-N bond to form N atoms which react with O_2 and HCN, which is oxidized to NO in a series of steps; (iii) and fuel-nitrogen NO, which is formed from the nitrogen present in coal and coal-derived fuels, usually entering the combustion process as HCN. NO can be dealt with through catalytic reduction, staged combustion (burning at a series of mixing ratios), reburning at a later stage with additional fuel, and injection of compounds such as NH_3 or cyanuric acid, $(HOCN)_3$. Progress has been made in the determination of the rate coefficients of many pertinent elementary reactions for each of these processes, but significant gaps remain. Examples are the full T, P dependence of the complex-forming reaction CH + N_2, and the complicated reaction pathways in the NH_2 + NO reaction important in de-NO_x through ammonia injection.

The mechanism of soot formation is a matter of dispute; unsaturated chains containing 4 or 5 carbons, aromatic rings, ions, and C_3H_3 radicals have all been suggested as precursors. Ignition (and engine knock) chemistry occurs at low temperatures and probably involves hydroperoxy (HO_2) and alkylperoxy (RO_2) radicals. The mechanism of production of toxic organic compounds such as aldehydes is poorly known, and few relevant rate coefficients are at hand.

88.2 Laser Combustion Diagnostics

The advent of the laser has altered much of experimental combustion science. Laser Doppler velocimetry, which employs particle scattering to measure directed flow velocity, is available in the form of commercial systems. Coherent anti-Stokes Raman scattering (CARS) is widely used in engineering laboratories to study practical combustors, including jet engines, industrial furnaces, internal combustion engines, and burning propellants. Laser-induced fluorescence (LIF) has come into prominence for the study of combustion systems and related chemical kinetics.

The spectroscopically based techniques of most interest in atomic, molecular, and optical physics are CARS, LIF, and the related recently emerging method of degenerate four wave mixing (DFWM). Each may be used to determine individual molecular species or temperature. Each is highly selective, due to its spectroscopic nature; they provide good spatial resolution, often on the order of a mm^3 or less; the use of pulsed lasers yields excellent time resolution, $\approx 10\,ns$. When used properly, they are nonintrusive, perturbing neither the flow dynamics nor flame chemistry. They can be used in very hostile environments (corrosive atmospheres, high temperatures) where physical probes would not survive.

LIF and CARS also have separate, complementary features. CARS is suitable for the determination of major species (fuel, O_2, CO_2, H_2O, and, in air-breathing combustion, N_2). It may be used for accurate temperature measurements, including measurements on a single laser shot when operated in a broadband mode. This makes CARS especially useful for problems in which fluid dynamics and flow patterns are most important. Because CARS produces a coherent, laser-like beam, spatial filtering can be used to discriminate against background radiation such as the intense thermal emission from hot combustor walls. LIF is highly sensitive and is useful for measurements on reaction intermediates present at low concentrations. It may be used in an imaging mode to obtain instantaneous two-dimensional patterns of radical species in rapidly time-varying flows.

Despite their utility, LIF and CARS are not user-friendly technologies. At present, considerable nurturing of the equipment, knowledge of the theory, and care in the data analysis are needed for accurate measurements, and nearly all such measurements are made by scientists trained in physics, chemistry, or physics-oriented engineering departments. Both methods (and DFWM) are examined here with an emphasis on the physics questions in their continued development and application.

88.2.1 Coherent Anti-Stokes Raman Scattering

CARS depends on the nonlinear polarization induced in molecules by the intense electric fields of lasers, exploiting resonant effects in the third order nonlinear susceptibility $\chi^{(3)}$. $\chi^{(3)}$ mixes the fields from a pump laser at frequency ω_1 and a Stokes beam at frequency ω_2 to form a third beam at $\omega_3 = 2\omega_1 - \omega_2$. $\chi^{(3)}$ is strongly enhanced when the difference $\omega_1 - \omega_2$ is tuned to a vibrationally or rotationally resonant frequency in the ground state of a molecule. This is usually accomplished using the intense green beam of a frequency doubled Nd:YAG laser for ω_1, and a dye laser Stokes shifted by the vibrational frequency of the molecule of interest for ω_2. In general, the signal in the visible region of the spectrum contains a combination of a nonresonant $\chi^{(3)}$ due to the electronic polarizability of the medium (mostly N_2 for air-breathing combustion) and the resonant part. For high concentrations c_A (e.g., $\sim 70\%$ N_2 in air based flames), the signal is a Lorentzian-shaped resonance on top of the flat nonresonant term, proportional to c_A^2/Γ where Γ is the linewidth. For lower concentrations (e.g., a few percent CO_2 in exhaust gases) the resonant and nonresonant amplitudes interfere, so that the signal is dispersion shaped with a magnitude $\propto c_A/\Gamma$.

A very useful and popular application of CARS operates the Stokes laser in a broadband mode. This generates the entire coherent anti-Stokes spectrum at once, which is then dispersed through a spectrometer and detected with an array. T may be determined from the rotational structure and the appearance of vibrational hot bands. These measurements are generally performed on the N_2, almost always present in large quantity in any combustion system of practical interest. This is the most reliable, accurate method for determination of pointwise temperatures with a single laser shot.

The nonlinear, coherent nature of CARS, in which interfering amplitudes are added, means that spectral details can have a significant influence on the observed spectrum. These include the concentration of all contributing species (including nonresonant background) and the widths of individual Raman lines. The linewidths must be dealt with correctly in order to deduce temperatures accurately from the spectral shapes. It has only been in the past five years that linewidths as a function of rotational level have been available for N_2, at the accuracy needed for application of the spectral fitting codes used for T determination. As P is increased, the lines broaden enough to begin to overlap; at high enough P (several atm for N_2) collisions occur so rapidly that the spectrum collapses to a single broad line. To make T measurements at high pressure, one must deal with both these P broadening and collisional narrowing effects; this requires an understanding of the underlying collision dynamics and energy transfer, generally on a rotational level state specific basis. Both energy transfer and dephasing collisions contribute to the linewidths. For molecules other than N_2, there is scant information on Raman linewidths even at atmospheric pressure, at accuracies needed for modeling CARS spectra. Molecules of particular interest are O_2, CO_2, H_2O, CO, and, in the future, hydrocarbon fuels. In some cases, especially H_2O, further spectral information for high rotational lines (populated at high temperature) is also needed.

Most of the treatments of CARS spectra assume a monochromatic laser. Finite bandwidth effects of real lasers must be taken into account for accurate theoretical treatments (and thus accurate T and concentration determinations). There are several nonequivalent theoretical treatments to convolute the frequency dependence of $\chi^{(3)}$ with the bandwidths of the pump and Stokes lasers.

88.2.2 Laser-Induced Fluorescence

LIF is much more sensitive than CARS, relying on resonant excitation to a real, emitting electronically excited state. The use of a tunable laser to scan the molecular excitation spectrum makes LIF also highly selective for small species that have characteristic, identifiable line spectra. Thus LIF is suitable for measurement of reactive intermediates of importance in understanding flame chemistry. Concentrations determined by LIF range from parts per billion to parts per thousand. Detectable species are diatomics such as OH and CH, triatomics including HCO, NCO, and NH_2, and a few larger molecules including CH_3O. Most of the pertinent molecular electronic transitions are in the UV or blue region of the spectrum. Because VUV radiation cannot penetrate flame gases, atoms (such as H and O) must be detected via two-photon absorption to a high-lying electronic state of the same symmetry as the ground state, which then radiates in the visible or near IR to another excited state.

A list of all molecules through teratomics composed of the five main atoms naturally occurring in combustion (H, C, O, N, and S), and detected to date by LIF, is given in Table 88.1. This forms a large fraction of the small reactive intermediates important in combustion chemistry, so that LIF has the ability to characterize fairly completely a combustion reaction mechanism. It would be particularly desirable to add certain other species; noteworthy are HO_2, C_2H, CH_3, and triplet CH_2, although suitable fluorescing electronic transitions for these molecules are not currently known. In addition to the species listed in Table 88.1, there are many other combustion-related compounds that can be seen with LIF. These include metal atoms and their compounds, species present in specialized combustion situations such as boron or chlorine containing radicals, and some polyatomic, partially oxidized hydrocarbon molecules.

LIF may be employed in several ways to study flames. With a pulsed laser, very rapid time resolution is achievable. Pointwise, single shot measurements (usually of OH) in turbulent flames may be interpreted on a statistical basis for comparison with flame models incorporating simplified chemistry. Two dimensional planar imaging (usually of OH, CH, or NO) is accomplished by focusing the laser into a sheet of radiation which passes through the flame. Imaging the illuminated region at right angles onto a two-dimensional array produces an instantaneous snapshot of the distribution of the radical throughout the flame. This may be used to understand flow patterns under conditions of rapid time variation, such as turbulence or flame spread following ignition.

On the other hand, turbulent flows are far too complex for an investigation of fine details of the combustion chemistry. This is best accomplished via experiments in laminar flames, where LIF is used to obtain spatial profiles (one- or two- dimensional) of reactive species, made as a function of height above a burner surface. Absolute or relative profiles provide highly constraining tests of detailed models of flame chemistry. Operation at reduced P spreads out the active flame front region, furnishing excellent spatial resolution for these profiles. The T profile, used as input to the model, is measured using rotational excitation scans. The OH radical, present throughout much of any given flame contain-

Table 88.1 Combustion chemistry intermediates detectable by laser-induced fluorescence

Molecule	Excitation wavelength (nm)	Molecule	Excitation wavelength (nm)
H*	205	NCO*	440
C*	280	HCO*	245
O*	226	HCN	189
N*	211	HNO	640
S	311	NH_2^*	598
OH*	309	C_3	405
CH*	413	C_2O	665
NH*	336	S_2O	340
SH*	324	SO_2^*	320
CN*	388	NO_2^*	590
CO*	280	HSO	585
CS	258	CS_2	320
NO*	226	N_3	272
NS*	231	NCN	329
SO*	267	CCN	470
O_2^*	217	NH_3^*	305
S_2^*	308	NO_3	570
C_2^*	516	$C_2H_2^*$	220
$^1CH_2^*$	537	CH_2O^*	320

An asterisk denotes that LIF detection has been performed in a flame

ing hydrogen and oxygen, is an ideal LIF thermometer. LIF may also be used in a semiquantitative way, so that simply the detection of some species at an approximate concentration can reveal new information about the chemical mechanism [88.7]. An example is LIF detection of NS in flames containing minor amounts of N and S compounds, indicating the importance of this radical in linking the chemistry of NO_x and SO_x formation.

LIF measurements involve quantitative knowledge of molecular spectroscopic characteristics and collisional behavior. Identification of the absorbing molecule requires assignment of the vibrational and rotational structure of the electronic transition in question. For weak signals, particularly in the presence of a strong absorber/fluorescor (e.g., CH_3O in the presence of OH near 310 nm), it may be necessary to scan many lines to ensure identification. The fluorescence spectrum is also valuable since selective detection of fluorescence may be used to discriminate between two molecules absorbing at the same wavelength. In many cases, the detailed spectroscopy of molecules of interest is well established in the literature; in other cases, it may be necessary to make studies via flow cells. A recent example is an LIF study of the B–X ultraviolet system of the HCO molecule [88.8], a radical whose reactions are important in controlling the hydrogen atom concentration in hydrocarbon flames. In some cases (e.g., NCO) LIF spectroscopic studies in flames themselves have been useful, in that high rotational and vibrational levels are populated and thus readily accessed. In addition to spectral line and band identification, rotational line strengths and vibrational band transition probabilities (including possible effects of an electronic transition moment that varies with internuclear distance) provide the Einstein A and B coefficients needed for the data analysis. Accurate values are especially important in the determination of T via excitation scans, since systematic errors in oscillator strengths can lead to errors of hundreds of degrees, but are not discernible through statistical goodness-of-fit criteria. Such large errors in T render meaningless any attempts to compare measurements with predictions that include detailed chemical kinetics.

Of considerable consequence for LIF measurements is an understanding of the collisional effects governing the fluorescence quantum yield. For example, in a flame at atmospheric pressure, approximately three of each thousand OH molecules excited by the laser will fluoresce; the remainder is removed nonradiatively by collisions with the ambient flame gases. Furthermore, to reduce interference or background, fluorescence detection is often accomplished using a filter or spectrometer with a bandpass encompassing only a portion of the total fluorescence. Vibrational and rotational energy transfer collisions of the excited radical with the surrounding gases may determine the fraction of emission into that particular bandpass.

Because of the extreme importance of the OH radical to combustion chemistry (and also in the atmosphere), its collisional behavior has been investigated extensively, although not yet completely. The results show definite quantum state and translational energy dependence of energy transfer and quenching cross sections. These are not only crucial to quantitative measurements of this radical, but they are also important in understanding molecular collision dynamics. OH is small enough that its interactions are amenable to quality ab initio calculations, at least with simple colliders like noble gases and H_2. Quantum scattering calculations may then be compared with experimental results; the state-to-state detail provided by this open shell radical furnishes valuable tests of those calculations. Furthermore, they may be compared with spectroscopic characteristics and half-collision dynamics of van der Waals complexes of OH with noble gases (and, in the future, molecular partners).

Quenching of the $A\ ^2\Sigma^+$ excited state of OH is the major factor determining the fluorescence quantum yield. Measurements have been made over a wide range of temperatures, from 200–2500 K, using cooled and room temperature flow cells, laser flash heated cells, flames, and shock tubes [88.9]. For nearly all collision partners the quenching cross section decreases with increasing T, i. e., increasing collision velocity. This shows that attractive forces are involved in the collision, and the form of the variation for most colliders can be explained in terms of a collision complex formation mechanism. The $A\ ^2\Sigma^+$ state of NO behaves similarly, but quenching of the $A\ ^2\Delta$ state of CH increases with T, showing that a barrier exists on the potential surfaces; the $A\ ^3\Pi$ state of NH exhibits more varied behavior.

Quenching also varies with rotational level for OH and NH, as does vibrational transfer ($v = 1 \rightarrow v' = 0$) for OH. This appears to be a dynamic, not energetic, effect, and has been ascribed to a rotational averaging of the effects of the highly anisotropic surface on which these polar hydrides interact with colliders. Vibrational transfer in OH has been studied in both the excited and ground electronic states. It is found to be much faster in the upper $A\ ^2\Sigma^+$ state for all colliders, by factors of 100–1000. This is perhaps related to the fact that, according to ab initio calculations on OH–Ar, there is a much deeper well for interaction of the excited radical than in the ground state. Rotational transfer in OH has been shown to have some definite propensities, including parity conservation for some collisions with He that is reproduced in quantum scattering calculations.

Such a wealth of detail is not yet available for other free radical combustion intermediates, although they are beginning to be investigated. Of particular current interest is NO and the major precursor to its formation, CH. The details of quenching by a variety of colliders are probably known well enough that quantum yields for OH, NO, and CH may be calculated reasonably accurately, given the local temperature and major species flame composition at the point of measurement. There are enough data to compute quenching rates to within about 30–50% for many flames; this is partly because H_2O, when present, accounts for the majority of the quenching for all three radicals. When the local environment (especially T) is not known, the situation is worse. This would arise for an instantaneous two-dimensional image of a radical in a flame. Even so, the net quenching rate for OH rarely varies more than threefold through the entire flame, so useful semiquantitative information can still be obtained.

If absolute LIF measurements are desired, some means of calibration is necessary. Stable compounds (NO, CO) can simply be introduced into the flames or corresponding cold flows. Free radicals pose a greater problem. OH is often present in sufficient quantity that absorption measurements in a stable, laminar one-dimensional flame can be performed, but this is generally not possible for other species. The recently introduced technique of cavity ring down spectroscopy may alter this situation [88.10]. Burnt gases, in thermal equilibrium at a measured T, furnish a reliable source of known concentrations of OH, H, and O but generally not other radicals (which are consumed earlier in the flame).

Two-photon excitation is used for detection of atomic species and the CO molecule. The high laser flux needed for two-photon absorption, together with the uv wavelengths required for the pertinent transitions, often leads to problems with photochemical interferences. Sometimes these produce the same species that is being measured, e.g., dissociation of vibrationally excited O_2 at the wavelength of atomic oxygen detection, with subsequent excitation of the spurious O atom. In other cases, especially in complex hydrocarbon fuels, a parent compound or partially oxidized fragment may be excited directly or photolyzed to yield an emitting product. Excitation scans and measurements as a function of laser power are needed to discern and avoid these interferences.

88.2.3 Degenerate Four-Wave Mixing

DFWM is a nonlinear process like CARS; but like LIF, it operates on real transitions. It combines some of the attributes of both methods: production of a coherent signal beam that can be spatially filtered to discriminate against background, and high sensitivity so that trace radical species may be detected [88.11]. It depends on $\chi^{(3)}$, as does CARS, and can be used in a broadband mode. Like LIF, it can produce two-dimensional images, with the proper laser sheet arrangement. DFWM adds some of its own advantages, especially a Doppler-free nature enhancing the spectral resolution and thus the molecular selectivity. Three laser beams are used to generate the DFWM signal, all the same frequency, and tuned to an absorption transition of the species of interest. Two pump beams create an interference pattern in the medium, and a probe beam scatters off the resulting grating to form the signal. The interference pattern may be in the ground/excited state populations or may exploit polarization phenomena. Rapid quenching of the excited state may produce a thermal grating in the medium; the probe beam can then scatter from the resulting variation in the index of refraction. DFWM depends only on absorption and may be used for sensitive detection in a nonfluorescing or poorly fluorescing case (e.g., a predissociative state or ir vibrational transition).

The interpretation of DFWM signals to obtain molecular concentrations or temperatures requires knowledge of the underlying physical phenomena. Because they are governed by $\chi^{(3)}$, the DFWM amplitudes interfere, and an understanding of molecular collisions, motion, and the effects of the laser power are needed for accurate modeling of DFWM spectra. Collisional effects are of particular interest due to the Doppler-free character of the technique. Signal intensities can be dramatically influenced by line broadening due to quenching, energy transfer, and dephasing collisions. The magnitude of the influence depends sensitively on the degree of optical saturation. A cohesive picture of these effects is currently being sought through a combination of measurement and theory. Furthermore, nearly all current theoretical treatments assume single-mode lasers; however, real lasers contain intensity and frequency fluctuations within each laser pulse. The effects of these fluctuations on the nonlinear wave-mixing process must be accounted for in a proper description of DFWM. Furthermore, because collisions affect the signal only during the laser pulse, rapid (ps) measurements may provide excellent sensitivity under very high P conditions.

The articles and books [88.7,9,11–18] are of a review or overview nature. In addition, archival articles illustrating the state of the combustion field can be found in the biennial International Symposia volumes published by the Combustion Institute, Pittsburgh, now referred to as the Proceedings of the Combustion Institute.

88.3 Recent Developments

In the last eight years, there have been but two important advances in fundamental combustion laser diagnostics. The first of these is the publication of a book [88.19] containing an in-depth discussion of state of the art laser diagnostics. It includes eight chapters on basic experimental methods, nine chapters on applications, which is presently the largest growing area, and nine chapters on perspectives, future needs, and emerging applications. The interested reader is referred to this comprehensive treatise on the topic.

In recent years, there has indeed been a number of new applications of LIF and CARS to practical combustors (see the Applications chapters of [88.19]). LIF has been extended to only a few new molecules but has been refined for several (see Chapt. 2 of [88.19] for a listing of all pertinent LIF molecules). DFWM has not found significant new applications, and efforts for nonfluorescing molecules have diverted to cavity ringdown spectroscopy. New laser sources are important, especially those in the infrared.

The major experimental advance has been the advent of cavity ring-down spectroscopy (see Sect. 43.2). This topic is covered in [88.20], and its particular applications to flame diagnostics are discussed in Chap. 4 of [88.19]. This method can be used on nonfluorescing molecules which do absorb available laser wavelengths. A recent article on CH in flames [88.21] discusses experimental methodology, references earlier papers, and compares this technique with LIF.

References

88.1 R. J. Kee, F. M. Rupley, J. A. Miller: Sandia National Laboratory Report SAND89-8009 (1989)

88.2 W. Gardiner, M. Frenklach, H. Wang, M. Goldenberg, C. Bowman, R. Hanson, D. Davidson, G. Smith, D. Golden, W. Gardiner, V. Lissianski: *Proceedings of the 1995 International Gas Research Conference* (Gas Research Institute, Chicago 1995)

88.3 A. H. Zewail: Science **242**, 1645 (1988)

88.4 H. Reisler, M. Noble, C. Wittig: Photodissociation processes in NO-containing molecules. In: *Molecular Photodissociation Dynamics*, ed. by M. N. R. Ashfold, J. E. Baggot (Royal Society of Chemistry, London 1987) p. 139

88.5 X. Yang, J. M. Price, J. A. Mack, C. G. Morgan, C. A. Rogaski, D. McGuire, E. H. Kim, A. M. Wodtke: J. Phys. Chem. **97**, 3944 (1993)

88.6 J. Warnatz: In: *Twenty-Fourth Symposium (International) on Combustion* (The Combustion Institute, Pittsburgh 1992) p. 553

88.7 D. R. Crosley: Comb. Flame **78**, 153 (1989)

88.8 A. D. Sappey, D. R. Crosley: J. Chem. Phys. **93**, 7601 (1990)

88.9 D. R. Crosley: J. Phys. Chem. **93**, 6273 (1989)

88.10 G. Meijer, M. G. H. Boogaarts, R. T. Jongma, D. H. Parker, A. M. Wodtke: Chem. Phys. Lett. **217**, 112 (1994)

88.11 R. L. Farrow, D. J. Rakestraw: Science **257**, 1894 (1992)

88.12 D. R. Crosley: Laser measurement of chemically reactive intermediates, in combustion. In: *Turbulence and Molecular Processes in Combustion*, ed. by T. Takeno (Elsevier, New York 1993) p. 221

88.13 A. C. Eckbreth: *Laser Diagnostics for Combustion Temperature and Species* (Abacus, Cambridge 1988)

88.14 W. C. Gardiner (Ed.): *Combustion Chemistry* (Springer, New York 1984)

88.15 L. P. Goss: CARS instrumentation for combustion applications. In: *Instrumentation for Flows with Combustion*, ed. by A. M. K. P. Taylor (Academic, New York 1993) p. 251

88.16 K. Kohse-Höinghaus: Prog. Energy Combust. Sci. **20**, 203 (1994)

88.17 A. Linan, F. A. Williams: *Fundamental Aspects of Combustion* (Oxford Univ. Press, Oxford, New York 1993)

88.18 J. A. Miller, R. J. Kee, C. K. Westbrook: Ann. Rev. Phys. Chem. **41**, 345 (1990)

88.19 K. Kohse-Höinghaus, J. B. Jeffries(Eds.): *Applied Combustion Diagnostics* (Taylor and Francis, New York 2002)

88.20 B. A. Paldus, A. A. Kachanov: Can. J. Phys. **83** (2005), in press

88.21 J. Luque, P. A. Berg, J. B. Jeffries, G. P. Smith, D. R. Crosley, J. J. Scherer: Appl. Phys. B **78**, 93 (2004)

89. Surface Physics

This chapter describes various applications of atomic and molecular physics to phenomena that occur at surfaces. Particular attention is placed on the application of electron- and photon-atom scattering processes to obtain surface specific structural and spectroscopic information.

The study of surfaces and interfaces touches on many fields of pure and applied science. In particular there are applications in the fields of semiconductor processing, thin film growth, catalysis, corrosion and fundamental physics in two-dimensions. A number of recent texts cover surface physics in general [89.1–4] as well as specific areas such as experimental techniques [89.5], surface electron spectroscopies [89.6,7], and the application of synchrotron radiation to surface science [89.8]. Also a number of book series that deal with areas of particular interest are published at regular intervals, such as surface chemistry [89.9], surface vibrations [89.10] and stimulated desorption processes [89.11].

89.1 Low Energy Electrons and Surface Science

To obtain information specific to the first few layers of atoms at the surface of a solid, techniques must be devised that discriminate between signals from the surface and from the bulk of the material. If a solid has dimensions of $\approx 1\,\text{cm}^3$, then only approximately one atom in 10^7 is located at the surface. This makes the application of bulk techniques to the study of surfaces (e.g. X-ray diffraction [89.12]) problematic.

There are several experimental approaches to achieving surface sensitivity. If the probe used is an atom or molecule that can be scattered from or desorbed from the surface, then these species can be analyzed by techniques such as mass spectrometry or resonant ionization. A more recent innovation has been the use of scanning probe microscopies [89.13] (e.g., the scanning tunneling microscope and its variants) which exploit a surface sensitivity such as the surface valence electron density. However the most common surface analytical techniques use the intrinsic surface sensitivity of low energy (10–2000 eV) electrons.

The surface specificity of low energy electrons arises from the very short inelastic mean free path (MFP) for these electrons in a solid. This property can be seen in Fig. 89.1 which plots measured values of the inelastic MFP for a number of materials and electron energies between 1 and 2000 eV. The curve is called "universal" because the same general trend of short inelastic MFP is observed for nearly all materials [89.14]. The dominant energy loss mechanisms are valence band excitations (plasmons and electronic excitations), and since most materials have similar valence electron densities, the resultant inelastic scattering MFP is to a good approximation, material independent.

The very short inelastic MFPs for low energy electrons has the result that any that escape from the solid without having undergone inelastic scattering can only

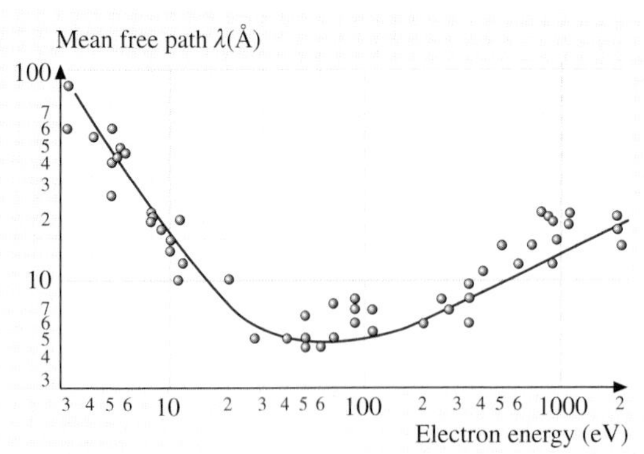

Fig. 89.1 The variation of the inelastic mean free path with electron kinetic energy ("universal curve"). Based on *Briggs* and *Seah* [89.14]

have originated very close to the surface, usually within a few atomic layer spacings. Many of the surface analysis techniques described in this chapter use this surface sensitivity to obtain surface structural and spectroscopic information. The same techniques are easily applicable to conducting and semiconducting materials, and can also be applied to insulating materials, if the charging effects are dealt with in some manner.

89.2 Electron–Atom Interactions

89.2.1 Elastic Scattering: Low Energy Electron Diffraction (LEED)

The elastic scattering of low energy (20–500 eV) electrons at surfaces is historically important in physics as the experiments of Davisson and Germer provided early experimental evidence of the wave nature of electrons. After these initial experiments, the technique was largely unused until the advent of cleaner ultra-high vacuum systems and surface preparation techniques in the 1960's. LEED is one of the most important and widely used surface characterization techniques due to its surface sensitivity and wide utility [89.5].

Diffraction from a two-dimensional net of scatterers results in a two-dimensional array of reciprocal lattice "rods" oriented normal to the surface. All kinematically allowed rods intersect the Ewald sphere at all energies — unlike the case in bulk X-ray diffraction. The short inelastic MFP for scattering low energy electrons yields the surface sensitivity of LEED for a crystal surface. The kinematic theory used in X-ray diffraction is not directly applicable to LEED since the low energy electrons can undergo several elastic collisions in the surface region. The elastic MFPs for low energy electrons are comparable in magnitude to the inelastic mean free paths shown in Fig. 89.1. This "multiple scattering" does not affect the positions of the diffraction beams, but does alter their intensities due to interference effects.

Kinematic analysis is sufficient to determine the diffraction beam positions for a proposed structure, and this is the most common use for LEED. A diffraction pattern is often sufficient to determine the surface periodicity and unit mesh size. However determining the surface crystal basis from the intensities of the diffracted LEED beams requires moderately sophisticated calculations to be performed for proposed structures.

Quantitative LEED generally compares measured "I(V) curves" (the intensity I of a particular diffraction beam as a function of electron energy measured in volts) with a calculated I(V) profile for a proposed surface crystal structure. These calculations determine the propagating electron wave functions Ψ_{LEED} that take into account electron-ion core scattering cross sections and phase shifts, available multiple scattering pathways, inelastic scattering cross sections and Debye-Waller effects. The structural parameters are varied systematically until agreement can be reached with the experimental I(V) curves [89.15, 16]. These calculations can yield surface atom positions to better than 0.1 Å vertically and 0.2 Å horizontally. Over 1000 surface structures have been determined using LEED and associated techniques [89.17].

Clean surfaces often have a different crystallography than simple termination of the bulk crystal structure. Due to the absence of neighbors above the surface, the surface atoms can undergo both relaxation (change in interlayer spacing) and reconstruction (changes in periodicity and bonding) [89.2]. While relaxation and reconstruction do occur on all types of surfaces, the reconstructions found on semiconductors are most striking. The strong covalent bonds that are broken when a semiconductor is cleaved give rise to high energy "dangling bonds". The surface energy is minimized by having the surface reconstruct to reduce the total number of these bonds. The resultant structure is formed through a balance between eliminating as many dangling bonds as possible and the resultant stress caused in other bonds due to the displacements of the surface atoms. One example of this is the clean Si(111) surface, in which 49 surface atoms form a new periodic arrangement to make up the reconstructed Si(111)–(7×7) unit cell [89.2]. The number of dangling bonds is thereby reduced from 49 to 19.

89.2.2 Inelastic Scattering: Electron Energy Loss Spectroscopy

Low energy surface excitations such as phonons, plasmons and electron-hole pair excitations can be studied using inelastic electron scattering. High resolution electron energy loss spectroscopy (HREELS) [89.18] uses incident electron beam energies of 1–300 eV with energy resolutions of 1–10 meV. Inelastic scattering phenomenon are divided into three types: dipole, impact and resonant scattering. Surface vibrations are also studied using infra-red absorption spectroscopy and inelastic atomic beam scattering techniques [89.19].

Dipole Scattering

In dipole scattering HREELS, the incident electron ($1 < E_\mathrm{i} < 20\,\mathrm{eV}$) undergoes a long range Coulomb interaction with dipole fields associated with a dynamic dipole moment. The main characteristic of dipole scattering is that the inelastically scattered electrons have an angular distribution that is strongly peaked (width $\Delta\theta \approx \hbar\omega/E_\mathrm{i}$) close to the specular direction ($\theta_\mathrm{i} = \theta_\mathrm{f}$). In this limit there is no momentum exchange with the surface vibrational mode (i. e., $\boldsymbol{k}_\parallel = 0$ in the bandstructure), so dipole scattering HREELS is similar to infrared spectroscopy in that only ir active vibrational modes at the zone center can be studied. At the surface of a conductor, dielectric screening guarantees that only ir active modes having a component perpendicular to the surface are visible — modes having dipole moments parallel to the surface are screened by the surface and so cannot be excited by dipole scattering. This "quasi-selection rule" allows the site symmetry for some systems to be decided [89.18]. Dipole scattering is most commonly used in the study of vibrations of molecules on surfaces, which often have dipole-active vibrational modes that are both intrinsic and caused by adsorption (i. e., frustrated translations and rotations).

Impact Scattering

In impact scattering, the incident electron samples the short-range interatomic potential, and is not restricted to dipole-active vibrational modes [89.18]. Higher electron energies are used ($30 < E_\mathrm{i} < 300\,\mathrm{eV}$) since the cross section for impact scattering generally increases with energy. The inelastically scattered electrons are distributed at all angles, and so there can be exchange of parallel momentum ($\boldsymbol{k}_\parallel$) with the surface, allowing the mapping of surface excitation bandstructure $\hbar\omega(\boldsymbol{k}_\parallel)$. The cross sections for impact scattering HREELS can be calculated using an approach similar to LEED, but now considering the normal displacements of atoms from equilibrium positions. The scattering potential $V(\boldsymbol{r}, [\boldsymbol{R}])$ (where $[\boldsymbol{R}]$ is the set of position vectors of the N atoms in the surface region) can be expanded in terms of the normal displacements $\boldsymbol{\mu}_i$ from equilibrium according to

$$V\big(\boldsymbol{r}, [\boldsymbol{R}]\big) = V\big(\boldsymbol{r}, [\boldsymbol{R}_0]\big) + \sum_{i=1}^{N} \Big(\nabla_{\mu_i} V\big(\boldsymbol{r}, [\boldsymbol{R}]\big)\Big)\Big|_{[\boldsymbol{R}_0]} \cdot \boldsymbol{\mu}_i + \cdots . \tag{89.1}$$

The first term is responsible for elastic scattering (LEED) and the second term is the dominant term for impact scattering. The inelastic cross section for mode i is then [89.20]

$$\frac{\mathrm{d}\sigma}{\mathrm{d}\Omega} = |\langle \Psi^*_{\mathrm{LEED}}(\boldsymbol{k}_f)| \frac{\partial V}{\partial \mu_i} |\Psi_{\mathrm{LEED}}(\boldsymbol{k}_i)\rangle|^2 \tag{89.2}$$

using electron wave functions calculated using the same formalism as LEED. The symmetry properties of this matrix element can be used to determine the polarization direction of surface vibrational modes by searching for systematic absences in the inelastic intensity in particular high symmetry direction of the surface [89.20]. Impact scattering studies are most commonly made in the study of surface phonon bandstructure and the vibrational modes of adsorbed molecules that are not dipole-active.

Resonant Scattering

Resonant electron scattering at surfaces [89.21, 22] is usually applied to the study of adsorbed molecules, and has much in common with resonant scattering from gas-phase atoms and molecules (see Chapt. 47). In resonant scattering, the incident electron combines with a target molecule to form a short-lived molecular ion, which subsequently decays and can leave the molecule vibrationally or electronically excited. The cross sections for this process have a typical profile. For example, in a shape resonance the formation of the temporary negative ion intermediate corresponds to adding the incident electron to a particular unoccupied orbital of the target molecule. The study of the angular dependence of resonance scattering cross sections [89.22] can yield information on the orientation of molecules on surfaces, since the electron capture and emission cross sections are fixed by the molecular orientation and the resonance symmetry.

89.2.3 Auger Electron Spectroscopy

Auger electron spectroscopy (AES) is one of the most widely used surface science techniques due to its chemical and surface sensitivity [89.5, 14, 23]. Core holes are created in near surface atoms using a high energy electron beam (2–5 keV, 1–100 μA) or less commonly, an X-ray source. For these low binding energy core holes, the Auger decay mode is highly probable (Sect. 61.2). Although the Auger energy and lineshape can give spectroscopic information, in surface science this is seldom used – rather it is the chemical fingerprint of the atoms which is of interest.

Auger electron spectroscopy is a surface sensitive technique by virtue of the low kinetic energy (50–1000 eV) of the emitted Auger electrons. Auger electrons from atoms more than a few Ångstroms below the surface are inelastically scattered and so not detected by the energy selective detector. The kinetic energy of the Auger electrons is

$$\mathcal{T}_e = E_A - E_B - E_C - U \qquad (89.3)$$

where E_A is the binding energy of the initial core electron and E_B, E_C are the binding energies of the other electrons (one or both are valence levels) involved in the Auger process. Energy shifts and relaxation are accounted in the term U which includes hole-hole interactions and atomic and solid state (dielectric) screening of the holes, and hence can be sensitive to the local chemical environment.

Auger spectra are typically obtained in derivative mode [i. e., $N' \mathrm{d}N(E)/\mathrm{d}E$] to separate the Auger transitions from the secondary electron background, and comparison is made to reference spectra [89.24]. The raw chemical sensitivity of AES is very high, and surface concentrations of $\approx 1\%$ of many common chemical species can be detected. Semiquantitative measurements may be made by comparison to these reference spectra, but such comparisons only give atomic concentrations within a factor of 2 (or worse) since the measured AES signal can be modified by a number of factors. More precise quantitative measurements can be made by calibration of the AES intensity for the atomic constituents at a surface [89.5]. The AES sensitivity to atoms A on a clean substrate (atoms B) can be determined if the absolute quantity of A can be established by some other means.

The chemical sensitivity of AES can also be exploited as a form of chemical microscopy (scanning Auger microscopy) since the exciting electron beam that is used may be focused to a very small size. The Auger signal from the small target volume can be analyzed for specific chemical components. By rastering the incident electron beam, a chemical map of the surface can be made.

89.3 Photon–Atom Interactions

89.3.1 Ultraviolet Photoelectron Spectroscopy (UPS)

Photoelectron spectroscopy (PES) (Sect. 61.1) has been historically divided into UV photoelectron spectroscopy for low photon energies (generally the study of valence electron states) and X-ray photoelectron spectroscopy (study of core electron levels). The use of low energy photons ($5 < h\nu < 50$ eV) for PES of solids has the advantage that the photons have a negligible momentum ($\boldsymbol{k} \approx 0$). This allows straightforward band mapping since the transitions are vertical in momentum space:

$$h\nu = E_f(\boldsymbol{k}+\boldsymbol{G}) - E_i(\boldsymbol{k}) \qquad (89.4)$$

where $\boldsymbol{k}$ is the electron state wavevector and $\boldsymbol{G}$ is a reciprocal lattice vector.

Most UPS studies are done using angle-resolved photoelectron detection (also called angle resolved photoelectron spectroscopy or ARPES). The kinetic energy $\mathcal{T}_e$ and emission angle θ of the photoelectrons are measured, allowing the initial state binding energy and momentum parallel to the surface ($\boldsymbol{k}_\parallel$) to be determined from

$$k_\parallel = \sqrt{\frac{2m\mathcal{T}_e}{\hbar^2}} \sin\theta \,. \qquad (89.5)$$

The mean free path of the photoelectrons from valence levels allows both surface and bulk electron states to be studied. Separation of the surface from bulk bands can be accomplished in several ways [89.5]. Since a crystal surface has two-dimensional symmetry, only the $\boldsymbol{k}_\parallel$ component of momentum is a good quantum number for the surface states. Also, in transporting the surface or bulk state photoelectron of momentum $\boldsymbol{k}$ through the surface to the detector, only the $\boldsymbol{k}_{\|}$ component is conserved. A bulk state disperses in three-dimensions, requiring knowledge of $\boldsymbol{k} = \boldsymbol{k}_{\|} + \boldsymbol{k}_{\perp}$. The magnitude of $\boldsymbol{k}_{\perp}$ cannot be determined unless the inner potential (change in potential normal to the surface) is known by some other means. This property allows bulk and surface states to be distinguished since surface states remain fixed in energy for a constant $\boldsymbol{k}_{\|}$, while bulk states disperse if $\boldsymbol{k}_{\perp}$ is changed while $\boldsymbol{k}_{\|}$ is kept constant. This is done by measuring the photoelectron spectrum for a range of UV photon energies at a fixed value of $\boldsymbol{k}_{\|}$ (often at $\theta = 0$ so $\boldsymbol{k}_{\|} = 0$). Bulk states disperse since $\boldsymbol{k}_{\perp}$ varies as the photon energy is changed, but the surface states remain at a fixed binding energy in the photoelectron spectrum.

The photoelectron transition matrix element between initial and final states $|i\rangle$ and $|f\rangle$ due to the incident photon vector potential $\boldsymbol{A}$ is

$$I \propto \left|\langle f|\boldsymbol{A}\cdot\boldsymbol{p}+\boldsymbol{p}\cdot\boldsymbol{A}|i\rangle\right|^2 \approx \left|\langle f|2\boldsymbol{A}\cdot\boldsymbol{p}|i\rangle\right|^2 . \quad (89.6)$$

The spatial variation of $\boldsymbol{A}$ near the surface (the surface photoeffect) can be neglected, although this is not strictly valid at these low photon energies [89.26].

Due to the low photoelectron kinetic energies in most UPS work, precise calculation of photoelectron spectrum intensities is rather difficult due to multiple scattering and phase shifts sensitive to valence electrons. However, the initial electron state symmetries can be determined using the symmetry properties of the matrix element (89.6). Selection rules allow the state symmetries to be determined by measuring the photoemission spectrum along high symmetry directions of the surface using polarized UV radiation [89.5].

The valence electron states of adsorbed molecules can also be studied using UPS. Peaks in the photoelectron spectrum due to valence levels of adsorbed molecules tend to have larger linewidths (≈ 1 eV typically) than in the gas phase due to solid state and instrumental effects, so vibrational structure is seldom resolved. The positions of the molecular valence states are shifted in energy due to the surface work function and solid state relaxation (dielectric screening) effects. In addition to these rigid shifts, chemical shifts are also observed due to bonding (chemisorption) interactions between the molecule and surface. The molecular character of the valence orbitals is often retained from the gas phase, so the shifted levels can be identified by using the symmetry properties of the photoemission matrix element. For example, photoemission from gas phase CO shows three valence states: 5σ, 1π and 4σ in order of increasing binding energy. A series of photoemission spectra [89.25] for gas phase CO, solid CO and CO chemisorbed on several transition metal surfaces (in order of increasing CO–surface binding energy) is shown in Fig. 89.2. For solid CO and weakly chemisorbed CO/Ag(111) all three valence orbitals of molecular CO are well resolved, though the adsorbed CO spectrum is shifted rigidly in energy by relaxation effects. Since CO is chemisorbed more strongly on Cu(111) and Pd(111), the 1π and 5σ valence states shift relative to one another. The strongly overlapping 1π and 5σ states observed for chemisorbed CO can be individually resolved by using the photoemission selection rules and linearly polarized UV radiation [89.25]. The origin of the chemical shift between the 5σ and 1π levels has been ascribed to bond formation involving the CO 5σ level and σ-symmetry d-electron states on the transition metal surfaces [89.2].

89.3.2 Inverse Photoemission Spectroscopy (IPES)

Inverse photoemission spectroscopy is properly classified as an incident electron technique but is included here

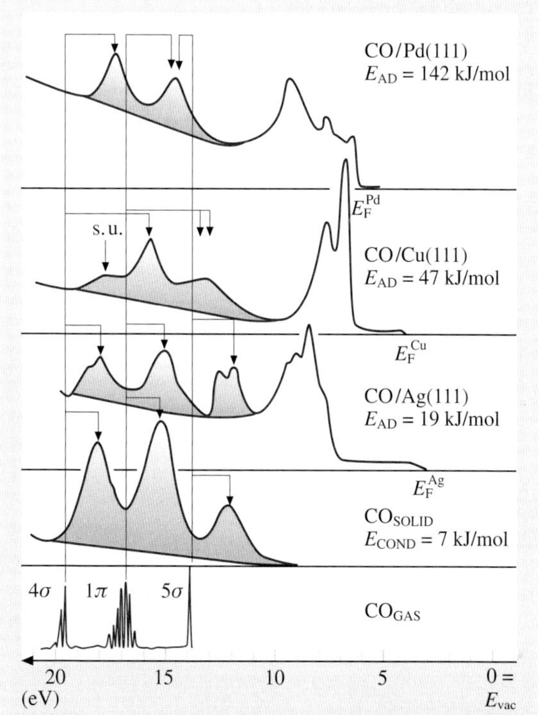

Fig. 89.2 A set of UPS spectra for CO adsorbed on various surfaces, as well as gas phase and solid CO. The peak labeled "su" is a shake-up satellite peak. After [89.25]

as a natural companion to UPS. IPES utilizes low energy electrons (5–50 eV) incident on a surface. Transitions from a high-lying initial electron state in the continuum to a lower unoccupied state cause a UV photon to be emitted. By detecting the emitted photon intensity as a function of energy, the joint density of states is measured [89.7]. IPES is used to study the unoccupied portion of surface and bulk bandstructure, particularly the region between the Fermi level and the vacuum level, which is difficult to access by other means. This allows the study of unoccupied states of adsorbed molecules (e.g., antibonding molecular levels) as well as intrinsic surface states such as the Rydberg-like states of electrons trapped in the image potential at the surface [89.7].

89.3.3 X-Ray Photoelectron Spectroscopy (XPS)

X-ray photoelectron spectroscopy allows the study of the energy of atomic core levels via the Einstein photoelectric equation $\mathcal{T}_e = h\nu - E_b$, where $\mathcal{T}_e$ is the kinetic energy of the photoemitted electron and E_b is the binding energy of the core level. Atoms bound in different chemical environments (e.g., at particular sites on a surface) experience different chemical shifts, and so are measured at slightly different binding energies. This sensitivity to chemical environment allows XPS to characterize the different types of binding sites for similar atoms, and measure their abundance by measuring the relative intensities of XPS emission from different species.

Shifts in $\mathcal{T}_e$ arise from two sources: intra-atomic and inter-atomic relaxation shifts. The intra-atomic relaxation E_a is due to screening of the core hole by other electrons in the emitter atom. The inter-atomic relaxation E_r is important in solids (particularly for metals) and is due to screening of the core hole by the dielectric response of the surrounding medium. These relaxation shifts (on the order of a few eV) tend to increase the kinetic energy of the emitted photoelectron, so the kinetic energy in the adiabatic limit is

$$\mathcal{T}_e = h\nu - E_b + E_a + E_r \,. \tag{89.7}$$

Since photoemission is a rapid process, nonadiabatic processes lead to shakeup and shakeoff features in which other electrons are excited to higher energy levels, causing the photoelectron to have lower kinetic energy [89.5] (see Sect. 62.4.4). It is also possible to excite discrete excitations of the solid, such as plasmons and a continuum of low energy electron-hole excitations, causing the XPS peak to have an asymmetric lineshape. The overall XPS distributions of $\mathcal{T}_e$ thus contain contributions from the adiabatic channel, and lower energy photoelectrons in a series of discrete peaks or a continuum from nonadiabatic processes.

The chemical abundance of a species can be determined from the sum rule that the total XPS cross section is proportional to the sum of the adiabatic peak and all the shake-up and shake-off components. However it is usually only convenient to measure the intensity of the adiabatic peak. If comparisons are made between chemical species in different chemical environments, the shake-up and shake-off intensities may differ. Hence the adiabatic channel intensities might not reflect the true abundance. Very often this problem can be minimized by careful calibration, and chemical analyses can be made to an accuracy of a few percent.

The surface sensitivity of XPS is due to the short inelastic MFP for the photoelectrons, which can be made to have energies in the range 10–1000 eV by appropriate choice of the X-ray wavelength. The XPS signal is proportional to $\exp[-z/(\lambda \cos\theta)]$, where z is the depth of the emitter, λ is the inelastic mean free path of the photoelectron and θ is the angle measured from the surface normal.

The popularity of XPS as a surface analysis technique is due to the availability of convenient and sufficiently intense monochromatic X-rays from lab sources (usually Al and Mg K_α X-ray lines at 1486.6 and 1253.6 eV). The increasing availability of continuously tunable X-rays from synchrotron radiation sources allows improved measurements due to the ability to tune the X-rays to slightly above threshold, where the XPS cross-sections are maximized, the inelastic MFP is shorter (hence more surface specific) and electron monochromators can operate with higher resolution [89.5].

As an example of the chemical sensitivity of XPS, photoelectron spectra of the $4f_{7/2}$ core levels of W(111) and Ta(111) are shown in Fig. 89.3. The spectra show discrete peaks from both bulk and surface atoms. The binding energies of the surface atoms are affected by the adsorption of hydrogen.

XPS can also be used to give local structural information due to elastic scattering of the photoemitted electron in the region near the emitter [89.5]. One form of this is the use of forward scattering from a buried emitter. A high energy XPS photoelectron is focused in the direction of nearby atoms due to a lower effective potential close to the atomic core. Angular scanning of the XPS detector will then detect a more intense signal along the bond axis. This method has proved useful in

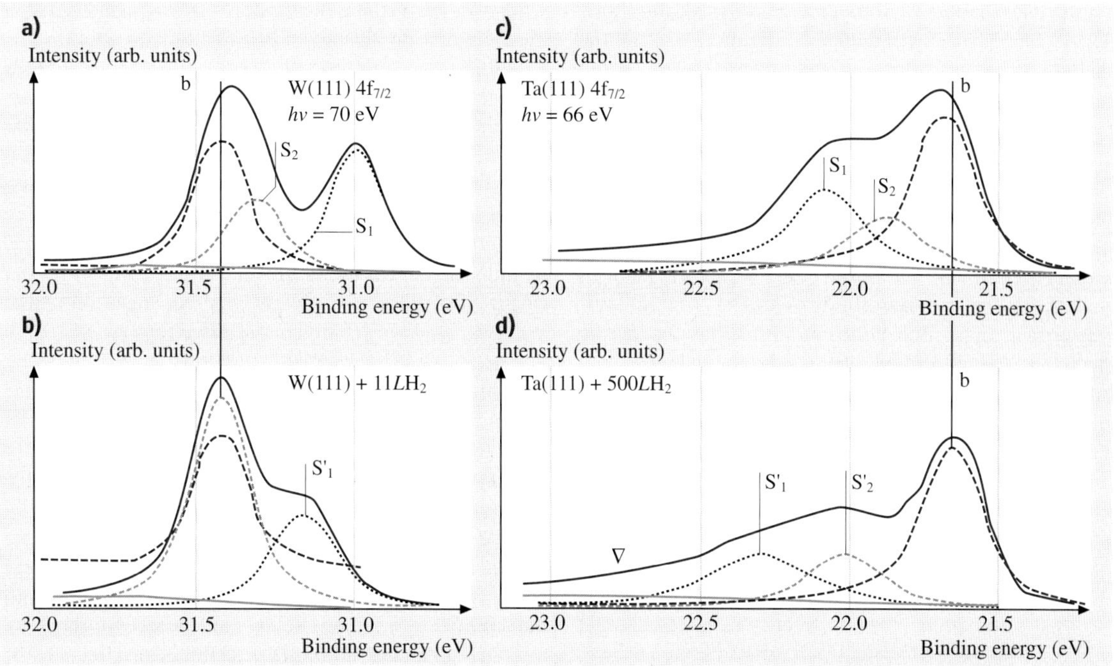

Fig. 89.3 XPS spectra of the $4f_{7/2}$ states of W(111) and Ta(111). Contributions from the bulk (b) and surface (S1,S2) atoms can be distinguished, and the surface atom peaks are shifted by the adsorption of hydrogen. After [89.28]

studying the orientation of adsorbed molecules and the structure of hetero-epitaxial thin films [89.27].

89.3.4 X-Ray Absorption Methods

X-ray absorption methods [89.8] measure the decay of the core hole rather than the intensity of emitted photoelectrons, since that XPS process is complicated by a number of possible final state processes, as discussed above. To obtain surface sensitivity, the X-ray absorption is measured using a low energy electron emitting channel such as an Auger electron emission or the total electron yield, which are proportional to the overall X-ray absorption. These methods require an intense and tunable source of monochromatic x-radiation near the excitation edge for the core level of a particular atom, and so are usually performed using synchrotron radiation. Synchrotron sources have the additional benefit of linearly polarized light, which is crucial for NEXAFS and useful for SEXAFS discussed next.

SEXAFS: Measurement of Bond Lengths

The surface extended X-ray absorption fine structure (SEXAFS) technique is most commonly used to measure the bond lengths for atoms adsorbed on a surface. SEXAFS utilizes the elastic backscattering of the emitted XPS photoelectron from nearby atoms that surround the emitter. Elastically scattered waves arrive back at the emitter and add coherently (constructively and destructively) to the outgoing wave, thus modifying the matrix element for the transition to the final state [89.29]. Experimentally, the cross section $\sigma(h\nu)$ for X-ray absorption above the threshold photon energy is modified by an oscillatory structure. The 'atomic' contribution to the absorption cross section can be removed using

$$\chi = \frac{\sigma - \sigma_0}{\sigma_0} , \qquad (89.8)$$

where σ and σ_0 are the measured surface and free atom X-ray absorption cross sections. If only single scattering events are involved in backscattering to the emitter then the fine structure function χ is

$$\chi(k) = \sum_{i=1}^{N} A_i(k) \sin\left[2kR_i + \phi_i(k)\right] , \qquad (89.9)$$

where k is the magnitude of the photoelectron wavevector and the sum is done for N 'shells' of atoms surrounding the emitter. The distance R_i is the radius of the ith shell and $2kR_i$ is the associated phase factor for the backscattered photoelectron, with an amplitude $A_i(k)$. The phase shift $\phi_i(k)$ is required due to the backscattering path through the potential surrounding the emitter and scattering from the atom at R_i. If the ϕ_i could be ignored, then a simple Fourier transform of $\chi(k)$ would reveal the radial distribution function and the bond lengths R_i [89.29]. In practice, the phase shifts cannot be neglected, but very often these can be found by studying chemically similar systems in which the bond lengths are known. The phase shifts for photoelectrons well above the absorption edge (having kinetic energies greater than $\approx$50 eV) are dominated by the atomic ion cores, and so are not sensitive to the valence electronic structure. The phase shifts can also be calculated in a straightforward way since this problem is essentially the same as done in LEED multiple scattering calculations.

The amplitudes $A_i(k)$ due to scattering from shell i at distance R_i from a point source emitter are [89.29]

$$A_i(k) = \frac{N_i^*}{kR_i^2}|f_i(\pi,k)|\exp\left(-2\langle u^2\rangle_i k^2\right) \times\exp\left(-2R_i/\lambda\right), \qquad (89.10)$$

where N_i^* is an effective number of atoms and $\langle u^2\rangle_i$ is the mean-square displacement of the atoms in shell i. The backscattering amplitude $|f_i(\pi,k)|$ has been separated from its phase factor $\phi_i(k)$ in (89.9). The inelastic mean free path for the photoelectrons reduces the contribution from successive shells by a factor $\exp(-2R_i/\lambda)$, and it is this term that allows the kinematic single-scattering approach to be used. Multiple scattering paths involve longer trajectories and so are more strongly attenuated by this exponential factor. If the near-edge energy region of the absorption cross section is not included in the analysis, then the scheme outlined in (89.9) and (89.10) is reasonable. The near-edge region (within $\approx$ 50 eV of the absorption edge) is troublesome, not only because of multiple scattering, but also because the phase shifts $\phi_i(k)$ for low energy photoelectrons are more sensitive to valence electron distributions, and so are more sensitive to the details of the local chemical environment.

The SEXAFS method is illustrated with the data of Fig. 89.4 for iodine adsorbed on Cu(111). Panel (a) shows the measured X-ray absorption $\sigma(h\nu)$ for both the surface system Cu(111)–I and bulk CuI. The extracted $\chi(k)$ and their Fourier transforms are shown in (b)

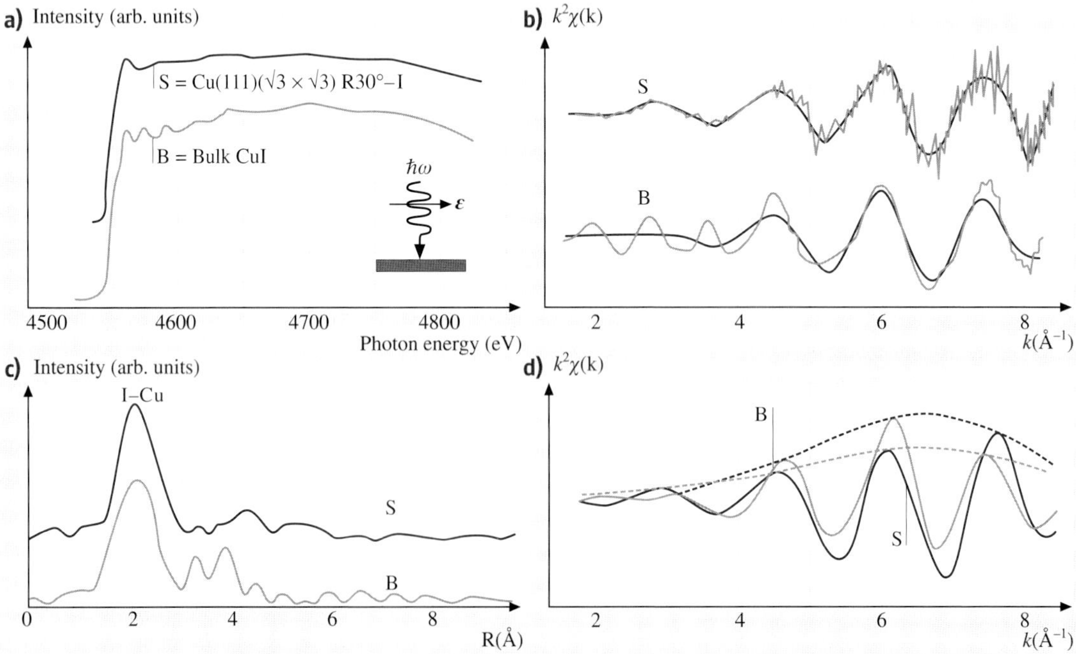

Fig. 89.4 (a) X-ray absorption data taken near the iodine edge for bulk CuI and Cu(111)–I. (b) The extracted fine structure function $\chi(k)$, shown multiplied by k^2 to enhance the high k structure. (c) The Fourier transform of $\chi(k)$ showing the location of the first shell of Cu atoms from the I emitters. (d) Back-transformed $k^2\chi(k)$ after applying a filter to extract the nearest neighbor data. After [89.30]

and (c). The Cu–I nearest neighbor bondlength is clearly shown by the peak in (c), and is found to be 0.07 Å longer than in bulk CuI. SEXAFS is most commonly applied to atomic adsorption systems since molecular systems are difficult to analyze, as they can contain several similar bond lengths, and the shells containing different atoms are difficult to model.

NEXAFS: Molecular Orientation at Surfaces

In SEXAFS, the X-ray absorption close to the excitation edge is avoided due to the problems of multiple scattering, and phase shifts that are chemically sensitive to valence electrons. The study of near-edge X-ray absorption fine structure (NEXAFS) can avoid these difficulties by using only the symmetry properties of the transition matrix element without concern for the absolute amplitudes [89.32]. For isolated molecules, unoccupied molecular states having σ or π symmetry are very commonly found close to or just below the threshold for photoemission. These 'molecular resonances' often remain for adsorbed molecules, and the symmetry properties of the absorption intensity $I \propto |\langle f|\boldsymbol{A}\cdot\boldsymbol{p}|i\rangle|^2$ can be used to determine the molecular orientation.

For the overall matrix element to be nonzero, it must be totally symmetric. Using linearly polarized synchrotron radiation, the direction of polarization $\boldsymbol{A}$ can be varied by rotating the crystal. For example, the K-edge X-ray absorption of CO on the Ni(100) surface in Fig. 89.5 shows final state resonances A (π-symmetry) and B (σ-symmetry). The intensity of these features depends of the direction of polarization of the incident X-rays. X-rays with a polarization vector parallel to the surface ($\theta = 90°$) strongly excite the π-symmetry absorption while the σ-symmetry absorption is absent. For grazing incidence X-rays polarized normal to the surface, the σ-resonance is prominent. From dipole selection rules, this polarization dependence of the X-ray absorption is evidence that the CO molecule is adsorbed

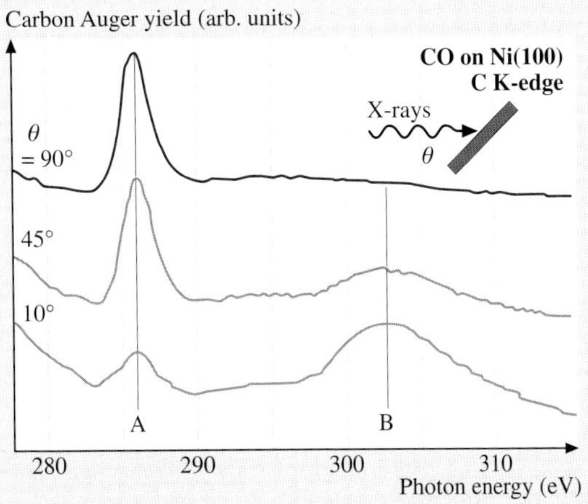

Fig. 89.5 Near-edge X-ray absorption data from the C K-edge from CO adsorbed on Ni(100) as function of incident photon angle θ. The molecular π (peak A) and σ (peak B) resonances are observed. The polarization dependence of the absorption allows the CO orientation to be determined. After [89.31]

with its bond perpendicular to the plane of the Ni(100) surface.

The σ-resonance final state in NEXAFS corresponds to multiple scattering of the photoelectron along the bond axis. The overall phase shift for this final state is approximately

$$\int \sqrt{E - V(r)}\, \mathrm{d}r \approx \sqrt{E - V(r)}\, R = \text{const.} \quad (89.11)$$

where R is the bond length. This sensitivity of the final state phase shift to bond lengths allows the σ-resonance energy to be used as a measure the molecular bond length [89.33]. A plot of the σ-resonance energy vs. $1/R^2$ shows a linear relationship, as is found in the case of simple hydrocarbons in the gas phase and adsorbed on a Cu(100) surface [89.32].

89.4 Atom–Surface Interactions

89.4.1 Physisorption

The binding interactions between atoms and surfaces can be classified as physisorption (long range attractive dispersion forces) and chemisorption, in which chemical bonds are formed. The long-range dispersion force between a polarizable atom and a conducting surface give rise to the leading term of the van der Waals potential $V(z) \propto -1/z^3$. At smaller atom-surface separations, the location of the reference image plane needs to be included, resulting in an attractive potential $V(z) = -C_v/|z - z_i|^3$ where z_i is the distance from the last atomic plane to the image plane, typically 2–3 Å. In principle, the constant C_v is calculable from the dielectric properties of the substrate and the atomic polarizability, but experiments have found values of C_v

40% less than expected [89.34]. The cause of this discrepancy has not been clarified, although contributions from surface roughness have been suggested.

At larger atom-surface separations, retardation effects must be taken into account in the dispersion interaction (the Casimir–Polder force, Sect. 79.2.4), and here theory predicts a $1/z^4$ interaction potential. This form for the potential has been confirmed experimentally [89.35].

As the atom approaches the surface, wave function overlap eventually causes repulsion. In the absence of chemical bond formation, the repulsive potential is simply proportional to the surface electron density $n(z)$, resulting in the overall physisorption potential

$$V(z) = Kn(z) - \frac{C_v}{|z - z_i|^3} . \tag{89.12}$$

The surface charge density $n(z)$ decreases exponentially above the surface [89.2], and the constant K can be determined to reasonable accuracy by an effective medium theoretical approach [89.36]. This shallow physisorption potential well has been studied experimentally in atomic beam (often He) scattering experiments. Under certain scattering conditions, it is possible for an incident atom to be 'selectively adsorbed' on the surface by making transitions into and out of bound states of the potential well of (89.12). The incident atom with wavevector $\boldsymbol{k}$ is diffracted by a surface reciprocal lattice vector $\boldsymbol{G}_{hk}$ into a bound state E_n of the potential well [89.2]. The diffracted atomic beam intensities due to scattering via selective adsorption show strong variations which can be related to the quantum numbers (n, h, k), and give information on the shape of the physisorption potential well.

89.4.2 Chemisorption

Many atoms and molecules will chemisorb when brought sufficiently close to a surface, forming chemical bonds (covalent or ionic) that are much stronger than the physisorption bond [89.3]. In chemisorption, there is charge transfer between the adsorbate and the surface, modifying the electronic structure of both. Valence electronic levels of the adsorbate are shifted in energy and also broadened by resonant interactions with the delocalized valence electrons at the surface (e.g., free electron-like s–p states). For many transition metal surfaces, interactions with the more localized d–states are also important.

The adsorbate-surface bond energies are most commonly studied by thermal desorption spectroscopy (TDS) [89.37, 38]. In this method, the adsorbate covered surface is heated using a linear temperature ramp, and the desorption rate of a particular species is measured using a mass spectrometer. By using a range of heating rates β, not only can the desorption energy be measured, but the kinetics governing the desorption process can be uncovered. For example, in the simplest case of a coverage-independent adsorption energy and first order kinetics, if the peak desorption rate occurs at a temperature T_0 then [89.39]

$$\frac{\nu}{\beta} = \frac{E_\mathrm{d}}{k_\mathrm{B} T_0^2} \exp\left(\frac{E_\mathrm{d}}{k_\mathrm{B} T_0}\right) \tag{89.13}$$

where E_d is the desorption energy and ν is the 'attempt frequency' for desorption. Since both ν and E_d are unknown, values for both can be found by measuring TDS spectra using two different heating rates β or by estimating ν (often $\nu \approx 10^{13}\,\mathrm{s}^{-1}$). More complex desorption kinetics are studied by utilizing the full desorption profile, and a range of heating rates and initial adsorbate coverages. These kinetic data are also applicable to adsorption if the adsorption and desorption are reversible processes. For irreversible adsorption systems, it is possible to measure the adsorption energies directly [89.40] by monitoring the ir radiation emitted from a surface as a submonolayer quantity of atoms or molecules is adsorbed.

89.5 Recent Developments

Surface Physics has both rapidly matured and expanded its connections with other disciplines in the last eight years. A number of notable review works have been published recently [89.41–45]. The application of surface science techniques has expanded rapidly in a number of technological areas, such as semiconductor devices, catalysis, and magnetic materials. The emerging field of nanotechnology has drawn heavily from the techniques of surface science. Important advances have been made in a large number of areas, including biosurfaces [89.46,47], cluster science [89.48], carbon fullerene materials [89.49, 50], electrochemistry [89.51], and surface photochemistry [89.52].

Continuing advances have been made in the application of light from synchrotron radiation sources — so much so that a complete catalog of applicable techniques would be difficult to compile. Many new facilities have been constructed, with higher brightness sources allowing the development of a large number of new techniques using synchrotron radiation from the ir through the soft and hard X-ray regimes [89.53, 54].

References

89.1 G. Ertl, J. Küppers: *Low Energy Electrons and Surface Chemistry*, 2nd edn. (VCH, Weinheim 1985)

89.2 A. Zangwill: *Physics at Surfaces* (Cambridge Univ. Press, Cambridge 1988)

89.3 V. Bortolani, N. H. March, M. P. Tosi (Eds.): *Interaction of Atoms and Molecules with Solid Surfaces* (Plenum, New York 1990)

89.4 H. Lüth: *Surfaces and Interfaces of Solids* (Springer, Berlin, Heidelberg 1993)

89.5 D. P. Woodruff, T. A. Delchar: *Modern Techniques of Surface Science*, 2nd edn. (Cambridge Univ. Press, Cambridge 1994)

89.6 S. D. Kevan (Ed.): *Angle-Resolved Photoemission: Theory and Current Applications* (Elsevier, Amsterdam 1992)

89.7 J. C. Fuggle, J. E. Inglesfield: *Unoccupied Electronic States* (Springer, Berlin, Heidelberg 1992)

89.8 R. Z. Bachrach (Ed.): *Synchrotron Radiation Research: Advances in Surface and Interface Science*, Vol. 1 (Plenum, New York 1992)

89.9 D. A. King, D. P. Woodruff (Eds.): *Chemical Physics of Solid Surfaces and Heterogeneous Catalysis* (Elsevier, New York 1990)

89.10 Y. J. Chabal, F. M. Hoffmann, G. P. Williams (Eds.): *Vibrations at Surfaces* (Elsevier, Amsterdam 1990)

89.11 A. R. Burns, E. R. Stechel, D. R. Jennison: *Desorption Induced by Electronic Transitions, DIET V* (Springer, Berlin, Heidelberg 1993)

89.12 P. H. Fuoss, K. S. Liang, P. Eisenberger: In: *Synchrotron Radiation Research: Advances in Surface and Interface Science*, Vol. 1, ed. by R. Z. Bachrach (Plenum, New York 1992)

89.13 R. Wiesendanger, H.-J. Güntherodt: *Scanning Tunneling Microscopy* (Springer, Berlin, Heidelberg 1993)

89.14 D. Briggs, M. P. Seah: *Practical Surface Analysis: Auger and X-Ray Photoelectron Spectroscopy*, Vol. 1 (Wiley, Chichester 1990)

89.15 P. J. Rous: Prog. Surface Science **39**, 3 (1992)

89.16 M. A. Van Hove: *Low-Energy Electron Diffraction* (Springer, Berlin, Heidelberg 1986)

89.17 J. M. MacLaren, J. B. Pendry, P. J. Rous, D. K. Saldin, G. A. Somorjai, M. A. Van Hove, D. Vvedensky: *Surface Crystallographic Information Service: A Handbook of Surface Structures* (Reidel, Dordrecht 1987)

89.18 H. Ibach, D. L. Mills: *Electron Energy Loss Spectroscopy and Surface Vibrations* (Academic, New York 1982)

89.19 H. Ibach: In: *Interaction of Atoms and Molecules with Solid Surfaces*, ed. by V. Bortolani, N. H. March, M. P. Tosi (Plenum, New York 1990)

89.20 S. Y. Tong, C. H. Li, D. L. Mills: Phys. Rev. B **24**, 806 (1981)

89.21 L. Sanche: J. Phys. B **23**, 1597 (1990)

89.22 R. E. Palmer, P. J. Rous: Rev. Mod. Phys. **64**, 383 (1992)

89.23 C. L. Bryant, R. P. Messmer (Eds.): *Auger Electron Spectroscopy* (Academic, New York 1988)

89.24 L. E. Davis, N. C. MacDonald, P. W. Palmberg, G. E. Riach, R. E. Weber: *Handbook of Auger Spectroscopy* (Physical Electronics Div., Perkin Elmer 1976)

89.25 H. J. Freund, M. Neumann: In: *Angle-Resolved Photoemission*, ed. by S. D. Kevan (Elsevier, Amsterdam 1992) p. 319

89.26 K. L. Kliewer: In: *Photoemission and the Electronic Properties of Surfaces*, ed. by B. Feuerbacher, B. Fitton, R. F. Willis (Wiley, Chichester 1978)

89.27 D. P. Woodruff: In: *Angle-Resolved Photoemission*, ed. by S. D. Kevan (Elsevier, Amsterdam 1992)

89.28 D. E. Eastman, F. J. Himpsel, J. F. van der Veen: J. Vac. Sci. Technol. **20**, 609 (1982)

89.29 J. E. Rowe: In: *Synchrotron Radiation Research*, Vol. 1, ed. by R. Z. Bachrach (Plenum, New York 1992) p. 117

89.30 P. H. Citrin, P. Eisenberger, R. C. Hewitt: Phys. Rev. Lett. **45**, 1948 (1980)

89.31 J. Stöhr, K. Baberschke, R. Jaeger, R. Treichler, S. Brennan: Phys. Rev. Lett. **47**, 381 (1981)

89.32 A. Bianconi, A. Marcelli: , Vol. 1, ed. by R. Z. Bachrach (Plenum, New York 1992) p. 63

89.33 J. Stöhr, F. Sette, A. L. Johnson: Phys. Rev. Lett. **53**, 1684 (1984)

89.34 A. Shih, V. A. Parsegian: Phys. Rev. A **12**, 835 (1975)

89.35 C. I. Sukenik, M. G. Boshier, D. Cho, V. Sandoghdar, E. A. Hinds: Phys. Rev. Lett. **70**, 560 (1993)

89.36 J. K. Norskov, N. D. Lang: Phys. Rev. B **21**, 2136 (1980)

89.37 D. Menzel: In: *Chemistry and Physics of Solid Surfaces IV*, ed. by R. Vanselow, R. Howe (Springer, Berlin, Heidelberg 1982) p. 102

89.38 H. J. Kreuzer, S. H. Payne: In: *Dynamics of Gas-Surface Interactions*, ed. by C. T. Rettner, M. N. R. Ashfold (Royal Society of Chemistry, Cambridge 1991) p. 220

89.39 P. A. Redhead: Vacuum **12**, 203 (1962)

89.40 J. T. Stuckless, N. Al-Sarraf, C. Wartnaby, D. A. King: J. Chem. Phys. **99**, 2202 (1993)

89.41 K. W. Kolasinski: *Surface Science: Foundations of Catalysis and Nanoscience* (John Wiley and Sons, New York 2002)

89.42 G. A. Somorjai: *Introduction to Surface Chemistry and Catalysis* (Wiley-Interscience, New York 1994)

89.43 C. B. Duke (Ed.): *Surface Science: The First Thirty Years* (Elsevier Science, New York 1994)

89.44 D. Bonnell (Ed.): *Scanning Probe Microscopy and Spectroscopy: Theory, Techniques, and Applications* (Wiley-VCH, Berlin 2000)

89.45 L. W. Bruch, M. W. Cole, E. Zaremba: *Physical Adsorption: Forces and Phenomena* (Oxford Univ. Press, Oxford 1997)

89.46 D. G. Castner, B. D. Ratner: Surface Science **500**, 28 (2002)

89.47 M. Tirrell, E. Kokkoli, M. Biesalski: Surface Science **500**, 61 (2002)

89.48 W. Eberhardt: Surface Science **500**, 242 (2002)

89.49 H. Dai: Surface Science **500**, 218 (2002)

89.50 M. Knupfer: Surface Science Reports **42**, 1 (2001)

89.51 D. M. Kolb: Surface Science **500**, 722 (2002)

89.52 H. L. Dai, W. Ho (Eds.): *Laser Spectroscopy and Photochemistry on Metal Surfaces* (World Scientific, Singapore 1995)

89.53 C. Lamberti: Surface Science Reports **53**, 1 (2004)

89.54 A. Baraldi et al.: Surface Science Reports **49**, 169 (2003)

90. Interface with Nuclear Physics

For an atom with a small to moderately large atomic number Z, the typical length scale a_0/Z of the innermost core orbitals is so much larger than typical nuclear length scales that the corrections to the energy levels and wave functions arising from the non-zero electric charge radius of the nucleus can accurately be computed using first-order perturbation theory, as is described in Sect. 90.1. Nonetheless, these relatively small shifts can sometimes have a profound effect on processes in atomic and/or nuclear physics, particularly if two or more energy levels are very close. For example, as is discussed in Sect. 90.2, the presence of the electron cloud makes energetically possible the β-decay of ^{187}Re to ^{187}Os, and significantly modifies the energy distribution of products in the β-decay of tritium in various chemical environments. Also, electronic screening can greatly enhance the cross-sections of low-energy nuclear reactions relative to what they would be for bare nuclei.

In isotopes of hydrogen, the replacement of an electron by a muon, with $m_\mu \approx 207 m_e$, results in a tiny neutral 'atom' which can closely approach another nucleus, thereby catalyzing nuclear fusion. For example, the rate of deuterium–tritium fusion is enhanced by 77 orders of magnitude if a single electron is replaced by a muon. A rich variety of bound-state properties and scattering processes for these exotic atoms and molecules has been extensively investigated, as is reviewed in Sect. 90.3.

A reader interested in the interface between atomic and nuclear physics should also consult Chapt. 27 (Quantum Electrodynamics), Chapt. 28 (Tests of Fundamental Physics), and Chapt. 29 (Parity Nonconserving Effects in Atoms) in this Handbook.

That nuclei are not infinitesimally small, structureless particles causes small but perceptible shifts in the electronic structure of atoms and molecules. Even for nuclei with small nuclear charge Z, the effects are readily detectable through modern high precision spectroscopy, and their magnitude grows as $Z^{14/3}$. Conversely, the presence of electrons tightly bound to atomic nuclei can alter the ordering of nuclear energy levels or make them unstable to β decay. Atomic effects can also influence nuclear branching ratios into the product channels. Nominally small atomic effects have been shown to affect the complicated chain of nuclear reactions responsible for the generation of energy in the sun. Setting bounds to the rest mass of the neutrino from the endpoint of the β-decay spectrum of tritium requires a precise understanding of atomic and molecular structure and scattering processes.

The great disparity between nuclear scales of energy (several MeV) and distance (10^{-5} to 10^{-4} Å) and the corresponding atomic scales (several eV and 1 Å, respectively) usually allows the separate treatment of nuclear and atomic effects. However, since not absolute energies but energy differences determine the magnitudes of perturbative effects, near coincidences in energy differences can greatly enhance the interplay between the two regimes. Such comparable differences of energies account for the important role of nuclear structure in the Lamb shift splitting between the $2s_{1/2}$ and $2p_{1/2}$ states of hydrogen (see Chapt. 27), and the influence of atomic structure on nuclear processes (see Sects. 90.1 and 90.2).

For the case of muonic atoms and molecules, the interplay is enhanced by the much larger mass of a muon relative to an electron. This decreases the distance scale by a factor of m_e/m_μ and increases the energy scale by a factor of m_μ/m_e. Small corrections such as the vacuum polarization part of the Lamb shift are amplified even more ($\approx (m_\mu/m_e)^3$ for low Z).

Besides the areas where atomic physics effects play an important role in nuclear physics, or vice versa, it is worth remembering that atomic and molecular physicists and nuclear physicists can benefit from knowing the theoretical techniques which have been developed in each others' fields. For example, it is well-known that group theoretical methods are widely employed in formulating and solving many-body problems in nuclear, atomic, and molecular physics. To take another case, the coupled-cluster method, which was first proposed in the late 1950s by *Coester* and *Kummel* in the context of nuclear theory [90.1–3], was applied a decade later to electronic structure problems in atomic and molecular physics and quantum chemistry by *Cizek*, *Paldus*, and *Shavitt* [90.4–6], and in the 1970s and 1980s was widely developed by Bartlett and coworkers at the University of Florida. Quite recently, 'quantum halos', which are very loosely bound states for which most of the probability density is spread diffusely over the classically forbidden region, have been treated in a unified manner for both nuclear and molecular systems [90.7].

90.1 Nuclear Size Effects in Atoms

90.1.1 Nuclear Size Effects on Nonrelativistic Energies

Interest in the influence of a finite nuclear charge distribution on the energy levels of the hydrogen atom goes back to the measurement of the Lamb shift [90.8–11], and even earlier indications that the fine structure of hydrogen did not quite agree with the predictions of the Dirac equation for a point nucleus [90.12–15]. The finite proton size does in fact raise the energy of the $2s_{1/2}$ state relative to $2p_{1/2}$, but the shift is only $\approx 0.012\%$ of the dominant electron self-energy contribution (Chapt. 27). It must nevertheless be taken into account in high precision tests of QED.

Early derivations were given by several authors [90.16–19] and generalized by *Zemach* [90.20] (see also [90.21]) to a form involving integrals over the nuclear electric and magnetic form factors. The basic result is illuminated by the following argument. Let $\rho(\boldsymbol{r})$ be the electron density, which may have no spatial symmetry properties in the particular case of a polyatomic molecule, and $\rho_n(\boldsymbol{r}_n)$ be the charge density of a nucleus, which obeys

$$\int d^3r_n\, \rho_n(\boldsymbol{r}_n) = Z\,. \tag{90.1}$$

Assume that $\rho_n(\boldsymbol{r}_n)$ has no permanent electric dipole moment, so that

$$\int d^3r_n\, \boldsymbol{r}_n\, \rho_n(\boldsymbol{r}_n) = \boldsymbol{0}\,. \tag{90.2}$$

By writing the Coulomb potential for a pointlike nucleus as

$$-\frac{Z}{r} = -\int d^3r_n \frac{Z\delta^{(3)}(\boldsymbol{r}_n)}{|\boldsymbol{r}-\boldsymbol{r}_n|}\,, \tag{90.3}$$

the first-order shift of the electronic energy due to the replacement of the pointlike nucleus by an extended nucleus is

$$\Delta E_{\text{nuc}} = \int d^3r \int d^3r_n \frac{Z\delta^{(3)}(\boldsymbol{r}_n)-\rho_n(\boldsymbol{r}_n)}{|\boldsymbol{r}-\boldsymbol{r}_n|}\,\rho(\boldsymbol{r})\,. \tag{90.4}$$

Since the Fourier transform, defined by

$$\hat{\rho}_n(\boldsymbol{k}) = \int d^3r_n\, e^{-i\boldsymbol{k}\cdot\boldsymbol{r}_n}\, \rho_n(\boldsymbol{r}_n)\,, \tag{90.5}$$

preserves inner products within a factor of $(2\pi)^3$ and maps convolutions to simple products, the integral in (90.4) reduces to

$$(2\pi)^3 \int d^3k\, [Z-\hat{\rho}_n(\boldsymbol{k})]\,\frac{4\pi}{k^2}\,\hat{\rho}(\boldsymbol{k})\,, \tag{90.6}$$

where the hats denote the Fourier transforms of the densities and $4\pi/k^2$ is the Fourier transform of the Coulomb potential $1/r$. Since $Z = \int \mathrm{d}^3 r_n\, \rho_n(\boldsymbol{r}_n)$, the energy shift can be reexpressed as

$$(2\pi)^3 \int \mathrm{d}^3 k \int \mathrm{d}^3 r_n\, (1 - \mathrm{e}^{-\mathrm{i}\boldsymbol{k}\cdot\boldsymbol{r}_n})\, \rho_n(\boldsymbol{r}_n)\, \frac{4\pi}{k^2}\, \hat{\rho}(\boldsymbol{k})\,, \tag{90.7}$$

which is still an exact first-order perturbation expression. Since typical nuclear length scales are much smaller than typical nonrelativistic atomic length scales, it is legitimate to expand the exponential in a Taylor series. The zeroth-order term, -1, is canceled by the +1. The linear term, $\mathrm{i}\boldsymbol{k}\cdot\boldsymbol{r}_n$, contributes nothing by the hypothesis that the nuclear charge distribution has no permanent electric dipole moment. The first nonvanishing term is

$$(2\pi)^3 \int \mathrm{d}^3 k \int \mathrm{d}^3 r_n\, \frac{1}{2} (\boldsymbol{k}\cdot\boldsymbol{r}_n)^2\, \rho_n(\boldsymbol{r}_n)\, \frac{4\pi}{k^2}\, \hat{\rho}(\boldsymbol{k})\,. \tag{90.8}$$

If $\rho(\boldsymbol{r})$ is nonzero at the nucleus, then for large k the leading behavior of $\hat{\rho}(\boldsymbol{k})$ is that of a spherically symmetric s-wave with a radial dependence proportional to k^{-4}. The angular integration in the variable $\boldsymbol{k}$ leads to the replacement of $(\boldsymbol{k}\cdot\boldsymbol{r}_n)^2$ by its average value $\frac{1}{3}k^2 r_n^2$, so that expression (90.8) reduces to

$$\frac{2\pi}{3}(2\pi)^3 \int \mathrm{d}^3 k\, \hat{\rho}(\boldsymbol{k}) \int \mathrm{d}^3 r_n\, r_n^2\, \rho_n(\boldsymbol{r}_n)\,, \tag{90.9}$$

which can be further simplified by observing that

$$(2\pi)^3 \int \mathrm{d}^3 k\, \hat{\rho}(\boldsymbol{k}) = (2\pi)^3 \int \mathrm{d}^3 \boldsymbol{k}\, \mathrm{e}^{\mathrm{i}\boldsymbol{0}\cdot\boldsymbol{k}} \hat{\rho}(\boldsymbol{k}) = \rho(0)\,, \tag{90.10}$$

and by definition

$$\int \mathrm{d}^3 r_n\, r_n^2\, \rho_n(\boldsymbol{r}_n) = Z\langle r_n^2\rangle\,, \tag{90.11}$$

thus yielding the final expression

$$\Delta E_{\mathrm{nuc}} = \frac{2\pi}{3} Z e^2 \rho(0)\, \langle r_n^2\rangle\,, \tag{90.12}$$

with

$$\rho(0) = \left(\frac{\mu}{m_\mathrm{e}}\right)^3 \frac{Z^3}{\pi n^3}\, a_0^{-3} \tag{90.13}$$

for a hydrogenic ion with reduced mass μ. This derivation is independent of the specific nuclear model or the assumption of spherical symmetry of the electron density. Since $\langle r_n^2\rangle$ scales as $Z^{2/3}$, ΔE_{nuc} then scales as $Z^{14/3}$. For a molecule with several nuclei, the contributions (90.12) from each nucleus should be summed.

For the helium atom, $\rho(0) = \langle \delta(\boldsymbol{r}_1) + \delta(\boldsymbol{r}_2)\rangle$ can be accurately calculated from high precision variational wave functions (Chapt. 11). For the $1s^2\,{}^1S_0$ ground state, $\rho(0) \simeq (\mu/m_\mathrm{e})^3[3.620\,8586 - 0.182\,37(\mu/M)]a_0^{-3}$ where M is the nuclear mass. Results for other states up to $n = 10$ are tabulated in [90.22]. Combined with high precision isotope shift measurements, the results can be used to extract differences in nuclear radii for pairs such as ${}^3\mathrm{He}/{}^4\mathrm{He}$, ${}^6\mathrm{Li}/{}^7\mathrm{Li}$, and H/D [90.23–26]. The method has recently been applied to the short-lived, neutron-rich nuclei ${}^6\mathrm{He}$, ${}^8\mathrm{Li}$, and ${}^9\mathrm{Li}$ [90.27, 28].

Expression (90.12) works well for atoms with small Z, since relativistic corrections to the electron density are small. However, it breaks down for heavier nuclei, for which relativistic wave functions are needed.

90.1.2 Nuclear Size Effects on Relativistic Energies

The preceding analysis breaks down for relativistic wave functions because they are singular at a point nucleus, making $\rho(\boldsymbol{0})$ infinite. In this case, the Dirac equations with Hamiltonians H_0 and H for the point nucleus and distributed nucleus cases, respectively, can be combined to obtain

$$(E - E_0)\Psi^\dagger \Psi_0 = \Psi^\dagger H \Psi_0 - \Psi^\dagger H_0 \Psi_0\,. \tag{90.14}$$

If a finite radius r_s is now chosen such that $H = H_0$ outside the sphere $r = r_\mathrm{s}$, then this equation can be integrated from r_s outward to yield [90.29]

$$\Delta E_{\mathrm{nuc}} = \frac{\hbar c\,(g f_0 - f g_0)_{r=r_\mathrm{s}}}{\int_{r_\mathrm{s}}^{\infty} (g g_0 + f f_0)\,\mathrm{d}r} \tag{90.15}$$

where f and g are the large and small radial components of Ψ (Chapt. 9), and the numerator is the surface term that remains after integrating by parts the $c\boldsymbol{\alpha}\cdot\boldsymbol{p}$ term in H. The units are $\hbar c/a_0 = \alpha m_\mathrm{e} c^2$. The solutions can be further expanded in terms of Bessel functions, or the Dirac equation can simply be integrated numerically.

For hydrogenic ions up to moderately large Z, the results are reasonably well represented by [90.30, 31]

$$\Delta E_{\mathrm{nuc}} = \frac{2}{3n^3}(Z\alpha)^2 m_\mathrm{e} c^2 \times \left[\delta_{\ell,0} + C_2 (Z\alpha)^2\right] \left(Z^2 \langle r_n^2\rangle / a_0^2\right)^{\gamma} \tag{90.16}$$

with $\gamma = \left[1-(Z\alpha)^2\right]^{1/2}$, and $C_2 \simeq 0.50$, 1.38, and 0.1875 for the $1S_{1/2}$, $2S_{1/2}$, and $2P_{1/2}$ states, respectively. Extensions to higher-order terms are discussed in [90.32]. The above formula was used in the tabulations of *Mohr* [90.31] for $10 \leq Z \leq 40$, while *Johnson* and *Soff* [90.33] used the numerical integration method for Z up to 110. The nuclear electric and magnetization density distributions are tabulated in [90.34], nuclear moments in [90.35], and nuclear masses in [90.36]. In the absence of better data, the rms nuclear radius can be estimated from $\langle r_n^2 \rangle^{1/2} \approx 0.777A^{1/3} + 0.778 \pm 0.06$ fm, where A is the atomic mass number.

90.1.3 Nuclear Size Effects on QED Corrections

Recent progress in the experimental study of transition energies in heavy ions stripped of most of their electrons [90.37–39] has inspired theoretical work on modifications of QED corrections due to an extended nuclear charge distribution. Calculations based on propagators expanded in terms of basis splines [90.40–42] (Sect. 8.1.1) have led to relatively rapid convergence with the number of angular functions.

90.2 Electronic Structure Effects in Nuclear Physics

90.2.1 Electronic Effects on Closely Spaced Nuclear Energy Levels

The presence of a nearby cloud of electrons can significantly affect nuclear processes involving closely spaced nuclear energy levels. One of the most dramatic cases involves the β-decay process $^{187}\mathrm{Re} \to {}^{187}\mathrm{Os} + e^- + \bar{\nu}_e$, which is energetically forbidden by about 12 keV for bare nuclei, but becomes allowed for the neutral atoms when the difference in electronic binding energies is included. The nuclear charges are $Z = 75$ for ^{187}Re and $Z = 76$ for ^{187}Os. There is also the possibility of the electron being captured into a bound state of ^{187}Os, as opposed to the continuum β-decay process.

The electronic binding energies of heavy atoms can be estimated from the Thomas–Fermi result $E_{\mathrm{TF}} \simeq -20.93\,Z^{7/3}$ eV for a neutral atom of charge Z. The difference between the energies of two atoms with nuclear charges $Z+1$ and Z, respectively, is then

$$E_{\mathrm{TF}}(Z+1) - E_{\mathrm{TF}}(Z) \simeq -48.83Z^{4/3}\,\mathrm{eV}, \quad (90.17)$$

which amounts to -15.4 keV at $Z = 75$. This is more than sufficient to overcome the 12 keV energy deficit in the otherwise energetically forbidden β-decay of ^{187}Re.

The general theory of bound state β-decay is discussed by *Bahcall* [90.43], who also calculated the ratio ρ of bound state β-decay to continuum β-decay for bare nuclei. In the case of $^{187}\mathrm{Re} \to {}^{187}\mathrm{Os}$, ρ is of importance in estimating changes in the half-life for β-decay of ^{187}Re under various conditions of ionization, since the measured isotope ratios $^{187}\mathrm{Re}/{}^{188}\mathrm{Re}$ and $^{187}\mathrm{Os}/{}^{188}\mathrm{Os}$ from terrestrial rocks and meteorites can be used to determine not only the age of the solar system, but also the age of our galaxy [90.44, 45]. Estimates based on a modified TF model [90.46] indicate that $\rho \simeq 0.01$, and further multiconfiguration Dirac–Fock calculations give $\rho = 0.005$ to 0.007 [90.47–49]. See [90.47–49] for further details and references.

90.2.2 Electronic Effects on Tritium Beta Decay

The mass of the neutrino, normally taken to be zero in the Standard Model, can be determined in principle from analysis of the β-decay process $^3\mathrm{T} \to {}^3\mathrm{He}^+ + e^- + \bar{\nu}_e$. An early measurement based on this method [90.50–52] yielded a neutrino mass of ≈ 25 eV. Several independent tests of this result were initiated soon thereafter. Since the experiments are performed not on bare tritons but on tritium gases and solids under various conditions, it is essential to understand quantitatively the atomic and molecular processes that affect the distribution of the highest-energy electrons produced from various initial states [90.53, 54].

Martin and *Cohen* [90.55] used a Stieltjes imaging technique to calculate shake-up and shake-off probabilities for the β-decay of T_2 into $^3\mathrm{HeT}^+$. Simultaneously, extensive calculations were carried out [90.56] using potential energy curves for the reactant T_2 and TH molecules and the product $^3\mathrm{HeT}^+$ and $^3\mathrm{HeH}^+$ molecules and accounting for the production of electronically and rovibrationally ground and excited final states, as well as resonant states. Nuclear motion was found to have a small but detectable effect on the results, and solid-state effects for frozen T_2 were also investigated and found to be small. These calculations played a crucial role in the interpretation of the experiments [90.57–61], which indicated that the neutrino mass is less than $\approx 10\,\text{eV}$. In 1998 there was published evidence from the super-Kamiokande experiment that the three flavors of neutrinos oscillate, as further confirmed by the Sudbury Neutrino Observatory. This implies that neutrinos have a nonzero rest mass [90.62]. Subsequently, upper bounds of the order of a few eV to the neutrino mass have been derived from measurements of tritium beta decay [90.63, 64] and from cosmological considerations [90.65].

90.2.3 Electronic Screening of Low Energy Nuclear Reactions

The cross section $\sigma(E)$ for a nuclear reaction involving charged reactants drops very rapidly for collision energies E below the Coulomb barrier. A WKB treatment shows that for low collision energies, the dependence of $\sigma(E)$ can be conveniently expressed as

$$\sigma(E) = S(E)\, E^{-1}\, \mathrm{e}^{-2\pi\eta}\,, \tag{90.18}$$

where $S(E)$ is the astrophysical factor, and

$$\eta = Z_1 Z_2 \alpha c\, (\mu/2E)^{1/2} \tag{90.19}$$

is the Sommerfeld parameter, which depends on the charge numbers Z_1 and Z_2 of the projectile and target nuclides, their reduced mass μ, and the cm energy E. For nuclear reactions involving light nuclei, it is found that $S(E)$ typically varies slowly with E except close to resonances. Thus the accurate determination of $S(E)$ at moderately low E can be used to extrapolate $\sigma(E)$ to much lower energies, which are beyond the reach of laboratory experiments but are of great relevance to the nuclear reactions that occur in stars.

However, electron screening effects can greatly enhance cross sections for nuclear reactions as measured in the laboratory at low energy [90.66], because at least the target nucleus is almost always surrounded by a cloud of electrons which screen the Coulomb repulsion between nuclei. The effect has been observed in various low-energy reactions such as $^3\text{He}(\text{d},\text{p})^4\text{He}$, $^6\text{Li}(\text{p},\alpha)^3\text{He}$, $^6\text{Li}(\text{d},\alpha)^4\text{He}$, and $^6\text{Li}(\text{p},\alpha)^4\text{He}$ [90.67–70]. Since reactions in stars involve bare nuclei, the laboratory data must be carefully corrected for screening effects.

Analysis of the data for the $^3\text{He}(\text{d,p})^4\text{He}$ reaction indicates that the effect of screening is always greater than that predicted in the adiabatic limit [90.71–73]. A more general theoretical treatment of the $\text{d}+{}^2\text{H}$ and $\text{d}+{}^3\text{He}$ reactions [90.74], using a time-dependent Hartree–Fock method for the electrons screening and classical motion for the nuclei found less enhancement than that observed. An improved treatment taking account of electron correlation and quantum-mechanical effects on the nuclear motion will likely be needed. This remains an important area of development for the future.

For some recent work on the subject of electronic screening of low-energy nuclear reactions, see [90.75–80].

90.2.4 Atomic and Molecular Effects in Relativistic Ion–Atom Collisions

High-energy accelerators can now produce beams of atomic ions partly or completely stripped of their electrons, even for Z as high as 92. The collisions of such beams of highly charged ions with fixed targets involve a broad array of atomic and molecular processes, such as excitation, ionization, charge transfer, and, in the extreme relativistic case, pair production. A similarly broad array of theoretical techniques is required to study these topics. A thorough review of them, including comparisons with experimental data where available, is given in [90.81, 82].

A topic of particular recent interest is the first experimental observation of the capture of electrons from electron-positron pair production in the extreme relativistic collision of a $0.96\,\text{GeV}$ / nucleon U^{92+} beam with gold, silver, copper, and Mylar targets [90.83]. The energy and angular distributions of the positrons were also measured. For the gold target, the cross section for capture was nearly as large as that for pair production without capture, and it was found to vary with the nuclear charge Z_t of the target nucleus roughly as $Z_\text{t}^{2.8(\pm 0.25)}$. Neither the dependence on Z_t nor the relatively great probability for capture is in agreement with perturbation theory, which highlights the need for further exploration of this exotic system.

90.3 Muon-Catalyzed Fusion

Exotic muonic atoms and molecules are more suitable subjects than electronic atoms and molecules for probing some physical effects. The muon μ^- is a leptonic elementary particle like the electron except that it is 206.768 times more massive and has a finite lifetime ($\tau_0 = 1/\lambda_0$, where λ_0 is the rate of decay) of $2.197\,\mu\text{s}$. This lifetime is amply long for most experiments. In normal atoms, the fine-structure splitting (due to $\boldsymbol{L} \cdot \boldsymbol{S}$ coupling) is much larger than the hyperfine splitting (due to $\boldsymbol{s}_{\text{nuc}} \cdot \boldsymbol{s}_{\text{e}}$ coupling); this relation is reversed in muonic atoms. Likewise, vacuum polarization, relativistic, finite-nuclear-size, and nonadiabatic effects are enhanced. (*Note*: the muonic Bohr radius $\hbar^2/m_\mu e^2 \approx 1/207\, a_0$ is similar in size to the Compton wavelength $\hbar/m_e c \approx 1/137\, a_0$.) Remarkably, muonic molecules make nuclear fusion possible at room temperature. In the phenomenon of muon-catalyzed fusion (μCF), there are both indirect and direct interactions between the atomic and molecular physics and the nuclear physics. Indirectly, the atomic and molecular densities and transition rates control the nuclear fusion rates, and, in turn, the kinetic energies of the fusion products affect the atomic and molecular kinetics. Directly, the nuclear structure affects some molecular energy levels that determine important resonant rates and the boundary condition on the muonic wave functions used to calculate the muon "sticking" loss.

In comparison with μCF, hot fusion schemes are made difficult by the electrostatic (Coulomb) repulsion between nuclei. In the two conventional approaches to controlled fusion, magnetic and inertial confinement, this barrier is partially surmounted by energetic collisions. (*Note*: the particle densities N and confinement times τ in the hot plasmas ($T \gtrsim 10^8$ K) are typically more than ten orders of magnitude different for these two schemes, but the product of the two required for d-t fusion is $N\tau \gtrsim 10^{14}\,\text{s/cm}^3$ in either case. For muon-catalyzed fusion, effectively $N\tau \approx 10^{25}\,\text{s/cm}^3$, but this criterion doesn't tell the real story.) On the other hand, in μCF the objective is to tunnel through the barrier without the benefit of kinetic energy. This feat is enabled by binding two hydrogenic nuclei (p, d, or t) in an exotic molecule like H_2^+ with the electron replaced by a negative muon.

Since the molecular size is inversely proportional to the mass of the binding particle, the average distance between nuclei in ppμ is $\approx 1/200$ Å (500 fm) instead of 1 Å as in ppe (i. e., H_2^+). This distance, which would be reached in a d + d collision at ≈ 3 keV ($\approx 3\times 10^7$ K), is still large compared with the separation of a few fm where the nuclear strong forces cause fusion, but fusion occurs rapidly because of the increased vibrational frequency and, more important, the increased probability of tunneling per vibration. The vibrational frequency is $(m_\mu/m_e)^{3/2} \approx 3\times 10^3$ times faster than for the corresponding electronic molecule. (*Note*: for comparison, the muonic/electronic energy scales as m_μ/m_e and the rotational energy scales as $(m_\mu/m_e)^2$. These relations [90.84] are based on the Born–Oppenheimer approximation, which is not very accurate for muonic molecules.) The effect on the tunneling probability depends on the nuclear masses; for dtμ, which has the largest nuclear matrix element (astrophysical S factor, Sect. 90.2.3), the increase is by a factor of $\approx 10^{77}$ compared with DT, and the consequential fusion rate is $\lambda^{\text{f}}_{\text{dt}\mu} \approx 10^{12}\,\text{s}^{-1}$.

Just on the basis of the fusion rate, one would expect a yield of $\lambda^{\text{f}}_{\text{dt}\mu}/\lambda_0 \approx 10^6$ muon-catalyzed d-t fusions for the average muon. While this number indeed provides an upper limit, the actual average number of fusions, ≈ 150 for dtμ, is much smaller and is determined by the atomic and molecular physics of the catalysis cycle (though the energy released in the nuclear fusion does play an important role here). Some of the atomic and molecular processes in the μCF cycle are quite ordinary, but others, like atomic capture and resonant molecular formation, have no counterpart with "normal" atoms.

Muon-catalyzed fusions of all pairs of hydrogen isotopes, except two protons, have been observed. Based on the experiments and theory, the reaction products are:

$$\text{pd}\mu \to \begin{cases} {}^3\text{He}(0.005) + \mu + \gamma(5.49) & (92\% - 89\%) \\ {}^3\text{He}(0.20) + \mu(5.29) & (8\% - 11\%) \end{cases} \tag{90.20}$$

$$\text{dd}\mu \to \begin{cases} \text{t}(1.01) + \text{p}(3.02) + \mu & (41\% - 53\%) \\ {}^3\text{He}(0.82) + \mu + \text{n}(2.45) & (59\% - 47\%) \end{cases} \tag{90.21}$$

$$\text{pt}\mu \to \begin{cases} {}^4\text{He}(0.05) + \mu + \gamma(19.76) & (60\% - 76\%) \\ {}^4\text{He}(0.59) + \mu(19.22) & (23\% - 14\%) \\ {}^4\text{He} + \mu + \text{e}^+ + \text{e}^-(19.81\text{total}) & \\ & (17\% - 10\%) \end{cases} \tag{90.22}$$

(*Note*: the last reaction of (90.22) has been theoretically predicted [90.85], but has not yet been observed.)

$$\text{dt}\mu \to {}^4\text{He}(3.54) + \mu + \text{n}(14.05) \tag{90.23}$$

$$\text{tt}\mu \to {}^4\text{He} + \mu + 2\text{n}(11.33\text{total})\ . \tag{90.24}$$

Here the product particle kinetic energies (in MeV) are given in parentheses. A μ without an energy designated is a spectator, i. e. serves to bring the nuclei together but plays no significant role in the kinematics of the reaction – such a μ may actually be bound (stuck) to one of the product nuclei. As indicated, the branching fractions

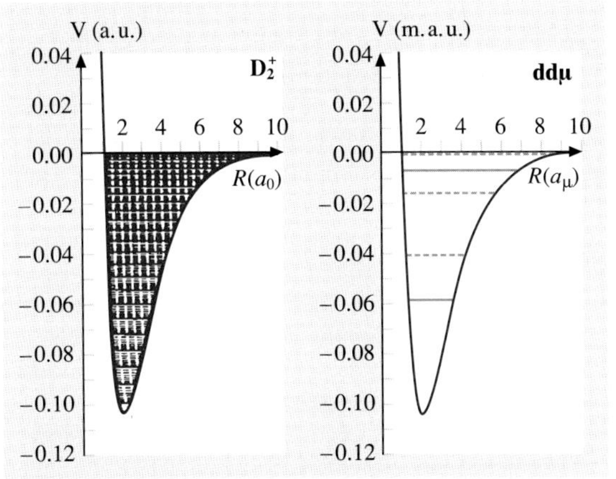

Fig. 90.1 Rovibrational energy levels for D_2^+ and ddμ. The $J = 0$ levels are shown as *solid lines* and the $J > 0$ levels are shown *dashed*. For D_2^+, all 28 vibrational levels are displayed, but the associated rotational levels are displayed only up to the next higher vibrational level. All levels of ddμ are displayed; the ($J = 1$, $v = 1$) level is barely discernible below the $V = 0$ axis

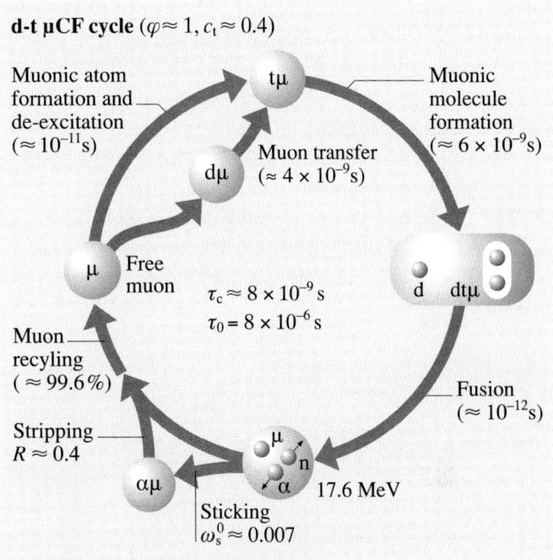

Fig. 90.2 The simplified d-t muon-catalyzed fusion cycle. The times are for density $\phi = 1$ (liquid hydrogen density) and tritium fraction $c_t = 0.4$. τ_c is the cycle time, and τ_0 is the muon-decay time

depend somewhat on the target parameters (isotopic composition, density, and temperature) [90.86–88].

Each reaction is of special interest in its own right: pdμ and ptμ for the contribution of μ conversion, and ttμ for the correlation of the two final state neutrons. Only the ddμ and dtμ molecular formations are resonant; i. e., their formation can occur in a one-body state because they, and only they, possess a loosely bound state such that the muonic binding energy can go into rovibrational energy of the electronic molecule. That the existence of such a state really is fortuitous can be seen in Fig. 90.1 where the bound rovibrational states of ddμ (1 m.a.u. = 5626.5 eV) are compared with the rovibrational states of D_2 (1 a.u. = 27.2 eV). Though both ddμ and dtμ can be formed resonantly, dtμ is unique in having a rapid (as compared with muon decay) formation rate and also in having a small sticking loss. The sticking loss is due to the possibility that the negatively charged muon may form a bound state with the positively charged fusion product. The relatively low branching fraction (< 1%) for $dt\mu \to {}^4He\mu + n$ is due simply to the high speed of the outgoing ^4He (α particle).

90.3.1 The Catalysis Cycle

A diagram of the μCF cycle for a d-t mixture is shown in Fig. 90.2. The basic steps in the cycle are

1. Atomic capture to form dμ or tμ (initially in a highly excited state, $n \gtrsim 14$).
2. Transfer of the μ from d to t, if necessary.
3. Resonant molecular formation, shown schematically. Here the dtμ is so small (in reality) that is can be considered to be a pseudo-nucleus in the electronic molecule.
4. Nuclear fusion.
5. Sticking (αμ formation) or recycling.

The reaction times shown are at liquid hydrogen density ($\phi = 1$ in the conventional LHD units) and a tritium fraction $c_t = 0.4$, which is close to the value that maximizes the number of cycles. The times for muonic atom formation and deexcitation, ≈ ps, are short compared with the times for muon transfer and molecular formation, ≈ ns, which in turn are short compared with the muon decay time ≈ μs.

Thus the time for a cycle is mainly given by the average time the μ spends as dμ waiting to transfer to t plus the average time it then spends as tμ waiting to form dtμ,

$$\tau_c \approx \tau_{d\mu} + \tau_{t\mu} , \tag{90.25}$$

or, in terms of rates,

$$\frac{1}{\lambda_c} \approx \frac{q_{1s} c_d}{\lambda_{dt} c_t} + \frac{1}{\lambda_{dt\mu} c_d} , \tag{90.26}$$

where c_d and c_t are the fractions of deuterium and tritium ($c_d + c_t = 1$), λ_{dt} is the d-to-t transfer rate in the 1s state, q_{1s} is the fraction of dμ atoms reaching the 1s state (before transfer), and $\lambda_{dt\mu}$ is the molecular formation rate. The factor q_{1s} takes into account the fact that any transfer in excited states is rapid (of necessity, since it must compete with the rapid deexcitation).

The cycle rate λ_c along with the sticking fraction ω_s constitute the two basic parameters of the catalysis cycle. The average number of fusions per muon Y is given by

$$\frac{1}{Y} \approx \frac{\lambda_0}{\lambda_c} + \omega_s \qquad (90.27)$$

where $\lambda_0 (\equiv 1/\tau_0)$ is the muon-decay rate. (*Note*: more precisely, W should appear in (90.27) in place of ω_s. W may include other losses, e.g., muon capture by impurities, but we will restrict the present discussion to ω_s, which is fundamental and normally dominant.) Coincidentally, the limits imposed on the yield by the cycling rate and by sticking are similar; for $\phi \approx 1$ and $T \approx 300$ K, $\lambda_c/\lambda_0 \approx 300$ and $1/\omega_s \approx 200$ for dtμ. More than 100 muon-catalyzed d-t fusions per muon have been observed. Similar considerations apply to the ddμ cycle, and, by further coincidence, the two limits are similar there as well, $\lambda_c^{dd\mu}/\lambda_0 \approx 7$ and $1/\omega_s^{\mathrm{eff}(dd\mu)} \approx 14$. (*Note*: the effective sticking probability (per cycle) in the ddμ cycle takes into account that in only 58% of the fusion reactions is a ^{3}He produced that can remove the μ by sticking. Sticking to t or p is possible but would facilitate rather than terminate the cycling.) The four experimental "knobs" are the temperature (T), density (ϕ), isotopic fractions (c_t, c_d, and c_p), as well as the molecular fractions (c_{D_2}, c_{DT}, c_{T_2}, $\cdots$) in the case of a target not in chemical equilibrium.

Each stage of the cycle is discussed in the following sections. The reader is referred to reviews [90.84, 89–92] for details of the theoretical and experimental methods and extensive values of the relevant parameters.

90.3.2 Muon Atomic Capture

The μCF process starts with a free muon, injected into a mixture of hydrogen isotopes, being stopped to form a muonic atom (stopping power is discussed in Sect. 91.1.1). The slowing and capture occur primarily by ionization, e.g.,

$$\mu + D \to \begin{cases} \mu + d + e \\ d\mu(n) + e\,. \end{cases} \qquad (90.28)$$

The muon is captured into an orbital with $n \gtrsim \sqrt{m_\mu/m_e} \approx 14$, which has about the same size and energy as that of the displaced electron.

Methods for hydrogen and helium atoms have been reviewed in [90.93]; the brief discussion here emphasizes the correct intuitive understanding. Until 1977 most calculations were done using the Born or Coulomb–Born approximation [90.94]. These methods are not very accurate for μ^- at velocities below 1 a.u., but, more importantly, their implementation treated slowing down and capture inconsistently. The upshot was prediction of capture of muons typically with kinetic energies of hundreds of eV, whereas it turns out that most captures actually occur at energies below 100 eV.

The perturbative methods fail because of the great electron charge redistribution that occurs during the capture process. Six other approaches have led to accurate treatment:

1. Adiabatic ionization with straight-line trajectories (AI-slt) [90.95],
2. Adiabatic ionization with curved trajectories (AI-ct) [90.93, 96],
3. Diabatic states (DS) [90.97, 98],
4. Classical-trajectory Monte Carlo (CTMC) [90.96],
5. Time-dependent Hartree–Fock (TDHF) [90.99],
6. Classical-quantal coupling (CQC) [90.100].

The first three are models tailored for the muon capture problem (the first two of these specialized for the hydrogen-atom target). The results for all six methods are shown in Fig. 90.3.

The early study by *Wightman* [90.95] shed a great deal of light on the capture process. His method, known as adiabatic ionization (AI), followed on the observation of *Fermi* and *Teller* [90.101] that there exists a critical strength of the dipole eR_c, formed by the negative muon and positive proton at distance $R_c = 0.639\,a_0$, for binding the electron. In collisions where the μ^- approaches closer than this distance, the electron is assumed to escape adiabatically, and, if the electron carries off more energy than the muon's initial kinetic energy, the pμ atom is formed. This cross section is thus

$$\sigma_{\mathrm{AI-slt}} = \pi R_c^2\,, \qquad (90.29)$$

and μ^- capture results if and only if $E < 0.5$ a.u., the target ionization energy.

The AI-slt model has three major shortcomings: (1) it does not take into account trajectory curvature, which is caused by the Coulomb attraction of μ^- toward the nucleus and can be large at the low trajectory velocities where capture usually occurs, (2) the adiabatically escaping electron takes off no kinetic energy, and (3) ionization occurs with unit probability if the approach is closer then R_c. The first failing is easy to remedy. The cross section with curved adiabatic trajectories (AI-ct) is just [90.96]

$$\sigma_{\text{AI-ct}} = \frac{\pi R_c^2}{E}\left(E + \frac{1}{R_c} - 0.5\,\text{a.u.}\right) \qquad (90.30)$$

as long as the collision energy E in the center-of-mass system is greater than 0.03 a.u. (0.8 eV). Hence trajectory curvature increases the capture cross section by a factor $1 + \frac{1.06}{E}$ (for E in a.u.), which is over a factor of three even at the highest collision energy (0.5 a.u.) where adiabatic capture can occur. For $E < 0.03$ a.u., the centrifugal barrier in the effective potential,

$$V_{a(\text{eff})}(R, b) = V_a(R) + \frac{b^2}{R^2}E\,, \qquad (90.31)$$

restricts penetration and reduces the cross section below the value given by (90.30) [90.93].

Cures for the second and third failings are less trivial. These two assumptions can be avoided by using the diabatic-states (DS) model [90.97, 98]. The adiabatic electronic energy no longer increases once it reaches the continuum ceiling; however, in view of the μ^- acceleration by the Coulomb attraction, the electron cloud actually does not have enough time to adjust adiabatically. In recognition of this situation the diabatic electronic energy crosses into the continuum at a distance larger than R_c and continues to rise smoothly. The concomitant probability of ionization is given by the ionization width, obtained by a Fermi-golden-rule-like formula. The first three approaches are somewhat specialized models, while the next three are general methods. The most economical in terms of computer time is the classical-trajectory Monte Carlo (CTMC) method [90.96] discussed in Chapt. 58. This method treats the dynamics of all particles completely classically. The time-dependent Hartree–Fock (TDHF) method discussed in Sect. 50.2 is purely quantum mechanical [90.99], but neglects correlation, which turns out to be important in the present problem. This deficiency is remedied by the classical-quantal coupling (CQC) method, which makes only the seemingly well-justified approximation of treating the muon classically while retaining the quantum treatment of the electron [90.100].

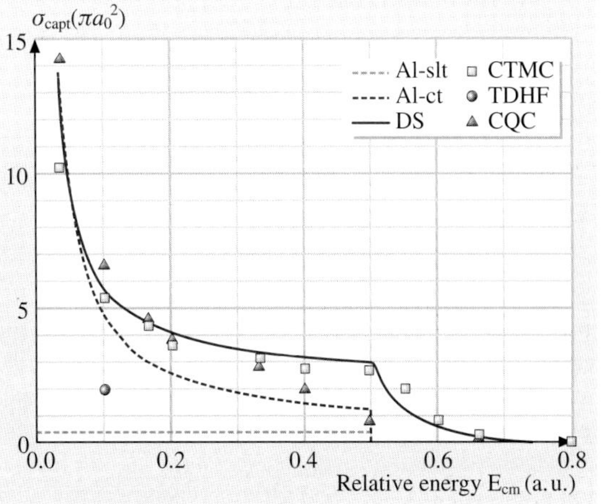

Fig. 90.3 Comparison of different capture cross sections for μ^-+H collisions: adiabatic ionization with straight-line trajectories(AI-slt), adiabatic ionization with curved trajectories (AI-ct), diabatic states with polarized orbital (DS), classical-trajectory Monte Carlo (CTMC), time-dependent Hartree-Fock (TDHF), and classical-quantal coupling (CQC)

Real μ^- capture experiments are done with molecules (H_2, DT, etc.). The naive notion that the H_2 cross section is simply twice that of H is quite unrealistic for slow ($v \ll 1$ a.u.) collisions. In the past decade there has been a major advance in the understanding of μ^- capture by hydrogen, which is the first step in μCF. The captures by the H atom and the H_2 molecule, previously thought to differ by less than a factor of two, have been shown theoretically to be quite different [90.102]. For a comprehensive review of all methods used to threat capture of negative particles see the review article [90.103]. This difference is primarily due to the vibrational degree of freedom, which enables the molecule to capture μ^- at collision energies up to ≈ 40 eV with $n \gtrsim 9$, whereas atomic capture cuts off above ≈ 14 eV with $n \gtrsim 14$. There is a corresponding isotope effect in the molecule, which is absent in the atom. Experiments may be conducted in the near future on the analogous capture of antiprotons by H and H_2 [90.104].

90.3.3 Muonic Atom Deexcitation and Transfer

The muon is captured in a highly excited state but normally must reach the 1s configuration of the heavier isotope (in case of mixtures like D/T) before the muonic molecule is formed. In the 1s configuration there are two hyperfine levels – the ground state with the nuclear and μ^- spins antiparallel and an excited state with spins parallel. Resonant molecular formation rates in the two states can be quite different and also depend strongly on the atom's kinetic energy. Thus there are several types of muonic atom collisions that must be taken into account: (1) elastic scattering in the ground and excited states, (2) isotopic transfer in excited states, (3) deexciting transitions (which may also occur radiatively), (4) isotopic transfer in the 1s state, and (5) hyperfine transitions. Cross sections for most of these processes have been calculated. The bulk of the calculations have been done by expanding in adiabatic (or modified adiabatic) eigenfunctions, but there also exist some calculations using the coupled-rearrangement-channel, Faddeev, and hyperspherical approaches (see [90.84] for references).

The cascade of the initially formed muonic atom, especially in mixtures, is a complicated process not yet completely characterized. It constitutes a crucial part of the $d-t$ μCF cycle in that it determines the parameter q_{1s} in (90.26). This parameter is essential to experimental analysis, but it was evident that early calculations yielded values of q_{1s} too small to be consistent with experiments. Recent calculations [90.105] suggest the explanation is that the excited muonic atoms are not thermalized. Epithermal atoms have three effects here: (1) the normal transfer rates are smaller, (2) the transfer is reversible down to lower principal quantum numbers n where E still exceeds the threshold for excitation of the next-higher level, and (3) excited-state [from $(t\mu)^*$] resonant formation of $(dt\mu)^*$ molecules that can predissociate back to dμ is enhanced [90.106]. (*Note*: the isotopic energy splittings are $134.7/n^2$ for dμ-pμ, $182.8/n^2$ for tμ-pμ, and $48.0/n^2$ for tμ-dμ.) The q_{1s} is determined by competition between transfer and deexcitation, which depend on the kinetic energies that result from further competition between superelastic deexcitation and thermalizing elastic collisions. It appears that the stage of the cascade most crucial to q_{1s} is $n \approx 4$ for normal muon transfer and $n = 2$ for the resonant sidepath.

For muons, the elastic cross sections are more difficult to calculate than the inelastic ones. The inelastic transitions occur at short range (a few a_μ) where the effects of electronic structure are negligible. However, electronic effects are not negligible for low energy (< 1 eV) elastic scattering where $\lambda_{dB} \approx 1\, a_0$. They have been taken into account for ground state but not yet excited state scattering. In doing so, it is not necessary to solve the general problem directly because of the following simplifications: (1) this energy is below the vibrational threshold so the molecular target can be taken as a rigid rotor and (2) the relative smallness of the muonic atoms makes the sudden approximation adequate.

If the 1s state is reached without muon transfer to the heavier isotope already having occurred, the transfer takes significant time and plays an important role in determining the tritium fraction c_t that optimizes the fusion yield. All of the 1s isotopic-exchange cross sections display the characteristic $\approx 1/v$ velocity dependence at thermal energies so that the corresponding rate $v\sigma$ is independent of temperature.

In muon-catalyzed d–d and d–t fusions, the resonant molecular formation rates in different hyperfine structure (hfs) states can differ by two or more orders of magnitude at low T due to their different energy levels. The hfs also has important effects on thermalization and diffusion via the different elastic cross sections. Under usual μCF experimental conditions, the hyperfine quenching (or "spin flip") is irreversible; the hfs splittings are 0.1820, 0.0485, and 0.2373 eV for pμ, dμ, and tμ, respectively. Theoretically, it is expected that transitions between hfs levels mainly occur in symmetric collisions since muon exchange suffices in such collisions [90.107]; e.g.,

$$t\mu(\uparrow\uparrow) + t(\downarrow) \rightarrow t(\uparrow) + t\mu(\downarrow\uparrow) \qquad (90.32)$$

(the usual terminology here is "muon exchange" although it might seem more logical to refer to the reaction as "triton exchange" since it is the identity of the tritons that enables the reaction). As in the case of the isotopic exchange cross sections, the behavior is $\approx 1/v$ at thermal energies, so the rates are nearly independent of temperature.

90.3.4 Muonic Molecule Formation

Until the prediction by *Vesman* [90.108] of a resonant formation for ddμ, it was thought that all muonic molecules were formed by an Auger process of the type

$$d\mu + D_2 \rightarrow [(dd\mu)de]^+ + e^- \,. \qquad (90.33)$$

Unlike the resonant process for ddμ and dtμ, the nonresonant process generally depends weakly on the temperature of the target, the hyperfine state of the muonic atom, and the "spectator" atom X in the molecule DX, where X can be H, D, or T. The nonresonant rate at low (liquid hydrogen) temperature for ddμ formation is about $3\times10^4\ s^{-1}$ and for dtμ is about $6\times10^5\ s^{-1}$. These rates are competitive with the resonant rates at low T for $d\mu(\uparrow\downarrow)+D_2$ and $t\mu(\uparrow\uparrow)+D_2$, but are 2 to 3 orders of magnitude smaller than the resonant rates for $d\mu(\uparrow\uparrow)+D_2$ and $t\mu(\uparrow\downarrow)+D_2$, respectively. The $(\uparrow\downarrow)$ state is the ground state; thus hfs quenching plays an important role in low-temperature experiments, especially for ddμ. At room temperature, resonant formation is dominant for both the ground and excited hyperfine states of dμ and tμ.

In the Vesman mechanism the binding energy of the muonic molecule goes into rovibrational excitation of the electronic host molecule instead of into ionization of a molecular electron. The process is resonant since the collision energy must be tuned to match the energy of the final discrete state. For the compound molecule formed, two sets of rovibrational quantum numbers are needed, e.g.,

$$(\mathrm{t}\mu)_F + [\mathrm{D}_2]_{K_i \nu_i} \rightarrow \left[(\mathrm{d}\mathrm{t}\mu)^{FS}_{J\upsilon}\,\mathrm{d}\,\mathrm{e}\,\mathrm{e}\right]_{K_f \nu_f}, \tag{90.34}$$

where (K_i, ν_i), and (K_f, ν_f) are the initial and final quantum numbers of the electronic molecule, (J, υ) are the quantum numbers of the muonic molecule, F is the spin of the muonic atom, and S is the total spin of the muonic molecule. The energetics of this process is shown in Fig. 90.4.

The resonant condition is achieved at the collision energy

$$\begin{aligned}\epsilon_{\mathrm{res}}\,(\mathrm{t}\mu + \mathrm{D}_2) = {} & \epsilon^{FS}_{11}[\mathrm{d}\mathrm{t}\mu] \\ & + E_{K_f \nu_f}[(\mathrm{d}\mathrm{t}\mu)\,\mathrm{d}\,\mathrm{e}\,\mathrm{e}] - E_{K_i 0}[\mathrm{D}_2]\,,\end{aligned} \tag{90.35}$$

where it is explicitly recognized that $(J, \upsilon) = (1, 1)$ is the only muonic level that can satisfy the resonant energy condition and that only $\nu_i = 0$ is populated at ordinary temperatures. Accurate calculations require values of ϵ_{res} to within about 0.1 meV. The rovibrational energies $E_{K\nu}$ of the electronic molecule, as well as the Coulomb contributions to the binding energy of the muonic molecule, ϵ^{FS}_{11}, are now known to this high accuracy. However, ϵ^{FS}_{11} is subject to corrections due to relativity, vacuum polarization, nuclear charge distributions and polarizabilities, the hyperfine interaction, and

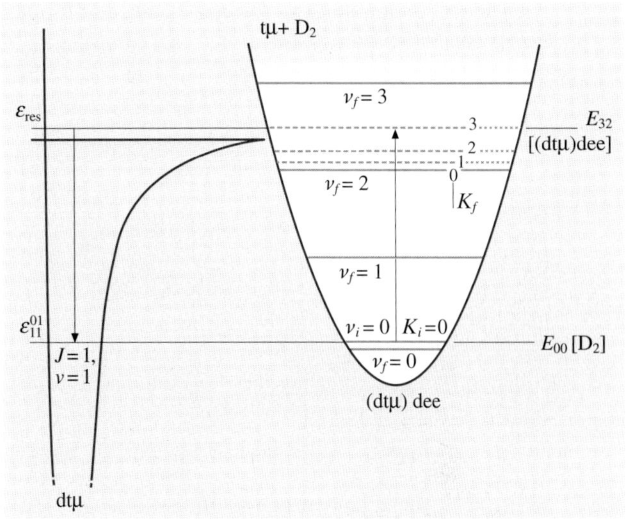

Fig. 90.4 Diagram of energy levels for the resonant reaction $\mathrm{t}\mu + [\mathrm{D}_2]_{K_i=0,\nu_i=0} \rightarrow [(\mathrm{d}\mathrm{t}\mu)^{F=0,S=1}_{J=1,\upsilon=1}\,\mathrm{d}\,\mathrm{e}\,\mathrm{e}]_{K_f\nu_f}$. The rovibrational quantum numbers are designated by (J, υ) for the muonic molecule and (K, ν) for the electronic molecules

the finite size and shape of the muonic molecule in the complex. The present overall accuracy is ≈ 1 meV. Some of the resulting values of ϵ_{res} are given in Table 90.1. The calculated cross section for reaction (90.34) is sharply peaked at E_{res}, but must be averaged over a kinetic energy distribution (e.g. Maxwellian) to obtain the observable rate. Still, the rate will display a characteristic resonant dependence on T.

Because the E_{res} are different for each target molecule (D_2, DT, T_2, ···), the effective molecular formation rate in a mixture depends on the molecular composition in addition to the isotopic fractions (c_{d}, c_{t}, ···) if the target is not in chemical equilibrium.

Table 90.1 Resonant (quasiresonant if negative) collision energies ϵ_{res} (in meV) calculated using (90.35)[a]

$\mathrm{d}\mu + \mathrm{D}_2$ $[(\mathrm{dd}\mu)_{11}\mathrm{dee}]_{\nu_f=7}$		$\mathrm{t}\mu + \mathrm{D}_2$ $[(\mathrm{dt}\mu)_{11}\mathrm{dee}]_{\nu_f=2}$		$[(\mathrm{dt}\mu)_{11}\mathrm{dee}]_{\nu_f=3}$		$\mathrm{t}\mu + \mathrm{DT}$ $[(\mathrm{dt}\mu)_{11}\mathrm{tee}]_{\nu_f=3}$	
F, S, K_i, K_f	ϵ_{res}	F, S, K_i, K_f	ϵ_{res}	F, S, K_i, K_f	ϵ_{res}	F, S, K_i, K_f	ϵ_{res}
$\frac{1}{2}, \frac{1}{2}$, 0, 1	52.7	0, 1, 0, 1	−14.0	0, 1, 0, 1	277.1	0, 1, 0, 1	163.8
$\frac{1}{2}, \frac{3}{2}$, 0, 1	76.9	0, 1, 0, 2	−4.3	1, 0, 0, 1	223.5	1, 0, 0, 1	110.2
$\frac{3}{2}, \frac{1}{2}$, 0, 1	4.2	0, 1, 0, 3	10.3	1, 1, 0, 1	226.9	1, 1, 0, 1	113.6
$\frac{3}{2}, \frac{3}{2}$, 0, 1	28.4	0, 1, 1, 2	−11.7	1, 2, 0, 1	233.3	1, 2, 0, 1	120.0
$\frac{1}{2}, \frac{1}{2}$, 1, 0	40.9	0, 1, 1, 3	2.9				
$\frac{1}{2}, \frac{3}{2}$, 1, 0	65.1						
$\frac{1}{2}, \frac{1}{2}$, 1, 2	54.1						
$\frac{1}{2}, \frac{3}{2}$, 1, 2	78.3						

[a] Note: $kT = 1$ meV for $T = 11.6$ K

The rate of ddμ and dtμ resonant molecular formation is calculated from

$$\lambda^{mf}(T) = N \sum_f \int \Big\{ \mathrm{d}\epsilon\, 2\pi\, |\langle i|H'|f\rangle|^2 \times f(\epsilon, T)\, I(\epsilon - \epsilon_{if}, T) \Big\} , \tag{90.36}$$

where N is the target density, $\langle i|H'|f\rangle$ is a transition matrix element, $f(\epsilon, T)$ is the collisional energy distribution, ϵ_{if} is the energy of the unperturbed resonance, and $I(\Delta\epsilon, T)$ is the intensity at energy $\Delta\epsilon$ relative to the unperturbed energy. H' has usually been taken to be the dipole interaction ("post" form of the rearrangement-collision Hamiltonian using the dtμ bound state as the zeroth-order Hamiltonian [90.109]),

$$H' = e^2 \boldsymbol{d} \cdot \boldsymbol{E} , \tag{90.37}$$

where $\boldsymbol{d}$ is the dipole operator of the dtμ (or ddμ) system and $\boldsymbol{E}$ is the electric field at the dtμ (or ddμ) center of mass due to the "spectator" nucleus and electrons [90.110]. Conservation of angular momentum thus requires [90.111]

$$\boldsymbol{L} + \boldsymbol{K}_i = \boldsymbol{J} + \boldsymbol{K}_f \tag{90.38}$$

where $\boldsymbol{L}$ is the orbital angular momentum of relative motion for $t\mu + D_2$ in reaction (90.34). At low T, $L = 0$ is predominant so that $K_f = K_i \pm 1$. This is simply the case for $d\mu + D_2 \rightarrow (dd\mu)dee$ where the most probable transition is $(K, \nu) = (0, 0) \rightarrow (1, 7)$. For dtμ the vibrational state of the electronic molecule changes by only $\Delta\nu_i = 2$ or 3 instead of 7, so the matrix element of (90.37) and the resulting rate are considerably larger than for ddμ. However, it can be seen in Fig. 90.4 that if D_2 is in its ground state ($K_i = 0$), the first level energetically accessible for $(dt\mu)dee$ has $K_f = 3$. If, as proves to be adequate in the case of ddμ, the intensity distribution I is taken to be a δ function, the lower levels are eliminated from (90.36). There are two possible solutions to this problem, whose importances have not been fully resolved: (1) the less likely $L > 0$ collisions contribute or (2) the levels with smaller K_f play a role even though they lie "below threshold".

The latter case is termed "quasiresonant". Theoretically the levels below threshold can contribute (1) directly if they are broadened so that they extend to positive energy [90.112–114] or (2) indirectly if configurations with different K_f are mixed [90.115]. Broadening can occur either inhomogeneously due to the finite lifetime (mainly with respect to Auger emission of an electron in the complex) or homogeneously due to collisions with neighboring molecules. Interactions with neighboring molecules also can mix the different K_f states, so the $K_f = 3$ state may "borrow" some intensity from the lower K_f states. Three-body molecular formation facilitated by neighboring molecules leads to a density dependence of the formation rate (normalized to LHD) that has been observed in experiments.

Experimentally, the resonant $dt\mu$ formation has now been observed directly [90.116]. Previously the experimental evidence for this mechanism derived from the magnitude and the temperature dependence of the μCF cycling rate. The new experiment, at TRIUMF, obtained the energy-dependent molecular-formation rate by measuring the time of flight between a cryogenic layer where the $t\mu$ atom was formed and a second cryogenic layer where the $dt\mu$ molecule was formed and fusion occurred.

90.3.5 Fusion

Usually the nuclear fusion rate in muonic molecules is calculated by a separable method; i. e., the united-atom limit of the molecular wave function is determined ignoring nuclear forces and then simply multiplied by a single number extracted from nuclear scattering experiments [90.117]. Fusion of $d - t$ is strongly dominated by the $I^\pi = \frac{3}{2}^+$ resonance of ^{5}He. For dtμ we have

$$\lambda^{f}_{dt\mu} = A \lim_{r_{dt} \rightarrow 0} \int |\psi_{dt\mu}|^2 \, \mathrm{d}^3 r_\mu , \tag{90.39}$$

where A is simply related to the low-energy limit of the astrophysical S factor by

$$A = \frac{\hbar}{\pi e^2 M_r} \lim_{E \rightarrow 0} S(E) , \tag{90.40}$$

where M_r is the reduced mass of the nuclei. $S(E)$ is usually obtained by fitting the $d + t$ fusion cross section observed in beam experiments to the form of (90.18).

The above formulation has yielded fusion rates in good agreement with another formulation – more accurate in principle – where a complex molecular wave function and energy are obtained directly incorporating the nuclear forces. In the latter approach,

$$\lambda^{f}_{dt\mu} = -2\, \mathrm{Im}(E_{dt\mu})/\hbar , \tag{90.41}$$

where $E_{dt\mu}$ is the complex eigenvalue (the imaginary part is $-\Gamma/2$, where Γ is the width). The nuclear effect in this formulation is taken into account by two different techniques: (1) using a complex optical potential [90.118, 119] (Sect. 48.2), and (2) using the nuclear R-matrix as an interior boundary condition [90.120–122] (Sect. 47.1.5). In the optical potential method, a short-range complex potential, determined by fitting experimental nuclear scattering data, is added to the three-body Coulomb potential; the real part describes elastic scattering and the imaginary (absorptive) part describes the fusion reaction. Then the eigenvalue problem is solved over all space with a regular boundary condition at the origin. In the R-matrix method, the same nuclear scattering data are used to determine a complex boundary condition at a distance characteristic of the nuclear forces, say $r_{dt} = a_{dt}$ where $a_{dt} \approx 5$ fm. The muonic eigenvalue problem is then formulated with the boundary condition at $r_{dt} = a_{dt}$ and solved over the space excluding $r_{dt} < a_{dt}$. The two methods can be used with similar basis sets to expand the wave function, which can also be used to calculate the sticking probability (Sect. 90.3.6).

The relation of other μCF cross sections to normal beam experiments is somewhat more complicated. In the cases of ddμ and ttμ, the fusion may occur in $J=1$ states, since the $J=1$ to $J=0$ transition is forbidden in molecules with identical nuclei. In this case the relevant information from beam experiments resides in the p-wave anisotropy, which is relatively small at low energies where σ is dominated by the s wave. It is then necessary to carry out the analysis in terms of the partial-wave transition amplitudes rather than the fit of the integrated σ via the S factor [90.123].

Fusions in pdμ and ptμ present a different complication [90.117]. There is a significant $E0$ contribution from muon conversion in addition to the $M1$ γ-ray contribution seen in p+d and p+t beam experiments (see Sect. 12.1 for discussion of multipole moments). The $E0$ contribution cannot be expressed through cross sections observed in the beam experiments, but has been determined using the bound-state nuclear wave functions of ^{3}He (or ^{4}He) and scattering wave functions of p+d (or p+t). For pdμ there is the additional complication that two different p−d spin states contribute significantly. For ptμ, theory [90.117] predicts that the probability of the fusion energy going into a e^+e^- pair is competitive with that of muon conversion, though the former has not yet been observed.

90.3.6 Sticking and Stripping

The fundamental mechanism of muon loss from the catalysis cycle, other than by particle decay, is via sticking to a helium nucleus, ^{4}He($\equiv \alpha$) or ^{3}He, produced in the fusion reaction. Especially in the case of dtμ, where the charged particle is fast and the sticking probability is already small, subsequent collisions may strip the muon. Thus the sticking probability is determined by two steps

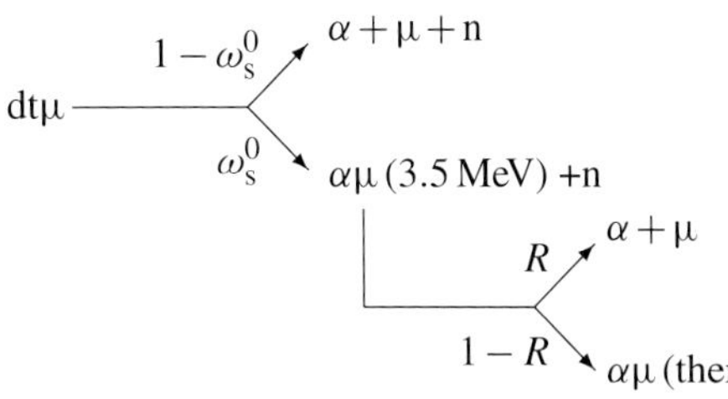

$$\text{dt}\mu \begin{array}{c} \xrightarrow{1-\omega_s^0} \alpha+\mu+n \\ \searrow_{\omega_s^0} \alpha\mu\,(3.5\,\text{MeV})+n \\ \Big|_R \xrightarrow{} \alpha+\mu \\ \searrow_{1-R} \alpha\mu\,(\text{thermal}). \end{array} \quad (90.42)$$

The initial sticking probability ω_s^0 depends only on intramolecular dynamics, but the stripping conditional probability depends on collisions. (Note that ω_s^0 is not the sticking in the zero-density limit since R is still finite in this limit.) The net sticking is then

$$\omega_s = \omega_s^0(1-R). \quad (90.43)$$

Since the nuclear reaction is very rapid compared with the atomic and molecular dynamics, the probability of sticking in a given state ν is given adequately by the sudden approximation,

$$P_\nu = \left|\left\langle \psi_\nu^{(f)} | \psi^{(i)} \right\rangle\right|^2, \quad (90.44)$$

where the initial wave function $\psi^{(i)}$ is the normalized molecular wave function in the limit $r_{dt} \to 0$ and the final wave function $\psi_\nu^{(f)}$ is given by

$$\psi_\nu^{(f)} = \phi_{n\ell m}(\mathbf{r})\,e^{i\mathbf{q}\cdot\mathbf{r}} \quad (90.45)$$

in which $\phi_{n\ell m}$ is an atomic wave function of $(^4\text{He}\mu)^+$ and the plane wave with momentum $\mathbf{q}$ represents its motion with respect to the initial molecule (recoil determined by conservation of energy and momentum). The total sticking is then

$$\omega_s^0 = \sum_\nu P_\nu. \quad (90.46)$$

The $(^4\text{He}\mu)^+$ wave function is known analytically since it is hydrogenic. Most of the labor goes into determination of the muonic molecule wave function. In the Born–Oppenheimer approximation this is simply "$(^5\text{He}\mu)^+$" and results in $(\omega_s^0)_{BO} = 1.20\%$ for dtμ [90.124]. More accurate nonadiabatic calculations show that the muonic motion lags behind that of the nuclei and reduces ω_s^0 to 0.886% [90.125, 126]. After inclusion of nuclear effects, the best current theoretical value of ω_s^0 is 0.912% [90.122, 127].

Since the ground state $(\text{He}\mu)^+$ ion is bound by 11 keV, it takes a quite energetic collision to strip off the muon. The reactivation fraction R is determined basically by competition between collisional processes that slow down the muonic ion and those that lead to stripping. Calculation of R requires a full kinetic treatment of the fast $(\text{He}\mu)^+$ ion, starting with its distribution among various states (1s, 2s, 2p, $\cdots$, $n \approx 10$). The most important processes are stopping power (due mainly to ionization of the medium) and muon ionization or transfer in collisions of the $(\text{He}\mu)^+$ ion with an isotope of H, but inelastic (excitation and deexcitation), Auger deexcitation, and ℓ-changing collisions as well as radiative deexcitation are also involved. The initial sticking occurs mostly in the 1s state (77% of the $\alpha\mu$'s from dtμ fusion), but the excited states have larger ionization cross sections. Most of the muons stripped from $\alpha\mu$ originally stuck in the 1s state, but a significant number are promoted to excited states before being ionized (so-called "ladder ionization"). The metastable 2s state is significant for its role in prolonging the excited state populations. The resulting values of R and ω_s are shown as a function of density in Fig. 90.5.

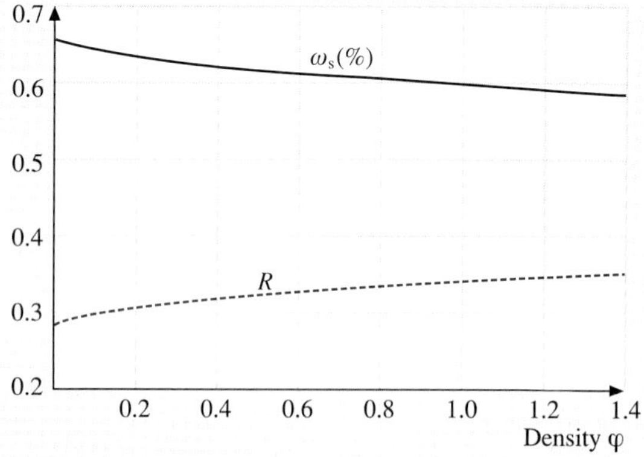

Fig. 90.5 Theoretical sticking fraction *(solid curve)* and reactivation probability *(dashed curve)* for d-t μCF

Most experiments on d-t μCF have been done with neutron detection [90.137–139], where λ_c and the muon loss probability per cycle W can be deduced from the time structure of the neutron emissions. The analysis is indirect and requires a theoretical model. What is actually measured is the product $W\phi\lambda_c$; the extraction of ω_s requires corrections for other loss mechanisms and separate determination of λ_c. Thus it is desirable to have other experimental diagnostics. Two types of corroborating experiments detect either X-rays from the $\alpha\mu$ formed by sticking or detect the species $(\alpha)^{2+}$ and $(\alpha\mu)^+$ by the different effects of their double and single electrical charges.

The theoretical sticking is compared in Table 90.2 with that from all three types of experiments. For a more meaningful comparison of measurements at different densities ϕ, the theoretical R has been used to convert all values to ω_s^0. The theoretical values are slightly, but significantly, higher than the observations. This discrepancy has not yet been resolved.

One intriguing explanation for the lower-than-predicted value of ω_s may be that a significant fraction of the fusions might occur in muonically excited bound

Table 90.2 Comparison of sticking values [a]

Source	ϕ	$\omega_s(\%)$	$\omega_s^0(\%)$
		Theory	
[90.122, 127]	1.2	0.59	0.91
		Neutron Experiments	
LANL [90.128]	≈ 1	0.43±0.05±0.06	0.66
PSI [90.129]	≈ 1	0.48±0.02±0.04	0.74
KEK [90.130]	1.2	0.51±0.004	0.78
KEK [90.131]	solid	0.421±0.008±0.029	0.65
		X-ray Experiments	
PSI [90.132]	1.2	0.39±0.10	0.60
KEK [90.130]	1.2	0.34±0.13	0.52
		$\alpha/\alpha\mu$ Experiments	
LANL [90.133]	0.001	–	0.80±0.15±0.12
PSI [90.134]	0.17	0.56±0.04	0.80

[a] Experimental values of ω_s^0 without error bars were obtained assuming the theoretical stripping [90.135, 136]
In cases of two error estimates, the first is statistical and the second is systematic
The extraction of ω_s from the X-ray experiments requires theoretical scaling

or resonant states for which the initial sticking is lower than in the ground state [90.140].

A recent experiment has systematically studied muon-catalyzed fusion in solid deuterium and tritium mixtures as a function of temperature and tritium concentration [90.141]. An unexpected decrease in the muon cycling rate (λ_c) *and* an increase in the muon loss (W) were observed. The former is likely due to the freezing out of phonons contributing to the resonance energy. The latter is especially intriguing. It is inconceivable that ω_s^0 for fusion in a given state of $dt\mu$ could depend on temperature, but this observation could imply either an unexpected effect of temperature on the muonic state in which fusion occurs *or* an unpredicted temperature dependence of the thermalization kinetics (e.g., due to ion channeling). It should be noted that the experimental analysis does not reject the possibility of some correlation between the extracted values of λ_c and W [90.140].

90.3.7 Prospectus

There is still a great deal to be learned from the less-studied μCF cycles, like those of p-d, p-t, and t-t, but more quantitative work is also needed on some key processes in the d-t μCF cycle: in particular, three-body effects on molecular formation at high densities, the excited-state cross sections and kinetics that go into the determination of the cascade factor q_{1s}, and the remaining discrepancy in the sticking factor ω_s which might have a theoretical or experimental resolution. Experimentally it is of interest to push on to higher temperatures and densities to see if more surprises lurk there. There have been a few schemes proposed to enhance stripping of stuck muons artificially, but none has been subjected to experiments yet.

The currently observed yield of about 150 d-t fusions (releasing 17.6 MeV each) per muon produces an energy return 25 times the rest-mass energy of the muon, but is only about one-third of that required for breakeven in a pure-fusion reactor. This conclusion is based on the estimated energy cost of producing a muon, ≈ 8 GeV [90.142, 143]. Other possible practical uses include a hybrid (fusion-fission) reactor [90.142, 143] or an intense 14 MeV neutron source [90.144].

Apart from such technological applications, the study of μCF is fruitful for a number of reasons including (1) bridging the gap between atomic and nuclear physics, (2) enabling nuclear reactions (including p-waves) at room temperature, (3) allowing precise studies under unusual physical conditions, (4) observing a compound electronic-muonic molecular environment, and (5) exhibiting phenomena spanning nine orders of magnitude in distance and energy. The experimental possibilities are far from exhausted even though the holy grail of pure fusion energy now appears just beyond reach.

References

90.1 F. Coester: Nucl. Phys. **7**, 421 (1958)
90.2 F. Coester, H. Kümmel: Nucl. Phys. **17**, 477 (1960)
90.3 H. Kümmel: Nucl. Phys. **22**, 177 (1969)
90.4 J. Cizek: J. Chem. Phys. **45**, 4256 (1966)
90.5 J. Cizek: Adv. Chem. Phys. **14**, 35 (1969)
90.6 J. Paldus, J. Cizek, I. Shavitt: Phys. Rev. A **5**, 50 (1972)
90.7 A. S. Jensen, K. Riisager, D. V. Fedorov, E. Garrido: Rev. Mod. Phys. **76**, 215 (2004)
90.8 W. E. Lamb, R. C. Retherford: Phys. Rev. **72**, 241 (1950)
90.9 W. E. Lamb, R. C. Retherford: Phys. Rev. **79**, 549 (1950)
90.10 W. E. Lamb, R. C. Retherford: Phys. Rev. **81**, 222 (1951)
90.11 W. E. Lamb, R. C. Retherford: Phys. Rev. **86**, 1014 (1952)
90.12 W. V. Houston, Y. M. Hsieh: Phys. Rev. **45**, 263 (1934)
90.13 W. V. Houston: Phys. Rev. **51**, 446 (1937)
90.14 R. C. Williams: Phys. Rev. **54**, 558 (1938)
90.15 S. Pasternack: Phys. Rev. **54**, 1113 (1938)
90.16 E. C. Kemble, R. D. Present: Phys. Rev. **44**, 1031 (1933)
90.17 R. Karplus, A. Klein, J. Schwinger: Phys. Rev. **86**, 301 (1951)
90.18 W. E. Lamb: Phys. Rev. **85**, 276 (1952)
90.19 E. E. Salpeter: Phys. Rev. **89**, 95 (1953)
90.20 A. C. Zemach: Phys. Rev. **104**, 1771 (1956)
90.21 S. Flügge: *Practical Quantum Mechanics*, Vol. I (Springer, New York 1971) pp. 191–192
90.22 G. W. F. Drake, Z.-C. Yan: Phys. Rev. A **46**, 2378 (1992)
90.23 D. Shiner, R. Dixson, V. Vedantham: Phys. Rev. Lett. **74**, 3553 (1995)
90.24 E. Riis, A. G. Sinclair, O. Poulsen, G. W. F. Drake, W. R. C. Rowley, A. P. Levick: Phys. Rev. A **49**, 207 (1994)
90.25 K. Pachucki, K. M. Weitz, T. W. Hänsch: Phys. Rev. A **49**, 2255 (1994)
90.26 A. Huber, T. Udem, B. Gross, J. Reichert, M. Kourogi, K. Pachucki, M. Weitz, T. W. Hänsch: Phys. Rev. Lett. **80**, 468 (1998)

90.27 L.-B. Wang, P. Mueller, K. Bailey, G.W.F. Drake, J.P. Greene, D. Henderson, R.J. Holt, R.V.F. Janssens, C.L. Jiang, Z.-T. Lu, T.P. O'Connor, R.C. Pardo, M. Paul, K.E. Rehm, J.P. Schiffer, X.D. Tang: Phys. Rev. Lett. **93**, 142501 (2004)

90.28 G. Ewald, W. Nördershäuser, A. Dax, S. Göte, R. Kirchner, H.-J. Kluge, Th. Kühl, R. Sanchez, A. Wojtaszek, B.A. Bushaw, G.W.F. Drake, Z.-C. Yan, C. Zimmermann: Phys. Rev. Lett. **93**, 113002 (2004)

90.29 A.L. Schawlow, C.H. Townes: Phys. Rev. **100**, 1273 (1955)

90.30 A.R. Bodmer: Proc. Phys. Soc. A **66**, 1041 (1953)

90.31 P.J. Mohr: At. Data Nucl. Data Tables **33**, 456 (1983)

90.32 J.L. Friar: Ann. Phys. (N.Y.) **122**, 151 (1979)

90.33 W.R. Johnson, G. Soff: At. Data Nucl. Data Tables **33**, 407 (1985)

90.34 C.W. de Jager, H. de Vries, C. de Vries: At. Data Nucl. Data Tables **14**, 479 (1974)

90.35 P. Raghaven: At. Data Nucl. Data Tables **42**, 189 (1989)

90.36 A.H. Wapstra, G. Audi: Nucl. Phys. A **432**, 1 (1985)

90.37 J.F. Seely, J.O. Ekberg, C.M. Brown, U. Feldman, W.E. Behring, J. Reader, M.C. Richardson: Phys. Rev. Lett. **57**, 2924 (1986)

90.38 T.E. Cowan, C.L. Bennett, D.D. Dietrich, J.V. Bixler, C.J. Hailey, J.R. Henderson, D.A. Knapp, M.A. Levine, R.E. Marrs, M.B. Schneider: Phys. Rev. Lett. **66**, 1150 (1991)

90.39 J. Schweppe, A. Belkacem, L. Blumenfeld, N. Claytor, B. Feinberg, H. Gould, V.E. Kostroune, L. Levy, S. Misawa, J.R. Mowat, M.H. Prior: Phys. Rev. Lett. **66**, 1434 (1991)

90.40 S.A. Blundell, N.J. Snyderman: Phys. Rev. A **44**, R1427 (1991)

90.41 S.A. Blundell: Phys. Rev. A **46**, 3762 (1992)

90.42 K.T. Cheng, W.R. Johnson, J. Sapirstein: Phys. Rev. A **47**, 1817 (1993)

90.43 J.N. Bahcall: Phys. Rev. **124**, 495 (1961)

90.44 D.D. Clayton: Ap. J. **139**, 637 (1964)

90.45 S.E. Woosley, W.A. Fowler: Ap. J. **233**, 411 (1979)

90.46 R.D. Williams, W.A. Fowler, S.E. Koonin: Astrophys. J. **281**, 363 (1984)

90.47 Z. Chen, L. Rosenberg, L. Spruch: Phys. Rev. A **35**, 1981 (1987)

90.48 Z. Chen, L. Rosenberg, L. Spruch: Adv. At. Mol. Opt. Phys. **26**, 297 (1989)

90.49 Z. Chen, L. Spruch: *AIP Conference Proceedings # 189, Relativistic, Quantum Electrodynamic, and Weak Interaction Effects in Atoms* (AIP, New York 1989) pp. 460–478

90.50 V.A. Lubimov, E.G. Novikov, V.Z. Nozik, E.F. Tretyakov, V.S. Kozik: Phys. Lett. B **94**, 266 (1980)

90.51 S. Boris, A. Golutvin, L. Laptin, V. Lubimov, V. Nagovizin, V. Nozik, E. Novikov, V. Soloshenko, I. Tihomirov, E. Tretjakov, N. Myasoedov: Phys. Rev. Lett. **58**, 2019 (1987)

90.52 S.D. Boris, A.I. Golutvin, L.P. Laptin, V.A. Lyubimov, N.F. Myasoedov, V.V. Nagovitsyn, V.Z. Nozik, E.G. Novikov, V.A. Soloshchenko, I.N. Tikhomirov, E.F. Tretyakov: Pis'ma Zh. Eksp. Teor. Fiz. **45**, 267 (1987) [Sov. Phys. JETP Lett. **45**, 333 (1987)]

90.53 K.-E. Bergkvist: Phys. Scr. **4**, 23 (1971)

90.54 K.-E. Bergkvist: Nucl. Phys. B **39**, 317, 371 (1972)

90.55 R.L. Martin, J.S. Cohen: Phys. Lett. A **110**, 95 (1985)

90.56 W. Kolos, B. Jeziorski, J. Rychlewski, K. Szalewicz, H.J. Monkhorst, O. Fackler: Phys. Rev. A **37**, 2297 (1988) and earlier references therein

90.57 M. Fritschi, E. Holzschuh, W. Kundig, J.W. Petersen, R.E. Pixley, H. Stussi: Phys. Lett. B **173**, 485 (1986)

90.58 J.F. Wilkerson, T.J. Bowles, J.C. Browne, M.P. Maley, R.G.H. Robertson, J.S. Cohen, R.L. Martin, D.A. Knapp, J.A. Helffrich: Phys. Rev. Lett. **58**, 2023 (1987)

90.59 R.G.H. Robinson et al.: Phys. Rev. Lett. **67**, 957 (1991)

90.60 E. Holzschuh, M. Fritschi, W. Kündig: Phys. Lett. B **287**, 381 (1992)

90.61 C. Weinheimer, M. Przyrembel, H. Backe, H. Barth, J. Bonn, B. Degen, T. Edling, H. Fischer, L. Fleischmann, J.U. Gross, R. Haid, A. Hermanni, G. Kube, P. Leiderer, T. Loeken, A. Molz, R.B. Moore, A. Osipowicz, E.W. Otten, A. Picard, M. Schrader, M. Steininger: Phys. Lett. B **300**, 210 (1993)

90.62 Y. Fukuda et al.: Phys. Rev. Lett. **81**, 1562 (1998)

90.63 Ch. Weinheimer, B. Degenddag, A. Bleile, J. Bonn, L. Bornschein, O. Kazachenko, A. Kovalik, E.W. Otten: Phys. Lett. B **460**, 219 (1999)

90.64 J. Bonn, B. Bornschein, L. Bornschein, L. Fickinger, B. Flatt, O. Kazachenko, A. Kovalik, C. Kraus, E.W. Otten, J.P. Schall, H. Ulrich, C. Weinheimer: Nucl. Phys. B (Proc. Suppl.) **91**, 273 (2001)

90.65 S. Hannestad: Phys. Rev. D **66**, 125011 (2002)

90.66 H.J. Assenbaum, K. Langanke, C. Rolfs: Z. Phys. A **327**, 461 (1987)

90.67 S. Engstler, A. Krauss, K. Neldner, C. Rolfs, U. Schröder, K. Langanke: Phys. Lett. B **202**, 179 (1988)

90.68 U. Schröder, S. Engstler, A. Krauss, K. Neldner, C. Rolfs, E. Somorjai, K. Langanke: Nucl. Instrum. Meth. B **40/41**, 466 (1989)

90.69 S. Engstler, G. Raimann, C. Angulo, U. Greife, C. Rolfs, U. Schröder, E. Somorjai, B. Kirch, K. Langanke: Phys. Lett. B **279**, 20 (1992)

90.70 S. Engstler, G. Raimann, C. Angulo, U. Greife, C. Rolfs, U. Schröder, E. Somorjai, B. Kirch, K. Langanke: Z. Phys. A **342**, 471 (1992)

90.71 G. Blüge, K. Langanke, H.G. Reusch, C. Rolfs: Z. Phys. A **333**, 219 (1989)

90.72 K. Langanke, D. Lukas: Ann. Phys. (Leipzig) **1**, 332 (1992)

90.73 K. Langanke: Adv. Nucl. Phys. **21**, 179 (1994)

90.74 T.D. Shoppa, S.E. Koonin, K. Langanke, R. Seki: Phys. Rev. C **48**, 837 (1993)

90.75 T. D. Shoppa, M. Jeng, S. E. Koonin, K. Langanke, D. Seki: Nucl. Phys. A **605**, 387 (1996)
90.76 J. N. Bahcall, X. Chen, M. Kamionkowski: Phys. Rev. C **57**, 2756 (1998)
90.77 F. Strieder, C. Rolfs, C. Spitaleri, P. Corvisiero: Naturwissenschaften **88**, 461 (2001)
90.78 S. Zavatarelli, P. Corvisiero, H. Costantini, P. G. P. Moroni, P. Prati, R. Bonetti, A. Guglielmetti, C. Broggini, L. Campajola, A. Formicola, L. Gialanella, G. Imbriani, A. Ordine, V. Roca, M. Romano, A. D'Onofrio, F. Terrasi, G. Gervino, C. Gustavino, M. Junker, D. Rogalla, C. Rolfs, F. Schumann, F. Strieder, H. P. Trautvetter: Nucl. Phys. A **688**, 514 (2001)
90.79 M. Aliotta, E. Raiola, G. Gyurky, A. Formicola, R. Bonetti, C. Broggini, L. Campajola, P. Corvisiero, H. Costantini, A. D'Onofrio, Z. Fulop, G. Gervino, L. Gialanella, A. Guglielmetti, C. Gustavino, G. Imbriani, M. Junker, P. G. Moroni, A. Ordine, P. Prati, V. Roca, D. Rogalla, C. Rolfs, M. Romano, F. Schumann, E. Somorjai, O. Straniero, F. Strieder, F. Terrasi, H. P. Trautvetter, S. Zavatarelli: Nucl. Phys. A **690**, 790 (2001)
90.80 S. Kimura, N. Takigawa, M. Abe, D. M. Brink: Phys. Rev. C **67**, 022801(R) (2003)
90.81 J. Eichler: Phys. Rep. **193**, 165 (1990)
90.82 R. Anholt, H. Gould: Adv. At. Mol. Phys. **22**, 315 (1986)
90.83 A. Belkacem, H. Gould, B. Feinberg, R. Bossingham, W. E. Meyerhof: Phys. Rev. Lett. **71**, 1514 (1993)
90.84 J. S. Cohen: In: *Review of Fundamental Processes and Applications of Atoms and Ions*, ed. by C. D. Lin (World Scientific, Singapore 1993) Chap. 2, pp. 61–110
90.85 L. Bogdanova, V. Markushin: Nucl. Phys. A **508**, 29c (1990)
90.86 P. Ackerbauer, W. H. Breunlich, M. Fuchs, S. Fussy, M. Jeitler, P. Kammel, B. Lauss, J. Marton, W. Prymas, J. Werner, J. Zmeskal, K. Lou, C. Petitjean, P. Baumann, H. Daniel, F. J. Hartmann, W. Schott, T. Vonegidy, P. Wojciechowski, D. Chatellard, J. P. Egger, E. Jeannet, T. Case, K. M. Crowe, R. H. Sherman, V. Markushin: Hyperfine Interact. **82**, 243 (1993)
90.87 D. V. Balin, V. N. Baturin, Yu. A. Chestnov, A. I. Ilyin, P. A. Kapinos, E. M. Maev, G. E. Petrov, L. B. Petrov, G. G. Semenchuk, Yu. A. Smirenin, A. A. Vorobyov, An. A. Vorobyov, N. I. Voropaev: Muon Catal. Fusion **5/6**, 163 (1990/91)
90.88 P. Baumann, H. Daniel, S. Grunewald, F. J. Hartmann, R. Lipowsky, E. Moser, W. Schott, T. Vonegidy, P. Ackerbauer, W. H. Breunlich, M. Fuchs, M. Jeitler, P. Kammel, J. Marton, N. Nagele, J. Werner, J. Zmeskal, H. Bossy, K. M. Crowe, R. H. Sherman, K. Lou, C. Petitjean, V. E. Markushin: Phys. Rev. Lett. **70**, 3720 (1993)
90.89 W. H. Breunlich, P. Kammel, J. S. Cohen, M. Leon: Ann. Rev. Nucl. Part. Sci. **39**, 311 (1989)
90.90 S. S. Gershteĭn, Yu. V. Petrov, L. I. Ponomarev: Usp. Fiz. Nauk **160**, 3 (1990) [Sov. Phys. Usp. **33**, 591 (1990)]
90.91 P. Froelich: Adv. Phys. **41**, 405 (1992)
90.92 C. Petitjean: Nucl. Phys. A **543**, 79 (1992)
90.93 J. S. Cohen: In: *Electromagnetic Cascade and Chemistry of Exotic Atoms*, Proceedings of the International School of Physics of Exotic Atoms, 5th Course, Erice, Sicily, 1989, ed. by L. M. Simons, D. Horvath, G. Torelli (Plenum, New York 1990) pp. 1–22
90.94 P. K. Haff, T. A. Tombrello: Ann. Phys. (N. Y.) **86**, 178 (1974)
90.95 A. S. Wightman: Phys. Rev. **77**, 521 (1950)
90.96 J. S. Cohen: Phys. Rev. A **27**, 167 (1983)
90.97 J. S. Cohen, R. L. Martin, W. R. Wadt: Phys. Rev. A **24**, 33 (1981)
90.98 J. S. Cohen, R. L. Martin, W. R. Wadt: Phys. Rev. A **27**, 1821 (1983)
90.99 J. D. Garcia, N. H. Kwong, J. S. Cohen: Phys. Rev. A **35**, 4068 (1987)
90.100 N. H. Kwong, J. D. Garcia, J. S. Cohen: J. Phys. B **22**, L633 (1989)
90.101 E. Fermi, E. Teller: Phys. Rev. **72**, 406 (1947)
90.102 J. S. Cohen: Phys. Rev. A **59**, 1160 (1999)
90.103 J. S. Cohen: Rep. Prog. Phys. **67**, 1769 (2004)
90.104 Y. Yamazaki: Hyperfine Interact. **138**, 141 (2001)
90.105 W. Czaplinski, A. Gula, A. Kravtsov, A. Mikhailov, N. Popov: Phys. Rev. A **50**, 525 (1994)
90.106 P. Froelich, J. Wallenius: Phys. Rev. Lett. **75**, 2108 (1995)
90.107 J. S. Cohen: Phys. Rev. A **43**, 4668 (1991)
90.108 E. A. Vesman: Pis'ma Zh. Eksp. Fiz. **5**, 113 (1967) [JETP Lett. **5**, 91 (1967)]
90.109 M. P. Faifman, L. I. Menshikov, T. A. Strizh: Muon Catal. Fusion **4**, 1 (1989)
90.110 J. S. Cohen, R. L. Martin: Phys. Rev. Lett. **53**, 738 (1984)
90.111 M. Leon: Phys. Rev. Lett. **52**, 605 (1984)
90.112 Yu. V. Petrov: Phys. Lett. B **163**, 28 (1985)
90.113 L. I. Menshikov, L. I. Ponomarev: Phys. Lett. B **167**, 141 (1986)
90.114 J. S. Cohen, M. Leon: Phys. Rev. A **39**, 946 (1989)
90.115 M. Leon: Phys. Rev. A **39**, 5554 (1989)
90.116 M. C. Fujiwara, A. Adamczak, J. M. Bailey, G. A. Beer, J. L. Beveridge, M. P. Faifman, T. M. Huber, P. Kammel, S. K. Kim, P. E. Knowles, A. R. Kunselman, M. Maier, V. E. Markushin, G. M. Marshall, C. J. Martoff, G. R. Mason, F. Mulhauser, A. Olin, C. Petitjean, T. A. Porcelli, J. Wozniak, J. Zmeskal: Phys. Rev. Lett. **85**, 1642 (2000)
90.117 L. N. Bogdanova: Muon Catal. Fusion **3**, 359 (1988)
90.118 L. N. Bogdanova, V. E. Markushin, V. S. Melezhik, L. I. Ponomarev: Yad. Phys. **34**, 1191 (1981) [Sov. J. Nucl. Phys. **34**, 662 (1981)]
90.119 M. Kamimura: In: *AIP Conference Proceedings 181, Muon-Catalyzed Fusion*, ed. by S. E. Jones, J. Rafelski, H. J. Monkhorst (AIP, New York 1989) p. 330

90.120 M. C. Struensee, G. M. Hale, R. T. Pack, J. S. Cohen: Phys. Rev. A **37**, 340 (1988)

90.121 K. Szalewicz, B. Jeziorski, A. Scrinzi, X. Zhao, R. Moszynski, W. Kolos, P. Froelich, H. J. Monkhorst, A. Velenik: Phys. Rev. A **42**, 3768 (1990)

90.122 C.-Y. Hu, G. M. Hale, J. S. Cohen: Phys. Rev. A **49**, 4481 (1994)

90.123 G. M. Hale: Muon Catal. Fusion **5/6**, 227 (1990/91)

90.124 L. Bracci, G. Fiorentini: Nucl. Phys. A **364**, 383 (1981)

90.125 S. E. Haywood, H. J. Monkhorst, S. A. Alexander: Phys. Rev. A **37**, 3393 (1988)

90.126 S. E. Haywood, H. J. Monkhorst, S. A. Alexander: Phys. Rev. A **43**, 5847 (1991)

90.127 B. Jeziorski, K. Szalewicz, A. Scrinzi, X. Zhao, R. Moszynski, W. Kolos, A. Velenik: Phys. Rev. A **43**, 1640 (1991)

90.128 S. E. Jones, S. F. Taylor, A. N. Anderson: Hyperfine Interact. **82**, 303 (1993)

90.129 C. Petitjean, D. V. Balin, V. N. Baturin, P. Baumann, W. H. Breunlich, T. Case, K. M. Crowe, H. Daniel, Y. S. Grigoriev, F. J. Hartmann, A. I. Ilyin, M. Jeitler, P. Kammel, B. Lauss, K. Lou, E. M. Maev, J. Marton, M. Muhlbauer, G. E. Petrov, W. Prymas, W. Schott, G. G. Semenchuk, Y. V. Smirenin, A. A. Vorobyov, N. I. Voropaev, P. Wojciechowski, J. Zmeskal: Hyperfine Interact. **82**, 273 (1993)

90.130 K. Nagamine, K. Ishida, S. Sakamoto, Y. Watanabe, T. Matsuzaki: Hyperfine Interact. **82**, 343 (1993)

90.131 K. Ishida, K. Nagamine, T. Matsuzaki, S. N. Nakamura, N. Kawamura, S. Sakamoto, M. Iwasaki, M. Tanase, M. Kato, K. Kurosawa, H. Sugai, I. Watanabe, K. Kudo, N. Takeda, G. H. Eaton: Hyperfine Interact. **118**, 203 (1999)

90.132 H. Bossy, H. Daniel, F. J. Hartmann, W. Neumann, H. S. Plendl, G. Schmidt, T. Vonegidy, W. H. Breunlich, M. Cargnelli, P. Kammel, J. Marton, N. Nagele, A. Scrinzi, J. Werner, J. Zmeskal, C. Petitjean: Phys. Rev. Lett. **59**, 2864 (1987)

90.133 M. A. Paciotti, O. K. Baker, J. N. Bradbury, J. S. Cohen, M. Leon, H. R. Maltrud, L. L. Sturgess, S. E. Jones, P. Li, L. M. Rees, E. V. Sheely, J. K. Shurtleff, S. F. Taylor, A. N. Anderson, A. J. Caffrey, J. M. Zabriskie, F. D. Brooks, W. A. Cilliers, J. D. Davies, J. B. A. England, G. J. Pyle, G. T. A. Squier, A. Bertin, M. Bruschi, M. Piccinini, A. Vitale, A. Zoccoli, V. R. Bom, C. W. E. van Eijk, H. de Haan, G. H. Eaton: In: *AIP Conference Proceedings 181, Muon-Catalyzed Fusion*, ed. by S. E. Jones, J. Rafelski, H. J. Monkhorst (AIP, New York 1989) p. 38

90.134 T. Case, K. M. Crowe, K. Lou, C. Petitjean, W. H. Breunlich, M. Jeitler, P. Kammel, B. Lauss, J. Marton, W. Prymas, J. Zmeskal, D. V. Balin, V. N. Baturin, Y. S. Grigoriev, A. I. Ilyin, E. M. Maev, G. E. Petrov, G. G. Semenchuk, Y. V. Smirenin, A. A. Vorobyov, N. I. Voropaev, P. Baumann, H. Daniel, F. J. Hartmann, M. Muhlbauer, W. Schott, P. Wojciechowski: Hyperfine Interact. **82**, 295 (1993)

90.135 M. C. Struensee, J. S. Cohen: Phys. Rev. A **38**, 44 (1988)

90.136 C. D. Stodden, H. J. Monkhorst, K. Szalewicz, T. G. Winter: Phys. Rev. A **41**, 1281 (1990)

90.137 S. E. Jones, A. N. Anderson, A. J. Caffrey, C. D. Vansiclen, K. D. Watts, J. N. Bradbury, J. S. Cohen, P. A. M. Gram, M. Leon, H. R. Maltrud, M. A. Paciotti: Phys. Rev. Lett. **56**, 588 (1986)

90.138 W. H. Breunlich, M. Cargnelli, P. Kammel, J. Marton, N. Naegele, P. Pawlek, A. Scrinzi, J. Werner, J. Zmeskal, J. Bistirlich, K. M. Crowe, M. Justice, J. Kurck, C. Petitjean, R. H. Sherman, H. Bossy, H. Daniel, F. J. Hartmann, W. Neumann, G. Schmidt: Phys. Rev. Lett. **58**, 329 (1987)

90.139 K. Nagamine, T. Matsuzaki, K. Ishida, Y. Hirata, Y. Watanabe, R. Kadono, Y. Miyake, K. Nishiyama, S. E. Jones, H. R. Maltrud: Muon Catal. Fusion **1**, 137 (1987)

90.140 P. Froelich, A. Flores-Riveros: Phys. Rev. Lett. **70**, 1595 (1993)

90.141 N. Kawamura, K. Nagamine, T. Matsuzaki, K. Ishida, S. N. Nakamura, Y. Matsuda, M. Tanase, M. Kato, H. Sugai, K. Kudo, N. Takeda, G. H. Eaton: Phys. Rev. Lett. **90**, 043401 (2003)

90.142 Yu. V. Petrov: Nature (London) **285**, 466 (1980)

90.143 Yu. V. Petrov: Muon Catal. Fusion **3**, 525 (1988)

90.144 T. Kase, K. Konashi, N. Sasao, H. Takahashi, Y.Hirao: Muon Catal. Fusion **5/6**, 521 (1990/91)

91. Charged-Particle–Matter Interactions

In the description of the interaction of fast charged particles with matter, two aspects can be distinguished: the effects on the particle, usually energy losses and deflections, and the spatial distribution of the energy lost by the particle in the absorber.

This Chapter discusses concepts needed in the operation of charged particle detectors and in describing radiation effects (Chapt. 92). Specifically, the radiation effects used for the instantaneous observation (i. e., within fractions of a millisecond) of the passage of a charged particle are described. Delayed effects, such as chemical reactions (e.g., biological effects, chemical dosimeters, photographic emulsions), metastable states, etc. are not discussed. It is assumed that particle speeds have been determined with, e.g., magnetic analyzers or by measurement of the time of flight. A measurement of particle ranges can also be used to determine the initial speed. The description is restricted to fast charged particles, defined by speeds $v > 6v_B$ (v_B is the Bohr speed), or $\beta = v/c > 0.04$. Interactions and cross sections at smaller speeds are discussed, e.g., by *Rudd* et al.[91.1] and in [91.2].

The present Chapter primarily considers energy loss straggling, rather than the stopping powers discussed by *Fano* [91.3].

It is important to understand that the mean energy-loss is not a suitable concept to use in the description of energy-loss spectra for thin absorbers. The most probable energy-loss should be used instead. The methods described here can be used to calculate reliable data for detector applications. No attempt is made to present a complete review. Anecdotal, qualitative examples of various effects are described. The following definitions are used:

1. The number of atoms or molecules per unit volume $N = N_A \varrho / A$, with A the molecular weight (in g/mole) of the absorber (with Z_2 electrons per molecule), ϱ its density, and N_A is Avogadro's number;
2. The relativistic factors β and γ for particles with rest mass M_0, speed $v = \beta c$ and kinetic energy $\mathcal{T}$:

$$\beta^2 = \left(\mathcal{T}/M_0c^2\right)\left(2+\mathcal{T}/M_0c^2\right) / \left(1+\mathcal{T}/M_0c^2\right)^2 ,$$

$$\gamma = M/M_0 = 1+\mathcal{T}/M_0c^2 ,$$

$$\gamma^2 = 1/\left(1-\beta^2\right) ,$$

$$\beta^2\gamma^2 = \gamma^2 - 1 ;$$

3. The coefficient of the Rutherford equation:

$$k_R = \frac{2\pi Z_1^2 e^4}{m_e c^2} = 2\pi r_0^2 m_e c^2 Z_1^2$$
$$= 2.549\,55 \times 10^{-19} Z_1^2 \text{ eV cm}^2 , \quad (91.1)$$

where m_e is the rest mass of an electron, $-e$ its charge, $Z_1 e$ the charge of the incident particle, and $r_0 = \alpha^2 a_0$ is the classical electron radius; and

4. The maximum energy loss of a heavy particle to an electron: $E_{max} \sim 2m_e c^2 \beta^2 \gamma^2$ [91.4].

91.1 Experimental Aspects

91.1.1 Energy Loss Experiments and Radiation Detectors

After passing through a thickness x of material, an initially monoenergetic beam of particles acquires a distribution of energies described by a probability density function

$$F(\Delta) = \frac{d\mathcal{N}(\Delta)}{d\Delta} , \quad (91.2)$$

where Δ is the energy loss per particle, and $d\mathcal{N}(\Delta)$ is the number of particles in the range Δ to $\Delta + d\Delta$. The straggling function $F(\Delta)$ represents the spectrum of energy losses Δ, such that $\int F(\Delta)\, d\Delta = \mathcal{N}$, the total number of particles observed. It can be characterized by the quantities

1. the most probable energy loss Δ_{mp},
2. the full width at half maximum Γ,
3. the moments [91.8]

$$\mu_\nu = \frac{1}{\mathcal{N}} \int F(\Delta)\Delta^\nu \, d\Delta , \quad (91.3)$$

and the central moments

$$C_\nu = \frac{1}{\mathcal{N}} \int F(\Delta)(\Delta - \mu_1)^\nu \, d\Delta . \quad (91.4)$$

Then $\mu_1 = \langle\Delta\rangle$ is the mean energy loss, $C_2 = \sigma^2$ is the variance and $\gamma_1 = C_3/C_2^{3/2}$ is the skewness. The fluence spectrum $\phi(\mathcal{T})$ is the complementary function describing the distribution of residual energies $\mathcal{T}$ of the particles.

As an example, Fig. 91.1 shows $F(\Delta)\, d\Delta$ for 1.27 GeV protons passing through a 32 μm thick silicon wafer [91.5]. The measured quantity is the ionization J resulting from the creation of electron–hole pairs in the silicon, so that $\Delta = JW$ is the energy deposited, where W is the energy required to create an electron–hole pair. For $\Delta < 15$ keV, the energy deposited differs little from the energy lost by the protons. For larger Δ, some of the secondary electrons may escape from the silicon, making the apparent energy deposited less than the energy lost [91.9]. The spectrum has a long tail extending up to a maximum energy loss of 4.6 MeV. This accounts for the large value of $\langle\Delta\rangle = 12.8$ keV, and the even larger value $C_2^{1/2} = 43$ keV for the standard deviation, relative to $\Delta_{mp} = 7.4$ keV and $\Gamma = 5.2$ keV.

In losing energy, the beam particles also suffer angular deflections, but for heavy particles, the deflections are usually small. For electrons, angular deflections are quite important, and are discussed in Chapts. 47, 64, and 65. Nuclear reactions cause large effects but are quite infrequent. Further examples of straggling spectra are given in [91.5, 10–12] for thin, extremely thin, moderately thick, and thick absorbers respectively.

Only for thick absorbers, the stopping power $S(\mathcal{T}) = -d\mathcal{T}/dx$ (i. e., the mean energy lost per unit thickness) provides a convenient measure of the energy loss process. For a beam with incident energy $\mathcal{T}_0$ and

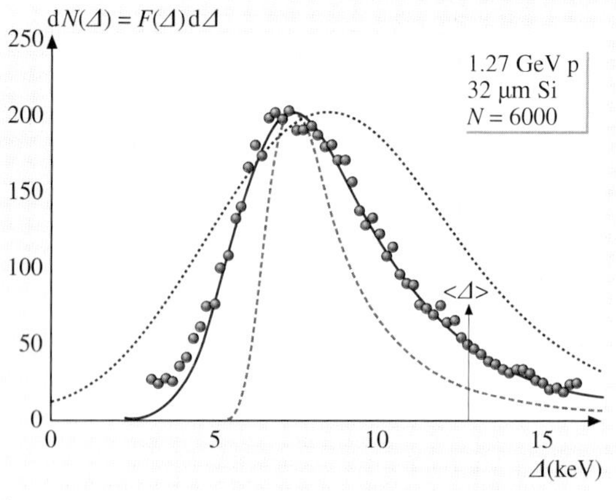

Fig. 91.1 Energy loss straggling functions $F(\Delta)d\Delta$ for 1.27 GeV protons traversing a 32 μm silicon detector, with $d\Delta = 0.21$ keV. The experimental data are shown by *circles*. Three calculated functions normalized to the same peak height are shown for comparison [91.5]. The *solid line* was calculated with the Bethe cross section Fig. 91.3 with $\Delta_{mp} = 7.4$ keV, $\Gamma = 5.2$ keV, and $\langle\Delta\rangle = 12.8$ keV. The *broken line* is the Landau function [91.6] and the *dotted line* is the Blunck–Leisegang modification [91.7]

$\langle \Delta \rangle < 0.1\mathcal{T}_0$, $S(\mathcal{T})$ is often approximated by

$$S(\mathcal{T}_0 - \langle \Delta \rangle/2) \simeq \langle \Delta \rangle / x\,, \qquad (91.5)$$

provided that $x \gg a_0$ (Sect. 91.4). The stopping power $S(\mathcal{T})$ depends on the absorber properties, especially the electron density NZ_2. The radiation dose $D(x)$ at a distance x into the material is given by

$$D(x) = \int \phi(\mathcal{T}, x) S(\mathcal{T})\, \mathrm{d}\mathcal{T}\,, \qquad (91.6)$$

where $\phi(\mathcal{T}, x)$ is the fluence spectrum. $D(x)$ is used in radiation dosimetry and protection [91.15, 16].

$S(\mathcal{T})$ ceases to be useful for very thin absorbers, such as microscopic biological specimens, micro- or nano-devices, or thin ionization chambers. Instead Δ_{mp} and Γ should be used. No simple equations can provide Δ_{mp} and Γ. They must be found from calculated or experimental $F(\Delta)$. Calculations of $F(\Delta)$ are described in 91.4.2 and 91.6.

A detailed theoretical description of the energy deposition process requires cross sections for the various scattering events in the target, the most important being collisions with electrons. The Rutherford cross section for the collision of a particle with a free electron is only useful for $\Delta \gtrsim 30\, U_Z$, where U_Z is the binding energy of a target electron (~ 500 eV for outer shell electrons). Various modifications to the Rutherford cross section for smaller Δ are discussed in Sect. 91.2.

91.1.2 Inelastic Scattering Events

If a beam of $\mathcal{N}$ monoenergetic particles with speed v passes through an absorber of infinitesimal thickness $\mathrm{d}x$, the average number of particles experiencing a collision with energy loss between E and $E + \mathrm{d}E$ is given by

$$\mathrm{d}\mathcal{N}(E) = \mathcal{N} N \sigma(E)\, \mathrm{d}E\, \mathrm{d}x\,, \qquad (91.7)$$

where $\sigma(E)$ is the collision cross section per molecule differential in E; it depends on β. Using this equation, $\sigma(E)$ can be obtained from measurements [91.17, 18] with extremely thin absorbers where particles on the average suffer less than one collision. If particles pass through a thicker absorber, they make several collisions with energy losses E_i; and the total energy loss is $\Delta = \sum E_i$ (Sect. 91.4.2).

Large differences exist between a gas and a solid of the same composition due to changes in the valence shell electrons. In isolated atoms [91.14], the smallest energy losses are to discrete excited electronic states. Ionizations occur for energy losses exceeding the binding energies U_l for each atomic shell l and the released electrons are given a kinetic energy $\delta = E - U_l$. For ionization, energy losses are continuous (Fig. 91.2). If atoms are brought closer together to form a liquid or a solid, their valence electrons come under the influence of the cores of surrounding atoms. A core is defined to consist of the nucleus and all the electrons inside the valence shell. For C, O, or H_2O, the core consists of two electrons in the K-shell, for Al or Si it contains two electrons in the K-shell and eight electrons in the L-shell. For metals, the valence electrons form a conduction band where they are nearly free. If a charged particle moves through the solid, the transfer of very small energies to the electron by Rutherford scattering is not observed. Instead, many thousands of them are excited at a time. For metals, this process is called plasmon excitation, for insulators, collective excitation (Fig. 91.2). For most substances, the plasmon excitation energy E_{p} is much larger than the energy of the lowest excited state of the atom E_1. For example, for Be, $E_1 = 3.6$ eV [91.13], while $E_{\mathrm{p}} = 19$ eV [91.17]. Similarly, for silicon, $E_1 = 3.6$ eV [91.13] and $E_{\mathrm{p}} = 16.7$ eV [91.5].

For molecules, measurements of electron energy losses [91.18] provide information about the difference in the structure of $\sigma(E)$ between gas and solid. For example, for benzene (C_6H_6) the vapor shows distinct structures for excitations to several discrete states, and

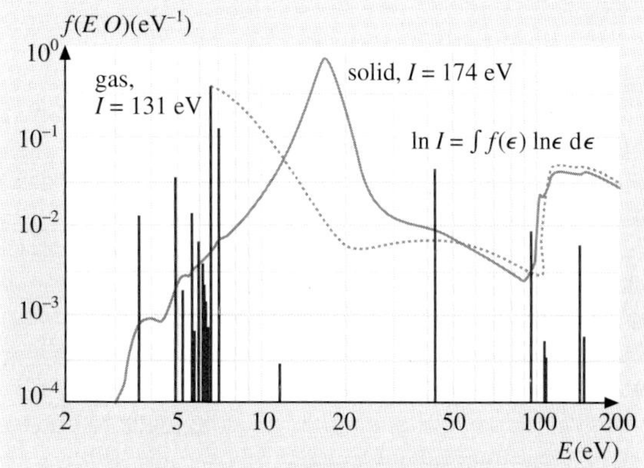

Fig. 91.2 Comparison of the dipole oscillator strength spectra for atomic and solid silicon. Discrete atomic excitations are shown by vertical lines, continuum excitations by the *dotted line*. Data for the atom are based on calculations [91.13]. Data for the solid are from [91.5]. The broad peak at ≈ 17 eV represents the plasmon excitations. Uncertainties may exceed 10%. The discrete atomic excitations disappear in the solid. The photoionization cross section is $\sigma(E) = 109.8\, f(E)$ Mb [91.14]

a broad peak at about 16 eV which appears to be equivalent to a collective excitation. For the solid, the structures are broadened, and the major peak for the collective excitations shifts to 21 eV. Data for water can be found in [91.19].

For energy losses well above E_p, $\sigma(E)$ decreases smoothly until the ionization energy of the next electron shell is reached, as shown in Fig. 91.2. In a molecular substance such as $CaCO_3$, several broad peaks appear between 7 and 40 eV, followed by narrow peaks at 280 eV (carbon K-shell), 345 eV (Ca L-shells), 530 eV (oxygen K-shell) and 4 keV (Ca K-shell) [91.17].

While there are large changes in the structure of the excitation of the valence electrons as the atoms are coalesced into a solid (Fig. 91.2), changes are less important for the inner shells because the binding energies of these shells are much larger than chemical energies. As an example, for an amorphous thin carbon film, the K-edge is at 284 eV [91.17], and K-shell excitation shows essentially just one peak at ≈ 296 eV. For the three biologically important molecules adenine, uracil and thymine, several peaks appear [91.20], all located between 284 and 300 eV. Thus there is at most a small change in the energy of the K-shell excitation, but the number of peaks as well as their locations changes considerably. For Al above the K-edge (1.56 keV), a structure with several peaks appears [91.21], with separations of about 40 eV. This structure, called extended X-ray absorption fine structure (EXAFS) [91.22], is caused by the presence of nearby atomic cores which backscatter the photoelectrons and thus change their wave functions. The discrete excitations of the atom below the K-edge disappear completely. Thus solid state and chemical effects are very important for valence shell electrons, and less so for inner-shell electrons.

91.2 Theory of Cross Sections

The cross section for the collision of two free charged particles is given by the Rutherford expression. If charged particles collide with electrons bound in atoms, molecules or solids, the cross section can be written as a modified Rutherford cross section. A plausible way of describing these interactions is to consider the emission of virtual photons by the fast particle, which then are absorbed by the material in the Fermi virtual photon method (FVP) [91.23]. The collision cross section then is proportional to the photo absorption cross section of the molecules. *Bohr* [91.24] described this as a "resonance" effect.

A variety of models has been used to obtain theoretical $\sigma(E)$ for bound electrons. Here, three of them will be described and compared with each other. Examples are shown in Fig. 91.3. Few analytic functions and methods are available to calculate cross sections (Chapt. 47). Usually, numerical calculations must be made to obtain reliable data for real absorbers.

91.2.1 Rutherford Cross Section

The cross section for close collisions of fast charged particles with loosely bound electrons is well approximated by the cross section for the collision of a charged particle with a free electron at rest. The nonrelativistic Rutherford cross section $\sigma_R(E)$ for an energy loss E in the collision of a charged particle with speed v in the laboratory frame is given by [91.25–28]

$$\sigma_R(E) = \frac{2\pi Z_1^2 e^4}{mv^2}\frac{1}{E^2} = \frac{k_R}{\beta^2}\frac{1}{E^2} . \tag{91.8}$$

Since the secondary electron receives all the energy E lost by the incident particle, the momentum transfer is $q = \sqrt{2mE}$. The cross section $\sigma_R(E)$ does not depend on particle mass M. The leading relativistic correction (For $\gamma > M/m$, further terms must be included [91.29–31]) is

$$\sigma_R'(E) = \sigma_R(E)\left(1 - \beta^2\frac{E}{E_{max}}x\right) . \tag{91.9}$$

91.2.2 Binary Encounter Approximation

A simple correction to $\sigma_R(E)$ can be achieved by taking into account the velocity of the bound electrons. With binding energy U_b, average kinetic energy $\mathcal{T}_e$, and average speed u of the electrons, the expression is

$$\sigma_L(E) = \sigma_R(E)\left(1 + \frac{4}{3}\frac{\mathcal{T}_e}{E}\right) , \quad E > U_b , \tag{91.10}$$

which is valid for $v \gg u$. The total cross section for a molecule includes contributions from each electron shell. Variants of this approach are described in [91.32, 33]. Figure 91.3 shows an example.

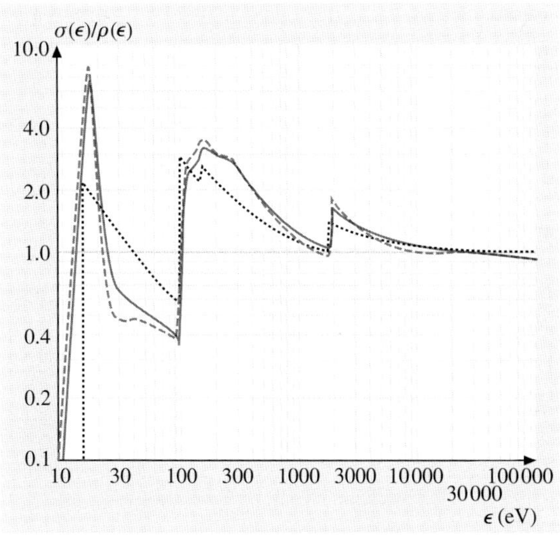

Fig. 91.3 Inelastic cross sections $\sigma(E)$ for single collisions in solid silicon, for incident protons with an energy $\mathcal{T} = 100\,\text{MeV}$, calculated using different theories. The *horizontal line* at 1.0 represents the Rutherford cross section from (91.8). The other curves are Bethe theory (*solid line*); FVP approximation (*dashed line*); binary encounter approximation from (91.10) [91.32] (*dotted line*). The curves extend to $E_{\max} = 230\,\text{keV}$

91.2.3 Bethe Model of Cross Section

The Bethe model [91.34] derives from the first Born approximation for inelastic scattering, which becomes essentially exact at high energies. In terms of the inelastic form factor $|\mathcal{F}(E, \boldsymbol{K})|^2$, the energy loss cross section is

$$d\sigma(E, Q) = \sigma_R(Q)|\mathcal{F}(E, \boldsymbol{K})|^2 \, dQ \,, \tag{91.11}$$

where $Q = q^2/2m_e$, and $\hbar \boldsymbol{K}$ is the momentum transfer vector. The generalized oscillator strength (GOS) is defined by

$$f(E, K) = E|\mathcal{F}(E, \boldsymbol{K})|^2/Q \,, \tag{91.12}$$

such that $f(E, K)$ reduces to the optical dipole oscillator strength $f(E, 0) \equiv f(E)$ in the limit $K \to 0$. Then

$$d\sigma(E, Q) = \sigma_R(E) E f(E, K) \, d\ln Q \,. \tag{91.13}$$

For hydrogenic atoms, $f(E, K)$ is well known [91.34–38], and a model spectrum is shown in Fig. 10 of [91.27, 28]. For more complicated atoms, many-electron effects introduce small corrections, as shown in Fig. 91.4 for Si using a Hartree–Fock approximation. Adding the contributions from all shells yields the Bethe result in Fig. 91.3 [91.5]. Similar results are given in [91.39] for Al. For the outermost electrons in metals, the electron gas model can be used to generate cross sections [91.40–42]. For semiconductors and insulators a model has been derived using the tight binding approximation for the ground state wave function and orthogonalized plane waves for excited states [91.41]. The dipole oscillator strength is derived from the data for optical absorption coefficients; for solids [91.43] and gases [91.44–46] see Chapt. 61.

91.2.4 Fermi Virtual Photon Method

In the Fermi virtual photon (FVP) method, $f(E, K)$ is approximated by $f(E, 0)$ for $Q < E$, with a delta function at $Q = E$ [91.23, 47–50] (Fig. 91.4). Then $\sigma(E)$ is given by [91.48]

$$\sigma(E) = \sigma_R(E)\left[E f(E, 0) \ln\left(\frac{2mv^2}{E}\right) + \int_0^E f(E', 0)\, dE' \right] , \tag{91.14}$$

and so only $f(E)$ need be known, or equivalently $\text{Im}(-1/\epsilon)$, where ϵ is the complex dielectric constant of the absorber. Data can be extracted from a variety of optical measurements [91.43, 44, 51], and from electron

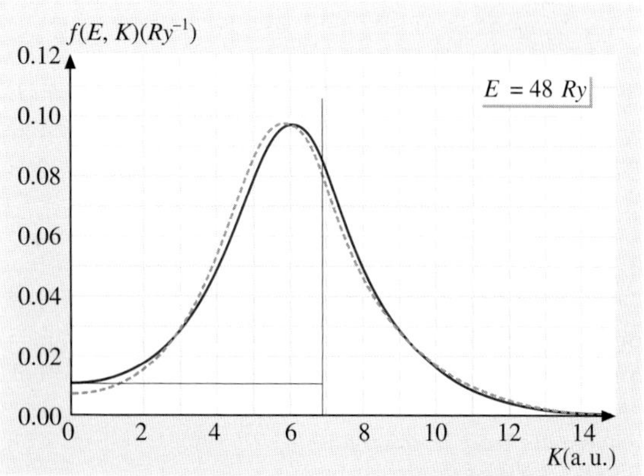

Fig. 91.4 Generalized oscillator strength (GOS) $f(E, K)$ (*solid line*) for longitudinal excitations of the 2p-shell of Si atoms (with a binding energy $U_L = 8\,\text{Ry}$), calculated with the Hartree–Fock–Slater potential. The energy transfer is $E = 650\,\text{eV}$. The hydrogenic approximation is given by the broken line; $f(E, K)$ peaks at $(Ka_0)^2 \approx E - U_L$. In the FVP model, (91.14), the GOS is replaced by a δ-function at $Ka_0 = E^{1/2}$ and by $f(E, 0)$ for $0 < Ka_0 < E^{1/2}$ (*straight lines*)

energy loss measurements [91.52]. A cross section calculated with this model is given in Fig. 91.3. The $\sigma(E)$ differ by as much as 50% from the Bethe result, but the moments differ by at most 10% (Table 91.1). It is evidently a better approximation than that given by the binary encounter approximation of (91.10).

Table 91.1 The coefficient $\tau(\beta) = M_0\beta^2/(NZk_R)$ for pions with $M_\pi = 139.567\,\text{MeV}/c^2$, calculated in the FVP approximation. For comparison, the Si (GOS) results were calculated from the complete $F(E, K)$ generalized oscillator strengths. The estimated accuracies are $\approx 1\%$ for Si (GOS) and $\sim 30\%$ for the others. The values are the same for all heavy particles with the same β; they differ slightly for electrons. For Si $\left(\varrho = 2.329\,\text{g/cm}^3\right)$, $NZk_R = 17.82\,\text{eV}/\mu\text{m}$, for the gases (at STP), $NZk_R = 6.3828\,Z\,\text{eV/cm}$. Units of $\tau(\beta)$ are eV^{-1}

$\mathcal{T}$ (MeV)	β	Si(GOS)	Si	He	Ar	Ethane	Butane
UT 1.	0.11907	0.1208	0.1333	0.1937	0.1081	0.2961	0.2855
3.	0.20406	0.1408	0.1528	0.2229	0.1242	0.3379	0.3259
10.	0.35950	0.1612	0.1738	0.2538	0.1412	0.3821	0.3686
30.	0.56791	0.1792	0.1917	0.2802	0.1557	0.4198	0.4050
100.	0.81276	0.1957	0.2082	0.3093	0.1717	0.4614	0.4451
300.	0.94825	0.2061	0.2186	0.3441	0.1908	0.5111	0.4931
1000.	0.99247	0.2127	0.2252	0.3958	0.2189	0.5843	0.5637
3000.	0.99901	0.2150	0.2275	0.4499	0.2468	0.6580	0.6333
10000.	0.99991	0.2155	0.2280	0.5057	0.2671	0.7034	0.6646
30000.	0.99999	0.2156	0.2281	0.5332	0.2737	0.7111	0.6697
100000.	1.00000	0.2156	0.2281	0.5407	0.2765	0.7128	0.6709

91.3 Moments of the Cross Section

Various moments of $\sigma(E)$ are defined by [91.5, 53]

$$M_\nu \equiv N \int E^\nu \sigma(E)\,\mathrm{d}E\,. \tag{91.15}$$

where the range of integration covers all non vanishing $\sigma(E)$, including a summation over discrete excited states. Then $M_0 = N\sigma_{\text{tot}} = 1/\lambda_{\text{mfp}}$, where σ_{tot} is the total collision cross section and λ_{mfp} the mean free path between collisions, and $M_1 = S$, the stopping power. M_2 and M_3 give the width and skewness of $F(\Delta)$.

Higher moments are not very useful [91.53] except for special applications [91.54–56]. For incident electrons, a further averaging of energy loss over different paths must be performed because of multiple angular scattering.

91.3.1 Total Collision Cross Section M_0

A simple result for M_0 is obtained with the Rutherford cross section

$$\begin{aligned} {}_{\mathrm{R}}M_0 &= NZ_2 \int \sigma_{\mathrm{R}}(E)\,\mathrm{d}E \\ &= NZ_2 \frac{k_{\mathrm{R}}}{\beta^2}\left(\frac{1}{E_{\min}} - \frac{1}{E_{\max}}\right)\,. \end{aligned} \tag{91.16}$$

Clearly, ${}_{\mathrm{R}}M_0$ is very sensitive to the choice of $E_{\min}$, and there is no simple prescription for choosing it, as is evident from Fig. 91.3. The same applies to the binary encounter approximation [91.32]. From the Bethe model, a good relativistic approximation to M_0 is given by [91.27, 28, 57]

$$\begin{aligned} M_0 &= NZ_2 \frac{k_{\mathrm{R}}}{\beta^2} g_0 \\ &\quad \times \left[\ln \beta^2\gamma^2 - \beta^2 + h_0 + 11.227\right] \\ &= NZ_2 \frac{k_{\mathrm{R}}}{\beta^2} \tau(\beta)\,. \end{aligned} \tag{91.17}$$

Values for g_0 and h_0 may be found in [91.57]. For Si, He, Ar, ethane, and butane, values of τ_β, calculated by numerical integration of $\sigma(E)$ obtained with the FVP method, are given in Table 91.1. For Si, comparison values calculated with the Bethe model are also given. The quantity $\tau(\beta)$ is more suitable for interpolation than M_0. An extensive description of cross sections are given for liquid water in [91.58].

91.3.2 Stopping Power M_1

The stopping power is usually written in the form

$$M_1 = S = NZ_2 \frac{k_R}{\beta^2} 2B \,, \tag{91.18}$$

where B is called the stopping number. M_1 calculated with the Rutherford cross section is

$$_R M_1 = NZ_2 \int_{E_1}^{E_{max}} E\sigma_R(E)\,dE = NZ_2 \frac{k_R}{\beta^2} \ln \frac{E_{max}}{E_1} \,. \tag{91.19}$$

To obtain realistic values, it is evident from Fig. 91.3 that a small value must be chosen for E_1 to compensate for the peaks in $\sigma(E)$. Note that for most applications in high energy physics (e.g. Time Projection Chambers [91.59, 60]) M_1 is not useful [91.4].

The Bethe Model

In the Bethe model of stopping powers, $d\sigma(E, K)$ from (91.13) must be integrated over both E and $Q = \hbar^2 K^2/2m$ [91.37,38]. Since $f(E, K)$ is nearly constant near $K = 0$ [91.61] (Fig. 91.4), the integral may be broken into four parts according to

$$S = \frac{\mathcal{N} k_R}{\beta^2} \int dE \left\{ \int_{Q_m}^{Q_1} f(E, 0) \frac{dQ}{Q} + \int_{Q_1}^{\infty} f(E, K) \frac{dQ}{Q} \right.$$
$$+ \int_0^{Q_1} [f(E, K) - f(E, 0)] \frac{dQ}{Q}$$
$$\left. - \int_0^{Q_m} [f(E, K) - f(E, 0)] \frac{dQ}{Q} \right\} , \tag{91.20}$$

where $Q_m \approx E^2/2mv^2$ [91.27, 28], and Q_1 is chosen such that $f(E, K)$ differs little from $f(E, 0)$ in the interval $0 \leq K \leq \sqrt{2mQ_1}/\hbar$ [91.36, 62]. Collisions with $Q < Q_1$ are "distant" and those with $Q > Q_1$ are "close". The integrals simplify by interchanging the order of integration in the second and third terms, and by using the sum rule [91.27, 28, 34]

$$\int_0^{\infty} f(E, K)\,dE = Z_2 \,, \tag{91.21}$$

for all K. The last term of (91.20) is then small and the second last vanishes exactly. The remaining two terms give

$$S = \frac{NZ_2 k_R}{\beta^2} \left(\ln \frac{2mv^2 Q_1}{\Im^2} + \ln \frac{Q_{max}}{Q_1} \right) , \tag{91.22}$$

where $\Im$ is the logarithmic mean excitation energy defined by

$$Z_2 \ln \Im = \int f(E) \ln E\,dE \,, \tag{91.23}$$

and $Q_{max} = 2m_e c^2 \beta^2 \gamma^2$ is the maximum energy loss (for electrons, $Q_{max} = \frac{1}{4} m_e c^2 \beta^2 \gamma^2$ [91.3]). Including relativistic and other correction terms, the final result for $M_1 \gg m_e$ is

$$S = \frac{\mathcal{N} Z_2 k_R}{\beta^2} 2 \left\{ B_0 - \frac{C(\beta)}{Z_2} + Z_1 L_1(\beta) + Z_1^2 L_2(\beta) \right.$$
$$\left. + \frac{1}{2} [G(M_1, \beta) - \delta(\beta)] \right\} , \tag{91.24}$$

where

$$B_0 = \ln \frac{2mc^2 \beta^2 \gamma^2}{\Im} - \beta^2 \tag{91.25}$$

is the uncorrected stopping number. In the limit $\beta \to 0$, $2B_0$ reduces to the terms in brackets in (91.22). The other correction terms are as follows [91.31].

The Shell Correction. $C(\beta)$ accounts for the last term in (91.20), together with modifications of $d\sigma(E, Q)$ near $Q = E_{max}$ [91.3]. It can be estimated on a shell-by-shell basis [91.63–66] using nonrelativistic hydrogenic calculations for the K- and L-shells [91.35, 37, 38], and rescaling methods for the outer shells which have not been calculated directly [91.63, 64, 67]. The effects are important for small β, but simple formulas are not known. A calculation for Al and Si based on a model more realistic than the hydrogenic one is given in [91.68].

The Barkas Term. $L_1(\beta)$ arises from polarization of the target electrons by the incident particle [91.69, 70]. Various approximating functions and fits to experimental data are described in [91.71].

The Bloch Term. $L_2(\beta)$ arises from corrections to the approximation that for close collisions, the electrons can be represented by plane waves [91.72]. Confinement of the electrons to the interior of a cylinder of atomic dimensions introduces transverse momentum components, resulting in the widely used correction

$$Z_1^2 L_2(\beta) = -q^2 \sum_{j=1}^{\infty} 1/\left[j \left(j^2 + y^2 \right) \right] \tag{91.26}$$

with $y = Z_1\alpha/\beta$. For $y = 0$, the sum is $\zeta(3) \simeq 1.202\,057$. A new approach can be found in [91.73, 74]. The need for the L_1 and L_2 terms was established experimentally in [91.75], and discussed in a more general context in [91.76].

The Mott Term. $G(M_1, \beta)$ is a kinematic recoil correction which becomes important for relativistic projectiles. In the limit of a point-like spinless nucleus, the correction is

$$G(M_1, \beta) = -\ln(1+2t) - t\beta^2/\gamma^2\,, \tag{91.27}$$

where $t = m\gamma/M_1$ [91.29]. The correction is negligible for $\gamma < 100$ and $M_1 > m_p$.

The Density Correction. $\delta(\beta)$ arises from the dielectric response of a solid absorber as a whole to the electric field of the projectile, and the work done by the interaction [91.3, 29]. Sternheimer's algorithm [91.77, 78] is usually used, as summarized in [91.29].

The remaining parameter to be discussed in (91.24) is $\Im$. If the oscillator strength distribution $f(E)$ is known, then $\Im$ can be calculated directly from (91.23). Results for many gases are given in [91.45]. The values give good agreement with experimental stopping powers [91.79]. For solids, the definition [91.3]

$$\ln \Im = \frac{2}{\pi(\hbar\omega_p)^2} \int_0^\infty E \,\mathrm{Im}\left[\frac{-1}{\epsilon(E)}\right] \ln E \, dE \tag{91.28}$$

may be used, where $\hbar\omega_p$ is the plasma energy for all the electrons, defined by $(\hbar\omega_p)^2 = 830.4\,\varrho Z/A$ with $\hbar\omega_p$ in eV. For metals, this is substantially larger than the plasmon energy associated with the conduction electrons. Only for Al [91.80] and for water [91.81] have sufficiently good measurements of $\mathrm{Im}[-1/\epsilon(E)]$ been available to permit the use of this method. For other materials, $\Im$ can be deduced from measurements of S, provided that the other corrections in (91.24) are known sufficiently well. A list of values for all elements and many compounds is given in [91.66]. See also [91.25, 63–65, 79, 82–87].

For rough estimates of $\Im$, the approximation [91.88]

$$\Im \approx \begin{cases} 11.7 + 11.2/Z_2\mathrm{eV}\,, & Z_2 \le 13 \\ 9.5 \pm 1\mathrm{eV}\,, & Z_2 > 13 \end{cases} \tag{91.29}$$

is useful, together with the Bragg rule

$$n_e \ln \Im = \sum_i n_i \ln \Im_i \tag{91.30}$$

for compounds and composite materials, where n_i is the electron density associated with element i. However, chemical shifts may be as large as 10% [91.89].

The Fermi Virtual Photon Method

From (91.14), the nonrelativistic FVP approximation to the stopping number $B(\beta)$ is

$$B(\beta) = \frac{1}{2} \ln\left(\frac{2m_e v^2}{\Im}\right) + \frac{1}{2Z_2} \int_{E_{min}}^{E_{max}} \frac{dE}{E} \int_0^E f(E')\, dE'\,. \tag{91.31}$$

Although the integrals must now be calculated numerically, the full $f(E, K)$ is not required, and in the case of silicon, the results are in close agreement with the corresponding Bethe model.

Stopping Power at Small Speeds

For small speeds, the various correction terms in the stopping number B, (91.24), become large compared with B_0 [91.63, 64, 68]. In particular, for $2mv^2 = \Im$, B_0 becomes zero. For example, for α-particles in U ($\Im = 840$ eV), $B_0 = 0$ at $T = 1.5$ MeV, and B then consists only of correction terms. For smaller energies, empirical approaches are used to describe S. Many of the tables referenced above give such data.

Mean Energy Loss per Collision

The quantity $\langle E \rangle = M_1/M_0$ is the mean energy loss per collision. For substances with $Z < 20$, $\langle E \rangle \sim 50$–$100$ eV. It changes at most by a factor of 1.5 for $\beta\gamma > 0.1$. In order to choose a suitable method for calculating straggling functions, it is useful to estimate the number of collisions in a thickness x of absorber. For less than 2000 collisions, the convolution method of Sect. 91.4.2 should be used (Fig. 15 in [91.5]).

91.3.3 Second Moment M_2

The relativistic result calculated with the Rutherford cross section is

$$\begin{aligned} {}_{R}M_2 &= NZ_2 \frac{k_R}{\beta^2}\left(1 - \beta^2/2\right) E_{max} \\ &= 2NZ_2 k_R mc^2 \frac{1-\beta^2/2}{1-\beta^2} \\ &= 0.156\,915 \frac{Z_1^2 Z_2}{A} \frac{1-\beta^2/2}{1-\beta^2}\ \mathrm{MeV}^2\,\mathrm{cm}^2/\mathrm{g}\,. \end{aligned} \tag{91.32}$$

For small β, ${}_{R}M_2$ is practically independent of β. A better approximation can be achieved with the binary encounter method:

$$_bM_2 = NZ_2\frac{k_R}{\beta^2}E_{max}\left[1-\frac{\beta^2}{2} + \sum_l \frac{4Z_lU_l}{3Z_2E_{max}}\left(\ln\frac{E_{max}}{_lE_{min}} - \beta^2\right)\right], \quad (91.33)$$

where the summation extends over atomic shells l. For $_lE_{min}$ the ionization energy J_l for shell l can be used, while U_l represents the kinetic energy of the electrons in shell l. This approximation is only useful for relatively large x [91.5]. By using sum rules, *Fano* [91.3, (72)] achieved a better approximation, which corresponds to the Bethe approximation for the stopping power [91.5,90]. Hydrogenic values of M_n are calculated in [91.53].

91.4 Energy Loss Straggling

91.4.1 Straggling Parameters

Parameters for Thick Absorbers

Straggling is due to the stochastic nature of the energy losses of the charged particles. Because the single collision spectrum is highly skewed (the most probable energy loss is ≈ 20 eV, the mean value ≈ 100 eV), straggling functions will also be skewed. Four parameters are useful in a preliminary study of straggling problems [91.47]. The parameter

$$\xi = NZ_2\frac{k_R}{\beta^2}x = 153.537\frac{Z_1^2}{\beta^2}\frac{Z_2}{A}x \text{ keV cm}^2 \quad (91.34)$$

gives the energy loss [calculated with $\sigma_R(E)$] which, on the average, is exceeded once for each particle in its passage through an absorber of thickness x, i. e., for $\Im \ll \xi \ll E_{max}$,

$$xNZ_2\int_\xi^{E_{max}} \sigma_R(E)\,dE = 1\,. \quad (91.35)$$

The parameter $\kappa = \xi/E_{max}$ is related to the skewness of the straggling function: $\gamma_1^2 = 1/4\kappa$. The mean energy loss is

$$\langle\Delta\rangle = xM_1 = 2\xi 2B = 2\kappa E_{max}B\,, \quad (91.36)$$

and the standard deviation of $F(\Delta)$ is

$$\omega^2 = xM_2 \sim xNZ_2\frac{k_R}{\beta^2}E_{max} = \xi E_{max} = \kappa E_{max}^2\,. \quad (91.37)$$

For thick absorbers, the straggling function becomes approximately Gaussian [91.5,24]. The requirement for this to occur is $\gamma_1 \to 0$, thus $\kappa \to \infty$, also $\langle\Delta\rangle \to \infty$, and $x \to \infty$.

Parameters for Thin Absorbers

The parameters described above are not suitable for describing $F(\Delta)$ for very thin absorbers. Instead, Δ_{mp} and Γ are used. Values for pions traversing Ar are given as a function of thickness x in Tables 91.2, 91.3. However, before comparing with experimental data, values of Δ must be converted into ionization values, and the detector and amplifier noise must be added [91.5] (Sect. 91.7). *Landau* [91.6] gave an expression for the most probable energy loss as a function of particle speed. It was modified in [91.5] to

$$\Delta_L = \xi\left(\ln\frac{2mc^2\beta^2\gamma^2}{I} + \ln\frac{\xi}{I} + 0.200 - \beta^2 - \delta\right). \quad (91.38)$$

Table 91.2 Calculated most probable energy loss Δ_{mp} of pions with $Z_1 = \pm 1$ and kinetic energy $\mathcal{T}$ passing through a distance x of argon gas (at 760 Torr, 293 K, $\varrho = 1.66\,g/dm^3$). For heavy ions (α, C^+), the values scale as $Z_1^{1.3}$ (for a better approximation, see Table V in [91.5]). Units of Δ_{mp} are keV. The quotient Δ_{mp}/x increases with x

	x (cm)				
$\mathcal{T}$ (MeV)	0.5	1	2	4	8
1	29.373	63.341	130.662	265.062	533.811
3	9.820	21.598	47.000	101.329	216.013
10	3.104	6.842	14.950	32.558	69.966
30	1.153	2.687	5.927	12.883	28.023
100	0.376	1.274	2.928	6.412	13.884
300	0.288	0.973	2.313	5.099	10.993
1×10^3	0.303	1.042	2.445	5.356	11.494
3×10^3	0.344	1.217	2.783	6.044	12.926
1×10^4	0.384	1.390	3.145	6.784	14.495
3×10^4	0.653	1.515	3.405	7.322	15.663
1×10^5	0.667	1.605	3.595	7.725	16.575

Table 91.3 Calculated values of Γ (fwhm) of the straggling function $F(\Delta)$ (see Table 91.2). As a rough approximation, $\Gamma \sim x^{0.8}$ (Table VI in [91.5])

	x(cm)				
$\mathcal{T}$(MeV)	**0.5**	**1**	**2**	**4**	**8**
1	16.014	24.961	35.503	51.327	69.120
3	6.283	12.012	22.928	43.393	68.589
10	2.507	4.563	8.659	16.153	30.026
30	1.360	2.279	3.922	7.441	13.881
100	0.781	1.434	2.350	4.178	7.705
300	0.574	1.239	2.030	3.410	6.232
1×10^3	0.621	1.286	2.085	3.397	6.340
3×10^3	0.752	1.391	2.244	3.687	6.844
1×10^4	0.905	1.510	2.432	3.999	7.502
3×10^4	0.983	1.615	2.548	4.314	7.956
1×10^5	1.040	1.701	2.689	4.666	8.399

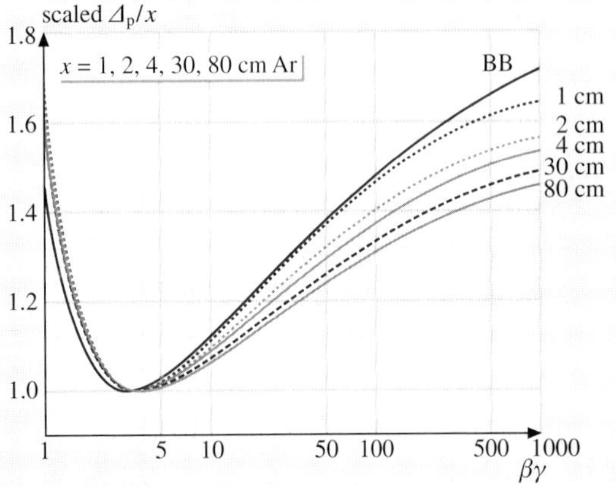

Fig. 91.5 The dependence of most probable energy-loss values $\Delta_{\mathrm{mp}}(\beta\gamma; x)/x >$ for segments of length x. *Solid line* BB: the Bethe–Bloch function $\mathrm{d}E/\mathrm{d}x(\beta\gamma)$ [91.4]. *Other lines* are for the segment lengths x marked at right. The functions are scaled with a factor $g(x)$ such that they concide at minimum ionization

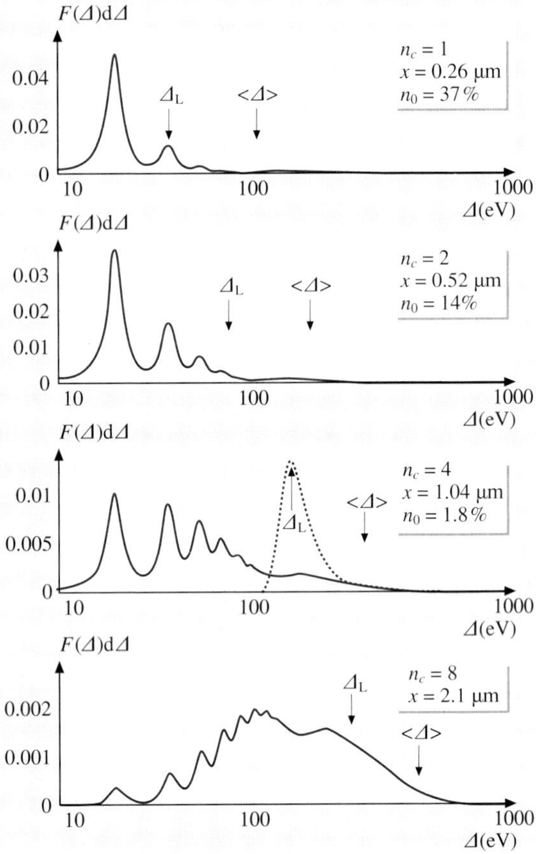

Fig. 91.6 Calculated straggling functions $F(\Delta)$ for 1 GeV pions traversing four thicknesses x of silicon. The average number of collisions is n_c, with $\langle E\rangle = 106\,\mathrm{eV}$, the mean energy loss is $\langle\Delta\rangle = n_c\langle E\rangle$, shown by an arrow. The number of particles traversing the absorber without a collision is n_0. The peak heights do not follow a Poisson distribution because successive convolutions give a broader distribution. The Landau function [91.6] for $n_c = 4$ is shown by the broken line. For all, the most probable energy loss calculated according to Landau, Δ_L (91.38), is shown

More accurate functions Δ_{mp} are obtained with the collision spectra of Fig. 91.3, and are called Bichsel functions [91.59]. For sufficiently large ξ, Δ_{mp}, and Δ_L agree within a few %. Examples are shown in Fig. 91.5, and for comparison the Bethe-Bloch function $\mathrm{d}E/\mathrm{d}x$ [91.4] is given.

Parameters for Extremely Thin Absorbers

If the number n_c of collisions in the absorber is less than about 16, $F(\Delta)$ still shows the details of $\sigma(E)$ (Fig. 91.6). The most probable energy losses are much less than $\langle\Delta\rangle$. Functions of this type have been observed with electron microscopes [91.10]. For comparison, the Landau straggling function [91.6] is also shown. This result, derived from $\sigma_R(E)$, does not show the structure of a realistic spectrum.

Detector noise and energy loss to ionization conversions introduce important changes [91.91–94] to the energy loss spectrum. In particular, if $n_c < 4$, a large fraction of the particles do not lose any energy in the absorber.

91.4.2 Analytic Methods for Calculating Energy Loss Straggling Function

Three principal methods are convolutions, Laplace transforms and the use of moments [91.5]. They are practical for different thicknesses of absorbers, as indicated.

Convolutions

For this method, a complete collision cross section must be available [91.47,50], and numerical calculations are required. The probability density function for particles having suffered n collisions is given by the n-fold convolution of the single collision spectrum [91.5,95]:

$$\sigma^{*n}(\Delta) = \int_0^{\Delta} \sigma(E)\sigma^{*(n-1)}(\Delta - E)\,\mathrm{d}E\,, \tag{91.39}$$

with

$$\sigma^{*0}(\Delta) = \delta(\Delta)\,, \quad \sigma^{*1}(\Delta) = \sigma(\Delta)\,. \tag{91.40}$$

Assuming that successive collisions are statistically independent, the number of collisions is described by the Poisson distribution

$$P_n^{n_c} = \frac{n_c^n}{n!}\mathrm{e}^{-n_c}\,, \tag{91.41}$$

where $P_n^{n_c}$ gives the fraction of particles suffering n collisions, and $n_c = xM_0$ is the average number of collisions for all particles. The complete straggling function is then

$$F(x,\Delta) = \sum_{n=0}^{\infty} P_n^{n_c}\sigma^{*n}(\Delta)\,. \tag{91.42}$$

Usually, this method is only used for absorbers thin enough that the single collision cross section changes negligibly during the passage through the absorber.

This approach would become quite complex for large n_c. The doubling method [91.96] uses the following procedure: the function for an absorber of thickness $2x$ is calculated by convoluting $F(x,\Delta)$ with itself, viz.

$$F(2x,\Delta) = \int_0^{\Delta} F(x,\Delta - g)F(x,g)\,\mathrm{d}g\,. \tag{91.43}$$

An initial distribution is calculated for an extremely thin absorber of thickness $\mathrm{d}x$ from

$$F(\mathrm{d}x,\Delta) = \delta(\Delta)(1 - M_0\,\mathrm{d}x) + \sigma(\Delta)\,\mathrm{d}x\,, \tag{91.44}$$

where $\delta(\Delta) = F(0,\Delta)$ and $M_0\,\mathrm{d}x \approx 0.001$.

For Si, the convolution method has been tested and compared with experiments for as many as 10^4 collisions [91.5]. The limitations of the algorithm are $\approx 10^5$ collisions. Since $P_0^{n_c} = \mathrm{e}^{-n_c}$, there will be particles that pass through an absorber without making a collision.

Laplace Transforms

The Laplace transform method is mathematically equivalent to the convolution method. The method, as implemented by *Landau* [91.6], is based on $\sigma_R(E)$. The implicit number of collisions is therefore too large, and the Landau function is too narrow, as shown in Figs. 91.1 and 91.6. A refinement proposed in [91.7] gives straggling functions which are too wide (Fig. 91.1).

Use of Moments

If the skewness γ_1 is not small, a Gaussian function is not a good approximation to a straggling function. This can be remedied by using a modified Gaussian of the form [91.55,56]

$$f(y) = \frac{\tilde{\delta}}{\sqrt{2\pi}}\frac{1}{\sqrt{y^2+1}} \times \exp\left\{-\frac{1}{2}\left[\tilde{\gamma} + \tilde{\delta}\ln\left(y + \sqrt{y^2+1}\right)\right]^2\right\}\,, \tag{91.45}$$

where $y = (\Delta - \langle\Delta\rangle)/\sigma$ and $\tilde{\gamma}$ and $\tilde{\delta}$ are related to M_2 and M_3 [91.55,56] such that the second and third moments are reproduced exactly.

The moments can be calculated as a function of particle speed, and thus the method is useful for mean energy losses up to about $0.7\mathcal{T}_0$. At these losses, some particles reach their full range and thus disappear from the beam. The moments are then distorted and other procedures are needed [91.11,97].

Thick Absorbers, Ranges and Range Straggling

In radiation therapy with charged particles, a parallel beam is directed at the body. Usually, the energy $\mathcal{T}_0$ of the particles is adjusted to penetrate a given distance into the body. A primary measure of this distance is the mean range R calculated with the continuous slowing down approximation (CSDA)

$$R(\mathcal{T}_0) = \int_0^{\mathcal{T}_0} \frac{\mathrm{d}T}{S(T)}\,. \tag{91.46}$$

Because of energy-loss straggling, there also is straggling in range amounting to a few percent of R [91.84].

In the practical application, a depth-dose curve is needed. At a given depth, the dose D is given by (91.6). For the first 50% of the range, the spread in energy of the beam is relatively small, and the dose is approximated closely by the stopping power $S(\mathcal{T})$ for the mean energy of the beam. For larger thicknesses, a detailed transport calculation must be made in order to find $\phi(\mathcal{T}, x)$ [91.11]. Range straggling functions can then also be obtained. Close agreement between measurement and calculation has been achieved [91.97].Similarly, the mean energy loss $\mathcal{T}_1$ in a thick absorber is given implicitly by

$$x = \int_{\mathcal{T}_1}^{\mathcal{T}_0} \frac{dT}{S(T)} \,. \qquad (91.47)$$

Usually, values are obtained from a range energy table. Such tables may be found in [91.25, 63–65, 79, 82–87].

91.4.3 Particle identification (PID)

In particle physics the determination of particle types produced in the collisions of fast particles is needed [91.59]. Since the cross sections in Sect. 91.2 only depend on particle speed β and charge $Z_1 e$, energy deposition measurements combined with the measurement of particle momentum p can be used to determine particle masses. This process is called particle identification. A large amount of empirical information about PID has been accumulated, but little effort has been made to correlate this information with the theory described here. A major problem is the dependence of Δ_{mp}/x, Table 91.2, on x. In addition, Δ_{mp}/x is less than $\mathrm{d}E/x$ and its dependence on particle speed, Fig. 91.5, differs from that of $\mathrm{d}E/\mathrm{d}x$, Fig. 91.5. Comparisons of calculated and measured functions $F(\Delta)$ can be used to diagnose problems of the performance of the particle detector.

91.5 Multiple Scattering and Nuclear Reactions

Coulomb scattering of particles heavier than electrons by nuclei and electrons usually produces many, but very small angular deflections [91.4, 98].

The Molière theory [91.99] giving the distribution $f(\theta)$ of angular deflections θ agrees well with experimental data [91.100]. As a first approximation, a Gaussian distribution $f(\theta) = \mathrm{e}^{-y^2}$, where $y = \theta/\theta_0$, can be used. The value θ_0 is given by $\theta_0^2 = \theta_1^2 B_M$, where

$$\theta_1^2 = 0.157 \frac{Z_2(Z_2+1)}{A} Z_1^2 \frac{x\gamma^2}{\left(Mc^2\beta^2\right)^2} \text{radians} \qquad (91.48)$$

and B_M has values of the order of 10. A more detailed description plus several tables is given in [91.84].

For a broad beam of particles, multiple scattering produces only a minor correction in the range. In a narrow "pencil" beam, it causes a broadening, e.g., a 320 MeV proton beam spreads to a width of about 2 cm after traveling through 40 cm of water [91.101].

It may be necessary to take into account the influence of nuclear interactions. As a first approximation, the total cross section is roughly [91.25]

$$\sigma_{\mathrm{tot}} = \pi \left(1.3 \times 10^{-13} A^{1/3} + \lambda\!\!\!^{-}\right)^2 \mathrm{cm}^2 \,, \qquad (91.49)$$

where $\lambda\!\!\!^{-} = \lambda_{\mathrm{d}B}/2\pi$.

91.6 Monte Carlo Calculations

In Monte Carlo calculations, the interactions occurring during the passage of the particles through matter can be simulated one at a time, collision by collision, and including those for the δ-rays [91.102, 103]. Particles travel random distances x_i between successive collisions, calculated by selecting a random number r_i and determining the distance to the next collision from

$$x_i = -\lambda_{\mathrm{mfp}} \ln r_i \,. \qquad (91.50)$$

The energy loss E_i is selected with a second random number from the integrated collision spectrum

$$Q(E) = \int_0^E \frac{\sigma(E)}{M_0} \mathrm{d}E \,. \qquad (91.51)$$

This process is repeated until $\sum x_i$ exceeds the absorber thickness. The total energy loss Δ of the particle is $\Delta = \sum_i E_i$.

91.7 Detector Conversion Factors

For all detectors, a study must be made of the process in which energy lost by the particles is converted into the observable signal, see Chapt. 92. The primary products of collisions are excited states, electron–ion and electron–hole pairs, and secondary electrons also called δ-rays. Secondary products are chemical species, fluorescent photons and Auger electrons. An important problem is the export of energy from the volume under observation by secondary radiations (δ-rays, mesons, X-rays, neutrons). Prominent among the effects [91.104] are ionization currents measured under applied electrical fields (in gaseous and solid state ionization chambers), photo effects by fluorescent radiation (scintillators), measurements of chemical yields (ferrous sulphate dosimeters, photographic emulsions, etc.), and the release of light by stored crystalline defects in thermoluminescent measurements.

The effect J can be related to the energy deposited D by a conversion factor W as $J = D/W$, where W represents the amount of energy needed to produce a unit of observable effect (e.g., an electron–ion pair). W depends on the absorber, particle type, and speed. For ionization, typical values are $W \approx 25\,\text{eV}$ gases, $W \approx 3.5\,\text{eV}$ semiconductors [91.84, 105].

For chemical effects [91.106], J is sometimes written as $J = YD$. Usually Y is given as the "yield per 100 eV". A typical value for a Fricke dosimeter is that, for fast electrons, 15 ferrous ions are converted into ferric ions per 100 eV of deposited energy. Usually, Y depends on time after the passage of the particles, and the full values are only reached after say 1 μs. For slow ions, Y is much smaller. Light emission (scintillations) has been observed from gases (typically, $W \approx 20\,\text{eV}$), liquids and solids (typically, $W \approx 1000\,\text{eV}$) [91.104].

The subsequent measurements of these radiation effects are done with current or charge amplifiers, proportional avalanche amplification, photo multipliers, channel plates, etc. [91.104].

References

91.1 M. E. Rudd, Y.-K. Kim, D. H. Madison, T. J. Gay: Rev. Mod. Phys. **64**, 441 (1992)

91.2 International Commission on Radiation Units and Measurements Report 55 (7910 Woodmont Ave., Bethesda, Maryland 1984)

91.3 U. Fano: Ann. Rev. Nucl. Sci. **13**, 1 (1963)

91.4 Particle Data Group: Review of Particle Physics, Phys. Lett. B **592**, 1 (2004)

91.5 H. Bichsel: Rev. Mod. Phys. **60**, 663 (1988)

91.6 L. Landau: J. Phys. (Moscow) **8**, 201 (1944)

91.7 O. Blunck, S. Leisegang: Z. Phys. **128**, 500 (1950)

91.8 H. Cramér: *The Elements of Probability Theory* (Wiley, New York 1955)

91.9 H. Bichsel: Microdosimetry, Rad. Protection Dosim. **13**, 91 (1985)

91.10 J.Ph. Perez, J. Sevely, B. Jouffrey: Phys. Rev. A **16**, 1061 (1977)

91.11 H. Bichsel, T. Hiraoka: Int. J. Quantum Chem. **23**, 565 (1989)

91.12 H. Bichsel, R. F. Mozley, W. Aron: Phys. Rev. **105**, 1788 (1957)

91.13 J. L. Dehmer, M. Inokuti, R. P. Saxon: Phys. Rev. A **12**, 102 (1975)

91.14 U. Fano, J. W. Cooper: Rev. Mod. Phys. **40**, 441 (1968)

91.15 AAPM report No. 16 *Protocol for heavy charged-particle therapy beam dosimetry*, published for the Amer. Assoc. of Physicists in Medicine by the A.I.P. (1986)

91.16 International Commission on Radiation Units and Measurements Report 59 (7910 Woodmont Ave., Bethesda, Maryland 1998)

91.17 C. C. Ahn, O. L. Krivanek, R. P. Burgner, M. M. Disko, P. R. Swann: *EELS Atlas*, HREM Facility (Arizona State University, Tempe 1980)

91.18 U. Killat: Z. Phys. **263**, 83 (1973)

91.19 C. D. Wilson, C. A. Dukes, R. A. Baragiola: Phys. Rev. B **63**, 121101 (R) (2001)

91.20 M. Isaacson: J. Chem. Phys. **56**, 1813 (1972)

91.21 P. Pianetta, T. W. Barbee Jr.: Nucl. Instrum. Meth. A **266**, 441 (1988)

91.22 J. J. Rehr, R. C. Albers: Rev. Mod. Phys. **72**, 621 (2000)

91.23 E. Fermi: Z. Phys. **29**, 315 (1924)

91.24 N. Bohr: Dan. Mat. Fys. Medd. **18**(8), 1 (1953)

91.25 H. Bichsel: Charged-particle interactions. In: *Radiation Dosimetry*, Vol. 1, 2nd edn., ed. by F. H. Attix, Wm. C. Roesch (Academic, New York 1968) Chap. 4, p. 157

91.26 R. D. Evans: *The Atomic Nucleus* (McGraw Hill, New York 1967) pp. 838–851

91.27 M. Inokuti: Rev. Mod. Phys. **43**, 297 (1971)
91.28 M. Inokuti: Rev. Mod. Phys. **50**, 23 (1978)
91.29 S. P. Ahlen: Rev. Mod. Phys. **52**, 121 (1980)
91.30 E. A. Uehling: Ann. Rev. Nucl. Sci. **4**, 315 (1954)
91.31 D. E. Groom, N. V. Mokhov, S. I. Striganov: Atomic Data and Nuclear Data Tables **78**, 183 (2001)
91.32 H. Bichsel: Scanning electron microscopy supplement, **4**, 147 (1990)
91.33 M. E. Rudd, J. H. Macek: Case Studies in Atomic Physics **3**, 47–136 (1972)
91.34 H. Bethe: Ann. Phys. **5**, 325 (1930)
91.35 H. Bichsel: Phys. Rev. A **28**, 1147 (1983); Calculations for L-shell electrons have not yet been published
91.36 G. S. Khandelwal: Phys. Rev. A **26**, 2983 (1982)
91.37 M. C. Walske: Phys. Rev. **88**, 1283 (1952)
91.38 M. C. Walske: Phys. Rev. **101**, 940 (1956)
91.39 S. T. Manson: Phys. Rev. A **6**, 1013 (1972)
91.40 J. Lindhard, A. Winther: Mat. Fys. Medd. Dan. Vid. Selsk. **34**(4), 1 (1964)
91.41 C. J. Tung, R. H. Ritchie, J. C. Ashley, V. E. Anderson: ORNL Report TM-5188 (1976)
91.42 R. E. Johnson, M. Inokuti: Comments At. Mol. Phys. **14**, 19 (1983)
91.43 E. D. Palik: *Handbook of Optical Constants of Solids* (Academic, Orlando 1985)
91.44 J. W. Gallagher, C. E. Brion, J. A. R. Samson, P. W. Langhoff: J. Phys. Chem. Ref. Data **17**, 9–153 (1988)
91.45 B. L. Jhanwar, Wm. J. Meath, J. C. F. MacDonald: Rad. Res. **96**, 20 (1983)
91.46 J. Berkowitz: *Atomic and Molecular Photoabsorption* (Academic, San Diego 2002)
91.47 H. Bichsel, R. P. Saxon: Phys. Rev. A. **11**, 1286 (1975)
91.48 W. W. M. Allison, J. H. Cobb: Ann. Rev. Nucl. Part. Sci. **30**, 253 (1980)
91.49 D. Liljequist: J. Phys. D **16**, 1567 (1983)
91.50 E. J. Williams: Proc. R. Soc. A **125**, 420 (1929)
91.51 J. H. Barkyoumb, D. Y. Smith: Phys. Rev. A **41**, 4863 (1990)
91.52 H. Raether: *Excitation of Plasmons and Interband Transit Ions by Electrons* (Springer, New York 1980)
91.53 H. Bichsel: Phys. Rev. B **1**, 2854 (1970)
91.54 P. Sigmund, U. Haagerup: Phys. Rev. A **34**, 892 (1986)
91.55 C. Tschalär: Nucl. Instrum. Meth. **61**, 141 (1968)
91.56 C. Tschalär: Nucl. Instrum. Meth. **64**, 237–243 (1968)
91.57 M. Inokuti, R. P. Saxon, J. L. Dehmer: Int. J. Radiat. Phys. Chem. **7**, 109 (1975)
91.58 M. Dingfelder, M. Inokuti, H. G. Paretzke: Radiation Physics and Chemistry **59**, 255 (2000)
91.59 STAR website: www.star.bnl.gov
91.60 STAR website: www.star.bnl.gov/~bichsel
91.61 H. A. Bethe, L. M. Brown, M. C. Walske: Phys. Rev. **79**, 413 (1950)
91.62 J. M. Peek: Phys. Rev. A **27**, 2384 (1983)
91.63 H. Bichsel: Phys. Rev. A **46**, 5761 (1992)
91.64 Stopping power of charged particles for heavy elements, Nat. Inst. Stand. Tech. Report, 4550 (1991)
91.65 J. F. Janni: At. Data Nucl. Data Tables **27**, 147 (1982)
91.66 International Commission on Radiation Units and Measurements Report No. 49 (7910 Woodmont Ave., Bethesda, Maryland 1993)
91.67 J. O. Hirschfelder, J. L. Magee: Phys. Rev. **73**, 207 (1948)
91.68 H. Bichsel: Phys. Rev. A **65**, 052709–1 (2002)
91.69 J. Lindhard: Nucl. Inst. Meth. **132**, 1 (1976)
91.70 J. Lindhard: Phys. Med. Biol. **37**, 2139 (1992)
91.71 H. Bichsel: Phys. Rev. A **41**, 3642 (1990)
91.72 F. Bloch: Ann. Phys. (5. Folge) **16**, 285 (1933)
91.73 J. Lindhard, A. H. Sœrensen: Phys. Rev. A **53**, 2443 (1996)
91.74 A. H. Sœrensen: Phys. Rev. A **55**, 2896 (1997)
91.75 H. H. Andersen, J. F. Bak, H. Knudsen, B. R. Nielsen: Phys. Rev. A **16**, 1929 (1977)
91.76 G. Basbas: Nucl. Instrum. Meth. B **4**, 227 (1984)
91.77 R. M. Sternheimer, S. T. Seltzer, M. J. Berger: Phys. Rev. B **26**, 6067 (1982)
91.78 R. M. Sternheimer, S. T. Seltzer, M. J. Berger: At. Data Nucl. Data Tables **30**, 261 (1984)
91.79 H. Bichsel, L. E. Porter: Phys. Rev. A **25**, 2499 (1982)
91.80 E. Shiles, T. Sasaki, M. Inokuti, D. Y. Smith: Phys. Rev. A **22**, 1612 (1980)
91.81 M. Dingfelder, D. Hantke, M. Inokuti, H. G. Paretzke: Radiation Physics, Chemistry **53**, 1 (1998)
91.82 International Commission on Radiation Units and Measurements Report 37 (7910 Woodmont Ave., Bethesda, Maryland 1984)
91.83 H. H. Andersen, J. F. Ziegler: *HYDROGEN Stopping Powers and Ranges in all Elements* (Pergamon, New York 1977)
91.84 H. Bichsel: *Am. Inst. Phys. Handbook*, ed. by D. E. Gray (McGraw Hill, New York 1972) pp. 8–189
91.85 F. Hubert, R. Bimbot, H. Gauvin: At. Data Nucl. Data Tables **46**, 1 (1990)
91.86 M. Inokuti (Ed.): *Coordinated Research Program on Atomic and Molecular Data for Radiotherapy and Related Research*, Report IAEA-TECDOC-799 (International Atomic Energy Agency, Vienna 1995) (Out of print)
91.87 J. F. Ziegler: *HELIUM Stopping Powers and Ranges in all Elements* (Pergamon, New York 1977) (Note that some of data in Table 1 of this reference are incorrect: apparently the "low energy electronic stopping" was used between 10 and 30 MeV)
91.88 P. Dalton, J. E. Turner: Health Phys. **15**, 257 (1968)
91.89 C. Tschalär, H. Bichsel: Phys. Rev. **175**, 476 (1968)
91.90 H. Bichsel: Phys. Rev. A **9**, 571 (1974)
91.91 J. Va'vra, P. Coyle, J. Kadyk, J. Wise: Nucl. Instrum. Meth. A **324**, 113 (1993)

91.92 D. Srdoc, B. Obelic, I. K. Bronic: J. Phys. B **20**, 4473 (1987)
91.93 G. W. Fraser, E. Mathieson: Nucl. Instrum. Meth. A **247**, 544 (1986)
91.94 B. Obelic: Nucl. Instrum. Meth. A **241**, 515 (1985)
91.95 J. R. Herring, E. Merzbacher: J. Elisha Mitchell Sci. Soc. **73**, 267 (1957)
91.96 A. M. Kellerer: *Mikrodosimetrie*, G. S. F. Bericht B-1 (Strahlenbiologisches Institut der Universität München, München 1968)
91.97 H. Bichsel, T. Hiraoka, K. Omata: Radiation Research **153**, 208 (2000)
91.98 Wm.T. Scott: Rev. Mod. Phys. **35**, 231 (1963)
91.99 G. Molière: Z. Natf. **6**, 1013 (1955)
91.100 H. Bichsel: Phys. Rev. **112**, 182 (1958)
91.101 H. Bichsel, K. M. Hanson, M. E. Schillaci: Phys. Med. Biol. **27**, 959 (1982)
91.102 H. Bichsel: Nucl. Instrum. Meth. B **52**, 136 (1990)
91.103 S. Agostinelli et al.: GEANT4 - a simulation toolkit, Nucl. Instrum. Meth. A **506**, 250 (2003)
91.104 G. F. Knoll: *Radiation Detection and Measurement*, 3rd edn. (Wiley, New York 1999)
91.105 International Commission on Radiation Units and Measurements Report No. 31 (7910 Woodmont Ave., Bethesda Maryland 1982)
91.106 Farhataziz, M. A. J. Rodgers: *Rad. Chem. Principles and Applications* (VCH Publishers, New York 1987)

92. Radiation Physics

Radiation physics entails studies of the interactions of ionizing radiation with matter. The term ionizing radiation refers to any energetic particles, either charged or uncharged, that can ionize atoms or molecules in matter. These particles include photons in the ultraviolet, X-ray, or γ-ray spectral region; electrons and positrons; mesons; protons and deuterons; α-particles; heavier ions including molecular ions; and neutrons. The matter under consideration includes substances in every phase of atomic aggregation (i. e., gas, liquid, solid, or plasma). Strictly speaking, the term "ionization" signifies an event in which at least one electron leaves an atom or molecule and eventually becomes free. This notion applies to a dilute gas, but not to condensed matter, for which it would be more precise to use the term "electronic activation" encompassing all modes of excitation, as sketched in Sect. 92.3.5.

92.1 General Overview

The motivation for studying the interactions of radiation with matter and their consequences is manifold. First, nuclear and particle physics experiments often require radiation detectors – devices that detect, score, or analyze energetic particles. The working principles of these detectors rest on knowledge of elementary processes of radiation interactions with matter [92.1]. Second, measurements of radiation fields, called dosimetry [92.2], are important to many purposes, including the protection of workers in industry and medicine, as well as of the general public; risk estimates for exposures to radiation from natural sources (such as cosmic rays and terrestrial radioactivity) or from artificial sources (such as accelerators, man-made radioisotopes, and nuclear-power facilities); and good performance of radiation diagnosis and therapy. The principles of dosimetry are largely based on radiation physics. Finally, ionizing radiation is useful for processing materials such as plastics and semiconductors, to endow desirable properties. To optimize methods of radiation processing, knowledge of radiation physics is highly valuable.

Atomic, molecular, and optical physics provides important underpinnings for radiation physics in two general contexts: instrumentation for the measurement of radiation and its effects on matter on the one hand, and elucidation of the atomic and molecular mechanisms leading to the effects on the other. The term instrumentation here means all devices that enable one to detect, identify, and quantify consequences of radiation interactions, such as particle kinetic energies and ionization, luminescence, and spectral analyses of atomic and molecular species produced. Conversely, needs in radiation physics have led to new atomic and molecular techniques such as resonance ionization spectroscopy [92.3], which was originally developed to determine the yield of excited atoms in gases exposed to radiation, but is now regarded as a branch of multiphoton spectroscopy valuable for studies of excited states of atoms and molecules (Chapt. 74). Thus, the instrumentation development constitutes a mutually beneficial interface between atomic, molecular, and optical physics and radiation physics.

The elucidation of microscopic mechanisms of radiation effects requires a grand synthesis of knowledge from various areas ranging from atomic, molecular, and optical physics to condensed matter physics, statistical physics, chemical kinetics, and molecular biology [92.4, 5]. Yet, the role of atomic, molecular, and optical physics is most fundamental [92.6, 7]. Interactions of radiation with matter lead to numerous collisions of energetic particles with atoms and molecules, and result in the production of various excited and ionized states, as well as electrons and other secondary particles. These have a wide range of kinetic energies, and in turn collide with atoms and molecules to generate further products. A detailed analysis of the chain of these early events demands knowledge in three topical areas: first, spectroscopic data (i. e., energy values and quantum numbers) of excited and ionized states; second, modes of decay and relaxation of these states, such as luminescence, molecular dissociation, and internal conversion and the branching ratios of each mode; and finally, the cross sections for all of the processes involved. All such knowledge is, in a broad sense, the fruit of work in atomic, molecular, and optical physics.

The following sections will concern key topics in basic radiation physics selected from a vast range of possibilities. The treatment will concentrate on basics and principles, and examples are meant to be illustrative rather than exhaustive; yet, the author hopes that the discussion will convey the charm and challenges of current research.

92.2 Radiation Absorption and its Consequences

92.2.1 Two Classes of Problems of Radiation Physics

It is useful to distinguish between two classes of problems in radiation physics [92.6], as illustrated in Fig. 92.1. Problems of Class I concern the fate of radiation after interactions with matter. Problems of Class II concern the fate of matter after interactions with radiation.

Problems of Class I are exemplified by the determination of the energy losses of particles penetrating matter, and the attenuation of a beam of photons in matter. Experimental studies on these problems are straightforward in principle; one only needs to analyze the kinetic energies of transmitted particles in the first problem and to count the number of transmitted photons in the second problem. Problems of Class I are also often simple to treat theoretically; for instance, in the Bethe theory of stopping power, discussed in Chapt. 91, the use of sum rules enables one to bypass detailed knowledge about excited and ionized states. As a consequence, many of the problems of Class I have been solved. Furthermore, many of the applications of radiation physics are based on the established knowledge of solutions of problems of Class I. For instance, diagnostic uses of X-ray photons in medicine and industry rest on the firm knowledge of the attenuation of photons in matter (more precisely, on its dependence on the atomic number).

In sharp contrast, problems of Class II are fundamentally difficult; indeed, none of them has been solved completely. Reasons for the difficulty are manifold. First, any piece of irradiated matter is a new material that needs to be fully characterized. Second, irradiated matter is in general in a nonequilibrium state that changes physically and chemically with time. Finally, existing tools and techniques, both experimental and theoretical, for material characterization are limited.

To illustrate the nature of Class II problems, it is useful to consider a prototype problem that has been treated reasonably well, but certainly not completely. Suppose a single electron with energy 10 keV enters a dilute hydrogen gas of a sufficiently large volume. The incident electron will collide with a hydrogen molecule. Possible outcomes include excitation of molecular states, ionization leading to H_2^+ in various vibrational and rotational states, and ionization leading to dissociation into $H^+ + H$ (in various excited states or in the ground state). An electron resulting from an ionizing event may have enough kinetic energy to cause further excitation and ionization. Eventually, all the liberated electrons will lose enough energy to become subexcitation electrons unable to excite even the lowest electronic level of H_2 at 9.5 eV. Table 92.1 summarizes the various atomic and molecular species produced. To produce this answer, it was necessary to survey all the spectroscopic and electron collision data on H_2, and to

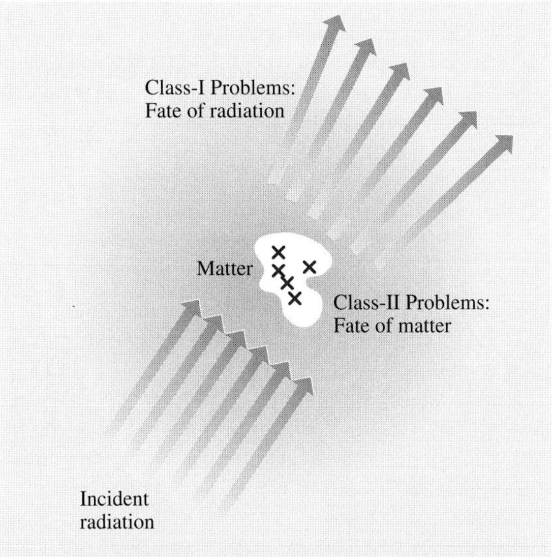

Fig. 92.1 Schematic diagram showing the definition of Class I and Class II problems. Incident radiation comes from the lower left, interacts with matter, and travels toward the upper right, having been scattered. The crosses symbolize interaction events taking place within matter

carry out an analysis of electron transport. Problems of this kind will be further discussed in Sect. 92.3.2 and Sect. 92.3.3.

92.2.2 Photons

High-energy photons (i. e. X-rays, γ-rays, or a major part of synchrotron radiation), are important as the ionizing radiation most frequently used in applications. Their initial interactions with matter result in a spectrum of energetic electrons. Detailed discussion is given in Chapts. 61 and 62.

92.2.3 Charged Particles

Energetic charged particles penetrating matter cause ionization and electronic excitation of atoms and molecules as they lose kinetic energy (Chapt. 65). The most fundamental index to characterize the energy loss process is the total cross section for inelastic collisions, which is the mean number of inelastic collisions per unit path length in a material of unit molecular density. The next fundamental index is the stopping power, which is the mean energy loss per unit path length. For many purposes, a more detailed description of particle penetration is necessary. An example is energy loss straggling, fully discussed in Chapt. 91.

Ionization is generally accompanied by the production of secondary electrons, which in general have a broad energy spectrum. This topic is reviewed by Rudd and coworkers [92.9–11].

92.2.4 Neutrons

Neutrons with kinetic energies in the MeV domain cause nuclear reactions, generally leading to the production of a number of charged particles. Neutrons of any kinetic energies collide with nuclei in matter elastically (viz., without causing nuclear reactions) and transfer substantial kinetic energy to nuclei. Consequently, neutrons may be regarded as a source of charged particles of various kinds, most importantly protons and light nuclei, all having broad energy spectra [92.12, 13].

Table 92.1 The mean number N_j of initial species produced in molecular hydrogen upon complete degradation of an incident electron at 10 keV, and the energy absorbed E_{abs} (in percent).[a] The left column of this table indicates kinds of species tersely. For instance, "Lyman" here means the $B^1\Sigma_u^+$ state, which emits the Lyman band. The term "Werner" means the $C^1\Pi_u$ state, which emits the Werner band. The designation "H(2p)" means the production of hydrogen atoms in the 2p state. The designation "Slow H(2s)" means the production of hydrogen atoms in the 2s state with no appreciable kinetic energy. The designation "Fast H(2s)" means the production of hydrogen atoms with kinetic energies of several eV. The designation "Triplet" means all the triplet states combined. The designation "Subexcitation electrons" means all the electrons that are not energetic enough to cause further electronic excitation

Initial species j	N_j	E_{abs}
$H_2(B^1\Sigma_u^+)$ (Lyman)	96.8	11.7
$H_2(C^1\Pi_u)$ (Werner)	112.1	14.0
H(2p)	5.6	0.8
Slow H(2s)	29.6	4.4
Fast H(2s)	5.3	1.7
H ($n = 3$)	3.2	0.5
Higher excited states	12.2	1.7
Ions	295.6	45.9
Triplets	102.6	10.0
Subexcitation electrons	295.6	9.4

[a] Adapted from *Douthat* [92.8]

92.3 Electron Transport and Degradation

92.3.1 The Dominant Role of Electrons

In many Class II problems (i. e., radiation actions on molecular substances such as gases, molecular liquids or solids, and biological cells), energetic electrons generated by ionizing radiation play a central role because they are numerous, and deliver to atoms and molecules a large fraction of the total energy of radiation. What follows is a resume of our current understanding.

92.3.2 Degradation Spectra and Yields of Products

The prototype H_2 problem discussed in Sect. 92.2.1 may be generalized as follows. Consider the total number N_j of a particular molecular species j produced as a result of complete slowing down of electrons in a chemically pure medium consisting of n molecules per unit volume. The product species j could be ions or excited states specified by a set of quantum numbers. Let $\sigma_j(T)$ be the cross section for the production of j in a collision of an electron of kinetic energy $\mathcal{T}$ with a molecule. During its passage over an infinitesimal path length $\mathrm{d}x$, an electron of kinetic energy $\mathcal{T}$ contributes the amount $\mathrm{d}N_j = n\sigma_j(\mathcal{T})\,\mathrm{d}x$ to N_j. Therefore, one may write

$$N_j = n \int \sigma_j(\mathcal{T})\,\mathrm{d}x\,. \tag{92.1}$$

To carry out this integration, one must have a relation between $\mathcal{T}$ and x. To the extent that one regards x and $\mathcal{T}$ as related through an analytic function, one may write

$$\mathrm{d}x = (\mathrm{d}\mathcal{T}/\mathrm{d}x)^{-1}\,\mathrm{d}\mathcal{T}\,, \tag{92.2}$$

where $\mathrm{d}\mathcal{T}/\mathrm{d}x$ is the mean energy loss per unit path length traversed, or the stopping power as defined in Chapt. 91. This treatment, called the continuous-slowing-down approximation (CSDA), is justified if $\mathcal{T}$ greatly exceeds the majority of energy losses upon individual collisions.

However, the CSDA is inadequate for electrons in general because the energy loss of an electron upon an individual collision is not always small compared with the current kinetic energy, and because the CSDA does not account for secondary electrons, which are abundantly produced. An appropriate formulation, introduced by *Spencer* and *Fano* [92.14], is obtained by writing

$$\mathrm{d}x = y(\mathcal{T})\,\mathrm{d}\mathcal{T} \tag{92.3}$$

in place of (92.2) and determining the function $y(\mathcal{T})$ through full analysis of electron transport, production, and slowing down. This function, called the degradation spectrum, has been an object of extensive studies [92.6, 15]. Once $y(\mathcal{T})$ is determined, the yield N_j of product species j is evaluated as

$$N_j = n \int_{E_{1,j}}^{\mathcal{T}_{\max}} \sigma_j(\mathcal{T})\,y(\mathcal{T})\,\mathrm{d}\mathcal{T}\,, \tag{92.4}$$

where $E_{1,j}$ is the threshold energy for the production of j, and $\mathcal{T}_{\max}$ is the highest kinetic energy of electrons in the medium.

Three basic properties of $y(\mathcal{T})$ are as follows [92.15]. First, $y(\mathcal{T})$ is proportional to the electron energy distribution (as treated by the Boltzmann equation) multiplied by electron speed v, or $(2\mathcal{T}/m)^{1/2}$ in the nonrelativistic case.

Second, $y(\mathcal{T})$ obeys an integral equation called the Spencer–Fano equation. This equation expresses the balance of the number of electrons arriving at kinetic energy $\mathcal{T}$ with the number of electrons departing from it. One may write the cross section for a collision process in which an electron of kinetic energy $\mathcal{T}$ collides with an atom or molecule and an electron of kinetic energy $\mathcal{T}'$ emerges as $\sigma(\mathcal{T} \to \mathcal{T}')$. Then, the expression $\int \sigma(\mathcal{T}' \to \mathcal{T})y(\mathcal{T}')\,\mathrm{d}\mathcal{T}' - \int \sigma(\mathcal{T} \to \mathcal{T}')\,\mathrm{d}\mathcal{T}'y(\mathcal{T})$ represents the change in the number of electrons at kinetic energy $\mathcal{T}$ due to all collisions in a medium of unit density. For convenience, one may write the above expression as $K_{\mathcal{T}}\,y(\mathcal{T})$ using a linear operator $K_{\mathcal{T}}$ called the collision operator. In the simple case where a source steadily generates $u(\mathcal{T})\,\mathrm{d}\mathcal{T}$ electrons having kinetic energies between $\mathcal{T}$ and $\mathcal{T} + \mathrm{d}\mathcal{T}$ in a medium consisting of n atoms or molecules of a single species, the Spencer–Fano equation takes the form

$$nK_{\mathcal{T}}\,y(\mathcal{T}) + u(\mathcal{T}) = 0\,. \tag{92.5}$$

Third, in the domain $E_1 < \mathcal{T} < \mathcal{T}_{\max}$, where E_1 is the lowest electronic-excitation threshold energy of the medium, $y(\mathcal{T})$ is invariably bimodal, as exemplified by Fig. 92.2. The behavior at high $\mathcal{T}$ is largely understandable from the CSDA (92.2), which approximately holds at $\mathcal{T} \gg E_1$; then, $y(\mathcal{T})$ is the reciprocal of the stopping power, which is given by the Bethe theory and corrections as explained in Chapt. 91. For example, for electrons with $\mathcal{T}$ between 1 keV and 10 keV, the stopping power of molecular hydrogen is a monotonically decreasing function of $\mathcal{T}$, and therefore $y(\mathcal{T})$ is a monotonically increasing function of $\mathcal{T}$, as seen in Fig. 92.2. In contrast, the steep rise of $y(\mathcal{T})$ at low

$\mathcal{T}$ occurs because most of the secondary electrons produced in an ionizing collision of any energetic charged particle (including an energetic electron) have low kinetic energies, corresponding to a few multiples of the ionization threshold energy, as fully documented by *Rudd* et al. [92.9–11].

As $\mathcal{T}$ varies, the principal contribution to the yield expression (92.4) comes from ranges where both $\sigma_j(\mathcal{T})$ and $y(\mathcal{T})$ are large. A large $y(\mathcal{T})$ means a long path length of electrons at that $\mathcal{T}$, and hence a modest energy loss per unit path length. From (92.2), the time interval $\mathrm{d}t$ during which electrons stay in the energy interval $\mathrm{d}\mathcal{T}$ is

$$\mathrm{d}t = (\mathrm{d}t/\mathrm{d}x)^{-1}\,\mathrm{d}x = v^{-1}y(\mathcal{T})\,\mathrm{d}\mathcal{T}\ . \qquad (92.6)$$

The quantity $v^{-1}y(\mathcal{T})\,\mathrm{d}\mathcal{T}$ represents the sojourn time of electrons at energy $\mathcal{T}$. Equation (92.4) expresses how N_j is determined by the competition between electron degradation, and the cross section $\sigma_j(\mathcal{T})$ for production of j upon individual collisions. The product $\sigma_j(\mathcal{T})y(\mathcal{T})$, called the yield spectrum for production of j, plays a central role in the theory.

Since $\mathcal{T}$ ranges from a few eV up to keV or MeV, it is convenient to rescale the yield spectrum to $\ln\mathcal{T}$. In fact, $\ln\mathcal{T}$ is more meaningful as a variable because it represents, in effect, the mean number of elastic collisions required to reduce $\mathcal{T}$ by a given amount. (In the neutron slowing down theory [92.16], $\ln\mathcal{T}$ is called lethargy by Fermi.) Equation (92.4) then becomes

$$N_j = n\int \mathcal{T}\sigma_j(\mathcal{T})y(\mathcal{T})\,\mathrm{d}(\ln\mathcal{T})\ , \qquad (92.7)$$

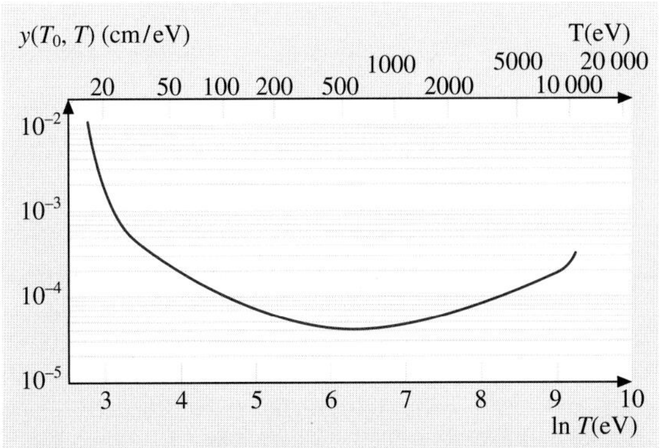

Fig. 92.2 Electron degradation spectrum in molecular hydrogen. Source electrons of kinetic energy $\mathcal{T}_{max} = 10\,\mathrm{keV}$ steadily enter the hydrogen gas at 0 °C and 1 atmosphere at the rate of 1 electron^{-1}cm^{-3}. Data are taken from *Douthat* [92.8]

with $\mathcal{T}\sigma_j(\mathcal{T})y(\mathcal{T})$ regarded as a function of $\ln\mathcal{T}$. The product $\mathcal{T}\sigma_j(\mathcal{T})$ is proportional to the collision strength in the theory of atomic collisions, as treated in Chapt. 47. It is also a key quantity in the Bethe theory [92.17].

Figure 92.3 is an example of a yield spectrum for ionization in molecular hydrogen. The area under the curve over any given interval of $\ln\mathcal{T}$ represents the number of ions produced in collisions of electrons in that interval. One sees a sizable contribution from high energies (between 1 keV and 10 keV). Below 1 keV, where the spectrum is roughly constant, each decade of $\mathcal{T}$ contributes roughly the same amount to the total ionization yield. This observation applies to any product j that results from dipole-allowed transitions.

In contrast, the yield spectrum for a product j resulting from forbidden transitions is dominated by values at low $\mathcal{T}$, where $\sigma_j(\mathcal{T})$ is appreciable, as Fig. 92.4 illustrates. An example is the production of two ground-state hydrogen atoms from a hydrogen molecule, which occurs solely from the lowest triplet repulsive state.

Finally, once electrons fall below E_1, they moderate much less rapidly than they do above E_1. These subexcitation electrons [92.18], lose energy through momentum transfer upon elastic collisions with molecules, rotational excitation, and vibrational excitation, as discussed in Chapt. 47. The behavior of the subexcitation electrons is treated by an extension of the Spencer–

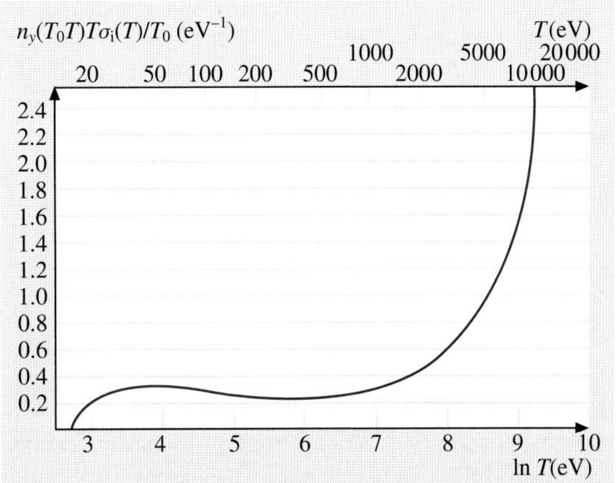

Fig. 92.3 The yield spectrum for ionization in molecular hydrogen. Source electrons are the same as for Fig. 92.2. The vertical axis represents $ny(\mathcal{T}_{max},\mathcal{T})\mathcal{T}\sigma_i(\mathcal{T})/\mathcal{T}_{max}$, where n is the number density of molecules, $\mathcal{T}_{max} = 10\,\mathrm{keV}$ is the source-electron energy, and $\sigma_i(\mathcal{T})$ is the ionization cross section of H_2 for electrons of energy $\mathcal{T}$. Data are taken from *Douthat* [92.8]

Fano equation [92.15], until they reach kinetic energies comparable with the thermal energy of the medium molecules. Then, the transport, moderation, and eventual energy gain from medium molecules become important, as well as the energy loss to them. The three main energy domains of electronic excitation, subexcitation, and thermal [92.19] are illustrated in Fig. 92.5. Because of the great disparity in the modes of interactions with the medium, electrons in each of the domains need different treatments.

Together with ions, molecular fragments, and other products of electron degradation, electrons in the thermal domain are important as precursors to subsequent chemical reactions, and therefore have been the subject of extensive study [92.4, 20, 21].

92.3.3 Quantities Expressing the Yields of Products

The total ionization yield is often expressed in terms of the quantity W, defined to be the mean energy required to produce an ion pair, or equivalently,

$$W = \mathcal{T}_{\text{abs}}/N_i \,, \tag{92.8}$$

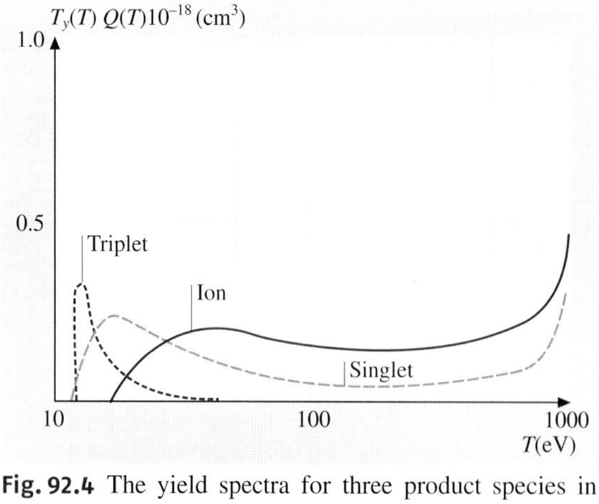

Fig. 92.4 The yield spectra for three product species in molecular hydrogen. Source electrons of kinetic energy $\mathcal{T}_{\max} = 1000$ eV steadily enter the hydrogen gas at 0°C and 1 atmosphere at the rate of 1 electron^{-1}cm^{-3}. The *solid curve* labeled "ION" represents ionization, the *long-broken curve* labeled "SINGLET" represents the production of the $B^1\Sigma_u^+$ state, and the *short-broken curve* labeled "TRIPLET" represents the production of the a $^3\Sigma_g^+$ state, which dissociates into two hydrogen atoms in the ground state (Courtesy: Mineo Kimura)

where $\mathcal{T}_{\text{abs}}$ is the mean enery absorbed and N_i is the mean number of ions observed over many events. $\mathcal{T}_{\text{abs}}$ is simply the initial injection energy if all the resulting electrons are degraded to kinetic energies below the ionization threshold. Since each event is stochastic, $\mathcal{T}_{\text{abs}}$ and N_i in general fluctuate from one event to the next.

A simple index for the statistical fluctuations in N_I is the variance V of N_I from its mean. The ratio $F = V/N_i$, called the Fano factor [92.22].

Both measurements and theory indicate that W and F are approximately the same for different kinds of ionizing radiations of sufficiently high energy, and that they depend primarily on the material of the medium [92.15, 23, 24]. This fact is important as a basis for radiation dosimetry through ionization measurements. The ratio W/I ranges from 1.7 to 3.2 depending on the nature of materials, where I is the ionization threshold energy. The ratio always exceeds unity because a fraction of radiation energy absorbed is expended to generate products other than ionization, such as discrete excited states, neutral molecular fragments, and subexcitation electrons.

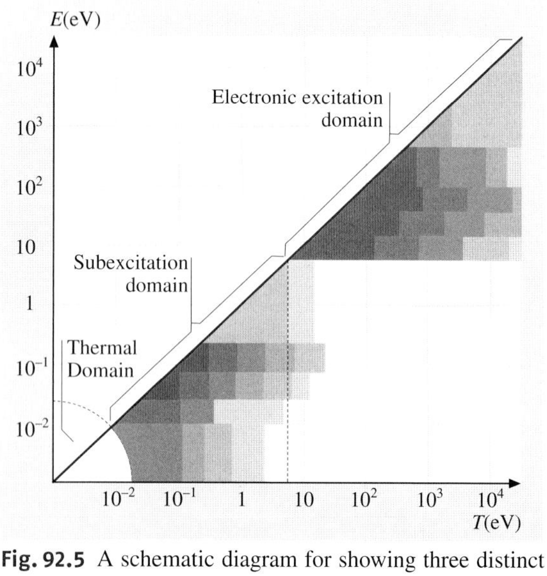

Fig. 92.5 A schematic diagram for showing three distinct domains of electron transport in a molecular substance. The horizontal axis represents the kinetic energy $\mathcal{T}$ of an electron. The vertical axis represents the energy loss E upon a single collision with a molecule. The vertical broken line indicates the first electronic excitation threshold E_1. The shade and fade schematically represent the magnitudes of cross sections

The Fano factor F is restricted to $0 < F < 1$ on general theoretical grounds. The maximum value $F = 1$ would occur if the ionization process were characterized by the Poisson statistics. In reality, the ionization process is subject to the constraints that it must compete with nonionizing events, and that the available total energy is fixed; therefore the fluctuations in the ionization yield must be smaller than those given by the Poisson statistics, viz., $F < 1$. Both theory and experimental data indicate that F and W tend to vary together from one material to another; more precisely, *Krajcar–Bronić* [92.24] found an approximate empirical relation

$$F = a(W/I) + b\,, \qquad (92.9)$$

where a and b are constant. This relation is valid for electrons of sufficiently high energies.

For molecular products other than ions, it is customary to use G_j defined by

$$G_j = N_j/\mathcal{T}_{\text{abs}}\,, \qquad (92.10)$$

where N_j is the mean number of molecular products of type j (see pp. 241–242 of *Kimura* et al. [92.15] for fuller discussion). Like W, G_j is often approximately the same for different kinds of ionizing radiations of sufficiently high energy, provided that the material, either liquid or solid, is homogeneous and that the absorbed energy is not extremely great. A counter example is the ferrous sulphate equeous solution, which is often used for dosimetry and in which the G value for the ferrous-ferric conversion depends appriciably on the kind of radiation (Chapt. 91). In the more general case of a heterogeneous material having complex molecular structure and aggregation, such as a biological cell, the yield of a product is not even approximately proportional to the absorbed energy, and the idea of a G value is useful only at the limit of vanishing absorbed energy.

In SI units, the G value has units of $\text{mol}\,\text{J}^{-1}$. This is related to the older unit, molecules $(100\,\text{eV}^{-1})$, by

$$1\,\text{mol}\,\text{J}^{-1} = 9.6485\times 10^{-6}\text{molecules}(100\,\text{eV})^{-1}\,. \qquad (92.11)$$

The meaning of the yield of a product depends on the process. The molecular species produced as a result of electron degradation (e.g., those given in Table 92.1) may be called initial products. Subsequently, they react with neighboring molecules or among themselves to form other species, which may be called secondary products, such as H_3^+, through the ion-molecule reaction

$$H_2^+ + H_2 \rightarrow H_3^+ + H\,. \qquad (92.12)$$

The yield of an initial product is seldom measured and is accessible by theory only. Most yield measurements concern a secondary product. To keep the distinction in mind, *Platzman* recommended [92.25] use of the symbol g for an initial product and the symbol G for a secondary product.

For data on G values, see *Tabata* et al. [92.4] and references cited therein.

92.3.4 Track Structures

The spatial distributions of initial products such as ions and excited states influence the kinetics of their subsequent chemical reactions, which lead to radiation effects. Spatial distribution is generally expected to have a role in a heterogeneous material; this role should be especially prominent when the scale characteristic of the distribution is comparable to the scale of the material inhomogeneity.

The simplest index of the distribution is the stopping power, which gives the linear density of energy lost from a particle along its path. From the point of view of radiation effects, or Class II problems, the linear density of energy imparted locally to matter is of greater interest. The distinction between the energy lost from a particle and the energy imparted locally to matter arises when one considers a volume having a linear scale less than the ranges of the majority of secondary electrons produced in the volume; sufficiently energetic secondary electrons will escape the volume and impart much of their kinetic energies elsewhere. This recognition led to the idea of the linear energy transfer (often abbreviated as LET), which is defined as the mean energy loss, excluding contributions from the production of secondary electrons having kinetic energies above a fixed value, such as 100 eV [92.26]. The LET, like the stopping power, depends on the particle charge and speed, as well as on the material, and is often used in radiation chemistry and biology to consider the role of the spatial distribution of initial products.

However, a given value of LET from particles of different charges and speeds does not necessarily lead to the same radiation effect. The aim of microdosimetry [92.27] is to provide a more detailed description of the energy imparted locally to a small volume of matter. For a fuller discussion see Chapt. 91.

The full representation of the spatial pattern of the initial products resulting from individual incident and secondary particles is called the track structure, because the pattern was first recognized in particle tracks visualized in cloud chambers, photographic emulsions, and other radiation detectors. Treatments of track structures and their consequences for the subsequent chemical reactions by Monte Carlo simulations are certainly informative and flexible for application to a wider range of problems, as exemplified by [92.28–33]. Such treatments are in principle sound, because they are independent of an assumption that the yield of a product is proportional or otherwise simply related to the energy lost or imparted.

The spatial distribution of the initial products and other consequences of radiation interactions with matter can be also studied by analytic methods using transport equations. An excellent exposition of the recent status of this approach is given in the NCRP Report No. 108 [92.34]. By solving relatively simple problems, the analytic methods are effective in elucidating principles and providing insights into the physics involved, while the Monte Carlo simulations are powerful in providing answers to complicated problems. In summary, the two approaches are complementary, rather than competitive.

Meaningful results from these treatments require a large volume of accurate cross section and other data as input. Various surveys of atomic and molecular cross section data are available to present recommended values and to identify needs for further studies, such as the report [92.35] sponsored by the International Atomic Energy Agency. A broader review of the current status of cross section determination, dissemination, and related topics is in [92.36].

92.3.5 Condensed Matter Effects

Beyond the interactions of ionizing particles with individual atoms and molecules in a gas or beam, the atomic or molecular aggregation in condensed matter has further important influences. A prelude to condensed matter effects is seen in studies of high-pressure gases [92.37] and in Chapt. 39.

Many phenomena are related to the complex dipole-response function $\epsilon(E)$, which is the electric displacement generated in matter by a spatially uniform electric field of unit strength oscillating at angular frequency $\omega = E/\hbar$ (see, e.g., *Landau* and *Lifshitz* [92.38]). For light at this frequency, the real part $\epsilon_1(E)$ describes the dispersion, and the imaginary part $\epsilon_2(E)$ describes the absorption. The probability that a glancing collision of a fast charged particle transfers energy E to matter is proportional to

$$\Pi(E) = \mathrm{Im}[-1/\epsilon(E)] = \frac{\epsilon_2}{\epsilon_1^2(E) + \epsilon_2^2(E)} . \tag{92.13}$$

In a low-density material, such as a dilute gas, $\epsilon_1(E)$ is close to unity, and $\epsilon_2(E)$ is much smaller than unity for all E. Then, $\Pi(E)$ is practically the same as $\epsilon_2(E)$; in other words, the spectrum of energy transfer in a glancing collision of a fast charged particle is effectively the same as the spectrum of photoabsortion. In a high-density material, $\Pi(E)$ differs appreciably from $\epsilon_2(E)$ at some E. An extreme case occurs at E at which $\epsilon_1(E)$ changes its sign; there $\Pi(E)$ has a peak. This is the well-known plasma excitation in metals.

A general criterion for determining the effect of atomic or molecular aggregation on the dipole oscillator strength spectrum, and hence on $\Pi(E)$, was given by *Fano* [92.39]. One may qualitatively summarize his criterion as follows: If the density of the dipole oscillator strength of an atom or molecule in both space and

Table 92.2 Condensed matter effects

Classification	Example	Remarks on characteristics
Shift of oscillator strength toward higher energies	Water, hydrocarbons, and organics	Occurs over wide ranges of exitation energies.
Excitation of special modes of motion	Plasmons in metals	Occurs at specific energies and carries considerable strength.
	Excitons in molecular crystals and ionic crystals	Occurs at specific energies and at minor strengths.
Interaction of fast ejected electrons with other atoms and molecules	EXAFS (extended X-ray absorption fine structure)	Occurs at energies above 100 eV, slightly above the pertinent threshold, and at weak strengths, chiefly due to interference of electrons emerging from different atoms.
Resonances in electron interactions	N_2, C_2H_4, C_2H_2, and other unsaturated hydrocarbons	Occurs at energies slightly above the excitation threshold or even lower; leads to inelasticity.
Diffraction of the electron de Broglie wave	Materials with periodic structure	Leads to band structure effects, such as transmission without energy loss at certain electron kinetic energies.

energy (or frequency) is sufficiently high, the effects of atomic or molecular aggregation will be appreciable. In other words, the effect will be seen in a sufficiently dense material and in a generally limited spectral region where the oscillator strength spectrum of an isolated atom or molecule is especially intense. With the use of the Fano criterion, one understands the occurrence of collective excitation (electronic excitation involving more than one atom or molecule), such as excitons, plasmons, and other elementary modes of excitation in condensed matter. For a fuller and more recent discussion of this topic, see *Fano* [92.40].

Many manifestations of condensed matter effects on the absorption of radiation energy have been identified, as discussed, for instance, in [92.7] and [92.41]. Table 92.2 presents a summary of these effects.

92.4 Connections with Related Fields of Research

92.4.1 Astrophysics and Space Physics

Some of the ideas and methods sketched in Sect. 92.3.1 and Sect. 92.3.2 are readily applicable to studies on terrestrial and planetary auroras [92.42–46] and on effects of cosmic rays or X-rays on interstellar clouds and other astronomical objects [92.47–53]. These studies treat consequences of the interactions of energetic charged particles or high-energy photons on voluminous gases, at low densities and often partially ionized. Often the presence of electric and magnetic fields influence radiation degradation phenomena.

92.4.2 Material Science

As fully discussed earlier, radiation effects on gases, liquids, and molecular substances, including the biological cell, result mainly from electronic excitation. Radiation effects on insulators, including crystalline solids, are also initiated by electronic excitation [92.54, 55] and are therefore closely related to atomic, molecular, and optical physics.

In metals, electronic excitation per se is of little consequence, because most of the electronic excitation energy is rapidly converted into phonons (thermal energy). Generally, a small fraction of electronic excitation energy causes appreciable atomic displacement. Radiation effects on these materials occur predominantly via direct energy transfer to atoms from neutrons, protons, and heavier charged particles, causing atomic displacement from a regular crystalline site [92.56–58].

More generally, the role of atoms and ions having kinetic energies far exceeding the thermal energy is important in chemistry. This topic, known as hot-atom chemistry [92.59, 60], represents another application of the knowledge of atomic, molecular, and optical physics as discussed in Chapts. 64 and 65, 66 and 67.

Radiation effects on the surface of a solid are a subject of extensive study, because of their importance as means for material-structure probing and also for material processing (for example, the ion implantation in the manufacture of semiconductors and other devices). A major phenomenon of interest is sputtering, which is the ejection of atoms and molecules from the surface by the action of ionizing radiation [92.61–64].

92.5 Supplement

A supplement to the foregoing sketch is provided by an essay [92.65] on the role of physics written in commemoration of the fiftieth anniversary of the Radiation Research Society, an organ devoted to physical, chemical, biological, and medical studies on the action of radiation. According to the essay, contributions by physicists have been steady, and show no sign of decline in the total volume only changing emphasis over years. Topics of the contributions concern radiation sources, dosimetry, instrumentation for measurements of radiation effects, fundamentals of radiation physics, mechanisms of radiation actions, and applications. The role of physics is most certain and decisive in the development of instrumentation in a broad sense, ranging from radiation sources to probes of radiation effects.

Finally, Ugo Fano was a giant in atomic, molecular, and optical physics and also a pioneer of radiation physics. After his death in 2001 many articles about his work and life naturally appeared in print. A special issue of *Physics Essays* in his honor [92.66] contains biographic materials, forty papers by his associates and friends, as well as his own memoir. A recent article [92.67] summarizes Fano's contributions to atomic, molecular, and optical physics, and includes the most accurate list of his publications.

References

92.1 G. S. Knoll: *Radiation Detection and Measurement*, 2nd edn. (Wiley, New York 1989)
92.2 K. R. Kase, B. E. Bjarngard, F. H. Attix: *The Dosimetry of Ionizing Radiation* (Academic, Orlando 1985–1987)
92.3 G. S. Hurst, M. G. Payne, S. D. Kramer, J. P. Young: Rev. Mod. Phys. **51**, 767 (1979)
92.4 Y. Tabata, Y. Ito, S. Tagawa (Eds.): *CRC Handbook of Radiation Chemistry* (CRC, Boca Raton 1991)
92.5 W. A. Glass, M. N. Varma (Eds.): *Physical and Chemical Mechanisms in Molecular Radiation Biology* (Plenum, New York 1991)
92.6 M. Inokuti: Radiation physics as a basis of radiation chemistry and biology. In: *Applied Atomic Collision Physics*, Vol. 4, ed. by S. Datz (Academic, Orlando 1983) p. 179
92.7 M. Inokuti: Atomic and Molecular Theory. In: *Physical and Chemical Mechanisms in Molecular Radiation Biology*, ed. by W. A. Glass, M. N. Varma (Plenum, New York 1991) p. 29
92.8 D. A. Douthat: J. Phys. B **12**, 663 (1979)
92.9 M. E. Rudd, Y.-K. Kim, D. H. Madison, J. W. Gallagher: Rev. Mod. Phys. **57**, 965 (1985)
92.10 M. E. Rudd: Nucl. Tracks Radiat. Meas. **16**, 213 (1989)
92.11 M. E. Rudd, Y.-K. Kim, D. H. Madison, T. J. Gay: Rev. Mod. Phys. **64**, 441 (1992)
92.12 R. S. Caswell: Radiat. Res. **27**, 92 (1966)
92.13 R. S. Caswell, J. J. Coyne, H. M. Gerstenberg, E. J. Axton: Radiat. Prot. Dosim. **23**, 41 (1988)
92.14 L. V. Spencer, U. Fano: Phys. Rev. **93**, 1172 (1954)
92.15 M. Kimura, M. Inokuti, M. A. Dillon: *Advances in Chemical Physics*, Vol. 84, ed. by I. Prigogine, S. A. Rice (Wiley, New York 1993) p. 193
92.16 A. M. Weinberg, E. P. Wigner: *The Physical Theory of Neutron Chain Reactors* (The Univ. Chicago, Chicago 1958)
92.17 M. Inokuti: Rev. Mod. Phys. **43**, 297 (1971)
92.18 R. L. Platzman: Radiat. Res. **2**, 1 (1955)
92.19 M. Inokuti: Appl. Radiat. Isot. **42**, 979 (1991)
92.20 C. Ferradini, J.-P. Jay-Gerin: *Excess Electrons in Dielectric Media* (CRC, Boca Raton 1991)
92.21 R. Holroyd: Ann. Rev. Phys. Chem. **40**, 439 (1989)
92.22 U. Fano: Phys. Rev. **72**, 26 (1947)
92.23 International Commission on Radiation Units and Measurements: *Average Energy Required to Produce an Ion Pair*, ICRU Report No. 31 (International Commission on Radiation Units and Measurements, Bethesda 1979)
92.24 I. Krajcar-Bronić: J. Phys. B **25**, L215 (1992)
92.25 R. L. Platzman: Energy spectrum of primary activations in the action of ionizing radiation. In: *Proceedings of the Third International Congress of Radiation Research, 1966*, ed. by G. Silini (North-Holland, Amsterdam 1967) p. 20
92.26 International Commission on Radiation Units and Measurements: *Linear Energy Transfer*, ICRU Report No. 16 (International Commission on Radiation Units and Measurements, Bethesda 1970)
92.27 International Commission on Radiation Units and Measurements: *Microdosimetry*, ICRU Report No. 36 (International Commission on Radiation Units and Measurements, Bethesda 1983)
92.28 H. G. Paretzke: Radiation track structure theory. In: *Kinetics of Non-Homogenous Processes*, ed. by G. R. Freeman (Wiley, New York 1987) p. 89
92.29 T. M. Jenkins, W. R. Nelson, A. Rindi (Eds.): *Monte Carlo Transport of Electrons and Photons* (Plenum, New York 1988)
92.30 R. H. Ritchie, R. N. Hamm, J. E. Turner, H. A. Wright, W. E. Bloch: Radiation interactions and energy transport in the condensed phase. In: *Physical and Chemical Mechanisms in Molecular Radiation Biology*, ed. by W. A. Glass, M. N. Varma (Plenum, New York 1991) p. 99
92.31 M. Zaider: Charged particle transport in condensed phase. In: *Physical and Chemical Mechanisms in Molecular Radiation Biology*, ed. by W. A. Glass, M. N. Varma (Plenum, New York 1991) p. 137
92.32 M. Terrissol, A. Beaudre: Radiat. Prot. Dosim. **31**, 175 (1990)
92.33 M. N. Varma, A. Chatterjee (Eds.): *Computational Approaches in Molecular Radiation Biology* (Plenum, New York 1994)
92.34 National Council on Radiation Protection and Measure ments: *Conceptual Basis for Calculations of Absorbed-Dose Distributions*, NCRP Report No. 108 (National Council on Radiation Protection and Measurements, Bethesda 1991)
92.35 International Atomic Energy Agency: *Atomic and Molecular Data for Radiotherapy and Radiation Research. Final Report of a Co-ordinated Research Programme*, Report IAEA-TECDOC-799 (International Atomic Energy Agency, Vienna 1995)

92.36 M. Inokuti (Ed.): *Advances in Atomic, Molecular, and Optical Physics*, Special Volume on Cross Section Data, Vol. 33 (Academic, San Diego 1994)

92.37 L. G. Christophorou: Radiation interactions in high-pressure gases. In: *Physical and Chemical Mechanisms in Molecular Radiation Biology*, ed. by W. A. Glass, M. N. Varma (Plenum, New York 1991) p. 183

92.38 L. D. Landau, E. M. Lifshitz: *Electrodynamics of Continuous Media* (Pergamon, London 1960) Chap. IX translated by J. B. Sykes and J. S. Bell

92.39 U. Fano: Phys. Rev. **118**, 451 (1960)

92.40 U. Fano: Rev. Mod. Phys. **64**, 313 (1993)

92.41 M. Inokuti: Radiat. Effects Defects **117**, 143 (1991)

92.42 K. Takayanagi, Y. Itikawa: Space Sci. Rev. **11**, 380 (1970)

92.43 J. L. Fox, G. A. Victor: Planet. Space Sci. **36**, 329 (1988)

92.44 S. P. Slinker, R. D. Taylor, A. W. Ali: J. Appl. Phys. **63**, 1 (1988)

92.45 S. P. Slinker, R. D. Taylor, A. W. Ali: J. Appl. Phys. **64**, 982 (1988)

92.46 K. Onda, M. Hayashi, K. Takayanagi: The Institute of Space, Astronautical Science Report **645**, 876 (1992)

92.47 T. E. Cravens, G. A. Victor, A. Dalgarno: Planet. Space Sci. **23**, 1059 (1970)

92.48 A. Dalgarno, R. A. McCray: Ann. Rev. Astron. Astrophys. **10**, 375 (1972)

92.49 A. E. Glassgold, W. D. Langer: Astrophys. J. **186**, 859 (1973)

92.50 T. E. Cravens, A. Dalgarno: Astrophys. J. **219**, 750 (1978)

92.51 R. Gredel, S. Lepp, A. Dalgarno: Astrophys. J. **323**, L137 (1987)

92.52 R. Gredel, S. Lepp, A. Dalgarno, E. Herbst: Astrophys. J. **347**, 289 (1989)

92.53 G. M. Vogt: Astrophys. J. **377**, 158 (1991)

92.54 N. Itoh (Ed.): *Direct Processes Induced by Electronic Excitation in Insulators* (World Scientific, Singapore 1989)

92.55 W. B. Fowler, N. Itoh (Eds.): *Atomic Processes Induced by Electronic Excitation in Non-Metallic Solids* (World Scientific, Singapore 1990)

92.56 F. Seitz, J. S. Koehler: Displacement of atoms during irradiation. In: *Solid State Physics. Advances in Research and Application*, Vol. 2, ed. by F. Seitz, D. Turnball (Academic, New York 1956) p. 305

92.57 M. W. Thompson: *Defects and Radiation Damage in Metals* (Cambridge Univ. Press, London 1969)

92.58 W. Schilling, H. Ullmaier: *Materials Science and Technology. A Comprehensive Treatment*, Vol. 10B, ed. by R. W. Cahn, P. Haasen, E. J. Kramer (VCH Verlagsgesellschaft, Weinheim 1994) p. 179

92.59 T. Matsuura: *Hot Atom Chemistry. Recent Trends and Applications in the Physical and Life Sciences and Technology* (Kodansha, Tokyo 1984)

92.60 J.-P. Adloff, P. P. Gasper, M. Imamura, A. G. Maddock, T. Matsuura, H. Sano, K. Yoshihara (Eds.): *Handbook of Hot Atom Chemistry* (Kodansha, Tokyo 1992)

92.61 R. Behrisch (Ed.): *Sputtering by Particle Bombardment I. Physical Sputtering of Single-Element Solids* (Springer, Berlin, Heidelberg 1981)

92.62 R. Behrisch (Ed.): *Sputtering by Particle Bombardment II. Sputtering of Alloys and Compounds, Electron and Neutron Sputtering, Surface Topography* (Springer, Berlin, Heidelberg 1983)

92.63 R. Behrisch, K. Wittmaack (Ed.): *Sputtering by Particle Bombardment III. Characteristics of Sputtered Particles, Technical Applications* (Springer, Berlin, Heidelberg 1991)

92.64 P. Sigmund (Ed.): Fundamental Processes in Sputtering of Atoms and Molecules, K. Dan. Vidensk. Selsk. Mat. Fys. Medd. **43**, 1 (1993)

92.65 M. Inokuti, S. M. Seltzer: Radiat. Res. **158**, 3 (2002)

92.66 M. Inokuti, A. R. P. Rau (Eds.): Phys. Essays **13**(2–3), 156 (2000)

92.67 M. Inokuti, A. R. P. Rau: Phys. Scr. T **68**, C96 (2003)

Acknowledgements

A.2 Angular Momentum Theory

by James D. Louck

This contribution on angular momentum theory is dedicated to Lawrence C. Biedenharn, whose tireless and continuing efforts in bringing understanding and structure to this complex subject is everywhere imprinted.

We also wish to acknowledge the many contributions of H. W. Galbraith and W. Y. C. Chen in sorting out the significance of results found in *Schwinger* [1.20]. The Supplement is dedicated to the memory of Brian G. Wybourne, whose contributions to symmetry techniques and angular momentum theory, both abstract and applied to physical systems, was monumental.

The author expresses his gratitude to Debi Erpenbeck, whose artful mastery of TEX and scrupulous attention to detail allowed the numerous complex relations to be displayed in two-column format.

Thanks are also given to Professors Brian Judd and Gordon Drake for the opportunity to make this contribution.

Author's note. It is quite impossible to attribute credits fairly in this subject because of its diverse origins across all areas of physics, chemistry, and mathematics. Any attempt to do so would likely be as misleading as it is informative. Most of the material is rooted in the very foundations of quantum theory itself, and the physical problems it addresses, making it still more difficult to assess unambiguous credit of ideas. Pragmatically, there is also the problem of confidence in the detailed correctness of complicated relationships, which prejudices one to cite those relationships personally checked. This accounts for the heavy use of formulas from [1.1], which is, by far, the most often used source. But most of that material itself is derived from other primary sources, and an inadequate attempt was made there to indicate the broad base of origins. While one might expect to find in a reference book a comprehensive list of credits for most of the formulas, it has been necessary to weigh the relative merits of presenting a mature subject from a viewpoint of conceptual unity versus credits for individual contributions. The first position was adopted. Nonetheless, there is an obligation to indicate the origins of a subject, noting those works that have been most influential in its developments. The list of textbooks and seminal articles given in the references is intended to serve this purpose, however inadequately.

Excerpts and Fig. 2.1 are reprinted from *Biedenharn* and *Louck* [1.1] with permission of Cambridge University Press. Tables 2.2–2.4 have been adapted from *Edmonds* [1.18] by permission of Princeton University Press. Thanks are given for this cooperation.

B.11 High Precision Calculations for Helium

by Gordon W. F. Drake

The author is grateful to R. N. Hill and J. D. Morgan III for suggesting some of the material at Sect. 11.1. This work was supported by the Natural Sciences and Engineering Research Council of Canada.

B.20 Thomas–Fermi and Other Density-Functional Theories

by John D. Morgan III

I am grateful to Cyrus Umrigar and Michael P. Teter of Cornell University for generously providing sabbatical support in the spring of 1995. I should also like to thank them, as well as Elliott Lieb and Mel Levy, for helpful discussions. This work was supported by my National Science Foundation grant PHY-9215442.

B.23 Many-Body Theory of Atomic Structure and Processes

by Miron Ya. Amusia

This work was supported by the Israeli Science Foundation under the grant 174/03.

B.28 Tests of Fundamental Physics

by Peter J. Mohr, Barry N. Taylor

The authors gratefully acknowledge helpful conversations with Prof. Michael Eides, Dr. Ulrich Jentschura, Prof. Toichiro Kinoshita, and Prof. Jonathan Sapirstein.

B.29 Parity Nonconserving Effects in Atoms

by Jonathan R. Sapirstein

The work described here was carried out in collaboration with S. A. Blundell and W. R. Johnson, and was supported in part by NSF grant PHY-92-04089.

B.30 Atomic Clocks and Constraints on Variations of Fundamental Constants

by Savely G. Karshenboim, Victor Flambaum, Ekkehard Peik

We are very grateful to our colleagues and to participants of the ACFC-2003 meeting for useful and stimulating discussions.

C.31 Molecular Structure

by David R. Yarkony

This work has been supported in part by NSF grant CHE 94-04193, AFOSR grant F49620-93-1-0067 and DOE grant DE-FG02-91ER14189

C.33 Radiative Transition Probabilities

by David L. Huestis

This work was supported by the NSF Atmospheric Chemistry Program, the NASA Stratospheric Chemistry Section, and the NASA Space Physics Division.

C.37 Gas Phase Reactions

by Eric Herbst

The author is grateful to Professor Anne B. McCoy, Department of Chemistry, The Ohio State University, for a critical reading of the manuscript.

C.38 Gas Phase Ionic Reactions

by Nigel G. Adams

The support of the National Science Foundation, NASA, and the Petroleum Research Fund-ACS for my experimental research program is gratefully acknowledged.

C.41 Laser Spectroscopy in the Submillimeter and Far-Infrared Regions

by Kenneth M. Evenson†, John M. Brown

We have benefitted invaluably from the help and collaboration of I. G. Nolt of NASA, Langley for his assistance with the detector technology, and of Kelly Chance for his line shape fitting program and his assistance with our studies of upper atmospheric species.

C.43 Spectroscopic Techniques: Cavity-Enhanced Methods

by Barbara A. Paldus, Alexander A. Kachanov

We have benefited invaluably from the collaboration with our co-workers at Picarro, and would especially like to acknowledge Eric Crosson, Bruce Richman, Sze Tan, Bernard Fidric, Ed Wahl, and Herb Burkard. We would like to extend our gratitude to Prof. Richard N. Zare and Dr. Marc Levenson for their unwavering support in helping make CRDS a commercial reality. We would also like to express our appreciation to all of our scientific collaborators worldwide, both industrial and academic, for their friendship, openness, and help over the years. Barb would like to dedicate this chapter to her father, Prof. Josef Paldus, who has been an inspiration to her throughout her career, and with whom it is an honor to share a publication in the same book.

D.45 Elastic Scattering: Classical, Quantal, and Semiclassical

by M. Raymond Flannery

This research is supported by the U.S. Air Force Office of Scientific Research under Grant No. F49620-94-1-0379. I wish to thank Dr. E. J. Mansky for numerous discussions on the content and form of this chapter and without whose expertise in computer typesetting this chapter would not have been possible.

D.54 Electron–Ion and Ion–Ion Recombination

by M. Raymond Flannery

This research is supported by the U.S. Air Force Office of Scientific Research under Grant No. F49620-94-1-0379. I wish to thank Dr. E. J. Mansky for numerous discussions on the content and form of this chapter and without whose expertise in computer typesetting this chapter would not have been possible.

D.56 Rydberg Collisions: Binary Encounter, Born and Impulse Approximations

by Edmund J. Mansky

The author thanks Prof. M. R. Flannery for many helpful discussions and Prof. E. W. McDaniel for access to his collection of reprints. The author would also like to thank Dr. D. R. Schultz of ORNL for the time necessary to complete this work.

This work was begun at the Georgia Institute of Technology (GIT) and completed at the Controlled Fusion Atomic Data Center at Oak Ridge National Laboratory (ORNL). The work at GIT was supported by US AFOSR grant No. F49620-94-1-0379. The work at ORNL was supported by the Office of Fusion Energy, US DOE contract No. DE-AC05-84OR21400 with Lockheed-Martin Energy Systems, Inc. and by ORNL Research Associates Program administered jointly by ORNL and ORISE.

E.61 Photon–Atom Interactions: Low Energy

by Denise Caldwell, Manfred O. Krause

This work was supported in part by the National Science Foundation under grant PHY-9207634 and in part by the US Department of Energy, Division of Chemical Sciences, under contract with Martin Marietta Energy Systems, Inc., DE-AC0584OR21400.

E.62 Photon–Atom Interactions: Intermediate Energies

by Bernd Crasemann

The author is indebted to Sue Mandeville for indefatigable assistance. This chapter is dedicated to the memory of Teijo Åberg.

E.65 Ion–Atom Collisions – High Energy

by Lew Cocke, Michael Schulz

This work was supported by the Chemical Sciences Division, Basic Energy Sciences, Office of Energy Research, U.S. Department to Energy and by the National Science Foundation.

F.74 Multiphoton and Strong-Field Processes

by Multiphoton and Strong-Field Processes

This work was supported in part by the U.S. DOE under contract number W-7405-ENG-48.

F.78 Quantized Field Effects

by Matthias Freyberger, Karl Vogel, Wolfgang P. Schleich, Robert F. O'Connell

We want to thank Prof. I. Bialynicki-Birula for stimulating discussions and a critical reading of the manuscript. The work of RFOC was supported in part by the U.S. Army Research Office under grant No. DAAH04-94-G-0333.

F.81 Quantum Information

by Sir Peter L. Knight, Stefan Scheel

We like to thank A. Beige, J. Eisert, E. A. Hinds, V. Kendon, W. J. Munro, M. B. Plenio, and many others for numerous stimulating discussions on this exciting subject. Funding by the UK Engineering and Physical Sciences Research Council (EPSRC), the Royal Society, the European Commission, and the Alexander von Humboldt foundation are gratefully acknowledged.

G.86 Atoms in Dense Plasmas

by Jon C. Weisheit, Michael S. Murillo

This work was performed under the auspices of the U.S. Department of Energy by the University of California, Los Alamos National Laboratory, under contract W-7405-Eng-36. LA-UR #04-7993. We wish to thank C. Iglesias for the results plotted in Fig. 86.2, and Los Alamos colleagues P. Bradley, J. Guzik, R. Peterson, and D. Saumon for providing the data plotted in Fig. 86.1.

G.87 Conduction of Electricity in Gases

by Alan Garscadden

This article is based on notes originally developed in cooperation with Dr. J. C. Ingraham. Dr. R. Nagpal assisted with an earlier version.

G.90 Interface with Nuclear Physics

by John D. Morgan III, James S. Cohen

J. D.M. is grateful to the Institute for Nuclear Theory of the University of Washington for making it possible for him to spend a productive semester there in the spring of 1993. This work has also been supported by National Science Foundation grant PHY-9215442.

G.92 Radiation Physics

by Mitio Inokuti

This work was supported by the U.S. Department of Energy, Office of Energy Research, Office of Health and Environmental Research, under Contract W-31-109-ENG-38.

About the Authors

Miron Ya. Amusia

The Hebrew University
Racah Institute of Physics
Jerusalem, Israel
amusia@vms.huji.ac.il

Chapter B.23

Dr. Miron Amusia is a Professor of Physics at the Hebrew University of Jerusalem, Israel, and Principal Scientist of the Ioffe Physical-Technical Institute, St-Petersburg, Russia. He is author and co-author of more than 400 referred papers and 9 books. His research is in many-body theory of atoms, nuclei, molecules, clusters, and condensed matter, but primarily in atomic physics. He is Fellow of American Physical Society and a member of several professional societies and editorial boards. He received the Humboldt research award and is a member of the Russian Academy of Natural Sciences.

Nils Andersen

University of Copenhagen
Niels Bohr Institute
Copenhagen, Denmark
noa@fys.ku.dk

Chapter D.46

Nils Andersen is Professor of Physics at the Niels Bohr Institute of the University of Copenhagen. His main activities include experimental and theoretical studies of atomic collisions involving optically prepared states. Recent research interests include cold and ultracold collisions.

Thomas Bartsch

Georgia Institute of Technology
School of Physics
Atlanta, GA, USA
bartsch@cns.physics.gatech.edu

Chapter B.15

Thomas Bartsch received his PhD from the University of Stuttgart, Germany, in 2002. He is currently a postdoctoral fellow at the Georgia Institute of Technology. His research is centred on applications of nonlinear dynamics to atomic and molecular physics.

Klaus Bartschat

Drake University
Department of Physics and Astronomy
Des Moines, IA, USA
klaus.bartschat@drake.edu

Chapter A.7

Dr. Bartschat is the Ellis & Nelle Levitt distinguished Professor of Physics at the Department of Physics and Astronomy at Drake University. His research in theoretical and computational atomic physics focuses on combining the general theory of measurement with highly accurate numerical calculations. He is a fellow of the american physical society and has published 2 books, 30 book chapters, 10 review articles, and more than 200 papers on electron and photon collisions with atoms and ions.

William E. Baylis

University of Windsor
Department of Physics
Windsor, ON, Canada
baylis@uwindsor.ca

Chapters .1, B.12

Professor Baylis earned degrees in physics from Duke (B.Sc.), the University of Illinois (M.Sc.), and the Technical University of Munich (D.Sc.). He has authored two books, edited or co-edited four more, contributed 28 chapters to other volumes, and published over a hundred journal articles. His publications are in theoretical physics and emphasize atomic and molecular structure, atomic collisions, and interactions with radiation. His most recent work concerns relativistic dynamics, the photon position operator and wave function, and applications of Clifford algebra, especially to the quantum - classical interface. He is a fellow of the American Physical Society, past chair of the Divisions of Atomic and Molecular Physics and of Theoretical Physics of the Canadian Association of Physicists, a member of the international editorial boards of the Springer Series of Atomic, Optical, and Plasma Physics and of the journal Advances in Applied Clifford Algebras. He is currently a University Professor at the University of Windsor.

Anand K. Bhatia

NASA Goddard Space Flight Center
Laboratory for Astronomy & Solar Physics
Greenbelt, MD, USA
anand.k.bhatia@nasa.gov

Chapter B.25

Dr. Bhatia received his Ph.D. in theoretical physics from the University of Maryland in 1963. Since then he has been at Goddard Space Flight Center. He has published a large number of papers in refereed journals on various topics in atomic and astrophysics: scattering of electrons and positrons from atoms, muonic fusion, polarizabilities of two-electron systems, Lamb shift, Rydberg states, excitation of ions etc. He is a Fellow of the American Physical Society.

Hans Bichsel

University of Washington
Center for Experimental Nuclear Physics and Astrophysics (CENPA)
Seattle, WA, USA
bichsel@npl.washington.edu

Chapter G.91

Professor Hans Bichsel has worked on the interactions of fast charged particles with matter for over 50 years. Some of his measurements are the most accurate of their type. At present he is studying the methods of particle identification for the time projection chambers at STAR and ALICE. Earlier he worked in nuclear physics and developed neutron radiation therapy in Seattle.

John M. Brown

University of Oxford
Physical and Theoretical Chemistry Laboratory
Oxford, England
john.m.brown@chem.ox.ac.uk

Chapter C.41

Professor Brown obtained his Ph.D. degree from the University of Cambridge in 1966. Before moving to Oxford in 1983, he was a Lecturer in the Department of Chemistry at Southampton University. He is a high-resolution, gas-phase spectroscopist with a special interest in free radical species. In addition to experimental studies at all wavelengths from microwave to the ultraviolet, he is interested in the development of theoretical models to describe the experimental results.

Henry Buijs

ABB Bomem Inc.
Québec, Canada
henry.l.buijs@ca.abb.com

Chapter C.40

Henry Buijs founded ABB Bomem Inc. in 1973 to bring to market state of the art Fourier Transform spectrometers. He received his Ph.D. from the University of British Columbia. He has interest in spectroscopic measurement in the atmosphere for Ozone chemistry, meteorological sounding and climate change assessment. ABB Bomem Inc. is leader in FT spectrometers for satellite based sensors and industrial process monitoring solutions.

Philip Burke

The Queen's University of Belfast
Department of Applied Mathematics and Theoretical Physics
Belfast, Northern Ireland, UK
p.burke@qub.ac.uk

Chapter D.47

Phil Burke is Emeritus Professor of Mathematical Physics at the Queen's University of Belfast, having been Professor at Queen's from 1967 until 1998. His research interests are the theory of atomic, molecular, and optical physics and their applications. He was awarded the Guthrie Medal and Prize in 1994 and the David Bates Prize in 2000. He is a Fellow of the Royal Society.

Denise Caldwell

National Science Foundation
Physics Division
Arlington, VA, USA
dcaldwel@nsf.gov

Chapter E.61

Dr. Caldwell is the Program Director for the Atomic, Molecular, Optical, and Plasma Physics program at the National Science Foundation. She was awarded her Ph.D. by Columbia University in 1976. She then held a postdoc at the University of Bielefeld and a junior faculty position at Yale University. In 1985 she joined the faculty at the University of Central Florida, where she maintained a research program on atomic photoionization using synchrotron radiation. In 1998 she left full-time academia to become a permanent staff member of the NSF. She is a Fellow of the American Physical Society.

Mark M. Cassar

University of Windsor
Department of Physics
Windsor, ON, Canada
cassar@uwindsor.ca

Chapter B.13

Mark M. Cassar received his Ph.D. from the University of Windsor, Canada in 2003. His research focuses on high-precision theoretical calculations for the energy level structure of three-body atomic and molecular systems.

Kelly Chance

Harvard-Smithsonian Center for Astrophysics
Cambridge, MA, USA
kchance@cfa.harvard.edu

Chapter G.85

Dr. Chance heads the Atomic and Molecular Physics Division of the Harvard-Smithsonian Center for Astrophysics. His current research applies molecular spectroscopy, structure and dynamics to studies of planetary atmospheres, with emphasis on satellite-based measurements of Earth's ozone layer composition and lower atmospheric pollution. Recent accomplishments include global measurements of tropospheric ozone, volatile organic compounds, and nitrogen oxides.

Raymond Y. Chiao

366 Leconte Hall
U.C. Berkeley
Berkeley, CA, USA
chiao@physics.berkeley.edu

Chapter F.80

Professor Chiao was awarded his Ph.D. by MIT in 1965. He has been Professor of Physics at Berkeley since 1977. His research interests are: Nonlinear and quantum optics; low temperature physics as applied to astrophysics; the relationship between general relativity and macroscopic quantum matter. He is writing a book with J. C. Garrison on Quantum Optics.

James S. Cohen

Los Alamos National Laboratory
Atomic and Optical Theory
Los Alamos, NM, USA
cohen@lanl.gov

Chapter G.90

Dr. Cohen is Group Leader of the Atomic and Optical Theory Group in the Theoretical Division of Los Alamos National Laboratory and a Fellow of the American Physical Society. He received a Ph.D. in Physics from Rice University in 1973. His general area of research is theoretical atomic and molecular physics, with special interest in exotic muonic and antiprotonic species.

Bernd Crasemann

University of Oregon
Department of Physics
Eugene, OR, USA
berndc@uoregon.edu

Chapter E.62

Bernd Crasemann is Professor Emeritus of Physics in the University of Oregon and Editor of Physical Review, Atomic, Molecular, and Optical Physics since 1993. He received his early education in Chile and a Ph.D. from the University of California at Berkeley. His work is in experimental and theoretical atomic inner-shell physics, particularly as explored with synchrotron radiation.

David R. Crosley

SRI International
Molecular Physics Laboratory
Menlo Park, CA, USA
david.crosley@sri.com

Chapter G.88

For most of his career, David R. Crosley has developed and used laser-induced fluorescence to study small free radicals. This research includes fundamental spectroscopic and energy transfer studies, as well as applications to combustion, atmospheric chemistry,and environmental monitoring. Notable among these are studies of OH, NH, and CH. He is a Fellow of the APS and AAAS.

Derrick Crothers

Queen's University Belfast
Department of Applied Mathematics and Theoretical Physics
Belfast, Northern Ireland, UK
d.crothers@qub.ac.uk

Chapter D.52

Derrick Crothers is Professor of Theoretical Physics (Personal chair). He researches in atomic, molecular, optical, and condensed matter physics. Topics include heavy-particle collisions, threshold phenomena, dielectrics and ferromagnetics. He was awarded an Honorary Professorship in Physics by St Petersburg State University in 2003.

Lorenzo J. Curtis

University of Toledo
Department of Physics and Astronomy
Toledo, OH, USA
ljc@physics.utoledo.edu

Chapter B.17

Lorenzo J. Curtis is a Distinguished University Professor of Physics at the University of Toledo. He received his Ph.D. from the University of Michigan in 1963 and was awarded the degree Philosophiae Doctorem Honoris Causa by the University of Lund in 1999. His research involves time-resolved atomic spectroscopy and the structure of highly ionized atoms. He is the author of over 200 scientific articles and a textbook on atomic structure,. He is an editor of Physica Scripta and a Member of the Editorial Board of Physical Review A.

Gordon W. F. Drake

University of Windsor
Department of Physics
Windsor, ON, Canada
gdrake@uwindsor.ca

Chapters .1, B.11

Dr. Gordon W.F. Drake is a Professor of Physics and Department Head at the University of Windsor, Canada. He received his Ph.D. degree from York University in Toronto. His research on high precision calculations and QED theory for helium and other few-body atomic systems has resulted in over 150 refereed journal articles, and numerous other review articles and book chapters. He is a Fellow of the American Physical Society and the Royal Society of Canada, and has been awarded numerous prizes and distinctions for his research. He is currently the Editor of the Canadian Journal of Physics, and an Associate Editor for Physical Review A, as well as Editor of the current volume, and Co-Editor-in-Chief of the Springer Series on Atomic, Optical, and Plasma Physics. He has served as President of the Canadian Association of Physicists, and as Chair of the Division of Atomic, Molecular, and Optical Physics of the American Physical Society.

Joseph H. Eberly

University of Rochester
Department of Physics and Astronomy
and Institute of Optics
Rochester, NY, USA
eberly@pas.rochester.edu

Chapter F.73

Joseph H. Eberly is Andrew Carnegie Professor of Physics and Professor of Optics at the University of Rochester. He earned his Ph.D. from Stanford University. He held the APS Chair of the Division of Laser Science from 1996–97 and was Divisional Councilor from 2003–2005. Eberly is OSA Vice President and its President in 2007. He is Foreign Member of the Academy of Science of Poland and received numerous awards such as the Charles Hard Townes Award in 1994, the Smoluchowski Medal in 1987, and the Humboldt Preis in 1984. He has published more than 300 research and review papers and several books in the areas of quantum optics, cavity QED and photon–atom interactions, evolution of coherence and quantum entanglement, high-field atomic physics, and nonlinear propagation of short optical pulses.

Guy T. Emery

Bowdoin College
Department of Physics
Brunswick, ME, USA
gemery@bowdoin.edu

Chapter B.16

Guy Emery was on the Brookhaven National Laboratory Staff, and taught physics at Indiana University and later Bowdoin College (Brunswick, ME). He was a visiting scientist at the Universities of Groningen and Osaka. His research has been in nuclear structure and reactions, the intersections of nuclear physics with atomic physics and particle physics, and in the history of physics.

Volker Engel

Universität Würzburg
Institut für Physikalische Chemie
Würzburg, Germany
voen@phys-chemie.uni-wuerzburg.de

Chapter C.35

Volker Engel studied Physics at the University of Göttingen and worked as a post-doctoral associate at the University of California, Santa Barbara. After his Habilitation in Physics (1993, University of Freiburg) he was appointed Professor in 1994 at the University of Würzburg. His research interests are in the time-dependent quantum theory of atomic and molecular dynamics in laser fields.

James M. Farrar

University of Rochester
Department of Chemistry
Rochester, NY, USA
farrar@chem.rochester.edu

Chapter E.67

James M. Farrar received his Ph.D. degree at the University of Chicago in 1974 working under the direction of Professor Yuan-Tseh Lee. Prior to joining the faculty at the University of Rochester, he was a postdoctoral fellow in the laboratory of Professor Bruce H. Mahan at the University of California at Berkeley. He has had a long-term interest in molecular beam studies of the dynamics of chemical reactions, and his current interests include low energy ion-molecule collisions and electronic spectroscopy of size-selected clusters ions. He is a Fellow of the American Physical Society.

Paul D. Feldman

The Johns Hopkins University
Department of Physics and Astronomy
Baltimore, MD, USA
pdf@pha.jhu.edu

Chapter G.83

Dr. Feldman is Professor of Physics and Astronomy at the Johns Hopkins University where he has been since 1967. He received his Ph.D. in physics from Columbia University in 1964. His recent work has been in space ultraviolet astronomy and spectroscopy with a focus on the study of the atmospheres of comets and planets and of the Earth's upper atmosphere.

Victor Flambaum

University of New South Wales
Department of Physics
Sydney, Australia
v.flambaum@unsw.edu.au

Chapter B.30

Dr. Victor Flambaum is a Professor of Physics and holds a Chair of Theoretical Physics. Ph.D., DSc. from the Institute of Nuclear Physics, Novosibirsk, Russia. He has about 200 publications in atomic, nuclear, particle, solid state, statistical physics, and astrophysics including works on violation of fundamental symmetries (parity, time reversal), test of unification theories, temporal and spatial variation of fundamental constants from Big Bang to present, many-body theory and high-precision atomic calculations, as well as statistical theory of finite chaotic Fermi systems and enhancement of weak interactions, high-temperature superconductivity, and conductance quantization.

David R. Flower

University of Durham
Department of Physics
Durham, United Kingdom
david.flower@durham.ac.uk

Chapter C.36

Professor Flower teaches at the University of Durham (UK). He was awarded his Ph.D. by the University of London in 1969. After working at the Observatoire de Paris (Meudon, France) and at the ETH (Zuerich, Switzerland), he joined the Physics Department of the University of Durham in 1978. He has been Professor of Physics since 1994. His research interests are in atomic and molecular physics related to astrophysics. He is currently preparing the second edition of his book on "Molecular Collisions in the Interstellar Medium".

A. Lewis Ford

Texas A&M University
Department of Physics
College Station, TX, USA
ford@physics.tamu.edu

Chapter D.50

Dr. Ford's research interests lie in theoretical atomic and molecular physics: inner-shell excitation, ionization, charge transfer, and electronic properties of diatomic molecules. Professor Ford joined the Texas A&M faculty in 1973. After receiving his B.A. degree from Rice University, he completed his Ph.D. at the University of Texas at Austin in 1972 and did post-doctoral work at Harvard. Professor Ford is a member of the American Physical Society, Division of Electron, Atomic,and Optical Physics.

Jane L. Fox

Wright State University
Department of Physics
Dayton, OH, USA
jane.fox@wright.edu

Chapter G.84

Jane Fox received her Ph.D. from Harvard University in Chemical Physics and has held positions at the State University of New York at Stony Brook, and the Harvard/Smithsonian Astrophysical Observatory. She has been elected a Fellow of the American Geophysical Union. Her research has focused on the chemistry, luminosity, heating of the thermospheres/ionospheres of the planets, and their evolution.

Matthias Freyberger

Universität Ulm
Abteilung für Quantenphysik
Ulm, Germany
matthias.freyberger@uni-ulm.de

Chapter F.78

Dr. Matthias Freyberger is extraordinary Professor at the Department of Quantum Physics at the University of Ulm, Germany. His research interests are in quantum optics, atom optics, quantum estimation theory, and the foundations of quantum mechanics.

Thomas F. Gallagher

University of Virginia
Department of Physics
Charlottesville, VA, USA
tfg@virginia.edu

Chapter B.14

Thomas F. Gallagher received his Ph.D. in physics in 1971 from Harvard University and is now the Jesse W. Beams Professor of Physics at the University of Virginia. His research is focused on the use of Rydberg atoms to realize novel physical systems.

Muriel Gargaud

Observatoire Aquitain des Sciences de l'Univers
Floirac, France
gargaud@obs.u-bordeaux1.fr

Chapter D.51

Muriel Gargaud is an astrophysicist at the "Observatoire Aquitain des Sciences de l'Univers" in Bordeaux, France. She studied for 2O years the physico-chemistry of the interstellar medium, her current research is now astrobiology. Astrobiolgy is an interdisciplinary research field (astronomy, geology, chemistry, biology) looking for the origins of life, its evolution and its development on Earth but also in and beyond the Solar System. She is the main scientific editor of "Lectures in Astrobiology" by Springer, Heidelberg 2005.

Alan Garscadden

Airforce Research Laboratory
Area B
Wright Patterson Air Force Base, OH, USA
alan.garscadden@wpafb.af.mil

Chapter G.87

Alan Garscadden received his B.Sc. and Ph.D. from Queen's University, Belfast, Northern Ireland. He is the chief Scientist, Propulsion Directorate, Air Force Research Laboratory. Wright-Patterson AFB, Ohio and Edwards AFB, California. Alan also performs basic and applied research in non-equilibrium plasmas and energized gas flows, lasers, mass spectroscopy measurements, and electron collision cross sections. He is a Fellow of the APS, IEEE, AIAA and of the UK Institute of Physics.

John Glass

British Telecommunications
Solution Design
Belfast, Northern Ireland, UK
john.glass@bt.com

Chapter D.52

John Glass earned his Ph.D. on Relativistic Ion-Atom Collisions from The Queen's University of Belfast in 1995. His Ph.D. focussed on distorted wave approximations in electron capture, in particular, the first fully symmetrical CDW solution via the Sommerfeld-Maue approximation. Dr. Glass now works in large-scale Business Support Systems, Solutions Design for British Telecommunications plc.

S. Pedro Goldman

The University of Western Ontario
Department of Physics & Astronomy
London, ON, Canada
goldman@uwo.ca

Chapter B.13

Professor Pedro Goldman completed a Ph.D. in Relativistic Atomic Physics at the University of Windsor. His work in atomic physics includes pioneering work on relativistic variational basis sets, relativistic calculations for many-electron atoms and diatomic molecules, accurate calculations for atoms in strong magnetic fields and accurate calculations of QED energy corrections and of the energy levels of Helium. Presently his research is directed to the optimization of the radiation therapy of tumours. He has as well received numerous teaching awards.

Ian P. Grant

University of Oxford
Mathematical Institute
Oxford, UK
ipg@maths.ox.ac.uk

Chapter B.22

Ian Grant is Emeritus Professor of Mathematical Physics, University of Oxford and a Fellow of the Royal Society. He graduated from Oxford with a degree in Mathematics and obtained his D. Phil. in Theoretical Physics in 1954. His interest in relativistic electronic structure of atoms arose whilst he was working for the UK Atomic Energy Authority at Aldermaston from 1957 to 1964 and the field has been a major component of his research ever since. He returned to Oxford to a research post in 1964 and was a full-time member of academic staff from 1969 until his retirement in 1998. He is the author of more than 220 research papers, many of them on relativistic quantum theory applied to atomic and molecular structure and processes.

William G. Harter

University of Arkansas
Department of Physics
Fayetteville, AR, USA
wharter@uark.edu

Chapter C.32

Professor Harter's research centers on theory of spectroscopy and what it reveals about quantum phenomena and symmetry principles of structure and dynamics. Current study focuses on how wave mechanics of light relates to matter waves and their relativistic symmetry ranging from intrinsic frames of floppy molecules to manifold dynamics of astrophysical objects. A strong educational effort is being developed to make modern theory more accessible. He is a Fellow of American Physical Society (DAMOP).

Carsten Henkel

Universität Potsdam
Institut für Physik
Potsdam, Germany
carsten.henkel
@quantum.physik.uni-potsdam.de

Chapter F.77

Carsten Henkel is Docteur en Sciences from the Université Paris-Sud Orsay. He habilitated in 2004 at Potsdam University where he is currently a Privatdozent. His research interests are in atom optics and nano optics. He is involved in several European projects on physical implementations of quantum information processing.

Eric Herbst

The Ohio State University
Departments of Physics
Columbus, OH, USA
herbst@mps.ohio-state.edu

Chapter C.37

Dr. Eric Herbst is Distinguished University Professor of Physics, Astronomy, and Chemistry at The Ohio State University. Herbst is a Fellow of both the American Physical Society and the Royal Society of Chemistry (UK). His specialty is the chemistry of molecules in interstellar clouds, which are large accumulations of gas and dust particles in our Galaxy and others in which star and planetary formation occur.

Robert N. Hill

Saint Paul, MN, USA
rnhill@fishnet.com

Chapter A.9

Professor Robert Nyden Hill received his Ph.D. from Yale University in 1962. In 1964, after postdoctoral fellowships at Princeton and Yale, he joined the faculty of the University of Delaware Physics Department. He retired in 1997, and moved to Saint Paul, Minnesota. He has published papers in relativistic dynamics, statistical mechanics, mathematical physics, and atomic and molecular physics.

David L. Huestis

SRI International
Molecular Physics Laboratory
Menlo Park, CA, USA
david.huestis@sri.com

Chapter C.33

David L. Huestis received his Ph.D. in Chemistry from the California Institute of Technology in 1973. He is a Fellow of the American Physical Society. His research activities include a wide range of experimental and theoretical investigations of fundamental kinetic and optical processes involving atoms, small molecules, liquids, and solids. Two major application areas have been chemical kinetics and optical physics of high-power visible and ultraviolet gas lasers and the optical emissions of terrestrial and planetary atmospheres.

Mitio Inokuti

Argonne National Laboratory
Physics Division
Argonne, IL, USA
inokuti@anl.gov

Chapter G.92

Dr. Mitio Inokuti earned his Ph.D. in Applied Physics from the University of Tokyo in 1962. From 1973–1995 he was Senior Physicist at Argonne National Laboratory. He is a Fellow of the American Physical Society and a member of the Radiation Research Society. Since 1985 he is a member of the International Commission on Radiation Units and Measurements, and since 1988 a member of the Editorial Board for Advances in Atomic, Molecular, and Optical Physics. He also is Associate Editor of the Journal of Applied Physics. His research interests focus on theoretical research in radiation physics and chemistry, and in atomic and molecular physics.

Juha Javanainen

University of Connecticut
Department of Physics Unit 3046
Storrs, CT, USA
jj@phys.uconn.edu

Chapters F.75, F.76

Juha Javanainen is Professor of Physics at the University of Connecticut. He has worked on a number of topics in theoretical quantum optics, and currently concentrates on quantum degenerate gases.

Erik T. Jensen

University of Northern British Columbia
Department of Physics
Prince George, BC, Canada
ejensen@unbc.ca

Chapter G.89

Erik Jensen is an Associate Professor of Physics at the University of Northern British Columbia (Canada). He obtained his Ph.D. in the Surface Physics Group at Cambridge University in 1990 and did post-Doctoral work with Prof. John Polanyi at the University of Toronto. His research interests are in low-energy electron and photon initiated dynamics for molecules at surfaces.

Brian R. Judd

The Johns Hopkins University
Department of Physics and Astronomy
Baltimore, MD, USA
juddbr@pha.jhu.edu

Chapters A.3, A.6

Brian Judd has had a life-long interest in applying group theory to the spectroscopic properties of the rare earths. He held appointments at Oxford, Chicago, Paris and Berkeley before joining the Physics Department of the Johns Hopkins University in 1966. He received the Spedding Award for Rare-Earth Research in 1988 and is an Honorary Fellow of Brasenose College, Oxford.

Alexander A. Kachanov

Research and Development
Picarro, Inc.
Sunnyvale, CA, USA
akachanov@picarro.com

Chapter C.43

Alexander Kachanov received the M.Sc. degree in physics from Moscow Institute of Physics and Technology in 1976, and the Ph.D. degree in physics from the Institute of Spectroscopy of the Russian Academy of Sciences in 1987. In 2001 he joined Picarro, Inc., where his research interests focus on ultra-sensitive gas detection and development of novel laser sources.

Savely G. Karshenboim

Chapter B.30

D.I.Mendeleev Institute for Metrology (VNIIM)
Quantum Metrology Department
St. Petersburg, Russia
sek@mpq.mpg.de

Dr. Savely G.Karshenboim was graduated in 1983 from St. Petersburg (then Leningrad) State University, Russia where he also received his Ph.D. in 1992 and habilitatatetd in 1999. He has been a member of D.I. Mendeleev Institute for Metrology since 1983 and is at present a head of Laboratory for Precision Physics and Metrology of simple atomic systems. Since 1994 until now he has enjoyed numerous visiting opportunities at Max-Planck-Institut für Quantenoptik. He is a member of the CODATA task group on fundamental constants and SUNAMCO commission of IUPAP. SUNAMCO is a commission on Symbols, Units, Nomenclature, Atomic Masses and Fundamental Constants.
Dr. Karshenboim's scientific interests include precision physics of simple atoms, quantum electrodynamics (QED), determination of fundamental constants and search for their variations.

Kate P. Kirby

Chapter G.85

Harvard-Smithsonian Center for Astrophysics
Cambridge, MA, USA
kkirby@cfa.havard.edu

Kate Kirby has a Ph.D. in Chemical Physics from the University of Chicago, and is currently director of the Institute for Theoretical Atomic, Molecular, and Optical Physics. Her research interests center on theoretical studies of ultracold molecule formation and atomic and molecular structure and processes which are of interest to astronomy and atmospheric physics. Such processes include: photoionization, photodissociation, radiative association, charge transfer, and line-broadening.

Sir Peter L. Knight

Chapter F.81

Imperial College London
Department of Physics Blackett Laboratory
London, UK
p.knight@imperial.ac.uk

Sir Peter Knight is Head of Physics at Imperial College. He is Chief Scientific Advisor to the National Physical Laboratory and past President of the Optical Society of America. He is a Fellow of the Royal Society and was knighted in 2005. He researches in strong field physics and quantum information and edits the Journal of Modern optics and contemporary physics.

Manfred O. Krause

Chapter E.61

Oak Ridge National Laboratory
Oak Ridge, TN, USA
mok@ornl.gov

Dr. Krause was a Senior Scientist at the Oak Ridge National Laboratory working primarily in the field of photoelectron spectrometry of atoms with the use of synchrotron radiation. He received his Dr. rer. nat. in physics at the Technische Universität and the Max Planck Institut für Metallforschung in Stuttgart in 1954. He joined the Oak Ridge National Laboratory in 1963 and retired in 1995. He is a Fellow of the American Physical Society, and was a Professeur d'Echange at the University of Paris in 1975 and an Alexander von Humboldt awardee at the University of Freiburg in 1976.

Paul G. Kwiat

Chapter F.80

University of Illinois at Urbana-Champaign
Department of Physics
Urbana, IL, USA
kwiat@uiuc.edu

Paul G. Kwiat is the Bardeen Chair in Physics, at the University of Illinois, in Urbana-Champaign. A Fellow of the American Physical Society and the Optical Society of America, he studies the phenomena of entanglement, quantum interrogation, quantum erasure, and optical implementations of quantum information protocols. He can't resist a good swing dance.

Maciej Lewenstein

ICFO–Institut de Ciéncies Fotóniques
Barcelona, Spain
maciej.lewenstein@icfo.es

Chapter F.74

Born in Warsaw Poland, Dr. Maciej Lewenstein worked for many years in the Center for Theoretical Physics in Warsaw. He graduated from the University of Essen, worked for several years at CEA, and the University of Hannover. Currently he leads the theoretical quantum optics group at ICFO, Barcelona, Spain. His interests include physics of ultracold gases, quantum information, and the physics of matter in strong fields. He is a Fellow of APS.

James D. Louck

Los Alamos National Laboratory
Retired Laboratory Fellow
Los Alamos, NM, USA
jimlouck@aol.com

Chapter A.2

James Louck is a Los Alamos National Laboratory Retired Fellow. He earned his Ph.D. in molecular physics from The Ohio State University in 1958, and is the co-author of three books. Except for the years 1960 - 1963 at Auburn University, his career was in the Theoretical Division at Los Alamos developing symmetry methods for physical systems. His current research is in the inter-relations between symmetry and combinatorics.

Joseph H. Macek

University of Tennessee and Oak Ridge National Laboratory
Department of Physics and Astronomy
Knoxville, TN, USA
jmacek@utk.edu

Chapter D.53

Dr. Joseph Macek is a Distinguished Professor at the University of Tennessee and a Distinguished Scientist at Oak Ridge National Laboratory. His currrent research concentrates on thetheory of atomic collisions. He has been assigned Co-Chair of the local committee for the annual meeting of the Division of Atomic and Molecular Physics of the American Physical Society, Knoxville, TN 2006.

Mary L. Mandich

Lucent Technologies Inc.
Bell Laboratories
Murray Hill, NJ, USA
mandich@lucent.com

Chapter C.39

Mary Mandich is a Technical Manager and Distinguished Member of Technical Staff at Bell Laboratories and currently leads research in high speed backplanes and optical remoting for next generation telecommunication networks. She obtained her Ph.D. degree in Physical Chemistry at Columbia University. She holds 6 U.S. Patents and has authored 2 book chapters and more than 55 scientific publications in chemistry, physics, and materials science.

Steven T. Manson

Georgia State University
Department of Physics and Astronomy
Atlanta, GA, USA
smanson@gsu.edu

Chapter D.53

Professor Manson is on the faculty at Georgia State University. He received the Ph.D. from Columbia University in 1966, and did a two-year post-doc at the NBS (now NIST) working with Ugo Fano and John Cooper. He started as a faculty member at Georgia State University in 1968 and has been Regents Professor since 1984. His research has been primarily in the area of theoretical studies of ionization of atoms and ions by charged particles and photons. He is a Fellow of the American Physical Society.

William C. Martin

National Institute of Standards and Technology
Atomic Physics Division
Gaithersburg, MD, USA
wmartin@nist.gov

Chapter B.10

Dr. Martin's research has included the measurement and energy-level analysis of atomic spectra. He has also published a number of critical compilations of atomic spectroscopic data, including a large volume for the rare-earth elements. In his current position as Scientist Emeritus at NIST, Dr. Martin is continuing work on internet-accessible atomic spectra databases.

Jim F. McCann

Queen's University Belfast
Dept. of Applied Mathematics and Theoretical Physics
Belfast, Northern Ireland, UK
j.f.mccann@qub.ac.uk

Chapter D.52

Jim McCann was a Ph.D. student of Prof. Derrick Crothers at Queen's University, Belfast. He is currently a Reader in Theoretical Physics at Queen's and works in the field of Quantum Optics and Quantum Information Processing.

Ronald McCarroll

Université Pierre et Marie Curie
Laboratoire de Chimie Physique
Paris Cedex 05, France
mccarrol@ccr.jussieu.fr

Chapter D.51

Ronald McCarroll is a Professor of Physics at the Université Pierre et Marie Curie in Paris. He obtained his Ph.D. degree in Theoretical Physics at Queen's University, Belfast. After a post-doctoral fellowship at the National Physics Laboratory, Teddington and a Lectureship at Queen's University, Belfast he was appointed as a Directeur de Recherche au CNRS at the Observatoire de Pari, Meudon. Later, he moved to the Université de Bordeaux I as Professor in Astrophysics and finally to Paris as Professor in Physics at the Univerité Pierre et Marie Curie. He has worked in the field atomic and molecular photodynamics, particularly in view of their application to astrophysics and the physics of fusion plasmas. He is the author of more then 130 papers in refereed journals and contributed more than 20 specialised reviews to books and other specialised publications.

Fiona McCausland

Northern Ireland Civil Service
Department of Enterprise Trade and Investment
Belfast, Northern Ireland, UK
fiona.mccausland@detini.gov.uk

Chapter D.52

Dr. Fiona McCausland gained her Ph.D. in Theoretical Physics in 1995 from the Queen's University of Belfast. Following a year spent as a Post Doctoral Research Assistant at the University, she joined the Northern Ireland Civil Service in September 1996. She currently holds the position of Project Manager in the Department of Enterprise, Trade and Investment.

William J. McConkey

University of Windsor
Department of Physics
Windsor, ON, Canada
mcconk@uwindsor.ca

Chapter E.63

Dr. Bill McConkey is a physicist with an extensive background in the measurement of absolute cross section data for the atomic, molecular, and optical physics community. His laboratory is recognised as a world leader in electron collisions research. He has been awarded the Gold Medal of the Canadian Association of Physicists (1999) and the Allis Prize of the American Physical Society (2004) for his work.

Robert P. McEachran

Australian National University
Atomic and Molecular Physics Laboratories Research School of Physical Sciences and Engineering
Canberra, Australia
robert.mceachran@anu.edu.au

Chapter D.48

Professor McEachran received his Ph.D. from the University of Western Ontario, Canada and then spent two years at the University College London (England) before joining York University in Toronto in 1964. In 1997 he accepted an Adjunct Professorship at the Australian National University. His current research interests are the theoretical treatment of electron/positron scattering from heavy atoms within a relativistic framework.

James H. McGuire

Tulane University
Department of Physics
New Orleans, LA, USA
mcguire@tulane.edu

Chapter D.57

Dr. McGuire is Murchison Mallory Chair and department chair at Tulane University. He is a past Chair of the Division of Atomic, Molecular and Optical Physics (DAMOP) of the American Physical Society. His research interests are in electron correlation dynamics. entanglement, complexity and correlation, and quantum time.

Dieter Meschede

Rheinische
Friedrich-Wilhelms-Universität Bonn
Institut für Angewandte Physik
Bonn, Germany
meschede@iap.uni-bonn.de

Chapter F.79

Professor Dieter Meschede teaches at the Institute for Applied Physics in Bonn. After his studies in Hanover and Cologne and having been awarded his Dr. rer. nat in Munich in 1984, he first worked at Yale University. Then he became senior scientist at the MPI for Quantum Optics, Garching. He has been Professor of Physics since 1990, first in Hanover, since 1994 in Bonn. Professor Meschede is author of "Optics, Light, and Laser", some 90 refereed articles, and, since 2001, editor of the "Gerthsen"textbook.

Pierre Meystre

University of Arizona
Department of Physics
Tucson, AZ, USA
meystre@physics.arizona

Chapter F.68

Pierre Meystre's research ranges from laser theory to cavity QED and to the physics of quantum-degenerate atomic and molecular systems. With Murray Sargent, he coauthored the textbook "Elements of Quantum Optics,"and he recently published the monograph "Atom Optics", both with Springer-erlag. He has been awarded the Senior Scientist Research Prize of the Humboldt Foundation and the R.W. Wood Prize of the Optical Society of America. He is currently a Regents Professor and the Head of the Physics Department at The University of Arizona.

Peter W. Milonni

Los Alamos, NM, USA
pwm@lanl.gov

Chapter F.70

Peter Milonni is a Laboratory Fellow (retired) at Los Alamos National Laboratory. His main interests are in theoretical physics, especially quantum optics and electrodynamics. He is an author of several books including Lasers (with J. H. Eberly), The Quantum Vacuum, and Fast Light, Slow Light, and Left-Handed Light. Previously he held positions with the U. S. Air Force, the Perkin-Elmer Corporation, and the University of Arkansas.

Peter J. Mohr

National Institute of Standards and Technology
Atomic Physics Division
Gaithersburg, MD, USA
mohr@nist.gov

Chapter B.28

Dr. Peter Mohr received his Ph.D. from the University of California at Berkeley in 1973 and spent some years at the Lawrence Berkeley Laboratory (1973–1978), at Yale University (1978–1985), at the National Science Foundation (1985–1987), and at the National Bureau of Standards/ National Institute of Standards and Technology from 1987 until now. He is a Fellow of the American Physical Society, and received the Alexander von Humboldt Senior Research Award in 1995. He held the Chair of the CODATA Task Group on Fundamental Constants from 1999 to 2006 and was Chair of the Precision Measurement and Fundamental Constants Topical Group of the American Physical Society from 2000–2001.

John D. Morgan III

University of Delaware
Department of Physics and Astronomy
Newark, DE, USA
jdmorgan@udel.edu

Chapters B.20, G.90

Dr. Morgan is Associate Professor and obtained his B.S. from The George Washington University, his M.Sc. in Theoretical Chemistry from Oxford University, and his Ph.D. in Chemistry from Berkeley. He has served on the editorial boards of the Journal of Mathematical Physics and the International Journal of Quantum Chemistry. His wide-ranging interests include the application of sophisticated mathematical techniques to assist the accurate calculation of properties of atoms and molecules.

Michael S. Murillo

Los Alamos National Laboratory
Theoretical Division
Los Alamos, NM, USA
murillo@lanl.gov

Chapter G.86

Dr. Murillo received his Ph.D. in theoretical atomic and plasma physics from Rice University. He then received a Director's Postdoctoral Felloship at Los Alamos, where he has remained since. His current research interests lie in the areas of dense and strongly coupled plasmas, including laser-produced plasmas, dusty plasmas, astrophysical plasmas, and ultracold plasmas. He applies both analytical and molecular dynamics methods to these systems.

Evgueni E. Nikitin

Technion–Israel Institute of Technology
Department of Chemistry
Haifa, Israel
nikitin@techunix.technion.ac.il

Chapter D.49

Professor, Nikitin Evgueni is a researcher, head of the research group, and Professor of Chemical Physics at the Institute of Chemical Physics, Moscow, since 1958. He is also Professor of Physical Chemistry, Technion, Haifa, since 1991. He is a member of the Deutsche Akademie der Naturforscher Leopoldina, the European Academy of Arts, Sciences and Humanities, and the International Academy of Quantum Molecular Sciences. His research concentrates on the theory of inelastic and reactive scattering, theory of nonadiabatic processes, statistical theory of chemical reactions, and atom-molecule processes at low energies. He authored 15 books and about 300 papers. Research awards: Alexander von Humboldt Award, Gauss Professorship, and Barecha Fellowship

Robert F. O'Connell

Louisiana State University
Department of Physics and Astronomy
Baton Rouge, LA, USA
oconnell@phys.lsu.edu

Chapter F.78

Professor O'Connell earned his Ph.D. in 1962 from the University of Notre Dame, Indiana. For many years , in collaboration with G. W. Ford , he has been studying dissipative and fluctuation phenomena in quantum mechanics and related applications. In addition, he is using the generalized quantum Langevin equation to explore recent topical questions in non-equilibrium statistical mechanics (particularly claims that the fundamental laws of thermodynamics may be violated in the quantum regime).

Francesca O'Rourke

Queen's University Belfast
Department of Applied Mathematics and Theoretical Physics
Belfast, UK
s.orourke@qub.ac.uk

Chapter D.52

Dr. O'Rourke obtained her Ph.D. in Ion-Atom Collisions from Queens University, Belfast, in 1991. She now lectures in Applied Mathematics and Theoretical Physics at Queens University, Belfast. Her current research interests include heavy particle collisions in atomic and molecular physics and more recently mathematical modelling in Biomedicine.

Ronald E. Olson

University of Missouri-Rolla
Physics Department
Rolla, MO, USA
olson@umr.edu

Chapter D.58

Ronald E. Olson, Curators' Professor of Physics earned his Ph.D. from Purdue University in 1967. He is a Fellow of the American Physics Society and a Fulbright Fellow to France. He was received the Humboldt Senior Prize Award, the University of Missouri system-wide Presidential Award for Research and Creativity. His research interests concentrate on theory of elastic and inelastic total and differential scattering cross sections: atom–atom, ion–atom, and ion–ion. Studies of multiply charged ion–atom collisions, Rydberg atom collisions, negative ion detachment mechanisms, and Penning and associative ionization.

Authors

Barbara A. Paldus

Skymoon Ventures
Palo Alto, CA, USA
bpaldus@skymoonventures.com

Chapter C.43

Dr. Barbara Paldus received her Ph.D. in electrical engineering from Stanford University. She is a partner at Skymoon Ventures, where she works with early stage photonics companies. Previously, she was CTO at Picarro, which she founded in 1998. She has received numerous research awards, most recently the Adolph Lomb Prize (2001) by the OSA for her work in cavity ring-down spectroscopy.

Josef Paldus

University of Waterloo
Department of Applied Mathematics
Waterloo, ON, Canada
paldus@scienide.uwaterloo.ca

Chapters A.4, A.5

Josef Paldus, FRSC, is a Distinguished Professor Emeritus in the Department of Applied Mathematics, Department of Chemistry, and Guelph-Waterloo Center for Graduate Work in Chemistry – Waterloo Campus, at the University of Waterloo, Waterloo, ON Canada. He is also an Adjunct Professor in the Department of Chemistry of the University of Florida in Gainesville, FL, USA. He received his Ph.D. degree from the Czechoslovak Academy of Sciences and his RNDr. and Dr.Sc. degrees from the Faculty of Mathematics and Physics of the Charles University in Prague, Czech Republic. His research interests are in the methodology of quantum chemistry, the many-electron correlation problem, and the electronic structure of molecular systems in general. On these topics he published about 300 papers, reviews, and monograph chapters. He is a member of several professional societies and editorial boards, and received various awards and international fellowships, notably a Killam Fellowship, Institute for Advanced Study in Berlin Fellowship, Alexander von Humboldt Senior Scientist Award, and most recently a Gold Medal of the Charles University. He is also a Fellow of the Royal Society of Canada and of the Fields Institute for Research in Mathematical Sciences.

Ruth T. Pedlow

Queen's University Belfast
Department of Applied Mathematics
and Theoretical Physics
Belfast, UK
r.pedlow@qub.ac.uk

Chapter D.52

Ruth Pedlow is working towards completion of her Ph.D. in heavy particle collisions in atomic and molecular physics at Queens University of Belfast.

David J. Pegg

University of Tennessee
Department of Physics
Knoxville, TN, USA
djpegg@utk.edu

Chapter E.60

Currently I am investigating the structure and dynamics of atomic and molecular negative ions by studying how they interact with photons and electrons. The threshold behaviour and resonance structure in detachment cross sections are used to measure correlation-sensitive parameters. Experiments on photo detachment involve the use of lasers or synchrotron radiation. Such measurements, for example, lead to information on the process of multiple electron detachment induced by the absorption of a single photon. Electron-impact detachment and dissociation processes are studied using a magnetic storage ring. These studies, for example, yield information on the production and decay of doubly negative charged molecular and cluster negative ions.

Ekkehard Peik

Physikalisch-Technische Bundesanstalt
Braunschweig, Germany
ekkehard.peik@ptb.de

Chapter B.30

Dr. Ekkehard Peik received his doctorate and the habilitation in physics at the University of Munich. His research interests are in the fields of laser-cooling and trapping of atoms and ions, precision laser spectroscopy and the application to optical time and frequency metrology and tests of fundamental physics. He is now head of the group 'Optical Clocks' at PTB and also a lecturer at the University of Hannover.

Ronald Phaneuf

University of Nevada
Department of Physics
Reno, NV, USA
phaneuf@unr.edu

Chapter E.64

Professor Phaneuf received a Ph.D. in atomic physics from the University of Windsor in 1973 and has since been engaged in experimental research on interactions of ions with electrons, atoms, molecules and photons using merged-beams and crossed-beams techniques. He was formerly at JILA and Oak Ridge National Laboratory. His current research emphasis is photon–ion interactions using synchrotron radiation.

Eric H. Pinnington

University of Alberta
Department of Physics
Edmonton, AB, Canada
pinning@phys.ualberta.ca

Chapter B.18

Eric Pinnington obtained his Ph.D. in Physics at Imperial College in 1962. Prior to joining the University of Alberta in 1965, he held an NRC postdoctoral fellowship at McMaster University in Hamilton, Ontario, and an Alexander von Humboldt Fellowship at the Max Planck Institute for Astrophysics in Munich. He was elected Fellow of the American Physical Society in 1995. He became Professor Emeritus of Physics in 1997.

Richard C. Powell

University of Arizona
Optical Sciences Center
Tuscon, AZ, USA
rcpowell@email.arizona.edu

Chapter F.71

Powell was educated in physics at the United States Naval Academy and Arizona State University. He has been a research scientist and professor at Air Force Cambridge Research Laboratories, Sandia National Laboratory, and Lawrence Livermore National Laboratory, Oklahoma State University and the University of Arizona. He has authored two textbooks and over 260 scientific papers in laser spectroscopy and solid-state laser development. Powell is an elected Fellow of both the American Physical Society and the Optical Society of America and has served a President of OSA. He has been elected to the Russian Academy of Engineering Science.

John F. Reading

Texas A&M University
Department of Physics
College Station, TX, USA
reading@physics.tamu.edu

Chapter D.50

Professor Reading earned his Ph.D. from the University of Birmingham, UK, in 1964. His current research interests are in theoretical calculations of cross sections for excitation and ionization following fast ion–atom collisions, the role of Pauli correlation in inner-shell vacancy production, and the role of dynamic electronic correlation. The latter especially in comparision of proton and anti-proton-induced single and double ionization of helium. He was named The Distinguished Texas Scientist of 1995 by the Texas Academy of Sciences and is Editor of the proceedings of several conferences on ion–atom collisions.

Jonathan R. Sapirstein

University of Notre Dame
Department of Physics
Notre Dame, IN, USA
jsapirst@nd.edu

Chapters B.27, B.29

Dr. Sapirstein earned his Ph.D. from Stanford University in 1979. He did postdoctoral work at UCLA and Cornell, and is at the University of Notre Dame, Indiana, since 1984. Current research interest in parity non-conservation in atoms, QED effects in highly charged many-electron ions, QED calculations in hydrogen, positronium, muonium, and helium. Dr. Sapirstein is a Fellow of the American Physical Society.

Stefan Scheel

Imperial College London
Blackett Laboratory
London, UK
s.scheel@imperial.ac.uk

Chapter F.81

Stefan Scheel received his Ph.D. (Dr. rer. nat.) from the Friedrich-Schiller-University Jena in 2001. He is an EPSRC Advanced Research Fellow in the Quantum Optics and Laser Science group in the Department of Physics at Imperial College London. His main research areas include QED in dielectric materials, quantum information processing using linear optics, and decoherence processes in atom chip experiments.

Axel Schenzle

Ludwig-Maximilians-Universität
Department für Physik
München, Germany
axel.schenzle@physik.uni-muenchen.de

Chapter F.79

Professor Schenzle has been working on various aspects of Theoretical Quantum Optics, the description of classical and quantummechanical noise in microscopic and mesoscopic systems, Bose–Einstein-Condensation, Quantum Information Theory, qunatum computing and decoherence. He has been Deputy Rector of the University of Munich and Dean for many years.

Reinhard Schinke

Max-Planck-Institut für Dynamik & Selbstorganisation
Göttingen, Germany
rschink@gwdg.de

Chapter C.34

Dr. Reinhard Schinke received his Ph.D. from the Physics department of the University of Kaiserslautern in 1976. His main area of research is molecular dynamics, in particular energy transfer in atomic collisions, chemical reactions, and photodissociation. He is author of the book Photodissociation Dynamics. In recent years his interest shifted to dynamical investigations of recombination processes with particular emphasis on the ozone isotope effect.

Wolfgang P. Schleich

Universität Ulm
Abteilung für Quantenphysik
Ulm, Germany
wolfgang.schleich@uni-ulm.de

Chapter F.78

Prof. Schleich studied physics and mathematics at the Ludwig-Maximilians-Universität München where he obtained his Diplom, Doktor, and Habilitation. He worked at the University of New Mexico (Albuquerque) and University of Texas (Austin) and the Max-Planck Institut für Quantenoptik in Garching. Since 1991 he has held a chair of theoretical physics at the Universität Ulm. He has more than 200 publications, is a Fellow of APS, IOP and OSA and an elected member of the Heidelberger Akademie der Wissenschaften and the Leopoldina, and has received numerous awards including the Leibniz Prize and the Max-Planck Prize.

Michael Schulz

University of Missouri-Rolla
Physics Department
Rolla, MO, USA
schulz@umr.edu

Chapter E.65

Professor Dr. Michael Schulz received his Ph.D. in Physics from the University of Heidelberg in 1987 to become a Teaching Assistant from 1981–1987. After positions at Oak Ridge National Laboratory and Kansas State University he joined the University of Missouri-Rolla as Assistant Professor in 1990. Since 2002 he is Professor of Physics and since 2003 Director of the Laboratory for Atomic, Molecular, and Optical Research. His scientific concentrate on experimental atomic physics, dynamics of many-body problem, correlation effects, and three-dimensional imaging of atomic break-up processes. He is a Fellow of the American Physical Society and was Mercator Scholar 2004–2005.

Peter L. Smith

Harvard University
Harvard-Smithsonian Center for Astrophysics
Cambridge, MA, USA
plsmith@cfa.havard.edu

Chapter C.44

Peter L. Smith received his Ph.D. degree in Physics from Caltech in 1972 and, after a year of teaching, came to and stayed at the Harvard-Smithsonian Center for Astrophysics. He is involved in measurements of fundamental atomic and molecular parameters at ultraviolet wavelengths for analysis of astronomical spectra, and design and calibration of instruments for ultraviolet spectroscopic and/or radiometric measurements, especially of the Sun, from earth-orbiting satellites.

Anthony F. Starace

The University of Nebraska
Department of Physics and Astronomy
Lincoln, NE, USA
astrace1@unl.edu

Chapter B.24

Dr. Starace earned his Ph.D. from the University of Chicago in 1971 and is George Holmes University Professor of Physics at the University of Nebraska since 2001. His primary research interests concern the interaction of intense laser light with atoms, especially single and multiphoton detachment and ionization processes. He is a Fellow of the American Physical Society and the American Association for the Advancement of Science, and is currently an Associate Editor of Reviews of Modern Physics.

Glenn Stark

Wellesley College
Department of Physics
Wellesley, MA, USA
gstark@wellesley.edu

Chapter C.44

Professor Stark's research interest is in the field of experimental molecular spectroscopy. His laboratory programs emphasize molecular transitions of interest to the astrophysics and aeronomy communities, primarily involving the measurement and interpretation of high-resolution absorption spectra of vacuum ultraviolet and extreme ultraviolet transitions. Related activities include Fourier transform spectroscopy of diatomic molecules, and laser spectroscopies of diatomics.

Allan Stauffer

Department of Physics and Astronomy
York University
Toronto, ON, Canada
stauffer@yorku.ca

Chapter D.48

Allan Stauffer has published numerous papers in the field of electron and positron scattering from atoms and simple molecules. In collaboration with numerous colleagues, he has been involved with extensive scattering calculations and developed methods to carry out these investigations and has worked closely with groups involved in measuring these processes.

Aephraim M. Steinberg

University of Toronto
Department of Physics
Toronto, ON, Canada
steinberg@physics.utoronto.ca

Chapter F.80

Aephraim Steinberg works on experimental quantum optics and laser cooling, with specific emphasis on foundational questions in quantum mechanics (esp. quantum measurement) and on quantum information. His obssession is with tunneling times; in 1994, he demonstrated (with Kwiat and Chiao) the superluminal tunneling of photons, and in 2005, he is starting an experiment to probe tunneling times for Bose-condensed atoms through optical barriers.

Stig Stenholm

Royal Institute of Technology
Physics Department
Stockholm, Sweden
stenholm@atom.kth.se

Chapter F.69

Stig Stenholm was Pprofessor of Laser Physics and Quantum Optics at the Royal Institute of Technology, Stockholm. He studied Technical Physics at the Helsinki Institute of Technology and Mathematics at the University of Helsinki. He worked at the Research Institute for Theoretical Physics in Helsinki until 1997, when moving to Stockholm. Theoretical research fields include spectroscopy, quantum optics, and informatics

Jack C. Straton

Portland State University
University Studies
Portland, OR, USA

Chapter D.57

Jack Straton earned a doctorate in quantum theory from the University of Oregon and served as both a volunteer and professional diversity trainer over the past 18 years. He is an Assistant Professor in Portland State University's interdisciplinary University Studies program, where his teaching blends science, art, diversity, and social responsibility. His research ranges from Quantum Scattering Theory to Anti-racist Pedagogy.

Carlos R. Stroud Jr.

University of Rochester
Institute of Optics
Rochester, NY, USA
stroud@optics.rochester.edu

Chapter F.73

Professor Stroud is Professor of Optics, Professor of Physics and Director of the Center for Quantum Information at the University of Rochester where he works in a variety of areas of experimental and theoretical quantum optics and atomic physics. His group pioneered the area of Rydberg electron wave packet physics observing localization, decays, revivals and interferometry with a single electron.

Barry N. Taylor

National Institute of Standards and Technology
Atom Physics Division
Gaithersburg, MD, USA
barry.taylor@nist.gov

Chapter B.28

Barry N. Taylor received his Ph.D. in Physics from the University of Pennsylvania in 1963. He remained at Penn as a faculty member until he joined RCA Laboratories in Princeton, NJ in 1966. He joined the National Bureau of Standards (now NIST) in 1970 as a Section Chief in the Electricity Division, becoming its Chief in 1974. In 1988 he became manager of the NIST Fundamental Constants Data Center, retiring from NIST and that position in 2001. Since then he has been a NIST Scientist Emeritus in the Data Center. Dr. Taylor has authored or co-authored over 100 publications, is a fellow of the APS and IEEE, and has received a number of awards. His current research focuses on the evaluation of data related to the fundamental constants and improving the International System of Units (SI).

Aaron Temkin

NASA Goddard Space Flight Center
Laboratory for Solar and Space Physics
Greenbelt, MD, USA
aaron.temkin-1@nasa.gov

Chapter B.25

Dr. Temkin is a research physicist (emeritus) at NASA/GSFC. He has specialized (primarily) in scattering problems of electrons from atoms and molecules, and associated processes (autoionization, in particular). He received his Ph.D. degree from the Massachusetts Institute of Technology in 1956, and has been at his present institution since 1960.

Sandor Trajmar

California Institute of Technology
Jet Propulsion Laboratory
Redwood City, USA
strajmar@comcast.net

Chapter E.63

Dr. Sandor Trajmar received his Ph.D. in physical chemistry from the University of California at Berkeley, California,. He was Head of the Electron collision Physics Group, Jet Propulsion Laboratory, California Institute of Technology, Pasadena, California. He retired in January 1997.

Elmar Träbert

Ruhr-Universität Bochum
Experimentalphysik III/NB3
Bochum, Germany
traebert@ep3.rub.de

Chapter B.18

Professor Elmar Träbert obtained his doctorate and professorial title at Ruhr-Universität Bochum. He has extensive experience in time-resolved spectroscopy and atomic lifetime measurements mainly from working with beam-foil spectroscopic techniques, a heavy-ion storage ring, as well as radio-frequency and electron beam ion traps in more than a dozen laboratories.

Turgay Uzer

Georgia Institute of Technology
School of Physics
Atlanta, GA, USA
turgay.uzer@physics.gatech.edu

Chapter B.15

Professor Turgay Uzer obtained his doctorate at Harvard and was a postdoctoral fellow at Caltech. Currently he is Regents' Professor in the School of Physics, Georgia Institute of Technology. His research interests include: Rydberg atoms and molecules, semiclassical theories, nonlinear dynamics/chaos, intramolecular energy transfer, and chemical reactivity.

Karl Vogel

Universität Ulm
Abteilung für Quantenphysik
Ulm, Germany
karl.vogel@uni-ulm.de

Chapter F.78

Dr. Vogel received his PhD from the Universität Ulm in 1989. His research area is theoretical quantum optics. In particular, he investigated how quantum states of the radiation field can be prepared and how they can be measured.

Jon C. Weisheit

Washington State University
Institute for Shock Physics
Pullman, WA, USA
weisheit@wsu.edu

Chapter G.86

Jon Weisheit recently joined Washington State Universtity's Intstitute for Shock Physics, where he holds appointments as Research Professor and Associate Director, and conducts research focused on understanding quantum phenomena in high energy density matter. He is a Fellow of the American Physical Society, and is a frequent advisor in government agencies on issues pertaining both to basic science and to national defense programs. Her received his graduate degrees in space science and in physics from Rice University.

Wolfgang L. Wiese

National Institute of Standards and Technology
Gaithersburg, MD, USA
wiese@nist.gov

Chapter B.10

Dr. Wolfgang Wiese is a physicist with extensive research background in atomic spectroscopy and in the critical tabulation of atomic reference data. He has worked at the National Institute of Standards and Technology for more than 40 years and has led the Atomic Physics Division from 1978 to 2004. He has authored 6 data volumes on Atomic Transition Probabilities, 15 book chapters and about 225 shorter research papers.

Martin Wilkens

Universität Potsdam
Institut für Physik
Potsdam, Germany
martin.wilkens@physik.uni-potsdam.de

Chapter F.77

Dr. Martin Wilkens received a Ph.D. In Physics from Essen University. He spent his post-doctoral years in Warsaw, Tucson, and Konstanz and has been appointed Professor for Theoretical Physics / Quantum Optics at Potsdam University in 1997. His current research areas are Bose-Einstein condensation, degenerate quantum gases, and quantum information processing and communication.

Authors

Detailed Contents

Detailed Cont.

Part C Molecules

Detailed Cont.

Part F Quantum Optics

68 Light–Matter Interaction

69 Absorption and Gain Spectra

Detailed Cont.

Detailed Cont.

Part G Applications

Subject Index

A

B

C

D

E

F

G

H

J

K

L

M

P

Q

R

S

Subject Index

Subject Index

U

V

W

X

Y

Z